Tenth Edition

Biology

Kenneth A. Mason

University of Iowa

Jonathan B. Losos

Harvard University

Susan R. Singer

Carleton College

based on the work of

Peter H. Raven
President Emeritus, Missouri Botanical Garden;
George Engelmann Professor of Botany Emeritus,
Washington University

George B. Johnson
Professor Emeritus of Biology, Washington University

Mc Graw Hill

Connect
Learn
Succeed™

BIOLOGY, TENTH EDITION

Published by McGraw-Hill, a business unit of The McGraw-Hill Companies, Inc., 1221 Avenue of the Americas, New York, NY 10020. Copyright © 2014 by The McGraw-Hill Companies, Inc. All rights reserved. Printed in the United States of America. Previous editions © 2011, 2008, and 2005. No part of this publication may be reproduced or distributed in any form or by any means, or stored in a database or retrieval system, without the prior written consent of The McGraw-Hill Companies, Inc., including, but not limited to, in any network or other electronic storage or transmission, or broadcast for distance learning.

Some ancillaries, including electronic and print components, may not be available to customers outside the United States.

This book is printed on acid-free paper.

1 2 3 4 5 6 7 8 9 0 DOW/DOW 1 0 9 8 7 6 5 4 3

ISBN 978-0-07-338307-1
MHID 0-07-338307-4

Senior Vice President, Products & Markets: *Kurt L. Strand*
Vice President, General Manager, Products & Markets: *Marty Lange*
Vice President, Content Production & Technology Services: *Kimberly Meriwether David*
Managing Director: *Michael S. Hackett*
Director, Biology: *Lynn Breithaupt*
Brand Manager: *Rebecca Olson*
Director of Development: *Elizabeth Sievers*
Director of Digital Content: *Tod Duncan, Ph.D.*
Executive Marketing Manager: *Patrick E. Reidy*
Lead Project Manager: *Sheila M. Frank*
Buyer: *Nicole Baumgartner*
Senior Media Project Manager: *Tammy Juran*
Designer: *Tara McDermott*
Cover Designer: *Elise Lansdon*
Cover Images: *Anolis gorgonae:* © D. Luke Mahler; DNA molecule: © Doug Struthers/Getty Images; *E. coli:* Drawing by Daniel Whitmire from data provided by Drs. Philip M. Silverman & Margaret B. Clarke, Oklahoma Medical Research Foundation; flower, *Centaurea americana* and spider, *Misumenops sp.:* Darrell S. Vodopich; lizard skin background: © Corbis/RF; orangutan on back: © Dave King/Getty Images
Senior Content Licensing Specialist: *Lori Hancock*
Photo Research: *Danny Muldung/Photoaffairs*
Art Studio: *Electronic Publishing Services Inc., NYC*
Compositor: *Electronic Publishing Services Inc., NYC*
Typeface: *10/12 Utopia Std*
Printer: *R. R. Donnelley*

All credits appearing on page or at the end of the book are considered to be an extension of the copyright page.

Library of Congress Cataloging-in-Publication Data

Cataloging-in-Publication Data has been requested from the Library of Congress.

The Internet addresses listed in the text were accurate at the time of publication. The inclusion of a website does not indicate an endorsement by the authors or McGraw-Hill, and McGraw-Hill does not guarantee the accuracy of the information presented at these sites.

www.mhhe.com

Brief Contents

About the Authors

Pictured left to right: Susan Rundell Singer, Jonathan Losos, Kenneth Mason

Kenneth Mason is a lecturer at the University of Iowa where he teaches introductory biology. He was formerly at Purdue University where for 6 years he was responsible for the largest introductory biology course on campus and collaborated with chemistry and physics faculty on an innovative new course supported by the National Science Foundation that combined biology, chemistry, and physics. Prior to Purdue, he was on the faculty at the University of Kansas for 11 years, where he did research on the genetics of pigmentation in amphibians, publishing both original work and reviews on the topic. While there he taught a variety of courses, was involved in curricular issues, and wrote the lab manual for an upper division genetics laboratory course. His latest move to the University of Iowa was precipitated by his wife's being named president of the University of Iowa.

Jonathan Losos is the Monique and Philip Lehner Professor for the Study of Latin America in the Department of Organismic and Evolutionary Biology and curator of herpetology at the Museum of Comparative Zoology at Harvard University. Losos's research has focused on studying patterns of adaptive radiation and evolutionary diversification in lizards. He is the recipient of several awards, including the prestigious Theodosius Dobzhansky and David Starr Jordan Prizes, the Edward Osborne Wilson Naturalist Award, and the Daniel Giraud Elliot Medal from the National Academy of Sciences. Losos has published more than 100 scientific articles.

Susan Rundell Singer is the Laurence McKinley Gould Professor of Natural Sciences in the department of biology at Carleton College in Northfield, Minnesota, where she has taught introductory biology, plant biology, genetics, and plant development for 26 years. Her research focuses on the development and evolution of flowering plants and genomics learning. Singer has authored numerous scientific publications on plant development and co-authored education reports including *Vision and Change* and "America's Lab Report." She received the American Society of Plant Biology's Excellence in Teaching Award, the Botanical Society's Bessey Award, is a AAAS fellow, served on the National Academies Board on Science Education, and chaired several National Research Council study committees including the committee that produced *Discipline-Based Education Research*.

The Learning Continues in the Digital Environment

The digital offerings for the study of biology have become a key component of both instructional and learning environments. In response to this, the author team welcomes Dr. Ian Quitadamo as Lead Digital Author for *Biology*, tenth edition. As Lead Digital Author, Ian oversaw the development of digital assessment tools in Connect. Ian's background makes him a unique and valuable addition to the tenth edition of *Biology*.

Ian Quitadamo Ian Quitadamo is an Associate Professor with a dual appointment in Biological Sciences and Science Education at Central Washington University in Ellensburg, WA. He teaches introductory and majors biology courses and cell biology, genetics, and biotechnology as well  as science teaching methods courses for future science teachers and interdisciplinary content courses in alternative energy and sustainability. Dr. Quitadamo was educated at Washington State University and holds a Bachelor's degree in biology, Master's degree in genetics and cell biology, and an interdisciplinary Ph.D. in science, education, and technology. Previously a researcher of tumor angiogenesis, he now investigates critical thinking and has published numerous studies of factors that affect student critical thinking performance. He has received the Crystal Apple award for teaching excellence, led various initiatives in critical thinking and assessment, and is active in training future and currently practicing science teachers. He served as a co-author on *Biology* , eleventh edition, by Mader and Windelspecht, copyright 2013, and is the lead digital author for *Biology*, third edition by Brooker, copyright 2014, both published by McGraw-Hill.

Committed to Excellence

With the release of the tenth edition, Raven & Johnson's *Biology* enters a new era. This edition provides an unmatched comprehensive text fully integrated with state-of-the-art digital resources. The material in the text is organized around learning outcomes keyed to major biological concepts, which provide a framework for understanding the ever-expanding universe of biological knowledge. An inquiry-based approach with robust, adaptive tools for discovery and assessment in both text and digital resources provide the intellectual challenge needed to promote student critical thinking and ensure academic success. A major strength of this hallmark tenth edition is assessment across multiple levels of Bloom's taxonomy that develops critical thinking and problem solving skill in addition to comprehensive factual knowledge. McGraw-Hill's Connect® platform offers a powerful suite of online tools including LearnSmart™ that adapts to student needs over time. The adaptive learning system helps students learn faster, study efficiently, and retain more knowledge of key concepts.

The tenth edition continues our tradition of providing the student with clear learning paths that emphasize data analysis and quantitative reasoning. Embedded eBook resources allow just-in-time exploration of major themes like evolution from directly within the eBook. Additional embedded eBook resources link to asides that delve more deeply into quantitative aspects.

The author team is experienced and fully committed to student learning and producing the best possible text for both students and faculty. Lead author Kenneth Mason (University of Iowa) has taught majors biology for more than 20 years at three major public universities. Jonathan Losos (Harvard University), a leading evolutionary biology researcher, has taught both undergraduate and graduate courses in biology for 20 years. Susan Rundell Singer (Carleton College) is a 26-year veteran science educator deeply involved in science education policy on a national level. As a team, we continually strive to improve the text by integrating the latest cognitive and best practices research with methods that are known to positively affect learning. We have multiple features that are focused on scientific inquiry, including an increased quantitative emphasis in the Scientific Thinking features. We continue to use the concise, accessible, and engaging writing style of past editions while maintaining the clear emphasis on evolution and scientific inquiry that have made this a leading textbook of choice for majors biology students. Our emphasis on evolution combined with integrated cell and molecular biology and genomics offers our readers a student-friendly text that is modern and well balanced.

The tenth edition continues to employ the aesthetically stunning art program that the Raven and Johnson *Biology* text is known for. Complex topics are represented clearly and succinctly, helping students to build the mental models needed to understanding biology.

Insights into the diversity of life that are provided by molecular tools have led to a reorganization of these topics in the tenth edition. This entire unit reflects the most current research on eukaryotic phylogenies, blending molecular, morphological, and development viewpoints. Nuclear reprogramming in stem cells, gene expression, and the importance of small RNAs in gene regulation continue to shape our treatment of these topics. These are just a few examples of the many changes in the tenth edition of *Biology* that provide students with scientifically accurate context, historical perspective, and relevant supporting details essential to a modern understanding of life science.

As the pace of scientific discovery continues to provide new insights into the foundation of life on Earth, our author team will continue to use every means possible to ensure students are as prepared as possible to engage in biological topics. Our goal now, as it has always been, is to ensure your success.

Our Consistent Themes

It is important to have consistent themes that organize and unify a text. A number of themes are used throughout the book to unify the broad-ranging material that makes up modern biology. This begins with the primary goal of this textbook to provide a comprehensive understanding of evolutionary theory and the scientific basis for this view. We use an experimental framework combining both historical and contemporary research examples to help students appreciate the progressive and integrated nature of science.

Biology Is Based on an Understanding of Evolution

When Peter Raven and George Johnson began work on *Biology* in 1982 they set out to write a text that presented biology the way they taught in their classrooms—as the product of evolution. We bear in mind always that all biology "only makes sense in the light of evolution"; so this text is enhanced by a consistent evolutionary theme that is woven throughout the text, and we have enhanced this theme in the tenth edition.

The enhanced evolutionary thread can be found in obvious examples such as the two chapters on molecular evolution, but can also be seen throughout the text. As each section considers the current state of knowledge, the "what" of biological phenomenona, they also consider how each system may have arisen by evolution, the "where it came from" of biological phenomenona.

We added an explicit phylogenetic perspective to the understanding of animal form and function. This is most obvious in the numerous figures containing phylogenies in the form and function chapters. The diversity material is supported by the most up-to-date approach to phylogenies of both animals and plants. Together these current approaches add even more evolutionary support to a text that set the standard for the integration of evolution in biology. Our approach allows evolution to be dealt with in the context in which it is relevant. The material throughout this book is considered not only in terms of present structure and function, but how that structure and function may have arisen via evolution by natural selection.

The emphasis on evolution is expanded in the eBook. Because a textbook limits content to a specific length, examples or additional information related to evolution cannot always be included in the printed book. The digital environment lifts these restrictions and has allowed us the opportunity to include interesting and instructional evolution material in the eBook. The topics that are supported by additional examples and information on the evolutionary aspects of a concept are highlighted with the inclusion of an "evolution" icon in the text. This icon is associated with a link in the eBook that reveals this additional material.

Biology Uses the Methods of Scientific Inquiry

Another unifying theme within the text is that knowledge arises from experimental work that moves us progressively forward. The use of historical and experimental approaches throughout allow the student not only to see where the field is now, but more importantly, how we arrived here. The incredible expansion of knowledge in biology has created challenges for authors in deciding what content to keep, and to what level an introductory text should strive. We have tried to keep as much historical context as possible and to provide this within an experimental framework consistently throughout the text.

We use a variety of approaches to expose the student to scientific inquiry. We use our Scientific Thinking figures to walk through an experiment and its implications. These figures always use material that is relevant to the story being told. Data are also provided throughout the text, and our new **Data Analysis questions** ask students to interpret these data. Students are also provided with **Inquiry questions** to stimulate critical thinking about the material throughout the book. The Data Analysis questions deal directly with data in figures or the text, while the Inquiry questions are more conceptual. This combination will allow the student experience in interpreting data, and lead the student to question the material in the text as well. Embedded eBook resources allow just-in-time exploration of quantitative aspects of the science. Quantitative Asides present in the eBook are indicated with a "quantitative" icon in the text. This icon in the eBook links to

 asides that delve more deeply into quantitative aspects of the topic under discussion.

Biology Is an Integrative Science

The explosion of molecular information has reverberated throughout all areas of biological study. Scientists are increasingly able to describe complicated processes in terms of the interaction of specific molecules, and this knowledge of life at the molecular level has illuminated relationships that were previously unknown. Using this cutting-edge information, we more strongly connect the different areas of biology in this edition.

One example of this integration concerns the structure and function of biological molecules—an emphasis of modern biology. This edition brings that focus to the entire book, using this as a theme to weave together the different aspects of content material with a modern perspective. Given the enormous amount of information that has accumulated in recent years, this emphasis on structure and function provides a necessary thread integrating these new perspectives into the fabric of the traditional biology text.

Although all current biology texts have added a genomics chapter, our text was one of the first to do so. This chapter has been updated, and we have an additional chapter on the evolution of genomes. More importantly, the results from the analysis of genomes and the proteomes they encode are presented throughout the book wherever this information is relevant. This allows a more modern perspective throughout the book rather than limiting it to a few chapters. Examples, for instance, can be found in the diversity chapters, where classification of some organisms were updated based on new findings revealed by molecular techniques.

This systems approach to biology also shows up at the level of chapter organization. We introduce genomes in the genetics section in the context of learning about DNA and genomics. We then come back to this topic with an entire chapter at the end of the evolution unit where we look at the evolution of genomes, followed by a chapter on the evolution of development, which leads into our unit on the diversity of organisms.

We're excited about the tenth edition of this quality textbook providing a learning path for a new generation of students. All of us have extensive experience teaching undergraduate biology, and we've used this knowledge as a guide in producing a text that is up to date, beautifully illustrated, and pedagogically sound for the student. We've also worked to provide clear explicit learning outcomes, and more closely integrate the text with its media support materials to provide instructors with an excellent complement to their teaching.

Ken Mason, Jonathan Losos, Susan Rundell Singer

Cutting Edge Science

Changes to the Tenth Edition

Part I: The Molecular Basis of Life

The material in this section does not change much with time. However, we have updated it to make it more friendly to the student. The Learning Outcomes have been analyzed and rewritten both for clarity and to increase linkage between Learning Outcomes and assessment in both the end-of-chapter material and online content.

In chapter 3, the material on nucleic acids has been rewritten to make it more modern. Our view of the role of RNA in particular has changed hugely in the last decade, and this introduction to these molecules has been rewritten to reflect this. Also in chapter 3 is the first Evolutionary Aside for the eBook.

Part II: Biology of the Cell

The Learning Outcomes have been analyzed and rewritten both for clarity and to increase linkage between Learning Outcomes and assessment in both end-of-chapter material and online content. Data Analysis questions were added, and some Evolutionary Aside and Quantitative Asides for the eBook were also included.

Chapter 4—New material on prokaryotic cytoskeleton was added. Material on the nuclear pore was updated and a new figure added to show our current view of this structure. The role of chromatin structure in gene expression is introduced earlier, and material on the ER and Golgi has been updated to present the most current view of these important organelles. The material on cell-to-cell connections has been updated and also given a more evolutionary perspective.

Chapter 5—Material on lipid rafts was reconsidered, and material on sphingolipids was added. These important lipids are often ignored, despite their importance in the nervous system of vertebrates. This material also includes a new figure of sphingolipids.

Chapter 7—The introduction to glycolysis was rewritten for clarity. This includes better integration of text and figures. The Krebs cycle overview figure was simplified, as was the explanatory text for greater clarity. The section on theoretical energy yield from chemiosmosis was completely rewritten to bring it up to the view of modern chemistry.

Chapter 9—The discussion of GPCR was updated to take into account new genetic data on their distribution. A new section on small ras-like G proteins was added, including a new figure showing their action. This both illustrates their importance in the control of cell division, and clarifies their connection to signaling by growth factors.

Chapter 10—Content on chromosome structure was updated and material on the behavior of chromosomes was rewritten for clarity. A figure from chapter 9 on role of growth factors was combined with a figure from chapter 10 for greater clarity and to reduce redundancy.

Part III: Genetic and Molecular Biology

The overall organization of this section remains the same. We have retained the split of transmission genetics into two chapters as it has proved successful for students.

Content changes in the molecular genetics portion of this section continue to update material that is the most rapidly changing in the entire book. We also continue to refine the idea that RNA plays a much greater role now than appreciated in the past. The more modern view of RNA continues to be under appreciated in introductory textbooks. New material continues to be put into historical context for greater student understanding. The Learning Outcomes have been analyzed and rewritten both for clarity and to increase linkage between learning outcomes and assessment in both the end-of-chapter material and online content. Data Analysis questions were added, and some Evolutionary Asides and Quantitative Asides for the eBook were also included.

Chapter 11—The behavior of chromosomes during meiosis was rewritten for clarity. This subject is one that causes great confusion for students, and two graphics—one figure and one in-text graphic—were updated to complement the new textual discussion. This section is now much easier for students to both appreciate the complex behavior of meiotic chromosomes, but also the molecular basis for this behavior.

Chapter 15—The definition of genes as one-gene/one-polypeptide was revised for clarity. The complexity of eukaryotic initiation is given greater appreciation. The idea of promoter-proximal pausing is introduced. This allows for a clearer view by students of the nature of the extensive transcription observed by whole-genome scans.

Chapter 16—Introductory material on the control of gene expression has been rewritten to reflect recent data and a more modern view of this control. Some material on DNA binding proteins was rewritten for clarity. The section on posttranscriptional control has been rewritten again as it is one of the most rapidly changing areas. This material is now on stronger conceptual ground.

Chapter 17—The chapter has been reorganized and revised to focus on biological concepts related to biotechnology, rather than using techniques as the organizational structure. Polymerase chain reaction is now clearly linked to student's prior learning about DNA replication. DNA sequencing was moved to chapter 18, which focuses on genomes so sequencing of single genes to entire genomes is explained in a coherent and cohesive way. Instead of a generic section on DNA analysis, a section on "Storing and Sorting DNA Fragments" has been introduced, followed by a section titled "Analyzing and Creating DNA Differences." We have revised the DNA fingerprinting section to include short tandem repeats. The applications sections have been updated to include, for example, sections on marker assisted breeding and transgenic salmon.

Chapter 18—A comprehensive approach to sequencing at all scales has been developed to frame the genomics chapter. A "Genes to Proteins" section also scales from individual genes and proteins to genomes and proteins. The section includes text and art explaining the yeast two-hybrid assay. A new section on comparing genomes has been added that provides the foundation for the comparative genomics in the Genome Evolution chapter placed after the principles of evolution. Text and artwork exploring genomic insights into human migration are now integrated into chapter 18.

Chapter 19—The material on nuclear reprogramming was rewritten for both clarity and to incorporate new data in this exciting area. New information on induced pluripotent stem cells is presented along with a better historical timeline of this topic. This is both of general interest to students, and is a source of controversy and misinformation. All material on plant development that was not used for direct comparisons to animals was removed or moved to chapter 41.

Part IV: Evolution

The evolution chapters were updated with new examples. A strong emphasis on the role of experimental approaches to studying evolutionary phenomena has been maintained and enhanced.

Chapter 20—This chapter has been reorganized to consolidate the discussion of selection acting on discrete and continuously distributed populations, before discussing the interaction among different evolutionary forces, which now is discussed toward the end of the chapter.

Chapter 21—The examples in chapter 21 have been updated in several important ways. First, research published in 2012 indicates that studies on selection on peppered moths by Kettlewell were completely correct in showing that bird predation favors those moths that contrasted their background. This section has been revised to clarify this previously controversial point. In addition, new data points on the decrease in the prevalence of black moths in recent years have been added, continuing to demonstrate that as air pollution has been alleviated, peppered moths are increasing in frequency. Material on dating the fossil record was moved to chapter 26. The discussion of the difference between "theory" and "hypothesis" in scientific terminology has been expanded.

Chapter 22—The discussion of hybrid inviability was expanded. The discussion of character displacement was enhanced with a detailed case study of stickleback fishes in northwestern lakes. The discussion of speciation and extinction through time has been moved to chapter 26 and chapter 59.

Chapter 23—Terminology concerning cladistic analysis and phylogenetic systematics was clarified. The example of using phylogenetic information to understand the spread of HIV from monkeys and apes to humans has been updated to reflect new discoveries.

Chapter 24—The increased number of sequenced genomes allowed us to increase the emphasis on comparative genomics informing our understanding of evolution, as represented in the new figure 24.1. The entire chapter was reorganized to present genome evolution in a conceptual rather than topical way, as reflected in the new section heads. New findings on the rapid rate of plant genome evolution are analyzed, as are the implications of additional primate sequences. A new section on how comparative genomics informs conservation biology was added.

Chapter 25—The seven sections in the previous edition have been consolidated into four sections focusing on core concepts in the evolution of development. The fully reorganized chapter now provides a more coherent and current overview of the maturing field of evolution of development. The chapter now emphasizes evolution of developmental patterns, how single-gene changes can alter form, and different ways to evolve the same structure. A new Scientific Thinking figure guides students through the research on how *Tbx4* and *Tbx 5* were co-opted for vertebrate limb development (figure 25.6). The role of *Hox* genes in digit development has been added, along with a supporting figure (figure 25.7). Our case study on the evolution of the eye has been updated, informed by new data on jellyfish and exemplified by the addition of figure 25.15.

Part V: Diversity of Life on Earth

You will notice some significant reorganization of material in Part V from the ninth edition.

Chapter 26—This is a new chapter for our tenth edition that sets the stage for the unit on Diversity of Life on Earth. This chapter focuses on the origins and diversity of life, beginning with an introduction to deep time. Understanding deep time is essential for student understanding of the origins and evidence for early life. "Earth's Changing Systems" presents our understanding of how Earth system changes have affected life and how life has affected Earth systems. The chapter then investigates the major innovations in the evolution of life, a springboard for the rest of the unit. New artwork supports student understanding of deep time and the changes that have occurred in Earth systems and life forms over geological time.

Chapter 27—New material has been added on the 2009 H1N1 pandemic. Material on bacteriophage life cycles was rewritten, and now is used as an example of a simple virus life cycle.

Chapter 28—Material on the origin of life that opened this chapter in the previous edition was moved to chapter 26 where all such material has been consolidated. A new introduction was written that looks at the history of the study of microbiology in brief.

Chapter 29—Reorganization of this chapter was guided by the newest phylogenies for the protists. The green algae presentation, including life cycles, was moved from the plant diversity chapter to the protist chapter in this edition to provide greater clarity about the algae in general for students.

Chapters 30 and 31—The Green Plant chapter in the ninth edition has been replaced with a chapter on Seedless Plants and a chapter on Seed Plants. This allowed us to move the information on the diversity of fruit and flower structure from the plant unit to the diversity unit where it is more appropriate. This approach reduces redundancy between the two units and keeps students focused on the most relevant concepts for understanding plant diversity.

Chapters 33–35—These chapters have been reorganized to reflect current understanding of phylogenetic relationships. In particular chapter 33 now discusses an overview of animal diversity and evolution, as well as the most basal members of animal phylogeny. Chapter 34 covers protostomes and chapter 35 covers deuterostomes.

Part VI: Plant Form and Function

This unit was reorganized at the chapter level in this edition. The Vegetative Plant Development chapter in the ninth edition was eliminated. Plant reproduction and development were consolidated in a single, coherent chapter 41. Information on the diversity of flowers and fruit was modified and integrated into the unit on diversity.

Chapter 37—Figure 37.2 was modified for clarity. To help students better understand mass flow, figure 37.19 was expanded to pull out details about critical events at both sources and sinks.

Chapter 41—A substantial reorganization of this chapter resulted in the incorporation of embryo development and germination into this chapter, following the section on pollination. To streamline the chapter, several figures addressing determination for flowering and the three-dimensional axes of embryo development were eliminated. The restructuring maintains student focus on what is truly core in understanding plant reproduction and development at the level of introductory biology.

Part VII: Animal Form and Function

The organizational changes made in the ninth edition have been maintained. This gives the student a system-level organization that is enhanced by the presentation of material that is both cellular and molecular in focus, and that puts this material into an evolutionary context. The Learning Outcomes have been analyzed and rewritten both for clarity and to increase linkage between Learning Outcomes and assessment in both the end-of-chapter material and online content. Data Analysis questions were added, and some Evolutionary Asides and Quantitative Asides for the eBook were also included.

Chapter 43—The material on the generation of a resting potential was rewritten for clarity. These changes emphasize membrane permeability and the role of ion channels. This provides a strong framework to understand how ion channels also function in graded and action potentials.

Chapter 50—This chapter was reorganized to make a more logical flow of topics. The section on nitrogenous wastes was moved up from section 4 to section 2. This places information on nitrogenous wastes immediately after the concept of osmoregulation, and consolidates all of the material on how various animals achieve this.

Chapter 51—The introduction and the material on innate immunity has been rewritten for clarity and to emphasize the connections between innate and adaptive immunity.

Chapter 52—The material on the development of the follicle has been updated to reflect a more accurate description of developmental timing.

Part VIII: Ecology and Behavior

This unit is rich in eBook Evolutionary Asides, for example, a number of case studies of the evolutionary significance of animal behavior are presented in chapter 54. In chapter 59, eBook Evolutionary Asides have been included exploring extinction through time, the evolutionary significance of biological "hot spots," and the evolutionary response of populations to overfishing. Befitting the nature of ecological science, the chapters are now also replete with Data Analysis questions accompanying the many graphical illustrations.

Chapter 55—The information on human population growth was updated using current statistics.

Chapter 56—The discussion of the definition of a biological community was revised.

Chapter 58—The information on human impacts on the environment and global warming was updated using the most current information available. New material discussing ocean acidification was added.

Chapter 59—Chapter 59 considers conservation biology, emphasizing the causes of species endangerment and what can be done. The information about species extinctions, including mass extinction (moved from chapter 22), was updated.

Committed to Preparing Students for the Future

Understand Biology With the Help of . . .

Integrated Learning Outcomes

Each section begins with specific Learning Outcomes that represent each major concept. At the end of each section, the Learning Outcomes Review serves as a check to help students confirm their understanding of the concepts in that section. Questions at the end of the Learning Outcomes Review ask students to think critically about what they have read.

> Any opportunity to identify "learning outcomes" is a welcome addition; we are forced more and more to identify these in learning assessments. I would use these as a guide for students to understand the minimum material they are expected to learn from each section.
>
> *Michael Lentz*
> *University of North Florida*

21.1 *The Beaks of Darwin's Finches: Evidence of Natural Selection*

Learning Outcomes

1. Describe how the species of Darwin's finches have adapted to feed in different ways.
2. Explain how climatic variation drives evolutionary change in the medium ground finch.

Upon Darwin's return to England, ornithologist John Gould informed Darwin that his collection was in fact a closely related group of distinct species, all similar to one another except for their beaks. In all, 14 species are now recognized.

Galápagos finches exhibit variation related to food gathering

The diversity of Darwin's finches is illustrated in figure 21.1. The ground finches feed on seeds that they crush in their powerful beaks; species with smaller and narrower beaks, such as the warbler finch, eat insects. Other species include fruit and bud eaters, and species that feed on cactus fruits and the insects they attract; some populations of the sharp-beaked ground finch even include "vampires" that sometimes creep up on seabirds and use their sharp beaks to pierce the seabirds' skin and drink their blood. Perhaps most remarkable are the tool users, woodpecker finches that pick up a twig, cactus spine, or leaf stalk, trim it into shape with their beaks, and then poke it into dead branches to pry out grubs.

The correspondence between the beaks of the finch species and their food source suggested to Darwin that natural selection had shaped them. In *The Voyage of the Beagle*, Darwin wrote, "Seeing this gradation and diversity of structure in one small, intimately related group of birds, one might really fancy that from an original paucity of birds in this archipelago, one species has been taken and modified for different ends."

21.3 *Artificial Selection: Human-Initiated Change*

Learning Outcomes

1. Contrast the processes of artificial and natural selection.
2. Explain what artificial selection demonstrates about the power of natural selection.

[...] eceding chapter, a variety of processes [...] y change. Most evolutionary biologists, [...] win's thinking that natural selection is [...] onsible for evolution. Although we can- [...] me, modern-day evidence allows us to [...] w evolution proceeds and confirms the [...] n as an agent of evolutionary change. [...] n both the field and the laboratory and [...] man-altered situations.

[...] re a classic example of evolution by [...] he visited the Galápagos Islands off [...] 1835, Darwin collected 31 specimens [...] ds. Darwin, not an expert on birds, [...] he specimens, believing by examining [...] ction contained wrens, "gross-beaks,"

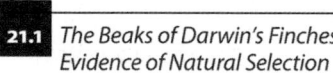

Woodpecker finch (*Cactospiza pallida*)   Large ground finch (*Geospiza magnirostris*)   Cactus finch (*Geospiza scandens*)

Figure 21.1 **Darwin's finches.** These species show differences in beaks and feeding habits among Darwin's finches. This diversity arose when an ancestral finch colonized the islands and diversified into habitats lacking other types of small birds. The beaks of several species resemble those of different families of birds on the mainland. For example, the warbler finch has a beak very similar t[...] related.

Warbler finch (*Certhidea olivacea*)   Vegetarian tree finch (*Platyspiza crassirostris*)

418   part **IV** Evolution

Learning Outcomes Review 21.3

In artificial selection, humans choose which plants or animals to mate in an attempt to conserve desirable traits. Rapid and substantial results can be obtained over a very short time, often in a few generations. From this we can see that natural selection is capable of producing major evolutionary change.

■ *In what circumstances might artificial selection fail to produce a desired change?*

The Learning Continues Online

The questions in Connect are tagged to the Learning Outcomes in the text so that the online assignments and tests can be more closely correlated with the material in the textbook. The online eBook in McGraw-Hill ConnectPlus™ provides students with clear understanding of concepts through a media-rich experience. Embedded animations bring key concepts to life. Also, the eBook provides an interactive experience with the Learning Outcome Review questions.

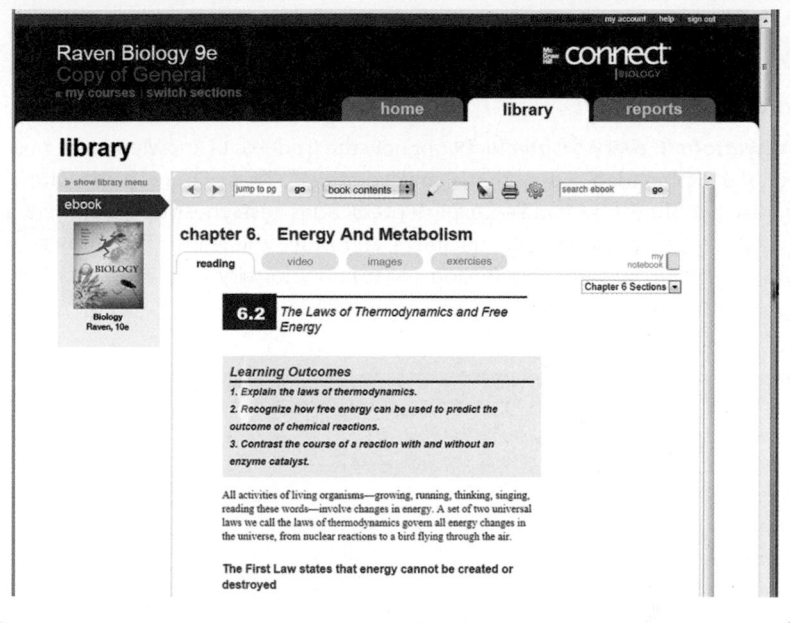

NEW! Evolutionary Asides are inserted at relevant places in the eBook. The student links to this online content with the Evolutionary Aside eBook icon found in the text. The Evolutionary Asides provide additional examples or discussions of evolutionary topics related to the textual discussion.

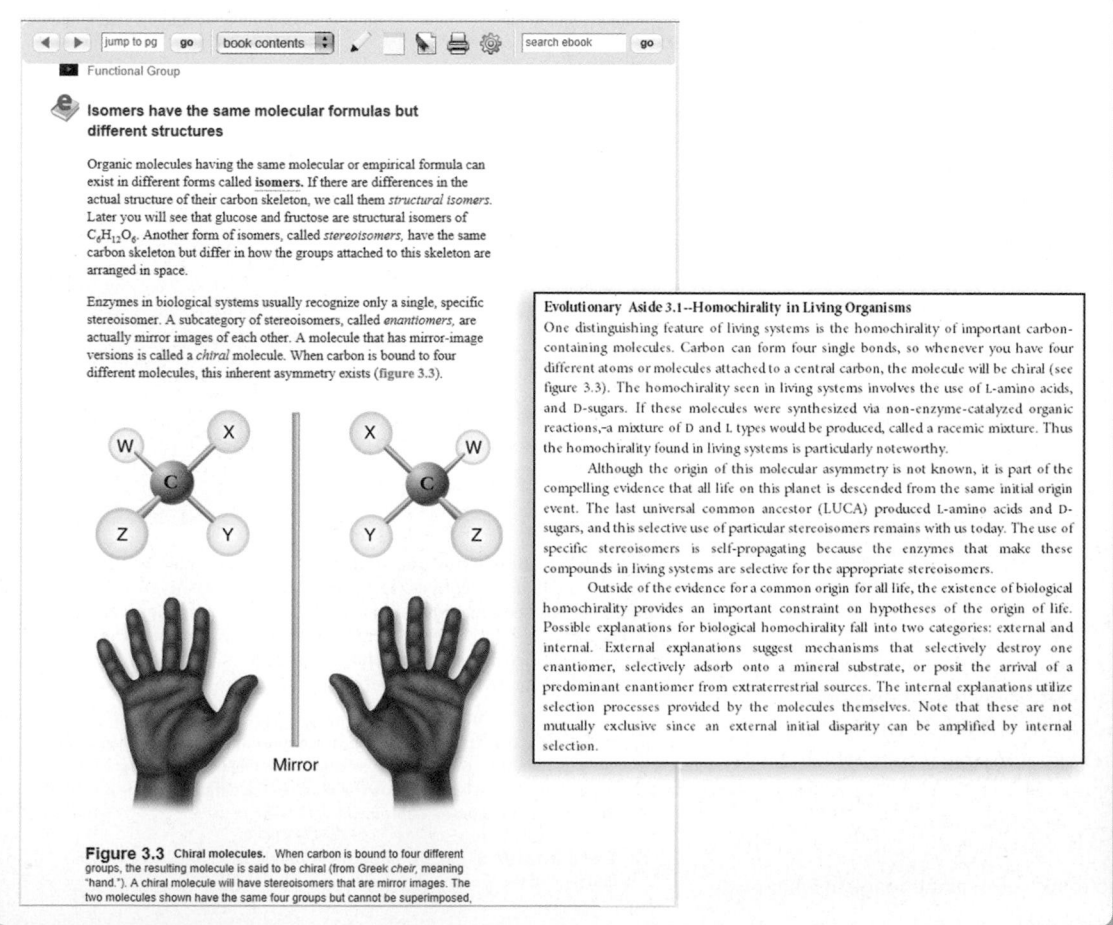

Figure 3.3 Chiral molecules. When carbon is bound to four different groups, the resulting molecule is said to be chiral (from Greek *cheir*, meaning "hand."). A chiral molecule will have stereoisomers that are mirror images. The two molecules shown have the same four groups but cannot be superimposed.

Apply Your Knowledge With...

Scientific Thinking Art

Key illustrations in every chapter highlight how the frontiers of knowledge are pushed forward by a combination of hypothesis and experiment. These figures begin with a hypothesis, then show how it makes explicit predictions, tests these by experiment and finally demonstrates what conclusions can be drawn, and where this leads. These provide a consistent framework to guide the student in the logic of scientific inquiry. Each illustration concludes with open-ended questions to promote scientific inquiry.

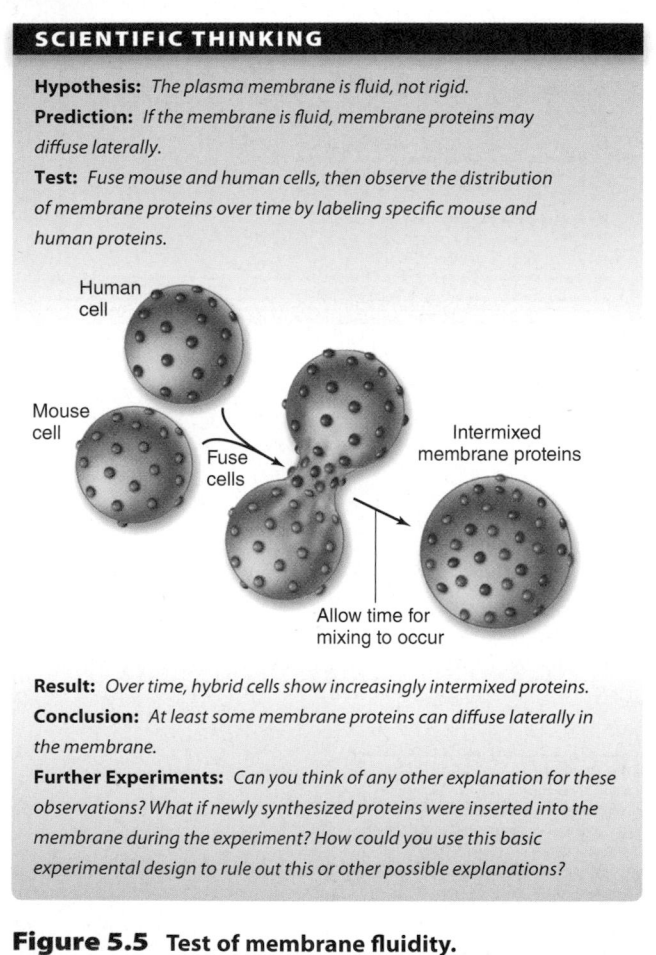

SCIENTIFIC THINKING

Hypothesis: *The plasma membrane is fluid, not rigid.*

Prediction: *If the membrane is fluid, membrane proteins may diffuse laterally.*

Test: *Fuse mouse and human cells, then observe the distribution of membrane proteins over time by labeling specific mouse and human proteins.*

Human cell

Mouse cell

Fuse cells

Intermixed membrane proteins

Allow time for mixing to occur

Result: *Over time, hybrid cells show increasingly intermixed proteins.*

Conclusion: *At least some membrane proteins can diffuse laterally in the membrane.*

Further Experiments: *Can you think of any other explanation for these observations? What if newly synthesized proteins were inserted into the membrane during the experiment? How could you use this basic experimental design to rule out this or other possible explanations?*

Figure 5.5 **Test of membrane fluidity.**

> Knowing how scientists solve problems, and then using this knowledge to solve a problem (as an example) drives home the concept of induction and deduction — I applaud this highly!
>
> *Marc LaBella*
> *Ocean County College*

NEW! Data Analysis Questions

It's not enough that students learn concepts and memorize scientific facts, a biologist needs to analyze data and apply that knowledge. Data Analysis questions inserted throughout the text challenge students to analyze data and Interpret experimental results, which shows a deeper level of understanding.

Inquiry Questions

Questions that challenge students to think about and engage in what they are reading at a more sophisticated level.

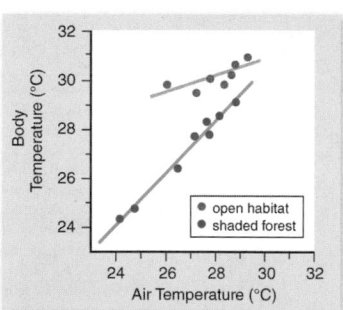

Figure 55.3 **Behavioral adaptation.** In open habitats, the Puerto Rican crested lizard *(Anolis cristatellus)* maintains a relatively constant temperature by seeking out and basking in patches of sunlight; as a result, it can maintain a relatively high temperature even when the air is cool. In contrast, in shaded forests, this behavior is not possible, and the lizard's body temperature conforms to that of its surroundings.

? **Inquiry question** When given the opportunity, lizards regulate their body temperature to maintain a temperature optimal for physiological functioning. Would lizards in open habitats exhibit different escape behaviors from lizards in shaded forest?

Data analysis Can the slope of the line tell us something about the behavior of the lizard?

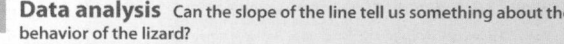

NEW! Quantitative Question Bank in Connect®

Developing quantitative reasoning skills is important to the success of today's students. In addition to the Question Bank and Test Bank in Connect a separate bank of quantitative questions is readily available for seamless use in homework/practice assignments, quizzes, and exams. These algorithmic-style questions provide an opportunity for students to more deeply explore quantitative concepts and to experience repeated practice that enables quantitative skill building over time.

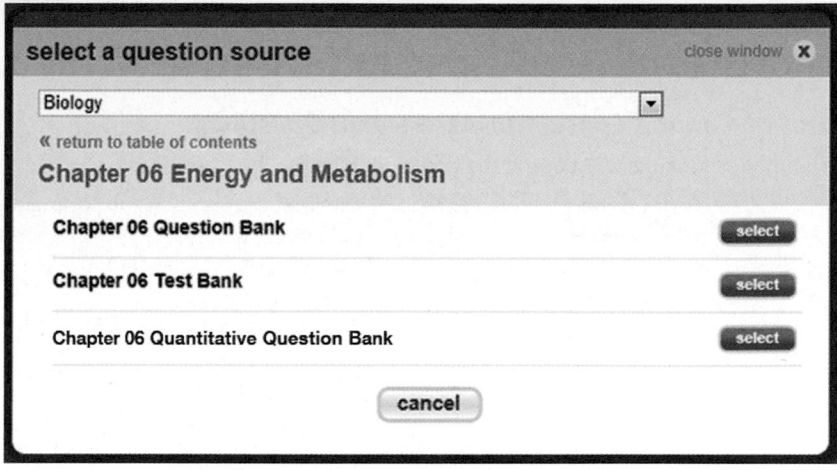

NEW! Quantitative Asides In the eBook

Quantitative Asides are inserted at relevant places in the eBook. The student links to this online content with the Quantitative Aside eBook icon found in the text. The Quantitative Asides provide additional examples or expanded discussions of a quantitative aspect of the topic under discussion.

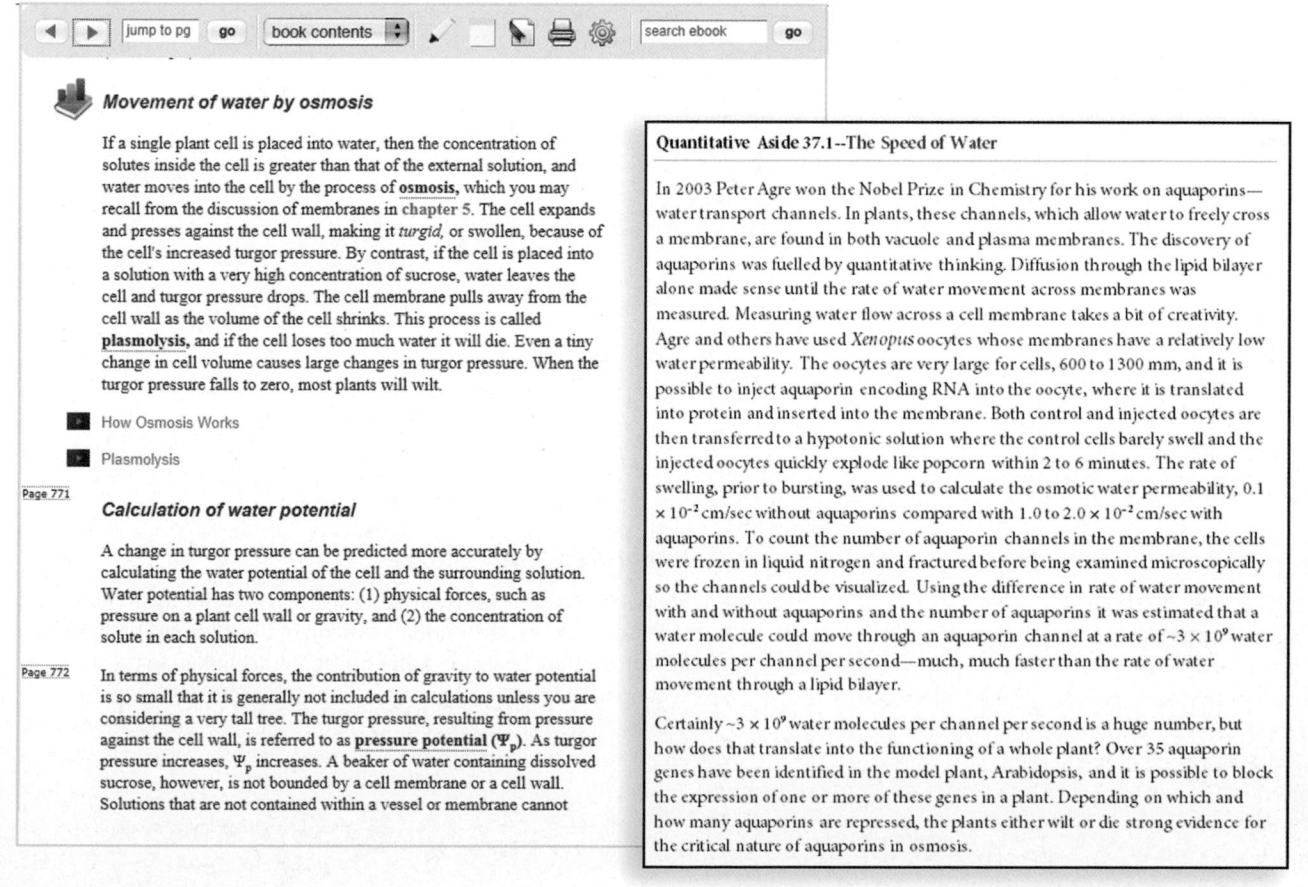

Synthesize and Tie It All Together With . . .

End-of-Chapter Conceptual Assessment Questions

Thought-provoking questions at the end of each chapter tie the concepts together by asking the student to go beyond the basics to achieve a higher level of cognitive thinking.

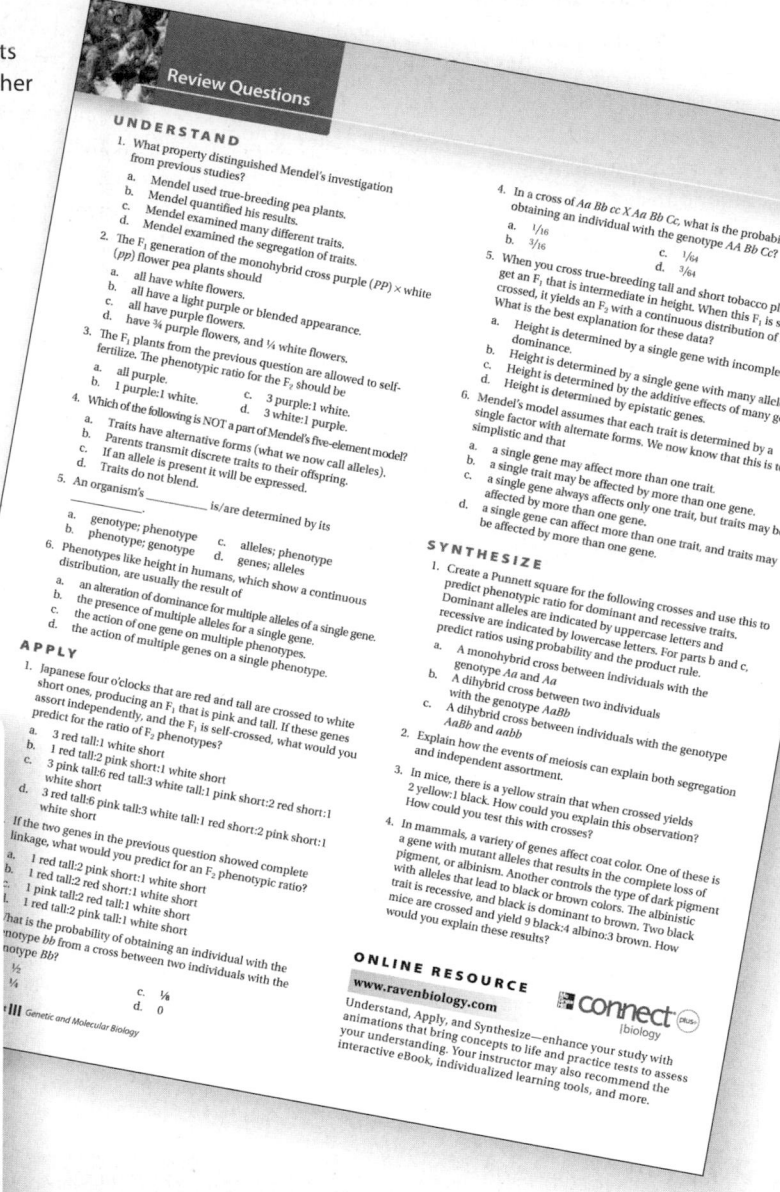

Review Questions

UNDERSTAND

1. What property distinguished Mendel's investigation from previous studies?
   a. Mendel used true-breeding pea plants.
   b. Mendel quantified his results.
   c. Mendel examined many different traits.
   d. Mendel examined the segregation of traits.
2. The F_1 generation of the monohybrid cross purple (PP) × white (pp) flower pea plants should
   a. all have white flowers.
   b. all have a light purple or blended appearance.
   c. all have purple flowers.
   d. have ¾ purple flowers, and ¼ white flowers.
3. The F_1 plants from the previous question are allowed to self-fertilize. The phenotypic ratio for the F_2 should be
   a. all purple.
   b. 1 purple:1 white.
   c. 3 purple:1 white.
   d. 3 white:1 purple.
4. Which of the following is NOT a part of Mendel's five-element model?
   a. Traits have alternative forms (what we now call alleles).
   b. Parents transmit discrete traits to their offspring.
   c. If an allele is present it will be expressed.
   d. Traits do not blend.
5. An organism's _____ is/are determined by its _____.
   a. genotype; phenotype
   b. phenotype; genotype
   c. alleles; phenotype
   d. genes; alleles
6. Phenotypes like height in humans, which show a continuous distribution, are usually the result of
   a. an alteration of dominance for multiple alleles of a single gene.
   b. the presence of multiple alleles for a single gene.
   c. the action of one gene on multiple phenotypes.
   d. the action of multiple genes on a single phenotype.

APPLY

1. Japanese four o'clocks that are red and tall are crossed to white short ones, producing an F_1 that is pink and tall. If these genes assort independently, and the F_1 is self-crossed, what would you predict for the ratio of F_2 phenotypes?
   a. 3 red tall:1 white short
   b. 1 red tall:2 pink short:1 white short
   c. 3 pink tall:6 red tall:3 white tall:1 pink short:2 red short:1 white short
   d. 3 red tall:6 pink tall:3 white tall:1 red short:2 pink short:1 white short
2. If the two genes in the previous question showed complete linkage, what would you predict for an F_2 phenotypic ratio?
   a. 1 red tall:2 pink short:1 white short
   b. 1 red tall:2 red short:1 white short
   c. 1 pink tall:2 red short:1 white short
   d. 1 red tall:2 pink tall:1 white short
3. What is the probability of obtaining an individual with the genotype bb from a cross between two individuals with the genotype Bb?
   a. ½
   b. ¼
   c. ⅛
   d. 0

4. In a cross of $Aa\ Bb\ cc \times Aa\ Bb\ Cc$, what is the probability of obtaining an individual with the genotype $AA\ Bb\ Cc$?
   a. 1/16
   b. 3/16
   c. 1/64
   d. 3/64
5. When you cross true-breeding tall and short tobacco plants you get an F_1 that is intermediate in height. When this F_1 is self-crossed, it yields an F_2 with a continuous distribution of heights. What is the best explanation for these data?
   a. Height is determined by a single gene with incomplete dominance.
   b. Height is determined by a single gene with many alleles.
   c. Height is determined by the additive effects of many genes.
   d. Height is determined by epistatic genes.
6. Mendel's model assumes that each trait is determined by a single factor with alternate forms. We now know that this is too simplistic and that
   a. a single gene may affect more than one trait.
   b. a single trait may be affected by more than one gene.
   c. a single gene always affects only one trait, but traits may be affected by more than one gene.
   d. a single gene can affect more than one trait, and traits may be affected by more than one gene.

SYNTHESIZE

1. Create a Punnett square for the following crosses and use this to predict phenotypic ratio for dominant and recessive traits. Dominant alleles are indicated by uppercase letters and recessive are indicated by lowercase letters. For parts b and c, predict ratios using probability and the product rule.
   a. A monohybrid cross between individuals with the genotype Aa and Aa
   b. A dihybrid cross between two individuals with the genotype $AaBb$
   c. A dihybrid cross between individuals with the genotype $AaBb$ and $aabb$
2. Explain how the events of meiosis can explain both segregation and independent assortment.
3. In mice, there is a yellow strain that when crossed yields 2 yellow:1 black. How could you explain this observation? How could you test this with crosses?
4. In mammals, a variety of genes affect coat color. One of these is a gene with mutant alleles that results in the complete loss of pigment, or albinism. Another controls the type of dark pigment with alleles that lead to black or brown colors. The albinistic trait is recessive, and black is dominant to brown. Two black mice are crossed and yield 9 black:4 albino:3 brown. How would you explain these results?

ONLINE RESOURCE
www.ravenbiology.com
Understand, Apply, and Synthesize—enhance your study with animations that bring concepts to life and practice tests to assess your understanding. Your instructor may also recommend the interactive eBook, individualized learning tools, and more.

connect |biology

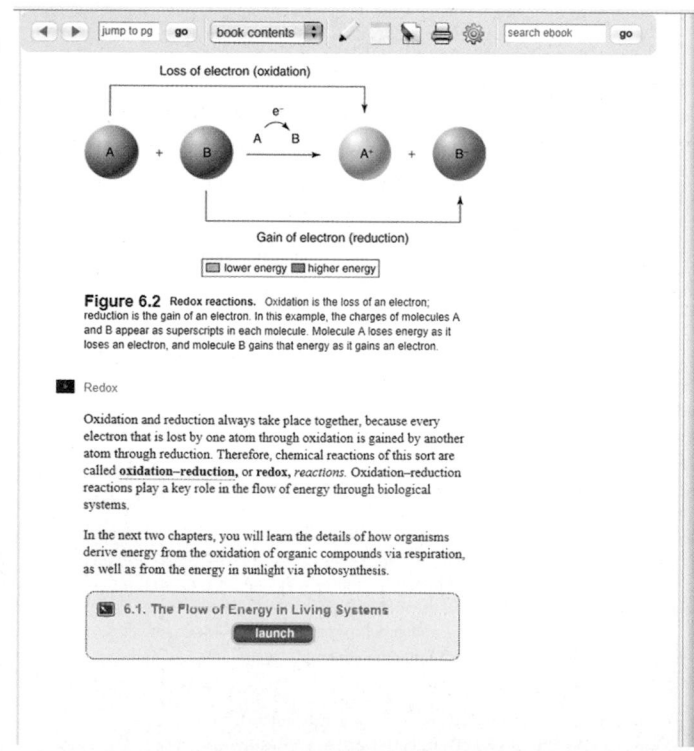

◀ ▶ | jump to pg | go | book contents ▾ | ✎ | ▭ | 🖊 | 🖨 | ⚙ | search ebook | go

Loss of electron (oxidation)

A + B → A⁺ + B⁻

Gain of electron (reduction)

▢ lower energy ▢ higher energy

Figure 6.2 Redox reactions. Oxidation is the loss of an electron; reduction is the gain of an electron. In this example, the charges of molecules A and B appear as superscripts in each molecule. Molecule A loses energy as it loses an electron, and molecule B gains that energy as it gains an electron.

■ Redox

Oxidation and reduction always take place together, because every electron that is lost by one atom through oxidation is gained by another atom through reduction. Therefore, chemical reactions of this sort are called **oxidation–reduction**, or **redox**, *reactions*. Oxidation–reduction reactions play a key role in the flow of energy through biological systems.

In the next two chapters, you will learn the details of how organisms derive energy from the oxidation of organic compounds via respiration, as well as from the energy in sunlight via photosynthesis.

▢ 6.1. The Flow of Energy in Living Systems
launch

Integrated Study Quizzes

Study quizzes have been integrated into the ConnectPlus eBook for students to assess their understanding of the information presented in each section. End-of-chapter questions are linked to the answer section of the text to provide for easy study. The notebook feature allows students to collect and manage notes and highlights from the eBook to create a custom study guide.

Committed to Biology Educators

McGraw-Hill Higher Education and Blackboard Have Teamed Up

Do More

Blackboard®, the Web-based course management system, has partnered with McGraw-Hill to better allow students and faculty to use online materials and activities to complement face-to-face teaching. Blackboard features exciting social learning and teaching tools that foster more logical, visually impactful, and active learning opportunities for students. You'll transform your closed-door classrooms into communities where students remain connected to their educational experience 24 hours a day.

This partnership allows you and your students access to McGraw-Hill's Connect and McGraw-Hill Create™ right from within your Blackboard course—all with one single sign-on. Not only do you get single sign-on with Connect and Create, you also get deep integration of McGraw-Hill content and content engines right in Blackboard. Whether you're choosing a book for your course or building Connect assignments, all the tools you need are right where you want them—inside of Blackboard.

Gradebooks are now seamless. When a student completes an integrated Connect assignment, the grade for that assignment automatically (and instantly) feeds your Blackboard grade center.

McGraw-Hill and Blackboard can now offer you easy access to industry leading technology and content, whether your campus hosts it or we do. Be sure to ask your local McGraw-Hill representative for details.

McGraw-Hill Connect® Biology

McGraw-Hill Connect Biology provides online presentation, assignment, and assessment solutions. It connects your students with the tools and resources they'll need to achieve success. With Connect Biology you can deliver assignments, quizzes, and tests online. A robust set of questions and activities are presented and aligned with the textbook's Learning Outcomes. As an instructor, you can edit existing questions and author entirely new problems. Track individual student performance—by question, assignment, or in relation to the class overall—with detailed grade reports. Integrate grade reports easily with Learning Management Systems (LMS), such as WebCT and Blackboard—and much more. ConnectPlus Biology provides students with all the advantages of Connect Biology plus 24/7 online access to an eBook. This media-rich version of the book is available through the McGraw-Hill Connect platform and allows seamless integration of text, media, and assessments.

To learn more, visit **www.mcgrawhillconnect.com**

My Lectures—Tegrity®

McGraw-Hill Tegrity records and distributes your class lecture with just a click of a button. Students can view anytime/anywhere via computer, iPod, or mobile device. It indexes as it records your PowerPoint® presentations and anything shown on your computer so students can use keywords to find exactly what they want to study. Tegrity is available as an integrated feature of McGraw-Hill Connect Biology and as a standalone.

Personalized and Adaptive Learning

LearnSmart™ **McGraw-Hill LearnSmart** is available as an integrated feature of McGraw-Hill Connect Biology. It is an adaptive learning system designed to help students learn faster, study more efficiently, and retain more knowledge for greater success. LearnSmart assesses a student's knowledge of course content through a series of adaptive questions. It pinpoints concepts the student does not understand and maps out a personalized study plan for success. This innovative study tool also has features that allow instructors to see exactly what students have accomplished and a built-in assessment tool for graded assignments.

Visit the following site for a demonstration: **www.mhlearnsmart.com**

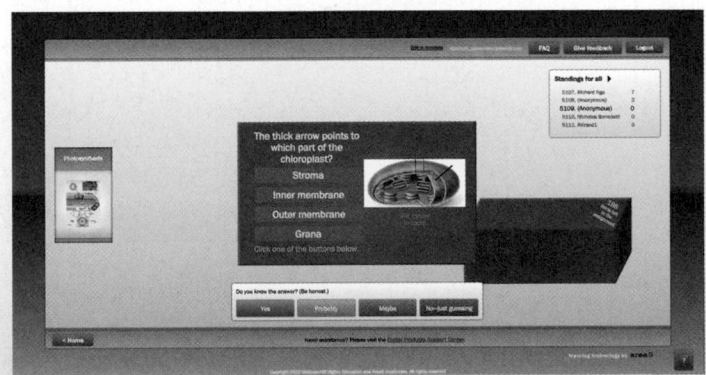

 **LabSmart**™ Based on the same world-class super-adaptive technology as LearnSmart, McGraw-Hill LabSmart is a must-see, outcomes-based lab simulation. It assesses a student's knowledge and adaptively corrects deficiencies, allowing the student to learn faster and retain more knowledge with greater success.

First, a student's knowledge is adaptively leveled on core learning outcomes: Questioning reveals knowledge deficiencies that are corrected by the delivery of content that is conditional on a student's response. Then, a simulated lab experience requires the student to think and act like a scientist: recording, interpreting, and analyzing data using simulated equipment found in labs and clinics. The student is allowed to make mistakes—a powerful part of the learning experience! A virtual coach provides subtle hints when needed; asks questions about the student's choices; and allows the student to reflect upon and correct those mistakes. Whether your need is to overcome the logistical challenges of a traditional lab, provide better lab prep, improve student performance, or make your online experience one that rivals the real world, LabSmart accomplishes it all.

Learn more at **www.mhlabsmart.com**

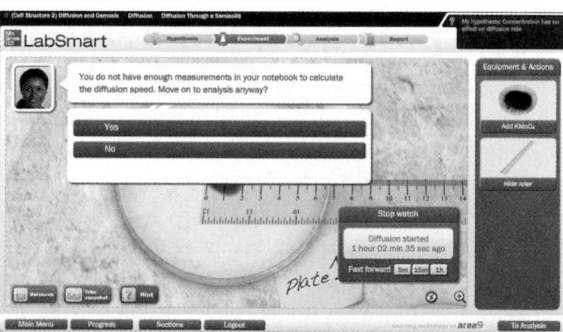

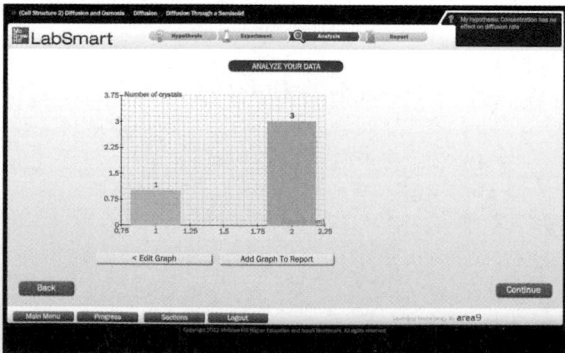

Preparing for Majors Biology

Do your majors biology students struggle the first few weeks of class, trying to get up to speed? McGraw-Hill can help.

McGraw-Hill has developed an adaptive learning tool designed to increase student success and aid retention through the first few weeks of class. Using this digital tool majors biology students can master some of the most fundamental and challenging principles of biology before they being to struggle in the first few weeks of class.

An initial diagnostic establishes a student's baseline comprehension and knowledge; then the program generates a learning plan tailored to the student's academic needs and schedule. As the student works through the learning plan, the program tracks the student's progress, delivering appropriate assessment and learning resources (e.g., tutorials, figures, animations, etc.) as needed. If a student incorrectly answers questions around a particular learning objective, they are asked to review learning resources around that objective before re-assessing their mastery of the objective.

Using this program, students can identify the content they don't understand, focus their time on content they need to know but don't, and therefore improve their chances of success in the majors biology course.

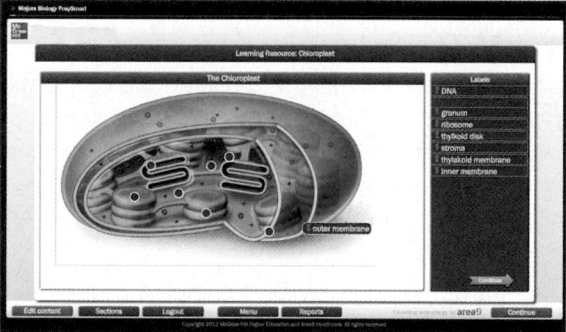

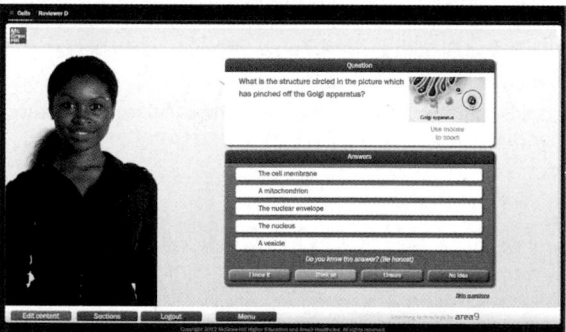

Powerful Presentation Tools

Everything you need for outstanding presentation in one place!

- FlexArt Image PowerPoints—including every piece of art that has been sized and cropped specifically for superior presentations as well as labels that you can edit, flexible art that can be picked up and moved, tables and photographs.
- Animation PowerPoints—Numerous full-color animations illustrating important processes. Harness the visual impact of concepts in motion by importing these slides into classroom presentations or online course materials.
- Lecture PowerPoints with animations fully embedded.
- Labeled and unlabeled JPEG images— Full-color digital files of all illustrations that can be readily incorporated into presentations, exams, or custom-made classroom materials.

Fully Developed Test Bank

The Digital Team (see page xviii) revised the Test Bank to fully align with the Learning Outcomes and complement questions written for the Question Bank intended for homework assignments. A thorough review process has been implemented to ensure accuracy. Provided within a computerized Test Bank powered by McGraw-Hill's flexible electronic testing program EZ Test Online, instructors can create paper and online tests or quizzes in this easy to use program! A tagging scheme allows you to sort questions by Learning Outcome, Bloom's level, topic, and section. Now, with EZ Test Online, instructors can select questions from multiple McGraw-Hill Test Banks or author their own, and then either print the test for paper distribution or give it online.

Contributors for other digital assets:

FlexArt Image PowerPoints—Carla Reinstadtler, *freelance content expert*

Lecture PowerPoints—Brian Shmaefsky, *Lone Star College*

eBook Quizzes—Amanda Rosenzweig, *Delgado Community College;* **Scott Cooper,** *University of Wisconsin–LaCrosse;* **Steven Clark,** *Lake-Sumter Community College;* **Lisa Bonneau,** *Mount Marty College*

Website—Kathleen Broomall, *University of Cincinnati–Clermont College* and **Carla Reinstadtler,** *freelance content expert*

LearnSmart—Lead: Peter Kourtev, *Central Michigan University* **Authors and reviewers: Isaac Barjis,** *New York City College of Technology;* **Anne Bullerjahn,** *Owens Community College;* **Elizabeth Drumm,** *Oakland Community College–Orchard Ridge Campus;* **Shelley Jansky,** *University of Wisconsin–Madison;* **Rita King,** *The College of New Jersey;* **Michelle Pass,** *University of North Carolina–Charlotte;* **Jennifer Warner,** *University of North Carolina, Charlotte*

Flexible Delivery Options

Raven et al. *Biology* is available in many formats in addition to the traditional textbook to give instructors and students more choices when deciding on the format of their biology text.

Foundations of Life—Chemistry, Cells, and Genetics
ISBN: 007-777580-5
Parts *1, 2, and 3*

Evolution, Diversity and Ecology
ISBN: 007-777581-3
Parts 4, 5, and 8

Plants and Animals
ISBN: 007-777582-1
Parts 6 and 7

Also available, customized versions for all of your course needs. You're in charge of your course, so why not be in control of the content of your textbook? At McGraw-Hill Custom Publishing, we can help you create the ideal text—the one you've always imagined. Quickly. Easily. With more than 20 years of experience in custom publishing, we're experts. But at McGraw Hill we're also innovators, leading the way with new methods and means for creating simplified value-added custom textbooks.

The options are never-ending when you work with McGraw Hill. You already know what will work best for you and your students. And here, you can choose it.

McGraw-Hill Create™

With **McGraw-Hill Create,** you can easily rearrange chapters, combine material from other content sources, and quickly upload content you have written, like your course syllabus or teaching notes. Find the content you need in Create by searching through thousands of leading McGraw-Hill textbooks. Arrange your book to fit your teaching style. Create even allows you to personalize your book's appearance by selecting the cover and adding your name, school, and course information. Order a Create book and you'll receive a complimentary print review copy in 3–5 business days or a complimentary electronic review copy (eComp) via e-mail in minutes. Go to www.mcgrawhillcreate.com today and register to experience how McGraw-Hill Create empowers you to teach *your* students *your* way. **www.mcgrawhillcreate.com**

Laboratory Manuals

Biology Laboratory Manual, Tenth Edition
Vodopich/Moore ISBN: 0-07-353225-8

This laboratory manual is designed for an introductory majors biology course with a broad survey of basic laboratory techniques. The experiments and procedures are simple, safe, easy to perform, and especially appropriate for large classes. Few experiments require a second class-meeting to complete the procedure. Each exercise includes many photographs, traditional topics, and experiments that help students  learn about life. Procedures within each exercise are numerous and discrete so that an exercise can be tailored to the needs of the students, the style of the instructor, and the facilities available.

Biological Investigations Lab Manual, Ninth Edition
Dolphin ISBN: 0-07-338305-8

This independent lab manual can be used for a one- or two-semester majors-level general biology lab and can be used with any majors-level general biology textbook. The labs are investigative and ask students to use more critical thinking and hands-on learning. The author emphasizes investigative, quantitative, and comparative approaches to studying the life sciences.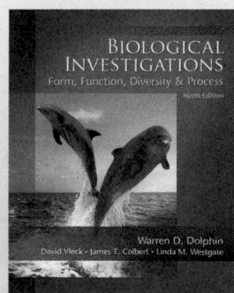

The Digital Story

Digital assessment is a major focus in higher education. Online tools promise anywhere, anytime access combined with the possibility of learning tailored to individual student needs. Digital assessments should span the spectrum of Bloom's taxonomy within the context of best-practice pedagogy. The increased challenge at higher Bloom's levels will help students grow intellectually and be better prepared to contribute to society.

Significant faculty demand for content at higher Bloom's levels led us to examine assessment quality and consistency of our Connect content, and to develop a scientific approach to systematically increase Bloom's levels and develop internally consistent and balanced digital assessments that promote student learning.

Our goal was to increase assessment quality of our Connect content to meet faculty and student needs. Our objective was to have 30% of all digital assessment questions in Connect at the Apply, Analyze, or Evaluate levels of Bloom's taxonomy. With thousands of existing questions, that is no small task. Consistent with best-practices research on how people learn, we took a comprehensive look at our existing digital assessments to determine Bloom's levels across our assignable content. Because this project was too extensive for a single person to accomplish, we assembled a team of faculty from research, comprehensive, liberal arts universities, and community colleges. Digital team members were selected based on commitment to student learning, biology content expertise, openness to a new vision for digital assessment and professional development, and question-writing skills.

Under the direction of lead digital author, Ian Quitadamo, team members were calibrated to a common perception of Bloom's taxonomy. The team then evaluated our existing Question Bank, Test Bank, Animation Quizzes, and Video Quizzes for appropriate level of Bloom's and compiled the results into a comprehensive database that was statistically analyzed. Results showed adequate coverage at the lower level of Bloom's taxonomy but less so at the higher levels of Bloom's. Knowing that assessment drives learning quality, we focused our efforts on "Blooming up" existing content and developing new assessments that examine students' problem-solving skills (see graphs for chapter 12). The end result of our team's scientific approach to developing digital content is a collection of engaging, diagnostic assessments that strengthen student ability to critically think, build connections across biology concepts, and develop quantitative reasoning skills that ultimately underlie student academic success and ability to contribute to society.

We would like to acknowledge our digital team and thank them for their tireless efforts:

Kerry Bohl, *University of South Florida*
David Bos, *Purdue University*
Scott Bowling, *Auburn University*
Scott Cooper, *University of Wisconsin–La Crosse*
Cynthia Dadmun, *Freelance content expert*
Jenny Dechaine, *Central Washington University*
Elizabeth Drumm, *Oakland Community College–Orchard Ridge Campus*
Susan Edwards, *Appalachian State University*

Julie Emerson, *Amherst College*
Brent Ewers, *University of Wyoming*
Chris Himes, *Massachusetts College of Liberal Arts*
Cintia Hongay, *Clarkson University*
Heather Jezorek, *University of South Florida*
Kristy Kappenman, *Central Washington University*
Jamie Kneitel, *California State University, Sacramento*
Marcy Lowenstein, *Florida International University*
Carolyn Martineau, *DePaul University*
Christin Munkittrick, *Freelance content expert*
Chris Osovitz, *University of South Florida*
Anneke Padolina, *Virginia Commonwealth University*
Marius Pfeiffer, *Tarrant County College*
Marceau Ratard, *Delgado Community College*
Nicolle Romero, *Freelance content expert*
Amanda Rosenzweig, *Delgado Community College*
Kathryn Spilios, *Boston University*
Jen Stanford, *Drexel University*
Martin St. Maurice, *Marquette University*
Salvatore Tavormina, *Austin Community College*
Sharon Thoma, *University of Wisconsin, Madison*
Gloriana Trujillo, *University of New Mexico*
Jennifer Wiatrowski, *Pasco-Hernando Community College*

Quantitative Question Bank

David Bos, *Purdue University*
Chris Osovitz, *University of South Florida*
Martin St. Maurice, *Marquette University*

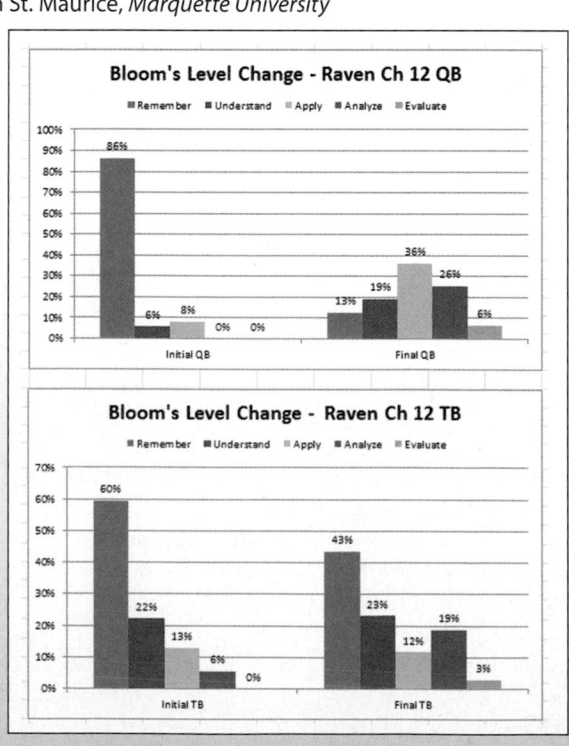

A Step Ahead in Quality

360° Development Process

McGraw-Hill's 360° Development Process is an ongoing, never-ending, education-oriented approach to building accurate and innovative print and digital products. It is dedicated to continual large-scale and incremental improvement, driven by multiple user feedback loops and checkpoints. This is initiated during the early planning stages of our new products, and intensifies during the development and production stages, then begins again upon publication in anticipation of the next edition.

This process is designed to provide a broad, comprehensive spectrum of feedback for refinement and innovation of our learning tools, for both student and instructor. The 360° Development Process includes market research, content reviews, course- and product-specific symposia, accuracy checks, and art reviews. We appreciate the expertise of the many individuals involved in this process.

General Biology Symposia

Every year McGraw-Hill conducts several General Biology Symposia, which are attended by instructors from across the country. These events are an opportunity for editors from McGraw-Hill to gather information about the needs and challenges of instructors teaching the majors biology course. It also offers a forum for the attendees to exchange ideas and experiences with colleagues they might not have otherwise met. The feedback we have received has been invaluable, and has contributed to the development of *Biology* and its supplements. A special thank you to recent attendees:

Thomas Abbott   *University of Connecticut*
Sylvester Allred   *Northern Arizona University*
Julie Anderson   *University of Wisconsin–Eau Claire*
Kim Baker   *University of Wisconsin–Green Bay*
Michael Bell   *Richland College*
Brian Berthelsen   *Iowa Western Community College*
Joe Beuchel   *Triton College*
Arlene Billock   *University of Louisiana–Lafayette*
Stephane Boissinot   *Queens College, the City University of New York*
David Bos   *Purdue University*
Scott Bowling   *Auburn University*
Jacqueline Bowman   *Arkansas Technical University*
Randy Brooks   *Florida Atlantic University*
Arthur Buikema   *Virginia Polytechnic Institute*
Anne Bullerjahn   *Owens Community College*
Helaine Burstein   *Ohio University*
Raymond Burton   *Germanna Community College*

Peter Busher   *Boston University*
Ruth Buskirk   *University of Texas–Austin*
Richard Cardullo   *University of California–Riverside*
Frank Cantelmo   *St. Johns University*
Jennifer Ciaccio   *Dixie State College*
Anne Barrett Clark   *Binghamton University*
Allison Cleveland   *University of South Florida–Tampa*
Clark Coffman   *Iowa State University*
Jennifer Coleman   *University of Massachusetts–Amherst*
Sehoya Cotner   *University of Minnesota*
Mitch Cruzan   *Portland State University*
Karen A. Curto   *University of Pittsburgh*
Rona Delay   *University of Vermont*
Mary Dettman   *Seminole State College of Florida*
Laura DiCaprio   *Ohio University*
Kathyrn Dickson   *California State College–Fullerton*
Cathy Donald-Whitney   *Collin County Community College*
Moon Draper   *University of Texas–Austin*

Tod Duncan   *University of Colorado–Denver*
Brent Ewers   *University of Wyoming*
Stanley Faeth   *Arizona State University*
Michael Ferrari   *University of Missouri–Kansas City*
David Fitch   *New York University*
Donald French   *Oklahoma State University*
Douglas Gaffin   *University of Oklahoma*
John Geiser   *Western Michigan University*
Karen Gerhart   *University of California–Davis*
Julie Gibbs   *College of DuPage*
Cynthia Giffen   *University of Wisconsin–Madison*
Sharon Gill   *Western Michigan University*
William Glider   *University of Nebraska–Lincoln*
Steven Gorsich   *Central Michigan University*
Christopher Gregg   *Louisiana State University*
Stan Guffey   *The University of Tennessee*
Sally Harmych   *University of Toledo*
Bernard Hauser   *University of Florida–Gainesville*
Jean Heitz   *Unversity of Wisconsin–Madison*
Mark Hens   *University of North Carolina–Greensboro*
Albert Herrera   *University of Southern California*
Ralph James Hickey   *Miami University of Ohio–Oxford*
Jodi Huggenvik   *Southern Illinois University–Carbondale*
Brad Hyman   *University of California–Riverside*
Rick Jellen   *Brigham Young University*
Michael Kempf   *University of Tennessee–Martin*
Kyoungtae Kim   *Missouri State University*
Sherry Krayesky   *University of Louisiana–Lafayette*
Jerry Kudenov   *University of Alaska–Anchorage*
Josephine Kurdziel   *University of Michigan*
Ellen Lamb   *University of North Carolina–Greensboro*
Brenda Leady   *University of Toledo*
Graeme Lindbeck   *Valencia Community College*
David Longstreth   *Louisiana State University*
Lucile McCook   *University of Mississippi*
Susan Meiers   *Western Illinois University*
Michael Meighan   *University of California–Berkeley*
John Merrill   *Michigan State University*
John Mersfelder   *Sinclair Community College*
Melissa Michael   *University of Illinois–Urbana-Champaign*
Michelle Mynlieff   *Marquette University*
Leonore Neary   *Joliet Junior College*
Shawn Nordell   *Saint Louis University*

John Osterman   *University of Nebraska–Lincoln*
Stephanie Pandolfi   *Michigan State University*
Anneke Padolina   *Virginia Commonwealth University*
C.O. Patterson   *Texas A&M University*
Nancy Pencoe   *University of West Georgia*
Roger Persell   *Hunter College*
Marius Pfeiffer   *Tarrant County College NE*
Steve Phelps   *University of Florida*
Debra Pires   *University of California–Los Angeles*
Thomas Pitzer   *Florida International University*
Steven Pomarico   *Louisiana State University*
Jo Anne Powell-Coffman   *Iowa State University*
Lynn Preston   *Tarrant County College*
Ian Quitadamo   *Central Washington University*
Rajinder Ranu   *Colorado State University*
Marceau Ratard   *Delgado Community College–City Park*
Melanie Rathburn   *Boston University*
Robin Richardson   *Winona State University*
Mike Robinson   *University of Miami*
Amanda Rosenzweig   *Delgado Community College–City Park*
Connie Russell   *Angelo State University*
Laurie Russell   *St. Louis University*
David Scicchitano   *New York University*
Timothy Shannon   *Francis Marion University*
Brian Shmaefsky   *Lone Star College–Kingwood*
Richard Showman   *University of South Carolina*
Allison Silveus   *Tarrant County College–Trinity River Campus*
Robert Simons   *University of California–Los Angeles*
Steve Skarda   *Linn Benton Community College*
Steven D. Skopik   *University of Delaware*
Phillip Sokolove   *University of Maryland–Baltimore County*
Martin St. Maurice   *Marquette University*
Brad Swanson   *Cental Michigan University*
David Thompson   *Northern Kentucky University*
Maureen Tubbiola   *St. Cloud State University*
Ashok Upadhyaya   *University of South Florida–Tampa*
Anthony Uzwiak   *Rutgers University*
Rani Vajravelu   *University of Central Florida*
Gary Walker   *Appalachian State University*
Pat Walsh   *University of Delaware*
Elizabeth Weiss-Kuziel   *University of Texas–Austin*
Clay White   *Lone Star College–CyFair*
Leslie Whiteman   *Virginia State University*

Jennifer Wiatrowski *Pasco–Hernando Community College*
David Williams *Valencia Community College, East Campus*

Holly Williams *Seminole Community College*
Michael Windelspecht *Appalachian State University*

Robert Winning *Eastern Michigan University*
Mary Wisgirda *San Jacinto College, South Campus*

Michelle Withers *West Virginia University*
Kevin Wolbach *University of the Sciences in Philadelphia*
Jay Zimmerman *St. John's University*

Tenth Edition Reviewers

Tamarah Adair *Baylor University*
Brian P. Ashburner *University of Toledo*
Suman Batish *Temple University*
Giacomo Bernardi *University of California, Santa Cruz*
Deborah Bielser *University of Illinois*
Helen Boswell *Southern Utah University*
Carolyn J.W. Bunde *Idaho State University*
Joseph C. Bundy, Jr. *The University of North Carolina at Greensboro*
Jason Carlson *St. Cloud Technical and Community College*
Rebekah Chapman *Georgia State University*
Jennifer Ciaccio *Dixie State College*
Hudson DeYoe *University of Texas Pan American*

Elizabeth Drumm *Oakland Community College*
Arundhati Ghosh *University of Pittsburgh*
Jennifer Hatchel *College of Coastal Georgia*
Margaret Horton *University of North Carolina at Greensboro*
David W Jones *Dixie State College of Utah*
Jason Knouft *Saint Louis University*
Ellen S. Lamb *The University of North Carolina at Greensboro*
Brenda Leady *University of Toledo*
Roger Lloyd *College of Coastal Georgia*
Janet Loxterman *Idaho State University*
Susan Mazer *University of California, Santa Barbara*

Bradley G. Mehrtens *University of Illinois at Urbana—Champaign*
Jamie Moon *University of North Florida*
Rajkumar Nathaniel *Nicholls State University*
Julie Nguyen *College of the Canyons*
Judith D. Ochrietor *University of North Florida*
Joanne Odden *Metropolitan State College of Denver*
Monique Ogletree *University of Houston*
Paul Pillitteri *Southern Utah University*
Nicola Plowes *Arizona State University*
Kumkum Prabhakar *Nassau Community College*
Marceau Ratard *Delgado Community College*

Melissa Reedy *University of Illinois at Urbana—Champaign*
Laurel Roberts *University of Pittsburgh*
Amanda Rosenzweig *Delgado Community College*
Benjamin Rowley *University of Central Arkansas*
Laurie Shornick *Saint Louis University*
Sonia Suri *Valencia Community College*
John-David Swanson *University of Central Arkansas*
Maureen Walter *Florida International University*
Chad Wayne *University of Houston*
Stacey Wild *East Tennessee State University*
Rebecca Yeomans *College of Coastal Georgia*

Previous Edition Reviewers

Tamarah Adair *Baylor University*
Gladys Alexandre-Jouline *University of Tennessee at Knoxville*
Gregory Andraso *Gannon University*
Jorge E. Arriagada *St. Cloud State University*
David Asch *Youngstown State University*
Jeffrey G. Baguley *University of Nevada – Reno*
Suman Batish *Temple University*
Donald Baud *University of Memphis*
Peter Berget *Carnegie Mellon University*
Randall Bernot *Ball State University*
Deborah Bielser *University of Illinois–Champaign*
Wendy Binder *Loyola Marymount University*
Todd A. Blackledge *University of Akron*
Andrew R. Blaustein *Oregon State University*
Dennis Bogyo *Valdosta State University*
David Bos *Purdue University*
Robert Boyd *Auburn University*
Graciela Brelles-Marino *California State Polytechnic University–Pomona*
Joanna Brooke *DePaul University*
Roxanne Brown *Blinn College*
Mark Browning *Purdue University*
Cedric O. Buckley *Jackson State University*
Arthur L. Buikema, Jr. *Virginia Tech*
Sharon Bullock *UNC – Charlotte*
Lisa Burgess *Broward College*
Scott Carlson *Luther College*
John L. Carr *University of Louisiana – Monroe*
Laura Carruth *Georgia State University*
Dale Cassamatta *University of North Florida*
Peter Chabora *Queens College–CUNY*
Tien-Hsien Chang *Ohio State University*
Genevieve Chung *Broward College*

Cynthia Church *Metropolitan State College of Denver*
William Cohen *University of Kentucky*
James Collins *Kilgore College*
Joanne Conover *University of Connecticut*
Iris Cook *Westchester Community College*
Erica Corbett *Southeastern Oklahoma State University*
Robert Corin *College of Staten Island – CUNY*
William G. R. Crampton *University of Central Florida*
Scott Crousillac *Louisiana State University–Baton Rouge*
Karen A. Curto *University of Pittsburgh*
Denise Deal *Nassau Community College*
Philias Denette *Delgado Community College*
Mary Dettman *Seminole Community College–Oviedo*
Ann Marie DiLorenzo *Montclair State University*
Ernest DuBrul *University of Toledo*
Richard Duhrkopf *Baylor University*
Susan Dunford *University of Cincinnati*
Andrew R. Dyer *University of South Carolina – Aiken*
Carmen Eilertson *Georgia State University*
Richard P. Elinson *Duquesne University*
William L. Ellis *Pasco-Hernando Community College*
Seema Endley *Blinn College*
Gary Ervin *Mississippi State University*
Karl Fath *Queens College–CUNY*
Zen Faulkes *The University of Texas– Pan American*
Myriam Feldman *Lake Washington Technical College*

Melissa Fierke *State University of New York*
Gary L. Firestone *University of California–Berkeley*
Jason Flores *UNC–Charlotte*
Markus Friedrich *Wayne State University*
Deborah Garrity *Colorado State University*
Christopher Gee *University of North Carolina-Charlotte*
John R. Geiser *Western Michigan University*
J.P. Gibson *University of Oklahoma*
Matthew Gilg *University of North Florida*
Teresa Golden *Southeastern Oklahoma State University*
Venkat Gopalan *Ohio State University*
Michael Groesbeck *Brigham Young University*
Theresa Grove *Valdosta State University*
David Hanson *University of New Mexico*
Paul Hapeman *University of Florida*
Nargess Hassanzadeh-Kiabi *California State University–Los Angeles*
Stephen K. Herbert *University of Wyoming*
Hon Ho *State University of New York at New Paltz*
Barbara Hunnicutt *Seminole Community College*
Steve Huskey *Western Kentucky University*
Cynthia Jacobs *Arkansas Tech University*
Jason B. Jennings *Southwest Tennessee Community College*
Frank J. Jochem *Florida International University–Miami*
Norman Johnson *University of Massachusetts*

Gregory A. Jones *Santa Fe Community College*
Jerry Kaster *University of Wisconsin–Milwaukee*
Mary Jane Keith *Wichita State University*
Mary Kelley *Wayne State University*
Scott Kight *Montclair State University*
Wendy Kimber *Stevenson University*
Jeff Klahn *University of Iowa*
David S. Koetje *Calvin College*
Olga Kopp *Utah Valley University*
John C. Krenetsky *Metropolitan State College of Denver*
Patrick J. Krug *California State University–LA*
Robert Kurt *Lafayette College*
Marc J. LaBella *Ocean County College*
Ellen S. Lamb *University of North Carolina–Greensboro*
David Lampe *Duquesne University*
Grace Lasker *Lake Washington Technical College*
Kari Lavalli *Boston University*
Shannon Erickson Lee *California Sate University–Northridge*
Zhiming Liu *Eastern New Mexico University*
J. Mitchell Lockhart *Valdosta State University*
David Logan *Clark Atlanta University*
Thomas A. Lonergan *University of New Orleans*
Andreas Madlung *University of Puget Sound*
Lynn Mahaffy *University of Delaware*
Jennifer Marcinkiewicz *Kent State University*
Henri Maurice *University of Southern Indiana*
Deanna McCullough *University of Houston–Downtown*

Dean McCurdy   *Albion College*
Richard Merritt   *Houston Community College–Northwest*
Stephanie Miller   *Jefferson State Community College*
Thomas Miller   *University of California, Riverside*
Hector C. Miranda, Jr.   *Texas Southern University*
Jasleen Mishra   *Houston Community College*
Randy Mogg   *Columbus State Community College*
Daniel Moon   *University of North Florida*
Janice Moore   *Colorado State University*
Richard C. Moore   *Miami University*
Juan Morata   *Miami Dade College–Wolfson*
Ellyn R. Mulcahy   *Johnson County Community College*
Kimberlyn Nelson   *Pennsylvania State University*
Howard Neufeld   *Appalachian State University*
Jacalyn Newman   *University of Pittsburgh*
Margaret N. Nsofor   *Southern Illinois University–Carbondale*
Judith D. Ochrietor   *University of North Florida*
Robert O'Donnell   *SUNY–Geneseo*
Olumide Ogunmosin   *Texas Southern University*
Nathan O. Okia   *Auburn University–Montgomery*
Stephanie Pandolfi   *Michigan State University*
Peter Pappas   *County College of Morris*
J. Payne   *Bergen Community College*
Andrew Pease   *Stevenson University*
Craig Peebles   *University of Pittsburgh*

David G. Pennock   *Miami University*
Beverly Perry   *Houston Community College*
John S. Peters   *College of Charleston, SC*
Stephanie Toering Peters   *Wartburg College*
Teresa Petrino-Lin   *Barry University*
Susan Phillips   *Brevard Community College–Palm Bay*
Paul Pillitteri   *Southern Utah University*
Thomas Pitzer   *Florida International University–Miami*
Uwe Pott   *University of Wisconsin–Green Bay*
Nimala Prabhu   *Edison State College*
Lynn Preston   *Tarrant County College–NW*
Kelli Prior   *Finger Lakes Community College*
Penny L. Ragland   *Auburn Montgomery*
Marceau Ratard   *Delgado Community College*
Michael Reagan   *College of St. Benedict/St. John's University*
Nancy A. Rice   *Western Kentucky University*
Linda Richardson   *Blinn College*
Amanda Rosenzweig   *Delgado Community College*
Cliff Ross   *University of North Florida*
John Roufaiel   *SUNY–Rockland Community College*
Kenneth Roux   *Florida State University*
Ann E. Rushing   *Baylor University*
Sangha Saha   *Harold Washington College*
Eric Saliim   *North Carolina Central University*
Thomas Sasek   *University of Louisiana–Monroe*

Leena Sawant   *Houston Community College*
Emily Schmitt   *Nova Southeastern University*
Mark Schneegurt   *Wichita State University*
Brenda Schoffstall   *Barry University*
Scott Schuette   *Southern Illinois University*
Pramila Sen   *Houston Community College*
Bin Shuai   *Wichita State University*
Susan Skambis   *Valencia Community College*
Michael Smith   *Western Kentucky University*
Ramona Smith   *Brevard Community College*
Nancy G. Solomon   *Miami University*
Sally K. Sommers Smith   *Boston University*
Melissa Spitler   *California State University–Northridge*
Ashley Spring   *Brevard Community College*
Moira Van Staaden   *Bowling Green State University*
Bruce Stallsmith   *University of Alabama–Huntsville*
Susan Stamler   *College of DuPage*
Nancy Staub   *Gonzaga University*
Stanley Stevens   *University of Memphis*
Ivan Still   *Arkansas Tech University*
Gregory W. Stunz   *Texas A&M University–Corpus Christi*
Ken D. Sumida   *Chapman University*
Rema Suniga   *Ohio Northern University*
Bradley Swanson   *Central Michigan University*
David Tam   *University of North Texas*

Franklyn Tan Te   *Miami Dade College–Wolfson*
William Terzaghi   *Wilkes University*
Melvin Thomson   *University of Wisconsin–Parkside*
Martin Tracey   *Florida International University*
James Traniello   *Boston University*
Bibit Halliday Traut   *City College of San Francisco*
Alexa Tullis   *University of Puget Sound*
Catherine Ueckert   *Northern Arizona University*
Mark VanCura   *Cape Fear CC/University of NC Pembroke*
Charles J. Venglarik   *Jefferson State Community College*
Diane Wagner   *University of Alaska–Fairbanks*
Maureen Walter   *Florida International University*
Wei Wan   *Texas A&M University*
James T. Warren, Jr.   *Penn State Erie*
Delon Washo-Krupps   *Arizona State University*
Frederick Wasserman   *Boston University*
Raymond R. White   *City College of San Francisco*
Stephen W. White   *Ozarks Technical Community College*
Kimberlyn Williams   *California State University–San Bernardino*
Martha Comstock Williams   *Southern Polytechnic State University*
David E. Wolfe   *American River College*
Amber Wyman   *Finger Lakes Community College*
Robert D. Young, Jr.   *Blinn College*

A Note From the Authors

A revision of this scope relies on the talents and efforts of many people working behind the scenes and we have benefited greatly from their assistance.

Linda Davoli has been the copyeditor for each edition under this author team. She has labored many hours and always improves the clarity and consistency of the text. She has made a tremendous contribution to the quality of the final product.

We were fortunate to again work with Electronic Publishing Services to update the art program and improve the layout of the pages. Our close collaboration resulted in a text that is pedagogically effective as well as more beautiful than any other biology text on the market.

We have the continued support of an excellent team at McGraw-Hill. Rebecca Olson, the Brand Manager for *Biology* has been a steady leader during a time of change. The Director of Development, Liz Sievers, provided support in so many ways it would be impossible to name them all. Sheila Frank, lead project manager, and Tara McDermott, designer, ensured our text was on time and elegantly designed. Patrick Reidy, marketing manager, is always a sounding board for more than just marketing, and many more people behind the scenes have all contributed to the success of our text. This includes the digital team, who are also credited elsewhere, but whom we owe a great deal for their efforts to help us move toward the future.

Throughout this edition we have had the support of spouses and children, who have seen less of us than they might have liked because of the pressures of getting this revision completed. They have adapted to the many hours this book draws us away from them, and, even more than us, looked forward to its completion.

In the end, the people we owe the most are the generations of students who have used the many editions of this text. They have taught us at least as much as we have taught them, and their questions and suggestions continue to improve the text and supplementary materials.

We would like to thank Darrell Vodopich of Baylor University for providing the the flower image on the front cover (basket flower—*Centaurea americana* and the crab spider—*Misumenops* sp.) and Luke Mahler of University of California-Davis for providing the Anolis image (*Anolis gorgonae*).

Finally, we need to thank our reviewers. Instructors from across the country are continually invited to share their knowledge and experience with us through reviews and focus groups. The feedback we received shaped this edition, resulting in a reorganization of the table of contents and expanded coverage in key areas. We would especially like to thank Allison Silveus of Tarrant County College-Trinity River Campus for helping to evaluate the reviews over the animal biology chapters. All of these people took time out of their already busy lives to help us build a better edition of *Biology* for the next generation of introductory biology students, and they have our heartfelt thanks.

Contents

Part III Genetic and Molecular Biology

Part IV Evolution

Part V Diversity of Life on Earth

Part VIII   Ecology and Behavior

Chapter 1

The Science of Biology

Chapter Contents

Part | The Molecular Basis of Life

Introduction

You are about to embark on a journey—a journey of discovery about the nature of life. Nearly 180 years ago, a young English naturalist named Charles Darwin set sail on a similar journey on board H.M.S. Beagle; a replica of this ship is pictured here. What Darwin learned on his five-year voyage led directly to his development of the theory of evolution by natural selection, a theory that has become the core of the science of biology. Darwin's voyage seems a fitting place to begin our exploration of biology—the scientific study of living organisms and how they have evolved. Before we begin, however, let's take a moment to think about what biology is and why it's important.

1.1 The Science of Life

Learning Outcomes

1. *Compare biology to other natural sciences.*
2. *Describe the characteristics of living systems.*
3. *Characterize the hierarchical organization of living systems.*

This is the most exciting time to be studying biology in the history of the field. The amount of information available about the natural world has exploded in the last 35 years, and we are now in a position to ask and answer questions that previously were only dreamed of.

We have determined the entire sequence of the human genome and are in the process of sequencing the genomes of other species at an ever-increasing pace. We are closing in on a description of the molecular workings of the cell in unprecedented detail, and we are in the process of finally unveiling the mystery of how a single cell can give rise to the complex organization seen in multicellular organisms. With robotics, advanced imaging, and analytical techniques, we have tools available that were formerly the stuff of science fiction.

In this text, we attempt to draw a contemporary picture of the science of biology, as well as provide some history and experimental perspective on this exciting time in the discipline. In this introductory chapter, we examine the nature of biology and the foundations of science in general to put into context the information presented in the rest of the text.

Biology unifies much of natural science

The study of biology is a point of convergence for the information and tools from all of the natural sciences. Biological systems are the most complex chemical systems on Earth, and their many functions are both determined and constrained by the principles of chemistry and physics. Put another way, no new laws of nature can be gleaned from the study of biology—but that study does illuminate and illustrate the workings of those natural laws.

The intricate chemical workings of cells are based on everything we have learned from the study of chemistry. And every level of biological organization is governed by the nature of energy transactions learned from the study of thermodynamics. Biological systems do not represent any new forms of matter, and yet they are the most complex organization of matter known. The complexity of living systems is made possible by a constant source of energy—the Sun. The conversion of this energy source into organic molecules by photosynthesis is one of the most beautiful and complex reactions known in chemistry and physics.

The way we do science is changing to grapple with increasingly difficult modern problems. Science is becoming more interdisciplinary, combining the expertise from a variety of traditional disciplines and emerging fields such as nanotechnology. Biology is at the heart of this multidisciplinary approach because biological problems often require many different approaches to arrive at solutions.

Life defies simple definition

In its broadest sense, biology is the study of living things—*the science of life*. Living things come in an astounding variety of shapes and forms, and biologists study life in many different ways. They live with gorillas, collect fossils, and listen to whales. They read the messages encoded in the long molecules of heredity and count how many times a hummingbird's wings beat each second.

What makes something "alive"? Anyone could deduce that a galloping horse is alive and a car is not, but why? We cannot say, "If it moves, it's alive," because a car can move, and gelatin can wiggle in a bowl. They certainly are not alive. Although we cannot define life with a single simple sentence, we can come up with a series of seven characteristics shared by living systems:

- **Cellular organization.** All organisms consist of one or more cells. Often too tiny to see, cells carry out the basic activities of living. Each cell is bounded by a membrane that separates it from its surroundings.
- **Ordered complexity.** All living things are both complex and highly ordered. Your body is composed of many different kinds of cells, each containing many complex

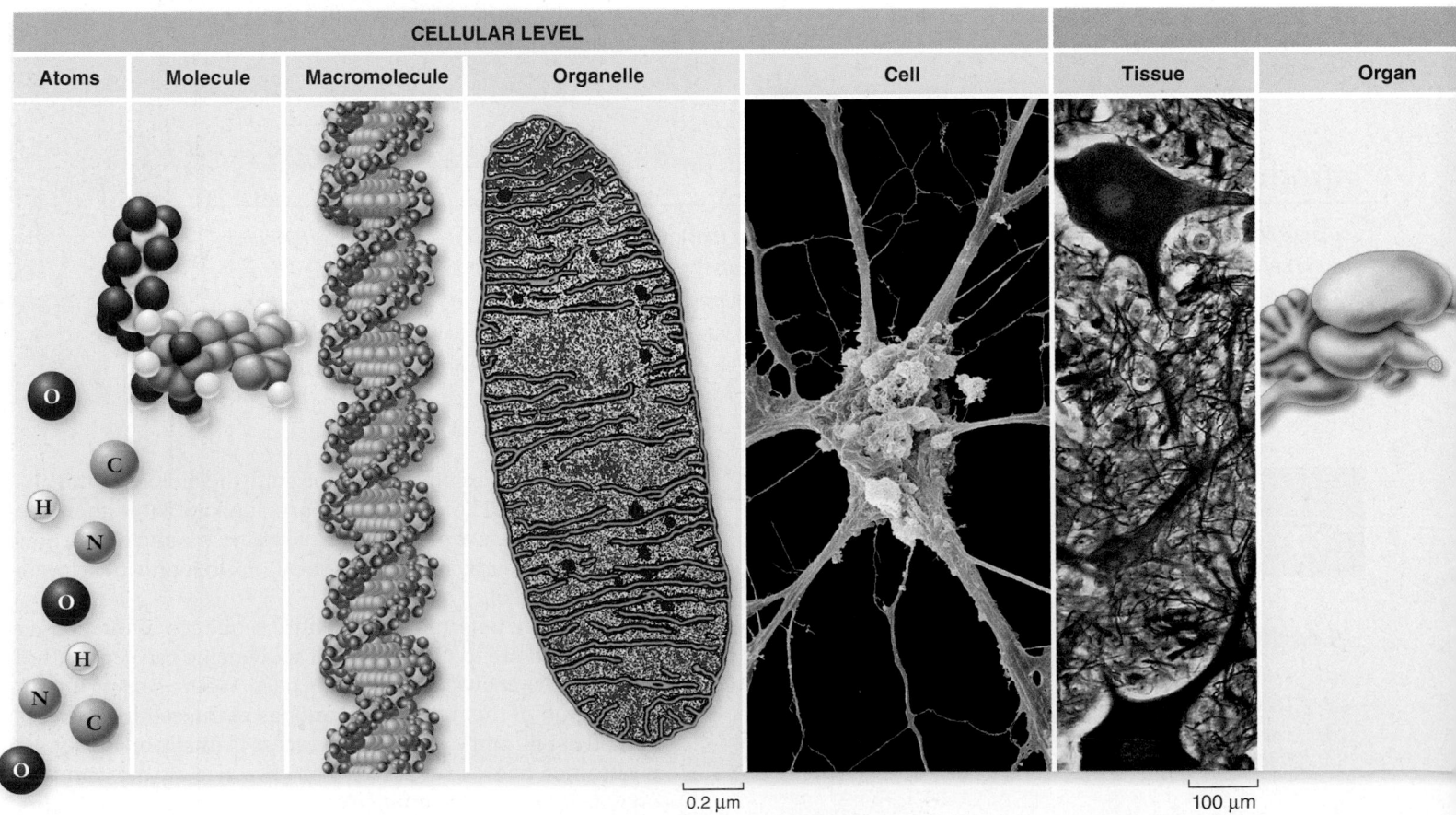

Atoms	Molecule	Macromolecule	Organelle	Cell	Tissue	Organ

CELLULAR LEVEL

O
C
H
N
O
H
N
C
O

0.2 μm

100 μm

molecular structures. Many nonliving things may also be complex, but they do not exhibit this degree of ordered complexity.

- **Sensitivity.** All organisms respond to stimuli. Plants grow toward a source of light, and the pupils of your eyes dilate when you walk into a dark room.
- **Growth, development, and reproduction.** All organisms are capable of growing and reproducing, and they all possess hereditary molecules that are passed to their offspring, ensuring that the offspring are of the same species.
- **Energy utilization.** All organisms take in energy and use it to perform many kinds of work. Every muscle in your body is powered with energy you obtain from your diet.
- **Homeostasis.** All organisms maintain relatively constant internal conditions that are different from their environment, a process called **homeostasis.** For example, your body temperature remains stable despite changes in outside temperatures.
- **Evolutionary adaptation.** All organisms interact with other organisms and the nonliving environment in ways that influence their survival, and as a consequence, organisms evolve adaptations to their environments.

Living systems show hierarchical organization

The organization of the biological world is hierarchical—that is, each level builds on the level below it.

1. **The Cellular Level.** At the cellular level (figure 1.1), **atoms,** the fundamental elements of matter, are joined together into clusters called **molecules.** Complex biological molecules are assembled into tiny structures called **organelles** within membrane-bounded units we call **cells.** The cell is the basic unit of life. Many independent organisms are composed only of single cells. Bacteria are single cells, for example. All animals and plants, as well as most fungi and algae, are multicellular—composed of more than one cell.
2. **The Organismal Level.** Cells in complex multicellular organisms exhibit three levels of organization. The most basic level is that of **tissues,** which are groups of similar cells that act as a functional unit. Tissues, in turn, are grouped into **organs**—body structures composed of several different tissues that act as a structural and functional unit. Your brain is an organ composed of nerve cells and a variety of associated tissues that form protective coverings and contribute blood. At the third level of organization, organs are grouped into **organ systems.** The nervous system, for example, consists of sensory organs, the brain and spinal cord, and neurons that convey signals.

Figure 1.1 Hierarchical organization of living systems. Life is highly organized from atoms to complex multicellular organisms. Along this hierarchy of structure, atoms form molecules that are used to form organelles, which in turn form the functional subsystems within cells. Cells are organized into tissues, then into organs and organ systems such as the goose's nervous system pictured. This organization extends beyond individual organisms to populations, communities, ecosystems, and finally the biosphere.

ORGANISMAL LEVEL		POPULATIONAL LEVEL				
Organ system	Organism	Population	Species	Community	Ecosystem	Biosphere

3. **The Populational Level.** Individual organisms can be categorized into several hierarchical levels within the living world. The most basic of these is the **population**—a group of organisms of the same species living in the same place. All populations of a particular kind of organism together form a **species,** its members similar in appearance and able to interbreed. At a higher level of biological organization, a **biological community** consists of all the populations of different species living together in one place.
4. **The Ecosystem Level.** At the highest tier of biological organization, a biological community and the physical habitat within which it lives together constitute an ecological system, or **ecosystem.** For example, the soil, water, and atmosphere of a mountain ecosystem interact with the biological community of a mountain meadow in many important ways.
5. **The Biosphere.** The entire planet can be thought of as an ecosystem that we call the biosphere.

As you move up this hierarchy, novel properties emerge. These **emergent properties** result from the way in which components interact, and they often cannot be deduced just from looking at the parts themselves. Examining individual cells, for example, gives little hint about the whole animal. You, and all humans, have the same array of cell types as a giraffe. It is because the living world exhibits many emergent properties that it is difficult to define "life."

The previous descriptions of the common features and organization of living systems begins to get at the nature of what it is to be alive. The rest of this book illustrates and expands on these basic ideas to try to provide a more complete account of living systems.

Learning Outcomes Review 1.1

Biology as a science brings together other natural sciences, such as chemistry and physics, to study living systems. Life does not have a simple definition, but living systems share a number of properties that together describe life. Living systems can be organized hierarchically, from the cellular level to the entire biosphere, with emerging properties that may exceed the sum of the parts.

■ *Can you study biology without studying other sciences?*

1.2 The Nature of Science

Learning Outcomes

1. *Compare the different types of reasoning used by biologists.*
2. *Demonstrate how to formulate a hypothesis.*

Much like life itself, the nature of science defies simple description. For many years scientists have written about the "scientific method" as though there is a single way of doing science. This oversimplification has contributed to confusion on the part of nonscientists about the nature of science.

At its core, science is concerned with developing an increasingly accurate understanding of the world around us using observation and reasoning. To begin with, we assume that natural forces acting now have always acted, that the fundamental nature of the universe has not changed since its inception, and that it is not changing now. A number of complementary approaches allow understanding of natural phenomena—there is no one "scientific method."

Scientists also attempt to be as objective as possible in the interpretation of the data and observations they have collected. Because scientists themselves are human, this is not completely possible, but because science is a collective endeavor subject to scrutiny, it is self-correcting. One person's results are verified by others, and if the results cannot be repeated, they are rejected.

Much of science is descriptive

The classic vision of the scientific method is that observations lead to hypotheses that in turn make experimentally testable predictions. In this way, we dispassionately evaluate new ideas to arrive at an increasingly accurate view of nature. We discuss this way of doing science later in this chapter, but it is important to understand that much of science is purely descriptive: In order to understand anything, the first step is to describe it completely. Much of biology is concerned with arriving at an increasingly accurate description of nature.

The study of biodiversity is an example of descriptive science that has implications for other aspects of biology in addition to societal implications. Efforts are currently underway to classify all life on Earth. This ambitious project is purely descriptive, but it will lead to a much greater understanding of biodiversity as well as the effect our species has on biodiversity.

One of the most important accomplishments of molecular biology at the dawn of the 21st century was the completion of the sequence of the human genome. Many new hypotheses about human biology will be generated by this knowledge, and many experiments will be needed to test these hypotheses, but the determination of the sequence itself was descriptive science.

Science uses both deductive and inductive reasoning

The study of logic recognizes two opposite ways of arriving at logical conclusions: deductive and inductive reasoning. Science makes use of both of these methods, although induction is the primary way of reasoning in hypothesis-driven science.

Deductive reasoning

Deductive reasoning applies general principles to predict specific results. More than 2200 years ago, the Greek scientist Eratosthenes used Euclidean geometry and deductive reasoning to accurately estimate the circumference of the Earth (figure 1.2). Deductive reasoning is the reasoning of

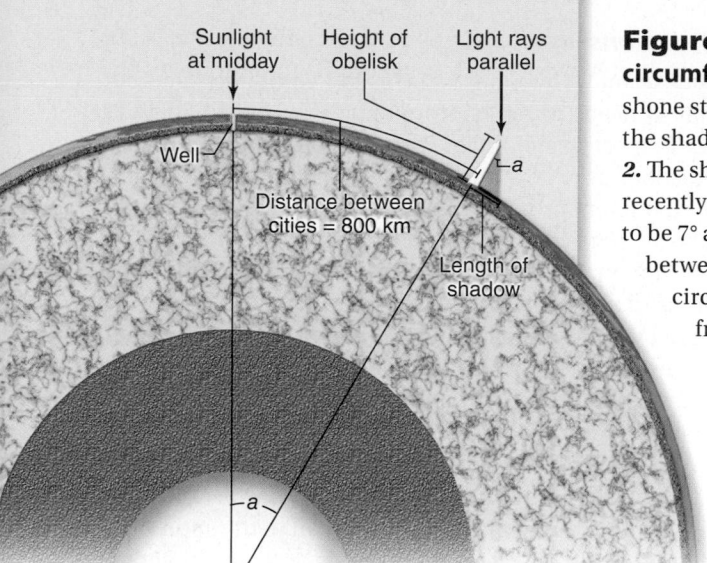

Figure 1.2 Deductive reasoning: how Eratosthenes estimated the circumference of the earth using deductive reasoning. *1.* On a day when sunlight shone straight down a deep well at Syene in Egypt, Eratosthenes measured the length of the shadow cast by a tall obelisk in the city of Alexandria, about 800 kilometers (km) away. *2.* The shadow's length and the obelisk's height formed two sides of a triangle. Using the recently developed principles of Euclidean geometry, Eratosthenes calculated the angle, *a*, to be 7° and 12´, exactly ¹⁄₅₀ of a circle (360°). *3.* If angle *a* is ¹⁄₅₀ of a circle, then the distance between the obelisk (in Alexandria) and the well (in Syene) must be equal to ¹⁄₅₀ the circumference of the Earth. *4.* Eratosthenes had heard that it was a 50-day camel trip from Alexandria to Syene. Assuming a camel travels about 18.5 km per day, he estimated the distance between obelisk and well as 925 km (using different units of measure, of course). *5.* Eratosthenes thus deduced the circumference of the Earth to be 50 × 925 = 46,250 km. Modern measurements put the distance from the well to the obelisk at just over 800 km. Using this distance Eratosthenes's value would have been 50 × 800 = 40,000 km. The actual circumference is 40,075 km.

mathematics and philosophy, and it is used to test the validity of general ideas in all branches of knowledge. For example, if all mammals by definition have hair, and you find an animal that does not have hair, then you may conclude that this animal is not a mammal. A biologist uses deductive reasoning to infer the species of a specimen from its characteristics.

Inductive reasoning

In **inductive reasoning,** the logic flows in the opposite direction, from the specific to the general. Inductive reasoning uses specific observations to construct general scientific principles. For example, if poodles have hair, and terriers have hair, and every dog that you observe has hair, then you may conclude that all dogs have hair. Inductive reasoning leads to generalizations that can then be tested. Inductive reasoning first became important to science in the 1600s in Europe, when Francis Bacon, Isaac Newton, and others began to use the results of particular experiments to infer general principles about how the world operates.

An example from modern biology is the role of homeobox genes in development. Studies in the fruit fly, *Drosophila melanogaster,* identified genes that could cause dramatic changes in developmental fate, such as a leg appearing in the place of an antenna. When the genes themselves were isolated and their DNA sequence determined, it was found that similar genes were found in many animals, including humans. This led to the general idea that the homeobox genes act as switches to control developmental fate.

Hypothesis-driven science makes and tests predictions

Scientists establish which general principles are true from among the many that might be true through the process of systematically testing alternative proposals. If these proposals prove inconsistent with experimental observations, they are rejected as untrue. Figure 1.3 illustrates the process.

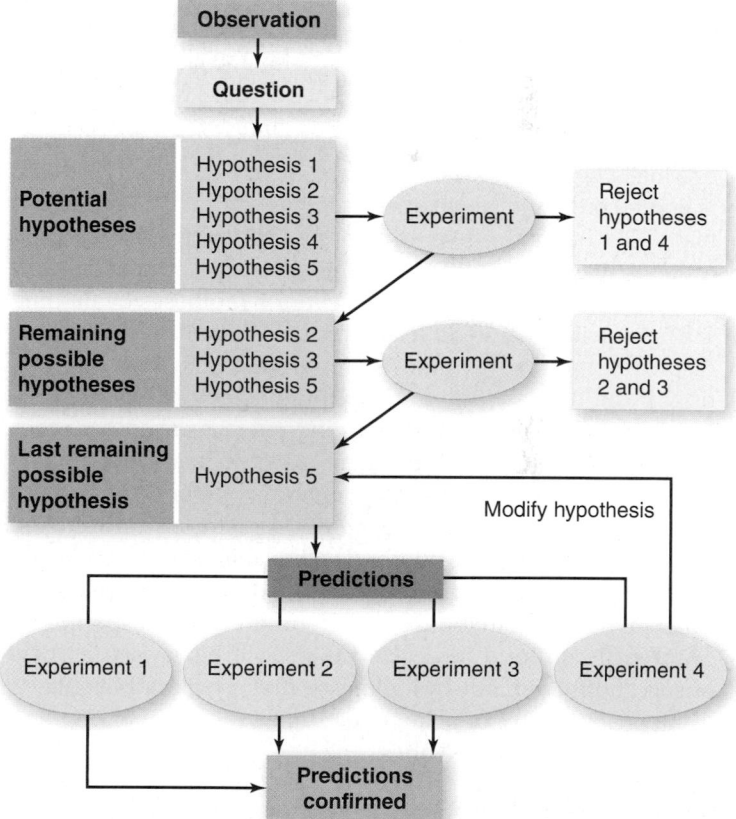

Figure 1.3 How science is done. This diagram illustrates how scientific investigations proceed. First, scientists make observations that raise a particular question. They develop a number of potential explanations (hypotheses) to answer the question. Next, they carry out experiments in an attempt to eliminate one or more of these hypotheses. Then, predictions are made based on the remaining hypotheses, and further experiments are carried out to test these predictions. The process can also be iterative. As experimental results are performed, the information can be used to modify the original hypothesis to fit each new observation.

After making careful observations, scientists construct a **hypothesis,** which is a suggested explanation that accounts for those observations. A hypothesis is a proposition that might be true. Those hypotheses that have not yet been disproved are retained. They are useful because they fit the known facts, but they are always subject to future rejection if, in the light of new information, they are found to be incorrect.

This process can also be *iterative,* that is, a hypothesis can be changed and refined with new data. For instance, geneticists George Beadle and Edward Tatum studied the nature of genetic information to arrive at their "one-gene/one-enzyme" hypothesis (see chapter 15). This hypothesis states that a gene represents the genetic information necessary to make a single enzyme. As investigators learned more about the molecular nature of genetic information, the hypothesis was refined to "one-gene/one-polypeptide" because enzymes can be made up of more than one polypeptide. With still more information about the nature of genetic information, other investigators found that a single gene can specify more than one polypeptide, and the hypothesis was refined again.

Testing hypotheses

We call the test of a hypothesis an **experiment.** Suppose that a room appears dark to you. To understand why it appears dark, you propose several hypotheses. The first might be, "There is no light in the room because the light switch is turned off." An alternative hypothesis might be, "There is no light in the room because the lightbulb is burned out." And yet another hypothesis might be, "I am going blind." To evaluate these hypotheses, you would conduct an experiment designed to eliminate one or more of the hypotheses.

For example, you might test your hypotheses by flipping the light switch. If you do so and the room is still dark, you have disproved the first hypothesis: Something other than the setting of the light switch must be the reason for the darkness. Note that a test such as this does not prove that any of the other hypotheses are true; it merely demonstrates that the one being tested is not. A successful experiment is one in which one or more of the alternative hypotheses is demonstrated to be inconsistent with the results and is thus rejected.

As you proceed through this text, you will encounter many hypotheses that have withstood the test of experiment. Many will continue to do so; others will be revised as new observations are made by biologists. Biology, like all science, is in a constant state of change, with new ideas appearing and replacing or refining old ones.

Establishing controls

Often scientists are interested in learning about processes that are influenced by many factors, or **variables.** To evaluate alternative hypotheses about one variable, all other variables must be kept constant. This is done by carrying out two experiments in parallel: a test experiment and a control experiment. In the **test experiment,** one variable is altered in a known way to test a particular hypothesis. In the **control experiment,** that variable is left unaltered. In all other respects the two experiments are identical, so any difference in the outcomes of the two experiments must result from the influence of the variable that was changed.

Much of the challenge of experimental science lies in designing control experiments that isolate a particular variable from other factors that might influence a process.

Using predictions

A successful scientific hypothesis needs to be not only valid but also useful—it needs to tell us something we want to know. A hypothesis is most useful when it makes predictions because those predictions provide a way to test the validity of the hypothesis. If an experiment produces results inconsistent with the predictions, the hypothesis must be rejected or modified. In contrast, if the predictions are supported by experimental testing, the hypothesis is supported. The more experimentally supported predictions a hypothesis makes, the more valid the hypothesis is.

As an example, in the early history of microbiology it was known that nutrient broth left sitting exposed to air becomes contaminated. Two hypotheses were proposed to explain this observation: spontaneous generation and the germ hypothesis. Spontaneous generation held that there was an inherent property in organic molecules that could lead to the spontaneous generation of life. The germ hypothesis proposed that preexisting microorganisms that were present in the air could contaminate the nutrient broth.

These competing hypotheses were tested by a number of experiments that involved filtering air and boiling the broth to kill any contaminating germs. The definitive experiment was performed by Louis Pasteur, who constructed flasks with curved necks that could be exposed to air, but that would trap any contaminating germs. When such flasks were boiled to sterilize them, they remained sterile, but if the curved neck was broken off, they became contaminated (figure 1.4).

SCIENTIFIC THINKING

Question: *What is the source of contamination that occurs in a flask of nutrient broth left exposed to the air?*

Germ Hypothesis: *Preexisting microorganisms present in the air contaminate nutrient broth.*

Prediction: *Sterilized broth will remain sterile if microorganisms are prevented from entering flask.*

Spontaneous Generation Hypothesis: *Living organisms will spontaneously generate from nonliving organic molecules in broth.*

Prediction: *Organisms will spontaneously generate from organic molecules in broth after sterilization.*

Test: *Use swan-necked flasks to prevent entry of microorganisms. To ensure that broth can still support life, break swan-neck after sterilization.*

  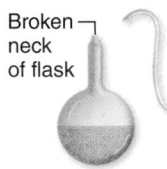

Broken neck of flask

Flask is sterilized by boiling the broth.

Unbroken flask remains sterile.

Broken flask becomes contaminated after exposure to germ-laden air.

Result: *No growth occurs in sterile swan-necked flasks. When the neck is broken off, and the broth is exposed to air, growth occurs.*

Conclusion: *Growth in broth is of preexisting microorganisms.*

Figure 1.4 **Experiment to test spontaneous generation versus germ hypothesis.**

This result was predicted by the germ hypothesis—that when the sterile flask is exposed to air, airborne germs are deposited in the broth and grow. The spontaneous generation hypothesis predicted no difference in results with exposure to air. This experiment disproved the hypothesis of spontaneous generation and supported the hypothesis of airborne germs under the conditions tested.

Reductionism breaks larger systems into their component parts

Scientists use the philosophical approach of **reductionism** to understand a complex system by reducing it to its working parts. Reductionism has been the general approach of biochemistry, which has been enormously successful at unraveling the complexity of cellular metabolism by concentrating on individual pathways and specific enzymes. By analyzing all of the pathways and their components, scientists now have an overall picture of the metabolism of cells.

Reductionism has limits when applied to living systems, however—one of which is that enzymes do not always behave exactly the same in isolation as they do in their normal cellular context. A larger problem is that the complex interworking of many interconnected functions leads to emergent properties that cannot be predicted based on the workings of the parts. For example, an examination of all of the proteins and RNAs in a ribosome in isolation would not lead to predictions about the nature of protein synthesis. On a higher level, understanding the physiology of a single Canada goose would not lead to predictions about flocking behavior. Biologists are just beginning to come to grips with this problem and to think about ways of dealing with the whole as well as the workings of the parts. The emerging field of systems biology focuses on this different approach.

Biologists construct models to explain living systems

Biologists construct models in many different ways for a variety of uses. Geneticists construct models of interacting networks of proteins that control gene expression, often even drawing cartoon figures to represent that which we cannot see. Population biologists build models of how evolutionary change occurs. Cell biologists build models of signal transduction pathways and the events leading from an external signal to internal events. Structural biologists build actual models of the structure of proteins and macromolecular complexes in cells.

Models provide a way to organize how we think about a problem. Models can also get us closer to the larger picture and away from the extreme reductionist approach. The working parts are provided by the reductionist analysis, but the model shows how they fit together. Often these models suggest other experiments that can be performed to refine or test the model.

As researchers gain more knowledge about the actual flow of molecules in living systems, more sophisticated kinetic models can be used to apply information about isolated enzymes to their cellular context. In systems biology, this modeling is being applied on a large scale to regulatory networks during development, and even to modeling an entire bacterial cell.

The nature of scientific theories

Scientists use the word **theory** in two main ways. The first meaning of *theory* is a proposed explanation for some natural phenomenon, often based on some general principle. Thus, we speak of the principle first proposed by Newton as the "theory of gravity." Such theories often bring together concepts that were previously thought to be unrelated.

The second meaning of *theory* is the body of interconnected concepts, supported by scientific reasoning and experimental evidence, that explains the facts in some area of study. Such a theory provides an indispensable framework for organizing a body of knowledge. For example, quantum theory in physics brings together a set of ideas about the nature of the universe, explains experimental facts, and serves as a guide to further questions and experiments.

To a scientist, theories are the solid ground of science, expressing ideas of which we are most certain. In contrast, to the general public, the word *theory* usually implies the opposite—a *lack* of knowledge, or a guess. Not surprisingly, this difference often results in confusion. In this text, *theory* will always be used in its scientific sense, in reference to an accepted general principle or body of knowledge.

Some critics outside of science attempt to discredit evolution by saying it is "just a theory." The hypothesis that evolution has occurred, however, is an accepted scientific fact—it is supported by overwhelming evidence. Modern evolutionary theory is a complex body of ideas, the importance of which spreads far beyond explaining evolution. Its ramifications permeate all areas of biology, and it provides the conceptual framework that unifies biology as a science. Again, the key is how well a hypothesis fits the observations. Evolutionary theory fits the observations very well.

Research can be basic or applied

In the past it was fashionable to speak of the "scientific method" as consisting of an orderly sequence of logical, either-or steps. Each step would reject one of two mutually incompatible alternatives, as though trial-and-error testing would inevitably lead a researcher through the maze of uncertainty to the ultimate scientific answer. If this were the case, a computer would make a good scientist. But science is not done this way.

As the British philosopher Karl Popper has pointed out, successful scientists without exception design their experiments with a pretty fair idea of how the results are going to come out. They have what Popper calls an "imaginative preconception" of what the truth might be. Because insight and imagination play such a large role in scientific progress, some scientists are better at science than others—just as Bruce Springsteen stands out among songwriters or Claude Monet stands out among Impressionist painters.

Some scientists perform *basic research*, which is intended to extend the boundaries of what we know. These individuals typically work at universities, and their research is usually supported by grants from various agencies and foundations.

The information generated by basic research contributes to the growing body of scientific knowledge, and it provides the scientific foundation utilized by *applied research*. Scientists who conduct applied research are often employed in some kind of industry. Their work may involve the manufacture of food additives, the creation of new drugs, or the testing of environmental quality.

Research results are written up and submitted for publication in scientific journals, where the experiments and conclusions are reviewed by other scientists. This process of careful evaluation, called *peer review,* lies at the heart of modern science. It helps to ensure that faulty research or false claims are not given the authority of scientific fact. It also provides other scientists with a starting point for testing the reproducibility of experimental results. Results that cannot be reproduced are not taken seriously for long.

Figure 1.5 Charles Darwin. This newly rediscovered photograph taken in 1881, the year before Darwin died, appears to be the last ever taken of the great biologist.

Learning Outcomes Review 1.2

Much of science is descriptive, amassing observations to gain an accurate view. Both deductive reasoning and inductive reasoning are used in science. Scientific hypotheses are suggested explanations for observed phenomena. Hypotheses need to make predictions that can be tested by controlled experiments. Theories are coherent explanations of observed data, but they may be modified by new information.

■ *How does a scientific theory differ from a hypothesis?*

1.3 An Example of Scientific Inquiry: Darwin and Evolution

Learning Outcomes

1. *Examine Darwin's theory of evolution by natural selection as a scientific theory.*
2. *Describe the evidence that supports the theory of evolution.*

Darwin's theory of evolution explains and describes how organisms on Earth have changed over time and acquired a diversity of new forms. This famous theory provides a good example of how a scientist develops a hypothesis and how a scientific theory grows and wins acceptance.

Charles Robert Darwin (1809–1882; figure 1.5) was an English naturalist who, after 30 years of study and observation, wrote one of the most famous and influential books of all time. This book, *On the Origin of Species by Means of Natural Selection,* created a sensation when it was published, and the ideas Darwin expressed in it have played a central role in the development of human thought ever since.

The idea of evolution existed prior to Darwin

In Darwin's time, most people believed that the different kinds of organisms and their individual structures resulted from direct actions of a Creator (many people still believe this). Species were thought to have been specially created and to be unchangeable over the course of time.

In contrast to these ideas, a number of earlier naturalists and philosophers had presented the view that living things must have changed during the history of life on Earth. That is, **evolution** has occurred, and living things are now different from how they began. Darwin's contribution was a concept he called *natural selection,* which he proposed as a coherent, logical explanation for this process, and he brought his ideas to wide public attention.

Darwin observed differences in related organisms

The story of Darwin and his theory begins in 1831, when he was 22 years old. He was part of a five-year navigational mapping expedition around the coasts of South America (figure 1.6), aboard H.M.S. *Beagle.* During this long voyage, Darwin had the chance to study a wide variety of plants and animals on continents and islands and in distant seas. Darwin observed a number of phenomena that were of central importance to his reaching his ultimate conclusion.

Repeatedly, Darwin saw that the characteristics of similar species varied somewhat from place to place. These geographical patterns suggested to him that lineages change gradually as species migrate from one area to another. On the Galápagos Islands, 960 km (600 miles) off the coast of Ecuador, Darwin encountered a variety of different finches on the various islands. The 14 species, although related, differed slightly in appearance, particularly in their beaks (figure 1.7).

Darwin thought it was reasonable to assume that all these birds had descended from a common ancestor arriving from the South American mainland several million years ago. Eating different foods on different islands, the finches' beaks had changed during their descent—"descent with modification," or evolution. (These finches are discussed in more detail in chapters 21 and 22.)

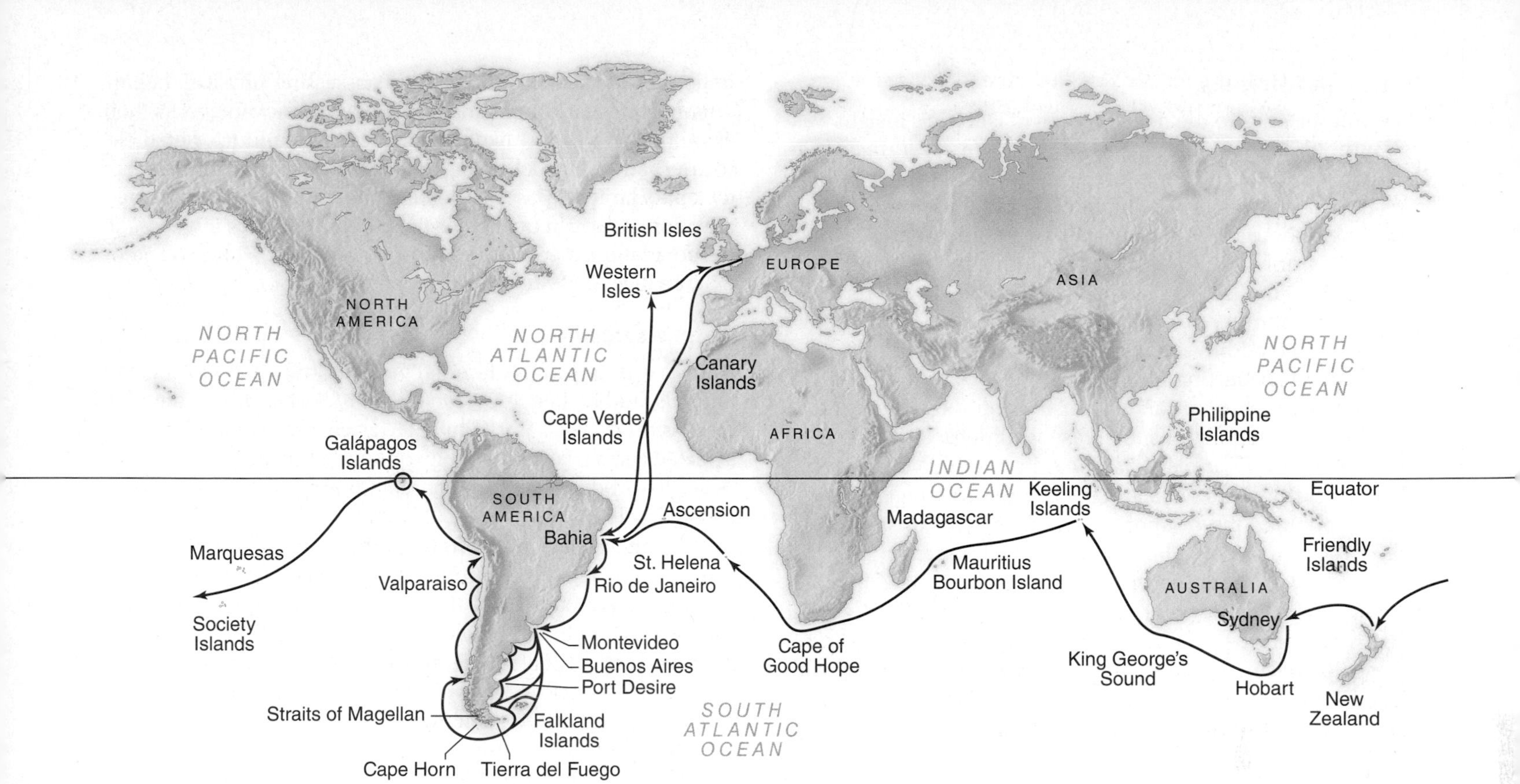

Figure 1.6 **The five-year voyage of H.M.S. *Beagle*.** Most of the time was spent exploring the coasts and coastal islands of South America, such as the Galápagos Islands. Darwin's studies of the animals of the Galápagos Islands played a key role in his eventual development of the concept of evolution by means of natural selection.

In a more general sense, Darwin was struck by the fact that the plants and animals on these relatively young volcanic islands resembled those on the nearby coast of South America. If each one of these plants and animals had been created independently and simply placed on the Galápagos Islands, why didn't they resemble the plants and animals of islands with similar climates—such as those off the coast of Africa, for example? Why did they resemble those of the adjacent South American coast instead?

Darwin proposed natural selection as a mechanism for evolution

It is one thing to observe the results of evolution, but quite another to understand how it happens. Darwin's great achievement lies in his ability to move beyond all the individual observations to formulate the hypothesis that evolution occurs because of natural selection.

Woodpecker Finch (*Cactospiza pallida*)

Large Ground Finch (*Geospiza magnirostris*)

Cactus Finch (*Geospiza scandens*)

Figure 1.7 **Three Galápagos finches and what they eat.** On the Galápagos Islands, Darwin observed 14 different species of finches differing mainly in their beaks and feeding habits. These three finches eat very different food items, and Darwin surmised that the different shapes of their bills represented evolutionary adaptations that improved their ability to eat the foods available in their specific habitats.

Darwin and Malthus

Of key importance to the development of Darwin's insight was his study of Thomas Malthus's *An Essay on the Principle of Population* (1798). In this book, Malthus stated that populations of plants and animals (including human beings) tend to increase geometrically, while humans are able to increase their food supply only arithmetically. Put another way, population increases by a multiplying factor—for example, in the series 2, 6, 18, 54, the starting number is multiplied by 3. Food supply increases by an additive factor—for example, the series 2, 4, 6, 8 adds 2 to each starting number. Figure 1.8 shows the difference that these two types of relationships produce over time.

Because populations increase geometrically, virtually any kind of animal or plant, if it could reproduce unchecked, would cover the entire surface of the world surprisingly quickly. Instead, populations of species remain fairly constant year after year, because death limits population numbers.

Sparked by Malthus's ideas, Darwin saw that although every organism has the potential to produce more offspring than can survive, only a limited number actually do survive and produce further offspring. Combining this observation with what he had seen on the voyage of the *Beagle,* as well as with his own experiences in breeding domestic animals, Darwin made an

important association: Individuals possessing physical, behavioral, or other attributes that give them an advantage in their environment are more likely to survive and reproduce than those with less advantageous traits. By surviving, these individuals gain the opportunity to pass on their favorable characteristics to their offspring. As the frequency of these characteristics increases in the population, the nature of the population as a whole will gradually change. Darwin called this process *selection.*

Natural selection

Darwin was thoroughly familiar with variation in domesticated animals, and he began *On the Origin of Species* with a detailed discussion of pigeon breeding. He knew that animal breeders selected certain varieties of pigeons and other animals, such as dogs, to produce certain characteristics, a process Darwin called **artificial selection**.

Artificial selection often produces a great variation in traits. Domestic pigeon breeds, for example, show much greater variety than all of the wild species found throughout the world. Darwin thought that this type of change could occur in nature, too. Surely if pigeon breeders could foster variation by artificial selection, nature could do the same—a process Darwin called **natural selection.**

Darwin drafts his argument

Darwin drafted the overall argument for evolution by natural selection in a preliminary manuscript in 1842. After showing the manuscript to a few of his closest scientific friends, however, Darwin put it in a drawer, and for 16 years turned to other research. No one knows for sure why Darwin did not publish his initial manuscript—it is very thorough and outlines his ideas in detail.

The stimulus that finally brought Darwin's hypothesis into print was an essay he received in 1858. A young English naturalist named Alfred Russel Wallace (1823–1913) sent the essay to Darwin from Indonesia; it concisely set forth the hypothesis of evolution by means of natural selection, a hypothesis Wallace had developed independently of Darwin. After receiving Wallace's essay, friends of Darwin arranged for a joint presentation of their ideas at a seminar in London. Darwin then completed his own book, expanding the 1842 manuscript he had written so long ago, and submitted it for publication.

The predictions of natural selection have been tested

More than 120 years have elapsed since Darwin's death in 1882. During this period, the evidence supporting his theory has grown progressively stronger. We briefly explore some of this evidence here; in chapter 21, we will return to the theory of evolution by natural selection and examine the evidence in more detail.

The fossil record

Darwin predicted that the fossil record would yield intermediate links between the great groups of organisms—for example, between fishes and the amphibians thought to have arisen from them, and between reptiles and birds. Furthermore, natural selection predicts the relative positions in time of such transitional forms. We now know the fossil record to a degree

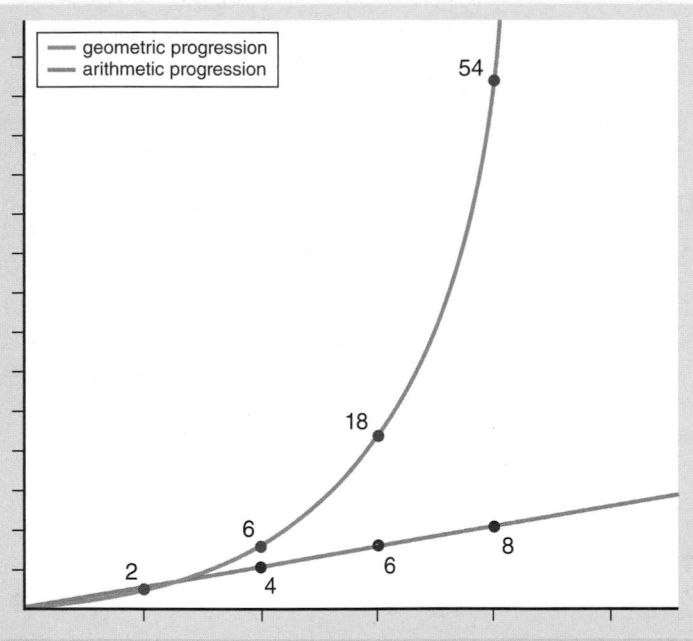

Figure 1.8 Geometric and arithmetic progressions. A geometric progression increases by a constant factor (for example, the curve shown increases ×3 for each step), whereas an arithmetic progression increases by a constant difference (for example, the line shown increases +2 for each step). Malthus contended that the human growth curve was geometric, but the human food production curve was only arithmetic.

🔍 **Data analysis** What is the effect of reducing the constant factor for a geometric progression? How would this change the curve in the figure?

❓ **Inquiry question** Might this effect be achieved with humans? How?

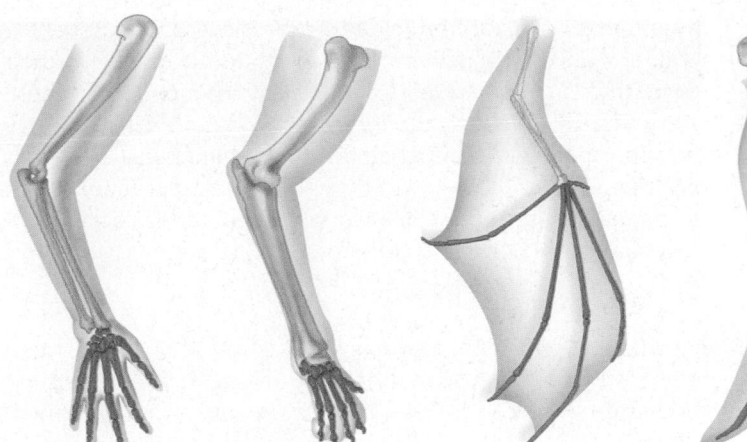

Human   Cat   Bat   Porpoise   Horse

Figure 1.9 Homology among vertebrate limbs. The forelimbs of these five vertebrates show the ways in which the relative proportions of the forelimb bones have changed in relation to the particular way of life of each organism.

that was unthinkable in the 19th century, and although truly "intermediate" organisms are hard to determine, paleontologists have found what appear to be transitional forms and found them at the predicted positions in time.

Recent discoveries of microscopic fossils have extended the known history of life on Earth back to about 3.5 billion years ago (BYA). The discovery of other fossils has supported Darwin's predictions and has shed light on how organisms have, over this enormous time span, evolved from the simple to the complex. For vertebrate animals especially, the fossil record is rich and exhibits a graded series of changes in form, with the evolutionary sequence visible for all to see.

The age of the Earth

Darwin's theory predicted the Earth must be very old, but some physicists argued that the Earth was only a few thousand years old. This bothered Darwin, because the evolution of all living things from some single original ancestor would have required a great deal more time. Using evidence obtained by studying the rates of radioactive decay, we now know that the physicists of Darwin's time were very wrong: The Earth was formed about 4.5 BYA.

The mechanism of heredity

Darwin received some of his sharpest criticism in the area of heredity. At that time, no one had any concept of genes or how heredity works, so it was not possible for Darwin to explain completely how evolution occurs.

Even though Gregor Mendel was performing his experiments with pea plants in Brünn, Austria (now Brno, the Czech Republic), during roughly the same period, genetics was established as a science only at the start of the 20th century. When scientists began to understand the laws of inheritance (discussed in chapters 12 and 13), this problem with Darwin's theory vanished.

Comparative anatomy

Comparative studies of animals have provided strong evidence for Darwin's theory. In many different types of vertebrates, for example, the same bones are present, indicating their evolutionary past. Thus, the forelimbs shown in figure 1.9 are all constructed from the same basic array of bones, modified for different purposes.

These bones are said to be **homologous** in the different vertebrates; that is, they have the same evolutionary origin, but they now differ in structure and function. They are contrasted with **analogous** structures, such as the wings of birds and butterflies, which have similar function but different evolutionary origins.

Molecular evidence

Evolutionary patterns are also revealed at the molecular level. By comparing the genomes (that is, the sequences of all the genes) of different groups of animals or plants, we can more precisely specify the degree of relationship among the groups. A series of evolutionary changes over time should involve a continual accumulation of genetic changes in the DNA.

This difference can be seen clearly in the protein hemoglobin (figure 1.10). Rhesus monkeys, which like

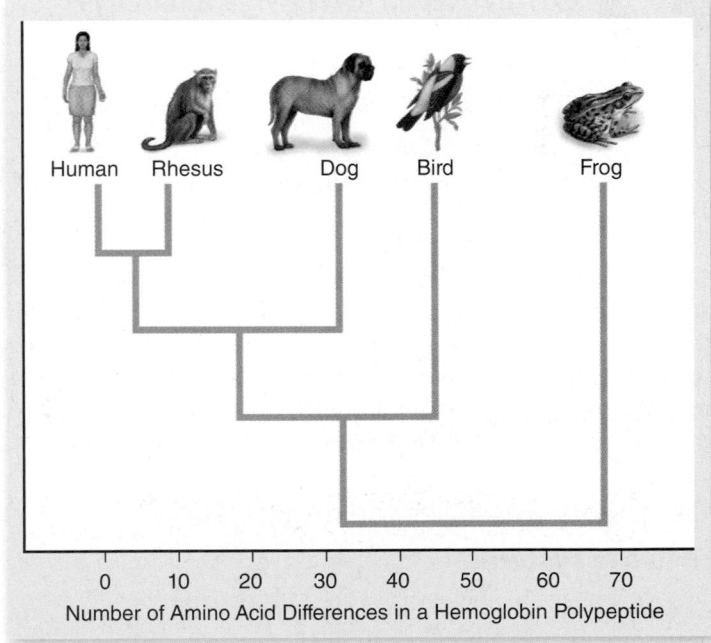

Human   Rhesus   Dog   Bird   Frog

Number of Amino Acid Differences in a Hemoglobin Polypeptide

Figure 1.10 Molecules reflect evolutionary patterns. Vertebrates that are more distantly related to humans have a greater number of amino acid differences in the hemoglobin polypeptide.

? Inquiry question Where do you imagine a snake might fall on the graph? Why?

humans are primates, have fewer differences from humans in the 146-amino-acid hemoglobin β-chain than do more distantly related mammals, such as dogs. Nonmammalian vertebrates, such as birds and frogs, differ even more.

The sequences of some genes, such as the ones specifying the hemoglobin proteins, have been determined in many organisms, and the entire time course of their evolution can be laid out with confidence by tracing the origins of particular nucleotide changes in the gene sequence. The pattern of descent obtained is called a **phylogenetic tree.** It represents the evolutionary history of the gene, its "family tree." Molecular phylogenetic trees agree well with those derived from the fossil record, which is strong direct evidence of evolution. The pattern of accumulating DNA changes represents, in a real sense, the footprints of evolutionary history.

a.   60 μm

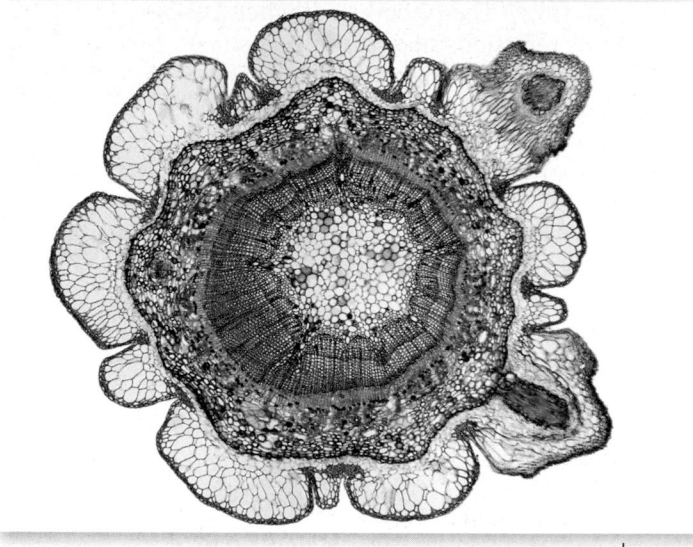

b.   500 μm

Learning Outcomes Review 1.3

Darwin observed differences in related organisms and proposed the hypothesis of evolution by natural selection to explain these differences. The predictions generated by natural selection have been tested and continue to be tested by analysis of the fossil record, genetics, comparative anatomy, and even the DNA of living organisms.

■ *Does Darwin's theory of evolution by natural selection explain the origin of life?*

1.4 Unifying Themes in Biology

Learning Outcomes

1. *Discuss the unifying themes in biology.*
2. *Contrast living and nonliving systems.*

The study of biology encompasses a large number of different subdisciplines, ranging from biochemistry to ecology. In all of these, however, unifying themes can be identified. Among these are cell theory, the molecular basis of inheritance, the relationship between structure and function, evolution, and the emergence of novel properties.

Cell theory describes the organization of living systems

As was stated at the beginning of this chapter, all organisms are composed of cells, life's basic units (figure 1.11). Cells were discovered by Robert Hooke in England in 1665, using one of the first microscopes, one that magnified 30 times. Not long after that, the Dutch scientist Anton van Leeuwenhoek used microscopes capable of magnifying 300 times and discovered an amazing world of single-celled life in a drop of pond water.

In 1839, the German biologists Matthias Schleiden and Theodor Schwann, summarizing a large number of observations by themselves and others, concluded that all living

Figure 1.11 Cellular basis of life. All organisms are composed of cells. Some organisms, including the protists, shown in part *(a)* are single-celled. Others, such as the plant shown in cross section in part *(b)* consist of many cells.

organisms consist of cells. Their conclusion has come to be known as the **cell theory.** Later, biologists added the idea that all cells come from preexisting cells. The cell theory, one of the basic ideas in biology, is the foundation for understanding the reproduction and growth of all organisms.

The molecular basis of inheritance explains the continuity of life

Even the simplest cell is incredibly complex—more intricate than any computer. The information that specifies what a cell is like—its detailed plan—is encoded in **deoxyribonucleic acid (DNA),** a long, cablelike molecule. Each DNA molecule is formed from two long chains of building blocks, called nucleotides, wound around each other (see chapter 14). Four different nucleotides are found in DNA, and the sequence in which they occur encodes the cell's information. Specific sequences of several hundred to many thousand nucleotides make up a **gene,** a discrete unit of information.

The continuity of life from one generation to the next—heredity—depends on the faithful copying of a cell's DNA into daughter cells. The entire set of DNA instructions that specifies a cell is called its *genome*. The sequence of the human genome, 3 billion nucleotides long, was decoded in rough draft form in 2001, a triumph of scientific investigation.

The relationship between structure and function underlies living systems

One of the unifying themes of molecular biology is the relationship between structure and function. Function in molecules, and larger macromolecular complexes, is dependent on their structure.

Although this observation may seem trivial, it has far-reaching implications. We study the structure of molecules and macromolecular complexes to learn about their function. When we know the function of a particular structure, we can infer the function of similar structures found in different contexts, such as in different organisms.

Biologists study both aspects, looking for the relationships between structure and function. On the one hand, this allows similar structures to be used to infer possible similar functions. On the other hand, this knowledge also gives clues as to what kinds of structures may be involved in a process if we know about the functionality.

For example, suppose that we know the structure of a human cell's surface receptor for insulin, the hormone that controls uptake of glucose. We then find a similar molecule in the membrane of a cell from a different species—perhaps even a very different organism, such as a worm. We might conclude that this membrane molecule acts as a receptor for an insulin-like molecule produced by the worm. In this way, we might be able to discern the evolutionary relationship between glucose uptake in worms and in humans.

The diversity of life arises by evolutionary change

The unity of life that we see in certain key characteristics shared by many related life-forms contrasts with the incredible diversity of living things in the varied environments of Earth. The underlying unity of biochemistry and genetics argues that all life has evolved from the same origin event. The diversity of life arises by evolutionary change leading to the present biodiversity we see.

Biologists divide life's great diversity into three great groups, called domains: Bacteria, Archaea, and Eukarya (figure 1.12). The domains Bacteria and Archaea are composed of single-celled organisms *(prokaryotes)* with little internal structure, and the domain Eukarya is made up of organisms *(eukaryotes)* composed of a complex, organized cell or multiple complex cells.

Within Eukarya are four main groups called kingdoms (see figure 1.12). Kingdom Protista consists of all the unicellular eukaryotes except yeasts (which are fungi), as well as the multicellular algae. Because of the great diversity among the protists, many biologists feel kingdom Protista should be split into several kingdoms.

Kingdom Plantae consists of organisms that have cell walls of cellulose and obtain energy by photosynthesis. Organisms in the kingdom Fungi have cell walls of chitin and obtain energy by secreting digestive enzymes and then absorbing the products they release from the external environment. Kingdom Animalia contains organisms that lack cell walls and obtain energy by first ingesting other organisms and then digesting them internally.

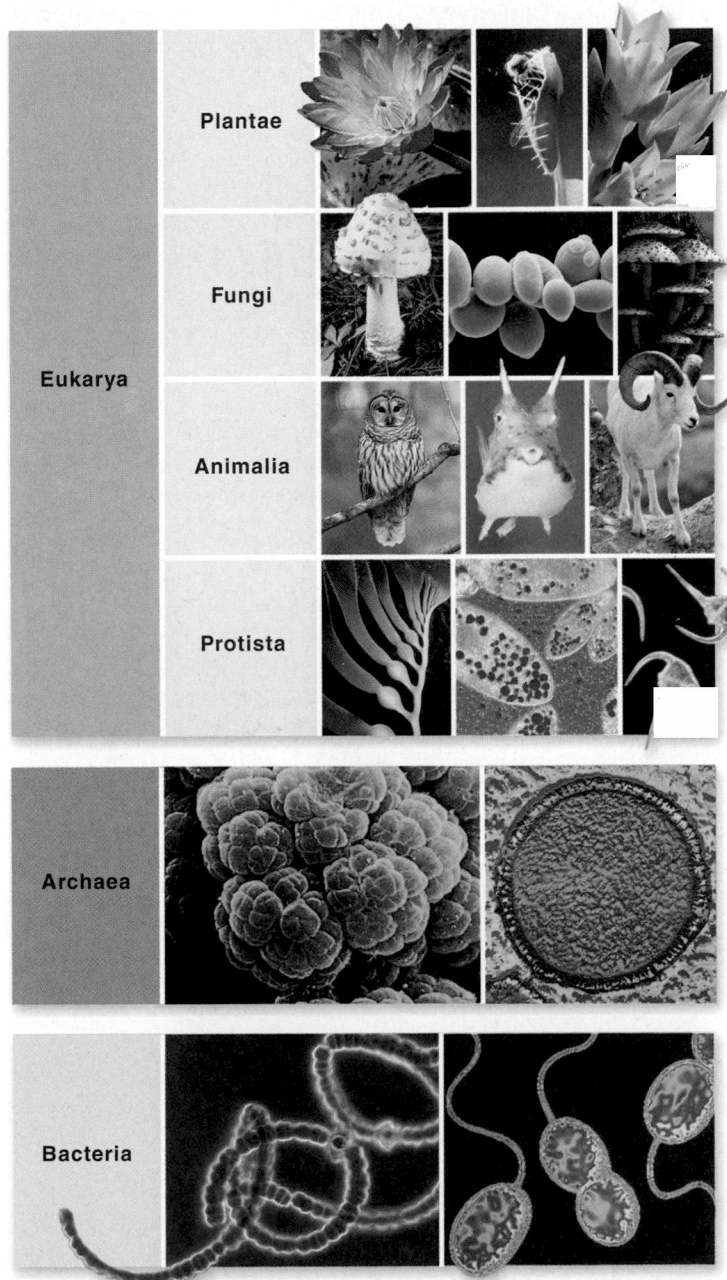

Figure 1.12 The diversity of life. Biologists categorize all living things into three overarching groups called domains: Bacteria, Archaea, and Eukarya. Domain Eukarya is composed of four kingdoms: Plantae, Fungi, Animalia, and Protista.

Evolutionary conservation explains the unity of living systems

Biologists agree that all organisms alive today have descended from some simple cellular creature that arose about 3.5 BYA. Some of the characteristics of that earliest organism have been preserved. The storage of hereditary information in DNA, for example, is common to all living things.

The retention of these conserved characteristics in a long line of descent usually reflects that they have a fundamental role in the biology of the organism—one not easily changed once adopted. A good example is provided by the homeodomain proteins, which play critical roles in early development in eukaryotes. Conserved characteristics can be seen in approximately 1850 homeodomain proteins, distributed among three different kingdoms of organisms (figure 1.13). The homeodomain proteins are powerful developmental tools that evolved early, and for which no better alternative has arisen.

Cells are information-processing systems

One way to think about cells is as highly complex nanomachines that process information. The information stored in DNA is used to direct the synthesis of cellular components, and the particular set of components can differ from cell to cell. The way that proteins fold in space is a form of information that is three-dimensional, and interesting properties emerge from the interaction of these shapes in macromolecular complexes. The control of gene expression allows differentiation of cell

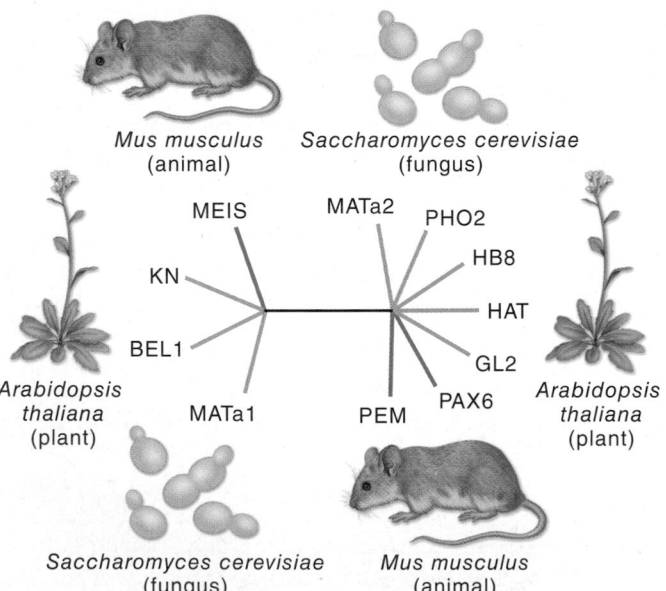

Figure 1.13 **Tree of homeodomain proteins.**
Homeodomain proteins are found in fungi (*brown*), plants (*green*), and animals (*blue*). Based on their sequence similarities, these 11 different homeodomain proteins (uppercase letters at the ends of branches) fall into two groups, with representatives from each kingdom in each group. That means, for example, the mouse homeodomain protein PAX6 is more closely related to fungal and flowering plant proteins, such as PHO2 and GL2, than it is to the mouse protein MEIS.

types in time and space, leading to changes over developmental time into different tissue types—even though all cells in an organism carry the same genetic information.

Cells also process information that they receive about the environment. Cells sense their environment through proteins in their membranes, and this information is transmitted across the membrane to elaborate signal-transduction chemical pathways that can change the functioning of a cell.

This ability of cells to sense and respond to their environment is critical to the function of tissues and organs in multicellular organisms. A multicellular organism can regulate its internal environment, maintaining constant temperature, pH, and concentrations of vital ions. This homeostasis is possible because of elaborate signaling networks that coordinate the activities of different cells in different tissues.

Living systems exist in a nonequilibrium state

A key feature of living systems is that they are open systems that function far from thermodynamic equilibrium. This has a number of implications for their behavior. A constant supply of energy is necessary to maintain a stable nonequilibrium state. Consider the state of the nucleic acids, and proteins in all of your cells: at equilibrium they are not polymers, they would all be hydrolyzed to monomer nucleotides and amino acids. Second, nonequilibrium systems exhibit self-organizing properties not seen in equilibrium systems.

These self-organizing properties of living systems show up at different levels of the hierarchical organization. At the cellular level, macromolecular complexes such as the spindle necessary for chromosome separation can self-organize. At the population level, a flock of birds, a school of fish, or the bacteria in a biofilm are all also self-organizing. This kind of interacting behavior of individual units leads to emergent properties that are not predictable from the nature of the units themselves.

Emergent properties are properties of collections of molecules, cells, individuals, that are distinct from the categorical properties that can be described by such statistics as mean and standard deviation. The mathematics necessary to describe these kind of interacting systems is nonlinear dynamics. The emerging field of systems biology is beginning to model biological systems in this way. The kinds of feedback and feedforward loops that exist between molecules in cells, or neurons in a nervous system, lead to emergent behaviors like human consciousness.

Learning Outcomes Review 1.4

Biology is a broad and complex field, but we can identify unifying themes in this complexity. Cells are the basic unit of life, and they are information-processing machines. The structures of molecules, macromolecular complexes, cells, and even higher levels of organization are related to their functions. The diversity of life can be classified and organized based on similar features; biologists identify three large domains that encompass six kingdoms. Living organisms are able to use energy to construct complex molecules from simple ones, and are thus not in a state of thermodynamic equilibrium.

■ *How do viruses fit into our definitions of living systems?*

Chapter Review

1.1 The Science of Life

Biology unifies much of natural science.

The study of biological systems is interdisciplinary because solutions require many different approaches to solve a problem.

Life defies simple definition.

Although life is difficult to define, living systems have seven characteristics in common. They are composed of one or more cells; are complex and highly ordered; can respond to stimuli; can grow, reproduce, and transmit genetic information to their offspring; need energy to accomplish work; can maintain relatively constant internal conditions (homeostasis); and are capable of evolutionary adaptation to the environment.

Living systems show hierarchical organization.

The hierarchical organization of living systems progresses from atoms to the biosphere. At each higher level, emergent properties arise that are greater than the sum of the parts.

1.2 The Nature of Science

At its core, science is concerned with understanding the nature of the world by using observation and reasoning.

Much of science is descriptive.

Science is concerned with developing an increasingly accurate description of nature through observation and experimentation.

Science uses both deductive and inductive reasoning.

Deductive reasoning applies general principles to predict specific results. Inductive reasoning uses specific observations to construct general scientific principles.

Hypothesis-driven science makes and tests predictions.

A hypothesis is constructed based on observations, and it must generate experimentally testable predictions. Experiments involve a test in which a variable is manipulated, and a control in which the variable is not manipulated. Hypotheses are rejected if their predictions cannot be verified by observation or experiment.

Reductionism breaks larger systems into their component parts.

Reductionism attempts to understand a complex system by breaking it down into its component parts. It is limited because parts may act differently when isolated from the larger system.

Biologists construct models to explain living systems.

A model provides a way of organizing our thinking about a problem; models may also suggest experimental approaches.

The nature of scientific theories.

Scientists use the word *theory* in two main ways: as a proposed explanation for some natural phenomenon and as a body of concepts that explains facts in an area of study.

Research can be basic or applied.

Basic research extends the boundaries of what we know; applied research seeks to use scientific findings in practical areas such as agriculture, medicine, and industry.

1.3 An Example of Scientific Inquiry: Darwin and Evolution

Darwin's theory of evolution shows how a scientist develops a hypothesis and sets forth evidence, as well as how a scientific theory grows and gains acceptance.

The idea of evolution existed prior to Darwin.

A number of naturalists and philosophers had suggested living things had changed during Earth's history. Darwin's contribution was the concept of natural selection as a mechanism for evolutionary change.

Darwin observed differences in related organisms.

During the voyage of the H.M.S. *Beagle,* Darwin had an opportunity to observe worldwide patterns of diversity.

Darwin proposed natural selection as a mechanism for evolution.

Darwin noted that species produce many offspring, but only a limited number survive and reproduce. He observed that the traits of offspring can be changed by artificial selection. Darwin proposed that individuals possessing traits that increase survival and reproductive success become more numerous in populations over time. This is the essence of descent with modification (natural selection). Alfred Russel Wallace independently came to the same conclusions from his own studies.

The predictions of natural selection have been tested.

Natural selection has been tested using data from many fields. Among these are the fossil record; the age of the Earth, determined by rates of radioactive decay to be 4.5 billion years; genetic experiments such as those of Gregor Mendel, showing that traits can be inherited as discrete units; comparative anatomy and the study of homologous structures; and molecular data that provides evidence for changes in DNA and proteins over time.

Taken together, these findings strongly support evolution by natural selection. No data to conclusively disprove evolution has been found.

1.4 Unifying Themes in Biology

Cell theory describes the organization of living systems.

The cell is the basic unit of life and is the foundation for understanding growth and reproduction in all organisms.

The molecular basis of inheritance explains the continuity of life.

Hereditary information, encoded in genes found in the DNA molecule, is passed on from one generation to the next.

The relationship between structure and function underlies living systems.

The function of macromolecules and their complexes is dictated by and dependent on their structure. Similarity of structure and function from one life-form to another may indicate an evolutionary relationship.

The diversity of life arises by evolutionary change.

Living organisms appear to have had a common origin from which a diversity of life arose by evolutionary change. They can be grouped into three domains comprising six kingdoms based on their differences.

Evolutionary conservation explains the unity of living systems.

The underlying similarities in biochemistry and genetics support the contention that all life evolved from a single source.

Cells are information-processing systems.

Cells can sense and respond to environmental changes through proteins located on their cell membranes. Differential expression of stored genetic information is the basis for different cell types.

Living systems exist in a nonequilibrium state.

Organisms are open systems that need a constant supply of energy to maintain their stable nonequilibrium state. Living things are able to self-organize, creating levels of complexity that may exhibit emergent properties.

Review Questions

UNDERSTAND

1. Which of the following is NOT a property of life?

   a. Energy utilization   c. Order
   b. Movement   d. Homeostasis

2. The process of inductive reasoning involves

   a. the use of general principles to predict a specific result.
   b. the generation of specific predictions based on a belief system.
   c. the use of specific observations to develop general principles.
   d. the use of general principles to support a hypothesis.

3. A hypothesis in biology is best described as

   a. a possible explanation of an observation.
   b. an observation that supports a theory.
   c. a general principle that explains some aspect of life.
   d. an unchanging statement that correctly predicts some aspect of life.

4. A scientific theory is

   a. a guess about how things work in the world.
   b. a statement of how the world works that is supported by experimental data.
   c. a belief held by many scientists.
   d. Both a and c are correct.

5. The cell theory states that

   a. cells are small.
   b. cells are highly organized.
   c. there is only one basic type of cell.
   d. all living things are made up of cells.

6. The molecule DNA is important to biological systems because

   a. it can be replicated.
   b. it encodes the information for making a new individual.
   c. it forms a complex, double-helical structure.
   d. nucleotides form genes.

7. The organization of living systems is

   a. linear with cells at one end and the biosphere at the other.
   b. circular with cells in the center.
   c. hierarchical with cells at the base, and the biosphere at the top.
   d. chaotic and beyond description.

8. The idea of evolution

   a. was original to Darwin.
   b. was original to Wallace.
   c. predated Darwin and Wallace.
   d. Both a and b are correct.

APPLY

1. What is the significance of Pasteur's experiment to test the germ hypothesis?

   a. It proved that heat can sterilize a broth.
   b. It demonstrated that cells can arise spontaneously.
   c. It demonstrated that some cells are germs.
   d. It demonstrated that cells can only arise from other cells.

2. Which of the following is NOT an example of reductionism?

   a. Analysis of an isolated enzyme's function in an experimental assay

   b. Investigation of the effect of a hormone on cell growth in a Petri dish
   c. Observation of the change in gene expression in response to specific stimulus
   d. An evaluation of the overall behavior of a cell

3. How is the process of natural selection different from that of artificial selection?

   a. Natural selection produces more variation.
   b. Natural selection makes an individual better adapted.
   c. Artificial selection is a result of human intervention.
   d. Artificial selection results in better adaptations.

4. If you found a fossil for a modern organism next to the fossil of a dinosaur, this would

   a. argue against evolution by natural selection.
   b. have no bearing on evolution by natural selection.
   c. indicate that dinosaurs may still exist.
   d. Both b and c are correct.

5. The theory of evolution by natural selection is a good example of how science proceeds because

   a. it rationalizes a large body of observations.
   b. it makes predictions that have been tested by a variety of approaches.
   c. it represents Darwin's belief of how life has changed over time.
   d. Both b and c are correct.

6. In which domain of life would you find only single-celled organisms?

   a. Eukarya   c. Archaea
   b. Bacteria   d. Both b and c are correct.

7. Evolutionary conservation occurs when a characteristic is

   a. important to the life of the organism.
   b. not influenced by evolution.
   c. no longer functionally important.
   d. found in more primitive organisms.

SYNTHESIZE

1. Exobiology is the study of life on other planets. In recent years, scientists have sent various spacecraft out into the galaxy in search for extraterrestrial life. Assuming that all life shares common properties, what should exobiologists be looking for as they explore other worlds?

2. The classic experiment by Pasteur (figure 1.4) tested the hypothesis that cells arise from other cells. In this experiment cell growth was measured following sterilization of broth in a swan-necked flask or in a flask with a broken neck.

   a. Which variables were kept the same in these two experiments?
   b. How does the shape of the flask affect the experiment?
   c. Predict the outcome of each experiment based on the two hypotheses.
   d. Some bacteria (germs) are capable of producing heat-resistant spores that protect the cell and allow it to continue to grow after the environment cools. How would the outcome of this experiment have been affected if spore-forming bacteria were present in the broth?

Chapter

2

The Nature of Molecules and the Properties of Water

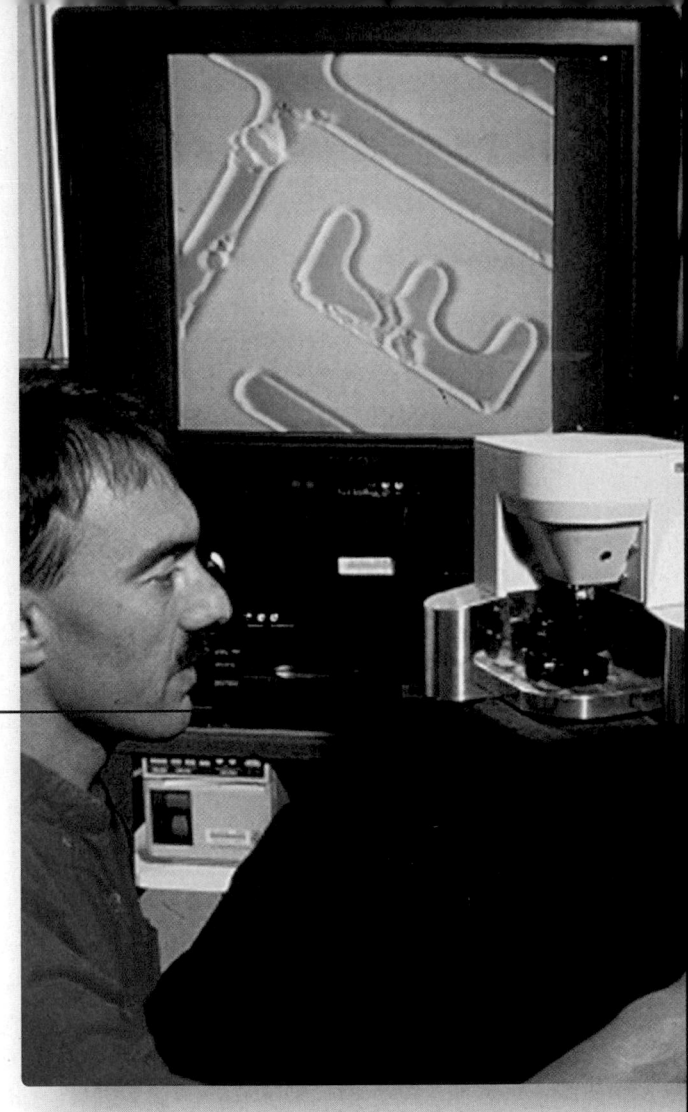

Chapter Contents

Introduction

About 12.5 billion years ago (BYA), an enormous explosion probably signaled the beginning of the universe. This explosion started a process of star building and planetary formation that eventually led to the formation of Earth, about 4.5 BYA. Around 3.5 BYA, life began on Earth and started to diversify. To understand the nature of life on Earth, we first need to understand the nature of the matter that forms the building blocks of all life.

The earliest speculations about the world around us included this most basic question, "What is it made of?" The ancient Greeks recognized that larger things may be built of smaller parts. This concept was formed into a solid experimental scientific idea in the early 20th century, when physicists began trying to break atoms apart. From those humble beginnings to the huge particle accelerators used by the modern physicists of today, the picture of the atomic world emerges as fundamentally different from the tangible, macroscopic world around us.

To understand how living systems are assembled, we must first understand a little about atomic structure, about how atoms can be linked together by chemical bonds to make molecules, and about the ways in which these small molecules are joined together to make larger molecules, until finally we arrive at the structures of cells and then of organisms. Our study of life on Earth therefore begins with physics and chemistry. For many of you, this chapter will be a review of material encountered in other courses.

2.1 The Nature of Atoms

Learning Outcomes

1. Define an element based on its composition.
2. Describe how atomic structure produces chemical properties.
3. Explain where electrons are found in an atom.

Any substance in the universe that has mass and occupies space is defined as *matter*. All matter is composed of extremely small particles called **atoms.** Because of their size, atoms are difficult to study. Not until early in the 20th century did scientists carry out the first experiments revealing the physical nature of atoms (figure 2.1).

SCIENTIFIC THINKING

Hypothesis: *Atoms are composed of diffuse positive charge with embedded negative charge (electrons).*

Prediction: *If alpha particles (α), which are helium nuclei, are shot at a thin foil of gold, the α-particles will not be deflected much by the diffuse positive charge or by the light electrons.*

Test: *α-Particles are shot at a thin sheet of gold foil surrounded by a detector screen, which shows flashes of light when hit by the particles.*

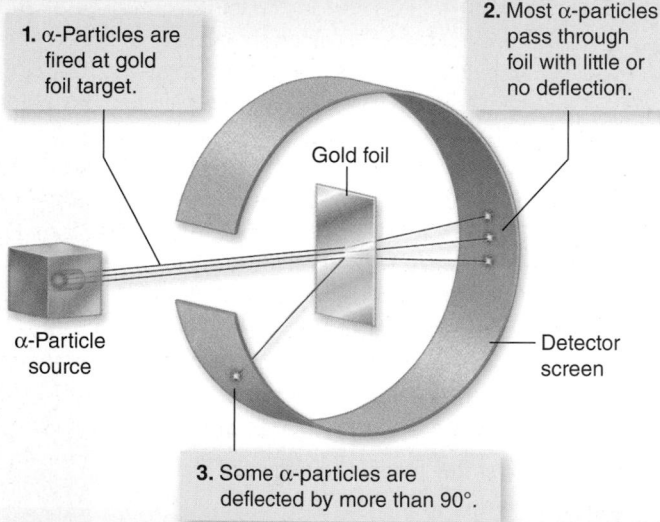

1. α-Particles are fired at gold foil target.

2. Most α-particles pass through foil with little or no deflection.

Gold foil

α-Particle source

Detector screen

3. Some α-particles are deflected by more than 90°.

Result: *Most particles are not deflected at all, but a small percentage of particles are deflected at angles of 90° or more.*

Conclusion: *The hypothesis is not supported. The large deflections observed led to a view of the atom as composed of a very small central region containing positive charge (the nucleus) surrounded by electrons.*

Further Experiments: *How does the Bohr atom with its quantized energy for electrons extend this model?*

Figure 2.1 Rutherford scattering experiment.
Large-angle scattering of α-particles led Rutherford to propose the existence of the nucleus.

Atomic structure includes a central nucleus and orbiting electrons

Objects as small as atoms can be "seen" only indirectly, by using complex technology such as tunneling microscopy (figure 2.2). We now know a great deal about the complexities of atomic structure, but the simple view put forth in 1913 by the Danish physicist Niels Bohr provides a good starting point for understanding atomic theory. Bohr proposed that every atom possesses an orbiting cloud of tiny subatomic particles called *electrons* whizzing around a core, like the planets of a miniature solar system. At the center of each atom is a small, very dense nucleus formed of two other kinds of subatomic particles: *protons* and *neutrons* (figure 2.3).

Atomic number and the elements

Within the nucleus, the cluster of protons and neutrons is held together by a force that works only over short, subatomic distances. Each proton carries a positive (+) charge, and each neutron has no charge. Each electron carries a negative (−) charge. Typically, an atom has one electron for each proton and is, thus, electrically neutral. Different atoms are defined by the number of protons, a quantity called the *atomic number.* The chemical behavior of an atom is due to the number and configuration of electrons, as we will see later in this chapter. Atoms with the same atomic number (that is, the same number of protons) have the same chemical properties and are said to belong to the same element. Formally speaking, an *element* is any substance that cannot be broken down to any other substance by ordinary chemical means.

Atomic mass

The terms *mass* and *weight* are often used interchangeably, but they have slightly different meanings. *Mass* refers to the amount of a substance, but *weight* refers to the force gravity exerts on a substance. An object has the same mass whether it is on the Earth or the Moon, but its weight will be greater on the Earth because the Earth's gravitational force is greater than the Moon's. The *atomic mass* of an atom is equal to the sum of the masses of its protons and neutrons. Atoms that occur naturally on Earth contain from 1 to 92 protons and up to 146 neutrons.

Figure 2.2 Scanning-tunneling microscope image. The scanning-tunneling microscope is a nonoptical way of imaging that allows atoms to be visualized. This image shows a lattice of oxygen atoms (dark blue) on a rhodium crystal (light blue).

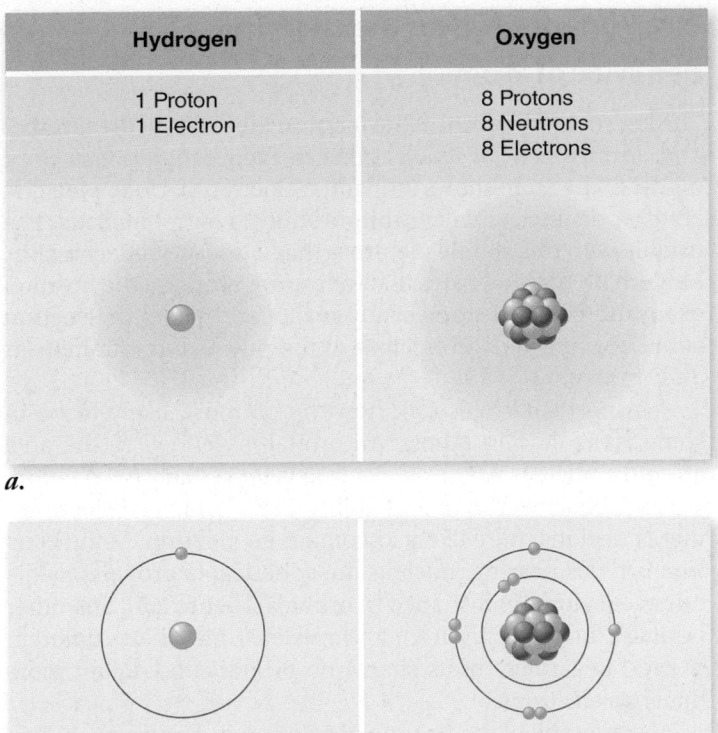

Hydrogen	Oxygen
1 Proton 1 Electron	8 Protons 8 Neutrons 8 Electrons

a.

b.

● proton (positive charge)	● electron (negative charge)	● neutron (no charge)

Figure 2.3 Basic structure of atoms. All atoms have a nucleus consisting of protons and neutrons, except hydrogen, the smallest atom, which usually has only one proton and no neutrons in its nucleus. Oxygen typically has eight protons and eight neutrons in its nucleus. In the simple "Bohr model" of atoms pictured here, electrons spin around the nucleus at a relatively far distance. *a.* Atoms are depicted as a nucleus with a cloud of electrons (not shown to scale). *b.* The electrons are shown in discrete energy levels. These are described in greater detail in the text.

The mass of atoms and subatomic particles is measured in units called *daltons.* To give you an idea of just how small these units are, note that it takes 602 million million billion (6.02×10^{23}) daltons to make 1 gram (g). A proton weighs approximately 1 dalton (actually 1.007 daltons), as does a neutron (1.009 daltons). In contrast, electrons weigh only 1/1840 of a dalton, so they contribute almost nothing to the overall mass of an atom.

Electrons

The positive charges in the nucleus of an atom are neutralized, or counterbalanced, by negatively charged electrons, which are located in regions called **orbitals** that lie at varying distances around the nucleus. Atoms with the same number of protons and electrons are electrically neutral; that is, they have no net charge, and are therefore called *neutral atoms.*

Electrons are maintained in their orbitals by their attraction to the positively charged nucleus. Sometimes other forces overcome this attraction, and an atom loses one or more electrons. In other cases, atoms gain additional electrons. Atoms in which the number of electrons does not equal the number of protons are known as *ions,* and they are charged particles. An atom having more protons than electrons has a net positive charge and is called a **cation.** For example, an atom of sodium (Na) that has lost one electron becomes a sodium ion (Na+), with a charge of +1. An atom having fewer protons than electrons carries a net negative charge and is called an **anion.** A chlorine atom (Cl) that has gained one electron becomes a chloride ion (Cl−), with a charge of −1.

Isotopes

Although all atoms of an element have the same number of protons, they may not all have the same number of neutrons. Atoms of a single element that possess different numbers of neutrons are called **isotopes** of that element.

Most elements in nature exist as mixtures of different isotopes. Carbon (C), for example, has three isotopes, all containing six protons (figure 2.4). Over 99% of the carbon found in nature exists as an isotope that also contains six neutrons. Because the total mass of this isotope is 12 daltons (6 from protons plus 6 from neutrons), it is referred to as carbon-12 and is symbolized ^{12}C. Most of the rest of the naturally occurring carbon is carbon-13, an isotope with seven neutrons. The rarest carbon isotope is carbon-14, with eight neutrons. Unlike the other two isotopes, carbon-14 is unstable: This means that its nucleus tends to break up into elements with lower atomic numbers. This nuclear breakup, which emits a significant amount of energy, is called *radioactive decay,* and isotopes that decay in this fashion are **radioactive isotopes.**

Some radioactive isotopes are more unstable than others, and therefore they decay more readily. For any given isotope, however, the rate of decay is constant. The decay time is usually expressed as the *half-life,* the time it takes for one-half

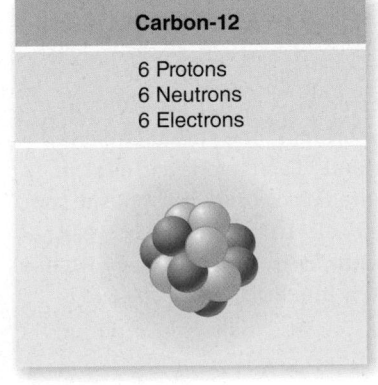

Carbon-12
6 Protons 6 Neutrons 6 Electrons

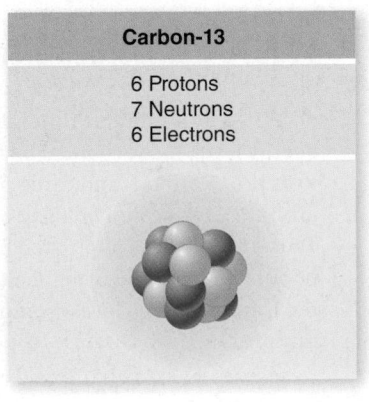

Carbon-13
6 Protons 7 Neutrons 6 Electrons

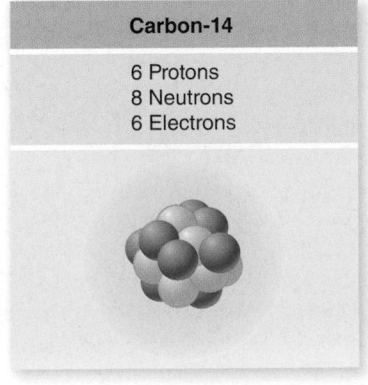

Carbon-14
6 Protons 8 Neutrons 6 Electrons

Figure 2.4 The three most abundant isotopes of carbon. Isotopes of a particular element have different numbers of neutrons.

of the atoms in a sample to decay. Carbon-14, for example, often used in the carbon dating of fossils and other materials, has a half-life of 5730 years. A sample of carbon containing 1 g of carbon-14 today would contain 0.5 g of carbon-14 after 5730 years, 0.25 g 11,460 years from now, 0.125 g 17,190 years from now, and so on. By determining the ratios of the different isotopes of carbon and other elements in biological samples and in rocks, scientists are able to accurately determine when these materials formed.

Radioactivity has many useful applications in modern biology. Radioactive isotopes are one way to label, or "tag," a specific molecule and then follow its progress, either in a chemical reaction or in living cells and tissue. The downside, however, is that the energetic subatomic particles emitted by radioactive substances have the potential to severely damage living cells, producing genetic mutations and, at high doses, cell death. Consequently, exposure to radiation is carefully controlled and regulated. Scientists who work with radioactivity follow strict handling protocols and wear radiation-sensitive badges to monitor their exposure over time to help ensure a safe level of exposure.

Electrons determine the chemical behavior of atoms

The key to the chemical behavior of an atom lies in the number and arrangement of its electrons in their orbitals. The Bohr model of the atom shows individual electrons as following discrete, or distinct, circular orbits around a central nucleus. The trouble with this simple picture is that it doesn't reflect reality. Modern physics indicates that we cannot pinpoint the position of any individual electron at any given time. In fact, an electron could be anywhere, from close to the nucleus to infinitely far away from it.

A particular electron, however, is more likely to be in some areas than in others. An orbital is defined as the area around a nucleus where an electron is most likely to be found. These orbitals represent probability distributions for electrons, that is, regions more likely to contain an electron. Some electron orbitals near the nucleus are spherical (s orbitals), while others are dumbbell-shaped (p orbitals) (figure 2.5). Still other orbitals, farther away from the nucleus, may have different shapes. Regardless of its shape, no orbital can contain more than two electrons.

Almost all of the volume of an atom is empty space. This is because the electrons are usually far away from the nucleus, relative to its size. If the nucleus of an atom were the size of a golf ball, the orbit of the nearest electron would be a mile away. Consequently, the nuclei of two atoms never come close enough in nature to interact with each other. It is for this reason that an atom's electrons, not its protons or neutrons, determine its chemical behavior, and it also explains why the isotopes of an element, all of which have the same arrangement of electrons, behave the same way chemically.

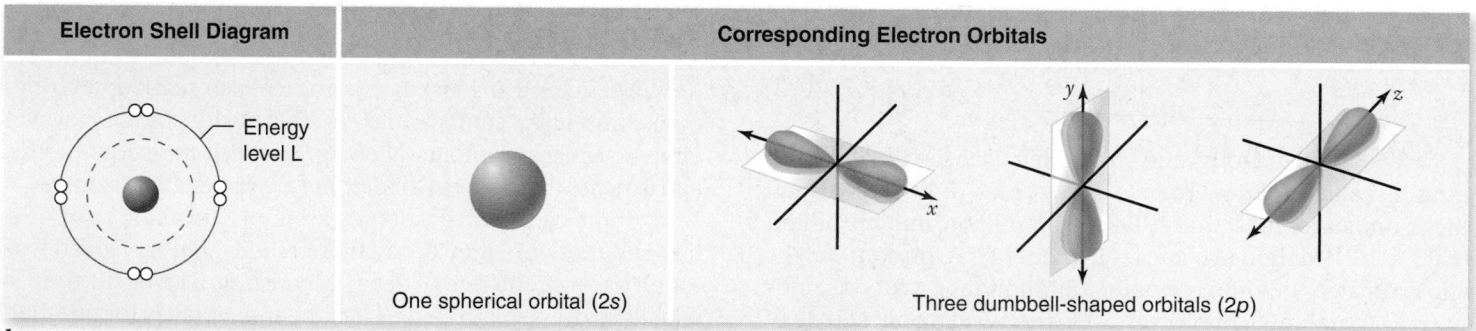

a.

b.

c.

Figure 2.5 Electron orbitals. *a.* The lowest energy level, or electron shell—the one nearest the nucleus—is level K. It is occupied by a single s orbital, referred to as 1s. *b.* The next highest energy level, L, is occupied by four orbitals: one s orbital (referred to as the 2s orbital) and three p orbitals (each referred to as a 2p orbital). Each orbital holds two paired electrons with opposite spin. Thus, the K level is populated by two electrons, and the L level is populated by a total of eight electrons. *c.* The neon atom shown has the L and K energy levels completely filled with electrons and is thus unreactive.

Atoms contain discrete energy levels

Because electrons are attracted to the positively charged nucleus, it takes work to keep them in their orbitals, just as it takes work to hold a grapefruit in your hand against the pull of gravity. The formal definition of energy is the ability to do work.

The grapefruit held above the ground is said to possess *potential energy* because of its position. If you release it, the grapefruit falls, and its potential energy is reduced. On the other hand, if you carried the grapefruit to the top of a building, you would increase its potential energy. Electrons also have a potential energy that is related to their position. To oppose the attraction of the nucleus and move the electron to a more distant orbital requires an input of energy, which results in an electron with greater potential energy. The chlorophyll that makes plants green captures energy from light during photosynthesis in this way. As you'll see in chapter 8—light energy excites electrons in the chlorophyll molecule. Moving an electron closer to the nucleus has the opposite effect: Energy is released, usually as radiant energy (heat or light), and the electron ends up with less potential energy (figure 2.6).

One of the initially surprising aspects of atomic structure is that electrons within the atom have discrete **energy levels.** These discrete levels correspond to quanta (sing., quantum), which means specific amount of energy. To use the grapefruit analogy again, it is as though a grapefruit could only be raised to particular floors of a building. Every atom exhibits a ladder of potential energy values, a discrete set of orbitals at particular energetic "distances" from the nucleus.

Because the amount of energy an electron possesses is related to its distance from the nucleus, electrons that are the same distance from the nucleus have the same energy, even if they occupy different orbitals. Such electrons are said to occupy the same energy level. The energy levels are denoted with letters K, L, M, and so on (see figure 2.6). Be careful not to confuse energy levels, which are drawn as rings to indicate an electron's *energy,* with orbitals, which have a variety of three-dimensional shapes and indicate an electron's most likely *location.* Electron orbitals are arranged so that as they are filled, this fills each energy level in successive order. This filling of

orbitals and energy levels is what is responsible for the chemical reactivity of elements.

During some chemical reactions, electrons are transferred from one atom to another. In such reactions, the loss of an electron is called **oxidation,** and the gain of an electron is called *reduction.*

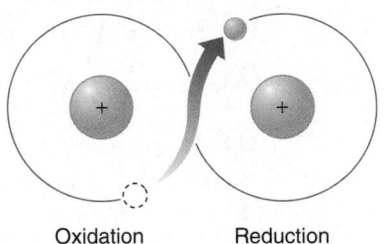

Oxidation   Reduction

Notice that when an electron is transferred in this way, it keeps its energy of position. In organisms, chemical energy is stored in high-energy electrons that are transferred from one atom to another in reactions involving oxidation and reduction (described in chapter 7). When the processes of oxidation and reduction are coupled, which often happens, one atom or molecule is oxidized while another is reduced in the same reaction. We call these combinations *redox reactions.*

Learning Outcomes Review 2.1

An atom consists of a nucleus of protons and neutrons surrounded by a cloud of electrons. For each atom, the number of protons is the atomic number; atoms with the same atomic number constitute an element. Atoms of a single element that have different numbers of neutrons are called isotopes. Electrons, which determine the chemical behavior of an element, are located about a nucleus in orbitals representing discrete energy levels. No orbital can contain more than two electrons, but each energy level consists of multiple orbitals, and thus contains many electrons with the same energy.

■ *If the number of protons exceeds the number of neutrons, is the charge on the atom positive or negative?*

■ *If the number of protons exceeds electrons?*

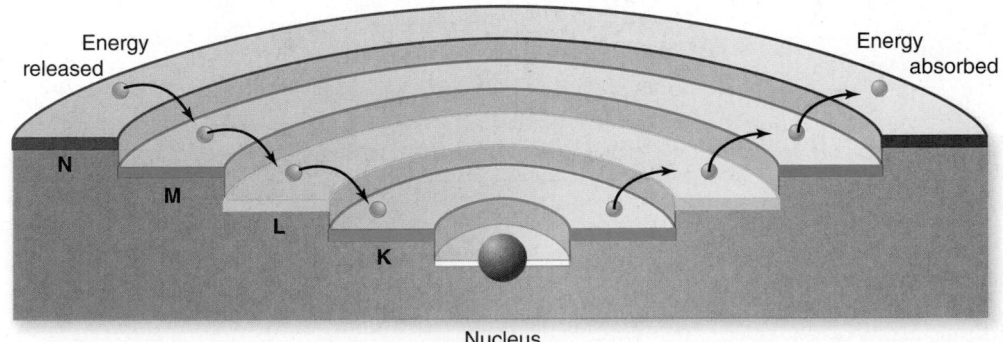

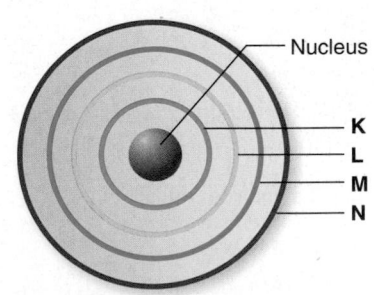

Figure 2.6 Atomic energy levels. Electrons have energy of position. When an atom absorbs energy, an electron moves to a higher energy level, farther from the nucleus. When an electron falls to lower energy levels, closer to the nucleus, energy is released. The first two energy levels are the same as shown in the previous figure.

Elements Found in Living Systems

Learning Outcomes

1. Relate atomic structure to the periodic table of the elements.
2. List the important elements found in living systems.

Ninety elements occur naturally, each with a different number of protons and a different arrangement of electrons. When the 19th-century Russian chemist Dmitri Mendeleev arranged the known elements in a table according to their atomic number, he discovered one of the great generalizations of science: The elements exhibit a pattern of chemical properties that repeats itself in groups of eight. This periodically repeating pattern lent the table its name: the periodic table of elements (figure 2.7).

The periodic table displays elements according to atomic number and properties

The eight-element periodicity that Mendeleev found is based on the interactions of the electrons in the outermost energy level of the different elements. These electrons are called **valence electrons,** and their interactions are the basis for the elements' differing chemical properties. For most of the atoms important to life, the outermost energy level can contain no more than eight electrons; the chemical behavior of an element reflects how many of the eight positions are filled.

Elements possessing all eight electrons in their outer energy level (two for helium) are *inert,* or nonreactive. These elements, which include helium (He), neon (Ne), argon (Ar), and so on, are termed the *noble gases.* In sharp contrast, elements with seven electrons (one fewer than the maximum number of eight) in their outer energy level, such as fluorine (F), chlorine (Cl), and bromine (Br), are highly reactive. They tend to gain the extra electron needed to fill the energy level. Elements with only one electron in their outer energy level, such as lithium (Li), sodium (Na), and potassium (K), are also very reactive. They tend to lose the single electron in their outer level.

Mendeleev's periodic table leads to a useful generalization, the **octet rule,** or *rule of eight* (Latin *octo,* "eight"): Atoms tend to establish completely full outer energy levels. For the main group elements of the periodic table, the rule of eight is accomplished by one filled *s* orbital and three filled *p* orbitals (figure 2.8). The exception to this is He, in the first row, which needs only two electrons to fill the 1*s* orbital. Most chemical behavior of biological interest can be predicted quite accurately from this simple rule, combined with the tendency of atoms to balance positive and negative charges. For instance, you read earlier that sodium ion (Na^+) has lost an electron, and chloride ion (Cl^-) has gained an electron. In the following section, we describe how these ions react to form table salt.

Of the 90 naturally occurring elements on Earth, only 12 (C, H, O, N, P, S, Na, K, Ca, Mg, Fe, Cl) are found in living systems in more than trace amounts (0.01% or higher). These elements all have atomic numbers less than 21, and thus, have low atomic masses. Of these 12, the first 4 elements (carbon, hydrogen, oxygen, and nitrogen) constitute 96.3% of the weight of your body. The majority of molecules that make up your body (other than water) are compounds of carbon, which we call

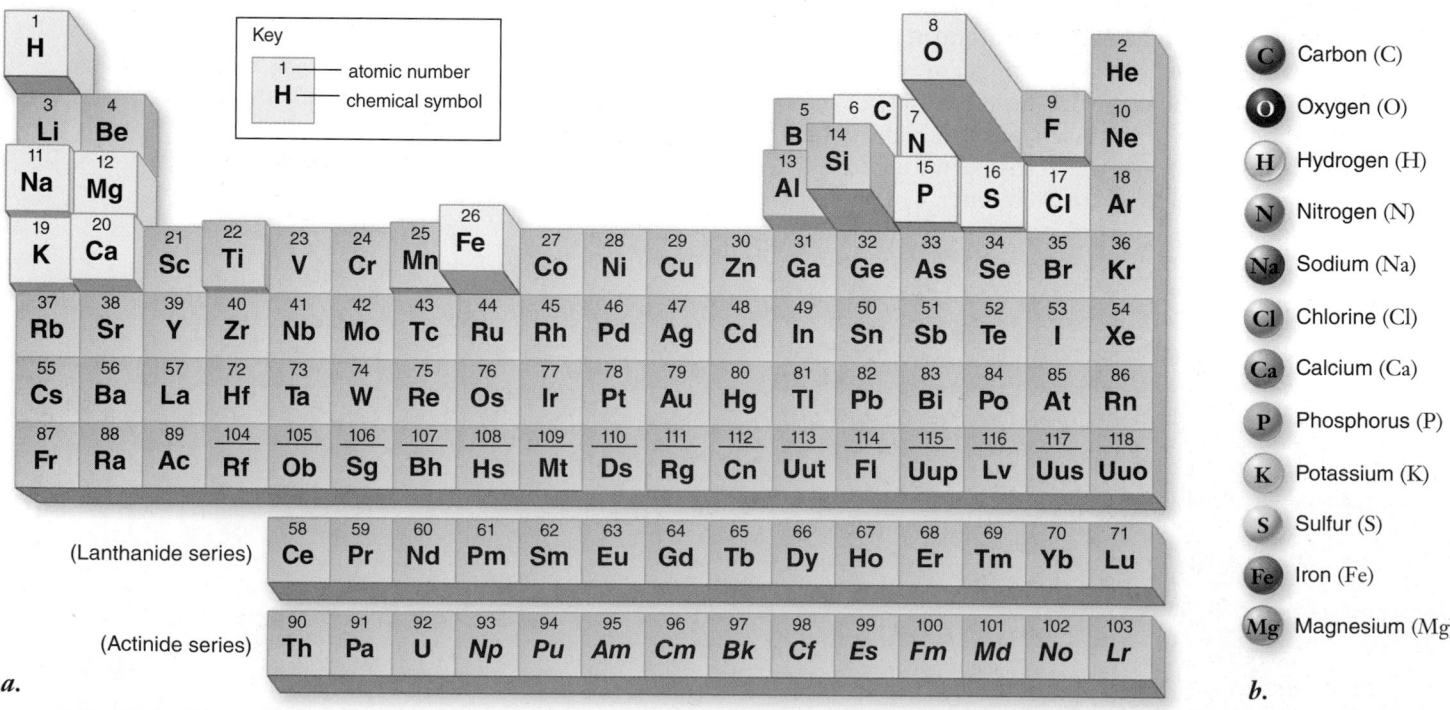

Figure 2.7 Periodic table of the elements. *a.* In this representation, the frequency of elements that occur in the Earth's crust is indicated by the height of the block. Elements shaded in green are found in living systems in more than trace amounts. *b.* Common elements found in living systems are shown in colors that will be used throughout the text.

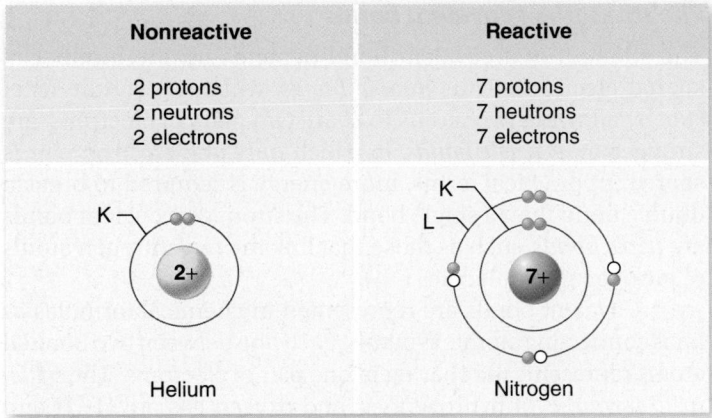

Nonreactive	Reactive
2 protons 2 neutrons 2 electrons	7 protons 7 neutrons 7 electrons
Helium	Nitrogen

Figure 2.8 Electron energy levels for helium and nitrogen. Green balls represent electrons, blue ball represents the nucleus with number of protons indicated by number of (+) charges. Note that the helium atom has a filled K shell and is thus unreactive, whereas the nitrogen atom has five electrons in the L shell, three of which are unpaired, making it reactive.

organic compounds. These organic compounds contain primarily these four elements (CHON), explaining their prevalence in living systems. Some trace elements, such as zinc (Zn) and iodine (I), play crucial roles in living processes even though they are present in tiny amounts. Iodine deficiency, for example, can lead to enlargement of the thyroid gland, causing a bulge at the neck called a goiter.

Learning Outcomes Review 2.2

The periodic table shows the elements in terms of atomic number and repeating chemical properties. Only 12 elements are found in significant amounts in living organisms: C, H, O, N, P, S, Na, K, Ca, Mg, Fe, and Cl.

■ *Why are the noble gases more stable than other elements in the periodic table?*

2.3 The Nature of Chemical Bonds

Learning Outcomes

1. *Predict which elements are likely to form ions.*
2. *Explain how molecules can be built from atoms joined by covalent bonds.*
3. *Contrast polar and nonpolar covalent bonds.*

A group of atoms held together by energy in a stable association is called a *molecule*. When a molecule contains atoms of more than one element, it is called a *compound*. The atoms in a molecule are joined by *chemical bonds;* these bonds can result when atoms with opposite charges attract each other (ionic bonds), when two atoms share one or more pairs of electrons

TABLE 2.1	Bonds and Interactions	
Name	**Basis of Interaction**	**Strength**
Covalent bond	Sharing of electron pairs	Strong
Ionic bond	Attraction of opposite charges	
Hydrogen bond	Sharing of H atom	
Hydrophobic interaction	Forcing of hydrophobic portions of molecules together in presence of polar substances	
van der Waals attraction	Weak attractions between atoms due to oppositely polarized electron clouds	Weak

(covalent bonds), or when atoms interact in other ways (table 2.1). We will start by examining *ionic bonds,* which form when atoms with opposite electrical charges (ions) attract.

Ionic bonds form crystals

Common table salt, the molecule sodium chloride (NaCl), is a lattice of ions in which the atoms are held together by ionic bonds (figure 2.9). Sodium has 11 electrons: 2 in the inner

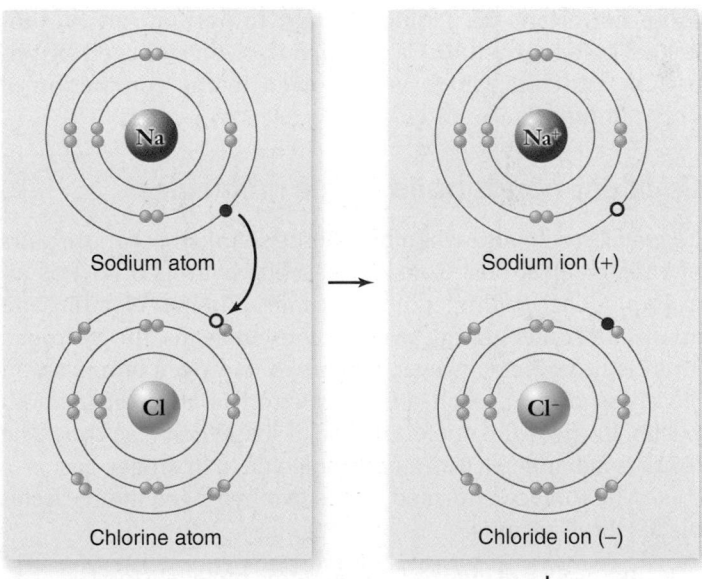

Sodium atom   Sodium ion (+)

Chlorine atom   Chloride ion (−)

a.

Figure 2.9 The formation of ionic bonds by sodium chloride. *a.* When a sodium atom donates an electron to a chlorine atom, the sodium atom becomes a positively charged sodium ion, and the chlorine atom becomes a negatively charged chloride ion. *b.* The electrostatic attraction of oppositely charged ions leads to the formation of a lattice of Na^+ and Cl^-.

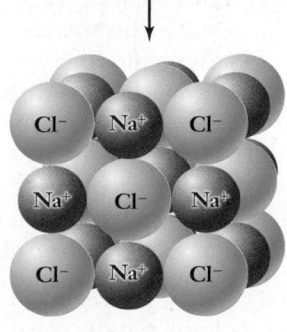

b. NaCl crystal

energy level (K), 8 in the next level (L), and 1 in the outer (valence) level (M). The single, unpaired valence electron has a strong tendency to join with another unpaired electron in another atom. A stable configuration can be achieved if the valence electron is lost to another atom that also has an unpaired electron. The loss of this electron results in the formation of a positively charged sodium ion, Na^+.

The chlorine atom has 17 electrons: 2 in the K level, 8 in the L level, and 7 in the M level. As you can see in the figure, one of the orbitals in the outer energy level has an unpaired electron (red circle). The addition of another electron fills that level and causes a negatively charged chloride ion, Cl^-, to form.

When placed together, metallic sodium and gaseous chlorine react swiftly and explosively, as the sodium atoms donate electrons to chlorine to form Na^+ and Cl^- ions. Because opposite charges attract, the Na^+ and Cl^- remain associated in an *ionic compound,* NaCl, which is electrically neutral. The electrical attractive force holding NaCl together, however, is not directed specifically between individual Na^+ and Cl^- ions, and no individual sodium chloride molecules form. Instead, the force exists between any one ion and *all* neighboring ions of the opposite charge. The ions aggregate in a crystal matrix with a precise geometry. Such aggregations are what we know as salt crystals. If a salt such as NaCl is placed in water, the electrical attraction of the water molecules, for reasons we will point out later in this chapter, disrupts the forces holding the ions in their crystal matrix, causing the salt to dissolve into a roughly equal mixture of free Na^+ and Cl^- ions.

Because living systems always include water, ions are more important than ionic crystals. Important ions in biological systems include Ca^{2+}, which is involved in cell signaling, K^+ and Na^+, which are involved in the conduction of nerve impulses.

Covalent bonds build stable molecules

Covalent bonds form when two atoms share one or more pairs of valence electrons. Consider gaseous hydrogen (H_2) as an example. Each hydrogen atom has an unpaired electron and an unfilled outer energy level; for these reasons, the hydrogen atom is unstable. However, when two hydrogen atoms are in close association, each atom's electron is attracted to both nuclei. In effect, the nuclei are able to share their electrons. The result is a diatomic (two-atom) molecule of hydrogen gas.

The molecule formed by the two hydrogen atoms is stable for three reasons:

1. **It has no net charge.** The diatomic molecule formed as a result of this sharing of electrons is not charged because it still contains two protons and two electrons.
2. **The octet rule is satisfied.** Each of the two hydrogen atoms can be considered to have two orbiting electrons in its outer energy level. This state satisfies the octet rule, because each shared electron orbits both nuclei and is included in the outer energy level of both atoms.
3. **It has no unpaired electrons.** The bond between the two atoms also pairs the two free electrons.

Unlike ionic bonds, covalent bonds are formed between two individual atoms, giving rise to true, discrete molecules.

The strength of covalent bonds

The strength of a covalent bond depends on the number of shared electrons. Thus *double bonds,* which satisfy the octet rule by allowing two atoms to share two pairs of electrons, are stronger than *single bonds,* in which only one electron pair is shared. In practical terms, more energy is required to break a double bond than a single bond. The strongest covalent bonds are *triple bonds,* such as those that link the two nitrogen atoms of nitrogen gas molecules (N_2).

Covalent bonds are represented in chemical formulas as lines connecting atomic symbols. Each line between two bonded atoms represents the sharing of one pair of electrons. The *structural formulas* of hydrogen gas and oxygen gas are H—H and O$=$O, respectively, and their *molecular formulas* are H_2 and O_2. The structural formula for N_2 is N$\equiv$N.

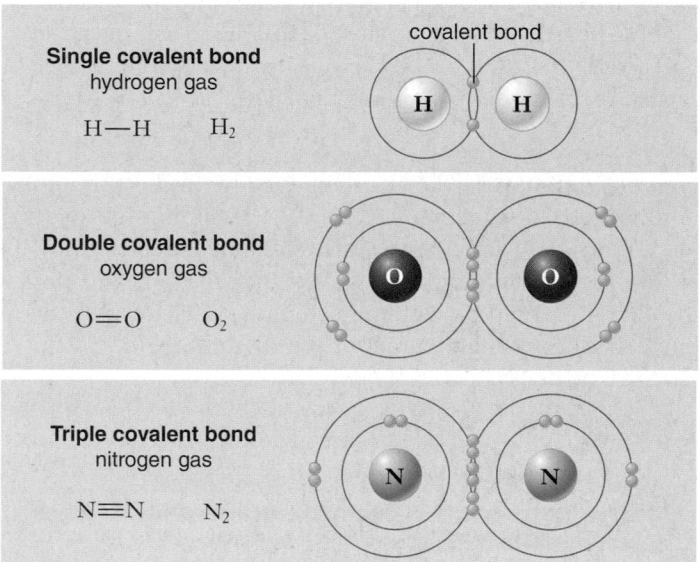

Molecules with several covalent bonds

A vast number of biological compounds are composed of more than two atoms. An atom that requires two, three, or four additional electrons to fill its outer energy level completely may acquire them by sharing its electrons with two or more other atoms.

For example, the carbon atom (C) contains six electrons, four of which are in its outer energy level and are unpaired. To satisfy the octet rule, a carbon atom must form four covalent bonds. Because four covalent bonds may form in many ways, carbon atoms are found in many different kinds of molecules. CO_2 (carbon dioxide), CH_4 (methane), and C_2H_5OH (ethanol) are just a few examples.

Polar and nonpolar covalent bonds

Atoms differ in their affinity for electrons, a property called **electronegativity.** In general, electronegativity increases left to right across a row of the periodic table and decreases down the column. Thus the elements in the upper-right corner have the highest electronegativity.

For bonds between identical atoms, for example, between two hydrogen or two oxygen atoms, the affinity for electrons is obviously the same, and the electrons are equally shared. Such bonds are termed **nonpolar.** The resulting compounds (H_2 or O_2) are also referred to as nonpolar.

For atoms that differ greatly in electronegativity, electrons are not shared equally. The shared electrons are more likely to be closer to the atom with greater electronegativity, and less likely to be near the atom of lower electronegativity. In this case, although the molecule is still electrically neutral (same number of protons as electrons), the distribution of charge is not uniform. This unequal distribution results in regions of partial negative charge near the more electronegative atom, and regions of partial positive charge near the less electronegative atom. Such bonds are termed **polar covalent bonds,** and the molecules polar molecules. When drawing polar molecules, these partial charges are usually symbolized by the lowercase Greek letter delta (δ). The partial charge seen in a polar covalent bond is relatively small—far less than the unit charge of an ion. For biological molecules, we can predict polarity of bonds by knowing the relative electronegativity of a small number of important atoms (table 2.2). Notice that although C and H differ slightly in electronegativity, this small difference is negligible, and C—H bonds are considered nonpolar.

Because of its importance in the chemistry of water, we will explore the nature of polar and nonpolar molecules in the following section on water. Water (H_2O) is a polar molecule with electrons more concentrated around the oxygen atom.

Chemical reactions alter bonds

The formation and breaking of chemical bonds, which is the essence of chemistry, is termed a *chemical reaction*. All chemical reactions involve the shifting of atoms from one molecule or ionic compound to another, without any change in the number or identity of the atoms. For convenience, we refer to the original molecules before the reaction starts as *reactants*, and the molecules resulting from the chemical reaction as *products*. For example:

$$6H_2O + 6CO_2 \longrightarrow C_6H_{12}O_6 + 6O_2$$
$$\text{reactants} \qquad \longrightarrow \qquad \text{products}$$

You may recognize this reaction as a simplified form of the photosynthesis reaction, in which water and carbon dioxide are combined to produce glucose and oxygen. Most animal life ultimately depends on this reaction, which takes place in plants. (Photosynthetic reactions will be discussed in detail in chapter 8.)

TABLE 2.2	Relative Electronegativities of Some Important Atoms
Atom	**Electronegativity**
O	3.5
N	3.0
C	2.5
H	2.1

The extent to which chemical reactions occur is influenced by three important factors:

1. **Temperature.** Heating the reactants increases the rate of a reaction because the reactants collide with one another more often. (Care must be taken that the temperature is not so high that it destroys the molecules.)
2. **Concentration of reactants and products.** Reactions proceed more quickly when more reactants are available, allowing more frequent collisions. An accumulation of products typically slows the reaction and, in reversible reactions, may speed the reaction in the reverse direction.
3. **Catalysts.** A catalyst is a substance that increases the rate of a reaction. It doesn't alter the reaction's equilibrium between reactants and products, but it does shorten the time needed to reach equilibrium, often dramatically. In living systems, proteins called enzymes catalyze almost every chemical reaction.

Many reactions in nature are reversible. This means that the products may themselves be reactants, allowing the reaction to proceed in reverse. We can write the preceding reaction in the reverse order:

$$C_6H_{12}O_6 + 6O_2 \longrightarrow 6H_2O + 6CO_2$$
$$\text{reactants} \qquad \longrightarrow \qquad \text{products}$$

This reaction is a simplified version of the oxidation of glucose by cellular respiration, in which glucose is broken down into water and carbon dioxide in the presence of oxygen. Virtually all organisms carry out forms of glucose oxidation; details are covered later, in chapter 7.

Learning Outcomes Review 2.3

An ionic bond is an attraction between ions of opposite charge in an ionic compound. A covalent bond is formed when two atoms share one or more pairs of electrons. Complex biological compounds are formed in large part by atoms that can form one or more covalent bonds: C, H, O, and N. A polar covalent bond is formed by unequal sharing of electrons. Nonpolar bonds exhibit equal sharing of electrons.

■ *How is a polar covalent bond different from an ionic bond?*

2.4 *Water: A Vital Compound*

Learning Outcomes

1. *Relate how the structure of water leads to hydrogen bonds.*
2. *Describe water's cohesive and adhesive properties.*

Of all the common molecules, only water exists as a liquid at the relatively low temperatures that prevail on the Earth's surface. Three-fourths of the Earth is covered by liquid water

a. Solid   ***b.*** Liquid   ***c.*** Gas

Figure 2.10 **Water takes many forms.** ***a.*** When water cools below 0°C, it forms beautiful crystals, familiar to us as snow and ice. ***b.*** Ice turns to liquid when the temperature is above 0°C. ***c.*** Liquid water becomes steam when the temperature rises above 100°C, as seen in this hot spring at Yellowstone National Park.

(figure 2.10). When life was beginning, water provided a medium in which other molecules could move around and interact, without being held in place by strong covalent or ionic bonds. Life evolved in water for 2 billion years before spreading to land. And even today, life is inextricably tied to water. About two-thirds of any organism's body is composed of water, and all organisms require a water-rich environment, either inside or outside it, for growth and reproduction. It is no accident that tropical rain forests are bursting with life, while dry deserts appear almost lifeless except when water becomes temporarily plentiful, such as after a rainstorm.

Water's structure facilitates hydrogen bonding

Water has a simple molecular structure, consisting of an oxygen atom bound to two hydrogen atoms by two single covalent bonds (figure 2.11). The resulting molecule is stable: It satisfies the octet rule, has no unpaired electrons, and carries no net electrical charge.

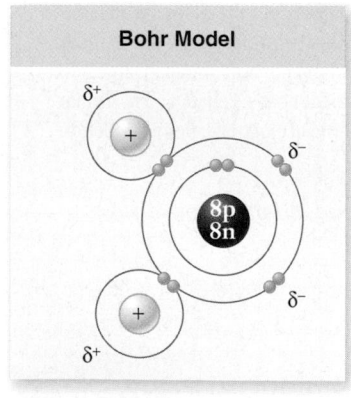

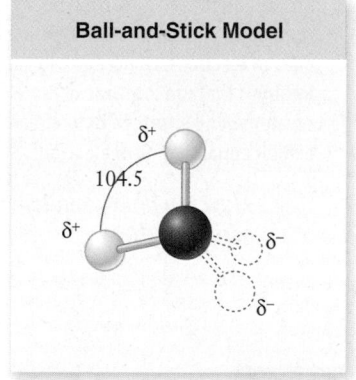

Figure 2.11 **Water has a simple molecular structure.** ***a.*** Each water molecule is composed of one oxygen atom and two hydrogen atoms. The oxygen atom shares one electron with each hydrogen atom. ***b.*** The greater electronegativity of the oxygen atom makes the water molecule polar: Water carries two partial negative charges (δ^-) near the oxygen atom and two partial positive charges (δ^+), one on each hydrogen atom. ***c.*** Space-filling model shows what the molecule would look like if it were visible.

The single most outstanding chemical property of water is its ability to form weak chemical associations, called **hydrogen bonds.** These bonds form between the partially negative O atoms and the partially positive H atoms of two water molecules. Although these bonds have only 5–10% of the strength of covalent bonds, they are important to DNA and protein structure, and thus responsible for much of the chemical organization of living systems.

The electronegativity of O is much greater than that of H (see table 2.2), and so the bonds between these atoms are highly polar. *The polarity of water underlies water's chemistry and the chemistry of life.*

If we consider the shape of a water molecule, we see that its two covalent bonds have a partial charge at each end: δ^- at the oxygen end and δ^+ at the hydrogen end. The most stable arrangement of these charges is a *tetrahedron (a pyramid with a triangle as its base),* in which the two negative and two positive charges are approximately equidistant from one another. The oxygen atom lies at the center of the tetrahedron, the hydrogen atoms occupy two of the apexes (corners), and the partial negative charges occupy the other two apexes (figure 2.11*b*). The bond angle between the two covalent oxygen–hydrogen bonds is 104.5°. This value is slightly less than the bond angle of a regular tetrahedron, which would be 109.5°. In water, the partial negative charges occupy more space than the partial positive regions, so the oxygen–hydrogen bond angle is slightly compressed.

Water molecules are cohesive

The polarity of water allows water molecules to be attracted to one another: that is, water is *cohesive.* The oxygen end of each water molecule, which is δ^-, is attracted to the hydrogen end, which is δ^+, of other molecules. The attraction produces hydrogen bonds among water molecules (figure 2.12). Each hydrogen bond is individually very weak and transient, lasting on average only a hundred-billionth (10^{-11}) of a second. The cumulative effects of large numbers of these bonds, however, can be enormous. Water forms an abundance of hydrogen bonds, which are responsible for many of its important physical properties (table 2.3).

Water's cohesion is responsible for its being a liquid, not a gas, at moderate temperatures. The cohesion of liquid water is also responsible for its **surface tension.** Small insects can walk on water (figure 2.13) because at the air–water interface, all the surface water molecules are hydrogen-bonded to molecules below them.

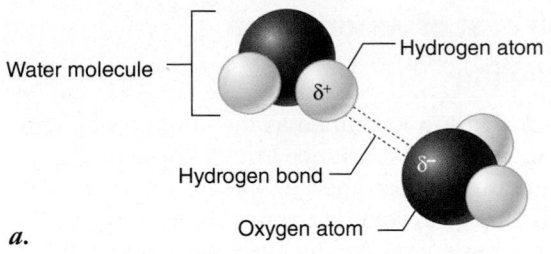

Water molecule
Hydrogen atom
Hydrogen bond
Oxygen atom

a.

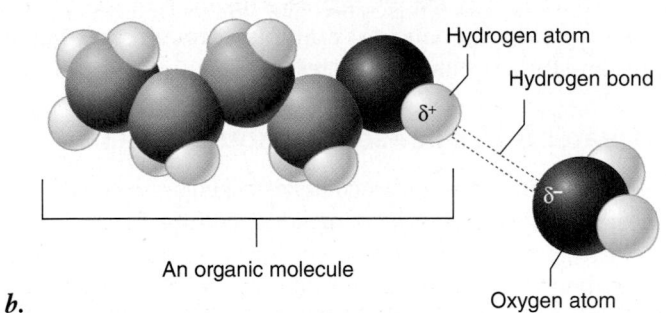

Hydrogen atom
Hydrogen bond
An organic molecule
Oxygen atom

b.

Figure 2.12 Structure of a hydrogen bond. *a.* Hydrogen bond between two water molecules. *b.* Hydrogen bond between an organic molecule (*n*-butanol) and water. H in *n*-butanol forms a hydrogen bond with oxygen in water. This kind of hydrogen bond is possible any time H is bound to a more electronegative atom (see table 2.2).

Water molecules are adhesive

The polarity of water causes it to be attracted to other polar molecules as well. This attraction for other polar substances is called *adhesion.* Water adheres to any substance with which it can form hydrogen bonds. This property explains why substances containing polar molecules get "wet" when they are immersed in water, but those that are composed of nonpolar molecules (such as oils) do not.

The attraction of water to substances that have electrical charges on their surface is responsible for capillary action. If a glass tube with a narrow diameter is lowered into a beaker of water, the water will rise in the tube above the level of the water in

Figure 2.13 Cohesion. Some insects, such as this water strider, literally walk on water. Because the surface tension of the water is greater than the force of one foot, the strider glides atop the surface of the water rather than sinking. The high surface tension of water is due to hydrogen bonding between water molecules.

the beaker, because the adhesion of water to the glass surface, drawing it upward, is stronger than the force of gravity, pulling it downward. The narrower the tube, the greater the electrostatic forces between the water and the glass, and the higher the water rises (figure 2.14).

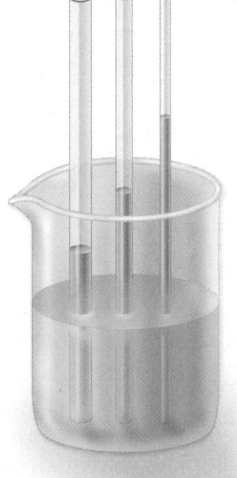

Figure 2.14 Adhesion. Capillary action causes the water within a narrow tube to rise above the surrounding water level; the adhesion of the water to the glass surface, which draws water upward, is stronger than the force of gravity, which tends to pull it down. The narrower the tube, the greater the surface area available for adhesion for a given volume of water, and the higher the water rises in the tube.

TABLE 2.3	The Properties of Water	
Property	**Explanation**	**Example of Benefit to Life**
Cohesion	Hydrogen bonds hold water molecules together.	Leaves pull water upward from the roots; seeds swell and germinate.
High specific heat	Hydrogen bonds absorb heat when they break and release heat when they form, minimizing temperature changes.	Water stabilizes the temperature of organisms and the environment.
High heat of vaporization	Many hydrogen bonds must be broken for water to evaporate.	Evaporation of water cools body surfaces.
Lower density of ice	Water molecules in an ice crystal are spaced relatively far apart because of hydrogen bonding.	Because ice is less dense than water, lakes do not freeze solid, allowing fish and other life in lakes to survive the winter.
Solubility	Polar water molecules are attracted to ions and polar compounds, making these compounds soluble.	Many kinds of molecules can move freely in cells, permitting a diverse array of chemical reactions.

2.5 Properties of Water

Water moderates temperature through two properties: its high specific heat and its high heat of vaporization. Water also has the unusual property of being less dense in its solid form, ice, than as a liquid. Water acts as a solvent for polar molecules and exerts an organizing effect on nonpolar molecules. All these properties result from its polar nature.

Water's high specific heat helps maintain temperature

The temperature of any substance is a measure of how rapidly its individual molecules are moving. In the case of water, a large input of thermal energy is required to break the many hydrogen bonds that keep individual water molecules from moving about. Therefore, water is said to have a high **specific heat,** which is defined as the amount of heat 1 g of a substance must absorb or lose to change its temperature by 1 degree Celsius (°C). Specific heat measures the extent to which a substance resists changing its temperature when it absorbs or loses heat. Because polar substances tend to form hydrogen bonds, the more polar it is, the higher is its specific heat. The specific heat of water (1 calorie/g/°C) is twice that of most carbon compounds and nine times that of iron. Only ammonia, which is more polar than water and forms very strong hydrogen bonds, has a higher specific heat than water (1.23 cal/g/°C). Still, only 20% of the hydrogen bonds are broken as water heats from 0° to 100°C.

Because of its high specific heat, water heats up more slowly than almost any other compound and holds its temperature longer. Because organisms have a high water content, water's high specific heat allows them to maintain a relatively constant internal temperature. The heat generated by the chemical reactions inside cells would destroy the cells if not for the absorption of this heat by the water within them.

Water's high heat of vaporization facilitates cooling

The **heat of vaporization** is defined as the amount of energy required to change 1 g of a substance from a liquid to a gas. A considerable amount of heat energy (586 cal) is required to accomplish this change in water. As water changes from a liquid to a gas it requires energy (in the form of heat) to break its many hydrogen bonds. The evaporation of water from a surface cools that surface. Many organisms dispose of excess body heat by evaporative cooling, for example, through sweating in humans and many other vertebrates.

Solid water is less dense than liquid water

At low temperatures, water molecules are locked into a crystal-like lattice of hydrogen bonds, forming solid ice (see figure 2.10a). Interestingly, ice is less dense than liquid water because the hydrogen bonds in ice space the water molecules relatively far apart. This unusual feature enables icebergs to float. If water did not have this property, nearly all bodies of water would be ice, with only the shallow surface melting every year. The buoyancy of ice is important ecologically because it means bodies of water freeze from the top down and not the bottom up. Because ice floats on the surface of lakes in the winter and the water beneath the ice remains liquid, fish and other animals keep from freezing.

The solvent properties of water help move ions and polar molecules

Water molecules gather closely around any substance that bears an electrical charge, whether that substance carries a full charge (ion) or a charge separation (polar molecule). For example, sucrose (table sugar) is composed of molecules that contain polar hydroxyl (OH) groups. A sugar crystal dissolves rapidly in water because water molecules can form hydrogen bonds with individual hydroxyl groups of the sucrose molecules. Therefore, sucrose is said to be *soluble* in water. Water is termed the *solvent,* and sugar is called the *solute.* Every time a sucrose molecule dissociates, or breaks away, from a solid sugar crystal, water molecules surround it in a cloud, forming a *hydration shell* that prevents it from associating with other sucrose molecules. Hydration shells also form around ions such as Na^+ and Cl^- (figure 2.15).

Water organizes nonpolar molecules

Water molecules always tend to form the maximum possible number of hydrogen bonds. When nonpolar molecules such as oils, which do not form hydrogen bonds, are placed in water, the water molecules act to exclude them. The nonpolar molecules aggregate, or clump together, thus minimizing their disruption of the hydrogen bonding of water. In effect, they shrink from contact with water, and for this reason they are referred to as **hydrophobic** (Greek *hydros,* "water," and *phobos,* "fearing"). In contrast, polar molecules, which readily form hydrogen bonds with water, are said to be **hydrophilic** ("water-loving").

The tendency of nonpolar molecules to aggregate in water is known as **hydrophobic exclusion.** By forcing the hydrophobic portions of molecules together, water causes these molecules to

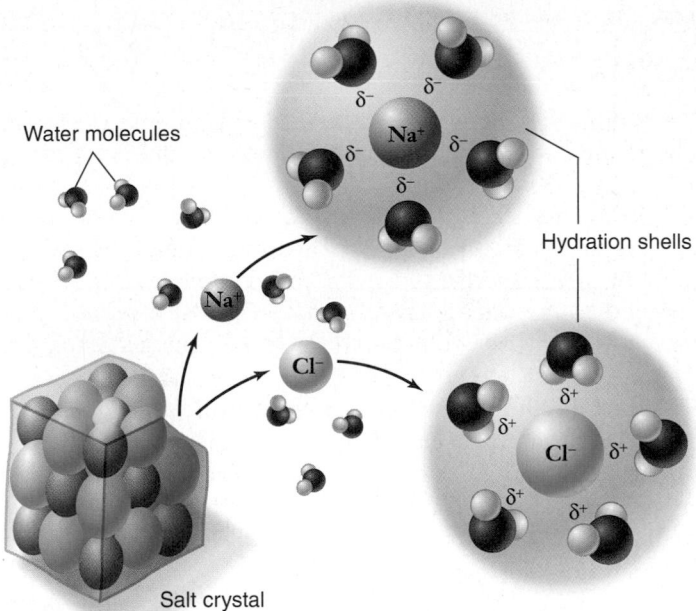

Figure 2.15 Why salt dissolves in water. When a crystal of table salt dissolves in water, individual Na^+ and Cl^- ions break away from the salt lattice and become surrounded by water molecules. Water molecules orient around Na^+ so that their partial negative poles face toward the positive Na^+; water molecules surrounding Cl^- orient in the opposite way, with their partial positive poles facing the negative Cl^-. Surrounded by hydration shells, Na^+ and Cl^- never reenter the salt lattice.

assume particular shapes. This property can also affect the structure of proteins, DNA, and biological membranes. In fact, the interaction of nonpolar molecules and water is critical to living systems.

Water can form ions

The covalent bonds of a water molecule sometimes break spontaneously. In pure water at 25°C, only 1 out of every 550 million water molecules undergoes this process. When it happens, a proton (hydrogen atom nucleus) dissociates from the molecule. Because the dissociated proton lacks the negatively charged electron it was sharing, its positive charge is no longer counterbalanced, and it becomes a hydrogen ion, H^+. The rest of the dissociated water molecule, which has retained the shared electron from the covalent bond, is negatively charged and forms a hydroxide ion, OH^-. This process of spontaneous ion formation is called *ionization*:

$$H_2O \longrightarrow OH^- + H^+$$
$$\text{water} \qquad \text{hydroxide ion} \qquad \text{hydrogen ion (proton)}$$

At 25°C, 1 liter (L) of water contains one ten-millionth (or 10^{-7}) mole of H^+ ions. A **mole** (mol) is defined as the weight of a substance in grams that corresponds to the atomic masses of all of the atoms in a molecule of that substance. In the case of H^+, the atomic mass is 1, and a mole of H^+ ions would weigh 1 g. One mole of any substance always contains 6.02×10^{23} molecules of the substance. Therefore, the **molar concentration** of hydrogen ions in pure water, represented as $[H^+]$, is 10^{-7} mol/L. (In reality, the H^+ usually associates with another water molecule to form a hydronium ion, H_3O^+.)

2.6 Acids and Bases

The concentration of hydrogen ions, and concurrently of hydroxide ions, in a solution is described by the terms *acidity* and *basicity*, respectively. Pure water, having an $[H^+]$ of 10^{-7} mol/L, is considered to be neutral, that is, neither acidic nor basic. Recall that for every H^+ ion formed when water dissociates, an OH^- ion is also formed, meaning that the dissociation of water produces H^+ and OH^- in equal amounts.

The pH scale measures hydrogen ion concentration

The *pH scale* (figure 2.16) is a more convenient way to express the hydrogen ion concentration of a solution. This scale defines *pH*, which stands for "partial hydrogen," as the negative logarithm of the hydrogen ion concentration in the solution:

$$pH = -\log [H^+]$$

Because the logarithm of the hydrogen ion concentration is simply the exponent of the molar concentration of H^+, the pH equals the exponent times –1. For water, therefore, an $[H^+]$ of 10^{-7} mol/L corresponds to a pH value of 7. This is the neutral point—a balance between H^+ and OH^-—on the pH scale. This balance occurs because the dissociation of water produces equal amounts of H^+ and OH^-.

Note that, because the pH scale is *logarithmic*, a difference of 1 on the scale represents a 10-fold change in $[H^+]$. A solution with a pH of 4 therefore has 10 times the $[H^+]$ of a solution with a pH of 5 and 100 times the $[H^+]$ of a solution with a pH of 6.

Acids

Any substance that dissociates in water to increase the $[H^+]$ (and lower the pH) is called an **acid.** The stronger an acid is, the more hydrogen ions it produces and the lower its pH. For example, hydrochloric acid (HCl), which is abundant in your stomach, ionizes completely in water. A dilution of 10^{-1} mol/L of HCl dissociates to form 10^{-1} mol/L of H^+, giving the solution

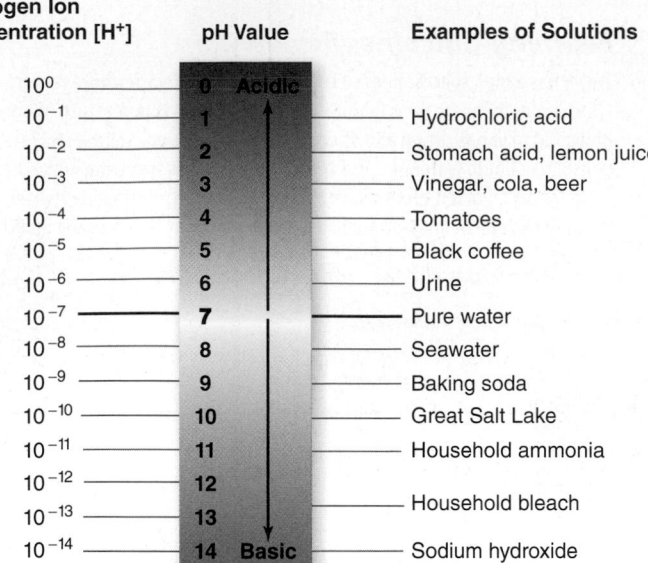

Hydrogen Ion Concentration [H+]	pH Value	Examples of Solutions
10^0	0 Acidic	
10^{-1}	1	Hydrochloric acid
10^{-2}	2	Stomach acid, lemon juice
10^{-3}	3	Vinegar, cola, beer
10^{-4}	4	Tomatoes
10^{-5}	5	Black coffee
10^{-6}	6	Urine
10^{-7}	7	Pure water
10^{-8}	8	Seawater
10^{-9}	9	Baking soda
10^{-10}	10	Great Salt Lake
10^{-11}	11	Household ammonia
10^{-12}	12	Household bleach
10^{-13}	13	
10^{-14}	14 Basic	Sodium hydroxide

Figure 2.16 **The pH scale.** The pH value of a solution indicates its concentration of hydrogen ions. Solutions with a pH less than 7 are acidic, whereas those with a pH greater than 7 are basic. The scale is logarithmic, which means that a pH change of 1 represents a 10-fold change in the concentration of hydrogen ions. Thus, lemon juice is 100 times more acidic than tomato juice, and seawater is 10 times more basic than pure water, which has a pH of 7.

a pH of 1. The pH of champagne, which bubbles because of the carbonic acid dissolved in it, is about 2.

Bases

A substance that combines with H+ when dissolved in water, and thus lowers the [H+], is called a **base.** Therefore, basic (or alkaline) solutions have pH values above 7. Very strong bases, such as sodium hydroxide (NaOH), have pH values of 12 or more. Many common cleaning substances, such as ammonia and bleach, accomplish their action because of their high pH.

Buffers help stabilize pH

The pH inside almost all living cells, and in the fluid surrounding cells in multicellular organisms, is fairly close to neutral, 7. Most of the enzymes in living systems are extremely sensitive to pH. Often even a small change in pH will alter their shape, thereby disrupting their activities. For this reason, it is important that a cell maintain a constant pH level.

But the chemical reactions of life constantly produce acids and bases within cells. Furthermore, many animals eat substances that are acidic or basic. Cola drinks, for example, are moderately strong (although dilute) acidic solutions. Despite such variations in the concentrations of H+ and OH−, the pH of an organism is kept at a relatively constant level by buffers (figure 2.17).

A **buffer** is a substance that resists changes in pH. Buffers act by releasing hydrogen ions when a base is added and absorbing hydrogen ions when acid is added, with the overall effect of keeping [H+] relatively constant.

Within organisms, most buffers consist of pairs of substances, one an acid and the other a base. The key buffer in human blood is an acid–base pair consisting of carbonic acid

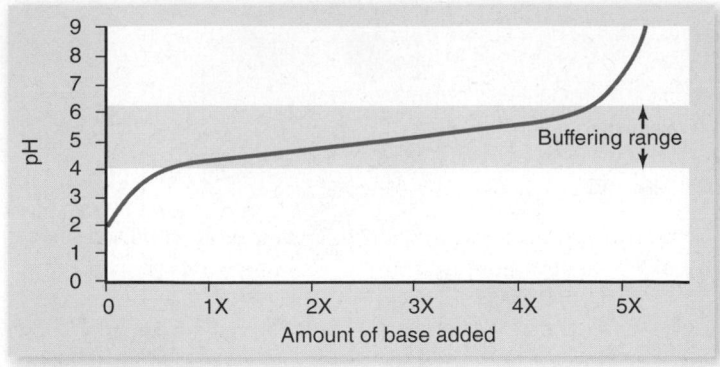

Figure 2.17 **Buffers minimize changes in pH.** Adding a base to a solution neutralizes some of the acid present, and so raises the pH. Thus, as the curve moves to the right, reflecting more and more base, it also rises to higher pH values. A buffer makes the curve rise or fall very slowly over a portion of the pH scale, called the "buffering range" of that buffer.

Data analysis If we call each step on the x-axis one volume of base, how many volumes of base must be added to change the pH from 4 to 6?

(acid) and bicarbonate (base). These two substances interact in a pair of reversible reactions. First, carbon dioxide (CO_2) and H_2O join to form carbonic acid (H_2CO_3), which in a second reaction dissociates to yield bicarbonate ion (HCO_3^-) and H+.

If some acid or other substance adds H+ to the blood, the HCO_3^- acts as a base and removes the excess H+ by forming H_2CO_3. Similarly, if a basic substance removes H+ from the blood, H_2CO_3 dissociates, releasing more H+ into the blood. The forward and reverse reactions that interconvert H_2CO_3 and HCO_3^- thus stabilize the blood's pH.

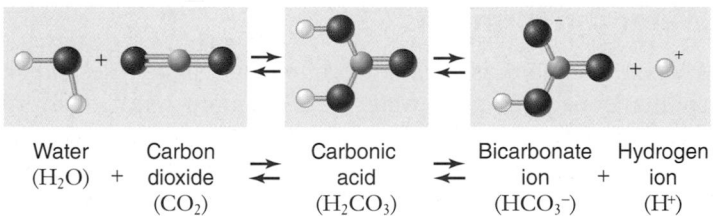

| Water (H_2O) | + | Carbon dioxide (CO_2) | ⇄ | Carbonic acid (H_2CO_3) | ⇄ | Bicarbonate ion (HCO_3^-) | + | Hydrogen ion (H+) |

The reaction of carbon dioxide and water to form carbonic acid is a crucial one because it permits carbon, essential to life, to enter water from the air. The Earth's oceans are rich in carbon because of the reaction of carbon dioxide with water.

In a condition called blood acidosis, human blood, which normally has a pH of about 7.4, drops to a pH of about 7.1. This condition is fatal if not treated immediately. The reverse condition, blood alkalosis, involves an increase in blood pH of a similar magnitude and is just as serious.

Learning Outcomes Review 2.6

Acid solutions have a high [H+], and basic solutions have a low [H+] (and therefore a high [OH−]). The pH of a solution is the negative logarithm of its [H+]. Low pH values indicate acids, and high pH values indicate bases. Even small changes in pH can be harmful to life. Buffer systems in organisms help to maintain pH within a narrow range.

■ *A change of 2 pH units indicates what change in [H+]?*

2.1 The Nature of Atoms

All matter is composed of atoms (figure 2.3).

Atomic structure includes a central nucleus and orbiting electrons.

Electrically neutral atoms have the same number of protons as electrons. Atoms that gain or lose electrons are called ions.

Each element is defined by its atomic number, the number of protons in the nucleus. Atomic mass is the sum of the mass of protons and neutrons in an atom. Isotopes are forms of a single element with different numbers of neutrons, and thus different atomic mass. Radioactive isotopes are unstable.

Electrons determine the chemical behavior of atoms.

The potential energy of electrons increases as distance from the nucleus increases. Electron orbitals are probability distributions. *s*-Orbitals are spherical; other orbitals have different shapes, such as the dumbbell-shaped *p*-orbitals.

Atoms contain discrete energy levels.

Energy levels correspond to quanta (sing. quantum) of energy, a "ladder" of energy levels that an electron may have.

The loss of electrons from an atom is called oxidation. The gain of electrons is called reduction. Electrons can be transferred from one atom to another in coupled redox reactions.

2.2 Elements Found in Living Systems

The periodic table displays elements according to atomic number and properties.

Atoms tend to establish completely full outer energy levels (the octet rule). Elements with filled outermost orbitals are inert.

Ninety elements occur naturally in the Earth's crust. Twelve of these elements are found in living organisms in greater than trace amounts: C, H, O, N, P, S, Na, K, Ca, Mg, Fe, and Cl.

Compounds of carbon are called organic compounds. The majority of molecules in living systems are composed of C bound to H, O, and N.

2.3 The Nature of Chemical Bonds

Molecules contain two or more atoms joined by chemical bonds. Compounds contain two or more different elements.

Ionic bonds form crystals.

Ions with opposite electrical charges form ionic bonds, such as NaCl (figure 2.9*b*).

Covalent bonds build stable molecules.

A molecule formed by a covalent bond is stable because it has no net charge, the octet rule is satisfied, and it has no unpaired electrons. Covalent bonds may be single, double, or triple, depending on the number of pairs of electrons shared. Nonpolar covalent bonds involve equal sharing of electrons between atoms. Polar covalent bonds involve unequal sharing of electrons.

Chemical reactions alter bonds.

Temperature, reactant concentration, and the presence of catalysts affect reaction rates. Most biological reactions are reversible, such as the conversion of carbon dioxide and water into carbohydrates.

2.4 Water: A Vital Compound

Water's structure facilitates hydrogen bonding.

Hydrogen bonds are weak interactions between a partially positive H in one molecule and a partially negative O in another molecule (figure 2.11).

Water molecules are cohesive.

Cohesion is the tendency of water molecules to adhere to one another due to hydrogen bonding. The cohesion of water is responsible for its surface tension.

Water molecules are adhesive.

Adhesion occurs when water molecules adhere to other polar molecules. Capillary action results from water's adhesion to the sides of narrow tubes, combined with its cohesion.

2.5 Properties of Water

Water's high specific heat helps maintain temperature.

The specific heat of water is high because it takes a considerable amount of energy to disrupt hydrogen bonds.

Water's high heat of vaporization facilitates cooling.

Breaking hydrogen bonds to turn liquid water into vapor takes a lot of energy. Many organisms lose excess heat through evaporative cooling, such as sweating.

Solid water is less dense than liquid water.

Hydrogen bonds are spaced farther apart in the solid phase of water than in the liquid phase. As a result, ice floats.

The solvent properties of water help move ions and polar molecules.

Water's polarity makes it a good solvent for polar substances and ions. Polar molecules or portions of molecules are attracted to water (hydrophilic). Molecules that are nonpolar are repelled by water (hydrophobic). Water makes nonpolar molecules clump together.

Water organizes nonpolar molecules.

Nonpolar molecules will aggregate to avoid water. This maximizes the hydrogen bonds that water can make. This hydrophobic exclusion can affect the structure of DNA, proteins and biological membranes.

Water can form ions.

Water dissociates into H^+ and OH^-. The concentration of H^+, shown as $[H^+]$, in pure water is 10^{-7} mol/L.

2.6 Acids and Bases (figure 2.16)

The pH scale measures hydrogen ion concentration.

pH is defined as the negative logarithm of $[H^+]$. Pure water has a pH of 7. A difference of 1 pH unit means a 10-fold change in $[H^+]$.

Acids have a greater $[H^+]$ and therefore a lower pH; bases have a lower $[H^+]$ and therefore a higher pH.

Buffers help stabilize pH.

Carbon dioxide and water react reversibly to form carbonic acid. A buffer resists changes in pH by absorbing or releasing H^+. The key buffer in the human blood is the carbonic acid/bicarbonate pair.

UNDERSTAND

1. The property that distinguishes an atom of one element (carbon, for example) from an atom of another element (oxygen, for example) is
   a. the number of electrons.
   b. the number of protons.
   c. the number of neutrons.
   d. the combined number of protons and neutrons.

2. If an atom has one valence electron, that is, a single electron in its outer energy level, it will most likely form
   a. one polar, covalent bond.
   b. two nonpolar, covalent bonds.
   c. two covalent bonds.
   d. an ionic bond.

3. An atom with a net positive charge must have more
   a. protons than neutrons.
   b. protons than electrons.
   c. electrons than neutrons.
   d. electrons than protons.

4. The isotopes carbon-12 and carbon-14 differ in
   a. the number of neutrons.
   b. the number of protons.
   c. the number of electrons.
   d. Both b and c are correct.

5. Which of the following is NOT a property of the elements most commonly found in living organisms?
   a. The elements have a low atomic mass.
   b. The elements have an atomic number less than 21.
   c. The elements possess eight electrons in their outer energy level.
   d. The elements are lacking one or more electrons from their outer energy level.

6. Ionic bonds arise from
   a. shared valence electrons.
   b. attractions between valence electrons.
   c. charge attractions between valence electrons.
   d. attractions between ions of opposite charge.

7. A solution with a high concentration of hydrogen ions
   a. is called a base.   c. has a high pH.
   b. is called an acid.   d. Both b and c are correct.

APPLY

1. Using the periodic table on page 22, which of the following atoms would you predict should form a positively charged ion (cation)?
   a. Fluorine (F)   c. Potassium (K)
   b. Neon (Ne)   d. Sulfur (S)

2. Refer to the element pictured. How many covalent bonds could this atom form?
   a. Two
   b. Three
   c. Four
   d. None

3. A molecule with polar covalent bonds would
   a. be soluble in water.
   b. not be soluble in water.
   c. contain atoms with very similar electronegativity.
   d. Both b and c are correct.

4. Hydrogen bonds are formed
   a. between any molecules that contain hydrogen.
   b. only between water molecules.
   c. when hydrogen is part of a polar bond.
   d. when two atoms of hydrogen share an electron.

5. If you shake a bottle of oil and vinegar then let it sit, it will separate into two phases because
   a. the nonpolar oil is soluble in water.
   b. water can form hydrogen bonds with the oil.
   c. polar oil is not soluble in water.
   d. nonpolar oil is not soluble in water.

6. The decay of radioactive isotopes involves changes to the nucleus of atoms. Explain how this differs from the changes in atoms that occur during chemical reactions.

SYNTHESIZE

1. Elements that form ions are important for a range of biological processes. You have learned something about the cations sodium (Na^+), calcium (Ca^{2+}) and potassium (K^+) in this chapter. Use your knowledge of the definition of a cation to identify other examples from the periodic table.

2. A popular theme in science fiction literature has been the idea of silicon-based life-forms in contrast to our carbon-based life. Evaluate the possibility of silicon-based life based on the chemical structure and potential for chemical bonding of a silicon atom.

3. Recent efforts by NASA to search for signs of life on Mars have focused on the search for evidence of liquid water rather than looking directly for biological organisms (living or fossilized). Use your knowledge of the influence of water on life on Earth to construct an argument justifying this approach.

ONLINE RESOURCE

www.ravenbiology.com

Understand, Apply, and Synthesize—enhance your study with animations that bring concepts to life and practice tests to assess your understanding. Your instructor may also recommend the interactive eBook, individualized learning tools, and more.

Chapter **3**

The Chemical Building Blocks of Life

Chapter Contents

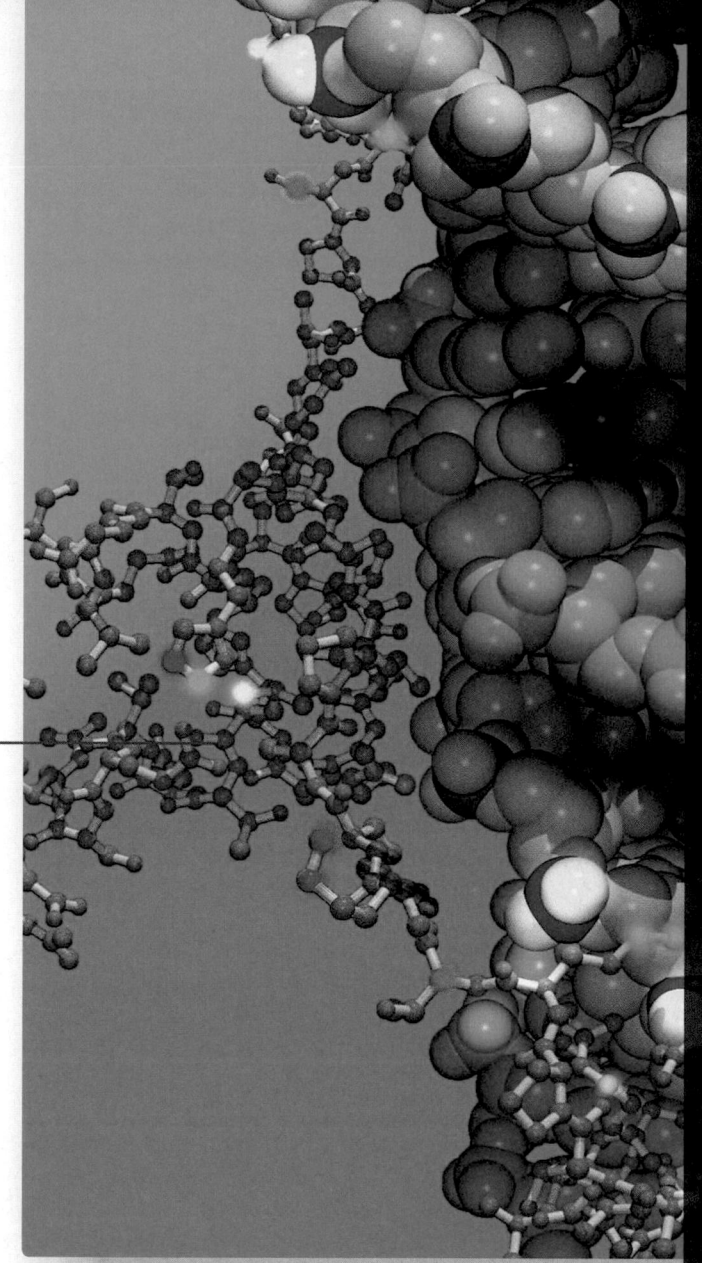

Introduction

A cup of water contains more molecules than there are stars in the sky. But many molecules are much larger than water molecules. Many thousands of distinct biological molecules are long chains made of thousands or even billions of atoms. These enormous assemblages, which are almost always synthesized by living things, are macromolecules. *As you may know, biological macromolecules can be divided into four categories: carbohydrates, nucleic acids, proteins, and lipids, and they are the basic chemical building blocks from which all organisms are composed.*

We take the existence of these classes of macromolecules for granted now, but as late as the 19th century many theories of "vital forces" were associated with living systems. One such theory held that cells contained a substance, protoplasm, that was responsible for the chemical reactions in living systems. Any disruption of cells was thought to disturb the protoplasm. Such a view makes studying the chemical reactions of cells in the lab (in vitro) impossible. The demonstration of fermentation in a cell-free system marked the beginning of modern biochemistry (figure 3.1). This approach involves studying biological molecules outside of cells to infer their role inside cells. Because these biological macromolecules all involve carbon-containing compounds, we begin with a brief summary of carbon and its chemistry.

Hypothesis: *Chemical reactions, such as the fermentation reaction in yeast, are controlled by enzymes and do not require living cells.*

Prediction: *If yeast cells are broken open, these enzymes should function outside of the cell.*

Test: *Yeast is mixed with quartz sand and diatomaceous earth and then ground in a mortar and pestle. The resulting paste is wrapped in canvas and subjected to 400–500 atm pressure in a press. Fermentable and nonfermentable substrates are added to the resulting fluid, with fermentation being measured by the production of CO_2.*

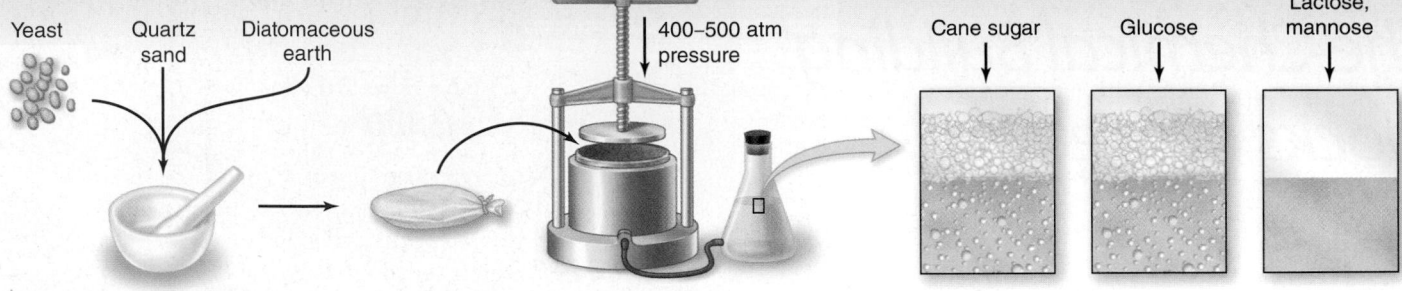

Grind in mortar/pestle.   Wrap in canvas and apply pressure in a press.

Result: *When a fermentable substrate (cane sugar, glucose) is used, CO_2 is produced; when a nonfermentable substrate (lactose, mannose) is used, no CO_2 is produced. In addition, visual inspection of the fluid shows no visible yeast cells.*

Conclusion: *The hypothesis is supported. The fermentation reaction can occur in the absence of live yeast.*

Historical Significance: *Although this is not precisely the intent of the original experiment, it represents the first use of a cell-free system. Such systems allow for the study of biochemical reactions in vitro and the purification of proteins involved. We now know that the "fermentation reaction" is actually a complex series of reactions. Would such a series of reactions be your first choice for this kind of demonstration?*

Figure 3.1 The demonstration of cell-free fermentation. The German chemist Eduard Buchner's (1860–1917) demonstration of fermentation by fluid produced from yeast, but not containing any live cells, both argued against the protoplasm theory and provided a method for future biochemists to examine the chemistry of life outside of cells.

3.1 Carbon: The Framework of Biological Molecules

Learning Outcomes

1. *Describe the relationship between functional groups and macromolecules.*
2. *Recognize the different kinds of isomers.*
3. *List the different kinds of biological macromolecules.*

In chapter 2, we reviewed the basics of chemistry. Biological systems obey all the laws of chemistry. Thus, chemistry forms the basis of living systems.

The framework of biological molecules consists predominantly of carbon atoms bonded to other carbon atoms or to atoms of oxygen, nitrogen, sulfur, phosphorus, or hydrogen. Because carbon atoms can form up to four covalent bonds, molecules containing carbon can form straight chains, branches, or even rings, balls, tubes, and coils.

Molecules consisting only of carbon and hydrogen are called *hydrocarbons*. Because carbon–hydrogen covalent bonds store considerable energy, hydrocarbons make good fuels. Gasoline, for example, is rich in hydrocarbons, and propane gas,

another hydrocarbon, consists of a chain of three carbon atoms, with eight hydrogen atoms bound to it. The chemical formula for propane is C_3H_8. Its structural formula is

$$\begin{array}{ccc} H & H & H \\ | & | & | \\ H-C-C-C-H \\ | & | & | \\ H & H & H \end{array}$$
Propane structural formula

Theoretically speaking, the length of a chain of carbon atoms is unlimited. As described in the rest of this chapter, the four main types of biological molecules often consist of huge chains of carbon-containing compounds.

Functional groups account for differences in molecular properties

Carbon and hydrogen atoms both have very similar electronegativities. Electrons in C—C and C—H bonds are therefore evenly distributed, with no significant differences in charge over the molecular surface. For this reason, hydrocarbons are nonpolar. Most biological molecules produced by cells, however, also contain other atoms. Because these other atoms frequently have different electronegativities, molecules containing them exhibit regions of partial positive or negative charge. They are polar. These molecules can be thought of as a

C—H core to which specific molecular groups, called **functional groups,** are attached. One such common functional group is —OH, called a *hydroxyl group.*

Functional groups have definite chemical properties that they retain no matter where they occur. Both the hydroxyl and carbonyl (C=O) groups, for example, are polar because of the electronegativity of the oxygen atoms (see chapter 2). Other common functional groups are the acidic carboxyl (COOH), phosphate (PO_4^-), and the basic amino (NH_2) group. Many of these functional groups can also participate in hydrogen bonding. Hydrogen bond donors and acceptors can be predicted based on their electronegativities shown in table 2.2. Figure 3.2 illustrates these biologically important functional groups and lists the macromolecules in which they are found.

Isomers have the same molecular formulas but different structures

Organic molecules having the same molecular or empirical formula can exist in different forms called **isomers.** If there are differences in the actual structure of their carbon skeleton, we call them *structural isomers.* Later you will see that glucose and fructose are structural isomers of $C_6H_{12}O_6$. Another form of isomers, called *stereoisomers,* have the same carbon skeleton but differ in how the groups attached to this skeleton are arranged in space.

Enzymes in biological systems usually recognize only a single, specific stereoisomer. A subcategory of stereoisomers, called *enantiomers,* are actually mirror images of each other. A molecule that has mirror-image versions is called a *chiral* molecule. When carbon is bound to four different molecules, this inherent asymmetry exists (figure 3.3).

Chiral compounds are characterized by their effect on polarized light. Polarized light has a single plane, and chiral molecules rotate this plane either to the right (Latin, *dextro*) or left (Latin, *levo*). We therefore call the two chiral forms *D* for *dextrorotatory* and *L* for *levorotatory.* Living systems tend to produce only a single enantiomer of the two possible forms; for example, in most organisms we find primarily D-sugars and L-amino acids.

Functional Group	Structural Formula	Example	Found In
Hydroxyl	—OH	Ethanol	carbohydrates, proteins, nucleic acids, lipids
Carbonyl	—C— (with O double bond)	Acetaldehyde	carbohydrates, nucleic acids
Carboxyl	—C(=O)OH	Acetic acid	proteins, lipids
Amino	—N(H)(H)	Alanine	proteins, nucleic acids
Sulfhydryl	—S—H	Cysteine	proteins
Phosphate	—O—P(O)(O⁻)—O⁻	Glycerol phosphate	nucleic acids
Methyl	—C(H)(H)—H	Alanine	proteins

Figure 3.2 The primary functional chemical groups. These groups tend to act as units during chemical reactions and give specific chemical properties to the molecules that possess them. Amino groups, for example, make a molecule more basic, and carboxyl groups make a molecule more acidic. These functional groups are also not limited to the examples in the "Found In" column but are widely distributed in biological molecules.

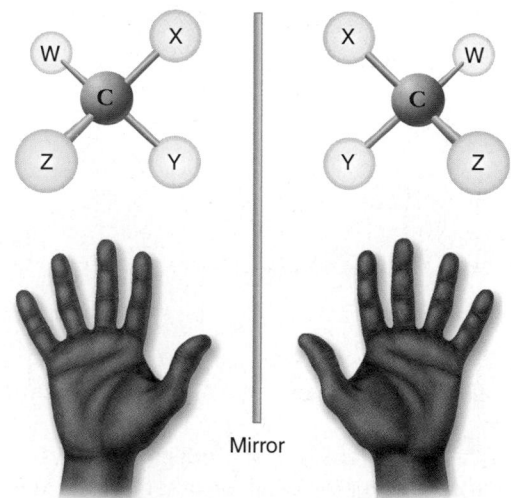

Figure 3.3 Chiral molecules. When carbon is bound to four different groups, the resulting molecule is said to be chiral (from Greek *cheir,* meaning "hand"). A chiral molecule will have stereoisomers that are mirror images. The two molecules shown have the same four groups but cannot be superimposed, much like your two hands cannot be superimposed but must be flipped to match. These types of stereoisomers are called *enantiomers.*

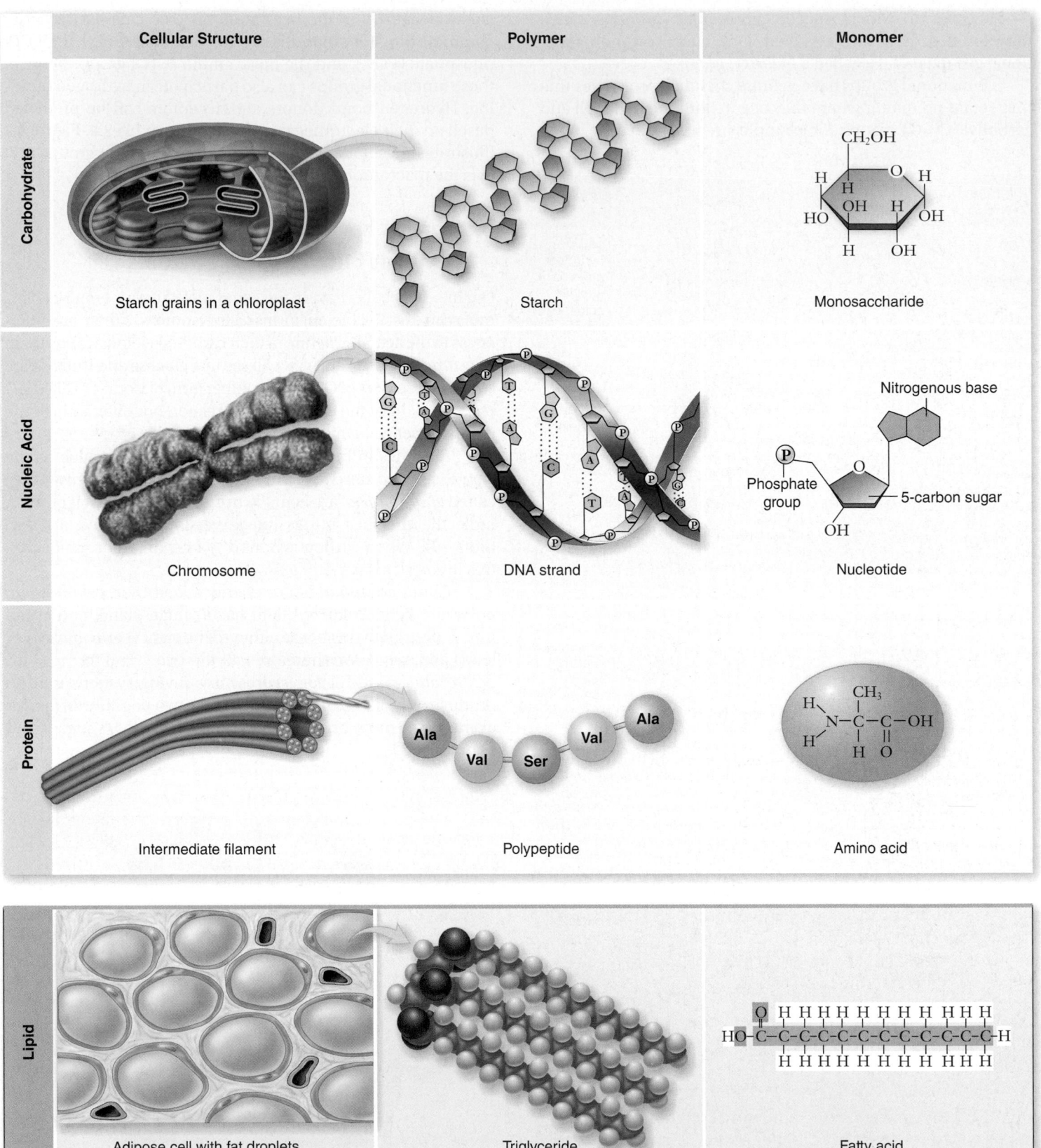

Cellular Structure	Polymer	Monomer

Carbohydrate

Starch grains in a chloroplast

Starch

Monosaccharide

Nucleic Acid

Chromosome

DNA strand

Nucleotide

Nitrogenous base

Phosphate group

5-carbon sugar

Protein

Intermediate filament

Polypeptide

Ala Val Ser Val Ala

Amino acid

Lipid

Adipose cell with fat droplets

Triglyceride

Fatty acid

Figure 3.4 Polymer macromolecules. The four major biological macromolecules are shown. Carbohydrates, nucleic acids, and proteins all form polymers and are shown with the monomers used to make them. Lipids do not fit this simple monomer–polymer relationship, however, because they are constructed from glycerol and fatty acids. All four types of macromolecules are also shown in their cellular context.

TABLE 3.1	Macromolecules		
Macromolecule	**Subunit**	**Function**	**Example**
C A R B O H Y D R A T E S			
Starch, glycogen	Glucose	Energy storage	Potatoes
Cellulose	Glucose	Structural support in plant cell walls	Paper; strings of celery
Chitin	Modified glucose	Structural support	Crab shells
N U C L E I C A C I D S			
DNA	Nucleotides	Encodes genes	Chromosomes
RNA	Nucleotides	Needed for gene expression	Messenger RNA
P R O T E I N S			
Functional	Amino acids	Catalysis; transport	Hemoglobin
Structural	Amino acids	Support	Hair; silk
L I P I D S			
Fats	Glycerol and three fatty acids	Energy storage	Butter; corn oil; soap
Phospholipids	Glycerol, two fatty acids, phosphate, and polar R groups	Cell membranes	Phosphatidylcholine
Prostaglandins	Five-carbon rings with two nonpolar tails	Chemical messengers	Prostaglandin E (PGE)
Steroids	Four fused carbon rings	Membranes; hormones	Cholesterol; estrogen
Terpenes	Long carbon chains	Pigments; structural support	Carotene; rubber

Biological macromolecules include carbohydrates, nucleic acids, proteins, and lipids

Remember that biological macromolecules are traditionally grouped into carbohydrates, nucleic acids, proteins, and lipids (table 3.1). In many cases, these macromolecules are polymers. A **polymer** is a long molecule built by linking together a large number of small, similar chemical subunits called **monomers.** They are like railroad cars coupled to form a train. The nature of a polymer is determined by the monomers used to build the polymer. Here are some examples. Complex carbohydrates such as starch are polymers composed of simple ring-shaped sugars. Nucleic acids (DNA and RNA) are polymers of nucleotides, and proteins are polymers of amino acids (figure 3.4). These long chains are built via chemical reactions termed *dehydration reactions* and are broken down by *hydrolysis reactions*. Lipids are macromolecules, but they really don't follow the monomer–polymer relationship. However, lipids are formed through dehydration reactions, which link the fatty acids to glycerol.

The dehydration reaction

Despite the differences between monomers of these major polymers, the basic chemistry of their synthesis is similar: To form a covalent bond between two monomers, an —OH group is removed from one monomer, and a hydrogen atom (H) is removed from the other (figure 3.5*a*). For example, this simple chemistry is the same for linking amino acids together to make a protein or assembling glucose units together to make starch. This reaction is also used to link fatty acids to glycerol in lipids. This chemical reaction is called condensation, or a **dehydration reaction,** because the removal of —OH and —H is the same as the removal of a molecule of water (H_2O). For every subunit added to a macromolecule, one water molecule is removed. These and other biochemical reactions require that the reacting substances are held close together and that the correct chemical bonds are stressed and broken. This process of positioning and stressing, termed *catalysis,* is carried out within cells by enzymes.

The hydrolysis reaction

Cells disassemble macromolecules into their constituent subunits through reactions that are the reverse of dehydration—a molecule of water is added instead of removed (figure 3.5*b*). In this process, called **hydrolysis,** a hydrogen atom is attached to one subunit and a hydroxyl group to the other, breaking a specific covalent bond in the macromolecule. When you eat a potato, which contains starch (discussed later), your body breaks the starch down into glucose units by hydrolysis. The potato plant built the starch molecules originally by dehydration reactions.

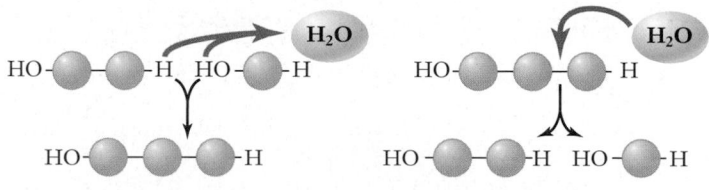

a. Dehydration reaction   *b.* Hydrolysis reaction

Figure 3.5 Making and breaking macromolecules.
a. Biological macromolecules are polymers formed by linking monomers together through dehydration reactions. This process releases a water molecule for every bond formed. *b.* Breaking the bond between subunits involves hydrolysis, which reverses the loss of a water molecule by dehydration.

3.2 Carbohydrates: Energy Storage and Structural Molecules

Monosaccharides are simple sugars

Carbohydrates are a loosely defined group of molecules that all contain carbon, hydrogen, and oxygen in the molar ratio 1:2:1. Their empirical formula (which lists the number of atoms in the molecule with subscripts) is $(CH_2O)_n$, where n is the number of carbon atoms. Because they contain many carbon–hydrogen (C—H) bonds, which release energy when oxidation occurs, carbohydrates are well suited for energy storage. Sugars are among the most important energy-storage molecules, and they exist in several different forms.

The simplest of the carbohydrates are the **monosaccharides** (Greek *mono,* "single," and Latin *saccharum,* "sugar"). Simple sugars contain as few as three carbon atoms, but those that play the central role in energy storage have six (figure 3.6). The empirical formula of six-carbon sugars is:

$$C_6H_{12}O_6 \qquad or \qquad (CH_2O)_6$$

Six-carbon sugars can exist in a straight-chain form, but dissolved in water (an aqueous environment) they almost always form rings.

The most important of the six-carbon monosaccharides for energy storage is glucose, which you first encountered in the examples of chemical reactions in chapter 2. Glucose has seven energy-storing C—H bonds (figure 3.7). Depending on the orientation of the carbonyl group (C=O) when the ring is closed, glucose can exist in two different forms: alpha (α) or beta (β).

Sugar isomers have structural differences

Glucose is not the only sugar with the formula $C_6H_{12}O_6$. Both structural isomers and stereoisomers of this simple six-carbon skeleton exist in nature. Fructose is a structural isomer that differs in the position of the carbonyl carbon (C=O); galactose is a stereoisomer that differs in the position of —OH and —H groups relative to the ring (figure 3.8). These differences often account for substantial functional differences between the isomers. Your taste buds can discern them: Fructose tastes much sweeter than glucose, despite the fact that both sugars have identical chemical composition. Enzymes that act on different sugars can distinguish both the structural and stereoisomers of this basic six-carbon skeleton. The different stereoisomers of glucose are also important in the polymers that can be made using glucose as a monomer, as you will see later in this chapter.

Disaccharides serve as transport molecules in plants and provide nutrition in animals

Most organisms transport sugars within their bodies. In humans, the glucose that circulates in the blood does so as a simple monosaccharide. In plants and many other organisms, however, glucose is converted into a transport form before it is moved from place to place within the organism. In such a form, it is less readily metabolized during transport.

Transport forms of sugars are commonly made by linking two monosaccharides together to form a **disaccharide** (Greek *di,* "two"). Disaccharides serve as effective reservoirs of glucose because the enzymes that normally use glucose in the organism cannot break the bond linking the two monosaccharide subunits. Enzymes that can do so are typically present only in the tissue that uses glucose.

3-carbon Sugar	5-carbon Sugars		6-carbon Sugars		
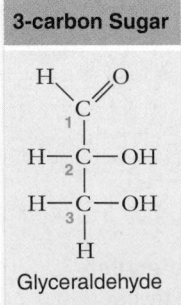	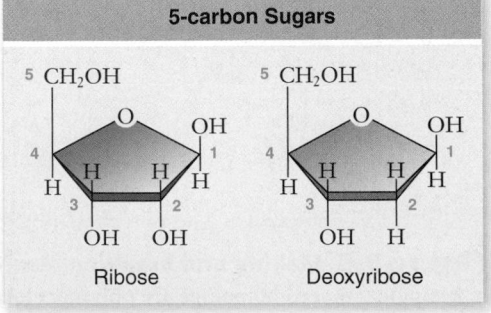		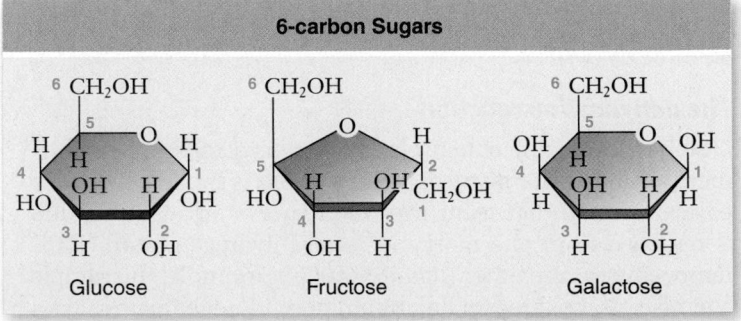		
Glyceraldehyde	Ribose	Deoxyribose	Glucose	Fructose	Galactose

Figure 3.6 Monosaccharides. Monosaccharides, or simple sugars, can contain as few as three carbon atoms and are often used as building blocks to form larger molecules. The five-carbon sugars ribose and deoxyribose are components of nucleic acids (see figure 3.15). The carbons are conventionally numbered (in *blue*) from the more oxidized end.

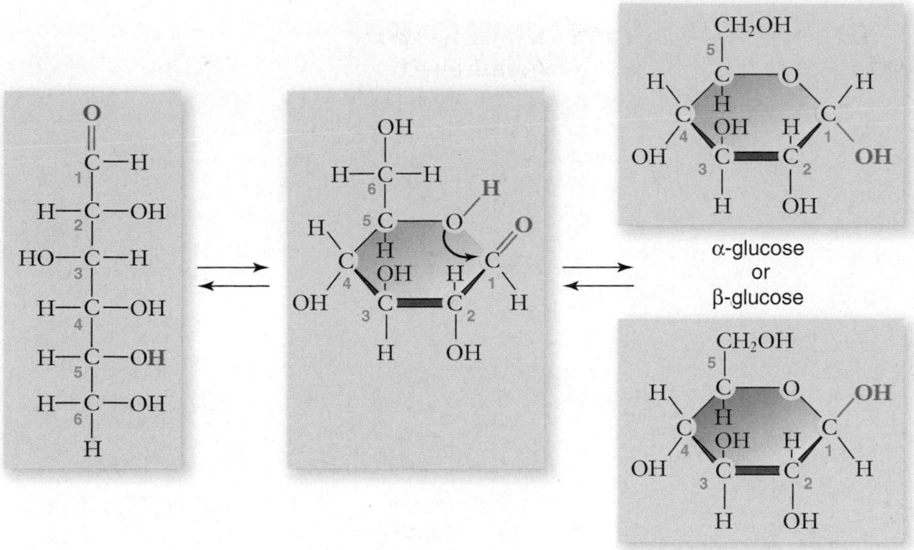

Figure 3.7 **Structure of the glucose molecule.** Glucose is a linear, six-carbon molecule that forms a six-membered ring in solution. Ring closure occurs such that two forms can result: α-glucose and β-glucose. These structures differ only in the position of the —OH bound to carbon 1. The structure of the ring can be represented in many ways; shown here are the most common, with the carbons conventionally numbered so that the forms can be compared easily. The heavy lines in the ring structures represent portions of the molecule that are projecting out of the page toward you.

Transport forms differ depending on which monosaccharides are linked to form the disaccharide. Glucose forms transport disaccharides with itself and with many other monosaccharides, including fructose and galactose. When glucose forms a disaccharide with the structural isomer fructose, the resulting disaccharide is *sucrose,* or table sugar (figure 3.9*a*). Sucrose is the form most plants use to transport glucose and is the sugar that most humans and other animals eat. Sugarcane and sugar beets are rich in sucrose.

When glucose is linked to the stereoisomer galactose, the resulting disaccharide is *lactose,* or milk sugar. Many mammals supply energy to their young in the form of lactose. Adults often have greatly reduced levels of lactase, the enzyme required to cleave lactose into its two monosaccharide components, and thus they cannot metabolize lactose efficiently. This can result in lactose intolerance in humans. Most of the energy that is channeled into lactose production is therefore reserved for offspring. For this reason, lactose as an energy source is primarily for offspring in mammals.

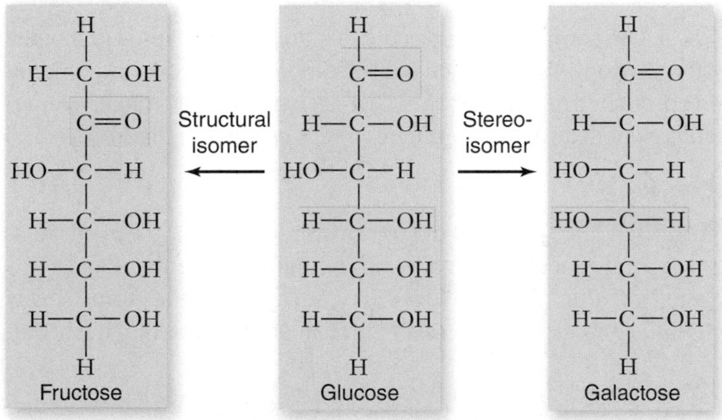

Polysaccharides provide energy storage and structural components

Polysaccharides are longer polymers made up of monosaccharides that have been joined through dehydration reactions. **Starch,** a storage polysaccharide, consists entirely of α-glucose molecules linked in long chains. **Cellulose,** a structural polysaccharide, also consists of glucose molecules linked in chains, but these molecules are β-glucose. Because starch is built from α-glucose we call the linkages α linkages; cellulose has β linkages.

Starches and glycogen

Organisms store the metabolic energy contained in monosaccharides by converting them into disaccharides, such as *maltose* (figure 3.9*b*). These are then linked together into the insoluble polysaccharides called *starches.* These polysaccharides differ mainly in how the polymers branch.

The starch with the simplest structure is *amylose.* It is composed of many hundreds of α-glucose molecules linked together in long, unbranched chains. Each linkage occurs between the

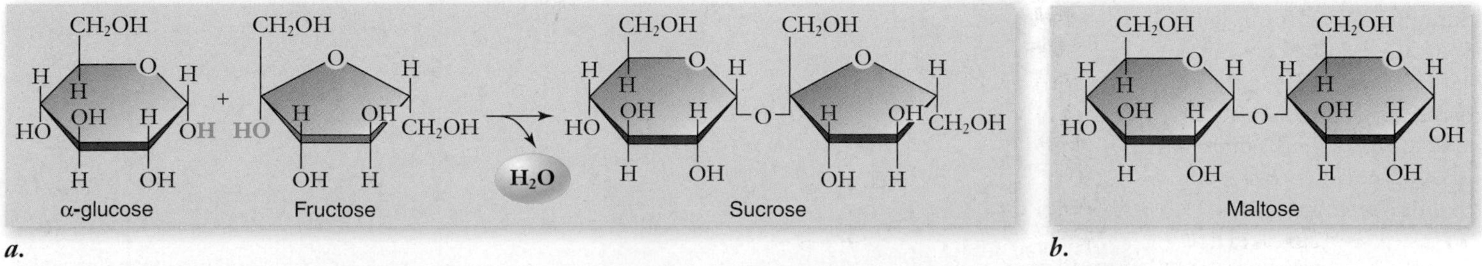

Figure 3.8 **Isomers and stereoisomers.** Glucose, fructose, and galactose are isomers with the empirical formula $C_6H_{12}O_6$. A structural isomer of glucose, such as fructose, has identical chemical groups bonded to different carbon atoms. Notice that this results in a five-membered ring in solution (see figure 3.6). A stereoisomer of glucose, such as galactose, has identical chemical groups bonded to the same carbon atoms but in different orientations (the —OH at carbon 4).

Figure 3.9 **How disaccharides form.** Some disaccharides are used to transport glucose from one part of an organism's body to another; one example is sucrose (*a*), which is found in sugarcane. Other disaccharides, such as maltose (*b*), are used in grain for storage.

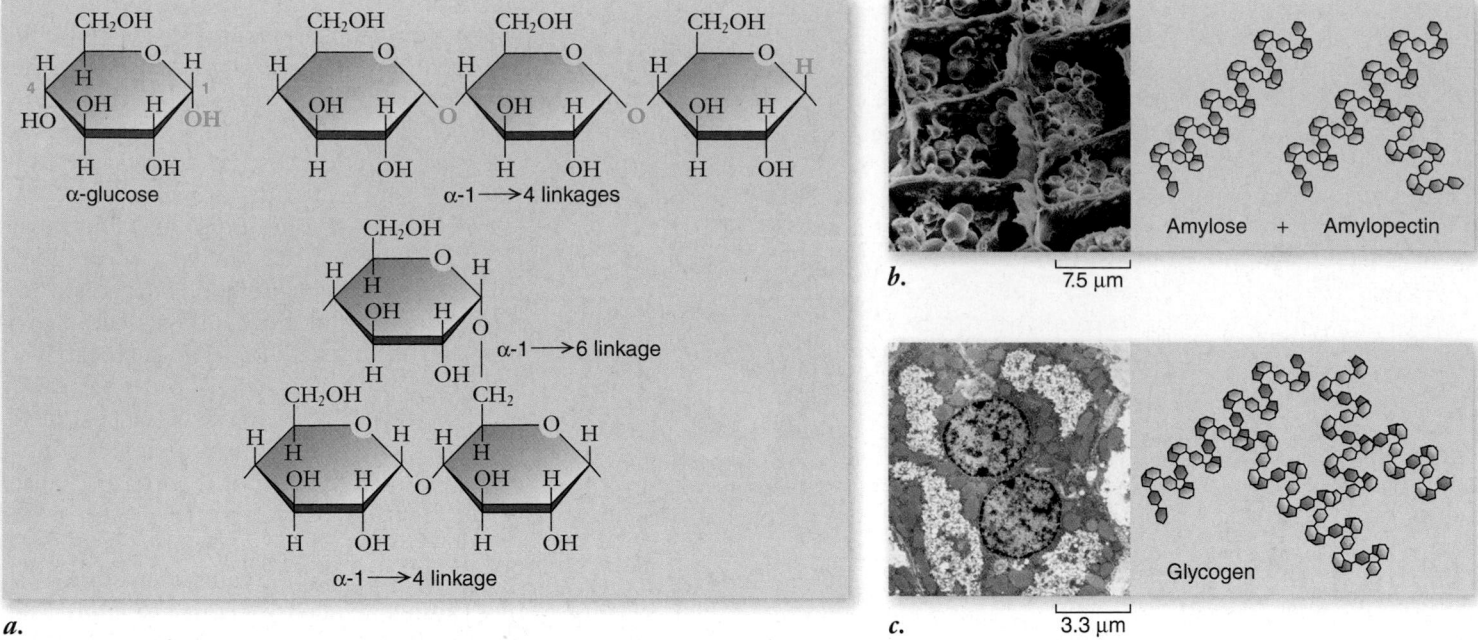

a.

Figure 3.10 Polymers of glucose: Starch and glycogen. *a.* Starch chains consist of polymers of α-glucose subunits joined by α-(1⟶4) glycosidic linkages. These chains can be branched by forming similar α-(1⟶6) glycosidic bonds. These storage polymers then differ primarily in their degree of branching. *b.* Starch is found in plants and is composed of amylose and amylopectin, which are unbranched and branched, respectively. The branched form is insoluble and forms starch granules in plant cells. *c.* Glycogen is found in animal cells and is highly branched and also insoluble, forming glycogen granules.

carbon 1 (C-1) of one glucose molecule and the C-4 of another, making them α-(1⟶4) linkages (figure 3.10*a*). The long chains of amylose tend to coil up in water, a property that renders amylose insoluble. Potato starch is about 20% amylose (figure 3.10*b*).

Most plant starch, including the remaining 80% of potato starch, is a somewhat more complicated variant of amylose called *amylopectin.* Pectins are branched polysaccharides with the branches occurring due to bonds between the C-1 of one molecule and the C-6 of another [α-(1⟶6) linkages]. These short amylose branches consist of 20 to 30 glucose subunits (figure 3.10*b*).

The comparable molecule to starch in animals is **glycogen.** Like amylopectin, glycogen is an insoluble polysaccharide containing branched amylose chains. Glycogen has a much longer average chain length and more branches than plant starch (figure 3.10*c*).

Cellulose

Although some chains of sugars store energy, others serve as structural material for cells. For two glucose molecules to link together, the glucose subunits must be of the same form. *Cellulose* is a polymer of β-glucose (figure 3.11). The bonds

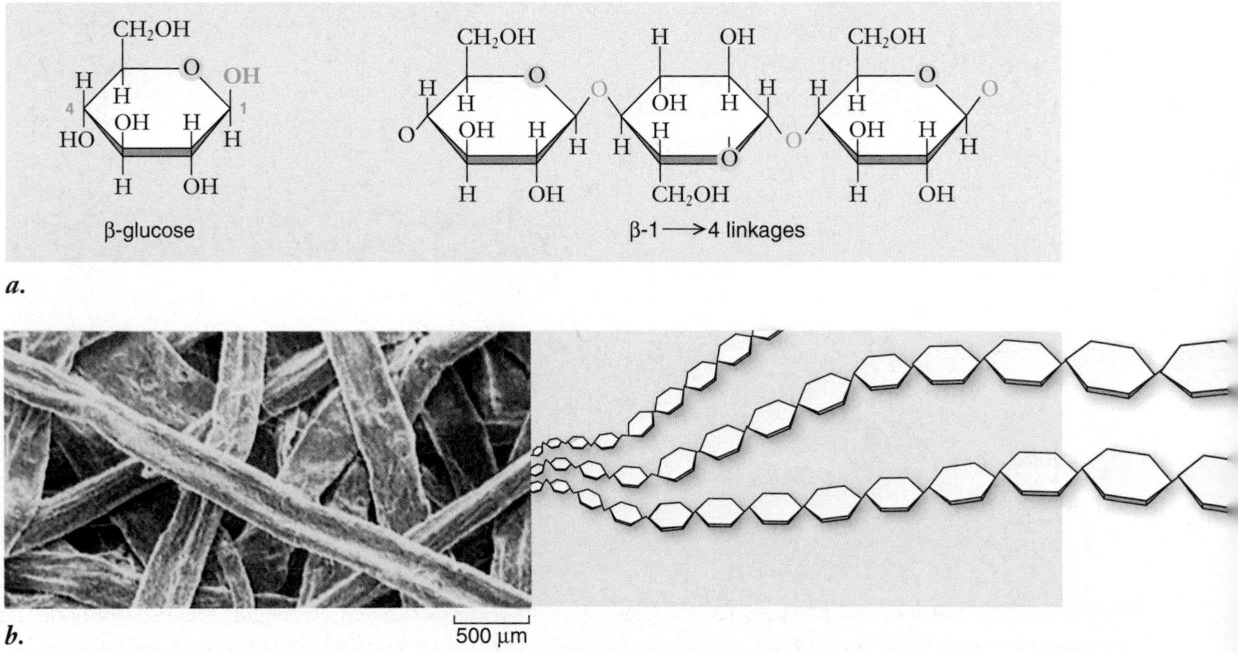

Figure 3.11 Polymers of glucose: Cellulose. Starch chains consist of α-glucose subunits, and cellulose chains consist of β-glucose subunits. *a.* Thus the bonds between adjacent glucose molecules in cellulose are β-(1⟶4) glycosidic linkages. *b.* Cellulose is unbranched and forms long fibers. Cellulose fibers can be very strong and are quite resistant to metabolic breakdown, which is one reason wood is such a good building material.

between adjacent glucose molecules still exist between the C-1 of the first glucose and the C-4 of the next glucose, but these are β-$(1 \longrightarrow 4)$ linkages.

The properties of a chain of glucose molecules consisting of all β-glucose are very different from those of starch. These long, unbranched β-linked chains make tough fibers. Cellulose is the chief component of plant cell walls (see figure 3.11b). It is chemically similar to amylose, with one important difference: The starch-hydrolyzing enzymes that occur in most organisms cannot break the bond between two β-glucose units because they only recognize α linkages.

Because cellulose cannot be broken down readily by most creatures, it works well as a biological structural material. But some animals, such as cows, are able to break down cellulose by means of symbiotic bacteria and protists in their digestive tracts. These organisms provide the necessary enzymes for cleaving the β-$(1 \longrightarrow 4)$ linkages, thus enabling access to a rich source of energy.

Chitin

Chitin, the structural material found in arthropods and many fungi, is a polymer of *N*-acetylglucosamine, a substituted version of glucose. When cross-linked by proteins, it forms a tough, resistant surface material that serves as the hard exoskeleton of insects and crustaceans (figure 3.12; see chapter 34). Few organisms are able to digest chitin, but most possess a chitinase enzyme, probably to protect against fungi.

Learning Outcomes Review 3.2

Monosaccharides have three to six or more carbon atoms typically arranged in a ring form. Disaccharides consist of two linked monosaccharides; polysaccharides are long chains of monosaccharides. Structural differences between sugar isomers can lead to functional differences. Starches are branched polymers of α-glucose used for energy storage. Cellulose in plants consists of unbranched chains of β-glucose that are not easily digested.

■ **How do the structures of starch, glycogen, and cellulose affect their function?**

Figure 3.12 Chitin. Chitin is the principal structural element in the external skeletons of many invertebrates, such as this lobster.

Nucleic Acids: Information Molecules

Learning Outcomes

1. *Describe the structure of nucleotides.*
2. *Contrast the structures of DNA and RNA.*
3. *Discuss the functions of DNA and RNA.*
4. *Recognize other nucleotides involved in energy metabolism.*

The biochemical activity of a cell depends on production of a large number of proteins, each with a specific sequence. The information necessary to produce the correct proteins is passed through generations of organisms, even though the proteins themselves are not inherited.

Nucleic acids carry information inside cells, just as disks contain the information in a computer or road maps display information needed by travelers. Two main varieties of nucleic acids are **deoxyribonucleic acid (DNA;** figure 3.13) and **ribonucleic acid (RNA).**

Genetic information is stored in DNA, and short-lived copies of this are made in the form of RNA, which is then used to direct the synthesis of proteins during the process of gene expression (as discussed in detail in chapter 15). Unique among macromolecules, nucleic acids are able to serve as templates for producing precise copies of themselves. This characteristic allows genetic information to be preserved during cell division and during the reproduction of organisms.

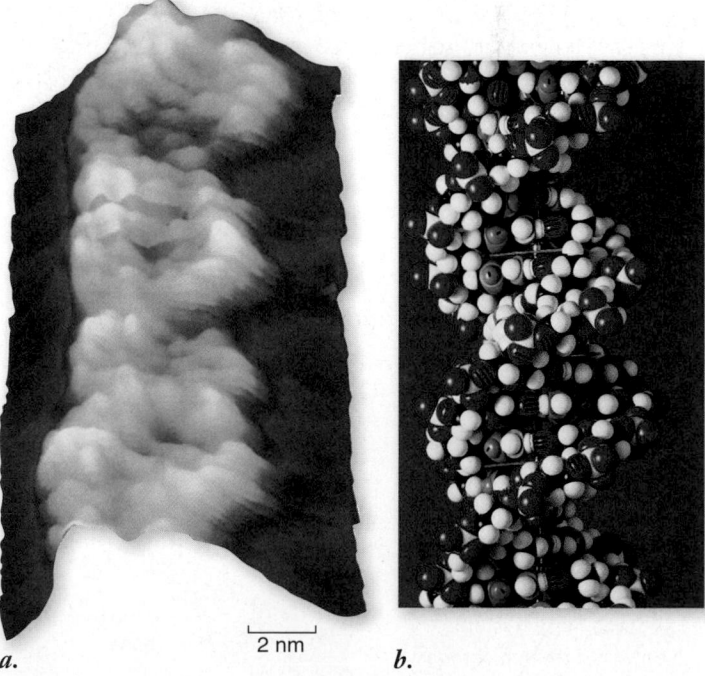

a.   2 nm   *b.*

Figure 3.13 Images of DNA. *a.* A scanning-tunneling micrograph of DNA (false color; 2,000,000×) showing approximately three turns of the DNA double helix. *b.* A space-filling model for comparison to the image of actual DNA in *(a)*.

The role of RNA in cells is much more complicated: RNA carries information, is part of the organelle responsible for protein synthesis, and recent work indicates it is also involved in the control of gene expression. As a carrier of information, the form of RNA called **messenger RNA (mRNA)** consists of transcribed single-stranded copies of portions of the DNA. These transcripts serve as blueprints specifying the amino acid sequences of proteins. This process will be described in detail in chapter 15.

Nucleic acids are nucleotide polymers

Nucleic acids are long polymers of repeating subunits called **nucleotides.** Each nucleotide consists of three components: a pentose, or five-carbon sugar (ribose in RNA and deoxyribose in DNA); a phosphate ($-PO_4^-$) group; and an organic nitrogenous (nitrogen-containing) base (figure 3.14). When a nucleic acid polymer forms, the phosphate group of one nucleotide binds to the hydroxyl group from the pentose sugar of another, releasing water and forming a *phosphodiester bond* by a dehydration reaction. A **nucleic acid,** then, is simply a chain of five-carbon sugars linked together by phosphodiester bonds with a nitrogenous base protruding from each sugar (see figure 3.15a). These chains of nucleotides, *polynucleotides,* have different ends: a phosphate on one end and an —OH from a sugar on the other end. We conventionally refer to these ends as 5′ ("five-prime," $-PO_4^-$) and 3′ ("three-prime," —OH) taken from the carbon numbering of the sugar (figure 3.15a).

Nucleotides have five types of nitrogenous bases (figure 3.15b). Two of these are large, double-ring molecules called *purines* that are each found in both DNA and RNA; the two purines are adenine (A) and guanine (G). The other three bases are single-ring molecules called *pyrimidines* that include

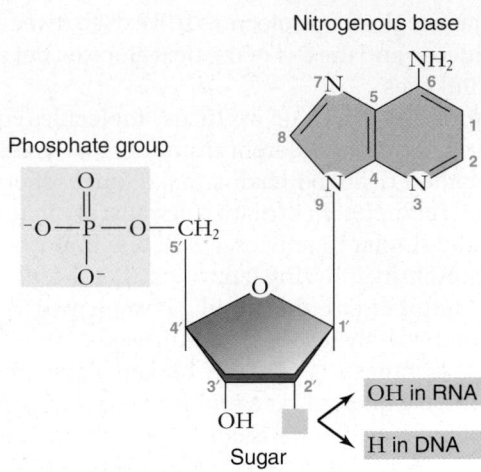

Figure 3.14 Structure of a nucleotide. The nucleotide subunits of DNA and RNA are made up of three elements: a five-carbon sugar (ribose or deoxyribose), an organic nitrogenous base (adenine is shown here), and a phosphate group. Notice that all the numbers on the sugar are given as "primes" (1′, 2′, etc.) to distinguish them from the numbering on the rings of the bases.

cytosine (C, in both DNA and RNA), thymine (T, in DNA only), and uracil (U, in RNA only).

DNA stores genetic information

Organisms use sequences of nucleotides in DNA to encode the information specifying the amino acid sequences of their proteins. This method of encoding information is very similar to the way in which sequences of letters encode information in a

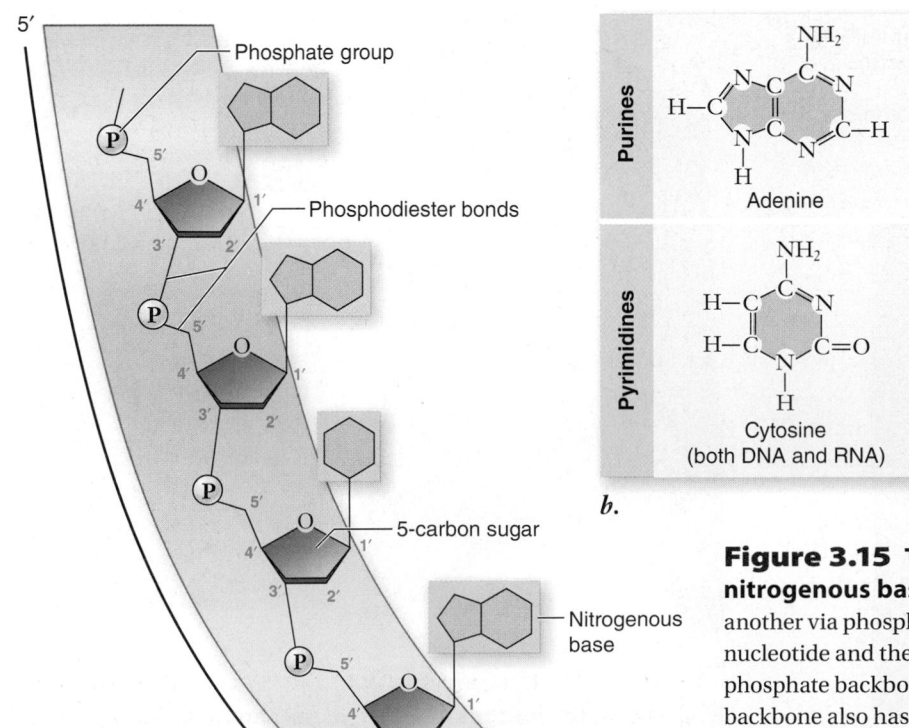

Figure 3.15 The structure of a nucleic acid and the organic nitrogenous bases. *a.* In a nucleic acid, nucleotides are linked to one another via phosphodiester bonds formed between the phosphate of one nucleotide and the sugar of the next nucleotide. We call this the sugar-phosphate backbone, and the organic bases protrude from this chain. The backbone also has different ends: a 5′ phosphate end and a 3′ hydroxyl end (the blue numbers come from the numbers in the sugars). *b.* The organic nitrogenous bases can be either purines or pyrimidines. The base thymine is found in DNA. The base uracil is found in RNA.

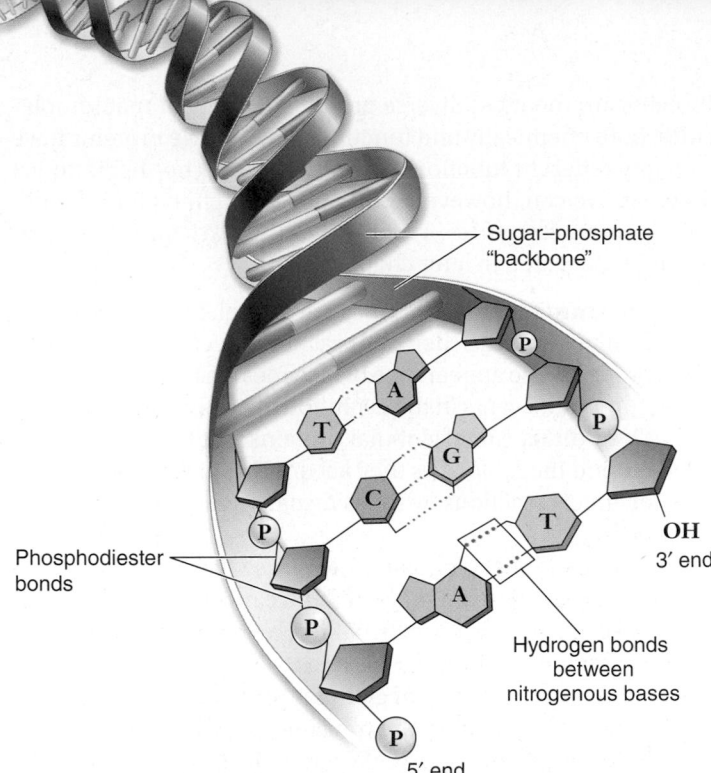

Figure 3.16 The structure of DNA. DNA consists of two polynucleotide chains running in opposite directions wrapped about a single helical axis. Hydrogen bond formation (dashed lines) between the nitrogenous bases, called base-pairing, causes the two chains of DNA to bind to each other and form a double helix.

sentence. A sentence written in English consists of a combination of the 26 different letters of the alphabet in a certain order; the code of a DNA molecule consists of different combinations of the four types of nucleotides in specific sequences, such as CGCTTACG.

DNA molecules in organisms exist as two chains wrapped about each other in a long linear molecule in eukaryotes, and a circular molecule in most prokaryotes. The two strands of a DNA polymer wind around each other like the outside and inside rails of a spiral staircase. Such a spiral shape is called a helix, and a helix composed of two chains is called a **double helix.** Each step of DNA's helical staircase is composed of a base-pair. The pair consists of a base in one chain attracted by hydrogen bonds to a base opposite it on the other chain (figure 3.16).

The base-pairing rules arise from the most stable hydrogen bonding configurations between the bases: Adenine pairs with thymine (in DNA) or with uracil (in RNA), and cytosine pairs with guanine. The bases that participate in base-pairing are said to be **complementary** to each other. Additional details of the structure of DNA and how it interacts with RNA in the production of proteins are presented in chapters 14 and 15.

In eukaryotic organisms, the DNA is further complexed with protein to form structures we call chromosomes. This actually forms a higher order structure that affects the function of DNA as it is involved in the control of gene expression (see chapter 16).

RNA has many roles in a cell

RNA is similar to DNA, but with two major chemical differences. First, RNA molecules contain ribose sugars, in which the

C-2 is bonded to a hydroxyl group. (In DNA, a hydrogen atom replaces this hydroxyl group.) Second, RNA molecules use uracil in place of thymine. Uracil has a similar structure to thymine, except that one of its carbons lacks a methyl ($-CH_3$) group.

RNA is produced by transcription (copying) from DNA, and is usually single-stranded (figure 3.17). The role of RNA in cells is quite varied: it carries information in the form of **mRNA,** it is part of the ribosome, in the form of **ribosomal RNA (rRNA),** and it carries amino acids in the form of **transfer RNA (tRNA).** There has been a revolution of late in how we view RNA since it has been found to function as an enzyme, and newly discovered forms of RNA are involved in regulating gene expression (explored in more detail in chapter 16).

Other nucleotides are vital components of energy reactions

In addition to serving as subunits of DNA and RNA, nucleotide bases play other critical roles in the life of a cell. For example,

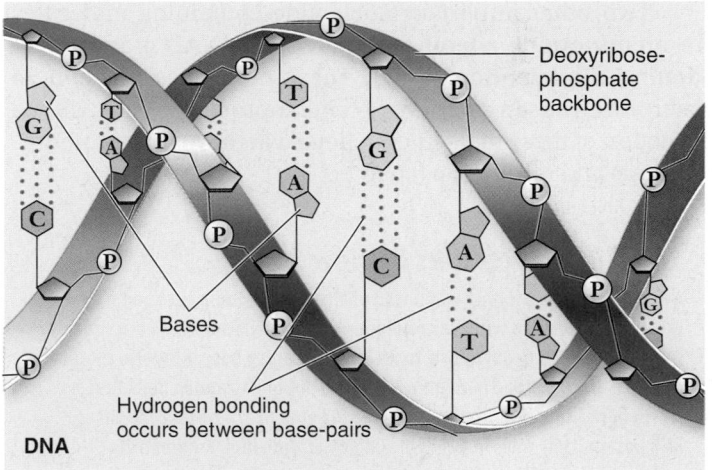

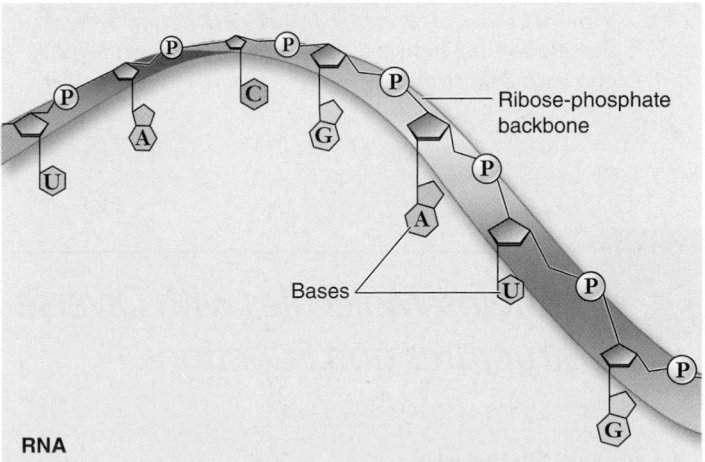

Figure 3.17 DNA versus RNA. DNA forms a double helix, uses deoxyribose as the sugar in its sugar–phosphate backbone, and uses thymine among its nitrogenous bases. RNA is usually single-stranded, uses ribose as the sugar in its sugar–phosphate backbone, and uses uracil in place of thymine.

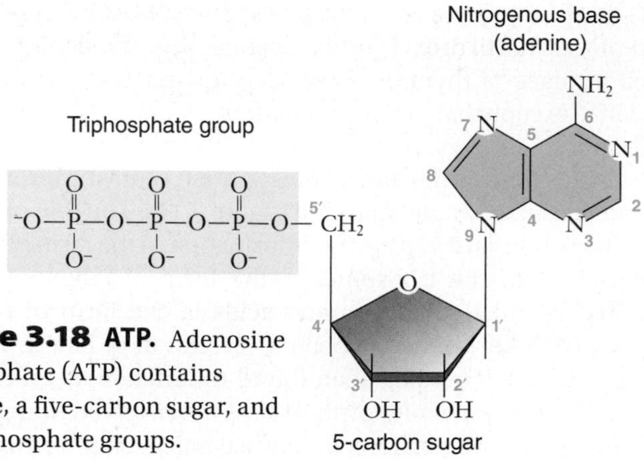

Triphosphate group

Nitrogenous base
(adenine)

Figure 3.18 ATP. Adenosine triphosphate (ATP) contains adenine, a five-carbon sugar, and three phosphate groups.

5-carbon sugar

adenine is a key component of the molecule **adenosine triphosphate** (**ATP;** figure 3.18)—the energy currency of the cell. Cells use ATP as energy in a variety of transactions, the way we use money in society. ATP is used to drive energetically unfavorable chemical reactions, to power transport across membranes, and to power the movement of cells.

Two other important nucleotide-containing molecules are **nicotinamide adenine dinucleotide (NAD⁺)** and **flavin adenine dinucleotide (FAD).** These molecules function as electron carriers in a variety of cellular processes. You will see the action of these molecules in detail when we discuss photosynthesis and respiration (chapters 7–8).

Learning Outcomes Review 3.3

A nucleic acid is a polymer composed of alternating phosphate and five-carbon sugar groups with a nitrogenous base protruding from each sugar. In DNA, this sugar is deoxyribose. In RNA, the sugar is ribose. RNA also contains the base uracil instead of thymine. DNA is a double-stranded helix that stores hereditary information as a specific sequence of nucleotide bases. RNA has multiple roles in a cell, including carrying information from DNA and forming part of the ribosome.

■ *If an RNA molecule is copied from a DNA strand, what is the relationship between the sequence of bases in RNA and each DNA strand?*

3.4 Proteins: Molecules with Diverse Structures and Functions

---

Learning Outcomes

1. Describe the possible levels of protein structure.
2. Explain how motifs and domains contribute to protein structure.
3. Understand the relationship between amino acid sequence and their three-dimensional structure.

Proteins are the most diverse group of biological macromolecules, both chemically and functionally. Because proteins have so many different functions in cells we could not begin to list them all. We can, however, group these functions into the following seven categories. This list is a summary only, however; details are covered in later chapters.

1. **Enzyme catalysis.** Enzymes are biological catalysts that facilitate specific chemical reactions. Because of this property, the appearance of enzymes was one of the most important events in the evolution of life. Enzymes are three-dimensional globular proteins that fit snugly around the molecules they act on. This fit facilitates chemical reactions by stressing particular chemical bonds.
2. **Defense.** Other globular proteins use their shapes to "recognize" foreign microbes and cancer cells. These cell-surface receptors form the core of the body's endocrine and immune systems.
3. **Transport.** A variety of globular proteins transport small molecules and ions. The transport protein hemoglobin, for example, transports oxygen in the blood. Membrane transport proteins help move ions and molecules across the membrane.
4. **Support.** Protein fibers play structural roles. These fibers include keratin in hair, fibrin in blood clots, and collagen. The last one, collagen, forms the matrix of skin, ligaments, tendons, and bones and is the most abundant protein in a vertebrate body.
5. **Motion.** Muscles contract through the sliding motion of two kinds of protein filaments: actin and myosin. Contractile proteins also play key roles in the cell's cytoskeleton and in moving materials within cells.
6. **Regulation.** Small proteins called hormones serve as intercellular messengers in animals. Proteins also play many regulatory roles within the cell—turning on and shutting off genes during development, for example. In addition, proteins receive information, acting as cell-surface receptors.
7. **Storage.** Calcium and iron are stored in the body by binding as ions to storage proteins.

Table 3.2 summarizes these functions and includes examples of the proteins that carry them out in the human body.

Proteins are polymers of amino acids

Proteins are linear polymers made with 20 different amino acids. **Amino acids,** as their name suggests, contain an amino group ($-NH_2$) and an acidic carboxyl group ($-COOH$). The specific order of amino acids determines the protein's structure and function. Many scientists believe amino acids were among the first molecules formed on the early Earth. It seems highly likely that the oceans that existed early in the history of the Earth contained a wide variety of amino acids.

Amino acid structure

The generalized structure of an amino acid is shown as amino and carboxyl groups bonded to a central carbon atom, with an additional hydrogen and a functional side group indicated

TABLE 3.2 The Many Functions of Protein

Function	Class of Protein	Examples	Examples of Use
Enzyme catalysis	Enzymes	Glycosidases	Cleave polysaccharides
		Proteases	Break down proteins
		Polymerases	Synthesize nucleic acids
		Kinases	Phosphorylate sugars and proteins
Defense	Immunoglobulins	Antibodies	Mark foreign proteins for elimination
	Toxins	Snake venom	Blocks nerve function
	Cell-surface antigens	MHC* proteins	"Self" recognition
Transport	Circulating transporters	Hemoglobin	Carries O_2 and CO_2 in blood
		Myoglobin	Carries O_2 and CO_2 in muscle
		Cytochromes	Electron transport
	Membrane transporters	Sodium–potassium pump	Excitable membranes
		Proton pump	Chemiosmosis
		Glucose transporter	Transports glucose into cells
Support	Fibers	Collagen	Forms cartilage
		Keratin	Forms hair, nails
		Fibrin	Forms blood clots
Motion	Muscle	Actin	Contraction of muscle fibers
		Myosin	Contraction of muscle fibers
Regulation	Osmotic proteins	Serum albumin	Maintains osmotic concentration of blood
	Gene regulators	*lac* Repressor	Regulates transcription
	Hormones	Insulin	Controls blood glucose levels
		Vasopressin	Increases water retention by kidneys
		Oxytocin	Regulates uterine contractions and milk production
Storage	Ion-binding	Ferritin	Stores iron, especially in spleen
		Casein	Stores ions in milk
		Calmodulin	Binds calcium ions

*MHC, major histocompatibility complex.

by R. These components completely fill the bonds of the central carbon:

$$H_2N—C—COOH$$

(with R above the central C and H below the central C)

The unique character of each amino acid is determined by the nature of the R group. Notice that unless the R group is an H atom, as in glycine, amino acids are chiral and can exist as two enantiomeric forms: D or L. In living systems, only the L-amino acids are found in proteins, and D-amino acids are rare.

The R group also determines the chemistry of amino acids. Serine, in which the R group is —CH_2OH, is a polar molecule. Alanine, which has —CH_3 as its R group, is nonpolar. The 20 common amino acids are grouped into five chemical classes, based on their R group:

1. Nonpolar amino acids, such as leucine, often have R groups that contain —CH_2 or —CH_3.
2. Polar uncharged amino acids, such as threonine, have R groups that contain oxygen (or —OH).
3. Charged amino acids, such as glutamic acid, have R groups that contain acids or bases that can ionize.
4. Aromatic amino acids, such as phenylalanine, have R groups that contain an organic (carbon) ring with alternating single and double bonds. These are also nonpolar.
5. Amino acids that have special functions have unique properties. Some examples are methionine, which is often the first amino acid in a chain of amino acids; proline, which causes kinks in chains; and cysteine, which links chains together.

Each amino acid affects the shape of a protein differently, depending on the chemical nature of its side group. For example, portions of a protein chain with numerous nonpolar amino acids tend to fold into the interior of the protein by hydrophobic exclusion.

Peptide bonds

In addition to its R group, each amino acid, when ionized, has a positive amino (NH_3^+) group at one end and a negative carboxyl (COO^-) group at the other. The amino and carboxyl groups on a pair of amino acids can undergo a dehydration reaction to form a covalent bond. The covalent bond that links two amino acids is called a **peptide bond** (figure 3.19). The two amino acids linked by such a bond are not free to rotate around the N—C linkage because the peptide bond has a partial double-bond character. This is different from the N—C and C—C bonds to the central carbon of the amino acid. This lack of rotation about the peptide bond is one factor that determines the structural character of the coils and other regular shapes formed by chains of amino acids.

A protein is composed of one or more long unbranched chains. Each chain is called a **polypeptide** and is composed of amino acids linked by peptide bonds. The terms *protein* and *polypeptide* tend to be used loosely and may be confusing. For proteins that include only a single polypeptide chain, the two terms are synonymous.

The pioneering work of Frederick Sanger in the early 1950s provided the evidence that each kind of protein has a specific amino acid sequence. Using chemical methods to remove successive amino acids and then identify them, Sanger succeeded in determining the amino acid sequence of insulin. In so doing he demonstrated clearly that this protein had a defined sequence, which was the same for all insulin molecules in the solution. Although many different amino acids occur in nature, only 20 commonly occur in proteins. Of these 20, 8 are called essential amino acids because humans cannot synthesize them and thus must get them from their diets. Figure 3.20 illustrates these 20 amino acids and their side groups.

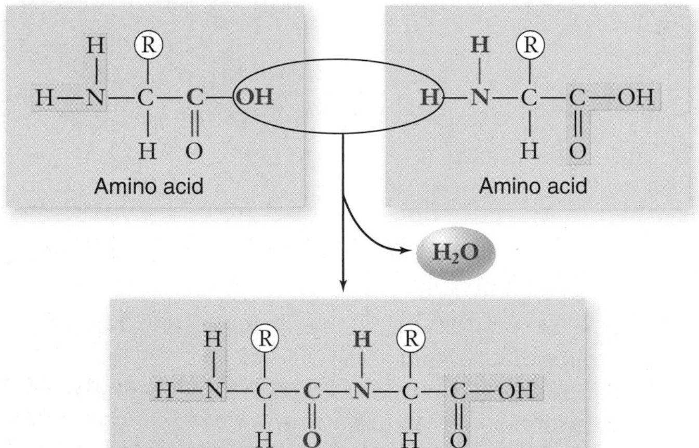

Proteins have levels of structure

The shape of a protein determines its function. One way to study the shape of something as small as a protein is to look at it with very short wavelength energy—in other words, with X-rays. X-rays can be passed through a crystal of protein to produce a diffraction pattern. This pattern can then be analyzed by a painstaking procedure that allows the investigator to build up a three-dimensional picture of the position of each atom. The first protein to be analyzed in this way was myoglobin, and the related protein hemoglobin was analyzed soon thereafter.

As more and more proteins were studied, a general principle became evident: In every protein studied, essentially all the internal amino acids are nonpolar ones—amino acids such as leucine, valine, and phenylalanine. Water's tendency to hydrophobically exclude nonpolar molecules literally shoves the nonpolar portions of the amino acid chain into the protein's interior (figure 3.21). This tendency forces the nonpolar amino acids into close contact with one another, leaving little empty space inside. Polar and charged amino acids are restricted to the surface of the protein, except for the few that play key functional roles.

The structure of proteins is usually discussed in terms of a hierarchy of four levels: *primary, secondary, tertiary,* and *quaternary* (figure 3.22). We will examine this view and then integrate it with a more modern approach arising from our increasing knowledge of protein structure.

Primary structure: amino acid sequence

The **primary structure** of a protein is its amino acid sequence. Because the R groups that distinguish the amino acids play no role in the peptide backbone of proteins, a protein can consist of any sequence of amino acids. Thus, because any of 20 different amino acids might appear at any position, a protein containing 100 amino acids could form any of 20^{100} different amino acid sequences (that's the same as 10^{130}, or 1 followed by 130 zeros—more than the number of atoms known in the universe). This important property of proteins permits great diversity.

Consider the protein hemoglobin, the protein your blood uses to transport oxygen. Hemoglobin is composed of two α-globin peptide chains and two β-globin peptide chains. The α-globin chains differ from the β-globin ones in the sequence of amino acids. Furthermore, any alteration in the normal sequence of either of the types of globin proteins, even by a single amino acid, can have drastic effects on how the protein functions.

Secondary structure: Hydrogen bonding patterns

The amino acid side groups are not the only portions of proteins that form hydrogen bonds. The peptide groups of the main chain can also do so. These hydrogen bonds can be with water or with other peptide groups. If the peptide groups formed too many hydrogen bonds with water, the proteins would tend to behave like a random coil and wouldn't produce

Figure 3.19 The peptide bond. A peptide bond forms when the amino end of one amino acid joins to the carboxyl end of another. Reacting amino and carboxyl groups are shown in red and nonreacting groups are highlighted in green. Notice that the resulting dipeptide still has an amino end and a carboxyl end. Because of the partial double-bond nature of peptide bonds, the resulting peptide chain cannot rotate freely around these bonds.

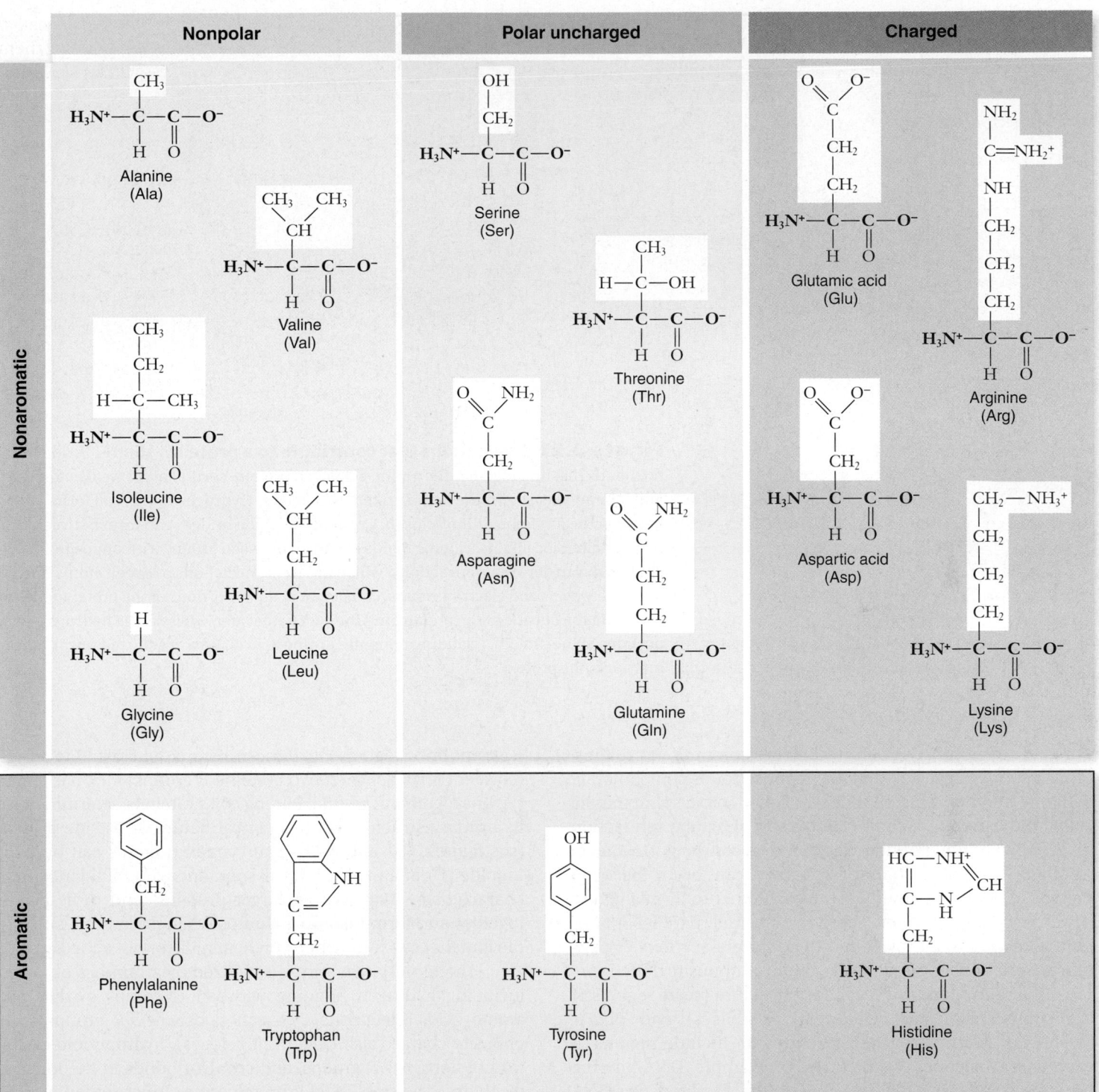

Figure 3.20 The 20 common amino acids. Each amino acid has the same chemical backbone, but differs in the side, or R, group. Seven of the amino acids are nonpolar because they have $-CH_2$ or $-CH_3$ in their R groups. Two of the seven contain ring structures with alternating double and single bonds, which classifies them also as aromatic. Another five are polar because they have oxygen or a hydroxyl group in their R groups. Five others are capable of ionizing to a charged form. The remaining three special-function amino acids have chemical properties that allow them to help form links between protein chains or kinks in proteins.

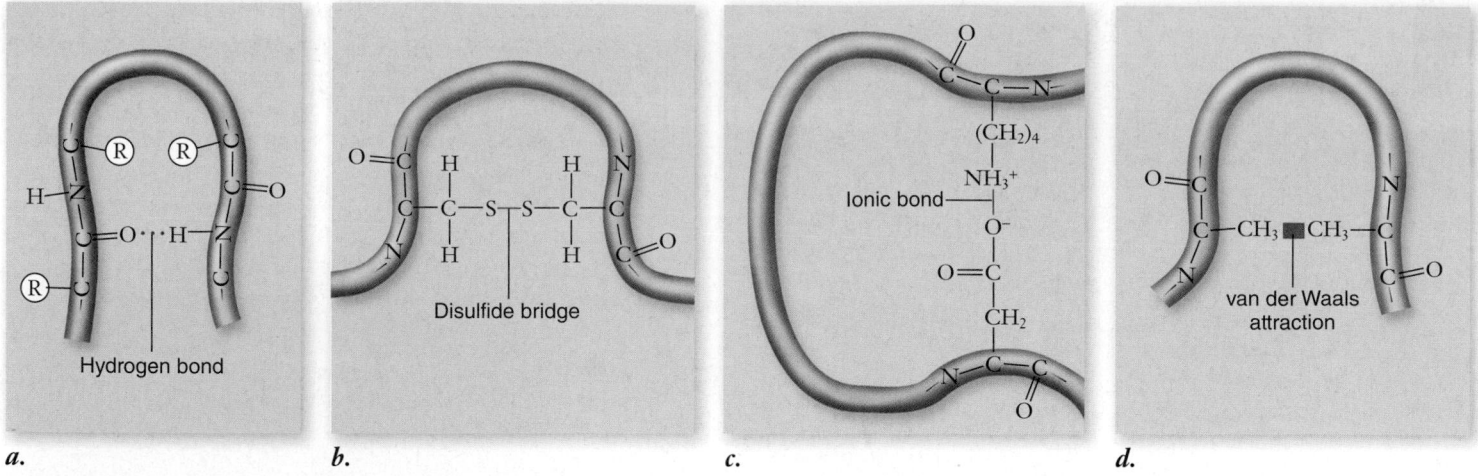

a.   b.   c.   d.

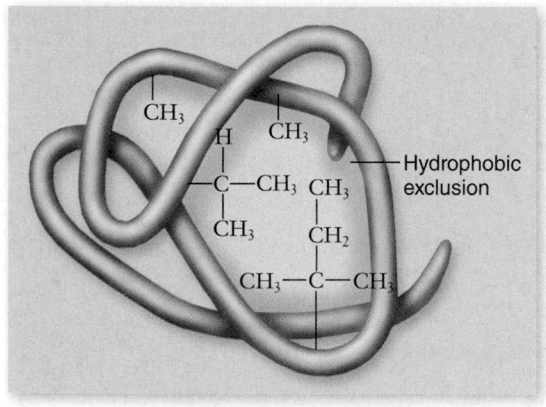

e.

Figure 3.21 Interactions that contribute to a protein's shape. Aside from the bonds that link together the amino acids in a protein, several other weaker forces and interactions determine how a protein will fold. ***a.*** Hydrogen bonds can form between the different amino acids. ***b.*** Covalent disulfide bridges can form between two cysteine side chains. ***c.*** Ionic bonds can form between groups with opposite charge. ***d.*** van der Waals attractions, which are weak attractions between atoms due to oppositely polarized electron clouds, can occur. ***e.*** Polar portions of the protein tend to gather on the outside of the protein and interact with water, whereas the hydrophobic portions of the protein, including nonpolar amino acid chains, are shoved toward the interior of the protein.

the kinds of globular structures that are common in proteins. Linus Pauling suggested that the peptide groups could interact with one another if the peptide was coiled into a spiral that he called the **α helix.** We now call this sort of regular interaction of groups in the peptide backbone **secondary structure.** Another form of secondary structure can occur between regions of peptide aligned next to each other to form a planar structure called a **β sheet.** These can be either parallel or anti-parallel depending on whether the adjacent sections of peptide are oriented in the same direction, or opposite direction.

These two kinds of secondary structure create regions of the protein that are cylindrical (α helices) and planar (β sheets). A protein's final structure can include regions of each type of secondary structure. For example, DNA-binding proteins usually have regions of α helix that can lay across DNA and interact directly with the bases of DNA. Porin proteins that form holes in membranes are composed of β sheets arranged to form a pore in the membrane. Finally in hemoglobin, the α- and β-globin peptide chains that make up the final molecule each have characteristic regions of secondary structure.

Tertiary structure: Folds and links

The final folded shape of a globular protein is called its **tertiary structure.** This tertiary structure contains regions that have secondary structure and determines how these are further arranged in space to produce the overall structure. A protein is initially driven into its tertiary structure by hydrophobic exclusion from water. Ionic bonds between oppositely charged R groups bring regions into close proximity, and disulfide bonds (covalent links between two cysteine R groups) lock particular regions together. The final folding of a protein is determined by its primary structure—the chemical nature of its side groups (see figures 3.21 and 3.22). Many small proteins can be fully unfolded ("denatured") and will spontaneously refold into their characteristic shape. Other larger proteins tend to associate together and form insoluble clumps when denatured, such as the film that can form when you heat milk for hot chocolate.

The tertiary structure is stabilized by a number of forces including hydrogen bonding between R groups of different amino acids, electrostatic attraction between R groups with opposite charge (also called salt bridges), hydrophobic exclusion of nonpolar R groups, and covalent bonds in the form of disulfides. The stability of a protein, once it has folded into its tertiary shape, is strongly influenced by how well its interior fits together. When two nonpolar chains in the interior are very close together, they experience a form of molecular attraction called van der Waals forces. Individually quite weak, these forces can add up to a strong attraction when many of them come into play, like the combined strength of hundreds of hooks and loops on a strip of Velcro. These forces are effective only over short distances, however. No "holes" or cavities exist in the interior of proteins. The variety of different nonpolar amino acids, with a different-sized R group with its own distinctive shape, allows nonpolar chains to fit very precisely within the protein interior.

It is therefore not surprising that changing a single amino acid can drastically alter the structure, and thus the function of a

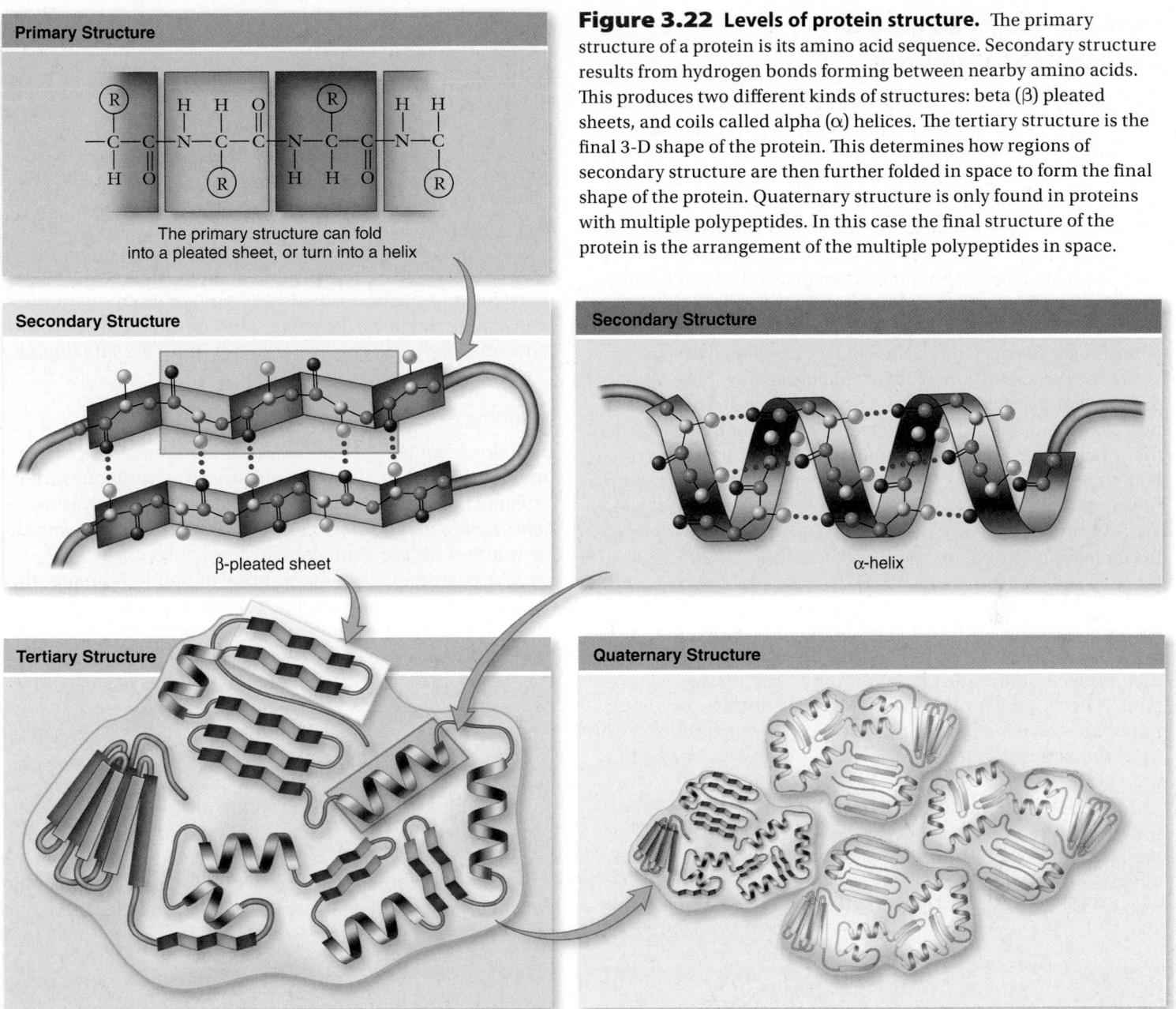

Primary Structure

The primary structure can fold
into a pleated sheet, or turn into a helix

Secondary Structure

β-pleated sheet

Secondary Structure

α-helix

Tertiary Structure

Quaternary Structure

Figure 3.22 Levels of protein structure. The primary structure of a protein is its amino acid sequence. Secondary structure results from hydrogen bonds forming between nearby amino acids. This produces two different kinds of structures: beta (β) pleated sheets, and coils called alpha (α) helices. The tertiary structure is the final 3-D shape of the protein. This determines how regions of secondary structure are then further folded in space to form the final shape of the protein. Quaternary structure is only found in proteins with multiple polypeptides. In this case the final structure of the protein is the arrangement of the multiple polypeptides in space.

protein. The sickle cell version of hemoglobin (HbS), for example, is a change of a single glutamic acid for a valine in the β-globin chain. This change substitutes a charged amino acid for a nonpolar one on the surface of the protein, leading the protein to become sticky and form clumps. Another variant of hemoglobin called HbE, actually the most common in human populations, causes a change from glutamic acid to lysine at a different site in the β-globin chain. In this case the structural change is not as dramatic, but it still impairs function, resulting in blood disorders called anemia and thalassemia. More than 700 structural variants of hemoglobin are known, with up to 7% of the world's population being carriers of forms that are medically important.

Quaternary structure: Subunit arrangements

When two or more polypeptide chains associate to form a functional protein, the individual chains are referred to as subunits of the protein. The arrangement of these subunits is termed its **quaternary structure**. In proteins composed of subunits, the interfaces where the subunits touch one another are often nonpolar, and they play a key role in transmitting information between the subunits about individual subunit activities.

Remember that the protein hemoglobin is composed of two α-chain subunits and two β-chain subunits. Each α- and β-globin chain has a primary structure consisting of a specific sequence of amino acids. This then assumes a characteristic secondary structure consisting of α helices and β sheets that are then arranged into a specific tertiary structure for each α- and β-globin subunit. Lastly, these subunits are then arranged into their final quaternary structure. This is the final structure of the protein. For proteins that consist of only a single peptide chain, the enzyme lysozyme for example, the tertiary structure is the final structure of the protein.

Motifs and domains are additional structural characteristics

To directly determine the sequence of amino acids in a protein is a laborious task. Although the process has been automated, it remains slow and difficult.

The ability to sequence DNA changed this situation rather suddenly. Sequencing DNA was a much simpler process, and even before it was automated, the number of known sequences rose quickly. With the advent of automation, the known sequences increased even more dramatically. Today the entire sequence of hundreds of bacterial genomes and more than a dozen animal genomes, including that of humans, has been determined. Because the DNA sequence is directly related to amino acid sequence in proteins, biologists now have a large database of protein sequences to compare and analyze. This new information has also stimulated thought about the logic of the genetic code and whether underlying patterns exist in protein structure. Our view of protein structure has evolved with this new information. Researchers still view the four-part hierarchical structure as important, but two new terms have entered the biologist's vocabulary: motif and domain.

Motifs

As biologists discovered the 3-D structure of proteins (an even more laborious task than determining the sequence), they noticed similarities between otherwise dissimilar proteins. These similar structures are called **motifs,** or sometimes "supersecondary structure." The term *motif* is borrowed from the arts and refers to a recurring thematic element in music or design.

One very common protein motif is the β-α-β motif, which creates a fold or crease; the so-called "Rossmann fold" at the core of nucleotide-binding sites in a wide variety of proteins. A second motif that occurs in many proteins is the β barrel, which is a β sheet folded around to form a tube. A third type of motif, the helix-turn-helix, consists of two α helices separated by a bend. This motif is important because many proteins use it to bind to the DNA double helix (figure 3.23; see also chapter 16).

Motifs indicate a logic to structure that investigators still do not understand. Do they simply represent a reuse by evolution of something that already works, or are they an optimal solution to a problem, such as how to bind a nucleotide? One way to think about it is that if amino acids are letters in the language of proteins, then motifs represent repeated words or phrases. Motifs have been useful in determining the function of unknown proteins. Databases of protein motifs are used to search new unknown proteins. Finding motifs with known functions may allow an investigator to infer the function of a new protein.

Domains

Domains of proteins are functional units within a larger structure. They can be thought of as substructure within the tertiary structure of a protein (see figure 3.23). To continue the metaphor: Amino acids are letters in the protein language, motifs are words or phrases, and domains are paragraphs.

Most proteins are made up of multiple domains that perform different parts of the protein's function. In many cases, these domains can be physically separated. For example, transcription factors (discussed in chapter 16) are proteins that bind to DNA and initiate its transcription. If the DNA-binding region is exchanged with a different transcription factor, then the specificity of the factor for DNA can be changed without changing its ability to stimulate transcription. Such "domain-swapping" experiments have been performed with many transcription factors, and they indicate, among other things, that the DNA-binding and activation domains are functionally separate.

These functional domains of proteins may also help the protein to fold into its proper shape. As a polypeptide chain

Motifs

β-α-β
motif

Helix-turn-Helix
motif

Figure 3.23 Motifs and domains. The elements of secondary structure can combine, fold, or crease to form motifs. These motifs are found in different proteins and can be used to predict function. Proteins also are made of larger domains, which are functionally distinct parts of a protein. The arrangement of these domains in space is the tertiary structure of a protein.

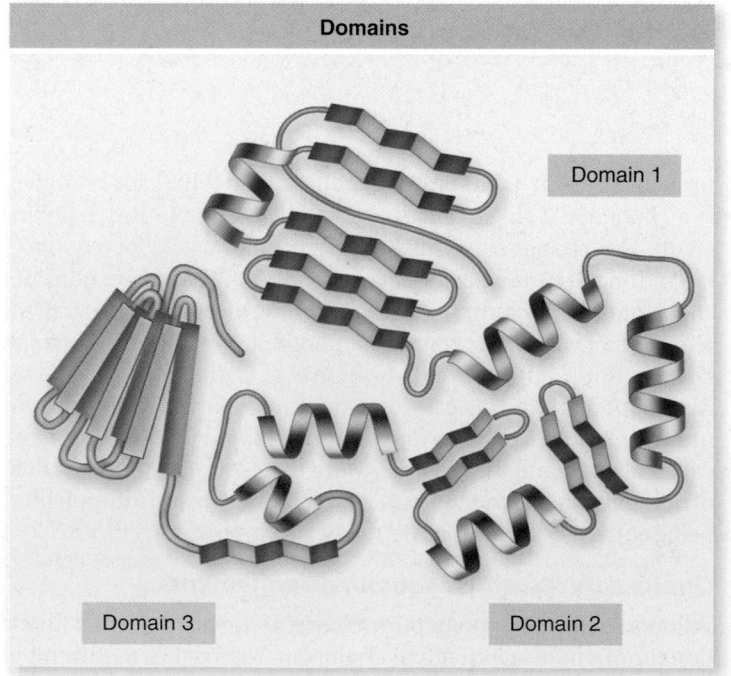

Domains

Domain 1

Domain 3

Domain 2

folds, the domains take their proper shape, each more or less independently of the others. This action can be demonstrated experimentally by artificially producing the fragment of a polypeptide that forms the domain in the intact protein, and showing that the fragment folds to form the same structure as it exhibits in the intact protein. A single polypeptide chain connects the domains of a protein, like a rope tied into several adjacent knots.

Domains can also correspond to the structure of the genes that encode them. Later, in chapter 15, you will see that genes in eukaryotes are often in pieces within the genome, and these pieces, called *exons,* sometimes encode the functional domains of a protein. This finding led to the idea of evolution acting by shuffling protein-encoding domains.

The process of folding relies on chaperone proteins

Until recently, scientific investigators thought that newly made proteins fold spontaneously, randomly trying out different configurations as hydrophobic interactions with water shoved nonpolar amino acids into the protein's interior until the final structure was arrived at. We now know this view is too simple. Protein chains can fold in so many different ways that trial and error would simply take too long. In addition, as the open chain folds its way toward its final form, nonpolar "sticky" interior portions are exposed during intermediate stages. If these intermediate forms are placed in a test tube in an environment identical to that inside a cell, they stick to other, unwanted protein partners, forming a gluey mess.

How do cells avoid having their proteins clump into a mass? A vital clue came in studies of unusual mutations that prevent viruses from replicating in bacterial cells. It turns out that the virus proteins produced inside the cells could not fold properly. Further study revealed that normal cells contain **chaperone proteins,** which help other proteins to fold correctly.

Molecular biologists have now identified many proteins that act as molecular chaperones. This class of proteins has multiple subclasses, and representatives have been found in essentially every organism that has been examined. Furthermore, these proteins seem to be essential for viability as well, illustrating their fundamental importance. Many are heat shock proteins, produced in greatly increased amounts when cells are exposed to elevated temperature. High temperatures cause proteins to unfold, and heat shock chaperone proteins help the cell's proteins to refold properly.

One class of these proteins, called chaperonins, has been extensively studied. In the bacterium *Escherichia coli* (*E. coli*), one example is the essential protein GroE chaperonin. In mutants in which the GroE chaperonin is inactivated, fully 30% of the bacterial proteins fail to fold properly. Chaperonins associate to form a large macromolecular complex that resembles a cylindrical container. Proteins can move into the container, and the container itself can change its shape considerably (figure 3.24). Experiments have shown that an improperly folded protein can enter the chaperonin and be refolded. Although we don't know exactly how this happens, it seems to involve changes in the hydrophobicity of the interior of the chamber.

The flexibility of the structure of chaperonins is amazing. We tend to think of proteins as being fixed structures, but this is clearly not the case for chaperonins and this flexibility is necessary for their function. It also illustrates that even domains that may be very widely separated in a very large protein are still functionally connected. The folding process within a chaperonin harnesses the hydrolysis of ATP to power these changes in structure necessary for function. This entire process can occur in a cyclic manner until the appropriate structure is achieved. Cells use these chaperonins both to accomplish the original folding of some proteins and to restore the structure of incorrectly folded ones.

Some diseases may result from improper folding

Chaperone protein deficiencies may be implicated in certain diseases in which key proteins are improperly folded. Cystic fibrosis is

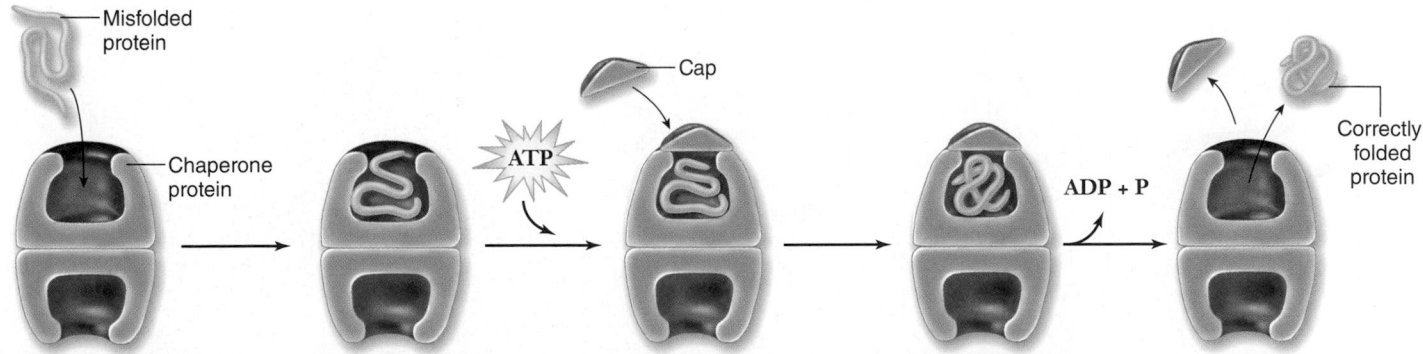

Chance for protein to refold

Figure 3.24 How one type of chaperone protein works. This barrel-shaped chaperonin is from the GroE family of chaperone proteins. It is composed of two identical rings each with seven identical subunits, each of which has three distinct domains. An incorrectly folded protein enters one chamber of the barrel, and a cap seals the chamber. Energy from the hydrolysis of ATP fuels structural alterations to the chamber, changing it from hydrophobic to hydrophilic. This change allows the protein to refold. After a short time, the protein is ejected, either folded or unfolded, and the cycle can repeat itself.

a hereditary disorder in which a mutation disables a vital protein that moves ions across cell membranes. As a result, people with cystic fibrosis have thicker than normal mucus. This results in breathing problems, lung disease, and digestive difficulties, among other things. One interesting feature of the molecular analysis of this disease has been the number of different mutations found in human populations. One diverse class of mutations all result in problems with protein folding. The number of different mutations that can result in improperly folded proteins may be related to the fact that the native protein often fails to fold properly.

Denaturation inactivates proteins

If a protein's environment is altered, the protein may change its shape or even unfold completely. This process is called **denaturation** (figure 3.25). Proteins can be denatured when the pH, temperature, or ionic concentration of the surrounding solution changes.

Denatured proteins are usually biologically inactive. This action is particularly significant in the case of enzymes. Because practically every chemical reaction in a living organism is catalyzed by a specific enzyme, it is vital that a cell's enzymes work properly.

The traditional methods of food preservation, salt curing and pickling, involve denaturation of proteins. Prior to the general availability of refrigerators and freezers, the only practical

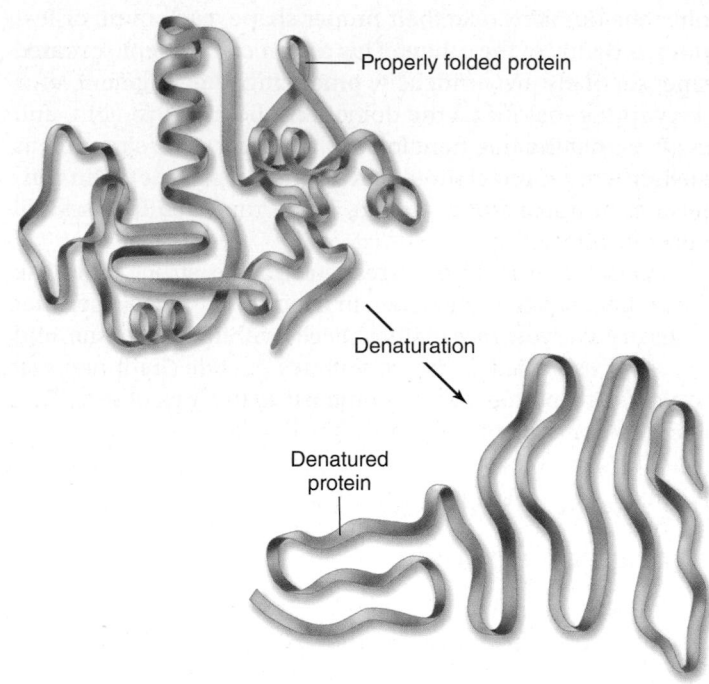

Figure 3.25 Protein denaturation. Changes in a protein's environment, such as variations in temperature or pH, can cause a protein to unfold and lose its shape in a process called denaturation. In this denatured state, proteins are biologically inactive.

SCIENTIFIC THINKING

Hypothesis: *The 3-D structure of a protein is the thermodynamically stable structure. It depends only on the primary structure of the protein and the solution conditions.*

Prediction: *If a protein is denatured and allowed to renature under native conditions, it will refold into the native structure.*

Test: *Ribonuclease is treated with a reducing agent to break disulfide bonds and is then treated with urea to completely unfold the protein. The disulfide bonds are reformed under nondenaturing conditions to see if the protein refolds properly.*

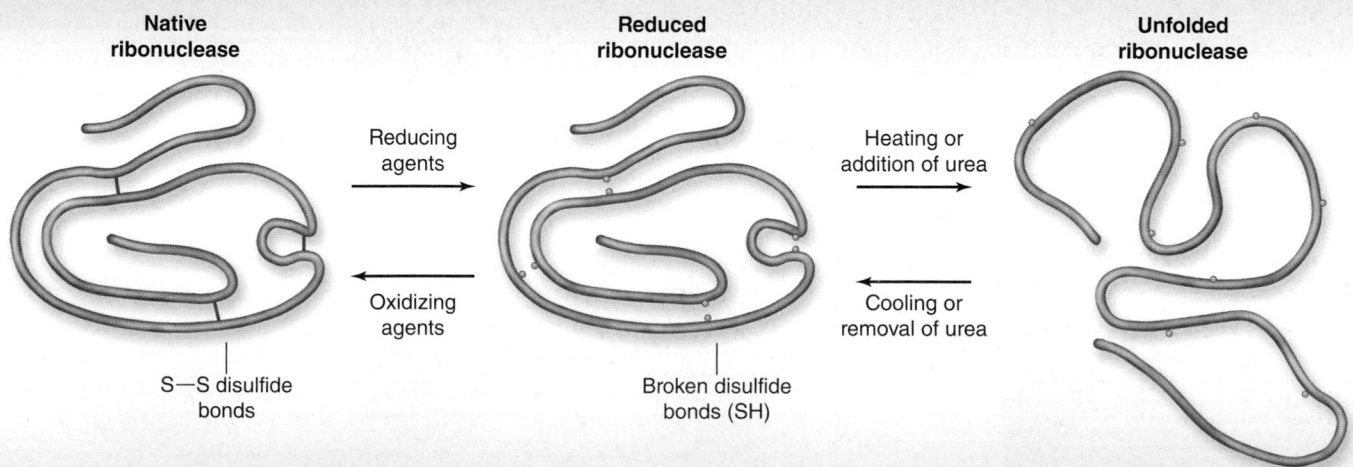

Result: *Denatured Ribonuclease refolds properly under nondenaturing conditions.*

Conclusion: *The hypothesis is supported. The information in the primary structure (amino acid sequence) is sufficient for refolding to occur. This implies that protein folding results in the thermodynamically stable structure.*

Further Experiments: *If the disulfide bonds were allowed to reform under denaturing conditions, would we get the same result? How can we rule out that the protein had not been completely denatured and therefore retained some structure?*

Figure 3.26 Primary structure determines tertiary structure.

way to keep microorganisms from growing in food was to keep the food in a solution containing a high concentration of salt or vinegar, which denatured the enzymes of most microorganisms and prevented them from growing on the food.

Most enzymes function within a very narrow range of environmental conditions. Blood-borne enzymes that course through a human body at a pH of about 7.4 would rapidly become denatured in the highly acidic environment of the stomach. Conversely, the protein-degrading enzymes that function at a pH of 2 or less in the stomach would be denatured in the relatively basic pH of the blood. Similarly, organisms that live near oceanic hydrothermal vents have enzymes that work well at these extremes of temperature (over 100°C). They cannot survive in cooler waters, because their enzymes do not function properly at lower temperatures. Any given organism usually has a tolerance range of pH, temperature, and salt concentration. Within that range, its enzymes maintain the proper shape to carry out their biological functions.

When a protein's normal environment is reestablished after denaturation, a small protein may spontaneously refold into its natural shape, driven by the interactions between its nonpolar amino acids and water (figure 3.26). This process is termed *renaturation,* and it was first established for the enzyme ribonuclease (RNase). The renaturation of RNase led to the doctrine that primary structure determines tertiary structure. Larger proteins can rarely refold spontaneously, however, because of the complex nature of their final shape, so this simple idea needs to be qualified.

The fact that some proteins can spontaneously renature implies that tertiary structure is strongly influenced by primary structure. In an extreme example, the *E. coli* ribosome can be taken apart and put back together experimentally. Although this process requires temperature and ion concentration shifts, it indicates an amazing degree of self-assembly. That complex structures can arise by self-assembly is a key idea in the study of modern biology.

It is important to distinguish denaturation from **dissociation.** For proteins with quaternary structure, the subunits may be dissociated without losing their individual tertiary structure. For example, the four subunits of hemoglobin may dissociate into four individual molecules (two α-globins and two β-globins) without denaturation of the folded globin proteins. They readily reassume their four-subunit quaternary structure.

Learning Outcomes Review 3.4

Proteins are molecules with diverse functions. They are constructed from 20 different kinds of amino acids. Protein structure can be viewed at four levels: (1) the amino acid sequence, or primary structure; (2) coils and sheets, called secondary structure; (3) the three-dimensional shape, called tertiary structure; and (4) individual polypeptide subunits associated in a quaternary structure. Different proteins often have similar substructures called motifs and can be broken down into functional domains. Proteins have a narrow range of conditions in which they fold properly; outside that range, proteins tend to unfold (denaturation). Under some conditions, denatured proteins can refold and become functional again (renaturation).

■ *How does our knowledge of protein structure help us to predict the function of unknown proteins?*

Learning Outcomes

1. *Describe the structure of triglycerides.*
2. *Explain how fats function as energy-storage molecules.*
3. *Apply knowledge of the structure of phospholipids to the formation of membranes.*

Lipids are a somewhat loosely defined group of molecules with one main chemical characteristic: They are insoluble in water. Storage fats such as animal fat are one kind of lipid. Oils such as those from olives, corn, and coconut are also lipids, as are waxes such as beeswax and earwax. Even some vitamins are lipids!

Lipids have a very high proportion of nonpolar carbon–hydrogen (C—H) bonds, and so long-chain lipids cannot fold up like a protein to confine their nonpolar portions away from the surrounding aqueous environment. Instead, when they are placed in water, many lipid molecules spontaneously cluster together and expose what polar (hydrophilic) groups they have to the surrounding water, while confining the nonpolar (hydrophobic) parts of the molecules together within the cluster. You may have noticed this effect when you add oil to a pan containing water, and the oil beads up into cohesive drops on the water's surface. This spontaneous assembly of lipids is of paramount importance to cells, as it underlies the structure of cellular membranes.

Fats consist of complex polymers of fatty acids attached to glycerol

Many lipids are built from a simple skeleton made up of two main kinds of molecules: fatty acids and glycerol. Fatty acids are long-chain hydrocarbons with a carboxylic acid (COOH) at one end. Glycerol is a three-carbon polyalcohol (three —OH groups). Many lipid molecules consist of a glycerol molecule with three fatty acids attached, one to each carbon of the glycerol backbone. Because it contains three fatty acids, a fat molecule is commonly called a **triglyceride** (the more accurate chemical name is *triacylglycerol*). This basic structure is depicted in figure 3.27. The three fatty acids of a triglyceride need not be identical, and often they are very different from one another. The hydrocarbon chains of fatty acids vary in length. The most common are even-numbered chains of 14 to 20 carbons. The many C—H bonds of fats serve as a form of long-term energy storage.

If all of the internal carbon atoms in the fatty acid chains are bonded to at least two hydrogen atoms, the fatty acid is said to be **saturated,** which refers to its having all the hydrogen atoms possible (see figure 3.27). A fatty acid that has double bonds between one or more pairs of successive carbon atoms is said to be **unsaturated.** Fatty acids with one double bond are called monounsaturated, and those with more than one double bond are termed **polyunsaturated.** Most naturally occurring unsaturated fatty acids have double bonds with a cis configuration where the carbon chain is on the same side before and after the double bond (double bonds in fatty acids in 3.27*b* are all cis).

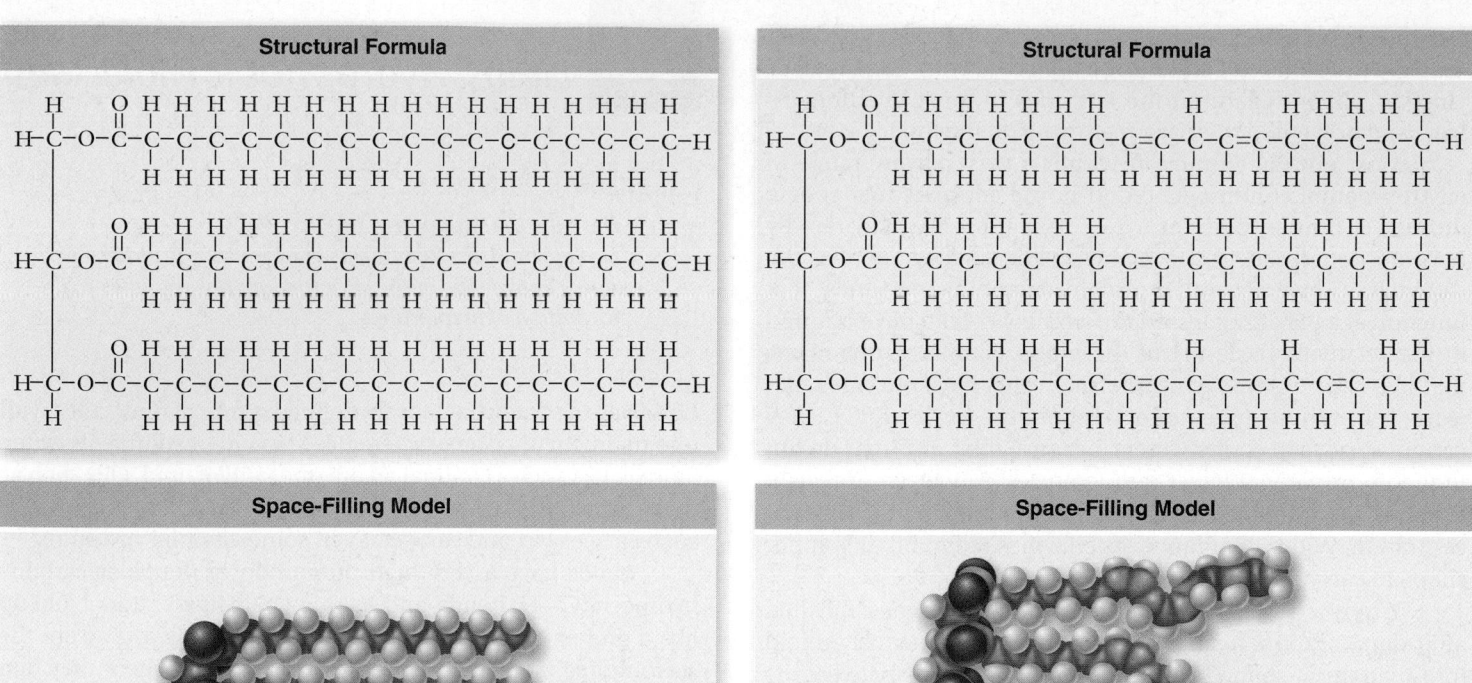

| Structural Formula | Structural Formula |

| Space-Filling Model | Space-Filling Model |

a. *b.*

Figure 3.27 Saturated and unsaturated fats. *a.* A saturated fat is composed of triglycerides that contain three saturated fatty acids (the kind that have no double bonds). A saturated fat therefore has the maximum number of hydrogen atoms bonded to its carbon chain. Most animal fats are saturated. *b.* Unsaturated fat is composed of triglycerides that contain three unsaturated fatty acids (the kind that have one or more double bonds). These have fewer than the maximum number of hydrogen atoms bonded to the carbon chain. This example includes both a monounsaturated and two polyunsaturated fatty acids. Plant fats are typically unsaturated. The many kinks of the double bonds prevent the triglyceride from closely aligning, which makes them liquid oils at room temperature.

When fats are partially hydrogenated industrially, this can produce double bonds with a trans configuration where the carbon chain is on opposite sides before and after the double bond. These are the so called trans fats. These have been linked to elevated levels of low-density lipoprotein (LDL) "bad cholesterol" and lowered levels of high-density lipoprotein (HDL) "good cholesterol." This condition is thought to be associated with an increased risk for coronary heart disease.

Having double bonds changes the behavior of the molecule because free rotation cannot occur about a C=C double bond as it can with a C—C single bond. This characteristic mainly affects melting point: that is, whether the fatty acid is a solid fat or a liquid oil at room temperature. Fats containing polyunsaturated fatty acids have low melting points because their fatty acid chains bend at the double bonds, preventing the fat molecules from aligning closely with one another. Most saturated fats, such as animal fat or those in butter, are solid at room temperature.

Placed in water, triglycerides spontaneously associate together, forming fat globules that can be very large relative to the size of the individual molecules. Because fats are insoluble in water, they can be deposited at specific locations within an organism, such as in vesicles of adipose tissue.

Organisms contain many other kinds of lipids besides fats (figure 3.28). *Terpenes* are long-chain lipids that are components of many biologically important pigments, such as chlorophyll and the visual pigment retinal. Rubber is also a terpene. *Steroids,* another class of lipid, are composed of four carbon rings. Most animal cell membranes contain the steroid cholesterol. Other steroids, such as testosterone and estrogen, function as hormones in multicellular animals. *Prostaglandins* are a group of about 20 lipids that are modified fatty acids, with two nonpolar "tails" attached to a five-carbon ring. Prostaglandins act as local chemical messengers in many vertebrate tissues. Later chapters explore the effects of some of these complex fatty acids.

Fats are excellent energy-storage molecules

Most fats contain over 40 carbon atoms. The ratio of energy-storing C—H bonds in fats is more than twice that of carbohydrates (see section 3.2), making fats much more efficient molecules for storing chemical energy. On average, fats yield about 9 kilocalories (kcal) of chemical energy per gram, as compared with about 4 kcal/g for carbohydrates.

Most fats produced by animals are saturated (except some fish oils), whereas most plant fats are unsaturated (see

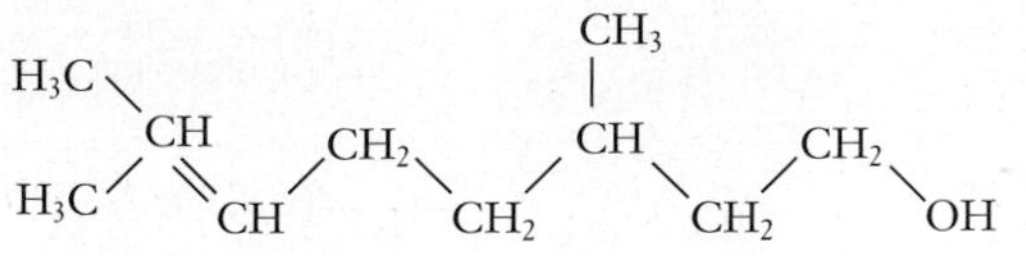

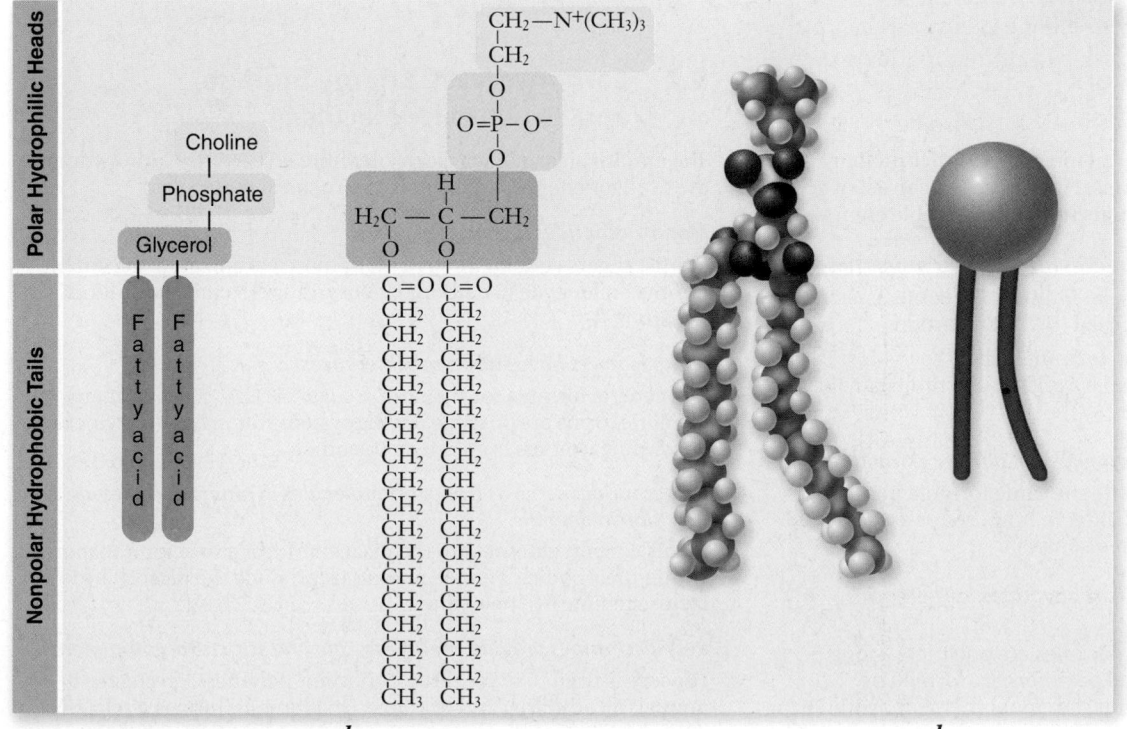

a. Terpene (citronellol)

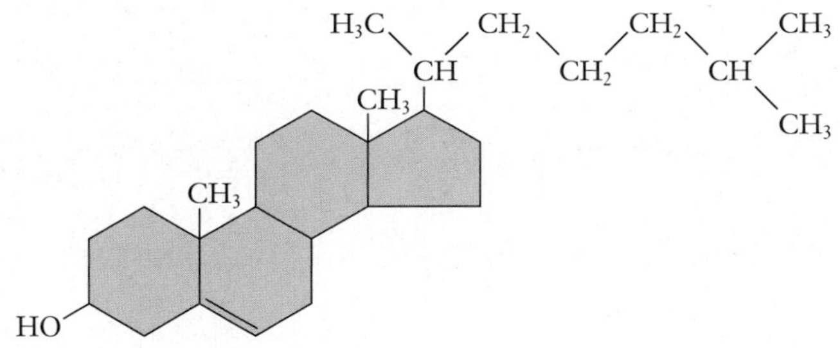

b. Steroid (cholesterol)

Figure 3.28 Other kinds of lipids. *a.* Terpenes are found in biological pigments, such as chlorophyll and retinal, and *(b)* steroids play important roles in membranes and as the basis for a class of hormones involved in chemical signaling.

figure 3.27). The exceptions are the tropical plant oils (palm oil and coconut oil), which are saturated even though they are liquid at room temperature. An oil may be converted into a solid fat by chemically adding hydrogen. Most peanut butter is usually artificially hydrogenated to make the peanut fats solidify, preventing them from separating out as oils while the jar sits on the store shelf. However, artificially hydrogenating unsaturated fats produces the *trans*-fatty acids described above.

When an organism consumes excess carbohydrate, it is converted into starch, glycogen, or fats reserved for future use. The reason that many humans in developed countries gain weight as they grow older is that the amount of energy they need decreases with age, but their intake of food does not. Thus, an increasing proportion of the carbohydrates they ingest is converted into fat.

A diet heavy in fats is one of several factors thought to contribute to heart disease, particularly atherosclerosis. In atherosclerosis, sometimes referred to as "hardening of the arteries," fatty substances called plaque adhere to the lining of blood vessels, blocking the flow of blood. Fragments of a plaque can break off from a deposit and clog arteries to the brain, causing a stroke.

Phospholipids form membranes

Complex lipid molecules called **phospholipids** are among the most important molecules of the cell because they form the core of all biological membranes. An individual phospholipid can be thought of as a substituted triglyceride, that is, a triglyceride with a phosphate replacing one of the fatty acids. The basic structure of a phospholipid includes three kinds of subunits:

1. *Glycerol,* a three-carbon alcohol, in which each carbon bears a hydroxyl group. Glycerol forms the backbone of the phospholipid molecule.
2. *Fatty acids,* long chains of $-CH_2$ groups (hydrocarbon chains) ending in a carboxyl ($-COOH$) group. Two fatty acids are attached to the glycerol backbone in a phospholipid molecule.
3. *A phosphate group* ($-PO_4^{2-}$) attached to one end of the glycerol. The charged phosphate group usually has a charged organic molecule linked to it, such as choline, ethanolamine, or the amino acid serine.

The phospholipid molecule can be thought of as having a polar "head" at one end (the phosphate group) and two long, very nonpolar "tails" at the other (figure 3.29). This structure is essential for how these molecules function, although it first

Figure 3.29 Phospholipids. The phospholipid phosphatidylcholine is shown as *(a)* a schematic, *(b)* a formula, *(c)* a space-filling model, and *(d)* an icon used in depictions of biological membranes.

appears paradoxical. Why would a molecule need to be soluble in water, but also not soluble in water? The formation of a membrane shows the unique properties of such a structure.

In water, the nonpolar tails of nearby lipid molecules aggregate away from the water, forming spherical *micelles,* with the tails facing inward (figure 3.30a). This is actually how detergent molecules work to make grease soluble in water. The grease is soluble within the nonpolar interior of the micelle and the polar surface of the micelle is soluble in water. With phospholipids, a more complex structure forms in which two layers of molecules line up, with the hydrophobic tails of each layer pointing toward one another, or inward, leaving the hydrophilic heads oriented outward, forming a bilayer (figure 3.30b). Lipid bilayers are the basic framework of biological membranes, discussed in detail in chapter 5.

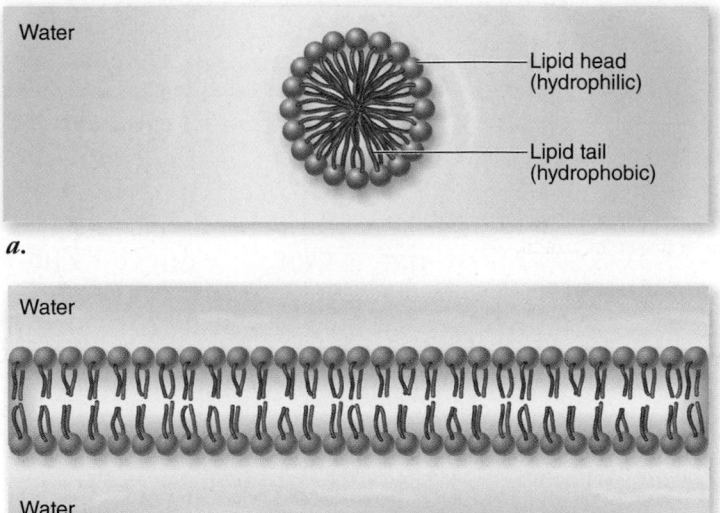

a.

b.

Figure 3.30 Lipids spontaneously form micelles or lipid bilayers in water. In an aqueous environment, lipid molecules orient so that their polar (hydrophilic) heads are in the polar medium, water, and their nonpolar (hydrophobic) tails are held away from the water. *a.* Droplets called micelles can form, or *(b)* phospholipid molecules can arrange themselves into two layers; in both structures, the hydrophilic heads extend outward and the hydrophobic tails inward. This second example is called a phospholipid bilayer.

Learning Outcomes Review 3.5

Triglycerides are made of fatty acids linked to glycerol. Fats can contain twice as many C—H bonds as carbohydrates and thus they store energy efficiently. Because the C—H bonds in lipids are nonpolar, they are not water-soluble and aggregate together in water. Phospholipids replace one fatty acid with a hydrophilic phosphate group. This allows them to spontaneously form bilayers, which are the basis of biological membranes.

■ *Why do phospholipids form membranes while triglycerides form insoluble droplets?*

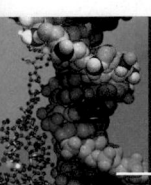

Chapter Review

3.1 Carbon: The Framework of Biological Molecules

Carbon, the backbone of all biological molecules, can form four covalent bonds and make long chains. Hydrocarbons consist of carbon and hydrogen, and their bonds store considerable energy.

Functional groups account for differences in molecular properties.
Functional groups are small molecular entities that confer specific chemical characteristics when attached to a hydrocarbon.

Carbon and hydrogen have similar electronegativity so C—H bonds are not polar. Oxygen and nitrogen have greater electronegativity, leading to polar bonds.

Isomers have the same molecular formulas but different structures.
Structural isomers are molecules with the same formula but different structures; stereoisomers differ in how groups are attached. Enantiomers are mirror-image stereoisomers.

Biological macromolecules include carbohydrates, nucleic acids, proteins, and lipids.
Most important biological macromolecules are polymers—long chains of monomer units. Biological polymers are formed by elimination of water (H and OH) from two monomers (dehydration reaction). They are broken down by adding water (hydrolysis).

3.2 Carbohydrates: Energy Storage and Structural Molecules

The empirical formula of a carbohydrate is $(CH_2O)_n$. Carbohydrates are used for energy storage and as structural molecules.

Monosaccharides are simple sugars.
Simple sugars contain three to six or more carbon atoms. Examples are glyceraldehyde (3 carbons), deoxyribose (5 carbons), and glucose (6 carbons).

Sugar isomers have structural differences.
The general formula for six-carbon sugars is $C_6H_{12}O_6$, and many isomeric forms are possible. Living systems often have enzymes for converting isomers from one to the other.

Disaccharides serve as transport molecules in plants and provide nutrition in animals.
Plants convert glucose into the disaccharide sucrose for transport within their bodies. Female mammals produce the disaccharide lactose to nourish their young.

Polysaccharides provide energy storage and structural components.
Glucose is used to make three important polymers: glycogen (in animals), and starch and cellulose (in plants). Chitin is a related structural material found in arthropods and many fungi.

3.3 Nucleic Acids: Information Molecules

Deoxyribonucleic acid (DNA) and ribonucleic acid (RNA) are polymers composed of nucleotide monomers. Cells use nucleic acids for information storage and transfer.

Nucleic acids are nucleotide polymers.

Nucleic acids contain four different nucleotide bases. In DNA these are adenine, guanine, cytosine, and thymine. In RNA, thymine is replaced by uracil.

DNA stores genetic information.

DNA exists as a double helix held together by specific base pairs: adenine with thymine and guanine with cytosine. The nucleic acid sequence constitutes the genetic code.

RNA has many roles in a cell.

RNA is made by copying DNA. RNA carries information from DNA and forms part of the ribosome. RNA can also be an enzyme and affect gene expression.

Other nucleotides are vital components of energy reactions.

Adenosine triphosphate (ATP) provides energy in cells; NAD⁺ and FAD transport electrons in cellular processes.

3.4 Proteins: Molecules with Diverse Structures and Functions

Most enzymes are proteins. Proteins also provide defense, transport, motion, and regulation, among many other roles.

Proteins are polymers of amino acids.

Amino acids are joined by peptide bonds to make polypeptides. The 20 common amino acids are characterized by R groups that determine their properties.

Proteins have levels of structure.

Protein structure is defined by the following hierarchy: primary (amino acid sequence), secondary (hydrogen bonding patterns), tertiary (three-dimensional folding), and quaternary (associations between two or more polypeptides).

Motifs and domains are additional structural characteristics.

Motifs are similar structural elements found in dissimilar proteins. They can create folds, creases, or barrel shapes. Domains are functional subunits or sites within a tertiary structure.

The process of folding relies on chaperone proteins.

Chaperone proteins assist in the folding of proteins. Heat shock proteins are an example of chaperone proteins.

Some diseases may result from improper folding.

Some forms of cystic fibrosis and Alzheimer disease are associated with misfolded proteins.

Denaturation inactivates proteins.

Denaturation refers to an unfolding of tertiary structure, which usually destroys function. Some denatured proteins may recover function when conditions are returned to normal. This implies that primary structure strongly influences tertiary structure.

Disassociation refers to separation of quaternary subunits with no changes to their tertiary structure.

3.5 Lipids: Hydrophobic Molecules

Lipids are insoluble in water because they have a high proportion of nonpolar C—H bonds.

Fats consist of complex polymers of fatty acids attached to glycerol.

Many lipids exist as triglycerides, three fatty acids connected to a glycerol molecule. Saturated fatty acids contain the maximum number of hydrogen atoms. Unsaturated fatty acids contain one or more double bonds between carbon atoms.

Fats are excellent energy-storage molecules.

The energy stored in the C—H bonds of fats is more than twice that of carbohydrates: 9 kcal/g compared with 4 kcal/g. For this reason, excess carbohydrate is converted to fat for storage.

Phospholipids form membranes.

Phospholipids contain two fatty acids and one phosphate attached to glycerol. In phospholipid-bilayer membranes, the phosphate heads are hydrophilic and cluster on the two faces of the membrane, and the hydrophobic tails are in the center.

Review Questions

UNDERSTAND

1. How is a polymer formed from multiple monomers?
   a. From the growth of the chain of carbon atoms
   b. By the removal of an —OH group and a hydrogen atom
   c. By the addition of an —OH group and a hydrogen atom
   d. Through hydrogen bonding

2. Why are carbohydrates important molecules for energy storage?
   a. The C—H bonds found in carbohydrates store energy.
   b. The double bonds between carbon and oxygen are very strong.
   c. The electronegativity of the oxygen atoms means that a carbohydrate is made up of many polar bonds.
   d. They can form ring structures in the aqueous environment of a cell.

3. Plant cells store energy in the form of _____, and animal cells store energy in the form of _____.
   a. fructose; glucose
   b. disaccharides; monosaccharides
   c. cellulose; chitin
   d. starch; glycogen

4. Which carbohydrate would you find as part of a molecule of RNA?
   a. Galactose
   b. Deoxyribose
   c. Ribose
   d. Glucose

5. A molecule of DNA or RNA is a polymer of
   a. monosaccharides.   c. amino acids.
   b. nucleotides.   d. fatty acids.

6. What makes cellulose different from starch?

    a. Starch is produced by plant cells, and cellulose is produced by animal cells.
    b. Cellulose forms long filaments, and starch is highly branched.
    c. Starch is insoluble, and cellulose is soluble.
    d. All of the choices are correct.

7. What monomers make up a protein?

    a. Monosaccharides   c. Amino acids
    b. Nucleotides   d. Fatty acids

8. A triglyceride is a form of _____ composed of _____.

    a. lipid; fatty acids and glucose
    b. lipid; fatty acids and glycerol
    c. carbohydrate; fatty acids
    d. lipid; cholesterol

APPLY

1. You can use starch or glycogen as an energy source, but not cellulose because

    a. starch and cellulose have similar structures.
    b. cellulose and glycogen have similar structures.
    c. starch and glycogen have similar structures.
    d. your body makes starch but not cellulose.

2. Which of the following is NOT a difference between DNA and RNA?

    a. Deoxyribose sugar versus ribose sugar
    b. Thymine versus uracil
    c. Double-stranded versus single-stranded
    d. Phosphodiester versus hydrogen bonds

3. Which part of an amino acid has the greatest influence on the overall structure of a protein?

    a. The (—NH₂) amino group
    b. The R group
    c. The (—COOH) carboxyl group
    d. Both a and c are correct.

4. A mutation that alters a single amino acid within a protein can alter

    a. the primary level of protein structure.
    b. the secondary level of protein structure.
    c. the tertiary level of protein structure.
    d. All of the choices are correct.

5. Two different proteins have the same domain in their structure. From this we can infer that they have

    a. the same primary structure.
    b. similar function.
    c. very different functions.
    d. the same primary structure but different function.

6. What aspect of triglyceride structure accounts for their insolubility in water?

    a. The COOH group of fatty acids
    b. The nonpolar C—H bonds in fatty acids
    c. The OH groups in glycerol
    d. The C=C bonds found in unsaturated fatty acids

7. The spontaneous formation of a lipid bilayer in an aqueous environment occurs because

    a. the polar head groups of the phospholipids can interact with water.
    b. the long fatty acid tails of the phospholipids can interact with water.
    c. the fatty acid tails of the phospholipids are hydrophobic.
    d. Both a and c are correct.

SYNTHESIZE

1. How do the four biological macromolecules differ from one another? How does the structure of each relate to its function?

2. Hydrogen bonds and hydrophobic interactions each play an important role in stabilizing and organizing biological macromolecules. Consider the four macromolecules discussed in this chapter. Describe how these affect the form and function of each type of macromolecule. Would a disruption in the hydrogen bonds affect form and function? Hydrophobic interactions?

3. Plants make both starch and cellulose. Would you predict that the enzymes involved in starch synthesis could also be used by the plant for cellulose synthesis? Construct an argument to explain this based on the structure and function of the enzymes and the polymers synthesized.

ONLINE RESOURCE

www.ravenbiology.com

Understand, Apply, and Synthesize—enhance your study with animations that bring concepts to life and practice tests to assess your understanding. Your instructor may also recommend the interactive eBook, individualized learning tools, and more.

Chapter 4

Cell Structure

Chapter Contents

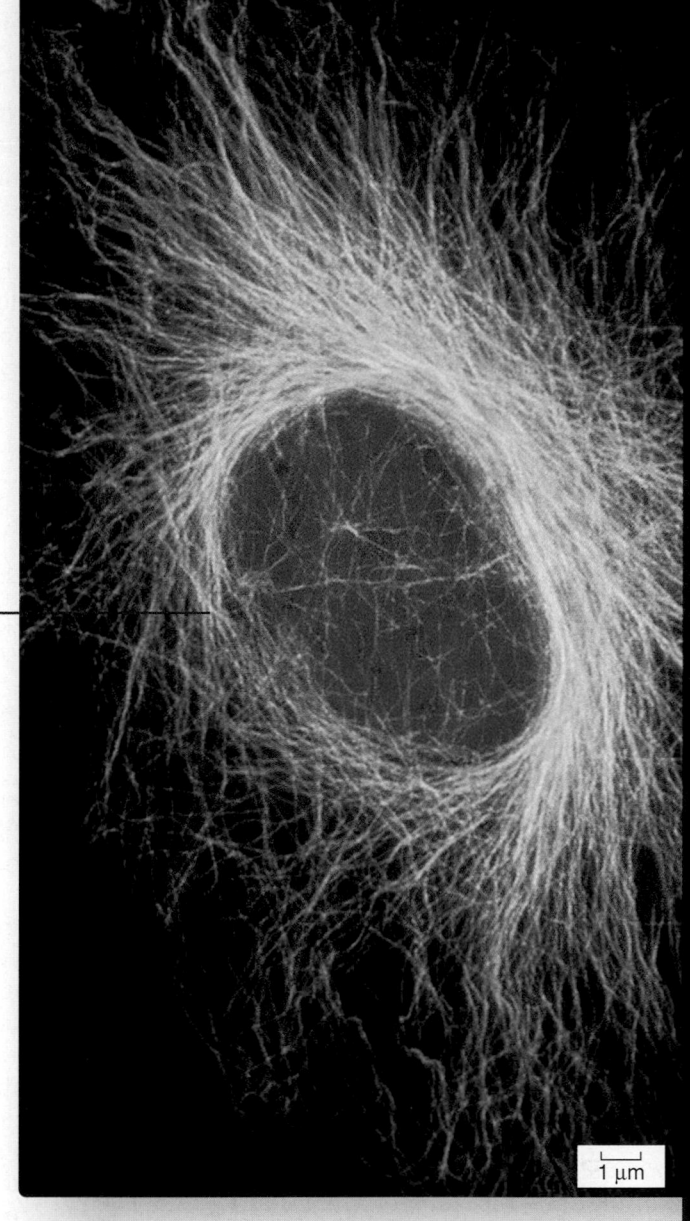

1 μm

Part II Biology of the Cell

Introduction

All organisms are composed of cells. The gossamer wing of a butterfly is a thin sheet of cells and so is the glistening outer layer of your eyes. The hamburger or tomato you eat is composed of cells, and its contents soon become part of your cells. Some organisms consist of a single cell too small to see with the unaided eye. Others, such as humans, are composed of many specialized cells, such as the fibroblast cell shown in the striking fluorescence micrograph on this page. Cells are so much a part of life that we cannot imagine an organism that is not cellular in nature. In this chapter, we take a close look at the internal structure of cells. In chapters 5 to 10, we will focus on cells in action—how they communicate with their environment, grow, and reproduce.

4.1   Cell Theory

Learning Outcomes

1. *Discuss the cell theory.*
2. *Describe the factors that limit cell size.*
3. *Categorize structural and functional similarities in cells.*

Cells are characteristically microscopic in size. Although there are exceptions, a typical eukaryotic cell is 10 to 100 micrometers (μm) (10–100 millionths of a meter) in diameter, although most prokaryotic cells are only 1 to 10 μm in diameter.

Because cells are so small, they were not discovered until the invention of the microscope in the 17th century. English natural philosopher Robert Hooke was the first to observe cells in 1665, naming the shapes he saw in cork *cellulae* (Latin, "small rooms"). This is known to us as *cells.* Another early microscopist, Dutch Anton van Leeuwenhoek, first observed living cells, which he termed "animalcules," or little animals.

After these early efforts, a century and a half passed before biologists fully recognized the importance of cells. In 1838, German botanist Matthias Schleiden stated that all plants "are aggregates of fully individualized, independent, separate beings, namely the cells themselves." In 1839, German physiologist Theodor Schwann reported that all animal tissues also consist of individual cells. Thus, the cell theory was born.

Cell theory is the unifying foundation of cell biology

The cell theory was proposed to explain the observation that all organisms are composed of cells. It sounds simple, but it is a far-reaching statement about the organization of life.

In its modern form, the *cell theory* includes the following three principles:

1. All organisms are composed of one or more cells, and the life processes of metabolism and heredity occur within these cells.
2. Cells are the smallest living things, the basic units of organization of all organisms.
3. Cells arise only by division of a previously existing cell.

Although life likely evolved spontaneously in the environment of early Earth, biologists have concluded that no additional cells are originating spontaneously at present. Rather, life on Earth represents a continuous line of descent from those early cells.

Cell size is limited

Most cells are relatively small for reasons related to the diffusion of substances into and out of them. The rate of diffusion is affected by a number of variables, including (1) surface area available for diffusion, (2) temperature, (3) concentration gradient of diffusing substance, and (4) the distance over which diffusion must occur. As the size of a cell increases, the length of time for diffusion from the outside membrane to the interior of the cell increases as well. Larger cells need to synthesize more macromolecules, have correspondingly higher energy requirements, and produce a greater quantity of waste. Molecules used for energy and biosynthesis must be transported through the membrane. Any metabolic waste produced must be removed, also passing through the membrane. The rate at which this transport occurs depends on both the distance to the membrane and the area of membrane available. For this reason, an organism made up of many relatively small cells has an advantage over one composed of fewer, larger cells.

The advantage of small cell size is readily apparent in terms of the **surface area-to-volume ratio.** As a cell's size increases, its volume increases much more rapidly than its surface area. For a spherical cell, the surface area is proportional to the square of the radius, whereas the volume is proportional to the cube of the radius. Thus, if the radii of two cells differ by a factor of 10, the larger cell will have 10^2, or 100 times, the surface area, but 10^3, or 1000 times, the volume of the smaller cell (figure 4.1).

The cell surface provides the only opportunity for interaction with the environment, because all substances enter and

Figure 4.1 Surface area-to-volume ratio. As a cell gets larger, its volume increases at a faster rate than its surface area. If the cell radius increases by 10 times, the surface area increases by 100 times, but the volume increases by 1000 times. A cell's surface area must be large enough to meet the metabolic needs of its volume.

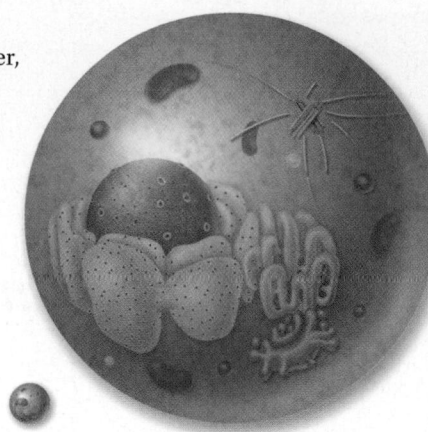

Cell radius (r)	1 unit	10 unit
Surface area ($4\pi r^2$)	12.57 unit^2	1257 unit^2
Volume ($\frac{4}{3}\pi r^3$)	4.189 unit^3	4189 unit^3
Surface Area / Volume	3	0.3

exit a cell via this surface. The membrane surrounding the cell plays a key role in controlling cell function. Because small cells have more surface area per unit of volume than large ones, control over cell contents is more effective when cells are relatively small.

Although most cells are small, some quite large cells do exist. These cells have apparently overcome the surface area-to-volume problem by one or more adaptive mechanisms. For example, some cells, such as skeletal muscle cells, have more than one nucleus, allowing genetic information to be spread around a large cell. Some other large cells, such as neurons, are long and skinny, so that any given point within the cell is close to the plasma membrane. This permits diffusion between the inside and outside of the cell to still be rapid.

Microscopes allow visualization of cells and components

Other than egg cells, not many cells are visible to the naked eye (figure 4.2). Most are less than 50 μm in diameter, far smaller than the period at the end of this sentence. So, to visualize cells we need the aid of technology. The development of microscopes and their refinement over the centuries has allowed us to continually explore cells in greater detail.

The resolution problem

How do we study cells if they are too small to see? The key is to understand why we can't see them. The reason we can't see such small objects is the limited resolution of the human eye. *Resolution* is the minimum distance two points can be apart and still be distinguished as two separate points. When two objects are closer together than about 100 μm, the light reflected from each strikes the same photoreceptor cell at the rear of the eye. Only when the objects are farther than 100 μm apart can the light from each strike different cells, allowing your eye to resolve them as two distinct objects rather than one.

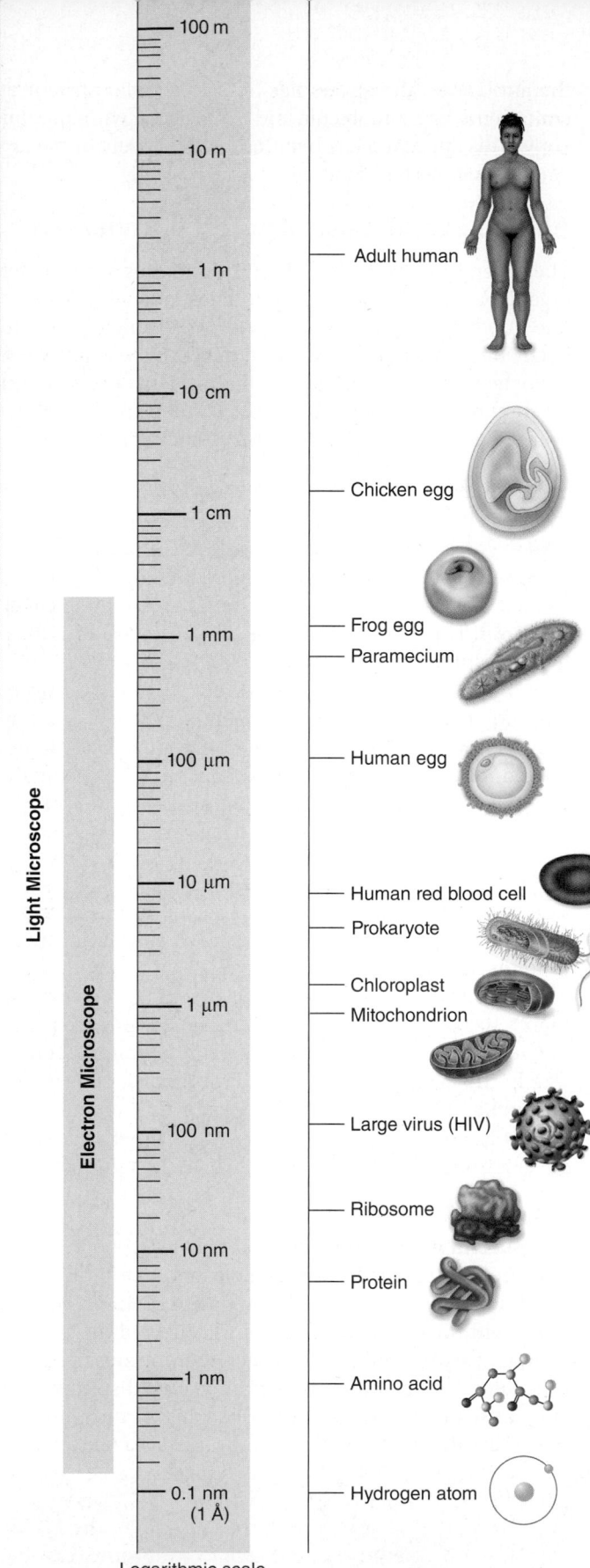

100 m

Human Eye

10 m

1 m — Adult human

10 cm

1 cm — Chicken egg

Light Microscope

1 mm — Frog egg
— Paramecium

100 μm — Human egg

10 μm — Human red blood cell
— Prokaryote

Electron Microscope

1 μm — Chloroplast
— Mitochondrion

100 nm — Large virus (HIV)

10 nm — Ribosome

1 nm — Protein

— Amino acid

0.1 nm — Hydrogen atom
(1 Å)

Logarithmic scale

Figure 4.2 The size of cells and their contents. Except for vertebrate eggs, which can typically be seen with the unaided eye, most cells are microscopic in size. Prokaryotic cells are generally 1 to 10 μm across.

$1 \text{ m} = 10^2 \text{ cm} = 10^3 \text{ mm} = 10^6 \text{ μm} = 10^9 \text{ nm}$

Types of microscopes

One way to overcome the limitations of our eyes is to increase magnification so that small objects appear larger. The first microscopists used glass lenses to magnify small cells and cause them to appear larger than the 100-μm limit imposed by the human eye. The glass lens increases focusing power. Because the glass lens makes the object appear closer, the image on the back of the eye is bigger than it would be without the lens.

Modern *light microscopes,* which operate with visible light, use two magnifying lenses (and a variety of correcting lenses) to achieve very high magnification and clarity (table 4.1). The first lens focuses the image of the object on the second lens, which magnifies it again and focuses it on the back of the eye. Microscopes that magnify in stages using several lenses are called *compound microscopes.* They can resolve structures that are separated by at least 200 nanometers (nm).

Light microscopes, even compound ones, are not powerful enough to resolve many of the structures within cells. For example, a cell membrane is only 5 nm thick. Why not just add another magnifying stage to the microscope to increase its resolving power? This doesn't work because when two objects are closer than a few hundred nanometers, the light beams reflecting from the two images start to overlap each other. The only way two light beams can get closer together and still be resolved is if their wavelengths are shorter. One way to avoid overlap is by using a beam of electrons rather than a beam of light. Electrons have a much shorter wavelength, and an *electron microscope,* employing electron beams, has 1000 times the resolving power of a light microscope. *Transmission electron microscopes,* so called because the electrons used to visualize the specimens are transmitted through the material, are capable of resolving objects only 0.2 nm apart—which is only twice the diameter of a hydrogen atom!

A second kind of electron microscope, the *scanning electron microscope,* beams electrons onto the surface of the specimen. The electrons reflected back from the surface, together with other electrons that the specimen itself emits as a result of the bombardment, are amplified and transmitted to a screen, where the image can be viewed and photographed. Scanning electron microscopy yields striking three-dimensional images. This technique has improved our understanding of many biological and physical phenomena (see table 4.1).

Using stains to view cell structure

Although resolution remains a physical limit, we can improve the images we see by altering the sample. Certain chemical stains increase the contrast between different cellular components. Structures within the cell absorb or exclude the stain differentially, producing contrast that aids resolution.

Stains that bind to specific types of molecules have made these techniques even more powerful. This method uses antibodies that bind, for example, to a particular protein. This process, called *immunohistochemistry,* uses antibodies generated in animals such as rabbits or mice. When these animals are injected with specific proteins, they produce antibodies that bind to the injected protein. The antibodies are then purified and chemically bonded to enzymes, to stains, or to fluorescent molecules. When cells are incubated in a solution containing

TABLE 4.1	Microscopes

LIGHT MICROSCOPES

Bright-field microscope:
Light is transmitted through a specimen, giving little contrast. Staining specimens improves contrast but requires that cells be fixed (not alive), which can distort or alter components.

28 μm

Dark-field microscope:
Light is directed at an angle toward the specimen. A condenser lens transmits only light reflected off the specimen. The field is dark, and the specimen is light against this dark background.

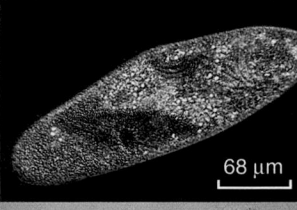

68 μm

Phase-contrast microscope:
Components of the microscope bring light waves out of phase, which produces differences in contrast and brightness when the light waves recombine.

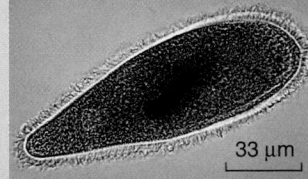

33 μm

Differential-interference–contrast microscope:
Polarized light is split into two beams that have slightly different paths through the sample. Combining these two beams produces greater contrast, especially at the edges of structures.

27 μm

Fluorescence microscope:
Fluorescent stains absorb light at one wavelength, then emit it at another. Filters transmit only the emitted light.

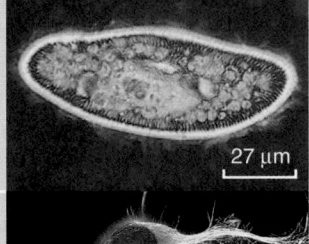

10 μm

Confocal microscope:
Light from a laser is focused to a point and scanned across the fluorescently stained specimen in two directions. This produces clear images of one plane of the specimen. Other planes of the specimen are excluded to prevent the blurring of the image. Multiple planes can be used to reconstruct a 3-D image.

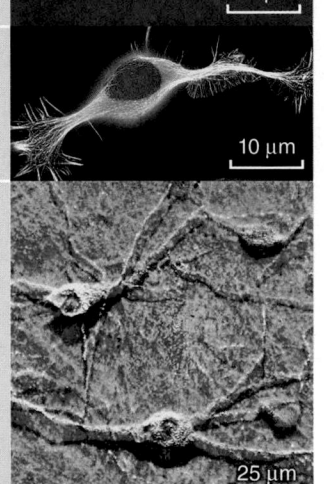

25 μm

ELECTRON MICROSCOPES

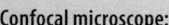

Transmission electron microscope:
A beam of electrons is passed through the specimen. Electrons that pass through are used to expose film. Areas of the specimen that scatter electrons appear dark. False coloring enhances the image.

3 μm

Scanning electron microscope:
An electron beam is scanned across the surface of the specimen, and electrons are knocked off the surface. Thus, the topography of the specimen determines the contrast and the content of the image. False coloring enhances the image.

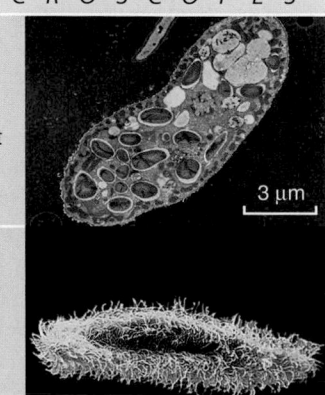

7 μm

the antibodies, the antibodies bind to cellular structures that contain the target molecule and can be seen with light microscopy. This approach has been used extensively in the analysis of cell structure and function.

All cells exhibit basic structural similarities

The general plan of cellular organization varies between different organisms, but despite these modifications, all cells resemble one another in certain fundamental ways. Before we begin a detailed examination of cell structure, let's first summarize four major features all cells have in common: (1) a nucleoid or nucleus where genetic material is located, (2) cytoplasm, (3) *ribosomes* to synthesize proteins, and (4) a plasma membrane.

Centrally located genetic material

Every cell contains DNA, the hereditary molecule. In **prokaryotes,** the simplest organisms, most of the genetic material lies in a single circular molecule of DNA. It typically resides near the center of the cell in an area called the **nucleoid.** This area is not segregated, however, from the rest of the cell's interior by membranes.

By contrast, the DNA of eukaryotes, which are more complex organisms, is contained in the nucleus, which is surrounded by a double-membrane structure called the **nuclear envelope.** In both types of organisms, the DNA contains the genes that code for the proteins synthesized by the cell. (Details of nucleus structure are described later in the chapter.)

The cytoplasm

A semifluid matrix called the **cytoplasm** fills the interior of the cell. The cytoplasm contains all of the sugars, amino acids, and proteins the cell uses to carry out its everyday activities. Although it is an aqueous medium, cytoplasm is more like jello than water due to the high concentration of proteins and other macromolecules. We call any discrete macromolecular structure in the cytoplasm specialized for a particular function an **organelle.** The part of the cytoplasm that contains organic molecules and ions in solution is called the **cytosol** to distinguish it from the larger organelles suspended in this fluid.

The plasma membrane

The **plasma membrane** encloses a cell and separates its contents from its surroundings. The plasma membrane is a phospholipid bilayer about 5 to 10 nm (5–10 billionths of a meter) thick, with proteins embedded in it. Viewed in cross section with the electron microscope, such membranes appear as two dark lines separated by a lighter area. This distinctive appearance arises from the tail-to-tail packing of the phospholipid molecules that make up the membrane (see chapter 5).

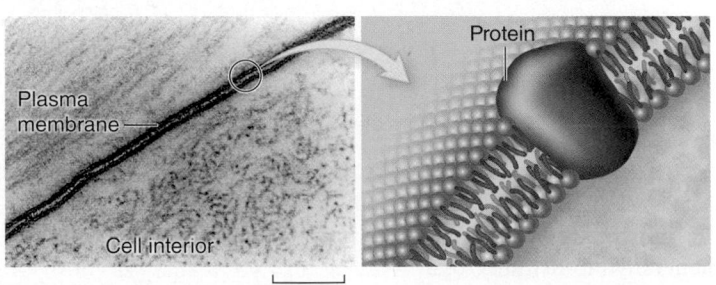

Plasma membrane

Cell interior

Protein

20 nm

The proteins of the plasma membrane are generally responsible for a cell's ability to interact with the environment. *Transport proteins* help molecules and ions move across the plasma membrane, either from the environment to the interior of the cell or vice versa. *Receptor proteins* induce changes within the cell when they come in contact with specific molecules in the environment, such as hormones, or with molecules on the surface of neighboring cells. These molecules can function as *markers* that identify the cell as a particular type. This interaction between cell surface molecules is especially important in multicellular organisms, whose cells must be able to recognize one another as they form tissues.

We'll examine the structure and function of cell membranes more thoroughly in chapter 5.

Learning Outcomes Review 4.1

All organisms are single cells or aggregates of cells, and all cells arise from preexisting cells. Cell size is limited primarily by the efficiency of diffusion across the plasma membrane. As a cell becomes larger, its volume increases more quickly than its surface area. Past a certain point, diffusion cannot support the cell's needs. All cells are bounded by a plasma membrane and filled with cytoplasm. The genetic material is found in the central portion of the cell; and in eukaryotic cells, it is contained in a membrane-bounded nucleus.

■ *Would finding life on Mars change our view of cell theory?*

4.2 Prokaryotic Cells

Learning Outcomes

1. Describe the organization of prokaryotic cells.
2. Distinguish between bacterial and archaeal cell types.

When cells were visualized with microscopes, two basic cellular architectures were recognized: eukaryotic and prokaryotic. These terms refer to the presence or absence, respectively, of a membrane-bounded nucleus that contains genetic material. We have already mentioned that in addition to lacking a nucleus, prokaryotic cells do not have an internal membrane system or numerous membrane-bounded organelles.

Prokaryotic cells have relatively simple organization

Prokaryotes are the simplest organisms. Prokaryotic cells are small. They consist of cytoplasm surrounded by a plasma membrane and are encased within a

rigid **cell wall.** They have no distinct interior compartments (figure 4.3). A prokaryotic cell is like a one-room cabin in which eating, sleeping, and watching TV all occur.

Prokaryotes are very important in the ecology of living organisms. Some harvest light by photosynthesis, others break down dead organisms and recycle their components. Still others cause disease or have uses in many important industrial processes. Prokaryotes have two main domains: archaea and bacteria. Chapter 28 covers prokaryotic diversity in more detail.

? Inquiry question What modifications would you include if you wanted to make a cell as large as possible?

Although prokaryotic cells do contain organelles like **ribosomes,** which carry out protein synthesis, most lack the membrane-bounded organelles characteristic of eukaryotic cells. It was long thought that prokaryotes also lack the elaborate cytoskeleton found in eukaryotes, but we have now found they have molecules related to both actin and tubulin, which form two of the cytoskeletal elements described later in the chapter. The strength and shape of the cell is determined by the cell wall and not these cytoskeletal elements (see figure 4.3). However, cell wall structure is influenced by the cytoskeleton. For instance, the presence of actin-like MreB fibers running the length of the cell lead to perpendicular cell-wall fibers that produce a rod-shaped cell. This can be seen when MreB protein is removed, cells become spherical rather than rod-shaped. During cell division, cell-wall deposition is influenced by the tubulin-like FtsZ protein (see chapter 10).

Pilus

Cytoplasm

Ribosomes

Nucleoid (DNA)

Plasma membrane

Cell wall

Capsule

Pili

Flagellum

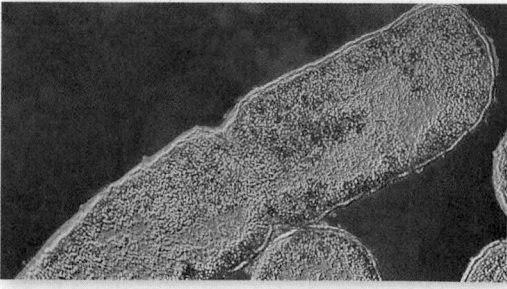

Figure 4.3 Structure of a prokaryotic cell. Generalized cell organization of a prokaryote. The nucleoid is visible as a dense central region segregated from the cytoplasm. Some prokaryotes have hairlike growths (called pili [singular, pilus]) on the outside of the cell.

0.3 μm

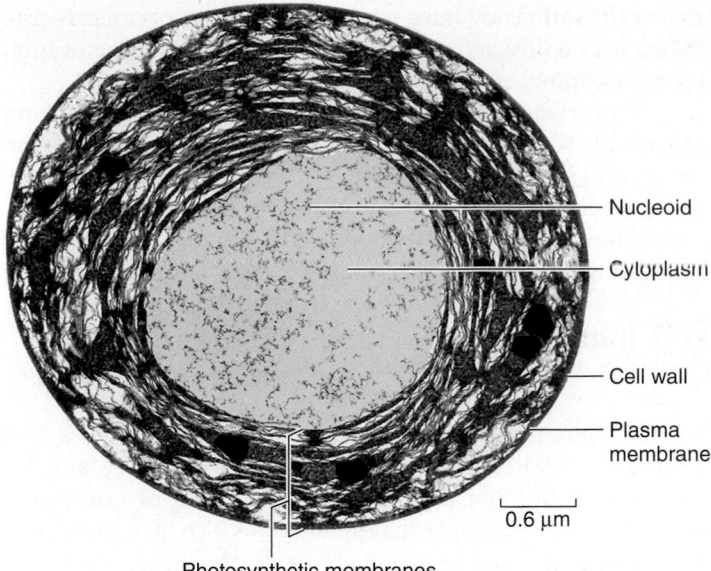

Figure 4.4 **Electron micrograph of a photosynthetic bacterial cell.** Extensive folded photosynthetic membranes are shown in green in this false colored electron micrograph of a *Prochloron* cell.

The plasma membrane of a prokaryotic cell carries out some of the functions organelles perform in eukaryotic cells. For example, some photosynthetic bacteria, such as the cyanobacterium *Prochloron* (figure 4.4), have an extensively folded plasma membrane, with the folds extending into the cell's interior. These membrane folds contain the bacterial pigments connected with photosynthesis. In eukaryotic plant cells, photosynthetic pigments are found in the inner membrane of the chloroplast.

Because a prokaryotic cell contains no membrane-bounded organelles, the DNA, enzymes, and other cytoplasmic constituents have access to all parts of the cell. Reactions are not compartmentalized as they are in eukaryotic cells, and the whole prokaryote operates as a single unit.

Bacterial cell walls consist of peptidoglycan

Most bacterial cells are encased by a strong **cell wall.** This cell wall is composed of *peptidoglycan,* which consists of a carbohydrate matrix (polymers of sugars) that is cross-linked by short polypeptide units. Details about the structure of this cell wall are discussed in chapter 28. Cell walls protect the cell, maintain its shape, and prevent excessive uptake or loss of water. The exception is the class Mollicutes, which includes the common genus *Mycoplasma,* which lack a cell wall. Plants, fungi, and most protists also have cell walls but with a chemical structure different from peptidoglycan.

The susceptibility of bacteria to antibiotics often depends on the structure of their cell walls. The drugs penicillin and vancomycin, for example, interfere with the ability of bacteria to cross-link the peptides in their peptidoglycan cell wall. Like removing all the nails from a wooden house, this destroys the integrity of the structural matrix, which can no longer prevent water from rushing in and swelling the cell to bursting.

Some bacteria also secrete a jelly-like protective capsule of polysaccharide around the cell. Many disease-causing bacteria have such a capsule, which enables them to adhere to teeth, skin, food—or to practically any surface that can support their growth.

Archaea lack peptidoglycan

We are still learning about the physiology and structure of archaea. Many of these organisms are difficult to culture in the laboratory, and so this group has not yet been studied in detail. More is known about their genetic makeup than about any other feature.

The cell walls of archaea are composed of various chemical compounds, including polysaccharides and proteins, and possibly even inorganic components. A common feature distinguishing archaea from bacteria is the nature of their membrane lipids. The chemical structure of archaeal lipids is distinctly different from that of lipids in bacteria and can include saturated hydrocarbons that are covalently attached to

Figure 4.5 **Some prokaryotes move by rotating their flagella.** *a.* The photograph shows *Vibrio cholerae,* the microbe that causes the serious disease cholera. *b.* The bacterial flagellum is a complex structure. The motor proteins, powered by a proton gradient, are anchored in the plasma membrane. Two rings are found in the cell wall. The motor proteins cause the entire structure to rotate. *c.* As the flagellum rotates it creates a spiral wave down the structure. This powers the cell forward.

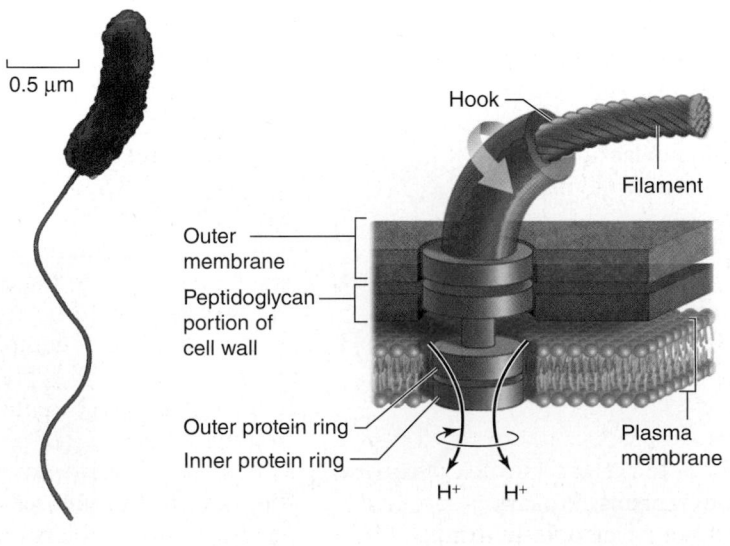

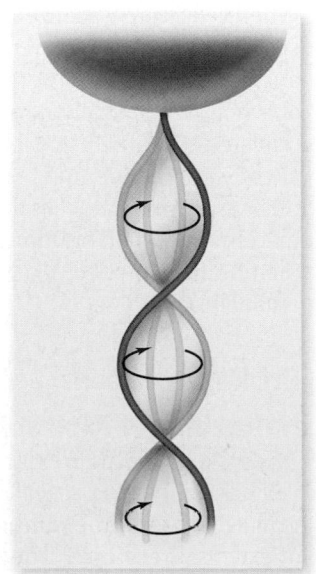

a.   *b.*   *c.*

glycerol at both ends, such that their membrane is a mono-layer. These features seem to confer greater thermal stability to archaeal membranes, although the trade-off seems to be an inability to alter the degree of saturation of the hydrocarbons—meaning that archaea with this characteristic cannot adapt to changing environmental temperatures.

The cellular machinery that replicates DNA and synthesized proteins in archaea is more closely related to eukaryotic systems than to bacterial systems. Even though they share a similar overall cellular architecture with prokaryotes, archaea appear to be more closely related on a molecular basis to eukaryotes.

Some prokaryotes move by means of rotating flagella

Flagella (singular, *flagellum*) are long, threadlike structures protruding from the surface of a cell that are used in locomotion. Prokaryotic flagella are protein fibers that extend out from the cell. There may be one or more per cell, or none, depending on the species. Bacteria can swim at speeds of up to 70 cell lengths per second by rotating their flagella like screws (figure 4.5). The rotary motor uses the energy stored in a gradient that transfers protons across the plasma membrane to power the movement of the flagellum. Interestingly, the same principle, in which a proton gradient powers the rotation of a molecule, is used in eukaryotic mitochondria and chloroplasts by an enzyme that synthesizes ATP (see chapters 7 and 8).

Learning Outcomes Review 4.2

Prokaryotes are small cells that lack complex interior organization. The two domains of prokaryotes are archaea and bacteria. The cell wall of bacteria is composed of peptidoglycan, which is not found in archaea. Archaea have cell walls made from a variety of polysaccharides and peptides, as well as membranes containing unusual lipids. Some bacteria move using a rotating flagellum.

■ *What features do bacteria and archaea share?*

4.3 Eukaryotic Cells

Learning Outcomes

1. *Compare the organization of eukaryotic and prokaryotic cells.*
2. *Discuss the role of the nucleus in eukaryotic cells.*
3. *Describe the role of ribosomes in protein synthesis.*

Eukaryotic cells (figures 4.6 and 4.7) are far more complex than prokaryotic cells. The hallmark of the eukaryotic cell is compartmentalization. This is achieved through a combination of an extensive **endomembrane system** that weaves through the cell interior and by numerous *organelles*. These organelles include membrane-bounded structures that form compartments within which multiple biochemical processes can proceed simultaneously and independently.

Plant cells often have a large, membrane-bounded sac called a **central vacuole,** which stores proteins, pigments, and waste materials. Both plant and animal cells contain **vesicles**—smaller sacs that store and transport a variety of materials. Inside the nucleus, the DNA is wound tightly around proteins and packaged into compact units called **chromosomes.**

All eukaryotic cells are supported by an internal protein scaffold, the **cytoskeleton.** Although the cells of animals and some protists lack cell walls, the cells of fungi, plants, and many protists have strong cell walls composed of cellulose or chitin fibers embedded in a matrix of other polysaccharides and proteins. Through the rest of this chapter, we will examine the internal components of eukaryotic cells in more detail.

The nucleus acts as the information center

The largest and most easily seen organelle within a eukaryotic cell is the **nucleus** (Latin, "kernel" or "nut"), first described by the Scottish botanist Robert Brown in 1831. Nuclei are roughly spherical in shape, and in animal cells, they are typically located in the central region of the cell (figure 4.8*a*). In some cells, a network of fine cytoplasmic filaments seems to cradle the nucleus in this position.

The nucleus is the repository of the genetic information that enables the synthesis of nearly all proteins of a living eukaryotic cell. Most eukaryotic cells possess a single nucleus, although the cells of fungi and some other groups may have from several to many nuclei. Mammalian erythrocytes (red blood cells) lose their nuclei when they mature. Many nuclei exhibit a dark-staining zone called the **nucleolus,** which is a region where intensive synthesis of ribosomal RNA is taking place.

The nuclear envelope

The surface of the nucleus is bounded by *two* phospholipid bilayer membranes, which together make up the **nuclear envelope** (see figure 4.8). The outer membrane of the nuclear envelope is continuous with the cytoplasm's interior membrane system, called the *endoplasmic reticulum* (described later).

Scattered over the surface of the nuclear envelope are what appear as shallow depressions in the electron micrograph but are in fact structures called **nuclear pores** (see figure 4.8*b, c*). These pores form 50 to 80 nm apart at locations where the two membrane layers of the nuclear envelope come together. The structure consists of a central framework with eightfold symmetry that is embedded in the nuclear envelope. This is bounded by a cytoplasmic face with eight fibers, and a nuclear face with a complex ring that forms a basket beneath the central ring. The pore allows ions and small molecules to diffuse freely between nucleoplasm and cytoplasm while controlling the passage of proteins and RNA–protein complexes. Transport across the pore is controlled and consists mainly of the import of proteins that function in the nucleus, and the export to the cytoplasm of RNA and RNA–protein complexes formed in the nucleus.

Figure 4.6 Structure of an animal cell.   In this generalized diagram of an animal cell, the plasma membrane encases the cell, which contains the cytoskeleton and various cell organelles and interior structures suspended in a semifluid matrix called the cytoplasm. Some kinds of animal cells possess finger-like projections called microvilli. Other types of eukaryotic cells—for example, many protist cells—may possess flagella, which aid in movement, or cilia, which can have many different functions.

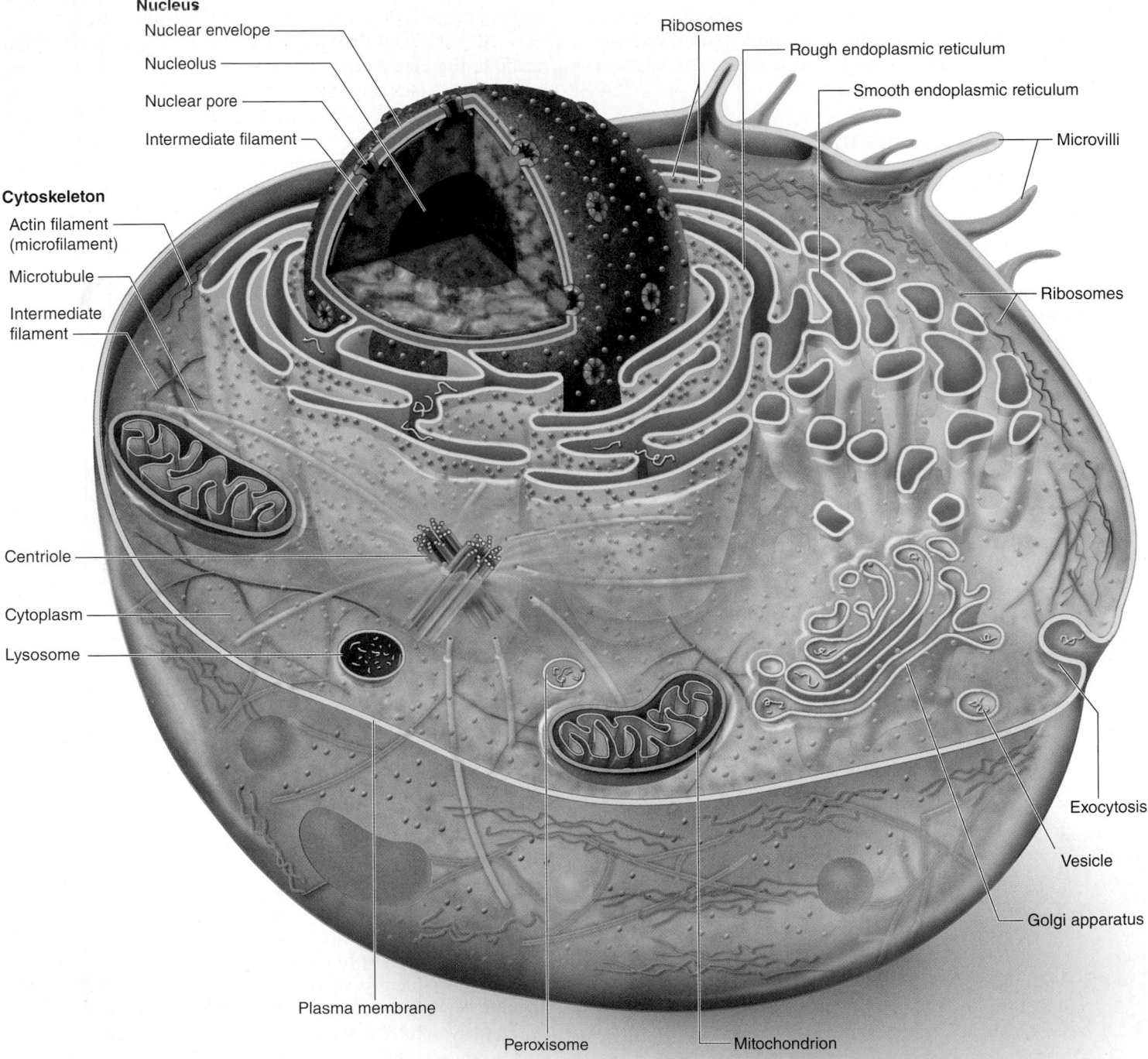

Nucleus
- Nuclear envelope
- Nucleolus
- Nuclear pore
- Intermediate filament

Cytoskeleton
- Actin filament (microfilament)
- Microtubule
- Intermediate filament

Centriole

Cytoplasm

Lysosome

Plasma membrane

Peroxisome

Mitochondrion

Ribosomes

Rough endoplasmic reticulum

Smooth endoplasmic reticulum

Microvilli

Ribosomes

Exocytosis

Vesicle

Golgi apparatus

Figure 4.7 **Structure of a plant cell.** Most mature plant cells contain a large central vacuole, which occupies a major portion of the internal volume of the cell, and organelles called chloroplasts, within which photosynthesis takes place. The cells of plants, fungi, and some protists have cell walls, although the composition of the walls varies among the groups. Plant cells have cytoplasmic connections to one another through openings in the cell wall called plasmodesmata. Flagella occur in sperm of a few plant species, but are otherwise absent from plant and fungal cells. Centrioles are also usually absent.

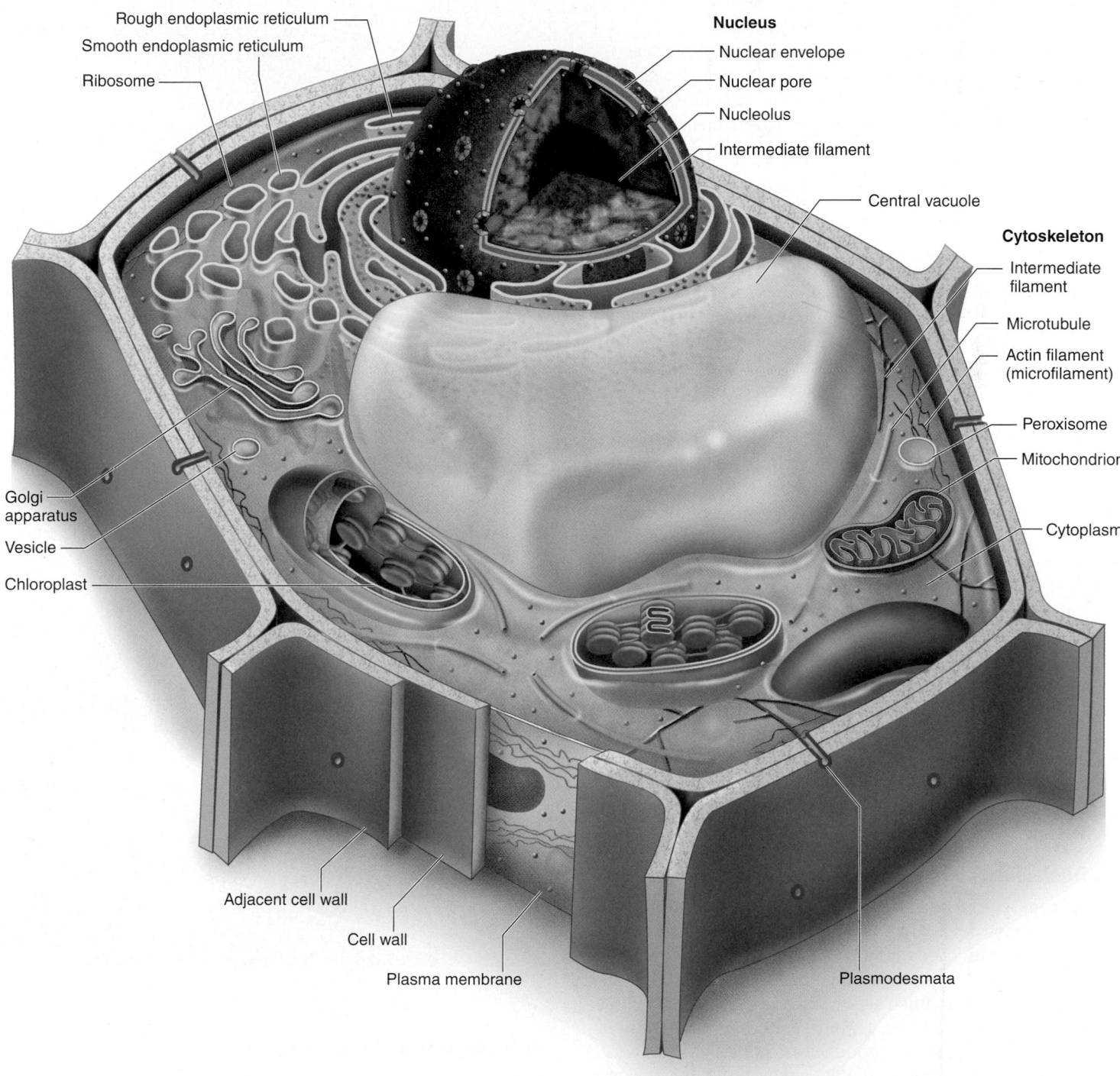

Rough endoplasmic reticulum

Smooth endoplasmic reticulum

Ribosome

Nucleus

Nuclear envelope

Nuclear pore

Nucleolus

Intermediate filament

Central vacuole

Cytoskeleton

Intermediate filament

Microtubule

Actin filament (microfilament)

Peroxisome

Mitochondrion

Cytoplasm

Golgi apparatus

Vesicle

Chloroplast

Adjacent cell wall

Cell wall

Plasma membrane

Plasmodesmata

Figure 4.8 The nucleus. *a.* The nucleus is composed of a double membrane called the nuclear envelope, enclosing a fluid-filled interior containing chromatin. The individual nuclear pores extend through the two membrane layers of the envelope. The close-up of the nuclear pore shows the central hub, cytoplasmic ring with fibers, and nuclear ring with basket. *b.* A freeze-fracture electron micrograph (see figure 5.4) of a cell nucleus, showing many nuclear pores. *c.* A transmission electron micrograph of the nuclear membrane showing a single nuclear pore. The dark material within the pore is protein, which acts to control access through the pore. *d.* The nuclear lamina is visible as a dense network of fibers made of intermediate filaments. The nucleus has been colored purple in the micrographs.

(b): © Dr. Richard Kessel & Dr. Gene Shih/Visuals Unlimited

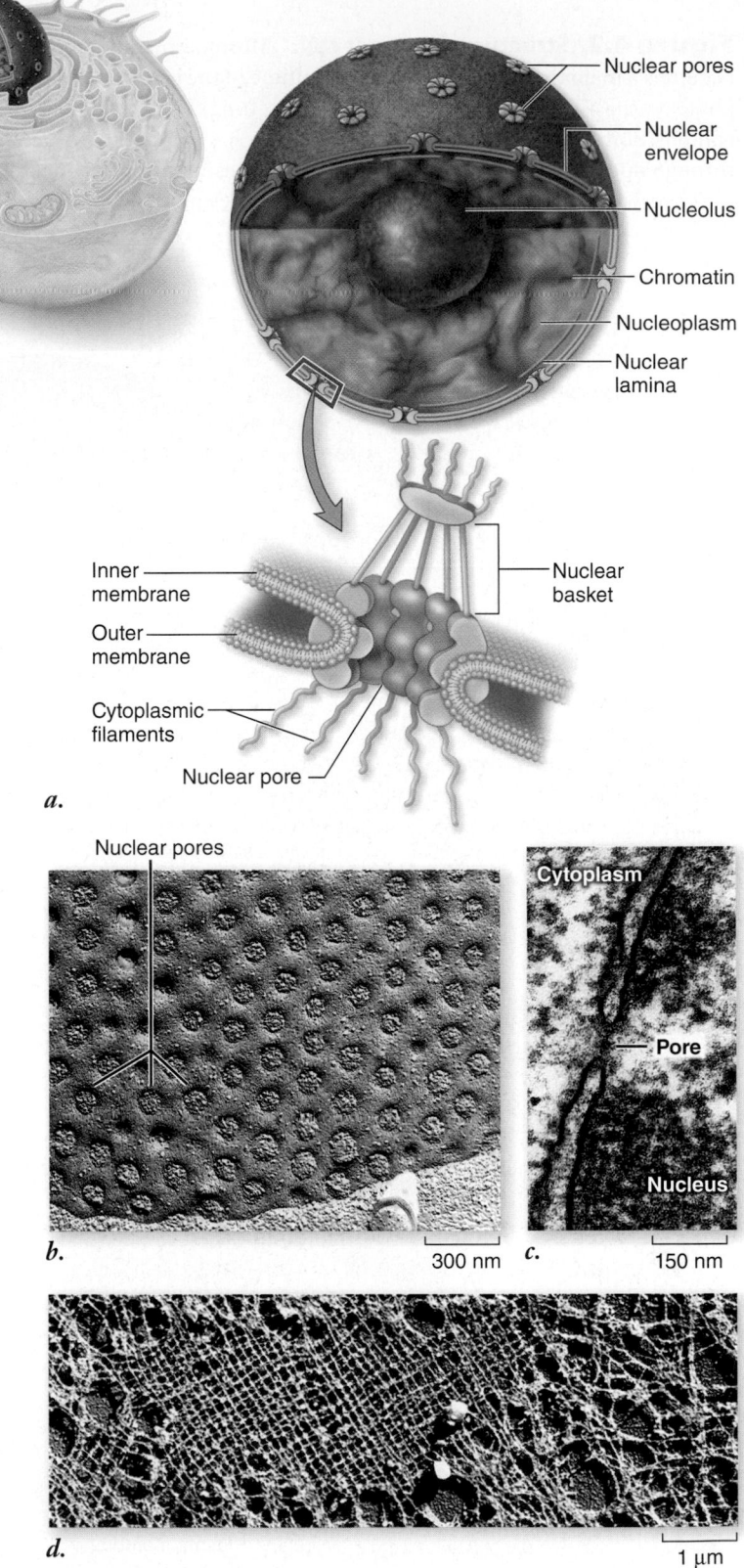

The inner surface of the nuclear envelope is covered with a network of fibers that make up the nuclear lamina (see figure 4.8*d*). This is composed of intermediate filament fibers called *nuclear lamins.* This structure gives the nucleus its shape and is also involved in the deconstruction and reconstruction of the nuclear envelope that accompanies cell division.

Chromatin: DNA packaging

In both prokaryotes and eukaryotes, DNA is the molecule that stores genetic information. In eukaryotes, the DNA is divided into multiple linear chromosomes, which are organized with proteins into a complex structure called **chromatin.** It is becoming clear that the very structure of chromatin affects the function of DNA. Changes in gene expression that do not involve changes in DNA sequence, so-called epigenetic changes, involve alterations in chromatin structure (see chapter 16). Although still not fully understood, this offers an exciting new view of many old ideas.

Chromatin is usually in a more extended form that is organized in the nucleus, although we still do not fully understand this organization. When cells divide, the chromatin must be further compacted into a more highly condensed state that forms the X-shaped chromosomes visible in the light microscope.

The nucleolus: Ribosomal subunit manufacturing

Before cells can synthesize proteins in large quantity, they must first construct a large number of ribosomes to carry out this synthesis. Hundreds of copies of the genes encoding the ribosomal RNAs are clustered together on the chromosome, facilitating ribosome construction. By transcribing RNA molecules from this cluster, the cell rapidly generates large numbers of the molecules needed to produce ribosomes.

The clusters of ribosomal RNA genes, the RNAs they produce, and the ribosomal proteins all come together within the nucleus during ribosome production. These ribosomal assembly areas are easily visible within the nucleus as one or more dark-staining regions called nucleoli (singular, *nucleolus*). Nucleoli can be seen under the light microscope even when the chromosomes are uncoiled.

Ribosomes are the cell's protein synthesis machinery

Although the DNA in a cell's nucleus encodes the amino acid sequence of each protein in the cell, the proteins are not assembled there. A simple experiment demonstrates this: If a brief pulse of radioactive amino acid is administered to a cell, the radioactivity shows up associated with newly made protein

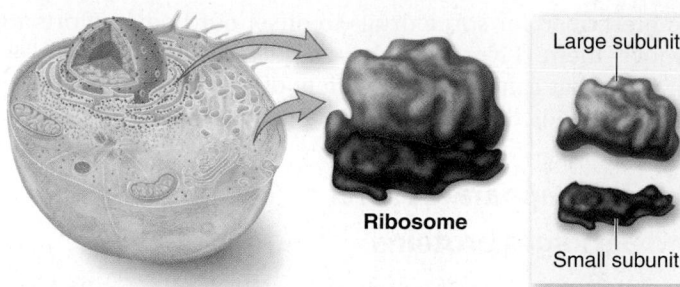

Figure 4.9 A ribosome. Ribosomes consist of a large and a small subunit composed of rRNA and protein. The individual subunits are synthesized in the nucleolus and then move through the nuclear pores to the cytoplasm, where they assemble to translate mRNA. Ribosomes serve as sites of protein synthesis.

in the cytoplasm, not in the nucleus. When investigators first carried out these experiments, they found that protein synthesis is associated with large RNA–protein complexes (called ribosomes) outside the nucleus.

Ribosomes are among the most complex molecular assemblies found in cells. Each ribosome is composed of two subunits (figure 4.9), each of which is composed of a combination of RNA, called **ribosomal RNA (rRNA),** and proteins. The subunits join to form a functional ribosome only when they are actively synthesizing proteins. This complicated process requires the two other main forms of RNA: **messenger RNA (mRNA),** which carries coding information from DNA, and **transfer RNA (tRNA),** which carries amino acids. Ribosomes use the information in mRNA to direct the synthesis of a protein. This process will be described in more detail in chapter 15.

Ribosomes are found either free in the cytoplasm or associated with internal membranes, as described in the following section. Free ribosomes synthesize proteins that are found in the cytoplasm, nuclear proteins, mitochondrial proteins, and proteins found in other organelles not derived from the endomembrane system. Membrane-associated ribosomes synthesize membrane proteins, proteins found in the endomembrane system, and proteins destined for export from the cell.

Ribosomes can be thought of as "universal organelles" because they are found in all cell types from all three domains of life. As we build a picture of the minimal essential functions for cellular life, ribosomes will be on the short list. Life is protein-based, and ribosomes are the factories that make proteins.

Learning Outcomes Review 4.3

In contrast to prokaryotic cells, eukaryotic cells exhibit compartmentalization. Eukaryotic cells contain an endomembrane system and organelles that carry out specialized functions. The nucleus, composed of a double membrane connected to the endomembrane system, contains the cell's genetic information. Material moves between the nucleus and cytoplasm through nuclear pores. Ribosomes translate mRNA, which is transcribed from DNA in the nucleus, into polypeptides that make up proteins. Ribosomes are a universal organelle found in all known cells.

■ **Would you expect cells in different organs in complex animals to have the same structure?**

Learning Outcomes

1. *Identify the different parts of the endomembrane system.*
2. *Contrast the different functions of internal membranes and compartments.*
3. *Evaluate the importance of each step in the protein-processing pathway.*

The interior of a eukaryotic cell is packed with membranes so thin that they are invisible under the low resolving power of light microscopes. This endomembrane system fills the cell, dividing it into compartments, channeling the passage of molecules through the interior of the cell, and providing surfaces for the synthesis of lipids and some proteins. The presence of these membranes in eukaryotic cells marks one of the fundamental distinctions between eukaryotes and prokaryotes.

The largest of the internal membranes is called the **endoplasmic reticulum (ER).** *Endoplasmic* means "within the cytoplasm," and *reticulum* is Latin for "a little net." The ER is composed of a phospholipid bilayer embedded with proteins. The ER has functional subdivisions, described below, and forms a variety of structures from folded sheets to complex tubular networks (figure 4.10). The two largest compartments in eukaryotic cells are the inner region of the ER, called the

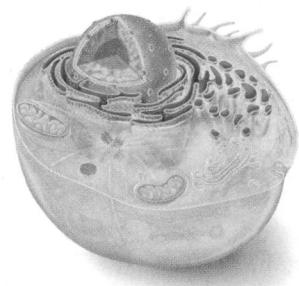
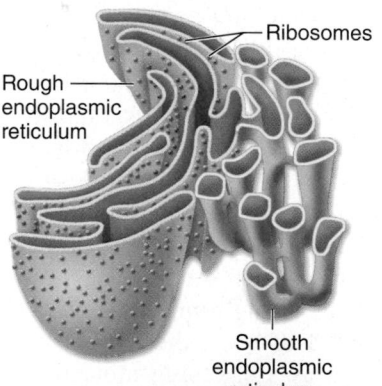

Figure 4.10 The endoplasmic reticulum. Rough ER (RER), blue in the drawing, is composed more of flattened sacs and forms a compartment throughout the cytoplasm. Ribosomes associated with the cytoplasmic face of the RER extrude newly made proteins into the interior, or lumen. The smooth ER (SER), green in the drawing, is a more tubelike structure connected to the RER. The micrograph has been colored to match the drawing.

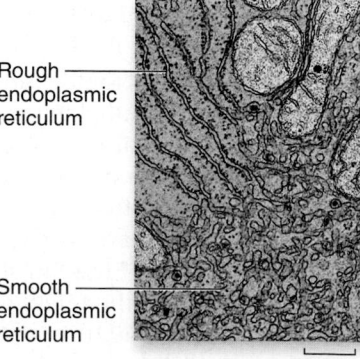

Ribosomes

Rough endoplasmic reticulum

Smooth endoplasmic reticulum

Rough endoplasmic reticulum

Smooth endoplasmic reticulum

80 nm

cisternal space, or lumen, and the region exterior to it, the cytosol, which is the fluid component of the cytoplasm containing dissolved organic molecules such as proteins and ions.

The rough ER is a site of protein synthesis

The **rough ER (RER)** gets its name from its pebbly surface appearance. The RER is not easily visible with a light microscope, but it can be seen using the electron microscope. It appears to be composed primarily of flattened sacs, the surfaces of which are bumpy with ribosomes (see figure 4.10).

The proteins synthesized on the surface of the RER are destined to be exported from the cell, sent to lysosomes or vacuoles (described in a later section), or embedded in the plasma membrane. These proteins enter the cisternal space as a first step in the pathway that will sort proteins to their eventual destinations. This pathway also involves vesicles and the Golgi apparatus, described later. The sequence of the protein being synthesized determines whether the ribosome will become associated with the ER or remain a cytoplasmic ribosome.

In the ER, newly synthesized proteins can be modified by the addition of short-chain carbohydrates to form **glycoproteins.** Those proteins destined for secretion are separated from other products and later packaged into vesicles that move to the Golgi for further modification and packaging for transport to other cellular locations.

The smooth ER has multiple roles

Regions of the ER with relatively few bound ribosomes are referred to as **smooth ER (SER).** The SER has a variety of structures ranging from a network of tubules, to flattened sacs, to higher order tubular arrays. The membranes of the SER contain many embedded enzymes. Enzymes anchored within the ER are involved in the synthesis of a variety of carbohydrates and lipids. Steroid hormones are synthesized in the SER as well. The majority of membrane lipids are assembled in the SER and then sent to whatever parts of the cell need membrane components. Membrane proteins in the plasma membrane and other cellular membrane are inserted by ribosomes on the RER.

An important function of the SER is to store intracellular Ca^{2+}. This keeps the cytoplasmic level low, allowing Ca^{2+} to be used as a signaling molecule. In muscle cells, for example, Ca^{2+} is used to trigger muscle contraction. In other cells, Ca^{2+} release from SER stores is involved in diverse signaling pathways.

The ratio of SER to RER is not fixed but depends on a cell's function. In multicellular animals such as ourselves, great variation exists in this ratio. Cells that carry out extensive lipid synthesis, such as those in the testes, intestine, and brain, have abundant SER. Cells that synthesize proteins that are secreted, such as antibodies, have much more extensive RER.

Another role of the SER is the modification of foreign substances to make them less toxic. In the liver, the enzymes of the SER carry out this detoxification. This action can include neutralizing substances that we have taken for a therapeutic reason, such as penicillin. Thus, relatively high doses are prescribed for some drugs to offset our body's efforts to remove them. Liver cells have extensive SER as well as enzymes that can process a variety of substances by chemically modifying them.

The Golgi apparatus sorts and packages proteins

Flattened stacks of membranes form a complex called the **Golgi body,** or **Golgi apparatus** (figure 4.11). These structures are named for Camillo Golgi, the 19th-century Italian physician who first identified them. The individual stacks of membrane are called **cisternae** (Latin, "collecting vessels"), and they vary in number within the Golgi body from 1 or a few in protists, to 20 or more in animal cells and to several hundred in plant cells. In vertebrates individual Golgi are linked to form a Golgi ribbon. They are especially abundant in glandular cells, which manufacture and secrete substances.

The Golgi apparatus functions in the collection, packaging, and distribution of molecules synthesized at one location and used at another within the cell or even outside of it. A Golgi body has a front and a back, with distinctly different membrane compositions at these opposite ends. The front, or receiving end, is called the *cis* face and is usually located near the ER. Materials arrive at the *cis* face in transport vesicles that bud off the ER and exit the *trans* face, where they are discharged in

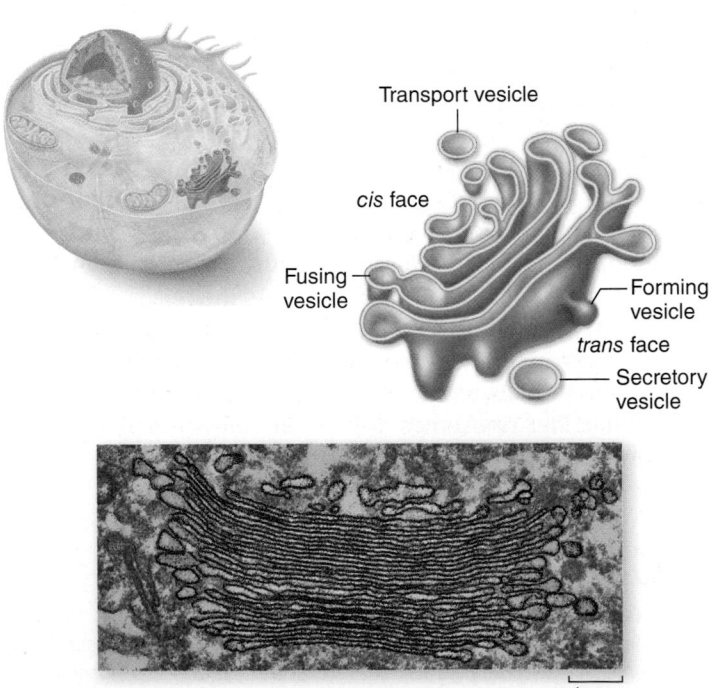

Transport vesicle

cis face

Fusing vesicle

Forming vesicle

trans face

Secretory vesicle

1 μm

Figure 4.11   The Golgi apparatus. The Golgi apparatus is a smooth, concave, membranous structure. It receives material for processing in transport vesicles on the *cis* face and sends the material packaged in transport or secretory vesicles off the *trans* face. The substance in a vesicle could be for export out of the cell or for distribution to another region within the same cell.

secretory vesicles (figure 4.12). How material transits through the Golgi has been a source of much contention. Models include maturation of the individual cisternae from *cis* to *trans,* transport between cisternae by vesicles, and direct tubular connections. Although there is probably transport of material by all of these, it now appears that the primary mechanism is cisternal maturation.

Proteins and lipids manufactured on the rough and smooth ER membranes are transported into the Golgi apparatus and modified as they pass through it. The most common alteration is the addition or modification of short sugar chains, forming glycoproteins and glycolipids. In many instances, enzymes in the Golgi apparatus modify existing glycoproteins and glycolipids made in the ER by cleaving a sugar from a chain or by modifying one or more of the sugars. These are then packaged into small, membrane-bounded vesicles that pinch off from the *trans* face of the Golgi. These vesicles then diffuse to other locations in the cell, distributing the newly synthesized molecules to their appropriate destinations.

Another function of the Golgi apparatus is the synthesis of cell-wall components. Noncellulose polysaccharides that form part of the cell wall of plants are synthesized in the Golgi apparatus and sent to the plasma membrane, where they can be added to the cellulose that is assembled on the exterior of the cell. Other polysaccharides secreted by plants are also synthesized in the Golgi apparatus.

Lysosomes contain digestive enzymes

Membrane-bounded digestive vesicles, called **lysosomes,** are also components of the endomembrane system. They arise from the Golgi apparatus. They contain high levels of degrading enzymes, which catalyze the rapid breakdown of proteins, nucleic acids, lipids, and carbohydrates. Throughout the lives of eukaryotic cells, lysosomal enzymes break down old organelles and recycle their component molecules. This makes room for newly formed organelles. For example, mitochondria are replaced in some tissues every 10 days.

The digestive enzymes in the lysosome are optimally active at acid pH. Lysosomes are activated by fusing with a food vesicle produced by *phagocytosis* (a specific type of endocytosis; see chapter 5) or by fusing with an old or worn-out organelle. The fusion event activates proton pumps in the lysosomal membrane, resulting in a lower internal pH. As the interior pH falls, the arsenal of digestive enzymes contained in the lysosome is activated. This leads to the degradation of macromolecules in the food vesicle or the destruction of the old organelle.

A number of human genetic disorders, collectively called lysosomal storage disorders, affect lysosomes. For example, the genetic abnormality called Tay–Sachs disease is caused by the loss of function of a single lysosomal enzyme (hexosaminidase). This enzyme is necessary to break down a membrane glycolipid found in nerve cells. Accumulation of glycolipid in lysosomes affects nerve cell function, leading to a variety of clinical symptoms such as seizures and muscle rigidity.

In addition to breaking down organelles and other structures within cells, lysosomes eliminate other cells that the cell

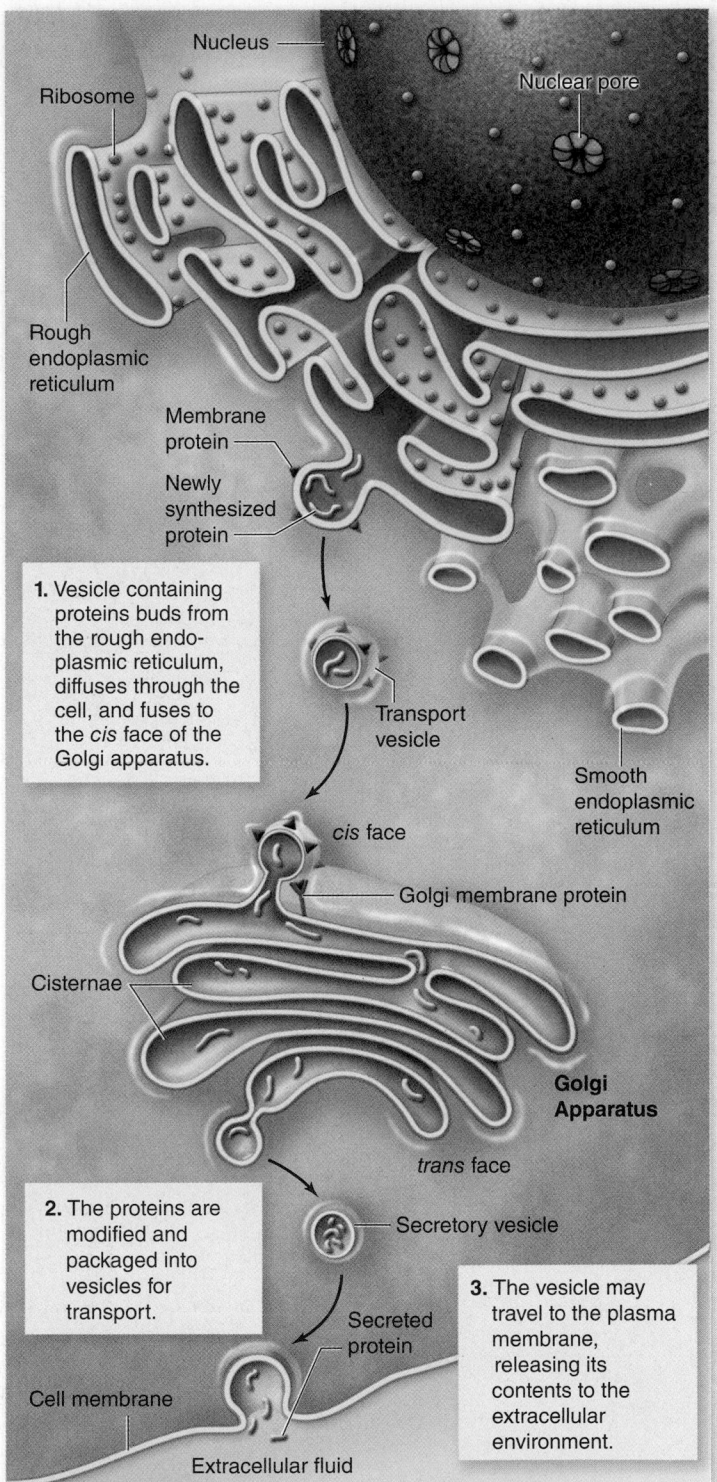

Figure 4.12 Protein transport through the endomembrane system. Proteins synthesized by ribosomes on the RER are translocated into the internal compartment of the ER. These proteins may be used at a distant location within the cell or secreted from the cell. They are transported within vesicles that bud off the RER. These transport vesicles travel to the *cis* face of the Golgi apparatus. There they can be modified and packaged into vesicles that bud off the *trans* face of the Golgi apparatus. Vesicles leaving the *trans* face transport proteins to other locations in the cell, or fuse with the plasma membrane, releasing their contents to the extracellular environment.

has engulfed by phagocytosis. When a white blood cell, for example, phagocytizes a passing pathogen, lysosomes fuse with the resulting "food vesicle," releasing their enzymes into the vesicle and degrading the material within (figure 4.13).

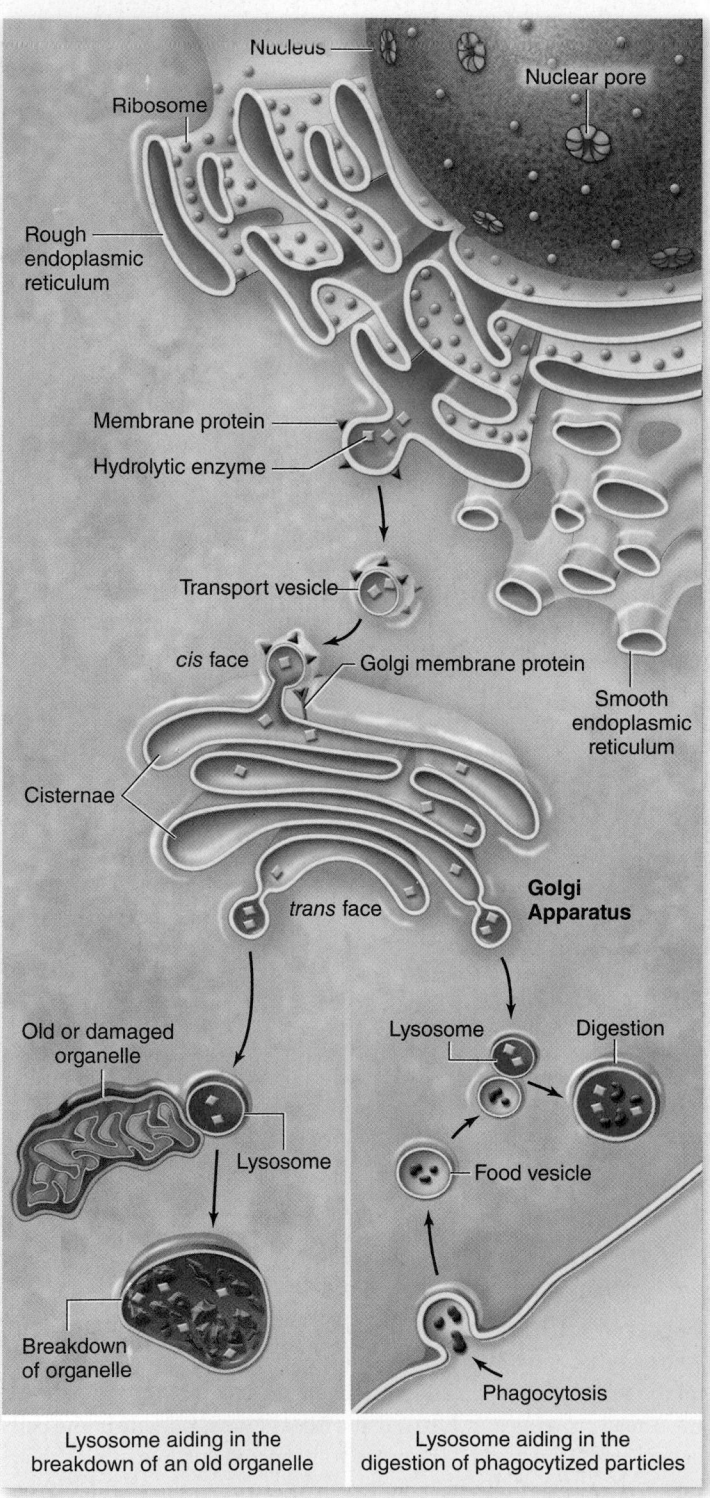

Figure 4.13 Lysosomes. Lysosomes are formed from vesicles budding off the Golgi. They contain hydrolytic enzymes that digest particles or cells taken into the cell by phagocytosis, and break down old organelles.

Microbodies are a diverse category of organelles

Eukaryotic cells contain a variety of enzyme-bearing, membrane-enclosed vesicles called **microbodies.** These are found in the cells of plants, animals, fungi, and protists. The distribution of enzymes into microbodies is one of the principal ways eukaryotic cells organize their metabolism.

Peroxisomes: Peroxide utilization

An important type of microbody is the **peroxisome** (figure 4.14), which contains enzymes involved in the oxidation of fatty acids. If these oxidative enzymes were not isolated within microbodies, they would tend to short-circuit the metabolism of the cytoplasm, which often involves adding hydrogen atoms to oxygen. Because many peroxisomal proteins are synthesized by cytoplasmic ribosomes, the organelles themselves were long thought to form by the addition of lipids and proteins, leading to growth. As they grow larger, they divide to produce new peroxisomes. Although division of peroxisomes still appears to occur, it is now clear that peroxisomes can form from the fusion of ER-derived vesicles. These vesicles then import peroxisomal proteins to form a mature peroxisome. Genetic screens have isolated some 32 genes that encode proteins involved in biogenesis and maintenance of peroxisomes. The human genetic diseases called peroxisome biogenesis disorders (PBDs) can be caused by mutations in some of these genes.

Peroxisomes get their name from the hydrogen peroxide produced as a by-product of the activities of oxidative enzymes.

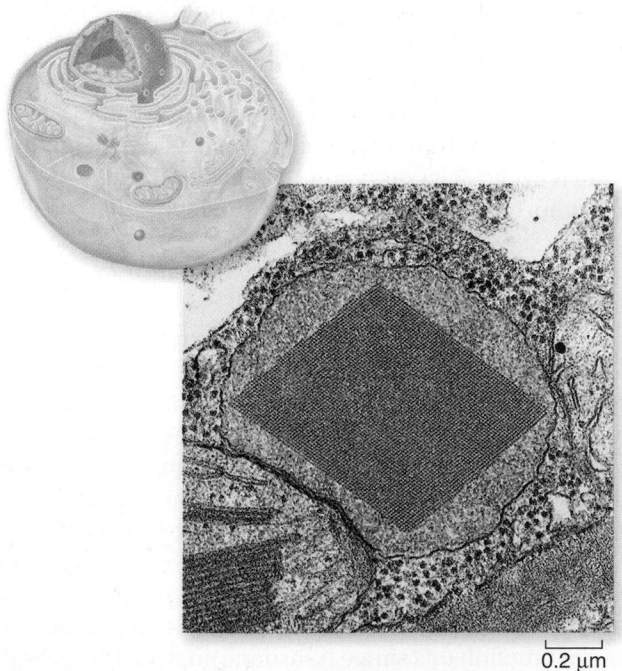

0.2 µm

Figure 4.14 A peroxisome. Peroxisomes are spherical organelles that may contain a large crystal structure composed of protein. Peroxisomes contain digestive and detoxifying enzymes that produce hydrogen peroxide as a by-product. A peroxisome has been colored green in the electron micrograph.

Hydrogen peroxide is dangerous to cells because of its violent chemical reactivity. However, peroxisomes also contain the enzyme catalase, which breaks down hydrogen peroxide into its harmless constituents—water and oxygen.

Plants use vacuoles for storage and water balance

Plant cells have specialized membrane-bounded structures called **vacuoles.** The most conspicuous example is the large central vacuole seen in most plant cells (figure 4.15). In fact, *vacuole* actually means blank space, referring to its appearance in the light microscope. The membrane surrounding this vacuole is called the **tonoplast** because it contains channels for water that are used to help the cell maintain its tonicity, or osmotic balance (see osmosis in chapter 5).

For many years biologists assumed that only one type of vacuole existed and that it served multiple functions. The functions assigned to this vacuole included water balance and storage of both useful molecules (such as sugars, ions, and pigments) and waste products. The vacuole was also thought to store enzymes involved in the breakdown of macromolecules and those used in detoxifying foreign substances. Old textbooks of plant physiology referred to vacuoles as the attic of the cell for the variety of substances thought to be stored there.

Studies of tonoplast transporters and the isolation of vacuoles from a variety of cell types have led to a more complex view of vacuoles. These studies have made it clear that different vacuolar types can be found in different cells. These vacuoles are specialized, depending on the function of the cell.

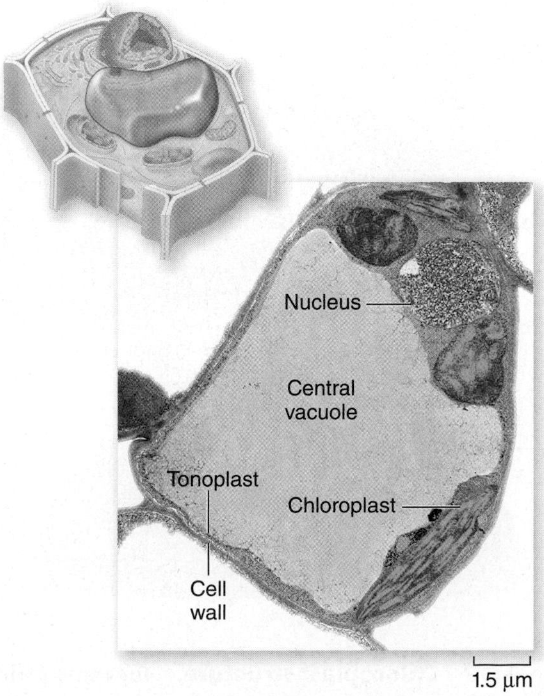

The central vacuole is clearly important for a number of roles in all plant cells. The central vacuole and the water channels of the tonoplast maintain the tonicity of the cell, allowing the cell to expand and contract, depending on conditions. The central vacuole is also involved in cell growth by occupying most of the volume of the cell. Plant cells grow by expanding the vacuole, rather than by increasing cytoplasmic volume.

Vacuoles with a variety of functions are also found in some types of fungi and protists. One form is the contractile vacuole, found in some protists, which can pump water and is used to maintain water balance in the cell. Other vacuoles are used for storage or to segregate toxic materials from the rest of the cytoplasm. The number and kind of vacuoles found in a cell depends on the needs of the particular cell type.

Learning Outcomes Review 4.4

The endoplasmic reticulum (ER) is an extensive system of folded membranes that spatially organize the cell's biosynthetic activities. Smooth ER (SER) is the site of lipid and membrane synthesis and is used to store Ca^{2+}. Rough ER (RER) is covered with ribosomes and is a site of protein synthesis. Proteins from the RER are transported by vesicles to the Golgi apparatus where they are modified, packaged, and distributed to their final location. Lysosomes are vesicles that contain digestive enzymes used to degrade materials such as invaders or worn-out components. Peroxisomes carry out oxidative metabolism that generates peroxides. Vacuoles are membrane-bounded structures with roles ranging from storage to cell growth in plants. They are also found in some fungi and protists.

■ **How do ribosomes on the RER differ from cytoplasmic ribosomes?**

Figure 4.15 The central vacuole. A plant's central vacuole stores dissolved substances and can expand in size to increase the tonicity of a plant cell. Micrograph shown with false color.

4.5 Mitochondria and Chloroplasts: Cellular Generators

Learning Outcomes

1. *Describe the structure of mitochondria and chloroplasts.*
2. *Compare the function of mitochondria and chloroplasts.*
3. *Explain the probable origin of mitochondria and chloroplasts.*

Mitochondria and chloroplasts share structural and functional similarities. Structurally, they are both surrounded by a double membrane, and both contain their own DNA and protein synthesis machinery. Functionally, they are both involved in energy metabolism, as we will explore in detail in later chapters on energy metabolism and photosynthesis.

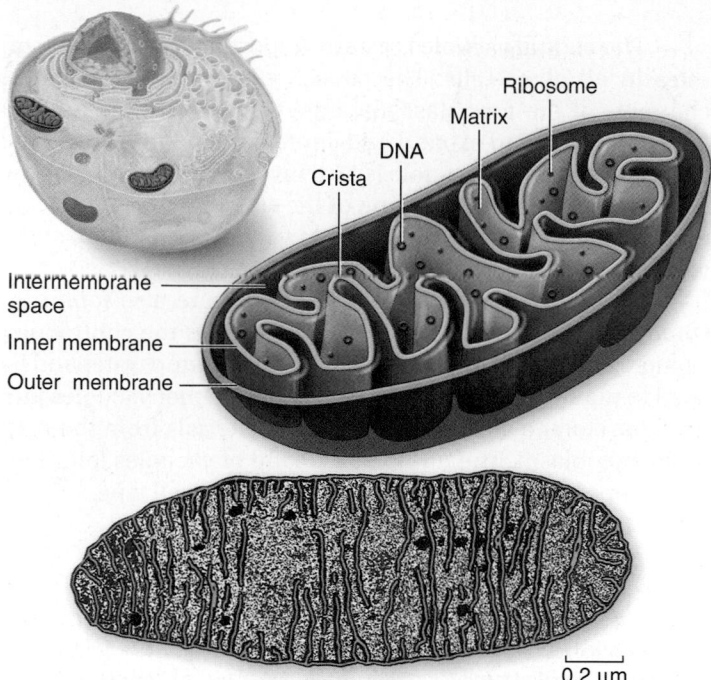

Figure 4.16 Mitochondria. The inner membrane of a mitochondrion is shaped into folds called cristae that greatly increase the surface area for oxidative metabolism. A mitochondrion in cross section and cut lengthwise is shown colored red in the micrograph.

Mitochondria metabolize sugar to generate ATP

Mitochondria (singular, *mitochondrion*) are typically tubular or sausage-shaped organelles about the size of bacteria that are found in all types of eukaryotic cells (figure 4.16). Mitochondria are bounded by two membranes: a smooth outer membrane, and an inner folded membrane with numerous contiguous layers called **cristae** (singular, *crista*).

The cristae partition the mitochondrion into two compartments: a **matrix,** lying inside the inner membrane; and an outer compartment, or **intermembrane space,** lying between the two mitochondrial membranes. On the surface of the inner membrane, and also embedded within it, are proteins that carry out oxidative metabolism, the oxygen-requiring process by which energy in macromolecules is used to produce ATP (chapter 7).

Mitochondria have their own DNA; this DNA contains several genes that produce proteins essential to the mitochondrion's role in oxidative metabolism. Thus, the mitochondrion, in many respects, acts as a cell within a cell, containing its own genetic information specifying proteins for its unique functions. The mitochondria are not fully autonomous, however, because most of the genes that encode the enzymes used in oxidative metabolism are located in the cell nucleus.

A eukaryotic cell does not produce brand-new mitochondria each time the cell divides. Instead, the mitochondria themselves divide in two, doubling in number, and these are partitioned between the new cells. Most of the components required for mitochondrial division are encoded by genes in the nucleus and are translated into proteins by cytoplasmic ribosomes. Mitochondrial replication is, therefore, impossible without nuclear participation, and mitochondria thus cannot be grown in a cell-free culture.

Chloroplasts use light to generate ATP and sugars

Plant cells and cells of other eukaryotic organisms that carry out photosynthesis typically contain from one to several hundred **chloroplasts.** Chloroplasts bestow an obvious advantage on the organisms that possess them: They can manufacture their own food. Chloroplasts contain the photosynthetic pigment chlorophyll that gives most plants their green color.

The chloroplast, like the mitochondrion, is surrounded by two membranes (figure 4.17). However, chloroplasts are larger and more complex than mitochondria. In addition to the outer and inner membranes, which lie in close association with each other, chloroplasts have closed compartments of stacked membranes called **grana** (singular, *granum*), which lie inside the inner membrane.

A chloroplast may contain a hundred or more grana, and each granum may contain from a few to several dozen disk-shaped structures called **thylakoids.** On the surface of the thylakoids are the light-capturing photosynthetic pigments, to be discussed in depth in chapter 8. Surrounding the thylakoid is a fluid matrix called the *stroma*. The enzymes used to synthesize glucose during photosynthesis are found in the stroma.

Like mitochondria, chloroplasts contain DNA, but many of the genes that specify chloroplast components are also located in the nucleus. Some of the elements used in photosynthesis, including the specific protein components

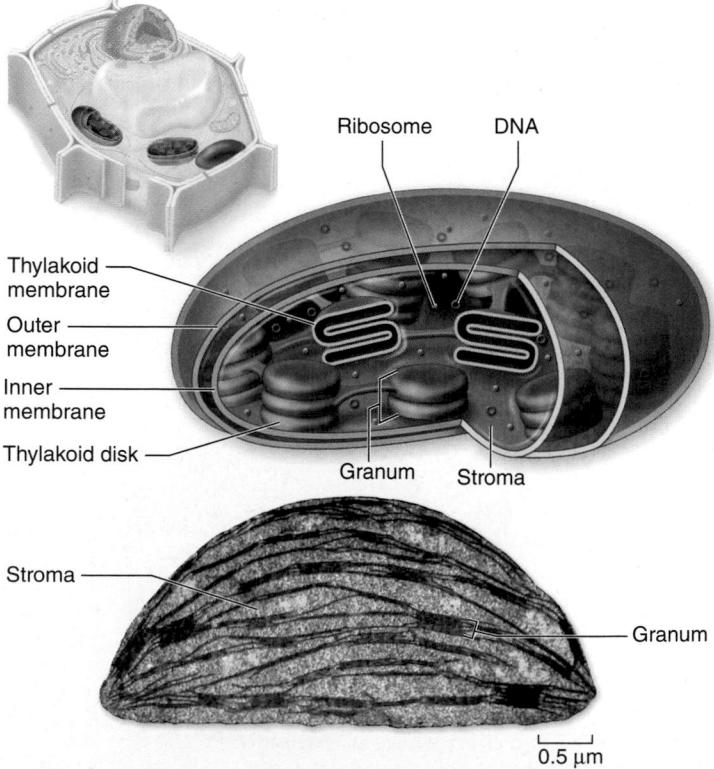

Figure 4.17 Chloroplast structure. The inner membrane of a chloroplast surrounds a membrane system of stacks of closed chlorophyll-containing vesicles called thylakoids, within which photosynthesis occurs. Thylakoids are typically stacked one on top of the other in columns called grana. The chloroplast has been colored green in the micrograph.

necessary to accomplish the reaction, are synthesized entirely within the chloroplast.

Other DNA-containing organelles in plants, called *leucoplasts,* lack pigment and a complex internal structure. In root cells and some other plant cells, leucoplasts may serve as starch-storage sites. A leucoplast that stores starch (amylose) is sometimes termed an **amyloplast.** These organelles—chloroplasts, leucoplasts, and amyloplasts—are collectively called **plastids.** All plastids are produced by the division of existing plastids.

> **? Inquiry question** Mitochondria and chloroplasts both generate ATP. What structural features do they share?

Mitochondria and chloroplasts arose by endosymbiosis

Symbiosis is a close relationship between organisms of different species that live together. As noted in chapter 29, the theory of **endosymbiosis** proposes that some of today's eukaryotic organelles evolved by a symbiosis arising between two cells that were each free-living. One cell, a prokaryote, was engulfed by and became part of another cell, which was the precursor of modern eukaryotes (figure 4.18).

According to the endosymbiont theory, the engulfed prokaryotes provided their hosts with certain advantages associated with their special metabolic abilities. Two key eukaryotic organelles are believed to be the descendants of these endosymbiotic prokaryotes: mitochondria, which are thought to have originated as bacteria capable of carrying out oxidative metabolism, and chloroplasts, which apparently arose from photosynthetic bacteria. This is discussed in detail in chapter 29.

> ### Learning Outcomes Review 4.5
>
> Mitochondria and chloroplasts have similar structures, with an outer membrane and an extensive inner membrane compartment. Both mitochondria and chloroplasts have their own DNA, but both also depend on nuclear genes for some functions. Mitochondria and chloroplasts are both involved in energy conversion: Mitochondria metabolize sugar to produce ATP, whereas chloroplasts harness light energy to produce ATP and synthesize sugars. Endosymbiosis theory proposes that both mitochondria and chloroplasts arose as prokaryotic cells were engulfed by a eukaryotic precursor.
>
> ■ *Many proteins in mitochondria and chloroplasts are encoded by nuclear genes. In light of the endosymbiont hypothesis, how might this come about?*

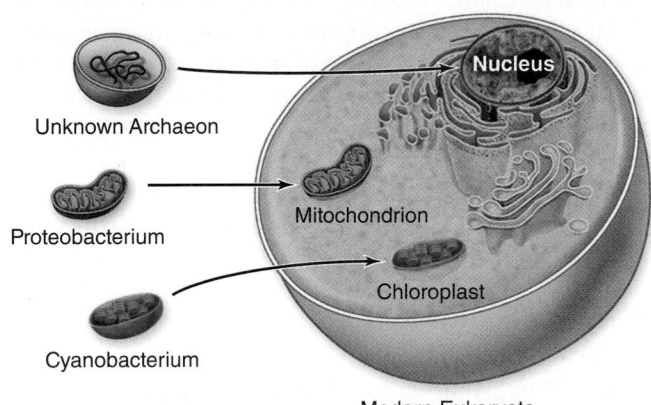

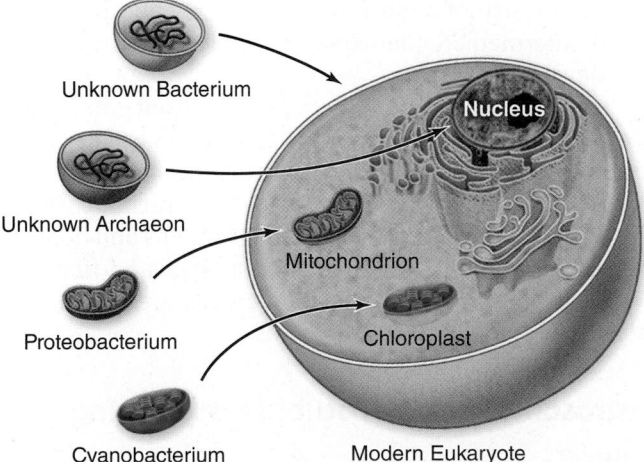

Figure 4.18 Possible origins of eukaryotic cells. Both mitochondria and chloroplasts are thought to have arisen by endosymbiosis when a free-living cell is taken up but not digested. The nature of the engulfing cell is unknown. Two possibilities are (1) the engulfing cell (top) is an archaeon that gave rise to the nuclear genome and cytoplasmic contents. (2) The engulfing cell (bottom) consists of a nucleus derived from an archaeon in a bacterial cell. This could arise by a fusion event or by engulfment of the archaeon by the bacterium.

4.6 The Cytoskeleton

> ### Learning Outcomes
>
> 1. *Contrast the structure and function of different fibers in the cytoskeleton.*
> 2. *Illustrate the role of microtubules in intracellular transport.*

The cytoplasm of all eukaryotic cells is crisscrossed by a network of protein fibers that supports the shape of the cell and anchors organelles to fixed locations. This network, called the cytoskeleton, is a dynamic system, constantly assembling and disassembling. Individual fibers consist of polymers of identical protein subunits that attract one another and spontaneously assemble into long chains. Fibers disassemble in the same way, as one subunit after another breaks away from one end of the chain.

Three types of fibers compose the cytoskeleton

Eukaryotic cells may contain the following three types of cytoskeletal fibers, each formed from a different kind of subunit: (1) actin filaments, sometimes called microfilaments, (2) microtubules, and (3) intermediate filaments.

Actin filaments (microfilaments)

Actin filaments are long fibers about 7 nm in diameter. Each filament is composed of two protein chains loosely twined together like two strands of pearls (figure 4.19). Each "pearl," or subunit, on the chain is the globular protein **actin.** Actin filaments exhibit polarity, that is, they have plus (+) and minus (–) ends. These designate the direction of growth of the filaments.

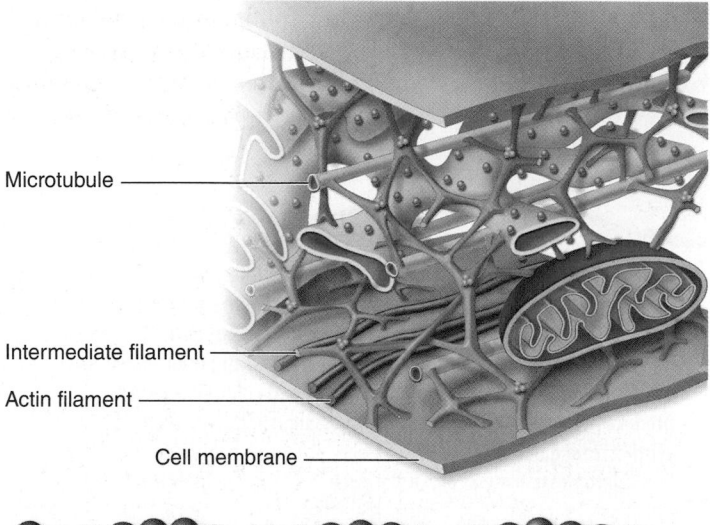

Microtubule

Intermediate filament

Actin filament

Cell membrane

a. Actin filaments

b. Microtubules

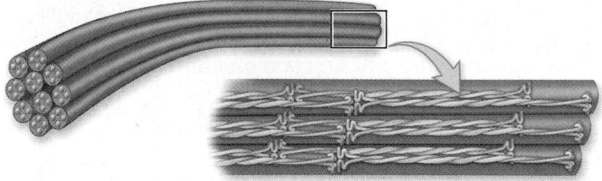

c. Intermediate filament

Figure 4.19 Molecules that make up the cytoskeleton.
a. *Actin filaments:* Actin filaments, also called *microfilaments,* are made of two strands of the globular protein actin twisted together. They are often found in bundles or in a branching network. Actin filaments in many cells are concentrated below the plasma membrane in bundles known as stress fibers, which may have a contractile function. ***b.*** *Microtubules:* Microtubules are composed of α- and β-tubulin protein subunits arranged side by side to form a tube. Microtubules are comparatively stiff cytoskeletal elements and have many functions in the cell including intracellular transport and the separation of chromosomes during mitosis. ***c.*** *Intermediate filaments:* Intermediate filaments are composed of overlapping staggered tetramers of protein. These tetramers are then bundled into cables. This molecular arrangement allows for a ropelike structure that imparts tremendous mechanical strength to the cell.

Actin molecules spontaneously form these filaments, even in a test tube.

Cells regulate the rate of actin polymerization through other proteins that act as switches, turning on polymerization when appropriate. Actin filaments are responsible for cellular movements such as contraction, crawling, "pinching" during division, and formation of cellular extensions.

Microtubules

Microtubules, the largest of the cytoskeletal elements, are hollow tubes about 25 nm in diameter, each composed of a ring of 13 protein protofilaments (see figure 4.19). Globular proteins consisting of dimers of α- and β-*tubulin* subunits polymerize to form the 13 protofilaments. The protofilaments are arrayed side by side around a central core, giving the microtubule its characteristic tube shape.

In many cells, microtubules form from nucleation centers near the center of the cell and radiate toward the periphery. They are in a constant state of flux, continually polymerizing and depolymerizing. The average half-life of a microtubule ranges from as long as 10 minutes in a nondividing animal cell to as short as 20 seconds in a dividing animal cell. The ends of the microtubule are designated as plus (+) (away from the nucleation center) or minus (–) (toward the nucleation center).

Along with facilitating cellular movement, microtubules organize the cytoplasm and are responsible for moving materials within the cell itself, as described shortly.

Intermediate filaments

The most durable element of the cytoskeleton in animal cells is a system of tough, fibrous protein molecules twined together in an overlapping arrangement (see figure 4.19). These *intermediate filaments* are characteristically 8 to 10 nm in diameter—between the size of actin filaments and microtubules. Once formed, intermediate filaments are stable and usually do not break down.

Intermediate filaments constitute a mixed group of cytoskeletal fibers. The most common type, composed of protein subunits called *vimentin,* provides structural stability for many kinds of cells. *Keratin,* another class of intermediate filament, is found in epithelial cells (cells that line organs and body cavities) and associated structures such as hair and fingernails. The intermediate filaments of nerve cells are called *neurofilaments.*

Centrosomes are microtubule-organizing centers

Centrioles are barrel-shaped organelles found in the cells of animals and most protists. They occur in pairs, usually located at right angles to each other near the nuclear membranes (figure 4.20). The region surrounding the pair in almost all animal cells is referred to as a *centrosome.* Surrounding the centrioles in the centrosome is the **pericentriolar material,** which contains ring-shaped structures composed of tubulin. The pericentriolar material can nucleate the assembly of microtubules in animal cells. Structures with this function are called

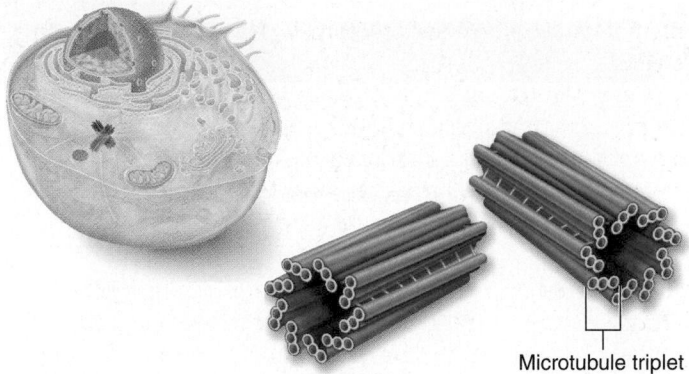

Figure 4.20 **Centrioles.** Each centriole is composed of nine triplets of microtubules. Centrioles are usually not found in plant cells. In animal cells they help to organize microtubules.

Microtubule triplet

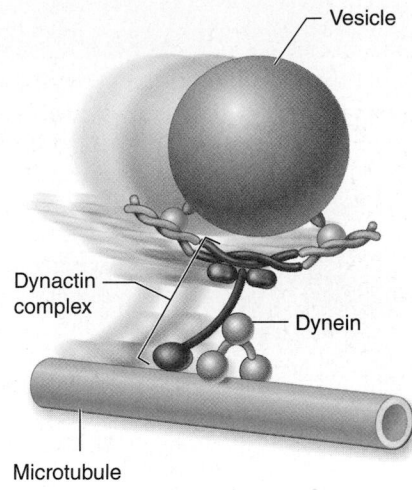

Vesicle

Dynactin complex

Dynein

Microtubule

Figure 4.21
Molecular motors.
Vesicles can be transported along microtubules using motor proteins that use ATP to generate force. The vesicles are attached to motor proteins by connector molecules, such as the dynactin complex shown here. The motor protein dynein moves the connected vesicle along microtubules.

microtubule-organizing centers. The centrosome is also responsible for the reorganization of microtubules that occurs during cell division. The centrosomes of plants and fungi lack centrioles, but still contain microtubule-organizing centers. You will learn more about the actions of the centrosomes when we describe the process of cell division in chapter 10.

The cytoskeleton helps move materials within cells

Actin filaments and microtubules often orchestrate their activities to affect cellular processes. For example, during cell reproduction (see chapter 10), newly replicated chromosomes move to opposite sides of a dividing cell because they are attached to shortening microtubules. Then, in animal cells, a belt of actin pinches the cell in two by contracting like a purse string.

Muscle cells also use actin filaments, which slide along filaments of the motor protein myosin when a muscle contracts. The fluttering of an eyelash, the flight of an eagle, and the awkward crawling of a baby all depend on these cytoskeletal movements within muscle cells.

Not only is the cytoskeleton responsible for the cell's shape and movement, but it also provides a scaffold that holds certain enzymes and other macromolecules in defined areas of the cytoplasm. For example, many of the enzymes involved in cell metabolism bind to actin filaments, as do ribosomes. By moving and anchoring particular enzymes near one another, the cytoskeleton, like the endoplasmic reticulum, helps organize the cell's activities.

Molecular motors

All eukaryotic cells must move materials from one place to another in the cytoplasm. One way cells do this is by using the channels of the endoplasmic reticulum as an intracellular highway. Material can also be moved using vesicles loaded with cargo that can move along the cytoskeleton like a railroad track. For example, in a nerve cell with an axon that may extend far from the cell body, vesicles can be moved along tracks of microtubules from the cell body to the end of the axon.

Four components are required to move material along microtubules: (1) a vesicle or organelle that is to be transported, (2) a motor protein that provides the energy-driven

motion, (3) a connector molecule that connects the vesicle to the motor molecule, and (4) microtubules on which the vesicle will ride like a train on a rail (figure 4.21).

The direction a vesicle is moved depends on the type of motor protein involved and the fact that microtubules are organized with their plus ends toward the periphery of the cell. In one case, a protein called kinectin binds vesicles to the motor protein *kinesin*. Kinesin uses ATP to power its movement toward the cell periphery, dragging the vesicle with it as it travels along the microtubule toward the plus end (figure 4.22). As nature's tiniest motors, these proteins pull the transport

SCIENTIFIC THINKING

Hypothesis: *Kinesin molecules can act as molecular motors and move along microtubules using energy from ATP.*

Test: *A microscope slide is covered with purified kinesin. Purified microtubules are added in a buffer containing ATP. The microtubules are monitored under a microscope using a video recorder to capture any movement.*

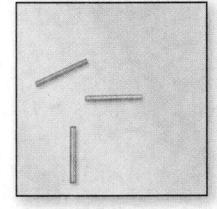 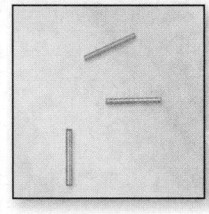 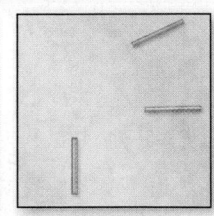

Frame 1   Frame 2   Frame 3

Result: *Over time, the movement of individual microtubules can be observed in the microscope. This is shown schematically in the figure by the movement of specific microtubules shown in color.*

Conclusion: *Kinesin acts as a molecular motor moving along (in this case actually moving) microtubules.*

Further Experiments: *Are there any further controls that are not shown in this experiment? What additional conclusions could be drawn by varying the amount of kinesin sticking to the slide?*

Figure 4.22 Demonstration of kinesin as molecular motor. Microtubules can be observed moving over a slide coated with kinesin.

TABLE 4.2

TABLE 4.2 Eukaryotic Cell Structures and their Functions

Structure		Description	Function
Plasma membrane		Phospholipid bilayer with embedded proteins	Regulates what passes into and out of cell; cell-to-cell recognition; connection and adhesion; cell communication
Nucleus		Structure (usually spherical) that contains chromosomes and is surrounded by double membrane	Instructions for protein synthesis and cell reproduction; contains genetic information
Chromosomes		Long threads of DNA that form a complex with protein	Contain hereditary information used to direct synthesis of proteins
Nucleolus		Site of genes for rRNA synthesis	Synthesis of rRNA and ribosome assembly
Ribosomes		Small, complex assemblies of protein and RNA, often bound to ER	Sites of protein synthesis
Endoplasmic reticulum (ER)		Network of internal membranes	Intracellular compartment forms transport vesicles; participates in lipid synthesis and synthesis of membrane or secreted proteins
Golgi apparatus		Stacks of flattened vesicles	Packages proteins for export from cell; forms secretory vesicles
Lysosomes		Vesicles derived from Golgi apparatus that contain hydrolytic digestive enzymes	Digest worn-out organelles and cell debris; digest material taken up by endocytosis
Microbodies		Vesicles that are formed from incorporation of lipids and proteins and that contain oxidative and other enzymes	Isolate particular chemical activities from rest of cell
Mitochondria		Bacteria-like elements with double membrane	"Power plants" of the cell; sites of oxidative metabolism
Chloroplasts		Bacteria-like elements with double membrane surrounding a third, thylakoid membrane containing chlorophyll, a photosynthetic pigment	Sites of photosynthesis
Cytoskeleton		Network of protein filaments	Structural support; cell movement; movement of vesicles within cells
Flagella (cilia)		Cellular extensions with 9 + 2 arrangement of pairs of microtubules	Motility or moving fluids over surfaces
Cell wall		Outer layer of cellulose or chitin; or absent	Protection; support

vesicles along the microtubular tracks. Another set of vesicle proteins, called the dynactin complex, binds vesicles to the motor protein *dynein* (see figure 4.22), which directs movement in the opposite direction along microtubules toward the minus end, inward toward the cell's center. (Dynein is also involved in the movement of eukaryotic flagella, as discussed later.) The destination of a particular transport vesicle and its content is thus determined by the nature of the linking protein embedded within the vesicle's membrane.

The major eukaryotic cell structures and their respective functions are summarized in table 4.2.

Learning Outcomes Review 4.6

The three principal fibers of the cytoskeleton are actin filaments (microfilaments), microtubules, and intermediate filaments. These fibers interact to modulate cell shape and permit cell movement. They also act to move materials within the cytoplasm. Material is also moved in large cells using vesicles and molecular motors. The motor proteins move vesicles along tracks of microtubules.

■ **What advantage does the cytoskeleton give to large eukaryotic cells?**

4.7 Extracellular Structures and Cell Movement

Learning Outcomes

1. Describe how cells move.
2. Identify the different cytoskeletal elements involved in cell movement.
3. Classify the elements of extracellular matrix in animal cells.

Essentially all cell motion is tied to the movement of actin filaments, microtubules, or both. Intermediate filaments act as intracellular tendons, preventing excessive stretching of cells. Actin filaments play a major role in determining the shape of cells. Because actin filaments can form and dissolve so readily, they enable some cells to change shape quickly.

Some cells crawl

The arrangement of actin filaments within the cell cytoplasm allows cells to crawl, literally! Crawling is a significant cellular phenomenon, essential to such diverse processes as inflammation, clotting, wound healing, and the spread of cancer. White blood cells in particular exhibit this ability. Produced in the bone marrow, these cells are released into the circulatory system and then eventually crawl out of venules and into the tissues to destroy potential pathogens.

At the leading edge of a crawling cell, actin filaments rapidly polymerize, and their extension forces the edge of the cell forward. This extended region is stabilized when microtubules polymerize into the newly formed region. Overall forward

movement of the cell is then achieved through the action of the protein **myosin,** which is best known for its role in muscle contraction. Myosin motors along the actin filaments contract, pulling the contents of the cell toward the newly extended front edge.

Cells crawl when these steps occur continuously, with a leading edge extending and stabilizing, and then motors contracting to pull the remaining cell contents along. Receptors on the cell surface can detect molecules outside the cell and stimulate extension in specific directions, allowing cells to move toward particular targets.

Flagella and cilia aid movement

Earlier in this chapter, we described the structure of prokaryotic flagella. Eukaryotic cells have a completely different kind of flagellum, consisting of a circle of nine microtubule pairs surrounding two central microtubules. This arrangement is referred to as the *9 + 2 structure* (figure 4.23).

As pairs of microtubules move past each other using arms composed of the motor protein dynein, the eukaryotic flagellum *undulates,* or waves up and down, rather than rotates. When examined carefully, each flagellum proves to be an outward projection of the cell's interior, containing cytoplasm and enclosed by the plasma membrane. The microtubules of the flagellum are derived from a **basal body,** situated just below the point where the flagellum protrudes from the surface of the cell.

The flagellum's complex microtubular apparatus evolved early in the history of eukaryotes. Today the cells of many multicellular and some unicellular eukaryotes no longer possess flagella and are nonmotile. Other structures, called

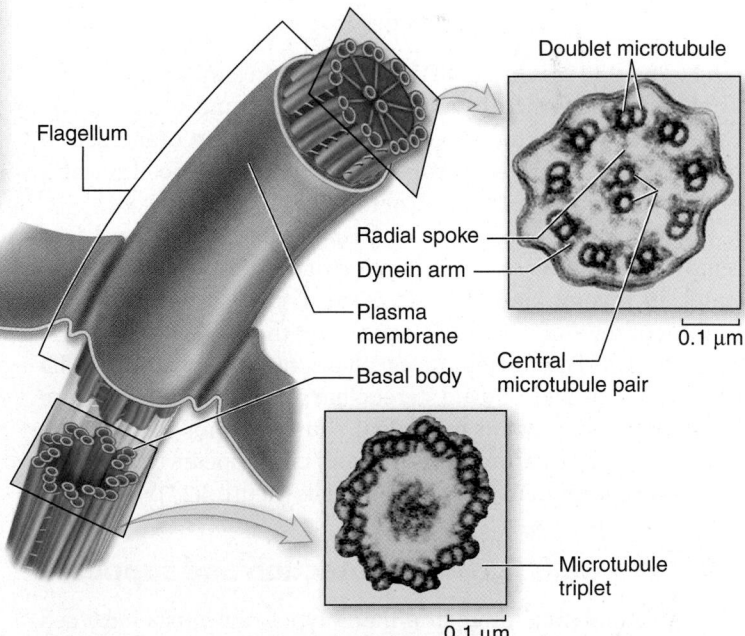

Figure 4.23 Flagella and cilia. A eukaryotic flagellum originates directly from a basal body. The flagellum has two microtubules in its core connected by radial spokes to an outer ring of nine paired microtubules with dynein arms (9 + 2 structure). The basal body consists of nine microtubule triplets connected by short protein segments. The structure of cilia is similar to that of flagella, but cilia are usually shorter.

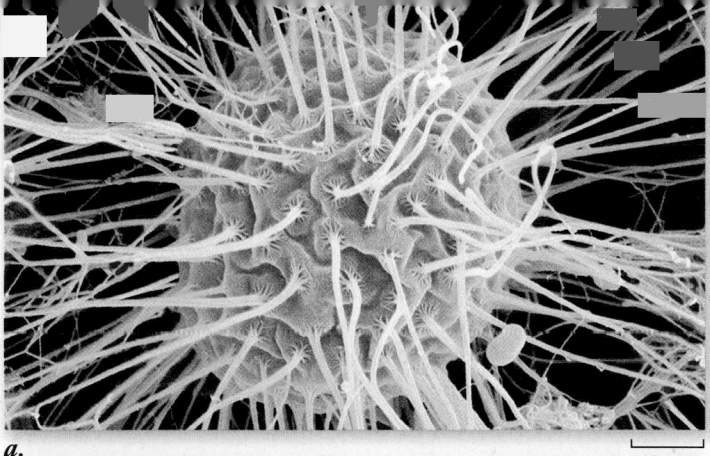

a.
40 µm

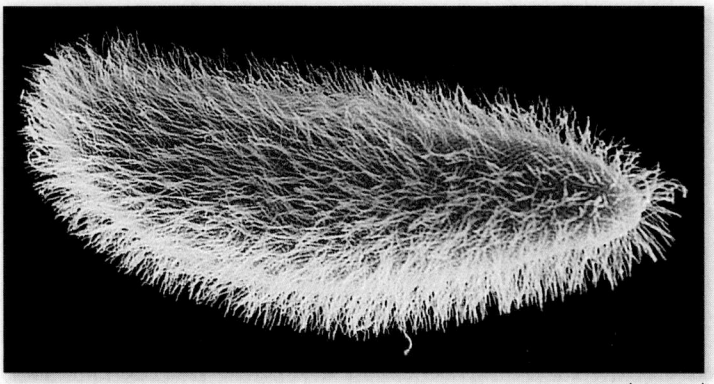

b.
67 µm

Figure 4.24 Flagella and cilia. *a.* A green alga with numerous flagella that allow it to move through the water. *b.* Paramecia are covered with many cilia, which beat in unison to move the cell. The cilia can also be used to move fluid into the paramecium's mouth to ingest material.

cilia (singular, *cilium*), with an organization similar to the 9 + 2 arrangement of microtubules can still be found within them. Cilia are short cellular projections that are often organized in rows. They are more numerous than flagella on the cell surface, but have the same internal structure.

In many multicellular organisms, cilia carry out tasks far removed from their original function of propelling cells through water. In several kinds of vertebrate tissues, for example, the beating of rows of cilia move water over the tissue surface. The sensory cells of the vertebrate ear also contain conventional cilia surrounded by actin-based stereocilia; sound waves bend these structures and provide the initial sensory input for hearing. Thus, the 9 + 2 structure of flagella and cilia appears to be a fundamental component of eukaryotic cells (figure 4.24).

Plant cell walls provide protection and support

The cells of plants, fungi, and many types of protists have cell walls, which protect and support the cells. The cell walls of these eukaryotes are chemically and structurally different from prokaryotic cell walls. In plants and protists, the cell walls are composed of fibers of the polysaccharide cellulose, whereas in fungi, the cell walls are composed of chitin.

In plants, **primary walls** are laid down when the cell is still growing. Between the walls of adjacent cells a sticky substance, called the **middle lamella,** glues the cells together (figure 4.25).

Some plant cells produce strong **secondary walls,** which are deposited inside the primary walls of fully expanded cells.

Animal cells secrete an extracellular matrix

Animal cells lack the cell walls that encase plants, fungi, and most protists. Instead, animal cells secrete an elaborate mixture of glycoproteins into the space around them, forming the *extracellular matrix (ECM)* (figure 4.26). The fibrous protein collagen, the same protein found in cartilage, tendons, and ligaments may be abundant in the ECM. Strong fibers of collagen and another fibrous protein, elastin, are embedded within a complex web of other glycoproteins, called proteoglycans, that form a protective layer over the cell surface.

> **?** **Inquiry question** The passageways of the human trachea (the path of airflow into and out of the lungs) are known to be lined with ciliated cells. What function could these cilia perform?

The ECM of some cells is attached to the plasma membrane by a third kind of glycoprotein, *fibronectin.* Fibronectin molecules bind not only to ECM glycoproteins but also to proteins called **integrins.** Integrins are an integral part of the

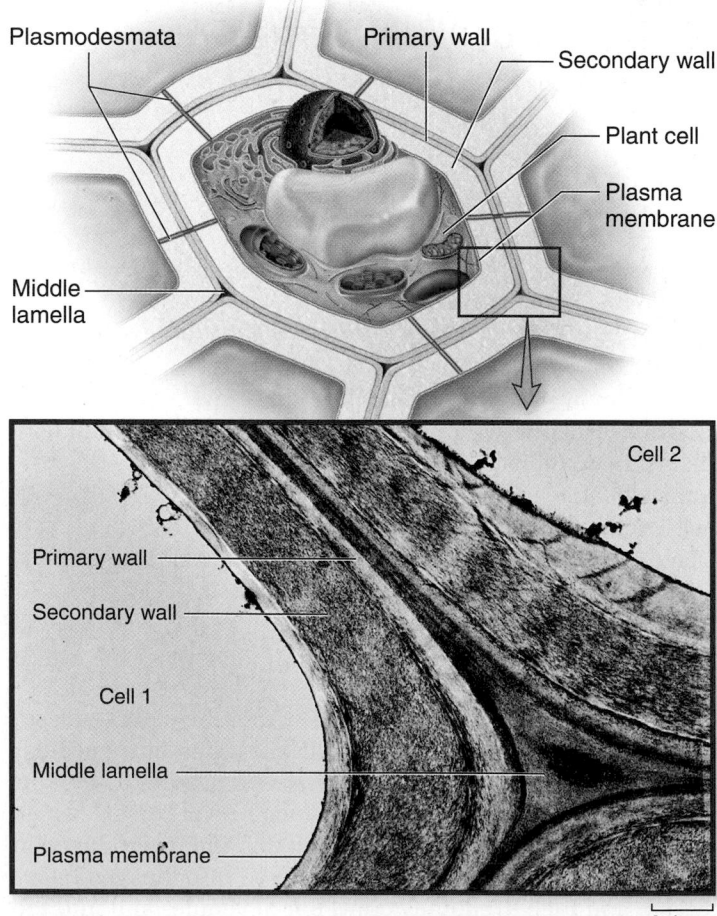

0.4 µm

Figure 4.25 Cell walls in plants. Plant cell walls are thick, strong, and rigid. Primary cell walls are laid down when the cell is young. Thicker secondary cell walls may be added later when the cell is fully grown.

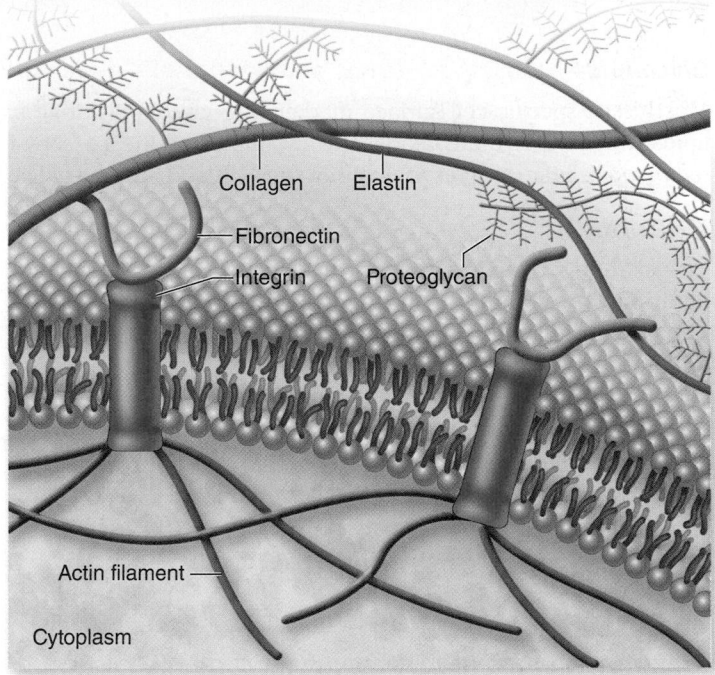

Collagen　Elastin

Fibronectin

Integrin　Proteoglycan

Actin filament

Cytoplasm

Figure 4.26　The extracellular matrix. Animal cells are surrounded by an extracellular matrix composed of various glycoproteins that give the cells support, strength, and resilience.

plasma membrane, extending into the cytoplasm, where they are attached to the microfilaments and intermediate filaments of the cytoskeleton. Linking ECM and cytoskeleton, integrins allow the ECM to influence cell behavior in important ways. They can alter gene expression and cell migration patterns by a combination of mechanical and chemical signaling pathways. In this way, the ECM can help coordinate the behavior of all the cells in a particular tissue.

Table 4.3 compares and reviews the features of three types of cells.

Learning Outcomes Review 4.7

Cell movement involves proteins. These can either be internal in the case of crawling cells that use actin and myosin, or external in the case of cells powered by cilia or flagella. Eukaryotic cilia and flagella are different from prokaryotic flagella because they are composed of bundles of microtubules in a 9 + 2 array. They undulate rather than rotate.

Plant cells have a cellulose-based cell wall. Animal cells lack a cell wall. In animal cells, the cytoskeleton is linked to a web of glycoproteins called the extracellular matrix.

■　*What cellular roles are performed by microtubules and microfilaments and not intermediate filaments?*

TABLE 4.3	A Comparison of Prokaryotic, Animal, and Plant Cells		
	Prokaryote	**Animal**	**Plant**
EXTERIOR STRUCTURES			
Cell wall	Present (protein-polysaccharide)	Absent	Present (cellulose)
Cell membrane	Present	Present	Present
Flagella/cilia	Flagella may be present	May be present (9 + 2 structure)	Absent except in sperm of a few species (9 + 2 structure)
INTERIOR STRUCTURES			
Endoplasmic reticulum	Absent	Usually present	Usually present
Ribosomes	Present	Present	Present
Microtubules	Absent	Present	Present
Centrioles	Absent	Present	Absent
Golgi apparatus	Absent	Present	Present
Nucleus	Absent	Present	Present
Mitochondria	Absent	Present	Present
Chloroplasts	Absent	Absent	Present
Chromosomes	Single; circle of DNA	Multiple; DNA–protein complex	Multiple; DNA–protein complex
Lysosomes	Absent	Usually present	Present
Vacuoles	Absent	Absent or small	Usually a large single vacuole

Learning Outcomes

1. *Differentiate between types of cell junctions.*
2. *Describe the roles of surface proteins.*

A basic feature of multicellular animals is the formation of diverse kinds of *tissue*, such as skin, blood, or muscle, where cells are organized in specific ways. Cells must also be able to communicate with each other and have markers of individual identity. All of these functions—connections between cells, markers of cellular identity, and cell communication—involve membrane proteins and proteins secreted by cells. As an organism develops, the cells acquire their identities by carefully controlling the *expression* of those genes, turning on the specific set of genes that encode the functions of each cell type. Table 4.4 provides a summary of the kinds of connections seen between cells that are explored in the following sections.

Surface proteins give cells identity

One key set of genes functions to mark the surfaces of cells, identifying them as being of a particular type. When cells make contact, they "read" each other's cell-surface markers and react accordingly. Cells that are part of the same tissue type recognize each other, and they frequently respond by forming connections between their surfaces to better coordinate their functions.

Glycolipids

Most tissue-specific cell-surface markers are glycolipids, that is, lipids with carbohydrate heads. The glycolipids on the surface of red blood cells are also responsible for the A, B, and O blood types.

MHC proteins

One example of the function of cell-surface markers is the recognition of "self" and "nonself" cells by the immune system. This function is vital for multicellular organisms, which need to defend themselves against invading or malignant cells. The immune system of vertebrates uses a particular set of markers to distinguish self from nonself cells, encoded by genes of the *major histocompatibility complex* (*MHC*). Cell recognition in the immune system is covered in chapter 51.

 ## Cell connections mediate cell-to-cell adhesion

The evolution of multicellularity required the acquisition of molecules that can connect cells to each other. It appears that multicellularity arose independently in different lineages, but the types of connections between cells are remarkably conserved, and many of the proteins involved are ancient.

The nature of the physical connections between the cells of a tissue in large measure determines what the tissue is like. Indeed, a tissue's proper functioning often depends critically on how the individual cells are arranged within it. Just as a house cannot maintain its structure without nails and cement, so a tissue cannot maintain its characteristic architecture without the appropriate cell junctions. Cell junctions can be

TABLE 4.4	Cell-to-Cell Connections and Cell Identity		
Type of Connection	**Structure**	**Function**	**Example**
Surface markers	Variable, integral proteins or glycolipids in plasma membrane	Identify the cell	MHC complexes, blood groups, antibodies
Septate junctions Tight junctions	Tightly bound, leakproof, fibrous claudin protein seal that surrounds cell	Holds cells together such that materials pass through but not between the cells	Junctions between epithelial cells in the gut
Adhesive junction (desmosome)	Variant cadherins, desmocollins, bind to intermediate filaments of cytoskeleton	Creates strong flexible connections between cells. Found in vertebrates	Epithelium
Adhesive junction (adherens junction)	Classical cadherins, bind to microfilaments of cytoskeleton	Connects cells together. Oldest form of cell junction, found in all multicellular organisms	Tissues with high mechanical stress, such as the skin
Adhesive junction (Hemidesmosome, focal adhesion)	Integrin proteins bind cell to extracellular matrix	Provide attachment to a substrate	Involved in cell movement and important during development
Communicating junction (gap junction)	Six transmembrane connexon/pannexin proteins creating a pore that connects cells	Allows passage of small molecules from cell to cell in a tissue	Excitable tissue such as heart muscle
Communicating junction (plasmodesmata)	Cytoplasmic connections between gaps in adjoining plant cell walls	Communicating junction between plant cells	Plant tissues

characterized by both their visible structure in the microscope, and the proteins involved in the junction.

Adhesive junctions

Adhesive junctions appear to have been the first to evolve. Primitive forms can even be found in sponges, and they are found in all animal species. They mechanically attach the cytoskeleton of a cell to the cytoskeletons of other cells or to the extracellular matrix. These junctions are found in tissues subject to mechanical stress, such as muscle and skin epithelium.

Adherens junctions are based on the protein **cadherin**, which is a Ca^{2+}-dependent adhesion molecule with very wide phylogenetic distribution. Cadherin is a single-pass transmembrane protein with an extracellular domain that can interact with the extracellular domain of a cadherin in an adjacent cell to join the cells together (figure 4.27). Adherens junctions are found in animals ranging from jellyfish to vertebrates. Cadherins found in these junctions are called classical cadherins and are broken down into types I and II. When cells bearing either type I or type II cadherins are mixed, they sort into populations joined by I to I or by II to II interactions. There is some evidence for interactions between type I and type II cadherins, but they are not as strong. On the cytoplasmic side, the cadherins interact indirectly through other proteins with actin to form flexible connections between cells (see figure 4.27).

Desmosomes are a cadherin-based junction unique to vertebrates. They contain the cadherins desmocollin and desmoglein, which interact with intermediate filaments of cytoskeletons instead of actin. Desmosomes join adjacent cells (figure 4.28b). These connections support tissues against mechanical stress.

Hemidesmosomes and focal adhesions connect cells to the basal lamina or other ECM. In this case the proteins that interact with the ECM are called integrins. The integrins are members of a large superfamily of cell-surface receptors that bind to a protein component of the extracellular matrix. At least 20 different integrins exist, each with a differently shaped binding domain. These junctions also connect to the cytoskeleton of cells: actin filaments at focal adhesions and intermediate filaments at hemidesmosomes.

Septate, or Tight, junctions

Septate junctions are found in both invertebrates and vertebrates and form a barrier that can seal off a sheet of cells. The proteins found at these junctions have been given different names in different systems; in *Drosophila*, the proteins include Discs Large and Neurexin. Their wide distribution indicates that they probably evolved soon after or with adherens junctions.

Tight junctions are unique to vertebrates and contain proteins called Claudins because of their ability to occlude or block substances from passing between cells. This form of junction between cells acts as a wall within the tissue, keeping molecules on one side or the other (see figure 4.28a).

Creating sheets of cells. The cells that line an animal's digestive tract are organized in a sheet only one cell thick. One surface of the sheet faces the inside of the tract, and the other faces the extracellular space, where blood vessels are located. Tight junctions encircle each cell in the sheet, like a belt cinched around a person's waist. The junctions between neighboring cells are so securely attached that there is no space between them for leakage. Hence, nutrients absorbed from the food in the digestive tract must pass directly through the cells in the sheet to enter the bloodstream because they cannot pass through spaces between cells.

The tight junctions between the cells lining the digestive tract also partition the plasma membranes of these cells into separate compartments. Transport proteins in the membrane facing the inside of the tract carry nutrients from that side to the cytoplasm of the cells. Other proteins, located in the membrane on the opposite side of the cells, transport those nutrients from the cytoplasm to the extracellular fluid, where they can enter the bloodstream. Tight junctions effectively segregate the proteins on opposite sides of the sheet, preventing them from drifting within the membrane from one side of the sheet to the other. When tight junctions are experimentally disrupted, just this sort of migration occurs.

Communicating junctions

The proteins involved in the junctions previously described can be found in some single-celled organisms as well. The evolution of multicellularity also led to a new form of cellular connection: the *communicating junctions*. These junctions allow communication between cells by diffusion through small openings. Communicating junctions permit small molecules or ions to pass from one cell to the other. In animals, these direct communication channels between cells are called *gap junctions,* and in plants, *plasmodesmata.*

Gap junctions in animals. **Gap junctions** are found in both invertebrates and vertebrates. In invertebrates they are formed by proteins known as pannexins. In vertebrates pannexin-base gap junctions exist, but there is an additional

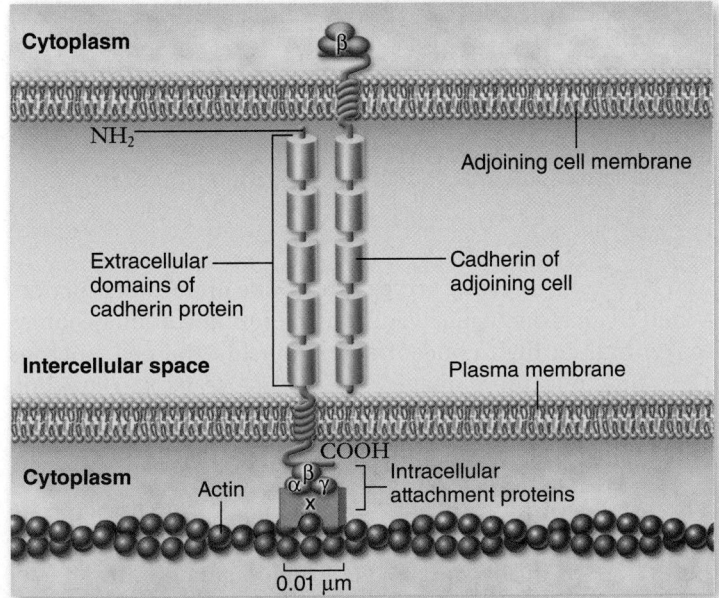

Figure 4.27 A cadherin-mediated junction. The cadherin molecule is anchored to actin in the cytoskeleton and passes through the membrane to interact with the cadherin of an adjoining cell.

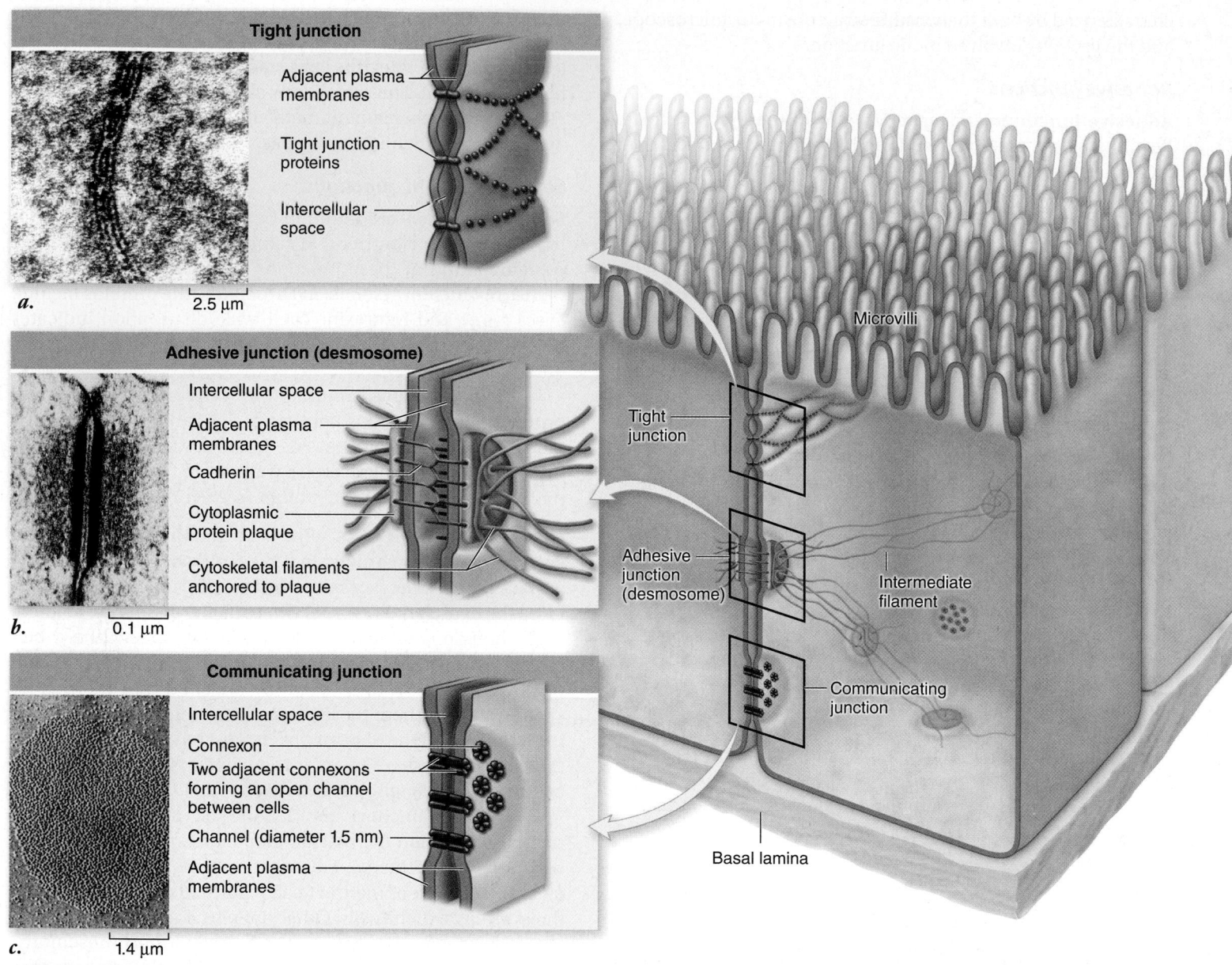

Tight junction

Adjacent plasma membranes

Tight junction proteins

Intercellular space

a. 2.5 μm

Adhesive junction (desmosome)

Intercellular space

Adjacent plasma membranes

Cadherin

Cytoplasmic protein plaque

Cytoskeletal filaments anchored to plaque

b. 0.1 μm

Communicating junction

Intercellular space

Connexon

Two adjacent connexons forming an open channel between cells

Channel (diameter 1.5 nm)

Adjacent plasma membranes

c. 1.4 μm

Microvilli

Tight junction

Adhesive junction (desmosome)

Intermediate filament

Communicating junction

Basal lamina

Figure 4.28 **Cell junction types in animal epithelium.** Here, the diagram of gut epithelial cells on the right illustrates the comparative structures and locations of common cell junctions. The detailed models on the left show the structures of the three major types of cell junctions: *(a)* tight junction; *(b)* adhesive junction, the example shown is a desmosome; *(c)* communicating junction, the example shown is a gap junction.

type based on similar proteins called connexons. In each case, a structure is formed by complexes of six identical transmembrane proteins (see figure 4.28c). The proteins are arranged in a circle to create a channel through the plasma membrane that protrudes several nanometers from the cell surface. A gap junction forms when the connexons/pannexins of two cells align perfectly, creating an open channel that spans the plasma membranes of both cells.

Gap junctions provide passageways large enough to permit small substances, such as simple sugars and amino acids, to pass from one cell to the next. Yet the passages are small enough to prevent the passage of larger molecules, such as proteins.

Gap junction channels are dynamic structures that can open or close in response to a variety of factors, including Ca^{2+}

and H^+ ions. This gating serves at least one important function. When a cell is damaged, its plasma membrane often becomes leaky. Ions in high concentrations outside the cell, such as Ca^{2+}, flow into the damaged cell and close its gap junction channels. This isolates the cell and prevents the damage from spreading.

Plasmodesmata in plants. In plants, cell walls separate every cell from all others. Cell–cell junctions occur only at holes or gaps in the walls, where the plasma membranes of adjacent cells can come into contact with one another. Cytoplasmic connections that form across the touching plasma membranes are called **plasmodesmata** (singular, *plasmodesma*) (figure 4.29). The majority of living cells within a

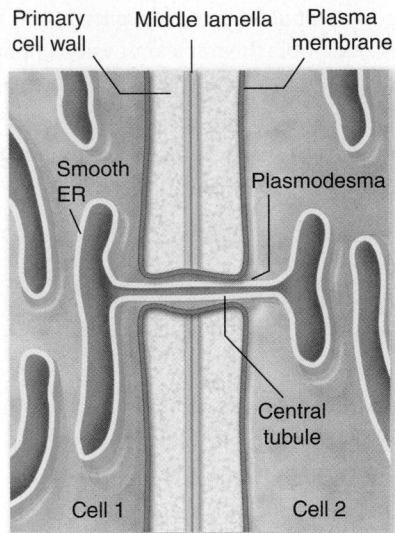

Primary cell wall   Middle lamella   Plasma membrane

Smooth ER

Plasmodesma

Central tubule

Cell 1   Cell 2

Figure 4.29 Plasmodesmata. Plant cells can communicate through specialized openings in their cell walls, called plasmodesmata, where the cytoplasm of adjoining cells are connected.

higher plant are connected to their neighbors by these junctions.

Plasmodesmata function much like gap junctions in animal cells, although their structure is more complex. Unlike gap junctions, plasmodesmata are lined with plasma membrane and contain a central tubule that connects the endoplasmic reticulum of the two cells.

Learning Outcomes Review 4.8

The evolution of multicellularity required the acquisition of cell adhesion molecules to connect cells together. Cell connections fall into three basic categories: (1) adhesive junctions provide strength and flexibility; (2) tight, or septate, junctions help to make sheets of cells that form watertight seals; and (3) communicating junctions, including gap junctions in animals and plasmodesmata in plants, allow passage of some materials between cells. Cells in multicellular organisms have distinct identity and connections. Cell identity is conferred by surface glycoproteins, which include the MHC proteins that are important in the immune system.

■ **How do cell junctions help to form tissues?**

Chapter Review

4.1 Cell Theory

Cell theory is the unifying foundation of cell biology.

All organisms are composed of one or more cells. Cells arise only by division of preexisting cells.

Cell size is limited.

Cell size is constrained by the diffusion distance. As cell size increases, diffusion becomes inefficient.

Microscopes allow visualization of cells and components.

Magnification gives better resolution than is possible with the naked eye. Staining with chemicals enhances contrast of structures.

All cells exhibit basic structural similarities.

All cells have centrally located DNA, a semifluid cytoplasm, and an enclosing plasma membrane.

4.2 Prokaryotic Cells (figure 4.3)

Prokaryotic cells have relatively simple organization.

Prokaryotic cells contain DNA and ribosomes, but they lack a nucleus, an internal membrane system, and membrane-bounded organelles. A rigid cell wall surrounds the plasma membrane.

Bacterial cell walls consist of peptidoglycan.

Peptidoglycan is composed of carbohydrate cross-linked with short peptides.

Archaea lack peptidoglycan.

Archaeal cell walls do not contain peptidoglycan, and they have unique plasma membranes.

Some prokaryotes move by means of rotating flagella.

Prokaryotic flagella rotate because of proton transfer across the plasma membrane.

4.3 Eukaryotic Cells (figures 4.6 and 4.7)

Eukaryotic cells have a membrane-bounded nucleus, an endomembrane system, and many different organelles.

The nucleus acts as the information center.

The nucleus is surrounded by an envelope of two phospholipid bilayers; the outer layer is contiguous with the ER. Pores allow exchange of small molecules. The nucleolus is a region of the nucleoplasm where rRNA is transcribed and ribosomes are assembled.

In most prokaryotes, DNA is organized into a single circular chromosome. In eukaryotes, numerous chromosomes are present.

Ribosomes are the cell's protein synthesis machinery.

Ribosomes translate mRNA to produce polypeptides. They are found in all cell types.

4.4 The Endomembrane System

The endoplasmic reticulum (ER) creates channels and passages within the cytoplasm (figure 4.10).

The rough ER is a site of protein synthesis.

The rough ER (RER), studded with ribosomes, synthesizes and modifies proteins and manufactures membranes.

The smooth ER has multiple roles.

The smooth endoplasmic reticulum (SER) lacks ribosomes; it is involved in carbohydrate and lipid synthesis and detoxification.

The Golgi apparatus sorts and packages proteins.

The Golgi apparatus receives vesicles from the ER, modifies and packages macromolecules, and transports them (figure 4.11).

Lysosomes contain digestive enzymes.

Lysosomes break down macromolecules and recycle the components of old organelles (figure 4.13).

Microbodies are a diverse category of organelles.

Plants use vacuoles for storage and water balance.

4.5 Mitochondria and Chloroplasts: Cellular Generators

Mitochondria and chloroplasts have a double-membrane structure, contain their own DNA, and can divide independently.

Mitochondria metabolize sugar to generate ATP.

The inner membrane of mitochondria is extensively folded into layers called cristae. Proteins on the surface and in the inner membrane carry out metabolism to produce ATP (figure 4.16).

Chloroplasts use light to generate ATP and sugars.

Chloroplasts capture light energy via thylakoid membranes arranged in stacks called grana, and use it to synthesize glucose (figure 4.17).

Mitochondria and chloroplasts arose by endosymbiosis.

The endosymbiont theory proposes that mitochondria and chloroplasts were once prokaryotes engulfed by another cell.

4.6 The Cytoskeleton

The cytoskeleton consists of crisscrossed protein fibers that support the shape of the cell and anchor organelles (figure 4.19).

Three types of fibers compose the cytoskeleton.

Actin filaments, or microfilaments, are long, thin polymers involved in cellular movement. Microtubules are hollow structures that move materials within a cell. Intermediate filaments serve a wide variety of functions.

Centrosomes are microtubule-organizing centers.

Centrosomes help assemble the nuclear division apparatus of animal cells (figure 4.20).

The cytoskeleton helps move materials within cells.

Molecular motors move vesicles along microtubules, like a train on a railroad track. Kinesin and dynein are two motor proteins.

4.7 Extracellular Structures and Cell Movement

Some cells crawl.

Cell crawling occurs as actin polymerization forces the cell membrane forward, while myosin pulls the cell body forward.

Flagella and cilia aid movement.

Eukaryotic flagella have a 9 + 2 structure and arise from a basal body. Cilia are shorter and more numerous than flagella.

Plant cell walls provide protection and support.

Plants have cell walls composed of cellulose fibers. The middle lamella, between cell walls, holds adjacent cells together.

Animal cells secrete an extracellular matrix.

Glycoproteins are the main component of the extracellular matrix (ECM) of animal cells.

4.8 Cell-to-Cell Interactions (figure 4.27)

Surface proteins give cells identity.

Glycolipids and MHC proteins on cell surfaces help distinguish self from nonself.

Cell connections mediate cell-to-cell adhesion.

Cell junctions include tight junctions, adhesive junctions, and communicating junctions. In animals, gap junctions allow the passage of small molecules between cells. In plants, plasmodesmata penetrate the cell wall and connect cells.

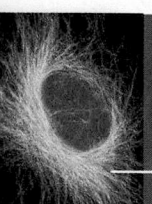

Review Questions

UNDERSTAND

1. Which of the following statements is NOT part of the cell theory?
   a. All organisms are composed of one or more cells.
   b. Cells come from other cells by division.
   c. Cells are the smallest living things.
   d. Eukaryotic cells have evolved from prokaryotic cells.

2. All cells have all of the following except
   a. plasma membrane.   c. cytoplasm.
   b. genetic material.   d. cell wall.

3. Eukaryotic cells are more complex than prokaryotic cells. Which of the following are found only in a eukaryotic cell?
   a. Cell wall
   b. Plasma membrane
   c. Endoplasmic reticulum
   d. Ribosomes

4. Which of the following are differences between bacteria and archaea?
   a. The molecular architecture of their cell walls
   b. The type of ribosomes found in each
   c. Archaea have an internal membrane system that bacteria lack.
   d. Both a and b are correct.

5. The cytoskeleton includes
   a. microtubules made of actin filaments.
   b. microfilaments made of tubulin.
   c. intermediate filaments made of twisted fibers of vimentin and keratin.
   d. smooth endoplasmic reticulum.

6. The smooth endoplasmic reticulum is
   a. involved in protein synthesis.
   b. a site of protein glycosylation.
   c. used to store a variety of ions.
   d. the site of lipid and membrane synthesis.

7. Plasmodesmata in plants and gap junctions in animals are functionally similar in that
    a. each is used to anchor layers of cells.
    b. they form channels between cells that allow diffusion of small molecules.
    c. they form tight junctions between cells.
    d. they are anchored to the extracellular matrix.

APPLY

1. The most important factor that limits the size of a cell is the
    a. quantity of proteins and organelles a cell can make.
    b. rate of diffusion of small molecules.
    c. surface area-to-volume ratio of the cell.
    d. amount of DNA in the cell.

2. All eukaryotic cells possess each of the following except
    a. mitochondria.   c. cytoskeleton.
    b. cell wall.   d. nucleus.

3. Adherens junctions, which contain cadherin, are found in all animals. Given this, which of the following predictions is most likely?
    a. Cadherins would not be found in the ancestor to all animals.
    b. Cadherins would be found in prokaryotes.
    c. Cadherins would be found in the ancestor to all animals.
    d. Cadherins would be found in vertebrates but not invertebrates.

4. Different motor proteins like kinesin and myosin are similar in that they can
    a. interact with microtubules.
    b. use energy from ATP to produce movement.
    c. interact with actin.
    d. do both a and b.

5. The protein sorting pathway involves the following organelles/compartments in order:
    a. SER, RER, transport vesicle, Golgi.
    b. RER, lysosome, Golgi.
    c. RER, transport vesicle, Golgi, final destination.
    d. Golgi, transport vesicle, RER, final destination.

6. Chloroplasts and mitochondria have many common features because both
    a. are present in plant cells.
    b. arose by endosymbiosis.
    c. function to oxidize glucose.
    d. function to produce glucose.

7. Eukaryotic cells are composed of three types of cytoskeletal filaments. How are these three filaments similar?
    a. They contribute to the shape of the cell.
    b. They are all made of the same type of protein.
    c. They are all the same size and shape.
    d. They are all equally dynamic and flexible.

SYNTHESIZE

1. The smooth endoplasmic reticulum is the site of synthesis of the phospholipids that make up all the membranes of a cell—especially the plasma membrane. Use the diagram of an animal cell (figure 4.6) to trace a pathway that would carry a phospholipid molecule from the SER to the plasma membrane. What endomembrane compartments would the phospholipids travel through? How can a phospholipid molecule move between membrane compartments?

2. Use the information provided in table 4.3 to develop a set of predictions about the properties of mitochondria and chloroplasts if these organelles were once free-living prokaryotic cells. How do your predictions match with the evidence for endosymbiosis?

3. In evolutionary theory, homologous traits are those with a similar structure and function derived from a common ancestor. Analogous traits represent adaptations to a similar environment, but from distantly related organisms. Consider the structure and function of the flagella found on eukaryotic and prokaryotic cells. Are the flagella an example of a homologous or analogous trait? Defend your answer.

4. The protist, *Giardia intestinalis,* is the organism associated with water-borne diarrheal diseases. *Giardia* is an unusual eukaryote because it seems to lack mitochondria. Provide two possible evolutionary scenarios for this in the context of the endosymbiotic theory.

ONLINE RESOURCE

www.ravenbiology.com

Understand, Apply, and Synthesize—enhance your study with animations that bring concepts to life and practice tests to assess your understanding. Your instructor may also recommend the interactive eBook, individualized learning tools, and more.

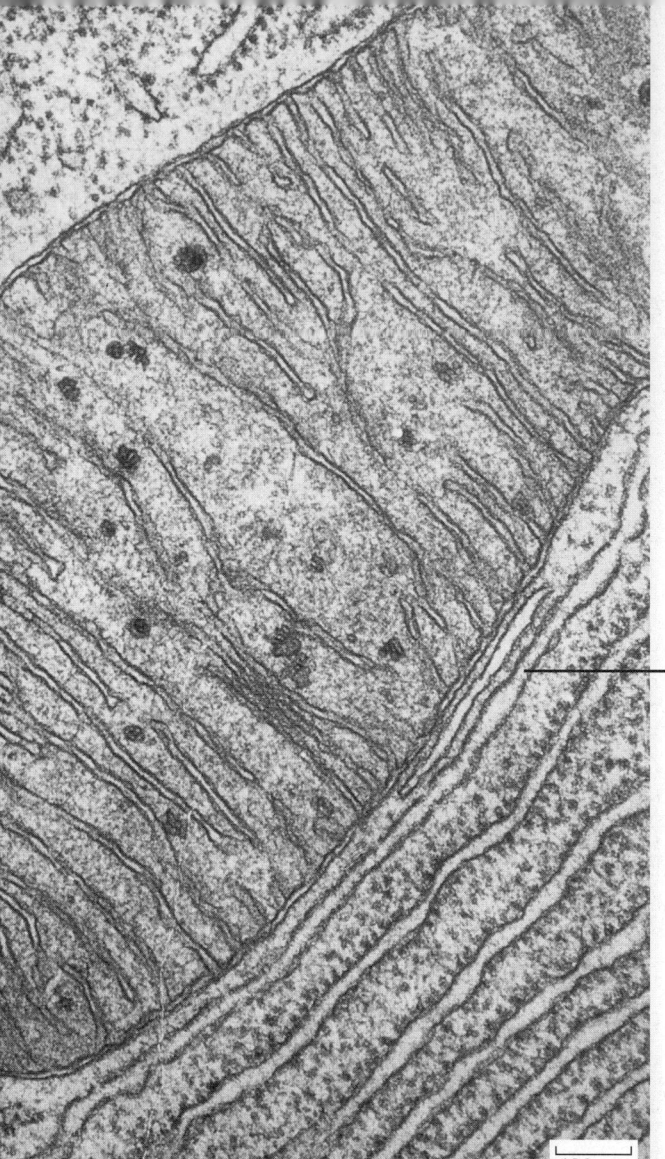

100 nm

Chapter 5

Membranes

Chapter Contents

Introduction

A cell's interactions with the environment are critical, a give-and-take that never ceases. Without it, life could not exist. Living cells are encased within a lipid membrane through which few water-soluble substances can pass. The membrane also contains protein passageways that permit specific substances to move into and out of the cell and allow the cell to exchange information with its environment. Eukaryotic cells also contain internal membranes like those of the mitochondrion and endoplasmic reticulum pictured here. We call the delicate skin of lipids with embedded protein molecules that encase the cell a plasma membrane. *This chapter examines the structure and function of this remarkable membrane.*

5.1 The Structure of Membranes

Learning Outcomes

1. *Describe the components of biological membranes.*
2. *Explain the fluid mosaic model of membrane structure.*

The membranes that encase all living cells are two phospholipid sheets that are only 5–10 nm thick; more than 10,000 of these sheets piled on one another would just equal the thickness of this sheet of paper. Biologists established the components of membranes—not only lipids, but also proteins and other molecules—through biochemical assays, but the organization of the membrane components remained elusive.

We begin by considering the theories that have been advanced about membrane structure. We then look at the individual components of membranes more closely.

The fluid mosaic model shows proteins embedded in a fluid lipid bilayer

The lipid layer that forms the foundation of a cell's membranes is a bilayer formed of **phospholipids.** These phospholipids include primarily the glycerol phospholipids (figure 5.1), and the sphingolipids such as sphingomyelin (figure 5.2). Note that although these look superficially similar, they are built on a different carbon skeleton. For many years, biologists thought that the protein components of the cell membrane covered the inner and outer surfaces of the phospholipid bilayer like a coat of paint. An early model portrayed the membrane as a sandwich; a phospholipid bilayer between two layers of globular protein.

In 1972, S. Jonathan Singer and Garth J. Nicolson revised the model in a simple but profound way: They proposed that the globular proteins are *inserted* into the lipid bilayer, with their nonpolar segments in contact with the nonpolar interior of the bilayer and their polar portions protruding out from the membrane surface. In this model, called the *fluid mosaic model,* a mosaic of proteins floats in or on the fluid lipid bilayer like boats on a pond (figure 5.3).

We now recognize two categories of membrane proteins based on their association with the membrane. *Integral membrane proteins* are embedded in the membrane, and *peripheral proteins* are associated with the surface of the membrane.

Cellular membranes consist of four component groups

A eukaryotic cell contains many membranes. Although they are not all identical, they share the same fundamental architecture. Cell membranes are assembled from four components (table 5.1):

1. **Phospholipid bilayer.** Every cell membrane is composed of phospholipids in a bilayer. The other components of the membrane are embedded within the bilayer, which provides a flexible matrix and, at the same time, imposes a barrier to permeability. Animal cell membranes also contain cholesterol, a steroid with a polar hydroxyl group (—OH). Plant cells have other sterols, but little or no cholesterol.

2. **Transmembrane proteins.** A major component of every membrane is a collection of proteins that float in the lipid bilayer. These proteins have a variety of functions, including transport and communication across the membrane. Many integral membrane proteins are not fixed in position. They can move about, just as the phospholipid molecules do. Some membranes are crowded with proteins, but in others, the proteins are more sparsely distributed.

3. **Interior protein network.** Membranes are structurally supported by intracellular proteins that reinforce the membrane's shape. For example, a red blood cell has a characteristic biconcave shape because a scaffold made

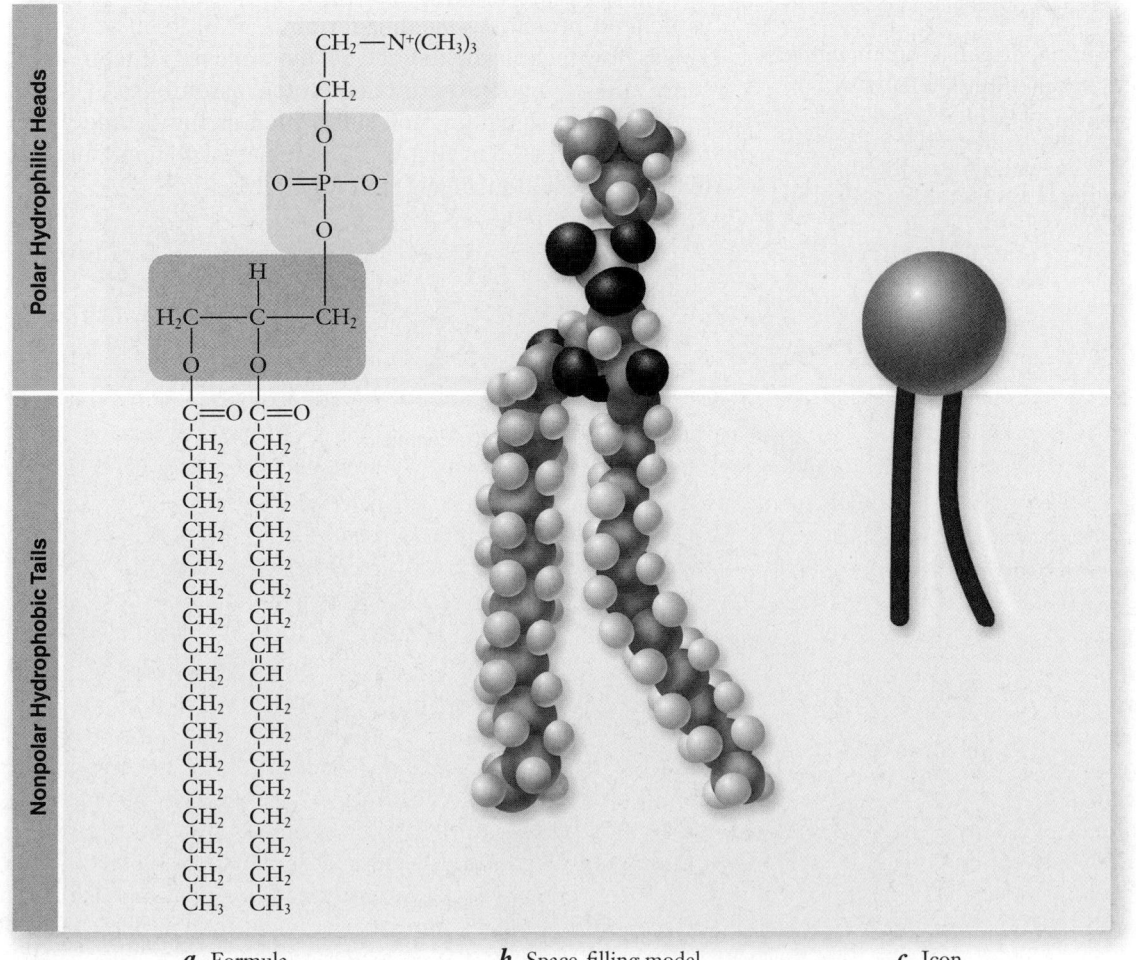

a. Formula   *b.* Space-filling model   *c.* Icon

Figure 5.1 Different views of phospholipid structure. Phospholipids are composed of glycerol *(pink)* linked to two fatty acids and a phosphate group. The phosphate group *(yellow)* can have additional molecules attached, such as the positively charged choline *(green)* shown. Phosphatidylcholine is a common component of membranes. It is shown in *(a)* with its chemical formula, *(b)* as a space-filling model, and *(c)* as the icon that is used in most of the figures in this chapter.

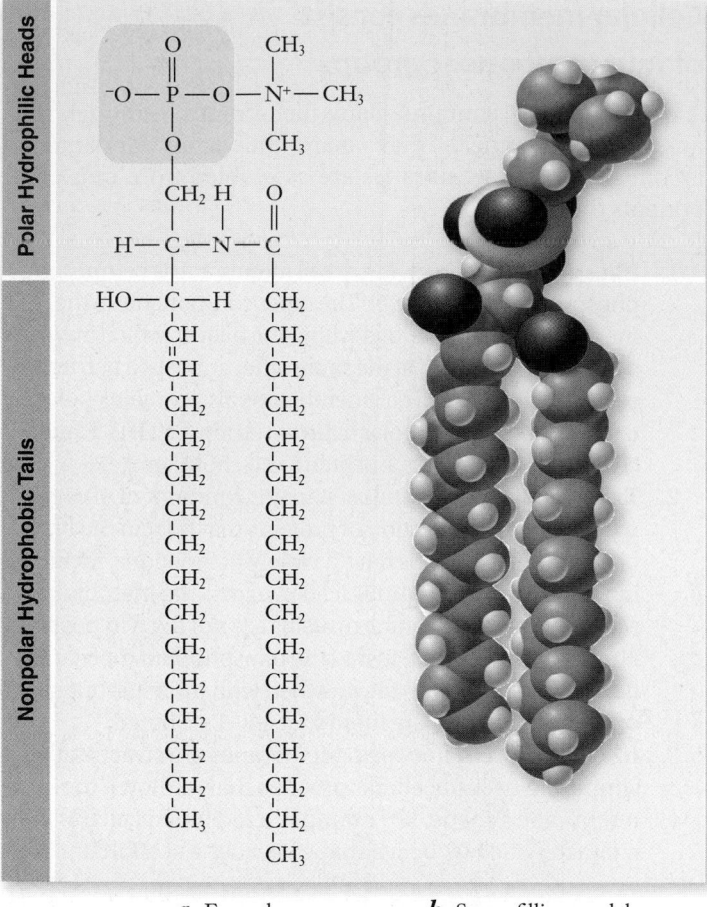

Polar Hydrophilic Heads

Nonpolar Hydrophobic Tails

a. Formula   *b.* Space-filling model

Figure 5.2 Sphingomyelin. Sphingomyelin is a sphingolipid found in animal cells. *a.* Formula. *b.* Space-filling model.

of a protein called spectrin links proteins in the plasma membrane with actin filaments in the cell's cytoskeleton.

Membranes use networks of other proteins to control the lateral movements of some key membrane proteins, anchoring them to specific sites.

4. **Cell-surface markers.** As you learned in the preceding chapter, membrane sections assemble in the endoplasmic reticulum, transfer to the Golgi apparatus, and then are transported to the plasma membrane. The ER adds chains of sugar molecules to membrane proteins and lipids, converting them into **glycoproteins** and **glycolipids.** Different cell types exhibit different varieties of these glycoproteins and glycolipids on their surfaces, which act as cell identity markers.

Cellular membranes have an organized substructure

Originally, it was believed that because of its fluidity, the plasma membrane was uniform, with lipids and proteins free to diffuse rapidly in the plane of the membrane. However, in the last decade evidence has accumulated suggesting the plasma membrane is not homogeneous and contains microdomains with distinct lipid and protein composition. This was first observed in epithelial cells in which the lipid composition of the apical and basal membranes was shown to be distinctly different. Theoretical work also showed that lipids can exist in either a disordered or an ordered phase within a bilayer.

This led to the idea of lipid microdomains called *lipid rafts* that are heavily enriched in cholesterol and sphingolipids. These lipids appear to interact with each other, and with raft-associated proteins—together forming an ordered structure. This is now technically defined as "dynamic nanometer-sized, sterol and sphingolipid-enriched protein assemblies." There is evidence that signaling molecules, such as the B- and T-cell receptors discussed in chapter 51, associate with lipid rafts and that this association affects their function.

Figure 5.3 The fluid mosaic model of cell membranes.
Integral proteins protrude through the plasma membrane, with nonpolar regions that tether them to the membrane's hydrophobic interior. Carbohydrate chains are often bound to the extracellular portion of these proteins, forming glycoproteins. Peripheral membrane proteins are associated with the surface of the membrane. Membrane phospholipids can be modified by the addition of carbohydrates to form glycolipids. Inside the cell, actin filaments and intermediate filaments interact with membrane proteins. Outside the cell, many animal cells have an elaborate extracellular matrix composed primarily of glycoproteins.

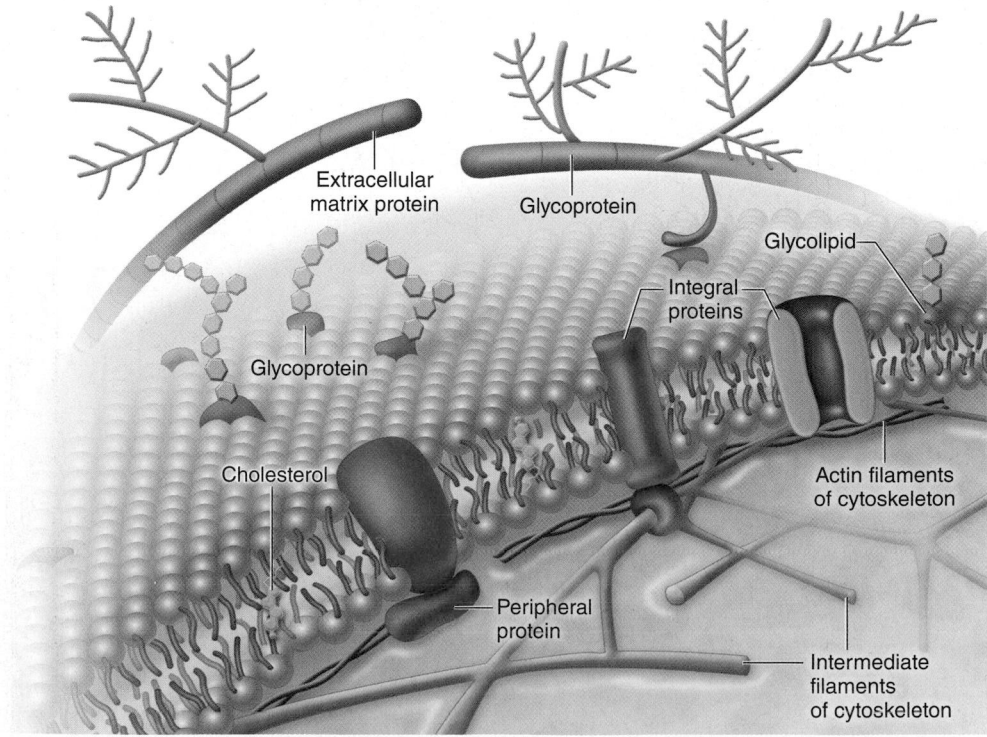

Extracellular matrix protein

Glycoprotein

Glycolipid

Integral proteins

Glycoprotein

Cholesterol

Actin filaments of cytoskeleton

Peripheral protein

Intermediate filaments of cytoskeleton

TABLE 5.1	Components of the Cell Membrane			
Component	**Composition**	**Function**	**How It Works**	**Example**
Phospholipid bilayer	Phospholipid molecules	Provides permeability barrier, matrix for proteins	Excludes water-soluble molecules from nonpolar interior of bilayer and cell	Bilayer of cell is impermeable to large water-soluble molecules, such as glucose
Transmembrane proteins	Carriers	Actively or passively transport molecules across membrane	Move specific molecules through the membrane in a series of conformational changes	Glycophorin carrier for sugar transport; sodium–potassium pump
	Channels	Passively transport molecules across membrane	Create a selective tunnel that acts as a passage through membrane	Sodium and potassium channels in nerve, heart, and muscle cells
	Receptors	Transmit information into cell	Signal molecules bind to cell-surface portion of the receptor protein. This alters the portion of the receptor protein within the cell, inducing activity	Specific receptors bind peptide hormones and neurotransmitters
Interior protein network	Spectrins	Determine shape of cell	Form supporting scaffold beneath membrane, anchored to both membrane and cytoskeleton	Red blood cell
	Clathrins	Anchor certain proteins to specific sites, especially on the exterior plasma membrane in receptor-mediated endocytosis	Proteins line coated pits and facilitate binding to specific molecules	Localization of low-density lipoprotein receptor within coated pits
Cell-surface markers	Glycoproteins	"Self" recognition	Create a protein/carbohydrate chain shape characteristic of individual	Major histocompatibility complex protein recognized by immune system
	Glycolipid	Tissue recognition	Create a lipid/carbohydrate chain shape characteristic of tissue	A, B, O blood group markers

In addition to these horizontal structures there is also vertical structure to the plasma membrane. That is, the distribution of membrane lipids in the plasma membrane is asymmetrical, with the outer leaflet enriched in the glycerol phospholipid phosphatidylcholine and in sphingolipids. This is despite being symmetrically distributed in the ER where they are synthesized. Some of this sorting occurs in the Golgi and is also affected by enzymes that transport lipids across the bilayer from one face to the other.

Electron microscopy has provided structural evidence

Electron microscopy allows biologists to examine the delicate, filmy structure of a cell membrane. We discussed two types of electron microscopes in chapter 4: the transmission electron microscope (TEM) and the scanning electron microscope (SEM). Both provide illuminating views of membrane structure.

When examining cell membranes with electron microscopy, specimens must be prepared for viewing. In one method of preparing a specimen, the tissue of choice is embedded in a hard epoxy matrix. The epoxy block is then cut with a microtome, a machine with a very sharp blade that makes incredibly thin, transparent "epoxy shavings" less than 1 μm thick that peel away from the block of tissue.

These shavings are placed on a grid, and a beam of electrons is directed through the grid with the TEM. At the high magnification an electron microscope provides, resolution is good enough to reveal the double layers of a membrane. False color can be added to the micrograph to enhance detail.

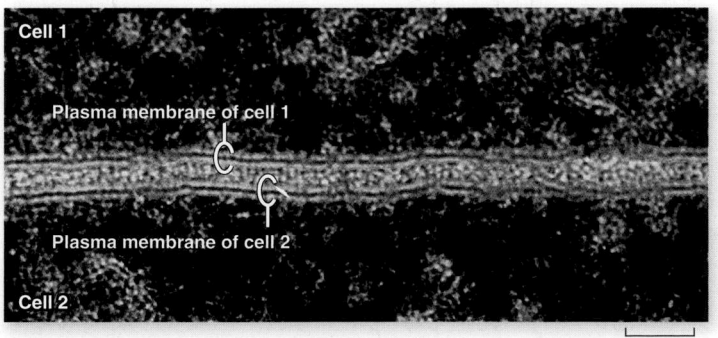

25 nm

Freeze-fracturing a specimen is another way to visualize the inside of the membrane (figure 5.4). The tissue is embedded in a medium and quick frozen with liquid nitrogen. The frozen tissue is then "tapped" with a knife, causing a crack between the phospholipid layers of membranes. Proteins, carbohydrates, pits, pores, channels, or any other structure affiliated with the membrane will pull apart (whole, usually) and stick with one or the other side of the split membrane.

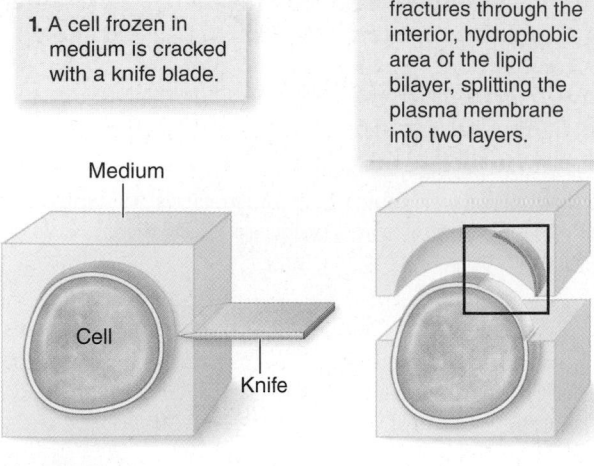

1. A cell frozen in medium is cracked with a knife blade.

Medium

Cell

Knife

2. The cell often fractures through the interior, hydrophobic area of the lipid bilayer, splitting the plasma membrane into two layers.

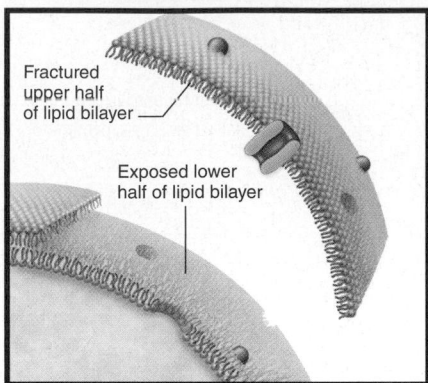

3. The plasma membrane separates such that proteins and other embedded membrane structures remain within one or the other layers of the membrane.

Fractured upper half of lipid bilayer

Exposed lower half of lipid bilayer

4. The exposed membrane is coated with platinum, which forms a replica of the membrane. The underlying membrane is dissolved away, and the replica is then viewed with electron microscopy.

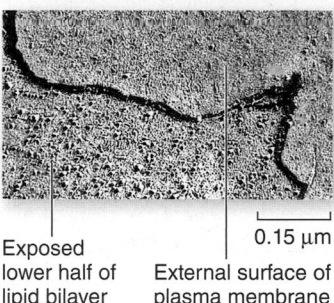

0.15 μm

Exposed lower half of lipid bilayer

External surface of plasma membrane

Figure 5.4 Viewing a plasma membrane with freeze-fracture microscopy.

Next, a very thin coating of platinum is evaporated onto the fractured surface, forming a replica or "cast" of the surface. After the topography of the membrane has been preserved in the cast, the actual tissue is dissolved away, and the cast is examined with electron microscopy, creating a textured and three-dimensional view of the membrane.

Learning Outcomes Review 5.1

Cellular membranes contain four components: (1) a phospholipid bilayer, (2) transmembrane proteins, (3) an internal protein network providing structural support, and (4) cell-surface markers composed of glycoproteins and glycolipids. The fluid mosaic model of membrane structure includes both the fluid nature of the membrane and the mosaic composition of proteins floating in the phospholipid bilayer. Transmission electron microscopy (TEM) and scanning electron microscopy (SEM) have provided evidence supporting the fluid mosaic model.

■ *If the plasma membrane were just a phospholipid bilayer, how would this affect its function?*

5.2 *Phospholipids: The Membrane's Foundation*

Learning Outcomes

1. *List the different components of phospholipids.*
2. *Explain how membranes form spontaneously.*
3. *Describe the factors involved in membrane fluidity.*

Like the fat molecules (triglycerides) described in chapter 3, glycerol phospholipids have a backbone derived from the three-carbon polyalcohol *glycerol.* Attached to this backbone are one to three fatty acids, long chains of carbon atoms ending in a car-boxyl (—COOH) group. A triglyceride molecule has three such chains, one attached to each carbon in the backbone. Because

these chains are nonpolar, they do not form hydrogen bonds with water, and triglycerides are not water-soluble.

A phospholipid, by contrast, has only two fatty acid chains attached to its backbone. The third carbon of the glycerol carries a phosphate group, thus the name *phospho*lipid. An additional polar organic molecule is often added to the phosphate group as well. By varying the polar organic group, and the fatty acid chains, a large variety of lipids can be constructed on this simple molecular framework.

In addition to the glycerol phospholipids, eukaryotic cell membranes have sphingolipids, which usually have saturated fatty acid chains that may aid in organizing the membrane into lipid rafts and other microstructures.

Phospholipids spontaneously form bilayers

The phosphate groups are charged, and other molecules attached to them are polar or charged. This creates a huge change in the molecule's physical properties compared with a triglyceride. The strongly polar phosphate end is hydrophilic, or "water-loving," while the fatty acid end is strongly nonpolar and hydrophobic, or "water-fearing." The two nonpolar fatty acids extend in one direction, roughly parallel to each other, and the polar phosphate group points in the other direction. To represent this structure, phospholipids are often diagrammed as a polar head with two dangling nonpolar tails, as in figure 5.1*c.*

What happens when a collection of phospholipid molecules is placed in water? The polar water molecules repel the long, nonpolar tails of the phospholipids while seeking partners for hydrogen bonding. Because of the polar nature of the water molecules, the nonpolar tails of the phospholipids end up packed closely together, sequestered as far as possible from water. Every phospholipid molecule is oriented with its polar head toward water and its nonpolar tails away. When *two* layers form with the tails facing each other, no tails ever come in contact with water. The resulting structure is the phospholipid bilayer. Phospholipid bilayers form spontaneously, driven by the tendency of water molecules to form the maximum number of hydrogen bonds.

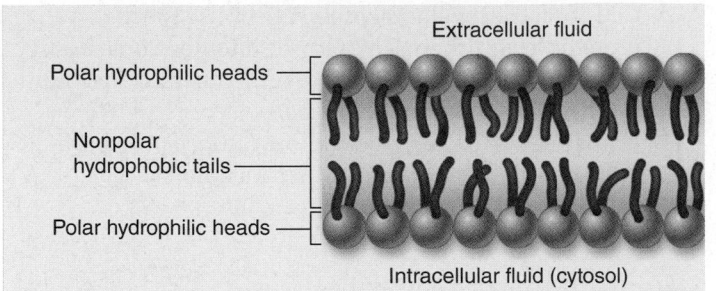

Extracellular fluid

Polar hydrophilic heads

Nonpolar hydrophobic tails

Polar hydrophilic heads

Intracellular fluid (cytosol)

The nonpolar interior of a lipid bilayer impedes the passage of any water-soluble substances through the bilayer, just as a layer of oil impedes the passage of a drop of water. This barrier to water-soluble substances is the key biological property of the lipid bilayer.

The phospholipid bilayer is fluid

A lipid bilayer is stable because water's affinity for hydrogen bonding never stops. Just as surface tension holds a soap bubble together, even though it is made of a liquid, so the hydrogen bonding of water holds a membrane together. Although water drives phospholipids into a bilayer configuration, it does not have any effect on the mobility of phospholipids and their nonlipid neighbors in the bilayer. Because phospholipids interact relatively weakly with one another, individual phospholipids

SCIENTIFIC THINKING

Hypothesis: *The plasma membrane is fluid, not rigid.*

Prediction: *If the membrane is fluid, membrane proteins may diffuse laterally.*

Test: *Fuse mouse and human cells, then observe the distribution of membrane proteins over time by labeling specific mouse and human proteins.*

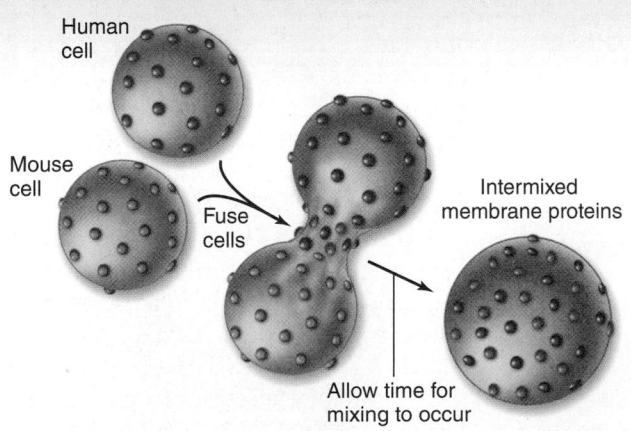

Human cell

Mouse cell

Fuse cells

Intermixed membrane proteins

Allow time for mixing to occur

Result: *Over time, hybrid cells show increasingly intermixed proteins.*

Conclusion: *At least some membrane proteins can diffuse laterally in the membrane.*

Further Experiments: *Can you think of any other explanation for these observations? What if newly synthesized proteins were inserted into the membrane during the experiment? How could you use this basic experimental design to rule out this or other possible explanations?*

Figure 5.5 Test of membrane fluidity.

and unanchored proteins are comparatively free to move about within the membrane. This can be demonstrated vividly by fusing cells and watching their proteins intermix with time (figure 5.5).

Membrane fluidity can change

The degree of membrane fluidity changes with the composition of the membrane itself. Much like triglycerides can be solid or liquid at room temperature, depending on their fatty acid composition, membrane fluidity can be altered by changing the membrane's fatty acid composition.

Saturated fats tend to make the membrane less fluid because they pack together well. Unsaturated fats make the membrane more fluid—the "kinks" introduced by the double bonds keep them from packing tightly. You saw this effect on fats and oils earlier in chapter 3.

In animal cells cholesterol may make up as much as 50% of membrane lipids in the outer leaflet. The cholesterol can fill gaps left by unsaturated fatty acids. This has the effect of decreasing membrane fluidity, but it increases the strength of the membrane. Overall this leads to a membrane with intermediate fluidity that is more durable and also less permeable.

Changes in the environment can have drastic effects on the membranes of single-celled organisms such as bacteria. Increasing temperature makes a membrane more fluid, and decreasing temperature makes it less fluid. Bacteria have evolved mechanisms to maintain a constant membrane fluidity despite fluctuating temperatures. Some bacteria contain enzymes called *fatty acid desaturases* that can introduce double bonds into fatty acids in membranes. Genetic studies, involving either the inactivation of these enzymes or the introduction of them into cells that normally lack them, indicate that the action of these enzymes confers cold tolerance. At colder temperatures, the double bonds introduced by fatty acid desaturase make the membrane more fluid, counteracting the environmental effect of reduced temperature.

Learning Outcomes Review 5.2

Biological membranes consist of a phospholipid bilayer. Each phospholipid has a hydrophilic (phosphate) head and a hydrophobic (lipid) tail. In water, phospholipid molecules spontaneously form a bilayer, with phosphate groups facing out toward the water and lipid tails facing in, where they are sequestered from water. Membrane fluidity varies with composition and conditions: unsaturated fats disturb packing of the lipid tails and make the membrane more fluid, as do higher temperatures.

■ *Would a phospholipid bilayer form in a nonpolar solvent?*

5.3 Proteins: Multifunctional Components

Learning Outcomes

1. *Illustrate the functions of membrane proteins.*
2. *Illustrate how proteins can associate with the membrane.*
3. *Identify a transmembrane domain.*

Cell membranes contain a complex assembly of proteins enmeshed in the fluid soup of phospholipid molecules. This very flexible organization permits a broad range of interactions with the environment, some directly involving membrane proteins.

Proteins and protein complexes perform key functions

Although cells interact with their environment through their plasma membranes in many ways, we will focus on six key classes of membrane protein in this chapter and in chapter 9 (figure 5.6).

1. **Transporters.** Membranes are very selective, allowing only certain solutes to enter or leave the cell, either through channels or carriers composed of proteins.
2. **Enzymes.** Cells carry out many chemical reactions on the interior surface of the plasma membrane, using enzymes attached to the membrane.
3. **Cell-surface receptors.** Membranes are exquisitely sensitive to chemical messages, which are detected by receptor proteins on their surfaces.
4. **Cell-surface identity markers.** Membranes carry cell-surface markers that identify them to other cells. Most cell types carry their own ID tags, specific combinations of cell-surface proteins and protein complexes such as glycoproteins that are characteristic of that cell type.

5. **Cell-to-cell adhesion proteins.** Cells use specific proteins to glue themselves to one another. Some act by forming temporary interactions, and others form a more permanent bond. (See chapter 4.)
6. **Attachments to the cytoskeleton.** Surface proteins that interact with other cells are often anchored to the cytoskeleton by linking proteins.

Structural features of membrane proteins relate to function

As we've just detailed, membrane proteins can serve a variety of functions. These diverse functions arise from the diverse structures of these proteins, yet they also have common structural features related to their role as membrane proteins.

The anchoring of proteins in the bilayer

Some membrane proteins are attached to the surface of the membrane by special molecules that associate strongly with phospholipids. Like a ship tied to a floating dock, these anchored proteins are free to move about on the surface of the membrane tethered to a phospholipid. The anchoring molecules are modified lipids that have (1) nonpolar regions that insert into the internal portion of the lipid bilayer and (2) chemical bonding domains that link directly to proteins.

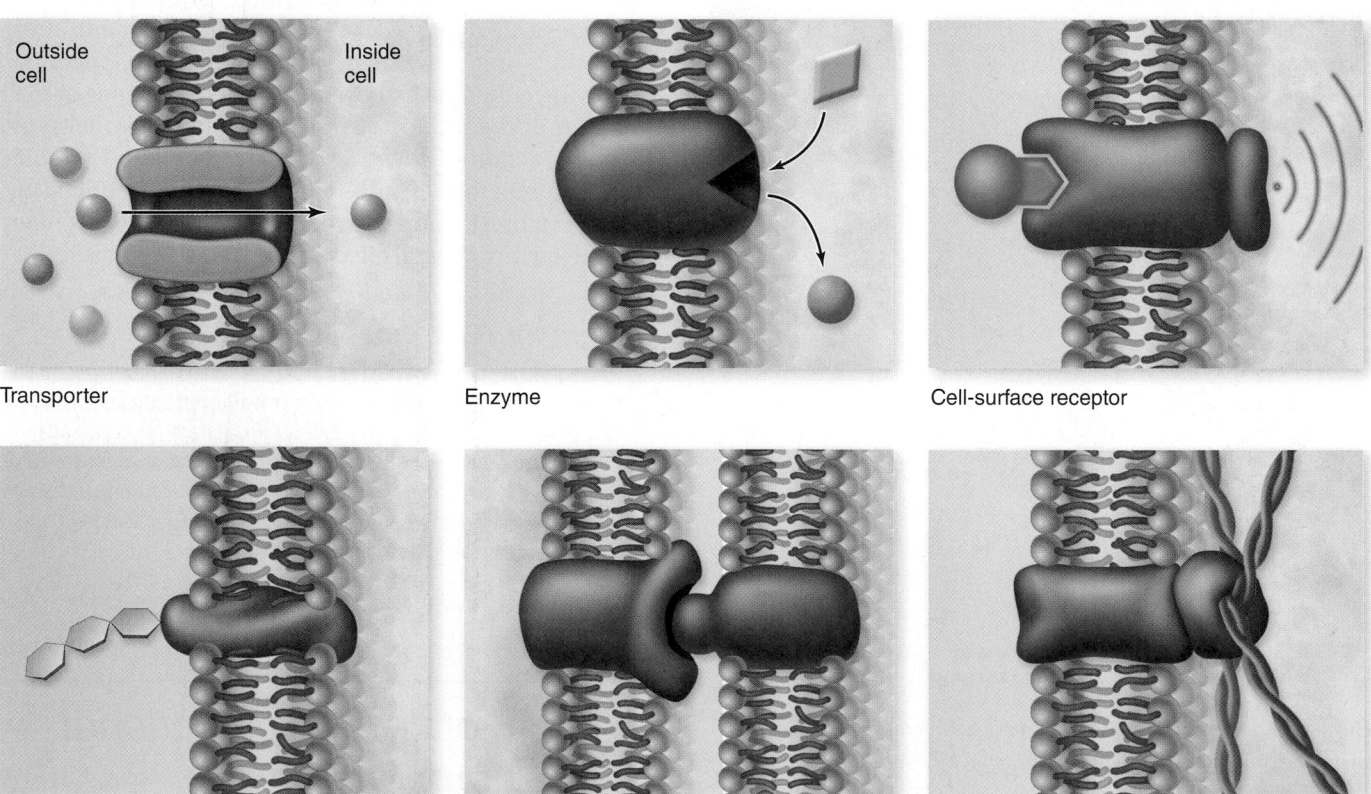

Figure 5.6 Functions of plasma membrane proteins. Membrane proteins act as transporters, enzymes, cell-surface receptors, and cell-surface identity markers, as well as aiding in cell-to-cell adhesion and securing the cytoskeleton.

? Inquiry question According to the fluid mosaic model, membranes are held together by hydrophobic interactions. Considering the forces that some cells may experience, why do membranes not break apart every time an animal moves?

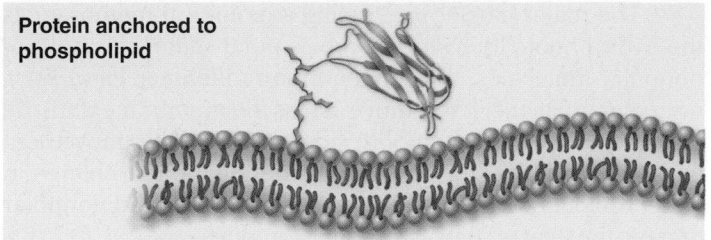

Protein anchored to phospholipid

In contrast, other proteins actually span the lipid bilayer (transmembrane proteins). The part of the protein that extends through the lipid bilayer and that is in contact with the nonpolar interior are α helices or β-pleated sheets (see chapter 3) that consist of nonpolar amino acids. Because water avoids nonpolar amino acids, these portions of the protein are held within the interior of the lipid bilayer. The polar ends protrude from both sides of the membrane. Any movement of the protein out of the membrane, in either direction, brings the nonpolar regions of the protein into contact with water, which "shoves" the protein back into the interior. These forces prevent the transmembrane proteins from simply popping out of the membrane and floating away.

Transmembrane domains

Cell membranes contain a variety of different transmembrane proteins, which differ in the way they traverse the lipid bilayer. The primary difference lies in the number of times that the protein crosses the membrane. Each membrane-spanning region is called a **transmembrane domain.** These domains are composed of hydrophobic amino acids usually arranged into α helices (figure 5.7).

Proteins need only a single transmembrane domain to be anchored in the membrane, but they often have more than one such domain. An example of a protein with a single transmembrane domain is the linking protein that attaches the spectrin network of the cytoskeleton to the interior of the plasma membrane.

Biologists classify some types of receptors based on the number of transmembrane domains they have, such as G protein–coupled receptors with seven membrane-spanning

domains (chapter 9). These receptors respond to external molecules, such as epinephrine, and initiate a cascade of events inside the cell.

Another example is bacteriorhodopsin, one of the key transmembrane proteins that carries out photosynthesis in halophilic (salt-loving) archaea. It contains seven nonpolar helical segments that traverse the membrane, forming a structure within the membrane through which protons pass during the light-driven pumping of protons.

Pores

Some transmembrane proteins have extensive nonpolar regions with secondary configurations of β-pleated sheets instead of α helices (chapter 3). The β sheets form a characteristic motif, folding back and forth in a cylinder so the sheets arrange themselves like a pipe through the membrane. This forms a polar environment in the interior of the β sheets spanning the membrane. This so-called *β barrel,* open on both ends, is a common feature of the porin class of proteins that are found within the outer membrane of some bacteria. The openings allow molecules to pass through the membrane (figure 5.8).

Learning Outcomes Review 5.3

Proteins in the membrane confer the main differences between membranes of different cells. Their functions include transport, enzymatic action, reception of extracellular signals, cell-to-cell interactions, and cell identity markers. Peripheral proteins can be anchored in the membrane by modified lipids. Integral membrane proteins span the membrane and have one or more hydrophobic regions, called transmembrane domains, that anchor them.

■ *Why are transmembrane domains hydrophobic?*

? Inquiry question Based only on amino acid sequence, how would you recognize an integral membrane protein?

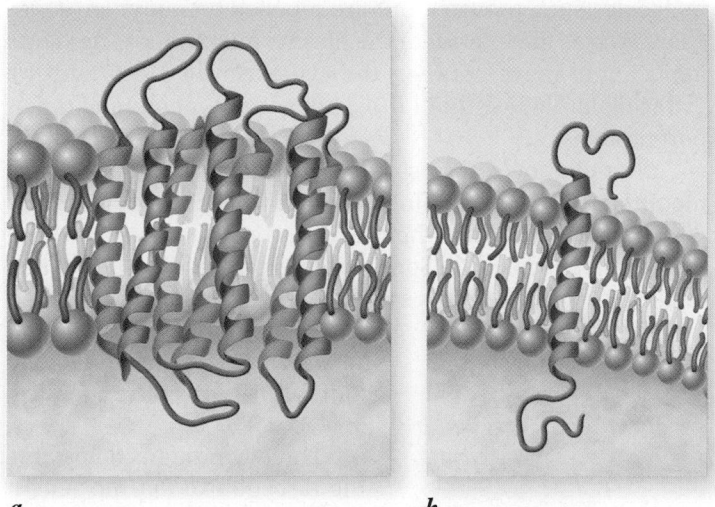

a.   *b.*

Figure 5.7 Transmembrane domains. Integral membrane proteins have at least one hydrophobic transmembrane domain (shown in blue) to anchor them in the membrane. *a.* Receptor protein with seven transmembrane domains. *b.* Protein with single transmembrane domain.

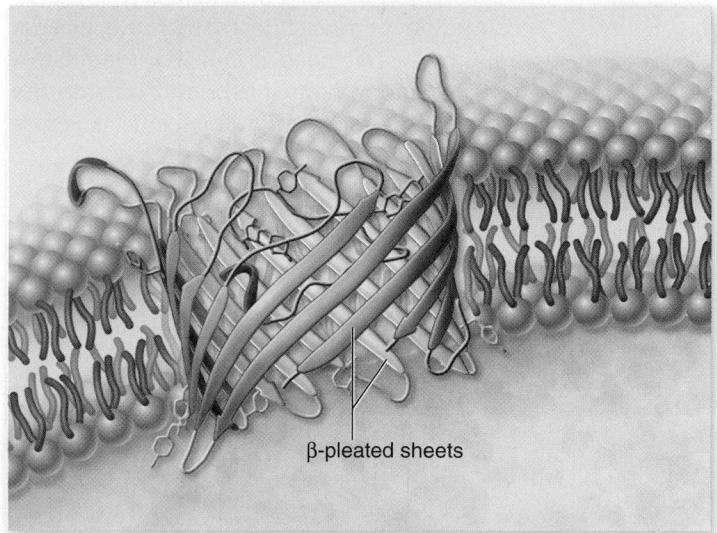

β-pleated sheets

Figure 5.8 A pore protein. The bacterial transmembrane protein porin creates large open tunnels called pores in the outer membrane of a bacterium. Sixteen strands of β-pleated sheets run antiparallel to one another, creating a so-called β barrel in the bacterial outer cell membrane. The tunnel allows water and other materials to pass through the membrane.

5.4 Passive Transport Across Membranes

Learning Outcomes

1. Compare simple diffusion and facilitated diffusion.
2. Differentiate between channel proteins and carrier proteins.
3. Predict the direction of water movement by osmosis.

Many substances can move in and out of the cell without the cell's having to expend energy. This type of movement is termed **passive transport.** Some ions and molecules can pass through the membrane fairly easily and do so because of a *concentration gradient*—a difference between the concentration on the inside of the membrane and that on the outside. Some substances also move in response to a gradient, but do so through specific channels formed by proteins in the membrane.

Transport can occur by simple diffusion

Molecules and ions dissolved in water are in constant random motion. This random motion causes a net movement of these substances from regions of high concentration to regions of lower concentration, a process called **diffusion** (figure 5.9).

Net movement driven by diffusion will continue until the concentration is the same in all regions. Consider what happens when you add a drop of colored ink to a bowl of water. Over time the ink becomes dispersed throughout the solution. This is due to diffusion of the ink molecules. In the context of cells, we are usually concerned with differences in concentration of molecules across the plasma membrane. We need to consider the relative concentrations both inside and outside the cell, as well as how readily a molecule can cross the membrane.

Figure 5.9 Diffusion. If a drop of colored ink is dropped into a beaker of water (*a*) its molecules dissolve (*b*) and diffuse (*c*). Eventually, diffusion results in an even distribution of ink molecules throughout the water (*d*).

The major barrier to crossing a biological membrane is the hydrophobic interior that repels polar molecules but not nonpolar molecules. If a concentration difference exists for a nonpolar molecule, it will move across the membrane until the concentration is equal on both sides. At this point, movement in both directions still occurs, but there is no net change in either direction. This includes molecules like O_2 and nonpolar organic molecules such as steroid hormones.

The plasma membrane has limited permeability to small polar molecules and very limited permeability to larger polar molecules and ions. The movement of water, one of the most important polar molecules, is discussed in its own section later on.

Proteins allow membrane diffusion to be selective

Many important molecules required by cells cannot easily cross the plasma membrane. These molecules can still enter the cell by diffusion through specific channel proteins or carrier proteins embedded in the plasma membrane, provided there is a higher concentration of the molecule outside the cell than inside. We call this process of diffusion mediated by a membrane protein **facilitated diffusion. Channel proteins** have a hydrophilic interior that provides an aqueous channel through which polar molecules can pass when the channel is open. **Carrier proteins,** in contrast to channels, bind specifically to the molecule they assist, much like an enzyme binds to its substrate. These channels and carriers are usually selective for one type of molecule, and thus the cell membrane is said to be **selectively permeable.**

Facilitated diffusion of ions through channels

You saw in chapter 2 that atoms with an unequal number of protons and electrons have an electric charge and are called ions. Those that carry a positive charge are called *cations* and those that carry a negative charge are called *anions.*

Because of their charge, ions interact well with polar molecules such as water, but are repelled by nonpolar molecules such as the interior of the plasma membrane. Therefore, ions cannot move between the cytoplasm of a cell and the extracellular fluid without the assistance of membrane transport proteins.

Ion channels possess a hydrated interior that spans the membrane. Ions can diffuse through the channel in either direction, depending on their relative concentration across the membrane (figure 5.10). Some channel proteins can be opened or closed in response to a stimulus. These channels are called *gated channels,* and depending on the nature of the channel, the stimulus can be either chemical or electrical.

Three conditions determine the direction of net movement of the ions: (1) their relative concentrations on either side of the membrane, (2) the voltage difference across the membrane and for the gated channels, and (3) the state of the gate (open or

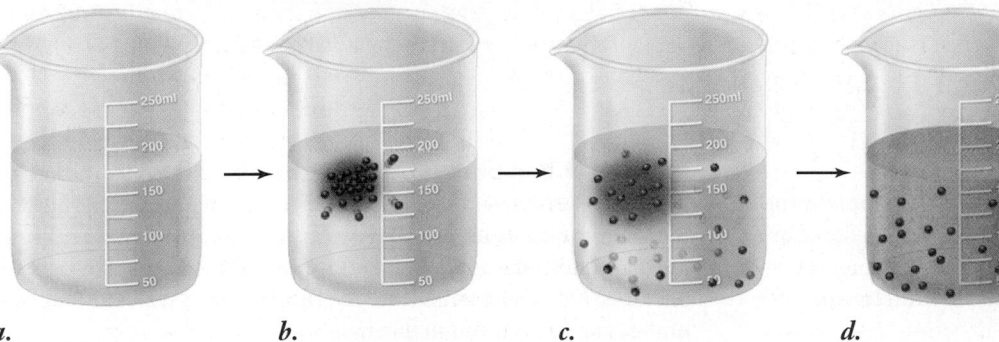

a.   *b.*   *c.*   *d.*

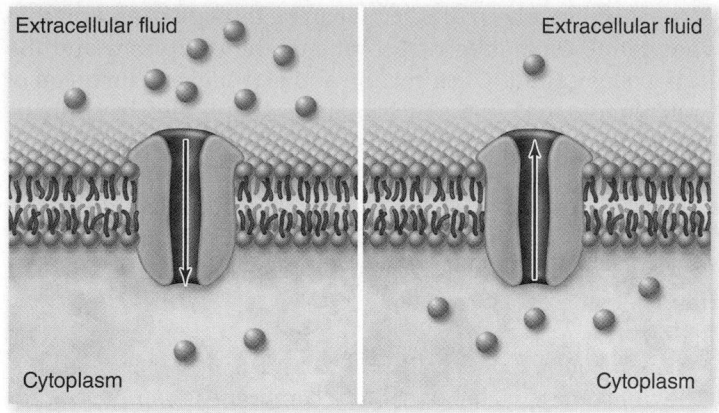

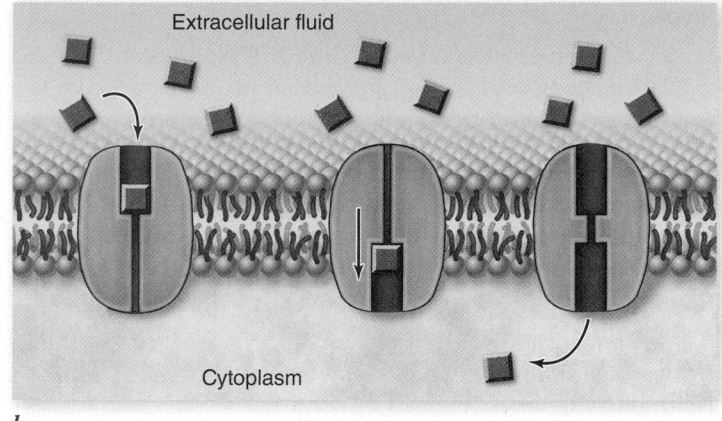

a.   *b.*

Figure 5.10 Facilitated diffusion. Diffusion can be facilitated by membrane proteins. *a.* The movement of ions through a channel is shown. On the left the concentration is higher outside the cell, so the ions move into the cell. On the right the situation is reversed. In both cases, transport continues until the concentration is equal on both sides of the membrane. At this point, ions continue to cross the membrane in both directions, but there is no net movement in either direction. *b.* Carrier proteins bind specifically to the molecules they transport. In this case, the concentration is higher outside the cell, so molecules bind to the carrier on the outside. The carrier's shape changes, allowing the molecule to cross the membrane. This is reversible, so net movement continues until the concentration is equal on both sides of the membrane.

closed). A voltage difference is an electrical potential difference across the membrane called a *membrane potential.* Changes in membrane potential form the basis for transmission of signals in the nervous system and some other tissues. (We discuss this topic in detail in chapter 43.) Each type of channel is specific for a particular ion, such as calcium (Ca^{2+}), sodium (Na^+), potassium (K^+), or chloride (Cl^-), or in some cases, for more than one cation or anion. Ion channels play an essential role in signaling by the nervous system.

Facilitated diffusion by carrier proteins

Carrier proteins can help transport both ions and other solutes, such as some sugars and amino acids, across the membrane. Transport through a carrier is still a form of diffusion and therefore requires a concentration difference across the membrane.

Carriers must bind to the molecule they transport, so the relationship between concentration and rate of transport differs from that due to simple diffusion. As concentration increases, transport by simple diffusion shows a linear increase in rate of transport. But when a carrier protein is involved, a concentration increase means that more of the carriers are bound to the transported molecule. At high enough concentrations all carriers will be occupied, and the rate of transport will be constant. This means that the carrier exhibits *saturation.*

This situation is somewhat like that of a stadium (the cell) where a crowd must pass through turnstiles to enter. If there are unoccupied turnstiles, you can go right through, but when all are occupied, you must wait. When ticket holders are passing through the gates at maximum speed, the rate at which they enter cannot increase, no matter how many are waiting outside.

Facilitated diffusion in red blood cells

Several examples of facilitated diffusion can be found in the plasma membrane of vertebrate red blood cells (RBCs). One RBC carrier protein, for example, transports a different molecule in each direction: chloride ion (Cl^-) in one direction and bicarbonate ion (HCO_3^-) in the opposite direction. As you will

learn in chapter 48, this carrier is important in the uptake and release of carbon dioxide.

The glucose transporter is a second vital facilitated diffusion carrier in RBCs. Red blood cells keep their internal concentration of glucose low through a chemical trick: They immediately add a phosphate group to any entering glucose molecule, converting it to a highly charged glucose phosphate that can no longer bind to the glucose transporter, and therefore cannot pass back across the membrane. This maintains a steep concentration gradient for unphosphorylated glucose, favoring its entry into the cell.

The glucose transporter that assists the entry of glucose into the cell does not appear to form a channel in the membrane. Instead, this transmembrane protein appears to bind to a glucose molecule and then to flip its shape, dragging the glucose through the bilayer and releasing it on the inside of the plasma membrane. After it releases the glucose, the transporter reverts to its original shape and is then available to bind the next glucose molecule that comes along outside the cell.

Osmosis is the movement of water across membranes

The cytoplasm of a cell contains ions and molecules, such as sugars and amino acids, dissolved in water. The mixture of these substances and water is called an *aqueous solution.* Water is termed the **solvent,** and the substances dissolved in the water are **solutes.** Both water and solutes tend to diffuse from regions of high concentration to ones of low concentration; that is, they diffuse down their concentration gradients.

When two regions are separated by a membrane, what happens depends on whether the solutes can pass freely through that membrane. Most solutes, including ions and sugars, are not lipid-soluble and, therefore, are unable to cross the lipid bilayer. The concentration gradient of these solutes can lead to the movement of water.

Osmosis

Water molecules interact with dissolved solutes by forming hydration shells around the charged solute molecules. When a membrane separates two solutions with different concentrations of solutes, the concentrations of *free* water molecules on the two sides of the membrane also differ. The side with higher solute concentration has tied up more water molecules in hydration shells and thus has fewer free water molecules.

As a consequence of this difference, free water molecules move down their concentration gradient, toward the higher solute concentration. This net diffusion of water across a membrane toward a higher solute concentration is called **osmosis** (figure 5.11).

The concentration of *all* solutes in a solution determines the **osmotic concentration** of the solution. If two solutions have unequal osmotic concentrations, the solution with the higher concentration is **hypertonic** (Greek *hyper*, "more than"), and the solution with the lower concentration is **hypotonic** (Greek *hypo*, "less than"). When two solutions have the same osmotic concentration, the solutions are **isotonic** (Greek *iso*, "equal"). The terms *hyperosmotic, hypoosmotic,* and *isosmotic* are also used to describe these conditions.

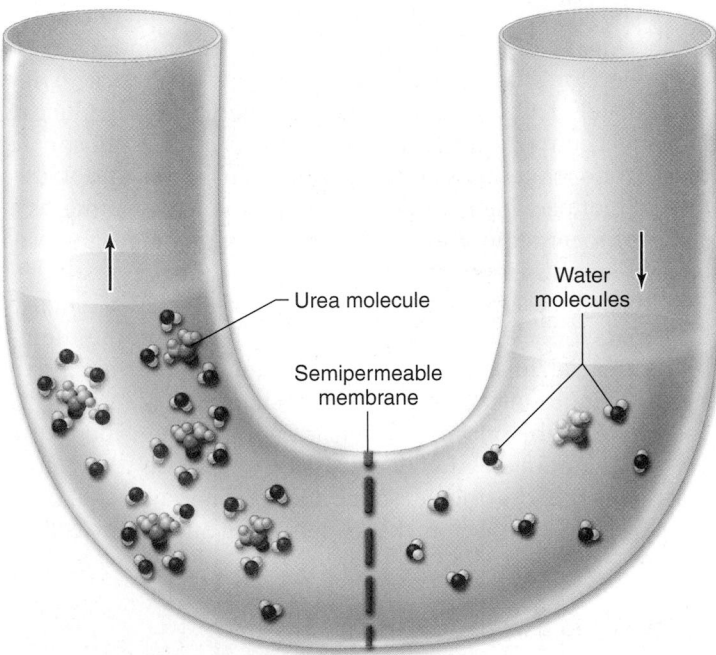

Figure 5.11 Osmosis. Concentration differences in charged or polar molecules that cannot cross a semipermeable membrane result in movement of water, which can cross the membrane. Water molecules form hydrogen bonds with charged or polar molecules creating a hydration shell around them in solution. A higher concentration of polar molecules (urea) shown on the left side of the membrane leads to water molecules gathering around each urea molecule. These water molecules are no longer free to diffuse across the membrane. The polar solute has reduced the concentration of free water molecules, creating a gradient. This causes a net movement of water by diffusion from right to left in the U-tube, raising the level on the left and lowering the level on the right.

A cell in any environment can be thought of as a plasma membrane separating two solutions: the cytoplasm and the extracellular fluid. The direction and extent of any diffusion of water across the plasma membrane is determined by comparing the osmotic strength of these solutions. Put another way, water diffuses out of a cell in a hypertonic solution (that is, the cytoplasm of the cell is hypotonic, compared with the extracellular fluid). This loss of water causes the cell to shrink until the osmotic concentrations of the cytoplasm and the extracellular fluid become equal.

Aquaporins: Water channels

The transport of water across the membrane is complex. Studies on artificial membranes show that water, despite its polarity, can cross the membrane, but this flow is limited. Water flow in living cells is facilitated by **aquaporins,** which are specialized channels for water.

A simple experiment demonstrates this. If an amphibian egg is placed in hypotonic spring water (the solute concentration in the cell is higher than that of the surrounding water), it does not swell. If aquaporin mRNA is then injected into the egg, the channel proteins are expressed and appear in the egg's plasma membrane. Water can now diffuse into the egg, causing it to swell.

More than 11 different kinds of aquaporins have been found in mammals. These fall into two general classes: those that are specific for only water, and those that allow other small hydrophilic molecules, such as glycerol or urea, to cross the membrane as well. This latter class explains how some membranes allow the easy passage of small hydrophilic substances.

The human genetic disease, hereditary (nephrogenic) diabetes insipidus (NDI), has been shown to be caused by a nonfunctional aquaporin protein. This disease causes the excretion of large volumes of dilute urine, illustrating the importance of aquaporins to our physiology.

Osmotic pressure

What happens to a cell in a hypotonic solution? (That is, the cell's cytoplasm is hypertonic relative to the extracellular fluid.) In this situation, water diffuses into the cell from the extracellular fluid, causing the cell to swell. The pressure of the cytoplasm pushing out against the cell membrane, or hydrostatic pressure, increases. The amount of water that enters the cell depends on the difference in solute concentration between the cell and the extracellular fluid. This is measured as **osmotic pressure,** defined as the force needed to stop osmotic flow.

If the membrane is strong enough, the cell reaches an equilibrium, at which the osmotic pressure, which tends to drive water into the cell, is exactly counterbalanced by the hydrostatic pressure, which tends to drive water back out of the cell. However, a plasma membrane by itself cannot withstand large internal pressures, and an isolated cell under such conditions would burst like an overinflated balloon (figure 5.12).

Accordingly, it is important for animal cells, which only have plasma membranes, to maintain osmotic balance. In contrast, the cells of prokaryotes, fungi, plants, and many protists are surrounded by strong cell walls, which can withstand high internal pressures without bursting.

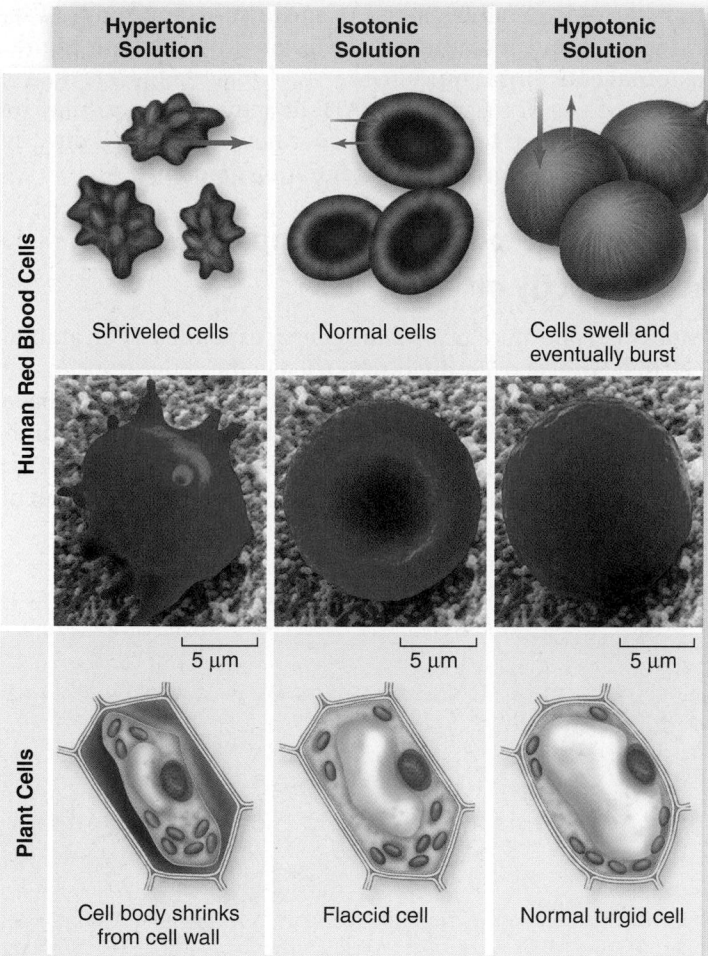

Hypertonic Solution	Isotonic Solution	Hypotonic Solution

Human Red Blood Cells

Shriveled cells | Normal cells | Cells swell and eventually burst

5 μm | 5 μm | 5 μm

Plant Cells

Cell body shrinks from cell wall | Flaccid cell | Normal turgid cell

Figure 5.12 How solutes create osmotic pressure. In a hypertonic solution, water moves out of the cell, causing the cell to shrivel. In an isotonic solution, water diffuses into and out of the cell at the same rate, with no change in cell size. In a hypotonic solution, water moves into the cell. Direction and amount of water movement is shown with blue arrows *(top)*. As water enters the cell from a hypotonic solution, pressure is applied to the plasma membrane until the cell ruptures. Water enters the cell due to osmotic pressure from the higher solute concentration in the cell. Osmotic pressure is measured as the force needed to stop osmosis. The strong cell wall of plant cells can withstand the hydrostatic pressure to keep the cell from rupturing. This is not the case with animal cells.

Maintaining osmotic balance

Organisms have developed many strategies for solving the dilemma posed by being hypertonic to their environment and therefore having a steady influx of water by osmosis.

Extrusion. Some single-celled eukaryotes, such as the protist *Paramecium,* use organelles called contractile vacuoles to remove water. Each vacuole collects water from various parts of the cytoplasm and transports it to the central part of the vacuole, near the cell surface. The vacuole possesses a small pore that opens to the outside of the cell. By contracting rhythmically, the vacuole pumps out

(extrudes) through this pore the water that is continuously drawn into the cell by osmotic forces.

Isosmotic Regulation. Some organisms that live in the ocean adjust their internal concentration of solutes to match that of the surrounding seawater. Because they are isosmotic with respect to their environment, no net flow of water occurs into or out of these cells.

Many terrestrial animals solve the problem in a similar way, by circulating a fluid through their bodies that bathes cells in an isotonic solution. The blood in your body, for example, contains a high concentration of the protein albumin, which elevates the solute concentration of the blood to match that of your cells' cytoplasm.

Turgor. Most plant cells are hypertonic to their immediate environment, containing a high concentration of solutes in their central vacuoles. The resulting internal hydrostatic pressure, known as **turgor pressure,** presses the plasma membrane firmly against the interior of the cell wall, making the cell rigid. Most green plants depend on turgor pressure to maintain their shape, and thus they wilt when they lack sufficient water.

Learning Outcomes Review 5.4

Passive transport involves diffusion, which requires a concentration gradient. Hydrophobic molecules can diffuse directly through the membrane (simple diffusion). Polar molecules and ions can also diffuse through the membrane, but only with the aid of a channel or carrier protein (facilitated diffusion). Channel proteins assist by forming a hydrophilic passageway through the membrane, whereas carrier proteins bind to the molecule they assist. Water passes through the membrane and through aquaporins in response to solute concentration differences inside and outside the cell. This process is called osmosis.

■ *If you require intravenous (IV) medication in the hospital, what should the concentration of solutes in the IV solution be relative to your blood cells?*

5.5 Active Transport Across Membranes

Learning Outcomes

1. *Differentiate between active transport and diffusion.*
2. *Describe the function of the Na⁺/K⁺ pump.*
3. *Explain the energetics of coupled transport.*

Diffusion, facilitated diffusion, and osmosis are passive transport processes that move materials down their concentration gradients, but cells can also actively move substances across a cell membrane *up* their concentration gradients. This process requires the expenditure of energy, typically from ATP, and is therefore called **active transport.**

Active transport uses energy to move materials against a concentration gradient

Like facilitated diffusion, active transport involves highly selective protein carriers within the membrane that bind to the transported substance, which could be an ion or a simple molecule, such as a sugar, an amino acid, or a nucleotide. These carrier proteins are called **uniporters** if they transport a single type of molecule and symporters or antiporters if they transport two different molecules together. **Symporters** transport two molecules in the same direction, and **antiporters** transport two molecules in opposite directions. These terms can also be used to describe facilitated diffusion carriers.

Active transport is one of the most important functions of any cell. It enables a cell to take up additional molecules of a substance that is already present in its cytoplasm in concentrations higher than in the extracellular fluid. Active transport also enables a cell to move substances out of its cytoplasm and into the extracellular fluid, despite higher external concentrations.

The use of energy from ATP in active transport may be direct or indirect. Let's first consider how ATP is used directly to move ions against their concentration gradients.

The sodium–potassium pump runs directly on ATP

More than one-third of all of the energy expended by an animal cell that is not actively dividing is used in the active transport of sodium (Na^+) and potassium (K^+) ions. Most animal cells have a low internal concentration of Na^+, relative to their surroundings, and a high internal concentration of K^+. They maintain these concentration differences by actively pumping Na^+ out of the cell and K^+ in.

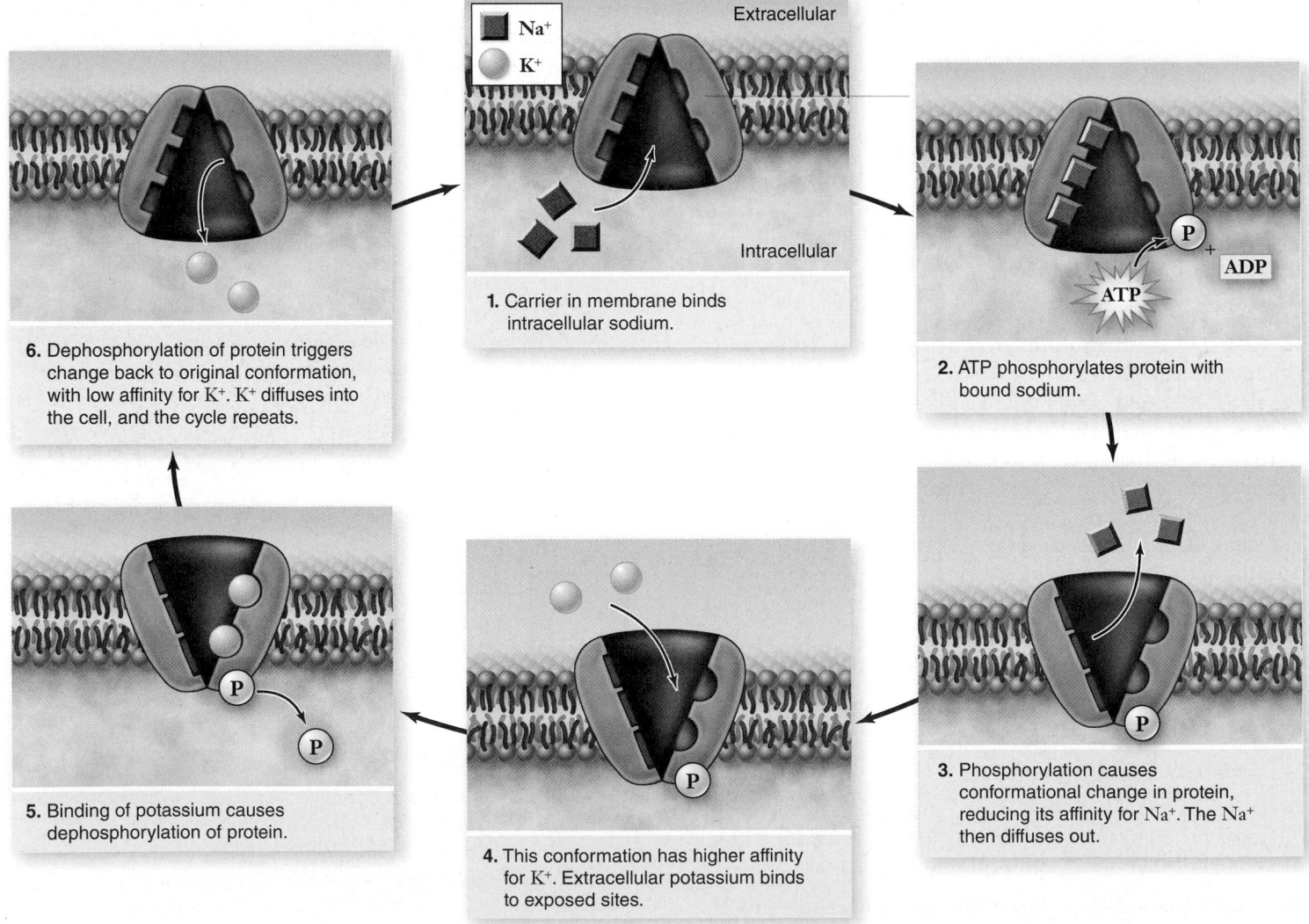

Figure 5.13 The sodium–potassium pump. The protein carrier known as the sodium–potassium pump transports sodium (Na^+) and potassium (K^+) across the plasma membrane. For every three Na^+ transported out of the cell, two K^+ are transported into it. The sodium–potassium pump is fueled by ATP hydrolysis. The affinity of the pump for Na^+ and K^+ is changed by adding or removing phosphate (P), which changes the conformation of the protein.

The remarkable protein that transports these two ions across the cell membrane is known as the **sodium–potassium pump** (Na$^+$/K$^+$ pump) (figure 5.13). This carrier protein uses the energy stored in ATP to move these two ions. In this case, the energy is used to change the conformation of the carrier protein, which changes its affinity for either Na$^+$ ions or K$^+$ ions. This is an excellent illustration of how subtle changes in the structure of a protein affect its function.

The important characteristic of the Na$^+$/K$^+$ pump is that it is an active transport mechanism, transporting Na$^+$ and K$^+$ from areas of low concentration to areas of high concentration. This transport is the opposite of passive transport by diffusion; it is achieved only by the constant expenditure of metabolic energy. The Na$^+$/K$^+$ pump works through the following series of conformational changes in the transmembrane protein (summarized in figure 5.13):

Step 1. Three Na$^+$ bind to the cytoplasmic side of the protein, causing the protein to change its conformation.

Step 2. In its new conformation, the protein binds a molecule of ATP and cleaves it into adenosine diphosphate (ADP) and phosphate (P$_i$). ADP is released, but the phosphate group is covalently linked to the protein. The protein is now phosphorylated.

Step 3. The phosphorylation of the protein induces a second conformational change in the protein. This change translocates the three Na$^+$ across the membrane, so they now face the exterior. In this new conformation, the protein has a low affinity for Na$^+$, and the three bound Na$^+$ break away from the protein and diffuse into the extracellular fluid.

Step 4. The new conformation has a high affinity for K$^+$, two of which bind to the extracellular side of the protein as soon as it is free of the Na$^+$.

Step 5. The binding of the K$^+$ causes another conformational change in the protein, this time resulting in the hydrolysis of the bound phosphate group.

Step 6. Freed of the phosphate group, the protein reverts to its original shape, exposing the two K$^+$ to the cytoplasm. This conformation has a low affinity for K$^+$, so the two bound K$^+$ dissociate from the protein and diffuse into the interior of the cell. The original conformation has a high affinity for Na$^+$. When these ions bind, they initiate another cycle.

In every cycle, three Na$^+$ leave the cell and two K$^+$ enter. The changes in protein conformation that occur during the cycle are rapid, enabling each carrier to transport as many as 300 Na$^+$ per second. The Na$^+$/K$^+$ pump appears to exist in all animal cells, although cells vary widely in the number of pump proteins they contain.

Coupled transport uses ATP indirectly

Some molecules are moved against their concentration gradient by using the energy stored in a gradient of a different molecule. In this process, called *coupled transport,* the energy released as one molecule moves down its concentration gradient is captured and used to move a different molecule against its gradient. As you just saw, the energy stored in ATP molecules can be used to create a gradient of Na$^+$ and K$^+$ across the membrane. These gradients can then be used to power the transport of other molecules across the membrane.

As one example, let's consider the active transport of glucose across the membrane in animal cells. Glucose is such an important molecule that there are a variety of transporters for it, one of which was discussed earlier under passive transport. In a multicellular organism, intestinal epithelial cells can have a higher concentration of glucose inside the cell than outside, so these cells need to be able to transport glucose against its concentration gradient. This requires energy and a different transporter than the one involved in facilitated diffusion of glucose.

The active glucose transporter uses the Na$^+$ gradient produced by the Na$^+$/K$^+$ pump as a source of energy to power the movement of glucose into the cell. In this system, both glucose and Na$^+$ bind to the transport protein, which allows Na$^+$ to pass into the cell down its concentration gradient, capturing the energy and using it to move glucose into the cell. In this kind of cotransport, both molecules are moving in the same direction across the membrane; therefore the transporter is a symporter (figure 5.14).

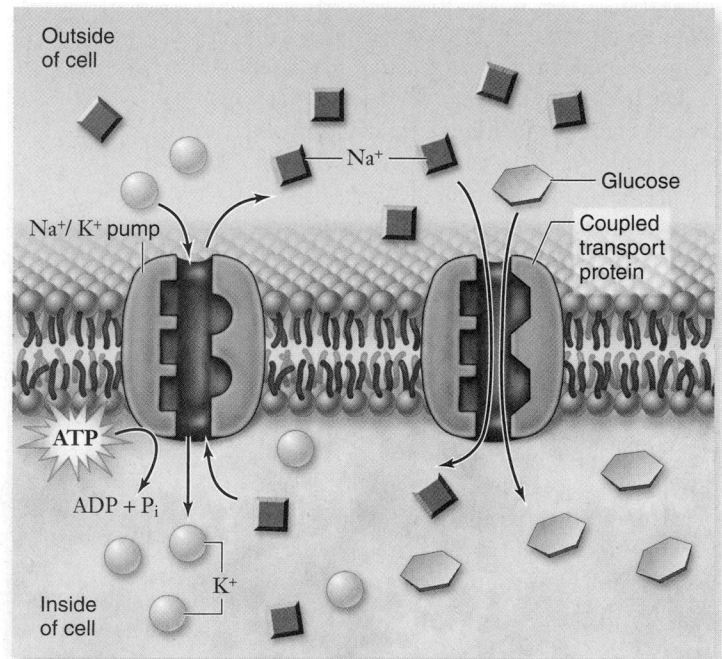

Figure 5.14 Coupled transport. A membrane protein transports Na$^+$ into the cell, down its concentration gradient, at the same time it transports a glucose molecule into the cell. The gradient driving the Na$^+$ entry allows sugar molecules to be transported against their concentration gradient. The Na$^+$ gradient is maintained by the Na$^+$/K$^+$ pump. ADP = adenosine diphosphate; ATP = adenosine triphosphate; P$_i$ = inorganic phosphate

In a related process, called *countertransport,* the inward movement of Na⁺ is coupled with the outward movement of another substance, such as Ca²⁺ or H⁺. As in cotransport, both Na⁺ and the other substance bind to the same transport protein, which in this case is an antiporter, as the substances bind on opposite sides of the membrane and are moved in opposite directions. In countertransport, the cell uses the energy released as Na⁺ moves down its concentration gradient into the cell to eject a substance against its concentration gradient. In both cotransport and countertransport, the potential energy in the concentration gradient of one molecule is used to transport another molecule against its concentration gradient. They differ only in the direction that the second molecule moves relative to the first.

Learning Outcomes Review 5.5

Active transport requires both a carrier protein and energy, usually in the form of ATP, to move molecules against a concentration gradient. The Na⁺/K⁺ pump uses ATP to moved Na⁺ in one direction and K⁺ in the other to create and maintain concentration differences of these ions. In coupled transport, a favorable concentration gradient of one molecule is used to move a different molecule against its gradient, such as in the transport of glucose by Na⁺.

■ *Can active transport involve a channel protein? Why or why not?*

5.6 Bulk Transport by Endocytosis and Exocytosis

Learning Outcomes

1. *Distinguish between endocytosis and exocytosis.*
2. *Illustrate how endocytosis can be specific.*

The lipid nature of cell plasma membranes raises a second problem. The substances cells require for growth are mostly large, polar molecules that cannot cross the hydrophobic barrier a lipid bilayer creates. How do these substances get into cells? Two processes are involved in this **bulk transport:** *endocytosis* and *exocytosis.*

Bulk material enters the cell in vesicles

In **endocytosis,** the plasma membrane envelops food particles and fluids. Cells use three major types of endocytosis: phagocytosis, pinocytosis, and receptor-mediated endocytosis (figure 5.15). Like active transport, these processes also require energy expenditure.

Figure 5.15 Endocytosis. Both (*a*) phagocytosis and (*b*) pinocytosis are forms of endocytosis. *c.* In receptor-mediated endocytosis, cells have pits coated with the protein clathrin that initiate endocytosis when target molecules bind to receptor proteins in the plasma membrane. Photo inserts (false color has been added to enhance distinction of structures): (*a*) A TEM of phagocytosis of a bacterium, *Rickettsia tsutsugamushi,* by a mouse peritoneal mesothelial cell. The bacterium enters the host cell by phagocytosis and replicates in the cytoplasm. (*b*) A TEM of pinocytosis in a smooth muscle cell. (*c*) A coated pit appears in the plasma membrane of a developing egg cell, covered with a layer of proteins. When an appropriate collection of molecules gathers in the coated pit, the pit deepens and will eventually seal off to form a vesicle.

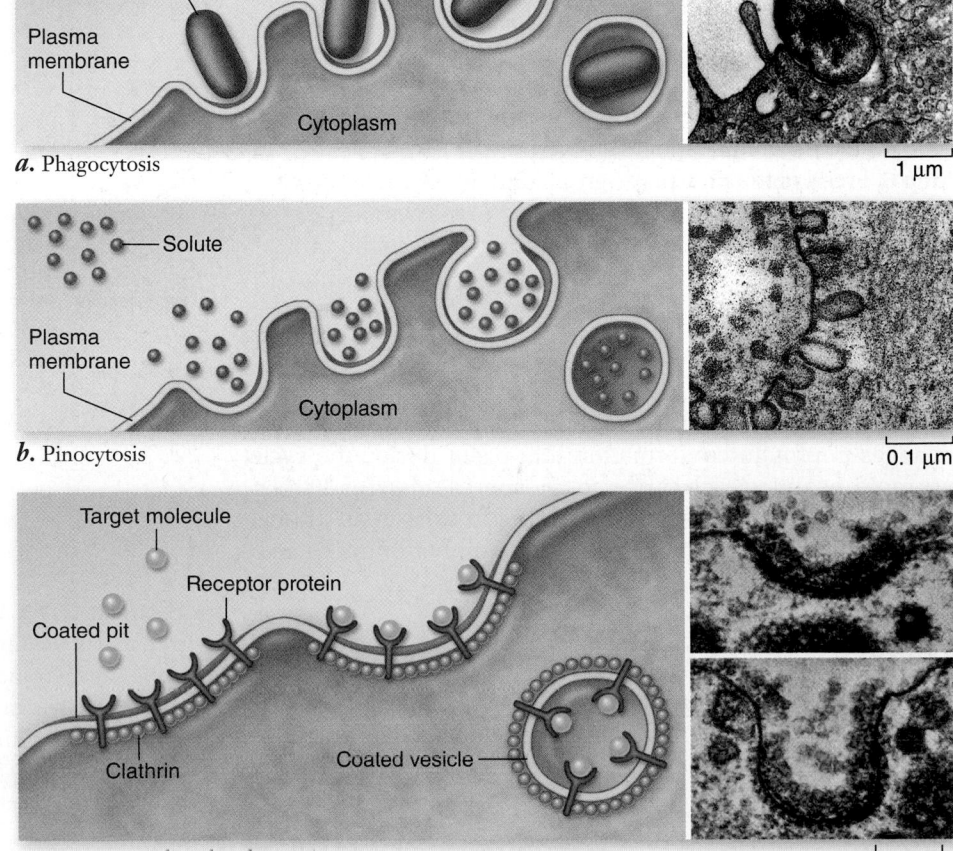

a. Phagocytosis — 1 µm

b. Pinocytosis — 0.1 µm

c. Receptor-mediated endocytosis — 90 nm

Phagocytosis and pinocytosis

If the material the cell takes in is particulate (made up of discrete particles), such as an organism or some other fragment of organic matter (figure 5.15*a*), the process is called **phagocytosis** (Greek *phagein*, "to eat," + *cytos*, "cell"). If the material the cell takes in is liquid (figure 5.15*b*), the process is called **pinocytosis** (Greek *pinein*, "to drink"). Pinocytosis is common among animal cells. Mammalian egg cells, for example, "nurse" from surrounding cells; the nearby cells secrete nutrients that the maturing egg cell takes up by pinocytosis.

Virtually all eukaryotic cells constantly carry out these kinds of endocytotic processes, trapping particles and extracellular fluid in vesicles and ingesting them. Endocytosis rates vary from one cell type to another. They can be surprisingly high; some types of white blood cells ingest up to 25% of their cell volume each hour.

Receptor-mediated endocytosis

Molecules are often transported into eukaryotic cells through **receptor-mediated endocytosis.** These molecules first bind to specific receptors in the plasma membrane—they have a conformation that fits snugly into the receptor. Different cell types contain a characteristic battery of receptor types, each for a different kind of molecule in their membranes.

The portion of the receptor molecule that lies inside the membrane is trapped in an indented pit coated on the cytoplasmic side with the protein *clathrin*. Each pit acts like a molecular mousetrap, closing over to form an internal vesicle when the right molecule enters the pit (figure 5.15*c*). The trigger that releases the trap is the binding of the properly fitted target molecule to the embedded receptor. When binding occurs, the cell reacts by initiating endocytosis; the process is highly specific and very fast. The vesicle is now inside the cell carrying its cargo.

One type of molecule that is taken up by receptor-mediated endocytosis is low-density lipoprotein (LDL). LDL molecules bring cholesterol into the cell where it can be incorporated into membranes. Cholesterol plays a key role in determining the stiffness of the body's membranes. In the human genetic disease familial hypercholesterolemia, the LDL receptors lack tails, so they are never fastened in the clathrin-coated pits and as a result, do not trigger vesicle formation. The cholesterol stays in the bloodstream of affected individuals, accumulating as plaques inside arteries and leading to heart attacks.

It is important to understand that endocytosis in itself does not bring substances directly into the cytoplasm of a cell. The material taken in is still separated from the cytoplasm by the membrane of the vesicle.

Material can leave the cell by exocytosis

The reverse of endocytosis is **exocytosis**, the discharge of material from vesicles at the cell surface (figure 5.16). In plant cells, exocytosis is an important means of exporting the materials needed to construct the cell wall through the plasma membrane. Among protists, contractile vacuole discharge is considered a form of exocytosis. In animal cells, exocytosis provides a mechanism for secreting many hormones, neurotransmitters, digestive enzymes, and other substances.

The mechanisms for transport across cell membranes are summarized in table 5.2.

Learning Outcomes Review 5.6

Large molecules and other bulky materials can enter a cell by endocytosis and leave the cell by exocytosis. These processes require energy. Endocytosis may be mediated by specific receptor proteins in the membrane that trigger the formation of vesicles.

■ *What feature unites transport by receptor-mediated endocytosis, transport by a carrier, and catalysis by an enzyme?*

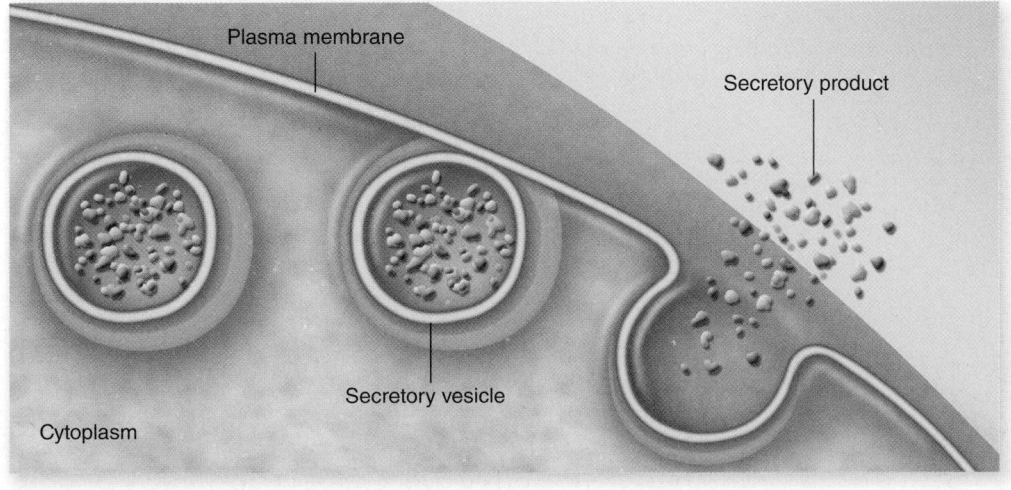
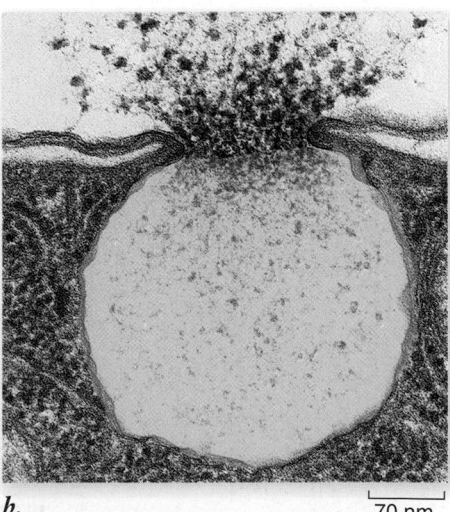

Plasma membrane

Secretory product

Secretory vesicle

Cytoplasm

a.

b.

70 nm

Figure 5.16 Exocytosis. *a.* Proteins and other molecules are secreted from cells in small packets called vesicles, whose membranes fuse with the plasma membrane, releasing their contents outside the cell. *b.* A false-colored transmission electron micrograph showing exocytosis.

TABLE 5.2	Mechanisms for Transport Across Cell Membranes		
Process		**How It Works**	**Example**
P A S S I V E *P R O C E S S E S*			
Diffusion			
Direct		Random molecular motion produces net migration of nonpolar molecules toward region of lower concentration	Movement of oxygen into cells
Facilitated Diffusion			
Protein channel		Polar molecules or ions move through a protein channel; net movement is toward region of lower concentration	Movement of ions in or out of cell
Protein carrier		Molecule binds to carrier protein in membrane and is transported across; net movement is toward region of lower concentration	Movement of glucose into cells
Osmosis			
Aquaporins		Diffusion of water across the membrane via osmosis; requires osmotic gradient	Movement of water into cells placed in a hypotonic solution
A C T I V E *P R O C E S S E S*			
Active Transport			
Protein carrier			
Na^+/K^+ pump		Carrier uses energy to move a substance across a membrane against its concentration gradient	Na^+ and K^+ against their concentration gradients
Coupled transport		Molecules are transported across a membrane against their concentration gradients by the cotransport of sodium ions or protons down their concentration gradients	Coupled uptake of glucose into cells against its concentration gradient using a Na^+ gradient
Endocytosis			
Membrane vesicle			
Phagocytosis		Particle is engulfed by membrane, which folds around it and forms a vesicle	Ingestion of bacteria by white blood cells
Pinocytosis		Fluid droplets are engulfed by membrane, which forms vesicles around them	"Nursing" of human egg cells
Receptor-mediated endocytosis		Endocytosis triggered by a specific receptor, forming clathrin-coated vesicles	Cholesterol uptake
Exocytosis			
Membrane vesicle		Vesicles fuse with plasma membrane and eject contents	Secretion of mucus; release of neurotransmitters

5.1 The Structure of Membranes

The fluid mosaic model shows proteins embedded in a fluid lipid bilayer.

Membranes are sheets of phospholipid bilayers with associated proteins (figure 5.3). Hydrophobic regions of a membrane are oriented inward and hydrophilic regions oriented outward. In the fluid mosaic model, proteins float on or in the lipid bilayer.

Cellular membranes consist of four component groups.

In eukaryotic cells, membranes have four components: a phospholipid bilayer, transmembrane proteins (integral membrane proteins), an interior protein network, and cell-surface markers. The interior protein network is composed of cytoskeletal filaments and peripheral membrane proteins, which are associated with the membrane but are not an integral part. Membranes contain glycoproteins and glycolipids on the surface that act as cell identity markers.

Cellular membranes have an organized substructure

Cholesterol and sphingolipid can associate to form microdomains. The two leaflets of the plasma membrane are also not identical.

Electron microscopy has provided structural evidence.

Transmission electron microscopy (TEM) and scanning electron microscopy (SEM) have confirmed the structure predicted by the fluid mosaic model.

5.2 Phospholipids: The Membrane's Foundation

Phospholipids are composed of two fatty acids and a phosphate group linked to a three-carbon glycerol molecule.

Phospholipids spontaneously form bilayers.

The phosphate group of a phospholipid is polar and hydrophilic; the fatty acids are nonpolar and hydrophobic, and they orient away from the polar head of the phospholipids. The nonpolar interior of the lipid bilayer impedes the passage of water and water-soluble substances.

The phospholipid bilayer is fluid.

Hydrogen bonding of water keeps the membrane in its bilayer configuration; however, phospholipids and unanchored proteins in the membrane are loosely associated and can diffuse laterally.

Membrane fluidity can change.

Membrane fluidity depends on the fatty acid composition of the membrane. Unsaturated fats tend to make the membrane more fluid because of the "kinks" of double bonds in the fatty acid tails. Temperature also affects fluidity. Some bacteria have enzymes that alter the fatty acids of the membrane to compensate for temperature changes.

5.3 Proteins: Multifunctional Components

Proteins and protein complexes perform key functions.

Transporters are integral membrane proteins that carry specific substances through the membrane. Enzymes often occur on the interior surface of the membrane. Cell-surface receptors respond to external chemical messages and change conditions inside the cell; cell identity markers on the surface allow recognition of the body's cells as "self." Cell-to-cell adhesion proteins glue cells together; surface proteins that interact with other cells anchor to the cytoskeleton.

Structural features of membrane proteins relate to function.

Surface proteins are attached to the surface by nonpolar regions that associate with polar regions of phospholipids. Transmembrane proteins may cross the bilayer a number of times, and each membrane-spanning region is called a transmembrane domain. Such a domain is composed of hydrophobic amino acids usually arranged in α helices. In certain proteins, β-pleated sheets in the nonpolar region form a pipelike passageway having a polar environment. An example is the porin class of proteins.

5.4 Passive Transport Across Membranes

Transport can occur by simple diffusion.

Simple diffusion is the passive movement of a substance along a chemical or electrical gradient. Biological membranes pose a barrier to hydrophilic polar molecules, while they allow hydrophobic substances to diffuse freely.

Proteins allow membrane diffusion to be selective.

Ions and large hydrophilic molecules cannot cross the phospholipid bilayer. Diffusion can still occur with the help of proteins, thus we call this facilitated diffusion. These proteins can be either channels, or carriers. Channels allow the diffusion of ions based on concentration and charge across the membrane. They are specific for different ions, but form an aqueous pore in the membrane. Carrier proteins bind to the molecules they transport, much like an enzyme. The rate of transport by a carrier is limited by the number of carriers in the membrane.

Osmosis is the movement of water across membranes.

The direction of movement due to osmosis depends on the solute concentration on either side of the membrane (figure 5.12). Solutions can be isotonic, hypotonic, or hypertonic. Cells in an isotonic solution are in osmotic balance; cells in a hypotonic solution will gain water; and cells in a hypertonic solution will lose water. Aquaporins are water channels that facilitate the diffusion of water.

5.5 Active Transport Across Membranes

Active transport uses energy to move materials against a concentration gradient.

Active transport uses specialized protein carriers that couple a source of energy to transport. They are classified based on the number of molecules and direction of transport. Uniporters transport a specific molecule in one direction; symporters transport two molecules in the same direction; and antiporters transport two molecules in opposite directions.

The sodium–potassium pump runs directly on ATP.

The sodium–potassium pump moves Na^+ out of the cell and K^+ into the cell against their concentration gradients using ATP. In every cycle of the pump, three Na^+ leave the cell and two K^+ enter it. This pump appears to be almost universal in animal cells.

Coupled transport uses ATP indirectly.

Coupled transport occurs when the energy released by a diffusing molecule is used to transport a different molecule against its concentration gradient in the same direction. Countertransport is similar to coupled transport, but the two molecules move in opposite directions.

5.6 Bulk Transport by Endocytosis and Exocytosis

Bulk transport moves large quantities of substances that cannot pass through the cell membrane.

Bulk material enters the cell in vesicles.

In endocytosis, the cell membrane surrounds material and pinches off to form a vesicle. In receptor-mediated endocytosis, specific molecules bind to receptors on the cell membrane.

Material can leave the cell by exocytosis.

In exocytosis, material in a vesicle is discharged when the vesicle fuses with the membrane.

UNDERSTAND

1. The fluid mosaic model of the membrane describes the membrane as
   a. containing a significant quantity of water in the interior.
   b. composed of fluid phospholipids on the outside and protein on the inside.
   c. composed of protein on the outside and fluid phospholipids on the inside.
   d. made of proteins and lipids that can freely move.

2. What chemical property characterizes the interior of the phospholipid bilayer?
   a. It is hydrophobic.   c. It is polar.
   b. It is hydrophilic.   d. It is saturated.

3. The transmembrane domain of an integral membrane protein
   a. is composed of hydrophobic amino acids.
   b. often forms an α-helical structure.
   c. can cross the membrane multiple times.
   d. All of the choices are correct.

4. The specific function of a membrane within a cell is determined by the
   a. degree of saturation of the fatty acids within the phospholipid bilayer.
   b. location of the membrane within the cell.
   c. presence of lipid rafts and cholesterol.
   d. type and number of membrane proteins.

5. The movement of water across a membrane is dependent on
   a. the solvent concentration.
   b. the solute concentration.
   c. the presence of carrier proteins.
   d. membrane potential.

6. If a cell is in an isotonic environment, then
   a. the cell will gain water and burst.
   b. no water will move across the membrane.
   c. the cell will lose water and shrink.
   d. osmosis still occurs, but there is no net gain or loss of cell volume.

7. Which of the following is NOT a mechanism for bringing material into a cell?
   a. Exocytosis   c. Pinocytosis
   b. Endocytosis   d. Phagocytosis

APPLY

1. A bacterial cell that can alter the composition of saturated and unsaturated fatty acids in its membrane lipids is adapted to a cold environment. If this cell is shifted to a warmer environment, it will react by
   a. increasing the amount of cholesterol in its membrane.
   b. altering the amount of protein present in the membrane.
   c. increasing the degree of saturated fatty acids in its membrane.
   d. increasing the percentage of unsaturated fatty acids in its membrane.

2. What variable(s) influence(s) whether a nonpolar molecule can move across a membrane by passive diffusion?
   a. The structure of the phospholipids bilayer
   b. The difference in concentration of the molecule across the membrane

   c. The presence of transport proteins in the membrane
   d. All of the choices are correct.

3. Which of the following does NOT contribute to the selective permeability of a biological membrane?
   a. Specificity of the carrier proteins in the membrane
   b. Selectivity of channel proteins in the membrane
   c. Hydrophobic barrier of the phospholipid bilayer
   d. Hydrogen bond formation between water and phosphate groups

4. How are *active* transport and *coupled* transport related?
   a. They both use ATP to move molecules.
   b. Active transport establishes a concentration gradient, but coupled transport doesn't.
   c. Coupled transport uses the concentration gradient established by active transport.
   d. Active transport moves one molecule, but coupled transport moves two.

5. A cell can use the process of facilitated diffusion to
   a. concentrate a molecule such as glucose inside a cell.
   b. remove all of a toxic molecule from a cell.
   c. move ions or large polar molecules across the membrane regardless of concentration.
   d. move ions or large polar molecules from a region of high concentration to a region of low concentration.

SYNTHESIZE

1. Figure 5.5 describes a classic experiment demonstrating the ability of proteins to move within the plane of the cell's plasma membrane. The following table outlines three different experiments using the fusion of labeled mouse and human cells.

Experiment	Conditions	Temperature (°C)	Result
1	Fuse human and mouse cells	37	Intermixed membrane proteins
2	Fuse human and mouse cells in presence of ATP inhibitors	37	Intermixed membrane proteins
3	Fuse human and mouse cells	4	No intermixing of membrane proteins

What conclusions can you reach about the movement of these proteins?

2. Each compartment of the endomembrane system of a cell is connected to the plasma membrane. Create a simple diagram of a cell including the RER, Golgi apparatus, vesicle, and the plasma membrane. Starting with the RER, use two different colors to represent the inner and outer halves of the bilayer for each of these membranes. What do you observe?

3. The distribution of lipids in the ER membrane is symmetric, that is, it is the same in both leaflets of the membrane. The Golgi apparatus and plasma membrane do not have symmetric distribution of membrane lipids. What kinds of processes could achieve this outcome?

ONLINE RESOURCE

www.ravenbiology.com

Understand, Apply, and Synthesize—enhance your study with animations that bring concepts to life and practice tests to assess your understanding. Your instructor may also recommend the interactive eBook, individualized learning tools, and more.

Chapter **6**

Energy and Metabolism

Chapter Contents

Introduction

Life can be viewed as a constant flow of energy, channeled by organisms to do the work of living. Each of the significant properties by which we define life—order, growth, reproduction, responsiveness, and internal regulation—requires a constant supply of energy. Both the lion and the giraffe need to eat to provide energy for a wide variety of cellular functions. Deprived of a source of energy, life stops. Therefore, a comprehensive study of life would be impossible without discussing bioenergetics, the analysis of how energy powers the activities of living systems. In this chapter, we focus on energy—what it is and how it changes during chemical reactions.

6.1 The Flow of Energy in Living Systems

Learning Outcomes

1. Differentiate between kinetic and potential energy.
2. Identify the source of energy for the biosphere.
3. Contrast oxidation and reduction reactions.

Thermodynamics is the branch of chemistry concerned with energy changes. Cells are governed by the laws of physics and chemistry, so we must understand these laws in order to understand how cells function.

Energy can take many forms

Energy is defined as the capacity to do work. We think of energy as existing in two states: kinetic energy and potential energy (figure 6.1). **Kinetic energy** is the energy of motion. Moving objects perform work by causing other matter to move. **Potential energy** is stored energy. Objects that are not actively moving but have the capacity to do so possess potential energy. A boulder perched on a hilltop has gravitational potential energy. As it begins to roll downhill, some of its potential energy is converted into kinetic energy. Much of the work that living organisms carry out involves transforming potential energy into kinetic energy.

Energy can take many forms: mechanical energy, heat, sound, electric current, light, or radioactivity. Because it can exist in so many forms, energy can be measured in many ways. Heat is the most convenient way of measuring energy because all other forms of energy can be converted into heat. In fact, the term *thermodynamics* means "heat changes."

The unit of heat most commonly employed in biology is the kilocalorie (kcal). One kilocalorie is equal to 1000 calories (cal). One calorie is the heat required to raise the temperature of one gram of water one degree Celsius (°C). (You are probably more used to seeing the term *Calorie* with a capital C. This is used on food labels and is actually the same as kilocalorie.) Another energy unit, often used in physics, is the *joule;* one joule equals 0.239 cal.

The Sun provides energy for living systems

Energy flows into the biological world from the Sun. It is estimated that sunlight provides the Earth with more than 13×10^{23} calories per year, or 40 million billion calories per second! Plants, algae, and certain kinds of bacteria capture a fraction of this energy through photosynthesis.

In photosynthesis, energy absorbed from sunlight is used to combine small molecules (water and carbon dioxide) into more complex ones (sugars). This process converts carbon from an inorganic to an organic form. In the process, energy from sunlight is stored as potential energy in the covalent bonds between atoms in the sugar molecules.

Breaking the bonds between atoms requires energy. In fact, the strength of a covalent bond is measured by the amount of energy required to break it. For example, it takes 98.8 kcal to break one mole (6.023×10^{23}) of the carbon–hydrogen (C—H) bonds found in organic molecules. Fat molecules have many C—H bonds, and breaking those bonds provides lots of energy.

a. Potential energy

b. Kinetic energy

Figure 6.1 Potential and kinetic energy. *a.* Objects that have the capacity to move but are not moving have potential energy. The energy required for the girl to climb to the top of the slide is stored as potential energy. *b.* Objects that are in motion have kinetic energy. The stored potential energy is released as kinetic energy as the girl slides down.

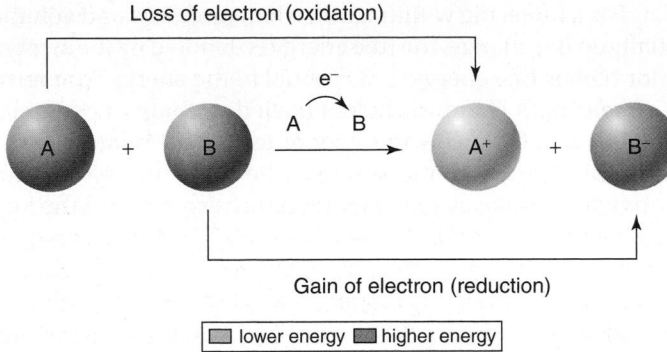

Loss of electron (oxidation)

$A + B$ e^- $A \quad B$ $A^+ + B^-$

Gain of electron (reduction)

☐ lower energy ☐ higher energy

Figure 6.2 Redox reactions. Oxidation is the loss of an electron; reduction is the gain of an electron. In this example, the charges of molecules A and B appear as superscripts in each molecule. Molecule A loses energy as it loses an electron, and molecule B gains that energy as it gains an electron.

This is one reason animals store fat. The oxidation of one mole of a 16-carbon fatty acid that is completely saturated with hydrogens yields 2340 kcal.

Oxidation–reduction reactions transfer electrons while bonds are made or broken

During a chemical reaction, the energy stored in chemical bonds may be used to make new bonds. In some of these reactions, electrons actually pass from one atom or molecule to another. An atom or molecule that loses an electron is said to be oxidized, and the process by which this occurs is called **oxidation.** The name comes from the fact that oxygen is the most common electron acceptor in biological systems. Conversely, an atom or molecule that gains an electron is said to be reduced, and the process is called *reduction.* The reduced form of a molecule has a higher level of energy than the oxidized form (figure 6.2).

Oxidation and reduction always take place together, because every electron that is lost by one atom through oxidation is gained by another atom through reduction. Therefore, chemical reactions of this sort are called **oxidation–reduction,** or **redox,** *reactions.* Oxidation–reduction reactions play a key role in the flow of energy through biological systems.

In the next two chapters, you will learn the details of how organisms derive energy from the oxidation of organic compounds via respiration, as well as from the energy in sunlight via photosynthesis.

Learning Outcomes Review 6.1

Energy is defined as the capacity to do work. The two forms of energy are kinetic energy, or energy of motion, and potential energy, or stored energy. The ultimate source of energy for living systems is the Sun. Organisms derive their energy from oxidation–reduction reactions. In oxidation, a molecule loses an electron; in reduction, a molecule gains an electron.

■ *What energy source might ecosystems at the bottom of the ocean use?*

6.2 The Laws of Thermodynamics and Free Energy

Learning Outcomes

1. *Explain the laws of thermodynamics.*
2. *Relate free energy changes to the outcome of chemical reactions.*
3. *Contrast the course of a reaction with and without an enzyme catalyst.*

All activities of living organisms—growing, running, thinking, singing, reading these words—involve changes in energy. A set of two universal laws we call the laws of thermodynamics govern all energy changes in the universe, from nuclear reactions to a bird flying through the air.

The First Law states that energy cannot be created or destroyed

The **First Law of Thermodynamics** concerns the amount of energy in the universe. Energy cannot be created or destroyed; it can only change from one form to another (from potential to kinetic, for example). The total amount of energy in the universe remains constant.

The lion eating a giraffe at the beginning of this chapter is acquiring energy. Rather than creating new energy or capturing the energy in sunlight, the lion is merely transferring some of the potential energy stored in the giraffe's tissues to its own body, just as the giraffe obtained the potential energy stored in the plants it ate while it was alive.

Within any living organism, chemical potential energy stored in some molecules can be shifted to other molecules and stored in different chemical bonds. It can also be converted into other forms, such as kinetic energy, light, or electricity. During each conversion, some of the energy dissipates into the environment as **heat,** which is a measure of the random motion of molecules (and therefore a measure of one form of kinetic energy). Energy continuously flows through the biological world in one direction, with new energy from the Sun constantly entering the system to replace the energy dissipated as heat.

Heat can be harnessed to do work only when there is a heat gradient—that is, a temperature difference between two areas. Cells are too small to maintain significant internal temperature differences, so heat energy is incapable of doing the work of cells. Instead, cells must rely on chemical reactions for energy.

Although the total amount of energy in the universe remains constant, the energy available to do work decreases as more of it is progressively lost as heat.

The Second Law states that some energy is lost as disorder increases

The **Second Law of Thermodynamics** concerns the transformation of potential energy into heat, or random molecular motion during any energy transaction. It states that the disorder in the universe, more formally called **entropy,** is continuously increasing. Put simply, disorder is more likely than order. For example, it is much more likely that a column of bricks will tumble over than that a pile of bricks will arrange themselves spontaneously to form a column.

In general, energy transformations proceed spontaneously to convert matter from a more ordered, less stable form to a less ordered, but more stable form. For this reason, the second law is sometimes called "time's arrow." Looking at the photographs in figure 6.3, you could put the pictures into correct sequence using the information that time had elapsed with only natural processes occurring. Although it might be great if our rooms would straighten themselves up, we know from experience how much work it takes to do so.

The Second Law of Thermodynamics can also be stated simply as "entropy increases." When the universe formed, it held all the potential energy it will ever have. It has become progressively more disordered ever since, with every energy exchange increasing the amount of entropy.

Chemical reactions can be predicted based on changes in free energy

It takes energy to break the chemical bonds that hold the atoms in a molecule together. Heat energy, because it increases atomic motion, makes it easier for the atoms to pull apart. Both chemical bonding and heat have a significant influence on a molecule. Chemical bonding reduces disorder; heat increases it. The net effect, the amount of energy actually available to break and subsequently form other chemical bonds, is called the *free energy* of that molecule. In a more general sense, **free energy** is defined as the energy available to do work in any system.

For a molecule within a cell, where pressure and volume usually do not change, the free energy is denoted by the symbol G (for "Gibbs free energy"). G is equal to the energy contained in a molecule's chemical bonds (called **enthalpy** and designated H) together with the energy term (TS) related to the degree of disorder in the system, where S is the symbol for *entropy* and T is the absolute temperature expressed in the Kelvin scale ($K = °C + 273$):

$$G = H - TS$$

Chemical reactions break some bonds in the reactants and form new ones in the products. Consequently, reactions can produce changes in free energy. When a chemical reaction occurs under conditions of constant temperature, pressure, and volume—as do most biological reactions—the change symbolized by the Greek capital letter delta, Δ, in free energy (ΔG) is simply:

$$\Delta G = \Delta H - T\Delta S$$

We can use the change in free energy, or ΔG, to predict whether a chemical reaction is spontaneous or not. For some reactions, the ΔG is positive, which means that the products of the reaction contain *more* free energy than the reactants; the bond energy (H) is higher, or the disorder (S) in the system is lower. Such reactions do not proceed spontaneously because they require an input of energy. Any reaction that requires an input of energy is said to be **endergonic** ("inward energy").

For other reactions, the ΔG is negative. In this case, the products of the reaction contain less free energy than the reactants; either the bond energy is lower, or the disorder is higher, or both. Such reactions tend to proceed spontaneously. These reactions release the excess free energy as heat and are thus said to be **exergonic** ("outward energy"). Any chemical reaction tends to proceed spontaneously if the difference in disorder $(T\Delta S)$ is *greater* than the difference in bond energies between reactants and products (ΔH).

Note that *spontaneous* does not mean the same thing as *instantaneous.* A spontaneous reaction may proceed very slowly. Figure 6.4 sums up endergonic and exergonic reactions.

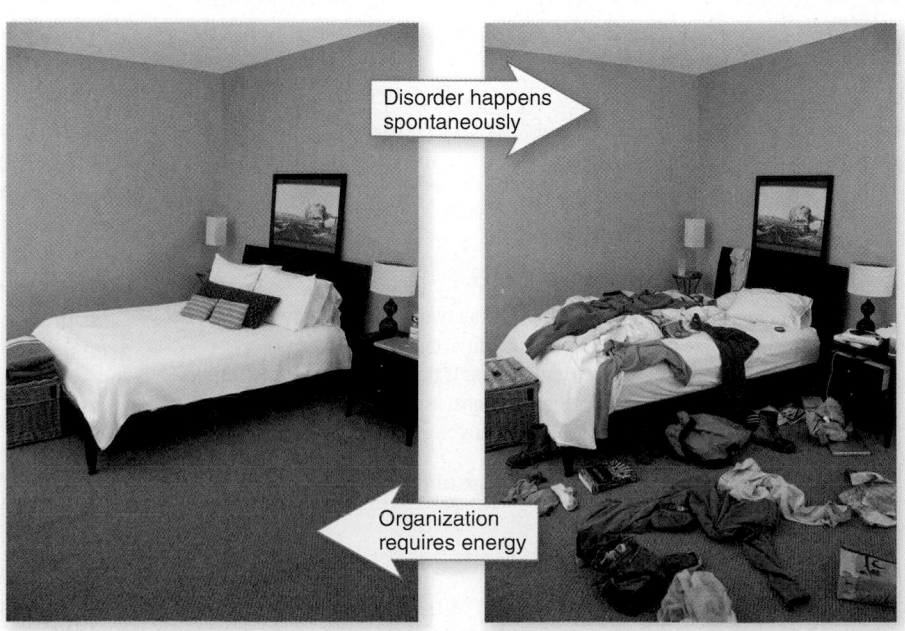

Figure 6.3 Entropy in action. As time elapses, the room shown at right becomes more disorganized. Entropy has increased in this room. It takes energy to restore it to the ordered state shown at left.

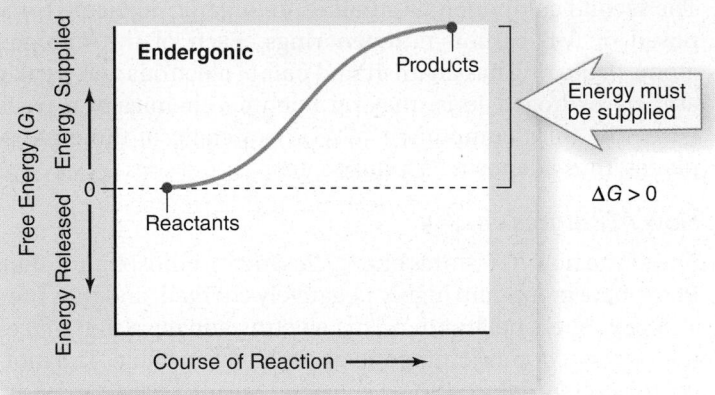

a.

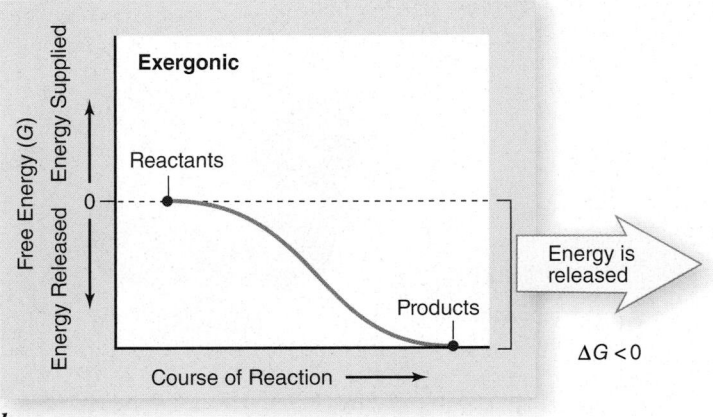

b.

Figure 6.4 Energy in chemical reactions. *a.* In an endergonic reaction, the products of the reaction contain more energy than the reactants, and the extra energy must be supplied for the reaction to proceed. ***b.*** In an exergonic reaction, the products contain less energy than the reactants, and the excess energy is released.

Because chemical reactions are reversible, a reaction that is exergonic in the forward direction will be endergonic in the reverse direction. For each reaction, an equilibrium exists at some point between the relative amounts of reactants and products. This equilibrium has a numeric value and is called the *equilibrium constant.* This characteristic of reactions provides us with another way to think about free energy changes: an exergonic reaction has an equilibrium favoring the products, and an endergonic reaction has an equilibrium favoring the reactants.

Spontaneous chemical reactions require activation energy

If all chemical reactions that release free energy tend to occur spontaneously, why haven't all such reactions already occurred? Consider the gasoline tank of your car: The oxidation of the hydrocarbons in gasoline is an exergonic reaction, but your gas tank does not spontaneously explode. One reason is that most reactions require an input of energy to get started. In the case of your car, this input consists of the electrical sparks in the engine's cylinders, producing a controlled explosion.

Activation energy

Before new chemical bonds can form, even bonds that contain less energy, existing bonds must first be broken, and that requires energy input. The extra energy needed to destabilize existing chemical bonds and initiate a chemical reaction is called **activation energy** (figure 6.5).

The rate of an exergonic reaction depends on the activation energy required for the reaction to begin. Reactions with larger activation energies tend to proceed more slowly because fewer molecules succeed in getting over the initial energy hurdle. The rate of reactions can be increased in two ways: (1) by increasing the energy of reacting molecules or (2) by lowering activation energy. Chemists often drive important industrial reactions by increasing the energy of the reacting molecules, which is frequently accomplished simply by heating up the reactants. The other strategy is to use a catalyst to lower the activation energy.

How catalysts work

Stressing particular chemical bonds can make them easier to break. The process of influencing chemical bonds in a way that lowers the activation energy needed to initiate a reaction is called **catalysis,** and substances that accomplish this are known as *catalysts* (see figure 6.5).

Catalysts exert their action by affecting an intermediate stage in a reaction—the transition state. The energy needed to reach this transition state is the activation energy. Catalysts stabilize this transition state, thus lowering activation energy.

Catalysts cannot violate the basic laws of thermodynamics; they cannot make an endergonic reaction proceed spontaneously. By reducing the activation energy, a catalyst accelerates both the forward and the reverse reactions by exactly the same amount. Therefore, a catalyst does not alter the proportion of reactant ultimately converted into product.

To understand this, imagine a bowling ball resting in a shallow depression on the side of a hill. Only a narrow rim of dirt below the ball prevents it from rolling down the hill. Now imagine digging away that rim of dirt. If you remove enough

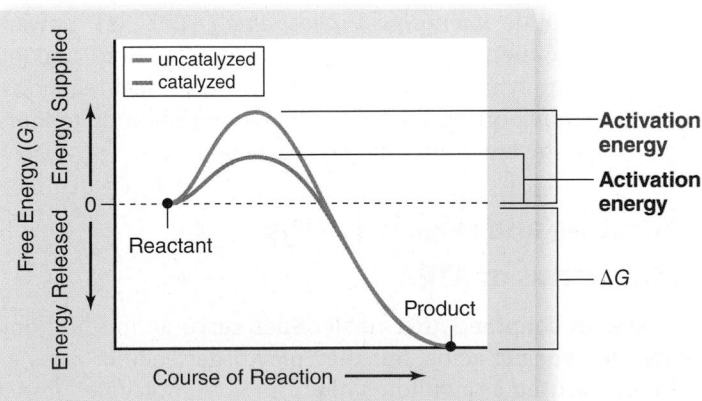

Figure 6.5 Activation energy and catalysis. Exergonic reactions do not necessarily proceed rapidly because activation energy must be supplied to destabilize existing chemical bonds. Catalysts accelerate particular reactions by lowering the amount of activation energy required to initiate the reaction. Catalysts do not alter the free-energy change produced by the reaction.

dirt from below the ball, it will start to roll down the hill—but removing dirt from below the ball will *never* cause the ball to roll up the hill. Removing the lip of dirt simply allows the ball to move freely; gravity determines the direction it then travels.

Similarly, the direction in which a chemical reaction proceeds is determined solely by the difference in free energy between reactants and products. Like digging away the soil below the bowling ball on the hill, catalysts reduce the energy barrier that is preventing the reaction from proceeding. Only exergonic reactions can proceed spontaneously, and catalysts cannot change that. What catalysts *can* do is make a reaction proceed much faster. In living systems, enzymes act as catalysts.

Learning Outcomes Review 6.2

The First Law of Thermodynamics states that energy cannot be created or destroyed. The Second Law states that disorder, or entropy, is increasing. Free-energy changes (ΔG) can predict whether chemical reactions take place. Reactions with a negative ΔG occur spontaneously, and those with a positive ΔG do not. Energy needed to initiate a reaction is termed activation energy. Catalysts stabilize an intermediate transition state, lowering activation energy and accelerating reactions.

■ *Can an enzyme make an endergonic reaction exergonic?*

6.3 ATP: The Energy Currency of Cells

Learning Outcomes

1. Describe the role of ATP in short-term energy storage.
2. Distinguish which bonds in ATP are "high energy."

The chief "currency" all cells use for their energy transactions is the nucleotide *adenosine triphosphate (ATP)*. ATP powers almost every energy-requiring process in cells, from making sugars, to supplying activation energy for chemical reactions, to actively transporting substances across membranes, to moving through the environment and growing.

Cells store and release energy in the bonds of ATP

You saw in chapter 3 that nucleotides serve as the building blocks for nucleic acids, but they play other cellular roles as well. ATP is used as a building block for RNA molecules, and it also has a critical function as a portable source of energy on demand for endergonic cellular processes.

The structure of ATP

ATP is composed of three smaller components (figure 6.6). The first component is a five-carbon sugar, ribose, which serves as the framework to which the other two subunits are attached.

The second component is adenine, an organic molecule composed of two carbon–nitrogen rings. Each of the nitrogen atoms in the ring has an unshared pair of electrons and weakly attracts hydrogen ions, making adenine chemically a weak base. The third component of ATP is a chain of three phosphates, thus adenosine *tri*phosphate.

How ATP stores energy

The key to how ATP stores energy lies in its triphosphate group. Phosphate groups are highly negatively charged, and thus they strongly repel one another. This electrostatic repulsion makes the covalent bonds joining the phosphates unstable. The molecule is often referred to as a "coiled spring," with the phosphates straining away from one another.

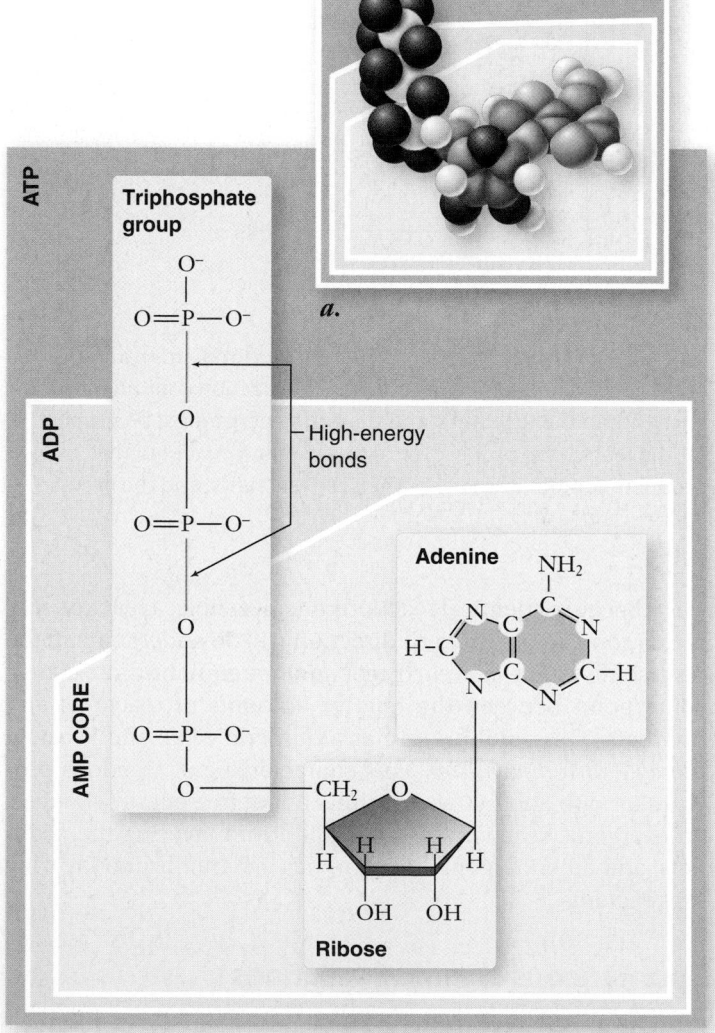

Figure 6.6 The ATP molecule. The model (*a*) and the structural diagram (*b*) both show that ATP has a core of AMP. Addition of one phosphate to AMP yields ADP, and addition of a second phosphate yields ATP. These two terminal phosphates are attached by high-energy bonds so that removing either by hydrolysis is an exergonic reaction that releases energy. ADP, adenosine diphosphate; AMP, adenosine monophosphate; ATP, adenosine triphosphate

The unstable bonds holding the phosphates together in the ATP molecule have a low activation energy and are easily broken by hydrolysis. When they break, they can transfer a considerable amount of energy. In other words, the hydrolysis of ATP has a negative ΔG, and the energy it releases can be used to perform work.

In most reactions involving ATP, only the outermost high-energy phosphate bond is hydrolyzed, cleaving off the phosphate group on the end. When this happens, ATP becomes *adenosine diphosphate (ADP)* plus an **inorganic phosphate (P_i),** and energy equal to 7.3 kcal/mol is released under standard conditions. The liberated phosphate group usually attaches temporarily to some intermediate molecule. When that molecule is dephosphorylated, the phosphate group is released as P_i.

Both of the two terminal phosphates can be hydrolyzed to release energy, leaving *adenosine monophosphate (AMP),* but the third phosphate is not attached by a high-energy bond. With only one phosphate group, AMP has no other phosphates to provide the electrostatic repulsion that makes the bonds holding the two terminal phosphate groups high-energy bonds.

ATP hydrolysis drives endergonic reactions

Cells use ATP to drive endergonic reactions. These reactions do not proceed spontaneously because their products possess more free energy than their reactants. However, if the cleavage of ATP's terminal high-energy bond releases more energy than the other reaction consumes, the two reactions can be coupled so that the energy released by the hydrolysis of ATP can be used to supply the endergonic reaction with the energy it needs. Coupled together, these reactions result in a net release of energy ($-\Delta G$) and are therefore exergonic and proceed spontaneously. Because almost all the endergonic reactions in cells require less energy than is released by the cleavage of ATP, ATP can provide most of the energy a cell needs.

 Data analysis Consider the reaction:
glutamate + $NH_3 \longrightarrow$ glutamine (ΔG = +3.4 kcal/mol). If this reaction is coupled to ATP hydrolysis (ΔG = –7.3 kcal/mol), what would be the overall ΔG? Would this process be endergonic or exergonic?

ATP cycles continuously

The same feature that makes ATP an effective energy donor—the instability of its phosphate bonds—prevents it from being a good long-term energy-storage molecule. Fats and carbohydrates serve that function better.

The use of ATP can be thought of as a cycle: Cells use exergonic reactions to provide the energy needed to synthesize ATP from ADP + P_i; they then use the hydrolysis of ATP to provide energy to drive the endergonic reactions they need (figure 6.7).

Most cells do not maintain large stockpiles of ATP. Instead, they typically have only a few seconds' supply of ATP at any given time, and they continually produce more from ADP and P_i. It is estimated that even a sedentary individual turns over an amount of ATP in one day roughly equal to his or her body weight. This statistic makes clear the importance of ATP synthesis. In the next two chapters we will explore in detail the cellular mechanisms for synthesizing ATP.

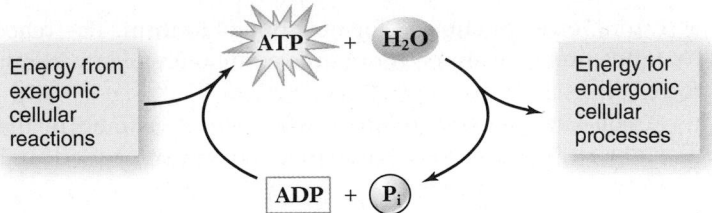

Figure 6.7 The ATP cycle. ATP is synthesized and hydrolyzed in a cyclic fashion. The synthesis of ATP from ADP + P_i is endergonic and is powered by exergonic cellular reactions. The hydrolysis of ATP to ADP + P_i is exergonic, and the energy released is used to power endergonic cellular functions such as muscle contraction. ADP, adenosine diphosphate; ATP, adenosine triphosphate; P_i, inorganic phosphate

Learning Outcomes Review 6.3

ATP is a nucleotide with three phosphate groups. Endergonic cellular processes can be driven by coupling to the exergonic hydrolysis of the two terminal phosphates. The bonds holding the terminal phosphate groups together are easily broken, releasing energy like a coiled spring. The cell is constantly building ATP using exergonic reactions and breaking it down to drive endergonic reactions.

■ *If the molecular weight of ATP is 507.18 g/mol, and the ΔG for hydrolysis is –7.3 kcal/mol, how much energy is released over the course of the day by a 100-kg man?*

6.4 Enzymes: Biological Catalysts

Learning Outcomes

1. *Discuss the specificity of enzymes.*
2. *Explain how enzymes bind to their substrates.*
3. *List the factors that influence the rate of enzyme-catalyzed reactions.*

The chemical reactions within living organisms are regulated by controlling the points at which catalysis takes place. Life itself, therefore, can be seen as regulated by catalysts. The agents that carry out most of the catalysis in living organisms are called enzymes. Most enzymes are proteins, although increasing evidence indicates that some enzymes are actually RNA molecules, as discussed later in this chapter.

An enzyme alters the activation energy of a reaction

The unique three-dimensional shape of an enzyme enables it to stabilize a temporary association between **substrates**—the molecules that will undergo the reaction. By bringing two substrates together in the correct orientation or by stressing particular chemical bonds of a substrate, an enzyme lowers the activation energy required for new bonds to form. The reaction thus proceeds much more quickly than it would without the enzyme.

The enzyme itself is not changed or consumed in the reaction, so only a small amount of an enzyme is needed, and it can be used over and over.

As an example of how an enzyme works, let's consider the reaction of carbon dioxide and water to form carbonic acid. This important enzyme-catalyzed reaction occurs in vertebrate red blood cells:

$$CO_2 + H_2O \rightleftharpoons H_2CO_3$$

carbon water   carbonic
dioxide   acid

This reaction may proceed in either direction, but because it has a large activation energy, the reaction is very slow in the absence of an enzyme: Perhaps 200 molecules of carbonic acid form in an hour in a cell in the absence of any enzyme. Reactions that proceed this slowly are of little use to a cell. Vertebrate red blood cells overcome this problem by employing an enzyme within their cytoplasm called *carbonic anhydrase* (enzyme names usually end in "–ase"). Under the same conditions, but in the presence of carbonic anhydrase, an estimated 600,000 molecules of carbonic acid form every *second!* Thus, the enzyme increases the reaction rate by more than one million times.

Thousands of different kinds of enzymes are known, each catalyzing one or a few specific chemical reactions. By facilitating particular chemical reactions, the enzymes in a cell determine the course of metabolism—the collection of all chemical reactions—in that cell.

Different types of cells contain different sets of enzymes, and this difference contributes to structural and functional variations among cell types. For example, the chemical reactions taking place within a red blood cell differ from those that occur within a nerve cell, in part because different cell types contain different arrays of enzymes.

Active sites of enzymes conform to fit the shape of substrates

Most enzymes are globular proteins with one or more pockets or clefts, called **active sites,** on their surface (figure 6.8). Substrates bind to the enzyme at these active sites, forming an **enzyme-substrate complex** (see figure 6.10). For catalysis to occur within the complex, a substrate molecule must fit precisely into an active site. When that happens, amino acid side groups of the enzyme end up very close to certain bonds of the substrate. These side groups interact chemically with the substrate, usually stressing or distorting a particular bond and consequently lowering the activation energy needed to break the bond. After the bonds of the substrates are broken, or new bonds are formed, the substrates have been converted to products. These products then dissociate from the enzyme, leaving the enzyme ready to bind its next substrate and begin the cycle again.

Proteins are not rigid. The binding of a substrate induces the enzyme to adjust its shape slightly, leading to a better *induced fit* between enzyme and substrate (figure 6.9). This

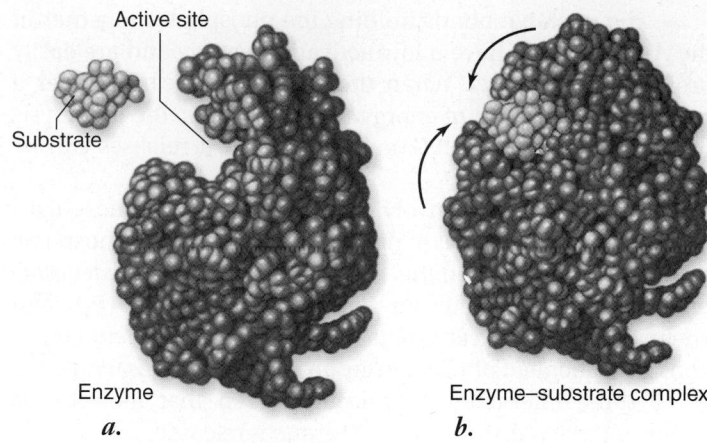

Enzyme
a.

Enzyme–substrate complex
b.

Figure 6.8 Enzyme binding its substrate. *a.* The active site of the enzyme lysozyme fits the shape of its substrate, a peptidoglycan that makes up bacterial cell walls. *b.* When the substrate, indicated in yellow, slides into the groove of the active site, the protein is induced to alter its shape slightly and bind the substrate more tightly. This alteration of the shape of the enzyme to better fit the substrate is called induced fit.

interaction may also facilitate the binding of other substrates; in such cases, one substrate "activates" the enzyme to receive other substrates.

Enzymes occur in many forms

Although many enzymes are suspended in the cytoplasm of cells, not attached to any structure, other enzymes function

Hypothesis: *Protein structure is flexible, not rigid.*

Prediction: *Antibody–antigen binding can involve a change in protein structure.*

Test: *Determine crystal structure of a fragment of a specific antibody with no antigen bound, and with antigen bound for comparison.*

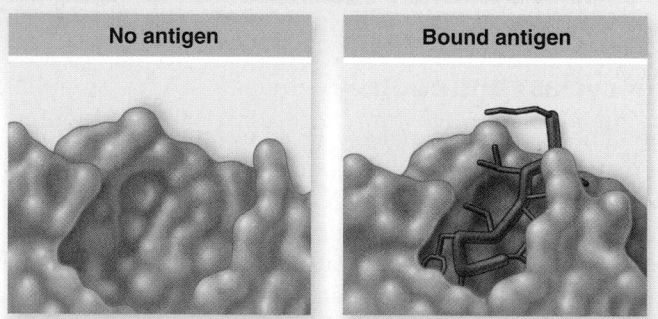

No antigen

Bound antigen

Result: *After binding, the antibody folds around the antigen forming a pocket.*

Conclusion: *In this case, binding involves an induced-fit kind of change in conformation.*

Further Experiments: *Why is this experiment easier to do with an antibody than with an enzyme? Can this experiment be done with an enzyme?*

Figure 6.9 Induced-fit binding of antibody to antigen.

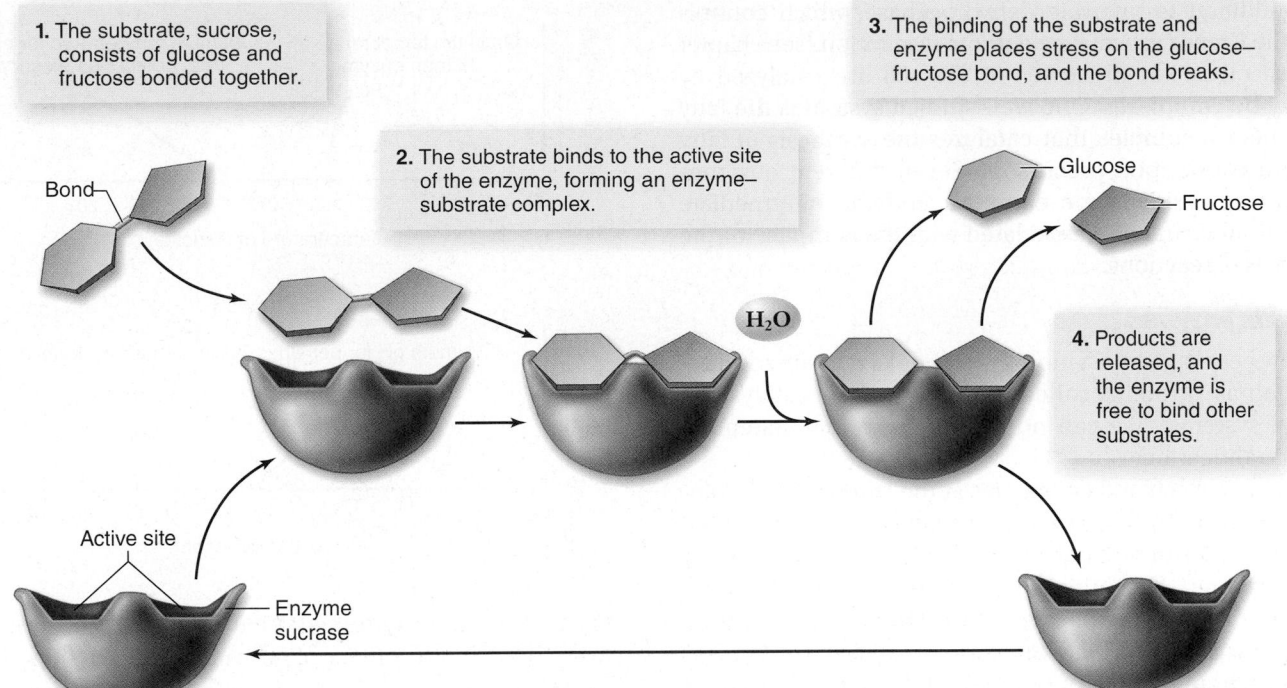

1. The substrate, sucrose, consists of glucose and fructose bonded together.

Bond

2. The substrate binds to the active site of the enzyme, forming an enzyme–substrate complex.

3. The binding of the substrate and enzyme places stress on the glucose–fructose bond, and the bond breaks.

Glucose

Fructose

H_2O

4. Products are released, and the enzyme is free to bind other substrates.

Active site

Enzyme sucrase

Figure 6.10 The catalytic cycle of an enzyme. Enzymes increase the speed at which chemical reactions occur, but they are not altered permanently themselves as they do so. In the reaction illustrated here, the enzyme sucrase is splitting the sugar sucrose into two simpler sugars: glucose and fructose.

as integral parts of cell membranes and organelles. Enzymes may also form associations called *multienzyme complexes* to carry out reaction sequences. And, as mentioned earlier, evidence exists that some enzymes may consist of RNA rather than being only protein.

Multienzyme complexes

Often several enzymes catalyzing different steps of a sequence of reactions are associated with one another in noncovalently bonded assemblies called **multienzyme complexes.** The bacterial pyruvate dehydrogenase multienzyme complex, shown in figure 6.11, contains enzymes that carry out three sequential reactions in oxidative metabolism. Each complex has multiple copies of each of the three enzymes—60 protein subunits in all. The many subunits work together to form a molecular machine that performs multiple functions.

Multienzyme complexes offer the following significant advantages in catalytic efficiency:

1. The rate of any enzyme reaction is limited by how often the enzyme collides with its substrate. If a series of sequential reactions occurs within a multienzyme complex, the product of one reaction can be delivered to the next enzyme without releasing it to diffuse away.
2. Because the reacting substrate doesn't leave the complex while it goes through the series of reactions, unwanted side reactions are prevented.
3. All of the reactions that take place within the multienzyme complex can be controlled as a unit.

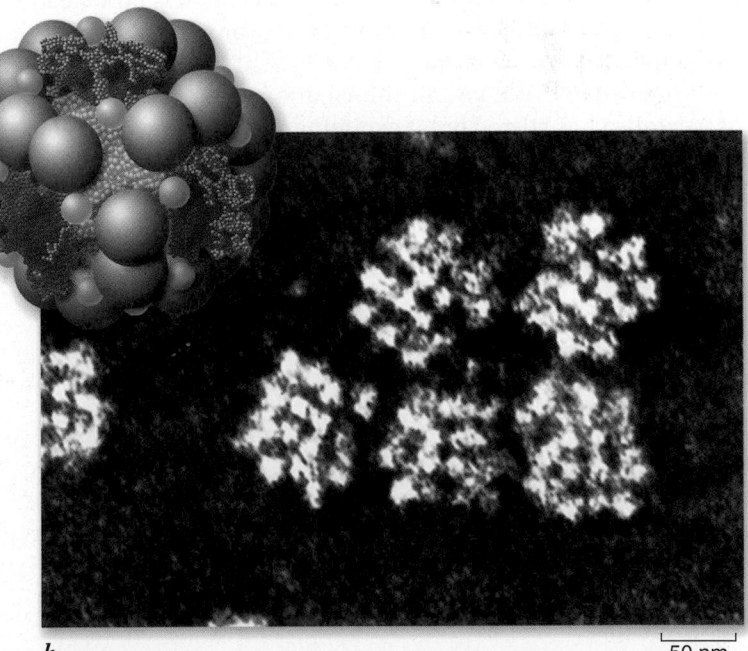

a.

b.

50 nm

Figure 6.11 A complex enzyme: pyruvate dehydrogenase. Pyruvate dehydrogenase, which catalyzes the oxidation of pyruvate, is one of the most complex enzymes known. *a.* A model of the enzyme showing the arrangement of the 60 protein subunits. *b.* Many of the protein subunits are clearly visible in the electron micrograph.

In addition to pyruvate dehydrogenase, which controls entry to the Krebs cycle during aerobic respiration (see chapter 7), several other key processes in the cell are catalyzed by multienzyme complexes. One well-studied system is the fatty acid synthetase complex that catalyzes the synthesis of fatty acids from two-carbon precursors. Seven different enzymes make up this multienzyme complex, and the intermediate reaction products remain associated with the complex for the entire series of reactions.

Nonprotein enzymes

Until a few years ago, most biology textbooks contained statements such as "Proteins called enzymes are the catalysts of biological systems." We can no longer make that statement without qualification.

Thomas R. Cech and colleagues at the University of Colorado reported in 1981 that certain reactions involving RNA molecules appear to be catalyzed in cells by RNA itself, rather than by enzymes. This initial observation has been corroborated by additional examples of RNA catalysis. Like enzymes, these RNA catalysts, which are loosely called "ribozymes," greatly accelerate the rate of particular biochemical reactions and show extraordinary substrate specificity.

Research has revealed at least two sorts of ribozymes. Some ribozymes have folded structures and catalyze reactions on themselves, a process called *intra*molecular catalysis. Other ribozymes act on other molecules without being changed themselves, a process called *inter*molecular catalysis.

The most striking example of the role of RNA as enzyme is emerging from recent work on the structure and function of the ribosome. For many years it was thought that RNA was a structural framework for this vital organelle, but it is now clear that ribosomal RNA plays a key role in ribosome function. The ribosome itself is a ribozyme.

The ability of RNA, an informational molecule, to act as a catalyst has stirred great excitement because it seems to answer the question—Which came first, the protein or the nucleic acid? It now seems at least possible that RNA evolved first and may have catalyzed the formation of the first proteins.

Environmental and other factors affect enzyme function

The rate of an enzyme-catalyzed reaction is affected by the concentrations of both the substrate and the enzyme that works on it. In addition, any chemical or physical factor that alters the enzyme's three-dimensional shape—such as temperature, pH, and the binding of regulatory molecules—can affect the enzyme's ability to catalyze the reaction.

Temperature

Increasing the temperature of an uncatalyzed reaction increases its rate because the additional heat increases random molecular movement. This motion can add stress to molecular bonds and affect the activation energy of a reaction.

The rate of an enzyme-catalyzed reaction also increases with temperature, but only up to a point called the *optimum*

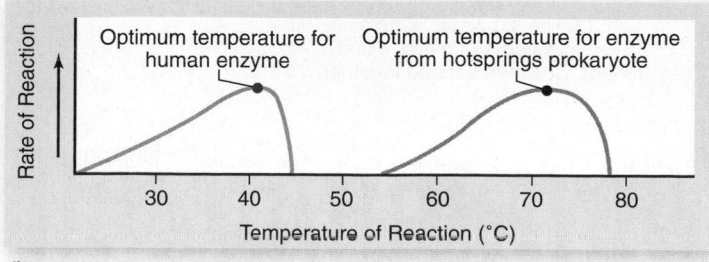

a.

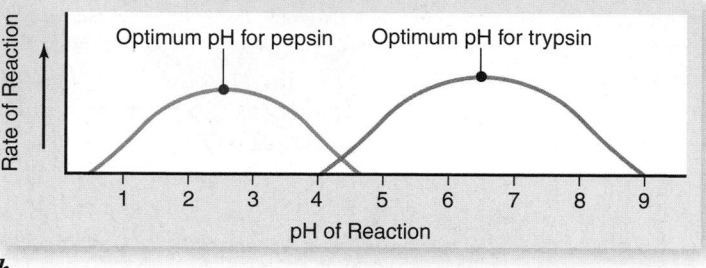

b.

Figure 6.12 Enzyme sensitivity to the environment. The activity of an enzyme is influenced by both **(a)** temperature and **(b)** pH. Most human enzymes, such as the protein-degrading enzyme trypsin, work best at temperatures of about 40°C and within a pH range of 6 to 8. The hot springs prokaryote tolerates a higher environmental temperature and a correspondingly higher temperature optimum for enzymes. Pepsin works in the acidic environment of the stomach and has a lower optimum pH.

temperature (figure 6.12*a*). Below this temperature, the hydrogen bonds and hydrophobic interactions that determine the enzyme's shape are not flexible enough to permit the induced fit that is optimum for catalysis. Above the optimum temperature, these forces are too weak to maintain the enzyme's shape against the increased random movement of the atoms in the enzyme. At higher temperatures, the enzyme denatures, as described in chapter 3.

Most human enzymes have an optimum temperature between 35°C and 40°C—a range that includes normal body temperature. Prokaryotes that live in hot springs have more stable enzymes (that is, enzymes held together more strongly), so the optimum temperature for those enzymes can be 70°C or higher. In each case the optimal temperature for the enzyme corresponds to the "normal" temperature usually encountered in the body or the environment, depending on the type of organism.

pH

Ionic interactions between oppositely charged amino acid residues, such as glutamic acid (−) and lysine (+), also hold enzymes together. These interactions are sensitive to the hydrogen ion concentration of the fluid in which the enzyme is dissolved, because changing that concentration shifts the balance between positively and negatively charged amino acid residues. For this reason, most enzymes have an *optimum pH* that usually ranges from pH 6 to 8.

Enzymes able to function in very acidic environments are proteins that maintain their three-dimensional shape even in the presence of high hydrogen ion concentrations. The enzyme pepsin, for example, digests proteins in the stomach at pH 2, a very acidic level (figure 6.12*b*).

Inhibitors and activators

Enzyme activity is also sensitive to the presence of specific substances that can bind to the enzyme and cause changes in its shape. Through these substances, a cell is able to regulate which of its enzymes are active and which are inactive at a particular time. This ability allows the cell to increase its efficiency and to control changes in its characteristics during development. A substance that binds to an enzyme and *decreases* its activity is called an **inhibitor.** Very often, the end product of a biochemical pathway acts as an inhibitor of an early reaction in the pathway, a process called *feedback inhibition* (discussed later in this chapter).

Enzyme inhibition occurs in two ways: **Competitive inhibitors** compete with the substrate for the same active site, occupying the active site and thus preventing substrates from binding; **noncompetitive inhibitors** bind to the enzyme in a location other than the active site, changing the shape of the enzyme and making it unable to bind to the substrate (figure 6.13).

Many enzymes can exist in either an active or inactive conformation; such enzymes are called *allosteric enzymes.* Most noncompetitive inhibitors bind to a specific portion of the enzyme called an **allosteric site.** These sites serve as chemical on/off switches; the binding of a substance to the site can switch the enzyme between its active and inactive configurations. A substance that binds to an allosteric site and reduces enzyme activity is called an **allosteric inhibitor** (figure 6.13*b*).

This kind of control is also used to activate enzymes. An **allosteric activator** binds to allosteric sites to keep an enzyme in its active configuration, thereby *increasing* enzyme activity.

Enzyme cofactors

Enzyme function is often assisted by additional chemical components known as **cofactors.** These can be metal ions that are often found in the active site participating directly in catalysis. For example, the metallic ion zinc is used by some enzymes, such as protein-digesting carboxypeptidase, to draw electrons away from their position in covalent bonds, making the bonds less stable and easier to break. Other metallic elements, such as molybdenum and manganese, are also used as cofactors. Like zinc, these substances are required in the diet in small amounts.

When the cofactor is a nonprotein organic molecule, it is called a **coenzyme.** Many of the small organic molecules essential in our diets that we call vitamins function as coenzymes. For example, the B vitamins B_6 and B_{12} both function as coenzymes for a number of different enzymes. Modified nucleotides are also used as coenzymes.

In numerous oxidation–reduction reactions that are catalyzed by enzymes, the electrons pass in pairs from the active site of the enzyme to a coenzyme that serves as the electron acceptor. The coenzyme then transfers the electrons to a different enzyme, which releases them (and the energy they bear) to the substrates in another reaction. Often, the electrons combine with protons (H^+) to form hydrogen atoms. In this way, coenzymes shuttle energy in the form of hydrogen atoms from one enzyme to another in a cell. The role of coenzymes and the specifics of their action will be explored in detail in the following two chapters.

Learning Outcomes Review 6.4

Enzymes are biological catalysts that accelerate chemical reactions inside the cell. Enzymes bind to their substrates based on molecular shape, which allows them to be highly specific. Enzyme activity is affected by conditions such as temperature and pH and the presence of inhibitors or activators. Some enzymes also require an inorganic cofactor or an organic coenzyme.

■ *Why do proteins and RNA function as enzymes but DNA does not?*

6.5 Metabolism: The Chemical Description of Cell Function

Learning Outcomes

1. *Explain the kinds of reactions that make up metabolism.*
2. *Discuss what is meant by a metabolic pathway.*
3. *Recognize that metabolism is a product of evolution.*

Living chemistry, the total of all chemical reactions carried out by an organism, is called **metabolism.** Those chemical reactions that expend energy to build up molecules are called *anabolic* reactions, or **anabolism.** Reactions that harvest energy by breaking down molecules are called *catabolic* reactions, or **catabolism.** This section presents a general overview of metabolic processes that will be described in much greater detail in later chapters.

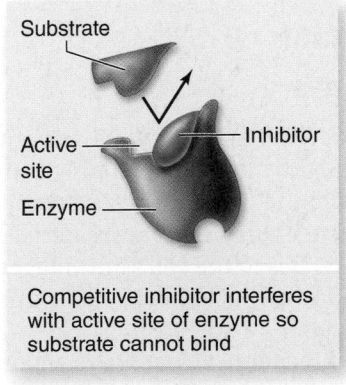
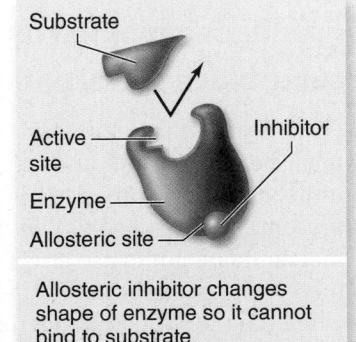

Substrate

Active site — Inhibitor

Enzyme

Competitive inhibitor interferes with active site of enzyme so substrate cannot bind

a. Competitive inhibition

Substrate

Active site — Inhibitor

Enzyme

Allosteric site

Allosteric inhibitor changes shape of enzyme so it cannot bind to substrate

b. Noncompetitive inhibition

Figure 6.13 How enzymes can be inhibited. *a.* In competitive inhibition, the inhibitor has a shape similar to the substrate and competes for the active site of the enzyme. *b.* In noncompetitive inhibition, the inhibitor binds to the enzyme at the allosteric site, a place away from the active site, effecting a conformational change in the enzyme, making it unable to bind to its substrate.

Biochemical pathways organize chemical reactions in cells

Organisms contain thousands of different kinds of enzymes that catalyze a bewildering variety of reactions. Many of these reactions in a cell occur in sequences called **biochemical pathways.** In such pathways, the product of one reaction becomes the substrate for the next (figure 6.14). Biochemical pathways are the organizational units of metabolism—the elements an organism controls to achieve coherent metabolic activity.

Many sequential enzyme steps in biochemical pathways take place in specific compartments of the cell; for example, the steps of the Krebs cycle (see chapter 7) occur in the matrix inside mitochondria in eukaryotes. By determining where many of the enzymes that catalyze these steps are located, we can "map out" a model of metabolic processes in the cell.

Biochemical pathways may have evolved in stepwise fashion

In the earliest cells, the first biochemical processes probably involved energy-rich molecules scavenged from the environment. Most of the molecules necessary for these processes are thought to have existed independently in the "organic soup" of the early oceans.

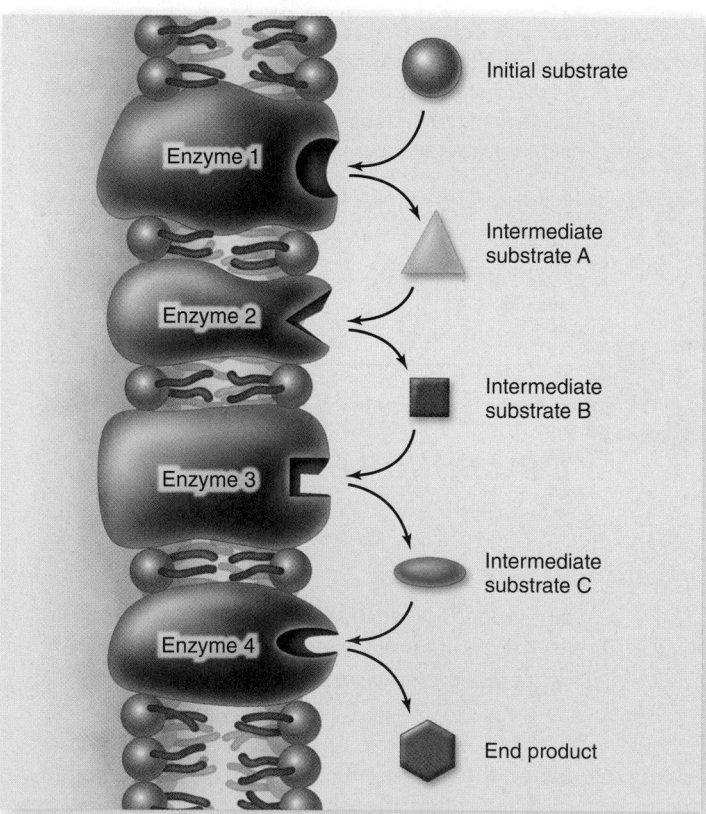

Figure 6.14 A biochemical pathway. The original substrate is acted on by enzyme 1, changing the substrate to a new intermediate, substrate A, recognized as a substrate by enzyme 2. Each enzyme in the pathway acts on the product of the previous stage. These enzymes may be either soluble or arranged in a membrane as shown.

The first catalyzed reactions were probably simple, one-step reactions that brought these molecules together in various combinations. Eventually, the energy-rich molecules became depleted in the external environment, and only organisms that had evolved some means of making those molecules from other substances could survive. Thus, a hypothetical reaction,

$$\begin{matrix} F \\ + \\ G \end{matrix} \longrightarrow H$$

where two energy-rich molecules (F and G) react to produce compound H and release energy, became more complex when the supply of F in the environment ran out.

A new reaction was added in which the depleted molecule, F, is made from another molecule, E, which was also present in the environment:

$$E \longrightarrow \begin{matrix} F \\ + \\ G \end{matrix} \longrightarrow H$$

When the supply of E was in turn exhausted, organisms that were able to make E from some other available precursor, D, survived. When D was depleted, those organisms in turn were replaced by ones able to synthesize D from another molecule, C:

$$C \longrightarrow D \longrightarrow E \longrightarrow \begin{matrix} F \\ + \\ G \end{matrix} \longrightarrow H$$

This hypothetical biochemical pathway would have evolved slowly through time, with the final reactions in the pathway evolving first and earlier reactions evolving later.

Looking at the pathway now, we would say that the "advanced" organism, starting with compound C, is able to synthesize H by means of a series of steps. This is how the biochemical pathways within organisms are thought to have evolved—not all at once, but one step at a time, backward.

Feedback inhibition regulates some biochemical pathways

For a biochemical pathway to operate efficiently, its activity must be coordinated and regulated by the cell. Not only is it unnecessary to synthesize a compound when plenty is already present, but doing so would waste energy and raw materials that could be put to use elsewhere. It is to the cell's advantage, therefore, to temporarily shut down biochemical pathways when their products are not needed.

The regulation of simple biochemical pathways often depends on an elegant feedback mechanism: The end-product of the pathway binds to an allosteric site on the enzyme that catalyzes the first reaction in the pathway. This mode of regulation is called **feedback inhibition** (figure 6.15).

In the hypothetical pathway we just described, the enzyme catalyzing the reaction C $\longrightarrow$ D would possess an allosteric site for H, the end-product of the pathway. As the pathway churned out its product and the amount of H in the cell increased, it would become more likely that an H molecule would encounter

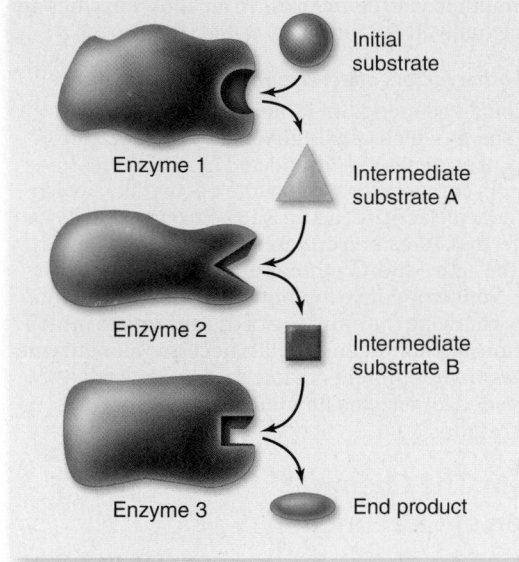

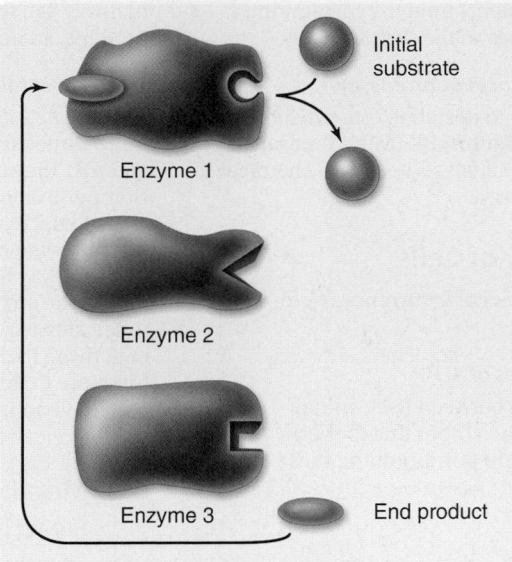

Figure 6.15 Feedback inhibition. *a.* A biochemical pathway with no feedback inhibition. *b.* A biochemical pathway in which the final end-product becomes the allosteric inhibitor for the first enzyme in the pathway. In other words, the formation of the pathway's final end-product stops the pathway. The pathway could be the synthesis of an amino acid, a nucleotide, or another important cellular molecule.

the allosteric site on the C $\longrightarrow$ D enzyme. Binding to the allosteric site would essentially shut down the reaction C $\longrightarrow$ D and in turn effectively shut down the whole pathway.

In this chapter we have reviewed the basics of energy and its transformations as carried out in living systems. Chemical bonds are the primary location of energy storage and release, and cells have developed elegant methods of making and breaking chemical bonds to create the molecules they need. Enzymes facilitate these reactions by serving as catalysts. In the following chapters you will learn the details of the mechanisms by which organisms harvest, store, and utilize energy.

Learning Outcomes Review 6.5

Metabolism is the sum of all chemical reactions in a cell. Anabolic reactions use energy to build up molecules. Catabolic reactions release energy by breaking down molecules. In a metabolic pathway, the end-product of one reaction is the substrate for the next reaction. Evolution may have favored organisms that could use precursor molecules to synthesize a nutrient. Over time, more reactions would be linked together as novel enzymes arose by mutation.

■ *Is a catabolic pathway likely to be subject to feedback inhibition?*

Chapter Review

6.1 The Flow of Energy in Living Systems

Thermodynamics is the study of energy changes.

Energy can take many forms.

Energy is the capacity to do work. Potential energy is stored energy, and kinetic energy is the energy of motion. Energy can take many forms: mechanical, heat, sound, electric current, light, or radioactive radiation. Energy is measured in units of heat known as kilocalories.

The Sun provides energy for living systems.

Photosynthesis stores light energy from the Sun as potential energy in the covalent bonds of sugar molecules. Breaking these bonds in living cells releases energy for use in other reactions.

Oxidation–reduction reactions transfer electrons while bonds are made or broken.

Oxidation is a reaction involving the loss of electrons. Reduction is the gain of electrons (figure 6.2). These two reactions take place together and are therefore termed redox reactions.

6.2 The Laws of Thermodynamics and Free Energy

The First Law states that energy cannot be created or destroyed.

Virtually all activities of living organisms require energy. Energy changes form as it moves through organisms and their biochemical systems, but it is not created or destroyed.

The Second Law states that some energy is lost as disorder increases.

The disorder, or entropy, of the universe is continuously increasing. In an open system like the Earth, which is receiving energy from the Sun, this may not be the case. To increase order, however, energy must be expended. In energy conversions, some energy is always lost as heat.

Chemical reactions can be predicted based on changes in free energy.

Free energy (G) is the energy available to do work in any system. Changes in free energy (ΔG) predict the direction of reactions. Reactions with a negative ΔG are spontaneous (exergonic) reactions, and reactions with a positive ΔG are not spontaneous (endergonic).

chapter **6** *Energy and Metabolism*

Endergonic chemical reactions absorb energy from the surroundings, whereas exergonic reactions release energy to the surroundings.

Spontaneous chemical reactions require activation energy.

Activation energy is the energy required to destabilize chemical bonds and initiate chemical reactions (figure 6.5). Even exergonic reactions require this activation energy. Catalysts speed up chemical reactions by lowering the activation energy.

6.3 ATP: The Energy Currency of Cells

Adenosine triphosphate (ATP) is the molecular currency used for cellular energy transactions.

Cells store and release energy in the bonds of ATP.

The energy of ATP is stored in the bonds between its terminal phosphate groups. These groups repel each other due to their negative charge and therefore the covalent bonds joining these phosphates are unstable.

ATP hydrolysis drives endergonic reactions.

Enzymes hydrolyze the terminal phosphate group of ATP to release energy for reactions. If ATP hydrolysis is coupled to an endergonic reaction with a positive ΔG with magnitude less than that for ATP hydrolysis, the two reactions together will be exergonic.

ATP cycles continuously.

ATP hydrolysis releases energy to drive endergonic reactions, and it is synthesized with energy from exergonic reactions (figure 6.7).

6.4 Enzymes: Biological Catalysts

An enzyme alters the activation energy of a reaction.

Enzymes lower the activation energy needed to initiate a chemical reaction.

Active sites of enzymes conform to fit the shape of substrates.

Substrates bind to the active site of an enzyme. Enzymes adjust their shape to the substrate so there is a better fit (figure 6.8).

Enzymes occur in many forms.

Enzymes can be free in the cytosol or exist as components bound to membranes and organelles. Enzymes involved in a biochemical

pathway can form multienzyme complexes. While most enzymes are proteins, some are actually RNA molecules, called ribozymes.

Environmental and other factors affect enzyme function.

An enzyme's functionality depends on its ability to maintain its three-dimensional shape, which can be affected by temperature and pH. The activity of enzymes can be affected by inhibitors. Competitive inhibitors compete for the enzyme's active site, which leads to decreased enzyme activity (figure 6.13). Enzyme activity can be controlled by effectors. Allosteric enzymes have a second site, located away from the active site, that binds effectors to activate or inhibit the enzyme. Noncompetitive inhibitors and activators bind to the allosteric site, changing the structure of the enzyme to inhibit or activate it. Cofactors are nonorganic metals necessary for enzyme function. Coenzymes are nonprotein organic molecules, such as certain vitamins, needed for enzyme function. Often coenzymes serve as electron acceptors.

6.5 Metabolism: The Chemical Description of Cell Function

Metabolism is the sum of all biochemical reactions in a cell. Anabolic reactions require energy to build up molecules, and catabolic reactions break down molecules and release energy.

Biochemical pathways organize chemical reactions in cells.

Chemical reactions in biochemical pathways use the product of one reaction as the substrate for the next.

Biochemical pathways may have evolved in stepwise fashion.

In the primordial "soup" of the early oceans, many reactions were probably single-step reactions combining two molecules. As one of the substrate molecules was depleted, organisms having an enzyme that could synthesize the substrate would have a selective advantage. In this manner, biochemical pathways are thought to have evolved "backward" with new reactions producing limiting substrates for existing reactions.

Feedback inhibition regulates some biochemical pathways.

Biosynthetic pathways are often regulated by the end product of the pathway. Feedback inhibition occurs when the end-product of a reaction combines with an enzyme's allosteric site to shut down the enzyme's activity (figure 6.15).

Review Questions

UNDERSTAND

1. A covalent bond between two atoms represents what kind of energy?
   a. Kinetic energy
   b. Potential energy
   c. Mechanical energy
   d. Solar energy

2. During a redox reaction the molecule that gains an electron has been
   a. reduced and now has a higher energy level.
   b. oxidized and now has a lower energy level.
   c. reduced and now has a lower energy level.
   d. oxidized and now has a higher energy level.

3. An endergonic reaction has the following properties
   a. $+\Delta G$ and the reaction is spontaneous.
   b. $+\Delta G$ and the reaction is not spontaneous.
   c. $-\Delta G$ and the reaction is spontaneous.
   d. $-\Delta G$ and the reaction is not spontaneous.

4. A spontaneous reaction is one in which
   a. the reactants have a higher free energy than the products.
   b. the products have a higher free energy than the reactants.
   c. an input of energy is required.
   d. entropy is decreased.

5. What is *activation energy*?

 a. The thermal energy associated with random movements of molecules
 b. The energy released through breaking chemical bonds
 c. The difference in free energy between reactants and products
 d. The energy required to initiate a chemical reaction

6. Which of the following is NOT a property of a catalyst?

 a. A catalyst reduces the activation energy of a reaction.
 b. A catalyst lowers the free energy of the reactants.
 c. A catalyst does not change as a result of the reaction.
 d. A catalyst works in both the forward and reverse directions of a reaction.

7. Where is the energy stored in a molecule of ATP?

 a. Within the bonds between nitrogen and carbon
 b. In the carbon-to-carbon bonds found in the ribose
 c. In the phosphorus-to-oxygen double bond
 d. In the bonds connecting the two terminal phosphate groups

APPLY

1. Cells use ATP to drive endergonic reactions because

 a. ATP is the universal catalyst.
 b. energy released by ATP hydrolysis makes ΔG for coupled reactions more negative.
 c. energy released by ATP hydrolysis makes ΔG for coupled reactions more positive.
 d. the conversion of ATP to ADP is also endergonic.

2. Which of the following statements is NOT true about enzymes?

 a. Enzymes use the three-dimensional shape of their active site to bind reactants.
 b. Enzymes lower the activation energy for a reaction.
 c. Enzymes make ΔG for a reaction more negative.
 d. Enzymes can catalyze the forward and reverse directions of a reaction.

3. ATP hydrolysis has a ΔG of –7.4 kcal/mol. Can an endergonic reaction with a ΔG of 12 kcal/mol be "driven" by ATP hydrolysis?

 a. No, the overall ΔG is still positive.
 b. Yes, the overall ΔG would now be negative.
 c. Yes, but only if an enzyme is used to lower ΔG.
 d. No, overall ΔG is now negative.

4. An online auction site offers a perpetual-motion machine. You decide not to bid on this because

 a. there is not enough energy in the universe to power this machine.
 b. the First Law says you cannot create energy.
 c. the Second Law says that energy loss due to entropy will not allow for perpetual motion.
 d. it could work, but would require a strong catalyst.

5. Enzymes have similar responses to both changes in temperature and pH. The effect of both is on the

 a. rate of movement of the substrate molecules.
 b. strength of the chemical bonds within the substrate.
 c. three-dimensional shape of the enzyme.
 d. rate of movement of the enzyme.

6. Feedback inhibition is an efficient way to control a metabolic pathway because the

 a. first enzyme in a pathway is inhibited by its own product.
 b. last enzyme in a pathway is inhibited by its own product.
 c. first enzyme in a pathway is inhibited by the end-product of the pathway.
 d. last enzyme in a pathway is inhibited by the end-product of the pathway.

SYNTHESIZE

1. Examine the graph showing the rate of reaction versus temperature for an enzyme–catalyzed reaction in a human.

 a. Describe what is happening to the enzyme at around 40°C.
 b. Explain why the line touches the *x*-axis at approximately 20°C and 45°C.
 c. Average body temperature for humans is 37°C. Suggest a reason why the temperature optimum of this enzyme is greater than 37°C.

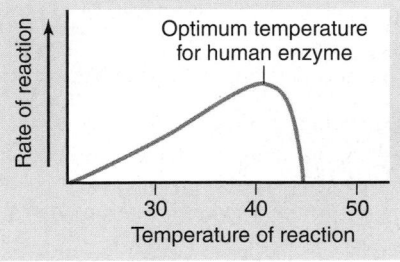

2. Phosphofructokinase functions to add a phosphate group to a molecule of fructose 6-phosphate. This enzyme functions early in glycolysis, an energy-yielding biochemical pathway discussed in chapter 7. The enzyme has an active site that binds fructose and ATP. An allosteric inhibitory site also binds ATP when cellular levels of ATP are very high.

 a. Predict the rate of the reaction if the levels of cellular ATP are low.
 b. Predict the rate of the reaction if levels of cellular ATP are very high.
 c. Describe what is happening to the enzyme when levels of ATP are very high.

ONLINE RESOURCE

www.ravenbiology.com

Understand, Apply, and Synthesize—enhance your study with animations that bring concepts to life and practice tests to assess your understanding. Your instructor may also recommend the interactive eBook, individualized learning tools, and more.

Chapter 7

How Cells Harvest Energy

Chapter Contents

Introduction

Life is driven by energy. All the activities organisms carry out—the swimming of bacteria, the purring of a cat, your thinking about these words—use energy. In this chapter, we discuss the processes all cells use to derive chemical energy from organic molecules and to convert that energy to ATP. Then, in chapter 8, we will examine photosynthesis, which uses light energy to make chemical energy. We consider the conversion of chemical energy to ATP first because all organisms, both the plant, a photosynthesizer, and the caterpillar feeding on the plant, pictured in the photo are capable of harvesting energy from chemical bonds. Energy harvest via respiration is a universal process.

Overview of Respiration

Plants, algae, and some bacteria harvest the energy of sunlight through photosynthesis, converting radiant energy into chemical energy. These organisms, along with a few others that use chemical energy in a similar way, are called **autotrophs** ("self-feeders"). All other organisms live on the organic compounds autotrophs produce, using them as food, and are called **heterotrophs** ("fed by others"). At least 95% of the kinds of organisms on Earth—all animals and fungi, and most protists and prokaryotes—are heterotrophs. Autotrophs also extract energy from organic compounds—they just have the additional capacity to use the energy from sunlight to synthesize these compounds. The process by which energy is harvested is **cellular respiration**—the oxidation of organic compounds to extract energy from chemical bonds.

Cells oxidize organic compounds to drive metabolism

Most foods contain a variety of carbohydrates, proteins, and fats, all rich in energy-laden chemical bonds. Carbohydrates and fats, as you recall from chapter 3, possess many carbon–hydrogen (C—H) bonds, as well as carbon–oxygen (C—O) bonds.

The job of extracting energy from the complex organic mixture in most foods is tackled in stages. First, enzymes break down the large molecules into smaller ones, a process called digestion (see chapter 47). Then, other enzymes dismantle these fragments a bit at a time, harvesting energy from C—H and other chemical bonds at each stage.

The reactions that break down these molecules share a common feature: They are oxidations. Energy metabolism is therefore concerned with redox reactions, and to understand the process we must follow the fate of the electrons lost from the food molecules.

These reactions are not the simple transfer of electrons, however; they are also **dehydrogenations.** That is, the electrons lost are accompanied by protons, so that what is really lost is a hydrogen atom, not just an electron.

Cellular respiration is the complete oxidation of glucose

In chapter 6, you learned that an atom that loses electrons is said to be *oxidized,* and an atom accepting electrons is said to be *reduced.* Oxidation reactions are often coupled with reduction reactions in living systems, and these paired reactions are called *redox reactions.* Cells utilize enzyme-facilitated redox reactions to take energy from food sources and convert it to ATP.

Redox reactions

Oxidation–reduction reactions play a key role in the flow of energy through biological systems because the electrons that pass from one atom to another carry energy with them. The amount of energy an electron possesses depends on its orbital position, or energy level, around the atom's nucleus. When this electron departs from one atom and moves to another in a redox reaction, the electron's energy is transferred with it.

Figure 7.1 shows how an enzyme catalyzes a redox reaction involving an energy-rich substrate molecule, with the help of a cofactor, **nicotinamide adenosine dinucleotide (NAD⁺).** In this reaction, NAD^+ accepts a pair of electrons from the substrate, along with a proton, to form **NADH** (this process is described in more detail shortly). The oxidized product is now released from the enzyme's active site, as is NADH.

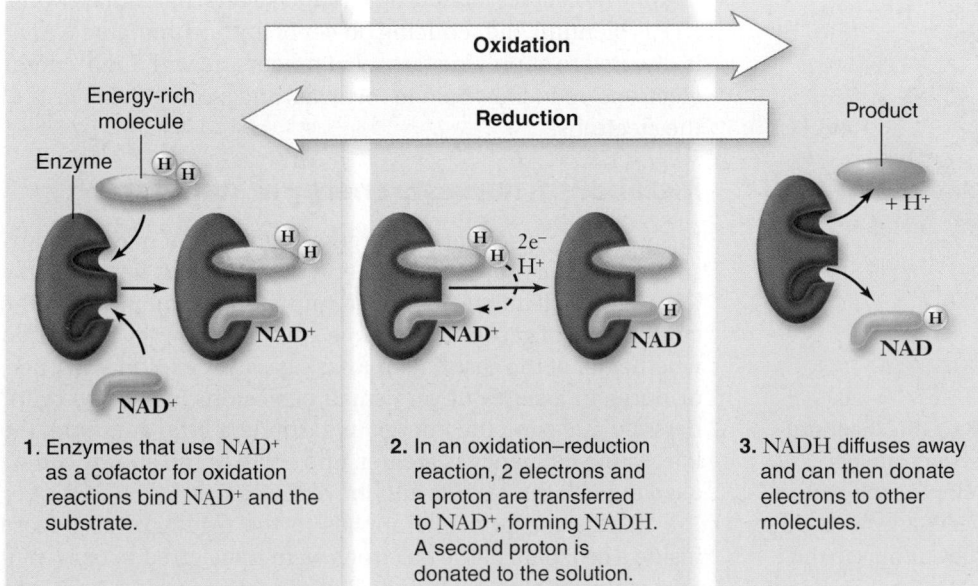

1. Enzymes that use NAD^+ as a cofactor for oxidation reactions bind NAD^+ and the substrate.

2. In an oxidation–reduction reaction, 2 electrons and a proton are transferred to NAD^+, forming NADH. A second proton is donated to the solution.

3. NADH diffuses away and can then donate electrons to other molecules.

Figure 7.1 Oxidation–reduction reactions often employ cofactors. Cells use a chemical cofactor called nicotinamide adenosine dinucleotide (NAD^+) to carry out many oxidation–reduction reactions. Two electrons and a proton are transferred to NAD^+ with another proton donated to the solution. Molecules that gain electrons are said to be reduced, and ones that lose energetic electrons are said to be oxidized. NAD^+ oxidizes energy-rich molecules by acquiring their electrons (in the figure, this proceeds $1 \longrightarrow 2 \longrightarrow 3$) and then reduces other molecules by giving the electrons to them (in the figure, this proceeds $3 \longrightarrow 2 \longrightarrow 1$). NADH is the reduced form of NAD^+.

In the overall process of cellular energy harvest dozens of redox reactions take place, and a number of molecules, including NAD⁺, act as electron acceptors. During each transfer of electrons energy is released. This energy may be captured and used to make ATP or to form other chemical bonds; the rest is lost as heat.

At the end of this process, high-energy electrons from the initial chemical bonds have lost much of their energy, and these depleted electrons are transferred to a final electron acceptor (figure 7.2). When this acceptor is oxygen, the process is called **aerobic respiration.** When the final electron acceptor is an inorganic molecule other than oxygen, the process is called **anaerobic respiration,** and when it is an organic molecule, the process is called **fermentation.**

"Burning" carbohydrates

Chemically, there is little difference between the catabolism of carbohydrates in a cell and the burning of wood in a fireplace. In both instances, the reactants are carbohydrates and oxygen, and the products are carbon dioxide, water, and energy:

$$C_6H_{12}O_6 + 6O_2 \longrightarrow 6CO_2 + 6H_2O + energy \text{ (heat and ATP)}$$
glucose   oxygen   carbon   water
                  dioxide

The change in free energy in this reaction is –686 kcal/mol (or –2870 kJ/mol) under standard conditions (that is, at room temperature, 1 atm pressure, and so forth). In the conditions that exist inside a cell, the energy released can be as high as –720 kcal/mol (–3012 kJ/mol) of glucose. This means that under actual cellular conditions, more energy is released than under standard conditions.

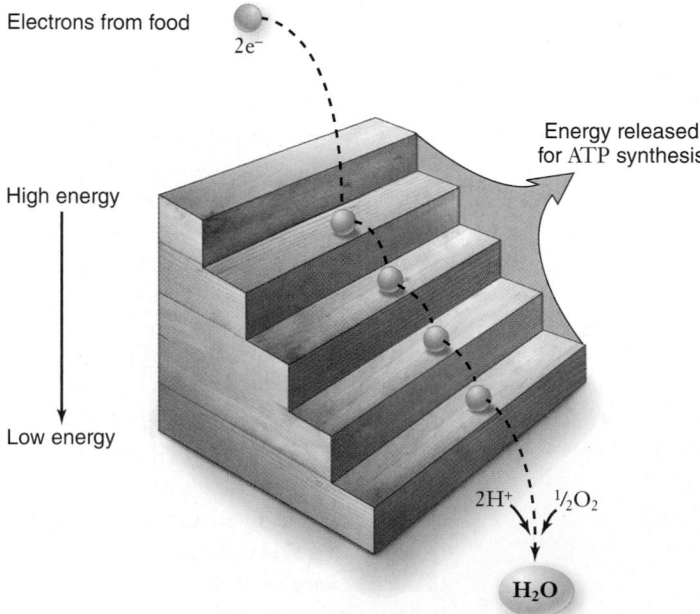

Figure 7.2 How electron transport works. This diagram shows how ATP is generated when electrons transfer from one energy level to another. Rather than releasing a single explosive burst of energy, electrons "fall" to lower and lower energy levels in steps, releasing stored energy with each fall as they tumble to the lowest (most electronegative) electron acceptor, O_2.

The same amount of energy is released whether glucose is catabolized or burned, but when it is burned, most of the energy is released as heat. Cells harvest useful energy from the catabolism of glucose by using a portion of the energy to drive the production of ATP.

Electron carriers play a critical role in energy metabolism

During respiration, glucose is oxidized to CO_2. If the electrons were given directly to O_2, the reaction would be combustion, and cells would burst into flames. Instead, as you have just seen, the cell transfers the electrons to intermediate electron carriers, then eventually to O_2.

Many forms of electron carriers are used in this process: (1) soluble carriers that move electrons from one molecule to another, (2) membrane-bound carriers that form a redox chain, and (3) carriers that move within the membrane. The common feature of all of these carriers is that they can be reversibly oxidized and reduced. Some of these carriers, such as the iron-containing cytochromes, can carry just electrons, and some carry both electrons and protons.

NAD⁺ is one of the most important electron (and proton) carriers. As shown on the left in figure 7.3, the NAD⁺ molecule is composed of two nucleotides bound together. The two nucleotides that make up NAD⁺, nicotinamide monophosphate (NMP) and adenosine monophosphate (AMP), are joined head-to-head by their phosphate groups. The two nucleotides serve different functions in the NAD⁺ molecule: AMP acts as the core, providing a shape recognized by many enzymes; NMP is the active part of the molecule, because it is readily reduced, that is, it easily accepts electrons.

When NAD⁺ acquires two electrons and a proton from the active site of an enzyme, it is reduced to NADH, shown on the right in figure 7.3. The NADH molecule now carries the two energetic electrons and can supply them to other molecules and reduce them.

This ability to supply high-energy electrons is critical to both energy metabolism and to the biosynthesis of many organic molecules, including fats and sugars. In animals, when ATP is plentiful, the reducing power of the accumulated NADH is diverted to supplying fatty acid precursors with high-energy electrons, reducing them to form fats and storing the energy of the electrons.

Metabolism harvests energy in stages

It is generally true that the larger the release of energy in any single step, the more of that energy is released as heat, and the less is available to be channeled into more useful paths. In the combustion of gasoline, the same amount of energy is released whether all of the gasoline in a car's gas tank explodes at once, or burns in a series of very small explosions inside the cylinders. By releasing the energy in gasoline a little at a time, the harvesting efficiency is greater, and more of the energy can be used to push the pistons and move the car.

The same principle applies to the oxidation of glucose inside a cell. If all of the electrons were transferred to oxygen in one explosive step, releasing all of the free energy at once, the

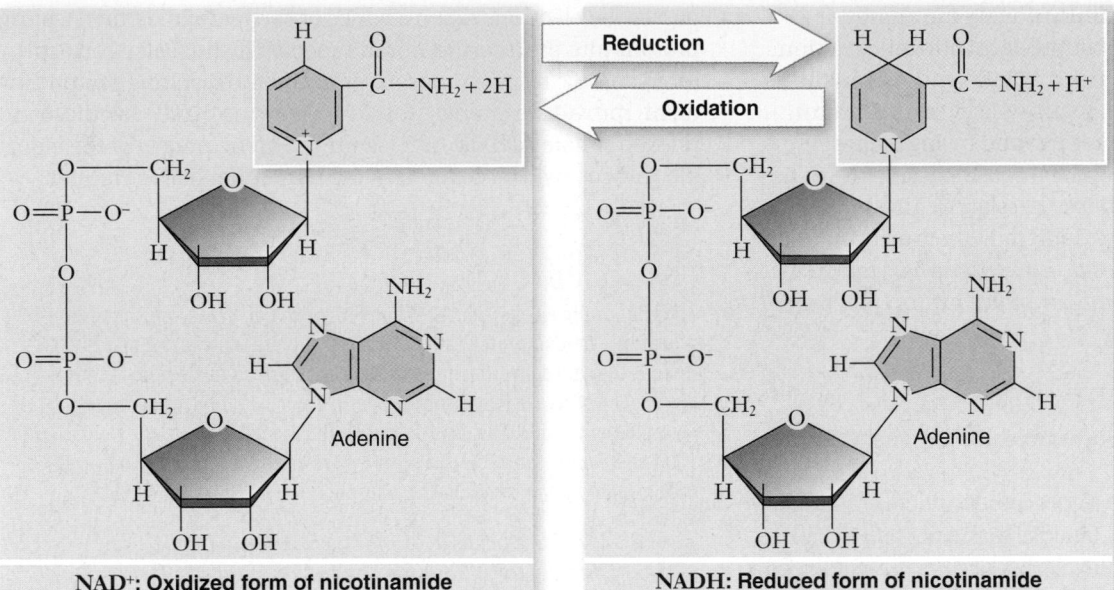

Figure 7.3 NAD⁺ and NADH. This dinucleotide serves as an "electron shuttle" during cellular respiration. NAD⁺ accepts a pair of electrons and a proton from catabolized macromolecules and is reduced to NADH.

NAD⁺: Oxidized form of nicotinamide

NADH: Reduced form of nicotinamide

cell would recover very little of that energy in a useful form. Instead, cells burn their fuel much as a car does, a little at a time.

The electrons in the C—H bonds of glucose are stripped off in stages in the series of enzyme-catalyzed reactions collectively referred to as glycolysis and the Krebs cycle. The electrons are removed by transferring them to NAD⁺, as described earlier, or to other electron carriers.

The energy released by all of these oxidation reactions is also not all released at once (see figure 7.2). The electrons are passed to another set of electron carriers called the **electron transport chain,** which is located in the mitochondrial inner membrane. Movement of electrons through this chain produces potential energy in the form of an electrochemical gradient. We examine this process in more detail later in this chapter.

ATP plays a central role in metabolism

The previous chapter introduced ATP as the energy currency of the cell. Cells use ATP to power most of those activities that require work—one of the most obvious of which is movement. Tiny fibers within muscle cells pull against one another when muscles contract. Mitochondria can move a meter or more along the narrow nerve cells that extend from your spine to your feet. Chromosomes are pulled apart by microtubules during cell division. All of these movements require the expenditure of energy by ATP hydrolysis. Cells also use ATP to drive endergonic reactions that would otherwise not occur spontaneously (see chapter 6).

How does ATP drive an endergonic reaction? The enzyme that catalyzes a particular reaction has two binding sites on its surface: one for the reactant and another for ATP. The ATP site splits the ATP molecule, liberating over 7 kcal ($\Delta G = -7.3$ kcal/mol) of chemical energy. This energy pushes the reactant at the second site "uphill," reaching the activation energy and driving the endergonic reaction. Thus endergonic reactions coupled to ATP hydrolysis become favorable.

The many steps of cellular respiration have as their ultimate goal the production of ATP. ATP synthesis is itself an endergonic reaction, which requires energy from cellular exergonic reactions to drive this synthesis.

Cells make ATP by two fundamentally different mechanisms

The synthesis of ATP can be accomplished by two distinct mechanisms: one that involves chemical coupling with an intermediate bound to phosphate, and another that relies on an electrochemical gradient of protons for the potential energy to phosphorylate ADP.

1. In *substrate-level phosphorylation,* ATP is formed by transferring a phosphate group directly to ADP from a phosphate-bearing intermediate, or substrate (figure 7.4). During **glycolysis,** the initial breakdown of glucose (discussed later), the chemical bonds of glucose are

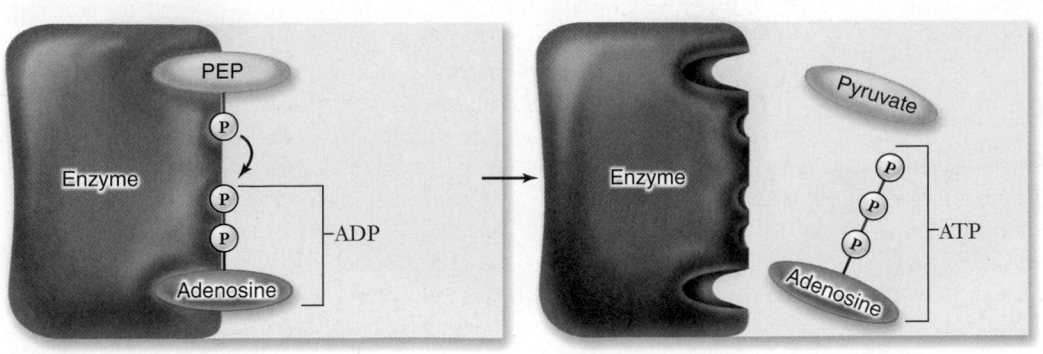

Figure 7.4 Substrate-level phosphorylation. Some molecules, such as phosphoenolpyruvate (PEP), possess a high-energy phosphate (P) bond similar to the bonds in ATP. When PEP's phosphate group is transferred enzymatically to ADP, the energy in the bond is conserved, and ATP is created.

shifted around in reactions that provide the energy required to form ATP by substrate-level phosphorylation.

2. In **oxidative phosphorylation,** ATP is synthesized by the enzyme **ATP synthase,** using energy from a proton (H^+) gradient. This gradient is formed by high-energy electrons from the oxidation of glucose passing down an electron transport chain (described later). These electrons, with their energy depleted, are then donated to oxygen, hence the term *oxidative phosphorylation*. ATP synthase uses the energy from the proton gradient to catalyze the reaction:

$$ADP + P_i \longrightarrow ATP$$

Eukaryotes and aerobic prokaryotes produce the vast majority of their ATP this way.

In most organisms, these two processes are combined. To harvest energy to make ATP from glucose in the presence of oxygen, the cell carries out a complex series of enzyme-catalyzed reactions that remove energetic electrons via oxidation reactions. These electrons are then used in an electron transport chain that passes the electrons down a series of carriers while translocating protons into the intermembrane space. The final electron acceptor in aerobic respiration is oxygen, and the resulting proton gradient provides energy for the enzyme ATP synthase to phosphorylate ADP to ATP (figure 7.5). The details of this complex process will be covered in the remainder of this chapter.

Learning Outcomes Review 7.1

Cells acquire energy from the complete oxidation of glucose. In these redox reactions, protons as well as electrons are transferred, and thus they are dehydrogenation reactions. Electron carriers aid in the gradual, stepwise release of the energy from oxidation, rather than rapid combustion. The result is the synthesis of ATP, a portable source of energy. ATP synthesis can occur by two mechanisms: substrate-level phosphorylation and oxidative phosphorylation.

■ *Why don't cells just link the oxidation of glucose directly to cellular functions that require the energy?*

Figure 7.5 An overview of aerobic respiration.

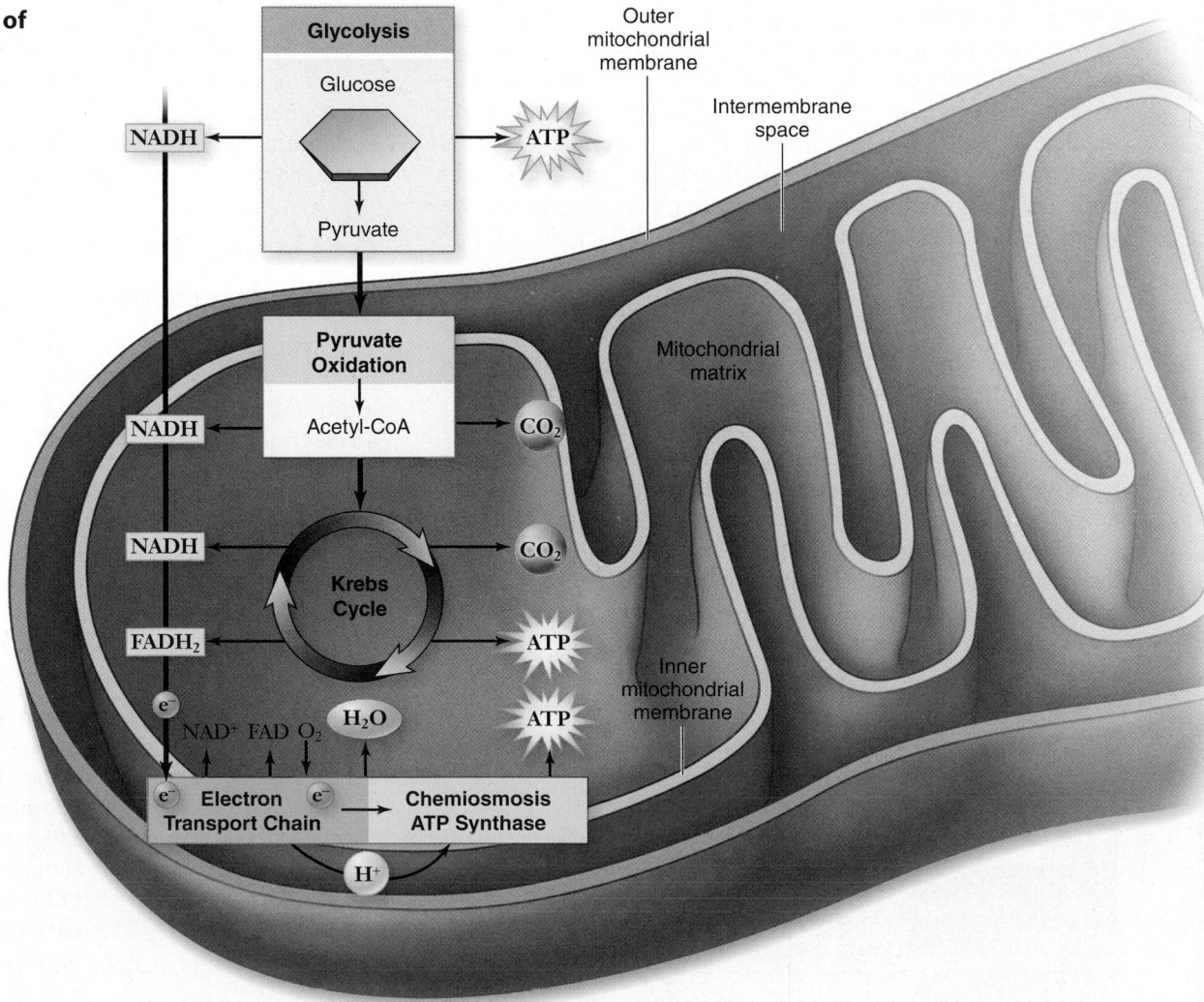

7.2 Glycolysis: Splitting Glucose

Glucose molecules can be dismantled in many ways, but primitive organisms evolved a glucose-catabolizing process that releases enough free energy to drive the synthesis of ATP in enzyme-coupled reactions. Glycolysis occurs in the cytoplasm and converts glucose into two 3-carbon molecules of pyruvate (figure 7.6). For each molecule of glucose that passes through this transformation, the cell nets two ATP molecules.

Glycolysis converts glucose into two pyruvate and yields two ATP and two NADH in the process

The first half of glycolysis consists of five sequential reactions that convert one molecule of glucose into two molecules of the 3-carbon compound **glyceraldehyde 3-phosphate (G3P).** These reactions require the expenditure of ATP, so they constitute an endergonic process. In the second half of glycolysis, five more reactions convert G3P into pyruvate in an energy-yielding process that generates ATP.

Priming reactions The first three reactions "prime" glucose by changing it into a compound that can be readily cleaved into two 3-carbon phosphorylated molecules. Two of these reactions transfer a phosphate from ATP, so this step requires the cell to use two ATP molecules.

Cleavage This 6-carbon diphosphate sugar is then split into two 3-carbon monophosphate sugars. One of these is G3P, and the other is converted into G3P. The G3P then undergoes a series of reactions that eventually yields more energy than was spent priming (figure 7.7).

Oxidation and ATP formation Each G3P is oxidized, transferring two electrons (and one proton) to NAD$^+$, thus forming NADH. A molecule of P$_i$ is also added to G3P to produce 1,3-bisphosphoglycerate (BPG). The phosphate incorporated can be transferred to ADP by substrate-level phosphorylation (see figure 7.4) to allow a positive yield of ATP at the end of the process.

Another four reactions convert BPG into pyruvate. In the process, the phosphates are transferred to ADP to yield two ATP per G3P. The entire process is shown in detail in figure 7.7.

Each glucose molecule is split into two G3P molecules, so the overall reaction sequence has a net yield of two molecules of ATP, as well as two molecules of NADH and two of pyruvate:

4 ATP (2 ATP for each of the 2 G3P molecules)
– 2 ATP (used in the two reactions in the first step)

2 ATP (net yield for entire process)

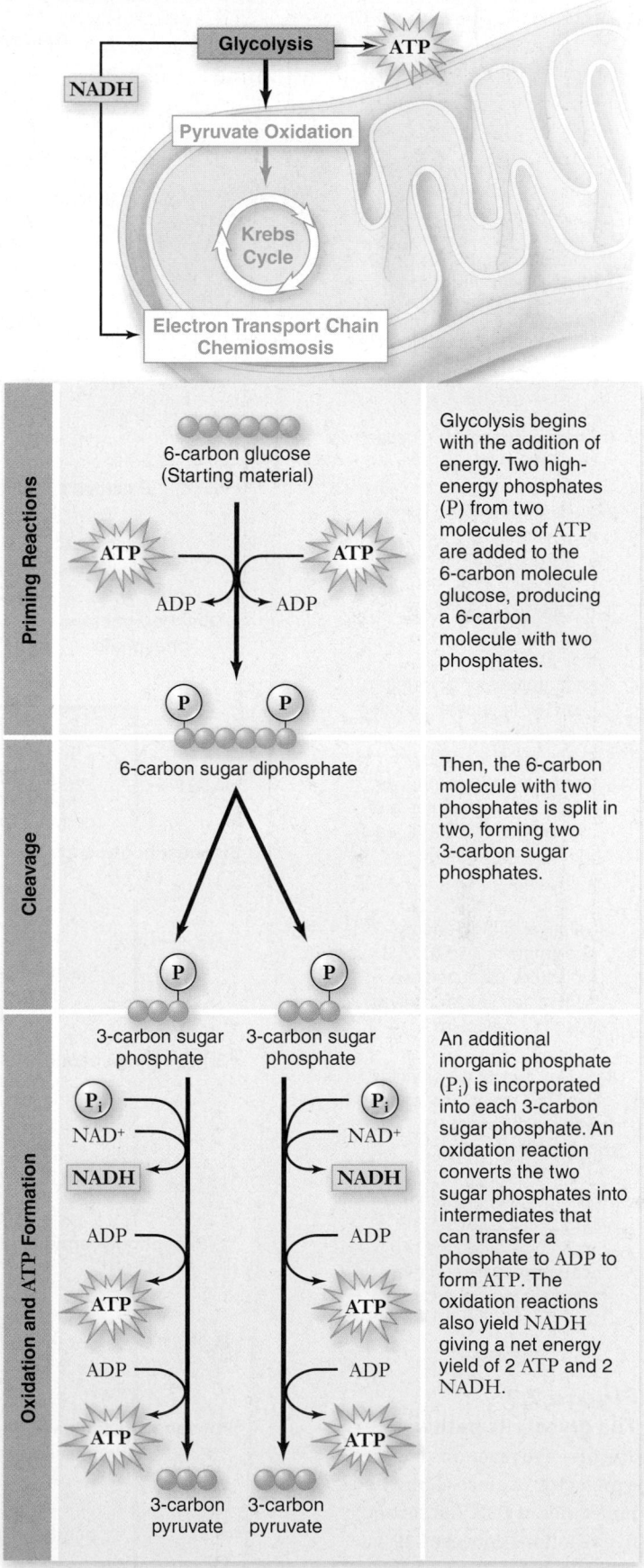

Figure 7.6 An overview of glycolysis.

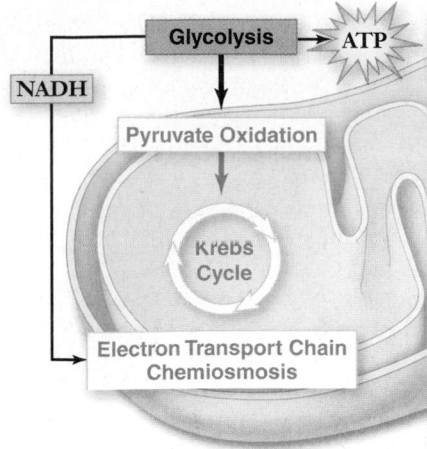

Glycolysis → ATP

NADH

Pyruvate Oxidation

Krebs Cycle

Electron Transport Chain Chemiosmosis

1. Phosphorylation of glucose by ATP.

2–3. Rearrangement, followed by a second ATP phosphorylation.

4–5. The 6-carbon molecule is split into two 3-carbon molecules—one G3P, another that is converted into G3P in another reaction.

6. Oxidation followed by phosphorylation produces two NADH molecules and two molecules of BPG, each with one high-energy phosphate bond.

7. Removal of high-energy phosphate by two ADP molecules produces two ATP molecules and leaves two 3PG molecules.

8–9. Removal of water yields two PEP molecules, each with a high-energy phosphate bond.

10. Removal of high-energy phosphate by two ADP molecules produces two ATP molecules and two pyruvate molecules.

Figure 7.7
The glycolytic pathway.
The first five reactions convert a molecule of glucose into two molecules of G3P. The second five reactions convert G3P into pyruvate.

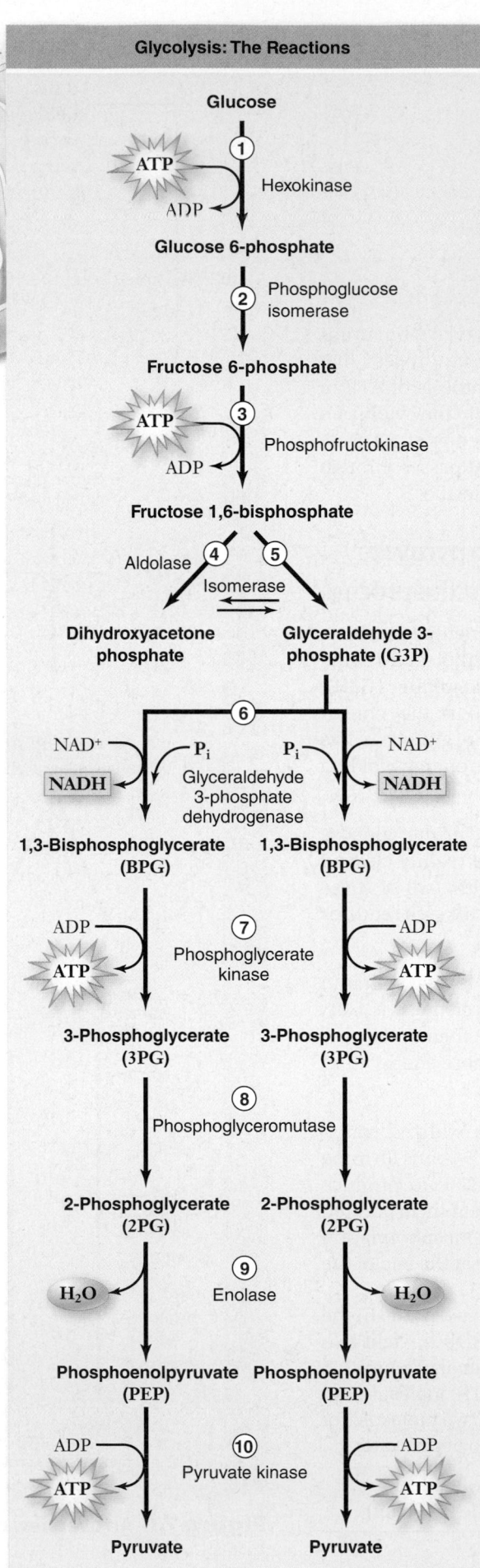

Glycolysis: The Reactions

Glucose

① ATP → ADP   Hexokinase

Glucose 6-phosphate

② Phosphoglucose isomerase

Fructose 6-phosphate

③ ATP → ADP   Phosphofructokinase

Fructose 1,6-bisphosphate

Aldolase ④   ⑤ Isomerase

Dihydroxyacetone phosphate   **Glyceraldehyde 3-phosphate (G3P)**

⑥   NAD⁺ → P_i P_i → NAD⁺
NADH   Glyceraldehyde 3-phosphate dehydrogenase   NADH

1,3-Bisphosphoglycerate (BPG)   **1,3-Bisphosphoglycerate (BPG)**

⑦ ADP → ATP   Phosphoglycerate kinase   ADP → ATP

3-Phosphoglycerate (3PG)   **3-Phosphoglycerate (3PG)**

⑧ Phosphoglyceromutase

2-Phosphoglycerate (2PG)   **2-Phosphoglycerate (2PG)**

⑨ H_2O ←   Enolase   → H_2O

Phosphoenolpyruvate (PEP)   **Phosphoenolpyruvate (PEP)**

⑩ ADP → ATP   Pyruvate kinase   ADP → ATP

Pyruvate   **Pyruvate**

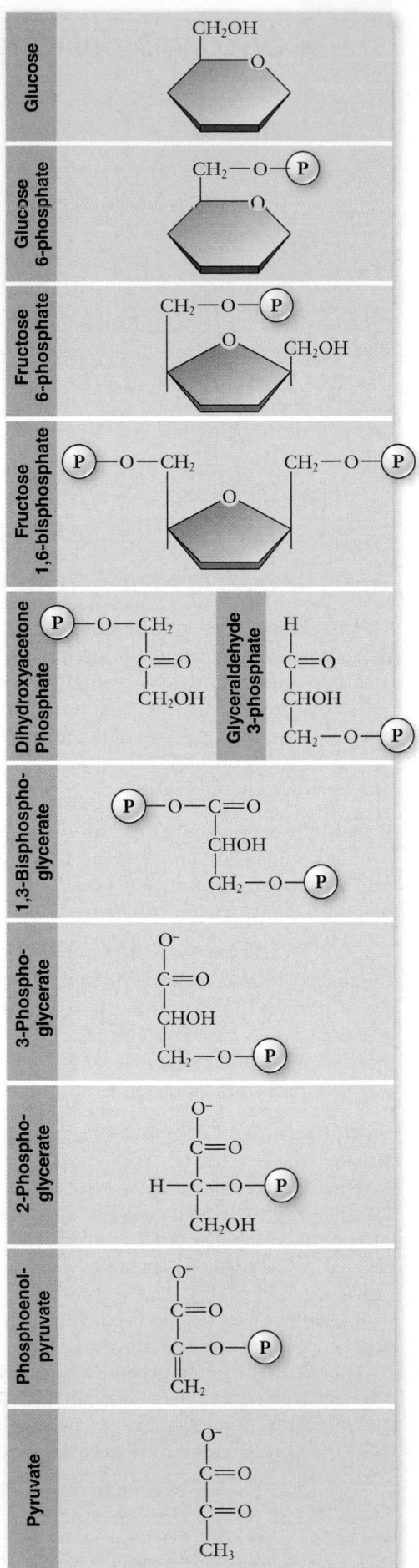

Glucose

Glucose 6-phosphate

Fructose 6-phosphate

Fructose 1,6-bisphosphate

Dihydroxyacetone Phosphate

Glyceraldehyde 3-phosphate

1,3-Bisphospho-glycerate

3-Phospho-glycerate

2-Phospho-glycerate

Phosphoenol-pyruvate

Pyruvate

The hydrolysis of one molecule of ATP yields a ΔG of –7.3 kcal/mol under standard conditions. Thus cells harvest a maximum of 14.6 kcal of energy per mole of glucose from glycolysis.

A brief history of glycolysis

Although the ATP yield from glycolysis is low, it is actually quite efficient, with just under 40% of the energy released being trapped as ATP. For more than a billion years during the anaerobic first stages of life on Earth, glycolysis was the primary way heterotrophic organisms generated ATP from organic molecules.

Like many biochemical pathways, glycolysis is believed to have evolved backward—the last steps in the process being the most ancient. Thus, the second half of glycolysis, the ATP-yielding breakdown of G3P, may have been the original process. The synthesis of G3P from glucose would have appeared later, perhaps when alternative sources of G3P were depleted.

Why does glycolysis take place in modern organisms, since its energy yield in the absence of oxygen is comparatively little? There are several possible answers. First, the process is energetically efficient, and better than the alternative—no ATP. Second, evolution is an incremental process: Change occurs by improving on past successes. In catabolic metabolism, glycolysis satisfied the one essential evolutionary criterion—it was an improvement. Cells that could not carry out glycolysis were at a competitive disadvantage, and only cells capable of glycolysis survived. Later improvements in catabolic metabolism built on this framework to increase the yield of ATP as oxygen became available as an oxidizing agent. Metabolism evolved as one layer of reactions added to another. Nearly every present-day organism carries out glycolysis, as a metabolic memory of its evolutionary past.

The last section of this chapter discusses the evolution of metabolism in more detail.

NADH must be recycled to continue respiration

Inspect for a moment the net reaction of the glycolytic sequence:

$$\text{glucose} + 2\ \text{ADP} + 2\ \text{P}_i + 2\ \text{NAD}^+ \longrightarrow 2\ \text{pyruvate} + 2\ \text{ATP} + 2\ \text{NADH} + 2\text{H}^+ + 2\text{H}_2\text{O}$$

You can see that three changes occur in glycolysis: (1) glucose is converted into two molecules of pyruvate; (2) two molecules of ADP are converted into ATP via substrate-level phosphorylation; and (3) two molecules of NAD^+ are reduced to NADH. This leaves the cell with two problems: extracting the energy that remains in the two pyruvate molecules, and regenerating NAD^+ to be able to continue glycolysis.

Recycling NADH

As long as food molecules that can be converted into glucose are available, a cell can continually churn out ATP to drive its activities. In doing so, however, it accumulates NADH and depletes the pool of NAD^+ molecules. A cell does not contain a large amount of NAD^+, and for glycolysis to continue, NADH must be recycled into NAD^+. Some molecule other than NAD^+ must ultimately accept the electrons taken from G3P and be reduced. Two processes can carry out this key task (figure 7.8):

1. **Aerobic respiration.** Oxygen is an excellent electron acceptor. Through a series of electron transfers, electrons taken from G3P can be donated to oxygen, forming water. This process occurs in the mitochondria of eukaryotic cells in the presence of oxygen. Because air is rich in oxygen, this process is also referred to as *aerobic metabolism*. A significant amount of ATP is also produced.
2. **Fermentation.** When oxygen is unavailable, an organic molecule can accept electrons. The organic molecules used are quite varied and include acetaldehyde in ethanolic fermentation or pyruvate itself in lactic acid fermentation. This reaction plays an important role in the metabolism of most organisms, even those capable of aerobic respiration.

The fate of pyruvate

The fate of the pyruvate that is produced by glycolysis depends on which of these two processes takes place. The aerobic

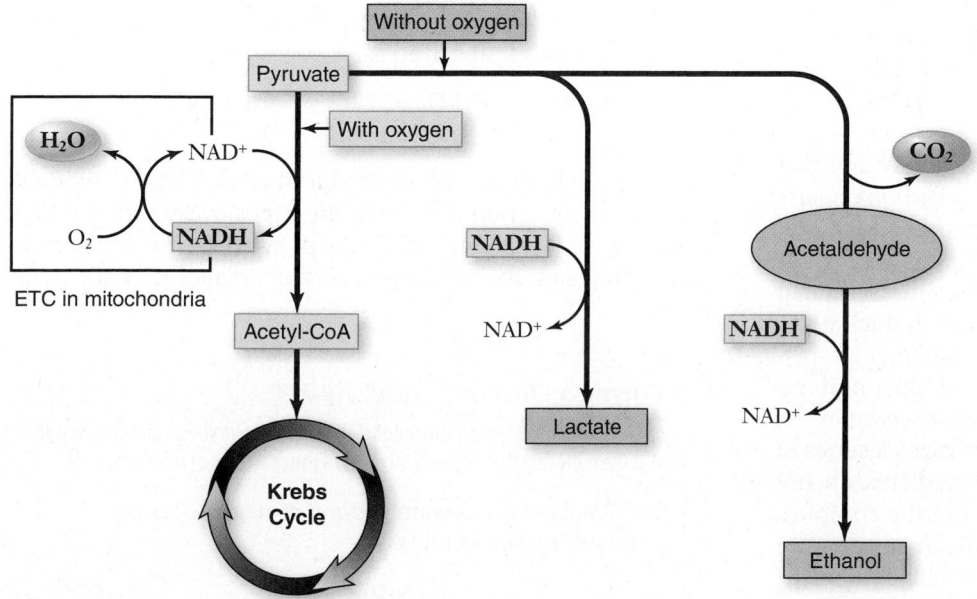

Figure 7.8 The fate of pyruvate and NADH produced by glycolysis. In the presence of oxygen, NADH is oxidized by the electron transport chain (ETC) in mitochondria using oxygen as the final electron acceptor. This regenerates NAD^+, allowing glycolysis to continue. The pyruvate produced by glycolysis is oxidized to acetyl-CoA, which enters the Krebs cycle. In the absence of oxygen, pyruvate is instead reduced, oxidizing NADH and regenerating NAD^+ thus allowing glycolysis to continue. Direct reduction of pyruvate, as in muscle cells, produces lactate. In yeast, carbon dioxide is first removed from pyruvate, producing acetaldehyde, which is then reduced to ethanol.

respiration path starts with the oxidation of pyruvate to produce acetyl coenzyme A (acetyl-CoA), which is then further oxidized in a series of reactions called the Krebs cycle. The fermentation path, by contrast, uses the reduction of all or part of pyruvate to oxidize NADH back to NAD⁺. We examine aerobic respiration next; fermentation is described in detail in a later section.

Learning Outcomes Review 7.2

Glycolysis splits the 6-carbon molecule glucose into two 3-carbon molecules of pyruvate. This process uses two ATP molecules in "priming" reactions and eventually produces four molecules of ATP per glucose for a net yield of two ATP. The oxidation reactions of glycolysis require NAD^+ and produce NADH. When oxygen is abundant, NAD^+ is regenerated in the electron transport chain, using O_2 as an acceptor. When oxygen is absent, NAD^+ is regenerated in a fermentation reaction using an organic molecule as an electron receptor.

■ *Does glycolysis taking place in the cytoplasm argue for or against the endosymbiotic origin of mitochondria?*

7.3 The Oxidation of Pyruvate to Produce Acetyl-CoA

Learning Outcome

1. *Diagram how the oxidation of pyruvate links glycolysis with the Krebs cycle.*

In the presence of oxygen, the oxidation of glucose that begins in glycolysis continues where glycolysis leaves off—with pyruvate. In eukaryotic organisms, the extraction of additional energy from pyruvate takes place exclusively inside mitochondria. In prokaryotes similar reactions take place in the cytoplasm and at the plasma membrane.

The cell harvests pyruvate's considerable energy in two steps. First, pyruvate is oxidized to produce a 2-carbon compound and CO_2, with the electrons transferred to NAD^+ to produce NADH. Next, the 2-carbon compound is oxidized to CO_2 by the reactions of the Krebs cycle.

Pyruvate is oxidized in a "decarboxylation" reaction that cleaves off one of pyruvate's three carbons. This carbon departs as CO_2 (figure 7.9). The remaining 2-carbon compound, called an acetyl group, is then attached to coenzyme A; this entire molecule is called *acetyl-CoA*. A pair of electrons and one associated proton is transferred to the electron carrier NAD^+, reducing it to NADH, with a second proton donated to the solution.

The reaction involves three intermediate stages, and it is catalyzed within mitochondria by a *multienzyme complex*. As chapter 6 noted, a multienzyme complex organizes a series of enzymatic steps so that the chemical intermediates do not diffuse away or undergo other reactions. Within the complex, component polypeptides pass the substrates from one enzyme to the next without releasing them. *Pyruvate dehydrogenase,*

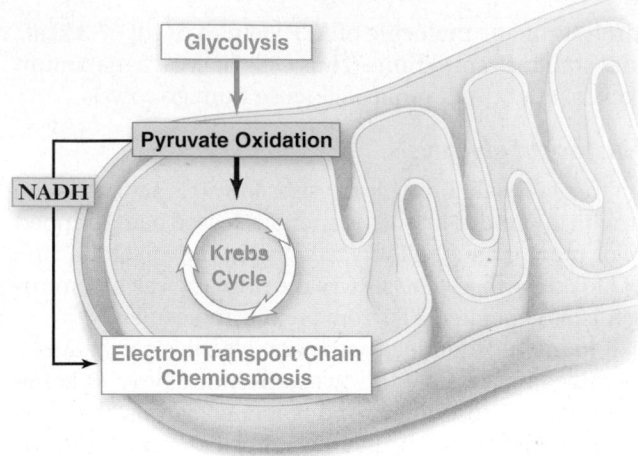

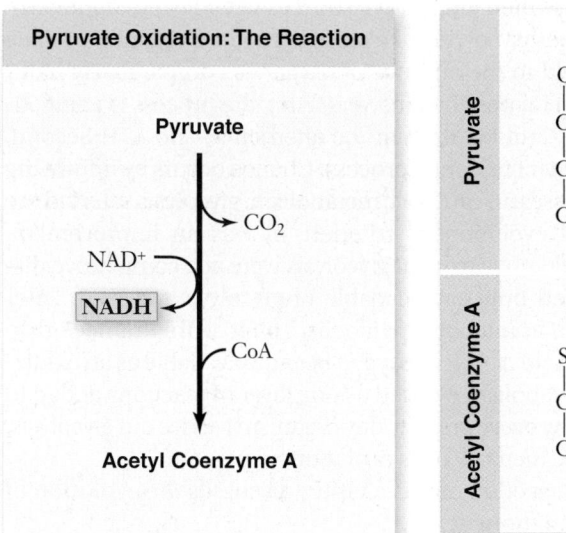

Figure 7.9 The oxidation of pyruvate. This complex reaction uses NAD^+ to accept electrons, reducing it to NADH. The product, acetyl coenzyme A (acetyl-CoA), feeds the acetyl unit into the Krebs cycle, and the CoA is recycled for another oxidation of pyruvate. NADH provides energetic electrons for the electron transport chain.

the complex of enzymes that removes CO_2 from pyruvate, is one of the largest enzymes known; it contains 60 subunits! The reaction can be summarized as:

$$pyruvate + NAD^+ + CoA \longrightarrow acetyl\text{-}CoA + NADH + CO_2 + H^+$$

The molecule of NADH produced is used later to produce ATP. The acetyl group is fed into the Krebs cycle, with the CoA being recycled for another oxidation of pyruvate. The Krebs cycle then completes the oxidation of the original carbons from glucose.

Learning Outcome Review 7.3

Pyruvate is oxidized in the mitochondria to produce acetyl-CoA and CO_2. Acetyl-CoA is the molecule that links glycolysis and the reactions of the Krebs cycle.

■ *What are the advantages and disadvantages of a multienzyme complex?*

In this third stage, the acetyl group from pyruvate is oxidized in a series of nine reactions called the *Krebs cycle.* These reactions occur in the matrix of mitochondria.

In this cycle, the 2-carbon acetyl group of acetyl-CoA combines with a 4-carbon molecule called oxaloacetate. The resulting 6-carbon molecule, citrate, then goes through a several-step sequence of electron-yielding oxidation reactions, during which two CO_2 molecules split off, restoring oxaloacetate. The regenerated oxaloacetate is used to bind to another acetyl group for the next round of the cycle.

In each turn of the cycle, a new acetyl group is added and two carbons are lost, as two CO_2 molecules and more electrons are transferred to electron carriers. These electrons are then used by the electron transport chain to drive *proton pumps* that generate ATP.

An overview of the Krebs cycle

The nine reactions of the Krebs cycle take in 2-carbon units in the form of acetyl-CoA and oxidize them, transferring electrons and protons to NADH and $FADH_2$ (figure 7.10).

The first reaction combines the 4-carbon oxaloacetate with the acetyl group to produce the 6-carbon citrate molecule. Five more steps, which have been simplified in figure 7.10, convert citrate to a 5-carbon intermediate and then to the 4-carbon succinate. During these reactions, two NADH and one ATP are produced.

Succinate undergoes three additional reactions, also simplified in the figure, to become oxaloacetate. During these reactions, one more NADH is produced; in addition, a molecule of flavin adenine dinucleotide (FAD), another cofactor, becomes reduced to $FADH_2$.

The specifics of each reaction are described next.

Figure 7.10 An overview of the Krebs cycle.

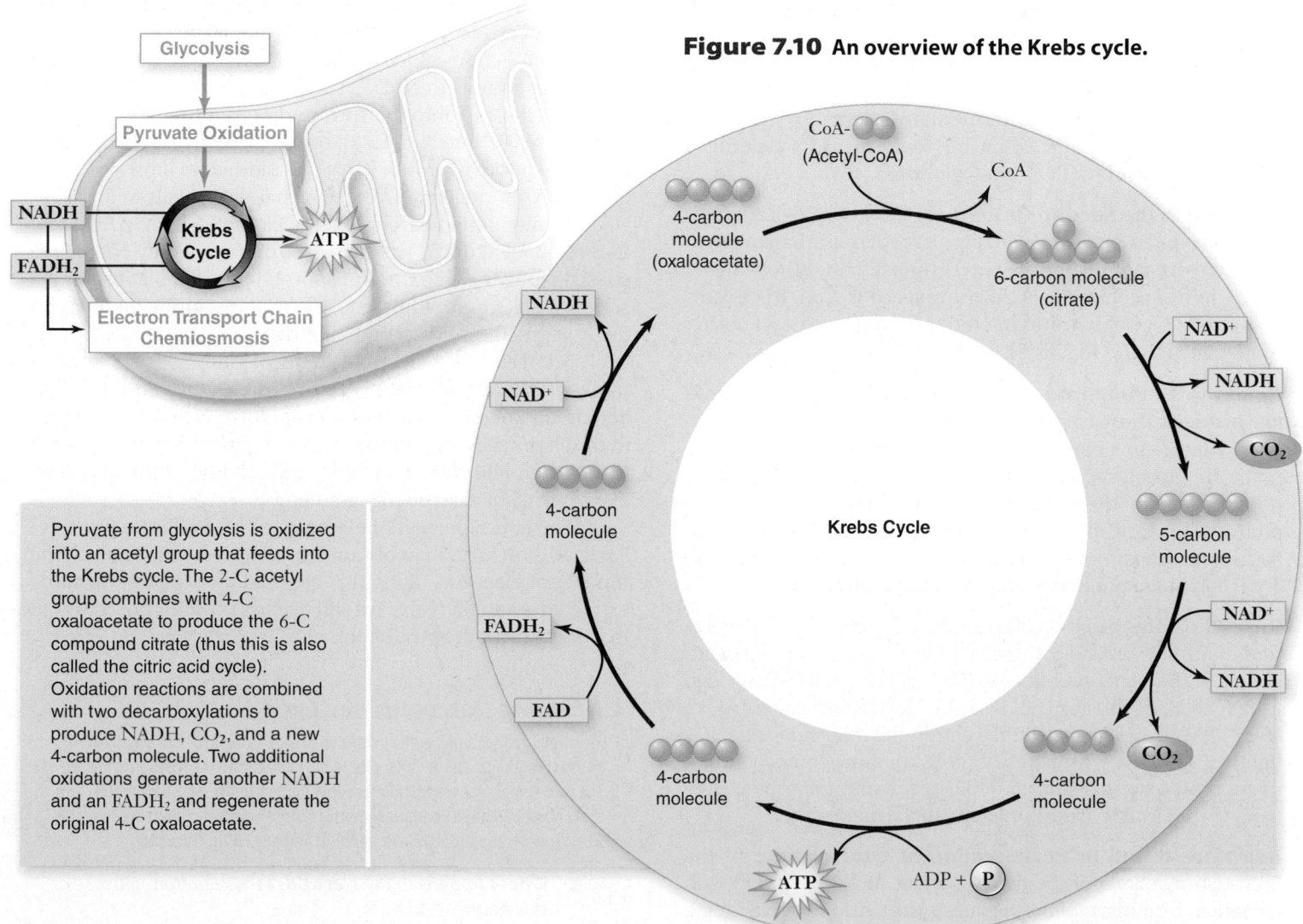

Pyruvate from glycolysis is oxidized into an acetyl group that feeds into the Krebs cycle. The 2-C acetyl group combines with 4-C oxaloacetate to produce the 6-C compound citrate (thus this is also called the citric acid cycle). Oxidation reactions are combined with two decarboxylations to produce NADH, CO_2, and a new 4-carbon molecule. Two additional oxidations generate another NADH and an $FADH_2$ and regenerate the original 4-C oxaloacetate.

The Krebs cycle extracts electrons and synthesizes one ATP

Figure 7.11 summarizes the sequence of the Krebs cycle reactions. A 2-carbon group from acetyl-CoA enters the cycle at the beginning, and two CO_2 molecules, one ATP, and four pairs of electrons are produced.

Reaction 1: Condensation Citrate is formed from acetyl-CoA and oxaloacetate. This condensation reaction is irreversible, committing the 2-carbon acetyl group to the Krebs cycle. The reaction is inhibited when the cell's ATP concentration is high and stimulated when it is low. The result is that when the cell possesses ample amounts of ATP, the Krebs cycle shuts down, and acetyl-CoA is channeled into fat synthesis.

Reactions 2 and 3: Isomerization Before the oxidation reactions can begin, the hydroxyl (—OH) group of citrate must be repositioned. This rearrangement is done in two steps: First, a water molecule is removed from one carbon; then water is added to a different carbon. As a result, an —H group and an —OH group change positions. The product is an isomer of citrate called *isocitrate*. This rearrangement facilitates the subsequent reactions.

Reaction 4: The First Oxidation In the first energy-yielding step of the cycle, isocitrate undergoes an oxidative decarboxylation reaction. First, isocitrate is oxidized, yielding a pair of electrons that reduce a molecule of NAD^+ to NADH. Then the oxidized intermediate is decarboxylated; the central carboxyl group splits off to form CO_2, yielding a 5-carbon molecule called α-*ketoglutarate*.

Reaction 5: The Second Oxidation Next, α-ketoglutarate is decarboxylated by a multienzyme complex similar to pyruvate dehydrogenase. The succinyl group left after the removal of CO_2 joins to coenzyme A, forming *succinyl-CoA*. In the process, two electrons are extracted, and they reduce another molecule of NAD^+ to NADH.

Reaction 6: Substrate-Level Phosphorylation The linkage between the 4-carbon succinyl group and CoA is a high-energy bond. In a coupled reaction similar to those that take place in glycolysis, this bond is cleaved, and the energy released drives the phosphorylation of guanosine diphosphate (GDP), forming guanosine triphosphate (GTP). GTP can transfer a phosphate to ADP converting it into ATP. The 4-carbon molecule that remains is called *succinate*.

Reaction 7: The Third Oxidation Next, succinate is oxidized to *fumarate* by an enzyme located in the inner mitochondrial membrane. The free-energy change in this reaction is not large enough to reduce NAD^+. Instead, FAD is the electron acceptor. Unlike NAD^+, FAD is not free to diffuse within the mitochondrion; it is tightly associated with its enzyme in the inner mitochondrial membrane. Its reduced form, $FADH_2$, can only contribute electrons to the electron transport chain in the membrane.

Reactions 8 and 9: Regeneration of Oxaloacetate In the final two reactions of the cycle, a water molecule is added to fumarate, forming *malate*. Malate is then oxidized, yielding a 4-carbon molecule of *oxaloacetate* and two electrons that reduce a molecule of NAD^+ to NADH. Oxaloacetate, the molecule that began the cycle, is now free to combine with another 2-carbon acetyl group from acetyl-CoA and begin the cycle again.

Glucose becomes CO_2 and potential energy

In the process of aerobic respiration, glucose is entirely consumed. The 6-carbon glucose molecule is cleaved into two 3-carbon pyruvate molecules during glycolysis. One of the carbons of each pyruvate is then lost as CO_2 in the conversion of pyruvate to acetyl-CoA. The two other carbons from acetyl-CoA are lost as CO_2 during the oxidations of the Krebs cycle.

All that is left to mark the passing of a glucose molecule into six CO_2 molecules is its energy, some of which is preserved in four ATP molecules and in the reduced state of 12 electron carriers. Ten of these carriers are NADH molecules; the other two are $FADH_2$.

Following the electrons in the reactions reveals the direction of transfer

As you examine the changes in electrical charge in the reactions that oxidize glucose, a good strategy for keeping the transfers clear is always to *follow the electrons*. For example, in glycolysis, an enzyme extracts two hydrogens—that is, two electrons and two protons—from glucose and transfers both electrons and one of the protons to NAD^+. The other proton is released as a hydrogen ion, H^+, into the surrounding solution. This transfer converts NAD^+ into NADH; that is, two negative electrons ($2e^-$) and one positive proton (H^+) are added to one positively charged NAD^+ to form NADH, which is electrically neutral.

As mentioned earlier, energy captured by NADH is not harvested all at once. The two electrons carried by NADH are passed along the electron transport chain, which consists of a series of electron carriers, mostly proteins, embedded within the inner membranes of mitochondria.

NADH delivers electrons to the beginning of the electron transport chain, and oxygen captures them at the end. The oxygen then joins with hydrogen ions to form water. At each step in the chain, the electrons move to a slightly more electronegative carrier, and their positions shift slightly. Thus, the electrons move *down* an energy gradient.

The entire process of electron transfer releases a total of 53 kcal/mol (222 kJ/mol) under standard conditions. The transfer of electrons along this chain allows the energy to be extracted gradually. Next, we will discuss how this energy is put to work to drive the production of ATP.

Learning Outcomes Review 7.4

The Krebs cycle completes the oxidation of glucose begun with glycolysis. In the first segment, acetyl-CoA is added to oxaloacetate to produce citrate. In the next segment, five reactions produce succinate, two NADH from NAD^+, and one ATP. Finally, succinate undergoes three more reactions to regenerate oxaloacetate, producing one more NADH and one $FADH_2$ from FAD.

■ *What happens to the electrons removed from glucose at this point?*

Glycolysis
Pyruvate Oxidation
NADH
FADH2
Krebs Cycle
ATP
Electron Transport Chain Chemiosmosis
1. Reaction 1: Condensation
2–3. Reactions 2 and 3: Isomerization
4. Reaction 4: The first oxidation
5. Reaction 5: The second oxidation
6. Reaction 6: Substrate-level phosphorylation
7. Reaction 7: The third oxidation
8–9. Reactions 8 and 9: Regeneration of oxaloacetate and the fourth oxidation

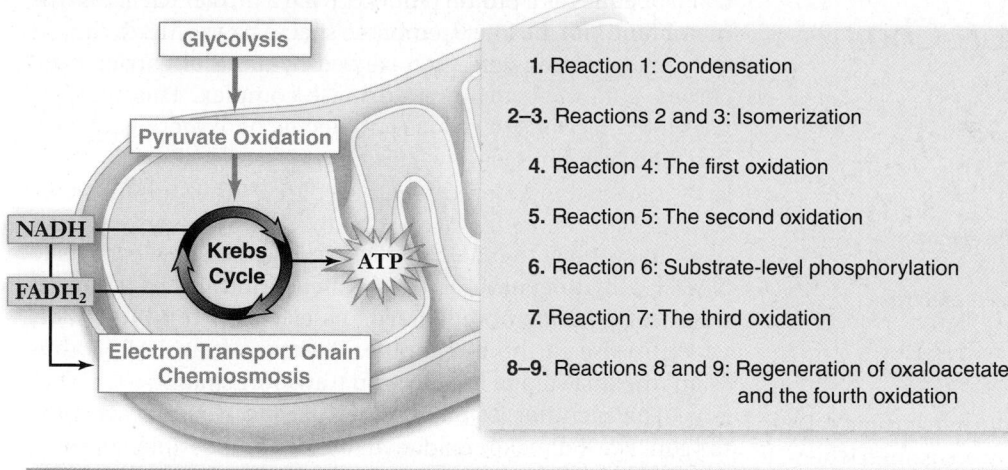

Figure 7.11 The Krebs cycle. This series of reactions takes place within the matrix of the mitochondrion. For the complete breakdown of a molecule of glucose, the two molecules of acetyl-CoA produced by glycolysis and pyruvate oxidation each have to make a trip around the Krebs cycle. Follow the different carbons through the cycle, and notice the changes that occur in the carbon skeletons of the molecules and where oxidation reactions take place as they proceed through the cycle.

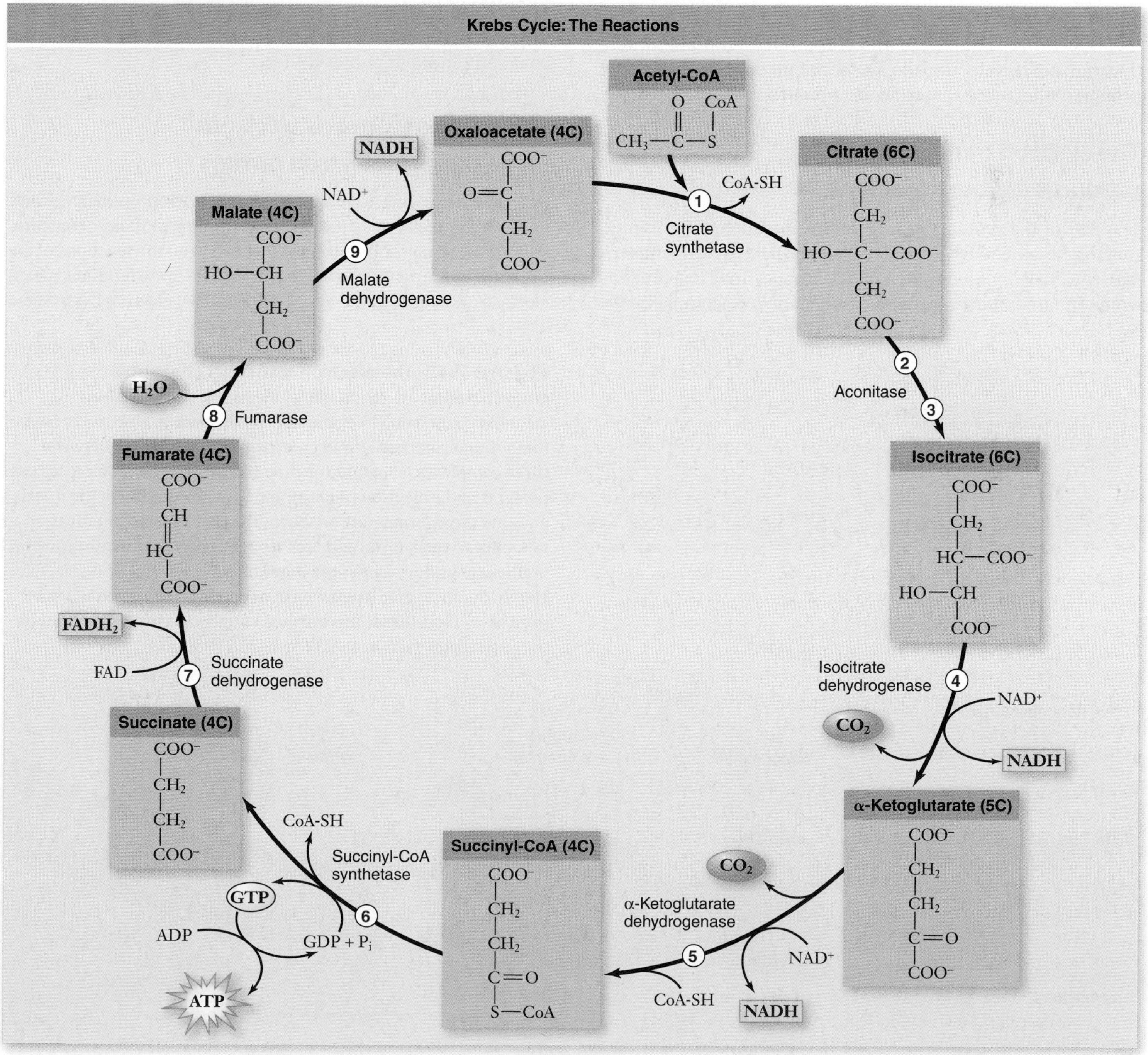

Krebs Cycle: The Reactions
Acetyl-CoA
Oxaloacetate (4C)
Citrate (6C)
CoA-SH
1 Citrate synthetase
NADH
NAD+
9 Malate dehydrogenase
Malate (4C)
2 Aconitase 3
Isocitrate (6C)
H2O
8 Fumarase
Fumarate (4C)
FADH2
FAD
7 Succinate dehydrogenase
4 Isocitrate dehydrogenase
CO2
NAD+
NADH
Succinate (4C)
α-Ketoglutarate (5C)
CoA-SH
Succinyl-CoA synthetase
Succinyl-CoA (4C)
CO2
α-Ketoglutarate dehydrogenase
GTP
ADP
6
GDP + Pi
5
NAD+
ATP
CoA-SH
NADH

The Electron Transport Chain and Chemiosmosis

Learning Outcomes

1. *Describe the structure and function of the electron transport chain.*
2. *Diagram how the proton gradient connects electron transport with ATP synthesis.*

The NADH and FADH$_2$ molecules formed during aerobic respiration each contain a pair of electrons that were gained when NAD$^+$ and FAD were reduced. The NADH and FADH$_2$ carry their electrons to the inner mitochondrial membrane, where they transfer the electrons to a series of membrane-associated proteins collectively called the *electron transport chain.*

The electron transport chain produces a proton gradient

The first of the proteins to receive the electrons is a complex, membrane-embedded enzyme called **NADH dehydrogenase.** A carrier called *ubiquinone* then passes the electrons to a protein–cytochrome complex called the *bc$_1$ complex.* Each complex in the

chain operates as a proton pump, driving a proton out across the membrane into the intermembrane space (figure 7.12*a*).

The electrons are then carried by another carrier, *cytochrome c,* to the cytochrome oxidase complex. This complex uses four electrons to reduce a molecule of oxygen. Each oxygen then combines with two protons to form water:

$$O_2 + 4H^+ + 4e^- \longrightarrow 2H_2O$$

In contrast to NADH, which contributes its electrons to NADH dehydrogenase, FADH$_2$, which is located in the inner mitochondrial membrane, feeds its electrons to ubiquinone, which is also in the membrane. Electrons from FADH$_2$ thus "skip" the first step in the electron transport chain.

The plentiful availability of a strong electron acceptor, oxygen, is what makes oxidative respiration possible. As you'll see in chapter 8, the electron transport chain used in aerobic respiration is similar to, and may well have evolved from, the chain employed in photosynthesis.

The gradient forms as electrons move through electron carriers

Respiration takes place within the mitochondria present in virtually all eukaryotic cells. The internal compartment, or matrix, of a mitochondrion contains the enzymes that carry out the reactions of the Krebs cycle. As mentioned earlier, protons (H$^+$) are produced when electrons are transferred to NAD$^+$. As the electrons harvested

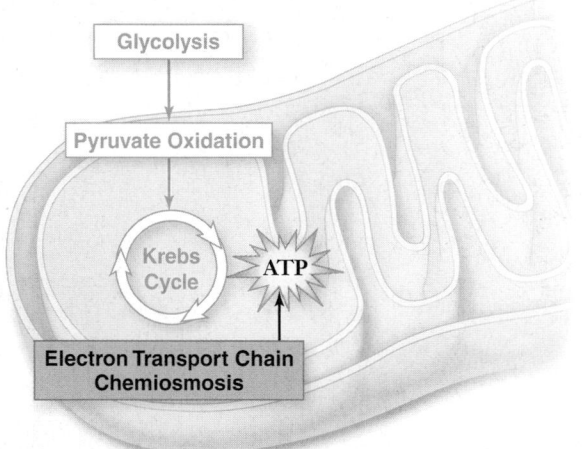

Figure 7.12 The electron transport chain and chemiosmosis. *a.* High-energy electrons harvested from catabolized molecules are transported by mobile electron carriers (ubiquinone, marked Q, and cytochrome c, marked C) between three complexes of membrane proteins. These three complexes use portions of the electrons' energy to pump protons out of the matrix and into the intermembrane space. The electrons are finally used to reduce oxygen, forming water. *b.* This creates a concentration gradient of protons across the inner membrane. This electrochemical gradient is a form of potential energy that can be used by ATP synthase. This enzyme couples the reentry of protons to the phosphorylation of ADP to form ATP.

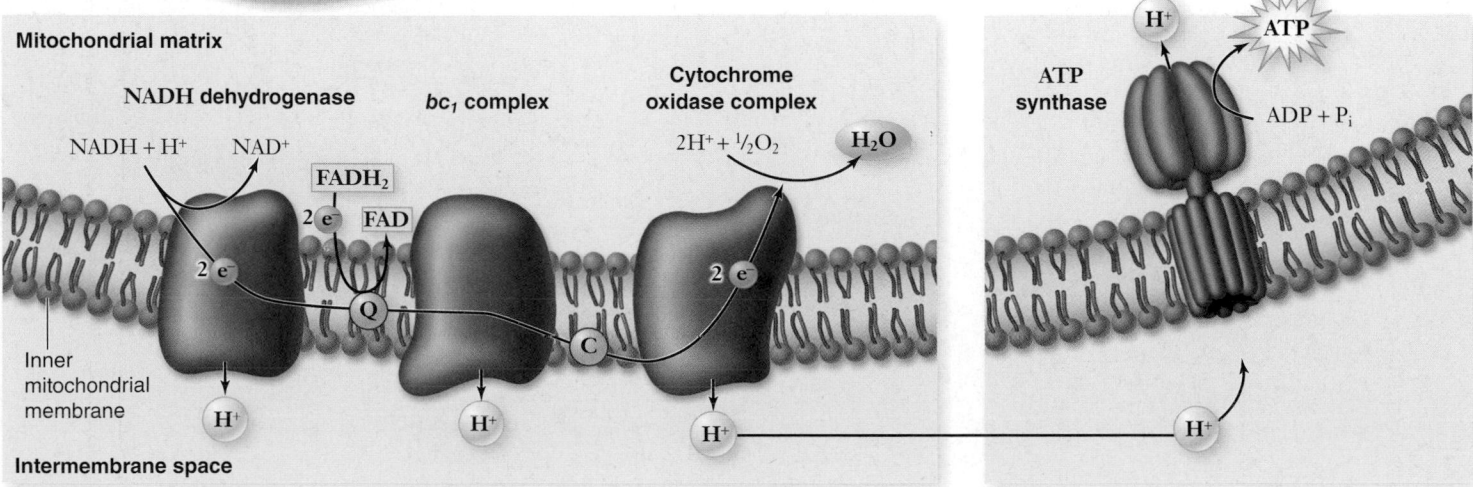

a. The electron transport chain

b. Chemiosmosis

by oxidative respiration are passed along the electron transport chain, the energy they release transports protons out of the matrix and into the outer compartment called the intermembrane space.

Three transmembrane complexes of the electron transport chain in the inner mitochondrial membrane actually accomplish the proton transport (see figure 7.12a). The flow of highly energetic electrons induces a change in the shape of pump proteins, which causes them to transport protons across the membrane. The electrons contributed by NADH activate all three of these proton pumps, whereas those contributed by $FADH_2$ activate only two because of where they enter the chain. In this way a proton gradient is formed between the intermembrane space and the matrix.

Chemiosmosis utilizes the electrochemical gradient to produce ATP

Because the mitochondrial matrix is negative compared with the intermembrane space, positively charged protons are attracted to the matrix. The higher outer concentration of protons also tends to drive protons back in by diffusion, but because membranes are relatively impermeable to ions, this process occurs only very slowly. Most of the protons that reenter the matrix instead pass through ATP synthase, an enzyme that uses the energy of the gradient to catalyze the synthesis of ATP from ADP and P_i. Because the chemical formation of ATP is driven by a diffusion force similar to osmosis, this process is referred to as *chemiosmosis* (figure 7.12b). The newly formed ATP is transported by facilitated diffusion to the many places in the cell where enzymes require energy to drive endergonic reactions. This chemiosmotic mechanism for the coupling of electron transport and ATP synthesis was controversial when it was proposed. Over the years, experimental evidence accumulated to support this hypothesis (figure 7.13).

The energy released by the reactions of cellular respiration ultimately drives the proton pumps that produce the

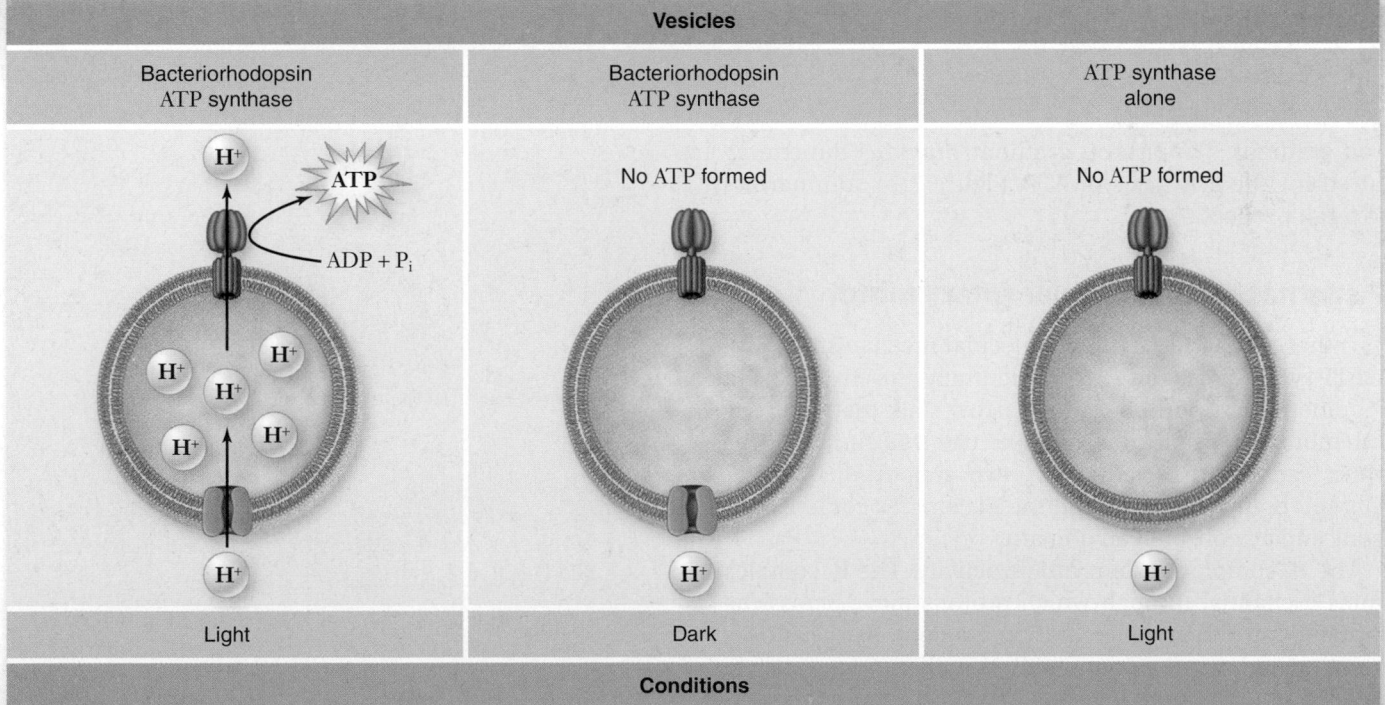

SCIENTIFIC THINKING

Hypothesis: *ATP synthase enzyme uses a proton gradient to provide energy for phosphorylation reaction.*

Prediction: *The source of the proton gradient should not matter. A proton gradient formed by the light-driven pump bacteriorhodopsin should power phosphorylation in the light but not in the dark.*

Test: *Artificial vesicles are made with bacteriorhodopsin and ATP synthase, and ATP synthase alone. These are illuminated with light and assessed for ATP production.*

Result: *The vesicle with both bacteriorhodopsin and ATP synthase can form ATP in the light but not in the dark. The vesicle with ATP synthase alone cannot form ATP in the light.*

Conclusion: *ATP synthase is able to utilize a proton gradient for energy to form ATP.*

Further Experiments: *What other controls would be appropriate for this type of experiment? Why is this experiment a better test of the chemiosmotic hypothesis than the acid bath experiment in Jangendorf/chapter 8 (see figure 8.16)?*

Figure 7.13 Evidence for the chemiosmotic synthesis of ATP by ATP synthase.

Figure 7.14 Aerobic respiration in the mitochondria. The entire process of aerobic respiration is shown in cellular context. Glycolysis occurs in the cytoplasm with the pyruvate and NADH produced entering the mitochondria. Here, pyruvate is oxidized and fed into the Krebs cycle to complete the oxidation process. All the energetic electrons harvested by oxidations in the overall process are transferred by NADH and FADH$_2$ to the electron transport chain. The electron transport chain uses the energy released during electron transport to pump protons across the inner membrane. This creates an electrochemical gradient that contains potential energy. The enzyme ATP synthase uses this gradient to phosphorylate ADP to form ATP.

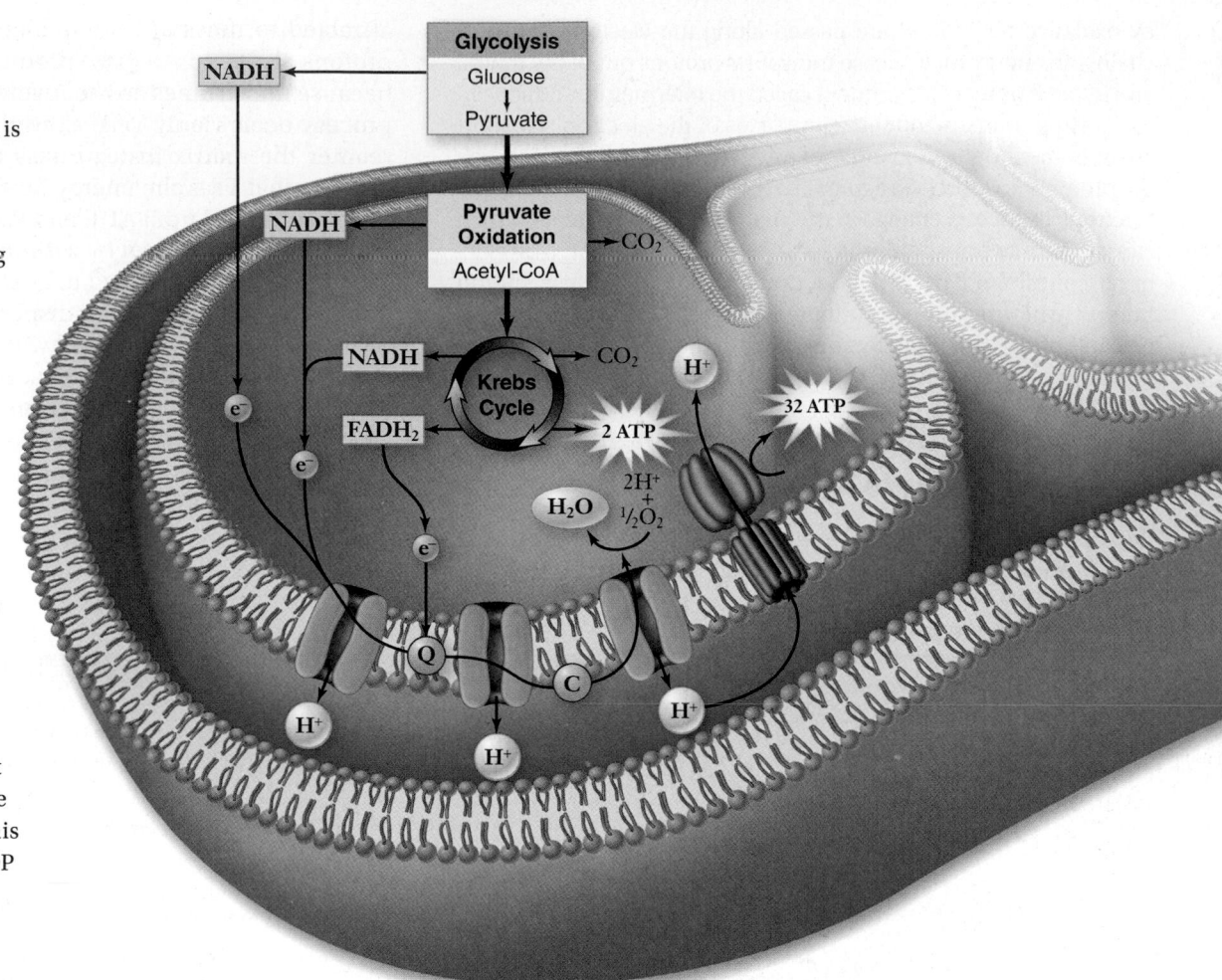

proton gradient. The proton gradient provides the energy required for the synthesis of ATP. Figure 7.14 summarizes the overall process.

ATP synthase is a molecular rotary motor

ATP synthase uses a fascinating molecular mechanism to perform ATP synthesis (figure 7.15). Structurally, the enzyme has a membrane-bound portion and a narrow stalk that connects the membrane portion to a knoblike catalytic portion. This complex can be dissociated into two subportions: the F$_0$ membrane-bound complex, and the F$_1$ complex composed of the stalk and a knob, or head domain.

The F$_1$ complex has enzymatic activity. The F$_0$ complex contains a channel through which protons move across the membrane down their concentration gradient. As they do so, their movement causes part of the F$_0$ complex and the stalk to rotate relative to the knob. The mechanical energy of this rotation is used to change the conformation of the catalytic domain in the F$_1$ complex.

Thus, the synthesis of ATP is achieved by a tiny rotary motor, the rotation of which is driven directly by a gradient of protons. The flow of protons is like that of water in a hydroelectric power plant. Like the flow of water driven by gravity causes a turbine to rotate and generate electrical current, the proton gradient produces the energy that drives the rotation of the ATP synthase generator.

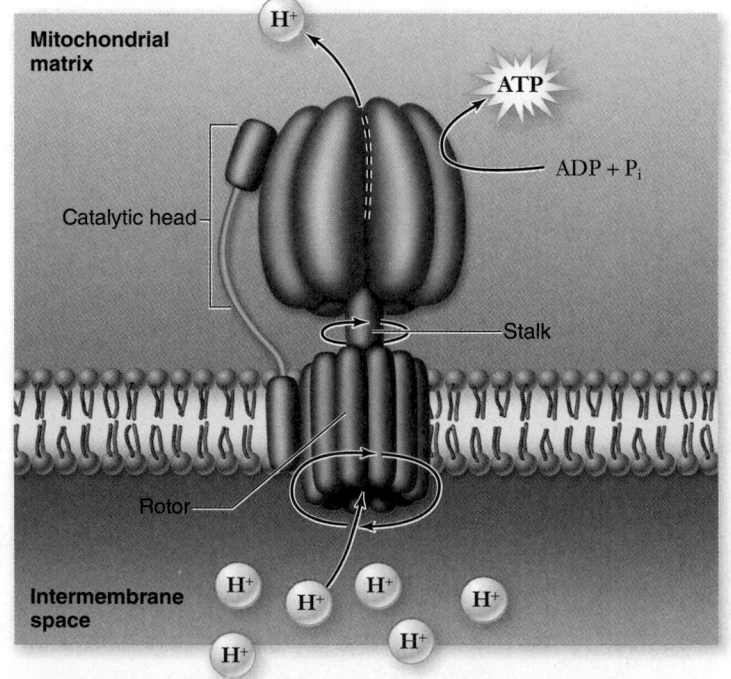

Figure 7.15 The ATP rotary engine. Protons move across the membrane down their concentration gradient. The energy released causes the rotor and stalk structures to rotate. This mechanical energy alters the conformation of the ATP synthase enzyme to catalyze the formation of ATP.

7.6 Energy Yield of Aerobic Respiration

Learning Outcome

1. *Calculate the number of ATP molecules produced by aerobic respiration.*

How much metabolic energy (in the form of ATP) does a cell gain from aerobic breakdown of glucose? This simple question has actually been a source of some controversy in biochemistry.

 The theoretical yield for eukaryotes is 30 molecules of ATP per glucose molecule

The number of molecules of ATP produced by ATP synthase per molecules of glucose depends on the number of protons transported across the inner membrane, and the number of protons needed per ATP synthesized. The number of protons transported per NADH and FADH₂ is 10 and 6 H⁺, respectively.

Each ATP synthesized requires 4 H⁺, leading to $10/4 = 2.5$ ATP/NADH, and $6/4 = 1.5$ ATP/FADH₂.

To finish the bookkeeping: oxidizing glucose to pyruvate via glycolysis yields 2 ATP directly, and $2 \times 2.5 = 5$ ATP from NADH. The oxidation of pyruvate to acetyl-CoA yields another $2 \times 2.5 = 5$ ATP from NADH. Lastly, the Krebs cycle produces 2 ATP directly, $6 \times 2.5 = 15$ ATP from NADH, and $2 \times 1.5 = 3$ ATP from FADH₂. Summing all of these leads to 32 ATP for respiration (figure 7.16).

This number is accurate for bacteria, but it does not hold for eukaryotes because the NADH produced in the cytoplasm by glycolysis needs to be transported into the mitochondria by active transport, which costs one ATP per NADH transported. This reduces the predicted yield for eukaryotes to 30 ATP.

Calculation of P/O ratios has changed over time

The value for the amount of ATP synthesized per O_2 molecule reduced is called the phosphate-to-oxygen ratio (P/O ratio). Both theoretical calculations, and direct measurement of this value, have been contentious issues. When theoretical calculations were first made, we lacked detailed knowledge of the respiratory chain, and the mechanism for coupling electron transport to ATP synthesis. Since redox reactions occur at three sites for NADH and two sites for FADH₂, it was assumed that three molecules of ATP were produced per NADH and two per FADH₂. We now know that assumption was overly simplistic.

Understanding that a proton gradient is the link between electron transport and ATP synthesis changed the nature of the calculations. We need to know the number of protons pumped during electron transport: 10 H⁺ per NADH, and 6 H⁺ per FADH₂. Then we need to know the number of protons needed per ATP. Since ATP synthase is a rotary motor, this calculation depends on the number of binding sites for ATP, and the number of protons required for rotation. We know that ATP synthase has three binding sites for ATP. If 12 protons are used per rotation, you get the value of 4 H⁺ per ATP used in the previous

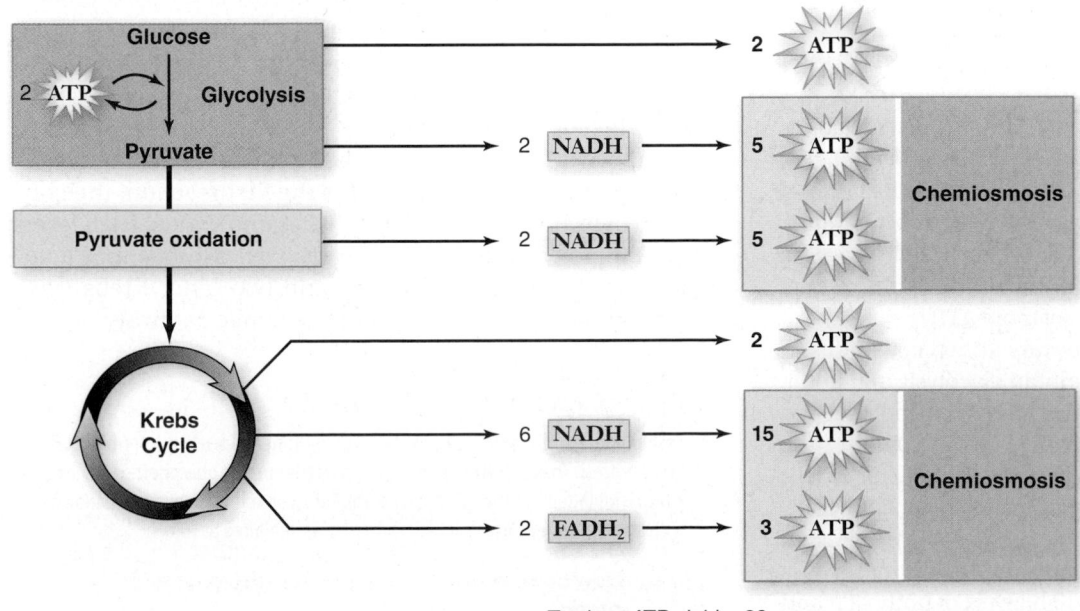

Total net ATP yield = 32
(30 in eukaryotes)

Figure 7.16 Theoretical ATP yield. The theoretical yield of ATP harvested from glucose by aerobic respiration totals 32 molecules. In eukaryotes this is reduced to 30 because it takes 1 ATP to transport each molecule of NADH that is generated by glycolysis in the cytoplasm into the mitochondria.

calculation. Actual measurements of the P/O ratio have been problematic, but now appear to be at most 2.5.

We can also calculate how efficiently respiration captures the free energy released by the oxidation of glucose in the form of ATP. The amount of free energy released by the oxidation of glucose is 686 kcal/mol, and the free energy stored in each ATP is 7.3 kcal/mol. Therefore, a eukaryotic cell harvests about $(7.3 \times 30)/686 = 32\%$ of the energy available in glucose. (By comparison, a typical car converts only about 25% of the energy in gasoline into useful energy.)

The higher energy yield of aerobic respiration was one of the key factors that fostered the evolution of heterotrophs. As this mechanism for producing ATP evolved, nonphotosynthetic organisms became more effective at using respiration to extract energy from molecules derived from other organisms. As long as some organisms captured energy by photosynthesis, others could exist solely by feeding on them.

Learning Outcome Review 7.6

Passage of electrons down the electron transport chain produces roughly 2.5 molecules of ATP per molecule of NADH (1.5 ATP per FADH$_2$). This process plus the ATP from substrate-level phosphorylation can yield a maximum of 32 ATP for the complete oxidation of glucose. NADH generated in the cytoplasm of eukaryotes yields only two ATP/NADH due to the cost of transport into the mitochondria, lowering the yield to 30 ATP.

■ *How does chemiosmosis allow for noninteger numbers of ATP/NADH?*

7.7 Regulation of Aerobic Respiration

Learning Outcome

1. *Understand the control points for cellular respiration.*

When cells possess plentiful amounts of ATP, the key reactions of glycolysis, the Krebs cycle, and fatty acid breakdown are inhibited, slowing ATP production. The regulation of these biochemical pathways by the level of ATP is an example of feedback inhibition. Conversely, when ATP levels in the cell are low, ADP levels are high, and ADP activates enzymes in the pathways of carbohydrate catabolism to stimulate the production of more ATP.

Control of glucose catabolism occurs at two key points in the catabolic pathway, namely at a point in glycolysis and at the beginning of the Krebs cycle (figure 7.17). The control point in glycolysis is the enzyme phosphofructokinase, which catalyzes the conversion of fructose phosphate to fructose bisphosphate. This is the first reaction of glycolysis that is not readily reversible, committing the substrate to the glycolytic sequence. ATP itself is an allosteric inhibitor (see chapter 6) of phosphofructokinase, as is the Krebs cycle intermediate citrate. High levels of both ATP and citrate inhibit phosphofructokinase. Thus, under conditions

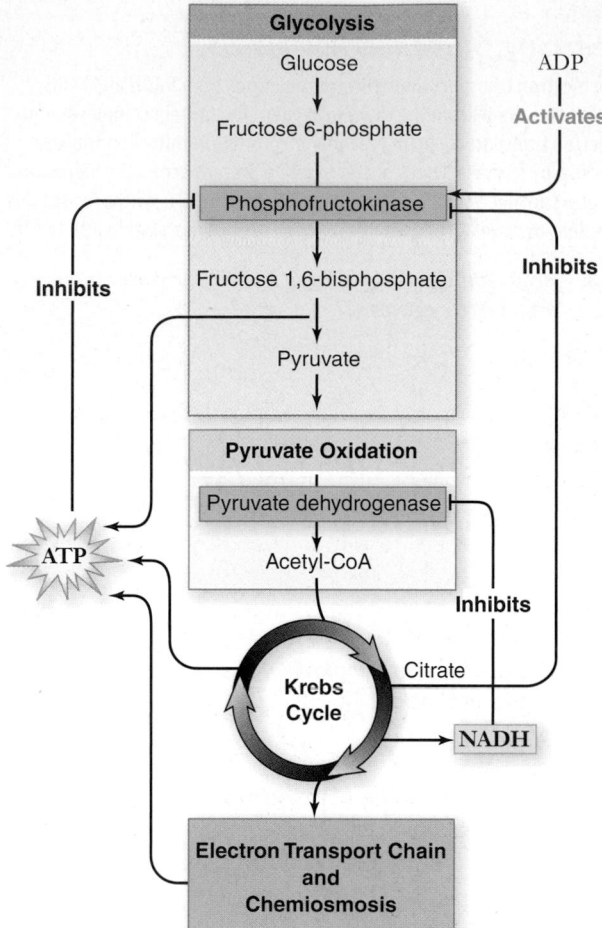

Figure 7.17 Control of glucose catabolism. The relative levels of ADP and ATP and key intermediates NADH and citrate control the catabolic pathway at two key points: the committing reactions of glycolysis and the Krebs cycle.

when ATP is in excess, or when the Krebs cycle is producing citrate faster than it is being consumed, glycolysis is slowed.

The main control point in the oxidation of pyruvate occurs at the committing step in the Krebs cycle with the enzyme pyruvate dehydrogenase, which converts pyruvate to acetyl-CoA. This enzyme is inhibited by high levels of NADH, a key product of the Krebs cycle.

Another control point in the Krebs cycle is the enzyme citrate synthetase, which catalyzes the first reaction, the conversion of oxaloacetate and acetyl-CoA into citrate. High levels of ATP inhibit citrate synthetase (as well as phosphofructokinase, pyruvate dehydrogenase, and two other Krebs cycle enzymes), slowing down the entire catabolic pathway.

Learning Outcome Review 7.7

Respiration is controlled by levels of ATP in the cell and levels of key intermediates in the process. The control point for glycolysis is the enzyme phosphofructokinase, which is inhibited by ATP or citrate (or both). The main control point in oxidation of pyruvate is the enzyme pyruvate dehydrogenase, inhibited by NADH.

■ *How does feedback inhibition ensure economic production of ATP?*

Oxidation Without O_2

In the presence of oxygen, cells can use oxygen to produce a large amount of ATP. But even when no oxygen is present to accept electrons, some organisms can still respire *anaerobically,* using inorganic molecules as final electron acceptors for an electron transport chain.

For example, many prokaryotes use sulfur, nitrate, carbon dioxide, or even inorganic metals as the final electron acceptor in place of oxygen (figure 7.18). The free energy released by using these other molecules as final electron acceptors is not as great as that using oxygen because they have a lower affinity for electrons. The amount of ATP produced is less, but the process is still respiration and not fermentation.

Methanogens use carbon dioxide

Among the heterotrophs that practice anaerobic respiration are Archaea such as thermophiles and methanogens. Methanogens use carbon dioxide (CO_2) as the electron acceptor, reducing CO_2 to CH_4 (methane). The hydrogens are derived from organic molecules produced by other organisms. Methanogens are found in diverse environments, including soil and the digestive systems of ruminants like cows.

Sulfur bacteria use sulfate

Evidence of a second anaerobic respiratory process among primitive bacteria is seen in a group of rocks about 2.7 BYA, known as the Woman River iron formation. Organic material in these rocks is enriched for the light isotope of sulfur, ^{32}S, relative to the heavier isotope, ^{34}S. No known geochemical process produces such enrichment, but biological sulfur reduction does, in a process still carried out today by certain prokaryotes.

In this sulfate respiration, the prokaryotes derive energy from the reduction of inorganic sulfates (SO_4) to hydrogen sulfide (H_2S). The hydrogen atoms are obtained from organic molecules other organisms produce. These prokaryotes thus are similar to methanogens, but they use SO_4 as the oxidizing (that is, electron-accepting) agent in place of CO_2.

The early sulfate reducers set the stage for the evolution of photosynthesis, creating an environment rich in H_2S. As discussed in chapter 8, the first form of photosynthesis obtained hydrogens from H_2S using the energy of sunlight.

Fermentation uses organic compounds as electron acceptors

In the absence of oxygen, cells that cannot utilize an alternative electron acceptor for respiration must rely exclusively on

a. |—| 0.625 μm   *b.*

Figure 7.18 Sulfur-respiring prokaryote. *a.* The micrograph shows the archaeal species *Thermoproteus tenax.* This organism can use elemental sulfur as a final electron acceptor for anaerobic respiration. *b. Thermoproteus* is often found in sulfur-containing hot springs such as the Norris Geyser Basin in Yellowstone National Park shown here.

glycolysis to produce ATP. Under these conditions, the electrons generated by glycolysis are donated to organic molecules in a process called *fermentation*. This process recycles NAD^+, the electron acceptor that allows glycolysis to proceed.

Bacteria carry out more than a dozen kinds of fermentation reactions, often using pyruvate or a derivative of pyruvate to accept the electrons from NADH. Organic molecules other than pyruvate and its derivatives can be used as well; the important point is that the process regenerates NAD^+:

$$\text{organic molecule} + NADH \longrightarrow \text{reduced organic} \\ \text{molecule} + NAD^+$$

Often the reduced organic compound is an organic acid—such as acetic acid, butyric acid, propionic acid, or lactic acid—or an alcohol.

Ethanol fermentation

Eukaryotic cells are capable of only a few types of fermentation. In one type, which occurs in yeast, the molecule that accepts electrons from NADH is derived from pyruvate, the end-product of glycolysis.

Yeast enzymes remove a terminal CO_2 group from pyruvate through decarboxylation, producing a 2-carbon molecule called acetaldehyde. The CO_2 released causes bread made with yeast to rise. The acetaldehyde accepts a pair of electrons from NADH, producing NAD^+ and ethanol (ethyl alcohol) (figure 7.19).

This particular type of fermentation is of great interest to humans, because it is the source of the ethanol in wine and beer. Ethanol is a by-product of fermentation that is actually toxic to yeast; as it approaches a concentration of about 12%, it begins to kill the yeast. That explains why naturally fermented wine contains only about 12% ethanol.

Lactic acid fermentation

Most animal cells regenerate NAD^+ without decarboxylation. Muscle cells, for example, use the enzyme lactate dehydrogenase to transfer electrons from NADH back to the pyruvate that is produced by glycolysis. This reaction converts pyruvate into lactic acid and regenerates NAD^+ from NADH (see figure 7.19). It therefore closes the metabolic circle, allowing glycolysis to continue as long as glucose is available.

Circulating blood removes excess lactate, the ionized form of lactic acid, from muscles, but when removal cannot keep pace with production, the accumulating lactic acid interferes with muscle function and contributes to muscle fatigue.

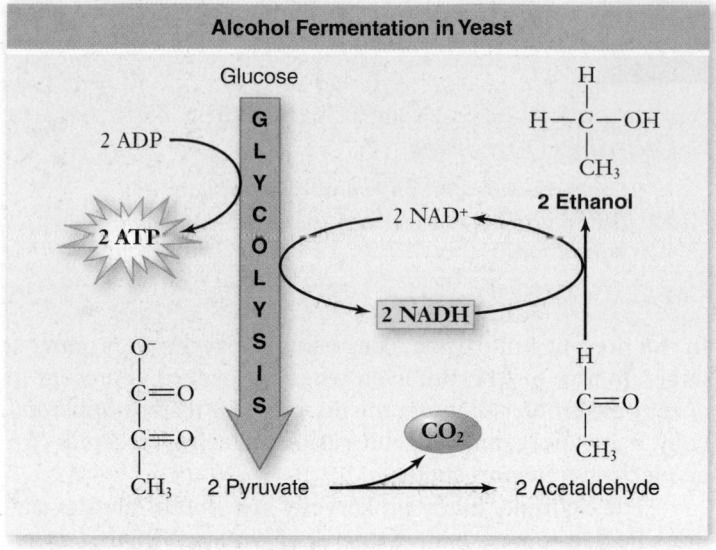

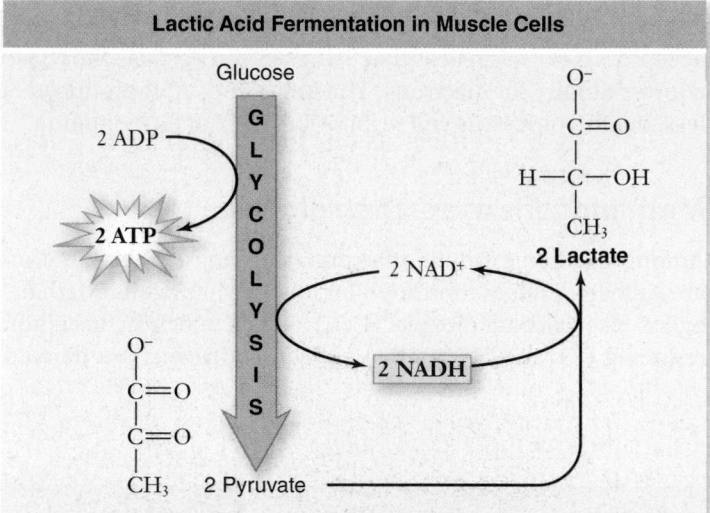

Figure 7.19 Fermentation. Yeasts carry out the conversion of pyruvate to ethanol. Muscle cells convert pyruvate into lactate, which is less toxic than ethanol. In each case, the reduction of a metabolite of glucose has oxidized NADH back to NAD^+ to allow glycolysis to continue under anaerobic conditions.

Learning Outcomes Review 7.8

Nitrate, sulfur, and CO_2 are all used as terminal electron acceptors in anaerobic respiration of different organisms. Organic molecules can also accept electrons in fermentation reactions that regenerate NAD^+. Fermentation reactions produce a variety of compounds, including ethanol in yeast and lactic acid in humans.

■ *In what kinds of ecosystems would you expect to find anaerobic respiration?*

7.9 Catabolism of Proteins and Fats

Learning Outcomes

1. *Identify the entry points for proteins and fats in energy metabolism.*
2. *Recognize the importance of key intermediates in metabolism.*

Thus far we have focused on the aerobic respiration of glucose, which organisms obtain from the digestion of carbohydrates or from photosynthesis. Organic molecules other than glucose,

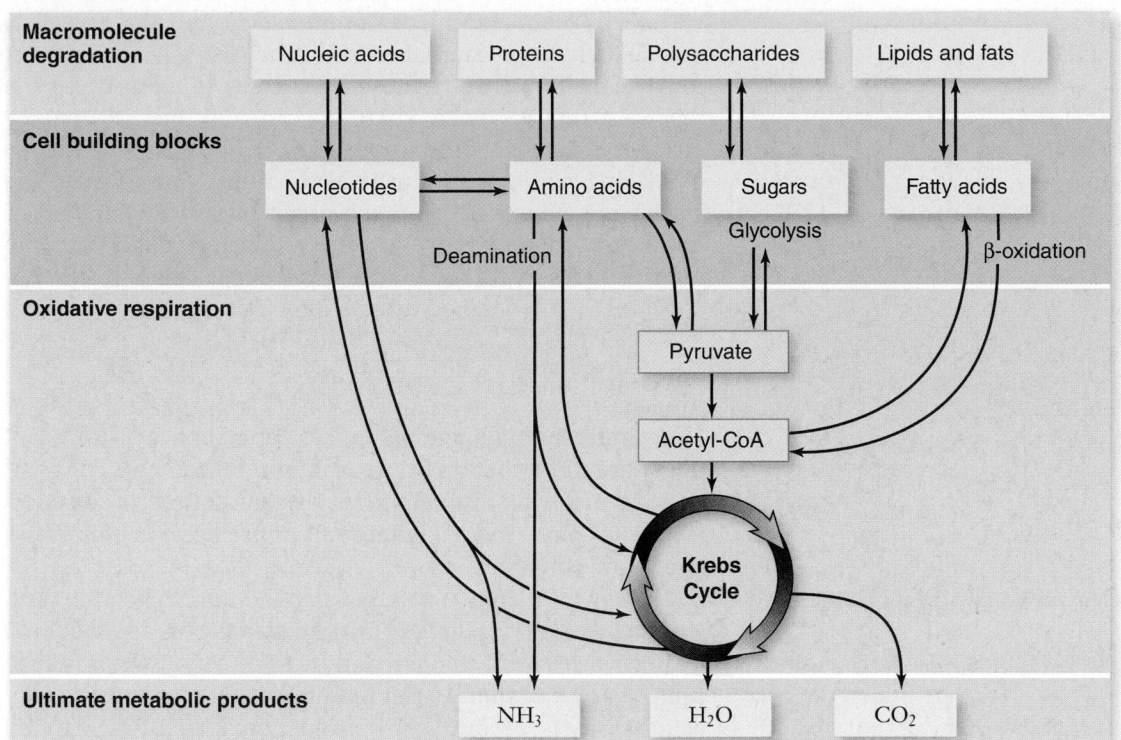

| Macromolecule degradation | Nucleic acids | Proteins | Polysaccharides | Lipids and fats |

Figure 7.20 How cells extract chemical energy. All eukaryotes and many prokaryotes extract energy from organic molecules by oxidizing them. The first stage of this process, breaking down macromolecules into their constituent parts, yields little energy. The second stage, oxidative or aerobic respiration, extracts energy, primarily in the form of high-energy electrons, and produces water and carbon dioxide. Key intermediates in these energy pathways are also used for biosynthetic pathways, shown by reverse arrows.

particularly proteins and fats, are also important sources of energy (figure 7.20).

Catabolism of proteins removes amino groups

Proteins are first broken down into their individual amino acids. The nitrogen-containing side group (the amino group) is then removed from each amino acid in a process called **deamination.** A series of reactions converts the carbon chain that remains into a molecule that enters glycolysis or the Krebs cycle. For example, alanine is converted into pyruvate, glutamate into α-ketoglutarate (figure 7.21), and aspartate into oxaloacetate. The reactions of glycolysis and the Krebs cycle then extract the high-energy electrons from these molecules and put them to work making ATP.

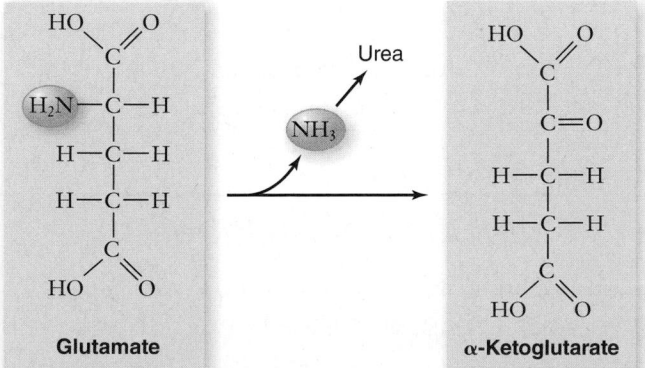

Figure 7.21 Deamination. After proteins are broken down into their amino acid constituents, the amino groups are removed from the amino acids to form molecules that participate in glycolysis and the Krebs cycle. For example, the amino acid glutamate becomes α-ketoglutarate, a Krebs cycle intermediate, when it loses its amino group.

Catabolism of fatty acids produces acetyl groups

Fats are broken down into fatty acids plus glycerol. Long-chain fatty acids typically have an even number of carbons, and the many C—H bonds provide a rich harvest of energy. Fatty acids are oxidized in the matrix of the mitochondrion. Enzymes remove the 2-carbon acetyl groups from the end of each fatty acid until the entire fatty acid is converted into acetyl groups (figure 7.22). Each acetyl group is combined with coenzyme A to form acetyl-CoA. This process is known as **β oxidation.** This process is oxygen-dependent, which explains why aerobic exercise burns fat, but anaerobic exercise does not.

How much ATP does the catabolism of fatty acids produce? Let's compare a hypothetical 6-carbon fatty acid with the 6-carbon glucose molecule, which we've said yields about 30 molecules of ATP in a eukaryotic cell. Two rounds of β oxidation would convert the fatty acid into three molecules of acetyl-CoA. Each round requires one molecule of ATP to prime the process, but it also produces one molecule of NADH and one of $FADH_2$. These molecules together yield four molecules of ATP (assuming 2.5 ATP per NADH, and 1.5 ATP per $FADH_2$).

The oxidation of each acetyl-CoA in the Krebs cycle ultimately produces an additional 10 molecules of ATP. Overall, then, the ATP yield of a 6-carbon fatty acid is approximately: 8 (from two rounds of β oxidation) − 2 (for priming those two rounds) + 30 (from oxidizing the three acetyl-CoAs) = 36 molecules of ATP. Therefore, the respiration of a 6-carbon fatty acid yields 20% more ATP than the respiration of glucose.

Moreover, a fatty acid of that size would weigh less than two thirds as much as glucose, so a gram of fatty acid contains more than twice as many kilocalories as a gram of glucose. You can see from this fact why fat is a storage molecule for excess

Figure 7.22

β oxidation. Through a series of reactions known as β oxidation, the last two carbons in a fatty acid combine with coenzyme A to form acetyl-CoA, which enters the Krebs cycle. The fatty acid, now two carbons shorter, enters the pathway again and keeps reentering until all its carbons have been used to form acetyl-CoA molecules. Each round of β oxidation uses one molecule of ATP and generates one molecule each of FADH$_2$ and NADH.

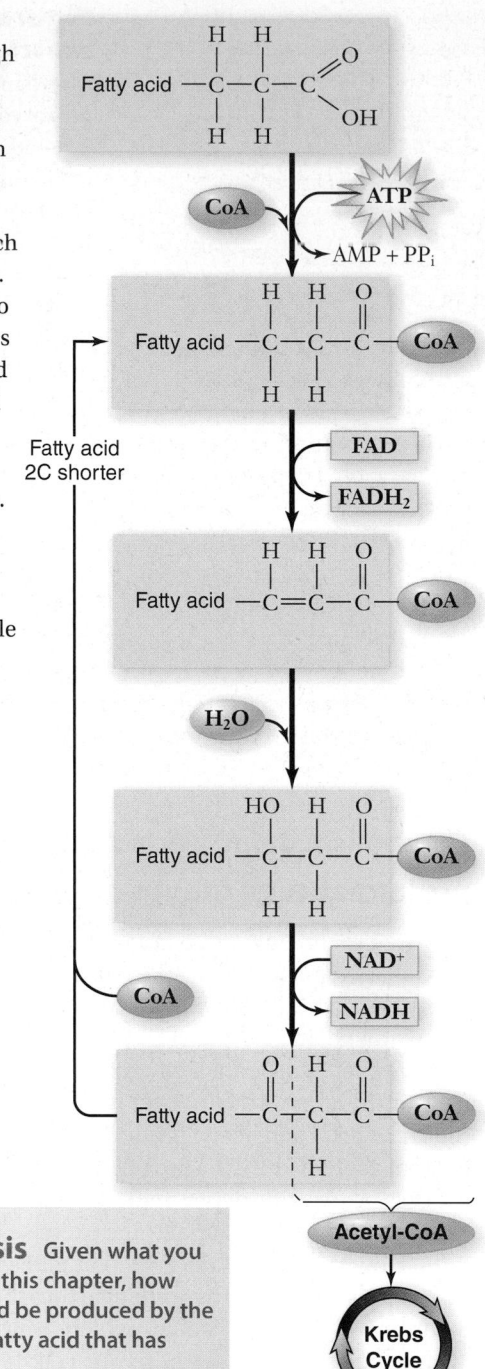

Data analysis Given what you have learned in this chapter, how many ATP would be produced by the oxidation of a fatty acid that has 16 carbons?

energy in many types of animals. If excess energy were stored instead as carbohydrate, as it is in plants, animal bodies would have to be much bulkier.

A small number of key intermediates connect metabolic pathways

Oxidation pathways of food molecules are interrelated in that a small number of key intermediates, such as pyruvate and acetyl-CoA, link the breakdown from different starting points. These key intermediates allow the interconversion of different types of molecules, such as sugars and amino acids (see figure 7.20).

Cells can make glucose, amino acids, and fats, as well as getting them from external sources. They use reactions similar to those that break down these substances. In many cases, the reverse pathways even share enzymes if the free-energy changes are small. For example, gluconeogenesis, the process of making new glucose, uses all but three enzymes of the glycolytic pathway. Thus, much of glycolysis runs forward or backward, depending on the concentrations of the intermediates—with only three key steps having different enzymes for forward and reverse directions.

Acetyl-CoA has many roles

Many different metabolic processes generate acetyl-CoA. Not only does the oxidation of pyruvate produce it, but the metabolic breakdown of proteins, fats, and other lipids also generates acetyl-CoA. Indeed, almost all molecules catabolized for energy are converted into acetyl-CoA.

Acetyl-CoA has a role in anabolic metabolism as well. Units of two carbons derived from acetyl-CoA are used to build up the hydrocarbon chains in fatty acids. Acetyl-CoA produced from a variety of sources can therefore be channeled into fatty acid synthesis or into ATP production, depending on the organism's energy requirements. Which of these two options is taken depends on the level of ATP in the cell.

When ATP levels are high, the oxidative pathway is inhibited, and acetyl-CoA is channeled into fatty acid synthesis. This explains why many animals (humans included) develop fat reserves when they consume more food than their activities require. Alternatively, when ATP levels are low, the oxidative pathway is stimulated, and acetyl-CoA flows into energy-producing oxidative metabolism.

Learning Outcomes Review 7.9

Proteins can be broken into their constituent amino acids, which are then deaminated and can enter metabolism at glycolysis or different steps of the Krebs cycle. Fats can be broken into units of acetyl-CoA by β oxidation and then fed into the Krebs cycle. Many metabolic processes can be used reversibly, to either build up (anabolism) or break down (catabolism) the major biological macromolecules. Key intermediates, such as pyruvate and acetyl-CoA, connect these processes.

■ *Can fats be oxidized in the absence of O$_2$?*

7.10 Evolution of Metabolism

Learning Outcome

1. Describe one possible hypothesis for the evolution of metabolism.

We talk about cellular respiration as a continuous series of stages, but it is important to note that these stages evolved over time, and metabolism has changed a great deal in that time.

Both anabolic processes and catabolic processes evolved in concert with each other. We do not know the details of this biochemical evolution, or the order of appearance of these processes. Therefore the following timeline is based on the available geochemical evidence and represents a hypothesis rather than a strict timeline.

The earliest life-forms degraded carbon-based molecules present in the environment

The most primitive forms of life are thought to have obtained chemical energy by degrading, or breaking down, organic molecules that were abiotically produced, that is, carbon-containing molecules formed by inorganic processes on the early Earth.

The first major event in the evolution of metabolism was the origin of the ability to harness chemical bond energy. At an early stage, organisms began to store this energy in the bonds of ATP.

The evolution of glycolysis also occurred early

The second major event in the evolution of metabolism was glycolysis, the initial breakdown of glucose. As proteins evolved diverse catalytic functions, it became possible to capture a larger fraction of the chemical bond energy in organic molecules by breaking chemical bonds in a series of steps.

Glycolysis undoubtedly evolved early in the history of life on Earth, because this biochemical pathway has been retained by all living organisms. It is a chemical process that does not appear to have changed for more than 2 billion years.

Anoxygenic photosynthesis allowed the capture of light energy

The third major event in the evolution of metabolism was anoxygenic photosynthesis. Early in the history of life, a different way of generating ATP evolved in some organisms. Instead of obtaining energy for ATP synthesis by reshuffling chemical bonds, as in glycolysis, these organisms developed the ability to use light to pump protons out of their cells and to use the resulting proton gradient to power the production of ATP through chemiosmosis.

Photosynthesis evolved in the absence of oxygen and works well without it. Dissolved H_2S, present in the oceans of the early Earth beneath an atmosphere free of oxygen gas, served as a ready source of hydrogen atoms for building organic molecules. Free sulfur was produced as a by-product of this reaction.

Oxygen-forming photosynthesis used a different source of hydrogen

The substitution of H_2O for H_2S in photosynthesis was the fourth major event in the history of metabolism. Oxygen-forming photosynthesis employs H_2O rather than H_2S as a source of hydrogen atoms and their associated electrons. Because it garners its electrons from reduced oxygen rather than from reduced sulfur, it generates oxygen gas rather than free sulfur.

More than 2 BYA, small cells capable of carrying out this oxygen-forming photosynthesis, such as cyanobacteria, became the dominant forms of life on Earth. Oxygen gas began to accumulate in the atmosphere. This was the beginning of a great transition that changed conditions on Earth permanently. Our atmosphere is now 20.9% oxygen, every molecule of which is derived from an oxygen-forming photosynthetic reaction.

Nitrogen fixation provided new organic nitrogen

Nitrogen is available from dead organic matter, and from chemical reactions that generated the original organic molecules. For life to expand, a new source of nitrogen was needed. Nitrogen fixation was the fifth major step in the evolution of metabolism. Proteins and nucleic acids cannot be synthesized from the products of photosynthesis because both of these biologically critical molecules contain nitrogen. Obtaining nitrogen atoms from N_2 gas, a process called *nitrogen fixation,* requires breaking an $N \equiv N$ triple bond.

This important reaction evolved in the hydrogen-rich atmosphere of the early Earth, where no oxygen was present. Oxygen acts as a poison to nitrogen fixation, which today occurs only in oxygen-free environments or in oxygen-free compartments within certain prokaryotes.

Aerobic respiration utilized oxygen

Respiration is the sixth and final event in the history of metabolism. Aerobic respiration employs the same kind of proton pumps as photosynthesis and is thought to have evolved as a modification of the basic photosynthetic machinery.

Biologists think that the ability to carry out photosynthesis without H_2S first evolved among purple nonsulfur bacteria, which obtain their hydrogens from organic compounds instead. It was perhaps inevitable that among the descendants of these respiring photosynthetic bacteria, some would eventually do without photosynthesis entirely, subsisting only on the energy and electrons derived from the breakdown of organic molecules. The mitochondria within all eukaryotic cells are thought to be descendants of these bacteria.

The complex process of aerobic metabolism developed over geological time, as natural selection favored organisms with more efficient methods of obtaining energy from organic molecules. The process of photosynthesis, as you have seen in this concluding section, has also developed over time, and the rise of photosynthesis changed life on Earth forever. The next chapter explores photosynthesis in detail.

Learning Outcome Review 7.10

Major milestones in the evolution of metabolism include the evolution of pathways to extract energy from organic compounds, the pathways of photosynthesis, and those of nitrogen fixation. Photosynthesis began as an anoxygenic process that later evolved to produce free oxygen, thus allowing the evolution of aerobic metabolism.

■ *What evidence can you cite for this hypothesis of the evolution of metabolism?*

7.1 Overview of Respiration (figure 7.5)

Cells oxidize organic compounds to drive metabolism.

Cellular respiration is the complete oxidation of glucose.

Aerobic respiration uses oxygen as the final electron acceptor for redox reactions. Anaerobic respiration utilizes inorganic molecules as acceptors, and fermentation uses organic molecules.

Electron carriers play a critical role in energy metabolism.

Electron carriers can be reversibly oxidized and reduced. For example, NAD^+ is reduced to NADH by acquiring two electrons; NADH supplies these electrons to other molecules to reduce them.

Metabolism harvests energy in stages.

Mitochondria of eukaryotic cells move electrons in steps via the electron transport chain to capture energy efficiently.

ATP plays a central role in metabolism.

The ultimate goal of cellular respiration is synthesis of ATP, which is used to power most of the cell's activities.

Cells make ATP by two fundamentally different mechanisms.

Substrate-level phosphorylation transfers a phosphate directly to ADP (figure 7.4). Oxidative phosphorylation generates ATP via the enzyme ATP synthase, powered by a proton gradient.

7.2 Glycolysis: Splitting Glucose (figures 7.6 & 7.7)

Glycolysis converts glucose into two pyruvate and yields two ATP and two NADH in the process.

Priming reactions add two phosphates to glucose; this is cleaved into two 3-carbon molecules of glyceraldehyde 3-phosphate (G3P). Oxidation of G3P transfers electrons to NAD^+, yielding NADH. After four more reactions, the final product is two molecules of pyruvate. Glycolysis produces 2 net ATP, 2 NADH, and 2 pyruvate.

Glycolysis is an ancient process with a low energy yield, but it can be efficient with up to 40% of available energy trapped as ATP. Glycolysis was probably the first catabolic reaction to evolve.

NADH must be recycled into NAD^+ to continue respiration.

In the presence of oxygen, pyruvate is oxidized to acetyl-CoA, which can be oxidized by the Krebs cycle. This process leads to a large amount of ATP. In the absence of oxygen, a fermentation reaction uses all or part of pyruvate to oxidize NADH.

In the presence of oxygen, NADH passes electrons to the electron transport chain. In the absence of oxygen, NADH passes the electrons to an organic molecule such as acetaldehyde (fermentation).

7.3 The Oxidation of Pyruvate to Produce Acetyl-CoA (figure 7.9)

Pyruvate is oxidized to yield 1 CO_2, 1 NADH, and 1 acetyl-CoA. Acetyl-CoA enters the Krebs cycle as 2-carbon acetyl units.

7.4 The Krebs Cycle (figures 7.10 & 7.11)

An overview of the Krebs cycle.

The Krebs cycle extracts electrons and synthesizes one ATP.

The first reaction is an irreversible condensation that produces citrate; it is inhibited when ATP is plentiful. The second and third reactions rearrange citrate to isocitrate. The fourth and fifth reactions are oxidations; in each reaction, one NAD^+ is reduced to NADH. The sixth reaction is a substrate-level phosphorylation producing

GTP, and from that ATP. The seventh reaction is another oxidation that reduces FAD to $FADH_2$. Reactions eight and nine regenerate oxaloacetate, including one final oxidation that reduces NAD^+ to NADH.

Glucose becomes CO_2 and potential energy.

As a glucose molecule is broken down to CO_2, some of its energy is preserved in 4 ATP, 10 NADH, and 2 $FADH_2$.

Following the electrons in the reactions reveals the direction of transfer.

7.5 The Electron Transport Chain and Chemiosmosis (figure 7.12)

The electron transport chain produces a proton gradient.

In the inner mitochondrial membrane, NADH is oxidized to NAD^+ by NADH dehydrogenase. Electrons move through ubiquinone and the bc_1 complex to cytochrome oxidase, where they join with H^+ and O_2 to form H_2O. This results in three protons being pumped into the intermembrane space. For $FADH_2$, electrons are passed directly to ubiquinone. Thus only two protons are pumped into the intermembrane space.

The gradient forms as electrons move through electron carriers.

Chemiosmosis utilizes the electrochemical gradient to produce ATP.

ATP synthase is a molecular rotary motor.

Protons diffuse back into the mitochondrial matrix via the ATP synthase channel. The enzyme uses this energy to synthesize ATP (figure 7.15).

7.6 Energy Yield of Aerobic Respiration

The theoretical yield for eukaryotes is 30 molecules of ATP per glucose molecule (figure 7.16).

Calculation of P/O ratios has changed over time.

7.7 Regulation of Aerobic Respiration

Glucose catabolism is controlled by the concentration of ATP molecules and intermediates in the Krebs cycle (figure 7.17).

7.8 Oxidation Without O_2 (figure 7.8)

In the absence of oxygen other final electron acceptors can be used for respiration.

Methanogens use carbon dioxide.

Sulfur bacteria use sulfate.

Fermentation uses organic compounds as electron acceptors (figure 7.19).

Fermentation is the regeneration of NAD^+ by oxidation of NADH and reduction of an organic molecule. In yeast, pyruvate is decarboxylated, then reduced to ethanol. In animals, pyruvate is reduced directly to lactate.

7.9 Catabolism of Proteins and Fats

Catabolism of proteins removes amino groups (figure 7.20).

Catabolism of fatty acids produces acetyl groups.

Fatty acids are converted to acetyl groups by successive rounds of β oxidation (figure 7.22). These acetyl groups feed into the Krebs cycle to be oxidized and generate NADH for electron transport.

A small number of key intermediates connect metabolic pathways.

Acetyl-CoA has many roles.
With high ATP, acetyl-CoA is converted into fatty acids.

7.10 Evolution of Metabolism

Major milestones are recognized in the evolution of metabolism; the order of events is hypothetical.

The earliest life-forms degraded carbon-based molecules present in the environment.

The evolution of glycolysis also occurred early.

Anoxygenic photosynthesis allowed the capture of light energy.

Oxygen-forming photosynthesis used a different source of hydrogen.

Nitrogen fixation provided new organic nitrogen.

Aerobic respiration utilized oxygen.

Review Questions

UNDERSTAND

1. An *autotroph* is an organism that
   a. extracts energy from organic sources.
   b. converts energy from sunlight into chemical energy.
   c. relies on the energy produced by other organisms as an energy source.
   d. does both a and b.

2. Which of the following processes is (are) required for the complete oxidation of glucose?
   a. The Krebs cycle
   b. Glycolysis
   c. Pyruvate oxidation
   d. All of the choices are correct.

3. Which of the following is NOT a product of glycolysis?
   a. ATP
   b. Pyruvate
   c. CO_2
   d. NADH

4. Glycolysis produces ATP by
   a. phosphorylating organic molecules in the priming reactions.
   b. the production of glyceraldehyde 3-phosphate.
   c. substrate-level phosphorylation.
   d. the reduction of NAD^+ to NADH.

5. What is the role of NAD^+ in the process of cellular respiration?
   a. It functions as an electron carrier.
   b. It functions as an enzyme.
   c. It is the final electron acceptor for anaerobic respiration.
   d. It is a nucleotide source for the synthesis of ATP.

6. The reactions of the Krebs cycle occur in the
   a. inner membrane of the mitochondria.
   b. intermembrane space of the mitochondria.
   c. cytoplasm.
   d. matrix of the mitochondria.

7. The electrons carried by NADH and $FADH_2$ can be
   a. pumped into the intermembrane space.
   b. transferred to the ATP synthase.
   c. moved between proteins in the inner membrane of the mitochondrion.
   d. transported into the matrix of the mitochondrion.

APPLY

1. Which of the following is NOT a true statement regarding cellular respiration?
   a. Enzymes catalyze reactions that transfer electrons.
   b. Electrons have a higher potential energy at the end of the process.
   c. Carbon dioxide gas is a by-product.
   d. The process involves multiple redox reactions.

2. The direct source of energy for the ATP produced by ATP synthase comes from
   a. the electron transport chain.
   b. a proton gradient.
   c. substrate-level phosphorylation.
   d. the oxidation reactions occurring during respiration.

3. Anaerobic respiration
   a. occurs in humans in the absence of O_2.
   b. occurs in yeast and is how we make beer and wine.
   c. yields less energy than aerobic respiration because other final electron acceptors have lower affinity for electrons than O_2.
   d. yields more energy than aerobic respiration because other final electron acceptors have higher affinity for electrons than O_2.

4. What is the importance of fermentation to cellular metabolism?
   a. It generates glucose for the cell in the absence of O_2.
   b. It oxidizes NADH to NAD^+ during electron transport.
   c. It oxidizes NADH to NAD^+ in the absence of O_2.
   d. It reduces NADH to NAD^+ in the absence of O_2.

5. The link between electron transport and ATP synthesis
   a. is a high-energy intermediate like phosphoenol pyruvate.
   b. is the transfer of electrons to ATP synthase.
   c. is a proton gradient.
   d. depends on the absence of oxygen.

6. A chemical agent that makes holes in the inner membrane of the mitochondria would
   a. stop the movement of electrons down the electron transport chain.
   b. stop ATP synthesis.
   c. stop the Krebs cycle.
   d. All of the choices are correct.

7. Yeast cells that have mutations in genes that encode enzymes in glycolysis can still grow on glycerol. They are able to utilize glycerol because it
   a. enters glycolysis after the step affected by the mutation.
   b. can feed into the Krebs cycle and generate ATP via electron transport and chemiosmosis.
   c. can be utilized by fermentation.
   d. can donate electrons directly to the electron transport chain.

SYNTHESIZE

1. Use the following table to outline the relationship between the molecules and the metabolic reactions.

Molecules	Glycolysis	Cellular Respiration
Glucose		
Pyruvate		
Oxygen		
ATP		
CO_2		

2. Human babies and hibernating or cold-adapted animals are able to maintain body temperature (a process called *thermogenesis*) due to the presence of brown fat. Brown fat is characterized by a high concentration of mitochondria. These brown fat mitochondria have a special protein located within their inner membranes. *Thermogenin* is a protein that functions as a passive proton transporter. Propose a likely explanation for the role of brown fat in thermogenesis based on your knowledge of metabolism, transport, and the structure and function of mitochondria.

3. Recent data indicate a link between colder temperatures and weight loss. If adults retain brown fat, how could this be explained?

ONLINE RESOURCE

www.ravenbiology.com

Understand, Apply, and Synthesize—enhance your study with animations that bring concepts to life and practice tests to assess your understanding. Your instructor may also recommend the interactive eBook, individualized learning tools, and more.

Chapter **8**

Photosynthesis

Chapter Contents

Introduction

The rich diversity of life that covers our Earth would be impossible without photosynthesis. Almost every oxygen atom in the air we breathe was once part of a water molecule, liberated by photosynthesis. All the energy released by the burning of coal, firewood, gasoline, and natural gas, and by our bodies' burning of all the food we eat—directly or indirectly—has been captured from sunlight by photosynthesis. It is vitally important, then, that we understand photosynthesis. Research may enable us to improve crop yields and land use, important goals in an increasingly crowded world. In chapter 7, we described how cells extract chemical energy from food molecules and use that energy to power their activities. In this chapter, we examine photosynthesis, the process by which organisms such as the aptly named sunflowers in the picture capture energy from sunlight and use it to build food molecules that are rich in chemical energy.

8.1 Overview of Photosynthesis

Learning Outcomes

1. *Explain the reaction for photosynthesis.*
2. *Describe the structure of the chloroplast.*

Life is powered by sunshine. The energy used by most living cells comes ultimately from the Sun and is captured by plants, algae, and bacteria through the process of photosynthesis.

The diversity of life is only possible because our planet is awash in energy streaming Earthward from the Sun. Each day, the radiant energy that reaches Earth equals the power from about 1 million Hiroshima-sized atomic bombs. Photosynthesis captures about 1% of this huge supply of energy (an amount equal to 10,000 Hiroshima bombs) and uses it to provide the energy that drives all life.

Photosynthesis combines CO₂ and H₂O, producing glucose and O₂

Photosynthesis occurs in a wide variety of organisms, and it comes in different forms. These include a form of photosynthesis that does not produce oxygen (anoxygenic) and a form that does (oxygenic). Anoxygenic photosynthesis is found in four different bacterial groups: purple bacteria, green sulfur bacteria, green nonsulfur bacteria, and heliobacteria. Oxygenic photosynthesis is found in cyanobacteria, seven groups of algae, and essentially all land plants. These two types of photosynthesis share similarities in the types of pigments they use to trap light energy, but they differ in the arrangement and action of these pigments.

In the case of plants, photosynthesis takes place primarily in the leaves. Figure 8.1 illustrates the levels of organization in a plant leaf. As you learned in chapter 4, the cells of plant leaves contain organelles called chloroplasts, which carry out the photosynthetic process. No other structure in a plant cell is able to carry out photosynthesis (figure 8.2). Photosynthesis takes place in three stages:

1. capturing energy from sunlight;
2. using the energy to make ATP and to reduce the compound NADP⁺, an electron carrier, to NADPH; and
3. using the ATP and NADPH to power the synthesis of organic molecules from CO₂ in the air.

The first two stages require light and are commonly called the **light-dependent reactions.**

The third stage, the formation of organic molecules from CO₂, is called **carbon fixation.** This process takes place via a cyclic series of reactions. As long as ATP and NADPH are available, the carbon fixation reactions can occur either in the presence or in the absence of light, and so these reactions are also called the **light-independent reactions.**

The following simple equation summarizes the overall process of photosynthesis:

$$6CO_2 + 12H_2O + light \longrightarrow C_6H_{12}O_6 + 6H_2O + 6O_2$$

$$\underset{\substack{\text{carbon}\\\text{dioxide}}}{} \quad \underset{\text{water}}{} \qquad\qquad \underset{\text{glucose}}{} \quad \underset{\text{water}}{} \quad \underset{\text{oxygen}}{}$$

You may notice that this equation is the reverse of the reaction for respiration. In respiration, glucose is oxidized to CO₂ using O₂ as an electron acceptor. In photosynthesis, CO₂ is reduced to glucose using electrons gained from the oxidation of water. The oxidation of H₂O and the reduction of CO₂ requires energy that is provided by light. Although this statement is an oversimplification, it provides a useful "global perspective."

In plants, photosynthesis takes place in chloroplasts

In the preceding chapter, you saw that a mitochondrion's complex structure of internal and external membranes contribute to its function. The same is true for the structure of the chloroplast.

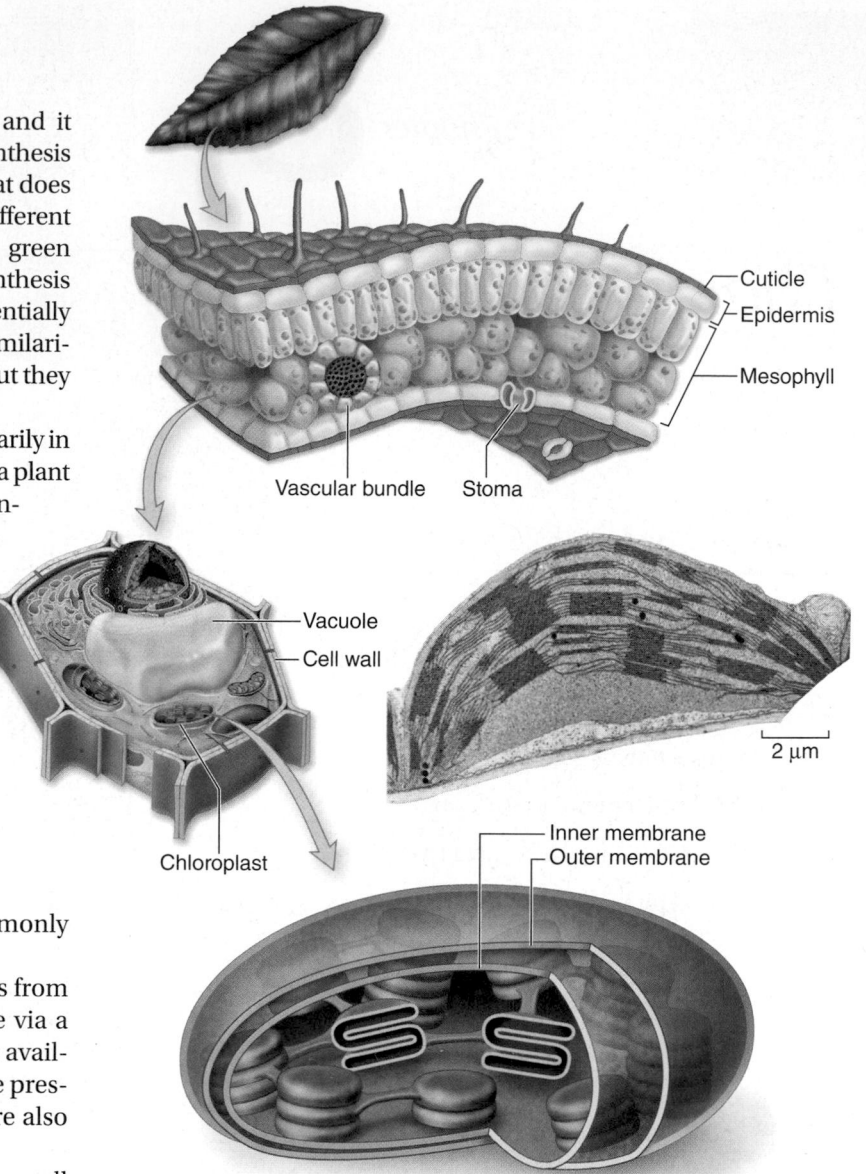

Figure 8.1 Journey into a leaf. A plant leaf possesses a thick layer of cells (the mesophyll) rich in chloroplasts. The inner membrane of the chloroplast is organized into flattened structures called thylakoid disks, which are stacked into columns called grana. The rest of the interior is filled with a semifluid substance called stroma.

The internal membrane of chloroplasts, called the *thylakoid membrane,* is a continuous phospholipid bilayer organized into flattened sacs that are found stacked on one another in columns called *grana* (singular, *granum*). The thylakoid membrane contains **chlorophyll** and other photosynthetic pigments for capturing light energy along with the machinery to make ATP. Connections between grana are termed *stroma lamella.*

Surrounding the thylakoid membrane system is a semi-liquid substance called **stroma.** The stroma houses the enzymes needed to assemble organic molecules from CO₂ using energy from ATP coupled with reduction via NADPH. In the thylakoid membrane, photosynthetic pigments are clustered together to form **photosystems,** which show distinct organization within the thylakoid.

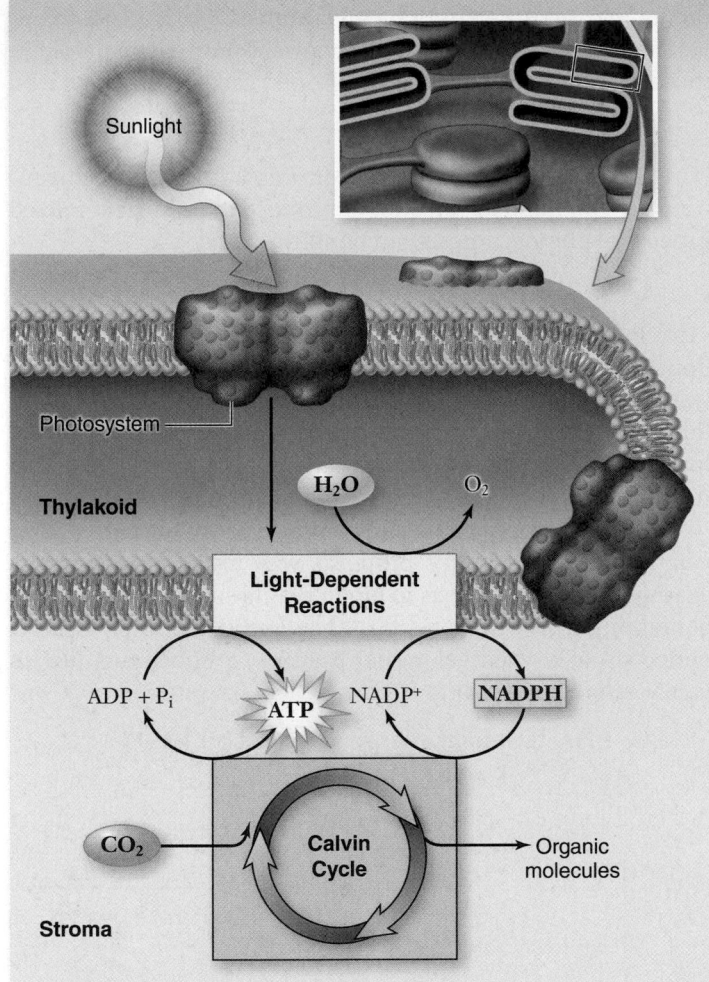

Figure 8.2 Overview of photosynthesis. In the light-dependent reactions, photosystems in the thylakoid absorb photons of light and use this energy to generate ATP and NADPH. Electrons lost from the photosystems are replaced by the oxidation of water, producing O_2 as a by-product. The ATP and NADPH produced by the light reactions is used during carbon fixation via the Calvin cycle in the stroma.

Each pigment molecule within the photosystem is capable of capturing photons, which are packets of energy. When light of a proper wavelength strikes a pigment molecule in the photosystem, the resulting excitation passes from one pigment molecule to another.

The excited electron is not transferred physically—rather, its *energy* passes from one molecule to another. The passage is similar to the transfer of kinetic energy along a row of upright dominoes. If you push the first one over, it falls against the next, and that one against the next, and so on, until all of the dominoes have fallen down.

Eventually, the energy arrives at a key chlorophyll molecule in contact with a membrane-bound protein that can accept an electron. The energy is transferred as an excited electron to that protein, which passes it on to a series of other membrane proteins that put the energy to work making ATP and NADPH.

These compounds are then used to build organic molecules. The photosystem thus acts as a large antenna, gathering the light energy harvested by many individual pigment molecules.

8.2 The Discovery of Photosynthetic Processes

The story of how we learned about photosynthesis begins over 300 years ago, and it continues to this day. It starts with curiosity about how plants manage to grow, often increasing their organic mass considerably.

Plants do not increase mass from soil and water alone

From the time of the Greeks, plants were thought to obtain their food from the soil, literally sucking it up with their roots. A Belgian doctor, Jan Baptista van Helmont (1580–1644) thought of a simple way to test this idea.

He planted a small willow tree in a pot of soil, after first weighing the tree and the soil. The tree grew in the pot for several years, during which time van Helmont added only water. At the end of five years, the tree was much larger, its weight having increased by 74.4 kg. However, the soil in the pot weighed only 57 g less than it had five years earlier. With this experiment, van Helmont demonstrated that the substance of the plant was not produced only from the soil. He incorrectly concluded, however, that the water he had been adding mainly accounted for the plant's increased biomass.

A hundred years passed before the story became clearer. The key clue was provided by the English scientist Joseph Priestly (1733–1804). On the 17th of August, 1771, Priestly put a living sprig of mint into air in which a wax candle had burnt out.

On the 27th of the same month, Priestly found that another candle could be burned in this same air. Somehow, the vegetation seemed to have restored the air. Priestly found that while a mouse could not breathe candle-exhausted air, air "restored" by vegetation was not "at all inconvenient to a mouse." The key clue was that *living vegetation adds something to the air.*

How does vegetation "restore" air? Twenty-five years later, the Dutch physician Jan Ingenhousz (1730–1799) solved the puzzle. He demonstrated that air was restored only in the presence of sunlight and only by a plant's green leaves, not by its roots. He proposed that the green parts of the plant carry out a process that uses sunlight to split carbon dioxide into carbon and oxygen. He suggested that the oxygen was released as O_2 gas into the air, while the carbon atom combined with water to form carbohydrates. Other research refined his conclusions, and by the end of the nineteenth century, the overall reaction for photosynthesis could be written as:

$$CO_2 + H_2O + \text{light energy} \longrightarrow (CH_2O) + O_2$$

It turns out, however, that there's more to it than that. When researchers began to examine the process in more detail in the twentieth century, the role of light proved to be unexpectedly complex.

Photosynthesis includes both light-dependent and light-independent reactions

At the beginning of the twentieth century, the English plant physiologist F. F. Blackman (1866–1947) came to the surprising conclusion that photosynthesis is in fact a multistage process, only one portion of which uses light directly.

Blackman measured the effects of different light intensities, CO_2 concentrations, and temperatures on photosynthesis. As long as light intensity was relatively low, he found photosynthesis could be accelerated by increasing the amount of light, but not by increasing the temperature or CO_2 concentration (figure 8.3). At high light intensities, however, an increase in temperature or CO_2 concentration greatly accelerated photosynthesis.

Blackman concluded that photosynthesis consists of an initial set of what he called "light" reactions, that are largely independent of temperature but depend on light, and a second set of "dark" reactions (more properly called light-independent reactions), that seemed to be independent of light but limited by CO_2.

Do not be confused by Blackman's labels—the so-called "dark" reactions occur in the light (in fact, they require the products of the light-dependent reactions); his use of the word *dark* simply indicates that light is not *directly* involved in those reactions.

Blackman found that increased temperature increased the rate of the light-independent reactions, but only up to about 35°C. Higher temperatures caused the rate to fall off rapidly. Because many plant enzymes begin to be denatured at 35°C, Blackman concluded that enzymes must carry out the light-independent reactions.

O_2 comes from water, not from CO_2

In the 1930s, C. B. van Niel (1897–1985) working at the Hopkins Marine Station at Stanford, discovered that purple sulfur bacteria

do not release oxygen during photosynthesis; instead, they convert hydrogen sulfide (H_2S) into globules of pure elemental sulfur that accumulate inside them. The process van Niel observed was:

$$CO_2 + 2H_2S + \text{light energy} \longrightarrow (CH_2O) + H_2O + 2S$$

The striking parallel between this equation and Ingenhousz's equation led van Niel to propose that the generalized process of photosynthesis can be shown as:

$$CO_2 + 2H_2A + \text{light energy} \longrightarrow (CH_2O) + H_2O + 2A$$

In this equation, the substance H_2A serves as an electron donor. In photosynthesis performed by green plants, H_2A is water, whereas in purple sulfur bacteria, H_2A is hydrogen sulfide. The product, A, comes from the splitting of H_2A. Therefore, the O_2 produced during green plant photosynthesis results from splitting water, not carbon dioxide.

When isotopes came into common use in the early 1950s, van Niel's revolutionary proposal was tested. Investigators examined photosynthesis in green plants supplied with water containing heavy oxygen (^{18}O); they found that the ^{18}O label ended up in oxygen gas rather than in carbohydrate, just as van Niel had predicted:

$$CO_2 + 2H_2^{18}O + \text{light energy} \longrightarrow (CH_2O) + H_2O + {}^{18}O_2$$

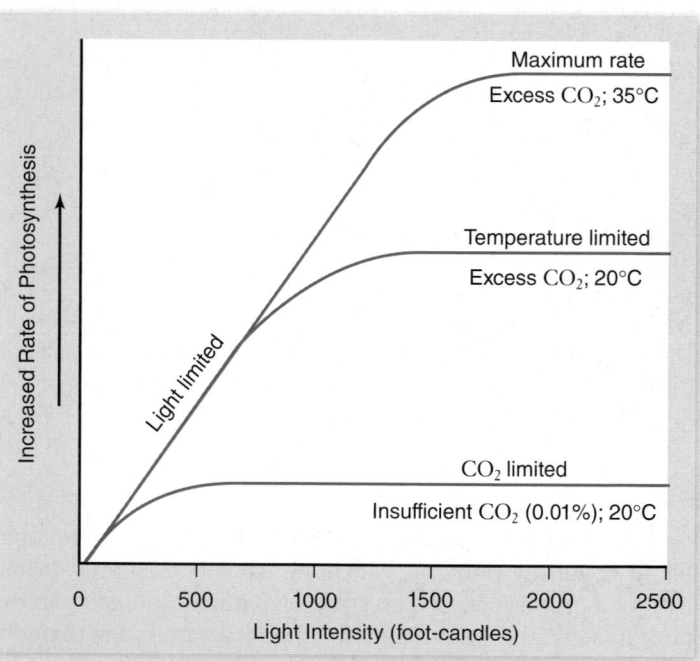

Figure 8.3 Discovery of the light-independent reactions. Blackman measured photosynthesis rates under differing light intensities, CO_2 concentrations, and temperatures. As this graph shows, light is the limiting factor at low light intensities, but temperature and CO_2 concentration are the limiting factors at higher light intensities. This implies the existence of reactions using CO_2 that involve enzymes.

 Data analysis Blackman found that increasing light intensity above 2000 foot-candles did not lead to any further increase in the rate of photosynthesis. Can you suggest a hypothesis that would explain this?

In algae and green plants, the carbohydrate typically produced by photosynthesis is glucose. The complete balanced equation for photosynthesis in these organisms thus becomes:

$$6CO_2 + 12H_2O + \text{light energy} \longrightarrow C_6H_{12}O_6 + 6H_2O + 6O_2$$

ATP and NADPH from light-dependent reactions reduce CO₂ to make sugars

In his pioneering work on the light-dependent reactions, van Niel proposed that the H^+ ions and electrons generated by the splitting of water were used to convert CO_2 into organic matter in a process he called *carbon fixation*. In the 1950s, Robin Hill (1899–1991) demonstrated that van Niel was right, light energy could be harvested and used in a reduction reaction. Chloroplasts isolated from leaf cells were able to reduce a dye and release oxygen in response to light. Later experiments showed that the electrons released from water were transferred to $NADP^+$ and that illuminated chloroplasts deprived of CO_2 accumulate ATP. If CO_2 is introduced, neither ATP nor NADPH accumulate, and the CO_2 is assimilated into organic molecules.

These experiments are important for three reasons: First, they firmly demonstrate that photosynthesis in plants occurs within chloroplasts. Second, they show that the light-dependent reactions use light energy to reduce $NADP^+$ and to manufacture ATP. Third, they confirm that the ATP and NADPH from this early stage of photosynthesis are then used in the subsequent reactions to reduce carbon dioxide, forming simple sugars.

Learning Outcomes Review 8.2

Early experiments indicated that plants "restore" air to usable form, that is, produce oxygen—but only in the presence of sunlight. Further experiments showed that there are both light-dependent and independent reactions. The light-dependent reactions produce O_2 from H_2O, and generate ATP and NADPH. The light-independent reactions synthesize organic compounds through carbon fixation.

■ *Where does the carbon in your body come from?*

Learning Outcomes

1. Discuss how pigments are important to photosynthesis.
2. Relate the absorption spectrum of a pigment to its color.

For plants to make use of the energy of sunlight, some biochemical structure must be present in chloroplasts and the thylakoids that can absorb this energy. Molecules that absorb light energy in the visible range are termed **pigments.** We are most familiar with them as dyes that impart a certain color to clothing or other materials. The color that we see is the color that is not absorbed—that is, it is reflected. To understand how plants use pigments to capture light energy, we must first review current knowledge about the nature of light.

Light is a form of energy

The wave nature of light produces an electromagnetic spectrum that differentiates light based on its wavelength (figure 8.4). We are most familiar with the visible range of this spectrum because we can actually see it, but visible light is only a small part of the entire spectrum. Visible light can be divided into its separate colors by the use of a prism, which separates light based on wavelength.

A particle of light, termed a **photon,** acts like a discrete bundle of energy. We use the wave concept of light to understand different colors of light and the particle nature of light to understand the energy transfers that occur during photosynthesis. Thus, we will refer both to wavelengths of light and to photons of light throughout the chapter.

The energy in photons

The energy content of a photon is inversely proportional to the wavelength of the light: Short-wavelength light contains photons of higher energy than long-wavelength light (see figure 8.4). X-rays, which contain a great deal of energy, have very short wavelengths—much shorter than those of visible light.

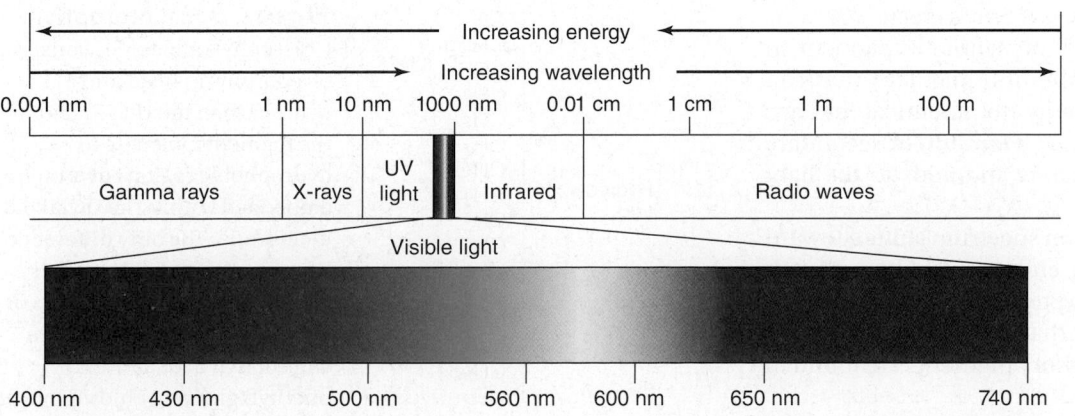

Figure 8.4 The electromagnetic spectrum. Light is a form of electromagnetic energy conveniently thought of as a wave. The shorter the wavelength of light, the greater its energy. Visible light represents only a small part of the electromagnetic spectrum between 400 and 740 nm.

A beam of light is able to remove electrons from certain molecules, creating an electrical current. This phenomenon is called the **photoelectric effect,** and it occurs when photons transfer energy to electrons. The strength of the photoelectric effect depends on the wavelength of light; that is, short wavelengths are much more effective than long ones in producing the photoelectric effect because they have more energy.

In photosynthesis, chloroplasts are acting as photoelectric devices: They absorb sunlight and transfer the excited electrons to a carrier. As we unravel the details of this process, it will become clear how this process traps energy and uses it to synthesize organic compounds.

Each pigment has a characteristic absorption spectrum

When a photon strikes a molecule with the amount of energy needed to excite an electron, then the molecule will absorb the photon raising the electron to a higher energy level. Whether the photon's energy is absorbed depends on how much energy it carries (defined by its wavelength), and also on the chemical nature of the molecule it hits.

As described in chapter 2, electrons occupy discrete energy levels in their orbits around atomic nuclei. To boost an electron into a different energy level requires just the right amount of energy, just as reaching the next rung on a ladder requires you to raise your foot just the right distance. A specific atom, therefore, can absorb only certain photons of light—namely, those that correspond to the atom's available energy levels. As a result, each molecule has a characteristic **absorption spectrum,** the range and efficiency of photons it is capable of absorbing.

As mentioned earlier, pigments are good absorbers of light in the visible range. Organisms have evolved a variety of different pigments, but only two general types are used in green plant photosynthesis: chlorophylls and carotenoids. In some organisms, other molecules also absorb light energy.

Chlorophyll absorption spectra

Chlorophylls absorb photons within narrow energy ranges. Two kinds of chlorophyll in plants, chlorophyll *a* and chlorophyll *b*, preferentially absorb violet-blue and red light (figure 8.5). Neither of these pigments absorbs photons with wavelengths between about 500 and 600 nm; light of these wavelengths is reflected. When these reflected photons are subsequently absorbed by the retinal pigment in our eyes, we perceive them as green.

Chlorophyll *a* is the main photosynthetic pigment in plants and cyanobacteria and is the only pigment that can act directly to convert light energy to chemical energy. **Chlorophyll *b*,** acting as an **accessory pigment,** or secondary light-absorbing pigment, complements and adds to the light absorption of chlorophyll *a*.

Chlorophyll *b* has an absorption spectrum shifted toward the green wavelengths. Therefore, chlorophyll *b* can absorb photons that chlorophyll *a* cannot, greatly increasing the proportion of the photons in sunlight that plants can harvest. In addition, a variety of different accessory pigments are found in plants, bacteria, and algae.

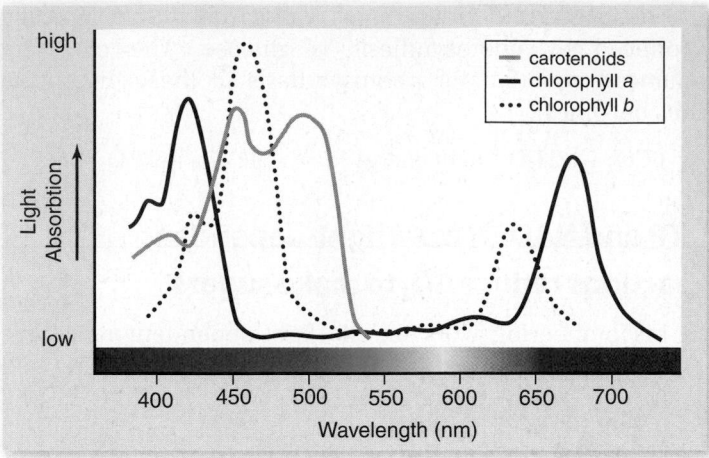

Figure 8.5 Absorption spectra for chlorophyll and carotenoids. The peaks represent wavelengths of light of sunlight absorbed by the two common forms of photosynthetic pigment, chlorophylls *a* and *b*, and the carotenoids. Chlorophylls absorb predominantly violet-blue and red light in two narrow bands of the spectrum and reflect green light in the middle of the spectrum. Carotenoids absorb mostly blue and green light and reflect orange and yellow light.

Structure of chlorophylls

Chlorophylls absorb photons by means of an excitation process analogous to the photoelectric effect. These pigments contain a complex ring structure, called a *porphyrin ring,* with alternating single and double bonds. At the center of the ring is a magnesium atom (figure 8.6).

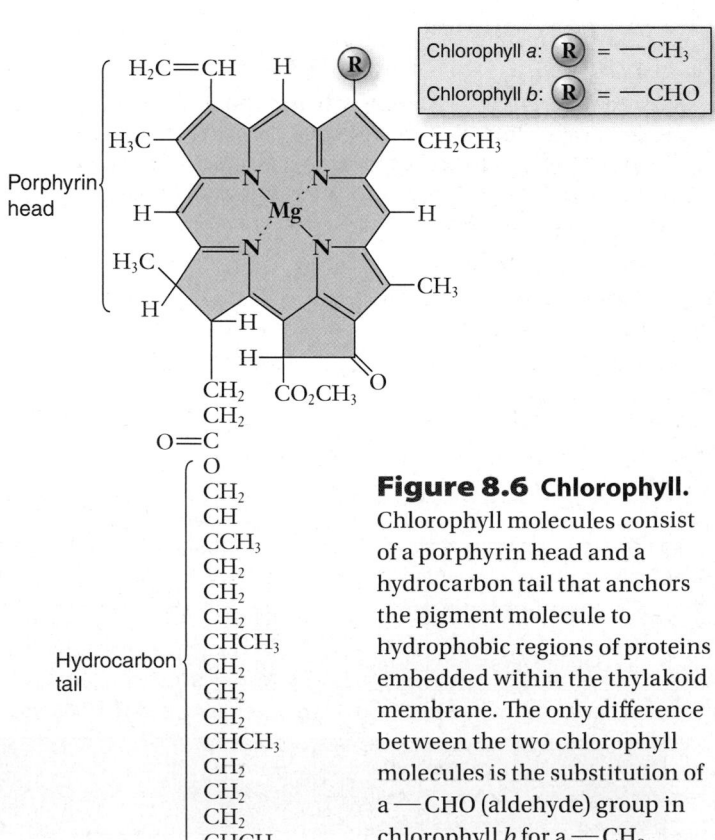

Figure 8.6 Chlorophyll. Chlorophyll molecules consist of a porphyrin head and a hydrocarbon tail that anchors the pigment molecule to hydrophobic regions of proteins embedded within the thylakoid membrane. The only difference between the two chlorophyll molecules is the substitution of a —CHO (aldehyde) group in chlorophyll *b* for a —CH₃ (methyl) group in chlorophyll *a*.

Hypothesis: *All wavelengths of light are equally effective in promoting photosynthesis.*

Prediction: *Illuminating plant cells with light broken into different wavelengths by a prism will produce the same amount of O_2 for all wavelengths.*

Test: *A filament of algae immobilized on a slide is illuminated by light that has passed through a prism. Motile bacteria that require O_2 for growth are added to the slide.*

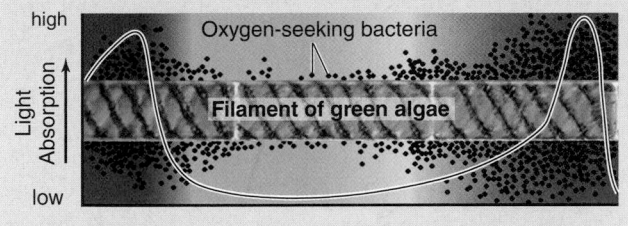

Result: *The bacteria move to regions of high O_2, or regions of most active photosynthesis. This is in the purple/blue and red regions of the spectrum.*

Conclusion: *All wavelengths are not equally effective at promoting photosynthesis. The most effective constitute the action spectrum for photosynthesis.*

Further Experiments: *How does the action spectrum relate to the various absorption spectra in figure 8.5?*

Figure 8.7 Determination of an action spectrum for photosynthesis.

Photons excite electrons in the porphyrin ring, which are then channeled away through the alternating carbon single- and double-bond system. Different small side groups attached to the outside of the ring alter the absorption properties of the molecule in the different kinds of chlorophyll (see figure 8.6). The precise absorption spectrum is also influenced by the local microenvironment created by the association of chlorophyll with different proteins.

The **action spectrum** of photosynthesis—that is, the relative effectiveness of different wavelengths of light in promoting photosynthesis—corresponds to the absorption spectrum for chlorophylls. This is demonstrated in the experiment in figure 8.7. All plants, algae, and cyanobacteria use chlorophyll *a* as their primary pigments.

It is reasonable to ask why these photosynthetic organisms do not use a pigment like retinal (the pigment in our eyes), which has a broad absorption spectrum that covers the range of 500 to 600 nm. The most likely hypothesis involves *photoefficiency.* Although retinal absorbs a broad range of wavelengths, it does so with relatively low efficiency. Chlorophyll, in contrast, absorbs in only two narrow bands, but does so with high efficiency. Therefore, plants and most other photosynthetic organisms achieve far higher overall energy capture rates with chlorophyll than with other pigments.

Carotenoids and other accessory pigments

Carotenoids consist of carbon rings linked to chains with alternating single and double bonds. They can absorb photons with a wide range of energies, although they are not always highly efficient in transferring this energy. Carotenoids assist in photosynthesis by capturing energy from light composed of wavelengths that are not efficiently absorbed by chlorophylls (figure 8.5; see also figure 8.8).

Carotenoids also perform a valuable role in scavenging free radicals. The oxidation–reduction reactions that occur in the chloroplast can generate destructive free radicals. Carotenoids can act as general-purpose antioxidants to lessen damage. Thus carotenoids have a protective role in addition to their role as light-absorbing molecules. This protective role is not surprising, because unlike the chlorophylls, carotenoids are found in many different kinds of organisms, including members of all three domains of life.

You may have heard that eating carrots can enhance vision. If this effect is real, it is probably due to the high content of β-carotene in carrots. This carotenoid consists of two molecules of vitamin A joined together. The oxidation of vitamin A produces retinal, the pigment used in vertebrate vision.

Figure 8.8 Fall colors are produced by carotenoids and other accessory pigments. During the spring and summer, chlorophyll in leaves masks the presence of carotenoids and other accessory pigments. When cool fall temperatures cause leaves to cease manufacturing chlorophyll, the chlorophyll is no longer present to reflect green light, and the leaves reflect the orange and yellow light that carotenoids and other pigments do not absorb.

Phycobiliproteins are accessory pigments found in cyanobacteria and some algae. These pigments contain a system of alternating double bonds similar to those found in other pigments and molecules that transfer electrons. Phycobiliproteins can be organized to form another light-harvesting complex that can absorb green light, which is reflected by chlorophyll. These complexes are probably ecologically important to cyanobacteria, helping them to exist in low-light situations in oceans. In this habitat, green light remains because red and blue light has been absorbed by green algae closer to the surface.

Learning Outcomes Review 8.3

A pigment is a molecule that can absorb light energy; its absorption spectrum shows the wavelengths at which it absorbs energy most efficiently. A pigment's color results from the wavelengths it does not absorb, which we then see. The main photosynthetic pigment is chlorophyll, which exists in several forms with slightly different absorption spectra. Many photosynthetic organisms have accessory pigments with absorption spectra different from chlorophyll; these increase light capture.

■ *What is the difference between an action spectrum and an absorption spectrum?*

8.4 *Photosystem Organization*

Learning Outcomes

1. *Describe the nature of photosystems.*
2. *Contrast the function of reaction center and antenna chlorophyll molecules.*

One way to study the role that pigments play in photosynthesis is to measure the correlation between the output of photosynthesis and the intensity of illumination—that is, how much photosynthesis is produced by how much light. Experiments on plants show that the output of photosynthesis increases linearly at low light intensities, but finally becomes saturated (no further increase) at high-intensity light. Saturation occurs because all of the light-absorbing capacity of the plant is in use.

Production of one O_2 molecule requires many chlorophyll molecules

Given the saturation observed with increasing light intensity, the next question is how many chlorophyll molecules have actually absorbed a photon. The question can be phrased this way: "Does saturation occur when all chlorophyll molecules have absorbed photons?" Finding an answer required being able to measure both photosynthetic output (on the basis of O_2 production) and the number of chlorophyll molecules present.

Using the unicellular algae *Chlorella,* investigators could obtain these values. Illuminating a *Chlorella* culture with pulses of light with increasing intensity should increase the

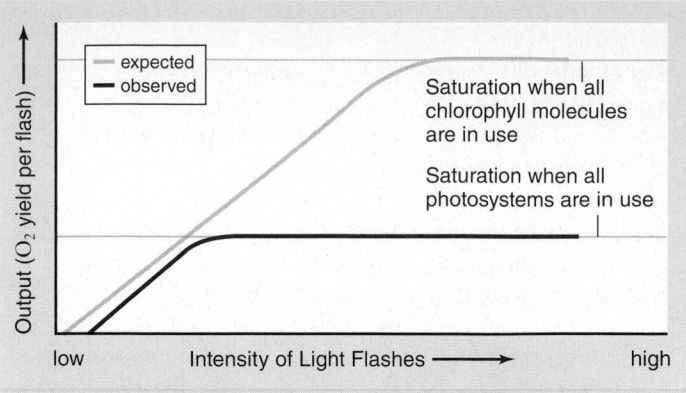

Figure 8.9 Saturation of photosynthesis. When photosynthetic saturation is achieved, further increases in intensity cause no increase in output. This saturation occurs far below the level expected for the number of individual chlorophyll molecules present. This led to the idea of organized photosystems, each containing many chlorophyll molecules. These photosystems saturate at a lower O_2 yield than that expected for the number of individual chlorophyll molecules.

 Data analysis Draw the curves for photosystems that have a greater or lesser number of chlorophyll molecules than the curve shown.

yield of O_2 per pulse until the system becomes saturated. Then O_2 production can be compared with the number of chlorophyll molecules present in the culture.

The observed level of O_2 per chlorophyll molecule at saturation, however, turned out to be only one molecule of O_2 per 2500 chlorophyll molecules (figure 8.9). This result was very different from what was expected, and it led to the idea that light is absorbed not by independent pigment molecules, but rather by clusters of chlorophyll and accessory pigment molecules (photosystems). Light is absorbed by any one of hundreds of pigment molecules in a photosystem, and each pigment molecule transfers its excitation energy to a single molecule with a lower energy level than the others.

A generalized photosystem contains an antenna complex and a reaction center

In chloroplasts and all but one class of photosynthetic prokaryotes, light is captured by photosystems. Each photosystem is a network of chlorophyll *a* molecules, accessory pigments, and associated proteins held within a protein matrix on the surface of the photosynthetic membrane. Like a magnifying glass focusing light on a precise point, a photosystem channels the excitation energy gathered by any one of its pigment molecules to a specific molecule, the reaction center chlorophyll. This molecule then passes the energy out of the photosystem as excited electrons that are put to work driving the synthesis of ATP and organic molecules.

A photosystem thus consists of two closely linked components: (1) an *antenna complex* of hundreds of pigment molecules that gather photons and feed the captured light energy to the reaction center; and (2) a *reaction center* consisting of one or more chlorophyll *a* molecules in a matrix of protein, that passes excited electrons out of the photosystem.

The antenna complex

The **antenna complex** is also called a light-harvesting complex, which accurately describes its role. This light-harvesting complex captures photons from sunlight (figure 8.10) and channels them to the reaction center chlorophylls.

In chloroplasts, light-harvesting complexes consist of a web of chlorophyll molecules linked together and held tightly in the thylakoid membrane by a matrix of proteins. Varying amounts of carotenoid accessory pigments may also be present. The protein matrix holds individual pigment molecules in orientations that are optimal for energy transfer.

The excitation energy resulting from the absorption of a photon passes from one pigment molecule to an adjacent molecule on its way to the reaction center. After the transfer, the excited electron in each molecule returns to the low-energy level it had before the photon was absorbed. Consequently, it is energy, not the excited electrons themselves, that passes from one pigment molecule to the next. The antenna complex funnels the energy from many electrons to the reaction center.

The reaction center

The **reaction center** is a transmembrane protein–pigment complex. The reaction center of purple photosynthetic bacteria is simpler than the one in chloroplasts but better understood. A pair of bacteriochlorophyll *a* molecules acts as a trap for photon energy, passing an excited electron to an acceptor precisely positioned as its neighbor. Note that here in the reaction center, the excited electron itself is transferred, and not just the energy, as was the case in the pigment–pigment transfers of the antenna complex. This difference allows the energy absorbed from

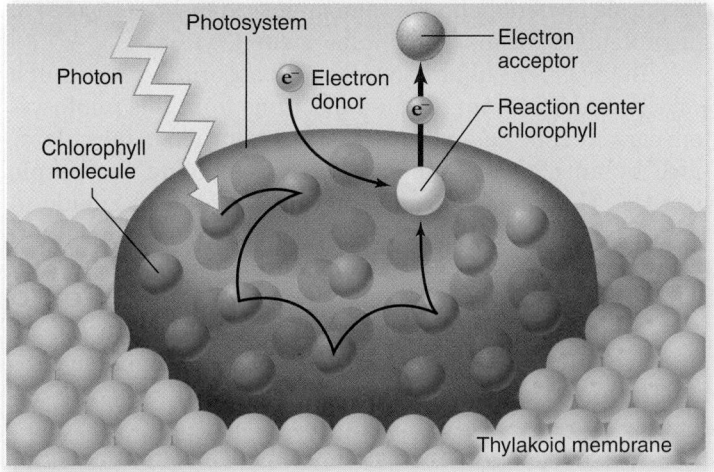

Figure 8.10 How the antenna complex works. When light of the proper wavelength strikes any pigment molecule within a photosystem, the light is absorbed by that pigment molecule. The excitation energy is then transferred from one molecule to another within the cluster of pigment molecules until it encounters the reaction center chlorophyll *a*. When excitation energy reaches the reaction center chlorophyll, electron transfer is initiated.

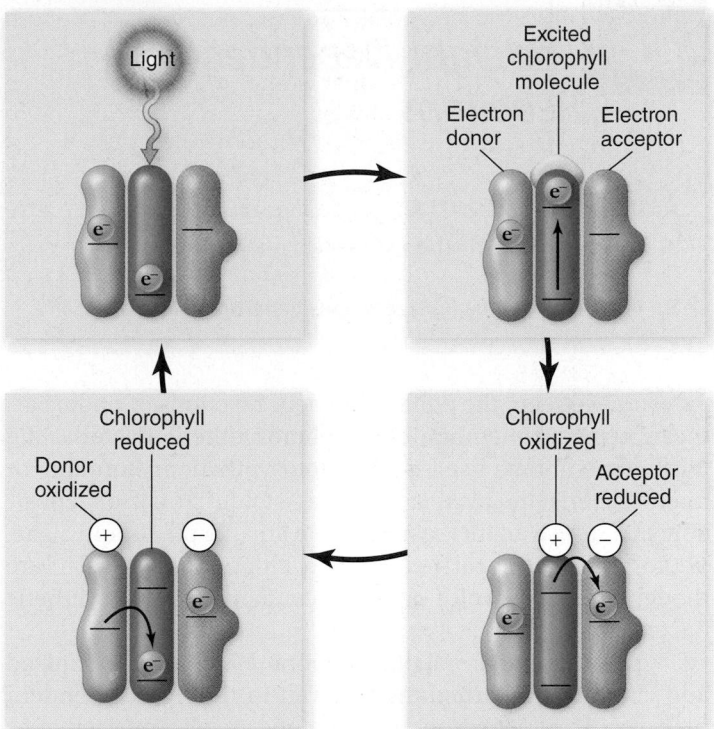

Figure 8.11 Converting light to chemical energy. When a chlorophyll in the reaction center absorbs a photon of light, an electron is excited to a higher energy level. This light-energized electron can be transferred to the primary electron acceptor, reducing it. The oxidized chlorophyll then fills its electron "hole" by oxidizing a donor molecule. The source of this donor varies with the photosystem as discussed in the text.

photons to move away from the chlorophylls, and it is the key conversion of light into chemical energy.

Figure 8.11 shows the transfer of excited electrons from the reaction center to the primary electron acceptor. By energizing an electron of the reaction center chlorophyll, light creates a strong electron donor where none existed before. The chlorophyll transfers the energized electron to the primary acceptor (a molecule of quinone), reducing the quinone and converting it to a strong electron donor. A nearby weak electron donor then passes a low-energy electron to the chlorophyll, restoring it to its original condition. The quinone transfers its electrons to another acceptor, and the process is repeated.

In plant chloroplasts, water serves as this weak electron donor. When water is oxidized in this way, oxygen is released along with two protons (H^+).

Learning Outcomes Review 8.4

Chlorophylls and accessory pigments are organized into photosystems found in the thylakoid membrane. The photosystem can be subdivided into an antenna complex, which is involved in light harvesting, and a reaction center, where the photochemical reactions occur. In the reaction center, an excited electron is passed to an acceptor; this transfers energy away from the chlorophylls and is key to the conversion of light into chemical energy.

■ *Why were photosystems an unexpected finding?*

8.5 The Light-Dependent Reactions

As you have seen, the light-dependent reactions of photosynthesis occur in membranes. In photosynthetic bacteria, the plasma membrane itself is the photosynthetic membrane. In many bacteria, the plasma membrane folds in on itself repeatedly to produce an increased surface area. In plants and algae, photosynthesis is carried out by chloroplasts, which are thought to be the evolutionary descendants of photosynthetic bacteria.

The internal thylakoid membrane is highly organized and contains the structures involved in the light-dependent reactions. For this reason, the reactions are also referred to as the thylakoid reactions. The thylakoid reactions take place in four stages:

1. **Primary photoevent.** A photon of light is captured by a pigment. This primary photoevent excites an electron within the pigment.
2. **Charge separation.** This excitation energy is transferred to the reaction center, which transfers an energetic electron to an acceptor molecule, initiating electron transport.
3. **Electron transport.** The excited electrons are shuttled along a series of electron carrier molecules embedded within the photosynthetic membrane. Several of them react by transporting protons across the membrane, generating a proton gradient. Eventually the electrons are used to reduce a final acceptor, NADPH.
4. **Chemiosmosis.** The protons that accumulate on one side of the membrane now flow back across the membrane through ATP synthase where chemiosmotic synthesis of ATP takes place, just as it does in aerobic respiration (see chapter 7).

These four processes make up the two stages of the light-dependent reactions mentioned at the beginning of this chapter. Steps 1 through 3 represent the stage of capturing energy from light; step 4 is the stage of producing ATP (and, as you'll see, NADPH). In the rest of this section we discuss the evolution of photosystems and the details of photosystem function in the light-dependent reactions.

Some bacteria use a single photosystem

Photosynthetic pigment arrays are thought to have evolved more than 2 BYA in bacteria similar to the purple and green bacteria alive today. In these bacteria, a single photosystem is used that generates ATP via electron transport. This process returns the electrons back to the reaction center. For this

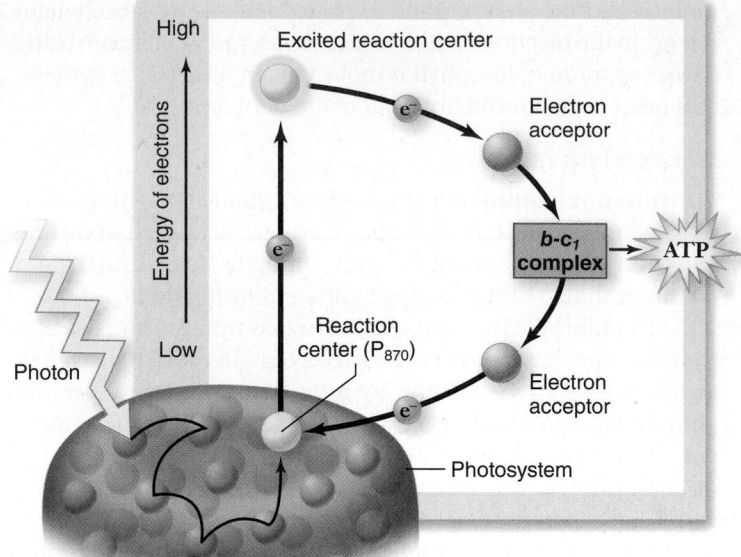

Figure 8.12 The path of an electron in purple nonsulfur bacteria. When a light-energized electron is ejected from the photosystem reaction center (P_{870}) it returns to the photosystem via a cyclic path that produces ATP but not NADPH.

reason, it is called cyclic photophosphorylation. These systems do not produce oxygen and so are also anoxygenic.

In the purple nonsulfur bacteria, peak absorption occurs at a wavelength of 870 nm (near infrared, not visible to the human eye), and thus the reaction center pigment is called P_{870}. Absorption of a photon by chlorophyll P_{870} does not raise an electron to a high enough level to be passed to NADP, so they must generate reducing power in a different way.

When the P_{870} reaction center absorbs a photon, the excited electron is passed to an electron transport chain that passes the electrons back to the reaction center, generating a proton gradient for ATP synthesis (figure 8.12). The proteins in the purple bacterial photosystem appear to be homologous to the proteins in the modern photosystem II.

In the green sulfur bacteria, peak absorption occurs at a wavelength of 840 nm. Excited electrons from this photosystem can either be passed to NADPH, or returned to the chlorophyll by an electron transport chain similar to the purple bacteria. They then use electrons from hydrogen sulfide to replace those passed to NADPH. The proteins in the green sulfur bacterial photosystem appear to be homologous to the proteins in the modern photosystem I.

Neither of these systems generates sufficient oxidizing power to oxidize H_2O. They are both anoxygenic and anaerobic. The linked photosystems of cyanobacteria and plant chloroplasts generate the oxidizing power necessary to oxidize H_2O, allowing it to serve as a source of both electrons and protons. This production of O_2 by oxygenic photosynthesis literally changed the atmosphere of the world.

Chloroplasts have two connected photosystems

In contrast to the sulfur bacteria, plants have two linked photosystems. This overcomes the limitations of cyclic photophosphorylation by providing an alternative source of electrons from the oxidation of water. The oxidation of water also generates O_2, thus oxygenic photosynthesis. The noncyclic transfer of electrons also produces NADPH, which can be used in the biosynthesis of carbohydrates.

One photosystem, called **photosystem I,** has an absorption peak of 700 nm, so its reaction center pigment is called P_{700}. This photosystem can pass electrons to NADPH similarly to the photosystem found in the sulfur bacteria discussed earlier. The other photosystem, called **photosystem II,** has an absorption peak of 680 nm, so its reaction center pigment is called P_{680}. This photosystem can generate an oxidation potential high enough to oxidize water. Working together, the two photosystems carry out a noncyclic transfer of electrons that generate both ATP and NADPH.

The photosystems were named I and II in the order of their discovery, and not in the order in which they operate in the light-dependent reactions. In plants and algae, the two photosystems are specialized for different roles in the overall process of oxygenic photosynthesis. Photosystem I transfers electrons ultimately to $NADP^+$, producing NADPH. The electrons lost from photosystem I are replaced by electrons from photosystem II. Photosystem II with its high oxidation potential can oxidize water to replace the electrons transferred to photosystem I. Thus there is an overall flow of electrons from water to NADPH.

These two photosystems are connected by a complex of electron carriers called the **cytochrome/b_6-f complex** (explained shortly). This complex can use the energy from the passage of electrons to move protons across the thylakoid membrane to generate the proton gradient used by an ATP synthase enzyme.

The two photosystems work together in noncyclic photophosphorylation

Evidence for the action of two photosystems came from experiments that measured the rate of photosynthesis using two light beams of different wavelengths: one red and the other far-red. Using both beams produced a rate greater than the sum of the rates using individual beams of these wavelengths (figure 8.13). This surprising result, called the *enhancement effect,* can be explained by a mechanism involving two photosystems acting in series (that is, one after the other), one photosystem absorbs preferentially in the red, the other in the far-red.

Plants use photosystems II and I in series, first one and then the other, to produce both ATP and NADPH. This two-stage process is called **noncyclic photophosphorylation** because the path of the electrons is not a circle—the electrons ejected from the photosystems do not return to them, but rather end up in NADPH. The photosystems are replenished with electrons obtained by splitting water.

The scheme shown in figure 8.14, called a *Z diagram,* illustrates the two electron-energizing steps, one catalyzed by each photosystem. The horizontal axis shows the progress of the light reactions and the relative positions of the complexes, and the vertical axis shows relative energy levels of electrons. The electrons originate from water, which holds onto its electrons very tightly (redox potential = +820 mV), and end up in NADPH, which holds its electrons much more loosely (redox potential = –320 mV).

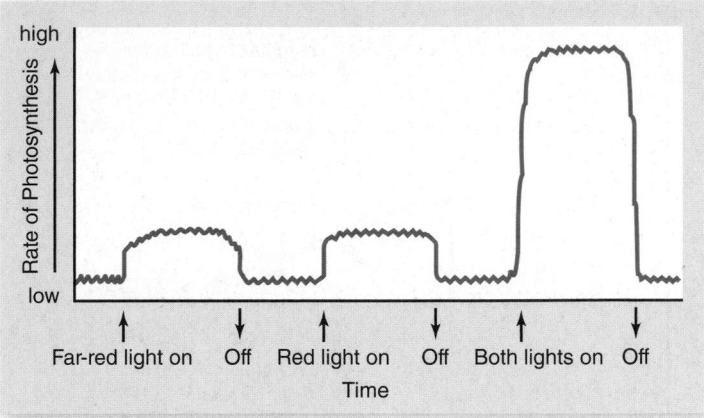

Figure 8.13 The enhancement effect. The rate of photosynthesis when red and far-red light are provided together is greater than the sum of the rates when each wavelength is provided individually. This result baffled researchers in the 1950s. Today, it provides key evidence that photosynthesis is carried out by two photochemical systems that act in series. One absorbs maximally in the far red, the other in the red portion of the spectrum.

 Data analysis If "both lights on" showed a rate equal to the sum of each light, what would you conclude?

Photosystem II acts first. High-energy electrons generated by photosystem II are used to synthesize ATP and are then passed to photosystem I to drive the production of NADPH. For every pair of electrons obtained from a molecule of water, one molecule of NADPH and slightly more than one molecule of ATP are produced.

Photosystem II

The reaction center of photosystem II closely resembles the reaction center of purple bacteria. It consists of a core of 10 transmembrane protein subunits with electron transfer components and two P_{680} chlorophyll molecules arranged around this core. The light-harvesting antenna complex consists of molecules of chlorophyll a and accessory pigments bound to several protein chains. The reaction center of photosystem II differs from the reaction center of the purple bacteria in that it also contains four manganese atoms. These manganese atoms are essential for the oxidation of water.

Although the chemical details of the oxidation of water are not entirely clear, the outline is emerging. Four manganese atoms are bound in a cluster to reaction center proteins. Two water molecules are also bound to this cluster of manganese atoms. When the reaction center of photosystem II absorbs a photon, an electron in a P_{680} chlorophyll molecule is excited, which transfers this electron to an acceptor. The oxidized P_{680} then removes an electron from a manganese atom. The oxidized manganese atoms, with the aid of reaction center proteins, remove electrons from oxygen atoms in the two water molecules. This process requires the reaction center to absorb four photons to complete the oxidation of two water molecules, producing one O_2 in the process.

chapter **8** *Photosynthesis* **157**

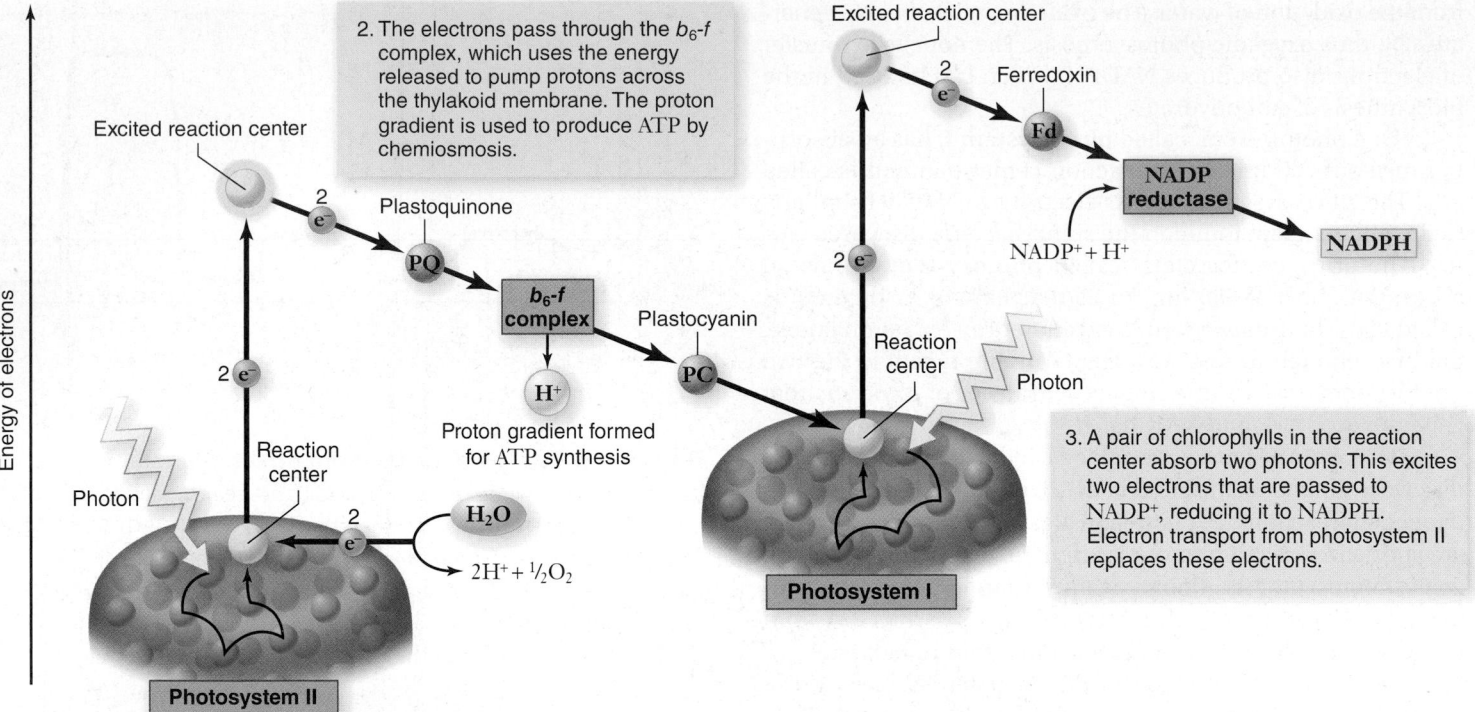

2. The electrons pass through the b_6-f complex, which uses the energy released to pump protons across the thylakoid membrane. The proton gradient is used to produce ATP by chemiosmosis.

Excited reaction center

2 e⁻

Plastoquinone

PQ

b_6-f complex

Plastocyanin

PC

Proton gradient formed for ATP synthesis

H⁺

2 e⁻

Reaction center

Photon

Photosystem II

H_2O

2 e⁻

$2H^+ + \frac{1}{2}O_2$

Energy of electrons

Excited reaction center

2 e⁻

Ferredoxin

Fd

NADP reductase

$NADP^+ + H^+$

NADPH

2 e⁻

Reaction center

Photon

Photosystem I

3. A pair of chlorophylls in the reaction center absorb two photons. This excites two electrons that are passed to $NADP^+$, reducing it to NADPH. Electron transport from photosystem II replaces these electrons.

1. A pair of chlorophylls in the reaction center absorb two photons of light. This excites two electrons that are transferred to plastoquinone (PQ). Loss of electrons from the reaction center produces an oxidation potential capable of oxidizing water.

Figure 8.14 Z diagram of photosystems I and II. Two photosystems work sequentially and have different roles. Photosystem II passes energetic electrons to photosystem I via an electron transport chain. The electrons lost are replaced by oxidizing water. Photosystem I uses energetic electrons to reduce $NADP^+$ to NADPH.

The role of the b_6-f complex

The primary electron acceptor for the light-energized electrons leaving photosystem II is a quinone molecule. The reduced quinone that results from accepting a pair of electrons (*plastoquinone*) is a strong electron donor; it passes the excited electron pair to a proton pump called the **b_6-f complex** embedded within the thylakoid membrane (figure 8.15). This complex closely resembles the bc_1 complex in the respiratory electron transport chain of mitochondria, discussed in chapter 7.

Arrival of the energetic electron pair causes the b_6-f complex to pump a proton into the thylakoid space. A small, copper-containing protein called *plastocyanin* then carries the electron pair to photosystem I.

Photosystem I

The reaction center of photosystem I consists of a core transmembrane complex consisting of 12 to 14 protein subunits with two bound P_{700} chlorophyll molecules. Energy is fed to it by an antenna complex consisting of chlorophyll *a* and accessory pigment molecules.

Photosystem I accepts an electron from plastocyanin into the "hole" created by the exit of a light-energized electron. The absorption of a photon by photosystem I boosts the electron leaving the reaction center to a very high energy level. The electrons are passed to an iron–sulfur protein called *ferredoxin*. Unlike photosystem II and the bacterial photosystem, the plant photosystem I does not rely on quinones as electron acceptors.

Making NADPH

Photosystem I passes electrons to ferredoxin on the stromal side of the membrane (outside the thylakoid). The reduced ferredoxin carries an electron with very high potential. Two of them, from two molecules of reduced ferredoxin, are then donated to a molecule of $NADP^+$ to form NADPH. The reaction is catalyzed by the membrane-bound enzyme *NADP reductase*.

Because the reaction occurs on the stromal side of the membrane and involves the uptake of a proton in forming NADPH, it contributes further to the proton gradient established during photosynthetic electron transport. The function of the two photosystems is summarized in figure 8.15.

ATP is generated by chemiosmosis

Protons are pumped from the stroma into the thylakoid compartment by the b_6-f complex. The splitting of water also produces added protons that contribute to the gradient. The thylakoid membrane is impermeable to protons, so this creates an electrochemical gradient that can be used to synthesize ATP.

ATP synthase

The chloroplast has ATP synthase enzymes in the thylakoid membrane that form a channel, allowing protons to cross back out into the stroma. These channels protrude like knobs on the external surface of the thylakoid membrane. As protons pass out of the thylakoid through the ATP synthase channel, ADP is phosphorylated to ATP and released into the stroma (see figure 8.15). The stroma contains the enzymes that catalyze the reactions of carbon fixation—the Calvin cycle reactions.

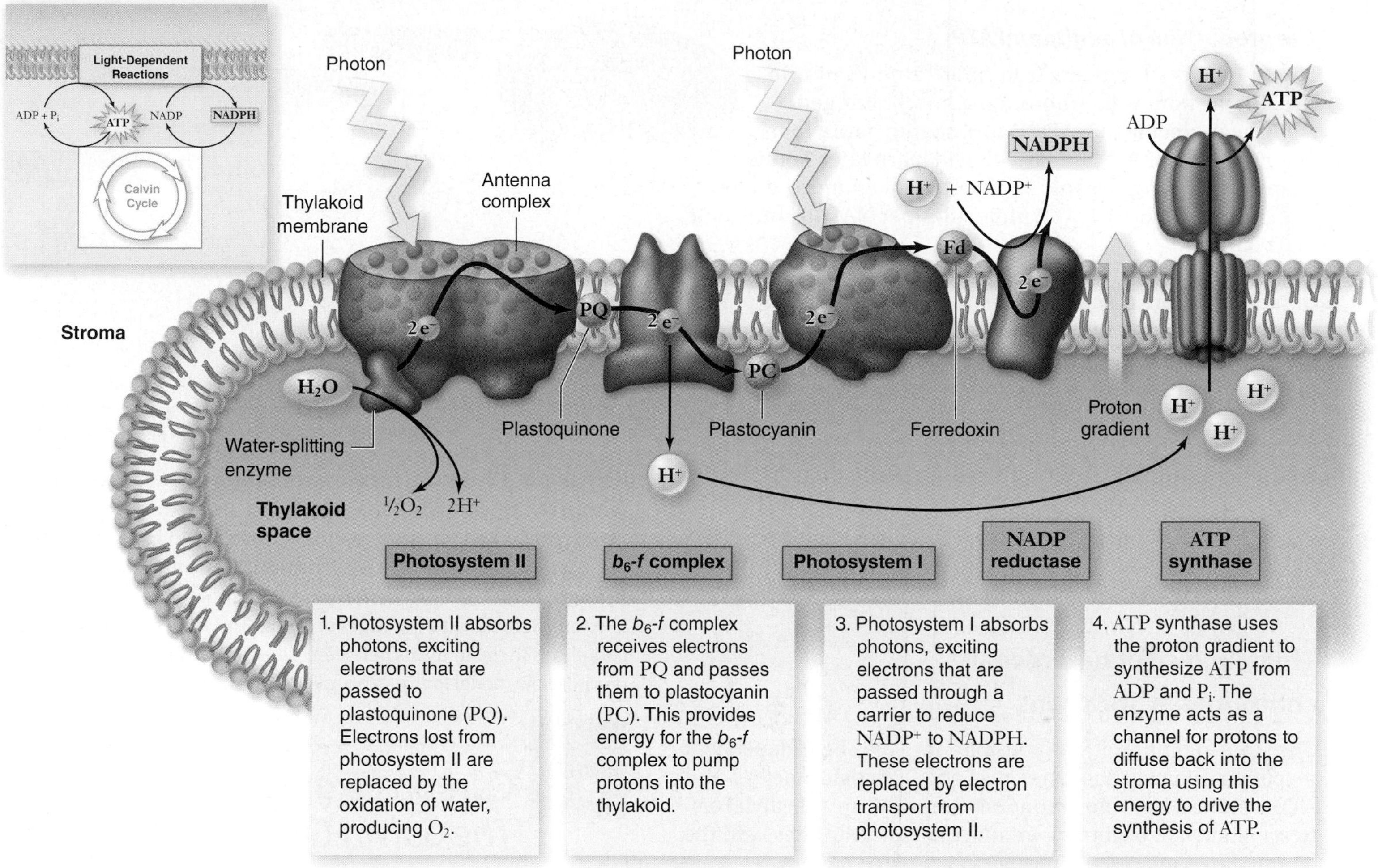

Figure 8.15

Light-Dependent Reactions

$ADP + P_i$ ATP   NADP   NADPH

Calvin Cycle

Photon

Photon

H^+ ATP

ADP

H^+ + NADP$^+$

NADPH

Fd

2e$^-$

Antenna complex

Thylakoid membrane

PQ

2e$^-$

2e$^-$

2e$^-$

PC

Stroma

H_2O

Water-splitting enzyme

Plastoquinone

Plastocyanin

Ferredoxin

Proton gradient

H^+

H^+

H^+

Thylakoid space

$^1/_2O_2$ 2H$^+$

H^+

Photosystem II	**b_6-f complex**	**Photosystem I**	**NADP reductase**	**ATP synthase**
1. Photosystem II absorbs photons, exciting electrons that are passed to plastoquinone (PQ). Electrons lost from photosystem II are replaced by the oxidation of water, producing O_2.	2. The b_6-f complex receives electrons from PQ and passes them to plastocyanin (PC). This provides energy for the b_6-f complex to pump protons into the thylakoid.	3. Photosystem I absorbs photons, exciting electrons that are passed through a carrier to reduce NADP$^+$ to NADPH. These electrons are replaced by electron transport from photosystem II.		4. ATP synthase uses the proton gradient to synthesize ATP from ADP and P_i. The enzyme acts as a channel for protons to diffuse back into the stroma using this energy to drive the synthesis of ATP.

Figure 8.15 The photosynthetic electron transport system and ATP synthase. The two photosystems are arranged in the thylakoid membrane joined by an electron transport system that includes the b_6-f complex. These function together to create a proton gradient that is used by ATP synthase to synthesize ATP.

This mechanism is the same as that seen in the mitochondrial ATP synthase, and, in fact, the two enzymes are evolutionarily related. This similarity in generating a proton gradient by electron transport and ATP by chemiosmosis illustrates the similarities in structure and function in mitochondria and chloroplasts. Evidence for this chemiosmotic mechanism for photophosphorylation was actually discovered earlier (figure 8.16) and formed the background for experiments using the mitochondrial ATP synthase.

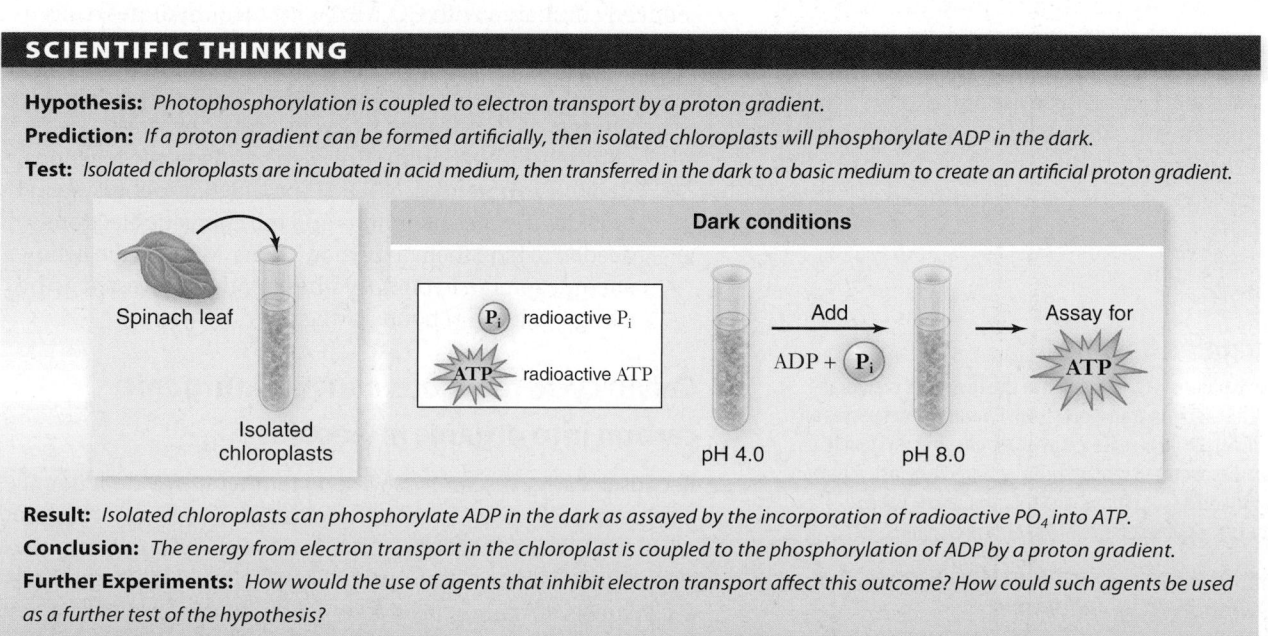

SCIENTIFIC THINKING

Figure 8.16 The Jagendorf acid bath experiment.

Hypothesis: *Photophosphorylation is coupled to electron transport by a proton gradient.*

Prediction: *If a proton gradient can be formed artificially, then isolated chloroplasts will phosphorylate ADP in the dark.*

Test: *Isolated chloroplasts are incubated in acid medium, then transferred in the dark to a basic medium to create an artificial proton gradient.*

Dark conditions

Spinach leaf

P_i radioactive P_i

ATP   radioactive ATP

Isolated chloroplasts

pH 4.0

Add
ADP + P_i

pH 8.0

Assay for ATP

Result: *Isolated chloroplasts can phosphorylate ADP in the dark as assayed by the incorporation of radioactive PO_4 into ATP.*

Conclusion: *The energy from electron transport in the chloroplast is coupled to the phosphorylation of ADP by a proton gradient.*

Further Experiments: *How would the use of agents that inhibit electron transport affect this outcome? How could such agents be used as a further test of the hypothesis?*

The production of additional ATP

The passage of an electron pair from water to NADPH in noncyclic photophosphorylation generates one molecule of NADPH and slightly more than one molecule of ATP. But as you will learn later in this chapter, building organic molecules takes more energy than that—it takes 1.5 ATP molecules per NADPH molecule to fix carbon.

To produce the extra ATP, many plant species are capable of short-circuiting photosystem I, switching photosynthesis into a *cyclic photophosphorylation* mode, so that the light-excited electron leaving photosystem I is used to make ATP instead of NADPH. The energetic electrons are simply passed back to the b_6-f complex, rather than passing on to NADP$^+$. The b_6-f complex pumps protons into the thylakoid space, adding to the proton gradient that drives the chemiosmotic synthesis of ATP. The relative proportions of cyclic and noncyclic photophosphorylation in these plants determine the relative amounts of ATP and NADPH available for building organic molecules.

Thylakoid structure reveals components' locations

The four complexes responsible for the light-dependent reactions—namely photosystems I and II, cytochrome b_6-f, and ATP synthase—are not randomly arranged in the thylakoid. Researchers are beginning to image these complexes with the atomic force microscope, which can resolve nanometer scale structures, and a picture is emerging in which photosystem II is found primarily in the grana, whereas photosystem I and ATP synthase are found primarily in the stroma lamella. Photosystem I and ATP synthase may also be found in the edges of the grana that are not stacked. The cytochrome b_6-f complex is found in the borders between grana and stroma lamella. One possible model for the arrangement of the complexes is shown in figure 8.17.

The thylakoid itself is no longer thought of only as stacked disks. Some models of the thylakoid, based on electron microscopy and other imaging, depict the grana as folds of the interconnecting stroma lamella. This kind of arrangement is more similar to the folds seen in bacterial photosynthesis, and it would therefore allow for more flexibility in how the various complexes are arranged relative to one another.

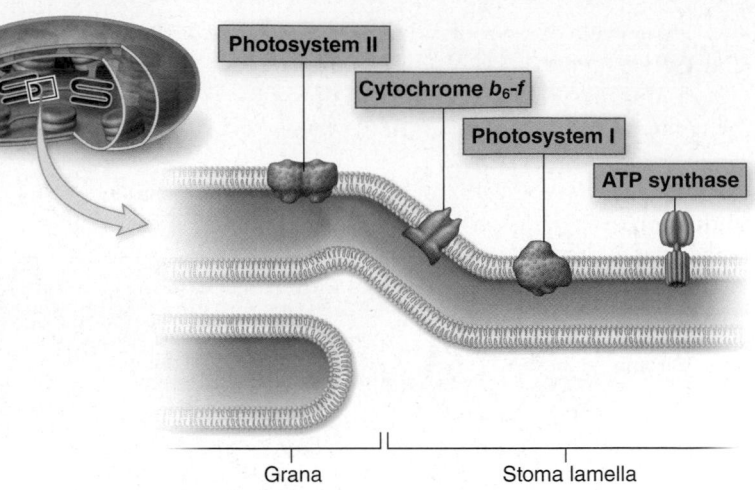

Figure 8.17 Model for the arrangement of complexes within the thylakoid. The arrangement of the two kinds of photosystems and the other complexes involved in photosynthesis is not random. Photosystem II is concentrated within grana, especially in stacked areas. Photosystem I and ATP synthase are concentrated in stroma lamella and the edges of grana. The cytochrome b_6-f complex is in the margins between grana and stroma lamella. This is one possible model for this arrangement.

8.6 Carbon Fixation: The Calvin Cycle

Learning Outcomes

1. Describe carbon fixation.
2. Demonstrate how six CO_2 molecules can be used to make one glucose.

Carbohydrates contain many C—H bonds and are highly reduced compared with CO_2. To build carbohydrates, cells use energy and a source of electrons produced by the light-dependent reactions of the thylakoids:

1. **Energy.** ATP (provided by cyclic and noncyclic photophosphorylation) drives the endergonic reactions.
2. **Reduction potential.** NADPH (provided by photosystem I) provides a source of protons and the energetic electrons needed to bind them to carbon atoms. Much of the light energy captured in photosynthesis ends up invested in the energy-rich C—H bonds of sugars.

Calvin cycle reactions convert inorganic carbon into organic molecules

Because early research showed temperature dependence, photosynthesis was predicted to involve enzyme-catalyzed reactions. These reactions form a cycle of enzyme-catalyzed steps much like the Krebs cycle of respiration. Unlike the Krebs cycle, however, carbon fixation is geared toward producing new compounds, so the nature of the cycles is quite different.

The cycle of reactions that allow carbon fixation is called the **Calvin cycle,** after its discoverer, Melvin Calvin (1911–1997). Because the first intermediate of the cycle, phosphoglycerate, contains three carbon atoms, this process is also called **C₃ photosynthesis.**

The key step in this process—the event that makes the reduction of CO_2 possible—is the attachment of CO_2 to a highly specialized organic molecule. Photosynthetic cells produce this molecule by reassembling the bonds of two intermediates in glycolysis—fructose 6-phosphate and glyceraldehyde 3-phosphate (G3P)—to form the energy-rich 5-carbon sugar **ribulose 1,5-bisphosphate (RuBP).**

CO_2 reacts with RuBP to form a transient 6-carbon intermediate that immediately splits into two molecules of the three-carbon *3-phosphoglycerate (PGA).* This overall reaction is called the *carbon fixation reaction* because inorganic carbon (CO_2) has been incorporated into an organic form: the acid PGA. The enzyme that carries out this reaction, **ribulose bisphosphate carboxylase/oxygenase** (usually abbreviated **rubisco**) is a large, 16-subunit enzyme found in the chloroplast stroma.

Carbon is transferred through cycle intermediates, eventually producing glucose

We will consider how the Calvin cycle can produce one molecule of glucose, although this glucose is not produced directly by the cycle (figure 8.18). In a series of reactions, six molecules of CO_2 are bound to six RuBP by rubisco to produce 12 molecules

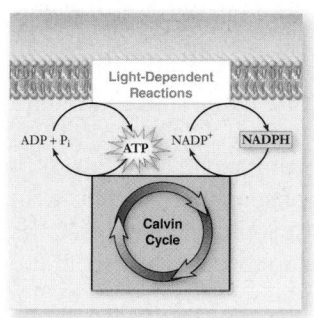

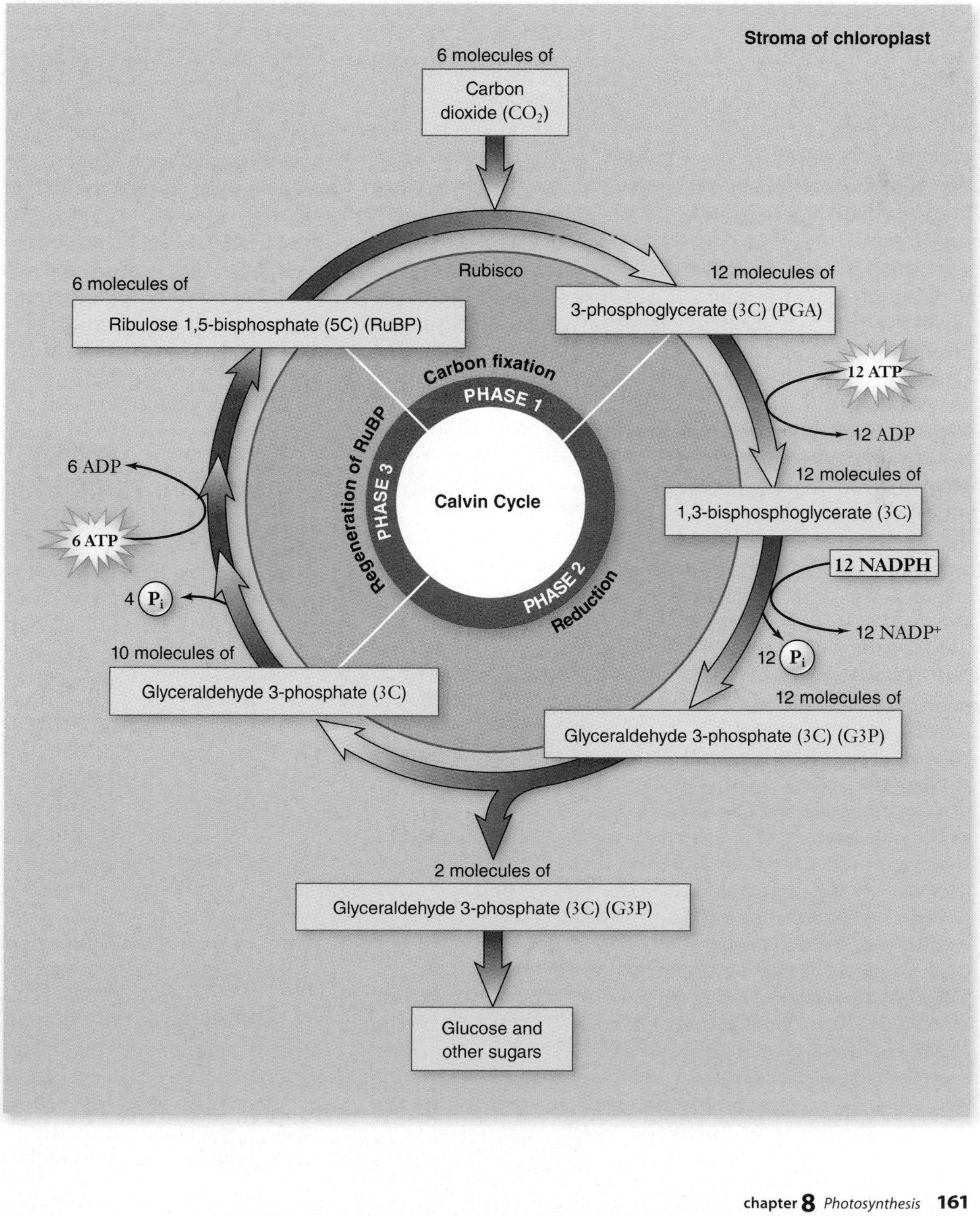

Figure 8.18 The Calvin cycle. The Calvin cycle accomplishes carbon fixation: converting inorganic carbon in the form of CO_2 into organic carbon in the form of carbohydrates. The cycle can be broken down into three phases: (1) carbon fixation, (2) reduction, and (3) regeneration of RuBP. For every six CO_2 molecules fixed by the cycle, a molecule of glucose can be synthesized from the products of the reduction reactions, G3P. The cycle uses the ATP and NADPH produced by the light reactions.

Stroma of chloroplast

6 molecules of
Carbon dioxide (CO_2)

Rubisco

12 molecules of
3-phosphoglycerate (3C) (PGA)

6 molecules of
Ribulose 1,5-bisphosphate (5C) (RuBP)

Carbon fixation
PHASE 1

12 ATP
12 ADP

Regeneration of RuBP
PHASE 3

Calvin Cycle

PHASE 2
Reduction

6 ADP

6 ATP

4 P_i

12 molecules of
1,3-bisphosphoglycerate (3C)

12 NADPH
12 NADP⁺

12 P_i

10 molecules of
Glyceraldehyde 3-phosphate (3C)

12 molecules of
Glyceraldehyde 3-phosphate (3C) (G3P)

2 molecules of
Glyceraldehyde 3-phosphate (3C) (G3P)

Glucose and other sugars

Light-Dependent Reactions

ADP + P_i ATP   NADP⁺   NADPH

Calvin Cycle

of PGA (containing $12 \times 3 = 36$ carbon atoms in all, 6 from CO_2 and 30 from RuBP). The 36 carbon atoms then undergo a cycle of reactions that regenerates the six molecules of RuBP used in the initial step (containing $6 \times 5 = 30$ carbon atoms). This leaves two molecules of *glyceraldehyde 3-phosphate (G3P)* (each with three carbon atoms) as the net gain. (You may recall G3P as also being the product of the first half of glycolysis, described in chapter 7.) These two molecules of G3P can then be used to make one molecule of glucose.

The net equation of the Calvin cycle is:

$$6CO_2 + 18 \text{ ATP} + 12 \text{ NADPH} + \text{water} \longrightarrow$$
$$2 \text{ glyceraldehyde 3-phosphate} + 16 \text{ P}_i + 18 \text{ ADP} + 12 \text{ NADP}^+$$

With six full turns of the cycle, six molecules of carbon dioxide enter, two molecules of G3P are produced, and six molecules of RuBP are regenerated. Thus six turns of the cycle produce two G3P that can be used to make a single glucose molecule. The six turns of the cycle also incorporated six CO_2 molecules, providing enough carbon to synthesize glucose, although the six carbon atoms do not all end up in this molecule of glucose.

Phases of the cycle

The Calvin cycle can be thought of as divided into three phases: (1) carbon fixation, (2) reduction, and (3) regeneration of RuBP. The carbon fixation reaction generates two molecules of the 3-carbon acid PGA; PGA is then reduced to G3P by reactions that are essentially a reverse of part of glycolysis; finally, the PGA is used to regenerate RuBP. Three turns around the cycle incorporate enough carbon to produce a new molecule of G3P, and six turns incorporate enough carbon to synthesize one glucose molecule.

We now know that light is required *indirectly* for different segments of the CO_2 reduction reactions. Five of the Calvin cycle enzymes—including rubisco—are light-activated; that is, they become functional or operate more efficiently in the presence of light. Light also promotes transport of required 3-carbon intermediates across chloroplast membranes. And finally, light promotes the influx of Mg^{2+} into the chloroplast stroma, which further activates the enzyme rubisco.

Output of the Calvin cycle

Glyceraldehyde 3-phosphate is a 3-carbon sugar, a key intermediate in glycolysis. Much of it is transported out of the chloroplast to the cytoplasm of the cell, where the reversal of several reactions in glycolysis allows it to be converted to fructose 6-phosphate and glucose 1-phosphate. These products can then be used to form sucrose, a major transport sugar in plants. (Sucrose, table sugar, is a disaccharide made of fructose and glucose.)

In times of intensive photosynthesis, G3P levels rise in the stroma of the chloroplast. As a consequence, some G3P in the chloroplast is converted to glucose 1-phosphate. This takes place in a set of reactions analogous to those occurring in the cytoplasm, by reversing several reactions similar to those of glycolysis. The glucose 1-phosphate is then combined into an insoluble polymer, forming long chains of starch stored as bulky starch grains in the cytoplasm. These starch grains represent stored glucose for later use.

Figure 8.19 Chloroplasts and mitochondria: completing an energy cycle. Water and O_2 cycle between chloroplasts and mitochondria within a plant cell, as do glucose and CO_2. Cells with chloroplasts take in CO_2 and H_2O and produce glucose and O_2. Cells without chloroplasts, such as animal cells, take in glucose and O_2 and produce CO_2 and H_2O. This leads to global cycling of carbon through photosynthesis and respiration (see figure 57.1).

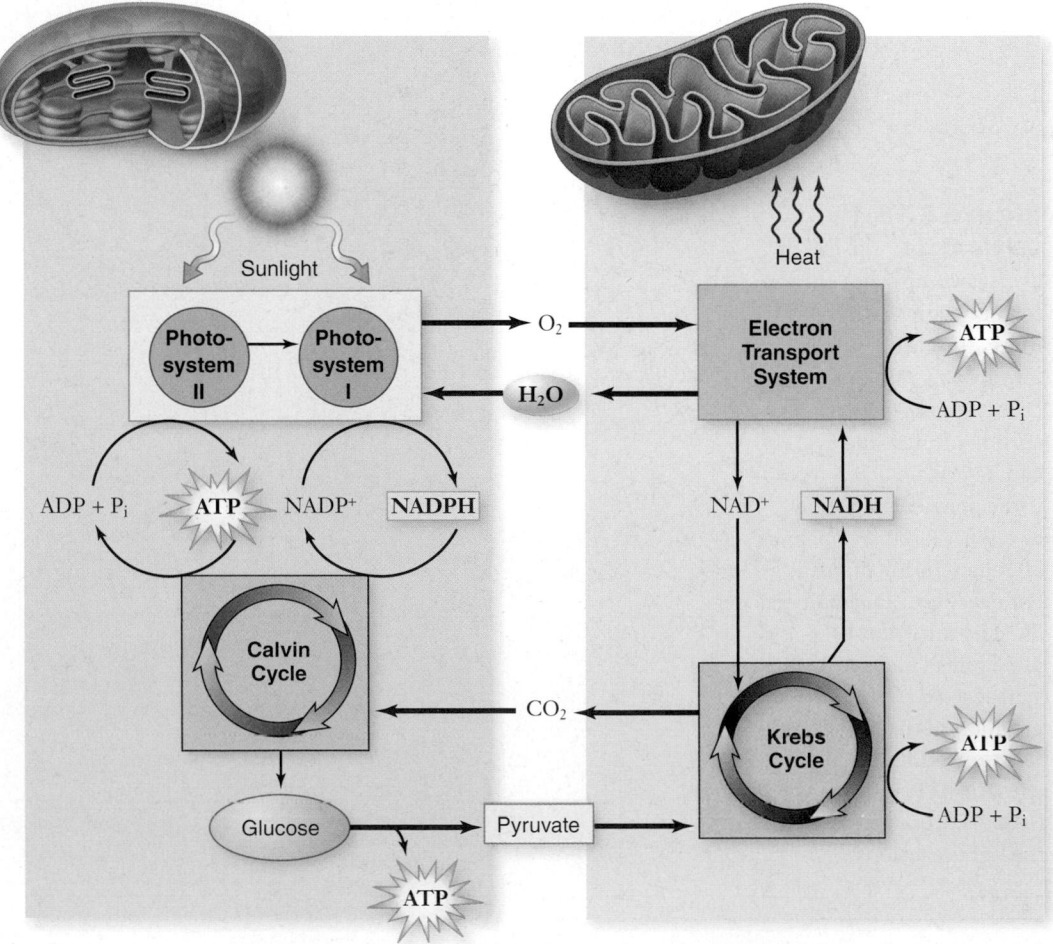

The energy cycle

The energy-capturing metabolisms of the chloroplasts studied in this chapter and the mitochondria studied in chapter 7 are intimately related (figure 8.19). Photosynthesis uses the products of respiration as starting substrates, and respiration uses the products of photosynthesis as starting substrates. The production of glucose from G3P even uses part of the ancient glycolytic pathway, run in reverse. Also, the principal proteins involved in electron transport and ATP production in plants are evolutionarily related to those in mitochondria.

Photosynthesis is but one aspect of plant biology, although it is an important one. In chapters 36 through 41, we examine plants in more detail. We have discussed photosynthesis as a part of cell biology because photosynthesis arose long before plants did, and because most organisms depend directly or indirectly on photosynthesis for the energy that powers their lives.

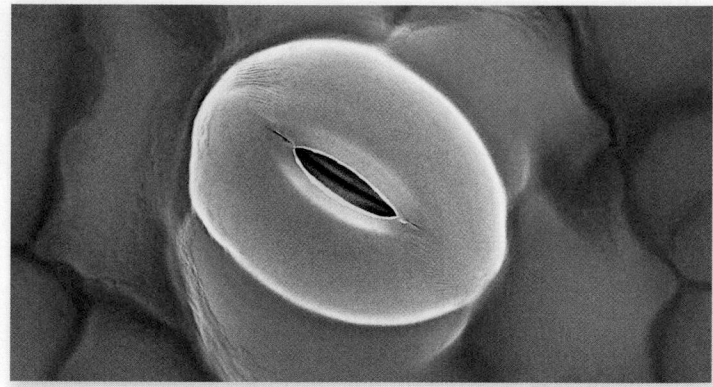

Figure 8.20 Stoma. A closed stoma in the leaf of a tobacco plant. Each stoma is formed from two guard cells whose shape changes with turgor pressure to open and close. Under dry conditions plants close their stomata to conserve water.

Learning Outcomes Review 8.6

Carbon fixation takes place in the stroma of the chloroplast, where inorganic CO_2 is incorporated into an organic molecule. The key intermediate is the 5-carbon sugar RuBP that combines with CO_2 in a reaction catalyzed by the enzyme rubisco. The cycle can be broken down into three stages: carbon fixation, reduction, and regeneration of RuBP. ATP and NADPH from the light reactions provide energy and electrons for the reduction reactions, which produce G3P. Glucose is synthesized when two molecules of G3P are combined.

■ How does the Calvin cycle compare with glycolysis?

8.7 Photorespiration

Learning Outcomes

1. Distinguish between how rubisco acts to make RuBP and how it oxidizes RuBP.
2. Compare the function of carbon fixation in the C_3, C_4, and CAM pathways.

Evolution does not necessarily result in optimum solutions. Rather, it favors workable solutions that can be derived from features that already exist. Photosynthesis is no exception. Rubisco, the enzyme that catalyzes the key carbon-fixing reaction of photosynthesis, provides a decidedly suboptimal solution. This enzyme has a second enzymatic activity that interferes with carbon fixation, namely that of *oxidizing* RuBP. In this process, called **photorespiration,** O_2 is incorporated into RuBP, which undergoes additional reactions that actually release CO_2. Hence, photorespiration releases CO_2, essentially undoing carbon fixation.

Photorespiration reduces the yield of photosynthesis

The carboxylation and oxidation of RuBP are catalyzed at the same active site on rubisco, and CO_2 and O_2 compete with each other at this site. Under normal conditions at 25°C, the rate of the carboxylation reaction is four times that of the oxidation reaction, meaning that 20% of photosynthetically fixed carbon is lost to photorespiration.

This loss rises substantially as temperature increases, because under hot, arid conditions, specialized openings in the leaf called *stomata* (singular, *stoma*) (figure 8.20) close to conserve water. This closing also cuts off the supply of CO_2 entering the leaf and does not allow O_2 to exit (figure 8.21). As a result, the low-CO_2 and high-O_2 conditions within the leaf favor photorespiration.

Figure 8.21 Conditions favoring photorespiration. In hot, arid environments, stomata close to conserve water, which also prevents CO_2 from entering and O_2 from exiting the leaf. The high-O_2/ low-CO_2 conditions favor photorespiration.

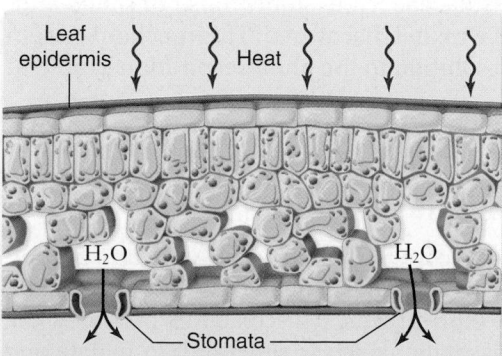

Under hot, arid conditions, leaves lose water by evaporation through openings in the leaves called stomata.

The stomata close to conserve water but as a result, O_2 builds up inside the leaves, and CO_2 cannot enter the leaves.

a. C₃ pathway

b. C₄ pathway

Figure 8.22 Comparison of C₃ and C₄ pathways of carbon fixation. *a.* The C₃ pathway uses the Calvin cycle to fix carbon. All reactions occur in mesophyll cells using CO_2 that diffuses in through stomata. *b.* The C₄ pathway incorporates CO_2 into a 4-carbon molecule of malate in mesophyll cells. This is transported to the bundle sheath cells where it is converted back into CO_2 and pyruvate, creating a high level of CO_2. This allows efficient carbon fixation by the Calvin cycle.

Plants that fix carbon using only C₃ photosynthesis (the Calvin cycle) are called **C₃ plants** (figure 8.22a). Other plants add CO_2 to phosphoenolpyruvate (PEP) to form a 4-carbon molecule. This reaction is catalyzed by the enzyme PEP *carboxylase.* This enzyme has two advantages over rubisco: it has a much greater affinity for CO_2 than rubisco, and it does not have oxidase activity.

The 4-carbon compound produced by PEP carboxylase undergoes further modification, only to be eventually decarboxylated. The CO_2 released by this decarboxylation is then used by rubisco in the Calvin cycle. This allows CO_2 to be pumped directly to the site of rubisco, which increases the local concentration of CO_2 relative to O_2, minimizing photorespiration. The 4-carbon compound produced by PEP carboxylase allows CO_2 to be stored in an organic form, to then be released in a different cell, or at a different time to keep the level of CO_2 high relative to O_2.

The reduction in the yield of carbohydrate as a result of photorespiration is not trivial. C₃ plants lose between 25% and 50% of their photosynthetically fixed carbon in this way. The rate depends largely on temperature. In tropical climates, especially those in which the temperature is often above 28°C, the problem is severe, and it has a major effect on tropical agriculture.

The two main groups of plants that initially capture CO_2 using PEP carboxylase differ in how they maintain high levels of CO_2 relative to O_2. In **C₄ plants** (figure 8.22b), the capture of CO_2 occurs in one cell and the decarboxylation occurs in an adjacent cell. This represents a spatial solution to the problem of photorespiration. The second group, **CAM plants,** perform both reactions in the same cell, but capture CO_2 using PEP carboxylase at night, then decarboxylate during the day. CAM stands for **crassulacean acid metabolism,** after the plant family Crassulaceae (the stonecrops, or hens-and-chicks), in which it was first discovered. This mechanism represents a temporal solution to the photorespiration problem.

C₄ plants have evolved to minimize photorespiration

The C₄ plants include corn, sugarcane, sorghum, and a number of other grasses. These plants initially fix carbon using PEP carboxylase in mesophyll cells. This reaction produces the organic acid oxaloacetate, which is converted to malate and transported to bundle-sheath cells that surround the leaf veins. Within the bundle-sheath cells, malate is decarboxylated to produce pyruvate and CO_2 (figure 8.23). Because the

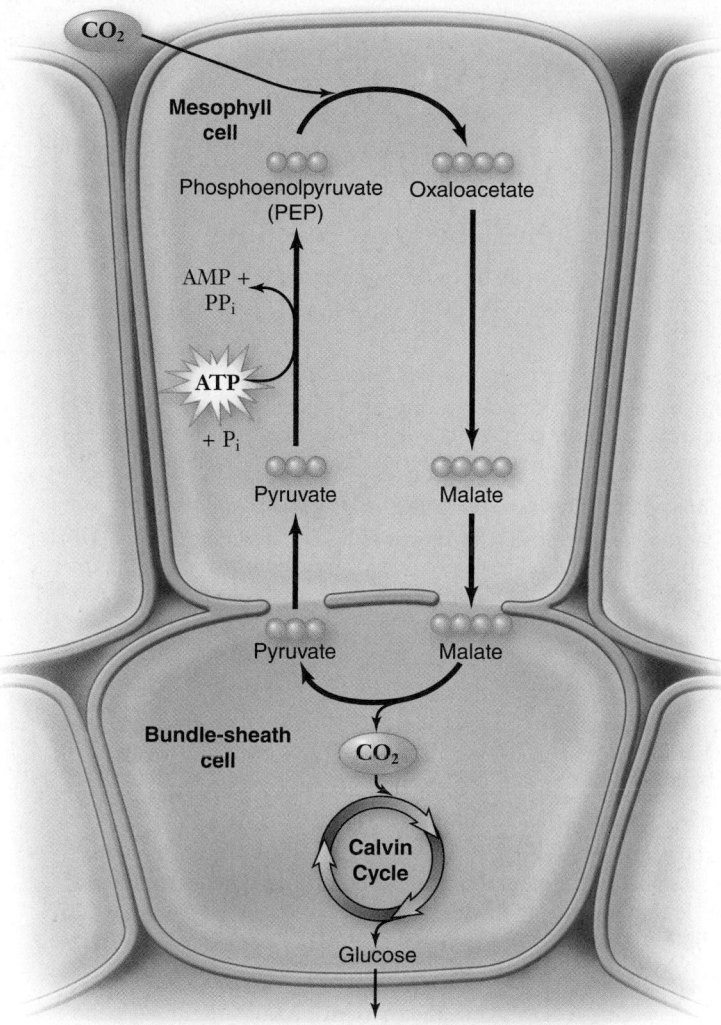

Figure 8.23 Carbon fixation in C₄ plants. This process is called the C₄ pathway because the first molecule formed, oxaloacetate, contains four carbons. The oxaloacetate is converted to malate, which moves into bundle-sheath cells where it is decarboxylated back to CO_2 and pyruvate. This produces a high level of CO_2 in the bundle-sheath cells that can be fixed by the usual C₃ Calvin cycle with little photorespiration. The pyruvate diffuses back into the mesophyll cells, where it is converted back to PEP to be used in another C₄ fixation reaction.

bundle-sheath cells are impermeable to CO_2, the local level of CO_2 is high and carbon fixation by rubisco and the Calvin cycle is efficient. The pyruvate produced by decarboxylation is transported back to the mesophyll cells, where it is converted back to PEP, thereby completing the cycle.

The C₄ pathway, although it overcomes the problems of photorespiration, does have a cost. The conversion of pyruvate back to PEP requires breaking two high-energy bonds in ATP. Thus each CO_2 transported into the bundle-sheath cells cost the equivalent of two ATP. To produce a single glucose, this requires 12 additional ATP compared with the Calvin cycle alone. Despite this additional cost, C₄ photosynthesis is advantageous in hot dry climates where photorespiration would remove more than half of the carbon fixed by the usual C₃ pathway alone.

Figure 8.24 Carbon fixation in CAM plants. CAM plants also use both C₄ and C₃ pathways to fix carbon and minimize photorespiration. In CAM plants, the two pathways occur in the same cell but are separated in time: The C₄ pathway is utilized to fix carbon at night, then CO_2 is released from these accumulated stores during the day to drive the C₃ pathway. This achieves the same effect of minimizing photorespiration while also minimizing loss of water by opening stomata at night when temperatures are lower.

The Crassulacean acid pathway splits photosynthesis into night and day

A second strategy to decrease photorespiration in hot regions has been adopted by the CAM plants. These include many succulent (water-storing) plants, such as cacti, pineapples, and some members of about two dozen other plant groups.

In these plants, the stomata open during the night and close during the day (figure 8.24). This pattern of stomatal opening and closing is the reverse of that in most plants. CAM plants initially fix CO_2 using PEP carboxylase to produce oxaloacetate. The oxaloacetate is often converted into other organic acids, depending on the particular CAM plant. These organic compounds accumulate during the night and are stored in the vacuole. Then during the day, when the stomata are closed, the organic acids are decarboxylated to yield high levels of CO_2. These high levels of CO_2 drive the Calvin cycle and minimize photorespiration.

Like C₄ plants, CAM plants use both C₃ and C₄ pathways. They differ in that they use both of these pathways in the same cell: the C₄ pathway at night and the C₃ pathway during the day. In C₄ plants the two pathways occur in different cells.

Learning Outcomes Review 8.7

Rubisco can also oxidize RuBP under conditions of high O_2 and low CO_2. In plants that use only C₃ metabolism (Calvin cycle), up to 20% of fixed carbon is lost to this photorespiration. Plants adapted to hot, dry environments are capable of storing CO_2 as a 4-carbon molecule and avoiding some of this loss; they are called C₄ plants. In CAM plants, CO_2 is fixed at night into a C₄ organic compound; in the daytime, this compound is used as a source of CO_2 C₃ metabolism when stomata are closed to prevent water loss.

■ *How do C₄ plants and CAM plants differ?*

Chapter Review

8.1 Overview of Photosynthesis

Photosynthesis is the conversion of light energy into chemical energy (figure 8.2).

Photosynthesis combines CO_2 and H_2O, producing glucose and O_2.
Photosynthesis has three stages: absorbing light energy, using this energy to synthesize ATP and NADPH, and using the ATP and NADPH to convert CO_2 to organic molecules. The first two stages consist of light-dependent reactions, and the third stage of light-independent reactions.

In plants, photosynthesis takes place in chloroplasts.
Chloroplasts contain internal thylakoid membranes and a fluid matrix called stroma. The photosystems involved in energy capture are found in the thylakoid membranes, and enzymes for assembling organic molecules are in the stroma.

8.2 The Discovery of Photosynthetic Processes

Plants do not increase mass from soil and water alone.
Early investigations revealed that plants produce O_2 from carbon dioxide and water in the presence of light.

Photosynthesis includes both light-dependent and light-independent reactions.
The light-dependent reactions require light; the light-independent reactions occur in both daylight and darkness. The rate of photosynthesis depends on the amount of light, the CO_2 concentration, and temperature.

O_2 comes from water, not from CO_2.
The use of isotopes revealed the individual origins and fates of different molecules in photosynthetic reactions.

ATP and NADPH from light-dependent reactions reduce CO_2 to make sugars.
Carbon fixation requires ATP and NADPH, which are products of the light-dependent reactions. As long as these are available, CO_2 is reduced by enzymes in the stroma to form simple sugars.

8.3 Pigments

Light is a form of energy.
Light exists both as a wave and as a particle (photon). Light can remove electrons from some metals by the photoelectric effect, and in photosynthesis, chloroplasts act as photoelectric devices.

Each pigment has a characteristic absorption spectrum (figure 8.5).
Chlorophyll a is the only pigment that can convert light energy into chemical energy. Chlorophyll b is an accessory pigment that increases the harvest of photons for photosynthesis.

Carotenoids and other accessory pigments further increase a plant's ability to harvest photons.

8.4 Photosystem Organization (figure 8.10)

Production of one O_2 molecule requires many chlorophyll molecules.
Measurement of O_2 output led to the idea of photosystems—clusters of pigment molecules that channel energy to a reaction center.

A generalized photosystem contains an antenna complex and a reaction center.
A photosystem is a network of chlorophyll a, accessory pigments, and proteins embedded in the thylakoid membrane. Pigment molecules of the antenna complex harvest photons and feed light energy to the reaction center. The reaction center is composed of two chlorophyll a molecules in a protein matrix that pass an excited electron to an electron acceptor.

8.5 The Light-Dependent Reactions

The light reactions can be broken down into four processes: primary photoevent, charge separation, electron transport, and chemiosmosis.

Some bacteria use a single photosystem (figure 8.12).
An excited electron moves along a transport chain and eventually returns to the photosystem. This cyclic process is used to generate a proton gradient. In some bacteria, this can also produce NADPH.

Chloroplasts have two connected photosystems (figure 8.14).
Photosystem I transfers electrons to $NADP^+$, reducing it to NADPH. Photosystem II replaces electrons lost by photosystem I. Electrons lost from photosystem II are replaced by electrons from oxidation of water, which also produces O_2.

The two photosystems work together in noncyclic photophosphorylation (figure 8.14).
Photosystem II and photosystem I are linked by an electron transport chain; the b_6-f complex in this chain pumps protons into the thylakoid space.

ATP is generated by chemiosmosis.
ATP synthase is a channel enzyme; as protons flow through the channel down their gradient, ADP is phosphorylated producing ATP, similar to the mechanism in mitochondria. Plants can make additional ATP by cyclic photophosphorylation.

Thylakoid structure reveals components' locations.
Imaging studies suggest that photosystem II is primarily found in the grana, while photosystem I and ATP synthase are found in the stroma lamella.

8.6 Carbon Fixation: The Calvin Cycle (figure 8.18)

Calvin cycle reactions convert inorganic carbon into organic molecules.
The Calvin cycle, also known as C_3 photosynthesis, uses CO_2, ATP, and NADPH to build simple sugars.

Carbon is transferred through cycle intermediates, eventually producing glucose.
The Calvin cycle occurs in three stages: carbon fixation via the enzyme rubisco's action on RuBP and CO_2; reduction of the resulting 3-carbon PGA to G3P, generating ATP and NADPH; and regeneration of RuBP. Six turns of the cycle fix enough carbon to produce two excess G3Ps used to make one molecule of glucose.

8.7 Photorespiration

Photorespiration reduces the yield of photosynthesis.
Rubisco can catalyze the oxidation of RuBP, reversing carbon fixation. Dry, hot conditions tend to increase this reaction.

C_4 plants have evolved to minimize photorespiration.
C_4 plants fix carbon by adding CO_2 to a 3-carbon molecule, forming oxaloacetate. Carbon is fixed in one cell by the C_4 pathway, then CO_2 is released in another cell for the Calvin cycle (figure 8.23).

The Crassulacean acid pathway splits photosynthesis into night and day.
CAM plants use the C_4 pathway during the day when stomata are closed, and the Calvin cycle at night in the same cell (figure 8.24).

UNDERSTAND

1. The *light-dependent* reactions of photosynthesis are responsible for the production of
   a. glucose.
   b. CO_2.
   c. ATP and NADPH.
   d. H_2O.

2. Which region of a chloroplast is associated with the capture of light energy?
   a. Thylakoid membrane
   b. Outer membrane
   c. Stroma
   d. Both a and c are correct.

3. The colors of light that are most effective for photosynthesis are
   a. red, blue, and violet.
   b. green, yellow, and orange.
   c. infrared and ultraviolet.
   d. All colors of light are equally effective.

4. During noncyclic photosynthesis, photosystem I functions to _____, and photosystem II functions to _____.
   a. synthesize ATP; produce O_2
   b. reduce $NADP^+$; oxidize H_2O
   c. reduce CO_2; oxidize NADPH
   d. restore an electron to its reaction center; gain an electron from water

5. How is a reaction center pigment in a photosystem different from a pigment in the antenna complex?
   a. The reaction center pigment is a chlorophyll molecule.
   b. The antenna complex pigment can only reflect light.
   c. The reaction center pigment loses an electron when it absorbs light energy.
   d. The antenna complex pigments are not attached to proteins.

6. The ATP and NADPH from the light reactions are used
   a. in glycolysis in roots.
   b. directly in most biochemical reactions in the cell.
   c. during the reactions of the Calvin cycle to produce glucose.
   d. to synthesize chlorophyll.

7. The carbon fixation reaction converts
   a. inorganic carbon into an organic acid.
   b. CO_2 into glucose.
   c. inactive rubisco into active rubisco.
   d. an organic acid into CO_2.

8. C_4 plants initially fix carbon by
   a. the same pathway as C_3 plants, but they modify this product.
   b. incorporating CO_2 into oxaloacetate, which is converted to malate.
   c. incorporating CO_2 into citrate via the Krebs cycle.
   d. incorporating CO_2 into glucose via reverse glycolysis.

APPLY

1. The overall flow of electrons in the light reactions is from
   a. antenna pigments to the reaction center.
   b. H_2O to CO_2.
   c. photosystem I to photosystem II.
   d. H_2O to NADPH.

2. If you could measure pH within a chloroplast, where would it be lowest?
   a. In the stroma
   b. In the lumen of the thylakoid
   c. In the cytoplasm immediately outside the chloroplast
   d. In the antenna complex

3. The excited electron from photosystem I
   a. can be returned to the reaction center to generate ATP by cyclic photophosphorylation.
   b. is replaced by oxidizing H_2O.
   c. is replaced by an electron from photosystem II.
   d. Both a and c are correct.

4. If the Calvin cycle runs through six turns
   a. all of the fixed carbon will end up in the same glucose molecule.
   b. 12 carbons will be fixed by the process.
   c. enough carbon will be fixed to make one glucose, but they will not all be in the same molecule.
   d. one glucose will be converted into six CO_2.

5. Which of the following are similarities between the structure and function of mitochondria and chloroplasts?
   a. They both create internal proton gradients by electron transport.
   b. They both generate CO_2 by oxidation reactions.
   c. They both have a double membrane system.
   d. Both a and c are correct.

6. Given that the C_4 pathway gets around the problems of photorespiration, why don't all plants use it?
   a. It is a more recent process, and many plants have not had time to evolve this pathway.
   b. It requires extra enzymes that many plants lack.
   c. It requires special transport tissues that many plants lack.
   d. It also has an energetic cost.

7. If the thylakoid membrane became leaky to ions, what would you predict to be the result on the light reactions?
   a. It would stop ATP production.
   b. It would stop NADPH production.
   c. It would stop the oxidation of H_2O.
   d. All of the choices are correct.

8. The overall process of photosynthesis
   a. results in the reduction of CO_2 and the oxidation of H_2O.
   b. results in the reduction of H_2O and the oxidation of CO_2.
   c. consumes O_2 and produces CO_2.
   d. produces O_2 from CO_2.

SYNTHESIZE

1. Compare and contrast the fixation of carbon in C_3, C_4, and CAM plants.

2. Diagram the relationship between the reactants and products of photosynthesis and respiration.

3. Do plant cells need mitochondria? Explain your answer.

ONLINE RESOURCE

www.ravenbiology.com

Understand, Apply, and Synthesize—enhance your study with animations that bring concepts to life and practice tests to assess your understanding. Your instructor may also recommend the interactive eBook, individualized learning tools, and more.

Chapter **9**

Cell Communication

Chapter Contents

Introduction

Springtime is a time of rebirth and renewal. Trees that have appeared dead produce new leaves and buds, and flowers sprout from the ground. For sufferers of seasonal allergy, this is not quite such a pleasant time. The pollen in the micrograph and other allergens produced stimulate the immune system to produce the molecule histamine and other molecules that form cellular signals. These signals cause inflammation, mucus secretion, vasodilation, and other responses that together cause the runny nose, itching, watery eyes, and other symptoms that make up the allergic reaction. We treat allergy symptoms by using drugs called antihistamines that interfere with this cellular signaling. The popular drug loratadine (better known as Claritin), for example, acts by blocking the receptor for histamine, thus preventing its action.

We will begin this chapter with a general overview of signaling, and the kinds of receptors cells use to respond to signals. Then we will look in more detail at how these different types of receptors can elicit a response from cells, and finally, how cells make connections with one another.

9.1 *Overview of Cell Communication*

Learning Outcomes

1. *Discriminate between methods of signaling based on distance from source to reception.*
2. *Describe how phosphorylation can affect protein function.*

Communication between cells is common in nature. Cell signaling occurs in all multicellular organisms, providing an indispensable mechanism for cells to influence one another. Effective signaling requires a signaling molecule, called a **ligand,** and a molecule to which the signal binds, called a **receptor protein.** The interaction of these two components initiates the process of *signal transduction,* which converts the information in the signal into a cellular response (figure 9.1).

The cells of multicellular organisms use a variety of molecules as signals, including but not limited to, peptides, large proteins, individual amino acids, nucleotides, and steroids and other lipids. Even dissolved gases such as NO (nitric oxide) are used as signals.

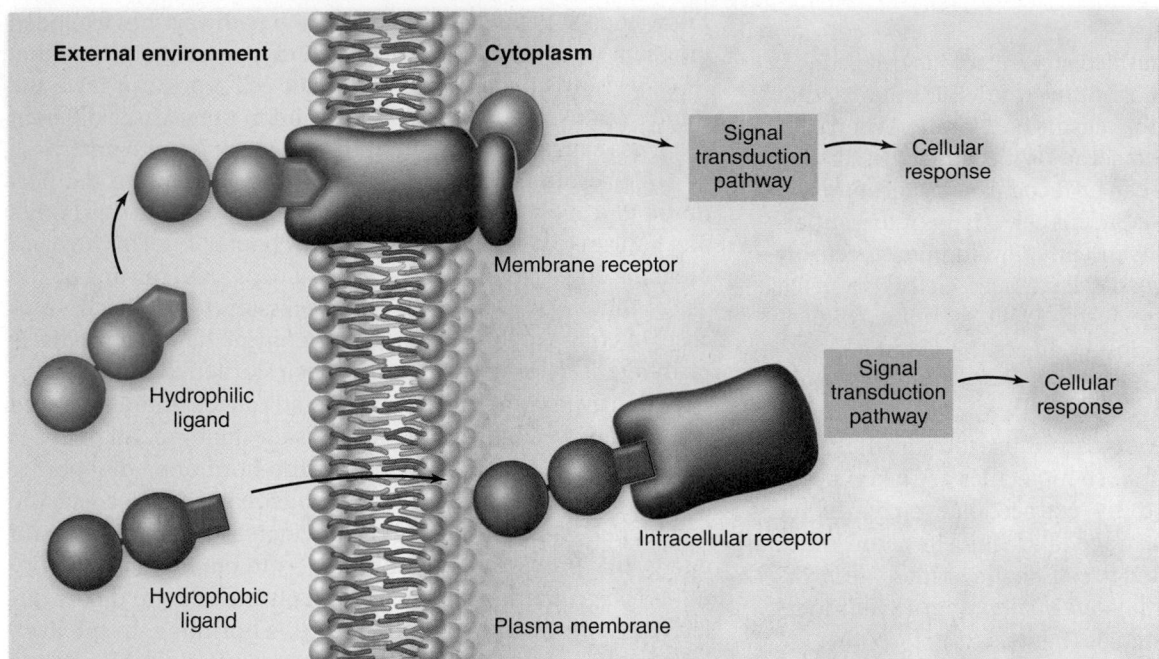

Figure 9.1 Overview of cell signaling. Cell signaling involves a signal molecule called a ligand, a receptor, and a signal transduction pathway that produces a cellular response. The location of the receptor can either be intracellular, for hydrophobic ligands that can cross the membrane, or in the plasma membrane, for hydrophilic ligands that cannot cross the membrane.

Any cell of a multicellular organism is exposed to a constant stream of signals. At any time, hundreds of different chemical signals may be present in the environment surrounding the cell. Each cell responds only to certain signals, however, and ignores the rest, like a person following the conversation of one or two individuals in a noisy, crowded room.

How does a cell "choose" which signals to respond to? The number and kind of receptor molecules determine this. When a ligand approaches a receptor protein that has a complementary shape, the two can bind, forming a complex. This binding induces a change in the receptor protein's shape, ultimately producing a response in the cell via a signal transduction pathway. In this way, a given cell responds to the signaling molecules that fit the particular set of receptor proteins it possesses and ignores those for which it lacks receptors.

Signaling is defined by the distance from source to receptor

Cells can communicate through any of four basic mechanisms, depending primarily on the distance between the signaling and responding cells (figure 9.2). These mechanisms are (1) direct contact, (2) paracrine signaling, (3) endocrine signaling, and (4) synaptic signaling.

In addition to using these four basic mechanisms, some cells actually send signals to themselves, secreting signals that bind to specific receptors on their own plasma membranes. This process, called *autocrine signaling,* is thought to play an important role in reinforcing developmental changes, and it is an important component of signaling in the immune system (chapter 51).

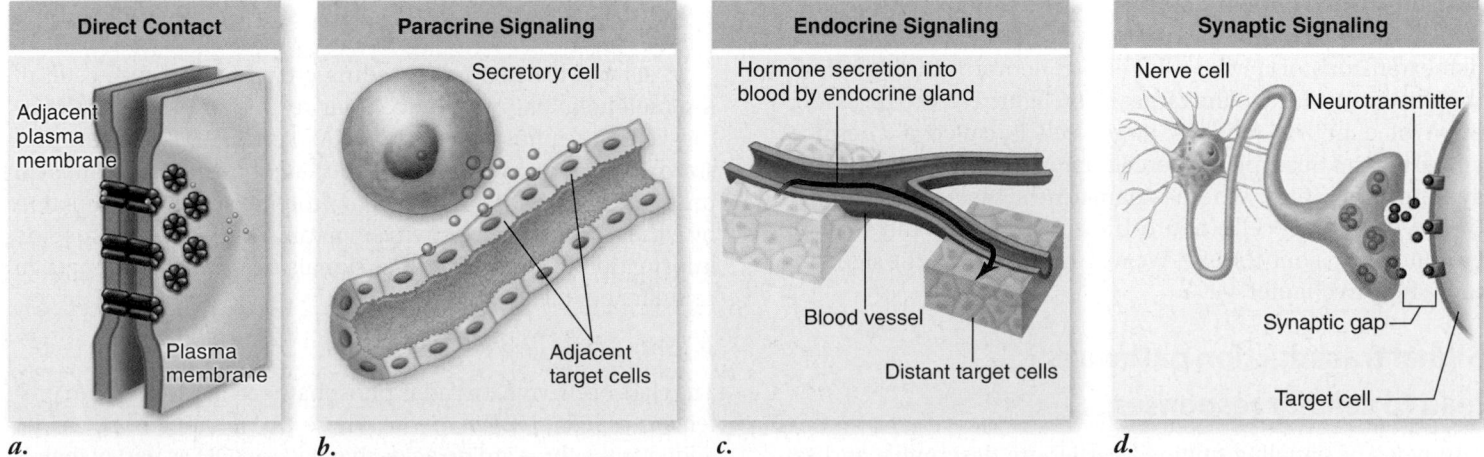

Figure 9.2 Four kinds of cell signaling. Cells communicate in several ways. *a.* Two cells in direct contact with each other may send signals across gap junctions. *b.* In paracrine signaling, secretions from one cell have an effect only on cells in the immediate area. *c.* In endocrine signaling, hormones are released into the organism's circulatory system, which carries them to the target cells. *d.* Chemical synapse signaling involves transmission of signal molecules, called neurotransmitters, from a neuron over a small synaptic gap to the target cell.

Direct contact

As you saw in chapter 5, the surface of a eukaryotic cell is richly populated with proteins, carbohydrates, and lipids attached to and extending outward from the plasma membrane. When cells are very close to one another, some of the molecules on the plasma membrane of one cell can be recognized by receptors on the plasma membrane of an adjacent cell. Many of the important interactions between cells in early development occur by means of direct contact between cell surfaces. Cells also signal through gap junctions (figure 9.2a). We'll examine contact-dependent interactions more closely later in this chapter.

Paracrine signaling

Signal molecules released by cells can diffuse through the extracellular fluid to other cells. If those molecules are taken up by neighboring cells, destroyed by extracellular enzymes, or quickly removed from the extracellular fluid in some other way, their influence is restricted to cells in the immediate vicinity of the releasing cell. Signals with such short-lived, local effects are called **paracrine** signals (figure 9.2b).

Like direct contact, paracrine signaling plays an important role in early development, coordinating the activities of clusters of neighboring cells. The immune response in vertebrates also involves paracrine signaling between immune cells (chapter 51).

Endocrine signaling

A released signal molecule that remains in the extracellular fluid may enter the organism's circulatory system and travel widely throughout the body. These longer-lived signal molecules, which may affect cells very distant from the releasing cell, are called **hormones,** and this type of intercellular communication is known as **endocrine signaling** (figure 9.2c). Chapter 45 discusses endocrine signaling in detail. Both animals and plants use this signaling mechanism extensively.

Synaptic signaling

In animals, the cells of the nervous system provide rapid communication with distant cells. Their signal molecules, **neurotransmitters,** do not travel to the distant cells through the circulatory system as hormones do. Rather, the long, fiber-like extensions of nerve cells release neurotransmitters from their tips very close to the target cells (figure 9.2d). The association of a neuron and its target cell is called a **chemical synapse,** and this type of intercellular communication is called **synaptic signaling.** Whereas paracrine signals move through the fluid between cells, neuro-transmitters cross the synaptic gap and persist only briefly. We will examine synaptic signaling more fully in chapter 44.

Signal transduction pathways lead to cellular responses

The types of signaling outlined earlier are descriptive and say nothing about how cells respond to signals. The events that occur within the cell on receipt of a signal are called **signal transduction.** These events form discrete pathways that lead to a cellular response to the signal received by receptors.

Knowledge of these signal transduction pathways has exploded in recent years and indicates a high degree of complexity that explains how in some cases different cell types can have the same response to different signals, and in other cases different cell types can have a different response to the same signal.

For example, a variety of cell types respond to the hormone glucagon by mobilizing glucose as part of the body's mechanism to control blood glucose (chapter 45). This involves breaking down stored glycogen into glucose and turning on the genes that encode the enzymes necessary to synthesize glucose. In contrast, the hormone epinephrine has diverse effects on different cell types. We have all been startled or frightened by a sudden event. Your heart beats faster, you feel more alert, and you can even feel the hairs on your skin stand up. All of this is due in part to your body releasing the hormone epinephrine (also called adrenaline) into the bloodstream. This leads to the heightened state of alertness and increased heart rate and energy that prepare us to respond to extreme situations.

These differing effects of epinephrine depend on the different cell types with receptors for this hormone. In the liver, cells are stimulated to mobilize glucose while in the heart muscle cells contract more forcefully to increase blood flow. In addition, blood vessels respond by expanding in some areas and contracting in others to redirect blood flow to the liver, heart, and skeletal muscles. These different reactions depend on the fact that each cell type has a receptor for epinephrine, but different sets of proteins that respond to this signal.

Phosphorylation is key in control of protein function

The function of a signal transduction pathway is to change the behavior or nature of a cell. This action may require changing the composition of proteins that make up a cell or altering the activity of cellular proteins. Many proteins are inactive or nonfunctional as they are initially synthesized and require modification after synthesis for activation. In other cases, a protein may be deactivated by modification. A major source of control for protein function is the addition or removal of phosphate groups, called **phosphorylation** or **dephosphorylation,** respectively.

As you learned in preceding chapters, the end result of the metabolic pathways of cellular respiration and photosynthesis was the phosphorylation of ADP to ATP. The ATP synthesized by these processes can donate phosphate groups to proteins. The phosphorylation of proteins alters their function by either turning their activity on or off. This is one way that the information from extracellular signals can result in changes in cellular activities.

Protein kinases

The class of enzyme that adds phosphate groups from ATP to proteins is called a *protein kinase.* These phosphate groups can be added to the three amino acids that have an OH as part of their R group, namely serine, threonine, and tyrosine. We categorize protein kinases as either serine–threonine or tyrosine kinases based on the amino acids they modify (figure 9.3). Most cytoplasmic protein kinases fall into the serine–threonine kinase class.

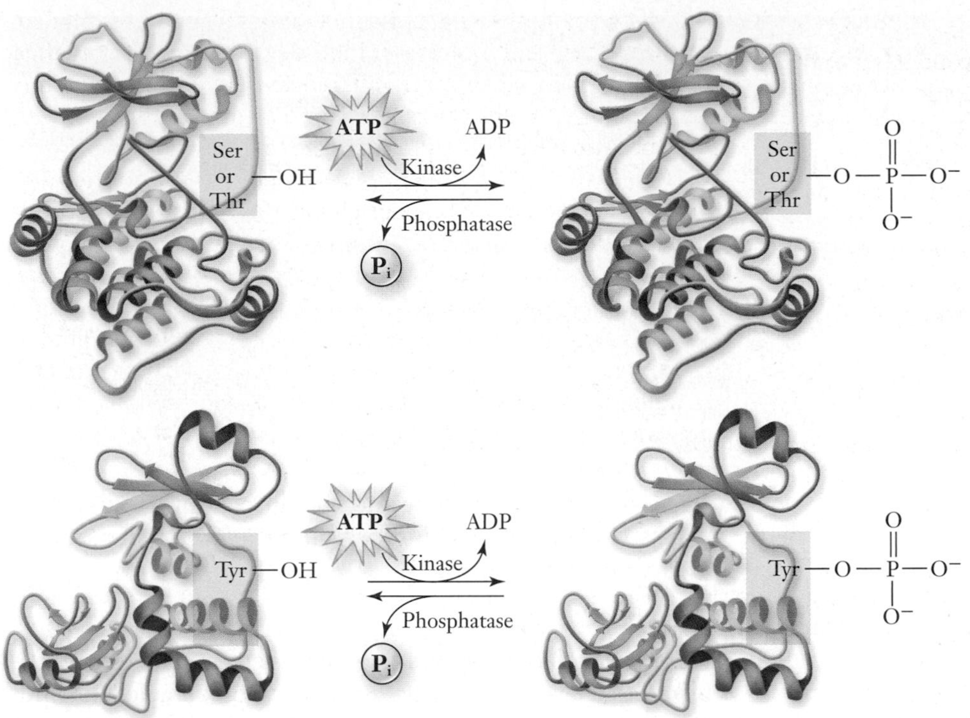

Figure 9.3 Phosphorylation of proteins. Many proteins are controlled by their phosphorylation state: that is, they are activated by phosphorylation and deactivated by dephosphorylation or the reverse. The enzymes that add phosphate groups are called kinases. These form two classes depending on the amino acid the phosphate is added to, either serine–threonine kinases or tyrosine kinases. The action of kinases is reversed by protein phosphatase enzymes.

Phosphatases

Part of the reason for the versatility of phosphorylation as a form of protein modification is that it is reversible. Another class of enzymes called **phosphatases** removes phosphate groups, reversing the action of kinases (see figure 9.3). Thus, a protein activated by a kinase will be deactivated by a phosphatase, and a protein deactivated by a kinase will be activated by a phosphatase.

Learning Outcomes Review 9.1

Cell communication involves chemical signals, or ligands, that bind to cellular receptors. Binding of ligand to receptor initiates signal transduction pathways that lead to a cellular response. Different cells may have the same response to one signal and the same signal can also elicit different responses in different cells. The phosphorylation–dephosphorylation of proteins is a common mechanism of controlling protein function found in signaling pathways.

■ *How are receptor ligand interactions similar to enzyme substrate interactions?*

9.2 Receptor Types

Learning Outcome

1. *Contrast the different types of receptors.*

The first step in understanding cell signaling is to consider the receptors themselves. Cells must have a specific receptor to be able to respond to a particular signaling molecule. The interaction of a receptor and its ligand is an example of molecular recognition, a process in which one molecule fits specifically based on its complementary shape with another molecule. This interaction causes subtle changes in the structure of the receptor, thereby activating it. This is the beginning of any signal transduction pathway.

Receptors are defined by location

The nature of these receptor molecules depends on their location and on the kind of ligands they bind. Intracellular receptors bind hydro-phobic ligands, which can easily cross the membrane, inside the cell. In contrast, cell surface or membrane receptors bind hydrophilic ligands, which cannot easily cross the membrane, outside the cell (see figure 9.1). Membrane receptors consist of transmembrane proteins that are in contact with both the cytoplasm and the extracellular environment. Table 9.1 summarizes the types of receptors and communication mechanisms discussed in this chapter.

Membrane receptors include three subclasses

When a receptor is a transmembrane protein, the ligand binds to the receptor outside of the cell and never actually crosses the plasma membrane. In this case, the receptor itself, and not the signaling molecule is responsible for information crossing the membrane. Membrane receptors can be categorized based on their structure and function.

Channel-linked receptors

Chemically gated ion channels are receptor proteins that allow the passage of ions (figure 9.4a). The receptor proteins that bind many neurotransmitters have the same basic structure. Each is a membrane protein with multiple transmembrane domains, meaning that the chain of amino acids threads back and forth across the plasma membrane several times. In the center of the protein is a pore that connects the extracellular fluid with the cytoplasm. The pore is big enough for ions to pass through, so the protein functions as an **ion channel.**

TABLE 9.1	Receptors Involved in Cell Signaling		
Receptor Type	**Structure**	**Function**	**Example**
Intracellular Receptors	No extracellular signal-binding site	Receives signals from lipid-soluble or noncharged, nonpolar small molecules	Receptors for NO, steroid hormone, vitamin D, and thyroid hormone
Cell Surface Receptors			
Chemically gated ion channels	Multipass transmembrane protein forming a central pore	Molecular "gates" triggered chemically to open or close	Neurons
Enzymatic receptors	Single-pass transmembrane protein	Binds signal extracellularly; catalyzes response intracellularly	Phosphorylation of protein kinases
G protein-coupled receptors	Seven-pass transmembrane protein with cytoplasmic binding site for G protein	Binding of signal to receptor causes GTP to bind a G protein; G protein, with attached GTP, detaches to deliver the signal inside the cell	Peptide hormones, rod cells in the eyes

The channel is said to be chemically gated because it opens only when a chemical (the neurotransmitter) binds to it. The type of ion that flows across the membrane when a chemically gated ion channel opens depends on the shape and charge structure of the channel. Sodium, potassium, calcium, and chloride ions all have specific ion channels.

The acetylcholine receptor found in muscle cell membranes functions as an Na$^+$ channel. When the receptor binds to its ligand, the neurotransmitter acetylcholine, the channel opens allowing Na$^+$ to flow into the muscle cell. This is a critical step linking the signal from a motor neuron to muscle cell contraction (chapter 46).

Enzymatic receptors

Many cell surface receptors either act as enzymes or are directly linked to enzymes (figure 9.4b). When a signal molecule binds to the receptor, it activates the enzyme. In almost all cases,

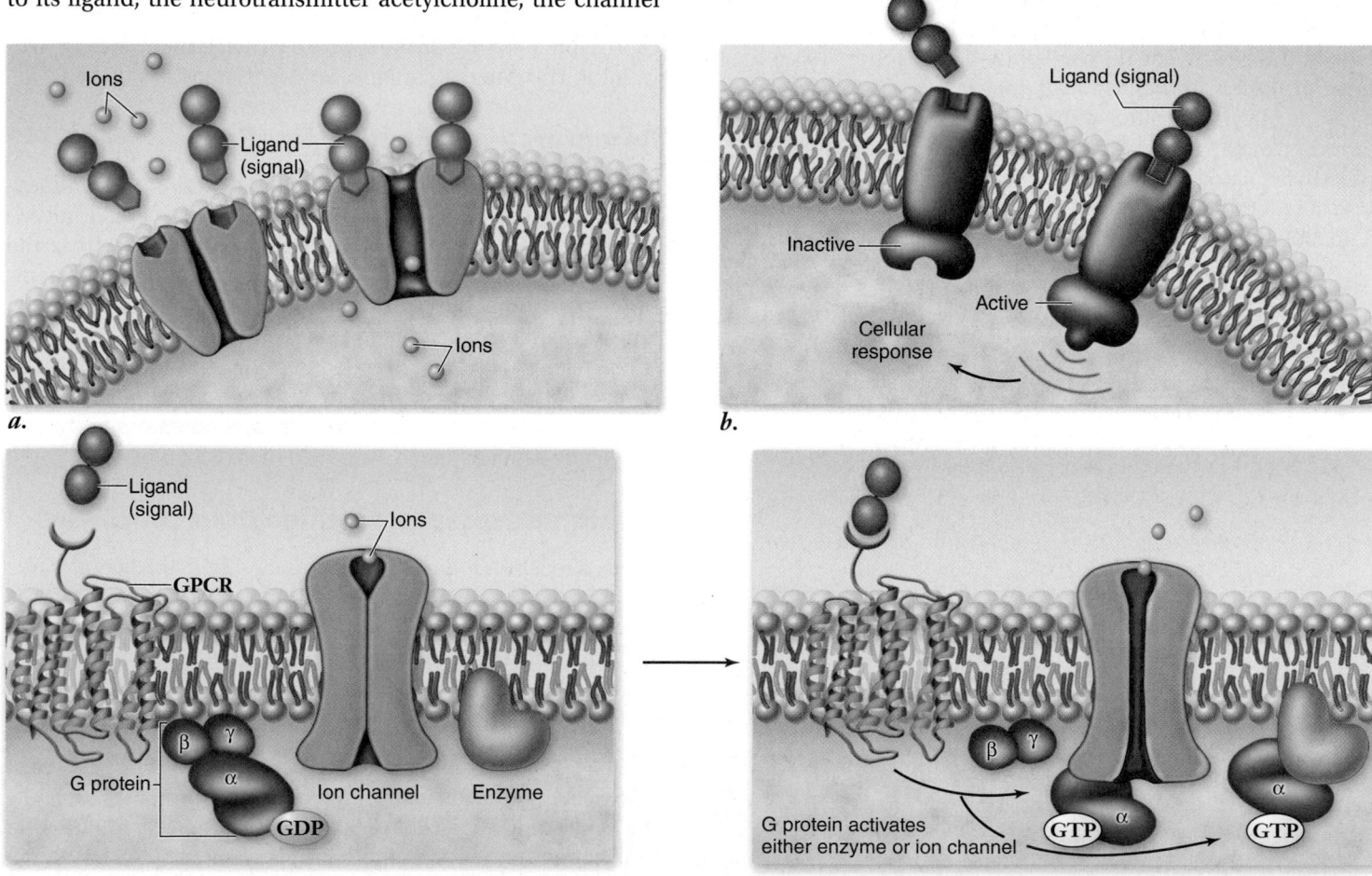

Figure 9.4 Cell surface receptors. *a.* Chemically gated ion channels form a pore in the plasma membrane that can be opened or closed by chemical signals. They are usually selective, allowing the passage of only one type of ion. *b.* Enzymatic receptors bind to ligands on the extracellular surface. A catalytic region on their cytoplasmic portion transmits the signal across the membrane by acting as an enzyme in the cytoplasm. *c.* G protein-coupled receptors (GPCR) bind to ligands outside the cell and to G proteins inside the cell. The G protein then activates an enzyme or ion channel, transmitting signals from the cell's surface to its interior.

these enzymes are **protein kinases,** enzymes that add phosphate groups to proteins. We discuss these receptors in detail in a later section of this chapter.

G Protein-coupled receptors

A third class of cell surface receptors acts indirectly on enzymes or ion channels in the plasma membrane with the aid of an assisting protein, called a **G protein.** The G protein, which is so named because it binds the nucleotide *guanosine triphosphate* (GTP), can be thought of as being inserted between the receptors and the enzyme (effector). That is, the ligand binds to the receptor, activating it, which activates the G protein, which in turn activates the effector protein (figure 9.4c). These receptors are also discussed in detail later on.

Membrane receptors can generate second messengers

Some enzymatic receptors and most G protein-coupled receptors utilize other substances to relay the message within the cytoplasm. These other substances, small molecules or ions called **second messengers,** alter the behavior of cellular proteins by binding to them and changing their shape. (The original signal molecule is considered the "first messenger.") Two common second messengers are **cyclic adenosine monophosphate (cyclic AMP, or cAMP)** and calcium ions. The role of these second messengers will be explored in more detail in a later section.

Learning Outcome Review 9.2

Receptors may be internal (intracellular receptors) or external (membrane receptors). Membrane receptors include channel-linked receptors, enzymatic receptors, and G protein-coupled receptors. Signal transduction through membrane receptors often involves the production of a second signaling molecule, or second messenger, inside the cell.

■ *Would a hydrophobic molecule be expected to have an internal or membrane receptor?*

Learning Outcomes
1. Describe the chemical nature of ligands for intracellular receptors.
2. Diagram the pathway of signal transduction through intracellular receptors.

Many cell signals are lipid-soluble or very small molecules that can readily pass through the plasma membrane of the target cell and into the cell, where they interact with an *intracellular receptor.* Some of these ligands bind to protein receptors located in the cytoplasm, others pass across the nuclear membrane as well and bind to receptors within the nucleus.

Steroid hormone receptors affect gene expression

Of all of the receptor types discussed in this chapter, the action of the steroid hormone receptors is the simplest and most direct.

Steroid hormones form a large class of compounds, including cortisol, estrogen, progesterone, and testosterone, that share a common nonpolar structure. Estrogen, progesterone, and testosterone are involved in sexual development and behavior (chapter 52). Other steroid hormones, such as cortisol, also have varied effects depending on the target tissue, ranging from the mobilization of glucose to the inhibition of white blood cells to control inflammation. Their anti-inflammatory action is the basis of their use in medicine.

The nonpolar structure allows these hormones to cross the membrane and bind to intracellular receptors. The location of steroid hormone receptors prior to hormone binding is cytoplasmic, but their primary site of action is in the nucleus. Binding of the hormone to the receptor causes the complex to

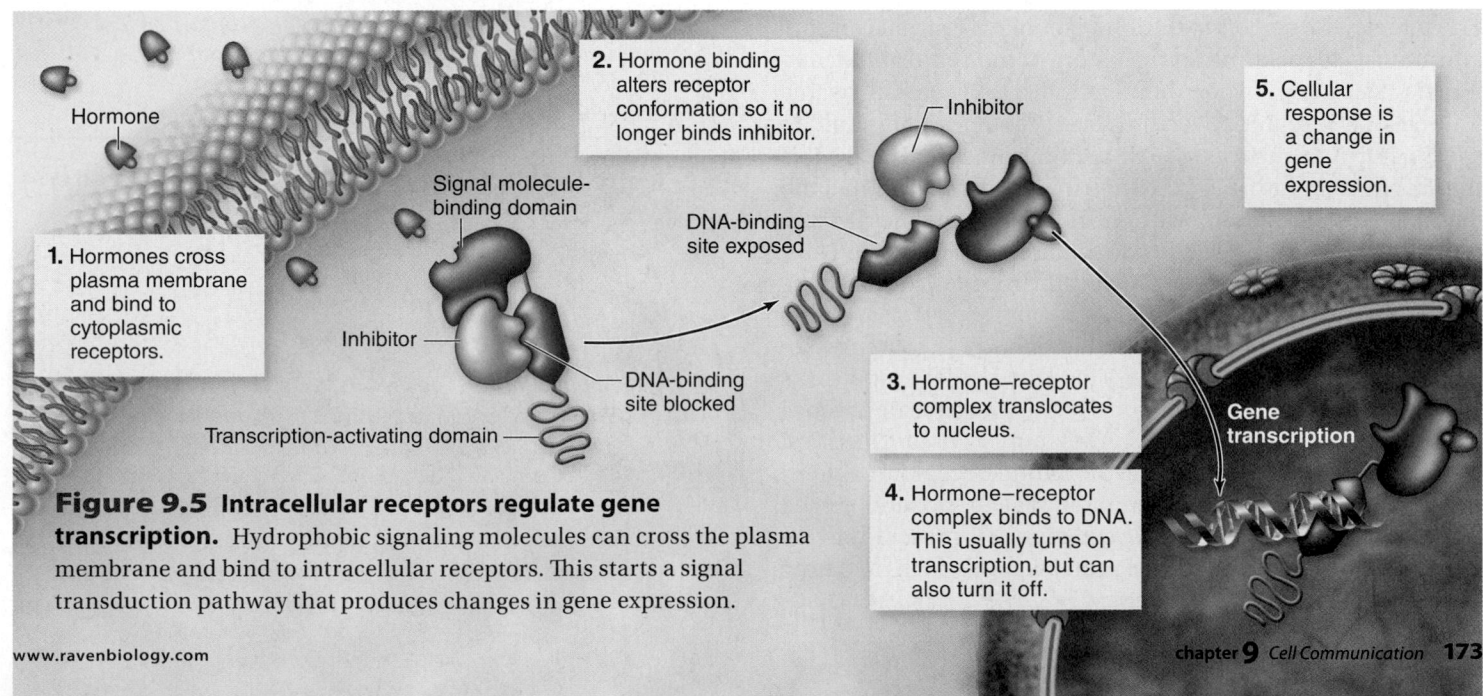

Figure 9.5 Intracellular receptors regulate gene transcription. Hydrophobic signaling molecules can cross the plasma membrane and bind to intracellular receptors. This starts a signal transduction pathway that produces changes in gene expression.

shift from the cytoplasm to the nucleus (figure 9.5). As the ligand–receptor complex makes it all the way to the nucleus of the cell, these receptors are often called **nuclear receptors.**

Steroid receptor action

The primary function of steroid hormone receptors, as well as receptors for a number of other small, lipid-soluble signal molecules such as vitamin D and thyroid hormone, is to act as regulators of gene expression (see chapter 16).

All of these receptors have similar structures; the genes that code for them appear to be the evolutionary descendants of a single ancestral gene. Because of their structural similarities, they are all part of the *nuclear receptor superfamily.*

Each of these receptors has three functional domains—

1. a hormone-binding domain,
2. a DNA-binding domain, and
3. a domain that can interact with coactivators to affect the level of gene transcription.

In its inactive state, the receptor typically cannot bind to DNA because an inhibitor protein occupies the DNA-binding site. When the signal molecule binds to the hormone-binding site, the conformation of the receptor changes, releasing the inhibitor and exposing the DNA-binding site, allowing the receptor to attach to specific nucleotide sequences on the DNA (see figure 9.5). This binding activates (or, in a few instances, suppresses) particular genes, usually located adjacent to the hormone-binding sequences. In the case of cortisol, which is a glucocorticoid hormone that can increase levels of glucose in cells, a number of different genes involved in the synthesis of glucose have binding sites for the hormone receptor complex.

The lipid-soluble ligands that intracellular receptors recognize tend to persist in the blood far longer than water-soluble signals. Most water-soluble hormones break down within minutes, and neurotransmitters break down within seconds or even milliseconds. In contrast, a steroid hormone such as cortisol or estrogen persists for hours.

Specificity and the role of coactivators

The target cell's response to a lipid-soluble cell signal can vary enormously, depending on the nature of the cell. This characteristic is true even when different target cells have the same intracellular receptor. Given that the receptor proteins bind to specific DNA sequences, which are the same in all cells, this may seem puzzling. It is explained in part by the fact that the receptors act in concert with **coactivators,** and the number and nature of these molecules can differ from cell to cell. Thus, a cell's response depends on not only the receptors but also the coactivators present.

The hormone estrogen has different effects in uterine tissue than in mammary tissue. This differential response is mediated by coactivators and not by the presence or absence of a receptor in the two tissues. In mammary tissue, a critical coactivator is lacking and the hormone–receptor complex instead interacts with another protein that acts to reduce gene expression. In uterine tissue, the coactivator is present, and the expression of genes that encode proteins involved in preparing the uterus for pregnancy are turned on.

Other intracellular receptors act as enzymes

A very interesting example of a receptor acting as an enzyme is found in the receptor for nitric oxide (NO). This small gas molecule diffuses readily out of the cells where it is produced and passes directly into neighboring cells, where it binds to the enzyme guanylyl cyclase. Binding of NO activates this enzyme, enabling it to catalyze the synthesis of *cyclic guanosine monophosphate (cGMP),* an intracellular messenger molecule that produces cell-specific responses such as the relaxation of smooth muscle cells.

When the brain sends a nerve signal to relax the smooth muscle cells lining the walls of vertebrate blood vessels, acetylcholine released by the nerve cell binds to receptors on epithelial cells. This causes an increase in intracellular Ca^{2+} in the epithelial cell that stimulates nitric oxide synthase to produce NO. The NO diffuses into the smooth muscle, where it increases the level of cGMP, leading to relaxation. This relaxation allows the vessel to expand and thereby increases blood flow. This explains the use of nitroglycerin to treat the pain of angina caused by constricted blood vessels to the heart. The nitroglycerin is converted by cells to NO, which then acts to relax the blood vessels.

The drug sildenafil (better known as Viagra) also functions via this signal transduction pathway by binding to and inhibiting the enzyme cGMP phosphodiesterase, which breaks down cGMP. This keeps levels of cGMP high, thereby stimulating production of NO. The reason for Viagra's selective effect is that it binds to a form of cGMP phosphodiesterase found in cells in the penis. This allows relaxation of smooth muscle in erectile tissue, thereby increasing blood flow.

Learning Outcomes Review 9.3

Hydrophobic signaling molecules can cross the membrane and bind to intracellular receptors. The steroid hormone receptors act by directly influencing gene expression. On binding hormone, the hormone–receptor complex moves into the nucleus to turn on (or sometimes turn off) gene expression. This may also require a coactivator that functions with the hormone–receptor complex. Thus, the cell's response to a hormone depends on the presence of a receptor and coactivators as well.

■ *Would these types of intracellular receptors be fast acting, or have effects of longer duration?*

9.4 *Signal Transduction Through Receptor Kinases*

Learning Outcomes

1. *Compare the function of RTKs to steroid hormone receptors.*
2. *Describe how information crosses the membrane in RTKs.*
3. *Explain the role of kinase cascades in signal transduction.*

Earlier you read that protein kinases phosphorylate proteins to alter protein function and that the most common kinases act on the amino acids serine, threonine, and tyrosine. The

receptor tyrosine kinases (RTKs) influence the cell cycle, cell migration, cell metabolism, and cell proliferation—virtually all aspects of the cell are affected by signaling through these receptors. Alterations to the function of these receptors and their signaling pathways can lead to cancers in humans and other animals.

Some of the earliest examples of cancer-causing genes, or oncogenes, involve RTK function (discussed in chapter 10). The avian erythroblastosis virus carries an altered form of the epidermal growth factor receptor that lacks most of its extracellular domain. When this virus infects a cell the altered receptors produced are stuck in the "on" state. The continuous signaling from this receptor leads to cells that have lost the normal controls over growth.

Receptor tyrosine kinases recognize hydrophilic ligands and form a large class of membrane receptors in animal cells. Plants possess receptors with a similar overall structure and function, but they are serine–threonine kinases. These plant receptors have been named **plant receptor kinases.**

Because these receptors are performing similar functions in plant and animal cells but differ in their substrates, the duplication and divergence of each kind of receptor kinase probably occurred after the plant–animal divergence. The proliferation of these types of signaling molecules is thought to coincide with the independent evolution of multicellularity in each group.

In this section, we will concentrate on the RTK family of receptors that has been extensively studied in a variety of animal cells.

RTKs are activated by autophosphorylation

Receptor tyrosine kinases have a relatively simple structure consisting of a single transmembrane domain that anchors them in the membrane, an extracellular ligand-binding domain, and an intracellular kinase domain. This kinase domain contains the catalytic site of the receptor, which acts as a protein kinase that adds phosphate groups to tyrosines. On ligand binding to a specific receptor, two of these receptor–ligand complexes associate together (often referred to as dimerization) and phosphorylate each other, a process called *autophosphorylation* (figure 9.6).

The autophosphorylation event transmits across the membrane the signal that began with the binding of the ligand to the receptor. The next step, propagation of the signal in the cytoplasm, can take a variety of different forms. These forms include activation of the tyrosine kinase domain to

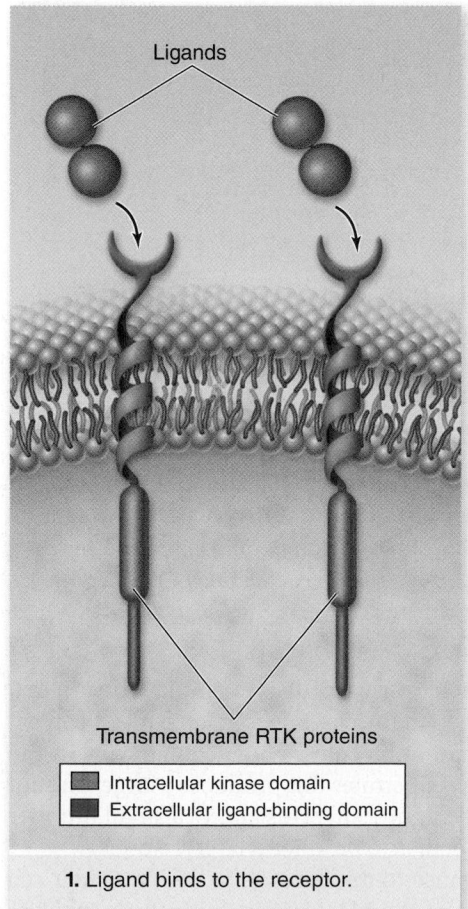

Ligands

Intracellular kinase domain
Extracellular ligand-binding domain

Transmembrane RTK proteins

1. Ligand binds to the receptor.

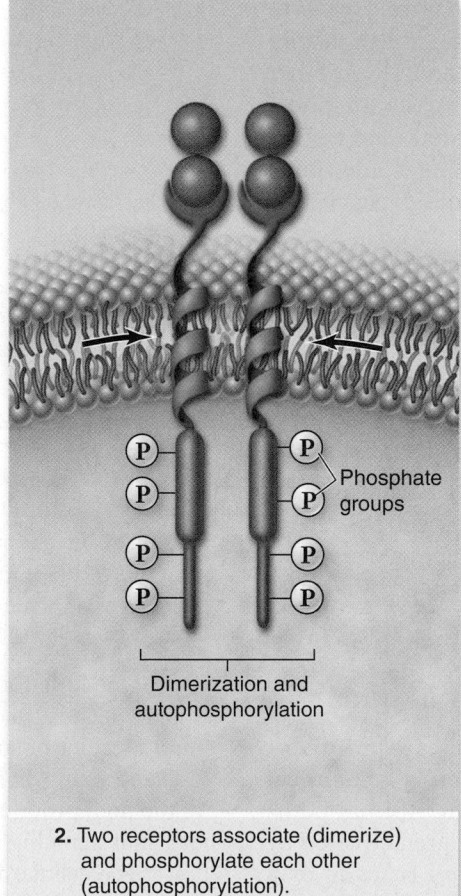

Phosphate groups

Dimerization and autophosphorylation

2. Two receptors associate (dimerize) and phosphorylate each other (autophosphorylation).

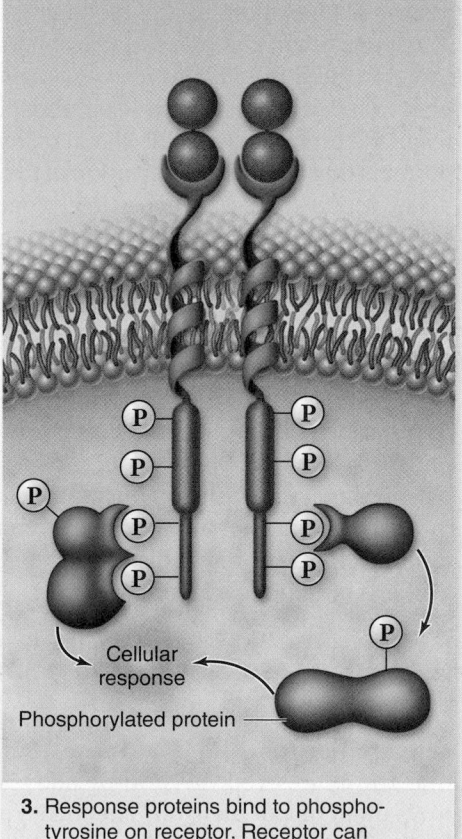

Cellular response

Phosphorylated protein

3. Response proteins bind to phospho-tyrosine on receptor. Receptor can phosphorylate other response proteins.

Figure 9.6 Activation of a receptor tyrosine kinase (RTK). These membrane receptors bind hormones or growth factors that are hydrophilic and cannot cross the membrane. The receptor is a transmembrane protein with an extracellular ligand binding domain and an intracellular kinase domain. Signal transduction pathways begin with response proteins binding to phosphotyrosine on receptor, and by receptor phosphorylation of response proteins.

phosphorylate other intracellular targets or interaction of other proteins with the phosphorylated receptor.

The cellular response after activation depends on the possible response proteins in the cell. Two different cells can have the same receptor yet a different response, depending on what response proteins are present in the cytoplasm. For example, fibroblast growth factor stimulates cell division in fibroblasts but stimulates nerve cells to differentiate rather than to divide.

Phosphotyrosine domains mediate protein–protein interactions

One way that the signal from the receptor can be propagated in the cytoplasm is via proteins that bind specifically to phosphorylated tyrosines in the receptor. When the receptor is activated, regions of the protein outside of the catalytic site are phosphorylated. This creates "docking" sites for proteins that bind specifically to phosphotyrosine. The proteins that bind to these phosphorylated tyrosines can initiate intracellular events to convert the signal from the ligand into a response (see figure 9.6).

The insulin receptor

The use of docking proteins is illustrated by the insulin receptor. The hormone insulin is part of the body's control system to maintain a constant level of blood glucose. The role of insulin is to lower blood glucose, acting by binding to an RTK. Another protein called the *insulin response protein* binds to the phosphorylated receptor and is itself phosphorylated. The insulin response protein passes the signal on by binding to additional proteins that lead to the activation of the enzyme glycogen synthase, which converts glucose to glycogen (figure 9.7), thereby lowering blood glucose. Other proteins activated by the insulin receptor act to inhibit the synthesis of enzymes involved in making glucose, and to increase the number of glucose transporter proteins in the plasma membrane.

Adapter proteins

Another class of proteins, **adapter proteins,** can also bind to phosphotyrosines. These proteins themselves do not participate in signal transduction but act as a link between the receptor and proteins that initiate downstream signaling events. For example, the Ras protein discussed later is activated by adapter proteins binding to a receptor.

Protein kinase cascades can amplify a signal

One important class of cytoplasmic kinases are **mitogen-activated protein (MAP) kinases.** A *mitogen* is a chemical that stimulates cell division by activating the normal pathways that control division. The MAP kinases are activated by a signaling module called a *phosphorylation cascade* or a **kinase cascade.** This module is a series of protein kinases that phosphorylate each other in succession. The final step in the cascade is the activation by phosphorylation of MAP kinase itself (figure 9.8).

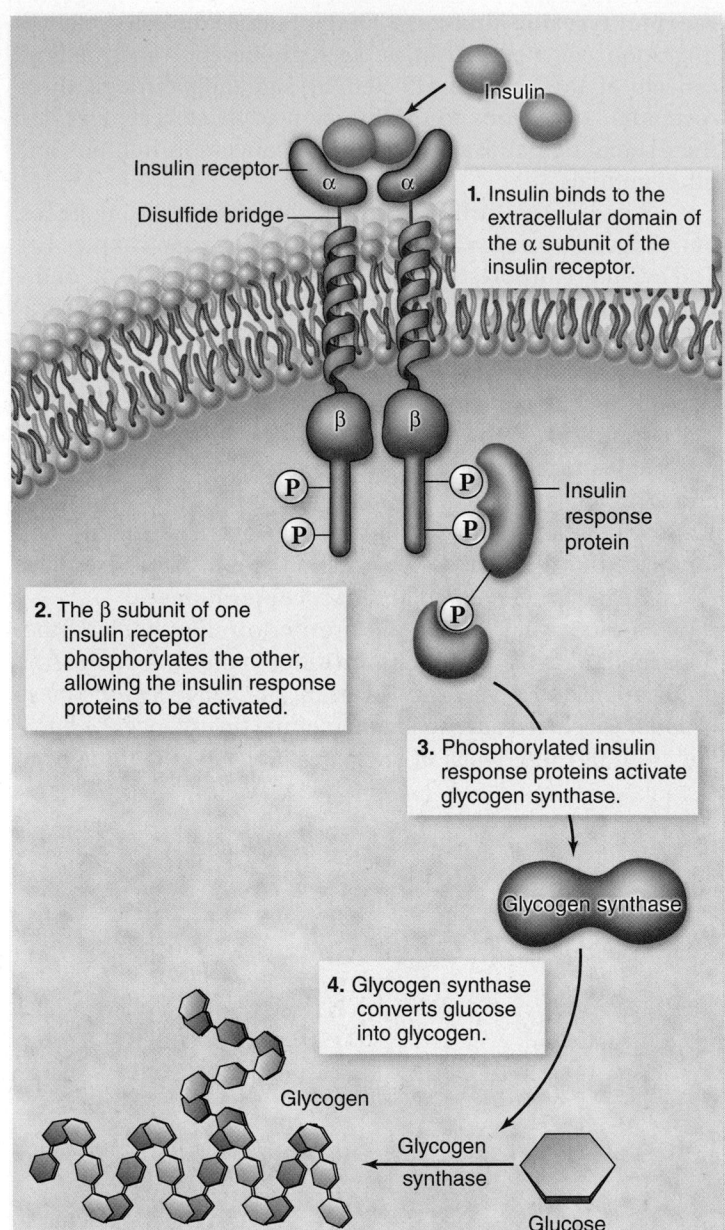

1. Insulin binds to the extracellular domain of the α subunit of the insulin receptor.

2. The β subunit of one insulin receptor phosphorylates the other, allowing the insulin response proteins to be activated.

3. Phosphorylated insulin response proteins activate glycogen synthase.

4. Glycogen synthase converts glucose into glycogen.

Figure 9.7 The insulin receptor. The insulin receptor is a receptor tyrosine kinase that initiates a variety of cellular responses related to glucose metabolism. One signal transduction pathway that this receptor mediates leads to the activation of the enzyme glycogen synthase. This enzyme converts glucose to glycogen.

One function of a kinase cascade is to amplify the original signal. Because each step in the cascade is an enzyme, it can act on a number of substrate molecules. With each enzyme in the cascade acting on many substrates this produces a large amount of the final product (see figure 9.8). This allows a small number of initial signaling molecules to produce a large response.

The cellular response to this cascade in any particular cell depends on the targets of the MAP kinase, but usually involves phosphorylating transcription factors that then activate gene expression (chapter 16). An example of this kind of signaling through growth factor receptors is provided in chapter 10 and

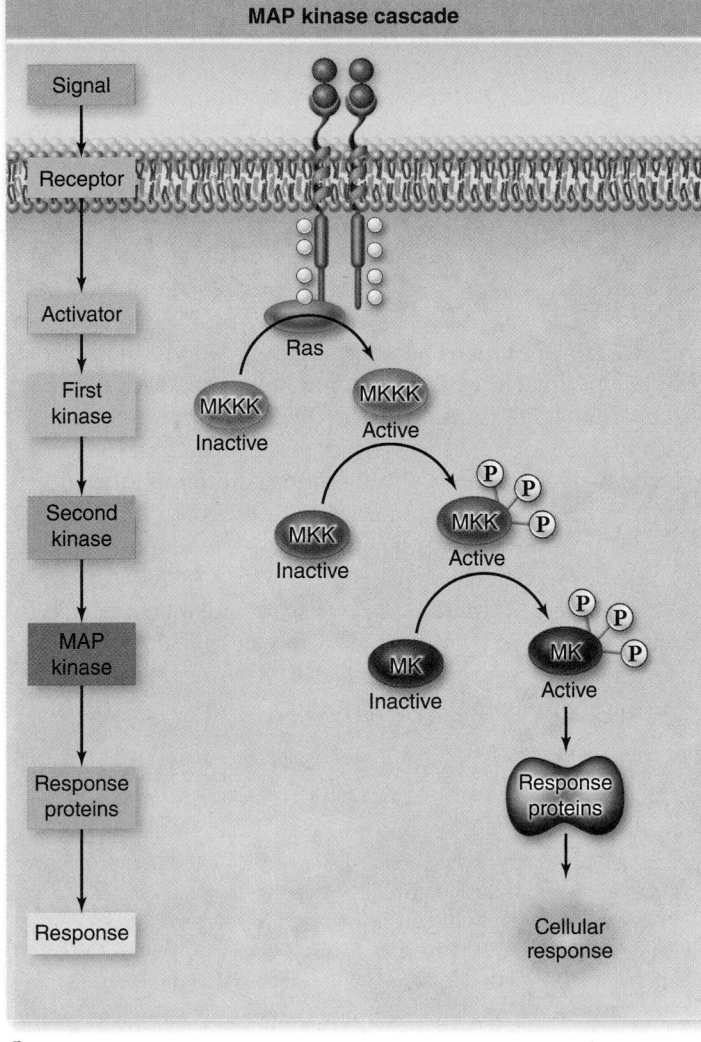

MAP kinase cascade

Signal

Receptor

Activator

First kinase

Second kinase

MAP kinase

Response proteins

Response

Ras

MKKK Inactive

MKKK Active

MKK Inactive

MKK Active

MK Inactive

MK Active

Response proteins

Cellular response

a.

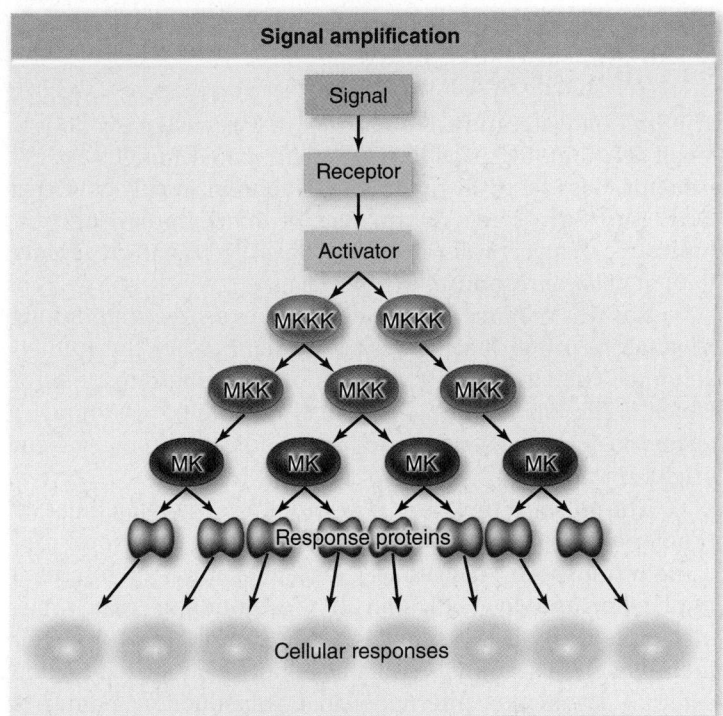

Signal amplification

Signal

Receptor

Activator

MKKK   MKKK

MKK   MKK   MKK

MK   MK   MK   MK

Response proteins

Cellular responses

b.

Figure 9.8 MAP kinase cascade leads to signal amplification. *a.* Phosphorylation cascade is shown as a flowchart on the left. The corresponding cellular events are shown on the right, beginning with the receptor in the plasma membrane. Each kinase is named starting with the last, the MAP kinase (MK), which is phosphorylated by a MAP kinase kinase (MKK), which is in turn phosphorylated by a MAP kinase kinase kinase (MKKK). The cascade is linked to the receptor protein by an activator protein. *b.* At each step the enzymatic action of the kinase on multiple substrates leads to amplification of the signal.

illustrates how signal transduction initiated by a growth factor can control the process of cell division through a kinase cascade.

Scaffold proteins organize kinase cascades

The proteins in a kinase cascade need to act sequentially to be effective. One way the efficiency of this process can be increased is to organize them in the cytoplasm. Proteins called *scaffold proteins* are thought to organize the components of a kinase cascade into a single protein complex, the ultimate in a signaling module. The scaffold protein binds to each individual kinase such that they are spatially organized for optimal function (figure 9.9).

The advantages of this kind of organization are many. A physically arranged sequence is clearly more efficient than one that depends on diffusion to produce the appropriate order of events. This organization also allows the segregation of signaling modules in different cytoplasmic locations.

The disadvantage of this kind of organization is that it reduces the amplification effect of the kinase cascade. Enzymes held in one place are not free to find new substrate molecules, but must rely on substrates being nearby.

The best studied example of a scaffold protein comes from mating behavior in budding yeast. Yeast cells respond to mating pheromones with changes in cell morphology and gene expression, mediated by a protein kinase cascade. A protein called Ste5 was originally identified as a protein required for mating behavior, but no enzymatic activity could be detected for this protein. It has now been shown that this protein interacts with all of the members of the kinase cascade and acts as a scaffold protein that organizes the cascade and insulates it from other signaling pathways.

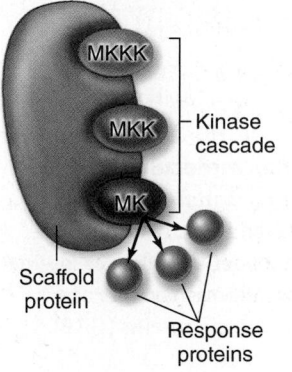

MKKK

MKK

MK

Kinase cascade

Scaffold protein

Response proteins

Figure 9.9 Kinase cascade can be organized by scaffold proteins. The scaffold protein binds to each kinase in the cascade, organizing them so each substrate is next to its enzyme. This organization also sequesters the kinases from other signaling pathways in the cytoplasm.

Ras is a small G protein that acts as a molecular switch

The link between the RTK and the MAP kinase cascade is a small GTP-binding protein (G protein) called **Ras.** Like all G proteins, Ras is actively bound to GTP, and inactively bound to GDP. The Ras protein is mutated in many human tumors, indicative of its central role in linking growth factor receptors to their cellular response.

Ras was the first protein identified in a large superfamily of small G proteins with over 150 members in the human genome. The superfamily consists of five subgroups, one of which is the Ras family. These small G proteins are found in eukaryotes from yeast to vertebrates, indicating their ancient origin.

The roles of these small G proteins vary, affecting cell proliferation, the cytoskeleton, membrane transport, and nuclear transport. They are an excellent example of how gene duplication and diversification allow evolution to create modular units with diverse functions. The common feature of all members of the family is to act as a molecular switch linking external signals to internal signal transduction pathways (figure 9.10).

The Ras switch is flipped by exchanging GDP for GTP, and by Ras hydrolyzing GTP to GDP. The link to outside signals comes from other proteins that affect the switch: guanine nucleotide exchange factors (GEFs) stimulate the exchange of GDP for GTP activating Ras. When a growth factor receptor is activated, it binds to an adapter protein that acts as a GEF. The activated Ras protein then activates the first kinase in the MAP kinase cascade (see figure 9.8 and chapter 10).

The action of Ras can be terminated by its intrinsic GTPase activity. This can be stimulated by a GAP protein, which provides the opportunity to fine-tune signaling based on the duration of Ras activity. The importance of these proteins is shown by mutations in GAP proteins that can lead to a predisposition for specific cancers such as neurofibromatosis.

RTKs are inactivated by internalization

It is important to cells that signaling pathways are only activated transiently. Continued activation could render the cell unable to respond to other signals or to respond inappropriately to a signal that is no longer relevant. Consequently, inactivation is as important for the control of signaling as activation. Receptor tyrosine kinases can be inactivated by two basic mechanisms—dephosphorylation and internalization. Internalization is by endocytosis, in which the receptor is taken up into the cytoplasm in a vesicle where it can be degraded or recycled.

The enzymes in the kinase cascade are all controlled by dephosphorylation by phosphatase enzymes. This leads to termination of the response at both the level of the receptor and the response proteins.

Learning Outcomes Review 9.4

Receptor tyrosine kinases (RTKs) are membrane receptors that can phosphorylate tyrosine. When activated, they autophosphorylate, creating binding domains for other proteins. These proteins transmit the signal inside the cell. One form of signaling pathway involves the MAP kinase cascade, a series of kinases that each activate the next in the series. This ends with a MAP kinase that activates transcription factors to alter gene expression.

■ *Ras protein is mutated in many human cancers. What are possible reasons for this?*

9.5 Signal Transduction Through G Protein-Coupled Receptors

Learning Outcomes

1. *Contrast signaling through GPCRs and RTKs.*
2. *Relate the function of second messengers to signal transduction pathways.*

The single largest category of receptor type in animal cells is **G protein-coupled receptors (GPCRs),** so named because the receptors act by coupling with a G protein. These receptors bind diverse ligands, including ions, organic odorants, peptides, proteins, and lipids. Light-sensing receptors are also part of this family, so we could even count photons as "ligands."

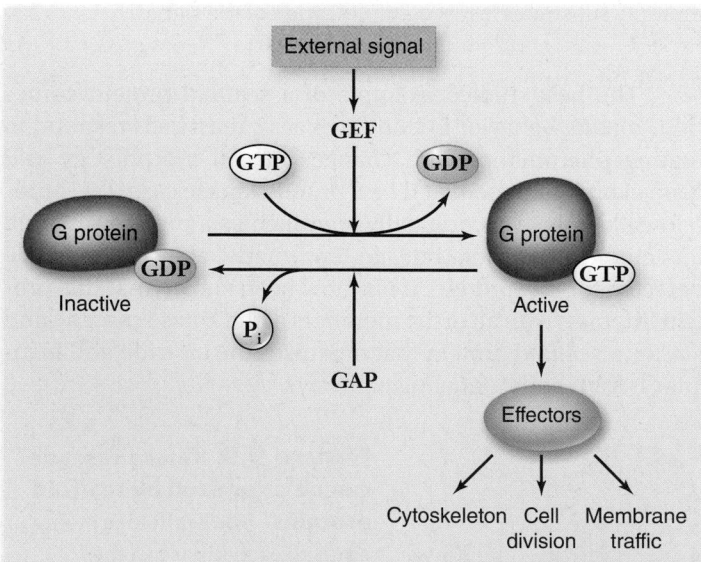

Figure 9.10 Small G proteins act as molecular switches. Small G proteins, such as Ras, link external signals to internal signal transduction pathways. External signals activate guanine nucleotide exchange proteins (GEF), which activate the G protein. The G protein can be inactivated by its weak intrinsic GTPase activity, which can be stimulated by activating proteins (GAP).

This superfamily of proteins also has a characteristic structure with seven transmembrane domains that anchor the receptors in the membrane. This arrangement of seven transmembrane domains is highly conserved and is used to search for new members in sequenced genomes. The analysis of many animal genomes indicates that GPCRs are the largest gene family in most animals. They have been found in virtually all types of eukaryotic organisms, indicating an ancient origin with duplication and divergence leading to a wide array of signaling pathways.

The latest count of genes encoding GPCRs in the human genome is 799, with about half of these encoding odorant receptors involved in the sense of taste and smell. In the mouse, over 1000 different odorant receptors are involved in the sense of smell. The family of GPCRs has been subdivided into five groups based on structure and function: Rhodopsin, Secretin, Adhesion, Glutamate, and Frizzled/Taste 2. The names refer to the first discovered member of each group; for example, Rhodopsin is the GPCR involved in light sensing in mammals. In this section, we will concentrate on the basic mechanism of activation and some of the possible signal transduction pathways.

G proteins link receptors with effector proteins

The function of the G protein in signaling by GPCRs is to provide a link between a receptor that receives signals and effector proteins that produce cellular responses. The G protein functions as a switch that is turned on by the receptor. In its "on" state, the G protein activates effector proteins to cause a cellular response.

All G proteins are active when bound to GTP and inactive when bound to GDP. The main difference between the G proteins in GPCRs and the small G proteins described earlier is that these G proteins are composed of three subunits, called α, β, and γ. As a result, they are often called *heterotrimeric G proteins*. When a ligand binds to a GPCR and activates its associated G protein, the G protein exchanges GDP for GTP and dissociates into two parts consisting of the G_α subunit bound to GTP, and the G_β and G_γ subunits together ($G_{\beta\gamma}$). The signal can then be propagated by either the G_α or the $G_{\beta\gamma}$ components, thereby acting to turn on effector proteins. The hydrolysis of bound GTP to GDP by G_α causes reassociation of the heterotrimer and restores the "off" state of the system (figure 9.11).

The effector proteins are usually enzymes. An effector protein might be a protein kinase that phosphorylates proteins to directly propagate the signal, or it may produce a second messenger to initiate a signal transduction pathway.

Effector proteins produce multiple second messengers

Often, the effector proteins activated by G proteins produce a second messenger. Two of the most common effectors are *adenylyl cyclase* and *phospholipase C,* which produce cAMP and IP$_3$ plus DAG, respectively.

Cyclic AMP

All animal cells studied thus far use cAMP as a second messenger (chapter 45). When a signaling molecule binds to a GPCR that uses the enzyme **adenylyl cyclase** as an effector, a large

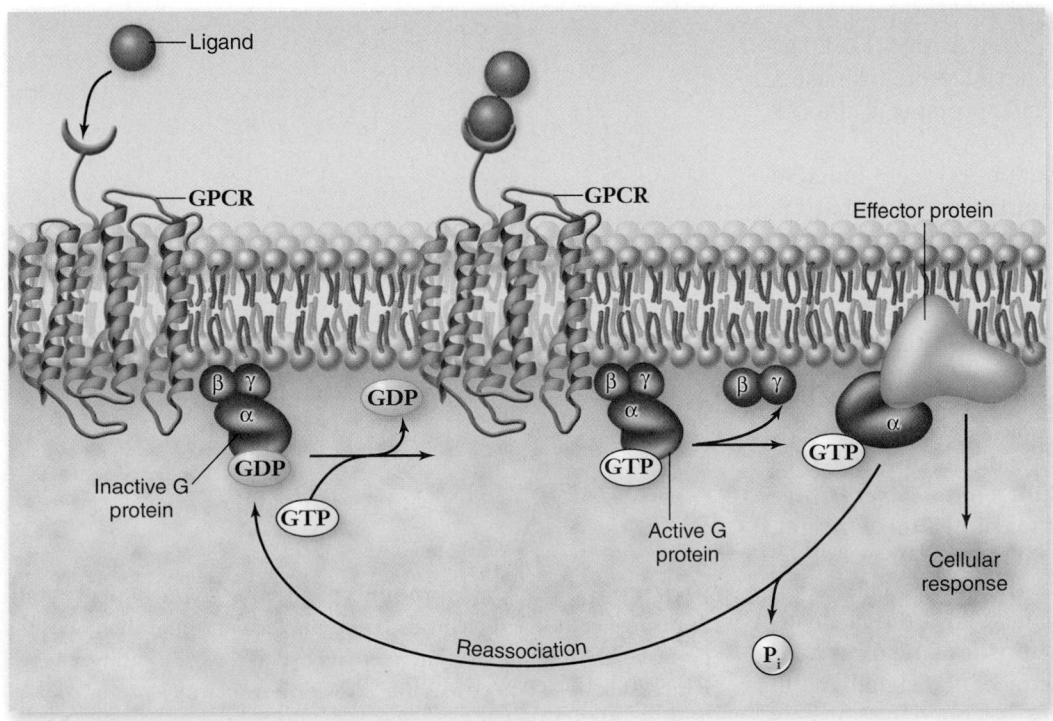

Figure 9.11 The action of G protein-coupled receptors. G protein-coupled receptors act through a heterotrimeric G protein that links the receptor to an effector protein. When ligand binds to the receptor, it activates an associated G protein, exchanging GDP for GTP. The active G protein complex dissociates into G_α and $G_{\beta\gamma}$. The G_α subunit (bound to GTP) is shown activating an effector protein. The effector protein may act directly on cellular proteins or produce a second messenger to cause a cellular response. G_α can hydrolyze GTP inactivating the system, then reassociate with $G_{\beta\gamma}$.

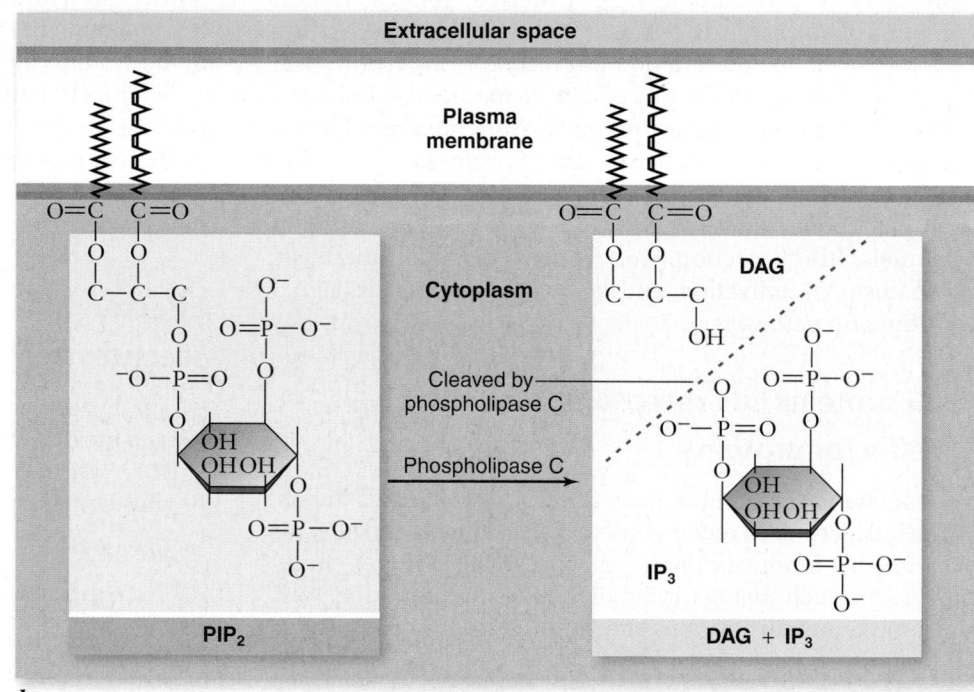

Figure 9.12 Production of second messengers. Second messengers are signaling molecules produced within the cell. *a.* The nucleotide ATP is converted by the enzyme adenylyl cyclase into cyclic AMP, or cAMP, and pyrophosphate (PP$_i$). *b.* The inositol phospholipid PIP$_2$ is composed of two lipids and a phosphate attached to glycerol. The phosphate is also attached to the sugar inositol. This molecule can be cleaved by the enzyme phospholipase C to produce two different second messengers: DAG, made up of the glycerol with the two lipids, and IP$_3$, inositol triphosphate.

amount of cAMP is produced within the cell (figure 9.12*a*). The cAMP then binds to and activates the enzyme protein kinase A (PKA), which adds phosphates to specific proteins in the cell (figure 9.13).

The effect of this phosphorylation on cell function depends on the identity of the cell and the proteins that are phosphorylated. In muscle cells, for example, PKA activates an enzyme necessary to break down glycogen and inhibits another enzyme necessary to synthesize glycogen. This leads to an increase in glucose available to the muscle. By contrast, in the kidney the action of PKA leads to the production of water channels that can increase the permeability of tubule cells to water.

Disruption of cAMP signaling can have a variety of effects. The symptoms of the disease cholera are due to altered cAMP levels in cells in the gut. The bacterium *Vibrio cholerae* produces a toxin that binds to a GPCR in the epithelium of the gut, causing it to be locked into an "on" state. This causes a large increase in intracellular cAMP that, in these cells, causes Cl⁻ ions to be transported out of the cell. Water follows the Cl⁻, leading to diarrhea and dehydration characteristic of the disease.

The molecule cAMP is also an extracellular signal. In the slime mold *Dictyostelium discoideum,* secreted cAMP acts as a signal for aggregation under conditions of starvation.

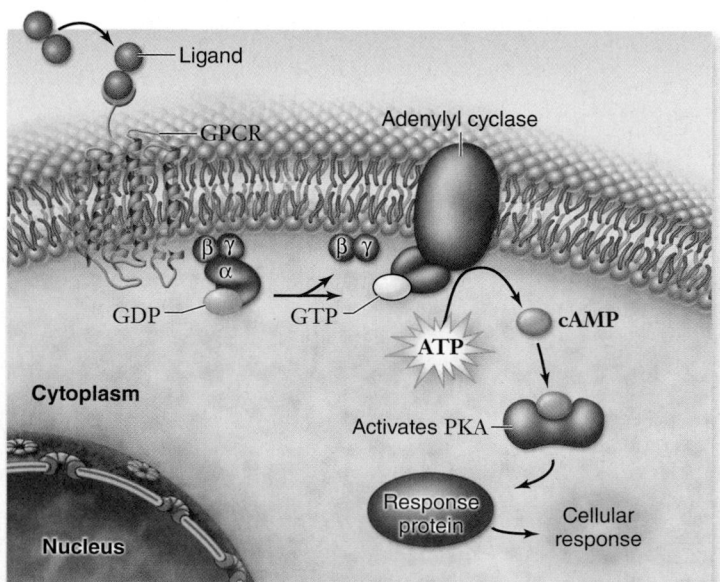

Figure 9.13 cAMP signaling pathway. Extracellular signal binds to a GPCR, activating a G protein. The G protein then activates the effector protein adenylyl cyclase, which catalyzes the conversion of ATP to cAMP. The cAMP then activates protein kinase A (PKA), which phosphorylates target proteins to cause a cellular response.

Hypothesis: *A previously identified G protein-coupled receptor is the cAMP receptor.*

Prediction: *If the function of the cAMP receptor is removed, then cells will not respond to starvation by aggregating.*

Test: *Use G protein-coupled receptor gene to direct synthesis of antisense RNA complementary to the normal mRNA. This will eliminate gene expression by the cellular copy of the G protein-coupled receptor.*

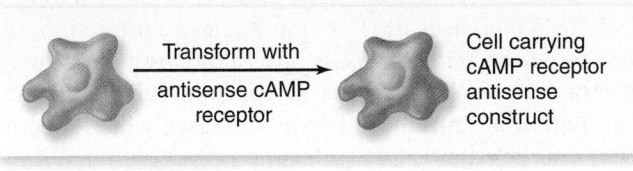

Transform with antisense cAMP receptor → Cell carrying cAMP receptor antisense construct

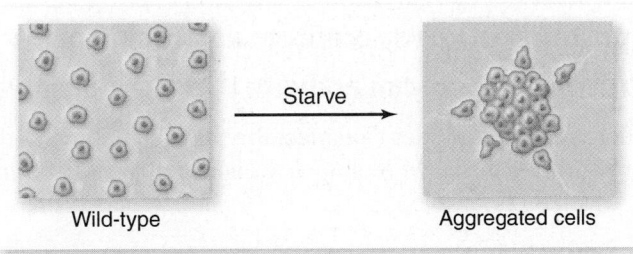

Wild-type — Starve → Aggregated cells

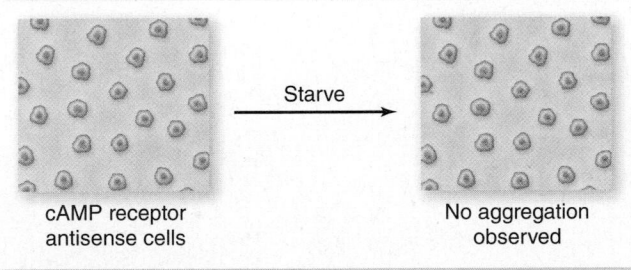

cAMP receptor antisense cells — Starve → No aggregation observed

Result: *Cells transformed with the antisense construct do not aggregate normally.*

Conclusion: *Previously identified G protein-coupled receptor is the cAMP receptor, which controls the aggregation response.*

Further Experiments: *How can this kind of experiment be used to unravel other aspects of this signaling system?*

Figure 9.14 The receptor for cAMP in *D. discoideum* is a GPCR.

Experiments have shown that the receptor for this signal is also a GPCR (figure 9.14).

Inositol phosphates

A common second messenger is produced from the molecules called inositol phospholipids. These are inserted into the plasma membrane by their lipid ends and have the *inositol phosphate* portion protruding into the cytoplasm. The most common inositol phospholipid is phosphatidylinositol-4,5-bisphosphate (PIP_2). This molecule is a substrate of the effector protein phospholipase C, which cleaves PIP_2 to yield **diacylglycerol (DAG)** and **inositol-1,4,5-trisphosphate (IP_3)** (see figure 9.12*b*).

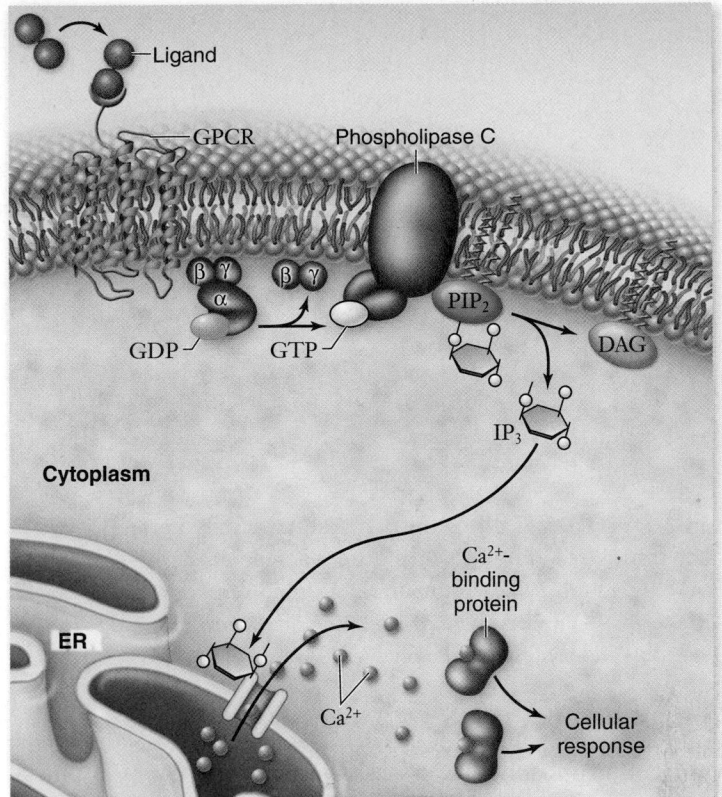

Figure 9.15 Inositol phospholipid and Ca²⁺ signaling. Extracellular signal binds to a GPCR activating a G protein. The G protein activates the effector protein phospholipase C, which converts PIP_2 to DAG and IP_3. IP_3 is then bound to a channel-linked receptor on the endoplasmic reticulum (ER) membrane, causing the ER to release stored Ca^{2+} into the cytoplasm. The Ca^{2+} then binds to Ca^{2+}-binding proteins such as calmodulin and PKC to cause a cellular response.

Both of these compounds then act as second messengers with a variety of cellular effects. DAG, like cAMP, can activate a protein kinase, in this case protein kinase C (PKC).

Calcium

Calcium ions (Ca^{2+}) serve widely as second messengers. Ca^{2+} levels inside the cytoplasm are normally very low (less than 10^{-7} M), whereas outside the cell and in the endoplasmic reticulum, Ca^{2+} levels are quite high (about 10^{-3} M). The endoplasmic reticulum has receptor proteins that act as ion channels to release Ca^{2+}. One of the most common of these receptors can bind the second messenger IP_3 to release Ca^{2+}, linking signaling through inositol phosphates with signaling by Ca^{2+} (figure 9.15).

The result of the outflow of Ca^{2+} from the endoplasmic reticulum depends on the cell type. For example, in skeletal muscle cells Ca^{2+} stimulates muscle contraction but in endocrine cells it stimulates the secretion of hormones.

Ca^{2+} initiates some cellular responses by binding to *calmodulin,* a 148-amino-acid cytoplasmic protein that contains four binding sites for Ca^{2+} (figure 9.16). When four Ca^{2+} ions are bound to calmodulin, the calmodulin/Ca^{2+} complex is able to bind to other proteins to activate them. These proteins include

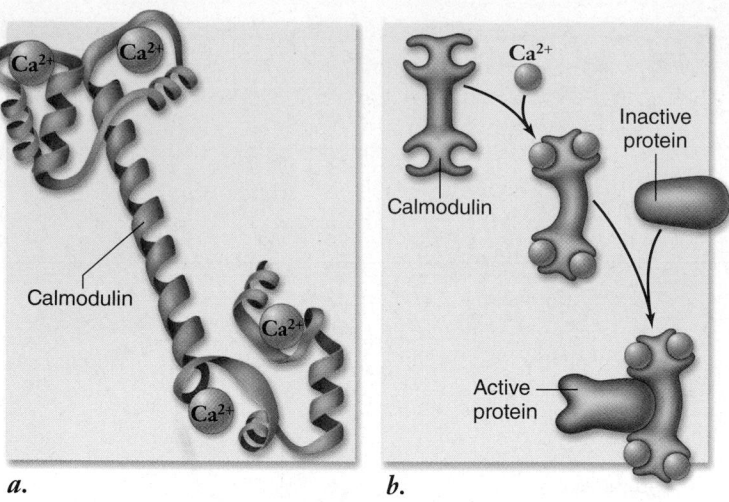

a.
Ca²⁺
Ca²⁺
Calmodulin
Ca²⁺
Ca²⁺
Calmodulin

b.
Ca²⁺
Calmodulin
Inactive protein
Active protein

Figure 9.16 Calmodulin. *a.* Calmodulin is a protein containing 148 amino acid residues that mediates Ca²⁺ function. *b.* When four Ca²⁺ are bound to the calmodulin molecule, it undergoes a conformational change that allows it to bind to other cytoplasmic proteins and effect cellular responses.

protein kinases, ion channels, receptor proteins, and cyclic nucleotide phosphodiesterases. These many uses of Ca²⁺ make it one of the most versatile second messengers in cells.

Different receptors can produce the same second messengers

As mentioned previously, the two hormones glucagon and epinephrine can both stimulate liver cells to mobilize glucose. The reason that these different signals have the same effect is that they both act by the same signal transduction pathway to stimulate the breakdown and inhibit the synthesis of glycogen.

The binding of either hormone to its receptor activates a G protein that simulates adenylyl cyclase. The production of cAMP leads to the activation of PKA, which in turn activates another protein kinase called phosphorylase kinase. Activated phosphorylase kinase then activates glycogen phosphorylase, which cleaves off units of glucose 6-phosphate from the glycogen polymer (figure 9.17). The action of multiple kinases again leads to amplification such that a few signaling molecules result in a large number of glucose molecules being released.

At the same time, PKA also phosphorylates the enzyme glycogen synthase, but in this case it inhibits the enzyme, thus preventing the synthesis of glycogen. In addition, PKA phosphorylates other proteins that activate the expression of genes encoding the enzymes needed to synthesize glucose. This convergence of signal transduction pathways from different receptors leads to the same result—glucose is mobilized.

Receptor subtypes can lead to different effects in different cells

We also saw earlier how a single signaling molecule, epinephrine, can have different effects in different cells. One way this happens is through the existence of multiple forms of the same

receptor. The receptor for epinephrine actually has nine different subtypes, or isoforms. These are encoded by different genes and are actually different receptor molecules. The sequences of these proteins are very similar, especially in the ligand-binding domain, which allows them to bind epinephrine. They differ mainly in their cytoplasmic domains, which interact with G proteins. This leads to different isoforms activating different G proteins, thereby leading to different signal transduction pathways.

Thus, in the heart, muscle cells have one isoform of the receptor that, when bound to epinephrine, activates a G protein that activates adenylyl cyclase, leading to increased cAMP. This increases the rate and force of contraction. In the intestine, smooth muscle cells have a different isoform of the receptor that, when bound to epinephrine, activates a different G protein that inhibits adenylyl cyclase, which decreases cAMP. This has the result of relaxing the muscle.

G protein-coupled receptors and receptor tyrosine kinases can activate the same pathways

Different receptor types can affect the same signaling module. For example, RTKs were shown to activate the MAP kinase

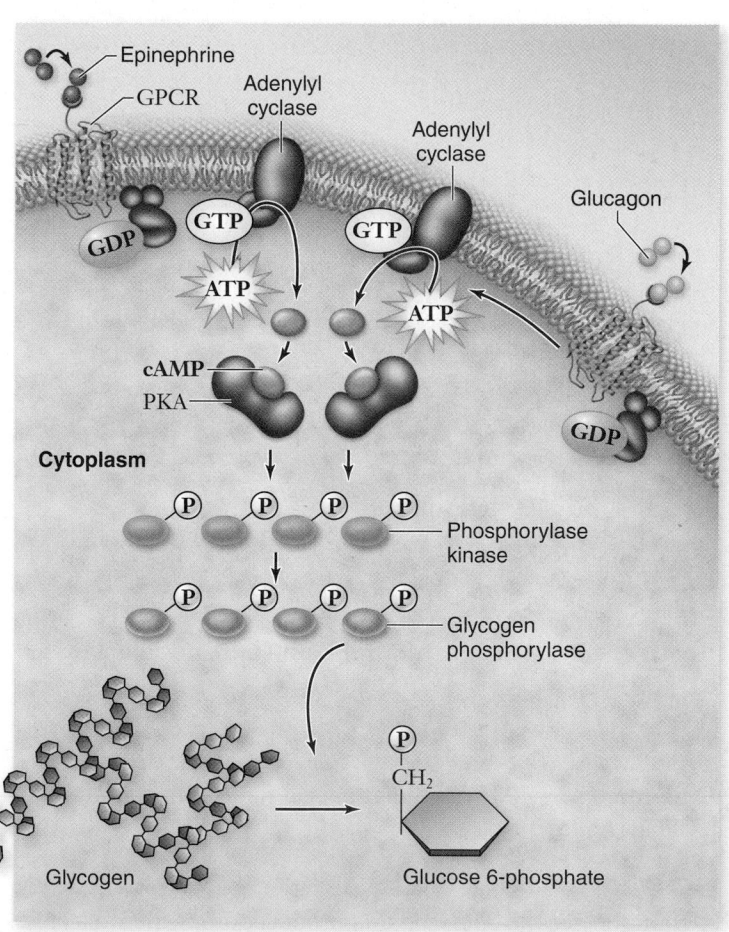

Figure 9.17 Different receptors can activate the same signaling pathway. The hormones glucagon and epinephrine both act through GPCRs. Each of these receptors acts via a G protein that activates adenylyl cyclase, producing cAMP. The activation of PKA begins a kinase cascade that leads to the breakdown of glycogen.

cascade, but GPCRs can also activate this same cascade. Similarly, the activation of phospholipase C was mentioned previously in the context of GPCR signaling, but it can also be activated by RTKs.

This cross-reactivity may appear to introduce complications into cell function, but in fact it provides the cell with an incredible amount of flexibility. Cells have a large, but limited number of intracellular signaling modules, which can be turned on and off by different kinds of membrane receptors. This leads to signaling networks that interconnect possible cellular effectors with multiple incoming signals.

The Internet represents an example of a network in which many different kinds of computers are connected globally. This network can be broken down into subnetworks that are connected to the overall network. Because of the nature of the connections, when you send an e-mail message across the Internet, it can reach its destination through many different pathways. Likewise, the cell has interconnected networks of signaling pathways in which many different signals, receptors, and response proteins are interconnected. Specific pathways

like the MAP kinase cascade, or signaling through second messengers like cAMP and Ca^{2+}, represent subnetworks within the global signaling network. A specific signal can activate different pathways in different cells, or different signals can activate the same pathway. We do not yet understand the cell at this level, but the field of systems biology is moving toward such global understanding of cell function.

Learning Outcomes Review 9.5

Signaling through GPCRs uses a three-part system—a receptor, a G protein, and an effector protein. G proteins are active when bound to GTP and inactive when bound to GDP. A ligand binding to the receptor activates the G protein, which then activates the effector protein. Effector proteins include adenylyl cyclase, which produces the second messenger cAMP. Another effector protein, phospholipase C, cleaves the inositol phosphates and results in the release of Ca^{2+} from the ER.

■ *There are far more GPCRs than any other receptor type. What is a possible explanation for this?*

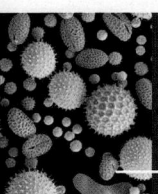

Chapter Review

9.1 Overview of Cell Communication (figure 9.1)

Cell communication requires signal molecules, called ligands, binding to specific receptor proteins producing a cellular response.

Signaling is defined by the distance from source to receptor (figure 9.2).

Direct contact—molecules on the plasma membrane of one cell contact the receptor molecules on an adjacent cell.

Paracrine signaling—short-lived signal molecules are released into the extracellular fluid and influence neighboring cells.

Endocrine signaling—long-lived hormones enter the circulatory system and are carried to target cells some distance away.

Synaptic signaling—short-lived neurotransmitters are released by neurons into the gap, called a synapse, between nerves and target cells.

Signal transduction pathways lead to cellular responses.

Intracellular events initiated by a signaling event are called signal transduction.

Phosphorylation is key in control of protein function.

Proteins can be controlled by phosphate added by kinase and removed by phosphatase enzymes.

9.2 Receptor Types (figure 9.4)

Receptors are defined by location.

Receptors are broadly defined as intracellular or cell-surface receptors (membrane receptors).

Membrane receptors are transmembrane proteins that transfer information across the membrane, but not the signal molecule.

Membrane receptors include three subclasses.

Channel-linked receptors are chemically gated ion channels that allow specific ions to pass through a central pore.

Enzymatic receptors are enzymes activated by binding a ligand; these enzymes are usually protein kinases.

G protein-coupled receptors interact with G proteins that control the function of effector proteins: enzymes or ion channels.

Membrane receptors can generate second messengers.

Some enzymatic and most G protein-coupled receptors produce second messengers, to relay messages in the cytoplasm.

9.3 Intracellular Receptors (figure 9.5)

Many cell signals are lipid-soluble and readily pass through the plasma membrane and bind to receptors in the cytoplasm or nucleus.

Steroid hormone receptors affect gene expression.

Steroid hormones bind cytoplasmic receptors, then are transported to the nucleus. Thus, they are called nuclear receptors. These can directly affect gene expression, usually activating transcription of the genes they control.

Nuclear receptors have three functional domains: hormone-binding, DNA-binding, and transcription-activating domains.

Ligand binding changes receptor shape, releasing an inhibitor occupying the DNA-binding site.

A cell's response to a lipid-soluble signal depends on the hormone–receptor complex and the other protein coactivators present.

Other intracellular receptors act as enzymes.

9.4 Signal Transduction Through Receptor Kinases

Receptor kinases in plants and animals recognize hydrophilic ligands and influence the cell cycle, cell migration, cell metabolism, and cell proliferation.

Because they are involved in growth control, alterations of receptor kinases and their signaling pathways can lead to cancer.

RTKs are activated by autophosphorylation.
The activated receptor can also phosphorylate other intracellular proteins.

Phosphotyrosine domains mediate protein–protein interactions.
Adapter proteins can bind to phosphotyrosine and act as links between the receptors and downstream signaling events.

Protein kinase cascades can amplify a signal.

Scaffold proteins organize kinase cascades.
Scaffold proteins and protein kinases form a single complex where the enzymes act sequentially and are optimally functional.

Internalized receptors are degraded or recycled.

Ras is a small G protein that acts as a molecular switch.
Small G proteins act as molecular switches linking external signals to signal transduction pathways.

RTKs are inactivated by internalization.

9.5 Signal Transduction Through G Protein-Coupled Receptors (figure 9.11)

G protein-coupled receptors function through activation of G proteins.

G proteins link receptors with effector proteins.
G proteins are active bound to GTP and inactive bound to GDP. Receptors promote exchange of GDP for GTP.

The activated G protein dissociates into two parts, G_α and $G_{\beta\gamma}$, each of which can act on effector proteins.

G_α also hydrolyzes GTP to GDP to inactivate the G protein.

Effector proteins produce multiple second messengers.
Two common effector proteins are adenylyl cyclase and phospholipase C, which produce second messengers known as cAMP, and DAG and IP_3, respectively.

Ca^{2+} is also a second messenger. Ca^{2+} release is triggered by IP_3 binding to channel-linked receptors in the ER.

Ca^{2+} can bind to a cytoplasmic protein calmodulin, which in turn activates other proteins, producing a variety of responses.

Different receptors can produce the same second messengers.
Different GPCR receptors can converge to activate the same effector enzyme and thus produce the same second messenger.

Receptor subtypes can lead to different effects in different cells.
Epinephrine causes increased contraction in heart muscle but relaxation in smooth muscle.

G protein-coupled receptors and receptor tyrosine kinases can activate the same pathways.
Both RTKs and GPCRs can activate MAP kinase cascades.

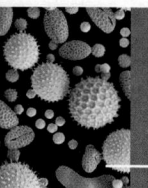

Review Questions

UNDERSTAND

1. Paracrine signaling is characterized by ligands that are
   a. produced by the cell itself.
   b. secreted by neighboring cells.
   c. present on the plasma membrane of neighboring cells.
   d. secreted by distant cells.

2. Signal transduction pathways
   a. are necessary for signals to cross the membrane.
   b. include the intracellular events stimulated by an extracellular signal.
   c. include the extracellular events stimulated by an intracellular signal.
   d. are only found in cases where the signal can cross the membrane.

3. The function of a _____ is to add phosphates to proteins, whereas a _____ functions to remove the phosphates.
   a. tyrosine; serine
   b. protein phosphatase; protein dephosphatase
   c. protein kinase; protein phosphatase
   d. receptor; ligand

4. Which of the following receptor types is not a membrane receptor?
   a. Channel-linked receptor
   b. Enzymatic receptor
   c. G protein-coupled receptor
   d. Steroid hormone receptors

5. How does the function of an intracellular receptor differ from that of a membrane receptor?
   a. The intracellular receptor binds a ligand.
   b. The intracellular receptor binds DNA.
   c. The intracellular receptor activates a kinase.
   d. The intracellular receptor functions as a second messenger.

6. Signaling through receptor tyrosine kinases often
   a. leads to the production of the second messenger cAMP.
   b. leads to the production of the second messenger IP_3.
   c. stimulates gene expression directly.
   d. leads to the activation of a cascade of kinase enzymes.

7. What is the function of Ras during tyrosine kinase cell signaling?

    a. It activates the opening of channel-linked receptors.
    b. It is an enzyme that synthesizes second messengers.
    c. It links the receptor protein to the MAP kinase pathway.
    d. It phosphorylates other enzymes as part of a pathway.

8. Which of the following best describes the immediate effect of ligand binding to a G protein-coupled receptor?

    a. The G protein trimer releases a GDP and binds a GTP.
    b. The G protein trimer dissociates from the receptor.
    c. The G protein trimer interacts with an effector protein.
    d. The α subunit of the G protein becomes phosphorylated.

APPLY

1. The action of steroid hormones is often longer-lived than that of peptide hormones. This is because they

    a. enter the cell and act like enzymes for a longer period of time.
    b. they turn on gene expression to produce proteins that persist in the cell.
    c. result in the production of second messengers that act directly on cellular processes.
    d. stimulate G proteins that act directly on cellular processes.

2. The ion Ca^{2+} can act as a second messenger because it is

    a. produced by the enzyme calcium synthase.
    b. normally at a high level in the cytoplasm.
    c. normally at a low level in the cytoplasm.
    d. stored in the cytoplasm.

3. Different receptors can have the same effect on a cell. One reason for this is that

    a. most receptors produce the same second messenger.
    b. different isoforms of receptors bind different ligands, but stimulate the same signaling pathway.
    c. signal transduction pathways intersect: the same pathway can be stimulated by different receptors.
    d. all receptors converge on the same signal transduction pathways.

4. In comparing small G proteins like Ras and GPCR proteins, we can say that

    a. both proteins have intrinsic GTPase activity that stops signaling.
    b. both proteins are active bound to GTP.
    c. Ras is active bound to GDP and GPCRs are active bound to GTP.
    d. both a and b are true.

5. The same signal can have different effects in different cells because there

    a. are different receptor subtypes that initiate different signal transduction pathways.
    b. may be different coactivators in different cells.
    c. may be different target proteins in different cells' signal transduction pathways.
    d. All of the choices are correct.

6. The receptors for steroid hormones and peptide hormones are fundamentally different because

    a. of the great difference in size of the molecule.
    b. peptides are one of the four major polymers and steroids are simple ringed structures.
    c. peptides are hydrophilic and steroids are hydrophobic.
    d. peptides are hydrophobic and steroids are hydrophilic.

SYNTHESIZE

1. Describe the common features found in all examples of cellular signaling discussed in this chapter. Provide examples to illustrate your answer.

2. The sheet of cells that form the gut epithelium folds into peaks called villi and valleys called crypts. The cells within the crypt region secrete a protein, Netrin-1, that becomes concentrated within the crypts. Netrin-1 is the ligand for a receptor protein that is found on the surface of all gut epithelial cells. Netrin-1 binding triggers a signal pathway that promotes cell growth. Gut epithelial cells undergo apoptosis (cell death) in the absence of Netrin-1 ligand binding.

    a. How would you characterize the type of signaling (autocrine, paracrine, endocrine) found in this system?
    b. Predict where the greatest amount of cell growth and cell death would occur in the epithelium.
    c. The loss of the Netrin-1 receptor is associated with some types of colon cancer. Suggest an explanation for the link between this signaling pathway and tumor formation.

ONLINE RESOURCE

www.ravenbiology.com

Understand, Apply, and Synthesize—enhance your study with animations that bring concepts to life and practice tests to assess your understanding. Your instructor may also recommend the interactive eBook, individualized learning tools, and more.

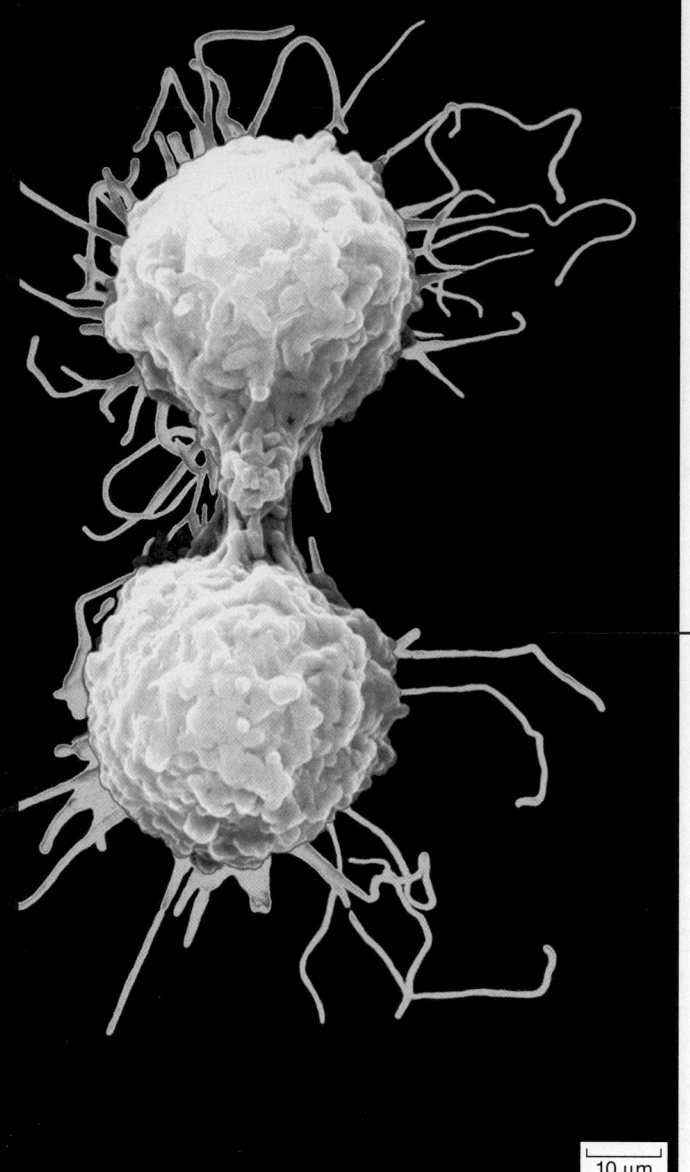

10 μm

Chapter **10**

How Cells Divide

Chapter Contents

Introduction

All species of organisms—bacteria, alligators, the weeds in a lawn—grow and reproduce. From the smallest creature to the largest, all species produce offspring like themselves and pass on the hereditary information that makes them what they are. In this chapter, we examine how cells, like the white blood cell shown in the figure, divide and reproduce. Cell division is necessary for the growth of organisms, for wound healing, and to replace cells that are lost regularly, such as those in your skin and in the lining of your gut. The mechanism of cell reproduction and its biological consequences have changed significantly during the evolution of life on Earth. The process is complex in eukaryotes, involving both the replication of chromosomes and their separation into daughter cells. Much of what we are learning about the causes of cancer relates to how cells control this process, and in particular their tendency to divide, a mechanism that in broad outline remains the same in all eukaryotes.

Bacterial Cell Division

Learning Outcome

1. Describe the process of binary fission.

Bacteria divide as a way of reproducing themselves. Although bacteria exchange DNA, they do not have a sexual cycle like eukaryotes. Thus all growth in a bacterial population is due to division to produce new cells. The reproduction of bacteria is clonal—that is, each cell produced by cell division is an identical copy of the original cell.

Binary fission is a simple form of cell division

Cell division in both bacterial and eukaryotic cells produces two new cells with the same genetic information as the original. Despite the differences in these cell types, the essentials of the process are the same: duplication and segregation of genetic information into daughter cells, and division of cellular contents. We begin by looking at the simpler process, **binary fission,** which occurs in bacteria.

Most bacteria have a genome made up of a single, circular DNA molecule. In spite of its apparent simplicity, the DNA molecule of the bacterium *Escherichia coli* is actually on the order of 500 times longer than the cell itself! Thus, this "simple" structure is actually packaged very tightly to fit into the cell. Although not found in a nucleus, the DNA is located in a region called the *nucleoid* that is distinct from the cytoplasm around it.

The compaction and organization of the nucleoid involves a class of proteins called structural maintenance of chromosome, or SMC, proteins. These are ancient proteins that have diversified over evolutionary time to fulfill a variety of roles related to DNA organization in different lineages. In eukaryotes the cohesin and condensin proteins discussed later in the chapter are SMC proteins.

During binary fission, the chromosome is replicated, and the two products are partitioned to each end of the cell prior to the actual division of the cell. One key feature of bacterial cell division is that replication and partitioning of the chromosome occur as a concerted process. In contrast, DNA replication in eukaryotic cells occurs early in division, and chromosome separation occurs much later.

Proteins control chromosome separation and septum formation

Binary fission begins with the replication of the bacterial DNA at a specific site—the origin of replication (see chapter 14)—and proceeds both directions around the circular DNA to a specific site of termination (figure 10.1). The cell grows by elongation, and division occurs roughly at midcell. For many years, it was thought that newly replicated *E. coli* DNA molecules were passively segregated by attachment to and growth of the membrane as the cell elongated. Experiments that follow the movement of the origin of replication show that it is at

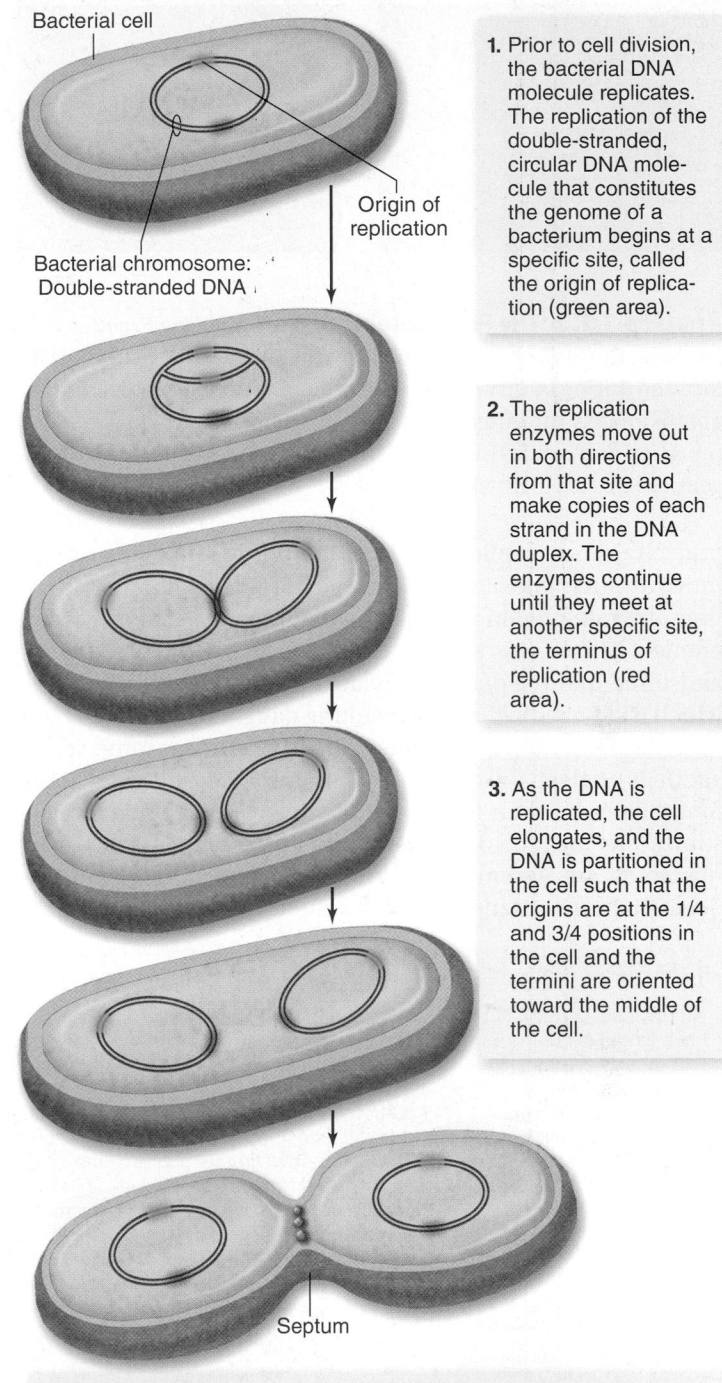

Bacterial cell

Bacterial chromosome: Double-stranded DNA

Origin of replication

1. Prior to cell division, the bacterial DNA molecule replicates. The replication of the double-stranded, circular DNA molecule that constitutes the genome of a bacterium begins at a specific site, called the origin of replication (green area).

2. The replication enzymes move out in both directions from that site and make copies of each strand in the DNA duplex. The enzymes continue until they meet at another specific site, the terminus of replication (red area).

3. As the DNA is replicated, the cell elongates, and the DNA is partitioned in the cell such that the origins are at the 1/4 and 3/4 positions in the cell and the termini are oriented toward the middle of the cell.

Septum

4. Septation then begins, in which new membrane and cell wall material begin to grow and form a septum at approximately the midpoint of the cell. A protein molecule called FtsZ (orange dots) facilitates this process.

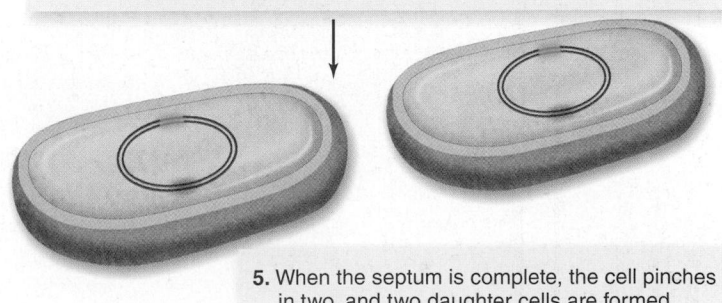

5. When the septum is complete, the cell pinches in two, and two daughter cells are formed, each containing a bacterial DNA molecule.

Figure 10.1 Binary fission.

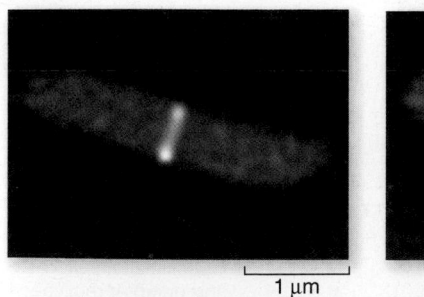 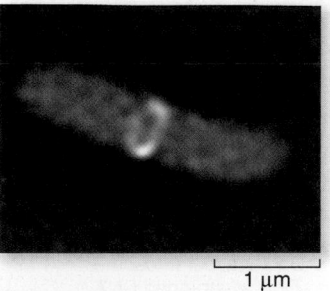

1 μm   1 μm

Figure 10.2 The FtsZ protein. In these dividing *E. coli* bacteria, the FtsZ protein is labeled with fluorescent dye to show its location during binary fission. The protein assembles into a ring at approximately the midpoint of the cell, where it facilitates septation and cell division. Bacteria carrying mutations in the *FtsZ* gene are unable to divide.

midcell prior to replication, then the newly replicated origins move toward opposite ends of the cell. This movement is faster than the rate of elongation, showing that growth alone is not enough. The origins appear to be captured at the one quarter and three quarter positions relative to the length of the cell, which will be midcell of the resulting daughter cells.

Although the actual mechanism of chromosome segregation is unclear, the order of events is not. During replication, first the origin, then the rest of the newly replicated chromosomes are moved to opposite ends of the cell as two new nucleoids are assembled. The final event of replication is decatenation (untangling) of the final replication products.

After replication and segregation, the midcell region is cleared of daughter nucleoids, and division occurs. The force behind chromosome segregation has been attributed to DNA replication itself, transcription, and the polymerization of actin-like molecules. At this point, no single model appears to explain the process, and it may involve more than one.

The cell's other components are partitioned by the growth of new membrane and production of the **septum** (see figure 10.1). This process, termed **septation,** usually occurs at the midpoint of the cell. It begins with the formation of a ring composed of many copies of the protein FtsZ (figure 10.2). Next, accumulation of a number of other proteins occurs, including ones embedded in the membrane. This structure contracts inward radially until the cell pinches off into two new cells. The midcell location of the FtsZ ring is caused by an oscillation between the two poles of an inhibitor of FtsZ formation.

The FtsZ protein is found in most prokaryotes, including archaea. It can form filaments and rings, and recent three-dimensional crystals show a high degree of similarity to eukaryotic tubulin. However, its role in bacterial division is quite different from the role of tubulin in mitosis in eukaryotes.

The evolution of eukaryotic cells included much more complex genomes composed of multiple linear chromosomes housed in a membrane-bounded nucleus. These complex genomes may be possible due to the evolution of mechanisms that delay chromosome separation after replication. Although it is unclear how this ability to keep chromosomes together evolved, it does seem more closely related to binary fission than we once thought (figure 10.3).

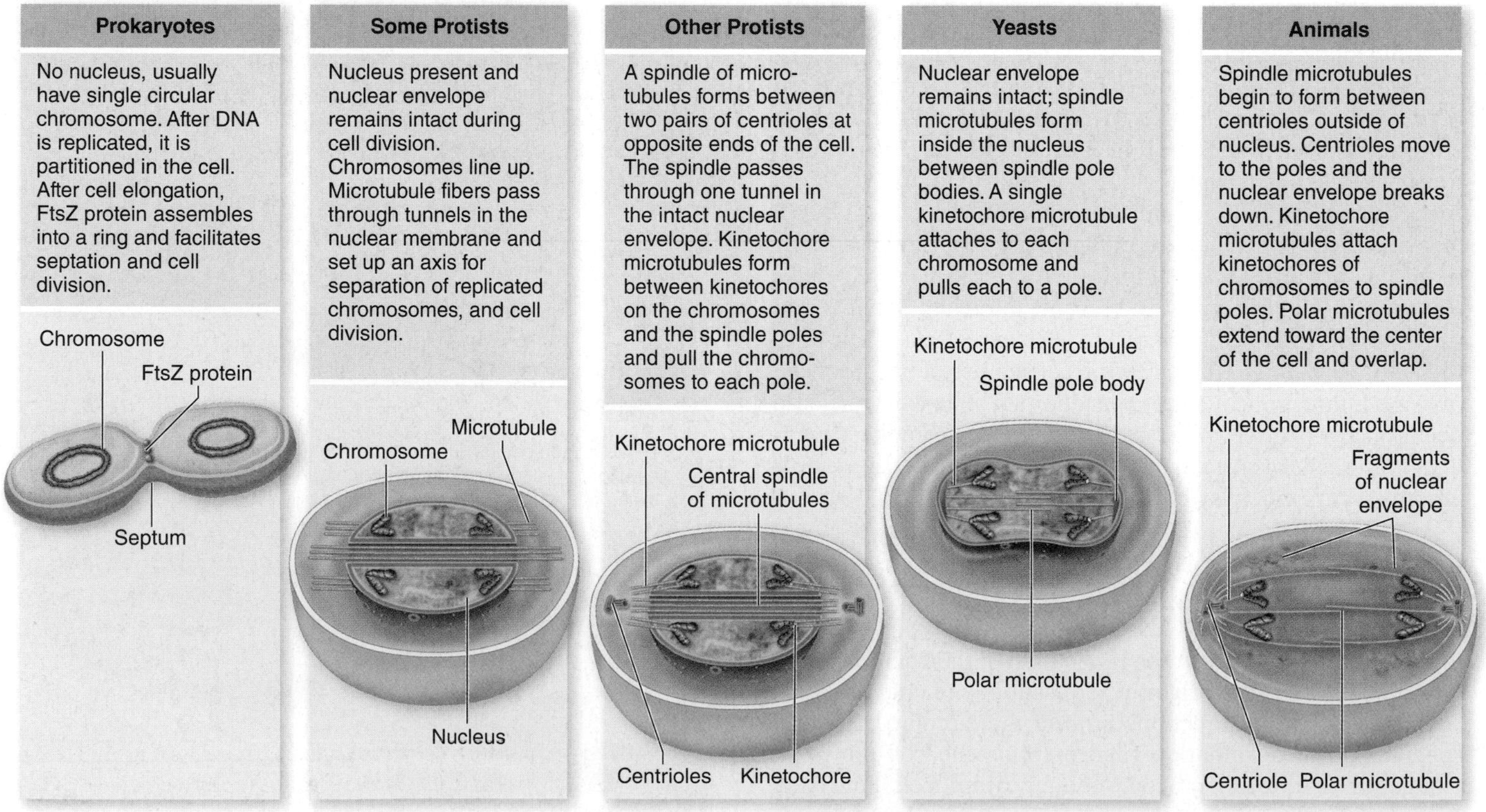

Prokaryotes	Some Protists	Other Protists	Yeasts	Animals
No nucleus, usually have single circular chromosome. After DNA is replicated, it is partitioned in the cell. After cell elongation, FtsZ protein assembles into a ring and facilitates septation and cell division.	Nucleus present and nuclear envelope remains intact during cell division. Chromosomes line up. Microtubule fibers pass through tunnels in the nuclear membrane and set up an axis for separation of replicated chromosomes, and cell division.	A spindle of microtubules forms between two pairs of centrioles at opposite ends of the cell. The spindle passes through one tunnel in the intact nuclear envelope. Kinetochore microtubules form between kinetochores on the chromosomes and the spindle poles and pull the chromosomes to each pole.	Nuclear envelope remains intact; spindle microtubules form inside the nucleus between spindle pole bodies. A single kinetochore microtubule attaches to each chromosome and pulls each to a pole.	Spindle microtubules begin to form between centrioles outside of nucleus. Centrioles move to the poles and the nuclear envelope breaks down. Kinetochore microtubules attach kinetochores of chromosomes to spindle poles. Polar microtubules extend toward the center of the cell and overlap.

Figure 10.3 A comparison of protein assemblies during cell division among different organisms. The prokaryotic protein FtsZ has a structure that is similar to that of the eukaryotic protein tubulin. Tubulin is the protein component of microtubules, which are fibers that eukaryotic cells use to construct the spindle apparatus that is used to separate chromosomes.

10.2 *Eukaryotic Chromosomes*

Learning Outcomes

1. Describe the structure of eukaryotic chromosomes.
2. Distinguish between homologues and sister chromatids.
3. Contrast replicated and nonreplicated chromosomes.

TABLE 10.1	Chromosome Number in Selected Eukaryotes
Group	**Total Number of Chromosomes**
F U N G I	
Neurospora (haploid)	7
Saccharomyces (a yeast)	16
I N S E C T S	
Mosquito	6
Drosophila	8
Honeybee	diploid females 32, haploid males 16
Silkworm	56
P L A N T S	
Haplopappus gracilis	2
Garden pea	14
Corn	20
Bread wheat	42
Sugarcane	80
Horsetail	216
Adder's tongue fern	1262
V E R T E B R A T E S	
Opossum	22
Frog	26
Mouse	40
Human	46
Chimpanzee	48
Horse	64
Chicken	78
Dog	78

Chromosomes were first observed by the German embryologist Walther Flemming (1843–1905) in 1879, while he was examining the rapidly dividing cells of salamander larvae. When Flemming looked at the cells through what would now be a rather primitive light microscope, he saw minute threads within their nuclei that appeared to be dividing lengthwise. Flemming called their division **mitosis,** based on the Greek word *mitos,* meaning "thread."

Chromosome number varies among species

Since their initial discovery, chromosomes have been found in the cells of all eukaryotes examined. Their number may vary enormously from one species to another. A few kinds of organisms have only a single pair of chromosomes, whereas some ferns have more than 500 pairs (table 10.1). Most eukaryotes have between 10 and 50 chromosomes in their body cells.

Human cells each have 46 chromosomes, consisting of 23 nearly identical pairs (figure 10.4). Each of these 46 chromosomes contains hundreds or thousands of genes that play important roles in determining how a person's body develops and functions. Human embryos missing even one chromosome, a condition called *monosomy,* do not survive in most cases. Having an extra copy of any one chromosome, a condition called *trisomy,* is usually fatal except where the smallest chromosomes are involved. (You'll learn more about human chromosome abnormalities in chapter 13.)

Eukaryotic chromosomes exhibit complex structure

Researchers have learned a great deal about chromosome structure and composition in the more than 125 years since

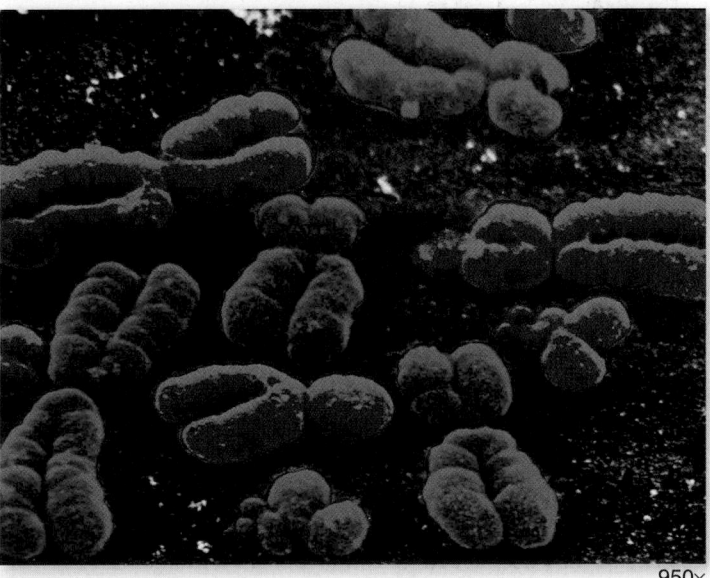

950×

Figure 10.4 Human chromosomes. This scanning electron micrograph shows human chromosomes as they appear immediately before nuclear division. Each DNA molecule has already replicated, forming identical copies held together at a visible constriction called the centromere. False color has been added to the chromosomes.

their discovery. But despite intense research, the exact structure of eukaryotic chromosomes during the cell cycle remains unclear. The structures described in this chapter represent the currently accepted model.

Composition of chromatin

Chromosomes are composed of **chromatin,** a complex of DNA and protein; most chromosomes are about 40% DNA and 60% protein. A significant amount of RNA is also associated with chromosomes because chromosomes are the sites of RNA synthesis.

Each chromosome contains a single DNA molecule that runs uninterrupted through the chromosome's entire length. A typical human chromosome contains about 140 million (1.4×10^8) nucleotides in its DNA. If we think of each nucleotide as a "word," then the amount of information an average chromosome contains would fill about 280 printed books of 1000 pages each, with 500 "words" per page.

If we could lay out the strand of DNA from a single chromosome in a straight line, it would be about 5 cm (2 in.) long. Fitting such a strand into a cell nucleus is like cramming a string the length of a football field into a baseball—and that's only 1 of 46 chromosomes! In the cell, however, the DNA is compacted, allowing it to fit into a much smaller space than would otherwise be possible.

The organization of chromatin in the nondividing nucleus is not well understood, but geneticists have recognized for years that some domains of chromatin, called **heterochromatin,** are not expressed, and other domains of chromatin, called **euchromatin,** are expressed. This genetically measurable state is also related to the physical state of chromatin, although researchers are just beginning to see the details.

Chromosome structure

If we gently disrupt a eukaryotic nucleus and examine the DNA with an electron microscope, we find that it resembles a string of beads (figure 10.5). Every 200 nucleotides (nt), the DNA duplex (double strand) is coiled around a core of eight **histone proteins.** Unlike most proteins, which have an overall negative charge, histones are positively charged because of an abundance of the basic amino acids arginine and lysine. Thus, they are strongly attracted to the negatively charged phosphate groups of the DNA, and the histone cores act as "magnetic forms" that promote and guide the coiling of the DNA. The complex of DNA and histone proteins is termed a **nucleosome.**

The DNA wrapped in nucleosomes is further coiled into an even more compact structure called the *solenoid*. The precise path of this higher order folding of chromatin is still a subject of debate, but it leads to a fiber with a diameter of 30 nm

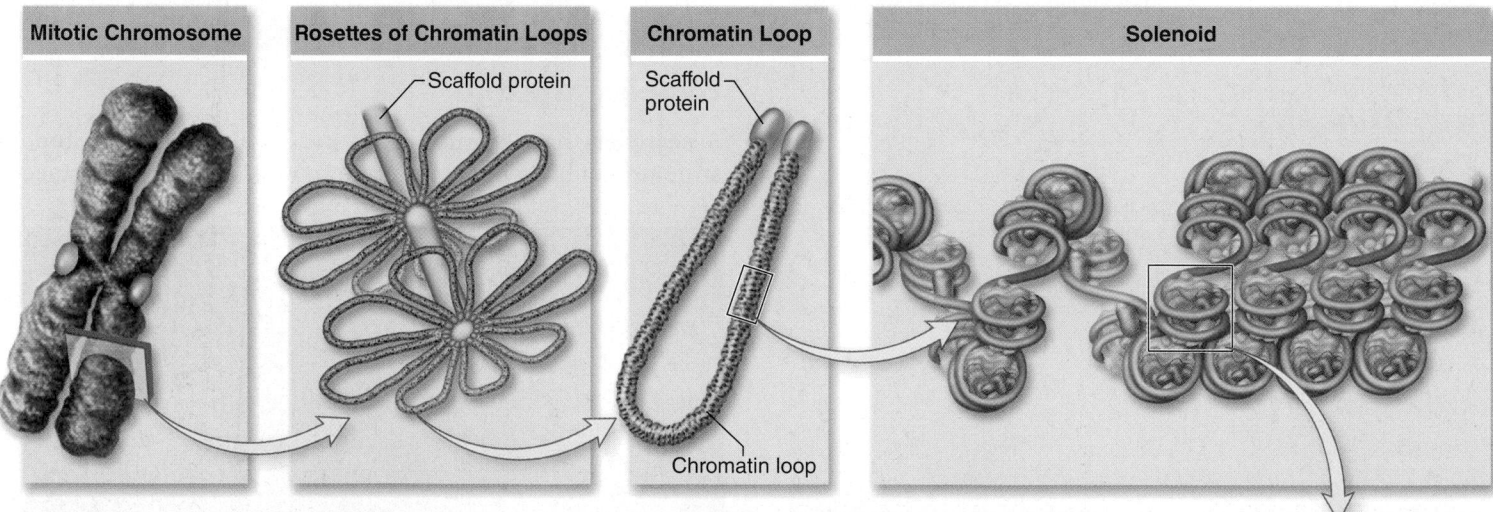

Mitotic Chromosome

Rosettes of Chromatin Loops
Scaffold protein

Chromatin Loop
Scaffold protein
Chromatin loop

Solenoid

Figure 10.5 Levels of eukaryotic chromosomal organization. Each chromosome consists of a long double-stranded DNA molecule. These strands require further packaging to fit into the cell nucleus. The DNA duplex is tightly bound to and wound around proteins called histones. The DNA-wrapped histones are called nucleosomes. The nucleosomes are further coiled into the solenoid. This solenoid is then organized into looped domains. The precise organization of mitotic chromosomes is unknown, but the solenoid is further condensed around a preexisting scaffolding of proteins. The arrangement shown is one of many possibilities.

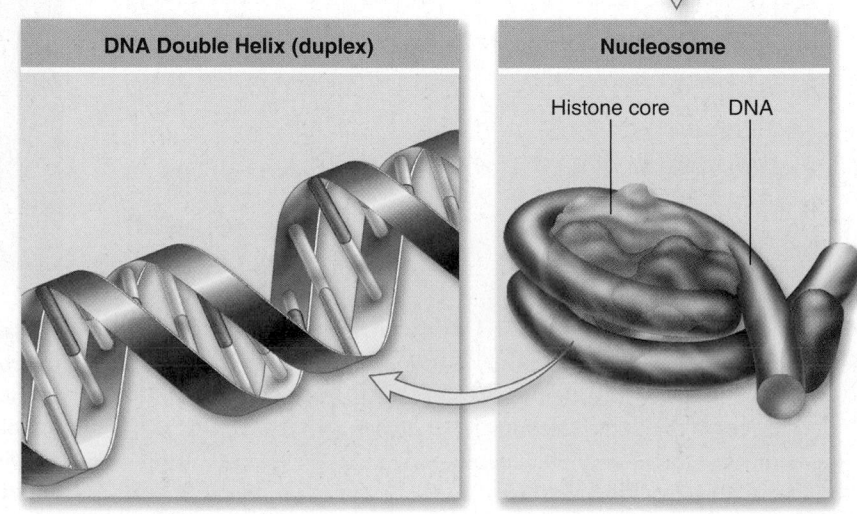

DNA Double Helix (duplex)

Nucleosome
Histone core   DNA

and thus is called the 30-nm fiber. This 30-nm fiber is the usual state of interphase (nondividing) chromatin.

During mitosis, a scaffold of proteins is assembled that allows the organization of even more compact chromosomes, which can be more readily separated by the mitotic machinery described later. The exact nature of the compaction is unknown, but one longstanding model involves looping of chromatin fibers from the scaffold like the fibers on a wire brush. This process is aided by a complex of proteins called condensin, which are evolutionarily related to the bacterial SMC that compact the nucleoid.

Chromosome karyotypes

Chromosomes vary in size, staining properties, the location of the centromere (a constriction found on all chromosomes, described shortly), the relative length of the two arms on either side of the centromere, and the positions of constricted regions along the arms. The particular array of chromosomes an individual organism possesses is called its **karyotype.** The karyotype in figure 10.6 shows the set of chromosomes from a normal human cell.

When defining the number of different chromosomes in a species, geneticists count the **haploid (n)** number of chromosomes. This refers to one complete set of chromosomes necessary to define an organism. For humans and many other species, the total number of chromosomes in a cell is called the

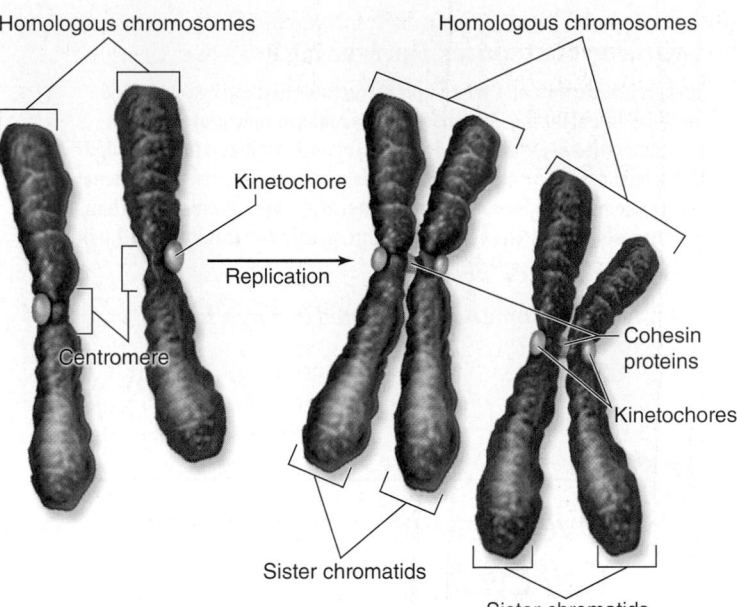

Figure 10.7 The difference between homologous chromosomes and sister chromatids. Homologous chromosomes are the maternal and paternal copies of the same chromosome—say, chromosome number 16. Sister chromatids are the two replicas of a single chromosome held together at their centromeres by cohesin proteins after DNA replication. The kinetochore (described later in the chapter) is composed of proteins found at the centromere that attach to microtubules during mitosis.

diploid (2n) number, which is twice the haploid number. For humans, the haploid number is 23 and the diploid number is 46. Diploid chromosomes reflect the equal genetic contribution that each parent makes to offspring. We refer to the maternal and paternal chromosomes as being **homologous,** and each one of the pair is termed a **homologue.**

Chromosome replication

Chromosomes as seen in a karyotype are only present for a brief period during cell division. Prior to replicating, each chromosome is composed of a single DNA molecule that is arranged into the 30-nm fiber described earlier. After replication, each chromosome is composed of two identical DNA molecules held together by a complex of proteins called **cohesins.** As the chromosomes become more condensed and arranged about the protein scaffold, they become visible as two strands that are held together at the centromere. At this point, we still call this one chromosome, but it is composed of two **sister chromatids** (figure 10.7).

The fact that the products of replication are held together is critical to the division process. One problem that a cell must solve is how to ensure that each new cell receives a complete set of chromosomes. If we were designing a system, we might use some kind of label to identify each chromosome, much like most of us use when we duplicate files on a computer. Instead of labeling chromosomes, the cell glues replication products together at the centromere. The process of mitosis then separates all of these copies at the same time, ensuring that each daughter cell gets one copy of each chromosome.

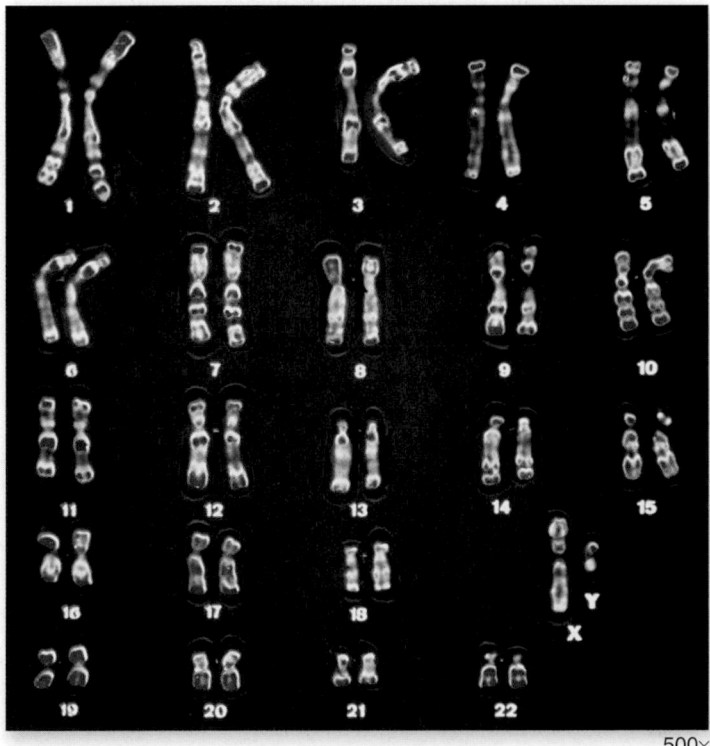

500×

Figure 10.6 A human karyotype. The individual chromosomes that make up the 23 pairs differ widely in size and in centromere position. In this preparation of a male karyotype, the chromosomes have been specifically stained to indicate differences in their composition and to distinguish them clearly from one another. Notice that members of a chromosome pair are very similar but not identical.

10.3 Overview of the Eukaryotic Cell Cycle

Compared with prokaryotes, the increased size and more complex organization of eukaryotic genomes required radical changes in the partitioning of replicated genomes into daughter cells. The **cell cycle** requires the duplication of the genome, its accurate segregation, and the division of cellular contents.

The cell cycle is divided into five phases

The cell cycle is divided into phases based on the key events of genome duplication and segregation. The cell cycle is usually diagrammed as in figure 10.8.

- **G_1 (gap phase 1)** is the primary growth phase of the cell. The term *gap phase* refers to its filling the gap between cytokinesis and DNA synthesis. For most cells, this is the longest phase.
- **S (synthesis)** is the phase in which the cell synthesizes a replica of the genome.
- **G_2 (gap phase 2)** is the second growth phase, and preparation for separation of the newly replicated genome. This phase fills the gap between DNA synthesis and the beginning of mitosis. During this phase microtubules begin to reorganize to form a spindle.

  G_1, S, and G_2 together constitute **interphase,** the portion of the cell cycle between cell divisions.
- **Mitosis** is the phase of the cell cycle in which the spindle apparatus assembles, binds to the chromosomes, and moves the sister chromatids apart. Mitosis is the essential step in the separation of the two daughter genomes. It is traditionally subdivided into five stages: prophase, prometaphase, metaphase, anaphase, and telophase.
- **Cytokinesis** is the phase of the cell cycle when the cytoplasm divides, creating two daughter cells. In animal

cells, the microtubule spindle helps position a contracting ring of actin that constricts like a drawstring to pinch the cell in two. In cells with a cell wall, such as plant cells, a plate forms between the dividing cells.

Mitosis and cytokinesis together are usually referred to collectively as M phase, to distinguish the dividing phase from interphase.

The duration of the cell cycle varies depending on cell type

The time it takes to complete a cell cycle varies greatly. Cells in animal embryos can complete their cell cycle in under 20 min; the shortest known animal nuclear division cycles occur in fruit fly embryos (8 min). These cells simply divide their nuclei as quickly as they can replicate their DNA, without cell growth. Half of their cycle is taken up by S, half by M, and essentially none by G_1 or G_2.

Because mature cells require time to grow, most of their cycles are much longer than those of embryonic tissue. Typically, a dividing mammalian cell completes its cell cycle in about 24 hr, but some cells, such as certain cells in the human liver, have cell cycles lasting more than a year. During the cycle, growth occurs throughout the G_1 and G_2 phases, as well as during the S phase. The M phase takes only about an hour, a small fraction of the entire cycle.

Most of the variation in the length of the cell cycle between organisms or cell types occurs in the G_1 phase. Cells often pause in G_1 before DNA replication and enter a resting state called the **G_0 phase;** cells may remain in this phase for

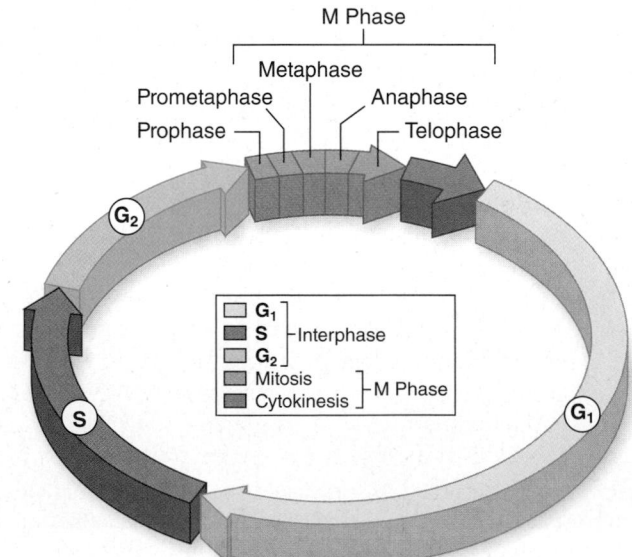

Figure 10.8 The cell cycle. The cell cycle is depicted as a circle. The first gap phase, G_1, involves growth and preparation for DNA synthesis. During S phase, a copy of the genome is synthesized. The second gap phase, G_2, prepares the cell for mitosis. During mitosis, replicated chromosomes are partitioned. Cytokinesis divides the cell into two cells with identical genomes.

days to years before resuming cell division. At any given time, most of the cells in an animal's body are in G_0 phase. Some, such as muscle and nerve cells, remain there permanently; others, such as liver cells, can resume G_1 phase in response to factors released during injury.

Learning Outcome Review 10.3

Cell division in eukaryotes is a complex process that involves five phases: a first gap phase (G_1); a DNA synthesis phase (S); a second gap phase (G_2); mitosis (M), during which chromatids are separated; and cytokinesis in which a cell becomes two separate cells.

- **When during the cycle is a cell irreversibly committed to dividing?**

10.4 Interphase: Preparation for Mitosis

Learning Outcomes

1. Describe the events that take place during interphase.
2. Illustrate the connection between sister chromatids after S phase.

The events that occur during interphase—the G_1, S, and G_2 phases—are very important for the successful completion of mitosis. During G_1, cells undergo the major portion of their growth. During the S phase, each chromosome replicates to produce two sister chromatids, which remain attached to each other at the centromere. In the G_2 phase, the chromosomes coil even more tightly.

The **centromere** is a point of constriction on the chromosome containing repeated DNA sequences that bind specific proteins. These proteins make up a disklike structure called the **kinetochore.** This disk functions as an attachment site for microtubules necessary to separate the chromosomes during cell division (figure 10.9). As seen in figure 10.6, each chromosome's centromere is located at a characteristic site along the length of the chromosome.

After the S phase, the sister chromatids appear to share a common centromere, but at the molecular level the DNA of the centromere has actually already replicated, so there are two complete DNA molecules. This means that two chromatids are held together by cohesin proteins at the centromere, and each chromatid has its own set of kinetochore proteins (figure 10.10). In multicellular animals, most of the cohesins that hold sister chromatids together after replication appear to be replaced by condensin as the chromosomes are condensed. This leaves the chromosomes still attached tightly at the centromere, but loosely attached elsewhere.

The cell grows throughout interphase. The G_1 and G_2 segments of interphase are periods of active growth, during which proteins are synthesized and cell organelles are

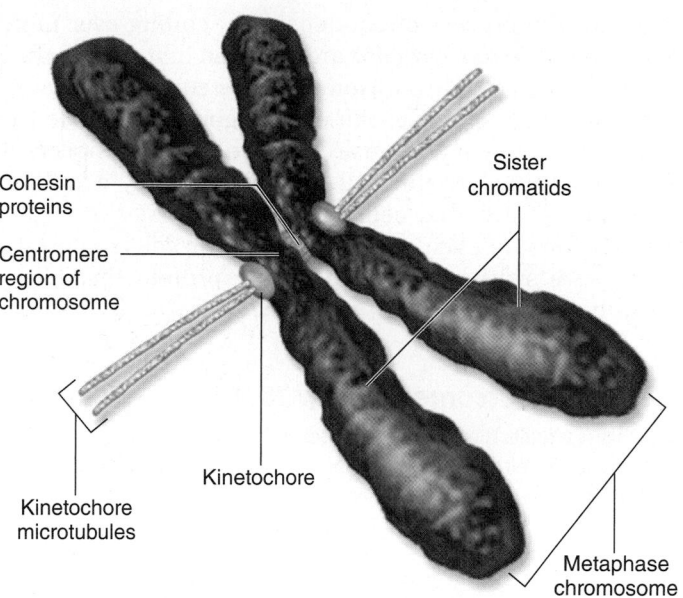

Figure 10.9 Kinetochores. Separation of sister chromatids during mitosis depends on microtubules attaching to proteins found in the kinetochore. These kinetochore proteins are assembled on the centromere of chromosomes. The centromeres of the two sister chromatids are held together by cohesin proteins.

produced. The cell's DNA replicates only during the S phase of the cell cycle.

After the chromosomes have replicated in S phase, they remain fully extended and uncoiled, although cohesin proteins are associated with them at this stage. In G_2 phase,

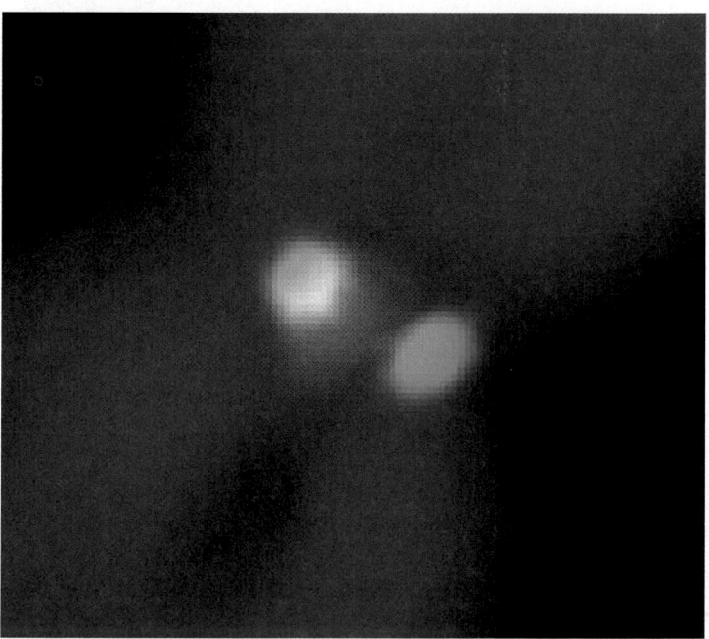

.25 μm

Figure 10.10 Proteins found at the centromere. In this image DNA, a cohesin protein, and a kinetochore protein have all been labeled with a different-colored fluorescent dye. Cohesin *(red)*, which holds centromeres together, lies between the sister chromatids *(blue)*. Each sister chromatid has its own separate kinetochore *(green)*.

they begin the process of condensation, coiling ever more tightly. Special *motor proteins* are involved in the rapid final condensation of the chromosomes that occurs early in mitosis. Also during G_2 phase, the cells begin to assemble the machinery they will later use to move the chromosomes to opposite poles of the cell. In animal cells, a pair of microtubule-organizing centers called *centrioles* replicate, producing one for each pole. All eukaryotic cells undertake an extensive synthesis of **tubulin,** the protein that forms microtubules.

Learning Outcomes Review 10.4

Interphase includes the G_1, S, and G_2 phases of the cell cycle. During interphase, the cell grows; replicates chromosomes, organelles, and centrioles; and synthesizes components needed for mitosis, including tubulin. Cohesin proteins hold chromatids together at the centromere of each chromosome.

■ *How would a mutation that deleted cohesin proteins affect cell division?*

10.5 M Phase: Chromosome Segregation and the Division of Cytoplasmic Contents

Learning Outcomes

1. **Describe the phases of mitosis.**
2. **Explain the importance of metaphase.**
3. **Compare cytokinesis in plants and animals.**

The process of mitosis is one of the most dramatic and beautiful biological processes that we can easily observe. In our attempts to understand this process, we have divided it into discrete phases but it should always be remembered that this is a dynamic, continuous process, not a set of discrete steps. This process is shown both schematically and in micrographs in figure 10.11.

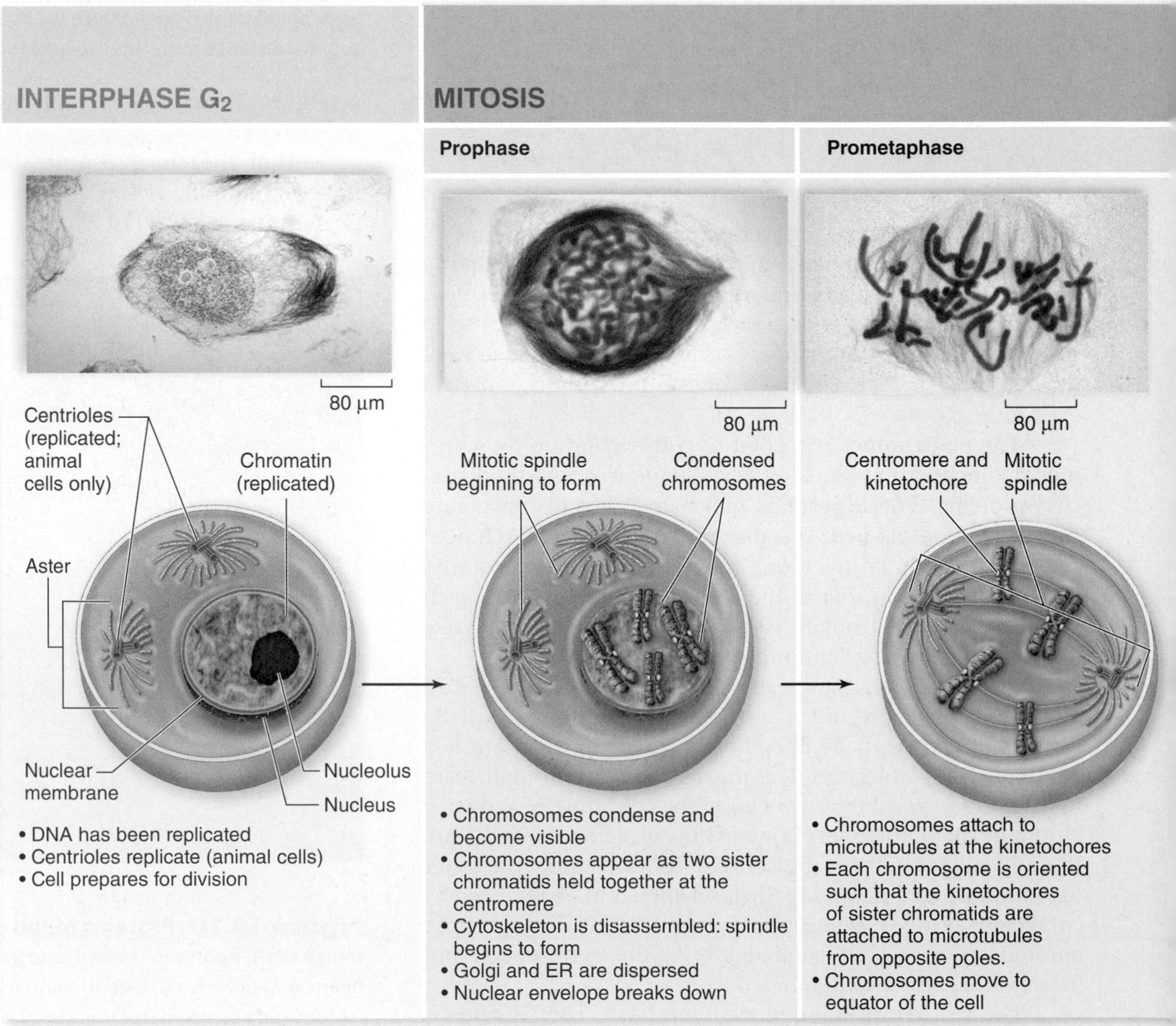

Figure 10.11 Mitosis and cytokinesis.
Mitosis is conventionally divided into five stages—prophase, prometaphase, metaphase, anaphase, and telophase—which together act to separate duplicated chromosomes. This is followed by cytokinesis, which divides the cell into two separate cells. Photos depict mitosis and cytokinesis in a plant, the African blood lily (*Haemanthus katharinae*), with chromosomes stained blue and microtubules stained red. Drawings depict mitosis and cytokinesis in animal cells.

INTERPHASE G_2

80 μm

Centrioles (replicated; animal cells only)
Chromatin (replicated)
Aster
Nuclear membrane
Nucleolus
Nucleus

- DNA has been replicated
- Centrioles replicate (animal cells)
- Cell prepares for division

MITOSIS

Prophase

80 μm

Mitotic spindle beginning to form
Condensed chromosomes

- Chromosomes condense and become visible
- Chromosomes appear as two sister chromatids held together at the centromere
- Cytoskeleton is disassembled: spindle begins to form
- Golgi and ER are dispersed
- Nuclear envelope breaks down

Prometaphase

80 μm

Centromere and kinetochore
Mitotic spindle

- Chromosomes attach to microtubules at the kinetochores
- Each chromosome is oriented such that the kinetochores of sister chromatids are attached to microtubules from opposite poles.
- Chromosomes move to equator of the cell

During prophase, the mitotic apparatus forms

When the chromosome condensation initiated in G_2 phase reaches the point at which individual condensed chromosomes first become visible with the light microscope, the first stage of mitosis, **prophase,** has begun. The condensation process continues throughout prophase; consequently, chromosomes that start prophase as minute threads appear quite bulky before its conclusion. Ribosomal RNA synthesis ceases when the portion of the chromosome bearing the rRNA genes is condensed.

The spindle and centrioles

The assembly of the **spindle** apparatus that will later separate the sister chromatids occurs during prophase. The normal microtubule structure in the cell disassembled in the G_2 phase is replaced by the spindle. In animal cells, the two centriole pairs formed during G_2 phase begin to move apart early in prophase, forming between them an axis of microtubules referred to as spindle fibers. By the time the centrioles reach the opposite poles of the cell, they have established a bridge of microtubules, called the spindle apparatus, between them. In plant cells, a similar bridge of microtubular fibers forms between opposite poles of the cell, although centrioles are absent in plant cells.

In animal cell mitosis, the centrioles extend a radial array of microtubules toward the nearby plasma membrane when they reach the poles of the cell. This arrangement of microtubules is called an **aster.** Although the aster's function is not fully understood, it probably braces the centrioles against the membrane and stiffens the point of microtubular attachment during the retraction of the spindle. Plant cells, which have rigid cell walls, do not form asters.

Breakdown of the nuclear envelope

During the formation of the spindle apparatus, the nuclear envelope breaks down, and the endoplasmic reticulum reabsorbs its components. At this point, the microtubular spindle fibers extend completely across the cell, from one pole to the other. Their orientation determines the plane in which the cell will subsequently divide, through the center of the cell at right angles to the spindle apparatus.

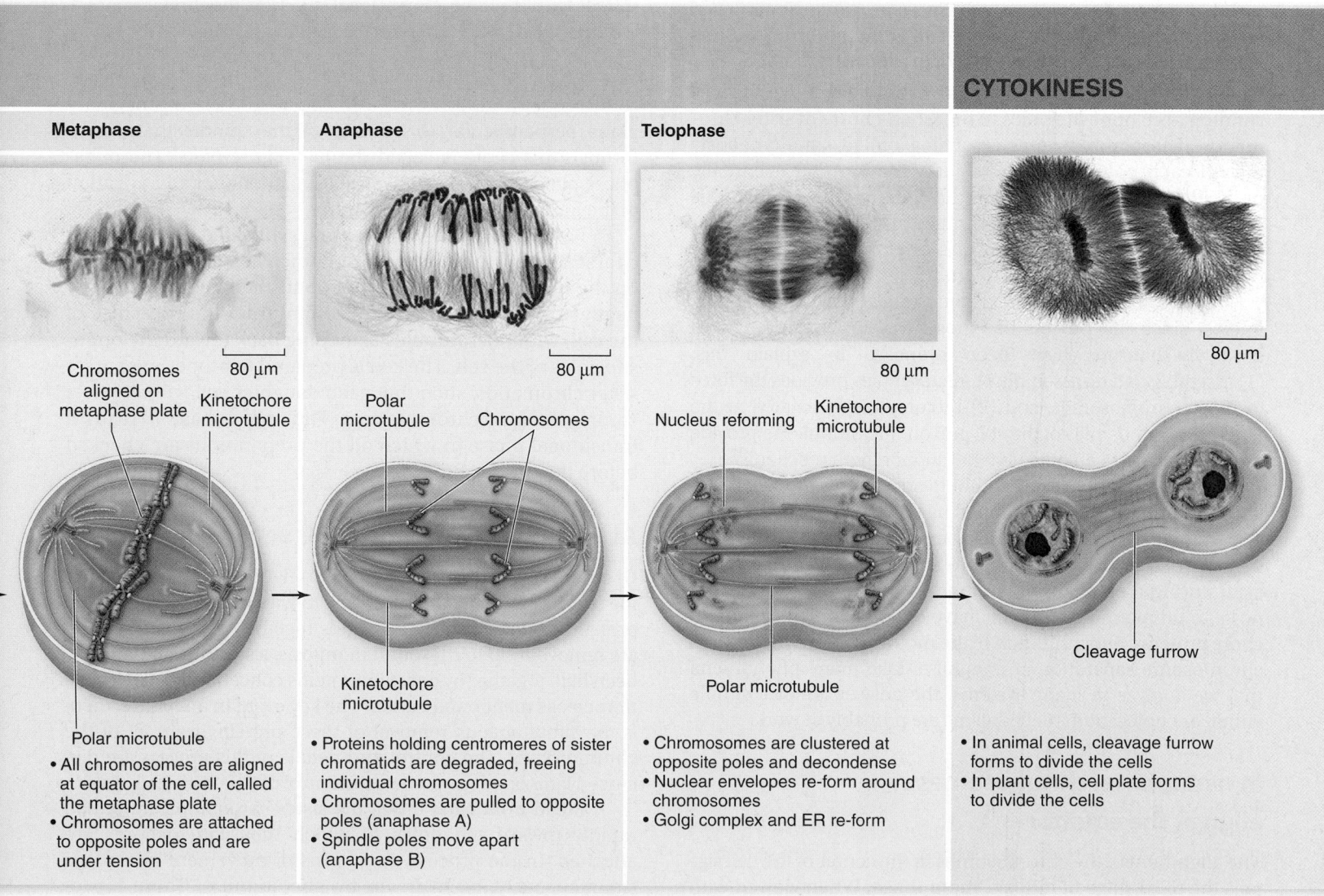

CYTOKINESIS

Metaphase

80 μm

Chromosomes aligned on metaphase plate
Kinetochore microtubule

Polar microtubule

- All chromosomes are aligned at equator of the cell, called the metaphase plate
- Chromosomes are attached to opposite poles and are under tension

Anaphase

80 μm

Polar microtubule
Chromosomes

Kinetochore microtubule

- Proteins holding centromeres of sister chromatids are degraded, freeing individual chromosomes
- Chromosomes are pulled to opposite poles (anaphase A)
- Spindle poles move apart (anaphase B)

Telophase

80 μm

Nucleus reforming
Kinetochore microtubule

Polar microtubule

- Chromosomes are clustered at opposite poles and decondense
- Nuclear envelopes re-form around chromosomes
- Golgi complex and ER re-form

80 μm

Cleavage furrow

- In animal cells, cleavage furrow forms to divide the cells
- In plant cells, cell plate forms to divide the cells

During prometaphase, chromosomes attach to the spindle

The transition from prophase to **prometaphase** occurs following the disassembly of the nuclear envelope. During prometaphase the condensed chromosomes become attached to the spindle by their kinetochores. Each chromosome possesses two kinetochores, one attached to the centromere region of each sister chromatid (see figure 10.9).

Microtubule attachment

As prometaphase continues, a second group of microtubules grow from the poles of the cell toward the centromeres. These microtubules are captured by the kinetochores on each pair of sister chromatids. This results in the kinetochores of each sister chromatid being connected to opposite poles of the spindle.

This bipolar attachment is critical to the process of mitosis; any mistakes in microtubule positioning can be disastrous. For example, the attachment of the kinetochores of both sister chromatids to the same pole leads to a failure of sister chromatid separation, and they will be pulled to the same pole ending up in the same daughter cell, with the other daughter cell missing that chromosome.

Movement of chromosomes to the cell center

Each chromosome is attached to the spindle by microtubules running from opposite poles to the kinetochores of sister chromatids. The chromosomes are being pulled simultaneously toward each pole, leading to a jerky motion that eventually pulls all of the chromosomes to the equator of the cell. At this point, the chromosomes are arranged at the equator, with sister chromatids under tension and oriented toward opposite poles by their kinetochore microtubules.

The force that moves chromosomes has been of great interest since the process of mitosis was first observed. Two basic mechanisms have been proposed to explain this: (1) assembly and disassembly of microtubules provides the force to move chromosomes, and (2) motor proteins located at the kinetochore and poles of the cell pull on microtubules to provide force. Data have been obtained that support both mechanisms.

In support of the microtubule-shortening proposal, isolated chromosomes can be pulled by microtubule disassembly. The spindle is a very dynamic structure, with microtubules being added to at the kinetochore and shortened at the poles, even during metaphase. In support of the motor protein proposal, multiple motor proteins have been identified as kinetochore proteins, and inhibition of the motor protein dynein slows chromosome separation at anaphase. Like many phenomena that we analyze in living systems, the answer is not a simple either–or choice; both mechanisms are probably at work.

In metaphase, chromosomes align at the equator

The alignment of the chromosomes in the center of the cell signals the third stage of mitosis, **metaphase.** When viewed with a light microscope, the chromosomes appear to array themselves in a circle along the inner circumference of the cell, just

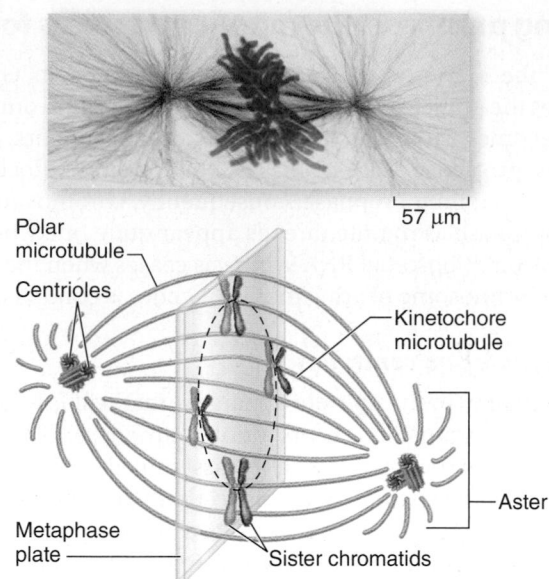

57 μm

Polar microtubule

Centrioles

Kinetochore microtubule

Aster

Metaphase plate

Sister chromatids

Figure 10.12   Metaphase. In metaphase, the chromosomes are arrayed at the midpoint of the cell. The imaginary plane through the equator of the cell is called the metaphase plate. As the spindle itself is a three-dimensional structure, the chromosomes are arrayed in a rough circle on the metaphase plate.

as the equator girdles the Earth (figure 10.12). An imaginary plane perpendicular to the axis of the spindle that passes through this circle is called the *metaphase plate*. The metaphase plate is not an actual structure, but rather an indication of the future axis of cell division.

Positioned by the microtubules attached to the kinetochores of their centromeres, all of the chromosomes line up on the metaphase plate. At this point their centromeres are neatly arrayed in a circle, equidistant from the two poles of the cell, with microtubules extending back toward the opposite poles of the cell. The cell is prepared to properly separate sister chromatids, such that each daughter cell will receive a complete set of chromosomes. Thus metaphase is really a transitional phase in which all the preparations are checked before the action continues.

At anaphase, the chromatids separate

Of all the stages of mitosis, shown in figure 10.11, **anaphase** is the shortest and the most amazing to watch. It begins when the proteins holding sister chromatids together at the centromere are removed. Up to this point in mitosis, sister chromatids have been held together by cohesin proteins concentrated at the centromere, as mentioned earlier. The key event in anaphase, then, is the simultaneous removal of these proteins from all of the chromosomes. The control and details of this process are discussed later on in the context of control of the entire cell cycle.

Freed from each other, the sister chromatids are pulled rapidly toward the poles to which their kinetochores are attached. In the process, two forms of movement take place simultaneously, each driven by microtubules. These movements are often called anaphase A and anaphase B to distinguish them.

First, during anaphase A, the *kinetochores are pulled toward the poles* as the microtubules that connect them to the poles shorten. This shortening process is not a contraction; the microtubules do not get any thicker. Instead, tubulin subunits are removed from the kinetochore ends of the microtubules. As more subunits are removed, the chromatid-bearing microtubules are progressively disassembled, and the chromatids are pulled ever closer to the poles of the cell.

Second, during anaphase B, the *poles move apart* as microtubular spindle fibers physically anchored to opposite poles slide past each other, away from the center of the cell (figure 10.13). Because another group of microtubules attach the chromosomes to the poles, the chromosomes move apart, too. If a flexible membrane surrounds the cell, it becomes visibly elongated.

When the sister chromatids separate in anaphase, the accurate partitioning of the replicated genome—the essential element of mitosis—is complete.

During telophase, the nucleus re-forms

In **telophase,** the spindle apparatus disassembles as the microtubules are broken down into tubulin monomers that can be used to construct the cytoskeletons of the daughter cells. A nuclear envelope forms around each set of sister chromatids, which can now be called chromosomes because they are no longer attached at the centromere. The chromosomes soon begin to uncoil into the more extended form that permits gene expression. One of the early group of genes expressed after mitosis is complete are the rRNA genes, resulting in the reappearance of the nucleolus.

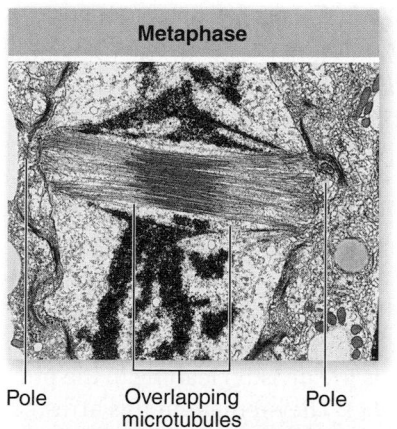

Metaphase

Pole — Overlapping microtubules — Pole

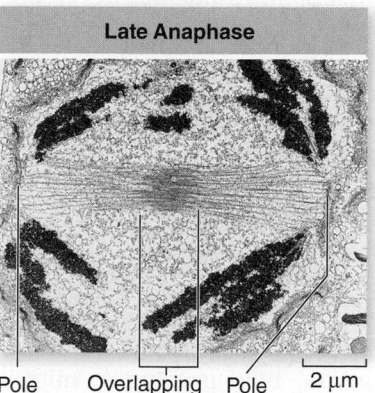

Late Anaphase

Pole — Overlapping microtubules — Pole — 2 μm

Figure 10.13 Microtubules slide past each other as the chromosomes separate. In these electron micrographs of dividing diatoms, the overlap of the microtubules lessens markedly during spindle elongation as the cell passes from metaphase to anaphase. During anaphase B the poles move farther apart as the chromosomes move toward the poles.

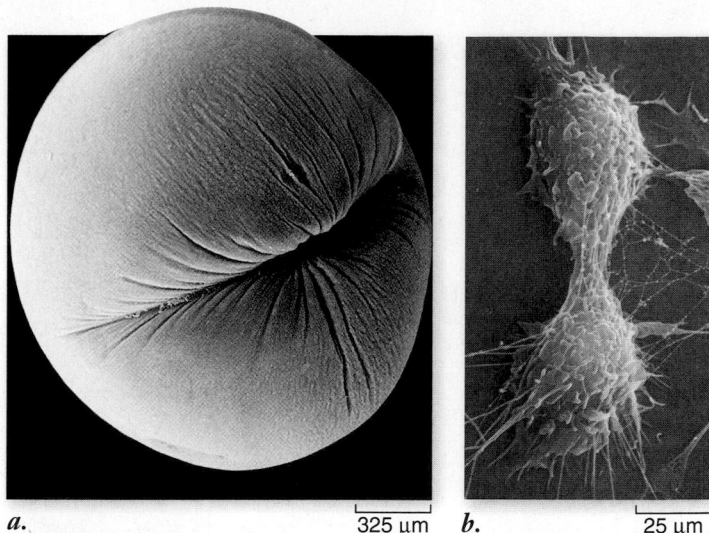

a.   325 μm   *b.*   25 μm

Figure 10.14 Cytokinesis in animal cells. *a.* A cleavage furrow forms around a dividing frog egg. *b.* The completion of cytokinesis in an animal cell. The two daughter cells are still joined by a thin band of cytoplasm occupied largely by microtubules.

Telophase can be viewed as a reversal of the process of prophase, bringing the cell back to the state of interphase. Mitosis is complete at the end of telophase. The eukaryotic cell has partitioned its replicated genome into two new nuclei positioned at opposite ends of the cell. Other cytoplasmic organelles, including mitochondria and chloroplasts (if present), were reassorted to areas that will separate and become the daughter cells.

Cell division is still not complete at the end of mitosis, however, because the division of the cell body proper has not yet begun. The phase of the cell cycle when the cell actually divides is called **cytokinesis.** It generally involves the cleavage of the cell into roughly equal halves.

In animal cells, a belt of actin pinches off the daughter cells

In animal cells and the cells of all other eukaryotes that lack cell walls, cytokinesis is achieved by means of a constricting belt of actin filaments. As these filaments slide past one another, the diameter of the belt decreases, pinching the cell and creating a **cleavage furrow** around the cell's circumference (figure 10.14*a*).

As constriction proceeds, the furrow deepens until it eventually slices all the way into the center of the cell. At this point, the cell is divided in two (figure 10.14*b*).

In plant cells, a cell plate divides the daughter cells

Plant cell walls are far too rigid to be squeezed in two by actin filaments. Instead, these cells assemble membrane components in their interior, at right angles to the spindle apparatus. This expanding membrane partition, called a **cell plate,** continues to grow outward until it reaches the interior surface of the plasma membrane and fuses with it, effectively dividing

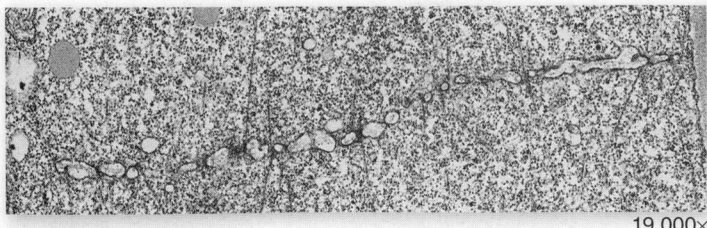

19,000×

Figure 10.15 Cytokinesis in plant cells. In this photomicrograph and companion drawing, a cell plate is forming between daughter nuclei. The cell plate forms from the fusion of Golgi-derived vesicles. Once the plate is complete, there will be two cells.

Vesicles containing membrane components fusing to form cell plate

Cell wall

the cell in two (figure 10.15). Cellulose is then laid down on the new membranes, creating two new cell walls. The space between the daughter cells becomes impregnated with pectins and is called a *middle lamella*.

In fungi and some protists, daughter nuclei are separated during cytokinesis

In most fungi and some groups of protists, the nuclear membrane does not dissolve, and as a result, all the events of mitosis occur entirely *within* the nucleus. Only after mitosis is complete in these organisms does the nucleus divide into two daughter nuclei; then, during cytokinesis, one nucleus goes to each daughter cell. This separate nuclear division phase of the cell cycle does not occur in plants, animals, or most protists.

After cytokinesis in any eukaryotic cell, the two daughter cells contain all the components of a complete cell. Whereas mitosis ensures that both daughter cells contain a full complement of chromosomes, no similar mechanism ensures that organelles such as mitochondria and chloroplasts are distributed equally between the daughter cells. But as long as at least one of each organelle is present in each cell, the organelles can replicate to reach the number appropriate for that cell.

Learning Outcomes Review 10.5

Mitosis is divided into phases: prophase, prometaphase, metaphase, anaphase, and telophase. The early phases involve restructuring the cell to create the microtubule spindle that pulls chromosomes to the equator of the cell in metaphase. Chromatids for each chromosome remain attached at the centromere by cohesin proteins. Chromatids are then pulled to opposite poles during anaphase when cohesin proteins are destroyed. The nucleus is re-formed in telophase, and cytokinesis then divides the cell cytoplasm and organelles. In animal cells, actin pinches the cell in two; in plant cells, a cell plate forms in the middle of the dividing cell.

■ *What would happen to a chromosome that loses cohesin protein between sister chromatids before metaphase?*

Learning Outcomes

1. *Distinguish the role of checkpoints in the control of the cell cycle.*
2. *Characterize the role of the anaphase-promoting complex/cyclosome in mitosis.*
3. *Describe cancer in terms of cell cycle control.*

Our knowledge of how the cell cycle is controlled, although still incomplete, has grown enormously in the past 30 years. Our current view integrates two basic concepts. First, the cell cycle has two irreversible points: the replication of genetic material and the separation of the sister chromatids. Second, the cell cycle can be put on hold at specific points called *checkpoints*. At any of these checkpoints, the process is checked for accuracy and can be halted if there are errors. This leads to extremely high fidelity overall for the entire process. The checkpoint organization also allows the cell cycle to respond to both the internal state of the cell, including nutritional state and integrity of genetic material, and to signals from the environment, which are integrated at major checkpoints.

Research uncovered cell cycle control factors

The history of investigation into control of the cell cycle is instructive in two ways. First, it allows us to place modern observations into context; second, we can see how biologists using very different approaches often end up at the same place. The following brief history introduces three observations and then shows how they can be integrated into a single mechanism.

Discovery of MPF

Research on the activation of frog oocytes led to the discovery of a substance that was first called *maturation-promoting factor (MPF)*. Frog oocytes, which go on to become egg cells, become arrested near the end of their development at the G_2 stage before meiosis I, which is the division leading to the production of gametes (chapter 11). They remain in this arrested state and await hormonal signaling to complete this division process.

Cytoplasm taken from a variety of actively dividing cells could prematurely induce cell division when injected into oocytes (figure 10.16). These experiments indicated the presence of a positive regulator of cell cycle progression in the cytoplasm of dividing cells: MPF. These experiments also fit well with cell fusion experiments done with mitotic and interphase cells that also indicated a cytoplasmic positive regulator that could induce mitosis (see figure 10.16).

Further studies highlighted two key aspects of MPF. First, MPF activity varied during the cell cycle: low in early G_2, rising throughout this phase, and then peaking in mitosis (figure 10.17). Second, the enzymatic activity of MPF involved

Hypothesis: *There are positive regulators of cell division.*

Prediction: *Frog oocytes are arrested in G_2 of meiosis I. They can be induced to mature (undergo meiosis) by progesterone treatment. If maturing oocytes contain a positive regulator of cell division, injection of cytoplasm should induce an immature oocyte to undergo meiosis.*

Test: *Oocytes are induced with progesterone, then cytoplasm from these maturing cells is injected into immature oocytes.*

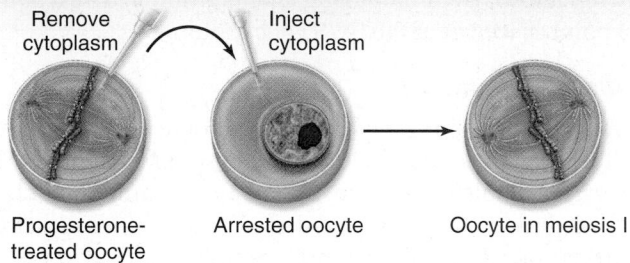

Remove cytoplasm   Inject cytoplasm

Progesterone-treated oocyte   Arrested oocyte   Oocyte in meiosis I

Result: *Injected oocytes progress from G_2 into meiosis I.*

Conclusion: *The progesterone treatment causes production of a positive regulator of maturation: Maturation Promoting Factor (MPF).*

Prediction: *If mitosis is driven by positive regulators, then cytoplasm from a mitotic cell should cause a G_1 cell to enter mitosis.*

Test: *M phase cells are fused with G_1 phase cells, then the nucleus from the G_1 phase cell is monitored microscopically.*

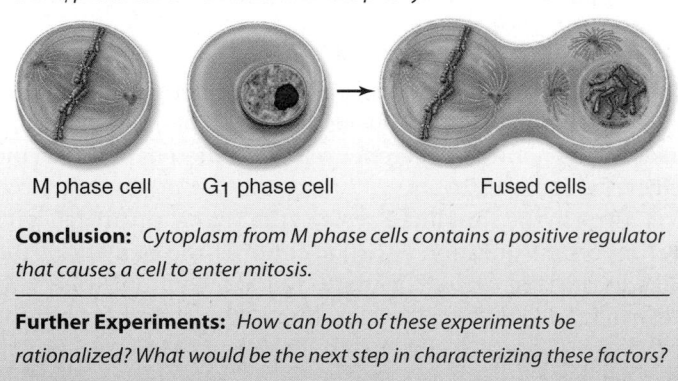

M phase cell G_1 phase cell   Fused cells

Conclusion: *Cytoplasm from M phase cells contains a positive regulator that causes a cell to enter mitosis.*

Further Experiments: *How can both of these experiments be rationalized? What would be the next step in characterizing these factors?*

Figure 10.16 Discovery of positive regulator of cell division.

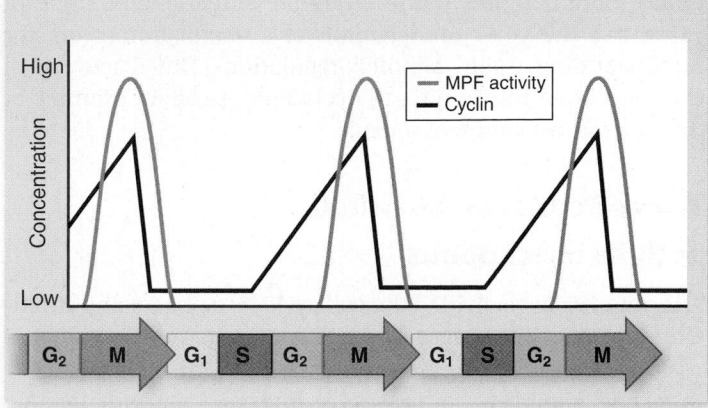

Figure 10.17 Correlation of MPF activity, amount of cyclin protein, and stages of the cell cycle. Cyclin concentration and MPF activity are shown plotted vs. stage of the cell cycle. MPF activity changes in a repeating pattern through the cell cycle. This also correlates with the level of mitotic cyclin in the cell, which shows a similar pattern. The reason for this correlation is that cyclin is actually one component of MPF, the other being a cyclin-dependent kinase (Cdk). Together these act as a positive regulator of cell division.

Despite much effort, no identified enzymatic activity was associated with these proteins. Their hallmark was the timing of their production and not any intrinsic activity.

Genetic analysis of the cell cycle

Geneticists using two different yeasts, budding yeast and fission yeast, as model systems set out to determine the genes necessary for control of the cell cycle. By isolating mutants that were halted during division, they identified genes that were necessary for cell cycle progression. These studies indicated that in yeast, there were two critical control points: the commitment to DNA synthesis, called START, as it meant committing to divide, and the commitment to mitosis. One particular gene, named *cdc2*, from fission yeast, was shown to be critical for passing both of these boundaries.

MPF is cyclin plus cdc2

All of these findings came together in an elegant fashion with the following observations. First, the protein encoded by the *cdc2* gene was shown to be a protein kinase. Second, the purification and identification of MPF showed that it was composed of both a cyclin component and a kinase component. Last, the kinase itself was the cdc2 protein!

The cdc2 protein was the first identified **cyclin-dependent kinase (Cdk),** that is, a protein kinase enzyme that is only active when complexed with cyclin. This finding led to the renaming of MPF as *mitosis*-promoting factor, as its role was clearly more general than simply promoting the maturation of frog oocytes.

These Cdk enzymes are the key positive drivers of the cell division cycle. They are often called the engine that drives cell division. The control of the cell cycle in higher eukaryotes is

---

the phosphorylation of proteins. This second point is not surprising given the importance of phosphorylation as a reversible switch on the activity of proteins (see chapter 9). The first observation indicated that MPF itself was not always active, but rather was being regulated with the cell cycle, and the second showed the possible enzymatic activity of MPF.

Discovery of cyclins

Other researchers examined proteins produced during the early divisions in sea urchin embryos. They identified proteins that were produced in synchrony with the cell cycle, and named them **cyclins** (see figure 10.17). These observations were extended in another marine invertebrate, the surf clam. Two forms of cyclin were found that cycled at slightly different times, reaching peaks at the G_1/S and G_2/M boundaries.

much more complex than the simple single-engine cycle of yeast, but the yeast model remains a useful framework for understanding more complex regulation. The discovery of Cdks and their role in the cell cycle is an excellent example of the progressive nature of science.

The cell cycle can be halted at three checkpoints

Although we have divided the cell cycle into phases and subdivided mitosis into stages, the cell recognizes three points at which the cycle can be delayed or halted. The cell uses these three checkpoints to both assess its internal state and integrate external signals (figure 10.18): G_1/S, G_2/M, and late metaphase (the spindle checkpoint). Passage through these checkpoints is controlled by the Cdk enzymes described earlier and also in the following section.

G_1/S checkpoint

The **G_1/S checkpoint** is the primary point at which the cell "decides" whether or not to divide. This checkpoint is therefore the primary point at which external signals can influence events of the cycle. It is the phase during which growth factors (discussed later on) affect the cycle and also the phase that links cell division to cell growth and nutrition.

In yeast systems, where the majority of the genetic analysis of the cell cycle has been performed, this checkpoint is called START. In animals, it is called the restriction point (R point). In all systems, once a cell has made this irreversible commitment to replicate its genome, it has committed to divide. Damage to DNA can halt the cycle at this point, as can starvation conditions or lack of growth factors.

G_2/M checkpoint

The **G_2/M checkpoint** has received a large amount of attention because of its complexity and its importance as the stimulus for the events of mitosis. Historically, Cdks active at this checkpoint were first identified as MPFs, a term that has now evolved into **M phase-promoting factor (MPF).**

Passage through this checkpoint represents the commitment to mitosis. This checkpoint assesses the success of DNA replication and can stall the cycle if DNA has not been accurately replicated. DNA-damaging agents result in arrest at this checkpoint as well as at the G_1/S checkpoint.

Spindle checkpoint

The **spindle checkpoint** ensures that all of the chromosomes are attached to the spindle in preparation for anaphase. The second irreversible step in the cycle is the separation of chromosomes during anaphase, and therefore it is critical that they are properly arrayed at the metaphase plate.

Cyclin-dependent kinases drive the cell cycle

The primary molecular mechanism of cell cycle control is phosphorylation, which you may recall is the addition of a phosphate group to the amino acids serine, threonine, and tyrosine in proteins (chapter 9). The enzymes that accomplish this phosphorylation are the Cdks (figure 10.19).

The action of Cdks

The first important cell cycle kinase was identified in fission yeast and named Cdc2 (now also called Cdk1). In yeast, this Cdk can partner with different cyclins at different points in the cell cycle (figure 10.20).

Even in the simplified cycle of the yeasts, we are left with the important question of what controls the activity of the Cdks during the cycle. For many years, a common view was that cyclins drove the cell cycle—that is, the periodic synthesis and destruction of cyclins acted as a clock. More recently, it has become clear that the Cdc2 kinase is also itself controlled

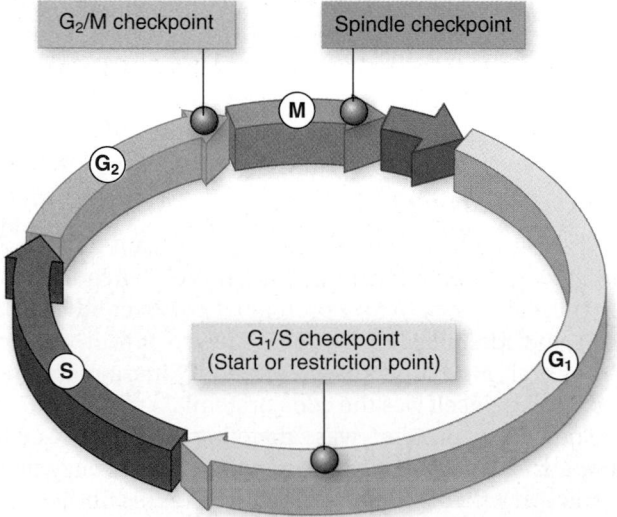

Figure 10.18 Control of the cell cycle. Cells use a centralized control system to check whether proper conditions have been achieved before passing three key checkpoints in the cell cycle.

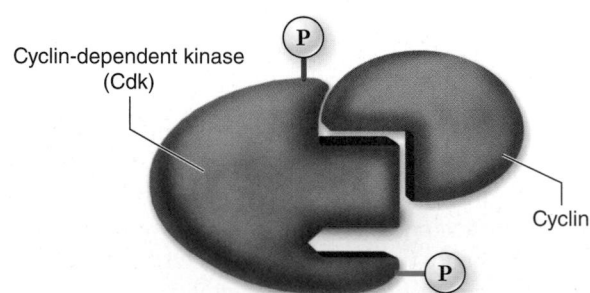

Figure 10.19 Cdk enzyme forms a complex with cyclin. Cdk is a protein kinase that activates numerous cell proteins by phosphorylating them. Cyclin is a regulatory protein required to activate Cdk. This complex is also called mitosis-promoting factor (MPF). The activity of Cdk is also controlled by the pattern of phosphorylation: phosphorylation at one site (represented by the red site) inactivates the Cdk, and phosphorylation at another site (represented by the green site) activates the Cdk.

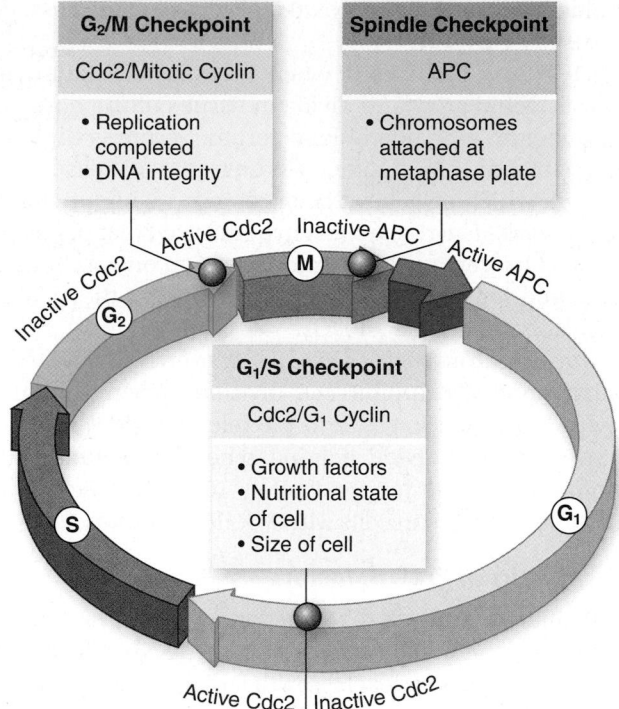

G₂/M Checkpoint

Cdc2/Mitotic Cyclin

• Replication completed
• DNA integrity

Spindle Checkpoint

APC

• Chromosomes attached at metaphase plate

Active Cdc2   Inactive APC   Active APC

Inactive Cdc2

G₂

M

G₁/S Checkpoint

Cdc2/G₁ Cyclin

• Growth factors
• Nutritional state of cell
• Size of cell

S

G₁

Active Cdc2 | Inactive Cdc2

Figure 10.20 Checkpoints of the yeast cell cycle.
The simplest cell cycle that has been studied in detail is the fission yeast. This is controlled by three main checkpoints and a single Cdk enzyme, called Cdc2. The Cdc2 enzyme partners with different cyclins to control the G_1/S and G_2/M checkpoints. The spindle checkpoint is controlled by the anaphase-promoting complex (APC).

by phosphorylation: Phosphorylation at one site activates Cdc2, and phosphorylation at another site inactivates it (see figure 10.19). Full activation of the Cdc2 kinase requires complexing with a cyclin and the appropriate pattern of phosphorylation.

As the G_1/S checkpoint is approached, the triggering signal in yeast appears to be the accumulation of G_1 cyclins. These form a complex with Cdc2 to create the active G_1/S Cdk, which phosphorylates a number of targets that bring about the increased enzyme activity for DNA replication.

The action of MPF

MPF and its role at the G_2/M checkpoint has been extensively analyzed in a number of different experimental systems. The control of MPF is sensitive to agents that disrupt or delay replication and to agents that damage DNA. It was once thought that MPF was controlled solely by the level of the M phase-specific cyclins, but it has now become clear that this is not the case.

Although M phase cyclin is necessary for MPF function, activity is controlled by inhibitory phosphorylation of the kinase component, Cdc2. The critical signal in this process is the removal of the inhibitory phosphates by a protein, phosphatase. This action forms a molecular switch based on positive feedback because the active MPF further activates its own activating phosphatase.

The checkpoint assesses the balance of the kinase that adds inhibitory phosphates with the phosphatase that removes them. Damage to DNA acts through a complex pathway that includes damage sensing and a response to tip the balance toward the inhibitory phosphorylation of MPF. Later on, we describe how some cancers overcome this inhibition.

The anaphase-promoting complex

The molecular details of the sensing system at the spindle checkpoint are not clear. The presence of all chromosomes at the metaphase plate and the tension on the microtubules between opposite poles are both important. The signal is transmitted through the **anaphase-promoting complex,** also called the *cyclosome (APC/C).*

The function of the APC/C is to trigger anaphase itself. As described earlier, the sister chromatids at metaphase are still held together by the protein complex cohesin. The APC does not act directly on cohesin, but rather acts by marking a protein called *securin* for destruction. The securin protein acts as an inhibitor of another protease called *separase* that is specific for one component of the cohesin complex. Once inhibition is lifted, separase destroys cohesin.

This process has been analyzed in detail in budding yeast, where it has been shown that the separase enzyme specifically degrades a component of cohesin called Scc1. This leads to the release of the sister chromatids and results in their sudden movement toward opposite poles during anaphase.

In vertebrates, most cohesin is removed from the sister chromatids during chromosome condensation, possibly with cohesin being replaced by condensin. At metaphase, the majority of the cohesin that remains on vertebrate chromatids is concentrated at the centromere (see figure 10.10). The destruction of this cohesin explains the anaphase movement of chromosomes and the apparent "division" of the centromeres.

The APC/C has two main roles in mitosis: it activates the protease that removes the cohesins holding sister chromatids together, and it is necessary for the destruction of mitotic cyclins to drive the cell out of mitosis. The APC/C complex marks proteins for destruction by the proteosome, the organelle responsible for the controlled degradation of proteins (chapter 16). The signal to degrade a protein is the addition of a molecule called *ubiquitin,* and the APC/C acts as a ubiquitin ligase. As we learn more about the APC/C and its functions, it is clear that the control of its activity is a key regulator of the cell cycle.

In multicellular eukaryotes, many Cdks and external signals act on the cell cycle

The major difference between more complex animals and single-celled eukaryotes such as fungi and protists is twofold: First, multiple Cdks control the cycle as opposed to the single Cdk in yeasts; and second, animal cells respond to a greater variety of external signals than do yeasts, which primarily respond to signals necessary for mating.

In higher eukaryotes there are more Cdk enzymes and more cyclins that can partner with these multiple Cdks, but their basic role is the same as in the yeast cycle. A more complex cell cycle is shown in figure 10.21. These more complex controls allow the integration of more input into control of the cycle. With the evolution of more complex forms of organization (tissues, organs, and organ systems), more complex forms of cell cycle control evolved as well.

A multicellular body's organization cannot be maintained without severely limiting cell proliferation—so that only certain cells divide, and only at appropriate times. The way cells inhibit individual growth of other cells is apparent in mammalian cells growing in tissue culture: A single layer of cells expands over a culture plate until the growing border of cells comes into contact with neighboring cells, and then the cells stop dividing. If a sector of cells is cleared away, neighboring cells rapidly refill that sector and then stop dividing again on cell contact.

How are cells able to sense the density of the cell culture around them? When cells come in contact with one another, receptor proteins in the plasma membrane activate a signal transduction pathway that acts to inhibit Cdk action. This prevents entry into the cell cycle.

Growth factors and the cell cycle

Growth factors act by triggering intracellular signaling systems. Fibroblasts, for example, possess numerous receptors on their plasma membranes for one of the first growth factors to be identified, **platelet-derived growth factor (PDGF).** The PDGF receptor is a receptor tyrosine kinase (RTK) that initiates a

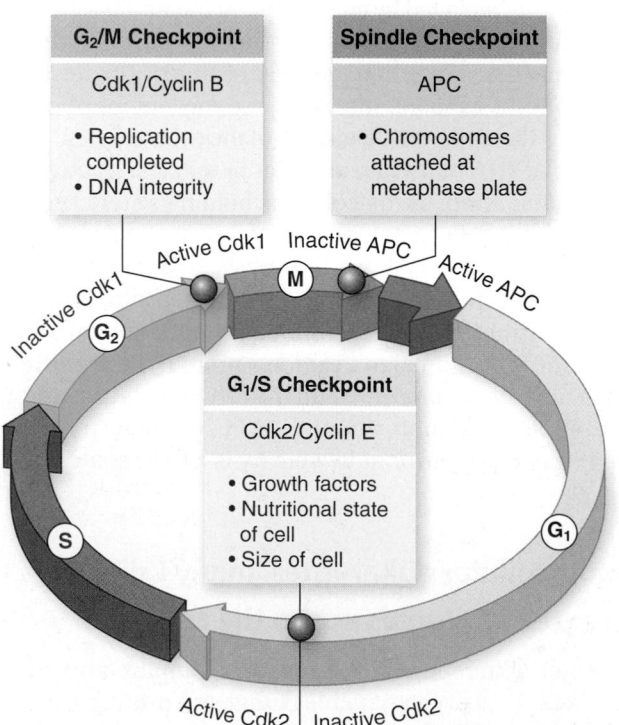

Figure 10.21 Checkpoints of the mammalian cell cycle. The more complex mammalian cell cycle is shown. This cycle is still controlled through three main checkpoints. These integrate internal and external signals to control progress through the cycle. These inputs control the state of two different Cdk–cyclin complexes and the anaphase-promoting complex (APC).

MAP kinase cascade to stimulate cell division (discussed in chapter 9).

PDGF was discovered when investigators found that fibroblasts would grow and divide in tissue culture only if the growth medium contained blood serum. Serum is the liquid that remains in blood after clotting; blood plasma, the liquid from which cells have been removed without clotting, would not work. The researchers hypothesized that platelets in the blood clots were releasing into the serum one or more factors required for fibroblast growth. Eventually, they isolated such a factor and named it PDGF.

Growth factors such as PDGF can override cellular controls that otherwise inhibit cell division. When a tissue is injured, a blood clot forms, and the release of PDGF triggers neighboring cells to divide, helping to heal the wound. Only a tiny amount of PDGF (approximately 10^{-10} M) is required to stimulate cell division in cells with PDGF receptors.

Characteristics of growth factors

Over 50 different proteins that function as growth factors have been isolated, and more undoubtedly exist. A specific cell surface receptor recognizes each growth factor, its binding site fitting that growth factor precisely. These growth factor receptors often initiate MAP kinase cascades in which the final kinase enters the nucleus and activates transcription factors by phosphorylation. These transcription factors stimulate the production of G_1 cyclins and the proteins that are necessary for cell cycle progression (figure 10.22).

The cellular selectivity of a particular growth factor depends on which target cells bear its unique receptor. Some growth factors, such as PDGF and epidermal growth factor (EGF), affect a broad range of cell types, but others affect only specific types. For example, nerve growth factor (NGF) promotes the growth of certain classes of neurons, and erythropoietin triggers cell division in red blood cell precursors. Most animal cells need a combination of several different growth factors to overcome the various controls that inhibit cell division.

The G_0 phase

If cells are deprived of appropriate growth factors, they stop at the G_1 checkpoint of the cell cycle. With their growth and division arrested, they remain in this dormant G_0 phase.

The ability to enter G_0 accounts for the incredible diversity seen in the length of the cell cycle in different tissues. Epithelial cells lining the human gut divide more than twice a day, constantly renewing this lining. By contrast, liver cells divide only once every year or two, spending most of their time in the G_0 phase. Mature neurons and muscle cells usually never leave G_0.

Cancer is a failure of cell cycle control

The unrestrained, uncontrolled growth of cells in humans leads to the disease called **cancer.** Cancer is essentially a disease of cell division—a failure of cell division control.

The p53 gene

One of the critical players in this control system has been identified. Officially dubbed *p53,* this gene plays a key role in the G_1 checkpoint of cell division.

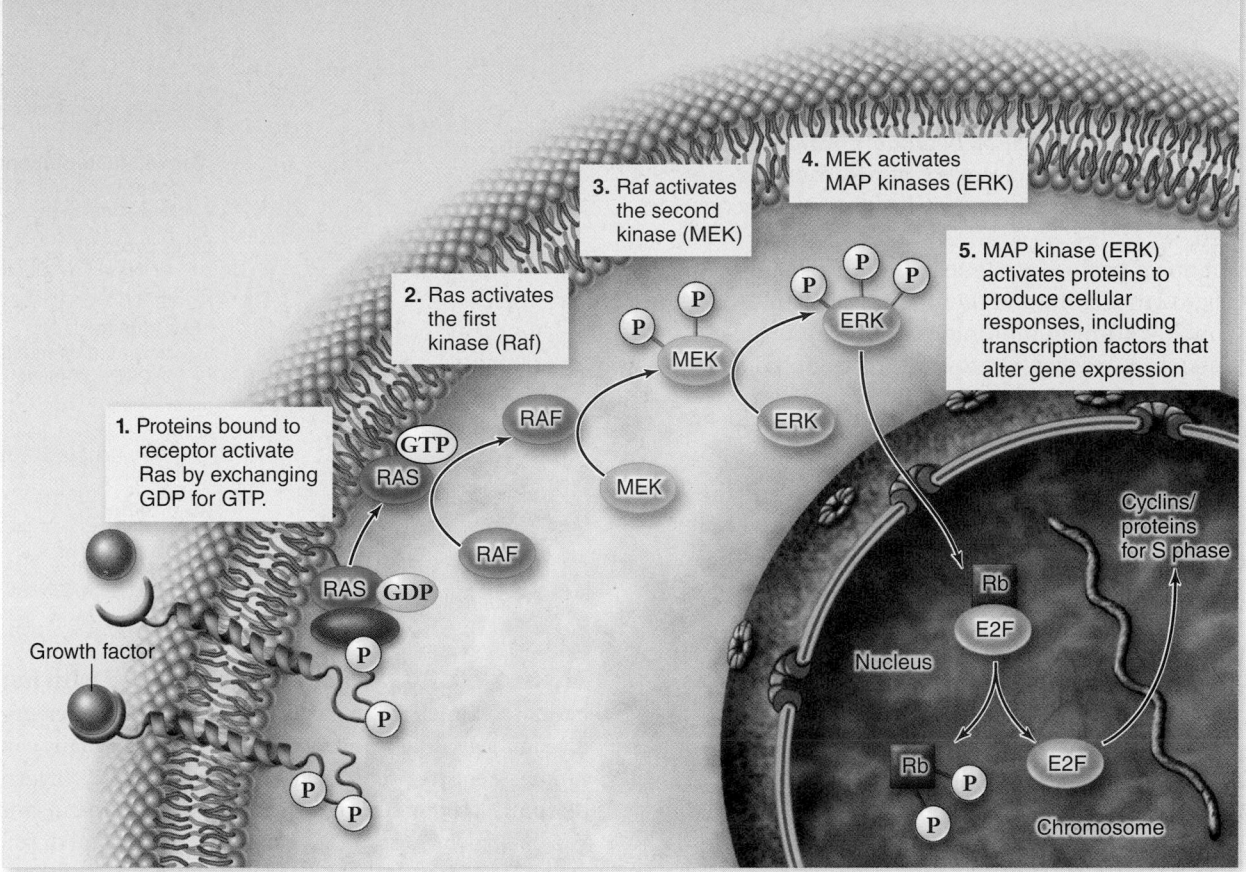

Figure 10.22 The cell proliferation-signaling pathway. Binding of a growth factor sets in motion a MAP kinase intracellular signaling pathway (described in chapter 9), which activates nuclear regulatory proteins that trigger cell division. In this example, when the nuclear retinoblastoma protein (Rb) is phosphorylated, another nuclear protein (the transcription factor E2F) is released and is then able to stimulate the production of cyclin and other proteins necessary for S phase.

The gene's product, the p53 protein, monitors the integrity of DNA, checking that it is undamaged. If the p53 protein detects damaged DNA, it halts cell division and stimulates the activity of special enzymes to repair the damage. Once the DNA has been repaired, p53 allows cell division to continue. In cases where the DNA damage is irreparable, p53 then directs the cell to kill itself.

By halting division in damaged cells, the *p53* gene prevents the development of many mutated cells, and it is therefore considered a **tumor-suppressor gene** although its activities are not limited to cancer prevention. Scientists have found that *p53* is entirely absent or damaged beyond use in the majority of cancerous cells they have examined. It is precisely because *p53* is nonfunctional that cancer cells are able to repeatedly undergo cell division without being halted at the G_1 checkpoint (figure 10.23).

Proto-oncogenes

The disease we call cancer is actually many different diseases, depending on the tissue affected. The common theme in all

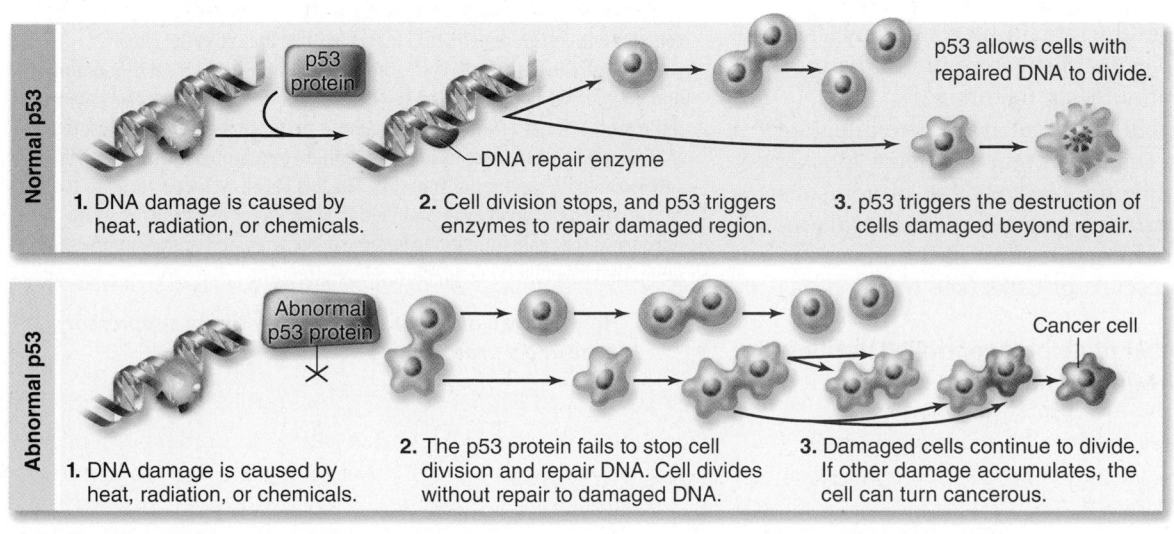

**Figure 10.23
Cell division, cancer, and p53 protein.**
Normal p53 protein monitors DNA, destroying cells that have irreparable damage to their DNA. Abnormal p53 protein fails to stop cell division and repair DNA. As damaged cells proliferate, cancer develops.

cases is the loss of control over the cell cycle. Research has identified numerous so-called **oncogenes,** genes that can, when introduced into a cell, cause it to become a cancer cell. This identification then led to the discovery of **proto-oncogenes,** which are normal cellular genes that become oncogenes when mutated.

The action of proto-oncogenes is often related to signalling by growth factors, and their mutation can lead to loss of growth control in multiple ways. Some proto-oncogenes encode receptors for growth factors, and others encode proteins involved in signal transduction that act after growth factor receptors. If a receptor for a growth factor becomes mutated such that it is permanently "on," the cell is no longer dependent on the presence of the growth factor for cell division. This is analogous to a light switch that is stuck on: The light will always be on. PDGF and EGF receptors both fall into the category of proto-oncogenes. Only one copy of a proto-oncogene needs to undergo this mutation for uncontrolled division to take place; thus, this change acts like a dominant mutation.

The number of proto-oncogenes identified has grown to more than 50 over the years. This line of research connects our understanding of cancer with our understanding of the molecular mechanisms governing cell cycle control.

Tumor-suppressor genes

After the discovery of proto-oncogenes, a second category of genes related to cancer was identified: the tumor-suppressor genes. We mentioned earlier that the *p53* gene acts as a tumor-suppressor gene, and a number of other such genes exist.

Both copies of a tumor-suppressor gene must lose function for the cancerous phenotype to develop, in contrast to the mutations in proto-oncogenes. Put another way, the proto-oncogenes act in a dominant fashion, and tumor suppressors act in a recessive fashion.

The first tumor-suppressor identified was the **retinoblastoma susceptibility gene (Rb),** which predisposes individuals for a rare form of cancer that affects the retina of the eye. Despite the fact that a cell heterozygous for a mutant *Rb* allele is normal, it is inherited as a dominant in families. The reason is that inheriting a single mutant copy of *Rb* means the individual has only one "good" copy left, and during the hundreds of thousands of divisions that occur to produce the retina, any error that damages the remaining good copy leads to a cancerous cell. A single cancerous cell in the retina then leads to the formation of a retinoblastoma tumor.

The role of the Rb protein in the cell cycle is to integrate signals from growth factors. The Rb protein is called a "pocket protein" because it has binding pockets for other proteins. Its role is therefore to bind important regulatory proteins and prevent them from stimulating the production of the necessary cell cycle proteins, such as cyclins or Cdks (see figure 10.21) discussed previously.

The binding of Rb to other proteins is controlled by phosphorylation: When it is dephosphorylated, it can bind a variety

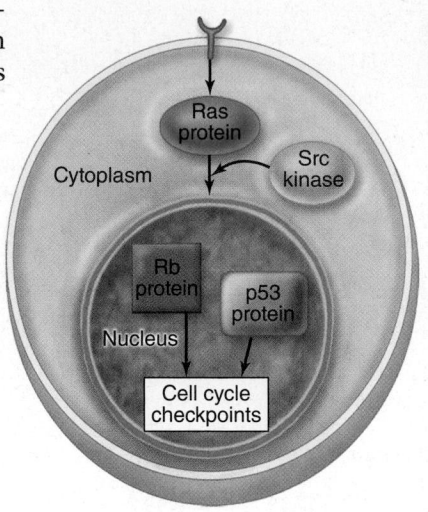

Proto-oncogenes
Growth factor receptor: more per cell in many breast cancers.
Ras protein: activated by mutations in 20–30% of all cancers.
Src kinase: activated by mutations in 2–5% of all cancers.

Tumor-suppressor Genes
Rb protein: mutated in 40% of all cancers.
p53 protein: mutated in 50% of all cancers.

Figure 10.24 Key proteins associated with human cancers. Mutations in genes encoding key components of the cell division-signaling pathway are responsible for many cancers. Among them are proto-oncogenes encoding growth factor receptors, protein relay switches such as Ras protein, and kinase enzymes such as Src, which act after Ras and growth factor receptors. Mutations that disrupt tumor-suppressor proteins, such as Rb and p53, also foster cancer development.

of regulatory proteins, but loses this capacity when phosphorylated. The action of growth factors results in the phosphorylation of Rb protein by a Cdk. This then brings us full circle, because the phosphorylation of Rb releases previously bound regulatory proteins, resulting in the production of S phase cyclins that are necessary for the cell to pass the G_1/S boundary and begin chromosome replication.

Figure 10.24 summarizes the types of genes that can cause cancer when mutated.

Learning Outcomes Review 10.6

Cyclin proteins are produced in synchrony with the cell cycle. These proteins complex with cyclin-dependent kinases to drive the cell cycle. Three checkpoints exist in the cell cycle: the G_1/S checkpoint, the G_2/M checkpoint, and the spindle checkpoint. The cell cycle can be halted at these checkpoints if the process is not accurate. The anaphase-promoting complex/cyclosome (APC/C) triggers anaphase by lifting inhibition on a protease that removes cohesin holding chromatids together. The loss of cell cycle control leads to cancer, which can occur by a combination of two basic mechanisms: proto-oncogenes that gain function to become oncogenes, and tumor-suppressor genes that lose function and allow cell proliferation.

■ *How can you distinguish between a tumor suppressor gene and a proto-oncogene?*

10.1 Bacterial Cell Division

Binary fission is a simple form of cell division.

Prokaryotic cell division is clonal, resulting in two identical cells. Bacterial DNA replication and partitioning of the chromosome are concerted processes.

Proteins control chromosome separation and septum formation.

DNA replication begins at a specific point, the origin, and proceeds bidirectionally to a specific termination site. Newly replicated chromosomes are segregated to opposite poles at the same time as they are replicated. New cells are separated by septation, which involves insertion of new cell membrane and other cellular materials at the midpoint of the cell. A ring of FtsZ and proteins embedded in the cell membrane expands radially inward, pinching the cell into two new cells.

10.2 Eukaryotic Chromosomes

Chromosome number varies among species.

The gain or loss of chromosomes is usually lethal.

Eukaryotic chromosomes exhibit complex structure.

Chromosomes are composed of chromatin, a complex of DNA, and protein. Heterochromatin is not expressed and euchromatin is expressed. The DNA of a single chromosome is a very long, double-stranded fiber. The DNA is wrapped around a core of eight histones to form a nucleosome, which can be further coiled into a 30-nm fiber in interphase cells. During mitosis, chromosomes are further condensed by arranging coiled 30-nm fibers radially around a protein scaffold.

Newly replicated chromosomes remain attached at a constricted area called a centromere, consisting of repeated DNA sequences. After replication, a chromosome consists of two sister chromatids held together at the centromere by a complex of proteins called cohesins (figure 10.7).

10.3 Overview of the Eukaryotic Cell Cycle (figure 10.8)

The cell cycle is divided into five phases.

The phases of the cell cycle are gap 1 (G_1), synthesis (S), gap 2 (G_2), mitosis, and cytokinesis (C). G_1, S, and G_2 are collectively called interphase, and mitosis and cytokinesis together are called M phase.

The duration of the cell cycle varies depending on cell type.

The length of a cell cycle varies with age, cell type, and species. Cells can exit G_1 and enter a nondividing phase called G_0; the G_0 phase can be temporary or permanent.

10.4 Interphase: Preparation for Mitosis

G_1, S, and G_2 are the three subphases of interphase. G_1 is the primary growth phase; during S phase, DNA synthesis occurs. G_2 phase occurs after S phase and before mitosis.

The centromere binds proteins assembled into a disklike structure called a kinetochore where microtubules attach during mitosis. The centromeric DNA is replicated, but the two DNA strands are held together by cohesin proteins.

10.5 M Phase: Chromosome Segregation and the Division of Cytoplasmic Contents (figure 10.11)

During prophase, the mitotic apparatus forms.

In prophase, chromosomes condense, the spindle is formed, and the nuclear envelope disintegrates. In animals cells, centriole pairs separate and migrate to opposite ends of the cell, establishing the axis of nuclear division.

During prometaphase, chromosomes attach to the spindle.

In metaphase, chromosomes align at the equator.

Chromatids of each chromosome are connected to opposite poles by kinetochore microtubules. They are held at the equator of the cell by the tension of being pulled toward opposite poles.

At anaphase, the chromatids separate.

At this point, cohesin proteins holding sister chromatids together at the centromeres are destroyed, and the chromatids are pulled to opposite poles. This movement is called anaphase A, and the movement of poles farther apart is called anaphase B.

During telophase, the nucleus re-forms.

Telophase reverses the events of prophase and prepares the cell for cytokinesis.

In animal cells, a belt of actin pinches off the daughter cells.

A contractile ring of actin under the membrane contracts during cytokinesis.

In plant cells, a cell plate divides the daughter cells.

Fusion of vesicles produces a new membrane in the middle of the cell to produce the cell plate.

In fungi and some protists, daughter nuclei are separated during cytokinesis.

10.6 Control of the Cell Cycle (figure 10.18)

Research uncovered cell cycle control factors.

Experiments showed that there are positive regulators of mitosis, and that there are proteins produced in synchrony with the cell cycle (cyclins). The positive regulators are cyclin-dependent kinases (Cdks). Cdks are complexes of a kinase and a regulatory molecule called cyclin. They phosphorylate proteins to drive the cell cycle.

The cell cycle can be halted at three checkpoints.

Checkpoints are points at which the cell can assess the accuracy of the process and stop if needed. The G_1/S checkpoint is a commitment to divide; the G_2/M checkpoint ensures DNA integrity; and the spindle checkpoint ensures that all chromosomes are attached to spindle fibers, with bipolar orientation.

Cyclin-dependent kinases drive the cell cycle.

The cycle progresses by the action of Cdks. Yeast have only one CDK enzyme; vertebrates have more than four enzymes. During the G_1 phase, G_1 cyclin combines with Cdc2 kinase to form the Cdk that triggers entry into S phase.

The anaphase-promoting complex/cyclosome (APC/C) activates a protease that removes cohesins holding the centromeres of sister chromatids together; the result is to trigger anaphase, separating the chromatids and drawing them to opposite poles. The APC/C also triggers destruction of mitotic cyclins to exit mitosis.

In multicellular eukaryotes, many Cdks and external signals act on the cell cycle.

Growth factors, like platelet-derived growth factor (PDGF), stimulate cell division. This acts through a MAP kinase cascade that results in the production of cyclins and activation of Cdks to stimulate cell division in fibroblasts after tissue injury.

Cancer is a failure of cell cycle control.

Mutations in proto-oncogenes have dominant, gain-of-function effects leading to cancer. Mutations in tumor-suppressor genes are recessive; loss of function of both copies leads to cancer.

UNDERSTAND

1. Binary fission in prokaryotes does not require the
   a. replication of DNA.
   b. elongation of the cell.
   c. separation of daughter cells by septum formation.
   d. assembly of the nuclear envelope.

2. Chromatin is composed of
   a. RNA and protein.
   b. DNA and protein.
   c. sister chromatids.
   d. chromosomes.

3. What is a nucleosome?
   a. A region in the cell's nucleus that contains euchromatin
   b. A region of DNA wound around histone proteins
   c. A region of a chromosome made up of multiple loops of chromatin
   d. A 30-nm fiber found in chromatin

4. What is the role of cohesin proteins in cell division?
   a. They organize the DNA of the chromosomes into highly condensed structures.
   b. They hold the DNA of the sister chromatids together.
   c. They help the cell divide into two daughter cells.
   d. They connect microtubules and chromosomes.

5. The kinetochore is a structure that functions to
   a. connect the centromere to microtubules.
   b. connect centrioles to microtubules.
   c. aid in chromosome condensation.
   d. aid in chromosomes cohesion.

6. Separation of the sister chromatids occurs during
   a. prophase.
   b. prometaphase.
   c. anaphase.
   d. telophase.

7. Why is cytokinesis an important part of cell division?
   a. It is responsible for the proper separation of genetic information.
   b. It is responsible for the proper separation of the cytoplasmic contents.
   c. It triggers the movement of a cell through the cell cycle.
   d. It allows cells to halt at checkpoints.

8. What steps in the cell cycle represent irreversible commitments?
   a. The S/G_2 checkpoint
   b. The G_1/S checkpoint
   c. Anaphase
   d. Both b and c are correct.

APPLY

1. Cyclin-dependent kinases (Cdks) are regulated by
   a. the periodic destruction of cyclins.
   b. bipolar attachment of chromosomes to the spindle.
   c. DNA synthesis.
   d. Both a and b are correct.

2. The bacterial SMC proteins, eukaryotic cohesin proteins, and condensin proteins share a similar structure. Functionally they all
   a. interact with microtubules.
   b. can act as kinase enzymes.
   c. interact with DNA to compact or hold strands together.
   d. connect chromosomes to cytoskeletal elements.

3. Genetically, proto-oncogenes act in a dominant fashion. This is because
   a. there is only one copy of each proto-oncogene in the genome.
   b. they act in a gain-of-function fashion to turn on the cell cycle.
   c. they act in a loss-of-function fashion to turn off the cell cycle.
   d. they require that both genomic copies are altered to affect function.

4. The metaphase to anaphase transition involves
   a. new force being generated to pull the chromatids apart.
   b. an increase in force on sister chromatids to pull them apart.
   c. completing DNA replication of centromeres allowing chromosomes to be pulled apart.
   d. loss of cohesion between sister chromatids.

5. The main difference between bacterial cell division and eukaryotic cell division is that
   a. since bacteria only have one chromosome, they can count the number of copies in the cell.
   b. eukaryotes mark their chromosomes to identify them and bacteria do not.
   c. bacterial DNA replication and chromosome segregation are concerted processes but in eukaryotes they are separated in time.
   d. None of the above is correct.

6. In animal cells, cytokinesis is accomplished by a contractile ring containing actin. The related process in bacteria is
   a. chromosome segregation, which also appears to use an actin-like protein.
   b. septation via a ring of FtsZ protein, which is an actin-like protein.
   c. cytokinesis, which requires formation of a cell plate via vesicular fusion.
   d. septation via a ring of FtsZ protein, which is a tubulin-like protein.

SYNTHESIZE

1. Regulation of the cell cycle is very complex and involves multiple proteins. In yeast, a complex of cdc2 and a mitotic cyclin is responsible for moving the cell past the G_2/M checkpoint. The activity of the cyclin-dependent kinase cdc2 is inhibited when it is phosphorylated by the kinase, Wee-1. What would you predict would be the phenotype of a Wee-1 mutant yeast? What other genes could be altered in a Wee-1 deficient mutant strain that would make the cells act normally?

2. Review your knowledge of signaling pathways (chapter 9). Create an outline illustrating how a growth factor (ligand) can lead to the production of a cyclin protein that would trigger S phase.

3. Compare and contrast how mutations in cellular proto-oncogenes and in tumor suppressor genes can lead to cancer cells.

ONLINE RESOURCE

www.ravenbiology.com

Understand, Apply, and Synthesize—enhance your study with animations that bring concepts to life and practice tests to assess your understanding. Your instructor may also recommend the interactive eBook, individualized learning tools, and more.

Chapter **11**

Sexual Reproduction and Meiosis

Chapter Contents

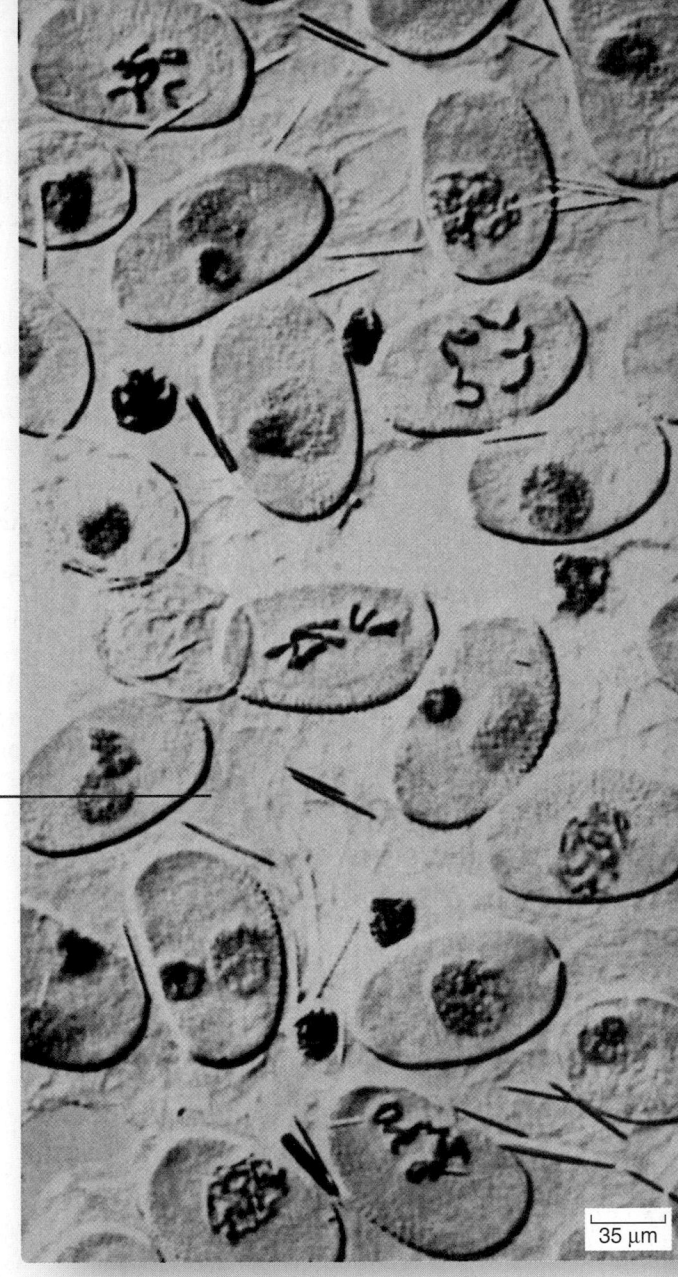

35 μm

Part IIII Genetic and Molecular Biology

Introduction

Most animals and plants reproduce sexually. Gametes of opposite sex unite to form a cell that, dividing repeatedly by mitosis, eventually gives rise to an adult body with some 100 trillion cells. The gametes that form the initial cell are the products of a special form of cell division called meiosis, *visible in the photo above, and the subject of this chapter. Meiosis is far more intricate than mitosis, and the details behind it are not as well understood. The basic process, however, is clear. Also clear are the profound consequences of sexual reproduction: It plays a key role in generating the tremendous genetic diversity that is the raw material of evolution.*

11.1 Sexual Reproduction Requires Meiosis

Learning Outcomes

1. *Characterize the function of meiosis in sexual reproduction.*
2. *Distinguish between germ-line and somatic cells.*

The essence of sexual reproduction is the genetic contribution of two cells. This mode of reproduction imposes difficulties for sexually reproducing organisms that biologists recognized early on. We are only recently making progress on the underlying mechanism for the elaborate behavior of chromosomes during meiosis. To begin, we briefly consider the history of meiosis and its relationship to sexual reproduction.

Meiosis reduces the number of chromosomes

Only a few years after Walther Flemming's discovery of chromosomes in 1879, Belgian cytologist Edouard van Beneden was surprised to find different numbers of chromosomes in different types of cells in the roundworm *Ascaris*. Specifically, he observed that the **gametes** (eggs and sperm) each contained two chromosomes, but all of the nonreproductive cells, or **somatic cells,** of embryos and mature individuals each contained four.

From his observations, van Beneden proposed in 1883 that an egg and a sperm, each containing half the complement of chromosomes found in other cells, fuse to produce a single cell called a **zygote.** The zygote, like all of the cells ultimately derived from it, contains two copies of each chromosome. The fusion of gametes to form a new cell is called **fertilization,** or **syngamy.**

It was clear even to early investigators that gamete formation must involve some mechanism that reduces the number of chromosomes to half the number found in other cells. If it did not, the chromosome number would double with each fertilization, and after only a few generations, the number of chromosomes in each cell would become impossibly large. For example, in just 10 generations, the 46 chromosomes present in human cells would increase to over 47,000 (46×2^{10}).

The number of chromosomes does not explode in this way because of a special reduction division, **meiosis.** Meiosis occurs during gamete formation, producing cells with half the normal number of chromosomes. The subsequent fusion of two of these cells ensures a consistent chromosome number from one generation to the next.

Sexual life cycles have both haploid and diploid stages

Meiosis and fertilization together constitute a cycle of reproduction. Two sets of chromosomes are present in the somatic cells of adult individuals, making them *diploid* cells, but only one set is present in the gametes, which are thus *haploid.* Reproduction that involves this alternation of meiosis and fertilization is called **sexual reproduction.** Its outstanding characteristic is that offspring inherit chromosomes from *two* parents (figure 11.1). You, for example, inherited 23 chromosomes from your mother (maternal homologue), and 23 from your father (paternal homologue).

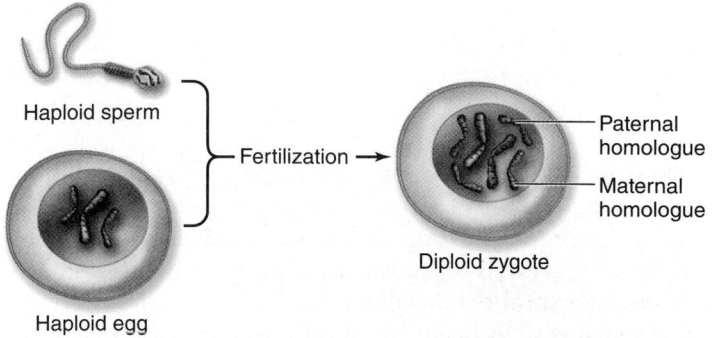

Figure 11.1 Diploid cells carry chromosomes from two parents. A diploid cell contains two versions of each chromosome, a maternal homologue contributed by the haploid egg of the mother, and a paternal homologue contributed by the haploid sperm of the father.

The life cycles of all sexually reproducing organisms follow a pattern of alternation between diploid and haploid chromosome numbers, but there is some variation in the life cycles. Many types of algae, for example, spend the majority of their life cycle in a haploid state. The zygote undergoing meiosis produces haploid cells that then undergo mitosis. Some plants and some algae alternate between a multicellular haploid phase and a multicellular diploid phase (specific examples can be found in chapters 30 and 31). In most animals, the diploid state dominates; the zygote first undergoes mitosis to produce diploid cells. Then later in the life cycle, some of these diploid cells undergo meiosis to produce haploid gametes (figure 11.2).

Germ-line cells are set aside early in animal development

In animals, the single diploid zygote undergoes mitosis to give rise to all of the cells in the adult body. The cells that will eventually undergo meiosis to produce gametes are set aside from somatic cells early in the course of development. These cells are referred to as **germ-line cells.**

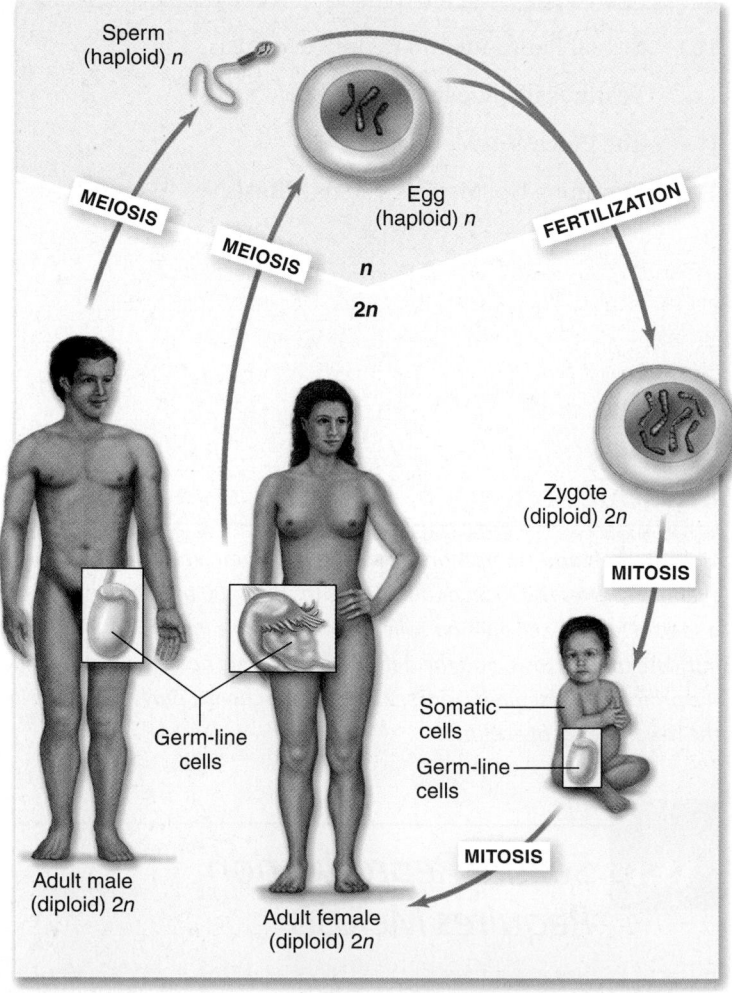

Figure 11.2 The sexual life cycle in animals. In animals, the zygote undergoes mitotic divisions and gives rise to all the cells of the adult body. Germ-line cells are set aside early in development and undergo meiosis to form the haploid gametes (eggs or sperm). The rest of the body cells are called somatic cells.

Both the somatic cells and the gamete-producing germ-line cells are diploid, but whereas somatic cells undergo mitosis to form genetically identical, diploid daughter cells, gamete-producing germ-line cells undergo meiosis to produce haploid gametes (see figure 11.2).

Learning Outcomes Review 11.1

Sexual reproduction involves the genetic contribution of two cells, each from a different individual. Meiosis produces haploid cells with half the number of chromosomes. Fertilization then unites these haploid cells to restore the diploid state of the next generation. Only germ-line cells are capable of meiosis. All other cells in the body, termed somatic cells, can undergo only mitotic division.

■ *Germ-line cells undergo meiosis, but how can the body maintain a constant supply of these cells?*

11.2 Features of Meiosis

Learning Outcomes

1. Describe how homologous chromosomes pair during meiosis.
2. Explain why meiosis I is called the reductive division.

The mechanism of meiotic cell division varies in important details in different organisms. These variations are particularly evident in the chromosomal separation mechanisms: Those found in protists and fungi are very different from those in plants and animals, which we describe here.

Meiosis in a diploid organism consists of two rounds of division, called **meiosis I** and **meiosis II,** with each round containing prophase, metaphase, anaphase, and telophase stages. Before describing the details of this process, we first examine the features of meiosis that distinguish it from mitosis.

Homologous chromosomes pair during meiosis

During early prophase I of meiosis, homologous chromosomes find each other and become closely associated, a process called pairing, or **synapsis** (figure 11.3a). Despite a long history of investigation, molecular details remain unclear.

Biologists have used electron microscopy, data from genetic crosses, and biochemical analysis to shed light on synapsis. Thus far the results of their investigations have not been integrated into a complete picture.

The synaptonemal complex

It is clear that homologous chromosomes find their proper partners and become intimately associated during prophase I. This process includes the formation in many species of an elaborate structure called the **synaptonemal complex,** consisting of the homologues paired closely along a lattice of proteins between them. The structure of the synaptonemal complex appears similar in all systems that have been examined, although its exact function is unclear. A representative example is shown in figure 11.3b. The result is that all four chromatids of the two homologues are closely associated during this phase of meiosis. This structure is also called a *tetrad* or *bivalent.*

The exchange of genetic material between homologues

While homologues are paired during prophase I, another process unique to meiosis occurs: genetic **recombination,** or **crossing over.** This process literally allows the homologues to exchange chromosomal material. The cytological observation of this phenomenon is called crossing over, and its detection genetically is called recombination—because alleles of genes

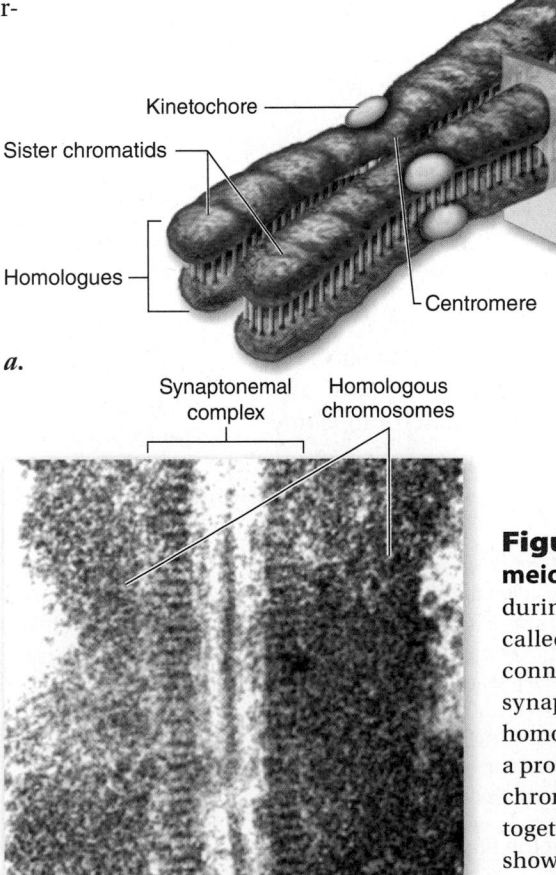

Figure 11.3 Unique features of meiosis. *a.* Homologous chromosomes pair during prophase I of meiosis. This process, called synapsis, produces homologues connected by a structure called the synaptonemal complex. The paired homologues can physically exchange parts, a process called crossing over. The sister chromatids of each homologue are also held together by cohesin proteins, which is not shown for clarity. *b.* The synaptonemal complex of the ascomycete *Neotiella rutilans,* a cup fungus.

that were formerly on separate homologues can now be found on the same homologue. (Genetic recombination is covered in detail in chapter 13.)

The sites of crossing over are called **chiasmata** (singular, *chiasma*), and these sites of contact are maintained until anaphase I. The physical connection of homologues due to crossing over and the continued connection of the sister chromatids lock homologues together.

Homologue association and separation

The association between the homologues persists throughout meiosis I and dictates the behavior of the chromosomes. During metaphase I, the paired homologues move to the metaphase plate and become oriented with homologues of each pair attached to opposite poles of the spindle. By contrast, in mitosis homologues behave independently of one another.

Then, during anaphase I, homologues are pulled to opposite poles for each pair of chromosomes. This again is in contrast to mitosis, in which sister chromatids, not homologues, are pulled to opposite poles.

You can now see why the first division is termed the "reduction division"—it results in daughter cells that contain one homologue from each chromosome pair. The second meiotic division does not further reduce the number of chromosomes; it will merely separate the sister chromatids for each homologue.

Meiosis features two divisions with one round of DNA replication

The most obvious distinction between meiosis and mitosis is the simple observation that meiosis involves two successive divisions with no replication of genetic material between them. One way to view this is that DNA replication must be suppressed between the two meiotic divisions. Because of the behavior of chromosomes during meiosis I, the resulting cells contain one replicated copy of each chromosome. A division that acts like mitosis, without DNA replication, converts these cells into ones with a single copy of each chromosome. This is the last key to understanding meiosis: The second meiotic division is like mitosis with no chromosome duplication.

Learning Outcomes Review 11.2

Meiosis is characterized by the pairing of homologous chromosomes during prophase I. In many species, an elaborate structure called the synaptonemal complex forms between homologues. During this pairing, homologues may exchange chromosomal material at sites called chiasmata. In meiosis I, the homologues separate from each other, reducing the chromosome number to the haploid state (thus the reductive division). It is followed by a second division without replication, during which sister chromatids become separated. The result of meiosis I and II is four haploid cells.

■ *If sister chromatids separated at the first division, would meiosis still work?*

Learning Outcomes

1. Describe the behavior of chromosomes through both meiotic divisions.
2. Explain the importance of monopolar attachment of homologous pairs at metaphase I.
3. Differentiate between the events of anaphase I and anaphase II of meiosis.

To understand meiosis, it is necessary to carefully follow the behavior of chromosomes during each division. The first meiotic division depends on each homologous pair behaving as a unit and not as individual chromosomes, as they do in mitosis. This is accomplished by a complex set of processes that together join homologues until anaphase I, when they are separated.

Prophase I sets the stage for the reductive division

Meiotic cells have an interphase period that is similar to mitosis with G_1, S, and G_2 phases. After interphase, germ-line cells enter meiosis I. In prophase I, the DNA coils tighter, and individual chromosomes first become visible under the light microscope as a matrix of fine threads. Because the DNA has already replicated before the onset of meiosis, each of these threads actually consists of two sister chromatids joined at their centromeres. In prophase I, homologous chromosomes become closely associated in synapsis, exchange segments by crossing over, and then separate.

Synapsis

During interphase in germ-line cells, the ends of the chromatids seem to be attached to the nuclear envelope at specific sites. The sites the homologues attach to are adjacent, so that during prophase I the members of each homologous pair of chromosomes are brought close together. Homologous pairs then align side by side, apparently guided by heterochromatin sequences, in the process of synapsis.

This association joins homologues along their entire length. The sister chromatids of each homologue are also joined by the cohesin complex in a process called *sister chromatid cohesion*. Sister chromatid cohesion also occurs in mitosis, but in meiosis the cohesin complex contains a meiosis-specific cohesin. This brings all four chromatids for each set of paired homologues into close association.

Crossing over

Along with the synaptonemal complex that forms during prophase I (see figure 11.3), another kind of structure appears at the same time that recombination occurs. These are called *recombination nodules,* and they are thought to contain the enzymatic machinery necessary to break and rejoin chromatids of homologous chromosomes.

Crossing over involves a complex series of events in which DNA segments are exchanged between nonsister chromatids. Reciprocal crossovers between nonsister chromatids are controlled such that each chromosome arm usually has one or a few crossovers per meiosis, no matter what the size of the chromosome. Human chromosomes typically have two or three.

When crossing over is complete, the synaptonemal complex breaks down, and the homologous chromosomes become less tightly associated but remain attached by chiasmata. At this point, there are four chromatids for each type of chromosome (two homologous chromosomes, each of which consists of two sister chromatids).

The four chromatids are held together in two ways: (1) The two sister chromatids of each homologue, the products of DNA replication, are held together by cohesin proteins (sister chromatid cohesion); and (2) exchange of material by crossing over between homologues locks all four chromatids together.

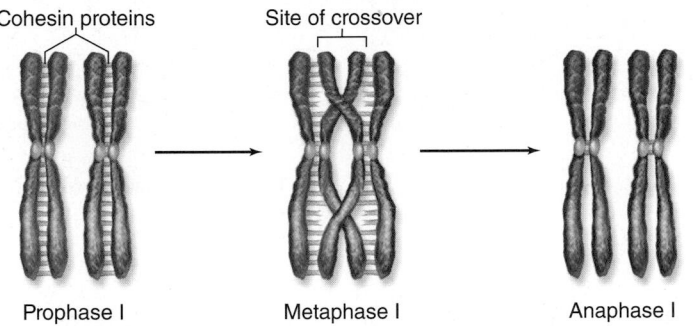

While the elaborate behavior of chromosome pairing is taking place, other events must occur during prophase I. The nuclear envelope must be dispersed, along with the interphase structure of microtubules. The microtubules re-form into a spindle, just as in mitosis.

During metaphase I, paired homologues align

Because of the events of prophase I, each of the paired homologues are locked together as a bivalent. As these bivalents capture spindle fibers, they move to the center of the cell, where

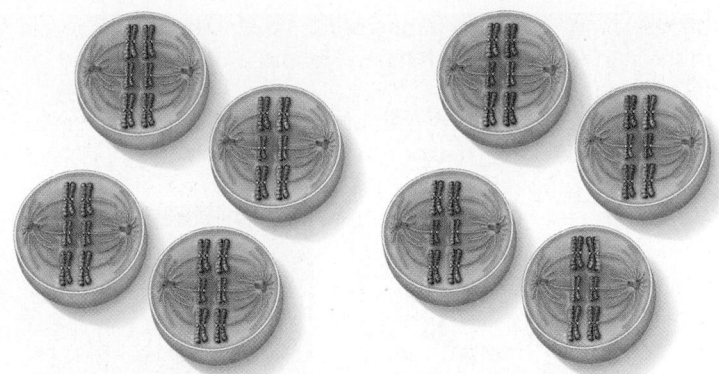

Figure 11.5 Random orientation of chromosomes on the metaphase plate. The number of possible chromosome orientations equals 2 raised to the power of the number of chromosome pairs. In this hypothetical cell with three chromosome pairs, eight (2^3) possible orientations exist. Each orientation produces gametes with different combinations of parental chromosomes.

they are aligned as paired homologues and not individual chromosomes.

The kinetochores of sister chromatids act as a unit to capture polar microtubules. This results in microtubules from opposite poles becoming attached to the kinetochores of *homologues,* and not to those of sister chromatids (figure 11.4).

The ability of sister centromeres to behave as a unit during meiosis I is not understood. It has been suggested, based on electron microscope data, that the centromere–kinetochore complex of sister chromatids is compacted during meiosis I, allowing them to function as a single unit.

The monopolar attachment of kinetochores of sister chromatids would be disastrous in mitosis, but it is critical to meiosis I. It produces tension on the paired homologues, pulling them to the equator of the cell. In this way, each joined pair of homologues lines up on the metaphase plate (see figure 11.4).

The orientation of each pair on the spindle axis is random; either the maternal or the paternal homologue may be oriented toward a given pole (figure 11.5; see also figure 11.6).

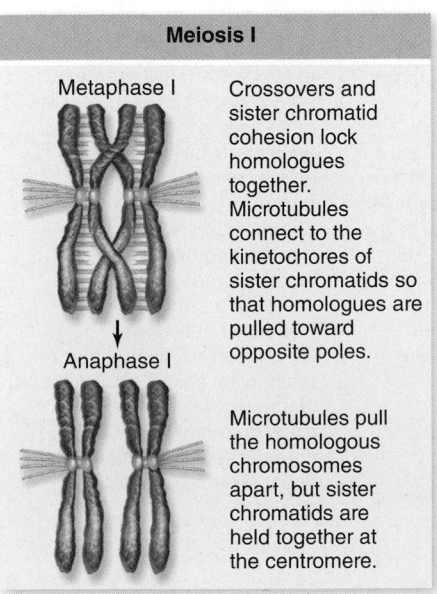

Meiosis I	
Metaphase I	Crossovers and sister chromatid cohesion lock homologues together. Microtubules connect to the kinetochores of sister chromatids so that homologues are pulled toward opposite poles.
Anaphase I	Microtubules pull the homologous chromosomes apart, but sister chromatids are held together at the centromere.

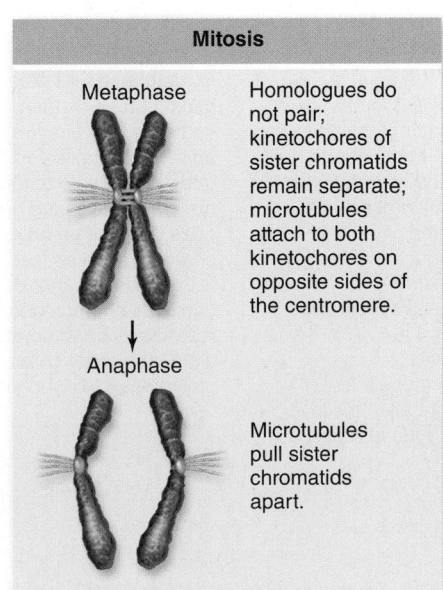

Mitosis	
Metaphase	Homologues do not pair; kinetochores of sister chromatids remain separate; microtubules attach to both kinetochores on opposite sides of the centromere.
Anaphase	Microtubules pull sister chromatids apart.

Figure 11.4 Alignment of chromosomes differs between meiosis I and mitosis. In metaphase I of meiosis I, the chiasmata and connections between sister chromatids hold homologous chromosomes together; paired kinetochores for sister chromatids of each homologue become attached to microtubules from one pole. By the end of meiosis I, connections between sister chromatid arms, but not centromeres, are broken as microtubules shorten, pulling the homologous chromosomes apart. In mitosis, microtubules from opposite poles attach to the kinetochore of each sister centromere; when the connections between sister centromeres are broken microtubules shorten, pulling the sister chromatids to opposite poles.

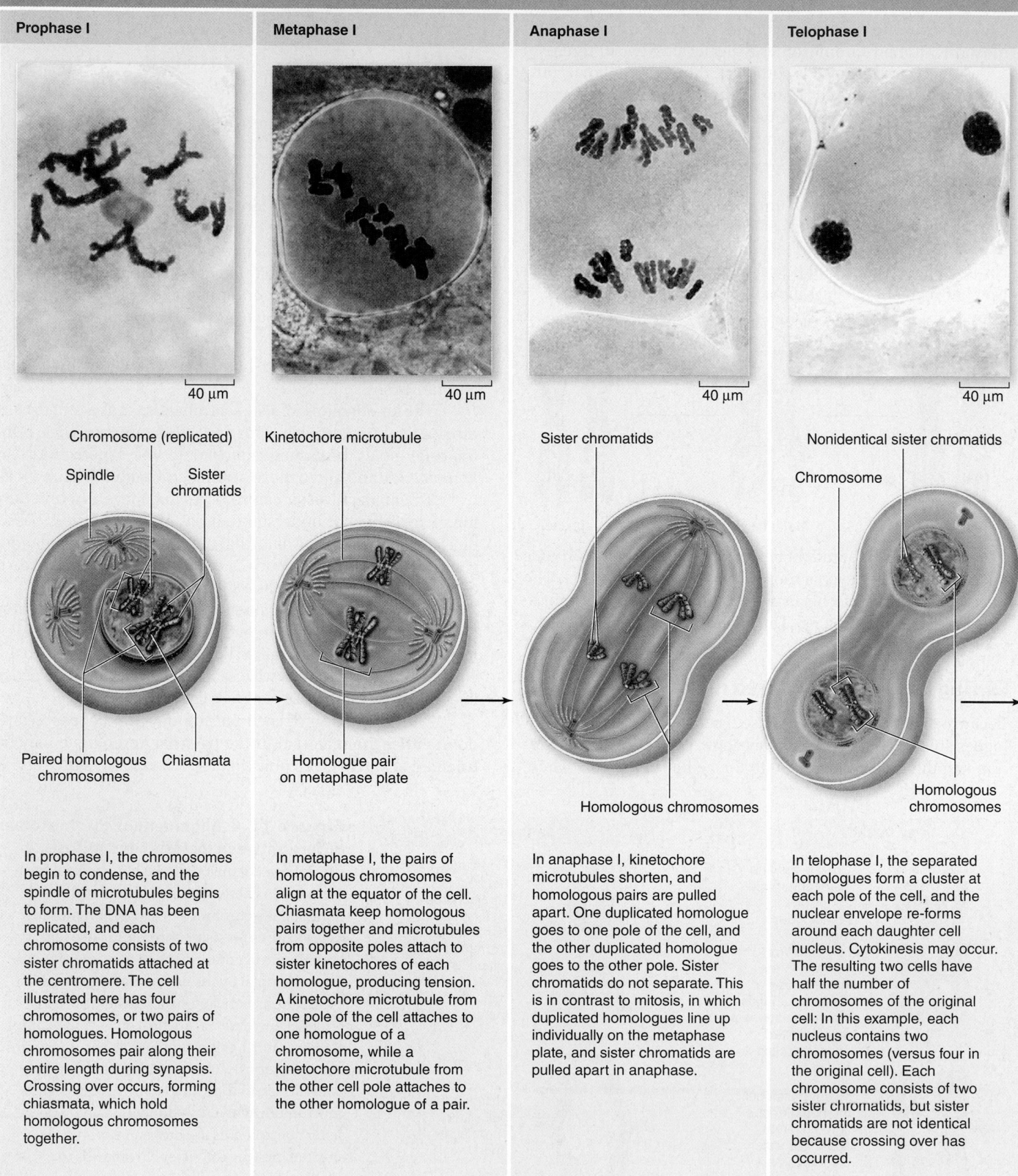

Prophase I

Chromosome (replicated)

Spindle   Sister
                 chromatids

Paired homologous   Chiasmata
chromosomes

In prophase I, the chromosomes begin to condense, and the spindle of microtubules begins to form. The DNA has been replicated, and each chromosome consists of two sister chromatids attached at the centromere. The cell illustrated here has four chromosomes, or two pairs of homologues. Homologous chromosomes pair along their entire length during synapsis. Crossing over occurs, forming chiasmata, which hold homologous chromosomes together.

Metaphase I

Kinetochore microtubule

Homologue pair
on metaphase plate

In metaphase I, the pairs of homologous chromosomes align at the equator of the cell. Chiasmata keep homologous pairs together and microtubules from opposite poles attach to sister kinetochores of each homologue, producing tension. A kinetochore microtubule from one pole of the cell attaches to one homologue of a chromosome, while a kinetochore microtubule from the other cell pole attaches to the other homologue of a pair.

Anaphase I

Sister chromatids

Homologous chromosomes

In anaphase I, kinetochore microtubules shorten, and homologous pairs are pulled apart. One duplicated homologue goes to one pole of the cell, and the other duplicated homologue goes to the other pole. Sister chromatids do not separate. This is in contrast to mitosis, in which duplicated homologues line up individually on the metaphase plate, and sister chromatids are pulled apart in anaphase.

Telophase I

Nonidentical sister chromatids

Chromosome

Homologous
chromosomes

In telophase I, the separated homologues form a cluster at each pole of the cell, and the nuclear envelope re-forms around each daughter cell nucleus. Cytokinesis may occur. The resulting two cells have half the number of chromosomes of the original cell: In this example, each nucleus contains two chromosomes (versus four in the original cell). Each chromosome consists of two sister chromatids, but sister chromatids are not identical because crossing over has occurred.

Figure 11.6 The stages of meiosis. Meiosis in plant cells (photos) and animal cells (drawings) is shown.

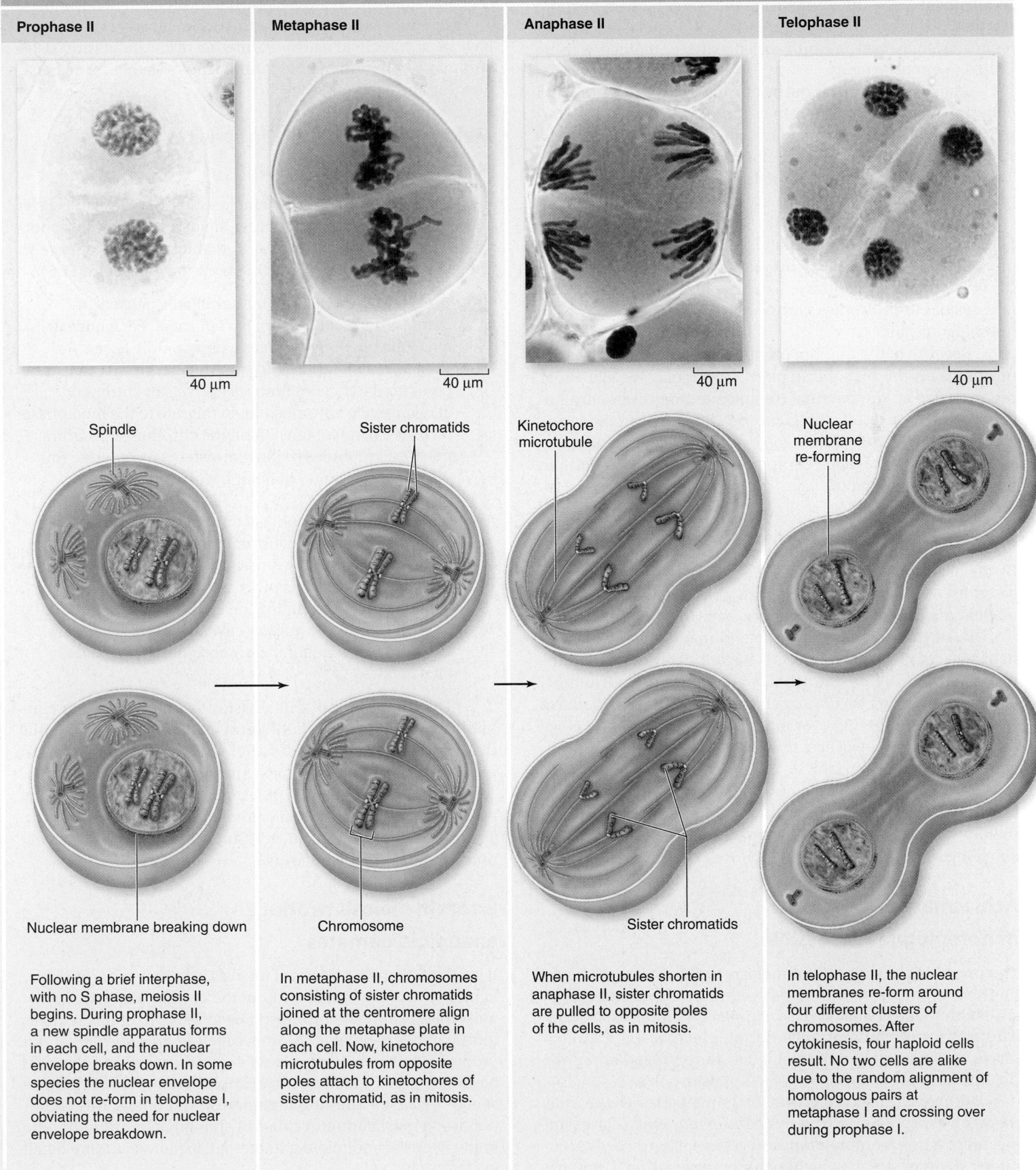

Prophase II

Spindle

Nuclear membrane breaking down

Following a brief interphase, with no S phase, meiosis II begins. During prophase II, a new spindle apparatus forms in each cell, and the nuclear envelope breaks down. In some species the nuclear envelope does not re-form in telophase I, obviating the need for nuclear envelope breakdown.

Metaphase II

Sister chromatids

Chromosome

In metaphase II, chromosomes consisting of sister chromatids joined at the centromere align along the metaphase plate in each cell. Now, kinetochore microtubules from opposite poles attach to kinetochores of sister chromatid, as in mitosis.

Anaphase II

Kinetochore microtubule

Sister chromatids

When microtubules shorten in anaphase II, sister chromatids are pulled to opposite poles of the cells, as in mitosis.

Telophase II

Nuclear membrane re-forming

In telophase II, the nuclear membranes re-form around four different clusters of chromosomes. After cytokinesis, four haploid cells result. No two cells are alike due to the random alignment of homologous pairs at metaphase I and crossing over during prophase I.

40 µm   40 µm   40 µm   40 µm

Anaphase I results from the differential loss of sister chromatid cohesion along the arms

In anaphase I, the microtubules of the spindle fibers begin to shorten. As they shorten, the connections between homologues at chiasmata are broken, allowing homologues to be pulled to opposite poles.

Anaphase I comes about by the release of sister chromatid cohesion along the chromosome arms, but not at the centromeres. This release is the result of the destruction of meiosis-specific cohesin in a process analogous to anaphase in mitosis. The difference is that the destruction is inhibited at the centromeres by a mechanism that is discussed later in the chapter.

As a result of this release, the homologues are pulled apart, but not the sister chromatids. Each homologue moves to one pole, taking both sister chromatids with it. When the spindle fibers have fully contracted, each pole has a complete haploid set of chromosomes consisting of one member of each homologous pair.

Because of the random orientation of homologous chromosomes on the metaphase plate, a pole may receive either the maternal or the paternal homologue from each chromosome pair. As a result, the genes on different chromosomes assort independently; that is, meiosis I results in the **independent assortment** of maternal and paternal chromosomes into the gametes (see chapter 12).

Telophase I completes meiosis I

By the beginning of telophase I, the chromosomes have segregated into two clusters, one at each pole of the cell. Now the nuclear membrane re-forms around each daughter nucleus.

Because each chromosome within a daughter nucleus had replicated before meiosis I began, each now contains two sister chromatids attached by a common centromere. Note that *the sister chromatids are no longer identical* because of the crossing over that occurred in prophase I (see figure 11.6); as you will see, this change has important implications for genetic variability.

Cytokinesis, the division of the cytoplasm and its contents, may or may not occur after telophase I. The second meiotic division, meiosis II, occurs after an interval of variable length.

Achiasmate segregation of homologues is possible

The preceding description of meiosis I relies on the observation that homologues are held together by chiasmata and by sister chromatid cohesion. This connection produces the critical behavior of chromosomes during metaphase I and anaphase I, when paired homologues move together to the metaphase plate and then move to opposite poles.

Although this connection of homologues is the rule, there are exceptions. In fruit fly *(Drosophila)* males for example, there is no recombination, and yet meiosis proceeds accurately, a process called **achiasmate segregation** ("without

chiasmata"). This seems to involve an alternative mechanism for joining homologues and then allowing their segregation during anaphase I. Telomeres and other heterochromatic sequences have been implicated, but the details are not known.

Despite these exceptions, the vast majority of species that have been examined use the formation of chiasmata and sister chromatid cohesion to hold homologues together for segregation during anaphase I.

Meiosis II is like a mitotic division without DNA replication

Typically, interphase between meiosis I and meiosis II is brief and does not include an S phase: Meiosis II resembles a normal mitotic division. Prophase II, metaphase II, anaphase II, and telophase II follow in quick succession (see figure 11.6).

Prophase II. At the two poles of the cell, the clusters of chromosomes enter a brief prophase II, each nuclear envelope breaking down as a new spindle forms.

Metaphase II. In metaphase II, spindle fibers from opposite poles bind to kinetochores of each sister chromatid, allowing each chromosome to migrate to the metaphase plate as a result of tension on the chromosomes from polar microtubules pulling on sister centromeres. This process is the same as metaphase during a mitotic division.

Anaphase II. The spindle fibers contract, and the cohesin complex joining the centromeres of sister chromatids is finally destroyed, allowing sister chromatids to be pulled to opposite poles. This process is essentially the same as anaphase during a mitotic division.

Telophase II. Finally, the nuclear envelope re-forms around the four sets of daughter chromosomes. Cytokinesis then follows.

The final result of this division is four cells, each containing a complete haploid set of chromosomes. The cells that contain these haploid nuclei may develop directly into gametes, as they do in animals. Alternatively, they may themselves divide mitotically, as they do in plants, fungi, and many protists, eventually producing greater numbers of gametes or, as in some plants and insects, adult individuals with varying numbers of chromosome sets.

Errors in meiosis produce aneuploid gametes

It is critical that the process of meiosis be accurate because any failure produces gametes without the correct number of chromosomes. Failure of chromosomes to move to opposite poles during either meiotic division is called *nondisjunction,* and it produces one gamete that lacks a chromosome and one that has two copies. Gametes with an improper number of chromosomes are called **aneuploid gametes.** In humans, this condition is the most common cause of spontaneous abortion. The implications of aneuploid gametes are explored in more detail in chapter 13.

11.4 Summing Up: Meiosis Versus Mitosis

The key to meiosis is understanding the differences between meiosis and mitosis. The basic machinery in both processes is the same, but the behavior of chromosomes is distinctly different during the first meiotic division (figure 11.7).

Meiosis is characterized by four distinct features:

1. Homologous pairing and crossing over joins maternal and paternal homologues during meiosis I.
2. Sister chromatids remain connected at the centromere and segregate together during anaphase I.
3. Kinetochores of sister chromatids are attached to the same pole in meiosis I and to opposite poles in mitosis.
4. DNA replication is suppressed between the two meiotic divisions.

Although the underlying molecular mechanisms are unclear, we will consider what we know of each of these features in the following sections.

Homologous pairing is specific to meiosis

The pairing of homologues during prophase I of meiosis is the first deviation from mitosis and sets the stage for all of the subsequent differences (see figure 11.7). How homologues find each other and become aligned is one of the great mysteries of meiosis. Some cytological evidence implicates telomeres and other specific sites as being necessary for pairing, but this finding does little to clarify the essential process.

The process of sister chromatid cohesion is similar to mitosis, but involves cohesin proteins that contain meiosis-specific subunits. In yeast, the protein Rec8 replaces the mitotic Scc1 protein as part of the cohesin complex. You saw in chapter 10 that Scc1 is destroyed during anaphase of mitosis to allow sister chromatids to be pulled to opposite poles. The replacement of this critical cohesin component with a meiosis-specific version seems to be a common feature in systems analyzed to date.

Synaptonemal complex proteins have been identified in diverse species, but these proteins show little sequence conservation, but do have similarities in structure where this has been analyzed. This may explain the similarity of structures observed cytologically. The transverse elements, while showing no sequence conservation, do share the feature of coiled-coil domains that promote protein–protein interactions.

The molecular details of the recombination process that produces crossing over are complex, but many of the proteins involved have been identified. The process is initiated with the introduction of a double-strand break in one homologue. This explains the similarity in the machinery necessary for meiotic recombination and the machinery involved in the repair of double-strand breaks in DNA. Recombination probably first evolved as a repair mechanism and was later co-opted for use in disjoining chromosomes. The importance of recombination for proper disjunction is clear from the observation in many organisms that loss of function for recombination proteins also results in higher levels of meiotic nondisjunction.

Sister chromatid cohesion is maintained through meiosis I but released in meiosis II

Meiosis I is characterized by the segregation of homologues, not sister chromatids, during anaphase. For this separation to occur, the centromeres of sister chromatids must cosegregate, or move to the same pole, during anaphase I. This means that meiosis-specific cohesin proteins must first be removed from the chromosome arms, then later from sister centromeres.

Homologues are joined by chiasmata, and sister chromatid cohesion around the site of exchange then holds homologues together. The destruction of Rec8 protein on the chromosome arms appears to be what allows homologues to be pulled apart at anaphase I.

This leaves the key distinction between meiosis and mitosis being the maintenance of sister chromatid cohesion at the centromere during all of meiosis I, but the loss of cohesion from the chromosome arms during anaphase I (see figure 11.7). Recently, some light was shed on this problem with the identification of conserved proteins, called Shugoshin (a Japanese term meaning "guardian spirit") required for cohesin protection from separase-mediated cleavage during meiosis I (figure 11.8). Mice have two Shugoshins: Sgo-1 and Sgo-2. Depletion of Sgo-2 results in early sister chromatid separation. This leaves the problem of why Sgo-2 acts only at anaphase I and not anaphase II. It has been suggested that the tension produced by anaphase II causes Sgo-2 to migrate from the centromere to the kinetochore.

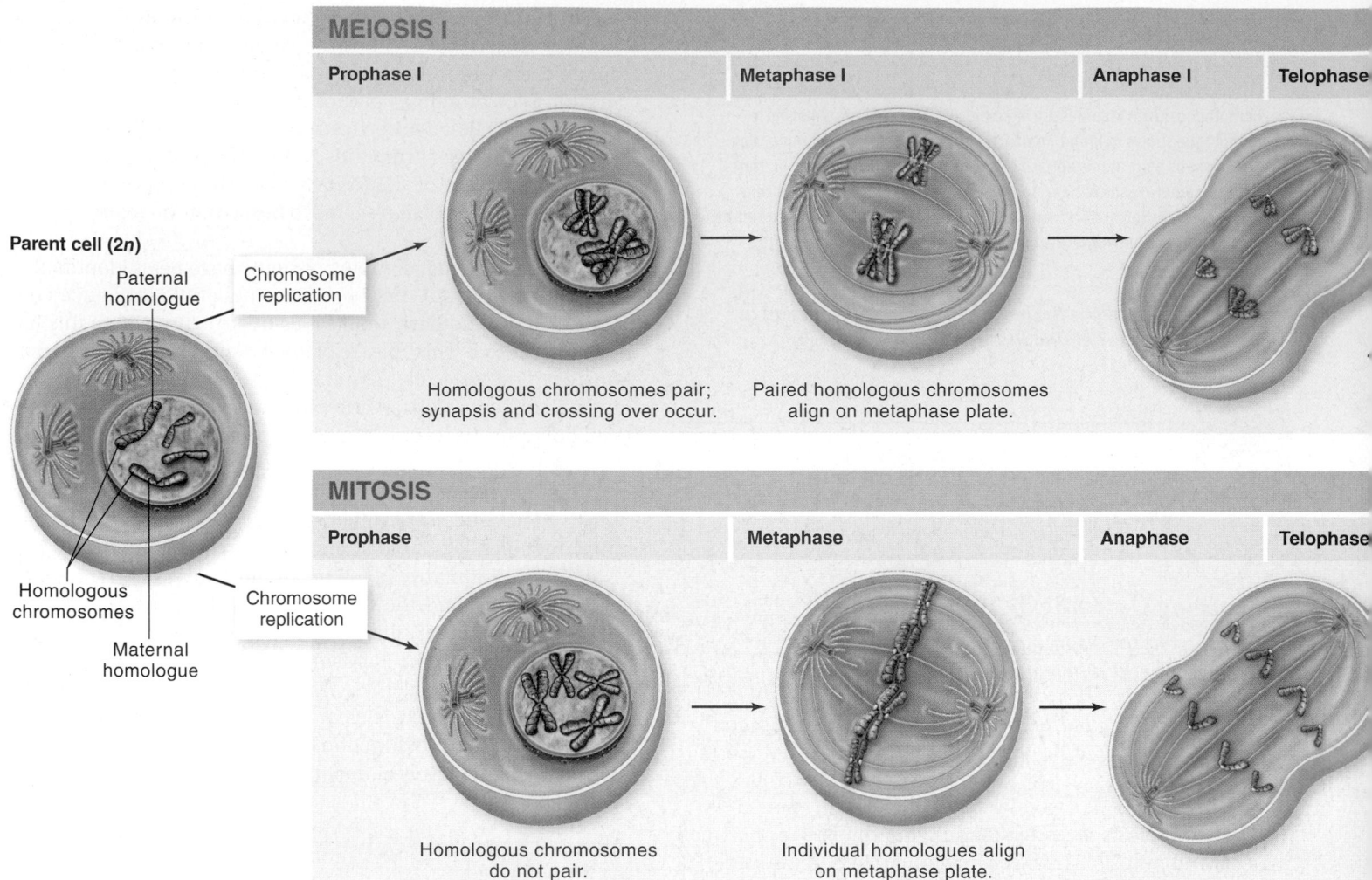

MEIOSIS I

| Prophase I | Metaphase I | Anaphase I | Telophase |

Homologous chromosomes pair; synapsis and crossing over occur.

Paired homologous chromosomes align on metaphase plate.

Parent cell (2*n*)

Paternal homologue

Chromosome replication

Homologous chromosomes

Maternal homologue

MITOSIS

| Prophase | Metaphase | Anaphase | Telophase |

Chromosome replication

Homologous chromosomes do not pair.

Individual homologues align on metaphase plate.

Sister kinetochores are attached to the same pole during meiosis I

The cosegregation of sister centromeres requires that the kinetochores of sister chromatids are attached to the same pole during meiosis I. This attachment is in contrast to both mitosis (see figure 11.7) and meiosis II, in which sister kinetochores must become attached to opposite poles.

The underlying basis of this monopolar attachment of sister kinetochores is unclear, but it seems to be based on structural differences between centromere–kinetochore complexes in meiosis I and in mitosis. Mitotic kinetochores visualized with the electron microscope appear to be recessed, making bipolar attachment more likely. Meiosis I kinetochores protrude more, making monopolar attachment easier.

It is clear that both the maintenance of sister chromatid cohesion at the centromere and monopolar attachment are required for the segregation of homologues that distinguishes meiosis I from mitosis.

Replication is suppressed between meiotic divisions

After a mitotic division, a new round of DNA replication must occur before the next division. For meiosis to succeed in halving the number of chromosomes, this replication must be suppressed between the two divisions. The detailed mechanism of suppression of replication between meiotic division is unknown. One clue is the observation that the level of one of the cyclins, cyclin B, is reduced between meiotic divisions, but is not lost completely, as it is between mitotic divisions.

During mitosis, the destruction of mitotic cyclin is necessary for a cell to enter another division cycle. The result of this maintenance of cyclin B between meiotic divisions in germ-line cells is the failure to form initiation complexes necessary for DNA replication to proceed. This failure to form initiation complexes appears to be critical to suppressing DNA replication.

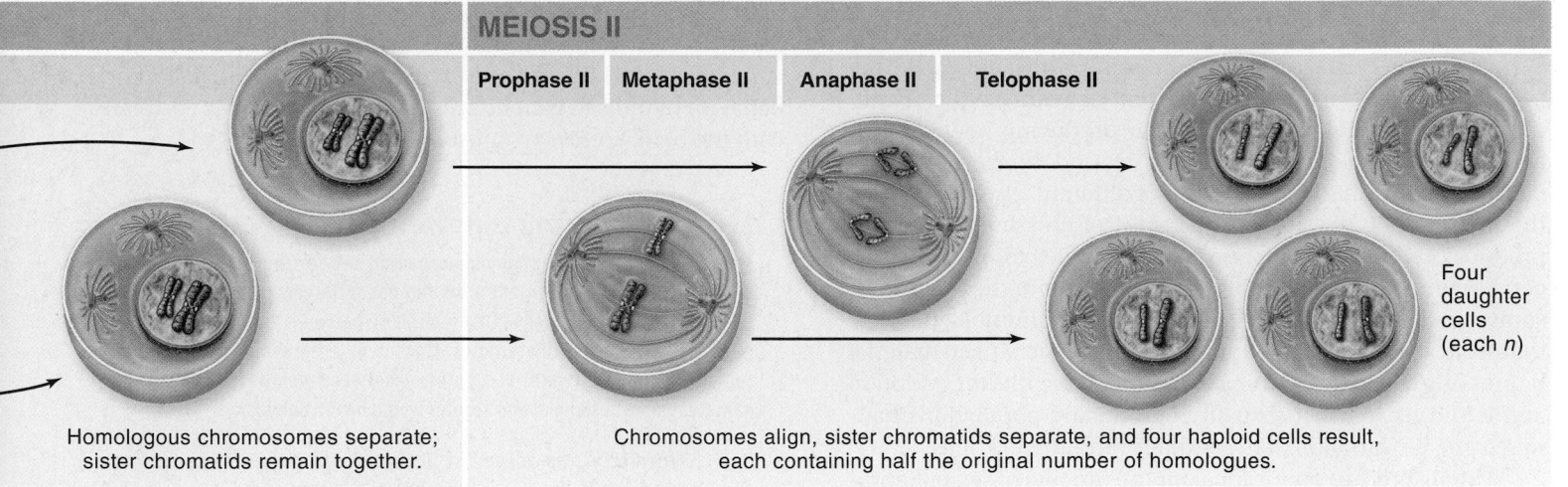

| Prophase II | Metaphase II | Anaphase II | Telophase II |

Four daughter cells (each *n*)

Homologous chromosomes separate; sister chromatids remain together.

Chromosomes align, sister chromatids separate, and four haploid cells result, each containing half the original number of homologues.

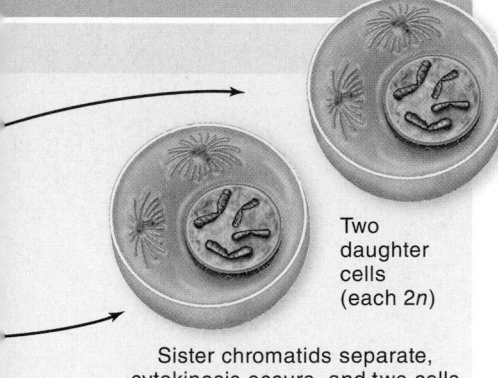

Two daughter cells (each 2*n*)

Sister chromatids separate, cytokinesis occurs, and two cells result, each containing the original number of homologues.

Figure 11.7 A comparison of meiosis and mitosis. Meiosis involves two nuclear divisions with no DNA replication between them. It thus produces four daughter cells, each with half the original number of chromosomes. Crossing over occurs in prophase I of meiosis. Mitosis involves a single nuclear division after DNA replication. It thus produces two daughter cells, each containing the original number of chromosomes.

? **Inquiry question** If the chromosomes of a mitotic cell behaved the same as chromosomes in meiosis I, would the resulting cells have the proper chromosomal constitution?

SCIENTIFIC THINKING

Question: *Why are cohesin proteins at the centromeres of sister chromatids not destroyed at anaphase I of meiosis?*

Hypothesis: *Meiosis-specific cohesin component Rec8 is protected by another protein at centromeres.*

Prediction: *If Rec8 and the centromere protecting protein are both expressed in mitotic cells, chromosome separation will be prevented. This is lethal to a dividing cell.*

Test: *Fission yeast strain is designed to produce Rec8 instead of normal mitotic cohesin. These cells are transformed with a cDNA library that expresses all cellular proteins. Transformed cells are duplicated onto media containing dye for dead cells (allows expression of Rec8 and cDNA), and media that will result in loss of plasmid cDNA (expresses only Rec8). Cells containing cDNA for protecting protein will be dead in presence of Rec8.*

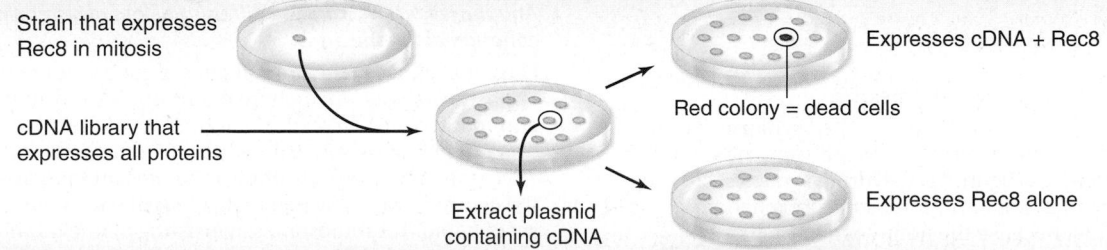

Strain that expresses Rec8 in mitosis

cDNA library that expresses all proteins

Extract plasmid containing cDNA

Expresses cDNA + Rec8

Red colony = dead cells

Expresses Rec8 alone

Result: *Transformed cells that die on the plates where Rec8 is coexpressed with cDNA identify the protecting protein. When the cDNA is extracted and analyzed, the encoded protein localizes to the centromeres of meiotic cells.*

Conclusion: *This screen identifies a protein with Rec8 protecting activity.*

Further Experiments: *If the gene encoding the protecting protein is deleted from cells, what would be the expected phenotype? In mitotic cells? In meiotic cells?*

Figure 11.8 Identification of meiosis-specific cohesin protector.

Meiosis produces cells that are not identical

The daughter cells produced by mitosis are identical to the parental cell, at least in terms of their chromosomal constitution. This exact copying is critical to producing new cells for growth, for development, and for wound healing. Meiosis, because of the random orientation of different chromosomes at the first meiotic division and because of crossing over, rarely produces cells that are identical. The gametes from meiosis all carry an entire haploid set of chromosomes, but these chromosomes are a mixture of maternal and paternal homologues; furthermore, the homologues themselves have exchanged material by crossing over. The resulting variation is essential for evolution and is the reason that sexually reproducing populations have much greater variation than asexually reproducing ones.

Meiosis is not only critical for the process of sexual reproduction, but is also the foundation for understanding the basis of heredity. The different cells produced by meiosis form the basis for understanding the behavior of observable traits in genetic crosses. In the next two chapters we will follow the behavior of traits in genetic crosses and see how this correlates with the behavior of chromosomes in meiosis.

Learning Outcomes Review 11.4

Meiosis is characterized by homologue pairing and crossing over; by loss of sister chromatid cohesion in the arms, but not at the centromere at the first division; by the suppression of DNA replication between the two meiotic divisions; and by sister kinetochores attachment to the same pole of the spindle. If replication were not suppressed between meiosis I and meiosis II, gametes would be diploid, and zygotes would be tetraploid.

■ **What features of meiosis lead to genetic variation in the products?**

Chapter Review

11.1 Sexual Reproduction Requires Meiosis (figure 11.2)

Meiosis reduces the number of chromosomes.
Eggs and sperm are haploid ($1n$) cells, which contain one set of all chromosomes, and products of meiotic division.

Sexual life cycles have both haploid and diploid stages.
During fertilization, or syngamy, the fusion of two haploid gametes results in a diploid ($2n$) zygote, which contains two sets of chromosomes. Meiosis and fertilization constitute a reproductive cycle in sexual organisms as they alternate between diploid and haploid chromosome numbers. Somatic cells divide by mitosis and form the body of an organism.

Germ-line cells are set aside early in animal development.
Cells that eventually will form haploid gametes by meiosis are called germ-line cells. These are set aside early in development in animals.

11.2 Features of Meiosis

Homologous chromosomes pair during meiosis.
The pairing of homologous chromosomes, called synapsis, occurs during early prophase I. Paired homologues are often joined by the synaptonemal complex (figure 11.3). During synapsis, crossing over occurs between homologous chromosomes, exchanging chromosomal material. Because the homologues are paired, they move as a unit to the metaphase plate during metaphase I. During anaphase I, homologues of each pair are pulled to opposite poles, producing two cells that each have one complete set of chromosomes.

Meiosis features two divisions with one round of DNA replication.
Meiosis II is like mitosis but without replication of DNA. Sister chromatids are pulled to opposite poles to yield four haploid cells.

11.3 The Process of Meiosis (figures 11.6 & 11.7)

Prophase I sets the stage for the reductive division.
Meiotic cells have an interphase period similar to mitosis with G_1, S, and G_2 phases. This is followed by prophase I in which homologous chromosomes align along their entire length. The sister chromatids are held together by cohesin proteins. Homologues exchange chromosomal material by crossing over, which assists in holding the homologues together during meiosis I. The nuclear envelope disperses and the spindle apparatus forms.

During metaphase I, paired homologues align.
Spindle fibers attach to the kinetochores of the homologues; the kinetochores of sister chromatids behave as a single unit. Homologues of each pair become attached by kinetochore microtubules to opposite poles, and homologous pairs move to the metaphase plate as a unit. The orientation of each homologous pair on the equator is random; either the maternal or paternal homologue may be oriented toward a given pole.

Anaphase I results from the differential loss of sister chromatid cohesion along the arms.
During anaphase I the homologues of each pair are pulled to opposite poles as kinetochore microtubules shorten. Loss of sister chromatid cohesion on the arms but not at the centromeres allows homologues to separate, but sister chromatids to stay together. This is due to the loss of cohesin proteins on the arms but not at the centromere. At the end of anaphase I each pole has a complete set of haploid chromosomes, consisting of one member of each homologous pair. Because of the random orientation of homologous pairs at metaphase I, meiosis I results in the independent assortment of maternal and paternal chromosomes in gametes.

Telophase I completes meiosis I.
During telophase I the nuclear envelope re-forms around each daughter nucleus. This phase does not occur in all species. Cytokinesis may or may not occur after telophase I.

Achiasmate segregation of homologues is possible.

Although homologues are usually held together by chiasmata, some systems are able to segregate chromosomes without this.

Meiosis II is like a mitotic division without DNA replication.

A brief interphase with no DNA replication occurs after meiosis I. During meiosis II, cohesin proteins at the centromeres that hold sister chromatids together are destroyed, allowing each to migrate to opposite poles of the cell. The result of meiosis I and II is four cells, each containing haploid sets of chromosomes that are not identical. Once completed, the haploid cells may produce gametes or divide mitotically to produce even more gametes or haploid adults.

Errors in meiosis produce aneuploid gametes.

Errors occur during meiosis because of nondisjunction, the failure of chromosomes to move to opposite poles. It may result in aneuploid gametes: one gamete with no chromosome, and another gamete with two copies of a chromosome.

11.4 Summing Up: Meiosis Versus Mitosis

Four distinct features of meiosis I are not found in mitosis: Maternal and paternal homologues pair, and exchange genetic information by crossing over; the kinetochores of sister chromatids function as a unit during meiosis I, allowing sister chromatids to cosegregate during anaphase I; kinetochores of sister chromatids are connected to a single pole in meiosis I and to opposite poles in mitosis; and DNA replication is suppressed between meiosis I and meiosis II.

Homologous pairing is specific to meiosis.

How homologues find each other during meiosis is not known. The proteins of the synaptonemal complex do not seem to be conserved in different species, but there are meiosis-specific cohesin proteins. These are involved in the differential destruction of cohesins on the arms versus the centromere during meiosis I. The recombination process that occurs between paired homologues is better known. This process uses proteins involved in DNA repair and starts with a double-stranded break in DNA.

Sister chromatid cohesion is maintained through meiosis I but released in meiosis II.

Shugoshin protein protects centromeric cohesin in anaphase I, so that sister chromatids remain connected. Cohesins on the arms are not protected and are thus degraded during anaphase I, allowing homologues to move to opposite poles.

Sister kinetochores are attached to the same pole during meiosis I.

Kinetochores of sister chromatids must be attached to the same spindle fibers (monopolar attachment) to segregate together.

Replication is suppressed between meiotic divisions.

Suppression of replication may be related to the maintenance of some cyclin proteins that are degraded at the end of mitosis.

Meiosis produces cells that are not identical.

Because of the independent assortment of homologues and the process of crossing over, gametes show great variation.

Review Questions

UNDERSTAND

1. In comparing somatic cells and gametes, somatic cells are
   a. diploid with half the number of chromosomes.
   b. haploid with half the number of chromosomes.
   c. diploid with twice the number of chromosomes.
   d. haploid with twice the number of chromosomes.

2. What are *homologous* chromosomes?
   a. The two halves of a replicated chromosome
   b. Two identical chromosomes from one parent
   c. Two genetically identical chromosomes, one from each parent
   d. Two genetically similar chromosomes, one from each parent

3. Chiasmata form
   a. between homologous chromosomes.
   b. sister chromatids.
   c. between replicated copies of the same chromosomes.
   d. sex chromosomes but not autosomes.

4. Crossing over involves each of the following with the exception of
   a. the transfer of DNA between two nonsister chromatids.
   b. the transfer of DNA between two sister chromatids.
   c. the formation of a synaptonemal complex.
   d. the alignment of homologous chromosomes.

5. During anaphase I
   a. sister chromatids separate and move to the poles.
   b. homologous chromosomes move to opposite poles.
   c. homologous chromosomes align at the middle of the cell.
   d. all the chromosomes align independently at the middle of the cell.

6. At metaphase I the kinetochores of sister chromatids are
   a. attached to microtubules from the same pole.
   b. attached to microtubules from opposite poles.
   c. held together with cohesin proteins.
   d. not attached to any microtubules.

7. What occurs during anaphase of meiosis II?
   a. The homologous chromosomes align.
   b. Sister chromatids are pulled to opposite poles.
   c. Homologous chromosomes are pulled to opposite poles.
   d. The haploid chromosomes line up.

APPLY

1. Which of the following does *not* contribute to genetic diversity?
   a. Independent assortment
   b. Recombination
   c. Metaphase of meiosis II
   d. Metaphase of meiosis I

2. How does DNA replication differ between mitosis and meiosis?

   a.   DNA replication takes less time in meiosis because the cells are haploid.

   b.   During meiosis, there is only one round of replication for two divisions.

   c.   During mitosis, there is only one round of replication every other division.

   d.   DNA replication is exactly the same in mitosis and meiosis.

3. Which of the following is NOT a distinct feature of meiosis?

   a.   Pairing and exchange of genetic material between homologous chromosomes

   b.   Attachment of sister kinetochores to spindle microtubules

   c.   Movement of sister chromatids to the same pole

   d.   Suppression of DNA replication

4. Which phase of meiosis I is most similar to the comparable phase in mitosis?

   a.   Prophase I   c.   Anaphase I

   b.   Metaphase I   d.   Telophase I

5. Structurally, meiotic cohesins have different components than mitotic cohesins. This leads to the following functional difference:

   a.   During metaphase I, the sister kinetochores become attached to the same pole.

   b.   Centromeres remain attached during anaphase I of meiosis.

   c.   Centromeres remain attached through both divisions.

   d.   Centromeric cohesins are destroyed at anaphase I, and cohesins along the arms are destroyed at anaphase II.

6. Mutations that affect DNA repair often also affect the accuracy of meiosis. This is because

   a.   the proteins involved in the repair of double-strand breaks are also involved in crossing over.

   b.   the proteins involved in DNA repair are also involved in sister chromatid cohesion.

   c.   DNA repair only occurs on condensed chromosomes such as those found in meiosis.

   d.   cohesin proteins are also necessary for DNA repair.

SYNTHESIZE

1. Diagram the process of meiosis for an imaginary cell with six chromosomes in a diploid cell.

   a.   How many homologous pairs are present in this cell? Create a drawing that distinguishes between homologous pairs.

   b.   Label each homologue to indicate whether it is maternal (M) or paternal (P).

   c.   Draw a new cell showing how these chromosomes would arrange themselves during metaphase of meiosis I. Do all the maternal homologues have to line up on the same side of the cell?

   d.   How would this picture differ if you were diagramming anaphase of meiosis II?

2. Mules are the offspring of the mating of a horse and a donkey. Mules are unable to reproduce. A horse has a total of 64 chromosomes, whereas donkeys have 62 chromosomes. Use your knowledge of meiosis to predict the diploid chromosome number of a mule. Propose a possible explanation for the inability of mules to reproduce.

3. Compare the processes of *independent assortment* and *crossing over*. Which process has the greatest influence on genetic diversity?

4. Aneuploid gametes are cells that contain the wrong number of chromosomes. Aneuploidy occurs as a result of *nondisjunction,* or lack of separation of the chromosomes during either phase of meiosis.

   a.   At what point in meiotic cell division would nondisjunction occur?

   b.   Imagine a cell had a diploid chromosome number of four. Create a diagram to illustrate the effects of nondisjunction of one pair of homologous chromosomes in meiosis I versus meiosis II.

ONLINE RESOURCE

www.ravenbiology.com

Understand, Apply, and Synthesize—enhance your study with animations that bring concepts to life and practice tests to assess your understanding. Your instructor may also recommend the interactive eBook, individualized learning tools, and more.

Chapter

12

Patterns of Inheritance

Chapter Contents

Introduction

Every living creature is a product of the long evolutionary history of life on Earth. All organisms share this history, but as far as we know, only humans wonder about the processes that led to their origin and investigate the possibilities. We are far from understanding everything about our origins, but we have learned a great deal. Like a partially completed jigsaw puzzle, the boundaries of this elaborate question have fallen into place, and much of the internal structure is becoming apparent. In this chapter, we discuss one piece of the puzzle—the enigma of heredity. Why do individuals, like the children in this picture, differ so much in appearance despite the fact that we are all members of the same species? And, why do members of a single family tend to resemble one another more than they resemble members of other families?

12.1 *The Mystery of Heredity*

Learning Outcomes

1. *Describe explanations for inheritance prior to Mendel.*
2. *Explain the advantages of Mendel's experimental system.*

As far back as written records go, patterns of resemblance among the members of particular families have been noted and commented on (figure 12.1), but there was no coherent model to explain these patterns. Before the 20th century, two concepts provided the basis for most thinking about heredity. The first was that heredity occurs within species. The second was that traits are transmitted directly from parents to offspring. Taken together, these ideas led to a view of inheritance as resulting from a blending of traits within fixed, unchanging species.

Figure 12.1 Heredity and family resemblance. Family resemblances are often strong—a visual manifestation of the mechanism of heredity.

Inheritance itself was viewed as traits being borne through fluid, usually identified as blood, that led to their blending in offspring. This older idea persists today in the use of the term "bloodlines" when referring to the breeding of domestic animals such as horses.

Taken together, however, these two classical assumptions led to a paradox. If no variation enters a species from outside, and if the variation within each species blends in every generation, then all members of a species should soon have the same appearance. It is clear that this does not happen—individuals within most species differ from one another, and they differ in characteristics that are transmitted from generation to generation.

Early plant biologists produced hybrids and saw puzzling results

The first investigator to achieve and document successful experimental **hybridizations** was Josef Kölreuter, who in 1760 cross-fertilized (or crossed, for short) different strains of tobacco and obtained fertile offspring. The hybrids differed in appearance from both parent strains. When individuals within the hybrid generation were crossed, their offspring were highly variable. Some of these offspring resembled plants of the hybrid generation (their parents), but a few resembled the original strains (their grandparents). The variation observed in second-generation offspring contradicts the theory of direct transmission. This can be seen as the beginning of modern genetics.

Over the next hundred years, other investigators elaborated on Kölreuter's work. T. A. Knight, an English landholder, in 1823 crossed two varieties of the garden pea, *Pisum sativum* (figure 12.2). One of these varieties had green seeds, and the other had yellow seeds. Both varieties were **true-breeding,** meaning that the offspring produced from self-fertilization remained uniform from one generation to the next. All of the progeny (offspring)

of the cross between the two varieties had yellow seeds. Among the offspring of these hybrids, however, some plants produced yellow seeds and others, less common, produced green seeds.

Other investigators made observations similar to Knight's, namely that alternative forms of observed traits were being distributed among the offspring. A modern geneticist would say the alternative forms of each trait were **segregating** among the progeny of a mating, meaning that some offspring exhibited one form of a trait (yellow seeds), and other offspring from the same mating exhibited a different form (green seeds). This segregation of alternative forms of a trait provided the clue that led Gregor Mendel to his understanding of the nature of heredity.

Within these deceptively simple results were the makings of a scientific revolution. Nevertheless, another century passed before the process of segregation was fully appreciated.

Mendel used mathematics to analyze his crosses

Born in 1822 to peasant parents in Austria, Gregor Mendel (figure 12.3) was educated in a monastery and went on to study science and mathematics at the University of Vienna, where he failed his examinations for a teaching certificate. He returned to the monastery and spent the rest of his life there, eventually becoming abbot. In the garden of the monastery, Mendel initiated his own series of experiments on plant hybridization. The results of these experiments would ultimately change our views of heredity irrevocably.

Practical considerations for use of the garden pea

For his experiments, Mendel chose the garden pea, the same plant Knight and others had studied. The choice was a good one for several reasons. First, many earlier investigators had produced hybrid peas by crossing different varieties, so Mendel knew that he could expect to observe segregation of traits among the offspring.

Figure 12.2 The garden pea, *Pisum sativum.* Easy to cultivate and able to produce many distinctive varieties, the garden pea was a popular experimental subject in investigations of heredity as long as a century before Gregor Mendel's experiments.

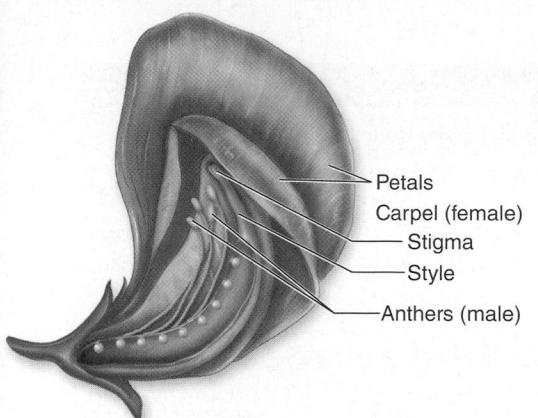

- Petals
- Carpel (female)
- Stigma
- Style
- Anthers (male)

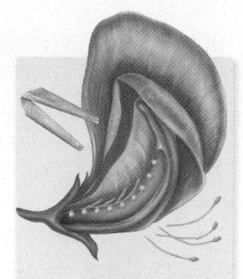

1. The anthers are cut away on the purple flower.

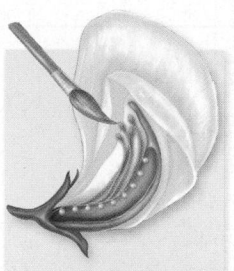

2. Pollen is obtained from the white flower.

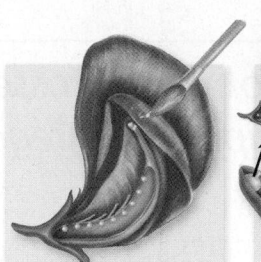

3. Pollen is transferred to the purple flower.

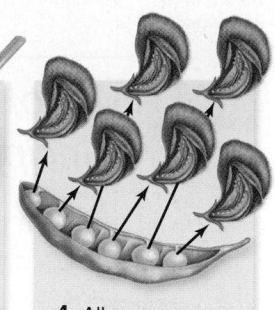

4. All progeny result in purple flowers.

Figure 12.3 How Mendel conducted his experiments. In a pea plant flower, petals enclose both the male anther (containing pollen grains, which give rise to haploid sperm) and the female carpel (containing ovules, which give rise to haploid eggs). This ensures self-fertilization will take place unless the flower is disturbed. Mendel collected pollen from the anthers of a white flower, then placed that pollen onto the stigma of a purple flower with anthers removed. This cross-fertilization yields all hybrid seeds that give rise to purple flowers. Using pollen from a white flower to fertilize a purple flower gives the same result.

? **Inquiry question** What confounding problems could have been seen if Mendel had chosen another plant with exposed male and female structures?

Second, a large number of pure varieties of peas were available. Mendel initially examined 34 varieties. Then, for further study, he selected lines that differed with respect to seven easily distinguishable traits, such as round versus wrinkled seeds and yellow versus green seeds, the latter a trait that Knight had studied.

Third, pea plants are small and easy to grow, and they have a relatively short generation time. A researcher can therefore conduct experiments involving numerous plants, grow several generations in a single year, and obtain results relatively quickly.

A fourth advantage of studying peas is that both the male and female sexual organs are enclosed within each pea flower (see figure 12.3), and gametes produced by the male and female parts of the same flower can fuse to form viable offspring, a process termed **self-fertilization.** This self-fertilization takes place automatically within an individual flower if it is not disturbed. It is also possible to prevent self-fertilization by removing a flower's male parts before fertilization occurs, then introduce pollen from a different strain, thus performing *cross-pollination* that results in *cross-fertilization* (see figure 12.3).

Mendel's experimental design

Mendel was careful to focus on only a few specific differences between the plants he was using and to ignore the countless other differences he must have seen. He also had the insight to realize that the differences he selected must be comparable. For example, he recognized that trying to study the inheritance of round seeds versus tall height would be useless.

Mendel usually conducted his experiments in three stages:

1. Mendel allowed plants of a given variety to self-cross for multiple generations to assure himself that the traits he was studying were indeed true-breeding, that is, transmitted unchanged from generation to generation.
2. Mendel then performed crosses between true-breeding varieties exhibiting alternative forms of traits. He also performed **reciprocal crosses:** using pollen from a white-flowered plant to fertilize a purple-flowered plant, then using pollen from a purple-flowered plant to fertilize a white-flowered plant.
3. Finally, Mendel permitted the hybrid offspring produced by these crosses to self-fertilize for several generations, allowing him to observe the inheritance of alternative forms of a trait. Most important, he counted the numbers of offspring exhibiting each trait in each succeeding generation.

This quantification of results is what distinguished Mendel's research from that of earlier investigators, who only noted differences in a qualitative way. Mendel's mathematical analysis of experimental results led to the inheritance model that we still use today.

Learning Outcomes Review 12.1

Prior to Mendel, concepts of inheritance did not form a consistent model. The dominant view was of blending inheritance, in which traits of parents were carried by fluid and "blended" in offspring. Plant hybridizers before Mendel, however, had already cast doubt on this model by observing characteristics in hybrids that seemed to change in second-generation offspring. Mendel's experiments with plants involved quantifying types of offspring and mathematically analyzing his observations.

■ *Which was more important to Mendel's success: his approach, or his choice of experimental material?*

12.2 Monohybrid Crosses: The Principle of Segregation

Learning Outcomes

1. Evaluate the outcome of a monohybrid cross.
2. Explain Mendel's principle of segregation.
3. Compare the segregation of alleles with the behavior of homologues in meiosis.

A *monohybrid cross* is a cross that follows only two variations on a single trait, such as white- and purple-colored flowers. This deceptively simple kind of cross can lead to important conclusions about the nature of inheritance.

The seven characteristics, or characters, Mendel studied in his experiments possessed two variants that differed from one another in ways that were easy to recognize and score (figure 12.4). We examine in detail Mendel's crosses with flower color. His experiments with other characters were similar, and they produced similar results.

The F$_1$ generation exhibits only one of two traits, without blending

When Mendel crossed white-flowered and purple-flowered plants, the hybrid offspring he obtained did not have flowers of intermediate color, as the hypothesis of blending inheritance would predict. Instead, in every case the flower color of the offspring resembled that of one of their parents. These offspring are customarily referred to as the **first filial generation,** or **F$_1$.** In a cross of white-flowered and purple-flowered plants, the F$_1$ offspring all had purple flowers, as other scientists had reported before Mendel.

Mendel referred to the form of each trait expressed in the F$_1$ plants as **dominant,** and to the alternative form that was not expressed in the F$_1$ plants as **recessive.** For each of the seven pairs of contrasting traits that Mendel examined, one of the pair proved to be dominant and the other recessive.

The F$_2$ generation exhibits both traits in a 3:1 ratio

After allowing individual F$_1$ plants to mature and self-fertilize, Mendel collected and planted the seeds from each plant to see what the offspring in the **second filial generation,** or **F$_2$,** would look like. He found that although most F$_2$ plants had purple flowers, some exhibited white flowers, the recessive trait. Although hidden in the F$_1$ generation, the recessive trait had reappeared among some F$_2$ individuals.

Believing the proportions of the F$_2$ types would provide some clue about the mechanism of heredity, Mendel counted the numbers of each type among the F$_2$ progeny. In the cross between the purple-flowered F$_1$ plants, he obtained a total of 929 F$_2$ individuals. Of these, 705 (75.9%) had purple flowers, and 224 (24.1%) had white flowers (see figure 12.4). Approximately ¼ of the F$_2$ individuals, therefore, exhibited the recessive form of the character.

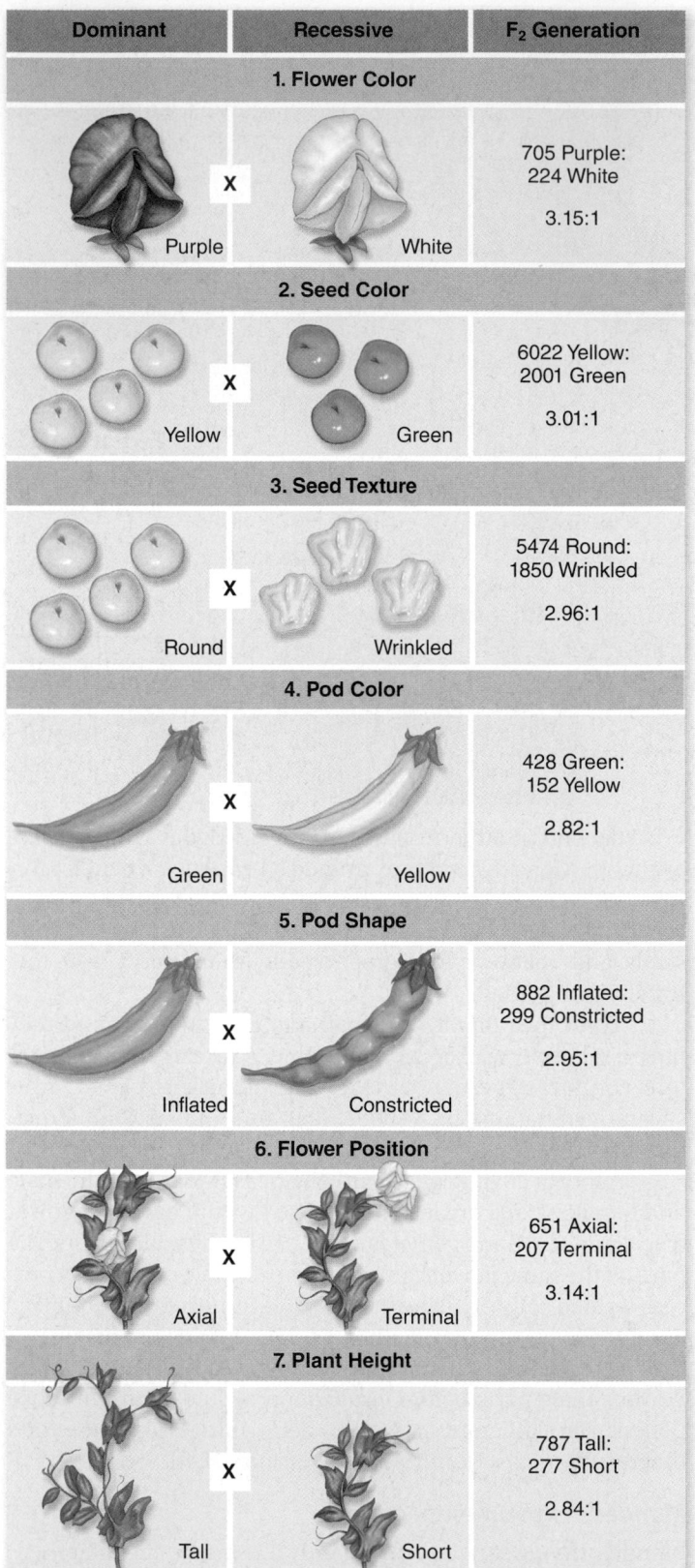

Dominant	Recessive	F$_2$ Generation
1. Flower Color		
Purple	White	705 Purple: 224 White 3.15:1
2. Seed Color		
Yellow	Green	6022 Yellow: 2001 Green 3.01:1
3. Seed Texture		
Round	Wrinkled	5474 Round: 1850 Wrinkled 2.96:1
4. Pod Color		
Green	Yellow	428 Green: 152 Yellow 2.82:1
5. Pod Shape		
Inflated	Constricted	882 Inflated: 299 Constricted 2.95:1
6. Flower Position		
Axial	Terminal	651 Axial: 207 Terminal 3.14:1
7. Plant Height		
Tall	Short	787 Tall: 277 Short 2.84:1

Figure 12.4 Mendel's seven traits. Mendel studied how differences among varieties of peas were inherited when the varieties were crossed. Similar experiments had been done before, but Mendel was the first to quantify the results and appreciate their significance. Results are shown for seven different monohybrid crosses. The F$_1$ generation is not shown in the table.

Mendel obtained the same numerical result with the other six characters he examined: Of the F_2 individuals, ¾ exhibited the dominant trait, and ¼ displayed the recessive trait (see figure 12.4). In other words, the dominant-to-recessive ratio among the F_2 plants was always close to 3:1.

The 3:1 ratio is actually 1:2:1

Mendel went on to examine how the F_2 plants passed traits to subsequent generations. He found that plants exhibiting the recessive trait were always true-breeding. For example, the white-flowered F_2 individuals reliably produced white-flowered offspring when they were allowed to self-fertilize. By contrast, only ⅓ of the dominant, purple-flowered F_2 individuals (¼ of all F_2 offspring) proved true-breeding, but ⅔ were not. This last class of plants produced dominant and recessive individuals in the third filial generation (F_3) in a 3:1 ratio.

This result suggested that, for the entire sample, the 3:1 ratio that Mendel observed in the F_2 generation was really a disguised 1:2:1 ratio: ¼ true-breeding dominant individuals, ½ not-true-breeding dominant individuals, and ¼ true-breeding recessive individuals (figure 12.5).

 Data analysis In the previous set of crosses, if the purple F_1 were backcrossed to the white parent, what would be the phenotypic ratio? The genotypic ratio?

Mendel's Principle of Segregation explains monohybrid observations

From his experiments, Mendel was able to understand four things about the nature of heredity:

- The plants he crossed did not produce progeny of intermediate appearance, as a hypothesis of blending inheritance would have predicted. Instead, different plants inherited each trait intact, as a discrete characteristic.
- For each pair of alternative forms of a trait, one alternative was not expressed in the F_1 hybrids, although it reappeared in some F_2 individuals. *The trait that "disappeared" must therefore be latent (present but not expressed) in the F_1 individuals.*
- The pairs of alternative traits examined were segregated among the progeny of a particular cross, some individuals exhibiting one trait and some the other.
- These alternative traits were expressed in the F_2 generation in the ratio of ¾ dominant to ¼ recessive. This characteristic 3:1 segregation is referred to as the **Mendelian ratio** for a monohybrid cross.

Mendel's five-element model

To explain these results, Mendel proposed a simple model that has become one of the most famous in the history of science, containing simple assumptions and making clear predictions. The model has five elements:

1. Parents do not transmit physiological traits directly to their offspring. Rather, they transmit discrete information for the traits, what Mendel called "factors." We now call these factors *genes.*

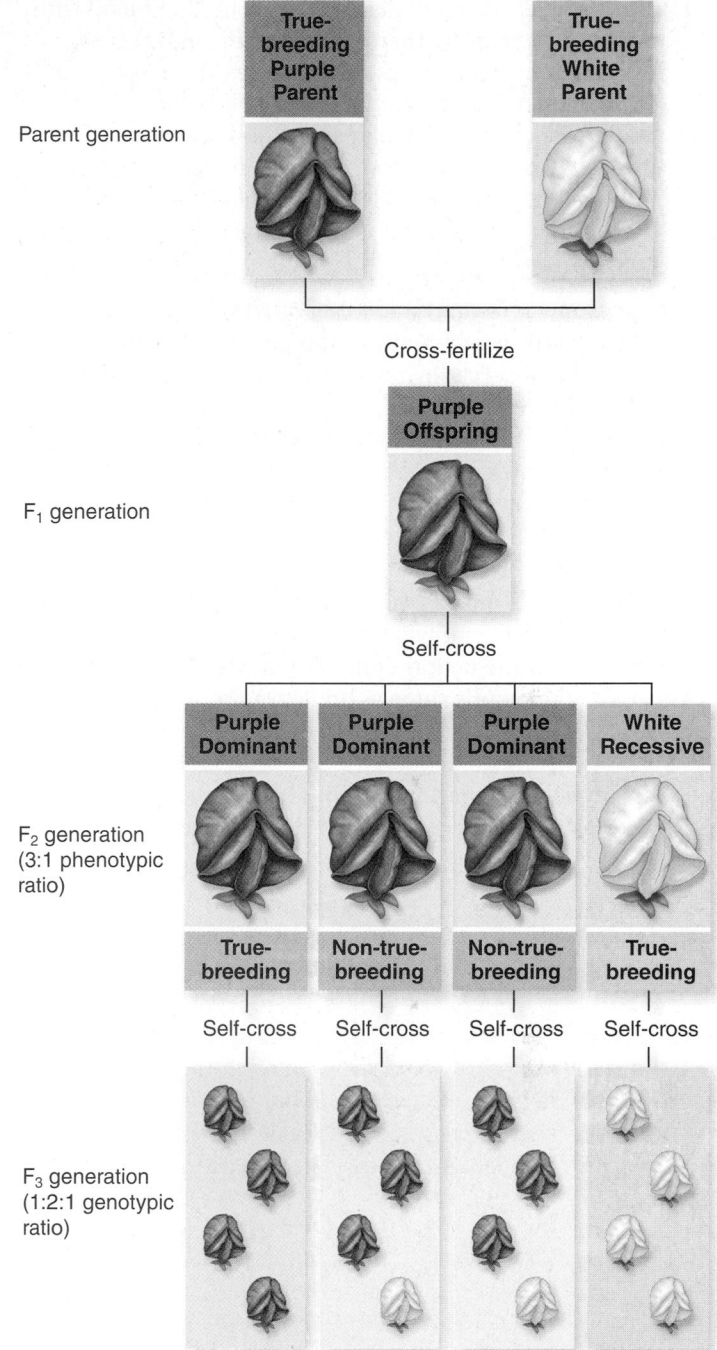

Figure 12.5 The F_2 generation is a disguised 1:2:1 ratio. By allowing the F_2 generation to self-fertilize, Mendel found from the offspring (F_3) that the ratio of F_2 plants was 1 true-breeding dominant: 2 not-true-breeding dominant: and 1 true-breeding recessive.

2. Each individual receives one copy of each gene from each parent. We now know that genes are carried on chromosomes, and each adult individual is diploid, with one set of chromosomes from each parent.
3. Not all copies of a gene are identical. The alternative forms of a gene are called **alleles.** When two haploid gametes containing the same allele fuse during fertilization, the resulting offspring is said to be **homozygous.** When the two haploid gametes contain different alleles, the resulting offspring is said to be **heterozygous.**

4. The two alleles remain discrete—they neither blend with nor alter each other. Therefore, when the individual matures and produces its own gametes, the alleles segregate randomly into these gametes.

5. The presence of a particular allele does not ensure that the trait it encodes will be expressed. In heterozygous individuals, only one allele is expressed (the dominant one), and the other allele is present but unexpressed (the recessive one).

Geneticists now refer to the total set of alleles that an individual contains as the individual's **genotype.** The physical appearance or other observable characteristics of that individual, which result from an allele's expression, is termed the individual's **phenotype.** In other words, the genotype is the blueprint, and the phenotype is the visible outcome in an individual.

This also allows us to present Mendel's ratios in more modern terms. The 3:1 ratio of dominant to recessive is the monohybrid phenotypic ratio. The 1:2:1 ratio of homozygous dominant to heterozygous to homozygous recessive is the monohybrid genotypic ratio. The genotypic ratio "collapses" into the phenotypic ratio due to the action of the dominant allele making the heterozygote appear the same as homozygous dominant.

The principle of segregation

Mendel's model accounts for the ratios he observed in a neat and satisfying way. His main conclusion—that alternative alleles for a character segregate from each other during gamete formation and remain distinct—has since been verified in many other organisms. It is commonly referred to as Mendel's first law of heredity, or the **Principle of Segregation.** It can be simply stated as: *The two alleles for a gene segregate during gamete formation and are rejoined at random, one from each parent, during fertilization.*

The physical basis for allele segregation is the behavior of chromosomes during meiosis. As you saw in chapter 11, homologues for each chromosome disjoin during anaphase I of meiosis. The second meiotic division then produces gametes that contain only one homologue for each chromosome.

It is a tribute to Mendel's intellect that his analysis arrived at the correct scheme, even though he had no knowledge of the cellular mechanisms of inheritance; neither chromosomes nor meiosis had yet been described.

The Punnett square allows symbolic analysis

To test his model, Mendel first expressed it in terms of a simple set of symbols. He then used the symbols to interpret his results.

Consider again Mendel's cross of purple-flowered with white-flowered plants. By convention, we assign the symbol *P* (uppercase) to the dominant allele, associated with the production of purple flowers, and the symbol *p* (lowercase) to the recessive allele, associated with the production of white flowers.

In this system, the genotype of an individual that is true-breeding for the recessive white-flowered trait would be designated

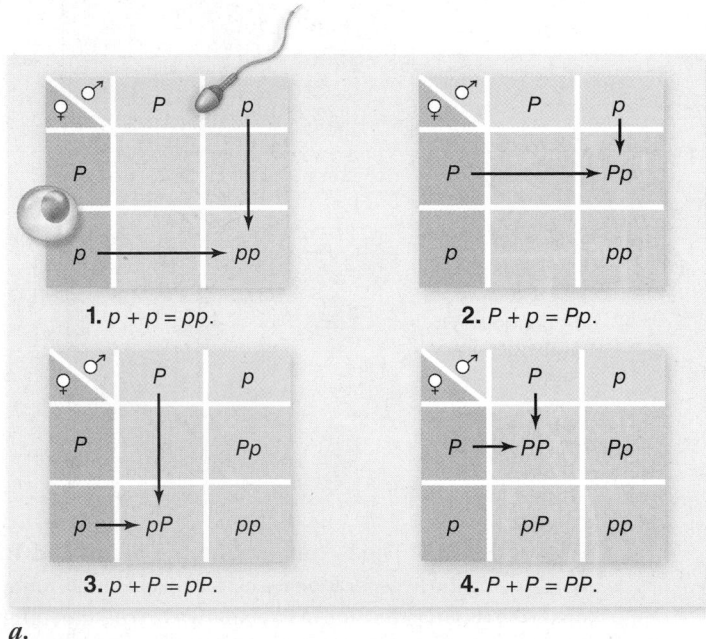

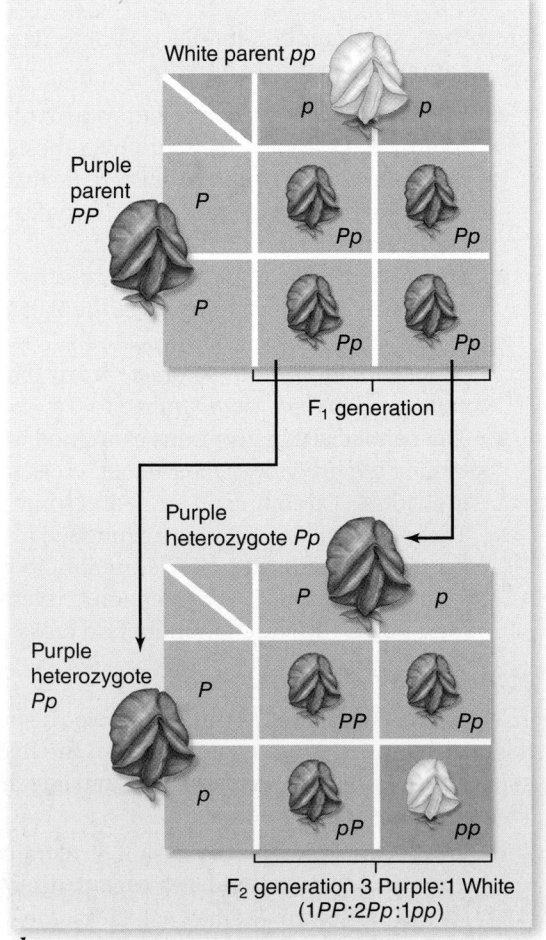

Figure 12.6 Using a Punnett square to analyze Mendel's cross.
a. To make a Punnett square, place the different female gametes along the side of a square and the different male gametes along the top. Each potential zygote is represented as the intersection of a vertical line and a horizontal line. *b.* In Mendel's cross of purple by white flowers, each parent makes only one type of gamete. The F₁ are all purple, *Pp,* heterozygotes. These F₁ offspring make two types of gametes that can be combined to produce three kinds of F₂ offspring: *PP* homozygous dominant (purple); *Pp* heterozygous (also purple); and *pp* homozygous recessive (white). The phenotypic ratio is 3 purple:1 white. The genotypic ratio is 1 *PP*:2 *Pp*:1 *pp.*

TABLE 12.1 — Some Dominant and Recessive Traits in Humans

Recessive Traits	Phenotypes	Dominant Traits	Phenotypes
Albinism	Lack of melanin pigmentation	Middigital hair	Presence of hair on middle segment of fingers
Alkaptonuria	Inability to metabolize homogentisic acid	Brachydactyly	Short fingers
Red-green color blindness	Inability to distinguish red or green wavelengths of light	Huntington disease	Degeneration of nervous system, starting in middle age
Cystic fibrosis	Abnormal gland secretion, leading to liver degeneration and lung failure	Phenylthiocarbamide (PTC) sensitivity	Ability to taste PTC as bitter
Duchenne muscular dystrophy	Wasting away of muscles during childhood	Camptodactyly	Inability to straighten the little finger
Hemophilia	Inability of blood to clot properly, some clots form but the process is delayed	Hypercholesterolemia (the most common human Mendelian disorder)	Elevated levels of blood cholesterol and risk of heart attack
Sickle cell anemia	Defective hemoglobin that causes red blood cells to curve and stick together	Polydactyly	Extra fingers and toes

pp. Similarly, the genotype of a true-breeding purple-flowered individual would be designated PP. In contrast, a heterozygote would be designated Pp (dominant allele first). Using these conventions and denoting a cross between two strains with ×, we can symbolize Mendel's original purple × white cross as $PP \times pp$.

Because a white-flowered parent (pp) can produce only p gametes, and a true-breeding purple-flowered parent (PP, *homozygous dominant*) can produce only P gametes, the union of these gametes can produce only heterozygous Pp offspring in the F$_1$ generation. Because the P allele is dominant, all of these F$_1$ individuals are expected to have purple flowers.

When F$_1$ individuals are allowed to self-fertilize, the P and p alleles segregate during gamete formation to produce both P gametes and p gametes. Their subsequent union at fertilization to form F$_2$ individuals is random.

The F$_2$ possibilities may be visualized in a simple diagram called a **Punnett square,** named after its originator, the English geneticist R. C. Punnett (figure 12.6a). Mendel's model, analyzed in terms of a Punnett square, clearly predicts that the F$_2$ generation should consist of ¾ purple-flowered plants and ¼ white-flowered plants, a phenotypic ratio of 3:1 (figure 12.6b).

Some human traits exhibit dominant/recessive inheritance

A number of human traits have been shown to display both dominant and recessive inheritance (table 12.1 provides a sample of these). Researchers cannot perform controlled crosses in humans the way Mendel did with pea plants; instead geneticists study crosses that have already been performed—in other words, family histories. The organized methodology we use is a **pedigree,** a consistent graphical representation of matings and offspring over multiple generations for a particular trait. The information in the pedigree may allow geneticists to deduce a model for the mode of inheritance of the trait. In analyzing these pedigrees, it is important to realize that disease-causing alleles are usually quite rare in the general population.

A dominant pedigree: Juvenile glaucoma

One of the most extensive pedigrees yet produced traced the inheritance of a form of blindness caused by a dominant allele. The disease allele causes a form of hereditary juvenile glaucoma. The disease causes degeneration of nerve fibers in the optic nerve, leading to blindness.

This pedigree followed inheritance over three centuries, following the origin back to a couple in a small town in northwestern France who died in 1495. A small portion of this pedigree is shown in figure 12.7. The dominant nature of the trait is

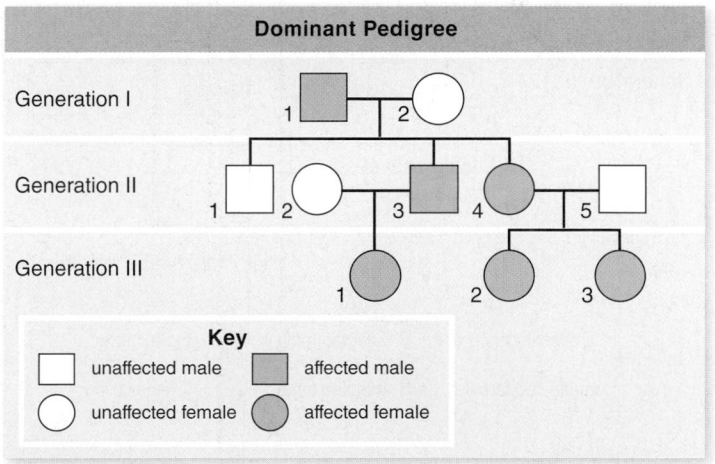

Figure 12.7 Dominant pedigree for hereditary juvenile glaucoma. Males are shown as squares and females are shown as circles. Affected individuals are shown shaded. The dominant nature of this trait can be seen in the trait appearing in every generation, a feature of dominant traits.

 Data analysis If one of the affected females in the third generation married an unaffected male, could she produce unaffected offspring? If so, what are the chances of having unaffected offspring?

obvious from the fact that every generation shows the trait. This is extremely unlikely for a recessive trait as it would require large numbers of unrelated individuals to be carrying the disease allele.

A recessive pedigree: Albinism

An example of inheritance of a recessive human trait is albinism, a condition in which the pigment melanin is not produced. Long thought to be due to a single gene, multiple genes are now known that lead to albinism; the common feature is the loss of pigment from hair, skin, and eyes. The loss of pigment makes albinistic individuals sensitive to the sun. The tanning effect we are all familiar with from exposure to the sun is due to increased numbers of pigment-producing cells, and increased production of pigment. This is lacking in albinistic individuals due to the lack of any pigment to begin with.

The pedigree in figure 12.8 is for a form of albinism due to a nonfunctional allele of the enzyme tyrosinase, which is required for the formation of melanin pigment. The genetic characteristics of this form of albinism are: females and males are affected equally, most affected individuals have unaffected parents, a single affected parent usually does not have affected offspring, and affected offspring are more frequent when parents are related. Each of these features can be seen in figure 12.8, and all of this fits a recessive mode of inheritance.

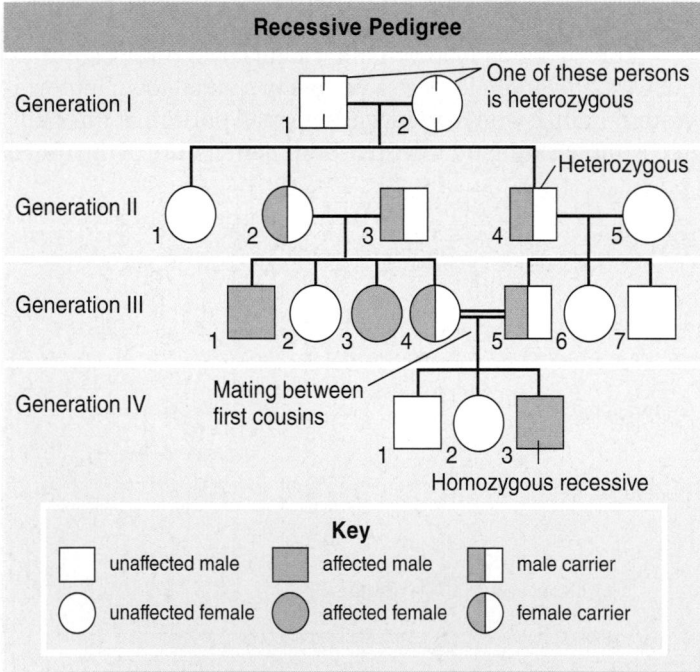

Figure 12.8 Recessive pedigree for albinism. One of the two individuals in the first generation must be heterozygous and individuals II-2 and II-4 must be heterozygous. Notice that for each affected individual, neither parent is affected, but both must be heterozygous (carriers). The double line indicates a consanguineous mating (between relatives) that, in this case, produced affected offspring.

? Inquiry question From the standpoint of genetic disease, why is it never advisable for close relatives to mate and have children?

Learning Outcomes Review 12.2

Mendel's monohybrid crosses refute the idea of blending. One trait disappears in the first generation (F_1), then reappears in a predictable ratio in the next (F_2). The trait observable in the F_1 is called dominant, and the other recessive. In the F_2, the ratio of observed dominant offspring to recessive is 3:1, and this represents a ratio of 1 homozygous dominant to 2 heterozygous to 1 homozygous recessive. The Principle of Segregation states that alleles segregate into different gametes, which randomly combine at fertilization. The physical basis for segregation is the separation of homologues during anaphase I of meiosis.

■ *What fraction of tall F_2 plants are true-breeding?*

12.3 Dihybrid Crosses: The Principle of Independent Assortment

Learning Outcomes

1. *Evaluate the outcome of a dihybrid cross.*
2. *Explain Mendel's Principle of Independent Assortment.*
3. *Compare the segregation of alleles for different genes with the behavior of different homologues in meiosis.*

The Principle of Segregation explains the behavior of alternative forms of a single trait in a monohybrid cross. The next step is to extend this to follow the behavior of two different traits in a single cross: a **dihybrid cross.**

With an understanding of the behavior of single traits, Mendel went on to ask if different traits behaved independently in hybrids. He first established a series of true-breeding lines of peas that differed in two of the seven characters he had studied. He then crossed contrasting pairs of the true-breeding lines to create heterozygotes. These heterozygotes are now doubly heterozygous, or dihybrid. Finally, he self-crossed the dihybrid F_1 plants to produce an F_2 generation, and counted all progeny types.

Traits in a dihybrid cross behave independently

Consider a cross involving different seed shape alleles (round, *R*, and wrinkled, *r*) and different seed color alleles (yellow, *Y*, and green, *y*). Crossing round yellow (*RR YY*) with wrinkled green (*rr yy*), produces heterozygous F_1 individuals having the same phenotype (namely round and yellow) and the same genotype (*Rr Yy*). Allowing these dihybrid F_1 individuals to self-fertilize produces an F_2 generation.

The F_2 generation exhibits four types of progeny in a 9:3:3:1 ratio

In analyzing these results, we first consider the number of possible phenotypes. We expect to see the two parental phenotypes: round yellow and wrinkled green. If the traits behave independently, then we can also expect one trait from each parent to produce plants with round green seeds and others with wrinkled yellow seeds.

Next consider what types of gametes the F_1 individuals can produce. Again, we expect the two types of gametes found in the parents: *RY* and *ry.* If the traits behave independently, then we can also expect the gametes *Ry* and *rY.* Using modern language, two genes each with two alleles can be combined four ways to produce these gametes: *RY, ry, Ry,* and *rY.*

A dihybrid Punnett square

We can then construct a Punnett square with these gametes to generate all possible progeny. This is a 4 × 4 square with 16 possible outcomes. Filling in the Punnett square produces all possible offspring (figure 12.9). From this we can see that there are 9 round yellow, 3 wrinkled yellow, 3 round green, and 1 wrinkled green. This predicts a phenotypic ratio of 9:3:3:1 for traits that behave independently.

Mendel's Principle of Independent Assortment explains dihybrid results

What did Mendel actually observe? From a total of 556 seeds from self-fertilized dihybrid plants, he observed the following results:

- 315 round yellow (signified *R__ Y__,* where the underscore indicates the presence of either allele),
- 108 round green *(R__ yy),*
- 101 wrinkled yellow *(rr Y__),* and
- 32 wrinkled green *(rr yy).*

These results are very close to a 9:3:3:1 ratio. (The expected 9:3:3:1 ratio from this many offspring would be 313:104:104:35.)

The alleles of two genes appeared to behave independently of each other. Mendel referred to this phenomenon as the traits assorting independently. Note that this *independent assortment* of different genes in no way alters the segregation of individual pairs of alleles for each gene. Round versus wrinkled seeds occur in a ratio of approximately 3:1 (423:133); so do yellow versus green seeds (416:140). Mendel obtained similar results for other pairs of traits.

We call this Mendel's second law of heredity, or the **Principle of Independent Assortment.** This can also be stated simply: *In a dihybrid cross, the alleles of each gene assort independently.* A more precise statement would be: *the segregation of different allele pairs is independent.* This statement more closely ties independent assortment to the behavior of chromosomes during meiosis (see chapter 11). The independent alignment of different homologous chromosome pairs during metaphase I leads to the independent segregation of the different allele pairs.

Learning Outcomes Review 12.3

Mendel's analysis of dihybrid crosses revealed that the segregation of allele pairs for different traits is independent; this finding is known as Mendel's Principle of Independent Assortment. When individuals that differ in two traits are crossed, and their progeny are intercrossed, the result is four different types that occur in a ratio of 9:3:3:1, Mendel's dihybrid ratio. This occurs because of the independent behavior of different homologous pairs of chromosomes during meiosis I.

- *Which is more important in terms of explaining Mendel's laws, meiosis I or meiosis II?*

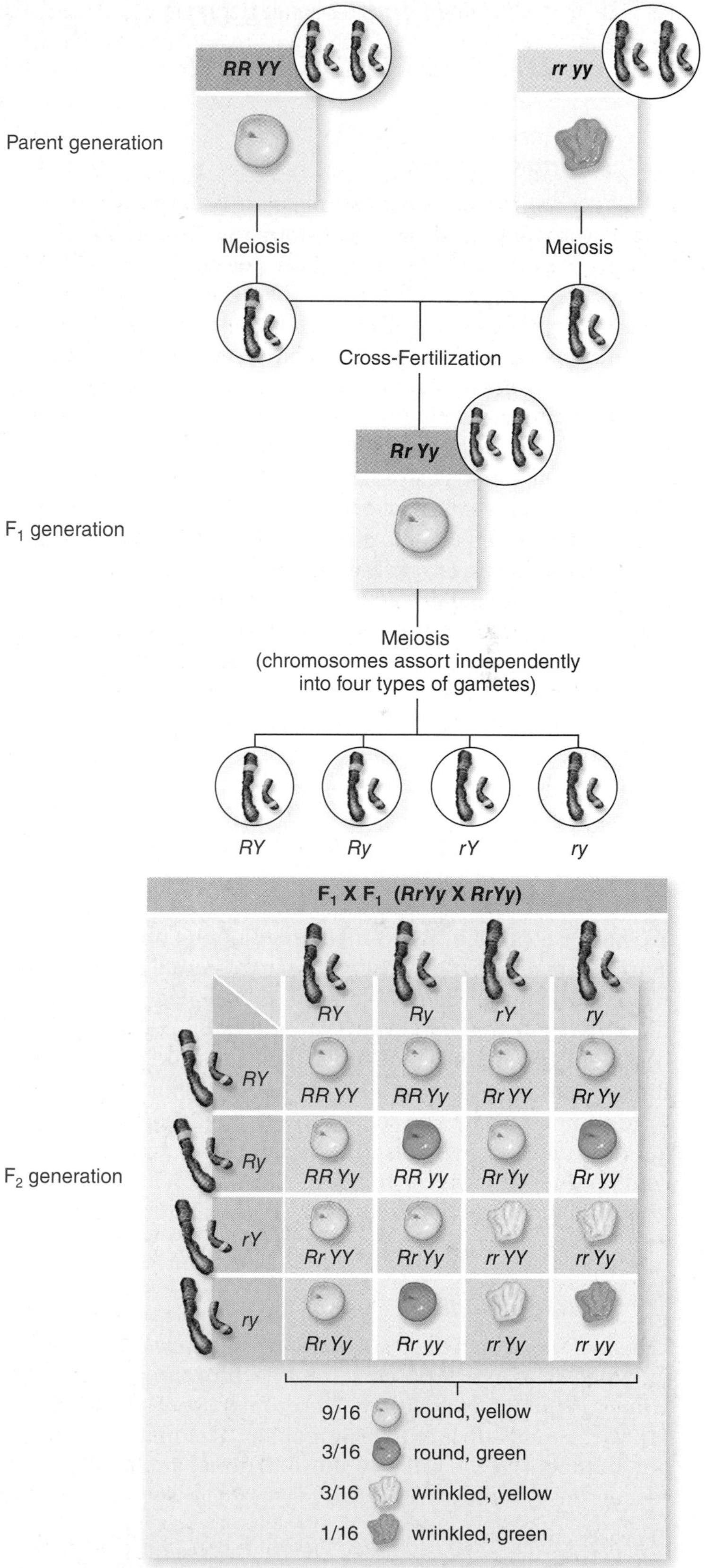

Figure 12.9 Analyzing a dihybrid cross. This Punnett square shows the results of Mendel's dihybrid cross between plants with round yellow seeds and plants with wrinkled green seeds. The ratio of the four possible combinations of phenotypes is predicted to be 9:3:3:1, the ratio that Mendel found.

Probability: Predicting the Results of Crosses

Probability allows us to predict the likelihood of the outcome of random events. Because the behavior of different chromosomes during meiosis is independent, we can use probability to predict the outcome of crosses. The probability of an event that is certain to happen is equal to 1. In contrast, an event that can never happen has a probability of 0. Therefore, probabilities for all other events have fractional values, between 0 and 1. For instance, when you flip a coin, two outcomes are possible; there is only one way to get the event "heads" so the probability of heads is one divided by two, or ½. In the case of genetics, consider a pea plant heterozygous for the flower color alleles P and p. This individual can produce two types of gametes in equal numbers, again due to the behavior of chromosomes during meiosis. There is one way to get a P gamete, so the probability of any particular gamete carrying a P allele is 1 divided by 2 or ½, just like the coin toss.

Two probability rules help predict monohybrid cross results

We can use probability to make predictions about the outcome of genetic crosses using only two simple rules. Before we describe these rules and their uses, we need another definition. We say that two events are *mutually exclusive* if both cannot happen at the same time. The heads and tails of a coin flip are examples of mutually exclusive events. Notice that this is different from two consecutive coin flips where you can get two heads or two tails. In this case, each coin flip represents an *independent event.* It is the distinction between independent and mutually exclusive events that forms the basis for our two rules.

The rule of addition

Consider a six-sided die instead of a coin: for any roll of the die, only one outcome is possible, and each of the possible outcomes are mutually exclusive. The probability of any particular number coming up is ⅙. The probability of either of two different numbers is the sum of the individual probabilities, or restated as the **rule of addition:**

For two mutually exclusive events, the probability of either event occurring is the sum of the individual probabilities.

$$\text{Probability of rolling either a 2 or a 6}$$
$$\text{is} = \frac{1}{6} + \frac{1}{6} = \frac{2}{6} = \frac{1}{3}$$

To apply this to our cross of heterozygous purple F_1, four mutually exclusive outcomes are possible: PP, Pp, pP, and pp. The probability of being heterozygous is the same as the probability of being either Pp or pP, or ¼ plus ¼, or ½.

$$\text{Probability of } F_2 \text{ heterozygote} = \frac{1}{4}Pp + \frac{1}{4}pP = \frac{1}{2}$$

In the previous example, of 379 total offspring, we would expect about 190 to be heterozygotes. (The actual number is 189.5.)

The rule of multiplication

The second rule, and by far the most useful for genetics, deals with the outcome of independent events. This is called the **product rule,** or **rule of multiplication,** and it states that the probability of two independent events both occurring is the *product* of their individual probabilities.

We can apply this to a monohybrid cross in which offspring are formed by gametes from each of two parents. For any particular outcome then, this is due to two independent events: the formation of two different gametes. Consider the purple F_1 parents from earlier. They are all Pp (heterozygotes), so the probability that a particular F_2 individual will be pp (homozygous recessive) is the probability of receiving a p gamete from the male (½) times the probability of receiving a p gamete from the female (½), or ¼:

$$\text{Probability of } pp \text{ homozygote} = \frac{1}{2}p \text{ (male parent)} \times \frac{1}{2}p$$
$$\text{(female parent)} = \frac{1}{4}pp$$

This is actually the basis for the Punnett square that we used before. Each cell in the square was the product of the probabilities of the gametes that contribute to the cell. We then use the addition rule to sum the probabilities of the mutually exclusive events that make up each cell.

We can use the result of a probability calculation to predict the number of homozygous recessive offspring in a cross between heterozygotes. For example, out of 379 total offspring, we would expect about 95 to exhibit the homozygous recessive phenotype. (The actual calculated number is 94.75.)

Dihybrid cross probabilities are based on monohybrid cross probabilities

Probability analysis can be extended to the dihybrid case. For our purple F_1 by F_1 cross, there are four possible outcomes, three of which show the dominant phenotype. Thus the probability of any offspring showing the dominant phenotype is ¾, and the probability of any offspring showing the recessive phenotype is ¼. Now we can use this and the product rule to predict the outcome of a dihybrid cross. We will use our example of seed shape and color from earlier, but now examine it using probability.

If the alleles affecting seed shape and seed color segregate independently, then the probability that a particular pair of alleles for seed shape would occur together with a particular pair of alleles for seed color is the product of the individual probabilities for each pair. For example, the probability that an individual with wrinkled green seeds ($rr\ yy$) would appear in the F_2 generation would be equal to the probability of obtaining wrinkled seeds (¼) times the probability of obtaining green seeds (¼), or ¹⁄₁₆.

$$\text{Probability of } rr\ yy = \frac{1}{4}\ rr \times \frac{1}{4}\ yy = \frac{1}{16}\ rr\ yy$$

Because of independent assortment, we can think of the dihybrid cross of consisting of two independent monohybrid crosses; since these are independent events, the product rule applies. So, we can calculate the probabilities for each dihybrid phenotype:

Probability of round yellow ($R\_\_ Y\_\_$) =
$$\tfrac{3}{4} R\_\_ \times \tfrac{3}{4} Y\_\_ = \tfrac{9}{16}$$

Probability of round green ($R\_\_ yy$) =
$$\tfrac{3}{4} R\_\_ \times \tfrac{1}{4} yy = \tfrac{3}{16}$$

Probability of wrinkled yellow ($rr\ Y\_\_$) =
$$\tfrac{1}{4} rr \times \tfrac{3}{4} Y\_\_ = \tfrac{3}{16}$$

Probability of wrinkled green ($rr\ yy$) =
$$\tfrac{1}{4} rr \times \tfrac{1}{4} yy = \tfrac{1}{16}$$

The hypothesis that color and shape genes are independently sorted thus predicts that the F_2 generation will display a 9:3:3:1 phenotypic ratio. These ratios can be applied to an observed total offspring to predict the expected number in each phenotypic group. The underlying logic and the results are the same as obtained using the Punnett square.

 Data analysis Purple-flowered, round, yellow peas are crossed to white-flowered, wrinkled, green peas to yield a purple-flowered, round, yellow F_1. If this F_1 is self-crossed, what proportion of progeny should be purple-flowered, round, yellow?

Learning Outcomes Review 12.4

The rule of addition states that the probability of either of two events occurring is the sum of their individual probabilities. The rule of multiplication states that the probability of two independent events both occurring is the product of their individual probabilities. These rules can be applied to genetic crosses to determine the probability of particular genotypes and phenotypes. Results can then be compared against these predictions.

■ *If genes* A *and* B *assort independently, in a cross of* Aa Bb *by* aa Bb, *what is the probability of having the dominant phenotype for both genes?*

12.5 The Testcross: Revealing Unknown Genotypes

Learning Outcome

1. *Interpret data from test crosses to infer unknown genotypes.*

To test his model further, Mendel devised a simple and powerful procedure called the **testcross.** In a testcross, an individual with unknown genotype is crossed with the homozygous recessive genotype—that is, the recessive parental variety. The contribution of the homozygous recessive parent can be ignored, because this parent can contribute only recessive alleles.

Consider a purple-flowered pea plant. It is impossible to tell whether such a plant is homozygous or heterozygous simply by looking at it. To learn its genotype, you can perform a testcross to a white-flowered plant. In this cross, the two possible test plant genotypes will give different results (figure 12.10):

Alternative 1: Unknown individual is homozygous dominant (PP) $PP \times pp$: All offspring have purple flowers (Pp).

Alternative 2: Unknown individual is heterozygous (Pp) $Pp \times pp$: ½ of offspring have white flowers (pp), and ½ have purple flowers (Pp).

Put simply, the appearance of the recessive phenotype in the offspring of a testcross indicates that the test individual's genotype is heterozygous.

For each pair of alleles Mendel investigated, he observed phenotypic F_2 ratios of 3:1 (see figure 12.4) and testcross ratios of 1:1, just as his model had predicted. Testcrosses can also be used to determine the genotype of an individual when two genes are involved. Mendel often performed testcrosses to verify the genotypes of dominant-appearing F_2 individuals.

An F_2 individual exhibiting both dominant traits ($A\_\_ B\_\_$) might have any of the following genotypes: *AABB,*

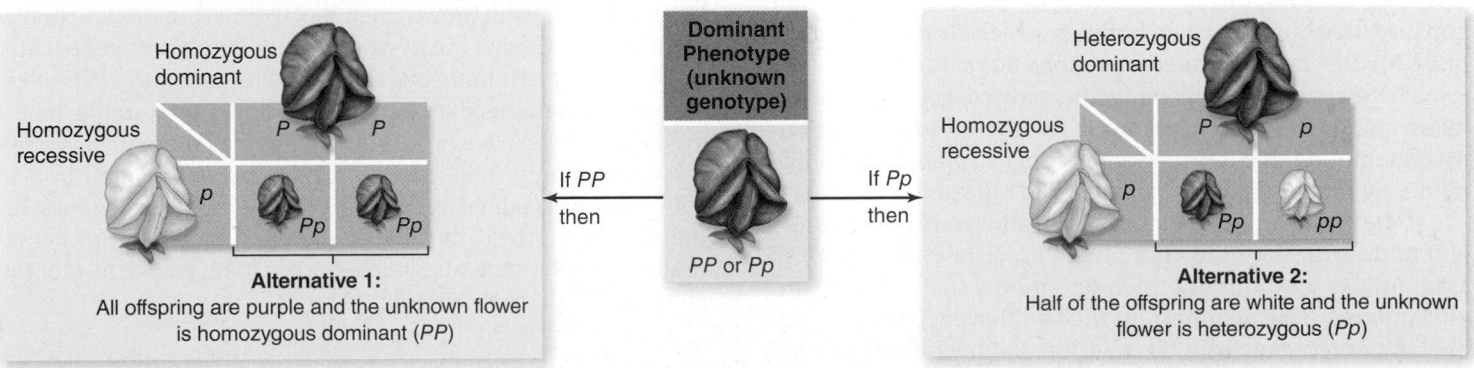

Figure 12.10 Using a testcross to determine unknown genotypes. Individuals with a dominant phenotype, such as purple flowers, can be either homozygous for the dominant allele, or heterozygous. Crossing an unknown purple plant to homozygous recessive *(white)* allows determination of its genotype. The Punnett "square" for each alternative shows only one row for the homozygous recessive white strain because they produce only *p*-bearing gametes.

AaBB, *AABb*, or *AaBb*. By crossing dominant-appearing F$_2$ individuals with homozygous recessive individuals (that is, *A__ B__* × *aabb*), Mendel was able to determine whether either or both of the traits bred true among the progeny, and so to determine the genotype of the F$_2$ parent.

 Data analysis Diagram all four possible testcrosses to determine genotypes for individuals that appear dominant for two traits.

Testcrossing is a powerful tool that simplifies genetic analysis. We will use this method of analysis in the next chapter, when we explore genetic mapping.

Learning Outcome Review 12.5

Individuals showing the dominant phenotype can be either homozygous dominant or heterozygous. Unknown genotypes can be revealed using a testcross, which is a cross to a homozygous recessive individual. Heterozygotes produce both dominant and recessive phenotypes in equal numbers as a result of the testcross.

■ *In a dihybrid testcross of a doubly heterozygous individual, what would be the expected phenotypic ratio?*

12.6 Extensions to Mendel

Learning Outcomes

1. *Describe how assumptions in Mendel's model result in oversimplification.*
2. *Discuss a genetic explanation for continuous variation.*
3. *Explain the genetic basis for observed alterations to Mendel's ratios.*

Although Mendel's results did not receive much notice during his lifetime, three different investigators independently rediscovered his pioneering paper in 1900, 16 years after his death. They came across it while searching the literature in preparation for publishing their own findings, which closely resembled those Mendel had presented more than 30 years earlier.

In the decades following the rediscovery of Mendel's ideas, many investigators set out to test them. However, scientists attempting to confirm Mendel's theory often had trouble obtaining the same simple ratios he had reported.

The reason that Mendel's simple ratios were not obtained had to do with the traits that others examined. A number of assumptions are built into Mendel's model that are oversimplifications. These assumptions include that each trait is specified by a single gene with two alternative alleles; that there are no environmental effects; and that gene products act independently. The idea of dominance also hides a wealth of biochemical complexity. In the following sections, you'll see how Mendel's simple ideas can be extended to provide a more complete view of genetics (table 12.2).

In polygenic inheritance, more than one gene can affect a single trait

Often, the relationship between genotype and phenotype is more complicated than a single allele producing a single trait. Most phenotypes also do not reflect simple two-state cases like purple or white flowers.

Consider Mendel's crosses between tall and short pea plants. In reality, the "tall" plants actually have normal height, and the "short" plants are dwarfed by an allele at a single gene. But in most species, including humans, height varies over a continuous range, rather than having discrete values. This continuous distribution of a phenotype has a simple genetic explanation: more than one gene is at work. The mode of inheritance operating in this case is often called **polygenic inheritance.**

In reality, few phenotypes result from the action of only one gene. Instead, most characters reflect multiple additive contributions to the phenotype by several genes. When multiple genes act jointly to influence a character, such as height or weight, the character often shows a range of small differences. When these genes segregate independently, a gradation in the degree of difference can be observed when a group consisting of many individuals is examined (figure 12.11). We call this gradation **continuous variation,** and we call such traits **quantitative traits.** The greater the number of genes influencing a character, the more continuous the expected distribution of the versions of that character.

This continuous variation in traits is similar to blending different colors of paint: Combining one part red with seven parts white, for example, produces a much lighter shade of pink than does combining five parts red with three parts white. Different ratios of red to white result in a continuum of shades, ranging from pure red to pure white.

Often, variations can be grouped into categories, such as different height ranges. Plotting the numbers in each height category produces a curve called a *histogram*, such as that shown in figure 12.11. The bell-shaped histogram approximates an idealized *normal distribution*, in which the central tendency is characterized by the mean, and the spread of the curve indicates the amount of variation.

Even simple-appearing traits can have this kind of polygenic basis. For example, human eye colors are often described in simple terms with brown dominant to blue, but this is actually incorrect. Extensive analysis indicates multiple genes are involved in determining eye color. This leads to more complex inheritance patterns than initially reported. For example, blue-eyed parents can have brown-eyed offspring, although it is rare. It is interesting to note that most phenotypic variation in eye color can be explained by the interaction of two to four genes. In the developing field of forensic genetics, attempts are being made to predict such phenotypes as eye, hair, and skin color by analysis of unknown samples.

In pleiotropy, a single gene can affect more than one trait

Not only can more than one gene affect a single trait, but a single gene can affect more than one trait. Considering the

TABLE 12.2	When Mendel's Laws/Results May Not Be Observed	
Genetic Occurrence	**Definition**	**Examples**
Polygenic inheritance	More than one gene can affect a single trait.	• Four genes are involved in determining eye color. • Human height
Pleiotropy	A single gene can affect more than one trait.	• A pleiotropic allele dominant for yellow fur in mice is recessive for a lethal developmental defect. • Cystic fibrosis • Sickle cell anemia
Multiple alleles for one gene	Genes may have more than two alleles.	ABO blood types in humans
Dominance is not always complete	• In incomplete dominance the heterozygote is intermediate. • In codominance no single allele is dominant, and the heterozygote shows some aspect of both homozygotes.	• Japanese four o'clocks • Human blood groups
Environmental factors	Genes may be affected by the environment.	Siamese cats
Gene interaction	Products of genes can interact to alter genetic ratios.	• The production of a purple pigment in corn • Coat color in mammals

complexity of biochemical pathways and the interdependent nature of organ systems in multicellular organisms, this should be no surprise.

An allele that has more than one effect on phenotype is said to be **pleiotropic.** The pioneering French geneticist Lucien Cuenot studied yellow fur in mice, a dominant trait, and

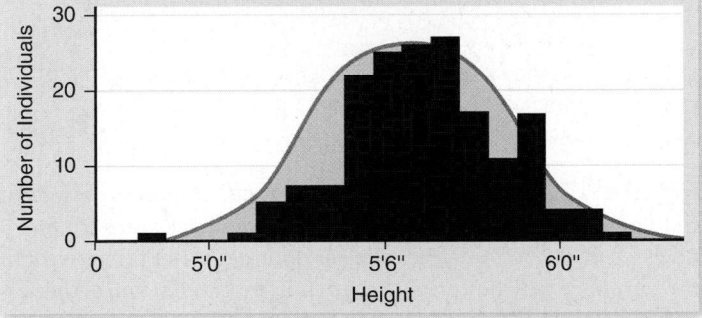

Figure 12.11 Height is a continuously varying trait. The photo and accompanying graph show variation in height among students of the 1914 class at the Connecticut Agricultural College. Because many genes contribute to height and tend to segregate independently of one another, the cumulative contribution of different combinations of alleles to height forms a *continuous* distribution of possible heights, in which the extremes are much rarer than the intermediate values. Variation can also arise due to environmental factors such as nutrition.

found he was unable to obtain a pure-breeding yellow strain by crossing individual yellow mice with each other. Individuals homozygous for the yellow allele died, because the yellow allele was pleiotropic: One effect was yellow coat color, but another was a lethal developmental defect.

 Data analysis When Cuenot crossed yellow mice, what ratio of yellow to wild-type mice did he observe?

A pleiotropic allele may be dominant with respect to one phenotypic consequence (yellow fur) and recessive with respect to another (lethal developmental defect). Pleiotropic effects are difficult to predict, because a gene that affects one trait often performs other, unknown functions.

Pleiotropic effects are characteristic of many inherited disorders in humans, including cystic fibrosis and sickle cell anemia (discussed in chapter 13). In these disorders, multiple symptoms (phenotypes) can be traced back to a single gene defect. Cystic fibrosis patients exhibit clogged blood vessels, overly sticky mucus, salty sweat, liver and pancreas failure, and several other symptoms. It is often difficult to deduce the nature of the primary defect from the range of a gene's pleiotropic effects. As it turns out, all these symptoms of cystic fibrosis are pleiotropic effects of a single defect, a mutation in a gene that encodes a chloride ion transmembrane channel.

Genes may have more than two alleles

Mendel always looked at genes with two alternative alleles. Although any diploid individual can carry only two alleles for a gene, there may be more than two alleles in a population. The example of ABO blood types in humans, described later on, involves an allelic series with three alleles.

If you think of a gene as a sequence of nucleotides in a DNA molecule, then the number of possible alleles is huge because even a single nucleotide change could produce a new allele. In reality, the number of alleles possible for any gene is constrained, but usually more than two alleles exist for any gene in an outbreeding population. The dominance relationships of these alleles cannot be predicted, but can be determined by observing the phenotypes for the various heterozygous combinations.

Dominance is not always complete

Mendel's idea of dominant and recessive traits can seem hard to explain in terms of modern biochemistry. For example, if a recessive trait is caused by the loss of function of an enzyme encoded by the recessive allele, then why should a heterozygote, with only half the activity of this enzyme, have the same appearance as a homozygous dominant individual?

The answer is that enzymes usually act in pathways and not alone. These pathways, as you have seen in earlier chapters, can be highly complex in terms of inputs and outputs, and they can sometimes tolerate large reductions in activity of single enzymes in the pathway without reductions in the level of the end-product. When this is the case, complete dominance will be observed; however, not all genes act in this way.

Incomplete dominance

In **incomplete dominance,** the heterozygote is intermediate in appearance between the two homozygotes. For example, in a cross between red- and white-flowering Japanese four o'clocks, described in figure 12.12, all the F_1 offspring have pink flowers—indicating that neither red nor white flower color was dominant. Looking only at the F_1, we might conclude that this is a case of blending inheritance. But when two of the F_1 pink flowers are crossed, they produce red-, pink-, and white-flowered plants in a 1:2:1 ratio. In this case the phenotypic ratio is the same as the genotypic ratio because all three genotypes can be distinguished.

Codominance

Most genes in a population possess several different alleles, and often no single allele is dominant; instead, each allele has its own effect, and the heterozygote shows some aspect of the phenotype of both homozygotes. The alleles are said to be **codominant.**

Codominance can be distinguished from incomplete dominance by the appearance of the heterozygote. In incomplete dominance, the heterozygote is intermediate between the two homozygotes, whereas in codominance, some aspect of both alleles is seen in the heterozygote. One of the clearest human examples is found in the human blood groups.

The different phenotypes of human blood groups are based on the response of the immune system to proteins on the surface of red blood cells. In homozygotes a single type of protein is found on the surface of cells, and in heterozygotes, two kinds of protein are found, leading to codominance.

The human ABO blood group system

The gene that determines ABO blood types encodes an enzyme that adds sugar molecules to proteins on the surface of red blood cells. These sugars act as recognition markers for the immune system (see chapter 51). The gene that encodes the enzyme, designated I, has three common alleles: I^A, whose product adds galactosamine; I^B, whose product adds galactose; and i, which codes for a protein that does not add a sugar.

Hypothesis: *The pink F_1 observed in a cross of red and white Japanese four o'clock flowers is due to failure of dominance and is not an example of blending inheritance.*

Prediction: *If pink F_1 are self-crossed, they will yield progeny the same as the Mendelian monohybrid genotypic ratio. This would be 1 red: 2 pink: 1 white.*

Test: *Perform the cross and count progeny.*

Result: *When this cross is performed, the expected outcome is observed.*

Conclusion: *Flower color in Japanese four o'clock plants exhibits incomplete dominance.*

Further Experiments: *How many offspring would you need to count to be confident in the observed ratio?*

Figure 12.12 Incomplete dominance. In a cross between a red-flowered (genotype $C^R C^R$) Japanese four o'clock and a white-flowered one ($C^W C^W$), neither allele is dominant. The heterozygous progeny have pink flowers and the genotype $C^R C^W$. If two of these heterozygotes are crossed, the phenotypes of their progeny occur in a ratio of 1:2:1 (red:pink:white).

The three alleles of the I gene can be combined to produce six different genotypes. An individual heterozygous for the I^A and I^B alleles produces both forms of the enzyme and exhibits both galactose and galactosamine on red blood cells. Because both alleles are expressed simultaneously in heterozygotes, the I^A and I^B alleles are codominant. Both I^A and I^B are dominant over the i allele, because both I^A and I^B alleles lead to sugar addition,

Alleles	Blood Type	Sugars Exhibited	Donates and Receives
I^AI^A, I^Ai (I^A dominant to i)	A	Galactosamine	Receives A and O Donates to A and AB
I^BI^B, I^Bi (I^B dominant to i)	B	Galactose	Receives B and O Donates to B and AB
I^AI^B (codominant)	AB	Both galactose and galactosamine	Universal receiver Donates to AB
ii (i is recessive)	O	None	Receives O Universal donor

Figure 12.13 ABO blood groups illustrate both codominance and multiple alleles. There are three alleles of the *I* gene: I^A, I^B, and *i*. I^A and I^B are both dominant to *i* (see types A and B), but codominant to each other (see type AB). The genotypes that give rise to each blood type are shown with the associated phenotypes in terms of sugars added to surface proteins and the body's reaction after a blood transfusion.

whereas the *i* allele does not. The different combinations of the three alleles produce four different phenotypes (figure 12.13):

1. Type A individuals add only galactosamine. They are either I^AI^A homozygotes or I^Ai heterozygotes (two genotypes).
2. Type B individuals add only galactose. They are either I^BI^B homozygotes or I^Bi heterozygotes (two genotypes).
3. Type AB individuals add both sugars and are I^AI^B heterozygotes (one genotype).
4. Type O individuals add neither sugar and are *ii* homozygotes (one genotype).

These four different cell-surface phenotypes are called the **ABO blood groups.**

A person's immune system can distinguish among these four phenotypes. If a type A individual receives a transfusion of type B blood, the recipient's immune system recognizes the "foreign" antigen (galactose) and attacks the donated blood cells, causing them to clump, or agglutinate. The same thing would happen if the donated blood is type AB. However, if the donated blood is type O, no immune attack occurs, because there are no galactose antigens.

In general, any individual's immune system can tolerate a transfusion of type O blood, and so type O is termed the "universal donor." Because neither galactose nor galactosamine is foreign to type AB individuals (whose red blood cells have both sugars), those individuals may receive any type of blood, and type AB is termed the "universal recipient." Nevertheless, matching blood is preferable for any transfusion.

Phenotypes may be affected by the environment

Another assumption, implicit in Mendel's work, is that the environment does not affect the relationship between genotype and phenotype. For example, the soil in the abbey yard where Mendel performed his experiments was probably not uniform, and yet its possible effect on the expression of traits was ignored. But in reality, although the expression of genotype produces phenotype, the environment can affect this relationship.

Environmental effects are not limited to the external environment. For example, the alleles of some genes encode heat-sensitive products that are affected by differences in internal body temperature. The *ch* allele in Himalayan rabbits and Siamese cats encodes a heat-sensitive version of the enzyme tyrosinase, which as you may recall is involved in albinism (figure 12.14). The Ch version of the enzyme is inactivated at temperatures above about 33°C. At the surface of the torso and head of these animals, the temperature is above 33°C and tyrosinase is inactive, producing a whitish coat. At the extremities, such as the tips of the ears and tail, the temperature is usually below 33°C and the enzyme is active, allowing production of melanin that turns the coat in these areas a dark color.

> **? Inquiry question** Many studies of identical twins separated at birth have revealed phenotypic differences in their development (height, weight, etc.). If these are identical twins, can you propose an explanation for these differences?

In epistasis, interactions of genes alter genetic ratios

The last simplistic assumption in Mendel's model is that the products of genes do not interact. But the products of genes may not act independently of one another, and the interconnected behavior of gene products can change the ratio expected by independent assortment, even if the genes are on different chromosomes that do exhibit independent assortment.

Given the interconnected nature of metabolism, it should not come as a surprise that many gene products are not independent. Genes that act in the same metabolic pathway, for example, should show some form of dependence at the level of function. In such cases, the ratio Mendel would predict is not readily observed, but it is still there in an altered form.

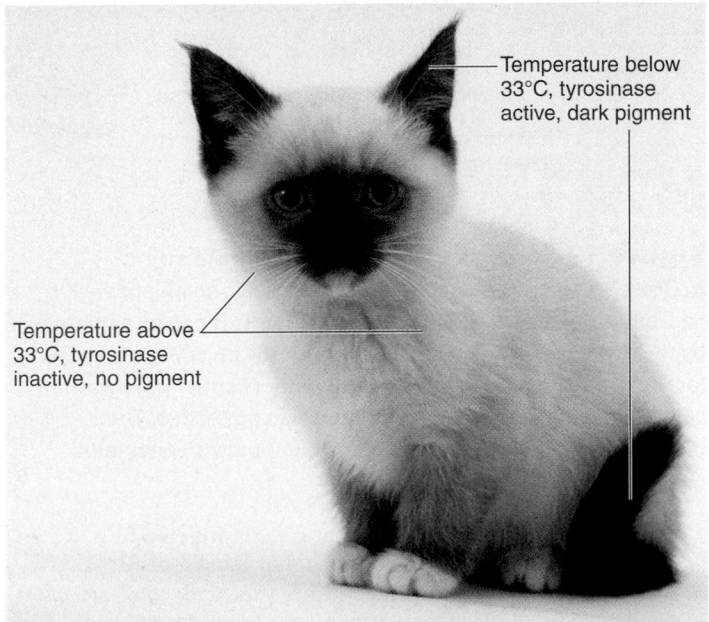

Temperature below 33°C, tyrosinase active, dark pigment

Temperature above 33°C, tyrosinase inactive, no pigment

Figure 12.14 Siamese cat. The pattern of coat color is due to an allele that encodes a temperature-sensitive form of the enzyme tyrosinase.

In the tests of Mendel's ideas that followed the rediscovery of his work, scientists had trouble obtaining Mendel's simple ratios, particularly with dihybrid crosses. Sometimes, it was not possible to identify successfully each of the four phenotypic classes expected, because two or more of the classes looked alike.

An example of this comes from the analysis of particular varieties of corn, *Zea mays*. Some commercial varieties exhibit a purple pigment called anthocyanin in their seed coats, whereas others do not. In 1918, geneticist R. A. Emerson crossed two true-breeding corn varieties, each lacking anthocyanin pigment. Surprisingly, all of the F_1 plants produced purple seeds.

When two of these pigment-producing F_1 plants were crossed to produce an F_2 generation, 56% were pigment producers and 44% were not. This is clearly not what Mendel's ideas would lead us to expect. Emerson correctly deduced that two genes were involved in producing pigment, and that the second cross had thus been a dihybrid cross. According to Mendel's theory, gametes in a dihybrid cross could combine in 16 equally possible ways—so the puzzle was to figure out how these 16 combinations could occur in the two phenotypic groups of progeny. Emerson multiplied the fraction that were pigment producers (0.56) by 16 to obtain 9, and multiplied the fraction that lacked pigment (0.44) by 16 to obtain 7. Emerson therefore had a *modified ratio* of 9:7 instead of the usual 9:3:3:1 ratio (figure 12.15).

This modified ratio is easily rationalized by considering the function of the products encoded by these genes. When gene products act sequentially, as in a biochemical pathway, an allele expressed as a defective enzyme early in the pathway blocks the flow of material through the rest of the pathway. In this case, it is impossible to judge whether the later steps of the pathway are functioning properly. This type of gene interaction, in which one gene can interfere with the expression of another, is the basis of the phenomenon called **epistasis.**

The pigment anthocyanin is the product of a two-step biochemical pathway:

$$\text{starting molecule} \xrightarrow{\text{enzyme 1}} \text{intermediate} \xrightarrow{\text{enzyme 2}} \text{anthocyanin}$$
$$\text{(colorless)} \qquad\qquad \text{(colorless)} \qquad\qquad \text{(purple)}$$

To produce pigment, a plant must possess at least one functional copy of each enzyme's gene. The dominant alleles encode functional enzymes, and the recessive alleles encode nonfunctional enzymes. Of the 16 genotypes predicted by random assortment, 9 contain at least one dominant allele of both genes; they therefore produce purple progeny. The remaining 7 genotypes lack dominant alleles at *either or both* loci (3 + 3 + 1 = 7) and so produce colorless progeny, giving the phenotypic ratio of 9:7 that Emerson observed (see figure 12.15).

You can see that although this ratio is not the expected dihybrid ratio, it is a modification of the expected ratio.

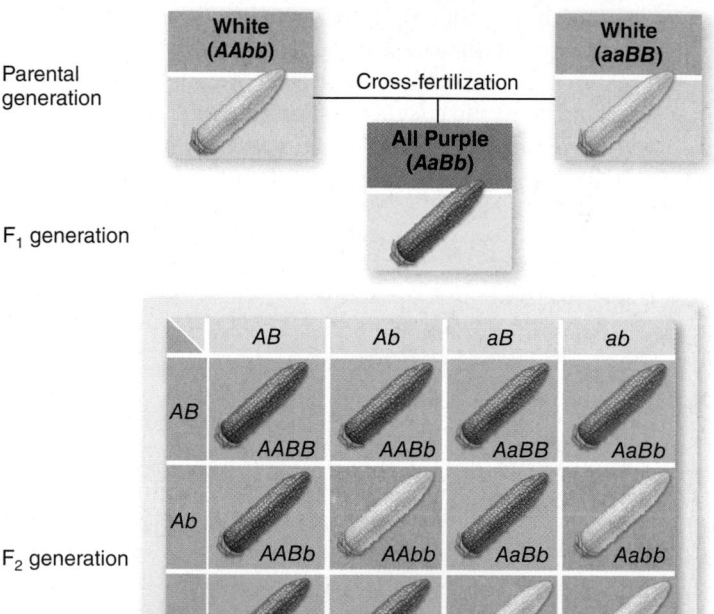

a.

b.

Figure 12.15 How epistasis affects grain color.
a. Crossing some white varieties of corn yields an all purple F_1. Self-crossing the F_1 yields 9 purple:7 white. This can be explained by the presence of two genes, each encoding an enzyme necessary for the production of purple pigment. Unless both enzymes are active (genotype is $A\_B\_$), no pigment is expressed. *b.* The biochemical pathway for pigment production with enzymes encoded by *A* and *B* genes.

Data analysis Mouse coat color is affected by a number of genes, including one that causes total loss of pigment (albinism) and another that leads to black/ brown fur. A black mouse is crossed to another black mouse, yielding progeny with a ratio of 9 black:3 brown:4 albino. How can you explain these data if these two genes segregate independently?

Chapter Review

12.1 The Mystery of Heredity

Early plant biologists produced hybrids and saw puzzling results.

Plant breeders noticed that some forms of a trait can disappear in one generation only to reappear later, that is, they segregate rather than blend.

Mendel used mathematics to analyze his crosses.

Mendel's experiments involved reciprocal crosses between true-breeding pea varieties followed by one or more generations of self-fertilization. His mathematical analysis of experimental results led to the present model of inheritance.

12.2 Monohybrid Crosses: The Principle of Segregation (figure 12.5)

The F_1 generation exhibits only one of two traits, without blending.

Mendel called the trait visible in the F_1 the dominant trait; the other he termed recessive.

The F_2 generation exhibits both traits in a 3:1 ratio.

When F_1 plants are self-fertilized, the F_2 shows a consistent ratio of 3 dominant:1 recessive. We call this 3:1 ratio the Mendelian monohybrid ratio.

The 3:1 ratio is actually 1:2:1.

Mendel then examined the F_2 and found the recessive F_2 plants always bred true, but only one out of three dominant F_2 bred true. This means the 3:1 ratio is actually 1 true-breeding dominant:2 non-true-breeding dominant:1 recessive.

Mendel's Principle of Segregation explains monohybrid observations.

Traits are determined by discrete factors we now call genes. These exist in alternative forms we call alleles. Individuals carrying two identical alleles for a gene are said to be homozygous, and individuals carrying different alleles are said to be heterozygous. The genotype is the entire set of alleles of all genes possessed by an individual. The phenotype is the individual's appearance due to these alleles.

The Principle of Segregation states that during gamete formation, the two alleles of a gene separate (segregate). Parental alleles then randomly come together to form the diploid zygote. The physical basis of segregation is the separation of homologues during anaphase of meiosis I.

The Punnett square allows symbolic analysis.

Punnett squares are formed by placing the gametes from one parent along the top of the square with the gametes from the other parent along the side. Zygotes formed from gamete combinations form the blocks of the square (figure 12.6).

Some human traits exhibit dominant/recessive inheritance.

Certain human traits have been found to have a Mendelian basis (table 12.1). Inheritance patterns in human families can be analyzed and inferred using a pedigree diagram of earlier generations.

12.3 Dihybrid Crosses: The Principle of Independent Assortment (figure 12.9)

Traits in a dihybrid cross behave independently.

If parents differing in two traits are crossed, the F_1 will be all dominant. Each F_1 parent can produce four different gametes that can be combined to produce 16 possible outcomes in the F_2. This yields a phenotypic ratio of 9:3:3:1 of the four possible phenotypes.

Mendel's Principle of Independent Assortment explains dihybrid results.

The Principle of Independent Assortment states that different traits segregate independently of one another. The physical basis of independent assortment is the independent behavior of different pairs of homologous chromosomes during meiosis I.

12.4 Probability: Predicting the Results of Crosses

Two probability rules help predict monohybrid cross results.

The rule of addition states that the probability of two independent events occurring is the sum of their individual probabilities. The rule of multiplication, or product rule, states that the probability of two independent events *both* occurring is the product of their individual probabilities.

Dihybrid cross probabilities are based on monohybrid cross probabilities.

A dihybrid cross is essentially two independent monohybrid crosses. The product rule applies and can be used to predict the cross's outcome.

12.5 The Testcross: Revealing Unknown Genotypes (figure 12.10)

In a testcross, an unknown genotype is crossed with a homozygous recessive genotype. The F_1 offspring will all be the same if the unknown genotype is homozygous dominant. The F_1 offspring will exhibit a 1:1 dominant:recessive ratio if the unknown genotype is heterozygous.

12.6 Extensions to Mendel

In polygenic inheritance, more than one gene can affect a single trait.

Many traits, such as human height, are due to multiple additive contributions by many genes, resulting in continuous variation.

In pleiotropy, a single gene can affect more than one trait.

A pleiotropic effect occurs when an allele affects more than one trait. These effects are difficult to predict.

Genes may have more than two alleles.

There may be more than two alleles of a gene in a population. Given the possible number of DNA sequences, this is not surprising.

Dominance is not always complete.

In incomplete dominance the heterozygote exhibits an intermediate phenotype; the monohybrid genotypic and phenotypic ratios are the same (figure 12.12). Codominant alleles each contribute to the phenotype of a heterozygote.

Phenotypes may be affected by the environment.

Genotype determines phenotype, but the environment will have an effect on this relationship. Environment means both external and internal factors. For example, in Siamese cats, a temperature-sensitive enzyme produces more pigment in the colder peripheral areas of the body.

In epistasis, interactions of genes alter genetic ratios.

Genes encoding enzymes that act in a single biochemical pathway are not independent. In corn, anthocyanin pigment production requires the action of two enzymes. Doubly heterozygous individuals for these enzymes yield a 9:7 ratio when self-crossed (figure 12.15).

Review Questions

UNDERSTAND

1. What property distinguished Mendel's investigation from previous studies?
   a. Mendel used true-breeding pea plants.
   b. Mendel quantified his results.
   c. Mendel examined many different traits.
   d. Mendel examined the segregation of traits.

2. The F_1 generation of the monohybrid cross purple (PP) × white (pp) flower pea plants should
   a. all have white flowers.
   b. all have a light purple or blended appearance.
   c. all have purple flowers.
   d. have ¾ purple flowers, and ¼ white flowers.

3. The F_1 plants from the previous question are allowed to self-fertilize. The phenotypic ratio for the F_2 should be
   a. all purple.   c. 3 purple:1 white.
   b. 1 purple:1 white. d. 3 white:1 purple.

4. Which of the following is NOT a part of Mendel's five-element model?
   a. Traits have alternative forms (what we now call alleles).
   b. Parents transmit discrete traits to their offspring.
   c. If an allele is present it will be expressed.
   d. Traits do not blend.

5. An organism's _____ is/are determined by its _____.
   a. genotype; phenotype   c. alleles; phenotype
   b. phenotype; genotype   d. genes; alleles

6. Phenotypes like height in humans, which show a continuous distribution, are usually the result of
   a. an alteration of dominance for multiple alleles of a single gene.
   b. the presence of multiple alleles for a single gene.
   c. the action of one gene on multiple phenotypes.
   d. the action of multiple genes on a single phenotype.

APPLY

1. Japanese four o'clocks that are red and tall are crossed to white short ones, producing an F_1 that is pink and tall. If these genes assort independently, and the F_1 is self-crossed, what would you predict for the ratio of F_2 phenotypes?
   a. 3 red tall:1 white short
   b. 1 red tall:2 pink short:1 white short
   c. 3 pink tall:6 red tall:3 white tall:1 pink short:2 red short:1 white short
   d. 3 red tall:6 pink tall:3 white tall:1 red short:2 pink short:1 white short

2. If the two genes in the previous question showed complete linkage, what would you predict for an F_2 phenotypic ratio?
   a. 1 red tall:2 pink short:1 white short
   b. 1 red tall:2 red short:1 white short
   c. 1 pink tall:2 red tall:1 white short
   d. 1 red tall:2 pink tall:1 white short

3. What is the probability of obtaining an individual with the genotype bb from a cross between two individuals with the genotype Bb?
   a. ½   c. ⅛
   b. ¼   d. 0

4. In a cross of $Aa\ Bb\ cc \times Aa\ Bb\ Cc$, what is the probability of obtaining an individual with the genotype $AA\ Bb\ Cc$?
   a. 1/16   c. 1/64
   b. 3/16   d. 3/64

5. When you cross true-breeding tall and short tobacco plants you get an F_1 that is intermediate in height. When this F_1 is self-crossed, it yields an F_2 with a continuous distribution of heights. What is the best explanation for these data?
   a. Height is determined by a single gene with incomplete dominance.
   b. Height is determined by a single gene with many alleles.
   c. Height is determined by the additive effects of many genes.
   d. Height is determined by epistatic genes.

6. Mendel's model assumes that each trait is determined by a single factor with alternate forms. We now know that this is too simplistic and that
   a. a single gene may affect more than one trait.
   b. a single trait may be affected by more than one gene.
   c. a single gene always affects only one trait, but traits may be affected by more than one gene.
   d. a single gene can affect more than one trait, and traits may be affected by more than one gene.

SYNTHESIZE

1. Create a Punnett square for the following crosses and use this to predict phenotypic ratio for dominant and recessive traits. Dominant alleles are indicated by uppercase letters and recessive are indicated by lowercase letters. For parts b and c, predict ratios using probability and the product rule.
   a. A monohybrid cross between individuals with the genotype Aa and Aa
   b. A dihybrid cross between two individuals with the genotype $AaBb$
   c. A dihybrid cross between individuals with the genotype $AaBb$ and $aabb$

2. Explain how the events of meiosis can explain both segregation and independent assortment.

3. In mice, there is a yellow strain that when crossed yields 2 yellow:1 black. How could you explain this observation? How could you test this with crosses?

4. In mammals, a variety of genes affect coat color. One of these is a gene with mutant alleles that results in the complete loss of pigment, or albinism. Another controls the type of dark pigment with alleles that lead to black or brown colors. The albinistic trait is recessive, and black is dominant to brown. Two black mice are crossed and yield 9 black:4 albino:3 brown. How would you explain these results?

ONLINE RESOURCE

www.ravenbiology.com

Understand, Apply, and Synthesize—enhance your study with animations that bring concepts to life and practice tests to assess your understanding. Your instructor may also recommend the interactive eBook, individualized learning tools, and more.

Chapter 13

Chromosomes, Mapping, and the Meiosis–Inheritance Connection

Chapter Contents

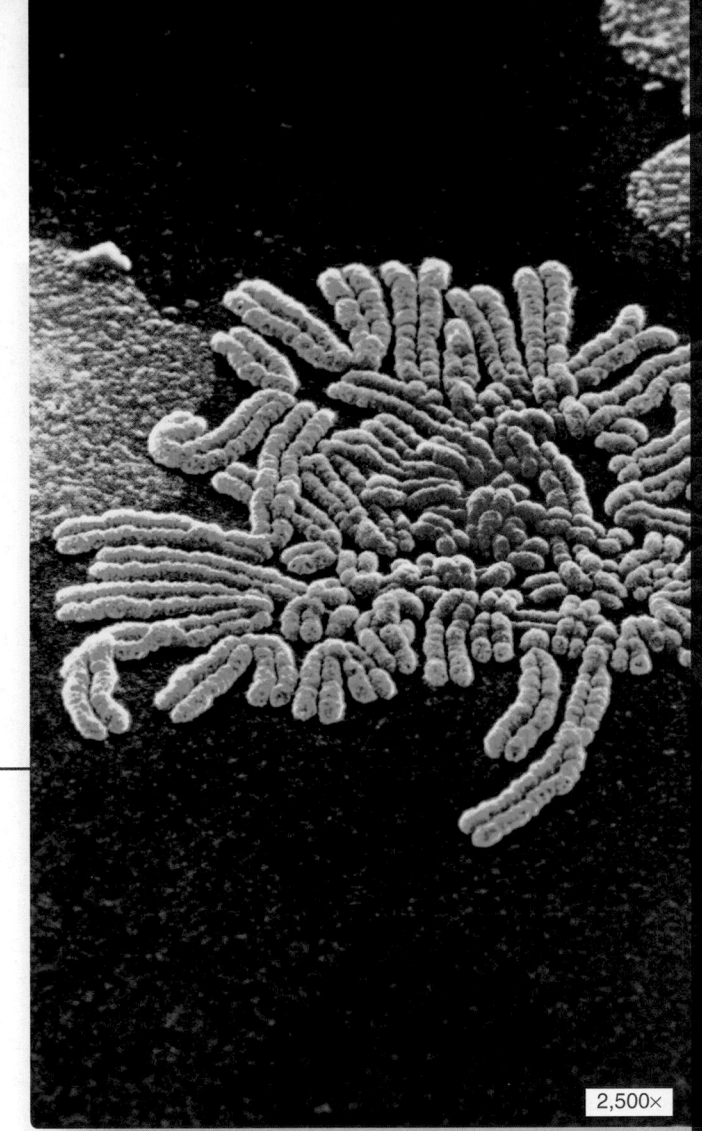

2,500×

Introduction

Mendel's experiments opened the door to understanding inheritance, but many questions remained. In the early part of the 20th century, we did not know the nature of the factors whose behavior Mendel had described. The next step, which involved many researchers in the early part of the century, was uniting information about the behavior of chromosomes, seen in the picture, and the inheritance of traits. The basis for Mendel's principles of segregation and independent assortment lie in events that occur during meiosis.

The behavior of chromosomes during meiosis not only explains Mendel's principles, but leads to new and different approaches to the study of heredity. The ability to construct genetic maps is one of the most powerful tools of classical genetic analysis. The tools of genetic mapping developed in flies and other organisms in combination with information from the Human Genome Project now allow us to determine the location of genes and isolate those that are involved in genetic diseases.

13.1 Sex Linkage and the Chromosomal Theory of Inheritance

Learning Outcomes

1. Describe sex-linked inheritance in fruit flies.
2. Explain the evidence for genes being on chromosomes.

A central role for chromosomes in heredity was first suggested in 1900 by the German geneticist Carl Correns, in one of the papers announcing the rediscovery of Mendel's work. Soon after, observations that similar chromosomes paired with one another during meiosis led directly to the **chromosomal theory of inheritance,** first formulated by the American Walter Sutton in 1902.

Morgan correlated the inheritance of a trait with sex chromosomes

In 1910, Thomas Hunt Morgan, studying the fruit fly *Drosophila melanogaster,* discovered a mutant male fly with white eyes instead of red (figure 13.1).

Morgan immediately set out to determine whether this new trait would be inherited in a Mendelian fashion. He first crossed the mutant male to a normal red-eyed female to see whether the red-eyed or white-eyed trait was dominant. All of the F_1 progeny had red eyes, so Morgan concluded that red eye color was dominant over white.

The F₁ cross

Following the experimental procedure that Mendel had established long ago, Morgan then crossed the red-eyed flies from the F_1 generation with each other. Of the 4252 F_2 progeny Morgan examined, 782 (18%) had white eyes. Although the ratio of red eyes to white eyes in the F_2 progeny was greater than 3:1, the results of the cross nevertheless provided clear evidence that eye color segregates. However, something about the outcome was strange and totally unpredicted by Mendel's theory— *all of the white-eyed F₂ flies were males!* (figure 13.2)

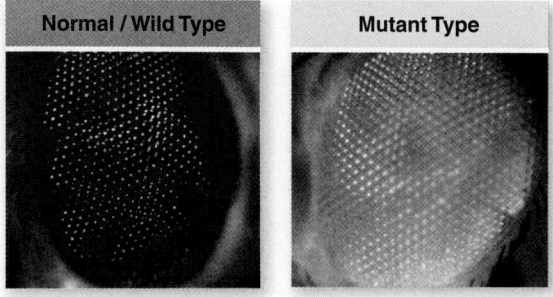

Figure 13.1 Red-eyed (wild-type) and white-eyed (mutant) *Drosophila*. Mutations are heritable alterations in genetic material. By studying the inheritance pattern of white and red alleles (located on the X chromosome), Morgan first demonstrated that genes are on chromosomes.

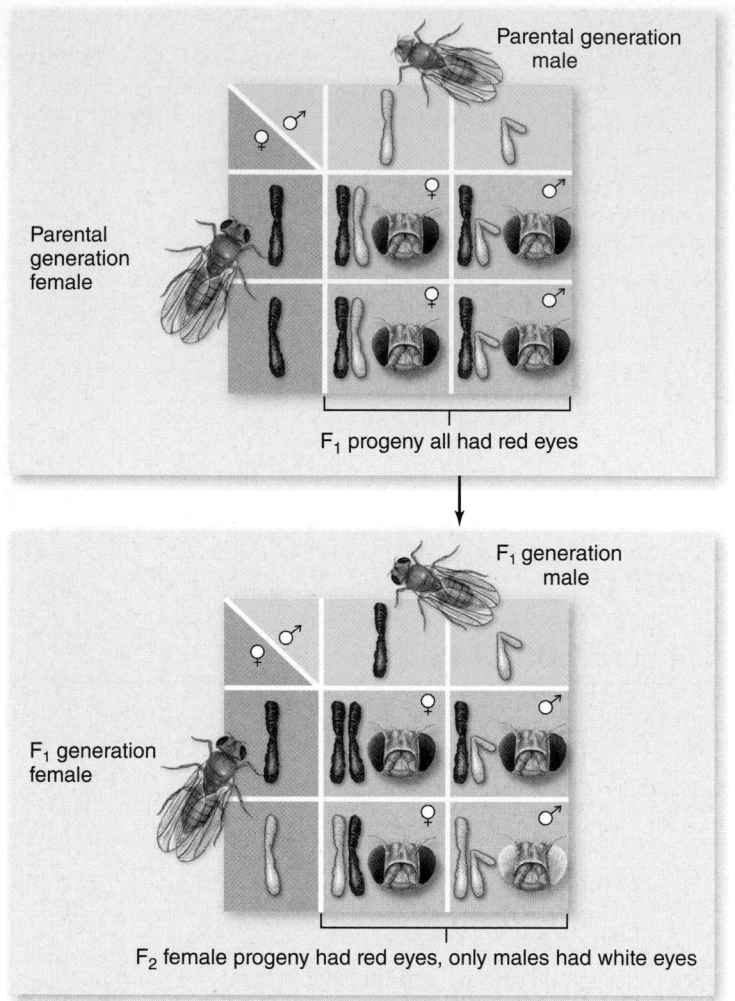

F₁ progeny all had red eyes

F₂ female progeny had red eyes, only males had white eyes

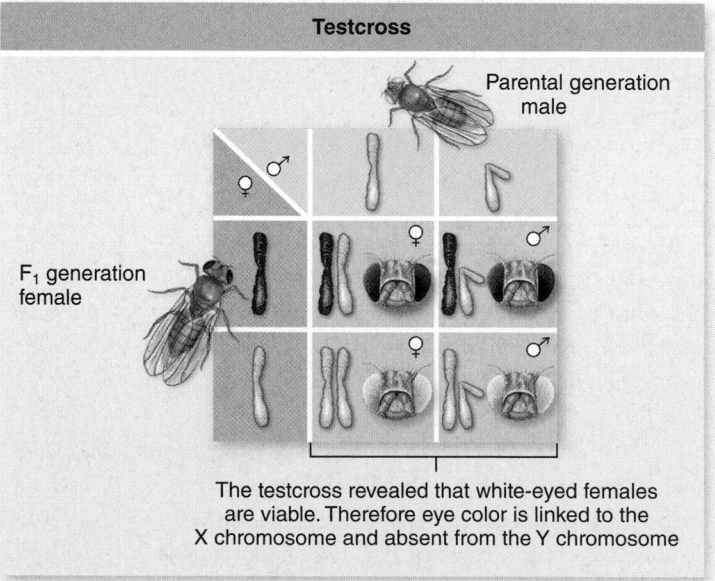

Testcross

The testcross revealed that white-eyed females are viable. Therefore eye color is linked to the X chromosome and absent from the Y chromosome

Figure 13.2 The chromosomal basis of sex linkage. White-eyed male flies are crossed to red-eyed females. The F_1 flies all have red eyes, as expected for a recessive white-eye allele. In the F_2, all of the white-eyed flies are male because the Y chromosome lacks the white-eye (*white*) gene. Inheritance of the sex chromosomes correlates with eye color, showing the *white* gene is on the X chromosome.

The testcross

Morgan sought an explanation for this result. One possibility was simply that white-eyed female flies don't exist; such individuals might not be viable for some unknown reason. To test this idea, Morgan testcrossed the female F_1 progeny with the original white-eyed male. He obtained white-eyed and red-eyed flies of both sexes in a 1:1:1:1 ratio, just as Mendel's theory had predicted (figure 13.2). Therefore, white-eyed female flies are viable. Given that white-eyed females can exist, Morgan turned to the nature of the chromosomes in males and females for an explanation.

The gene for eye color lies on the X chromosome

In *Drosophila*, the sex of an individual is determined by the number of copies it has of a particular chromosome, the **X chromosome.** Observations of *Drosophila* chromosomes revealed that female flies have two X chromosomes, but male flies have only one. In males, the single X chromosome pairs in meiosis with a dissimilar partner called the **Y chromosome.** These two chromosomes are termed **sex chromosomes** because of their association with sex.

During meiosis, a female produces only X-bearing gametes, but a male produces both X-bearing and Y-bearing gametes. When fertilization involves an X sperm, the result is an XX zygote, which develops into a female; when fertilization involves a Y sperm, the result is an XY zygote, which develops into a male.

The solution to Morgan's puzzle is that the gene causing the white-eye trait in *Drosophila* resides only on the X chromosome—it is absent from the Y chromosome. (We now know that the Y chromosome in flies carries almost no functional genes.) A trait determined by a gene on the X chromosome is said to be **sex-linked,** or X-linked, because it is associated with the sex of the individual. Knowing the white-eye trait is recessive to the red-eye trait, we can now see that Morgan's result was a natural consequence of the Mendelian segregation of chromosomes (see figure 13.2).

Morgan's experiment was one of the most important in the history of genetics because it presented the first clear evidence that the genes determining Mendelian traits do indeed reside on the chromosomes, as Sutton had proposed. Mendelian traits segregate in genetic crosses because homologues separate during gamete formation.

Learning Outcomes Review 13.1

Morgan showed that the trait for white eyes in *Drosophila* segregated with the sex of offspring. X and Y chromosomes also segregate with sex. This correlates the behavior of a trait with the behavior of chromosomes. This finding supported the chromosomal theory of inheritance, which states that traits are carried on chromosomes.

■ *What are the expectations for a cross of white-eyed females to red-eyed males?*

Learning Outcomes

1. *Describe the relationship between sex chromosomes and sex determination.*
2. *Explain the genetic consequences of dosage compensation in mammals.*

The structure and number of sex chromosomes vary in different species (table 13.1). In the fruit fly, *Drosophila*, females are XX and males XY, which is also the case for humans and other mammals. However, in birds, the male has two Z chromosomes, and the female has a Z and a W chromosome. Some insects, such as grasshoppers, have no Y chromosome—females are XX and males are characterized as XO (O indicates the absence of a chromosome).

In humans, the Y chromosome generally determines maleness

In chapter 10, you learned that humans have 46 chromosomes (23 pairs). Twenty-two of these pairs are perfectly matched in both males and females and are called **autosomes.** The remaining pair are the sex chromosomes: XX in females, and XY in males.

X chromosome

Y chromosome

35,000×

TABLE 13.1 Sex Determination in Some Organisms		Female	Male
Humans, *Drosophila*		XX	XY
Birds		ZW	ZZ
Grasshoppers		XX	XO
Honeybees		Diploid	Haploid

The Y chromosome in males is highly condensed. Because few genes on the Y chromosome are expressed, recessive alleles on a male's single X chromosome have no *active* counterpart on the Y chromosome.

The "default" setting in human embryonic development leads to female development. Some of the active genes on the Y chromosome, notably the *SRY* gene, are responsible for the masculinization of genitalia and secondary sex organs, producing features associated with "maleness" in humans. Consequently, any individual with *at least one Y chromosome* is normally a male.

The exceptions to this rule actually provide support for this mechanism of sex determination. For example, movement of part of the Y chromosome to the X chromosome can cause otherwise XX individuals to develop as male. There is also a genetic disorder that causes a failure to respond to the androgen hormones (androgen insensitivity syndrome) that causes XY individuals to develop as female. Lastly, mutations in *SRY* itself can cause XY individuals to develop as females.

This form of sex determination seen in humans is shared among mammals, but is not universal in vertebrates. Among fishes and some species of reptiles, environmental factors can cause changes in the expression of this sex-determining gene, and thus in the sex of the adult individual.

Some human genetic disorders display sex linkage

From ancient times, people have noted conditions that seem to affect males to a greater degree than females. Red-green color blindness is one well-known condition that is more common in males because the gene affected is carried on the X chromosome.

Another example is hemophilia, a disease that affects a single protein in a cascade of proteins involved in the formation of blood clots. Thus, in an untreated hemophiliac, even minor cuts will not stop bleeding. This form of hemophilia is caused by an X-linked recessive allele; women who are heterozygous for the allele are asymptomatic carriers, and men who receive an X chromosome with the recessive allele exhibit the disease.

The allele for hemophilia was introduced into a number of different European royal families by Queen Victoria of England. Because these families kept careful genealogical records, we have an extensive pedigree for this condition. In the five generations after Victoria, ten of her male descendants have had hemophilia as shown in the pedigree in figure 13.3.

The Russian house of Romanov inherited this condition through Alexandra Feodorovna, a granddaughter of Queen

Figure 13.3 **The royal hemophilia pedigree.** Queen Victoria, shown at the bottom center of the photo, was a carrier for hemophilia. Two of Victoria's four daughters, Alice and Beatrice, inherited the hemophilia allele from Victoria. Two of Alice's daughters are standing behind Victoria (wearing feathered boas): Princess Irene of Prussia *(right)* and Alexandra *(left)*, who would soon become czarina of Russia. Both Irene and Alexandra were also carriers of hemophilia. From the pedigree, it is clear that Alice introduced hemophilia into the Russian and Prussian royal houses, and Victoria's daughter Beatrice introduced it into the Spanish royal house. Victoria's son Leopold, himself a victim, also transmitted the disorder in a third line of descent. Half-shaded symbols represent carriers with one normal allele and one defective allele; fully shaded symbols represent affected individuals.

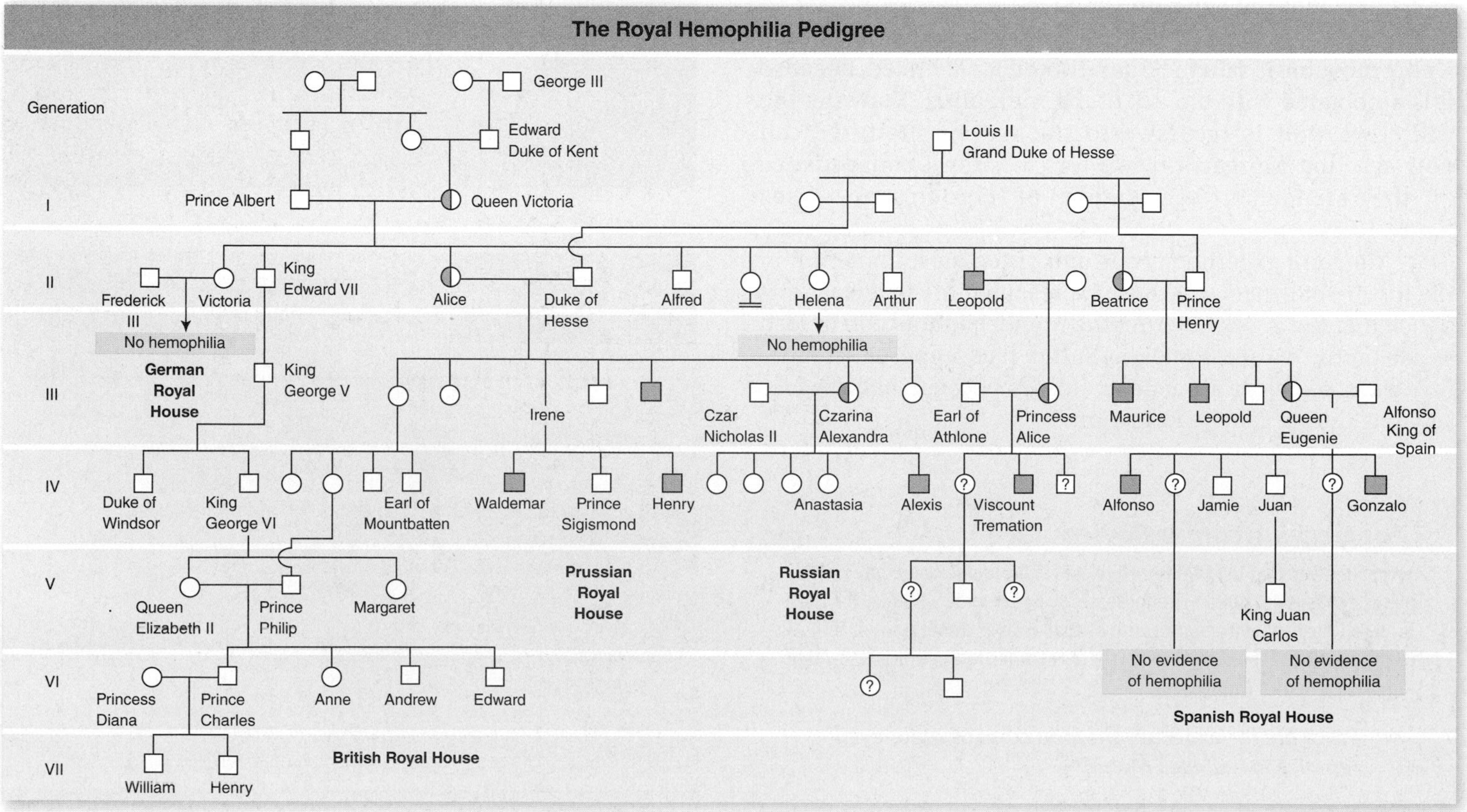

The Royal Hemophilia Pedigree

Generation

George III

Edward
Duke of Kent

Louis II
Grand Duke of Hesse

I — Prince Albert — Queen Victoria

II — Frederick III / Victoria — King Edward VII — Alice — Duke of Hesse — Alfred — Helena — Arthur — Leopold — Beatrice — Prince Henry

No hemophilia

German Royal House

III — King George V — Irene — Czar Nicholas II — Czarina Alexandra — Earl of Athlone — Princess Alice — Maurice — Leopold — Queen Eugenie — Alfonso King of Spain

No hemophilia

IV — Duke of Windsor — King George VI — Earl of Mountbatten — Waldemar — Prince Sigismond — Henry — Anastasia — Alexis — Viscount Tremation — Alfonso — Jamie — Juan — Gonzalo

V — Queen Elizabeth II — Prince Philip — Margaret

Prussian Royal House

Russian Royal House

King Juan Carlos

No evidence of hemophilia   No evidence of hemophilia

Spanish Royal House

VI — Princess Diana — Prince Charles — Anne — Andrew — Edward

British Royal House

VII — William — Henry

Victoria. She married Czar Nicholas II, and their only son, Alexis, was afflicted with the disease. The entire family was executed during the Russian revolution. (Recently, a woman who had long claimed to be Anastasia, a surviving daughter, was shown not to be a Romanov using modern genetic techniques to test her remains.)

Ironically, this condition has not affected the current British royal family, because Victoria's son Edward, who became King Edward VII, did not receive the hemophilia allele. All of the subsequent rulers of England are his descendants.

Dosage compensation prevents doubling of sex-linked gene products

Although males have only one copy of the X chromosome and females have two, female cells do not produce twice as much of the proteins encoded by genes on the X chromosome. Instead, one of the X chromosomes in females is inactivated early in embryonic development, shortly after the embryo's sex is determined. This inactivation is an example of **dosage compensation,** which ensures an equal level of expression from the sex chromosomes despite a differing number of sex chromosomes in males and females. (In *Drosophila,* by contrast, dosage compensation is achieved by increasing the level of expression on the male X chromosome.)

Which X chromosome is inactivated in females varies randomly from cell to cell. If a woman is heterozygous for a sex-linked trait, some of her cells will express one allele and some the other. The inactivated X chromosome is highly condensed, making it visible as an intensely staining **Barr body,** seen below, attached to the nuclear membrane.

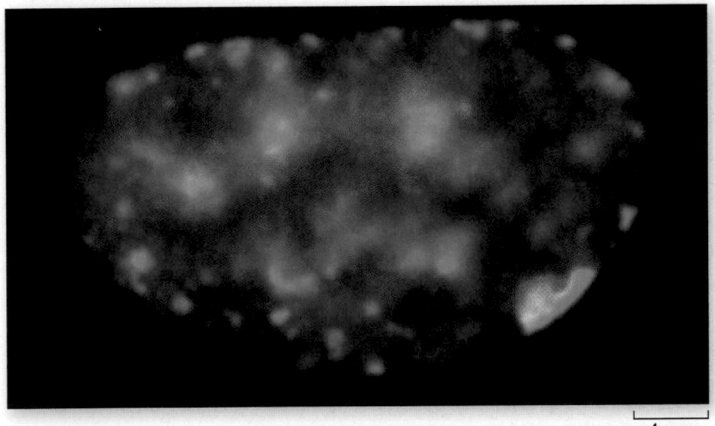

4 μm

X-chromosome inactivation can lead to genetic mosaics

X-chromosome inactivation to produce dosage compensation is not unique to humans but is true of all mammals. Females that are heterozygous for X-chromosome alleles are **genetic mosaics:** Their individual cells may express different alleles, depending on which chromosome is inactivated.

One example is the calico cat, a female that has a patchy distribution of dark fur, orange fur, and white fur (figure 13.4). The dark fur and orange fur are due to heterozygosity for a gene on the

Second gene causes patchy distribution of pigment: white fur = no pigment, orange or black fur = pigment

Allele for black fur is inactivated

X-chromosome allele for orange fur

Inactivated X chromosome becomes Barr body

Nucleus

Allele for orange fur is inactivated

X-chromosome allele for black fur

Inactivated X chromosome becomes Barr body

Nucleus

Figure 13.4 A calico cat. The cat is heterozygous for alleles of a coat color gene that produce either black fur or orange fur. This gene is on the X chromosome, so the different-colored fur is due to inactivation of one X chromosome. The patchy distribution and white color is due to a second gene that is epistatic to the coat color gene and thus masks its effects.

X chromosome that determines pigment color. One allele results in dark fur, and another allele results in orange fur. Which of these colors is observed in any particular patch is due to inactivation of one X chromosome: If the chromosome containing the orange allele is inactivated, then the fur will be dark, and vice versa.

The patchy distribution of color, and the presence of white fur, is due to a second gene that is epistatic to the fur color gene (see chapter 12). That is, the presence of this second gene produces a patchy distribution of pigment, with some areas totally lacking pigment. In the areas that lack pigment, the effect of either fur color allele is masked. Thus, in this one animal we can see an excellent example of both epistasis and X inactivation.

Learning Outcomes Review 13.2

Sex determination begins with the presence or absence of certain chromosomes termed the sex chromosomes. Additional factors may influence sex determination in different species. In humans, males are XY, and therefore they exhibit recessive traits for alleles on the X chromosome. In mammalian females, one X chromosome in each cell becomes inactivated to balance the levels of gene expression. This random inactivation can lead to genetic mosaics.

■ *Would you expect an XXX individual to be viable? If so, would that individual be male or female?*

Exceptions to the Chromosomal Theory of Inheritance

Learning Outcome

1. Describe the inheritance pattern for genes contained in a chloroplast or mitochondrion DNA.

Although the chromosomal theory explains most inheritance, there are exceptions. Primarily, these are due to the presence of DNA in organelle genomes, specifically in mitochondria and chloroplasts. Non-Mendelian inheritance via organelles was studied in depth by Ruth Sager, who in the face of universal skepticism constructed the first map of chloroplast genes in *Chlamydomonas,* a unicellular green alga, in the 1960s and 1970s.

Mitochondria and chloroplasts are not partitioned with the nuclear genome by the process of meiosis. Thus any trait that is due to the action of genes in these organelles will not show Mendelian inheritance.

Mitochondrial genes are inherited from the female parent

Organelles are usually inherited from only one parent, generally the mother. When a zygote is formed, it receives an equal contribution of the nuclear genome from each parent, but it gets all of its mitochondria from the egg cell, which contains a great deal more cytoplasm (and thus organelles). As the zygote divides, these original mitochondria divide as well and are partitioned randomly.

As a result, the mitochondria in every cell of an adult organism can be traced back to the original maternal mitochondria present in the egg. This mode of uniparental (one-parent) inheritance from the mother is called **maternal inheritance.**

In humans, the disease Leber's hereditary optic neuropathy (LHON) shows maternal inheritance. The genetic basis of this disease is a mutant allele for a subunit of NADH dehydrogenase. The mutant allele reduces the efficiency of electron flow in the electron transport chain in mitochondria (see chapter 7), in turn reducing overall ATP production. Some nerve cells in the optic system are particularly sensitive to reduction in ATP production, resulting in neural degeneration.

A mother with this disease will pass it on to all of her progeny, whereas a father with the disease will not pass it on to any of his progeny. Note that this condition differs from sex-linked inheritance because males and females are equally affected.

Chloroplast genes may also be passed on uniparentally

The inheritance pattern of chloroplasts is also usually maternal, although both paternal and biparental inheritance of chloroplasts may be observed in some species. Carl Correns first hypothesized in 1909 that chloroplasts were responsible for

inheritance of variegation (mixed green and white leaves) in the plant commonly known as the four o'clock *(Mirabilis jalapa).* The offspring exhibited the phenotype of the female parent, regardless of the male's phenotype.

In Sager's work on *Chlamydomonas,* resistance to the antibiotic streptomycin was shown to be transmitted via the chloroplast DNA from only the mt$^+$ mating type. The mt$^-$ mating type does not contribute chloroplast DNA to the zygote formed by fusion of mt$^+$ and mt$^-$ gametes.

Learning Outcome Review 13.3

The genomes of mitochondria and chloroplasts divide independently of the nucleus. These organelles are carried in the cytoplasm of the egg cell, so any traits determined by these genomes are maternally inherited and thus do not follow Mendelian rules. In some species, however, chloroplasts may be passed on paternally or biparentally.

- **How can you explain the lack of mt$^-$ chloroplast DNA in Chlamydomonas *zygotes from mt$^-$ by mt$^+$ crosses?***

Genetic Mapping

Learning Outcomes

1. Describe how genes on the same chromosome will segregate.
2. Explain the relationship between frequency of recombinant progeny and map distance.
3. Calculate map distances from the frequency of recombinants in testcrosses.

We have seen that Mendelian traits are determined by genes located on chromosomes and that the independent assortment of Mendelian traits reflects the independent assortment of chromosomes in meiosis. This is fine as far as it goes, but it is still incomplete. Of Mendel's seven traits in figure 12.4, six are on different chromosomes and two are on the same chromosome, yet all show independent assortment with one another. The two on the same chromosome should not behave the same as those that are on different chromosomes. In fact, organisms generally have many more genes that assort independently than the number of chromosomes. This means that independent assortment cannot be due only to the random alignment of different chromosomes during meiosis.

? Inquiry question Mendel did not examine plant height and pod shape in his dihybrid crosses. The genes for these traits are very close together on the same chromosome. How would this have changed Mendel's results?

The solution to this problem is found in an observation that was introduced in chapter 11: the crossing over of homologues during meiosis. In prophase I of meiosis, homologues appear to physically exchange material by crossing over (figure 13.5). In chapter 11, you saw how this was part of the mechanism that allows homologues, and not sister chromatids, to disjoin at anaphase I.

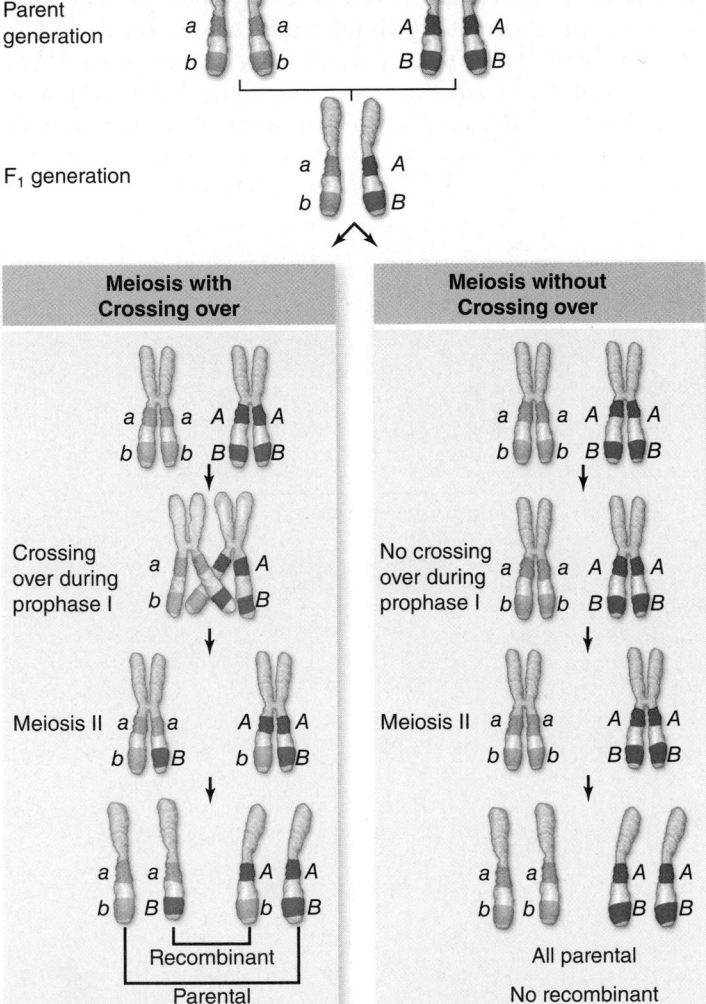

Figure 13.5 Crossing over exchanges alleles on homologues. When a crossover occurs between two loci, it leads to the production of recombinant chromosomes. If no crossover occurs, then the chromosomes will carry the parental combination of alleles.

Genetic recombination exchanges alleles on homologues

Consider a dihybrid cross performed using the Mendelian framework. Two true-breeding parents that each differ with respect to two traits are crossed, producing doubly heterozygous F₁ progeny. If the genes for the two traits are on a single chromosome, then during meiosis we would expect alleles for both loci to segregate together and produce only gametes that resemble the two parental types. But if a crossover occurs between the two loci, then each homologue would carry one allele from each parent and produce gametes that combine these parental traits (see figure 13.5). We call gametes with this new combination of alleles *recombinant* gametes as they are formed by recombining the parental alleles.

The first investigator to provide evidence for this was Morgan, who studied three genes on the X chromosome of *Drosophila*. He found an excess of parental types, which he explained as due to the genes all being on the X chromosome and therefore coinherited (inherited together). He went

further, suggesting that the recombinant genotypes were due to crossing over between homologues during meiosis.

Experiments performed independently by Barbara McClintock and Harriet Creighton in maize and by Curt Stern in *Drosophila* provided evidence for this physical exchange of genetic material. The experiment done by Creighton and McClintock is detailed in figure 13.6. In this experiment, they used a chromosome with two alterations visible under a microscope: a knob on one end of the chromosome and an extension of the other end making it longer. In addition to these visible markers, this chromosome also carried a gene that determines kernel color (colored or colorless) and a gene that determines kernel texture (waxy or starchy).

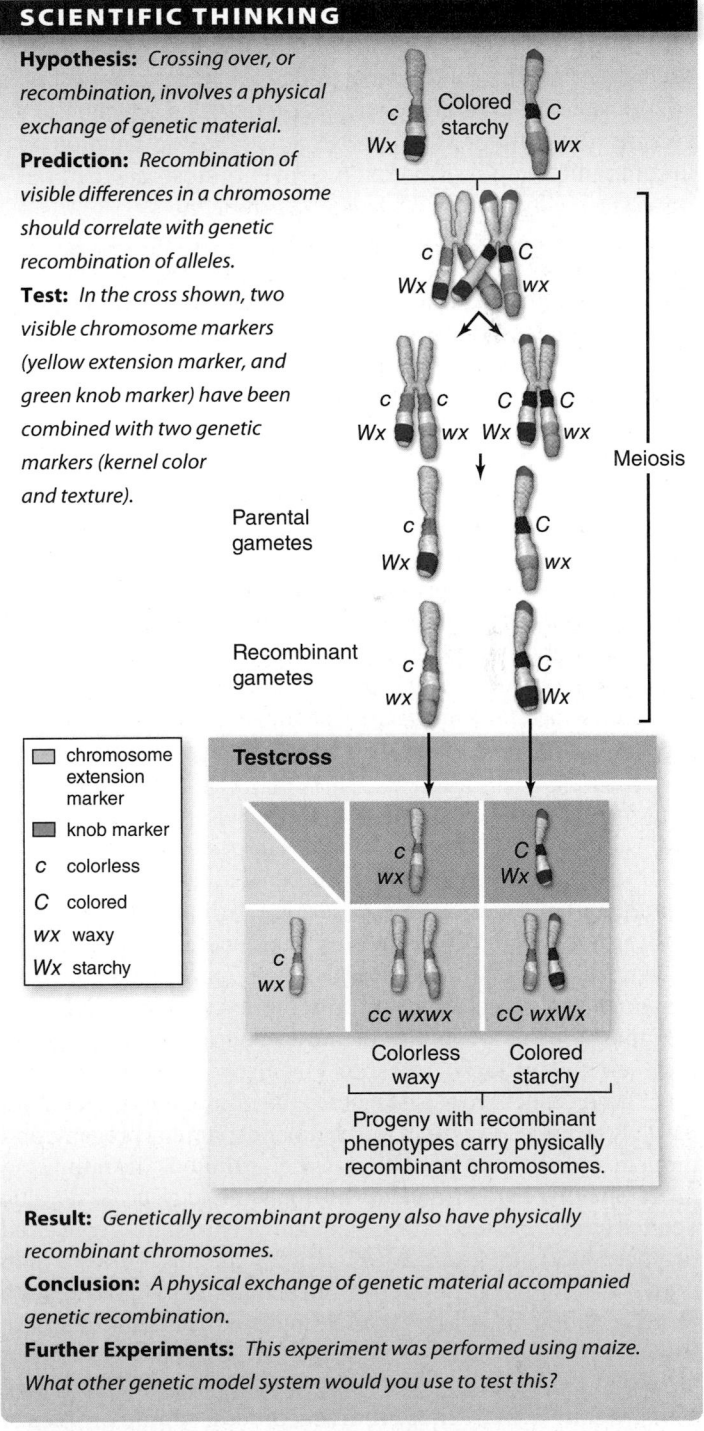

SCIENTIFIC THINKING

Hypothesis: *Crossing over, or recombination, involves a physical exchange of genetic material.*

Prediction: *Recombination of visible differences in a chromosome should correlate with genetic recombination of alleles.*

Test: *In the cross shown, two visible chromosome markers (yellow extension marker, and green knob marker) have been combined with two genetic markers (kernel color and texture).*

Result: *Genetically recombinant progeny also have physically recombinant chromosomes.*

Conclusion: *A physical exchange of genetic material accompanied genetic recombination.*

Further Experiments: *This experiment was performed using maize. What other genetic model system would you use to test this?*

Figure 13.6 The Creighton and McClintock experiment.

The long chromosome, which also had the knob, carried the dominant colored allele for kernel color *(C)* and the recessive waxy allele for kernel texture *(wx)*. Heterozygotes were constructed with this chromosome paired with a visibly normal chromosome carrying the recessive colorless allele for kernel color *(c)* and the dominant starchy allele for kernel texture *(Wx)* (see figure 13.6). These plants appeared colored and starchy because they were heterozygous for both loci, and they were also heterozygous for the two visibly distinct chromosomes.

These plants, heterozygous for both chromosomal and genetic markers, were testcrossed to colorless waxy plants with normal appearing chromosomes. The progeny were analyzed for both physical recombination (using a microscope to observe chromosome appearance) and genetic recombination (by examining the phenotype of progeny). The results were striking: All of the progeny that were genetically recombinant (appear colored starchy or colorless waxy) also now had only one of the chromosomal markers. That is, genetic recombination was accompanied by physical exchange of chromosomal material.

Recombination is the basis for genetic maps

The ability to map the location of genes on chromosomes using data from genetic crosses is one of the most powerful tools of genetics. The insight that allowed this technique, like many great insights, is so simple as to seem obvious in retrospect.

Morgan had already suggested that the frequency with which a particular group of recombinant progeny appeared was a reflection of the relative location of genes on the chromosome. An undergraduate in Morgan's laboratory, Alfred Sturtevant put this observation on a quantitative basis. Sturtevant reasoned that the frequency of recombination observed in crosses could be used as a measure of genetic distance. That is, as physical distance on a chromosome increases, so does the probability of recombination (crossover) occurring between the gene loci. Using this logic, the frequency of recombinant gametes produced is a measure of their distance apart on a chromosome.

Linkage data

To be able to measure recombination frequency easily, investigators use a testcross instead of intercrossing the F_1 progeny as Mendel did. In a testcross, as described earlier, the phenotypes of the progeny reflect the gametes produced by the doubly heterozygous F_1 individual. In the case of recombination, progeny that appear parental have not undergone crossover, and progeny that appear recombinant have experienced a crossover between the two loci in question (see figure 13.5).

When genes are close together, the number of recombinant progeny is much lower than the number of parental progeny, and the genes are defined on this basis as being **linked.** The number of recombinant progeny divided by total progeny gives a value defined as the **recombination frequency.** This value is converted to a percentage, and each 1% of recombination is termed a **map unit.** This unit has been named the centimorgan (cM) for T. H. Morgan, although it is also called simply a map unit (m.u.) as well.

Constructing maps

Constructing genetic maps then becomes a simple process of performing testcrosses with doubly heterozygous individuals

and counting progeny to determine percent recombination. This is best shown with an example using a two-point cross.

Drosophila homozygous for two mutations, vestigial wings *(vg)* and black body *(b)*, are crossed to flies homozygous for the wild type, or normal alleles, of these genes *(vg⁺ b⁺)*. The doubly heterozygous F_1 progeny are then testcrossed to homozygous

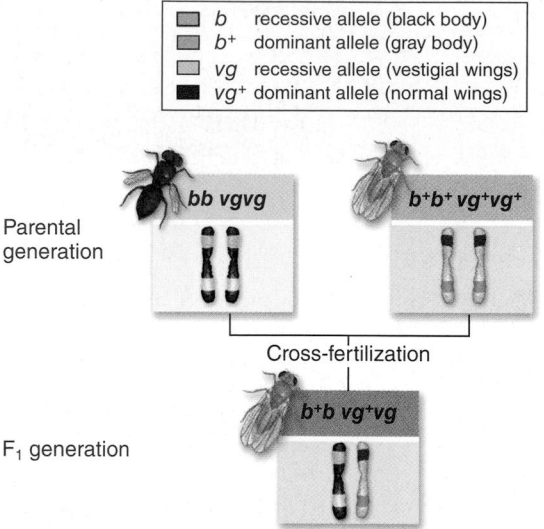

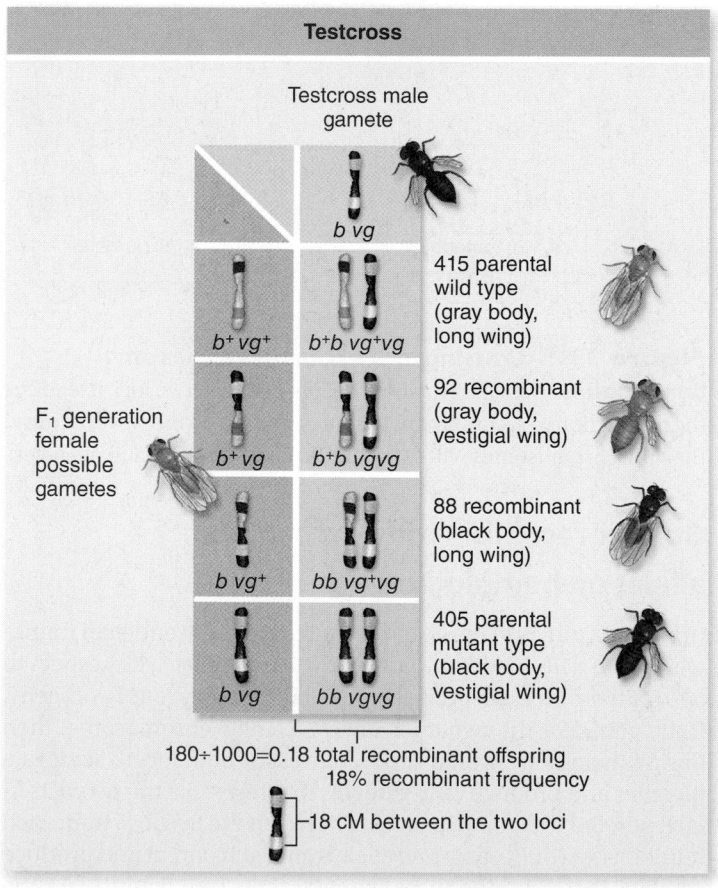

Figure 13.7 Two-point cross to map genes. Flies homozygous for long wings *(vg⁺)* and gray bodies *(b⁺)* are crossed to flies homozygous for vestigial wings *(vg)* and black bodies *(b)*. Both vestigial wing and black body are recessive to the normal (wild-type) long wing and gray body. The F_1 progeny are then testcrossed to homozygous vestigial black to produce the progeny for mapping. Data are analyzed in the text.

recessive individuals *(vg b/vg b)*, and progeny are counted (figure 13.7). The data are shown here:

vestigial wings, black body *(vg b)*   405   (parental)
long wings, gray body *(vg⁺ b⁺)*   415   (parental)
vestigial wings, gray body *(vg b⁺)*   92   (recombinant)
long wings, black body *(vg⁺ b)*   88   (recombinant)
Total Progeny   1000

The numbers of recombinant progeny are added together, and this sum is divided by total progeny to produce the recombination frequency. The recombination frequency is 92 + 88 divided by 1000, or 0.18. Converting this number to a percentage yields 18 cM as the map distance between these two loci.

Multiple crossovers can yield independent assortment results

As the distance separating loci increases, the probability of recombination occurring between them during meiosis also increases. What happens when more than one recombination event occurs?

If homologues undergo two crossovers between loci, then the parental combination is restored. This leads to an underestimate of the true genetic distance because not all events can be noted. As a result, the relationship between true distance on a chromosome and the recombination frequency is not linear. It begins as a straight line, but the slope decreases; the curve levels off at a recombination frequency of 0.5 (figure 13.8).

At long distances, multiple events between loci become frequent. In this case, odd numbers of crossovers (1, 3, 5) produce recombinant gametes, and no crossover or even numbers of crossovers (0, 2, 4) produce parental gametes. At large enough distances, these frequencies are about equal, leading to the number of recombinant gametes being equal to the number of parental gametes, and the loci exhibit independent assortment! This is how Mendel could use two loci on the same chromosome and have them assort independently.

> **Data analysis**   What would Mendel have observed in a dihybrid cross if the two loci were 10 cM apart on the same chromosome? Is this likely to have led him to the idea of independent assortment?

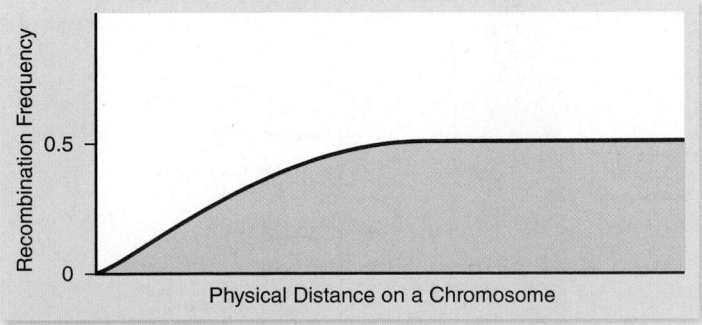

Figure 13.8 Relationship between true distance and recombination frequency. As distance on a chromosome increases, the recombinants are not all detected due to double crossovers. This leads to a curve that levels off at 0.5.

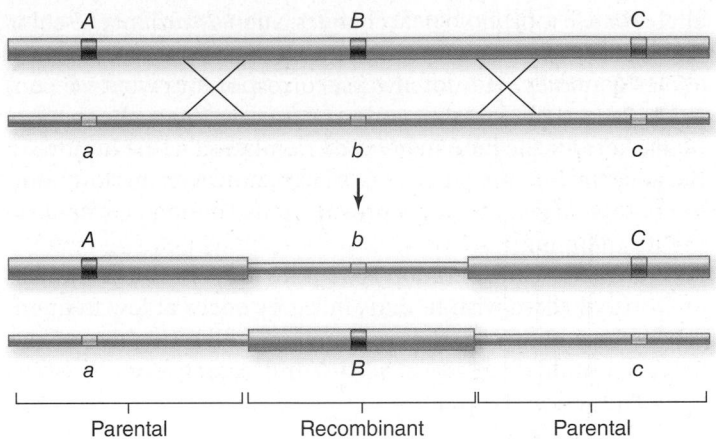

Figure 13.9 Use of a three-point cross to order genes. In a two-point cross, the outside loci appear parental for double crossovers. With the addition of a third locus, the two crossovers can still be detected because the middle locus will be recombinant. This double crossover class should be the least frequent, so whatever locus has recombinant alleles in this class must be in the middle.

 ## Three-point crosses can be used to put genes in order

Because multiple crossovers reduce the number of observed recombinant progeny, longer map distances are not accurate. As a result, when geneticists try to construct maps from a series of two-point crosses, determining the order of genes is problematic. Using three loci instead of two, or a three-point cross, can help solve the problem.

In a three-point cross, the gene in the middle allows us to see recombination events on either side. For example, a double crossover for the two outside loci is actually a single crossover between the middle locus and each outside locus (figure 13.9).

The probability of two crossovers is equal to the product of the probability of each individual crossover, each of which is relatively low. Therefore, in any three-point cross, the class of offspring with two crossovers is the least frequent class. Analyzing these individuals to see which locus is recombinant identifies the locus that lies in the middle of the three loci in the cross (see figure 13.9).

In practice, geneticists use three-point crosses to determine the order of genes, then use data from the closest two-point crosses to determine distances. Longer distances are generated by simple addition of shorter distances. This avoids using inaccurate measures from two-point crosses between distant loci.

Genetic maps can be constructed for the human genome

Human genes can be mapped, but the data must be derived from historical pedigrees, such as those of the royal families of Europe mentioned earlier. The principle is the same—genetic distance is still proportional to recombination frequency—but the analysis requires the use of complex statistics and summing data from many families.

The difficulty of mapping in humans

Looking at nonhuman animals with extensive genetic maps, the majority of genetic markers have been found at loci where alleles cause morphological changes, such as variant eye color, body color, or wing morphology in flies. In humans, such alleles generally, but not always, correspond to what we consider disease states. As recently as the early 1980s, the number of markers for the human genome numbered in the hundreds. Because the human genome is so large, however, this low number of markers would never provide dense enough coverage to use for mapping.

Another consideration is that the disease-causing alleles are those that we wish to map, but they occur at low frequencies in the population. Any one family would be highly unlikely to carry multiple disease alleles, the segregation of which would allow for mapping.

Anonymous markers

This situation changed with the development of **anonymous markers,** genetic markers that can be detected using molecular techniques, but that do not cause a detectable phenotype. The nature of these markers has evolved with technology, leading to a standardized set of markers scattered throughout the genome. These markers, which have a relatively high density, can be detected using techniques that are easy to automate. As a result of analysis, geneticists now have several thousand markers to work with, instead of hundreds, and have produced a human genetic map that would have been unthinkable 25 years ago (figure 13.10). (In the following chapters of this unit, you'll learn about some of the molecular techniques that have been developed for use with genomes.)

Single-nucleotide polymorphisms (SNPs)

The information developed from sequencing the human genome can then be used to identify and map single bases that differ between individuals. Any differences between individuals in populations are termed *polymorphisms;* polymorphisms affecting a single base of a gene locus are called **single-nucleotide polymorphisms (SNPs).** Over 3.1 million such differences have been identified and are being placed on both the genetic map and the human genome sequence. This confluence of techniques will enable the ultimate resolution of genetic analysis.

The recent progress in gene mapping applies to more than just the relatively small number of genes that show simple Mendelian inheritance. The development of a high-resolution genetic map, and the characterization of millions of SNPs, opens up the possibility of being able to characterize complex quantitative traits in humans as well.

On a more practical level, the types of molecular markers described earlier are used in forensic analysis. Although not quite as rapid as some television programs would have you believe, this does allow rapid DNA testing of crime scene samples to help eliminate or confirm crime suspects and for paternity testing.

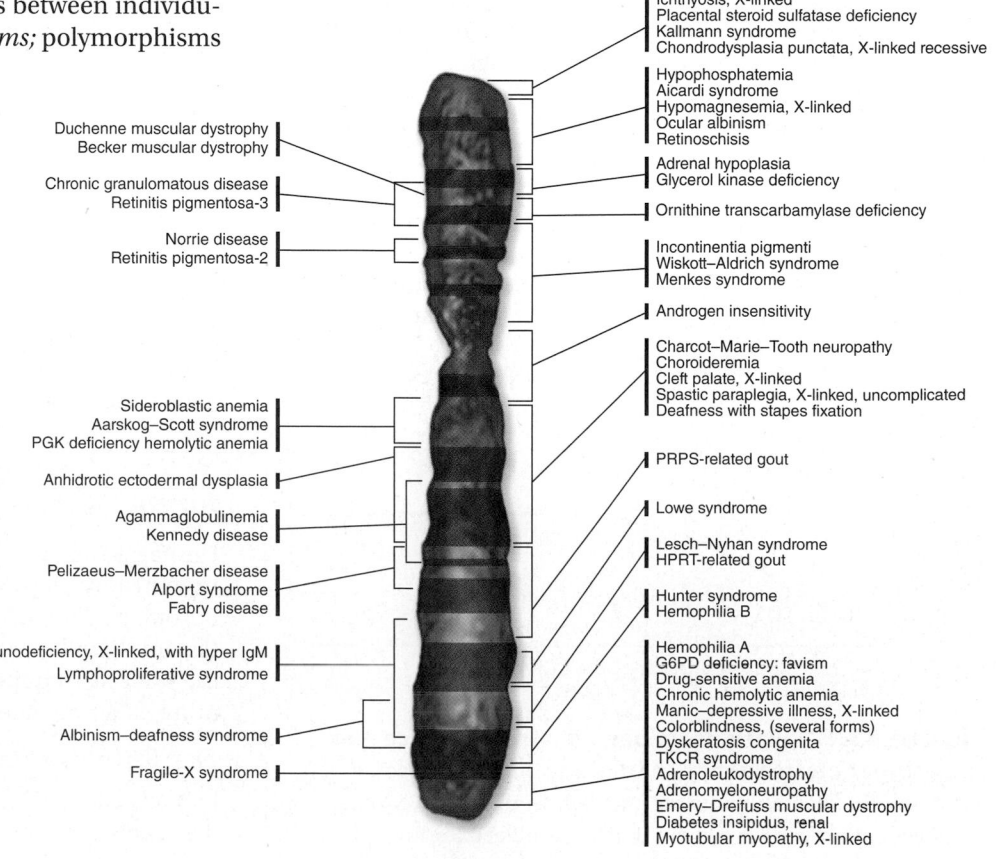

Figure 13.10 The human X-chromosome gene map. Only a partial map for the human X chromosome is presented here; a more detailed map would require a much larger figure. The black bands represent staining patterns that can be seen under the microscope, and the constriction represents the centromere. Analysis of the sequence of the X chromosome indicates 1098 genes on the X chromosome. Many of these may have mutant alleles that can affect disease states. By analyzing inheritance patterns of affected and unaffected individuals, the 59 diseases shown have been traced to specific segments of the X chromosome, indicated by brackets.

13.5 Selected Human Genetic Disorders

Learning Outcomes

1. Explain how mutations can cause disease.
2. Describe the consequences of nondisjunction in humans.
3. Recognize how genomic imprinting can lead to non-Mendelian inheritance.

Diseases that run in families have been known for many years. These can be nonlife-threatening like albinism, or may result in premature death like Huntington's, which were used as examples of recessive and dominant traits in humans previously. A small sample of diseases due to alterations of alleles of a single gene is provided in table 13.2. We will discuss the nature of these genetic changes later in chapter 15. In this section we discuss some of the genetic disorders that have been found in human populations.

Sickle cell anemia is due to altered hemoglobin

The first human disease shown to be the result of a mutation in a protein was sickle cell anemia. It is caused by a defect in the oxygen carrier molecule, hemoglobin, that leads to impaired oxygen delivery to tissues. The defective hemoglobin molecules stick to one another, leading to stiff, rodlike structures that alter the shape of the red blood cells that carry them. These

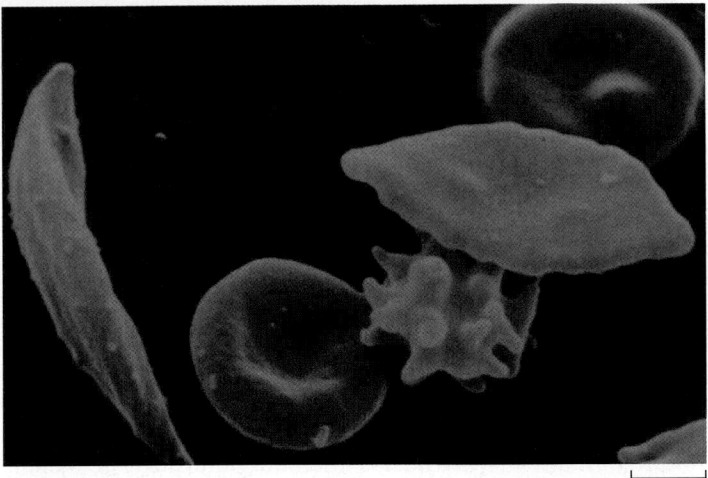

1 μm

Figure 13.11 Sickle cell anemia. In individuals homozygous for the sickle cell trait, many of the red blood cells have sickled or irregular shapes, such as the cell on the far left.

red blood cells take on a characteristic shape that led to the name "sickle cell" (figure 13.11).

Individuals homozygous for the sickle cell allele exhibit intermittent illness and reduced life span. Individuals heterozygous for the sickle cell allele are indistinguishable from normal individuals in a normal oxygen environment, although their red cells do exhibit reduced ability to carry oxygen.

The sickle cell allele is particularly prevalent in people of African descent. In some regions of Africa, up to 45% of the population is heterozygous for the trait, and 6% are

TABLE 13.2	Some Important Genetic Disorders			
Disorder	**Symptom**	**Defect**	**Dominant/Recessive**	**Frequency Among Human Births**
Cystic fibrosis	Mucus clogs lungs, liver, and pancreas	Failure of chloride ion transport mechanism	Recessive	1/2500 (Caucasians)
Sickle cell anemia	Blood circulation is poor	Abnormal hemoglobin molecules	Recessive	1/600 (African Americans)
Tay–Sachs disease	Central nervous system deteriorates in infancy	Defective enzyme (hexosaminidase A)	Recessive	1/3500 (Ashkenazi Jews)
Phenylketonuria	Brain fails to develop in infancy, treatable with dietary restriction	Defective enzyme (phenylalanine hydroxylase)	Recessive	1/12,000
Hemophilia	Blood fails to clot	Defective blood-clotting factor VIII	X-linked recessive	1/10,000 (Caucasian males)
Huntington disease	Brain tissue gradually deteriorates in middle age	Production of an inhibitor of brain cell metabolism	Dominant	1/24,000
Muscular dystrophy (Duchenne)	Muscles waste away	Degradation of myelin coating of nerves stimulating muscles	X-linked recessive	1/3700 (males)
Hypercholesterolemia	Excessive cholesterol levels in blood lead to heart disease	Abnormal form of cholesterol cell surface receptor	Dominant	1/500

homozygous. This proportion of heterozygotes is higher than would be expected on the basis of chance alone. It turns out that heterozygosity confers a greater resistance to the blood-borne parasite that causes malaria. In regions of central Africa where malaria is endemic, the sickle cell allele also occurs at a high frequency.

The sickle cell allele is not the end of the story for the β-globin gene; a large number of other alterations of this gene have been observed that lead to anemias. In fact, for hemoglobin, which is composed of two α-globins and two β-globins, over 700 structural variants have been cataloged. It is estimated that 7% of the human population worldwide are carriers for different inherited hemoglobin disorders.

The Human Gene Mutation Database has cataloged the nature of many disease alleles, including the sickle cell allele. The majority of alleles seem to be simple changes. Almost 57% of the close to 85,000 alleles in the Human Gene Mutation Database are single-base substitutions. Another 24% are due to small insertions or deletions of less than 20 bases. This careful survey of disease-causing alterations has also revealed more large-scale changes. This is now termed *copy number variation (CNV)* and appears more important than previously thought.

Nondisjunction of chromosomes changes chromosome number

The failure of homologues or sister chromatids to separate properly during meiosis is called **nondisjunction.** This failure leads to the gain or loss of a chromosome, a condition called **aneuploidy.** The frequency of aneuploidy in humans is surprisingly high, being estimated to occur in 5% of conceptions.

Nondisjunction of autosomes

Humans who have lost even one copy of an autosome are called **monosomics,** and generally do not survive embryonic development. In all but a few cases, humans who have gained an extra autosome (called **trisomics**) also do not survive. Data from clinically recognized spontaneous abortions indicate levels of aneuploidy as high as 35%.

Five of the smallest human autosomes—those numbered 13, 15, 18, 21, and 22—can be present as three copies and still allow the individual to survive, at least for a time. The presence of an extra chromosome 13, 15, or 18 causes severe developmental defects, and infants with such a genetic makeup die within a few months. In contrast, individuals who have an extra copy of chromosome 21 or, more rarely, chromosome 22, usually survive to adulthood. In these people, the maturation of the skeletal system is delayed, so they generally are short and have poor muscle tone. Their mental development is also affected, and children with trisomy 21 show some degree of intellectual disability.

The developmental defect produced by trisomy 21 (figure 13.12) was first described in 1866 by J. Langdon Down; for this reason, it is called Down syndrome. About 1 in every 750 children exhibits Down syndrome, and the frequency is comparable in all racial groups. Similar conditions also occur in chimpanzees and other related primates.

In humans, the defect occurs when a particular small portion of chromosome 21 is present in three copies instead of two. In 97% of the cases examined, all of chromosome 21 is

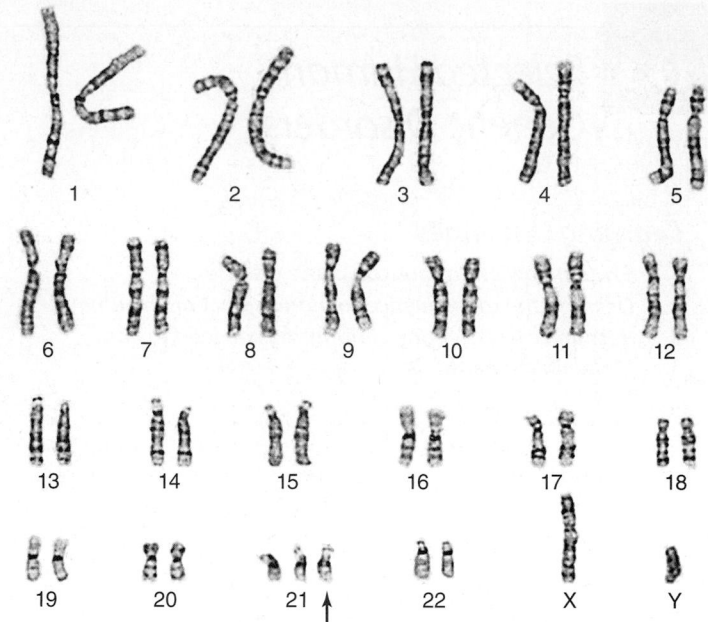

Figure 13.12 Down syndrome. As shown in this male karyotype, Down syndrome is associated with trisomy of chromosome 21 (arrow shows third copy of chromosome 21).

present in three copies. In the other 3%, a small portion of chromosome 21 containing the critical segment has been added to another chromosome by a process called *translocation* (see chapter 15); it exists along with the normal two copies of chromosome 21. This latter condition is known as *translocation Down syndrome.*

In mothers younger than 20 years of age, the risk of giving birth to a child with Down syndrome is about 1 in 1700; in mothers 20 to 30 years old, the risk is only about 1 in 1400. However, in mothers 30 to 35 years old, the risk rises to 1 in 750, and by age 45, the risk is as high as 1 in 16 (figure 13.13).

Primary nondisjunctions are far more common in women than in men because all of the eggs a woman will ever produce have developed to the point of prophase in meiosis I by the time she is born. By the time a woman has children, her eggs are as old as she is. Therefore, there is a much greater chance for cell-division problems of various kinds, including those that cause primary nondisjunction, to accumulate over time in female gametes. In contrast, men produce new sperm daily. For this reason, the age of the mother is more critical than that of the father for couples contemplating childbearing.

Nondisjunction of sex chromosomes

Individuals who gain or lose a sex chromosome do not generally experience the severe developmental abnormalities caused by similar changes in autosomes. Although such individuals have somewhat abnormal features, they often reach maturity and in some cases may be fertile.

X chromosome nondisjunction. When X chromosomes fail to separate during meiosis, some of the gametes produced possess both X chromosomes, and so are XX gametes; the other gametes have no sex chromosome and are designated "O" (figure 13.14).

If an XX gamete combines with an X gamete, the resulting XXX zygote develops into a female with one functional

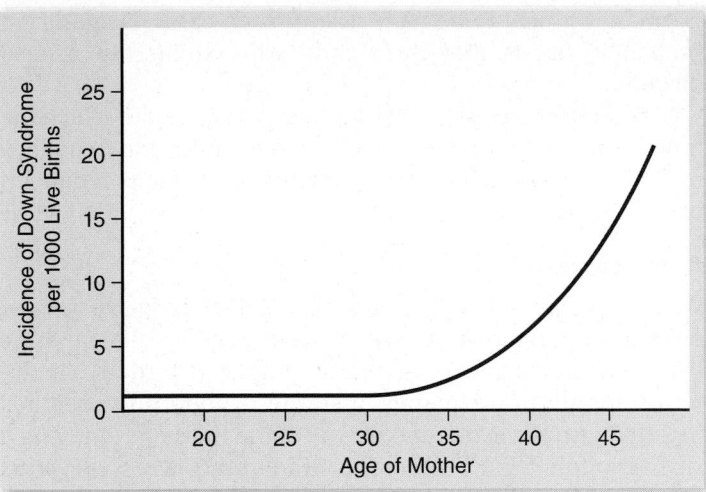

Figure 13.13 Correlation between maternal age and the incidence of Down syndrome. As women age, the chances they will bear a child with Down syndrome increase. After a woman reaches 35, the frequency of Down syndrome rises rapidly.

Data analysis Over a five-year period between ages 20 and 25, the incidence of Down syndrome increases 0.1 per thousand; over a five-year period between ages 35 and 40, the incidence increases to 8.0 per thousand, 80 times as great. The period of time is the same in both instances. What has changed?

X chromosome and two Barr bodies. She may be taller in stature but is otherwise normal in appearance.

If an XX gamete instead combines with a Y gamete, the effects are more serious. The resulting XXY zygote develops into a male who has many female body characteristics and, in some cases but not all, diminished mental capacity. This condition, called *Klinefelter syndrome,* occurs in about 1 out of every 500 male births.

If an O gamete fuses with a Y gamete, the resulting OY zygote is nonviable and fails to develop further; humans cannot survive when they lack the genes on the X chromosome. But if an O gamete fuses with an X gamete, the XO zygote develops into a sterile female of short stature, with a webbed neck and sex organs that never fully mature during puberty. The mental abilities of an XO individual are in the low-normal range. This condition, called *Turner syndrome,* occurs roughly once in every 5000 female births.

Y chromosome nondisjunction. The Y chromosome can also fail to separate in meiosis, leading to the formation of YY gametes. When these gametes combine with X gametes, the XYY zygotes develop into fertile males of normal appearance. The frequency of the XYY genotype *(Jacob syndrome)* is about 1 per 1000 newborn males.

Genomic imprinting depends on the parental origin of alleles

By the late 20th century, geneticists were confident that they understood the basic mechanisms governing inheritance. It

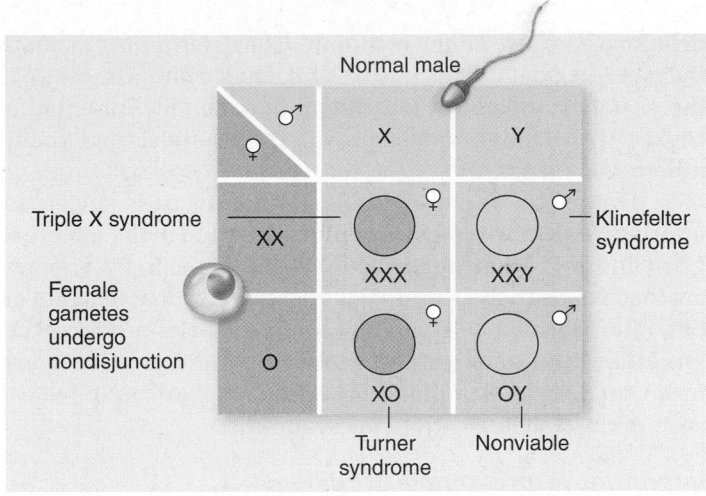

Figure 13.14 How nondisjunction can produce abnormalities in the number of sex chromosomes. When nondisjunction occurs in the production of female gametes, the gamete with two X chromosomes (XX) produces Klinefelter males (XXY) and triple-X females (XXX). The gamete with no X chromosome (O) produces Turner females (XO) and nonviable OY males lacking any X chromosome.

Inquiry question Can you think of two nondisjunction scenarios that would produce an XXY male?

came as quite a surprise to find that for some genes inheritance depends on the parent of origin. Even stranger, the only two groups that show this pattern are flowering plants and mammals. In **genomic imprinting,** the phenotype caused by a specific allele is exhibited when the allele comes from one parent, but not from the other.

The basis for genomic imprinting is the expression of a gene depending on passage through maternal or paternal germ lines. Some genes are inactivated in the paternal germ line and therefore are not expressed in the zygote. Other genes are inactivated in the maternal germ line, with the same result. This condition makes the zygote effectively haploid for an imprinted gene. The expression of variant alleles of imprinted genes depends on the parent of origin. Furthermore, imprinted genes seem to be concentrated in particular regions of the genome. These regions include genes that are both maternally and paternally imprinted.

Prader–Willi and Angelman syndromes

An example of genomic imprinting in humans involves the two diseases Prader–Willi syndrome (PWS) and Angelman syndrome (AS). The effects of PWS include respiratory distress, obesity, short stature, mild intellectual disability, and obsessive–compulsive behavior. The effects of AS include developmental delay, severe intellectual disability, hyperactivity, aggressive behavior, and inappropriate laughter.

Genetic studies have implicated genes on chromosome 15 for both disorders, but the pattern of inheritance is complementary. The most common cause of both syndromes is a deletion of material on chromosome 15 and, in fact, the same deletion can cause either syndrome. The determining factor is the parental origin of the normal and deleted chromosomes. If the chromosome with the deletion is paternally inherited it causes PWS; if the chromosome with the deletion is maternally inherited it causes AS.

The region of chromosome 15 that is lost is subject to imprinting, with some genes being inactivated in the maternal germ line, and others in the paternal germ line. In PWS, genes are inactivated in the maternal germ line, such that deletion or other functional loss of paternally derived alleles produces the syndrome. The opposite is true for AS syndrome: Genes are inactivated in the paternal germ line, such that loss of maternally derived alleles leads to the syndrome.

Imprinting Is an Example of Epigenetics

Genomic imprinting is actually an example of a more general phenomenon: **epigenetic inheritance.** An epigenetic trait is defined as a stably heritable phenotype resulting from changes in a chromosome without alteration in the DNA sequence. This seems contradictory, but it illustrates the point that the sequence of bases in genes is not the end of the story. Another example from this chapter is X-chromosome inactivation, a phenomenon in which an entire chromosome is silenced. This is inherited through mitotic divisions.

As we will see in chapter 16, the control of gene expression involves the interaction of regulatory proteins with DNA and also with the proteins that are involved in chromosome structure. In some well studied cases, the pattern of imprinting that occurs in the male and female germ line is due to male- and female-specific patterns of DNA methylation and alterations to the proteins that are involved in chromosome structure.

Some genetic defects can be detected early in pregnancy

Although most genetic disorders cannot yet be cured, we are learning a great deal about them, and progress toward successful therapy is being made in many cases. In the absence of a cure, however, the only recourse is to try to avoid producing children with these conditions. The process of identifying parents at risk for having children with genetic defects and of assessing the genetic state of early embryos is called **genetic counseling.**

Pedigree analysis

One way of assessing risks is through pedigree analysis, often employed as an aid in genetic counseling. By analyzing a person's pedigree, it is sometimes possible to estimate the likelihood that the person is a carrier for certain disorders. For example, if a counseling client's family history reveals that a relative has been afflicted with a recessive genetic disorder, such as cystic fibrosis, it is possible that the client is a heterozygous carrier of the recessive allele for that disorder.

When a couple is expecting a child, and pedigree analysis indicates that both of them have a significant chance of being heterozygous carriers of a deleterious recessive allele, the pregnancy is said to be high-risk. In such cases, a significant probability exists that their child will exhibit the clinical disorder.

Another class of high-risk pregnancy is that in which the mothers are older than 35. As discussed earlier, the frequency of Down syndrome increases dramatically in the pregnancies of older women (see figure 13.13).

Amniocentesis

When a pregnancy is diagnosed as high-risk, many women elect to undergo **amniocentesis,** a procedure that permits the prenatal diagnosis of many genetic disorders. In the fourth month of pregnancy, a sterile hypodermic needle is inserted into the expanded uterus of the mother, removing a small sample of the amniotic fluid that bathes the fetus (figure 13.15). Within the fluid are free-floating cells derived from the fetus; once removed, these cells can be grown in cultures in the laboratory.

During amniocentesis, the position of the needle and that of the fetus are usually observed by means of *ultrasound.* The sound waves used in ultrasound are not harmful to mother or fetus, and they permit the person withdrawing the amniotic fluid to do so without damaging the fetus. In addition, ultrasound can be used to examine the fetus for signs of major abnormalities. However, about 1 out of 200 amniocentesis procedures may result in fetal death and miscarriage.

Chorionic villi sampling

In recent years, physicians have increasingly turned to a new, less invasive procedure for genetic screening called **chorionic villi sampling (CVS).** Using this method, the physician removes cells from the chorion, a membranous part of the placenta that nourishes the fetus (figure 13.15). This procedure can be used earlier in pregnancy (by the eighth week) and yields results much more rapidly than does amniocentesis. Risks from chorionic villi sampling are comparable to those for amniocentesis.

To test for certain genetic disorders, genetic counselors look for three characteristics in the cultures of cells obtained from amniocentesis or chorionic villi sampling. First, analysis of the karyotype can reveal aneuploidy (extra or missing chromosomes) and gross chromosomal alterations. Second, in many cases it is possible to test directly for the proper functioning of enzymes involved in genetic disorders. The lack of normal enzymatic activity signals the presence of the disorder. As examples, the lack of the enzyme responsible for breaking down phenylalanine indicates phenylketonuria (PKU); the absence of the enzyme responsible for the breakdown of gangliosides indicates Tay–Sachs disease; and so forth. Additionally, with information from the Human Genome Project, more disease alleles for genetic disorders are known. If there are a small number of alleles for a specific disease in the population, these can be identified as well.

With the changes in human genetics brought about by the Human Genome Project (see chapter 18), it is possible to design tests for many more diseases. Difficulties still exist in

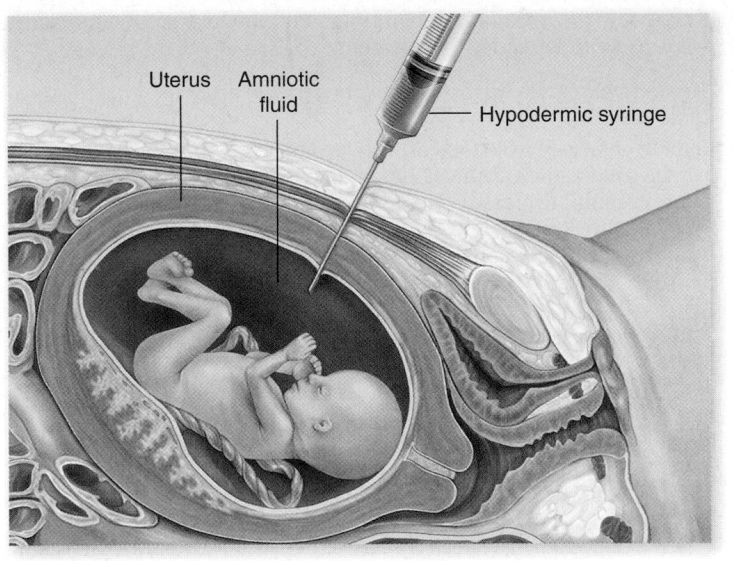
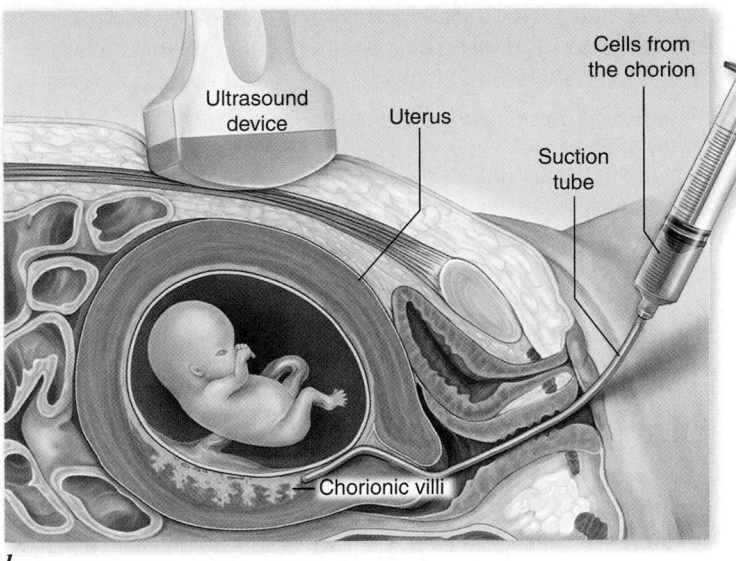

a.   *b.*

Figure 13.15 Two ways to obtain fetal cells. *a.* In amniocentesis, a needle is inserted into the amniotic cavity, and a sample of amniotic fluid containing some free cells derived from the fetus is withdrawn into a syringe. *b.* In chorionic villi sampling, cells are removed by suction with a tube inserted through the cervix. This can be done as early as the eighth to tenth week. In each case, the cells can be grown in culture, then examined for karyotypes, and used in biochemical and genetic tests.

discerning the number and frequency of disease-causing alleles, but these problems are not insurmountable. At present, tests for at least 13 genes with alleles that lead to clinical syndromes are available. This number is bound to rise and to be expanded to include alleles that do not directly lead to disease states but that predispose a person for a particular disease.

? Inquiry question Based on what you read in this chapter, what reasons could a mother have to undergo CVS, considering its small but potential risks?

Learning Outcomes Review 13.5

Mutations in DNA that result in altered proteins can cause hereditary diseases. Pedigree studies and genetic testing may clarify the risk of disease. At the chromosome level, nondisjunction during meiosis can result in gametes with too few or too many chromosomes, most of which produce inviable offspring. Imprinting refers to inactivation of alleles depending on which parent the alleles come from; offspring in whom imprinting occurs appear haploid for the affected gene even though they are diploid.

■ ***During spermatogenesis, is there any difference in outcome between first- and second-division nondisjunction?***

Chapter Review

13.1 Sex Linkage and the Chromosomal Theory of Inheritance

Morgan correlated the inheritance of a trait with sex chromosomes (figure 13.2).
Morgan crossed red-eyed and white-eyed flies and found differences in inheritance based on the sex of offspring. All white-eyed offspring were males, but testcrosses showed that white-eyed females were possible, supporting the idea that the white-eye gene was on the X chromosome.

The gene for eye color lies on the X chromosome.
The inheritance of eye color in *Drosophila* segregates with the X chromosome, a phenomenon termed sex-linked inheritance.

13.2 Sex Chromosomes and Sex Determination

Sex determination in animals is usually associated with a chromosomal difference. In some animals, females have two similar sex chromosomes and males have sex chromosomes that differ. In other species, females have sex chromosomes that differ (table 13.1).

In humans, the Y chromosome generally determines maleness.
The Y chromosome is highly condensed and does not have active counterparts to most genes on the X chromosome. The *SRY* gene on the Y chromosome is responsible for the masculinization of genitalia and secondary sex organs. An XY individual can develop into a sterile female due to mutations in the *SRY* gene or the failure of the embryo to respond to androgens.

Some human genetic disorders display sex linkage (figure 13.3).

Human genetic disorders show sex linkage when the relevant gene is on the X chromosome; hemophilia is an example.

Dosage compensation prevents doubling of sex-linked gene products.

In fruit flies, males double the gene expression from their single X chromosome. In mammals, one of the X chromosomes in a female is randomly inactivated during development.

X-chromosome inactivation can lead to genetic mosaics.

In a mammalian female that is heterozygous for X-chromosome alleles, X inactivation produces a mosaic pattern, as shown in the coat color of calico cats (figure 13.4).

13.3 Exceptions to the Chromosomal Theory of Inheritance

Mitochondrial genes are inherited from the female parent.

Mitochondria have their own genomes and divide independently; they are passed to offspring in the cytoplasm of the egg cell.

Chloroplast genes may also be passed on uniparentally.

Chloroplasts also reside in the cytoplasm, have their own genomes, and divide independently. They are usually inherited maternally.

13.4 Genetic Mapping

Mendel's independent assortment is too simplistic. Genes on the same chromosome may or may not segregate independently.

Genetic recombination exchanges alleles on homologues.

Homologous chromosomes may exchange alleles by crossing over (figure 13.5). This occurs by breakage and rejoining of chromosomes as shown by crosses in which chromosomes carry both visible and genetic markers (figure 13.6).

Recombination is the basis for genetic maps.

Genes close together on a single chromosome are said to be linked. The farther apart two linked genes are, the greater the frequency of recombination. This allows genetic maps to be constructed based on recombination frequency. A map unit is expressed as the percentage of recombinant progeny.

Multiple crossovers can yield independent assortment results.

The probability of multiple crossovers increases with distance between two genes and results in an underestimate of recombination frequency. The maximum recombination frequency is 50%, the same value as for independent assortment.

Three-point crosses can be used to put genes in order (figure 13.9).

If three genes are used instead of two, data from multiple crossovers can be used to order genes. Longer map distances fail to reflect the effect of multiple crossovers and thus underestimate true distance. By evaluating intervening genes with less separation, more accurate distances can be obtained.

Genetic maps can be constructed for the human genome.

Human genetic mapping was difficult because it required multiple disease-causing alleles segregating in a family. The process has been made easier by the use of anonymous markers, identifiable molecular markers that do not cause a phenotype. Single-nucleotide polymorphisms (SNPs) can be used to detect differences between individuals for identification.

13.5 Selected Human Genetic Disorders

Sickle cell anemia is due to altered hemoglobin.

The phenotypes in sickle cell anemia can all be traced to alterations in the structure of hemoglobin that affect the shape of red blood cells. Over 700 variants of hemoglobin structure have been characterized, some of which also cause disorders.

Nondisjunction of chromosomes changes chromosome number.

Nondisjunction is the failure of homologues or sister chromatids to separate during meiosis. The result is aneuploidy: monosomy or trisomy of a chromosome in the zygote. Most aneuploidies are lethal, but some, such as trisomy 21 in humans (Down syndrome), can result in viable offspring. X-chromosome nondisjunction occurs when X chromosomes fail to separate during meiosis. The resulting gamete carries either XX or O (zero sex chromosomes) (figure 13.14). Y-chromosome nondisjunction results in YY gametes.

Genomic imprinting depends on the parental origin of alleles.

In genomic imprinting, the expression of a gene depends on whether it passes through the maternal or paternal germ line. Imprinted genes appear to be inactivated by methylation. Imprinting produces a haploid phenotype.

Some genetic defects can be detected early in pregnancy.

Genetic defects in humans can be determined by pedigree analysis, amniocentesis, or chorionic villi sampling.

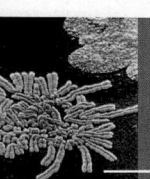

Review Questions

UNDERSTAND

1. Why is the white-eye phenotype always observed in males carrying the white-eye allele?
   a. Because the trait is dominant
   b. Because the trait is recessive
   c. Because the allele is located on the X chromosome and males only have one X
   d. Because the allele is located on the Y chromosome and only males have Y chromosomes

2. In an organism's genome, *autosomes* are
   a. the chromosomes that differ between the sexes.
   b. chromosomes that are involved in sex determination.
   c. only inherited from the mother (maternal inheritance).
   d. all of the chromosomes other than sex chromosomes.

3. What cellular process is responsible for genetic recombination?
   a. The independent alignment of homologous pairs during meiosis I
   b. Separation of the homologues in meiosis I
   c. Separation of the chromatids during meiosis II
   d. Crossing over between homologues

4. The map distance between two genes is determined by the
   a. recombination frequency.
   b. frequency of parental types.
   c. ratio of genes to length of a chromosome.
   d. ratio of parental to recombinant progeny.

5. How many map units separate two alleles if the recombination frequency is 0.07?
   a. 700 cM   b. 70 cM   c. 7 cM   d. 0.7 cM

6. How does maternal inheritance of mitochondrial genes differ from sex linkage?
   a. Mitochondrial genes do not contribute to the phenotype of an individual.
   b. Because mitochondria are inherited from the mother, only females are affected.
   c. Since mitochondria are inherited from the mother, females and males are equally affected.
   d. Mitochondrial genes must be dominant. Sex-linked traits are typically recessive.

7. Which of the following genotypes due to nondisjunction of sex chromosomes is lethal?
   a. XXX   b. XXY   c. OY   d. XO

APPLY

1. A recessive sex-linked gene in humans leads to a loss of sweat glands. A woman heterozygous for this will
   a. have no sweat glands.
   b. have normal sweat glands.
   c. have patches of skin with and without sweat glands.
   d. have an excess of sweat glands.

2. As real genetic distance increases, the distance calculated by recombination frequency becomes an
   a. overestimate due to multiple crossovers that cannot be scored.
   b. underestimate due to multiple crossovers that cannot be scored.
   c. underestimate due to multiple crossovers adding to recombination frequency.
   d. overestimate due to multiple crossovers adding to recombination frequency.

3. Down syndrome is the result of trisomy for chromosome 21. Why is this trisomy viable and trisomy for most other chromosomes is not?
   a. Chromosome 21 is a large chromosome and excess genetic material is less harmful.
   b. Chromosome 21 behaves differently in meiosis I than the other chromosomes.
   c. Chromosome 21 is a small chromosome with few genes so this does less to disrupt the genome.
   d. Chromosome 21 is less prone to nondisjunction than other chromosomes.

4. Genes that are on the same chromosome can show independent assortment
   a. when they are far enough apart for two crossovers to occur.
   b. when they are far enough apart that odd numbers of crossovers is about equal to even.
   c. only if recombination is low for that chromosome.
   d. only if the genes show genomic imprinting.

5. The A and B gene are 10 cM apart on a chromosome. If an A B/a b heterozygote is testcrossed to a b/a b, how many of each progeny class would you expect out of 100 total progeny?
   a. 25 A B, 25 a b, 25 A b, 25 a B
   b. 10 A B, 10 a b
   b. 45 A B, 45 a b
   d. 45 A B, 45 a b, 5 A b, 5 a B

6. During the process of spermatogenesis, a nondisjunction event that occurs during the second division would be
   a. worse than the first division because all four meiotic products would be aneuploid.
   b. better than the first division because only two of the four meiotic products would be aneuploid.
   c. the same outcome as the first division with all four products aneuploid.
   d. the same outcome as the first division as only two products would be aneuploid.

SYNTHESIZE

1. Color blindness is caused by a sex-linked, recessive gene. If a woman, whose father was color blind, marries a man with normal color vision, what percentage of their children will be color blind? What percentage of male children? Of female children?

2. Assume that the genes for seed color and seed shape are located on the same chromosome. A plant heterozygous for both genes is testcrossed wrinkled green with the following results:

green, wrinkled	645
green, round	36
yellow, wrinkled	29
yellow, round	590

   What were the genotypes of the parents, and how far apart are these genes?

3. Is it possible to have a calico cat that is male? Why or why not?

ONLINE RESOURCE

www.ravenbiology.com

Understand, Apply, and Synthesize—enhance your study with animations that bring concepts to life and practice tests to assess your understanding. Your instructor may also recommend the interactive eBook, individualized learning tools, and more.

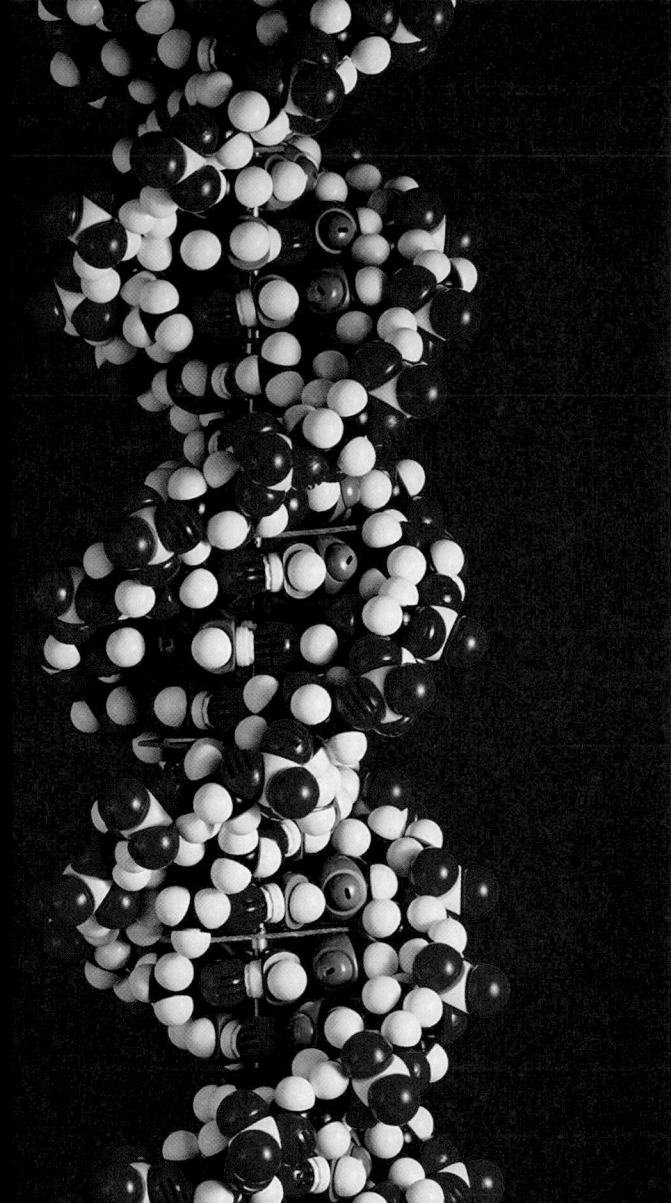

Chapter 14

DNA: The Genetic Material

Chapter Contents

Introduction

The rediscovery of Mendel at the turn of the 20th century led to a period of rapid discovery of genetic mechanisms detailed in the previous two chapters. One question not directly answered for more than 50 years was perhaps the most simple: What are genes actually made of? Genes were known to be on chromosomes, but they are complex structures composed of DNA, RNA, and protein. This chapter describes the chain of experiments that led to our current understanding of DNA, modeled in the picture above, and of the molecular mechanisms of heredity. These experiments are among the most elegant in science. The elucidation of the structure of DNA was the beginning of a molecular era, whose pace is only accelerating today.

14.1 The Nature of the Genetic Material

Learning Outcomes

1. *Describe the experiments of Griffith and Avery.*
2. *Evaluate the evidence for DNA as genetic material.*

In the previous two chapters, you learned about the nature of inheritance and how genes, which contain the information to specify traits, are located on chromosomes. This finding led to the question of what part of the chromosome actually contains the genetic information. Specifically, biologists wondered about the chemical identity of the genetic information. They knew that chromosomes are composed primarily of both protein and DNA. Which of these organic molecules actually makes up the genes?

Starting in the late 1920s and continuing for about 30 years, a series of investigations addressed this question.

DNA consists of four chemically similar nucleotides. In contrast, protein contains 20 different amino acids that are much more chemically diverse than nucleotides. These characteristics seemed initially to indicate greater informational capacity in protein than in DNA.

However, experiments began to reveal evidence in favor of DNA. We describe three of those major findings in this section.

Griffith finds that bacterial cells can be transformed

The first clue came in 1928 with the work of the British microbiologist Frederick Griffith. Griffith was trying to make a vaccine that would protect against influenza, which was thought at the time to be caused by the bacteria *Streptococcus pneumoniae*. There are two forms of this bacteria: The normal virulent form that causes pneumonia, and a mutant, nonvirulent form that does not. The normal virulent form of this bacterium is referred to as the S form because it forms smooth colonies on a culture dish. The mutant, nonvirulent form, which lacks an enzyme needed to manufacture the polysaccharide coat, is called the R form because it forms rough colonies.

Griffith performed a series of simple experiments in which mice were infected with these bacteria, then monitored for disease symptoms (figure 14.1). Mice infected with the virulent S form died from pneumonia, whereas infection with the nonvirulent R form had no effect. This result shows that the polysaccharide coat is necessary for virulence. If the virulent S form is first heat-killed, infection does not harm the mice, showing that the coat itself is not sufficient to cause disease. Lastly, infecting mice with a mixture of heat-killed S form with live R form caused pneumonia and death in the mice. This was unexpected as neither treatment alone caused disease. Furthermore, high levels of live S form bacteria were found in the lungs of the dead mice.

Somehow, the information specifying the polysaccharide coat had passed from the dead, virulent S bacteria to the live, coatless R bacteria in the mixture, permanently altering the coatless R bacteria into the virulent S variety. Griffith called this transfer of virulence from one cell to another **transformation.** Our modern interpretation is that genetic material was actually transferred between the cells.

Avery, MacLeod, and McCarty identify the transforming principle

The agent responsible for transforming *Streptococcus* went undiscovered until 1944. In a classic series of experiments, Oswald Avery and his coworkers Colin MacLeod and Maclyn McCarty identified the substance responsible for transformation in Griffith's experiment.

They first prepared the mixture of dead S *Streptococcus* and live R *Streptococcus* that Griffith had used. Then they removed as much of the protein as they could from their preparation, eventually achieving 99.98% purity. They found that despite the removal of nearly all protein, the transforming activity was not reduced.

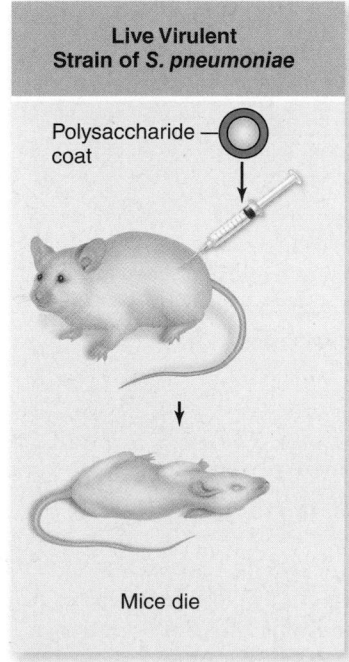
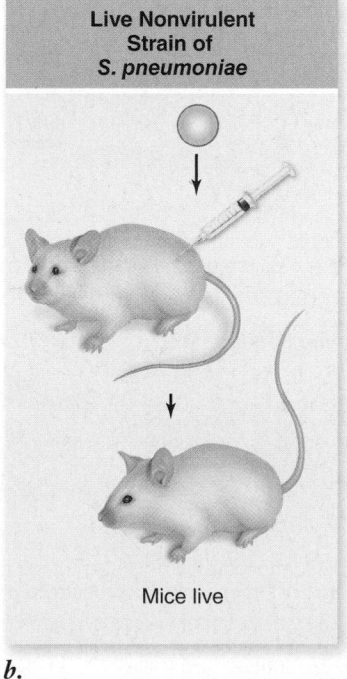
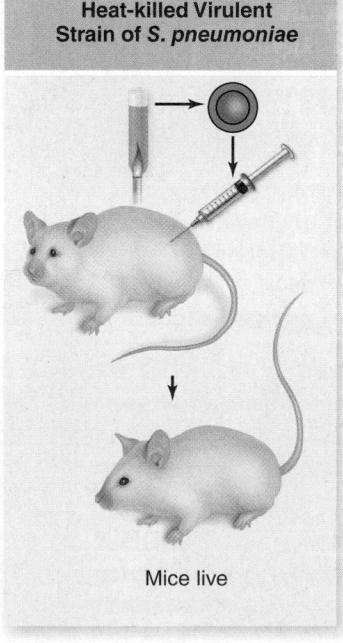
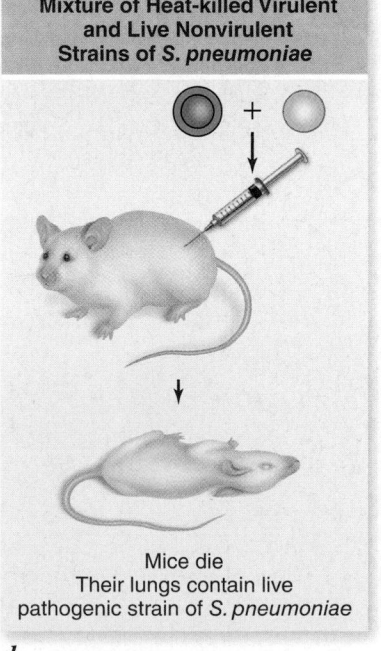

Figure 14.1 Griffith's experiment. Griffith was trying to make a vaccine against pneumonia and instead discovered transformation. *a.* Injecting live virulent bacteria into mice produces pneumonia. Injection of nonvirulent bacteria (*b*) or heat-killed virulent bacteria (*c*) had no effect. *d.* However, a mixture of heat-killed virulent and live nonvirulent bacteria produced pneumonia in the mice. This indicates the genetic information for virulence was transferred from dead, virulent cells to live, nonvirulent cells, transforming them from nonvirulent to virulent.

Moreover, the properties of this substance resembled those of DNA in several ways:

1. The elemental composition agreed closely with that of DNA.
2. When spun at high speeds in an ultracentrifuge, it migrated to the same level (density) as DNA.
3. Extracting lipids and proteins did not reduce transforming activity.
4. Protein-digesting enzymes did not affect transforming activity, nor did RNA-digesting enzymes.
5. DNA-digesting enzymes destroyed all transforming activity.

These experiments supported the identity of DNA as the substance transferred between cells by transformation and indicated that the genetic material, at least in this bacterial species, is DNA.

Hershey and Chase demonstrate that phage genetic material is DNA

Avery's results were not widely accepted at first because many biologists continued to believe that proteins were the repository of hereditary information. But additional evidence supporting Avery's conclusion was provided in 1952 by Alfred Hershey and Martha Chase, who experimented with viruses that infect bacteria. These viruses are called **bacteriophages,** or more simply, **phages.**

Viruses, described in more detail in chapter 27, are much simpler than cells; they generally consist of genetic material (DNA or RNA) surrounded by a protein coat. The phage used in these experiments is called a *lytic* phage because infection causes the cell to burst, or lyse. When such a phage infects a bacterial cell, it first binds to the cell's outer surface and then injects its genetic information into the cell. There, the viral genetic information is expressed by the bacterial cell's machinery, leading to production of thousands of new viruses. The buildup of viruses eventually causes the cell to lyse, releasing progeny phage.

The phage used by Hershey and Chase contains only DNA and protein, and therefore it provides the simplest possible system to differentiate the roles of DNA and protein. Hershey and Chase set out to identify the molecule that the phage injects into the bacterial cells. To do this, they needed a method to label both DNA and protein in unique ways that would allow them to be distinguished. Nucleotides contain

SCIENTIFIC THINKING

Hypothesis: *DNA is the genetic material in bacteriophage.*

Prediction: *The phage life cycle requires reprogramming the cell to make phage proteins. The information for this must be introduced into the cell during infection.*

Test: *DNA can be specifically labeled using radioactive phosphate (^{32}P), and protein can be specifically labeled using radioactive sulfur (^{35}S). Phage are grown on either ^{35}S or ^{32}P, then used to infect cells in two experiments. The phage heads remain attached to the outside of the cell and can be removed by brief agitation in a blender. The cell suspension can be collected by centrifugation, leaving the phage heads in the supernatant.*

^{35}S-Labeled Bacteriophages

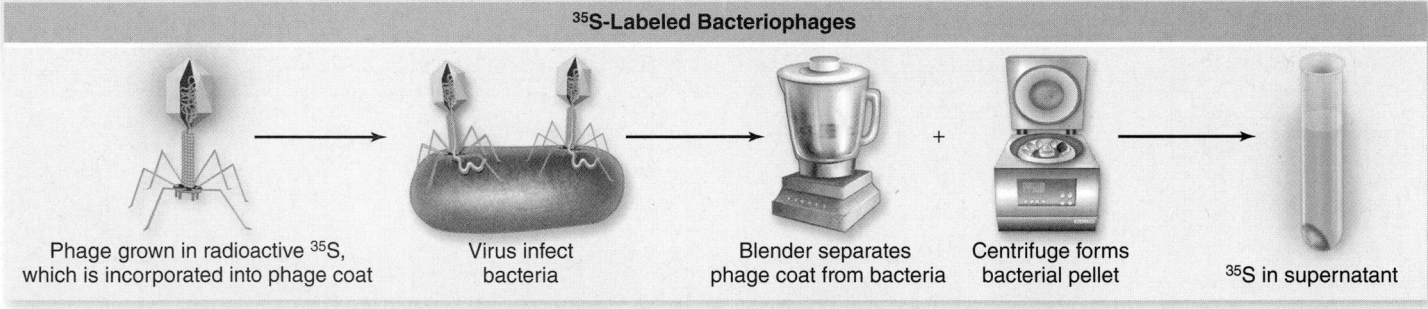

| Phage grown in radioactive ^{35}S, which is incorporated into phage coat | Virus infect bacteria | Blender separates phage coat from bacteria | Centrifuge forms bacterial pellet | ^{35}S in supernatant |

^{32}P-Labeled Bacteriophages

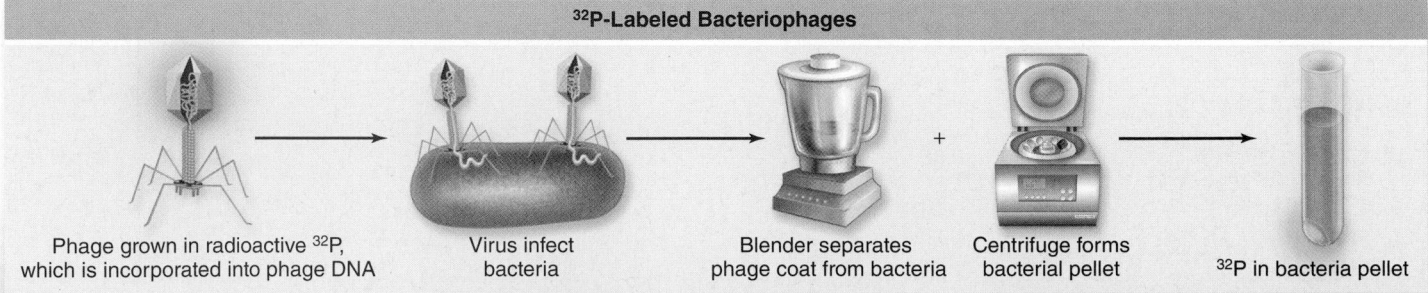

| Phage grown in radioactive ^{32}P, which is incorporated into phage DNA | Virus infect bacteria | Blender separates phage coat from bacteria | Centrifuge forms bacterial pellet | ^{32}P in bacteria pellet |

Result: *When the experiment is done, only ^{32}P makes it into the cell in any significant quantity.*

Conclusion: *Thus, DNA must be the molecule that is used to reprogram the cell.*

Further Experiments: *How does this experiment complement or extend the work of Avery on the identity of the transforming principle?*

Figure 14.2 Hershey–Chase experiment showed DNA is genetic material for phage.

phosphorus, but proteins do not, and some amino acids contain sulfur, but DNA does not. Thus, the radioactive ^{32}P isotope can be used to label DNA specifically, and the isotope ^{35}S can be used to label proteins specifically. The two isotopes are easily distinguished based on the particles they emit when they decay.

Two experiments were performed (figure 14.2). In one, viruses were grown on a medium containing ^{32}P, which was incorporated into DNA; in the other, viruses were grown on medium containing ^{35}S, which was incorporated into coat proteins. Each group of labeled viruses was then allowed to infect separate bacterial cultures.

After infection, the bacterial cell suspension was agitated in a blender to remove the infecting viral particles from the surfaces of the bacteria. This step ensured that only the part of the virus that had been injected into the bacterial cells—that is, the genetic material—would be detected.

Each bacterial suspension was then centrifuged to produce a pellet of cells for analysis. In the ^{32}P experiment, a large amount of radioactive phosphorus was found in the cell pellet, but in the ^{35}S experiment, very little radioactive sulfur was found in the pellet (see figure 14.2). Hershey and Chase deduced that DNA, and not protein, constituted the genetic information that viruses inject into bacteria.

Learning Outcomes Review 14.1

Experiments with pneumonia-causing bacteria showed that virulence could be passed from one cell to another, a phenomenon termed transformation. When the factor responsible for transformation was purified, it was shown to be DNA. Labeling experiments with phage also indicated that the genetic material was DNA and not protein.

■ **Why was protein an attractive candidate for the genetic material?**

Learning Outcomes

1. **Explain how the Watson–Crick structure rationalized the data available to them.**
2. **Evaluate the significance of complementarity for DNA structure and function.**

A Swiss chemist, Friedrich Miescher, discovered DNA in 1869, only four years after Mendel's work was published—although it is unlikely that Miescher knew of Mendel's experiments.

Miescher extracted a white substance from the nuclei of human cells and fish sperm. The proportion of nitrogen and phosphorus in the substance was different from that found in any other known constituent of cells, which convinced Miescher that he had discovered a new biological substance. He called this substance "nuclein" because it seemed to be specifically associated with the nucleus. Because Miescher's nuclein was slightly acidic, it came to be called *nucleic acid.*

DNA's components were known, but its three-dimensional structure was a mystery

Although the three-dimensional structure of the DNA molecule was not elucidated until Watson and Crick, it was known that it contained three main components (figure 14.3):

1. a five-carbon sugar
2. a phosphate (PO_4) group

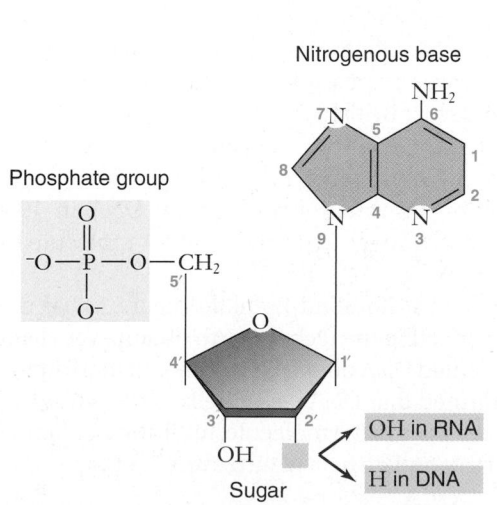

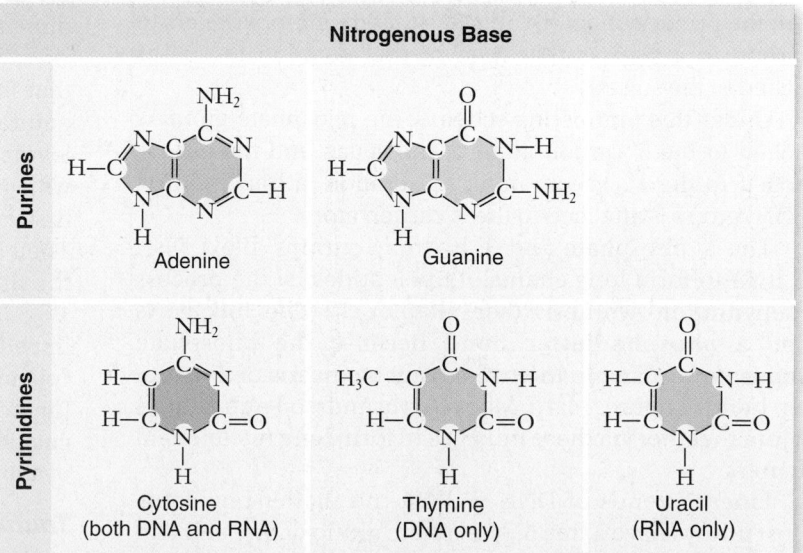

Figure 14.3 Nucleotide subunits of DNA and RNA. The nucleotide subunits of DNA and RNA are composed of three components: a five-carbon sugar (deoxyribose in DNA and ribose in RNA); a phosphate group; and a nitrogenous base (either a purine or a pyrimidine).

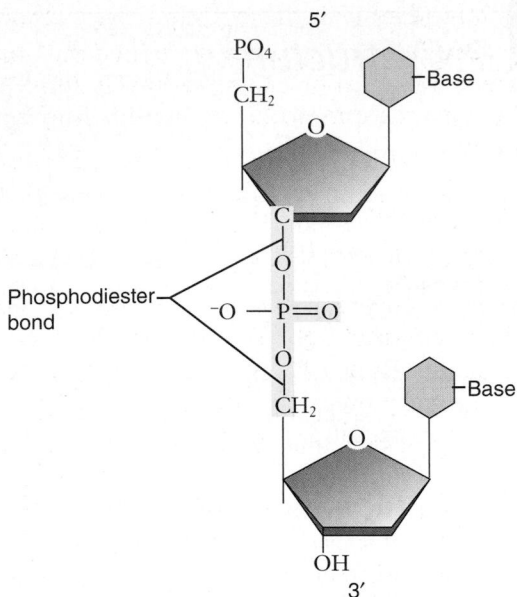

Figure 14.4 A phosphodiester bond.

3. a nitrogen-containing (nitrogenous) base. The base may be a **purine** (adenine, A, or guanine, G), a two-ringed structure; or a **pyrimidine** (thymine, T, or cytosine, C), a single-ringed structure. RNA contains the pyrimidine uracil (U) in place of thymine.

The convention in organic chemistry is to number the carbon atoms of a molecule and then to use these numbers to refer to any functional group attached to a carbon atom (see chapter 3). In the ribose sugars found in nucleic acids, four of the carbon atoms together with an oxygen atom form a five-membered ring. As illustrated in figure 14.3, the carbon atoms are numbered 1′ to 5′, proceeding clockwise from the oxygen atom; the prime symbol (′) indicates that the number refers to a carbon in a sugar rather than to the atoms in the bases attached to the sugars.

Under this numbering scheme, the phosphate group is attached to the 5′ carbon atom of the sugar, and the base is attached to the 1′ carbon atom. In addition, a free hydroxyl (—OH) group is attached to the 3′ carbon atom.

The 5′ phosphate and 3′ hydroxyl groups allow DNA and RNA to form long chains of nucleotides by the process of dehydration synthesis (see chapter 3). The linkage is called a **phosphodiester bond** because the phosphate group is now linked to the two sugars by means of a pair of ester bonds (figure 14.4). Many thousands of nucleotides can join together via these linkages to form long nucleic acid polymers.

Linear strands of DNA or RNA, no matter how long, almost always have a free 5′ phosphate group at one end and a free 3′ hydroxyl group at the other. Therefore, every DNA and RNA molecule has an intrinsic polarity, and we can refer unambiguously to each end of the molecule. By convention, the sequence of bases is usually written in the 5′-to-3′ direction.

Chargaff, Franklin, and Wilkins obtained some structural evidence

To understand the model that Watson and Crick proposed, we need to review the evidence that they had available to construct their model.

Chargaff's rules

A careful study carried out by Erwin Chargaff showed that the nucleotide composition of DNA molecules varied in complex ways, depending on the source of the DNA. This strongly suggested that DNA was not a simple repeating polymer and that it might have the information-encoding properties genetic material requires. Despite DNA's complexity, however, Chargaff observed an important underlying regularity in the ratios of the bases found in native DNA: *The amount of adenine present in DNA always equals the amount of thymine, and the amount of guanine always equals the amount of cytosine.* Two important findings from this work are often called *Chargaff's rules:*

1. The proportion of A always equals that of T, and the proportion of G always equals that of C, or: A = T, and G = C.
2. The ratio of G–C to A–T varies with different species.

As mounting evidence indicated that DNA stored the hereditary information, investigators began to puzzle over how such a seemingly simple molecule could carry out such a complex coding function.

X-ray diffraction patterns of DNA

The technique of X-ray diffraction provided more direct information about the possible structure of DNA. In X-ray diffraction, crystals of a molecule are bombarded with a beam of X-rays. The rays are bent, or diffracted, by the molecules they encounter, and the diffraction pattern is recorded on photographic film. The patterns resemble the ripples created by tossing a rock into a smooth lake. When analyzed mathematically, the diffraction pattern can yield information about the three-dimensional structure of a molecule.

The problem with using this technique with DNA was that in the 1950s, it was impossible to obtain true crystals of natural DNA. However, British researcher Maurice Wilkins learned how to prepare uniformly oriented DNA fibers, and with graduate student Ray Gosling succeeded in obtaining the first crude diffraction information on natural DNA in 1950. Their early X-ray photos suggested that the DNA molecule has the shape of a helix.

The British chemist Rosalind Franklin (figure 14.5a) continued this work, perfecting the technique to obtain ever clearer "pictures" of the oriented DNA fibers. The clearest of these images (figure 14.5b) confirmed that DNA was a helix, and allowed calculation of the dimensions of the molecule, indicating a diameter of about 2 nm and a complete helical turn every 3.4 nm.

Tautomeric forms of bases

One piece of evidence important to Watson and Crick was the form of the bases themselves. Because of the alternating double and single bonds in the bases, they actually exist in equilibrium between two different forms when in solution. The different forms have to do with keto (C=O) versus enol

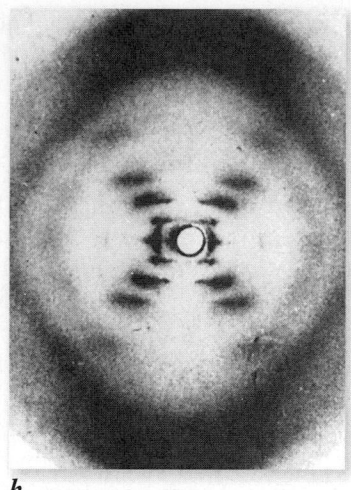

a. **b.**

Figure 14.5 Rosalind Franklin's X-ray diffraction patterns. *a.* Rosalind Franklin. ***b.*** This X-ray diffraction photograph of DNA fibers, made in 1953 by Rosalind Franklin, was interpreted to show the helical structure of DNA.

(C—OH) groups and amino (—NH₂) versus imino (=NH) groups that are attached to the bases. These structural forms are called *tautomers.*

The importance of this distinction is that the two forms exhibit very different hydrogen-bonding possibilities. The predominant forms of the bases contain the keto

and amino groups (see figure 14.3), but a prominent biochemistry text of the time actually contained the opposite, and incorrect, information. Legend has it that Watson learned the correct forms while having lunch with a biochemist friend.

The Watson–Crick model fits the available evidence

Learning informally of Franklin's results before they were published in 1953, American chemist James Watson and English molecular biologist Francis Crick, two young investigators at Cambridge University, quickly worked out a likely structure for the DNA molecule (figure 14.6), which we now know was substantially correct. Watson and Crick did not perform a single experiment themselves related to DNA structure; rather, they built detailed molecular models based on the information available.

The key to the model was their understanding that each DNA molecule is actually made up of *two* chains of nucleotides that are intertwined—the double helix.

The phosphodiester backbone

The two strands of the double helix are made up of long polymers of nucleotides, and as described earlier, each strand is made up of repeating sugar and phosphate units joined by phosphodiester bonds (figure 14.7). We call this the *phosphodiester backbone* of the molecule. The two strands of the

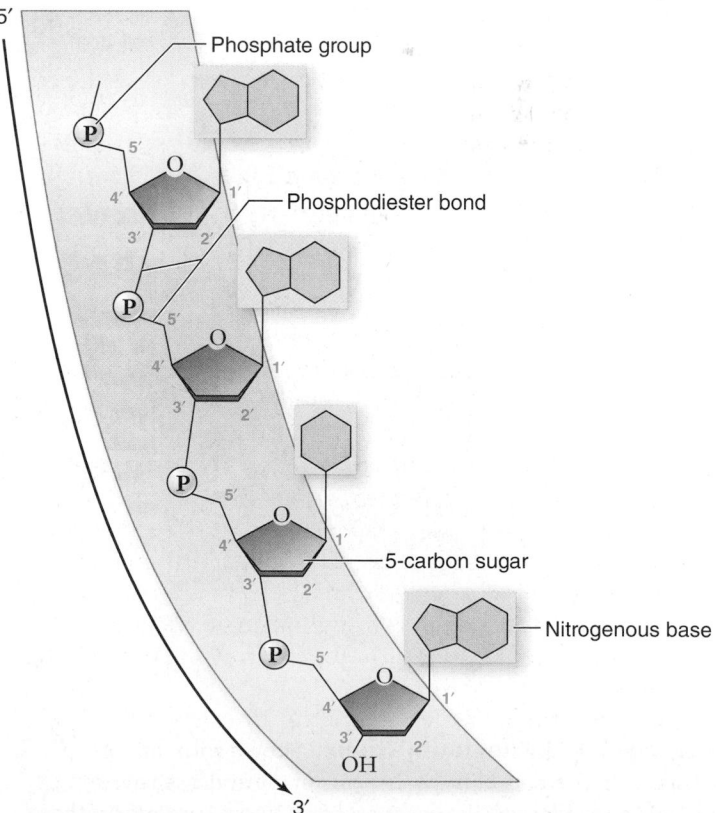

Figure 14.7 Structure of a single strand of DNA. The phosphodiester backbone is composed of alternating sugar and phosphate groups. The bases are attached to each sugar.

Figure 14.6 The DNA double helix. James Watson *(left)* and Francis Crick *(right)* deduced the structure of DNA in 1953 from Chargaff's rules, knowing the proper tautomeric forms of the bases and using Franklin's diffraction studies.

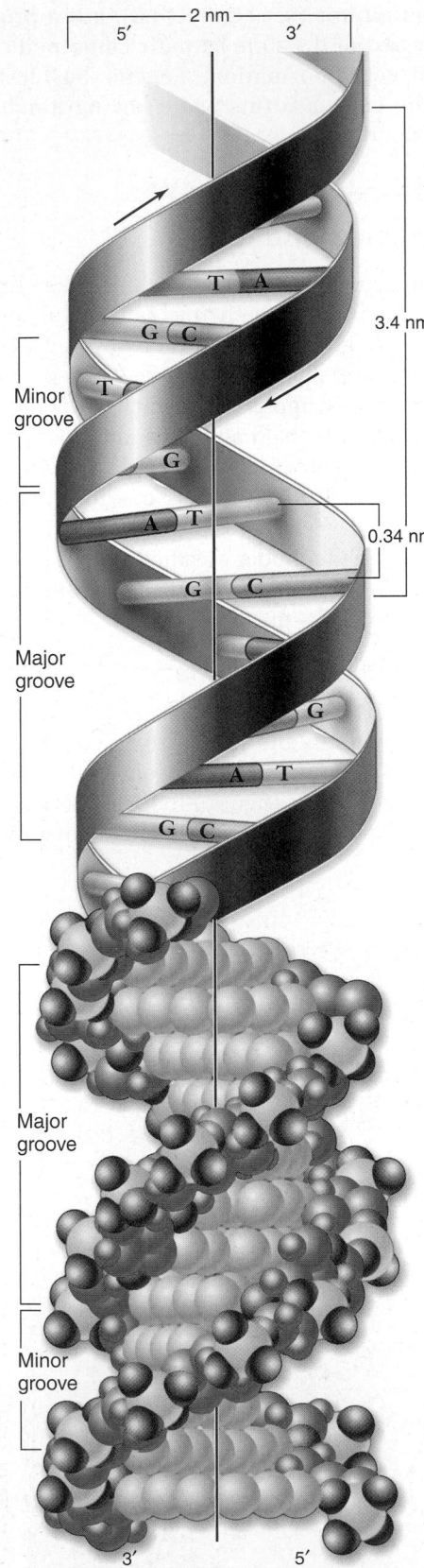

Figure 14.8 The double helix. Shown with the phosphodiester backbone as a ribbon on top and a space-filling model on the bottom. The bases protrude into the interior of the helix where they hold it together by base-pairing. The backbone forms two grooves, the larger major groove and the smaller minor groove.

backbone are then wrapped about a common axis forming a double helix (figure 14.8). The helix is often compared to a spiral staircase, in which the two strands of the double helix are the handrails on the staircase.

Complementarity of bases

Watson and Crick proposed that the two strands were held together by formation of hydrogen bonds between bases on opposite strands. These bonds would result in specific **base-pairs:** Adenine (A) can form two hydrogen bonds with thymine (T) to form an A–T base-pair, and guanine (G) can form three hydrogen bonds with cytosine (C) to form a G–C base-pair (figure 14.9).

Note that this configuration also pairs a two-ringed purine with a single-ringed pyrimidine in each case, so that the diameter of each base-pair is the same. This consistent diameter is indicated by the X-ray diffraction data.

We refer to this pattern of base-pairing as *complementary,* which means that although the strands are not identical, they each can be used to specify the other by base-pairing. If the sequence of one strand is ATGC, then the complementary strand sequence must be TACG. This characteristic becomes critical for DNA replication and expression, as you will see later in this chapter.

The Watson–Crick model also explained Chargaff's results: In a double helix, adenine forms two hydrogen bonds with thymine, but it will not form hydrogen bonds properly with cytosine. Similarly, guanine forms three hydrogen bonds with cytosine, but it will not form hydrogen bonds properly with thymine. Because of this base-pairing, adenine and thymine always occur in the same proportions in any DNA molecule, as do guanine and cytosine.

Figure 14.9 Base-pairing holds strands together. The hydrogen bonds that form between A and T and between G and C are shown with dashed lines. These produce AT and GC base-pairs that hold the two strands together. This always pairs a purine with a pyrimidine, keeping the diameter of the double helix constant.

 Data analysis Explain how the Watson–Crick model accounts for the data discussed in the text.

Antiparallel configuration

As stated earlier, a single phosphodiester strand has an inherent polarity, meaning that one end terminates in a 3′ OH and the other end terminates in a 5′ PO₄. Strands are thus referred to as having either a 5′-to-3′ or a 3′-to-5′ polarity. Two strands could be put together in two ways: with the polarity the same in each (parallel) or with the polarity opposite (antiparallel). Native double-stranded DNA always has the antiparallel configuration, with one strand running 5′ to 3′ and the other running 3′ to 5′ (see figure 14.8). In addition to its complementarity, this antiparallel nature also has important implications for DNA replication.

The Watson–Crick DNA molecule

In the Watson and Crick model, each DNA molecule is composed of two complementary phosphodiester strands that each form a helix with a common axis. These strands are antiparallel, with the bases extending into the interior of the helix. The bases from opposite strands form base-pairs with each other to join the two complementary strands (see figures 14.8 and 14.9).

Although the hydrogen bonds between each individual base-pair are low-energy bonds, the sum of bonds between the many base-pairs of the polymer has enough energy that the entire molecule is stable. To return to our spiral staircase analogy—the backbone is the handrails, the base-pairs are the steps.

Although the Watson–Crick model provided a rational structural for DNA, researchers had to answer further questions about how DNA could be replicated, a crucial step in cell division, and also about how cells could repair damaged or otherwise altered DNA. We explore these questions in the rest of this chapter. (In the following chapter, we continue with the genetic code and the connection between the code and protein synthesis.)

Learning Outcomes Review 14.2

Chargaff showed that in DNA, the amount of adenine was equal to the amount of thymine, and the amount of guanosine was equal to that of cytosine. X-ray diffraction studies by Franklin and Wilkins indicated that DNA formed a helix. Watson and Crick built a model consisting of two antiparallel strands wrapped in a helix about a common axis. The two strands are held together by hydrogen bonds between the bases: adenine pairs with thymine and guanine pairs with cytosine. The two strands are thus complementary to each other.

■ **Why was information about the proper tautomeric form of the bases critical?**

14.3 Basic Characteristics of DNA Replication

Learning Outcomes

1. Illustrate the products of semiconservative replication.
2. Describe the requirements for DNA replication.

The accurate replication of DNA prior to cell division is a basic and crucial function. Research has revealed that this complex process requires the participation of a large number of cellular proteins. Before geneticists could look for these details, however, they needed to perform some groundwork on the general mechanisms.

Meselson and Stahl demonstrate the semiconservative mechanism

The Watson–Crick model immediately suggested that the basis for copying the genetic information is complementarity. One chain of the DNA molecule may have any conceivable base sequence, but this sequence completely determines the sequence of its partner in the duplex.

In replication, the sequence of parental strands must be duplicated in daughter strands. That is, one parental helix with two strands must yield two daughter helices with four strands. The two daughter molecules are then separated during the course of cell division.

Three models of DNA replication are possible (figure 14.10):

1. In a *conservative model,* both strands of the parental duplex would remain intact (conserved), and new DNA copies would consist of all-new molecules. Both daughter strands would contain all-new molecules.

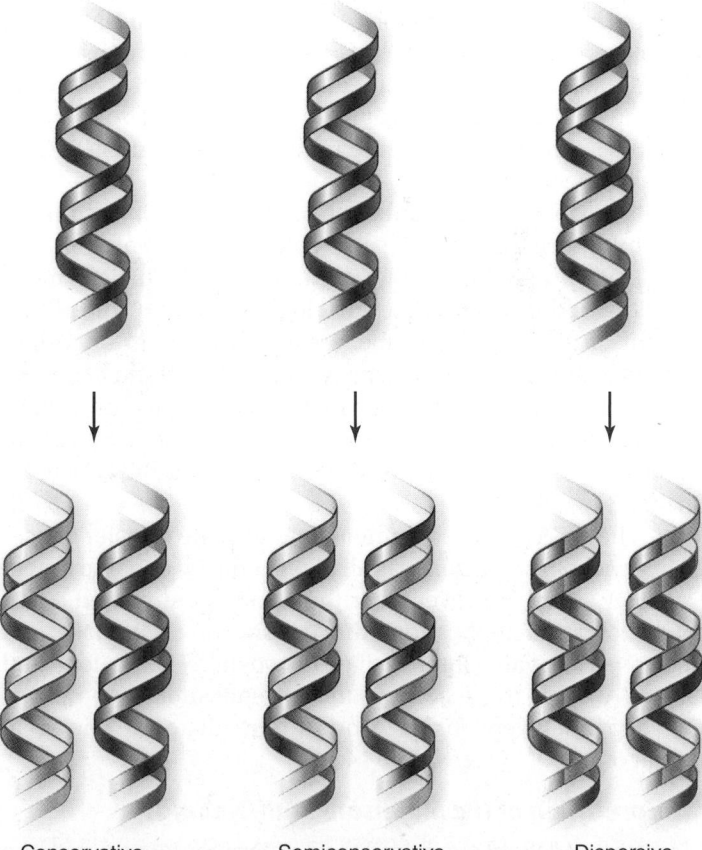

Conservative   Semiconservative   Dispersive

Figure 14.10 Three possible models for DNA replication. The conservative model produces one entirely new molecule and conserves the old. The semiconservative model produces two hybrid molecules of old and new strands. The dispersive model produces hybrid molecules with each strand a mixture of old and new.

2. In a **semiconservative model,** one strand of the parental duplex remains intact in daughter strands (semiconserved); a new complementary strand is built for each parental strand consisting of new molecules. Daughter strands would consist of one parental strand and one newly synthesized strand.

3. In a *dispersive model,* copies of DNA would consist of mixtures of parental and newly synthesized strands; that is, the new DNA would be dispersed throughout each strand of both daughter molecules after replication.

Notice that these three models suggest general mechanisms of replication, without specifying any molecular details of the process.

The Meselson–Stahl experiment

The three models for DNA replication were evaluated in 1958 by Matthew Meselson and Franklin Stahl. To distinguish between these models, they labeled DNA and then followed the labeled DNA through two rounds of replication (figure 14.11).

The label Meselson and Stahl used was a heavy isotope of nitrogen (^{15}N), not a radioactive label. Molecules containing ^{15}N have a greater density than those containing the common ^{14}N isotope. Ultracentrifugation can be used to separate molecules that have different densities.

Bacteria were grown in a medium containing ^{15}N, which became incorporated into the bases of the bacterial DNA. After several generations, the DNA of these bacteria was denser than that of bacteria grown in a medium containing the normally available ^{14}N. Meselson and Stahl then transferred the bacteria from the ^{15}N medium to ^{14}N medium and collected the DNA at various time intervals.

The DNA for each interval was dissolved in a solution containing a heavy salt, cesium chloride. This solution was spun at very high speeds in an ultracentrifuge. The enormous centrifugal forces caused cesium ions to migrate toward the bottom of the centrifuge tube, creating a gradient of cesium concentration, and thus of density. Each DNA strand floated or sank in the gradient until it reached the point at which its density exactly matched the density of the cesium at that location. Because ^{15}N strands are denser than ^{14}N strands, they migrated farther down the tube.

The DNA collected immediately after the transfer of bacteria to new ^{14}N medium was all of one density equal to that of ^{15}N DNA alone. However, after the bacteria completed a first round of DNA replication, the density of their DNA had decreased to a value intermediate between ^{14}N DNA alone and ^{15}N DNA. After the second round of replication, two density classes of DNA were observed: one intermediate and one equal to that of ^{14}N DNA (see figure 14.11).

Interpretation of the Meselson–Stahl findings

Meselson and Stahl compared their experimental data with the results that would be predicted on the basis of the three models.

1. The conservative model was not consistent with the data because after one round of replication, two densities should have been observed: DNA strands would either be all-heavy (parental) or all-light (daughter). This model is rejected.

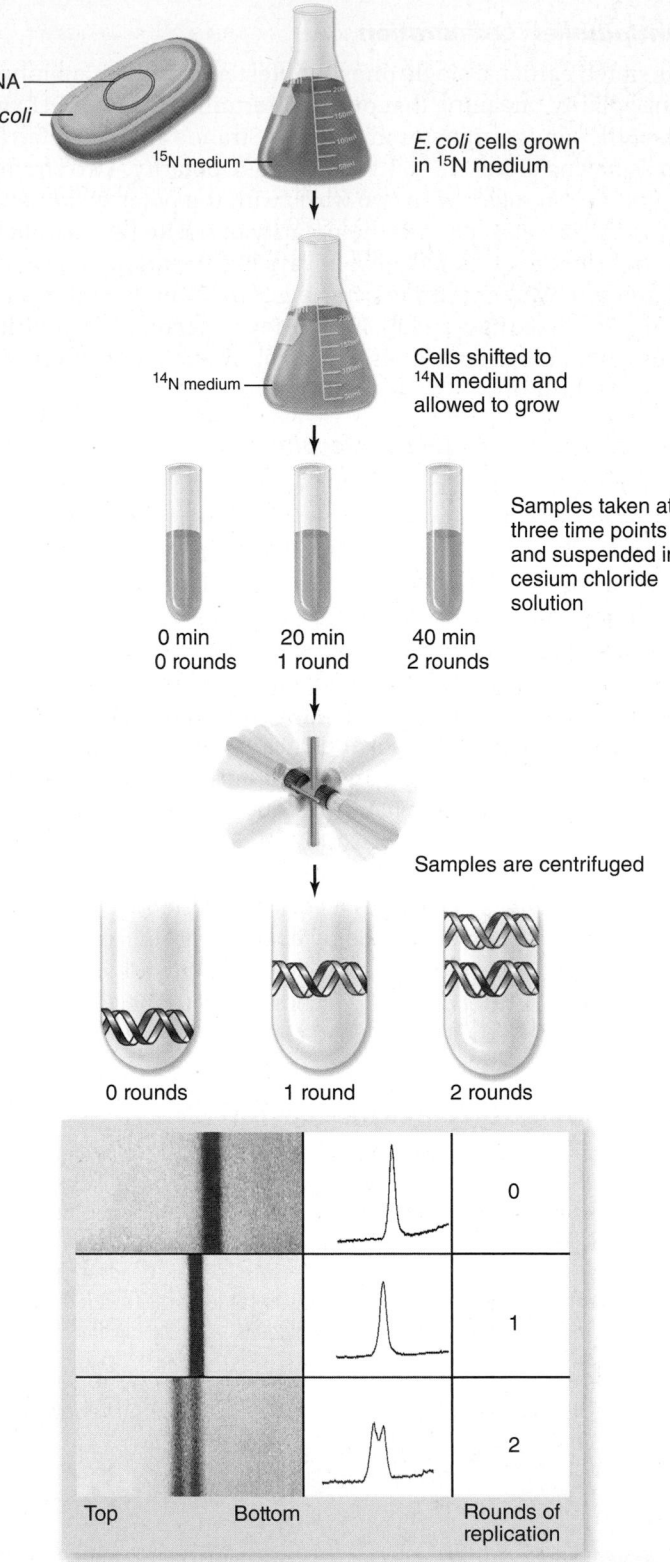

Figure 14.11 The Meselson–Stahl experiment. Bacteria grown in heavy ^{15}N medium are shifted to light ^{14}N medium and grown for two rounds of replication. Samples are taken at time points corresponding to zero, one, and two rounds of replication and centrifuged in cesium chloride to form a gradient. The actual data are shown at the bottom with the interpretation of semiconservative replication shown schematically.

 Data analysis What would you predict for the products of a third round of replication?

2. The semiconservative model is consistent with all observations: After one round of replication, a single density would be predicted because all DNA molecules would have a light strand and a heavy strand. After two rounds of replication, half of the molecules would have two light strands, and half would have a light strand and a heavy strand—and so two densities would be observed. Therefore, the results support the semiconservative model.

3. The dispersive model was consistent with the data from the first round of replication, because in this model, every DNA helix would consist of strands that are mixtures of ½ light (new) and ½ heavy (old) molecules. But after two rounds of replication, the dispersive model would still yield only a single density; DNA strands would be composed of ¾ light and ¼ heavy molecules. Instead, two densities were observed. Therefore, this model is also rejected.

The basic mechanism of DNA replication is semiconservative. At the simplest level, then, DNA is replicated by opening up a DNA helix and making copies of both strands to produce two daughter helices, each consisting of one old strand and one new strand.

DNA replication requires a template, nucleotides, and a polymerase enzyme

Replication requires three things: something to copy, something to do the copying, and the building blocks to make the copy. The parental DNA molecules serve as a template, enzymes perform the actions of copying the template, and the building blocks are nucleoside triphosphates.

The process of replication can be thought of as having a beginning where the process starts; a middle where the majority of building blocks are added; and an end where the process is finished. We use the terms *initiation, elongation,* and *termination* to describe a biochemical process. Although this may seem overly simplistic, in fact, discrete functions are usually required for initiation and termination that are not necessary for elongation.

A number of enzymes work together to accomplish the task of assembling a new strand, but the enzyme that actually matches the existing DNA bases with complementary nucleotides and then links the nucleotides together to make the new strand is **DNA polymerase** (figure 14.12). All DNA polymerases that have been examined have several common features. They all add new bases to the 3′ end of existing

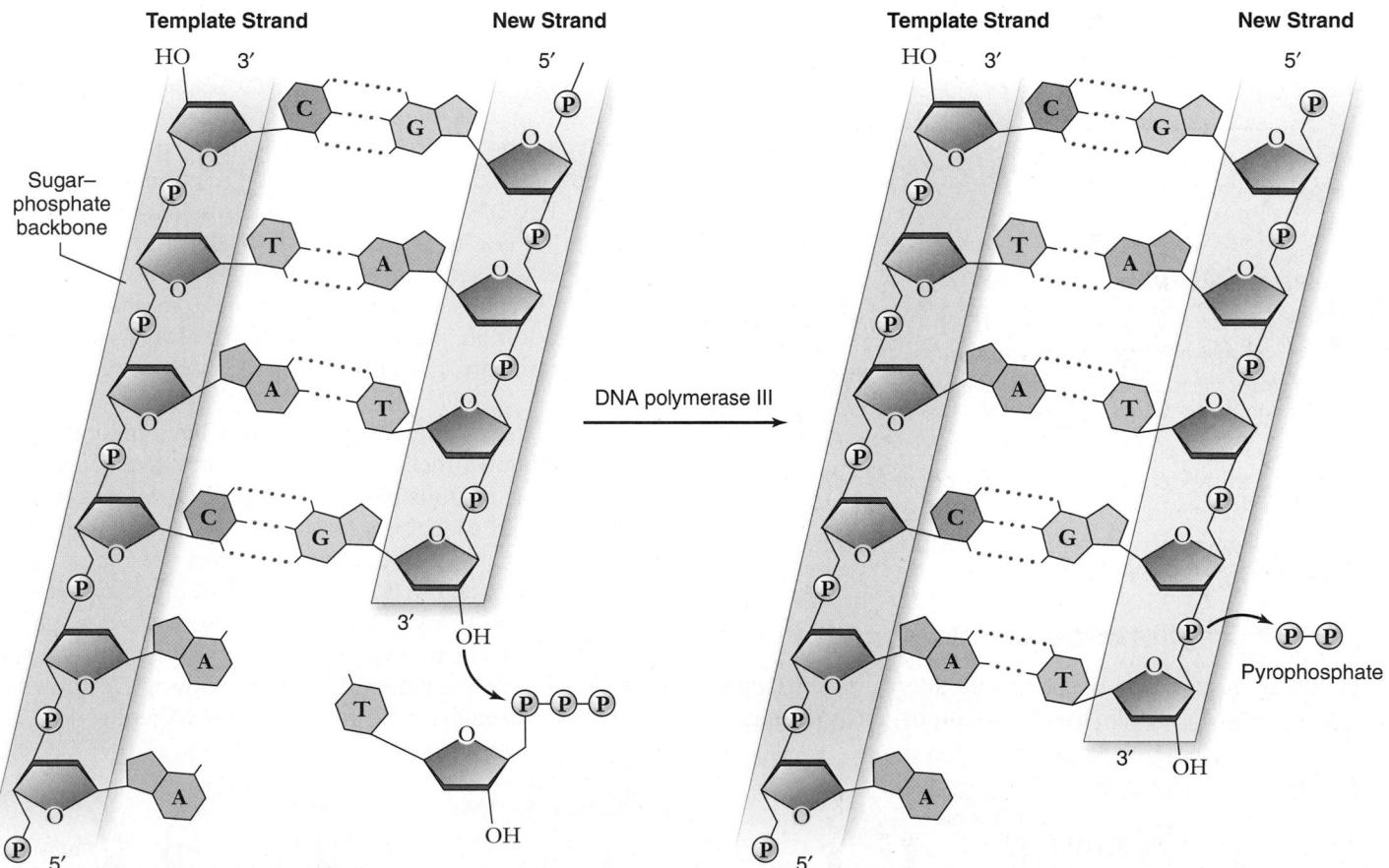

Figure 14.12 Action of DNA polymerase. DNA polymerases add nucleotides to the 3′ end of a growing chain. The nucleotide added depends on the base that is in the template strand. Each new base must be complementary to the base in the template strand. With the addition of each new nucleoside triphosphate, two of its phosphates are cleaved off as pyrophosphate.

? Inquiry question Why do you think it is important that the sugar–phosphate backbone of DNA is held together by covalent bonds, and the cross-bridges between the two strands are held together by hydrogen bonds?

strands. That is, they synthesize in a 5′-to-3′ direction by extending a strand base-paired to the template. All DNA polymerases also require a *primer* to begin synthesis; they cannot begin without a strand of RNA or DNA base-paired to the template. RNA polymerases do not have this requirement, so they usually synthesize the primers.

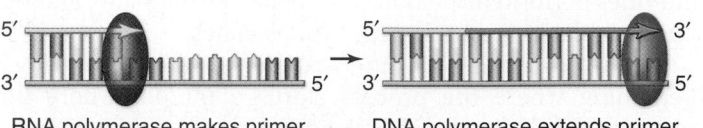

RNA polymerase makes primer   DNA polymerase extends primer

14.4 Prokaryotic Replication

To build up a more detailed picture of replication, we first concentrate on prokaryotic replication using *E. coli* as a model. We can then look at eukaryotic replication primarily in how it differs from the prokaryotic system.

Prokaryotic replication starts at a single origin

Replication in *E. coli* initiates at a specific site, the origin (called *oriC*), and ends at a specific site, the terminus. The sequence of

oriC consists of repeated nucleotides that bind an initiator protein and an AT-rich sequence that can be opened easily during initiation of replication. (A–T base-pairs have only two hydrogen bonds, compared with the three hydrogen bonds in G–C base-pairs.)

After initiation, replication proceeds bidirectionally from this unique origin to the unique terminus (figure 14.13). We call the DNA controlled by an origin a **replicon.** In this case, the chromosome plus the origin forms a single replicon.

 ### E. coli has at least three different DNA polymerases

As mentioned earlier, DNA polymerase refers to a group of enzymes responsible for the building of a new DNA strand from the template. The first DNA polymerase isolated in *E. coli* was given the name **DNA polymerase I (Pol I).** At first, investigators assumed this polymerase was responsible for the bulk synthesis of DNA during replication. A mutant was isolated, however, that had no Pol I activity, but could still replicate its chromosome. Two additional polymerases were isolated from this strain of *E. coli* and were named **DNA polymerase II (Pol II)** and **DNA polymerase III (Pol III).** As with all other known polymerases, all three of these enzymes synthesize polynucleotide strands only in the 5′-to-3′ direction and require a primer.

Many DNA polymerases have additional enzymatic activity that aids their function. This activity is a nuclease activity, or the ability to break phosphodiester bonds between nucleotides. Nucleases are classified as either **endonucleases** (which cut DNA internally) or **exonucleases** (which chew away at an end of DNA). DNA Pol I, Pol II, and Pol III have 3′-to-5′ exonuclease activity, which serves as a proofreading function because it allows the enzyme to remove a mispaired base. In addition, the DNA Pol I enzyme also has a 5′-to-3′ exonuclease activity, the importance of which will become clear shortly.

The three different polymerases have different roles in the replication process. DNA Pol III is the main replication enzyme; it is responsible for the bulk of DNA synthesis. DNA Pol I acts on the lagging strand to remove primers and replace them with DNA. The Pol II enzyme does not appear to play a role in replication but is involved in DNA repair processes.

For many years, these three polymerases were thought to be the only DNA polymerases in *E. coli,* but recently several new ones have been identified. There are now five known polymerases, although not all are active in DNA replication.

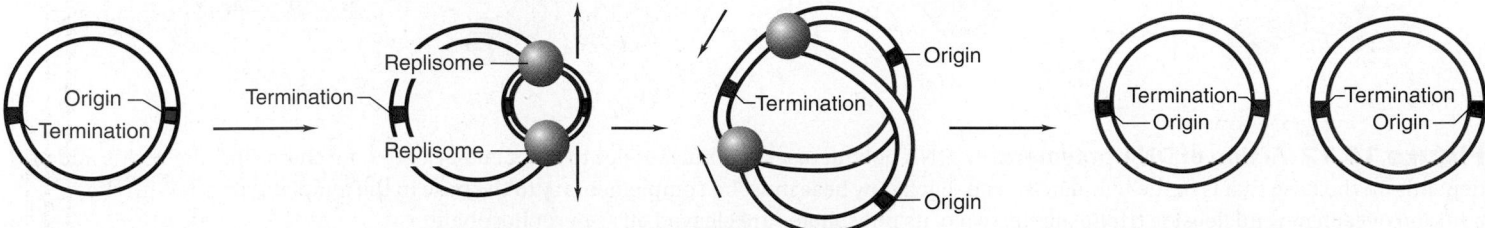

Figure 14.13 Replication is bidirectional from a unique origin. Replication initiates from a unique origin. Two separate replisomes are loaded onto the origin and initiate synthesis in the opposite directions on the chromosome. These two replisomes continue in opposite directions until they come to a unique termination site.

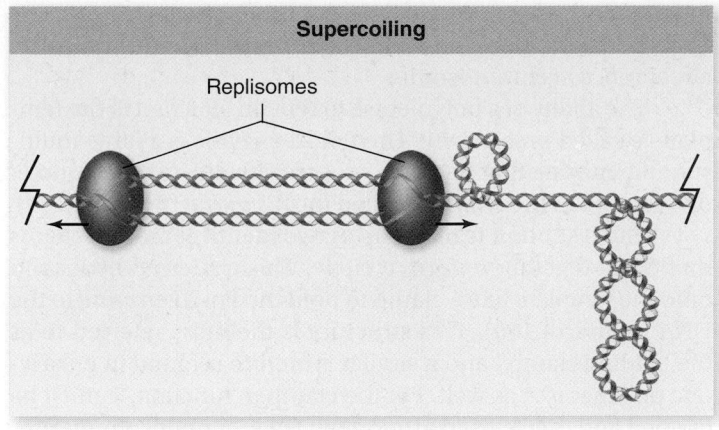

Supercoiling

Replisomes

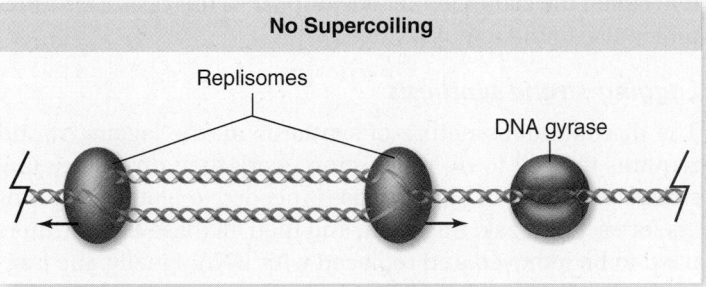

No Supercoiling

Replisomes

DNA gyrase

Figure 14.14 Unwinding the helix causes torsional strain. If the ends of a linear DNA molecule are constrained, as they are in the cell, unwinding the helix produces torsional strain. This can cause the double helix to further coil in space (supercoiling). The enzyme DNA gyrase can relieve supercoiling.

Unwinding DNA requires energy and causes torsional strain

Although some DNA polymerases can unwind DNA as they synthesize new DNA, another class of enzymes has the single function of unwinding DNA strands to make this process more efficient. Enzymes that use energy from ATP to unwind the DNA template are called **helicases.**

The single strands of DNA produced by helicase action are unstable because the process exposes the hydrophobic bases to water. Cells solve this problem by using a protein, called single-strand-binding protein (SSB), to coat exposed single strands.

The unwinding of the two strands introduces torsional strain in the DNA molecule. Imagine two rubber bands twisted together. If you now unwind the rubber bands, what happens? The rubber bands, already twisted about each other, will further coil in space. When this happens with a DNA molecule it is called **supercoiling** (figure 14.14). The branch of mathematics that studies how forms twist and coil in space is called *topology,* and therefore we describe this coiling of the double helix as the *topological state* of DNA. This state describes how the double helix itself coils in space. You have already seen an example of this coiling with DNA wrapped about histone proteins in the nucleosomes of eukaryotic chromosomes (see chapter 10).

Enzymes that can alter the topological state of DNA are called **topoisomerases.** Topoisomerase enzymes act to relieve the torsional strain caused by unwinding and to prevent this supercoiling from happening. **DNA gyrase** is the topoisomerase involved in DNA replication (see figure 14.14).

Replication is semidiscontinuous

Earlier, DNA was described as being antiparallel—meaning that one strand runs in the 3′-to-5′ direction, and its complementary strand runs in the 5′-to-3′ direction. The antiparallel nature of DNA combined with the nature of the polymerase enzymes puts constraints on the replication process. Because polymerases can synthesize DNA in only one direction, and the two DNA strands run in opposite directions, polymerases on the two strands must be synthesizing DNA in opposite directions (figure 14.15).

The requirement of DNA polymerases for a primer means that on one strand primers need to be added as the helix is

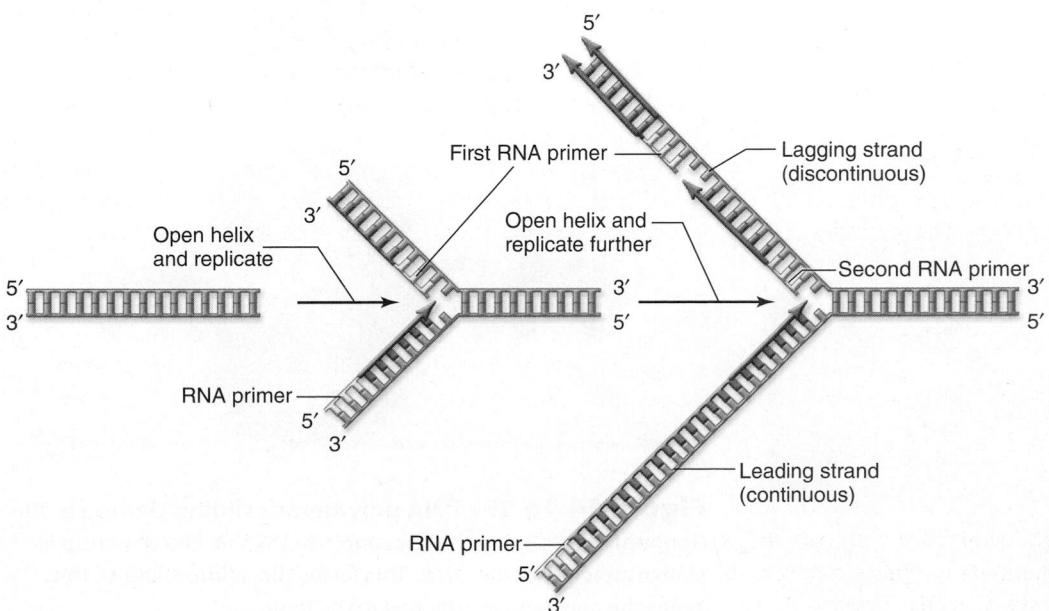

Figure 14.15 Replication is semidiscontinuous. The 5′-to-3′ synthesis of the polymerase and the antiparallel nature of DNA mean that only one strand, the leading strand, can be synthesized continuously. The other strand, the lagging strand, must be made in pieces, each with its own primer.

opened up (see figure 14.15). This means that one strand can be synthesized in a continuous fashion from an initial primer, but the other strand must be synthesized in a discontinuous fashion with multiple priming events and short sections of DNA being assembled. The strand that is continuous is called the **leading strand,** and the strand that is discontinuous is the **lagging strand.** DNA fragments synthesized on the lagging strand are named **Okazaki fragments** in honor of the man who first experimentally demonstrated discontinuous synthesis. They introduce a need for even more enzymatic activity on the lagging strand, as is described next.

Synthesis occurs at the replication fork

The partial opening of a DNA helix to form two single strands has a forked appearance, and is thus called the **replication fork.** All of the enzymatic activities that we have discussed plus a few more are found at the replication fork (table 14.1). Synthesis on the leading strand and on the lagging strand proceed in different ways, however.

Priming

The primers required by DNA polymerases during replication are synthesized by the enzyme *DNA primase.* This enzyme is an RNA polymerase that synthesizes short stretches of RNA 10–20 bp (base-pairs) long that function as primers for DNA polymerase. Later on, the RNA primer is removed and replaced with DNA.

Leading-strand synthesis

Synthesis on the leading strand is relatively simple. A single priming event is required, and then the strand can be extended indefinitely by the action of DNA Pol III. If the enzyme remains attached to the template, it can synthesize around the entire circular *E. coli* chromosome.

The ability of a polymerase to remain attached to the template is called *processivity.* The Pol III enzyme is a large multisubunit enzyme that has high processivity due to the action of one subunit of the enzyme, called the β *subunit* (figure 14.16a).

The β subunit is made up of two identical protein chains that come together to form a circle. This circle can be loaded onto the template like a clamp to hold the Pol III enzyme to the DNA (figure 14.16b). This structure is therefore referred to as the "sliding clamp," and a similar structure is found in eukaryotic polymerases as well. For the clamp to function, it must be opened and then closed around the DNA. A multisubunit protein called the clamp loader accomplishes this task. This function is also found in eukaryotes.

Lagging-strand synthesis

The discontinuous nature of synthesis on the lagging strand requires the cell to do much more work than on the leading strand (see figure 14.15). Primase is needed to synthesize primers for each Okazaki fragment, and then all these RNA primers need to be removed and replaced with DNA. Finally, the fragments need to be stitched together.

DNA Pol III accomplishes the synthesis of Okazaki fragments. The removal and replacement of primer segments, however, is accomplished by DNA Pol I. Using its 5′-to-3′ exonuclease activity, it can remove primers in front and then replace them by using its usual 5′-to-3′ polymerase activity. The synthesis is

TABLE 14.1	DNA Replication Enzymes of *E. coli*		
Protein	**Role**	**Size (kDa)**	**Molecules per Cell**
Helicase	Unwinds the double helix	300	20
Primase	Synthesizes RNA primers	60	50
Single-strand binding protein	Stabilizes single-stranded regions	74	300
DNA gyrase	Relieves torque	400	250
DNA polymerase III	Synthesizes DNA	≈900	20
DNA polymerase I	Erases primer and fills gaps	103	300
DNA ligase	Joins the ends of DNA segments; DNA repair	74	300

a.

b.

Figure 14.16 The DNA polymerase sliding clamp. *a.* The β subunit forms a ring that can encircle DNA. *b.* The β subunit is shown attached to the DNA. This forms the "sliding clamp" that keeps the polymerase attached to the template.

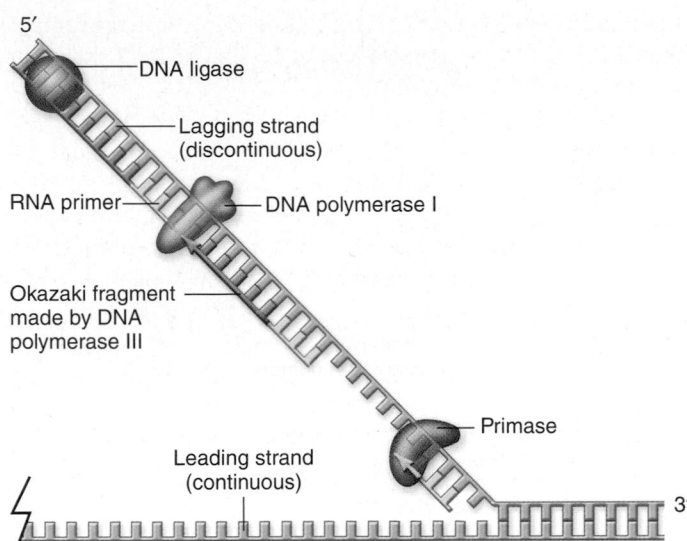

Figure 14.17 Lagging-strand synthesis. The action of primase synthesizes the primers needed by DNA polymerase III (not shown). These primers are removed by DNA polymerase I using its 5′-to-3′ exonuclease activity, then extending the previous Okazaki fragment to replace the RNA. The nick between Okazaki fragments after primer removal is sealed by DNA ligase.

primed by the previous Okazaki fragment, which is composed of DNA and has a free 3′ OH that can be extended.

This leaves only the last phosphodiester bond to be formed where synthesis by Pol I ends. This is done by **DNA ligase,** which seals this "nick," eventually joining the Okazaki fragments into complete strands. All of this activity on the lagging strand is summarized in figure 14.17.

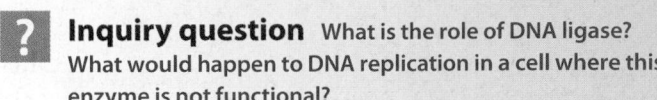
Termination

Termination occurs at a specific site located roughly opposite *oriC* on the circular chromosome. The last stages of replication produce two daughter molecules that are intertwined like two rings in a chain. These intertwined molecules are unlinked by the same enzyme that relieves torsional strain at the replication fork—DNA gyrase.

The replisome contains all the necessary enzymes for replication

The enzymes involved in DNA replication form a macromolecular assembly called the **replisome.** This assembly can be thought of as the "replication organelle," just as the ribosome is the organelle that synthesizes protein. The replisome has two main subcomponents: the *primosome,* and a complex of two DNA Pol III enzymes, one for each strand. The primosome is composed of primase and helicase, along with a number of accessory proteins. The need for constant priming on the lagging strand explains the need for the primosome complex as part of the replisome.

The two Pol III complexes include two synthetic core subunits, each with its own β subunit. The entire replisome complex is held together by a number of proteins that includes the clamp loader. The clamp loader is required to periodically load a β subunit on the lagging strand and to transfer the Pol III to this new β subunit (figure 14.18).

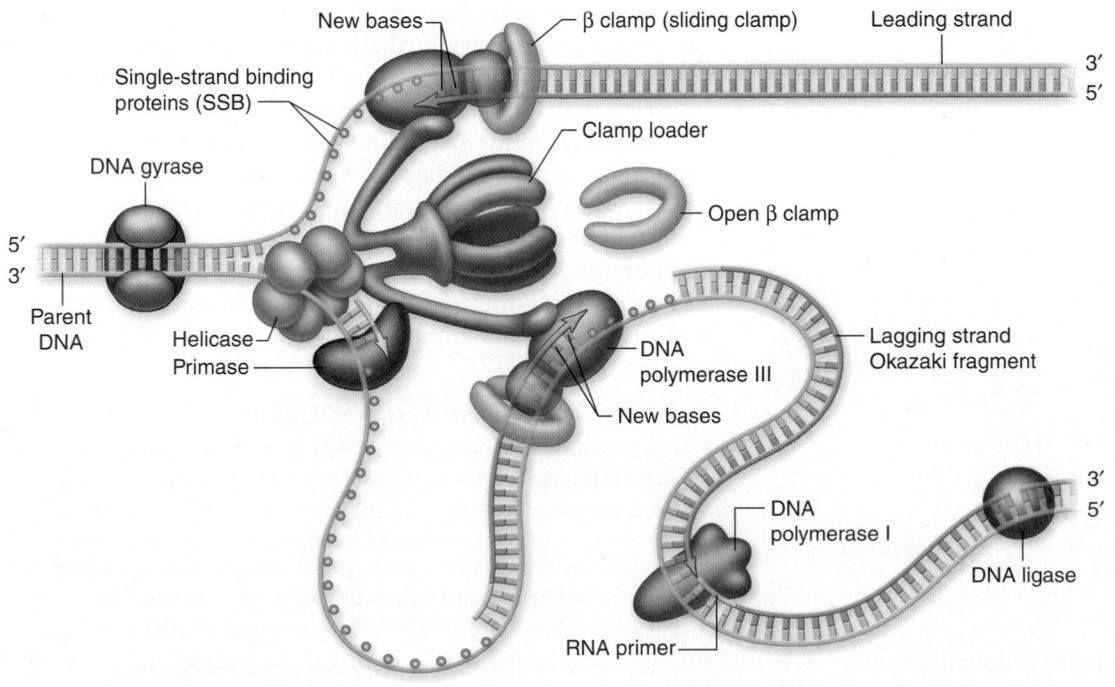

Figure 14.18 The replication fork. A model for the structure of the replication fork with two polymerase III enzymes held together by a large complex of accessory proteins. These include the "clamp loader," which loads the β subunit sliding clamp periodically on the lagging strand. The polymerase III on the lagging strand periodically releases its template and reassociates along with the β clamp. The loop in the lagging-strand template allows both polymerases to move in the same direction despite DNA being antiparallel. Primase, which makes primers for the lagging-strand fragments, and helicase are also associated with the central complex. Polymerase I removes primers and ligase joins the fragments together.

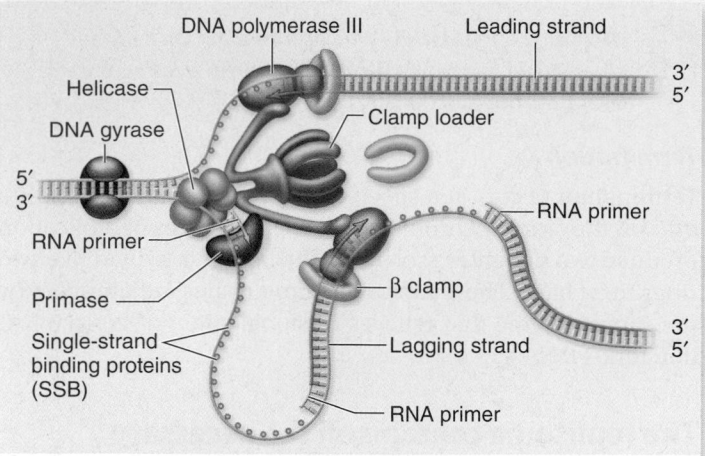

Helicase
DNA polymerase III
DNA gyrase
Leading strand
Clamp loader
RNA primer
RNA primer
Primase
β clamp
Single-strand binding proteins (SSB)
Lagging strand
RNA primer
3′ 5′
5′ 3′
3′ 5′

1. A DNA polymerase III enzyme is active on each strand. Primase synthesizes new primers for the lagging strand.

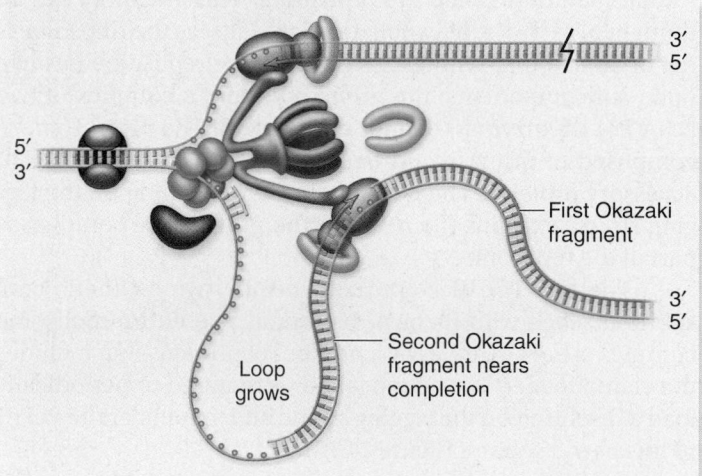

First Okazaki fragment
Second Okazaki fragment nears completion
Loop grows
5′ 3′
3′ 5′
3′ 5′

2. The "loop" in the lagging-strand template allows replication to occur 5′-to-3′ on both strands, with the complex moving to the left.

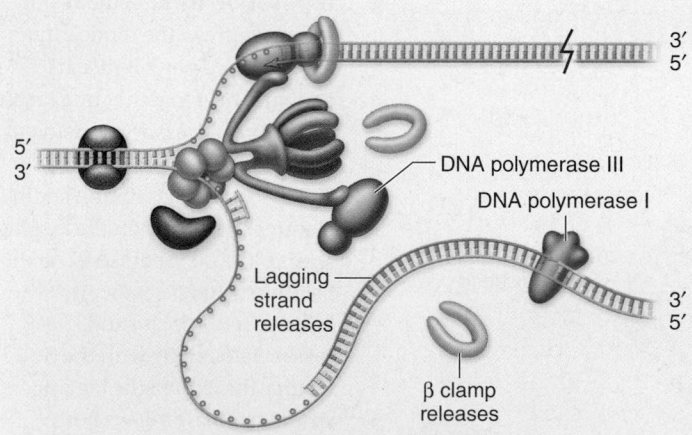

DNA polymerase III
DNA polymerase I
Lagging strand releases
β clamp releases
5′ 3′
3′ 5′
3′ 5′

3. When the polymerase III on the lagging strand hits the previously synthesized fragment, it releases the β clamp and the template strand. DNA polymerase I attaches to remove the primer.

Figure 14.19 DNA synthesis by the replisome. The semidiscontinuous synthesis of DNA is illustrated in stages using the model from figure 14.18.

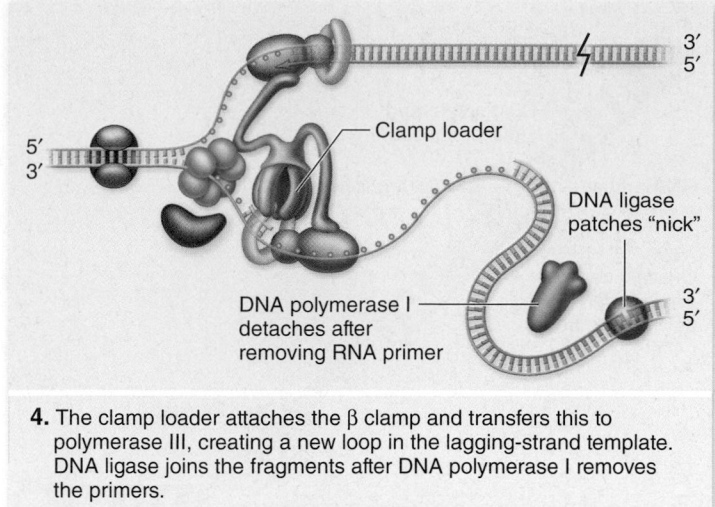

Clamp loader
DNA ligase patches "nick"
DNA polymerase I detaches after removing RNA primer
5′ 3′
3′ 5′
3′ 5′

4. The clamp loader attaches the β clamp and transfers this to polymerase III, creating a new loop in the lagging-strand template. DNA ligase joins the fragments after DNA polymerase I removes the primers.

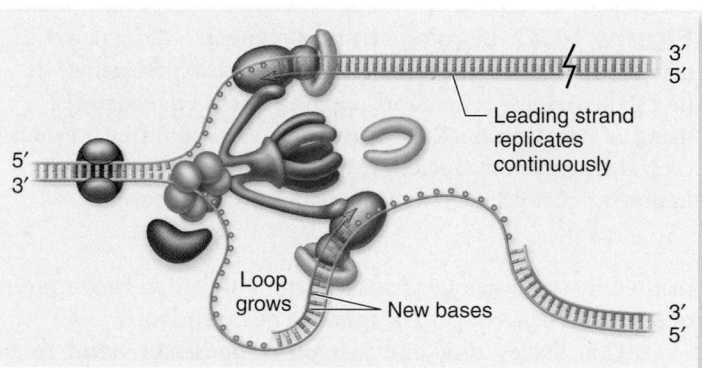

Leading strand replicates continuously
Loop grows
New bases
5′ 3′
3′ 5′
3′ 5′

5. After the β clamp is loaded, the DNA polymerase III on the lagging strand adds bases to the next Okazaki fragment.

Even given the difficulties with lagging-strand synthesis, the two Pol III enzymes in the replisome are active on both leading and lagging strands simultaneously. How can the two strands be synthesized in the same direction when the strands are antiparallel? The model first proposed, still with us in some form, involves a loop formed in the lagging strand, so that the polymerases can move in the same direction (see figure 14.18). Current evidence also indicates that this replication complex is probably stationary, with the DNA strand moving through it like thread in a sewing machine, rather than the complex moving along the DNA strands. This stationary complex also pushes the newly synthesized DNA outward, which may aid in chromosome segregation. This process is summarized in figure 14.19.

Learning Outcomes Review 14.4

E. coli has three DNA polymerases: DNA Pol I, II, and III. Synthesis on one strand is discontinuous because DNA is antiparallel, and polymerases only synthesize in the 5′-to-3′ direction. Replication occurs at the replication fork, where the two strands are separated. Assembled here is a massive complex, the replisome, containing DNA polymerase III, primase, helicase, and other proteins. The lagging strand requires DNA polymerase I to remove the primers and replace them with DNA, and ligase to join Okazaki fragments.

■ *How are the nuclease functions of the different polymerases used during replication?*

Eukaryotic Replication

Learning Outcomes

1. *Compare eukaryotic replication with prokaryotic.*
2. *Explain the function of telomeres.*
3. *Evaluate the role of telomerase in cell division.*

Eukaryotic replication is complicated by two main factors: the larger amount of DNA organized into multiple chromosomes, and the linear structure of the chromosomes. This process requires new enzymatic activities only for dealing with the ends of chromosomes; otherwise the basic enzymology is the same.

Eukaryotic replication requires multiple origins

The sheer amount of DNA and how it is packaged constitute a problem for eukaryotes (figure 14.20). Eukaryotes usually have multiple chromosomes that are each larger than the *E. coli*

110,000×

Figure 14.20 DNA of a single human chromosome. This chromosome has been relieved of most of its packaging proteins, leaving the DNA in its native form. The residual protein scaffolding appears as the dark material in the lower part of the micrograph.

chromosome. If only a single unique origin existed for each chromosome, the length of time necessary for replication would be prohibitive. This problem is solved by the use of multiple origins of replication for each chromosome, resulting in multiple *replicons*.

The origins are not as sequence-specific as *oriC,* and their recognition seems to depend on chromatin structure as well as on sequence. The number of origins used can also be adjusted during the course of development, so that early on, when cell divisions need to be rapid, more origins are activated. Each origin must be used only once per cell cycle.

The enzymology of eukaryotic replication is more complex

The replication machinery of eukaryotes is similar to that found in bacteria, but it is larger and more complex. The initiation phase of replication requires more factors to assemble both helicase and primase complexes onto the template, then load the polymerase with its sliding clamp unit.

The eukaryotic primase is interesting in that it is a complex of both an RNA polymerase and a DNA polymerase. It first makes short RNA primers, then extends these with DNA to produce the final primer. The reason for this added complexity is unclear.

The main replication polymerase itself is also a complex of two different enzymes that work together. One is called *DNA polymerase epsilon* (pol ε) and the other *DNA polymerase delta* (pol δ). The sliding clamp subunit that allows the enzyme complex to stay attached to the template is called PCNA (for proliferating cell nuclear antigen). This unusual name reflects the fact that PCNA was first identified as an antibody-inducing protein in proliferating (dividing) cells. The PCNA sliding clamp forms a trimer, but this structure is similar to the β subunit sliding clamp. The clamp loader is also similar to the bacterial structure. Despite the additional complexity, the action of the replisome is similar to that described earlier for *E. coli,* and the replication fork has essentially the same components.

Archaeal replication proteins are similar to eukaryotic proteins

Despite their lack of a membrane-bounded nucleus, Archaeal replication proteins are more similar to eukaryotes than to bacterial. The main replication polymerase is most similar to eukaryotic pol δ, and the sliding clamp is similar to the PCNA protein. The clamp loading complex is also more similar to eukaryotic than bacterial. The most interesting conclusion from all of these data are that all three domains of life have similar functions involved in replicating chromosomes. All three domains assemble similar protein complexes with clamp loader, sliding clamp, two polymerases, helicase, and primase at the replication fork.

Linear chromosomes have specialized ends

The specialized structures found on the ends of eukaryotic chromosomes are called **telomeres.** These structures protect

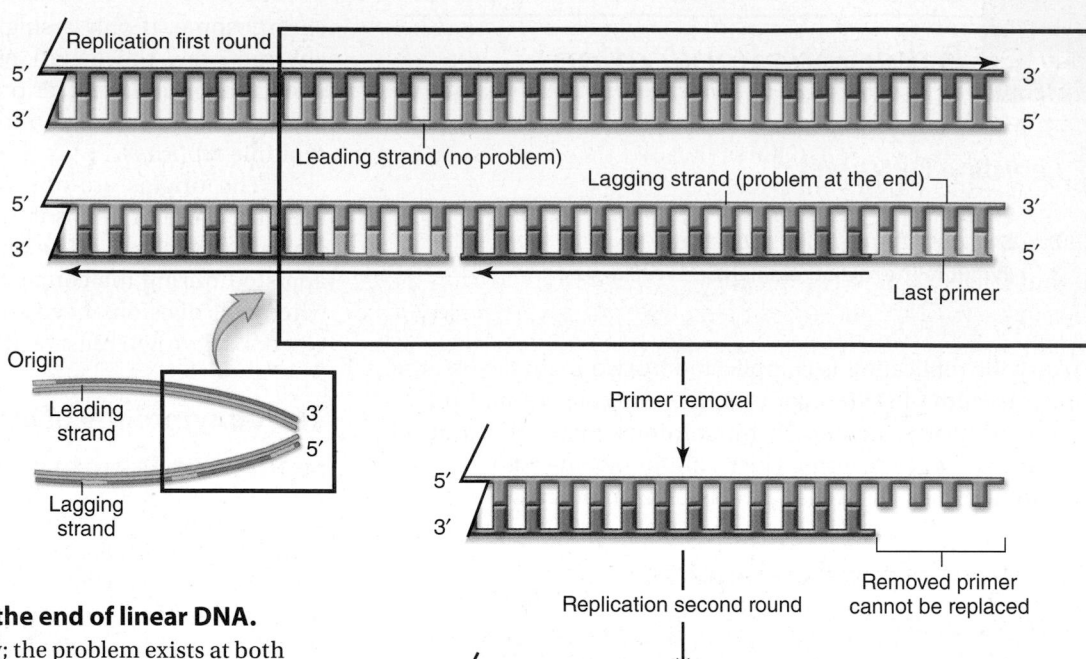

Figure 14.21 Replication of the end of linear DNA.
Only one end is shown for simplicity; the problem exists at both ends. The leading strand can be completely replicated, but the lagging strand cannot be finished. When the last primer is removed, it cannot be replaced. During the next round of replication, when this shortened template is replicated, it will produce a shorter chromosome.

the ends of chromosomes from nucleases and maintain the integrity of linear chromosomes. These telomeres are composed of specific DNA sequences, but they are not made by the replication complex.

Replicating ends

The very structure of a linear chromosome causes a cell problems in replicating the ends. The directionality of polymerases, combined with their requirement for a primer, create this problem.

Consider a simple linear molecule like the one in figure 14.21. Replication of one end of each template strand is simple, namely the 5′ end of the leading-strand template. When the polymerase reaches this end, synthesizing in the 5′-to-3′ direction, it eventually runs out of template and is finished.

But on the other strand's end, the 3′ end of the lagging strand, removal of the last primer on this end leaves a gap. This gap cannot be primed, meaning that the polymerase complex cannot finish this end properly. The result would be a gradual shortening of chromosomes with each round of cell division (see figure 14.21).

The action of telomerase

When the sequence of telomeres was determined, they were found to be composed of short repeated sequences of DNA. This repeating nature is easily explained by their synthesis. They are made by an enzyme called **telomerase,** which uses an internal RNA as a template and not the DNA itself (figure 14.22).

Figure 14.22 Action of telomerase. Telomerase contains an internal RNA that the enzyme uses as a template to extend the DNA of the chromosome end. Multiple rounds of synthesis by telomerase produce repeated sequences. This single strand is completed by normal synthesis using it as a template (not shown).

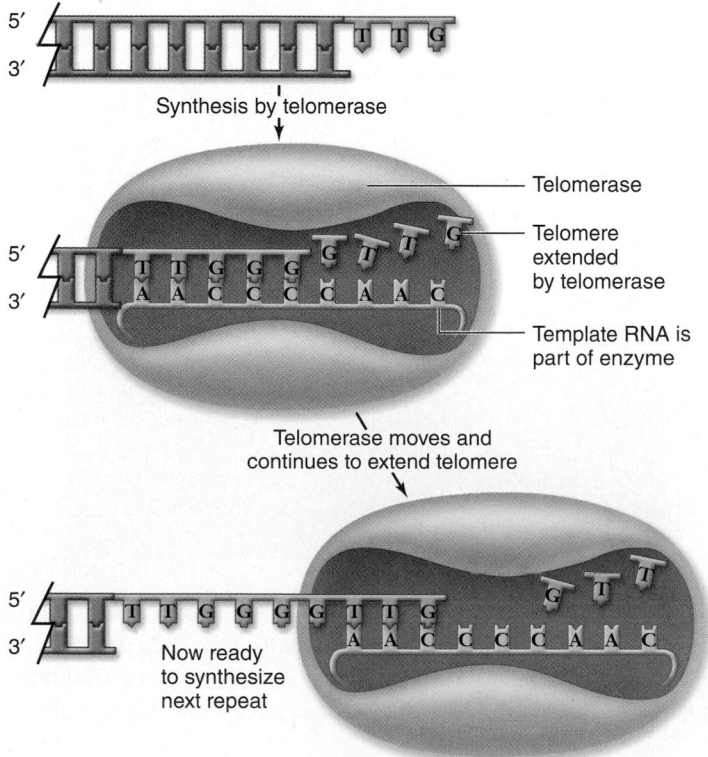

The use of the internal RNA template allows short stretches of DNA to be synthesized, composed of repeated nucleotide sequences complementary to the RNA of the enzyme. The other strand of these repeated units is synthesized by the usual action of the replication machinery copying the strand made by telomerase.

Telomerase, aging, and cancer

A gradual shortening of the ends of chromosomes occurs in the absence of telomerase activity. During embryonic and childhood development in humans, telomerase activity is high, but it is low in most somatic cells of the adult. The exceptions are cells that must divide as part of their function, such as lymphocytes. The activity of telomerase in somatic cells is kept low by preventing the expression of the gene encoding this enzyme.

Evidence for the shortening of chromosomes in the absence of telomerase was obtained by producing mice with no telomerase activity. These mice appear to be normal for up to six generations, but they show steadily decreasing telomere length that eventually leads to nonviable offspring.

This finding indicates a relationship between cell senescence (aging) and telomere length. Normal cells undergo only a specified number of divisions when grown in culture. This limit is at least partially based on telomere length.

Support for the relationship between senescence and telomere length comes from experiments in which telomerase was introduced into fibroblasts in culture. These cells have their lifespan increased relative to controls that have no added telomerase. Interestingly, these cells do not show the hallmarks of malignant cells, indicating that activation of telomerase alone does not make cells malignant.

A relationship has been found, however, between telomerase and cancer. Cancer cells do continue to divide indefinitely, and this would not be possible if their chromosomes were being continually shortened. Cancer cells generally show activation of telomerase, which allows them to maintain telomere length; but this is clearly only one aspect of conditions that allow them to escape normal growth controls.

> **? Inquiry question** How does the structure of eukaryotic genomes affect replication? Does this introduce problems that are not faced by prokaryotes?

Learning Outcomes Review 14.5

Eukaryotic replication is complicated by a large amount of DNA organized into chromosomes, and by the linear nature of chromosomes. Eukaryotes replicate a large amount of DNA in a short time by using multiple origins of replication. Linear chromosomes end in telomeres, and the length of telomeres is correlated with the ability of cells to divide. The enzyme telomerase synthesizes the telomeres. Cancer cells show activation of telomerase, which extends the ability of the cells to divide.

- *What might be the result of abnormal shortening of telomeres or a lack of telomerase activity?*

14.6 DNA Repair

Learning Outcomes

1. **Explain why DNA repair is critical for cells.**
2. **Describe the different forms of DNA repair.**

As you learned earlier, many DNA polymerases have 3′-to-5′ exonuclease activity that allows "proofreading" of added bases. This action increases the accuracy of replication, but errors still occur. Without error correction mechanisms, cells would accumulate errors at an unacceptable rate, leading to high levels of deleterious or lethal mutations. A balance must exist between the introduction of new variation by mutation, and the effects of deleterious mutations on the individual.

Cells are constantly exposed to DNA-damaging agents

In addition to errors in DNA replication, cells are constantly exposed to agents that can damage DNA. These agents include radiation, such as UV light and X-rays, and chemicals in the environment. Agents that damage DNA can lead to mutations, and any agent that increases the number of mutations above background levels is called a **mutagen.**

The number of potentially mutagenic agents that organisms encounter is huge. Sunlight itself includes radiation in the UV range and is thus mutagenic. Ozone normally screens out much of the harmful UV radiation in sunlight, but some remains. The relationship between sunlight and mutations is shown clearly by the increase in skin cancer in regions of the southern hemisphere that are underneath a seasonal "ozone hole."

Organisms also may encounter mutagens in their diet in the form of either contaminants in food or natural plant products that can damage DNA. When a simple test was designed to detect mutagens, screening of possible sources indicated an amazing diversity of mutagens in the environment and in natural sources. As a result, consumer products are now screened to reduce the load of mutagens we are exposed to, but we cannot escape natural sources.

DNA repair restores damaged DNA

Cells cannot escape exposure to mutagens, but systems have evolved that enable cells to repair some damage. These DNA repair systems are vital to continued existence, whether a cell is a free-living, single-celled organism or part of a complex multicellular organism.

The importance of DNA repair is indicated by the multiplicity of repair systems that have been discovered and characterized. All cells that have been examined show multiple pathways for repairing damaged DNA and for reversing errors that occur during replication. These systems are not perfect, but they do reduce the mutational load on organisms to an acceptable level. In the rest of this section, we illustrate the action of DNA repair by concentrating on two examples drawn from these multiple repair pathways.

Repair can be either specific or nonspecific

DNA repair falls into two general categories: specific and nonspecific. Specific repair systems target a single kind of lesion in DNA and repair only that damage. Nonspecific forms of repair use a single mechanism to repair multiple kinds of lesions in DNA.

Photorepair: A specific repair mechanism

Photorepair is specific for one particular form of damage caused by UV light, namely the *thymine dimer*. Thymine dimers are formed by a photochemical reaction of UV light and adjacent thymine bases in DNA. The UV radiation causes the thymines to react, covalently linking them together: a thymine dimer (figure 14.23).

Repair of these thymine dimers can be accomplished by multiple pathways, including photorepair. In photorepair, an enzyme called a *photolyase* absorbs light in the visible range and uses this energy to cleave the thymine dimer. This action restores the two thymines to their original state (see figure 14.23). It is interesting that sunlight in the UV range can cause this damage, and sunlight in the visible range can be used to repair the damage. Photorepair does not occur in cells deprived of visible light.

The photolyase enzyme has been found in many different species, ranging from bacteria, to single-celled eukaryotes, to humans. The ubiquitous nature of this enzyme illustrates the importance of this form of repair. For as long as cells have existed on Earth, they have been exposed to UV light and its potential to damage DNA.

Excision repair: A nonspecific repair mechanism

A common form of nonspecific repair is **excision repair.** In this pathway, a damaged region is removed, or excised, and is then replaced by DNA synthesis (figure 14.24). In *E. coli*, this action is accomplished by proteins encoded by the *uvr A, B,*

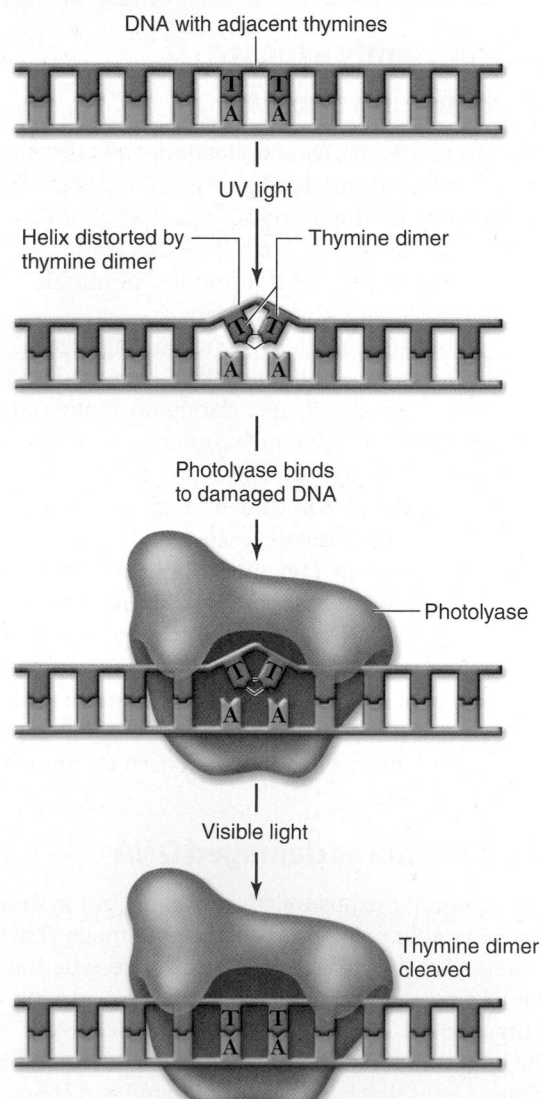

Figure 14.23 Repair of thymine dimer by photorepair.
UV light can catalyze a photochemical reaction to form a covalent bond between two adjacent thymines, thereby creating a thymine dimer. A photolyase enzyme recognizes the damage and binds to the thymine dimer. The enzyme absorbs visible light and uses the energy to cleave the thymine dimer.

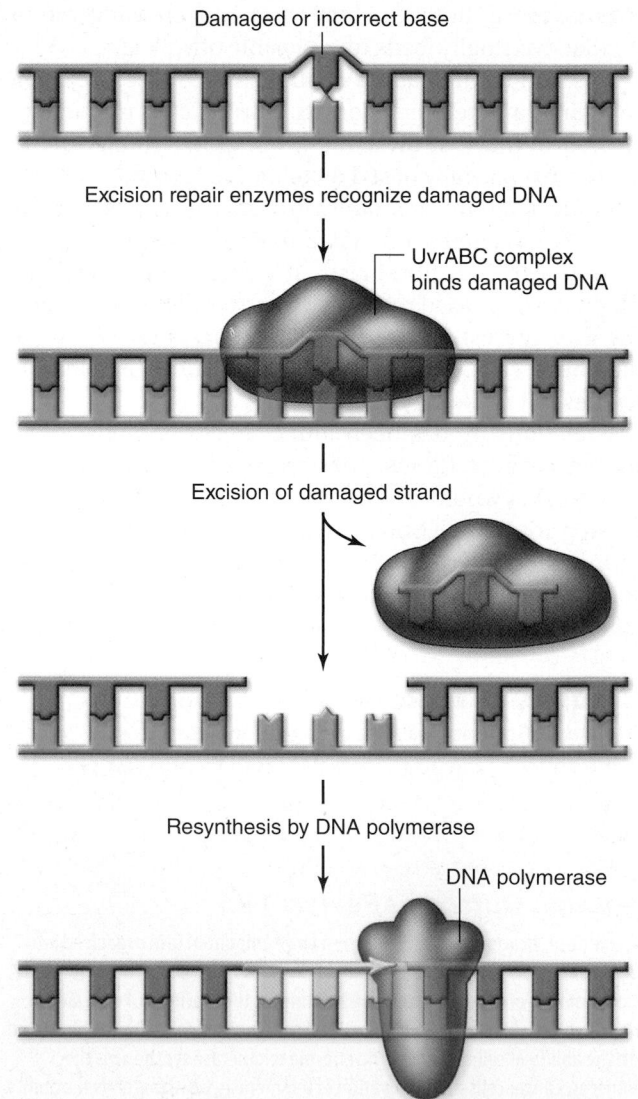

Figure 14.24 Repair of damaged DNA by excision repair. Damaged DNA is recognized by the uvr complex, which binds to the damaged region and removes it. Synthesis by DNA polymerase replaces the damaged region. DNA ligase finishes the process (not shown).

and *C* genes. Although these genes were identified based on mutations that increased sensitivity of the cell to UV light (hence the "uvr" in their names), their proteins can act on damage due to other mutagens.

Excision repair follows three steps: (1) recognition of damage, (2) removal of the damaged region, and (3) resynthesis using the information on the undamaged strand as a template (see figure 14.24). Recognition and excision are accomplished by the UvrABC complex. The UvrABC complex binds to damaged DNA and then cleaves a single strand on either side of the damage, removing it. In the synthesis stage, DNA pol I or pol II replaces the damaged DNA. This restores the original information in the damaged strand by using the information in the complementary strand.

Other repair pathways

Cells have other forms of nonspecific repair, and these fall into two categories: error-free and error-prone. It may seem strange to have an error-prone pathway, but it can be thought of as a last-ditch effort to save a cell that has been exposed to such massive damage that it has overwhelmed the error-free systems. In fact, this system in *E. coli* is part of what is called the "SOS response."

Cells can also repair damage that produces breaks in DNA. These systems use enzymes related to those that are involved in recombination during meiosis (see chapter 11). It is thought that recombination uses enzymes that originally evolved for DNA repair.

The number of different systems and the wide spectrum of damage that can be repaired illustrate the importance of maintaining the integrity of the genome. Accurate replication of the genome is useless if a cell cannot reverse errors that can occur during this process or repair damage due to environmental causes.

> **? Inquiry question** Cells are constantly exposed to DNA-damaging agents, ranging from UV light to by-products of oxidative metabolism. How does the cell deal with this, and what would happen if the cell had no way of dealing with this?

Learning Outcomes Review 14.6

The ability to repair DNA is critical because of replication errors and the constant presence of damaging agents that can cause mutation. Cells have multiple repair pathways; some of these systems are specific for a single type of damage, such as photorepair that reverses thymine dimers caused by UV light. Other systems are nonspecific, such as excision repair that removes and replaces damaged regions.

■ *Could a cell survive with no form of DNA repair?*

Chapter Review

14.1 The Nature of the Genetic Material

Griffith finds that bacterial cells can be transformed.
Nonvirulent *S. pneumoniae* could take up an unknown substance from a virulent strain and become virulent.

Avery, MacLeod, and McCarty identify the transforming principle.
The transforming substance could be inactivated by DNA-digesting enzymes, but not by protein-digesting enzymes.

Hershey and Chase demonstrate that phage genetic material is DNA.
Radioactive labeling showed that the infectious agent of phage is its DNA, and not its protein.

14.2 DNA Structure

DNA's components were known, but its three-dimensional structure was a mystery.
The nucleotide building blocks for DNA contain deoxyribose and the bases adenine (A), guanine (G), cytosine (C), and thymine (T). Phosphodiester bonds are formed between the 5′ phosphate of one nucleotide and the 3′ hydroxyl of another nucleotide (figure 14.4).

Chargaff, Franklin, and Wilkins obtained some structural evidence.
Chargaff found equal amounts of adenine and thymine, and of cytosine and guanine, in DNA. The bases exist primarily in keto and enol forms that exhibit hydrogen bonding. X-ray diffraction studies by Franklin and Wilkins indicated that DNA had a helical structure.

The Watson–Crick model fits the available evidence (figures 14.8 & 14.9).
DNA consists of two antiparallel polynucleotide strands wrapped about a common helical axis. These strands are held together by hydrogen bonds forming specific base pairs (A/T and G/C). The two strands are complementary; one strand can specify the other.

14.3 Basic Characteristics of DNA Replication

Meselson and Stahl demonstrate the semiconservative mechanism (figure 14.11).
Semiconservative replication uses each strand of a DNA molecule to specify the synthesis of a new strand. Meselson and Stahl showed this by using a heavy isotope of nitrogen and separating the replication products. Replication produces two new molecules each composed of one new strand and one old strand.

DNA replication requires a template, nucleotides, and a polymerase enzyme.
All new DNA molecules are produced by DNA polymerase copying a template. All known polymerases synthesize new DNA in the 5′-to-3′ direction. These enzymes also require a primer. The building blocks used in replication are deoxynucleotide triphosphates with high-energy bonds; they do not require any additional energy.

14.4 Prokaryotic Replication

Prokaryotic replication starts at a single origin.

The *E. coli* origin has AT-rich sequences that are easily opened. The chromosome and its origin form a replicon.

E. coli *has at least three different DNA polymerases.*

Some DNA polymerases can also degrade DNA from one end, called exonuclease activity. Pol I, II, and III all have 3′-to-5′ exonuclease activity that can remove mispaired bases. Pol I can remove bases in the 5′-to-3′ direction, important to removing RNA primers.

Unwinding DNA requires energy and causes torsional strain.

DNA helicase uses energy from ATP to unwind DNA. The torsional strain introduced is removed by the enzyme DNA gyrase.

Replication is semidiscontinuous.

Replication is discontinuous on one strand (figure 14.15). The continuous strand is called the leading strand, and the discontinuous strand is called the lagging strand.

Synthesis occurs at the replication fork.

The partial opening of a DNA strand forms two single-stranded regions called the replication fork. At the fork, synthesis on the leading strand requires a single primer, and the polymerase stays attached to the template because of the β subunit that acts as a sliding clamp. On the lagging strand, DNA primase adds primers periodically, and DNA Pol III synthesizes the Okazaki fragments. DNA Pol I removes primer segments, and DNA ligase joins the fragments.

The replisome contains all the necessary enzymes for replication.

The replisome consists of two copies of Pol III, DNA primase, DNA helicase, and a number of accessory proteins. It moves in one direction by creating a loop in the lagging strand, allowing the antiparallel template strands to be copied in the same direction (figures 14.18 & 14.19).

14.5 Eukaryotic Replication

Eukaryotic replication requires multiple origins.

The sheer size and organization of eukaryotic chromosomes requires multiple origins of replication to be able to replicate DNA in the time available in S phase.

The enzymology of eukaryotic replication is more complex.

The eukaryotic primase synthesizes a short stretch of RNA and then switches to making DNA. This primer is extended by the main replication polymerase, which is a complex of two enzymes. The sliding clamp subunit was originally identified as protein produced by proliferating cells and is called PCNA.

Archaeal replication proteins are similar to eukaryotic proteins.

The replication proteins of archaea, including the sliding clamp, clamp loader, and DNA polymerases, are more similar to those of eukaryotes than to prokaryotes.

Linear chromosomes have specialized ends.

The ends of linear chromosomes are called telomeres. They are made by telomerase, not by the replication complex. Telomerase contains an internal RNA that acts as a template to extend the DNA of the chromosome end. Adult cells lack telomerase activity, and telomere shortening correlates with senescence.

14.6 DNA Repair

Cells are constantly exposed to DNA-damaging agents.

Errors from replication and damage induced by agents such as UV light and chemical mutagens can lead to mutations.

DNA repair restores damaged DNA.

Without repair mechanisms, cells would accumulate mutations until inviability occurred.

Repair can be either specific or nonspecific.

The enzyme photolyase uses energy from visible light to cleave thymine dimers caused by UV light. Excision repair is nonspecific. In prokaryotes, the uvr system can remove a damaged region of DNA.

Review Questions

UNDERSTAND

1. What was the key finding from Griffith's experiments using live and heat-killed pathogenic bacteria?
   a. Bacteria with a smooth coat could kill mice.
   b. Bacteria with a rough coat are not lethal.
   c. DNA is the genetic material.
   d. Genetic material can be transferred from dead to live bacteria.

2. Which of the following is NOT a component of DNA?
   a. The pyrimidine uracil   c. The purine adenine
   b. Five-carbon sugars   d. Phosphate groups

3. Chargaff studied the composition of DNA from different sources and found that
   a. the number of phosphate groups always equals the number of five-carbon sugars.
   b. the proportions of A equal that of C and G equals T.
   c. the proportions of A equal that of T and G equals C.
   d. purines bind to pyrimidines.

4. The bonds that hold two complementary strands of DNA together are
   a. hydrogen bonds.   c. ionic bonds.
   b. peptide bonds.   d. phosphodiester bonds.

5. The basic mechanism of DNA replication is semiconservative with two new molecules,
   a. each with new strands.
   b. one with all new strands and one with all old strands.
   c. each with one new and one old strand.
   d. each with a mixture of old and new strands.

6. One common feature of all DNA polymerases is that they
   a. synthesize DNA in the 3′-to-5′ direction.
   b. synthesize DNA in the 5′-to-3′ direction.
   c. synthesize DNA in both directions by switching strands.
   d. do not require a primer.

7. Which of the following is *not* part of the Watson–Crick model of the structure of DNA?
   a. DNA is composed of two strands.
   b. The two DNA strands are oriented in parallel (5′-to-3′).
   c. Purines bind to pyrimidines.
   d. DNA forms a double helix.

APPLY

1. If one strand of a DNA is 5′ ATCGTTAAGCGAGTCA 3′, then the complementary strand would be:
   a. 5′ TAGCAATTCGCTCAGT 3′.
   b. 5′ ACTGAGCGAATTGCTA 3′.
   c. 5′ TGACTCGCTTAACGAT 3′.
   d. 5′ ATCGTTAAGCGAGTCA 3′.

2. Hershey and Chase used radioactive phosphorus and sulfur to
   a. label DNA and protein uniformly.
   b. differentially label DNA and protein.
   c. identify the transforming principle.
   d. Both b and c are correct.

3. The Meselson and Stahl experiment used a density label to be able to
   a. determine the directionality of DNA replication.
   b. differentially label DNA and protein.
   c. distinguish between newly replicated and old strands.
   d. distinguish between replicated DNA and RNA primers.

4. The difference in leading- versus lagging-strand synthesis is a consequence of
   a. only the physical structure of DNA.
   b. only the activity of DNA polymerase enzymes.
   c. both the physical structure of DNA and the action of polymerase enzyme.
   d. the larger size of the lagging strand.

5. If the activity of DNA ligase was removed from replication, this would have a greater affect on
   a. synthesis on the lagging strand versus the leading strand.
   b. synthesis on the leading strand versus the lagging strand.
   c. priming of DNA synthesis versus actual DNA synthesis.
   d. photorepair of DNA versus DNA replication.

6. Successful DNA synthesis requires all of the following *except*
   a. helicase.   b. endonuclease.
   c. DNA primase.   d. DNA ligase.

7. The synthesis of telomeres
   a. uses DNA polymerase, but without the sliding clamp.
   b. uses enzymes involved in DNA repair.
   c. requires telomerase, which does not use a template.
   d. requires telomerase, which uses an internal RNA as a template.

8. When mutations that affected DNA replication were isolated, two kinds were found. In cultures that were not synchronized (that is, not all dividing at the same time), one class put an immediate halt to replication, whereas the other put a much slower stop to the process. The first class affects functions at the replication fork like polymerase and primase. The second class affects functions necessary for
   a. elongation on the lagging but not the leading strand.
   b. elongation on the leading but not the lagging strand.
   c. initiation: cells complete replication but cannot start a new round.
   d. the sliding clamp: loss makes the polymerase slower.

SYNTHESIZE

1. The work by Griffith provided the first indication that DNA was the genetic material. Review the four experiments outlined in figure 14.1. Predict the likely outcome for the following variations on this classic research.
   a. Heat-killed pathogenic and heat-killed nonpathogenic
   b. Heat-killed pathogenic and live nonpathogenic in the presence of an enzyme that digests proteins (proteases)
   c. Heat-killed pathogenic and live nonpathogenic in the presence of an enzyme that digests DNA (endonuclease)

2. In the Meselson–Stahl experiment, a control experiment was done to show that the hybrid bands after one round of replication were in fact two complete strands, one heavy and one light. Using the same experimental setup as detailed in the text, how can this be addressed?

3. Enzyme function is critically important for the proper replication of DNA. Predict the consequence of a loss of function for each of the following enzymes.
   a. DNA gyrase   c. DNA ligase
   b. DNA polymerase III   d. DNA polymerase I

ONLINE RESOURCE

www.ravenbiology.com

Understand, Apply, and Synthesize—enhance your study with animations that bring concepts to life and practice tests to assess your understanding. Your instructor may also recommend the interactive eBook, individualized learning tools, and more.

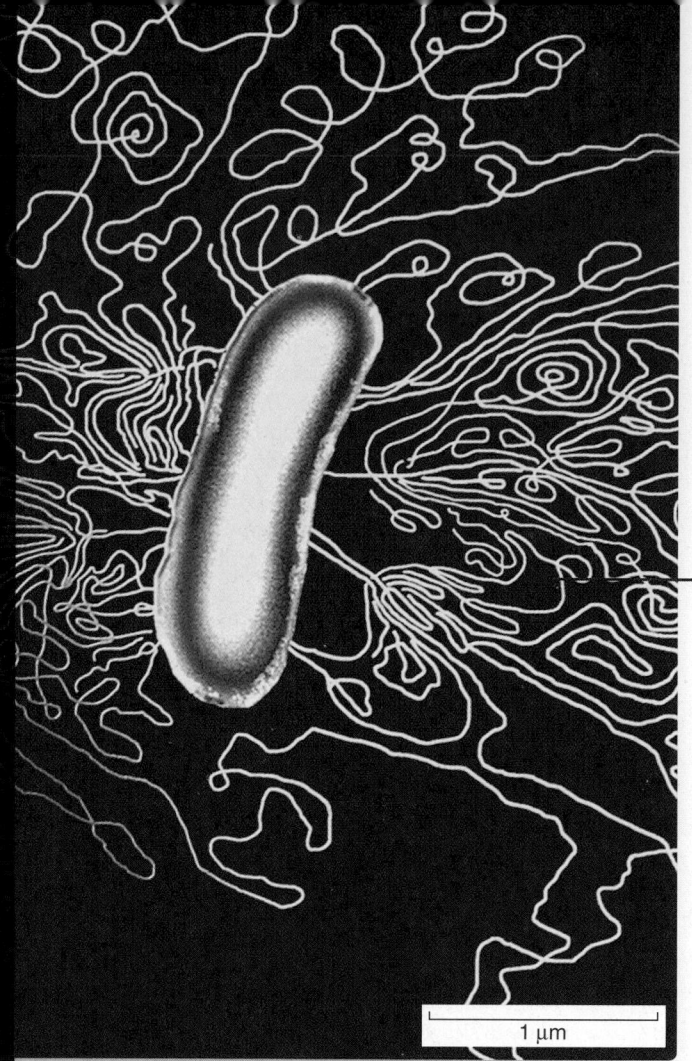

1 μm

Chapter 15

Genes and How They Work

Chapter Contents

Introduction

You've seen how genes specify traits and how those traits can be followed in genetic crosses. You've also seen that the information in genes resides in the DNA molecule; the picture above shows all the DNA within the entire E. coli chromosome. Information in DNA is replicated by the cell and then partitioned equally during the process of cell division. The information in DNA is much like a blueprint for a building. The construction of the building uses the information in the blueprint, but requires building materials and carpenters and other skilled laborers using a variety of tools working together to actually construct the building. Similarly, the information in DNA requires nucleotide and amino acid building blocks, multiple forms of RNA, and many proteins acting in a coordinated fashion to make up the structure of a cell.

We now turn to the nature of the genes themselves and how cells extract the information in DNA in the process of gene expression. Gene expression can be thought of as the conversion of genotype into the phenotype.

15.1   The Nature of Genes

Learning Outcomes

1. *Evaluate the evidence for the one-gene/one-polypeptide hypothesis.*
2. *Distinguish between transcription and translation.*
3. *List the roles played by RNA in gene expression.*

We know that DNA encodes proteins, but this knowledge alone tells us little about how the information in DNA can control cellular functions. Researchers had evidence that genetic mutations affected proteins, and in particular enzymes, long before the structure and code of DNA was known. In this section we review the evidence of the link between genes and enzymes.

Garrod concluded that inherited disorders can involve specific enzymes

In 1902, the British physician Archibald Garrod noted that certain diseases among his patients seemed to be more prevalent in particular families. By examining several generations of these families, he found that some of the diseases behaved as though they were the product of simple recessive alleles. Garrod concluded that these disorders were Mendelian traits, and that they had resulted from changes in the hereditary information in an ancestor of the affected families.

Garrod investigated several of these disorders in detail. In alkaptonuria, patients produced urine that contained homogentisic acid (alkapton). This substance oxidized rapidly when exposed to air, turning the urine black. In normal individuals, homogentisic acid is broken down into simpler substances. With considerable insight, Garrod concluded that patients suffering from alkaptonuria lack the enzyme necessary to catalyze this breakdown. He speculated that many other inherited diseases might also reflect enzyme deficiencies.

Beadle and Tatum showed that genes specify enzymes

From Garrod's finding, it took but a short leap of intuition to surmise that the information encoded within the DNA of chromosomes acts to specify particular enzymes. This point was not actually established, however, until 1941, when a series of experiments by George Beadle and Edward Tatum at Stanford University provided definitive evidence. Beadle and Tatum deliberately set out to create mutations in chromosomes and verified that they behaved in a Mendelian fashion in crosses. These alterations to single genes were analyzed for their effects on the organism (figure 15.1).

Neurospora crassa, *the bread mold*

One of the reasons Beadle and Tatum's experiments produced clear-cut results was their choice of experimental organism, the bread mold *Neurospora crassa*. This fungus can be grown readily in the laboratory on a defined medium consisting of only a carbon source (glucose), a vitamin (biotin), and inorganic salts. This type of medium is called "minimal" because it represents the minimal requirements to support growth. Any cells that can grow on minimal medium must be able to synthesize all necessary biological molecules.

Beadle and Tatum exposed *Neurospora* spores to X-rays, expecting that the DNA in some of the spores would experience damage in regions encoding the ability to make compounds needed for normal growth (see figure 15.1). Such a mutation would cause cells to be unable to grow on minimal medium. Such mutations are called **nutritional mutations** because cells carrying them grow only if the medium is supplemented with additional nutrients.

Nutritional mutants

To identify mutations causing metabolic deficiencies, Beadle and Tatum placed subcultures of individual fungal cells grown on a rich medium onto minimal medium. Any cells that had lost the ability to make compounds necessary for growth would not grow on minimal medium. Using this approach, Beadle and Tatum succeeded in isolating and identifying many nutritional mutants.

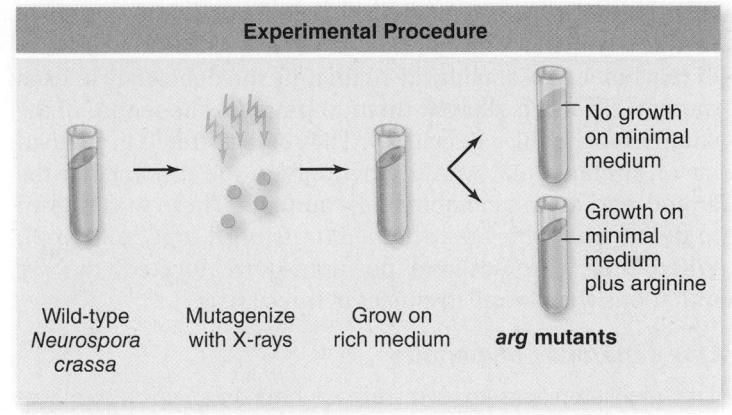

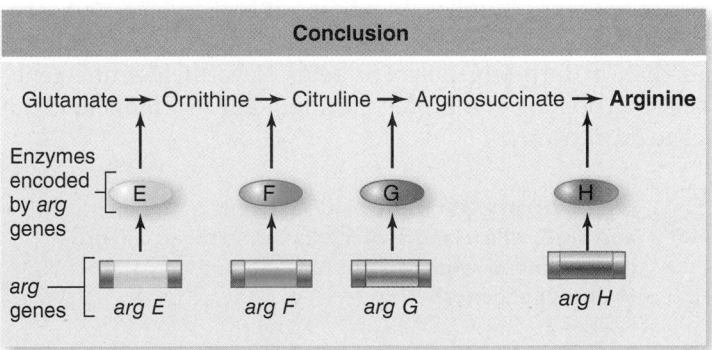

Figure 15.1 The Beadle and Tatum experiment. Wild-type *Neurospora* were mutagenized with X-rays to produce mutants deficient in the synthesis of arginine (top panel). The specific defect in each mutant was identified by growing on medium supplemented with intermediates in the biosynthetic pathway for arginine (middle panel). A mutant will grow only on media supplemented with an intermediate produced after the defective enzyme in the pathway for each mutant. The enzymes in the pathway can then be correlated with genes on chromosomes (bottom panel).

Next, the researchers supplemented the minimal medium with different compounds known to be intermediates in this biochemical pathway to identify the deficiency in each mutant. This step allowed them to pinpoint the nature of the strain's biochemical deficiency. They concentrated in particular on mutants that would grow only in the presence of the amino acid arginine, dubbed *arg* mutants. These were shown to define four genes they named *argE, argF, argG,* and *argH.* When their chromosomal positions were located, the *arg* mutations were found to cluster in three areas.

One gene/one polypeptide

The next step was to determine where each mutation was blocked in the biochemical pathway for arginine biosynthesis. To do this, they supplemented the medium with each intermediate in the pathway to see which would support each mutant's growth. If the mutation affects an enzyme in the pathway that acts prior to the supplement, then growth should be supported—but not if the mutation affects a step after the intermediate used (see figure 15.1). Beadle and Tatum were able to isolate a mutant strain defective for each enzyme in the biosynthetic pathway. The mutation was always located at a specific chromosomal sites and each mutation had a unique location. Thus, each of the mutants they examined had a defect in a single enzyme, caused by a mutation at a single site on a chromosome.

Beadle and Tatum concluded that genes specify the structure of enzymes, and that each gene encodes the structure of one enzyme (see figure 15.1). They called this relationship the *one-gene/one-enzyme hypothesis.* Today, because many enzymes contain multiple polypeptide subunits, each encoded by a separate gene, the relationship is more commonly referred to as the **one-gene/one-polypeptide hypothesis.** This hypothesis clearly states the molecular relationship between genotype and phenotype.

As you learn more about genomes and gene expression, this clear relationship becomes overly simple. Eukaryotic genes are more complex than those of prokaryotes, and some enzymes are composed, at least in part, of RNA, itself an intermediate in the production of proteins. Nevertheless, one-gene/one-polypeptide is a useful starting point for thinking about gene expression.

 Data analysis If you made a double mutant with *argE* and *argG*, which kind(s) of media would this strain grow on? In general, what can a double mutant tell you about the order of genes?

The central dogma describes information flow in cells as DNA to RNA to protein

The conversion of genotype to phenotype requires information stored in DNA to be converted to protein. The nature of information flow in cells was first described by Francis Crick as the **central dogma of molecular biology.** Information passes in one direction from the gene (DNA) to an RNA copy of the gene, and the RNA copy directs the sequential assembly

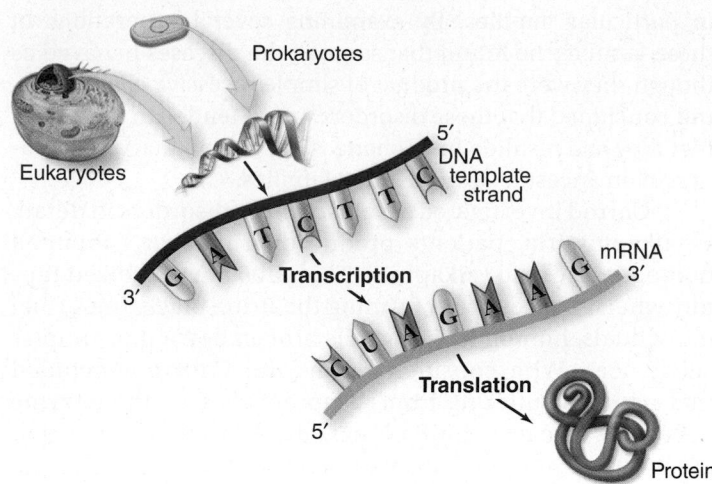

Figure 15.2 The central dogma of molecular biology. DNA is transcribed to make mRNA, which is translated to make a protein.

of a chain of amino acids into a protein (figure 15.2). Stated briefly,

$$DNA \longrightarrow RNA \longrightarrow protein$$

The central dogma provides an intellectual framework that describes information flow in biological systems. We call the DNA-to-RNA step **transcription** because it produces an exact copy of the DNA, much as a legal transcription contains the exact words of a court proceeding. The RNA-to-protein step is termed **translation** because it requires translating from the nucleic acid to the protein "languages."

Since the original formulation of the central dogma, a class of viruses called **retroviruses** was discovered that can convert their RNA genome into a DNA copy, using the viral enzyme **reverse transcriptase.** This conversion violates the direction of information flow of the central dogma, and the discovery forced an updating of the possible flow of information to include this "reverse" flow from RNA to DNA.

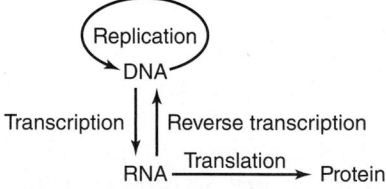

Transcription makes an RNA copy of DNA

The process of transcription produces an RNA copy of the information in DNA. That is, transcription is the DNA-directed synthesis of RNA by the enzyme RNA polymerase (figure 15.3). This process uses the principle of complementarity, described in chapter 14, to use DNA as a template to make RNA.

Because DNA is double-stranded and RNA is single-stranded, only one of the two DNA strands needs to be copied. We call the strand that is copied the **template strand.** The RNA transcript's sequence is complementary to the template strand.

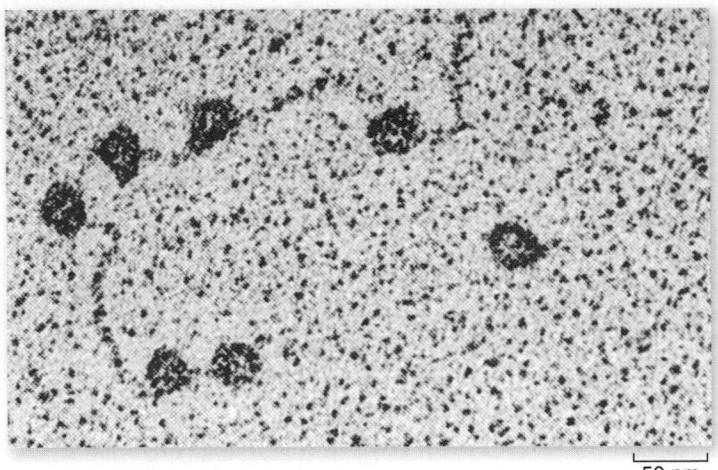

50 nm

Figure 15.3 RNA polymerase. In this electron micrograph, the dark circles are RNA polymerase molecules synthesizing RNA from a DNA template.

The strand of DNA not used as a template is called the **coding strand.** It has the same sequence as the RNA transcript, except that U (uracil) in the RNA is T (thymine) in the DNA-coding strand. Another naming convention for the two strands of the DNA is to call the coding strand the sense strand, as it has the same "sense" as the RNA. The template strand would then be the antisense strand.

```
        Coding (sense)  5′–TCAGCCGTCAGCT–3′ ⎤
                                            ⎬ DNA
   Template (antisense) 3′–AGTCGGCAGTCGA–5′ ⎦
                             |
                        Transcription
                             ↓
           Coding  5′–UCAGCCGUCAGCU– 3′   mRNA
```

The RNA transcript used to direct the synthesis of polypeptides is termed *messenger RNA (mRNA).* Its name reflects its use by the cell to carry the DNA message to the ribosome for processing.

? Inquiry question RNA polymerase has no proofreading capacity. How does this affect the error rate in transcription compared with DNA replication? Why do you think it is more important for DNA polymerase than for RNA polymerase to proofread?

Translation uses information in RNA to synthesize proteins

The process of translation is by necessity much more complex than transcription. In this case, RNA cannot be used as a direct template for a protein because there is no complementarity—that is, a sequence of amino acids cannot be aligned to an RNA template based on any kind of "chemical fit." Molecular geneticists suggested that some kind of adapter molecule must exist that can interact with both RNA and amino acids, and *transfer RNA (tRNA)* was found to fill this role. This need for an intermediary adds a level of complexity to the process that is not seen in either DNA replication or transcription of RNA.

Translation takes place on the ribosome, the cellular protein-synthesis machinery, and it requires the participation of multiple kinds of RNA and many proteins. Here we provide an outline of the processes; all are described in detail in the sections that follow.

RNA has multiple roles in gene expression

All RNAs are synthesized from a DNA template by transcription. Gene expression requires the participation of multiple kinds of RNA, each with different roles in the overall process. Here is a brief summary of these roles, which are described in detail later.

Messenger RNA. Even before the details of gene expression were unraveled, geneticists recognized that there must be an intermediate form of the information in DNA that can be transported out of the eukaryotic nucleus to the cytoplasm for ribosomal processing. This hypothesis was called the "messenger hypothesis," and we retain this language in the name *messenger RNA* (mRNA).

Ribosomal RNA. The class of RNA found in ribosomes is called **ribosomal RNA (rRNA).** There are multiple forms of rRNA, and rRNA is found in both ribosomal subunits. This rRNA is critical to the function of the ribosome.

Transfer RNA. The intermediary adapter molecule between mRNA and amino acids, as mentioned earlier, is **transfer RNA (tRNA).** Transfer RNA molecules have amino acids covalently attached to one end and an anticodon that can base-pair with an mRNA codon at the other. The tRNAs act to interpret information in mRNA and to help position the amino acids on the ribosome.

Small nuclear RNA. Small nuclear RNAs (snRNAs) are part of the machinery that is involved in nuclear processing of eukaryotic "pre-mRNA." We discuss this splicing reaction later in the chapter.

SRP RNA. In eukaryotes, where some proteins are synthesized by ribosomes on the rough endoplasmic reticulum (RER), this process is mediated by the **signal recognition particle,** or **SRP,** described later in the chapter. The SRP contains both RNA and proteins.

Small RNAs. This class of RNA includes both **micro-RNA (miRNA)** and **small interfering RNA (siRNA).** These are involved in the control of gene expression discussed in chapter 16.

Learning Outcomes Review 15.1

Garrod showed that altered enzymes can cause metabolic disorders. Beadle and Tatum demonstrated that each gene encodes a unique enzyme. Genetic information flows from DNA (genes) to protein (enzymes) using messenger RNA as an intermediate. Transcription converts information in DNA into an RNA transcript, and translation converts this information into protein. RNA comes in several varieties having different functions; these include mRNA (the transcript), tRNA (the intermediary), and rRNA (in ribosomes), as well as snRNA, SNP RNA, and small RNAs (miRNA, siRNA).

■ *Why do cells need an adapter molecule like tRNA between RNA and protein?*

15.2 *The Genetic Code*

How does a sequence of nucleotides in a DNA molecule specify the sequence of amino acids in a polypeptide? The answer to this essential question came in 1961, through an experiment led by Francis Crick and Sydney Brenner. That experiment was so elegant and the result so critical to understanding the genetic code that we describe it here in detail.

The code is read in groups of three

Crick and Brenner reasoned that the genetic code most likely consisted of a series of blocks of information called **codons,** each corresponding to an amino acid in the encoded protein. They further hypothesized that the information within one codon was probably a sequence of three nucleotides. With four DNA nucleotides (G, C, T, and A), using two in each codon can produce only 4^2, or 16, different codons—not enough to code for 20 amino acids. However, three nucleotides results in 4^3, or 64, different combinations of three, more than enough.

Spaced or unspaced codons?

In theory, the sequence of codons in a gene could be punctuated with nucleotides between the codons that are not used, like the spaces that separate the words in this sentence. Alternatively, the codons could lie immediately adjacent to each other, forming a continuous sequence of nucleotides.

If the information in the genetic message is separated by spaces, then altering any single word would not affect the entire sentence. In contrast, if all of the words are run together but read in groups of three, then any alteration that is not in groups of three would alter the entire sentence. These two ways of using information in DNA imply different methods of translating the information into protein.

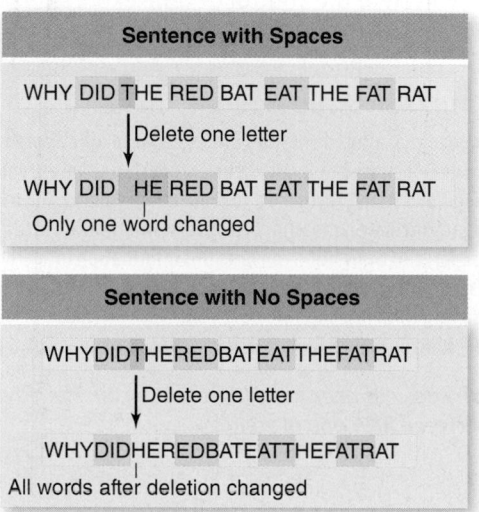

Determining that codons are unspaced

To choose between these alternative mechanisms, Crick and his colleagues used a chemical to create mutations that caused single-base insertions or deletions from a viral DNA molecule. They then showed that combining an insertion with a deletion restored function even though either one individually displayed loss of function. In this case, only the region between the insertion or deletion would be altered. By choosing a region of the gene that encoded a part of the protein not critical to function, this small change did not cause a change in phenotype.

When they combined a single deletion or two deletions near each other, the genetic message shifted, altering all of the amino acids after the deletion. When they made three deletions, however, the protein after the deletions was normal. They

SCIENTIFIC THINKING

Hypothesis: *The genetic code is read in groups of three bases.*

Prediction: *If the genetic code is read in groups of three, then deletion of one or two bases would shift the reading frame after the deletion. Deletion of three bases, however, would produce a protein with a single amino acid deleted but no change downstream.*

Test: *Single-base deletion mutants are collected, each of which exhibits a mutant phenotype. Three of these deletions in a single region are combined to assess the effect of deletion of three bases.*

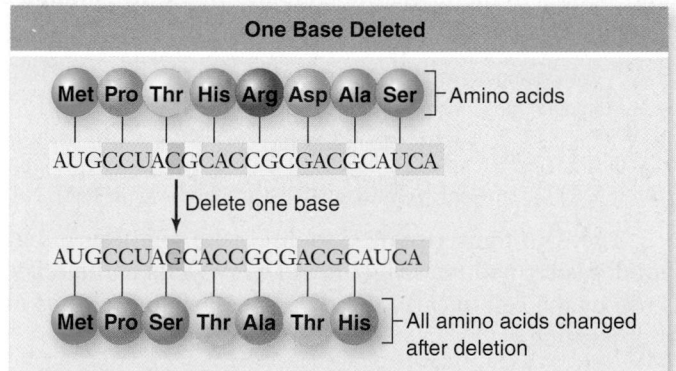

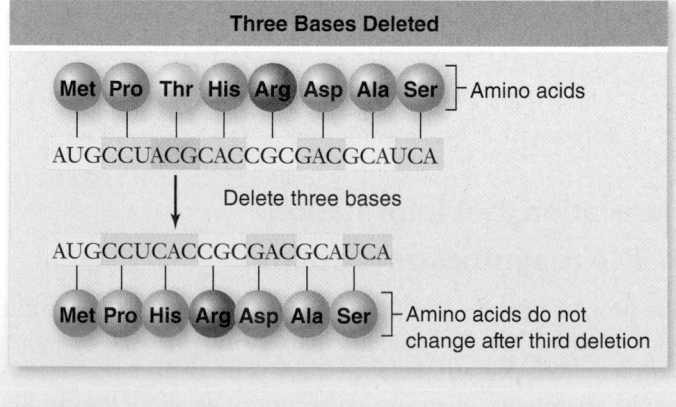

Result: *The combination of three deletions does not have the same drastic effect as the loss of one or two bases.*

Conclusion: *The genetic code is read in groups of three.*

Further Experiments: *If you also had mutants with single-base additions, what would be the effect of combining a deletion and an addition?*

Figure 15.4 **The genetic code is triplet.**

obtained the same results when they made additions to the DNA consisting of 1, 2, or 3 nt (nucleotides).

Thus, Crick and Brenner concluded that the genetic code is read in increments of three nucleotides (in other words, it is a triplet code), and that reading occurs continuously without punctuation between the 3-nt units (figure 15.4).

These experiments indicate the importance of the **reading frame** for the genetic message. Because there is no punctuation, the reading frame established by the first codon in the sequence determines how all subsequent codons are read. We now call the kinds of mutations that Crick and Brenner used **frameshift mutations** because they alter the reading frame of the genetic message.

Nirenberg and others deciphered the code

The determination of which of the 64 possible codons encoded each particular amino acids was one of the greatest triumphs of 20th-century biochemistry. Accomplishing this decryption depended on two related technologies: (1) cell-free biochemical systems that would support protein synthesis from a defined RNA and (2) the ability to produce synthetic, defined RNAs that could be used in the cell-free system.

During a five-year period from 1961 to 1966, work performed primarily in American biochemist Marshall Nirenberg's laboratory led to the elucidation of the genetic code. Nirenberg's group first showed that adding the synthetic RNA molecule polyU (an RNA molecule consisting of a string of uracil nucleotides) to their cell-free systems produced the polypeptide polyphenylalanine (a string of phenylalanine amino acids). Therefore, UUU encodes phenylalanine.

Next they used enzymes to produce RNA polymers with more than one nucleotide. These polymers allowed them to infer the composition of many of the possible codons, but not the order of bases in each codon.

The researchers then were able to use enzymes to synthesize defined 3-base sequences that could be tested for binding to the protein-synthesis machinery. This so-called *triplet-binding assay* allowed them to identify 54 of the 64 possible triplets.

The organic chemist H. Gobind Khorana provided the final piece of the puzzle by using organic synthesis to produce artificial RNA molecules of defined sequence, and then examining what polypeptides they directed in cell-free systems. The combination of all of these methods allowed the determination of all 64 possible 3-nt sequences, and the full genetic code was determined (table 15.1).

The code is degenerate but specific

Some obvious features of the code jump out of table 15.1. First, 61 of the 64 possible codons are used to specify amino acids. Three codons, UAA, UGA, and UAG, are reserved for another function: they signal "stop" and are known as **stop codons.** The only other form of "punctuation" in the code is that AUG is used to signal "start" and is therefore the **start codon.** In this case the codon has a dual function because it also encodes the amino acid methionine (Met).

| TABLE 15.1 | The Genetic Code |

First Letter	SECOND LETTER								Third Letter
	U		C		A		G		
U	UUU	Phe Phenylalanine	UCU	Ser Serine	UAU	Tyr Tyrosine	UGU	Cys Cysteine	U
	UUC		UCC		UAC		UGC		C
	UUA	Leu Leucine	UCA		UAA	"Stop"	UGA	"Stop"	A
	UUG		UCG		UAG	"Stop"	UGG	Trp Tryptophan	G
C	CUU	Leu Leucine	CCU	Pro Proline	CAU	His Histidine	CGU	Arg Arginine	U
	CUC		CCC		CAC		CGC		C
	CUA		CCA		CAA	Gln Glutamine	CGA		A
	CUG		CCG		CAG		CGG		G
A	AUU	Ile Isoleucine	ACU	Thr Threonine	AAU	Asn Asparagine	AGU	Ser Serine	U
	AUC		ACC		AAC		AGC		C
	AUA		ACA		AAA	Lys Lysine	AGA	Arg Arginine	A
	AUG	Met Methionine; "Start"	ACG		AAG		AGG		G
G	GUU	Val Valine	GCU	Ala Alanine	GAU	Asp Aspartate	GGU	Gly Glycine	U
	GUC		GCC		GAC		GGC		C
	GUA		GCA		GAA	Glu Glutamate	GGA		A
	GUG		GCG		GAG		GGG		G

A codon consists of three nucleotides read in the sequence shown. For example, ACU codes for threonine. The first letter, A, is in the First Letter column; the second letter, C, is in the Second Letter column; and the third letter, U, is in the Third Letter column. Each of the mRNA codons is recognized by a corresponding anticodon sequence on a tRNA molecule. Many amino acids are specified by more than one codon. For example, threonine is specified by four codons, which differ only in the third nucleotide (ACU, ACC, ACA, and ACG).

Figure 15.5 Transgenic pig. The piglet on the right is a conventional piglet. The piglet on the left was engineered to express a gene from jellyfish that encodes green fluorescent protein. The color of this piglet's nose is due to expression of this introduced gene. Such transgenic animals indicate the universal nature of the genetic code.

You can see that 61 codons are more than enough to encode 20 amino acids. That leaves lots of extra codons. One way to deal with this abundance would be to use only 20 of the 61 codons, but this is not what cells do. In reality, all 61 codons are used, making the code **degenerate,** which means that some amino acids are specified by more than one codon. The reverse, however, in which a single codon would specify more than one amino acid, is never found.

This degeneracy is not uniform. Some amino acids have only one codon, and some have up to six. In addition, the degenerate base usually occurs in position 3 of a codon, such that the first two positions are the same, and two or four of the possible nucleotides at position 3 encode the same amino acid. (The nature of protein synthesis on ribosomes explains how this codon usage works, and it is discussed later.)

The code is practically universal, but not quite

The genetic code is the same in almost all organisms. The universality of the genetic code is among the strongest evidence that all living things share a common evolutionary heritage. Because the code is universal, genes can be transferred from one organism to another and can be successfully expressed in their new host (figure 15.5). This universality of gene expression is central to many of the advances of genetic engineering discussed in chapter 17.

In 1979, investigators began to determine the complete nucleotide sequences of the mitochondrial genomes in humans, cattle, and mice. It came as something of a shock when these investigators learned that the genetic code used by these mammalian mitochondria was not quite the same as the "universal code" that has become so familiar to biologists.

In the mitochondrial genomes, what should have been a stop codon, UGA, was instead read as the amino acid tryptophan; AUA was read as methionine rather than as isoleucine; and AGA and AGG were read as stop codons rather than as arginine. Furthermore, minor differences from the universal code have also been found in the genomes of chloroplasts and in ciliates (certain types of protists).

Thus, it appears that the genetic code is not quite universal. Some time ago, presumably after they began their endosymbiotic existence, mitochondria and chloroplasts began to read the code differently, particularly the portion associated with "stop" signals.

> **? Inquiry question** The genetic code is almost universal. Why do you think it is nearly universal?

Learning Outcomes Review 15.2

The genetic code was shown to be nucleotide base triplets with two forms of punctuation and no spaces: three bases code for an amino acid, and the groups of three are read in order. Sixty-one codons specify amino acids, one of which also codes for "start," and three codons indicate "stop," for 64 total. Because some amino acids are specified by more than one codon, the code is termed degenerate. All codons encode only one amino acid, however.

■ *What would be the outcome if a codon specified more than one amino acid?*

15.3 *Prokaryotic Transcription*

Learning Outcomes

1. *Describe the transcription process in bacteria.*
2. *Differentiate among initiation, elongation, and termination of transcription.*
3. *Define the unique features of prokaryotic transcription.*

We begin an examination of gene expression by describing the process of transcription in prokaryotes. The later description of eukaryotic transcription will concentrate on their differences from prokaryotes.

Figure 15.6 Bacterial RNA polymerase and transcription initiation. *a.* RNA polymerase has two forms: core polymerase and holoenzyme. *b.* The σ subunit of the holoenzyme recognizes promoter elements at –35 and –10 and binds to the DNA. The helix is opened at the –10 region, and transcription begins at the start site at +1.

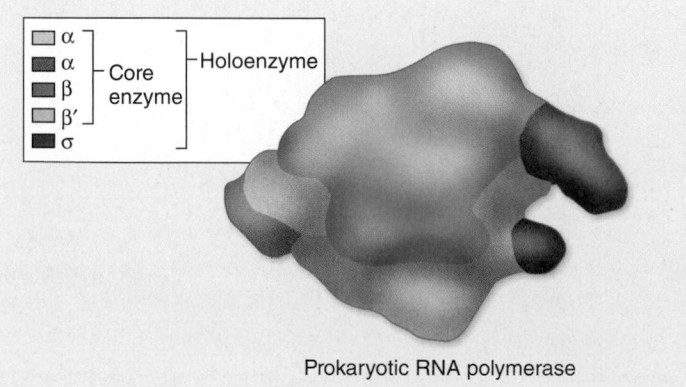

Prokaryotic RNA polymerase

a.

Prokaryotes have a single RNA polymerase

The single **RNA polymerase** of prokaryotes exists in two forms: the *core polymerase* and the *holoenzyme.* The core polymerase can synthesize RNA using a DNA template, but it cannot initiate synthesis accurately. The holoenzyme can accurately initiate synthesis.

The core polymerase is composed of four subunits: two identical α subunits, a β subunit, and a β' subunit (figure 15.6*a*). The two α subunits help to hold the complex together and can bind to regulatory molecules. The active site of the enzyme is formed by the β and β' subunits, which bind to the DNA template and the ribonucleotide triphosphate precursors.

The *holoenzyme,* which can properly initiate synthesis, is formed by the addition of a σ (sigma) subunit to the core polymerase (see figure 15.6*a*). Its ability to recognize specific signals in DNA allows RNA polymerase to locate the beginning of genes, which is critical to its function. Note that initiation of mRNA synthesis does not require a primer, in contrast to DNA replication.

Initiation occurs at promoters

Accurate initiation of transcription requires two sites in DNA: one called a **promoter** that forms a recognition and binding site for the RNA polymerase, and the actual **start site.** The polymerase also needs a signal to end transcription, which we call a **terminator.** We then refer to the region from promoter to terminator as a **transcription unit.**

The action of the polymerase moving along the DNA can be thought of as analogous to water flowing in a stream. We can speak of sites on the DNA as being "upstream" or "downstream" of the start site. We can also use this comparison to form a simple system for numbering bases in DNA to refer to positions in the transcription unit. The first base transcribed is called +**1,** and this numbering continues downstream until the last base is transcribed. Any bases upstream of the start site receive negative numbers, starting at –**1.**

The promoter is a short sequence found upstream of the start site and is therefore not transcribed by the polymerase. Two 6-base sequences are common to bacterial promoters: One

is located 35 nt upstream of the start site (–35), and the other is located 10 nt upstream of the start site (–10) (figure 15.6*b*). These two sites provide the promoter with asymmetry; they indicate not only the site of initiation, but also the direction of transcription.

The binding of RNA polymerase to the promoter is the first step in transcription. Promoter binding is controlled by the σ subunit of the RNA polymerase holoenzyme, which recognizes the –35 sequence in the promoter and positions the RNA polymerase at the correct start site, oriented to transcribe in the correct direction.

Once bound to the promoter, the RNA polymerase begins to unwind the DNA helix at the –10 site (see figure 15.6*b*). The polymerase covers a region of about 75 bp but only unwinds about 12–14 bp.

> **? Inquiry question** The prokaryotic promoter has two distinct elements that are not identical. How is this important to the initiation of transcription?

Elongation adds successive nucleotides

In prokaryotes, the transcription of the RNA chain usually starts with ATP or GTP. One of these forms the 5' end of the chain, which grows in the 5'-to-3' direction as ribonucleotides are added. As the RNA polymerase molecule leaves the promoter region, the σ factor is no longer required, although it may remain in association with the enzyme.

This process of leaving the promoter, called *clearance,* or *escape,* involves more than just synthesizing the first few nucleotides of the transcript and moving on, because the enzyme has made strong contacts to the DNA during initiation. It is necessary to break these contacts with the promoter region to be able to move progressively down the template. The enzyme goes through conformational changes during this clearance stage, and subsequently contacts less of the DNA than it does during the initial promoter binding.

The region containing the RNA polymerase, the DNA template, and the growing RNA transcript is called the **transcription bubble** because it contains a locally unwound

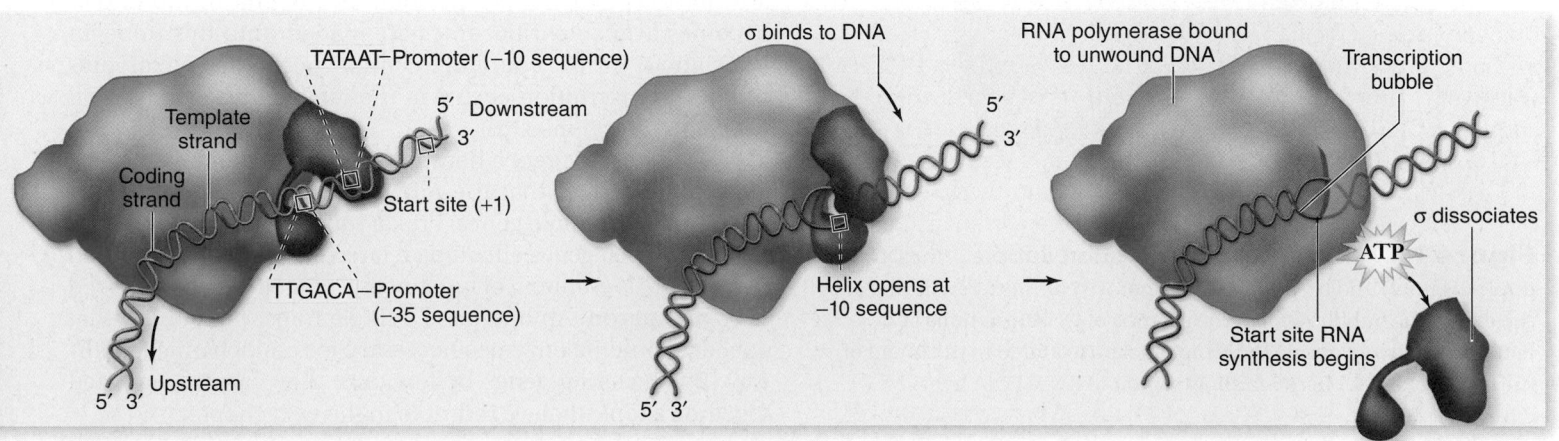

b.

"bubble" of DNA (figure 15.7). Within the bubble, the first 9 bases of the newly synthesized RNA strand temporarily form a helix with the template DNA strand. This stabilizes the positioning of the 3′ end of the RNA so it can interact with an incoming ribonucleotide triphosphate. The enzyme itself covers about 50 bp of DNA around this transcription bubble.

The transcription bubble created by RNA polymerase moves down the bacterial DNA at a constant rate, about 50 nt/sec, with the growing RNA strand protruding from the bubble. After the transcription bubble passes, the now-transcribed DNA is rewound as it leaves the bubble.

Termination occurs at specific sites

The end of a bacterial transcription unit is marked by terminator sequences that signal "stop" to the polymerase. Reaching these sequences causes the formation of phosphodiester bonds to cease, the RNA–DNA hybrid within the transcription bubble to dissociate, the RNA polymerase to release the DNA, and the DNA within the transcription bubble to rewind.

The simplest terminators consist of a series of G–C base-pairs followed by a series of A–T base-pairs. The RNA transcript of this stop region can form a double-stranded structure in the GC region called a *hairpin,* which is followed by four or more uracil (U) ribonucleotides (figure 15.8). Formation of the hairpin causes the RNA polymerase to pause,

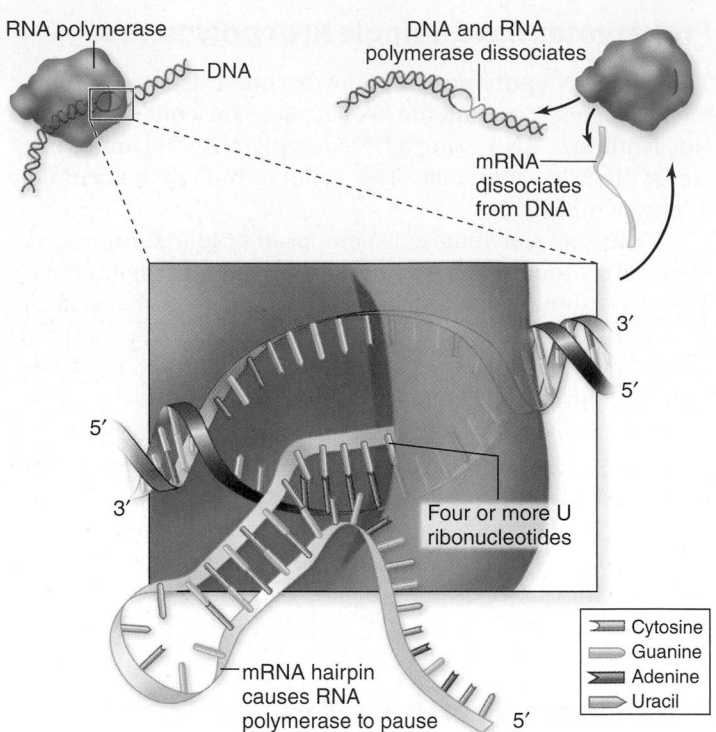

Figure 15.8 **Bacterial transcription terminator.** The self-complementary G–C region forms a double-stranded stem with a single-stranded loop called a hairpin. The stretch of U's forms a less stable RNA–DNA hybrid that falls off the enzyme.

placing it directly over the run of four uracils. The pairing of U with the DNA's A is the weakest of the four hybrid base-pairs, and it is not strong enough to hold the hybrid strands when the polymerase pauses. Instead, the RNA strand dissociates from the DNA within the transcription bubble, and transcription stops. A variety of protein factors also act at these terminators to aid in terminating transcription.

Prokaryotic transcription is coupled to translation

In prokaryotes, the mRNA produced by transcription begins to be translated before transcription is finished—that is, they are *coupled* (figure 15.9). As soon as a 5′ end of the mRNA becomes available, ribosomes are loaded onto this to begin translation. (This coupling cannot occur in eukaryotes because transcription occurs in the nucleus, and translation occurs in the cytoplasm.)

Another difference between prokaryotic and eukaryotic gene expression is that the mRNA produced in prokaryotes may contain multiple genes. Prokaryotic genes are often organized such that genes encoding related functions are clustered together. This grouping of functionally related genes is referred to as an **operon.** An operon is a single transcription unit that encodes multiple enzymes necessary for a biochemical pathway. By clustering genes by function, they can be regulated together, a topic that we return to in the next chapter.

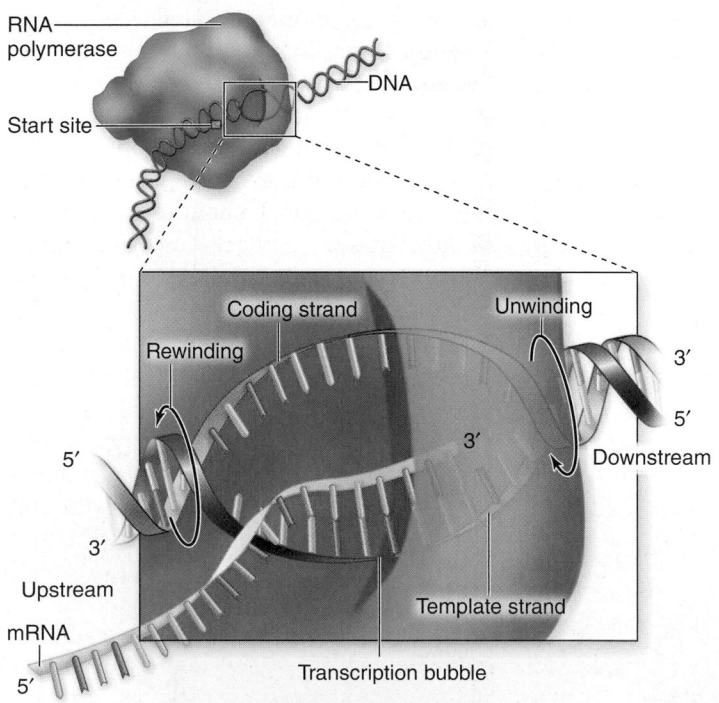

Figure 15.7 **Model of a transcription bubble.** The DNA duplex is unwound by the RNA polymerase complex, rewinding at the end of the bubble. One of the strands of DNA functions as a template, and nucleotide building blocks are added to the 3′ end of the growing RNA. There is a short region of RNA–DNA hybrid within the bubble.

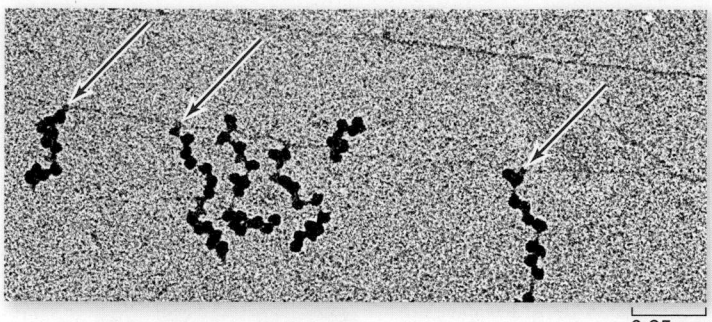

0.25 μm

RNA polymerase

DNA

mRNA

Ribosomes

Polyribosome

Polypeptide
chains

Figure 15.9 Transcription and translation are coupled in prokaryotes. In this micrograph of gene expression in *E. coli*, translation is occurring during transcription. The arrows point to RNA polymerase enzymes, and ribosomes are attached to the mRNAs extending from the polymerase. Polypeptides being synthesized by ribosomes, which are not visible in the micrograph, have been added to the last mRNA in the drawing.

Learning Outcomes Review 15.3

Transcription in bacteria is accomplished by RNA polymerase, which has two forms: a core polymerase and a holoenzyme. Initiation is accomplished by the holoenzyme form, which can accurately recognize promoter sequences. Elongation consists of RNA synthesis by the core enzyme, which adds RNA nucleotides in sequence until it reaches a terminator where synthesis stops, then the transcript is released. In prokaryotes, translation of the RNA transcript begins before transcription is finished, making the processes coupled.

■ *Yeast are unicellular organisms like bacteria; would you expect them to have the same transcription/ translation coupling?*

15.4 Eukaryotic Transcription

Learning Outcomes

1. List the different eukaryotic RNA polymerases.
2. Differentiate promoters for the three polymerases.
3. Describe the processing of eukaryotic transcripts.

The basic mechanism of transcription by RNA polymerase is the same in eukaryotes as in prokaryotes; however, the details of the two processes differ enough that it is necessary to consider them separately. Here we concentrate only on how eukaryotic systems differ from prokaryotic systems, such as the bacterial system just discussed. All other features may be assumed to be the same.

Eukaryotes have three RNA polymerases

Unlike prokaryotes, which have a single RNA polymerase enzyme, eukaryotes have three different RNA polymerases, which are distinguished in both structure and function. The enzyme **RNA polymerase I** transcribes rRNA, **RNA polymerase II** transcribes mRNA and some small nuclear RNAs, and **RNA polymerase III** transcribes tRNA and some other small RNAs. Together, these three enzymes accomplish all transcription in the nucleus of eukaryotic cells.

Each polymerase has its own promoter

The existence of three different RNA polymerases requires different signals in the DNA to allow each polymerase to recognize where to begin transcription. Each polymerase recognizes a different promoter structure.

RNA polymerase I promoters

RNA polymerase I promoters at first puzzled biologists, because comparisons of rRNA genes between species showed no similarities outside the coding region. The current view is that these promoters are also specific for each species, and for this reason, cross-species comparisons do not yield similarities.

RNA polymerase II promoters

The RNA polymerase II promoters are the most complex of the three types, probably a reflection of the huge diversity of genes that are transcribed by this polymerase. When the first eukaryotic genes were isolated, many had a sequence called the **TATA box** upstream of the start site. This sequence was similar to the prokaryotic –10 sequence, and it was assumed that the TATA box was the primary promoter element. With the sequencing of entire genomes, many more genes have been analyzed, and this assumption has proved too simple. It has been replaced by the idea of a "core promoter" that can be composed of a number of different elements, including the TATA box. Additional control elements allow for tissue-specific and developmental time–specific expression (see chapter 16).

RNA polymerase III promoters

Promoters for RNA polymerase III also were a source of surprise for biologists in the early days of molecular biology who were examining the control of eukaryotic gene expression. A common technique for analyzing regulatory regions was to make successive deletions from the 5′ end of genes until enough was deleted to abolish specific transcription. The logic followed experiences with prokaryotes, in which the regulatory regions had been found at the 5′ end of genes. But in the case of tRNA genes, the 5′ deletions had no effect on

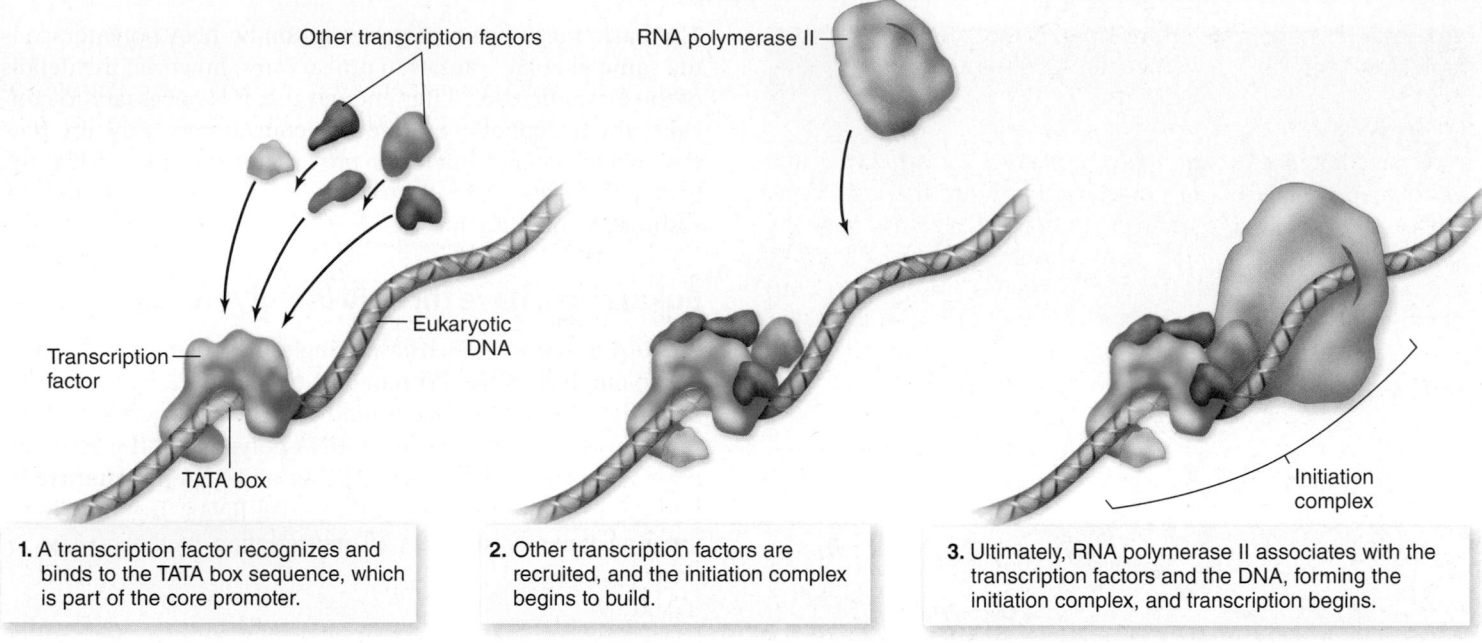

1. A transcription factor recognizes and binds to the TATA box sequence, which is part of the core promoter.

2. Other transcription factors are recruited, and the initiation complex begins to build.

3. Ultimately, RNA polymerase II associates with the transcription factors and the DNA, forming the initiation complex, and transcription begins.

Figure 15.10 Eukaryotic initiation complex. Unlike transcription in prokaryotic cells, in which the RNA polymerase recognizes and binds to the promoter, eukaryotic transcription requires the binding of transcription factors to the promoter before RNA polymerase II binds to the DNA. The association of transcription factors and RNA polymerase II at the promoter is called the initiation complex.

expression! The promoters were found to actually be internal to the gene itself. This has not proved to be the case for all polymerase III genes, but appears to be for most.

Initiation and termination differ from that in prokaryotes

The initiation at RNA polymerase II promoters is analogous to prokaryotic initiation but instead of a single factor allowing promoter recognition, eukaryotes use a host of **transcription factors.** These proteins are necessary to assemble RNA polymerase II on a promoter. The transcription factors interact with RNA polymerase II to form an initiation complex at the promoter (figure 15.10). We explore this complex in detail in chapter 16 when we describe the control of gene expression.

Curiously, it now appears that recruitment to the promoter is not the end of the story. Recent global analyses showed that 30% of human genes have Pol II paused 20–50 bp

downstream of the promoter. This promoter-proximal pausing can be relieved by elongation factors, and allows another level of control on transcription.

The termination of transcription for RNA polymerase II also differs from prokaryotes. Although termination sites exist, they are not well defined, and the end of the mRNA is not even formed by RNA polymerase II because the primary transcript is modified.

Eukaryotic transcripts are modified

A primary difference between prokaryotes and eukaryotes is the fate of the transcript itself. Prokaryotes translate the mRNA during transcription, but eukaryotes extensively modify the transcript in the nucleus before its translation in the cytoplasm. We call the RNA synthesized by RNA polymerase II the **primary transcript,** which is processed to produce the **mature mRNA.**

The 5′ cap

The first base in the transcript is usually an adenine (A) or a guanine (G), and this is modified by the addition of GTP to the 5′ PO_4 group, forming what is known as the **5′ cap** (figure 15.11). This cap is joined to the transcript by its 5′ end; the only such 5′-to-5′ bond found in nucleic acids. The G in the GTP is also modified by the addition of a methyl group, so it is often called a **methyl-G cap.** The cap is added while transcription is still in

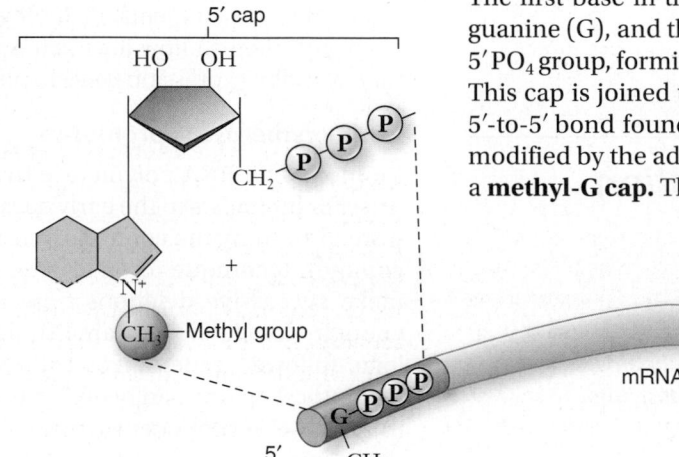

Figure 15.11 Posttranscriptional modifications to 5′ and 3′ ends. Eukaryotic mRNA molecules are modified in the nucleus with the addition of a methylated GTP to the 5′ end of the transcript, called the 5′ cap, and a long chain of adenine residues to the 3′ end of the transcript, called the 3′ poly-A tail.

progress. This cap protects the 5′ end of the mRNA from degradation and participates in translation initiation.

The 3′ poly-A tail

A major difference between prokaryotes and eukaryotes is that in eukaryotes, the end of the transcript is not the end of the mRNA. The eukaryotic transcript is cleaved downstream of a specific site (AAUAAA) prior to the termination site. A series of adenine (A) residues, called the **3′ poly-A tail,** is added after this cleavage by the enzyme poly-A polymerase. Thus the end of the mRNA is not created by RNA polymerase II (see figure 15.11).

The enzyme poly-A polymerase is part of a complex that recognizes the poly-A site, cleaves the transcript, then adds 100–200 A's to the end. The poly-A tail appears to play a role in the stability of mRNAs by protecting them from degradation (see chapter 16).

Splicing of primary transcripts

Eukaryotic genes may contain noncoding sequences that have to be removed to produce the final mRNA. This process, called pre-mRNA splicing, is accomplished by an organelle called the **spliceosome.** This complex topic is discussed in the next section.

Learning Outcomes Review 15.4

Eukaryotes have three RNA polymerases called polymerase I, II, and III. Each synthesizes a different RNA and recognizes its own promoter. The RNA polymerase I promoter is species-specific. The polymerase II promoter is complex, but often includes a sequence called the TATA box. The polymerase III promoter is internal to the gene, rather than close to the 5′ end. Polymerase II is responsible for mRNA synthesis. The primary mRNA transcript is modified by addition of a 5′ cap and a 3′ poly-A tail consisting of 100–200 adenines. Noncoding regions are removed by splicing.

■ *Does the complexity of the eukaryotic genome require three polymerases?*

15.5 Eukaryotic pre-mRNA Splicing

Learning Outcomes

1. *Explain the relationship between genes and proteins in prokaryotes and eukaryotes.*
2. *Describe the splicing reaction for pre-mRNA.*
3. *Illustrate how splicing changes the nature of genes.*

The first genes isolated were prokaryotic genes found in *E. coli* and its viruses. A clear picture of the nature and some of the control of gene expression emerged from these systems before any eukaryotic genes were isolated. It was assumed that although details would differ, the outline of gene expression in eukaryotes would be similar. The world of biology was in for a shock with the isolation of the first genes from eukaryotic organisms.

Eukaryotic genes may contain interruptions

Many eukaryotic genes appeared to contain sequences that were not represented in the mRNA. It is hard to exaggerate how unexpected this finding was. A basic tenet of molecular biology based on *E. coli* was that a gene was *colinear* with its protein product, that is, the sequence of bases in the gene corresponds to the sequence of bases in the mRNA, which in turn corresponds to the sequence of amino acids in the protein.

In the case of eukaryotes, genes can be interrupted by sequences that are not represented in the mRNA and the protein. The term "split genes" was used at the time, but the nomenclature that has stuck describes the unexpected nature of these sequences. We call the noncoding DNA that interrupts the sequence of the gene "intervening sequences," or **introns,** and we call the coding sequences **exons** because they are expressed (figure 15.12).

The spliceosome is the splicing organelle

It is still true that the mature eukaryotic mRNA is colinear with its protein product, but a gene that contains introns is not. Imagine looking at an interstate highway from a satellite. Scattered randomly along the thread of concrete would be cars, some moving in clusters, others individually; most of the road

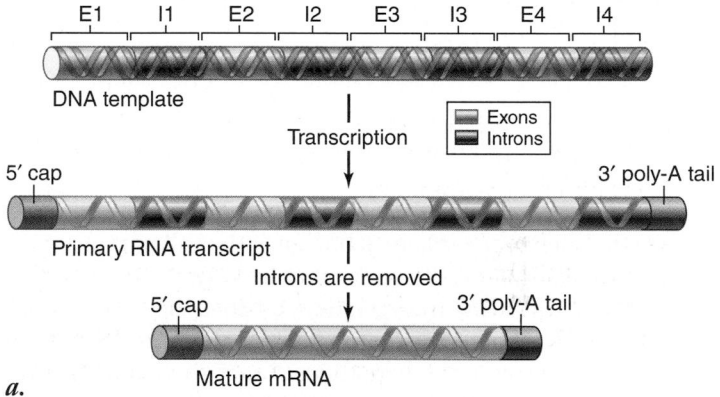

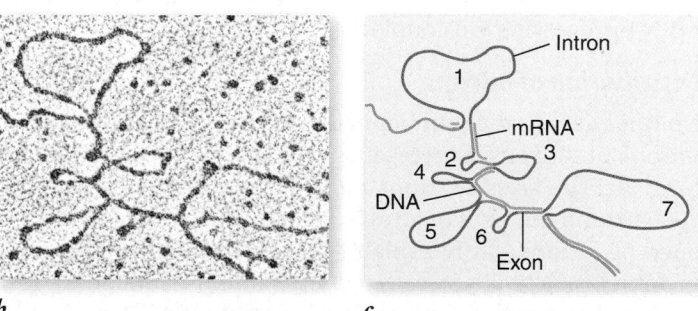

Figure 15.12 Eukaryotic genes contain introns and exons. *a.* Eukaryotic genes contain sequences that form the coding sequence called exons and intervening sequences called introns. *b.* An electron micrograph showing hybrids formed with the mRNA and the DNA of the ovalbumin gene, which has seven introns. Introns within the DNA sequence have no corresponding sequence in the mRNA and thus appear as seven loops. *c.* A schematic drawing of the micrograph.

 Data analysis Illustrate how the gene in figure 15.12 could encode multiple transcripts.

would be bare. That is what a eukaryotic gene is like—scattered exons embedded within much longer sequences of introns.

In humans, only 1 to 1.5% of the genome is devoted to the exons that encode proteins; 24% is devoted to the noncoding introns within which these exons are embedded.

The splicing reaction

The obvious question is—How do eukaryotic cells deal with the noncoding introns? The answer is that the primary transcript is cut and put back together to produce the mature mRNA. The latter process is referred to as **pre-mRNA splicing,** and it occurs in the nucleus prior to the export of the mRNA to the cytoplasm.

The intron-exon junctions are recognized by **small nuclear ribonucleoprotein particles,** called **snRNPs** (pronounced "snurps"). The snRNPs are complexes composed of snRNA and protein. These snRNPs then cluster together with other associated proteins to form a larger complex called the **spliceosome,** which is responsible for the splicing, or removal, of the introns.

For splicing to occur accurately, the spliceosome must be able to recognize intron-exon junctions. Introns all begin with the same 2-base sequence and end with another 2-base sequence that tags them for removal. In addition, within the intron there is a conserved A nucleotide, called the *branch point,* which is important for the splicing reaction (figure 15.13).

The splicing process begins with cleavage of the 5′ end of the intron. This 5′ end becomes attached to the 2′ OH of the branch point A, forming a branched structure called a *lariat* due to its resemblance to a cowboy's lariat in a rope (see figure 15.13). The 3′ end of the first exon is then used to displace the 3′ end of the intron, joining the two exons together and releasing the intron as a lariat.

The processes of transcription and RNA processing do not occur in a linear sequence, but are rather all part of a concerted process that produces the mature mRNA. The capping reaction occurs during transcription, as does the splicing process. The RNA polymerase II enzyme itself helps to recruit the other factors necessary for modification of the primary transcript, and in this way the process of transcription and pre-mRNA processing are coupled.

Distribution of introns

No rules govern the number of introns per gene or the sizes of introns and exons. Some genes have no introns; others may have 50. The sizes of exons range from a few nucleotides to 7500 nt, and the sizes of introns are equally variable. The presence of introns partly explains why so little of a eukaryotic genome is actually composed of "coding sequences" (see chapter 18 for results from the Human Genome Project).

One explanation for the existence of introns suggests that exons represent functional domains of proteins, and that the intron–exon arrangements found in genes represent the shuffling of these functional units over long periods of evolutionary time. This hypothesis, called *exon shuffling,* was proposed soon after the discovery of introns and has been the subject of much debate over the years.

The recent flood of genomic data has shed light on this issue by allowing statistical analysis of the placement of introns and on intron–exon structure. This analysis has provided support for the exon shuffling hypothesis for many genes; however, it is also clearly

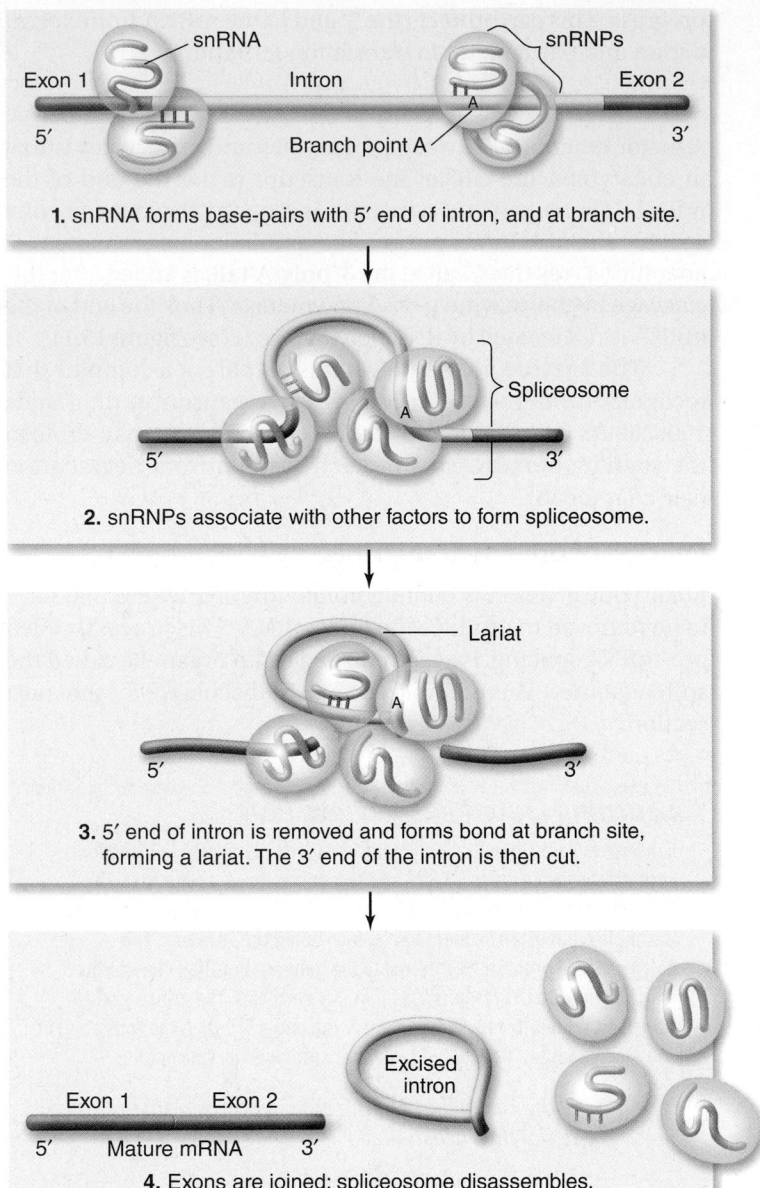

1. snRNA forms base-pairs with 5′ end of intron, and at branch site.

2. snRNPs associate with other factors to form spliceosome.

3. 5′ end of intron is removed and forms bond at branch site, forming a lariat. The 3′ end of the intron is then cut.

4. Exons are joined; spliceosome disassembles.

Figure 15.13 Pre-mRNA splicing by the spliceosome.
Particles called snRNPs contain snRNA that interacts with the 5′ end of an intron and with a branch site internal to the intron. Several snRNPs come together with other proteins to form the spliceosome. As the intron forms a loop, the 5′ end is cut and linked to a site near the 3′ end of the intron. The intron forms a lariat that is excised, and the exons are spliced together. The spliceosome then disassembles and releases the spliced mRNA.

not universal, because all proteins do not show this kind of pattern. It is possible that introns do not have a single origin, and therefore cannot be explained by a single hypothesis.

Splicing can produce multiple transcripts from the same gene

One consequence of the splicing process is greater complexity in gene expression in eukaryotes. A single primary transcript can be spliced into different mRNAs by the inclusion of different sets of exons, a process called **alternative splicing.**

Evidence indicates that the normal pattern of splicing is important to an organism's function. Up to half of known human genetic disorders may be due to altered splicing. Mutations in the signals for splicing can introduce new splice sites or can abolish normal patterns of splicing. (In chapter 16 we consider how alternative splicing can be used to regulate gene expression.)

Although many specific cases of alternative splicing have been documented, the availability of the human genome and high throughput systems to analyze transcription have led to a flood of data comparing transcripts from different tissues to the genome. This has led to continuously increasing estimates of the frequency of alternative transcripts. If the latest estimates hold up, the conclusion is that alternative splicing is essentially universal. Two different groups arrived at estimates of more than 90% of human genes being alternatively spliced with more than 80% of these having minor isoforms that make up more than 15% of mRNAs for the gene.

It is important to note that these analyses are global surveys using next generation sequencing methods to analyze RNA populations from different tissues. The possible functions of the protein products of these splice variants have been investigated for only a small fraction of the potentially spliced genes. These analyses, however, do explain how the 25,000 genes of the human genome can encode the more than 100,000 different proteins reported to exist in human cells. The emerging field of proteomics addresses the number and functioning of proteins encoded by the human genome.

Learning Outcomes Review 15.5

In prokaryotes, genes appear to be colinear with their protein products. Eukaryotic genes, by contrast, contain exon regions, which are expressed, and intron sequences, which interrupt the exons. The introns are removed by the spliceosome in a process that leaves the exons joined together. Alternative splicing can generate different mRNAs, and thus different proteins, from the same gene. Recent estimates are that as many as half of human genes may be alternatively spliced.

■ *What advantages would alternative splicing confer on an organism?*

15.6 The Structure of tRNA and Ribosomes

Learning Outcomes

1. **Explain why the tRNA charging reaction is critical to translation.**
2. **Identify the tRNA-binding sites in the ribosome.**

The ribosome is the key organelle in translation, but it also requires the participation of mRNA, tRNA, and a host of other factors. Critical to this process is the interaction of the ribosomes with tRNA and mRNA. To understand this, we first examine the structure of the tRNA adapter molecule and the ribosome itself.

Aminoacyl-tRNA synthetases attach amino acids to tRNA

Each amino acid must be attached to a tRNA with the correct anticodon for protein synthesis to proceed. This covalent attachment is accomplished by the action of activating enzymes called **aminoacyl-tRNA synthetases.** One of these enzymes is present for each of the 20 common amino acids.

tRNA structure

Transfer RNA is a bifunctional molecule that must be able to interact with mRNA and with amino acids. The structure of tRNAs is highly conserved in all living systems, and it can be formed into a cloverleaf type of structure based on intramolecular base-pairing that produces double-stranded regions. This primary structure is then folded in space to form an L-shaped molecule that has two functional ends: the **acceptor stem** and the **anticodon loop** (figure 15.14).

2D "Cloverleaf" Model	3D Ribbon-like Model	3D Space-filled Model	Icon
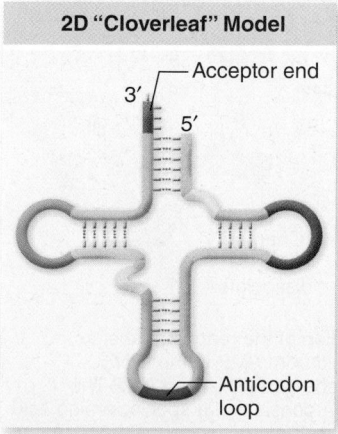	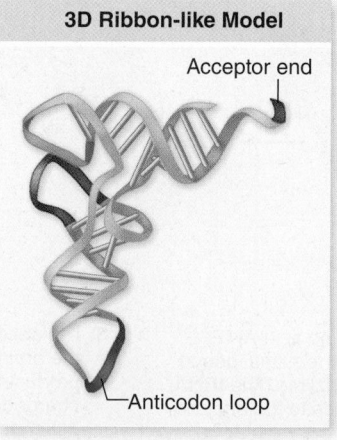	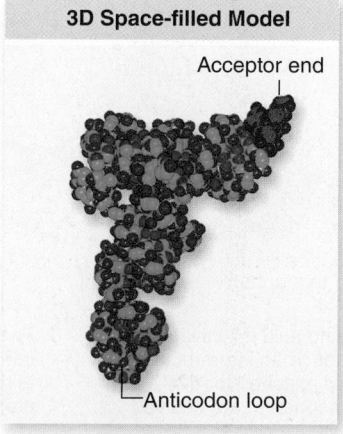	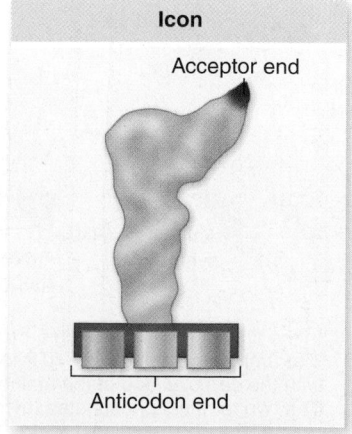

Figure 15.14 The structure of tRNA. Base-pairing within the molecule creates three stem and loop structures in a characteristic cloverleaf shape. The loop at the bottom of the cloverleaf contains the anticodon sequence, which can base-pair with codons in the mRNA. Amino acids are attached to the free, single-stranded —OH end of the acceptor stem. In its final three-dimensional structure, the loops of tRNA are folded into the final L-shaped structure.

The acceptor stem is the 3′ end of the molecule, which always ends in 5′ CCA 3′. The amino acid is attached to this end of the molecule. The anticodon loop is the bottom loop of the cloverleaf, and it can base-pair with codons in mRNA.

The charging reaction

The aminoacyl-tRNA synthetases must be able to recognize specific tRNA molecules as well as their corresponding amino acids. Although 61 codons code for amino acids, there are actually not 61 tRNAs in cells, although the number varies from species to species. Therefore, some aminoacyl-tRNA synthetases must be able to recognize more than one tRNA—but each recognizes only a single amino acid.

The reaction catalyzed by the enzymes is called the tRNA **charging reaction,** and the product is an amino acid joined to a tRNA, now called a *charged tRNA.* An ATP molecule provides energy for this endergonic reaction. The charged tRNA produced by the reaction is an activated intermediate that can undergo the peptide bond-forming reaction without an additional input of energy.

The charging reaction joins the acceptor stem to the carboxyl terminus of an amino acid (figure 15.15). Keeping this directionality in mind is critical to understanding the function of the ribosome, because each peptide bond will be formed between the amino group of one amino acid and the carboxyl group of another amino acid.

The correct attachment of amino acids to tRNAs is important because the ribosome does not verify this attachment. Ribosomes can only ensure that the codon–anticodon pairing is correct. In an elegant experiment, cysteine was converted chemically to alanine after the charging reaction, when the amino acid was already attached to tRNA. When this charged tRNA was used in an in vitro protein synthesis system, alanine was incor-

porated in the place of cysteine, showing that the ribosome cannot "proofread" the amino acids attached to tRNA.

In a very real sense, therefore, the charging reaction is the actual translation step; amino acids are incorporated into a peptide based solely on the tRNA anticodon and its interaction with the mRNA.

The ribosome has multiple tRNA-binding sites

The synthesis of any biopolymer can be broken down into initiation, elongation, and termination—you have seen this division for DNA replication as well as for transcription. In the case of translation, or protein synthesis, all three of these steps take place on the ribosome, a large macromolecular assembly consisting of rRNA and proteins. Details of the process by which the two ribosome subunits are assembled during initiation are described shortly.

For the ribosome to function it must be able to bind to at least two charged tRNAs at once so that a peptide bond can be formed between their amino acids, as described in the previous overview. The bacterial ribosome contains three binding sites, summarized in figure 15.16:

- The **P site** (peptidyl) binds to the tRNA attached to the growing peptide chain.
- The **A site** (aminoacyl) binds to the tRNA carrying the next amino acid to be added.
- The **E site** (exit) binds the tRNA that carried the previous amino acid added (see figure 15.16).

Transfer RNAs move through these sites successively during the process of elongation. Relative to the mRNA, the sites are arranged 5′ to 3′ in the order E, P, and A. The incoming

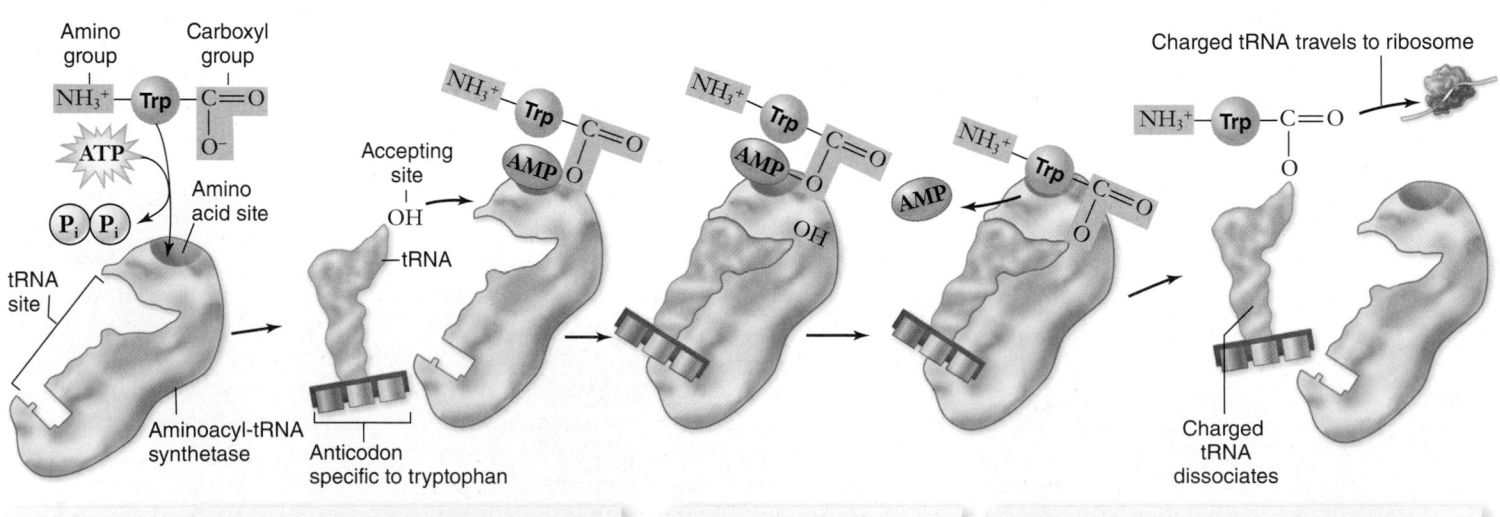

1. In the first step of the reaction, the amino acid is activated. The amino acid reacts with ATP to produce an intermediate with the carboxyl end of the amino acid attached to AMP. The two terminal phosphates (pyrophosphates) are cleaved from ATP in this reaction.

2. The amino acid-AMP complex remains bound to the enzyme. The tRNA next binds to the enzyme.

3. The second step of the reaction transfers the amino acid from AMP to the tRNA, producing a charged tRNA and AMP. The charged tRNA consists of a specific amino acid attached to the 3′ acceptor stem of its RNA.

Figure 15.15 tRNA charging reaction. There are 20 different aminoacyl-tRNA synthetase enzymes each specific for one amino acid, such as tryptophan (Trp). The enzyme must also recognize and bind to the tRNA molecules with anticodons specifying that amino acid, ACC for tryptophan. The reaction uses ATP and produces an activated intermediate that will not require further energy for peptide bond formation.

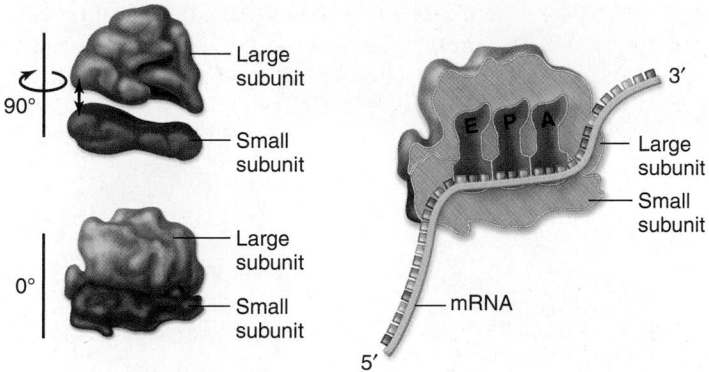

Large subunit

Small subunit

90°

Large subunit

Small subunit

0°

E P A

Large subunit

Small subunit

mRNA

3′

5′

Figure 15.16 Ribosomes have two subunits. Ribosome subunits come together and apart as part of a ribosome cycle. The smaller subunit fits into a depression on the surface of the larger one. Ribosomes have three tRNA-binding sites: aminoacyl site (A), peptidyl site (P), and empty site (E).

charged tRNAs enter the ribosome at the A site, transit through the P site, and then leave via the E site.

The ribosome has both decoding and enzymatic functions

The two functions of the ribosome involve decoding the transcribed message and forming peptide bonds. The decoding function resides primarily in the small subunit of the ribosome. The formation of peptide bonds requires the enzyme **peptidyl transferase,** which resides in the large subunit.

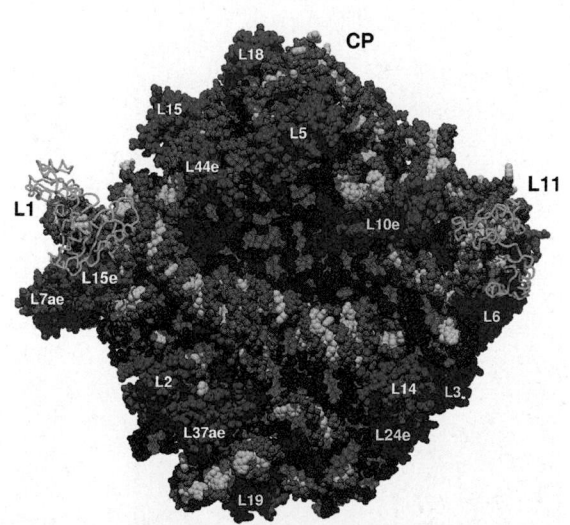

CP

L18

L15

L5

L44e

L11

L1

L10e

L15e

L7ae

L6

L2

L14 L3

L37ae

L24e

L19

Figure 15.17 3-D structure of prokaryotic ribosome. The complete atomic structure of a prokaryotic large ribosomal subunit has been determined at 2.4-Å resolution. Bases of RNA are white, the polynucleotide backbone is red, and proteins are blue. The faces of each ribosomal subunit are lined with rRNA such that their interaction with tRNAs, amino acids, and mRNA all involve rRNA. Proteins are absent from the active site but abundant everywhere on the surface. The proteins stabilize the structure by interacting with adjacent RNA strands.

Our view of the ribosome has changed dramatically over time. Initially, molecular biologists assumed that the proteins in the ribosome carried out its function and that the rRNA was a structural scaffold necessary to hold the proteins in the correct position. Now this view has mostly been reversed; the ribosome is seen instead as rRNAs that are held in place by proteins. The faces of the two subunits that interact with each other are lined with rRNA, and the parts of both subunits that interact with mRNA, tRNA, and amino acids are also primarily rRNA (figure 15.17). It is now thought that the peptidyl transferase activity resides in an rRNA in the large subunit.

Learning Outcomes Review 15.6

Transfer RNA has two functional regions, one that bonds with an amino acid, and the other that can base-pair with mRNA. The tRNA charging reaction joins the carboxyl end of an amino acid to the 3′ acceptor stem of its tRNA; without charged tRNAs, translation cannot take place. This reaction is catalyzed by 20 different aminoacyl-tRNA synthetases, one for each amino acid. The ribosome has three different binding sites for tRNA, one for the tRNA adding to the growing peptide chain (P site), one for the next charged tRNA (A site), and one for the previous tRNA, which is now without an amino acid (E site). The ribosome can be thought of as having both a decoding function and an enzymatic function.

■ *What would be the effect on translation of a mutant tRNA that has an anticodon complementary to a STOP codon?*

15.7 The Process of Translation

Learning Outcomes

1. *Describe the process of translation initiation.*
2. *Explain the elongation cycle.*
3. *Compare translation on the RER and in the cytoplasm.*

The process of translation is one of the most complex and energy-expensive tasks that cells perform. An overview of the process, as you saw earlier, is perhaps deceptively simple: The mRNA is threaded through the ribosome, while tRNAs carrying amino acids bind to the ribosome, where they interact with mRNA by base-pairing with the mRNA's codons. The ribosome and tRNAs position the amino acids such that peptide bonds can be formed between each new amino acid and the growing polypeptide.

Initiation requires accessory factors

As mentioned earlier, the start codon is AUG, which also encodes the amino acid methionine. The ribosome usually uses the first AUG it encounters in an mRNA strand to signal the start of translation.

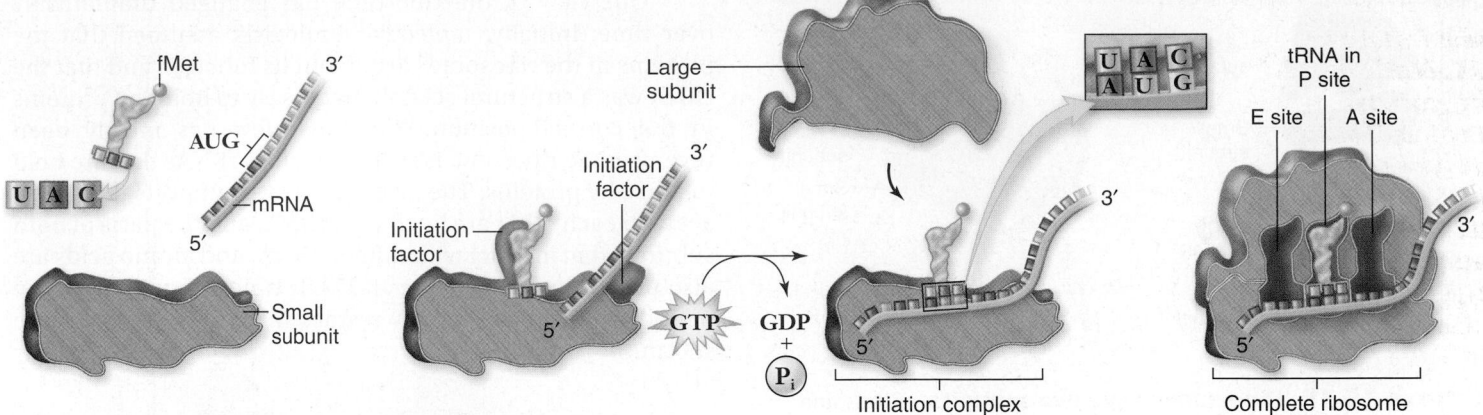

Figure 15.18 Initiation of translation. In prokaryotes, initiation factors play key roles in positioning the small ribosomal subunit, the initiator tRNA^fMet, and the mRNA. When the tRNA^fMet is positioned over the first AUG codon of the mRNA, the large ribosomal subunit binds, forming the E, P, and A sites where successive tRNA molecules bind to the ribosomes, and polypeptide synthesis begins. Ribosomal subunits are shown as a cutaway sectioned through the middle.

Prokaryotic initiation

In prokaryotes, the **initiation complex** includes a special **initiator tRNA** molecule charged with a chemically modified methionine, *N-formylmethionine*. The initiator tRNA is shown as tRNA^fMet. The initiation complex also includes the small ribosomal subunit and the mRNA strand (figure 15.18). The small subunit is positioned correctly on the mRNA due to a conserved sequence in the 5′ end of the mRNA called the **ribosome-binding sequence (RBS)** that is complementary to the 3′ end of a small subunit rRNA.

A number of initiation factors mediate this interaction of the ribosome, mRNA, and tRNA^fMet to form the initiation complex. These factors are involved in initiation only and are not part of the ribosome.

Once the complex of mRNA, initiator tRNA, and small ribosomal subunit is formed, the large subunit is added, and

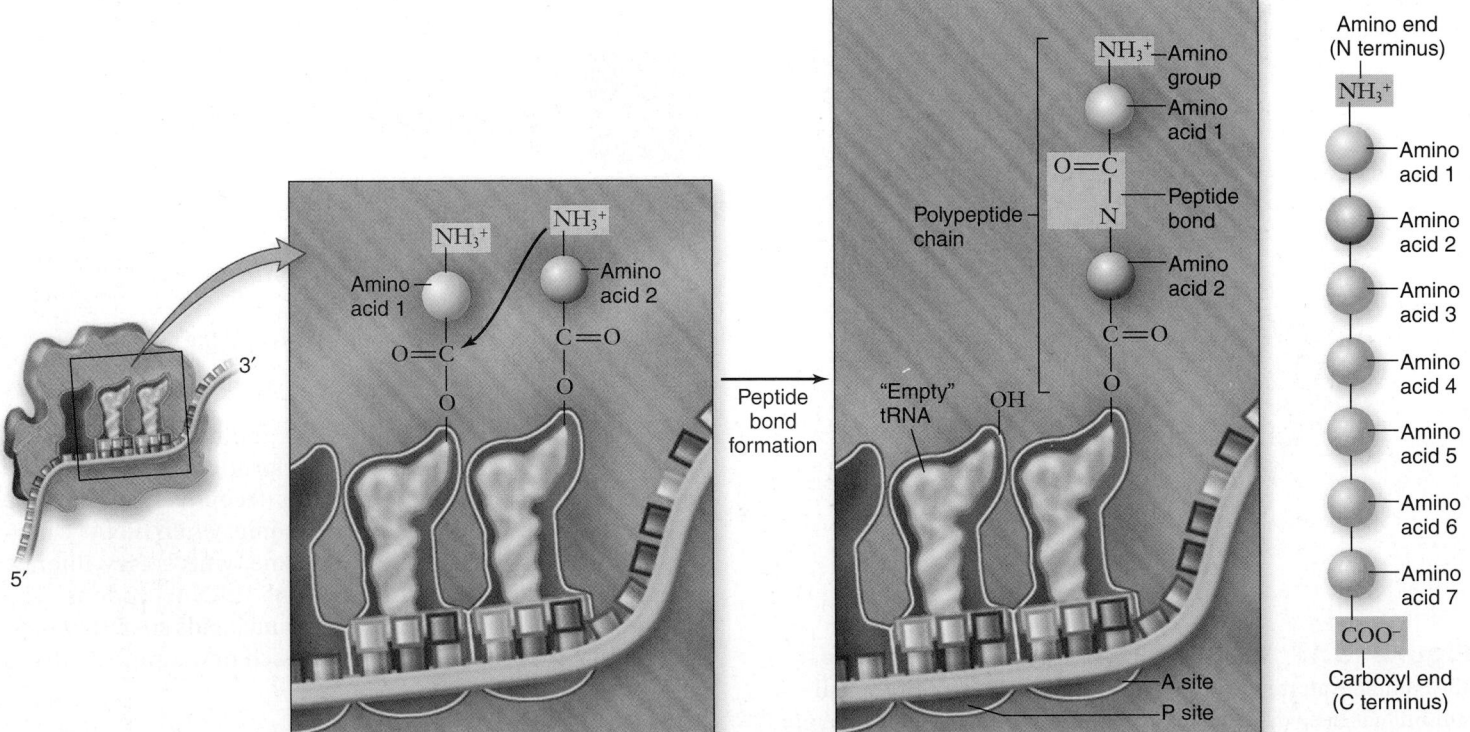

Figure 15.19 Peptide bond formation. Peptide bonds are formed between a "new" charged tRNA in the A site and the growing chain attached to the tRNA in the P site. The bond forms between the amino group of the new amino acid and the carboxyl group of the growing chain. This breaks the bond between the growing chain and its tRNA, transferring it to the A site as the new amino acid remains attached to its tRNA.

translation can begin. With the formation of the complete ribosome, the initiator tRNA is bound to the P site with the A site empty.

Eukaryotic initiation

Initiation in eukaryotes is similar, although it differs in two important ways. First, in eukaryotes, the initiating amino acid is methionine rather than *N*-formylmethionine. Second, the initiation complex is far more complicated than in prokaryotes, containing nine or more protein factors, many consisting of several subunits. Eukaryotic mRNAs also lack an RBS. The small subunit binds to the mRNA initially by binding to the 5′ cap of the mRNA.

Elongation adds successive amino acids

When the entire ribosome is assembled around the initiator tRNA and mRNA, the second charged tRNA can be brought to the ribosome and bind to the empty A site. This requires an **elongation factor** called **EF-Tu,** which binds to the charged tRNA and to GTP.

A peptide bond can then form between the amino acid of the initiator tRNA and the newly arrived charged tRNA in the A site. The geometry of this bond relative to the two charged tRNAs is critical to understanding the process. Remember that an amino acid is attached to a tRNA by its carboxyl terminus. The peptide bond is formed between the amino end of the incoming amino acid (in the A site) and the carboxyl end of the growing chain (in the P site) (figure 15.19).

The addition of successive amino acids is a series of events that occur in a cyclic fashion. Figure 15.20 shows the details of the elongation cycle.

1. **Matching tRNA anticodon with mRNA codon.** Each new charged tRNA comes to the ribosome bound to EF-Tu and GTP. The charged tRNA binds to the A site if its anticodon is complementary to the mRNA codon in the A site.

   After binding, GTP is hydrolyzed, and EF-Tu–GDP dissociates from the ribosome where it is recycled by another factor. This two-step binding and hydrolysis of GTP is thought to increase the accuracy of translation.

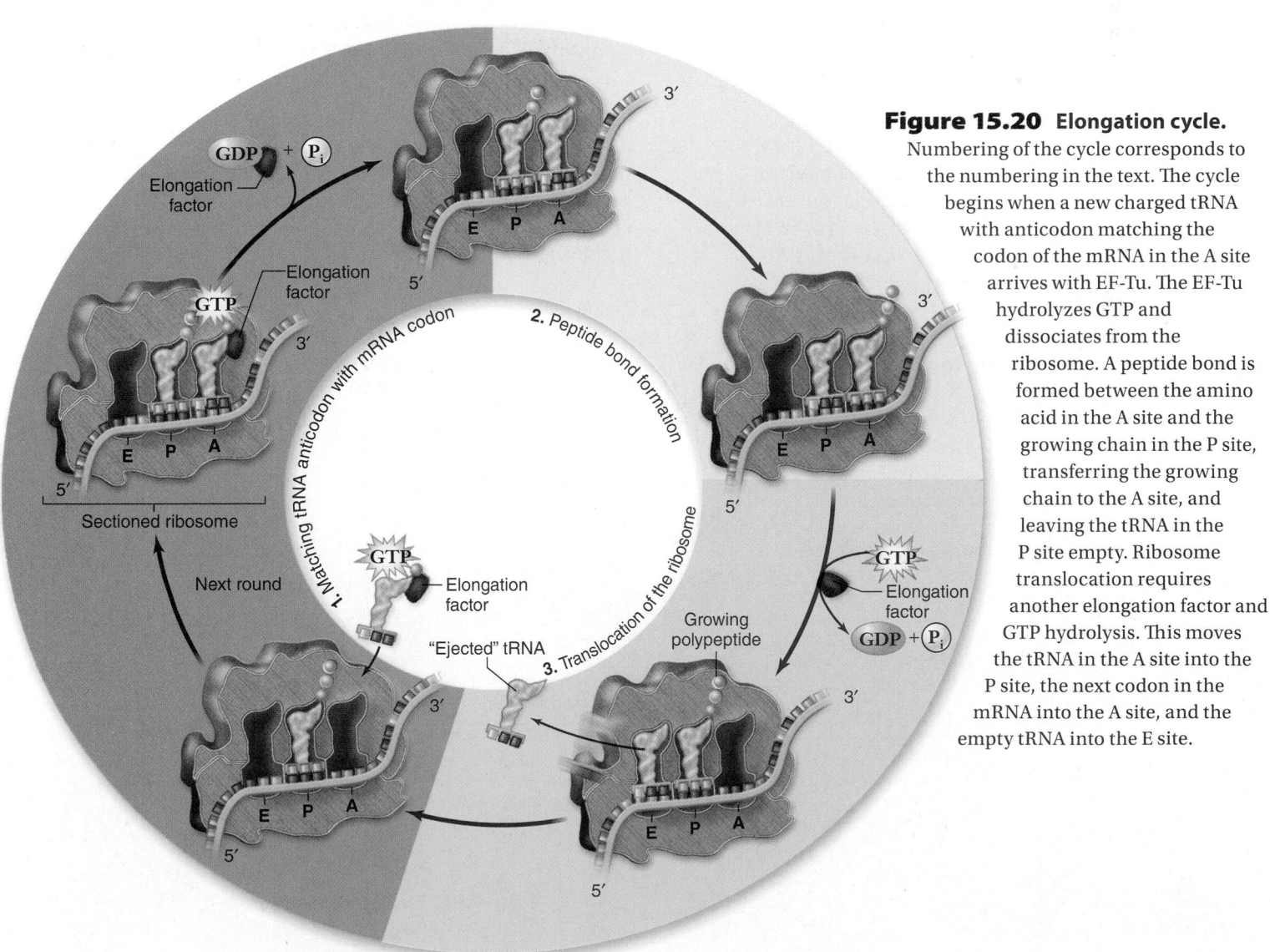

Figure 15.20 Elongation cycle. Numbering of the cycle corresponds to the numbering in the text. The cycle begins when a new charged tRNA with anticodon matching the codon of the mRNA in the A site arrives with EF-Tu. The EF-Tu hydrolyzes GTP and dissociates from the ribosome. A peptide bond is formed between the amino acid in the A site and the growing chain in the P site, transferring the growing chain to the A site, and leaving the tRNA in the P site empty. Ribosome translocation requires another elongation factor and GTP hydrolysis. This moves the tRNA in the A site into the P site, the next codon in the mRNA into the A site, and the empty tRNA into the E site.

chapter **15** *Genes and How They Work* **295**

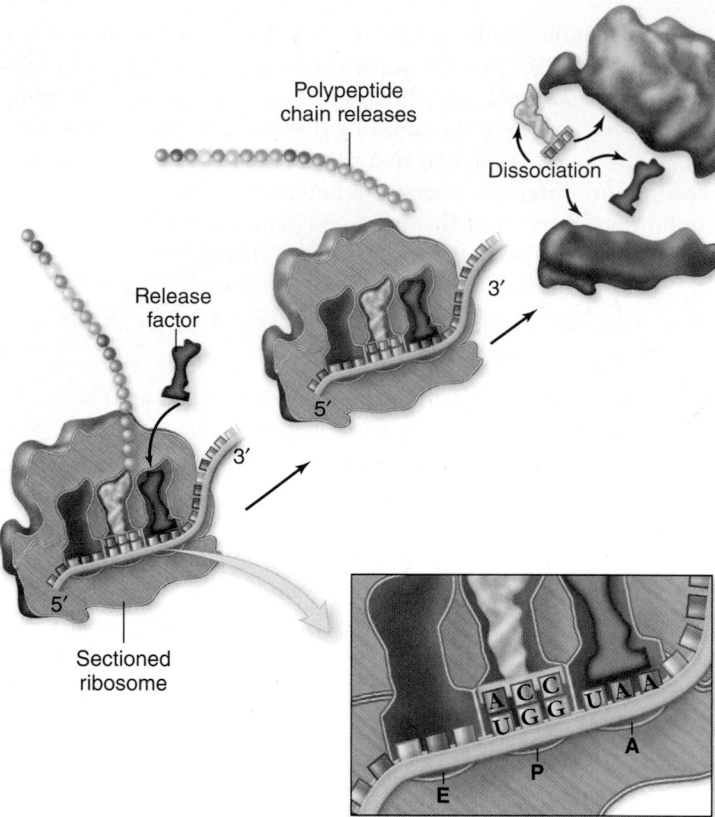

Figure 15.21 Termination of protein synthesis. There is no tRNA with an anticodon complementary to any of the three termination signal codons. When a ribosome encounters a termination codon, it stops translocating. A specific protein release factor facilitates the release of the polypeptide chain by breaking the covalent bond that links the polypeptide to the P site tRNA.

2. **Peptide bond formation.** Peptidyl transferase, located in the large subunit, catalyzes the formation of a peptide bond between the amino group of the amino acid in the A site and the carboxyl group of the growing chain. This also breaks the bond between the growing chain and the tRNA in the P site leaving it empty (no longer charged). The overall result of this is to transfer the growing chain to the tRNA in the A site.

3. **Translocation of the ribosome.** After the peptide bond has been formed, the ribosome moves relative to the mRNA and the tRNAs. The next codon in the mRNA shifts into the A site, and the tRNA with the growing chain moves to the P site. The uncharged tRNA formerly in the P site is now in the E site, and it will be ejected in the next cycle. This translocation step requires the accessory factor EF-G and the hydrolysis of another GTP.

This elongation cycle continues with each new amino acid added. The ribosome moves down the mRNA in a 5′-to-3′ direction, reading successive codons. The tRNAs move through the ribosome in the opposite direction, from the A site to the P site and finally the E site, before they are ejected as empty tRNAs, which can be charged with another amino acid and then used again.

Wobble pairing

As mentioned, there are fewer tRNAs than codons. This situation is easily rationalized because the pairing between the 3′ base of the codon and the 5′ base of the anticodon is less stringent than normal. In some tRNAs, the presence of modified bases with less accurate pairing in the 5′ position of the

Figure 15.22 Synthesis of proteins on RER. Proteins that are synthesized on RER arrive at the ER because of sequences in the peptide itself. A signal sequence in the amino terminus of the polypeptide is recognized by the signal recognition particle (SRP). This complex docks with a receptor associated with a channel in the ER. The peptide passes through the channel into the lumen of the ER as it is synthesized.

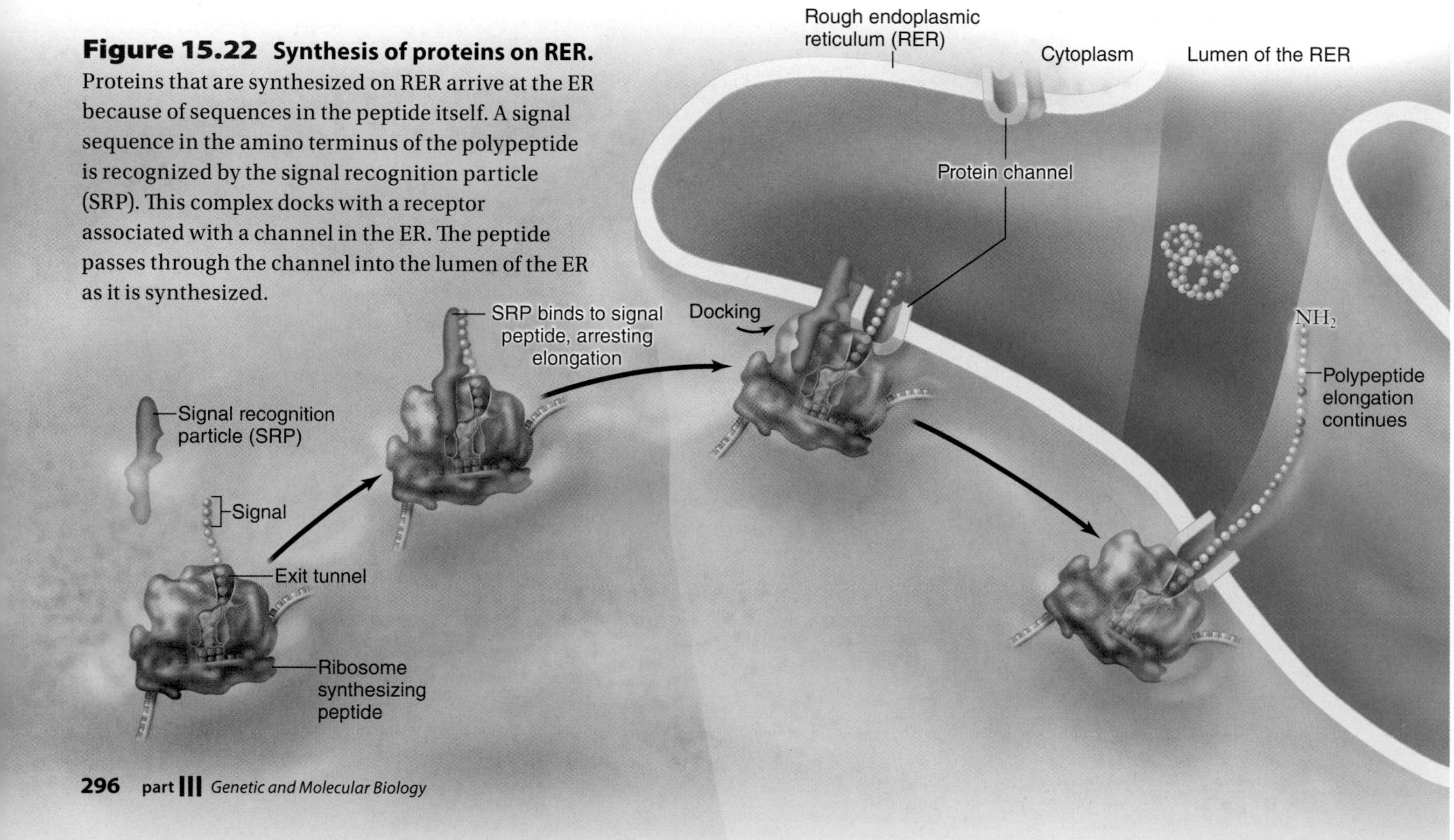

anticodon enhances this flexibility. This effect is referred to as **wobble pairing** because these tRNAs can "wobble" a bit on the mRNA, so that a single tRNA can "read" more than one codon in the mRNA.

> **? Inquiry question** How is the wobble phenomenon related to the number of tRNAs and the degeneracy of the genetic code?

Termination requires accessory factors

Elongation continues in this fashion until a chain-terminating stop codon is reached (for example, UAA in figure 15.21). These stop codons do not bind to tRNA; instead, they are recognized by release factors, proteins that release the newly made polypeptide from the ribosome.

Proteins may be targeted to the ER

In eukaryotes, translation can occur either in the cytoplasm or on the RER. Proteins that are translated on the RER are targeted there based on their own initial amino acid sequence. The ribosomes found on the RER are actively translating and are not permanently bound to the ER.

A polypeptide that starts with a short series of amino acids called a **signal sequence** is specifically recognized and bound by a cytoplasmic complex of proteins called the **signal recognition particle (SRP).** The complex of signal sequence and SRP is in turn recognized by a receptor protein in the ER membrane. The binding of the ER receptor to the signal sequence/SRP complex holds the ribosome engaged in translation of the protein on the ER membrane, a process called *docking* (figure 15.22).

As the protein is assembled, it passes through a channel formed by the docking complex and into the interior ER compartment, the cisternal space. This is the basis for the docking metaphor—the ribosome is not actually bound to the ER itself, but with the newly synthesized protein entering the ER, the ribosome is like a boat tied to a dock with a rope.

The basic mechanism of protein translocation across membranes by the SRP and its receptor and channel complex has been conserved across all three cell types: eukaryotes, bacteria, and archaea. Given that only eukaryotic cells have an endomembrane system, this universality may seem curious; however, bacteria and archaea both export proteins through their plasma membrane, and the mechanism used is similar to the way in which eukaryotes move proteins into the cisternal space of the ER.

Once within the ER cisternal space, or lumen, the newly synthesized protein can be modified by the addition of sugars (glycosylation) and transported by vesicles to the Golgi apparatus (see chapter 4). This is the beginning of the protein-trafficking pathway that can lead to other intracellular targets, to incorporation into the plasma membrane, or to release outside of the cell itself.

Learning Outcomes Review 15.7

Translation initiation involves the interaction of the small ribosomal subunit with mRNA and a charged initiator tRNA. The elongation cycle involves bringing in new charged tRNAs to the ribosome's A site, forming peptide bonds between amino acids, and translocating the ribosome along the mRNA chain. The tRNAs transit through the ribosome from A to P to E sites during the process. In eukaryotes, signal sequences of a newly forming polypeptide may target it and its ribosome to be moved to the RER. Polypeptides formed on the RER enter the cisternal space rather than being released into the cytoplasm.

■ *What stages of translation require energy?*

15.8 Summarizing Gene Expression

Because of the complexity of the process of gene expression, it is worth stepping back to summarize some key points:

■ The process of gene expression converts information in the genotype into the phenotype.

■ A copy of the gene in the form of mRNA is produced by transcription, and the mRNA is used to direct the synthesis of a protein by translation.

■ Both transcription and translation can be broken down into initiation, an elongation cycle, and termination—processes that produce their respective polymers. (The same is true for DNA replication.)

■ Eukaryotic gene expression is much more complex than that of prokaryotes.

The structure of eukaryotic genes with interrupted coding sequences complicates both the process of gene expression and the nature of genetic information. It means that processing must occur between transcription and translation, and that one gene can produce multiple messages. Transcription in eukaryotes also takes place in the nucleus, whereas translation takes place in the cytoplasm. This necessitates that the mRNA be transported through nuclear pores to the cytoplasm prior to translation. The entire eukaryotic process is summarized in figure 15.23, and differences in gene expression between prokaryotes and in eukaryotes are summarized in table 15.2.

Learning Outcome Review 15.8

The greater complexity of eukaryotic gene expression is related to the functional organization of the cell, with DNA in the nucleus and ribosomes in the cytoplasm. The differences in gene expression between prokaryotes and eukaryotes is mainly in detail, but some differences have functional significance.

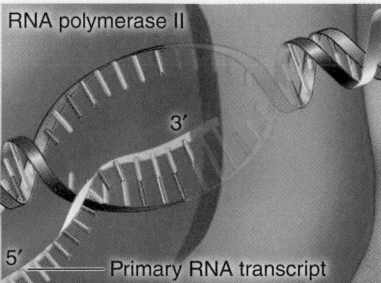

1. RNA polymerase II in the nucleus copies one strand of the DNA to produce the primary transcript.

RNA polymerase II

3′

5′ — Primary RNA transcript

Figure 15.23 An overview of gene expression in eukaryotes.

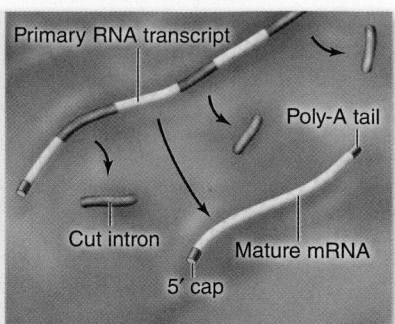

2. The primary transcript is processed by addition of a 5′ methyl-G cap, cleavage and polyadenylation of the 3′ end, and removal of introns. The mature mRNA is then exported through nuclear pores to the cytoplasm.

Primary RNA transcript

Poly-A tail

Cut intron

5′ cap

Mature mRNA

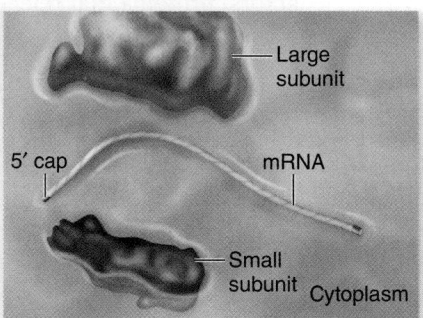

3. The 5′ cap of the mRNA associates with the small subunit of the ribosome. The initiator tRNA and large subunit are added to form an initiation complex.

Large subunit

5′ cap

mRNA

Small subunit

Cytoplasm

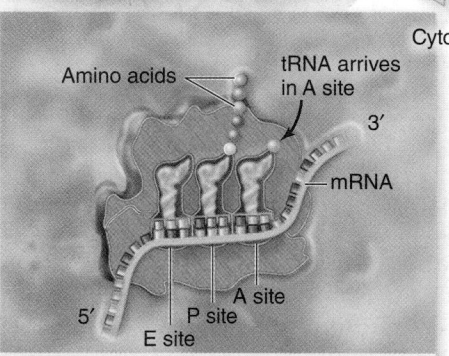

Cytoplasm

Amino acids

tRNA arrives in A site

3′

mRNA

A site

P site

E site

5′

4. The ribosome cycle begins with the growing peptide attached to the tRNA in the P site. The next charged tRNA binds to the A site with its anticodon complementary to the codon in the mRNA in this site.

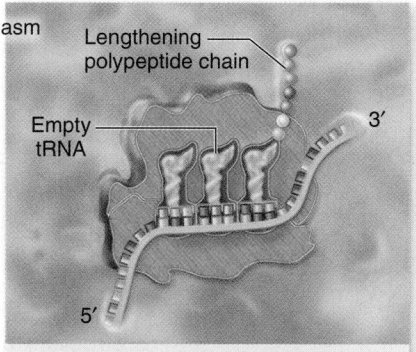

Lengthening polypeptide chain

Empty tRNA

3′

5′

5. Peptide bonds form between the amino terminus of the next amino acid and the carboxyl terminus of the growing peptide. This transfers the growing peptide to the tRNA in the A site, leaving the tRNA in the P site empty.

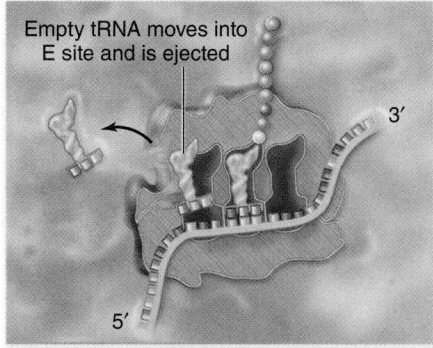

Empty tRNA moves into E site and is ejected

3′

5′

6. Ribosome translocation moves the ribosome relative to the mRNA and its bound tRNAs. This moves the growing chain into the P site, leaving the empty tRNA in the E site and the A site ready to bind the next charged tRNA.

TABLE 15.2	Differences Between Prokaryotic and Eukaryotic Gene Expression	
Characteristic	**Prokaryotes**	**Eukaryotes**
Introns	No introns, although some archaeal genes possess them.	Most genes contain introns.
Number of genes in mRNA	Several genes may be transcribed into a single mRNA molecule. Often these have related functions and form an operon, which helps coordinate regulation of biochemical pathways.	Only one gene per mRNA molecule; regulation of pathways accomplished in other ways.
Site of transcription and translation	No membrane-bounded nucleus, transcription and translation are coupled.	Transcription in nucleus; mRNA is transported to the cytoplasm for translation.
Initiation of translation	Begins at AUG codon preceded by special sequence that binds the ribosome.	Begins at AUG codon preceded by the 5′ cap (methylated GTP) that binds the ribosome.
Modification of mRNA after transcription	None; translation begins before transcription is completed. Transcription and translation are coupled.	A number of modifications while the mRNA is in the nucleus: Introns are removed and exons are spliced together; a 5′ cap is added; a poly-A tail is added.

15.9 Mutation: Altered Genes

Learning Outcomes

1. *Describe the effects of different point mutations.*
2. *Explain the nature of triplet repeat expansion.*
3. *List the different chromosomal mutations and their effects.*

Geneticists analyze the function of genes by looking at altered forms; that is, by finding or inducing mutations. The behavior of the protein produced from the altered gene provides information about how the normal protein functions. In terms of the organism, however, mutations are usually negative; most mutations have deleterious effects on the phenotype of the organism. In chapter 13, you saw how a number of genetic diseases, such as sickle cell anemia, are due to single-base changes. We now take this one step farther by considering how the DNA itself is altered. This ranges from the alteration of a single base, to the loss of genetic material (deletion), to the loss of an entire chromosome. Changing a single base can result in an amino acid substitution that can lead to a debilitating clinical phenotype. This is illustrated for the case of sickle cell anemia in figure 15.24. In the sickle cell allele, a single A is changed to a T, resulting in a glutamic acid being replaced with a valine. The substitution of nonpolar valine causes the β-chains to aggregate into polymers, which consequently alters the shape of the cells, leading to the disease state.

Point mutations affect a single site in the DNA

A mutation that alters a single base is termed a **point mutation.** The mutation can be either the substitution of one base for another, or the deletion or addition of a single base (or a small number of bases) (figure 15.25).

Base substitution

The substitution of one base pair for another in DNA is called a **base substitution mutation.** Because of the degenerate nature of the genetic code, base substitution may or may not alter the amino acid encoded. If the new codon from the base substitution still encodes the same amino acid, we say the mutation is *silent* (figure 15.25*b*). When base substitution changes an amino acid in a protein, it is also called a **missense mutation** as the "sense" of the codon produced after transcription of the mutant gene will be altered (figure 15.25*c*). These fall into two classes, *transitions* and *transversions*. A transition does not change the type of bases in the base pair, that is, a pyrimidine is substituted for a pyrimidine, or purine for purine. In contrast, a transversion does change the type of bases in a base pair, that is, pyrimidine to purine or the reverse. A variety of human genetic diseases, including sickle cell anemia, are caused by base substitutions.

Nonsense mutations

A special category of base substitution arises when a base is changed such that the transcribed codon is converted to a stop codon (see figure 15.25*d*). We call these **nonsense mutations** because the mutation does not make "sense" to the translation apparatus. The stop codon results in premature termination of translation and leads to a truncated protein. How short the resulting protein is depends on where a stop codon has been introduced in the gene.

Frameshift mutations

The addition or deletion of a single base has much more profound consequences than does the substitution of one base for another. These mutations are called *frameshift mutations* because they alter the reading frame in the mRNA downstream of the mutation. This class of mutations was used by Crick and Brenner, as described earlier in the chapter, to infer the nature of the genetic code.

Changing the reading frame early in a gene, and thus in its mRNA transcript, means that the majority of the protein will

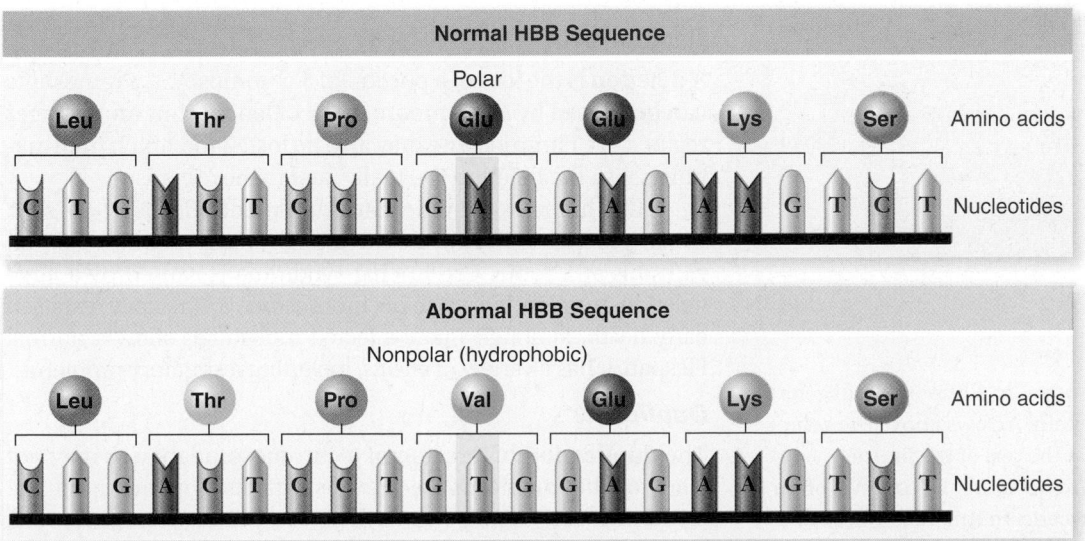
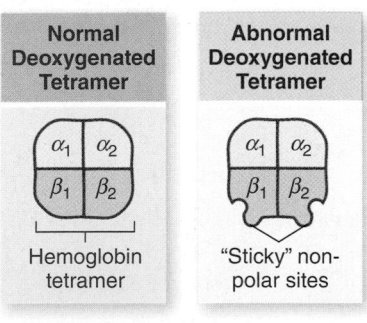
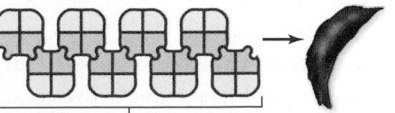

Figure 15.24 Sickle cell anemia is caused by an altered protein. Hemoglobin is composed of a tetramer of two α-globin and two β-globin chains. The sickle cell allele of the β-globin gene contains a single base change resulting in the substitution of Val for Glu. This creates a hydrophobic region on the surface of the protein that is "sticky" leading to their association into long chains that distort the shape of the red blood cells.

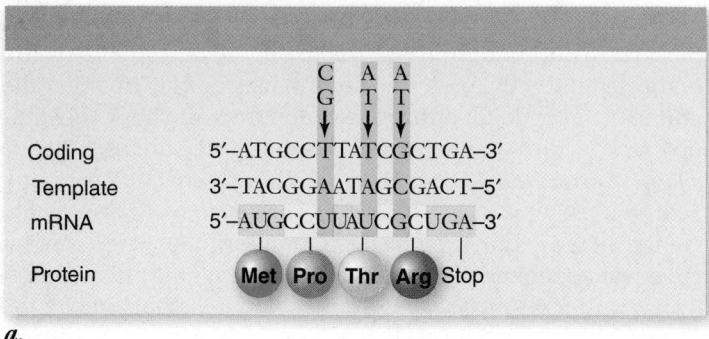

a.

Silent Mutation

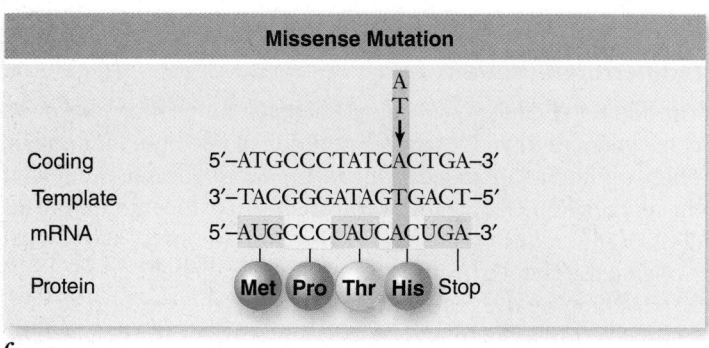

b.

Missense Mutation

Coding 5′–ATGCCCTATCACTGA–3′
Template 3′–TACGGGATAGTGACT–5′
mRNA   5′–AUGCCCUAUCACUGA–3′

Protein Met Pro Thr His Stop

c.

Nonsense Mutation

Coding 5′–ATGCCCTAACGCTGA–3′
Template 3′–TACGGGATTGCGACT–5′
mRNA   5′–AUGCCCUAACGCUGA–3′

Protein Met Pro Stop

d.

Figure 15.25 Types of mutations. *a.* A hypothetical gene is shown with encoded mRNA and protein. Arrows above the gene indicate sites of mutations described in the rest of the figure. *b.* Silent mutation. A change in the third position of a codon is often silent due to degeneracy in the genetic code. In this case T/A to C/G mutation does not change the amino acid encoded (proline). *c.* Missense mutation. The G/C to A/T mutation changes the amino acid encoded from arginine to histidine. *d.* Nonsense mutation. The T/A to A/T mutation produces a UAA stop codon in the mRNA.

be altered. Frameshifts also can cause premature termination of translation because 3 in 64 codons are stop codons, which represents a high probability in the sequence that has been randomized by the frameshift.

Triplet repeat expansion mutations

Given the long history of molecular genetics, and the relatively short time that molecular analysis has been possible on humans, it is surprising that a new kind of mutation was discovered in humans. However, one of the first genes isolated that was associated with human disease, the gene for *Huntington disease,* provided a new kind of mutation. The gene for Huntington contains a triplet sequence of DNA that is repeated, and this repeat unit is expanded in the disease allele relative to the normal allele. Since this initial discovery, at least 20 other human genetic diseases appear to be due to this mechanism. The prevalence of this kind of mutation is unknown, but at present humans and mice are the only organisms in which they have been observed, implying that they may be limited to vertebrates, or even mammals. No such mutation has ever been found in *Drosophila,* for example.

The expansion of the triplet can occur in the coding region or in noncoding transcribed DNA. In the case of Huntington disease, the repeat unit is actually in the coding region of the gene where the triplet encodes glutamine, and expansion results in a polyglutamine region in the protein. A number of other neurodegenerative disorders also show this kind of mutation. In the case of fragile-X syndrome, an inherited form of intellectual disability, the repeat is in noncoding DNA.

Chromosomal mutations change the structure of chromosomes

Point mutations affect a single site in a chromosome, but more extensive changes can alter the structure of the chromosome itself, resulting in **chromosomal mutations.** Many human cancers are associated with chromosomal abnormalities, so these are of great clinical relevance. We briefly consider possible alterations to chromosomal structure, all of which are summarized in figure 15.26.

Deletions

A **deletion** is the loss of a portion of a chromosome. Frameshifts can be caused by one or more small deletions, but much larger regions of a chromosome may also be lost. If too much information is lost, the deletion is usually fatal to the organism.

One human syndrome that is due to a deletion is *cri-du-chat,* which is French for "cry of the cat" after the noise made by children with this syndrome. Cri-du-chat syndrome is caused by a large deletion from the short arm of chromosome 5. It usually results in early death, although many affected individuals show a normal lifespan. It has a variety of effects, including respiratory problems.

Duplications

The **duplication** of a region of a chromosome may or may not lead to phenotypic consequences. Effects depend upon the location of the "breakpoints" where the duplication occurred. If the duplicated region does not lie within a gene, there may be no effect. If the duplication occurs next to the original region, it is termed a *tandem duplication.* These tandem duplications are important in the evolution of families of related genes, such as the globin family that encode the protein hemoglobin.

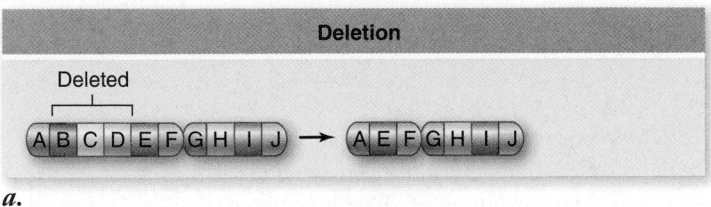

Deletion

Deleted

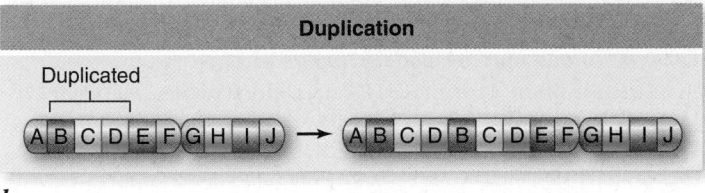

Duplication

Duplicated

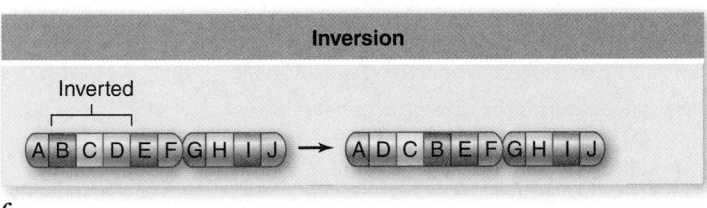

Inversion

Inverted

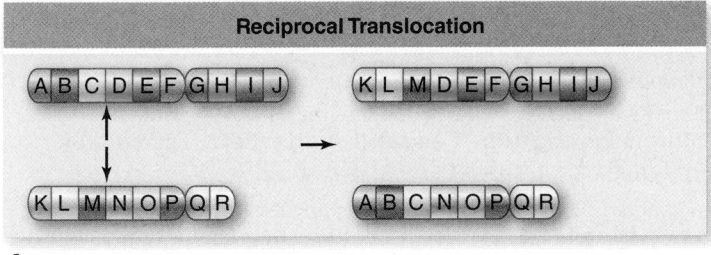

Reciprocal Translocation

a.

b.

c.

d.

Figure 15.26 Chromosomal mutations. Larger-scale changes in chromosomes are also possible. Material can be deleted *(a)*, duplicated *(b)*, and inverted *(c)*. Translocations occur when one chromosome is broken and becomes part of another chromosome. This often occurs where both chromosomes are broken and exchange material, an event called a reciprocal translocation *(d)*.

Inversions

An **inversion** results when a segment of a chromosome is broken in two places, reversed, and put back together. An inversion may not have an effect on phenotype if the sites where the inversion occurs do not break within a gene. In fact, although humans all have the "same" genome, the order of genes in all individuals in a population is not precisely the same due to inversions that occur in different lineages.

Translocations

If a piece of one chromosome is broken off and joined to another chromosome, we call this a **translocation.** Translocations are complex because they can cause problems during meiosis, particularly when two different chromosomes try to pair with each other during meiosis I.

Translocations can also move genes from one chromosomal region to another in a manner that changes the expression of genes in the region involved. Two forms of leukemia

have been shown to be associated with translocations that move oncogenes into regions of a chromosome where they are expressed inappropriately in blood cells.

Mutations are the starting point of evolution

If no changes occurred in genomes over time, then there could be no evolution. Too much change, however, is harmful to the individual with a greatly altered genome. Thus a delicate balance must exist between the amount of new variation that arises in a species and the health of individuals in the species. This topic is explored in more detail later in the book when we consider evolution and population genetics (chapter 20).

The larger scale alteration of chromosomes has also been important in evolution, although its role is poorly understood. It is clear that gene families arise by the duplication of an ancestral gene, followed by the functional divergence of the duplicated copies. It is also clear that even among closely related species, the number and arrangements of genes on chromosomes can differ. Large-scale rearrangements may have occurred.

Our view of the nature of genes has changed with new information

In this and the preceding chapters, we have seen multiple views of genes. Mendel used crosses to follow traits determined by what we now call genes. The behavior of these genes can be predicted based on the behavior of chromosomes during meiosis. Morgan and others learned to map the location of genes on chromosomes. These findings led to the view of genes as abstract entities that could be followed through generations and mapped to chromosomal locations like "beads on a string," with the beads being genes and the string the chromosome.

The original molecular analysis of genes led to the simple one-gene/one-polypeptide paradigm. This oversimplification was changed when geneticists observed the alternative splicing of eukaryotic genes, which can lead to multiple protein products from the same genetic information. Furthermore, some genes do not encode proteins at all, but only RNA, which can either be a part of the gene expression machinery (rRNA, tRNA, and other forms) or can itself act as an enzyme. Other stretches of DNA are important for regulating genes but are not expressed. All of these findings make a simple definition of genes difficult.

We are left with the rich complexity of the nature of genes, which defies simple definition. To truly understand the nature of genes we must consider both their molecular nature as well as their phenotypic expression. This brings us full circle, back to the relationship between genotype and phenotype, with a much greater appreciation for the complexity of this relationship.

Learning Outcomes Review 15.9

Point mutations (single-base changes, additions, or deletions) include missense mutations that cause substitution of one amino acid for another, nonsense mutations that halt transcription, and frameshift mutations that throw off the correct reading of codons. Triplet repeat expansion is the abnormal duplication of a codon with each round of cell division. Mutations affecting chromosomes include deletions, duplications, inversions, and translocations.

■ *Would an inversion or duplication always be expected to have a phenotype?*

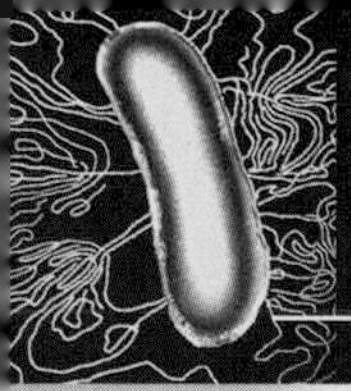

Chapter Review

15.1 The Nature of Genes

Garrod concluded that inherited disorders can involve specific enzymes.

Garrod found that alkaptonuria is due to an altered enzyme.

Beadle and Tatum showed that genes specify enzymes.

Neurospora mutants unable to synthesize arginine were found to lack specific enzymes. Beadle and Tatum advanced the "one gene/one polypeptide" hypothesis (figure 15.1).

The central dogma describes information flow in cells as DNA to RNA to protein (figure 15.2).

We call the DNA strand copied to mRNA the template (antisense) strand; the other the coding (sense) strand.

Transcription makes an RNA copy of DNA.

Translation uses information in RNA to synthesize proteins.

An adapter molecule, tRNA, is required to connect the information in mRNA into the sequence of amino acids.

RNA has multiple roles in gene expression.

15.2 The Genetic Code

The code is read in groups of three.

Crick and Brenner showed that the code is nonoverlapping and is read in groups of three. This finding established the concept of reading frame.

Nirenberg and others deciphered the code.

A codon consists of 3 nucleotides, so there are 64 possible codons. Three codons signal "stop," and one codon signals "start" and also encodes methionine. Thus 61 codons encode the 20 amino acids.

The code is degenerate but specific.

Many amino acids have more than one codon, but each codon specifies only a single amino acid.

The code is practically universal, but not quite.

In some mitochondrial and protist genomes, a STOP codon is read as an amino acid; otherwise the code is universal.

15.3 Prokaryotic Transcription

Prokaryotes have a single RNA polymerase.

Prokaryotic RNA polymerase exists in two forms: core polymerase, which can synthesize mRNA; and holoenzyme, core plus σ factor, which can accurately initiate synthesis (figure 15.6).

Initiation occurs at promoters.

Initiation requires a start site and a promoter. The promoter is upstream of the start site, and binding of RNA polymerase holoenzyme to its –35 region positions the polymerase properly.

Elongation adds successive nucleotides.

Transcription proceeds in the 5′-to-3′ direction. The transcription bubble contains RNA polymerase, the locally unwound DNA template, and the growing mRNA transcript (figure 15.7).

Termination occurs at specific sites.

Terminators consist of complementary sequences that form a double-stranded hairpin loop where the polymerase pauses (figure 15.8).

Prokaryotic transcription is coupled to translation.

Translation begins while mRNAs are still being transcribed.

15.4 Eukaryotic Transcription

Eukaryotes have three RNA polymerases.

RNA polymerase I transcribes rRNA; polymerase II transcribes mRNA and some snRNAs; polymerase III transcribes tRNA.

Each polymerase has its own promoter.

Initiation and termination differ from that in prokaryotes.

Unlike prokaryotic promoters, RNA polymerase II promoters require a host of transcription factors. Although termination sites exist, the end of the mRNA is modified after transcription.

Eukaryotic transcripts are modified (figure 15.11).

After transcription, a methyl-GTP cap is added to the 5′ end of the transcript. A poly-A tail is added to the 3′ end. Noncoding internal regions are also removed by splicing.

15.5 Eukaryotic pre-mRNA Splicing

Eukaryotic genes may contain interruptions.

Coding DNA (an exon) is interrupted by noncoding introns. These introns are removed by splicing (figure 15.13).

The spliceosome is the splicing organelle.

snRNPs recognize intron–exon junctions and recruit spliceosomes. The spliceosome ultimately joins the 3′ end of the first exon to the 5′ end of the next exon.

Splicing can produce multiple transcripts from the same gene.

15.6 The Structure of tRNA and Ribosomes

Aminoacyl-tRNA synthetases attach amino acids to tRNA.

The tRNA charging reaction attaches the carboxyl terminus of an amino acid to the 3′ end of the correct tRNA (figure 15.15).

The ribosome has multiple tRNA-binding sites (figure 15.16).

A charged tRNA first binds to the A site, then moves to the P site where its amino acid is bonded to the peptide chain, and finally, without its amino acid, moves to the E site from which it is released.

The ribosome has both decoding and enzymatic functions.

Ribosomes hold tRNAs and mRNA in position for a ribosomal enzyme to form peptide bonds.

15.7 The Process of Translation

Initiation requires accessory factors.

In prokaryotes, initiation-complex formation is aided by the ribosome-binding sequence (RBS) of mRNA, complementary to a small subunit. Eukaryotes use the 5′ cap for the same function.

Elongation adds successive amino acids (figure 15.20).

As the ribosome moves along the mRNA, new amino acids from charged tRNAs are added to the growing peptide.

Termination requires accessory factors.

Stop codons are recognized by termination factors.

Proteins may be targeted to the ER.

In eukaryotes, proteins with a signal sequence in their amino terminus bind to the SRP, and this complex docks on the ER.

(15.8 Summary is omitted.)

15.9 Mutation: Altered Genes

Point mutations affect a single site in the DNA .

Base substitutions exchange one base for another, and frameshift mutations involve the addition or deletion of a base. Triplet repeat expansion mutations can cause genetic diseases.

Chromosomal mutations change the structure of chromosomes.

Chromosomal mutations include additions, deletions, inversions, or translocations.

Mutations are the starting point of evolution.

Our view of the nature of genes has changed with new information.

UNDERSTAND

1. The experiments with nutritional mutants in *Neurospora* by Beadle and Tatum provided evidence that
   a. bread mold can be grown in a lab on minimal media.
   b. X-rays can damage DNA.
   c. cells need enzymes.
   d. genes specify enzymes.

2. What is the *central dogma* of molecular biology?
   a. DNA is the genetic material.
   b. Information passes from DNA directly to protein.
   c. Information passes from DNA to RNA to protein.
   d. One gene encodes only one polypeptide.

3. In the genetic code, one codon
   a. consists of three bases.
   b. specifies a single amino acid.
   c. specifies more than one amino acid.
   d. Both a and b are correct.

4. Eukaryotic transcription differs from prokaryotic in that
   a. eukaryotes have only one RNA polymerase.
   b. eukaryotes have three RNA polymerases.
   c. prokaryotes have three RNA polymerases.
   d. Both a and c are correct.

5. An anticodon would be found on which of the following types of RNA?
   a. snRNA (small nuclear RNA)
   b. mRNA (messenger RNA)
   c. tRNA (transfer RNA)
   d. rRNA (ribosomal RNA)

6. RNA polymerase binds to a _____ to initiate _____.
   a. mRNA; translation
   b. promoter; transcription
   c. primer; transcription
   d. transcription factor; translation

7. During translation, the codon in mRNA is actually "read" by
   a. the A site in the ribosome.
   b. the P site in the ribosome.
   c. the anticodon in a tRNA.
   d. the anticodon in an amino acid.

APPLY

1. You have mutants that all affect the same biochemical pathway. If feeding an intermediate in the pathway supports growth, this tells you that the enzyme encoded by the affected gene
   a. acts after the intermediate used.
   b. acts before the intermediate used.
   c. must act to produce the intermediate.
   d. must not act to produce the intermediate.

2. The splicing process
   a. occurs in prokaryotes.
   b. joins introns together.
   c. can produce multiple mRNAs from the same transcript.
   d. only joins exons for each gene in one way.

3. The enzyme that forms peptide bonds is called peptidyl *transferase* because it transfers
   a. a new amino acid from a tRNA to the growing peptide.
   b. the growing peptide from a tRNA to the next amino acid.
   c. the peptide from one amino acid to another.
   d. the peptide from the ribosome to a charged tRNA.

4. In comparing gene expression in prokaryotes and eukaryotes
   a. eukaryotic genes can produce more than one protein.
   b. prokaryotic genes can produce more than one protein.
   c. both produce mRNAs that are colinear with the protein.
   d. Both a and c are correct.

5. The codon CCA could be mutated to produce
   a. a silent mutation.　　c. a Stop codon.
   b. a codon for Lys.　　d. Both a and b are correct.

6. An inversion will
   a. necessarily cause a mutant phenotype.
   b. only cause a mutant phenotype if the inversion breakpoints fall within a gene.
   c. halt transcription in the inverted region because the chromosome is now backward.
   d. interfere with translation of genes in the inverted region.

7. What is the relationship between mutations and evolution?
   a. Mutations make genes better.
   b. Mutations can create new alleles.
   c. Mutations happened early in evolution, but not now.
   d. There is no relationship between evolution and genetic mutations.

SYNTHESIZE

1. A template strand of DNA has the following sequence:
   3′ – CGTTACCCGAGCCGTACGATTAGG – 5′
   Use the sequence information to determine
   a. the predicted sequence of the mRNA for this gene.
   b. the predicted amino acid sequence of the protein.

2. Frameshift mutations often result in truncated proteins. Explain this observation based on the genetic code.

3. Describe how each of the following mutations will affect the final protein product (protein begins with START codon). Name the type of mutation.
   Original template strand:
   3′ – CGTTACCCGAGCCGTACGATTAGG – 5′
   a. 3′ – CGTTACCCGAGCCGTAACGATTAGG – 5′
   b. 3′ – CGTTACCCGATCCGTACGATTAGG – 5′
   c. 3′ – CGTTACCCGAGCCGTTCGATTAGG – 5′

4. There are a number of features that are unique to bacteria, and others that are unique to eukaryotes. Could any of these features offer the possibility to control gene expression in a way that is unique to either eukaryotes or bacteria?

ONLINE RESOURCE

www.ravenbiology.com

Understand, Apply, and Synthesize—enhance your study with animations that bring concepts to life and practice tests to assess your understanding. Your instructor may also recommend the interactive eBook, individualized learning tools, and more.

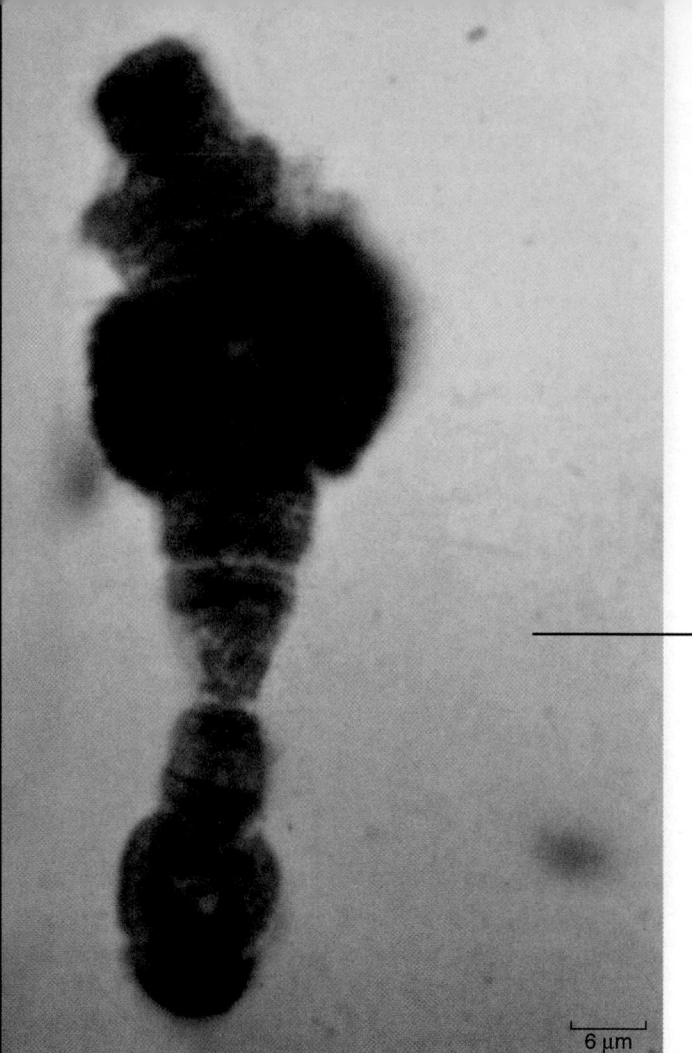

6 μm

Chapter 16

Control of Gene Expression

Chapter Contents

Introduction

In a symphony, various instruments play their own parts at different times; the musical score determines which instruments play when. Similarly, in an organism, different genes are expressed at different times, with a "genetic score," written in regulatory regions of the DNA, determining which genes are active when. The picture shows the expanded "puff" of this Drosophila *chromosome, which represents genes that are being actively expressed. Gene expression and how it is controlled is our topic in this chapter.*

16.1 Control of Gene Expression

Learning Outcomes

1. *Identify when gene expression is usually controlled.*
2. *Describe the usual action of regulatory proteins.*
3. *List differences between control of gene expression in prokaryotes and eukaryotes.*

Control of gene expression is essential to all organisms. In prokaryotes, cells adapt to environmental changes by changing gene expression. In multicellular eukaryotes, it is critical for directing development and maintaining homeostasis. In fact, different cell types in a multicellular organism have distinct functions based on the proteins they express. All cells in a multicellular organism share a set of "housekeeping" genes that sustain the cell, but they also express a unique set of proteins that distinguish each cell type.

Control can occur at all levels of gene expression

You learned in the previous chapter that gene expression is the conversion of genotype to phenotype—the flow of information from DNA to produce functional proteins that control cellular activities. This traditional view of gene expression includes controlling the process primarily at the level of transcription initiation. Although this view remains valid, evidence is accumulating that both the extent of the genome transcribed, and the control of this transcription in multicellular organisms is even more complex than previously expected.

In this chapter, we will develop the control of initiation of transcription in some detail because of its importance, and because it is still the best studied mechanism for the control of gene expression. With this framework in place, we will also consider how chromatin structure affects gene expression and how control can be exerted posttranscriptionally as well. The latter topic will lead us into the exciting new world of regulatory RNA molecules.

RNA polymerase is key to transcription, and it must have access to the DNA helix and must be capable of binding to the gene's promoter for transcription to begin. **Regulatory proteins** act by modulating the ability of RNA polymerase to bind to the promoter. This idea of controlling the access of RNA polymerase to a promoter is common to both prokaryotes and eukaryotes, but the details differ greatly, as you will see.

These regulatory proteins bind to specific nucleotide sequences on the DNA that are usually only 10–15 nt in length. (Even a large regulatory protein has a "footprint," or binding area, of only about 20 nt.) Hundreds of these regulatory sequences have been characterized, and each provides a binding site for a specific protein that is able to recognize the sequence. Binding of the protein either *blocks* transcription by getting in the way of RNA polymerase or *stimulates* transcription by facilitating the binding of RNA polymerase to the promoter.

Control strategies in prokaryotes are geared to adjust to environmental changes

Control of gene expression is accomplished very differently in prokaryotes than it is in eukaryotes. Prokaryotic cells have been shaped by evolution to grow and divide as rapidly as possible, enabling them to exploit transient resources. Proteins in prokaryotes turn over rapidly, allowing these organisms to respond quickly to changes in their external environment by changing patterns of gene expression.

In prokaryotes, the primary function of gene control is to adjust the cell's activities to its immediate environment. Changes in gene expression alter which enzymes are present in response to the quantity and type of nutrients and the amount of oxygen available. Almost all of these changes are fully reversible, allowing the cell to adjust its enzyme levels up or down in response to environment changes.

Control strategies in eukaryotes maintain homeostasis and drive development

The cells of multicellular organisms, in contrast, have been shaped by evolution to be protected from transient changes in their immediate environment. Most of them experience fairly constant conditions. Indeed, *homeostasis*—the maintenance of a constant internal environment—is a hallmark of multicellular organisms. Cells in such organisms respond to signals in

their immediate environment (such as growth factors and hormones) by altering gene expression, and in doing so they participate in regulating the body as a whole.

Perhaps the most important role for controlled changes in gene expression is the development of the organism itself. This occurs by coordinated changes in gene expression that occur both over developmental time and in different tissues. Understanding this complex program has been a major goal of developmental genetics, and this work has uncovered circuits of gene expression that have been preserved over very long evolutionary time (see chapter 19). The same regulatory circuit may also be used within an organism to pattern different structures in different places at different developmental times. For example, the gene *sonic hedgehog* acts to pattern the neural tube, then later to pattern digits in developing limbs.

Unicellular eukaryotes also use different control mechanisms from those of prokaryotes. All eukaryotes have a membrane-bounded nucleus, use similar mechanisms to condense DNA into chromosomes, and have the same gene expression machinery, all of which differ from those of prokaryotes.

Learning Outcomes Review 16.1

Gene expression is usually controlled at the level of transcription initiation. Regulatory proteins bind to specific DNA sequences and affect the binding of RNA polymerase to promoters. Individual proteins may either prevent or stimulate transcription. In prokaryotes, regulation is focused on adjusting the cell's activities to the environment to ensure viability. In multicellular eukaryotes, regulation is geared to maintaining internal homeostasis, and even in unicellular forms, this control has mechanisms to deal with a bounded nucleus and multiple chromosomes.

■ *Would you expect the control of gene expression in a unicellular eukaryote like yeast to be more like that of humans or* E. coli?

16.2 Regulatory Proteins

Learning Outcomes

1. *Explain how proteins can interact with base pairs without unwinding the helix.*
2. *Describe the common features of DNA-binding motifs.*

The ability of certain proteins to bind to *specific* DNA regulatory sequences and either block transcription or facilitate the binding of RNA polymerase is the basis for transcriptional control. To understand how cells control gene expression, it is first necessary to gain a clear picture of this molecular recognition process.

Proteins can interact with DNA through the major groove

In the past, molecular biologists thought that the DNA helix had to unwind before proteins could distinguish one DNA sequence from another; only in this way, they reasoned, could regulatory proteins gain access to the hydrogen bonds between base-pairs. We now know it is unnecessary for the helix to unwind because proteins can bind to its outside surface, where the edges of the base-pairs are exposed.

Careful inspection of a DNA molecule reveals two helical grooves winding around the molecule, one deeper than the other. Within the deeper groove, called the **major groove,** the nucleotides' hydrogen bond donors and acceptors are accessible. The pattern created by these chemical groups is unique for each of the four possible base-pair arrangements, providing a ready way for a protein nestled in the groove to read the sequence of bases (figure 16.1).

DNA-binding domains interact with specific DNA sequences

Protein–DNA recognition is an area of active research; so far, the structures of over 30 regulatory proteins have been analyzed. Although each protein is unique in its fine details, the part of the protein that actually binds to the DNA is much less variable. Almost all of these proteins employ one of a small set of **DNA-binding motifs.** A motif, as described in chapter 3, is a form of three-dimensional substructure that is found in many proteins. These DNA-binding motifs share the property of interacting with specific sequences of bases, usually through the major groove of the DNA helix.

DNA-binding motifs are the key structure within the DNA-binding domain of these proteins. This domain is a functionally distinct part of the protein necessary to bind to DNA in a sequence-specific manner. Regulatory proteins also need to be able to interact with the transcription apparatus, which is accomplished by a different regulatory domain.

Note that two proteins that share the same DNA-binding domain do not necessarily bind to the same DNA sequence. The similarities in the DNA-binding motifs appear in their three-dimensional structure, and not in the specific contacts that they make with DNA.

Several common DNA-binding motifs are shared by many proteins

A limited number of common DNA-binding motifs are found in a wide variety of different proteins. Four of the best known are detailed in the following sections to give the sense of how DNA-binding proteins interact with DNA.

The helix-turn-helix motif

The most common DNA-binding motif is the **helix-turn-helix,** constructed from two α-helical segments of the protein linked by a short, nonhelical segment, the "turn" (figure 16.2*a*). As the first motif recognized, the helix-turn-helix motif has since been identified in hundreds of DNA-binding proteins.

A close look at the structure of a helix-turn-helix motif reveals how proteins containing such motifs interact with the major groove of DNA. The helical segments of the motif interact with one another, so that they are held at roughly right angles. When this motif is pressed against DNA, one of the

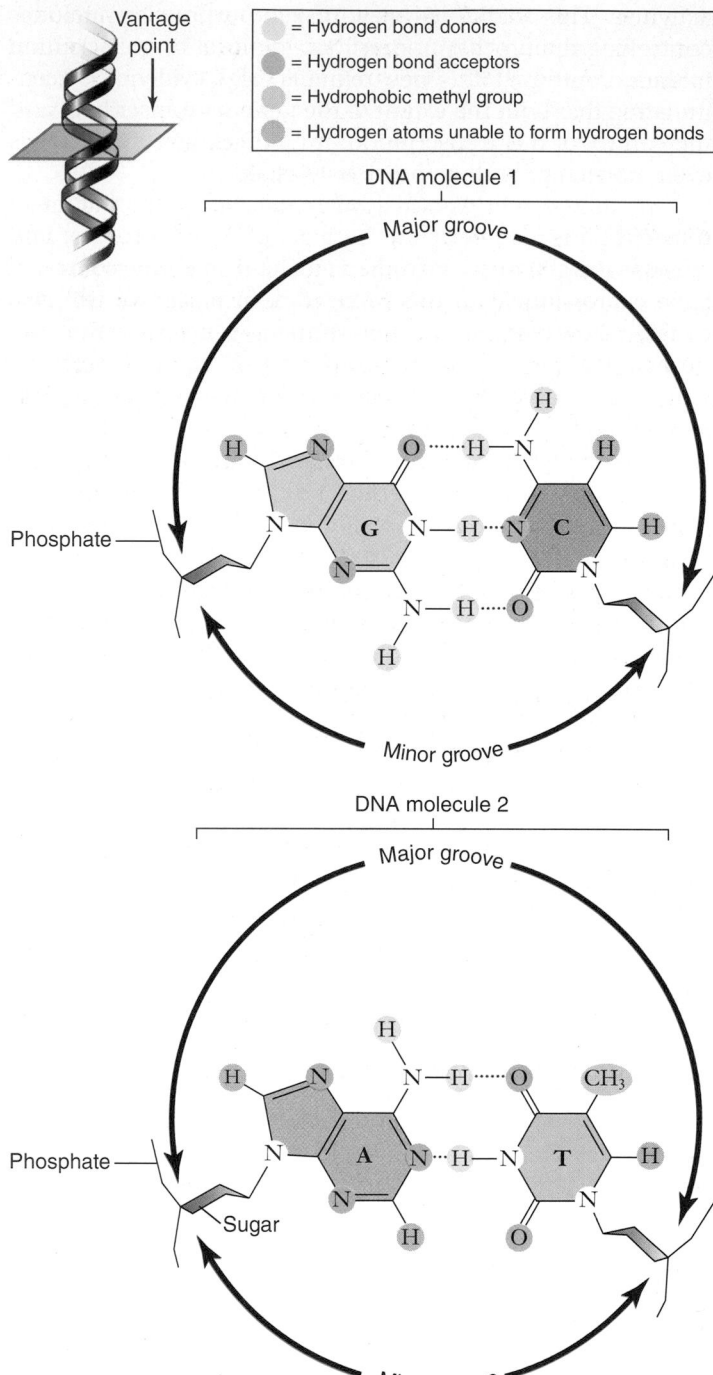

Figure 16.1 Reading the major groove of DNA. Looking down into the major groove of a DNA helix, we can see the edges of the bases protruding into the groove. Each of the four possible base-pair arrangements (two are shown here) extends a unique set of chemical groups into the groove, indicated in this diagram by differently colored circles. A regulatory protein can identify the base-pair arrangement by this characteristic signature.

helical segments (called the *recognition helix*) fits snugly in the major groove of the DNA molecule, and the other butts up against the outside of the DNA molecule, helping to ensure the proper positioning of the recognition helix.

Most DNA-regulatory sequences recognized by helix-turn-helix motifs occur in symmetrical pairs. Such sequences are bound by proteins containing two helix-turn-helix motifs separated by 3.4 nanometers (nm), the distance required for one turn of the DNA helix (see figure 16.2a). Having *two* protein–DNA-binding sites doubles the zone of contact between protein and DNA and greatly strengthens the affinity between them.

The homeodomain motif

A special class of helix-turn-helix motifs, the **homeodomain,** plays a critical role in development in a wide variety of eukaryotic organisms, including humans. These motifs were discovered when researchers began to characterize a set of homeotic mutations in *Drosophila* (mutations that cause one body part to be replaced by another). They found that the mutant genes encoded regulatory proteins. Normally these proteins would initiate key stages of development by binding to developmental switch-point genes. More than 50 of these regulatory proteins have been analyzed, and they all contain a nearly identical sequence of 60 amino acids, which was termed the *homeodomain*. The most conserved part of the homeodomain contains a recognition helix of a helix-turn-helix motif. The rest of the homeodomain forms the other two helices of this motif.

The zinc finger motif

A different kind of DNA-binding motif uses one or more zinc atoms to coordinate its binding to DNA. Called **zinc fingers**, these motifs exist in several forms. In one form, a zinc atom links an α-helical segment to a β-sheet segment so that the helical segment fits into the major groove of DNA.

This sort of motif often occurs in clusters, the β sheets spacing the helical segments so that each helix contacts the major groove. The effect is like a hand wrapped around the DNA with the fingers lying in the major groove. The more zinc fingers in the cluster, the more the protein associates with DNA.

The leucine zipper motif

In another DNA-binding motif, the name actually refers to a dimerization motif that allows different subunits of a protein to associate with the DNA. This so-called **leucine zipper** is created where a region on one subunit containing several hydrophobic amino acids (usually leucines) interacts with a similar region on the other subunit. This interaction holds the subunits together and creates a Y-shaped structure where the two arms of the Y are helical regions that each fit into the major groove of DNA but on opposite sides of the helix (figure 16.2b), holding the DNA like a pair of tongs.

Learning Outcomes Review 16.2

A DNA helix exhibits a major groove and a minor groove; regulatory proteins interact with DNA by accessing bases along the major groove. These proteins all contain DNA-binding motifs, and they often include one or two α-helical segments. These motifs form the active part of the DNA-binding domain, and another domain of the protein interacts with the transcription apparatus.

■ *What would be the effect of a mutation in a helix-turn-helix protein that altered the spacing of the two helices?*

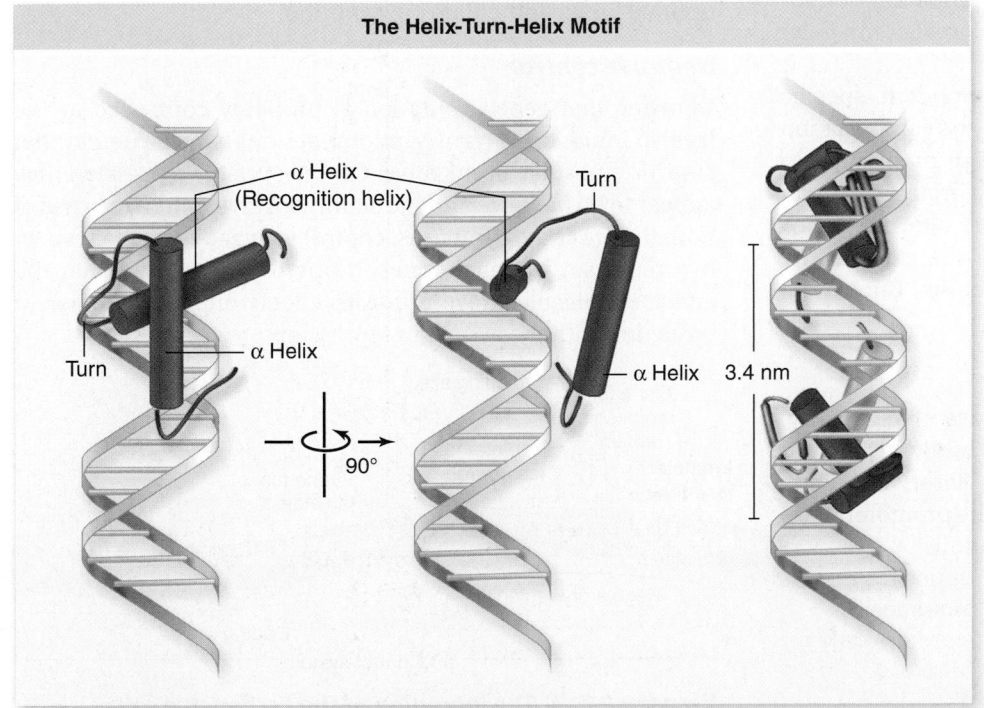

The Helix-Turn-Helix Motif

α Helix (Recognition helix)

Turn

α Helix

Turn

90°

α Helix   3.4 nm

a.

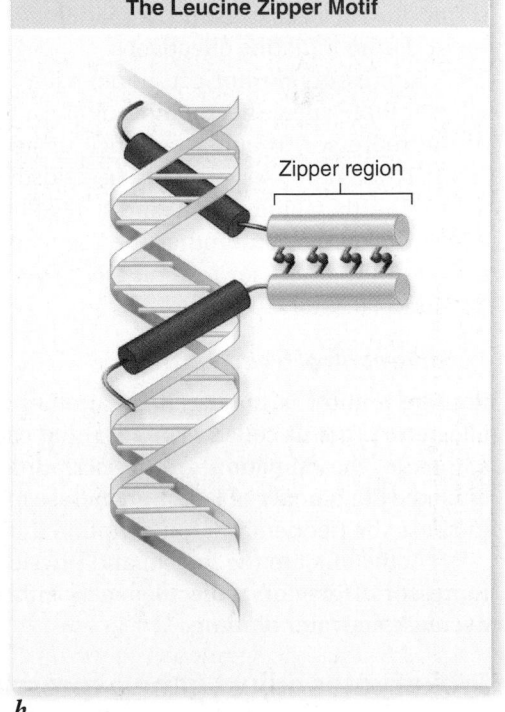

The Leucine Zipper Motif

Zipper region

b.

Figure 16.2 Major DNA-binding motifs. Two different DNA-binding motifs are pictured interacting with DNA. *a.* The helix-turn-helix motif binds to DNA using one α helix, the recognition helix, to interact with the major groove. The other helix positions the recognition helix. Proteins with this motif are usually dimers, with two identical subunits, each containing the DNA-binding motif. The two copies of the motif *(red)* are separated by 3.4 nm, precisely the spacing of one turn of the DNA helix. *b.* The leucine zipper acts to hold two subunits in a multisubunit protein together, thereby allowing α-helical regions to interact with DNA.

16.3 *Prokaryotic Regulation*

The details of regulation can be revealed by examining mechanisms used by prokaryotes to control the initiation of transcription. Prokaryotes and eukaryotes share some common themes, but they have some profound differences as well. Later on we discuss eukaryotic systems and concentrate on how they differ from the simpler prokaryotic systems.

Control of transcription can be either positive or negative

Control at the level of transcription initiation can be either positive or negative. **Positive control** increases the frequency of initiation, and **negative control** decreases the frequency of initiation. Each of these forms of control are mediated by regulatory proteins, but the proteins have opposite effects.

Negative control by repressors

Negative control is mediated by proteins called **repressors.** Repressors are proteins that bind to regulatory sites on DNA called **operators** to prevent or decrease the initiation of transcription. They act as a kind of roadblock to prevent the polymerase from initiating effectively.

Repressors do not act alone; each responds to specific effector molecules. Effector binding can alter the conformation of the repressor to either enhance or abolish its binding to DNA. These repressor proteins are allosteric proteins with an active site that binds DNA and a regulatory site that binds effectors. Effector binding at the regulatory site changes the ability of the repressor to bind DNA (see chapter 6 for more details on allosteric proteins).

Positive control by activators

Positive control is mediated by another class of regulatory, allosteric proteins called *activators* that can bind to DNA and stimulate the initiation of transcription. These activators enhance the binding of RNA polymerase to the promoter to increase the frequency of transcription initiation.

Activators are the logical and physical opposites of repressors. Effector molecules can either enhance or decrease activator binding.

Prokaryotes adjust gene expression in response to environmental conditions

Changes in the environments that bacteria and archaea encounter often result in changes in gene expression. In general, genes encoding proteins involved in catabolic pathways (breaking down molecules) respond oppositely from genes encoding proteins involved in anabolic pathways (building up molecules). In the discussion that follows, we describe enzymes in the catabolic pathway that transports and utilizes the sugar lactose. Later we describe the anabolic pathway that synthesizes the amino acid tryptophan.

As mentioned in the preceding chapter, prokaryotic genes are often organized into operons, multiple genes that are part of a single transcription unit having a single promoter. Genes that are involved in the same metabolic pathway are often organized in this fashion. The proteins necessary for the utilization of lactose are encoded by the *lac* **operon,** and the proteins necessary for the synthesis of tryptophan are encoded by the *trp* **operon.**

> **? Inquiry question** What advantage might a bacterium get by linking into a single operon several genes, all of the products of which contribute to a single biochemical pathway?

Induction and repression

If a bacterium encounters lactose, it begins to make the enzymes necessary to utilize lactose. When lactose is not present, however, there is no need to make these proteins. Thus, we say that the synthesis of the proteins is *induced* by the presence of lactose. **Induction** therefore occurs when enzymes for a certain pathway are produced in response to a substrate.

When tryptophan is available in the environment, a bacterium will not synthesize the enzymes necessary to make tryptophan. If tryptophan ceases to be available, then the bacterium begins to make these enzymes. **Repression** occurs when bacteria capable of making biosynthetic enzymes do not produce them. In the case of both induction and repression, the bacterium is adjusting to produce the enzymes that are optimal for its immediate environment.

Negative control

Knowing that gene expression is probably controlled at the level of initiation of transcription does not tell us whether that control is positive or negative. On the surface, repression may appear to be negative and induction positive; but in the case of both the *lac* and *trp* operons, control is negative by the respective repressor proteins for each operon. The key is that the effector molecules have opposite effects on the repressor in induction with those seen in repression.

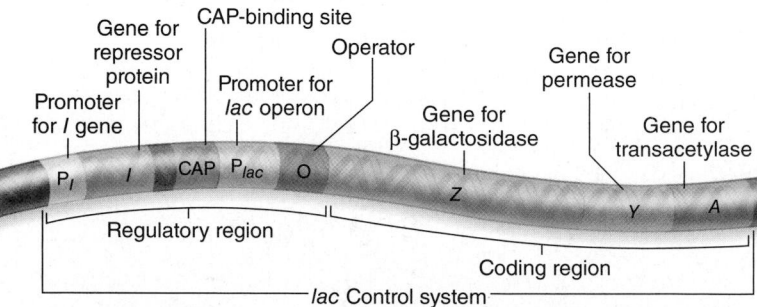

Figure 16.3 **The *lac* region of the *Escherichia coli* chromosome.** The *lac* operon consists of a promoter, an operator, and three genes (*lac Z, Y,* and *A*) that encode proteins required for the metabolism of lactose. In addition, there is a binding site for the catabolite activator protein (CAP), which affects RNA polymerase binding to the promoter. The *I* gene encodes the repressor protein, which can bind the operator and block transcription of the *lac* operon.

For either mechanism to work, the molecule in the environment, such as lactose or tryptophan, must produce the proper effect on the gene being regulated. In the case of *lac* induction, the presence of lactose must *prevent* a repressor protein from binding to its regulatory sequence. In the case of *trp* repression, by contrast, the presence of tryptophan must *cause* a repressor protein to bind to its regulatory sequence.

These responses are opposite because the needs of the cell are opposite in anabolic versus catabolic pathways. Each pathway is examined in detail in the following sections to show how protein–DNA interactions allow the cell to respond to environmental conditions.

The *lac* operon is negatively regulated by the *lac* repressor

The control of gene expression in the *lac* operon was elucidated by the pioneering work of Jacques Monod and François Jacob.

The *lac* operon consists of the genes that encode functions necessary to utilize lactose: β-galactosidase *(lacZ)*, lactose permease *(lacY)*, and lactose transacetylase *(lacA)*, plus the regulatory regions necessary to control the expression of these genes (figure 16.3). In addition, the gene for the *lac* repressor *(lacI)* is linked to the rest of the *lac* operon and is thus considered part of the operon although it has its own promoter. The arrangement of the control regions upstream of the coding region is typical of most prokaryotic operons, although the linked repressor is not.

Action of the repressor

Initiation of transcription of the *lac* operon is controlled by the *lac* repressor. The repressor binds to the operator, which is adjacent to the promoter (figure 16.4*a*). This binding prevents RNA polymerase from binding to the promoter. This DNA binding is sensitive to the presence of lactose: The

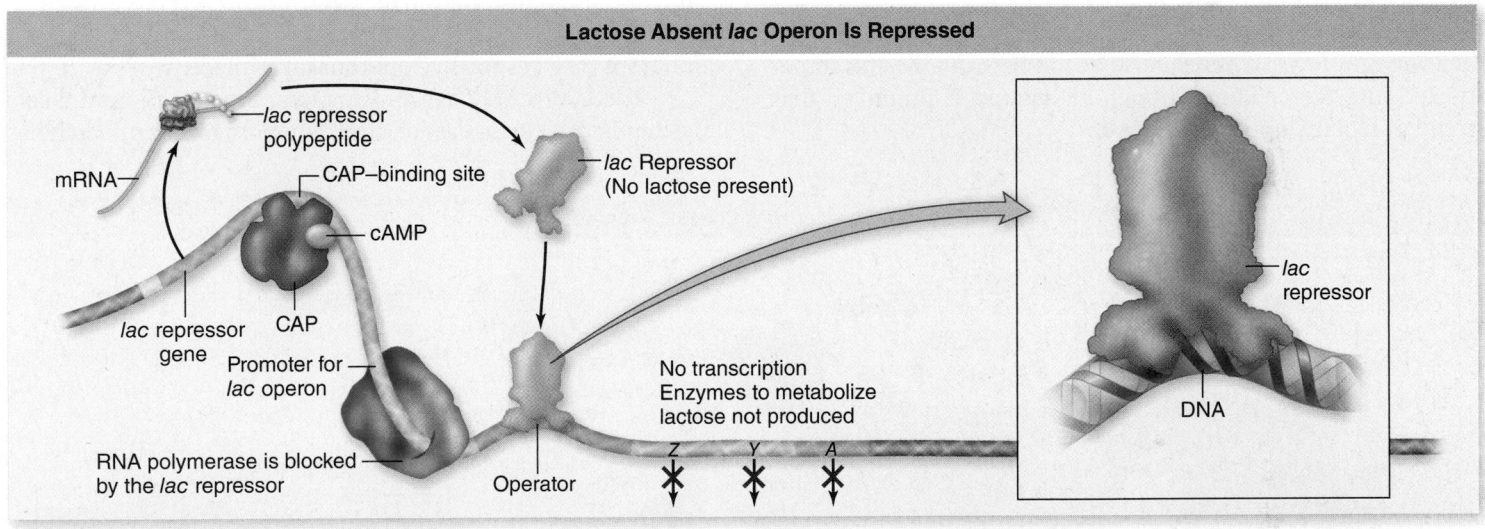

Figure 16.4 Induction of the *lac* operon. *a.* In the absence of lactose the *lac* repressor binds to DNA at the operator site, thus preventing transcription of the operon. When the repressor protein is bound to the operator site, the *lac* operon is shut down (repressed). *b.* The *lac* operon is transcribed (induced) when CAP is bound and when the repressor is not bound. Allolactose binding to the repressor alters the repressor's shape so it cannot bind to the operator site and block RNA polymerase activity.

🔍 **Data analysis** What would the phenotype be of a repressor mutation that prevents DNA binding? Inducer binding? How do these compare with a mutation in the operator that prevents repressor binding?

repressor binds DNA in the absence of lactose, but not in the presence of lactose.

Interaction of repressor and inducer

In the absence of lactose, the *lac* repressor binds to the operator, and the operon is repressed (see figure 16.4*a*). The effector that controls the DNA binding of the repressor is a metabolite of lactose, allolactose, which is produced when lactose is available. Allolactose binds to the repressor, altering its conformation so that it no longer can bind to the operator (figure 16.4*b*). The operon is now induced. Since allolactose allows induction of the operon, it is usually called the inducer.

As the level of lactose falls, allolactose concentrations decrease, making it no longer available to bind to the repressor and allowing the repressor to bind to DNA again. Thus this system of negative control by the *lac* repressor and its inducer, allolactose, allows the cell to respond to changing levels of lactose in the environment.

Even in the absence of lactose, the *lac* operon is expressed at a very low level. When lactose becomes available, it is transported into the cell and enough allolactose is produced that induction of the operon can occur.

The presence of glucose prevents induction of the *lac* operon

Glucose repression is a mechanism for the preferential use of glucose in the presence of other sugars such as lactose. If bacteria are grown in the presence of both glucose and lactose, the *lac* operon is not induced. As glucose is used up, the *lac* operon is induced, allowing lactose to be used as an energy source.

Despite the name *glucose repression,* this mechanism involves an activator protein that can stimulate transcription from multiple catabolic operons, including the *lac* operon. This activator, **catabolite activator protein (CAP),** is an allosteric protein with cAMP as an effector. This protein is also called **cAMP response protein (CRP)** because it binds cAMP, but we will use the name CAP to emphasize its role as a positive regulator. CAP alone does not bind to DNA, but binding of the effector cAMP to CAP changes its conformation such that it can bind to DNA (figure 16.5). The level of cAMP in cells is reduced in the presence of glucose so that no stimulation of transcription from CAP-responsive operons takes place.

The CAP–cAMP system was long thought to be the sole mechanism of glucose repression. But more recent research has

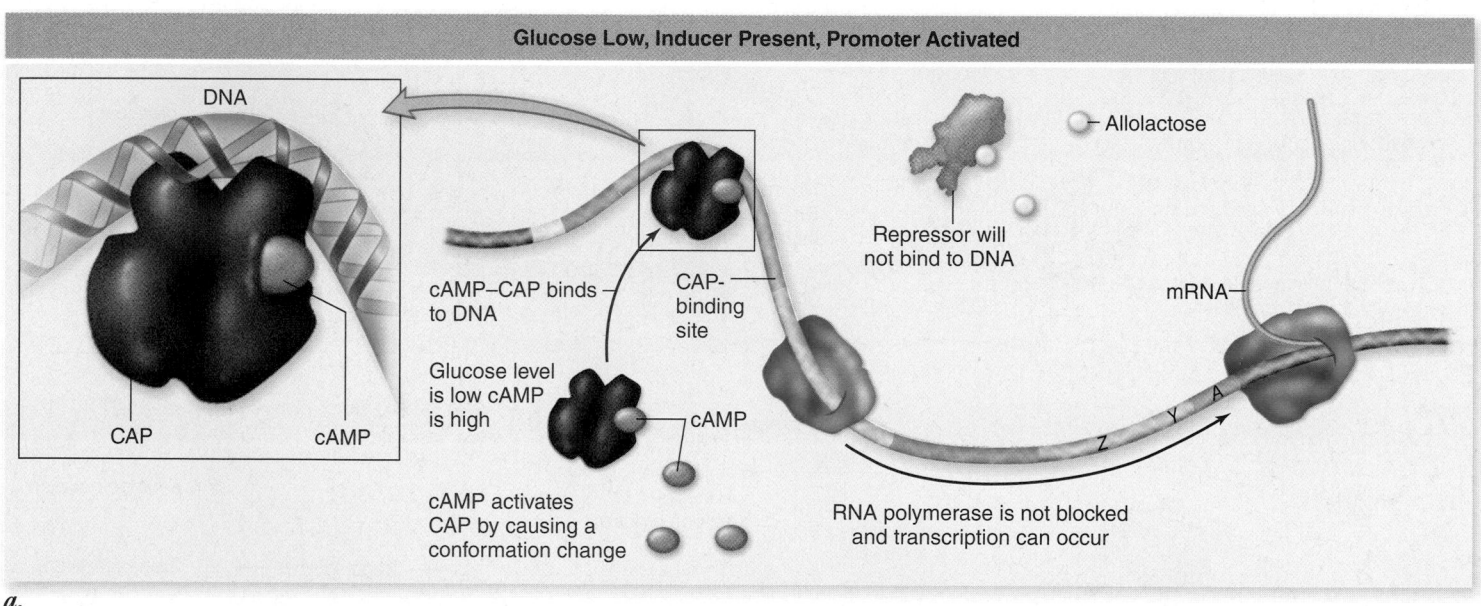

Glucose Low, Inducer Present, Promoter Activated

DNA

CAP   cAMP

cAMP–CAP binds to DNA

CAP-binding site

Glucose level is low cAMP is high

cAMP

cAMP activates CAP by causing a conformation change

Allolactose

Repressor will not bind to DNA

mRNA

Z   Y   A

RNA polymerase is not blocked and transcription can occur

a.

Figure 16.5 Effect of glucose on the *lac* operon. Expression of the *lac* operon is controlled by a negative regulator (repressor) and a positive regulator (CAP). The action of CAP is sensitive to glucose levels. *a.* For CAP to bind to DNA, it must bind to cAMP. When glucose levels are low, cAMP is abundant and binds to CAP. The CAP–cAMP complex causes the DNA to bend around it. This brings CAP into contact with RNA polymerase (not shown) making polymerase binding to the promoter more efficient. *b.* High glucose levels produce two effects: cAMP is scarce so CAP is unable to activate the promoter, and the transport of lactose is blocked (inducer exclusion).

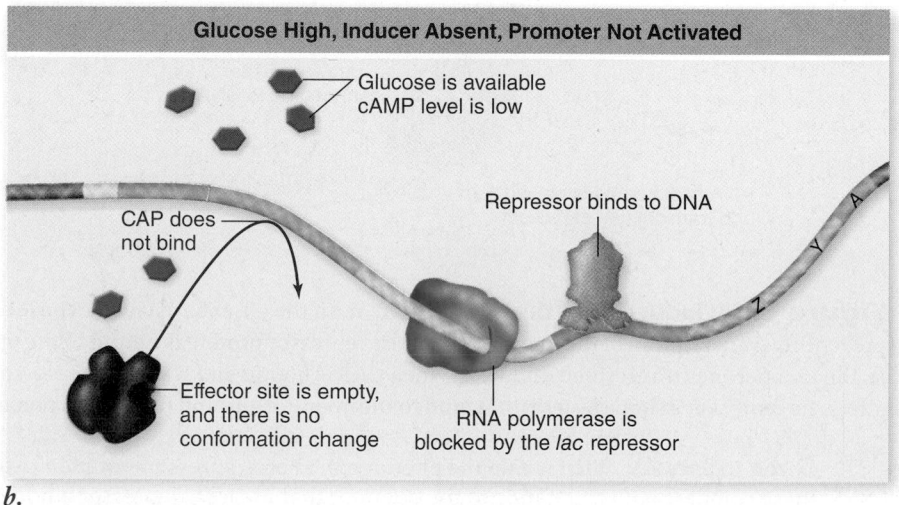

Glucose High, Inducer Absent, Promoter Not Activated

Glucose is available cAMP level is low

CAP does not bind

Repressor binds to DNA

Effector site is empty, and there is no conformation change

RNA polymerase is blocked by the *lac* repressor

Z   Y   A

b.

indicated that the presence of glucose inhibits the transport of lactose into the cell. This deprives the cell of the *lac* operon inducer, allolactose, allowing the repressor to bind to the operator. This mechanism, called **inducer exclusion,** is now thought to be the main form of glucose repression of the *lac* operon.

Given that inducer exclusion occurs, the role of CAP in the absence of glucose seems superfluous. But in fact, the positive control of CAP–cAMP is necessary because the promoter of the *lac* operon alone is not efficient in binding RNA polymerase. This inefficiency is overcome by the action of the positive control of the CAP–cAMP activator. Thus, the highest levels of expression occur in the absence of glucose and in the presence of lactose. In this case the presence of the activator, and absence of the repressor combine to produce the highest levels of expression (see figure 16.5).

The *trp* operon is controlled by the *trp* repressor

Like the *lac* operon, the *trp* operon consists of a series of genes that encode enzymes involved in the same biochemical pathway. In the case of the *trp* operon these enzymes are necessary for synthesizing tryptophan. The regulatory region that controls transcription of these genes is located upstream of the genes. The *trp* operon is controlled by a repressor encoded by a gene located outside the *trp* operon. The *trp* operon is continuously expressed in the absence of tryptophan and is not expressed in the presence of tryptophan.

The *trp* repressor is a helix-turn-helix protein that binds to the operator site located adjacent to the *trp* promoter (figure 16.6). In the absence of tryptophan, the *trp* repressor does not bind to its operator, allowing expression

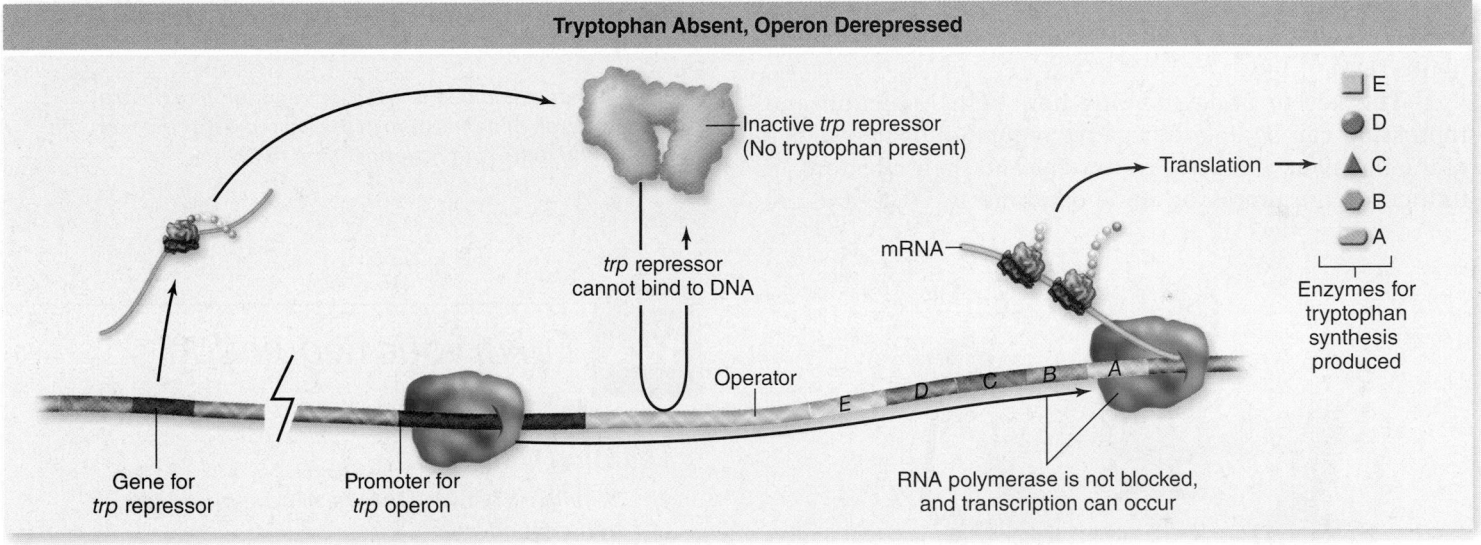

a.

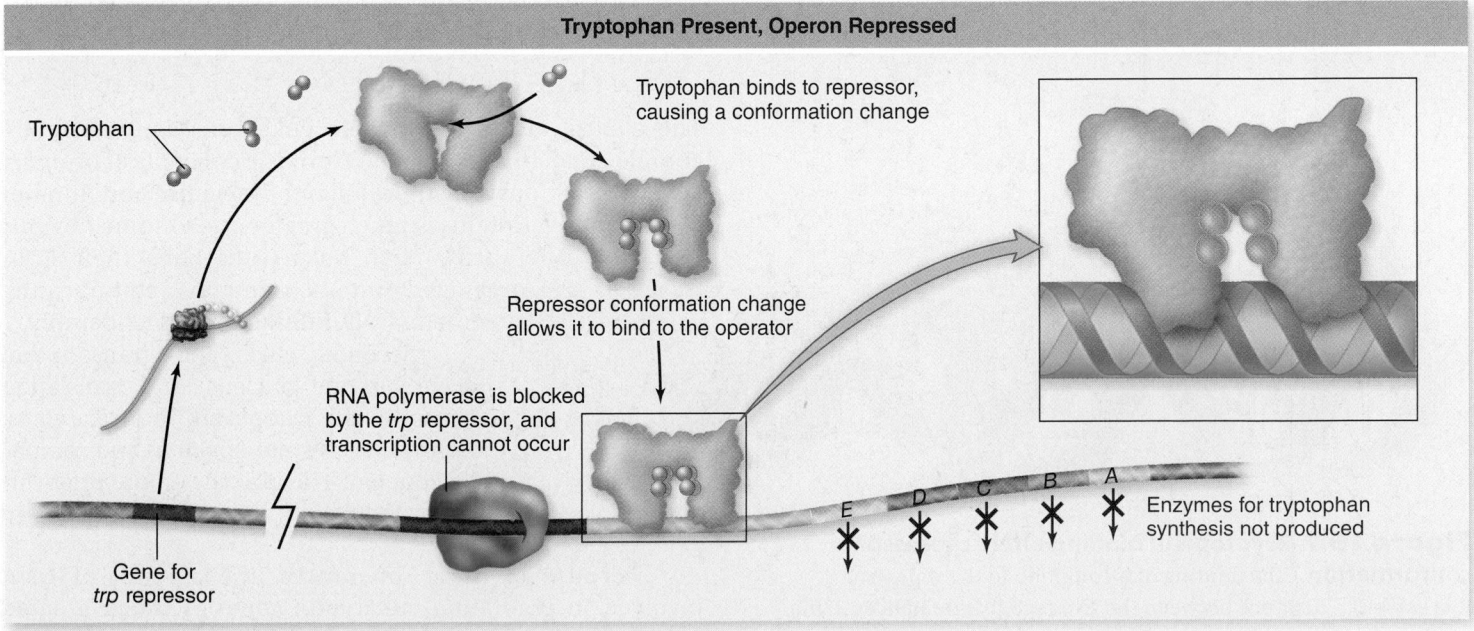

b.

Figure 16.6 How the *trp* operon is controlled. The tryptophan operon encodes the enzymes necessary to synthesize tryptophan.
a. The tryptophan repressor alone cannot bind to DNA. The promoter is free to function, and RNA polymerase transcribes the operon.
b. When tryptophan is present, it binds to the repressor, altering its conformation so it now binds DNA. The tryptophan–repressor complex binds tightly to the operator, preventing RNA polymerase from initiating transcription.

of the operon, and production of the enzymes necessary to make tryptophan.

When levels of tryptophan rise, then tryptophan (the *corepressor*) binds to the repressor and alters its conformation, allowing it to bind to its operator. Binding of the repressor–corepressor complex to the operator prevents RNA polymerase from binding to the promoter. The actual change in repressor structure due to tryptophan binding is an alteration of the orientation of a pair of helix-turn-helix motifs that allows their recognition helices to fit into adjacent major grooves of the DNA (figure 16.7).

When tryptophan is present and bound to the repressor and this complex is bound to the operator, the operon is said to be *repressed*. As tryptophan levels fall, the repressor alone cannot bind to the operator, allowing expression of the operon. In this state, the operon is said to be **derepressed,** distinguishing this state from induction (see figure 16.6).

The key to understanding how both induction and repression can be due to negative regulation is knowledge of the behavior of repressor proteins and their effectors. In induction, the repressor alone can bind to DNA, and the inducer prevents DNA binding. In the case of repression, the repressor only binds DNA when bound to the corepressor. Induction and repression are excellent examples of how interactions of molecules can affect their structures, and how molecular structure is critical to function.

Learning Outcomes Review 16.3

Induction occurs when expression of genes in a pathway is turned on in response to a substrate; repression occurs when expression is prevented in response to a substrate. The *lac* operon is negatively controlled by a repressor protein that binds to DNA, thus preventing transcription. When lactose is present, the operon is turned on; allolactose binds to the repressor, which then no longer binds to DNA. This operon is also positively regulated by an activator protein. The *trp* operon is negatively controlled by a repressor protein that must be bound to tryptophan in order to bind to DNA. In the absence of tryptophan, the repressor cannot bind DNA, and the operon is derepressed.

■ *What would be the effect on regulation of the* trp *operon of a mutation in the* trp *repressor that can still bind to* trp, *but no longer bind to DNA?*

16.4 Eukaryotic Regulation

Learning Outcomes

1. Distinguish between the role of general and specific transcription factors.
2. Describe the formation of a Pol II initiation complex.
3. Explain how transcription factors can have an effect from a distance in the DNA.

The control of transcription in eukaryotes is much more complex than in prokaryotes. The basic concepts of protein–DNA interactions are still valid, but the nature and number of interacting proteins is much greater due to some obvious differences. First, eukaryotes have their DNA organized into chromatin, complicating protein–DNA interactions considerably.

Second, eukaryotic transcription occurs in the nucleus, and translation occurs in the cytoplasm; in prokaryotes, these processes are spatially and temporally coupled. This provides more opportunities for regulation in eukaryotes than in prokaryotes.

Because of these differences, the amount of DNA involved in regulating eukaryotic genes is much greater. The need for a fine degree of flexible control is especially important for multicellular eukaryotes, with their complex

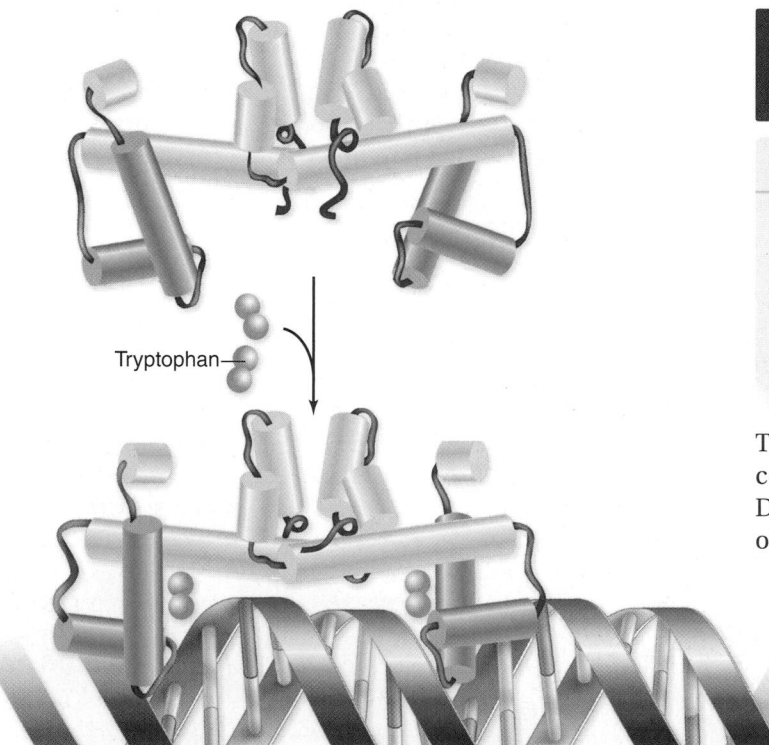

Tryptophan

⊢———— 3.4 nm ————⊣

Figure 16.7 Tryptophan binding alters repressor conformation. The binding of tryptophan to the repressor increases the distance between the two recognition helices in the repressor, allowing the repressor to fit snugly into two adjacent portions of the major groove in DNA.

Transcription factors can be either general or specific

In the preceding chapter we introduced the concept of transcription factors. Eukaryotic transcription requires a variety of these protein factors, which fall into two categories: *general transcription factors* and *specific transcription factors*. General factors are necessary for the assembly of a transcription apparatus and recruitment of RNA polymerase II to a promoter. Specific factors increase the level of transcription in certain cell types or in response to signals.

General transcription factors

Transcription of RNA polymerase II templates (the majority being genes that encode protein products) requires more than just RNA polymerase II to initiate transcription. A host of **general transcription factors** are also necessary to establish productive initiation. These factors are required for transcription to occur, but they do not increase the rate above this basal rate.

General transcription factors are named with letter designations that follow the abbreviation TFII, for "transcription factor RNA polymerase II." The most important of these factors, TFIID, contains the TATA-binding protein that

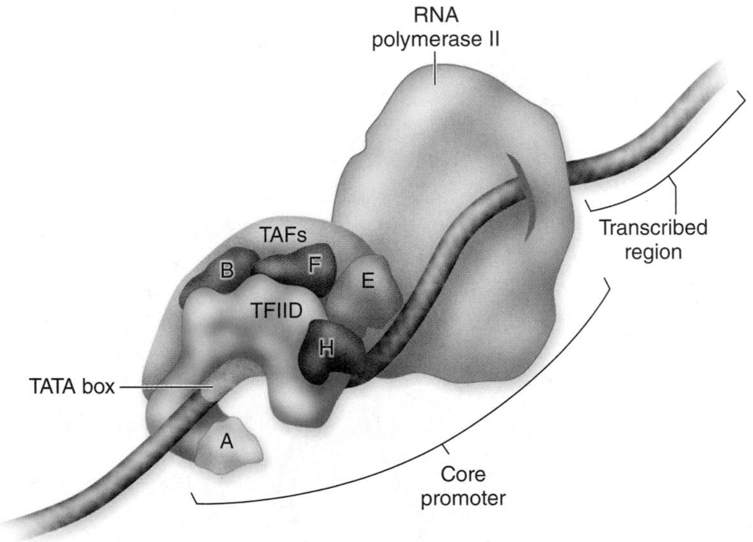

Figure 16.8 Formation of a eukaryotic initiation complex. The general transcription factor, TFIID, binds to the TATA box and is joined by the other general factors, TFIIE, TFIIF, TFIIA, TFIIB, and TFIIH. This complex is added to by a number of transcription-associated factors (TAFs) that together recruit the RNA Pol II molecule to the core promoter.

recognizes the TATA box sequence found in many eukaryotic promoters.

Binding of TFIID is followed by binding of TFIIE, TFIIF, TFIIA, TFIIB, and TFIIH and a host of accessory factors called *transcription-associated factors,* TAFs. The *initiation complex* that results (figure 16.8) is clearly much more complex than the bacterial RNA polymerase holoenzyme binding to a promoter. And there is yet another level of complexity: The initiation complex, although capable of initiating synthesis at a basal level, does not achieve transcription at a high level without the participation of other, specific factors.

Specific transcription factors

Specific transcription factors act in a tissue- or time-dependent manner to stimulate higher levels of transcription than the basal level. The number and diversity of these factors are overwhelming. Some sense can be made of this proliferation of factors by concentrating on the DNA-binding motif, as opposed to the specific factors.

A key common theme that emerges from the study of these factors is that specific transcription factors, called *activators,* have a domain organization. Each factor consists of a DNA-binding domain and a separate activating domain that interacts with the transcription apparatus, and these domains are essentially independent in the protein. If the DNA-binding domains are "swapped" between different factors the binding specificity for the factors is switched without affecting their ability to activate transcription.

Promoters and enhancers are binding sites for transcription factors

Promoters, as mentioned in the preceding chapter, contain DNA-binding sites for general transcription factors. These factors then mediate the binding of RNA polymerase II to the promoter (and also the binding of RNA polymerases I and III to their specific promoters). In contrast, the holoenzyme portion of the RNA polymerase of prokaryotes can directly recognize a promoter and bind to it.

Enhancers were originally defined as DNA sequences necessary for high levels of transcription that can act independently of position or orientation. At first, this concept seemed counterintuitive, especially since molecular biologists had been conditioned by prokaryotic systems to expect control regions to be immediately upstream of the coding region. It turns out that enhancers are the binding site of the specific transcription factors. The ability of enhancers to act over large distances was at first puzzling, but investigators now think this action is accomplished by DNA bending to form a loop, positioning the enhancer closer to the promoter.

Although more important in eukaryotic systems, this looping was first demonstrated using prokaryotic

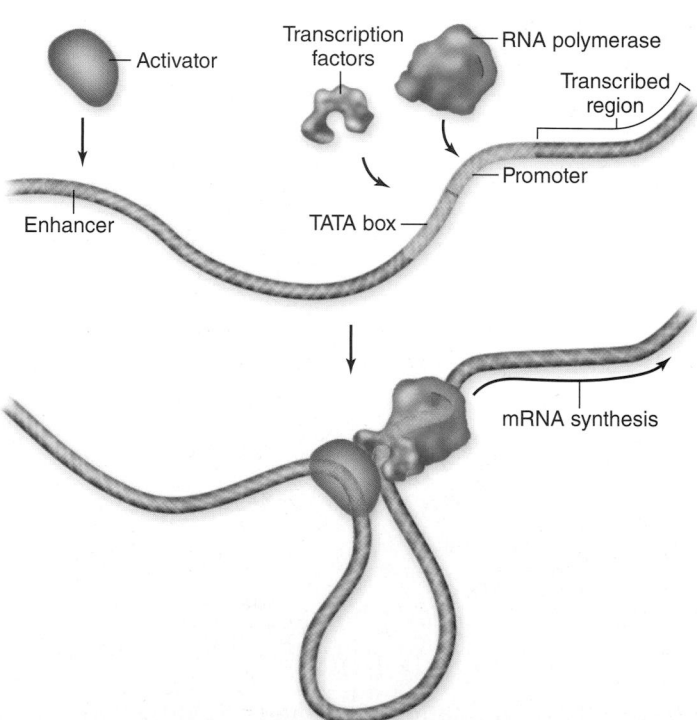

DNA-binding proteins (figure 16.9). The important point is that the linear distance separating two sites on the chromosome does not have to translate to great physical distance, because the flexibility of DNA allows bending and looping. An activator bound to an enhancer can thus be brought into contact with the transcription factors bound to a distant promoter (figure 16.10).

Coactivators and mediators link transcription factors to RNA polymerase II

Other factors specifically mediate the action of transcription factors. These *coactivators* and *mediators* are also necessary for activation of transcription by the transcription factor. They act by binding the transcription factor and then binding to another part of the transcription apparatus. Mediators are essential to the function of some transcription factors, but not all transcription factors require them. The number of coactivators is much smaller than the number of transcription factors because the same coactivator can be used with multiple transcription factors.

NtrC (activator)
Enhancer
5 nm
Promoter
RNA polymerase

Bacterial RNA polymerase is loosely bound to the promoter. The activator (NtrC) binds at the enhancer.

ATP

ADP

5 nm

DNA loops around so that the activator comes into contact with the RNA polymerase.

RNA polymerase

Activator

mRNA synthesis

The activator triggers RNA polymerase activation, and transcription begins. DNA unloops.

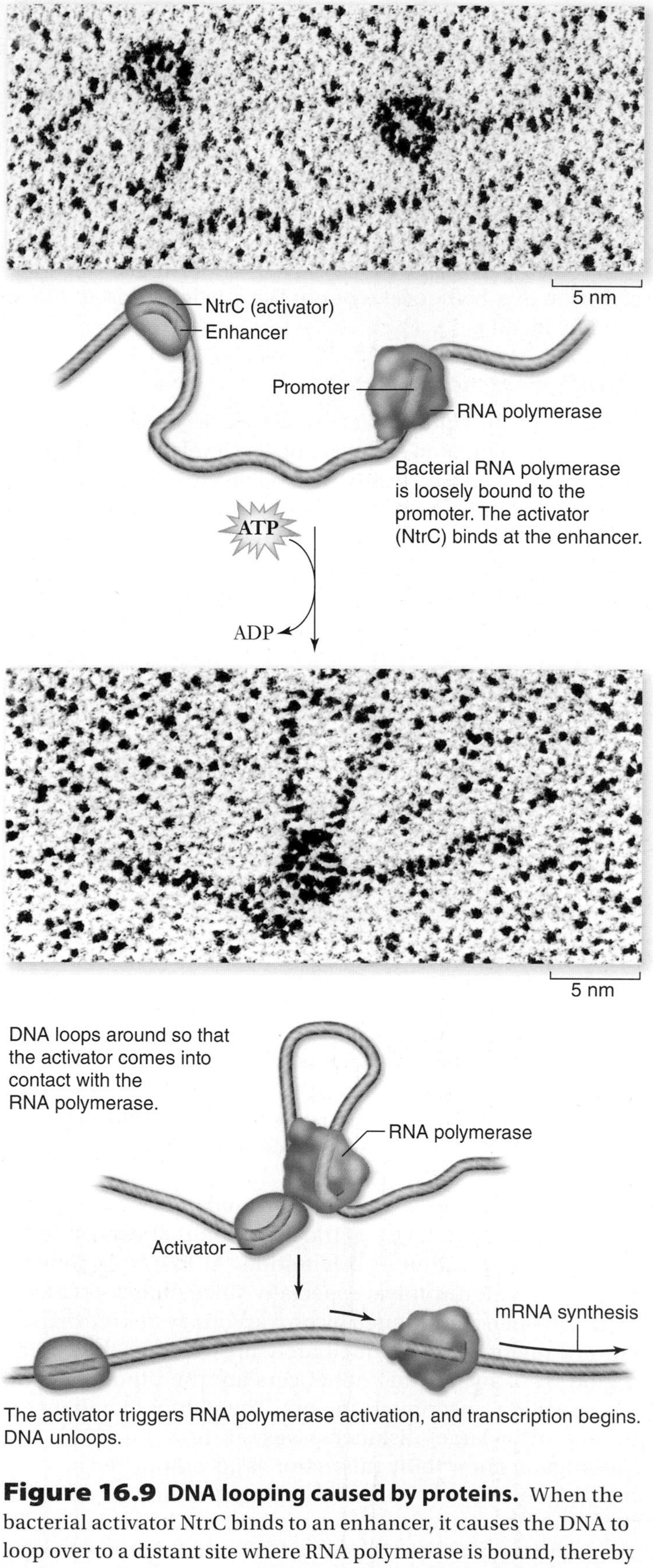

Figure 16.9 DNA looping caused by proteins. When the bacterial activator NtrC binds to an enhancer, it causes the DNA to loop over to a distant site where RNA polymerase is bound, thereby activating transcription. Although such enhancers are rare in prokaryotes, they are common in eukaryotes.

Activator
Transcription factors
RNA polymerase
Transcribed region
Enhancer
TATA box
Promoter

mRNA synthesis

Figure 16.10 How enhancers work. The enhancer site is located far away from the gene being regulated. Binding of an activator *(gray)* to the enhancer allows the activator to interact with the transcription factors *(blue)* associated with RNA polymerase, stimulating transcription.

The transcription complex brings things together

Although a few general principles apply to a broad range of situations, nearly every eukaryotic gene—or group of genes with coordinated regulation—represents a unique case. Virtually all genes that are transcribed by RNA polymerase II need the same suite of general factors to assemble an initiation complex, but the assembly of this complex and its ultimate level of transcription depend on specific transcription factors that in combination make up the **transcription complex** (figure 16.11).

The makeup of eukaryotic promoters, therefore, is either very simple, if we consider only what is needed for the initiation complex, or very complicated, if we consider all factors that may bind in a complex and affect transcription. This kind of combinatorial gene regulation leads to great flexibility because it can respond to the many signals a cell may receive affecting transcription, allowing integration of these signals.

? Inquiry question   How do eukaryotes coordinate the activation of many genes whose transcription must occur at the same time?

Learning Outcomes Review 16.4

In eukaryotes, initiation requires general transcription factors that bind to the promoter and recruit RNA polymerase II to form an initiation complex. General factors produce the basal level of transcription. Specific transcription factors, which bind to enhancer sequences, can increase the level of transcription. Enhancers can act at a distance because DNA can loop, bringing an enhancer and a promoter closer together. Additional coactivators and mediators link certain specific transcription factors to RNA polymerase II.

■ *What would be the effect of a mutation that results in the loss of a general transcription factor versus the loss of a specific factor?*

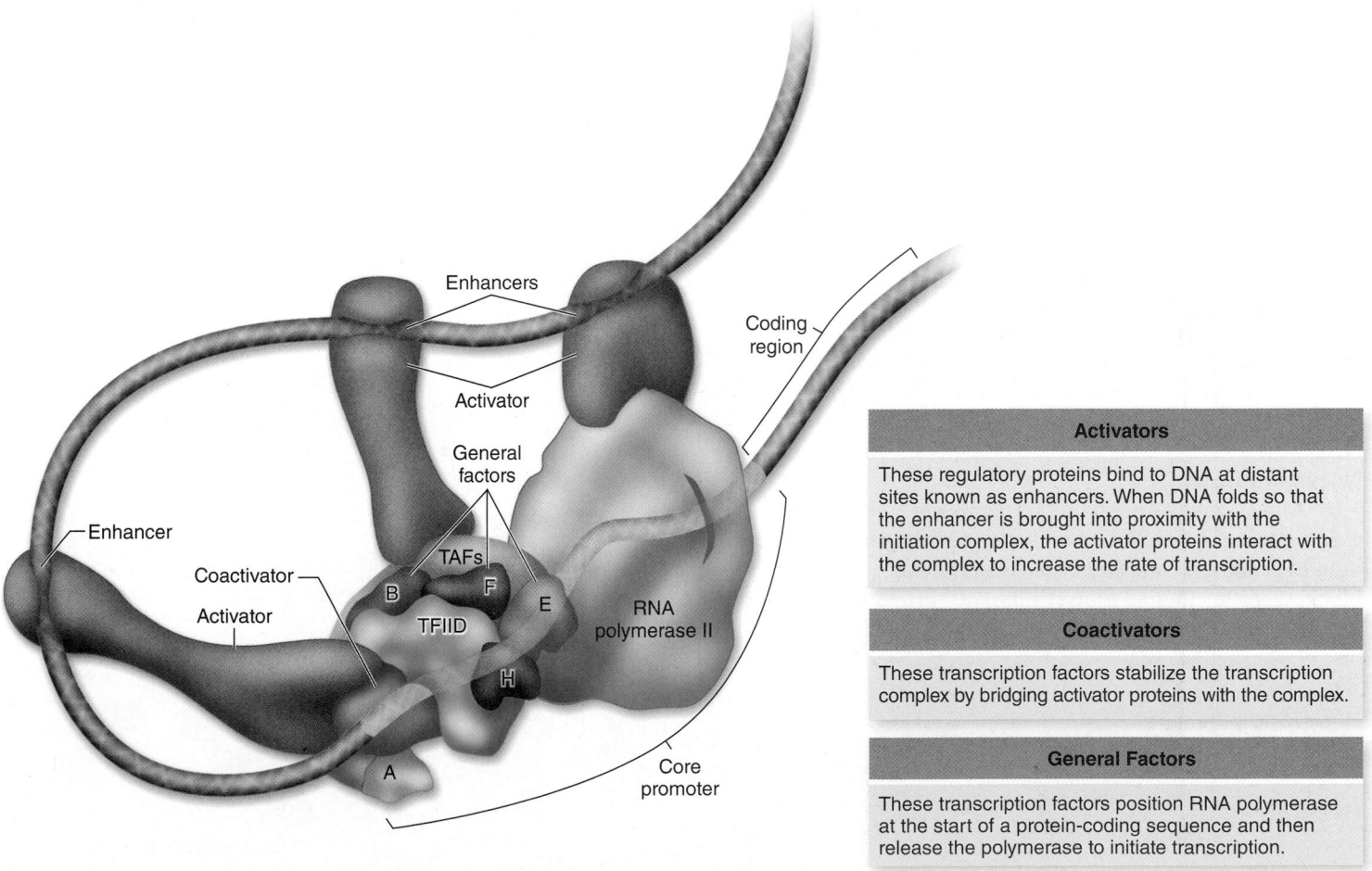

Activators

These regulatory proteins bind to DNA at distant sites known as enhancers. When DNA folds so that the enhancer is brought into proximity with the initiation complex, the activator proteins interact with the complex to increase the rate of transcription.

Coactivators

These transcription factors stabilize the transcription complex by bridging activator proteins with the complex.

General Factors

These transcription factors position RNA polymerase at the start of a protein-coding sequence and then release the polymerase to initiate transcription.

Figure 16.11 Interactions of various factors within the transcription complex. All specific transcription factors bind to enhancer sequences that may be distant from the promoter. These proteins can then interact with the initiation complex by DNA looping to bring the factors into proximity with the initiation complex. As detailed in the text, some transcription factors, called activators, can directly interact with the RNA polymerase II or the initiation complex, whereas others require additional coactivators.

16.5 *Eukaryotic Chromatin Structure*

Learning Outcomes

1. *Describe how chromatin structure can affect gene expression.*
2. *Explain the function of chromatin remodeling complexes.*

Eukaryotes have the additional gene expression hurdle of possessing DNA that is packaged into chromatin. The packaging of DNA first into nucleosomes and then into higher order chromatin structures is now thought to be directly related to the control of gene expression.

Chromatin structure at its lowest level is the organization of DNA and histone proteins into *nucleosomes* (see chapter 10). These nucleosomes may block binding of transcription factors and RNA polymerase II at the promoter.

The higher order organization of chromatin, which is not completely understood, appears to depend on the state of the histones in nucleosomes. Histones can be modified to result in a greater condensation of chromatin, making promoters even less accessible for protein–DNA interactions. A chromatin remodeling complex exists that can make DNA more accessible.

Both DNA and histone proteins can be modified

Chemical *methylation* of the DNA was once thought to play a major role in gene regulation in vertebrate cells. The addition of a methyl group to cytosine creates 5-methylcytosine, but this change has no effect on its base-pairing with guanine (figure 16.12). Similarly, the addition of a methyl group to uracil produces thymine, which clearly does not affect base-pairing with adenine.

Many inactive mammalian genes are methylated, and it was tempting to conclude that methylation caused the inactivation. But methylation is now viewed as having a less direct role, blocking the accidental transcription of "turned-off" genes. Vertebrate cells apparently possess a protein that binds to clusters of 5-methylcytosine, preventing transcriptional activators from gaining access to the DNA. DNA methylation in vertebrates thus ensures that once a gene is turned off, it stays off.

The histone proteins that form the core of the nucleosome (chapter 10) can also be modified. This modification is correlated with active versus inactive regions of chromatin, similar to the methylation of DNA just described. Histones can also be methylated, and this alteration is generally found in inactive regions of chromatin. Finally, histones can be modified by the addition of an acetyl group, and this addition is correlated with active regions of chromatin.

Some transcription activators alter chromatin structure

The control of eukaryotic transcription requires the presence of many different factors to activate transcription. Some activators seem to interact directly with the initiation complex or with coactivators that themselves interact with the initiation complex, as described earlier. Other cases are not so clear. The emerging consensus is that some coactivators have been shown to be histone acetylases. In these cases, it appears that transcription is increased by removing higher order chromatin structure that would prevent transcription (figure 16.13). Some corepressors have been shown to be histone deacetylases as well.

These observations have led to the suggestion that a "histone code" might exist, analogous to the genetic code.

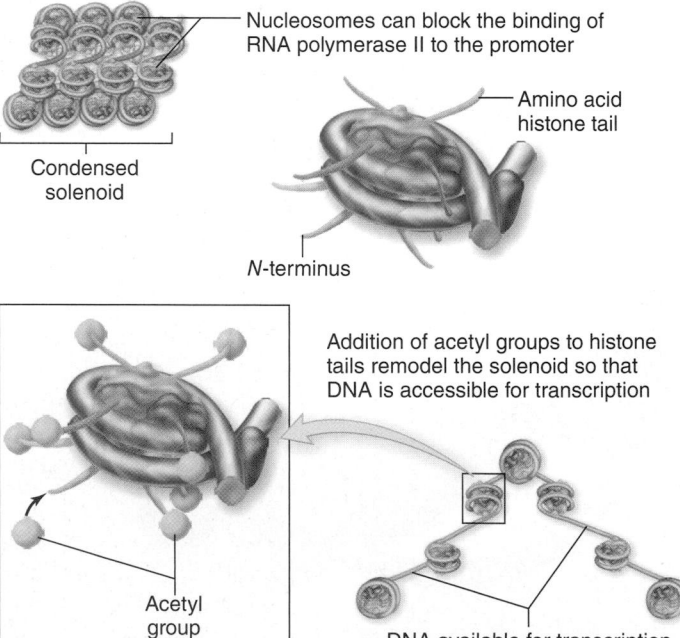

Figure 16.13 Histone modification affects chromatin structure. DNA in eukaryotes is organized first into nucleosomes and then into higher order chromatin structures. The histones that make up the nucleosome core have amino tails that protrude. These amino tails can be modified by the addition of acetyl groups. The acetylation alters the structure of chromatin, making it accessible to the transcription apparatus.

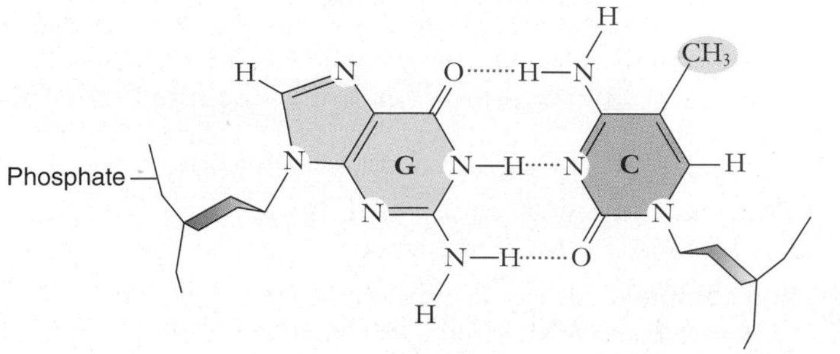

Figure 16.12 DNA methylation. Cytosine is methylated, creating 5-methylcytosine. Because the methyl group *(green)* is positioned to the side, it does not interfere with the hydrogen bonds of a G–C base-pair, but it can be recognized by proteins.

This histone code is postulated to underlie the control of chromatin structure and, thus, of access of the transcription machinery to DNA.

Chromatin-remodeling complexes also change chromatin structure

The outline of how alterations to chromatin structure can regulate gene expression are beginning to emerge. A key discovery is the existence of so-called **chromatin-remodeling complexes.** These large complexes of proteins include enzymes that modify histones and DNA and that also change chromatin structure itself.

One class of these remodeling factors, ATP-dependent chromatin remodeling factors, function as molecular motors that affect DNA and histones. These ATP-dependent remodeling factors use energy from ATP to alter the relationships between histones and DNA. They can catalyze four different changes in histone/DNA binding (figure 16.14): (1) nucleosome sliding along DNA, which changes the position of a nucleosome on the DNA; (2) create a remodeled state where

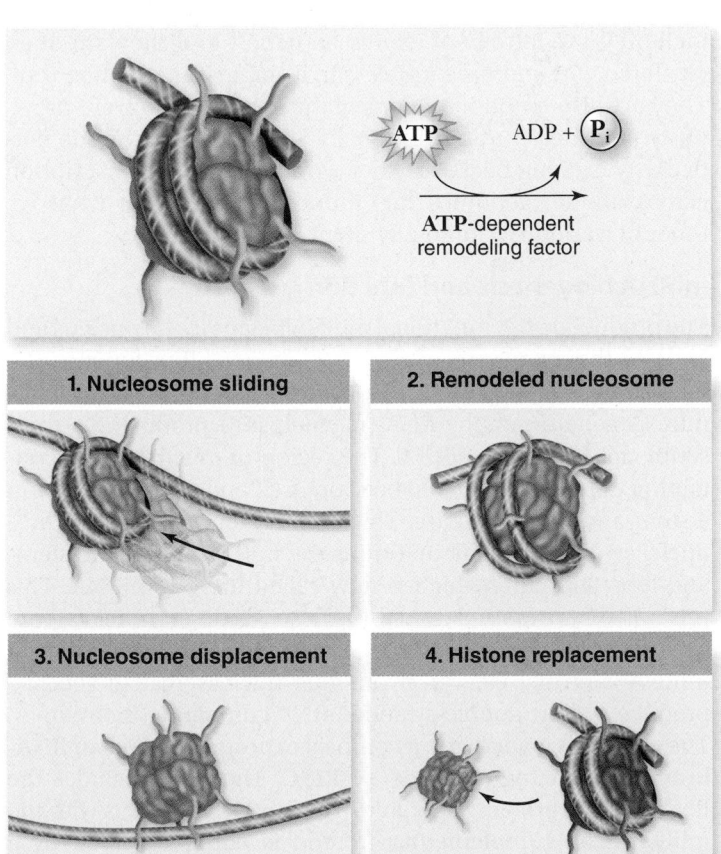

Figure 16.14 Function of ATP-dependent remodeling factors. ATP-dependent remodeling factors use the energy from ATP to alter chromatin structure. They can *(1)* slide nucleosomes along DNA to reveal binding sites for proteins; *(2)* create a remodeled state of chromatin where the DNA is more accessible; *(3)* completely remove nucleosomes from DNA; and *(4)* replace histones in nucleosomes with variant histones.

DNA is more accessible; (3) removal of nucleosomes from DNA; and (4) replacement of histones with variant histones. These functions all act to make DNA more accessible to regulatory proteins that in turn, affect gene expression.

Learning Outcomes Review 16.5

Eukaryotic DNA is packaged into chromatin, adding another structural challenge to transcription. Changes in chromatin structure correlate with modification of DNA and histones, and access to DNA by transcriptional regulators requires changes in chromatin structure. Some transcriptional activators modify histones by acetylation. Large chromatin-remodeling complexes include enzymes that alter the structure of chromatin, making DNA more accessible to regulatory proteins.

■ **Genes that are turned on in all cells are called "housekeeping" genes. Explain the idea behind this name.**

16.6 Eukaryotic Posttranscriptional Regulation

Learning Outcomes

1. **Explain how small RNAs can affect gene expression.**
2. **Differentiate between the different kinds of posttranscriptional regulation.**

The separation of transcription in the nucleus and translation in the cytoplasm in eukaryotes provides possible points of regulation that do not exist in prokaryotes. For many years we thought of this as "alternative" forms of regulation, but it now appears that they play a much more central role than previously suspected. In this section we will consider several of these mechanisms for controlling gene expression, beginning with the exciting new area of regulation by small RNAs.

Small RNAs act after transcription to control gene expression

Developmental genetics has provided important insights into the regulation of gene expression. A striking example is the discovery of small RNAs that affect gene expression. The mutant *lin-4* was known to alter developmental timing in the worm *C. elegans,* and genetic studies had shown that *lin-4* regulated another gene, *lin-14*. When Ambros, Lee, and Feinbaum isolated the *lin-4* gene in 1992, they found it did not encode a protein product. Instead, the *lin-4* gene encoded only two small RNA molecules, one of 22 nt and the other of 61 nt. Furthermore, the 22-nt RNA was derived from the longer 61-nt RNA. Further work showed that this small RNA was complementary to a region in *lin-14*. A model was developed in which the lin-4 RNA acted as a translational repressor of the lin-14 mRNA (figure 16.15). Although not called that at the time, this was the first identified **micro-RNA**, or **miRNA.**

A completely different line of inquiry involved the use of double-stranded RNAs to turn off gene expression. This has been shown to act via another class of small RNA called **small interfering RNAs**, or **siRNAs.** These may be experimentally introduced, derived from invading viruses, or even encoded in the genome. The use of siRNA to control gene expression revealed the existence of cellular mechanisms for the control of gene expression via small RNAs.

Since its discovery, gene silencing by small RNAs has been a source of great interest for both its experimental uses, and as an explanation for posttranslational control of gene expression. Recent research has uncovered a wealth of new types of small RNAs, but we will confine ourselves to the two classes of miRNA and siRNA, as these are well established and illustrate the RNA silencing machinery.

miRNA genes

The discovery of the role of miRNAs in gene expression initially appeared to be confined to nematodes because the *lin-4* gene did not have any obvious homologues in other systems. Seven years later, a second gene, *let-7*, was discovered in the same pathway in *C. elegans*. The *let-7* gene also encoded a 22-nt RNA that could influence translation. In this case, homologues for *let-7* were immediately found in both *Drosophila* and humans.

As an increasing number of miRNAs were discovered in different organisms, miRNA gene discovery has turned to computer searching and high-throughput methods such as microarrays and new next-generation sequencing. A database devoted to miRNAs currently lists 695 known human miRNA sequences.

Genes for miRNA are found in a variety of locations, including the introns of expressed genes, and they are often clustered with multiple miRNAs in a single transcription unit. They are also found in regions of the genome that were previously considered transcriptionally silent. This finding is particularly exciting because other work looking at transcription across animal genomes has found that much of what we thought was transcriptionally silent is actually not.

miRNA biogenesis and function

The production of a functional miRNA begins in the nucleus and ends in the cytoplasm with an ~22-nt RNA that functions to repress gene expression (figure 16.16). The initial transcript of a miRNA gene occurs by RNA polymerase II producing a transcript called the pri-miRNA. The region of this transcript containing the miRNA can fold back on itself and base-pair to form a stem-and-loop structure. This is cleaved in the nucleus by a nuclease called Drosha that trims the miRNA to just the stem-and-loop structure, which is now called the pre-miRNA. This pre-miRNA is exported from the nucleus through a nuclear pore bound to the protein exportin 5. Once in the cytoplasm, the pre-miRNA is further cleaved by another nuclease called Dicer to produce a short double-stranded RNA containing the miRNA. The miRNA is loaded into a complex of proteins called an **RNA-induced silencing complex,** or **RISC.** The RISC includes the RNA-binding protein Argonaute (Ago), which interacts with the miRNA. The complementary strand is either removed by a nuclease or is removed during the loading process.

At this point, the RISC is targeted to repress the expression of other genes based on sequence complementarity to the miRNA. The complementary region is usually in the 3′ untranslated region of genes, and the result can be cleavage of the mRNA or inhibition of translation. It appears that in animals, the inhibition of translation is more common than the cleavage of the mRNA, although the precise mechanism of this inhibition is still unclear. In plants, the cleavage of the mRNA by the

---

Hypothesis: *The region of the* lin-14 *gene complementary to the* lin-4 *miRNA controls* lin-14 *expression.*

Prediction: *If the* lin-4 *complementary region of the* lin-14 *gene is spliced into a reporter gene, then this reporter gene should show regulation similar to* lin-14.

Test: *Recombinant DNA is used to make two versions of a reporter gene (β-galactosidase). In transgenic worms (C. elegans), expression of the reporter gene produces a blue color.*

1. The β-galactosidase gene with the lin-14 *3′ untranslated region containing the* lin-4 *complementary region (shown below)*

2. The β-galactosidase gene with a control 3′ untranslated region with no lin-4 *complementary region (not shown)*

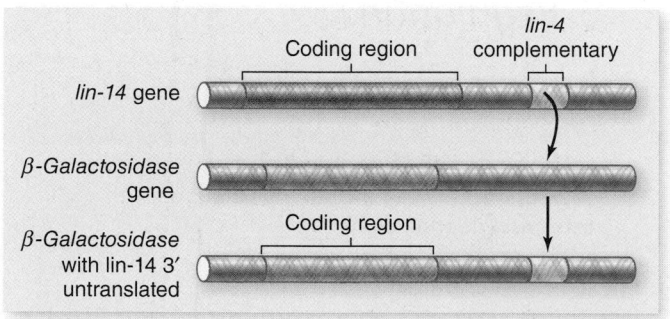

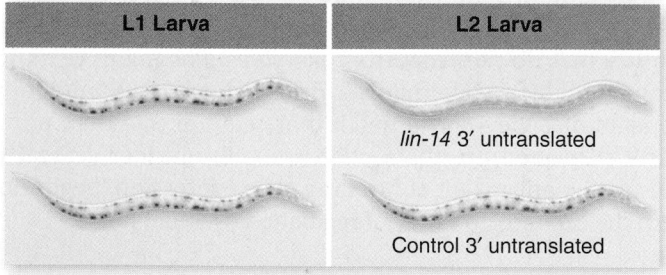

Result:

1. Transgenic worms with reporter gene plus lin-14 *3′ untranslated region show expression in L1 but not L2 stage larvae. This is the pattern expected for the* lin-14 *gene, which is controlled by* lin-4.

2. Transgenic worms with reporter gene with control 3′ untranslated region do not show expression pattern expected for control by lin-4.

Conclusion: *The 3′ untranslated region from* lin-14 *is sufficient to turn off gene expression in L2 larvae.*

Further Experiments: *What expression pattern would you predict for these constructs in a mutant that lacks* lin-4 *function?*

Figure 16.15 Control of *lin-14* gene expression. The *lin-14* gene is controlled by the *lin-4* gene. This is mediated by a region of the 3′ untranslated region of the *lin-14* mRNA that is complementary to *lin-4* miRNA.

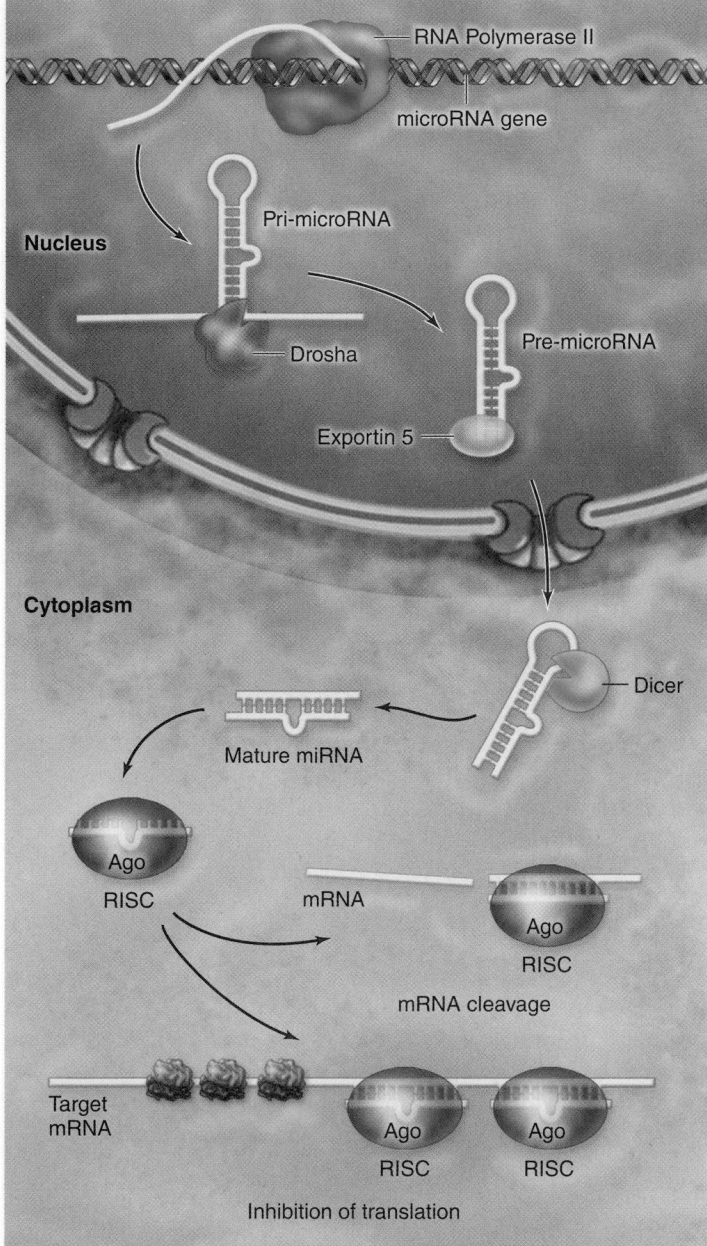

Figure 16.16 Biogenesis and function of miRNA. Genes for miRNAs are transcribed by RNA polymerase II to produce a pri-miRNA. This is processed by the Drosha nuclease to produce the pre-miRNA, which is exported from the nucleus bound to export factor Exportin 5. Once in the cytoplasm, the pre-miRNA is processed by Dicer nuclease to produce the mature miRNA. The miRNA is loaded into a RISC, which can act to either cleave target mRNAs, or to inhibit translation of target mRNAs.

RISC is common and seems to be related to the more precise complementarity found between plant miRNAs and their targets than that found in animal systems.

RNA interference

Small RNA-mediated gene silencing has been known for a number of years. Some confusion arose in the nomenclature in this area because work in different systems led to a profusion of names. However, RNA interference, cosuppression,

and posttranscriptional gene silencing all act through similar biochemical mechanisms. The term **RNA interference** is currently the most commonly used and involves the production of siRNAs.

The production of siRNAs is similar to that of miRNAs, except that they arise from a long piece of double-stranded RNA (figure 16.17). This can be either a very long region of self complementarity, or from two complementary RNAs. These long double-stranded RNAs are processed by Dicer to yield multiple siRNAs that are loaded into an Ago containing RISC. The siRNAs usually have near-perfect complementarity to their target mRNAs, and the result is cleavage of the mRNA by the siRNA containing RISC.

The source of the double-stranded RNA to produce siRNAs can be either from the cell or from outside the cell. From the cell itself, genes can produce RNAs with long regions of self-complementarity that fold back to produce a substrate for Dicer in the cytoplasm. They can also arise from repeated regions of the genome that contain transposable elements. Exogenous double-stranded RNAs can be introduced experimentally or by infection with a virus.

Small RNAs may have evolved to protect the genome

The observation that viral RNA can be degraded via the RNA-silencing pathway may point toward the evolutionary origins of small RNAs. A related observation is that RNA silencing can

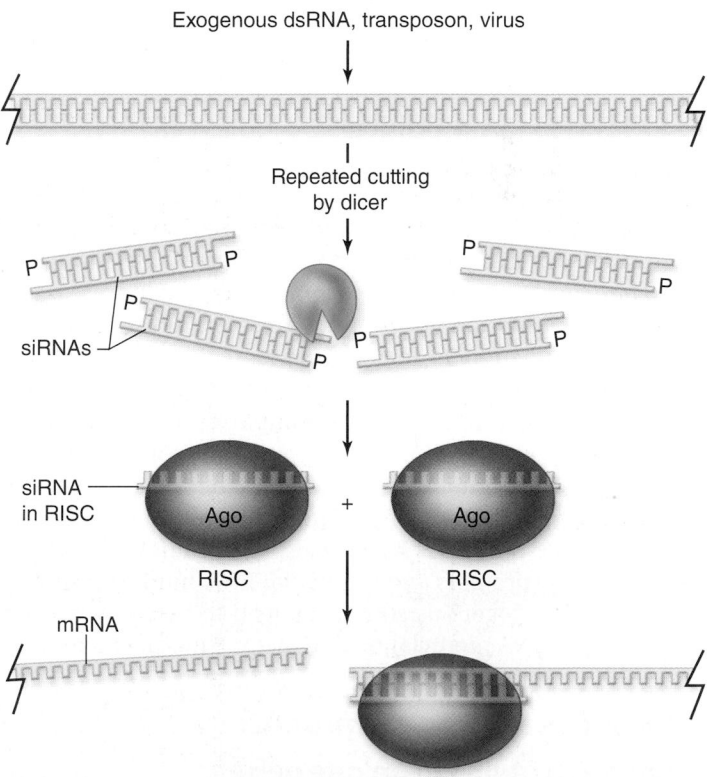

Figure 16.17 Biogenesis and function of siRNA. SiRNAs can arise from a variety of sources that all produce long double-stranded regions of RNA. The double-stranded RNA is processed by Dicer nuclease to produce a number of siRNAs that are each loaded onto their own RISC. The RISC then cleaves target mRNA.

control the action of transposons as well. In both mice and fruit flies, genetic evidence supports the involvement of the RNA interference machinery in the germ line where a specific class of small RNA appears to be involved in silencing transposons during spermatogenesis and oogenesis. Thus the origins of this mechanism may be an ancient pathway for protection of the genome from assault from both within and without. The conservation of key proteins suggests that the ancestor to all eukaryotes had some form of RNA-silencing pathway.

Distinguishing miRNAs and siRNAs

The biogenesis of both miRNA and siRNA involves cleavage by Dicer and incorporation into a RISC complex. The main thing that distinguishes these small RNAs is their targets: miRNAs tend to repress genes different from their origin, whereas endogenous siRNAs tend to repress the genes they were derived from. Additionally, siRNAs are used experimentally to turn off the expression of genes. This technique uses the cells' RNA-silencing machinery to turn off a gene by introducing a double-stranded RNA complementary to the gene.

The two classes of small RNA have other differences. When multiple species are examined, miRNAs tend to be evolutionarily conserved, but siRNAs do not. Although the biogenesis for both is similar in terms of the nucleases involved, the actual structure of the double-stranded RNAs is not the same. The transcript of miRNA genes form stem-loop structures containing the miRNA, but the double-stranded RNAs generating siRNAs may be bimolecular, or very long stem-loops. These longer double-stranded regions lead to multiple siRNAs, whereas only a single miRNA is generated from a pre-miRNA.

Small RNAs can mediate heterochromatin formation

RNA-silencing pathways have also been implicated in the formation of heterochromatin in fission yeast, plants, and *Drosophila*. In fission yeast, centromeric heterochromatin formation is driven by siRNAs produced by the action of the Dicer nuclease. This heterochromatin formation also involves modification of histone proteins and thus connects RNA interference with chromatin-remodeling complexes in this system. It is not yet clear how widespread this phenomenon is.

Plants are an interesting case in that they have a variety of small RNA species. The RNA interference pathway in plants is more complex than that in animals, with multiple forms of Dicer nuclease proteins and Argonaute RNA-binding proteins. One class of endogenous siRNA can lead to heterochromatin formation by DNA methylation and histone modification.

Alternative splicing can produce multiple proteins from one gene

The latest estimates of the frequency of alternative splicing mentioned in the preceding chapter emphasize its importance. However, the functional significance of these data is still not clear. Here we will consider some well characterized examples.

Alternative splicing can change the splicing events that occur during different stages of development or in different tissues. An example of developmental differences is found in *Drosophila,* in which sex determination is the result of a complex series of alternative splicing events that differ in males and females.

An excellent example of tissue-specific alternative splicing in action is found in two different human organs: the thyroid gland and the hypothalamus. The thyroid gland is responsible for producing hormones that control processes such as metabolic rate. The hypothalamus, located in the brain, collects information from the body (for example, salt balance) and releases hormones that in turn regulate the release of hormones from other glands, such as the pituitary gland. (You'll learn more about these glands in chapter 45.)

These two organs produce two distinct hormones: *calcitonin* and *CGRP* (calcitonin-gene-related peptide) as part of their function. Calcitonin controls calcium uptake and the balance of calcium in tissues such as bones and teeth. CGRP is involved in a number of neural and endocrine functions. Although these two hormones are used for very different physiological purposes, they are produced from the same transcript (figure 16.18).

The synthesis of one product versus another is determined by tissue-specific factors that regulate the processing of the primary transcript. In the case of calcitonin and CGRP, pre-mRNA splicing is controlled by different factors that are present in the thyroid and in the hypothalamus.

RNA editing alters mRNA after transcription

In some cases, the editing of mature mRNA transcripts can produce an altered mRNA that is not truly encoded in the genome—an unexpected possibility. RNA editing was first discovered as the insertion of uracil residues into some RNA transcripts in protozoa, and it was thought to be an anomaly.

RNA editing of a different sort has since been found in mammalian species, including humans. In this case, the editing involves chemical modification of a base to change its base-pairing properties, usually by deamination. For example, both deamination of cytosine to uracil and deamination of adenine to inosine have been observed (inosine pairs as G would during translation).

Apolipoprotein B

The human protein apolipoprotein B is involved in the transport of cholesterol and triglycerides. The gene that encodes this protein, *apoB,* is large and complex, consisting of 29 exons scattered across almost 50 kilobases (kb) of DNA.

The protein exists in two isoforms: a full-length APOB100 form and a truncated APOB48 form. The truncated form is due to an alteration of the mRNA that changes a codon for glutamine to one that is a stop codon. Furthermore, this editing occurs in a tissue-specific manner; the edited form appears only in the intestine, whereas the liver makes only the full-length form. The full-length APOB100 form is part of the low-density lipoprotein (LDL) particle that carries cholesterol. High levels of serum LDL are thought to be a

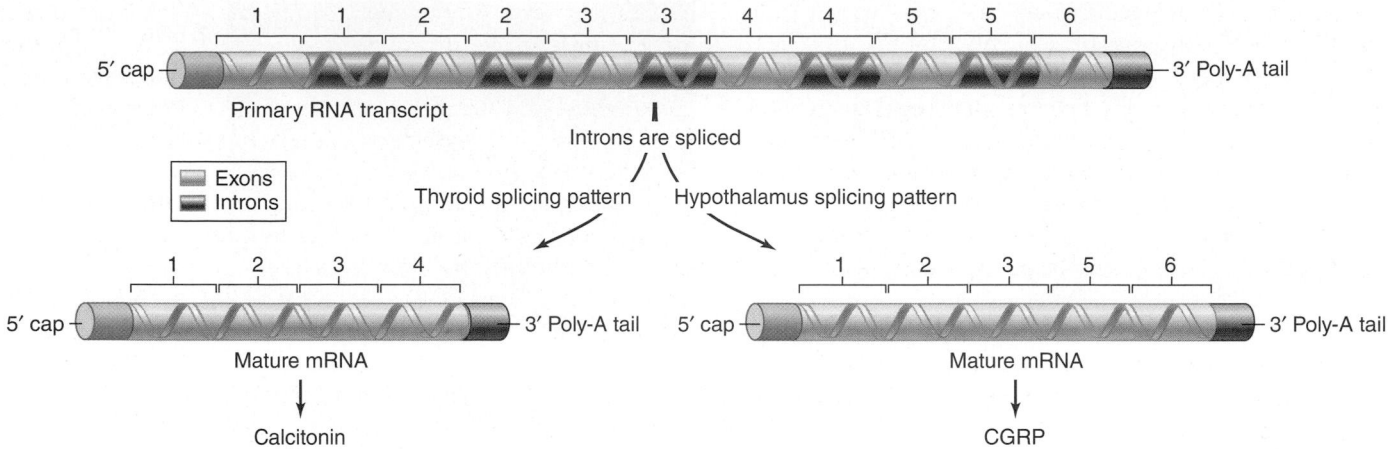

Figure 16.18 Alternative splicing. Many primary transcripts can be spliced in different ways to give rise to multiple mRNAs. In this example, in the thyroid the primary transcript is spliced to contain four exons encoding the protein calcitonin. In the hypothalamus the fourth exon, which contains the poly-A site used in the thyroid, is skipped and two additional exons are added to encode the protein calcitonin-gene-related peptide (CGRP).

major predictor of atherosclerosis in humans. It does not appear that editing has any effect on the levels of the intestine-specific transcript.

The 5-HT serotonin receptor

RNA editing has also been observed in some brain receptors for opiates in humans. One of these receptors, the serotonin (5-HT) receptor, is edited at multiple sites to produce a total of 12 different isoforms of the protein.

It is unclear how widespread these forms of RNA editing are, but they are further evidence that the information encoded within genes is not the end of the story for protein production.

mRNA must be transported out of the nucleus for translation

Processed mRNA transcripts exit the nucleus through the nuclear pores (described in chapter 4). The passage of a transcript across the nuclear membrane is an active process that requires the transcript to be recognized by receptors lining the interior of the pores. Specific portions of the transcript, such as the poly-A tail, appear to play a role in this recognition.

There is little hard evidence that gene expression is regulated at this point, although it could be. On average, about 10% of primary transcripts consists of exons that will make up mRNA sequences, but only about 5% of the total mRNA produced as primary transcript ever reaches the cytoplasm. This observation suggests that about half of the exons in primary transcripts never leave the nucleus, but it is unclear whether the disappearance of this mRNA is selective.

Initiation of translation can be controlled

The translation of a processed mRNA transcript by ribosomes in the cytoplasm involves a complex of proteins called *translation factors*. In at least some cases, gene expression is regulated by modification of one or more of these factors. In other instances, **translation repressor proteins** shut down transla-

tion by binding to the beginning of the transcript, so that it cannot attach to the ribosome.

In humans, the production of ferritin (an iron-storing protein) is normally shut off by a translation repressor protein called aconitase. Aconitase binds to a 30-nt sequence at the beginning of the ferritin mRNA, forming a stable loop to which ribosomes cannot bind. When iron enters the cell, the binding of iron to aconitase causes the aconitase to dissociate from the ferritin mRNA, freeing the mRNA to be translated and increasing ferritin production 100-fold.

The degradation of mRNA is controlled

Another aspect that affects gene expression is the stability of mRNA transcripts in the cell cytoplasm. Unlike prokaryotic mRNA transcripts, which typically have a half-life of about 3 min, eukaryotic mRNA transcripts are very stable. For example, β-globin gene transcripts have a half-life of over 10 hr, an eternity in the fast-moving metabolic life of a cell.

The transcripts encoding regulatory proteins and growth factors, however, are usually much less stable, with half-lives of less than 1 hr. What makes these particular transcripts so unstable? In many cases, they contain specific sequences near their 3′ ends that make them targets for enzymes that degrade mRNA. A sequence of A and U nucleotides near the 3′ poly-A tail of a transcript promotes removal of the tail, which destabilizes the mRNA.

Loss of the poly-A tail leads to rapid degradation by 3′ to 5′ RNA exonucleases. Another consequence of this loss is the stimulation of decapping enzymes that remove the 5′ cap leading to degradation by 5′ to 3′ RNA exonucleases.

Other mRNA transcripts contain sequences near their 3′ ends that are recognition sites for endonucleases, which cause these transcripts to be digested quickly. The short half-lives of the mRNA transcripts of many regulatory genes are critical to the function of those genes because they enable the levels of regulatory proteins in the cell to be altered rapidly.

A review of various methods of posttranscriptional control of gene expression is provided in figure 16.19.

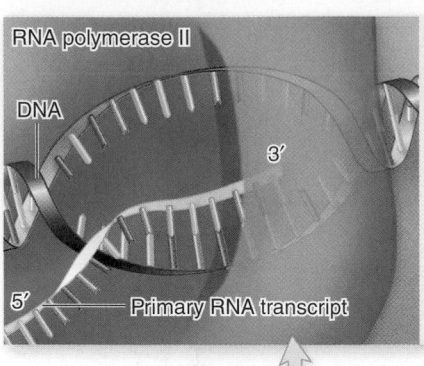

1. Initiation of transcription
Transcription is controlled by the frequency of initiation. This involves transcription factors that bind to promoters and enhancers.

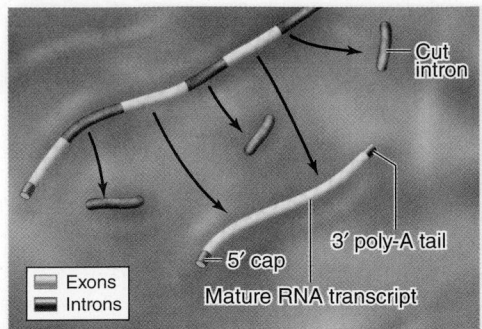

2. RNA splicing
Gene expression can be controlled by altering the rate of splicing in eukaryotes. Alternative splicing can produce multiple mRNAs from one gene.

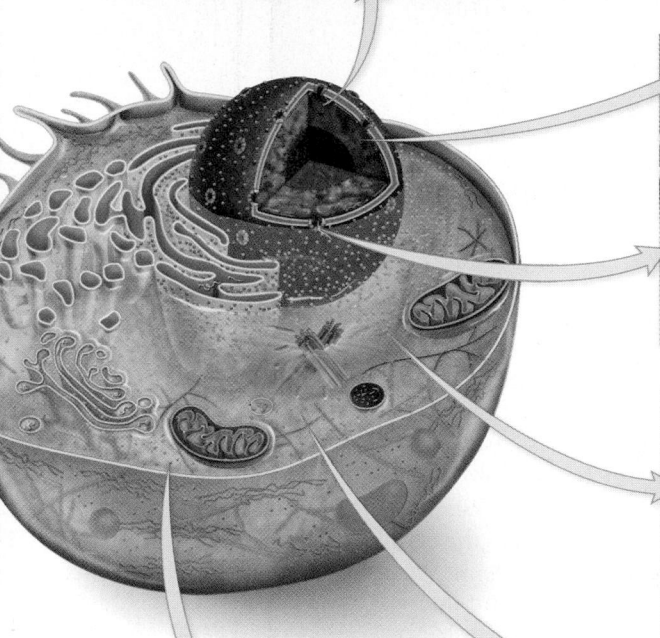

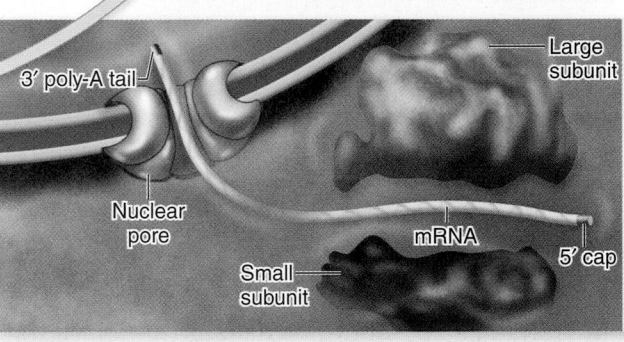

3. Passage through the nuclear membrane
Gene expression can be regulated by controlling access to or efficiency of transport channels.

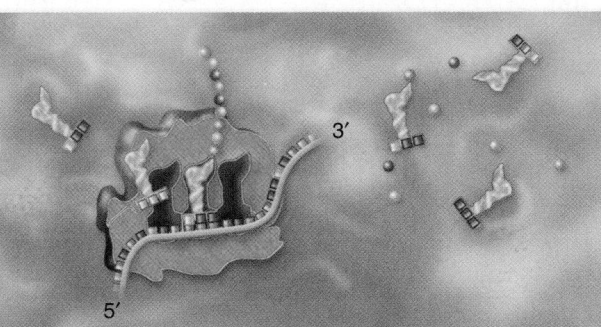

4. Protein synthesis
Many proteins take part in the translation process, and regulation of the availability of any of them alters the rate of gene expression by speeding or slowing protein synthesis.

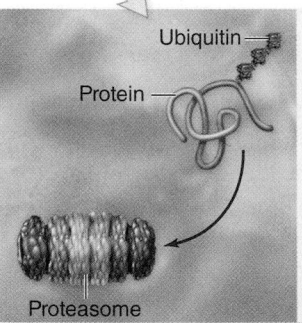

6. Protein degradation
Proteins to be degraded are labeled with ubiquitin, then destroyed by the proteasome.

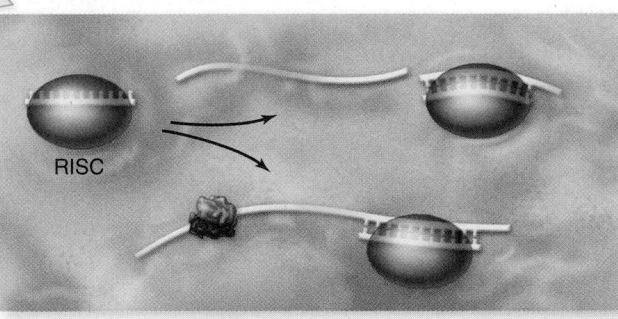

5. RNA interference
Gene expression is regulated by small RNAs. Protein complexes containing siRNA and miRNA target specific mRNAs for destruction or inhibit their translation.

Figure 16.19 Mechanisms for control of gene expression in eukaryotes.

Learning Outcomes Review 16.6

Small RNAs control gene expression by either selective degradation of mRNA, inhibition of translation, or alteration of chromatin structure. Multiple mRNAs can be formed from a single gene via alternative splicing, which can be tissue- and developmentally specific. The sequence of an mRNA transcript can also be altered by RNA editing.

■ *How could the phenomenon of RNA interference be used in drug design?*

16.7 Protein Degradation

Learning Outcomes

1. *Describe the role of ubiquitin in the degradation of proteins.*
2. *Explain the function of the proteasome.*

If all of the proteins produced by a cell during its lifetime remained in the cell, serious problems would arise. Protein labeling studies in the 1970s indicated that eukaryotic cells turn over proteins in a controlled manner. That is, proteins are continually being synthesized and degraded. Although this protein turnover is not as rapid as in prokaryotes, it indicates that a system regulating protein turnover is important.

Proteins can become altered chemically, rendering them nonfunctional; in addition, the need for any particular protein may be transient. Proteins also do not always fold correctly, or they may become improperly folded over time. These changes can lead to loss of function or other chemical behaviors, such as aggregating into insoluble complexes. In fact, a number of neurodegenerative diseases, such as Alzheimer dementia, Parkinson disease, and mad cow disease, are related to proteins that aggregate, forming characteristic plaques in brain cells. Thus, in addition to normal turnover of proteins, cells need a mechanism to get rid of old, unused, and incorrectly folded proteins.

Enzymes called **proteases** can degrade proteins by breaking peptide bonds, converting a protein into its constituent amino acids. Although there is an obvious need for these enzymes, they clearly cannot be floating around in the cytoplasm active at all times.

One way that eukaryotic cells handle such problems is to confine destructive enzymes to a specific cellular compartment. You may recall from chapter 4 that lysosomes are vesicles that contain digestive enzymes, including proteases. Lysosomes are used to remove proteins and old or nonfunctional organelles, but this system is not specific for particular proteins. Cells need another regulated pathway to remove proteins that are old or unused, but leave the rest of cellular proteins intact.

Addition of ubiquitin marks proteins for destruction

Eukaryotic cells solve this problem by marking proteins for destruction, then selectively degrading them. The mark that cells use is the attachment of a **ubiquitin** molecule. Ubiquitin, so named because it is found in essentially all eukaryotic cells (that is, it is ubiquitous), is a 76–amino-acid protein that can exist as an isolated molecule or in longer chains that are attached to other proteins.

The longer chains are added to proteins in a stepwise fashion by an enzyme called *ubiquitin ligase* (figure 16.20). This reaction requires ATP and other proteins, and it takes place in a multistep, regulated process. Proteins that have a ubiquitin chain attached are called *polyubiquitinated,* and this state is a signal to the cell to destroy this protein.

Two basic categories of proteins become ubiquitinated: those that need to be removed because they are improperly folded or nonfunctional, and those that are produced and degraded in a controlled fashion by the cell. An example of the latter are the cyclin proteins that help to drive the cell cycle (chapter 10). When these proteins have fulfilled their role in active division of the cell, they become polyubiquitinated and

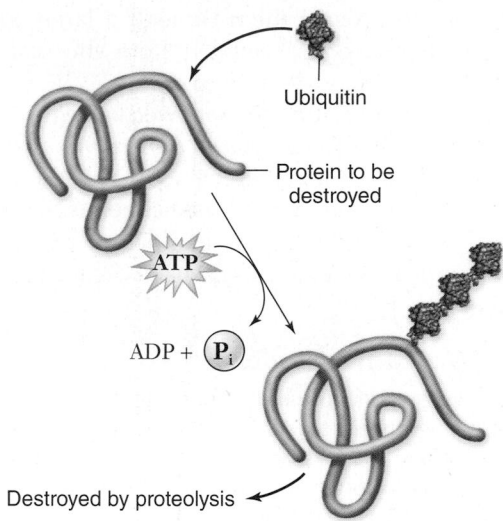

Figure 16.20 Ubiquitination of proteins. Proteins that are to be degraded are marked with ubiquitin. The enzyme ubiquitin ligase uses ATP to add ubiquitin to a protein. When a series of these have been added, the polyubiquitinated protein is destroyed.

are removed. In this way, a cell can control entry into cell division or maintain a nondividing state.

The proteasome degrades polyubiquitinated proteins

The cellular organelle that degrades proteins marked with ubiquitin is the **proteasome,** a large cylindrical complex that proteins enter at one end and exit the other as amino acids or peptide fragments (figure 16.21).

The proteasome complex contains a central region that has protease activity and regulatory components at each end. Although not membrane-bounded, this organelle can be thought of as a form of compartmentalization on a very small scale. By using a two-step process, first to mark proteins for

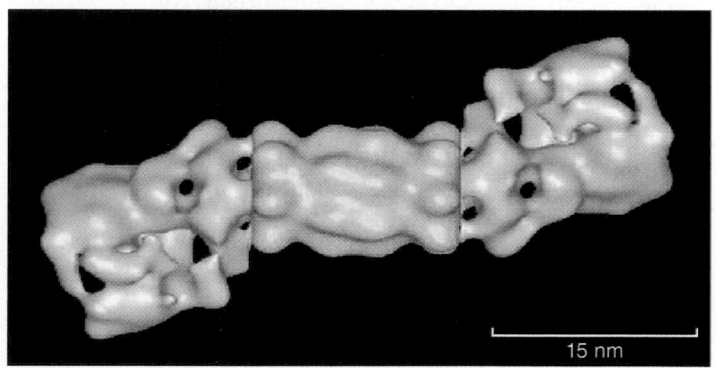

Figure 16.21 The *Drosophila* proteasome. The central complex contains the proteolytic activity, and the flanking regions act as regulators. Proteins enter one end of the cylinder and are cleaved to peptide fragments that exit the other end.

destruction, then to process them through a large complex, proteins to be degraded are isolated from the rest of the cytoplasm.

The process of ubiquitination followed by degradation by the proteasome is called the *ubiquitin–proteasome pathway*. It can be thought of as a cycle in that the ubiquitin added to proteins is not itself destroyed in the proteasome. As the proteins are degraded, the ubiquitin chain itself is simply cleaved back into ubiquitin units that can then be reused (figure 16.22).

Learning Outcomes Review 16.7

Control of protein degradation in eukaryotes involves addition of the protein ubiquitin, which marks the protein for destruction. The proteasome, a cylindrical complex with protease activity in its center, recognizes ubiquitinated proteins and breaks them down, much like a shredder destroys documents. Ubiquitin is recycled unchanged.

■ *If the ubiquitination process was not tightly controlled, what effect would this have on a cell?*

? **Inquiry question** What are two reasons a cell would polyubiquitinate a polypeptide?

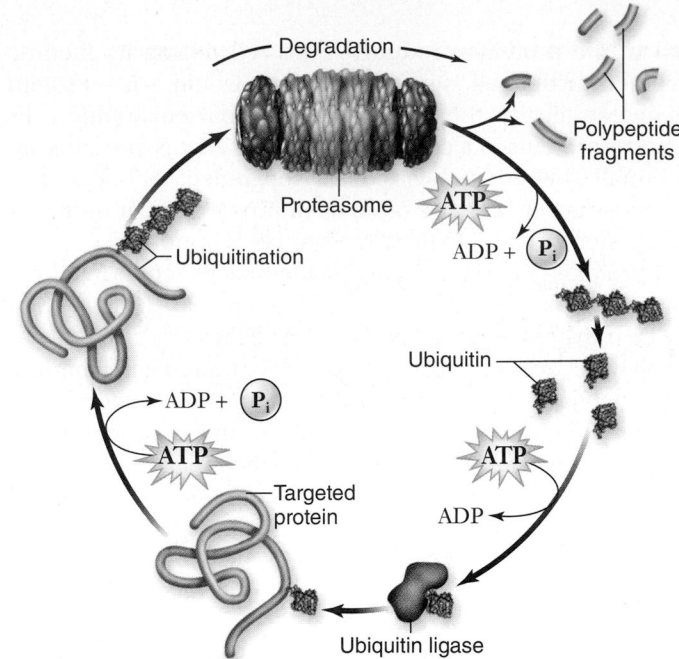

Figure 16.22 Degradation by the ubiquitin–proteasome pathway. Proteins are first ubiquitinated, then enter the proteasome to be degraded. In the proteasome, the polyubiquitin is removed and then is later "deubiquitinated" to produce single ubiquitin molecules that can be reused.

Chapter Review

16.1 Control of Gene Expression

Control can occur at all levels of gene expression.
Transcription is controlled by regulatory proteins that modulate the ability of RNA polymerase to bind to the promoter. These may either block transcription or stimulate it.

Control strategies in prokaryotes are geared to adjust to environmental changes.

Control strategies in eukaryotes maintain homeostasis and drive development.

16.2 Regulatory Proteins

Proteins can interact with DNA through the major groove.
A DNA double helix exhibits a major groove and a minor groove; bases in the major groove are accessible to regulatory proteins.

DNA-binding domains interact with specific DNA sequences.
A region of the regulatory protein that can bind to the DNA is termed a DNA-binding motif (figure 16.2).

Several common DNA-binding motifs are shared by many proteins.
Common motifs include the helix-turn-helix motif, the homeodomain motif, the zinc finger motif, and the leucine zipper.

16.3 Prokaryotic Regulation

Control of transcription can be either positive or negative.
Negative control is mediated by proteins called repressors that interfere with transcription. Positive control is mediated by a class of regulatory proteins called activators that stimulate transcription.

Prokaryotes adjust gene expression in response to environmental conditions.
The *lac* operon is induced in the presence of lactose; that is, the enzymes to utilize lactose are only produced when lactose is present. The *trp* operon is repressed; that is, the enzymes needed to produce tryptophan are turned off when tryptophan is present.

The lac *operon is negatively regulated by the* lac *repressor.*
The *lac* operon is induced when the effector (allolactose) binds to the repressor, altering its conformation such that it no longer binds DNA (figure 16.4).

The presence of glucose prevents induction of the lac *operon.*
Maximal expression of the *lac* operon requires positive control by catabolite activator protein (CAP) complexed with cAMP. When glucose is low, cAMP is high. Glucose repression involves both inducer exclusion, in which lactose is prevented from entering the cell, and the control of CAP function by the level of glucose.

The trp *operon is controlled by the* trp *repressor.*
The *trp* operon is repressed when tryptophan, acting as a corepressor, binds to the repressor, altering its conformation such

that it can bind to DNA and turn off the operon. This prevents expression in the presence of excess *trp.*

16.4 Eukaryotic Regulation

Transcription factors can be either general or specific.

General transcription factors are needed to assemble the transcription apparatus and recruit RNA polymerase II at the promoter. Specific factors act in a tissue- or time-dependent manner to stimulate higher rates of transcription.

Promoters and enhancers are binding sites for transcription factors.

General factors bind to the promoter to recruit RNA polymerase. Specific factors bind to enhancers, which may be distant from the promoter but can be brought closer by DNA looping.

Coactivators and mediators link transcription factors to RNA polymerase II (figure 16.11).

Some, but not all, transcription factors require a mediator. The number of coactivators is small because a single coactivator can be used with multiple transcription factors.

The transcription complex brings things together.

16.5 Eukaryotic Chromatin Structure

In eukaryotes, DNA is wrapped around proteins called histones, forming nucleosomes. These may block binding of transcription factors to promoters and enhancers.

Both DNA and histone proteins can be modified.

Methylation of DNA bases, primarily cytosine, correlates with genes that have been "turned off." Methylation is associated with inactive regions of chromatin.

Some transcription activators alter chromatin structure.

Acetylation of histones results in active regions of chromatin.

Chromatin-remodeling complexes also change chromatin structure.

Chromatin-remodeling complexes contain enzymes that move, reposition, and transfer nucleosomes.

16.6 Eukaryotic Posttranscriptional Regulation

Small RNAs act after transcription to control gene expression.

RNA interference is mediated by siRNAs formed by cleavage of double-stranded RNA by the Dicer nuclease. The siRNA is bound to a protein, Argonaute, in an RNA-induced silencing complex (RISC). The RISC can cleave mRNA or inhibit translation. Another class of small RNA, miRNA, is formed by the action of two nucleases, Drosha and Dicer, on RNA stem-and-loop structures. These also form a RISC that can either degrade mRNA or stop translation.

Small RNAs may have evolved to protect the genome.

Viral RNAs are degraded and transposons are silenced in the germ line by RNA interference. The origins of this machinery are ancient.

Small RNAs can mediate heterochromatin formation.

In fission yeast, *Drosophila,* and plants, RNA interference pathways lead to the formation of heterochromatin.

Alternative splicing can produce multiple proteins from one gene.

In response to tissue-specific factors, alternative splicing of pre-mRNA from one gene can result in multiple proteins.

RNA editing alters mRNA after transcription.

mRNA must be transported out of the nucleus for translation.

Initiation of translation can be controlled.

Translation factors may be modified to control initiation; translation repressor proteins can bind to the beginning of a transcript so that it cannot attach to the ribosome.

The degradation of mRNA is controlled.

An mRNA transcript is relatively stable, but it may carry targets for enzymes that degrade it more quickly as needed by the cell.

16.7 Protein Degradation

Addition of ubiquitin marks proteins for destruction.

In eukaryotes, proteins targeted for destruction have ubiquitin added to them as a marker.

The proteasome degrades polyubiquitinated proteins.

A cell organelle—the cylindrical proteasome—degrades ubiquitinated proteins that pass through it.

Review Questions

UNDERSTAND

1. In prokaryotes, control of gene expression usually occurs at the
   a. splicing of pre-mRNA into mature mRNA.
   b. initiation of translation.
   c. initiation of transcription.
   d. All of the choices are correct.

2. Regulatory proteins interact with DNA by
   a. unwinding the helix and changing the pattern of base-pairing.
   b. binding to the sugar–phosphate backbone of the double helix.
   c. unwinding the helix and disrupting base-pairing.
   d. binding to the major groove of the double helix and interacting with base-pairs.

3. In *E. coli,* induction in the *lac* operon and repression in the *trp* operon are both examples of
   a. negative control by a repressor.
   b. positive control by a repressor.
   c. negative control by an activator.
   d. positive control by a repressor.

4. The *lac* operon is controlled by two main proteins. These proteins
   a. both act in a negative fashion.
   b. both act in a positive fashion.
   c. act in the opposite fashion, one negative and one positive.
   d. act at the level of translation.

5. In eukaryotes, binding of RNA polymerase to a promoter requires the action of

   a. specific transcription factors.
   b. general transcription factors.
   c. repressor proteins.
   d. inducer proteins.

6. In eukaryotes, the regulation of gene expression occurs

   a. only at the level of transcription.
   b. only at the level of translation.
   c. at the level of transcription initiation, or posttranscriptionally.
   d. only posttranscriptionally.

7. In the *trp* operon, the repressor binds to DNA

   a. in the absence of *trp.*
   b. in the presence of *trp.*
   c. in either the presence or absence of *trp.*
   d. only when *trp* is needed in the cell.

APPLY

1. The *lac* repressor, the *trp* repressor and CAP are all

   a. negative regulators of transcription.
   b. positive regulators of transcription.
   c. allosteric proteins that bind to DNA and an effector.
   d. proteins that can bind DNA or other proteins.

2. Specific transcription factors in eukaryotes interact with enhancers, which may be a long distance from the promoter. These transcription factors then

   a. alter the structure of the DNA between enhancer and promoter.
   b. do not interact with the transcription apparatus.
   c. can interact with the transcription apparatus via DNA looping.
   d. can interact with the transcription apparatus by removing the intervening DNA.

3. Repression in the *trp* operon and induction in the *lac* operon are both mechanisms that

   a. would only be possible with positive regulation.
   b. allow the cell to control the level of enzymes to fit environmental conditions.
   c. would only be possible with negative regulation.
   d. cause the cell to make the enzymes from these two operons all the time.

4. Regulation by small RNAs and alternative splicing are similar in that both

   a. act after transcription.
   b. act via RNA/protein complexes.
   c. regulate the transcription machinery.
   d. Both a and b are correct.

5. Eukaryotic mRNAs differ from prokaryotic mRNAs in that they

   a. usually contain more than one gene.
   b. are colinear with the genes that encode them.
   c. are not colinear with the genes that encode them.
   d. Both a and c are correct.

6. In the cell cycle, cyclin proteins are produced in concert with the cycle. This likely involves

   a. control of initiation of transcription of cyclin genes, and ubiquitination of cyclin proteins.
   b. alternative splicing of cyclin genes to produce different cyclin proteins.
   c. RNA editing to produce the different cyclin proteins.
   d. transcription/translation coupling.

7. A mechanism of control in *E. coli* not discussed in this chapter involves pausing of ribosomes allowing a transcription terminator to form in the mRNA. In eukaryotic fission yeast, this mechanism should

   a. be common since they are unicellular.
   b. not be common since they are unicellular.
   c. not occur as transcription occurs in the nucleus and translation in the cytoplasm.
   d. not occur due to possibility of alternative splicing.

SYNTHESIZE

1. You have isolated a series of mutants affecting regulation of the *lac* operon. All of these are constitutive, that is, they express the *lac* operon all the time. You also have both mutant and wild-type alleles for each mutant in all combinations, and on F′ plasmids, which can be introduced into cells to make the cell diploid for the relevant genes. How would you use these tools to determine which mutants affect DNA binding sites on DNA, and which affect proteins that bind to DNA?

2. Examples of positive and negative control of transcription can be found in the regulation of expression of the bacterial operons *lac* and *trp*. Use these two operon systems to describe the difference between positive and negative regulation.

3. What forms of eukaryotic control of gene expression are unique to eukaryotes? Could prokaryotes use the mechanisms, or are they due to differences in these cell types?

4. The number and type of proteins found in a cell can be influenced by genetic mutation and regulation of gene expression. Discuss how these two processes differ.

ONLINE RESOURCE

www.ravenbiology.com

Understand, Apply, and Synthesize—enhance your study with animations that bring concepts to life and practice tests to assess your understanding. Your instructor may also recommend the interactive eBook, individualized learning tools, and more.

Chapter

17

Biotechnology

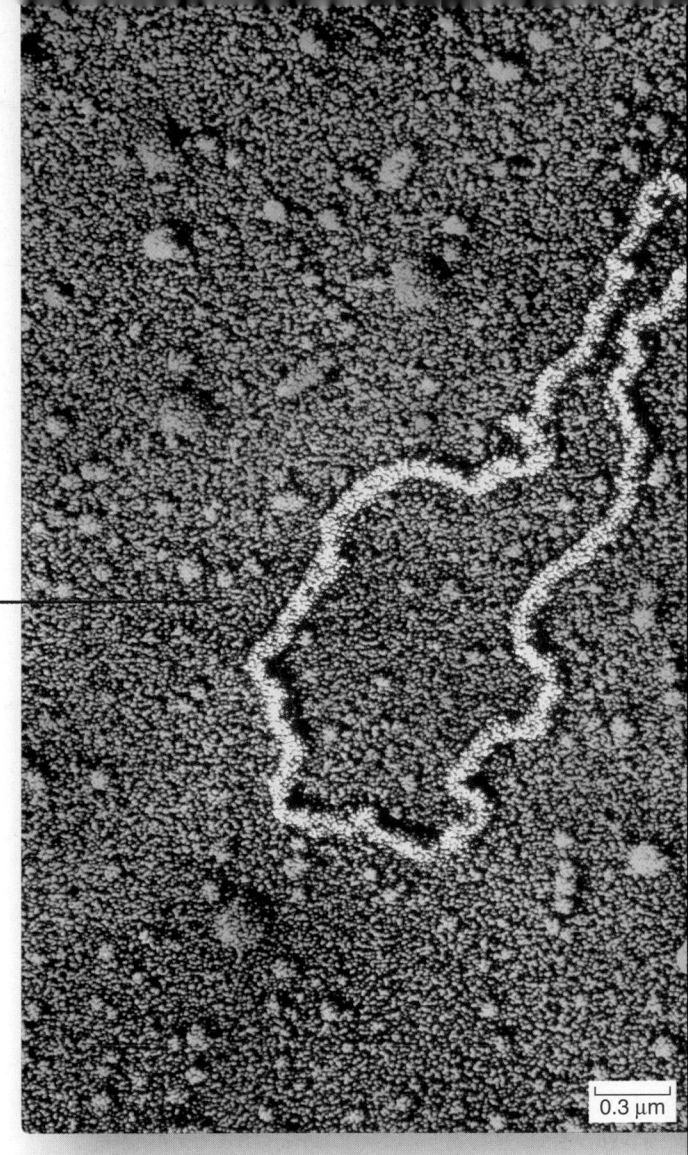

Chapter Contents

0.3 μm

Introduction

Biotechnology is the manipulation of living organisms resulting in a desired product or process. Over the past three decades, the development of new and powerful techniques for studying and manipulating DNA has revolutionized biology. The knowledge gained in the last 30 years is greater than that accrued during the history of biology. Biotechnology also affects more aspects of everyday life than any other area of biology. From the food on your table to the future of medicine, biotechnology touches your life.

Biotechnology includes the application of molecular biology principles you have studied to numerous aspects of life. The ability to isolate specific DNA sequences arose from the study and use of small DNA molecules found in bacteria, such as the plasmid pictured here. In this chapter, we explore these technologies and consider how they apply to specific problems of practical importance.

17.1 Recombinant DNA

Learning Outcomes

1. *Contrast the potential results of using breeding versus recombinant DNA.*
2. *Relate endogenous roles of enzymes to their recombinant DNA applications.*
3. *Explain why DNA fragments can be separated with gel electrophoresis and why it is useful.*

Over 15,000 years ago humans began domesticating dogs from wolves, and since that time selective breeding has led to the 155 existing breeds of dogs. Crop domestication followed, and the development of hybrid corn in the early 20th century substantially increased yield. Although creating new gene combinations through genetic crosses is not new, the ability to directly isolate and manipulate genetic material was one of the most profound changes to come about in the field of biology in the late 20th century. The construction of **recombinant DNA** molecules, that is, a single DNA molecule made from two different sources, began in the mid-1970s. The development of this technology, which has led to the entire field of biotechnology, is based on enzymes that can be used to manipulate DNA.

In the 1970s, the Indian-born American microbiologist Ananda Chakrabarty, at General Electric's Research and Development Center, was developing bacteria to consume oil spills. Four *Pseudomonas* strains had genes coding for single, distinct enzymes that digested the hydrocarbons in oil into methane. The genes were encoded in plasmids (small, circular DNA) in the bacteria. Together the four enzymes were more effective in dissipating an oil spill than any one alone. Chakrabarty combined the genes encoding the enzymes into a single plasmid, creating the first recombinant DNA. He sought a patent for the bacteria containing the recombinant DNA. In 1980, in the case of *Diamond versus Chakrabarty,* the Supreme Court ruled that Chakrabarty had the right to patent his oil-consuming bacteria because such bacteria did not occur in nature and existed only because of his ingenuity.

Restriction endonucleases cleave DNA at specific sites

Restriction endonucleases are the key to inserting a sequence of DNA from one organism into a piece of DNA, such as a plasmid, from another. The use of restriction enzymes revolutionized molecular biology because of their ability to cleave DNA at specific sites. As described in chapter 14, nucleases are enzymes that degrade DNA, and many were known prior to the isolation of the first restriction enzyme (*Hind*II) in 1970. If a DNA sequence were a rope, then restriction enzymes would be a knife that always cut that rope into specific pieces.

Discovery and significance of restriction endonucleases

This site-specific cleavage activity, long sought by molecular biologists, was discovered from basic research into why bacterial viruses can infect some cells but not others. This phenomenon was termed host restriction. The bacteria produce enzymes that can cleave the invading viral DNA at specific sequences. The host cells protect their own DNA from cleavage by modifying bases at the cleavage sites; the restriction enzymes do not cleave the modified bacterial DNA. Since the initial discovery of these restriction endonucleases, hundreds more have been isolated that recognize and cleave different **restriction sites.**

The ability to cut DNA at specific places is significant in two ways: First, it allows physical maps to be constructed based on the positioning of cleavage sites for restriction enzymes. These restriction maps provide crucial data for identifying and working with DNA molecules (as discussed in detail in chapter 18).

Second, restriction endonuclease cleavage allows for the creation of recombinant molecules. The ability to construct recombinant molecules is critical to research because many steps in the process of cloning and manipulating DNA require the ability to combine molecules from different sources.

How restriction enzymes work

There are three types of restriction enzymes, but only type II cleaves at precise locations. Types I and III cleave with less precision and are not often used in cloning and manipulating DNA.

Type II enzymes enable creation of recombinant molecules; these enzymes recognize a specific DNA sequence,

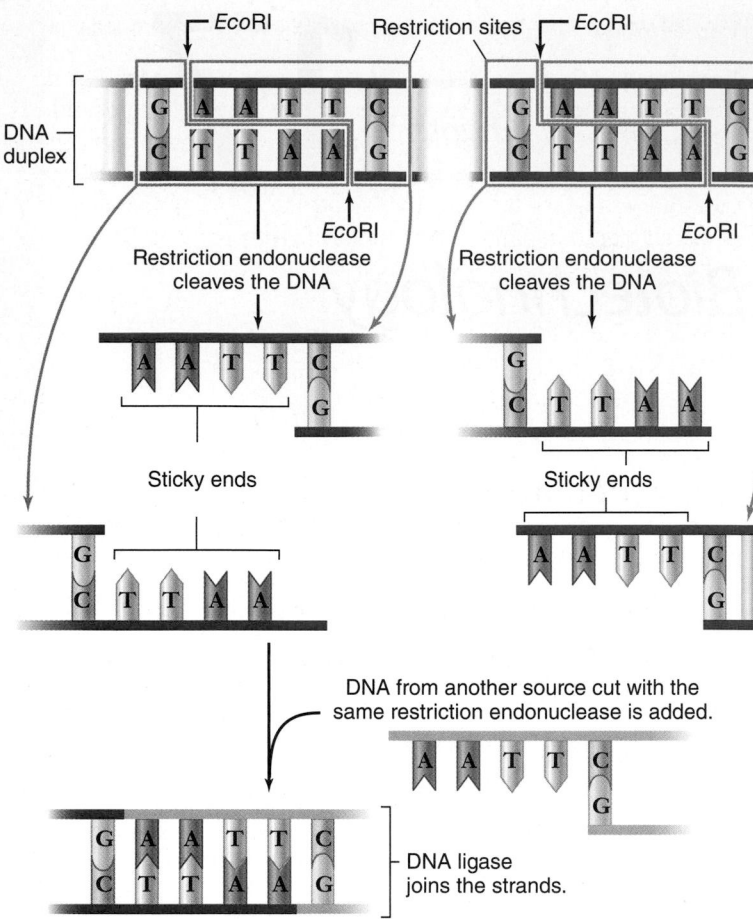

Figure 17.1 Many restriction endonucleases produce DNA fragments with sticky ends. The restriction endonuclease *Eco*RI always cleaves the sequence 5′–GAATTC–3′ between G and A. Because the same sequence occurs on both strands, both are cut. However, the two sequences run in opposite directions on the two strands. As a result, single-stranded tails called "sticky ends" are produced that are complementary to each other. These complementary ends can then be joined to a fragment from another DNA that is cut with the same enzyme. These two molecules can then be joined by DNA ligase to produce a recombinant molecule.

ranging from 4 bases to 12 bases, and cleave the DNA at a specific base within this sequence (figure 17.1). The recognition sites for most type II enzymes are palindromes. A linguistic **palindrome** is a word or phrase that reads the same forward and in reverse, such as the sentence: "Madam I'm Adam." The palindromic DNA sequence reads the same from 5′ to 3′ on one strand as it does on the complementary strand (see figure 17.1).

Given this kind of sequence, cutting the DNA at the same base on either strand can lead to staggered cuts that produce "sticky ends." These short, unpaired sequences are the same for any DNA that is cut by this enzyme. Thus, these sticky ends allow DNA molecules from different sources to be easily joined together (see figure 17.1). Although less common, some type II restriction enzymes, including *Pvu*II, can cut both strands in the same position, producing blunt, not sticky, ends. Blunt-cut ends can be joined with other blunt-cut ends.

Gel electrophoresis separates DNA fragments

Restriction endonucleases cut DNA into fragments of different sizes. Separating the fragments based on size makes it possible to select the DNA sequence of interest. The most common separation technique used is gel electrophoresis. This technique takes advantage of the negative charge on DNA molecules by using an electrical field to provide the force necessary to separate DNA molecules based on size.

The gel, which is made of either agarose or polyacrylamide and spread thinly on supporting material, provides a three-dimensional matrix that separates molecules based on size (figure 17.2). The gel is submerged in a buffer solution containing ions that can carry current and is subjected to an electrical field.

The strong negative charges from the phosphate groups in the DNA backbone cause it to migrate toward the positive pole (figure 17.2b). The gel acts as a sieve to separate DNA molecules based on size: The larger the molecule, the slower it will move through the gel matrix. Over a given period, smaller molecules migrate farther than larger ones. The DNA in gels can be visualized using a fluorescent dye that binds to DNA (figure 17.2c, d).

Electrophoresis is one of the most important methods in the toolbox of modern molecular biology, with uses ranging from DNA fingerprinting to DNA sequencing, both of which are described later on.

DNA ligase allows construction of recombinant plasmids

Once a specific restriction fragment is isolated, it can be spliced into a plasmid that has been cut with the same restriction enzyme. Another enzyme is needed, however, to join the two fragments together to create a stable DNA molecule. The

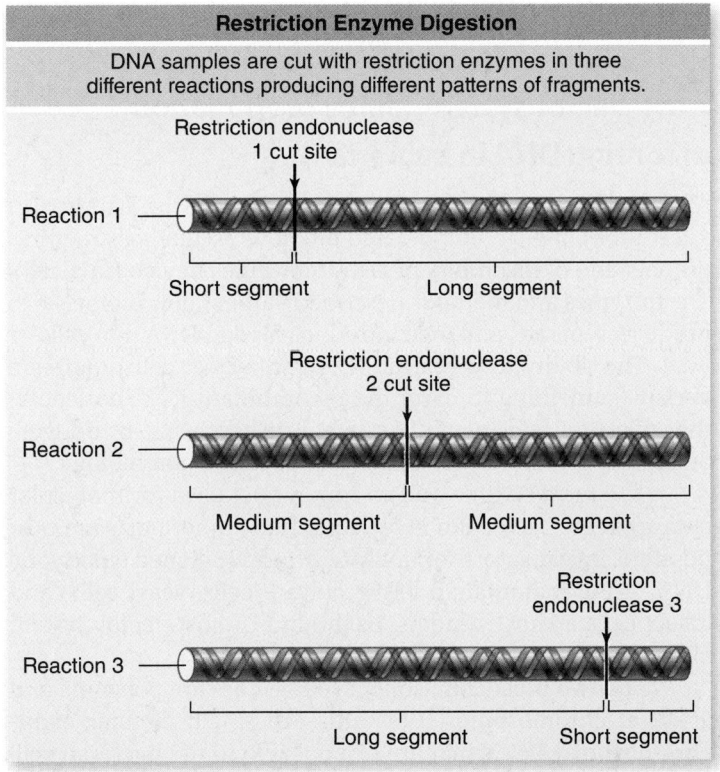

a.

b.

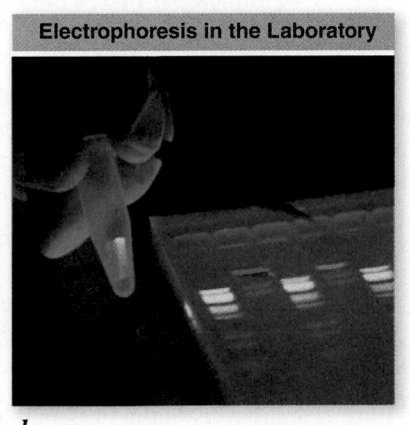

c.

d.

Figure 17.2 Gel electrophoresis. *a*. Three restriction enzymes are used to cut DNA into specific pieces depending on each enzyme's recognition sequence. ***b*.** The fragments are loaded into a gel (agarose or polyacrylamide), and an electrical current is applied. The DNA fragments migrate through the gel based on size, with larger ones moving more slowly. ***c*.** This results in a pattern of fragments separated based on size, with the smaller fragments migrating farther than larger ones. ***d*.** The fragments can be visualized by staining with the dye ethidium bromide. When the gel is exposed to UV light, the DNA with bound dye fluoresces, appearing as pink bands in the gel. In the photograph, one band of DNA has been excised from the gel for further analysis and can be seen glowing in the tube the technician holds.

enzyme DNA ligase accomplishes this by catalyzing the formation of a phosphodiester bond between adjacent phosphate and hydroxyl groups of DNA nucleotides. The action of ligase is to seal nicks in one or both strands (see figure 17.1). This is the same enzyme that joins Okazaki fragments on the lagging strand during DNA replication (see chapter 14).

Learning Outcomes Review 17.1

Restriction endonucleases are part of bacterial cells' strategies for fighting viral infection. Type II endonucleases cleave DNA at specific sites. DNA ligase can be used to link together fragments following action of restriction endonucleases. Gel electrophoresis employs electrical charge to separate DNA fragments according to size.

- Assess the effect of recombinant DNA on biotechnology.
- Compare and contrast the endogenous roles of EcoRI and ligase in E. coli with their use in a molecular biology lab.

17.2 Introducing Foreign DNA into Cells

Learning Outcomes

1. Explain the role of a vector in molecular cloning.
2. Compare recombinant technology methods in bacteria and eukaryotes.

Obtaining a fragment of DNA that encodes a gene of interest is the first step toward genetic engineering. Restriction enzymes are used to insert a DNA fragment into a plasmid, but a plasmid in isolation cannot replicate. To be useful, many identical copies of the fragment must be generated and stably maintained. The term **clone** refers to a genetically identical copy and is used in several different contexts. Here we explore the idea of molecular cloning, amplifying a sequence of DNA.

Molecular cloning involves the isolation of a specific sequence of DNA, usually one that encodes a particular protein product. This is sometimes called *gene cloning,* but the term *molecular cloning* is more accurate. In this section the focus is on the use of cells in molecular cloning.

Transformation allows introduction of foreign DNA into cells

In chapter 14 you learned that Frederick Griffith demonstrated that genetic material could be transferred between bacterial cells. This process, called *transformation,* is a natural process in the cells that Griffith was studying. The bacterium *E. coli,* used routinely in molecular biology laboratories, does not undergo natural transformation; but artificial transformation techniques have been developed to allow introduction of foreign DNA into *E. coli.* Through temperature shifts or application of an electrical charge, the *E. coli* membrane becomes transiently permeable to the foreign DNA. In this way, recombinant molecules can be propagated in a cell that will make many copies of the constructed molecules.

In general, the introduction of DNA from an outside source into a cell is referred to as transformation. This process is important in *E. coli* for molecular cloning and the propagation of cloned DNA. Researchers also want to be able to reintroduce DNA into the original cells from which it was isolated, including eukaryotic cells. A transformed cell can also be used to form all or part of a multicellular, eukaryotic organism, called a **transgenic organism.**

Here we focus on amplifying recombinant DNA sequences in transformed cells. Amplification and expression are not the same. For example, a eukaryotic gene can be amplified if it is inserted in a plasmid in *E. coli,* but if it has a eukaryotic promoter and/or eukaryotic introns, the bacterial cell will not be able to correctly transcribe and translate the gene. This distinction will be taken into consideration when we look at applications of molecular cloning later in the chapter.

Host–vector systems allow propagation of foreign DNA in bacteria

Although short sequences of DNA can be synthesized in vitro (in a test tube), the cloning of large unknown sequences requires propagation of recombinant DNA molecules in vivo (in a cell). The enzymes and methods described earlier allow biologists to produce, separate, and then introduce foreign DNA into cells.

The ability to propagate DNA in a host cell requires a **vector** (something to carry the recombinant DNA molecule) that can replicate in the host when it has been introduced. Such host–vector systems are crucial to molecular biology.

The most flexible and common host used for molecular cloning is the bacterium *E. coli,* but many other hosts are now possible. Investigators routinely reintroduce cloned eukaryotic DNA, using mammalian tissue culture cells, yeast cells, and insect cells as host systems. Each kind of host–vector system allows particular uses of the cloned DNA.

The two most commonly used vectors are plasmids and artificial chromosomes. **Plasmids** are small, circular extrachromosomal DNAs that are dispensable to the bacterial cell. Bacterial and eukaryotic artificial chromosomes are used to clone larger pieces of DNA.

Plasmid vectors

Plasmid vectors are typically used to clone relatively small pieces of DNA, up to a maximum of about 10 kilobases (kb). Larger inserts are not stable and are lost from the plasmid over time. A plasmid vector must have three components:

1. An *origin of replication* to allow it to be replicated in *E. coli* independently of the host chromosome,
2. A *selectable marker,* usually antibiotic resistance, and
3. *One or more unique restriction sites* where foreign DNA can be added.

The selectable marker allows the presence of the plasmid to be easily identified through genetic selection. For example, cells that contain a plasmid with an antibiotic resistance gene continue to live when plated on antibiotic-containing growth media, whereas cells that lack the plasmid will die (they are killed by the antibiotic).

A fragment of DNA is inserted by the techniques previously described into a region of the plasmid with restriction sites called the multiple-cloning site (MCS). This region contains a number of unique restriction sites such that when the plasmid is cut with the relevant restriction enzymes, a linear plasmid results. When DNA of interest is cut with the same restriction enzyme, it can then be ligated into this site. The plasmid is then introduced into cells by transformation (figure 17.3), and the fragment of DNA is amplified as the bacterial cells divide.

This region of the vector often has been engineered to contain another gene that becomes inactivated, a process called *insertional inactivation,* because it is now interrupted by the inserted DNA. One of the first cloning vectors, pBR322, used another antibiotic resistance gene for insertional activation; resistance to one antibiotic and sensitivity to the other indicated the presence of inserted DNA.

More recent vectors use the gene for β-galactosidase, an enzyme that cleaves galactoside sugars such as lactose. When the enzyme cleaves the artificial substrate X-gal, a blue color is produced. In these plasmids, insertion of foreign DNA interrupts the β-galactosidase gene, preventing a functional enzyme from being produced. When transformed cells are plated on a medium containing both antibiotic (to select for plasmid-containing cells) and X-gal, they remain white, whereas transformed cells with no inserted DNA are blue (see figure 17.3).

Artificial chromosomes

The size of DNA molecules that can be cloned in plasmid vectors has limited the large-scale analysis of genomes. To deal with this, geneticists decided to follow the strategy of cells and construct chromosomes, leading to the development of yeast artificial chromosomes (YACs) and bacterial artificial chromosomes (BACs). Progress has also been made on creating mammalian artificial chromosomes.

> **? Inquiry question** An investigator wishes to clone a 32-kb recombinant molecule. What do you think is the best vector to use?

Eukaryotic cell transformation is challenging

Transforming eukaryotic cells is more challenging than transforming bacteria. The primary experimental difficulty has been identifying a suitable vector for introducing recombinant DNA. Eukaryotic cells do not possess the many plasmids that bacteria have, so the choice of potential vectors is limited.

Two successful approaches for transforming animal cells are injecting DNA into the pronucleus of a fertilized egg or injecting embryonic stem cells into an embryo. Specific applications are explored in chapter 19. General approaches to transforming plant cells include electrically charging cells and employing "the gene gun," to bombard cells with tiny gold or tungsten particles coated with DNA. This technique has the advantage of being usable for any species, but it does not provide as precise engineering because the copy number of introduced genes is much harder to control. However, a bacterial pathogen has been quite effective in transforming plants.

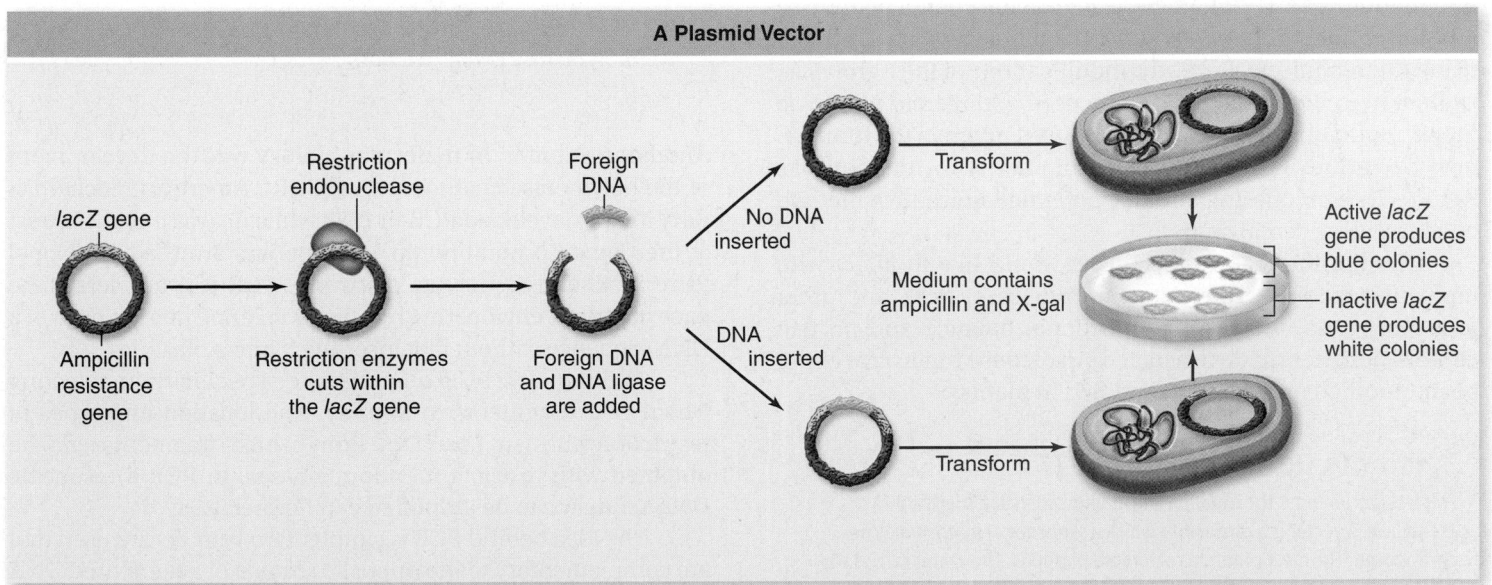

A Plasmid Vector

Figure 17.3 Molecular cloning with vectors. Plasmids are cut within the β-galactosidase gene *(lacZ),* and foreign DNA and DNA ligase are added. Foreign DNA inserted into *lacZ* interrupts the coding sequence, thus inactivating the gene. Plating cells on a medium containing the antibiotic ampicillin selects for plasmid-containing cells. The medium also contains X-gal, and when *lacZ* is intact *(top),* the expressed enzyme cleaves the X-gal, producing blue colonies. When *lacZ* is inactivated *(bottom),* X-gal is not cleaved, and colonies remain white.

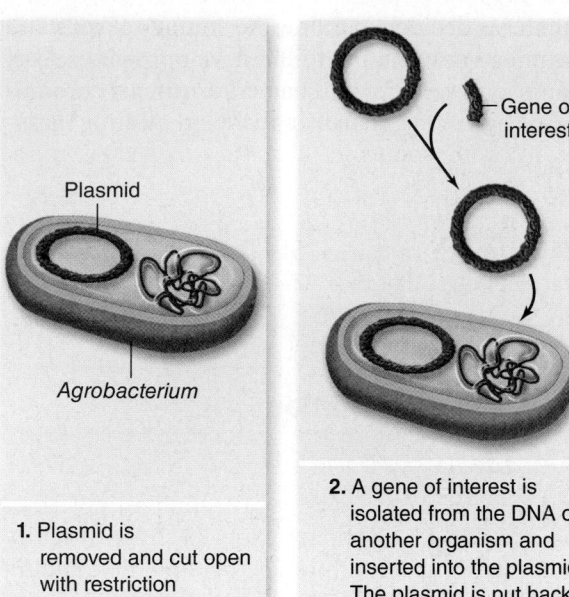
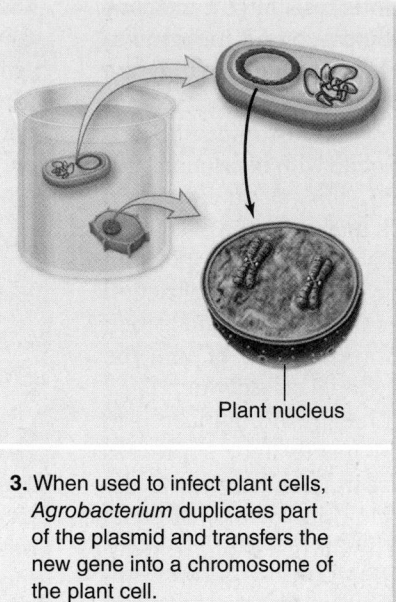
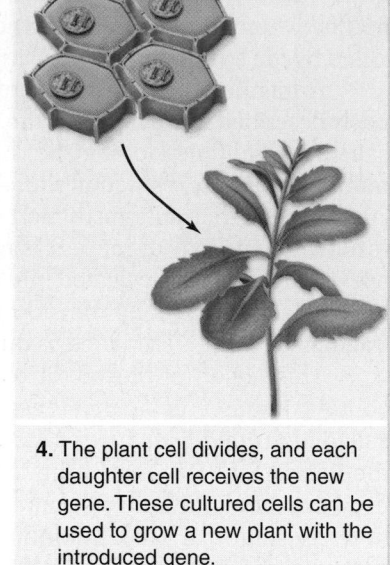

1. Plasmid is removed and cut open with restriction endonuclease.

2. A gene of interest is isolated from the DNA of another organism and inserted into the plasmid. The plasmid is put back into the *Agrobacterium*.

3. When used to infect plant cells, *Agrobacterium* duplicates part of the plasmid and transfers the new gene into a chromosome of the plant cell.

4. The plant cell divides, and each daughter cell receives the new gene. These cultured cells can be used to grow a new plant with the introduced gene.

Plasmid

Agrobacterium

Gene of interest

Plant nucleus

Figure 17.4 The Ti plasmid. This *Agrobacterium tumefaciens* plasmid is used in plant genetic engineering.

Transforming plants with a bacterial pathogen

The most successful results thus far have been obtained with the **Ti (tumor-inducing) plasmid** of the plant bacterium *Agrobacterium tumefaciens,* which normally infects broadleaf plants such as tomato, tobacco, and soybean. This plant pathogen causes tumors to form on the plant, but the tumor-inducing genes in the tumor-inducing (Ti) plasmid in *Agrobacterium* can be removed. The **transfer DNA (T-DNA)** within the Ti plasmid integrates into the plant DNA, and researchers have succeeded in attaching other genes to this portion of the plasmid (figure 17.4). The characteristics of a number of plants that are susceptible to *Agrobacterium* infections have been altered using this technique. Recently, modifications of the *Agrobacterium* system have allowed it to be used with cereal plants, so the gene gun technology may not be used much in the future. A new bacterium has also been manipulated to function like *Agrobacterium,* offering another potential alternative method of engineering cereal crops.

Among the features scientists would like to affect with molecular cloning are resistance to disease, frost, and other forms of stress; increase in nutritional balance and protein content; and herbicide resistance. All of these traits have either been modified or are being modified in plants.

Learning Outcomes Review 17.2

Molecular cloning is the isolation and amplification of a specific DNA sequence. A vector is a carrier into which a sequence of interest may be introduced. The most common vectors are plasmids. The vector carries the DNA sequence into a cell, which then multiplies, copying its own DNA along with that of the vector.

■ *Review the essential steps needed to make multiple copies of a restriction digest DNA fragment that you have isolated using gel electrophoresis.*

17.3 Amplifying DNA Without a Vector—The Polymerase Chain Reaction

Learning Outcomes

1. *Explain how DNA can be amplified in the absence of a vector.*
2. *Compare and contrast DNA replication and the polymerase chain reaction.*

Another revolution in molecular biology was the development of the polymerase chain reaction (PCR). American biochemist Kary Mullis developed PCR in 1983 while he was a staff chemist at the Cetus Corporation; in 1993, he was awarded the Nobel Prize in Chemistry for his discovery. PCR can accelerate the pace of genetic engineering by quickly creating many clones of a DNA sequence without first inserting it into a plasmid.

Additionally, PCR offers greater specificity in selecting which DNA fragments to amplify. The location and types of restriction sites in the DNA limit which fragments can be obtained with restriction endonucleases. In PCR the specific DNA sequence to be amplified can be specified.

The idea behind PCR is simple: Two primers are used that are complementary to the opposite strands of a denatured DNA sequence. When DNA polymerase acts on these primers and the sequence of interest, the polymerase produces complementary strands, each containing the other primer. If this procedure is done cyclically, the result is a large quantity of a sequence corresponding to the DNA that lies between the two primers (figure 17.5). Simply put, PCR is DNA replication in vitro.

PCR mimics DNA replication

Two developments turned this simple concept into a powerful technique. First, each cycle requires denaturing the DNA after each round of synthesis, which is easily done by raising the temperature; however, this destroys most polymerase enzymes. The solution was to isolate a DNA polymerase from a thermophilic, or heat-loving bacterium. *Thermus aquaticus,* found in the hot springs in Yellowstone National Park, was an ideal candidate. The enzyme produced from the bacterium, called **Taq polymerase** for the first few letters of the bacterium's name, allows the reaction mixture to be repeatedly heated without destroying enzyme activity.

The second innovation was the development of machines with heating blocks that can be rapidly cycled over large temperature ranges with very accurate temperature control.

Thus each cycle of PCR involves three steps:

1. Denaturation (high temperature)
2. Annealing of primers (low temperature)
3. Synthesis (intermediate temperature)

Steps 1 to 3 are repeated, and the two copies become four. It is not necessary to add any more polymerase, because the heating step does not harm Taq polymerase. Each complete cycle, which takes only 1–2 min, doubles the number of DNA molecules. After 20 cycles, a single fragment produces more than a million (2^{20}) copies!

In this way, the process of PCR allows the **amplification** of a single DNA fragment from a small amount of a complex mixture of DNA. This result is similar to what is isolated using molecular cloning, but in the case of PCR, the DNA cannot be reintroduced directly into a cell. The PCR product can be analyzed using electrophoresis, cloned into a vector for other manipulations, or directly sequenced. The size of the fragment that can be synthesized in this way is limited, but it has been adapted for an amazing number of uses.

PCR has many applications

PCR, now fully automated, has revolutionized many aspects of science and medicine because it allows for the investigation of minute samples of DNA. In criminal investigations, DNA fingerprints can now be prepared from the cells in a tiny speck of dried blood or from the tissue at the base of a single human hair. In medicine, physicians can detect genetic defects in very early embryos by collecting a single cell and amplifying its DNA. Due to its sensitivity, speed, and ease of use, technicians now routinely use PCR methods for these applications.

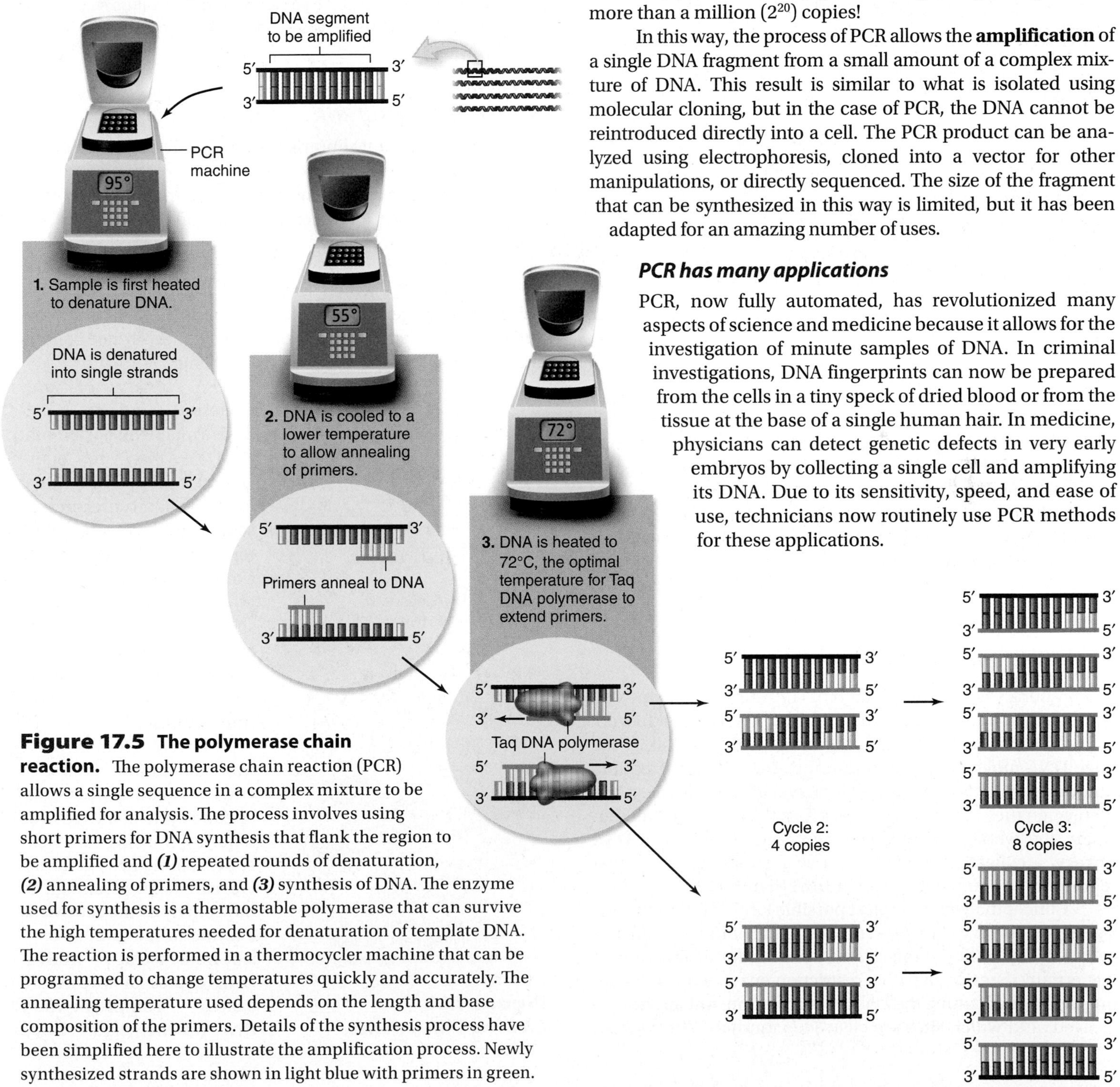

Figure 17.5 The polymerase chain reaction. The polymerase chain reaction (PCR) allows a single sequence in a complex mixture to be amplified for analysis. The process involves using short primers for DNA synthesis that flank the region to be amplified and *(1)* repeated rounds of denaturation, *(2)* annealing of primers, and *(3)* synthesis of DNA. The enzyme used for synthesis is a thermostable polymerase that can survive the high temperatures needed for denaturation of template DNA. The reaction is performed in a thermocycler machine that can be programmed to change temperatures quickly and accurately. The annealing temperature used depends on the length and base composition of the primers. Details of the synthesis process have been simplified here to illustrate the amplification process. Newly synthesized strands are shown in light blue with primers in green.

PCR has even been used to analyze mitochondrial DNA from the early human species *Homo neanderthalensis.* This application provided the first glimpse of data from extinct related species.

17.4 Storing and Sorting DNA Fragments

So far we have considered single fragments of DNA that can be isolated and amplified using either restriction endonuclease digests that are inserted into plasmids and introduced into cells or PCR. Using these two methods, a large number of different DNA fragments can be isolated. Storing and sorting different DNA fragments requires a DNA library. It is also possible to isolate all the RNA in a specific tissue at a specific time and convert it to DNA fragments, which represent the expressed genes. These DNA libraries contain only DNA that is expressed as RNA in the cell.

DNA libraries may contain the entire genome of an organism

A DNA library is a representation of very complex mixtures in DNA, such as an entire genome, in a form that is easier to work with than the enormous chromosomes within a cell. If the huge DNA molecules in chromosomes can be converted into random fragments and inserted into a vector, such as plasmids, then when they are propagated in a host they will together represent the whole genome. This aggregate is termed a **DNA library,** a collection of DNAs in a vector that taken together represent the complex mixture of DNA (figure 17.6).

Conceptually the simplest possible kind of DNA library is a **genomic library**—a representation of the entire genome in a vector. Many copies of this genome are randomly fragmented by partially digesting it with a restriction enzyme that cuts frequently. By not cutting the DNA to completion, not all sites are cleaved, and which sites are cleaved is random. The random

fragments are then inserted into vectors and introduced into host cells. Genomic libraries are usually constructed in BACs.

A variety of different kinds of libraries can be made depending on the source DNA used. Any particular clone in the library contains only a single DNA, and all of them together make up the library. Keep in mind that unlike a library full of books, which is organized and catalogued, a DNA library is a random collection of overlapping DNA fragments. We explore how to find a sequence of interest in this random collection later in the chapter.

 Data analysis The human genome contains 3 billion base-pairs. If you construct a BAC library of the human genome with 500-bp fragments of DNA, what is the minimum number of clones in your library?

Using reverse transcriptase to make a DNA copy of RNA

In addition to genomic libraries, investigators often wish to isolate only the *expressed* part of genes. The structure of eukaryotic genes is such that the mRNA may be much smaller than the gene itself due to the presence of introns in the gene. After transcription by RNA polymerase II, the primary transcript is spliced to produce the mRNA (see chapter 15). Because of this, genomic libraries, which are crucial to understanding the structure of the gene, are not of much use if we want to express the gene in a bacterial species, whose genes do not contain introns and have no mechanism for splicing.

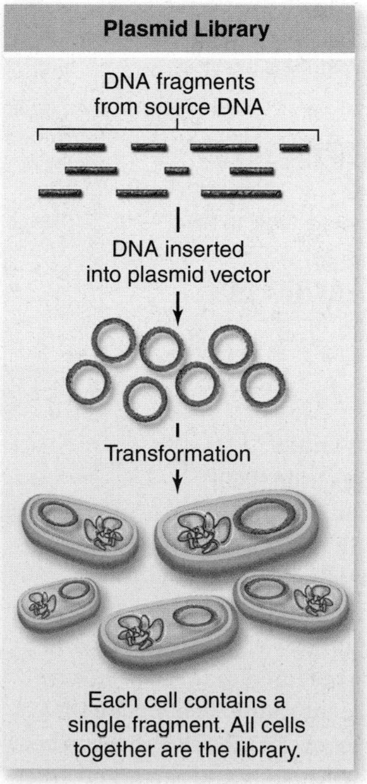

A library of only expressed sequences represents a much smaller amount of DNA than the entire genome. The starting point for a DNA library of expressed sequences is isolated mRNA representing the genes expressed in a specific tissue at a specific developmental stage. Such a library of expressed sequences is made possible by the use of another enzyme: **reverse transcriptase.**

Reverse transcriptase was isolated from a class of viruses called retroviruses. The life cycle of a retrovirus requires making a DNA copy from its RNA genome. We can take advantage of the activity of the retrovirus enzyme to make DNA copies from isolated mRNA. DNA copies of mRNA are called **complementary DNA (cDNA)** (figure 17.7).

Plasmid Library

DNA fragments from source DNA

↓

DNA inserted into plasmid vector

↓

Transformation

↓

Each cell contains a single fragment. All cells together are the library.

Figure 17.6
Creating DNA libraries.

A cDNA library is made by first isolating mRNA from genes being expressed and then using the reverse transcriptase enzyme to make cDNA from the mRNA. The cDNA is then used to make a library, as mentioned earlier. These cDNA libraries are commonly made to represent the genes expressed in many different tissues or cells. Although all genomic libraries made from an individual will be identical, cDNA libraries from the same cells at different developmental stages or cells from different tissues will each be distinct. Questions about which genes influence physiological responses to the environment or

development in an organism can be answered by comparing cDNA libraries.

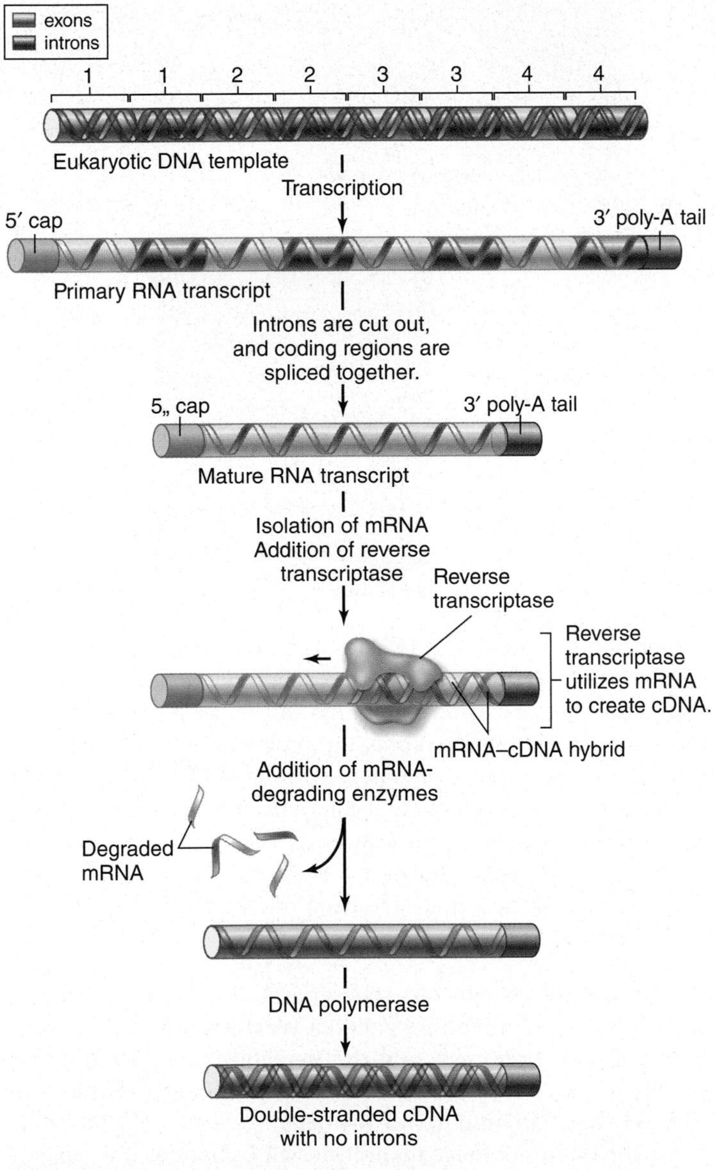

17.5 Analyzing and Creating DNA Differences

Although all members of a species have the same genes, the alleles of these genes differ, contributing to variation among individuals. Advances in studying and manipulating DNA have led to new approaches to differentiating among alleles, which can be useful in identifying individuals. Closely related species have very similar genes that can also be distinguished with DNA analysis. In addition, DNA technology is used to create mutant alleles of a gene in an individual to better understand gene function. In this section we explore DNA analysis methods with a broad range of applications.

DNA differences can be detected using blotting methods

Once a gene has been cloned, it may be used as a probe to identify the same gene in a DNA library. Any single-stranded nucleic acid (DNA or RNA) can be tagged with a radioactive label or with another detectable label, such as a fluorescent dye. This can then be used as a probe to identify its complement in a complex mixture of DNA or RNA. This renaturing is termed *hybridization* because the combination of labeled probe and unlabeled DNA form a hybrid molecule through base-pairing.

A probe may also be used to identify the same or a similar gene in DNA isolated from a cell or tissue in the same or related

Figure 17.7 The formation of cDNA. A mature mRNA transcript is usually much smaller than the gene due to the loss of intron sequences by splicing. mRNA is isolated from the cytoplasm of a cell, which the enzyme reverse transcriptase uses as a template to make a DNA strand complementary to the mRNA. That newly made strand of DNA is the template for the enzyme DNA polymerase, which assembles a complementary DNA strand along it, producing cDNA—a double-stranded DNA version of the intron-free mRNA.

Figure labels:
- exons
- introns
- Eukaryotic DNA template
- Transcription
- 5′ cap / 3′ poly-A tail
- Primary RNA transcript
- Introns are cut out, and coding regions are spliced together.
- 5„ cap / 3′ poly-A tail
- Mature RNA transcript
- Isolation of mRNA Addition of reverse transcriptase
- Reverse transcriptase
- Reverse transcriptase utilizes mRNA to create cDNA.
- mRNA–cDNA hybrid
- Addition of mRNA-degrading enzymes
- Degraded mRNA
- DNA polymerase
- Double-stranded cDNA with no introns

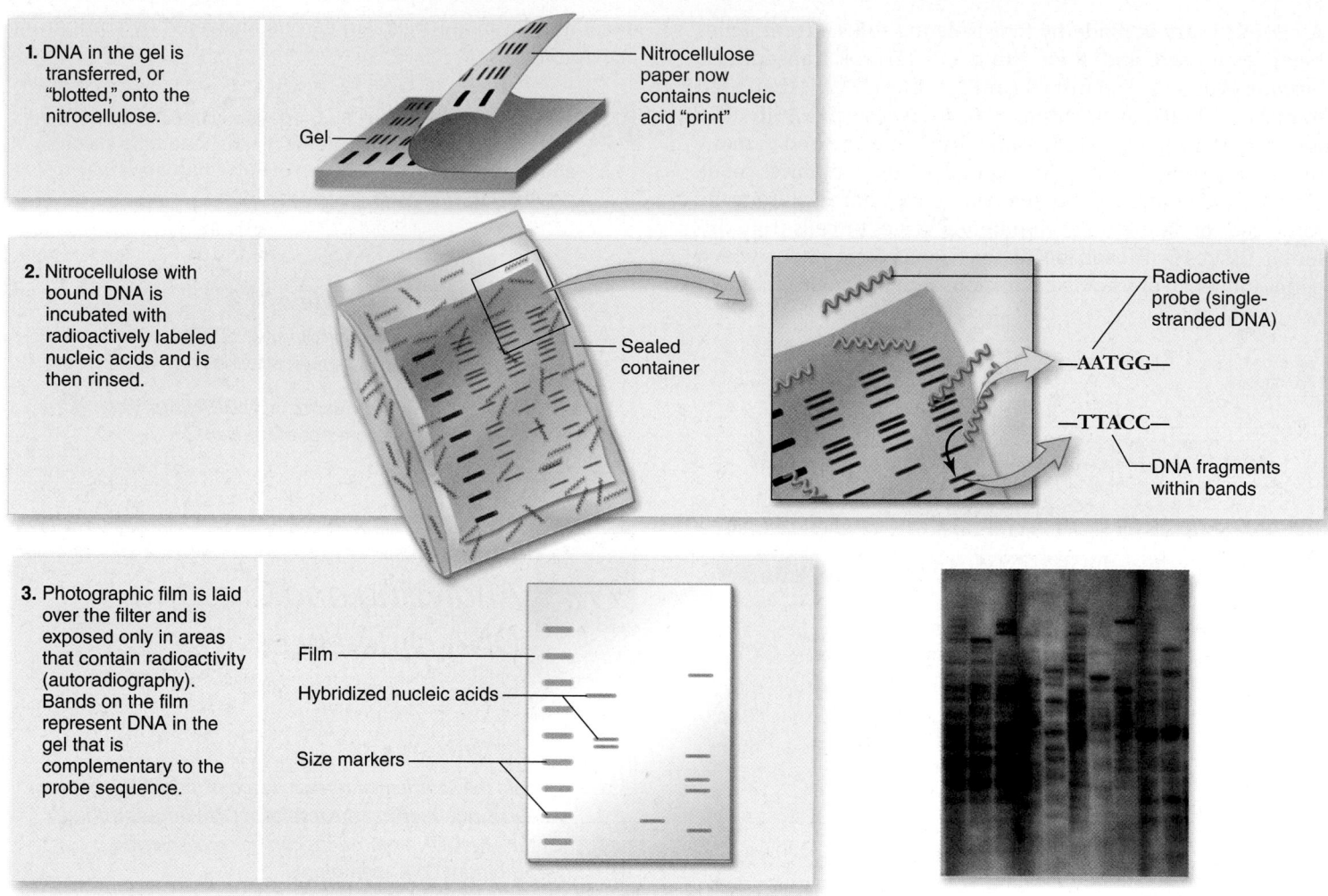

1. DNA in the gel is transferred, or "blotted," onto the nitrocellulose.

Nitrocellulose paper now contains nucleic acid "print"

Gel

2. Nitrocellulose with bound DNA is incubated with radioactively labeled nucleic acids and is then rinsed.

Sealed container

Radioactive probe (single-stranded DNA)

—AATGG—

—TTACC—

DNA fragments within bands

3. Photographic film is laid over the filter and is exposed only in areas that contain radioactivity (autoradiography). Bands on the film represent DNA in the gel that is complementary to the probe sequence.

Film

Hybridized nucleic acids

Size markers

Figure 17.8 The Southern blot procedure. Edwin M. Southern developed this procedure in 1975 to enable DNA fragments of interest to be visualized in a complex sample containing many other fragments of similar size. In steps 1–3, the DNA is separated on a gel, and then transferred ("blotted") onto a solid support medium such as nitrocellulose paper or a nylon membrane and denatured with alkaline chemicals into single strands. Sequences of interest can be detected by using a radioactively labeled probe. This probe (usually several hundred nucleotides in length) of single-stranded DNA (or an mRNA complementary to the gene of interest) is incubated with the filter containing the DNA fragments. All DNA fragments that contain nucleotide sequences complementary to the probe will form hybrids with the probe. Only a short segment of the probe and the complementary sequence are shown in step 3. The fragments differ in size, with the smallest moving the farthest in the gel. The fragments of interest are then detected using photographic film. A representative image is shown. The use of film for detection is being replaced by phosphor imagers, computer-controlled devices that have electronic sensors for light or radioactive emissions.

species (figure 17.8). In this procedure, called a **Southern blot,** DNA from the sample is cleaved into fragments with a restriction endonuclease, and the fragments are separated by gel electrophoresis. The double-stranded helix of each DNA fragment is then denatured into single strands by making the pH of the gel basic. Then the gel is "blotted" with a sheet of filter paper, transferring some of the DNA strands to the sheet.

Next, the filter is incubated with a labeled probe consisting of purified, single-stranded DNA corresponding to a specific gene (or mRNA transcribed from that gene). Any fragment that has a nucleotide sequence complementary to the probe's sequence hybridizes with the probe (see figure 17.8).

This kind of blotting technique has also been adapted for use with RNA and proteins. When mRNA is separated by electrophoresis, the technique is called a **Northern blot.** The methodology is the same except for the starting material (mRNA instead of DNA) and that no denaturation step is

required. Proteins can also be separated by electrophoresis and blotted by a procedure called a **Western blot.** In this case both the electrophoresis and the detection step are different from Southern blotting. The detection, in this case, requires an antibody that can bind to one protein.

The names of these techniques all go back to the original investigator, the British biologist Edwin M. Southern; the Northern and Western blotting names were word play on Southern's name using the cardinal points of the compass.

DNA fingerprinting is used to identify particular genomes

In some cases, an investigator wants to do more than find a specific gene, but instead is looking for variation in the genes of different individuals. Identifying differences can be important in diagnosing a genetic disease, establishing biological relationship, identifying

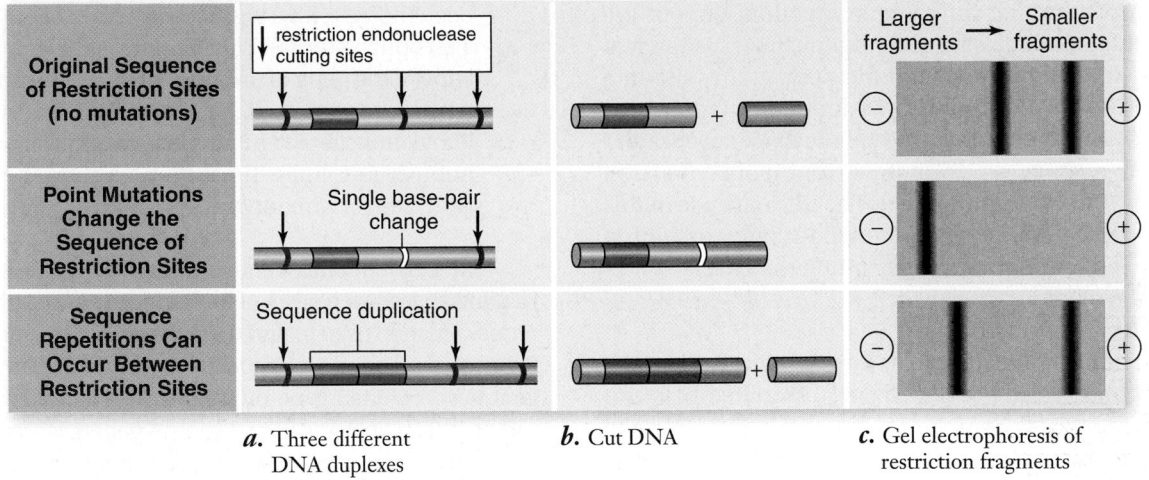

	restriction endonuclease cutting sites			Larger fragments → Smaller fragments

Original Sequence of Restriction Sites (no mutations)

Point Mutations Change the Sequence of Restriction Sites — Single base-pair change

Sequence Repetitions Can Occur Between Restriction Sites — Sequence duplication

a. Three different DNA duplexes

b. Cut DNA

c. Gel electrophoresis of restriction fragments

Figure 17.9 Restriction fragment length polymorphism (RFLP) analysis. *a.* Three samples of DNA differ in their restriction sites due to a single base-pair substitution in one case and a sequence duplication in another case. *b.* When the samples are cut with a restriction endonuclease, different numbers and sizes of fragments are produced. *c.* Gel electrophoresis separates the fragments, and different banding patterns result.

the remains of missing persons or victims of a mass disaster, or convicting or exonerating an individual accused of a crime. Early on, **restriction fragment length polymorphisms, or RFLPs,** were most commonly used to distinguish individuals (figure 17.9).

Point mutations that change the sequence of DNA can eliminate sequences recognized by restriction enzymes or create new recognition sequences. For example, Huntington disease, cystic fibrosis, and sickle cell anemia are all caused by a single base-pair mutation and have associated RFLPs that have been used as molecular markers for diagnosis.

More commonly, short, repeated sequences vary among individuals. These variations are particularly useful in **DNA fingerprinting,** a process that compares DNA from several different regions of individual genomes for identification purposes. Short tandem repeats (STRs), typically 2–4 nt long, are not part of coding or regulatory regions of genes and mutate over generations so that the length of the repeats varies. We say that the population is **polymorphic** for these molecular markers. These markers can be used as DNA "fingerprints" in criminal investigations and other identification applications (figure 17.10).

STRs can be detected using RFLPs and Southern blotting, but currently PCR and automated DNA sequencing is the preferred method (see chapter 18 for DNA sequencing approaches). Since 1997, 13 STRs have been established as the standard of evidence for identification in court, and approved identification kits are available commercially. The 13 STRs form the basis of a federal profiling database called CODIS (Combined DNA Index System). New DNA fingerprints can be compared with those of known individuals in CODIS.

Since 1972, approximately 3000 people have been reported missing in California, and researchers in the California Department of Justice use STR fingerprints to try and match unidentified remains with samples from missing persons cases. This is done by comparing a database with the DNA fingerprints of relatives with a database with STR data obtained from all unidentified remains examined by California coroners. The goal is to provide closure for families and information for law enforcement officers.

DNA fingerprinting has also been used to exonerate wrongly convicted individuals who had been imprisoned for years before DNA analysis was widely available, as well as to identify individuals after catastrophes. After the September 11, 2001, attacks on the World Trade Centers in New York, DNA fingerprinting was the only means for identifying some of the victims. The DNA Shoah Project uses DNA fingerprinting to reunite families separated during the Holocaust. After a devastating earthquake in the Republic of Haiti in 2010, DNA fingerprinting was used to reunite 13 children with their families.

DNA differences can be created to determine gene function

In addition to identifying differences in DNA, recombinant technologies can be used to create differences. Creating

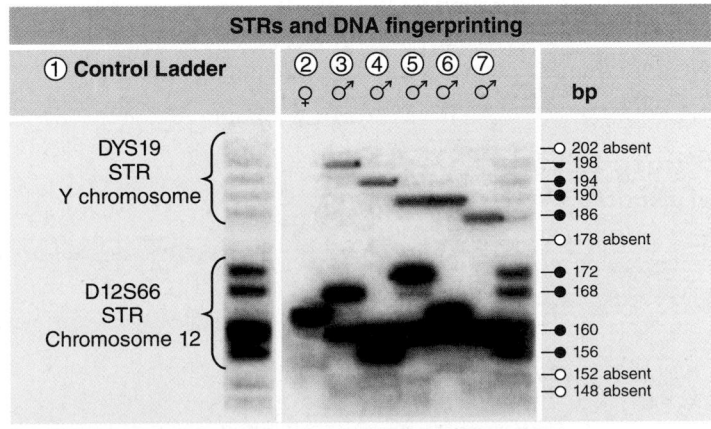

Figure 17.10 Using STRs and DNA fingerprinting to identify individuals. A Y-chromosome STR distinguishes between men and women (*top*) because it is absent in women (*column 2*). The STR is different lengths in men, depending on the number of repeats (columns 3–7). A second STR found on chromosome 12 appears in both men and women (*bottom*).

differences is a way to determine gene function. One of the most important technologies for research purposes is **in vitro mutagenesis**—the ability to create mutations at any site in a cloned gene to examine their effect on function. Rather than depending on mutations induced by chemical agents or radiation in intact organisms, which is time- and labor-intensive, the DNA itself is directly manipulated. The ultimate use of this approach is to be able to replace the wild-type gene with a mutant copy to test the function of the mutated gene.

"Knockout" mice

In mice, in vitro mutagenesis produces **knockout mice** in which a known gene is inactivated ("knocked out"). The effect of loss of this function is then assessed in the adult mouse, or if it is lethal, the stage of development at which function fails can be determined. The idea is simple, but the technology is quite complex. A streamlined description of the steps in this type of experiment are outlined as follows and illustrated in figure 17.11:

1. The cloned gene is disrupted by replacing it with a marker gene using recombinant DNA techniques. The marker gene codes for resistance to the antibiotic neomycin in bacteria, which allows mouse cells to survive when grown in a medium containing the related drug G418. The construction is done such that the marker gene is flanked by the DNA normally flanking the gene of interest in the chromosome.

2. The interrupted gene is introduced into **embryonic stem cells (ES cells).** These cells are derived from early embryos and can develop into different adult tissues. In these cells, the gene can recombine with the chromosomal copy of the gene based on the flanking DNA. This is the same kind of recombination used to map genes (see chapter 13). The knockout gene with the drug resistance gene does not have an origin of replication, and thus it will be lost if no recombination occurs. Cells are grown in medium containing G418 to select for recombination events. (Only those containing the marker gene can grow in the presence of G418.)

3. The ES cells containing the knocked-out gene are injected into an embryo early in its development, which is then implanted into a pseudopregnant female (one that has

been mated with a vasectomized male and as a result has a receptive uterus). The pups from this female have one copy of the gene of interest knocked out, and the phenotype is characterized. If the mutation is recessive and not lethal, transgenic animals can then be crossed to generate homozygous lines. These homozygous lines can be analyzed for phenotypes.

In conventional genetics, genes are identified based on mutants that show a particular phenotype. Molecular genetic techniques are then used to find the gene and isolate a molecular clone for analysis. The use of knockout mice is an example of **reverse genetics:** A cloned gene of unknown function is used to make a mutant that is deficient in that gene. A geneticist can then assess the effect on the entire organism of eliminating a single gene.

Sometimes this approach leads to surprises, such as happened when the gene for the p53 tumor suppressor was knocked out. Because this protein is found mutated in many human cancers and plays a key role in the regulation of the cell cycle (see chapter 10), it was thought to be essential—the knockout was expected to be lethal in the embryo. Instead, the mice were born normal; that is, development had proceeded normally. These mice do have a phenotype, however; they exhibit an increased incidence of tumors in a variety of tissues as they age.

RNA interference—RNAi

RNA interference (RNAi) provides another experimental approach to knockdown or knockout expression of a gene of interest. Unlike knockout mutations, which permanently alter the DNA, RNAi can target a specific RNA sequence so it is degraded and not translated into protein. Details of the mechanism of RNAi are presented in chapter 15. RNAi is used to determine the function of a gene in many organisms, including the nematode *Caenorhabditis elegans*. Silencing of the expression of a gene is initiated when double-stranded RNA (dsRNA) is inserted into cells. Three methods are successfully used to insert dsRNA into *C. elegans* worms; injecting dsRNA, soaking worms in dsRNA, and feeding worms *E. coli* that express the desired dsRNA.

The *C. elegans* genome has been fully sequenced. Coupling the feeding approach with library construction, it is now possible to test most of the genes in the genome. A library of

Figure 17.11 Construction of a knockout mouse. Steps in the construction of a knockout mouse. Some technical details have been omitted, but the basic concept is shown.

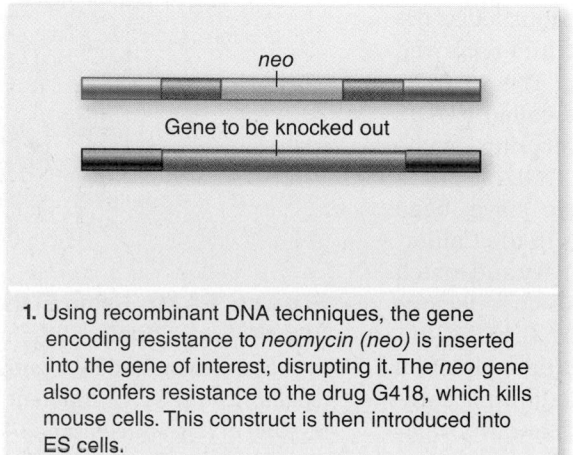

1. Using recombinant DNA techniques, the gene encoding resistance to *neomycin (neo)* is inserted into the gene of interest, disrupting it. The *neo* gene also confers resistance to the drug G418, which kills mouse cells. This construct is then introduced into ES cells.

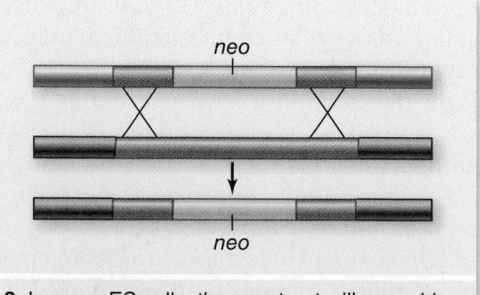

2. In some ES cells, the construct will recombine with the chromosomal copy of the gene to be knocked out. This replaces the chromosomal copy with the *neo* disrupted construct. This is the equivalent to a double crossover event in a genetic cross.

16,757 *E. coli* colonies contains 86% of the *C. elegans* genes that can be used to feed worms. Unlike knockout mutations, RNAi most often knocks down, but does not completely turn off a gene. The advantage to RNAi is that it works very quickly, and changes in phenotype provide information about the function of the gene that is knocked down.

Learning Outcomes Review 17.5

The Southern blotting technique allows identification of a target DNA sequence by separating single-stranded DNA fragments and hybridizing fragments of interest with a labeled probe. Genetic variation among individuals, including variation in STRs, can be used to differentiate individuals by separating RFLPs using gel electrophoresis. Conversely, variation can be created among individuals using recombinant DNA.

■ *Propose a strategy for distinguishing your DNA from your neighbor's, given restriction endonucleases, probes for the relevant RFLPs, and a well-equipped molecular biology lab.*

17.6 Medical Applications

Learning Outcomes

1. *Explain how recombinant DNA can be used in vaccine production.*
2. *Evaluate potential problems of gene therapy.*

The ability to clone individual genes for analysis ushered in an era of unprecedented advancement in research. The early days of genetic engineering led to a rash of start-up companies, many of which are no longer in business. At the same time, all of the major pharmaceutical companies either began research in this area or actively sought to acquire smaller companies with promising technology. The number of medical applications of this technology are far too numerous to mention here, but we highlight a few. Agricultural applications are discussed in the following section.

Eukaryotic genes can be expressed in bacterial cells

Bacterial cells can produce eukaryotic proteins if the differences in the control of transcription and translation discussed in chapter 16 are carefully considered and controlled. In 1982, the human insulin gene was first successfully expressed in *E. coli*. This was a technological breakthrough because individuals with diabetes who required insulin injections to regulate their blood glucose levels could now obtain human insulin more easily and cheaply via bacterial production than from the older method of extraction from pigs' pancreases. To successfully produce insulin in cultured *E. coli* cells, three challenges had to be met (figure 17.12):

1. A eukaryotic promoter is not recognized by the bacterial cellular machinery.
   The eukaryotic insulin promoter gene was inserted in a bacterial plasmid with a bacterial promoter at the 5′ end of the gene.
2. Eukaryotic genes have introns that bacteria cannot excise because they lack posttranscriptional modification.
   The regions of the insulin gene that coded for introns were removed before the gene was spliced into the bacterial plasmid.
3. The insulin peptide undergoes posttranslational modification that excises part of the peptide, creating two peptide sequences held together by a disulfide bond.
   Two DNA sequences that code for only the final, modified peptides were cloned into different plasmids and inserted into two different *E. coli*. The peptides are produced in two separate cultures, purified, and then combined to make human insulin.

Recombinant DNA may simplify vaccine production

Another area of potential significance involves the use of genetic engineering to produce vaccines against communicable diseases. Three types of vaccines are under investigation: *subunit vaccines*, *DNA vaccines*, and *plant-based vaccines*.

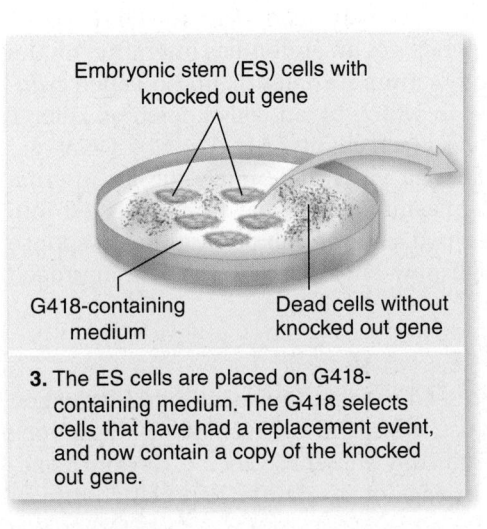

Embryonic stem (ES) cells with knocked out gene

G418-containing medium

Dead cells without knocked out gene

3. The ES cells are placed on G418-containing medium. The G418 selects cells that have had a replacement event, and now contain a copy of the knocked out gene.

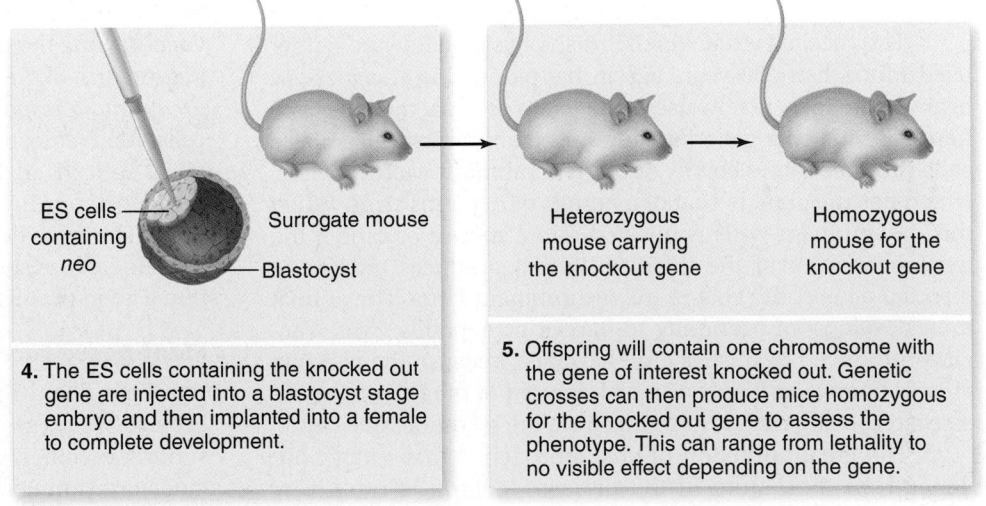

ES cells containing *neo*

Blastocyst

Surrogate mouse

Heterozygous mouse carrying the knockout gene

Homozygous mouse for the knockout gene

4. The ES cells containing the knocked out gene are injected into a blastocyst stage embryo and then implanted into a female to complete development.

5. Offspring will contain one chromosome with the gene of interest knocked out. Genetic crosses can then produce mice homozygous for the knocked out gene to assess the phenotype. This can range from lethality to no visible effect depending on the gene.

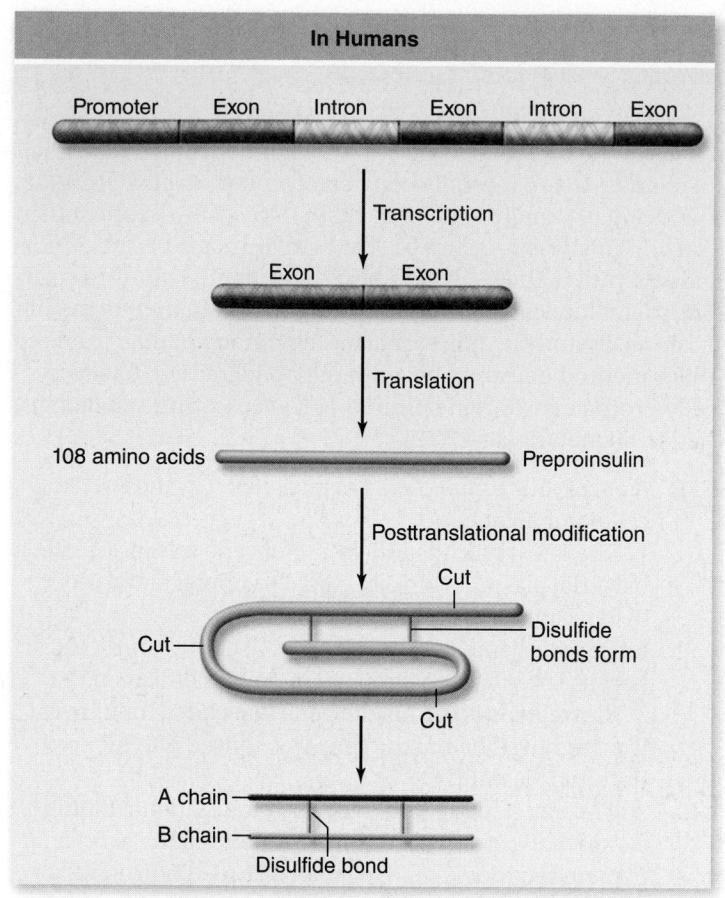

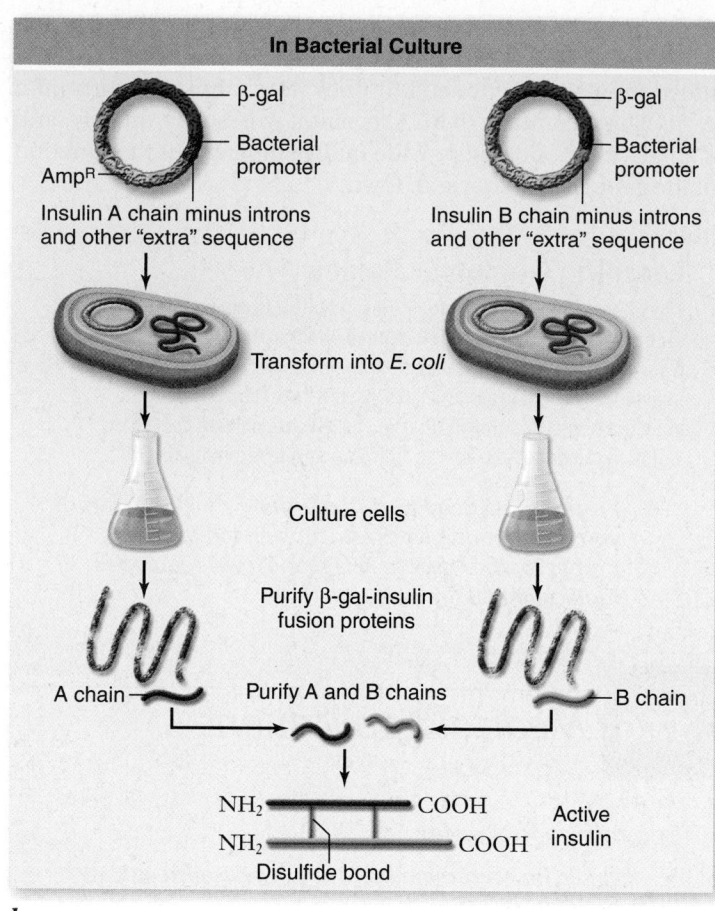

Figure 17.12 Genetically engineering *E. coli* to manufacture human insulin. Three modifications were required in bacterial culture (b): (1) eukaryotic promoters do not work in bacteria, (2) bacteria lack posttranscriptional modification and cannot remove introns, and (3) bacteria lack posttranscriptional modification so the single polypeptide could not be cleaved into two polypeptides.

Subunit vaccines

Subunit vaccines may be developed against pathogens, including the viruses that cause herpes and hepatitis and the parasitic protist that causes malaria. Genes encoding a part, or subunit, of a protein on the surface of the pathogen are cloned, inserted into a vector, and used to produce the vaccine. For example, genes encoding part of the polysaccharide coat of the herpes simplex virus or hepatitis B virus are spliced into a fragment of the vaccinia (cowpox) virus genome (figure 17.13).

The vaccinia virus, which British physician Edward Jenner used more than 200 years ago in his pioneering vaccinations against smallpox, is now used as a vector to carry the herpes or hepatitis viral coat gene into cultured mammalian cells. These cells produce many copies of the recombinant vaccinia virus, which has the outside coat of a herpes or hepatitis virus. When this recombinant virus is injected into a mouse or rabbit, the immune system of the infected animal produces antibodies directed against the coat of the recombinant virus. The animal then develops an immunity to herpes or hepatitis virus. Vaccines produced in this way are harmless because the vaccinia virus is benign, and only a small fragment of the DNA from the disease-causing virus is introduced via the recombinant virus.

The great attraction of this approach is that it does not depend on the nature of the disease. Malaria is caused by *Plasmodium falciparum,* a mosquito-borne protist, and has been difficult to eradicate in developing countries. The subunit vaccine RTS,S/AS01E is the first successful vaccine for malaria and reduced the incidence of malaria in vaccinated individuals by almost a half. A wide variety of subunit vaccines are now being developed.

DNA vaccines

In 1995, the first clinical trials began to test a novel kind of **DNA vaccine,** one that depends not on antibodies but rather on the second arm of the body's immune defense, the so-called *cellular immune response,* in which blood cells known as killer T cells attack infected cells (see chapter 51). The first DNA vaccines spliced an influenza virus gene encoding an internal nucleoprotein into a plasmid, which was then injected into mice. The mice developed a strong cellular immune response to influenza. Although new and controversial, the approach offers great promise.

Plant-based vaccines

The Texas Plant-Expressed Vaccine Consortium has launched Project GreenVax, to use transgenic tobacco plants to produce subunit vaccines, specifically influenza vaccine. Currently vaccines are produced in eggs or in mammalian cell culture in

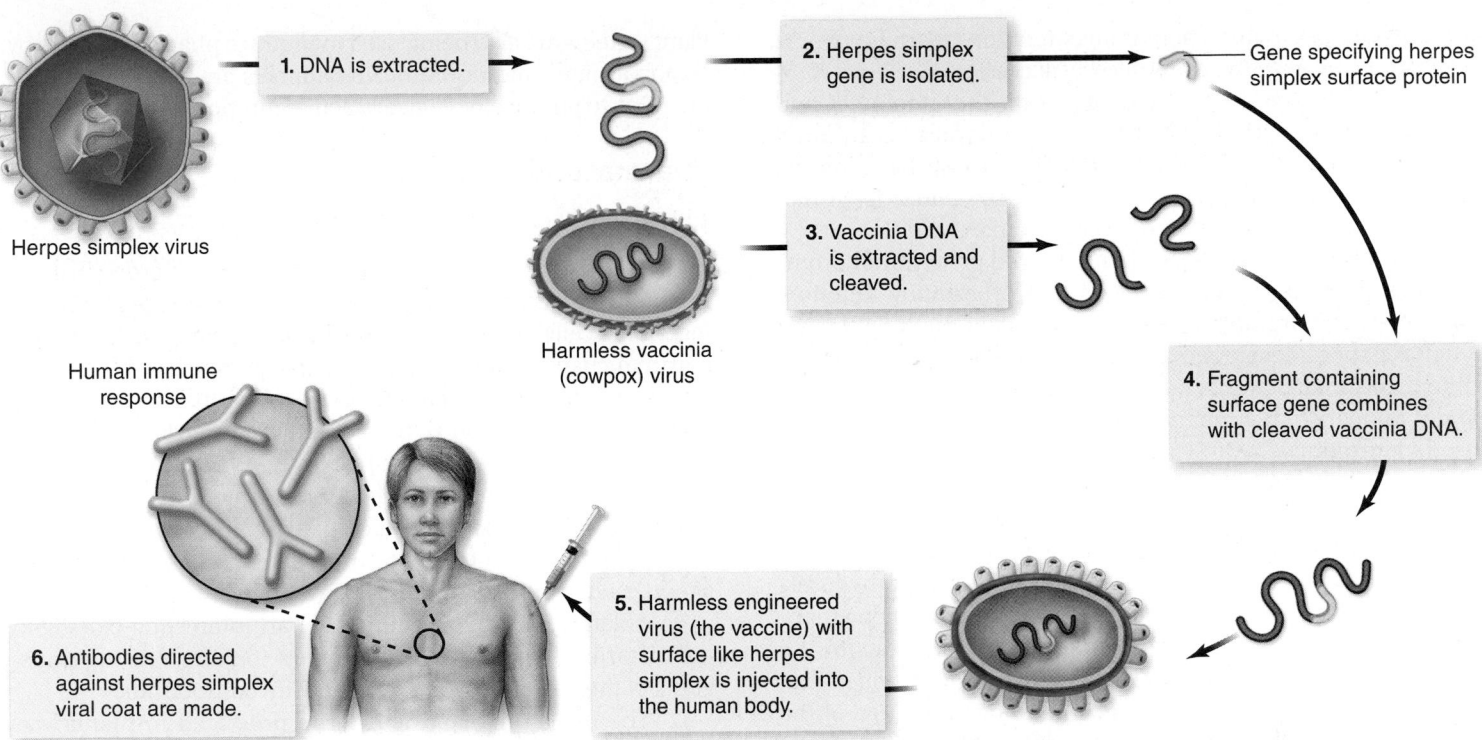

Figure 17.13 Strategy for constructing a subunit vaccine against herpes simplex. Recombinant DNA techniques can be used to construct vaccines for a single protein from a virus or bacterium. In this example, the protein is a surface protein from the herpes simplex virus.

bioreactors (large fermentation tanks), which are costly and relatively slow processes. In contrast, large-scale, low-cost production is possible in plants that replace the bioreactors. Advantages include being able to quickly produce vaccines in response to new strains of influenza and to provide vaccines to parts of the world where availability is currently limited because of cost.

Project GreenVax facilities include processing areas to purify the vaccine produced by the plants. At some point, it may be possible to eliminate the purification processes and use edible plants so the vaccine could be delivered orally.

Gene therapy can treat genetic diseases directly

In 1990, researchers first attempted to combat genetic defects by the transfer of human genes. When a hereditary disorder is the result of a single defective gene, an obvious way to cure the disorder would be to add a working copy of the gene. This approach is being used in an attempt to combat cystic fibrosis, and it offers the potential of treating muscular dystrophy and a variety of other disorders (table 17.1).

Clinical trials for treating macular degeneration, a genetic eye disease that is a leading cause of blindness among adults in the United States, using an RNAi vector (see chapter 16) are promising. Individuals with a certain type of macular degeneration lose their sight because of the uncontrolled proliferation of blood vessels under the retina. For the patient, it is a lot like looking through a car windshield with broken wipers in the middle of a thunderstorm. RNAi gene therapy involves injection of double-stranded RNA coding for a gene necessary for blood vessel proliferation. The RNAi mechanism has the counterintuitive effect of suppressing production of the protein needed for blood vessel development, preventing progression of the disease. Other applications of RNAi to treat genetic disease have not developed as quickly. Despite the initial promise of RNAi, one of the key challenges is delivering the dsRNA to

TABLE 17.1	Diseases Being Treated in Clinical Trials of Gene Therapy
Disease
Cancer (melanoma, renal cell, ovarian, neuroblastoma, brain, head and neck, lung, liver, breast, colon, prostate, mesothelioma, leukemia, lymphoma, multiple myeloma)
SCID (severe combined immunodeficiency)
Cystic fibrosis
Gaucher disease
Familial hypercholesterolemia
Hemophilia
Purine nucleoside phosphorylase deficiency
α_1-Antitrypsin deficiency
Fanconi anemia
Hunter syndrome
Chronic granulomatous disease
Rheumatoid arthritis
Peripheral vascular disease
Acquired immunodeficiency syndrome (AIDS)
Duchenne muscular dystrophy
Macular degeneration (wet variety)
Batten disease (neurological disorder)

the targeted cells. The macular degeneration treatment works beautifully because the eye is readily accessible.

One disease that illustrates both the potential and the problems with gene therapy is **severe combined immunodeficiency disease (SCID).** This disease has multiple forms, including an X-linked form (X-SCID) and a form that lacks the enzyme adenosine deaminase (ADA-SCID).

Trials for both of these forms showed great initial promise, with patients exhibiting restoration of immune function. Unfortunately, problems arose in the case of the X-SCID trial when a patient developed a rare form of leukemia. Since that time, two other patients have developed the same leukemia, and it appears to be due to the gene therapy itself. The vector used to introduce the X-SCID gene integrated into the genome next to a proto-oncogene called *LMO2* in all three cases. Activation of this gene can cause childhood leukemia.

The insertion of a gene during gene therapy has always been a random event, and it has been a concern that the insertion could inactivate an essential gene, or turn on a gene inappropriately. That effect had not been observed prior to the X-SCID trial, despite a large number of genes introduced into blood cells in particular. For leukemia to occur in 15% of the patients treated implies that some influence of the genetic background associated with X-SCID potentiates this development. This possibility is supported by the observation that the ADA-SCID patients treated have not been affected thus far.

On the positive side, 15 children treated successfully are still alive, several of them after more than a decade after gene therapy, with functioning immune systems. On the negative side, three other children treated have developed leukemia. New approaches are being explored, and clinical trials for SCID have resumed.

Learning Outcomes Review 17.6

Recombinant DNA technology has allowed genes from eukaryotes, such as humans, to be isolated, inserted into vectors, and recombined into bacterial genomes, where the genes' products can be mass-produced. Gene therapy is the process of using genetic engineering to replace defective genes; however, in some cases unwanted effects result from random gene insertion.

■ *Create a flow chart to illustrate how a gene coding for an influenza coat protein can be used to create a plant-based influenza vaccine.*

17.7 Agricultural Applications

Learning Outcomes

1. *Compare recombinant technology techniques used in plants with those used in bacteria.*
2. *Describe the controversial issues in the use of GM plants.*

Perhaps no area of genetic engineering touches all of us so directly as the applications being used in agriculture today. Crops are being modified to resist disease, to be tolerant of herbicides, and for changes in nutritional and other content in a variety of ways.

Plant systems are also being used to produce pharmaceuticals by "biopharming," and domesticated animals are being genetically modified to produce biologically active compounds.

Pharmaceuticals can be produced by "biopharming"

The medicinal use of plants goes back as far as recorded history. In modern times, the pharmaceutical industry began by isolating biologically active compounds from plants. This approach began to change when in 1897, the Bayer company introduced acetylsalicylic acid, otherwise known as aspirin. This compound was a synthetic version of the compound salicylic acid, which was isolated from the bark of the white willow. The production of pharmaceuticals has since been dominated more by organic synthesis and less by the isolation of plant products.

One exception to this trend is cancer chemotherapeutic agents such as taxol, vinblastine, and vincristine, all of which were isolated from plant sources. In an interesting closing of the historical loop, the industry is now looking at using transgenic plants for the production of useful compounds.

The first human protein to be produced in plants was human serum albumin, which was produced in 1990 by both genetically engineered tobacco and potato plants. Since that time more than 20 proteins have been produced in transgenic plants. This first crop of transgenic pharmaceuticals is now undergoing regulatory examination.

Herbicide-resistant crops allow for no-till planting

Recently, broadleaf plants have been genetically engineered to be resistant to **glyphosate,** a powerful, biodegradable herbicide that kills most actively growing plants (figure 17.14). Glyphosate works by inhibiting an enzyme called 5-enolpyruvylshikimate-3-phosphate (EPSP) synthetase, which plants require to produce aromatic amino acids.

Humans do not make aromatic amino acids; we get them from our diet, so we are unaffected by glyphosate. To make glyphosate-resistant plants, scientists used a Ti plasmid to insert extra copies of the EPSP synthetase gene into plants. These engineered plants produce 20 times the normal level of EPSP synthetase, enabling them to synthesize proteins and grow despite glyphosate's suppression of the enzyme. In later experiments, a bacterial form of the EPSP synthetase gene that differs from the plant form by a single nucleotide was introduced into plants via Ti plasmids; the bacterial enzyme is not inhibited by glyphosate (see figure 17.14).

These advances are of great interest to farmers because a crop resistant to glyphosate does not have to be weeded through tilling—the field could simply be treated with the herbicide. Because glyphosate is a broad-spectrum herbicide, farmers no longer need to employ a variety of different herbicides, most of which kill only a few kinds of weeds. Furthermore, glyphosate breaks down readily in the environment, unlike many other herbicides commonly used in agriculture. A plasmid is actively being sought for the introduction of the EPSP synthetase gene into cereal plants that would also make them glyphosate-resistant.

At this point four important crop plants have been modified to be glyphosate-resistant: maize (corn), cotton, soybeans,

Hypothesis: *Petunias can acquire tolerance to the herbicide glyphosate by overexpressing EPSP synthase.*

Prediction: *Transgenic petunia plants with a chimeric EPSP synthase gene with strong promoter will be glyphosate tolerant.*

Test:

1. Use restriction enzymes and ligase to "paste" the cauliflower mosaic virus promoter (35S) to the EPSP synthase gene and insert the construct in Ti plasmids.

2. Transform Agrobacterium with the recombinant plasmid.

3. Infect petunia cells and regenerate plants. Regenerate uninfected plants as controls.

4. Challenge plants with glyphosate.

Result: *Glyphosate kills control plants, but not transgenic plants.*

Conclusion: *Additional EPSP synthase provides glyphosate tolerance.*

Further Experiments: *The transgenic plants are tolerant, but not resistant (note bleaching at shoot tip). How could you determine if additional copies of the gene would increase tolerance? Can you think of any downsides to expressing too much EPSP synthase in petunia?*

Figure 17.14 Genetically engineered herbicide resistance.

and canola. The use of glyphosate-resistant soy has been especially popular, accounting for 60% of the global area of genetically modified (GM) crops grown in nine countries worldwide. In the United States, 90% of soy currently grown is GM soy. Use of GM crops varies globally, with the Americas, led by the United States, the largest adopter. The area currently with the largest growth in the use of GM crops is Asia, and Europe has been the slowest to adopt their use.

Bt crops are resistant to some insect pests

Many commercially important plants are attacked by insects, and the usual defense against such attacks has been to apply insecticides. Over 40% of the chemical insecticides used today are targeted against boll weevils, bollworms, and other insects that eat cotton plants. Scientists have produced plants that are resistant to insect pests, removing the need to use many externally applied insecticides.

The approach is to insert into crop plants genes encoding proteins that are harmful to the insects that feed on the plants, but harmless to other organisms. The most commonly used protein is a toxin produced by the soil bacterium *Bacillus thuringiensis* (Bt toxin). When insects ingest Bt toxin, endogenous enzymes convert it into an insect-specific toxin, causing paralysis and death. Because these enzymes are not found in other animals, the protein is harmless to them.

The same four crops that have been modified for herbicide resistance have also been modified for insect resistance using the Bt toxin. The use of Bt maize is the second most common GM crop, representing 14% of global area of GM crops in

nine countries. The global distribution of these crops is also similar to the herbicide-resistant relatives.

Given the popularity of both of these types of crop modifications, it is not surprising that they have also been combined, so-called *stacked GM crops,* in both maize and cotton. Stacked crops now represent 9% of global area of GM crops.

Golden Rice shows the potential of GM crops

One of the successes of GM crops is the development of Golden Rice. The World Health Organization (WHO) estimates that vitamin A deficiency affects between 140 and 250 million preschool children worldwide, 250,000–500,000 of whom become blind. The deficiency is especially severe in developing countries where the major staple food is rice. Golden Rice has been genetically modified to produce β-carotene (provitamin A), which can be converted by enzymes in the body to vitamin A, thus alleviating the deficiency.

Golden Rice is named for its distinctive color imparted by the presence of β-carotene in the endosperm (the outer layer of rice that has been milled). Rice does not normally make β-carotene in endosperm tissue, but does produce a precursor, geranyl geranyl diphosphate, that can be converted by three enzymes, phytoene synthase, phytoene desaturase, and lycopene β-cyclase, to β-carotene. These three genes were engineered to be expressed in endosperm and introduced into rice to complete the biosynthetic pathway producing β-carotene in endosperm (figure 17.15).

This case of genetic engineering is interesting for two reasons. First, it introduces a new biochemical pathway in tissue of the transgenic plants. Second, it could not have been done by conventional breeding as no rice cultivar known produces these

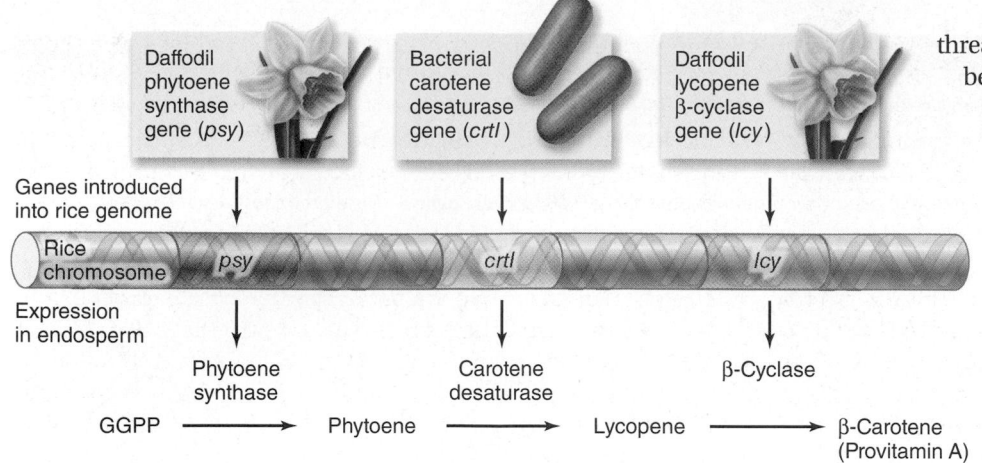

Figure 17.15 Construction of Golden Rice. Rice does not normally express the enzymes needed to synthesize β-carotene in endosperm. Three genes were added to the rice genome to allow expression of the pathway for β-carotene in endosperm. The source of the genes and the pathway for synthesis of β-carotene is shown. The result is Golden Rice, which contains enriched levels of β-carotene in endosperm.

enzymes in endosperm. The original constructs used two genes from daffodil and one from a bacterium (see figure 17.15). There are many reasons to expect failure in the introduction of a biochemical pathway without disrupting normal metabolism. That the original form of Golden Rice makes significant amounts of β-carotene in an otherwise healthy plant is impressive. A second-generation version that makes much higher levels of β-carotene has also been produced by using the gene for phytoene synthase from maize in place of the original daffodil gene.

Golden Rice was originally constructed in a public facility in Switzerland and made available for free with no commercial entanglements. Since its inception, Golden Rice has been improved on both by public groups and by industry scientists, and these improved versions are also being made available without commercial strings attached.

Marker assisted breeding accelerates breeding programs

High-yielding crops have a multitude of desirable traits. Breeders cross different plants to maximize desired gene combinations, but this is a very time-consuming process. **Marker assisted breeding** (MAB) combines classic plant breeding with molecular biology tools. Instead of screening for phenotypically obvious traits at maturity, breeders extract DNA from leaf tissue of young seedlings and screen for molecular markers (DNA sequences) that are associated with agriculturally important traits. Screening relies on the DNA fingerprinting techniques (described earlier in this chapter). Seedlings with the desired combination of molecular markers are raised to maturity and the remaining plants are discarded.

Molecular markers are often repetitive sequences, similar to the STRs used for fingerprinting. They are variants that tend to cosegregate with desirable traits, but the variation in the DNA is not necessarily in a gene known to code for that trait. For example, variation in the number of repeats in an STR near a corn gene coding for an enzyme that converts sugar to starch could be an indicator of the relative sweetness of the corn.

MAB is being used to breed for nematode resistance in soybeans. Fields of soybeans are being destroyed by this microscopic, threadlike worm that bores into the roots of soybeans and use nutrients from the plant. Peking landraces of soybean from China are resistant to nematodes, but lack many of the traits in commercial lines that enhance yield and make it possible to mechanically harvest the soybeans. These plants are being bred with commercial lines, and molecular markers (MAB) are used to sort out promising offspring, saving vast amounts of time and growing space.

MAB is a tool that allows breeders to take advantage of the diversity in traditional breeding programs. The Food and Agricultural Organization of the United Nations proposes that MAB will play a significant role in breeding programs for crops as well as livestock, forestry, and fisheries globally.

GM crops raise a number of social issues

The adoption of GM crops has been resisted in some places for a variety of reasons. Some people have wondered about the safety of these crops for human consumption, the likelihood of introduced genes moving into wild relatives, and the possible loss of biodiversity associated with these crops.

Powerful forces have aligned on opposing sides in this debate. On the side in favor of the use of GM crops are the multinational companies that are utilizing this technology to produce seeds for the various GM crops and groups noting the benefits of GM crops like Golden Rice in feeding the developing world. On the other side are a variety of political organizations that are opposed to genetically modified foods. Scientists can be found on both sides of the controversy.

The controversy originally centered on the safety of introduced genes for human consumption. In the United States, this issue has been "settled" for the crops already mentioned, and a large amount of GM soy and maize is consumed in this country. Although some opponents still raise the issue of long-term use and allergic reactions, no negative effects have been documented so far. Existing crops will be monitored for adverse effects, and each new modification will require regulatory approval for human consumption.

Another concern is that genes of the GM crops will spread outside into non-GM crop, through gene flow. The amount of cross-pollination between plants of the same species varies. Soybeans tend to self-pollinate with limited outcrossing; but corn freely outcrosses, and so genes from GM plants more frequently move to non-GM corn plants. This is a concern for organic farmers with fields close to GM corn fields, as current regulations for organic certification exclude GM crops. The amount of outcrossing depends on the proximity of crops, wind speed, and direction, as well as temperature and humidity. For plants that depend on pollinators for reproduction, the range of species the pollinator visits and the distance it travels are also factors.

Gene flow can also occur between related species through hybridization. In the United States, at least 15 weedy species have hybridized with crop plants. For two of the major U.S. crops, corn and soybean, no closely related weedy species reproduce with them, so this is an unlikely mechanism for acquiring herbicide resistance. Sunflowers, however, were first domesticated in the United States, and the risk of a transgene moving into a weedy population is a concern. One study revealed that 10–33% of wild sunflowers had hybridized with domesticated varieties.

In the case of herbicide-resistant GM crop plants, the repeated application of a single herbicide over a number of years creates selective pressure on weedy species. Over time, weeds with resistance to the herbicide will increase in frequency, and the herbicide becomes ineffectual. Almost 200 weed species worldwide have acquired resistance to one or more herbicides. Management practices that reduce herbicide use and vary the herbicides used can slow the evolution of herbicide resistance.

Domesticated animals can also be genetically modified

Humans have been breeding and selecting domestic animals for thousands of years. With the advent of genetic engineering, this process can be accelerated, and genes can be introduced from other species.

The production of transgenic livestock is in an early stage, and it is hard to predict where it will go. At this point, one of the uses of biotechnology is not to construct transgenic animals, but to use DNA markers to identify animals and to map genes that are involved in such traits as palatability in food animals, texture of hair or fur, and other features of animal products. This information is then used in MAB. Molecular techniques combined with the ability to clone domestic animals (see chapter 19) could produce improved animals for economically desirable traits.

Transgenic animal technology has not been as successful as initially predicted. To date, no transgenic animal has been approved for human consumption. Early on, pigs were engineered to overproduce growth hormone in the hope that this would lead to increased and faster growth. These animals proved to have only slightly increased growth, and they had lower fat levels, which reduces flavor, as well as showing other deleterious effects.

One interesting idea for transgenics is the "Enviropig." This animal has been engineered with the gene for phytase under the control of a salivary gland-specific promoter. The enzyme phytase breaks down phosphorus in the feed and can reduce phosphate excretion by up to 65%. Phosphate is a major problem in pig waste, and reducing the amount of phosphate runoff into rivers and streams reduces the overgrowth of algae. Enviropigs can now be raised in Canada for research purposes, but it will likely be years before they make it to the consumer's table.

For over a decade, AquaBounty Technologies has been seeking approval from the Food and Drug Administration (FDA) to market its GM salmon that grows more rapidly than non-GM salmon because of a transgenic growth hormone gene (figure 17.16). The hormone gene was isolated from a related fish species, and there is no evidence that the protein could harm humans. The fish are all female triploids so they could not reproduce with wild relatives if they escaped into the wild, but they could compete with wild salmon for resources. The focus at the FDA was on whether or not to label the salmon as genetically modified in the grocery store, when in June 2011 the U.S. House of Representatives voted to ban the FDA from approving any GM salmon for human consumption. As with the Enviropig, GM salmon will not be on the menu any time soon.

Learning Outcomes Review 17.7

To date, herbicide resistance, pathogen protection, nutritional enhancement, and drug production have been targets of agricultural genetic engineering. Controversy regarding the use of GM plants has centered on the potential of unforeseen effects on human health and on the environment.

■ *Compare and contrast molecular assisted breeding (MAB) and the development of GM plants for pest resistance.*

a.

b.

Figure 17.16 Transgenic salmon grow more rapidly than wild salmon. *a.* The rate of growth in transgenic salmon compared with that for wild salmon. *b.* The size of the transgenic salmon is larger than a wild salmon of the same age.

Data analysis Based on the information in this figure, evaluate the claim that transgenic salmon are huge compared with wild salmon.

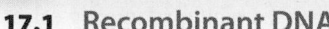

17.1 Recombinant DNA

Restriction endonucleases cleave DNA at specific sites.

DNA molecules fragmented by known type II restriction endonucleases can be recombined with other DNA cut with the same restriction endonucleases (figure 17.1).

Gel electrophoresis separates DNA fragments.

An electric field applied to a gel matrix causes DNA to migrate through the matrix, separating the DNA fragment by size. Smaller fragments migrate farther than large fragments (figure 17.2).

DNA ligase allows construction of recombinant molecules.

Just as in DNA replication, DNA ligase catalyzes formation of a phosphodiester bond between nucleotides, forming a recombinant molecule.

17.2 Introducing Foreign DNA into Cells

Transformation allows introduction of foreign DNA into cells.

Artificial transformation techniques introduce foreign DNA into bacterial and eukaryotic cells, which are then termed transgenic.

Host–vector systems allow propagation of foreign DNA in bacteria.

Plasmids and artificial chromosomes can be used as vectors. Foreign DNA is inserted using restriction enzymes and DNA ligase; once the vector is inside the host cell, it is replicated during the cell cycle.

Eukaryotic cell transformation is challenging.

Eukaryotic cells lack plasmids and cannot be transformed in the same way as bacteria. Other strategies have been developed and include injecting DNA into animal cells and using a bacterial pathogen to insert DNA into some plant cells (figure 17.4).

17.3 Amplifying DNA Without a Vector—The Polymerase Chain Reaction

PCR mimics DNA replication (figure 17.5).

The polymerase chain reaction (PCR) amplifies a single small DNA fragment using two short primers that flank the region to be amplified. Cyclic replication is accomplished via heating and cooling; a key factor is Taq polymerase, which is not denatured at high temperature.

PCR has many applications.

Because PCR can amplify even tiny amounts of DNA, it has many applications, ranging from DNA fingerprinting for forensics to genome sequencing of extinct species.

17.4 Storing and Sorting DNA Fragments

DNA libraries may contain the entire genome of an organism.

A DNA library is a complex mixture of DNAs collected into vectors that are randomly organized (figure 17.6).

A library can also be created for the expressed parts of the genome by isolating RNA and converting it to cDNA using the enzyme reverse transcriptase (figure 17.7).

17.5 Analyzing and Creating DNA Differences

DNA differences can be detected using blotting methods.

DNA can be reversibly denatured and renatured, resulting in single- and then double-stranded DNA. Renaturation of complementary strands from different sources is called hybridization. Known DNA can be labeled to identify complementary strands.

In Southern blotting, a complex mixture is separated by electrophoresis and transferred to filter paper (figure 17.8). Specific sequences of DNA can then be identified by hybridization. Similar techniques can identify mRNA (Northern blotting) and proteins (Western blotting).

DNA fingerprinting is used to identify particular genomes.

Regions of repetitive DNA sequences, including STRs, vary among individuals in terms of the number of repeats. This variation can be assessed using RFLPs to create distinct DNA fingerprints, which are used to identify individuals in criminal and missing person cases, as well as major disasters.

DNA differences can be created to determine gene function.

Reverse genetics approaches begin with a sequence of DNA and aim to find the function of that sequence. In knockout mice, a gene is inactivated by replacing the wild-type version with a mutant copy (figure 17.11).

RNAi can be used to knock down gene expression by inserting dsRNA into an organism. In this way, the function of the gene can be analyzed and clarified.

17.6 Medical Applications

Eukaryotic genes can be expressed in bacterial cells.

Bacterial production of human proteins, such as insulin, has allowed better results and has increased production to treat disease.

Recombinant DNA may simplify vaccine production.

Subunit vaccines produced in cultured cells have been shown to be effective in animals.

DNA vaccines, which alter the cellular immune response, are also promising. Both these approaches require further testing.

Gene therapy can treat genetic diseases directly.

Gene therapy involves inserting a normal gene to replace a defective one. Clinical trials for some therapies are underway, but the many challenges include safely inserting a foreign gene into the appropriate cells.

17.7 Agricultural Applications

Pharmaceuticals can be produced by "biopharming."

More than 20 proteins that are used as pharmaceuticals are being made by recombinant plants.

Herbicide-resistant crops allow for no-till planting.

Herbicide-resistant crops are widely used in the United States, reducing both the need for tilling and the associated fossil fuel consumption.

Bt crops are resistant to some insect pests.

Recombinant crops that produce Bt are resistant to insect damage.

Golden Rice shows the potential of GM crops.

Golden rice produces provitamin A, reducing blindness in developing countries where rice is a staple crop. Golden rice is an example of a not-for-profit effort to provide food security in the developing world.

Marker assisted breeding accelerates breeding programs.

By isolating DNA from the leaves of young seedlings in a breeding program and using DNA fingerprinting to compare individuals, only those with promising molecular markers are raised to maturity. This combination of conventional breeding with the use of molecular tools is called MAB and accelerates the rate of breeding programs substantially.

GM crops raise a number of social issues.

Concerns about GM plants include unintended allergic reactions to proteins inserted from a different organism, although no adverse affects have been documented yet. The spread of foreign genes into noncultivated plants in the environment through hybridization is being followed closely, and to date almost 200 weed species worldwide have acquired resistance to one or more herbicides.

Domesticated animals can also be genetically modified.

To date, results with transgenic animals have been mixed, and no recombinant meat or fish is currently available on grocery shelves.

UNDERSTAND

1. A recombinant DNA molecule is one that is
   a. produced through the process of crossing over that occurs in meiosis.
   b. constructed from DNA from different sources.
   c. constructed from novel combinations of DNA from the same source.
   d. produced through mitotic cell division.

2. What is the basis of separation of different DNA fragments by gel electrophoresis?
   a. The negative charge on DNA
   b. The size of the DNA fragments
   c. The sequence of the fragments
   d. The presence of a dye

3. The basic logic of organizing a genome into BAC libraries is to create
   a. a nested set of DNA fragments produced by restriction enzymes.
   b. a nested set of DNA fragments that each begin with different bases.
   c. primers to allow PCR amplification of the region between the primers.
   d. a nested set of DNA fragments that end with known bases.

4. A DNA library is
   a. an orderly array of all the genes within an organism.
   b. a collection of vectors.
   c. the collection of plasmids found within a single *E. coli*.
   d. a collection of DNA fragments representing the entire genome of an organism.

5. Molecular hybridization is used to
   a. generate cDNA from mRNA.
   b. introduce a vector into a bacterial cell.
   c. screen a DNA library.
   d. introduce mutations into genes.

6. In vitro mutagenesis is used to
   a. produce large quantities of mutant proteins.
   b. create mutations at specific sites within a gene.
   c. create random mutations within multiple genes.
   d. create organisms that carry foreign genes.

7. Insertion of a gene for a surface protein from a medically important virus such as herpes into a harmless virus is an example of
   a. a DNA vaccine.   c. gene therapy.
   b. reverse genetics.   d. a subunit vaccine.

8. What is a Ti plasmid?
   a. A vector that can transfer recombinant genes into plant genomes
   b. A vector that can be used to produce recombinant proteins in yeast
   c. A vector that is specific to cereal plants like rice and corn
   d. A vector that is specific to embryonic stem cells

APPLY

1. How is the gene for β-galactosidase used in the construction of a plasmid?
   a. The gene is a promoter that is sensitive to the presence of the sugar galactose.
   b. It is an origin of replication.
   c. It is a cloning site.
   d. It is a marker for insertion of DNA.

2. Which of the following statements is accurate for DNA replication in your cells, but not PCR?
   a. DNA primers are required.
   b. DNA polymerase is stable at high temperatures.
   c. Ligase is essential.
   d. dNTPs are necessary.

3. What potential problems must be considered in creating a transgenic bacterium with the human insulin gene to produce insulin?
   a. Introns in the human gene will not be processed after transcription.
   b. The bacterial cell will be unable to posttranslationally process the insulin peptide sequence.
   c. There is no way to get the bacterium to transcribe high levels of a human gene.
   d. Both a and b present problems.

SYNTHESIZE

1. Many human proteins, such as hemoglobin, are only functional as an assembly of multiple subunits. Assembly of these functional units occurs within the endoplasmic reticulum and Golgi apparatus of a eukaryotic cell. Discuss what limitations, if any, exist to the large-scale production of genetically engineered hemoglobin.

2. As a plant breeder your assignment is to integrate the resistance to root nematodes in a Chinese soybean into an otherwise robust soybean grown in Iowa. You have a set of STRs for both soybeans and information about which STRs have alleles that correspond with root nematode resistance. Using your knowledge of MAB and DNA fingerprinting, write a short proposal to your supervisor outlining a plan to breed for nematode resistance over the next three years.

ONLINE RESOURCE

www.ravenbiology.com

Understand, Apply, and Synthesize—enhance your study with animations that bring concepts to life and practice tests to assess your understanding. Your instructor may also recommend the interactive eBook, individualized learning tools, and more.

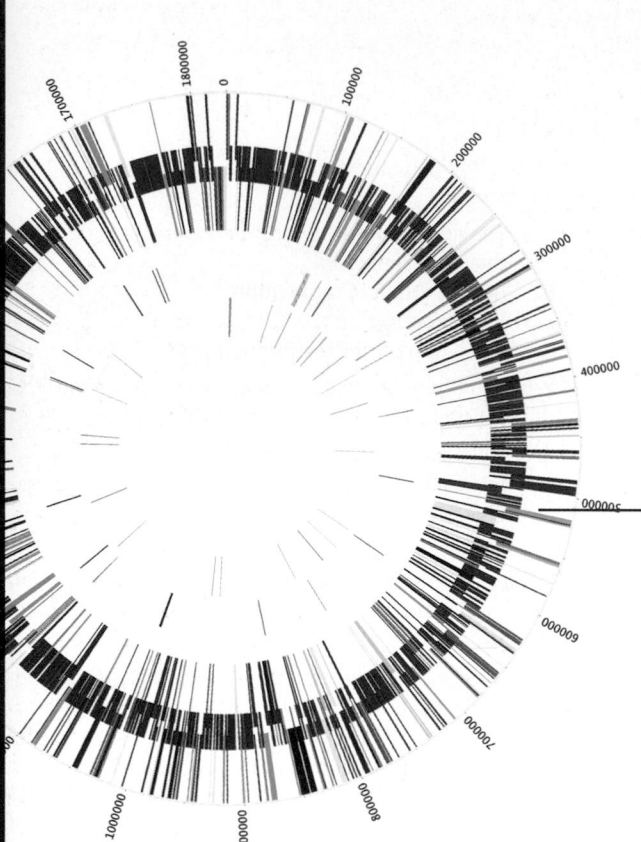

Chapter 18

Genomics

Chapter Contents

Introduction

The pace of discovery in biology in the last 30 years has been like the exponential growth of a population. Starting with the isolation of the first genes in the mid-1970s, researchers had accomplished the first complete genome sequence by the mid-1990s—that of the bacterial species Haemophilus influenzae, *shown in the picture (genes with similar functions are shown in the same color). By the turn of the 21st century, the molecular biology community had completed a draft sequence of the human genome. Put another way, scientific accomplishments moved from cloning a single gene, to determining the sequence of a million base-pairs in 20 years, then determining the sequence of 3 billion base-pairs in another 5 years, and now sequencing 20 billion base-pairs at one time. In the previous chapter you learned about the basic techniques of molecular biology. In this chapter you will see how those techniques have been applied to the analysis of whole genomes. This analysis integrates ideas from classical and molecular genetics with biotechnology, scaled and applied to whole genomes.*

18.1 Mapping Genomes

Learning Outcomes

1. *Distinguish between a genetic map and a physical map.*
2. *Explain how genetic and physical maps can be linked.*

We use maps to find our location, and depending on how accurately we wish to do this, we may use multiple maps with different resolutions. In genomics, we can locate a gene on a chromosome, in a subregion of a chromosome, and finally its precise location in the chromosome's DNA sequence. The DNA sequence level requires knowing the entire sequence of the genome, something that was once out of our reach technologically. Knowing the entire sequence is useless, however, without other kinds of maps; finding a single gene within the sequence of the human genome is like trying to find your house on a map of the world.

To overcome this difficulty, maps of genomes are constructed at different levels of resolution and using different kinds of information. We can distinguish between *genetic maps* and *physical maps*. **Genetic maps** are abstract maps that place the relative location of genes on chromosomes based on recombination frequency (see chapter 13). **Physical maps** use landmarks within DNA sequences, ranging from restriction sites (described in the preceding chapter) to the ultimate level of detail: the actual DNA sequence.

Different kinds of physical maps can be generated

To make sense of genome mapping, it is important to have physical landmarks on the genome that are at a lower level of resolution than the entire sequence. In fact, long before the Human Genome Project was even conceived, physical maps of DNA were needed as landmarks on cloned DNA. Two types of physical maps are (1) restriction maps, constructed using restriction enzymes and (2) chromosome-banding patterns, generated by cytological dye methods.

Restriction maps

Distances between "landmarks" on a physical map are measured in base-pairs (1000 base-pairs [bp] equal 1 kilobase [kb]). It is not necessary to know the DNA sequence of a segment of DNA in order to create a physical map, or to know whether the DNA encompasses information for a specific gene.

The first physical maps were created by cutting genomic DNA with different restriction enzymes, both singly and with combinations of enzymes (figure 18.1). The analysis of the patterns of fragments generated was used to generate a map.

In terms of larger pieces of DNA, this process is repeated and then used to put the pieces back together, based on size and overlap, into a contiguous segment of the genome, called a **contig**. Coincidentally, the very first restriction enzymes to be isolated came from *Haemophilus*, which was also the first free-living genome to be completely sequenced.

Chromosome-banding patterns

Cytologists studying chromosomes with light microscopes found that by using different stains, they could produce reproducible patterns of bands on the chromosomes. In this way, they could identify all of the chromosomes and divide them into subregions based on banding pattern.

The use of different stains allows for the construction of a cytological map of the entire genome. These large-scale physical maps are like a map of an entire country, in that they encompass the whole genome, but at low resolution.

Cytological maps are used to characterize chromosomal abnormalities associated with human diseases, such as chronic myelogenous leukemia. In this disease, a reciprocal translocation occurs between chromosome 9 and

1. Multiple copies of a segment of DNA are cut with restriction enzymes.

2. The fragments produced by enzyme A only, by enzyme B only, and by enzymes A and B together are run side-by-side on a gel, which separates them according to size.

3. The fragments are arranged so that the smaller ones produced by the simultaneous cut can be grouped to generate the larger ones produced by the individual enzymes.

4. A physical map is constructed.

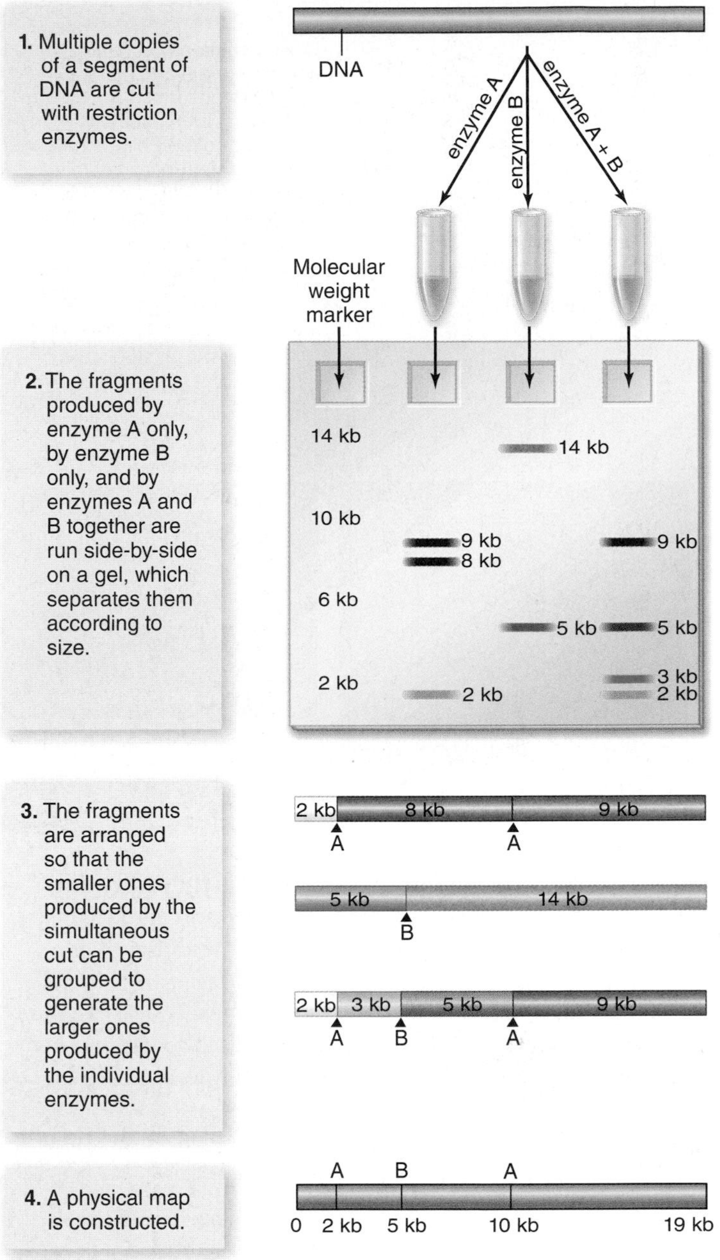

Figure 18.1 Restriction enzymes can be used to create a physical map. DNA is digested with two different restriction enzymes singly and in combination, then electrophoresed to separate the fragments. The location of sites can be deduced by comparing the sizes of fragments from the individual reactions with the combined reaction.

chromosome 22 (figure 18.2*a*), resulting in an altered form of tyrosine kinase that is always turned on, causing white blood cell proliferation.

The use of hybridization with cloned DNA has added to the utility of chromosome-banding analysis. In this case, because the hybridization involves whole chromosomes, it is called *in situ hybridization*. It is done using fluorescently labeled probes, and so its complete name is **fluorescence in situ hybridization (FISH)** (figure 18.2*b*).

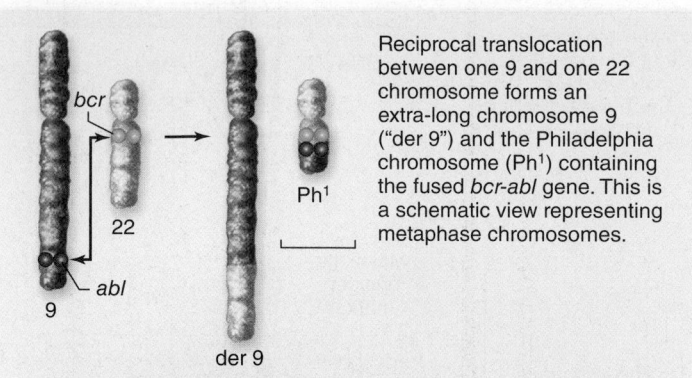

Reciprocal translocation between one 9 and one 22 chromosome forms an extra-long chromosome 9 ("der 9") and the Philadelphia chromosome (Ph¹) containing the fused *bcr-abl* gene. This is a schematic view representing metaphase chromosomes.

a.

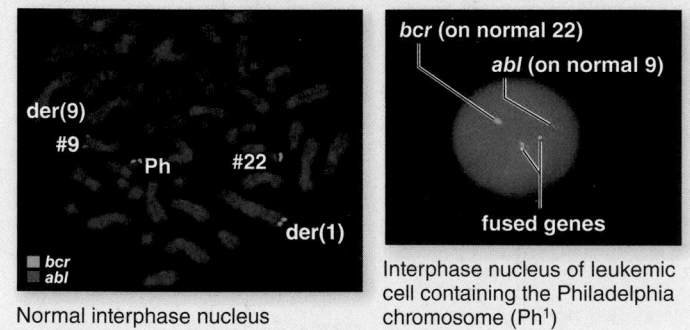

Normal interphase nucleus

Interphase nucleus of leukemic cell containing the Philadelphia chromosome (Ph¹)

b.

Figure 18.2 Use of fluorescence in situ hybridization (FISH) to correlate cloned DNA with cytological maps. *a.* Karyotype of human chromosomes showing the translocation between chromosomes 9 and 22. *b.* FISH using a *bcr* (*green*) and *abl* (*red*) probe. The yellow color indicates the fused genes (*red* plus *green* fluorescence combined). The *abl* gene and the fused *bcr-abl* gene both encode a tyrosine kinase, but the fused gene is always expressed.

? Inquiry question Why are there only three colored spots on the karyotype for two different genes?

Sequence-tagged sites provide a common language for physical maps

The construction of a physical map for a large genome requires the efforts of many laboratories in different locations. A variety of difficulties arose in comparing data from different labs, as well as integrating different types of landmarks used on physical and genetic maps.

In the early days of the Human Genome Project, this problem was addressed by the creation of a common molecular language that could be used to describe the different types of landmarks.

Defining common markers

Since all genetic information is ultimately based on DNA sequence, it was important for this common language to be sequence-based, but not to require generating a large amount of sequence for any landmark. The solution was the **sequence-tagged site,** or **STS.** This site is a small stretch of DNA that is unique in the genome, that is, it only occurs once.

The boundary of the STS is defined by PCR primers, so the presence of the STS can be identified by PCR using any DNA as a template (see chapter 17). These sites need to be only 200–500 bp long, an amount of sequence that can be determined easily. The STS can contain any other kind of landmark—for example, part of a cloned gene that has been genetically mapped, or a restriction site that is polymorphic. Any marker that has been mapped can be converted to an STS by sequencing only 200–500 bp of the entire marker.

The use of STSs

As maps are generated, new STSs are identified and added to a database that indicates the sequence of the STS, its location in the genome, and the PCR primers needed to identify it. Any researcher is then able to identify the presence or absence of any STS in the DNA that he or she is analyzing.

Fragments of DNA can be pieced together using STSs by identifying overlapping regions in fragments. Because of the high density of STSs in the human genome and the relative ease of identifying an STS in a DNA clone, investigators were able to develop physical maps on the huge scale of the 3.2-gigabase genome in the mid-1990s (figure 18.3). STSs provide a scaffold for assembling genome sequences.

Genetic maps provide a link to phenotypes

The first genetic (linkage) map was made in 1911 when Alfred Sturtevant mapped five genes in *Drosophila,* as described in chapter 13. Distances on a genetic map reflect the frequency of recombination between genes and are measured in centimorgans (cM) in honor of the geneticist Thomas Hunt Morgan. One centimorgan corresponds to 1% recombination frequency between two loci. Over 15,000 genes have been mapped on the *Drosophila* genome.

Linkage mapping can be done without knowing the DNA sequence of a gene. Computer programs make it possible to create a linkage map for a thousand genes at a time. But a few limitations to genetic maps still exist. One is that distances between genes determined by recombination frequencies do not directly correspond to physical distance on a chromosome. The sequence and conformation of DNA between genes varies, as does the frequency of recombination. Another limitation is that not all genes have obvious phenotypes that can be followed in segregating crosses.

The human genetic map is quite dense, with a marker roughly every 1 cM. This level of detail would have been unheard of 20 years ago, and it was made possible by development of molecular markers that do not cause a phenotype change.

The most common type of markers are short repeated sequences, called short tandem repeats, or STR loci, that differ in repeat length between individuals. These repeats are identified by using PCR to amplify the region containing the repeat, then analyzing the products using electrophoresis. Once a map is constructed using these markers, genes with alleles that cause a disease state can be mapped relative to the

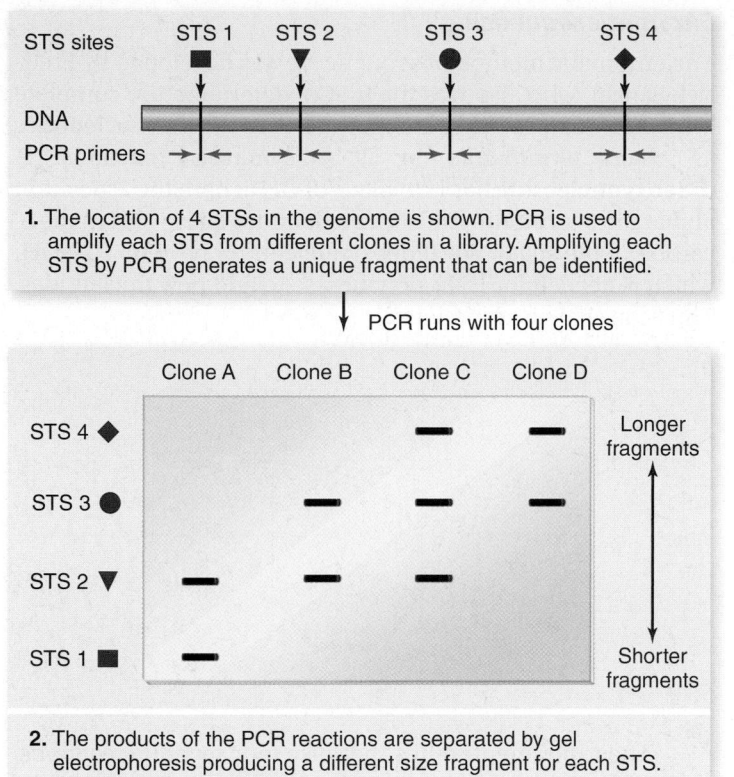

1. The location of 4 STSs in the genome is shown. PCR is used to amplify each STS from different clones in a library. Amplifying each STS by PCR generates a unique fragment that can be identified.

PCR runs with four clones

2. The products of the PCR reactions are separated by gel electrophoresis producing a different size fragment for each STS.

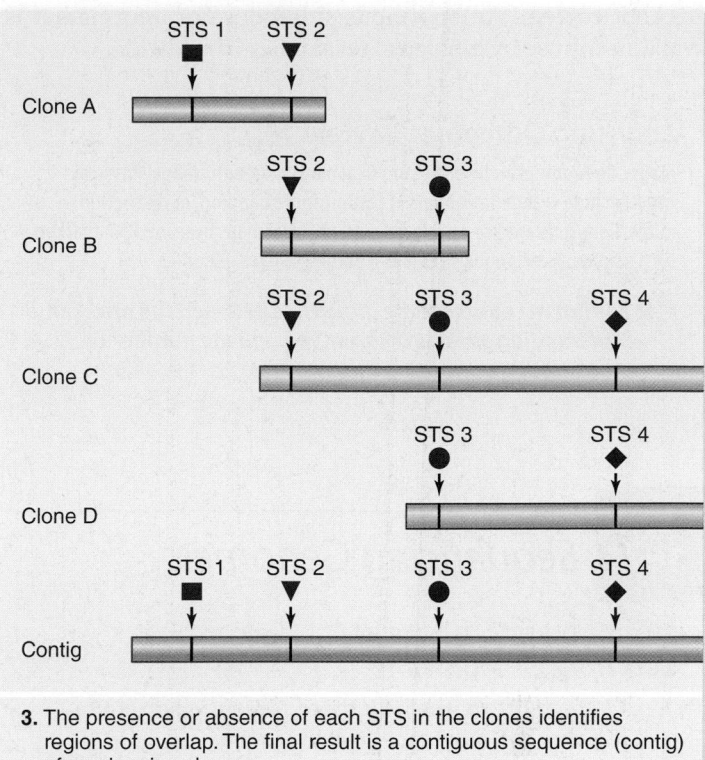

3. The presence or absence of each STS in the clones identifies regions of overlap. The final result is a contiguous sequence (contig) of overlapping clones.

Figure 18.3 Creating a physical map with sequence-tagged sites. The presence of landmarks called sequence-tagged sites, or STSs, in the human genome made it possible to begin creating a physical map large enough in scale to provide a foundation for sequencing the entire genome. *(1)* Primers *(green arrows)* that recognize unique STSs are added to cloned DNA, followed by DNA amplification via polymerase chain reaction (PCR). *(2)* PCR products are separated based on size on a DNA gel, and the STSs contained in each clone are identified. *(3)* Cloned DNA segments are aligned based on STSs to create a contig.

 Data analysis You could construct the physical map using a subset of the clones in panel 2. List the possible combinations that would allow you to construct the physical map with the fewest number of clones.

molecular landmarks. As described in chapter 17, databases of STR loci provide DNA fingerprints for identifying individuals in missing persons' cases, establishing relationship, identifying remains in mass tragedies, and for court proceedings.

Physical maps can be correlated with genetic maps

We need to be able to correlate genetic maps with physical maps, particularly genome sequences, to aid in finding physical sequences for genes that have been mapped genetically.

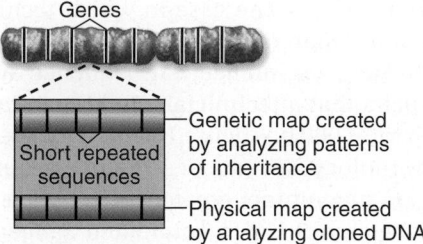

The problem in finding genes is that the resolution of genetic maps at present is not nearly as fine-grained as the genome sequence. Markers that are 1 cM apart may be as much as a million base-pairs apart.

Since the markers used to construct genetic maps are now primarily molecular markers, they can be easily located within a genome sequence. Similarly, any gene that has been cloned can be placed within the genome sequence and can also be mapped genetically. This provides an automatic correlation between the two maps. Genes that have been mapped genetically, but not isolated as molecular clones, present a problem because of the nature of genetic maps. Distances measured on genetic maps are not uniform due to variation in recombination frequency along the chromosome. So 1 cM of genetic distance translates to different numbers of base-pairs in different regions.

Different kinds of maps are stored in databases so they can be aligned and viewed. The National Center for Biotechnology Information (NCBI) is a branch of the National Library of Medicine, and it serves as the U.S. repository for these data and more. Similar databases exist in Europe and Japan, and all are kept current. An enormous storehouse of information is available for use by biological researchers worldwide.

Learning Outcomes Review 18.1

Maps of genomes can be either physical maps or genetic maps. Physical maps include cytogenetic maps of chromosome banding, or restriction maps. Genetic maps are correlated with physical maps by using DNA markers such as sequence-tagged sites (STSs) unique to each genome.

■ *What accounts for the difference between the proximity of banding sites on a karyotype and the number of base-pairs separating the two sites?*

18.2 Sequencing Genomes

Learning Outcomes

1. *Characterize the main hurdle to sequencing an entire genome and how it has been overcome.*
2. *Differentiate between clone-by-clone sequencing and shotgun sequencing.*

The ultimate physical map is the base-pair sequence of an entire genome. Large-scale DNA sequencing made genomics possible. Conceptually, all approaches to sequencing adapt DNA replication to an in vitro environment, using modified nucleotides to stop or pause replication and limiting the replication to just one strand of the DNA. The rapid advances in genomics have been enabled by newer and much faster automated sequencing technologies.

DNA sequencing provides information about genes and genomes

The development of sequencing technology has paralleled the advancement of molecular biology. The earliest approach was to generate a set of nested fragments that each begin with the same sequence and end in a specific base. When this set of fragments is separated by high-resolution gel electrophoresis, the result is a "ladder" of fragments in which each band consists of fragments that end in a specific base. By starting with the shortest fragment, one can then read the sequence by moving up the ladder.

The problem then became how to generate the sets of fragments that end in specific bases. In the early days of sequencing, both a chemical method and an enzymatic method were utilized. The chemical method involved organic reactions specific for the different bases that made breaks in the DNA chains at specific bases. The enzymatic method used DNA polymerase to synthesize chains, but it also included in the reaction modified nucleotides that could be incorporated but not extended: so-called *chain terminators*. The enzymatic method has proved more versatile, and it is easier to adapt to different uses.

Enzymatic sequencing

The enzymatic method of sequencing was developed by Fredrick Sanger, who also was the first to determine the complete sequence of a protein. This method uses dideoxynucleotides as chain terminators in DNA synthesis reactions. A **dideoxynucleotide** has H in place of OH at both the 2′ position and at the 3′ position. All DNA nucleotides lack —OH at the 2′ carbon of the sugar, but dideoxynucleotides have no 3′ —OH, which is needed for DNA polymerase to add new nucleotides. Thus, the chain is terminated.

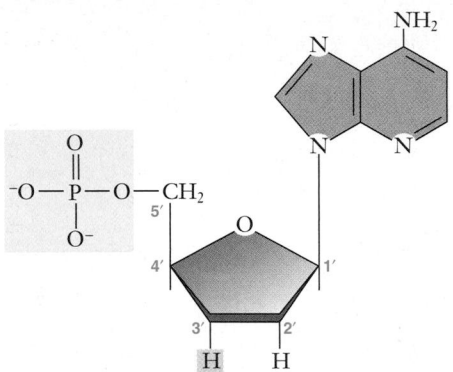

The experimenter must perform four separate reactions, each with a single dideoxynucleotide, to generate a set of fragments that terminate in specific bases. Thus all of the fragments produced in the A reaction incorporate dideoxyadenosine and must end in A, and the same for the other three reactions with different terminators. When these fragments are separated by high-resolution gel electrophoresis, each reaction is run in a different track, or lane, to generate a pattern of nested fragments that can be read from the smallest fragment to fragments that are each longer by one base (figure 18.4*a*).

Remember that all DNA polymerase enzymes require a primer to begin synthesis (see chapter 14). With cloned DNA in a vector, the site of insertion is known, and short DNA molecules that are complementary to regions adjacent to the cloned DNA can be chemically synthesized for use as primers. This serves the dual purposes of providing a primer and ensuring that the first few bases sequenced are known because they are in the vector itself. This provides a landmark for determining where the sequence of interest begins. Using the sequence determined with the first set of reactions, new complementary primers can be designed near the end of this sequence and chemically synthesized to use in the next set of reactions to extend the region sequenced.

Automated sequencing

Large-scale genome sequencing requires the use of high-throughput automated methods and computer analysis. Genome sequencing is one case in which technology drove the science, rather than the other way around. In a few hours, an automated Sanger sequencer can sequence the same number of base-pairs that a technician could manually sequence in a year—up to 50,000 bp (figure 18.4*b*). With the current generation of technology, the rate of sequence generation is now five orders of magnitude greater than when the human genome was sequenced with automated Sanger sequencers. Without the automation of sequencing, it would have been

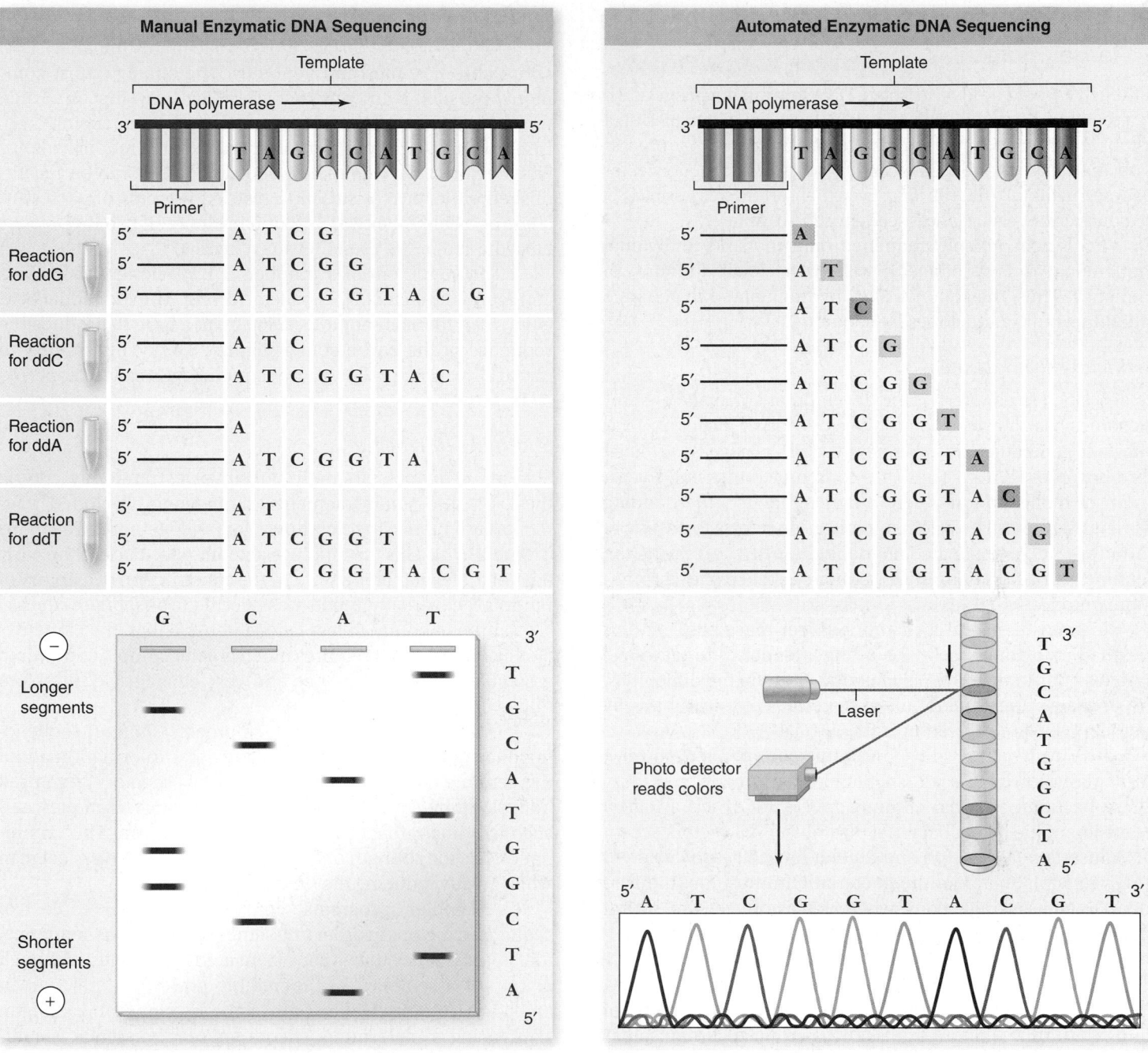

Figure 18.4 Manual and automated enzymatic DNA sequencing. The sequence to be determined is shown at the top as a template strand for DNA polymerase with a primer attached. ***a.*** In the manual method, four reactions were done, one for each nucleotide. For example, the A tube would contain dATP, dGTP, dCTP, dTTP, and ddATP. This leads to fragments that end in A due to the dideoxy terminator. The fragments generated in each reaction are shown along with the results of gel electrophoresis. ***b.*** In automated sequencing, each ddNTP is labeled with a different color fluorescent dye, which allows the reaction to be done in a single tube. The fragments generated by the reactions are shown. When these are electrophoresed in a capillary tube, a laser at the bottom of the tube excites the dyes, and each will emit a different color that is detected by a photodetector.

impossible to sequence large, eukaryotic genomes like that of humans.

New automated sequencing technology is even faster than automated Sanger sequencing. Modified, fluorescent nucleotides are still used in some sequencers, but they allow for a controlled pause rather than a complete stop to elongating the new strand of DNA. High-resolution cameras can capture the identity of each new base-pair as it is added so a sequence can be discerned from individual sequencing events, rather than analyzing a pool of terminated fragments. The new approach is termed sequencing by synthesis, but it still adapts the basics of DNA replication with the use of one, rather than two, primers to drive replication in a single direction.

Genome sequencing requires larger molecular clones

Although it would be ideal to isolate DNA from an organism, add it to a sequencer, and then come back in a week or two to pick up a computer-generated printout of the genome sequence, the process is not quite that simple. Sequencers provide accurate sequences for DNA segments up to 800 bp long. Even then, errors are possible. So, to reduce errors, each clone is sequenced 5–10 times.

Even with reliable sequence data in hand, individual sequencing runs produce a relatively small amount of sequence. Thus, the genome must be fragmented to generate individual molecular clones for sequencing

Artificial chromosomes

As described in chapter 17, the development of artificial chromosomes has allowed scientists to clone larger pieces of DNA. The first generation of these new vectors was yeast artificial chromosomes (YACs). These are constructed by using a yeast origin of replication and centromere sequence, then adding foreign DNA to it. The origin of replication allows the artificial chromosome to replicate independently of the rest of the genome, and the centromere sequences make the chromosome mitotically stable.

YACs were useful for cloning larger pieces of DNA but they had many drawbacks, including a tendency to rearrange or to lose portions of DNA by deletion. Despite the difficulties, the YACs were used early on to construct physical maps by restriction enzyme digestion of the YAC DNA.

The artificial chromosomes most commonly used now, particularly for large-scale sequencing, are made in *E. coli.* These bacterial artificial chromosomes (BACs) are a logical extension of the use of bacterial plasmids. BAC vectors accept DNA inserts between 100 and 200 kb long. The downside of BAC vectors is that, like the bacterial chromosome, they are maintained as a single copy, whereas plasmid vectors exist at high copy numbers.

Human artificial chromosomes

Human artificial chromosomes can introduce large segments of human DNA into cultured cells. These artificial chromosomes are usually constructed by fragmentation of chromosomes with centromere sequence. Although circular, some can still segregate correctly during mitosis up to 98% of the time. Construction of linear human artificial chromosomes is not yet possible.

Whole-genome sequencing is approached in two ways: clone-by-clone and shotgun

Sequencing an entire genome is an enormous task. Multiple copies of a genome are fragmented into chunks of overlapping sequences that must be assembled by matching the overlapping regions after sequencing. The shorter the sequences, the harder it is to piece together the puzzle. Two strategies are employed: assemble portions of a chromosome first and then figure out how the bigger pieces fit together (**clone-by-clone sequencing**) or try to assemble all the pieces at once, instead of in a stepwise fashion (**shotgun sequencing**).

Clone-by-clone sequencing

The cloning of large inserts in BACs facilitates the analysis of entire chromosomes and genomes. The strategy most commonly pursued is to construct a physical map first, and then use it to place the site of BAC clones for later sequencing.

Aligning large portions of a chromosome requires identifying regions that overlap between clones. This can be accomplished either by constructing restriction maps of each BAC clone or by identifying STSs found in clones. If two BAC clones have the same STS, then they must overlap.

The alignment of a number of BAC clones results in a contiguous stretch of DNA called a *contig*. The individual BAC clones can then be sequenced 500 bp at a time to produce the sequence of the entire contig (figure 18.5*a*). This strategy of physical mapping followed by sequencing is called clone-by-clone sequencing.

Shotgun sequencing

Shotgun sequencing, simply put, involves randomly cutting the DNA into small fragments, sequencing all cloned fragments, and then using a computer to put together the overlaps (figure 18.5*b*). This terminology actually goes back to the early days of molecular cloning when the construction of a library of randomly cloned fragments was referred to as *shotgun cloning.* This approach is much less labor-intensive than the clone-by-clone method, but it requires much greater computer power to assemble the final sequence and very efficient algorithms to find overlaps.

Unlike the clone-by-clone approach, shotgun sequencing does not tie the sequence to any other information about the genome (see figure 18.5*b*). Many investigators have used both clone-by-clone and shotgun-sequencing techniques, and such hybrid approaches are becoming the norm. This combination has the strength of tying the sequence to a physical map while greatly reducing the time involved.

Assembler programs compare multiple copies of sequenced regions in order to assemble a **consensus sequence,** that is, a sequence that is consistent across all copies. Although computer assemblers are incredibly powerful, final human analysis is required after both clone-by-clone and shotgun sequencing to determine when a genome sequence is sufficiently accurate to be useful to researchers.

The Human Genome Project used both sequencing methods

The vast scale of genomics ushered in a new way of doing biological research involving large teams. Although a single individual can isolate and manually sequence a molecular clone for a single gene, a huge genome like the human genome requires the collaborative efforts of hundreds of researchers.

The Human Genome Project originated in 1990 when a group of American scientists formed the International Human Genome Sequencing Consortium. The goal of this publicly funded effort was to use a clone-by-clone approach to sequence the human genome. Genetic and physical maps were used as scaffolding to sequence each chromosome.

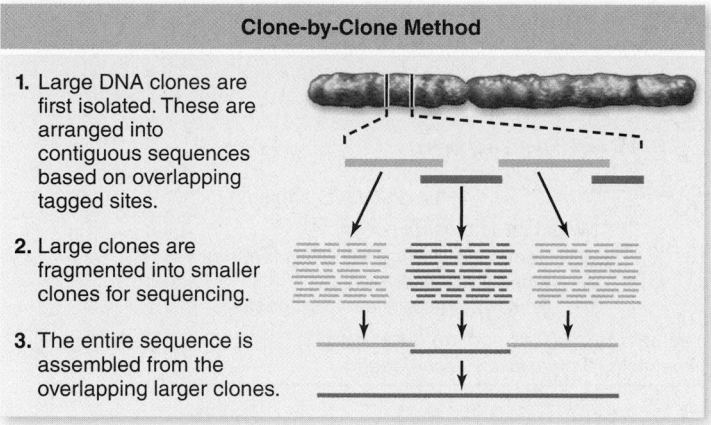

Clone-by-Clone Method

1. Large DNA clones are first isolated. These are arranged into contiguous sequences based on overlapping tagged sites.

2. Large clones are fragmented into smaller clones for sequencing.

3. The entire sequence is assembled from the overlapping larger clones.

a.

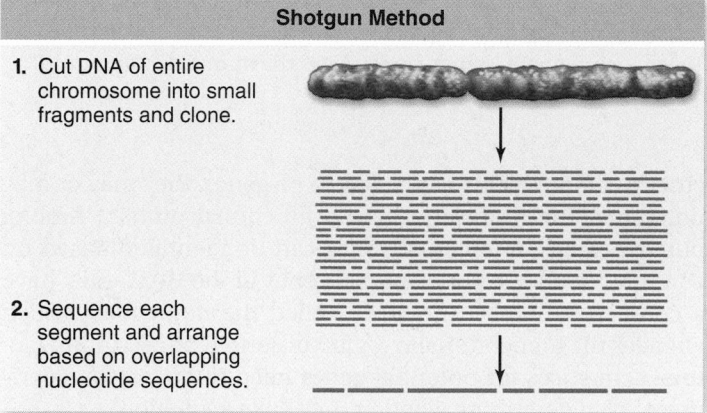

Shotgun Method

1. Cut DNA of entire chromosome into small fragments and clone.

2. Sequence each segment and arrange based on overlapping nucleotide sequences.

b.

Figure 18.5 Comparison of sequencing methods.
a. The clone-by-clone method uses large clones assembled into overlapping regions by STSs. Once assembled, these can be fragmented into smaller clones for sequencing. *b.* In the shotgun method the entire genome is fragmented into small clones and sequenced. Computer algorithms assemble the final DNA sequence based on overlapping nucleotide sequences.

In May 1998, American biologist Craig Venter, whose research group had sequenced *Haemophilus influenzae*, announced that his private company (Celera Genomics) would also endeavor to sequence the human genome. He proposed to shotgun-sequence the 3.2-gigabase genome in only two years. The Consortium rose to the challenge, and the race to sequence the human genome began. The upshot was a tie of sorts. On June 26, 2000, the groups jointly announced success, and each published its findings simultaneously in 2001. The Consortium's draft alone included 248 authors.

The draft sequence of the human genome was just the beginning. Gaps in the sequence are still being filled, and the map is constantly being refined. The most recent "finished" human sequence is down to only 260 gaps, a 400-fold reduction in gaps, and it now includes 99% of the euchromatic sequence, up from 95%. The reference sequence has an error rate of 1 per 100,000 bases. Newer sequencing technologies are being used to close the remaining gaps. A few individuals, including James Watson (who codiscovered the structure of

DNA) and both Venter and his colleague Francis Collins (who lead the U.S. team of the International Consortium), have now had their personal genomes sequenced. The cost for having one's genome sequenced is predicted to fall to $1000 in the next few years, making the process much more affordable and accessible. Many more people seeking personal sequencing raises many questions about genome privacy, including how private the information will remain.

Research on the whole genome can move ahead. Now that the ultimate physical map is in place and is being integrated with the genetic map, diseases that result from changes in more than one gene, such as diabetes, can be addressed. Comparisons with other genomes are already changing our understanding of genome evolution (see chapter 24).

Learning Outcomes Review 18.2

Because of the enormous size of genomes, sequencing requires the use of automated sequencers running many samples in parallel. Two approaches have been developed for whole-genome sequencing: one that uses clones already aligned by physical mapping (clone-by-clone sequencing), and one that involves sequencing random clones and using a computer to assemble the final sequence (shotgun sequencing). In either case, significant computing power is necessary to assemble a final sequence.

■ **Compare and contrast DNA replication and DNA sequencing.**

18.3 *Characterizing Genomes*

Learning Outcomes

1. *Describe the classes of DNA found in a genome.*
2. *Explain what an SNP is and why SNPs are helpful in characterizing genomes.*

Automated sequencing technology has produced huge amounts of sequence data, eventually sequencing entire genomes. This has allowed researchers studying complex problems to move beyond approaches restricted to the analysis of individual genes. However, sequencing projects in themselves are descriptive analyses that tell us nothing about the organization of genomes, let alone the function of gene products and how they may be interrelated. Additional research and evaluation has given us both answers and new puzzles.

The Human Genome Project found fewer genes than expected

For many years, geneticists had estimated the number of human genes to be around 100,000. This estimate, although based on some data, was really just a guess. Imagine researchers' surprise when the number turned out to be less than 25,000! This represents only about twice as many genes as

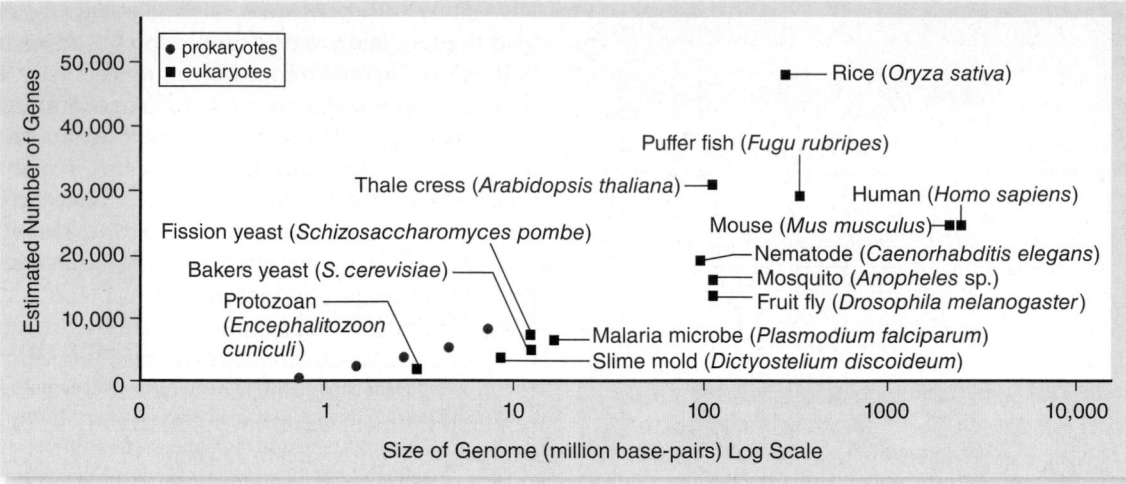

Figure 18.6 Size and complexity of genomes. In general, eukaryotic genomes are larger and have more genes than prokaryotic genomes, although the size of the organism is not the determining factor. The mouse genome is nearly as large as the human genome, and the rice genome contains more genes than both.

Drosophila and fewer genes than rice (figure 18.6). Clearly, the complexity of an organism is not a simple function of the number of genes in its genome.

The human transcriptome and proteome are far more complex than might be inferred from under 25,000 coding genes. As discussed in chapter 16, alternative splicing can produce multiple mRNAs and thus multiple proteins from a single gene. An analysis of the human genome revealed that about 95% of genes with multiple exons undergo alternative splicing. An estimated 100,000 alternative splicing events occur in the major tissues of humans. Understanding genome sequences is just the beginning of understanding how the blueprint in the genome is used to build an organism.

 Data analysis Estimate the number of splice variants that occur, on average, on each of the 25,000 genes in the human genome.

Finding genes in sequence data requires computer searches

Once a genome has been sequenced, the next step is to determine which regions of the genome contain which genes, and what those genes do. A lot of information can be mined from the sequence data. Using markers from physical maps and information from genetic maps, it is possible to find the sequence of the small percentage of genes that are identified by mutations with an observable (phenotypic) effect. Genes can also be found by comparing expressed sequences to genomic sequences. The analysis of expressed sequences is discussed later in this section.

Locating starts and stops

Information in the nucleotide sequence itself can also be used in the search for genes. A protein-coding gene begins with a start codon, such as ATG, and it contains no stop codons (TAA, TGA, or TAG) for a distance long enough to encode a protein. This coding region is referred to as an **open reading frame**

(ORF). Although ORFs are likely to be genes, they may or may not actually be translated into a functional protein. Among putative genes, families of genes can be identified based on common domains. For example, genes in the Hox family have a conserved, 180-bp sequence called the homeobox, which encodes the homeodomain region of certain transcription factors. Sequences for potential genes need to be tested experimentally to determine whether they have a function.

The addition of information to the basic sequence information, like identifying ORFs, is called sequence **annotation.** This process is what converts simple sequence data into something that we can recognize based on landmarks such as regions that are transcribed and regions that are known or thought to encode proteins.

 Inquiry question Given the sequence of a genome, how might you predict the number of genes in the genome? Why might your estimate be inaccurate?

Inferring function across species: the BLAST algorithm

It is also possible to search genome databases for sequences that are homologous to known genes in other species. A researcher who has isolated a molecular clone for a gene of unknown function can search the database for similar sequences to infer function. The tool that makes this possible is a search algorithm called BLAST (which stands for Basic Local Alignment Search Tool). Using a networked computer, one can submit a sequence to the BLAST server and get back a reply with all possible similar sequences contained in the sequence database.

Using these techniques, sequences that are not part of ORFs have been identified that have been conserved over millions of years of evolution. These sequences may be important for the regulation of the genes contained in the genome.

Using computer programs to search for genes, to compare genomes, and to assemble genomes are only a few of the new genomics approaches falling under the heading of **bioinformatics.**

Genomes contain both coding and noncoding DNA

When genome sequences are analyzed, regions that encode proteins and other regions that do not encode proteins are revealed. For many years investigators had known of the latter, but they did not know the extent and nature of the noncoding DNA. We first consider the types of coding DNA that have been found, then move on to look at types of noncoding DNA.

Protein-encoding DNA in eukaryotes

Four different classes of protein-encoding genes are found in eukaryotic genomes, differing largely in gene copy number.

Single-copy genes. Many genes exist as single copies on a particular chromosome. Most mutations in these genes result in recessive Mendelian inheritance.

Segmental duplications. Sometimes whole blocks of genes are copied from one chromosome to another, resulting in *segmental duplication.* Blocks of similar genes in the same order are found throughout the human genome. Chromosome 19 seems to have been the biggest borrower, sharing blocks of genes with 16 other chromosomes.

Multigene families. As more has been learned about eukaryotic genomes, many genes have been found to exist as parts of *multigene families,* groups of related but distinctly different genes that often occur together in clusters. About 10,000 multigene families with two or more genes are found in the human genome. Comparisons of mammalian gene families show that 164 of these gene families are evolving at accelerated rates. Biological functions of genes in these families include immune response and brain development. These multigene families may include silent copies called *pseudogenes,* which are inactivated by mutation.

Tandem clusters. Identical copies of genes can also be found in *tandem clusters.* These genes are transcribed simultaneously, increasing the amount of mRNA available for protein production. Tandem clusters also include genes that do not encode proteins, such as clusters of rRNA genes.

Noncoding DNA in eukaryotes

One of the most notable characteristics is the amount of non-coding DNA they possess. The Human Genome Project has revealed a particularly startling picture. Each of your cells has about 6 feet of DNA stuffed into it, but of that, less than 1 inch is devoted to genes! Nearly 99% of the DNA in your cells is non-protein-coding DNA.

True genes are scattered about the human genome in clumps among the much larger amount of noncoding DNA, like isolated oases in a desert. Seven major sorts of noncoding human DNA have been described. (Table 18.1 shows the composition of the human genome, including noncoding DNA.)

Noncoding DNA within genes. As discussed in chapter 15, a human gene is not simply a stretch of DNA, like the letters of a word. Instead, a human gene is made up of numerous fragments of protein-encoding information (exons) embedded within a much larger matrix of noncoding DNA (introns). Together, introns make up about 24% of the human genome and exons less than 1.5%.

Structural DNA. Some regions of the chromosomes remain highly condensed, tightly coiled, and untranscribed throughout the cell cycle. Called *constitutive heterochromatin,* these portions tend to be localized around the centromere or located near the ends of the chromosome, at the telomeres.

Simple sequence repeats. Scattered about chromosomes are **simple sequence repeats (SSRs).** An SSR is a 1- to 6-nt sequence such as CA or CGG, repeated like a broken record thousands and thousands of times. SSRs can arise from DNA replication errors. SSRs make up about 3% of the human genome.

Segmental duplications. Blocks of genomic sequences composed of from 10,000 to 300,000 bp have duplicated and moved either within a chromosome or to a nonhomologous chromosome.

Pseudogenes. These inactive genes may have lost function because of mutation.

Transposable elements. Fully 45% of the human genome consists of mobile bits of DNA called *transposable elements.*

TABLE 18.1	Classes of DNA Sequences Found in the Human Genome
Class	**Description**
Protein-encoding genes	Translated portions of the 25,000 genes scattered about the chromosomes
Introns	Noncoding DNA that constitutes the great majority of each human gene
Segmental duplications	Regions of the genome that have been duplicated
Pseudogenes (inactive genes)	Sequence that has characteristics of a gene but is not a functional gene
Structural DNA	Constitutive heterochromatin, localized near centromeres and telomeres
Simple sequence repeats	Stuttering repeats of a few nucleotides such as CGG, repeated thousands of times
Transposable elements	21%: Long interspersed elements (LINEs), which are active transposons 13%: Short interspersed elements (SINEs), which are active transposons 8%: Retrotransposons, which contain long terminal repeats (LTRs) at each end 3%: DNA transposon fossils
Noncoding RNA	RNAs that do not encode a protein but have regulatory functions, many of which are currently unknown

Some of these elements code for proteins, but many do not. Because of the significance of these elements, we describe them more fully in the following section.

microRNA genes. Hidden within the non-protein-coding DNA lies an extraordinary mechanism for controlling gene expression and development. Compact regulatory RNAs have a much larger role in directing development in complex organisms than we imagined even a few years ago. Specifically, DNA that was once considered "junk" has been shown to encode microRNAs, or miRNAs, which are processed after transcription to lengths of 21 to 23 nt, but never translated. About 10,000 unique miRNAs have been identified that are complementary to one or more mature mRNAs.

Long, noncoding RNA. In addition to the many small RNAs such as microRNAs that are not translated into protein but serve a regulatory role, tens of thousands of longer noncoding RNAs likely regulate gene expression. This recently discovered, hidden world of regulatory networks reveals a new level of complexity in the precise control of gene expression. Long, noncoding RNAs have important roles in physiology and development and are only beginning to be characterized.

Transposable elements: mobile DNA

Discovered by Barbara McClintock in 1950, **transposable elements,** also termed *transposons* and *mobile genetic elements,* are bits of DNA that are able to move from one location on a chromosome to another. Barbara McClintock received the 1983 Nobel Prize in physiology or medicine for discovery of these elements and their unexpected ability to change location.

Transposable elements move around in different ways. In some cases, the transposon is duplicated, and the duplicated DNA moves to a new place in the genome, so the number of copies of the transposon increases. Other types of transposons are excised without duplication and insert themselves elsewhere in the genome. The role of transposons in genome evolution is discussed in chapter 24.

Human chromosomes contain four sorts of transposable elements. Fully 21% of the genome consists of **long interspersed elements (LINEs).** These ancient and very successful elements are about 6000 bp long, and they contain all the equipment needed for transposition. LINEs encode a reverse transcriptase enzyme that can make a cDNA copy of the transcribed LINE RNA. The result is a double-stranded segment that can reinsert into the genome rather than undergo translation into a protein. Since these elements use an RNA intermediate, they are termed *retrotransposons.*

Short interspersed elements (SINEs) are similar to LINEs, but they cannot transpose without using the transposition machinery of LINEs. Nested within the genome's LINEs are over half a million copies of a SINE element called Alu (named for a restriction enzyme that cuts within the sequence). The Alu SINE is 300 bp and represents 10% of the human genome. Like a flea on a dog, Alu moves with the LINE it resides within. Just as a flea sometimes jumps to a different dog, so Alu sometimes uses the enzymes of its LINE to move to a new chromosome location. Alu can also jump right into genes, causing harmful mutations.

Two other sorts of transposable elements are also found in the human genome: 8% of the human genome is devoted to retrotransposons called **long terminal repeats (LTRs).** Although the transposition mechanism is a bit different from that of LINEs, LTRs also use reverse transcriptase to ensure that copies are double-stranded and can reintegrate into the genome.

Some 3% of the genome is devoted to dead transposons, elements that have lost the signals for replication and can no longer move.

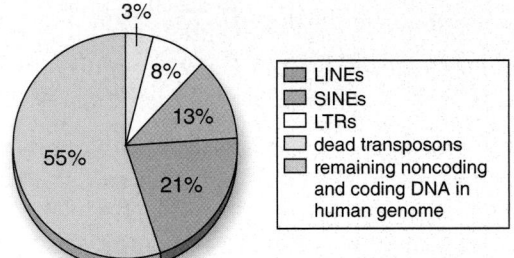

3%
8%
13%
55%
21%

- LINEs
- SINEs
- LTRs
- dead transposons
- remaining noncoding and coding DNA in human genome

? Inquiry question How do you think these repetitive elements would affect the determination of gene order?

Expressed sequence tags identify genes that are transcribed

Given the complexity of coding and noncoding DNA, it is important to be able to recognize regions of the genome that are actually expressed—that is, transcribed and then translated.

Because DNA is easier to work with than protein, one approach is to isolate mRNA, use this to make cDNA, then sequence one or both ends of as many cDNAs as possible. With automated sequencing, this task is not difficult, and these short sections of cDNA have been named **expressed sequence tags (ESTs).** An EST is another form of STS, and thus it can be included in physical maps. This technique does not tell us anything about the function of any particular EST, but it does provide one view, at the whole-genome level, of what genes are expressed, at least as mRNAs.

ESTs have been used to identify 87,000 cDNAs in different human tissues. About 80% of these cDNAs were previously unknown. The estimated 25,000 genes of the human genome can result in these 87,000 different cDNAs because of *alternative splicing* (figure 18.7).

SNPs are single-base differences between individuals

One fact becoming clear from analysis of the human genome is that a huge amount of genetic variation exists in our species. This information has practical use.

Single-nucleotide polymorphisms (SNPs) are sites where individuals differ by only a single nucleotide. To be classified as a polymorphism, an SNP must be present in at least 1% of the population. SNPs occur about every 100 to 300 bp in the 3 billion base-pair human genome. As of July 2011, 4.4 million nonredundant human SNPs had been identified and validated, representing about 30% of the variation available. These SNPs are being used to look for associations between genes. We

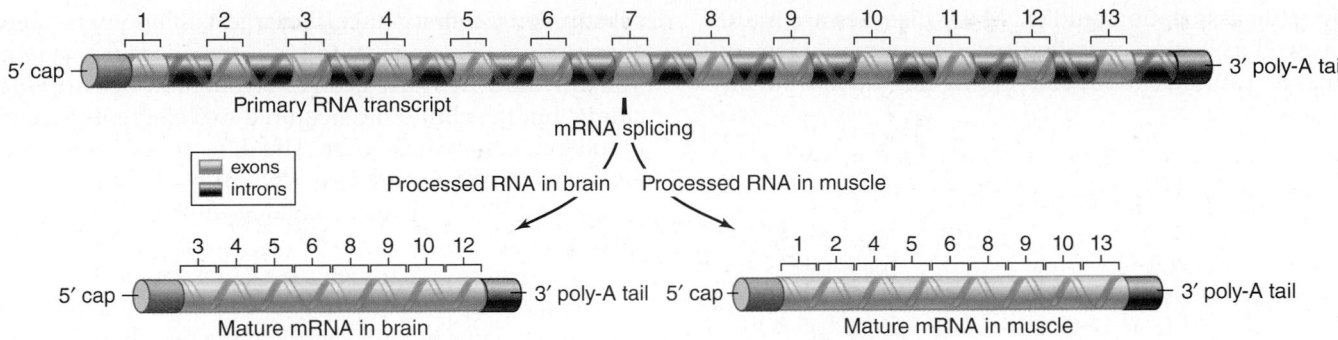

Figure 18.7 Alternative splicing can result in the production of different mRNAs from the same coding sequence. In some cells, exons can be excised along with neighboring introns, resulting in different proteins. Alternative splicing explains why 25,000 human genes can code for three to four times as many proteins.

expect that the genetic recombination occurring during meiosis randomizes all but the most tightly linked genes. We call the tendency for genes *not* to be randomized **linkage disequilibrium.** This kind of association can be used to map genes and when done at the level of the whole genome is called a **genome-wide association study**.

Analysis of SNPs shows that many cosegregate with specific genomic regions at almost 100% frequencies. This unexpected result has led to the idea of genomic **haplotypes,** or regions of chromosomes that are not being exchanged by recombination. The existence of haplotypes allows the genetic characterization of genomic regions by describing a small number of SNPs (figure 18.8). The SNPs have been mapped onto the human haplotypes, creating a haplotype map. A large number of SNPs have now been identified that are associated with disease phenotypes through

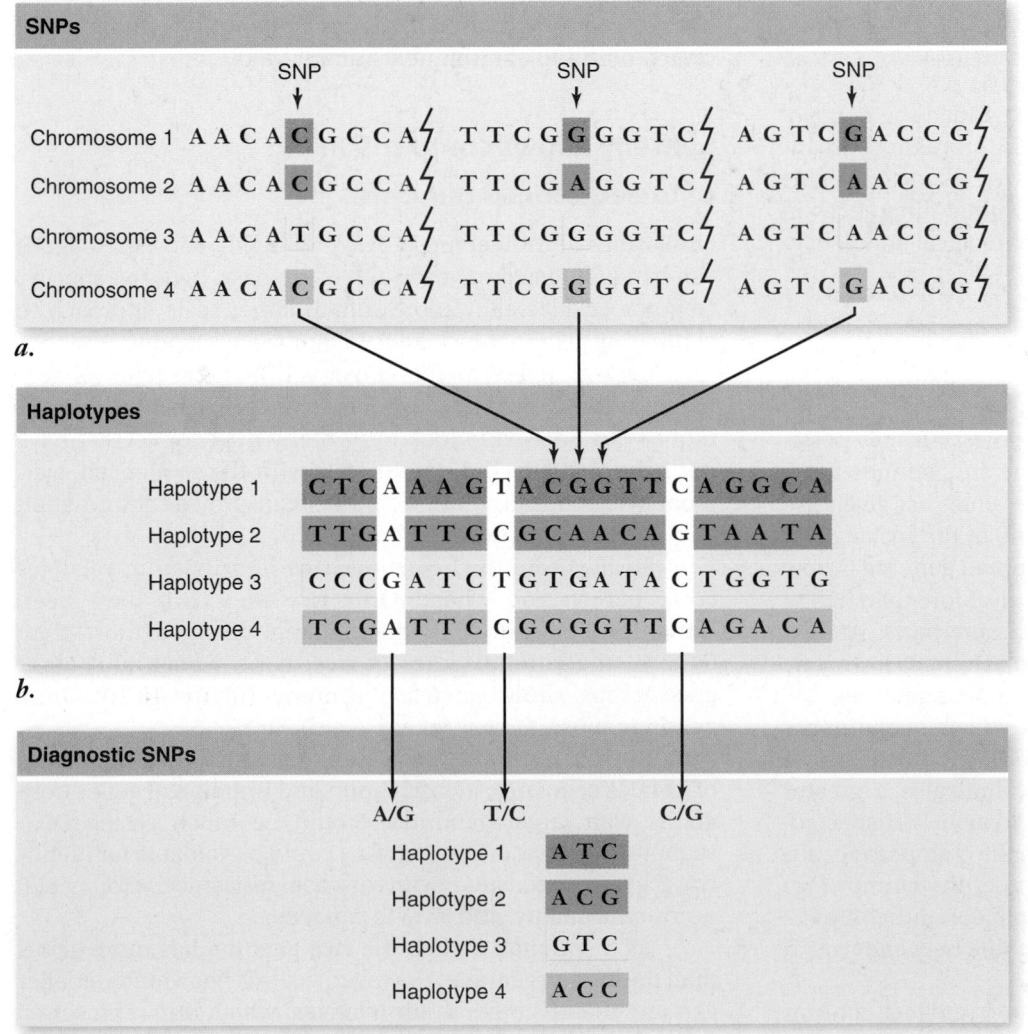

Figure 18.8 Construction of a haplotype map. Single-nucleotide polymorphisms (SNPs) are single-base differences between individuals. Sections of DNA sequences from four individuals are shown in *(a),* with three SNPs indicated by arrows. *b.* These three SNPs are shown aligned along with 17 other SNPs from this chromosomal region. This represents a haplotype map for this region of the chromosome. Haplotypes are regions of the genome that are not exchanged by recombination during meiosis. *c.* Haplotypes can be identified using a small number of diagnostic SNPs that differ between the different haplotypes. In this case, 3 SNPs out of the 20 in this region are all that are needed to uniquely identify each haplotype. This greatly facilitates locating disease-causing genes, as haplotypes represent large regions of the genome that behave as a single site during meiosis.

genome-wide association studies. Many diseases are multi-genic, as well as having contributing environmental factors. One must be cautious about concluding too much from any given SNP variant.

Learning Outcomes Review 18.3

Coding sequences in a genome can be found as a single copy, as repeated clusters, as part of segmental duplications, or as part of a gene family. A significant amount of noncoding DNA is found in all eukaryotic organisms. Transposable elements are capable of movement in the genome and are found in all eukaryotic genomes. Single-nucleotide polymorphisms (SNPs) provide a way of identifying individual variation, and they have also revealed cases of nonrandom recombination (genomic haplotypes).

■ *What explanation could you suggest, based on principles of natural selection, for the many repeated transposable elements in the human genome?*

18.4 Comparing Genomes

Learning Outcome

1. *Describe the advances that have come from comparative genomics.*

Sequencing whole genomes has vastly expanded the range of traits available for evolutionary analysis. Comparing organisms at the level of whole genomes is giving us fresh insights into our shared genomes and the often subtle differences in DNA that correspond with notable morphological and behavioral differences.

Comparative genomics reveals conserved regions in genomes

With the large number of sequenced genomes, it is now possible to make comparisons at both the gene and genome level. The flood of information from different genomes has given rise to a new field: *comparative genomics.* One of the striking lessons learned from the sequence of the human genome is how very similar humans are to other organisms. More than half of the genes of *Drosophila* have human counterparts. Among mammals, the similarities are even greater. Humans have only 300 genes that have no counterpart in the mouse genome.

The use of comparative genomics to ask evolutionary questions is also a field of great promise. The comparison of the many prokaryotic genomes already indicates a greater degree of lateral gene transfer than was previously suspected. Sequenced animal genomes now include the chimpanzee, our closest living relative. The draft sequence of the chimp *(Pan troglodytes)* genome has just been completed, and comparisons between the chimp and human genome may allow us to unravel what makes us uniquely human.

The early returns from this sequencing effort confirm that our genomes differ by only 1.23% in terms of nucleotide substitutions. At first glance, the largest difference between our genomes actually appears to be in transposable elements. In humans, the SINEs have been threefold more active than in the chimp, but the chimp has acquired two elements that are not found in the human genome. The differences due to insertion and deletion of bases are fewer than substitutions but account for about 1.5% of the euchromatic sequence being unique in each genome.

The genome of an even closer, but extinct, relative, *Homo neanderthalensis* (Neandertal) is remarkably similar to our own. About half a million years ago the two shared a common ancestor in Africa. Then Neandertals headed north, eventually settling in Europe and Asia and dying out about 28,000 years ago. Humans migrated from Africa later and, for about 100,000 years, overlapped with Neanderthals in Europe. The most compelling evidence that humans and Neandertals interbred in Europe is the finding that Europeans and Asians, but not African's, share 1% to 4% of their genome with Neandertals (figure 18.9).

Enough DNA remained in 38,000- and 44,000-year-old Neandertal bones from Croatia that the genome could be sequenced and reconstructed. Of the nearly 10 million amino acids encoded in our genome, only 78 differences that could alter the shape and/or function of a protein were consistently found in humans and not Neanderthals. More than 200 other regions within the genome appear to have evolved since the human and Neanderthal branches split, but it is nothing that clearly points to our uniquely human traits.

Synteny allows comparison of unsequenced genomes

Similarities and differences between highly conserved genes can be investigated on a gene-by-gene basis between species. Genome science allows for a much larger scale approach to comparing genomes by taking advantage of synteny.

Synteny refers to the conserved arrangements of segments of DNA in related genomes. Physical mapping techniques can be used to look for synteny in genomes that have not been sequenced. Comparisons with the sequenced, syntenous segment in another species can provide information about the unsequenced genome.

To illustrate this, consider rice and its grain relatives corn, barley, and wheat. Only rice and corn have been sequenced. Even though these plants diverged more than 50 MYA, the chromosomes of rice, corn, wheat, and other grass crops show extensive synteny (figure 18.10). In a genomic sense, "rice is wheat."

By understanding the rice and corn genomes at the level of its DNA sequence, identification and isolation of genes from grains with larger genomes should be much easier. DNA sequence analysis of cereal grains could be valuable for identifying genes associated with disease resistance, crop yield, nutritional quality, and growth capacity.

As mentioned earlier, the rice genome has more genes than the human genome. However, rice still has a much smaller genome than its other grain relatives, which also represent a major food source for humans.

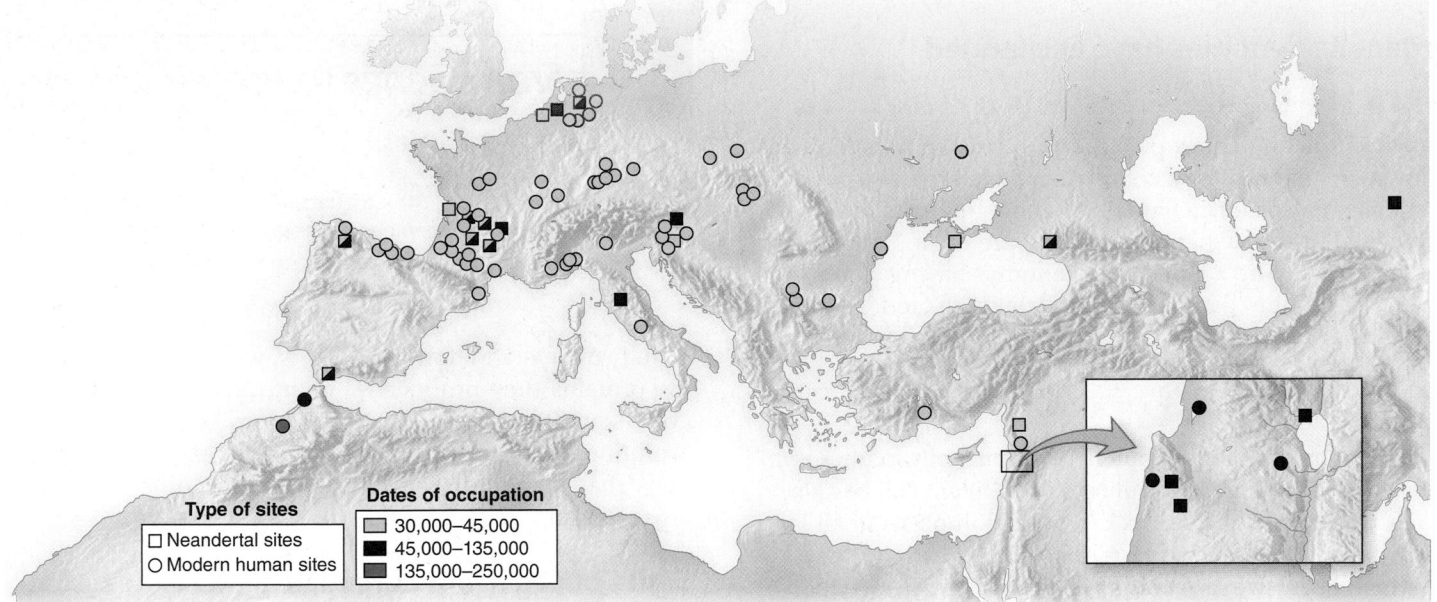

Figure 18.9 Europeans and Asians, but not Africans, share between 1% and 4% of their genomes with Neandertals. From 30,000 to 45,000 years ago humans and Neandertals coexisted in Europe, and possibly in the Middle East as early as 80,000 years ago. Interbreeding likely occurred after European and Asian ancestors left Africa.

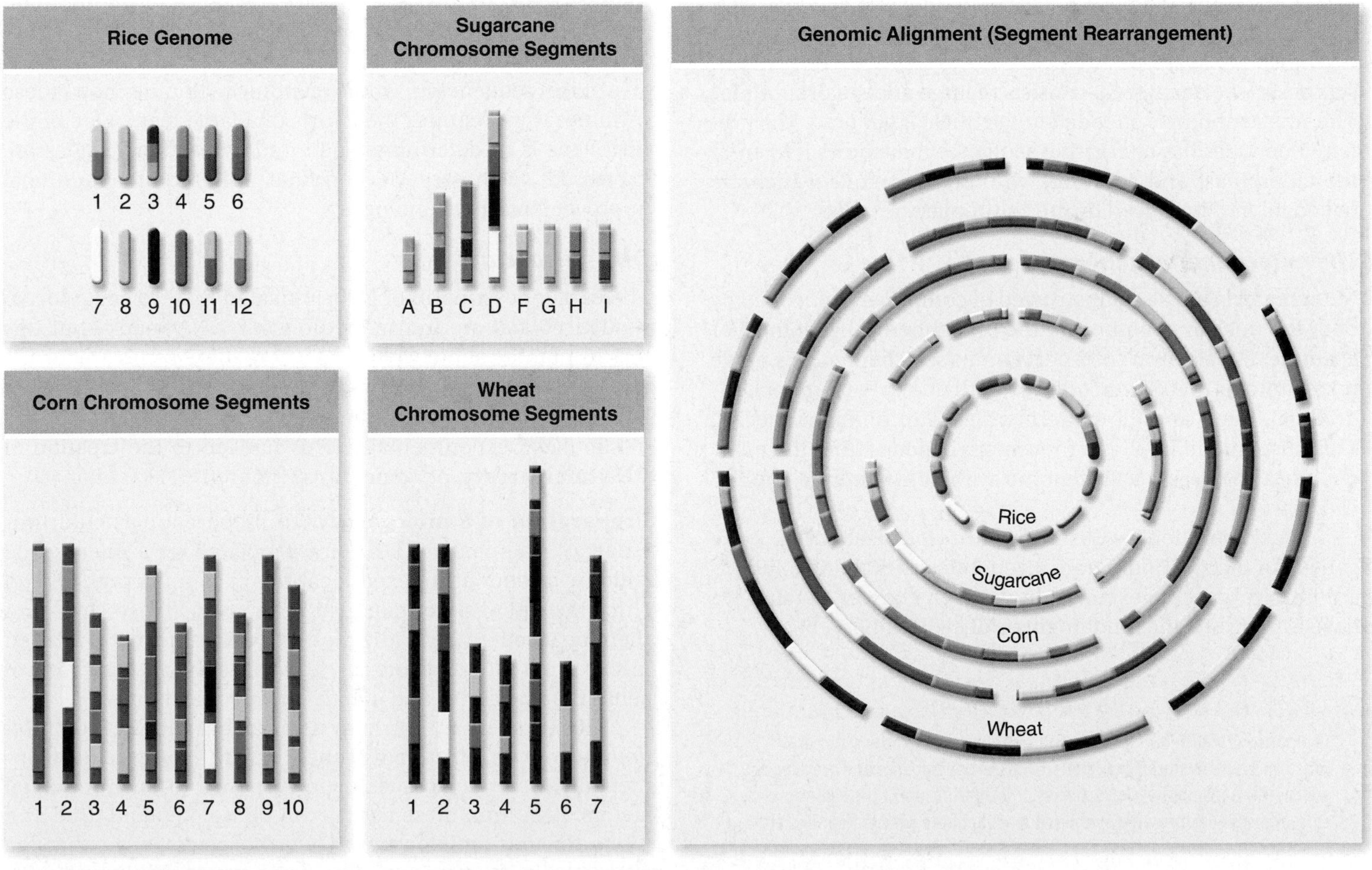

Rice Genome

1 2 3 4 5 6

7 8 9 10 11 12

Sugarcane Chromosome Segments

A B C D F G H I

Corn Chromosome Segments

1 2 3 4 5 6 7 8 9 10

Wheat Chromosome Segments

1 2 3 4 5 6 7

Genomic Alignment (Segment Rearrangement)

Rice

Sugarcane

Corn

Wheat

Figure 18.10 Grain genomes are rearrangements of similar chromosome segments. Shades of the same color represent pieces of DNA that are conserved among the different species but have been rearranged. By splitting the individual chromosomes of major grass species into segments and rearranging the segments, researchers have found that the genome components of rice, sugarcane, corn, and wheat are highly conserved. This implies that the order of the segments in the ancestral grass genome has been rearranged by recombination as the grasses have evolved.

Organelle genomes have exchanged genes with the nuclear genome

Mitochondria and chloroplasts are considered to be descendants of ancient bacterial cells living in eukaryotes as a result of endosymbiosis (see chapter 4). Their genomes have been sequenced in some species, and they are most like prokaryotic genomes. The chloroplast genome, having about 100 genes, is minute compared with the rice genome, with 32,000–55,000 genes.

The chloroplast genome

The chloroplast, a plant organelle that functions in photosynthesis, can independently replicate in the plant cell because it has its own genome. The DNA in the chloroplasts of all land plants have about the same number of genes, and they are present in about the same order. In contrast to the evolution of the DNA in the plant cell nucleus, chloroplast DNA has evolved at a more conservative pace and therefore shows a more easily interpretable evolutionary pattern when scientists study DNA sequence similarities. Chloroplast DNA is also not subject to modification caused by transposable elements or to mutations due to recombination.

Over time, some genetic exchange appears to have occurred between the nuclear and chloroplast genomes. For example, Rubisco, the key enzyme in the Calvin cycle of photosynthesis (see chapter 8), consists of large and small subunits. The small subunit is encoded in the nuclear genome. The protein it encodes has a targeting sequence that allows it to enter the chloroplast and combine with large subunits, which are coded for and produced by the chloroplast.

The mitochondrial genome

Mitochondria are also constructed of components encoded by both the nuclear genome and the mitochondrial genome. For example, the electron transport chain (see chapter 7) is made up of proteins that are encoded by both nuclear and mitochondrial genomes, and the pattern varies with different species. This observation implies a movement of genes from the mitochondria to the nuclear genome with some lineage-specific variation.

The evolutionary history of the localization of these genes is a puzzle. Comparative genomics and their evolutionary implications are explored in detail in chapter 24, after we have established the fundamentals of evolutionary theory.

Learning Outcome Review 18.4

Comparisons of different genomes allow geneticists to infer structural, functional, and evolutionary relationships between genes and proteins as well as relationships between species. Information about the evolution of humans is now informed by the sequencing of the extinct Neandertal genome.

■ *The human and Neandertal genomes are so similar that it is difficult to identify sequences that make us uniquely human. Are there other ways the two genomes could differ (consider the cereal synteny example)?*

18.5 *From Genes to Proteins*

Learning Outcomes

1. *Distinguish between genomics and proteomics.*
2. *Explain how gene expression can be analyzed.*

To fully understand how genes work, we need to characterize the proteins they produce. This information is essential to understanding cell biology, physiology, development, and evolution. In many ways, we continue to ask the same questions that Mendel asked, but at a much different level of organization.

Functional genomics reveals gene function at the genome level

Bioinformatics takes advantage of high-end computer technology to analyze the growing gene databases, look for relationships among genomes, and then hypothesize functions of genes based on sequence. Genomics is now shifting gears and moving back to hypothesis-driven science, to **functional genomics,** the study of the function of genes and their products.

Like sequencing whole genomes, finding how these genomes work requires the efforts of a large team. One of the first steps is to determine when and where these genes are expressed. Each step beyond that will require additional improvements in technology.

DNA microarrays

The earlier description of ESTs indicated that we could locate sequences that are transcribed on our DNA maps—but this tells us nothing about when and where these genes are turned on. To be able to analyze gene expression at the whole-genome level requires a representation of the genome that can be manipulated experimentally. This has led to the creation of **DNA microarrays,** or "gene chips" (figure 18.11).

Preparation of a microarray. To prepare a particular microarray, fragments of DNA are deposited on a microscope slide by a robot at indexed locations (i.e., an array). Silicon chips instead of slides can also be arrayed. These chips can then be used in hybridization experiments with labeled mRNA from different sources. This gives a high-level view of genes that are active and inactive in specific tissues.

Researchers are currently using a chip with 24,000 *Arabidopsis* genes on it to identify genes that are expressed developmentally in certain tissues or in response to environmental factors. RNA from these tissues can be isolated and used as a probe for these microarrays. Only those sequences that are expressed in the tissues will be present and will hybridize to the microarray.

Microarray analysis and cancer. One of the most exciting uses of microarrays has been the profiling of gene expression patterns in human cancers. Microarray analysis has revealed

Hypothesis: *Flowers and leaves will express some of the same genes.*

Prediction: *When mRNAs isolated from* Arabidopsis *flowers and from leaves are used as probes on an* Arabidopsis *genome microarray, the two different probe sets will hybridize to both common and unique sequences.*

Test:

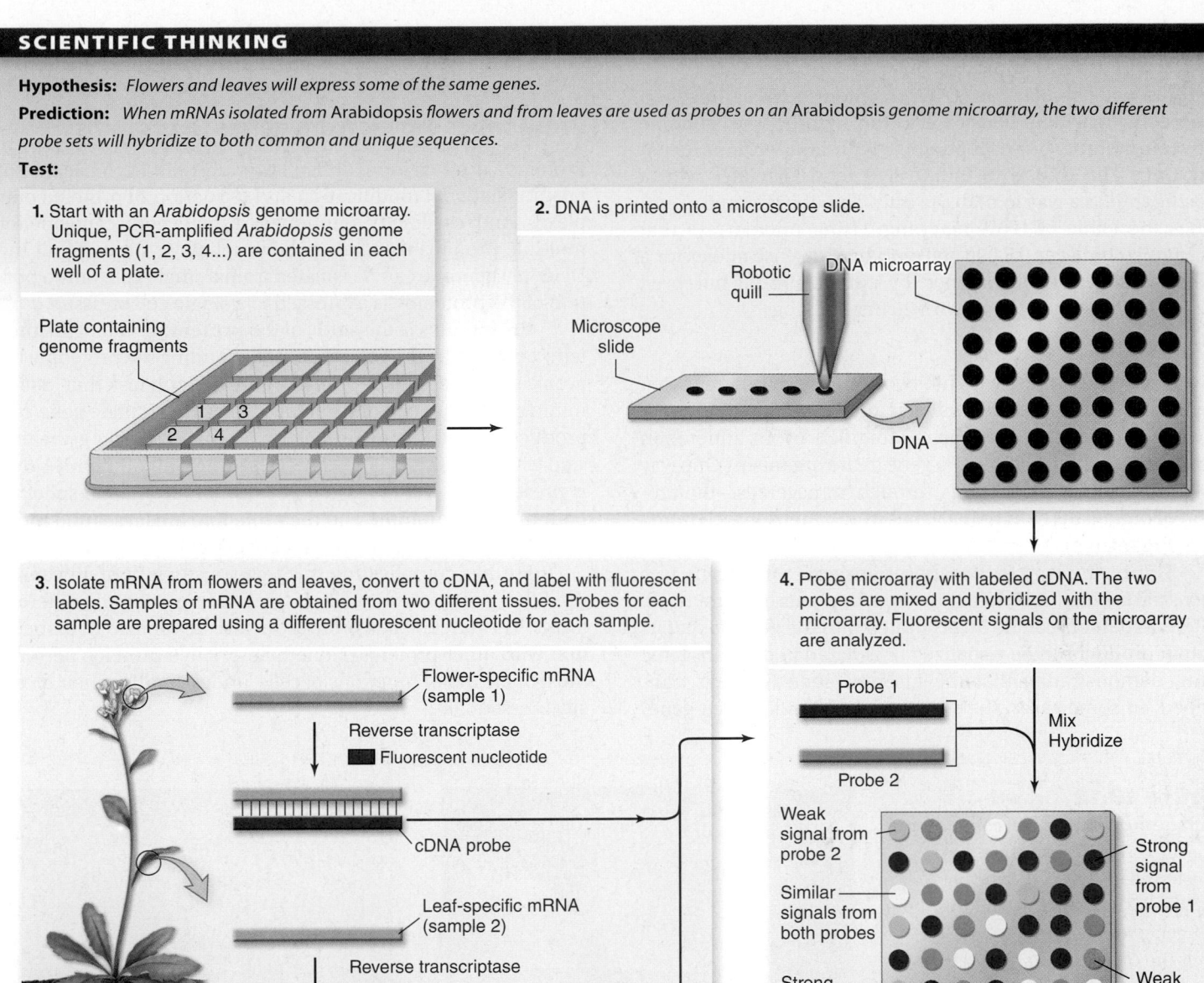

1. Start with an *Arabidopsis* genome microarray. Unique, PCR-amplified *Arabidopsis* genome fragments (1, 2, 3, 4...) are contained in each well of a plate.

Plate containing genome fragments

2. DNA is printed onto a microscope slide.

Robotic quill
DNA microarray
Microscope slide
DNA

3. Isolate mRNA from flowers and leaves, convert to cDNA, and label with fluorescent labels. Samples of mRNA are obtained from two different tissues. Probes for each sample are prepared using a different fluorescent nucleotide for each sample.

Flower-specific mRNA (sample 1)
Reverse transcriptase
Fluorescent nucleotide
cDNA probe
Leaf-specific mRNA (sample 2)
Reverse transcriptase
Different fluorescent nucleotide
cDNA probe

4. Probe microarray with labeled cDNA. The two probes are mixed and hybridized with the microarray. Fluorescent signals on the microarray are analyzed.

Probe 1
Mix Hybridize
Probe 2
Weak signal from probe 2
Similar signals from both probes
Strong signal from probe 2
Strong signal from probe 1
Weak signal from probe 1

Result: *Yellow spots represent sequences that hybridized to cDNA from both flowers and leaves. Red spots represent genes expressed only in flowers. Green spots represent genes expressed only in leaves.*

Conclusion: *Some* Arabidopsis *genes are expressed in both flowers and leaves, but there are genes expressed in flowers but not leaves and leaves but not flowers.*

Further Experiments: *How could you use microarrays to determine whether the genes expressed in both flowers and leaves are housekeeping genes or are unique to flowers and leaves?*

Figure 18.11 Microarrays.

that different cancers have different gene expression patterns. These findings are already being used to diagnose and design specific treatments for particular cancers.

From a large body of data, several patterns emerge:

1. Specific cancer types can be reliably distinguished from other cancer types and from normal tissue based on microarray data.

2. Subtypes of particular cancers often have different gene expression patterns in microarray data.

3. Gene expression patterns from microarray data can be used to predict disease recurrence, tendency to metastasize, and treatment response.

This represents an important step forward in both the diagnosis and treatment of human cancers.

Microarray analysis and genome-wide association mapping

Genome-wide association (GWA), as mentioned earlier, is an approach that compares SNPs throughout the genome between members in a population with and without a specific trait. The goal is to find a SNP that correlates with a specific trait as a way to map the trait. The dog genome exemplifies the value of GWA mapping. Using microarrays that distinguish between 15,000 SNP variants, disease alleles for a recessive trait can be mapped by comparing 20 purebred dogs exhibiting the disease with 20 healthy dogs.

Transgenics

How can we determine whether two genes from different species having similar sequences have the same function? And, how can we be sure that a gene identified by an annotation program actually functions as a gene in the organism? One way to address these questions is through transgenics—the creation of organisms containing genes from other species (transgenic organisms).

The technology for creating transgenic organisms was discussed in chapter 17; it is illustrated for plants in figure 18.12. Different markers can be incorporated into the gene so that its protein product can be visualized or isolated in the transgenic plant, demonstrating that the inserted gene is being transcribed. In some cases, the transgene (inserted foreign gene) may affect a visible phenotype. Of course, transgenics are but one of many ways to address questions about gene function.

Proteomics moves from genes to proteins

Proteins are much more difficult to study than DNA because of posttranslational modification and formation of protein complexes. And, as already mentioned, a single gene can code for multiple proteins using alternative splicing. Although all the DNA in a genome can be isolated from a single cell, only a portion of the proteome is expressed in a single cell or tissue.

Proteomics is the study of the **proteome**—all of the proteins encoded by the genome. Understanding the proteome for even a single cell will be a much more difficult task than determining the sequence of a genome. Because a single gene can produce more than one protein by alternative splicing, the first step is to characterize the **transcriptome**—all of the RNA that is present in a cell or tissue. Because of alternative splicing, both the transcriptome and the proteome are larger and more complex than the simple number of genes in the genome.

Further complicating this issue, a single protein can be modified posttranslationally to produce functionally different forms. The function of a protein can also depend on its association with other proteins. Nonetheless, since proteins perform most of the major functions of cells, understanding their diversity is essential.

Figure 18.12 Growth of a transgenic plant. DNA containing a gene for herbicide resistance was transferred into wheat *(Triticum aestivum)*. The DNA also contains the *GUS* gene, which is used as a tag or label. The *GUS* gene produces an enzyme that catalyzes the conversion of a staining solution from clear to blue. *a.* Embryonic tissue just prior to insertion of foreign DNA. *b.* Following DNA transfer, callus cells containing the foreign DNA are indicated by color from the *GUS* gene *(dark blue spots)*. *c.* Shoot formation in the transgenic plants growing on a selective medium. Here, the gene for herbicide resistance in the transgenic plants allows growth on the selective medium containing the herbicide. *d.* Comparison of growth on the selection medium for transgenic plants bearing the herbicide resistance gene *(left)* and a nontransgenic plant *(right)*.

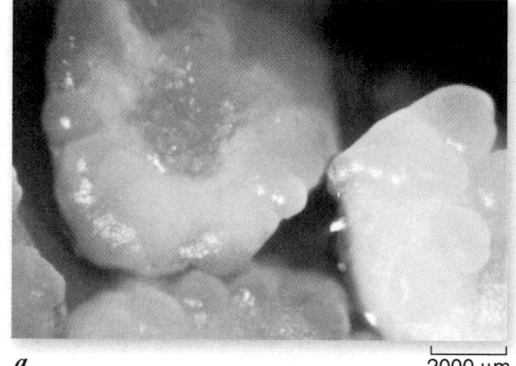
a. 2000 μm

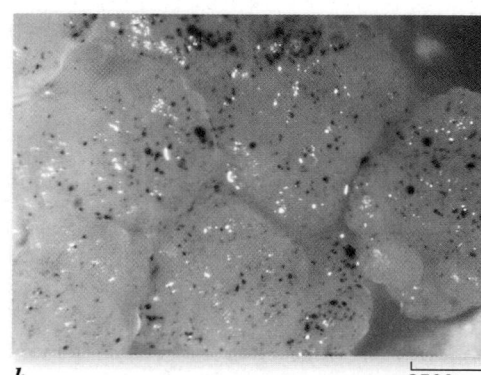
b. 2500 μm

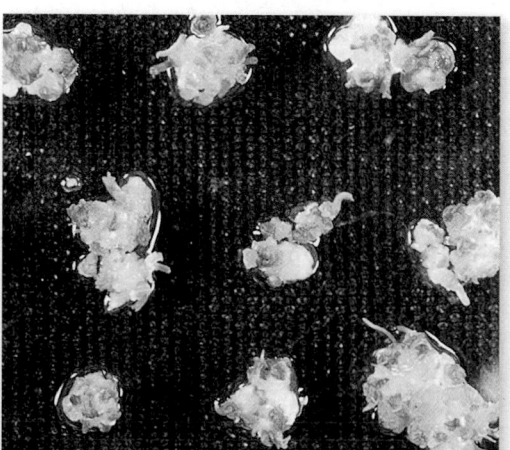
c.

d.

Inquiry question Why is the "proteome" likely to be different from simply the predicted protein products found in the complete genome sequence?

Predicting protein function

The use of new methods to quickly identify and characterize large numbers of proteins is the distinguishing feature between traditional protein biochemistry and proteomics. As with genomics, the challenge is one of scale.

Ideally, a researcher would like to be able to examine a nucleotide sequence and know what sort of functional protein the sequence specifies. Databases of protein structures in different organisms can be searched to predict the structure and function of genes known only by sequence, as identified in genome projects. Analysis of these data provides a clearer picture of how gene sequence relates to protein structure and function. Having a greater number of DNA sequences available allows for more extensive comparisons as well as identification of common structural patterns as groups of proteins continue to emerge.

Although there may be as many as a million different proteins, most are just variations on a handful of themes. The same shared structural motifs—barrels, helices, molecular zippers—are found in the proteins of plants, insects, and humans (figure 18.13; also see chapter 3 for more information on protein motifs). The maximum number of distinct motifs has been estimated to be fewer than 5000. About 1000 of these motifs have already been cataloged. Efforts are now under way to detail the shapes of all the common motifs.

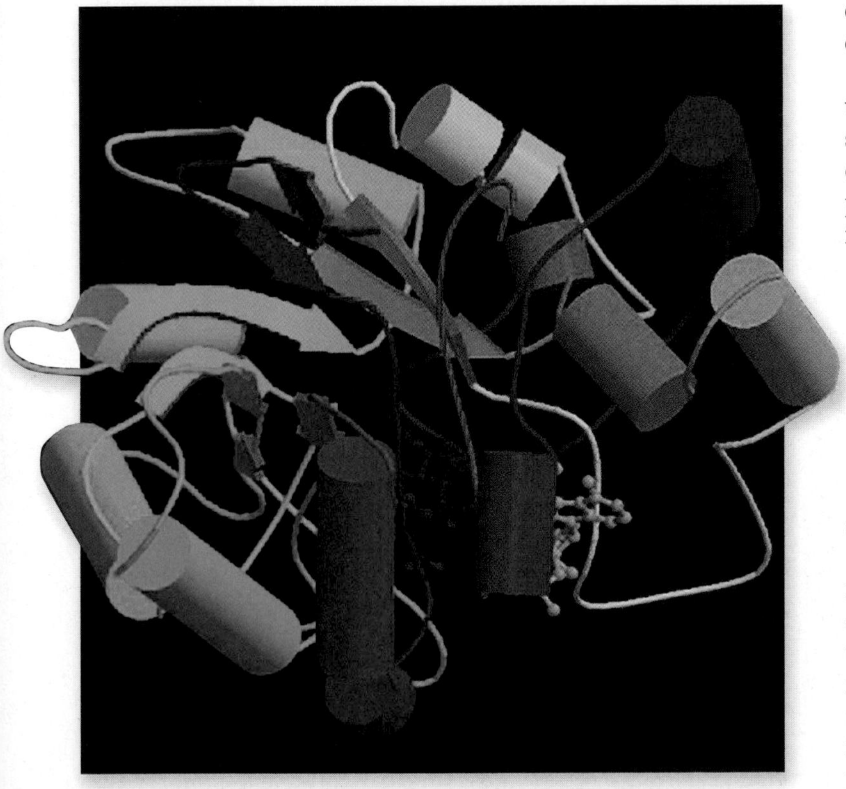

Figure 18.13 Computer-generated model of an enzyme. Searchable databases contain known protein structures, including human aldose reductase shown here. Secondary structural motifs are shown in different colors.

Large-scale screens reveal protein–protein interactions

We often study proteins in isolation, compared with their normal cellular context. This approach is obviously artificial. One immediate goal of proteomics, therefore, is to map all the physical interactions between proteins in a cell. This is a daunting task that requires tools that can be automated, similarly to the way that genome sequencing was automated.

One approach is to use the yeast two-hybrid system, which integrates much of the technology discussed in this chapter. It takes advantage of one feature of eukaryotic gene regulation, namely that the structure of proteins that turn on eukaryotic gene expression—transcription factors—have a modular structure.

The *Gal4* gene of yeast encodes a transcriptional activator with modular structure consisting of a DNA-binding domain that binds sequences in *Gal4*-responsive promoters and an activation domain that interacts with the transcription apparatus to turn on transcription. The system uses two vectors: one containing a fragment of the *Gal4* gene that encodes the DNA-binding domain, and another containing a fragment of the *Gal4* gene that encodes the transcription activation domain. Neither of these alone can activate transcription.

When cDNAs are inserted into each of these two vectors in the proper reading frame, they are expressed as a single protein consisting of the protein of interest and part of the Gal4 activator protein (figure 18.14). These hybrid proteins are called *fusion proteins* since they are literally fused in the same polypeptide chain. The DNA-binding hybrid is called the *bait,* and the activating domain hybrid is called the *prey.*

These vectors are inserted into cells of different mating types that can be crossed. One of these vectors also contains a so-called *reporter gene* encoding a protein that can be assayed for enzymatic activity. The reporter gene is under control of a *Gal4*-responsive regulatory region, so that when active *Gal4* is present, the reporter gene is expressed and can be detected by an enzymatic assay.

The DNA-binding hybrid binds to DNA adjacent to the reporter gene. When the two proteins in bait and prey interact, the prey hybrid brings the activating domain into position to turn on gene expression from the reporter gene (see figure 18.13).

The beauty of this system is that it is both simple and flexible. It can be used with two known proteins or with a known protein in the bait vector and entire cDNA libraries in the prey vector. In the latter case, all of the possible interactions in a cell type can be mapped.

It is already clear that even more protein interactions occur in cells than anticipated. In the future these data will

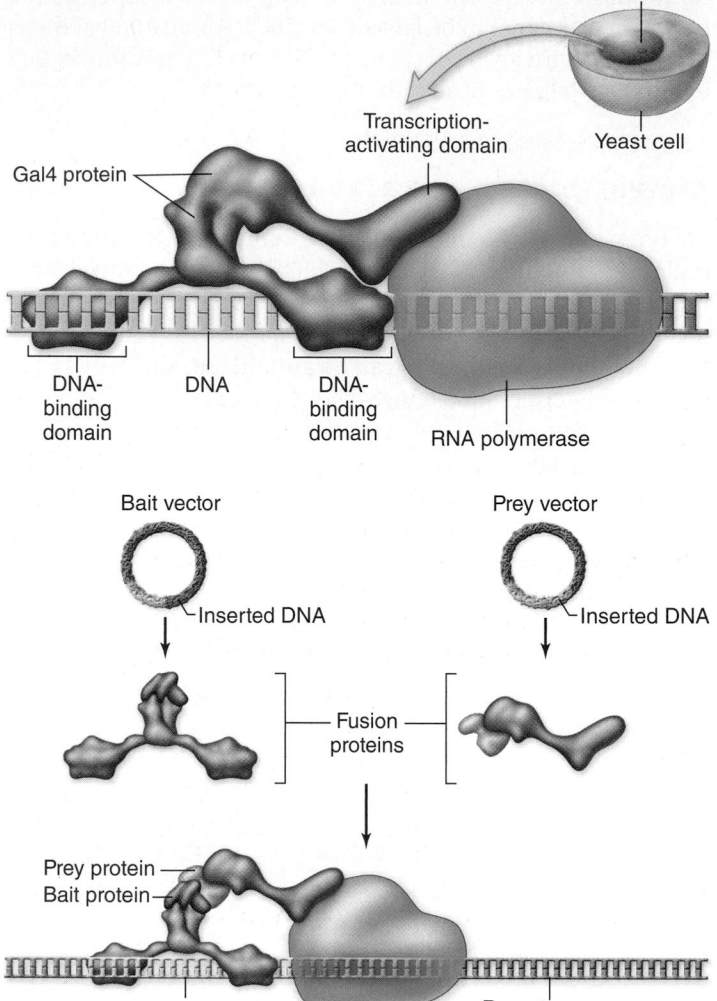

Figure 18.14 The yeast two-hybrid system detects interacting proteins. The Gal4 protein is a transcriptional activator *(top)*. The *Gal4* gene has been split and engineered into two different vectors such that one will encode only the DNA-binding domain (bait vector) and the other the transcription-activating domain (prey vector). When other genes are spliced into these vectors, they produce fusion proteins containing part of Gal4 and the proteins to be tested. If the proteins being tested interact, this will restore *Gal4* function and activate expression of a reporter gene.

form the basis for understanding the networks of protein interactions that make up the normal activities of a cell.

The yeast two-hybrid system can be automated once libraries of known cDNAs are available in each of the two vectors used. The use of two-hybrid screens has been applied to budding yeast to generate a map of all possible interacting proteins. This method is difficult to apply to more complex multicellular organisms, but in a technical tour-de-force, it has been applied to *Drosophila melanogaster* as well.

For vertebrates, the two-hybrid system is being applied more selectively, by concentrating on a biologically significant

process, such as signal transduction. The technique can then be used to map all of the interacting proteins in a specific signaling pathway.

? Inquiry question What is the relationship among genome, transcriptome, and proteome?

Learning Outcomes Review 18.5

Functional genomics provides tools to begin to understand the functions of the genes in a genome. Microarrays enable evaluation of gene expression for many genes at once. Proteomics involves similar analysis of all the proteins coded by a genome, that is, an organism's proteome. Because of alternative splicing, this task is much more complex.

■ *Why is establishment of a species' transcriptome an important step in studying its proteome?*

18.6 Applications of Genomics

Learning Outcomes

1. *List ways in which genomics could be applied to infectious disease research.*
2. *Explain how genomics could enhance crop production and nutritional yield.*
3. *Evaluate the issues of genome ownership and privacy.*

Space allows us to highlight only a few of the myriad applications of genomics to show the possibilities. The tools being developed truly represent a revolution in biology that will likely have a lasting influence on the way that we think about living systems.

Genomics can help to identify infectious diseases

The genomics revolution has yielded millions of new genes to be investigated. The potential of genomics to improve human health is enormous. Mutations in a single gene can explain some, but not most, hereditary diseases. With entire genomes to search, the probability of unraveling human, animal, and plant diseases is greatly improved.

Although proteomics will likely lead to new pharmaceuticals, the immediate effect of genomics is being seen in diagnostics. Both improved technology and gene discovery are enhancing the diagnosis of genetic abnormalities.

Diagnostics are also being used to identify individuals. For example, short tandem repeats (STRs), discovered through genomic research, were among the forensic diagnostic tools used to identify remains of victims of the September 11, 2001, terrorist attack on the World Trade Center in New York City.

The September 11 attacks were followed by an increased awareness and concern about biological weapons. Five people died and 17 more were infected with anthrax after envelopes containing anthrax spores were sent through the U.S. mail. A massive FBI investigation initially focused on the wrong individual, Steven J. Hatfill, a government scientist. Genome sequencing allowed exploration of possible sources of the deadly bacteria. A difference of only 10 bp between strains allowed the FBI to trace the source to a single vial of the bacteria used in a vaccine research program at U.S. Army Medical Research Institute for Infectious Diseases. By 2008, Hatfield was exonerated. Another researcher, Bruce E. Ivins, committed suicide just before being formally charged by the FBI with criminal activity in the 2001 anthrax attacks. Ivins had been working on vaccine development. In addition, substantial effort has been turned toward the use of genomic tools to distinguish between naturally occurring infections and intentional outbreaks of disease. The Centers for Disease Control and Prevention (CDC) have ranked bacteria and viruses that are likely targets for bioterrorism (table 18.2).

Genomics can help improve agricultural crops

Globally speaking, poor nutrition is the greatest impediment to human health. Much of the excitement about the rice genome project is based on its potential for improving the yield and nutritional quality of rice and other cereals worldwide. The development of Golden Rice (see chapter 17) is an

TABLE 18.2	High-Priority Pathogens for Genomic Research	
Pathogen	**Disease**	**Genome***
Variola major	Smallpox	Complete
Bacillus anthracis	Anthrax	Complete
Yersinia pestis	Plague	Complete
Clostridium botulinum	Botulism	Complete
Francisella tularensis	Tularemia	Complete
Filoviruses	Ebola and Marburg hemorrhagic fever	Both are complete
Arenaviruses	Lassa fever and Argentine hemorrhagic fever	Both are complete

* These viruses and bacteria have multiple strains. "Complete" indicates that at least one has been sequenced. For example, the Florida strain of anthrax was the first to be sequenced.

example of improved nutrition through genetic approaches, and access to the entire genome could potentially lead to more improvements. About one third of the world population obtains half its calories from rice (figure 18.15). In some regions, individuals consume up to 1.5 kg of rice daily. More than 500 million tons of rice is produced each year, but this may not be adequate to provide enough rice for the world in the future.

Due in large part to scientific advances in crop breeding and farming techniques, in the last 50 years world grain production has more than doubled, with an increase in cropland of only 1%. The world now farms a total area the size of South America, but without the scientific advances of the past 50 years, an area equal to the entire western hemisphere would need to be farmed to produce enough food for the world.

Unfortunately, water usage for crops has tripled in that time period, and quality farmland is being lost to soil erosion. Scientists are also concerned about the effects of global climate change on agriculture worldwide. Increasing the yield and quality of crops, especially on more marginal farmland, will depend on many factors—but genetic engineering, built on the findings of genomics projects, can contribute significantly to the solution.

Most crops grown in the United States produce less than half of their genetic potential because of environmental stresses (salt, water, and temperature), herbivores, and pathogens (figure 18.16). Identifying genes that can provide resistance to stress and pests is the focus of many current genomics research projects. Having access to entire genomic sequences will enhance the probability of identifying critical genes.

Figure 18.15 Rice field. Most of the rice grown globally is directly consumed by humans and is the dietary mainstay of 2 billion people.

Figure 18.16 Corn crop productivity well below its genetic potential due to drought stress. Corn production can be limited by water deficiencies due to the drought that occurs during the growing season in dry climates. Global climate change may increase drought stress in areas where corn is the major crop.

Synthetic biology extends the potential of genetic engineering

Synthetic biology is the next frontier beyond genetic engineering approaches described in chapter 17. It is now possible to synthetically construct an entire bacterial genome and insert it into a bacterium. Although this accomplishment has been touted as a synthetic cell or synthetic life, creating all the other cell components de novo is still on the horizon.

Synthetic biology can be used to creatively engineer a vast number of solutions. Biosensors are cells that respond to toxins or other molecules with a visible or otherwise measurable signal. Biofuel production, therapeutic treatment for disease, and creation of organisms to degrade environmental hazards are among the many promises of synthetic biology. Along with the huge benefits, the potential for misuse of the technology is leading to extensive ethical discussions.

Genomics raises ethical issues over ownership of genomic information

Genome science is also a source of ethical challenges and dilemmas. One example is the issue of gene patents. Actually, it is the use of a gene, not the gene itself, that is patentable. For a patent to be granted for a gene's use, the product and its function must be known.

The public genome consortia, supported by federal funding, have been driven by the belief that the sequence of genomes should be freely available to all and should not be patented. Private companies patent gene functions, but they often make sequence data available with certain restrictions. The physical sciences have negotiated the landscape of public and for-profit research for decades, but this is relatively new territory for biologists.

A March 2010 ruling in New York state's Southern District Court invalidated Myriad Genetics patents on the *BRCA1* and *BRCA2* genes. The molecular diagnostics company had a monopoly on these two genes, which are used in diagnosing breast cancer risk. From a patient perspective, the patents prevented individuals from seeking a second independent analysis of their *BRCA1* and *BRCA2* allele. Appeals are underway, and the controversy over gene patents continues.

Another ethical issue involves privacy. How sequence data are used is the focus of thoughtful and ongoing discussions. The Universal Declaration on the Human Genome and Human Rights states, "The human genome underlies the fundamental unity of all members of the human family, as well as the recognition of their inherent dignity and diversity. In a symbolic sense, it is the heritage of humanity."

Although we talk about "the" human genome, each of us has subtly different genomes that can be used to identify us. Genetic disorders such as cystic fibrosis and Huntington disease can already be identified by screening, but genomics will greatly increase the number of identifiable traits. The Genetic Information Nondiscrimination Act (GINA) was signed into law in 2008 to prevent discrimination based on genotype. Employers and health insurance companies may not request genetic tests or discriminate based on someone's genetic code. Life, disability, and long-term care insurance coverage are not covered by GINA, however. Members of the military are also excluded from GINA's privacy protection. The U.S. Armed Forces require DNA samples from members of the military for possible casualty identification. The genome privacy debate continues.

Behavioral genomics is an area that is also rich with possibilities and dilemmas. Very few behavioral traits can be accounted for by single genes. Two genes have been associated with fragile-X syndrome, and three with early-onset Alzheimer disease. Comparisons of multiple genomes will likely lead to the identification of multiple genes controlling a range of behaviors. Will this change the way we view acceptable behavior?

In Iceland, the parliament has voted to have a private company create a database from pooled medical, genetic, and genealogical information about all Icelanders, a particularly fascinating population from a genetic perspective. Because minimal migration or immigration has occurred there over the last 800 years, the information that can be mined from the Icelandic database is phenomenal. Ultimately, the value of that information has to be weighed, however, against any possible discrimination or stigmatization of individuals or groups.

? **Inquiry question** As of February 2008 a draft version of the corn genome had been sequenced. How could you use information from the corn and rice genome sequences to try to improve drought tolerance in corn?

Learning Outcomes Review 18.6

Genomics is one approach to better diagnosis, based on knowledge of infectious agents' genetic makeup; it also allows identification of individual disease strains. Genomics has enhanced DNA identification of remains. Agricultural crop yields and nutritional content could be improved if genes that confer disease resistance or increased synthesis can be identified.

■ *You are assigned to develop a biosensor for a toxin and are working with a bioluminescent gene. What part of the gene will you work with and why?*

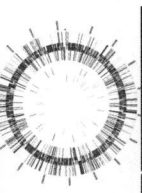

Chapter Review

18.1 Mapping Genomes

Different kinds of physical maps can be generated.
Physical genetic maps include fully sequenced genomes, restriction maps, and maps of chromosome-banding patterns.

Sequence-tagged sites provide a common language for physical maps.
Any physical site can be used as a sequence-tagged site (STS), based on a small stretch of a unique DNA sequence that allows unambiguous identification of a fragment (figure 18.3).

Genetic maps provide a link to phenotypes.
Short tandem repeats (STRs) are the most common type of markers for distinguishing regions of the genome and assessing its phenotypic effects.

Physical maps can be correlated with genetic maps.
Physical and genetic maps can be correlated. Any gene that can be cloned can be placed within the genome sequence and mapped. However, absolute correspondence of distances cannot be accomplished.

18.2 Sequencing Genomes

DNA sequencing provides information about genes and genomes.
DNA sequencing technology is an adaptation of DNA replication in vitro that uses modified nucleotides and copies a single strand of the DNA. Modified nucleotides are labeled and stop or pause the extension of a new strand. A whole-genome sequence is the ultimate physical map.

Genome sequencing requires larger molecular clones.
Yeast artificial chromosomes (YACs) have allowed cloning of larger pieces of DNA, although their use has some drawbacks. Bacterial artificial chromosomes (BACs) are most commonly used now.

Whole-genome sequencing is approached in two ways: clone-by-clone and shotgun.
Clone-by-clone sequencing starts with known clones, often in BACs that can be aligned with each other.

Shotgun sequencing involves sequencing random clones, then using a computer to assemble the finished sequence.

The Human Genome Project used both sequencing methods.
By 2004, the "finished" sequence was announced, and it includes 99% of the euchromatic human DNA sequence.

18.3 Characterizing Genomes

The Human Genome Project found fewer genes than expected.
Although eukaryotic genomes are larger and have more genes than those of prokaryotes, the size of the organism is not always correlated with the size of the genome. The human genome contains only around 25,000 genes, fewer than found in rice (figure 18.7).

Finding genes in sequence data requires computer searches.
In a sequenced genome, protein-coding genes are identified by looking for open-reading frames (ORFs). An ORF begins with a start codon and contains no stop codon for a distance long enough to encode a protein. Genes are then grouped based on conserved regions.

Genomes contain both coding and noncoding DNA.
Protein-encoding DNA includes single-copy genes, segmental duplications, multigene families, and tandem clusters. Noncoding DNA in eukaryotes makes up about 99% of DNA. Approximately 45% of the human genome is composed of mobile transposable elements, including LINEs, SINEs, and LTRs.

Expressed sequence tags identify genes that are transcribed.
The number and location of expressed genes can be estimated by sequencing the ends of randomly selected cDNAs to produce expressed sequence tags (ESTs).

SNPs are single-base differences between individuals.
Single-nucleotide difference between individuals are called single-nucleotide polymorphisms (SNPs). To be classified as a polymorphism, an SNP must be present in at least 1% of the population. At least 50,000 SNPs are currently known in coding regions.

Genomic haplotypes are regions of chromosomes that are not exchanged by recombination. These regions can be used to map genes by association (figure 18.8).

18.4 Comparing Genomes

Comparative genomics reveals conserved regions in genomes.

More than half of the genes of *Drosophila* have human counterparts. The biggest difference between our genome and the chimpanzee genome is in transposable elements.

Synteny allows comparison of unsequenced genomes.

Synteny refers to the conserved arrangements of segments of DNA in related genomes (figure 18.9). Many separate species have been found to have large regions of synteny.

Organelle genomes have exchanged genes with the nuclear genome.

Both chloroplasts and mitochondria contain components that indicate exchange of genetic material with the nuclear genome.

18.5 From Genes to Proteins

Functional genomics reveals gene function at the genome level.

Functional genomics uses high-end computer technology to analyze gene function and gene products. DNA microarrays allow the expression of all of the genes in a cell to be monitored at once (figure 18.10).

Proteomics moves from genes to proteins.

Proteomics characterizes all of the proteins produced by a cell. The transcriptome is all the mRNAs present in a cell at a specific time. Protein microarrays can identify and characterize large numbers of proteins.

Large-scale screens reveal protein–protein interactions.

The yeast two-hybrid system is used to generate large-scale maps of interacting proteins; however, the scope of this task is daunting in humans, mice, and other vertebrates. Selective applications in specific areas, such as signal transduction, have been undertaken.

18.6 Applications of Genomics

Genomics can help to identify infectious diseases.

Genomics can help identify naturally occurring and intentional outbreaks of infectious diseases and tracing of disease strains.

Genomics can help improve agricultural crops.

Genomics can potentially increase the nutritional value of crops and alter their responses to environmental stresses, potentially helping to feed a growing population.

Synthetic biology extends the potential of genetic engineering.

The ability to synthesize entire genomes de novo makes it possible to engineer organisms to function as biosensors and chemical factories, clean up contaminated environments, and perform other yet to be imagined roles.

Genomics raises ethical issues over ownership of genomic information.

Questions regarding profit and ownership of genomic data provide ongoing challenges for the ethical use of scientific knowledge.

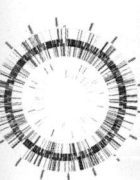

Review Questions

UNDERSTAND

1. A genetic map is based on the
   a. sequence of the DNA.
   b. relative position of genes on chromosomes.
   c. location of sites of restriction enzyme cleavage.
   d. banding pattern on a chromosome.

2. What is an STS?
   a. A unique sequence within the DNA that can be used for mapping
   b. A repeated sequence within the DNA that can be used for mapping
   c. An upstream element that allows for mapping of the 3′ region of a gene
   d. Both b and c are correct.

3. Which number represents the total number of genes in the human genome?
   a. 2500   c. 25,000
   b. 10,000   d. 100,000

4. An open reading frame (ORF) is distinguished by the presence of
   a. a stop codon.
   b. a start codon.

   c. a sequence of DNA long enough to encode a protein.
   d. All of the choices are correct.

5. What is a BLAST search?
   a. A mechanism for aligning consensus regions during whole-genome sequencing
   b. A search for similar gene sequences from other species
   c. A method of screening a DNA library
   d. A method for identifying ORFs

6. Which of the following is NOT an example of a protein-encoding gene?
   a. Single-copy gene   c. Pseudogene
   b. Tandem clusters   d. Multigene family

7. What is a proteome?
   a. The collection of all genes encoding proteins
   b. The collection of all proteins encoded by the genome
   c. The collection of all proteins present in a cell
   d. The amino acid sequence of a protein

8. Which of the following is NOT an example of noncoding DNA?
   a. Promoter   c. Pseudogene
   b. Intron   d. Exon

APPLY

1. An artificial chromosome is useful because it
   a. produces more consistent results than a natural chromosome.
   b. allows for the isolation of larger DNA sequences.
   c. provides a high copy number of a DNA sequence.
   d. is linear.

2. Comparisons between genomes is made easier because of
   a. synteny.
   b. haplotypes.
   c. transposons.
   d. expressed sequence tags.

3. Which of the following techniques relies on prior knowledge of overlapping sequences?
   a. Yeast two-hybrid system
   b. Shotgun method of genome sequencing
   c. FISH
   d. Clone-by-clone method of genome sequencing

4. The duplication of a gene due to uneven meiotic crossing over is thought to lead to the production of a
   a. segmental duplication.
   b. tandem duplication.
   c. simple sequence repeat.
   d. multigene family.

5. What information can be obtained from a DNA microarray?
   a. The sequence of a particular gene
   b. The presence of genes within a specific tissue
   c. The pattern of gene expression
   d. Differences between genomes

6. Which of the following is true regarding microarray technology and cancer?
   a. A DNA microarray can determine the type of cancer.
   b. A DNA microarray can measure the response of a cancer to therapy.
   c. A DNA microarray can be used to predict whether the cancer will metastasize.
   d. All of the choices are correct.

7. Which of the following techniques could be used to examine protein–protein interactions in a cell?
   a. Two-hybrid screens
   b. Protein structure databases
   c. Protein microarrays
   d. Both a and c are correct.

SYNTHESIZE

1. You are in the early stages of a genome-sequencing project. You have isolated a number of clones from a BAC library and mapped the inserts in these clones using STSs. Use the STSs shown to align the clones into a contiguous sequence of the genome (a contig).

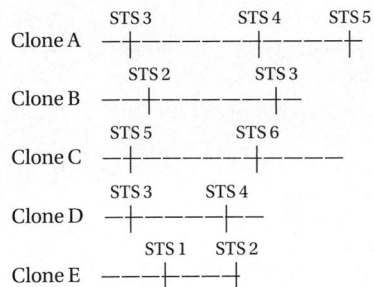

2. Genomic research can be used to determine if an outbreak of an infectious disease is natural or "intentional." Explain what a genomic researcher would be looking for in a suspected intentional outbreak of a disease like anthrax.

ONLINE RESOURCE

www.ravenbiology.com

Understand, Apply, and Synthesize—enhance your study with animations that bring concepts to life and practice tests to assess your understanding. Your instructor may also recommend the interactive eBook, individualized learning tools, and more.

Chapter 19

Cellular Mechanisms of Development

5.5 μm

Chapter Contents

Introduction

Recent work with different kinds of stem cells, like those pictured, have captured the hopes and imagination of the public. For thousands of years, humans have wondered how organisms arise, grow, change, and mature. We are now in an era when long-standing questions may be answered, and new possibilities for regenerative medicine seem on the horizon.

We have explored gene expression from the perspective of individual cells, examining the diverse mechanisms cells employ to control the transcription of particular genes. Now we broaden our perspective and look at the unique challenge posed by the development of a single cell, the fertilized egg, into a multicellular organism. In the course of this developmental journey, a pattern of gene expression takes place that causes particular lines of cells to proceed along different paths, spinning an incredibly complex web of cause and effect. Yet, for all its complexity, this developmental program works with impressive precision. In this chapter, we explore the mechanisms of development at the cellular and molecular level.

19.1 The Process of Development

Development can be defined as the process of systematic, gene-directed changes through which an organism forms the successive stages of its life cycle. Development is a continuum, and explorations of development can be focused on any point along this continuum. The study of development plays a central role in unifying the understanding of both the similarities and diversity of life on Earth.

We can divide the overall process of development into four subprocesses:

- **Cell Division.** A developing plant or animal begins as a fertilized egg, or zygote, that must undergo cell division to produce the new individual. In all cases early development involves extensive cell division, but in many cases it does not include much growth as the egg cell itself is quite large.

- **Differentiation.** As cells divide, orchestrated changes in gene expression result in differences between cells that ultimately result in cell specialization. In differentiated cells, certain genes are expressed at particular times, but other genes may not be expressed at all.

■ **Pattern Formation.** Cells in a developing embryo must become oriented to the body plan of the organism the embryo will become. Pattern formation involves cells' abilities to detect positional information that guides their ultimate fate.

■ **Morphogenesis.** As development proceeds, the form of the body—its organs and anatomical features—is generated. Morphogenesis may involve cell death as well as cell division and differentiation.

Despite the overt differences between groups of plants and animals, most multicellular organisms develop according to molecular mechanisms that are fundamentally very similar. This observation suggests that these mechanisms evolved very early in the history of multicellular life.

19.2 *Cell Division*

Learning Outcomes

1. *Characterize the role of cell division in early development.*
2. *Describe the use of* C. elegans *to track cell lineages.*
3. *Distinguish differences in cell division between animals and plants.*

When a frog tadpole hatches out of its protective coats, it is roughly the same overall mass as the fertilized egg from which it came. Instead of being made up of just one cell, however, the tadpole consists of about a million cells, which are organized into tissues and organs with different functions. Thus, the very first process that must occur during embryogenesis is cell division.

Immediately following fertilization, the diploid zygote undergoes a period of rapid mitotic divisions that ultimately result in an early embryo comprised of dozens to thousands of diploid cells. In animal embryos, the timing and number of these divisions are species-specific and are controlled by a set of molecules that we examined in chapter 10: the *cyclins* and *cyclin-dependent kinases (Cdks)*. These molecules exert control over checkpoints in the cycle of mitosis.

Development begins with cell division

In animal embryos, the period of rapid cell division following fertilization is called **cleavage.** During cleavage, the enormous mass of the zygote is subdivided into a larger and larger number of smaller and smaller cells, called **blastomeres** (figure 19.1). Hence, cleavage is not accompanied by any increase in the overall size of the embryo. The G_1 and G_2 phases of the cell cycle, during which a cell increases its mass and size, are extremely shortened or eliminated during cleavage (figure 19.2).

Because of the absence of the two gap/growth phases, the rapid rate of mitotic divisions during cleavage is never again approached in the lifetime of any animal. For example, zebrafish blastomeres divide once every several minutes during cleavage, to create an embryo with a thousand cells in just under 3 hr! In contrast, cycling adult human intestinal epithelial cells divide on average only once every 19 hr. A comparison of the different patterns of cleavage can be found in chapter 53.

When external sources of nutrients become available— for example, during larval feeding stages or after implantation of mammalian embryos—daughter cells can increase in size following cytokinesis, and an overall increase in the size of the organism occurs as more cells are produced.

Every cell division is known in the development of *C. elegans*

One of the most completely described models of development is the tiny nematode *Caenorhabditis elegans*. Only about 1 mm long, the adult worm consists of 959 somatic cells.

Because *C. elegans* is transparent, individual cells can be followed as they divide. By observing them, researchers have learned how each of the cells that make up the adult worm is derived from the fertilized egg. As shown on the lineage map in figure 19.3*a*, the egg divides into two cells, and these daughter cells continue to divide. Each horizontal line on the map represents one round of cell division. The length of each vertical line represents the time between cell divisions, and the end of each vertical line represents one fully differentiated cell. In figure 19.3*b*, the major organs of the worm are color-coded to match the colors of the corresponding groups of cells on the lineage map.

Some of these differentiated cells, such as some cells that generate the worm's external cuticle, are "born" after only 8 rounds of cell division; other cuticle cells require as many as 14 rounds. The cells that make up the worm's pharynx, or feeding organ, are born after 9 to 11 rounds of division, whereas cells in the gonads require up to 17 divisions.

Exactly 302 nerve cells are destined for the worm's nervous system. Exactly 131 cells are programmed to die, mostly within minutes of their "birth." The fate of each cell is the same in every *C. elegans* individual, except for the cells that will become eggs and sperm.

Figure 19.1 Cleavage divisions in a frog embryo. *a.* The first cleavage division divides the egg into two large blastomeres. *b.* After two more divisions, four small blastomeres sit on top of four large blastomeres, each of which continues to divide to produce *(c)* a compact mass of cells.

a. 0.8 mm *b.* 0.8 mm

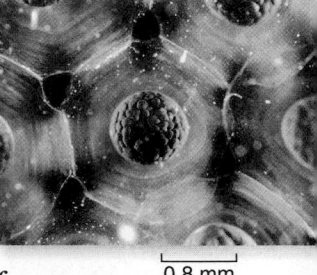

c. 0.8 mm

chapter **19** *Cellular Mechanisms of Development*   **373**

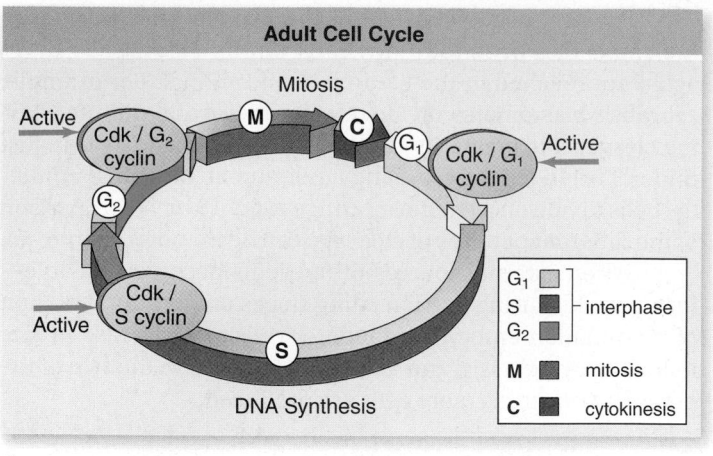

Adult Cell Cycle

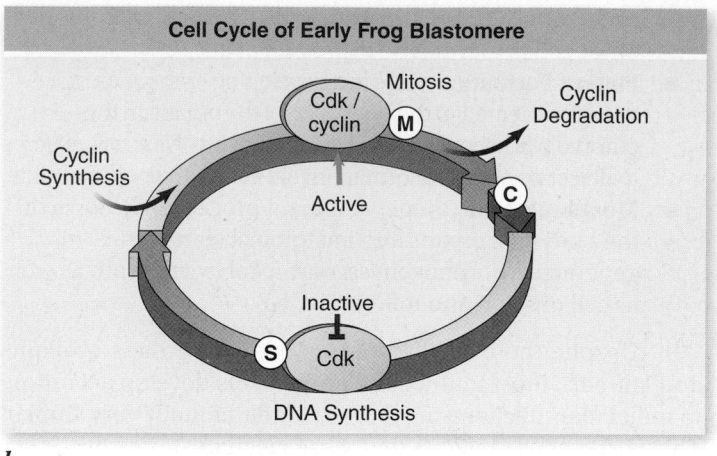

Cell Cycle of Early Frog Blastomere

a. *b.*

Figure 19.2 Cell cycle of adult cell and embryonic cell. In contrast to the cell cycle of adult somatic cells (*a*), the dividing cells of early frog embryos lack G₁ and G₂ stages (*b*), enabling the cleavage stage nuclei to rapidly cycle between DNA synthesis and mitosis. Large stores of cyclin mRNA are present in the unfertilized egg. Periodic degradation of cyclin proteins correlates with exiting from mitosis. Cyclin degradation and Cdk inactivation allow the cell to complete mitosis and initiate the next round of DNA synthesis.

Plant growth occurs in specific areas called meristems

A major difference between animals and plants is that most animals are mobile, at least in some phase of their life cycles, and therefore they can move away from unfavorable circumstances. Plants, in contrast, are anchored in position and must simply endure whatever environment they experience. Plants compensate for this restriction by allowing development to accommodate local circumstances.

Instead of creating a body in which every part is specified to have a fixed size and location, a plant assembles its body throughout its life span from a few types of modules, such as leaves, roots, branch nodes, and flowers. Each module has a rigidly controlled structure and organization, but how the modules are utilized is quite flexible—they can be adjusted to environmental conditions.

Figure 19.3 Studying embryonic cell division and development in the nematode.
Development in *C. elegans* has been mapped out such that the fate of each cell from the single egg cell has been determined. *a.* The lineage map shows the number of cell divisions from the egg, and the color coding links their placement in (*b*) the adult organism.

M. E. Challinor illustration. From Howard Hughes Medical Institute © as published in *From Egg to Adult,* 1992. Reprinted by permission.

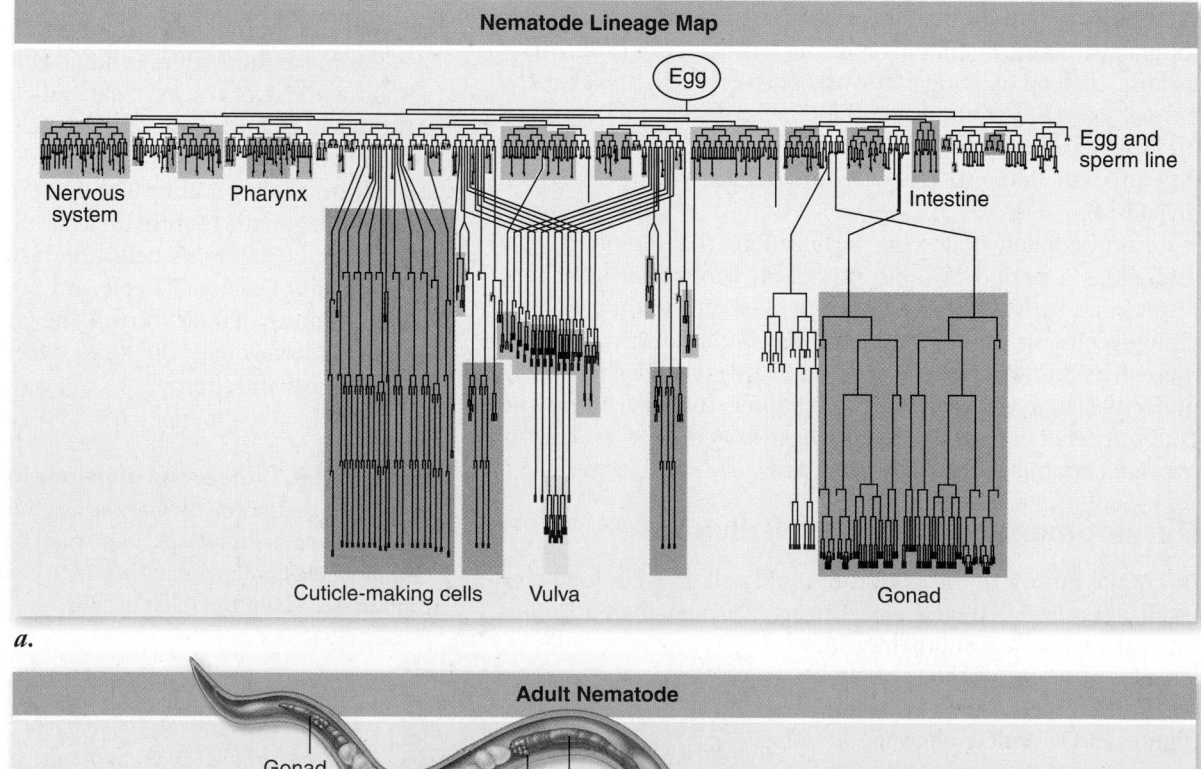

a.

b.

Plants develop by building their bodies outward, creating new parts from groups of stem cells that are contained in structures called **meristems.** As meristematic stem cells continually divide, they produce cells that can differentiate into the tissues of the plant.

This simple scheme indicates a need to control the process of cell division. We know that cell-cycle control genes are present in both yeast (fungi) and animal cells, implying that these are a eukaryotic innovation—and in fact, the plant cell cycle is regulated by the same mechanisms, namely through cyclins and cyclin-dependent kinases. In one experiment, overexpression of a Cdk inhibitor in transgenic *Arabidopsis thaliana* plants resulted in strong inhibition of cell division in leaf meristems, leading to significant changes in leaf size and shape.

Learning Outcomes Review 19.2

In animal embryos, a series of rapid cell divisions that skip the G_1 and G_2 phases convert the fertilized egg into many cells with no change in size. In the nematode *C. elegans,* every cell division leading to the adult form is known, and this pattern is invariant, allowing biologists to trace development in a cell-by-cell fashion. In plants, growth is restricted to specific areas called meristems, where undifferentiated stem cells are retained.

- How are early cell divisions in an embryo different from in an adult organism?

19.3 Cell Differentiation

Learning Outcomes

1. Describe the progressive nature of determination.
2. Illustrate with examples how cells become committed to developmental pathways.
3. Differentiate between the different types of stem cells.

In chapter 16, we examined the mechanisms that control eukaryotic gene expression. These processes are critical for the development of multicellular organisms, in which life functions are carried out by different tissues and organs. In the course of development, cells become different from one another because of the differential expression of subsets of genes—not only at different times, but in different locations of the growing embryo. We now explore some of the mechanisms that lead to differential gene expression during development.

Cells become determined prior to differentiation

A human body contains more than 210 major types of differentiated cells. These differentiated cells are distinguishable from one another by the particular proteins that they synthesize, their morphologies, and their specific functions. A molecular decision to become a particular type of differentiated cell occurs prior to any overt changes in the cell. This molecular decision-making process is called **cell determination,** and it commits a cell to a particular developmental pathway.

Tracking determination

Determination is often not visible in the cell and can only be "seen" by experiment. The standard experiment to test whether a cell or group of cells is determined is to move the donor cell(s) to a different location in a host (recipient) embryo. If the cells of the transplant develop into the same type of cell as they would have if left undisturbed, then they are judged to be already determined (figure 19.4).

Determination has a time course; it depends on a series of intrinsic or extrinsic events, or both. For example, a cell in the prospective brain region of an amphibian embryo at the early gastrula stage (see chapter 53) has not yet been determined; if transplanted elsewhere in the embryo, it will develop according to the site of transplant. By the late gastrula stage,

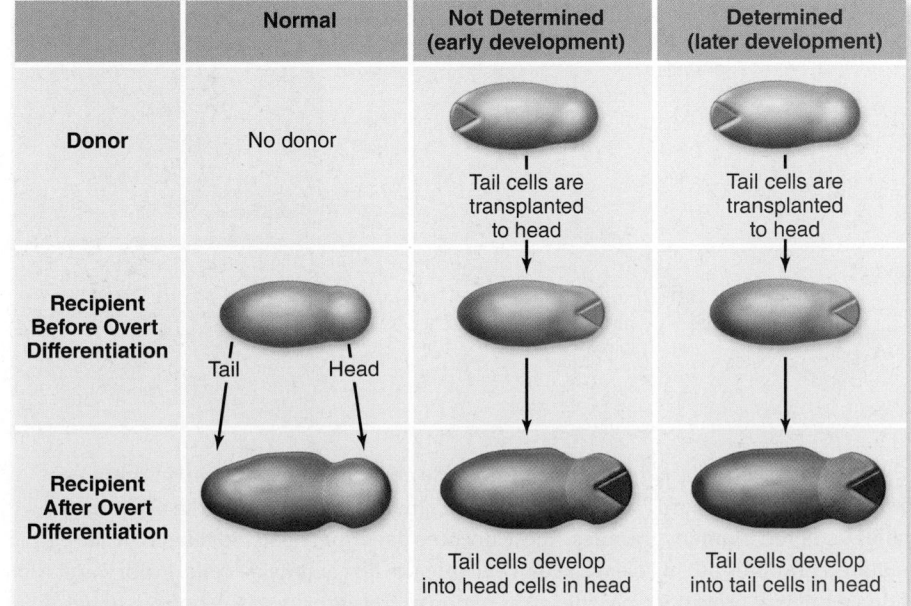

	Normal	Not Determined (early development)	Determined (later development)
Donor	No donor	Tail cells are transplanted to head	Tail cells are transplanted to head
Recipient Before Overt Differentiation	Tail Head		
Recipient After Overt Differentiation		Tail cells develop into head cells in head	Tail cells develop into tail cells in head

Figure 19.4 The standard test for determination. The gray ovals represent embryos at early stages of development. The cells to the right normally develop into head structures, whereas the cells to the left usually form tail structures. If prospective tail cells from an early embryo are transplanted to the opposite end of a host embryo, they develop according to their new position into head structures. These cells are not determined. At later stages of development, the tail cells are determined since they now develop into tail structures after transplantation into the opposite end of a host embryo!

however, additional cell interactions have occurred, determination has taken place, and the cell will develop as neural tissue no matter where it is transplanted.

Determination often takes place in stages, with a cell first becoming partially committed, acquiring positional labels that reflect its location in the embryo. These labels can have a great influence on how the pattern of the body subsequently develops. In a chicken embryo, tissue at the base of the leg bud normally gives rise to the thigh. If this tissue is transplanted to the tip of the identical-looking wing bud, which would normally give rise to the wing tip, the transplanted tissue will develop into a toe rather than a thigh. The tissue has already been determined as leg, but it is not yet committed to being a particular part of the leg. Therefore, it can be influenced by the positional signaling at the tip of the wing bud to form a tip (but in this case, a tip of leg).

The molecular basis of determination

Cells initiate developmental changes by using transcription factors to change patterns of gene expression. When genes encoding these transcription factors are activated, one of their effects is to reinforce their own activation. This reinforcement makes the developmental switch deterministic, initiating a chain of events that leads down a particular developmental pathway.

Cells in which a set of regulatory genes have been activated may not actually undergo differentiation until some time later, when other factors interact with the regulatory protein and cause it to activate still other genes. Nevertheless, once the initial "switch" is thrown, the cell is fully committed to its future developmental path.

Cells become committed to follow a particular developmental pathway in one of two ways:

1. via the differential inheritance of cytoplasmic determinants, which are maternally produced and deposited into the egg during oogenesis; or
2. via cell–cell interactions.

The first situation can be likened to a person's social status being determined by who his or her parents are and what he or she has inherited. In the second situation, the person's social standing is determined by interactions with his or her neighbors. Clearly both can be powerful factors in the development and maturation of that individual.

Determination can be due to cytoplasmic determinants

Many invertebrate embryos provide good visual examples of cell determination through the differential inheritance of

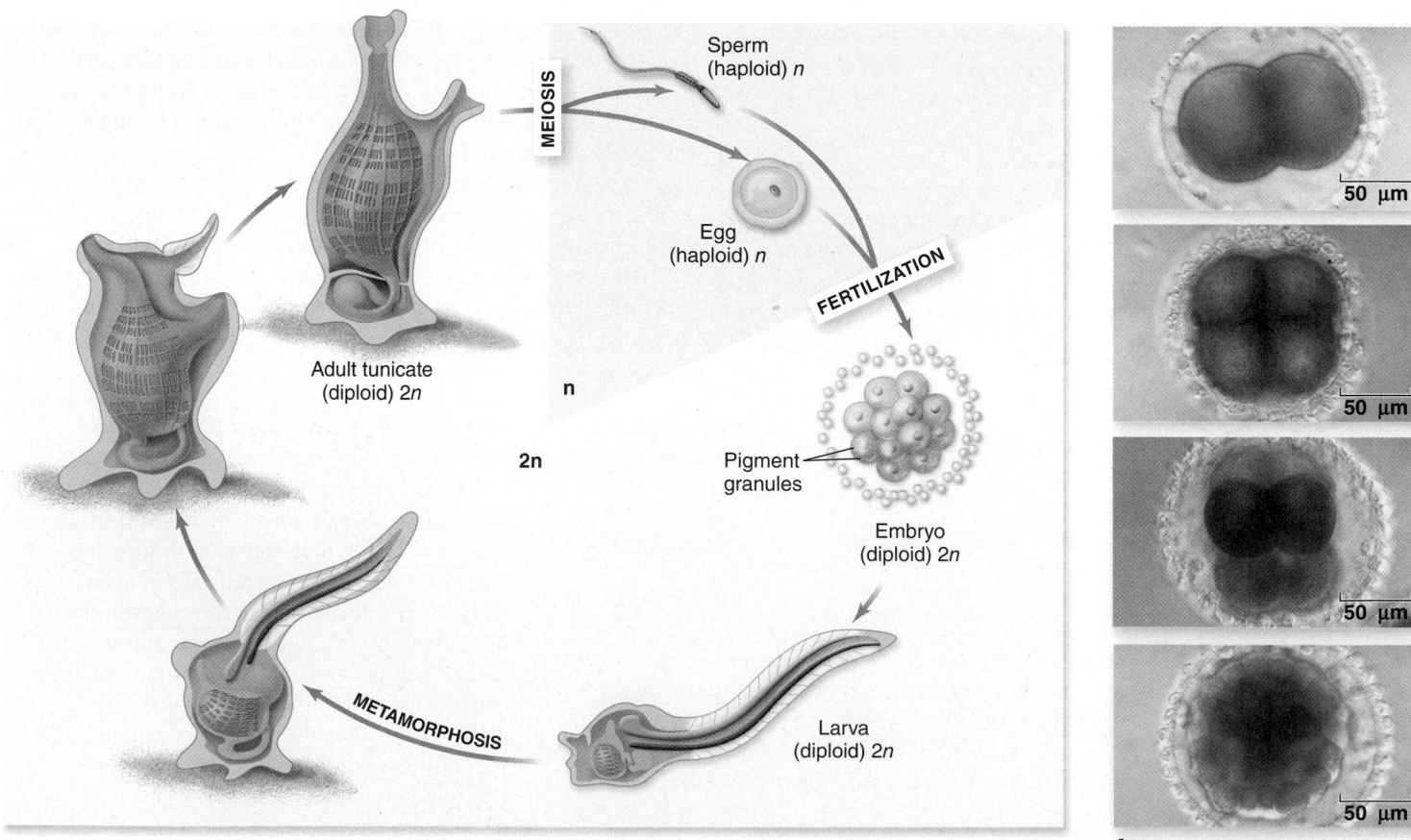

a.

Figure 19.5 Muscle determinants in tunicates. *a.* The life cycle of a solitary tunicate. Muscle cells that move the tail of the swimming tadpole are arranged on either side of the notochord and nerve cord. The tail is lost during metamorphosis into the sedentary adult. *b.* The egg of the tunicate *Styela* contains bright yellow-colored pigment granules. These become asymmetrically localized in the egg following fertilization, and cells that inherit the yellow-colored granules during cleavage will become the larval muscle cells. Embryos at the 2-cell, 4-cell, 8-cell, and 64-cell stages are shown. The tadpole tail will grow out from the lower region of the embryo in the bottom panel.

cytoplasmic determinants. Tunicates are marine invertebrates (see chapter 35), and most adults have simple, saclike bodies that are attached to the underlying substratum. Tunicates are placed in the phylum Chordata, however, due to the characteristics of their swimming, tadpolelike larval stage, which has a dorsal nerve cord and notochord (figure 19.5a). The muscles that move the tail develop on either side of the notochord.

In many tunicate species, yellow-colored pigment granules become asymmetrically localized in the egg following fertilization and subsequently segregate to the tail muscle cell progenitors during cleavage (figure 19.5b). When these pigment granules are shifted experimentally into other cells that normally do not develop into muscle, their fate is changed and they become muscle cells. Thus, the molecules that flip the switch for muscle development appear to be associated with the pigment granules.

The next step is to determine the identity of the molecules involved. Experiments indicate that the female parent provides the egg with mRNA encoded by the *macho-1* gene. The elimination of *macho-1* function leads to a loss of tail muscle in the tadpole, and the misexpression of *macho-1* mRNA leads to the formation of additional (ectopic) muscle cells from nonmuscle lineage cells. The *macho-1* gene product has been shown to be a transcription factor that can activate the expression of several muscle-specific genes.

Induction can lead to cell differentiation

In chapter 9, we examined a variety of ways by which cells communicate with one another. We can demonstrate the importance of cell–cell interactions in development by separating the cells of an early frog embryo and allowing them to develop independently.

Under these conditions, blastomeres from one pole of the embryo (the "animal pole") develop features of ectoderm, and blastomeres from the opposite pole of the embryo (the "vegetal pole") develop features of endoderm. None of the two separated groups of cells ever develop features characteristic of mesoderm, the third main cell type. If animal-pole cells and vegetal-pole cells are placed next to each other, however, some of the animal-pole cells develop as mesoderm. The interaction between the two cell types triggers a switch in the developmental path of these cells. This change in cell fate due to interaction with an adjacent cell is called **induction.** Signaling molecules act to alter gene expression in the target cells, in this case, some of the animal-pole cells.

Another example of inductive cell interactions is the formation of the notochord and mesenchyme, a specific tissue, in tunicate embryos. Muscle, notochord, and mesenchyme all arise from mesodermal cells that form at the vegetal margin of the 32-cell stage embryo. These prospective mesodermal cells receive signals from the underlying endodermal precursor cells that lead to the formation of notochord and mesenchyme (figure 19.6).

The chemical signal is a member of the *fibroblast growth factor (FGF)* family of signaling molecules. It induces the overlying marginal zone cells to differentiate into either notochord (anterior) or mesenchyme (posterior). The FGF receptor on the marginal zone cells is a receptor tyrosine kinase that

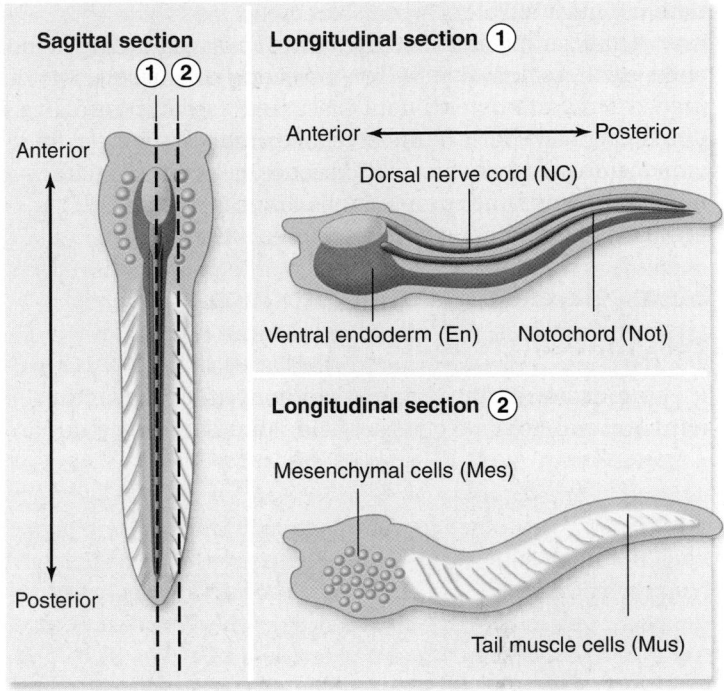

a.

b.   *c.*

Figure 19.6 Inductive interactions contribute to cell fate specification in tunicate embryos. *a.* Internal structures of a tunicate larva. To the left is a sagittal section through the larva with dotted lines indicating two longitudinal sections. Section 1, through the midline of a tadpole, shows the dorsal nerve cord (NC), the underlying notochord (Not) and the ventral endoderm cells (En). Section 2, a more lateral section, shows the mesenchymal cells (Mes) and the tail muscle cells (Mus). *b.* View of the 32-cell stage looking up at the endoderm precursor cells. Fibroblast growth factor (FGF) secreted by these cells is indicated with light-green arrows. Only the surfaces of the marginal cells that directly border the endoderm precursor cells bind FGF signal molecules. Note that the posterior vegetal blastomeres also contain the *macho-1* determinants (red and white stripes). *c.* Cell fates have been fixed by the 64-cell stage. Colors are as in *(a).* Cells on the anterior margin of the endoderm precursor cells become notochord and nerve cord, respectively, whereas cells that border the posterior margin of the endoderm cells become mesenchyme and muscle cells, respectively.

signals through a MAP kinase cascade to activate a transcription factor that turns on gene expression resulting in differentiation (figure 19.7).

This example is also a case of two cells responding differently to the same signal. The presence or absence of the *macho-1* muscle determinant discussed earlier controls this difference in cell fate. In the presence of *macho-1*, cells differentiate into mesenchyme; in its absence, cells differentiate into notochord. Thus, the combination of *macho-1* and FGF signaling leads to four different cell types (see figure 19.7)

Stem cells can divide and produce cells that differentiate

It is important, both during development, and even in the adult animal, to have cells set aside that can divide but are not determined for only a single cell fate. We call cells that are capable of continued division but that can also give rise to differentiated cells, **stem cells.** These cells can be characterized based on the degree to which they have become determined. At one extreme, we call a cell that can give rise to any tissue in an organism **totipotent.** In mammals, the only cells that can give rise to both the embryo and the extraembryonic membranes are the zygote and early blastomeres from the first few cell divisions. Cells that can give rise to all of the cells in the organism's body are called **pluripotent.** A stem cell that can give rise to a limited number of cell types, such as the cells that give rise to the different blood cell types, are called **multipotent.** Then at the other extreme, **unipotent** stem cells give rise to only a single cell type, such as the cells that give rise to sperm cells in males.

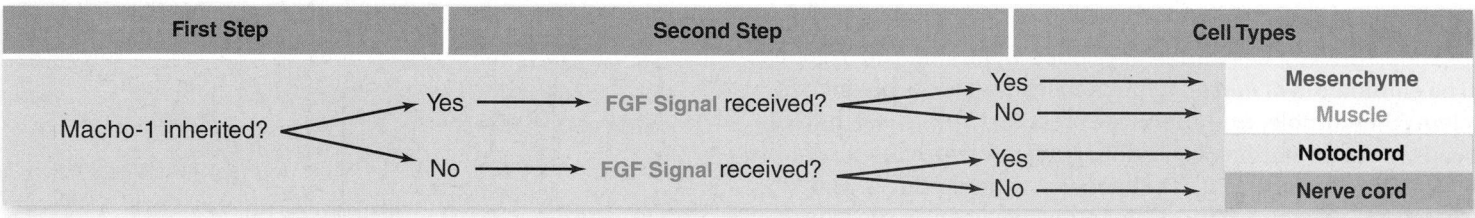

a.

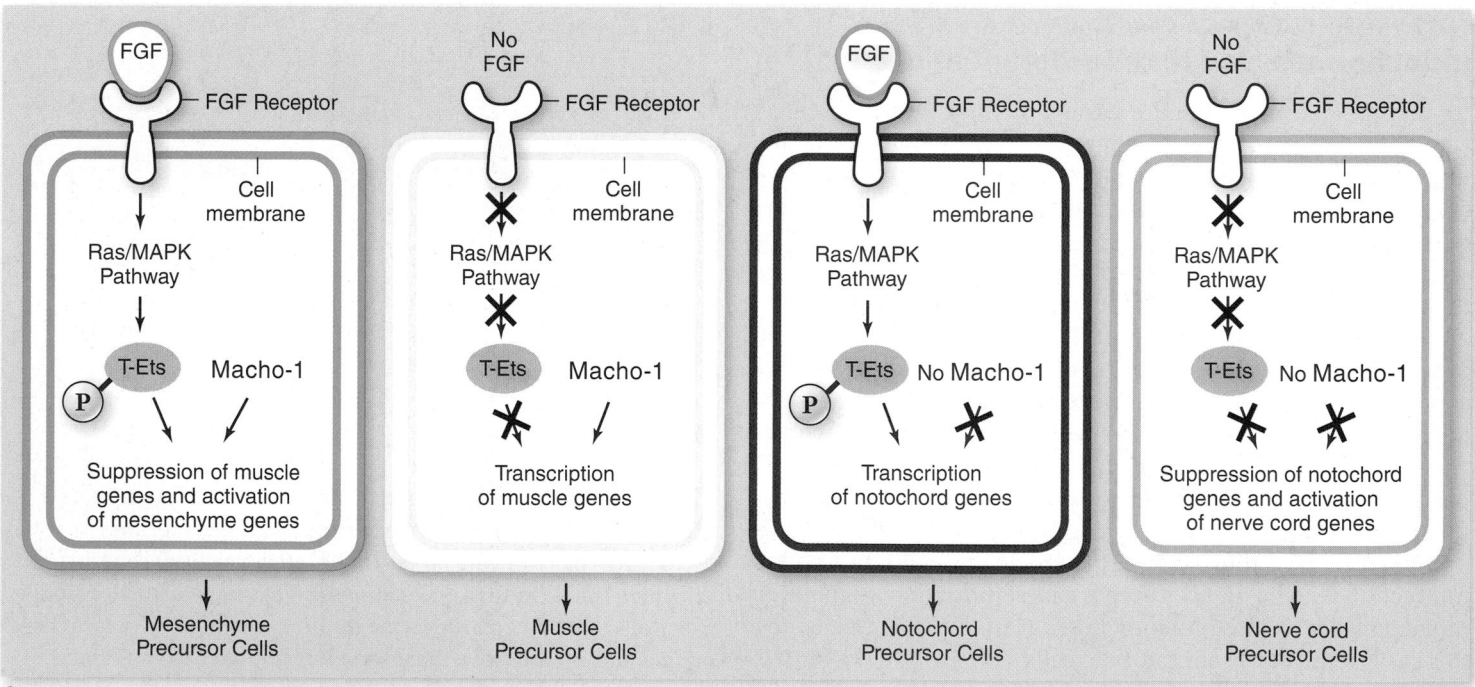

b.

Figure 19.7 **Model for cell fate specification by Macho-1 muscle determinant and FGF signaling.** *a.* Two-step model of cell fate specification in vegetal marginal cells of the tunicate embryo. The first step is inheritance (or not) of muscle *macho-1* mRNA. The second step is FGF signaling from the underlying endoderm precursor cells. *b.* Posterior vegetal margin cells inherit *macho-1* mRNA. Signaling by FGF activates a Ras/MAP kinase pathway that produces the transcription factor T-Ets. Macho-1 protein and T-Ets suppress muscle-specific genes and turn on mesenchyme specific genes *(green cells).* In cells with Macho-1 that do not receive the FGF signal, Macho-1 alone turns on muscle-specific cells *(yellow cells).* Anterior vegetal margin cells do not inherit *macho-1* mRNA. If these cells receive the FGF signal, T-Ets turns on notochord-specific genes *(purple cells).* In cells that lack Macho-1 and FGF, notochord-specific genes are suppressed and nerve cord-specific genes are activated *(gray cells).*

Data analysis What type of cells would develop if you injected embryos with a reagent that blocked the FGF receptor, thus preventing its signaling? What about with a reagent that turned on the FGF receptor, thereby causing it to be always on?

Embryonic stem cells are pluripotent cells derived from embryos

A form of pluripotent stem cells that has been derived in the laboratory are called embryonic stem cells (ES cells). These cells are made from mammalian embryos that have undergone the cleavage stage of development to produce a ball of cells called a blastocyst. The blastocyst consists of an outer ball of cells, the trophectoderm, which will become the placenta, and the inner cell mass that will go on to form the embryo (see chapter 53 for details). Embryonic stem cells can be isolated from the inner cell mass and grown in culture (figure 19.8). In mice, these cells have been studied extensively and have been shown to be able to develop into any type of cell in the tissues of the adult. However, these cells cannot give rise to the extra-embryonic tissues that arise during development, so they are pluripotent, but not totipotent.

Once these cells were found in mice, it was only a matter of time before human ES cells were derived as well. In 1998, the first human ES cells (hES cells) were isolated and grown in culture. While there are differences between human and mouse ES cells, there are also substantial similarities. These embryonic stem cells hold great promise for regenerative medicine based on their potential to produce any cell type as described below. These cells have also been the source of much controversy and ethical discussion due to their embryonic origin.

Differentiation in culture

In addition to their possible therapeutic uses, ES cells offer a way to study the differentiation process in culture. The manipulation of these cells by additions to the culture media will allow us to tease out the factors involved in differentiation at the level of the actual cell undergoing the process. Early attempts at assessing differentiation in culture was plagued by the culture conditions. The medium in early experiments contained fetal calf serum (common in tissue culture), which is ill-defined, and varies lot-to-lot. More recently, more defined culture conditions have been found that allow greater reproducibility in controlling differentiation in culture.

Using more defined media, ES cells have been used to recapitulate in culture the early events in mouse development. Thus mouse ES cells can be used to first give rise to ectoderm, endoderm, and mesoderm, then these three cell types will give rise to the different cells each germ layer is determined to become. This work is in early stages but is tremendously exciting as it offers the promise of understanding the molecular cues that are involved in the stepwise determination of different cell types.

In humans, ES cells have been used to give rise to a variety of cell types in culture. For example, human ES cells have been shown to give rise to different kinds of blood cells in culture. Work is underway to produce hematopoietic stem cells in culture, which could be used to replace such cells in patients with diseases that affect blood cells. Human ES cells have also been used to produce cardiomyocytes in culture. These cells could be used to replace damaged heart tissue after heart attacks.

Learning Outcomes Review 19.3

Cell differentiation is preceded by determination, where the cell becomes committed to a developmental pathway, but has not yet differentiated. Differential inheritance of cytoplasmic factors can cause determination and differentiation, as can interactions between neighboring cells (induction). Inductive changes are mediated by signaling molecules that trigger transduction pathways. Stem cells are able to divide indefinitely, and they can give rise to differentiated cells. Embryonic stem cells are pluripotent cells that can give rise to all adult structures.

■ *How could you distinguish whether a cell becomes determined by induction or because of cytoplasmic factors?*

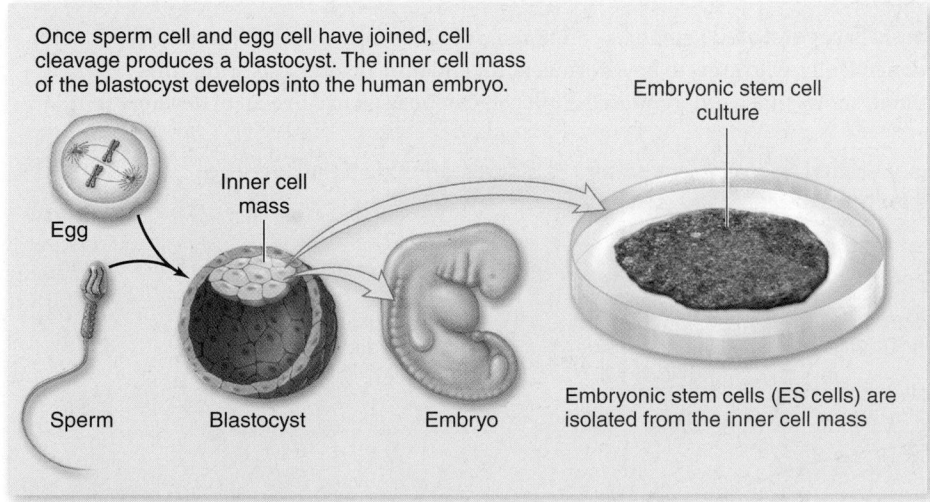

Once sperm cell and egg cell have joined, cell cleavage produces a blastocyst. The inner cell mass of the blastocyst develops into the human embryo.

Egg

Sperm

Inner cell mass

Blastocyst

Embryo

Embryonic stem cell culture

Embryonic stem cells (ES cells) are isolated from the inner cell mass

a.

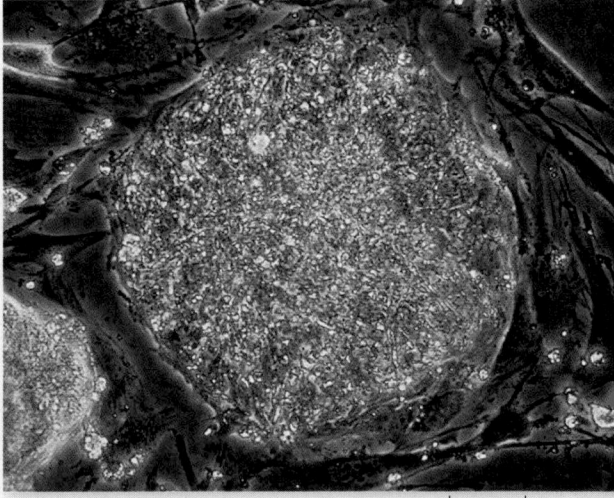

500 μm

b.

Figure 19.8 Isolation of embryonic stem cells. *a.* Early cell divisions lead to the blastocyst stage that consists of an outer layer and an inner cell mass, which will go on to form the embryo. Embryonic stem cells (ES cells) can be isolated from this stage by disrupting the embryo and plating the cells. Stem cells removed from a six-day blastocyst can be established in culture and maintained indefinitely in an undifferentiated state. *b.* Human embryonic stem cells. This mass in the photograph is a colony of undifferentiated human embryonic stem cells being studied in the developmental biologist James Thomson's research lab at the University of Wisconsin–Madison.

19.4 Nuclear Reprogramming

Learning Outcomes

1. Contrast different methods of nuclear reprogramming.
2. Differentiate between reproductive and therapeutic cloning.

The study of the process of determination and differentiation leads quite naturally to questions about whether this process can be reversed. This is of interest both in terms of the experimental possibilities to understand the basic process, and the prospect of creating patient-specific populations of specific cell types to replace cells lost to disease or trauma. This has led to a fascinating path with many twists and turns that has accelerated in the recent past. We will briefly consider the history of this topic, then look at the most recent results available.

Reversal of determination has allowed cloning

Experiments carried out in the 1950s showed that single cells from fully differentiated tissue of an adult plant could develop into entire, mature plants. The cells of an early cleavage stage mammalian embryo are also totipotent. When mammalian embryos naturally split in two, identical twins result. If individual blastomeres are separated from one another, any one of them can produce a completely normal individual. In fact, this type of procedure has been used to produce sets of four or eight identical offspring in the commercial breeding of particularly valuable lines of cattle.

Early research in amphibians

An early question in developmental biology was whether the production of differentiated cells during development involved irreversible changes to cells. Experiments carried out in the 1950s by Robert Briggs and Thomas King, and by John Gurdon in the 1960s and 1970s showed nuclei could be transplanted between cells. Using very fine pipettes (hollow glass tubes), these researchers sucked the nucleus out of a frog or toad egg and replaced the egg nucleus with a nucleus sucked out of a body cell taken from another individual.

The conclusions from these experiments are somewhat contradictory. On the one hand, cells do not appear to undergo any truly irreversible changes, such as loss of genes. On the other hand, the more differentiated the cell type, the less successful the nucleus in directing development when transplanted. This led to the concept of *nuclear reprogramming,* that is, a nucleus from a differentiated cell undergoes **epigenetic** changes that must be reversed to allow the nucleus to direct development. Epigenetic changes do not change a cell's DNA but are stable through cell divisions. The early work on amphibians showed that tadpoles' intestinal cell nuclei could be reprogrammed to produce viable adult frogs. These animals not only can be considered clones, but they show that tadpole nuclei can be completely reprogrammed. However, nuclei from adult differentiated cells could only be reprogrammed to produce tadpoles, but not viable, fertile adults. Thus this work showed that adult nuclei have remarkable developmental potential, but cannot be reprogrammed to be totipotent.

Early research in mammals

Given the work done in amphibians, much effort was put into nuclear transfer in mammals, primarily mice and cattle. Not only did this not result in reproducible production of cloned

Figure 19.9 Proof that determination in animals is reversible. Scientists combined a nucleus from an adult mammary cell with an enucleated egg cell to successfully clone a sheep, named Dolly, who grew to be a normal adult and bore healthy offspring. This experiment, the first successful cloning of an adult animal, shows that a differentiated adult cell can be used to drive all of development.

 Data analysis The sheep used for the donor nucleus had a different pattern of pigmentation than the donor egg. Why is this important, and which animal should Dolly resemble?

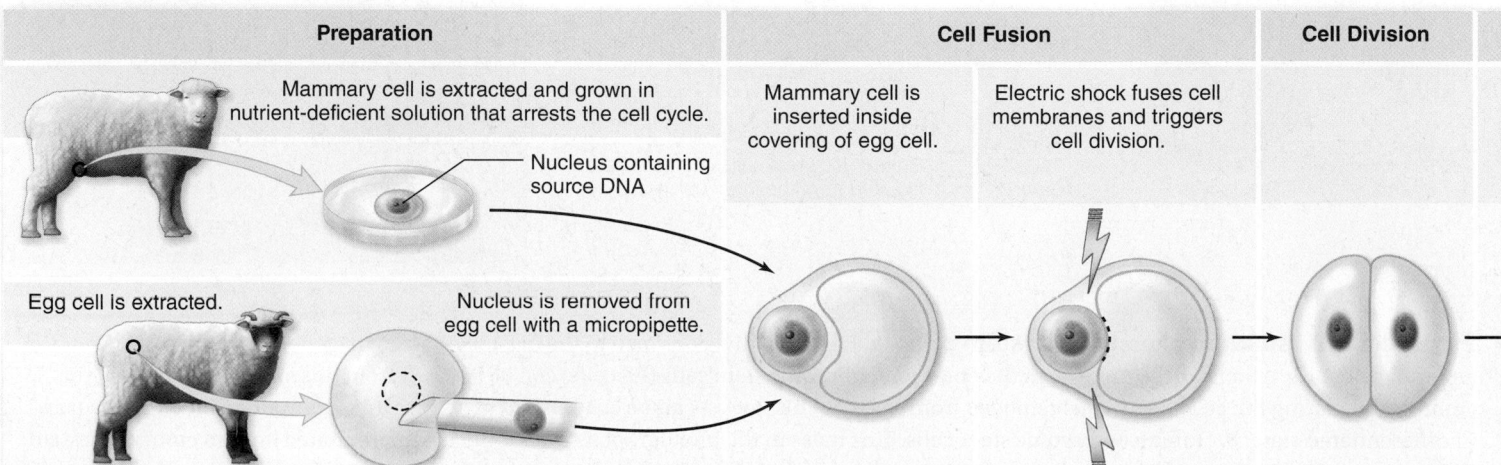

animals, but this work led to the discovery of imprinting through the production of embryos with only maternal or paternal input (see chapter 13 for more information on imprinting). These embryos never developed, and showed different kinds of defects depending on whether the maternal or paternal genome was the sole contributor.

Successful nuclear transplant in mammals

These results stood until a sheep was cloned using the nucleus from a cell of an early embryo in 1984. The key to this success was in picking a donor cell very early in development. This exciting result was soon replicated by others in a host of other organisms, including pigs and monkeys. Only early embryo cells seemed to work, however.

Geneticists at the Roslin Institute in Scotland reasoned that the egg and donated nucleus would need to be at the same stage of the cell cycle for successful development. To test this idea, they performed the following procedure (figure 19.9):

1. They removed differentiated mammary cells from the udder of a six-year-old sheep. The cells were grown in tissue culture, and then the concentration of serum nutrients was substantially reduced for five days, causing them to pause at the beginning of the cell cycle.
2. In parallel preparation, eggs obtained from a ewe were enucleated.
3. Mammary cells and egg cells were surgically combined in a process called **somatic cell nuclear transfer (SCNT)** in January of 1996. Mammary cells and eggs were fused to introduce the mammary nucleus into egg.
4. Twenty-nine of 277 fused couplets developed into embryos, which were then placed into the reproductive tracts of surrogate mothers.
5. A little over five months later, on July 5, 1996, one sheep gave birth to a lamb named Dolly, the first clone generated from a fully differentiated animal cell.

Dolly matured into an adult ewe, and she was able to reproduce the old-fashioned way, producing six lambs. Thus, Dolly established beyond all dispute that determination in animals is reversible—that with the right techniques, the nucleus of a fully differentiated cell *can* be reprogrammed to be totipotent.

Reproductive cloning has inherent problems

The term **reproductive cloning** refers to the process just described, in which scientists use SCNT to create an animal that is genetically identical to another animal. Since Dolly's birth in 1997, scientists have successfully cloned one or more cats, dogs, rabbits, rats, mice, cattle, goats, pigs, and mules. All of these procedures used some form of adult cell.

Low success rate and age-associated diseases

The efficiency in all reproductive cloning is quite low—only 3–5% of adult nuclei transferred to donor eggs result in live births. In addition, many clones that are born usually die soon thereafter of liver failure or infections. Many become oversized, a condition known as *large offspring syndrome (LOS)*. In 2003, three of four cloned piglets developed to adulthood, but all three suddenly died of heart failure at less than 6 months of age.

Dolly herself was euthanized at the relatively young age of six. Although she was put down because of virally induced lung cancer, she had been diagnosed with advanced-stage arthritis a year earlier. Thus, one difficulty in using genetic engineering and cloning to improve livestock is production of enough healthy animals.

Lack of imprinting

The reason for these problems lies in a phenomenon discussed in chapter 13: *genomic imprinting*. Imprinted genes are expressed differently depending on parental origin—that is, they are turned off in either egg or sperm, and this "setting" continues through development into the adult. Normal mammalian development depends on precise genomic imprinting.

The chemical reprogramming of the DNA, which occurs in adult reproductive tissue, takes months for sperm and years for eggs. During cloning, by contrast, the reprogramming of the donor DNA must occur within a few hours. The organization of the chromatin in a somatic cell is also quite different from that in a newly fertilized egg. Significant chromatin remodeling of the transferred donor nucleus must also occur if the cloned embryo is to survive. Cloning fails because there is likely not enough time in these few hours to get the remodeling and reprogramming jobs done properly.

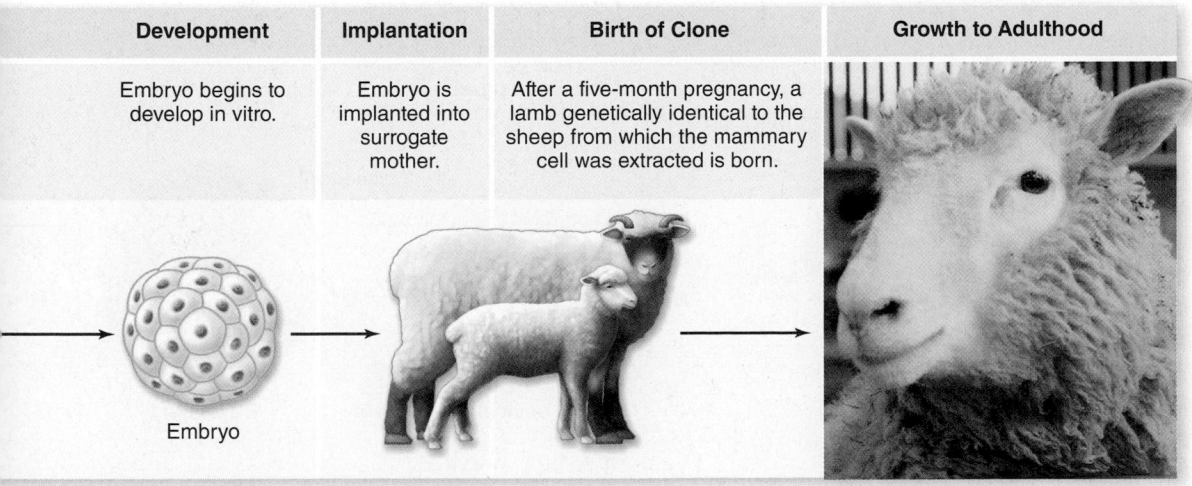

Development	Implantation	Birth of Clone	Growth to Adulthood
Embryo begins to develop in vitro.	Embryo is implanted into surrogate mother.	After a five-month pregnancy, a lamb genetically identical to the sheep from which the mammary cell was extracted is born.	

Embryo

chapter **19** *Cellular Mechanisms of Development* **381**

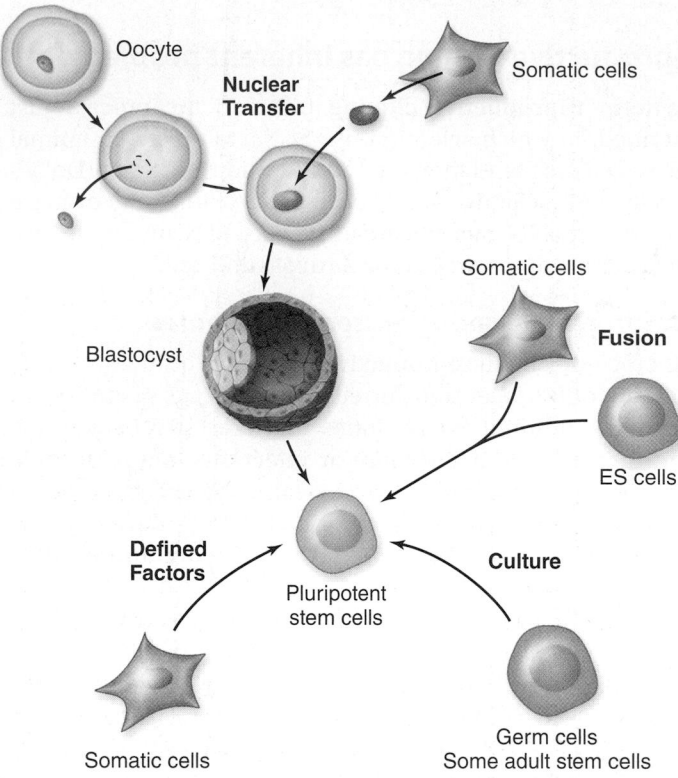

Figure 19.10 Methods to reprogram adult cell nuclei.
Cells taken from adult organisms can be reprogrammed to pluripotent cells in a number of different ways. Nuclei from somatic cells can be transplanted into oocytes as during cloning. Somatic cells can be fused to ES cells created by some other means. Germ cells, and some adult stem cells, after prolonged culture appear to be reprogrammed. Recent work has shown that somatic cells in culture can be reprogrammed by introduction of defined factors.

Nuclear reprogramming has been accomplished by use of defined factors

Stimulated by the discovery of ES cells and success in the reproductive cloning of mammals, work turned to finding ways to reprogram adult cells to pluripotency without the use of embryos (figure 19.10). One approach was to fuse an ES cell to a differentiated cell. These fusion experiments showed that the nucleus of the differentiated cell could be reprogrammed by exposure to ES cell cytoplasm. Of course, the resulting cells are tetraploid (four copies of the genome), which limits their experimental and practical utility. Another line of research showed that primordial germ cells explanted into culture can give rise to cells that act similar to ES cells after extended time in culture.

All of these different lines of inquiry showed that reprogramming of somatic nuclei was possible. Investigations into the characteristics of pluripotency identified a set of transcription factors that were active in ES cells. Then in 2006 it was shown that introducing genes that encode four of these transcription factors, Oct4, Sox2, c-Myc, and Klf4 could reprogram fibroblast cells in culture. Following introduction of the transcription factors genes, cells were selected that express a target gene regulated by Oct4 and Sox2, and these cells appeared to be pluripotent. These were named induced pluripotent stem cells, or iPS cells.

The protocol has been refined by selection for another target gene known to be critical to the pluripotent state: *Nanog*. These *Nanog*-expressing iPS cells appear to be similar to ES cells in terms of developmental potential, as well as gene expression pattern. There is some indication that their chromatin structure, and thus their epigenetic state, may not be the same as that of ES cells.

It is worth asking what this work has taught us about the pluripotent state and the differentiated state. It is becoming clear that reprogramming is a multistep process. When this is done in culture, only a subset of cells make each step, thus explaining why the entire process is inefficient. Starting from a fibroblast, cells first change shape, becoming more spherical, and divide more rapidly. They then reverse part of their developmental program, becoming more like epithelial cells, a so-called mesenchyme-to-epithelial transition.

Lastly, the stable expression of the core pluripotency regulatory factors Oct4, Sox2, and Nanog is established. The pluripotent state is maintained by a combination of transcription factors and chromatin structure (epigenetic changes).

This technology has now been used to construct ES cells from patients with the inherited neurological disorder spinal muscular atrophy. These ES cells differentiate in culture into motor neurons that show the phenotype of the disease. The ability to derive

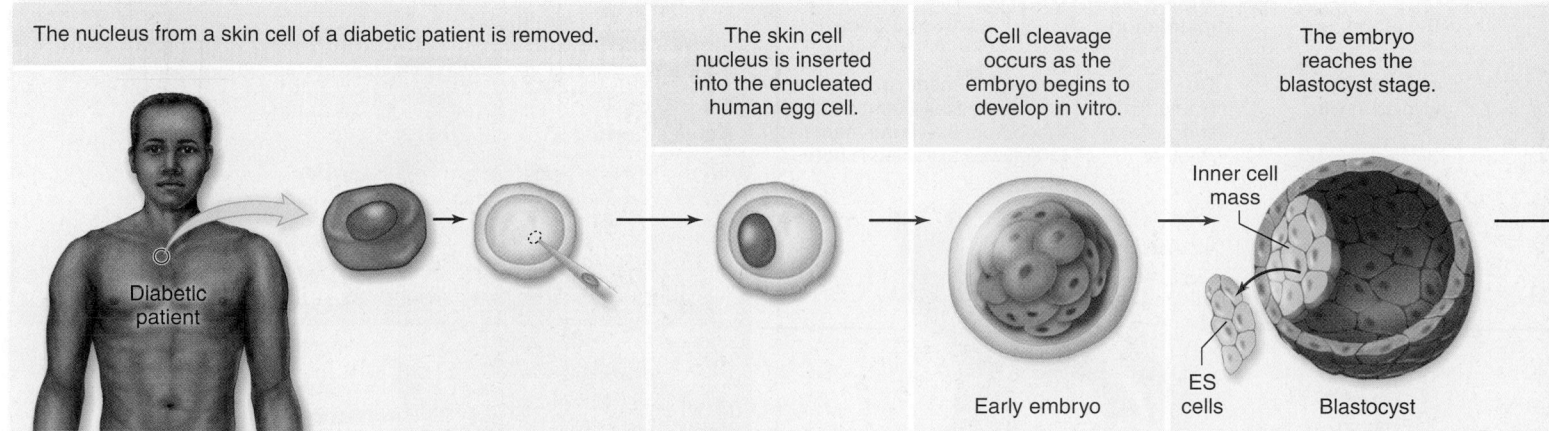

The nucleus from a skin cell of a diabetic patient is removed.

The skin cell nucleus is inserted into the enucleated human egg cell.

Cell cleavage occurs as the embryo begins to develop in vitro.

The embryo reaches the blastocyst stage.

Diabetic patient

Early embryo

Inner cell mass

ES cells

Blastocyst

disease-specific stem cells is an incredible advance for research on such diseases. This will allow us to study the cells affected by genetic diseases and to screen for possible therapeutics.

Pluripotent cell types themselves have potential for therapeutic applications. One way to solve the problem of graft rejection, such as in skin grafts in severe burn cases, is to produce patient-specific lines of ES cells. The first method to accomplish this was called **therapeutic cloning** and it uses the same SCNT procedure that created Dolly to assemble an embryo. The nucleus is removed from a skin cell and inserted into an egg whose nucleus has already been removed. The egg with its skin cell nucleus is allowed to form a blastocyst-stage embryo. This artificial embryo is then used to derive ES cells for transfer to injured tissue (figure 19.11).

Therapeutic cloning successfully addresses one key problem that must be solved before stem cells can be used to repair human tissues damaged by heart attack, nerve injury, diabetes, or Parkinson disease—the problem of immune acceptance. Since stem cells are cloned from a person's own tissues, they pass the immune system's "self" identity check, and the body readily accepts them. The first human trials using this technology were recently halted by the company Geron, reportedly for financial reasons. There is still great interest in this technology and the use of iPS cells would remove the ethical problems of embryo destruction, and the practical problem of the requirement for oocytes.

Learning Outcomes Review 19.4

Cloning has long been practiced in plants. In animals, cells from early-stage embryos are also totipotent, but attempts to use adult nuclei for cloning led to mixed results. The nucleus of a differentiated cell requires reprogramming to be totipotent. This appears to be necessary at least in part because of genomic imprinting. Nuclei may be reprogrammed by fusion with an embryonic stem cell, which produces a tetraploid cell, or through the introduction of four important transcription factors. That reprogramming is possible was shown by reproductive cloning via somatic cell nuclear transfer (SCNT). In therapeutic cloning, the goal is to produce replacement tissue using a patient's own cells.

■ **What changes must occur to produce a totipotent cell from a differentiated nucleus?**

19.5 Pattern Formation

Learning Outcomes

1. Describe A/P axis formation in Drosophila.
2. Describe D/V axis formation in Drosophila.
3. Explain the importance of homeobox-containing genes in development.

For cells in multicellular organisms to differentiate into appropriate cell types, they must gain information about their relative locations in the body. All multicellular organisms seem to use positional information to determine the basic pattern of body compartments and, thus, the overall architecture of the adult body. This positional information then leads to intrinsic changes in gene activity, so that cells ultimately adopt a fate appropriate for their location.

Pattern formation is an unfolding process. In the later stages, it may involve morphogenesis of organs (to be discussed later), but during the earliest events of development, the basic body plan is laid down, along with the establishment of the anterior–posterior (A/P, head-to-tail) axis and the dorsal–ventral (D/V, back-to-front) axis. Thus, pattern formation can be considered the process of taking a radially symmetrical cell and imposing two perpendicular axes to define the basic body plan, which in this way becomes bilaterally symmetrical. Developmental biologists use the term **polarity** to refer to the acquisition of axial differences in developing structures.

The fruit fly *Drosophila melanogaster* is the best understood animal in terms of the genetic control of early patterning. We will concentrate on the *Drosophila* system here, and later in chapter 53 we will examine axis formation in vertebrates in the context of their overall development.

A hierarchy of gene expression that begins with maternally expressed genes controls the development of *Drosophila*. To understand the details of these gene interactions, we first need to briefly review the stages of *Drosophila* development.

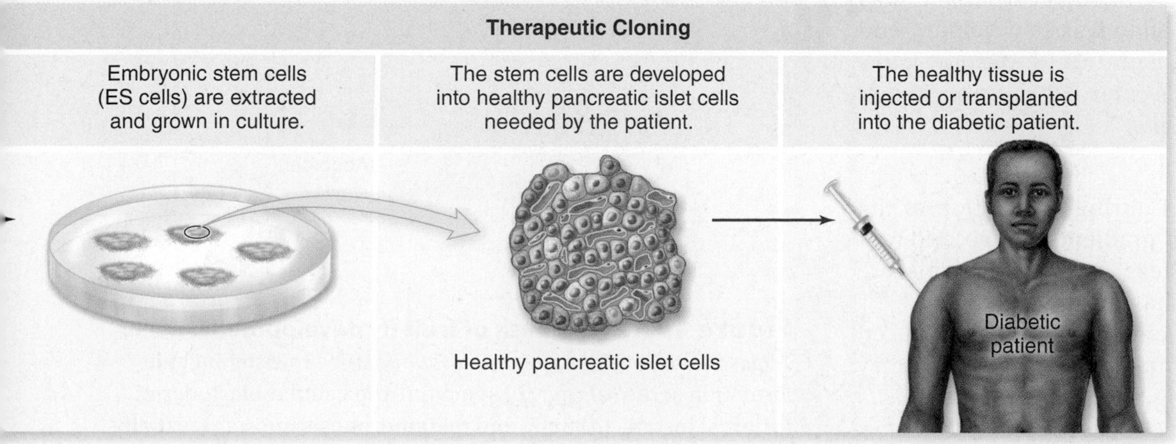

Therapeutic Cloning

Embryonic stem cells (ES cells) are extracted and grown in culture.

The stem cells are developed into healthy pancreatic islet cells needed by the patient.

Healthy pancreatic islet cells

The healthy tissue is injected or transplanted into the diabetic patient.

Diabetic patient

Figure 19.11 How human embryos might be used for therapeutic cloning. In therapeutic cloning, after initial stages to reproductive cloning, the embryo is broken apart and its embryonic stem cells are extracted. These are grown in culture and used to replace the diseased tissue of the individual who provided the DNA. This is useful only if the disease in question is not genetic, as the stem cells are genetically identical to the patient.

Drosophila embryogenesis produces a segmented larva

Drosophila and many other insects produce two different kinds of bodies during their development: the first, a tubular eating machine called a **larva,** and the second, an adult flying sex machine with legs and wings. The passage from one body form to the other, called **metamorphosis,** involves a radical shift in development (figure 19.12). In this chapter, we concentrate on the process of going from a fertilized egg to a larva, which is termed *embryogenesis.*

Prefertilization maternal contribution

The development of an insect like *Drosophila* begins before fertilization, with the construction of the egg. Specialized *nurse cells* that help the egg grow move some of their own maternally encoded mRNAs into the maturing oocyte (figure 19.12*a*).

Following fertilization, the maternal mRNAs are transcribed into proteins, which initiate a cascade of sequential gene activations. Embryonic nuclei do not begin to function (that is, to direct new transcription of genes) until approximately 10 nuclear divisions have occurred. Therefore, the action of maternal, rather than zygotic, genes determines the initial course of *Drosophila* development.

Postfertilization events

After fertilization, 12 rounds of nuclear division without cytokinesis produce about 4000 nuclei, all within a single cytoplasm. All of the nuclei within this **syncytial blastoderm** (figure 19.12*b*) can freely communicate with one another, but nuclei located in different sectors of the egg encounter different maternal products.

Once the nuclei have spaced themselves evenly along the surface of the blastoderm, membranes grow between them to form the **cellular blastoderm.** Embryonic folding and primary tissue development soon follow, in a process fundamentally similar to that seen in vertebrate development. Within a day of fertilization, embryogenesis creates a segmented, tubular body—which is destined to hatch out of the protective coats of the egg as a larva.

Morphogen gradients form the basic body axes in *Drosophila*

Pattern formation in the early *Drosophila* embryo requires positional information encoded in labels that can be read by cells. The unraveling of this puzzle, work that earned the 1995 Nobel Prize for researchers Christiane Nüsslein-Volhard and Eric Wieschaus, is summarized in figure 19.13. We now know that two different genetic pathways control the establishment of A/P and D/V polarity in *Drosophila.*

Anterior–posterior axis

Formation of the A/P axis begins during maturation of the oocyte and is based on opposing gradients of two different proteins: **Bicoid** and **Nanos.** These protein gradients are established by an interesting mechanism.

Nurse cells in the ovary secrete maternally produced *bicoid* and *nanos* mRNAs into the maturing oocyte where they are differentially transported along microtubules to opposite poles of the oocyte (figure 19.14*a*). This differential transport comes about due to the use of different motor proteins to move

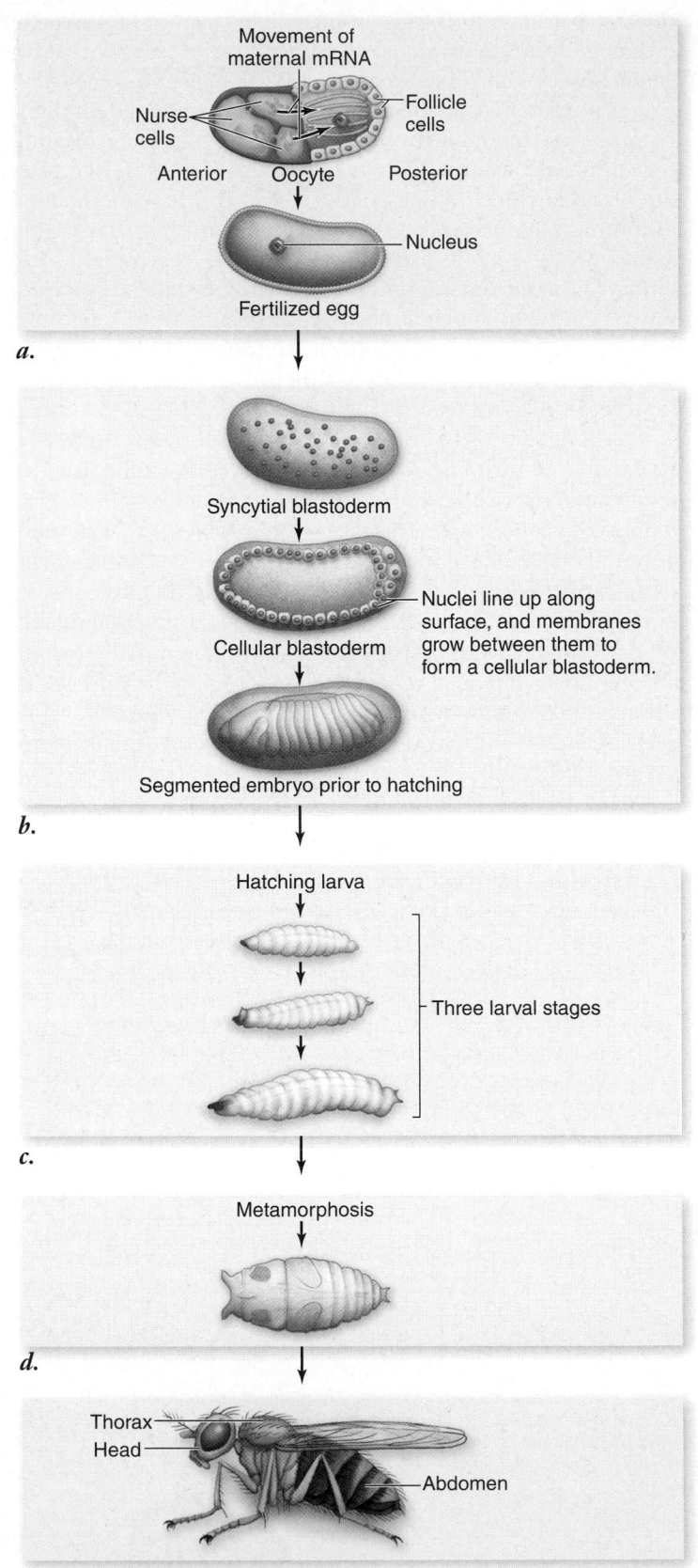

Figure 19.12 The path of fruit fly development. Major stages in the development of *Drosophila melanogaster* include formation of the (*a*) egg, (*b*) syncytial and cellular blastoderm, (*c*) larval instars, (*d*) pupa and metamorphosis into a (*e*) sexually mature adult.

Establishing the Polarity of the Embryo

Fertilization of the egg triggers the production of Bicoid protein from maternal RNA in the egg. The Bicoid protein diffuses through the egg, forming a gradient. This gradient determines the polarity of the embryo, with the head and thorax developing in the zone of high concentration (*green* fluorescent dye in antibodies that bind bicoid protein allows visualization of the gradient).

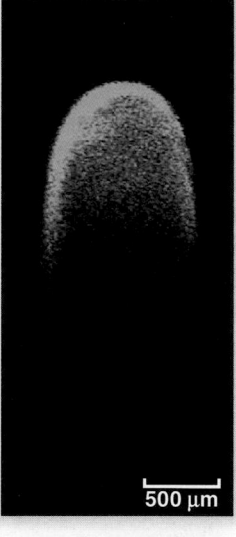

500 μm

Setting the Stage for Segmentation

About 2 1/2 hours after fertilization, Bicoid protein turns on a series of brief signals from so-called gap genes. The gap proteins act to divide the embryo into large blocks. In this photo, fluorescent dyes in antibodies that bind to the gap proteins Krüppel (*orange*) and Hunchback (*green*) make the blocks visible; the region of overlap is yellow.

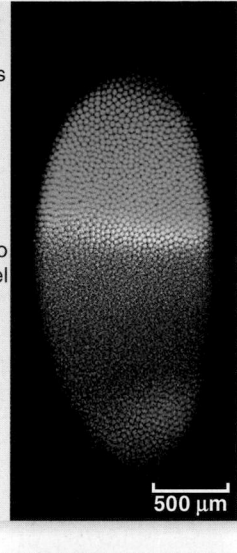

500 μm

Figure 19.13 Body organization in an early *Drosophila* embryo. In these fluorescent microscope images by 1995 Nobel laureate Christiane Nüsslein-Volhard and Sean Carroll, we watch a *Drosophila* egg pass through the early stages of development, in which the basic segmentation pattern of the embryo is established. The proteins in the photographs were made visible by binding fluorescent antibodies to each specific protein.

Laying Down the Fundamental Regions

About 0.5 hr later, the gap genes switch on the "pair-rule" genes, which are each expressed in seven stripes. This is shown for the pair-rule gene *hairy*. Some pair-rule genes are only required for even-numbered segments while others are only required for odd numbered segments.

500 μm

Forming the Segments

The final stage of segmentation occurs when a "segment-polarity" gene called *engrailed* divides each of the seven regions into anterior and posterior compartments of the future segments. This occurs a little after the formation of the cellular blastoderm (see figure 19.12). The curved appearance of the embryo at this stage is because of a phenomenon called germ band extension that causes the embryo to fold over itself.

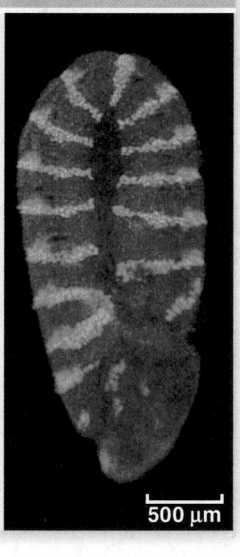

500 μm

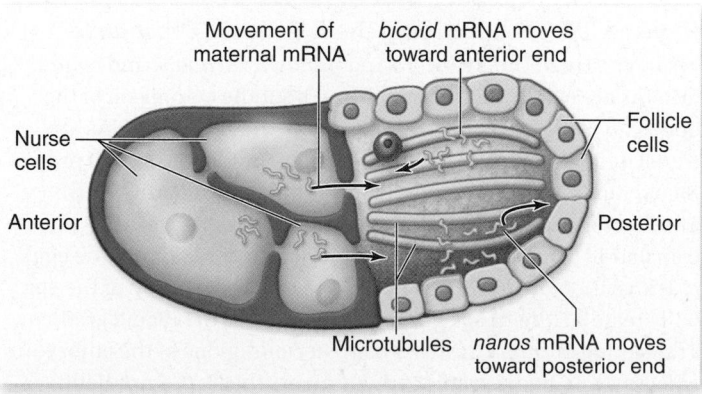

a.

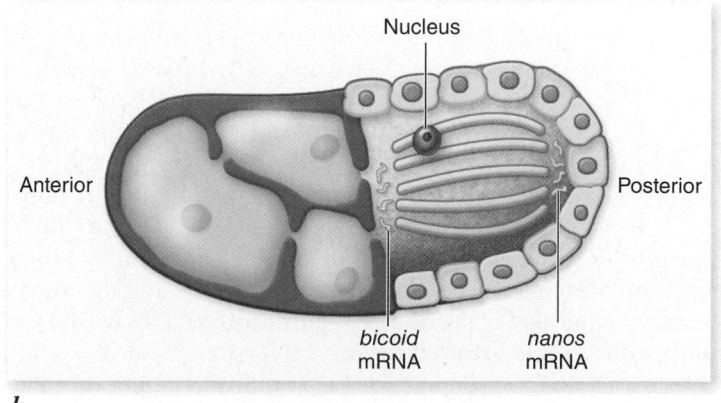

b.

Figure 19.14 Specifying the A/P axis in *Drosophila* embryos I. *a.* In the ovary, nurse cells secrete maternal mRNAs into the cytoplasm of the oocyte. Clusters of microtubules direct oocyte growth and maturation. Motor proteins travel along the microtubules transporting molecules in two directions. *Bicoid* mRNAs are transported toward the anterior pole of the oocyte, *nanos* mRNA is transported toward the posterior pole of the oocyte. *b.* A mature oocyte, showing localization of *bicoid* mRNAs to the anterior pole and *nanos* mRNAs to the posterior pole.

the two mRNAs. The *bicoid* mRNA then becomes anchored in the cytoplasm at the end of the oocyte closest to the nurse cells, and this end will develop into the anterior end of the embryo. *Nanos* mRNA becomes anchored to the opposite end of the oocyte, which will become the posterior end of the embryo. Thus, by the end of oogenesis, the *bicoid* and *nanos* mRNAs are already set to function as cytoplasmic determinants in the fertilized egg (figure 19.14*b*).

Following fertilization, translation of the anchored mRNA and diffusion of the proteins away from their respective sites of synthesis create opposing gradients of each protein: Highest levels of Bicoid protein are at the anterior pole of the embryo (figure 19.15*a*), and highest levels of the Nanos protein are at the posterior pole. Concentration gradients of soluble molecules can specify different cell fates along an axis, and proteins that act in this way, like Bicoid and Nanos, are called **morphogens.** The importance of these morphogens can be seen by the effects of loss-of-function mutants: loss of Bicoid produces an embryo with only posterior sides, and loss of Nanos protein produces an embryo with only anterior sides.

The Bicoid and Nanos proteins control the translation of two other maternal messages, *hunchback* and *caudal*, that encode transcription factors. **Hunchback** activates genes required for the formation of anterior structures, and **Caudal** activates genes required for the development of posterior (abdominal) structures. The *hunchback* and *caudal* mRNAs are evenly distributed across the egg (figure 19.15*b*), so how is it that proteins translated from these mRNAs become localized?

The answer is that Bicoid protein binds to and inhibits translation of *caudal* mRNA. Therefore, *caudal* is only translated in the posterior regions of the egg where Bicoid is absent. Similarly, Nanos protein binds to and prevents translation of the *hunchback* mRNA. As a result, *hunchback* is only translated in the anterior regions of the egg (figure 19.15*c*). Thus, shortly after fertilization, four protein gradients exist in the embryo: anterior–posterior gradients of Bicoid and Hunchback proteins, and posterior–anterior gradients of Nanos and Caudal proteins (figure 19.15*c*).

Dorsal–ventral axis

The dorsal–ventral axis in *Drosophila* is established by actions of the *dorsal* gene product. Once again the process begins in the ovary, when maternal transcripts of the *dorsal* gene are put into the oocyte. However, unlike *bicoid* or *nanos,* the *dorsal* mRNA does not become asymmetrically localized. Instead, a series of steps are required for Dorsal to carry out its function.

First, the oocyte nucleus, which is located to one side of the oocyte, synthesizes *gurken* mRNA. The *gurken* mRNA then accumulates in a crescent between the nucleus and the membrane on that side of the oocyte (figure 19.16*a*). This will be the future dorsal side of the embryo.

The Gurken protein is a soluble cell-signaling molecule, and when it is translated and released from the oocyte, it binds to receptors in the membranes of the overlying follicle cells (figure 19.16*b*). These cells then differentiate into a dorsal morphology. Meanwhile, no Gurken signal is released from the other side of the oocyte, and the follicle cells on that side of the oocyte adopt a ventral fate.

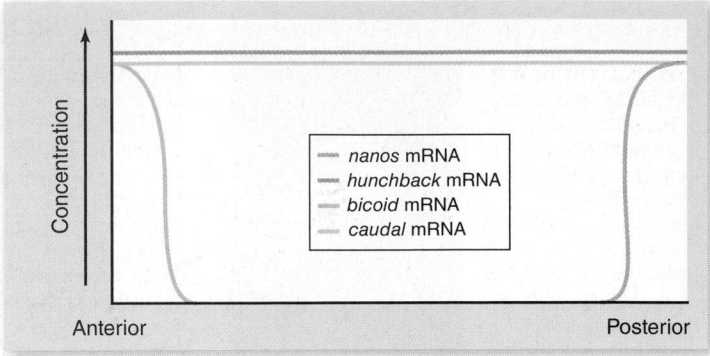

a. Oocyte mRNAs

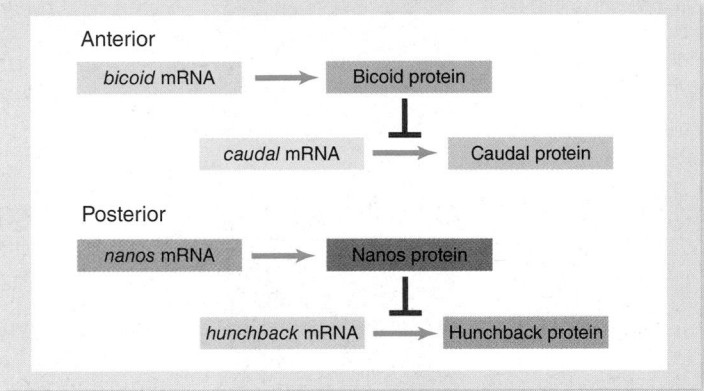

b. After fertilization

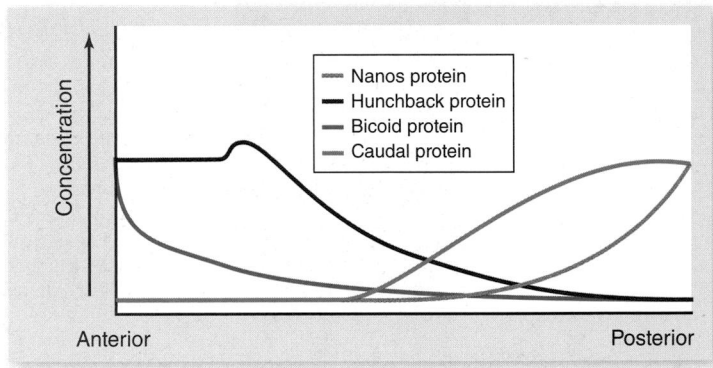

c. Early cleavage embryo proteins

Figure 19.15 Specifying the A/P axis in *Drosophila* embryos II. *a.* Unlike *bicoid* and *nanos, hunchback* and *caudal* mRNAs are evenly distributed throughout the cytoplasm of the oocyte. *b.* Following fertilization, *bicoid* and *nanos* mRNAs are translated into protein, making opposing gradients of each protein. Bicoid binds to and represses translation of *caudal* mRNAs (in anterior regions of the egg). Nanos binds to and represses translation of *hunchback* mRNAs (in posterior regions of the egg). *c.* Translation of *hunchback* mRNAs in anterior regions of the egg will create a Hunchback gradient that mirrors the Bicoid gradient. Translation of *caudal* mRNAs in posterior regions of the embryo will create a Caudal gradient that mirrors the Nanos gradient.

Following fertilization, a signaling molecule is differentially activated on the ventral surface of the embryo in a complex sequence of steps. This signaling molecule then binds to a

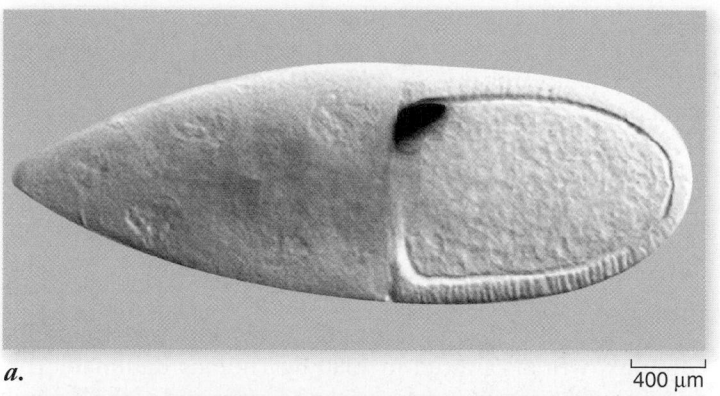

a.   400 μm

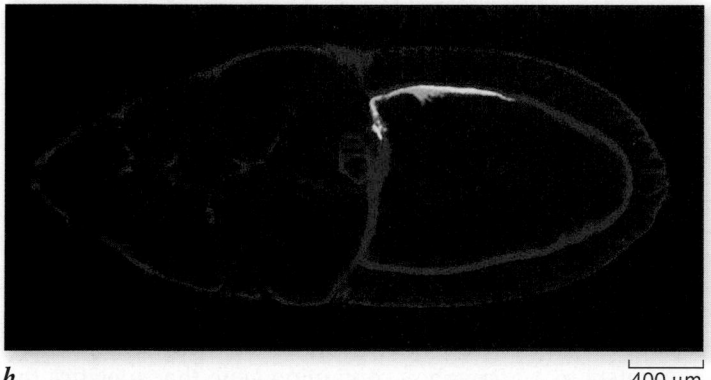

b.   400 μm

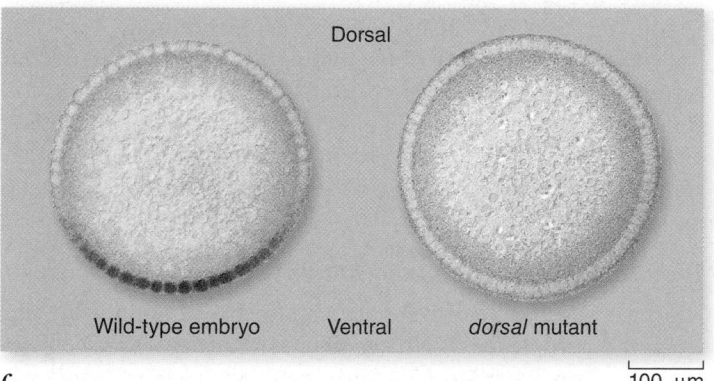

Dorsal

Wild-type embryo   Ventral   *dorsal* mutant

c.   100 μm

Figure 19.16 Specifying the D/V axis in *Drosophila* embryos. *a.* The *gurken* mRNA *(dark stain)* is concentrated between the oocyte nucleus (not visible) and the dorsal, anterior surface of the oocyte. *b.* In a more mature oocyte, Gurken protein *(yellow stain)* is secreted from the dorsal anterior surface of the oocyte, forming a gradient along the dorsal surface of the egg. Gurken then binds to membrane receptors in the overlying follicle cells. Double staining for actin *(red)* shows the cell boundaries of the oocyte, nurse cells, and follicle cells. *c.* For these images, cellular blastoderm stage embryos were cut in cross section to visualize the nuclei of cells around the perimeter of the embryos. Dorsal protein *(dark stain)* is localized in nuclei on the ventral surface of the blastoderm in a wild-type embryo *(left).* The *dorsal* mutant on the right will not form ventral structures, and Dorsal is not present in ventral nuclei of this embryo.

membrane receptor in the ventral cells of the embryo and activates a signal transduction pathway in those cells. Activation of this pathway results in the selected transport of the Dorsal

protein (which is everywhere) into ventral nuclei, forming a gradient along the D/V axis. The Dorsal protein levels are highest in the nuclei of ventral cells (figure 19.16*c*).

(Note that many *Drosophila* genes are named for the mutant phenotype that results from a loss of function in that gene. A lack of *dorsal* function produces dorsalized embryos with no ventral structures.)

The Dorsal protein is a transcription factor, and once it is transported into nuclei, it activates genes required for the proper development of ventral structures, simultaneously repressing genes that specify dorsal structures. Hence, the product of the *dorsal* gene ultimately directs the development of ventral structures.

Although profoundly different mechanisms are involved, the unifying factor controlling the establishment of both A/P and D/V polarity in *Drosophila* is that *bicoid, nanos, gurken,* and *dorsal* are all maternally expressed genes. The polarity of the future embryo in both instances is therefore laid down in the oocyte using information coming from the maternal genome.

The preceding discussion simplifies events, but the outline is clear: Polarity is established by the creation of morphogen gradients in the embryo based on maternal information in the egg. These gradients then drive the expression of the zygotic genes that will actually pattern the embryo. This reliance on a hierarchy of regulatory genes is a unifying theme for all of development.

The body plan is produced by sequential activation of genes

Let us now return to the process of pattern formation in *Drosophila* along the A/P axis. Determination of structures is accomplished by the sequential activation of three classes of **segmentation genes.** These genes create the hallmark segmented body plan of a fly, which consists of three fused head segments, three thoracic segments, and eight abdominal segments (see figure 19.12*e*).

To begin, Bicoid protein exerts its profound effect on the organization of the embryo by activating the translation and transcription of *hunchback* mRNA (which is the first mRNA to be transcribed after fertilization). *Hunchback* is a member of a group of nine genes called the **gap genes.** These genes map out the initial subdivision of the embryo along the A/P axis (see figure 19.13).

All of the gap genes encode transcription factors, which, in turn, regulate the expression of eight or more **pair-rule genes.** Each of the pair-rule genes, such as *hairy,* produces seven distinct bands of protein, which appear as stripes when visualized with fluorescent reagents (see figure 19.13). These bands subdivide the broad gap regions and establish boundaries that divide the embryo into seven zones. When mutated, each of the pair-rule genes alters every other body segment.

All of the pair-rule genes also encode transcription factors, and they, in turn, regulate the expression of each other and of a group of nine or more **segment polarity genes.** The segment polarity genes are each expressed in 14 distinct bands of cells, which subdivide each of the seven zones specified by the pair-rule genes (see figure 19.13). The *engrailed* gene, for example, divides each of the seven zones established by *hairy* into anterior and posterior compartments. The segment polarity genes encode proteins that function in cell–cell signaling pathways. Thus, they function in inductive events—which occur *after* the syncytial blastoderm is divided into cells—to fix the anterior and posterior fates of cells within each segment.

In summary, within 3 hr after fertilization, a highly orchestrated cascade of segmentation gene activity transforms the broad gradients of the early embryo into a periodic, segmented structure with A/P and D/V polarity. The activation of the segmentation genes depends on the free diffusion of maternally encoded morphogens, which is only possible within the syncytial blastoderm of the early *Drosophila* embryo.

Segment identity arises from the action of homeotic genes

With the basic body plan laid down, the next step is to give identity to the segments of the embryo. A highly interesting class of *Drosophila* mutants has provided the starting point for understanding the creation of segment identity.

In these mutants, a particular segment seems to have changed its identity—that is, it has characteristics of a different segment. In wild-type flies, a pair of legs emerges from each of the three thoracic segments, but only the second thoracic segment has wings. Mutations in the *Ultrabithorax* gene cause a fly to grow an extra pair of wings, as though it has two second thoracic segments (figure 19.17). Even more bizarre are mutations in *Antennapedia,* which cause legs to grow out of the head in place of antennae!

Thus, mutations in these genes lead to the appearance of perfectly normal body parts in inappropriate places. Such mutants are termed *homeotic mutants* because the transformed body part looks similar (homeotic) to another. The genes in which such mutants occur are therefore called **homeotic genes.**

Homeotic gene complexes

In the early 1950s, geneticist and Nobel laureate Edward Lewis discovered that several homeotic genes, including *Ultrabithorax,* map together on the third chromosome of *Drosophila* in a tight cluster called the **bithorax complex.** Mutations in these genes all affect body parts of the thoracic and abdominal segments, and Lewis concluded that the genes of the bithorax complex control the development of body parts in the rear half of the thorax and all of the abdomen.

Interestingly, the order of the genes in the bithorax complex mirrors the order of the body parts they control, as though the genes are activated serially. Genes at the beginning of the cluster switch on development of the thorax; those in the middle control the anterior part of the abdomen; and those at the end affect the posterior tip of the abdomen.

A second cluster of homeotic genes, the **Antennapedia complex,** was discovered in 1980 by Thomas Kaufman. The Antennapedia complex governs the anterior end of the fly, and the order of genes in this complex also corresponds to the order of segments they control (figure 19.18*a*).

The homeobox

An interesting relationship was discovered after the genes of the bithorax and Antennapedia complexes were cloned and sequenced. These genes all contain a conserved sequence of 180 nucleotides that codes for a 60-amino-acid, DNA-binding domain. Because this domain was found in all of the homeotic genes, it was named the *homeodomain,* and the DNA that encodes it is called the homeobox. Thus, the term **Hox gene** now refers to a homeobox-containing gene that specifies the identity of a body part. These genes function as transcription factors that bind DNA using their homeobox domain.

Clearly, the homeobox distinguishes portions of the genome that are devoted to pattern formation. How the *Hox* genes do this is the subject of much current research. Scientists believe that the ultimate targets of *Hox* gene function must be genes that control cell behaviors associated with organ morphogenesis.

Evolution of homeobox-containing genes

A large amount of research has been devoted to analyzing the clustered complexes of *Hox* genes in other organisms. These investigations have led to a fairly coherent view of homeotic gene evolution.

It is now clear that the *Drosophila* bithorax and Antennapedia complexes represent two parts of a single cluster of genes. In vertebrates, there are four copies of *Hox* gene clusters. As in *Drosophila,* the spatial domains of *Hox* gene expression correlate with the order of the genes on the chromosome (figure 19.18*b*). The existence of four *Hox* clusters in vertebrates is viewed by many as evidence that two duplication events of the entire genome have occurred in the vertebrate lineage.

This idea raises the issue of when the original cluster arose. To answer this question, researchers have turned to more primitive organisms, such as *Amphioxus* (now called *Branchiostoma*), a lancelet chordate (see chapter 35). The finding of only one cluster of *Hox* genes in *Amphioxus* implies that indeed there

Figure 19.17 Mutations in homeotic genes. Three separate mutations in the bithorax complex caused this fruit fly to develop an additional second thoracic segment, with accompanying wings.

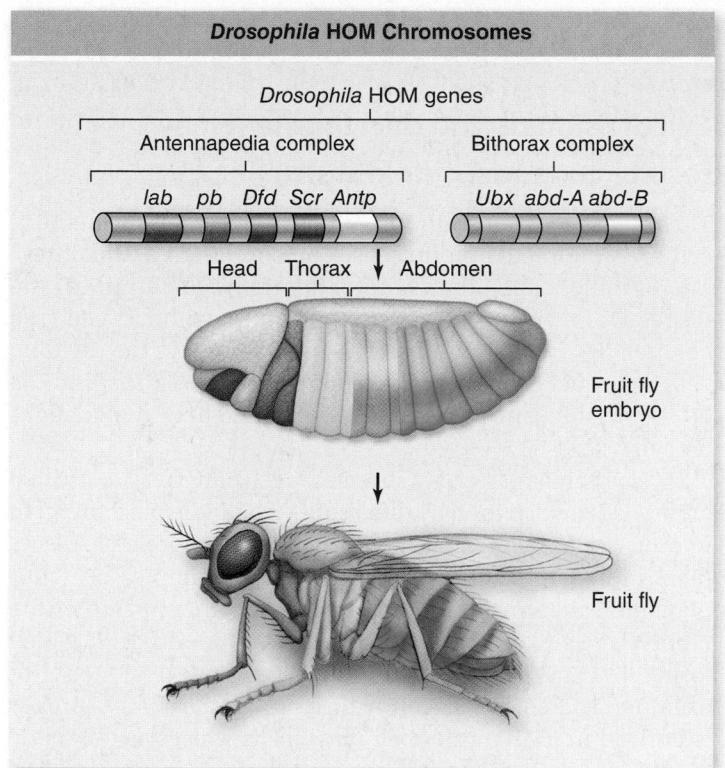

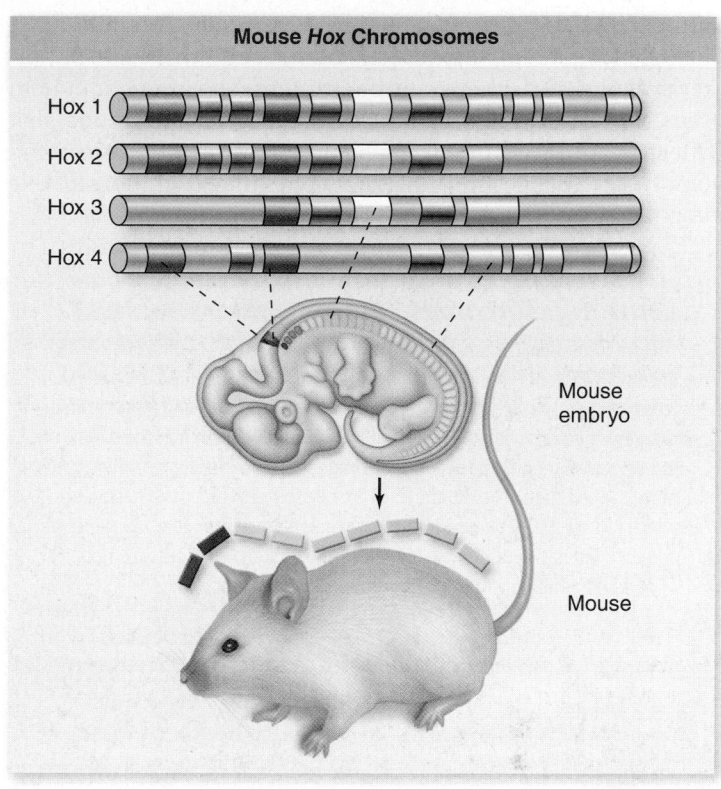

Figure 19.18 A comparison of homeotic gene clusters in the fruit fly *Drosophila melanogaster* and the mouse *Mus musculus*.
a. *Drosophila* homeotic genes. Called the homeotic gene complex, or HOM complex, the genes are grouped into two clusters: the Antennapedia complex (anterior) and the bithorax complex (posterior). **b.** The *Drosophila* HOM genes and the mouse *Hox* genes are related genes that control the regional differentiation of body parts in both animals. These genes are located on a single chromosome in the fly and on four separate chromosomes in mammals. In this illustration, the genes are color-coded to match the parts of the body along the A/P axis in which they are expressed. Note that the order of the genes along the chromosome(s) is mirrored by their pattern of expression in the embryo and in structures in the adult fly.

have been two duplications in the vertebrate lineage, at least of the *Hox* cluster. Given the single cluster in arthropods, this finding implies that the common ancestor to all animals with bilateral symmetry had a single *Hox* cluster as well.

The next logical step is to look at even more-primitive animals: the radially symmetrical cnidarians such as *Hydra* (see chapter 34). Thus far, *Hox* genes have been found in a number of cnidarian species, and recent sequence analyses suggest that cnidarian *Hox* genes are also arranged into clusters. Thus, the appearance of the ancestral *Hox* cluster likely preceded the divergence between radial and bilateral symmetries in animal evolution.

Pattern formation in plants is also under genetic control

The evolutionary split between plant and animal cell lineages occurred about 1.6 BYA, before the appearance of multicellular organisms with defined body plans. The implication is that multicellularity evolved independently in plants and animals. Because of the activity of meristems, additional modules can be added to plant bodies throughout their lifetimes. In addition, plant flowers and roots have a radial organization, in contrast to the bilateral symmetry of most animals. We may

therefore expect that the genetic control of pattern formation in plants is fundamentally different from that of animals.

Although plants have homeobox-containing genes, they are not organized into complexes of *Hox* genes similar to the ones that determine regional identity of developing structures in animals. Instead, the predominant homeotic gene family in plants appears to be the **MADS-box** genes.

MADS-box genes are a family of transcriptional regulators found in most eukaryotic organisms, including plants, animals, and fungi. The MADS-box is a conserved DNA-binding and dimerization domain, named after the first five genes to be discovered with this domain. Only a small number of *MADS*-box genes are found in animals, where their functions include the control of cell proliferation and tissue-specific gene expression in postmitotic muscle cells. They do not appear to play a role in the patterning of animal embryos.

In contrast, the number and functional diversity of *MADS*-box genes increased considerably during the evolution of land plants, and there are more than 100 *MADS*-box genes in the *Arabidopsis* genome. In flowering plants, the *MADS*-box genes dominate the control of development, regulating such processes as the transition from vegetative to reproductive growth, root development, and floral organ identity.

Although distinct from genes in the *Hox* clusters of animals, plant *Hox* genes encode transcription factors that have important developmental functions. One such example is the family of *knottedlike homeobox (knox)* genes, which are important regulators of shoot apical meristem development in both seed-bearing and nonseed-bearing plants. Mutations that affect expression of *knox* genes produce changes in leaf and petal shape, suggesting that these genes play an important role in generating leaf form.

Learning Outcomes Review 19.5

Pattern formation in animals involves the coordinated expression of a hierarchy of genes. Gradients of morphogens in *Drosophila* specify A/P and D/V axes, then lead to sequential activation of segmentation genes. Bicoid and Nanos protein gradients determine the A/P axis. The protein Dorsal determines the D/V axis, but activation requires a series of steps beginning with the oocyte's Gurken protein. The action of homeotic genes provide segment identity. These genes, which include a DNA-binding homeodomain sequence, are called *Hox* genes (for *homeobox* genes), and they are organized into clusters. Plants use a different set of developmental control genes called *MADS*-box genes.

■ *Why would you expect homeotic genes to be conserved across species evolution?*

19.6 Morphogenesis

Learning Outcomes

1. *Discuss the importance of cell shape changes and cell migration in development.*
2. *Explain how cell death can contribute to morphogenesis.*
3. *Describe the role of the extracellular matrix in cell migration.*

At the end of cleavage, the *Drosophila* embryo still has a relatively simple structure: It comprises several thousand identical-looking cells, which are present in a single layer surrounding a central yolky region. The next step in embryonic development is **morphogenesis**—the generation of ordered form and structure.

Morphogenesis is the product of changes in cell structure and cell behavior. Animals regulate the following processes to achieve morphogenesis:

■ The number, timing, and orientation of cell divisions;
■ Cell growth and expansion;
■ Changes in cell shape;
■ Cell migration; and
■ Cell death.

Plant and animal cells are fundamentally different in that animal cells have flexible surfaces and can move, but plant cells are immotile and encased within stiff cellulose walls. Each cell in a plant is fixed into position when it is created. Thus, animal cells use cell migration extensively during development while plants use the other four mechanisms but lack cell migration. We consider the morphogenetic changes in animals here, and plant morphogenesis is detailed in chapter 41.

Cell division during development may result in unequal cytokinesis

The orientation of the mitotic spindle determines the plane of cell division in eukaryotic cells. The coordinated function of microtubules and their motor proteins determines the respective position of the mitotic spindle within a cell (see chapter 10). If the spindle is centrally located in the dividing cell, two equal-sized daughter cells will result. If the spindle is off to one side, one large daughter cell and one small daughter cell will result.

The great diversity of cleavage patterns in animal embryos is determined by differences in spindle placement. In many cases, the fate of a cell is determined by its relative placement in the embryo during cleavage. For example, in preimplantation mammalian embryos, cells on the outside of the embryo usually differentiate into trophectoderm cells, which form only extraembryonic structures later in development (for example, a part of the placenta). In contrast, the embryo proper is derived from the inner cell mass, cells which, as the name implies, are in the interior of the embryo.

Cells change shape and size as morphogenesis proceeds

In animals, cell differentiation is often accompanied by profound changes in cell size and shape. For example, the large nerve cells that connect your spinal cord to the muscles in your big toe develop long processes called *axons* that span this entire distance. The cytoplasm of an axon contains microtubules, which are used for motor-driven transport of materials along the length of the axon.

As another example, muscle cells begin as *myoblasts,* undifferentiated muscle precursor cells. They eventually undergo conversion into the large, multinucleated *muscle fibers* that make up mammalian skeletal muscles. These changes begin with the expression of the *MyoD1* gene, which encodes a transcription factor that binds to the promoters of muscle-determining genes to initiate these changes.

Programmed cell death is a necessary part of development

Not every cell produced during development is destined to survive. For example, human embryos have webbed fingers and toes at an early stage of development. The cells that make up the webbing die in the normal course of morphogenesis. As another example, vertebrate embryos produce a very large number of neurons, ensuring that enough neurons are available to make the necessary synaptic connections, but over half of these neurons never make connections and die in an orderly way as the nervous system develops.

Unlike accidental cell deaths due to injury, these cell deaths are planned—and indeed required—for proper development and morphogenesis. Cells that die due to injury typically swell and burst, releasing their contents into the extracellular fluid. This form of cell death is called necrosis. In contrast, cells programmed to die shrivel and shrink in a process called apoptosis, which means "falling away," and their remains are taken up by surrounding cells.

Genetic control of apoptosis

Apoptosis occurs when a "death program" is activated. All animal cells appear to possess such programs. In *C. elegans,* the same 131 cells always die during development in a predictable and reproducible pattern.

Work on *C. elegans* showed that three genes are central to this process. Two (*ced-3* and *ced-4*) activate the death program itself; if either is mutant, those 131 cells do not die, and go on instead to form nervous tissue and other tissue. The third gene (*ced-9*) represses the death program encoded by the other two: All 1090 cells of the *C. elegans* embryo die in *ced-9* mutants. In *ced-9/ced-3* double mutants, all 1090 cells live, which suggests that *ced-9* inhibits cell death by functioning prior to *ced-3* in the apoptotic pathway (figure 19.19*a*).

The mechanism of apoptosis appears to have been highly conserved during the course of animal evolution. In human nerve cells, the *Apaf1* gene is similar to *ced-4* of *C. elegans* and activates the cell death program, and the human *bcl-2* gene acts similarly to *ced-9* to repress apoptosis. If a copy of the human *bcl-2* gene is transferred into a nematode with a defective *ced-9* gene, *bcl-2* suppresses the cell death program of *ced-3* and *ced-4*.

The mechanism of apoptosis

The product of the *C. elegans ced-4* gene is a protease that activates the product of the *ced-3* gene, which is also a protease. The human *Apaf1* gene is actually named for its role: *A*poptotic *p*rotease *a*ctivating *f*actor. It activates two proteases called caspases that have a role similar to the Ced-3 protease in *C. elegans* (figure 19.19*b*). When the final proteases are activated, they chew up proteins in important cellular structures such as the cytoskeleton and the nuclear lamina, leading to cell fragmentation.

The role of Ced-9/Bcl-2 is to inhibit this program. Specifically, it inhibits the activating protease, preventing the activation of the destructive proteases. The entire process is thus controlled by an inhibitor of the death program.

Both internal and external signals control the state of the Ced-9/Bcl-2 inhibitor. For example, in the human nervous system, neurons have a cytoplasmic inhibitor of Bcl-2 that allows the death program to proceed (see figure 19.19*b*). In the presence of nerve growth factor, a signal transduction pathway leads to the cytoplasmic inhibitor being inactivated, allowing Bcl-2 to inhibit apoptosis and the nerve cell to survive.

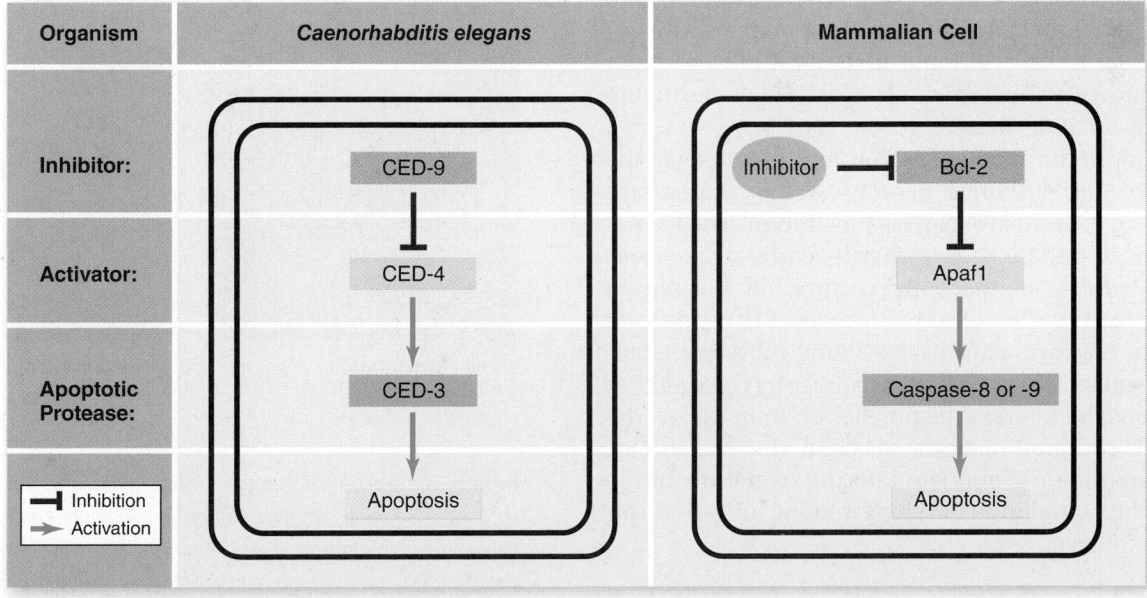

a.   *b.*

Figure 19.19 Programmed cell death pathway. Apoptosis, or programmed cell death, is necessary for the normal development of all animals. *a.* In the developing nematode, for example, two genes, *ced-3* and *ced-4*, code for proteins that cause the programmed cell death of 131 specific cells. In the other (surviving) cells of the developing nematode, the product of a third gene, *ced-9*, represses the death program encoded by *ced-3* and *ced-4*. *b.* The mammalian homologues of the apoptotic genes in *C. elegans* are *bcl-2* (*ced-9* homologue), *Apaf1* (*ced-4* homologue), and *caspase-8* or *-9* (*ced-3* homologues). In the absence of any cell survival factor, Bcl-2 is inhibited and apoptosis occurs. In the presence of nerve growth factor (NGF) and NGF receptor binding, Bcl-2 is activated, thereby inhibiting apoptosis.

Cell migration gets the right cells to the right places

The migration of cells is important during many stages of animal development. The movement of cells involves both adhesion and the loss of adhesion. Adhesion is necessary for cells to get "traction," but cells that are initially attached to others must lose this adhesion to be able to leave a site.

Cell movement also involves cell-to-substrate interactions, and the extracellular matrix may control the extent or route of cell migration. The central paradigm of morphogenetic cell movements in animals is a change in cell adhesiveness, which is mediated by changes in the composition of macromolecules in the plasma membranes of cells or in the extracellular matrix. Cell-to-cell interactions are often mediated through cadherins, but cell-to-substrate interactions often involve integrin-to-extracellular-matrix (ECM) interactions.

Cadherins

Cadherins are a large gene family, with over 80 members identified in humans. In the genomes of *Drosophila, C. elegans,* and humans, the cadherins can be sorted into several subfamilies that exist in all three genomes.

The cadherin proteins are all transmembrane proteins that share a common motif, the *cadherin domain,* a 110-amino-acid domain in the extracellular portion of the protein that mediates Ca^{2+}-dependent binding between like cadherins (homophilic binding).

Experiments in which cells are allowed to sort in vitro illustrate the function of cadherins. Cells with the same cadherins adhere specifically to one another, while not adhering to other cells with different cadherins. If cell populations with different cadherins are dispersed and then allowed to reaggregate, they sort into two populations of cells based on the nature of the cadherins on their surface.

An example of the action of cadherins can be seen in the development of the vertebrate nervous system. All surface ectoderm cells of the embryo express E-cadherin. The formation of the nervous system begins when a central strip of cells on the dorsal surface of the embryo turns off E-cadherin expression and turns on N-cadherin expression. In the process of **neurulation,** the formation of the neural tube (see chapter 53), the central strip of N-cadherin-expressing cells folds up to form the tube. The neural tube pinches off from the overlying cells, which continue to express E-cadherin. The surface cells outside the tube differentiate into the epidermis of the skin, whereas the neural tube develops into the brain and spinal cord of the embryo.

Integrins

In some tissues, such as connective tissue, much of the volume of the tissue is taken up by the spaces *between* cells. These spaces are filled with a network of molecules secreted by surrounding cells, termed a *matrix.* In connective tissue such as cartilage, long polysaccharide chains are covalently linked to proteins (proteoglycans), within which are embedded strands of fibrous protein (collagen, elastin, and fibronectin). Migrating cells traverse this matrix by binding to it with cell surface proteins called integrins.

Integrins are attached to actin filaments of the cytoskeleton and protrude out from the cell surface in pairs, like two hands. The "hands" grasp a specific component of the matrix, such as collagen or fibronectin, thus linking the cytoskeleton to the fibers of the matrix. In addition to providing an anchor, this binding can initiate changes within the cell, alter the growth of the cytoskeleton, and activate gene expression and the production of new proteins.

The process of **gastrulation** (described in detail in chapter 53), during which the hollow ball of animal embryonic cells folds in on itself to form a multilayered structure, depends on fibronectin–integrin interactions. For example, injection of antibodies against either fibronectin or integrins into salamander embryos blocks binding of cells to fibronectin in the ECM and inhibits gastrulation. The result is like a huge traffic jam following a major accident on a freeway: Cells (cars) keep coming, but they get backed up since they cannot get beyond the area of inhibition (accident site) (figure 19.20). Similarly, a

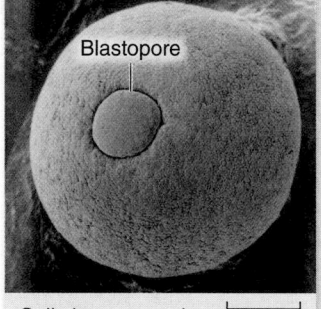

SCIENTIFIC THINKING

Hypothesis: *Fibronectin is required for cell migration during gastrulation.*

Prediction: *Blocking fibronectin with antifibronectin antibodies before gastrulation should prevent cell movement.*

Test: *Staged salamander embryos were injected either with antifibronectin antibody, or with preimmune serum as a control, prior to gastrulation. Cell movements were then monitored photographically.*

Treated with Preimmune

Blastopore

Cells have moved into the interior   285 µm

a.

Treated with Antifibronectin

Blastopore

Cells pile up on the surface   285 µm

b.

Result: *The experimental embryos injected with antifibronectin antibody show extremely aberrant gastrulation where cells pile up and do not enter the interior of the embryo. Control embryos gastrulate normally.*

Conclusion: *Fibronectin is required for cells to migrate into the interior of the embryo during gastrulation.*

Further Experiments: *How can this same system be used to analyze the role of fibronectin in other early morphogenetic events?*

Figure 19.20 Fibronectin is necessary for cell migration during gastrulation.

targeted knockout of the fibronectin gene in mice resulted in gross defects in the migration, proliferation, and differentiation of embryonic mesoderm cells.

Thus, cell migration is largely a matter of changing patterns of cell adhesion. As a migrating cell travels, it continually extends projections that probe the nature of its environment. Tugged this way and that by different tentative attachments, the cell literally feels its way toward its ultimate target site.

Chapter Review

19.1 The Process of Development

Development is the sequence of systematic, gene-directed changes throughout a life cycle. The four subprocesses of development are growth, cell differentiation, pattern formation, and morphogenesis.

19.2 Cell Division

Development begins with cell division.

In animals, cleavage stage divisions divide the fertilized egg into numerous smaller cells called blastomeres. During cleavage the G_1 and G_2 phases of the cell cycle are shortened or eliminated (figure 19.2).

Every cell division is known in the development of C. elegans.

The lineage of 959 adult somatic *Caenorhabditis elegans* cells is invariant. Knowledge of the differentiation sequence and outcome allows study of developmental mechanisms.

Plant growth occurs in specific areas called meristems.

Plant growth continues throughout the life span from meristematic stem cells that can divide and differentiate into any plant tissue.

19.3 Cell Differentiation

Cells become determined prior to differentiation.

The process of determination commits a cell to a particular developmental pathway prior to its differentiation. This is not visible but can be tracked experimentally. Determination is due to differential inheritance of cytoplasmic factors or cell-to-cell interactions.

Determination can be due to cytoplasmic determinants.

In tunicates, determination of tail muscle cells depends on the presence of mRNA for the Macho-1 transcription factor, which is deposited in the egg cytoplasm during gamete formation.

Induction can lead to cell differentiation.

Induction occurs when one cell type produces signal molecules that induce gene expression in neighboring target cells.

In frogs, cells from animal and vegetal poles do not develop into mesoderm when isolated. In tunicates, signaling by the growth factor FGF induces mesoderm development.

Stem cells can divide and produce cells that differentiate.

Stem cells replace themselves by division and produce cells that differentiate. Totipotent stem cells can give rise to any cell type including extraembryonic tissues; pluripotent cells can give rise to

all cells of an organism; and multipotent stem cells can give rise to many kinds of cells.

Embryonic stem cells are pluripotent cells derived from embryos.

Embryonic stem cells are derived from the inner cell mass of the blastocyst (figure 19.8). They can differentiate into any adult tissue in a mouse.

19.4 Nuclear Reprogramming

Reversal of determination has allowed cloning.

Cells undergo no irreversible changes during development. However, transplanted nuclei from older donors are less able to direct complete development. The cloning of the sheep Dolly showed that the nucleus of an adult cell can be reprogrammed to be totipotent (figure 19.9).

Reproductive cloning has inherent problems.

Reproductive cloning has a low success rate, and clones often develop age-associated diseases.

Nuclear reprogramming has been accomplished by use of defined factors.

Adult cells can be converted into pluripotent cells by introduction of four genes for transcription factors. These induced pluripotent cells appear to be similar to ES cells.

The use of cells cloned from a patient's cells to replace damaged tissue could avoid the problem of transplant rejection.

19.5 Pattern Formation

Drosophila embryogenesis produces a segmented larva.

The maternal contribution of mRNA along with the postfertilization events of cellular blastoderm formation produce a segmented embryo.

Morphogen gradients form the basic body axes in Drosophila.

Pattern formation produces two perpendicular axes in a bilaterally symmetrical organism. Positional information leads to changes in gene activity so cells adopt a fate appropriate for their location.

Formation of the anterior/posterior (A/P) axis is based on opposing gradients of morphogens, Bicoid and Nanos, synthesized from maternal mRNA (figures 19.14, 19.15).

The dorsal/ventral (D/V) axis is established by a gradient of the Dorsal transcription factor. Successive action of transcription factors divides the embryo into segments.

The body plan is produced by sequential activation of genes.

Segment identity arises from the action of homeotic genes.

Homeotic genes, called *Hox* genes because they contain a DNA sequence called the homeobox, give identity to embryo segments.

Hox genes are found in four clusters in vertebrates.

Pattern formation in plants is also under genetic control.

Plants have *MADS*-box genes that control the transition from vegetative to reproductive growth, root development, and floral organ identity.

19.6 Morphogenesis

Cell division during development may result in unequal cytokinesis.

Cells change shape and size as morphogenesis proceeds.

Depending on the orientation of the mitotic spindle, cells of equal or different sizes can arise. Morphogenesis involves changes in cell shape and size and cell migration.

Programmed cell death is a necessary part of development.

Apoptosis, the programmed death of cells, removes structures once they are no longer needed (figure 19.19).

Cell migration gets the right cells to the right places.

The migration of cells requires both adhesion and loss of adhesion between cells and their substrate.

Cell-to-cell interactions are often mediated by cadherin proteins, whereas cell-to-substrate interactions may involve integrin-to-extracellular-matrix interactions.

Integrins bind to fibers found in the extracellular matrix. This action can alter the cytoskeleton and activate gene expression.

Review Questions

UNDERSTAND

1. During development, cells become
   a. differentiated before they become determined.
   b. determined before they become differentiated.
   c. determined by the loss of genetic material.
   d. differentiated by the loss of genetic material.

2. Determination can occur by
   a. the action of cytoplasmic determinants.
   b. induction by other cells.
   c. the loss of chromosomes during cell division.
   d. Both a and b are correct.

3. The rapid divisions that occur early in development are made possible by shortening
   a. M phase.
   b. S phase.
   c. G_1 and G_2 phases.
   d. All of the choices are correct.

4. A pluripotent cell is one that can
   a. become any cell type in an organism.
   b. produce an indefinite supply of a single cell type.
   c. produce a limited amount of a specific cell type.
   d. produce multiple cell types.

5. Plant meristems
   a. are only present during development.
   b. contain stem cells.
   c. undergo meiosis.
   d. All of the choices are correct.

6. Pattern formation involves cells determining their position in the embryo. One mechanism that can accomplish this is
   a. the loss of genetic material.
   b. alterations of chromosome structure.
   c. gradients of morphogens.
   d. changes in the cell cycle.

7. The process of nuclear reprogramming
   a. is a normal part of pattern formation.
   b. reverses the changes that occur during differentiation.
   c. requires the introduction of new DNA.
   d. is not possible with mammalian cells.

APPLY

1. What is the common theme in cell determination by induction or cytoplasmic determinants?
   a. The activation of transcription factors
   b. The activation of cell division
   c. A change in gene expression
   d. Both a and c are correct.

2. The process of reproductive cloning
   a. shows that nuclear reprogramming is possible.
   b. is very efficient in mammals.
   c. always produces adult animals that are identical to the donor.
   d. Both a and b are correct.

3. Production of anterior–posterior and dorsal–ventral axes in the fruit fly *Drosophila*
   a. both use gradients of mRNA.
   b. are conceptually similar but mechanistically different.
   c. use the exact same mechanisms.
   d. both use gradients of protein.

4. For pattern formation to occur, the cells in the developing embryo must
   a. "know" their position in the embryo.
   b. be determined during the earliest divisions.
   c. differentiate as they are "born."
   d. must all be reprogrammed after each cell division.

5. The genes that encode the morphogen gradients in *Drosophila* were all identified in mutant screens. A mutation that removes the gradient necessary for the A/P morphogen gradient would be expected to
   a. affect the larvae but not the adult.
   b. affect the adult but not the larvae.
   c. be lethal and lead to an abnormal embryo.
   d. produce replacement of one adult structure with another.

6. What would be the likely result of a mutation of the *bcl-2* gene on the level of apoptosis?
   a. No change
   b. A decrease in apoptosis
   c. An increase in apoptosis
   d. An initial decrease, followed by an increase in apoptosis

7. *MADS*-box, and *Hox* genes are
   a. found only in plants and animals, respectively.
   b. found only in animals and plants, respectively.
   c. have similar roles in development in plants and animals, respectively.
   d. have similar roles in development in animals and plants, respectively.

SYNTHESIZE

1. The fate map for *C. elegans* (refer to figure 19.3) diagrams development of a multicellular organism from a single cell. Use this fate map to determine the number of cell divisions required to establish the population of cells that will become (a) the nervous system and (b) the gonads.

2. Carefully examine the *C. elegans* fate map in figure 19.3. Notice that some of the branchpoints (daughter cells) do *not* go on to produce more cells. What is the cellular mechanism underlying this pattern?

3. You have generated a set of mutant embryonic mouse cells. Predict the developmental consequences for each of the following mutations.
   a. Knockout mutation for N-cadherin
   b. Knockout mutation for integrin
   c. Deletion of the cytoplasmic domain of integrin

4. Assume you have the factors in hand necessary to reprogram an adult cell, and the factors necessary to induce differentiation to any cell type. How could these be used to replace a specific damaged tissue in a human patient?

ONLINE RESOURCE

www.ravenbiology.com

Understand, Apply, and Synthesize—enhance your study with animations that bring concepts to life and practice tests to assess your understanding. Your instructor may also recommend the interactive eBook, individualized learning tools, and more.

Chapter **20**

Genes Within Populations

Chapter Contents

Introduction

No other human being is exactly like you (unless you have an identical twin). Often the particular characteristics of an individual have an important bearing on its survival, on its chances to reproduce, and on the success of its offspring. Evolution is driven by such factors, as different alleles rise and fall in populations. These deceptively simple matters lie at the core of evolutionary biology, which is the topic of this chapter and chapters 21 through 25.

20.1 Genetic Variation and Evolution

Learning Outcomes

1. *Define* evolution *and* population genetics.
2. *Explain the difference between evolution by natural selection and the inheritance of acquired characteristics.*

Genetic variation, that is, differences in alleles of genes found within individuals of a population, provides the raw material for natural selection, which will be described shortly. Natural populations contain a wealth of such variation. In this chapter, we explore genetic variation in natural populations and consider the evolutionary forces that cause allele frequencies in natural populations to change.

The word *evolution* is widely used in the natural and social sciences. It refers to how an entity—be it a social system, a gas, or a planet—changes through time. Although development of the modern concept of evolution in biology can be traced to Darwin's landmark work, *On the Origin of Species,* the first five editions of his book never actually used the term. Rather, Darwin used the phrase "descent with modification."

Although many more complicated definitions have been proposed, Darwin's words probably best capture the essence of biological evolution: Through time, species accumulate differences; as a result, descendants differ from their ancestors. In this way, new species arise from existing ones.

Many processes can lead to evolutionary change

You have already learned about the development of Darwin's ideas in chapter 1. Darwin was not the first to propose a theory of evolution. Rather, he followed a long line of earlier philosophers and naturalists who deduced that the many kinds of organisms around us were produced by a process of evolution.

Unlike his predecessors, however, Darwin proposed natural selection as the mechanism of evolution. Natural selection produces evolutionary change when some individuals in a population possess certain inherited characteristics and then produce more surviving offspring than individuals lacking these characteristics. As a result, the population gradually comes to include more and more individuals with the advantageous characteristics. In this way, the population evolves and becomes better adapted to its local circumstances.

A rival theory, championed by the prominent biologist Jean-Baptiste Lamarck, was that evolution occurred by the **inheritance of acquired characteristics.** According to Lamarck, changes that individuals acquired during their lives were passed on to their offspring. For example, Lamarck proposed that ancestral giraffes with short necks tended to stretch their necks to feed on tree leaves, and this extension of the neck was passed on to subsequent generations, leading to the long-necked giraffe (figure 20.1*a*). In Darwin's theory, by contrast, the variation is not created by experience, but is the result of preexisting genetic differences among individuals (figure 20.1*b*).

One way to monitor how populations change through time is to look at changes in the frequencies of alleles of a gene from one generation to the next. Natural selection, by favoring individuals with certain alleles, can lead to change in such *allele frequencies,* but it is not the only process that can do so. Allele frequencies can also change when mutations occur repeatedly, changing one allele to another, and when migrants bring alleles into a population. In addition, when populations are small, the frequencies of alleles can change randomly as the result of chance events. Often, natural selection overwhelms the effects of these other processes, but as you will see later in this chapter, this is not always the case.

Evolution can result from any process that causes a change in the genetic composition of a population. We cannot talk about evolution, therefore, without also considering **population genetics,** the study of the properties of genes in populations.

Populations contain ample genetic variation

Biologists have always wanted to know how much genetic variation exists in natural populations. The ability to ask this question has been limited by the techniques available to analyze variation at different levels: proteins, genes, and now genomes. The story of genetic variation in natural populations is told using increasingly sophisticated tools for detecting differences.

Initial approaches to understanding variation examined the most obvious differences—the morphological. At this level, natural populations usually show substantial genetic variation,

Some individuals born happen to have longer necks due to genetic differences.

Individuals pass on their traits to next generation.

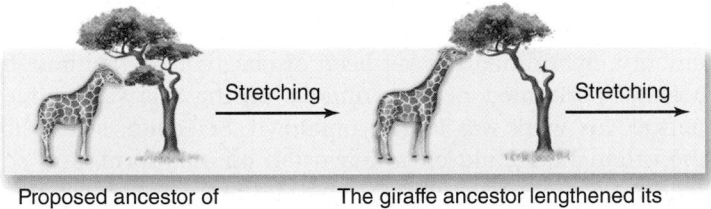
Proposed ancestor of giraffes has characteristics of modern-day okapi.

The giraffe ancestor lengthened its neck by stretching to reach tree leaves, then passed the change to offspring.

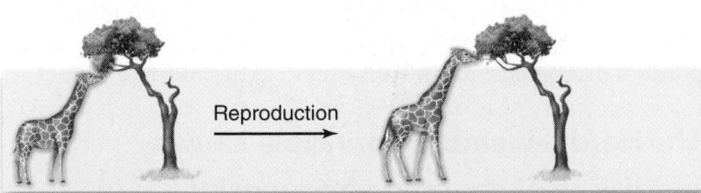

a. **Lamarck's theory: acquired variation is passed on to descendants.**

Figure 20.1 Two ideas of how giraffes might have evolved long necks.

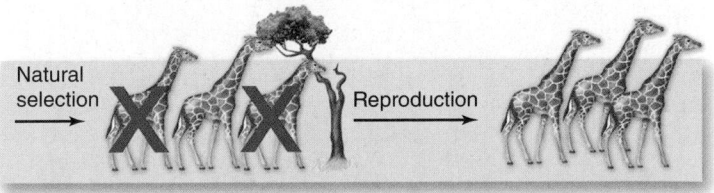
Over many generations, longer-necked individuals are more successful, perhaps because they can feed on taller trees, and pass the long-neck trait on to their offspring.

b. **Darwin's theory: natural selection or genetically-based variation leads to evolutionary change.**

as figure 20.2 illustrates. An example closer to home are the genes that influence blood groups in humans. Chemical analysis has revealed the existence of more than 30 blood group genes in humans, in addition to the ABO locus. At least one third of these genes are routinely found in several alternative allelic forms in human populations. In addition to these, more than 45 variable genes encode other proteins in human blood cells and plasma that are not considered blood groups. In short, many genetically variable genes are present in this one system alone.

The first approach to directly assay genetic variation within populations was the use of electrophoresis to separate proteins produced by alternative alleles of enzyme-encoding genes (see chapter 17). As DNA analysis tools were developed in the laboratory, they were adapted for use with samples collected from natural populations. This led to, in order of increasing detail, examining restriction fragment length polymorphisms (RFLPs, see chapter 17), sequencing specific genes, and, most recently, sequencing entire genomes.

One of the most useful tools, both for genetic mapping and for the analysis of population-level variation, has been single-nucleotide polymorphisms (SNPs). These are defined as single-base differences between individuals that exist in the population at more than 1% (see chapter 24). A large-scale international effort has identified 3.1 million SNPs in the human genome, which translates to a density of about 1 per kilobase of DNA.

Figure 20.2 Polymorphic variation. This natural population of lupines, *Lupinus,* exhibits considerable variation in flower color. Individual differences are inherited and passed on to offspring.

Such variation is also being assayed in many other species of interest. The results for most species are similar to those in humans: the closer you look, the more variation you see. One reason for the appeal of SNPs is that we have been able to automate the analysis of multiple samples. This approach is being applied to many species of scientific and economic interest. The U.S. storehouse for such information, the National Center for Biotechnology Information (NCBI), now has a database with SNPs found in over 100 species, revealing the near ubiquity of genetic variation and providing tools for assessing patterns in human and natural populations.

Learning Outcomes Review 20.1

Evolution can be described as descent with modification. Natural selection occurs when individuals carrying certain alleles leave more offspring than those without the alleles. Natural populations generally contain considerable amounts of genetic variation. Population genetics studies this variability through statistical analyses.

■ *Why is genetic variation in a population necessary for evolution to occur?*

20.2 Changes in Allele Frequency

Learning Outcomes

1. *Explain the Hardy–Weinberg principle.*
2. *Describe the characteristics of a population that is in Hardy–Weinberg equilibrium.*
3. *Demonstrate how the operation of evolutionary processes can be detected.*

Genetic variation within natural populations was a puzzle to Darwin and his contemporaries in the mid-1800s. The way in which meiosis produces genetic segregation among the progeny of a hybrid had not yet been discovered. And, although Mendel performed his experiments during this same time period, his work was largely unknown. Selection, scientists then thought, should always favor an optimal form, and so tend to eliminate variation. Moreover, the theory of *blending inheritance*—in which offspring were expected to be phenotypically intermediate relative to their parents—was widely accepted. If blending inheritance were correct, then the effect of any new genetic variant would quickly be diluted to the point of disappearance in subsequent generations.

The Hardy–Weinberg principle allows prediction of genotype frequencies

Following the rediscovery of Mendel's research, two people in 1908 solved the puzzle of why genetic variation persists— Godfrey H. Hardy, an English mathematician, and Wilhelm

Weinberg, a German physician. These workers were initially confused about why, after many generations, a population didn't come to be composed solely of individuals with the dominant phenotype. The conclusion they independently came to was that the original proportions of the genotypes in a population will remain constant from generation to generation, as long as the following assumptions are met:

1. No mutation takes place.
2. No genes are transferred to or from other sources (no immigration or emigration takes place).
3. Mating is random (individuals do not choose mates based on their phenotype or genotype).
4. The population size is very large.
5. No selection occurs.

Because the genotypes' proportions do not change, they are said to be in **Hardy–Weinberg equilibrium.**

The Hardy–Weinberg equation with two alleles: A binomial expansion

In algebraic terms, the Hardy–Weinberg principle is written as an equation. Consider a population of 100 cats in which 84 are black and 16 are white. The frequencies of the two phenotypes would be 0.84 (or 84%) black and 0.16 (or 16%) white. Based on these phenotypic frequencies, can we deduce the underlying frequency of genotypes?

If we assume that the white cats are homozygous recessive for an allele we designate as *b,* and the black cats are either homozygous dominant *BB* or heterozygous *Bb,* we can calculate the **allele frequencies** of the two alleles in the population from the proportion of black and white individuals, assuming that the population is in Hardy–Weinberg equilibrium.

Let the letter *p* designate the frequency of the *B* allele and the letter *q* the frequency of the alternative allele. Because there are only two alleles, *p* plus *q* must always equal 1 (that is, the total population). In addition, we know that the sum of the three genotype frequencies must also equal 1. If the frequency of the *B* allele is *p,* then the probability that an individual will have two *B* alleles is simply the probability that each of its alleles is a *B.* The probability of two events happening independently is the product of the probability of each event; in this case, the probability that the individual received a *B* allele from its father is *p,* and the probability the individual received a *B* allele from its mother is also *p,* so the probability that both happened is $p * p = p^2$ (figure 20.3). By the same reasoning, the probability that an individual will have two *b* alleles is q^2.

What about the probability that an individual will be a heterozygote? There are two ways this could happen: The individual could receive a *B* from its father and a *b* from its mother, or vice versa. The probability of the first case is $p * q$ and the probability of the second case is $q * p$. Because the result in either case is that the individual is a heterozygote, the probability of that outcome is the sum of the two probabilities, or $2pq$.

So, to summarize, if a population is in Hardy–Weinberg equilibrium with allele frequencies of *p* and *q,* then the probability that an individual will have each of the three possible genotypes is $p^2 + 2pq + q^2$. You may recognize this as the *binomial expansion:*

$$(p + q)^2 = p^2 + 2pq + q^2$$

Finally, we may use these probabilities to predict the distribution of genotypes in the population, again assuming that the population is in Hardy–Weinberg equilibrium. If the probability that any individual is a heterozygote is $2pq$, then we would expect the proportion of heterozygous individuals in the population to be $2pq$; similarly, the frequency of *BB* and *bb* homozygotes would be expected to be p^2 and q^2.

Let us return to our example. Remember that 16% of the cats are white. If white is a recessive trait, then this means that such individuals must have the genotype *bb.* If the frequency of this genotype is $q^2 = 0.16$ (the frequency of white cats), then *q* (the frequency of the *b* allele) = 0.4. Because $p + q = 1$, therefore, *p,* the frequency of allele *B,* would be 1.0 − 0.4 = 0.6

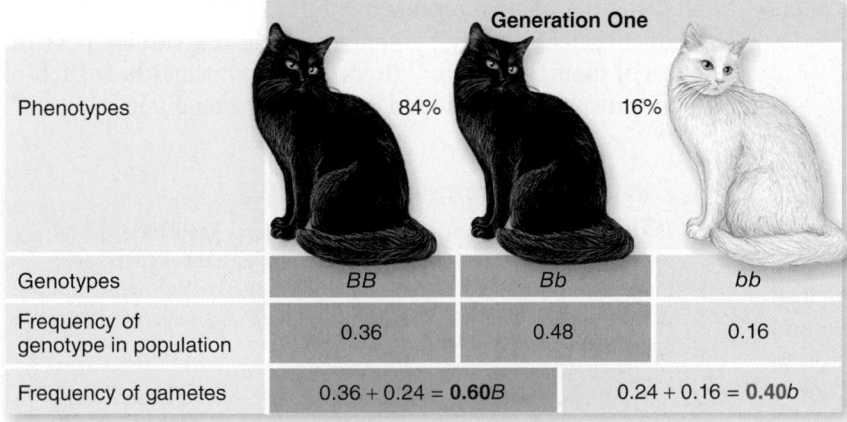
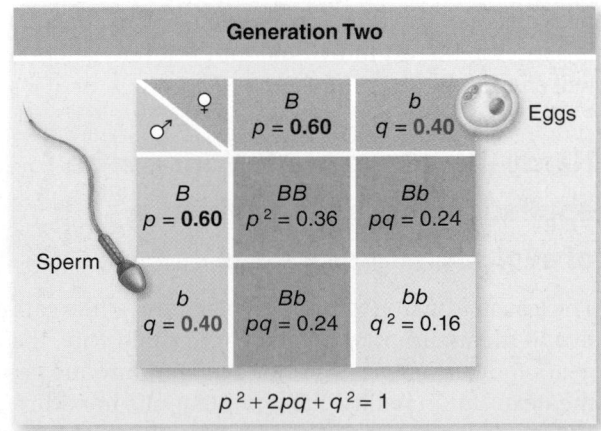

Figure 20.3 The Hardy–Weinberg equilibrium. In the absence of factors that alter them, the frequencies of gametes, genotypes, and phenotypes remain constant generation after generation.

Data analysis If all white cats died, what proportion of the kittens in the next generation would be white?

(remember, the frequencies must add up to 1). We can now easily calculate the expected **genotype frequencies:** homozygous dominant *BB* cats would make up the p^2 group, and the value of $p^2 = (0.6)^2 = 0.36$, or 36 homozygous dominant *BB* individuals in a population of 100 cats. The heterozygous cats have the *Bb* genotype and would have the frequency corresponding to $2pq$, or $(2 * 0.6 * 0.4) = 0.48$, or 48 heterozygous *Bb* individuals.

Using the Hardy–Weinberg equation to predict frequencies in subsequent generations

The Hardy–Weinberg equation is another way of expressing the Punnett square described in chapter 12, with two alleles assigned frequencies, p and q. Figure 20.3 allows you to trace genetic reassortment during sexual reproduction and see how it affects the frequencies of the *B* and *b* alleles during the next generation.

In constructing this diagram, we have assumed that the union of sperm and egg in these cats is random, so that all combinations of *b* and *B* alleles occur. The alleles are therefore mixed randomly and are represented in the next generation in proportion to their original occurrence. Each individual egg or sperm in each generation has a 0.6 chance of receiving a *B* allele ($p = 0.6$) and a 0.4 chance of receiving a *b* allele ($q = 0.4$).

In the next generation, therefore, the chance of combining two *B* alleles is p^2, or 0.36 (that is, 0.6 * 0.6), and approximately 36% of the individuals in the population will continue to have the *BB* genotype. The frequency of *bb* individuals is q^2 (0.4 * 0.4) and so will continue to be about 16%, and the frequency of *Bb* individuals will be $2pq$ (2 * 0.6 * 0.4), or on average, 48%.

Phenotypically, if the population size remains at 100 cats, we would still see approximately 84 black individuals (with either *BB* or *Bb* genotypes) and 16 white individuals (with the *bb* genotype). Allele, genotype, and phenotype frequencies have remained unchanged from one generation to the next, despite the reshuffling of genes that occurs during meiosis and sexual reproduction. Dominance and recessiveness of alleles can therefore be seen only to affect how an allele is expressed in an individual and not how allele frequencies will change through time.

Hardy–Weinberg predictions can be applied to data to find evidence of evolutionary processes

The lesson from the example of black and white cats is that if all five of the assumptions listed earlier hold true, the allele and genotype frequencies will not change from one generation to the next. But in reality, most populations in nature will not fit all five assumptions. The primary utility of this method is to determine whether some evolutionary process or processes are operating in a population and, if so, to suggest hypotheses about what they may be.

Suppose, for example, that the observed frequencies of the *BB, bb,* and *Bb* genotypes in a different population of cats

were 0.6, 0.2, and 0.2, respectively. We can calculate the allele frequencies for *B* as follows: 60% (0.6) of the cats have two *B* alleles, 20% have one, and 20% have none. This means that the average number of *B* alleles per cat is 1.4 [(0.6 × 2) + (0.2 × 1) + (0.2 × 0) = 1.4]. Because each cat has two alleles for this gene, the frequency is 1.4/2.0 = 0.7. Similarly, you should be able to calculate that the frequency of the *b* allele = 0.3.

If the population were in Hardy–Weinberg equilibrium, then, according to the equation earlier in this section, the frequency of the *BB* genotype would be $0.7^2 = 0.49$, lower than it really is. Similarly, you can calculate that there are fewer heterozygotes and more *bb* homozygotes than expected; then clearly, the population is not in Hardy–Weinberg equilibrium.

What could cause such an excess of homozygotes and deficit of heterozygotes? A number of possibilities exist, including (1) natural selection favoring homozygotes over heterozygotes, (2) individuals choosing to mate with genetically similar individuals (because *BB* * *BB* and *bb* * *bb* matings always produce homozygous offspring, but only half of *Bb* * *Bb* produce heterozygous offspring, such mating patterns would lead to an excess of homozygotes), or (3) an influx of homozygous individuals from outside populations (or conversely, emigration of heterozygotes to other populations). By detecting a lack of Hardy–Weinberg equilibrium, we can generate potential hypotheses that we can then investigate directly.

The operation of evolutionary processes can be detected in a second way. As discussed previously, if all of the Hardy–Weinberg assumptions are met, then allele frequencies will stay the same from one generation to the next. Changes in allele frequencies between generations would indicate that one of the assumptions is not met.

Suppose, for example, that the frequency of *b* was 0.53 in one generation and 0.61 in the next. Again, there are a number of possible explanations: For example, (1) selection favoring individuals with *b* over *B*, (2) immigration of *b* into the population or emigration of *B* out of the population, or (3) high rates of mutation that more commonly occur from *B* to *b* than vice versa. Another possibility is that the population is a small one, and that the change represents the random fluctuations that result because, simply by chance, some individuals pass on more of their genes than others. We will discuss how each of these processes is studied in the rest of the chapter.

Learning Outcomes Review 20.2

The Hardy–Weinberg principle states that in a large population with no selection and random mating, the proportion of alleles does not change through the generations. Finding that a population is not in Hardy–Weinberg equilibrium indicates that one or more evolutionary agents are operating.

- *If you know the genotype frequencies in a population, how can you determine whether the population is in Hardy–Weinberg equilibrium?*
- *What would you conclude if you found a population not in Hardy–Weinberg equilibrium? What would be your next step?*

Five Agents of Evolutionary Change

The five assumptions of the Hardy–Weinberg principle also indicate the five agents that can lead to evolutionary change in populations. They are mutation, gene flow, nonrandom mating, genetic drift in small populations, and the pressures of natural selection. Any one of these may bring about changes in allele or genotype proportions.

Mutation changes alleles

Mutation from one allele to another can obviously change the proportions of particular alleles in a population. Mutation rates are generally so low that they have little effect on the Hardy–Weinberg proportions of common alleles. A typical gene mutates about once per 100,000 cell divisions. Because this rate is so low, other evolutionary processes are usually more important in determining how allele frequencies change.

Nonetheless, mutation is the ultimate source of genetic variation and thus makes evolution possible (figure 20.4*a*). It is important to remember, however, that the likelihood of a particular mutation occurring is not affected by natural selection; that is, mutations do not occur more frequently in situations in which they would be favored by natural selection.

Gene flow occurs when alleles move between populations

Gene flow is the movement of alleles from one population to another. It can be a powerful agent of change. Sometimes gene flow is obvious, as when an animal physically moves from one place to another. If the characteristics of the newly arrived individual differ from those of the animals already there, and if the newcomer is adapted well enough to the new area to survive and mate successfully, the genetic composition of the receiving population may be altered.

Other important kinds of gene flow are not as obvious. These subtler movements include the drifting of gametes or the immature stages of plants or marine animals from one place to another (figure 20.4*b*). Pollen, the male gamete of flowering plants, is often carried great distances by insects and other animals that visit flowers. Seeds may also blow in the wind or be carried by animals to new populations far from their place of origin. In addition, gene flow may also result from the mating of individuals belonging to adjacent populations.

Consider two populations initially different in allele frequencies: In population 1, $p = 0.2$ and $q = 0.8$; in population 2, $p = 0.8$ and $q = 0.2$. Gene flow will tend to bring the rarer allele into each population. Thus, allele frequencies will change from generation to generation, and the populations will not be in Hardy–Weinberg equilibrium. Only when allele frequencies reach 0.5 for both alleles in both populations will equilibrium be attained. This example also indicates that gene flow tends to homogenize allele frequencies among populations.

Nonrandom mating shifts genotype frequencies

Individuals with certain genotypes sometimes mate with one another more commonly than would be expected on a

Mutation	Gene Flow	Nonrandom Mating	Genetic Drift	Selection
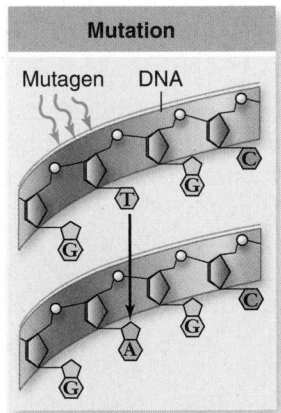		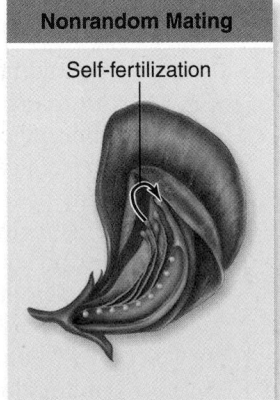		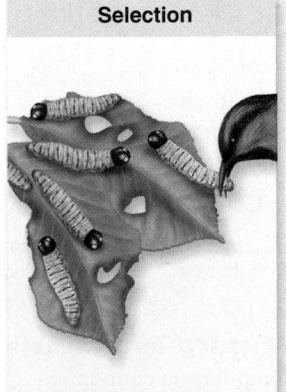

a. The ultimate source of variation. Individual mutations occur so rarely that mutation alone usually does not change allele frequency much.

b. A very potent agent of change. Individuals or gametes move from one population to another.

c. Inbreeding is the most common form. It does not alter allele frequency but reduces the proportion of heterozygotes.

d. Statistical accidents. The random fluctuation in allele frequencies increases as population size decreases.

e. The only agent that produces *adaptive* evolutionary changes.

Figure 20.4 Five agents of evolutionary change. *a.* Mutation, *b.* gene flow, *c.* nonrandom mating, *d.* genetic drift, and *e.* selection.

random basis, a phenomenon known as *nonrandom mating* (figure 20.4*c*). **Assortative mating,** in which phenotypically similar individuals mate, is a type of nonrandom mating that causes the frequencies of particular genotypes to differ greatly from those predicted by the Hardy–Weinberg principle.

Assortative mating does not change the frequency of the individual alleles, but rather increases the proportion of homozygous individuals because phenotypically similar individuals are likely to be genetically similar and thus are also more likely to produce offspring with two copies of the same allele. This is why populations of self-fertilizing plants consist primarily of homozygous individuals.

By contrast, **disassortative mating,** in which phenotypically different individuals mate, produces an excess of heterozygotes.

Genetic drift may alter allele frequencies in small populations

In small populations, frequencies of particular alleles may change drastically by chance alone. Such changes in allele frequencies occur randomly, as if the frequencies were drifting from their values. These changes are thus known as **genetic drift** (figure 20.4*d*). For this reason, a population must be large to be in Hardy–Weinberg equilibrium.

If the gametes of only a few individuals form the next generation, the alleles they carry may by chance not be representative of the parent population from which they were drawn, as illustrated in figure 20.5. In this example, a small number of individuals are removed from a bottle. By chance, most of the individuals removed are green, so the new population has a much higher population of green individuals than the parent generation had.

A set of small populations that are isolated from one another may come to

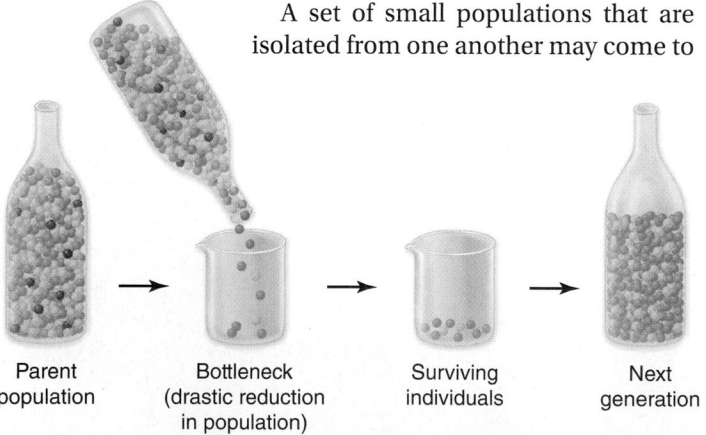

Figure 20.5 Genetic drift: a bottleneck effect. The parent population contains roughly equal numbers of green and yellow individuals and a small number of red individuals. By chance, the few remaining individuals that contribute to the next generation are mostly green. The bottleneck occurs because so few individuals form the next generation, as might happen after an epidemic or a catastrophic storm.

? **Inquiry question** Why are rare alleles particularly likely to be lost in a population bottleneck?

differ strongly as a result of genetic drift, even if the forces of natural selection are the same for both. Because of genetic drift, sometimes harmful alleles may increase in frequency in small populations, despite selective disadvantage, and favorable alleles may be lost even though they are selectively advantageous. It is interesting to realize that humans have lived in small groups for much of the course of their evolution; consequently, genetic drift may have been a particularly important factor in the evolution of our species.

Larger populations also experience the effect of genetic drift, but to a lesser extent than smaller populations—the magnitude of genetic drift is inversely related to population size. However, large populations may have been much smaller in the past, and genetic drift may have greatly altered allele frequencies at that time. Imagine a population containing only two alleles of a gene, *B* and *b*, in equal frequency (that is, $p = q = 0.50$). In a large Hardy–Weinberg population, the genotype frequencies are expected to be 0.25 *BB*, 0.50 *Bb*, and 0.25 *bb*. If only a small sample of individuals produces the next generation, large deviations in these genotype frequencies can occur simply by chance.

Suppose, for example, that four individuals form the next generation, and that by chance they are two *Bb* heterozygotes and two *BB* homozygotes—that is, the allele frequencies in the next generation would be $p = 0.75$ and $q = 0.25$. In fact, if you were to replicate this experiment 1000 times, each time randomly drawing four individuals from the parental population, then in about 8 of the 1000 experiments, one of the two alleles would be missing entirely.

This result leads to an important conclusion: Genetic drift can lead to the loss of alleles in isolated populations. Alleles that initially are uncommon are particularly vulnerable (see figure 20.5).

Although genetic drift occurs in any population, it is particularly likely in populations that were founded by a few individuals or in which the population was reduced to a very small number at some time in the past.

The founder effect

Sometimes one or a few individuals disperse and become the founders of a new, isolated population at some distance from their place of origin. These pioneers are not likely to carry all the alleles present in the source population. Thus, some alleles may be lost from the new population, and others may change drastically in frequency. In some cases, previously rare alleles in the source population may be a significant fraction of the new population's genetic endowment. This phenomenon is called the **founder effect.**

Founder effects are not rare in nature. Many self-pollinating plants start new populations from a single seed. Founder effects have been particularly important in the evolution of organisms on distant oceanic islands, such as the Hawaiian and Galápagos Islands. Most of the organisms in such areas probably derive from one or a few initial founders. Although rare, such events are occasionally observed, such as when a mass of vegetation carrying several iguanas washed up on the shore of the Caribbean island of Anguilla in 1996, leading to the establishment of a population that still occurs there to this day.

In a similar way, isolated human populations begun by relatively few individuals are often dominated by genetic features characteristic of their founders. Amish populations in the United States, for example, have unusually high frequencies of a number of conditions, such as polydactylism (the presence of a sixth finger).

The bottleneck effect

Even if organisms do not move from place to place, occasionally their populations may be drastically reduced in size. This may result from flooding, drought, epidemic disease, and other natural forces, or from changes in the environment. The few surviving individuals may constitute a random genetic sample of the original population (unless some individuals survive specifically because of their genetic makeup). The resulting alterations and loss of genetic variability have been termed the **bottleneck effect.**

The genetic variation of some living species appears to be severely depleted, probably as the result of a bottleneck effect in the past. For example, the northern elephant seal, which breeds on the western coast of North America and nearby islands, was nearly hunted to extinction in the nineteenth century and was reduced to a single population containing perhaps no more than 20 individuals on the island of Guadalupe off the coast of Baja, California (figure 20.6). As a result of this bottleneck, the species has lost almost all of its genetic variation, even though the seal populations have rebounded and now number in the tens of thousands and breed in locations as far north as near San Francisco.

Any time a population becomes drastically reduced in numbers, such as in endangered species, the bottleneck effect is a potential problem. Even if population size rebounds, the lack of variability may mean that the species remains vulnerable to extinction—a topic to which we will return in chapter 58.

Selection favors some genotypes over others

As Darwin pointed out, some individuals leave behind more progeny than others, and the rate at which they do so is affected by phenotype and behavior. We describe the results of this process as **selection** (see figure 20.4e). In *artificial selection,* a breeder selects for the desired characteristics. In *natural selection,* environmental conditions determine which individuals in a population produce the most offspring.

Evolution by natural selection occurs when the following conditions are met:

1. **Phenotypic variation must exist among individuals in a population.** Natural selection works by favoring individuals with some traits over individuals with alternative traits. If no variation exists, natural selection cannot operate.

2. **Variation among individuals must result in differences in the number of offspring surviving in the next generation.** This is natural selection. Because of their phenotype or behavior, some individuals are more successful than others in producing offspring. Although many traits are phenotypically variable, individuals exhibiting variation do not always differ in survival and reproductive success.

3. **Phenotypic variation must have a genetic basis.** For natural selection (see item 2) to result in evolutionary change, the selected differences must have a genetic basis. Not all variation has a genetic basis—even genetically identical individuals may be phenotypically quite distinctive if they grow up in different environments. Such environmental effects are common in nature. In many turtles, for example, individuals that hatch from eggs laid in moist soil are heavier, with longer and wider shells, than individuals from nests in drier areas.

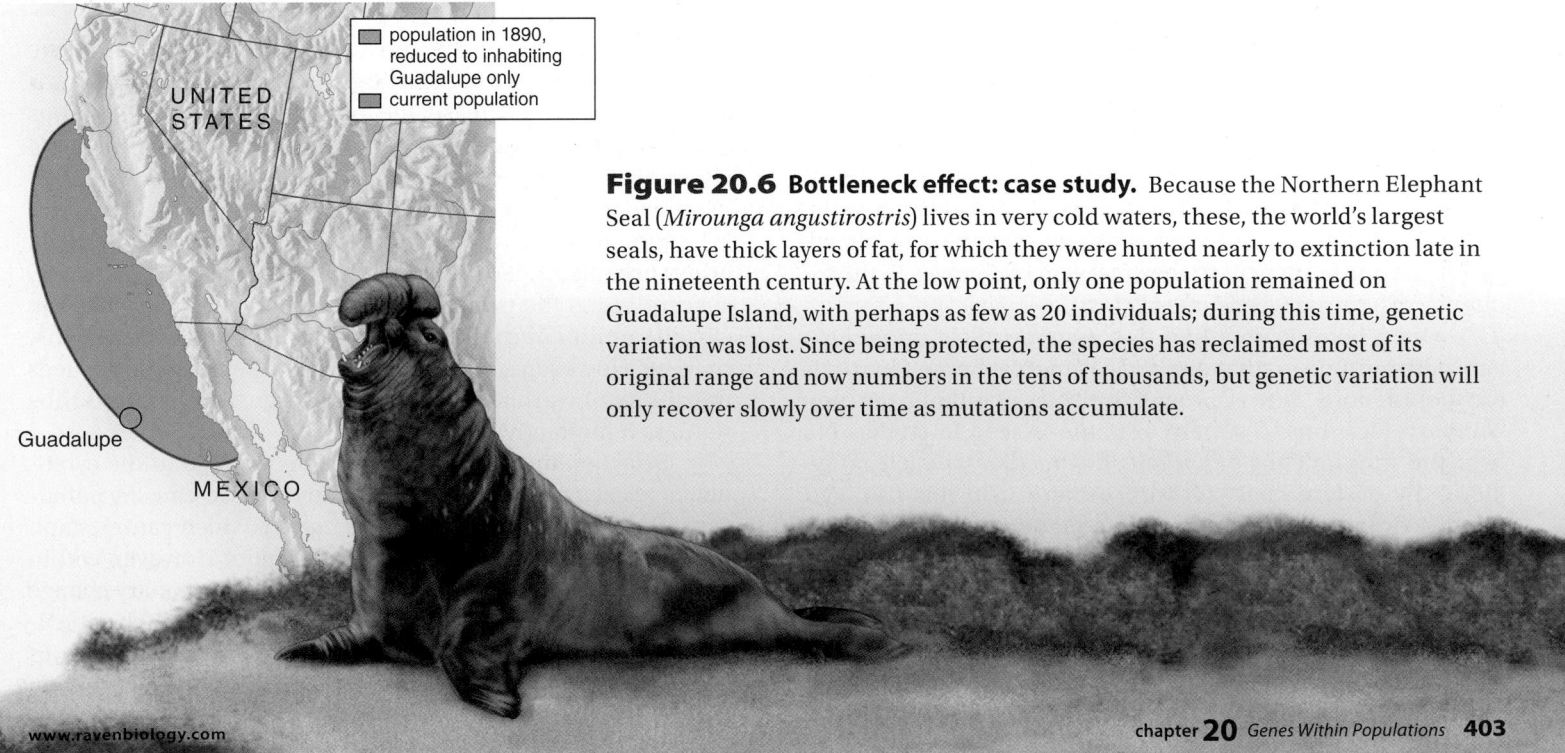

Figure 20.6 Bottleneck effect: case study. Because the Northern Elephant Seal (*Mirounga angustirostris*) lives in very cold waters, these, the world's largest seals, have thick layers of fat, for which they were hunted nearly to extinction late in the nineteenth century. At the low point, only one population remained on Guadalupe Island, with perhaps as few as 20 individuals; during this time, genetic variation was lost. Since being protected, the species has reclaimed most of its original range and now numbers in the tens of thousands, but genetic variation will only recover slowly over time as mutations accumulate.

population in 1890, reduced to inhabiting Guadalupe only
current population

UNITED STATES

Guadalupe

MEXICO

When phenotypically different individuals do not differ genetically, then differences in the number of their offspring will not alter the genetic composition of the population in the next generation, and thus, no evolutionary change will have occurred.

It is important to remember that natural selection and evolution are not the same—the two concepts often are incorrectly equated. Natural selection is a process, whereas evolution is the historical record, or outcome, of change through time. Natural selection (the process) can lead to evolution (the outcome), but natural selection is only one of several processes that can result in evolutionary change. Moreover, natural selection can occur without producing evolutionary change; only if variation is genetically based will natural selection lead to evolution.

Selection to avoid predators

The result of evolution driven by natural selection is that populations become better adapted to their environment. Many of the most dramatic documented instances of adaptation involve genetic changes that decrease the probability of capture by a predator. The caterpillar larvae of the common sulphur butterfly *Colias eurytheme* usually exhibit a pale green color, providing excellent camouflage against the alfalfa plants on which they feed. An alternative bright yellow color morph is reduced to very low frequency because this color renders the larvae highly visible on the food plant, making it easier for bird predators to see them (see figure 20.4e).

One of the most dramatic examples of background matching involves ancient lava flows in the deserts of the American Southwest. In these areas, the black rock formations produced when the lava cooled contrast starkly with the surrounding bright glare of the desert sand. Populations of many species of animals occurring on these rocks—including lizards, rodents, and a variety of insects—are dark in color, whereas sand-dwelling populations in surrounding areas are much lighter (figure 20.7).

Predation is the likely cause for these differences in color. Laboratory studies have confirmed that predatory birds such as owls are adept at picking out individuals occurring on backgrounds to which they are not adapted.

Selection to match climatic conditions

Many studies of selection have focused on genes encoding enzymes, because in such cases the investigator can directly assess the consequences to the organism of changes in the frequency of alternative enzyme alleles.

Often investigators find that enzyme allele frequencies vary with latitude, so that one allele is more common in northern populations, but is progressively less common at more southern locations. A superb example is seen in studies of a fish, the mummichog *(Fundulus heteroclitus),* which ranges along the eastern coast of North America. In this fish, geographic variation occurs in allele frequencies for the gene that produces the enzyme lactate dehydrogenase, which catalyzes the conversion of pyruvate to lactate (see section 7.8).

Biochemical studies show that the enzymes formed by these alleles function differently at different temperatures, thus explaining their geographic distributions. The form of the

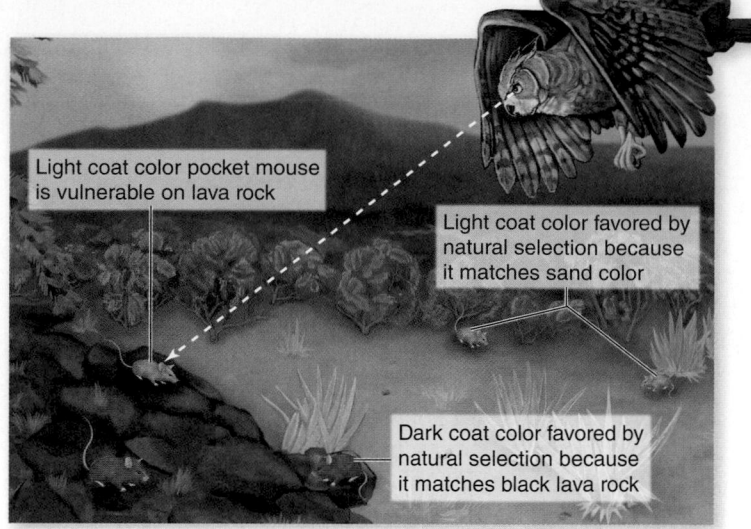

Figure 20.7 Pocket mice from the Tularosa Basin of New Mexico whose color matches their background. Black lava formations are surrounded by desert, and selection favors coat color in pocket mice that matches their surroundings. Genetic studies indicate that the differences in coat color are the result of small differences in the DNA of alleles of a single gene.

enzyme more frequent in the north is a better catalyst at low temperatures than is the enzyme from the south. Moreover, studies indicate that at low temperatures, individuals with the northern allele swim faster, and presumably survive better, than individuals with the alternative allele.

Selection for pesticide and microbial resistance

A particularly clear example of selection in natural populations is provided by studies of pesticide resistance in insects. The widespread use of insecticides has led to the rapid evolution of resistance in more than 500 pest species. The cost of this evolution, in terms of crop losses and increased pesticide use, has been estimated at $3–8 billion per year.

In the housefly, the resistance allele at the *pen* gene decreases the uptake of insecticide, whereas alleles at the *kdr* and *dld-r* genes decrease the number of target sites, thus decreasing the binding ability of the insecticide (figure 20.8). Other alleles enhance the ability of the insects' enzymes to identify and detoxify insecticide molecules.

Single genes are also responsible for resistance in other organisms. For example, Norway rats are normally susceptible to the pesticide warfarin, which diminishes the clotting ability of the rat's blood and leads to fatal hemorrhaging. However, a resistance allele at a single gene reduces the ability of warfarin to bind to its target enzyme and thus renders it ineffective.

Selection imposed by humans has also led to the evolution of resistance to antibiotics in many disease-causing pathogens. For example, *Staphylococcus aureus,* which causes staph infections, was initially treated by penicillin. However, within four years of mass-production of the drug, evolutionary change in *S. aureus* modified an enzyme so that it would attack penicillin and render it inactive. Since that time, several other drugs have been developed to attack the microbe, and each time

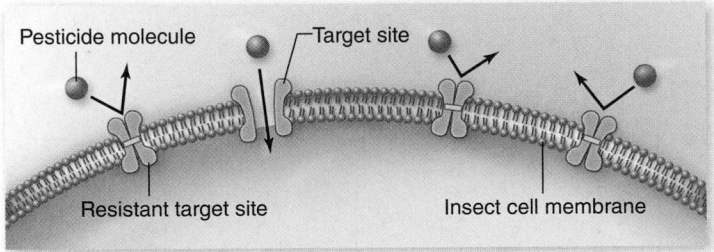

a. Insect cells with resistance allele at *pen* gene: decreased uptake of the pesticide.

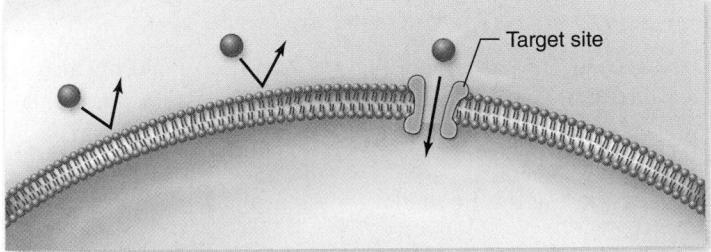

b. Insect cells with resistance allele at *kdr* gene: decreased number of target sites for the pesticide.

Figure 20.8 Selection for pesticide resistance. Resistance alleles at genes such as *pen* and *kdr* allow insects to be more resistant to pesticides. Insects that possess these resistance alleles have become more common through selection.

resistance has evolved. As a result, staph infections have re-emerged as a major health threat. The speed by which adaptation occurs may be surprising, but remember that microbes have enormous population sizes. Even though mutation rates are low, the vast size of these populations guarantees a steady supply of new mutations for natural selection to utilize.

Learning Outcomes Review 20.3

Five factors can bring about deviation from the predicted Hardy–Weinberg genotype frequencies. Of these, only selection regularly produces adaptive evolutionary change, but the genetic constitution of populations, and thus the course of evolution, can also be affected by mutation, gene flow, nonrandom mating, and genetic drift.

■ *How do each of these processes cause populations to vary from Hardy–Weinberg equilibrium?*

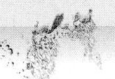

20.4 *Quantifying Natural Selection*

Learning Outcomes

1. *Define evolutionary fitness.*
2. *Explain the different components of fitness.*
3. *Demonstrate how the success of different phenotypes can be compared by calculating their relative fitness.*

Selection occurs when individuals with one phenotype leave more surviving offspring in the next generation than individuals with an alternative phenotype. Evolutionary biologists quantify reproductive success as **fitness,** the number of surviving offspring left in the next generation.

Fitness is a relative concept; the most fit phenotype is simply the one that produces, on average, the greatest number of offspring.

A phenotype with greater fitness usually increases in frequency

Suppose, for example, that in a population of toads, two phenotypes exist: green and brown. Suppose, further, that green toads leave, on average, 4.0 offspring in the next generation, but brown toads leave only 2.5. By custom, the most fit phenotype is assigned a fitness value of 1.0, and other phenotypes are expressed as relative proportions. In this case, the fitness of the green phenotype would be 4.0/4.0 = 1.000, and the fitness of the brown phenotype would be 2.5/4.0 = 0.625. The difference in fitness would therefore be 1.000 – 0.625 = 0.375. A difference in fitness of 0.375 is quite large; natural selection in this case strongly favors the green phenotype.

If differences in color have a genetic basis, then we would expect evolutionary change to occur; the frequency of green toads should be substantially greater in the next generation. Further, if the fitness of two phenotypes remained unchanged, we would expect alleles for the brown phenotype eventually to disappear from the population.

 Inquiry question Why might the frequency of green toads not increase in the next generation, even if color differences have a genetic basis?

Fitness may consist of many components

Although selection is often characterized as "survival of the fittest," differences in survival are only one component of fitness.

Even if no differences in survival occur, selection may operate if some individuals are more successful than others in attracting mates. In many territorial animal species, for example, large males mate with many females, and small males rarely get to mate. Selection with respect to mating success is termed *sexual selection;* we describe this topic more fully in the discussion of behavioral biology in chapter 54.

In addition, the number of offspring produced per mating is also important. Large female frogs and fish lay more eggs than do smaller females, and thus they may leave more offspring in the next generation.

Fitness is therefore a combination of survival, mating success, and number of offspring per mating. Selection favors phenotypes with the greatest fitness, but predicting fitness from a single component can be tricky because traits favored for one component of fitness may be at a disadvantage for others. As an example, in water striders, larger females lay more eggs per day (figure 20.9). Thus, natural selection at this stage

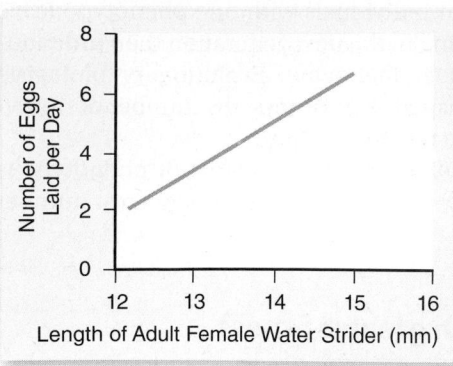
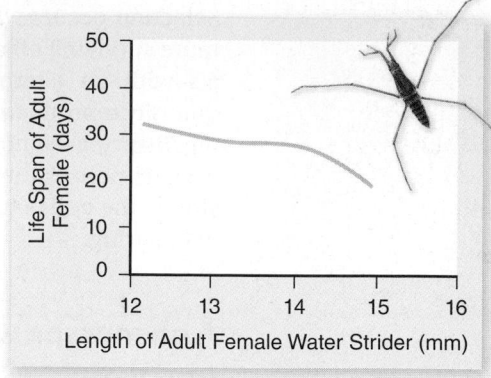
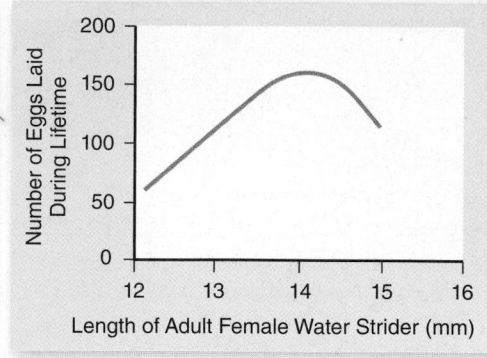

Figure 20.9 Body size and egg-laying in water striders. Larger female water striders lay more eggs per day (left panel), but also survive for a shorter period of time (center panel). As a result, intermediate-sized females produce the most offspring over the course of their entire lives and thus have the highest fitness (right panel).

 Inquiry question What evolutionary change in body size might you expect? If the number of eggs laid per day was not affected by body size, would your prediction change?

 Data analysis Assuming that the values on the *x*-axis represent the final body size of different animals and that the *y*-axis represents egg-laying rate and survival once they reach that size, how many eggs would you expect a 12-mm water strider to lay? And a 15-mm strider?

favors large size. However, larger females also die at a younger age and thus have fewer opportunities to reproduce than smaller females. Overall, the two opposing directions of selection cancel each other out, and the intermediate-sized females leave the most offspring in the next generation.

Learning Outcomes Review 20.4

Fitness is defined by an organism's reproductive success relative to other members of its population. This success is determined by how long it survives, how often it mates, and how many offspring it produces per mating. Relative fitness assigns numerical values to different phenotypes relative to the most fit phenotype.

■ *Is one of these factors always the most important in determining reproductive success? Explain.*

20.5 Natural Selection's Role in Maintaining Variation

Learning Outcomes

1. Define frequency-dependent selection, oscillating selection, and heterozygote advantage.
2. Explain how these processes affect the amount of genetic variation in a population.

In the previous pages, natural selection has been discussed as a process that removes variation from a population by favoring one allele over others at a gene locus. However, in some

circumstances, selection can do exactly the opposite and actually maintain population variation.

Frequency-dependent selection may favor either rare or common phenotypes

In some circumstances, the fitness of a phenotype depends on its frequency within the population, a phenomenon termed **frequency-dependent selection.** This type of selection favors certain phenotypes depending on how commonly or uncommonly they occur.

Negative frequency-dependent selection

In negative frequency-dependent selection, rare phenotypes are favored by selection. Assuming a genetic basis for phenotypic variation, such selection will have the effect of making rare alleles more common, thus maintaining variation.

Negative frequency-dependent selection can occur for many reasons. For example, it is well known that animals or people searching for something form a "search image." That is, they become particularly adept at picking out certain objects. Consequently, predators may form a search image for common prey phenotypes. Rare forms may thus be preyed upon less frequently.

An example is fish predation on an insect, the water boatman, which occurs in three different colors. Experiments indicate that each of the color types is preyed upon disproportionately when it is the most common one; fish eat more of the common-colored insects than would occur by chance alone (figure 20.10).

Another cause of negative frequency dependence is resource competition. If genotypes differ in their resource requirements, as occurs in many plants, then the rarer genotype will have fewer competitors. When the different resource

Question: *Does negative frequency-dependent selection maintain variation in a population?*

Hypothesis: *Fish may disproportionately capture water boatmen (a type of aquatic insect) with the most common color.*

Experiment: *Place predatory fish in different aquaria with the different frequencies of the color types in each aquarium.*

Result: *Fish prey disproportionately on the common color in each aquarium. The rare color in each aquarium generally survives best.*

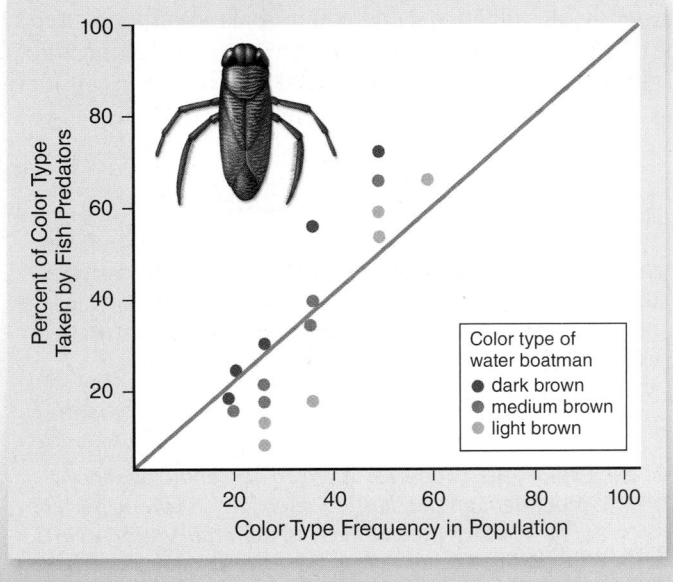

Figure 20.10 Frequency-dependent selection.

 Data analysis How can this experiment distinguish between frequency-dependent and directional selection?

types are equally abundant, the rarer genotype will be at an advantage relative to the more common genotype.

Positive frequency-dependent selection

Positive frequency-dependent selection has the opposite effect; by favoring common forms, it tends to eliminate variation from a population. For example, predators don't always select common individuals. In some cases, "oddballs" stand out from the rest and attract attention (figure 20.11).

The strength of selection should change through time as a result of frequency-dependent selection. In negative frequency-dependent selection, rare genotypes should become increasingly common, and their selective advantage will decrease correspondingly. Conversely, in positive frequency dependence, the rarer a genotype becomes, the greater the chance it will be selected against.

In oscillating selection, the favored phenotype changes as the environment changes

In some cases, selection favors one phenotype at one time and another phenotype at another time, a phenomenon called

oscillating selection. If selection repeatedly oscillates in this fashion, the effect will be to maintain genetic variation in the population.

One example, discussed in chapter 21, concerns the medium ground finch of the Galápagos Islands. In times of drought, the supply of small, soft seeds is depleted, but there are still enough large seeds around. Consequently, birds with big bills are favored. However, when wet conditions return, the ensuing abundance of small seeds favors birds with smaller bills.

Oscillating selection and frequency-dependent selection are similar because in both cases the form of selection changes through time. But it is important to recognize that they are not the same: In oscillating selection, the fitness of a phenotype does not depend on its frequency; rather, environmental changes lead to the oscillation in selection. In contrast, in frequency-dependent selection, it is the change in frequencies themselves that leads to the changes in fitness of the different phenotypes.

In some cases, heterozygotes may exhibit greater fitness than homozygotes

If heterozygotes are favored over homozygotes, then natural selection actually tends to maintain variation in the population. This **heterozygote advantage** favors individuals with copies of both alleles, and thus works to maintain both alleles in the population. Some evolutionary biologists believe that heterozygote advantage is pervasive and can explain the high levels of polymorphism observed in natural populations. Others, however, believe that it is relatively rare.

The best documented example of heterozygote advantage is sickle cell anemia, a hereditary disease affecting hemoglobin in humans. Individuals with sickle cell anemia exhibit symptoms of severe anemia and abnormal red blood cells that are irregular in shape, with a great number of long, sickle-shaped

Figure 20.11 Positive frequency-dependent selection. In some cases, rare individuals stand out from the rest and draw the attention of predators; thus, in these cases, common phenotypes have the advantage (positive frequency-dependent selection).

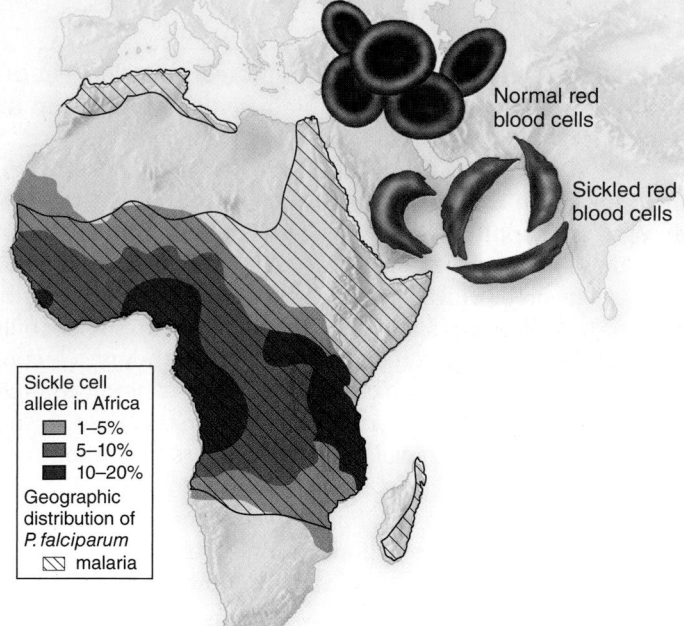

Normal red blood cells

Sickled red blood cells

Sickle cell allele in Africa
■ 1–5%
■ 5–10%
■ 10–20%
Geographic distribution of *P. falciparum*
▨ malaria

Figure 20.12 Frequency of sickle cell allele and distribution of *Plasmodium falciparum* malaria. The red blood cells of people homozygous for the sickle cell allele collapse into sickled shapes when the oxygen level in the blood is low. The distribution of the sickle cell allele in Africa coincides closely with that of *P. falciparum* malaria.

cells (figure 20.12). Chapter 13 discusses why the sickle cell mutation (*S*) causes red blood cells to sickle.

The average incidence of the *S* allele in central African populations is about 0.12, far higher than that found among African Americans. From the Hardy–Weinberg principle, you can calculate that 1 in 5 central African individuals is heterozygous at the *S* allele, and 1 in 100 is homozygous and develops the fatal form of the disorder. People who are homozygous for the sickle cell allele almost never reproduce because they usually die before they reach reproductive age.

Why, then, is the *S* allele not eliminated from the central African population by selection rather than being maintained at such high levels? As it turns out, one of the leading causes of illness and death in central Africa, especially among young children, is malaria. People who are heterozygous for the sickle cell allele (and thus do not suffer from sickle cell anemia) are much less susceptible to malaria. The reason is that when the parasite that causes malaria, *Plasmodium falciparum,* enters a red blood cell, it causes extremely low oxygen tension in the cell, which leads to sickling in cells of individuals either homozygous or heterozygous for the sickle cell allele (but not in individuals that do not have the sickle cell allele). Such cells are quickly filtered out of the bloodstream by the spleen, thus eliminating the parasite. (The spleen's filtering effect is what leads to anemia in persons homozygous for the sickle cell allele because large numbers of red blood cells become sickle-shaped and are removed; in the case of heterozygotes, only those cells containing the *Plasmodium* parasite sickle, whereas the remaining cells are not affected, and thus anemia does not occur.)

Consequently, even though most homozygous recessive individuals die at a young age, the sickle cell allele is maintained at high levels in these populations because it is

associated with resistance to malaria in heterozygotes and also, for reasons not yet fully understood, with increased fertility in female heterozygotes. Figure 20.12 shows the overlap between regions where sickle cell anemia is found and where malaria is prevalent.

For people living in areas where malaria is common, having the sickle cell allele in the heterozygous condition has adaptive value (see figure 20.12). Among African Americans, however, many of whose ancestors have lived for many generations in a country where malaria is now essentially absent, the environment does not place a premium on resistance to malaria. Consequently, no adaptive value counterbalances the ill effects of the disease; in this nonmalarial environment, selection is acting to eliminate the *S* allele. Only 1 in 375 African Americans develops sickle cell anemia, far fewer than in central Africa.

Learning Outcomes Review 20.5

Selection can maintain variation within populations in a number of ways. Negative frequency-dependent selection tends to favor rare phenotypes. Oscillating selection favors different phenotypes at different times. In some cases, heterozygotes have a selective advantage that may act to retain alleles that are deleterious in the homozygous state.

- **How would genetic variation in a population change if heterozygotes had the lowest fitness?**
- **Explain the difference between negative frequency-dependent and oscillating selection. Over long periods of time, which is more likely to maintain variation in a population?**

20.6 Selection Acting on Traits Affected by Multiple Genes

Learning Outcomes

1. *Define and contrast disruptive, stabilizing, and directional selection.*
2. *Explain the evolutionary outcome of each of these types of selection.*

In nature, many traits—perhaps most—are affected by more than one gene. The interactions between genes are typically complex, as you saw in chapter 12. For example, alleles of many different genes play a role in determining human height (see figure 12.13). In such cases, selection operates on all the genes, influencing most strongly those that make the greatest contribution to the phenotype. How selection changes the population depends on which genotypes are favored.

Disruptive selection removes intermediates

In some situations, selection acts to eliminate intermediate types, a phenomenon called **disruptive selection** (figure 20.13*a*). A clear example is the different beak sizes of the African

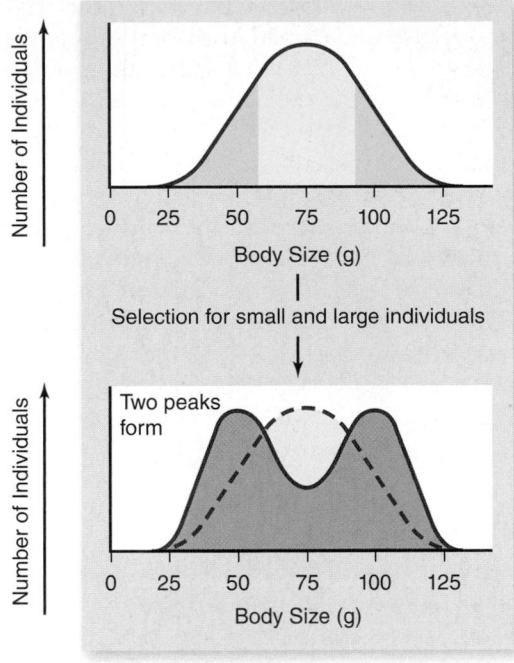

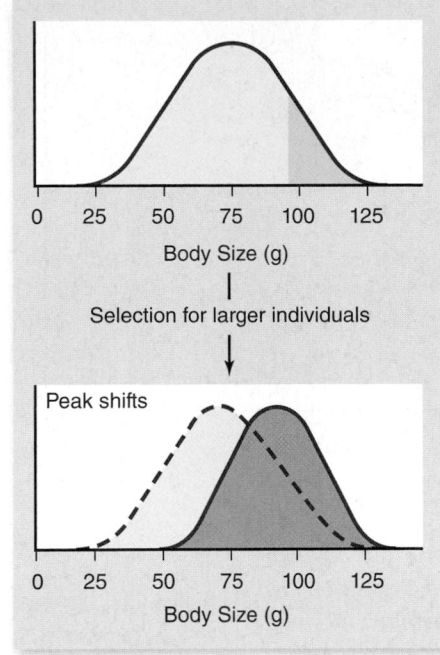

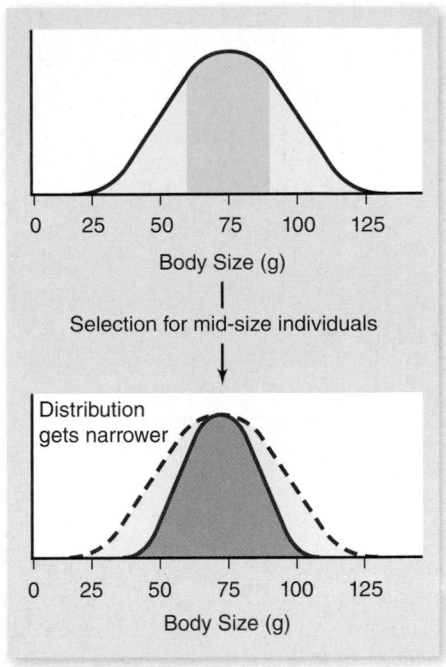

a. Disruptive selection

b. Directional selection

c. Stabilizing selection

Figure 20.13 Three kinds of selection. The top panels show the populations before selection has occurred (under the solid red line). Within the population, those favored by selection are shown in light brown. The bottom panels indicate what the populations would look like in the next generation. The dashed red lines are the distribution of the original population and the solid, dark brown lines are the true distribution of the population in the next generation. *a.* In disruptive selection, individuals in the middle of the range of phenotypes of a certain trait are selected against, and the extreme forms of the trait are favored. *b.* In directional selection, individuals concentrated toward one extreme of the array of phenotypes are favored. *c.* In stabilizing selection, individuals with midrange phenotypes are favored, with selection acting against both ends of the range of phenotypes.

black-bellied seedcracker finch (figure 20.14). Populations of these birds contain individuals with large and small beaks, but very few individuals with intermediate-sized beaks.

As their name implies, these birds feed on seeds, and the available seeds fall into two size categories: large and small. Only large-beaked birds can open the tough shells of large seeds, whereas birds with the smaller beaks are more adept at handling small seeds. Birds with intermediate-sized beaks are at a disadvantage with both seed types—they are unable to open large seeds and too clumsy to efficiently process small seeds. Consequently, selection acts to eliminate the intermediate phenotypes, in effect partitioning (or "disrupting") the population into two phenotypically distinct groups.

Directional selection eliminates phenotypes on one end of a range

When selection acts to eliminate one extreme from an array of phenotypes, the genes promoting this extreme become less frequent in the population and may eventually disappear. This form of selection is called **directional selection** (see figure 20.13*b*). Thus, in the *Drosophila* population illustrated in figure 20.15, eliminating flies that move toward light causes the population over time to contain fewer individuals with alleles promoting such behavior. If you were to pick an individual at random from a later generation of flies, there is a smaller chance that the fly would spontaneously move toward light than if you had

SCIENTIFIC THINKING

Question: *Does disruptive selection promote differences in beak size in the African Black-bellied Seedcracker Finches (Pyrenestes ostrinus)?*

Field Study: *Capture, measure, and release birds in a population. Follow the birds through time to determine how long each lives.*

Result: *Large- and small-beaked birds have higher survival rates than birds with intermediate-sized beaks.*

Interpretation: *What would happen if the distribution of seed size and hardness in the environment changed?*

Figure 20.14 Disruptive selection for large and small beaks. Differences in beak size in the black-bellied seedcracker finch of west Africa are the result of disruptive selection.

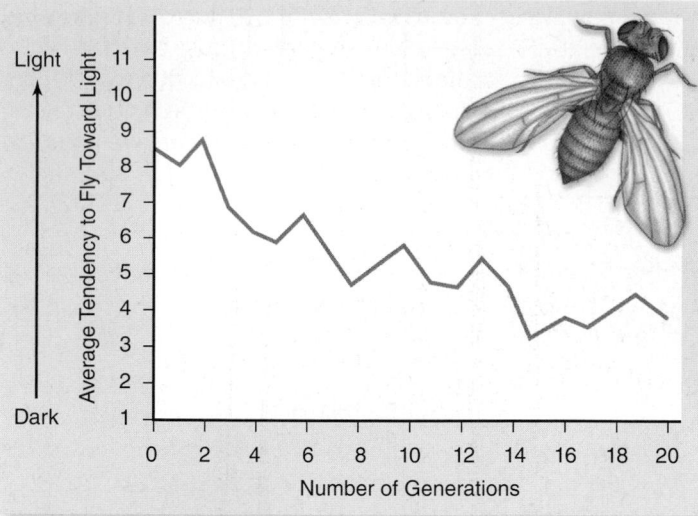

Figure 20.15 Directional selection for negative phototropism in *Drosophila*. Flies that moved toward light were discarded, and only flies that moved away from light were used as parents for the next generation. This procedure was repeated for 20 generations, producing substantial evolutionary change.

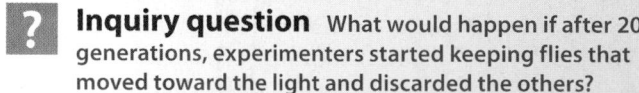

? Inquiry question What would happen if after 20 generations, experimenters started keeping flies that moved toward the light and discarded the others?

selected a fly from the original population. Artificial selection has changed the population in the direction of being less attracted to light. Directional selection often occurs in nature when the environment changes; one example is the widespread evolution of pesticide resistance discussed earlier in this chapter.

Stabilizing selection favors individuals with intermediate phenotypes

When selection acts to eliminate both extremes from an array of phenotypes, the result is to increase the frequency of the already common intermediate type. This form of selection is called **stabilizing selection** (see figure 20.13c). In effect, selection is operating to prevent change away from this middle range of values. Selection does not change the most common phenotype of the population, but rather makes it even more common by eliminating extremes. Many examples are known. In humans, infants with intermediate weight at birth have the highest survival rate (figure 20.16). In ducks and chickens, eggs of intermediate weight have the highest hatching success.

Learning Outcomes Review 20.6

In disruptive selection, intermediate forms of a trait diminish; in stabilizing selection, intermediates increase, whereas in disruptive selection they decrease. Directional selection shifts frequencies toward one end or the other and may eventually eliminate alleles entirely.

■ *How does directional selection differ from frequency-dependent selection?*

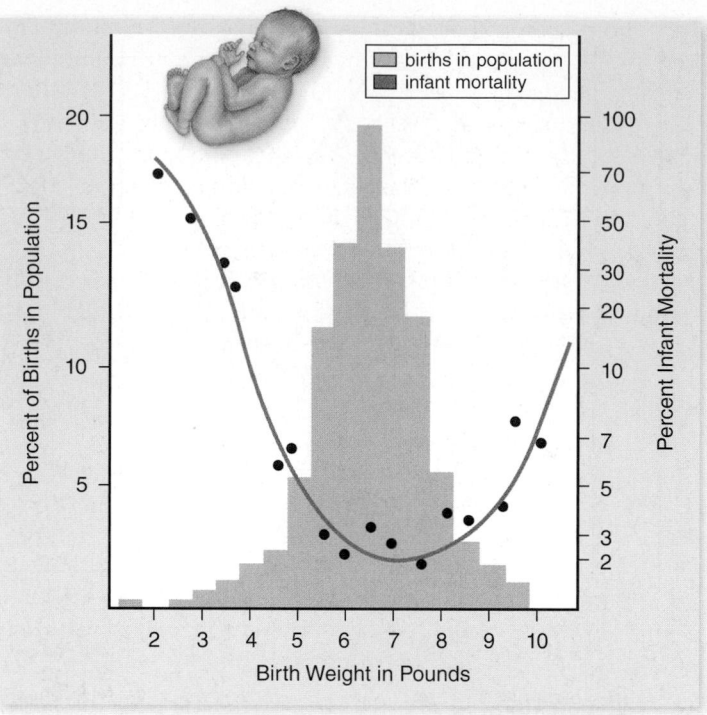

Figure 20.16 Stabilizing selection for birth weight in humans. The death rate among babies (red curve; right *y*-axis) is lowest at an intermediate birth weight; both smaller and larger babies have a greater tendency to die than those around the most frequent weight (tan area; left *y*-axis) of between 7 and 8 pounds. Recent medical advances have reduced mortality rates for small and large babies.

? Inquiry question As improved medical technology leads to decreased infant mortality rates, how would you expect the distribution of birth weights in the population to change?

Data analysis Sketch what a graph would look like that showed the absolute number of infant mortality deaths vs. birth weight.

20.7 Experimental Studies of Natural Selection

Learning Outcome

1. *Explain how experiments can be used to test evolutionary hypotheses.*

To study evolution, biologists have traditionally investigated what has happened in the past, sometimes many millions of years ago. To learn about dinosaurs, a paleontologist looks at dinosaur fossils. To study human evolution, an anthropologist looks at human fossils and, increasingly, examines the "family tree" of mutations that have accumulated in human DNA over millions of years. In this traditional approach, evolutionary biology is similar to astronomy and history, relying on observation rather than experimentation to examine ideas about past events.

Nonetheless, evolutionary biology is not entirely an observational science. Darwin was right about many things, but one area in which he was mistaken concerns the pace at which evolution occurs. Darwin thought that evolution occurred at a very slow, almost imperceptible pace. But in recent years many case studies have demonstrated that in some circumstances, evolutionary change can occur rapidly. Consequently, experimental studies can be devised to test evolutionary hypotheses.

Although laboratory studies on fruit flies and other organisms have been common for more than 50 years, scientists have only recently started conducting experimental studies of evolution in nature. One excellent example of how observations of the natural world can be combined with rigorous experiments in the lab and in the field concerns research on guppies.

Guppy color variation in different environments suggests natural selection at work

The guppy is a popular aquarium fish because of its bright coloration and prolific reproduction. In nature, guppies are found in small streams in northeastern South America and in many mountain streams on the nearby island of Trinidad. One interesting feature of several of the streams is that they have waterfalls. Amazingly, guppies and some other fish are capable of colonizing portions of the stream above the waterfall.

The killifish is a particularly good colonizer; apparently on rainy nights, it will wriggle out of the stream and move through the damp leaf litter. Guppies are not so proficient, but they are good at swimming upstream. During flood seasons, rivers sometimes overflow their banks, creating secondary channels that move through the forest. On these occasions, guppies may be able to swim upstream in the secondary channels and invade the pools above waterfalls.

By contrast, some species are not capable of such dispersal and thus are only found in streams below the first waterfall. One species whose distribution is restricted by waterfalls is the pike cichlid, a voracious predator that feeds on other fish, including guppies.

Because of these barriers to dispersal, guppies can be found in two very different environments. In pools just below the waterfalls, predation by the pike cichlid is a substantial risk, and rates of survival are relatively low. But in similar pools just above the waterfall, the only predator present is the killifish, which rarely preys on guppies.

Guppy populations above and below waterfalls exhibit many differences. In the high-predation pools, guppies exhibit drab coloration. Moreover, they tend to reproduce at a younger age and attain relatively smaller adult sizes. Male fish above the waterfall, in contrast, are colorful (figure 20.17), mature later, and grow to larger sizes.

These differences suggest the operation of natural selection. In the low-predation environment, males display gaudy colors and spots that they use to court females. Moreover, larger males are most successful at holding territories and mating with females, and larger females lay more eggs. Thus, in the absence of predators, larger and more colorful fish may have produced more offspring, leading to the evolution of those traits.

In pools below the waterfall, natural selection would favor different traits. Colorful males are likely to attract the attention of the pike cichlid, and high predation rates mean that most fish live short lives. Individuals that are more drab and shunt energy into early reproduction, rather than growth to a larger size, are therefore likely to be favored by natural selection.

Experimentation reveals the agent of selection

Although the differences between guppies living above and below the waterfalls suggest evolutionary responses to differences in the strength of predation, alternative explanations are possible. Perhaps, for example, only very large fish are capable of crawling past the waterfall to colonize pools. If this were the case, then a founder effect would occur in which the new population was established solely by individuals with genes for large size. The only way to rule out such alternative possibilities is to conduct a controlled experiment.

The laboratory experiment

The first experiments were conducted in large pools in laboratory greenhouses. At the start of the experiment, a group of

Figure 20.17 The evolution of protective coloration in guppies. In pools below waterfalls where predation is high, male guppies are drab in color. In the absence of the highly predatory pike cichlid *(Crenicichla alta)* in pools above waterfalls, male guppies are much more colorful and attractive to females. The killifish is also a predator, but it only rarely eats guppies. The evolution of these differences in guppies can be experimentally tested.

2000 guppies was divided equally among 10 large pools. Six months later, pike cichlids were added to four of the pools and killifish to another four, with the remaining two pools left to serve as "no-predation" controls.

Fourteen months later (which corresponds to 10 guppy generations), the scientists compared the populations. The guppies in the killifish and control pools were indistinguishable—brightly colored and large. In contrast, the guppies in the pike cichlid pools were smaller and drab in coloration (figure 20.18).

These results established that predation can lead to rapid evolutionary change, but do these laboratory experiments reflect what occurs in nature?

SCIENTIFIC THINKING

Question: *Does the presence of predators affect the evolution of guppy color?*

Hypothesis: *Predation on the most colorful individuals will cause a population to become increasingly dull through time. Conversely, in populations with few or no predators, increased color will evolve.*

Experiment: *Establish laboratory populations of guppies in large pools with or without predators.*

Result: *The populations with predators evolved to have fewer spots, while the populations in pools without predators evolved more spots.*

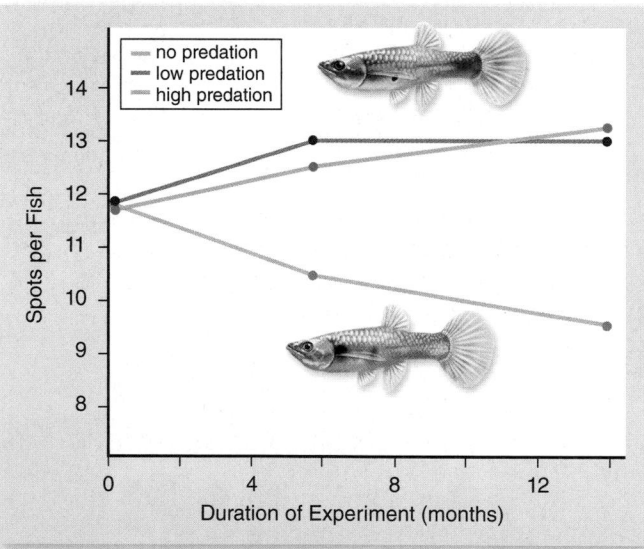

Interpretation: *Why does color increase in the absence of predators? How would you test your hypothesis?*

Figure 20.18 Evolutionary change in spot number.
Guppy populations raised for 10 generations in low-predation or no-predation environments in laboratory greenhouses evolved a greater number of spots, whereas selection in more dangerous environments, such as the pools with the highly predatory pike cichlid, led to less conspicuous fish. The same results are seen in field experiments conducted in pools above and below waterfalls.

? Inquiry question How do these results depend on the manner by which the guppy predators locate their prey?

The field experiment
To find out whether the laboratory results were an accurate reflection of natural processes, the scientists located two streams that had guppies in pools below a waterfall, but not above it. As in other Trinidadian streams, the pike cichlid was present in the lower pools, but only the killifish was found above the waterfalls.

The scientists then transplanted guppies to the upper pools and returned at several-year intervals to monitor the populations. Despite originating from populations in which predation levels were high, the transplanted populations rapidly evolved the traits characteristic of low-predation guppies: they matured late, attained greater size, and had brighter colors. The control populations in the lower pools, by contrast, continued to be drab and to mature early and at a smaller size. Laboratory analysis confirmed that the variations between the populations were the result of genetic differences.

These results demonstrate that substantial evolutionary change can occur in less than 12 years. More generally, these studies indicate how scientists can formulate hypotheses about how evolution occurs and then test these hypotheses in natural conditions. The results give strong support to the theory of evolution by natural selection.

Learning Outcome Review 20.7
Although much of evolutionary theory is derived from observation, experiments are sometimes possible in natural settings. Studies have revealed that traits can shift in populations in a relatively short time. The data obtained from evolutionary experiments can be used to refine theoretical assumptions.

■ *What experiments could you design to test other examples of natural selection, such as the evolution of pesticide resistance or background color matching?*

20.8 Interactions Among Evolutionary Forces

Learning Outcomes
1. *Discuss how evolutionary processes can work simultaneously, but in opposing ways.*
2. *Evaluate what determines the evolutionary outcome when multiple processes are operating simultaneously.*

The amount of genetic variation in a population may be determined by the relative strength of different evolutionary processes. Sometimes these processes act together, and in other cases they work in opposition.

Mutation and genetic drift may counter selection

In theory, if allele *B* mutates to allele *b* at a high enough rate, allele *b* could be maintained in the population, even if natural selection strongly favored allele *B*. In nature, however, mutation rates are rarely high enough to counter the effects of natural selection.

The effect of natural selection also may be countered by genetic drift. Both of these processes may act to remove variation from a population. But selection is a nonrandom process that operates to increase the representation of alleles that enhance survival and reproductive success, whereas genetic drift is a random process in which any allele may increase. Thus, in some cases, drift may lead to a decrease in the frequency of an allele that is favored by selection. In some extreme cases, drift may even lead to the loss of a favored allele from a population.

Remember, however, that the magnitude of drift is inversely related to population size; consequently, natural selection is expected to overwhelm drift, except when populations are very small.

Gene flow may promote or constrain evolutionary change

Gene flow can be either a constructive or a constraining force. On one hand, gene flow can spread a beneficial mutation that arises in one population to other populations. On the other hand, gene flow can impede adaptation within a population by the continual flow of inferior alleles from other populations.

Consider two populations of a species that live in different environments. In this situation, natural selection might favor different alleles—*B* and *b*—in the two populations. In the absence of other evolutionary processes such as gene flow, the frequency of *B* would be expected to reach 100% in one population and 0% in the other. However, if gene flow occurred between the two populations, then the less favored allele would continually be reintroduced into each population. As a result, the frequency of the alleles in the populations would reflect a balance between the rate at which gene flow brings the inferior allele into a population, and the rate at which natural selection removes it.

A classic example of gene flow opposing natural selection occurs on abandoned mine sites in Great Britain. Although mining activities ceased hundreds of years ago, the concentration of metal ions in the soil is still much greater than in surrounding areas. Large concentrations of heavy metals are generally toxic to plants, but alleles at certain genes confer the ability to grow on soils high in heavy metals. The ability to tolerate heavy metals comes at a price, however; individuals with the resistance allele exhibit lower growth rates on nonpolluted soil. Consequently, we would expect the resistance allele to occur with a frequency of 100% on mine sites and 0% elsewhere.

Heavy-metal tolerance has been studied intensively in the slender bent grass *Agrostis tenuis,* in which the resistance allele occurs at intermediate levels in many areas (figure 20.19). The explanation relates to the reproductive system of this grass, in which pollen, the floral equivalent of sperm, is dispersed by the wind. As a result, pollen grains—and the alleles they carry—can move great distances, leading to levels of gene flow between mine sites and unpolluted areas high enough to counteract the effects of natural selection.

In general, the extent to which gene flow can hinder the effects of natural selection should depend on the relative strengths of the two processes. In species in which gene flow is generally strong, such as in birds and wind-pollinated plants,

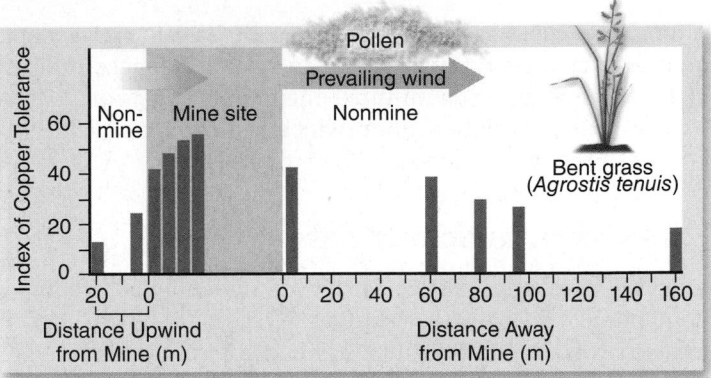

Figure 20.19 Degree of copper tolerance in grass plants on and near ancient mine sites. Individuals with tolerant alleles have decreased growth rates on unpolluted soil. Thus, we would expect copper tolerance to be 100% on mine sites and 0% on nonmine sites. However, prevailing winds blow pollen containing nontolerant alleles onto the mine site and tolerant alleles beyond the site's borders. The amount of pollen received decreases with distance, which explains the changes in levels of tolerance. The index of copper tolerance is calculated as the growth rate of a plant on soil with high concentrations of copper relative to growth rate on soils with low levels of copper; the higher the index, the more tolerant the plant is of heavy metal pollution.

Data analysis Examine the index of copper tolerance on nonmine areas. What does it suggest about the factor responsible? In particular, how does the index change as a function of distance from the mine on the right-hand side? And how does the relationship between distance and index value differ on the two sides of the mine? What process might be responsible for such patterns?

the frequency of the allele less favored by natural selection may be relatively high. In more sedentary species that exhibit low levels of gene flow, such as salamanders, the favored allele should occur at a frequency near 100%.

Learning Outcomes Review 20.8

Allele frequencies sometimes reflect a balance between opposing processes. Gene flow, for example, may increase some alleles while natural selection decreases them. Where several processes are involved, observed frequencies depend on the relative strength of the processes.

■ *Under what circumstances might evolutionary processes operate in the same direction, and what would be the outcome?*

20.9 The Limits of Selection

Learning Outcomes

1. *Define pleiotropy and epistasis.*
2. *Explain how these phenomena may affect the evolutionary response to selective pressure.*

Although selection is the most powerful of the principal agents of genetic change, there are limits to what it can accomplish. These limits result from multiple phenotypic effects of alleles, lack of genetic variation upon which selection can act, and interactions between genes.

Genes have multiple effects

Alleles often affect multiple aspects of a phenotype (the phenomenon of *pleiotropy;* see chapter 12). These multiple effects tend to set limits on how much a phenotype can be altered.

For example, selecting for large clutch size in chickens eventually leads to eggs with thinner shells that break more easily. For this reason, we could never produce chickens that lay eggs twice as large as the best layers do now. Likewise, we cannot produce gigantic cattle that yield twice as much meat as our leading breeds, or corn with an ear at the base of every leaf, instead of just at the bases of a few leaves.

Evolution requires genetic variation

Over 80% of the gene pool of the thoroughbred horses racing today goes back to 31 ancestors from the late eighteenth century. Despite intense directional selection on thoroughbreds, their performance times have not improved for more than 50 years (figure 20.20). Decades of intense selection presumably have removed variation from the population at a rate greater than mutation can replenish it, such that little genetic variation now remains, and evolutionary change is not possible.

In some cases, phenotypic variation for a trait may never have had a genetic basis. The compound eyes of insects are made up of hundreds of visual units, termed ommatidia (described in chapter 34). In some individuals, the left eye contains more ommatidia than the right. In other individuals, the right eye contains more than the left (figure 20.21). However, despite intense selection experiments in the laboratory, scientists have never been able to produce a line of

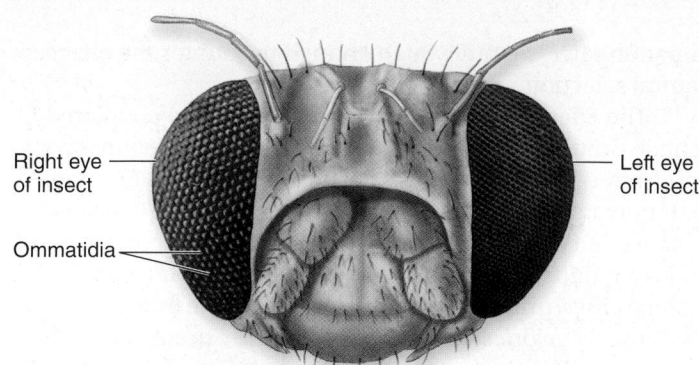

Figure 20.21 Phenotypic variation in insect ommatidia. In some individuals, the number of ommatidia in the left eye is greater than the number in the right.

fruit flies that consistently has more ommatidia in the left eye than in the right.

The reason is that separate genes do not exist for the left and right eyes. Rather, the same genes affect both eyes, and differences in the number of ommatidia result from differences that occur as the eyes are formed in the development process. Thus, despite the existence of phenotypic variation, no underlying genetic variation is available for selection to favor.

Gene interactions affect fitness of alleles

As discussed in chapter 12, *epistasis* is the phenomenon in which an allele for one gene may have different effects, depending on alleles present at other genes. Because of epistasis, the selective advantage of an allele at one gene may vary from one genotype to another. If a population is polymorphic for a second gene, then selection on the first gene may be constrained because different alleles are favored in different individuals of the same population.

Studies on bacteria illustrate how selection on alleles for one gene can depend on which alleles are present at other genes. In *E. coli,* two biochemical pathways exist to break down gluconate, each using enzymes produced by different genes. One gene produces the enzyme 6-PGD, for which there are several alleles. When the common allele for the second gene, which codes for the other biochemical pathway, is present, selection does not favor one allele over another at the 6-PGD gene. In some *E. coli,* however, an alternative allele at the second gene occurs that is not functional. The bacteria with this alternative allele are forced to rely only on the 6-PGD pathway, and in this case, selection favors one 6-PGD allele over another. Thus, epistatic interactions exist between the two genes, and the outcome of natural selection on the 6-PGD gene depends on which alleles are present at the second gene.

Figure 20.20 Selection for increased speed in racehorses is no longer effective. Kentucky Derby winning speeds have not improved significantly since 1950.

? Inquiry question What might explain the lack of change in winning speeds?

Learning Outcomes Review 20.9

In pleiotropy, a single gene affects multiple traits; in epistasis, interaction between alleles of different genes affects a single trait. Both these conditions can constrain the effects of natural selection.

■ *How can epistasis and pleiotropy constrain the evolutionary response to natural selection?*

20.1 Genetic Variation and Evolution

Many processes can lead to evolutionary change.

Darwin proposed that evolution of species occurs by the process of natural selection. Other processes can also lead to evolutionary change.

Populations contain ample genetic variation.

For a population to be able to evolve, it must contain genetic variation. DNA testing shows that natural populations generally have substantial variation.

20.2 Changes in Allele Frequency (figure 20.3)

The Hardy–Weinberg principle allows prediction of genotype frequencies.

Hardy–Weinberg equilibrium exists when observed genotype frequencies match the prediction from calculated frequencies. It occurs only when evolutionary processes are not acting to shift the distribution of alleles or genotypes in the population.

Hardy–Weinberg predictions can be applied to data to find evidence of evolutionary processes.

If genotype frequencies are not in Hardy–Weinberg equilibrium, then evolutionary processes must be at work.

20.3 Five Agents of Evolutionary Change (figure 20.4)

Mutation changes alleles.

Mutations are the ultimate source of genetic variation. Because mutation rates are low, mutation usually is not responsible for deviations from Hardy–Weinberg equilibrium.

Gene flow occurs when alleles move between populations.

Gene flow is the migration of new alleles into a population. It can introduce genetic variation and can homogenize allele frequencies between populations.

Nonrandom mating shifts genotype frequencies.

Assortative mating, in which similar individuals tend to mate, increases homozygosity; disassortative mating increases the frequency of heterozygotes.

Genetic drift may alter allele frequencies in small populations.

Genetic drift refers to random shifts in allele frequency. Its effects may be severe in small populations.

Selection favors some genotypes over others.

For evolution by natural selection to occur, genetic variation must exist, it must result in differential reproductive success, and it must be inheritable.

20.4 Quantifying Natural Selection

A phenotype with greater fitness usually increases in frequency.

Fitness is defined as the reproductive success of an individual. Relative fitness refers to the success of one genotype relative to others in a population. Usually, the genotype with highest relative fitness increases in frequency in the next generation.

Fitness may consist of many components.

Reproductive success is determined by how long an individual survives, how often it mates, and how many offspring it has per reproductive event.

20.5 Natural Selection's Role in Maintaining Variation

Frequency-dependent selection may favor either rare or common phenotypes.

Negative frequency-dependent selection favors rare phenotypes and maintains variation within a population. Positive frequency-dependent selection favors the common phenotype and leads to decreased variation.

In oscillating selection, the favored phenotype changes as the environment changes.

If environmental change is cyclical, selection would favor first one phenotype, then another, maintaining variation.

In some cases, heterozygotes may exhibit greater fitness than homozygotes.

Heterozygote advantage favors individuals with both alleles.

20.6 Selection Acting on Traits Affected by Multiple Genes (figure 20.12)

Disruptive selection removes intermediates.

When intermediate phenotypes are at a disadvantage, a population may exhibit a bimodal trait distribution.

Directional selection eliminates phenotypes at one end of a range.

Directional selection tends to shift the mean value of the population toward the favored end of the distribution.

Stabilizing selection favors individuals with intermediate phenotypes.

Stabilizing selection eliminates both extremes and increases the frequency of an intermediate type. The population may have the same mean value, but with decreased variation.

20.7 Experimental Studies of Natural Selection

The hypothesis that natural selection leads to evolutionary change can be tested experimentally.

Guppy color variation in different environments suggests natural selection at work.

Experimentation reveals the agent of selection.

Guppies in natural populations subject to different predators were shown to undergo color change over generations.

20.8 Interactions Among Evolutionary Forces

Mutation and genetic drift may counter selection.

In theory, a high rate of mutation could oppose natural selection, but this rarely happens. Genetic drift also can work counter to natural selection.

Gene flow may promote or constrain evolutionary change.

Gene flow can spread a beneficial mutation to other populations, but it can also impede adaptation due to influx of alleles with low fitness in a population's environment.

20.9 The Limits of Selection

Genes have multiple effects.

Pleiotropic genes, which have multiple effects, set limits on how much a phenotype can be altered. Even if one affected trait is favored, other affected traits may not be.

Evolution requires genetic variation.

Intense selection pressure may remove genetic variation.

Gene interactions affect fitness of alleles.

In epistasis, fitness of one allele may vary depending on the genotype of a second gene.

UNDERSTAND

1. Assortative mating
   a. affects genotype frequencies expected under Hardy–Weinberg equilibrium.
   b. affects allele frequencies expected under Hardy–Weinberg equilibrium.
   c. has no effect on the genotypic frequencies expected under Hardy–Weinberg equilibrium because it does not affect the relative proportion of alleles in a population.
   d. increases the frequency of heterozygous individuals above Hardy–Weinberg expectations.

2. When the environment changes from year to year and different phenotypes have different fitness in different environments
   a. natural selection will operate in a frequency-dependent manner.
   b. the effect of natural selection may oscillate from year to year, favoring alternative phenotypes in different years.
   c. genetic variation is not required to get evolutionary change by natural selection.
   d. None of the choices is correct.

3. Many factors can limit the ability of natural selection to cause evolutionary change, including
   a. a conflict between reproduction and survival as seen in Trinidadian guppies.
   b. lack of genetic variation.
   c. pleiotropy.
   d. All of the choices are correct.

4. Stabilizing selection differs from directional selection because
   a. in the former, phenotypic variation is reduced but the average phenotype stays the same, whereas in the latter both the variation and the mean phenotype change.
   b. the former requires genetic variation, but the latter does not.
   c. intermediate phenotypes are favored in directional selection.
   d. None of the choices is correct.

5. Founder effects and bottlenecks are
   a. expected only in large populations.
   b. mechanisms that increase genetic variation in a population.
   c. two different modes of natural selection.
   d. forms of genetic drift.

6. *Relative fitness*
   a. refers to the survival rate of one phenotype compared to that of another.
   b. is the physical condition of an individual's siblings and cousins.
   c. refers to the reproductive success of a phenotype.
   d. None of the choices is correct.

7. For natural selection to result in evolutionary change
   a. variation must exist in a population.
   b. reproductive success of different phenotypes must differ.
   c. variation must be inherited from one generation to the next.
   d. All of the choices are correct.

APPLY

1. In a population of red (dominant allele) or white flowers in Hardy–Weinberg equilibrium, the frequency of red flowers is 91%. What is the frequency of the red allele?
   a. 9%   c. 91%
   b. 30%   d. 70%

2. Genetic drift and natural selection can both lead to rapid rates of evolution. However,
   a. genetic drift works fastest in large populations.
   b. only drift leads to adaptation.
   c. natural selection requires genetic drift to produce new variation in populations.
   d. both processes of evolution can be slowed by gene flow.

3. Suppose that the relationship between birth weight and infant mortality, instead of being at a minimum at intermediate sizes, changed such that babies born at 5 or 10 pounds had the highest survival, with a valley in between such that 7.5-pound babies had low survival rate. How would you expect the distribution of birth weights to change over time?
   a. It would not change.
   b. The distribution would shift to the right.
   c. The distribution would become bimodal, with two peaks and the mean value unchanged.
   d. The distribution would become bimodal, with two peaks and the mean value shifted to the right.

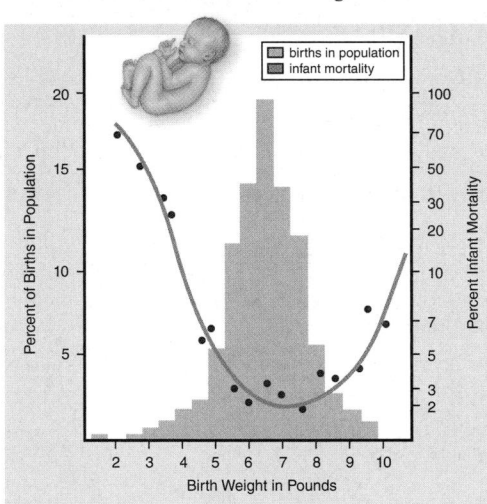

SYNTHESIZE

1. In Trinidadian guppies a combination of elegant laboratory and field experiments builds a very compelling case for predator-induced evolutionary changes in color and life history traits. It is still possible, though not likely, that there are other differences between the sites above and below the falls aside from whether predators are present. What additional studies could strengthen the interpretation of the results?

2. On large, black lava flows in the deserts of the southwestern United States, populations of many types of animals are composed primarily of black individuals. By contrast, on small lava flows, populations often have a relatively high proportion of light-colored individuals. How can you explain this difference?

3. Based on a consideration of how strong artificial selection has helped eliminate genetic variation for speed in thoroughbred horses, we are left with the question of why, for many traits like speed (continuous traits), there is usually abundant genetic variation. This is true even for traits we know are under strong selection. Where does genetic variation ultimately come from, and how does the rate of production compare with the strength of natural selection? What other mechanisms can maintain and increase genetic variation in natural populations?

Chapter **21**

The Evidence for Evolution

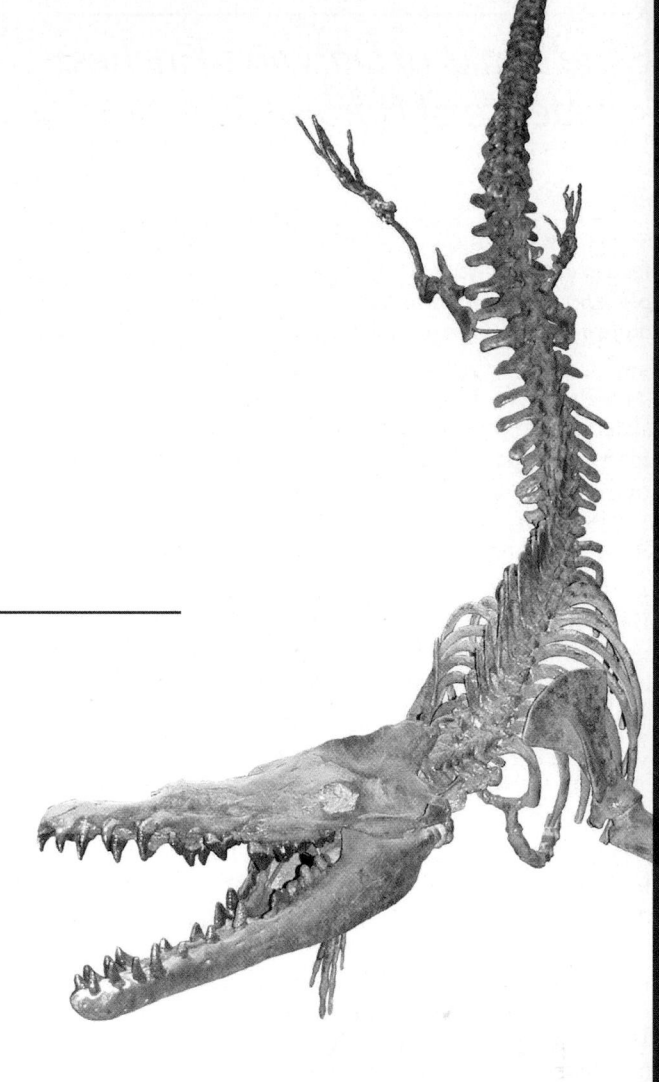

Chapter Contents

Introduction

As we discussed in chapter 1, when Darwin proposed his revolutionary theory of evolution by natural selection, little actual evidence existed to bolster his case. Instead, Darwin relied on observations of the natural world, logic, and results obtained by breeders working with domestic animals. Since his day, however, the evidence for Darwin's theory has become overwhelming.

The case is built upon two pillars: first, evidence that natural selection can produce evolutionary change, and second, evidence from the fossil record that evolution has occurred. The whale skeleton pictured here is the Vogtle whale (Georgiacetus vogtlensis), which is the oldest whale fossil found in North America (40 million years old). The pelvic bones and hindlimbs reveal a link between land mammals and whales. In addition to evidence from studies of natural selection and fossils, information from many different areas of biology—fields as different as anatomy, molecular biology, and biogeography—is only interpretable scientifically as being the outcome of evolution.

The Beaks of Darwin's Finches: Evidence of Natural Selection

Learning Outcomes

1. *Describe how the species of Darwin's finches have adapted to feed in different ways.*
2. *Explain how climatic variation drives evolutionary change in the medium ground finch.*

As you learned in the preceding chapter, a variety of processes can produce evolutionary change. Most evolutionary biologists, however, agree with Darwin's thinking that natural selection is the primary process responsible for evolution. Although we cannot travel back through time, modern-day evidence allows us to test hypotheses about how evolution proceeds and confirms the power of natural selection as an agent of evolutionary change. This evidence comes from both the field and the laboratory and from both natural and human-altered situations.

Darwin's finches are a classic example of evolution by natural selection. When he visited the Galápagos Islands off the coast of Ecuador in 1835, Darwin collected 31 specimens of finches from three islands. Darwin, not an expert on birds, had trouble identifying the specimens, believing by examining their beaks that his collection contained wrens, "gross-beaks," and blackbirds.

Upon Darwin's return to England, ornithologist John Gould informed Darwin that his collection was in fact a closely related group of distinct species, all similar to one another except for their beaks. In all, 14 species are now recognized.

Galápagos finches exhibit variation related to food gathering

The diversity of Darwin's finches is illustrated in figure 21.1. The ground finches feed on seeds that they crush in their powerful beaks; species with smaller and narrower beaks, such as the warbler finch, eat insects. Other species include fruit and bud eaters, and species that feed on cactus fruits and the insects they attract; some populations of the sharp-beaked ground finch even include "vampires" that sometimes creep up on seabirds and use their sharp beaks to pierce the seabirds' skin and drink their blood. Perhaps most remarkable are the tool users, woodpecker finches that pick up a twig, cactus spine, or leaf stalk, trim it into shape with their beaks, and then poke it into dead branches to pry out grubs.

The correspondence between the beaks of the finch species and their food source suggested to Darwin that natural selection had shaped them. In *The Voyage of the Beagle,* Darwin wrote, "Seeing this gradation and diversity of structure in one small, intimately related group of birds, one might really fancy that from an original paucity of birds in this archipelago, one species has been taken and modified for different ends."

Woodpecker finch (*Cactospiza pallida*)

Large ground finch (*Geospiza magnirostris*)

Cactus finch (*Geospiza scandens*)

Warbler finch (*Certhidea olivacea*)

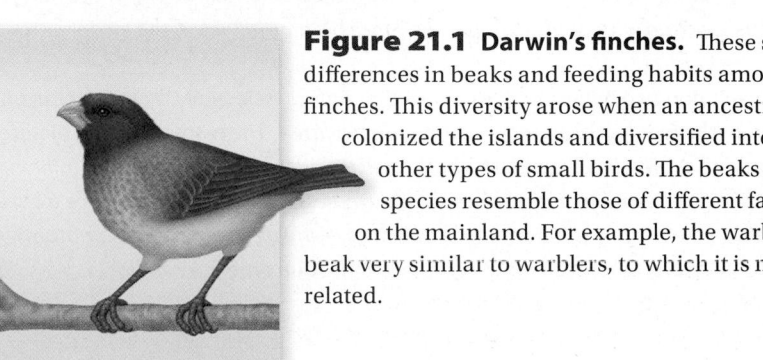

Vegetarian tree finch (*Platyspiza crassirostris*)

Figure 21.1 Darwin's finches. These species show differences in beaks and feeding habits among Darwin's finches. This diversity arose when an ancestral finch colonized the islands and diversified into habitats lacking other types of small birds. The beaks of several species resemble those of different families of birds on the mainland. For example, the warbler finch has a beak very similar to warblers, to which it is not closely related.

Modern research has verified Darwin's selection hypothesis

Darwin's observations suggest that differences among species in beak size and shape have evolved as the species adapted to use different food resources, but can this hypothesis be tested? In chapter 20, you read that the theory of evolution by natural selection requires that three conditions be met:

1. Phenotypic variation must exist in the population.
2. This variation must lead to differences among individuals in lifetime reproductive success.
3. Phenotypic variation among individuals must be genetically transmissible to the next generation.

The key to successfully testing Darwin's proposal proved to be patience. For more than 40 years, starting in 1973, Peter and Rosemary Grant of Princeton University and their students have studied the medium ground finch on a tiny island in the center of the Galápagos called Daphne Major. These finches feed preferentially on small, tender seeds, produced in abundance by plants in wet years. The birds resort to larger, drier seeds, which are harder to crush, only when small seeds become depleted during long periods of dry weather, when plants produce few seeds.

The Grants quantified beak shape among the medium ground finches of Daphne Major by carefully measuring beak depth (height of beak, from top to bottom, at its base) on individual birds. Measuring many birds every year, they were able to assemble for the first time a detailed portrait of evolution in action. The Grants found that not only did a great deal of variation in beak depth exist among members of the population, but the average beak depth changed from one year to the next in a predictable fashion.

During droughts, plants produced few seeds, and all available small seeds were quickly eaten, leaving large seeds as the major remaining source of food. As a result, birds with deeper, more powerful beaks survived better, because they were better able to break open these large seeds. Consequently, the average beak depth of birds in the population increased the next year. Then, when normal rains returned, average beak depth of the population decreased to its original size (figure 21.2a).

Conversely, in particularly wet years, plants flourished, producing an abundance of small seeds; as a result, small-beaked birds were favored, and beak depth decreased greatly.

Could these changes in beak dimension reflect the action of natural selection? An alternative possibility might be that the changes in beak depth do not reflect changes in gene frequencies, but rather are simply a response to diet—for example, perhaps crushing large seeds causes a growing bird to develop a larger beak.

To rule out this possibility, the Grants measured the relationship of parent beak size to offspring beak size, examining many broods over several years. The depth of the beak was very similar between parents and offspring regardless of environmental conditions (figure 21.2b), suggesting that the differences among individuals in beak size reflect genetic differences, and therefore that the year-to-year changes in average beak depth represent evolutionary change resulting from natural selection.

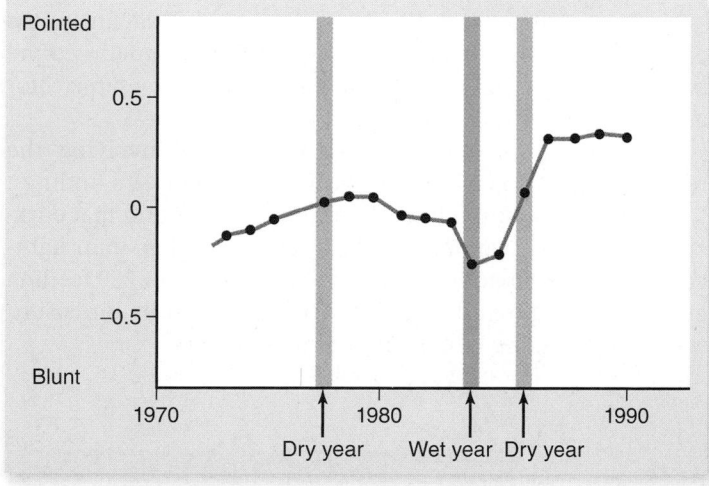

a.

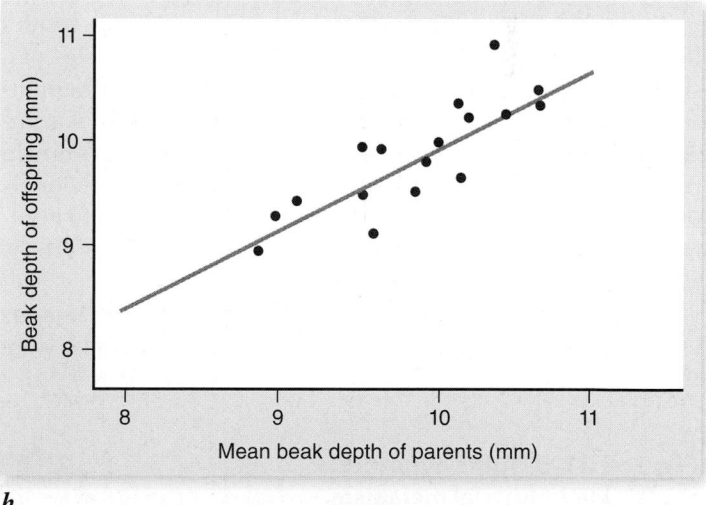

b.

Figure 21.2 **Evidence that natural selection alters beak shape in the medium ground finch (*Geospiza fortis*).** *a.* In dry years, when only large, tough seeds are available, the mean beak depth increases. (The *y*-axis measures beak shape relative to the average beak shape across all years.) In wet years, when many small seeds are available, mean beak depth decreases. *b.* Beak depth is inherited from parents to offspring.

? Inquiry question What would the relationship in figure 21.2b look like if beak shape were not determined genetically, but rather by an environmental factor, such as what a nestling bird ate during its growth period?

🔍 Data analysis Suppose that a male with a beak depth of 10 mm mated with a female with a beak depth of 8 mm. What would the expected beak depth of the offspring be? Would it matter if the female's beak was 10 mm and the male's 6 mm?

21.2 Peppered Moths and Industrial Melanism: More Evidence of Selection

When the environment changes, natural selection often may favor different traits in a species. One classic example concerns the peppered moth, *Biston betularia*. Adults come in a range of shades, from light gray with black speckling (hence the name "peppered" moth) to jet black (melanic).

Extensive genetic analysis has shown that the moth's body color is a genetic trait that reflects different alleles of a single gene. Recent molecular genetic studies have demonstrated that all-black individuals are the descendants of a single mutation. This dominant allele was present but very rare in populations before 1850. From that time on, dark individuals increased in frequency in moth populations near industrialized centers until they made up almost 100% of these populations.

Biologists soon noticed that in industrialized regions where the dark moths were common, the tree trunks were darkened almost black by the soot of pollution, which also killed many of the light-colored lichens on tree trunks.

Light-colored moths decreased in polluted areas

Why did dark moths gain a survival advantage around 1850? In 1896, a British amateur moth collector named J. W. Tutt proposed what became the most commonly accepted hypothesis explaining the decline of the light-colored moths. He suggested that peppered forms were more visible to predators on sooty trees that have lost their lichens. Consequently, birds ate the peppered moths resting on the trunks of trees during the day. The black forms, in contrast, had an advantage because they were camouflaged (figure 21.3).

Although Tutt initially had no evidence, British ecologist Bernard Kettlewell tested the hypothesis in the 1950s by releasing equal numbers of dark and light individuals into two sets of woods: one near heavily polluted Birmingham, and the other in unpolluted Dorset. Kettlewell then set up lights in the woods to attract moths to traps to see how many of both kinds of moths survived. To evaluate his results, he had marked the released moths with a dot of paint on the underside of their wings, where birds could not see it.

In the polluted area near Birmingham, Kettlewell recaptured only 19% of the light moths, but 40% of the dark ones. This indicated that dark moths had a far better chance of surviving in these polluted woods, where tree trunks were dark. In the relatively unpolluted Dorset woods, Kettlewell recovered 12.5% of the light moths but only 6% of the dark ones. This result indicated that where the tree trunks were still light-colored, light moths had a much better chance of survival.

Kettlewell later solidified his argument by placing moths on trees and filming birds looking for food. Sometimes the birds actually passed right over a moth that was the same color as its background.

Recently, an enormous six-year study involving the release of nearly 5,000 moths confirmed Kettlewell's findings. Conducted in an unpolluted forest, the study found that dark-colored moths disappeared at a rate 10% higher than light-colored moths. In addition, direct observations of 250 feeding events revealed that dark moths were captured by birds substantially more often than light-colored moths.

Figure 21.3 Tutt's hypothesis explaining industrial melanism. These photographs show preserved specimens of the peppered moth *(Biston betularia)* placed on trees. Tutt proposed that the dark melanic variant of the moth is more visible to predators on unpolluted trees *(left)*, while the light "peppered" moth is more visible to predators on bark blackened by industrial pollution *(right)*.

When environmental conditions reverse, so does selection pressure

In industrialized areas throughout Eurasia and North America, dozens of other species of moths have evolved in the same way as the peppered moth. The term **industrial melanism** refers to the phenomenon in which darker individuals come to predominate over lighter ones. In the second half of the 20th century, with the widespread implementation of pollution controls, the trend toward melanism began reversing for many species of moths throughout the northern continents.

In Great Britain, the air pollution that promoted industrial melanism began to reverse following enactment of the Clean Air Act in 1956. Beginning in 1959, the *Biston* population at Caldy Common outside Liverpool has been sampled each year. The frequency of the melanic (dark) form has dropped from a high of 93% in 1959 to less than 5% in 2002 (figure 21.4).

The drop correlates well with a significant drop in air pollution, particularly with a lowering of the levels of sulfur dioxide and suspended particulates, both of which act to darken trees. The drop is consistent with a 15% selective disadvantage acting against moths with the dominant melanic allele.

Interestingly, the same reversal of melanism occurred in the United States. Of 576 peppered moths collected at a field station near Detroit from 1959 to 1961, 515 were melanic, a frequency of 89%. The American Clean Air Act, passed in 1963, led to significant reductions in air pollution. Resampled in 2001, the Detroit field station peppered moth population also had less than 5% melanic moths; a similar decreasing trend was also observed in Pennsylvania (see figure 21.4). The moth populations in Liverpool and Detroit, both part of the same natural experiment, exhibit strong evidence for natural selection.

The agent of selection may be difficult to pin down

Although the evidence for natural selection in the case of the peppered moth is strong, Tutt's hypothesis about the agent of selection is currently being reevaluated. Researchers have noted that the recent selection against melanism does not appear to correlate with changes in tree lichens.

At Caldy Common, the light form of the peppered moth began to increase in frequency long before lichens began to reappear on the trees. At the Detroit field station, the lichens never changed significantly as the dark moths first became dominant and then declined over a 30-year period. In fact, investigators have not been able to find peppered moths on Detroit trees at all, whether covered with lichens or not. Some evidence suggests the moths rest on leaves in the treetops during the day, but no one is sure. Could poisoning by pollution rather than predation by birds be the agent of natural selection on the moths? Perhaps—but to date, only selection resulting from bird predation has been demonstrated.

Researchers supporting the bird predation hypothesis point out that a bird's ability to detect moths may depend less on the presence or absence of lichens, and more on other ways in which the environment is darkened by industrial pollution. Pollution tends to cover all objects in the environment with a

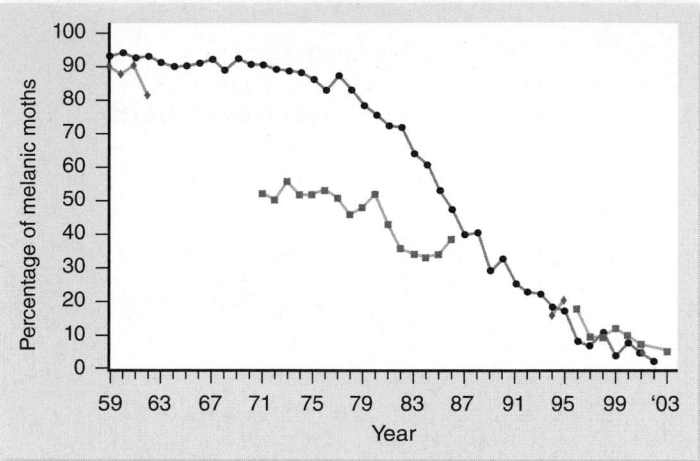

Figure 21.4 Selection against melanism. The red circles indicate the frequency of melanic *Biston betularia* moths at Caldy Common in Great Britain. Green diamonds indicate frequencies of melanic *B. betularia* in Michigan, and the blue squares indicate corresponding frequencies in Pennsylvania.

? **Inquiry question** What can you conclude from the fact that the frequency of melanic moths decreased to the same degree in the two locations?

fine layer of particulate dust, which tends to decrease how much light surfaces reflect. In addition, pollution has a particularly severe effect on birch trees, which are light in color. Both effects would tend to make the environment darker, and thus would favor darker moths by protecting them from predation by birds.

Despite this uncertainty over the agent of selection, the overall pattern is clear. Kettlewell's experiments established indisputably that selection favors dark moths in polluted habitats and light moths in pristine areas. The increase and subsequent decrease in the frequency of melanic moths, correlated with levels of pollution independently on two continents, demonstrates clearly that this selection drives evolutionary change.

The current reconsideration of the agent of natural selection illustrates well the way in which scientific progress is achieved: Hypotheses, such as Tutt's, are put forth and then tested. If rejected, new hypotheses are formulated, and the process begins anew.

Learning Outcomes Review 21.2

Natural selection has favored the dark form of the peppered moth in areas subject to severe air pollution, perhaps because on darkened trees they are less easily seen by moth-eating birds. As pollution has abated, selection has in turn shifted to favor the light form. Although selection is clearly occurring, further research is required to understand whether predation by birds is the agent of selection.

■ *How would you test the idea that predation by birds is the agent of selection on moth coloration?*

21.3 Artificial Selection: Human-Initiated Change

Learning Outcomes

1. Contrast the processes of artificial and natural selection.
2. Explain what artificial selection demonstrates about the power of natural selection.

Humans have imposed selection upon plants and animals since the dawn of civilization. Just as in natural selection, such **artificial selection** operates by favoring individuals with certain phenotypic traits, allowing them to reproduce and pass their genes on to the next generation. Assuming that phenotypic differences are genetically determined, this directional selection should lead to evolutionary change, and indeed it has.

Artificial selection, imposed in laboratory experiments, agriculture, and the domestication process, has produced substantial change in almost every case in which it has been applied. This success is strong proof that selection is an effective evolutionary process.

Experimental selection produces changes in populations

With the rise of genetics as a field of science in the 1920s and 1930s, researchers began conducting experiments to test the hypothesis that selection can produce evolutionary change. A favorite subject was the laboratory fruit fly, *Drosophila melanogaster*. Geneticists have imposed selection on just about every conceivable aspect of the fruit fly—including body size, eye color, growth rate, life span, and exploratory behavior—with a consistent result: Selection for a trait leads to strong and predictable evolutionary response.

In one classic experiment, scientists selected for fruit flies with many bristles (stiff, hairlike structures) on their abdomens. At the start of the experiment, the average number of bristles was 9.5. Each generation, scientists picked out the 20% of the population with the greatest number of bristles and allowed them to reproduce, thus establishing the next generation. After 86 generations of this directional selection, the average number of bristles had quadrupled, to nearly 40! In another experiment, fruit flies in one population were selected for high numbers of bristles, while fruit flies in the other cage were selected for low numbers of bristles. Within 35 generations, the populations did not overlap at all in range of variation (figure 21.5).

Similar experiments have been conducted on a wide variety of other laboratory organisms. For example, by selecting for rats that were resistant to tooth decay, in less than 20 generations scientists were able to increase the average time for onset of decay from barely over 100 days to greater than 500 days.

SCIENTIFIC THINKING

Question: *Can artificial selection lead to substantial evolutionary change?*

Hypothesis: *Strong directional selection will quickly lead to a large shift in the mean value of the population.*

Experiment: *In one population, every generation pick out the 20% of the population with the most bristles and allow them to reproduce to form the next generation. In the other population, do the same with the 20% with the fewest number of bristles.*

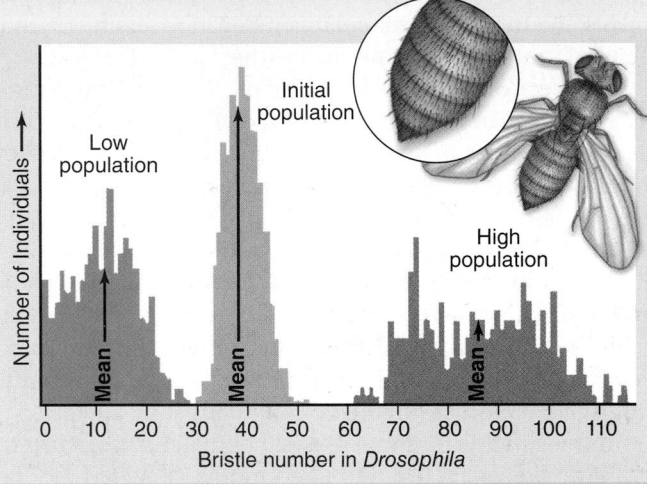

Result: *After 35 generations, mean number of bristles has changed substantially in both populations.*

Interpretation: *Note that at the end of the experiment, the range of variation lies outside the range seen in the initial population. Selection can move a population beyond its original range because mutation and recombination continuously introduce new variation into populations.*

Figure 21.5 Artificial selection can lead to rapid and substantial evolutionary change.

 Inquiry question What would happen if, within a population, both small and large individuals were allowed to breed, but middle-sized ones were not?

Agricultural selection has led to extensive modification of crops and livestock

Familiar livestock, such as cattle and pigs, and crops, such as corn and strawberries, are greatly different from their wild ancestors (figure 21.6). These differences have resulted from generations of human selection for desirable traits, such as greater milk production and larger corn ear size.

An experiment with corn demonstrates the ability of artificial selection to rapidly produce major change in crop plants. In 1896, agricultural scientists began selecting for the oil content of corn kernels, which initially was 4.5%. Just as in the fruit fly experiments, the top 20% of all individuals were allowed to reproduce. By 1986, at which time 90 generations had passed, average oil content of the corn kernels had increased approximately 450%.

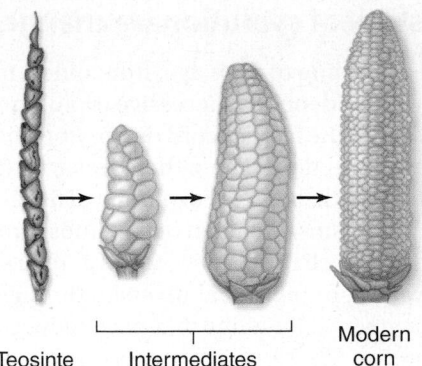

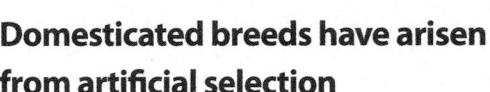

Figure 21.6 Corn looks very different from its ancestor. Teosinte, which can be found today in a remote part of Mexico, is very similar to the ancestor of modern corn. Artificial selection has transformed it into the form we know today.

Domesticated breeds have arisen from artificial selection

Human-imposed selection has produced a great variety of breeds of cats, dogs (figure 21.7), pigeons, and other domestic animals. In some cases, breeds have been developed for particular purposes. Greyhound dogs, for example, resulted from selection for maximal running ability, resulting in an animal with long legs, a long tail for balance, an arched back to increase stride length, and great muscle mass. By contrast, the odd proportions of the ungainly dachshund resulted from selection for dogs that could enter narrow holes in pursuit of badgers. In other cases, varieties have been selected primarily for their appearance, such as the many colorful breeds of pigeons or cats.

Domestication also has led to unintentional selection for some traits. In recent years, as part of an attempt to domesticate the silver fox, Russian scientists have chosen the most docile animals in each generation and allowed them to reproduce. Within 40 years, most foxes were exceptionally tame, not only allowing themselves to be petted, but also whimpering to get attention and sniffing and licking their caretakers (figure 21.8). In many respects, they had become no different from domestic dogs.

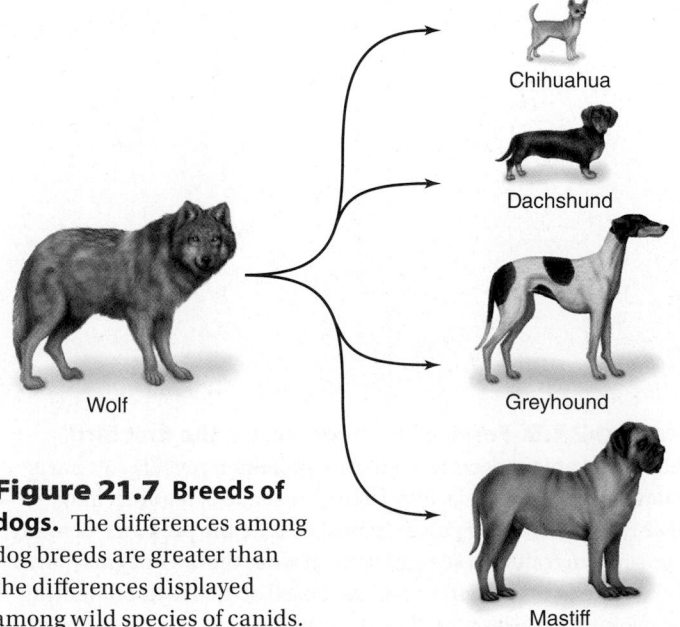

Chihuahua

Dachshund

Greyhound

Wolf

Mastiff

Figure 21.7 Breeds of dogs. The differences among dog breeds are greater than the differences displayed among wild species of canids.

Figure 21.8 Domesticated foxes. After 40 years of selectively breeding the tamest individuals, artificial selection has produced silver foxes that are not only as friendly as domestic dogs, but also exhibit many physical traits seen in dog breeds.

It was not only their behavior that changed, however. These foxes also began to exhibit other traits seen in some dog breeds, such as different color patterns, floppy ears, curled tails, and shorter legs and tails. Presumably, the genes responsible for docile behavior either affect these traits as well or are closely linked to the genes for these other traits (the phenomena of pleiotropy and linkage, which are discussed in chapters 12 and 13).

Can selection produce major evolutionary changes?

Given that we can observe the results of selection operating over a relatively short time, most scientists think that natural selection is the process responsible for the evolutionary changes documented in the fossil record. Some critics of evolution accept that selection can lead to changes within a species, but contend that such changes are relatively minor in scope and not equivalent to the substantial changes documented in the fossil record. In other words, it is one thing to change the number of bristles on a fruit fly or the size of an ear of corn, and quite another to produce an entirely new species.

This argument does not fully appreciate the extent of change produced by artificial selection. Consider, for example, the existing breeds of dogs, all of which have been produced since wolves were first domesticated, perhaps 10,000 years ago. If the various dog breeds did not exist and a paleontologist found fossils of animals similar to dachshunds, greyhounds, mastiffs, and chihuahuas, there is no question that they would be considered different species. Indeed, the differences in size and shape exhibited by these breeds are greater than those between members of different genera in the family Canidae—such as coyotes, jackals, foxes, and wolves—which have been evolving separately for 5 to 10 million years. Consequently, the claim that artificial selection produces only minor changes is clearly incorrect. If selection operating over a period of only 10,000 years can produce such substantial differences, it should be powerful enough, over the course of many millions of years, to produce the diversity of life we see around us today.

21.4 *Fossil Evidence of Evolution*

Learning Outcomes

1. *Describe how fossils are formed.*
2. *Explain the importance of the discovery of transitional fossils.*
3. *Name the evolutionary trends revealed by the study of horse evolution.*

The most direct evidence that evolution has occurred is found in the fossil record. Today we have a far more complete understanding of this record than was available in Darwin's time.

Fossils are the preserved remains of once-living organisms. They include specimens preserved in amber, Siberian permafrost, and dry caves, as well as the more common fossils preserved as rocks.

Rock fossils are created when three events occur. First, the organism must become buried in sediment; then, the calcium in bone or other hard tissue must mineralize; and finally, the surrounding sediment must eventually harden to form rock.

The process of fossilization occurs only rarely. Usually, animal or plant remains decay or are scavenged before the process can begin. In addition, many fossils occur in rocks that are inaccessible to scientists. When they do become available, they are often destroyed by erosion and other natural processes before they can be collected. As a result, only a very small fraction of the species that have ever existed (estimated by some to be as many as 500 million) are known from fossils. Nonetheless, the fossils that have been discovered are sufficient to provide detailed information on the course of evolution through time.

The age of fossils can be estimated

By dating the rocks in which fossils occur, we can get an accurate idea of how old the fossils are. In Darwin's day, rocks were dated by their position with respect to one another *(relative dating)*; rocks in deeper strata are generally older. Knowing the relative positions of sedimentary rocks and the rates of erosion of different kinds of sedimentary rocks in different environments, geologists of the 19th century derived a fairly accurate idea of the relative ages of rocks.

Today, geologists can determine the absolute age of rocks using isotopic dating. At the time a rock forms, some elements exist as different isotopes. Over time the less stable isotope is converted into the other isotope and the ratio of the two forms changes. Details on absolute dating are found in chapter 26.

Fossils present a history of evolutionary change

When fossils are arrayed according to their age, from oldest to youngest, they often provide evidence of successive evolutionary change. At the largest scale, the fossil record documents the course of life through time, from the origin of first prokaryotic and then eukaryotic organisms, through the evolution of fishes, the rise of land-dwelling organisms, the reign of the dinosaurs, and on to the origin of humans. In addition, the fossil record shows the waxing and waning of biological diversity through time, such as the periodic mass extinctions that have reduced the number of living species. These topics are discussed at greater length in chapter 26.

Fossils document evolutionary transitions

Given the low likelihood of fossil preservation and recovery, it is not surprising that there are gaps in the fossil record. Nonetheless, intermediate forms are often available to illustrate how the major transitions in life occurred.

Undoubtedly the most famous of these is the oldest known bird, *Archaeopteryx* (meaning "ancient feather") which lived around 165 million years ago (MYA) (figure 21.9). This specimen is clearly intermediate between birds and dinosaurs. Its feathers, similar in many respects to those of birds today, clearly reveal that it is a bird. Nonetheless, in many other respects—for example, possession of teeth, a bony tail, and other anatomical characteristics—it is indistinguishable from some carnivorous dinosaurs. Indeed, it is so similar to these dinosaurs that several specimens lacking preserved feathers were misidentified as dinosaurs and lay in the wrong natural history museum cabinet for several decades before the mistake was discovered!

Archaeopteryx reveals a pattern commonly seen in intermediate fossils—rather than being intermediate in every trait, such fossils usually exhibit some traits like their ancestors and

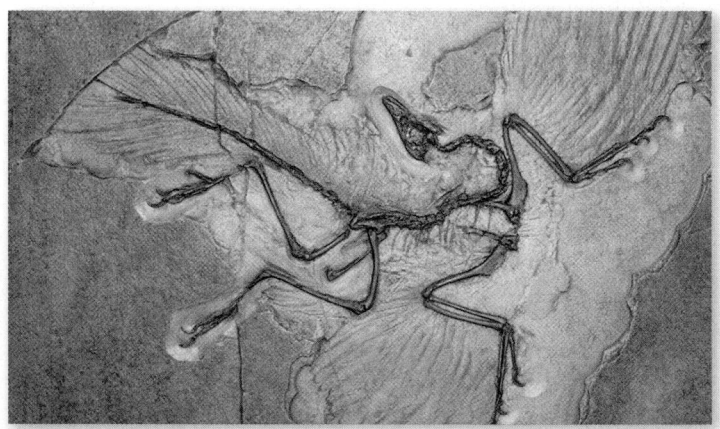

Figure 21.9 Fossil of *Archaeopteryx,* the first bird. The remarkable preservation of this specimen reveals soft parts usually not preserved in fossils; the presence of feathers makes clear that *Archaeopteryx* was a bird, despite the presence of many dinosaurian traits. To see a picture of what this animal may have looked like in life, see figure 35.29. Fossil evidence documenting the evolutionary descent of birds from dinosaurs is considered in greater detail in chapter 23 (see figure 23.11).

others like their descendants. In other words, traits evolve at different rates and different times; expecting an intermediate form to be intermediate in every trait would not be correct.

The first *Archaeopteryx* fossil was discovered in 1859, the year Darwin published *On the Origin of Species*. Since then, paleontologists have continued to fill in the gaps in the fossil record. Today, the fossil record is far more complete, particularly among the vertebrates; fossils have been found linking all the major groups.

Recent years have seen spectacular discoveries, closing some of the major remaining gaps in our understanding of vertebrate evolution. For example, a four-legged aquatic mammal was discovered only recently that provides important insights concerning the evolution of whales and dolphins from land-dwelling, hoofed ancestors (figure 21.10). Similarly, a fossil snake with legs has shed light on the evolution of snakes, which are descended from lizards that gradually became more and more elongated with the simultaneous reduction and eventual

disappearance of the limbs. In chapter 35, we discuss the most recent such discovery, *Tiktaalik*, a species that bridged the gap between fish and the first amphibians (see figure 35.14).

On a finer scale, evolutionary change within some types of animals is known in exceptional detail. For example, about 200 MYA, oysters underwent a change from small, curved shells to larger, flatter ones, with progressively flatter fossils seen in the fossil record over a period of 12 million years. A host of other examples illustrate similar records of successive change. The demonstration of this successive change is one of the strongest lines of evidence that evolution has occurred.

The evolution of horses is a prime example of evidence from fossils

One of the most studied cases in the fossil record concerns the evolution of horses. Modern-day members of the family Equidae include horses, zebras, donkeys, and asses, all of which are large, long-legged, fast-running animals adapted to living on open grasslands. These species, all classified in the genus *Equus,* are the last living descendants of a long lineage that has produced 34 genera since its origin in the Eocene period, approximately 55 MYA. Examination of these fossils has provided a particularly well-documented case of how evolution has proceeded through adaptation to changing environments.

The first horse

The earliest known members of the horse family, species in the genus *Hyracotherium,* didn't look much like modern-day horses at all. Small, with short legs and broad feet, these species occurred in wooded habitats, where they probably browsed on leaves and herbs and escaped predators by dodging through openings in the forest vegetation. The evolutionary path from these diminutive creatures to the workhorses of today has involved changes in a variety of traits, including size, toe reduction, and tooth size and shape (figure 21.11).

Changes in size

The first species of horses were as big as a large house cat or a medium-sized dog. By contrast, modern equids can weigh more than 500 kg. Examination of the fossil record reveals that horses changed little in size for their first 30 million years, but since then, a number of different lineages have exhibited rapid and substantial increases. However, evolution has not been unidirectional and trends toward decreased size were also exhibited in some branches of the equid evolutionary tree.

Toe reduction

The feet of modern horses have a single toe enclosed in a tough, bony hoof. By contrast, *Hyracotherium* had four toes on its front feet and three on its hind feet. Rather than hooves, these toes were encased in fleshy pads like those of dogs and cats.

Examination of fossils clearly shows the transition through time: a general increase in length of the central toe, development of the bony hoof, and reduction and loss of the

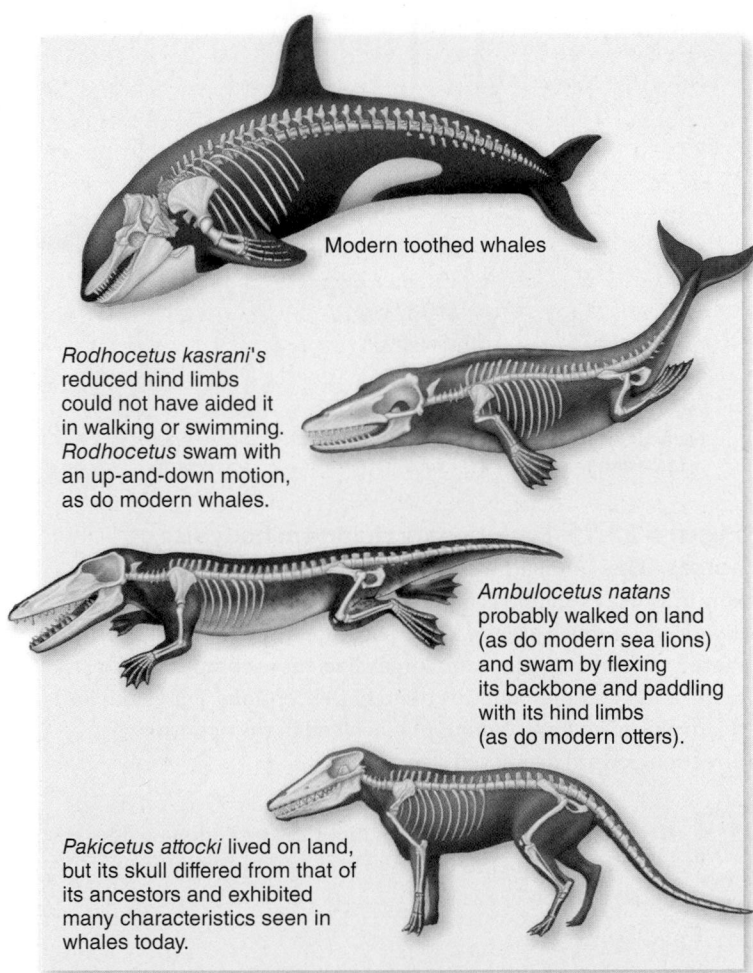

Figure 21.10 Whale "missing links." The recent discoveries of *Ambulocetus, Rodhocetus,* and *Pakicetus* have filled in the gaps between whales and their hoofed mammal ancestors. The features of *Pakicetus* illustrate that intermediate forms are not intermediate in all characteristics; rather, some traits evolve before others. In the case of the evolution of whales, changes occurred in the skull prior to evolutionary modification of the limbs. All three fossil forms occurred in the Eocene period, 45–55 MYA.

Modern toothed whales

Rodhocetus kasrani's reduced hind limbs could not have aided it in walking or swimming. *Rodhocetus* swam with an up-and-down motion, as do modern whales.

Ambulocetus natans probably walked on land (as do modern sea lions) and swam by flexing its backbone and paddling with its hind limbs (as do modern otters).

Pakicetus attocki lived on land, but its skull differed from that of its ancestors and exhibited many characteristics seen in whales today.

other toes (see figure 21.11). As with body size, these trends occurred concurrently on several different branches of the horse evolutionary tree and were not exhibited by all lineages.

At the same time as toe reduction was occurring, these horse lineages were evolving changes in the length and skeletal structure of their limbs, leading to animals capable of running long distances at high speeds.

Tooth size and shape

The teeth of *Hyracotherium* were small and relatively simple in shape. Through time, horse teeth have increased greatly in length and have developed a complex pattern of ridges on their molars and premolars. The effect of these changes is to produce teeth better capable of chewing tough and gritty vegetation, such as grass, which tends to wear teeth down.

Accompanying these changes have been alterations in the shape of the skull that strengthened its ability to withstand the stresses imposed by continual chewing. As with body size, evolutionary change has not been constant through time. Rather, much of the change in tooth shape has occurred within the past 20 million years, and changes have not been constant among all horse lineages.

All of these changes may be understood as adaptations to changing global climates. In particular, during the late Miocene and early Oligocene epochs (approximately 20 to 25 MYA), grasslands became widespread in North America, where much of horse evolution occurred. As horses adapted to these habitats, high-speed locomotion probably became more important to escape predators. By contrast, the greater flexibility provided by multiple toes and shorter limbs, which was advantageous for ducking through complex forest vegetation, was no longer beneficial. At the same time, horses were eating grasses and other vegetation that contained more grit and other hard substances, thus favoring teeth and skulls better suited for withstanding such materials.

Evolutionary trends

For many years, horse evolution was held up as an example of constant evolutionary change through time. Some even saw in the record of horse evolution evidence for a progressive, guiding force, consistently pushing evolution in a single direction. We now know that such views are misguided, and that the course of evolutionary change over millions of years is rarely so simple.

Rather, the fossils demonstrate that even though overall trends have been evident in a variety of characteristics, evolutionary change has been far from constant and uniform through time. Instead, rates of evolution have varied widely, with long periods of little observable change and some periods of great change. Moreover, when changes happen, they often occur simultaneously in different lineages of the horse evolutionary tree.

Finally, even when a trend exists, exceptions, such as the evolutionary decrease in body size exhibited by some lineages, are not uncommon. These patterns are usually discovered for any group of plants and animals for which we have an extensive fossil record, as you will see when we discuss human evolution in chapter 35.

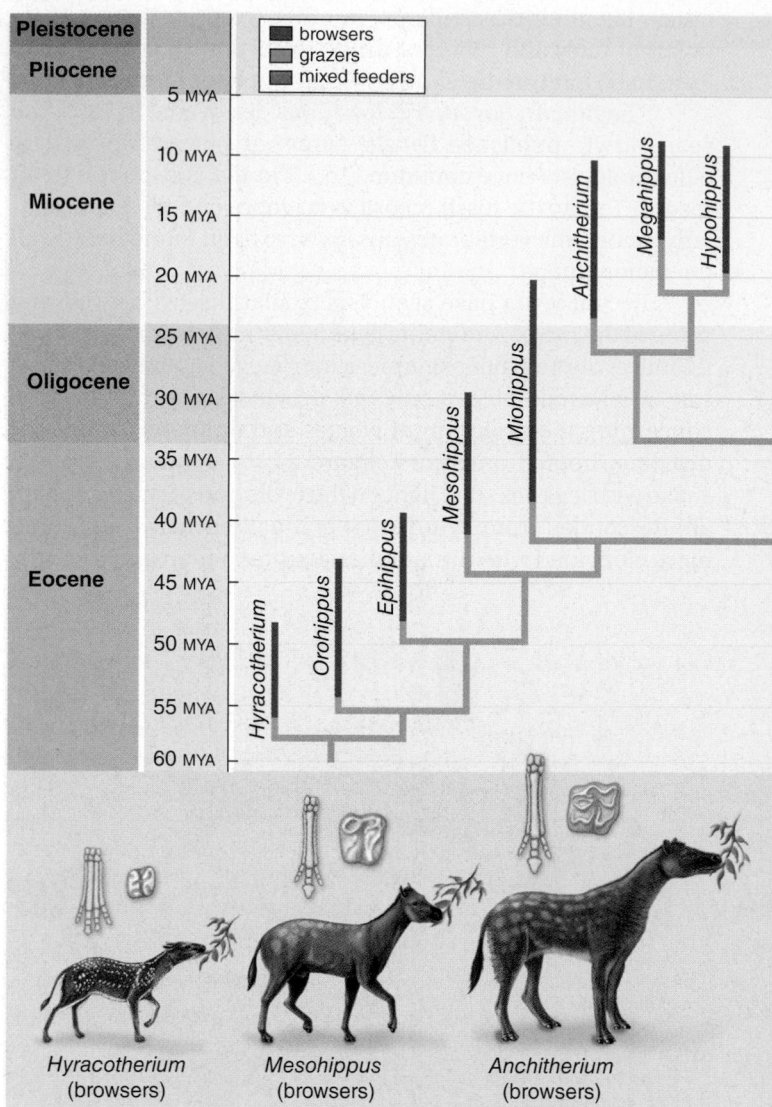

Figure 21.11 Evolutionary change in body size of horses. Lines indicate evolutionary relationships of the horse family. Horse evolution is more like a bush than a single-trunk tree; diversity was much greater in the past than it is today. In general, there has been a trend toward larger size, more complex molar teeth, and fewer toes, but this trend has exceptions. For example, a relatively recent form, *Nannippus,* evolved in the opposite direction, toward decreased size.

 Inquiry question Why might the evolutionary line leading to *Nannippus* have experienced an evolutionary decrease in body size?

Horse diversity

One reason that horse evolution was originally conceived of as linear through time may be that modern horse diversity is relatively limited. For this reason it is easy to mentally picture a straight line from *Hyracotherium* to modern-day *Equus.* But today's limited horse diversity—only one surviving genus—is unusual. In fact, at the peak of horse diversity in the Miocene

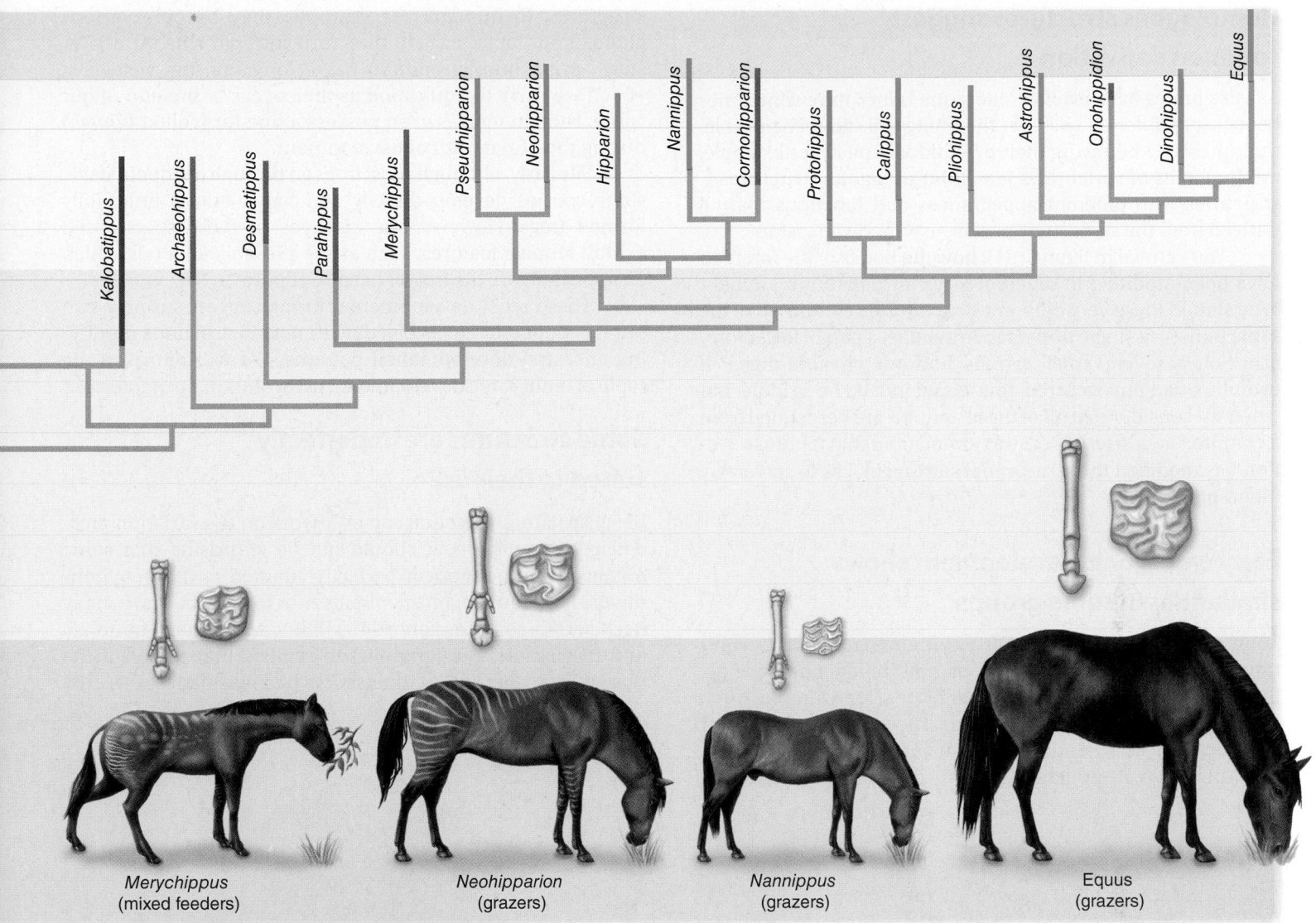

Merychippus
(mixed feeders)

Neohipparion
(grazers)

Nannippus
(grazers)

Equus
(grazers)

epoch, 13 genera of horses could be found in North America alone. These species differed in body size and in a wide variety of other characteristics. Presumably, they lived in different habitats and exhibited different dietary preferences. Had this diversity existed to modern times, early evolutionary biologists would likely have had a different outlook on horse evolution.

Learning Outcomes Review 21.4

Fossils form when an organism is preserved in a matrix such as amber, permafrost, or rock. They can be used to construct a record of evolutionary transitions over long periods of time, which allows us to understand how major changes in evolution occur. The extensive fossil record for horses provides a detailed view of evolutionary diversification of this group, although trends are not constant and uniform and may include exceptions.

■ **Why might rates and direction of evolutionary change vary through time?**

21.5 Anatomical Evidence for Evolution

Learning Outcomes

1. Explain the evolutionary significance of homologous and vestigial structures.
2. Describe how patterns of early development provide evidence for evolution.

Much of the power of the theory of evolution is its ability to provide a sensible framework for understanding the diversity of life. Many observations from throughout biology simply cannot be understood in any meaningful way except as a result of evolution.

Homologous structures suggest common derivation

As vertebrates have evolved, the same bones have sometimes been put to different uses. Yet the bones are still recognizable, their presence betraying their evolutionary past. For example, the forelimbs of vertebrates are all **homologous structures**—structures with different appearances and functions that all derived from the same body part in a common ancestor.

You can see in figure 21.12 how the bones of the forelimb have been modified in different ways for different mammals. Why should these very different structures be composed of the same bones—a single upper forearm bone, a pair of lower forearm bones, several small carpals, and one or more digits? If evolution had not occurred, this would indeed be a riddle. But when we consider that all of these animals are descended from a common ancestor, it is easy to understand that natural selection has modified the same initial starting blocks to serve very different purposes.

Early embryonic development shows similarities in some groups

Some of the strongest anatomical evidence supporting evolution comes from comparisons of how organisms develop. Embryos of different types of vertebrates, for example, often are similar early on, but become more different as they develop. Early in their development vertebrate embryos possess pharyngeal pouches, which develop into different structures. In humans, for example, they become various glands and ducts; in fish, they turn into gill slits. At a later stage, every human embryo has a long tail, the vestige of which we carry to adulthood as the coccyx at the end of our spine. Human fetuses even possess a fine fur (called *lanugo*) during the fifth month of development.

Similarly, although most frogs go through a tadpole stage, some species develop directly and hatch out as little, fully formed frogs. However, the embryos of these species still exhibit tadpole features, such as the presence of a tail, which disappear before the froglet hatches (figure 21.13).

These relict developmental forms suggest strongly that our development has evolved, with new instructions modifying ancestral developmental patterns. We will return to the topic of embryonic development and evolution in chapter 25.

Some structures are imperfectly suited to their use

Because natural selection can only work on the variation present in a population, it should not be surprising that some organisms do not appear perfectly adapted to their environments. For example, most animals with long necks have many neck vertebrae for enhanced flexibility: Geese have up to 25, and plesiosaurs, the long-necked reptiles that patrolled the seas during the age of dinosaurs, had as many as 76. By

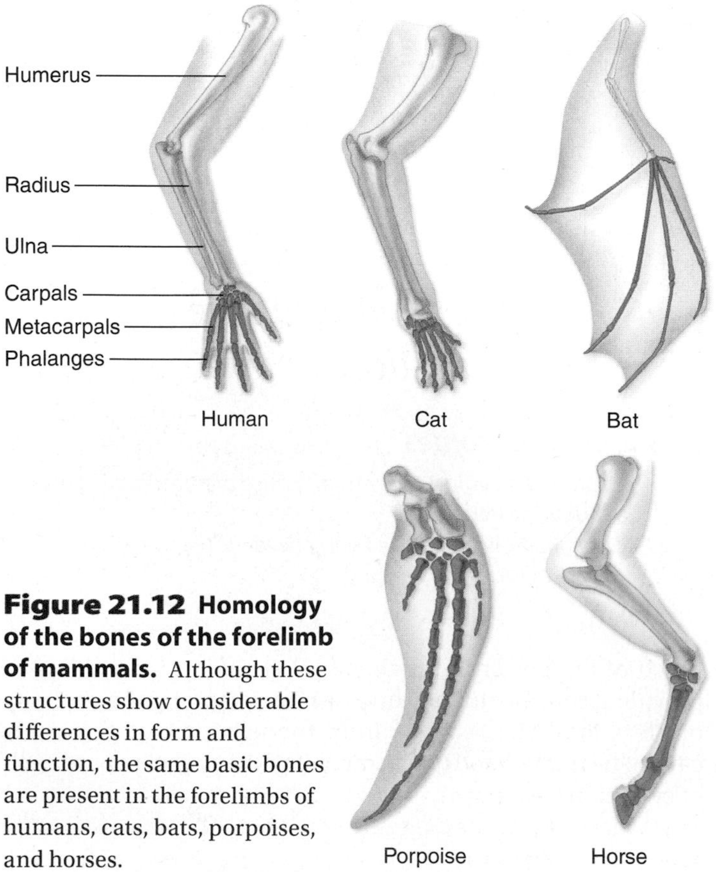

Humerus
Radius
Ulna
Carpals
Metacarpals
Phalanges

Human   Cat   Bat

Porpoise   Horse

Figure 21.12 Homology of the bones of the forelimb of mammals. Although these structures show considerable differences in form and function, the same basic bones are present in the forelimbs of humans, cats, bats, porpoises, and horses.

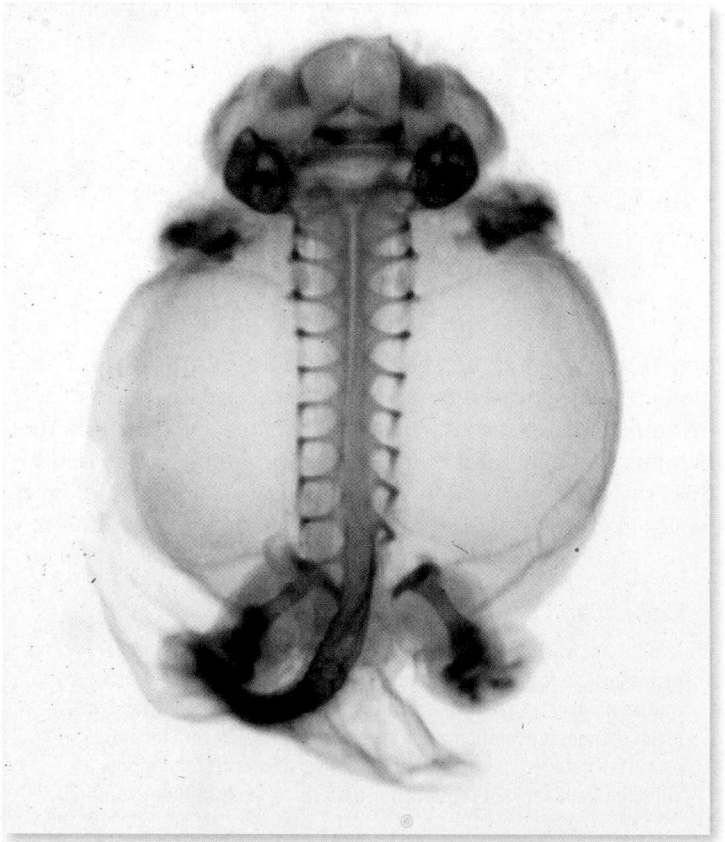

Figure 21.13 Developmental features reflect evolutionary ancestry. Some species of frogs have lost the tadpole stage. Nonetheless, tadpole features first appear and then disappear during development in the egg.

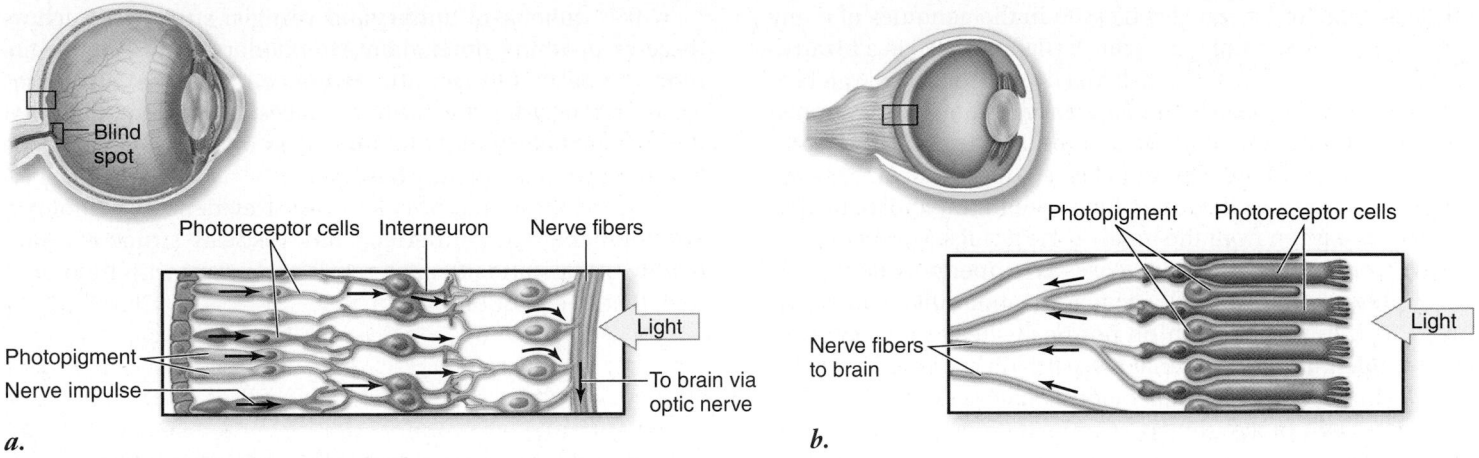

a.

b.

Figure 21.14 The eyes of vertebrates and mollusks. *a.* Photoreceptors of vertebrates point backward, whereas *(b)* those of mollusks face forward. As a result, vertebrate nerve fibers pass in front of the photoreceptor; and where they bundle together and exit the eye, a blind spot is created. Mollusks' eyes have neither of these problems.

contrast, almost all mammals have only 7 neck vertebrae, even the giraffe. In the absence of variation in vertebrae number, selection led to an evolutionary increase in vertebra size to produce the long neck of the giraffe.

An excellent example of an imperfect design is the eye of vertebrate animals, in which the photoreceptors face backward, toward the wall of the eye (figure 21.14*a*). As a result, the nerve fibers extend not backward, toward the brain, but forward into the eye chamber, where they slightly obstruct light. Moreover, these fibers bundle together to form the optic nerve, which exits through a hole at the back of the eye, creating a blind spot.

By contrast, the eye of mollusks—such as squid and octopuses—are more optimally designed: The photoreceptors face forward, and the nerve fibers exit at the back, neither obstructing light nor creating a blind spot (figure 21.14*b*).

Such examples illustrate that natural selection is like a tinkerer, working with whatever material is available to craft a workable solution, rather than like an engineer, who can design and build the best possible structure for a given task. Workable, but imperfect, structures such as the vertebrate eye are an expected outcome of evolution by natural selection.

Vestigial structures can be explained as holdovers from the past

Many organisms possess **vestigial structures** that have no apparent function, but resemble structures their ancestors possessed. Humans, for example, possess a complete set of muscles for wiggling their ears, just like many other mammals do. Although these muscles allow other mammals to move their ears to pinpoint sounds such as the movements or growl of a predator, they have little purpose in humans other than amusement.

As other examples, boa constrictors have hip bones and rudimentary hind legs. Manatees (a type of aquatic mammal

often referred to as "sea cows") have fingernails on their fins, which evolved from legs. Blind cave fish, which never see the light of day, have small, nonfunctional eyes. Figure 21.15 illustrates the skeleton of a baleen whale, which contains pelvic bones, as other mammal skeletons do, even though such bones serve no known function in the whale.

The human vermiform appendix is apparently vestigial; it represents the degenerate terminal part of the cecum, the blind pouch or sac in which the large intestine begins. In other mammals, such as mice, the cecum is the largest part of the large intestine and functions in storage—usually of bulk cellulose in herbivores. Although some functions have been suggested, it is difficult to assign any current function to the human vermiform appendix. In many respects, it can be a dangerous organ: appendicitis, which results from infection of the appendix, can be fatal.

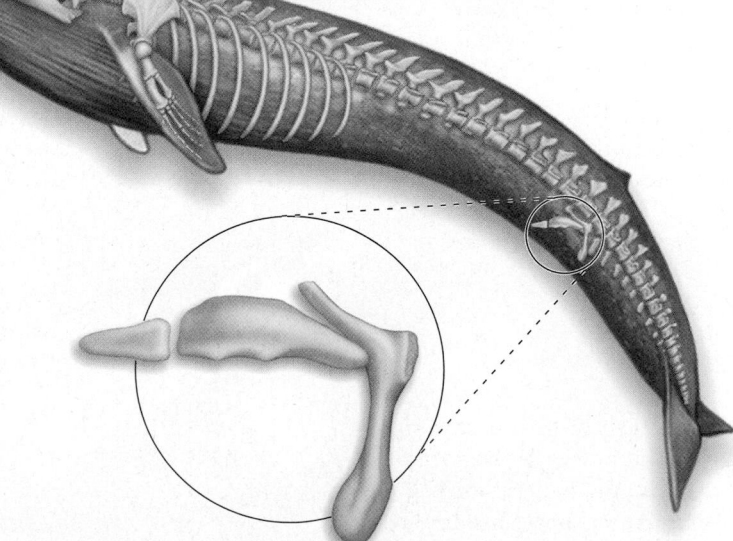

Figure 21.15 Vestigial structures. The skeleton of a whale reveals the presence of pelvic bones. These bones resemble those of other mammals, but are only weakly developed in the whale and have no apparent function.

Vestigial traits can also be seen in the genomes of many organisms. For example, the icefish (figure 21.16) is a bizarre-looking, nearly see-through fish that lives in the frigid waters of the Antarctic. The icefish's transparency results not only from a lack of pigment in its body structures, but also from the near invisibility of its blood. Our blood is red due to the presence of red blood cells, which contain hemoglobin, the molecule that transports oxygen from the lungs to the tissues. However, oxygen concentration in water increases as temperature decreases. The waters of the Antarctic, which are about 0°C, contain so much oxygen that the fish do not need special molecules to carry oxygen. The result is that these fish do not have hemoglobin, and consequently their blood is colorless. Nonetheless, when the DNA of icefish was examined, scientists discovered that they have the same gene that produces hemoglobin as that found in other vertebrates. However, the icefish hemoglobin gene has a variety of mutations that render it nonfunctional, and thus the icefish does not produce hemoglobin. The presence of this inoperative version of the hemoglobin gene in icefish means that its ancestors had hemoglobin; however, once the icefish's progenitors occupied the cold waters of the Antarctic and lost the need for hemoglobin, mutations that would be harmful and thus would be filtered out by natural selection in other species were able to persist in the population. Just by chance, some of these mutations increased in frequency in the population through time, eventually becoming established in all individuals and knocking out the fish's ability to produce hemoglobin.

Fossil genes, or **pseudogenes,** such as the hemoglobin gene in the icefish, are actually quite common in the genomes of most organisms and are discussed in chapter 24: when a trait disappears, the gene does not just vanish from the genome; rather, some mutation renders it inactive, and once that occurs, other mutations can accumulate.

Figure 21.16 Photo of an icefish. This nearly transparent fish is found in the frigid waters of the Antarctic.

It is difficult to understand vestigial structures such as these as anything other than evolutionary relicts, holdovers from the past. However, the existence of vestigial structures argues strongly for the common ancestry of the members of the groups that share them, regardless of how different those groups have subsequently become.

All of these anatomical lines of evidence—homology, development, and imperfect and vestigial structures—are readily understandable as a result of descent with modification, that is, evolution.

Learning Outcomes Review 21.5

Comparisons of the anatomy of different living animals often reveal evidence of shared ancestry. In cases of homology, the same organ has evolved to carry out different functions. In other cases, an organ is still present, usually in diminished form, even though it has lost its function altogether; such an organ or structure is termed vestigial.

- *How might homologous and vestigial structures be explained other than as a result of evolutionary descent with modification?*

21.6 Convergent Evolution and the Biogeographical Record

Learning Outcomes

1. *Explain the principle of convergent evolution.*
2. *Demonstrate how the biogeographical distribution of plant and animal species on islands provides evidence of evolutionary diversification.*

Biogeography, the study of the geographic distribution of species, reveals that different geographical areas sometimes exhibit groups of plants and animals of strikingly similar appearance, even though the organisms may be only distantly related.

It is difficult to explain so many similarities as the result of coincidence. Instead, natural selection appears to have favored parallel evolutionary adaptations in similar environments. Because selection in these instances has tended to favor changes that made the two groups more alike, their phenotypes have converged. This form of evolutionary change is referred to as **convergent evolution.**

Marsupials and placentals demonstrate convergence

In the best known case of convergent evolution, two major groups of mammals—marsupials and placentals—have evolved

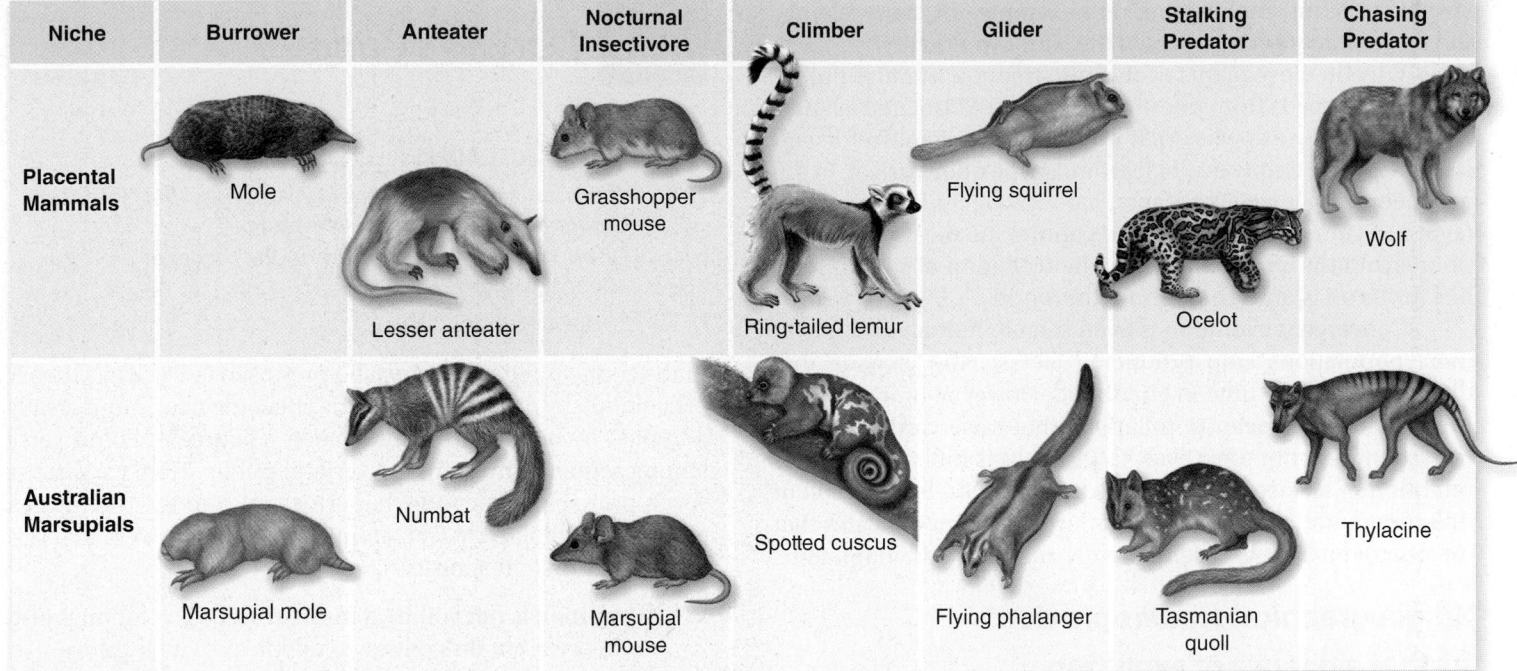

Niche	Burrower	Anteater	Nocturnal Insectivore	Climber	Glider	Stalking Predator	Chasing Predator
Placental Mammals	Mole	Lesser anteater	Grasshopper mouse	Ring-tailed lemur	Flying squirrel	Ocelot	Wolf
Australian Marsupials	Marsupial mole	Numbat	Marsupial mouse	Spotted cuscus	Flying phalanger	Tasmanian quoll	Thylacine

Figure 21.17 Convergent evolution. Many marsupial species in Australia resemble placental mammals occupying similar ecological niches elsewhere in the rest of the world. Marsupials evolved in isolation after Australia separated from other continents.

in very similar ways in different parts of the world. Marsupials are a group in which the young are born in a very immature condition and held in a pouch until they are ready to emerge into the outside world. In placentals, by contrast, offspring are not born until they can safely survive in the external environment (with varying degrees of parental care).

Australia separated from the other continents more than 70 MYA; at that time, both marsupials and placental mammals had evolved, but in different places. In particular, only marsupials occurred in Australia. As a result of this separation, the only placental mammals in Australia today are bats and a few colonizing rodents (which arrived relatively recently), and the continent is dominated by marsupials.

What are the Australian marsupials like? To an astonishing degree, they resemble the placental mammals living today on the other continents (figure 21.17). The similarity between some individual members of these two sets of mammals argues strongly that they are the result of convergent evolution, similar forms having evolved in different, isolated areas because of similar selective pressures in similar environments.

Convergent evolution is a widespread phenomenon

When species interact with the environment in similar ways, they often are exposed to similar selective pressures, and they therefore frequently develop the same evolutionary adaptations. Consider, for example, fast-moving marine predators (figure 21.18). The hydrodynamics of moving through water require a streamlined body shape to minimize friction. It is no coincidence that dolphins, sharks, and tuna—among the fastest of marine species—have all evolved to have the same basic shape. We can infer as well that ichthyosaurs—marine reptiles that lived during the Age of the Dinosaurs—exhibited a similar lifestyle.

Island trees exhibit a similar phenomenon. Most islands are covered by trees (or were until the arrival of humans). Careful inspection of these trees, however, reveals that they are not closely related to the trees with which we are familiar. Although they have all the characteristics of trees, such as being tall and having a tough outer covering, in many cases island trees are members of plant families that elsewhere exist only as flowers,

Figure 21.18 Convergence among fast-swimming predators. Fast movement through water requires a streamlined body form, which has evolved numerous times.

chapter **21** *The Evidence for Evolution*

shrubs, or other small bushes. For example, on many islands, the native trees are members of the sunflower family.

Why do these plants evolve into trees on islands? Probably because seeds from trees rarely make it to isolated islands. As a result, those species that do manage to colonize distant islands face an empty ecological landscape upon arrival. In the absence of other treelike plants, natural selection often would favor individual plants that could capture the most sunlight for photosynthesis, and the result is the evolution of similar treelike forms on islands throughout the world.

Convergent evolution is even seen in humans. People in most populations stop producing lactase, the enzyme that digests milk, some time in childhood. However, individuals in African and European populations that raise cattle produce lactase throughout their lives. DNA analysis indicates that the retention of lactase production into adulthood is the result of different mutations in Africa and Europe, which indicates that the populations have independently acquired this adaptation.

Biogeographical studies provide further evidence of evolution

Darwin made several important observations during his voyage around the world. He noted that islands often are missing plants and animals common on continents, such as frogs and land mammals. Accidental human introductions have proved that these species can survive if they are released on islands, so lack of suitable habitat is not the cause. In addition, those species that are present on islands often have diverged from their continental relatives and sometimes—as with Darwin's finches and the island trees just discussed—occupy ecological niches used by other species on continents. Lastly, island species usually are more closely related to species on nearby continents, even though the environment on continents and nearby islands often is not very similar.

Darwin deduced the explanation for these phenomena. Many islands have never been connected to continental areas. The species that occur there arrived by dispersing across the water; dispersal from nearby areas is more likely than from more distant sources, though long-distance colonization does occur occasionally. Some species, those that can fly, float, or swim are more likely to get to the island than others. Some, like frogs, are particularly vulnerable to dehydration in saltwater and have almost no chance of island colonization.

The absence of some types of plants and animals provides opportunity to those that do arrive; as a result, colonizers often evolve into many species exhibiting great ecological and morphological diversity. This phenomenon, termed adaptive radiation, is discussed in the next chapter.

Learning Outcomes Review 21.6

Convergence is the evolution of similar forms in different lineages when exposed to similar selective pressures. The biogeographical distribution of species often reflects the outcome of evolutionary diversification with closely related species in nearby areas.

- **Why does convergent evolution occur and why might species occupying similar environments in different localities sometimes not exhibit it?**

Learning Outcomes

1. **Characterize the criticisms of evolutionary theory and list counterarguments that can be made.**
2. **Distinguish between hypothesis and theory in scientific usage.**

In the century and a half since he proposed it, Darwin's theory of evolution by natural selection has become nearly universally accepted by biologists, but has been a source of controversy among some members of the general public. Here we discuss seven principle objections that critics raise to the teaching of evolution as biological fact, along with some answers that scientists present in response.

1. **Evolution is not solidly demonstrated.** "Evolution is just a theory," Darwin's critics point out, as though *theory* meant a lack of knowledge, or some kind of guess.

   Scientists, however, use the word *theory* in a very different sense than the general public does. They use *theory* only for those ideas that are strongly supported by many lines of evidence, like the theories of gravity and evolution. It is important to recognize that both evolution and gravitation are "just theories." The term *theory* is not the equivalent of a "hunch" or a "notion," or even an initial hypothesis. Rather, it is a well-tested phenomenon that rationalizes the data available.

2. **There are no fossil intermediates.** "No one ever saw a fin on the way to becoming a leg," critics claim, pointing to the many gaps in the fossil record in Darwin's day.

   Since that time, however, many fossil intermediates in vertebrate evolution have indeed been found. A clear line of fossils now traces the transition between hoofed mammals and whales, between reptiles and mammals, between dinosaurs and birds, and between apes and humans. The fossil evidence of evolution between major forms is compelling.

3. **The intelligent design argument.** "The organs of living creatures are too complex for a random process to have produced—the existence of a clock is evidence of the existence of a clockmaker."

   Evolution by natural selection is not a random process. Quite the contrary, by favoring those variations that lead to the highest reproductive fitness, natural selection is a nonrandom process that can construct highly complex organs by incrementally improving them from one generation to the next.

   For example, the intermediates in the evolution of the mammalian ear can be seen in fossils, and many intermediate "eyes" are known in various invertebrates. These intermediate forms arose because they have value—being able to detect light slightly is better than not being able to detect it at all. Complex structures such as eyes evolved as a progression of slight improvements. Moreover, inefficiencies of certain designs, such as the vertebrate eye and the existence of vestigial structures, do not support the idea of an intelligent designer.

4. **Evolution violates the Second Law of Thermodynamics.** "A jumble of soda cans doesn't by itself jump neatly into a stack—things become more disorganized due to random events, not more organized."

   Biologists point out that this argument ignores what the second law really says: Disorder increases in a closed system, which the Earth most certainly is not. Energy continually enters the biosphere from the Sun, fueling life and all the processes that organize it.

5. **Proteins are too improbable.** "Hemoglobin has 141 amino acids. The probability that the first one would be leucine is 1/20, and that all 141 would be the ones they are by chance is $(1/20)^{141}$, an impossibly rare event."

   This argument illustrates a lack of understanding of probability and statistics—probability cannot be used to argue backward. The probability that a student in a classroom has a particular birthdate is 1/365; arguing this way, the probability that everyone in a class of 50 would have the birthdates that they do is $(1/365)^{50}$, and yet there the class sits, all with their actual birthdates.

6. **Natural selection cannot account for major changes in evolution.** "No scientist has come up with an experiment in which fish evolve into frogs and leap away from predators."

   Can we extrapolate from our understanding that natural selection produces relatively small changes that are observable in populations *within* species to explain the major differences observed *between* species? Most biologists who have studied the problem think so. The differences between breeds produced by artificial selection—such as chihuahuas, mastiffs, and greyhounds—are more distinctive than the differences between some wild species, and laboratory selection experiments sometimes create forms that cannot interbreed and thus would in nature be considered different species. Thus, production of radically different forms has indeed been observed, repeatedly. These changes take millions of years, and they are seen clearly in the fossil record.

7. **The irreducible complexity argument.** Because each part of a complex cellular mechanism such as blood clotting is essential to the overall process, the intricate machinery of the cell cannot be explained by evolution from simpler stages.

   What's wrong with this argument is that each part of a complex molecular machine evolves as part of the whole system. Natural selection can act on a complex system because at every stage of its evolution, the system functions. Parts that improved function are added.

Subsequently, other parts may be modified or even lost, so that parts that were not essential when they first evolved become essential. In this way, an "irreducible complex" structure can evolve by natural selection. The same process works at the molecular level.

For example, snake venom initially evolved as enzymes to increase the ability of snakes to digest large prey items, which were captured by biting the prey and then constricting them with coils. Subsequently, the digestive enzymes evolved to become increasingly lethal. Rattlesnakes kill large prey by injecting them with venom, letting them go, and then tracking them down and eating them after they die. To do so, they have evolved extremely toxic venom, highly modified syringe-like front teeth, and many other characteristics. Take away the fangs or the venom and the rattlesnakes can't feed—what initially evolved as nonessential parts are now indispensable; irreducible complexity has evolved by natural selection.

The mammalian blood clotting system similarly has evolved from much simpler systems. The core clotting system evolved at the dawn of the vertebrates more than 500 MYA, and it is found today in primitive fishes such as lampreys. One hundred million years later, as vertebrates continued to evolve, proteins were added to the clotting system, making it sensitive to substances released from damaged tissues. Fifty million years later, a third component was added, triggering clotting by contact with the jagged surfaces produced by injury. At each stage, as the clotting system evolved to become more complex, its overall performance came to depend on the added elements. Thus, blood clotting has become "irreducibly complex" as the result of Darwinian evolution.

Statements that various structures could not have been built by natural selection have repeatedly been made over the past 150 years. In many cases, after detailed scientific study, the likely path by which such structures have evolved has been discovered.

Learning Outcomes Review 21.7

Darwin's theory of evolution is controversial to some in the general public. Objections are often based on a misunderstanding of the theory. In scientific usage, a hypothesis is an educated guess, whereas a theory is an explanation that fits available evidence and has withstood rigorous testing.

■ *Suppose someone suggests that humans originally came from Mars. Would this be a hypothesis or a theory, and how could it be tested?*

chapter **21** *The Evidence for Evolution*

Chapter Review

21.1 The Beaks of Darwin's Finches: Evidence of Natural Selection

Galápagos finches exhibit variation related to food gathering.

The correspondence between beak shape and its use in obtaining food suggested to Darwin that finch species had diversified and adapted to eat different foods.

Modern research has verified Darwin's selection hypothesis.

Natural selection acts on variation in beak morphology, favoring larger-beaked birds during extended droughts and smaller-beaked birds during long periods of heavy rains.

Because this variation is heritable, evolutionary change occurs in the frequencies of beak sizes in subsequent generations.

21.2 Peppered Moths and Industrial Melanism: More Evidence of Selection

Light-colored moths decreased in polluted areas.

In polluted areas where soot built up on tree trunks, the dark-colored form of the peppered moth became more common. In unpolluted areas, light-colored forms remained predominant.

Experiments suggested that predation by birds was the cause; light-colored moths stand out on dark trunks, and vice versa.

When environmental conditions reverse, so does selection pressure.

In the last 50 years, pollution has decreased in many areas and the frequency of light-colored moths has rebounded.

The agent of selection may be difficult to pin down.

Recent research has questioned whether bird predation is the agent of selection. Regardless, the observation that the dark-colored form has increased during times of pollution and then declined as pollution abates indicates that natural selection has acted on moth coloration.

21.3 Artificial Selection: Human-Initiated Change
(figure 21.5)

Experimental selection produces changes in populations.

Laboratory experiments in directional selection have shown that substantial evolutionary change can occur in these controlled populations.

Agricultural selection has led to extensive modification of crops and livestock.

Domesticated breeds have arisen from artificial selection.

Crop plants and domesticated animal breeds are often substantially different from their wild ancestors.

If artificial selection can rapidly create substantial change over short periods of time, then it is reasonable to assume that natural selection could have created the Earth's diversity of life over millions of years.

21.4 Fossil Evidence of Evolution

The age of fossils can be estimated.

Specimens become fossilized in different ways. Fossils in rock can be dated by calculating the extent of radioactive decay based on half-lives of known isotopes.

Fossils present a history of evolutionary change.

Fossils document evolutionary transitions.

The history of life on Earth can be traced through the fossil record. In recent years, new fossil discoveries have provided more detailed understanding of major evolutionary transitions.

The evolution of horses is a prime example of evidence from fossils.

The fossil record indicates that horses have evolved from small, forest-dwelling animals to the large and fast plains-dwelling species alive today.

Over the course of 50 million years, evolution has not been constant and uniform. Rather, change has been rapid at some times, slow at others. Although a general trend toward increase in size is evident, some species evolved to smaller sizes.

21.5 Anatomical Evidence for Evolution

Homologous structures suggest common derivation.

Homologous structures may have different appearances and functions even though derived from the same common ancestral body part.

Early embryonic development shows similarities in some groups.

Embryonic development shows similarity in developmental patterns among species whose adult phenotypes are very different.

Species that have lost a feature that was present in an ancestral form often develop and then lose that feature during embryological development.

Some structures are imperfectly suited to their use.

Natural selection can influence only the variation present in a population; because of this, evolution often results in workable, but imperfect structures, such as the vertebrate eye.

Vestigial structures can be explained as holdovers from the past.

The existence of vestigial structures supports the concept of common ancestry among organisms that share them.

21.6 Convergent Evolution and the Biogeographical Record

Marsupials and placentals demonstrate convergence.

Convergent evolution may occur in species or populations exposed to similar selective pressures. Marsupial mammals in Australia have converged upon features of their placental counterparts elsewhere.

Convergent evolution is a widespread phenomenon.

Examples include hydrodynamic streamlining in marine species and the evolution of tree species on islands from ancestral forms that were not treelike.

Biogeographical studies provide further evidence of evolution.

Island species usually are closely related to species on nearby continents even if the environments are different. Early island colonizers often evolve into diverse species because other, competing species are scarce.

21.7 Darwin's Critics

Darwin's theory of evolution by natural selection is almost universally accepted by biologists. Many criticisms have been made both historically and recently, but most stem from a lack of understanding of scientific principles, the theory's actual content, or the time spans involved in evolution.

UNDERSTAND

1. Artificial selection is different from natural selection because
   a. artificial selection is not capable of producing large changes.
   b. artificial selection does not require genetic variation.
   c. natural selection cannot produce new species.
   d. breeders (people) choose which individuals reproduce based on desirability of traits.

2. Gaps in the fossil record
   a. demonstrate our inability to date geological sediments.
   b. are expected since the probability that any organism will fossilize is extremely low.
   c. have not been filled in as new fossils have been discovered.
   d. weaken the theory of evolution.

3. The evolution of modern horses *(Equus)* is best described as
   a. the constant change and replacement of one species by another over time.
   b. a complex history of lineages that changed over time, with many going extinct.
   c. a simple history of lineages that have always resembled extant horses.
   d. None of the choices is correct.

4. Homologous structures
   a. are structures in two or more species that originate as the same structure in a common ancestor.
   b. are structures that look the same in different species.
   c. cannot serve different functions in different species.
   d. must serve different functions in different species.

5. Convergent evolution
   a. is an example of stabilizing selection.
   b. depends on natural selection to independently produce similar phenotypic responses in different species or populations.
   c. occurs only on islands.
   d. is expected when different lineages are exposed to vastly different selective environments.

6. Darwin's finches are a noteworthy case study of evolution by natural selection because evidence suggests
   a. they are descendants of many different species that colonized the Galápagos.
   b. they radiated from a single species that colonized the Galápagos.
   c. they are more closely related to mainland species than to one another.
   d. None of the choices is correct.

7. The possession of fine fur in 5-month human embryos indicates
   a. that the womb is cold at that point in pregnancy.
   b. humans evolved from a hairy ancestor.
   c. hair is a defining feature of mammals.
   d. some parts of the embryo grow faster than others.

APPLY

1. In Darwin's finches,
   a. occurrence of wet and dry years preserves genetic variation for beak size.
   b. increasing beak size over time proves that beak size is inherited.

c. large beak size is always favored.
   d. All of the choices are correct.

2. Artificial selection experiments in the laboratory such as in figure 21.5 are an example of
   a. stabilizing selection.
   b. negative frequency-dependent selection.
   c. directional selection.
   d. disruptive selection.

3. Convergent evolution is often seen among species on different islands because
   a. island populations are usually smaller and more affected by genetic drift.
   b. disruptive selection occurs commonly on islands.
   c. island species are usually most closely related to species in similar habitats elsewhere.
   d. when islands are first colonized, many ecological resources are unused, allowing descendants of a colonizing species to diversify and adapt to many different parts of the environment.

SYNTHESIZE

1. What conditions are necessary for evolution by natural selection?

Refer to figure 21.2 for the following two questions.

2. Explain how data shown in figure 21.2*a* and *b* relate to the conditions identified by you in question 1.

3. On figure 21.2*b*, draw the relationship between offspring beak depth and parent beak depth, assuming that there is no genetic basis to beak depth in the medium ground finch.

4. Refer to figure 21.5, artificial selection in the laboratory. In this experiment, one population of *Drosophila* was selected for low numbers of bristles and the other for high numbers. Note that not only did the means of the populations change greatly in 35 generations, but also all individuals in both experimental populations lie outside the range of the initial population. What would happen if the direction of selection were reversed, such that a greater number of bristles was selected for in the low-bristle population, and vice versa? How would the rate of evolutionary change compare with that in the initial part of the experiment before selection was reversed?

5. The ancestor of horses was a small, many-toed animal that lived in forests, whereas today's horses are large animals with a single hoof that live on open plains. A series of intermediate fossils illustrate how this transition has occurred, and for this reason, many old treatments of horse evolution portrayed it as a steady increase through time in body size accompanied by a steady decrease in toe number. Why is this interpretation incorrect?

ONLINE RESOURCE

www.ravenbiology.com

Understand, Apply, and Synthesize—enhance your study with animations that bring concepts to life and practice tests to assess your understanding. Your instructor may also recommend the interactive eBook, individualized learning tools, and more.

Chapter **22**

The Origin of Species

Chapter Contents

A Introduction

Although Darwin titled his book On the Origin of Species, *he never actually discussed what he referred to as that "mystery of mysteries"—how one species gives rise to another. Rather, his argument concerned evolution by natural selection; that is, how one species evolves through time to adapt to its changing environment. Although an important mechanism of evolutionary change, the process of adaptation does not explain how one species becomes another, a process we call speciation. As we shall see, adaptation may be involved in the speciation process, but it does not have to be.*

Before we can discuss how one species gives rise to another, we need to understand exactly what a species is. Even though the definition of a species is of fundamental importance to evolutionary biology, this issue has still not been completely settled and is currently the subject of considerable research and debate.

22.1 The Nature of Species and the Biological Species Concept

Learning Outcomes

1. Distinguish between the biological species concept and the ecological species concept.
2. Define the two kinds of reproductive isolating mechanisms.
3. Describe the relationship of reproductive isolating mechanisms to the biological species concept.

Any concept of a species must account for two phenomena: the distinctiveness of species that occur together at a single locality, and the connection that exists among different populations belonging to the same species.

Sympatric species inhabit the same locale but remain distinct

Put out birdfeeders on your balcony or in your back yard, and you will attract a wide variety of birds (especially if you include different kinds of foods). In the midwestern United States, for example, you might routinely see cardinals, blue jays, downy woodpeckers, house finches—even hummingbirds in the summer.

Although it might take a few days of careful observation, you would soon be able to readily distinguish the many different species. The reason is that species that occur together (termed **sympatric**) are distinctive entities that are phenotypically different, utilize different parts of the habitat, and behave differently. This observation is generally true not only for birds, but also for most other types of organisms.

Occasionally, two species occur together that appear to be nearly identical. In such cases, we need to go beyond visual similarities. When other aspects of the phenotype are examined, such as the mating calls or the chemicals exuded by each species, they usually reveal great differences. In other words, even though we might have trouble distinguishing them, the organisms themselves have no such difficulties.

Populations of a species exhibit geographic variation

Within a single species, individuals in populations that occur in different areas may be distinct from one another. Such groups of distinctive individuals may be classified as **subspecies** (the vague term *race* has a similar connotation, but is no longer commonly used). In areas where these populations occur close to one another, individuals often exhibit combinations of features characteristic of both populations (figure 22.1). In other words, even though geographically distant populations may appear distinct, they are usually connected by intervening populations that are intermediate in their characteristics.

The biological species concept focuses on the ability to exchange genes

What can account for both the distinctiveness of sympatric species and the connectedness of geographically separate populations of the same species? One obvious possibility is that each species exchanges genetic material only with other members of its species. If sympatric species commonly exchanged genes, which they generally do not, we might expect such species to rapidly lose their distinctions, as the **gene pools** (that is, all of the alleles present in a species) of the different species became homogenized. Conversely, the ability of geographically distant populations of a single species to share genes through the process of gene flow may keep these populations integrated as members of the same species.

Based on these ideas, in 1942 the evolutionary biologist Ernst Mayr set forth the **biological species concept,** which defines *species* as ". . . groups of actually or potentially interbreeding natural populations which are reproductively isolated from other such groups."

In other words, the biological species concept says that a species is composed of populations whose members mate with each other and produce fertile offspring—or would do so if they came into contact. Conversely, populations whose members do not mate with each other or who cannot produce fertile offspring are said to be **reproductively isolated** and, therefore, are members of different species.

What causes reproductive isolation? If organisms cannot interbreed or cannot produce fertile offspring, they clearly belong to different species. However, some populations that are considered separate species can interbreed and produce

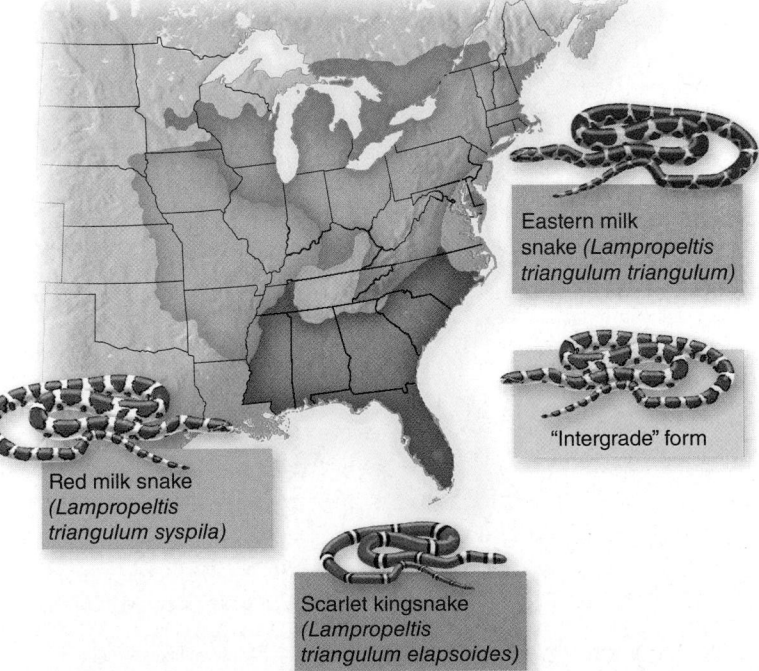

Eastern milk snake (*Lampropeltis triangulum triangulum*)

"Intergrade" form

Red milk snake (*Lampropeltis triangulum syspila*)

Scarlet kingsnake (*Lampropeltis triangulum elapsoides*)

Figure 22.1 Geographic variation in the milk snake, *Lampropeltis triangulum.* Although subspecies appear phenotypically quite distinctive from one another, they are connected by populations that are phenotypically intermediate.

fertile offspring, but they ordinarily do not do so under natural conditions. They are still considered reproductively isolated in that genes from one species generally will not enter the gene pool of the other.

Table 22.1 summarizes the steps at which barriers to successful reproduction may occur. Such barriers are termed **reproductive isolating mechanisms** because they prevent genetic exchange between species. We will discuss examples of these next, beginning with those that prevent the formation of zygotes, which are called **prezygotic isolating mechanisms.** Mechanisms that prevent the proper functioning of zygotes after they form are called **postzygotic isolating mechanisms.**

Prezygotic isolating mechanisms prevent the formation of a zygote

Mechanisms that prevent formation of a zygote include ecological or environmental isolation, behavioral isolation, temporal isolation, mechanical isolation, and prevention of gamete fusion.

Ecological isolation

Even if two species occur in the same area, they may utilize different portions of the environment and thus not hybridize because they do not encounter each other. For example, in India, the ranges of lions and tigers overlapped until about 150 years ago. Even so, there were no records of natural hybrids. Lions stayed mainly in the open grassland and hunted in groups called prides; tigers tended to be solitary creatures of the forest (figure 22.2). Because of their ecological and behavioral differences, lions and tigers rarely came into direct contact with each other, even though their ranges overlapped over thousands of square kilometers.

In another example, the ranges of two toads, *Bufo woodhousei* and *B. americanus,* overlap in some areas. Although these two species can produce viable hybrids, they usually do not interbreed because they utilize different portions of the habitat for breeding. *B. woodhousei* prefers to breed in streams, and *B. americanus* breeds in rainwater puddles.

Similar situations occur among plants. Two species of oaks occur widely in California: the valley oak, *Quercus lobata,* and the scrub oak, *Q. dumosa.* The valley oak, a graceful deciduous tree that can be as tall as 35 m, occurs in the fertile soils of open grassland on gentle slopes and valley floors. In contrast, the scrub oak is an evergreen shrub, usually only 1 to 3 m tall, which often forms the kind of dense scrub known as chaparral. The scrub oak is found on steep slopes in less fertile soils. Hybrids between these different oaks do occur and are fully fertile, but they are rare. The sharply distinct habitats of their parents limit their occurrence together, and there is little intermediate habitat where the hybrids might flourish.

Behavioral isolation

Chapter 54 describes the often elaborate courtship and mating rituals of some groups of animals. Related species of organisms

TABLE 22.1	Reproductive Isolating Mechanisms	
Mechanism		**Description**
PREZYGOTIC ISOLATING MECHANISMS		
Ecological isolation		Species occur in the same area, but they occupy different habitats and rarely encounter each other.
Behavioral isolation		Species differ in their mating rituals.
Temporal isolation		Species reproduce in different seasons or at different times of the day.
Mechanical isolation		Structural differences between species prevent mating.
Prevention of gamete fusion		Gametes of one species function poorly with the gametes of another species or within the reproductive tract of another species.
POSTZYGOTIC ISOLATING MECHANISMS		
Hybrid inviability or infertility		Hybrid embryos do not develop properly, hybrid adults do not survive in nature, or hybrid adults are sterile or have reduced fertility.

Figure 22.2 Lions and tigers are ecologically isolated. The ranges of lions and tigers overlap in India. However, lions and tigers do not hybridize in the wild because they utilize different portions of the habitat. Hybrids, such as this tiglon, have been successfully produced in captivity, but hybridization does not occur in the wild.

Figure 22.3 Differences in courtship rituals can isolate related bird species. These Galápagos blue-footed boobies select their mates only after an elaborate courtship display. This male is lifting his feet in a ritualized high-step that shows off his bright blue feet. The display behavior of the two other species of boobies that occur in the Galápagos is very different, as is the color of their feet.

such as birds often differ in their courtship rituals, which tends to keep these species distinct in nature even if they inhabit the same places (figure 22.3). For example, mallard and pintail ducks are perhaps the two most common freshwater ducks in North America. In captivity, they produce completely fertile hybrid offspring, but in nature they nest side by side and only rarely hybridize.

Sympatric species avoid mating with members of the wrong species in a variety of ways; every mode of communication imaginable appears to be used by some species. Differences in visual signals, as just discussed, are common; however, other types of animals rely more on other sensory modes for communication. Many species, such as frogs, birds, and a variety of insects, use sound to attract mates. Predictably, sympatric species of these animals produce different calls. Similarly, the "songs" of lacewings are produced when they vibrate their abdomens against the surface on which they are sitting, and sympatric species produce different vibration patterns (figure 22.4).

Other species rely on the detection of chemical signals, called **pheromones.** The use of pheromones in moths has been particularly well studied. When female moths are ready to mate, they emit a pheromone that males can detect at great distances. Sympatric species differ in the pheromone they produce: Either they use different chemical compounds, or, if using the same compounds, the proportions used are different. Laboratory studies indicate that males are remarkably adept at distinguishing the pheromones of their own species from those of other species or even from synthetic compounds similar, but not identical, to that of their own species.

Some species even use electroreception. African and South Asian electric fish independently have evolved specialized organs in their tails that produce electrical discharges and electroreceptors on their skins to detect them. These discharges are used to communicate in social interactions; field experiments indicate that males can distinguish between signals produced by their own and other species, probably on the basis of differences in the timing of the electrical pulses.

Temporal isolation

Two species of wild lettuce, *Lactuca graminifolia* and *L. canadensis,* grow together along roadsides throughout the southeastern United States. Hybrids between these two species are easily made experimentally and are completely fertile. But these hybrids are rare in nature because *L. graminifolia* flowers in early spring and *L. canadensis* flowers in summer. When their blooming periods overlap, as happens occasionally, the two species do form hybrids, which may become locally abundant.

Many species of closely related amphibians have different breeding seasons that prevent hybridization. For example, five species of frogs of the genus *Rana* occur together in most of the eastern United States, but hybrids are rare because the peak breeding time is different for each of them.

Mechanical isolation

Structural differences prevent mating between some related species of animals. Aside from such obvious features as size, the structure of the male and female copulatory organs may be incompatible. In many insect and other arthropod groups, the sexual organs, particularly those of the male, are so diverse that they are used as a primary basis for distinguishing species.

Similarly, flowers of related species of plants often differ significantly in their proportions and structures. Some of these differences limit the transfer of pollen from one plant species to another. For example, bees may carry the pollen of one species on a certain place on their bodies; if this area does not come into contact with the receptive structures of the flowers of another plant species, the pollen is not transferred.

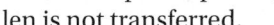

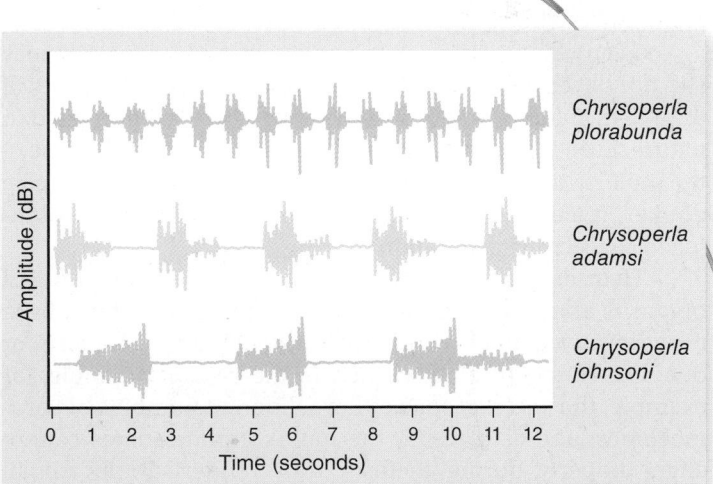

Figure 22.4 Differences in courtship song of sympatric species of lacewings. Lacewings are small insects that rely on auditory signals produced by moving their abdomens to vibrate the surface on which they are sitting to attract mates. As these recordings indicate, the vibration patterns produced by sympatric species differ greatly. Females, which detect the calls as they are transmitted through solid surfaces such as branches, are able to distinguish calls of different species and only respond to individuals producing their own species' call.

chapter **22** *The Origin of Species*

Prevention of gamete fusion

In animals that shed gametes directly into water, the eggs and sperm derived from different species may not attract or fuse with one another. Many animals that have internal fertilization may not hybridize successfully because the sperm of one species functions so poorly within the reproductive tract of another that fertilization never takes place. In plants, the growth of pollen tubes may be impeded in hybrids between different species. In both plants and animals, isolating mechanisms such as these prevent the union of gametes, even following successful mating.

Postzygotic isolating mechanisms prevent normal development into reproducing adults

All of the factors we have discussed so far tend to prevent hybridization. If hybrid matings do occur and zygotes are produced, many factors may still prevent those zygotes from developing into normally functioning, fertile individuals.

As you saw in chapter 19, development is a complex process. In hybrids, the genetic complements of two species may be so different that they cannot function together normally in embryonic development. For example, hybridization between sheep and goats usually produces embryos that die in the earliest developmental stages.

The leopard frogs (*Rana pipiens* complex) of the eastern United States are a group of similar species, assumed for a long time to constitute a single species (figure 22.5). Careful examination, however, revealed that although the frogs appear similar, successful mating between them is rare because of problems that occur as the fertilized eggs develop. Many of the hybrid combinations cannot be produced even in the laboratory. Examples of this kind, in which similar species have been recognized only as a result of hybridization experiments, are common in plants.

Even when hybrids survive the embryo stage, they may still not develop normally. If the hybrids are less physically fit than their parents, they will almost certainly be eliminated in nature. Even if a hybrid is vigorous and strong, as in the case of the mule, which is a hybrid between a female horse and a male donkey, it may still be sterile and thus incapable of contributing to succeeding generations.

Hybrids may be sterile because the development of sex organs is abnormal, because the chromosomes derived from the respective parents cannot pair properly during meiosis, or due to a variety of other causes. In the case of the mule, for example, the parental species have different numbers of chromosomes, and as a result, the mule's chromosomes cannot align properly during meiosis, thus rendering the animal infertile.

The biological species concept does not explain all observations

The biological species concept has proved to be an effective way of understanding the existence of species in nature. Nonetheless, it fails to take into account all observations, leading some biologists to propose alternative species concepts.

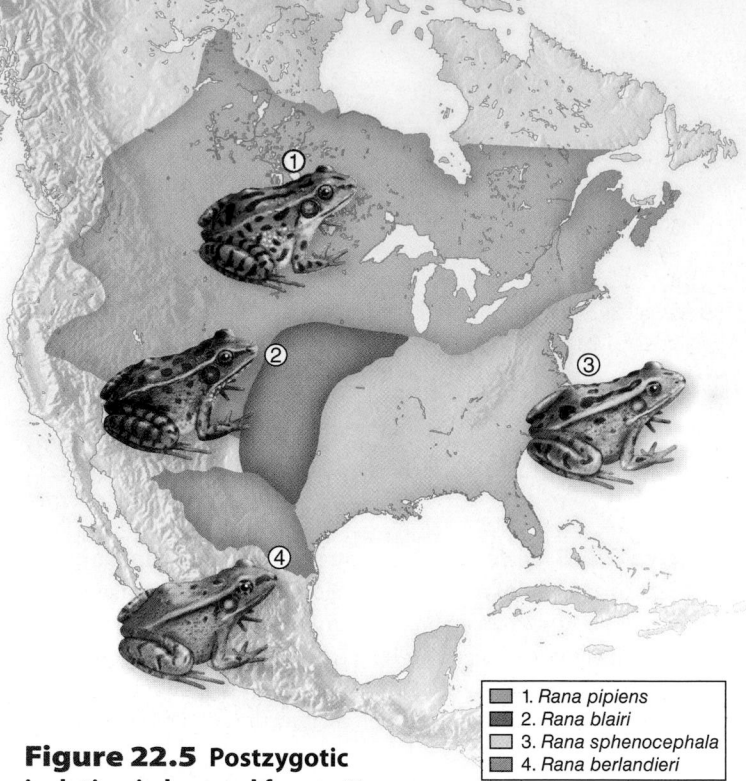

Figure 22.5 Postzygotic isolation in leopard frogs. These four species resemble one another closely in their external features. Their status as separate species first was suspected when hybrids between some pairs of these species were found to produce defective embryos in the laboratory. Subsequent research revealed that the mating calls of the four species differ substantially, indicating that the species have both pre- and postzygotic isolating mechanisms.

1. *Rana pipiens*
2. *Rana blairi*
3. *Rana sphenocephala*
4. *Rana berlandieri*

One criticism of the biological species concept concerns the extent to which all species truly are reproductively isolated. By definition, under the biological species concept, species should not interbreed and produce fertile offspring. But in recent years, biologists have detected much greater amounts of interspecies hybridization than was previously thought to occur between populations that seem to coexist as distinct biological entities.

Botanists have always been aware that plant species often undergo substantial amounts of hybridization. More than 50% of California plant species included in one study, for example, were not well defined by genetic isolation. This coexistence without genetic isolation can be long-lasting: Fossil data show that balsam poplars and cottonwoods have been phenotypically distinct for 12 million years, but they also have routinely produced hybrids throughout this time. Consequently, many botanists have long felt that the biological species concept is misnamed and only applies to animals.

New evidence, however, increasingly indicates that hybridization is not all that uncommon in animals, either. In recent years, many cases of substantial hybridization between animal species have been documented. One survey indicated that almost 10% of the world's 9500 bird species are known to have hybridized in nature.

The Galápagos finches provide a particularly well-studied example. Three species on the island of Daphne

Major—the medium ground finch, the cactus finch, and the small ground finch—are clearly distinct morphologically, and they occupy different ecological niches. Studies over the past 20 years by Peter and Rosemary Grant found that, on average, 2% of the medium ground finches and 1% of the cactus ground finches mated with other species every year. Furthermore, hybrid offspring appeared to be at no disadvantage in terms of survival or subsequent reproduction. This is not a trivial amount of genetic exchange, and one might expect to see the species coalesce into one genetically variable population—but the species are maintaining their distinctiveness.

Hybridization is not rampant throughout the animal world, however. Most bird species do not hybridize, and probably even fewer experience significant amounts of hybridization. Still, hybridization is common enough to cast doubt on whether reproductive isolation is the only force maintaining the integrity of species.

Natural selection and the ecological species concept

An alternative hypothesis proposes that the distinctions among species are maintained by natural selection. The idea is that each species has adapted to its own specific part of the environment. Stabilizing selection, described in chapter 20, then maintains the species' adaptations. Hybridization has little effect because alleles introduced into one species' gene pool from other species are quickly eliminated by natural selection.

You probably recall from chapter 20 that the interaction between gene flow and natural selection can have many outcomes. In some cases, strong selection can overwhelm any effects of gene flow—but in other situations, gene flow can prevent populations from eliminating less successful alleles from a population.

As a general explanation, then, an ecological species concept is not likely to have any fewer exceptions than does the biological species concept, although it might prove to be a more successful description for certain types of organisms or habitats.

Other weaknesses of the biological species concept

The biological species concept has been criticized for other reasons as well. For example, it can be difficult to apply the concept to populations that are geographically separated in nature. Because individuals of these populations do not encounter each other, it is not possible to observe whether they would interbreed naturally.

Although experiments can determine whether fertile hybrids can be produced, this information is not enough. Many species that coexist without interbreeding in nature will readily hybridize in the artificial settings of the laboratory or zoo. Consequently, evaluating whether such populations constitute different species is ultimately a judgment call. In addition, the concept is more limited than its name would imply. Many organisms are asexual and reproduce without mating. Reproductive isolation therefore has no meaning for such organisms.

For these reasons, a variety of other ideas have been put forward to establish criteria for defining species. Many of these are specific to a particular type of organism, and none has universal applicability. In reality, there may be no single

explanation for what maintains the identity of species. Given the incredible variation evident in plants, animals, and microorganisms in all aspects of their biology, it would not be surprising to find that different processes are operating in different organisms.

In addition, some scientists have turned from emphasizing the processes that maintain species distinctions to examining the evolutionary history of populations. These genealogical species concepts are currently a topic of great debate and are discussed further in chapter 23.

Learning Outcomes Review 22.1

Species are populations of organisms that are distinct from other, co-occurring species, and are interconnected geographically. The biological species concept therefore defines species based on their ability to interbreed. Reproductive isolating mechanisms prevent successful interbreeding between species. The ecological species concept relies on adaptation and natural selection as a force for maintaining separation of species.

- **How does the ability to exchange genes explain why sympatric species remain distinct and geographic populations of one species remain connected?**
- **How does the ecological species concept explain this phenomenon?**

22.2 Natural Selection and Reproductive Isolation

Learning Outcomes

1. *Define reinforcement in the context of reproductive isolation.*
2. *Explain the possible outcomes when two populations that are partially reproductively isolated become sympatric.*

One of the oldest questions in the field of evolution is: How does one ancestral species become divided into two descendant species? If species are defined by the existence of reproductive isolation, then the process of speciation is identical to the evolution of reproductive isolating mechanisms.

Selection may reinforce isolating mechanisms

The formation of species is a continuous process, and as a result, two populations may only be partially reproductively isolated. For example, because of behavioral or ecological differences, individuals of two populations may be more likely to mate with members of their own population, and yet between-population matings may still occur. If mating occurs and fertilization produces a zygote, postzygotic barriers may also be incomplete: developmental problems may result in lower embryo survival or reduced fertility, but some individuals may survive and reproduce.

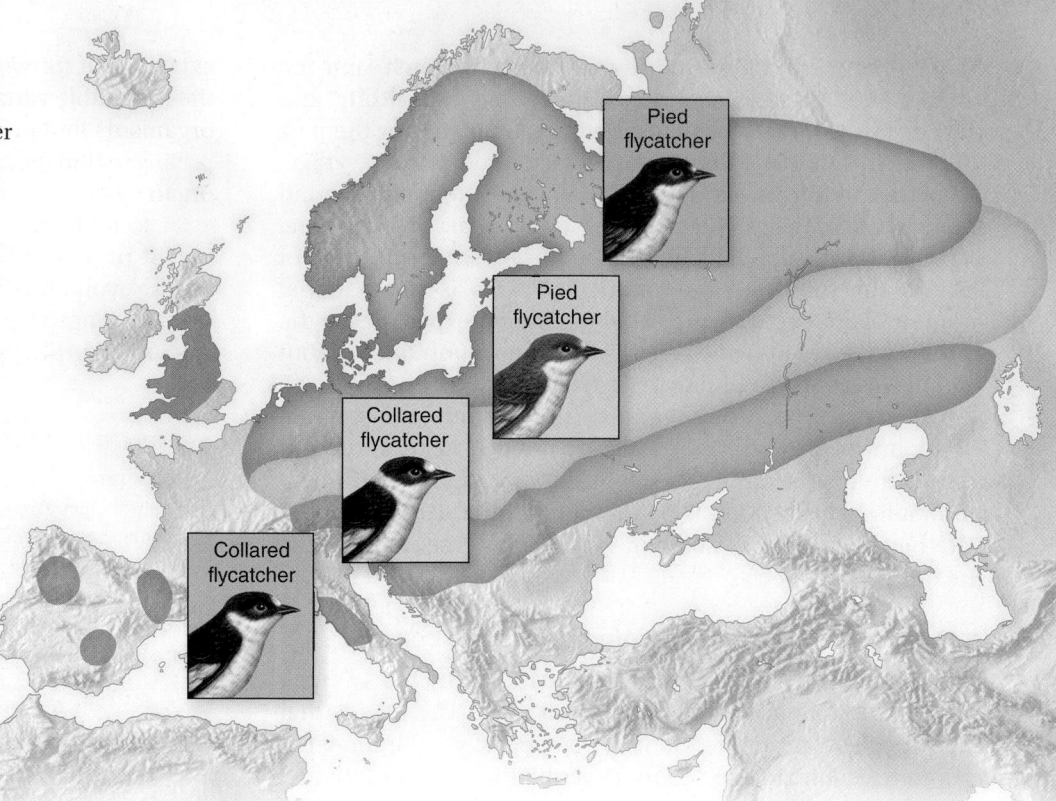

Figure 22.6 Reinforcement in European flycatchers. The pied flycatcher *(Ficedula hypoleuca)* and the collared flycatcher *(F. albicollis)* appear very similar when they occur alone. However, in places where the two species occur sympatrically (indicated by the yellow color on the map), they have evolved differences in color and pattern, which allow individuals to choose mates from their own species and thus avoid hybridizing.

What happens when two populations come into contact thus depends on the extent to which isolating mechanisms have already evolved. If isolating mechanisms have not evolved at all, then the two populations will interbreed freely, and whatever other differences have evolved between them should disappear over the course of time, as genetic exchange homogenizes the populations. Conversely, if the populations are completely reproductively isolated, then no genetic exchange will occur, and the two populations will remain different species.

How reinforcement can complete the speciation process

The intermediate state, in which reproductive isolation has partially evolved but is not complete, is perhaps the most interesting situation. If the hybrids are partly sterile, or not as well adapted to the existing habitats as their parents, they will be at a disadvantage. Selection would favor any alleles in the parental populations that prevented hybridization, because individuals that did not engage in hybridization would produce more successful offspring.

The result would be the continual improvement of prezygotic isolating mechanisms until the two populations were completely reproductively isolated. This process is termed **reinforcement** because initially incomplete isolating mechanisms are reinforced by natural selection until they are completely effective.

An example of reinforcement is provided by pied and collared flycatchers. Throughout much of Europe, these two bird species are geographically separated **(allopatric)** and are very similar in color (figure 22.6). However, in central Europe, the two species occur together and occasionally hybridize, producing offspring that usually have very low fertility. At those sites, the species have evolved to look very different from each other, and birds prefer to mate with individuals with their own species' coloration. In contrast, birds from the allopatric populations prefer the allopatric color pattern. As a consequence of

the color differences where the species are sympatric, the rate of hybridization is extremely low. These results indicate that when populations of the two species came into contact, natural selection led to the evolution of differences in color patterns, thus enabling birds of one species to distinguish individuals of their own species from members of the other species, producing behavioral, prezygotic isolation.

How gene flow may counter speciation

Reinforcement is not inevitable, however. When incompletely isolated populations come together, gene flow immediately begins to occur between them. Although hybrids may be inferior, they are not completely inviable or infertile—if they were, the species would already be completely reproductively isolated. When these surviving hybrids reproduce with members of either population, they will serve as a conduit of genetic exchange from one population to the other, and the two populations will tend to lose their genetic distinctiveness. Thus, a race ensues: Can complete reproductive isolation evolve before gene flow erases the differences between the populations? Experts disagree on the likely outcome, but many consider reinforcement to be the less common outcome.

Learning Outcomes Review 22.2

Natural selection may favor the evolution of increased prezygotic reproductive isolation between sympatric populations. This phenomenon is termed reinforcement, and it may lead to populations becoming completely reproductively isolated. In contrast, however, genetic exchange between populations may decrease genetic differences among populations, thus preventing speciation from occurring.

■ *How might the initial degree of reproductive isolation affect the probability that reinforcement will occur when two populations come into sympatry?*

What role does natural selection play in the speciation process? Certainly, the process of reinforcement is driven by natural selection, favoring the evolution of complete reproductive isolation. But reinforcement may not be common. In situations other than reinforcement, does natural selection play a role in the evolution of reproductive isolating mechanisms?

Random changes may cause reproductive isolation

As mentioned in chapter 20, populations may diverge for purely random reasons. Genetic drift in small populations, founder effects, and population bottlenecks all may lead to changes in traits that cause reproductive isolation.

For example, in the Hawaiian Islands, closely related species of *Drosophila* often differ greatly in their courtship behavior. Colonization of new islands by these fruit flies probably involved a founder effect, in which one or a few flies—perhaps only a single pregnant female—were blown by strong winds to the new island. Changes in courtship behavior between ancestor and descendant populations may be the result of such founder events.

Given enough time, any two isolated populations will diverge because of genetic drift (remember that even large populations experience drift, but at a lower rate than in small populations). In some cases, this random divergence may affect traits responsible for reproductive isolation, and speciation may occur.

Adaptation can lead to speciation

Although random processes may sometimes be responsible, in many cases natural selection probably plays a role in the speciation process. As populations of a species adapt to different circumstances, they likely accumulate many differences that may lead to reproductive isolation. For example, if one population of flies adapts to wet habitats and another to dry areas, then natural selection will favor a variety of corresponding differences in physiological and sensory traits. These differences may produce ecological and behavioral isolation and may cause any hybrids the two populations produce to be poorly adapted to either habitat.

Selection might also act directly on mating behavior. Male *Anolis* lizards, for example, court females by extending a colorful flap of skin, called a *dewlap,* located under their throats (figure 22.7). The ability of one lizard to see the dewlap of another lizard depends not only on the color of the dewlap, but on the environment in which the lizards occur. A light-colored dewlap, for example, is most effective in reflecting light in a dim forest, whereas dark colors are more apparent in the bright glare of open habitats. As a result, when these lizards occupy new habitats, natural selection favors evolutionary change in dewlap color because males whose dewlaps cannot be seen will attract few mates. But the lizards also distinguish members of their own species from other species by the color of the dewlap. Adaptive change in mating signals in new environments could therefore have the incidental consequence of producing reproductive isolation from populations in the ancestral environment.

Laboratory scientists have conducted experiments on fruit flies and other fast-reproducing organisms in which they isolate populations in different laboratory chambers and measure how much reproductive isolation evolves. These experiments indicate that genetic drift by itself can lead to some degree of reproductive isolation, but in general, reproductive isolation evolves more rapidly when the populations are forced to adapt to different laboratory environments (such as temperature or food type). Although natural selection in the experiment does not directly favor traits because they lead to reproductive isolation, the incidental effect of adaptive divergence is that populations in different environments become reproductively isolated. For this reason, some biologists believe that the term *isolating mechanisms* is misguided, because it implies that the traits evolved specifically for the purpose of genetically isolating a species, which in most cases—except reinforcement—is probably incorrect.

Figure 22.7 Dewlaps of different species of Caribbean *Anolis* lizards. Males use their dewlaps in both territorial and courtship displays. Coexisting species almost always differ in their dewlaps, which are used in species recognition. Darker-colored dewlaps, such as those of the two species on the left, are easier to see in open habitats, whereas lighter-colored dewlaps, like those of the two species on the right, are more visible in shaded environments.

a. *b.* *c.*

Figure 22.8 Populations can become geographically isolated for a variety of reasons. *a.* Colonization of remote areas by one or a few individuals can establish populations in a distant place. *b.* Barriers to movement can split an ancestral population into two isolated populations. *c.* Extinction of intermediate populations can leave the remaining populations isolated from one another.

22.4 The Geography of Speciation

Learning Outcomes

1. Compare and contrast sympatric and allopatric speciation.
2. Explain the conditions required for sympatric speciation to occur.

Speciation is a two-part process. First, initially identical populations must diverge, and second, reproductive isolation must evolve to maintain these differences. The difficulty with this process, as we have seen, is that the homogenizing effect of gene flow between populations is constantly acting to erase any differences that may arise, either by genetic drift or natural selection. Gene flow only occurs between populations that are in contact, however, and populations can become geographically isolated for a variety of reasons (figure 22.8). Consequently, evolutionary biologists have long recognized that speciation is much more likely in geographically isolated populations.

Allopatric speciation takes place when populations are geographically isolated

Ernst Mayr was the first biologist to demonstrate that geographically separated, or *allopatric*, populations appear much more likely to have evolved substantial differences leading to speciation. Marshalling data from a wide variety of organisms and localities, Mayr made a strong case for allopatric speciation as the primary means of speciation.

For example, the little paradise kingfisher varies little throughout its wide range in New Guinea, despite the great variation in the island's topography and climate. By contrast, isolated populations on nearby islands are strikingly different from one another and from the mainland population (figure 22.9). Thus, geographic isolation seems to have been an important prerequisite for the evolution of differences between populations.

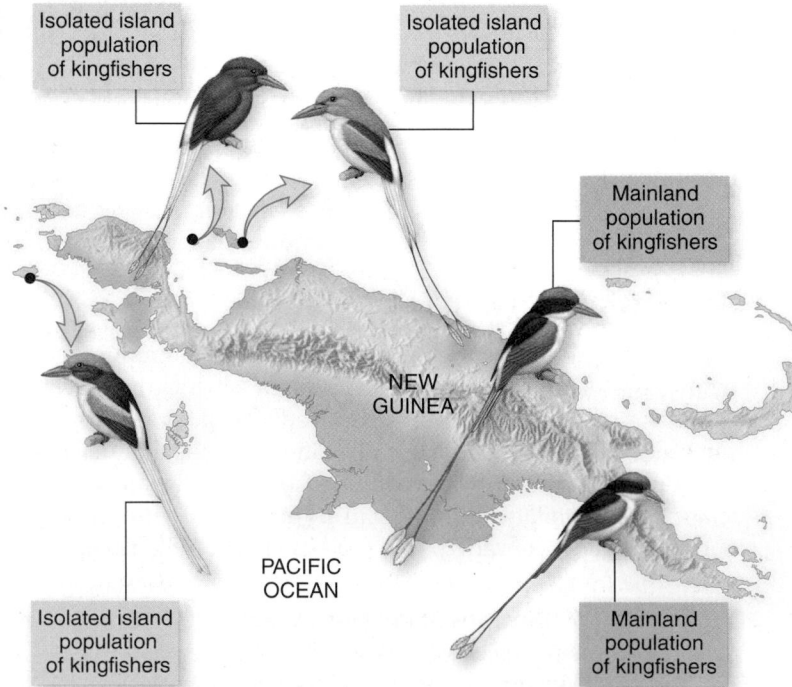

Figure 22.9 Phenotypic differentiation in the little paradise kingfisher *Tanysiptera hydrocharis* in New Guinea. Isolated island populations *(left)* are quite distinctive, showing variation in tail feather structure and length, plumage coloration, and bill size, whereas kingfishers on the mainland *(right)* show little variation.

Many other examples indicate that speciation can occur under allopatric conditions. Because we would expect isolated populations to diverge over time, by either drift or selection, this result is not surprising. Rather, the more intriguing question becomes: Is geographic isolation *required* for speciation to occur?

Sympatric speciation occurs without geographic separation

For decades, biologists have debated whether one species can split into two at a single locality, without the two new species ever having been geographically separated. Investigators have suggested that this sympatric speciation could occur either instantaneously or over the course of multiple generations. Although most of the hypotheses suggested so far are highly controversial, one type of instantaneous sympatric speciation is known to occur commonly, as the result of polyploidy.

Instantaneous speciation through polyploidy

Instantaneous sympatric speciation occurs when an individual is born that is reproductively isolated from all other members of its species. In most cases, a mutation that would cause an individual to be greatly different from others of its species would have many adverse pleiotropic side effects, and the individual would not survive. One exception often seen in plants, however, occurs through the process of **polyploidy,** which produces individuals that have more than two sets of chromosomes.

Polyploid individuals can arise in two ways. In **autopolyploidy,** all of the chromosomes may arise from a single species. This might happen, for example, due to an error in cell division that causes a doubling of chromosomes. Such individuals, termed *tetraploids* because they have four sets of chromosomes, can self fertilize or mate with other tetraploids, but cannot mate and produce fertile offspring with normal diploids. The reason is that the tetraploid species produce "diploid" gametes that produce triploid offspring (having three sets of chromosomes) when combined with haploid gametes from normal diploids. Triploids are sterile because the odd number of chromosomes prevent proper pairing during meiosis.

A more common type of polyploid speciation is **allopolyploidy,** which may happen when two species hybridize (figure 22.10). The resulting offspring, having one copy of the chromosomes of each species, is usually infertile because the chromosomes do not pair correctly in meiosis. However, such individuals are often otherwise healthy, can reproduce asexually, and can even become fertile through a variety of events. For example, if the chromosomes of such an individual were to spontaneously double, as just described, the resulting tetraploid would have two copies of each set of chromosomes. Consequently, pairing would no longer be a problem in meiosis. As a result, such tetraploids would be able to interbreed, and a new species would have been created.

It is estimated that about half of the approximately 260,000 species of plants have a polyploid episode in their history, including many of great commercial importance, such as bread wheat, cotton, tobacco, sugarcane, bananas, and potatoes. Speciation by polyploidy is also known to occur in a variety of animals, including insects, fish, and salamanders, although much more rarely than in plants.

Sympatric speciation by disruptive selection

Some investigators believe that sympatric speciation can occur over the course of multiple generations through the process of disruptive selection. As noted in chapter 20, disruptive

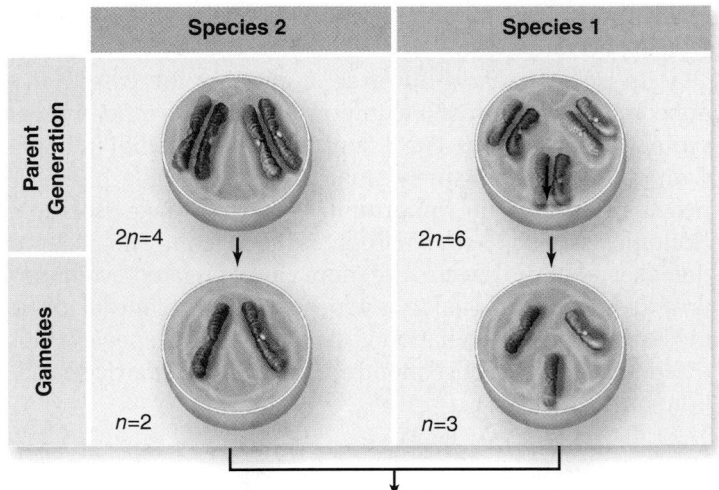

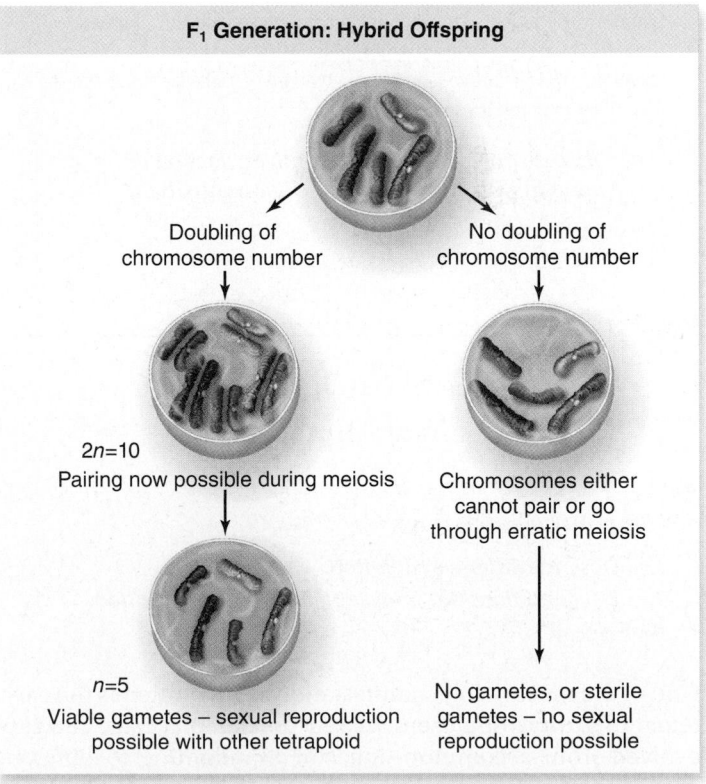

Figure 22.10 Allopolyploid speciation. Hybrid offspring from parents with different numbers of chromosomes often cannot reproduce sexually. Sometimes, the number of chromosomes in such hybrids doubles to produce a tetraploid individual that can undergo meiosis and reproduce with similar tetraploid individuals.

selection can cause a population to contain individuals exhibiting two different phenotypes.

One might think that if selection is strong enough, these two phenotypes would evolve over a number of generations into different species. But before the two phenotypes could become different species, they would have to evolve reproductive isolating mechanisms. Initially, the two phenotypes would not be reproductively isolated at all, and genetic exchange between individuals of the two phenotypes would tend to prevent genetic divergence in mating preferences or other isolating mechanisms. As a result, the two phenotypes would be retained as polymorphisms within a single population. For this reason, most biologists consider sympatric speciation of this type to be a rare event.

In recent years, however, a number of cases have appeared that are difficult to interpret in any way other than as sympatric speciation. For example, Lake Barombi Mbo in Cameroon is an extremely small and ecologically homogeneous lake, with no opportunity for within-lake isolation. Nonetheless, 11 species of closely related cichlid fish occur in the lake; all of the species are more closely related evolutionarily to one another than to any species outside of the lake. The most reasonable explanation is that an ancestral species colonized the lake and subsequently underwent sympatric speciation multiple times.

Learning Outcomes Review 22.4

Sympatric speciation occurs without geographic separation, whereas allopatric speciation occurs in geographically isolated populations. Polyploidy and disruptive selection are two ways by which a single species may undergo sympatric speciation.

■ *How do polyploidy and disruptive selection differ as ways in which sympatric speciation can occur?*

22.5 Adaptive Radiation and Biological Diversity

Learning Outcomes

1. *Describe adaptive radiation.*
2. *List conditions that may lead to adaptive radiation.*

One of the most visible manifestations of evolution is the existence of groups of closely related species that have recently evolved from a common ancestor by adapting to different parts of the environment. These **adaptive radiations** are particularly common in situations in which a species occurs in an environment with few other species and many available resources. One example is the creation of new islands through volcanic activity, such as the Hawaiian and Galápagos Islands.

Another example is a catastrophic event leading to the extinction of most other species, a phenomenon termed mass extinction that we discuss in chapter 59.

Adaptive radiation can also result when a new trait, called a **key innovation,** evolves within a species allowing it to use

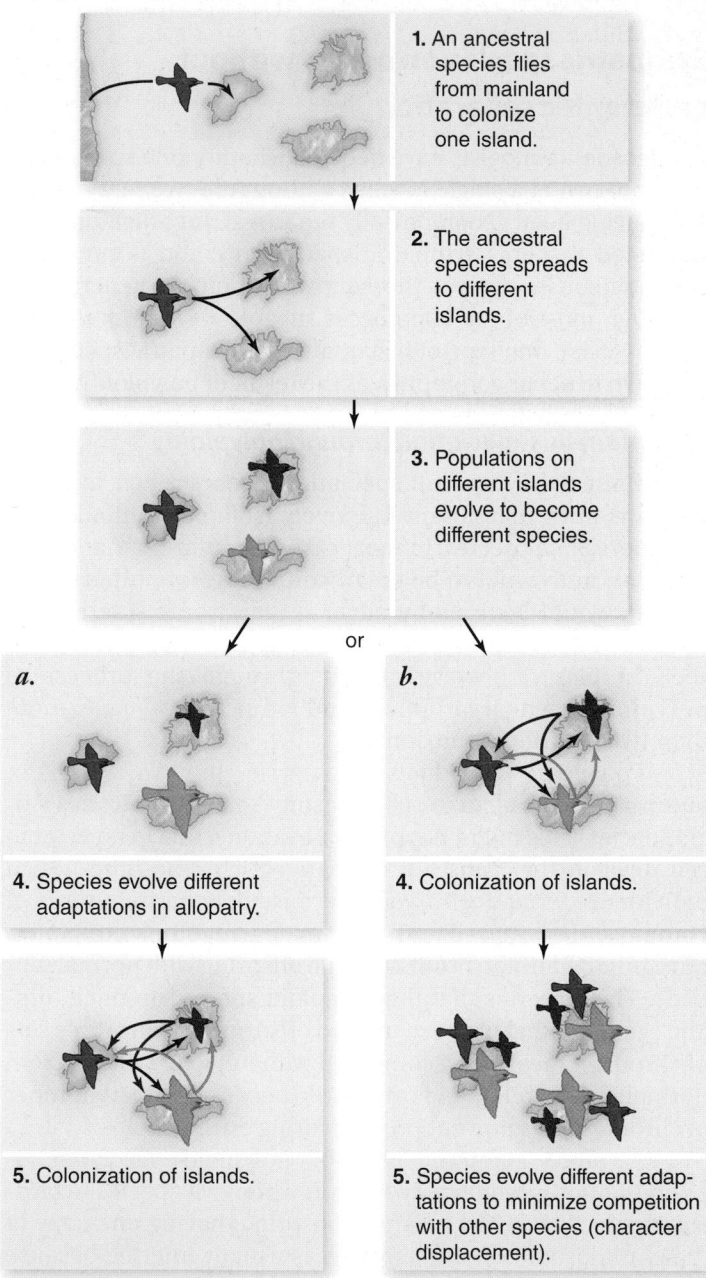

Figure 22.11 Classic model of adaptive radiation on island archipelagoes. *(1)* An ancestral species colonizes an island in an archipelago. Subsequently, the population colonizes other islands *(2),* after which the populations on the different islands speciate in allopatry *(3).* Then some of these new species colonize other islands, leading to local communities of two or more species. Adaptive differences can either evolve when species are in allopatry in response to different environmental conditions *(a)* or as the result of ecological interactions between species *(b)* by the process of character displacement.

resources or other aspects of the environment that were previously inaccessible. Classic examples of key innovation leading to adaptive radiation are the evolution of lungs in fish and of wings in birds and insects, both of which allowed descendant species to diversify and adapt to many newly available parts of the environment.

Adaptive radiation requires both speciation and adaptation to different habitats. A classic model postulates that a species colonizes multiple islands in an archipelago. Speciation subsequently occurs allopatrically, and then the newly arisen species colonize other islands, producing multiple species per island (figure 22.11).

Adaptation to new habitats can occur either during the allopatric phase, as the species respond to different environments on the different islands, or after two species become sympatric. In the latter case, this adaptation may be driven by the need to minimize competition for available resources with other species. Populations on different islands evolve to become different species. In this process, termed **character displacement,** natural selection in each species favors those individuals that use resources not used by the other species. Because those individuals will have greater fitness, whatever traits cause the differences in resource use will increase in frequency (assuming that a genetic basis exists for these differences), and, over time, the species will diverge (figure 22.12).

One example of character displacement involves three-spined sticklebacks, small fish found in temperate lakes throughout the northern hemisphere. In British Columbia, many lakes were created in the past 12,000 years when the glaciers melted at the end of the last Ice Age. Some of these lakes near the ocean have been invaded by marine populations of the stickleback. These fish tend to occur in habitats throughout the lake and evolved to be different from their marine ancestors. So different, in fact, that in the few lakes in which a second colonization occurred, the two stickleback populations had evolved a high degree of reproductive isolation. As a result, natural selection led to the evolution of differences in the two populations to minimize competition for food. One stickleback adapted to using open water, and evolved a streamlined body form for greater speed and more effective gill structures for straining out the zooplankton that occur in the open, whereas the other population adapted to foraging near the margins and bottom of the lake, evolving a larger and stouter body and altered the shape of the mouth to better feed on large invertebrates from the lake floor (figure 22.13). Similar character displacement occurred in seven doubly colonized lakes.

SCIENTIFIC THINKING

Question: *Does competition for resources cause character displacement?*

Hypothesis: *Competition with similar species will cause natural selection to promote evolutionary divergence.*

Experiment: *Place a species of fish in a pond with another, similar fish species and measure the form of selection. As a control, place a population of the same species in a pond without the second species. Note that the size of food that these fish eat is related to the size of the fish.*

Result: *In the pond with two species, directional selection favors those individuals which have phenotypes most dissimilar from the other species, and thus are most different in resource use. Directional selection does not occur in the control population.*

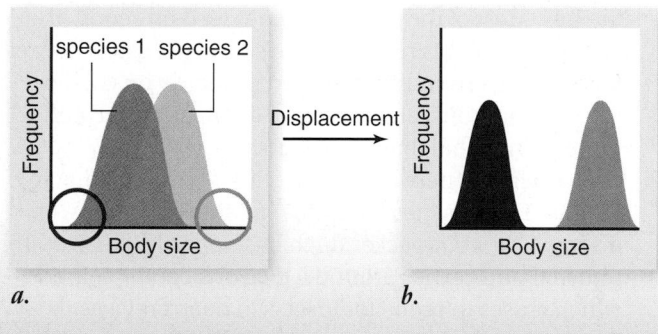

Interpretation: *Would you expect character displacement to occur if resources were unlimited?*

Figure 22.12 Character displacement. *a.* Two species are initially similar and thus overlap greatly in resource use, as might happen if the two species were similar in size (in many species, body size and food size are closely related). Individuals in each species that are most different from the other species (circled) will be favored by natural selection, because they will not have to compete with the other species. For example, the smallest individuals of one species and the largest of the other would not compete with the other species for food and thus would be favored. *b.* As a result, the species will diverge in resource use and minimize competition between the species.

a.

b.

Figure 22.13 Stickleback fish adapted to using different habitats. Fish living in open water *(a)* are more streamlined in body form, whereas those that occur near the substrate *(b)* are larger, bulkier, and have a mouth better equipped to pick invertebrates off the lake bottom.

An alternative possibility is that adaptive radiation occurs through repeated instances of sympatric speciation, producing a suite of species adapted to different habitats. As discussed earlier such scenarios are hotly debated.

In the following sections we discuss four exemplary cases of adaptive radiation.

> **?** **Inquiry question** How would the scenario for adaptive radiation differ depending on whether speciation is allopatric or sympatric? What is the relationship between character displacement and sympatric speciation?

Hawaiian *Drosophila* exploited a rich, diverse habitat

More than 1000 species in the fly genus *Drosophila* occur on the Hawaiian Islands. New species of *Drosophila* are still being discovered in Hawaii, although the rapid destruction of the native vegetation is making the search more difficult.

Aside from their sheer number, Hawaiian *Drosophila* species are unusual because of their incredible diversity of morphological and behavioral traits (figure 22.14). Evidently, when their ancestors first reached these islands, they encountered many "empty" habitats that other kinds of insects and other animals occupied elsewhere. As a result, the species have adapted to all manners of fruit fly life and include predators, parasites, and herbivores, as well as species specialized for eating the detritus in leaf litter and the nectar of flowers. The larvae of various species live in rotting stems, fruits, bark, leaves, or roots, or feed on sap. No comparable diversity of *Drosophila* species is found anywhere else in the world.

The great diversity of Hawaiian species is a result of the geological history of these islands. New islands have continually arisen from the sea in this region. As they have done so, they appear to have been invaded successively by the various *Drosophila* groups present on the older islands. New species thus have evolved as new islands have been colonized.

In addition, the Hawaiian Islands are among the most volcanically active islands in the world. Periodic lava flows often have created patches of habitat within an island surrounded by a "sea" of barren rock. These land islands are termed *kipukas. Drosophila* populations isolated in these kipukas often undergo speciation. In these ways, rampant speciation combined with ecological opportunity has led to an unparalleled diversity of insect life.

Darwin's finch species adapted to use different food types

The diversity of Darwin's finches on the Galápagos Islands was first mentioned in chapter 21. Presumably, the ancestor of Darwin's finches reached these islands before other land birds, and many of the types of habitats that other types of birds use on the mainland were unoccupied.

As the new arrivals moved into these vacant ecological niches and adopted new lifestyles, they were subjected to many different sets of selective pressures. Under these circumstances, and aided by the geographic isolation afforded by the many islands of the Galápagos archipelago, the ancestral finches rapidly split into a series of diverse populations, some of which evolved into separate species. These species now occupy many different habitats on the Galápagos Islands, which are comparable to the habitats several distinct groups of birds occupy on the mainland. As illustrated in figure 22.15, the 14 species fall into four groups:

1. **Ground finches.** There are six species of *Geospiza* ground finches. Most of the ground finches feed on seeds. The size of their bills is related to the size of the seeds they eat. Some of the ground finches feed primarily on cactus flowers and fruits, and they have a longer, larger, and more pointed bill than the others.
2. **Tree finches.** There are five species of insect-eating tree finches. Four species have bills suitable for feeding on insects. The woodpecker finch has a chisel-like beak. This unusual bird carries around a twig or a cactus spine, which it uses to probe for insects in deep crevices.
3. **Vegetarian finch.** The very heavy bill of this species is used to wrench buds from branches.
4. **Warbler finches.** These unusual birds play the same ecological role in the Galápagos woods that warblers play on the mainland, searching continuously over the leaves and branches for insects. They have slender, warblerlike beaks.

Recently, scientists have examined the DNA of Darwin's finches to study their evolutionary history. These studies suggest that the deepest branches in the finch evolutionary tree lead to warbler finches, which implies that warbler finches were among the first types to evolve after colonization of the islands. All of the ground species are closely related to one another, as are all of the tree finches. Nonetheless, within each

a.   *b.*

Figure 22.14 Hawaiian *Drosophila*. The hundreds of species that have evolved on the Hawaiian Islands are extremely variable in appearance, although genetically almost identical. *a. Drosophila heteroneura. b. Drosophila silvestris.*

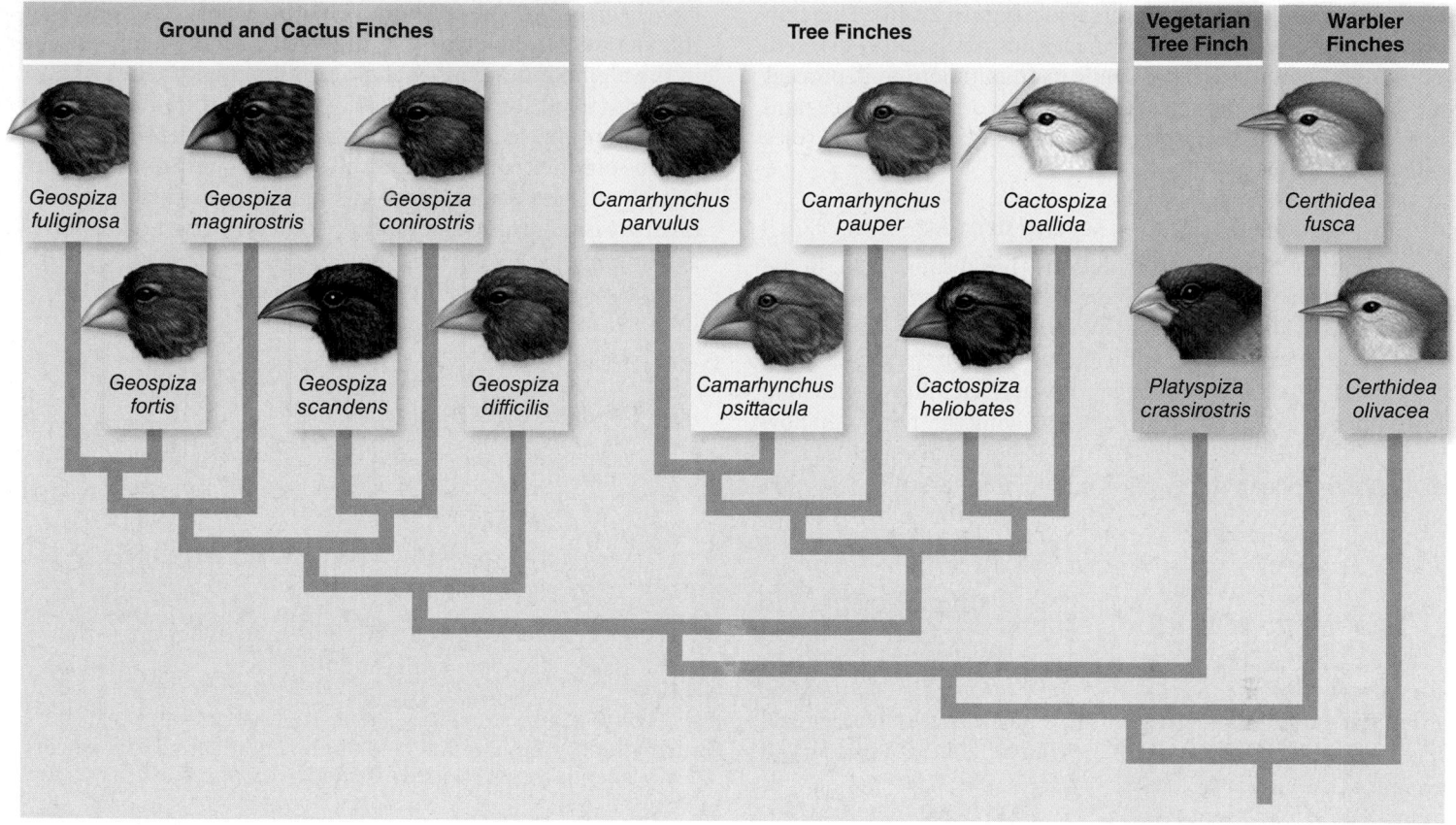

Figure 22.15 An evolutionary tree of Darwin's finches. This evolutionary tree, derived from examination of DNA sequences, suggests that warbler finches are an early offshoot. Ground and tree finches subsequently diverged, and then species within each group specialized to use different resources. Recent studies have shown, surprisingly, that the two warbler finches are not each other's closest relatives. Rather, *Certhidea fusca* is more closely related to the remaining Darwin's finches than it is to *C. olivacea*.

group, species differ in beak size and other attributes, as well as in resource use.

Field studies, conducted in conjunction with those discussed in chapter 21, demonstrate that ground species compete for resources; the differences between species likely resulted from character displacement as initially similar species diverged to minimize competitive pressures.

Lake Victoria cichlid fishes diversified very rapidly

Lake Victoria is an immense, shallow, freshwater sea about the size of Switzerland in the heart of equatorial East Africa. Until recently, the lake was home to an incredibly diverse collection of over 450 species of cichlid fishes.

Geologically recent radiation

The cluster of cichlid species appears to have evolved recently and quite rapidly. By sequencing the cytochrome *b* gene in many of the lake's fish, scientists have been able to estimate that the first cichlids entered Lake Victoria only 200,000 years ago.

Dramatic changes in water level encouraged species formation. As the lake rose, it flooded new areas and opened up new habitats. Many of the species may have originated after the lake dried down 14,000 years ago, isolating local populations in small lakes until the water level rose again.

Cichlid diversity

Cichlids are small, perchlike fishes ranging from 5 to 25 centimeters (cm) in length, and the males come in endless varieties of colors. The ecological and morphological diversity of these fish is remarkable, particularly given the short span of time over which they have evolved.

We can gain some sense of the vast range of types by looking at how different species eat. There are mud biters, algae scrapers, leaf chewers, snail crushers, zooplankton eaters, insect eaters, prawn eaters, and fish eaters. Snail shellers pounce on slow-crawling snails and spear their soft parts with long, curved teeth before the snail can retreat into its shell. Scale scrapers rasp slices of scales off other fish. There are even cichlid species that are "pedophages," eating the young of other cichlids.

Cichlid fish have a remarkable key innovation that may have been instrumental in their evolutionary radiation: They carry a second set of functioning jaws (figure 22.16). This trait occurs in many other fish, but in cichlids it is greatly enlarged. The ability of these second jaws to manipulate and process food has freed the oral jaws to evolve for other purposes, and the result has been the incredible diversity of ecological roles filled by these fish.

Abrupt extinction in the last several decades

Recently, much of the cichlid diversity has disappeared. In the 1950s, the Nile perch, a large commercial fish with a voracious appetite, was introduced to Lake Victoria. Since then, it has spread through the lake, eating its way through the cichlids.

By 1990, many of the open-water cichlid species had become extinct, as well as others living in rocky shallow regions. Over 70% of all the named Lake Victoria cichlid species had disappeared, as well as untold numbers of species that had yet to be described. We will revisit the story of Lake Victoria when we discuss conservation biology in chapter 59.

New Zealand alpine buttercups underwent speciation in glacial habitats

Adaptive radiations such as those we have described in Hawaiian *Drosophila,* Galápagos finches, and cichlid fishes seem to have been favored by periodic isolation. A clear example of the role periodic isolation plays in species formation can be seen in the alpine buttercups that grow among the glaciers of New Zealand (figure 22.17).

More species of alpine buttercups grow on the two main islands of New Zealand than in all of North and South America combined. The evolutionary mechanism responsible for this diversity is recurrent isolation associated with the recession of glaciers.

The 14 species of alpine buttercups occupy five distinctive habitats within glacial areas:

- *snowfields*—rocky crevices among outcrops in permanent snowfields at 2130- to 2740-m elevation;
- *snowline fringe*—rocks at lower margin of snowfields between 1220 and 2130 m;
- *stony debris*—slopes of exposed loose rocks at 610 to 1830 m;
- *sheltered situations*—shaded by rock or shrubs at 305 to 1830 m; and
- *boggy habitats*—sheltered slopes and hollows, poorly drained tussocks at elevations between 760 and 1525 m.

Buttercup speciation and diversification have been promoted by repeated cycles of glacial advance and retreat.

Figure 22.16 Cichlid fishes of Lake Victoria. These fishes have evolved adaptations to use a variety of different habitats. The enlarged second set of jaws located in the throat of these fish has provided evolutionary flexibility, allowing oral jaws to be modified in many ways.

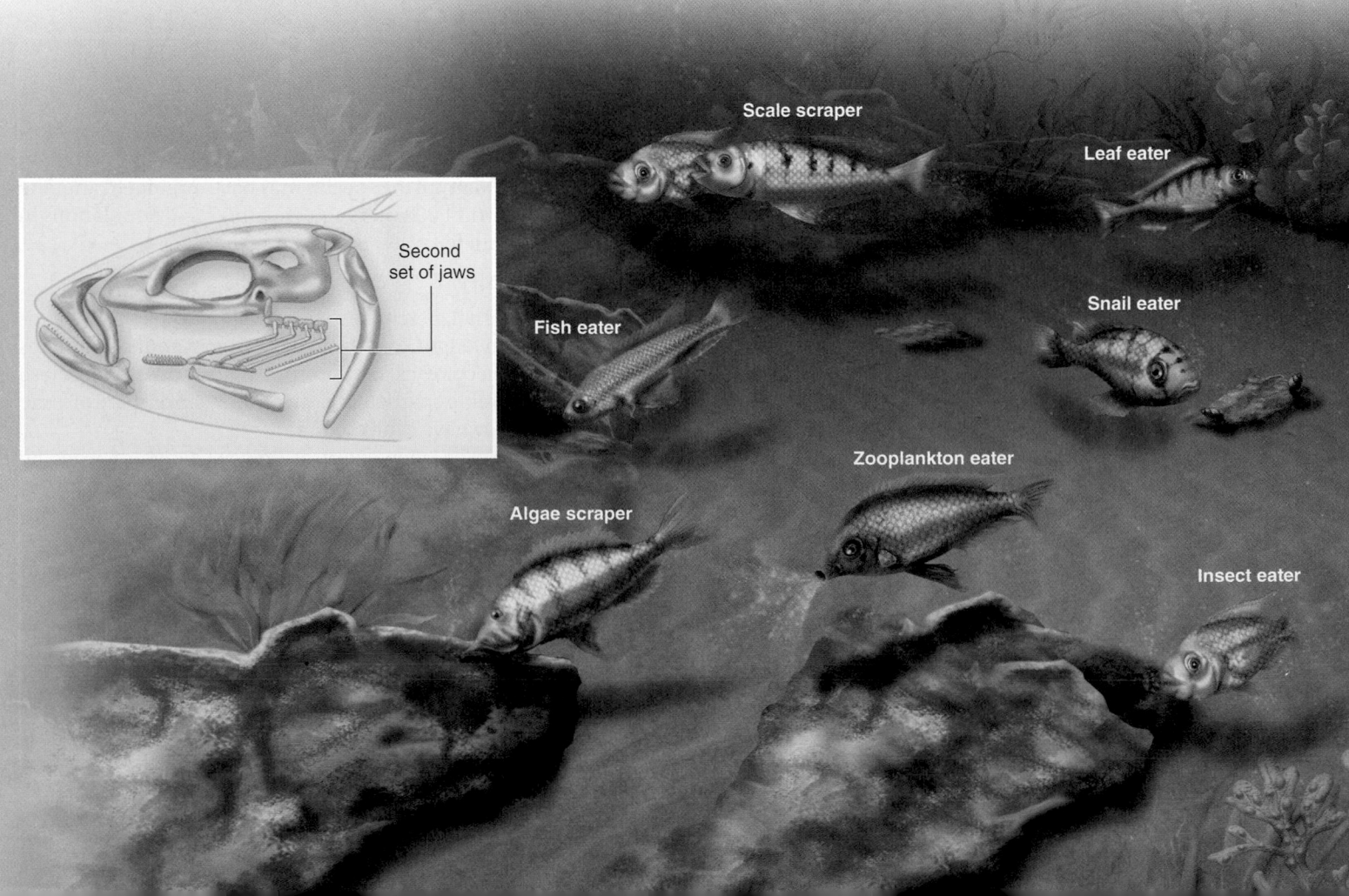

snowfield snowline fringe stony debris sheltered boggy

a.

Glaciers link alpine zones into one continuous range.

Glaciers recede →

Mountain populations become isolated, permitting divergence and speciation.

Glaciation →

Alpine zones are reconnected. Separately evolved species come back into contact.

b.

Figure 22.17 **New Zealand alpine buttercups (genus *Ranunculus*).** Periodic glaciation encouraged species formation among alpine buttercups in New Zealand. *a.* Fourteen species of alpine *Ranunculus* grow among the glaciers and mountains of New Zealand. *b.* The formation of extensive glaciers during the Pleistocene epoch linked the alpine zones *(white)* of many mountains together. When the glaciers receded, these alpine zones were isolated from one another, only to become reconnected with the advent of the next glacial period. During periods of isolation, populations of alpine buttercups diverged in the isolated habitats.

As the glaciers retreat up the mountains, populations become isolated on mountain peaks, permitting speciation (see figure 22.17). In the next glacial advances, these new species can expand throughout the mountain range, coming into contact with their close relatives. In this way, one initial species could give rise to many descendants. Moreover, on isolated mountaintops during glacial retreats, species have convergently evolved to occupy similar habitats; these distantly related but ecologically similar species have then been brought back into contact in subsequent glacial advances.

Learning Outcomes Review 22.5

Adaptive radiation occurs when a species diversifies, producing descendant species that are adapted to use many different parts of the environment. Adaptive radiation may occur under conditions of recurrent isolation, which increases the rate at which speciation occurs, and by occupation of areas with few competitors and many types of available resources, such as on volcanic islands. The evolution of a key innovation may also allow adaptation to parts of the environment that previously couldn't be utilized.

■ *In contrast to the archipelago model, how might an adaptive radiation proceed in a case of sympatric speciation by disruptive selection?*

22.6 *The Pace of Evolution*

Learning Outcomes

1. *Define stasis and compare it to gradual evolutionary change.*
2. *Explain the components of the punctuated equilibrium hypothesis.*

We have discussed the manner in which speciation may occur, but we haven't yet considered the relationship between speciation and the evolutionary change that occurs within a species. Two hypotheses, *gradualism* and *punctuated equilibrium,* have been advanced to explain the relationship.

Gradualism is the accumulation of small changes

For more than a century after the publication of *On the Origin of Species,* the standard view was that evolution occurred very

a. Gradualism   *b.* Punctuated equilibrium

Figure 22.18 Two views of the pace of macroevolution.
a. Gradualism suggests that evolutionary change occurs slowly through time and is not linked to speciation, whereas *(b)* punctuated equilibrium requires that phenotypic change occurs in bursts associated with speciation, separated by long periods of little or no change.

slowly. Such change would be nearly imperceptible from generation to generation, but would accumulate such that, over the course of thousands and millions of years, major changes could occur. This view is termed **gradualism** (figure 22.18*a*).

Punctuated equilibrium is long periods of stasis followed by relatively rapid change

Gradualism was challenged in 1972 by paleontologists Niles Eldredge of the American Museum of Natural History in New York and Stephen Jay Gould of Harvard University, who argued that species experience long periods of little or no evolutionary change (termed **stasis**), punctuated by bursts of evolutionary change occurring over geologically short time intervals. They called this phenomenon **punctuated equilibrium** (figure 22.18*b*) and argued that these periods of rapid change occurred only during the speciation process.

Initial criticism of the punctuated equilibrium hypothesis focused on whether rapid change could occur over short periods of time. As we have seen in the last two chapters, however, when natural selection is strong, rapid and substantial evolutionary change can occur. A more difficult question involves the long periods of stasis: Why would species exist for thousands, or even millions, of years without changing?

Although a number of possible reasons have been suggested, most researchers now believe that a combination of stabilizing and oscillating selection is responsible for stasis. If the environment does not change over long periods of time, or if environmental changes oscillate back and forth, then stasis may occur for long periods. One factor that may enhance this stasis is the ability of species to shift their ranges; for example, during the ice ages, when the global climate cooled, the geographic ranges of many species shifted southward, so that the species continued to experience similar environmental conditions.

> **? Inquiry question** Why would changes in geographic ranges promote evolutionary stasis?

Evolution may include both types of change

Eldredge and Gould's proposal prompted a great deal of research. Some well-documented groups, such as African mammals, clearly have evolved gradually, not in spurts. Other groups, such as marine bryozoa, seem to show the irregular pattern of evolutionary change predicted by the punctuated equilibrium model. It appears, in fact, that gradualism and punctuated equilibrium are two ends of a continuum. Although some groups have evolved solely in a gradual manner and others only in a punctuated mode, many other groups show evidence of both gradual and punctuated episodes at different times in their evolutionary history.

The idea that speciation is necessarily linked to phenotypic change has not been supported, however. On one hand, it is now clear that speciation can occur without substantial phenotypic change. For example, many closely related salamander species are nearly indistinguishable. On the other, it is also clear that phenotypic change can occur within species in the absence of speciation.

> ### Learning Outcomes Review 22.6
> Gradualism is the accumulation of almost imperceptible changes that eventually results in major differences. Punctuated equilibrium proposes that long periods of stasis are interrupted (punctuated) by periods of rapid change. Evidence for both gradualism and punctuated equilibrium has been found in different groups. Stasis refers to a period in which little or no evolutionary change occurs. Stasis may result from stabilizing or oscillating selection.
>
> ■ *Could evolutionary change be punctuated in time (that is, rapid and episodic), but not linked to speciation?*

22.1 The Nature of Species and the Biological Species Concept

Sympatric species inhabit the same locale but remain distinct.

Most sympatric species are readily distinguishable phenotypically and ecologically. Others usually can be identified by careful study.

Populations of a species exhibit geographic variation.

Populations that differ greatly in phenotype or ecologically are usually connected by geographically intermediate populations.

The biological species concept focuses on the ability to exchange genes.

In the biological species concept, species are defined as those populations that interbreed, or have the potential to do so, and produce fertile offspring. Reproductive isolating mechanisms prevent genetic exchange between species.

Prezygotic isolating mechanisms prevent the formation of a zygote.

Prezygotic isolating mechanisms prevent a viable zygote from being created. These mechanisms include ecological, behavioral, temporal, and mechanical isolation, as well as failure of gametes to unite to form a zygote.

Postzygotic isolating mechanisms prevent normal development into reproducing adults.

Postzygotic isolating mechanisms prevent a zygote from developing into a viable and fertile individual.

The biological species concept does not explain all observations.

The ecological species concept, one alternative explanation, focuses on the role of natural selection and differences among species in their ecological requirements.

In reality, no one species concept can explain all the diversity of life.

22.2 Natural Selection and Reproductive Isolation

Selection may reinforce isolating mechanisms.

Populations may evolve complete reproductive isolation in allopatry. If populations that have evolved only partial reproductive isolation come into contact, natural selection can lead to increased reproductive isolation, a process termed "reinforcement."

Alternatively, genetic exchange between the populations can lead to homogenization and thus prevent speciation from occurring.

22.3 The Role of Genetic Drift and Natural Selection in Speciation

Random changes may cause reproductive isolation.

In small populations, genetic drift may cause populations to diverge; the differences that evolve may cause the populations to become reproductively isolated.

Adaptation can lead to speciation.

Adaptation to different situations or environments may incidentally lead to reproductive isolation. In contrast to reinforcement, these differences are not directly favored by natural selection because they prevent hybridization.

22.4 The Geography of Speciation (figure 22.9)

Allopatric speciation takes place when populations are geographically isolated.

Allopatric, or geographically isolated, populations may evolve into separate species because no gene flow occurs between them. Most speciation probably occurs in allopatry.

Sympatric speciation occurs without geographic separation.

Sympatric speciation can occur in two ways. One is polyploidy, which instantly creates a new species. Disruptive selection also can cause one species to divide into two, but scientists debate how commonly it occurs.

22.5 Adaptive Radiation and Biological Diversity

Adaptive radiation occurs when a species finds itself in a new or suddenly changed environment with many resources and few competing species (figure 22.11).

The evolution of a new trait that allows individuals to use previously inaccessible parts of the environment, termed a "key innovation," may also trigger an adaptive radiation.

Hawaiian Drosophila *exploited a rich, diverse habitat.*

At least 1000 species of Hawaiian *Drosophila* have been identified.

Darwin's finch species adapted to use different food types.

Fourteen species in four genera have evolved to exploit four different habitats based on type of food.

Lake Victoria cichlid fishes diversified very rapidly.

Cichlids underwent rapid radiation to form more than 450 species, although today more than 70% of these species are now extinct.

New Zealand alpine buttercups underwent speciation in glacial habitats.

Periodic isolation by glaciers has led to 14 species in distinct habitats.

22.6 The Pace of Evolution

Gradualism is the accumulation of small changes.

Historically, scientists took the view that speciation occurred gradually through very small cumulative changes.

Punctuated equilibrium is long periods of stasis followed by relatively rapid change.

The punctuated equilibrium hypothesis contends that not only is change rapid and episodic, but that it is only associated with the speciation process.

Evolution may include both types of change.

Scientists generally agree that evolutionary change occurs on a continuum, with gradualism and punctuated change being the extremes.

UNDERSTAND

1. Prezygotic isolating mechanisms include all of the following except
   a. hybrid sterility.
   b. courtship rituals.
   c. habitat separation.
   d. seasonal reproduction.

2. Reproductive isolation is
   a. a result of individuals not mating with each other.
   b. a specific type of postzygotic isolating mechanism.
   c. required by the biological species concept.
   d. None of the choices is correct.

3. Problems with the biological species concept include the fact that
   a. many species reproduce asexually.
   b. postzygotic isolating mechanisms decrease hybrid viability.
   c. prezygotic isolating mechanisms are extremely rare.
   d. All of the choices are correct.

4. Allopatric speciation
   a. is less common than sympatric speciation.
   b. involves geographic isolation of some kind.
   c. is the only kind of speciation that occurs in plants.
   d. requires polyploidy.

5. Gradualism and punctuated equilibrium are
   a. two ends of the continuum of the rate of evolutionary change over time.
   b. mutually exclusive views about how all evolutionary change takes place.
   c. mechanisms of reproductive isolation.
   d. None of the choices is correct.

6. Prezygotic isolation
   a. always involves mechanisms that prevent interbreeding of members of different species.
   b. includes the death of a zygote shortly after fertilization.
   c. only occurs in plants.
   d. becomes stronger as the result of reinforcement.

7. Speciation by allopolyploidy
   a. takes a long time.
   b. is common in birds.
   c. leads to reduced numbers of chromosomes.
   d. occurs after hybridization between two species.

8. Adaptive radiation
   a. is the result of enriched uranium used in power plants.
   b. is the evolution of closely related species adapted to use different parts of the environment.
   c. results from genetic drift.
   d. is the outcome of stabilizing selection favoring the maintenance of adaptive traits.

9. Leopard frogs from different geographic populations of the *Rana pipiens* complex
   a. are members of a single species because they look very similar to one another.
   b. are different species shown to have pre- and postzygotic isolating mechanisms.
   c. frequently interbreed to produce viable hybrids.
   d. are genetically identical due to effective reproductive isolation.

10. Character displacement
    a. arises through competition and natural selection, favoring divergence in resource use.
    b. arises through competition and natural selection, favoring convergence in resource use.
    c. does not promote speciation.
    d. reduced speciation rates in Galápagos finches.

APPLY

1. If reinforcement is weak and hybrids are not completely infertile,
   a. genetic divergence between populations may be overcome by gene flow.
   b. speciation will occur 100% of the time.
   c. gene flow between populations will be impossible.
   d. the speciation will be more likely than if hybrids were completely infertile.

2. Natural selection can
   a. enhance the probability of speciation.
   b. enhance reproductive isolation.
   c. act against hybrid survival and reproduction.
   d. All of the choices are correct.

3. Hybridization between incompletely isolated populations
   a. always leads to reinforcement due to the inferiority of hybrids.
   b. can serve as a mechanism for preserving gene flow between populations, thus preventing speciation.
   c. only occurs in plants.
   d. never affects rates of speciation.

4. Natural selection can lead to speciation
   a. by causing small populations to diverge more than large populations.
   b. because the evolutionary changes that two populations acquire while adapting to different habitats may have the effect of making them reproductively isolated.
   c. by favoring the same evolutionary change in multiple populations.
   d. by favoring intermediate phenotypes.

SYNTHESIZE

1. Natural selection can lead to the evolution of prezygotic isolating mechanisms, but not postzygotic isolating mechanisms. Explain.

2. If there is no universally accepted definition of a species, what good is the term? Will the idea of and need for a "species concept" be eliminated in the future?

3. Refer to figure 22.6. In Europe, pied and collared flycatchers are dissimilar in sympatry, but very similar in allopatry, consistent with character divergence in coloration. In this case, there is no competition for ecological resources as in other cases of character divergence discussed. Why have differences in color evolved in sympatric populations of the two species?

4. Refer to figure 22.15. *Geospiza fuliginosa* and *Geospiza fortis* are found in sympatry on at least one island in the Galápagos and in allopatry on several islands in the same archipelago. Compare your expectations about degree of morphological similarity of the two species in these two contexts, given the hypothesis that competition for food played a large role in the adaptive radiation of this group. Would your expectations be the same for a pair of finch species that are not as closely related? Explain.

Chapter

23

Systematics, Phylogenies, and Comparative Biology

Chapter Contents

Introduction

All organisms share many biological characteristics. They are composed of one or more cells, carry out metabolism and transfer energy with ATP, and encode hereditary information in DNA. Yet, there is also a tremendous diversity of life, ranging from bacteria and amoebas to blue whales and sequoia trees. For generations, biologists have tried to group organisms based on shared characteristics. The most meaningful groupings are based on the study of evolutionary relationships among organisms. New methods for constructing evolutionary trees and a sea of molecular sequence data are leading to improved evolutionary hypotheses to explain life's diversification.

23.1 Systematics

Learning Outcomes

1. *Understand what a phylogeny represents.*
2. *Explain why phenotypic similarity does not necessarily indicate close evolutionary relationship.*

One of the great challenges of modern science is to understand the history of ancestor–descendant relationships that unites all forms of life on Earth, from the earliest single-celled organisms to the complex organisms we see around us today. If the fossil record were perfect, we could trace the evolutionary history of species and examine how each arose and proliferated; however, as discussed in chapter 21, the fossil record is far from complete. Although it answers many questions about life's diversification, it leaves many others unsettled.

Consequently, scientists must rely on other types of evidence to establish the best hypothesis of evolutionary relationships. Bear in mind that the outcomes of such studies *are* hypotheses, and as such, they require further testing. All hypotheses may be disproved by new data, leading to the formation of better, more accurate scientific ideas.

The reconstruction and study of evolutionary relationships is called **systematics.** By looking at the similarities and differences between species, systematists can construct an evolutionary tree, or **phylogeny,** which represents a hypothesis about patterns of relationship among species.

Branching diagrams depict evolutionary relationships

Darwin envisioned that all species were descended from a single common ancestor, and that the history of life could be depicted as a branching tree (figure 23.1). In Darwin's view, the twigs of the tree represent existing species. As one works down the tree, the joining of twigs and branches reflects the pattern of common ancestry back in time to the single common ancestor of all life. The process of descent with modification from common ancestry results in all species being related in this branching, hierarchical fashion, and their evolutionary history can be depicted using branching diagrams or phylogenetic trees. Figure 23.1b shows how evolutionary relationships are depicted with a branching diagram. Humans and chimpanzees are descended from a common ancestor and are each other's closest living relative (the position of this common ancestor is indicated by the node labeled 1). Humans, chimps, and gorillas share an older common ancestor (node 2), and all great apes share a more distant common ancestor (node 3).

One key to interpreting a phylogeny is to look at how recently species share a common ancestor, rather than looking at the arrangement of species across the top of the tree. If you compare the three versions of the phylogeny of figure 23.1b, you can see that the relationships are the same: Regardless of where they are positioned, chimpanzees and humans are still more closely related to each other than to any other species.

Moreover, even though humans are placed next to gibbons in version 1 of figure 23.1b, the pattern of relationships still indicates that humans are more closely related (that is, share a more recent common ancestor) with gorillas and orangutans than with gibbons. Phylogenies are also sometimes displayed on their side, rather than upright (figure 23.1b, version 3), but this arrangement also does not affect its interpretation.

Similarity may not accurately predict evolutionary relationships

We might expect that the greater the time since two species diverged from a common ancestor, the more different they would be. Early systematists relied on this reasoning and constructed phylogenies based on overall similarity. If, in fact, species evolved at a constant rate, then the amount of divergence between two species would be a function of how long they had been diverging, and thus phylogenies based on degree of similarity would be accurate. As a result, we might think that chimps and gorillas are more closely related to each other than either is to humans.

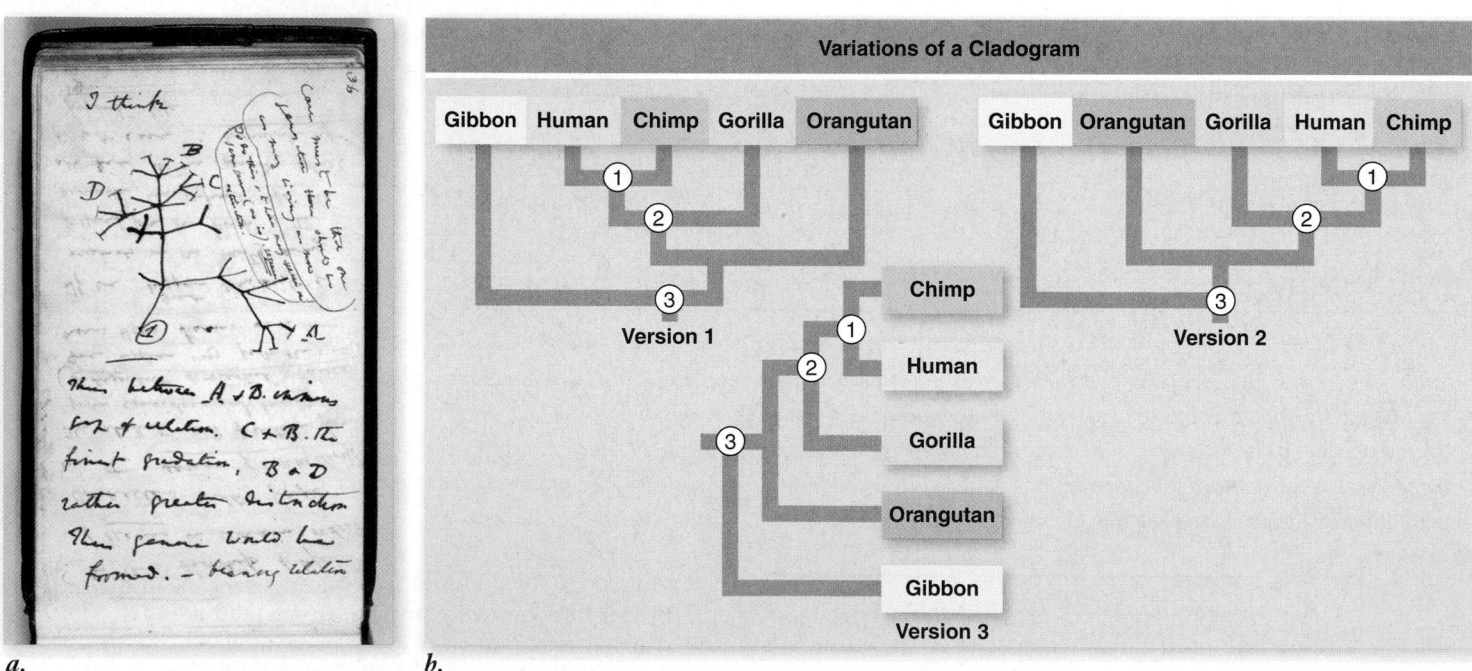

Figure 23.1 Phylogenies depict evolutionary relationships. *a.* A drawing from one of Darwin's notebooks, written in 1837 as he developed his ideas that led to *On the Origin of Species*. Darwin viewed life as a branching process akin to a tree, with species on the twigs, and evolutionary change represented by the branching pattern displayed by a tree as it grows. *b.* An example of a phylogeny. Note that these three versions convey the same information despite the differences in arrangement of species and orientation. Humans and chimpanzees are more closely related to each other than they are to any other living species. This is apparent because they share a common ancestor (the node labeled 1) that was not an ancestor of other species. Similarly, humans, chimpanzees, and gorillas are more closely related to one another than any of them is to orangutans because they share a common ancestor (node 2) that was not ancestral to orangutans. Node 3 represents the common ancestor of all apes. At each node, the two descendants can be rotated without changing the meaning. For example, one difference between versions 1 and 2 is that the descendants of node 2 have been rotated so that gorilla branches to the right in version 1 and to the left in version 2. However, this does not affect the interpretation that human and chimp are more closely related to each other than either is to gorilla.

But as chapter 22 revealed, evolution can occur very rapidly at some times and very slowly at others. In addition, evolution is not unidirectional—sometimes species' traits evolve in one direction, and then back the other way (a result of oscillating selection; see chapter 20). Species invading new habitats are likely to experience new selective pressures and may change greatly; those staying in the same habitats as their ancestors may change only a little. For this reason, similarity is not necessarily a good predictor of how long it has been since two species shared a common ancestor.

A second fundamental problem exists as well: Evolution is not always divergent. In chapter 21, we discussed convergent evolution, in which two species independently evolve the same features. Often, species evolve convergently because they use similar habitats, in which similar adaptations are favored. As a result, two species that are not closely related may end up more similar to each other than they are to their close relatives. Evolutionary reversal, the process in which a species re-evolves the characteristics of an ancestral species, also has this effect.

Learning Outcomes Review 23.1

Systematics is the study of evolutionary relationships. Phylogenies, or phylogenetic trees, are graphic representations of relationships among species. Similarity of organisms alone does not necessarily correlate with their relatedness because evolutionary change is not constant in rate and direction.

■ *Why might a species be most phenotypically similar to a species that is not its closest evolutionary relative?*

23.2 Cladistics

Learning Outcomes

1. *Describe the difference between ancestral and derived similarities.*
2. *Explain why only shared, derived characters indicate close evolutionary relationship.*
3. *Demonstrate how a cladogram is constructed.*

Because phenotypic similarity may be misleading, most systematists no longer construct their phylogenetic hypotheses solely on this basis. Rather, they distinguish similarity among species that is inherited from the most recent common ancestor of an entire group, which is called **derived,** from similarity that arose prior to the common ancestor of the group, which is termed *ancestral*. In this approach, termed **cladistics,** only **shared derived characters** are considered informative in determining evolutionary relationships.

The cladistic method requires that character variation be identified as ancestral or derived

To employ the method of cladistics, systematists first gather data on a number of characters for all the species in the analysis. Characters can be any aspect of the phenotype, including morphology, physiology, behavior, and DNA. As chapters 18 and 24 show, the revolution in genomics should soon provide a vast body of data that may revolutionize our ability to identify and study character variation.

To be useful, the characters should exist in recognizable **character states.** For example, consider the character "teeth" in amniote vertebrates (namely birds, reptiles, and mammals; see chapter 35). This character has two states: presence in most mammals and reptiles and absence in birds and a few other groups such as turtles.

A cladistic analysis begins by determining the states for a number of characters for each **taxon** (taxa are species or higher level groups, such as genera or families) in the analysis. In the example in figure 23.2, six characters are used for seven taxa. The first character, presence of jaws, is scored as absent in lamprey (denoted by 0) and present in the other species (denoted with a 1). Another character, hair, is absent in all of the taxa except the three mammal species.

Examples of ancestral versus derived characters

The presence of hair is a shared derived feature of mammals (figure 23.2); in contrast, the presence of lungs in mammals is an ancestral feature because it is also present in amphibians and reptiles (represented by a salamander and a lizard) and therefore presumably evolved prior to the common ancestor of mammals (see figure 23.2). The presence of lungs, therefore, does not tell us that mammal species are all more closely related to one another than to reptiles or amphibians, but the shared, derived feature of hair suggests that all mammal species share a common ancestor that existed more recently than the common ancestor of mammals, amphibians, and reptiles.

To return to the question concerning the relationships of humans, chimps, and gorillas, a number of morphological and DNA characters exist that are derived and shared by chimps and humans, but not by gorillas or other great apes. These characters suggest that chimps and humans diverged from a common ancestor (see figure 23.1*b,* node 1) that existed more recently than the common ancestor of gorillas, chimps, and humans (node 2).

Determination of ancestral versus derived

Once the data are assembled, the first step in a manual cladistic analysis is to **polarize** the characters—that is, to determine whether particular character states are ancestral or derived. To polarize the character "teeth," for example, systematists must determine which state—presence or absence—was exhibited by the most recent common ancestor of this group.

Usually, the fossils available do not represent the most recent common ancestor—or we cannot be confident that they do. As a result, the method of *outgroup comparison* is used to assign character polarity. To use this method, a species or group of species that is closely related to, but not a member of, the group under study is designated as the **outgroup.** When the group under study exhibits multiple character states, and one of those states is exhibited by the outgroup, then that state is considered to be ancestral and other states are considered to be derived. However, outgroup species also evolve from their ancestors, so the outgroup species will not always exhibit the ancestral condition.

Polarity assignments are most reliable when the same character state is exhibited by several different outgroups. In

Traits: Organism	Jaws	Lungs	Amniotic Membrane	Hair	No Tail	Bipedal
Lamprey	0	0	0	0	0	0
Shark	1	0	0	0	0	0
Salamander	1	1	0	0	0	0
Lizard	1	1	1	0	0	0
Tiger	1	1	1	1	0	0
Gorilla	1	1	1	1	1	0
Human	1	1	1	1	1	1

a.

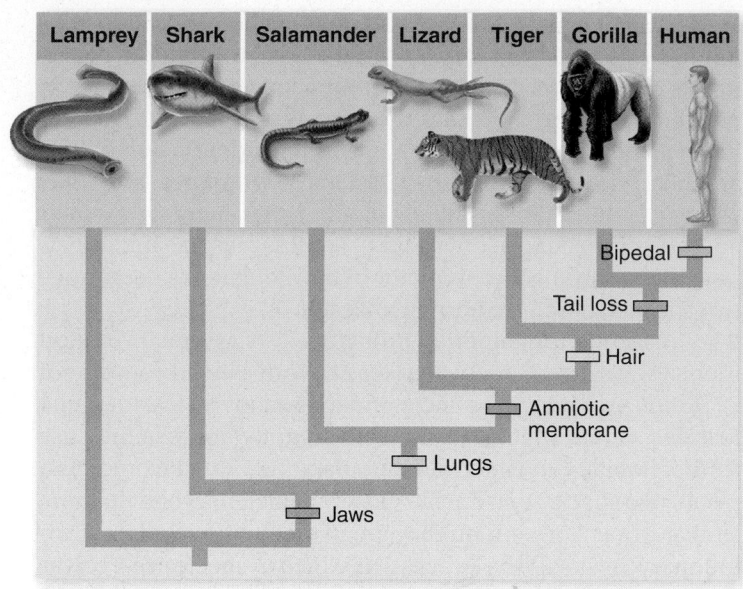

b.

Figure 23.2 A cladogram. *a.* Morphological data for a group of seven vertebrates are tabulated. A "1" indicates possession of the derived character state, and a "0" indicates possession of the ancestral character state (note that the derived state for character "no tail" is the absence of a tail; for all other traits, absence of the trait is the ancestral character state). *b.* A tree, or cladogram, diagrams the relationships among the organisms based on the presence of derived characters. The derived characters between the cladogram branch points are shared by all organisms above the branch points and are not present in any below them. The outgroup (in this case, the lamprey) does not possess any of the derived characters.

the preceding example, teeth are generally present in the nearest outgroups of amniotes—amphibians and fish—as well as in many species of amniotes themselves. Consequently, the presence of teeth in mammals and reptiles is considered ancestral, and their absence in birds and turtles is considered derived.

Construction of a cladogram

Once all characters have been polarized, systematists use this information to construct a **cladogram,** which depicts a hypothesis of evolutionary relationships. Species that share a common ancestor, as indicated by the possession of shared derived characters, are said to belong to a **clade.** Clades are thus evolutionary units and refer to a common ancestor and all of its descendants. A derived character shared by clade members is called a **synapomorphy** of that clade. Figure 23.2*b* illustrates that a simple cladogram is a nested set of clades, each characterized by its own synapomorphies. For example, amniotes are a clade for which the evolution of an amniotic membrane is a synapomorphy. Within that clade, mammals are a clade, with hair as a synapomorphy, and so on.

Ancestral states are also called **plesiomorphies,** and shared ancestral states are called **symplesiomorphies.** In contrast to synapomorphies, symplesiomorphies are not informative about phylogenetic relationships.

Consider, for example, the character state "presence of a tail," which is exhibited by lampreys, sharks, salamanders, lizards, and tigers. Does this mean that tigers are more closely related to—and shared a more recent common ancestor with—lizards and sharks than to apes and humans, their fellow mammals? The answer, of course, is no: Because symplesiomorphies reflect character states inherited from a distant ancestor, they do not imply that species exhibiting that state are closely related.

Homoplasy complicates cladistic analysis

In real-world cases, phylogenetic studies are rarely as simple as the examples we have shown so far. The reason is that in some cases, the same character has evolved independently in several species. These characters would be categorized as shared derived characters, but they would be false signals of a close evolutionary relationship. In addition, derived characters may sometimes be lost as species within a clade re-evolve to the ancestral state.

Homoplasy refers to a shared character state that has not been inherited from a common ancestor exhibiting that character state. Homoplasy can result from convergent evolution or from evolutionary reversal. For example, adult frogs do not have a tail. Thus, absence of a tail is a synapomorphy that unites not only gorillas and humans, but also frogs. However, frogs are not closely related to gorillas and humans; they have neither an amniotic membrane nor hair, both of which are synapomorphies for clades that contain gorillas and humans.

In cases such as this, when there are conflicts among the characters, systematists rely on the **principle of parsimony,** which favors the hypothesis that requires the fewest assumptions. As a result, the phylogeny that requires the fewest evolutionary events is considered the best hypothesis of phylogenetic relationships (figure 23.3). In the example just stated, therefore, grouping frogs with salamanders is favored because it requires only one instance of homoplasy (the multiple origins of taillessness), whereas a phylogeny in which frogs were most closely related to humans and gorillas would require two

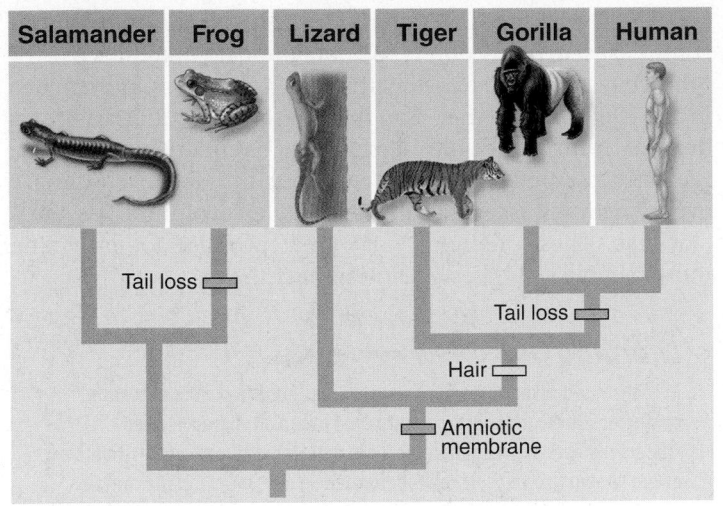

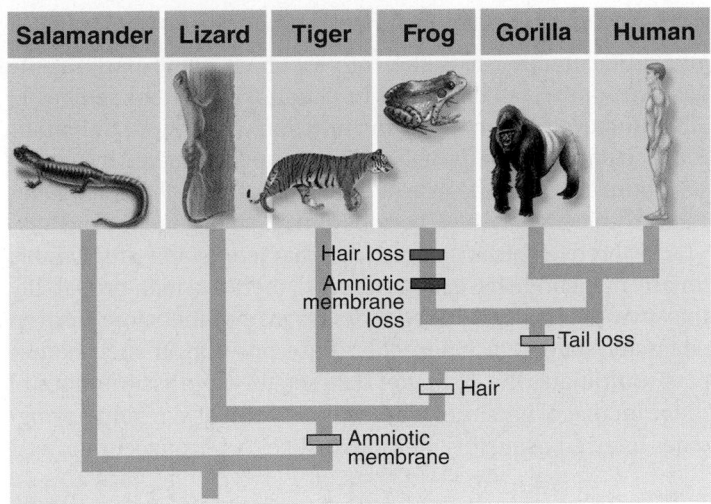

a.

b.

Figure 23.3 Parsimony and homoplasy. *a.* The placement of frogs as closely related to salamanders requires that tail loss evolved twice, an example of homoplasy. *b.* If frogs are closely related to gorillas and humans, then tail loss only had to evolve once. However, this arrangement would require two additional evolutionary changes: Frogs would have had to have lost the amniotic membrane and hair (alternatively, hair could have evolved independently in tigers and the clade of humans and gorillas; this interpretation would require two evolutionary changes in the hair character, just like the interpretation shown in the figure, in which hair evolved only once, but then was lost in frogs). Based on the principle of parsimony, the cladogram that requires the fewest number of evolutionary changes is favored; in this case the cladogram in *(a)* requires four changes, whereas that in *(b)* requires five, so *(a)* is considered the preferred hypothesis of evolutionary relationships.

 Data analysis Construct a data matrix like in figure 23.2, showing the character states for all six species for the traits hair, amniotic membrane, and tail.

homoplastic evolutionary events (the loss of both amniotic membranes and hair in frogs).

The examples presented so far have all involved morphological characters, but systematists increasingly use DNA sequence data to construct phylogenies because of the large number of characters that can be obtained through sequencing. Cladistics analyzes sequence data in the same manner as any other type of data: Character states are polarized by reference to the sequence of an outgroup, and a cladogram is constructed that minimizes the amount of character evolution required (figure 23.4).

Other phylogenetic methods work better than cladistics in some situations

If characters evolve from one state to another at a slow rate compared with the frequency of speciation events, then the principle of parsimony works well in reconstructing evolutionary relationships. In this situation, the principle's underlying assumption—that shared derived similarity is indicative of recent common ancestry—is usually correct. In recent years, however, systematists have realized that some characters evolve so rapidly that the principle of parsimony may be misleading.

	DNA Sequence									
Site	1	2	3	4	5	6	7	8	9	10
Species A	G	C	A	T	A	G	G	C	G	T
Species B	A	C	A	G	C	C	G	C	A	T
Species C	G	C	A	T	A	G	G	T	G	T
Species D	A	C	A	T	C	G	G	T	G	G
Outgroup	A	T	A	T	C	C	G	T	A	T

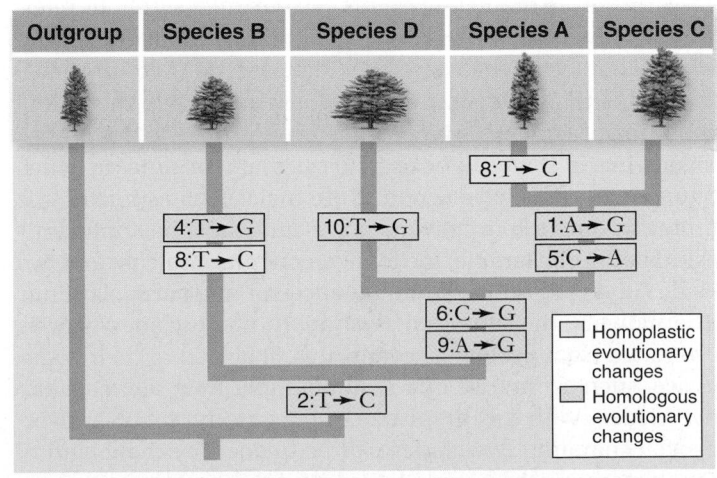

Figure 23.4 Cladistic analysis of DNA sequence data. Sequence data are analyzed just like any other data. The most parsimonious interpretation of the DNA sequence data requires nine evolutionary changes. Each of these changes is indicated on the phylogeny. Change in site 8 is homoplastic: Species A and B independently evolved from thymine to cytosine at that site.

Rapid rates of evolutionary change and homoplasy

Of particular interest is the rate at which some parts of the genome evolve. As discussed in chapter 18, some stretches of DNA do not appear to have any function. As a result, mutations that occur in these parts of the DNA are not eliminated by natural selection, and thus the rate of evolution of new character states can be quite high in these regions as a result of genetic drift.

Moreover, because only four character states are possible for any nucleotide base (A, C, G, or T), there is a high probability that two species will independently evolve the same derived character state at any particular base position. If such homoplasy dominates the character data set, then the assumptions of the principle of parsimony are violated, and as a result, phylogenies inferred using this method are likely to be inaccurate.

 Inquiry question Why do high rates of evolutionary change and a limited number of character states cause problems for parsimony analyses?

Statistical approaches

Because evolution can sometimes proceed rapidly, systematists in recent years have been exploring other methods based on statistical approaches, such as maximum likelihood, to infer phylogenies. These methods start with an assumption about the rate at which characters evolve and then fit the data to these models to derive the phylogeny that best accords (i.e., "maximally likely") with these assumptions.

One advantage of these methods is that different assumptions of rate of evolution can be used for different characters. If some DNA characters evolve more slowly than others—for example, because they are constrained by natural selection—then the methods can employ different models of evolution for the different characters. This approach is more effective than parsimony in dealing with homoplasy when rates of evolutionary changes are high.

The molecular clock

In general, cladograms such as the one in figure 23.2 only indicate the order of evolutionary branching events; they do not contain information about the timing of these events. In some cases, however, branching events can be timed, either by reference to fossils, or by making assumptions about the rate at which characters change. One widely used but controversial method is the **molecular clock,** which states that the rate of evolution of a molecule is constant through time. In this model, divergence in DNA can be used to calculate the times at which branching events have occurred. To make such estimates, the timing of one or more divergence events must be confidently estimated. For example, the fossil record may indicate that two clades diverged from a common ancestor at a particular time. Alternatively, the timing of separation of two clades may be estimated from geological events that likely led to their divergence, such as the rise of a mountain that now separates the two clades. With this information, the amount of DNA divergence separating two clades can be divided by the length of time separating the two clades, which produces an estimate of the rate of DNA divergence per unit of time (usually, per million years). Assuming a molecular clock, this rate can then be used to date other divergence events in a cladogram.

Although the molecular clock appears to hold true in some cases, in many others the data indicate that rates of evolution have not been constant through time across all branches in an evolutionary tree. For this reason, evolutionary dates derived from molecular data must be treated cautiously. Recently, methods have been developed to date evolutionary events without assuming that molecular evolution has been clocklike. These methods hold great promise for providing more reliable estimates of evolutionary timing.

Learning Outcomes Review 23.2

In cladistics, derived character states are distinguished from ancestral character states, and species are grouped based on shared derived character states. Derived characters are determined from comparison to a group known to be closely related, termed an outgroup. A clade contains all descendants of a common ancestor. A cladogram is a hypothetical representation of evolutionary relationships based on derived character states. Homoplasies may give a false picture of relationships.

- Why is cladistics more successful at inferring phylogenetic relationships in some cases than in others?
- Why are only shared derived, instead of all derived, characters useful in cladistics for reconstructing phylogenies?

23.3 Systematics and Classification

Learning Outcomes

1. Differentiate among monophyletic, paraphyletic, and polyphyletic groups.
2. Explain the meaning of the phylogenetic species concept and why it is controversial.

Whereas systematics is the reconstruction and study of evolutionary relationships, **classification** refers to how we place species and higher groups—genus, family, class, and so forth—into the taxonomic hierarchy (a topic we discuss in greater detail in chapter 26).

Current classification sometimes does not reflect evolutionary relationships

Systematics and traditional classification are not always congruent; to understand why, we need to consider how species may be grouped based on their phylogenetic relationships. A **monophyletic** group includes the most recent common ancestor of the group and all of its descendants. By definition, a clade is a monophyletic group. A **paraphyletic** group includes the most recent common ancestor of the group, but not all its descendants, and a **polyphyletic** group does not include the most recent common ancestor of all members of the group (figure 23.5).

Taxonomic hierarchies are based on shared traits, and ideally they should reflect evolutionary relationships. Traditional taxonomic groups, however, do not always fit well with

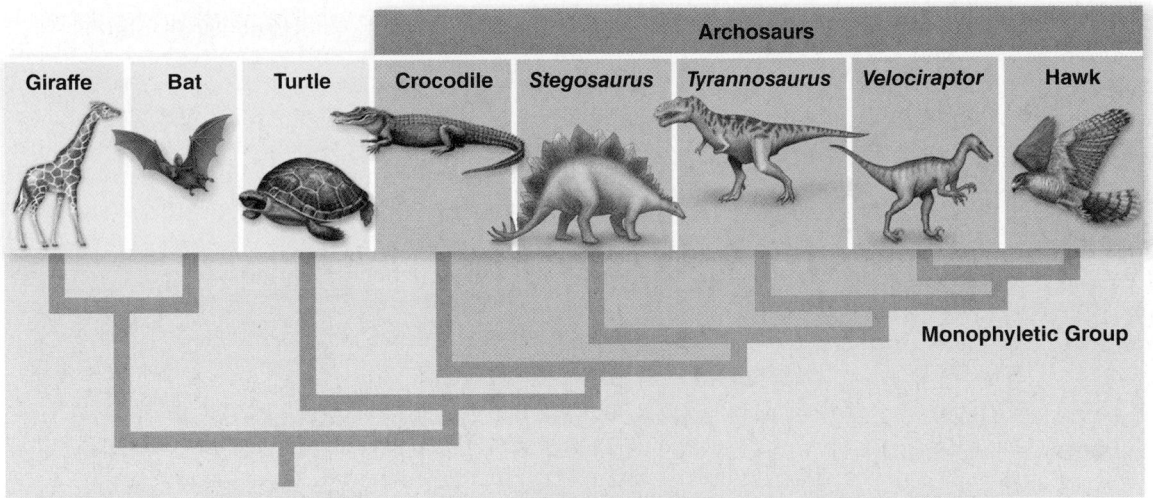

Giraffe Bat Turtle Crocodile *Stegosaurus* *Tyrannosaurus* *Velociraptor* Hawk

Archosaurs

Monophyletic Group

a.

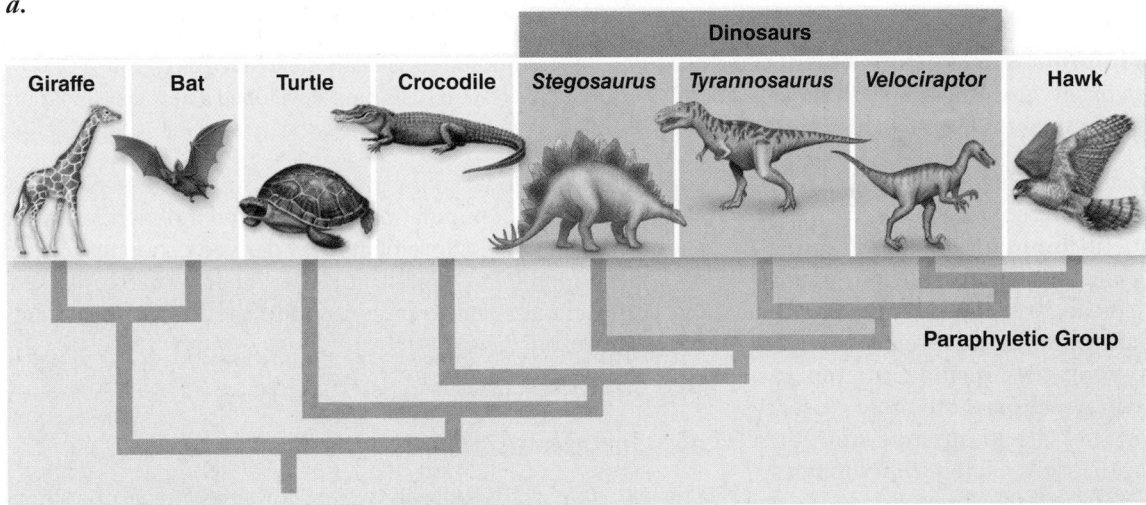

Giraffe Bat Turtle Crocodile *Stegosaurus* *Tyrannosaurus* *Velociraptor* Hawk

Dinosaurs

Paraphyletic Group

b.

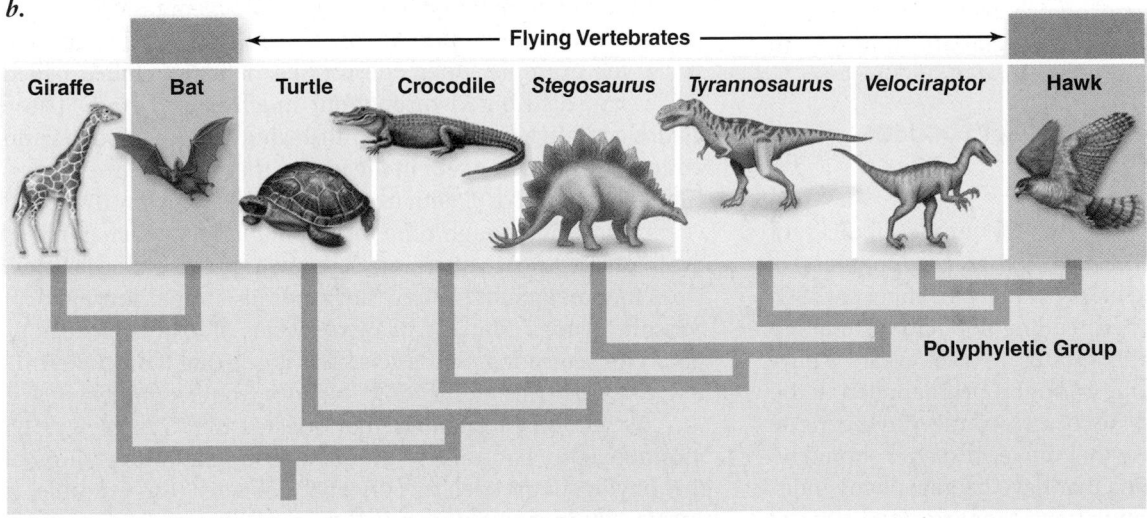

Giraffe Bat Turtle Crocodile *Stegosaurus* *Tyrannosaurus* *Velociraptor* Hawk

Flying Vertebrates

Polyphyletic Group

c.

Figure 23.5
Monophyletic, paraphyletic, and polyphyletic groups.
a. A monophyletic group consists of the most recent common ancestor and all of its descendants. For example, the name "Archosaurs" is given to the monophyletic group that includes a crocodile, *Stegosaurus*, *Tyrannosaurus*, *Velociraptor*, and a hawk.
b. A paraphyletic group consists of the most recent common ancestor and some of its descendants. For example, some, but not all, taxonomists traditionally give the name "dinosaurs" to the paraphyletic group that includes *Stegosaurus*, *Tyrannosaurus*, and *Velociraptor*. This group is paraphyletic because one descendant of the most recent ancestor of these species, birds, is not included in the group. Other taxonomists include birds within the Dinosauria because *Tyrannosaurus* and *Velociraptor* are more closely related to birds than to other dinosaurs.
c. A polyphyletic group does not contain the most recent common ancestor of the group. For example, bats and birds could be classified in the same group, which we might call "flying vertebrates," because they have similar shapes, anatomical features, and habitats. However, their similarities reflect convergent evolution, not common ancestry.

? Inquiry question Based on this phylogeny, are there any alternatives to convergence to explain the presence of wings in birds and bats? What types of data might be used to test these hypotheses?

new understanding of phylogenetic relationships. For example, birds have historically been placed in the class Aves, and dinosaurs have been considered part of the class Reptilia. But recent phylogenetic advances make clear that birds evolved from dinosaurs. The last common ancestor of all birds and a dinosaur was a meat-eating dinosaur (see figure 23.5).

Therefore, having two separate monophyletic groups, one for birds and one for reptiles (including dinosaurs and crocodiles, as well as lizards, snakes, and turtles), is not possible based on phylogeny. And yet the terms Aves and Reptilia are so familiar and well established that suddenly referring to birds as a type of dinosaur, and thus a type of reptile, is difficult

chapter **23** *Systematics, Phylogenies, and Comparative Biology* **461**

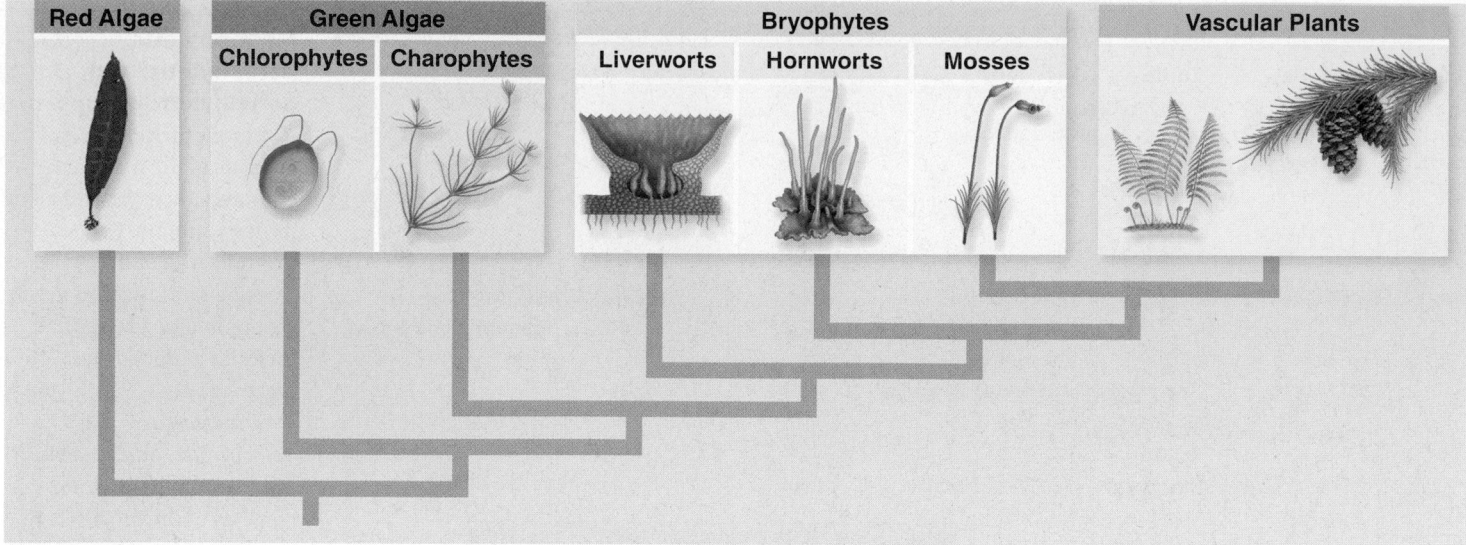

Figure 23.6 Phylogenetic information transforms plant classification. The traditional classification included two groups that we now realize are not monophyletic: the green algae and bryophytes. For this reason, plant systematists have developed a new classification of plants that does not include these groups (discussed in chapter 30).

for some. Nonetheless, biologists increasingly refer to birds as a type of dinosaur and hence a type of reptile.

Situations like this are not uncommon. Another example concerns the classification of plants. Traditionally, three major groups were recognized: green algae, bryophytes, and vascular plants (figure 23.6). However, recent research reveals that neither the green algae nor the bryophytes constitute monophyletic groups. Rather, some bryophyte groups are more closely related to vascular plants than they are to other bryophytes, and some green algae are more closely related to bryophytes and vascular plants than they are to other green algae. As a result, systematists no longer recognize green algae or bryophytes as evolutionary groups, and the classification system has been changed to reflect evolutionary relationships.

The phylogenetic species concept focuses on shared derived characters

In the preceding chapter, you read about a number of different ideas concerning what determines whether two populations belong to the same species. The biological species concept (BSC) defines species as groups of interbreeding populations that are reproductively isolated from other groups. In recent years, a phylogenetic perspective has emerged and has been applied to the question of species concepts. Advocates of the **phylogenetic species concept (PSC)** propose that the term *species* should be applied to groups of populations that have been evolving independently of other groups of populations. Moreover, they suggest that phylogenetic analysis is the way to identify such species. In this view, a species is a population or set of populations characterized by one or more shared derived characters.

This approach solves two of the problems with the BSC that were discussed in chapter 22. First, the BSC cannot be applied to allopatric populations because scientists cannot determine whether individuals of the populations would interbreed and produce fertile offspring if they ever came together. The PSC solves this problem: Instead of trying to predict what will happen in the future if allopatric populations ever come into contact, the PSC looks to the past to determine whether a population (or groups of populations) has evolved independently for a long enough time to develop its own derived characters.

Second, the PSC can be applied equally well to both sexual and asexual species, in contrast to the BSC, which deals only with sexual forms.

The phylogenetic species concept also has drawbacks

The PSC is controversial, however, for several reasons. First, some critics contend that it will lead to the recognition of every slightly differentiated population as a distinct species. In Missouri, for example, open, desert-like habitat patches called glades are distributed throughout much of the state. These glades contain a variety of warmth-loving species of plants and animals that do not occur in the forests that separate the glades. Glades have been isolated from one another for a few thousand years, allowing enough time for populations on each glade to evolve differences in some rapidly evolving parts of the genome. Does that mean that each of the hundreds, if not thousands, of Missouri glades contains its own species of lizards, grasshoppers, and scorpions? Some scientists argue that if one takes the PSC to its logical extreme, that is exactly what would result.

A second problem is that species may not always be monophyletic, contrary to the definition of some versions of the phylogenetic species concept. Consider, for example, a species composed of five populations, with evolutionary relationships like those indicated in figure 23.7. Suppose that population C becomes isolated and evolves differences that make it qualify as a species by any concept (for example, reproductively isolated, ecologically differentiated). But this distinction would mean that the remaining populations, which might still be perfectly capable of exchanging genes, would be paraphyletic, rather than monophyletic. Such situations probably occur often in the natural world.

Phylogenetic species concepts, of which there are many different permutations, are increasingly used, but are

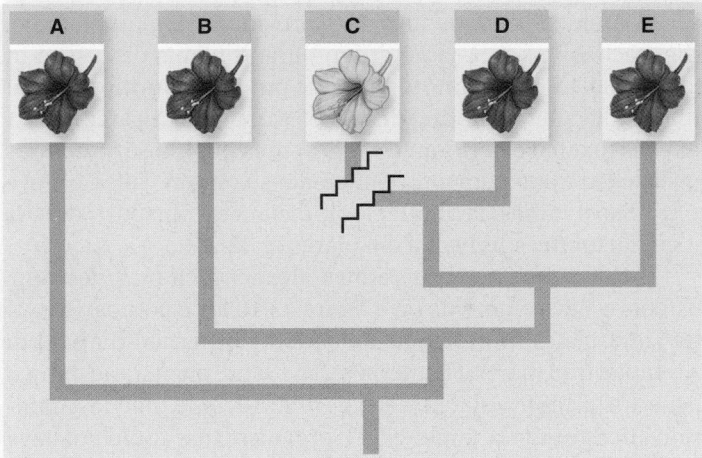

Figure 23.7 Paraphyly and the phylogenetic species concept. The five populations initially were all members of the same species, with their historical relationships indicated by the cladogram. Then, population C evolved in some ways to become greatly differentiated ecologically and reproductively from the other populations. By all species concepts, this population would qualify as a different species. However, the remaining four species do not form a clade; they are paraphyletic because population C has been removed and placed in a different species. This scenario may occur commonly in nature, but most versions of the phylogenetic species concept do not recognize paraphyletic species.

also contentious for the reasons just discussed. Evolutionary biologists are trying to find ways to reconcile the historical perspective of the PSC with the process-oriented perspective of the BSC and other species concepts.

Learning Outcomes Review 23.3

By definition, a clade is monophyletic. A paraphyletic group contains the most recent common ancestor, but not all its descendants; a polyphyletic group does not contain the most recent common ancestor of all members. The phylogenetic species concept focuses on the possession of shared derived characters, in contrast to the biological species concept, which emphasizes reproductive isolation. The PSC solves some problems of the BSC but has difficulties of its own.

■ *Under the biological species concept, is it possible for a species to be polyphyletic?*

Learning Outcomes

1. Explain the importance of homoplasy for interpreting patterns of evolutionary change.
2. Describe how phylogenetic trees can reveal the existence of homoplasy.
3. Discuss how a phylogenetic tree can indicate the timing of species diversification.

Phylogenies not only provide information about evolutionary relationships among species, but they are also indispensable for understanding how evolution has occurred. By examining the distribution of traits among species in the context of their phylogenetic relationships, much can be learned about how and why evolution may have proceeded. In this way, phylogenetics is the basis of all comparative biology.

Homologous features are derived from the same ancestral source; homoplastic features are not

In chapter 21, we pointed out that homologous structures are those that are derived from the same body part in a common ancestor. Thus, the forelegs of a dolphin (flipper) and of a horse (leg) are homologous because they are derived from the same bones in an ancestral vertebrate. By contrast, the wings of birds and those of dragonflies are homoplastic structures because they are derived from different ancestral structures. Phylogenetic analysis can help determine whether structures are homologous or homoplastic.

Homologous parental care in dinosaurs, crocodiles, and birds

Recent fossil discoveries have revealed that many species of dinosaurs exhibited parental care. They incubated eggs laid in nests and took care of growing baby dinosaurs, many of which could not have fended for themselves. Some recent fossils show dinosaurs sitting on a nest in exactly the same posture used by birds today (figure 23.8a)! Initially, these discoveries were treated as remarkable and unexpected—dinosaurs apparently had independently evolved behaviors similar to those of

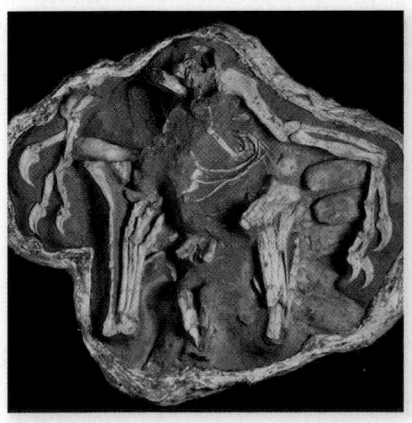

a.   *b.*

Figure 23.8 Parental care in dinosaurs and crocodiles *a.* Fossil dinosaur incubating its eggs. This remarkable fossil of *Oviraptor* shows the dinosaur sitting on its nest of eggs just as chickens do today. Not only is the dinosaur squatting on the nest, but its forelimbs are outstretched, perhaps to shade the eggs. *b.* Crocodile exhibiting parental care. Female crocodilians build nests and then remain nearby, guarding them, while the eggs incubate. When they are ready to hatch, the baby crocodiles vocalize; females respond by digging up the eggs and carrying the babies to the water.

modern-day organisms. But examination of the phylogenetic position of dinosaurs (see figure 23.5) indicates that they are most closely related to two living groups of animals—crocodiles and birds—both of which exhibit parental care (figure 23.8*b*).

It appears likely, therefore, that the parental care exhibited by crocodiles, dinosaurs, and birds did not evolve convergently from different ancestors that did not exhibit parental care; rather, the behaviors are homologous, inherited by each of these groups from their common ancestor that cared for its young.

Homoplastic convergence: Saber teeth and plant conducting tubes

In other cases, by contrast, phylogenetic analysis can indicate that similar traits have evolved independently in different clades. This convergent evolution from different ancestral sources indicates that such traits represent homoplasies. As one example, the fossil record reveals that extremely elongated canine teeth (saber teeth) occurred in a number of different groups of extinct carnivorous mammals. Although how these teeth were actually used is still debated, all saber-toothed carnivores had body proportions similar to those of cats, which suggests that these different types of carnivores all evolved into a similar predatory lifestyle. Examination of the saber-toothed character state in a phylogenetic context reveals that it most likely evolved independently at least three times (figure 23.9).

Conducting tubes in plants provide a similar example. The tracheophytes, a large group of land plants discussed in chapter 30, transport photosynthetic products, hormones, and other molecules over long distances through elongated, tubular cells that have perforated walls at the end. These structures are stacked upon each other to create a conduit called a sieve tube. Sieve tubes facilitate long-distance transport that is essential for the survival of tall plants on land.

Most members of the brown algae, which includes kelp, also have sieve elements (see figure 23.10 for a comparison of the sieve plates in brown algae and angiosperms) that aid in the rapid transport of materials. The land plants and brown algae are distantly related (see figure 23.10), and their last common ancestor was a single-celled organism that could not have had a multicellular transport system. This indicates that the strong structural and functional similarity of sieve elements in these plant groups is an example of convergent evolution.

Complex characters evolve through a sequence of evolutionary changes

Most complex characters do not evolve, fully formed, in one step. Rather, they are often built up, step-by-step, in a series of evolutionary transitions. Phylogenetic analysis can help discover these evolutionary sequences.

SCIENTIFIC THINKING

Question: *How many times have saber teeth evolved in mammals?*

Hypothesis: *Saber teeth are homologous and have only evolved once in mammals (or, conversely, saber teeth are convergent and have evolved multiple times in mammals).*

Phylogenic Analysis: *Examine the distribution of saber teeth on a phylogeny of mammals, and use parsimony to infer the history of saber tooth evolution (note that not all branches within marsupials and placentals are shown on the phylogeny).*

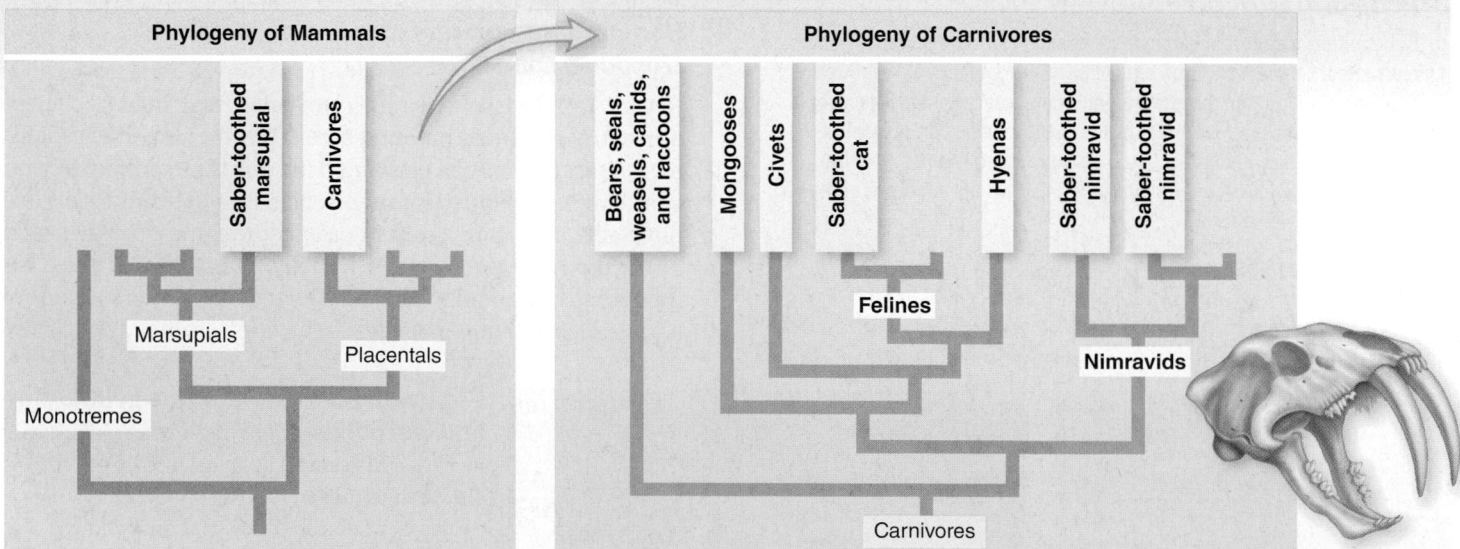

Result: *Saber teeth have evolved at least three times in mammals: once within marsupials, once in felines, and at least once in a group of now-extinct cat-like carnivores alled nimravids.*

Interpretation: *Note that it is possible that saber teeth evolved twice in nimravids, but another possibility that requires the same number of evolutionary changes (and thus is equally parsimonious) is that saber teeth evolved only once in the ancestor of nimravids and then were subsequently lost in one group of nimravids. (Note that for clarity, not all branches within marsupials and placentals are shown in this illustration.)*

Figure 23.9 Distribution of saber-toothed mammals.

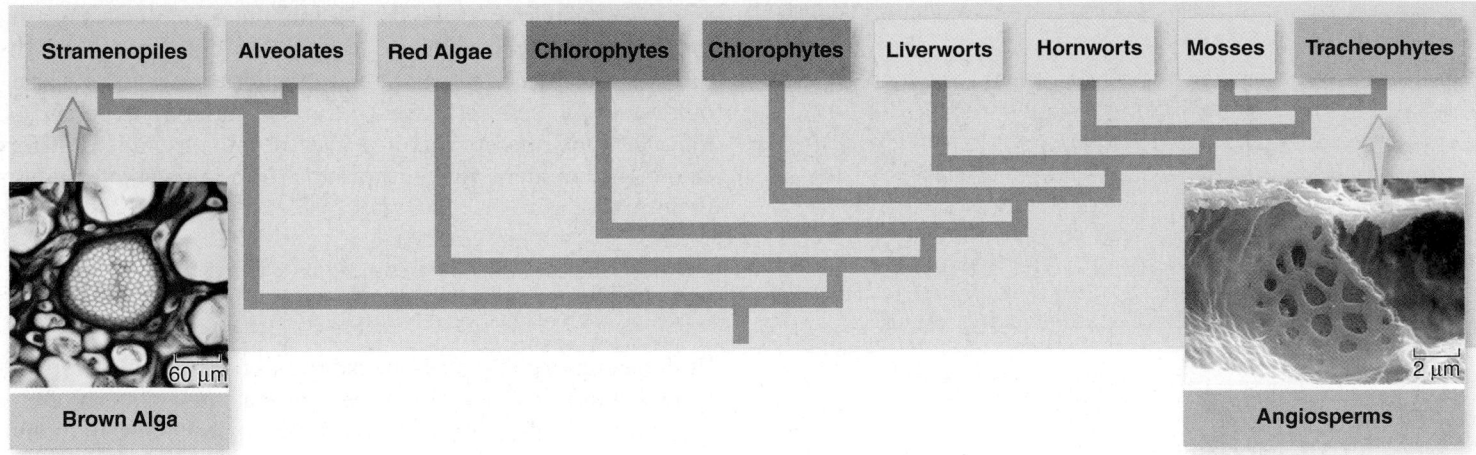

Figure 23.10 Convergent evolution of conducting tubes. Sieve tubes, which transport hormones and other substances throughout the plant, have evolved in two distantly related plant groups (brown algae are stramenopiles and angiosperms are tracheophytes).

© Dr. Richard Kessel & Dr. Gene Shih/Visuals Unlimited

Modern-day birds—with their wings, feathers, light bones, and breastbone—are exquisitely adapted flying machines. Fossil discoveries in recent years now allow us to reconstruct the evolution of these features. When the fossils are arranged phylogenetically, it becomes clear that the features characterizing living birds did not evolve simultaneously. Figure 23.11 shows how the features important to flight evolved

sequentially, probably over a long period of time, in the ancestors of modern birds.

One important finding often revealed by studies of the evolution of complex characters is that the initial stages of a character evolved as an adaptation to some environmental selective pressure different from that for which the character is currently adapted. Examination of figure 23.11 reveals that the first

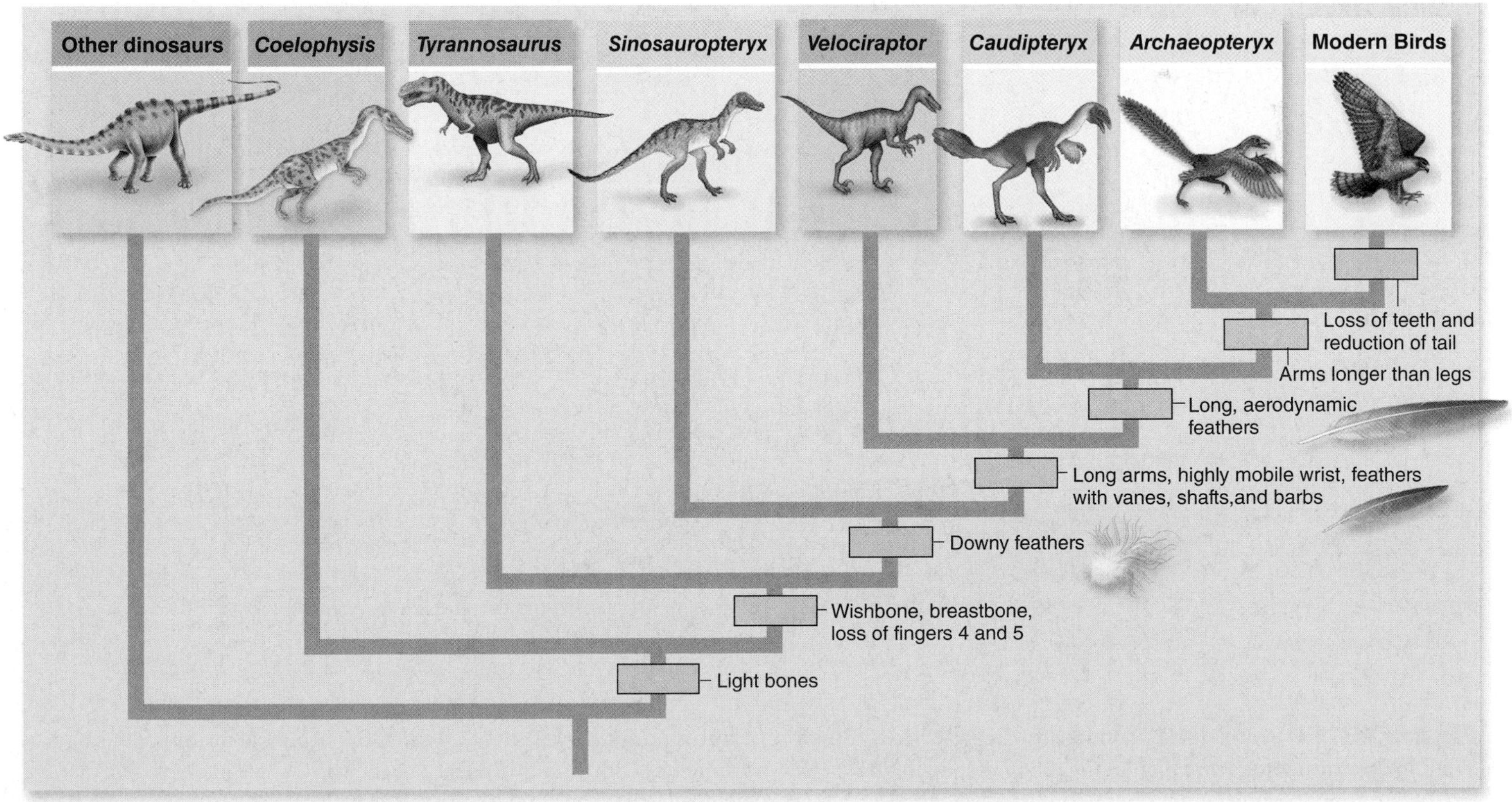

Figure 23.11 The evolution of birds. The traits we think of as characteristic of modern birds have evolved in stages over many millions of years.

From Richard O. Prum and Alan H. Brush, "The Evolutionary Origin and Diversification of Feathers," *Quarterly Review of Biology*, September 2002. Reprinted with permission of the University of Chicago Press.

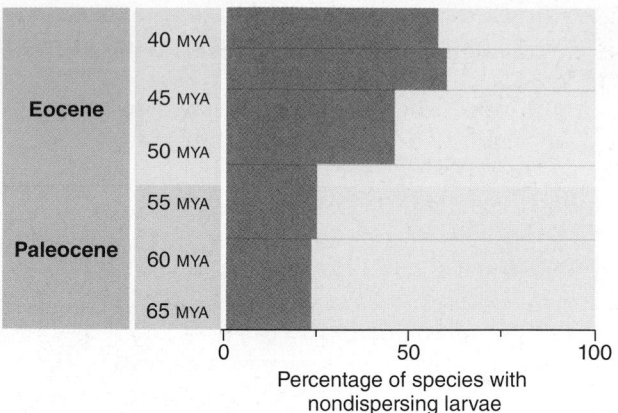

Figure 23.12 Larvae dispersal. Increase through time in the proportion of species whose larvae do not disperse far from their place of birth.

feathery structures evolved deep in the theropod phylogeny, in animals with forearms clearly not modified for flight. Therefore, the initial feather-like structures must have evolved for some other reason, perhaps serving as insulation or decoration. Through time, these structures have become modified to the extent that modern feathers produce excellent aerodynamic performance.

Phylogenetic methods can be used to distinguish between competing hypotheses

Understanding the causes of patterns of biological diversity observed today can be difficult because a single pattern often could have resulted from several different processes. In many cases, scientists can use phylogenies to distinguish between competing hypotheses.

Larval dispersal in marine snails

An example of this use of phylogenetic analysis concerns the evolution of larval forms in marine snails. Most species of

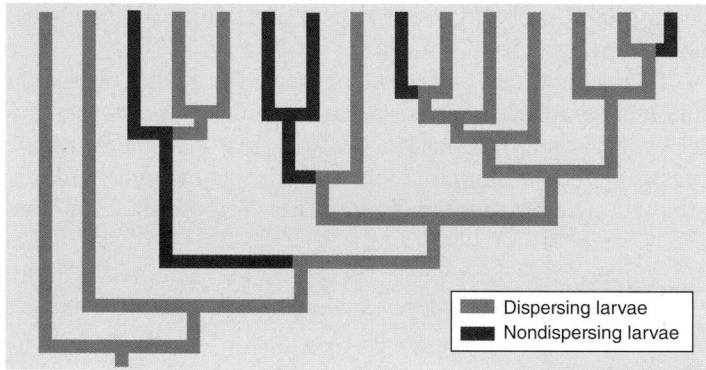

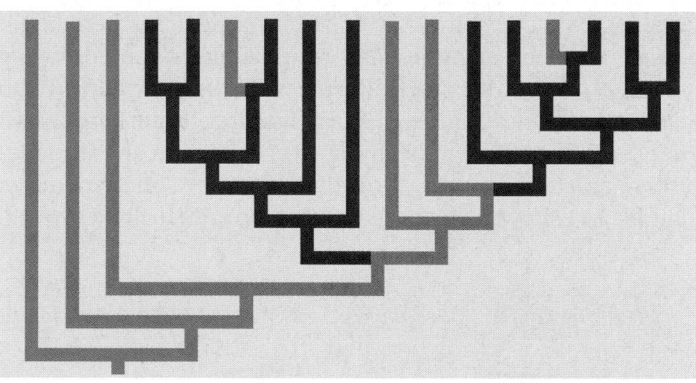

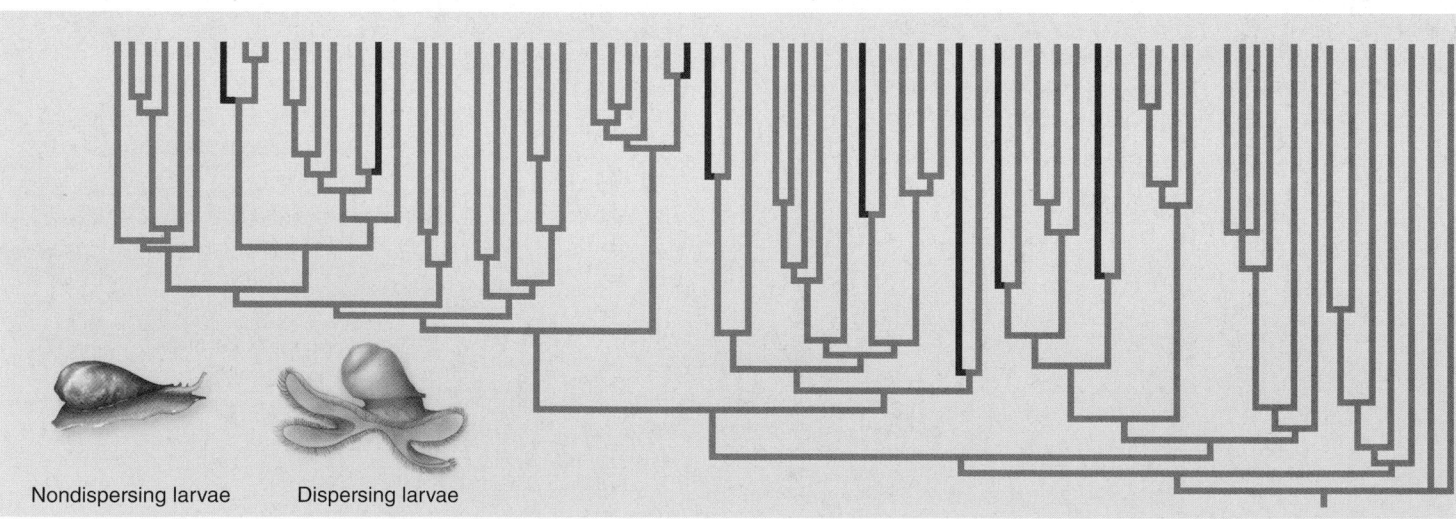

Figure 23.13 Phylogenetic investigation of the evolution of nondispersing larvae. *a.* In this hypothetical example, the evolutionary transition from dispersing to nondispersing larvae occurs more frequently (four times) than the converse (once). By contrast, in *(b)* the number of evolutionary changes in each direction is the same, but clades that have nondispersing larvae diversify to a greater extent due to higher rates of speciation or lower rates of extinction (assuming that extinct forms are not shown). *c.* Phylogeny for *Conus,* a genus of marine snails. Nondispersing larvae have evolved eight separate times from dispersing larvae, with no instances of evolution in the reverse direction. This phylogeny does not show all species, however; nondispersing clades contain on average 3.5 times as many species as dispersing clades.

snails produce microscopic larvae that drift in the ocean currents, sometimes traveling hundreds or thousands of miles before becoming established and transforming into adults. Some species, however, have evolved larvae that settle to the ocean bottom very quickly and thus don't disperse far from their place of origin. Studies of fossil snails indicate that the proportion of species that produce nondispersing larvae has increased through geological time (figure 23.12).

Two processes could produce an increase in nondispersing larvae through time. First, if evolutionary change from dispersing to nondispersing occurs more often than change in the opposite direction, then the proportion of species that are nondispersing would increase through time.

Alternatively, if species that are nondispersing speciate more frequently, or become extinct less frequently, than dispersing species, then through time the proportion of nondispersers would also increase (assuming that the descendants of nondispersing species also were nondispersing). This latter case is a reasonable hypothesis because nondispersing species probably have lower amounts of gene flow than dispersing species, and thus might more easily become geographically isolated, increasing the likelihood of allopatric speciation (see chapter 22).

These two processes would result in different phylogenetic patterns. If evolution from a dispersing ancestor to a nondispersing descendant occurred more often than the reverse, then an excess of such changes should be evident in the phylogeny, as shown by more dispersing $\longrightarrow$ nondispersing branchpoints in figure 23.13a. In contrast, if nondispersing species underwent greater speciation, then clades of nondispersing species would contain more species than clades of dispersing species, as shown in figure 23.13b.

Evidence for both processes was revealed in an examination of the phylogeny of marine snails in the genus *Conus,* in which 30% of species are nondispersing (figure 23.13c). The phylogeny indicates that possession of dispersing larvae was

the ancestral state; nondispersing larvae are inferred to have evolved eight times, with no evidence for evolutionary reversal from nondispersing to dispersing larvae.

At the same time, clades of nondispersing larvae tend to have on average 3.5 times as many species as dispersing larvae, which suggests that in nondispersing species, rates of speciation are higher, rates of extinction are lower, or both.

This analysis therefore indicates that the evolutionary increase in nondispersing larvae through time may be a result both of a bias in the direction in which evolution proceeds plus an increase in rate of diversification (that is, speciation rate minus extinction rate) in nondispersing clades.

The lack of evolutionary reversal is not surprising because when larvae evolve to become nondispersing, they often lose a variety of structures used for feeding while drifting in the ocean current. In most cases, once a structure is lost, it rarely re-evolves, and thus the standard view is that the evolution of nondispersing larvae is a one-way street, with few examples of re-evolution of dispersing larvae.

Loss of the larval stage in marine invertebrates

A related phenomenon in many marine invertebrates is the loss of the larval stage entirely. Most marine invertebrates—in groups as diverse as snails, sea stars, and anemones—pass through a larval stage as they develop. But in a number of different types of organisms, the larval stage is omitted, and the eggs develop directly into adults.

The evolutionary loss of the larval stage has been suggested as another example of a nonreversible evolutionary change because once the larval stages are lost, it is difficult for them to re-evolve—or so the argument goes. A recent study on one group of marine limpets, shelled marine organisms related to snails, shows that this is not necessarily the case. Among these limpets, direct development has evolved many times; however, in three cases, the phylogeny strongly suggests that evolution reversed and a larval stage re-evolved (figure 23.14a).

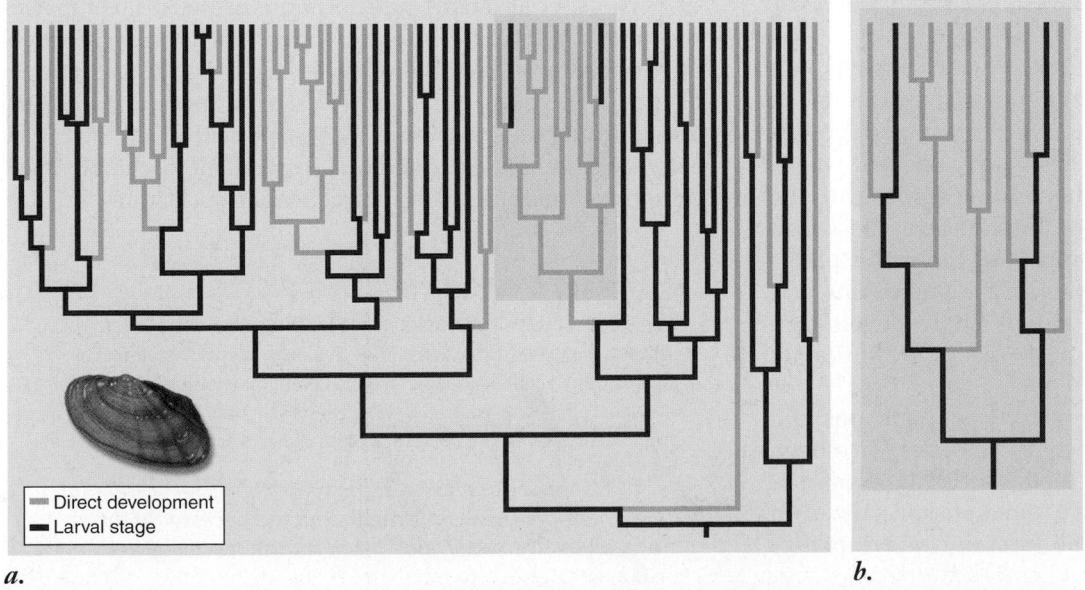

Legend:
— Direct development
— Larval stage

a.

b.

Figure 23.14 Evolution of direct development in a family of limpets. *a.* Direct development evolved many times (indicated by beige lines stemming from a red ancestor), and three instances of reversed evolution from direct development to larval development are indicated (red lines from a beige ancestor). *b.* A less parsimonious interpretation of evolution in the clade in the light blue box is that, rather than two evolutionary reversals, six instances of the evolution of development occurred without any evolutionary reversal.

? Inquiry question How might you distinguish between the hypotheses in parts *a* and *b* of this figure?

It is important to remember that patterns of evolution suggested by phylogenetic analysis are not always correct—evolution does not necessarily occur parsimoniously. In the limpet study, for example, it is possible that within the clade in the light blue box, presence of a larva was retained as the ancestral state, and direct development evolved independently six times (figure 23.14*b*). Phylogenetic analysis cannot rule out this possibility, even if it is less phylogenetically parsimonious.

If the re-evolution of lost traits seems unlikely, then the alternative hypothesis that direct development evolved six times—rather than only once at the base of the clade, with two instances of evolutionary reversal—should be considered. For example, studies of the morphology or embryology of direct-developing species might shed light on whether such structures are homologous or convergent. In some cases, artificial selection experiments in the laboratory or genetic manipulations can test the hypothesis that it is difficult for lost structures to re-evolve. Conclusions from phylogenetic analyses are always stronger when supported by results of other types of studies.

Phylogenetics helps explain species diversification

One of the central goals of evolutionary biology is to explain patterns of species diversity: Why do some types of plants and animals exhibit more **species richness**—a greater number of species per clade—than others? Phylogenetic analysis can be used both to suggest and to test hypotheses about such differences.

Species richness in beetles

Beetles (order Coleoptera) are the most diverse group of animals. Approximately 60% of all animal species are insects, and approximately 80% of all insect species are beetles. Among beetles, families that are herbivorous are particularly species-rich.

Examination of the phylogeny provides insight into beetle evolutionary diversification (figure 23.15). Among the Phytophaga, the clade which contains most herbivorous beetle species, the deepest branches belong to beetle families that specialize on conifers. This finding agrees with the fossil record because conifers were among the earliest seed plant groups to evolve. By contrast, the flowering plants (angiosperms) evolved more recently, in the Cretaceous, and beetle families specializing on them have shorter evolutionary branches, indicating their more recent evolutionary appearance.

This correspondence between phylogenetic position and timing of plant origins suggests that beetles have been remarkably conservative in their diet. The family Nemonychidae, for example, appears to have remained specialized on conifers since the beginning of the Jurassic, approximately 210 MYA.

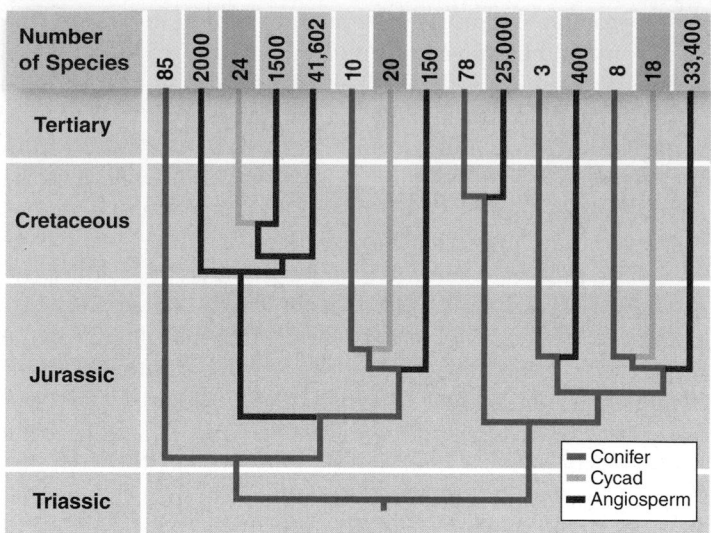

Figure 23.15 Evolutionary diversification of the Phytophaga, the largest clade of herbivorous beetles. Clades that originated deep in the phylogenetic tree feed on conifers; clades that feed on angiosperms, which evolved more recently, originated more recently. Age of clades is established by examination of fossil beetles.

Phylogenetic explanations for beetle diversification

The phylogenetic perspective suggests factors that may be responsible for the incredible diversity of beetles. The phylogeny for the Phytophaga indicates that it is not the evolution of herbivory itself that is linked to great species richness. Rather, specialization on angiosperms seems to have been a prerequisite for great species diversification. Specialization on angiosperms appears to have arisen five times independently within herbivorous beetles; in each case, the angiosperm-specializing clade is substantially more species-rich than the clade to which it is most closely related (termed a *sister clade)* and which specializes on some other type of plant.

Why specialization on angiosperms has led to great species diversity is not yet clear and is the focus of much current research. One possibility is that this diversity is linked to the great species-richness of angiosperms themselves. With more than 250,000 species of angiosperms, beetle clades specializing on them may have had a multitude of opportunities to adapt to feed on different species, thus promoting divergence and speciation.

Learning Outcomes Review 23.4

Homologous traits are derived from the same ancestral character states, whereas homoplastic traits are not, even though they may have similar function. Phylogenetic analysis can help determine whether homology or homoplasy has occurred. By correlating phylogenetic branching with known evolutionary events, the timing and cause of diversification can be inferred.

- *Does the possession of the same character state by all members of a clade mean that the ancestor of that clade necessarily possessed that character state?*

Phylogenetics and Disease Evolution

The examples so far have illustrated the use of phylogenetic analysis to examine relationships among species. Such analyses can also be conducted on virtually any group of biological entities, as long as evolutionary divergence in these groups occurs by a branching process, with little or no genetic exchange between different groups. No example illustrates this better than recent attempts to understand the evolution of the virus that causes autoimmune deficiency syndrome (AIDS).

HIV has evolved from a simian viral counterpart

AIDS was first recognized in the early 1980s, and it rapidly became epidemic in the human population. Current estimates are that more than 33 million people are infected with the human immunodeficiency virus (HIV), of whom more than 2 million die each year.

At first, scientists were perplexed about where HIV had originated and how it had infected humans. In the mid-1980s, however, scientists discovered a related virus in laboratory monkeys, termed simian immunodeficiency virus (SIV). In biochemical terms, the viruses are very similar, although genetic differences exist. At last count, SIV has been detected in 36 species of primates, but only in species found in sub-Saharan Africa. Interestingly, SIV—which appears to be transmitted sexually—does not appear to cause illness in some of these species.

Based on the degree of genetic differentiation among strains of SIV, scientists estimate that SIV may have been around for more than a million years in these primates, perhaps providing enough time for these species to adapt to the virus and thus prevent it from having adverse effects.

Phylogenetic analysis identifies the path of transmission

Phylogenetic analysis of strains of HIV and SIV reveals three clear findings. First, HIV obviously descended from SIV. All strains of HIV are phylogenetically nested within clades of SIV strains, indicating that HIV is derived from SIV (figure 23.16).

Second, a number of different strains of HIV exist, and they appear to represent independent transfers from different primate species. Each of the human strains is more closely related to a strain of SIV than it is to other HIV strains, indicating separate origins of the HIV strains.

Finally, humans have acquired HIV from different host species. HIV-1, which is the virus responsible for the global epidemic, has four subtypes. Two of these subtypes are most closely related to strains in chimpanzees, whereas a third is most closely related to a strain in gorillas, indicating transmission from both ape species. The origin of the fourth subtype is not clear; further sampling will likely reveal it to be the sister taxon to a currently undiscovered chimp or a gorilla strain.

By contrast, subtypes of HIV-2, which is much less widespread (in some cases known from only one individual), are related to SIV found in West African monkeys, primarily the sooty mangabey *(Cercocebus atys)*. Moreover, the subtypes of HIV-2 also appear to represent multiple cross-species transmissions to humans.

Transmission from other primates to humans

Several hypotheses have been proposed to explain how SIV jumped from chimps and monkeys to humans. The most likely idea is that transmission occurred as the result of

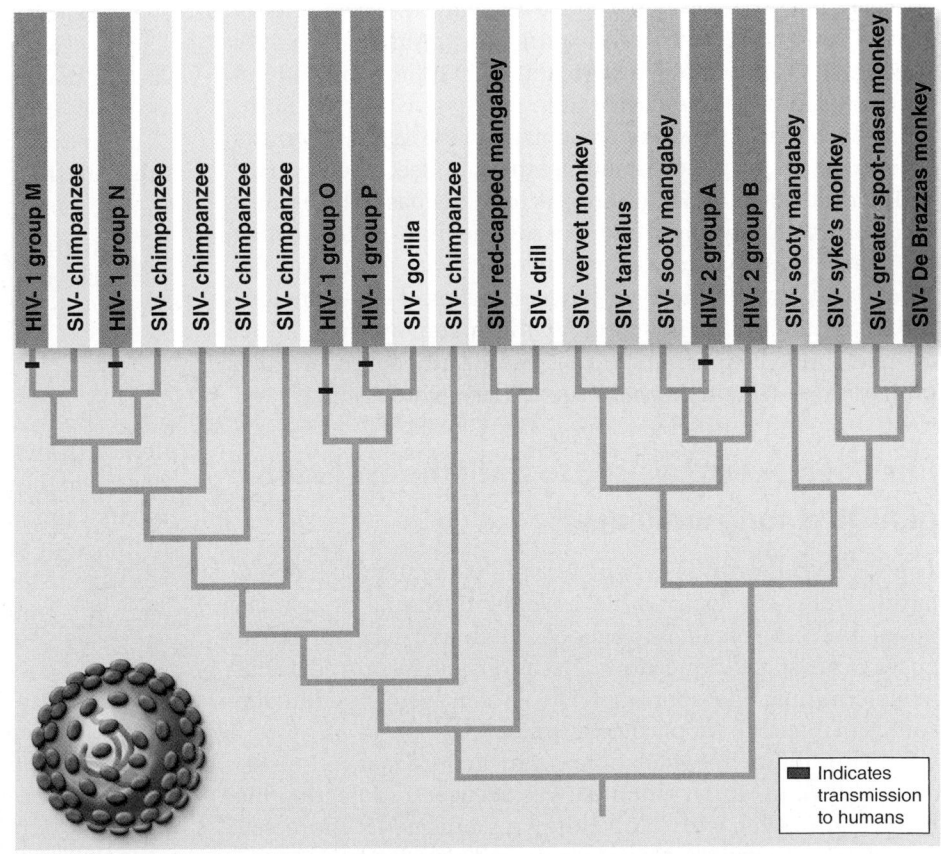

Figure 23.16 Evolution of HIV and SIV. HIV has evolved multiple times and from strains of SIV in different primate species (each primate species indicated by a different color; drill and tantalus are species of monkey). Boxes indicate transmission to humans from other primate species.

blood-to-blood contact that may occur when humans kill and butcher monkeys and apes. Recent years have seen a huge increase in the rate at which primates are hunted for the "bush-meat" market, particularly in central and western Africa. This increase has resulted from a combination of increased human populations desiring ever greater amounts of protein, combined with increased access to the habitats in which these primates live as the result of road building and economic development. The unfortunate result is that population sizes of many primate species, including our closest relatives, are plummeting toward extinction. A second consequence of this hunting is that humans are increasingly brought into contact with bodily fluids of these animals, and it is easy to imagine how during the butchering process, blood from a recently killed animal might enter the human bloodstream through cuts in the skin, perhaps obtained during the hunting process.

Establishing the crossover time line and location

Where and when did this cross-species transmission occur? HIV strains are most diverse in Africa, and the incidence of HIV is higher there than elsewhere in the world. Combined with the evidence that HIV is related to SIV in African primates, it seems certain that AIDS appeared first in Africa.

As for when the jump from other primates to humans occurred, the fact that AIDS was not recognized until the 1980s suggests that HIV probably arose recently. Descendants of slaves brought to North America from West Africa in the 19th century lacked the disease, indicating that it probably did not occur at the time of the slave trade.

Once the disease was recognized in the 1980s, scientists scoured repositories of blood samples to see whether HIV could be detected in blood samples from the past. The earliest HIV-positive result was found in a sample from 1959, pushing the date of origin back at least two decades. Based on the amount of genetic difference between strains of HIV-1, including the 1959 sample, and assuming the operation of a molecular clock, scientists estimate that the deadly strain of AIDS probably crossed into humans some time before 1940.

Phylogenies can be used to track the evolution of AIDS among individuals

The AIDS virus evolves extremely rapidly, so much so that different strains can exist within a single individual in a single population. As a result, phylogenetic analysis can be applied to answer very specific questions; just as phylogeny proved useful in determining the source of HIV, it can also pinpoint the source of infection for particular individuals.

This ability became apparent in a court case in Louisiana in 1998, in which a dentist was accused of injecting his former girlfriend with blood drawn from an HIV-infected patient. The dentist's records revealed that he had drawn blood from the patient and had done so in a suspicious manner. Scientists sequenced the viral strains from the victim, the patient, and from a large number of HIV-infected people in the local community. The phylogenetic analysis clearly

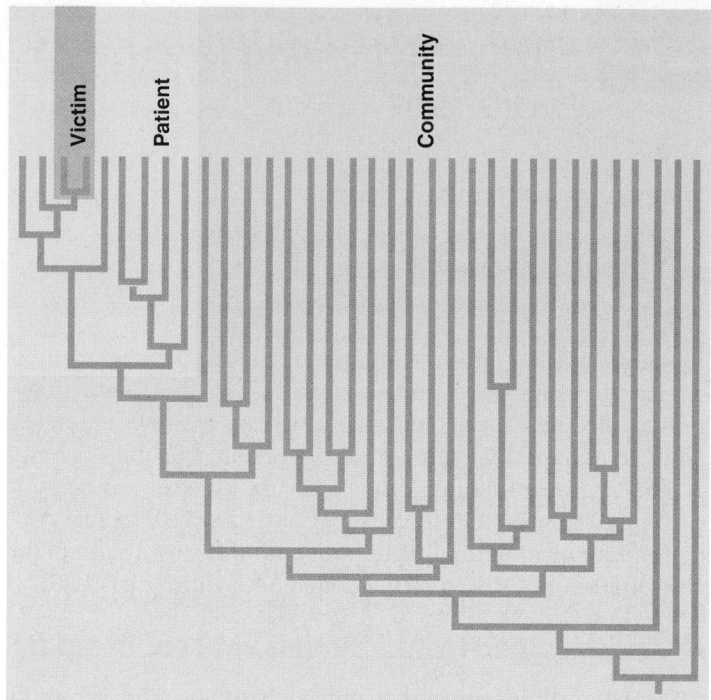

Figure 23.17 Evolution of HIV strains reveals the source of infection. HIV mutates so rapidly that a single HIV-infected individual often contains multiple genotypes in his or her body. As a result, it is possible to create a phylogeny of HIV strains and to identify the source of infection of a particular individual. In this case, the HIV strains of the victim clearly are derived from strains in the body of another individual, the patient. Other HIV strains are from HIV-infected individuals in the local community.

? Inquiry question What would the phylogeny look like if the victim had not gotten HIV from the patient?

demonstrated that the victim's viral strain was most closely related to the patient's (figure 23.17). This analysis, which for the first time established phylogenetics as a legally admissible form of evidence in courts in the United States, helped convict the dentist, who is now serving a 50-year sentence for attempted murder.

Learning Outcome Review 23.5

Modern phylogenetic techniques and analysis can track the evolution of disease strains, uncovering sources and progression. The HIV virus provides a prime example: Analysis of viral strains has shown that the progression from simian immunodeficiency virus (SIV) into human hosts has occurred several times. Phylogenetic analysis is also used to track the transmission of human disease.

■ *Could HIV have arisen in humans and then have been transmitted to other primate species?*

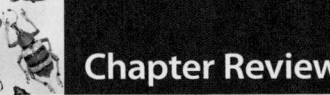

23.1 Systematics

Branching diagrams depict evolutionary relationships.

Systematics is the study of evolutionary relationships, which are depicted on branching evolutionary trees, called phylogenies.

Similarity may not accurately predict evolutionary relationships.

The rate of evolution can vary among species and can even reverse direction. Closely related species can therefore be dissimilar in phenotypic characteristics.

Conversely, convergent evolution results in distantly related species being phenotypically similar.

23.2 Cladistics

The cladistic method requires that character variation be identified as ancestral or derived.

Derived character states are those that differ from the ancestral condition. Only shared derived characters are useful for inferring phylogenies.

Character polarity is established using an outgroup comparison in which the outgroup consists of one or a group of species, relative to the group under study.

Character states exhibited by the outgroup are assumed to be ancestral, and other character states are considered derived.

A cladogram is a graphically represented hypothesis of evolutionary relationships.

Homoplasy complicates cladistic analysis.

Homoplasy refers to a shared character state, such as wings of birds and wings of insects, that has not been inherited from a common ancestor.

Cladograms are constructed based on the principle of parsimony, which indicates that the phylogeny requiring the fewest evolutionary changes is accepted as the best working hypothesis.

Other phylogenetic methods work better than cladistics in some situations.

When evolutionary change is rapid, other methods, such as statistical approaches and the use of the molecular clock, are sometimes more useful.

23.3 Systematics and Classification

Current classification sometimes does not reflect evolutionary relationships.

A monophyletic group consists of the most recent common ancestor and all of its descendants.

A paraphyletic group consists of the most recent common ancestor and some of its descendants.

A polyphyletic group does not contain the most recent ancestor of the group.

Some currently recognized taxa are not monophyletic, such as reptiles, which are paraphyletic with respect to birds.

The phylogenetic species concept focuses on shared derived characters.

The phylogenetic species concept (PSC) emphasizes the possession of shared derived characters, whereas the biological species concept focuses on reproductive isolation. Many versions of this concept recognize species that are monophyletic.

The phylogenetic species concept also has drawbacks.

Among criticisms of the PSC are that it subdivides groups too far via impractical distinctions, and that the PSC definition of a group may not always apply as selection proceeds.

23.4 Phylogenetics and Comparative Biology

Homologous features are derived from the same ancestral source; homoplastic features are not.

Homologous structures can be identified by phylogenetic analysis, establishing whether or not different structures have been built from the same ancestral structure.

Complex characters evolve through a sequence of evolutionary changes.

Most complex features do not evolve in a single step but include stages of transition. They may have begun as an adaptation to a selective pressure different from the one for which the feature is currently adapted.

Phylogenetic methods can be used to distinguish between competing hypotheses.

Different evolutionary scenarios can be distinguished by phylogenetic analysis. The minimum number of times a trait may have evolved can be established, and the direction of trait evolution, the timing, and the cause of diversification can be inferred.

Phylogenetics helps explain species diversification.

Questions regarding the causes of species richness may be addressed with phylogenetic analysis.

23.5 Phylogenetics and Disease Evolution

HIV has evolved from a simian viral counterpart.

Phylogenetic methods have indicated that HIV is related to SIV.

Phylogenetic analysis identifies the path of transmission.

It is clear that HIV has descended from SIV, and that independent transfers from simians to humans have occurred several times.

Phylogenies can be used to track the evolution of AIDS among individuals.

Even though HIV evolves rapidly, phylogenetic analysis can trace the origin of a current strain to a specific source of infection.

UNDERSTAND

1. Overall similarity of phenotypes may not always reflect evolutionary relationships
   a. due to convergent evolution.
   b. because of variation in rates of evolutionary change of different kinds of characters.
   c. due to homoplasy.
   d. All of the choices are correct.

2. Cladistics
   a. is based on overall similarity of phenotypes.
   b. requires distinguishing similarity due to inheritance from a common ancestor from other reasons for similarity.
   c. is not affected by homoplasy.
   d. None of the choices is correct.

3. The principle of parsimony
   a. helps evolutionary biologists distinguish among competing phylogenetic hypotheses.
   b. does not require that the polarity of traits be determined.
   c. is a way to avoid having to use outgroups in a phylogenetic analysis.
   d. cannot be applied to molecular traits.

4. Parsimony suggests that parental care in birds, crocodiles, and some dinosaurs
   a. evolved independently multiple times by convergent evolution.
   b. evolved once in an ancestor common to all three groups.
   c. is a homoplastic trait.
   d. is not a homologous trait.

5. The forelimb of a bird and the forelimb of a rhinoceros
   a. are homologous and symplesiomorphic.
   b. are not homologous but are symplesiomorphic.
   c. are homologous and synapomorphic.
   d. are not homologous but are synapomorphic.

6. In order to determine polarity for different states of a character
   a. there must be a fossil record of the groups in question.
   b. genetic sequence data must be available.
   c. an appropriate name for the taxonomic group must be selected.
   d. an outgroup must be identified.

7. In a paraphyletic group
   a. all species are more closely related to each other than they are to a species outside the group.
   b. evolutionary reversal is common.
   c. polyphyly also usually occurs.
   d. some species are more closely related to species outside the group than they are to some species within the group.

8. A paraphyletic group includes
   a. an ancestor and all of its descendants.
   b. an ancestor and some of its descendants.
   c. descendants of more than one common ancestor.
   d. All of the choices are correct.

9. Sieve tubes and sieve elements are
   a. homoplastic because they have different function.
   b. homologous because they have similar function.
   c. homoplastic because their common ancestor was single-celled.
   d. structures involved in transport within animals.

APPLY

1. A taxonomic group that contains species that have similar phenotypes due to convergent evolution is
   a. paraphyletic.   c. polyphyletic.
   b. monophyletic.   d. a good cladistic group.

2. Rapid rates of character change relative to the rate of speciation pose a problem for cladistics because
   a. the frequency with which distantly related species evolve the same derived character state may be high.
   b. evolutionary reversals may occur frequently.
   c. homoplasy will be common.
   d. All of the choices are correct.

3. Species recognized by the phylogenetic species concept
   a. sometimes also would be recognized as species by the biological species concept.
   b. are sometimes paraphyletic.
   c. are characterized by symplesiomorphies.
   d. are more frequent in plants than in animals.

SYNTHESIZE

1. List the synapomorphy and the taxa defined by that synapomorphy for the groups pictured in figure 23.2. Name each group defined by a set of synapomorphies in a way that might be construed as informative about what kind of characters define the group.

2. Identifying "outgroups" is a central component of cladistic analysis. As described on page 457, a group is chosen that is closely related to, but not a part of the group under study. If one does not know the relationships of members of the group under study, how can one be certain that an appropriate outgroup is chosen? Can you think of any approaches that would minimize the effect of a poor choice of outgroup?

3. As noted in your reading, cladistics is a widely utilized method of systematics, and our classification system (taxonomy) is increasingly becoming reflective of our knowledge of evolutionary relationships. Using birds as an example, discuss the advantages and disadvantages of recognizing them as reptiles versus as a group separate and equal to reptiles.

4. Across many species of limpets, loss of larval development and reversal from direct development appears to have occurred multiple times. Under the simple principle of parsimony, are changes in either direction merely counted equally in evaluating the most parsimonious hypothesis? If it is much more likely to lose a larval mode than to re-evolve it from direct development, should that be taken into account? If so, how?

5. Birds, pterosaurs (a type of flying reptile that lived in the Cretaceous period), and bats all have modified their forelimbs to serve as wings, but they have done so in different ways. Are these structures homologous? If so, how can they be both homologous and convergent? Do any organisms possess wings that are convergent, but not homologous?

6. In what sense does the biological species concept focus on evolutionary mechanisms and the phylogenetic species concept on evolutionary patterns? Which, if either, is correct?

ONLINE RESOURCE

www.ravenbiology.com

Understand, Apply, and Synthesize—enhance your study with animations that bring concepts to life and practice tests to assess your understanding. Your instructor may also recommend the interactive eBook, individualized learning tools, and more.

Chapter **24**

Genome Evolution

Chapter Contents

Introduction

Genomes contain the raw material for evolution, and many clues to evolution are hidden in the ever-changing nature of genomes. As more genomes have been sequenced, the new and exciting field of comparative genomics has emerged and has yielded some surprising results as well as many, many questions. Comparing whole genomes, not just individual genes, enhances our ability to understand the workings of evolution, to improve crops, and to identify the genetic basis of disease so that we might develop more effective treatments with minimal side effects. The focus of this chapter is the role of comparative genomics in enhancing our understanding of genome evolution and how this new knowledge can be applied to improve our lives.

24.1 Comparative Genomics

Learning Outcomes

1. *Describe the kinds of differences that can be found between genomes.*
2. *Relate timescale to genome evolution.*
3. *Explain why genomes may evolve at different rates.*

A key challenge of modern evolutionary biology is finding a way to link changes in DNA sequences, which we are now able to study in great detail, with the evolution of the complex morphological characters used to construct a traditional phylogeny. Many different genes contribute to complex characters. Making the connection between a specific change in a gene and a modification in a morphological character is particularly difficult.

Comparing genomes (entire DNA sequences) provides a powerful tool for exploring evolutionary divergence among organisms to connect DNA level changes with morphological

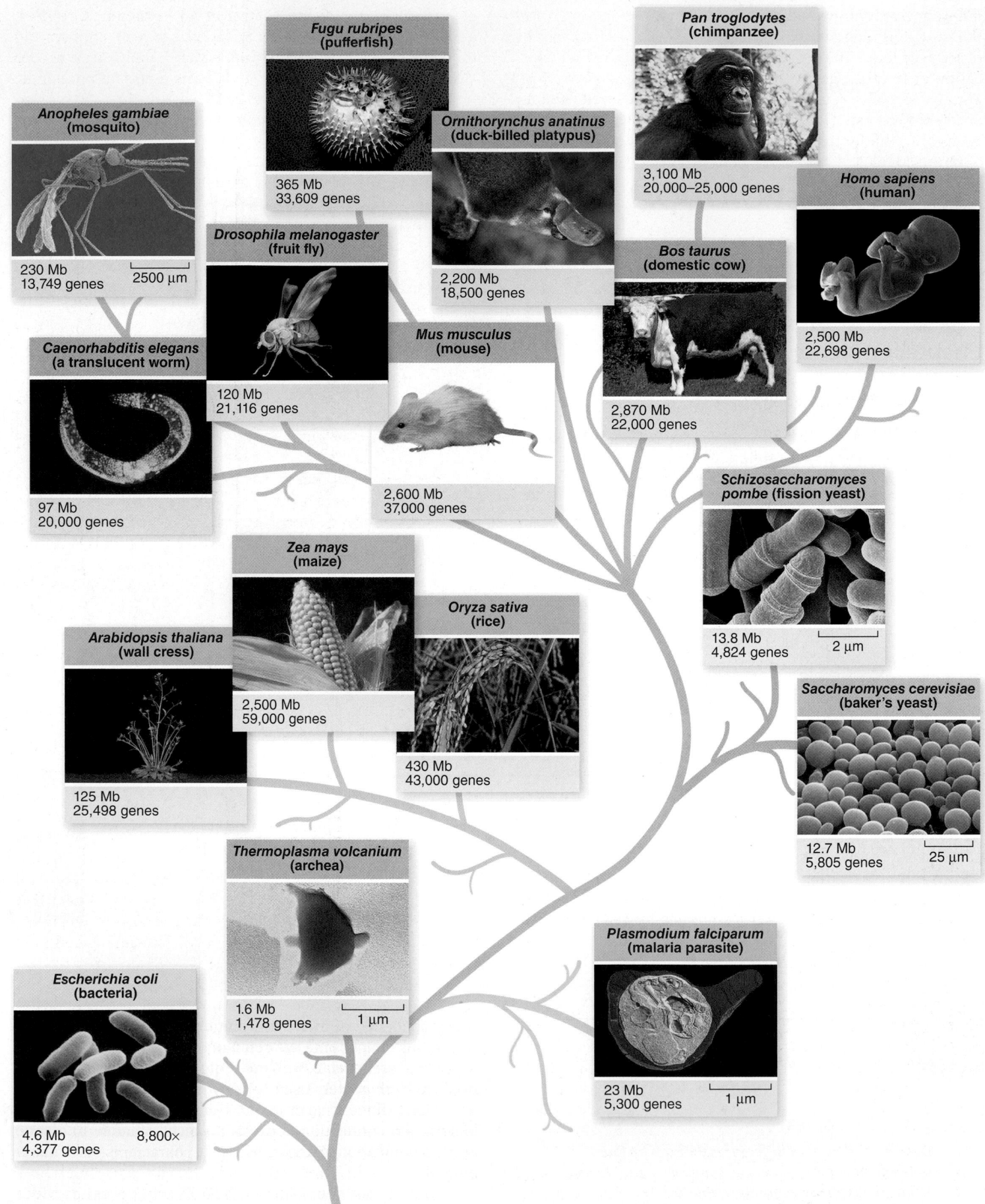

Figure 24.1 Milestones of comparative eukaryotic genomics.

Anopheles gambiae
(mosquito)

230 Mb
13,749 genes
2500 µm

Fugu rubripes
(pufferfish)

365 Mb
33,609 genes

Ornithorynchus anatinus
(duck-billed platypus)

2,200 Mb
18,500 genes

Pan troglodytes
(chimpanzee)

3,100 Mb
20,000–25,000 genes

Homo sapiens
(human)

2,500 Mb
22,698 genes

Drosophila melanogaster
(fruit fly)

120 Mb
21,116 genes

Bos taurus
(domestic cow)

2,870 Mb
22,000 genes

Mus musculus
(mouse)

2,600 Mb
37,000 genes

Caenorhabditis elegans
(a translucent worm)

97 Mb
20,000 genes

*Schizosaccharomyces
pombe* (fission yeast)

13.8 Mb
4,824 genes
2 µm

Zea mays
(maize)

2,500 Mb
59,000 genes

Oryza sativa
(rice)

430 Mb
43,000 genes

Arabidopsis thaliana
(wall cress)

125 Mb
25,498 genes

Saccharomyces cerevisiae
(baker's yeast)

12.7 Mb
5,805 genes
25 µm

Thermoplasma volcanium
(archea)

1.6 Mb
1,478 genes
1 µm

Plasmodium falciparum
(malaria parasite)

23 Mb
5,300 genes
1 µm

Escherichia coli
(bacteria)

4.6 Mb
4,377 genes
8,800×

differences. Genomes are more than instruction books for building and maintaining an organism; they also record the history of life. The growing number of fully sequenced genomes in all kingdoms is leading to a revolution in comparative evolutionary biology (figure 24.1). Genetic differences between species can be explored in a very direct way, examining the footprints on the evolutionary path between different species.

Evolutionary differences accumulate over long periods

Genomes of viruses and bacteria can evolve in a matter of days, whereas complex eukaryotic species evolve over millions of years. To illustrate this point, we compare four vertebrate genomes: human, tiger pufferfish *(Fugu rubripes)*, mouse *(Mus musculus)*, and chimpanzee *(Pan troglodytes)*.

Comparison between human and pufferfish genomes

The draft (preliminary) sequence of the tiger pufferfish was completed in 2002; it was only the second vertebrate genome to be sequenced. For the first time, we were able to compare the genomes of two vertebrates: humans and pufferfish. These two animals last shared a common ancestor 450 MYA.

Some human and pufferfish genes have been conserved during evolution, but others are unique to each species. About 25% of human genes have no counterparts in *Fugu*. Also, extensive genome rearrangements have occurred during the 450 million years since the mammal lineage and the teleost fish diverged, indicating a considerable scrambling of gene order. Finally, the human genome is 97% repetitive DNA (chapter 18), but repetitive DNA accounts for less than one-sixth of the *Fugu* sequence.

Comparison between human and mouse genomes

Later in 2002, a draft sequence of the mouse genome was completed by an international consortium of investigators, allowing for the first time a comparison of two mammalian genomes. In contrast to the human–pufferfish genome comparison, the differences between these two mammalian genomes are minuscule.

The human genome has about 400 million more nucleotides than that of the mouse. A comparison of the genomes reveals that both have about 25,000 genes, and that they share the bulk of them; in fact, the human genome shares 99% of its genes with mice. Humans and mice diverged about 75 MYA, approximately one-sixth of the amount of time that separates humans from pufferfish. There are only 300 genes unique to either human or mouse, constituting about 1% of the genome. Even 450 million years after last sharing a common ancestor, 75% of the genes found in humans have counterparts in pufferfish.

Comparison between human and chimpanzee genomes

Humans and chimpanzees, *Pan troglodytes,* diverged only about 4.1 MYA, leaving even less time for their genomes to accumulate mutational differences. The chimp genome was sequenced in 2005, providing a comparative window between us and our closest living relative. A 1.5% difference in insertions and deletions (indels) is found between chimps and humans. Comparing chimp and human indels to outgroups can allow a determination of whether the indel was ancestrally present, and thus can identify which species has the derived condition. Fifty-three of the potentially human-specific indels lead to loss-of-function changes that might correlate with some of the traits that distinguish us from chimps, including a larger cranium and lack of body hair. As will be discussed later in this chapter, mutations leading to differences in the patterns of gene expression are particularly important in understanding why chimps are chimps and humans are humans.

Comparisons of single-nucleotide substitutions reveal that only 2.7% of the two genomes have consistent differences in single nucleotides. Mutations in coding DNA are classified into two groups: those that alter the amino acids coded for in the sequence (nonsynonymous changes) and those that do not (synonymous changes). For example, a synonymous mutation that changed UUU to UUC would still code for phenylalanine (use the genetic code in table 15.1 to see if you can identify examples of possible synonymous and nonsynonymous changes).

With the addition of the sequenced orangutan *(Pongo pygmaeus)* genome, even more can be learned about the often subtle changes in genomes that distinguish members of the great apes. Like humans and chimps, orangutans are also susceptible to cardiovascular disease and diabetes. A three-way comparison can enhance our understanding of how these diseases evolved in the great apes.

 ## Genomes evolve at different rates

As noted earlier, the genomes of viruses and bacteria can evolve in a matter of days. Evolution of insect genomes occurs more rapidly than mammalian genomes, which evolve over millions of years. A short generation time may be responsible for these rate differences. However, even in some complex eukaryotic organisms, genome evolution can occur more rapidly.

Comparing the 30 plant genomes that have been sequenced with sequenced animal genomes yielded a surprising finding. Plant genomes change much more rapidly than animal genomes. This is especially evident in noncoding DNA, which tends to change more rapidly than protein-coding DNA. Many conserved, noncoding regions of DNA can be identified when a fish and a primate genome are aligned, even though the two diverged 400 MYA. But, very few similarities are found between the noncoding DNA of *Arabidopsis* and rice, *Oryza sativa,* although the two diverged only 200 MYA. The sequencing of the corn genome has revealed that two varieties can differ by as much as 20%. Contrast this with the differences between human and chimp genomes.

Genomic change in plants is so fast that geneticist Detlef Weigel has concluded that plants practice "genomic anarchy." What accounts for all the variation in plant genomes? Transposable elements and other mobile elements frequently remodel the genome. Different bits of different genes come together in new combinations, and, as a result, new regulatory elements are created.

Plant, fungal, and animal genomes have unique and shared genes

We now step back farther and consider genomic differences among the eukaryotic kingdoms that diverged long before the examples just discussed. You have already seen that many genes are highly conserved in animals. Plant genes are also highly conserved, but new families of genes have emerged at each stage of plant evolution.

Evolution of plant-specific genes

Sequenced plant genomes offer a window into a billion years of evolution. A total of 3814 gene families are shared by every plant, from the algae to grapes and corn. All the essentials for photosynthesis are found in the genome of the single-celled alga, *Chlamydomonas reinhardtii*, along with a gene that is now used to make flowers in the flowering plants. Although multicellularity was a huge evolutionary step, the multicellular alga *Volvox* has just about the same number of genes as *Chlamydomonas*. Lots of extra genes were not required for this major innovation.

The move onto land is represented by the genome of the moss *Physcomitrella patens* that arose 450 MYA. New genes that allowed the plant to survive in an arid environment arose, resulting in 3006 new gene families. Although moss are very different from flowering plants, making spores instead of seed, for example, they share 80% of the toolkit of developmental genes found in the flowering plant *Arabidoposis*. The shift from mosses to plants with internal transport systems required another 516 gene families. The transition to flowering plants, which now dominate land-based flora, required another 1350 gene families. It is perhaps not surprising that the biggest genomic change in the billion years of plant history was in the gene families required for colonizing land, a completely new environment.

Comparison of plants with animals and fungi

About one-third of the genes in *Arabidopsis* and rice appear to be in some sense "plant" genes—that is, genes not found in any animal or fungal genome sequenced so far. These include the many thousands of genes involved in photosynthesis and photosynthetic anatomy. Few plant genomes have been sequenced to date, however.

Among the remaining genes found in plants, many are very similar to those found in animal and fungal genomes, particularly the genes involved in basic intermediary metabolism, in genome replication and repair, and in transcription and protein synthesis. Prior to the availability of whole-genome sequences, assessment of the extent of genetic similarity and difference among diverse organisms had been difficult at best.

Learning Outcomes Review 24.1

Genomes vary in number of similar genes, arrangement of genes, total number of base-pairs, and base-pair differences. Closely related animals such as humans and chimps exhibit highly similar genomes, but greater variation is found in related plant genomes.

- *Would you expect a high degree of similarity between genomes of a bony fish, such as a swordfish, and a cartilaginous fish, such as a shark? What genes might be different?*

24.2 *Genome Size*

Learning Outcomes

1. *Differentiate between autopolyploidy and allopolyploidy.*
2. *Explain why most crosses between two species do not result in a new polyploid species.*
3. *Explain why the genome of a polyploid is not identical to the sum of the two parental genomes.*
4. *Explain why genome size and genome number do not correlate.*

Both genome size and gene number vary greatly among the eukaryotes species (see figure 24.1). One contributing factor is whole-genome duplication, which results in polyploidy, the focus of this section. Genome duplication alone, however, cannot explain why rice has 43,000 genes distributed among 430 Mb of DNA, but within the 3200 Mb of the human genome, only 25,000 genes are found. To more fully understand the relationship between gene number and genome size in eukaryotes, the noncoding DNA must also be considered.

 Data analysis Estimate and compare the number of genes per mega base-pair of genomic DNA in humans and rice. Which genomes in figure 24.1 are most like humans in terms of the ratio of genes to base-pairs?

Ancient and newly created polyploids guide studies of genome evolution

Polyploidy (three or more chromosome sets) can give rise to new species, as you learned in chapter 22. Polyploidy can result from either genome duplication in one species or from hybridization of two different species. Genome duplication is more prevalent in plants than in animals. In **autopolyploids**, the genome of one species is duplicated through a meiotic error, resulting in four copies of each chromosome. **Allopolyploids** result from the hybridization and subsequent duplication of the genomes of two different species (figure 24.2). The origins of wheat, illustrated in figure 24.3, involve two successive allopolyploid events.

Two avenues of research have lead to intriguing insights into genome alterations following polyploidization. The first approach studies ancient polyploids, called **paleopolyploids.** Sequence comparisons and phylogenetic tools establish the time and patterns of polyploidy events (figure 24.4 on page 478). Sequence divergence between homologues, as well as the presence or absence of duplicated gene pairs from the hybridization, can be used for historical reconstructions of genome evolution. All copies of duplicate gene pairs arising through polyploidy are not necessarily found thousands or millions of years after polyploidization. The loss of duplicate genes is considered later in this section.

The second approach is to create **synthetic polyploids** by crossing plants most closely related to the ancestral species and then chemically inducing chromosome doubling. Unless the hybrid genome becomes doubled, the plant will be sterile

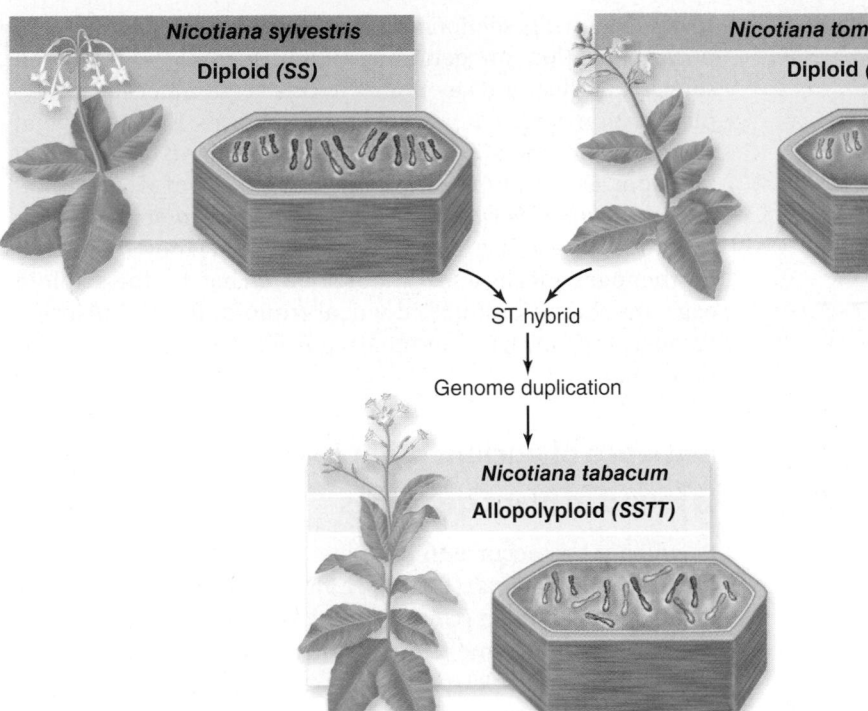

Figure 24.2 **Allopolyploidy** occurred in tobacco 5 MYA, but can be approximated by crossing the progenitor species and initiating a doubling of chromosomes, often through tissue culture followed by plant regeneration, which can lead to chromosome doubling. Tobacco species have many chromosomes, not all of which are visible in a single plane of a cell.

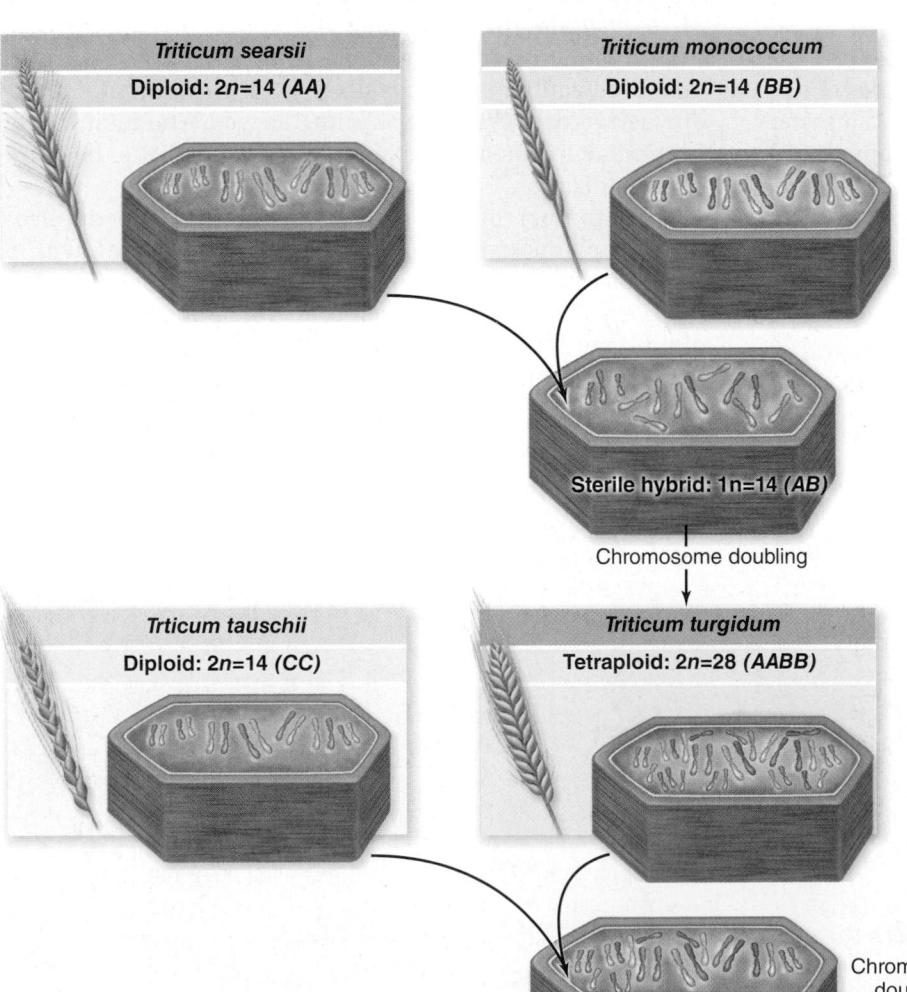

Figure 24.3 **Evolutionary history of wheat.** Domestic wheat arose in southwestern Asia in the hilly country of what is now Iraq. This region contains a rich assembly of grasses of the genus *Triticum*. Domestic wheat *(T. aestivum)* is a polyploid species of *Triticum* that arose through two so-called allopolyploid events. (1) Two different diploid genomes, symbolized here as *AA* and *BB*, hybridized to form an *AB* species hybrid; the species were so different that A and B chromosomes could not pair in meiosis, so the *AB* hybrid was sterile. However, in some plants the chromosome number spontaneously doubled due to a failure of chromosomes to separate in meiosis, producing a fertile tetraploid species, *AABB*. This wheat (durum semolina) is used in the production of pasta. (2) In a similar fashion, the tetraploid species *AABB* hybridized much more recently, within the last 10,000 years, with a different diploid species, with genome *CC*. The second hybridization produced, after another doubling event, the hexaploid *T. aestivum, AABBCC*. This bread wheat is one of the most important food plants in the world. Not all chromosomes appear in a single plane of any cell.

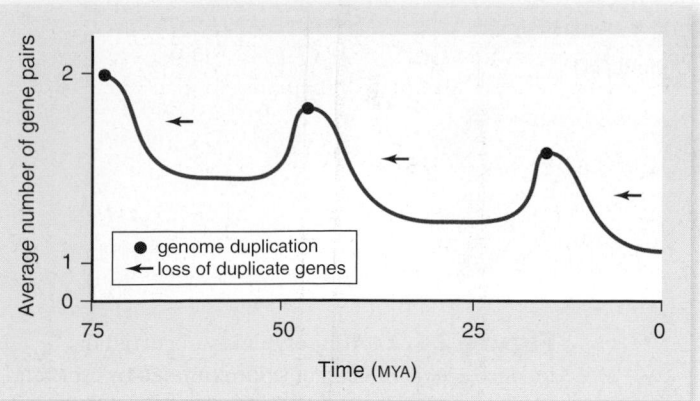

Figure 24.4 **Sequence comparisons of numerous genes in a polyploid genome tell us how long ago allopolyploidy or autopolyploidy events occurred.** Complex analyses of sequence divergence among duplicate gene pairs and presence or absence of duplicate gene pairs provide information about when both genome duplication and gene loss occurred. The graph reveals multiple polyploidy events over evolutionary time.

because it will lack homologous chromosomes that need to pair during metaphase I of meiosis.

Because meiosis requires an even number of chromosome sets, species with ploidy levels that are multiples of two can reproduce sexually. However, meiosis would be a disaster in a 3*n* organism, such as asexually propagated commercial bananas, since three sets of chromosomes can't be evenly divided between two cells. Triploid bananas are seedless. The aborted ovules appear as the little brown dots in the center of any cross section of a banana.

> **?** **Inquiry question** Sketch out what would happen in meiosis in a 3*n* banana cell, referring back to chapter 11 if necessary.

In the following sections we examine further the effect of polyploidization on genomes. Plant examples have been selected to illustrate key points in this section because polyploidization occurs more frequently in plants. The somewhat surprising findings, however, are not limited to the plant kingdom. For example, triploid populations of whiptail lizards *(Cnemidophorus tesselatus)* are found in southeastern Colorado. These lizards are all female, and they reproduce parthenogenetically (see chapter 52). That is, the triploid eggs are able to produce identical triploid, female offspring through mitosis, with no meiosis or fertilization needed for reproduction.

Evidence of ancient polyploidy is found in plant genomes

Polyploidy has occurred numerous times in the evolution of the flowering plants (figure 24.5). The legume clade that includes the soybean *(Glycine max),* the plant *Medicago truncatula* (a forage legume often used in research), and the garden pea *(Pisum sativum)* underwent a major polyploidization event 44–58 MYA and again 15 MYA.

A quick comparison of the genomes of soybean and *M. truncatula* reveal a huge difference in genome size (figure 24.6). In addition to increasing genome size through polyploidization, genomes like *M. truncatula* definitely downsized over evolutionary time as well. The total size of a genome cannot be explained solely on the basis of polyploidization.

A number of independent whole-genome duplications in plants cluster around 65 MYA, coinciding with a mass extinction event caused by a catastrophic change due to an asteroid impact or increased volcanic activity. Polyploids may have had a better chance of surviving the extinction event with increased genes and alleles available for selection.

Figure 24.5 **Polyploidy has occurred numerous times in the evolution of the flowering plants (the numbers indicate millions of years ago—MYA).**

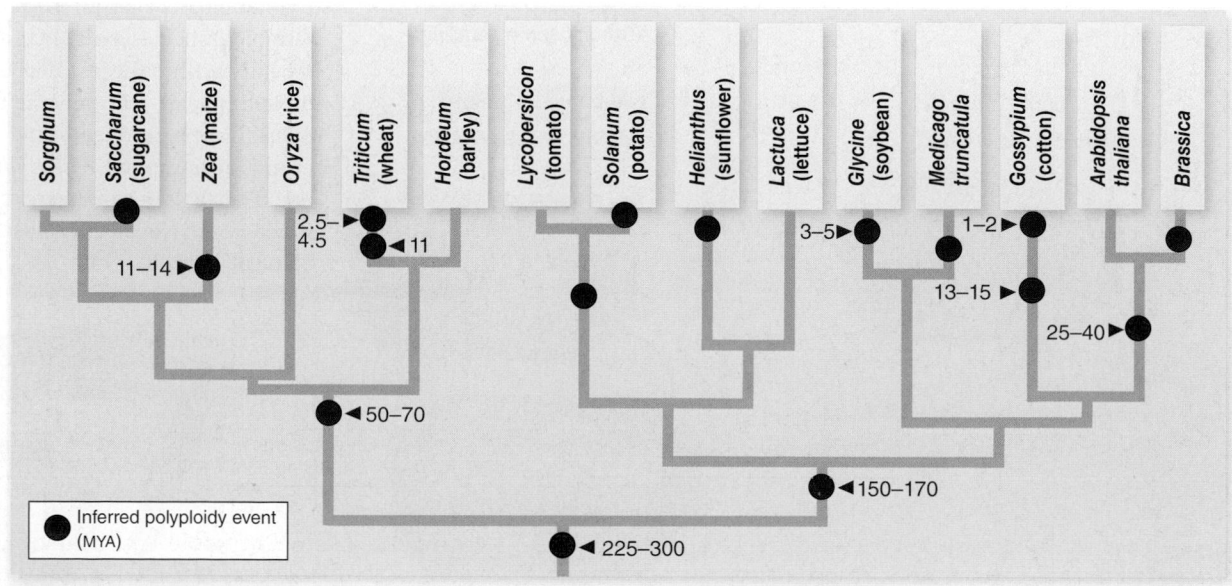

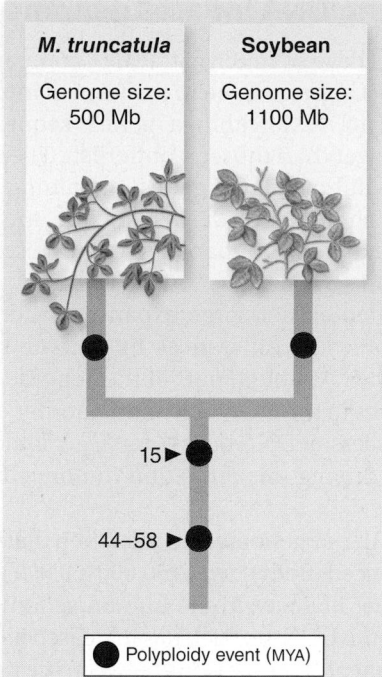

Figure 24.6 Genome downsizing. Genome downsizing must have occurred in *Medicago truncatula* (the numbers indicate millions of years ago—MYA).

Polyploidy induces elimination of duplicated genes

Formation of an allopolyploid from two different species is often followed by a rapid loss of genes (figure 24.7) or even whole chromosomes. In some polyploids, however, loss of one copy of many duplicated genes arises over a much longer period. In some species a great deal of gene loss occurs in the first few generations after polyploidization.

Modern tobacco, *Nicotiana tabacum,* arose from the hybridization and genome duplication of a cross between *Nicotiana sylvestris* (female parent) and *N. tomentosiformis* (male parent) (see figure 24.2). To complement the analysis based on a cross that occurred over 5 MYA, researchers constructed synthetic *N. tabacum* and observed the chromosome loss that followed. Curiously, the loss of chromosomes is not even. More *N. tomentosiformis* chromosomes were jettisoned than those of *N. sylvestris*. Similar unequal chromosome loss has been observed in synthetic wheat hybrids, in which 13% of the genome of one parent is lost in contrast to 0.5% of the other parental genome. A comparison of closely related *Arabidopsis thaliana* and *Arabidopsis lyrata* found that *A. lyrata* has 500 more genes than *A. thaliana*. In the 10,000 years since the two species diverged, hundreds of thousands of small regions of the *A. thaliana* genome have been deleted. It is possible that different rates of genome replication could explain the differential loss, as is true for synthetic human–mouse hybrid cultured cells.

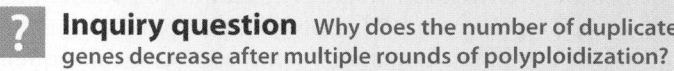

? Inquiry question Why does the number of duplicate genes decrease after multiple rounds of polyploidization?

Polyploidy can alter gene expression

A striking discovery is the change in gene expression that occurs in the early generations after polyploidization. Some of this may be connected to an increase in the methylation of

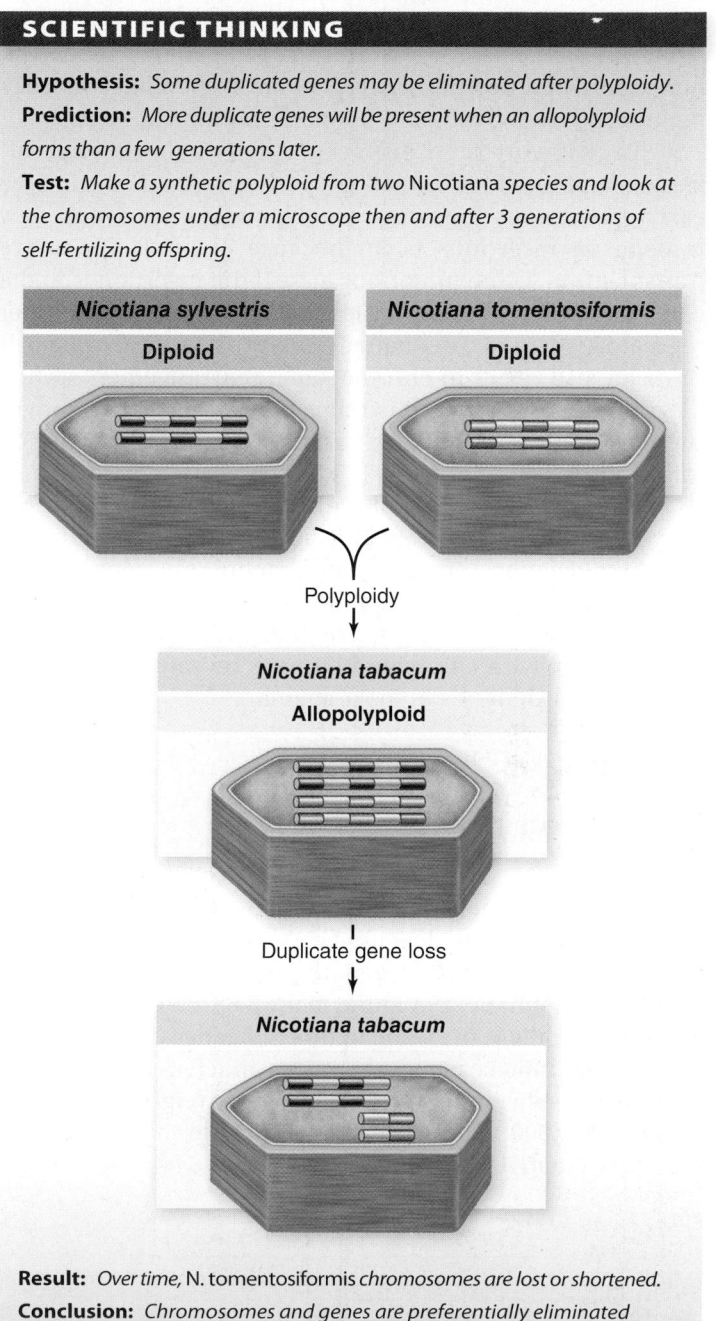

SCIENTIFIC THINKING

Hypothesis: *Some duplicated genes may be eliminated after polyploidy.*

Prediction: *More duplicate genes will be present when an allopolyploid forms than a few generations later.*

Test: *Make a synthetic polyploid from two* Nicotiana *species and look at the chromosomes under a microscope then and after 3 generations of self-fertilizing offspring.*

Nicotiana sylvestris — Diploid

Nicotiana tomentosiformis — Diploid

Polyploidy

Nicotiana tabacum — Allopolyploid

Duplicate gene loss

Nicotiana tabacum

Result: *Over time,* N. tomentosiformis *chromosomes are lost or shortened.*

Conclusion: *Chromosomes and genes are preferentially eliminated following polyploidy.*

Further Experiments: *Why might the chromosomes and genes of one species be preferentially eliminated? How could you test your explanation?*

Figure 24.7 Polyploidy may be followed by the unequal loss of duplicate genes from the combined genomes.

cytosines in the DNA. Methylated genes cannot be transcribed, as described in chapter 16. Simply put, polyploidization can lead to a short-term silencing of some genes. In subsequent generations, methylation decreases.

Transposons jump around following polyploidization

Barbara McClintock, in her Nobel Prize-winning work on transposable (mobile) genetic elements, referred to these jumping DNA regions as *controlling elements.* She hypothesized that transposons could respond to genome shock and jump into a new position in the genome. Depending on where the transposon moved, new phenotypes could emerge.

Recent work on transposon activity following hybridization supports McClintock's hypothesis. Again, during the early generations following a polyploidization event, new transposon insertions occur because of unusually active transposition. Gene mutations arise if the insertion is in the coding region of a gene. Insertions in promoter and other regulatory regions can change gene expression. Transposition can also result in chromosomal rearrangements when a portion of the genome is relocated. All of these changes provide additional genetic variation on which evolution might act.

The relative number of transposable elements in a species may influence the rate of genome evolution whether or not hybridization occurs. The orangutan's genome has evolved more slowly than either of its close relatives—humans and chimps. Comparisons of the recently sequenced orangutan genome with chimp and humans revealed many fewer transposable elements in the orangutan.

Polyploidy alone cannot account for variation in genome size

It is difficult to account for all genome size variation with polyploidy. Even with gene or chromosome elimination following hybridization, polyploidy is unlikely to result in such a range of ratios of gene number to genome size. Humans have nine times the amount of DNA found in the 3.65×10^8-bp pufferfish genome, but about the same number of genes. Plants have an even greater range of genome sizes. As much as a 200-fold difference has been found, yet all these plants weigh in with about 24,000 to 60,000 genes. Tulips, for example, have 170 times more DNA than *Arabidopsis.*

Noncoding DNA inflates genome size

Why do humans have so much extra DNA compared with pufferfish? Much of it appears to be in the form of introns, noncoding segments (ncDNA) within a gene's sequence, that are substantially bigger than those in pufferfish. The *Fugu* genome has only a handful of "giant" genes containing long introns; studying them should provide insight into the evolutionary forces that have driven the change in genome size during vertebrate evolution.

As described earlier, large expanses of retrotransposon DNA contribute to the differences in genome size from one species to another. Although part of the genome, ncDNA does not contain genes in the usual sense. As another example, *Drosophila* exhibits less ncDNA than *Anopheles,* although the evolutionary force driving this reduction in noncoding regions is unclear.

The 430-Mb rice genome and the 2-bbp maize genome are now fully sequenced. Both encode 50,000–60,000 genes, roughly twice the number of genes in the human genome. Maize contains lots of repetitive DNA, which has increased its DNA content, but not gene content. Comparisons between the rice, maize, and wheat genomes should provide clues about the genome of their common ancestor and the dynamic evolutionary balance between opposing forces that increase genome size (polyploidy, transposable element proliferation, and gene duplication) and those that decrease genome size (mutational loss).

Learning Outcomes Review 24.2

Autopolyploids arise from duplication of a species' genome; allopolyploids occur from hybridization between two species. Unless the number of chromosomes doubles, an allopolyploid will likely be sterile because of lack of homologous pairs at meiosis. Polyploidization can lead to major changes in genome structure, including loss of genes, alteration of gene expression, increased transposon hopping, and chromosomal rearrangements. Polyploidy is considered important in the generation of biodiversity and adaptation, especially in plants, but it cannot explain why increases or decreases in genome size do not correlate with the number of genes. Evidently DNA content is not the same as gene content. Polyploidy in plants does not by itself explain differences in genome size. Often a greater amount of DNA is explained by the presence of introns and non-protein-coding sequences than by gene duplicates.

■ *How might a genome with a small number of genes and a small number of total base-pairs evolve into a genome with the same small number of genes and a thousand-fold larger genome?*

Learning Outcomes

1. *Define the terms segmental duplication, genome rearrangement, and pseudogene.*
2. *Explain why horizontal gene transfer can complicate evolutionary hypotheses.*

Genome evolution can occur at the level of whole-genome duplication as evidenced by polyploidy organisms, but within genomes, evolutionary changes are seen at all levels, from individual genes to whole chromosomes and including duplications of portions of genomes. Duplications provide opportunities for genes with the same function to diverge because a "backup pair" of genes is in place. After duplication, one gene can lose function, because the duplicate compensates for it. The mutated gene is not selected against because another functioning gene, not just an allele, exists. The duplicated gene can also diverge and acquire new functions because a backup copy exists.

Individual chromosomes may be duplicated

Aneuploidy refers to the duplication or loss of an individual chromosome rather than of an entire genome. Failure of a pair of homologous chromosomes or sister chromatids to separate during meiosis is the most common way that aneuploidy occurs. In general, plants are better able to tolerate aneuploidy than animals, but the explanation for this difference is elusive.

DNA segments may be duplicated

One of the greatest sources of novel traits in genomes is duplication of segments of DNA. Two genes within an organism that have arisen from the duplication of a single gene in an ancestor are called paralogues. In contrast, orthologues reflect the inheritance of a single gene from a common ancestor—a conserved gene.

When a gene is duplicated, the most likely fates of the duplicate gene are (1) losing function through subsequent mutation (becomes a pseudogene), (2) gaining a novel function through subsequent mutation (neofunctionalization), and (3) having the total function of the ancestral gene partitioned into the two duplicates (subfunctionalization). Gene families grow through gene duplication. In reality, however, most duplicate genes lose function.

How then can researchers claim that gene duplication is a major evolutionary force for gene innovation, that is, genes gaining new function? One piece of the answer can be found by noting where gene duplication is most likely to occur in the genome. In humans, the highest rates of duplication have occurred in the three most gene-rich chromosomes of the genome. The seven chromosomes with the fewest genes also show the least amount of duplication. (Remember, having fewer genes does not mean that there is less total DNA.)

Even more compelling, certain types of human genes appear to be more likely to be duplicated: growth and development genes, immune system genes, and cell-surface receptors. About 5% of the human genome consists of segmental duplications (figure 24.8). Finally, and importantly, gene duplication is thought to be a major evolutionary force for gene innovation because the duplicated genes are found to have different patterns of gene expression (see chapter 25 for examples). For example, the two duplicated copies may be expressed in different or overlapping sets of tissues or organs during development.

Segmental duplications may account for differences between human and chimp. Species-specific segmental duplications tend to contain genes that are differentially expressed between the species. This is true for genes expressed in liver, kidney, or heart. A simple explanation is that there are more copies of the gene being expressed, but experimental evidence shows that it is more likely that the regulatory DNA near the duplicate is different.

Both rice and *Arabidopsis* have higher *copy numbers* for gene families (multiple slightly divergent copies of a gene) than are seen in animals or fungi, suggesting that these plants may have undergone numerous episodes of segmental duplication during the 150–200 million years since rice and *Arabidopsis* diverged from a common ancestor. As whole-genome duplications have also occurred since they diverged, additional analysis will be needed to determine the relative effects of polyploidy and segmental duplication in plant evolution.

Figure 24.8 Segmental duplication on the human Y chromosome. Each orange region has 98% sequence similarity with a sequence on a different human chromosome. Each dark blue region has 98% sequence similarity with a sequence elsewhere on the Y chromosome.

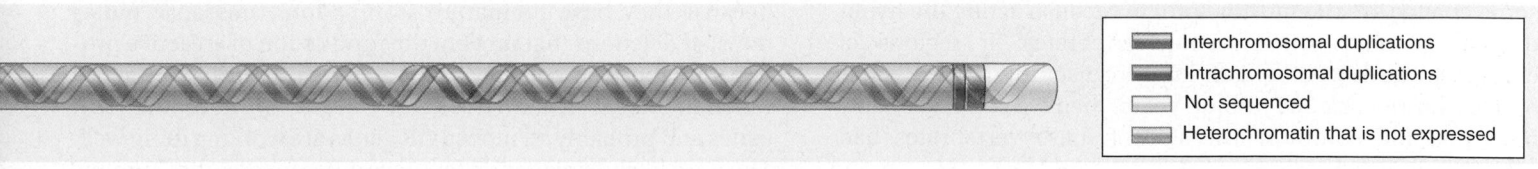

- Interchromosomal duplications
- Intrachromosomal duplications
- Not sequenced
- Heterochromatin that is not expressed

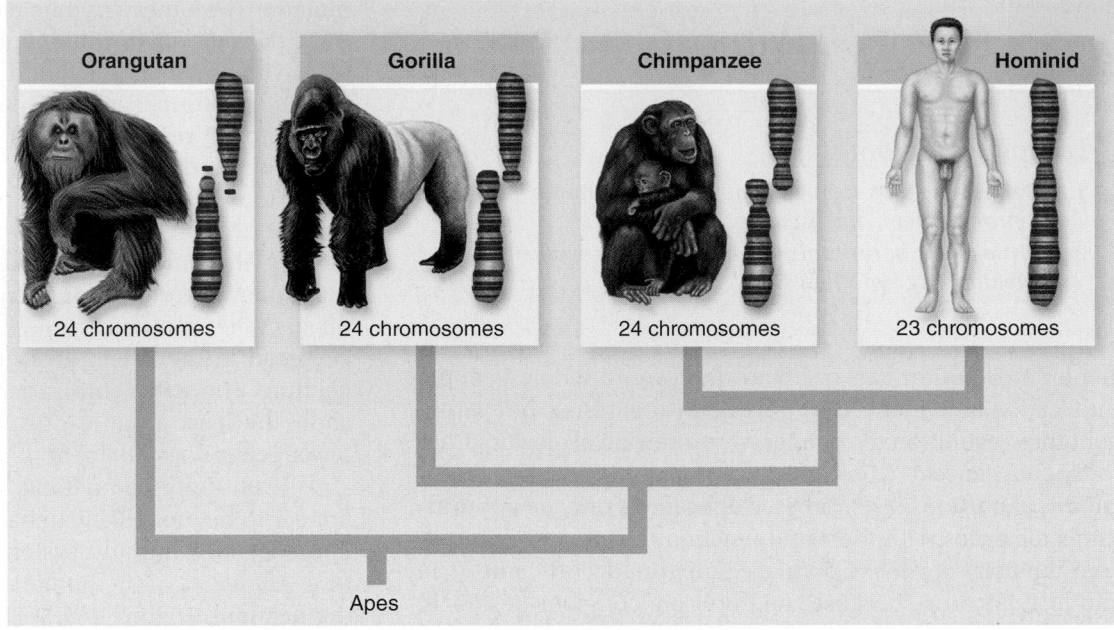

Figure 24.9 Living great apes. All living great apes, with the exception of humans, have a haploid chromosome number of 24. Humans have not lost a chromosome; rather, two smaller chromosomes fused to make a single chromosome.

Genomes may become rearranged

Humans have one fewer chromosome than chimpanzees, gorillas, and orangutans (figure 24.9). It's not that we have lost a chromosome. Rather, at some point in time, two midsized ape chromosomes fused to make what is now human chromosome 2, the second largest chromosome in our genome.

The fusion leading to human chromosome 2 is an example of the sort of genome reorganization that has occurred in many species. Rearrangements like this can provide evolutionary clues, but they are not always definitive proof of how closely related two species are.

Consider the organization of known orthologues shared by humans, chickens, and mice. One study estimated that 72 chromosome rearrangements had occurred since the chicken and human last shared a common ancestor. This number is substantially less than the estimated 128 rearrangements between chicken and mouse, or 171 between mouse and human.

This does not mean that chickens and humans are more closely related than mice and humans or mice and chickens. What these data actually show is that chromosome rearrangements have occurred at a much lower frequency in the lineages that led to humans and to chickens than in the lineages leading to mice. Chromosomal rearrangements in mouse ancestors seem to have occurred at twice the rate seen in the human line. These different rates of change help counter the notion that humans existed hundreds of millions of years ago.

Genomes that have undergone relatively slow chromosome change are the most helpful in reconstructing the hypothetical genomes of ancestral vertebrates. If regions of chromosomes have changed little in distantly related vertebrates over the last 300 million years, then we can reasonably infer that the common ancestor of these vertebrates had genomic similarities.

Variation in the organization of genomes is as intriguing as gene sequence differences. Chromosome rearrangements are common, yet over long segments of chromosomes, the linear order of mouse and human genes is the same—the common ancestral sequence has been preserved in both species. This **conservation of synteny** (see chapter 18) was anticipated from earlier gene-mapping studies, and it provides strong evidence that evolution actively shapes the organization of the eukaryotic genome. As seen in figure 24.10, the conservation of synteny allows researchers to more readily locate a gene in a different species using information about synteny, thus underscoring the power of a comparative genomic approach.

Gene inactivation results in pseudogenes

The loss of gene function is another important way genomes evolve. Consider the olfactory receptor (OR) genes that are responsible for our sense of smell. These genes code for receptors that bind odorants, initiating a cascade of signaling events that eventually lead to our perception of scents.

Gene inactivation seems to be the best explanation for our reduced sense of smell relative to that of the great apes and other mammals. Mice have the largest mammalian, OR gene family, with about 1500 OR genes—about 50% more OR genes than humans. Only 20% of the mouse OR genes are pseudogenes, in contrast to about 60% in humans, which are inactive **pseudogenes** (sequences of DNA that are very similar to functional genes but that do not produce a functional product because they have premature stop codons, missense mutations, or deletions that prevent the production of an active protein). In contrast, half the chimpanzee and gorilla OR genes function effectively, and over 95% of New World monkey OR genes and probably all mouse OR genes are working quite well. The most likely explanation for these differences is that humans

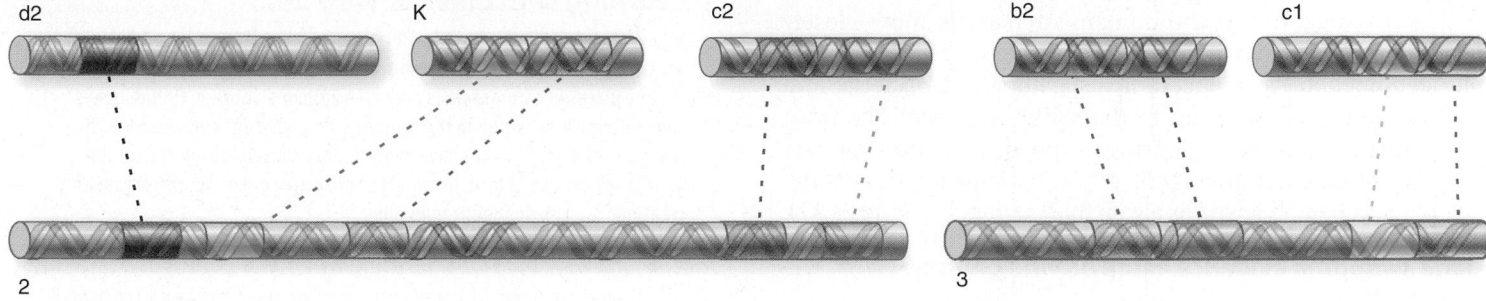

d2   K   c2   b2   c1

2   3

M. truncatula

Figure 24.10 **Synteny and gene identification.** Genes sequenced in the model legume, *Medicago truncatula,* can be used to identify homologous genes in soybean, *Glycine max,* because large regions of the genomes are syntenic as illustrated for some of the linkage groups (chromosomes) of the two species. Regions of the same color represent homologous genes.

came to rely on other senses, reducing the selection pressure against loss of OR gene function by random mutation.

An older question about the possibility of positive selection for OR genes in chimps was resolved with the completion of the chimp genome. A careful analysis indicated that both humans and chimps are gradually losing OR genes to pseudogenes and that there is no evidence to support positive selection for any of the OR genes in the chimp.

Rearranged DNA can acquire new functions

Errors in meiosis that rearrange parts of genes most often create pseudogenes, but occasionally a broken piece of a gene can end up in a new spot in the genome where it acquires a new function. One of the most intriguing examples occurred within a family of fish in the suborder Notothenioidae found in the Antarctic ocean. These fish are called icefish because they survive the frigid temperatures in the Antarctic, in part because of a protein in their blood that works like antifreeze. Reconstructing evolutionary history using comparative genomics reveals that 9-bp fragment of a gene coding for a digestive enzyme moved to a new location where it evolved to encode part of an antifreeze protein. The series of errors that gave rise to the new protein persisted only because the change coincided with a massive cooling of the Antarctic waters. Natural selection acted on this mutation over millions of years.

 ## Noncoding DNA can acquire regulatory functions

So far, we have primarily compared genes that code for proteins. As more genomes are sequenced, we learn that much of the genome is composed of ncDNA. The repetitive DNA is often retrotransposon DNA, contributing to as much as 30% of animal genomes and 40–80% of plant genomes. (Refer to chapter 18 for more information on repetitive DNA in genomes.) Yet there are conserved noncoding regions (CNCs) that evolve

much more slowly than expected, assuming there was no associated function. An analysis of CNCs in 40 vertebrates led to the conclusion that as the rate of base-pair substitution slowed down over evolutionary time, these CNCs had begun to regulate the expression of transcription factors, genes that specifically influence development, and genes that encode receptor-binding proteins that are important in signaling.

Horizontal gene transfer complicates matters

Evolutionary biologists build phylogenies on the assumption that genes are passed from generation to generation, a process called **vertical gene transfer (VGT).** Hitchhiking genes from other species, a process referred to as **horizontal gene transfer (HGT)** and sometimes called *lateral gene transfer,* can lead to phylogenetic complexity. HGT was likely most prevalent very early in the history of life, when the boundaries between individual cells and species seem to have been less firm than they are now and DNA more readily moved among different organisms. Earlier in the history of life, gene swapping between species was rampant. Today HGT continues in prokaryotes and eukaryotes, but less frequently. An intriguing example of more recent HGT between moss and a flowering plant is described in chapter 31.

Gene swapping in early lineages

The extensive gene swapping among early organisms has caused many researchers to reexamine the base of the tree of life. Early phylogenies based on ribosomal RNA (rRNA) sequences indicate that an early prokaryote gave rise to two major domains: the Bacteria and the Archaea. From one of these lineages, the domain Eukarya emerged; its organelles originated as unicellular organisms engulfed specialized prokaryotes.

This rRNA phylogeny is being revised as more microbial genomes are sequenced. By 2011, the Microbial Genome Program of the U.S. Department of Energy had sequenced more than 541 microbial genomes and microbial communities. With new sequencing technology, microbial genomes can be sequenced in less than a day. Phylogenies built with rRNA sequences suggest that the domain Archaea is more closely related to the Eukarya than to the Bacteria. But as more microbial genomes are sequenced, investigators find bacterial and archaeal genes showing up in the same organism! The most likely conclusion is that organisms swapped genes, even absorbing DNA obtained from a food source. Perhaps the base of the tree of life is better viewed as a web than a branch (figure 24.11).

Gene swapping evidence in the human genome

Let's move closer to home and look at the human genome, which is riddled with foreign DNA, often in the form of transposons. The many transposons of the human genome provide a paleontological record over several hundred million years.

Comparisons of versions of a transposon that has duplicated many times allow researchers to construct a "family tree" to identify the ancestral form of the transposon. The percent of sequence divergence found in duplicates allows an estimate of the time at which that particular transposon originally invaded the human genome. In humans, most of the DNA hitchhiking seems to have occurred millions of years ago in very distant ancestor genomes.

Our genome carries many more ancient transposons than do the genomes of *Drosophila, C. elegans,* and *Arabidopsis.* One explanation for the observed low level of transposons in *Drosophila* is that fruit flies somehow eliminate unnecessary DNA from their genome 75 times faster than humans do. Our genome has simply hung on to hitchhiking DNA more often.

The human genome has had minimal transposon activity in the past 50 million years; mice, by contrast, are continuing to acquire new transposable elements. This difference may explain in part the more rapid change in chromosome organization in mice than in humans.

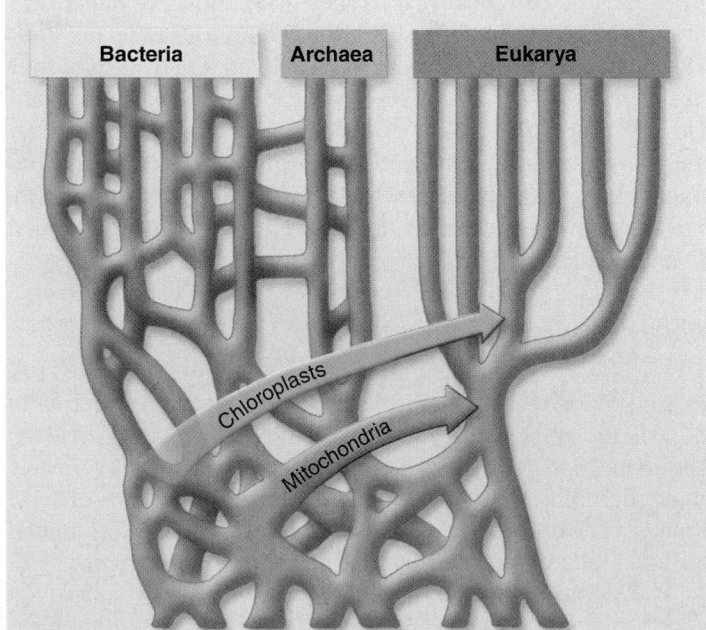

Figure 24.11 Horizontal gene transfer. Early in the history of life, organisms may have freely exchanged genes beyond multiple endosymbiotic events. To a lesser extent, this transfer continues today. The tree of life may be more like a web or a net.

Learning Outcomes Review 24.3

In segmental duplication, part of a chromosome and the genes it contains are duplicated. In genome rearrangement, segments of chromosomes may change places or chromosomes may fuse with one another. Pseudogenes have become inactivated in the course of evolution but still persist in the genome. All these changes have evolutionary consequences. Horizontal gene transfer has led to an unexpected mixing of genes among organisms, creating many phylogenetic questions.

■ *How would you determine whether a gene was a pseudogene or an example of horizontal gene transfer?*

24.4 Gene Function and Expression Patterns

Learning Outcomes

1. *Explain how species with nearly identical genes can look very different.*
2. *Describe the action of the* FOXP2 *gene across species.*

Gene function can be inferred by comparing genes in different species. You saw earlier that the function of 1000 human genes was understood once the mouse genome was sequenced. One of the major puzzles arising from comparative genomics is that organisms with very different forms can share so many conserved genes in their genomic toolkit.

The best explanation for why a mouse develops into a mouse and not a human is that the same or similar genes are expressed at different times, in different tissues, and in different amounts and combinations. For example, the cystic fibrosis gene (cystic fibrosis transmembrane conductance regulator, *CFTR*), which has been identified in both species and affects a chloride ion channel, illustrates this point. Defects in the human *CFTR* gene cause especially devastating effects in the lungs, but mice with the mutant *CFTR* gene do not have lung symptoms. Possibly variations in expression of *CFTR* between mouse and human explain the difference in lung symptoms when *CFTR* is defective.

Chimp and human gene transcription patterns differ

Humans and chimps diverged from a common ancestor only about 4.1 MYA—too little time for much genetic differentiation to evolve, but enough for significant morphological and behavioral differences to have developed. Sequence comparisons indicate that chimp DNA is 98.7% identical to human DNA. If just the gene sequences encoding proteins are considered, the similarity increases to 99.2%. How could two species differ so much in body and behavior, and yet have almost equivalent sets of genes?

One potential answer to this question is based on the observation that chimp and human genomes show very different patterns of gene transcription activity, at least in brain cells. Investigators used microarrays containing up to 18,000 human genes to analyze RNA isolated from cells in the fluid extracted from several regions of living brains of chimps and humans. (See figure 18.12 for a summary of this technique.) The RNA was linked with a fluorescent tag and then incubated with the microarray under conditions that allow the formation of DNA–RNA hybrids if the sequences are complementary. If the transcript of a particular gene is present in the cells, then the microarray spot corresponding to that gene lights up under UV light. The more copies of the RNA, the more intense the signal.

Because the chimp genome is so similar to that of humans, the microarray detects the activity of chimp genes reasonably well. Although the same genes were transcribed in chimp and human brain cells, the patterns and levels of transcription varied. Much of the difference between human and chimp brains lies in which genes are transcribed and when and where that transcription occurs.

? Inquiry question You are given a microarray of ape genes and RNA from both human and ape brain cells. Using the experimental technique described for comparison of humans and chimps, what would you expect to find in terms of genes being transcribed? What about levels of transcription?

Posttranscriptional differences may also play a role in building distinct organisms from similar genomes. As research continues to push the frontiers of proteomics and functional genomics, a more detailed picture of the subtle differences in the developmental and physiological processes of closely related species will be revealed. The integration of development and genome evolution is explored in depth in the following chapter.

Speech is uniquely human: An example of complex expression

Development of human culture is closely tied to the capacity to control the larynx and mouth to produce speech. Humans with a single point mutation in the transcription factor gene *FOXP2* have impaired speech and grammar but not impaired language comprehension.

The *FOXP2* gene is also found in chimpanzees, gorillas, orangutans, rhesus macaques, and even the mouse, yet none of these mammals speak (figure 24.12). The gene is expressed in areas of the brain that affect motor function, including the complex coordination needed to create words.

FOXP2 protein in mice and humans differs by only three amino acids. There is only a single amino acid difference between mouse and chimp, gorilla, and rhesus macaque, which all have identical amino acid sequences for FOXP2. Two more amino acid differences exist between humans and the sequence shared by chimp, gorilla, and macaque. The difference of only two amino acids between human and other primate FOXP2 may be linked to language acquisition in humans. Evidence points to strong selective pressure for the two *FOXP2* mutations that allow brain, larynx, and mouth to coordinate to produce speech.

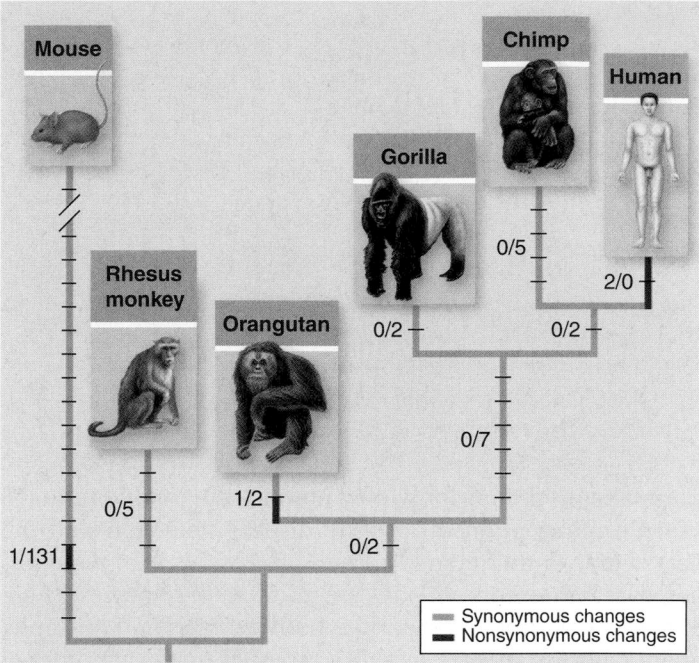

Figure 24.12 Evolution of *FOXP2*. Comparisons of synonymous and nonsynonymous changes in mouse and primate *FOXP2* genes indicate that changing two amino acids in the gene corresponds to the emergence of human language. Beige bars represent synonymous changes and brown bars represent nonsynonymous changes.

Is it possible that two amino acid changes lead to speech, language, and ultimately human culture? This box of mysteries will take a long time to unpack, but hints indicate that the changes are linked to signaling and gene expression. The two altered amino acids may change the ability of FOXP2 transcription factor to be phosphorylated. One way signaling pathways operate is through activation or inactivation of an existing transcription factor by phosphorylation.

Comparative genomics efforts are now extending beyond primates. A role for *FOXP2* in songbird singing and vocal learning has been proposed. Mice communicate via squeaks, with lost young mice emitting high-pitched squeaks. *FOXP2* mutations leave mice squeakless and transgenic baby mice with the human *FOXP2* squeak differently. For both mice and songbirds, it is a stretch to claim that *FOXP2* is a language gene—but it likely is needed in the neuromuscular pathway to make sounds.

Learning Outcomes Review 24.4

To understand functional differences between genes shared by species, one must look beyond sequence similarity. Alterations in the time and place of gene expression can lead to marked differences in phenotype. As an example, the FOXP2 factor appears to be involved in sound production in mice, chimps, gorillas, macaques, and humans, and only very small differences may have led to human speech.

■ *How can a single-nucleotide difference in a gene lead to a noticeably different phenotype? Give examples.*

24.5 Applying Comparative Genomics

Learning Outcomes

1. Describe how comparative genomics can reveal the genetic basis for disease.
2. Explain how genome comparisons between a pathogen and its host can aid drug development.
3. Describe how genome comparisons can be useful when working with endangered species.

Comparisons among individual human genomes continue to provide information on genetic disease detection and the best course of treatment. An even broader array of possibilities arises when comparisons are made among species. There are advantages to comparing both closely and distantly related pairs of species, as well as comparing the genomes of a pathogen and its host. Genome-level comparisons are also assisting conservation biologists in developing breeding programs. Examples of the benefits of each type of genome comparison follow.

Distantly related genomes offer clues for causes of disease

Sequences that are conserved between humans and pufferfish provide valuable clues for understanding the genetic basis of many human diseases. Amino acids critical to protein function tend to be preserved over the course of evolution, and changes at such sites within genes are more likely to cause disease.

It is difficult to distinguish functionally conserved sites when comparing human proteins with those of other mammals because not enough time has elapsed for sufficient changes to accumulate at nonconserved sites. A promising exception is the duck-billed platypus *(Ornithorhynchus anatinus),* which diverged from other mammals about 166 MYA and whose genome provides clues to the evolution of the immune system. Because the pufferfish genome is only distantly related to humans, conserved sequences are far more easily distinguished than even in the platypus.

Closely related organisms enhance medical research

Because studies in humans must be strictly regulated for ethical reasons, it is much easier to design experiments to identify gene function in an experimental system like that of the mouse. Comparing mouse and human genomes quickly revealed the function of 1000 previously unidentified human genes. The effects of these genes can be studied in mice, and the results can be used in potential treatments for human diseases.

A draft of the rat genome has been completed, and even more exciting news about the evolution of mammalian genomes may emerge from comparisons of these species. One of the most exciting aspects of comparing rat and mice genomes is the potential to capitalize on the extensive research on rat physiology, especially heart disease, and the long history of genetics in mice. Linking genes to disease has become much easier.

Humans have been found to contain segmental duplications that are absent in the chimp. Some of these duplications correspond to human disease. These differences can aid medical researchers in developing treatments for genetic disease. For example, some of the regions duplicated only in humans correspond to regions of the human genome that have been implicated in Prader–Willi syndrome and spinal muscular atrophy. Children with Prader–Willi lack muscle tone and struggle with life-threatening obesity because of an insatiable appetite. Spinal muscular atrophy affects all muscles in the body, but especially those nearest the trunk. Affected individuals can have difficulty swallowing and breathing, as well as experiencing weakness in the legs.

Pathogen–host genome differences reveal drug targets

With genome sequences in hand, pharmaceutical researchers are more likely to find suitable drug targets to eliminate pathogens without harming the host. Diseases—including malaria and Chagas disease—in many developing countries have both human and insect hosts. Both these infections are caused by protists (see chapter 29), and the value of comparative genomics in drug discovery to treat them is illustrated in the following sections.

Malaria

Anopheles gambiae, the malaria-carrying mosquito, along with *Plasmodium falciparum,* the protistan parasite it transmits, together have an enormous effect on human health, resulting in 1.7–2.5 million deaths each year from malaria. The genomes of both *Anopheles* and *Plasmodium* were sequenced in 2002.

P. falciparum, which causes malaria, has a relatively small genome of 2.46×10^7 bp that proved very difficult to sequence. It has an unusually high proportion of adenine and thymine, making it hard to distinguish one portion of the genome from the next. The project took five years to complete. *P. falciparum* appears to have about 5300 genes, with those of related function clustered together, suggesting that they might share the same regulatory DNA.

P. falciparum is a particularly crafty organism that hides from our immune system inside red blood cells, regularly changing the proteins it presents on the surface of the red blood cell. This "cloaking" has made developing a vaccine or other treatment for malaria particularly difficult.

Recently, a link to chloroplast-like structures in *P. falciparum* has raised other possibilities for treatment. An odd

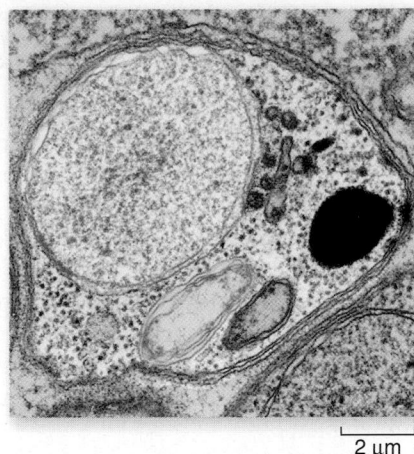

Figure 24.13 *Plasmodium* apicoplast. Drugs targeting enzymes used for fatty acid biosynthesis within *Plasmodium* apicoplasts (colored dark green) offer hope for treating malaria.

subcellular component called the *apicoplast,* found only in *Plasmodium* and its relatives, appears to be derived from a chloroplast appropriated from algae engulfed by the parasite's ancestor (figure 24.13).

Analysis of the *Plasmodium* genome reveals that about 12% of all the parasite's proteins, encoded by the nuclear genome, head for the apicoplast. These proteins act there to produce fatty acids. The apicoplast is the only location in which the parasite makes the fatty acids, suggesting that drugs targeted at this biochemical pathway might be very effective against malaria.

Another disease-prevention possibility is to look at chloroplast-specific herbicides, which might kill *Plasmodium* by targeting the chloroplast-derived apicoplast. Coupled with a newer vaccine, this treatment could substantially reduce the incidence of malaria.

Chagas disease

Trypanosoma cruzi, an insect-borne protozoan, kills about 21,000 people in Central and South America each year. As many as 18 million suffer from this infection, called Chagas disease, the symptoms of which include damage to the heart and other internal organs. Genome sequencing of *T. cruzi* was completed in 2005.

A surprising and hopeful finding is that a common core of 6200 genes is shared among *T. cruzi* and two other insect-borne pathogens: *Trypanosoma brucei* and *Leishmania major.* *T. brucei* causes African sleeping sickness, and *L. major* infections result in lesions of the skin of the limbs and face. These core genes are being considered as possible targets for drug treatments.

Currently, no effective vaccines and only a few drugs with limited effectiveness are available to treat any of these diseases. The genomic similarities may aid not only in targeting drug development, but perhaps also result in a treatment or vaccine that is effective against all three devastating illnesses (figure 24.14).

Genome comparisons inform conservation biology

Genomes of species on the brink of extinction are being mined for information that could contribute to disease reduction and conservation efforts. Here we consider the application of genomics to three endangered species, the Tasmanian devil *(Sarcophilus harrisii),* the giant panda *(Ailuropoda melanoleura),* and the polar bear *(Ursus maritimus).*

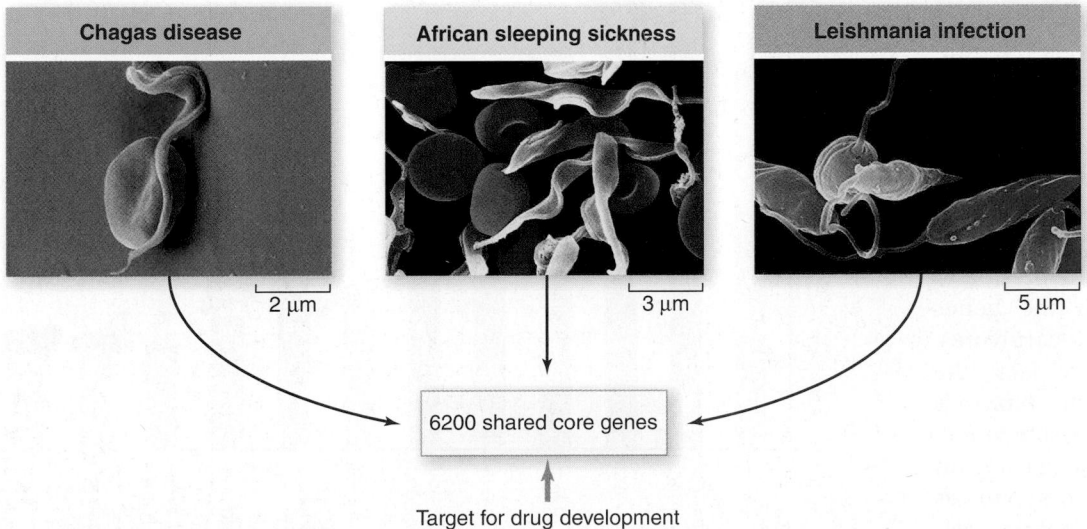

Figure 24.14 Comparative genomics may aid in drug development. The organisms that cause Chagas disease, African sleeping sickness, and leishmaniasis, which claim millions of lives in developing nations each year, share 6200 core genes. Drug development targeted at proteins encoded by the shared core genes could yield a single treatment for all three diseases.

Figure 24.15 DFTD is a fatal condition for Tasmanian devils. Tumors interfere with eating, and devils die within months of the first appearance of lesions.

An unexpected finding arose when the mitochondrial genomes of modern and fossil polar and brown bears (including DNA from the jaw of a 110,000 to 130,000-year-old polar bear) were sequenced. Polar bears evolved about 150,000 years ago. Geneticists were shocked to find that the entire maternal line of polar bears can be traced back to a brown bear living in Ireland between 20,000 and 50,000 years ago. Despite the close relation to brown bears and the ability to interbreed, this finding does not offer much hope for the polar bears facing extinction as warmer temperatures melt their sea ice homes. Average temperatures for the next 50 years are predicted to be much warmer than anything polar bears have experienced in their 150,000 years on Earth or the last 20,000 or so years since they mated with brown bears.

The Tasmanian devil, a marsupial on the Australian island of Tasmania, faces decimation from devil facial tumor disease (DFTD, figure 24.15). Close to 90% of the population is affected, and, since 1996, 60% of the devils have been wiped out due to DFTD. A comparison of genomes from two devils, named Cedric and Spirit, from distant corners of Tasmania revealed extremely low genetic diversity that can be traced back 100 years before the DFTD outbreak. Only the now extinct Tasmanian tiger had less genetic diversity (figure 24.16). Breeding programs can use the genomic information to preserve what diversity there is in the population.

Sequencing of the genome of the giant panda offered more promising news about population diversity (see figure 24.16). Almost three times as many SNPs (2.7 million) were identified in the panda than for the devil. Bamboo is the mainstay of a panda diet, and habitat destruction is a factor in the decline of pandas. Curiously, the panda's relatives are carnivores, and pandas maintain the genes associated with carnivores. Furthermore, they lack the genes coding for enzymes needed to fully digest bamboo. Attention is now focused on the microbes in the panda's gut that digest bamboo. Collectively, this information may assist conservationists in keeping the population from dipping below its current level of 2500–3000 individuals.

Learning Outcomes Review 24.5

DNA sequences conserved over evolutionary time tend to be those critical for protein function and survival. Variations in these conserved sequences may provide clues to diseases with a hereditary component. Knowledge of a pathogenic organism's genome and its differences from a host's genome may allow targeting of drugs and vaccines that affect the invader but leave the host unharmed. Comparative genomics within species can provide information of population diversity of endangered species. Conservation biologists can use this information to develop breeding strategies.

■ *A pathogen makes a critical protein that differs from the human version by only seven amino acids. What approaches might lead to an effective drug against the pathogen? What drawbacks might be encountered?*

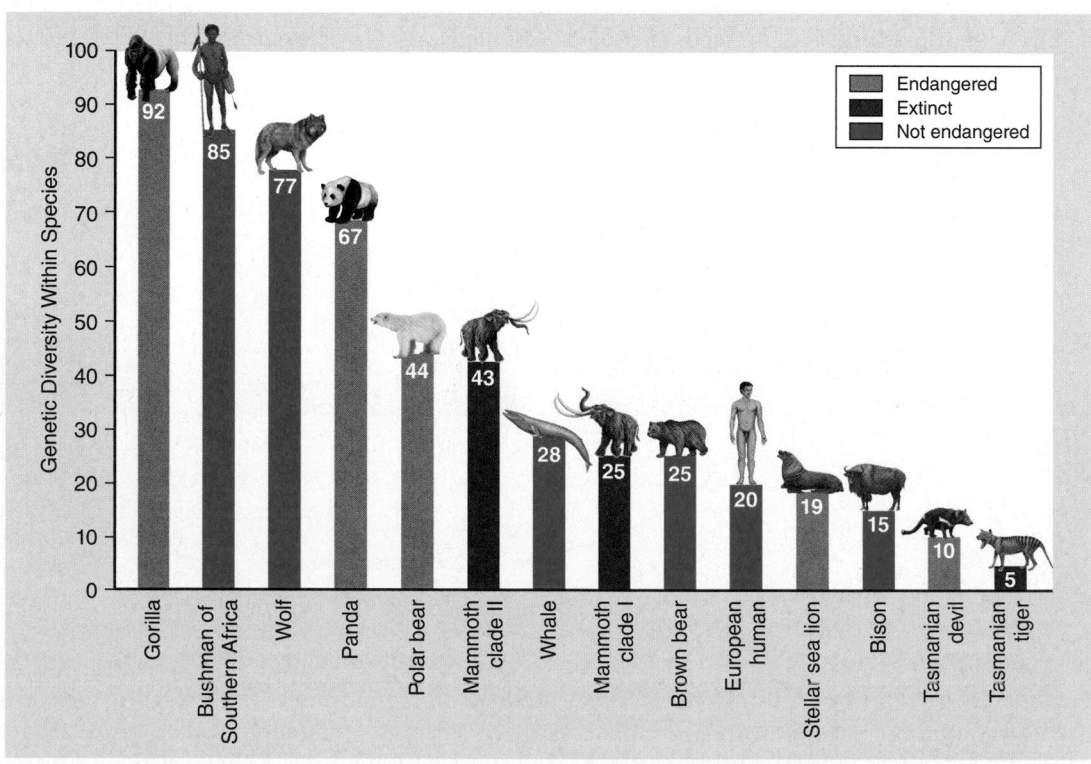

Figure 24.16 Genetic diversity of mitochondrial genomes based on average number of differences between pairs of individuals in the populations indicated.

24.1 Comparative Genomics

Evolutionary differences accumulate over long periods.

Even distantly related species may often have many genes in common. Changes in DNA codons that do not alter the amino acid specified are termed synonymous changes.

Genomes evolve at different rates.

Viral, bacterial, and even insect genomes evolve more rapidly than mammalian genomes. Plant genomes change more quickly than animal genomes, possibly as the result of massive genome remodeling from extensive transposition of mobile elements.

Plant, fungal, and animal genomes have unique and shared genes.

Plant genomes have changed much more rapidly than animal genomes. It appears about one-third of plant genes are unique to plants. Of the remaining plant genes, many are also found in animal and fungal genomes and are required for metabolism and gene expression.

24.2 Genome Size

Ancient and newly created polyploids guide studies of genome evolution.

Autopolyploidy results from an error in meiosis that leads to a duplicated genome; allopolyploidy is the result of hybridization between species (figure 24.2).

Evidence of ancient polyploidy is found in plant genomes.

Polyploidy has occurred numerous times in the evolution of flowering plants, and downsizing of genomes is common.

Polyploidy induces elimination of duplicated genes.

Downsizing of a polyploid genome can be caused by unequal loss of duplicated genes (figure 24.7).

Polyploidy can alter gene expression.

Polyploidization can lead to short-term silencing of genes via methylation of cytosines in the DNA.

Transposons jump around following polyploidization.

Transposons become highly active after polyploidization; their insertions into new positions may lead to new phenotypes.

Polyploidy alone cannot account for variation in genome size.

The vast range of ratios of gene number to genome size, even in closely related species, cannot be explained by polyploidy alone.

Noncoding DNA inflates genome size.

Genome size is most often inflated due to the presence of introns and non-protein-coding sequences. Genome size does not correlate with the number of genes. Some species have very little noncoding DNA, and others have extensive amounts of ncDNA, often retrotransposon DNA. Two species, such as rice and maize, can vary substantially in the amount of ncDNA and yet have similar numbers of genes.

24.3 Evolution Within Genomes

Individual chromosomes may be duplicated.

Aneuploidy, the duplication or loss of individual chromosomes, results from errors in meiosis. It is tolerated better in plants than in animals.

DNA segments may be duplicated (figure 24.8).

Paralogues are duplicated ancestral genes; orthologues are conserved ancestral genes. Duplicated DNA is common for genes associated with growth and development, immunity, and cell-surface receptors.

Genomes may become rearranged.

Genomes may be rearranged by moving gene locations within a chromosome or by the fusion of two chromosomes.

Conservation of synteny refers to preservation of long segments of ancestral chromosome sequences identifiable in related species (figure 24.10).

Gene inactivation results in pseudogenes.

Some ancestral genes become inactivated as they acquire mutations and are termed pseudogenes.

Rearranged DNA can acquire new functions.

Occasionally part of a gene can end up in a new spot in the genome where its function changes.

Noncoding DNA can acquire regulatory functions.

DNA that has no function in an organism can acquire mutations without any detrimental effect. However, conserved noncoding regions (CNCs) evolved more slowly than expected. These CNC regions appear to have gained functions that regulate the expression of neighboring protein-coding genes.

Horizontal gene transfer complicates matters.

Horizontal gene transfer creates many phylogenetic questions, such as the origins of the three major domains (figure 24.11).

24.4 Gene Function and Expression Patterns

Chimp and human gene transcription patterns differ.

Even when species have highly similar genes, expression of these genes may vary greatly. Posttranscriptional differences may also contribute to species differences.

Speech is uniquely human: An example of complex expression.

Small evolutionary changes in the FOXP2 protein and its expression may have led to human speech (figure 24.12).

24.5 Applying Comparative Genomics

Distantly related genomes offer clues for causes of disease.

Changes in amino acid sequences of critical proteins are a likely cause of diseases, and these differences can be identified by genome comparison.

Closely related organisms enhance medical research.

By comparing related organisms, researchers can focus on genes that cause diseases and devise possible treatments.

Pathogen–host genome differences reveal drug targets.

Analysis of the genomes of pathogenic organisms may provide new avenues of treatment and prevention (figure 24.14).

Genome comparisons inform conservation biology.

Comparative genomics within species can provide information of population diversity of endangered species. Conservation biologists can use this information to develop breeding strategies.

UNDERSTAND

1. Humans and pufferfish diverged from a common ancestor about 450 MYA, and these two genomes have

   a. very few of the same genes in common.
   b. all the same genes.
   c. a large proportion of the genes in common.
   d. no nucleotide divergence.

2. Genome comparisons have suggested that mouse DNA has mutated about twice as fast as human DNA. What is a possible explanation for this discrepancy?

   a. Mice are much smaller than humans.
   b. Mice live in much less sanitary conditions than humans and are therefore exposed to a wider range of mutation-causing substances.
   c. Mice have a smaller genome size.
   d. Mice have a much shorter generation time.

3. Polyploidy in plants

   a. has only arisen once and therefore is very rare.
   b. only occurs naturally when there is a hybridization event between two species.
   c. is common, but never occurs in animals.
   d. is common, and does occur in some animals.

4. Homologous genes in distantly related organisms can often be easily located on chromosomes due to

   a. horizontal gene transfer.
   b. conservation of synteny.
   c. gene inactivation.
   d. pseudogenes.

5. All of the following are believed to contribute to genomic diversity among various species, *except*

   a. gene duplication.
   b. gene transcription.
   c. lateral gene transfer.
   d. chromosomal rearrangements.

6. What is the fate of *most* duplicated genes?

   a. Gene inactivation
   b. Gain of a novel function through subsequent mutation
   c. They are transferred to a new organism using lateral gene transfer.
   d. They become orthologues.

APPLY

1. Chimp and human DNA whole-genome sequences differ by about 2.7%. Determine which of the following explanations is most consistent with the substantial differences in morphology and behavior between the two species.

   a. It must be due largely to gene expression.
   b. It must be due exclusively to environmental differences.
   c. It cannot be explained with current genetic theory.
   d. The differences are caused by random effects during development.

2. You are offered a summer research opportunity to investigate a region of ncDNA in maize. A friend politely smiles and says that only graduate students get to work on the coding regions of DNA. How would you critique your friend's statement?

   a. The friend has a point; ncDNA is "junk" DNA and therefore not very important.
   b. The ncDNA produces protein through mechanisms other than transcription.
   c. Most ncDNA is usually translated.
   d. Often ncDNA produces RNA transcripts that themselves have regulatory function.

3. Analyze the conclusion that the *Medicago truncatula* genome has been downsized relative to its ancestral legume, and circle the evidence that is consistent with this conclusion.

   a. *Medicago* has a proportional decrease in the number of genes.
   b. *Medicago* has a proportional increase in the number of genes.
   c. *Medicago* has an increase in the amount of DNA.
   d. *Medicago* has a decrease in the amount of DNA.

4. Analyze why an herbicide that targets the chloroplast is effective against malaria.

   a. Because *Plasmodium* needs a functional apicoplast
   b. Because the main vector for malaria is a plant
   c. Because mosquitoes require plant leaves for food
   d. Because *Plasmodium* mitochondria are very similar to chloroplasts

SYNTHESIZE

1. The *FOXP2* gene is associated with speech in humans. It is also found in chimpanzees, gorillas, orangutans, rhesus macaques, and even the mouse, yet none of these mammals speak. Develop a hypothesis that explains why *FOXP2* supports speech in humans but not other mammals.

2. One of the common misconceptions about sequencing projects (especially the high-profile Human Genome Project) is that creating a complete road map of the DNA will lead directly to cures for genetically based diseases. Given the percentage similarity in DNA between humans and chimps, is this simplistic view justified? Explain.

3. How does horizontal gene transfer (HGT) complicate phylogenetic analysis?

ONLINE RESOURCE

www.ravenbiology.com

Understand, Apply, and Synthesize—enhance your study with animations that bring concepts to life and practice tests to assess your understanding. Your instructor may also recommend the interactive eBook, individualized learning tools, and more.

Chapter **25**

Evolution
of Development

Chapter Contents

Introduction

How is it that closely related species of frogs can have completely different patterns of development? One frog goes from fertilized egg to adult frog with no intermediate tadpole stage. The sister species has an extra developmental stage neatly slipped in between early development and the formation of limbs—the tadpole stage. The answer to this and other such evolutionary differences in development that yield novel phenotypes are now being investigated with modern genetic and genomic tools. Research findings are accentuating the biological puzzle that many developmental genes are highly conserved, and a tremendous diversity of life shares this basic toolkit of developmental genes. In this chapter, we explore the emerging field of evolution of development, a field that brings together previously distinct fields of biology.

25.1 Evolution of Developmental Patterns

Learning Outcomes

1. *Explain how the same gene can produce different morphologies in different species.*
2. *Identify types of genes most likely to affect morphology.*
3. *Evaluate the limitations of comparative genomics in exploring the evolution of development.*

Ultimately, to explain the differences among species, we need to look at changes in developmental processes. Heritable changes in development are the result of changes in genes or gene expression that produce a different phenotype.

Phenotypic diversity could either result from changes in protein-coding regions of many different genes across species or could be explained by how a much smaller set of genes is deployed and regulated. In some cases, changes in the protein-coding region have been implicated in novel phenotypes. In other cases, a set of genes conserved across species appears to be responsible for the basic body plan of organisms with changes in regulation of gene expression, accounting for phenotypic differences. The latter is true for two sea urchin species.

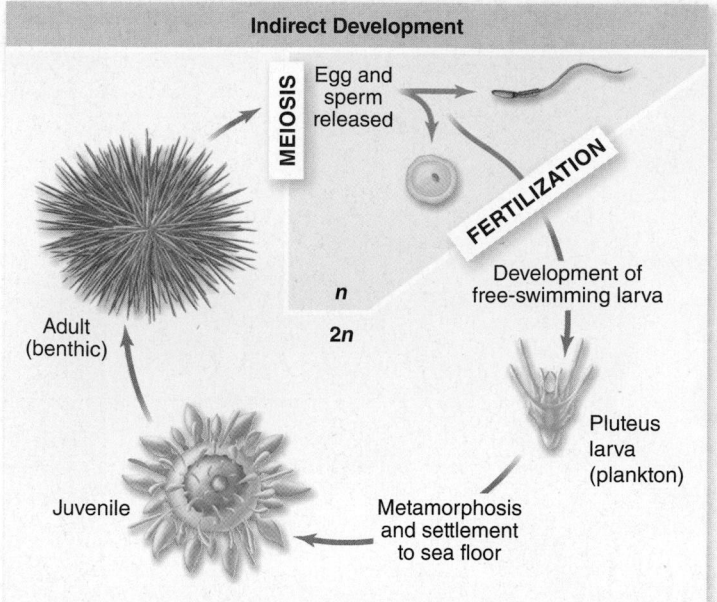
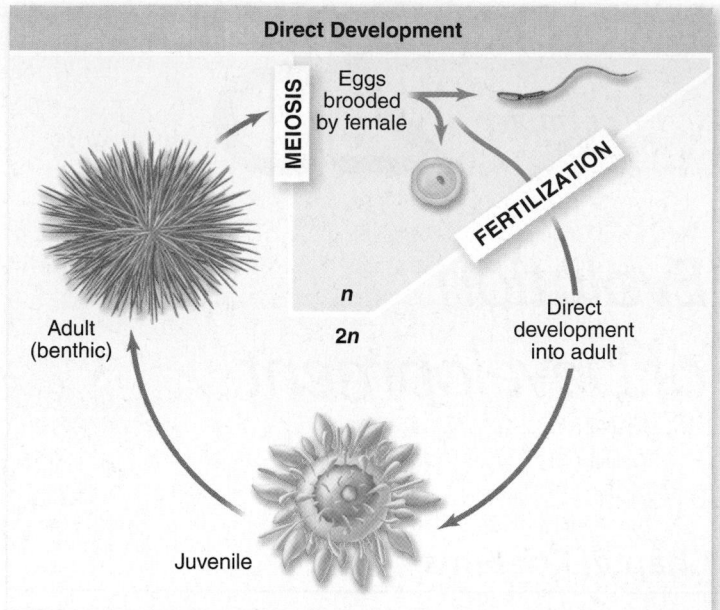

Figure 25.1 Direct and indirect sea urchin development. Phylogenetic analysis shows that indirect development (pluteus larva) was the ancestral state. Direct-developing sea urchins have lost an intermediate stage of development.

Closely related sea urchins have been discovered that have very distinctive developmental patterns (figure 25.1). One species exhibits indirect development, involving a pluteus (free-swimming) larval stage. The direct-developing urchin never makes a pluteus larva—it just jumps ahead to its adult form. One explanation is that the two forms have different developmental genes, but it turns out that this is not the case. Instead, the two forms have undergone dramatic changes in patterns of developmental gene expression, even though their adult form is nearly the same. In this case, patterns of expression have changed.

Highly conserved genes produce diverse morphologies

Transcription factors and genes involved in signaling pathways are responsible for coordination of development. As you saw in chapter 9, key elements of kinase and G protein-signaling pathways are also highly conserved among organisms. Even subtle changes in a signaling pathway can alter the enzyme that is activated or repressed, the transcription factor that is activated or repressed, or the activation or repression of gene expression. Any of these changes can have dramatic effects on the development of an organism.

A relatively small number of gene families, about two dozen, regulate animal and plant development. The developmental roles of several of these families, including *Hox* gene transcription factors, are described in chapter 19.

Hox (homeobox) genes appeared before the divergence of plants and animals; in plants, they have a role in shoot growth and leaf development, and in animals they establish body plans. These genes code for proteins with a highly conserved homeodomain that binds to the regulatory region of other genes to activate or repress these genes' expression. *Hox* genes specify when and where genes are expressed.

Another family of transcription factors, *MADS* box genes, are found throughout the eukaryotes. The *MADS* box also codes for a DNA-binding motif. Large numbers of *MADS* box genes establish the body plan of plants, especially the flowers. Although the *MADS* box region is highly conserved, variation exists in other regions of the coding sequence. Later in this chapter, we consider how there came to be so many *MADS* box genes in plants and how such similar genes can have very different functions.

Developmental mechanisms evolve

Understanding how development evolves requires integration of knowledge about genes, gene expression, development, and evolution. Either transcription factors or signaling molecules can be modified during evolution, changing the timing or position of gene expression and, as a result, gene function.

Heterochrony

Alterations in timing of developmental events due to a genetic change are called **heterochrony.** A heterochronic mutation could affect a gene that controls when a plant transitions from the juvenile to the adult stage, at which point it can produce reproductive organs. A mutation in a gene that delays flowering in plants can result in a small plant that flowers quickly rather than requiring months or years of growth.

Most mutations that affect developmental regulatory genes are lethal, but every so often a novel phenotype emerges that persists because of increased fitness. If a mutation leading to early flowering increased the fitness of a plant, the new phenotype will persist. For example, a tundra plant that flowers earlier, enabling it to be fertilized and set seed, could have increased fitness over an individual of the same species that flowers later, just as the short summer comes to a close.

Homeosis

Alternations in the spatial pattern of gene expression can result in **homeosis**. A four-winged *Drosophila* fly is an example of a change in gene expression pattern with a dramatic effect on morphology. Mutations in three genes in the *Bithorax* complex are required to produce this phenotype, which resembles more ancestral insects with four rather than two wings (see figure 19.17).

The *Drosophila Antennapedia* mutant, which has a leg where an antenna should be, is another example of a homeotic mutation. Mutations in genes such as *Antennapedia* can arise spontaneously in the natural world or by mutagenesis in the laboratory, but their bizarre phenotypes would have little survival value in nature.

Changes in transcription

The timing or location of gene expression can be modified in several ways, giving rise to heterochrony or homeosis. The coding sequence of a gene can contain multiple regions with different functions (figure 25.2). The DNA-binding motifs, exemplified by *MADS* box and *Hox* genes, could be altered so that they no longer bind to their target genes; as a result, that developmental pathway would cease to function. Alternatively, the modified transcription factor might bind to a different target and initiate a new sequence of developmental events.

The regulatory region of a gene encoding a transcription factor may also be altered. A sequence change in the promoter could prevent transcription of the transcription factor gene. Alternatively a changed regulatory region might bind a different transcription factor. In this case, the downstream targets would be the same, but the cells that express the target genes or the time at which these target genes are expressed could change.

Changes in signaling pathways

Signaling pathways help cells coordinate information about neighboring cells and the external environment that is essential for successful development. If the structure of a ligand changes, it may no longer bind to its target receptor; it could bind to a different receptor or no receptor at all. If, as a result of a genetic change, a receptor is produced in a different cell type, a homeotic phenotype may appear. And, as mentioned earlier, small changes in signaling molecules can alter their targets.

The sections that follow use specific examples of the evolution of diverse morphologies. For each example, consider how the mechanism of development has been altered and what the outcome is. Keep in mind that these are the successful examples; most morphological novelties that arise quickly, but do not improve fitness, go extinct.

Understanding evolution of development requires functional analysis

Comparative genomics is amazingly useful for understanding morphological diversity. Limitations exist, however, in the

SCIENTIFIC THINKING

Hypothesis: *A transcription factor can affect the expression of more than one gene.*

Prediction: *The protein encoded by a transcription factor gene will have multiple DNA- or protein-binding sites.*

Test: *Experimentally identify molecules that bind to the transcription factor.*

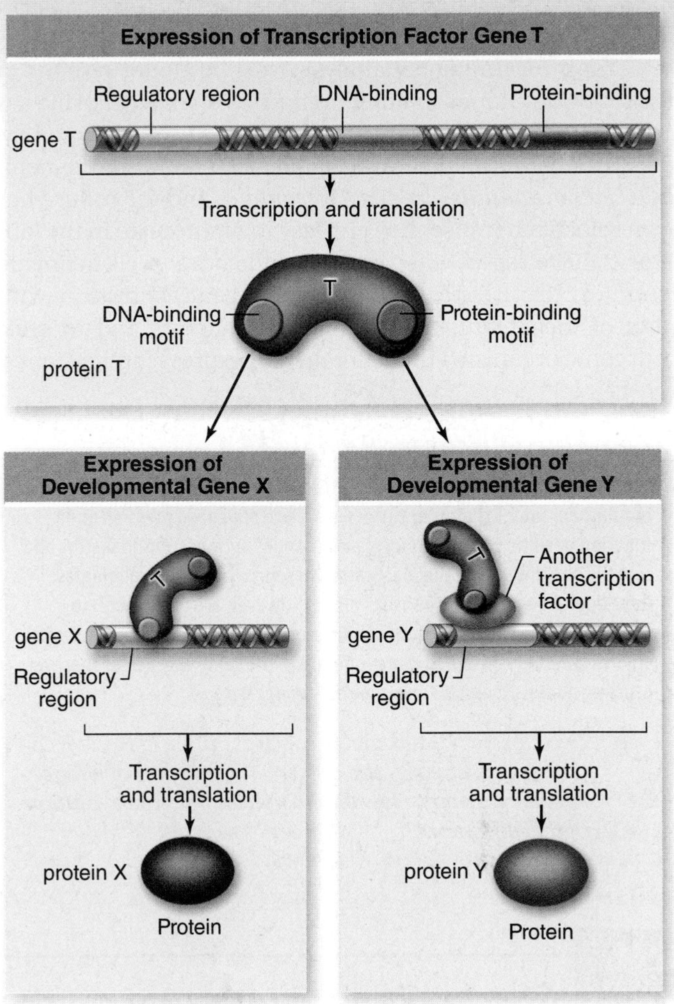

Result: *This transcription factor has a site that binds to the regulatory region of a gene and a site that binds to a transcription factor that regulates expression of a second gene.*

Conclusion: *A single transcription factor can regulate the expression of more than one gene.*

Further Experiments: *Determine the specific developmental role of each binding domain by creating mutations in regions of the gene that code for specific binding sites in the protein.*

Figure 25.2 **Transcription factors have a key role in the evolution of development.**

inferences about evolution of development that we can draw from sequence comparison alone. *Functional genomics* includes a range of experiments designed to test the actual function of a gene in different species as explained in chapter 18.

Sequence comparisons among organisms are essential for both phylogenetic and comparative developmental studies. Careful analysis is needed to distinguish paralogues from orthologues. Rapidly evolving research using bioinformatics, which utilizes computer programming to analyze DNA and protein data, leads to hypotheses that can be tested experimentally. A single base mutation can change an active gene into an inactive pseudogene, and experiments are necessary to demonstrate the actual function of the gene.

Tools for functional analysis exist in model systems but need to be developed in other organisms on the tree of life if we are going to piece together evolutionary history. Model systems such as yeast, the flowering plant *Arabidopsis,* the nematode *Caenorhabditis elegans, Drosophila,* and the mouse have been selected because they are easy to manipulate in the laboratory, have short life cycles, and have well-delineated genomes. Also, it is possible to visualize gene expression within parts of the organism using labeled markers and to create transgenic organisms that contain and express foreign genes.

Learning Outcomes Review 25.1

Highly conserved genes can undergo small changes in their coding or regulatory regions that alter the place or time of gene expression and function, resulting in new body plans. Changes in transcription factors and signaling pathways are the most common source of new morphologies. Genetic and genomic comparisons alone, however, are not enough to determine the function of genes in different species; functional genomics studies whether conserved genes operate in the same way across species utilizing model organisms and genetic engineering.

- **Two closely related species of** Drosophila *in Hawaii can be distinguished by the presence of one pair of wings versus two pairs. How would you explain the evolution of this difference?*

25.2 Single-Gene Changes and the Alteration of Form and Function

Learning Outcomes

1. *Explain how a small number of mutations can give rise to new morphologies and even new species.*
2. *Explain how a gene could acquire a new function.*
3. *Describe how duplicated genes could give rise to new functions in an organism.*

In the preceding chapter, we discussed the similarity between human and mouse genomes. If all but 300 of the 20,000–25,000 human genes are shared with mice, why are mice and humans so different? Part of the answer is that genes with similar sequences in two different species may work in slightly or even dramatically different ways. Here we explore several examples of single-gene mutations that altered the form and function of plants and animals to better understand how such small changes can sometimes have such dramatic effects on the overall body plan of an organism.

Cauliflower and broccoli began with a stop codon

The species *Brassica oleracea* is particularly fascinating because individual members can have extraordinarily diverse phenotypes (figure 25.3).Wild cabbage, kale, tree kale, red cabbage, green cabbage, brussels sprouts, broccoli, and cauliflower are all members of the same species. Some flower early, some late. Some have long stems, others have short ones. Some form a few flowers, and others, like broccoli and cauliflower, initiate many flowers, but development of the flowers is arrested. Curiously, these plants with such different appearances are very closely related.

One piece of the puzzle lies with the gene *CAL* (*Cauliflower*), which was first cloned in a close *Brassica* relative, *Arabidopsis*. In combination with another mutation, *Apetala1, Arabidopsis* plants can be turned from plants with a limited number of simple flowers into miniature broccoli or cauliflower plants with masses of arrested flower meristems or flower buds. These two genes are needed for the transition to making flowers and arose through duplication of a single ancestral gene within the brassica group. When they are absent, meristems continue to make branches, but are delayed in producing flowers.

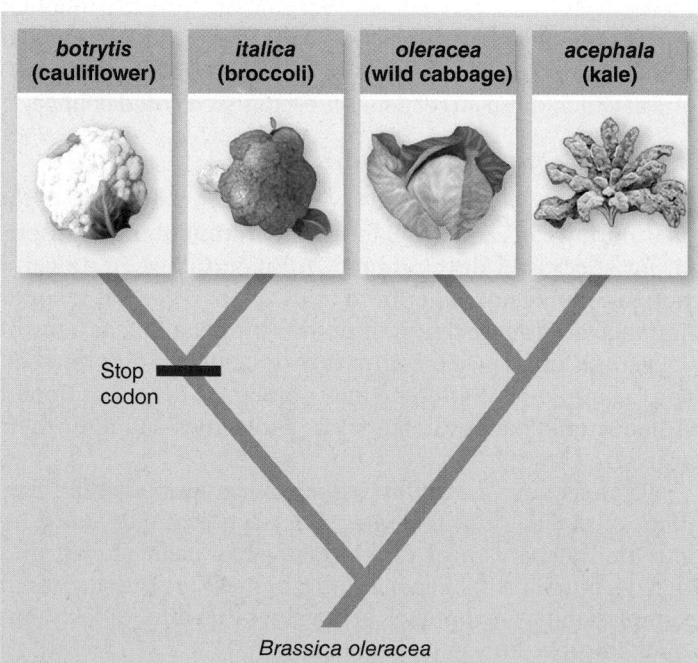

Figure 25.3 Evolution of cauliflower and broccoli.
A point mutation that converted an amino acid–coding region into a stop codon resulted in the extensive reproductive branching pattern that was artificially selected for in two crop plants that are both *Brassica oleracea.*

The *CAL* gene was cloned from large numbers of *B. oleracea* subspecies, and a stop codon, TAG, was found in the middle of the *CAL* coding sequences of broccoli and cauliflower. A phylogenetic analysis of *B. oleracea* coupled with the *CAL* sequence analysis leads to the conclusion that this stop codon appeared after the ancestors of broccoli and cauliflower diverged from other subspecies members, but before broccoli and cauliflower diverged from each other (see figure 25.3).

> **?** **Inquiry question** Knowing that cauliflower and broccoli have a stop codon in the middle of the *CAL* gene-coding sequence, predict the wild-type function of *CAL*. What additional evolutionary events may have occurred since broccoli and cauliflower diverged?

A second, somewhat unusual feature of this example is that the driving selective force for this diversity was artificial. Wild relatives are still found scattered along the rocky coasts of Spain and the Mediterranean region. The most likely scenario is that humans found a *cal* mutant and selected for that phenotype through cultivation. The large heads of broccoli and cauliflower offer a larger amount of a vegetable material than the wild kale plants and a tasty alternative to *Brassica* leaves.

Cichlid fish jaws demonstrate morphological diversity

Single-gene mutations can lead to rapid speciation. Here we consider modifications in form and function from adaptive radiation of cichlid fish in Lake Malawi in East Africa. In less than a few million years, hundreds of species have evolved in the lake from a common ancestor. The rapid speciation of cichlids is illustrated in figure 22.15.

One possible explanation for the evolutionary exuberance of these fish is that different species' feeding habits have led to different morphologies. There are bottom feeders, biters, and rammers. The rammers have particularly long snouts with which to ram their prey; biters have an intermediate snout; and the bottom feeders have short snouts adapted to scrounging for food at the base of the lake (figure 25.4).

How did these fish acquire such different snout forms? An extensive genetic analysis revealed that two genes, of yet unknown function, are likely responsible for the shape and size of the jaw. The results of crossing long- and short-snouted cichlids indicate the importance of one of these two genes in specifically determining jaw length and height.

Regulating the length versus the height of the jaw may well be an early and important developmental event. The overall size of the fish and the extent of muscle development both hinge on the form of the jaw. The range of jaw forms appears to have persisted because the cichlids establish unique niches for feeding within the lake.

Stickleback fish lose their armor with a single mutation

Not all mutations that give rise to new forms are passed down through generations. There is a critical link between persistence of a new mutation and increased fitness. The freshwater, three-spine stickleback fish, *Gasterousteus aculeatus,* originated after the last ice age from marine populations with bony plates that protect the fish from predators. Freshwater populations, subject to less predation, have lost their bony armor. The *Ectodysplasin (Eda)* gene is one of a few associated with reduced armor in freshwater three-spine sticklebacks. The *Eda* allele that causes reduced armor originated about 2 MYA in marine sticklebacks and persists with a frequency of about 1% in marine environments. The frequency is much higher in freshwater populations. To test the fitness of the *Eda* allele in freshwater, marine sticklebacks that were heterozygous for the *Eda* allele were moved to four freshwater environments and allowed to breed. Positive selection for the reduced armor allele was observed and correlated with longer length in juvenile fish, likely because fewer resources were allocated to armor development. Although the reduced armor allele has persisted in marine populations for 2 million years as a rare genetic variant, increased frequency of the allele and phenotype are only seen under conditions in which they are adaptive.

Figure 25.4 Diversity of cichlid fish jaws. A difference in one gene is responsible for a short snout in *Labeotropheus fuelleborni* and a long snout in *Metriaclima zebra*. Genes that affect jaw length can affect body shape as well because of the constraints the size of the jaw places on muscle development.

chapter **25** *Evolution of Development*

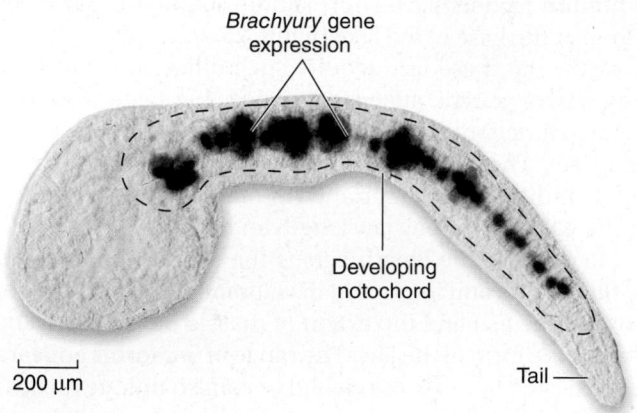

Figure 25.5 **Co-opting a gene for a new function.**
Brachyury is a gene found in invertebrates that has been used for notochord development in this ascidian, a basal chordate. By attaching the *Brachyury* promoter to a gene with a protein product that stains blue, it is possible to see that *Brachyury* gene expression in ascidians is associated with the development of the notochord, a novel function compared with its function in organisms lacking a notochord. The orthologue in the nematode *Caenorhabditis elegans* is important for hind gut and male tail development, but there is no evidence of a notochord precursor.

Ancestral genes may be co-opted for new functions

The evolution of chordates can partially be explained by the co-option of an existing gene for a new function. Ascidians are basal chordates that have a notochord but no vertebrae (see chapter 34). The *Brachyury* gene of ascidians encodes a transcription factor, and it is expressed in the developing notochord (figure 25.5).

Brachyury is not a novel gene that appeared as vertebrates evolved. It is also found in invertebrates. For example, a mollusk homologue of *Brachyury* is associated with anterior–posterior axis specification. Most likely, an ancestral *Brachyury* gene was co-opted for a new role in notochord development.

Brachyury is a member of a gene family with a specific sequence motif, that is, a conserved sequence of base-pairs within the gene. A region of *Brachyury* encodes a protein domain called the **T box,** which is a transcription factor. So, *Brachyury*-encoded protein turns on a gene or genes. The details of which genes are regulated by *Brachyury* are only now being discovered.

How a single genetic toolkit can be used to build an insect, a bird, a bat, a whale, or a human is an intriguing puzzle for evolutionary developmental biologists, exemplified with the *Brachyury* gene. One explanation is that *Brachyury* turns on different genes or combinations of genes in different animals. Although there are not yet enough data to sort out the details of *Brachyury*, we can look at limb formation for an explanation of how such a change could have evolved.

Limbs have developed through modification of transcriptional regulation

Most tetrapods have four limbs—two hindlimbs and two forelimbs—although two or more limbs have been lost many times in lizards, including snakes (which in evolutionary terms are lizards). The wing in a bird is actually a forelimb. Our arm is a forelimb. Clearly, these are two very different structures, but they have a common evolutionary origin. As you learned in chapter 23, these are termed *homologous structures.*

At the genetic level, humans and birds both express the *Tbx5* gene in developing forelimb buds and *Tbx4* in hindlimbs. Like *Brachyury, Tbx5* is a member of a transcription factor gene family with T box motif, that is, a conserved sequence of base-pairs within the gene. *Tbx5*-encoded protein turns on *Fibroblast growth factor-10* (Fgf10) that is needed to make a limb. Mutations in the human *Tbx5* gene cause Holt–Oram syndrome, resulting in forelimb and heart abnormalities.

The link between *Tbx5*, which initiates forelimb development, and human heart development can be traced back to limbless amphioxus, a chordate that lacks vertebrates. Amphioxus has a homologue, *AmphiTbx4/5*, expressed in the heart region (figure 25.6). The evolutionary tale of *Tbx5* is one of both gene co-option and duplication, which we come back to later in this chapter.

Two whole-genome duplications accompanied the emergence of the vertebrates, and the duplicated *AmphiTbx4/5* gave rise to both *Tbx5* and *Tbx4,* which were co-opted for forelimb and hindlimb development, respectively. Two scenarios for the evolution of *Tbx5* are possible. The coding region could be modified so the transcription factor interacted with other genes. Or, the regulatory region could be altered. To distinguish between the two possibilities, transgenic mice were made using the *AmphiTbx4/5* gene.

If the regulatory, not the coding sequence, evolved a new function, then swapping the mouse *Tbx5* coding region with the amphioxus *AmphiTbx4/5* coding region should not affect forelimb development in the mouse. Transgenic mice with a *Tbx5* regulatory region and an *AmphiTbx4/5* coding region form forelimbs (see figure 25.6). A second experiment is needed to determine if the regulatory region changed in the 520 million years since amphioxus and mice last shared a common ancestor. When the amphioxus *AmphiTbx4/5* regulatory region is swapped with the mouse *Tbx5* regulatory region, no forelimbs develop although the *Tbx5* coding region is present (see figure 25.6). This experiment demonstrates that the key vertebrate innovation was a new regulatory region, specific to the forelimb area in the case of *Tbx5.*

Changes in gene regulation also explain the evolution of digit formation of limbs. The two whole genome duplications in early vertebrate evolution gave rise to four clusters of *Hox* genes, with *Hoxc* and *Hoxd* arising from a common ancestor in the second duplication. Of the two, only *Hoxd* contributes to limb development, specifically digit formation. The HOXD12 protein is expressed in the forelimb, but not the HOXC12 protein. To sort out the role of regulatory elements, the entire upstream chromosomal regions of mouse *Hoxc* and *Hoxd* were swapped. In mice with the shuffled genes, the HOXC12 protein

Background: *AmphiTbx4/5 gave rise to the vertebrate Tbx4 and Tbx5 after whole-genome duplication. The three DNA sequences are very similar.*

Hypothesis 1: *Tbx4 and Tbx5 were co-opted for limb development in vertebrates.*

Hypothesis 2: *Tbx5 underwent changes in its regulatory rather than coding DNA.*

Prediction 1: *AmphiTbx4/5 will be expressed in the heart region of amphioxus and Tbx5 and Tbx4 in the forelimb and hindlimb, respectively, of a mouse.*

Experiment 1: *Transgenic amphioxi and mice are made with AmphiTbx4/5, Tbx5, and Tbx4 promoters placed in front of a β-galactosidase gene that will enzymatically convert an added stain blue in cells where it is expressed.*

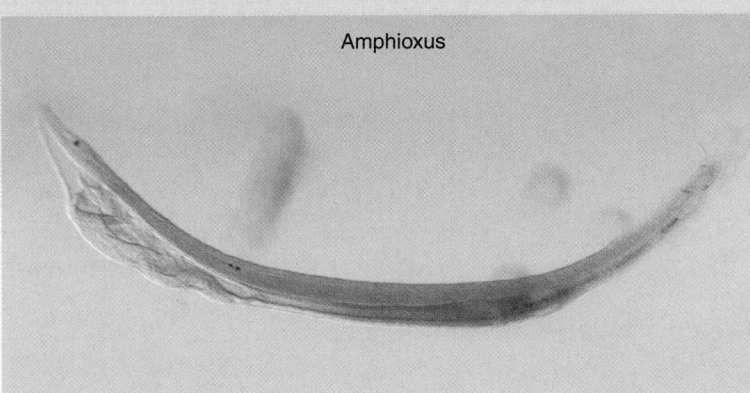
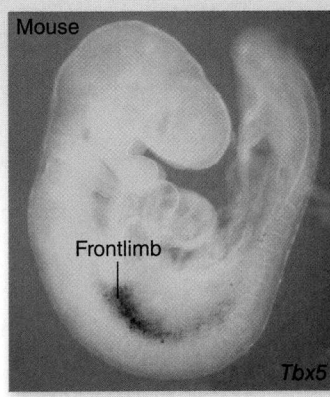
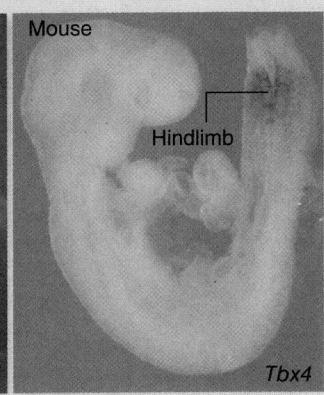

Result 1: *Expression patterns are as predicted.*

Prediction 2: *AmphiTbx4/5 will support forelimb bud development in mice missing the Tbx5k coding region, if it is under regulatory control of Tbx5 but not AmphiTbx4/5.*

Experiment 2: *Transgenic mice containing AmphiTbx4/5 and the regulatory but not coding region of Tbx5 and transgenic mice with AmphiTbx4/5 and its 70-kb upstream and downstream regulatory regions, but no Tbx5 sequences were constructed and allowed to develop.*

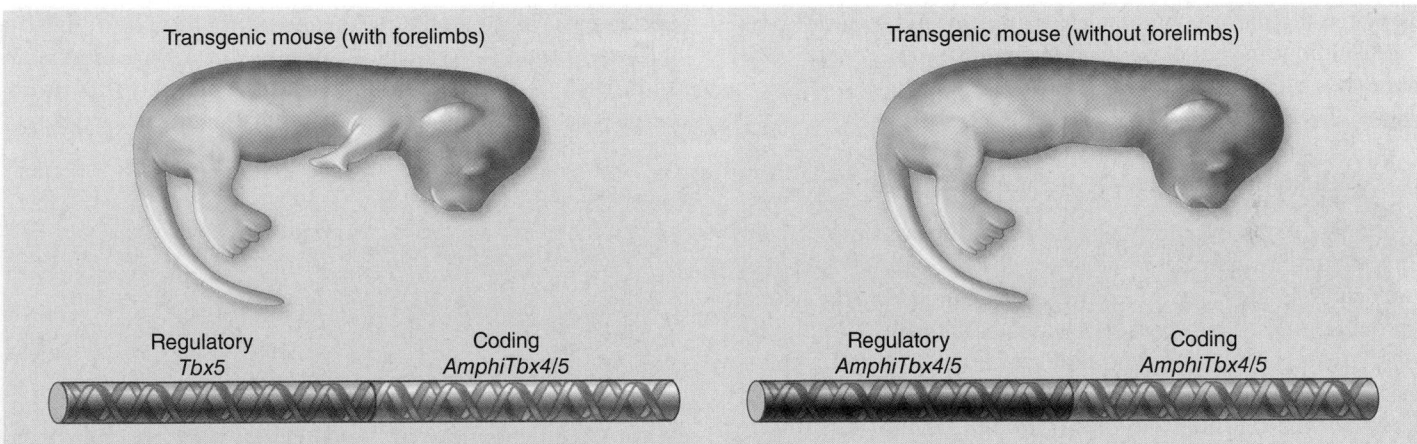

Result 2: *AmphiTbx4/5 supports limb development under Tbx5, but not AmphiTbx4/5 regulatory control.*

Conclusion: *New regulatory regions for Tbx5 and Tbx4 lead to limb bud formation.*

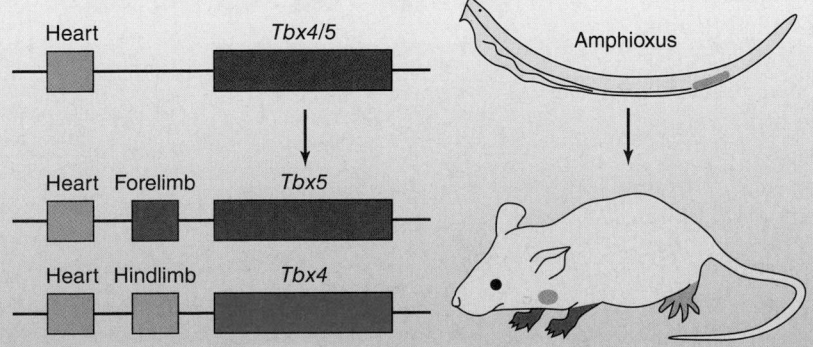

Further Experiments: *Identify the regulatory region that results in Tbx5 expression in the forelimb. How would you design these experiments?*

Figure 25.6 Amphioxus heart gene co-opted for limb development in vertebrates through changes in gene regulation.

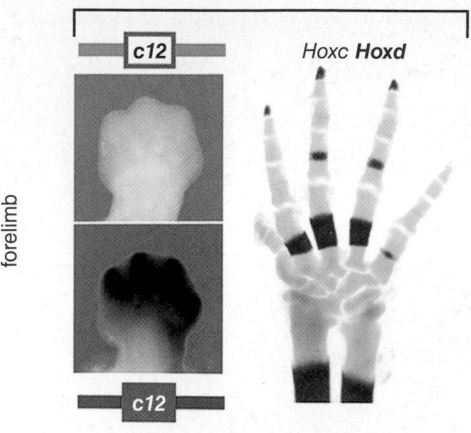
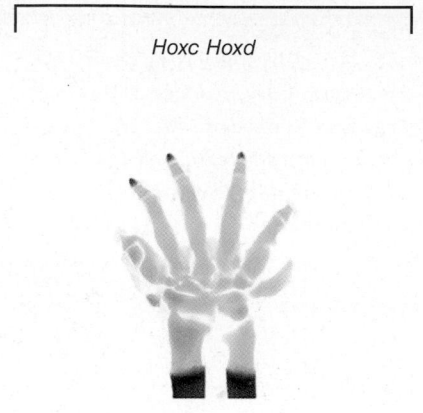
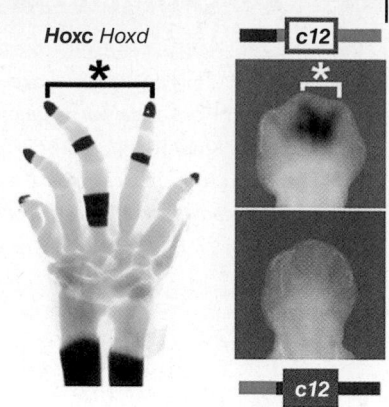

a. **Wild type mouse:** *Hoxd12* supports forelimb digit development and *Hoxc12* genes are not expressed.

b. Mouse forelimb digits are aberrant when *Hoxc* and *Hoxd* genes are absent.

c. *Hoxd* and *Hoxc* regulatory regions are swapped. Hoxc12 can partially rescue Hoxd12 activity in the forelimb.

Figure 25.7 *Hoxc12* **can partially substitute for** *Hoxd12* **in forelimb digit development.**

was able to replace the function of HOXD12 in digit formation and rescue forelimb development (figure 25.7).

In the case of both *Tbx5* and *Hoxd12,* the whole-genome duplications in early vertebrate development yielded duplicate genes with redundant function that evolution could act on. In both cases, regulatory regions rather than coding regions experienced change essential for limb formation.

Gene duplications provide opportunities for new gene functions

The analysis of the evolution of limb development emphasized how changes in gene regulatory sequences can lead to new gene functions. It is also an example of the importance of gene duplication in the evolution of new gene functions. A duplicate gene provides a backup gene that can mutate without being lethal to the organism. In this section, we explore a specific example of the evolution of development through gene duplication and divergence in flower form.

Gene duplications of *paleoAP3* **and flowering-plant morphology.** Before the flowering plants originated, a *MADS* box gene duplicated, giving rise to genes called *PI* and *paleoAP3.* In ancestral flowering plants, these genes affected stamen development, and this function has been retained. (Stamens are the male reproductive structures of flowering plants.)

The *paleoAP3* gene duplicated to produce *AP3* and an *AP3* duplicate some time after members of the poppy family last shared a common ancestor with the clade of plants called the eudicots (plants like apple, tomato, and *Arabidopsis*). This clade of eudicots is distinguished on the genome level by both the duplication of *paleoAP3* and the origins of a precise pattern of petal development in their last common ancestor (figure 25.8). The phylogenetic inference is that *AP3* gained a role in petal development.

Alteration in gene divergence of *AP3* **function and controlling petal development.** Although the occurrence of *AP3* through duplication corresponds with a uniform developmental process for specifying petal development, the correlation could simply be coincidental. Experiments that mix and match parts of the *AP3* and *PI* genes, and then introduce them into *ap3* mutant plants, confirm that the phylogenetic correspondence is not a coincidence. The *ap3* plants do not produce either petals or stamens. A summary of the experiments is shown in figure 25.9.

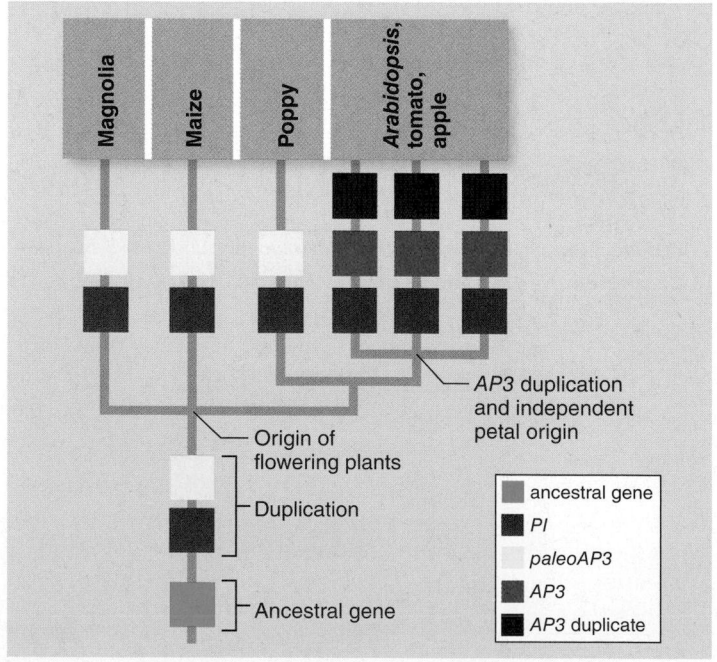

Figure 25.8 **Petal evolution through gene duplication.**
Two gene duplications resulted in the *AP3* gene in the eudicots that has acquired a role in petal development.

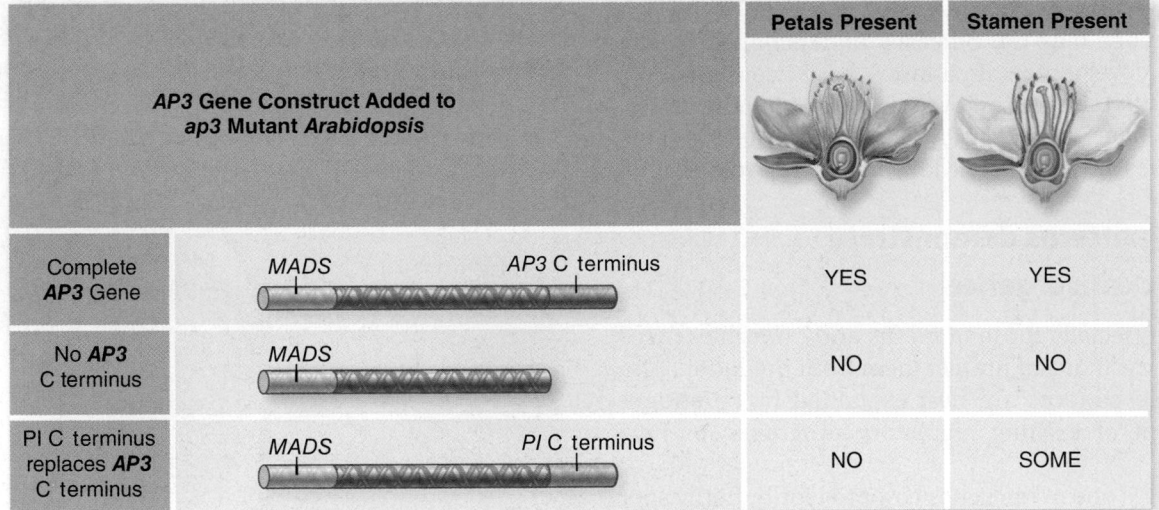

AP3 Gene Construct Added to ap3 Mutant Arabidopsis		Petals Present	Stamen Present
Complete **AP3** Gene	MADS *AP3* C terminus	YES	YES
No **AP3** C terminus	MADS	NO	NO
PI C terminus replaces **AP3** C terminus	MADS *PI* C terminus	NO	SOME

Figure 25.9 *AP3* **has acquired a domain necessary for petal development.** The *AP3* gene includes MADS box encoding a DNA-binding domain and a highly specific sequence near the C terminus. Without the 3′ region of the *AP3* gene, the *Arabidopsis* plant will not make petals.

Earlier in this chapter, the *MADS* box transcription factor gene family was introduced. One region of a *MADS* gene codes for a DNA-binding motif; other regions code for different functions, including protein–protein binding. The PI and AP3 proteins can bind to each other and, as a result, can regulate the transcription of genes needed for stamen and petal formation.

Both AP3 and PI have distinct sequences at the C (carboxy) protein terminus (coded for by the 3′ end of the genes). The C-terminus sequence of the AP3 protein is essential for specifying petal function, and it contains a conserved sequence shared among the eudicots. The *AP3* C-terminus DNA sequence was deleted from the wild-type gene, and the new construct was inserted into *ap3* plants to create a transgenic plant. Other transgenic plants were created by inserting the complete *AP3* sequence into *ap3* plants. The complete *AP3* sequence rescued the mutant, and petals were produced. No petals formed when the C-terminus motif was absent.

AP3 is also needed for stamen development, an ancestral trait found in *paleoAP3*. Plants lacking *AP3* fail to produce either stamens or petals. Transgenic plants with the C-terminus deletion construct also failed to produce stamens.

The *pi* mutant phenotype also lacks stamens and petals. To test whether the *PI* C terminus could substitute for the *AP3* C terminus in specifying petal formation, the *PI* C terminus was added to the truncated *AP3* gene. No petals formed, but stamen development was partially rescued. These experiments demonstrate that *AP3* has acquired an essential role in petal development, encoded in a sequence at the 3′ end of the gene.

? Inquiry question Explain how functional analysis was used to support the claim that petal development evolved through the acquisition of petal function in the *AP3* gene of *Arabidopsis*.

Learning Outcomes Review 25.2

Although most mutations are lethal, some confer a fitness advantage. These may consist of very small mutations, such as a change to a single codon, that have large effects on development and morphology. During the long course of evolution, genes have been co-opted for new functions. A change in the protein-coding region of a transcription factor can change the genes it can bind to and regulate. A change in the regulatory region of a gene can change where or when that gene is expressed, which can lead to altered morphology. Gene duplication allows for divergence that can lead to novel function.

■ *A marine three-spine stickleback fish is mated with a freshwater three-spine with reduced armor, and all the offspring have reduced armor. Both populations have identical Eda coding regions. Could this difference in the Eda gene be the cause of the difference in armor? How could you test this?*

25.3 Different Ways to Evolve the Same Structure

Learning Outcomes

1. Differentiate between homologous structures and homoplastic structures.
2. Explain how two very similar morphologies can arise from different developmental pathways.

Homoplastic structures, also known as analogous structures, have the same or similar functions, but arose independently—unlike homologous structures that arose once from a common ancestor. Phylogenies reveal convergent events, but the origin of the convergence may not be easily understood. In many cases different developmental pathways have been modified, as is the case with the spots on butterfly wings. In other cases, such as flower shape, it is not always as clear whether the same or different genes are responsible for convergent evolution.

Insect wing patterns demonstrate homoplastic convergence

Insect wings, especially those of moths and butterflies, have beautiful patterns that can protect them from predation. The origins of these patterns are best explained by co-option, the recruitment of existing regulatory programs for new functions.

Distal-less is one of the genes co-opted for butterfly spot development. Limb development in insects and arthropods requires *Distal-less,* but the expression of this gene also predicts where spots will form on butterfly wings (figure 27.10). *Distal-less* determines the center of the spot, but several other genes have been co-opted to determine the overall size and pattern of different spots.

Not all insects have co-opted the same sets of genes as *Precis coenia* for these new functions, but all the evolutionary pathways have converged around production of these novel, highly patterned wings.

Flower shapes also demonstrate convergence

Flowers exhibit two types of symmetry. Looking down on a **radially symmetrical** flower, you see a circle. No matter how you cut that flower, as long as you have a straight line that intersects with the center, you end up with two identical parts. Examples of radially symmetrical flowers are daisies, roses, tulips, and many other flowers.

Bilaterally symmetrical flowers have mirror-image halves on each side of a single central axis. If they are cut in any other direction, two nonsimilar shapes result. Plants with bilaterally symmetrical flowers include snapdragons, mints, and peas. Bilaterally symmetrical flowers are attractive to their pollinators, and the shape may have been an important factor in their evolutionary success.

At the crossroads of evolution and development, two questions arise: first, what genes are involved in bilateral symmetry? And second, are the same genes involved in the numerous, independent origins of asymmetrical flowers?

Cycloidia (CYC) is a snapdragon gene responsible for the bilateral symmetry of the flower. Snapdragons with mutations in *CYC* have radially symmetrical flowers (see figure 41.17*b*). Beginning with robust phylogenies, researchers have selected flowers that evolved bilateral symmetry independently from snapdragons and cloned the *CYC* gene. The *CYC* gene in closely related symmetrical flowers has also been sequenced.

Comparisons of the *CYC* gene sequence among phylogenetically diverse flowers indicate that both radial symmetry

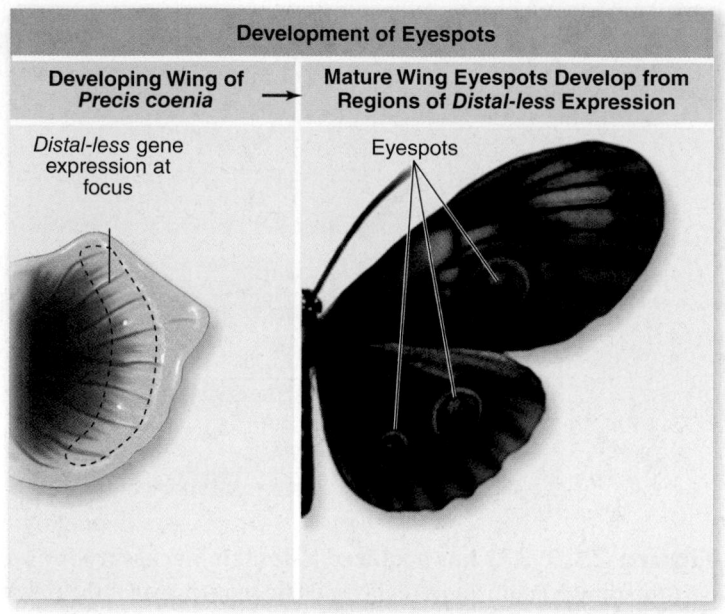

Development of Eyespots

| Developing Wing of *Precis coenia* | → | Mature Wing Eyespots Develop from Regions of *Distal-less* Expression |

Distal-less gene expression at focus

Eyespots

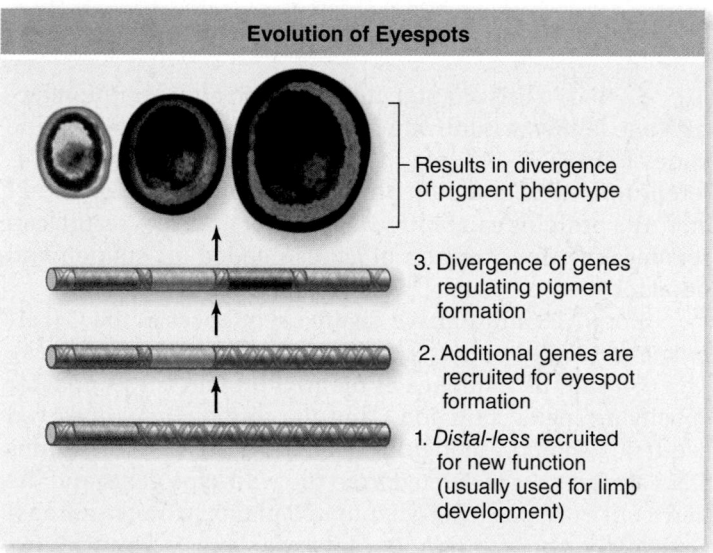

Evolution of Eyespots

Results in divergence of pigment phenotype

3. Divergence of genes regulating pigment formation

2. Additional genes are recruited for eyespot formation

1. *Distal-less* recruited for new function (usually used for limb development)

Figure 25.10 Butterfly eyespot evolution. The *Distal-less* gene, usually used for limb development, was recruited for eyespot development on butterfly wings. *Distal-less* initiates the development of different-colored spots in different butterfly species by regulating different pigment genes in different species. Eyespots can protect butterflies by startling predators.

and bilateral symmetry evolved in multiple ways in flowers. Although radial symmetry is the ancestral condition, some radially symmetrical flowers have a bilaterally symmetrical ancestor. Loss of *CYC* function accounts for the loss of bilateral symmetry in some of these plants.

Gain of bilateral symmetry arose independently among some species because of the *CYC* gene. This change is an example of convergent evolution through mutations of the same gene. In other cases, *CYC* is not clearly responsible for the bilateral symmetry. Other genes also played a role in the convergent evolution of bilaterally symmetrical flowers.

25.4 Diversity of Eyes in the Natural World: A Case Study

The eye is one of the most complex organs. Biologists have studied it for centuries. Indeed, explaining how such a complicated structure could evolve was one of the great challenges facing Darwin. If all parts of a structure such as an eye are required for proper functioning, how could natural selection build such a structure?

Darwin's response was that even intermediate structures—which provide, for example, the ability to distinguish light from dark—would be advantageous compared with the ancestral state of no visual capability whatsoever, and thus these structures would be favored by natural selection. In this way, by incremental improvements in function, natural selection could build a complicated structure.

Morphological evidence indicates eyes evolved at least twenty times

Comparative anatomists long have noted that the structures of the eyes of different types of animals are quite different. Consider, for example, the difference in the eyes of a vertebrate, an insect, a mollusk (octopus), and a jellyfish (figure 25.11, also see figure 45.15). The eyes of these organisms are extremely different in many ways, ranging from compound eyes, to simple eyes, to mere eyespots.

Consequently, morphologists concluded that eyes are examples of convergent evolution and are homoplastic. For this reason, evolutionary biologists traditionally viewed the eyes of different organisms as having independently evolved, perhaps as many as 20 times. Moreover, this view holds that the most recent common ancestor of all these forms was a primitive animal with no ability to detect light. Although this was the conclusion of morphologists, molecular studies are pointing to a different conclusion.

The same gene, *Pax6*, initiates fly and mouse eye development

In the early 1990s, biologists studied the development of the eye in both vertebrates and insects. In each case, a gene was discovered that codes for a transcription factor important in lens formation; the mouse gene was given the name **Pax6**, whereas the fly gene was called *eyeless*. A mutation in the *eyeless* gene led to a lack of production of the transcription factor, and thus the complete absence of eye development, giving the gene its name.

Vertebrate	Insect	Mollusk	Jellyfish
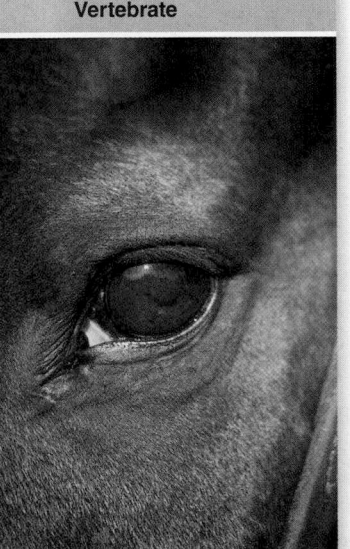		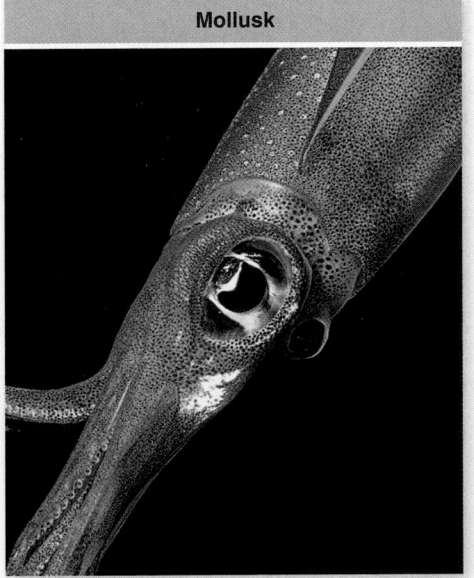	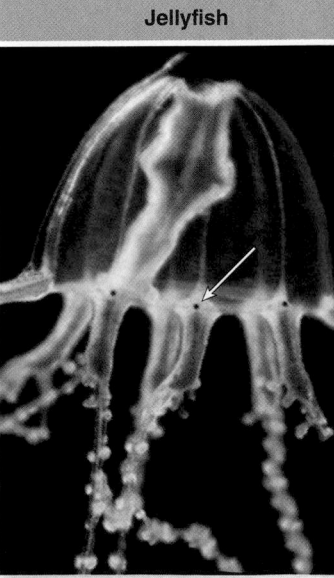

Figure 25.11 A diversity of eyes. Morphological and anatomical comparisons of eyes are consistent with the hypothesis of independent, convergent evolution of eyes in diverse species such as flies and humans.

100 µm

Figure 25.12 Mouse *Pax6* makes an eye on the leg of a fly. *Pax6* and *eyeless* are functional homologues. The *Pax6* master regulator gene can initiate compound eye development in a fruit fly or simple eye development in a mouse.

When these genes were sequenced, it became apparent that they were highly similar; in essence, the homologous *Pax6* gene was responsible for triggering lens formation in both insects and vertebrates. A stunning demonstration of this homology was conducted by the Swiss biologist Walter Gehring, who inserted the mouse version of the *Pax6* into the genome of a fruit fly, creating a transgenic fly. In this fly, the *Pax6* gene was turned on by regulatory factors in the fly's leg and an eye formed on the leg of the fly (figure 25.12)!

These results were truly shocking to the evolutionary biology community. Insects and vertebrates diverged from a common ancestor more than 500 MYA. Moreover, given the large differences in structure of the vertebrate eye and insect eye, the standard assumption was that the eyes evolved independently, and thus that their development would be controlled by completely different genes. That eye development was affected by the same homologous gene, and that these genes were so similar that the vertebrate gene seemed to function normally in the insect genome, was completely unexpected.

The *Pax6* story extends to eyeless fish found in caves (figure 25.13). Fish that live in dark caves need to rely on senses other than sight. In cavefish, *Pax6* gene expression is greatly reduced. Eyes start to develop, but then degenerate.

Ribbon worms, but not planaria, use *Pax6* for eye development

Research on other bilaterians (animals with mirror-image symmetry) yielded further surprises about the *Pax6* gene. Even the very simple ribbon worm, *Lineus sanguineus,* relies on *Pax6* for development of its eyespots. A *Pax6* homologue has been cloned and has been shown to express at the sites where eyespots develop.

Ribbon worms can regenerate their head region if it is removed. In an elegant experiment, the head of a ribbon worm was removed, and biologists followed the regeneration of eyespots. At the same time, the expression of the *Pax6* homologue

was observed using in situ hybridization. To observe *Pax6* gene expression, an antisense RNA sequence of the *Pax6* was made and labeled with a color marker. When the regenerating ribbon worms were exposed to the antisense *Pax6* probe, the antisense RNA paired with expressed *Pax6* RNA transcripts and could be seen as colored spots under the microscope (figure 25.14).

Surface Dweller — Has Pax6

a.

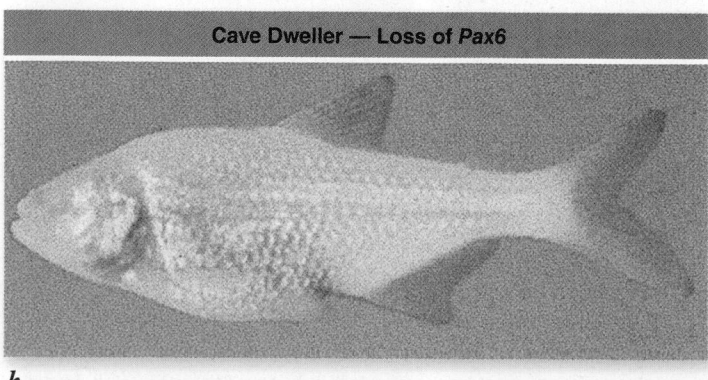

Cave Dweller — Loss of Pax6

b.

Figure 25.13 Cavefish have lost their sight. Mexican tetras (*Astyanax mexicanus*) have (*a*) surface-dwelling members and (*b*) cave-dwelling members of the same species. The cavefish have very tiny eyes, partly because of reduced expression of *Pax6.*

Hypothesis: Pax6 *is necessary for eyespot regeneration in ribbon worms.*

Prediction: *Eyespots will regenerate where* Pax6 *is expressed.*

Test:

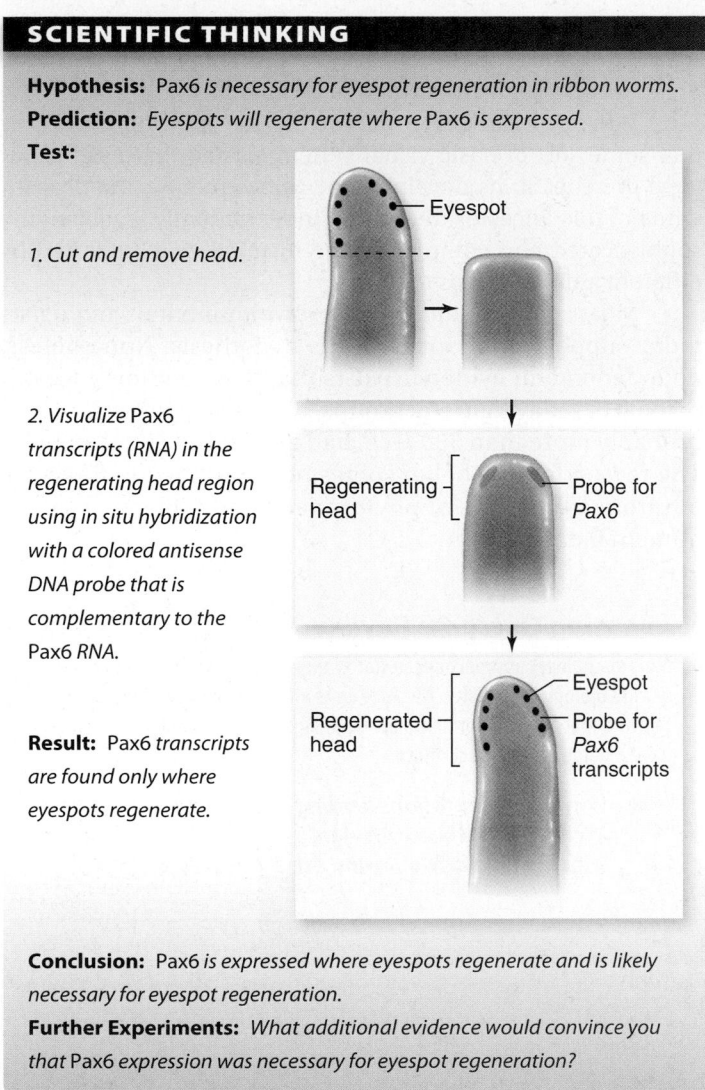

1. Cut and remove head.

Eyespot

2. Visualize Pax6 *transcripts (RNA) in the regenerating head region using in situ hybridization with a colored antisense DNA probe that is complementary to the* Pax6 *RNA.*

Regenerating head — Probe for *Pax6*

Result: Pax6 *transcripts are found only where eyespots regenerate.*

Regenerated head — Eyespot — Probe for *Pax6* transcripts

Conclusion: Pax6 *is expressed where eyespots regenerate and is likely necessary for eyespot regeneration.*

Further Experiments: *What additional evidence would convince you that* Pax6 *expression was necessary for eyespot regeneration?*

Figure 25.14 *Pax6* **expression correlates with ribbon worm eyespot regeneration.**

Although the evidence supporting the role of *Pax6* for eye development in bilaterians continues to grow, there are exceptions. The flatworm planaria can also regenerate eyespots when it is cut in half lengthwise, but no *Pax6* gene expression is associated with regenerating the eyespots. These *Pax6*-related genes are, however, expressed in the central nervous system. A *Pax6*-responsive element, P3-enhancer, has also been identified and shown to be active in planaria.

Cnidarians use other *Pax* genes for eye development

Pax genes were involved in eye development much earlier in evolutionary history than was imagined. In the radially symmetrical cnidarians, which include the jellyfish, two different groups of jellyfish, the hydrozoans and cubozoans, rely on *PaxA* and *PaxB* genes, respectively, for eye development. Walter Gehring repeated his startling transgenic fly experiment (see figure 25.12) by misexpressing a jellyfish *PaxA* gene in a fly, which made a compound eye on its leg again!

As more *Pax* family genes are analyzed, an intriguing phylogeny has emerged (figure 25.15). Different *Pax* genes members have been recruited by different animal lineages for eye development after early gene duplications. Learning more about *Pax* genes in the simplest animals—sponges—should help solve the mystery of the *Pax* genes and the origins of the eye.

The initiation of eye development may have evolved just once

Several explanations are possible for these findings. One is that eyes in different types of animals evolved truly independently, as originally believed. But if this is the case, why are the *Pax* genes so structurally similar and able to play a similar role in so many different groups? Opponents of single evolution of the eye point out that *Pax6* is involved not only in development of the eye, but also in development of the entire forehead region

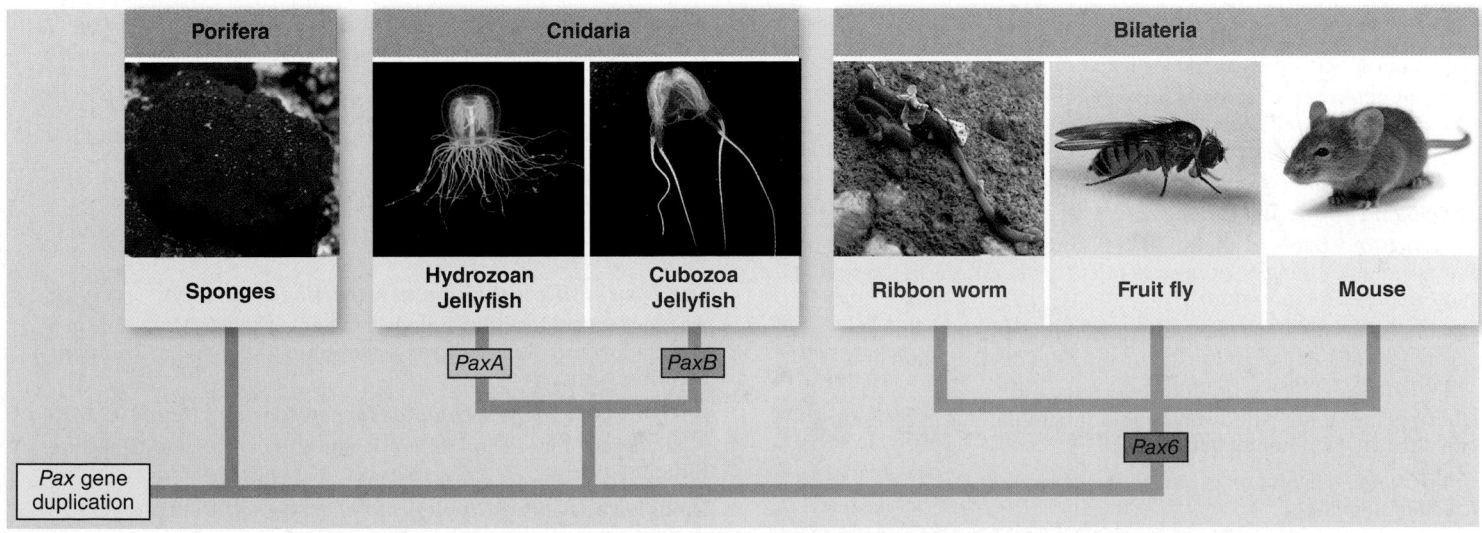

Figure 25.15 **Three different** *Pax* **gene family members were recruited by different animal lineages for eye development.**

of many organisms. Consequently, it is possible that if *Pax6* had a regulatory role in the forehead of early bilaterians, perhaps it has been independently co-opted time and time again to serve a role in eye development. This role would be consistent with the data on planarians and the use of *PaxA* and *PaxB* in jellyfish eye development.

Many other biologists find this interpretation unlikely. The consistent use of *Pax6* in eye development in so many organisms, the fact that it functions in the same role in each case, and the great similarity in DNA sequence and even functional replaceability suggest to many that *Pax6* acquired its evolutionary role in eye development only a single time, in the common ancestor of all extant organisms that use *Pax6* in eye development. The replaceability of *Pax6* with jellyfish *PaxA* in fly eye development further strengthens the case.

The use of different *Pax* family member for eye development is consistent with single origins also. It can be argued that after the early duplications, all the *Pax* genes served redundant functions, including eyespot or light perception functions, and then individual genes became specialized for specific functions. In different lineages, this subfunctionalization evolved differently, but had a common origin. Support for this conclusion comes from findings that other genes are conserved in eye development.

Given the great dissimilarity among eyes of different groups, how can this be? One hypothesis is that the common ancestor of these groups was not completely blind, as traditionally assumed. Rather, that organism may have had some sort of rudimentary visual system—maybe no more than a pigmented photoreceptor cell, maybe a slightly more elaborate organ that could distinguish light from dark.

Whatever the exact phenotype, the important point is that some sort of basic visual system existed that used a *Pax* gene or genes in its development. Subsequently, the descendants of this ancestor diversified independently, evolving the sophisticated and complex image-forming eyes exhibited by different animal groups today.

Most evolutionary and developmental biologists today support some form of this hypothesis. Nonetheless, no independent evidence exists that the common ancestor of most of today's animal groups, a primitive form that lived probably more than 500 MYA, had any ability to detect light. The reason for this belief comes not from the fossil record, but from a synthesis of phylogenetic and molecular developmental data.

Learning Outcome Review 25.4

Multidisciplinary approaches can clarify the evolutionary history of the world's biological diversity. The *Pax6* gene and its many homologues indicate that eye development, although highly diverse in outcome, may have a single evolutionary origin.

■ *Why would mutations leading to defective* **Pax6** *persist in cavefish? If these fish were introduced into a habitat with light, what would you expect to occur?*

Chapter Review

25.1 Evolution of Developmental Patterns

Highly conserved genes produce diverse morphologies.

The *Hox* genes establish body form in animals; *MADS* box genes have a similar function in plants. Changes in these transcription factors and in genes involved in signaling pathways are responsible for new morphologies.

Developmental mechanisms evolve.

Heterochrony refers to alteration of timing of developmental events due to genetic changes; homeosis refers to alterations in the spatial pattern of gene expression.

Modifications of different parts of the coding and regulatory sequences of a transcription factor can alter development and phenotypic expression (figure 25.2).

Changes in signaling pathways, including mutations in receptors, can alter developmental patterns.

Understanding evolution of development requires functional analysis.

Sequence comparisons are essential for both phylogenetic and comparative development studies, but we can only infer function from this information.

25.2 Single-Gene Changes and the Alteration of Form and Function

Cauliflower and broccoli began with a stop codon.

The wide diversity of cabbage subspecies is due to a simple mutation of one gene (figure 25.3).

Cichlid fish jaws demonstrate morphological diversity.

Jaw morphology in fish has also been modified by mutations in one or a few genes.

Stickleback fish lose their armor with a single mutation.

An allele that alters morphology can persist at very low levels in a population and rapidly increase in frequency when environmental conditions change.

Ancestral genes may be co-opted for new functions (figure 25.6).

A single gene may act on different genes or combinations of genes in different species.

Limbs have developed through modification of transcriptional regulation.

Species differences in limbs have resulted from evolutionary changes in gene expression and timing of expression.

Gene duplications provide opportunities for new gene functions.

Duplication of the *AP3* gene and subsequent divergence produced flower petals, whereas the ancestral form affected only stamen development.

Studies have narrowed the active region in *AP3* to the C terminus, which acts differently from the C terminus of the related *PI* gene; only *AP3* can produce petals (figure 25.9).

25.3 Different Ways to Evolve the Same Structure

Insect wing patterns demonstrate homoplastic convergence.

Wing patterns in butterflies have evolved as wing scales developed from ancestral sensory bristles (figure 25.10).

Flower shapes also demonstrate convergence.

Both radial and bilateral symmetry have arisen in multiple ways in flowers, even though radial symmetry is considered ancestral.

25.4 Diversity of Eyes in the Natural World: A Case Study

Morphological evidence indicates eyes evolved at least twenty times.

Homoplasy and convergent evolution are supported as explaining the diversity of eyes found in the animal kingdom.

The same gene, Pax6, initiates fly and mouse eye development.

Transgenic experiments showed that the *Pax6* gene from a mouse, inserted into the genome of *Drosophila*, could cause development of an eye (figure 25.12).

Ribbon worms, but not planaria, use Pax6 for eye development.

Further evidence for the use of *Pax6* in eye development in the bilaterians comes from ribbon worms, but there are exceptions, including the planaria (figure 25.14).

Cnidarians use other Pax genes for eye development.

Jellyfish rely on *PaxA* and *PaxB* genes, rather than *Pax6* for eye development and also used other conserved genes for eye development (figure 25.15).

The initiation of eye development may have evolved just once.

It appears that at some distant point in evolutionary time, *Pax* genes were part of a visual system that later diverged many times, with specific *Pax* family members diverging in terms of their functions.

Review Questions

UNDERSTAND

1. Heterochrony is
   a. the alteration of the spatial pattern of gene expression.
   b. a change in the relative position of a body part.
   c. a change in the relative timing of developmental events.
   d. a change in a signaling pathway.

2. Vast differences in the phenotypes of organisms as different as fruit flies and humans
   a. must result from differences among many thousands of genes controlling development.
   b. have apparently arisen largely through manipulation of the timing and regulation of expression of probably less than 100 highly conserved genes.
   c. can be entirely explained by heterochrony.
   d. can be entirely explained by homeotic factors.

3. Homoplastic structures
   a. can involve convergence of completely unrelated developmental pathways.
   b. always are morphologically distinct.
   c. are produced by divergent evolution of homologous structures.
   d. are derived from the same structure in a shared common ancestor.

4. *Hox* genes are
   a. found in both plants and animals.
   b. found only in animals.
   c. found only in plants.
   d. only associated with genes in the *MADS* complex.

5. The *Brachyury* and *Tbx5* in vertebrates and the *Ap3* gene in flowering plants
   a. are examples of *Hox* genes.
   b. are examples of co-opting a gene for a new function.
   c. are homologues for determining the body plan of eukaryotes.
   d. help regulate the formation of appendages.

6. Which of the following statements about *Pax6* is false?
   a. *Pax6* has a similar function in mice and flies.
   b. *Pax6* is involved in eyespot formation in ribbon worms.
   c. *Pax6* is required for eye formation in *Drosophila*.
   d. *Pax6* is required for eyespot formation in planaria.

7. Which of the following statements about *Tbx5* is true?
   a. *Tbx5, Tbx4,* and *AmphiTbx4/5* have very similar coding regions.
   b. *Tbx5* is involved in tail development in vertebrates.
   c. *Tbx5, Tbx4,* and *AmphiTbx4/5* have very similar regulatory regions.
   d. *Tbx5* initiates hindlimb development.

8. Homeosis
   a. refers to a maintained and unchanging genetic environment.
   b. is a temporal change in gene expression.
   c. is a spatial change in gene expression.
   d. is not an important genetic mechanism in development.

9. Transcription factors are
   a. genes.
   b. sequences of RNA.
   c. proteins that affect the expression of genes.
   d. None of the choices are correct.

10. Independently derived mutations of the *CYC* gene in plants
   a. suggests bilateral floral symmetry among all plants is homologous.
   b. establishes that radial floral symmetry is preferred by pollinators.
   c. establishes that radial floral symmetry is derived for all plants.
   d. None of the choices are correct.

APPLY

1. Choose the statement that best explains how the *Tbx5* protein can be responsible for the development of the heart in amphioxus but initiates forelimb development in mice.
   a. *Tbx5* is a key component of bone.
   b. *Tbx5* is a transcription factor that binds to Fgf-10.
   c. *Tbx5*'s regulatory region has evolved since the two vertebrate whole-genome duplications.
   d. *Tbx5* is a signaling molecule involved in the signaling pathway for limb development in both species.

2. Analyze why it was important to create transgenic plants to determine the role of *AP3* in petal formation and choose the most compelling reason.
   a. It provided a functional test of the role of *AP3* in petal development.
   b. Duplication of *AP3* could not be resolved on the phylogeny.
   c. To check if the phylogenetic position of *AP3* is really derived.
   d. Because tests already established the role of *paleoAP3* in stamen development.

3. The *Eda* allele that causes reduced armor originated about 2 MYA in marine stickleback fish and persists with a frequency of about 1% in marine environments. The frequency is much higher in freshwater populations. Apply your understanding of evolution to determine the most likely reason for the difference in *Eda* allele frequency.
   a. There is positive selection for armor in marine environments because fish are more buoyant in salt water.
   b. Fish have many predators in the marine environment and fish with reduced armor have reduced fitness.
   c. There is negative selection for armor in freshwater because building armor is energetically expensive.
   d. Both b and c are valid.

4. In *Drosophila* species, the yellow *(y)* gene is responsible for the patterning of black pigment on the body and the wings. A comparison of two species reveals that one has a black spot on each wing, and the other species lacks black pigmentation on the wing. The sequence of the coding region for *y* is identical in both species. Critique the following explanations and choose the most plausible one.
   a. The protein coded for by the *y* gene has a different structure in species with the black spots than the species without.
   b. The species without black pigmentation on the wing has a mutation in a regulatory region of the *y* gene.

   c. A deletion mutation is present in one of the exons of the *y* gene in the species without black pigment in the wings.
   d. Convergent evolution explains why both species develop black wing spots.

SYNTHESIZE

1. The *paired-like homeodomain transcription factor 1 (pitx1)* is expressed in the hindlimbs of developing mouse embryos, and its homologue is expressed in the pelvic region of the nine-spine stickleback fish *(Pungitius pungitius)*. You are beginning a research project on *pitx1*, and your supervisor is convinced that *pitx1* has been co-opted to make armor around the pelvic region of this species of stickleback. You begin by isolating RNA from a freshwater *Pungitius pungitius* that has very reduced armor. You convert the RNA to cDNA and amplify the *pitx1* cDNA using PCR. You have your *pitx1* PCR product sequenced and find it has exactly the same sequence as the marine *Pungitius pungitius* with armor. In light of this evidence, analyze your supervisor's hypothesis.

2. From the chapter on evolution of development it would seem that the generation of new developmental patterns would be fairly easy and fast, leading to the ability of organisms to adapt quickly to environmental changes. Construct an explanation for why it can take millions of years, typically, for many of the traits examined to evolve. (*Hint:* Consider the differences in *Eda* allele frequency in marine and freshwater three-spine stickleback fish.)

3. Phenotypic diversity among major groups of organisms can be explained in several ways. On one end of the spectrum, such differences could arise out of differences in many genes that control development. On the other end, small sets of genes might differ in how they regulate the expression of various parts of the genome. Evaluate which view represents our current understanding.

4. Critique the argument that eyes have multiple evolutionary origins.

5. Based on the information in figure 25.8, construct an explanation for the evolutionary differences in genes related to *AP3* in maize and tomato. Be sure to consider what happened before and after the two species diverged from a common ancestor.

6. Having read all of this chapter, return to the claim that the difference between direct and indirect development in sea urchins is caused by a change in gene expression, not differences in genes. Starting at the level of DNA sequence, formulate an argument in support of the claim.

ONLINE RESOURCE

www.ravenbiology.com

Understand, Apply, and Synthesize—enhance your study with animations that bring concepts to life and practice tests to assess your understanding. Your instructor may also recommend the interactive eBook, individualized learning tools, and more.

Chapter 26

The Origin and Diversity of Life

Chapter Contents

Part **V** Diversity of Life on Earth

Introduction

Different life-forms descend from a common ancestor and have many things in common: they are composed of one or more cells, they carry out metabolism and transfer energy with ATP, and they encode hereditary information in DNA. But living things are also highly diverse, ranging from bacteria and amoebas to blue whales and sequoia trees. Coral reefs, such as the one pictured here, are microcosms of diversity, comprising many life-forms and sheltering an enormous array of life. The origins and history of life on Earth are intertwined with Earth's ever-changing geology, climate, and atmosphere. To understand the diversity of life, questions must be asked about the history of Earth on a geological time scale and about the effects life itself has had on Earth's systems.

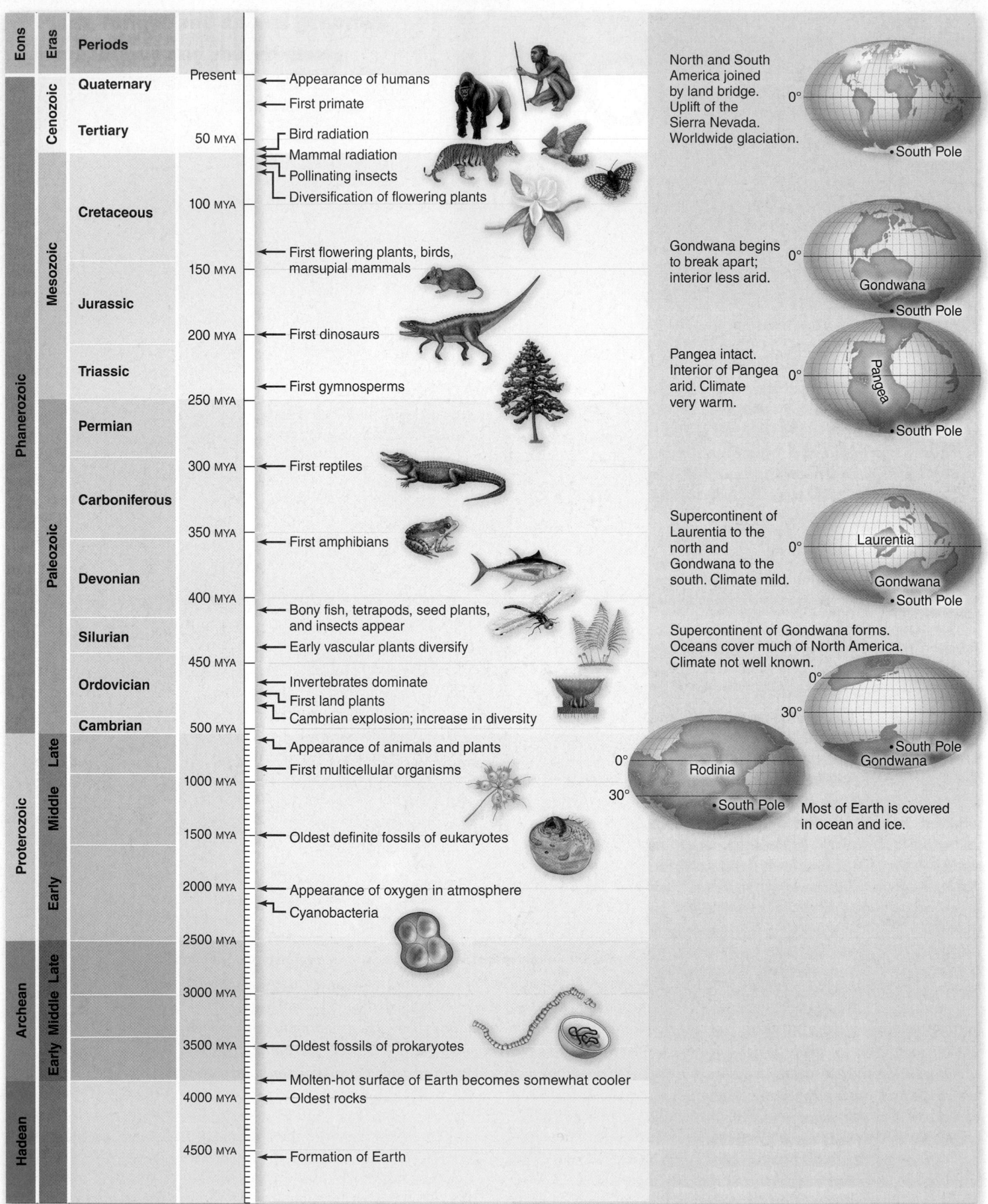

Figure 26.1 Geological timescale and the evolution of life on Earth.

Learning Outcomes

1. Calculate the percent of the Earth's history that includes a given species.
2. Given two fossils, propose a method to determine their ages.

Earth changed over geological time

Many of the fundamental questions about the history of Earth are geological. To explore the origins and diversification of life over billions of years, it is important to understand geological, or deep, time. Geologic time is divided into four eons, stretching over 4.6 billion years (figure 26.1). Eons are subdivided into eras, which are further subdivided into periods.

So much change has occurred since the Earth formed, that no rocks exist from the first 500–700 million years of Earth's history (called the Hadean eon) that preceded the earliest fossils. Although it is impossible to be certain what early Earth was like, geological evidence is consistent with a meteor hitting the Earth almost 4.6 BYA with such force that that debris from the impact formed the Moon. The rocky mantle of the Earth literally melted as atmospheric temperatures exceeded 2000°C.

Hadean Earth was also pummeled by asteroids, which could potentially vaporize entire oceans. It was a wildly dynamic environment, shifting between a fiery and sometimes frozen Earth, that was unlikely to have supported life.

CO$_2$ levels shifted, affecting temperature

The early atmosphere likely had high levels of CO$_2$, and water slowly vaporized from the molten rock. The Earth cooled over a 2-million-year period. As the temperature cooled, clouds made of silicate condensed in the atmosphere and rained down, forming a warm ocean under a CO$_2$ atmosphere. CO$_2$ levels dropped and the Earth cooled, and for a period of time the ocean froze (figure 26.2).

Decreases in CO$_2$ contributed to the decrease in temperature because less radiant energy was absorbed in the atmosphere. What changed the CO$_2$ level in the atmosphere? The ocean and atmosphere were in equilibrium in terms of CO$_2$ levels. Volcanic eruptions added CO$_2$ to the atmosphere and ocean, while the weathering of rock decreased CO$_2$ levels.

Weathering occurred more rapidly under hot, wet conditions than cold, dry conditions. Weathering is the conversion of silicate rock to soil. It occurs when the CO$_2$ in the atmosphere combines with water (H$_2$O) to create a carbonic acid (H$_2$CO$_3$) rain. The carbonic acid interacted with the rock, releasing bicarbonate ions (HCO$_3^-$) and Ca^{2+}. The weathered solutes moved through rivers and oceans and formed calcium carbonate (CaCO$_3$), which precipitated and sequestered the CO$_2$ in the ocean (figure 26.3).

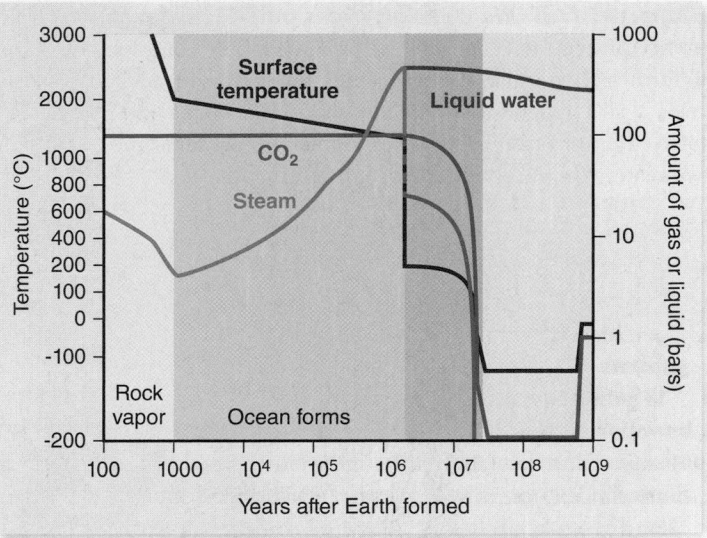

Figure 26.2 A rapid decrease in atmospheric CO$_2$ during the Hadean led to a corresponding decrease in temperature and a frozen ocean.

Continents moved over geological time

Earth's crust formed rigid slabs of rock called plates under both continents and oceans. These huge slabs shift a few centimeters each year, a process called plate tectonics. The term *tectonics* comes from the Greek word for build or builder, and plate movements built and continue to build the geological features of Earth. Most major earthquakes and volcanoes occur at the edges of the plates when they move.

Although the plates move slowly, over deep geological time the cumulative effects are astounding. Several times in Earth's history, all the continents have come together to form a single supercontinent (see figure 26.1). Two proposed supercontinents, Rodinia (all the continents) and Gondwana (composed of all the current Southern Hemisphere continents), existed from 1100 MYA to 650 MYA and 500 MYA, respectively, and both occupied the Southern Hemisphere. Gondwana contributed to the supercontinent Pangea which was fully formed 225 MYA and began to separate 25 to 50 million years later. Geologists have the greatest confidence in plate tectonic evidence from the last 200 million years, beginning with the breakup of Pangea.

Life emerged in the Archean

At some point, life emerged. Some fossil evidence exists from the Archean eon that followed the Hadean. Two billion years into Earth's history, the Proterozoic (early life) eon occurred. It was characterized by the formation of the supercontinent Rodinia, which 650 MYA broke up into several continents before the start of the Phanerozoic (visible life) eon. Collectively, the Hadean, Archean, and Proterozoic eons are referred to as the Precambrian. With the start of the Phanerozoic's Paleozoic era, marked first by the Cambrian period, a remarkable diversification of multicellular organisms took place. Beginning with the Phanerozoic both geologists and biologists begin to focus on chunks of time on the order of periods and shorter.

Figure 26.3 **Weathering rocks pull CO₂ from the atmosphere.** H_2O and CO_2 in the atmosphere combine to form carbonic acid (H_2CO_3) which interacts with rock to release HCO_3^- and Ca^{2+}, which wash into the ocean and form calcium carbonate ($CaCO_3$), which precipitates and sequesters the carbon in the ocean sediment.

CO₂ in the atmosphere

CO_2 combines with H_2O to form carbonic acid
$$CO_2 + H_2O \rightleftharpoons H_2CO_3$$

Carbonic acid reacts with rocks

Ocean

Carbon carried by rivers

Calcium and bicarbonate form calcium carbonate which precipitates
$$Ca^{2+} + HCO_3^- \rightleftharpoons CaCO_3$$

The Phanerozoic eon represents only 12% of Earth's history, yet it contains most of the biological history of the diversification of life (see figure 26.1). Birds and mammals have existed for 4% of Earth's existence, whereas humans have existed for 0.2% of the history of Earth.

The past can be reconstructed from the fossil record

Much of what we know about the history of life on Earth comes from the fossil record. As Earth changed, layers of rock were deposited upon each other and fossils became embedded within these layers (see chapter 21 for a more thorough discussion of the formation of fossils). The relative age of fossils is determined based on the geology of the rocks in which they are found. For a long time it was only possible to use a relative time scale to determine when organisms lived.

Isotopic age dating provides a means of calculating the absolute versus the relative age of fossils. Isotope analysis uses the isotopes found in different rocks. For example, potassium is one of the most common atoms in organisms. All potassium

(K) atoms have the same number of protons, but the isotopes of K vary in the number of neutrons they have. K^{40} has 19 protons and 21 neutrons and is less stable than K^{39} or K^{41}. K^{40} is converted (decays) over time and forms Ar^{40} (argon). The K^{40} half-life is 1.25 billion years. That is, it takes 1.25 billion years for the amount of K^{40} to decrease by 50%. The long half-life makes it useful for dating ancient fossils by determining the ratio of K^{40} to Ar^{40} (figure 26.4).

For events that occurred more recently, radiocarbon dating can be used. Carbon in the form of atmospheric CO_2, with a mix of isotopes C^{14} and C^{12}, is incorporated into plants via photosynthesis. The relative amount of C^{14} to C^{12} decreases with a half-life of about 5700 years. Other isotopes can be used for more intermediate dates.

 Data analysis Starting with 100% as the amount of K^{40} in a fossil, plot a graph of percentage K^{40} as a function of time over four half-lives. How many half-lives will it take for less than 1% of the original isotope to remain? Will you ever end up with 0% K^{40}? Explain your response.

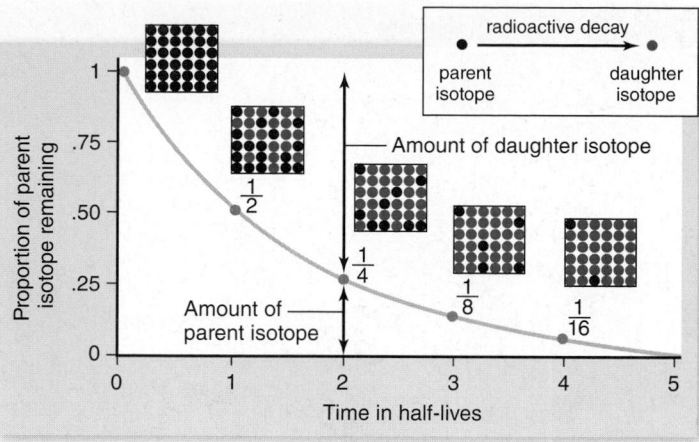

Figure 26.4 Isotopic decay. Isotopes decay at a known rate, called their half-life. After 1 half-life, one-half of the original amount of parent isotope has transformed into a daughter isotope. After each successive half-life, one-half of the remaining amount of the parent isotope is transformed.

Learning Outcomes Review 26.1

The history of Earth extends over 4.6 billion years, but the geological record of what occurred during the Hadean eon before life emerged is limited. The fossil record provides clues to the timing and sequence of the evolution of the diversity of life. Isotopic dating provides information about the age of a fossil.

- **Given an opportunity to study the diversity of life in the fossil record, which eon would you choose? Support your choice.**

26.2 Origins of Life

Learning Outcomes

1. *Contrast today's atmosphere with the likely atmosphere at the end of the Hadean eon.*
2. *Describe key steps necessary for life to originate.*

We really don't know how life began on Earth. Since we cannot recreate the process now, we have to use various lines of scientific exploration to piece together the puzzle of life's origins, beginning with the geology of early Earth. Hadean Earth was a hot mass of molten rock about 4.6 BYA. As it cooled, much of the water vapor present in Earth's atmosphere condensed into liquid water that accumulated on the surface in chemically rich oceans. One scenario for the origin of life is that it originated in this dilute, hot, smelly soup of ammonia, formaldehyde, formic acid, cyanide, methane, hydrogen sulfide, and organic hydrocarbons. Whether at the oceans' edges, in hydrothermal deep-sea vents, or elsewhere, the consensus among researchers is that life arose spontaneously from these early waters. Although the way in which this happened remains a

puzzle, we cannot escape a certain curiosity about the earliest steps that eventually led to the origin of all living things on Earth, including ourselves. How did organisms evolve from the complex molecules that swirled in the early oceans?

Long before there were cells with the properties of life, organic (carbon-based) molecules formed from inorganic molecules. The formation of proteins, nucleic acids, carbohydrates, and lipids were essential, but not sufficient for life. The evolution of cells required early organic molecules to assemble into a functional, interdependent unit.

Early organic molecules may have originated in various ways

Organic molecules may have had extraterrestrial origins

Organic molecules are the basis of all living organisms. How the first organic molecules formed is not known, and some could have extraterrestrial origins. Hundreds of thousands of meteorites and comets are known to have slammed into the early Earth, and recent findings suggest that at least some may have carried organic materials. For example, chemical analysis of the Tagish Lake meteorite, a rocky, carbon-based meteorite, found that nearly 3% of the weight was organic matter. Soluble organic compounds in the meteorite included carboxylic and sulfonic acids, along with trace levels of amino acids. Glycine is the most abundant amino acid in the meteorite and its carbon isotope ratios are inconsistent with rocks found on Earth, supporting the claim that some organic molecules may have had extraterrestrial origins.

Organic molecules may have originated on early Earth

Very few geochemists agree on the exact composition of the early atmosphere. One popular view is that it contained principally carbon dioxide (CO_2) and nitrogen gas (N_2), along with significant amounts of water vapor (H_2O). It is possible that the early atmosphere also contained hydrogen gas (H_2) and compounds in which hydrogen atoms were bonded to the other light elements (sulfur, nitrogen, and carbon), producing hydrogen sulfide (H_2S), ammonia (NH_3), and methane (CH_4).

We refer to such an atmosphere as a *reducing atmosphere* because of the ample availability of hydrogen atoms and their electrons. Because a reducing atmosphere would not have required as much energy to drive chemical reactions as it would today, it would have made it easier to form the carbon-rich molecules from which life evolved.

An early attempt to determine what kinds of organic molecules might have been produced on the early Earth was carried out in 1953 by American chemists Stanley L. Miller and Harold C. Urey. In what has become a classic experiment, they attempted to reproduce the conditions in the Earth's primitive oceans under a reducing atmosphere. Even if their hypothesis proves incorrect—the jury is still out on this—this experiment is critically important because it ushered in the whole new field of prebiotic chemistry.

To carry out their experiment, Miller and Urey (1) assembled a reducing atmosphere rich in hydrogen and excluding gaseous oxygen; (2) placed this atmosphere over liquid water; (3) maintained this mixture at a temperature somewhat below

Figure 26.5 The Miller–Urey experiment. The apparatus consisted of a closed tube connecting two chambers. The upper chamber contained a mixture of gases thought to resemble the primitive Earth's atmosphere. Electrodes discharged sparks through this mixture, simulating lightning. Condensers then cooled the gases, causing water droplets to form, which passed into the second heated chamber, the "ocean." Any complex molecules formed in the atmosphere chamber would be dissolved in these droplets and carried to the ocean chamber, from which samples were withdrawn for analysis.

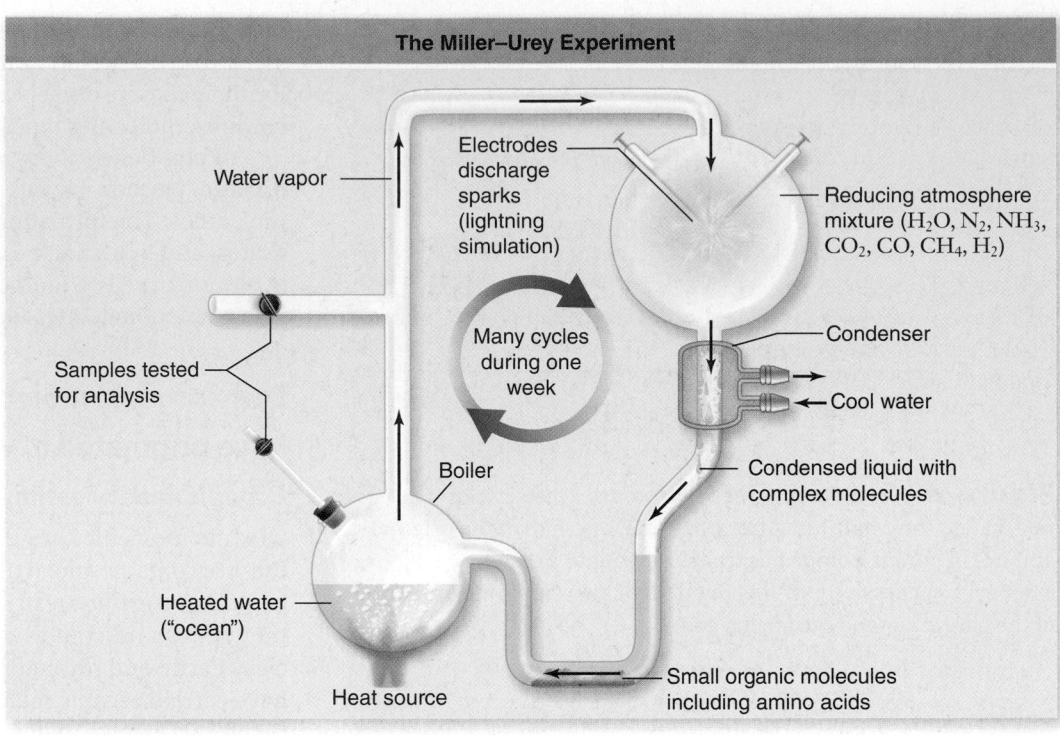

The Miller–Urey Experiment

100°C; and (4) simulated lightning by bombarding it with energy in the form of sparks (figure 26.5).

They found that within a week, 15% of the carbon originally present as methane gas (CH_4) had converted into other simple carbon compounds. Among these compounds were formaldehyde (CH_2O) and hydrogen cyanide (HCN). These compounds then combined to form simple molecules, such as formic acid (HCOOH) and urea (NH_2CONH_2), and more complex molecules containing carbon–carbon bonds, including the amino acids glycine and alanine.

In similar experiments performed later by other scientists, more than 30 different carbon compounds were identified, including the amino acids glycine and alanine, but also glutamic acid, valine, proline, and aspartic acid. As we saw in chapter 3, amino acids are the basic building blocks of proteins, and proteins are one of the major kinds of molecules of which organisms are composed. Other biologically important molecules were also formed in these experiments. For example, hydrogen cyanide contributed to the production of a complex ring-shaped molecule called adenine—one of the bases found in DNA and RNA. Thus, the key molecules of life could have formed in the reducing atmosphere of the early Earth.

Metabolic pathways may have emerged in various ways

Many hypotheses for the emergence of metabolic pathways exist. One scenario assumes that primitive organisms were autotrophic, building all the complex organic molecules they required from simple inorganic compounds, rather than heterotrophic and acquiring all organic compounds from the surrounding environment. For example, glucose may have been synthesized from formaldehyde, CH_2O, in the alkaline conditions that could have existed on early Earth. Glycolysis and a version of the Krebs cycle (see chapter 7) that functioned without enzymes are proposed to be the core from which other metabolic pathways emerged. Early autotrophs could have made, stored, and later used glucose as an energy source.

Enzymes to catalyze metabolic pathways also emerged. Although most enzymes are proteins, RNA can catalyze reactions as well as store genetic information. According to the hypothesis of an RNA world, RNA, rather than DNA, was the first nucleic acid that permitted self-replication, an important step toward life. Later DNA, which is more stable than RNA, took over the information storage function. Proteins that have a greater variety of building blocks (amino acids) gained the enzymatic function.

Ribozymes are RNA sequences with an enzymatic function. Strong evidence supporting the RNA world hypothesis comes from the ribosome, which is used in cells for translation of RNA into proteins. Although the ribosome is composed of both protein and RNA, it is an RNA sequence that is involved in the central mechanism for translation. This is consistent with the hypothesis that early cells used RNA to catalyze the synthesis of peptides from an RNA sequence.

Although much evidence supports an early RNA world, there have been some questions about the hypothesis. With what is known about prebiotic Earth, it is unlikely that there was much ribose sugar then, which is essential for the sugar–phosphate backbone of RNA. Research has shown that it is possible to synthesize RNA nucleotides without pure ribose under the conditions postulated to have existed on prebiotic Earth.

Another challenge to the hypothesis was the puzzle of how long chains of RNA nucleotides formed. Evidence suggests that nucleotides could have concentrated on clay surfaces and that bonds would have formed, linking the concentrated nucleotides. Increased concentrations of RNA nucleotides, fostering bond formation, may also have occurred in ice crystals in salty water. These newer findings support the RNA world hypothesis.

Single cells were the first life-forms

In addition to metabolism, cells require membranes. Constraining organic molecules to a physical space within a lipid or protein bubble could lead to an increased concentration of specific molecules. This in turn could increase the probability of metabolic reactions occurring.

Although modern membranes are made up of a bilayer of phospholipids (see chapter 5), early membranes may have been composed of fatty acids. These are simpler molecules than phospholipids and were more likely to form under prebiotic conditions. Just like phospholipids, they have hydrophilic heads and hydrophobic tails so they can form bilayers and enclosed cell-like structures.

At some point, these bubbles became living cells with cell membranes and all the properties of life described earlier. For most of the history of life on Earth, these single-celled organisms were the only life-forms. We don't know exactly how cells formed because we can't recreate that process, but at some point simple cellular life evolved.

Learning Outcomes Review 26.2

Whether all the organic molecules necessary for life formed on Earth or some formed elsewhere and came to Earth within meteors remains an open question. Although conditions on early Earth cannot be completely reconstructed, it is likely that the temperatures were extreme and that the atmosphere had a very different gaseous composition than it does today that allowed organic molecules, metabolic pathways, and cells to evolve.

- *If you could time travel back to early Earth and return with a primordial pool sample, what types of molecules would you look for to unpack the origins of early life? Construct an argument for the origins of life based on your predicted molecules.*

26.3 *Evidence for Early Life*

Learning Outcome

1. *Evaluate the strength of fossil evidence dating the origins of life.*

The more we learn about the Earth's early history, the more likely it seems that Earth's first organisms emerged and lived at very high temperatures. By about 3.8 BYA, ocean temperatures are thought to have dropped to a hot 49° to 88°C (120°–190°F). Around 3.8 BYA, life first appeared, promptly after the Earth was habitable. Thus, as intolerable as early Earth's infernal temperatures seem to us today, they gave birth to life. Here we explore the evidence for life dating back 3.2 to 3.8 billion years.

Fossil evidence indicates life may have originated 3.2 BYA

Early life may have arisen during the Archean, but evidence of life in the form of microfossils is difficult both to find and to interpret. Nonbiological processes can produce microfossil-like structures, and rocks older than 3 billion years are rarely unchanged by geologic action over time. Two main formations of 3.5- to 3.8-billion-year-old rocks have been found that are mostly intact: the Kaapvaal craton in South Africa and the Pilbara craton in western Australia. (A *craton* is a rock layer of undisturbed continental crust.) Structures have been found in each of these formations and others that are interpreted to be biological in origin. Although this interpretation has been controversial, the accumulation of evidence over time favors these structures as being true fossil cells.

Microfossils are fossilized forms of microscopic life. Many microfossils are small (1–2 μm in diameter) and appear to be single-celled, lack external appendages, and have little evidence of internal structure. Thus, microfossils seem to resemble present-day prokaryotes.

Currently, the oldest microfossils are 3.5 billion years old. The claim that these microfossils are the remains of living organisms is supported by isotopic data and by spectroscopic analysis that indicates they do contain complex carbon molecules. Whether these microscopic structures are true fossil cells is still controversial, and the identity of the prokaryotic groups represented by the various microfossils is still unclear.

More compelling evidence comes from microfossils that are 300 μm in diameter and were part of what were microbial mats in shallow marine environments in South Africa 3.2 BYA (figure 26.6). Wrinkled organic walls (160 nm thick) surrounding these carbonaceous structures can be seen with scanning electron microscopy. Transmission electron microscopy revealed hollow organic-walled vesicles between compressed walls. The features of the walls are similar to those observed in Proterozoic microfossils that are well documented to be biotic in origin. The size of 3.2-billion-year-old microfossils is consistent with eukaryotic cells, but they are more likely to be cyanobacteria (discussed shortly).

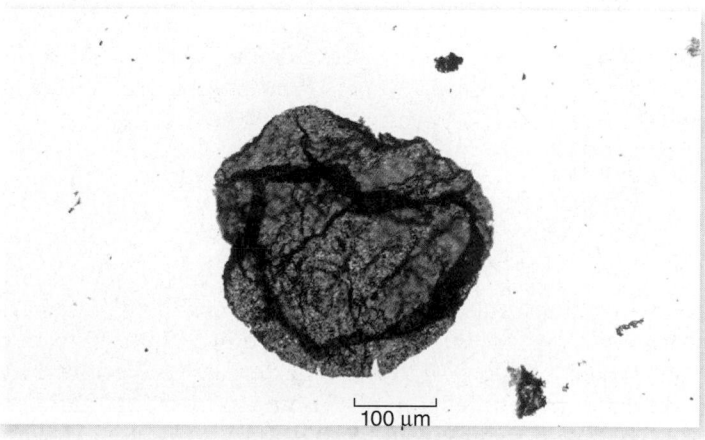

100 μm

Figure 26.6 A 3.2-billion-year-old microfossil from South Africa.

In addition to these microfossils, indirect evidence for ancient life can be found in the form of sedimentary deposits called **stromatolites.** These structures are commonly interpreted as a combination of sedimentary deposits and precipitated material that are held in place by mats of microorganisms. The microorganisms that make up the mats are thought to be cyanobacteria. Formations of stromatolites are as old as 2.7 billion years. Because relatively modern stromatolites are also known, the formation and biological nature of these structures is less contentious (figure 26.7).

Isotopic data indicate that carbon fixation is an ancient process

Another way to ask when life began is to look for the signature of living systems in the geological record. Living systems alter their environments, and sometimes this change can be detected. The most obvious change is that living systems are selective in the isotopes of carbon in compounds they use. Living organisms incorporate C^{12} into their cells before any other carbon isotope, and thus they can alter the ratios of these isotopes in the atmosphere. They also have a higher level of C^{12} in their fossilized bodies than does the nonorganic rock around them.

Much work has been done on dating and analyzing carbon compounds in the oldest rocks, looking for signatures of life. Analysis of carbon signatures indicates carbon fixation, the incorporation of inorganic carbon into organic form, was active as long as 3.8 BYA, consistent with the dating of the oldest microfossils.

The ancient fixation of carbon happened via two main pathways. The most common pathway for carbon fixation is the Calvin cycle (see chapter 8). This is the pathway used by cyanobacteria, algae, and modern land plants that perform oxygenic photosynthesis using two photosystems. The Calvin cycle is also active in green and purple sulfur bacteria that perform anoxygenic photosynthesis using a single photosystem. This anoxygenic form of photosynthesis could account for ancient carbon fixation.

Figure 26.7 Stromatolites. Mats of bacterial cells that trap mineral deposits and form the characteristic dome shapes seen here.

To date, the entire Calvin cycle has not been demonstrated in the archaea, a group of prokaryotes, although the key enzyme for this pathway has been identified in a few archaeal isolates. Instead, some archaea use a reductive version of the Krebs cycle (see chapter 7). This pathway of carbon fixation is also used by some lithotrophic bacteria, which derive energy from the oxidation of inorganic compounds, and by the green sulfur bacteria. Two other pathways may also occur in the lithotrophs, archaea, and the green nonsulfur bacteria. Evidence suggests that the ability to fix carbon has evolved more than once over the course of evolution.

Some hydrocarbons found in ancient rocks may have biological origins

Another way to look for evidence of ancient life is to look for organic molecules, which are clearly of biological origin; such molecules are called *biomarkers*. Although the process sounds simple, it has proved difficult to find such markers. One type of biomarker molecule is hydrocarbons, which are derived from the fatty acid tails of lipids. These can be analyzed for their carbon isotope ratios to indicate biological origin. The analysis of extractable hydrocarbons from the Pilbara formation in Australia found lipids that are indicative of cyanobacteria as long ago as 2.7 billion years. The search for definitive chemical markers for living systems in the oldest rocks and in meteorites is an area of intense interest.

Learning Outcome Review 26.3

Evidence for the earliest cells exists in microfossils. The earliest microfossils are controversial, but they are at least 3.5 billion years old. Other evidence for early life includes isotopic ratios that are skewed by biological activity. The Calvin cycle and a reductive version of the Krebs cycle, as well as other pathways, appear to have led to carbon fixation in ancient life. Some hydrocarbons appear to be biomarkers and may therefore also indicate ancient life-forms.

■ *You have discovered a fossil that may be an early bacterial cell. What evidence would be sufficient to convince you that this is indeed an early bacterial cell?*

26.4 Earth's Changing System

Learning Outcomes

1. *Construct an explanation for relationship between CO_2 levels and glaciation.*
2. *Argue that plate tectonics has affected the evolution of life on Earth.*

The climate (temperature and water availability) and atmosphere (including levels of CO_2 and O_2) are among the many factors that affect the ability of organisms to survive and reproduce. Over the course of Earth's history, repeated and dramatic shifts in all these factors have led to mass extinctions and otherwise influenced the course of evolution (the phenomenon of mass extinction is discussed in chapter 59). Shifting plates have led to volcanic eruptions that alter the atmosphere, including simply blocking sunlight. Oscillating CO_2 levels over geological time correlate with temperature changes as increased atmospheric concentrations of CO_2 trap the heat radiating from the Earth, creating a greenhouse effect. Some changes affecting the climate and atmosphere are strictly geological, but living organisms account for other changes. For example, the evolution of photosynthesis increased O_2 in the atmosphere. In this section we explore how shifts in climate and the atmosphere have affected Earth and life on Earth over geological time.

Earth's climate has been ever-changing

The range of temperatures and rainfall over Earth's history is astounding. Early Earth experienced temperatures over 2000°C and global mean temperatures of –50°C. Temperature shifts immediately affect terrestrial, aquatic, and marine organisms. Over a slightly longer time frame, the level of oceans is affected, further disrupting life.

Earth has been cooling since its formation, but several sudden drops in temperature, including three exceptionally sharp drops that occurred both early and late in the Proterozoic, have decimated life. These extreme drops in temperature resulted in glacial ice covering Earth from pole to pole, a phenomenon called snowball Earth (figure 26.8). Under such conditions the ice reflected back most of the radiant energy from the Sun, maintaining the cold temperatures. Even at the

equator, temperatures would have climbed no higher than –20°C, equivalent to current Antarctic temperatures. With frozen oceans, temperatures were less moderated, and the shifts in temperature were more marked than on Earth today. The glacial ice could still flow and with it move sediments. The resulting sedimentary deposits leave clues to periods of snowball Earth.

Glaciation results in massive extinctions of species. It is difficult to imagine life surviving pole-to-pole glaciation, yet only 80 million years after the last snowball Earth event, bilaterally symmetrical animals are found in the fossil record. Although we don't fully understand the factors initiating snowball Earth or leading to its recovery to a more moderate climate, changes in CO_2 levels and plate tectonics are currently the best supported explanations and are discussed shortly.

Geologic changes and living organisms explain shifts in the atmosphere

Geologic changes can explain many but not all the shifts in the composition of the atmosphere. Living organisms have also substantially altered the atmosphere, as discussed in the next section. Here we focus on CO_2 as one example, recognizing that changes in concentration of other atmospheric gases, including O_2, N_2, and CH_4, have affected the evolution of life on Earth.

At the time of the last two snowball Earth events in the late Proterozoic, most of the continents were in the tropics (see figure 26.1). The hot, wet climate accelerated weathering, described earlier, which led to a significant decline in the concentration of atmospheric CO_2 (see figure 26.3). The accompanying decrease in temperature slowed the weathering, stabilizing both temperature and CO_2 levels. Weathering is hypothesized to have led to the late Proterozoic glaciations.

In addition to changes in weathering associated with a hot, wet climate, plate tectonics can also affect weathering and thus atmospheric levels of CO_2. As a large continent breaks into smaller pieces, more area is exposed to the ocean and becomes wetter. The increased moisture leads to increased weathering and decreased atmospheric CO_2 concentrations. Thus increased weathering from continental shifts during the late Proterozoic also contributed to glaciation.

Continental motion affected evolution

Although they contribute to a changing atmosphere and climate, shifting plates also affect evolution by reproductively isolating populations or allowing previously separate populations to interbreed (see chapter 22). The continents sit on submerged plates that are in motion. During the Proterozoic eon, all the landmass existed as a single supercontinent named Rodinia that began to break up into smaller continents about 700 MYA. This and other shifts have continued throughout Earth's history (see figure 26.1).

The Paleozoic era began with the expansive diversification of life of the Cambrian period and a number of separate continents. But, as the Paleozoic came to a close, Earth once

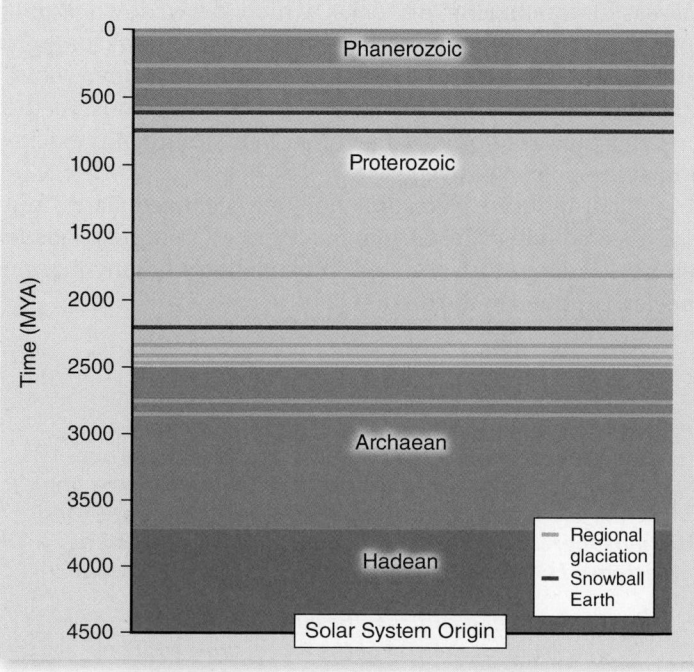

Figure 26.8 Three global glaciation events occurred during the Proterozoic.

chapter **26** *The Origin and Diversity of Life* **515**

again had a single supercontinent—Pangea. During the Carboniferous and Permian periods of the Paleozoic, major collisions of continents complete the development of the Appalachian mountains of North America, when eastern North America and northwestern Africa smashed together. Pangea lasted for 100 million years and began to break up during the Late Triassic and Early Jurassic periods of the Mesozoic era.

The current Cenozoic era began 65 MYA. Australia and Antarctica separated, as did Greenland and North America. The Atlantic Ocean continued to grow as plates in the mid-Atlantic spread. Greenhouse conditions during the Cretaceous period led to a rise in sea level and extensive continental areas were submerged.

During the last 2 million years, cold periods led to glaciation and a drop in sea level. As a result, a land bridge formed between Asia and North America. Humans and other animals migrated between the once disconnected continents. A connection between Australia and southeast Asia also allowed for migration. Understanding when landmasses were connected and which species existed at the time is critical in explaining the evolution of the diversity of life on Earth.

Life changes Earth

 ### Oxygenic photosynthesis produced atmospheric O_2

The early atmosphere contained CO_2, but, unlike the modern atmosphere, it lacked oxygen gas (O_2). The evolution of photosynthesis added O_2 to the ocean and atmosphere, providing an environment conducive to the evolution of cellular respiration (figure 26.9).

Geologic evidence shows a 200-million-year lag between the origins of photosynthesis and substantial concentrations of O_2 in the environment. One explanation for the lag is the formation and precipitation of iron oxide in the ocean. Much of the O_2 released in the ocean during the first 200 million years reacted with elemental iron to form iron oxide, which prevented an increase in atmospheric O_2 concentrations.

As O_2 increased in the atmosphere, some of it interacted with ultraviolet (UV) radiation from the Sun and formed O_3 (ozone). The ozone layer protects terrestrial Earth from UV radiation, reducing the rate of mutations and making life on land possible.

Did plants contribute to glaciations?

There is growing evidence that plants contributed to two glaciations. The initial colonization of land by plants was also followed by gradual cooling and abrupt glaciation 488 to 444 MYA in the Ordovician period. The CO_2 levels just prior to the cooling were in the range of 20 times higher than present day concentrations, and climate models predict the levels would need to be reduced by 50% to trigger glaciation. Geological weathering could explain some cooling through decreased CO_2, but the decrease would not be sufficient to trigger glaciation.

The early land plants lacked roots, but studies using extant relatives show they released organic acids that can increase rock weathering. Although additional weathering caused by plants could accelerate decreases in CO_2 and temperature, it would still be insufficient to trigger glaciation. The most convincing proposal is that phosphorous, an essential nutrient for plant growth, was released from the rocks through weathering and entered the ocean where it supported extensive algal growth. The rapid growth of the photosynthetic algae drew down atmospheric CO_2, triggering glaciation. Increased phosphorous in sedimentary rock from that time supports this conclusion. The entire system equilibrated after the initial release of phosphorous and plants began recycling phosphorous in the soil, eliminating the more massive runoff into aquatic environments.

A second glaciation was concurrent with vascular plants diversification and spread 400 to 360 MYA during the Devonian period. Vascular plants' extensive root systems increased weathering of rocks, which drew down atmospheric CO_2 levels. Plant roots release the same organic acids as the rootless early land plants that weather rocks, releasing essential nutrients including phosphorous. As the vascular plants colonized Earth, global cooling and glaciation at the poles followed.

Plant-induced glaciation not only changed Earth, but also affected other life. Glaciation often led to rapid drops in sea level. This in turn resulted in extinction of many marine species, captured in the fossil record.

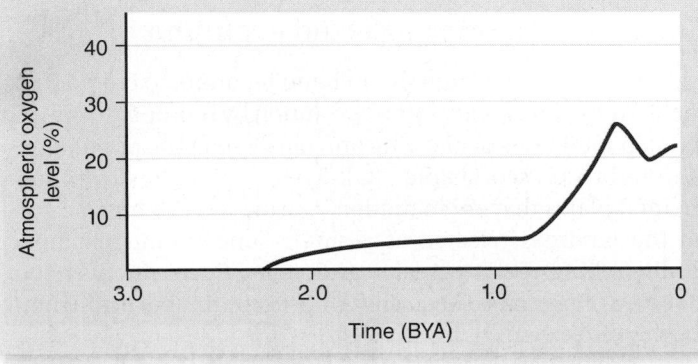

Figure 26.9 Atmospheric O_2 levels over time.

Learning Outcomes Review 26.4

Changes in climate, atmosphere, and location of continents has supported the diversification of life and led to extinctions, as a result of glaciation. In concert, life has changed the Earth's system including land, water, and atmosphere. Both abiotic and biotic factors can lead to decreased CO_2 levels, which can precipitate glaciation.

■ *Compare and contrast the events that triggered the glaciation at the end of the Proterozoic with the glaciation early in the Ordovician period.*

Ever-Changing Life on Earth

Beginning with a single cell in the Archean eon, life has evolved into three monophyletic clades called domains: Eubacteria, Archaea, and Eukaryotes. Six supergroups have been identified within the eukaryotes, based on their phylogenetic relationships: Excavata (organisms lacking typical mitochondria), Chromalveolata (organisms with chloroplasts obtained through secondary endosymbiosis), Archaeplastida (organisms with chloroplasts for photosynthesis), Rhizaria (organisms with slender pseudopods used for movement), Amoebozoans (organisms with blunt pseudopods used for movement), and Opisthokonts (fungi, animal ancestors, and animals) (figure 26.10). Here we will consider the key evolutionary events supporting this incredible diversity of life evolving in response to an ever-changing environment.

Compartmentalization of cells enabled the advent of eukaryotes

For at least 1 billion years, bacteria and archaea ruled the Earth. No other types of organisms existed to eat them or compete with them, and their tiny cells formed the world's oldest fossils.

Archaea are more closely related to eukaryotes and may be found in environments that are extreme by today's standards, including high temperature, high pressure, or high salt. Both bacteria and archaea are distinct from eukaryotes in that they lack compartmentalization of their cells.

Members of the third great domain of life, the eukaryotes, appear in the fossil record much later, only about 1.5 BYA. But despite the metabolic similarity of eukaryotic cells to prokaryotic cells, their structure and function enabled these cells to be larger, and eventually, allowed multicellular life to evolve.

Evolution of the endomembrane system

The hallmark of eukaryotes is complex cellular organization, highlighted by an extensive endomembrane system that subdivides the eukaryotic cell into functional compartments, including the nucleus (figure 26.11, chapter 4). The evolution of a nuclear membrane, not found in bacteria and archaea, accounts for increased complexity in eukaryotes. In eukaryotes, RNA transcripts from nuclear DNA are processed and transported across the nuclear membrane into the cytosol, where translation occurs. The physical separation of transcription and translation in eukaryotes adds additional levels of control to the process of gene expression.

The Golgi apparatus and endoplasmic reticulum are key innovations that facilitate intracellular transport and the localization of proteins in specific regions of the cell (see chapter 4). These membrane systems, as well as the nuclear membrane, arose through the infolding of the cellular membrane. Not all

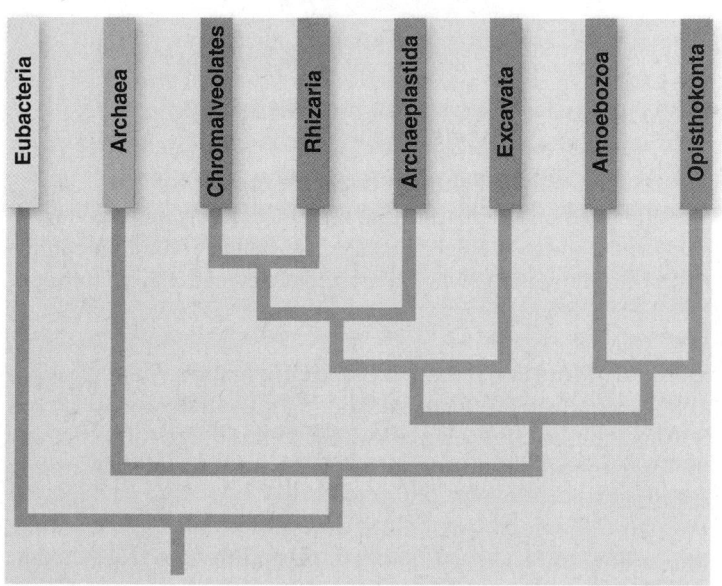

Figure 26.10 Six supergroups have been identified within the Eukaryote domain, one of three domains of life on Earth.

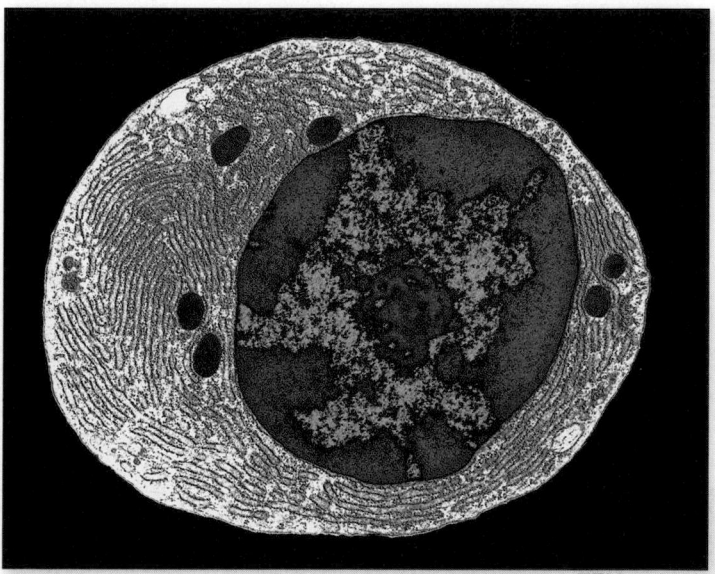

TEM 15,500×

Figure 26.11 Cellular compartmentalization. These complex, single-celled organisms called Paramecia are classified as protists. The yeasts, stained red in this photograph, have been consumed and are enclosed within membrane-bounded sacs called digestive vacuoles.

cellular compartments, however, are derived from the endo-membrane system.

Endosymbiosis and the origin of eukaryotes

With few exceptions, modern eukaryotic cells possess the energy-producing organelles termed *mitochondria,* and photosynthetic eukaryotic cells possess *chloroplasts,* the energy-harvesting organelles. Mitochondria and chloroplasts are both believed to have entered early eukaryotic cells by a process called **endosymbiosis,** which is discussed in more detail in chapter 29.

Mitochondria are the descendants of relatives of purple sulfur bacteria and the parasite *Rickettsia* that were incorporated into eukaryotic cells early in the history of the group. Chloroplasts are derived from cyanobacteria. The red and green algae acquired their chloroplasts by directly engulfing a cyanobacterium. The brown algae most likely engulfed red algae to obtain chloroplasts (figure 26.12).

Molecular phylogenetic data indicate that eukaryotes arose early in the Proterozoic and diversified later in that eon. Microfossil evidence from the Proterozoic supports the existence of eukaryotes as early as 1.5 BYA.

Multicellularity leads to cell specialization

The unicellular body plan has been tremendously successful, with unicellular prokaryotes and eukaryotes constituting about half of the biomass on Earth. But a single cell has limits, even with the within cell specialization provided by compartmentalization in eukaryotes. The evolution of multicellularity allowed organisms to deal with their environments in novel ways through differentiation of cell types into tissues and organs.

True multicellularity, in which the activities of individual cells are coordinated and the cells themselves are in contact, occurs only in eukaryotes and is one of their major characteristics. Bacteria and many single-celled eukaryotes form colonial aggregates of many cells, but the cells in the aggregates have little differentiation or integration of function (figure 26.13).

Multicellularity has arisen independently in different eukaryotic supergroups. For example, multicellularity arose independently in the red, brown, and green algae. One lineage of multicellular green algae was the ancestor of the plants (see chapters 29). A different unicellular ancestor in the Opisthokonts gave rise to all multicellular animals.

Multicellularity required that cells connect to each other and communicate. Although each cell has identical genetic information, gene expression varies among cells to allow specialization. Mechanisms for coordinating gene expression and cell differentiation evolved.

Sexual reproduction increases genetic diversity

Another major characteristic of eukaryotic species as a group is sexual reproduction. Although some interchange of genetic

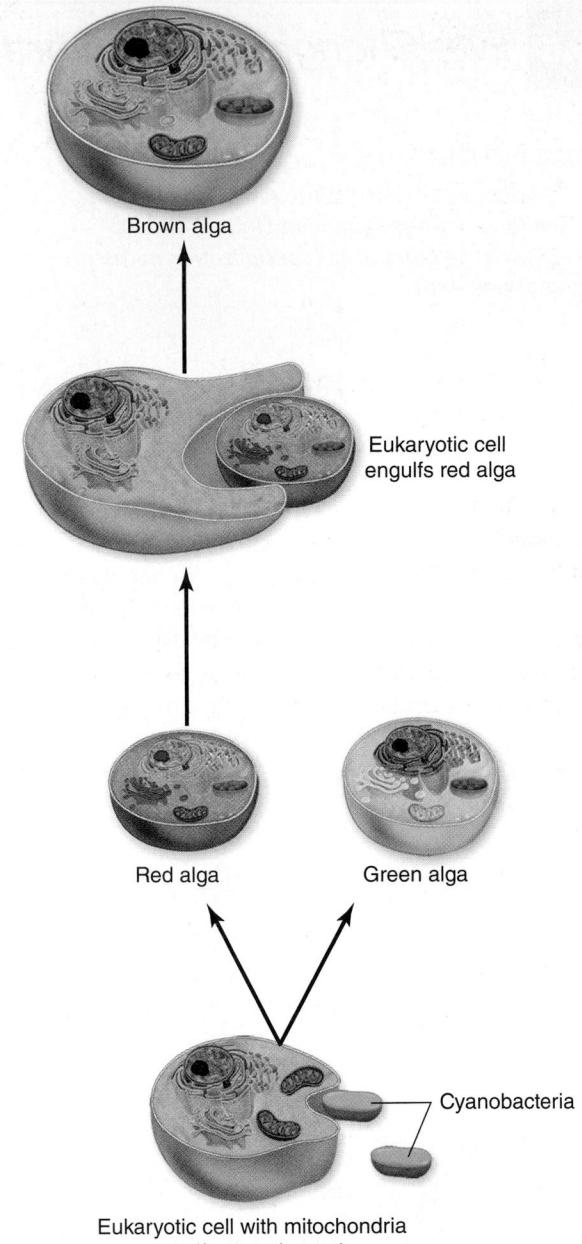

Figure 26.12 All chloroplasts are monophyletic.
The same cyanobacteria were engulfed by multiple hosts that were ancestral to the red and green algae. Brown algae share the same ancestral chloroplast DNA, but most likely gained it by engulfing red algae.

material occurs in bacteria, it is certainly not a regular, predictable mechanism in the same sense that sex is in eukaryotes. Sexual reproduction allows greater genetic diversity through the processes of meiosis and crossing over, as you learned in chapter 13.

In some of the eukaryotic supergroups, sexual reproduction occurs only occasionally. The first eukaryotes were probably haploid; diploids seem to have arisen on a number of separate occasions by the fusion of haploid cells, which then eventually divided by mitosis.

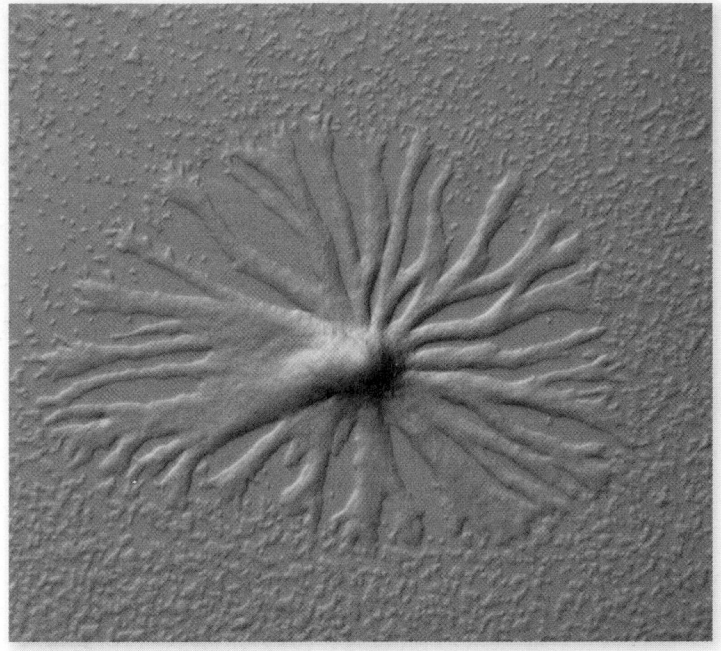

Figure 26.13 **Aggregation of** *Dictyostelium discoideum* **to form a colonial organism.**

Figure 26.14 **Fossil from the Cambrian explosion.** An unusually large number of fossils with soft body parts, such as this *Marrella splendens* fossil from the Burgess Shale in Yoho National Park in British Columbia, provide evidence for the rapid diversification of animal life during this period.

Rapid diversification occurred during the Cambrian

The evolutionary innovations in cells occurred while life was primarily aquatic and established the foundations that led to the tremendous diversity of life on Earth today. As the Proterozoic came to a close with the end of the third snowball Earth event, the Paleozoic began with the Cambrian period lasting from 542 to 488 MYA. This was a period of extremely rapid expansion of life referred to as the Cambrian explosion. We know a great deal about this period because not only the hard parts of organisms, but also the soft parts were preserved in the fossil record in three sites in British Columbia, Greenland, and China (figure 26.14).

For 3 billion years, life, with the exception of a few groups of algae, had been unicellular. In the period leading up to the Cambrian radiation, the first multicellular animals appeared. During the 50 million years that followed, ancestors of almost every group of animals evolved.

Major innovations allowed for the move onto land

The Cambrian radiation was confined to the ocean. Shortly after, plants and then animals colonized terrestrial environments. The evolution of photosynthesis, which resulted in an O_2-rich atmosphere, also resulted in the ozone layer, which protects life on the surface from UV radiation.

Successfully moving from a watery to a terrestrial environment required innovations to prevent desiccation and to obtain water. Gas exchange also required new strategies. In animals, lungs were more effective than gills or gas exchange through moist skin. Plants evolved stomata, regulated openings in the surface of the plant, to facilitate gas exchange and prevent water loss. These and other major innovations are discussed in later chapters.

Naming diverse organisms is essential in biology

An in-depth exploration of the diversity of life requires a shared system for naming organisms. Biology has shifted from an emphasis on identifying and naming organisms to constructing evolutionary hypotheses to explain the relatedness of species as described in chapter 23, but scientific names for species are still essential for biologists to communicate with each other. By convention, a species is given a binomial name. The first part of the name identifies the genus, and the second part the individual species.

Historically, genera with similar characters were grouped into a cluster called a **family,** and similar families were placed into the same **order** (figure 26.15). Orders with common properties were placed into the same **class,** and classes with similar characteristics into the same **phylum** (plural, *phyla*). Finally, the phyla were assigned to one of several great groups, the **kingdoms.** As better information and

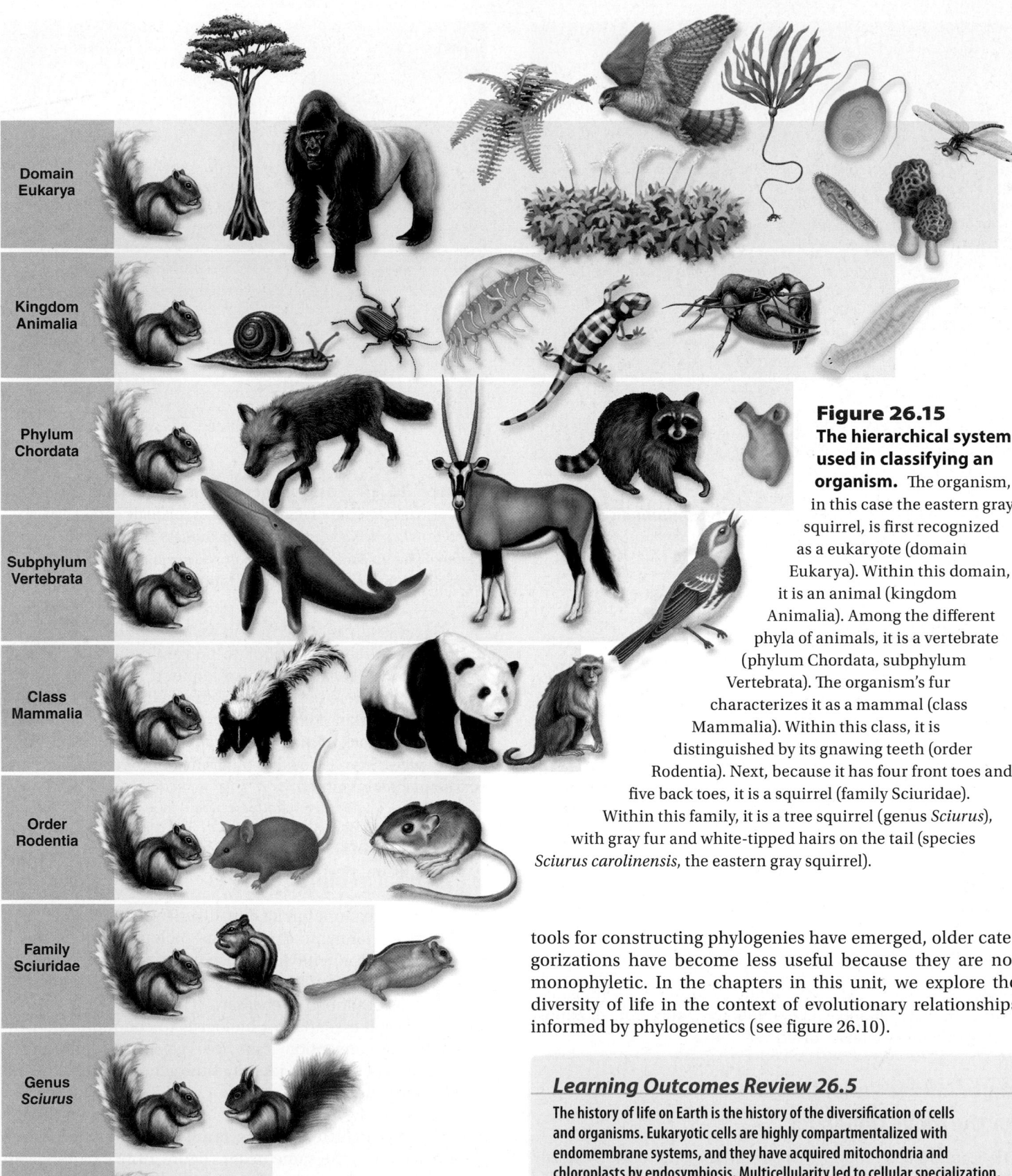

**Figure 26.15
The hierarchical system used in classifying an organism.** The organism, in this case the eastern gray squirrel, is first recognized as a eukaryote (domain Eukarya). Within this domain, it is an animal (kingdom Animalia). Among the different phyla of animals, it is a vertebrate (phylum Chordata, subphylum Vertebrata). The organism's fur characterizes it as a mammal (class Mammalia). Within this class, it is distinguished by its gnawing teeth (order Rodentia). Next, because it has four front toes and five back toes, it is a squirrel (family Sciuridae). Within this family, it is a tree squirrel (genus *Sciurus*), with gray fur and white-tipped hairs on the tail (species *Sciurus carolinensis*, the eastern gray squirrel).

Domain Eukarya

Kingdom Animalia

Phylum Chordata

Subphylum Vertebrata

Class Mammalia

Order Rodentia

Family Sciuridae

Genus *Sciurus*

Species *Sciurus carolinensis*

Sciurus carolinensis

tools for constructing phylogenies have emerged, older categorizations have become less useful because they are not monophyletic. In the chapters in this unit, we explore the diversity of life in the context of evolutionary relationships informed by phylogenetics (see figure 26.10).

Learning Outcomes Review 26.5

The history of life on Earth is the history of the diversification of cells and organisms. Eukaryotic cells are highly compartmentalized with endomembrane systems, and they have acquired mitochondria and chloroplasts by endosymbiosis. Multicellularity led to cellular specialization, and meiosis and sexual reproduction increased genetic diversity. Bursts of rapid radiation of species, as well as periodic extinctions, have led to the diversity of marine and terrestrial life in existence today.

■ *Analyze ways evolutionary innovations prior to the Cambrian may have contributed to the rapid radiation.*

26.1 Deep Time

Earth changed over geological time (figure 26.1).

During the 4.6-billion-year history of life on Earth, Earth has experienced extreme shifts in temperature that correspond with shifting CO_2 levels. Continents formed, collided to make supercontinents, and separated again multiple times. Most of the biological history of life on Earth has occurred in the last 12% of Earth's history.

The past can be reconstructed from the fossil record.

The position of fossils within layers of rocks can provide information about the relative age of fossils, and isotopic dating can provide absolute age.

26.2 Origins of Life

Early organic molecules may have originated in various ways.

Some organic molecules may have had extraterrestrial origins, but most likely formed under the reducing environment of early Earth.

Metabolic pathways may have emerged in various ways.

We are not quite sure how metabolic pathways arose. Autotrophic organisms may have evolved ways to incorporate inorganic molecules in the environment into organic molecules. Early pathways for utilizing energy did not require enzymes.

Single cells were the first life-forms.

At some point, simple cellular life evolved as membranes formed and bounded metabolic and self-replicating activities.

26.3 Evidence for Early Life

Fossil evidence indicates life may have originated 3.2 BYA.

Microfossils with organic wells have been studied with scanning electron microscopy and date back at least 3.2 billion years. More substantial evidence comes from 2.7-billion-year-old stromatolites, sedimentary deposits of microorganisms.

Isotopic data indicate that carbon fixation is an ancient process.

Carbon dating revealed that organisms were sequestering carbon using an ancient form of photosynthesis as early as 3.8 BYA.

Some hydrocarbons found in ancient rocks may have biological origins.

Organic molecules that clearly have biological origins, including lipids, have been identified in 2.7-billion-year-old rock formations in Australia.

26.4 Earth's Changing System

Earth's climate has been ever-changing.

Earth has been cooling since its formation, but extreme shifts in temperature, correlating with changes in CO_2 levels, have been associated with glaciation events, including three glaciations that covered the entire globe. Glaciations can lead to mass extinction and affect the course of evolution.

Geologic changes and living organisms explain shifts in the atmosphere.

Weathering in hot, wet climates and on increased moist surface areas resulting from the breakup of supercontinents resulted in the sequestration of CO_2 in the oceans. In some cases, the drop in CO_2 concentrations can be sufficient to trigger glaciations.

Continental motion affected evolution.

Plate tectonics explains the gradual movement of continents that can change the climate and affect whether or not populations are able to mix and interbreed.

Life changes Earth.

Plant life has contributed to two glaciation events by pulling down CO_2 from the atmosphere through both photosynthesis and the weathering of rock to release phosphorous, a necessary plant nutrient.

26.5 Ever-Changing Life on Earth

Compartmentalization of cells enabled the advent of eukaryotes.

Through both the infolding of membranes to form endomembrane systems and endosymbiosis, eukaryotic cells have gained compartments that enhance cellular function (figure 26.12).

Multicellularity leads to cell specialization.

Multicellularity arose independently in eukaryotic supergroups.

Sexual reproduction increases genetic diversity.

The evolution of meiosis increased the genetic diversity available for natural selection.

Rapid diversification occurred during the Cambrian.

The evolution of compartmentalization, multicellularity, and sexual reproduction established the foundations for the rapid evolution of animal life during the Cambrian period (542–488 MYA).

Major innovations allowed for the move onto land.

Moving to a terrestrial environment required adaptations to prevent desiccation, obtain water, and facilitate gas exchange in a dry environment.

Naming diverse organisms is essential in biology (figure 26.15).

Although biology has shifted from naming organisms to constructing evolutionary hypotheses of relatedness, scientific names are still essential for biologists to communicate with each other. Species are given a genus and species name.

UNDERSTAND

1. The Miller–Urey experiment demonstrated that
   a. life originated on Earth.
   b. organic molecules could have originated in the early atmosphere.
   c. the early genetic material on the planet was DNA.
   d. the early atmosphere contained large amounts of oxygen.

2. Plate tectonics can contribute to
   a. volcanoes and earthquakes.
   b. formation of supercontinents.
   c. increased weathering and CO_2 sequestration.
   d. All of the choices are correct.

3. Identify which of the following statements is false and correct the statement.
   a. Brown and red algae are not closely related phylogenetically.
   b. Chloroplasts in brown and red algae are monophyletic.
   c. Brown algae gained chloroplasts by engulfing green algae (endosymbiosis).
   d. None of the statements are false.

4. Which of the following events occurred first in eukaryotic evolution?
   a. Endosymbiosis and mitochondria evolution
   b. Endosymbiosis and chloroplast evolution
   c. Compartmentalization and formation of the nucleus
   d. Formation of multicellular organisms

5. Isotopic decay of C^{14}
   a. can be used to date fossils from the Archaean.
   b. can support the claim that carbon was fixed by organisms as early as 3.8 BYA.
   c. has a half-life of 1.25 billion years.
   d. is useful in establishing the extraterrestrial origins of organic molecules.

APPLY

1. A global glaciation would be unlikely to occur if
   a. a supercontinent formed near the equator and there was extensive rainfall.
   b. millions of acres of forest were cleared.
   c. vast amounts of phosphorous found its way into aquatic and oceanic environments.
   d. there was a rapid expansion of algal populations in the ocean.

2. Although we do not know how early life arose, the following likely happened:
   a. All organic molecules were transported to Earth by meteors.
   b. High levels of oxygen were essential for glycolysis.
   c. Lipids organized to form cell membranes.
   d. Organic molecules formed once temperatures reached moderate levels, equivalent to today's environment.

3. Which bits of evidence would convince you that life originated as early as 3.2 BYA?
   a. You look at a microfossil under a scanning electron microscope and it is the same shape as a cell.
   b. A high-quality transmission electron micrograph of a fossil cell reveals cellular compartments, including a possible nucleus.
   c. Potassium dating of a fossil containing a possible cell indicates that the fossil is 3.2 billion years old.
   d. Using transmission and scanning electron microscopy, you find evidence of a carbon-based material in what appears to be a cell wall of a fossil isotopically dated as 3.2 billion years old.

4. During which times would you expect that geographic isolation would be particularly important in the evolution of life?
   a. Cambrian period
   b. End of the Paleozoic era
   c. The beginning of the Cenozoic era
   d. Both a and c are correct.

5. The chloroplasts of brown algae
   a. have a chromosome with a very different DNA sequence than the chromosome of a red alga.
   b. are surrounded by four membranes.
   c. are surrounded by two membranes.
   d. have a chromosome with a DNA sequence that is similar to red algae, but very different from green algae.

SYNTHESIZE

1. Vascular plants sequestered large amounts of CO_2 through photosynthesis when they first expanded across terrestrial environment, leading to a glaciation. Evaluate the claim that a similar glaciation event occurred when the first plants colonized land. Be sure to consider whether or not photosynthesis by these new species alone could cause a glaciation.

2. Synthesizing what you have learned in this chapter, analyze the multiple effects the collision of two plates could have on the evolution of life on Earth.

3. Analyze the factors that may have contributed to the Cambrian explosion.

ONLINE RESOURCE

www.ravenbiology.com

Understand, Apply, and Synthesize—enhance your study with animations that bring concepts to life and practice tests to assess your understanding. Your instructor may also recommend the interactive eBook, individualized learning tools, and more.

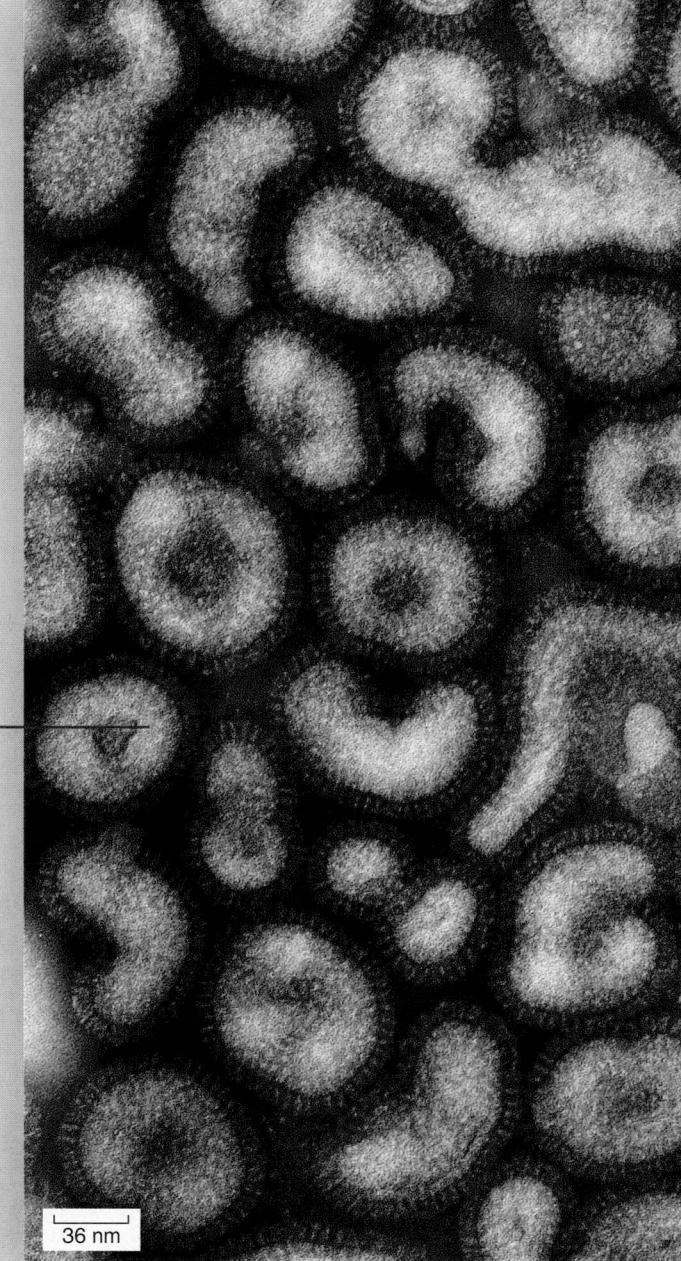

36 nm

Chapter 27

Viruses

Chapter Contents

Introduction

We begin our exploration of the diversity of life with viruses. Viruses are genetic elements enclosed in protein; they are not considered organisms since they lack many of the features associated with life, including cellular structure, and independent metabolism or replication. For this reason viral particles are not called viral cells, but virions, *and they are generally not described as living or dead but as active or inactive. Because of their disease-producing potential, however, viruses are important biological entities. The virus particles pictured here are responsible for causing influenza—flu for short. In the flu season of 1918 to 1919, an influenza pandemic killed an estimated 50 to 100 million people worldwide, twice as many as were killed in combat during World War I. Other viruses cause such diseases as AIDS, SARS, and hemorrhagic fever, and some cause certain forms of cancer.*

For more than four decades, viral studies have been thoroughly intertwined with those of genetics and molecular biology. Classic studies using viruses that infect bacteria (known as bacteriophage) *have led to the discovery of restriction enzymes and the identification of nucleic acid, not protein, as the hereditary material. Currently, viruses are one of the principal tools used to experimentally carry genes from one organism to another. Applications of this technology could include treating genetic illnesses and fighting cancer.*

The Nature of Viruses

All viruses have the same basic structure—a core of nucleic acid surrounded by protein. This structure lacks cytoplasm, and it is not a cell. Individual viruses contain only a single type of nucleic acid, either DNA or RNA. The DNA or RNA genome may be linear or circular; single-stranded or double-stranded.

RNA viruses may be segmented, with multiple RNA molecules within a virion, or nonsegmented, with a single RNA molecule. Viruses are classified, in part, by the nature of their genomes: RNA viruses, DNA viruses, or retroviruses.

Viruses are strands of nucleic acids encased in a protein coat

Nearly all viruses form a protein sheath, or **capsid,** around their nucleic acid core (figure 27.1). The capsid is composed of one to a few different protein molecules repeated many times. The repeating units are called capsomeres.

In several viruses, specialized enzymes are stored with the nucleic acid, inside the capsid. One example is reverse transcriptase, which is required for retroviruses to complete their cycle and is not found in the host. This enzyme is needed early in the infection process and is carried within each virion.

Many animal viruses have an *envelope* around the capsid that is rich in proteins, lipids, and glycoprotein molecules. The lipids found in the envelope are derived from the host cell; however, the proteins found in a viral envelope are generally virally encoded.

Viral hosts include virtually every kind of organism

Viruses occur as obligate intracellular parasites in every kind of organism that has been investigated for their presence. Viruses infect fungal cells, bacterial cells, and protists as well as cells of plants and animals; however, each type of virus can replicate in only a very limited number of cell types. A virus that infects bacteria would be ill-equipped to infect a human or plant cell.

The suitable cells for a particular virus are collectively referred to as its **host range.** Once inside a multicellular host, many viruses also exhibit **tissue tropism,** targeting only a specific set of cells. For example, rabies virus grows within neurons, and hepatitis virus replicates within liver cells. Once inside a host cell, some viruses, such as the highly dangerous Ebola virus, wreak havoc on the cells they infect; others produce little or no damage. Still other viruses remain dormant until a specific signal or event triggers their expression.

As one example, a person can get chicken pox as a child, recover, and develop the disease shingles decades later. Both chicken pox and shingles are caused by the same virus, varicella zoster. This virus can remain dormant, or *latent,* for years. Stresses to the immune system may trigger an outbreak of shingles in people who have had chicken pox in the past. This is caused by the same virus, but the infection may be called herpes zoster because the virus is actually a herpesvirus.

Any given organism may often be susceptible to more than one kind of virus. This observation suggests that many more kinds of viruses may exist than there are kinds of organisms—perhaps trillions of different viruses. Only a few thousand viruses have been described at this point.

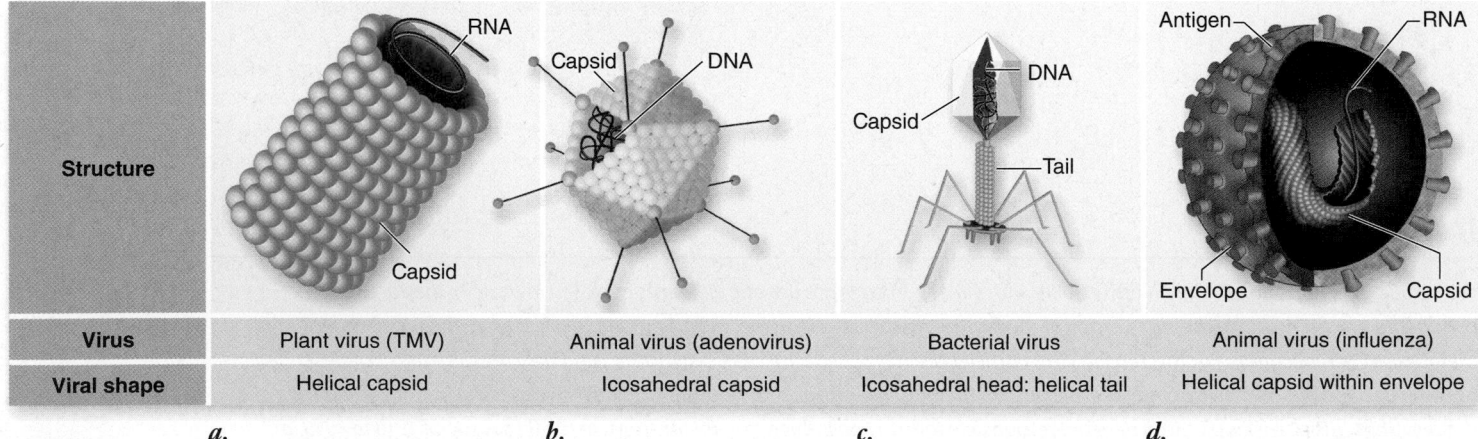

Structure				
Virus	Plant virus (TMV)	Animal virus (adenovirus)	Bacterial virus	Animal virus (influenza)
Viral shape	Helical capsid	Icosahedral capsid	Icosahedral head: helical tail	Helical capsid within envelope
	a.	*b.*	*c.*	*d.*

Figure 27.1 Structure of virions. Viruses are characterized as helical, icosahedral, binal, or polymorphic, depending on their symmetry. *a.* The capsid may have helical symmetry such as the tobacco mosaic virus (TMV). TMV infects plants and consists of 2130 identical protein molecules *(green)* that form a cylindrical coat around the single strand of RNA *(red). b.* The capsid of icosahedral viruses has 20 facets made of equilateral triangles. These viruses can come in many different sizes all based on the same basic shape. *c.* Bacteriophage come in a variety of shapes, but binal symmetry is exclusively seen in phages such as the T4 phage of *E. coli.* This form of symmetry is characterized by an icosahedral head, which contains the viral genome, and a helical tail. *d.* Viruses can also have an envelope surrounding the capsid such as the influenza virus. This gives the virus a polymorphic shape. This virus has eight RNA segments, each within a helical capsid.

Viruses replicate by taking over host machinery

An infecting virus can be thought of as a set of instructions, not unlike a computer program. A cell is normally directed by chromosomal DNA-encoded instructions, just as a computer's operation is controlled by the instructions in its operating system. A virus is simply a set of instructions, the viral genome, that can trick the cell's replication and metabolic enzymes into making copies of the virus. Computer viruses get their name because they perform similar actions, taking over a computer and directing its activities. Like a computer with a virus, a cell with a virus is often damaged by infection.

Viruses can reproduce only when they enter cells. When they are outside of a cell, viral particles are called *virions* and are metabolically inert. Viruses lack ribosomes and the enzymes necessary for protein synthesis and most, if not all, of the enzymes for nucleic acid replication. Inside cells, the virus hijacks the transcription and translation systems to produce viral proteins from *early genes,* which are the genes in the viral genome expressed first. This is followed by the expression of *middle genes* and eventually *late genes.* This cascade of gene expression leads to replication of viral nucleic acid and production of viral capsid proteins. The late genes generally code for proteins important in assembly and release of viral particles from a host cell.

Most viruses come in two simple shapes

Most viruses have an overall structure that is either *helical* or *icosahedral.* Helical viruses, such as the tobacco mosaic virus in figure 27.1a, have a rodlike or threadlike appearance. Icosahedral viruses have a soccer ball shape, the geometry of which is revealed only under the highest magnification with an electron microscope.

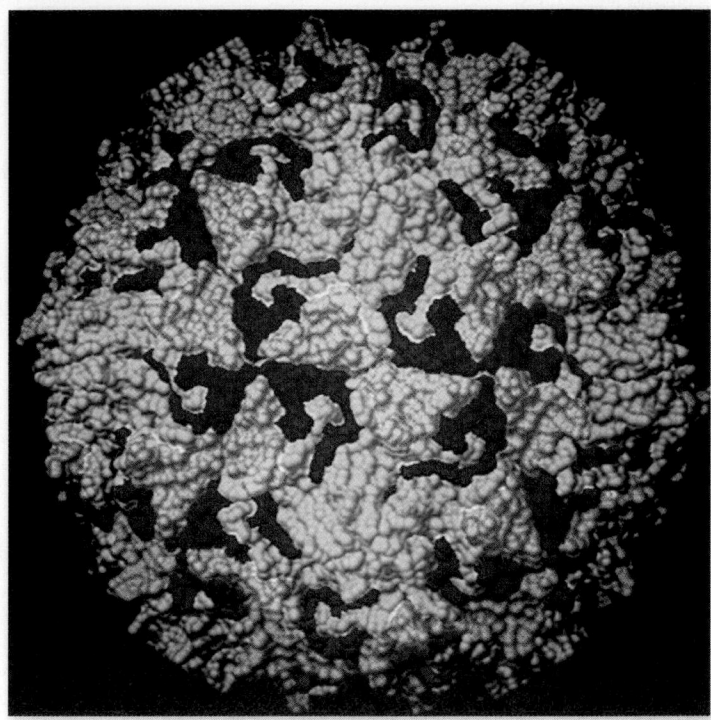

Figure 27.2 Icosahedral virion. The poliovirus has icosahedral symmetry. The capsid is formed from multiple copies of four different proteins shown in different colors. (One protein is internal and cannot be seen.)

The **icosahedron** is a structure with 20 equilateral triangular facets. Most animal viruses are icosahedral in basic structure (figure 27.1b). The icosahedron is the basic design of the geodesic dome, and it is the most efficient symmetrical arrangement that subunits can take to form a shell with maximum internal capacity (figure 27.2).

Some viruses, such as the T-even bacteriophage shown in figure 27.3, are complex. Complex viruses have a *binal,* or

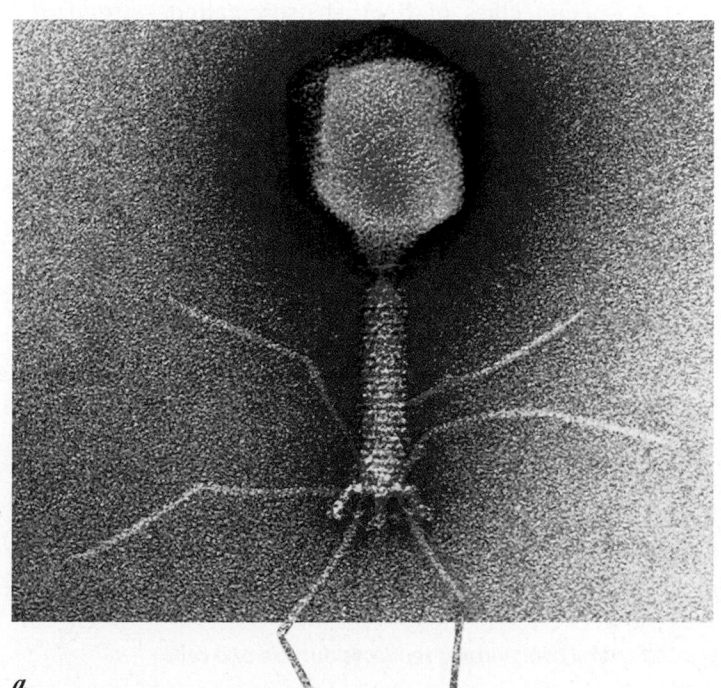

a.

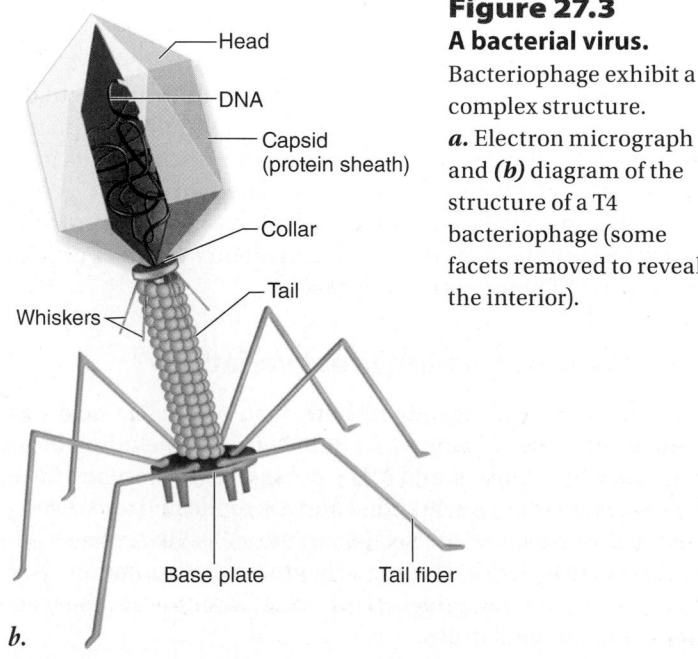

b.

**Figure 27.3
A bacterial virus.**
Bacteriophage exhibit a complex structure.
a. Electron micrograph and *(b)* diagram of the structure of a T4 bacteriophage (some facets removed to reveal the interior).

Labels: Head, DNA, Capsid (protein sheath), Collar, Whiskers, Tail, Base plate, Tail fiber

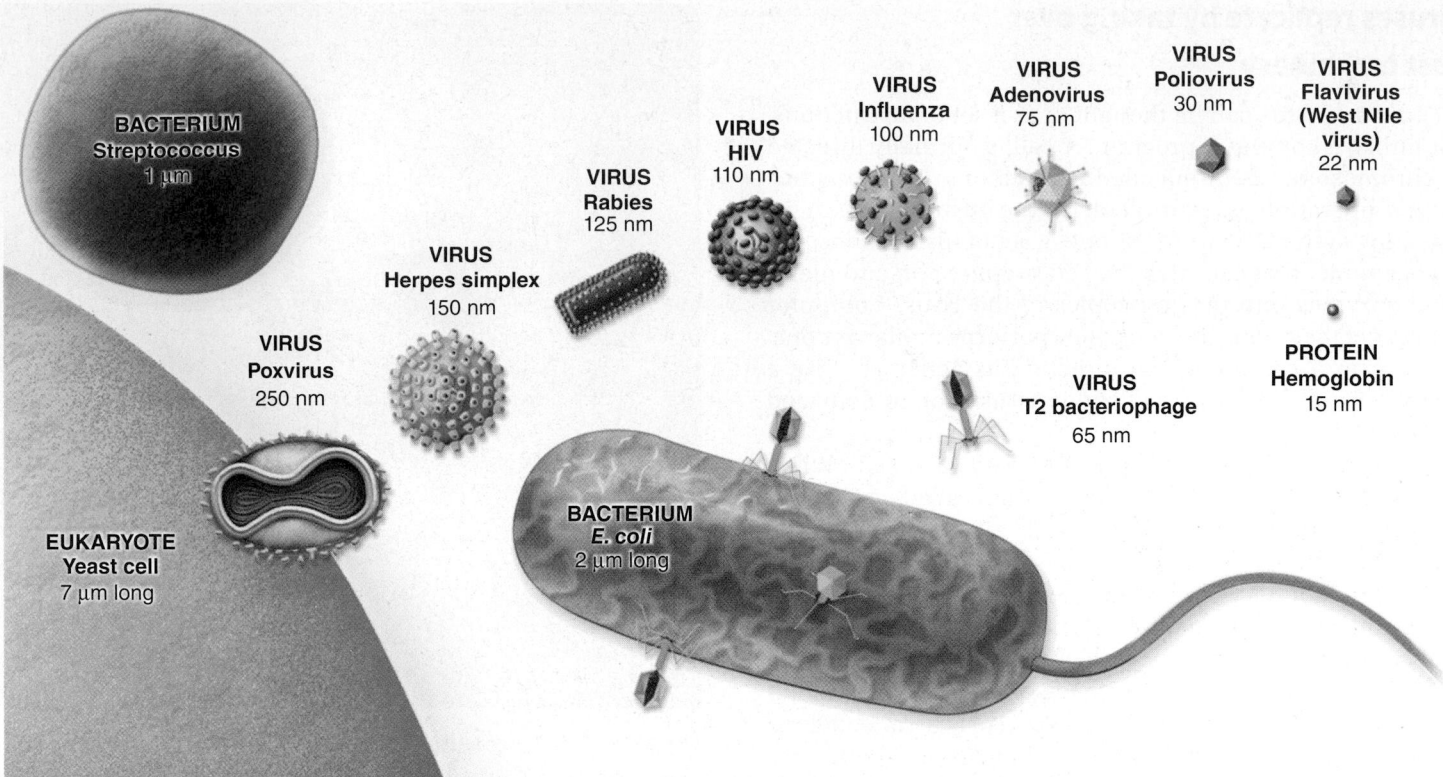

Figure 27.4 **Viruses vary in size and shape.** Note the dramatic differences in the size of a eukaryotic yeast cell, prokaryotic bacterial cells, and the many different viruses.

twofold, symmetry that is not either purely icosahedral or helical. The T-even phage shown has a head structure that is an elongated icosahedron. A collar connects the head to a hollow tube with helical symmetry that ends in a complex baseplate with tail fibers. Although animal viruses do not have this binal symmetry, some, such as the poxviruses, do have a complex multilayered capsid structure. Some enveloped viruses, such as influenza, are *polymorphic*, having no distinctive symmetry.

Viruses also vary greatly in size. As shown in figure 27.4, the very smallest viruses, such as the poliovirus, have actually been synthesized in a lab using nothing more than sequence data and a machine capable of synthesizing nucleic acids from nucleotides. The larger viruses, such as the poxviruses, generally carry more genes, have more complex structures, and tend to have a very short cycle time between entry of viral particles and release of newly formed virions.

Viral genomes exhibit great variation

Viral genomes vary greatly in both type of nucleic acid and number of strands (table 27.1). Some viruses, including those that cause flu, measles, and AIDS, possess RNA genomes. Most RNA viruses are single-stranded and are replicated and assembled in the cytosol of infected eukaryotic cells. RNA virus replication is error-prone, leading to high rates of mutation. This makes them difficult targets for the host immune system, vaccines, and antiviral drugs.

In single-stranded RNA viruses, if the genome has the same base sequence as the mRNA used to produce viral proteins, then the genomic RNA can serve as the mRNA. Such viruses are called *positive-strand viruses.* In contrast, if the genome is complementary to the viral mRNA, then the virus is called a *negative-strand virus.*

A special class of RNA viruses, called *retroviruses,* have an RNA genome that is reverse-transcribed into DNA by the enzyme *reverse transcriptase.* The DNA fragments produced by reverse transcription are often integrated into a host's chromosomal DNA. *Human immunodeficiency virus (HIV),* the agent that causes *acquired immune deficiency syndrome (AIDS),* is a retrovirus. (We describe HIV in detail later on.)

Other viruses, such as the viruses causing smallpox and herpes, have DNA genomes. Most DNA viruses are double-stranded, and their DNA is replicated in the nucleus of eukaryotic host cells.

Learning Outcomes Review 27.1

Viruses have a very simple structure that includes a nucleic acid genome encased in a protein coat. Viruses replicate by taking over a host's cell systems and are thus obligate intracellular parasites. Viruses show diverse genomes that are composed of DNA or RNA, which may be single- or double-stranded; most DNA viruses are double-stranded.

■ *Why can't viruses replicate outside of a cell?*

TABLE 27.1 | Important Human Viral Diseases

Disease	Pathogen	Genome	Vector/Epidemiology
Chicken pox	Varicella-zoster virus	Double-stranded DNA	Spread through contact with infected individuals. No cure. Rarely fatal. Vaccine approved in U.S. in early 1995. May exhibit latency leading to shingles. A vaccine for shingles that consists of a higher dose of the '95 vaccine was approved in 2006.
Hepatitis B (viral)	Hepadnavirus	Double-stranded DNA	Highly infectious through contact with infected body fluids. Approximately 1% of U.S. population infected. Vaccine available. No cure. Can be fatal.
Herpes	Herpes simplex virus	Double-stranded DNA	Blisters; spread primarily through skin-to-skin contact with cold sores/blisters. Very prevalent worldwide. No cure. Exhibits latency—the disease can be dormant for several years.
Mononucleosis	Epstein–Barr virus	Double-stranded DNA	Spread through contact with infected saliva. May last several weeks; common in young adults. No cure. Rarely fatal.
Smallpox	Variola virus	Double-stranded DNA	Historically a major killer; the last recorded case of smallpox was in 1977. A worldwide vaccination campaign wiped out the disease completely.
AIDS	HIV	(+) Single-stranded RNA (two copies)	Destroys immune defenses, resulting in death by opportunistic infection or cancer. For the year 2010, WHO estimated that 34 million people are living with AIDS, with an estimated 2.7 million new HIV infections and an estimated 1.8 million deaths.
Polio	Enterovirus	(+) Single-stranded RNA	Acute viral infection of the CNS that can lead to paralysis and is often fatal. Prior to the development of Salk's vaccine in 1954, 60,000 people a year contracted the disease in the U.S. alone.
West Nile fever	Flavivirus	(+) Single-stranded RNA	Spread by mosquitoes and can be amplified in bird hosts. Can lead to neurological problems. Present in U.S. since 1999.
Ebola	Filoviruses	(–) Single-stranded RNA	Acute hemorrhagic fever; virus attacks connective tissue, leading to massive hemorrhaging and death. Peak mortality is 50–90% if untreated. Outbreaks confined to local regions of central Africa.
Influenza	Influenza viruses	(–) Single-stranded RNA (eight segments)	Historically a major killer (20–50 million died during 18 months in 1918–1919); wild Asian ducks, chickens, and pigs are major reservoirs. The ducks are not affected by the flu virus, which shuffles its antigen genes while multiplying within them, leading to new flu strains. Vaccines are available.
Measles	Paramyxoviruses	(–) Single-stranded RNA	Extremely contagious through contact with infected individuals. Vaccine available. Usually contracted in childhood, when it is not serious; more dangerous to adults.
SARS	Coronavirus	(–) Single-stranded RNA	Acute respiratory infection; an emerging disease, can be fatal, especially in the elderly. Commonly infected animals include bats, foxes, skunks, and raccoons. Domestic animals can be infected.
Rabies	Rhabdovirus	(–) Single-stranded RNA	An acute viral encephalomyelitis transmitted by the bite of an infected animal. Fatal if untreated. Commonly infected animals include bats, foxes, skunks, and raccoons. Domestic animals can be infected.

Bacteriophage: Bacterial Viruses

Bacteriophage (both singular and plural) are viruses that infect bacteria. They are diverse, both structurally and functionally, and are united solely by their occurrence in bacterial hosts. Many of these types of bacteriophage, called *phage* for short, are large and complex, with relatively large amounts of DNA and proteins.

E. coli-infecting viruses were among the first bacteriophage to be discovered and are still some of the best studied. Some of these viruses that infect *E. coli* have been named as members of a "T" series (T1, T2, and so forth); others have been given different types of names. To illustrate the diversity of these viruses, T3 and T7 phage are icosahedral and have short tails. In contrast, the so-called T-even phage (T2, T4, and T6) have an icosahedral head, a capsid that consists primarily of three proteins, a connecting neck with a collar and long "whiskers," a long tail, and a complex base plate (see figure 27.3).

Archaeal viruses have diverse morphologies

Archaeal viruses were initially thought to be similar to bacterial viruses, but recent evidence argues against this. Surveys of viruses in several extreme environments dominated by archaeal species have uncovered an unexpected diversity of viral forms. In addition to the viral types described in the previous section, viruses with a two-tailed structure, with a bottle-shaped structure, and with a spindle-shaped structure have all been observed. All of these viruses have double-stranded DNA genomes, and most appear to be unrelated to any bacteriophage. The characterization of these viruses is in the early stages, so we will not discuss them further.

Bacterial viruses exhibit two reproductive cycles

The usual result of viral infection is production of new virus particles, which are released, usually killing the cell. This release of viruses then allows infection of new cells, or horizontal transmission of the virus. This kind of infection cycle is called a lytic cycle because the virus usually causes the cells to rupture, or lyse. Some bacterial viruses, however, can also enter a latent phase, called the lysogenic cycle, after the initial infection. These latent viruses are then transmitted vertically, that is, through cell division.

The lytic cycle

The lytic cycle of virus reproduction is illustrated in figure 27.5. The basic steps of a lytic bacteriophage cycle are similar to those of a nonenveloped animal virus. We will use the lytic cycle as an example of a viral life cycle. Although the basic outline of this infective cycle is common to most viruses, the details vary as do the viruses themselves.

The first step is called **attachment** (or absorption), in which the virus contacts the cell and becomes specifically bound to the cell. This step limits the host range of the virus as they bind to specific proteins on the surface of the cell. Different phages may target different parts of the outer surface of a bacterial cell. The next step is called **penetration** and results in the release of the viral genome into the host. This has been studied in detail in the binal phage, such as T4. Once contact is established, the tail contracts, and the tail tube passes through an opening that appears in the base plate, piercing the bacterial cell wall. The viral genome is literally injected into the host cytoplasm.

Once inside the bacterial cell, in the **synthesis** phase, the virus takes over the cell's replication and protein synthesis machinery in order to synthesize viral components. These components are then assembled (**assembly** phase) to produce mature virus particles.

In the **release** phase, mature virus particles are released, either through the action of enzymes that lyse the host cell or by budding through the host cell wall. The time between adsorption and the formation of new viral particles is called an eclipse period because if a cell is lysed at this point, few if any active virions can be released.

The lysogenic cycle

Bacteriophage that are capable of latent infection do this by integrating their nucleic acid into the genome of the infected host cell. This integration allows a virus to be replicated along with the host cell's DNA as the host divides. These viruses are called **temperate,** or **lysogenic,** phage. The DNA segment that is integrated into a host cell's genome is called a **prophage,** and the resulting cell is called a **lysogen.**

Among the bacteriophage that do this is the binal phage lambda (λ) of *E. coli*. Lambda may be the best studied biological entity on the planet. When phage λ infects a cell, the early events constitute a genetic switch that will determine whether the virus will replicate and destroy the cell or become a lysogen and be passively replicated with the cell's genome. This lysis/lysogeny "decision" depends on the expression of early genes. Early on, two regulatory proteins are produced that compete for binding to sites on the phage's DNA. Depending on which protein "wins," either the genes necessary for replication of the genome will be expressed beginning the lytic cycle, or the enzymes necessary for integrating the viral genome into the chromosome will be expressed and the lysogenic cycle initiated (figure 27.5, *right*).

In a lysogenic phage the expression of its genome is repressed (see chapter 16) by one of the two viral regulatory proteins mentioned earlier. This is not a permanent state, however; in times of cell stress, the prophage can be derepressed, and the enzymes necessary for excision of the genome expressed. The viral genome then is in the same state as for the initial stage of infection, and the lytic cycle can commence, leading to formation of viral particles and lysis of the cell.

The switch from a lysogenic prophage to a lytic cycle is called induction because it requires turning on the gene expression necessary for the lytic cycle. It can be stimulated in

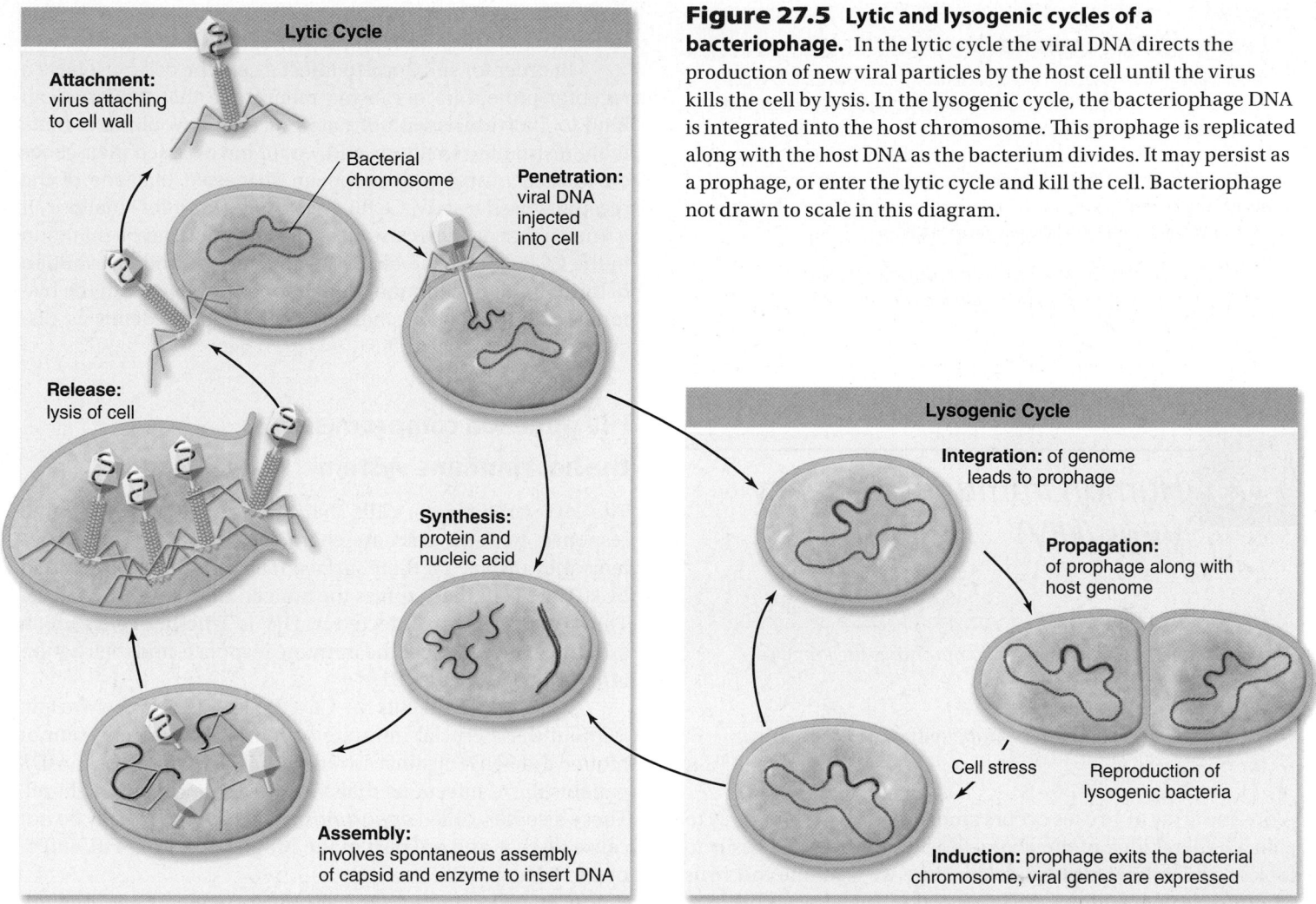

Figure 27.5 Lytic and lysogenic cycles of a bacteriophage. In the lytic cycle the viral DNA directs the production of new viral particles by the host cell until the virus kills the cell by lysis. In the lysogenic cycle, the bacteriophage DNA is integrated into the host chromosome. This prophage is replicated along with the host DNA as the bacterium divides. It may persist as a prophage, or enter the lytic cycle and kill the cell. Bacteriophage not drawn to scale in this diagram.

Lytic Cycle

Attachment: virus attaching to cell wall

Bacterial chromosome

Penetration: viral DNA injected into cell

Release: lysis of cell

Synthesis: protein and nucleic acid

Assembly: involves spontaneous assembly of capsid and enzyme to insert DNA

Lysogenic Cycle

Integration: of genome leads to prophage

Propagation: of prophage along with host genome

Cell stress

Reproduction of lysogenic bacteria

Induction: prophage exits the bacterial chromosome, viral genes are expressed

the laboratory by stressors such as starvation or ultraviolet radiation. The molecular events of induction take advantage of host proteins that respond to stress to produce a protease that can destroy the repressor protein that is keeping the viral genome silent. The normal function of this protease is to degrade a host repressor that controls DNA repair genes. The two repressor proteins are similar enough that both are degraded by the protease.

Bacteriophage can contribute genes to the host genome

During the integrated portion of a lysogenic reproductive cycle, a few viral genes may be expressed at the same time as host cell genes. Sometimes the expression of these genes has an important effect on the host cell, altering it in novel ways. When the phenotype or characteristics of the lysogenic bacterium is altered by the prophage, the alteration is called **phage conversion.**

Phage conversion of the cholera-causing bacterium

The bacterium *Vibrio cholerae* usually exists in a harmless form, but a second, disease-causing form also occurs. In this latter form, the bacterium is responsible for the deadly disease cholera, but how the bacteria changed from harmless to deadly was not known until recently.

Research now shows that a lysogenic bacteriophage that infects *V. cholerae* introduces into the host bacterial cell a gene that codes for the cholera toxin. This gene, along with the rest of the phage genome, becomes incorporated into the bacterial chromosome. The toxin gene is expressed along with the other host genes, thereby converting the benign bacterium to a disease-causing agent.

The receptors used by this toxin-encoding phage are pili (hairlike projections) found on the outer surface of *V. cholerae* (chapter 28); in recent experiments, it was determined that mutant bacteria that did not have pili were resistant to infection by the bacteriophage. This discovery has important implications in efforts to develop vaccines against cholera, which have been unsuccessful up to this point. Phage conversion could change any pili-expressing, nontoxigenic *V. cholerae* into a toxin-producing, potentially deadly form.

Another example involved in human disease is the toxin found in *Corynebacterium diphtheriae*. This toxin is the product of phage conversion, as are the changes to the outer surface of certain infectious *Salmonella* species.

27.3 Human Immunodeficiency Virus (HIV)

Learning Outcomes

1. *Explain how the HIV virus compromises the immune system.*
2. *Describe the disease AIDS.*
3. *Illustrate the different therapeutic options for AIDS.*

A diverse array of viruses occurs among animals. A good way to gain a general idea of the characteristics of these viruses is to look at one animal virus in detail. Here we examine the virus responsible for a comparatively new and fatal viral disease, *acquired immune deficiency syndrome (AIDS).*

AIDS is caused by HIV

The disease now known as AIDS was first reported in the United States in 1981, although a few dozen people in the United States had likely died of AIDS prior to that time and had not been diagnosed. Frozen plasma samples and estimates based on evolutionary speed and current diversity of HIV strains trace the origins of HIV in the human population to Africa in the 1950s. It was not long before the infectious agent, a retrovirus, was identified by laboratories in France. Study of HIV revealed it to be closely related to a chimpanzee virus (simian immunodeficiency virus, SIV), suggesting a recent host expansion to humans from chimpanzees in central Africa.

Infected humans have varying degrees of resistance to HIV. Some have little resistance to infection and rapidly progress from having HIV-positive status to developing AIDS and eventually die. Others, even after repeated exposure, fail to become HIV-positive or may become HIV-positive without developing AIDS.

A relatively recent hypothesis to explain this great variability in susceptibility is genetic variation among these groups due to the selective pressure put on the human population by the smallpox virus (variola major) over the centuries. Because of successful vaccination and immunization, smallpox has been eradicated from the human population; however, before its eradication, it caused billions of deaths worldwide.

In order for smallpox to infect a cell, the cell must have a receptor protein in its plasma membrane that the virus can bind to. Individuals with mutated receptors would have been more resistant to smallpox and would have passed their genes on to their offspring. It has been suggested that one of the receptors used by HIV, CCR5, is also a receptor for smallpox. It is known that people resistant to HIV infection have a mutation in the CCR5 gene. The historical appearance and distribution of this mutation in human populations correlates with the historical distribution of smallpox. The AIDS epidemic is discussed further in chapter 51.

HIV infection compromises the host immune system

The HIV virus targets cells that are critical to the immune response. Immune cells are characterized based on the proteins they display on their surface. The relevant cells targeted by HIV are cells that express the antigen CD4, thus **CD4+ cells.** The specific cell type infected by HIV is **T-helper cells,** which are critical to regulating the immune response, and their action is described in chapter 51.

HIV infects and kills the CD4+ cells until very few are left. Without these crucial immune system cells, the body cannot mount a defense against invading bacteria or viruses. AIDS patients die of infections that a healthy person could fight off. These diseases, called *opportunistic infections,* normally do not cause disease and are part of the progression from HIV infection to having AIDS.

Clinical symptoms typically do not begin to develop until after a long latency period, generally 8 to 10 years after the initial infection with HIV. Some individuals, however, may develop symptoms in as few as two years. During latency, HIV particles are not in circulation, but the virus can be found integrated within the genome of macrophages and CD4+ T cells as a provirus (equivalent to a prophage in bacteria).

HIV testing

HIV tests do not test for the presence of circulating virus but rather for the presence of antibody against HIV. Because only those people exposed to HIV in their bloodstream at one time or another would have anti-HIV antibodies, this screening provides an effective way to determine whether further testing is needed to confirm HIV-positive status.

The spread of AIDS

Although carriers of HIV have no clinical symptoms during the long latency period, they are apparently fully infectious, which makes the spread of HIV very difficult to control. The reason HIV remains hidden for so long seems to be that its infection cycle continues throughout the 8- to 10-year latency period without doing serious harm to the infected person because of an effective immune response. Eventually, however, a random mutational event in the virus or a failure of the immune response allows the virus to quickly overcome the immune defense, beginning the course of AIDS.

HIV infects key immune-system cells

The way in which HIV infects humans provides a good example of how animal viruses replicate (figure 27.6). Most other viral infections follow a similar course, although the details of entry and replication differ in individual cases.

Attachment

When HIV is introduced into the human bloodstream, the virus particles circulate throughout the body but only infect CD4$^+$ cells. Most other animal viruses are similarly narrow in their requirements; hepatitis goes only to the liver, and rabies to the brain. This tissue tropism is determined by the proteins found on a cell surface and on a viral surface.

For example, the common cold virus uses the ICAM-1 membrane protein as a receptor to enter cells. ICAM-1 is a protein that is up-regulated (increased) in times of immune activation and stress. So, the more inflammation and stress in an area, the more receptors exist for the virus to enter a cell and continue the disease process.

How does a virus such as HIV recognize a target cell? Recall from chapter 4 that every kind of cell in the human body

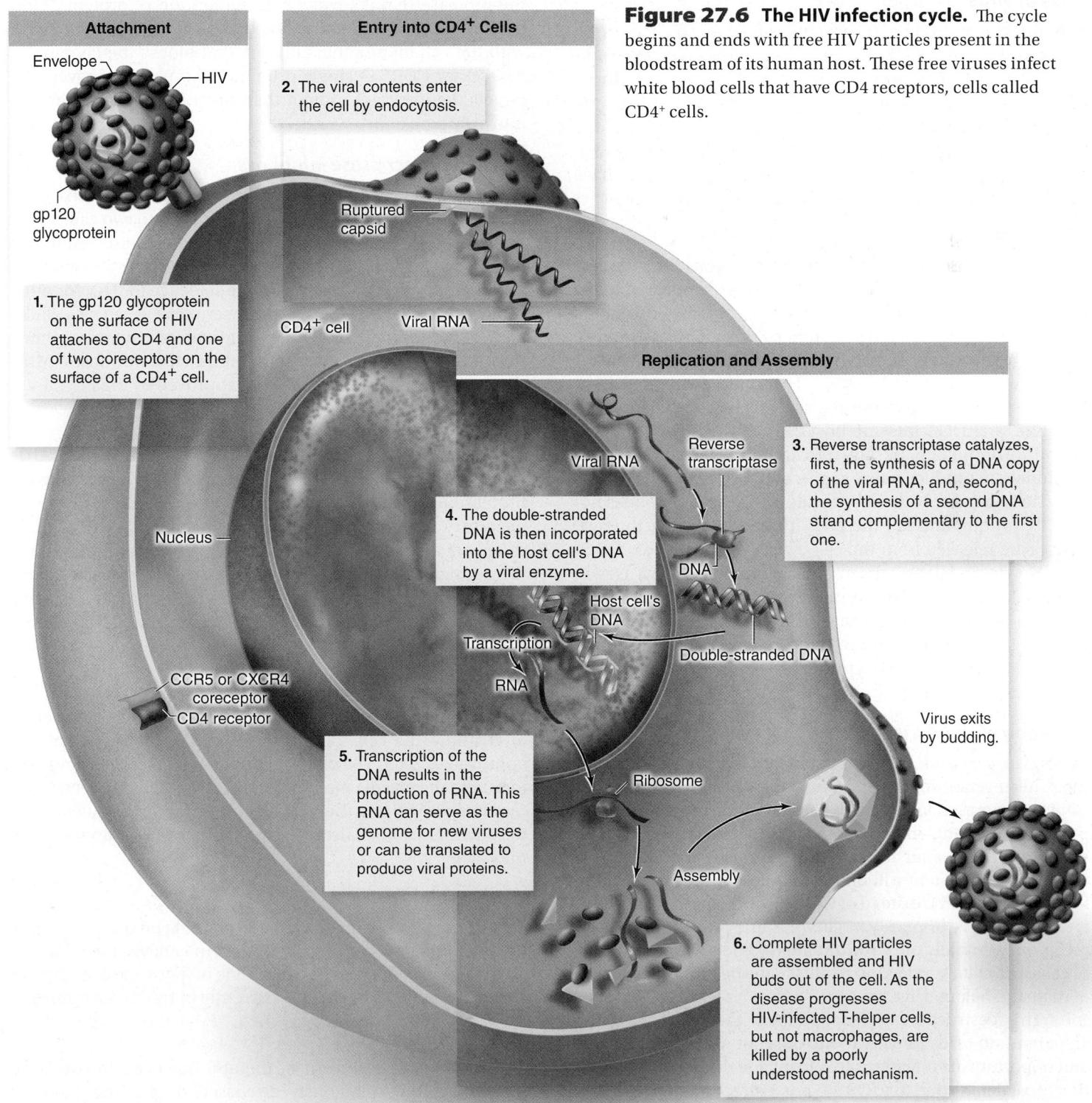

Figure 27.6 The HIV infection cycle. The cycle begins and ends with free HIV particles present in the bloodstream of its human host. These free viruses infect white blood cells that have CD4 receptors, cells called CD4$^+$ cells.

Attachment

Envelope
HIV
gp120 glycoprotein

1. The gp120 glycoprotein on the surface of HIV attaches to CD4 and one of two coreceptors on the surface of a CD4$^+$ cell.

Entry into CD4$^+$ Cells

2. The viral contents enter the cell by endocytosis.

Ruptured capsid

CD4$^+$ cell

Viral RNA

Nucleus

CCR5 or CXCR4 coreceptor
CD4 receptor

Replication and Assembly

Viral RNA
Reverse transcriptase

3. Reverse transcriptase catalyzes, first, the synthesis of a DNA copy of the viral RNA, and, second, the synthesis of a second DNA strand complementary to the first one.

DNA

4. The double-stranded DNA is then incorporated into the host cell's DNA by a viral enzyme.

Host cell's DNA

Transcription

RNA

Double-stranded DNA

5. Transcription of the DNA results in the production of RNA. This RNA can serve as the genome for new viruses or can be translated to produce viral proteins.

Ribosome

Assembly

Virus exits by budding.

6. Complete HIV particles are assembled and HIV buds out of the cell. As the disease progresses HIV-infected T-helper cells, but not macrophages, are killed by a poorly understood mechanism.

has a specific array of cell-surface glycoprotein markers that serve to identify them to other, similar cells. Invading viruses take advantage of this to bind to specific cell types. Each HIV particle possesses a glycoprotein called gp120 on its surface that precisely fits the cell-surface marker protein CD4 on the surfaces of the immune system macrophages and T cells. Macrophages, another type of white blood cell, are infected first. Because macrophages commonly interact with CD4⁺ T cells, this may be one way that the T cells are infected. Several coreceptors also significantly affect the likelihood of viral entry into cells, including the CCR5 receptor, which is mutated in HIV-immune individuals.

Entry of virus

After docking onto the CD4 receptor of a cell, HIV requires a coreceptor such as CCR5, to pull itself across the cell membrane. After gp120 binds to CD4, it goes through a conformational change that allows it to then bind the coreceptor. Receptor binding is thought to ultimately result in fusion of the viral and target cell membranes and entry of the virus through a fusion pore. The coreceptor, CCR5, is hypothesized to have been used by the smallpox virus as was mentioned earlier.

Replication

Once inside the host cell, the HIV particle sheds its protective coat. This leaves viral RNA floating in the cytoplasm, along with the reverse transcriptase enzyme that was also within the virion. Reverse transcriptase synthesizes a double strand of DNA complementary to the virus RNA, often making mistakes and introducing new mutations. This double-stranded DNA then enters the nucleus along with a viral enzyme that incorporates the viral DNA into the host cell's DNA. After a variable period of dormancy the HIV provirus directs the host cell's machinery to produce many copies of the virus.

As is the case with most enveloped viruses, HIV does not directly rupture and kill the cells it infects. Instead, the new viruses are released from the cell by *budding*, a process much like exocytosis. HIV synthesizes large numbers of viruses in this way, challenging the immune system over a period of years. In contrast, naked viruses, those lacking an envelope, generally lyse the host cell in order to exit. Some enveloped viruses may produce enzymes that damage the host cell enough to kill it or may produce lytic enzymes as well.

Evolution of HIV during infection

During an infection, HIV is constantly replicating and mutating. The reverse transcriptase enzyme is less accurate than DNA polymerases, leading to a high mutation rate. Eventually, by chance, variants in the gene for gp120 arise that cause the gp120 protein to alter its second-receptor partner. This new form of gp120 protein will bind to a different second receptor, for example CXCR4, instead of CCR5. During the early phase of an infection, HIV primarily targets immune cells with the CCR5 receptor. Eventually the virus mutates to infect a broader range of cells. Ultimately, infection results in the destruction and loss of critical T-helper cells.

This destruction of T cells blocks the body's immune response and leads directly to the onset of AIDS, with cancers and opportunistic infections free to invade the defenseless victim. Most deaths due to AIDS are not a direct result of HIV, but are from other diseases that normally do not harm a host with a normal immune system.

AIDS treatment targets different phases of the HIV life cycle

The federal Food and Drug Administration (FDA) currently lists 35 antiretroviral drugs that are used in AIDS therapy. These target four aspects of the HIV life cycle: viral entry, genome replication, integration of viral DNA, and maturation of HIV proteins (figure 27.7). Of these, the vast majority are inhibitors of the replication enzyme, reverse transcriptase, and the protease that is involved in maturation of proteins. Only two drugs block viral entry: one that blocks the fusion of virus with the cell membrane, and one that blocks the chemokine coreceptor CCR5. A single drug has also been approved that targets the integrase protein that integrates the viral genome into a chromosome.

Reverse transcriptase inhibitors

The first drug licensed for clinical use was AZT, a reverse transcriptase inhibitor. This class of drugs falls into two categories: nucleotide or nucleoside reverse transcriptase inhibitors (NRTI) such as AZT, and nonnucleoside reverse transcriptase inhibitors (NNRTI). These drugs are selective for HIV (or other retroviruses) because reverse transcriptase is not a cellular enzyme. So although the NRTI drugs affect cellular enzymes, they inhibit reverse transcriptase at much lower doses. More than 17 RT inhibitors are currently approved by the FDA.

Protease inhibitors

The second class of drugs discovered to be effective in AIDS treatment were the protease inhibitors. These drugs target a protease that cleaves a polyprotein into the smaller proteins necessary for viral replication and assembly. Some of these drugs are actually a good example of the concept of rational drug design. Drug designers started with the protease enzyme, then targeted drugs at transition state analogues of this enzyme. There are now more than 10 of these drugs that have received FDA approval.

Blocking viral entry

Two drugs have been approved by the FDA that block the entry of HIV into the cell. One of these, the fusion inhibitor, was approved in 2003. The drug blocks the fusion of the viral envelope with the plasma membrane of a target cell. This entry also requires recognizing the CD4 receptor protein, and a coreceptor such as CCR5. The coreceptor blocker was approved in mid-2007.

Integrase inhibitors

A number of companies have been working on drugs targeting the viral integrase protein. This protein catalyzes the integration reaction of the viral genome. One of these was approved in late 2007. At least one other is currently in testing for approval.

Combination therapy

The most successful form of therapy has been to use combinations of the previously discussed drugs. The standard

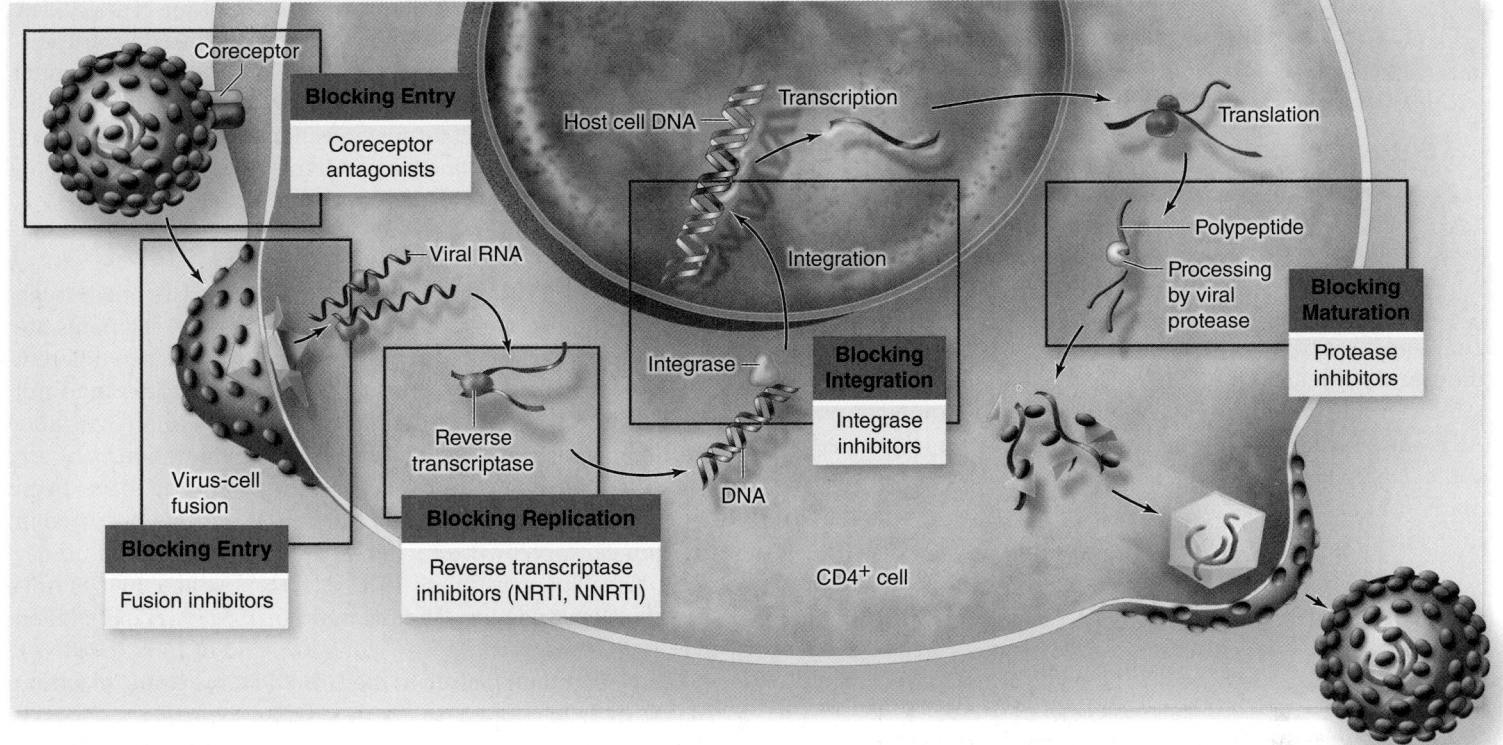

Figure 27.7 **Viral targets for therapeutic drugs.** A simplified version of the HIV infection cycle is shown with the steps targeted for drug intervention. Four parts of the life cycle have been targeted: viral entry, genome replication, integration of the genome, and maturation of viral proteins. NRTI, nucleoside reverse transcriptase inhibitor; NNRTI, nonnucleoside reverse transcriptase inhibitor.

treatment regimen consists of a minimum of three active drugs: an NNRTI or protease inhibitor combined with two different NRTIs. This form of combination therapy has entirely eliminated the HIV virus from many patients' bloodstreams. All of these patients began to receive the combination drug therapy within three months of contracting the virus, before their bodies had an opportunity to develop tolerance to any single drug. Widespread use of this *combination therapy,* otherwise known as *highly active antiretroviral therapy (HAART),* has cut the U.S. AIDS death rate by three-fourths since its introduction in the mid-1990s.

Unfortunately, this sort of combination therapy does not appear to actually eliminate HIV from the body. Although the virus disappears from the bloodstream, traces of it can still be detected in the patient's lymph tissue. When combination therapy is discontinued, virus levels in the bloodstream once again rise.

In addition to the search for new drugs to add to the mix of HAART, much effort has been put into simplifying the drug regimens. The more complex the regimen of drugs, the less likely that patients will stick with it. New drugs that combine NRTI and NNRTI compounds into a single pill to reduce this complexity have been approved by the FDA.

Vaccine development for HIV has been unsuccessful

A large amount of effort has been put toward developing an anti-HIV vaccine. Thus far, this has been totally unsuccessful. A recent large-scale international HIV vaccine trial was halted when an early examination of data indicated that the vaccine was essentially useless for preventing infection or lowering viral load. This has led to a retooling of clinical trials to have periodic examination of data instead of waiting for endpoints, but has not helped in terms of the vaccine itself. This vaccine was a subunit vaccine in which a specific HIV protein was engineered into an adenovirus vector.

While the high mutation rate of HIV has always been seen as a problem for vaccine development, it appears that the reason for vaccine failures may be more basic, and harder to surmount. A vaccine needs to produce a strong cellular immune response, and thus far no trial HIV subunit vaccine has done this. The only type of vaccine in an animal system that has been shown to provide protection against infection was a vaccine made from attenuated SIV. Unfortunately over time, the attenuated virus was able to mutate into an infective virus, and the experimental animals eventually developed simian AIDS.

Learning Outcomes Review 27.3

HIV is a retrovirus that enters cells via membrane fusion. The virus primarily infects host CD4+ T cells, ultimately resulting in massive death of these cells, which thereby compromises the host's immune system. Most deaths result from cancer and infections that typically do not harm hosts with normal immune systems. Combination drug therapy is a treatment modality in developed countries, and much research is being done to develop vaccines or to find agents that can prevent infection.

■ *Does combination therapy such as HAART represent a cure for AIDS?*

Other Viral Diseases

Humans have known and feared diseases caused by viruses for thousands of years. Among the diseases that viruses cause (see table 27.1) are influenza, smallpox, hepatitis, yellow fever, polio, AIDS, and SARS. In addition, viruses have been implicated in some cancers including leukemias. Viruses not only cause many human diseases, but also cause major losses in agriculture, forestry, and the productivity of natural ecosystems.

The flu is caused by influenza virus

Perhaps the most lethal virus in human history has been the influenza virus. The influenza pandemic of 1918 and 1919 is thought to have infected a third of the world's population.

Types and subtypes

Flu viruses are enveloped segmented RNA viruses that infect animals. An individual flu virus resembles a rod studded with spikes composed of two kinds of protein. The three general "types" of flu virus are distinguished by their capsid protein, which surrounds the viral RNA segments and is different for each type: **Type A flu virus** causes most of the serious flu epidemics in humans and also occurs in mammals and birds. Type B and type C viruses are restricted to humans and rarely cause serious health problems.

Different strains of flu virus, called subtypes, differ in their protein spikes. One of these proteins, hemagglutinin (H), aids the virus in gaining access to the cell interior. The other, neuraminidase (N), helps the daughter viruses break free of the host cell once virus replication has been completed.

Parts of the H molecule contain "hotspots" that display an unusual tendency to change as a result of mutation of the viral RNA during imprecise replication. Point mutations cause changes in these spike proteins in 1 of 100,000 viruses during the course of each generation. These highly variable segments of the H molecule are targets against which the body's antibodies are directed. These constantly changing H-molecule regions improve the reproductive capacity of the virus and hinder our ability to make effective vaccines.

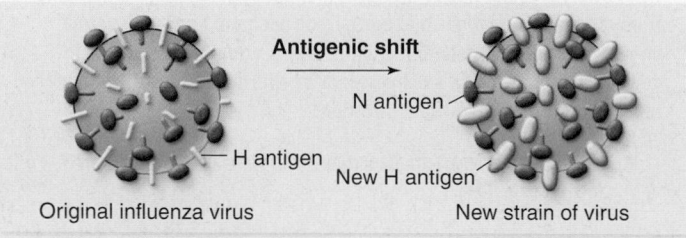

Antigenic shift

N antigen

H antigen

New H antigen

Original influenza virus

New strain of virus

Because of accumulating changes in the H and N molecules, different flu vaccines are required to protect against different subtypes. Type A flu viruses are currently classified into 13 distinct *H subtypes* and 9 distinct *N subtypes*, each of which requires a different vaccine to protect against infection. Thus, the type A virus that caused the 1918 influenza pandemic (that is, worldwide epidemic) has type 1H and type 1N and is designated A(H1N1).

The importance of recombination

The greatest problem in combating flu viruses arises not through mutation, but through recombination. Viral RNA segments are readily reassorted by genetic recombination when two different subtypes simultaneously infect the same cell. This may put together novel combinations of H and N spikes unrecognizable by human antibodies specific for the old configuration.

Viral recombination of this kind seems to have been responsible for the three major flu pandemics that occurred in the 20th century, by producing drastic shifts in H–N combinations. The "Spanish flu" of 1918, A(H1N1), killed 50–100 million people worldwide. The Asian flu of 1957, A(H2N2), killed over 100,000 Americans. The Hong Kong flu of 1968, A(H3N2), infected 50 million people in the United States alone, of whom 70,000 died.

Origin of new strains

It is no accident that new strains of flu usually originate in the Far East. The most common hosts of influenza virus are ducks, chickens, and pigs, which in Asia often live in close proximity to each other and to humans. Pigs are subject to infection by both bird and human strains of the virus, and individual animals are often simultaneously infected with multiple strains. This creates conditions favoring genetic recombination between strains, producing new combinations of H and N subtypes.

In 1997, a form of avian influenza, A(H5N1), was discovered that could infect humans. Avian influenza, or "bird flu," is highly contagious and deadly among domestic bird populations, and this H5N1 strain is transmitted between domestic birds, which live in close contact with humans, and wild birds, which migrate worldwide. This new influenza variant caused concern because, although the number of infections worldwide is low, the mortality rate in these rare infections is quite high. However, this strain does not appear to spread by human-to-human contact.

While H5N1 captured the interest of the press worldwide, another viral reassortment of avian, human, and swine viruses occurred. This led to the H1N1 pandemic of 2009. The virus first appeared in an outbreak of influenza in Veracruz, Mexico, and spread worldwide over 2009–10. The WHO raised the event to pandemic status in May 2009. This pandemic is interesting on many levels, both scientific and social. Events of this type are instructive in terms of how societies deal with a crisis, whether it ends up being as severe as expected or not. In the end, the infection rate was similar to previous pandemics, but this is not much different from seasonal forms. The mortality rate was not high (around 1.5%), but the age distribution was unusual affecting more younger people with older people exhibiting some immunity. In the end, the death toll was similar to seasonal influenza, but if measured in terms of life expectancy, the greater toll on the young raises this number.

New viruses emerge by infecting new hosts

Sometimes viruses that originate in one organism pass to another, thus expanding their host range. Often, this expansion is deadly to the new host. HIV, for example, is thought to have arisen in chimpanzees and relatively recently passed to humans. Influenza is fundamentally a bird virus. Viruses that originate in one organism and then pass to another and cause disease are called **emerging viruses.** They represent a considerable threat in an age when airplane travel potentially allows infected individuals to move about the world quickly, spreading an infection.

Hantavirus

An emerging virus caused a sudden outbreak of a deadly pneumonia in the southwestern United States in 1993. This disease was traced to a species of **hantavirus** and was called the *sin nombre,* or *no-name, virus.* Hantavirus is a single-stranded RNA virus associated with rodents. This virus was eventually traced to deer mice. The deer mouse hantavirus is transmitted to humans through fecal and urine contamination in areas of human habitation. Controlling the deer mouse population has limited the disease.

Hemorrhagic fever: Ebola

Sometimes the origin of an emerging virus is unknown, making an outbreak more difficult to control. Among the most lethal of emerging viruses are a collection of filamentous viruses arising in central Africa that cause severe hemorrhagic fever. With lethality rates in excess of 50%, these so-called **filoviruses** are among the most lethal infectious diseases known. One, **Ebola virus** (figure 27.8), has exhibited lethality rates in excess of 90% in isolated outbreaks in central Africa. The outbreak of Ebola virus in the summer of 1995 in Zaire killed 245 people out of 316 infected—a mortality rate of 78%. A recent (2004) outbreak of Ebola in Yambio, southern Sudan, caused 17 infections and 7 deaths. This outbreak was rapidly contained by isolating patients from family members as soon as symptoms appeared. The natural host of Ebola is unknown.

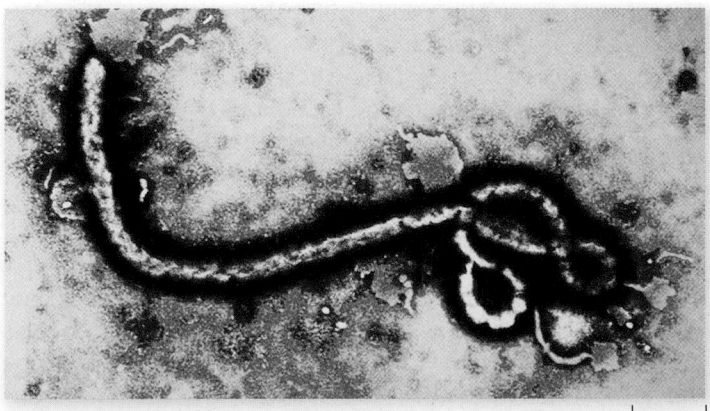

0.3 µm

Figure 27.8 The Ebola virus. This virus, with a fatality rate that can exceed 90%, appears sporadically in central Africa. The natural host of the virus currently is unknown.

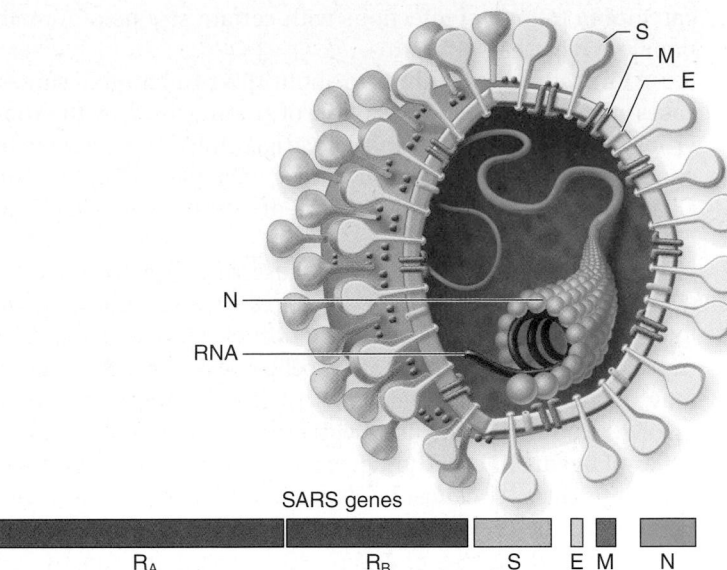

SARS genes

R_A | R_B | S | E M | N

Figure 27.9 SARS coronavirus. The 29,751-nucleotide SARS genome is composed of RNA and contains six principal genes: R_A and R_B, replicases; S, spike proteins; E, envelope glycoproteins; M, membrane glycoprotein; N, nucleocapsid protein.

SARS

A recently emerged species of coronavirus (figure 27.9) was responsible for the 2003 worldwide outbreak of **severe acute respiratory syndrome (SARS),** a respiratory infection with pneumonia-like symptoms that is fatal in over 8% of cases. When the 29,751-nucleotide RNA genome of the SARS coronavirus was sequenced, it proved to be a completely new form of coronavirus, not closely related to any of the three previously described forms.

Virologists suspect that the SARS coronavirus most likely came from civets (a weasel-like mammal) and possibly other wild animals that live in China and are eaten as delicacies. If the SARS virus indeed exists in natural populations, future outbreaks will be difficult to prevent without an effective vaccine. Recent data have implicated bats as the natural reservoir for SARS virus. The significance of this finding for control of the virus is unclear at present.

Genome sequences have been analyzed from SARS patients at various stages of the outbreak, and these analyses indicate that the virus's mutation rate is low, in marked contrast to HIV, another RNA virus. The stable genome of the SARS virus should make development of a SARS vaccine practical. The lessons learned from developing antiviral agents against other RNA viruses, such as HIV and influenza, have helped develop drugs to treat SARS. Several anti-SARS agents and vaccines are currently being tested in laboratories across the world.

Viruses can cause cancer

Through epidemiological studies and research, scientists have established a link between some viral infections and the subsequent development of cancer. Examples include the association between chronic hepatitis B infections and the development of liver cancer, and the development of cervical carcinoma following infections with certain strains of human papillomaviruses (HPV).

Viruses may contribute to about 15% of all human cancer cases worldwide. They are capable of altering the growth properties of human cells they infect by triggering the expression of cancer-causing genes called oncogenes (see chapter 10). Changes in the normal function of these genes leads to cancer.

These changes can occur because viral proteins interfere with the regulation of oncogene expression. Alternatively, the integration of a viral genome into a host chromosome may disrupt a gene required to control the cell cycle. Viruses themselves may encode these oncogenes as well. Virus-induced cancer involves complex interactions with cellular genes and requires a series of events in order to develop. The association of viruses with some forms of cancer has led to research on vaccine development for the prevention of these cancers. In June 2006, the FDA approved the use of a new HPV vaccine in women and young girls from the age of 11 to prevent cervical cancer.

Learning Outcomes Review 27.4

Many types of viruses have caused disease in humans for as long as we have recorded history. Some of these, such as influenza, have caused pandemics responsible for millions of deaths worldwide. Recombination is common in the influenza virus, making natural immunity and development of vaccines problematic. Emerging diseases can occur when viruses switch hosts, that is, jump from another species to humans. Hantavirus, Ebola, and SARS all fall into this category. Virus infection has also been linked to the development of certain cancers.

■ *Why does the effectiveness of flu shots vary from year to year?*

27.5 Prions and Viroids: Subviral Particles

Learning Outcomes

1. Explain why prion replication was a "heretical" concept.
2. Describe the mechanism of prion transmission.

For decades, scientists have been fascinated by a peculiar group of fatal brain diseases. These diseases have an unusual property: Years and often decades pass after infection before the disease is detected in infected individuals. The brains of infected individuals develop numerous small cavities as neurons die, producing a marked spongy appearance. Called *transmissible spongiform encephalopathies (TSEs),* these diseases include scrapie in sheep; bovine spongiform encephalopathy (BSE), or "mad cow" disease in cattle; chronic wasting disease in deer and elk; and kuru, Creutzfeldt–Jakob disease (CJD), and variant Creutzfeldt–Jakob disease (vCJD) in humans.

TSEs can be transmitted experimentally by injecting infected brain tissue into a recipient animal's brain. TSEs can also spread via tissue transplants and, apparently, tainted food. The disease kuru was once common in the Fore people of Papua New Guinea, because they practiced ritual cannibalism, eating the brains of infected individuals. Mad cow disease spread widely among the cattle herds of England in the 1990s because cows were fed bonemeal prepared from sheep and cattle carcasses to increase the protein content of their diet. Like the Fore, the British cattle were eating the tissue of cattle that had died of the disease.

In the years following the outbreak of BSE, there has been a significant increase in CJD incidence in England. Some cases appear to be genetic. Mysteriously, patients with no family history of CJD were being diagnosed with the disease. This led to the discovery of a new form of CJD called variant CJD, or vCJD, that is acquired from eating meat of BSE-infected animals. Concern exists that vCJD may be transmitted from person to person through blood products, similar to the transmission of HIV through blood and blood products.

Prion replication was a heretical suggestion

In the 1960s, British researchers Tikvah Alper and John Stanley Griffith noted that infectious TSE preparations remained infectious even after exposure to radiation that would destroy DNA or RNA. They suggested that the infectious agent was a protein. They speculated that the protein could sometimes misfold, and then catalyze other proteins to do the same, the misfolding spreading like a chain reaction. This "heretical" suggestion was not accepted by the scientific community, because it violated a key tenet of molecular biology: Only DNA or RNA act as hereditary material, transmitting information from one generation to the next.

Evidence has accumulated that prions cause TSEs

In the early 1970s, American physician Stanley Prusiner began to study TSEs. Try as he might, Prusiner could find no evidence of nucleic acids or viruses in the infectious TSE preparations. He concluded, as Alper and Griffith had, that the infectious agent was a *protein,* which in a 1982 paper he named a **prion,** for "proteinaceous infectious particle."

Prusiner went on to isolate a distinctive prion protein, and to amass evidence that prions play a key role in triggering TSEs. Every host tested to date expresses a normal prion protein (PrP^c) in their cells. The disease-causing prions are the same protein, but folded differently (PrP^{sc}). These misfolded proteins have been shown in vitro to serve as a template for normal PrP to misfold. The misfolded PrP proteins are very resistant to degradation, making it possible for them to pass through the acidic digestive tract intact and therefore to be transmitted by ingesting tainted food.

Experimental evidence has accumulated to support this idea. Injection of prions with different abnormal conformations into hosts leads to the same abnormal conformations as the parent prions. Mice genetically engineered to lack PrP^c are

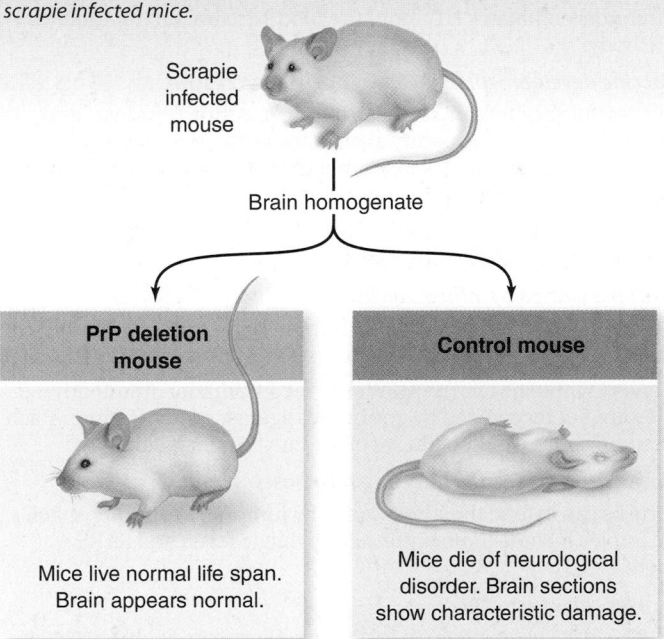
Figure 27.10 Demonstration that normal prion protein is necessary for scrapie infectivity.

immune to TSE infection (figure 27.10). If brain tissue with the prion protein is grafted into the mice, the grafted tissue—but not the rest of the brain—can then be infected with TSE. However, infectious PrP has not been generated in vitro in quantities that would produce disease in nature. The mechanism of pathogenesis also remains controversial.

Prions have also been found in yeast and other fungi. Three different "genes" that behave in a non-Mendelian fashion have all been shown to be prions in yeast. There has been greater progress in the in vitro conversion of normal cellular proteins to the prion form in the yeast system.

Viroids are infectious RNA with no protein coat

Viroids are tiny, naked molecules of circular RNA, only a few hundred nucleotides long, that are important infectious disease agents in plants. A recent viroid outbreak killed over 10 million coconut palms in the Philippines. Despite their small size, viroids autonomously replicate without a helper virus, although they obviously use host proteins. Despite being nucleic acids, the important information they carry appears to be in their three-dimensional structure. There are also some intriguing hints that they might use the plant siRNA machinery to affect gene expression.

Learning Outcomes Review 27.5

Prions and viroids are smaller and simpler than viruses. Prions are infectious particles that do not seem to contain any nucleic acid. They appear to be misfolded proteins that cause related cellular proteins to also misfold. Prions are the causative agent of TSEs. Viroids are infectious RNAs that are implicated in some plant diseases.

■ *If prions are infectious proteins, then what is the form of genetic information that they carry?*

Chapter Review

27.1 The Nature of Viruses

Viruses are strands of nucleic acids encased in a protein coat.

Viral genomes can consist of either DNA or RNA and can be classified as DNA viruses, RNA viruses, or retroviruses.

Most viruses have a protein sheath or capsid around their nucleic acid core. Many animal viruses have an envelope around the capsid composed of proteins that are virally encoded, lipids from the host cell, and glycoproteins.

Some viruses also have enzymes inside their capsid that are important in early infection.

Viral hosts include virtually every kind of organism.

Each virus has a limited host range, and many also exhibit tissue tropism.

Viruses replicate by taking over host machinery.

Viruses are obligate intracellular parasites lacking ribosomes and proteins needed for replication. Viruses take over host machinery and direct their own nucleic acid and protein synthesis.

Most viruses come in two simple shapes.

Viruses vary in size and come in two simple shapes: helical (rodlike) or icosahedral (spherical) (figures 27.1 and 27.4).

Viral genomes exhibit great variation.

The DNA or RNA viral genome may be linear or circular, single- or double-stranded. RNA viruses may have multiple RNA molecules (segmented), or only one RNA molecule (nonsegmented). Retroviruses contain RNA that is transcribed into DNA by reverse transcriptase.

27.2 Bacteriophage: Bacterial Viruses

Archaeal viruses have diverse morphologies.

Bacterial viruses exhibit two reproductive cycles.

The lytic cycle kills the host cell, whereas the lysogenic cycle incorporates the virus into the host genome as a prophage (figure 27.5). A cell containing a prophage is called a lysogen.

The prophage can be induced by DNA damage and other environmental cues to reenter the lytic cycle.

For most phage, steps in infection include attachment, injection of DNA (penetration), macromolecular synthesis, assembly of new phage, and release of progeny phage.

Bacteriophage can contribute genes to the host genome.

Phage conversion occurs when foreign DNA is contributed to the host by a bacterial virus.

27.3 Human Immunodeficiency Virus (HIV)

AIDS is caused by HIV.

Human immunodeficiency virus (HIV) causes acquired immunodeficiency syndrome (AIDS) (figure 27.6).

HIV infection compromises the host immune system.

HIV specifically targets macrophages and CD4[+] cells, a type of helper T-lymphocyte cell. With the loss of these cells the body cannot fight off opportunistic infections, which ultimately lead to death.

HIV infects key immune-system cells.

The viral glycoprotein gp120 precisely fits on the cell-surface marker protein CD4[+] on macrophages and T cells. When HIV attaches to two receptors, CD4[+] and CCR5, receptor-mediated endocytosis is activated, bringing the HIV particle into the cell.

Once inside the cell, the protective coat is shed, releasing viral RNA and reverse transcriptase into the cytoplasm. Reverse transcriptase makes double-stranded DNA complementary to the viral RNA. This DNA may be incorporated into the host DNA as a provirus.

Replicated viruses are budded off the host cell by exocytosis.

HIV has a high mutation rate because the reverse transcriptase enzyme is much less accurate than DNA polymerases. Mutations lead to an altered glycoprotein gp120, which now binds instead

to the CXCR4 receptor found only on the surface of CD4[+] cells. Incorporation of the altered HIV particle leads to a rapid decline in T cells and immune response.

AIDS treatment targets different phases of the HIV life cycle.

Drugs target reverse transcriptase, a protease involved in protein maturation, viral entry, and the integration of the genome. Most approved drugs are reverse transcriptase and protease inhibitors.

Combination drug therapy using nucleoside analogues and protease inhibitors eliminates HIV from the bloodstream but not totally from the body.

Vaccine development for HIV has been unsuccessful.

Successful vaccines must elicit a strong immune response, and attempts to create immunity against a specific HIV subunit have failed. Attenuated viruses in animal tests have also proved capable of mutating into infectious forms.

27.4 Other Viral Diseases

The flu is caused by influenza virus.

One of the most lethal viruses in human history is type A influenza virus. Influenza can also infect other mammals and birds.

Genes in influenza viruses undergo recombination frequently, so they are not recognized by antibodies against past infections. Each year the composition of flu vaccines must be changed.

New viruses emerge by infecting new hosts.

Viruses can extend their host range by jumping to another species. Examples include hantavirus, hemorrhagic fever, and SARS (figure 27.9).

Viruses can cause cancer.

Viruses have been linked to formation of cancers, including liver cancer and cervical papillomas.

27.5 Prions and Viroids: Subviral Particles

Prion replication was a heretical suggestion.

Prions are proteinaceous infectious particles consisting of a misfolded form of a protein. This misfolding catalyzes a chain reaction of misfolding in normal proteins, causing disease.

Evidence has accumulated that prions cause TSEs (figure 27.10).

TSE disease-causing prions (PrP[sc]) are the same protein as the normal version (PrP[c]) but misfolded.

Viroids are infectious RNA with no protein coat.

Viroids are circular, naked molecules of RNA that infect plants. They use host protein to replicate.

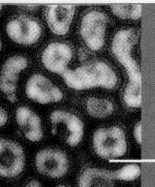

Review Questions

UNDERSTAND

1. The reverse transcriptase enzyme is active in which class of viruses?
   a. Positive-strand RNA viruses
   b. Double-stranded DNA viruses
   c. Retroviruses
   d. Negative-strand RNA viruses

2. Which of the following is not part of a virus?
   a. Capsid
   b. Ribosomes
   c. Genetic material
   d. All of the choices are found in viruses.

3. Which of the following is common in animal viruses but not in bacteriophage?

   a. DNA
   b. Capsid
   c. Envelope
   d. Icosahedral shape

4. Which of the following would not be part of the life cycle of a lytic virus?

   a. Macromolecular synthesis
   b. Attachment to host cell
   c. Assembly of progeny virus
   d. Integration into the host genome

5. A process by which a virus may change a benign bacteria into a virulent strain is called

   a. induction.
   b. phage conversion.
   c. lysogeny.
   d. replication.

6. Prior to entry, the _____ glycoprotein of the HIV virus recognizes the _____ receptor on the surface of the macrophage.

   a. CCR5; gp120
   b. CXCR4; CCR5
   c. CD4; CCR5
   d. gp120; CD4

7. The use of multiple drugs in HAART to treat AIDS has

   a. completely removed the virus from infected individuals.
   b. reduced the viral level in the bloodstream to undetectable levels.
   c. been a complete failure.
   d. been supplanted by the new HIV vaccine.

APPLY

1. The varying degrees of resistance to HIV in populations has been suggested to be related to the patterns of smallpox outbreaks over human history. This explanation hinges on

   a. the similarity in the genomes of the two viruses.
   b. the fact that both viruses use reverse transcriptase.
   c. both viruses using the same receptor to bind to host cells.
   d. the fact that both viruses compromise the immune system.

2. The idea of a protein that was an infectious agent was heretical because

   a. proteins are not that important in cells.
   b. proteins are not the informational molecules in cells.
   c. the function of proteins does not depend on their structure.
   d. proteins require nucleic acids for their function.

3. Bacterial viruses and animal viruses are similar in that they both

   a. have only DNA as genetic material.
   b. have only RNA as genetic material.
   c. require host functions for some aspect of their life cycle.
   d. do not require any host proteins.

4. The drugs used against HIV in AIDS therapy are not effective against the flu because

   a. HIV is an RNA virus and influenza is a DNA virus.
   b. HIV is a DNA virus and influenza is an RNA virus.
   c. the two viruses have different-sized genomes.
   d. the proteins targeted by HIV drugs are not found in influenza.

5. Phage conversion in which viruses add genes to a bacterial cell can be considered to be a form of

   a. standard inheritance.
   b. horizontal gene transfer.
   c. vertical gene transfer.
   d. parasitism.

6. According to the prion hypothesis, the infectious agent for scrapie must have "genetic" information in

   a. the sequence of amino acids in the scrapie protein.
   b. the sequence of bases in the scrapie gene.
   c. the three-dimensional structure of the scrapie prion.
   d. All of the choices are correct.

7. The difficulty designing a single flu vaccine that will work forever is that influenza

   a. is an RNA virus.
   b. is a DNA virus.
   c. both mutates and can be recombined to form new viruses.
   d. infects only humans.

8. The SARS outbreak is an example of

   a. a virus jumping from one species to another.
   b. mutation of a virus that only infects humans.
   c. how viruses can disable the human immune system.
   d. two viruses combining to form a new virus.

SYNTHESIZE

1. *E. coli* lysogens derived from infection by phage λ can be induced to form progeny viruses by exposure to radiation. The inductive event is the destruction of a repressor protein that keeps the prophage genome unexpressed. What might be the normal role of the protein that recognizes and destroys the λ repressor?

2. Most biologists believe that viruses evolved following the origin of the first cells. Defend or critique this concept.

3. Much effort has been expended to produce a vaccine for HIV. To date, this has not been successful. Why has this been such a difficult task? Are any other viruses equally resistant to a vaccine, and are the reasons the same? Why do you think that we could make a vaccine against the smallpox virus that allowed us to completely eradicate this virus?

4. What do we mean by the term "emerging virus"? How is this a medical problem, and what is a recent example?

5. How might phage λ be used to transfer *E. coli* genes between different bacterial cells? Could this be used to transfer any gene?

ONLINE RESOURCE

www.ravenbiology.com

Understand, Apply, and Synthesize—enhance your study with animations that bring concepts to life and practice tests to assess your understanding. Your instructor may also recommend the interactive eBook, individualized learning tools, and more.

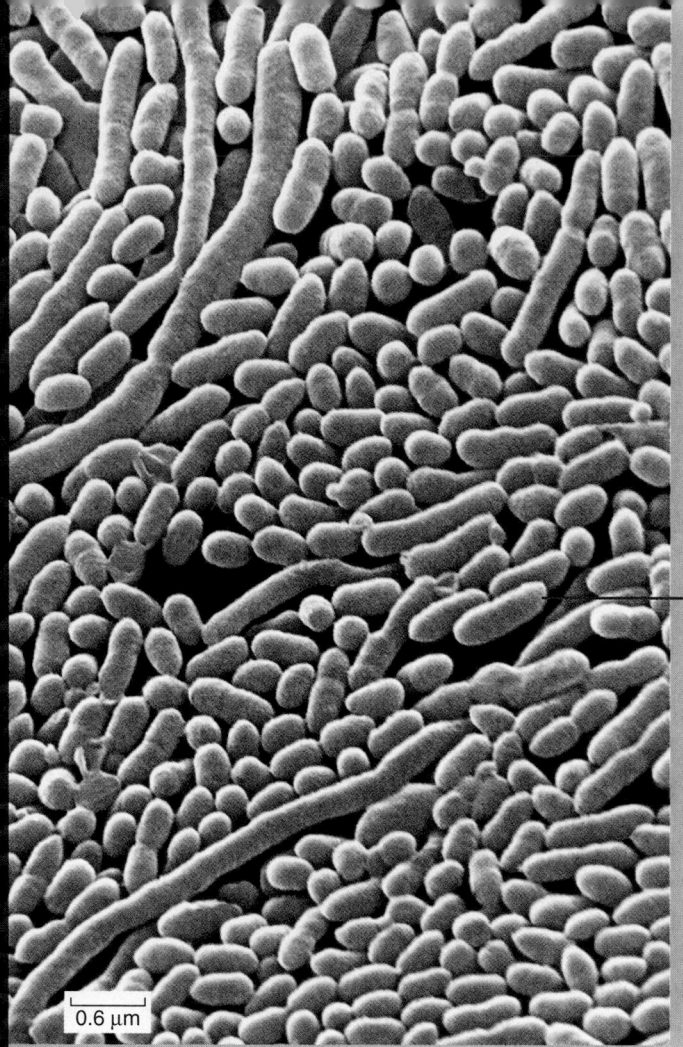

0.6 μm

Chapter 28

Prokaryotes

Chapter Contents

Introduction

One of the hallmarks of living organisms is their cellular organization. You learned earlier that living things come in two basic cell types: prokaryotes and eukaryotes. It might surprise you to learn that prokaryotes, despite being considerably smaller, are more numerous than their eukaryotic counterparts. Based on sheer numbers, prokaryotes are the dominant life-forms on the planet. If we examined a human being closely, we would discover that for every single cell of the human body there are approximately 10 prokaryotic cells—and there are trillions of human cells.

Prokaryotic microbes play an important role in global ecology as well. We no longer think of geochemistry, but now of biogeochemistry. Most biologists think prokaryotes were the first organisms to evolve. The diversity of eukaryotic organisms that currently live on Earth could not exist without prokaryotes because they make possible many of the essential functions of ecosystems. Prokaryotic photosynthesis, for example, is thought to have been the source of the oxygen in ancient Earth's atmosphere, and it still contributes significantly to oxygen production today. An understanding of prokaryotes is essential to understanding all life on Earth, past and present.

28.1 Prokaryotic Diversity

Learning Outcomes

1. Differentiate among archaea, bacteria, and eukarya.
2. Describe the basic features of bacteria and archaea.
3. Explain classification methods for prokaryotes.

A brief history of microbiology

The size of prokaryotic cells led to their being undiscovered for most of human history. Despite this, as early as 1546, the Italian physician Girolamo Fracastoro suggested that disease was caused by unseen organisms. The history of microbiology entwines the twin strands of technology for visualizing the unseen (microscopy) and the investigation of the nature of infectious diseases. This history also includes the prolonged debate about the very nature of how living things arise: from other living things or by spontaneous generation from nonliving matter.

Although he did not invent the microscope, Dutch scientist Antony van Leeuwenhoek was the first to observe and accurately describe microbial life. From these earliest observations we have built more and more sophisticated microscopes to study this unseen realm (see chapter 4). The modern electron microscope allows us to see extensive substructure within cells.

The controversy over spontaneous generation continued into the 19th century, eventually being refuted by French microbiologist Louis Pasteur (see figure 1.4) and others. During this same period in the 19th century, evidence accumulated that microorganisms could cause disease: the so-called germ theory of disease. Ultimately, German physician Robert Koch studied the disease anthrax in detail and was able to culture the causative agent. He proposed four postulates to prove a causal relationship between a microorganism and a disease:

1. The microorganism must be present in every case of the disease and absent from healthy individuals.
2. The putative causative agent must be isolated and grown in pure culture.
3. The same disease must result when the cultured microorganism is used to infect a healthy host.
4. The same microorganism must be isolated again from the diseased host.

The study of microorganisms and disease also led to advances in knowledge about the immune system, and how challenge with a disease-causing agent can prevent future infection (see chapter 51). These many different historical strands produced a vibrant field—microbiology—which is still relevant today. In this chapter we will concentrate on the study of prokaryotic microorganisms, and later in chapter 29 we will consider their eukaryotic counterparts.

Given the rich history of microbiology, which was barely touched upon here, it is amazing that at the end of the 20th century, new techniques that allow us to survey microbial diversity without having to grow cells in culture led to the conclusion that we have barely begun to understand the diversity of prokaryotic life. Although thousands of different kinds of prokaryotes are currently recognized, it is estimated that only between 1 and 10% of all prokaryotic species are known and characterized, leaving between 90 and 99% unknown and undescribed. Every place microbiologists look, new species are being discovered, often altering the way we think about prokaryotes.

Archaea and bacteria are the oldest, structurally simplest, and most abundant forms of life. They are also the only organisms with prokaryotic cellular organization. Prokaryotes were abundant for over a billion years before eukaryotes appeared in the world. Early photosynthetic bacteria (cyanobacteria) altered the Earth's atmosphere by producing oxygen, which stimulated extreme bacterial and eukaryotic diversity.

Prokaryotes are ubiquitous and live everywhere eukaryotes do; they are also able to thrive in places no eukaryote could live. Bacteria and archaea have been found in deep-sea caves, volcanic rims, and deep within glaciers. Some of the extreme environments in which prokaryotes can be found would be lethal to any other life-form.

Many archaea are extremophiles. They live in hot springs that would cook other organisms (figure 28.1), in hypersaline

Figure 28.1 Hot springs such as these in Yellowstone National Park contain thermophilic prokaryotes.

environments that would dehydrate other cells, and in atmospheres rich in otherwise-toxic gases such as methane or hydrogen sulfide. They have even been recovered living beneath 435 m of ice in Antarctica!

These harsh environments may be similar to the conditions present on the early Earth when life first began. It is likely that prokaryotes evolved to dwell in these harsh conditions early on and have retained the ability to exploit these areas as the rest of the atmosphere has changed.

Prokaryotes are fundamentally different from eukaryotes

Prokaryotes differ from eukaryotes in numerous important features. These differences represent some of the most fundamental distinctions that separate any groups of organisms.

Unicellularity. With a few exceptions, prokaryotes are fundamentally single-celled. In some types, individual cells adhere to one another within a matrix and form filaments; however, the cells retain their individuality. Even in this case, the characteristics of the organism are the characteristics of the individual cells, unlike in a multicellular organism.

In their natural environments, most bacteria appear to be capable of forming a complex community of different species called a **biofilm** (figure 28.2). These

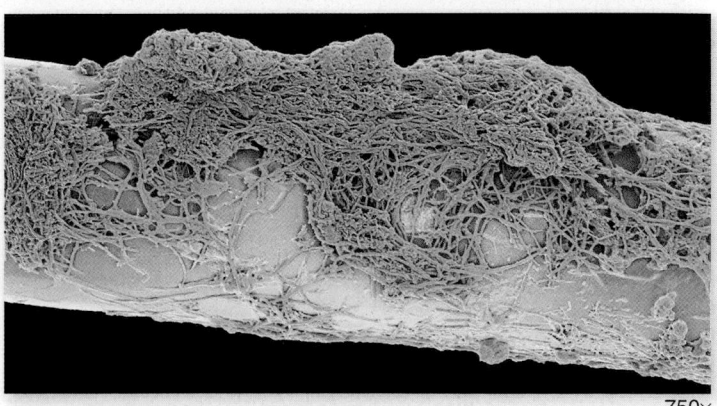

750×

Figure 28.2 Used toothbrush bristle shows a biofilm of the bacteria that cause dental plaque.

associations are more resistant to antibiotics, desiccation, and other environmental stressors than is a simple colony of a single type of microbe, such as laboratory culture.

Cell size. As new species of prokaryotes are discovered, investigators are finding that the size of prokaryotic cells varies tremendously, by as much as five orders of magnitude. The largest bacterial cells currently characterized are from *Thiomargarita namibiensis* (figure 28.3). A single cell from this species is up to 750 µm across, which is visible to the naked eye and is roughly the size of the eye of a bumblebee. Most prokaryotic cells, however, are only 1 µm or less in diameter, whereas most eukaryotic cells are well over 10 times bigger. This generality is misleading, however, because there are very small eukaryotes as well as very large prokaryotes.

Nucleoid. Prokaryotes lack a membrane-bounded nucleus; instead they usually have a single circular chromosome made up of DNA and histone-like proteins in a nucleoid region of the cell. The exceptions to this are *Vibrio cholerae*, which has two circular chromosomes, and *Borrelia* species, which have multiple linear chromosomes. Prokaryotic cells often have accessory DNA molecules called plasmids as well.

Cell division and genetic recombination. Cell division in eukaryotes takes place by mitosis and involves spindles made up of microtubules. Cell division in prokaryotes takes place mainly by binary fission (see chapter 10), which does not use a spindle. Prokaryotes also do not have a sexual cycle (see chapter 11), although we now know that they exchange genetic material extensively. These mechanisms are collectively called horizontal gene transfer and are not related to reproduction.

Internal compartmentalization. The cytoplasm of prokaryotes, unlike that of eukaryotes, does not have extensive internal compartments and no membrane-bounded organelles (see chapter 4). Instead the plasma membrane of prokaryotes can be extensively infolded, and both respiration and photosynthesis take place on organized portions of the plasma membrane.

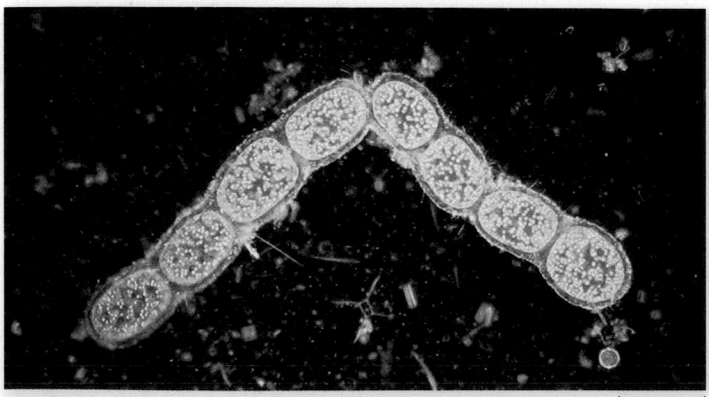

Figure 28.3 Cells of the bacterium *Thiomargarita namibiensis* can be as large as 750 µm across.

0.2 mm

Flagella. Prokaryotic flagella do not show the 9 + 2 architecture found in eukaryotes. Instead, they are composed of a single fiber of the protein flagellin. Bacterial flagella also are rigid and spin like propellers, whereas eukaryotic flagella have a whiplike motion (described in more detail later and in figure 28.9).

Metabolic diversity. Photosynthetic bacteria have two basic patterns of photosynthesis: (1) oxygenic, producing oxygen, and (2) anoxygenic, nonoxygen producing. Anoxygenic photosynthesis involves the formation of products such as sulfur and sulfate instead of oxygen. Prokaryotic cells are also the only chemolithotrophic organisms, meaning that they use the energy stored in chemical bonds of inorganic molecules to synthesize carbohydrates (we will explore this metabolic diversity in section 28.4).

Despite similarities, bacteria and archaea differ fundamentally

Archaea and bacteria are similar in that both have a prokaryotic cellular structure, but they vary considerably at the biochemical and molecular levels. They differ in four key areas: plasma membranes, cell walls, DNA replication, and gene expression.

Plasma membranes. All prokaryotes have plasma membranes with a fluid mosaic architecture (see chapter 5). The plasma membranes of archaea differ from both bacteria and eukaryotes. Archaean membrane lipids are composed of glycerol linked to hydrocarbon chains by ether linkages, not the ester linkages seen in bacteria and eukaryotes (figure 28.4a). These hydrocarbons may also be branched, and they may be organized as tetraethers that form a monolayer instead of a bilayer (figure 28.4b).

In the case of some hyperthermophiles, the majority of the membrane may be this tetraether monolayer. This structural feature is part of what allows these archaeans to withstand high temperatures.

Cell wall. Both kinds of prokaryotes typically have cell walls covering the plasma membrane that strengthen the cell. The cell walls of bacteria are constructed, minimally, of **peptidoglycan,** which is formed from carbohydrate polymers linked together by peptide cross-bridges. The peptide cross-bridges also contain D-amino acids, which are never found in cellular protein. The cell walls of archaea lack peptidoglycan, although some have **pseudomurein,** which is similar to peptidoglycan in structure and function. This wall layer is also a carbohydrate polymer with peptide cross-bridges, but the carbohydrates are different, and the peptide cross-bridge structure also differs. Other archaeal cell walls have been found to be composed of a variety of proteins and carbohydrates, making generalizations difficult.

DNA replication. Although both archaea and bacteria have a single replication origin, the nature of this origin and the proteins that act there are quite different. Archaeal

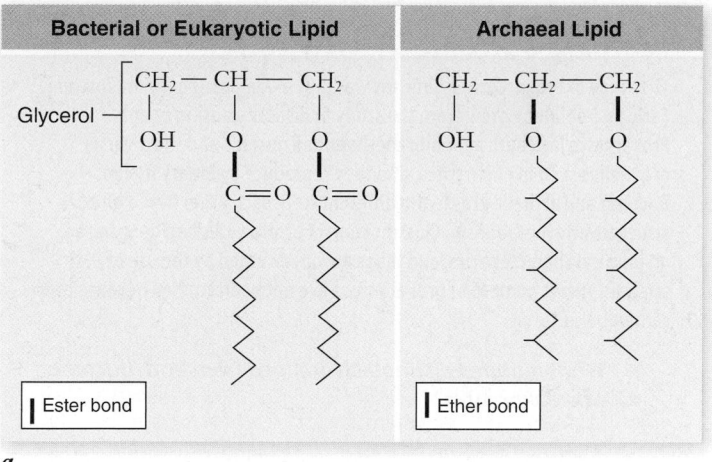

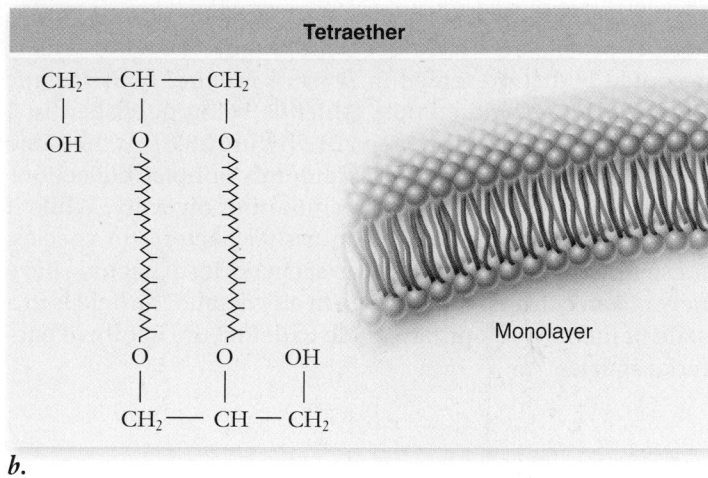

a.

b.

Figure 28.4 Archaea membrane lipids. *a.* Archaea membrane lipids are formed on a glycerol skeleton similar to bacterial and eukaryotic lipids, but the hydrocarbon chains are connected to the glycerol by ether linkages not ester linkages. The hydrocarbons can also be branched and even contain rings. *b.* These lipids can also form as tetraethers instead of diethers. The tetraether forms a monolayer as it includes two polar regions connected by hydrophobic hydrocarbons.

initiation of DNA replication is more similar to that of eukaryotes (see chapter 14).

Gene expression. The machinery used for gene expression also differs between archaea and bacteria. The archaea may have more than one RNA polymerase, and these enzymes more closely resemble the eukaryotic RNA polymerases than they do the single bacterial RNA polymerase. Some of the translation machinery is also more similar to that of eukaryotes (see chapter 15).

Most prokaryotes have not been characterized

Prokaryotes are not easily classified according to their forms, and only recently has enough been learned about their biochemical and metabolic characteristics to develop a satisfactory overall classification scheme comparable to that used for other organisms.

Early classification characteristics

Early systems for classifying prokaryotes relied on differential stains such as the Gram stain and differences in the observable phenotype of the organism. Key characteristics once used in classifying prokaryotes were

1. photosynthetic or nonphotosynthetic
2. motile or nonmotile
3. unicellular or colony-forming or filamentous
4. formation of spores or division by transverse binary fission
5. importance as human pathogens or not

Molecular approaches to classification

With the development of genetic and molecular approaches, prokaryotic classifications may help reflect true evolutionary relatedness. Molecular approaches include

1. the analysis of the amino acid sequences of key proteins

2. the analysis of nucleic acid–base sequences by establishing the percent of guanine (G) and cytosine (C)
3. nucleic acid hybridization, which is essentially the mixing of single-stranded DNA from two species and determining the amount of base-pairing (closely related species will have more bases pairing)
4. gene and RNA sequencing, especially looking at ribosomal RNA
5. whole-genome sequencing

The three-domain system of phylogeny, originated by Carl Woese, (figure 28.5 and chapter 26) relies on all of these molecular methods, but emphasizes the comparison of rRNA sequences to establish the evolutionary relatedness of all organisms. The rRNA sequences were chosen for their high degree of evolutionary conservation to ask questions about these most ancient splits in the tree of life.

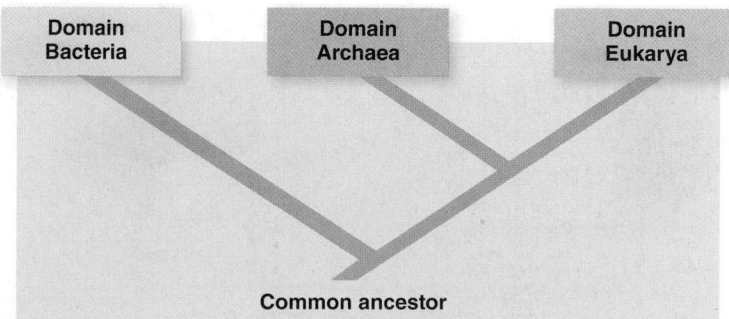

Figure 28.5 The three domains of life. The two prokaryotic domains, Archaea and Bacteria, are not closely related, though both are prokaryotes. In many ways (see text), archaea more closely resemble eukaryotes than bacteria. This tree is based on rRNA sequences.

Based on these sorts of molecular data, several groupings of prokaryotes have been proposed. The most widely accepted is that presented in *Bergey's Manual of Systematic Bacteriology*, second edition, which is being published in 5 volumes, the last published in 2012 (figure 28.6). At the same time, large scale sequencing of randomly sampled collections of bacteria show an incredible amount of diversity. While it has always been challenging to assign bacteria to species, these new data indicate that the vast majority of bacteria have never been cultured and studied in any detail. The field is in a state of flux as attempts are made to define the nature of bacterial species.

Learning Outcomes Review 28.1

The study of prokaryotic organisms was first made possible by microscopes. Early microbiology grew from the study of disease-causing agents. Prokaryotes lack both a membrane-bounded nucleus and the diverse organelles seen in eukaryotes, and they reproduce by binary fission. Bacteria and archaea are clearly different from each other based on both structure and metabolism. Classification of prokaryotes had been based on physical characteristics, and it has now been aided by the use of DNA analysis. A vast number of prokaryotes have not been studied because they cannot be cultured.

■ **What features distinguish archaea from both bacteria and eukaryotes?**

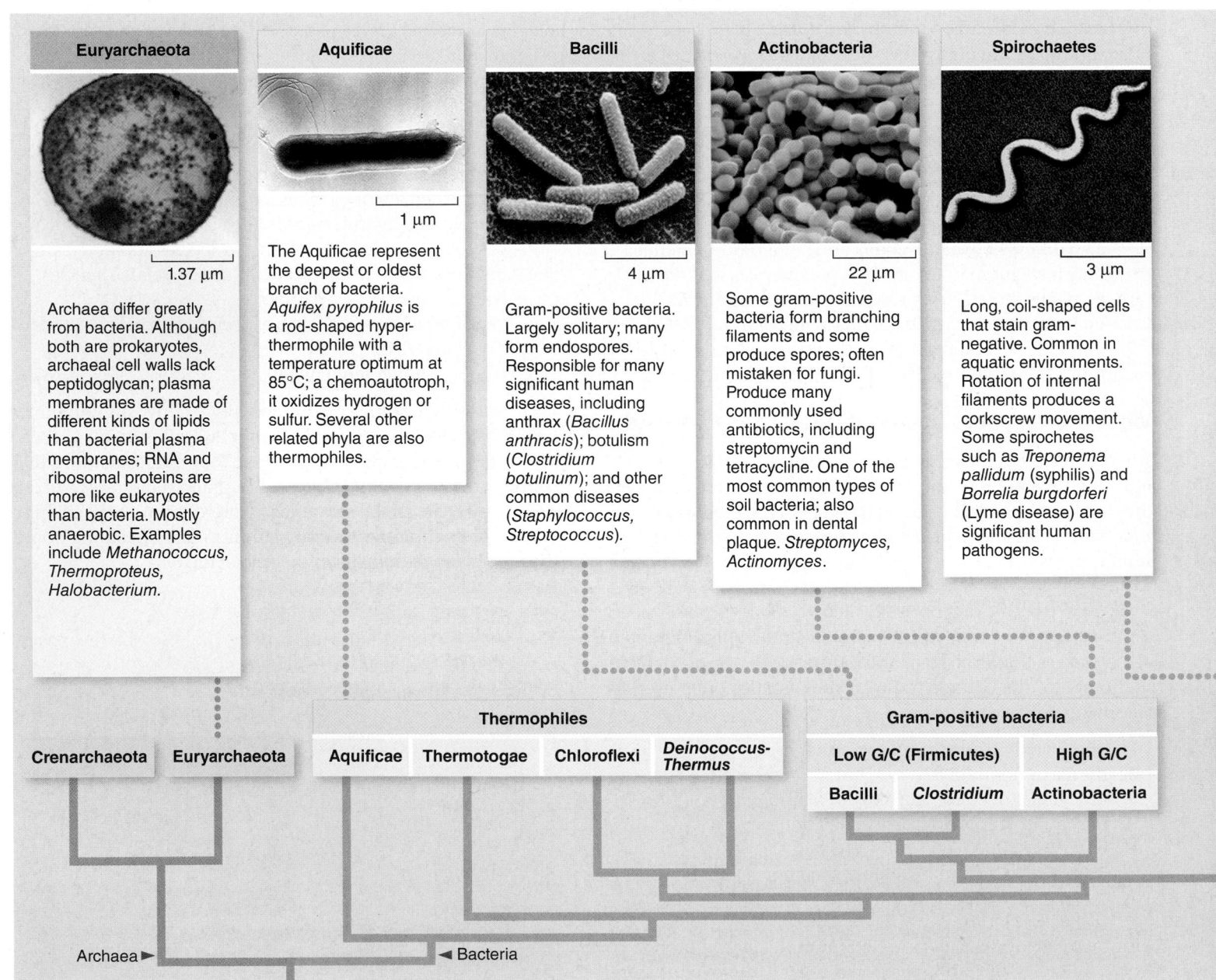

Figure 28.6 Some major clades of prokaryotes. The classification adopted here is that of *Bergey's Manual of Systematic Bacteriology*, second edition, 2001. G/C refers to %G/C in genome.

28.2 Prokaryotic Cell Structure

Prokaryotic cells are relatively simple, but they can be categorized based on cell shape. They also have some variations in structure that give them different staining properties for certain dyes. Other features are found in some types of cells but not in others.

Learning Outcomes

1. Describe the general features common to all prokaryotic cells.
2. Explain the differences between gram-positive and gram-negative bacterial cells.
3. Describe the features that distinguish different kinds of prokaryotic cells.

Prokaryotes have three basic forms: rods, cocci, and spirals

Although it is an oversimplification, it is useful to divide bacteria based on easily definable morphologies. Most prokaryotes exhibit one of three basic shapes: rod-shaped, often called a

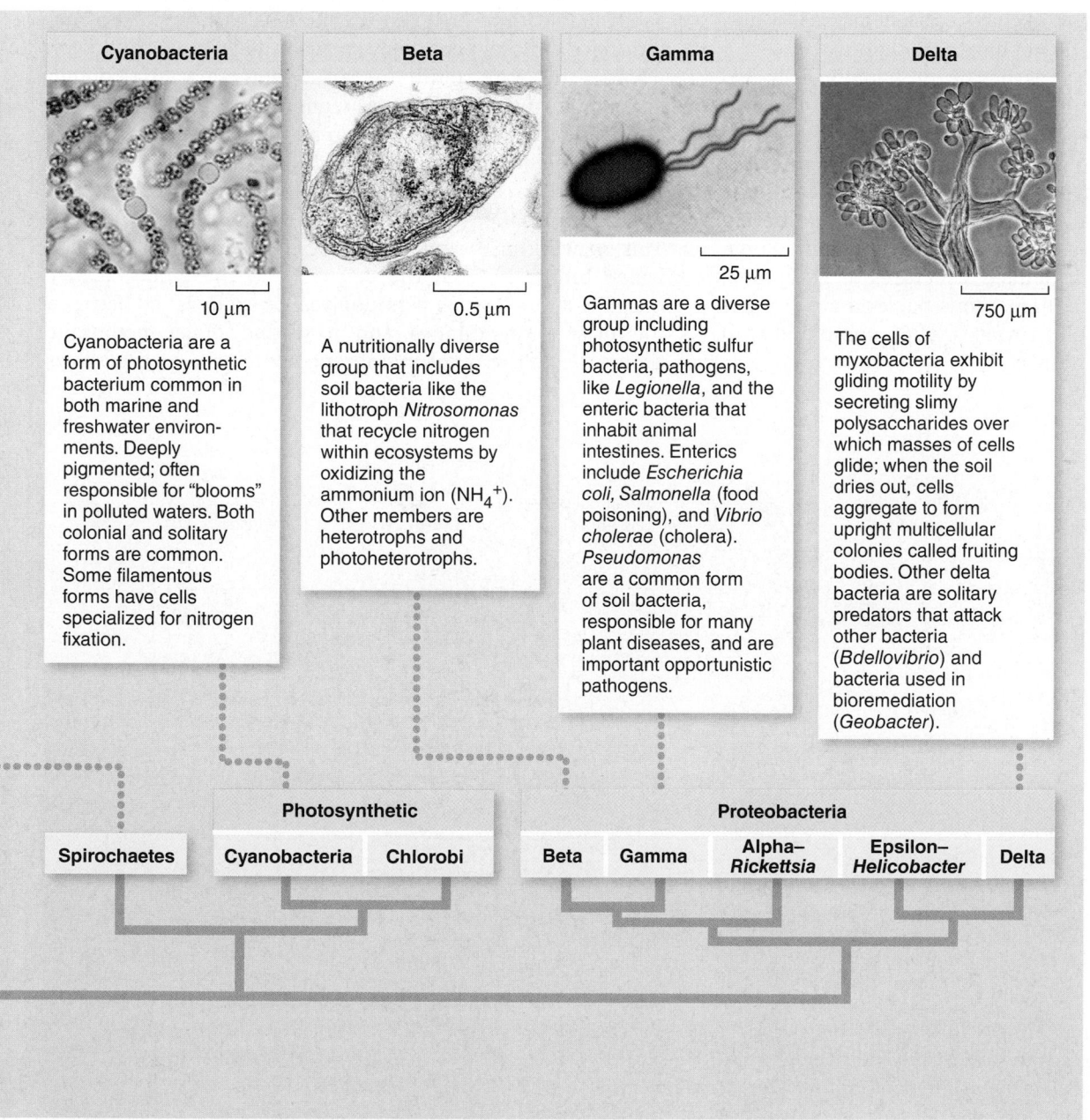

Cyanobacteria

10 μm

Cyanobacteria are a form of photosynthetic bacterium common in both marine and freshwater environments. Deeply pigmented; often responsible for "blooms" in polluted waters. Both colonial and solitary forms are common. Some filamentous forms have cells specialized for nitrogen fixation.

Beta

0.5 μm

A nutritionally diverse group that includes soil bacteria like the lithotroph *Nitrosomonas* that recycle nitrogen within ecosystems by oxidizing the ammonium ion (NH_4^+). Other members are heterotrophs and photoheterotrophs.

Gamma

25 μm

Gammas are a diverse group including photosynthetic sulfur bacteria, pathogens, like *Legionella*, and the enteric bacteria that inhabit animal intestines. Enterics include *Escherichia coli*, *Salmonella* (food poisoning), and *Vibrio cholerae* (cholera). *Pseudomonas* are a common form of soil bacteria, responsible for many plant diseases, and are important opportunistic pathogens.

Delta

750 μm

The cells of myxobacteria exhibit gliding motility by secreting slimy polysaccharides over which masses of cells glide; when the soil dries out, cells aggregate to form upright multicellular colonies called fruiting bodies. Other delta bacteria are solitary predators that attack other bacteria (*Bdellovibrio*) and bacteria used in bioremediation (*Geobacter*).

Spirochaetes	Photosynthetic		Proteobacteria				
	Cyanobacteria	Chlorobi	Beta	Gamma	Alpha–*Rickettsia*	Epsilon–*Helicobacter*	Delta

bacillus (plural, *bacilli*); *coccus* (plural, *cocci*), spherical- or ovoid-shaped; and *spirillum* (plural, *spirilla*), long and helical-shaped; these bacteria are also called *spirochetes*.

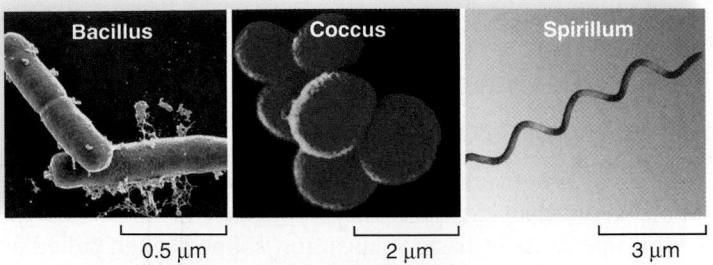

0.5 μm   2 μm   3 μm

© Dr. Richard Kessel & Dr. Gene Shih/Visuals Unlimited

The bacterial cell wall is the single most important contributor to cell shape. Bacteria that normally lack cell walls, such as the mycoplasmas, do not have a set shape.

As diverse as their shapes may be, prokaryotic cells also have many different methods to move through their environment. A *flagellum* or several flagella may be found on the outer surface of many prokaryotic cells. These structures are used to propel the organisms in a fluid environment. Some rod-shaped and spherical bacteria form colonies, adhering end-to-end after they have divided, forming chains. Some bacterial cells change into stalked structures or grow long, branched filaments. Some filamentous bacteria are capable of a gliding motion on solid surfaces, often combined with rotation around a longitudinal axis.

Prokaryotes have a tough cell wall and other external structures

The prokaryotic cell wall is often complex, consisting of many layers. Minimally it consists of peptidoglycan, a polymer unique to bacteria. This polymer forms a rigid network of polysaccharide strands cross-linked by peptide side chains. It is an important structure because it maintains the shape of the cell and protects the cell from swelling and rupturing in hypotonic solutions, which are most commonly found in the environment. The archaea do not possess peptidoglycan, but some have a similar structure called pseudomurein, or pseudopeptidoglycan.

Gram-positive and gram-negative bacteria

Two types of bacteria can be identified using a staining process called the **Gram stain,** hence their names. **Gram-positive** bacteria have a thicker peptidoglycan wall and stain a purple color, whereas the more common **gram-negative** bacteria contain less peptidoglycan and do not retain the purple-colored dye. These gram-negative bacteria can be stained with a red counterstain and then appear dark pink (figure 28.7).

In the gram-positive bacteria, the peptidoglycan forms a thick, complex network around the outer surface of the cell. This network also contains lipoteichoic and teichoic acid, which protrudes from the cell wall. In the gram-negative bacteria, a thin layer of peptidoglycan is sandwiched between the plasma membranes and a second outer membrane (figure 28.8). The outer membrane contains large molecules

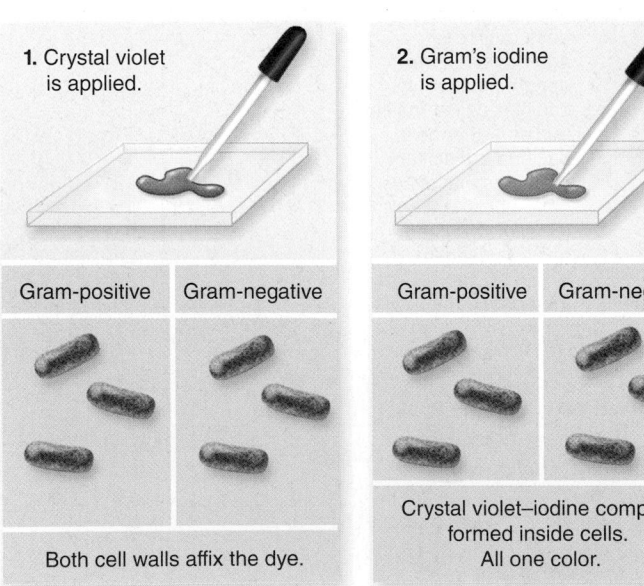

1. Crystal violet is applied.

Gram-positive | Gram-negative

Both cell walls affix the dye.

2. Gram's iodine is applied.

Gram-positive | Gram-negative

Crystal violet–iodine complex formed inside cells. All one color.

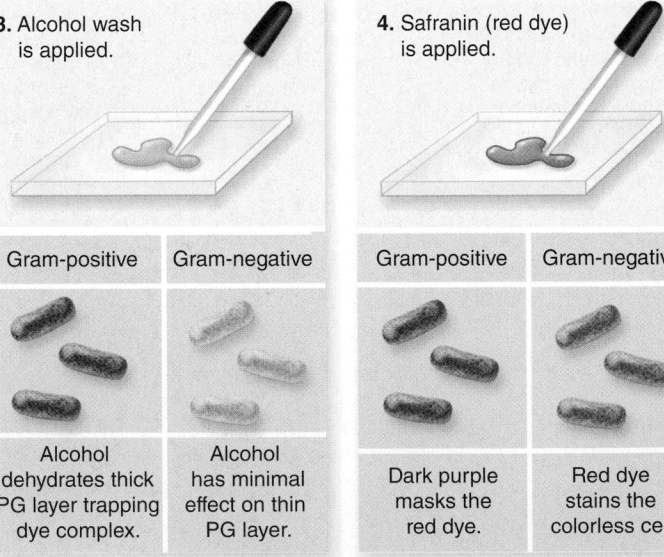

3. Alcohol wash is applied.

Gram-positive | Gram-negative

Alcohol dehydrates thick PG layer trapping dye complex. | Alcohol has minimal effect on thin PG layer.

4. Safranin (red dye) is applied.

Gram-positive | Gram-negative

Dark purple masks the red dye. | Red dye stains the colorless cell.

a.

Figure 28.7 The Gram stain. *a.* The thick peptidoglycan (PG) layer encasing gram-positive bacteria traps crystal violet dye, so the bacteria appear purple in a gram-stained smear (named after Hans Christian Gram—Danish bacteriologist, 1853–1938—who developed the technique). Because gram-negative bacteria have much less peptidoglycan (located between the plasma membrane and an outer membrane), they do not retain the crystal violet dye and so exhibit the red counterstain (usually a safranin dye). *b.* A micrograph showing the results of a Gram stain with both gram-positive and gram-negative cells.

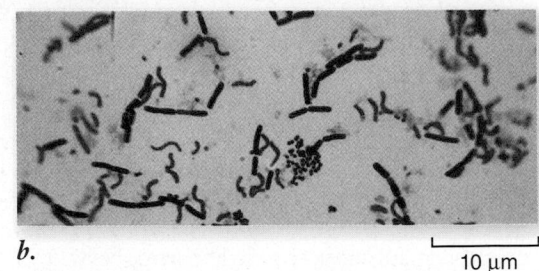

b.   10 μm

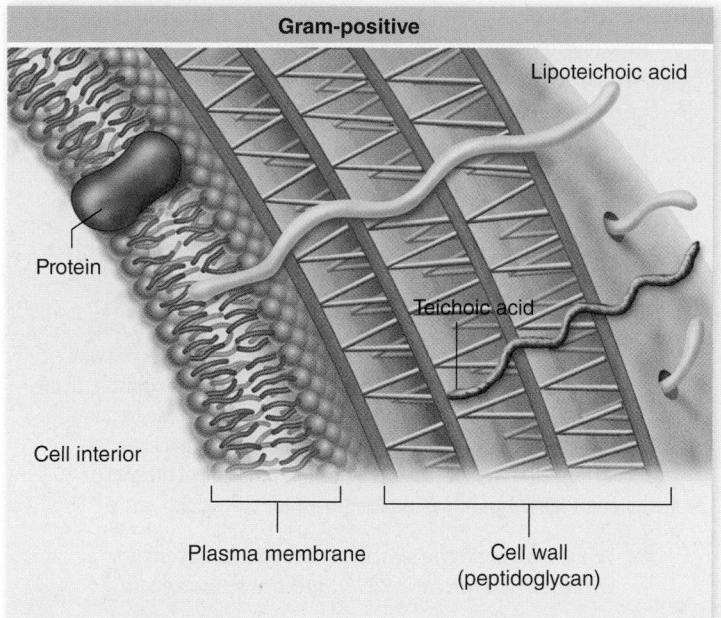

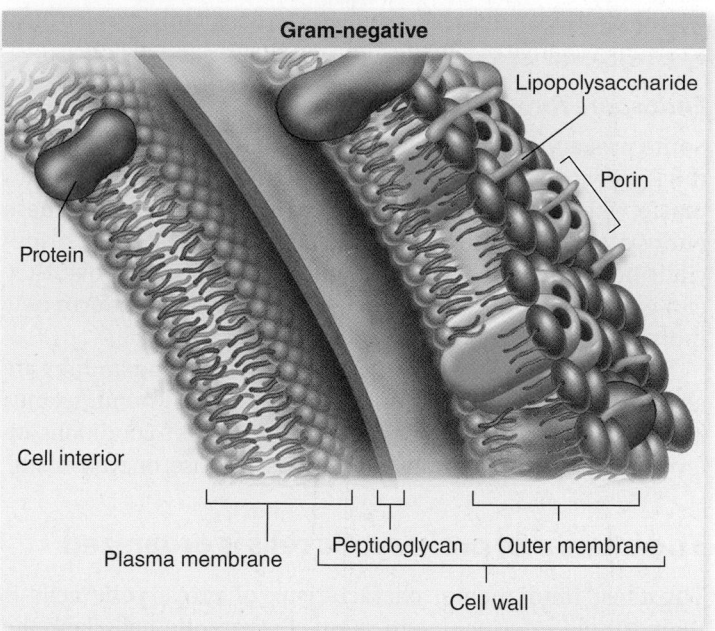

Figure 28.8 The structure of gram-positive and gram-negative cell walls. The gram-positive cell wall is much simpler, composed of a thick layer of cross-linked peptidoglycan chains. Molecules of lipoteichoic acid and teichoic acid are also embedded in the wall and exposed on the surface of the cell. The gram-negative cell wall is composed of multiple layers. The peptidoglycan layer is thinner than in gram-positive bacteria and is surrounded by an additional membrane composed of lipopolysaccharide. Porin proteins form aqueous pores in the outer membrane. The space between the outer membrane and peptidoglycan is called the periplasmic space.

of **lipopolysaccharide,** lipids with polysaccharide chains attached. The outer membrane layer makes gram-negative bacteria resistant to many antibiotics that interfere with cell-wall synthesis in gram-positive bacteria. For example, penicillin acts to inhibit the cross-linking of peptidoglycan in a gram-positive cell wall, killing growing bacterial populations.

S-layer

In some bacteria and archaea, an additional protein or glycoprotein layer forms a rigid paracrystalline surface called an *S-layer* outside of the peptidoglycan or outer membrane layers of gram-positive and gram-negative bacteria, respectively. Among the archaea, the S-layer is almost universal and can be found outside of a pseudopeptidoglycan layer or, in contrast to the bacteria, may be the only rigid layer surrounding the cell. The functions of S-layers are diverse and variable but often involve adhesion to surfaces or protection.

The capsule

In some bacteria, an additional gelatinous layer, the **capsule,** surrounds the other wall layers. A capsule enables a prokaryotic cell to adhere to surfaces and to other cells, and, most important, to evade an immune response by interfering with recognition by phagocytic cells. Therefore, a capsule often contributes to the ability of bacteria to cause disease.

Bacterial flagella and pili

Many kinds of prokaryotes have slender, rigid, helical flagella composed of the protein **flagellin** (figure 28.9). These flagella range from 3 to 12 μm in length and are very thin—only 10 to 20 nm thick. They are anchored in the cell wall and spin like a propeller, moving the cell through a liquid environment.

Bacterial cells that have lost the genes for flagellin are not able to swim.

Pili (singular, *pilus*) are other hairlike structures that occur on the cells of some gram-negative prokaryotes. They are shorter than prokaryotic flagella and about 7.5 to 10 nm thick. Pili are

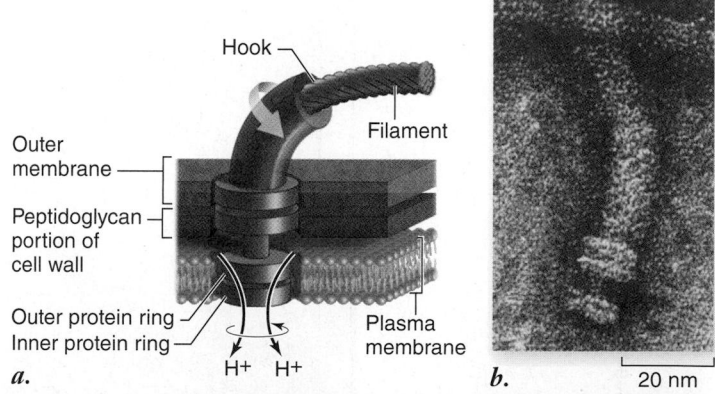

Figure 28.9 The flagellar motor of a gram-negative bacterium. *a.* A protein filament, composed of the protein flagellin, is attached to a protein rod that passes through a sleeve in the outer membrane and through a hole in the peptidoglycan layer to rings of protein anchored in the cell wall and plasma membrane, like rings of ball bearings. The rod rotates when the inner protein ring attached to the rod turns with respect to the outer ring fixed to the cell wall. The inner ring is an H⁺ ion channel, a proton pump that uses the flow of protons into the cell to power the movement of the inner ring past the outer one. The membrane wall anchor of the flagellum is called the basal body. *b.* Electron micrograph of bacterial flagellum.

more important in adhesion than movement, and they also have a role in exchange of genetic information (discussed later).

Endospore formation

Some prokaryotes are able to form **endospores,** developing a thick wall around their genome and a small portion of the cytoplasm when they are exposed to environmental stress. These endospores are highly resistant to environmental stress, especially heat, and when environmental conditions improve, they can germinate and return to normal cell division to form new individuals after decades or even centuries.

The bacteria that cause tetanus, botulism, and anthrax are all capable of forming spores. With a puncture wound, tetanus endospores may be driven deep into the skin where conditions are favorable for them to germinate and cause disease, or even death.

The interior of prokaryotic cells is organized

The most fundamental characteristic of prokaryotic cells is their simple interior organization. Prokaryotic cells lack the extensive functional compartmentalization seen within eukaryotic cells, but they do have the following structures:

Internal membranes. Many prokaryotes possess invaginated regions of the plasma membrane that function in respiration or photosynthesis (figure 28.10).

Nucleoid region. Prokaryotes lack nuclei and generally do not possess linear chromosomes. Instead, their genes are encoded within a single double-stranded ring of DNA that is highly condensed to form a visible region of the cell known as the **nucleoid region.** Many prokaryotic cells also possess plasmids, which as described earlier are small, independently replicating circles of DNA. Plasmids contain only a few genes, and although these genes may confer a selective advantage, they are not essential for the cell's survival.

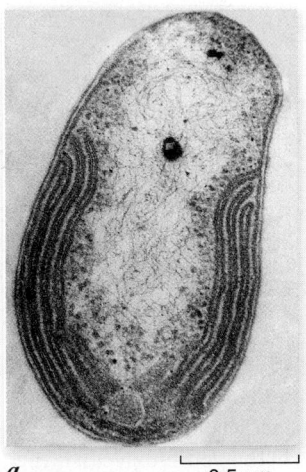
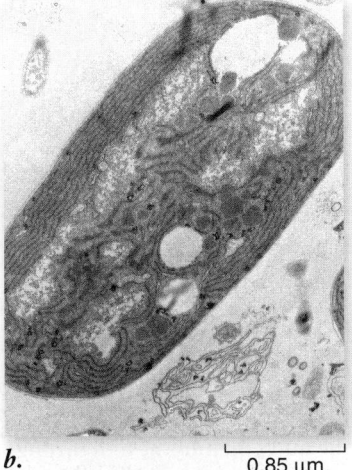

a.   0.5 μm   *b.*   0.85 μm

Figure 28.10 Prokaryotic cells often have complex internal membranes. *a.* This aerobic bacterium exhibits extensive respiratory membranes (long dark curves that hug the cell wall) within its cytoplasm, not unlike those seen in mitochondria. *b.* This cyanobacterium has thylakoid-like membranes (ripple-like shapes along the edges and in the center) that provide a site for photosynthesis.

Ribosomes. Prokaryotic ribosomes are smaller than those of eukaryotes and differ in protein and RNA content. Antibiotics such as tetracycline and chloramphenicol can tell the difference, however—they bind to prokaryotic ribosomes and block protein synthesis, but they do not bind to eukaryotic ribosomes.

Learning Outcomes Review 28.2

The three basic shapes of prokaryotes are rod-shaped, spherical, and spiral-shaped. Bacteria have a cell wall containing peptidoglycan, which is the basis for the Gram stain. Gram-positive bacteria have a thick cell wall, relative to gram-negative species. Many also have an external capsule. Some bacteria have flagella and pili. Some can form heat-resistant endospores. Although prokaryotes do not have membrane-bounded organelles, the interior of the cell is organized and may include infolding of the plasma membrane. Prokaryotic DNA is localized in a nucleoid region.

■   *What would be the simplest method to determine whether two bacteria belong to the same species?*

28.3 Prokaryotic Genetics

Learning Outcomes

1. *Contrast the mechanisms of DNA exchange in prokaryotes.*
2. *Explain genetic mapping in* E. coli.
3. *Describe how genetics explains the spread of antibiotic resistance.*

In sexually reproducing populations, traits are transferred vertically from parent to child. Prokaryotes do not reproduce sexually, but they can exchange DNA between different cells. This horizontal gene transfer occurs when genes move from one cell to another by **conjugation,** requiring cell-to-cell contact, or by means of viruses *(transduction)*. Some species of bacteria can also pick up genetic material directly from the environment *(transformation)*.

All of these processes have been observed in archaea, but the study of archaeal genetics is still in its infancy because of the difficulty in culturing most species. We concentrate here on bacterial systems, primarily *E. coli,* which has been studied extensively.

Conjugation depends on the presence of a conjugative plasmid

Plasmids may encode functions that can confer an advantage to the cell, such as antibiotic resistance, on which natural selection can operate—but they are not required for normal function. In some cases, plasmids can be transferred from one cell to another via conjugation. The best known plasmid capable of transfer is called the **F plasmid,** for fertility factor; cells containing F plasmids are termed **F⁺** cells, and cells that lack the F plasmid are **F⁻** cells. The F plasmid occurs in *E. coli* and, like all plasmids, acts as an independent genetic entity that nevertheless depends on the cell for replication. Studies involving the F plasmid were

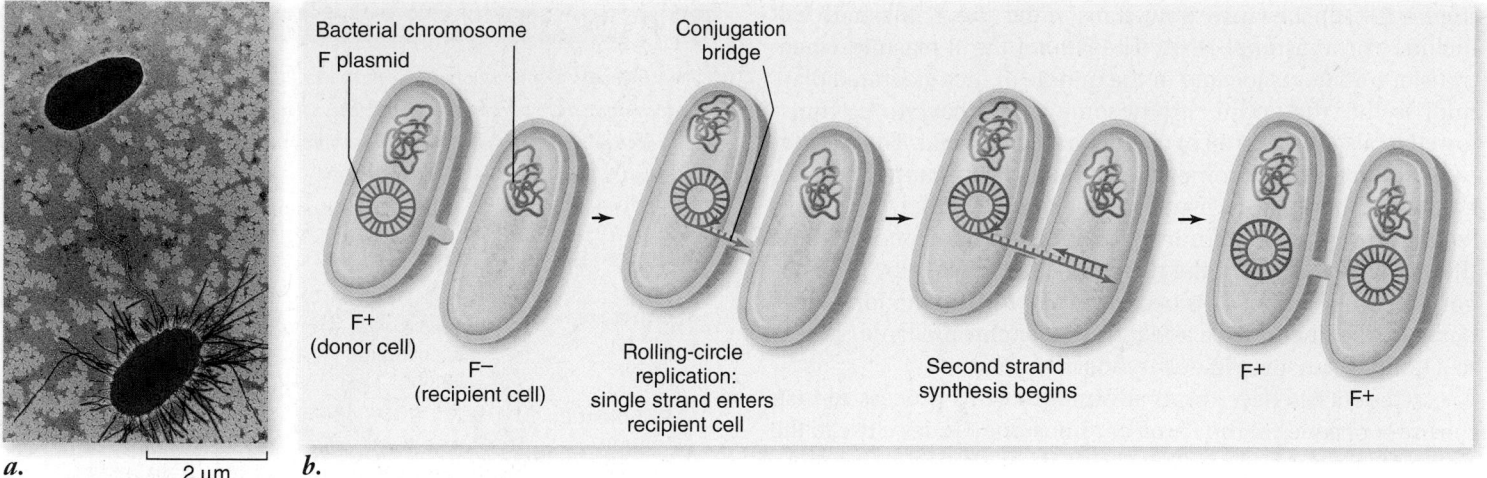

Figure 28.11 **Conjugation bridge and transfer of F plasmid between F⁺ and F⁻ cell.** *a.* The electron micrograph shows two *E. coli* cells caught in the act of conjugation. The connection between the cells is the extended F pilus. *b.* F⁻ cells are converted to F⁺ cells by the transfer of the F plasmid. The cells are joined by a conjugation bridge and the plasmid is replicated in the donor cell, displacing one parental strand. The displaced strand is transferred to the recipient cell, then replicated. After successful transfer, the recipient cell becomes an F⁺ cell capable of expressing genes for the F pilus and acting as a donor.

Labels in figure: Bacterial chromosome; F plasmid; Conjugation bridge; F⁺ (donor cell); F⁻ (recipient cell); Rolling-circle replication: single strand enters recipient cell; Second strand synthesis begins; F⁺; F⁺; *a.*; 2 μm; *b.*

critical to our current understanding of bacterial genetics and the organization of the *E. coli* chromosome.

F plasmid transfer

The F plasmid contains a DNA replication origin and several genes that promote its transfer to other cells. These genes encode protein subunits that assemble on the surface of the bacterial cell, forming a hollow pilus that is necessary for the transfer process (figure 28.11*a*).

First, the F plasmid binds to a site on the interior of the F⁺ cell just beneath the pilus, now called a *conjugation bridge.* Then, by a process called *rolling-circle replication,* the F plasmid begins to copy its DNA at the binding point. As it is replicated, the displaced single strand of the plasmid passes into the other cell. There, a complementary strand is added, creating a new, stable F plasmid (figure 28.11*b*).

Recombination between the F plasmid and host chromosome

The F plasmid can integrate into the host chromosome by recombining with it (see chapter 13). The molecular events in this process are similar to events during meiosis in eukaryotes when crossing over (recombination) exchanges material between chromosomes. This process is also called homologous recombination. In the case of the F plasmid and the *E. coli* chromosome, a single recombination event between two circles produces a larger circle, consisting of the chromosome and the integrated plasmid. This integration is actually mediated by host-encoded proteins, but it takes advantage of regions in the F plasmid called insertion sequences (IS) that also exist in the *E. coli* chromosome. These IS elements are actually transposable elements that probably moved from the chromosome to the F plasmid.

When the F plasmid is integrated into the chromosome, the cell is called an **Hfr cell** for high frequency of recombination

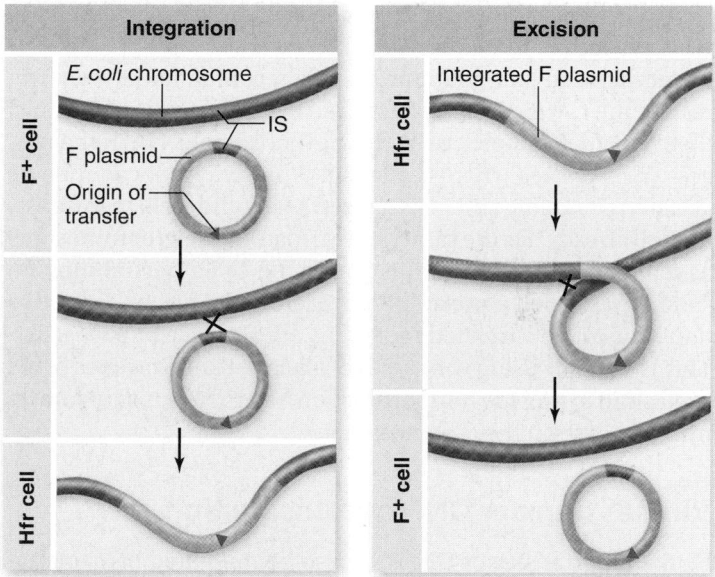

Figure 28.12 **Integration and excision of F plasmid.** The F plasmid contains short insertion sequences (IS) that also exist in the chromosome. This allows the plasmid to pair with the chromosome, and a single recombination event between two circles leads to a larger circle. This integrates the plasmid into the chromosome, creating an Hfr cell, as shown on the left. The process is reversible because the IS elements in the integrated plasmid can pair, and now a recombination event will return the two circles and convert the Hfr back to an F⁺ cell as shown on the right.

Labels in figure: Integration; Excision; F⁺ cell; *E. coli* chromosome; IS; F plasmid; Origin of transfer; Hfr cell; Integrated F plasmid; Hfr cell; F⁺ cell.

Data analysis If the excision of an F plasmid is not precise, and some *E. coli* DNA is added to the F plasmid, what are the consequences when this plasmid is transferred to a new cell?

(figure 28.12), because now transfer by the F plasmid will include chromosomal DNA. The site on the F plasmid where transfer initiates is located in the middle of the integrated plasmid, so that the entire chromosome would have to be transferred to also transfer all of the integrated plasmid. The transfer of the entire chromosome takes around 100 minutes, and the conjugation bridge is usually broken before that time. This leads to transfer of portions of donor chromosome that can then replace regions of the recipient chromosome by homologous recombination. This occurs by *two* recombination events between the linear piece and the circular chromosome, similar to a double crossover in eukaryotic meiosis.

Geneticists have taken advantage of this process to map the order of genes in the *E. coli* chromosome. Genes close to the origin of transfer are transferred early in the process, and those far from the origin are transferred later. If the process of mating is experimentally interrupted at different times, then gene order can be mapped based on time of entry of each gene (figure 28.13). The entry of genes can be detected by using a donor with wild-type alleles that can replace mutant alleles in the recipient by homologous recombination as described. These experiments have shown that the *E. coli* chromosome is indeed circular, and the genetic map is therefore circular. The units of the map are minutes, and the entire map is 100 minutes long.

The F plasmid can also excise itself by reversing the integration process. In this case, the IS elements bounding the integrated plasmid pair and now a single recombination event will restore the two circles (see figure 28.12). If excision is inaccurate, the F plasmid can pick up some chromosomal DNA in the process. This creates what is called an F plasmid that can then be transferred rapidly and in its entirety to another cell. In this case, the cell already has the same genetic material in its chromosome as that carried by the F′. This makes the cell a **partial diploid,** sometimes called a **merodiploid.** Merodiploids can be used to determine if new isolated mutations are alleles of known genes. This is done by using wild types of alleles of known genes of the F′ plasmid to provide normal function heterozygous to unknown mutant alleles in the chromosome.

Viruses transfer DNA by transduction

Horizontal transfer of DNA can also be mediated by bacteriophage. In **generalized transduction,** virtually any gene can be transferred between cells; in **specialized transduction,** only a few genes are transferred.

Generalized transduction

Generalized transduction can be thought of as an accident of the biology of some types of lytic phage (see chapter 27). In these viruses, after the viral genome is replicated and the phage head is constructed, the phage packaging machinery stuffs DNA into the phage head until no more fits, so-called headfull packaging. Sometimes the phage begins with bacterial DNA instead of phage DNA and packages this DNA into a phage head (figure 28.14). When this viral particle goes on to infect another cell, it injects the bacterial DNA into the infected cell instead of viral DNA. This DNA can then be incorporated into the recipient chromosome by homologous recombination. Similar to transfer by Hfr cells described earlier, two

Hypothesis: *Conjugation using Hfr strains involves the linear transfer of information from donor to recipient cell.*

Prediction: *If there is a linear transfer of information, then different markers should appear in a time sequence.*

Test: *Mating strains are agitated at time points to break the conjugation bridge, then plated to determine genotype.*

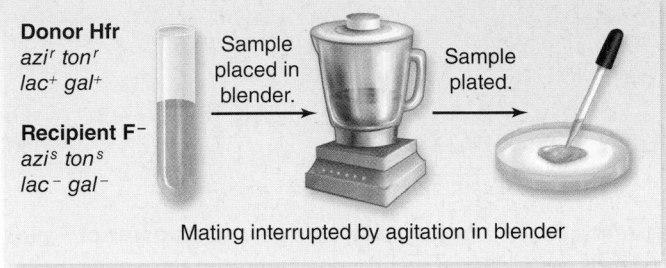

Mating interrupted by agitation in blender

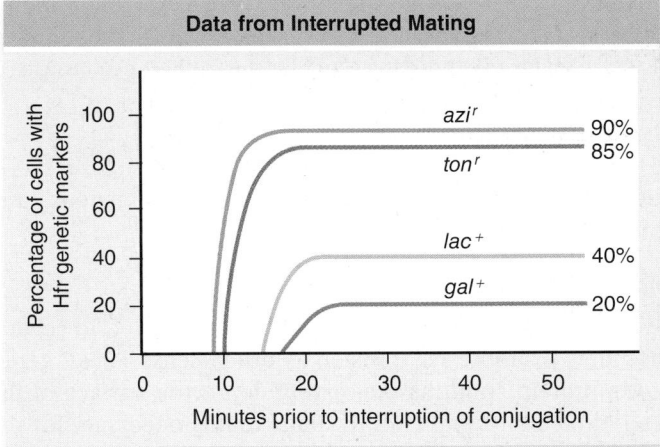

Data from Interrupted Mating

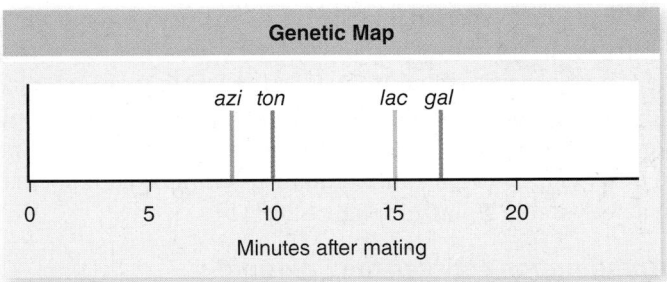

Genetic Map

Result: *The different genes from the donor strain appear in a linear time sequence.*

Conclusion: *The transfer of genetic information is linear. This sequence can be used to construct a genetic map ordering the genes on the chromosomes.*

Further Experiments: *Can other methods of DNA exchange also be used for genetic mapping?*

Figure 28.13 **Interrupted mating experiment allows construction of genetic map.**

recombination events are necessary to integrate the linear piece of DNA into the circular chromosome (see figure 28.14).

Generalized transduction has also been used for mapping purposes in *E. coli,* although the logic is different from

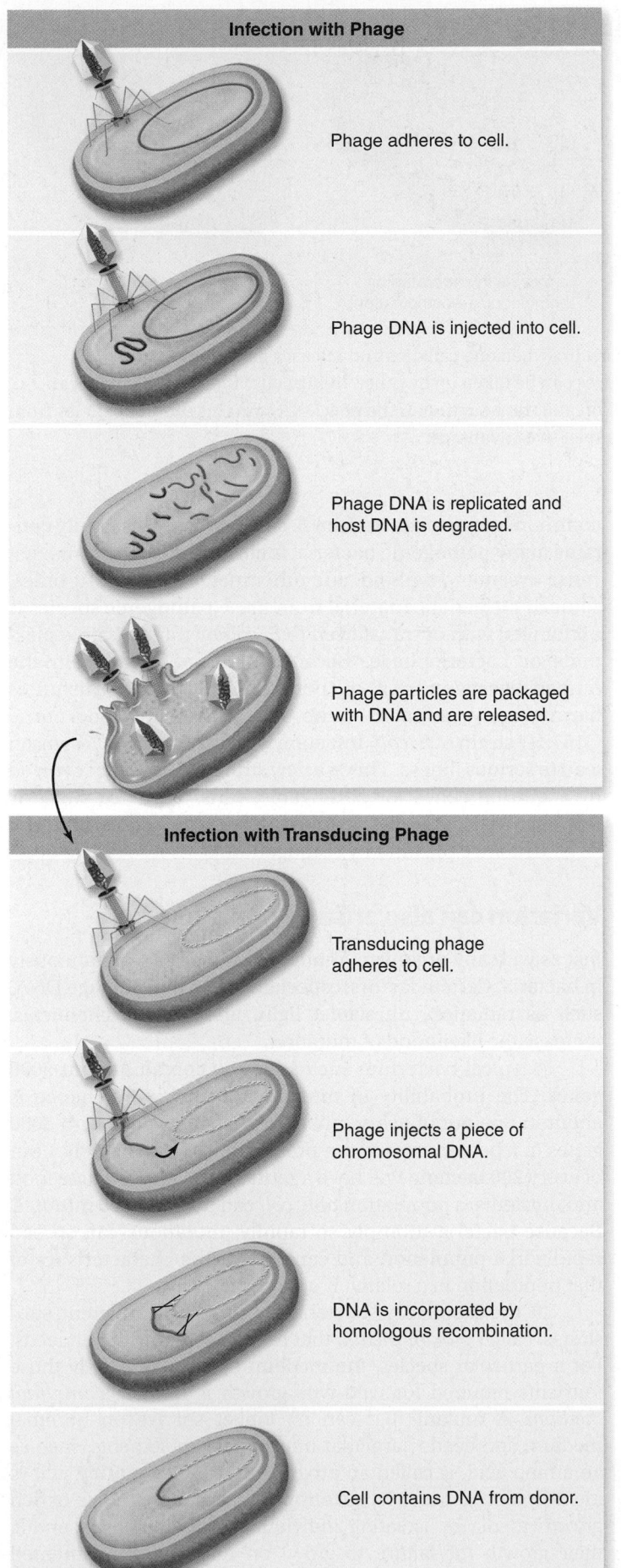

Infection with Phage

Phage adheres to cell.

Phage DNA is injected into cell.

Phage DNA is replicated and host DNA is degraded.

Phage particles are packaged with DNA and are released.

Infection with Transducing Phage

Transducing phage adheres to cell.

Phage injects a piece of chromosomal DNA.

DNA is incorporated by homologous recombination.

Cell contains DNA from donor.

Figure 28.14 Transduction by generalized transducing phage. When some phage infect cells, they degrade the host DNA into pieces. When the phage package their DNA, they can package host DNA in place of phage DNA to produce a transducing phage as shown on the top. When a transducing phage infects a cell, it injects host DNA that can then be integrated into the host genome by homologous recombination. With a linear piece of DNA, it requires two recombination events, which replace the chromosomal DNA with the transducing DNA as shown on the bottom. If the new allele is different from the old, the cell's phenotype will change.

that in conjugation. In transduction, the closer together two genes are, the more likely it is that they will be transferred in a single transduction event. This can be expressed mathematically as the *cotransduction frequency*. Correlation of maps from the two methods allows an empirical conversion between cotransduction frequency and minutes in the genetic map.

Specialized transduction

Specialized transduction is limited to phage that exhibit a lysogenic life cycle (see chapter 27). The prototype for this is phage λ from *E. coli*. When λ infects a cell and its genome integrates into the host chromosome, it does not destroy the cell but is passed on by cell division. This integration event is similar to the integration of the F plasmid, except that in the case of λ the recombination is a site-specific event mediated by phage-encoded proteins.

In this lysogenic state, the phage is called a prophage and it is dormant. The prophage encodes the functions necessary to eventually excise itself and undergo lytic growth, leading to the death of the cell. If this excision event is imprecise, it may take some chromosomal DNA with it, in the process making a specialized transducing phage. These phage carry both phage genes and chromosomal genes, unlike generalized transducing phage that carry only chromosomal DNA.

Because the phage head can carry only as much DNA as is found in the phage genome, imprecise excision results in deletion of phage genes. Thus specialized transducing phage may be defective if genes necessary for phage growth are lost in the process.

Specialized transducing phage particles can then integrate into the chromosome, just like wild-type phage, also making the cell diploid for the genes carried by the phage. Phage particles that can integrate as prophages may become trapped in the host genome if the genes necessary for excision become defective by mutation or are lost. The *E. coli* genome contains a number of such cryptic prophage, some of which encode functions important to the cell and must now be considered part of the host genome.

Transformation is the uptake of DNA directly from the environment

Transformation is a naturally occurring process in some species, such as the bacteria that were studied by Frederick Griffith (see chapter 14). Griffith discovered the process despite not knowing what chemical component was transferred. Transformation

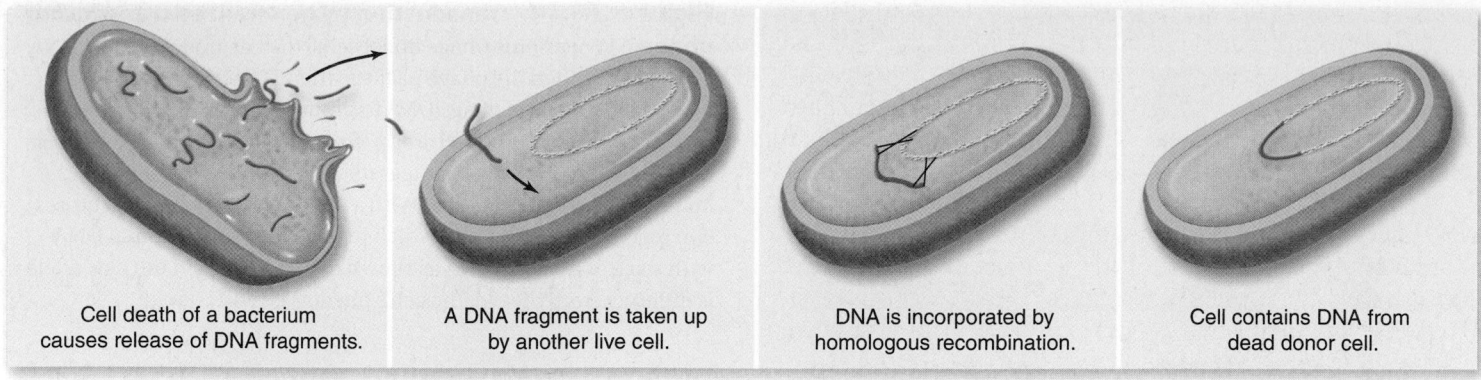

| Cell death of a bacterium causes release of DNA fragments. | A DNA fragment is taken up by another live cell. | DNA is incorporated by homologous recombination. | Cell contains DNA from dead donor cell. |

Figure 28.15 **Natural transformation.** Natural transformation occurs when one cell dies and releases its contents to the surrounding environment. The DNA is usually fragmented, and small pieces can be taken up by other, living cells. The DNA taken up can replace chromosomal DNA by homologous recombination as in conjugation and transduction. If the new DNA contains different alleles from the chromosome, the phenotype of the cell changes, possibly providing a selective advantage.

occurs when one bacterial cell has died and ruptured, spilling its fragmented DNA into the surrounding environment. This DNA can be taken up by another cell and incorporated into its genome, thereby transforming it (figure 28.15). When the uptake occurs under natural conditions, it is termed natural transformation. Some species of both gram-positive and gram-negative bacteria exhibit natural transformation, although the mechanisms seem to differ between the groups.

The proteins involved in the process of natural transformation are all encoded by the bacterial chromosome. The implication is that natural transformation may be the only one of the mechanisms of DNA exchange that evolved as part of normal cellular machinery. The transfer of chromosomal DNA by either conjugation or transduction can be thought of as accidents of plasmid or phage biology, respectively.

Transformation is also important in molecular cloning, but *E. coli* does not exhibit natural transformation. When transformation is accomplished in the laboratory it is called artificial transformation. Artificial transformation is useful for cloning and DNA manipulation (see chapter 17).

Antibiotic resistance can be transferred by resistance plasmids

Some conjugative plasmids pick up antibiotic resistance genes, becoming resistance plasmids, or **R plasmids.** The rapid transfer of newly acquired, antibiotic resistance genes by plasmids has been an important factor in the appearance of the resistant strains of the pathogen *Staphylococcus aureus* discussed in the next section.

The means by which resistance plasmids acquire antibiotic resistance genes is often through transposable elements, which were described in chapter 18. These elements can move from chromosome to chromosome or from plasmid to chromosome and back again, and they can transfer antibiotic resistance genes in the process. If a conjugative plasmid picks up these genes, then the bacterium carrying it has a selective advantage in the presence of those antibiotics.

An important example in terms of human health involves the Enterobacteriaceae, the family of bacteria to which the common intestinal bacterium *E. coli* belongs. This family contains many pathogenic bacteria, including the organisms that cause dysentery, typhoid, and other major diseases. At times, some of the genetic material from these pathogenic species is exchanged with or transferred to *E. coli* by transmissible plasmids or bacteriophage. Because of its abundance in the human digestive tract, *E. coli* poses a special threat if it acquires harmful traits, as seen by the outbreaks of the food-borne O157:H7 strain of *E. coli*. Infection with this strain of *E. coli* can lead to serious illness. This is a new strain of *E. coli* that evolved by acquiring genes for pathogenic traits. Evidence suggests this occurred by both transduction and the acquisition of a large virulence plasmid by conjugation.

Variation can also arise by mutation

Just as with any organism, mutations can arise spontaneously in bacteria. Certain factors, especially those that damage DNA, such as radiation, ultraviolet light, and various chemicals, increase the likelihood of mutation.

A typical bacterium such as *E. coli* contains about 5000 genes. The probability of mutation occurring by chance is about in one out of every million copies of a gene. With 5000 genes in a bacterium, we can predict that approximately 1 out of every 200 bacteria will have a mutation. With adequate food and nutrients, a population of *E. coli* can double in 20 minutes. Because bacteria multiply so rapidly, mutations can spread rapidly in a population and can change the characteristics of that population in a relatively short time.

In the laboratory, bacteria are grown on different substrates, called *growth media,* that reflect their nutritional needs. For a particular species, the medium that contains only those nutrients required for wild-type growth is termed a *minimal medium.* A mutant that can no longer survive on minimal medium and needs particular nutritional supplements, such as an amino acid, is called an **auxotroph. Replica plating** allows identification of bacterial auxotrophs from a master plate of rich growth media by isolating individual colonies and observing their growth (or failure to grow) on different supplemented media. The technique is somewhat like using a rubber stamp—an impression of colonies growing in a Petri plate is made on a velvet surface, and then this surface is pressed onto different media in other plates. The impression contains many thousands if not millions of cells from each colony—and each colony has grown from a single cell. In this way, a bacterium with a highly specific mutation can be isolated, identified, and grown.

The ability of prokaryotes to change rapidly in response to new challenges often has adverse effects on humans. A number of antibiotic-resistant strains of the bacterium *Staphylococcus aureus* (termed methicillin-resistant *Staphylococcus aureus,* or MRSA) had been known in hospital settings for some time. More recently these have been observed in infections out of the hospital setting, so-called community acquired MRSA.

Of most concern among these strains is **vancomycin-resistant *Staphylococcus aureus*** (VRSA). This appears to have arisen rapidly by mutation and is alarming because vancomycin is the drug of last resort, making these strains and the infections they cause very difficult to stop. *Staphylococcus* infections, or "staph" infections for short, provide an excellent example of the way in which mutation and intensive selection can bring about rapid change in bacterial populations.

Learning Outcomes Review 28.3

Prokaryotic DNA exchange is horizontal, from donor cell to recipient cell. DNA can be exchanged by conjugation via plasmids, by transduction via viruses, and by transformation through the direct uptake of DNA from the environment. These forms of DNA exchange can be used experimentally to map genes. Variation in prokaryotes also arises by mutation. Extensive use of antibiotics has led to selection for resistant organisms. Resistance genes can be transferred, rapidly spreading resistance.

■ *How does transfer of genetic information in bacteria differ from eukaryotic sex?*

28.4 Prokaryotic Metabolism

Learning Outcomes

1. *Describe the different ways that prokaryotes acquire energy and carbon.*
2. *Explain how bacterial proteins can cause disease in humans.*

The variation seen in prokaryotes manifests itself most noticeably in biochemical rather than morphological diversity. Wide variation has been found in the types of metabolism prokaryotes exhibit, especially in the means by which they acquire energy and carbon.

Prokaryotes acquire carbon and energy in four basic ways

Prokaryotes have evolved many mechanisms to acquire the energy and carbon they need for growth and reproduction.

Many are *autotrophs* that obtain their carbon from inorganic CO_2. Other prokaryotes are *heterotrophs* that obtain at least some of their carbon from organic molecules, such as glucose. Depending on the method by which they acquire energy, autotrophs and heterotrophs are categorized as follows:

Photoautotrophs. Many bacteria carry out photosynthesis, using the energy of sunlight to build organic molecules from carbon dioxide. The **cyanobacteria** use chlorophyll *a* as the key light-capturing pigment and H_2O as an electron donor, releasing oxygen gas as a by-product. They are therefore oxygenic, and their method of photosynthesis is very similar to that found in algae and plants.

Other bacteria use bacteriochlorophyll as their light-capturing pigment and H_2S as an electron donor, leaving elemental sulfur as the by-product. These bacteria do not produce oxygen (anoxygenic) and have a simpler method of photosynthesis. These are the purple and green sulfur bacteria.

Archaeal species also carry out photosynthesis, the simplest form known. This involves a single protein, bacteriorhodopsin, that uses energy from light to translocate protons across a membrane. This then provides a proton motive force for ATP synthesis. Recent surveys of microbial diversity in marine ecosystems using DNA sequencing have found a new relative of the rhodopsin family called proteorhodopsin. First found in a bacterial species, these proteorhodopsins are quite widespread, found in bacterial, archaeal, and even algal species. This raises the possibility that photosynthesis in marine systems may be more widespread and complex than previously thought.

Chemolithoautotrophs. Some prokaryotes obtain energy by oxidizing inorganic substances. Nitrifiers, for example, oxidize ammonia or nitrite to obtain energy, producing the nitrate that is taken up by plants. This process is called **nitrification,** and it is essential in terrestrial ecosystems because plants primarily absorb nitrogen in the form of nitrate.

Other chemolithoautotrophs oxidize sulfur, hydrogen gas, and other inorganic molecules. On the dark ocean floor at depths of 2500 m, entire ecosystems subsist on prokaryotes that oxidize hydrogen sulfide as it escapes from thermal vents.

Photoheterotrophs. The so-called purple and green nonsulfur bacteria use light as their source of energy but obtain carbon from organic molecules, such as carbohydrates or alcohols that have been produced by other organisms.

Chemoheterotrophs. The majority of prokaryotes obtain both carbon atoms and energy from organic molecules. These include decomposers and most pathogens. Human beings and all nonphotosynthetic eukaryotes are chemoheterotrophs as well.

Some bacteria can attack other cells directly

Invading pathogens of the genera *Yersinia* can introduce proteins directly into host cells by a specialized form of secretion. (*Yersinia pestis* is the bacterial species responsible for bubonic plague.) Most proteins secreted by gram-negative bacteria have special signal sequences that allow them to pass through the bacterium's double membrane. The proteins secreted by *Yersinia* lacked a key signal sequence that two known secretion mechanisms require for transport. The proteins must therefore have been secreted by means of a third type of system, which researchers called the *type III system*. This kind of system acts like a kind of molecular syringe allowing the pathogen to inject proteins directly into the cytoplasm of host cells.

As more bacterial species are studied, the genes coding for the type III system are turning up in other gram-negative animal pathogens, and even in more distantly related plant pathogens. The genes seem more closely related to one another than are the bacteria. Furthermore, the genes are similar to those that code for bacterial flagella.

These proteins are used to transfer other virulence proteins, such as toxins, into nearby eukaryotic cells. Given the similarity of the type III genes to the genes that code for flagella, the transfer proteins may form a flagellum-like structure that shoots virulence proteins into the host cells. Once in the eukaryotic cells, the virulence proteins affect the host's response to the pathogen.

In *Yersinia,* proteins secreted by the type III system are injected into macrophages; the proteins disrupt signals that tell the macrophages to engulf bacteria. *Salmonella* and *Shigella* use their type III proteins to enter the cytoplasm of eukaryotic cells, and thus they are protected from the immune system of their host. The proteins secreted by certain strains of *E. coli* alter the cytoskeleton of nearby intestinal eukaryotic cells, resulting in a bulge onto which the bacterial cells can tightly bind.

Bacteria are costly plant pathogens

Although the majority of commercially relevant plant pathogens are fungi, many diseases of plants are associated with particular heterotrophic bacteria. Almost every kind of plant is susceptible to one or more kinds of bacterial disease, including blights, soft rots, and wilts. Fire blight, which destroys pear and apple trees and related plants, is a well-known example of bacterial disease.

The early symptoms of these plant diseases vary, but they are commonly manifested as spots of various sizes on the stems, leaves, flowers, or fruits. Most bacteria that cause plant diseases are members of the group of rod-shaped gram-negative bacteria known as pseudomonads.

Learning Outcomes Review 28.4

Prokaryotes exhibit amazing metabolic diversity with both autotrophic and heterotrophic species. Photoautotrophs use light as an energy source; chemolithoautotrophs oxidize inorganic compounds. Photoheterotrophs use light as an energy source and organic compounds as carbon sources. Chemoheterotrophs use organic compounds for both energy and carbon. Bacterial animal pathogens attack host cells with toxic proteins that disrupt the host's immune response, among other effects.

■ *Why is metabolism a better way than morphology to characterize prokaryotes?*

Learning Outcomes

1. *Describe common human bacterial pathogens.*
2. *Explain how bacteria can cause ulcers.*
3. *Identify sexually transmitted diseases caused by bacteria.*

In the early 20th century, before the discovery and widespread use of antibiotics, infectious diseases killed nearly 20% of all U.S. children before they reached the age of five. Sanitation and antibiotics considerably improved the situation. In recent years, however, we have seen the appearance or reappearance of many bacterial diseases, including cholera, leprosy, tetanus, bacterial pneumonia, whooping cough, diphtheria, and Lyme disease (table 28.1). Members of the genus *Streptococcus* are associated with scarlet fever, rheumatic fever, pneumonia, "flesh-eating disease," and other infections. Tuberculosis, another bacterial disease, is still a leading cause of death in humans worldwide.

Bacteria have many different methods to spread through a susceptible population. Tuberculosis and many other bacterial diseases of the respiratory tract are mostly spread through the air in droplets of mucus or saliva. Diseases such as typhoid fever, paratyphoid fever, and bacillary dysentery are spread by fecal contamination of food or water. Lyme disease and Rocky Mountain spotted fever are spread to humans by tick vectors.

Tuberculosis has infected humans for all of recorded history

Tuberculosis (TB) has been a scourge to humanity for thousands of years. There is evidence that peoples from ancient Egypt and pre-Columbian South America died from TB; the TB bacillus (*Mycobacterium tuberculosis*) has been identified in prehistoric mummies. TB afflicts the respiratory system, thwarts the immune system, and is easily transmitted from person to person through the air.

The spread of tuberculosis

Currently, about one-third of all people worldwide are regularly exposed to *Mycobacterium tuberculosis.* An estimated 9.27 million new cases were diagnosed, and 1.8 million deaths occurred in 2007. In 2006, the World Health Organization reported the incidence of TB falling in five of six WHO regions, but the numbers continue to rise in Africa driven by the spread of HIV.

Since the mid-1980s, the United States has experienced a resurgence of TB. This peaked in the mid-1990s and has been declining since, although the rate of decline is leveling off. The latest statistics from the CDC indicate 11,182 TB cases in 2010, down from 13,754 cases in 2006.

Tuberculosis treatment

Most TB patients are placed on multiple, expensive antibiotics for six to twelve months. Alarming outbreaks of

TABLE 28.1

TABLE 28.1 Important Human Bacterial Diseases

Disease	Pathogen	Vector/Reservoir	Epidemiology
Anthrax	*Bacillus anthracis*	Animals, including processed skins	Bacterial infection that can be transmitted through contact or ingestion. Rare except in sporadic outbreaks. May be fatal.
Botulism	*Clostridium botulinum*	Improperly prepared food	Contracted through ingestion or contact with wound. Produces acute toxic poison; can be fatal.
Chlamydia	*Chlamydia trachomatis*	Humans, sexually transmitted disease (STD)	Urogenital infections with possible spread to eyes and respiratory tract. Increasingly common over past 20 years.
Cholera	*Vibrio cholerae*	Human feces, plankton	Causes severe diarrhea that can lead to death by dehydration; 50% peak mortality rate if untreated. A major killer in times of crowding and poor sanitation; after the 2010 earthquake in Haiti, 180,000 were infected in just 70 days.
Dental caries	*Streptococcus mutans, Streptococcus sobrinus*	Humans	A dense collection of these bacteria on the surface of teeth leads to secretion of acids that destroy minerals in tooth enamel; sugar alone does not cause caries.
Diphtheria	*Corynebacterium diphtheriae*	Humans	Acute inflammation and lesions of respiratory mucous membranes. Spread through respiratory droplets. Vaccine available.
Gonorrhea	*Neisseria gonorrhoeae*	Humans only	STD, on the increase worldwide. Usually not fatal.
Hansen disease (leprosy)	*Mycobacterium leprae*	Humans, feral armadillos	Chronic infection of the skin; worldwide incidence about 10–12 million, especially in southeast Asia. Spread through contact with infected individuals.
Lyme disease	*Borrelia burgdorferi*	Ticks, deer, small rodents	Spread through bite of infected tick. Lesion followed by malaise, fever, fatigue, pain, stiff neck, and headache.
Peptic ulcers	*Helicobacter pylori*	Humans	Originally thought to be caused by stress or diet, most peptic ulcers now appear to be caused by this bacterium; good news for ulcer sufferers because it can be treated with antibiotics.
Plague	*Yersinia pestis*	Fleas of wild rodents: rats and squirrels	Killed one-fourth of the population of Europe in the 14th century; endemic in wild rodent populations of the western United States today.
Pneumonia	*Streptococcus, Mycoplasma, Chlamydia, Haemophilus*	Humans	Acute infection of the lungs; often fatal without treatment. Vaccine for streptococcal pneumonia available.
Tuberculosis	*Mycobacterium tuberculosis*	Humans	An acute bacterial infection of the lungs, lymph, and meninges. Its incidence is on the rise, complicated by the development of new strains of the bacterium that are resistant to antibiotics.
Typhoid fever	*Salmonella typhi*	Humans	A systemic bacterial disease of worldwide incidence. Fewer than 500 cases a year are reported in the United States. Spread through contaminated water or foods (such as improperly washed fruits and vegetables). Vaccines are available for travelers.
Typhus	*Rickettsia typhi*	Lice, rat fleas, humans	Historically a major killer in times of crowding and poor sanitation; transmitted from human to human through the bite of infected lice and fleas. Peak untreated mortality rate of 70%.

multidrug-resistant (MDR) strains of TB have occurred, however, in the United States and worldwide. These MDR strains are resistant to most of the best available anti-TB medications. MDR TB is of particular concern because it requires much more time and is more expensive to treat. Also, it is more likely to prove fatal.

This spread of MDR TB is likely due to the extremely long course of antibiotics required to treat the disease. Patients often quit taking the antibiotics before completing the course, setting up conditions in their bodies to allow drug-resistant bacteria to thrive.

The basic principles of TB treatment and control are to make sure all patients complete a full course of medication, so that all of the bacteria causing the infection are killed and drug-resistant strains do not develop. Great efforts are being made to ensure that high-risk individuals who are infected but not yet sick receive preventive therapy under observation. Such programs are approximately 90% effective in reducing the likelihood of developing active TB and spreading it to others. The efforts are having a significant effect. From 2000–2008 the TB rate decreased almost 4% annually, and in 2009 the case rate decreased 11%.

Bacterial biofilms are involved in tooth decay

Bacteria and other organisms may form mixed cultures on certain surfaces that are extremely difficult to treat. On teeth, this biofilm, or plaque, consists largely of bacterial cells surrounded by a polysaccharide matrix. Most of the bacteria in plaque are filaments of rod-shaped cells classified as various species of *Actinomyces,* which extend out perpendicular to the surface of the tooth. Many other bacterial species are also present in plaque.

Tooth decay, or dental caries, is caused by the bacteria present in the plaque, which persist especially in places that are difficult to reach with a toothbrush. Diets that are high in simple sugars are especially harmful to teeth because certain bacteria, notably *Streptococcus sobrinus* and *S. mutans,* ferment the sugars to lactic acid. This acid production reduces the pH in the area around the plaque, breaking down the structure of the hydroxyapatite that makes tooth enamel hard. As the enamel degenerates, the remaining soft matrix of the tooth becomes vulnerable to bacterial attack.

Bacteria can cause ulcers

Bacteria can also be the cause of disease states that on the surface appear to have no infectious basis. Peptic ulcer disease is due to craterlike lesions in the gastrointestinal tract that are exposed to peptic acid. Ulcers can be caused by drugs, such as nonsteroidal anti-inflammatory drugs, and also by some tumors of the pancreas that cause an oversecretion of peptic acid. In 1982, a bacterium named *Campylobacter pylori* (now named *Helicobacter pylori*) was isolated from gastric juices. Over the years evidence has accumulated that this bacterium is actually the causative agent in the majority of cases of peptic ulcer disease.

Antibiotic therapy can now eliminate *H. pylori,* treating the cause of the disease, and not just the symptoms. The discovery of the action of this bacterial species illustrates how even disease states that appear to be unrelated to infectious disease may actually be caused by cryptic (unknown) infection.

Many sexually transmitted diseases are bacterial

A number of bacteria cause sexually transmitted diseases (STDs), three particularly important examples of which are gonorrhea, syphilis, and chlamydia (figure 28.16).

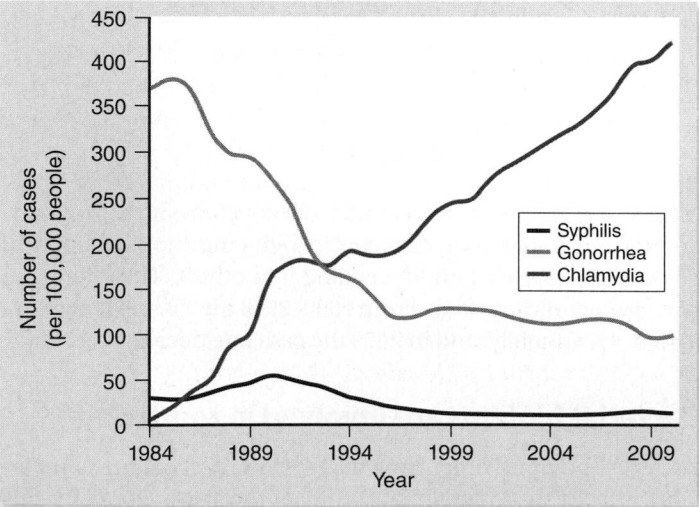

Figure 28.16 **Trends in sexually transmitted diseases in the United States.**

? Inquiry question How is it possible for the incidence of one STD (chlamydia) to rise as another (gonorrhea) falls?

Gonorrhea

Gonorrhea is one of the most prevalent communicable diseases in North America. Caused by the bacterium *Neisseria gonorrhoeae,* gonorrhea can be transmitted through sexual intercourse or any other sexual contact in which body fluids are exchanged, such as oral or anal intercourse. It can also pass from mother to baby during delivery through the birth canal.

The incidence of gonorrhea has been on the decline in the United States, but it remains a serious threat worldwide. Of particular concern is the appearance of antibiotic-resistant strains of *N. gonorrhoeae.*

Syphilis

Syphilis, a very destructive STD, was once prevalent and deadly but is now less common due to the advent of blood-screening procedures and antibiotics. Syphilis is caused by a spirochete bacterium, *Treponema pallidum,* transmitted during sexual intercourse or through direct contact with an open syphilis chancre sore. The bacterium can also be transmitted from a mother to her fetus, often causing damage to the heart, eyes, and nervous system of the baby.

Once inside the body, the disease progresses in four distinct stages. The first, or primary stage, is characterized by the appearance of a small, painless, often unnoticed sore called a *chancre.* The chancre resembles a blister and occurs at the location where the bacterium entered the body about three weeks following exposure. This stage of the disease is highly infectious, and an infected person may unwittingly transmit the disease to others. This sore heals without treatment in approximately four weeks, deceptively indicating a "cure" of the disease, although the bacterium remains in the body.

The second stage of syphilis, or secondary syphilis, is marked by a rash, a sore throat, and sores in the mouth. The bacteria can be transmitted at this stage through kissing or contact with an open sore. Commonly at this point, the disease enters the third stage, a latent period. This latent stage of syphilis is symptomless and may last for several years. At this point, the person is no longer infectious, but the bacteria are still present in the body, attacking the internal organs.

The final stage of syphilis is the most debilitating, as the damage done by the bacteria in the third stage becomes evident. Sufferers at this stage of syphilis experience heart disease, mental deficiency, and nerve damage, which may include loss of motor functions or blindness.

Chlamydia

Chlamydia is caused by an unusual bacterium. *Chlamydia trachomatis* is genetically a bacterium but is an obligate intracellular parasite, much like a virus in this respect. It is susceptible to antibiotics but it depends on its host to replicate its genetic material. The bacterium is transmitted through vaginal, anal, or oral intercourse with an infected person.

Chlamydia is called the "silent STD" because women usually experience no symptoms until after the infection has become established. In part because of this symptomless nature, the incidence of chlamydia has skyrocketed, increasing from 6.5 cases per 100,000 population in 1984 to 426 cases per 100,000 population in 2010. The 1,307,893 total cases reported in 2010 was the largest number of cases ever reported to the CDC for any condition.

The effects of an established chlamydia infection on the female body are extremely serious. Chlamydia can cause pelvic inflammatory disease (PID), which can lead to sterility and sometimes death.

It has recently been established that infection of the male or female reproductive tract by chlamydia can cause heart disease. Chlamydiae produce a peptide similar to one produced by cardiac muscle. As the body's immune system tries to fight off the infection, it recognizes and reacts to this peptide. The similarity between the bacterial and cardiac peptides confuses the immune system, and T cells attack cardiac muscle fibers, inadvertently causing inflammation of the heart and other problems.

The treatment for the disease is antibiotics, usually tetracycline, which can penetrate the eukaryotic plasma membrane to attack the bacterium. Any woman who experiences the symptoms associated with this STD or who is at risk of developing an STD should be tested for the presence of the chlamydia bacterium; otherwise, her fertility may be at risk.

Learning Outcomes Review 28.5

Many human diseases are due to bacterial infection, including tuberculosis, streptococcal and staphylococcal infection, and sexually transmitted diseases. The causative agent of most peptic ulcers is *Helicobacter pylori,* an inhabitant of the digestive tract. Bacteria are responsible for many STDs, including gonorrhea, syphilis, and chlamydia. In many cases symptoms of infection disappear although the disease is still present, and all can have serious consequences if untreated, especially for women.

■ *Why is infection by most pathogens not fatal?*

28.6 Beneficial Prokaryotes

Learning Outcomes

1. *Recognize the role of prokaryotes in the global cycling of elements.*
2. *Describe examples of bacterial/eukaryotic symbiosis.*
3. *Explain how bacteria can be used for bioremediation.*

Prokaryotes were largely responsible for creating the current properties of the atmosphere and the soil through billions of years of their activity. Today, they still affect the Earth and human life in many important ways.

Prokaryotes are involved in cycling important elements

Life on Earth is critically dependent on the cycling of chemical elements between organisms and the physical environments in which they live—that is, between the living and nonliving elements of ecosystems. Prokaryotes, algae, and fungi play many key roles in this chemical cycling, a process discussed in detail in chapter 57.

Decomposition

The carbon, nitrogen, phosphorus, sulfur, and other atoms of biological systems all have come from the physical environment, and when organisms die and decay, these elements all return to it. The prokaryotes and fungi that carry out the decomposition portion of chemical cycles, releasing a dead organism's atoms to the environment, are called *decomposers.*

Fixation

Other prokaryotes play important roles in fixation, the other half of chemical cycles, helping to return elements from inorganic forms to organic forms that heterotrophic organisms can use.

Carbon. The role of photosynthetic prokaryotes in fixing carbon is obvious. The organic compounds that plants, algae, and photosynthetic prokaryotes produce from CO_2 pass up through food chains to form the bodies of all the ecosystem's heterotrophs. Ancient cyanobacteria are thought to have added oxygen to the Earth's atmosphere as a by-product of their photosynthesis. Modern photosynthetic prokaryotes continue to contribute to the production of oxygen.

Nitrogen. Less obvious, but no less critical to life, is the role of prokaryotes in recycling nitrogen. The nitrogen in the Earth's atmosphere is in the form of N_2 gas. A triple covalent bond links the two nitrogen atoms and is not easy to break. Among the Earth's organisms, only a very few species of prokaryotes are able to accomplish this feat, reducing N_2 to ammonia (NH_3), which is used to build amino acids and other nitrogen-containing biological molecules. When the organisms that contain these molecules die, decomposers return nitrogen to the soil as ammonia. This is then converted to nitrate (NO^{3-}) by nitrifying bacteria, making nitrogen available for plants. The nitrate can also be converted back into molecular nitrogen by *denitrifiers* that return the nitrogen to the atmosphere, completing the cycle.

To fix atmospheric nitrogen, prokaryotes employ an enzyme complex called nitrogenase, encoded by a set of genes called *nif* ("nitrogen fixation") genes. The nitrogenase complex is extremely sensitive to oxygen and is found in a wide range of free-living prokaryotes.

In aquatic environments, nitrogen fixation is carried out largely by cyanobacteria such as *Anabaena,* which forms long chains of cells. Because the nitrogen fixation process is strictly anaerobic, individual cyanobacteria cells may develop into *heterocysts,* specialized nitrogen-fixing cells impermeable to oxygen.

In soil, nitrogen fixation occurs in the roots of plants that harbor symbiotic colonies of nitrogen-fixing bacteria. These associations include *Rhizobium* (a genus of proteobacteria; see figure 28.6) with legumes, *Frankia* (an actinomycete) with many woody shrubs, and *Anabaena* with water ferns.

Prokaryotes may live in symbiotic associations with eukaryotes

Many prokaryotes live in symbiotic association with eukaryotes. **Symbiosis** refers to the ecological relationship between different species that live in direct contact with each other. The symbiotic association of nitrogen-fixing bacteria with plant roots is an example of *mutualism,* a form of symbiosis in which both parties benefit. The bacteria supply the plant with useful nitrogen, and the plant supplies the bacteria with sugars and other organic nutrients (see chapter 38).

Many bacteria live symbiotically within the digestive tracts of animals, providing nutrients to their hosts. Cattle and other grazing mammals are unable to digest cellulose in the grass and plants they eat because they lack the required cellulase enzyme. Colonies of cellulase-producing bacteria inhabiting the gut allow cattle to digest their food (see chapter 47 for a fuller account). Similarly, humans maintain large colonies of bacteria in the large intestine that produce vitamins—particularly B$_{12}$ and K—that the body cannot make.

Many bacteria inhabit the outer surfaces of animals and plants without doing damage. These associations are examples of *commensalism,* in which one organism (the bacterium) receives benefits while the animal or plant is neither benefited nor harmed.

Parasitism is a form of symbiosis in which one member (in this case, the bacterium) benefits, and the other (the infected animal or plant) is harmed. Infection might be considered a form of parasitism.

Bacteria are used in genetic engineering

Because the genetic code is universal, a gene from a human can be inserted into a bacterial cell, and the bacterium produces a human protein. The use of bacteria in genetic engineering was discussed in chapter 17, and it is a large part of modern molecular biology.

In addition to the production of pharmaceutical agents such as insulin, applying genetic engineering methods to produce improved strains of bacteria for commercial use holds promise for the future. Bacteria are now widely used as "biofactories" in the commercial production of a variety of enzymes, vitamins, and antibiotics. Immense cultures of bacteria, often genetically modified to enhance performance, are used to produce commercial acetone and other industrially important compounds.

Bacteria can be used for bioremediation

The use of organisms to remove pollutants from water, air, and soil is called *bioremediation.* The normal functioning of sewage treatment plants depends on the activity of microorganisms. In sewage treatment plants, the solid matter from raw sewage is broken down by bacteria and archaea naturally present in the sewage. The end product, methane gas (CH_4) is often used as an energy source to heat the treatment plant.

Biostimulation, that is, the addition of nutrients such as nitrogen and phosphorus sources, has been used to encourage the growth of naturally occurring microbes that can degrade crude oil spills. This approach was used successfully to clean up the Alaskan shoreline after the crude oil spill of the *Exxon Valdez* in 1998. Similarly, biostimulation has been used to encourage the growth of naturally occurring microbial flora in contaminated groundwater. Current efforts include those concentrated on the use of endogenous microbes such as *Geobacter* (see figure 28.6) to eliminate radioactive uranium from groundwater contaminated during the cold war.

Chlorinated compounds released into the environment by a variety of sources are another serious pollutant. Some bacteria can actually use these compounds for energy by performing reductive dehalogenation that is linked to electron transport, a process termed *halorespiration.* Although still at the development stage, the use of such bacteria to remove halogenated compounds from toxic waste holds great promise.

Learning Outcomes Review 28.6

Prokaryotes are vital to ecosystems for both recycling elements and fixation, or making elements available in organic form. Bacteria are involved in fixation of both carbon and nitrogen and are the only organisms that can fix nitrogen. These nitrogen-fixing bacteria may live in symbiotic association with plants. Bacteria are a key component of waste treatment, and they are also being used in bioremediation to remove toxic compounds introduced into the environment.

■ *Does the information about nitrogen fixation shed any light on the practice of crop rotation?*

Chapter Review

28.1 Prokaryotic Diversity

A brief history of microbiology.
Microbiology grew out of the study of infectious disease, and was aided by technology to view the unseen world. Spontaneous generation was disproved by experimentation, and Koch's postulates provide guidelines to assign a causative agent to a disease.

Prokaryotes are fundamentally different from eukaryotes.
Prokaryotic features include unicellularity, small circular DNA, division by binary fission, lack of internal compartmentalization, a singular flagellum, and metabolic diversity.

Despite similarities, bacteria and archaea differ fundamentally.
Bacteria and archaea differ in four key areas: plasma membranes, cell walls, DNA replication, and gene expression.

Archaeal lipids have ether instead of ester linkages and can form tetraether monolayers. The cell walls of bacteria contain peptidoglycans, but those of archaea do not.

Both bacterial and archaeal DNA have a single replication origin, but the origin and the replication proteins are different. Archaeal initiation of DNA replication and RNA polymerases are more like those of eukaryotes.

Most prokaryotes have not been characterized.

Nine clades of prokaryotes have been found so far, but many bacteria have not been studied (figure 28.6).

28.2 Prokaryotic Cell Structure

Prokaryotes have three basic forms: rods, cocci, and spirals.

Prokaryotes have a tough cell wall and other external structures.

Bacteria are classified as gram-positive or gram-negative based on the Gram stain (figure 28.7). Gram-positive bacteria have a thick peptidoglycan layer in the cell wall that contains teichoic acid (figure 28.8). Gram-negative bacteria have a thin peptidoglycan layer and an outer membrane containing lipopolysaccharides in their cell wall (figure 28.8).

Some bacteria have a gelatinous layer, the capsule, enabling the bacterium to adhere to surfaces and evade an immune response.

Many bacteria have a slender, rigid, helical flagellum composed of flagellin, which can rotate to drive movement (figure 28.9). Some bacteria have hairlike pili that have roles in adhesion and exchange of genetic information.

Some bacteria form highly resistant endospores in response to environmental stress.

The interior of prokaryotic cells is organized.

In prokaryotes, invaginated regions of the plasma membrane function in respiration and photosynthesis. The nucleoid region contains a compacted circular DNA with no bounding membrane.

Prokaryotic ribosomes are smaller than those of eukaryotes and some antibiotics work by binding to these ribosomes, blocking protein synthesis.

28.3 Prokaryotic Genetics

Conjugation depends on the presence of a conjugative plasmid.

DNA can be exchanged by conjugation (figure 28.11), which depends on the presence of conjugative plasmids like the F plasmid in *E. coli*. The F+ donor cell transfers the F plasmid to the F- recipient cell.

The F plasmid can also integrate into the bacterial genome. Excision may be imprecise, so that the F plasmid carries genetic information from the host.

Viruses transfer DNA by transduction (figure 28.14).

Generalized transduction occurs when viruses package host DNA and transfer it on subsequent infection. Specialized transduction is limited to lysogenic phage.

Transformation is the uptake of DNA directly from the environment (figure 28.15).

Transformation occurs when cells take up DNA from the surrounding medium. It can be induced artificially in the laboratory.

Antibiotic resistance can be transferred by resistance plasmids.

R plasmids have played a significant role in the appearance of strains resistant to antibiotics, such as *S. aureus* and *E. coli* O157:H7.

Variation can also arise by mutation.

Mutations can occur spontaneously in bacteria due to radiation, UV, and various chemicals.

28.4 Prokaryotic Metabolism

Prokaryotes acquire carbon and energy in four basic ways.

Photoautotrophs carry out photosynthesis and obtain carbon from carbon dioxide. Chemolithoautotrophs obtain energy by oxidizing inorganic substances. Photoheterotrophs use light for energy but obtain carbon from organic molecules. Chemoheterotrophs, the largest group, obtain carbon and energy from organic molecules.

Some bacteria can attack other cells directly.

Some bacteria release proteins through their cell walls, and these proteins may transfer other, virulent proteins into eukaryotic cells.

Bacteria are costly plant pathogens.

Gram-negative bacteria known as pseudomonads are responsible for most plant diseases.

28.5 Human Bacterial Disease (table 28.1)

Bacterial diseases are spread through mucus or saliva droplets, contaminated food and water, and insect vectors.

Tuberculosis has infected humans for all of recorded history.

Tuberculosis continues to be a major public health problem. Treatment requires a long course of antibiotics.

Bacterial biofilms are involved in tooth decay.

Bacteria can cause ulcers.

Most stomach ulcers are caused by infection with *Helicobacter pylori*.

Many sexually transmitted diseases are bacterial.

The potentially dangerous sexually transmitted diseases gonorrhea, syphilis, and chlamydia are caused by bacteria.

28.6 Beneficial Prokaryotes

Prokaryotes are involved in cycling important elements.

Prokaryotes are involved in the recycling of carbon and nitrogen; only bacteria can fix nitrogen.

Prokaryotes may live in symbiotic associations with eukaryotes.

Bacteria are used in genetic engineering.

Genetically engineered prokaryotes can be used to produce human pharmaceutical agents and other useful products.

Bacteria can be used for bioremediation.

Review Questions

UNDERSTAND

1. Which of the following would be an example of a biomarker?
   a. A microfossil found in a meteorite
   b. A hydrocarbon found in an ancient rock layer
   c. An area that is high in carbon-12 concentration in a rock layer
   d. A newly discovered formation of stromatolites

2. A cell that can use energy from the sun, and CO_2 as a carbon source is a
   a. photoautotroph.   c. photoheterotroph.
   b. chemoautotroph.   d. chemoheterotroph.

3. Gram-positive (+) and gram-negative (–) bacteria are characterized by differences in
   a. the cell wall: gram+ have peptidoglycan, gram– have pseudo-peptidoglycan.
   b. the plasma membrane: gram+ have ester-linked lipids, gram– have ether-linked lipids.
   c. the cell wall: gram+ have a thick layer of peptidoglycan and gram– have an outer membrane.
   d. chromosomal structure: gram+ have circular chromosomes, gram– have linear chromosomes.

4. Which of the following characteristics is unique to the archaea?
   a. A fluid mosaic model of plasma membrane structure
   b. The use of an RNA polymerase during gene expression
   c. Ether-linked phospholipids
   d. A single origin of DNA replication

5. The horizontal transfer of DNA using a plasmid is an example of
   a. generalized transduction.
   b. binary fission.
   c. transformation.
   d. conjugation.

6. The disease tuberculosis is
   a. caused by a bacterial pathogen.
   b. an emerging disease that is now worldwide.
   c. caused by a viral pathogen.
   d. not treatable with antibiotics.

7. Prokaryotes participate in the global cycling of
   a. proteins and nucleic acids.
   b. carbon and nitrogen.
   c. carbohydrates and lipids.
   d. All of the choices are correct.

APPLY

1. Which of the following is typically not associated with a prokaryote?
   a. Horizontal transfer of genetic information
   b. A lack of internal compartmentalization
   c. Multiple, linear chromosomes
   d. A cell size of 1 μm

2. The mechanisms of DNA exchange in prokaryotes share the feature of
   a. vertical transmission of information.
   b. horizontal transfer of information.
   c. requiring cell contact.
   d. the presence of a plasmid in one cell.

3. The cell wall in both gram-positive and gram-negative cells is
   a. composed of phospholipids.
   b. a target for antibiotics that affect peptidoglycan synthesis.
   c. composed of peptidoglycan.
   d. surrounded by a membrane.

4. The three domains of life
   a. represent variations of the same basic cell type.
   b. include two different basic cell types.
   c. consist of three different basic cell types.
   d. describe current cells but say nothing about their history.

5. Ulcers and tooth decay do not appear related, but in fact both
   a. are due to eating particular kinds of foods.
   b. are caused by viral infection.
   c. are caused by environmental factors.
   d. can be due to bacterial infection.

6. Bacteria lack independent internal membrane systems, but are able to perform photosynthesis and respiration, both of which use membranes. They are able to perform these functions because
   a. they actually have internal membranes, but only for these functions.
   b. invaginations of the plasma membrane can provide an internal membrane surface.
   c. they take place outside of the cell between the membrane and the cell wall.
   d. they use protein-based structures to take the place of internal membranes.

7. Plants cannot fix nitrogen, yet some plants do not need nitrogen from the soil. This is because
   a. of a symbiotic association with a bacterium that can fix nitrogen.
   b. these plants are the exceptions that can fix nitrogen.
   c. they have been infected by a parasitic virus that can fix nitrogen.
   d. they are able to obtain nitrogen from the air.

SYNTHESIZE

1. If a new form of carbon fixation was discovered that was not biased toward carbon-12, would this affect our analysis of the earliest evidence for life?

2. Frederick Griffith's experiments (see chapter 14) played an important role in showing that DNA is the genetic material. Griffith showed that dead virulent bacteria mixed with live nonvirulent bacteria could cause pneumonia in mice. Live rough bacteria could also be cultured from the infected mice. The difference between the two strains is a polysaccharide capsule found in the smooth strain. Given what you have learned in this chapter, how would you explain these observations?

3. In the 1960s, it was common practice to prescribe multiple antibiotics to fight bacterial infections. It is also often the case that patients do not always take the entire "course" of their antibiotics. Antibiotic resistance genes are often found on conjugative plasmids. How do these factors affect the evolution of antibiotic resistance and of resistance to multiple antibiotics in particular?

4. Soil-based nitrogen-fixing bacteria appear to be highly vulnerable to exposure to UV radiation. Suppose that the ozone level continues to be depleted, what are the long-term effects on the planet?

ONLINE RESOURCE

www.ravenbiology.com

Understand, Apply, and Synthesize—enhance your study with animations that bring concepts to life and practice tests to assess your understanding. Your instructor may also recommend the interactive eBook, individualized learning tools, and more.

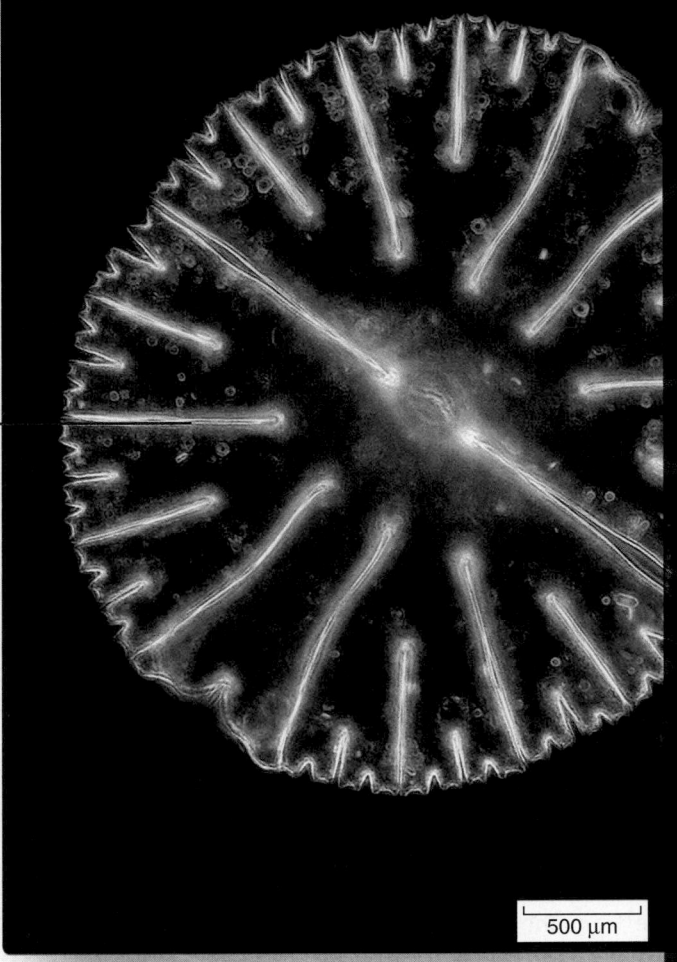

500 μm

Chapter *29*

29

Protists

Chapter Contents

Introduction

For more than half of the long history of life on Earth, all life was microscopic. The biggest organisms that existed for over 2 billion years were single-celled bacteria fewer than 6 μm thick. These prokaryotes lacked internal membranes, except for invaginations of surface membranes in photosynthetic bacteria.

The first evidence of a different kind of organism is found in tiny fossils in rock 1.5 billion years old. These fossil cells are much larger than bacteria (up to 10 times larger) and contain internal membranes and what appear to be small, membrane-bounded structures. The complexity and diversity of form among these single cells is astonishing. The step from relatively simple to quite complex cells marks one of the most important events in the evolution of life—the appearance of a new kind of organism, the eukaryote. Eukaryotes that are clearly not animals, plants, or fungi are collectively referred to as protists.

29.1 *Eukaryotic Origins and Endosymbiosis*

Protists were the first eukaryotes. Eukaryotic cells are distinguished from prokaryotes by the presence of a cytoskeleton and compartmentalization that includes a nuclear envelope and organelles. The exact sequence of events that led to large, complex eukaryotic cells is unknown, but several key events are agreed upon. Loss of a rigid cell wall allowed membranes to fold inward, increasing surface area. Membrane flexibility also made it possible for one cell to engulf another.

Fossil evidence dates the origins of eukaryotes

Indirect chemical traces hint that eukaryotes may go as far back as 2.7 billion years, but no fossils as yet support such an early appearance. In rocks about 1.5 billion years old, we begin to see the first microfossils that are noticeably different in appearance from the earlier, simpler forms, none of which were more than 6 μm in diameter (figure 29.1). These cells are much larger than those of prokaryotes and have internal membranes and thicker walls.

These early fossils mark a major event in the evolution of life: A new kind of organism had appeared. These new cells are called eukaryotes, from the Greek words meaning "true nucleus," because they possess an internal structure called a nucleus. All organisms other than prokaryotes are eukaryotes.

In the sections that follow, the origins of eukaryotic internal structure are considered. Keep in mind that, as discussed in chapter 24, horizontal gene transfer occurred frequently while eukaryotic cells were evolving. Eukaryotic cells evolved not only through horizontal gene transfer, but through infolding of membranes and engulfing other cells. Today's eukaryotic cell is the result of cutting and pasting of DNA and organelles from different species.

The nucleus and ER arose from membrane infoldings

Many prokaryotes have infoldings of their outer membranes extending into the cytoplasm that serve as passageways to the surface. The network of internal membranes in eukaryotes is called the endoplasmic reticulum (ER), and the nuclear envelope, an extension of the ER network that isolates and protects the nucleus, is thought to have evolved from such infoldings (figure 29.2).

Mitochondria evolved from engulfed aerobic bacteria

Bacteria that live within other cells and perform specific functions for their host cells are called *endosymbiotic bacteria.* Their widespread presence in nature led biologist Lynn Margulis in the early 1970s to champion the theory of endosymbiosis, which was first proposed by Konstantin Mereschkowsky in 1905. Endosymbiosis means living together in close association.

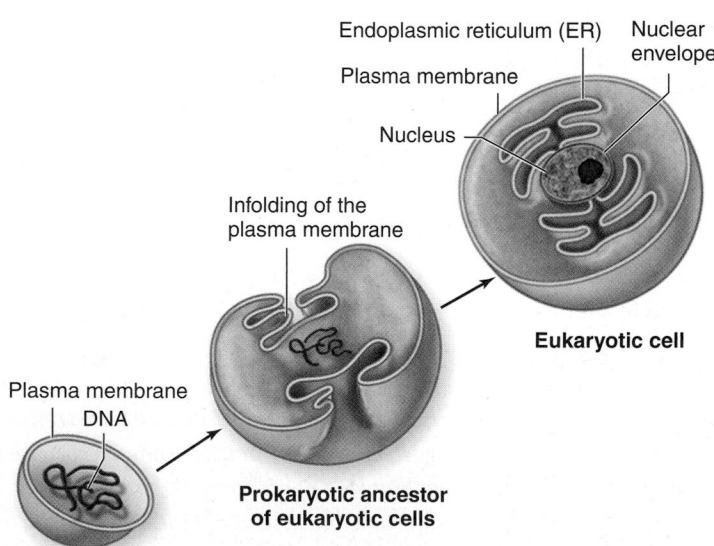

Figure 29.2 Origin of the nucleus and endoplasmic reticulum. Many prokaryotes today have infoldings of the plasma membrane (see also figure 28.10). The eukaryotic internal membrane system, called the endoplasmic reticulum (ER), and the nuclear envelope may have evolved from such infoldings of the plasma membrane, encasing the DNA of prokaryotic cells that gave rise to eukaryotic cells.

50 μm

Figure 29.1 Early eukaryotic fossil. Fossil algae that lived in Siberia 1 BYA.

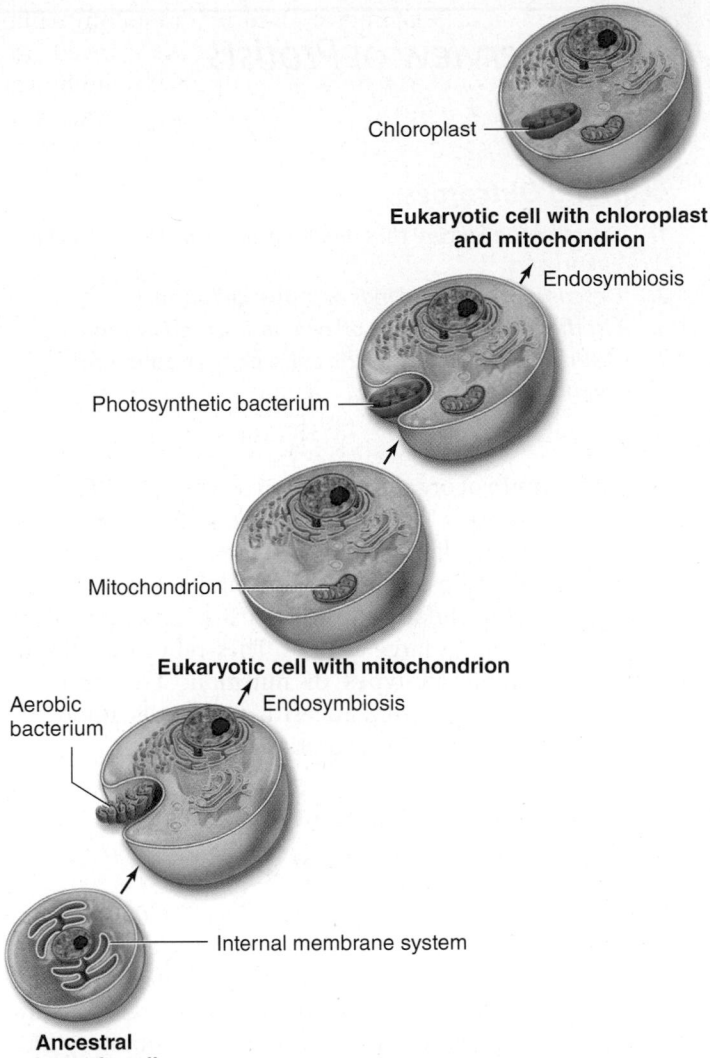

Chloroplast

Eukaryotic cell with chloroplast and mitochondrion

↗ Endosymbiosis

Photosynthetic bacterium

Mitochondrion

Eukaryotic cell with mitochondrion

Aerobic bacterium

↗ Endosymbiosis

Internal membrane system

Ancestral eukaryotic cell

Figure 29.3 The theory of endosymbiosis. Scientists propose that ancestral eukaryotic cells, which already had an internal system of membranes, engulfed aerobic bacteria, which then became mitochondria within the eukaryotic cell. Chloroplasts also originated this way, with eukaryotic cells engulfing photosynthetic bacteria.

Endosymbiosis, a concept that is now widely accepted, suggests that a critical stage in the evolution of eukaryotic cells involved endosymbiotic relationships with prokaryotic organisms. According to this theory, energy-producing bacteria may have come to reside within larger bacteria, eventually evolving into what we now know as mitochondria (figure 29.3). Possibly the original host cell was anaerobic with hydrogen-dependent metabolic pathways. The symbiont had a form of respiration that produced H_2. The host depended on the symbiont for H_2 under anaerobic conditions and was able later to adapt to an O_2-rich atmosphere using the symbiont's respiratory pathways.

Chloroplasts evolved from engulfed photosynthetic bacteria

Photosynthetic bacteria may have come to live within other larger bacteria, leading to the evolution of chloroplasts, the photosynthetic organelles of plants and algae (see figure 29.3). The history of chloroplast evolution is an example of the care that must be taken in phylogenetic studies. All chloroplasts are likely derived from a single line of cyanobacteria, but the organisms that host these chloroplasts are not monophyletic. This apparent paradox is resolved by considering the possibility of secondary, and even tertiary endosymbiosis. Figure 26.12 explains how red and green algae both obtained their chloroplasts by engulfing photosynthetic cyanobacteria. The brown algae most likely obtained their chloroplasts by engulfing one or more red algae, a process called **secondary endosymbiosis** (figure 29.4).

A phylogenetic tree based only on chloroplast gene sequences from brown, red, and green algae reveals a close evolutionary relationship. This tree is misleading, however, because it is not possible to tell just from these data how much the algal lines had diverged at the time they engulfed the same line of cyanobacteria. Nuclear gene sequences, as well as morphological and chemical traits, are more helpful than chloroplast gene sequences in sorting out algal relations.

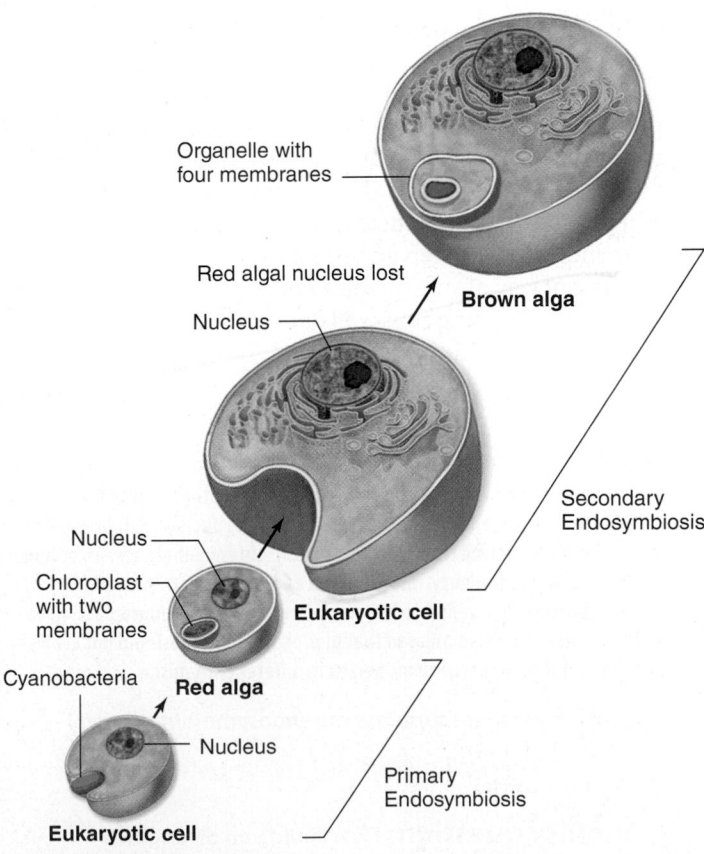

Organelle with four membranes

Red algal nucleus lost

Brown alga

Nucleus

Secondary Endosymbiosis

Nucleus

Chloroplast with two membranes

Eukaryotic cell

Cyanobacteria

Red alga

Nucleus

Primary Endosymbiosis

Eukaryotic cell

Figure 29.4 Endosymbiotic origins of chloroplasts in red and brown algae.

Endosymbiosis is supported by a range of evidence

The fact that we now witness so many symbiotic relationships lends general support to the endosymbiotic theory. Even stronger support comes from the observation that present-day organelles such as mitochondria and chloroplasts contain their own DNA, which is remarkably similar to the DNA of bacteria in size and character. During the billion and a half years in which mitochondria have existed as endosymbionts within eukaryotic cells, most of their genes have been transferred to the chromosomes of the host cells—but not all. Each mitochondrion still has its own genome, a circular, closed molecule of DNA similar to that found in bacteria, on which are located genes encoding the essential proteins of oxidative metabolism. These genes are transcribed within the mitochondrion, using mitochondrial ribosomes that are smaller than those of eukaryotic cells, very much like bacterial ribosomes in size and structure. Many antibiotics that inhibit protein synthesis in bacteria also inhibit protein synthesis in mitochondria and chloroplasts, but not in the cytoplasm. Chloroplasts and mitochondria replicate via binary fission, not mitosis, further supporting bacterial origins.

Mitosis evolved in eukaryotes

The mechanisms of mitosis and cytokinesis, now so common among eukaryotes, did not evolve all at once. Traces of very different, and possibly intermediate, mechanisms survive today in some of the eukaryotes. In fungi and in some groups of protists, for example, the nuclear membrane does not dissolve, as it does in plants, animals, and most other protists, and mitosis is confined to the nucleus. When mitosis is complete in these organisms, the nucleus divides into two daughter nuclei, and only then does the rest of the cell divide. We do not know whether mitosis without nuclear membrane dissolution represents an intermediate step on the evolutionary journey, or simply a different way of solving the same problem. We cannot see the interiors of dividing cells well enough in fossils to be able to trace the history of mitosis.

Learning Outcomes Review 29.1

Eukaryotes are organisms that contain a nucleus and other membrane-bounded organelles. Endoplasmic reticulum and the nuclear membrane are believed to have evolved from infoldings of the outer membranes. According to the endosymbiont theory, mitochondria and chloroplasts evolved from engulfed bacteria that remained intact. Mitochondria and chloroplasts have their own DNA, which is similar to that of prokaryotes. Mitosis did not evolve all at once; different mechanisms persist in different organisms.

■ *What evidence supports the endosymbiont theory?*

? **Inquiry question** How could you distinguish between primary and secondary endosymbiosis by looking at micrographs of cells with chloroplasts?

Learning Outcomes

1. *Describe the feature that distinguishes protists from other eukaryotes.*
2. *Describe the various kinds of protist cell surfaces.*
3. *List the two main means of locomotion used by protists.*
4. *Distinguish between phototrophs, phagotrophs, and osmotrophs.*

Protists are a group of organisms united on the basis of a single negative characteristic: They are eukaryotes that are not fungi, plants, or animals. In all other respects, they vary considerably, with no uniting features. Many are unicellular, but numerous colonial and multicellular groups also exist. Most are microscopic, but some are as large as trees. They represent all symmetries and exhibit all types of nutrition. The origin of eukaryotes, which began with ancestral protists, is among the most significant events in the evolution of life.

Eukaryotes are organized into six supergroups that all contain protists

Unlike plants, fungi, and animals, the over 200,000 different protists are not monophyletic. Eukaryotes diverged rapidly in a world that was shifting from anaerobic to aerobic conditions. We may never be able to completely sort out the relationships among different lineages during this major evolutionary transition. Applications of a variety of molecular methods are providing insights into the evolutionary relationships among protists. Molecular systematics is helping to sort out the protists which are now grouped into six supergroups: Excavata, Chromalveolata, Archaeplastida, Rhizaria, Amoebozoans, and Ophisthokonts. With the exception of Chromalveolata—the supergroups are monophyletic.

In this chapter, we explore the diverse and fascinating world of the protists based on our current understanding of phylogeny (figure 29.5). Understanding the evolution of protists is key to understanding the origins of plants, fungi, and animals. The supergroups encompass all of eukaryotic diversity and many protists are closely related to plants, animals, or fungi. The remaining chapters in this diversity unit focus on the monophyletic fungi, plants, and animals that trace their ancestry back to common ancestors shared with different protist lineages.

Protist cell surfaces vary widely

Protists possess a varied array of cell surfaces. Some protists, such as amoebas, are surrounded only by their plasma membrane. All other protists have a plasma membrane with an extracellular matrix (ECM) deposited on the outside of the

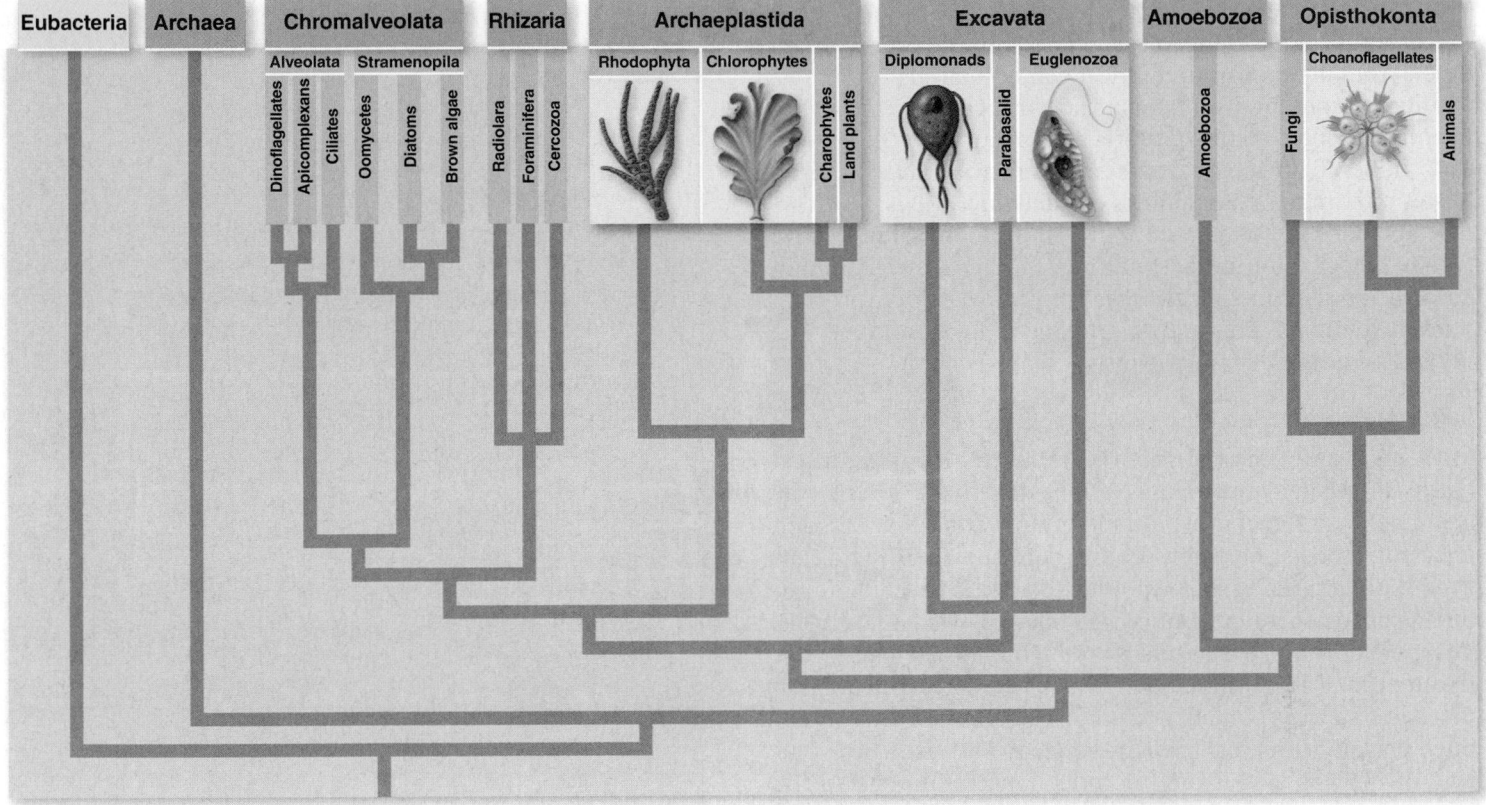

Figure 29.5 Eukaryotic evolutionary relationships. Analyses of current comparative molecular data group the eukaryotes into six supergroups shown here. Within the eukaryotes, plants, fungi, and animals are monophyletic clades, but protists are polyphyletic. Protist lineages are shaded in blue.

membrane. Some ECMs form strong cell walls; for instance, diatoms and foraminifera secrete glassy shells of silica.

Many protists with delicate surfaces are capable of surviving unfavorable environmental conditions. How do they manage to survive so well? They form cysts, which are dormant forms with resistant outer coverings in which cell metabolism is more or less completely shut down. Not all cysts are so sturdy, however. Vertebrate parasitic amoebas, for example, form cysts that are quite resistant to gastric acidity, but will not tolerate desiccation or high temperature.

Protists have several means of locomotion

Movement in protists is also accomplished by diverse mechanisms. Protist movement is chiefly by either flagellar rotation or pseudopods. Many protists wave one or more flagella to propel themselves through the water, and others use banks of short, flagella-like structures called cilia to create water currents for their feeding or propulsion. Pseudopods (Greek, meaning "false feet") are the chief means of locomotion among amoebas, whose pseudopods are large, blunt extensions of the cell body called lobopodia. Other related protists extend thin, branching protrusions called filopodia. Still other protists extend long, thin pseudopods called axopodia supported by axial rods of microtubules. Axopodia can be extended or

retracted. Because the tips can adhere to adjacent surfaces, the cell can move by a rolling motion, shortening the axopodia in front and extending those in the rear.

Protists have a range of nutritional strategies

Protists can be heterotrophic or autotrophic. Some autotrophic protists are photosynthetic and are called **phototrophs.** Others are heterotrophs that obtain energy from organic molecules synthesized by other organisms.

Among the heterotrophic protists are some called *phagotrophs,* which ingest visible particles of food by pulling them into intracellular vesicles called food vacuoles or phagosomes. Lysosomes fuse with the food vacuoles, introducing enzymes that digest the food particles within. Digested molecules are absorbed across the vacuolar membrane.

Protists that ingest food in soluble form are called *osmotrophs.* Another example of the protists' tremendous nutritional flexibility is seen in *mixotrophs,* protists that are both phototrophic and heterotrophic.

Protists reproduce asexually and sexually

Protists typically reproduce asexually, although some have an obligate sexual reproductive phase and others undergo sexual reproduction at times of stress, including food shortages.

Asexual reproduction

Asexual reproduction involves mitosis, but the process often differs from the mitosis in multicellular animals. For example, the nuclear membrane often persists throughout mitosis, with the microtubular spindle forming within it.

In some species, a cell simply splits into nearly equal halves after mitosis. Sometimes the daughter cell is considerably smaller than its parent and then grows to adult size—a type of cell division called **budding.** In *schizogony,* common among some protists, cell division is preceded by several nuclear divisions. This allows cytokinesis to produce several individuals almost simultaneously.

Sexual reproduction

Most eukaryotic cells also possess the ability to reproduce sexually, something prokaryotes cannot do at all. Meiosis (see chapter 11) is a major evolutionary innovation that arose in ancestral protists and allows for the production of haploid cells from diploid cells. Sexual reproduction is the process of producing offspring by fertilization, the union of two haploid cells. The great advantage of sexual reproduction is that it allows for frequent genetic recombination, which generates the variation that is the starting point of evolution. Not all eukaryotes reproduce sexually, but most have the capacity to do so. The evolution of meiosis and sexual reproduction contributed to the tremendous explosion of diversity among the eukaryotes.

Protists are the bridge to multicellularity

Diversity was also promoted by the development of *multicellularity.* Some single eukaryotic cells began living in association with others, in colonies. Eventually, individual members of the colony began to assume different duties, and the colony began to take on the characteristics of a single individual. Multicellularity has arisen many times among the eukaryotes. Practically every organism big enough to be seen with the unaided eye, including all animals and plants, is multicellular. The great advantage of multicellularity is that it fosters specialization; some cells devote all of their energies to one task, other cells to another. Few innovations have had as great an influence on the history of life as the specialization made possible by multicellularity.

Learning Outcomes Review 29.2

Eukaryotes are grouped into six supergroups based on evolutionary relationships: Excavata, Chromalveolata, Archaeplastida, Rhizaria, Amoebozoans, and Ophisthokonts. All the supergroups include protists. All protists have plasma membranes, but other cell-surface components, such as deposited extracellular material (ECM), are highly variable. Protists mainly use flagella or pseudopodial movement to propel themselves. Phototrophic protists carry out photosynthesis; phagotrophs ingest food particles; and osmotrophs ingest dissolved nutrients. Sexual reproduction is common, but asexual reproduction also occurs in many groups. Multicellular organisms likely arose from colonial protists.

■ *Explain how the evolution of meiosis might be linked to the tremendous diversity of protists.*

29.3 Lack of Typical Mitochondria in Excavata

Learning Outcomes

1. List the main features of diplomonads and parabasalids.
2. Give examples of diplomonads and parabasalids.
3. Explain why euglenozoa cannot be classified as either plants or animals.
4. Describe the distinguishing feature of kinetoplastids.

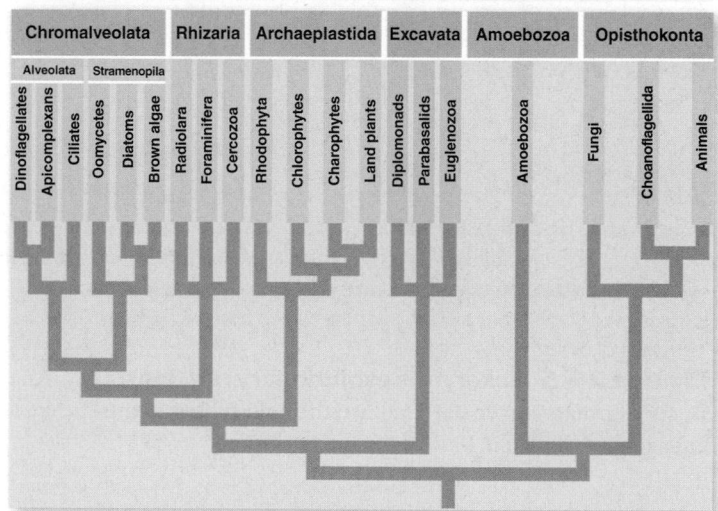

The *diplomonads, parabasalids,* and euglenozoans are grouped as Excavata (the excavates) based on cytoskeletal and DNA sequence similarities showing evolutionary relatedness. Many have a characteristic groove on the side of their bodies that they use for feeding, hence the name excavate. Although these groups have similar features, differences separate them into three distinct groups. Parabasalids and diplomonads have two nuclei, multiple flagella, and modified mitochondria; euglenozoans have flagella with a distinctive structure relative to flagella in other protists. A number of the euglenozoa have acquired chloroplasts through endosymbiosis. None of the algae are closely related to euglenozoa, a reminder that endosymbiosis is widespread.

Diplomonads have two nuclei

Diplomonads are unicellular and move with multiple flagella. This group lacks mitochondria, but has two nuclei. The parasite, *Giardia intestinalis,* is an example of a diplomonad (figure 29.6). *Giardia* is a parasite that can pass from human to human via contaminated water and causes diarrhea. Mitochondrial genes are found in their nuclei, leading to the conclusion that *Giardia* evolved from aerobic ancestors. Electron micrographs of *Giardia* cells stained with mitochondrial-specific antibodies reveal degenerate mitochondria called mitosomes that do not participate in cellular respiration.

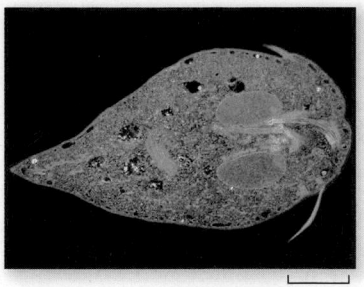

Figure 29.6 *Giardia intestinalis.* This parasitic diplomonad lacks a mitochondrion.

0.6 μm

Parabasalids have undulating membranes

Parabasalids contain an intriguing array of species. Some live in the gut of termites and digest cellulose, the main component of the termite's wood-based diet. The symbiotic relationship is one layer more complex because these parabasalids have a symbiotic relationship with bacteria that also aid in the digestion of cellulose. The persistent activity of these three symbiotic organisms from three different kingdoms can lead to the collapse of a home built of wood or recycle tons of fallen trees in a forest. Another parabasalid, *Trichomonas vaginalis,* causes a sexually transmitted disease in humans.

Parabasalids have undulating membranes that assist in locomotion (figure 29.7). Like diplomonads, parabasalids also use flagella to move and they lack mitochondria. The lack of mitochondria in both groups is now believed to be a derived rather than an ancestral trait.

Euglenozoa include free-living and parasitic groups

Euglenozoa diverged early and were among the earliest free-living eukaryotes to possess mitochondria. Both free-living euglenids and parasitic kinetoplastids are considered euglenozoans.

Euglenids

Euglenids clearly illustrate the impossibility of distinguishing plant-like protists from animal-like protists. About one-third of the approximately 40 genera of euglenids have chloroplasts and are fully autotrophic; the others lack chloroplasts, ingest their food, and are heterotrophic.

Some euglenids with chloroplasts may become heterotrophic in the dark; the chloroplasts become small and nonfunctional. If they are put back in the light, they may become green within a few hours. Photosynthetic euglenids may sometimes feed on dissolved or particulate food.

Individual euglenids range from 10 to 500 μm long and vary greatly in form. Interlocking proteinaceous strips arranged in a helical pattern form a flexible structure called the *pellicle,* which lies within the plasma membrane of the euglenids. Because its pellicle is flexible, a euglenid is able to change its shape.

Reproduction occurs by mitotic cell division. The nuclear envelope remains intact throughout the process of mitosis. No sexual reproduction is known to occur in this group.

In *Euglena* (figure 29.8), the genus for which the euglenid phylum is named, two flagella are attached at the base of a flask-shaped opening called the *reservoir,* which is located at the anterior end of the cell. One of the flagella is long and has a row of very fine, short, hairlike projections along one side. A second, shorter flagellum is located within the reservoir but does not emerge from it. Contractile vacuoles collect excess water from all parts of the organism and empty it into the reservoir, which apparently helps regulate the osmotic pressure within the organism. A light-sensitive stigma, which also occurs in the green algae (Chlorophyta), helps these photosynthetic organisms move toward light.

Cells of *Euglena* contain numerous small chloroplasts. These chloroplasts, like those of the green algae and plants,

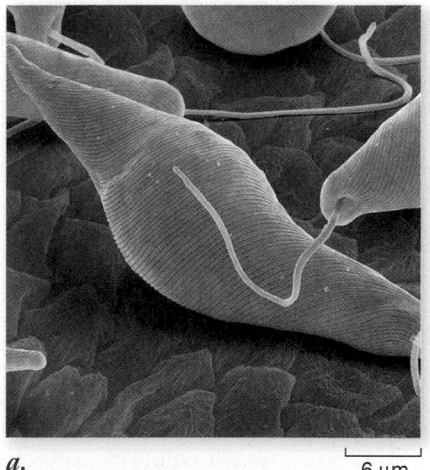

Figure 29.8 Euglenids.
a. Micrograph of *Euglena gracilis.*
b. Diagram of *Euglena.* Paramylon granules are areas where food reserves are stored.

a.

6 μm

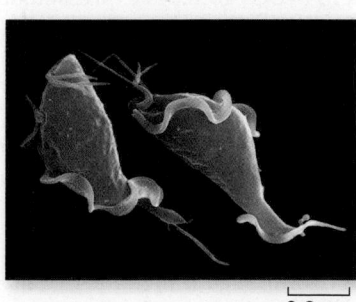

Figure 29.7 Undulating membrane characteristic of parabasalids. Vaginitis can be caused by this parasite species, *Trichomonas vaginalis.*

0.8 μm

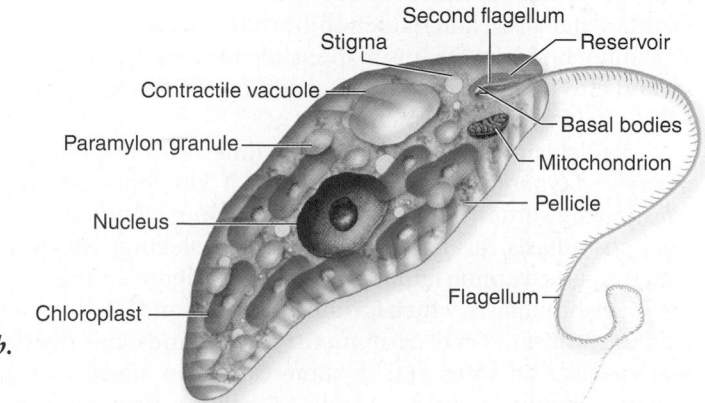

b.

Stigma
Second flagellum
Reservoir
Contractile vacuole
Paramylon granule
Basal bodies
Mitochondrion
Pellicle
Nucleus
Chloroplast
Flagellum

Hypothesis: Euglena *cells do not retain photosynthetic pigments in a dark environment.*

Prediction: *Photosynthetic pigments will be degraded when light-grown* Euglena *cells are transferred to the dark and new pigment will not be produced.*

Test: *Grow* Euglena *under normal light conditions. Transfer the culture to two flasks. Take a sample from each flask and measure the amount of photosynthetic pigments in each. Maintain one flask in the light and transfer the other to the dark. After several days, extract the photosynthetic pigments from each flask, and compare amounts with each other and with initial levels.*

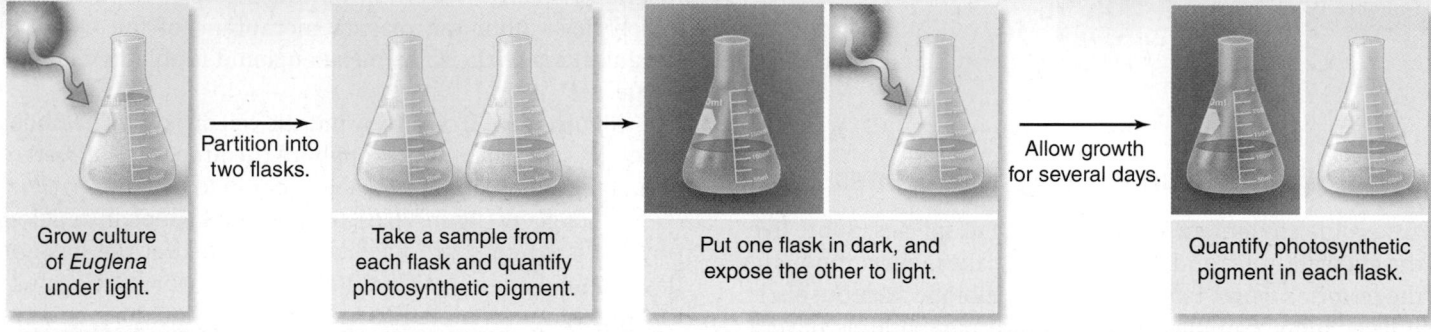

Grow culture of *Euglena* under light.

Partition into two flasks.

Take a sample from each flask and quantify photosynthetic pigment.

Put one flask in dark, and expose the other to light.

Allow growth for several days.

Quantify photosynthetic pigment in each flask.

Result: *Photosynthetic pigment levels are lower in the dark-grown flask than in the light-grown one. Pigment levels in the dark-grown flask are lower than at the beginning of the experiment. Pigment levels in the light-grown flask are unchanged.*

Conclusion: *The hypothesis is supported. Maintenance of* Euglena *in the dark resulted in a loss of photosynthetic pigment. Pigments were degraded in the dark-grown flask.*

Further Experiments: *Transfer dark-grown flasks back to the light and measure changes in pigment levels over time. Are original pigment levels restored after growth in light?*

Figure 29.9 Effect of light on *Euglena* photosynthetic pigments.

contain chlorophylls *a* and *b,* together with carotenoids. Although the chloroplasts of euglenids differ somewhat in structure from those of green algae, they probably had a common origin. *Euglena's* photosynthetic pigments are light-sensitive (figure 29.9). It seems likely that euglenid chloroplasts ultimately evolved from a symbiotic relationship through ingestion of green algae. Recent phylogenetic evidence indicates that *Euglena* had multiple origins within the euglenids, and the concept of a single *Euglena* genus is now being debated.

Parasitic Kinetoplastids

A second major group within the euglenozoa is the *kinetoplastids.* The name kinetoplastid refers to a unique, single mitochondrion in each cell. The mitochondria have two types of DNA: minicircles and maxicircles. (Remember that prokaryotes have circular DNA, and mitochondria had prokaryotic origins.) This mitochondrial DNA is responsible for very rapid glycolysis and also for an unusual kind of editing of the RNA by guide RNAs encoded in the minicircles.

Parasitism has evolved multiple times within the kinetoplastids. Trypanosomes are a group of kinetoplastids that cause many serious human diseases, the most familiar being trypanosomiasis, also known as African sleeping sickness, which causes extreme lethargy and fatigue (figure 29.10).

Leishmaniasis, which is transmitted by sand flies infected with the protist parasite *Leishmania,* is a trypanosomic disease that causes skin sores and in some cases can affect internal organs, leading to death. About 1.5 million new cases are reported each year. The rise in leishmaniasis in South America correlates with the move of infected individuals from rural to urban environments, where there is a greater chance of spreading the parasite.

Chagas disease is caused by *Trypanosoma cruzi.* At least 90 million people, from the southern United States to Argentina, are at risk of contracting *T. cruzi* from small wild mammals that carry the parasite and can spread it to other mammals and humans through skin contact with urine and feces. Blood transfusions have also increased the spread of the infection. Chagas disease can lead to severe cardiac and digestive problems in humans and domestic animals, but it appears to be tolerated in the wild mammals.

Control is especially difficult because of the unique attributes of these parasites. For example, tsetse fly-transmitted trypanosomes have evolved an elaborate genetic mechanism for repeatedly changing the antigenic nature of their protective glycoprotein coat, thus dodging the antibodies their hosts produce against them (see chapter 51). Only a single gene out of some 1000 variable-surface glycoprotein (VSG) genes is expressed at a time. A VSG gene is usually duplicated and moved to 1 of about 20 expression sites near the telomere where it is transcribed. Only one expression site is transcribed at a time.

In the guts of the flies that spread them, trypanosomes are noninfective. When they are ready to transfer to the skin or bloodstream of their host, trypanosomes migrate to the salivary glands and acquire the thick coat of glycoprotein antigens that protect them from the host's antibodies. Later, when they are taken up by a tsetse fly, the trypanosomes again shed their coats.

The production of vaccines against such a system is complex, but tests of a protein kinase inhibitor hold promise.

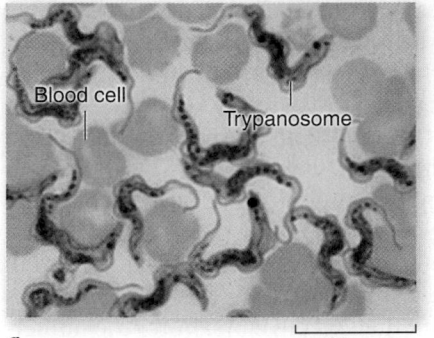

Figure 29.10 A kinetoplastid. *a. Trypanosoma* among red blood cells. The nuclei (dark-staining bodies), anterior flagella, and undulating, changeable shape of the trypanosomes are visible in this photomicrograph. *b.* The tsetse fly, shown here sucking blood from a human arm, can carry trypanosomes.

Releasing sterilized flies to impede the reproduction of populations is another technique being tried to control the fly population. Traps made of dark cloth and scented like cows, but poisoned with insecticides, have likewise proved effective.

The sequencing of the genomes of the three kinetoplastids described earlier revealed a core of common genes in all three, as described in chapter 24. The devastating toll all three take on human life could be alleviated by the development of a single drug targeted at one or more of the core proteins shared by the three parasites.

Learning Outcomes Review 29.3

Excavata are grouped based on cytoskeletal structure and DNA sequence similarities. Within the excavates, diplomonads lack mitochondria but may contain mitochondrial genes. They are unicellular, have two nuclei, and move with flagella; an example is *Giardia*. Parabasalids also lack mitochondria and use flagella and undulating membranes for locomotion; an example is *Trichomonas*. The euglenids includes phototrophs and heterotrophs. Some members have chloroplasts that remain nonfunctional unless light is present, and some phototrophs may feed if food particles are present. The kinetoplastids contain a single mitochondrion with two types of DNA and the ability to edit RNA with RNA guides. Trypanosomes are disease-causing kinetoplastids.

- *In what type of habitat would it be useful to use undulating membranes for locomotion?*
- *How does a contractile vacuole regulate osmotic pressure in a Euglena cell?*

29.4 Secondary Endosymbiosis in Chromalveolata

Learning Outcomes

1. *Identify the distinguishing feature of the members of alveolates.*
2. *Explain the function of the apical complex in apicomplexans.*
3. *Describe the characteristic features of the stramenopiles.*
4. *Explain how the oomycetes are distinguished from other protists.*

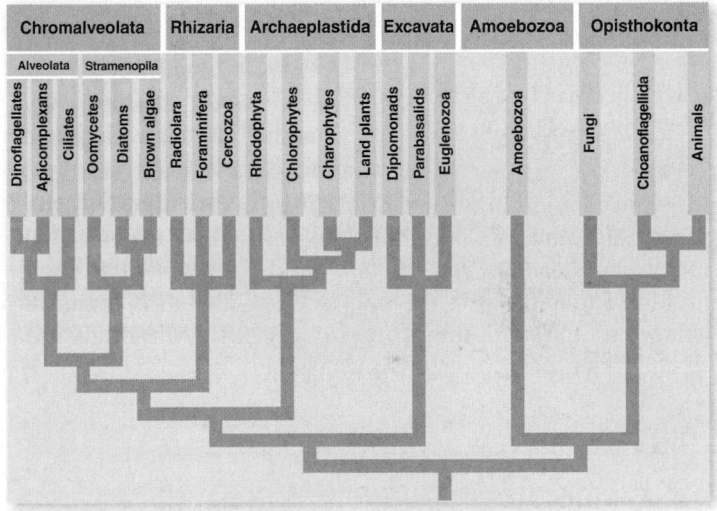

Chromalveolates constitute a supergroup that may have arisen by one or more secondary endosymbiosis events. Phylogenies based on sequence data do not support a monophyletic Chromalveolate supergroup, but the exact evolutionary history remains to be resolved. One branch is the **alveolates,** including the *dinoflagellates, apicomplexans,* and *ciliates,* all of which have a common lineage but diverse modes of locomotion. A common trait is the presence of flattened vesicles called alveoli (hence the name Alveolata) stacked in a continuous layer below their plasma membranes (figure 29.11). The alveoli may

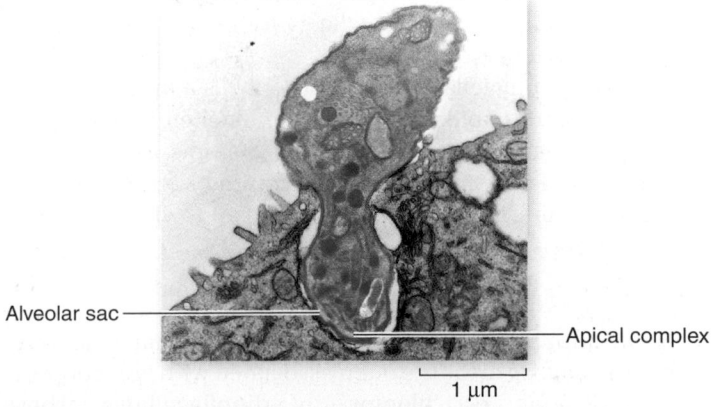

Figure 29.11 Alveoli are a continuum of vesicles just below the plasma membrane of dinoflagellates, apicomplexans, and ciliates. The apical complex of apicomplexans forces the parasite into host cells.

18,500×

Figure 29.12 Stramenopiles have very fine hairs on their flagella.

function in membrane transport, similar to Golgi bodies. A second branch of Chromalveolates is the stramenopiles, including *brown algae, diatoms,* and the *oomycetes* (water molds). The name refers to unique, fine hairs found on the flagella of members of this group (figure 29.12), although a few species have lost their hairs during evolution.

Dinoflagellates are photosynthesizers with distinctive features

Most dinoflagellates are photosynthetic unicells with two flagella. Dinoflagellates live in both marine and freshwater environments. Some dinoflagellates are luminous and contribute to the twinkling or flashing effects seen in the sea at night, especially in the tropics.

The flagella, protective coats, and biochemistry of dinoflagellates are distinctive, and the dinoflagellates do not appear to be directly related to any other phylum. Plates made of a cellulose-like material, often encrusted with silica, encase the dinoflagellate cells (figure 29.13). Grooves at the junctures of these plates usually house the flagella, one encircling the cell like a belt, and the other perpendicular to it. By beating in their grooves, these flagella cause the dinoflagellate to spin as it moves.

Most dinoflagellates have chlorophylls *a* and *c,* in addition to carotenoids, so that in the biochemistry of their chloroplasts, they resemble the diatoms and the brown algae. Possibly this lineage acquired such chloroplasts by forming endosymbiotic relationships with members of those groups.

Red tide: Overgrowth of dinoflagellates

The poisonous and destructive "red tides" that occur frequently in coastal areas are often associated with great population explosions, or "blooms," of dinoflagellates, whose pigments color the water (figure 29.14). The blooms are most often triggered by excess nutrients from agricultural and other human activity. Red tides have a profound, detrimental effect on the fishing industry worldwide. Some 20 species of dinoflagellates produce powerful neurotoxins that inhibit the diaphragm, causing respiratory failure in many vertebrates. When the toxic dinoflagellates are abundant, many fishes, birds, and marine mammals may die.

Although sexual reproduction does occur under starvation conditions, dinoflagellates reproduce primarily by asexual cell division. Asexual cell division relies on a unique form of mitosis in which the permanently condensed chromosomes divide within a permanent nuclear envelope. After the numerous chromosomes duplicate, the nucleus divides into two daughter nuclei.

Also, the dinoflagellate chromosome is unique among eukaryotes in that the DNA is not generally complexed with histone proteins. In all other eukaryotes, the chromosomal DNA is complexed with histones to form nucleosomes, structures that represent the first order of DNA packaging in the nucleus (chapter 10). How dinoflagellates maintain distinct chromosomes with a small amount of histones remains a mystery.

Apicomplexans include the malaria parasite

Apicomplexans are spore-forming parasites of animals. They are called apicomplexans because of a unique arrangement of fibrils, microtubules, vacuoles, and other cell organelles at one end of the cell, termed an *apical complex* (see figure 29.11). The apical complex is a cytoskeletal and secretory complex that enables the apicomplexan to invade its host. The best known apicomplexan is the malarial parasite *Plasmodium.* (The use of the genome sequence of the parasite and the mosquito that carries it is discussed in chapter 24.)

Plasmodium and malaria

Plasmodium glides inside the red blood cells of its host with amoeboid-like contractility. Like other apicomplexans, *Plasmodium* has a complex life cycle involving sexual and asexual

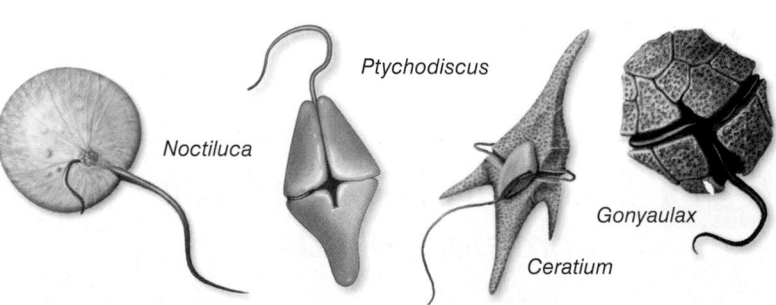

Figure 29.13 Some dinoflagellates. *Noctiluca,* which lacks the heavy cellulose armor characteristic of most dinoflagellates, is one of the bioluminescent organisms that cause the waves to sparkle in warm seas. In the other three genera, the shorter, encircling flagellum is seen in its groove, with the longer one projecting away from the body of the dinoflagellate. (Not drawn to scale.)

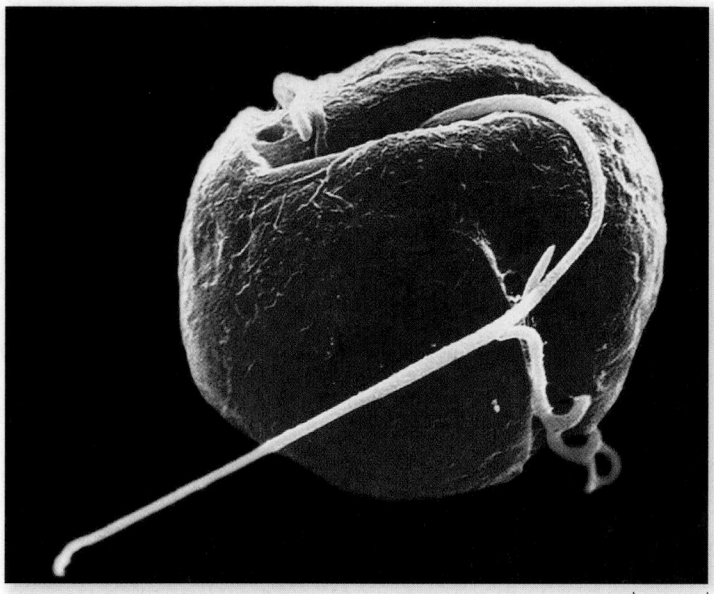

Figure 29.14 Red tide. Although small in size, huge populations of dinoflagellates, including this *Gymnopodium* species, can color the sea red and release toxins into the water.

0.8 μm

phases and alternation between different hosts, in this case mosquitoes *(Anopheles gambiae)* and humans (figure 29.15). Even though *Plasmodium* has mitochondria, it grows best in a low-O_2, high-CO_2 environment.

Efforts to eradicate malaria have focused on (1) eliminating the mosquito vectors; (2) developing drugs to poison the parasites that have entered the human body; and (3) developing vaccines. From the 1940s to the 1960s, wide-scale applications of dichlorodiphenyltrichloroethane (DDT) killed mosquitoes in the United States, Italy, Greece, and certain areas of Latin America. For a time, the worldwide elimination of malaria appeared possible. But this hope was soon crushed by the development of DDT-resistant mosquitoes in many regions. Furthermore, the use of DDT has had serious environmental consequences. In addition to the problems with resistant strains of mosquitoes, strains of *Plasmodium* have appeared that are resistant to the drugs historically used to kill them, including quinine.

An experimental vaccine containing a surface protein of one malaria-causing parasite, *P. falciparum,* seems to induce the immune system to defend against future infections. In tests, six out of seven vaccinated people did not get malaria after being bitten by mosquitoes that carried *P. falciparum.* Many are hopeful that this new vaccine may be able to fight malaria. (Chapter 24 contains a discussion of the genome sequences of both *Plasmodium* and its mosquito vector.)

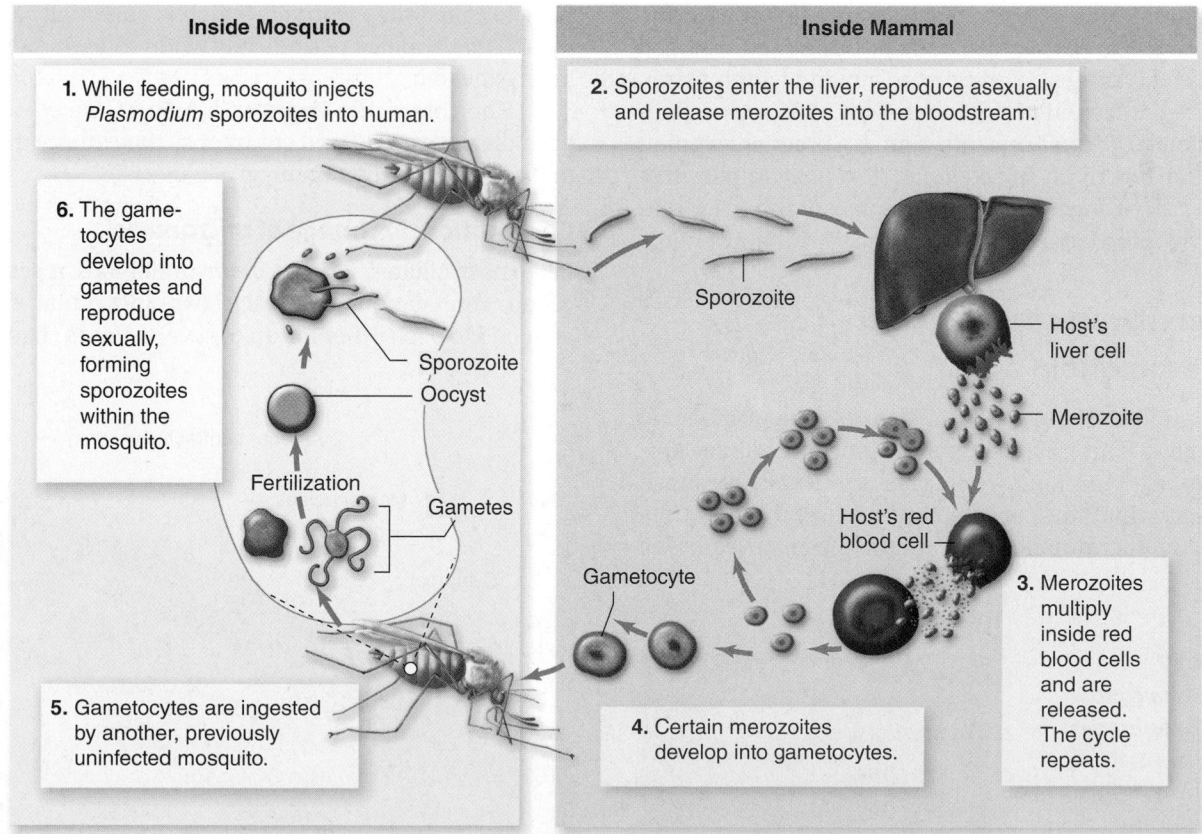

Figure 29.15 The life cycle of *Plasmodium*. *Plasmodium,* the apicomplexan that causes malaria, has a complex life cycle that alternates between mosquitoes and mammals.

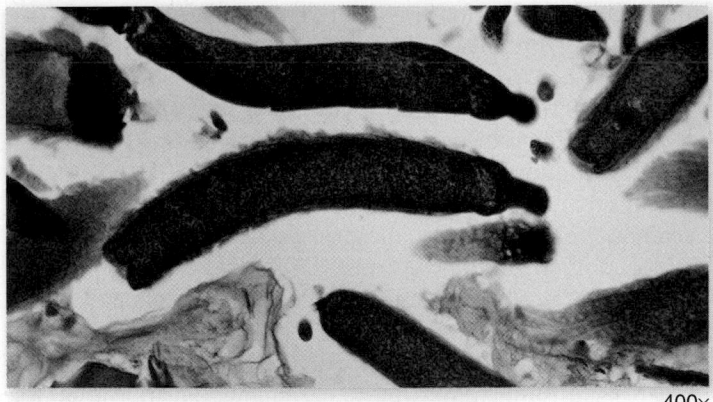

Figure 29.16 **Gregarine entering a cell.**

400×

Gregarines

Gregarines are another group of apicomplexans that use their distinctive apical complex to attach themselves in the intestinal epithelium of arthropods, annelids, and mollusks. Most of the gregarine body, aside from the apical complex, is in the intestinal cavity, and nutrients appear to be obtained through the apicomplexan attachment to the cell (figure 29.16).

Toxoplasma

Using its apical complex, *Toxoplasma gondii* invades the epithelial cells of the human gut. Most individuals infected with the parasite mount an immune response, preventing any permanent damage. In the absence of a fully functional immune system, however, *Toxoplasma* can damage brain (figure 29.17), heart, and skeletal tissues, in addition to gut and lymph tissue, during extended infections. Individuals with AIDS are particularly susceptible to *Toxoplasma* infection. If a pregnant women touches a litter box of an infected cat, *Toxoplasma* parasites from the cat can, if ingested, cross the placental barrier and harm the developing fetus with an immature immune system.

Ciliates are characterized by their mode of locomotion

As the name indicates, most ciliates feature large numbers of cilia (tiny beating hairs). These heterotrophic, unicellular protists are 10 to 3000 μm long. Their cilia are usually arranged either in longitudinal rows or in spirals around the cell. Cilia are anchored to microtubules beneath the plasma membrane

Figure 29.17
Micrograph of a cyst filled with *Toxoplasma*.
Toxoplasma can enter the brain and form cysts filled with slowly replicating parasites.

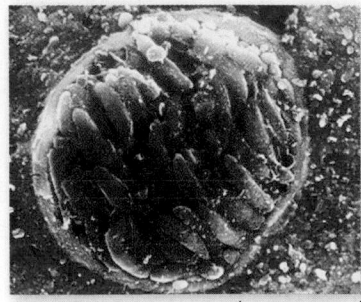

10 μm

(see chapter 5), and they beat in a coordinated fashion. In some groups, the cilia have specialized functions, becoming fused into sheets, spikes, and rods that may then function as mouths, paddles, teeth, or feet.

The ciliates have a pellicle, a tough but flexible outer covering, that enables them to squeeze through or move around obstacles.

Micronucleus and macronucleus

All known ciliates have two different types of nuclei within their cells: a small **micronucleus** and a larger **macronucleus** (figure 29.18). Macronuclei divide by mitosis and are essential for the physiological function of the well-known ciliate *Paramecium*. The micronucleus of some individuals of *Tetrahymena pyriformis,* a common laboratory species, was experimentally removed in the 1930s, and their descendants continue to reproduce asexually to this day! *Paramecium,* however, is not immortal. The cells divide asexually for about 700 generations and then die if sexual reproduction has not occurred. The micronucleus in ciliates is evidently needed only for sexual reproduction.

Vacuoles

Ciliates form vacuoles for ingesting food and regulating water balance. Food first enters the gullet, which in *Paramecium* is lined with cilia fused into a membrane (see figure 29.18). From the gullet, the food passes into food vacuoles, where enzymes and hydrochloric acid aid in its digestion. Afterward, the vacuole empties its waste contents through a special pore in the pellicle called the *cytoproct,* which is essentially an exocytotic vesicle that appears periodically when solid particles are ready to be expelled.

The contractile vacuoles, which regulate water balance, periodically expand and contract as they empty their contents to the outside of the organism.

Conjugation: Exchange of micronuclei

Like most ciliates, *Paramecium* undergoes a sexual process called conjugation, in which two individual cells remain attached to each other for up to several hours (figure 29.19).

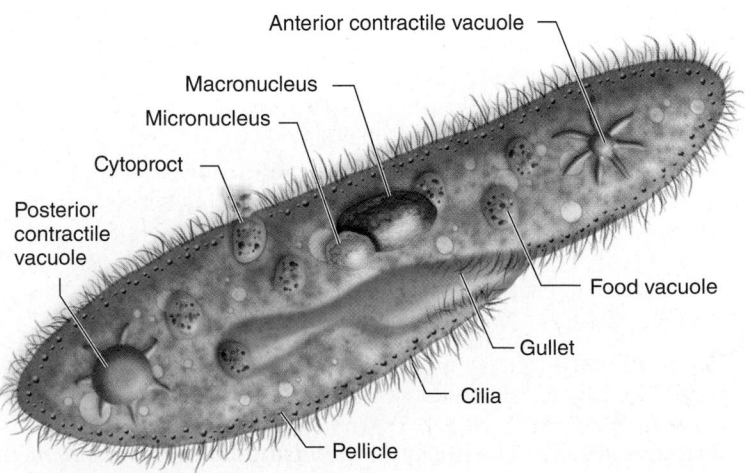

Anterior contractile vacuole
Macronucleus
Micronucleus
Cytoproct
Posterior contractile vacuole
Food vacuole
Gullet
Cilia
Pellicle

Figure 29.18 *Paramecium.* The main features of this ciliate include cilia, two nuclei, and numerous specialized organelles.

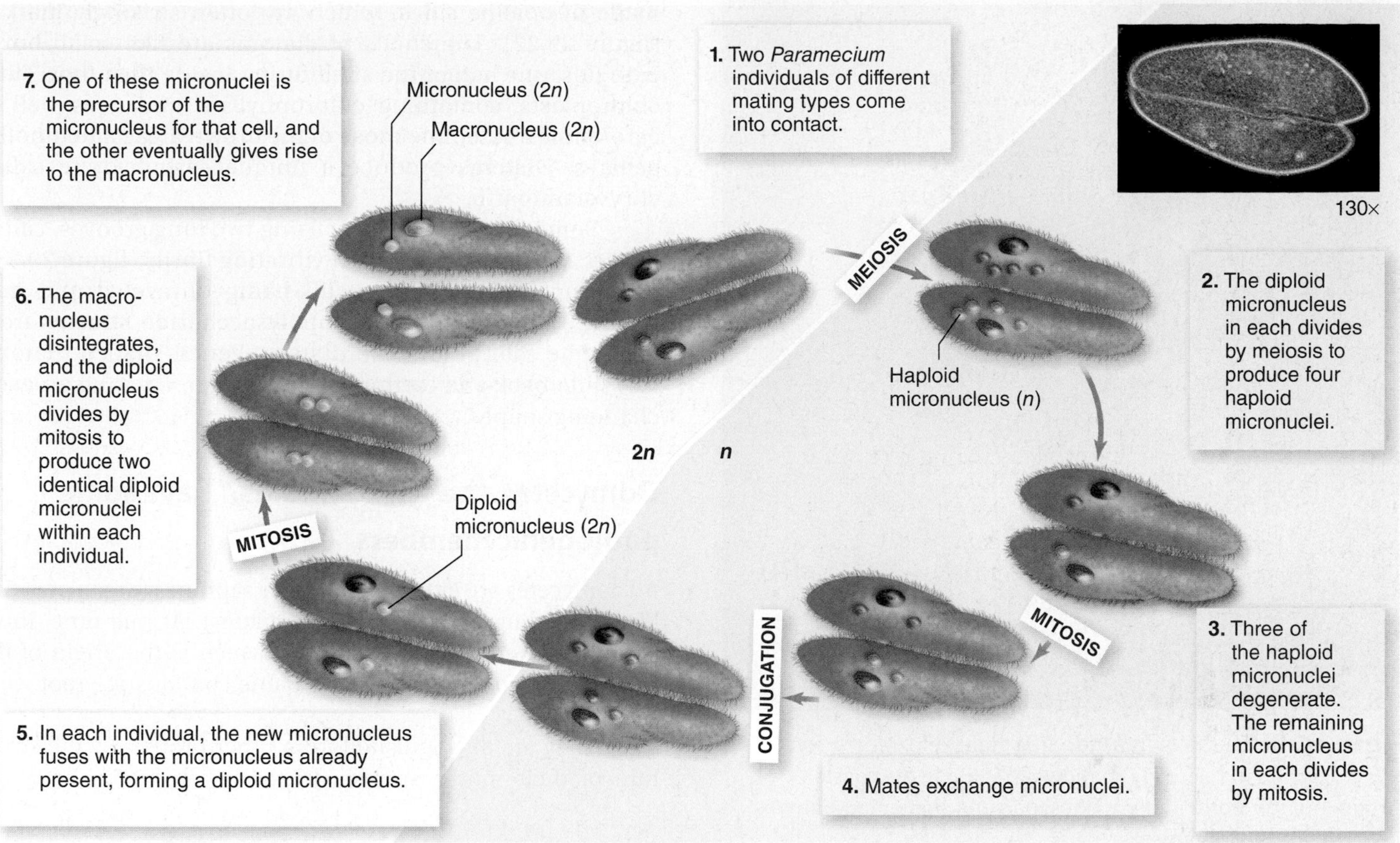

Life cycle of Paramecium

7. One of these micronuclei is the precursor of the micronucleus for that cell, and the other eventually gives rise to the macronucleus.

Micronucleus (2n)

Macronucleus (2n)

1. Two *Paramecium* individuals of different mating types come into contact.

130×

6. The macro-nucleus disintegrates, and the diploid micronucleus divides by mitosis to produce two identical diploid micronuclei within each individual.

MEIOSIS

Haploid micronucleus (n)

2. The diploid micronucleus in each divides by meiosis to produce four haploid micronuclei.

2n   n

Diploid micronucleus (2n)

MITOSIS

MITOSIS

CONJUGATION

3. Three of the haploid micronuclei degenerate. The remaining micronucleus in each divides by mitosis.

5. In each individual, the new micronucleus fuses with the micronucleus already present, forming a diploid micronucleus.

4. Mates exchange micronuclei.

Figure 29.19 Life cycle of *Paramecium*. In sexual reproduction, two mature cells fuse in a process called conjugation.

Paramecia have multiple mating types. Only cells of two different genetically determined mating types can conjugate. Meiosis in the micronuclei produces several haploid micronuclei, and the two partners exchange a pair of their micronuclei through a cytoplasmic bridge between them.

In each conjugating individual, the new micronucleus fuses with one of the micronuclei already present in that individual, resulting in the production of a new diploid micronucleus. After conjugation, the macronucleus in each cell disintegrates, and the new diploid micronucleus undergoes mitosis, thus giving rise to two new identical diploid micronuclei in each individual.

One of these micronuclei becomes the precursor of the future micronuclei of that cell, while the other micronucleus undergoes multiple rounds of DNA replication, becoming the new macronucleus. This complete segregation of the genetic material is unique to the ciliates and makes them ideal organisms for the study of certain aspects of genetics.

"Killer" strains

Paramecium strains that kill other, sensitive strains of *Paramecium* long puzzled researchers. Initially, killer strains were believed to have genes coding for a substance toxic to sensitive strains. The true source of the toxin turned out to be an endosymbiotic bacterium in the "killer" strains. If this bacterium is engulfed by a "nonkiller" strain, the toxin is released, and the sensitive *Paramecium* dies.

Brown algae include large seaweeds

Brown algae, along with diatoms and oomycetes ("water molds") make up the Stramenopiles within the Chromalveolates. Brown algae are the most conspicuous seaweeds in many northern regions (figure 29.20). The **haplodiplontic** life cycle of the brown algae is marked by alternation of generations between a multicellular sporophyte (diploid) and a multicellular gametophyte (haploid) (figure 29.21). Some sporophyte cells go through meiosis and produce spores. These spores germinate and undergo mitosis to produce the large individuals we recognize, such as the kelps. The gametophytes are often much smaller, filamentous individuals, perhaps a few centimeters in width.

Even in an aquatic environment, transport can be a challenge for the very large brown algal species. Distinctive transport cells that stack one upon the other enhance transport within some species (see figure 23.10). However, even though the large kelp look like plants, it is important to realize that they do not contain the complex tissues such as xylem that are found in plants.

Diatoms are unicellular organisms with double shells

Diatoms, members of the phylum Chrysophyta, are photosynthetic, unicellular organisms with unique double shells

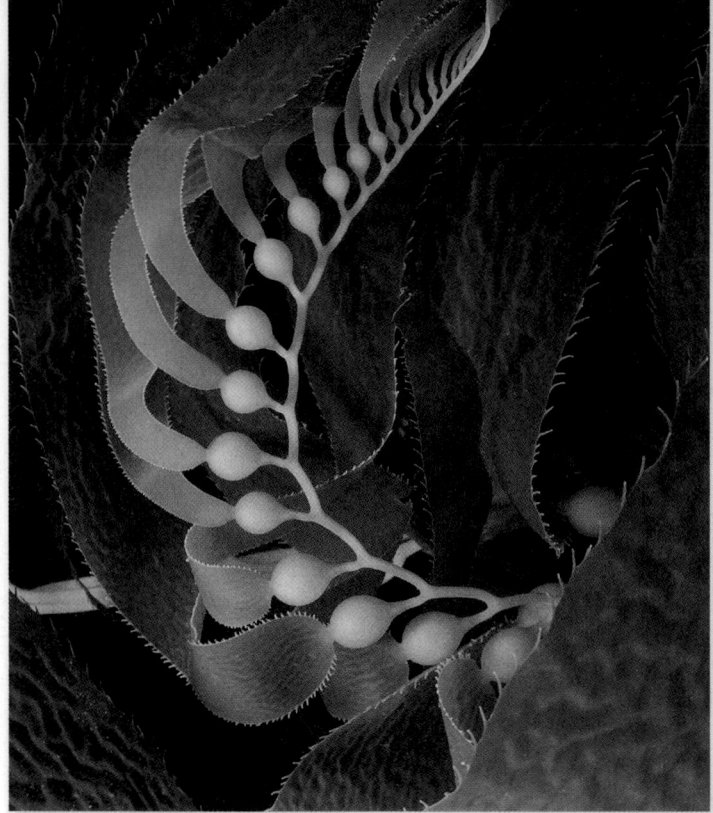

Figure 29.20 Brown alga. The giant kelp, *Macrocystis pyrifera*, grows in relatively shallow water along the coasts throughout the world and provides food and shelter for many different kinds of organisms.

made of opaline silica, which are often strikingly marked (figure 29.22). The shells of diatoms are like small boxes with lids, one half of the shell fitting inside the other. Their chloroplasts, containing chlorophylls *a* and *c,* as well as carotenoids, resemble those of the brown algae and dinoflagellates. Diatoms produce a unique carbohydrate called chrysolaminarin.

Some diatoms move by using two long grooves, called *raphes,* which are lined with vibrating fibrils (figure 29.23). The exact mechanism is still being unraveled and may involve the ejection of mucopolysaccharide streams from the raphe that propel the diatom. Pencil-shaped diatoms can slide back and forth over each other, creating an ever-changing shape.

Oomycetes, the "water molds," have some pathogenic members

All oomycetes are either parasites or saprobes (organisms that live by feeding on dead organic matter). At one time, these organisms were considered fungi, which is the origin of the term *water mold* and why their name contains the root word *-mycetes.*

They are distinguished from other protists by the structure of their motile spores, or zoospores, which bear two

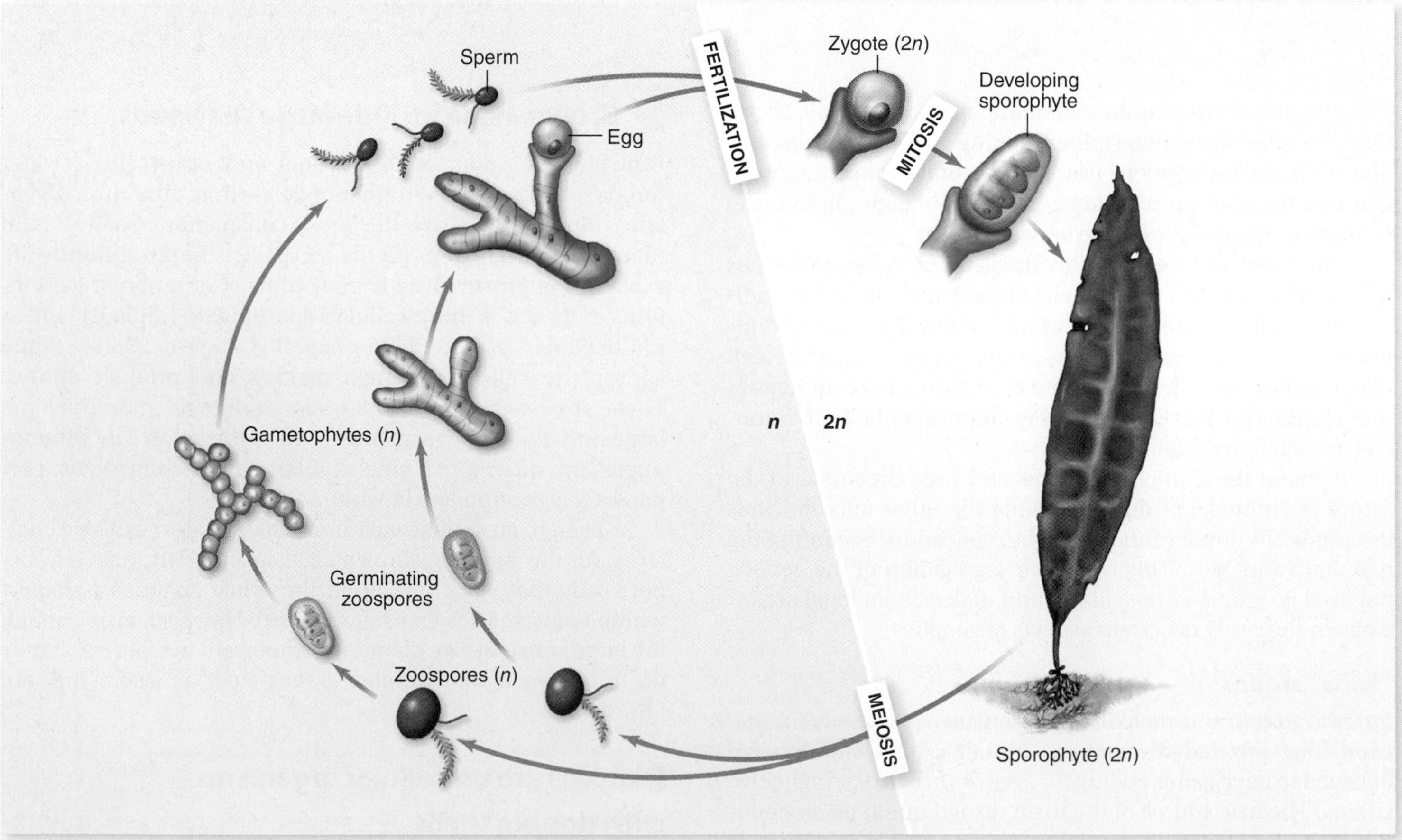

Figure 29.21 Haplodiplontic cycle of *Laminaria,* a brown alga. Multicellular haploid and diploid stages are found in this life cycle, although the male and female gametophytes are quite small.

90 μm

Figure 29.22 Diatoms. These different radially symmetrical diatoms have unique silica, two-part shells.

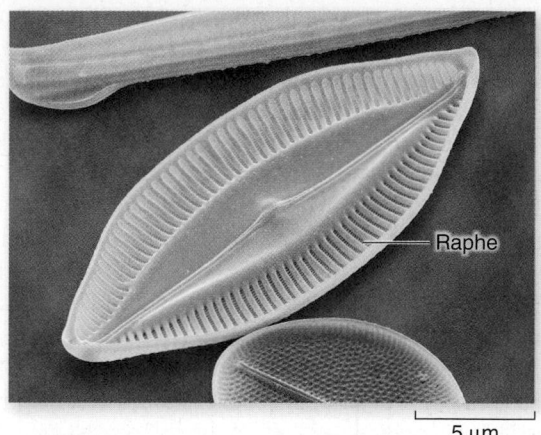

Raphe

5 μm

Figure 29.23 Diatom raphes are lined with fibrils that aid in locomotion.

unequal flagella, one pointed forward and the other backward. Zoospores are produced asexually in a sporangium. Sexual reproduction involves the formation of male and female reproductive organs that produce gametes. Most oomycetes are found in water, but their terrestrial relatives are plant pathogens.

Phytophthora infestans, which causes late blight of potatoes, was one of multiple factors, including weather and oppressive social conditions that contributed to the Irish potato famine of 1845 and 1847. During the famine, about 400,000 people starved to death or died of diseases complicated by starvation, and about 2 million Irish immigrated to the United States and elsewhere.

Another oomycete, *Saprolegnia,* is a fish pathogen that can cause serious losses in fish hatcheries. When these fish are released into lakes, the pathogen can infect amphibians and kill millions of amphibian eggs at a time at certain locations. This pathogen is thought to contribute to the phenomenon of amphibian decline.

Learning Outcomes Review 29.4

All members of the Alveolata contain flattened vesicles called alveoli. Dinoflagellates have pairs of flagella arranged perpendicularly to each other, which causes them to swim with a spinning motion. Blooms of dinoflagellates result in red tides. Apicomplexans are animal parasites that produce a structure called an apical complex, which is composed of cytoskeleton and secretory structures and aids in penetrating their host. The ciliates are unicellular, heterotrophic protists with cilia used for feeding and propulsion. Most members of the stramenopiles have fine hairs on their flagella. Brown algae are large seaweeds that provide food and habitat for marine organisms. They undergo an alternation of generations. Diatoms are unicellular with silica in their cell walls, which forms a shell with two halves. Some can propel themselves. Oomycetes are unique in the production of zoospores that bear two unequal flagella.

- **What would be a major difficulty in finding a poison to fight the malaria-causing protist Plasmodium?**
- **How could you distinguish between the sporophyte and the gametophyte of a brown alga?**

29.5 Chloroplasts in Archaeplastida

Learning Outcomes

1. List the major characteristics of red algae.
2. Describe how humans use red algae.
3. Explain why charophytes are considered the closest relatives of land plants.

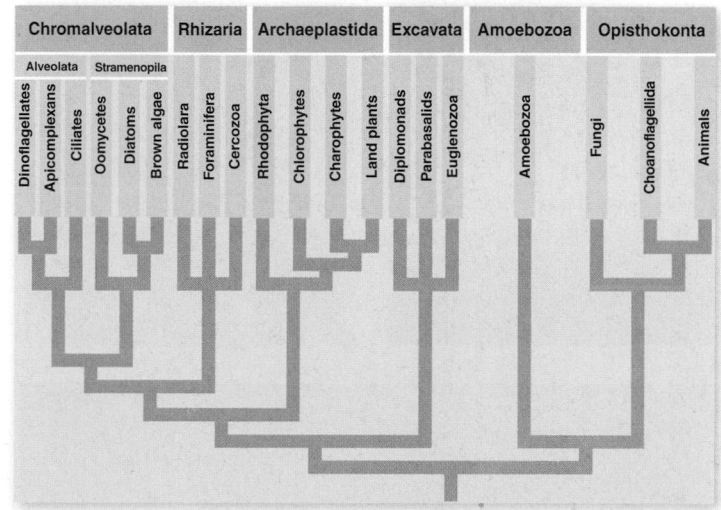

Archaeplastids acquired their chloroplasts through primary endosymbiosis, unlike the brown algae in which secondary endosymbiosis resulted in photosynthetic cells (see figures 29.3 and 29.4). As noted earlier, the common origins of the chloroplast genome in all the algae complicates phylogenetic research, especially since some of the chloroplast genes moved into the nuclear genome. Extensive comparisons of brown, red, and green algal DNA sequences are consistent with a closer evolutionary relationship between the reds and the greens than the browns, which belong to the Chromalveolates. A smaller group of algae, the glaucophytes, are also grouped with the Archaeplastida. Glaucophyte plastids were acquired through primary endosymbiosis and have a peptidoglycan (sugar and amino acid polymer) layer that harks back to the cyanobacterial endosymbiont.

The green algae have two groups—Chlorophyta and Charophyta—with the Charophyta sharing a common ancestor with the land plants. Land plants are the focus of chapters 30 and 31, with other Archaeplastida discussed here.

 Data analysis To construct a phylogeny with brown, red, and green algae, look for the amount of sequence variation in several genes from the chloroplast genomes of species in each of the groups and nuclear genomes. Few nucleotide differences are to be found among the chloroplast genes. More nucleotide differences occur between the brown algal and either the red or green algal nuclear genes than between the red and green nuclear genes. Draw a phylogeny and explain your reasoning.

Rhodophyta are Red Algae

The red algae, formally known as **Rhodophyta,** include more than 7000 species, ranging in size from microscopic organisms to *Schizymenia borealis* with blades as long as 2 m (figure 29.24). Sushi rolls are wrapped in nori, a red alga. Red algal polysaccharides are used commercially to thicken ice cream and cosmetics.

This lineage lacks flagella and centrioles and has the accessory photosynthetic pigments phycoerythrin, phycocyanin, and allophycocyanin, which are arranged within structures called *phycobilisomes.* As with the brown algae, red algae have both haploid and diploid phases in their life cycles.

Chlorophytes and Charophytes are Green Algae

Green algae have two distinct lineages: the chlorophytes, discussed here, and another lineage, the charophytes, that gave rise to the land plants (see figure 29.5). The chlorophytes are of special interest here because of their unusual diversity and lines of specialization. The chlorophytes have an extensive fossil record dating back 900 million years. Modern chlorophytes closely resemble land plants, especially in their chloroplasts, which are biochemically similar to those of the plants. They contain chlorophylls *a* and *b,* as well as carotenoids.

Unicellular chlorophytes

Early green algae probably resembled *Chlamydomonas reinhardtii,* diverging from land plants over 1 BYA (figure 29.25).

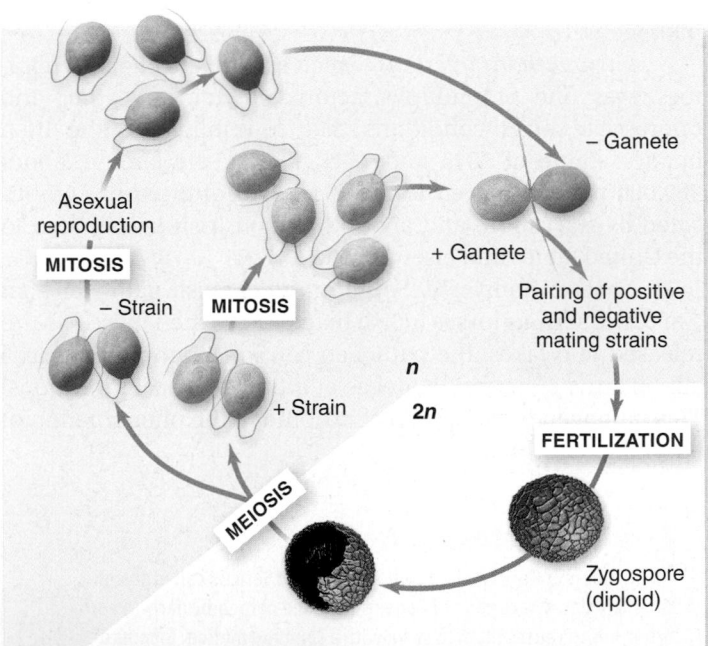

Figure 29.25 *Chlamydomonas* **life cycle.** This single-celled chlorophyte has both asexual and sexual reproduction. Unlike multicellular green plants, gamete fusion is not followed by mitosis.
© Dr. Richard Kessel & Dr. Gene Shih/Visuals Unlimited

Figure 29.24 **Red algae come in many forms and sizes.**

Individuals are microscopic (usually less than 25 μm long), green, and rounded, and they have two flagella at the anterior end. They are soil dwellers that move rapidly in water by beating their flagella in opposite directions. Most individuals of *Chlamydomonas* are haploid. *Chlamydomonas* reproduces asexually as well as sexually, but because it is always unicellular, the life cycle is not haplodiplontic (see figure 29.25).

Several lines of evolutionary specialization have been derived from organisms such as *Chlamydomonas,* including the evolution of nonmotile, unicellular green algae. *Chlamydomonas* is capable of retracting its flagella and settling down as an immobile unicellular organism if the pond in which it lives dries out. Some common algae found in soil and bark, such as *Chlorella,* are essentially like *Chlamydomonas* in this trait, but they do not have the ability to form flagella.

Genome-sequencing projects are providing new insights into the evolution of the green algae and land plants. A comparison of the 6968 protein families predicted by the *Chlamydomonas* genome were compared with a red algal genome and two land plant genomes (moss and *Arabidopsis*). Of these proteins, 172 are not found in red algae. Comparing these conserved proteins in the green algae and land plants will provide new information about the evolution of the land plants, which is discussed in chapter 30.

Cell specialization in colonial chlorophytes

Multicellularity arose many times in the eukaryotes. Colonial chlorophytes provide examples of cellular specialization, an aspect of multicellularity. A line of specialization from cells like those of *Chlamydomonas* concerns the formation of motile, colonial organisms. In these genera of green algae, the *Chlamydomonas*-like cells retain some of their individuality.

The most elaborate of these organisms is *Volvox* (figure 29.26), a hollow sphere made up of a single layer of 500 to 60,000 individual cells, each cell having two flagella. Only a small number of the cells are reproductive. Some reproductive cells may divide asexually, bulge inward, and give rise to new colonies that initially remain within the parent colony. Others produce gametes.

Haplodiplontic life cycles in multicellular chlorophytes

Haplodiplontic life cycles are found in some chlorophytes and the streptophytes, which include both charophytes and land plants. *Ulva,* a multicellular chlorophyte, has identical gametophyte and sporophyte generations that consist of flattened sheets two cells thick (figure 29.27). Unlike the charophytes, none of the ancestral chlorophytes gave rise to land plants.

 Inquiry question Are *Ulva* gametes formed by meiosis? Explain your response.

Charophytes: Closest Relatives to Land Plants

Charophytes are also green algae, and they are distinguished from chlorophytes by their close phylogenetic relationship to the land plants. Currently, the molecular evidence from rRNA and DNA sequences favors the charophytes as the green algal

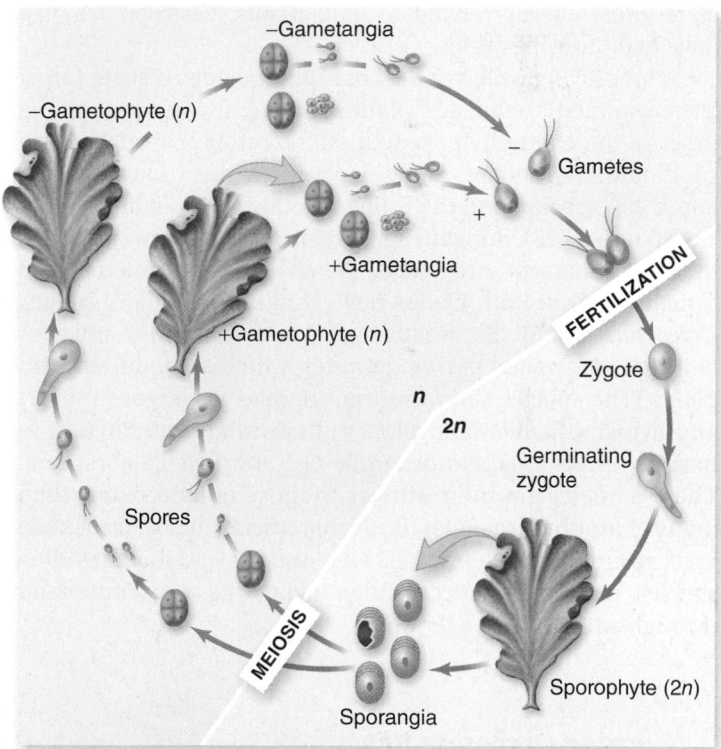

Figure 29.27 Life cycle of *Ulva.* This chlorophyte alga has a haplodiplontic life cycle. The gametophyte and sporophyte are multicellular and identical in appearance.

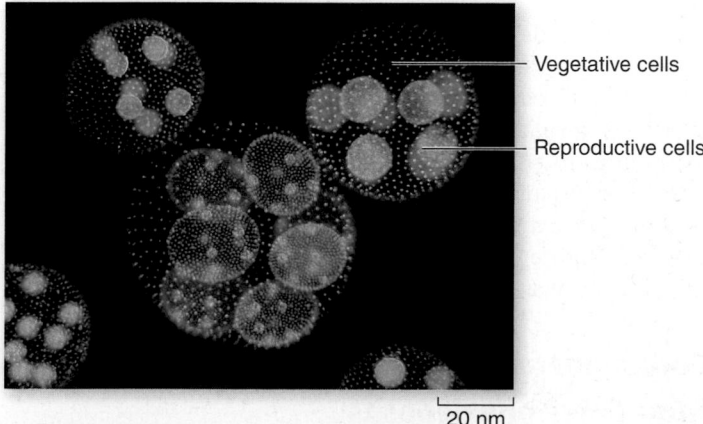

20 nm

Figure 29.26 *Volvox.* This chlorophyte forms a colony in which some cells are specialized for reproduction.

45 μm

Figure 29.28 *Chara,* a member of the Charales, and *Coleochaete,* a member of the Coleochaetales, represent the two clades most closely related to land plants.

clade most closely related to land plants. Charophytes also have haplontic life cycles.

Identifying which of the charophyte clades is sister (most closely related) to the land plants puzzled biologists for a long time, as the charophyte algae fossil record is scarce. The two candidate charophyte clades have been the Charales, with about 300 species, and the Coleochaetales, with about 30 species (figure 29.28). Both lineages are primarily freshwater algae, but the Charales are huge, relative to the microscopic Coleochaetales. Both clades have similarities to land plants. *Coleochaete* and its relatives have cytoplasmic linkages between cells called *plasmodesmata,* which are found in land plants. The species *Chara* in the Charales undergoes mitosis and cytokinesis like land plant cells. Sexual reproduction in both relies on a large, nonmotile egg and flagellated sperm. These gametes are more similar to those of land plants than many charophyte relatives. Both charophyte clades form green mats around the edges of freshwater ponds and marshes. One species must have successfully inched its way onto land through adaptations to drying.

Learning Outcomes Review 29.5

Red algae vary greatly in size and produce accessory pigments that may give them a red color. They lack centrioles and flagella, and they reproduce using an alternation of generations. Humans use red algae as a food and a thickening agent. The chlorophytes have chloroplasts very similar to those of land plants. Specializations in this group include the evolution of nonmotile, unicellular species that can tolerate drying and formation of colonial organisms that exhibit a degree of cell specialization. Charophytes are the green algae that are most likely sister to the land plants based on molecular evidence, morphology, and reproduction

■ *Why would you expect to get different results from analysis of nuclear DNA versus plastid DNA?*

■ *What major barrier must be overcome for sexual reproduction of land-based organisms?*

Learning Outcomes

1. *Distinguish between the shells of most foraminifera and those of diatoms.*
2. *Distinguish between cellular and plasmodial slime molds.*

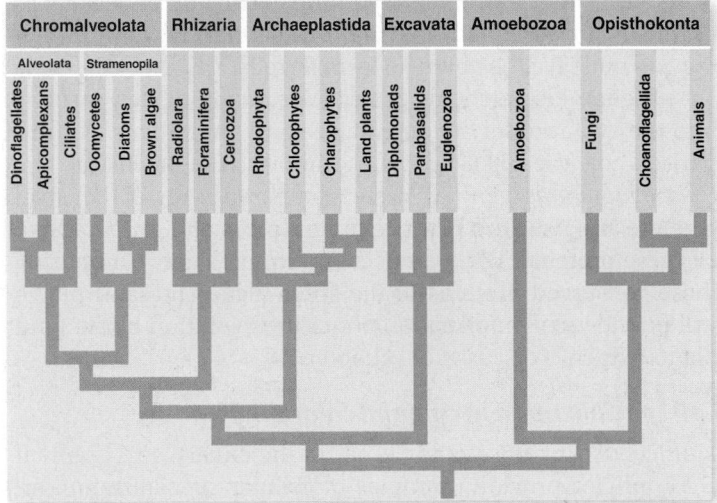

Rhizaria use pseudopods for locomotion, but are distinct from the supergroup Amoebozoans. Amoeboid motion was once used as a trait to group protists, but the inclusion of molecular data in phylogenetic reconstructions led to the realization that locomotion alone was not a useful trait in evolutionary analyses. Within the Rhizaria, three distinct monophyletic groups have been identified. Radiolara are common marine phytoplankton. Formanifera are also marine phytoplankton, and their calcium carbonate shells contribute to chalky deposits over time. Cercozoa are found in the soil.

Radiolarians have silica exoskeletons

The pseudopods of amoeboid cells give them truly amorphous bodies. One group, however, has more distinct structures. *Radiolarians* secrete glassy exoskeletons made of silica. These skeletons give the unicellular organisms a distinct shape, exhibiting either bilateral or radial symmetry. The shells of different species form many elaborate and beautiful shapes, with pseudopods extruding outward along spiky projections of the skeleton (figure 29.29). Microtubules support these cytoplasmic projections.

Foraminifera fossils created huge limestone deposits

Members of the phylum Foraminifera are heterotrophic marine protists. They range in diameter from about 20 μm to

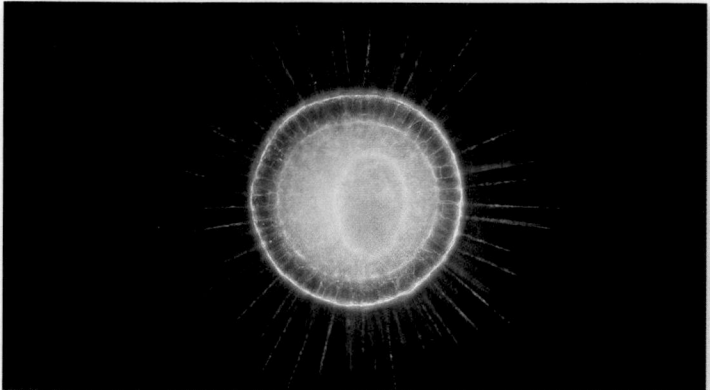

Figure 29.29 *Actinosphaerium* with needle-like pseudopods.

10×

several centimeters. They resemble tiny snails and can form 3-m-deep layers in marine sediments. Characteristic of the group are pore-studded shells (called *tests*) composed of organic materials usually reinforced with grains of calcium carbonate, sand, or even plates from shells of echinoderms or spicules (minute needles of calcium carbonate) from sponge skeletons.

Depending on the building materials they use, foraminifera may have shells of very different appearance. Some of them are brilliantly colored red, salmon, or yellow-brown.

Most foraminifera live in sand or are attached to other organisms, but two families consist of free-floating planktonic organisms. Their tests may be single-chambered, but are more often multichambered, and they sometimes have a spiral shape resembling that of a tiny snail. Thin cytoplasmic projections called *podia* emerge through openings in the tests (figure 29.30). Podia are used for swimming, gathering materials for the tests, and feeding. Foraminifera eat a wide variety of small organisms.

The life cycles of foraminifera are extremely complex, involving alternation between haploid and diploid generations. Foraminifera have contributed massive accumulations of their tests to the fossil record for more than 200 million years. Because

Figure 29.31 White Cliffs of Dover. The limestone that forms these cliffs is composed almost entirely of fossil shells of protists, including foraminifera.

of the excellent preservation of their tests and the striking differences among them, forams are very important as geological markers. The pattern of occurrence of different forams is often used as a guide in searching for oil-bearing strata. Limestones all over the world, including the famous White Cliffs of Dover in southern Great Britain, are often rich in forams (figure 29.31).

Cercozoa locomote with flagella or pseudopods

Cercozoa are a morphologically diverse group of primarily soil protists. Some rely on flagella for locomotion, but others extend pseudopods for amoeboid movement. Some have silica-based shells made up of scales or plates. A cercozoan, *Paulinella chromatophora*, may have ingested a green alga as recently as 60 MYA, establishing an endosymbiotic relationship (figure 29.32). This species is being investigated for clues to the evolution of endosymbiosis.

8 μm

Figure 29.30 A representative of the foraminifera. *Podia*, thin cytoplasmic projections, extend through pores in the calcareous test, or shell, of this living foram.

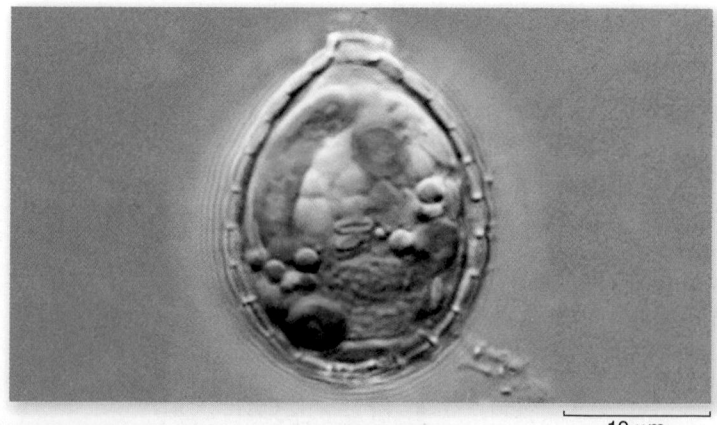

10 μm

Figure 29.32 The cercozoan *Paulinella chromatophora*. Cercozoans may be a more recent endosymbiont, offering insights into the evolution of endosymbiosis.

29.7 Blunt Pseudopods in Amoebozoans

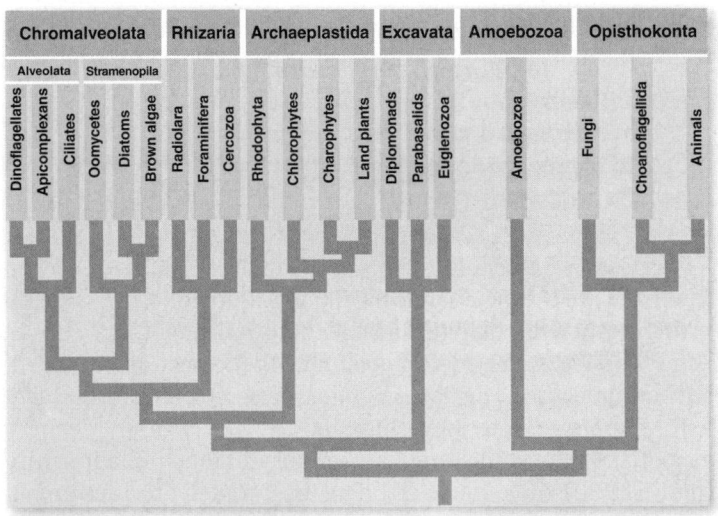

Amoebas move from place to place by means of their pseudopods. Pseudopods are flowing projections of cytoplasm that extend and pull the amoeba forward or engulf food particles. An amoeba puts a pseudopod forward and then flows into it (figure 29.33). Microfilaments of actin and myosin similar to those

found in muscles are associated with these movements. The pseudopods can form at any point on the cell body so that it can move in any direction. The amoebas in the Amoebozoan supergroup are most closely related to the Opisthokonts, which include the close relatives of fungi and animals. Members of these two supergroups have a single flagellum, in contrast with two or more in the other supergroups. Amoebozoans and Opisthokonts are collectively referred to as Unikonts, with "uni" referring to the single flagellum; however, they are distinct supergroups.

Most amoeba are free living, but some are parasitic

Found in the soil, as well as freshwater, most of these Amoebozoans are free living and important to the soil ecosystem. A few rare cases of a species acting as a human pathogen have been reported. In immune-suppressed individuals, the *Acanthamoeba* enters the body through a wound and is able to cross the blood-brain barrier into the brain. Once in the brain, inflammation and death follow. Intestinal illness is caused by a different species, *Entamoeba histolytica*.

Plasmodial slime molds are multinucleate

Plasmodial slime molds stream along as a **plasmodium,** a nonwalled, multinucleate mass of cytoplasm that resembles a moving mass of slime (figure 29.34). This form is called the *feeding phase,* and the plasmodia may be orange, yellow, or another color.

Plasmodia show a back-and-forth streaming of cytoplasm that is very conspicuous, especially under a microscope. They are able to pass through the mesh in cloth or simply to flow around or through other obstacles. As they move, they engulf and digest bacteria, yeasts, and other small particles of organic matter.

A multinucleated *Plasmodium* cell undergoes mitosis synchronously, with the nuclear envelope breaking down, but only at late anaphase or telophase. Centrioles are absent.

When either food or moisture is in short supply, the plasmodium migrates relatively rapidly to a new area. Here it stops moving and either forms a mass in which spores differentiate

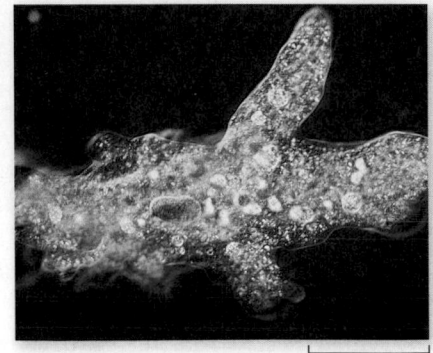

Figure 29.33
Amoeba proteus.
The projections are pseudopods; an amoeba moves by flowing into them.

60 μm

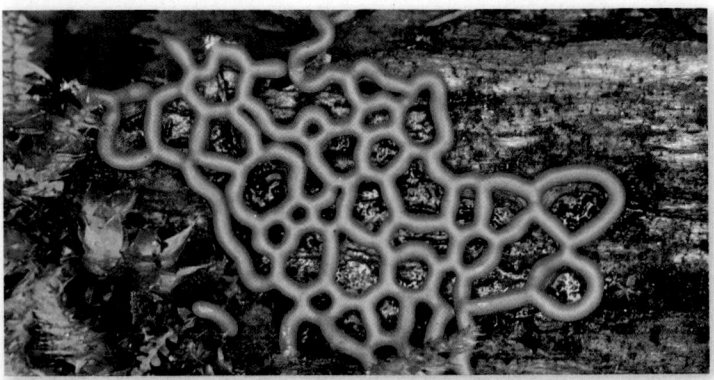

Figure 29.34 A plasmodial protist. This multinucleate pretzel slime mold, *Hemitrichia serpula,* moves about in search of the bacteria and other organic particles that it ingests.

Figure 29.35 Sporangia of a plasmodial slime mold.
These *Arcyria* sporangia are found in the phylum Myxomycota.

or divides into a large number of small mounds, each of which produces a single, mature **sporangium,** the structure in which spores are produced. These sporangia are often beautiful and extremely complex in form (figure 29.35). The spores are highly resistant to unfavorable environmental influences and may last for years if kept dry.

Cellular slime molds exhibit cell differentiation

The cellular slime molds have become an important group for the study of cell differentiation because of their relatively simple developmental systems (figure 29.36). The individual organisms behave as separate amoebas, moving through the soil and ingesting bacteria. When food becomes scarce, the individuals aggregate to form a moving "slug." Cyclic adenosine monophosphate (cAMP) is sent out in pulses by some of the cells, and other cells move in the direction of the cAMP to form the slug. In the cellular slime mold *Dictyostelium discoideum,* this slug goes through morphogenesis to make stalk and spore cells. The spores then go on to form a new amoeba if they land in a moist habitat.

Learning Outcome Review 29.7

At least three lineages of slime mold exist. Cellular slime molds are multicellular, and plasmodial slime molds consist of large multinucleate single cells.

■ *Would you say that cellular slime molds are closely related to plasmodial slime molds?*

29.8 *Propulsion via a Single Posterior Flagellum in Opisthokonts*

Learning Outcome

1. *Describe the evolutionary significance of the choanoflagellates.*

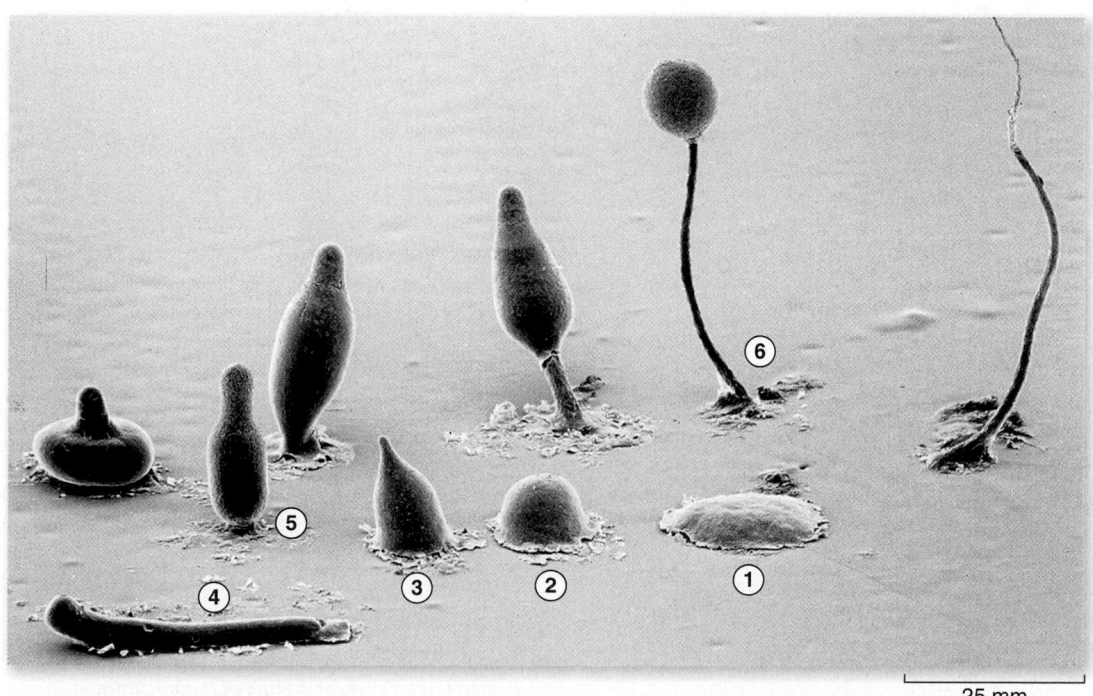

.25 mm

Figure 29.36 Development in *Dictyostelium discoideum*, a cellular slime mold. *1.* First, a spore germinates, forming an amoeba that feeds and reproduces until the food runs out. At that point, amoebas aggregate and move toward a fixed center. *2.* The aggregated amoebas begin to form a mound. *3.* The mound produces a tip and begins to fall sideways. *4.* Next, the aggregate forms a multicellular "slug," 2–3 mm long, that migrates toward light. *5.* The slug stops moving, and a process called culmination begins. Cells differentiate into stalk and spore cells. *6.* In the mature fruiting body, amoebas become encysted as spores.

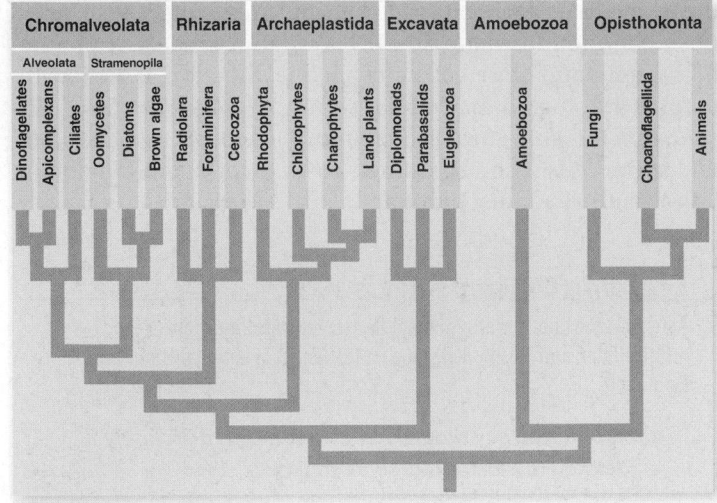

Chromalveolata						Rhizaria			Archaeplastida				Excavata			Amoebozoa	Opisthokonta		
Alveolata			Stramenopila																
Dinoflagellates	Apicomplexans	Ciliates	Oomycetes	Diatoms	Brown algae	Radiolara	Foraminifera	Cercozoa	Rhodophyta	Chlorophytes	Charophytes	Land plants	Diplomonads	Parabasalids	Euglenozoa	Amoebozoa	Fungi	Choanoflagellida	Animals

Fungi and animals are more closely related to each other than to plants because they share a common ancestor, which leads them to be grouped as Opisthokonts. Of particular interest in the Opisthokonts are the *Choanoflagellates,* unicellular organisms that are most like the common ancestor of the sponges and, indeed, all animals. Choanoflagellates have a single emergent flagellum surrounded by a funnel-shaped, contractile collar composed of closely placed filaments, a structure that is exactly matched in the sponges, which are animals. These protists feed on bacteria strained out of the water by their collar. Colonial forms resemble freshwater sponges (figure 29.37).

Figure 29.37
Colonial choanoflagellates resemble their close animal relatives, the sponges.

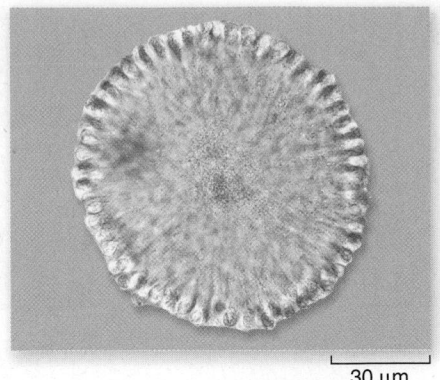

30 μm

The close relationship of choanoflagellates to animals was further demonstrated by the strong homology between a surface receptor (a tyrosine kinase receptor) found in choanoflagellates and sponges. This surface receptor initiates a signaling pathway involving phosphorylation (see chapter 9).

Learning Outcome Review 29.8

The choanoflagellates are believed to be the closest relatives of animals. Colonial forms are similar to freshwater sponges, and both organisms have a homologous cell-surface receptor.

- *What other types of studies might connect choanoflagellates with sponges?*

Chapter Review

29.1 Eukaryotic Origins and Endosymbiosis

Fossil evidence dates the origins of eukaryotes.

Although eukaryotes may have arisen earlier, the fossil evidence of their appearance dates back to 1.5 BYA.

The nucleus and ER arose from membrane infoldings (figure 29.2).

Mitochondria evolved from engulfed aerobic bacteria.

According to the theory of endosymbiosis, ancestral eukaryotic cells engulfed aerobic bacteria, which then became mitochondria (figure 29.3).

Chloroplasts evolved from engulfed photosynthetic bacteria.

Chloroplasts are believed to have arisen when ancestral eukaryotic cells engulfed photosynthetic bacteria (figure 29.4). Brown algae arose through a secondary endosymbiotic event when a red alga was engulfed by a single-celled organism.

Endosymbiosis is supported by a range of evidence.

In support of endosymbiosis, several organelles are found to contain their own DNA, which closely resembles that of prokaryotes. Over a billion and a half years, many of the chloroplast and mitochondrial genes have migrated to the nuclear genome.

Mitosis evolved in eukaryotes.

Mechanisms of mitosis vary among organisms, suggesting that the process did not evolve all at once.

29.2 Overview of Protists

Eukaryotes are organized into six supergroups that all contain protists.

Molecular systematics is helping to sort out the protists, which are now grouped into six supergroups: Excavata, Chromalveolata, Archaeplastida, Rhizaria, Amoebozoans, and Ophisthokonts (figure 29.5). With the exception of Chromalveolata, the supergroups are monophyletic.

Monophyletic clades have been identified among the protists.

Protist cell surfaces vary widely.

Extracellular material (ECM) may cover the plasma membrane.

Protists have several means of locomotion.

Protists mainly use flagella or pseudopods for locomotion, although many other means of propulsion are found.

Protists have a range of nutritional strategies.

Protists include phototrophs, heterotrophs, (phagotrophs or osmotrophs), and mixotrophs capable of both modes.

Protists reproduce asexually and sexually.

Protists can reproduce asexually by mitosis, budding, or schizogony. They may also carry out sexual reproduction.

Protists are the bridge to multicellularity.

Colonial protists may be the precursors of multicellular organisms.

29.3 Lack of Typical Mitochondria in Excavata

Diplomonads have two nuclei.

Diplomonads are unicellular, move with flagella, and have two nuclei.

Parabasalids have undulating membranes.

Parabasalids use flagella and undulating membranes for locomotion.

Euglenozoa include free-living and parasitic groups.

Free-living euglenids, exemplified by *Euglena*, can produce chloroplasts to carry out photosynthesis in the light. They contain a pellicle and move via anterior flagella. Kinetoplastids are parasitic and are distinctive in having a single, unique mitochondrion with two types of circular DNA.

29.4 Secondary Endosymbiosis in Chromalveolata

Dinoflagellates are photosynthesizers with distinctive features.

Dinoflagellates have pairs of flagella arranged so that they swim with a spinning motion. Blooms of dinoflagellates cause red tides (figure 29.13).

Apicomplexans include the malaria parasite.

Apicomplexans are spore-forming animal parasites (figure 29.15). They have a unique arrangement of organelles at one end of the cell, called the apical complex, which is used to invade the host.

Ciliates are characterized by their mode of locomotion.

Ciliates are unicellular, heterotrophic protists that use numerous cilia for feeding and propulsion. Each cell has a macronucleus and a micronucleus. Micronuclei are exchanged during conjugation. (figure 29.19)

Brown algae include large seaweeds.

Brown algae are typically large seaweeds that have a haplodiplontic life cycle, producing gametophyte and sporophyte stages (figure 29.21).

Diatoms are unicellular organisms with double shells.

Diatoms have silica in their cell walls. Each diatom produces two overlapping glassy shells that fit like a box and lid.

Oomycetes, the "water molds," have some pathogenic members.

Oomycetes are parasitic and are unique in the production of asexual spores (zoospores) that bear two unequal flagella

29.5 Chloroplasts in Archaeplastida

Rhodophyta are red algae.

Red algae produce accessory pigments that may give them a red color. They lack centrioles and flagella, and they reproduce using an alternation of generations.

Chlorophytes and charophytes are green algae (figures 29.25 and 29.27).

Unicellular chlorophytes include *Chlamydomonas,* which has two flagella, and *Chlorella,* which has no flagella and reproduces asexually.

Volvox is an example of a colonial green alga; some cells are specialized for producing gametes or for asexual reproduction. It may represent a step on the way to multicellularity. Multicellular chlorophytes can have haplodiplontic life cycles.

Ulva has sporophyte and gametophyte generations; however, the chlorophytes did not give rise to land plants although they are the closest relatives to land plants.

Both groups within the Charophytes, Charales and Coleochaetales, have plasmodesmata, cytoplasmic links between cells. They also undergo mitosis and cytokinesis like terrestrial plants. Charophytes are most closely related to the land plants of all the algae.

29.6 Slender Pseudopods in Rhizaria

Radiolarians have silica exoskeletons.

Glassy exoskeletons made of silica give radiolarians a distinct shape. Microtubules support pseudopods that extrude toward the tips of the spiky exoskeleton.

Foraminifera fossils created huge limestone deposits.

The Foraminifera are heterotrophic marine protists with pore-studded shells primarily formed by deposit of calcium carbonate.

Cercozoa locomote with flagella or pseudopods.

Like the marine radiolarians, cercozoans have silica exoskeletons, but most are found in soils. Cercozoans ingest green algae and may provide clues to the evolution of endosymbiosis.

29.7 Blunt Pseudopods in Amoebozoans

Most amoeba are freeliving, but some are parasitic.

Many protists are freeliving and found in the soil and freshwater ecosystems, but a few species are pathogenic to humans.

Plasmodial slime molds are multinucleate.

Plasmodia in their feeding phase are macroscopically visible masses of oozing slime (figure 29.34). These colorful large cells undergo repeated rounds of mitosis without cell division.

Cellular slime molds exhibit cell differentiation (figure 29.36).

Cellular slime molds such as *Dictyostelium discoideum* can signal and interact with neighboring cells, resulting in differentiated cell types in an aggregate organism.

29.8 Propulsion via a Single Posterior Flagellum in Opisthokonts

Colonial choanoflagellates are structurally similar to freshwater sponges, and molecular similarities have been found.

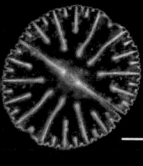

Review Questions

UNDERSTAND

1. Fossil evidence of eukaryote dates back to
   a. 2.5 BYA. c. 2.5 MYA.
   b. 1.5 BYA. d. 1.5 MYA.

2. DNA is not found in this organelle.
   a. Endoplasmic reticulum
   b. Nucleus
   c. Chloroplast
   d. Mitochondrion

3. The products of budding are
   a. two cells of equal size.
   b. two cells, one of which is smaller than the other.
   c. many cells of equal size.
   d. many cells of variable size.

4. Both diplomonads and parabasalids
   a. contain chloroplasts.
   b. have multinucleate cells.
   c. lack mitochondria.
   d. have silica in their cell walls.

5. Trypanosomes are examples of
   a. euglenoids. c. parabasalids.
   b. diplomonads. d. kinetoplastids.

6. The function of the apical complex in Apicomplexans is to
   a. propel the cell through water.
   b. penetrate host tissue.
   c. absorb food.
   d. detect light.

7. If a cell contains a pellicle, it
   a. can change shape readily.
   b. is shaped like a sphere.
   c. is shaped like a torpedo.
   d. must have a contractile vacuole.

8. Stramenopila are
   a. tiny flagella. c. small hairs on flagella.
   b. large cilia. d. pairs of large flagella.

9. Choose all of the following that exhibit an alternation of multicellular generations.
   a. Dinoflagellates c. Red algae
   b. Brown algae d. Diatoms

10. Choose all of the following that are photosynthetic.
    a. Diatoms c. Apicomplexans
    b. Ciliates d. Dinoflagellates

11. Which is most likely the ancestor of animals?
    a. Trypanosomes c. Ciliates
    b. Diplomonads d. Choanoflagellates

12. When food is scarce, cells of this organism communicate with each other to form a multicellular slug.
    a. Cellular slime molds c. Foraminifera
    b. True amoebas d. Diatoms

APPLY

1. Analyze the following statements and choose the one that most accurately supports the endosymbiotic theory.
   a. Mitochondria rely on mitosis for replication.
   b. Chloroplasts contain DNA but translation does not occur in chloroplasts.
   c. Vacuoles have double membranes.
   d. Antibiotics that inhibit protein synthesis in bacteria can have the same effect on mitochondria.

2. Determine which feature of the choanoflagellates was likely the most significant for the evolution of animals?
   a. Flagellum with a funnel-shaped, contractile collar also found in sponges
   b. A tyrosine kinase receptor on the surface of choanoflagellates that has strong homology to fungi
   c. A colonial form that resembles some fungi
   d. Eyespots that are similar to ribbon worms

3. Examine the life cycle of cellular slime molds, and determine which feature affords the greatest advantage for surviving food shortages.
   a. Cellular slime molds produce spores when starved.
   b. Cellular slime molds are saprobes.
   c. A diet of bacteria ensures there will never be a shortage of food.
   d. Cellular slime molds use cAMP to guide each other to food sources.

SYNTHESIZE

1. Modern taxonomic treatments rely heavily on phylogenetic data to classify organisms. In the past, taxonomists often used a morphological species concept, in which species were defined based on similarities in growth form. Give an example of how a morphological species concept would group a set of protists differently from how a phylogenetic species concept would.

2. Three methods have been used to try to eradicate malaria. One is to eliminate the mosquito vectors of the parasite, a second is to kill the parasites after they entered the human body, and the third is to develop a vaccine against the parasite, allowing the human immune system to provide protection from the disease. Which do you suppose is the most promising in the long run? Why? Think about both the biology of the disease and the efficacy of carrying out each of the methods on a large scale.

3. Design an experiment to demonstrate that cells of cellular slime molds are attracted to cyclic AMP. Then, design a follow-up experiment to determine whether they are always attracted to cAMP or only when resources are scarce.

ONLINE RESOURCE

www.ravenbiology.com

Understand, Apply, and Synthesize—enhance your study with animations that bring concepts to life and practice tests to assess your understanding. Your instructor may also recommend the interactive eBook, individualized learning tools, and more.

Chapter 30

Seedless Plants

Chapter Contents

Introduction

Colonization of land by plants fundamentally altered the history of life on Earth. A terrestrial environment offers abundant CO_2 and solar radiation for photosynthesis. But for at least 500 million years, the lack of water and higher ultraviolet (UV) radiation on land confined green algal ancestors to an aquatic environment. Evolutionary innovations for reproduction, structural support, and prevention of water loss are key in the story of plant adaptation to land. The evolutionary shift on land to life cycles dominated by a diploid generation masks recessive mutations arising from higher UV exposure. As a result, larger numbers of alleles persist in the gene pool, creating greater genetic diversity. Long before seeds and flowers evolved, the seedless plants covered the Earth. In this chapter we consider the evolutionary innovations in seedless plants during the first 100 million years of terrestrial life.

30.1 Origin of Land Plants

Learning Outcomes

1. *Explain the relationship between the different algae clades and plants.*
2. *Describe the haplodiplontic life cycle.*
3. *Distinguish between a sporophyte and a gametophyte.*
4. *Identify two major environmental challenges for land plants and associated adaptations.*

Green algae and the land plants shared a common ancestor a little over 1 BYA and are collectively referred to as the green plants. DNA sequence data are consistent with the claim that a single individual gave rise to all green plants. The green plants are photoautotrophic, but not all photoautotrophs are plants. The definition of a green plant is broad, but it excludes the red and brown algae. All algae—red, brown, and green—shared a primary endosymbiotic event 1.5 BYA. But sharing an ancestral chloroplast lineage is not the same as being monophyletic, as discussed in chapter 29. Red and green algae last shared a common ancestor about 1.4 BYA. Brown algae became photosynthetic through endosymbiosis with a eukaryotic red alga that had itself already acquired a photosynthetic cyanobacterium, as described in the preceding chapter.

Plants are also not fungi, which are more closely related to metazoan animals (see chapters 32 and 33). Fungi, however, were essential to the colonization of land by plants, enhancing plants' nutrient uptake from the soil.

One of the most significant evolutionary events in the billion-year-old history of the green plants is the adaptation to terrestrial living. Thus this chapter and the next focus on the land plants.

Land plants evolved from freshwater algae

Some saltwater algae evolved to thrive in a freshwater environment. Just a single species of freshwater green algae gave rise to the entire terrestrial plant lineage, from mosses through the flowering plants (angiosperms). Given the incredibly harsh conditions of life on land, it is not surprising that all land plants share a single common ancestor. Exactly what this ancestral alga was is still a mystery, but close relatives, members of the charophytes, exist in freshwater lakes today.

The green algae split into two major clades: the chlorophytes, which never made it to land, and the charophytes, which are sister clade to all the land plants (figure 30.1). Together charophytes and land plants are referred to as streptophytes. Land plants, although diverse, have certain characteristics in common. Unlike the charophytes, land plants have multicellular haploid and diploid stages. Diploid embryos are also land plant innovations. Over time, the trend has been toward more embryo protection and a smaller haploid stage in the life cycle.

Land plants have adapted to terrestrial life

Unlike their freshwater ancestors, most land plants have only limited amounts of water available to them. As an adaptation to living on land, most plants are protected from desiccation—the tendency of organisms to lose water to the air—by a waxy surface material called the cuticle that is secreted onto their exposed surfaces. The cuticle is relatively impermeable, preventing water loss. This solution, however, limits the gas exchange essential for respiration and photosynthesis. Gas diffusion into and out of a plant occurs through tiny mouth-shaped openings called **stomata** (singular, *stoma*), which allow water to diffuse out at the same time. Chapter 37 describes how stomata can be closed at times to limit water loss.

Moving water within plants is a challenge that increases with plant size. Members of the land plants can be distinguished based on the presence or absence of **tracheids,** specialized cells that facilitate the transport of water and minerals (see chapter 37). Tracheophytes have specialized transport cells called tracheids and have evolved highly efficient transport systems: water-conducting xylem and food-conducting phloem strands of tissues in their stems, roots, and leaves. Some plants that grow in aquatic environments, including water lilies, have tracheids. Aquatic tracheophytes had terrestrial ancestors that adapted back to a watery environment.

Terrestrial plants are exposed to higher intensities of UV irradiation than aquatic algae, increasing the chance of mutation. Diploid genomes mask the effect of a single, deleterious allele. All land plants have both haploid and diploid generations, and the evolutionary shift toward a dominant diploid

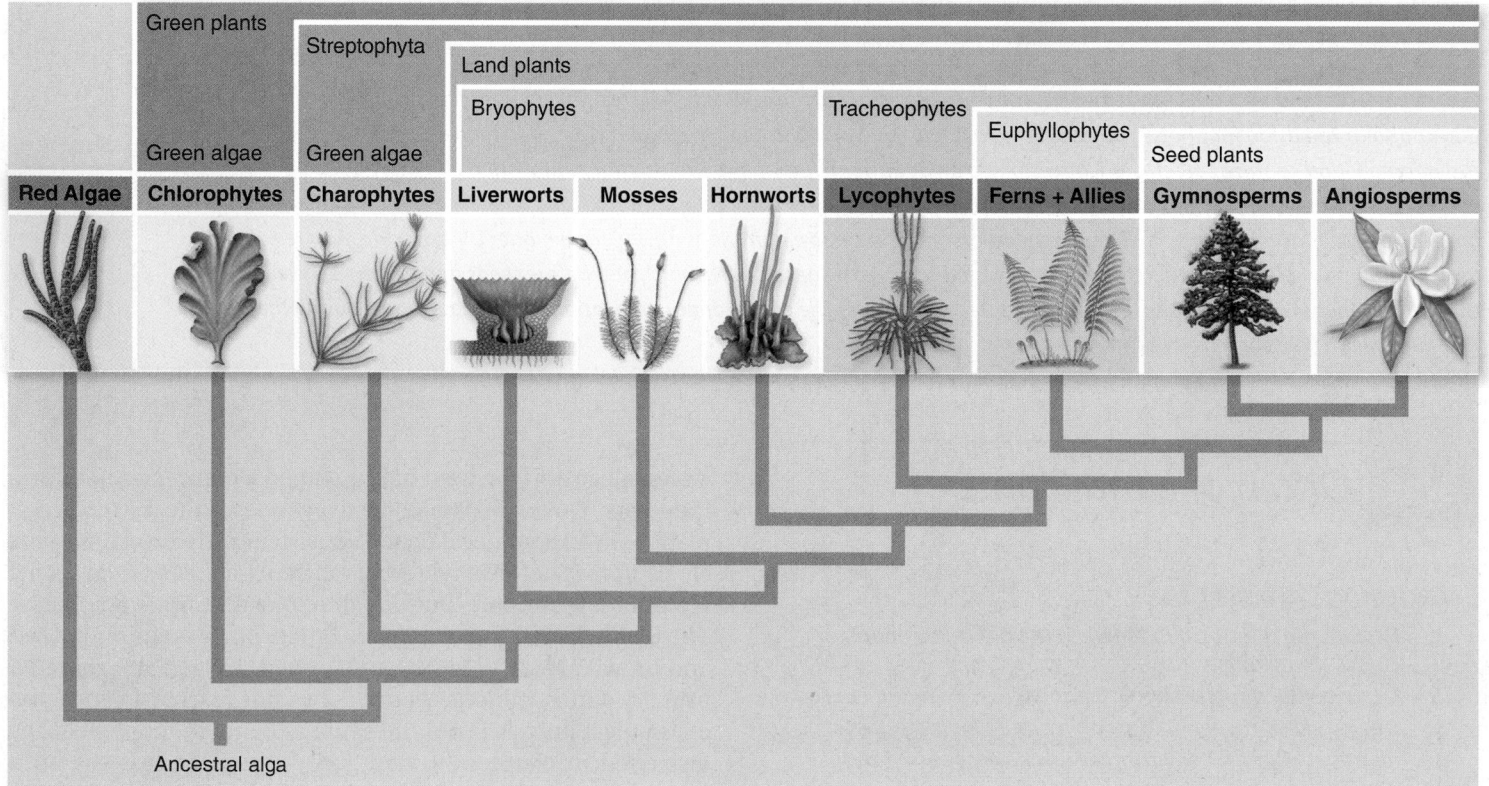

Figure 30.1 Green plant phylogeny.

generation allows for greater genetic variability to persist in terrestrial plants.

The haplodiplontic cycle produces alternation of generations

Humans have a **diplontic** life cycle, meaning that only the diploid stage is multicellular; by contrast, the land plant life cycle is **haplodiplontic,** having multicellular haploid and diploid stages. Most multicellular green plants have this haplodiplontic life cycle. Many multicellular green algae and all land plants have haplodiplontic life cycles and undergo mitosis after both gamete fusion and meiosis. The result is a multicellular haploid individual and a multicellular diploid individual—unlike in the human life cycle, in which gamete fusion directly follows meiosis. The basic haplodiplontic cycle is summarized in figure 30.2.

Many brown, red, and green algae are also haplodiplontic. Humans produce gametes via meiosis, but land plants actually produce gametes by *mitosis* in a multicellular, haploid individual. The diploid generation, or **sporophyte,** alternates with the haploid generation, or **gametophyte.** Sporophyte means "spore plant," and gametophyte means "gamete plant." These terms indicate the kinds of reproductive cells the respective generations produce.

The diploid sporophyte produces haploid spores (not gametes) by meiosis. Meiosis takes place in structures called sporangia, where diploid **spore mother cells (sporocytes)** undergo meiosis, each producing four haploid **spores.** Spores are the first cells of the gametophyte generation. Spores divide by mitosis, producing a multicellular, haploid gametophyte.

The haploid gametophyte is the source of gametes. When the gametes fuse, the zygote they form is diploid and is the first cell of the next sporophyte generation. The zygote grows into a diploid sporophyte by mitosis and produces sporangia in which meiosis ultimately occurs.

The relative sizes of haploid and diploid generations vary

All land plants are haplodiplontic; however, the haploid generation consumes a much larger portion of the life cycle in mosses and ferns than it does in the seed plants—the gymnosperms and angiosperms. In mosses, liverworts, and ferns, the gametophyte is photosynthetic and free living. When you look at mosses, what you see is largely gametophyte tissue; the sporophytes are usually smaller, brownish or yellowish structures attached to the tissues of the gametophyte. In other plants, the gametophyte is usually nutritionally dependent on the sporophyte. When you look at a gymnosperm or angiosperm, such as most trees, the largest, most visible portion is a sporophyte.

Although the sporophyte generation can get very large, the size of the gametophyte is limited in all plants. The gametophyte generation of mosses produces gametes at its tips. The egg is stationary, and sperm lands near the egg in a droplet of water. If the moss were the height of a sequoia, not only would vascular tissue be needed for conduction and support, but the sperm would have to swim up the tree! In contrast, the small gametophyte of the fern develops on the forest floor where gametes can meet. Tree ferns are especially abundant in Australia; the haploid spores that the sporophyte trees produce fall to the ground and develop into gametophytes.

Having completed an overview of plant life cycles, we next consider the major groups of seedless land plants. As we proceed, you will see a reduction of the gametophyte from group to group, a loss of multicellular **gametangia** (structures in which gametes are produced), and increasing specialization for life on land.

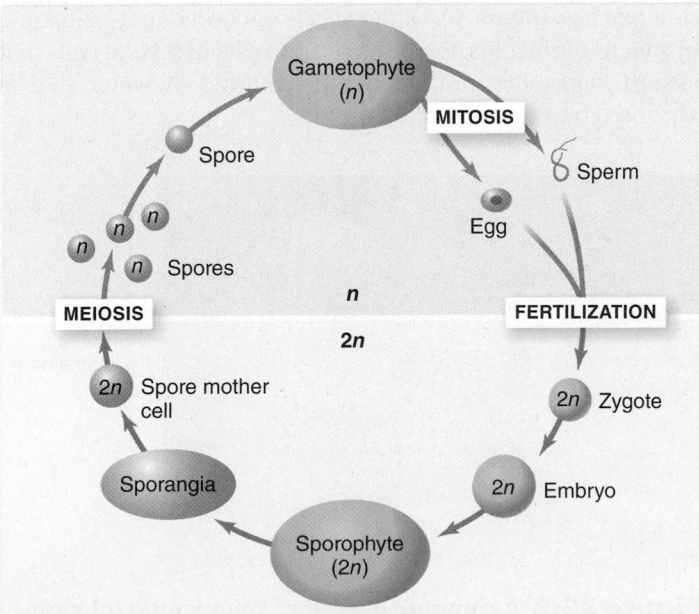

Figure 30.2 A generalized multicellular plant life cycle. Note that both haploid and diploid individuals can be multicellular. Also, spores are produced by meiosis, but gametes are produced by mitosis.

Learning Outcomes Review 30.1

All algae acquired chloroplasts necessary for photosynthesis, but green algae diverged from red algae after that event. A single freshwater green alga successfully invaded land; its descendants eventually developed reproductive strategies, conducting systems, stomata, and cuticles as adaptations. Green plants include all green algae and the land plants, whereas the streptophytes include only the land plants and their sister clade, the charophytes. Most plants have a haplodiplontic life cycle, a haploid form alternates with a diploid form in a single organism. Diploid sporophytes produce haploid spores by meiosis. Each spore can develop into a haploid gametophyte by mitosis; the gametophyte form produces haploid gametes, again by mitosis. When the gametes fuse, the diploid sporophyte is formed once more.

- How would you distinguish a small aquatic tracheophyte from a freshwater alga?
- What distinguishes gamete formation in plants from gamete formation in humans?

30.2 Bryophytes: Dominant Gametophyte Generation

Learning Outcome

1. Describe adaptations of bryophytes for terrestrial environments.

Land plants began diverging 450 MYA. Bryophytes are the closest living descendants of these first land plants. Plants in this group are also called nontracheophytes because they lack the derived transport cell called a *tracheid.*

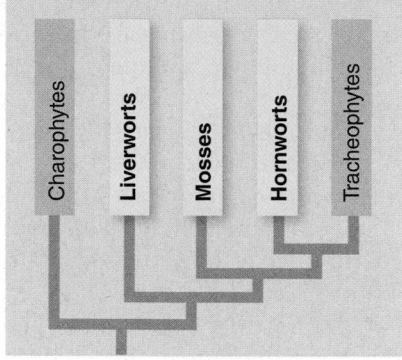

Fossil evidence and molecular systematics can be used to reconstruct early terrestrial plant life. Water and gas availability were limiting factors. These plants likely had little ability to regulate internal water levels and likely tolerated desiccation, traits found in most extant mosses, although some are aquatic.

Algae, including the Charales, lack roots. Fungi and early land plants cohabited, and the fungi formed close associations with the plants that enhanced water uptake. The tight symbiotic relationship between fungi and plants, called **mycorrhizal associations,** are also found in many existing bryophytes. More information on mycorrhizal fungi is found in chapter 32.

Bryophytes are unspecialized but successful in many environments

The approximately 24,700 species of bryophytes are simple but highly adapted to a diversity of terrestrial environments, even deserts. Most bryophytes are small; few exceed 7 cm in height. Bryophytes have conducting cells other than tracheids for water and nutrients. The tracheid is a derived trait that characterizes the tracheophytes, all land plants but the bryophytes. Bryophytes are sometimes called nonvascular plants, but *nontracheophyte* is a more accurate term because they do have conducting cells of different types.

Scientists now agree that bryophytes consist of three quite distinct clades of relatively unspecialized plants: liverworts, mosses, and hornworts. Their gametophytes are photosynthetic and are more conspicuous than the sporophytes. Sporophytes are attached to the gametophytes and depend on them nutritionally in varying degrees. Some of the sporophytes are completely enclosed within gametophyte tissue; others are

not and usually turn brownish or straw-colored at maturity. Like ferns and certain other vascular (tracheophyte) plants, bryophytes require water (such as rainwater) to reproduce sexually, tracing back to their aquatic origins. It is not surprising that they are especially common in moist places, both in the tropics and temperate regions.

Liverworts are an ancient phylum

The Old English word *wyrt* means "plant" or "herb." Some common liverworts (phylum Hepaticophyta) have flattened gametophytes with lobes resembling those of liver—hence the name "liverwort." Although the lobed liverworts are the best-known representatives of this phylum, they constitute only about 20% of the species (figure 30.3). The other 80% are leafy and superficially resemble mosses. The gametophytes are prostrate instead of erect, with single celled rhizoids that aid in absorption like roots but are not organs.

Some liverworts have air chambers containing upright, branching rows of photosynthetic cells, each chamber having a pore at the top to facilitate gas exchange. Unlike stomata, the pores are fixed open and cannot close.

Sexual reproduction in liverworts is similar to that in mosses. Lobed liverworts may form gametangia in umbrella-like structures. Asexual reproduction occurs when lens-shaped pieces of tissue that are released from the gametophyte grow to form new gametophytes.

Mosses have rhizoids and water-conducting tissue

Unlike other bryophytes, the gametophytes of mosses typically consist of small, leaflike structures (not true leaves, which contain vascular tissue) arranged spirally or alternately around a stemlike axis (figure 30.4); the axis is anchored to its substrate by means of rhizoids. Each rhizoid consists of several cells that absorb water, but not nearly the volume of water that is absorbed by a vascular plant root.

Figure 30.3 A common liverwort, *Marchantia* (phylum Hepaticophyta). The microscopic sporophytes are formed by fertilization within the tissues of the umbrella-shaped structures that arise from the surface of the flat, green, creeping gametophyte.

Figure 30.4 A hair-cup moss, *Polytrichum* (phylum Bryophyta). The leaflike structures belong to the gametophyte. Each of the yellowish brown stalks with a capsule, or sporangium, at its summit is a sporophyte.

Moss leaflike structures have little in common with leaves of vascular plants, except for the superficial appearance of the green, flattened blade and slightly thickened midrib that runs lengthwise down the middle. Only one cell layer thick (except at the midrib), they lack vascular strands and stomata, and all the cells are haploid. However, mosses do have stomata on the capsule portion of the sporophyte generation and because of that are the basal land group with stomata.

Water may rise up a strand of specialized cells in the center of a moss gametophyte axis. Some mosses also have specialized food-conducting cells surrounding those that conduct water.

Moss reproduction

Multicellular gametangia are formed at the tips of the leafy gametophytes (figure 30.5). Female gametangia (**archegonia**) may develop either on the same gametophyte as the male gametangia (**antheridia**) or on separate plants. A single egg is produced in the swollen lower part of an archegonium, whereas numerous sperm are produced in an antheridium.

When sperm are released from an antheridium, they swim with the aid of flagella through a film of dew or rainwater to the archegonia. One sperm (which is haploid) unites with an egg (also haploid), forming a diploid zygote. The zygote divides by mitosis and develops into the sporophyte, a slender, basal stalk with a swollen capsule, the *sporangium,* at its tip. As the sporophyte develops, its base is embedded in gametophyte tissue, its nutritional source.

The sporangium is often cylindrical or club-shaped. Spore mother cells within the sporangium undergo meiosis, each producing four haploid spores. In many mosses at

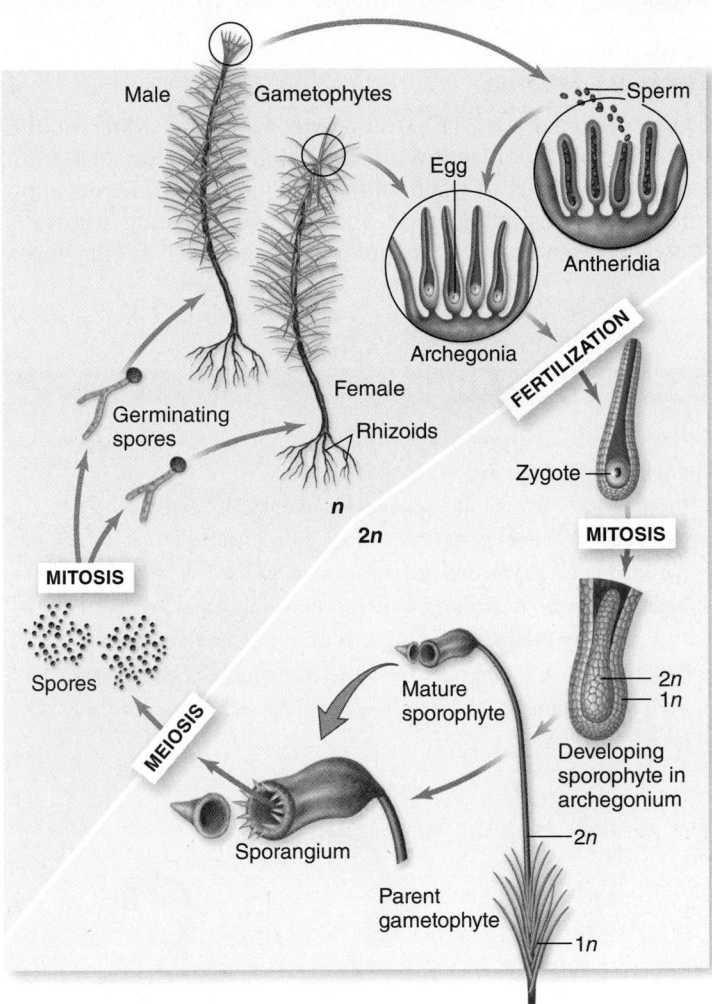

Figure 30.5 Life cycle of a typical moss. The majority of the life cycle of a moss is in the haploid state. The leafy gametophyte is photosynthetic, but the smaller sporophyte is not and is nutritionally dependent on the gametophyte. Water is required to carry sperm to the egg.

maturity, the top of the sporangium pops off, and the spores are released. A spore that lands in a suitable damp location may germinate and grow, using mitosis, into a threadlike structure, which branches to form rhizoids and "buds" that grow upright. Each bud develops into a new gametophyte plant consisting of a leafy axis.

Moss distribution

In the Arctic and the Antarctic, mosses are the most abundant plants. The greatest diversity of moss species, however, is found in the tropics. Many mosses are able to withstand prolonged periods of drought, although mosses are not common in deserts.

Most mosses are highly sensitive to air pollution and are rarely found in abundance in or near cities or other areas with high levels of air pollution. Some mosses, such as the peat mosses (*Sphagnum*), can absorb up to 25 times their weight in water and are valuable commercially as a soil conditioner or as a fuel when dry.

The moss genome

Moss plants can survive extreme water loss—an adaptive trait in the early colonization of land that has been lost from vegetative tissues of tracheophytes (figure 30.6). Desiccation tolerance and phylogenetic position were among the traits that led researchers to sequence the genome of the moss

Figure 30.7 Hornworts (phylum Anthocerotophyta). Hornwort sporophytes are seen in this photo. Unlike the sporophytes of other bryophytes, most hornwort sporophytes are photosynthetic.

Physcomitrella patens as being the first land plant that is not a tracheophyte. Although the moss genome is a single genome bracketed by *Chlamydomonas* and the tracheophytes, many evolutionary hints are hidden within it. Evidence indicates the loss of genes associated with a watery life, including flagellar arms, were lost in the last common ancestor of the land plants. Genes associated with tolerance of terrestrial stresses, including temperature and water availability, are absent in *Chlamydomonas* and present in moss. For example, the plant hormone abscisic acid (ABA) is important in stress responses in moss and other land plants and genes needed for ABA signaling are not found in algae.

Hornworts developed stomata

The origin of hornworts (phylum Anthocerotophyta) is a puzzle. They are most likely among the earliest land plants, yet the earliest hornwort fossil spores date from the Cretaceous period (65–145 MYA), when angiosperms were emerging.

The small hornwort sporophytes resemble tiny green broom handles or horns, rising from filmy gametophytes usually less than 2 cm in diameter (figure 30.7). The sporophyte base is embedded in gametophyte tissue, from which it derives some of its nutrition. However, the sporophyte has stomata to regulate gas exchange, is photosynthetic, and provides much of the energy needed for growth and reproduction. Hornwort cells usually have a single large chloroplast.

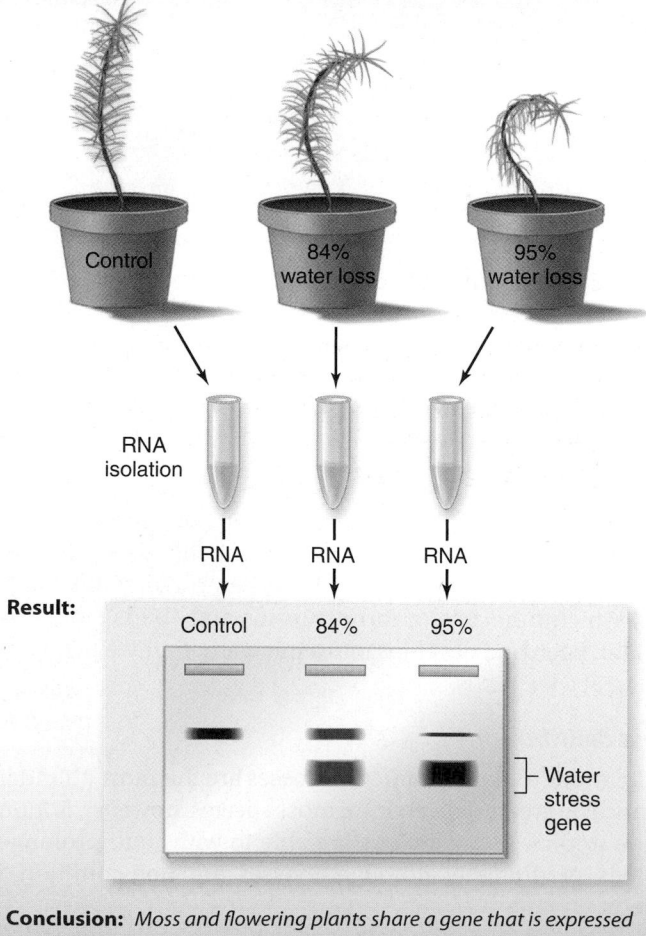

SCIENTIFIC THINKING

Hypothesis: *Desiccation tolerance genes in moss and flowering plants first appeared in a common ancestor.*

Prediction: *The late embryogenesis abundant (LEA) protein gene, a desiccation tolerance gene, from flowering plants will be expressed in moss plants when they experience severe water loss.*

Test: *Isolate RNA from moss plants that have not been water stressed (control), have been dehydrated to 84% water loss, and have been dehydrated to 95% water loss. Load a gel with equal amounts of RNA from each treatment. Probe the gel with a cDNA sequence for the LEA gene that is labeled.*

Control   84% water loss   95% water loss

RNA isolation

RNA   RNA   RNA

Result:

Control   84%   95%

Water stress gene

Conclusion: *Moss and flowering plants share a gene that is expressed under water stress conditions.*

Further Experiments: *Are other stress genes shared by bryophytes and flowering plants? Repeat the experiment with other stress-induced genes.*

Figure 30.6 Moss and flowering plants share desiccation tolerance genes.

Learning Outcome Review 30.2

The bryophytes exhibit adaptations to terrestrial life. Moss adaptations include rhizoids to anchor the moss body and to absorb water, and water-conducting tissues. Mosses are found in a variety of habitats, and some can survive droughts. Hornworts developed stomata that can open and close to regulate gas exchange.

■ *What might account for the abundance of mosses in the Arctic and Antarctic?*

30.3 Tracheophyte Plants: Roots, Stems, and Leaves

Learning Outcomes

1. Explain the evolutionary significance of tracheids.
2. Analyze the claim that roots, stems, and leaves are evolutionary innovations unique to tracheophytes.

Tracheophytes, also known as vascular plants, first appeared about 410 MYA. The first tracheophytes with a relatively complete record belonged to the phylum Rhyniophyta. We are not certain what the earliest of these vascular plants looked like, but fossils of *Cooksonia* provide some insight into their characteristics (figure 30.8).

Cooksonia, the first known vascular land plant, appeared in the late Silurian period about 420 MYA, but is now extinct. It was successful partly because it encountered little competition as it spread out over vast tracts of land. The plants were only a few centimeters tall and had no roots or leaves. They consisted of little more than a branching axis, the branches forking evenly and expanding slightly toward the tips. They were **homosporous** (producing only one type of spore). Sporangia formed at branch tips. Other ancient vascular plants that followed evolved more complex arrangements of sporangia.

Vascular tissue allows for distribution of nutrients

Cooksonia and the other early plants that followed it became successful colonizers of the land by developing efficient water- and food-conducting systems called *vascular tissues.* These tissues consist of strands of specialized cylindrical or elongated cells that form a network throughout a plant, extending from near the tips of the roots, through the stems, and into true leaves, defined by the presence of vascular tissue in the blade. One type of vascular tissue, **xylem,** conducts water and dissolved minerals upward from the roots; another type of tissue, **phloem,** conducts sucrose and hormones throughout the plant. Tracheids are the cells in the early vascular plants that conducted water in xylem tissue. Vascular tissue enables enhanced height and size in the tracheophytes. It develops in the sporophyte, but (with a few exceptions) not in the gametophyte. (Vascular tissue structure is discussed more fully in chapter 37.) A cuticle and stomata are also characteristic of vascular plants.

 Inquiry question Explain why tracheophytes may have had a selective advantage during the evolution of land plants.

Tracheophytes are grouped in three clades

Three clades of vascular plants exist today: (1) lycophytes (club mosses), (2) pterophytes (ferns and their relatives), and

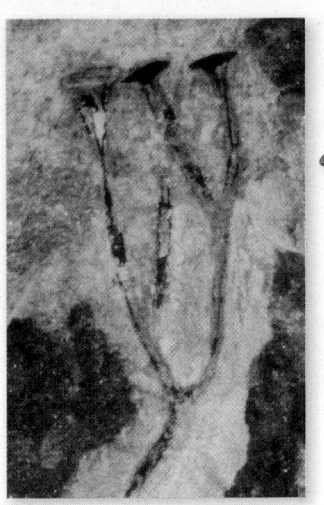
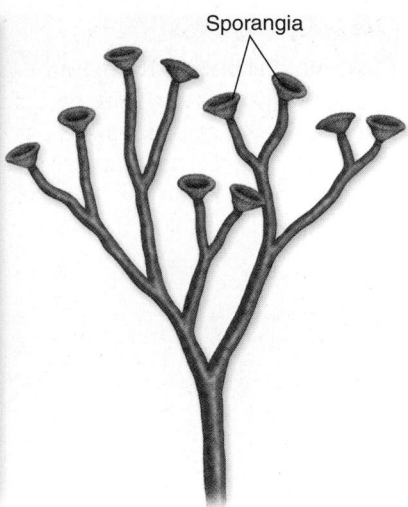

Sporangia

Figure 30.8 *Cooksonia,* **the first known vascular land plant.** This fossil represents a plant that lived some 420 MYA. *Cooksonia* belongs to phylum Rhyniophyta, consisting entirely of extinct plants. Its upright, branched stems, which were no more than a few centimeters tall, terminated in sporangia, as seen here. It probably lived in moist environments such as mudflats, had a resistant cuticle, and produced spores typical of vascular plants.

(3) seed plants. The first two clades are explored in this chapter (table 30.1). Advances in molecular systematics have changed the way we view the evolutionary history of vascular plants. Whisk ferns and horsetails were long believed to be distinct phyla that were transitional between bryophytes and vascular plants. Phylogenetic evidence now shows they are the closest living relatives to ferns, and they are grouped as pterophytes.

Tracheophytes dominate terrestrial habitats everywhere, except for the highest mountains and the tundra. The haplo-diplontic life cycle persists, but the gametophyte has been reduced in size relative to the sporophyte during the evolution of tracheophytes. A similar reduction in multicellular gametangia has occurred as well.

Stems evolved prior to roots

Fossils of early vascular plants reveal stems, but no roots or leaves. The earliest vascular plants, including *Cooksonia,* had transport cells in their stems, but the lack of roots limited the size of these plants.

Roots provide structural support and transport capability

True roots are found only in the tracheophytes. Other, somewhat similar structures enhance either transport or support in nontracheophytes, but only roots have a dual function: providing both transport and support. Lycophytes diverged from other tracheophytes before roots appeared, based on fossil evidence. It appears that roots evolved at least two separate times.

TABLE 30.1

The Phyla of Extant Seedless Vascular Plants

Phylum	Examples		Key Characteristics	Approximate Number of Living Species
Lycophyta	Club mosses		Homosporous or heterosporous. Sperm motile. External water necessary for fertilization. About 12–13 genera.	1,150
Pterophyta	Ferns		Primarily homosporous (a few heterosporous). Sperm motile. External water necessary for fertilization. Leaves uncoil as they mature. Sporophytes and virtually all gametophytes are photosynthetic. About 365 genera.	11,000
	Horsetails		Homosporous. Sperm motile. External water necessary for fertilization. Stems ribbed, jointed, either photosynthetic or nonphotosynthetic. Leaves scalelike, in whorls; nonphotosynthetic at maturity. One genus.	15
	Whisk ferns		Homosporous. Sperm motile. External water necessary for fertilization. No differentiation between root and shoot. No leaves; one of the two genera has scalelike extensions and the other leaflike appendages.	6

Leaves evolved more than once

Leaves increase surface area of the sporophyte, enhancing photosynthetic capacity. Lycophytes have single vascular strands supporting relatively small leaves called lycophylls. True leaves, called euphylls, are found only in ferns and seed plants, having distinct origins from lycophylls (figure 30.9). Lycophylls may have resulted from vascular tissue penetrating small, leafy protuberances on stems. Euphylls most likely arose from branching stems that became webbed with leaf tissue.

About 40 million years separates the appearance of vascular tissue and the wide euphyll leaf—a curiously long time. The current hypothesis is that a 90% drop in atmospheric CO_2 360 MYA allowed for the increase in leaf size because of an increase in the number of stomata on a leaf. Large, horizontal leaves capture 200% more radiation than thin, axial leaves. Although beneficial for photosynthesis, larger leaves correspondingly increase leaf temperature, which can be lethal. Stomatal openings in the leaf enhance the movement of water out of the leaf, thereby cooling it. The density of stomata on leaf surfaces correlates with CO_2 concentration, as the stomatal openings are essential for gas exchange. As the atmospheric CO_2 levels dropped, plants could not obtain sufficient CO_2 for photosynthesis. In the low-CO_2 atmosphere, natural selection favored plants with higher stomatal densities. Higher stomatal densities favored larger leaves with a photosynthetic advantage that did not overheat. Leaves up to 120 mm wide and 160 mm long have been identified in the fossil record from that time period.

 Data analysis Compare the amount of time that passed between the early diversification of land plants, the origins of tracheophytes, and the origins of leaves.

Seeds are another innovation in some tracheophyte phyla

Seeds are highly resistant structures well suited to protecting a plant embryo from drought and to some extent from predators. In addition, almost all seeds contain a supply of food for the young plant. Lycophytes and pterophytes do not have seeds.

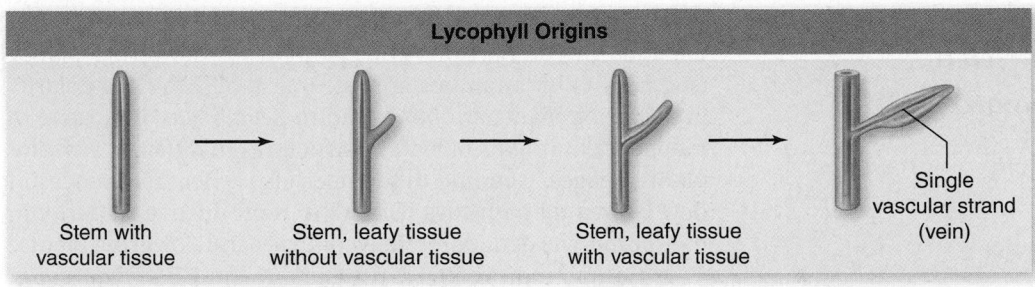

Figure 30.9 Evolution of leaves.

Lycophyll Origins

Stem with vascular tissue → Stem, leafy tissue without vascular tissue → Stem, leafy tissue with vascular tissue → Single vascular strand (vein)

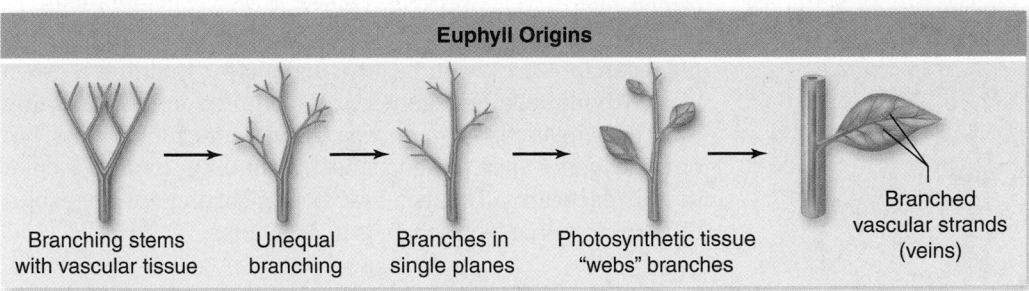

Euphyll Origins

Branching stems with vascular tissue → Unequal branching → Branches in single planes → Photosynthetic tissue "webs" branches → Branched vascular strands (veins)

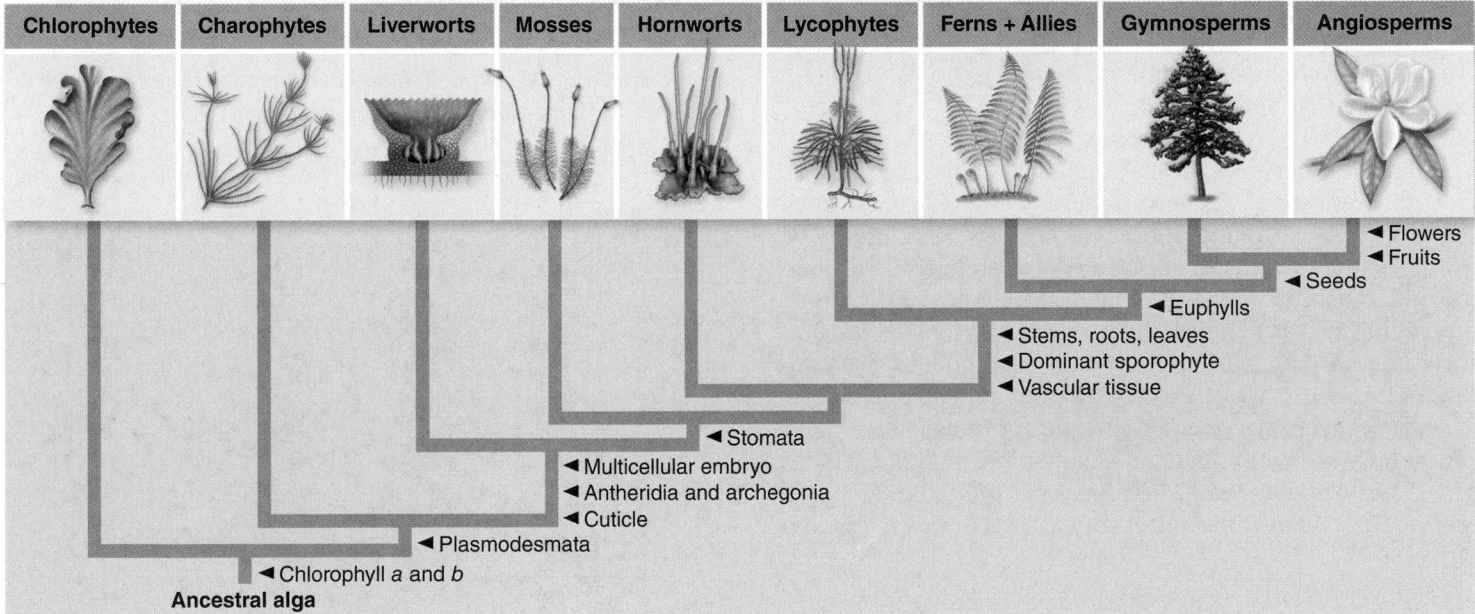

Figure 30.10 Land plant innovations.

Fruits in the flowering plants (angiosperms) add a layer of protection to seeds and have adaptations that assist in seed dispersal, expanding the potential range of the species. Flowers allow plants to secure the benefits of wide outcrossing in promoting genetic diversity, as discussed in chapter 31. Before moving on to the specifics of lycophytes and pterophytes, review the evolutionary history of terrestrial innovations in the land plants illustrated in figure 30.10.

Learning Outcomes Review 30.3

Most tracheophytes have well-developed vascular tissues, including tracheids, that enable efficient delivery of water and nutrients throughout the organism. They also exhibit specialized roots, stems, leaves, cuticles, and stomata. Many produce seeds, which protect and nourish embryos.

■ *Why would vascular tissue be prevalent in the sporophyte, but not the gametophyte, generation?*

30.4 Lycophytes: Dominant Sporophyte Generation and Vascular Tissue

The earliest vascular plants lacked seeds. Members of four phyla of living vascular plants also lack seeds, as do at least three other phyla known only from fossils. As we explore the adaptations of the vascular plants, we focus on both reproductive strategies and the advantages of increasingly complex transport systems.

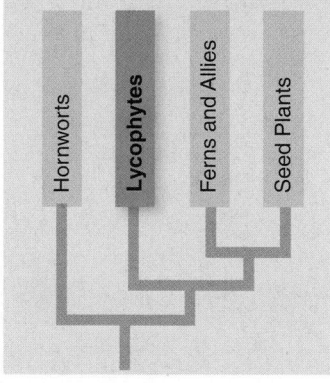

The lycophytes (club mosses) are relic species of an ancient past when vascular plants first evolved (figure 30.11). They are the sister group to all vascular plants. Several genera of club mosses, some of them treelike, became extinct about 270 MYA. Today, club mosses are worldwide in distribution but are most abundant in the tropics and moist temperate regions.

Members of the 12 to 13 genera and about 1150 living species of club mosses superficially resemble true mosses, but once their internal vascular structure and reproductive processes became known, it was clear that they are unrelated to mosses. The sporophyte stage is the dominant (obvious) stage; sporophytes have leafy stems that are seldom more than 30 cm long.

The lycophyte *Selaginella moellendorffii* is the first seedless vascular plant with a fully sequenced genome. A few clues to the evolution of vascular plants, hidden in the genome, emerged in comparisons with genomes of flowering plants. Genes that play an important role in establishing leaf polarity in flowering plants are not found in *Selaginella,* indicative of independent origins of leaflike structures in different vascular plant lineages. Genome differences also reflect differences in developmental pathways leading to reproductive maturity in the sporophyte generation in lycophytes and flowering plants.

Comparing predicted proteins in the *Chlamydomonas* (green alga), *Physcomitrella* (moss), and *Selaginella* with 15 angiosperms revealed 3814 gene families that all the green plants share—the essential instructions for building a green plant. About 3000 new genes were acquired in the transition from the single-celled green alga to the multicellular moss, but only 516 genes were added in the transition from nonvascular to vascular plants. This is a first step in sorting out the evolutionary steps that led to the vascular plants.

30.5 Pterophytes: Ferns and Their Relatives

The phylogenetic relationships among ferns and their near relations is intriguing. A common ancestor gave rise to two clades: One clade diverged to produce a line of ferns and horsetails; the other diverged to yield another line of ferns and whisk ferns— ancient-looking plants.

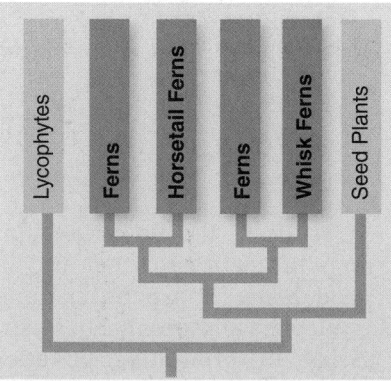

Whisk ferns and horsetails are close relatives of ferns. Like lycophytes and bryophytes, they all form antheridia and archegonia. Free water is required for the process of fertilization, during which the sperm, which have flagella, swim to and unite with the eggs. In contrast, most seed plants have nonflagellated sperm.

Figure 30.11
A club moss. *Selaginella moellendorffii's* sporophyte generation grows on moist forest floors.

Whisk ferns lost their roots and leaves secondarily

In whisk ferns, which occur in the tropics and subtropics, the sporophytic generation consists merely of evenly forking green stems without roots (figure 30.12). The two or three species of the genus *Psilotum* do, however, have tiny, green, spirally arranged flaps of tissue lacking veins and stomata. Another genus, *Tmesipteris,* has more leaflike appendages. Currently, systematists believe that whisk ferns lost leaves and roots when they diverged from others in the fern lineage.

Given the simple structure of whisk ferns, it was particularly surprising to discover that they are monophyletic with ferns. The gametophytes of whisk ferns are essentially colorless and are less than 2 mm in diameter, but they can be up to 18 mm long. They form symbiotic associations with fungi, which furnish their nutrients. Some develop elements of vascular tissue and have the distinction of being the only gametophytes known to do so.

Horsetails have jointed stems with brushlike leaves

The 15 living species of horsetails are all homosporous. They constitute a single genus, *Equisetum.* Fossil forms of *Equisetum* extend back 300 million years to an era when some of their relatives were treelike. Today, they are widely scattered around the world, mostly in damp places. Some that grow among the coastal redwoods of California may reach a height of 3 m, but most are less than a meter tall (figure 30.13).

Horsetail sporophytes consist of ribbed, jointed, photosynthetic stems that arise from branching underground *rhizomes* with roots at their nodes. A whorl of nonphotosynthetic, scalelike leaves emerges at each node. The hollow stems have silica deposits in the epidermal cells of the ribs, and the interior parts of the stems have two sets of vertical, tubular canals. The larger outer canals, which alternate with the ribs, contain air, and the smaller inner canals opposite the ribs contain water. Horsetails are also called scouring rushes because pioneers of the American West used them to scrub pans.

Figure 30.13 A horsetail, *Equisetum telmateia.* This species forms two kinds of erect stems; one is green and photosynthetic, and the other, which terminates in a spore-producing "cone," is mostly light brown.

Ferns have fronds that bear sori

Ferns are the most abundant group of seedless vascular plants, with about 11,000 living species. Recent research indicates that they may be the closest relatives to the seed plants.

The fossil record indicates that ferns originated during the Devonian period about 350 MYA and became abundant and varied in form during the next 50 million years. Their apparent ancestors were established on land as much as 375 MYA. Rainforests and swamps of lycopsid and fern trees growing in the Eastern United States and Europe over 300 MYA formed the coal currently being mined. Today, ferns flourish in a wide range of habitats throughout the world; however, about 75% of the species occur in the tropics.

The conspicuous sporophytes may be less than a centimeter in diameter (as in small aquatic ferns such as *Azolla*), or more than 24 m tall, with leaves up to 5 m or longer in the tree ferns (figure 30.14). The sporophytes and the much smaller gametophytes, which rarely reach 6 mm in diameter, are both photosynthetic.

Figure 30.12 **A whisk fern.** Whisk ferns have no roots or leaves. The green, photosynthetic stems have yellow sporangia attached.

Figure 30.14 **A tree fern (phylum Pterophyta) in the forests of Malaysia.** The ferns are by far the largest group of seedless vascular plants.

The fern life cycle (figure 30.15) differs from that of a moss primarily in the much greater development, independence, and dominance of the fern's sporophyte. The fern sporophyte is structurally more complex than the moss sporophyte, having vascular tissue and well-differentiated roots, stems, and leaves. The gametophyte, however, lacks the vascular tissue found in the sporophyte.

Fern morphology

Fern sporophytes, like horsetails, have rhizomes. Leaves, referred to as *fronds,* usually develop at the tip of the rhizome as tightly rolled-up coils ("fiddleheads") that unroll and expand (figure 30.16). Fiddleheads are considered a delicacy in several cuisines, but some species contain secondary compounds linked to stomach cancer.

Many fronds are highly dissected and feathery, making the ferns that produce them prized as ornamental garden plants. Some ferns, such as *Marsilea,* have fronds that resemble a four-leaf clover, but *Marsilea* fronds still begin as coiled fiddleheads. Other ferns produce a mixture of photosynthetic fronds and nonphotosynthetic reproductive fronds that tend to be brownish in color.

Fern reproduction

Ferns produce distinctive sporangia, usually in clusters called **sori** (singular, *sorus*), typically on the underside of the fronds. Sori are often protected during their development by a transparent, umbrella-like covering. (At first glance, one might mistake the sori for an infection on the plant.) Diploid spore mother cells in each sporangium undergo meiosis, producing haploid spores.

At maturity, the spores are catapulted from the sporangium by a snapping action, and those that land in suitable damp locations may germinate, producing gametophytes that are often heart-shaped, are only one cell layer thick (except in the center), and have rhizoids that anchor them to their substrate. These rhizoids are not true roots because they lack vascular tissue, but they do aid in transporting water and nutrients from the soil. Flask-shaped archegonia and globular antheridia are produced on either the same or a different gametophyte.

Figure 30.15
Life cycle of a typical fern. Both the gametophyte and sporophyte are photosynthetic and can live independently. Water is necessary for fertilization. Sperm are released on the underside of the gametophyte and swim in moist soil to neighboring gametophytes. Spores are dispersed by wind.

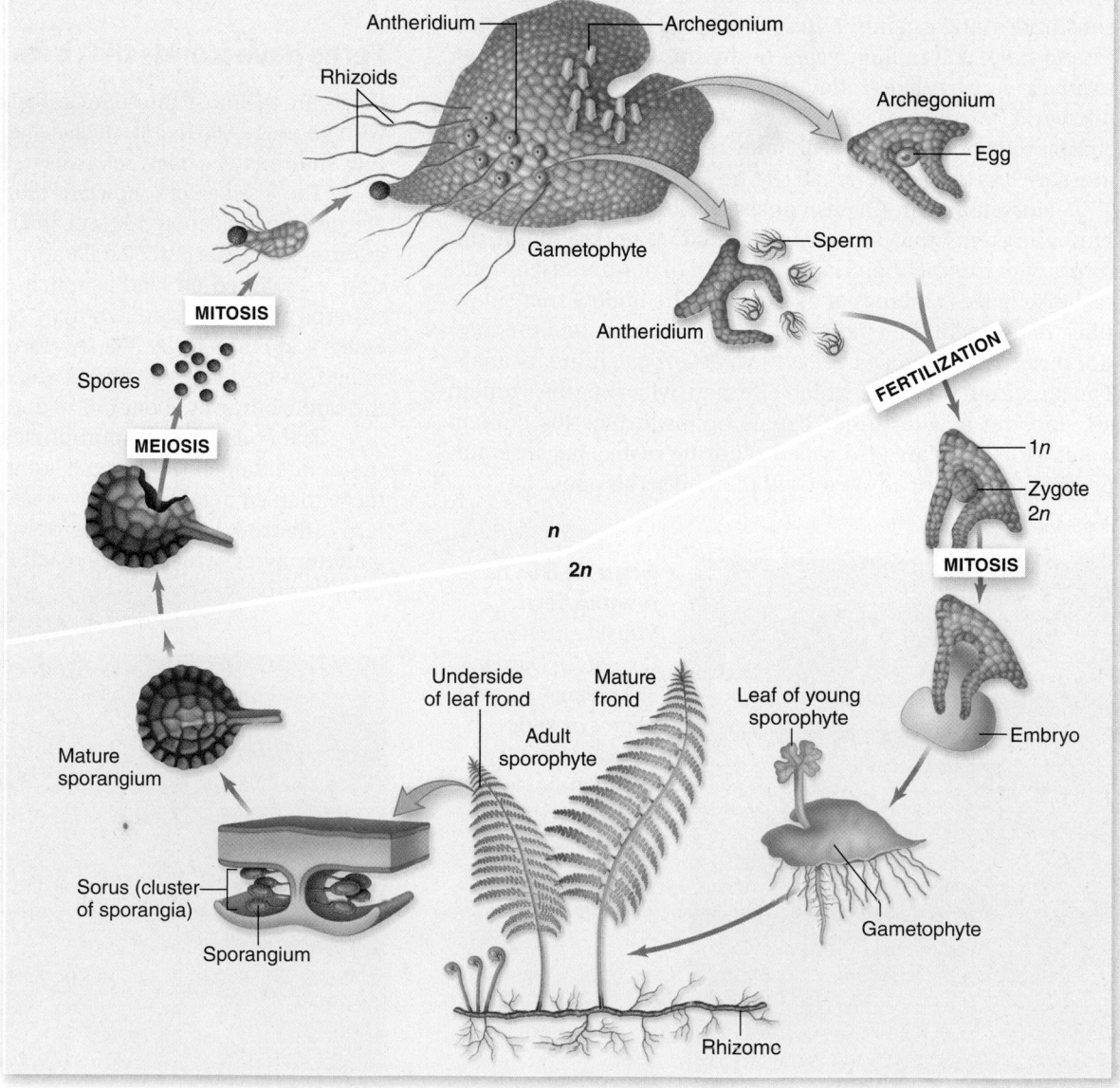

Tightly Coiled Fern | Uncoiling Fern

Figure 30.16 Fern "fiddlehead." Fronds develop in a coil and slowly unfold on ferns, including the tree fern fronds in these photos.

The multicellular archegonia provide some protection for the developing embryo.

The sperm formed in the antheridia have flagella, with which they swim toward the archegonia when water is present, often in response to a chemical signal secreted by the archegonia. One sperm unites with the single egg toward the base of an archegonium, forming a zygote. The zygote then develops into a new sporophyte, completing the life cycle (see figure 30.15).

The developing fern embryo has substantially more protection from the environment than a charophyte zygote, but it cannot enter a dormant phase to survive a harsh winter the way a seed plant embryo can. Although extant ferns do not produce seeds, seed fern fossils have been found that date back 365 million years. The seed ferns are not actually pterophytes, but gymnosperms. The evolution of seed plants is the topic of chapter 31.

Learning Outcomes Review 30.5

Ferns and their relatives have a large and conspicuous sporophyte with vascular tissue. Many have well-differentiated roots, stems, and leaves (fronds). The gametophyte generation is small and lacks vascular tissue.

■ *What would be the advantage of silica deposits in stems, as is found in horsetails?*

Chapter Review

30.1 Origin of Land Plants

Land plants evolved from freshwater algae.

Green algae and the land plants are called green plants.

All green plants arose from a single freshwater green algal species (figure 30.1). The charophytes are the sister clade of the land plants and together these groups are called the streptophytes.

Land plants have adapted to terrestrial life.

Land plants have two major characteristics: protected embryos and multicellular haploid and diploid phases. A waxy cuticle, stomata, and specialized cells for transport of water and minerals enhance survival.

The haplodiplontic cycle produces alternation of generations.

Most multicellular green plants have haplodiplontic life cycles.

Plants have a haplodiplontic life cycle with multicellular diploid sporophytes and haploid gametophytes (figure 30.2).

The relative sizes of haploid and diploid generations vary.

As some plants became more complex, the sporophyte stage became the dominant phase.

30.2 Bryophytes: Dominant Gametophyte Generation

Bryophytes are unspecialized but successful in many environments.

Bryophytes consist of three distinct clades: liverworts, mosses, and hornworts. Bryophytes do not have true roots or tracheids, but do have conducting cells for movement of water and nutrients.

In liverworts and mosses, the nonphotosynthetic sporophyte is nutritionally dependent on the gametophyte.

Liverworts are an ancient phylum.

The gametophyte of some liverworts is flattened and has lobes that resemble those of the liver. They produce upright structures that contain the gametangia.

Mosses have rhizoids and water-conducting tissue.

Mosses exhibit alternation of generations and have widespread distribution. Many are able to withstand droughts.

Hornworts developed stomata.

Stomata in the sporophyte can open and close to regulate gas exchange. The sporophyte is also photosynthetic.

30.3 Tracheophyte Plants: Roots, Stems, and Leaves (table 30.1)

Vascular tissue allows for distribution of nutrients.

The evolution of tracheids allowed more efficient vascular systems to develop. This vascular tissue develops in the sporophyte.

Vascular plants have a much reduced gametophyte.

Tracheophytes are grouped in three clades.

The vascular plants found today exist in three clades: lycophytes, pterophytes, and seed plants (figure 30.10).

Stems evolved prior to roots.

Roots provide structural support and transport capability.

Leaves evolved more than once.

Lycophytes have small leaves called lycophylls that lack vascularization.

True leaves (euphylls) are found only in ferns and seed plants and have origins different from lycophylls.

Seeds are another innovation in some tracheophyte phyla.

Seeds are resistant structures that protect the embryo from desiccation and to some extent from predators.

30.4 Lycophytes: Dominant Sporophyte Generation and Vascular Tissue

Lycophyte ancestors were the earliest vascular plants and were among the first plants to have a dominant sporophyte generation.

30.5 Pterophytes: Ferns and Their Relatives

Ancestors of the pterophytes gave rise to two clades: one line of ferns and horsetails, and a second line of ferns and whisk ferns.

Pterophytes require water for fertilization and are seedless.

Whisk ferns lost their roots and leaves secondarily.

The sporophyte of a whisk fern consists of evenly forking green stems without roots.

Horsetails have jointed stems with brushlike leaves.

Scalelike leaves of horsetail sporophytes emerge in a whorl. The stems have silica deposits in epidermal cells of their ribs.

Ferns have fronds that bear sori.

The leaves of ferns, called fronds, develop as tightly rolled coils that unwind to expand. Sporangia called sori develop on the underside of the fronds. The gametophyte is often heart shaped and can live independently.

Review Questions

UNDERSTAND

1. Which of the following plant structures is not matched to its correct function?

   a. Stomata—allow gas transfer
   b. Tracheids—allow the movement of water and minerals
   c. Cuticle—prevents desiccation
   d. All of the choices are matched correctly.

2. Which of the following genera most likely directly gave rise to the land plants?

   a. *Volvox*   c. *Ulva*
   b. *Chlamydomonas*   d. *Chara*

3. Which of the following would not be found in a bryophyte?

   a. Mycorrhizal associations
   b. Rhizoids
   c. Tracheid cells
   d. Photosynthetic gametophytes

4. Which of the following statements is correct regarding the bryophytes?

   a. The bryophytes represent a monophyletic clade.
   b. The sporophyte stage of all bryophytes is photosynthetic.
   c. Archegonium and antheridium represent haploid structures that produce reproductive cells.
   d. Stomata are common to all bryophytes.

5. Evolutionary innovations that increase desiccation tolerance include

   a. waxy cuticles.
   b. abscisic acid–signaling pathways.
   c. rhizoids.
   d. All of the choices are correct.

6. Which of the following statements about the pterophytes is accurate?

   a. Horsetails and whisk ferns form a single clade.
   b. Ferns form a single clade.
   c. Whisk ferns have euphylls.
   d. All pterophytes have a dominant sporophyte generation.

APPLY

1. Compare what happens to a spore mother cell as it gives rise to a spore with what happens to a spore as it gives rise to a gametophyte.

   a. The spore mother cell and the spore both go through meiosis.
   b. The spore mother cell and the spore both go through mitosis.
   c. The spore mother cell goes through mitosis, and the spore goes through meiosis.
   d. The spore mother cell goes through meiosis, and the spore goes through mitosis.

2. How could a plant without roots obtain sufficient nutrients from the soil?

   a. It cannot; all land plants have roots.
   b. Mychorrizal fungi associate with the plant and assist with the transfer of nutrients.
   c. Charophytes associate with the plant and assist with the transfer of nutrients.
   d. It relies on its xylem in the absence of a root.

3. A major innovation of land plants is embryo protection. How is a moss embryo protected from desiccation?

   a. By the seed
   b. By the antheridium
   c. By the archegonium
   d. By the lycophyll

4. In comparing the *Selaginella* and *Physcomitrella* genomes, you would expect to find that

   a. both have genes needed for flagellar arms.
   b. *Selaginella* has 3000 new genes not found in *Physcomitrella*.
   c. *Selaginella*, but not *Physcomitrella*, has abscisic acid genes and other abiotic stress genes.
   d. some of the novel genes in *Selaginella* encode proteins necessary for tracheid development.

5. The following evolutionary trends are seen in the seedless land plants.

   a. Gametophytes became photosynthetic.
   b. A key innovation accompanying the appearance of pterophytes was the haplodiplontic life cycle.
   c. The gametophyte generation became the dominant generation.
   d. There is increased protection of the sporophyte generation in its early developmental stages.

6. Identify which of the following statements is true.

   a. Moss and lycophytes have leaves.
   b. Leaves of lycophytes and pterophytes have different origins.
   c. Leaves were a bryophyte innovation.
   d. Leaf polarity genes in *Selaginella* support the hypothesis that leaves evolved one time in all the land plants.

SYNTHESIZE

1. You have access to the sequenced genomes for moss and the lycophyte *Selaginella*. Your goal in analyzing the data is to write a ground-breaking paper that answers an important question about the evolution of land plants. What question would you try to answer?

2. Would you expect the sporophyte generation of a moss or a fern to have more mitotic divisions? Why?

3. Imagine hypothetical moss and fern trees, each 10 m tall. Which would face the greater barriers to sexual reproduction? Why?

ONLINE RESOURCE

www.ravenbiology.com

Understand, Apply, and Synthesize—enhance your study with animations that bring concepts to life and practice tests to assess your understanding. Your instructor may also recommend the interactive eBook, individualized learning tools, and more.

Chapter **31**

Seed Plants

Chapter Contents

Introduction

The history of the land plants is replete with evolutionary innovations allowing the ancestors of aquatic algae to colonize the harsh and varied terrestrial terrains. Early innovations made survival on land possible, later followed by an explosion of plant life that continues to change the land and atmosphere, and support terrestrial animal life. Seed-producing plants have come to dominate the terrestrial landscape over the last several hundred million years. Much of the remarkable success of seed plants can be attributed to the evolution of the seed, an innovation that protects and provides food for delicate embryos. Seeds allow embryos to "stop the clock" and germinate after a harsh winter or extremely dry season has passed. Fruits, a later innovation, enhanced the dispersal of embryos across a broader landscape. Within the seed plants, the flowering plants have coevolved intricate relationships with animals to enhance sexual reproduction, as well as dispersal of the next generation.

31.1 The Evolution of Seed Plants

Learning Outcomes

1. *List the evolutionary advantages of seeds.*
2. *Distinguish between pollen and sperm in seed plants.*

Numerous evolutionary solutions to terrestrial challenges have resulted in well over 300,000 species of seed plants dominating all terrestrial communities today. Collectively these seed plants affect almost every aspect of our lives, from improving environmental quality to providing pharmaceuticals, food, fuels, building materials, and clothing.

Seed plants, which have additional embryo protection, were the ancestors of gymnosperms and angiosperms. Seed plants appear to have evolved from spore-bearing plants known as progymnosperms. Progymnosperms shared several features with modern gymnosperms, including secondary vascular tissues (which allow for an increase in girth later in development). Some progymnosperms had leaves. Their reproduction was very simple, and it is not certain which particular group of progymnosperms gave rise to seed plants.

The rapid success and diversity of seed plants has long intrigued biologists, particularly the diversification of the flowering plants (angiosperms). Likely the diversity can be attributed to a confluence of features, with the evolution of the seed being one important event. Phylogenetic analysis of seedless and seed plants with sequenced genomes revealed vast numbers of gene duplications around 319 MYA, when the early seed plants (ancestral to gymnosperms and angiosperms) began to diversify from their seedless plant ancestor, and again 192 MYA when the earliest angiosperms were appearing. The gene duplications are at such a scale that the most logical explanation is that whole-genome duplications occurred at the origins of both seed plants and angiosperms. As discussed in chapter 24, whole-genome duplications can accelerate evolution with the spare genes that can tolerate a range of mutations.

The seed protects the embryo

From an evolutionary and ecological perspective, the seed represents an important advance. The embryo is protected by an extra layer or two of sporophyte tissue called the **integument,** creating the **ovule** (figure 31.1). Within the ovule, the megasporangium divides meiotically, producing a haploid megaspore. The megaspore produces the egg that combines with the sperm, resulting in the zygote. Seeds also contain a food supply for the developing embryo.

During development, the integuments harden to produce the seed coat. In addition to protecting the embryo from drought, the seed can be easily dispersed. Perhaps even more significantly, the presence of seeds introduces into the life cycle a dormant phase that allows the embryo to survive until environmental conditions are favorable for further growth.

Water is not required to transport the male gametophyte to the female gametophyte

Seed plants produce two kinds of gametophytes—male and female—each of which consists of just a few cells. The greatly reduced gametophyte follows the trend of increasing sporophyte dominance seen in the evolution of the seedless plants.

Pollen grains, multicellular male gametophytes, are conveyed to the egg in the female gametophyte by wind or by a pollinator. In some seed plants, the sperm moves toward the egg through a growing **pollen tube.** This eliminates the need for external water. In contrast to the seedless plants, the whole male gametophyte, rather than just the sperm, moves to the female gametophyte.

A female gametophyte forms within the protection of the integuments, collectively forming the ovule. In angiosperms, the ovules are completely enclosed within additional diploid sporophyte tissue. The ovule and the surrounding protective tissue are called the ovary. The ovary develops into the fruit.

Learning Outcomes Review 31.1

A common ancestor that had seeds gave rise to the gymnosperms and the angiosperms. Whole-genome duplication may have contributed to the rise and dominance of seed plants, including the angiosperms. Seeds protect the embryo, aid in dispersal, and can allow for an extended pause in the life cycle. Seed plants produce male and female gametophytes; the male gametophyte is a pollen grain, which is carried to the female gametophyte by wind or other means. The sperm is within the pollen grain.

■ *Why is water not essential for fertilization in seed plants?*

31.2 Gymnosperms: Plants with "Naked Seeds"

Learning Outcomes

1. *Describe the distinguishing features of a gymnosperm.*
2. *List the four groups of living gymnosperms.*

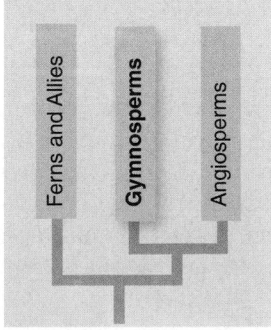

Seeds distinguish the gymnosperms from the pterophytes. There are four groups of living **gymnosperms:** coniferophytes, cycadophytes, gnetophytes, and ginkgophytes, all of which lack the flowers and fruits of angiosperms (table 31.1). In all of them, the ovule, which becomes a seed, rests exposed on a scale (a modified shoot or leaf) and is not completely enclosed by sporophyte tissues at the time of pollination. The name *gymnosperm* literally means "naked seed." Although the ovules are naked at the time of pollination, the seeds of gymnosperms are sometimes enclosed by other sporophyte tissues by the time they are mature.

Details of reproduction vary somewhat in gymnosperms, and their forms vary greatly. For example, cycads and *Ginkgo* have motile sperm, whereas conifers and gnetophytes have sperm with no flagella. All sperm are carried within a pollen

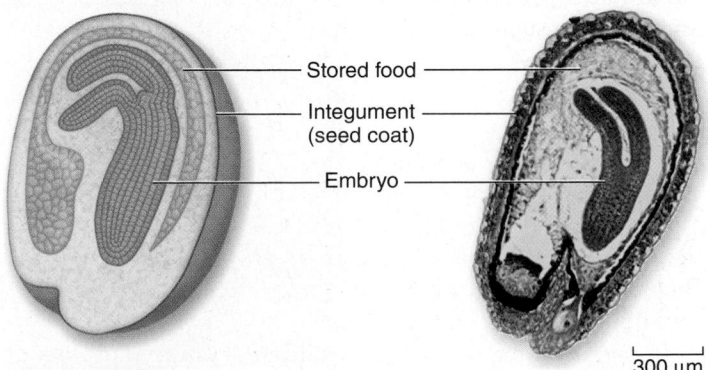

Figure 31.1 Cross-section of an ovule.

300 μm

- Stored food
- Integument (seed coat)
- Embryo

TABLE 31.1	The Five Phyla of Extant Seed Plants			
Phylum	**Examples**		**Key Characteristics**	**Approximate Number of Living Species**
Coniferophyta	Conifers (including pines, spruces, firs, yews, redwoods, and others)		Heterosporous seed plants. Sperm not motile; conducted to egg by a pollen tube. Leaves mostly needle-like or scalelike. Trees, shrubs. About 50 genera. Many produce seeds in cones.	601
Cycadophyta	Cycads		Heterosporous. Sperm flagellated and motile but confined within a pollen tube that grows to the vicinity of the egg. Palmlike plants with pinnate leaves. Secondary growth slow compared with that of the conifers. Ten genera. Seeds in cones.	206
Gnetophyta	Gnetophytes		Heterosporous. Sperm not motile; conducted to egg by a pollen tube. The only gymnosperms with vessels. Trees, shrubs, vines. Three very diverse genera (*Ephedra, Gnetum, Welwitschia*).	65
Ginkgophyta	*Ginkgo*		Heterosporous. Sperm flagellated and motile but conducted to the vicinity of the egg by a pollen tube. Deciduous tree with fan-shaped leaves that have evenly forking veins. Seeds resemble a small plum with fleshy, foul-smelling outer covering. One genus.	1
Anthophyta	Flowering plants (angiosperms)		Heterosporous. Sperm not motile; conducted to egg by a pollen tube. Seeds enclosed within a fruit. Leaves greatly varied in size and form. Herbs, vines, shrubs, trees. About 14,000 genera.	>300,000

tube. The female cones range from tiny, woody structures weighing less than 25 g and having a diameter of a few millimeters, to massive structures produced in some cycads, weighing more than 45 kg and growing to lengths of more than a meter.

Conifers are the largest gymnosperm phylum

The most familiar gymnosperms are **conifers** (phylum Coniferophyta), which include pines (figure 31.2), spruces, firs, cedars, hemlocks, yews, larches, cypresses, and others. The coastal redwood *(Sequoia sempervirens),* a conifer native to northwestern California and southwestern Oregon, is the tallest living vascular plant; it may attain a height of nearly 100 m (300 ft). Another conifer, the bristlecone pine *(Pinus longaeva)* of the White Mountains of California, is the oldest living tree; one specimen is 4900 years old.

Conifers are found in the colder temperate and sometimes drier regions of the world. Various species are sources of timber, biofuel feedstock, paper, resin, taxol (used to treat cancer), and other economically important products. Genomes of three conifers—loblolly pine, sugar pine, and Douglas fir—are currently being sequenced because of their economic importance. With genomes 10-fold larger than the human genome this is a substantial challenge.

Pines are an exemplary conifer genus

More than 100 species of pines exist today, all native to the northern hemisphere, although the range of one species does extend a

Figure 31.2 Conifers. Loblolly pine *(Pinus taeda)* is found on 58 million acres of land in the southeastern part of the United States. Worldwide 16 percent of the annual timber supply comes from this conifer.

little south of the equator. Pines and spruces, which belong to the same family, are members of the vast coniferous forests that lie between the arctic tundra and the temperate deciduous forests and prairies to their south. During the past century, pines have been extensively planted in the southern hemisphere.

Pine morphology

Pines have tough, needle-like leaves produced mostly in clusters of two to five. Among the conifers, only pines have clustered leaves. The leaves, which have a thick cuticle and recessed stomata, represent an evolutionary adaptation for retarding water loss. This strategy is important because many of the trees grow in areas where the topsoil is frozen for part of the year, making it difficult for the roots to obtain water.

The leaves and other parts of the sporophyte have canals into which surrounding cells secrete resin. The resin deters insect and fungal attacks. The resin of certain pines is harvested commercially for its volatile liquid portion, called *turpentine,* and for the solid *rosin,* which is used on bowed stringed instruments. The wood of pines lacks some of the more rigid cell types found in other trees, and it is considered a "soft" rather than a "hard" wood. The thick bark of pines is an adaptation for surviving fires

and subzero temperatures. Some cones actually depend on fire to open them, releasing seeds to reforest burned areas.

Reproductive structures

All seed plants produce two types of spores that give rise to two types of gametophytes (figure 31.3). The male gametophytes (pollen grains) of pines develop from microspores, which are produced in male cones that develop in clusters of 30 to 70, typically at the tips of the lower branches; there may be hundreds of such clusters on any single tree.

The male pine cones generally are 1 to 4 cm long and consist of small, papery scales arranged in a spiral or in whorls. A pair of microsporangia form as sacs within each scale. Numerous microspore mother cells in the microsporangia undergo meiosis, each becoming four microspores. The microspores develop into four-celled pollen grains with a pair of air sacs that give them added buoyancy when released into the air. A single cluster of male pine cones may produce more than a million pollen grains.

Female pine cones typically are produced on the upper branches of the same tree that produces male cones. Female cones are larger than male cones, and their scales become woody.

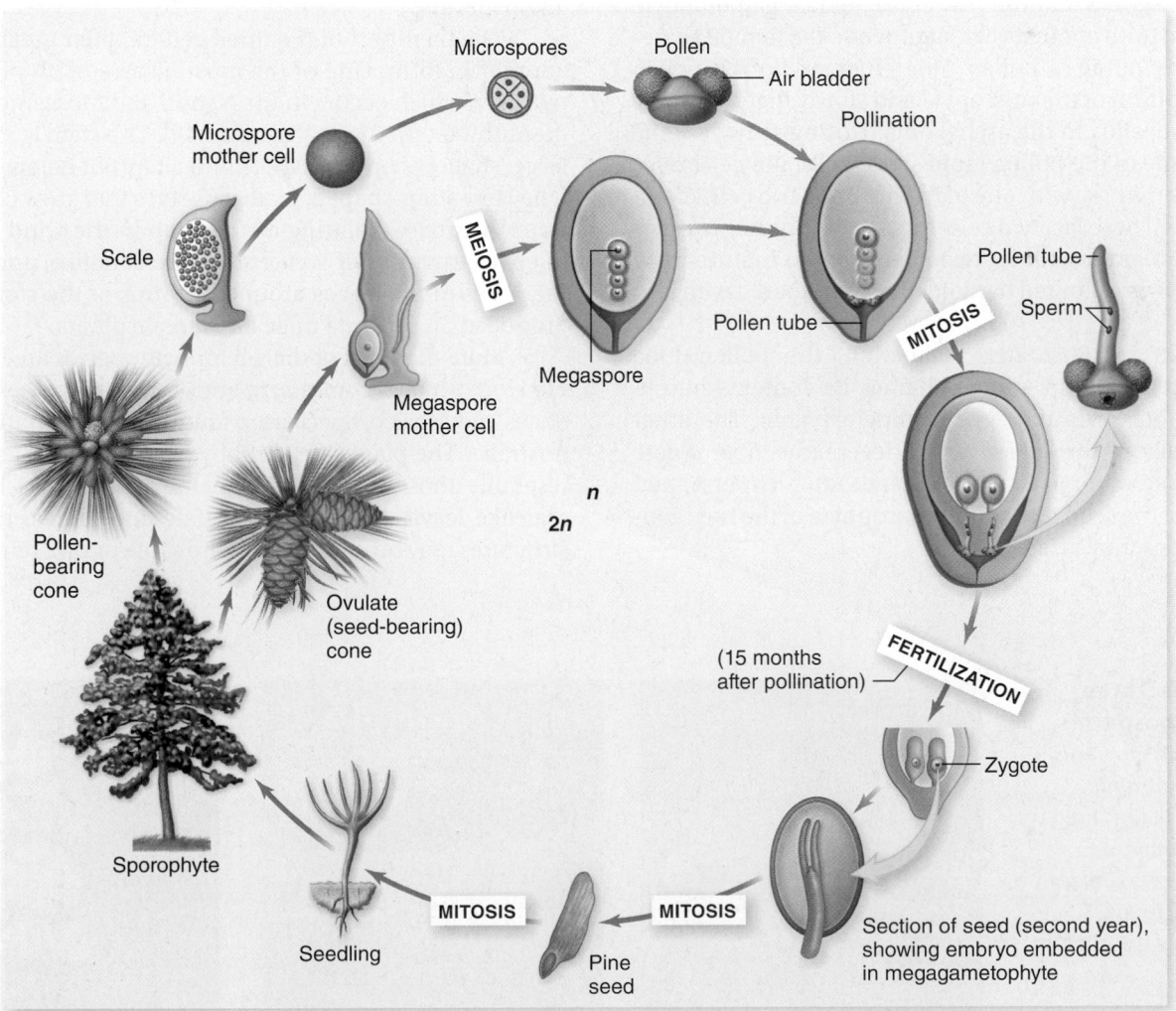

Figure 31.3 Life cycle of a typical pine. The male and female gametophytes are dramatically reduced in size in these plants. Wind generally disperses the male gametophyte (pollen), which produces sperm. Pollen tube growth delivers the sperm to the egg on the female cone. Additional protection for the embryo is provided by the integument, which develops into the seed coat.

Two ovules develop toward the base of each scale. Each ovule contains a megasporangium called the **nucellus.** The nucellus itself is completely surrounded by a thick layer of cells called the integument that has a small opening (the **micropyle**) toward one end. One of the layers of the integument later becomes the seed coat. A single megaspore mother cell within each megasporangium undergoes meiosis, becoming a row of four megaspores. Three of the megaspores break down, but the remaining one, over the better part of a year, slowly develops into a female gametophyte. The female gametophyte at maturity may consist of thousands of cells, with two to six archegonia formed at the micropylar end. Each archegonium contains an egg so large it can be seen without a microscope.

Fertilization and seed formation

Female cones usually take two or more seasons to mature. At first they may be reddish or purplish in color, but they soon turn green, and during the first spring, the scales spread apart. While the scales are open, pollen grains carried by the wind drift down between them, some catching in sticky fluid oozing out of the micropyle. The pollen grains within the sticky fluid are slowly drawn down through the micropyle to the top of the nucellus, and the scales close shortly thereafter.

The archegonia and the rest of the female gametophyte are not mature until about a year later. While the female gametophyte is developing, a pollen tube emerges from a pollen grain at the bottom of the micropyle and slowly digests its way through the nucellus to the archegonia. During growth of the pollen tube, one of the pollen grain's four cells, the *generative cell,* divides by mitosis, with one of the resulting two cells dividing once more. These last two cells function as sperm. The germinated pollen grain with its two sperm is the mature male gametophyte, a very limited haploid phase compared with fern gametophytes.

About 15 months after pollination, the pollen tube reaches an archegonium and discharges its contents into it. One sperm unites with the egg, forming a zygote. The other sperm and cells of the pollen grain degenerate. The zygote develops into an embryo within the seed. After dispersal and germination of the seed, the young sporophyte of the next generation develops into a tree.

Cycads resemble palms, but are not flowering plants

Cycads (phylum Cycadophyta) are slow-growing gymnosperms of tropical and subtropical regions. The sporophytes of most of the 100 known species resemble palm trees (figure 31.4*a*) with trunks that can attain heights of 15 m or more. Unlike palm trees, which are flowering plants, cycads produce cones and have a life cycle similar to that of pines.

The female cones, which develop upright among the leaf bases, are huge in some species and can weigh up to 45 kg. The sperm of cycads, although formed within a pollen tube, are released within the ovule to swim to an archegonium. These sperm are the largest sperm cells among all living organisms. Several species of cycads are facing extinction in the wild and soon may exist only in botanical gardens.

Gnetophytes have xylem vessels

There are three genera and about 65 living species of gnetophytes (phylum Gnetophyta). They are the only gymnosperms with vessels in their xylem. **Vessels** are a particularly efficient conducting cell type that is a common feature in angiosperms.

The members of the three genera differ greatly from one another in form. One of the most bizarre of all plants is *Welwitschia,* which occurs in the Namib and Mossamedes deserts of southwestern Africa (figure 31.4*b*). The stem is shaped like a large, shallow cup that tapers into a taproot below the surface. It has two strap-shaped, leathery leaves that grow continuously from their base, splitting as they flap in the wind. The reproductive structures of *Welwitschia* are conelike, appear toward the bases of the leaves around the rims of the stems, and are produced on separate male and female plants.

More than half of the gnetophyte species are in the genus *Ephedra,* which is common in arid regions of the western United States and Mexico. Species are found on every continent except Australia. The plants are shrubby, with stems that superficially resemble those of horsetails, being jointed and having tiny, scalelike leaves at each node. Male and female reproductive structures may be produced on the same or different plants.

Figure 31.4 Three phyla of gymnosperms.
a. A cycad, *Cycas circinalis.*
b. Welwitschia mirabilis represents one of the three genera of gnetophytes.
c. Maidenhair tree, *Ginkgo biloba,* the only living representative of the phylum Ginkgophyta.

a.

b.

c.

The drug ephedrine, widely used in the treatment of respiratory problems, was in the past extracted from the Chinese species of *Ephedra,* but it has now been largely replaced with synthetic preparations (pseudoephedrine). Because ephedrine found in herbal remedies for weight loss was linked to strokes and heart attacks, it was withdrawn from the market in April of 2004. Sales restrictions were placed on pseudoephedrine-containing products in 2006 because it can be used to manufacture the illegal drug methamphetamine.

The best-known species of the third genus, *Gnetum,* is a tropical tree, but most species are vinelike. All species have broad leaves similar to those of angiosperms. One *Gnetum* species is cultivated in Java for its tender shoots, which are cooked as a vegetable.

Only one species of the ginkgophytes remains extant

The fossil record indicates that members of the ginkgophytes (phylum Ginkgophyta) were once widely distributed, particularly in the northern hemisphere; today, only one living species, *Ginkgo biloba,* remains (figure 31.4c). This tree, which sheds its leaves in the fall, was first encountered by Europeans in cultivation in Japan and China; it apparently no longer exists in the wild.

Like the sperm of cycads, those of *Ginkgo* have flagella. The ginkgo is **dioecious**—that is, the male and female reproductive structures are produced on separate trees. The fleshy outer coverings of the seeds of female ginkgo plants exude the foul smell of rancid butter, caused by the presence of butyric and isobutyric acids, which are also found in butter and vomit. As a result, male plants vegetatively propagated from shoots are preferred for cultivation. Because of its beauty and resistance to air pollution, *Ginkgo* is commonly planted along city streets.

Learning Outcomes Review 31.2

Gymnosperms are mostly cone-bearing seed plants. In gymnosperms, the ovules are not completely enclosed by sporophyte tissue at pollination, and thus have "naked seeds." The four groups of gymnosperms are conifers, cycads, gnetophytes, and ginkgophytes.

■ *What adaptation do conifers exhibit to capture wind-borne pollen?*

31.3 Angiosperms: The Flowering Plants

Learning Outcomes

1. List the defining features of angiosperms.
2. Describe the roles of some animals in the angiosperm life cycle.
3. Explain double fertilization and its outcome.

Over 300,000 known species of flowering plants are called **angiosperms** because their ovules, unlike those of gymnosperms, are enclosed within diploid tissues at the time of pollination. The *carpel,* a modified leaf that encapsulates seeds, develops into the fruit, a unique angiosperm feature (figure 31.5). Although some

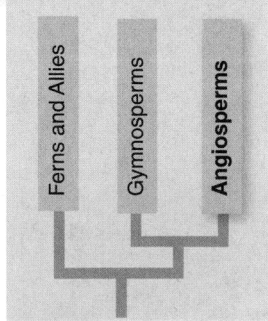

Figure 31.5
Evolution of fruit. Unlike gymnosperms, the ovule of angiosperms is surrounded by diploid tissue derived from leaves. This tissue is called the carpel and develops into the fruit. The leaf origins of carpels are still visible in the pea pod.

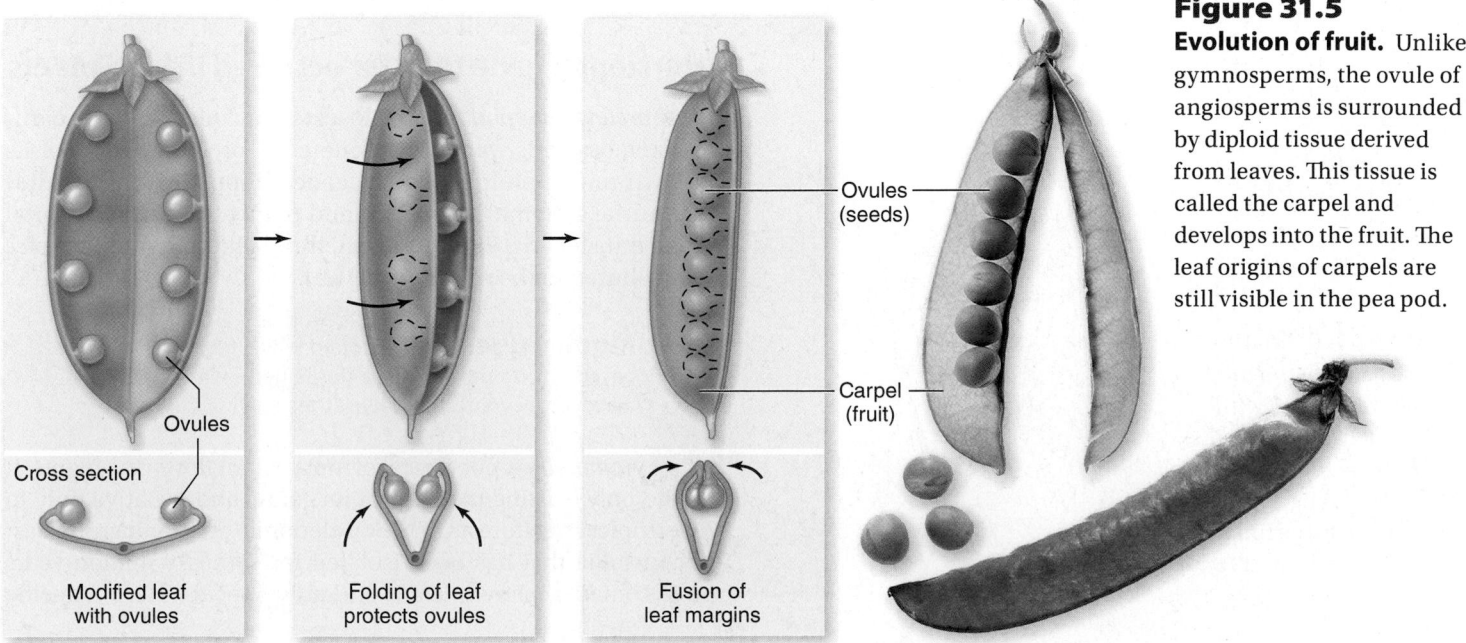

Ovules

Cross section

Modified leaf with ovules

Folding of leaf protects ovules

Fusion of leaf margins

Ovules (seeds)

Carpel (fruit)

gymnosperms, including the yew (*Taxus* spp.), have fleshlike tissue around their seeds, it is of a different origin and not a true fruit.

Sorting out angiosperm origins is complicated

The origins of the angiosperms puzzled even Darwin who referred to their origin as an "abominable mystery." What particularly puzzled Darwin was the very rapid diversification and success of the angiosperms. Fossil pollen and plants accompanied by molecular sequence data provide exciting clues about basal angiosperms, indicating origins as early as 167 to 199 MYA.

In the remote Liaoning province of China, a complete angiosperm fossil that is at least 125 million years old has been found (figure 31.6). The fossil may represent a new, basal, and extinct angiosperm family, Archaefructaceae, with two species: *Archaefructus liaoningensis* and *A. sinensis*. *Archaefructus* was an herbaceous, aquatic plant. This family is proposed to be the sister clade to all other angiosperms, but there is a lively debate about the validity of this claim.

Archaefructus fossils have both male and female reproductive structures; however, they lack the sepals and petals that evolved in later angiosperms to attract pollinators. The fossils were so well preserved that fossil pollen could be examined using scanning electron microscopy. Although *Archaefructus* is

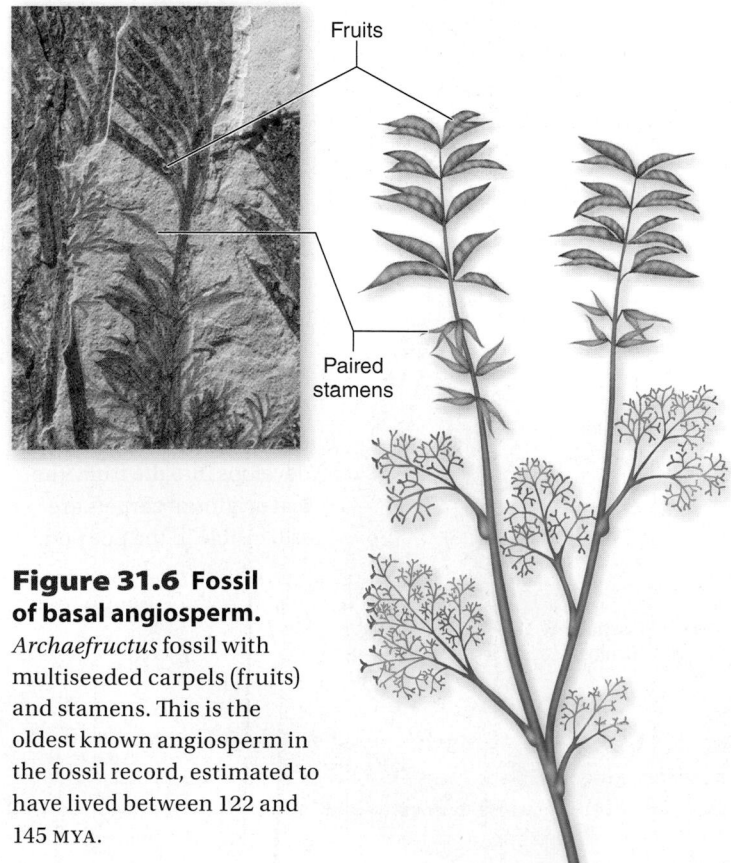

Figure 31.6 Fossil of basal angiosperm. *Archaefructus* fossil with multiseeded carpels (fruits) and stamens. This is the oldest known angiosperm in the fossil record, estimated to have lived between 122 and 145 MYA.

Fruits

Paired stamens

Figure 31.7 An ancient living angiosperm, *Amborella trichopoda.* This plant is believed to be the closest living relative to the original angiosperm.

carpel

petal

sepal

ancient, it is unlikely to be the very first angiosperm. Still, the incredibly well-preserved fossils provide valuable detail on angiosperms in the Upper Jurassic to Lower Cretaceous period, when dinosaurs roamed the Earth.

Consensus has also been growing on the most basal living angiosperm—*Amborella trichopoda* (figure 31.7). *Amborella,* with small, cream-colored flowers, is even more primitive than water lilies. This small shrub, found only on the island of New Caledonia in the South Pacific, is the last remaining species of the earliest extant lineage of the angiosperms that arose about 135 MYA.

Although *Amborella* is not the original angiosperm, it is sufficiently close that studying its reproductive biology may help us understand the early radiation of the angiosperms. The angiosperm phylogeny reflects an evolutionary hypothesis that is driving new research on angiosperm origins (figure 31.8).

Horizontal gene transfer occurred in land plants

Amborella trichopoda is the closest living relative to the earliest angiosperms, yet at least one copy of 20 out of 31 of its known mitochondrial protein genes hopped into the mitochondrial genome from other land plants through horizontal gene transfer (HGT). In addition, three different moss species contributed to the mix (figure 31.9).

? Inquiry question Explain why a phylogenetic tree based on comparisons of a single gene could result in an inaccurate evolutionary hypothesis.

Amborella is not typical of most extant flowering plants. It is the only existing member of its genus and is native only to the tropical rain forests of New Caledonia, an island group east of Australia that has been isolated for some 70 million years and contains many ancient endemic species. Here parasitic

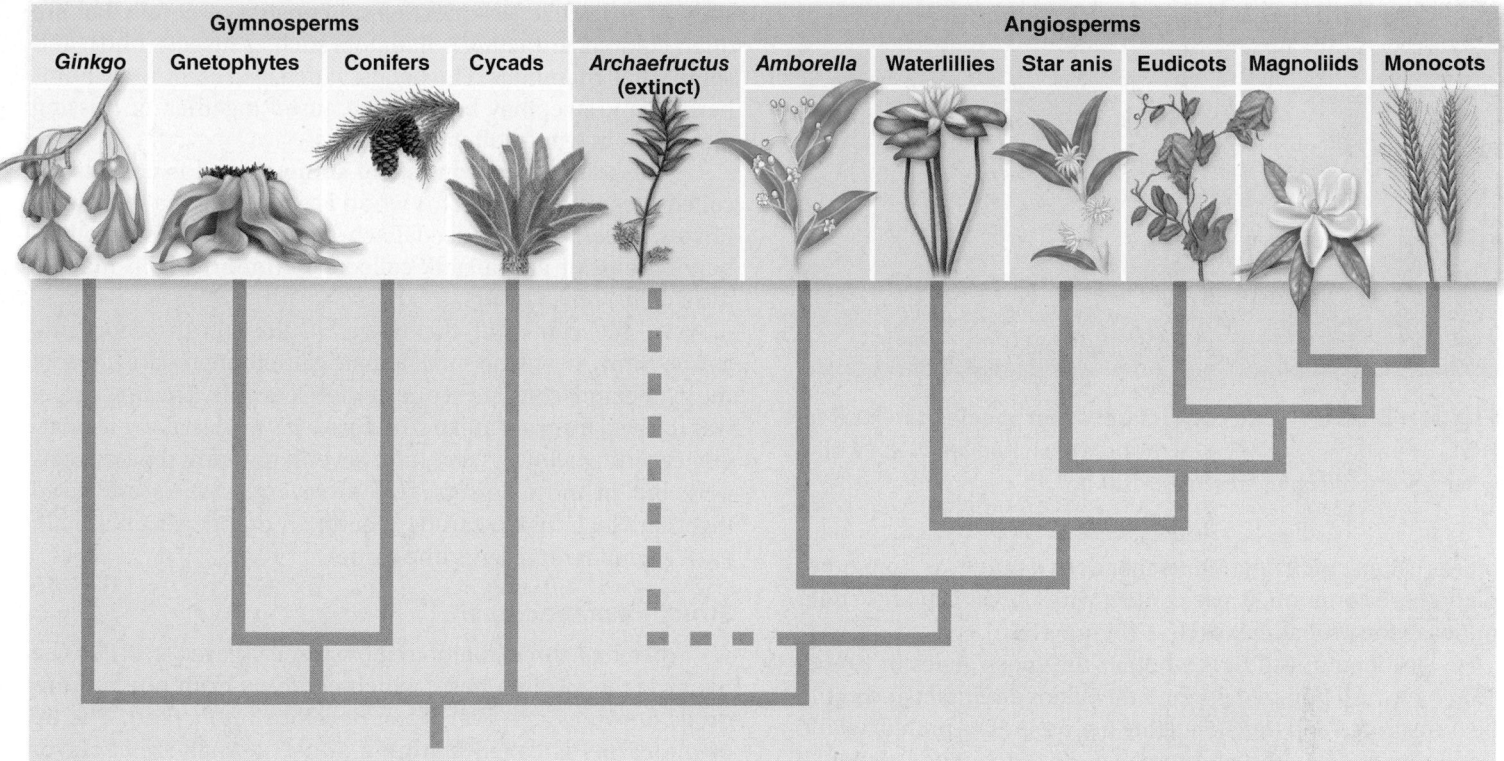

Figure 31.8 *Archaefructus* **may be the sister clade to all other angiosperms.** All members of the *Archaefructus* lineage are extinct, leaving *Amborella* as the most basal, living angiosperm. Gymnosperm species' labels are shaded green.

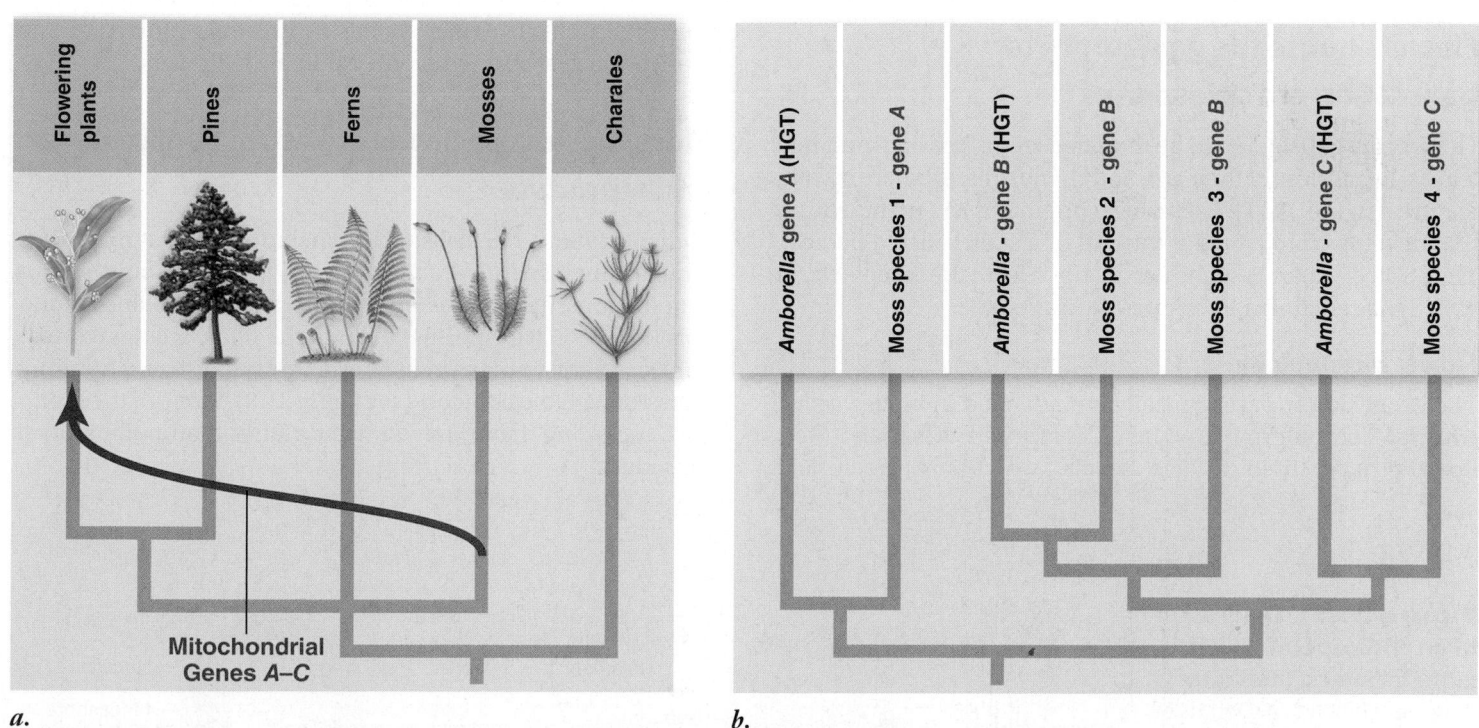

a.

b.

Figure 31.9 **The flowering plant** *Amborella* **acquired three moss genes through horizontal gene transfer (HGT).**
a. Phylogenetic relationship of *Amborella* to other land plants. As shown by the arrow connecting moss and the flowering plants, HGT is the only plausible explanation for the presence of moss mitochondrial genes in *Amborella*. *b.* Phylogenetic relationships among the horizontally transferred gene.

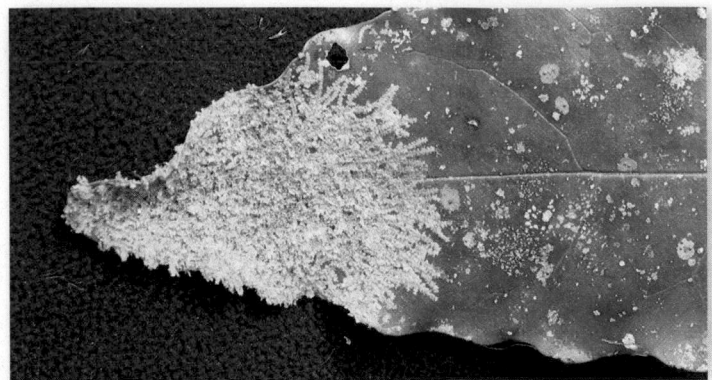

Figure 31.10 Close contact between species can lead to HGT. Here moss is growing on the base of an *Amborella* leaf with lichens scattered on the rest of the leaf.

plants called *epiphytes* (plants that derive nutrients from other plants) are common. Close contact with parasitic plants could increase the probability of HGT (figure 31.10).

An open question is whether the moss genes in *Amborella* have functions. About half the genes are intact and could be transcribed and translated into a protein. The protein would be similar to an existing protein in the plant, but its function, if any, remains to be determined.

? Inquiry question How would you determine if a moss gene in *Amborella* had a function? (*Hint:* Refer to chapter 24.)

Flowers house the gametophyte generation of angiosperms

Flowers are considered to be modified stems bearing modified leaves. Regardless of their size and shape, they all share certain features (figure 31.11). Each flower originates as a **primordium** that develops into a bud at the end of a stalk called a pedicel. The pedicel expands slightly at the tip to form the receptacle, to which the remaining flower parts are attached.

Flower morphology

The other flower parts typically are attached in circles called **whorls.** The outermost whorl is composed of **sepals.** Most flowers have three to five sepals, which are green and somewhat leaflike. The next whorl consists of **petals** that are often colored, attracting pollinators such as insects, birds, and some small mammals. The petals, which also commonly number three to five, may be separate, fused together, or missing altogether in wind-pollinated flowers.

The third whorl consists of **stamens** and is collectively called the androecium. This whorl is where the male gametophytes, pollen, are produced. Each stamen consists of a pollen-bearing **anther** and a stalk called a **filament,** which may be missing in some flowers.

At the center of the flower is the fourth whorl, the **gynoecium,** where the small female gametophytes are housed; the gynoecium consists of one or more **carpels.** The first carpel was derived from a leaflike structure with ovules along its margins. Primitive flowers can have several to many separate carpels, but in most flowers, two to several carpels are fused together. Such fusion can be seen in an orange sliced in half; each segment represents one carpel.

Structure of the carpel

A carpel has three major regions (see figure 31.11*a*). The **ovary** is the swollen base, which contains from one to hundreds of ovules; the ovary later develops into a **fruit.** The tip of the carpel is called a **stigma.** Most stigmas are sticky or feathery, causing pollen grains that land on them to adhere. Typically, a neck or stalk called a **style** connects the stigma and the ovary; in some flowers, the style may be very short or even missing.

Many flowers have nectar-secreting glands called *nectaries,* often located toward the base of the ovary. Nectar is a fluid containing sugars, amino acids, and other molecules that attracts insects, birds, and other animals to flowers.

Most species use flowers to attract pollinators and reproduce

Eudicots (about 175,000 species) include the great majority of familiar angiosperms—almost all kinds of trees and shrubs, snapdragons, mints, peas, sunflowers, and other plants. Monocots (about 65,000 species) include the lilies, grasses, cattails, palms, agaves, yuccas, orchids, and irises and share a common ancestor with the eudicots (see figure 31.8). Some of the monocots, including maize, rely on wind rather than pollinators to reproduce.

Figure 31.11 Diagram of an angiosperm flower.
a. The main structures of the flower are labeled. *b.* Details of an ovule. The ovary as it matures will become a fruit; as the ovule's outer layers (integuments) mature, they will become a seed coat.

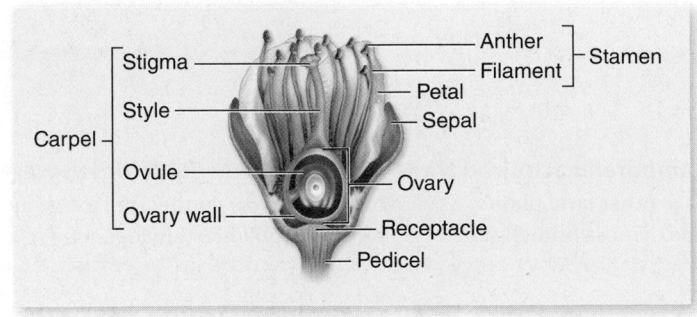

a.

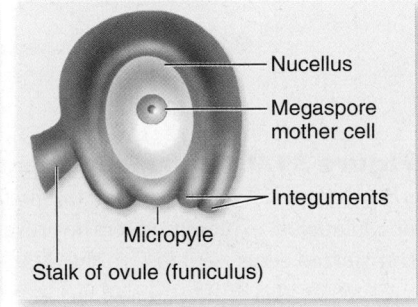

b.

The angiosperm life cycle includes double fertilization

During development of a flower bud, a single megaspore mother cell in the ovule undergoes meiosis, producing four megaspores (figure 31.12). In most flowering plants, three of the megaspores soon disappear; the nucleus of the remaining megaspore divides mitotically, and the cell slowly expands until it becomes many times its original size.

The female gametophyte

While the expansion of the megaspore is occurring, each of the daughter nuclei divides twice, resulting in eight haploid nuclei arranged in two groups of four. At the same time, two layers of the ovule, the integuments, differentiate and become the *seed coat* of a seed. The integuments, as they develop, form the micropyle, a small gap or pore at one end that was described earlier (see figure 31.11*b*).

One nucleus from each group of four migrates toward the center, where they function as polar nuclei. Polar nuclei may fuse together, forming a single diploid nucleus, or they may form a single cell with two haploid nuclei. Cell walls also form around the remaining nuclei. In the group closest to the micropyle, one cell functions as the egg; the other two nuclei are called synergids. At the other end, the three cells are now called antipodals; they have no apparent function and eventually break down and disappear.

The large sac with eight nuclei in seven cells is called an embryo sac; it constitutes the female gametophyte. Although it is completely dependent on the sporophyte for nutrition, it is a multicellular, haploid individual.

Pollen production

While the female gametophyte is developing, a similar but less complex process takes place in the anthers (see figure 31.12). Most anthers have patches of tissue (usually four) that

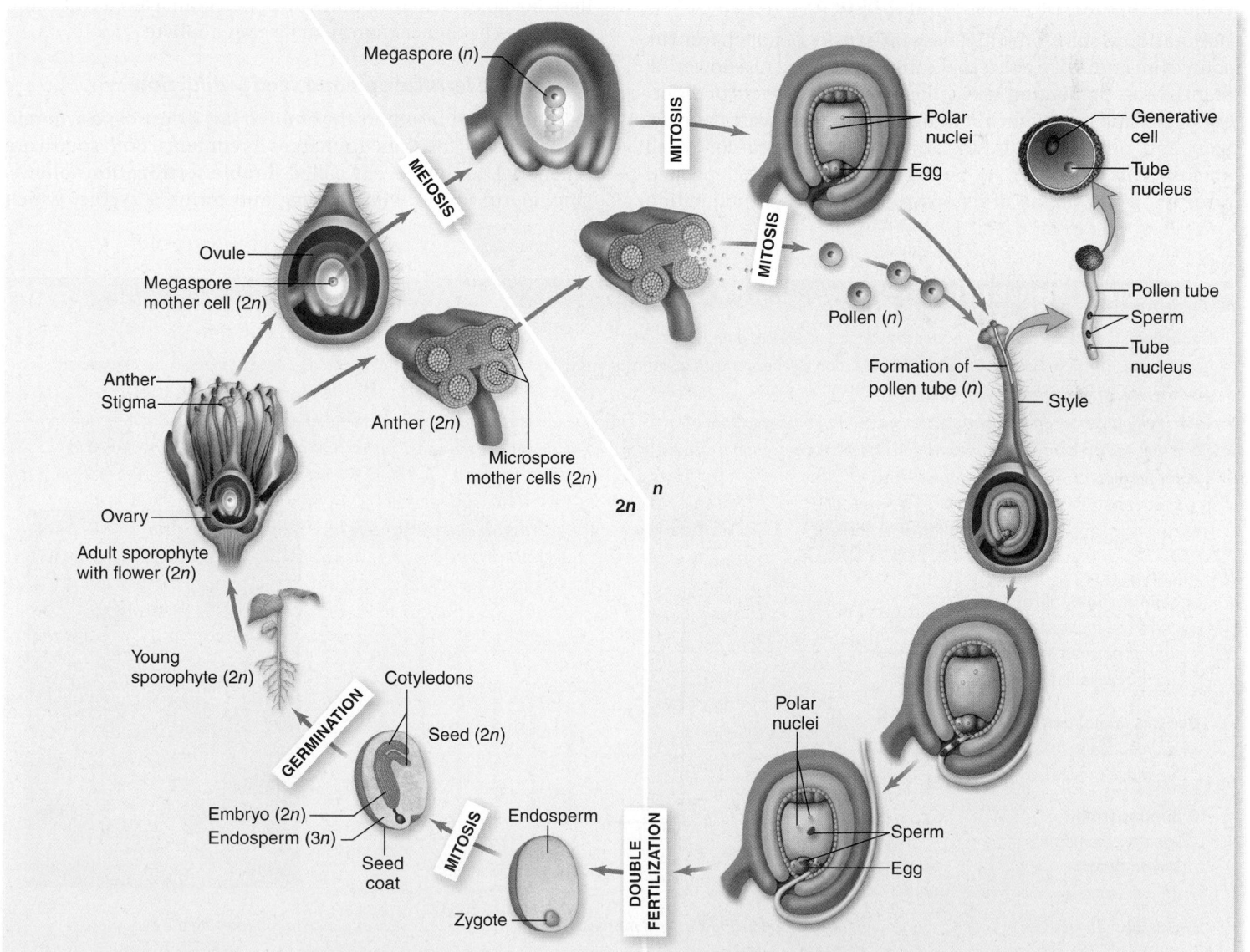

Figure 31.12 Life cycle of a typical angiosperm. As in pines, external water is no longer required for fertilization. In most species of angiosperms, animals carry pollen to the carpel. The outer wall of the carpel forms the fruit, which often entices animals to disperse the seed.

eventually become chambers lined with nutritive cells. The tissue in each patch is composed of many diploid microspore mother cells that undergo meiosis more or less simultaneously, each producing four microspores.

The four microspores at first remain together as a quartet, or tetrad, and the nucleus of each microspore divides once; in most species, the microspores of each quartet then separate. At the same time, a two-layered wall develops around each microspore. As the microspore-containing anther continues to mature, the wall between adjacent pairs of chambers breaks down, leaving two larger sacs. At this point, the binucleate microspores have become pollen grains.

The outer pollen grain wall layer often becomes beautifully sculptured, and it contains chemicals that may react with others in a stigma to signal whether development of the male gametophyte should proceed to completion. The pollen grain has areas called *apertures,* through which a pollen tube may later emerge.

Pollination and the male gametophyte

Pollination is simply the mechanical transfer of pollen from its source (an anther) to a receptive area (the stigma of a flowering plant). Most pollination takes place between flowers of different plants and is brought about by insects, wind, water, gravity, bats, and other animals. In as many as one-quarter of all angiosperms, however, a pollen grain may be deposited directly on the stigma of its own flower, and self-pollination

occurs. Pollination may or may not be followed by *fertilization,* depending on the genetic compatibility of the pollen grain and the flower on whose stigma it has landed.

If the stigma is receptive, the pollen grain's dense cytoplasm absorbs substances from the stigma and bulges through an aperture. The bulge develops into a pollen tube that responds to chemical and mechanical stimuli that guide it to the embryo sac. It follows a diffusion gradient of the chemicals and grows down through the style and into the micropyle. The pollen tube usually takes several hours to two days to reach the micropyle, but in a few instances, the journey may take up to a year. Pollen tube growth is more rapid in angiosperms than gymnosperms. Rapid pollen tube growth rate is an innovation that is believed to have preceded fruit development and been essential in the origins of angiosperms (figure 31.13)

One of the pollen grain's two cells, the *generative cell,* lags behind. Its nucleus divides in the pollen grain or in the pollen tube, producing two sperm cells. Unlike sperm in mosses, ferns, and some gymnosperms, the sperm of flowering plants have no flagella. At this point, the pollen grain with its tube and sperm has become a mature male gametophyte.

Double fertilization and seed production

As the pollen tube enters the embryo sac, it destroys a synergid in the process and then discharges its contents. Both sperm are functional, and an event called **double fertilization** follows. One sperm unites with the egg and forms a zygote, which

---

SCIENTIFIC THINKING

Hypothesis: *An increase in pollen tube growth accompanied angiosperm origins.*

Prediction: *The time from pollination to fertilization will be greater in gymnosperms than angiosperms that are sister to the ancestral angiosperm and also more derived angiosperms.*

Test: *Pollinate three angiosperm species, including* Amborella, *that are most closely related to the first angiosperm. At different time intervals, cut sections of the carpel and observe pollen tube position under a fluorescent microscope. Calculate and compare the rate of pollen tube growth with results published for gymnosperms and more derived angiosperms.*

Species	Pollen Tube Growth Rate (µm/hour)
Close relatives of ancestral angiosperm	
Amborella trichopoda	80
Nuphar polysepala	589
Austrobaileya scandens	271
Derived angiosperms	
Petunia inflata	900
Zea mays (maize)	10,000
Gymnosperms	
Gnetum gnemon	6
Ginkgo biloba	0.31

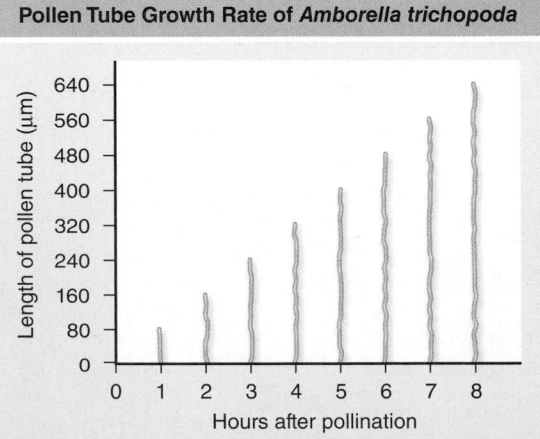

Pollen Tube Growth Rate of *Amborella trichopoda*

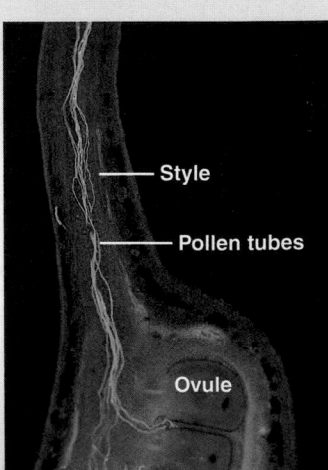

Conclusion: *An increase in pollen tube growth rate is found in angiosperms most closely related to the ancestral angiosperm, supporting the hypothesis.*

Further Experiments: *Develop a hypothesis to explain the difference in pollen tube growth rates among the tested angiosperms. Is there a correlation between the distance a pollen tube must travel to fertilize an egg and the rate of growth? How could you test your hypothesis?*

Figure 31.13 Rapid pollen tube growth accompanied the origin of angiosperms.

develops into an embryo sporophyte plant. The other sperm and the two polar nuclei unite, forming a triploid primary endosperm nucleus.

The primary endosperm nucleus begins dividing rapidly and repeatedly, becoming triploid endosperm tissue that may soon consist of thousands of cells. Endosperm tissue can become an extensive part of the seed in grasses such as corn, and it provides nutrients for the embryo in most flowering plants (see figure 41.35).

Until recently, the nutritional, triploid endosperm was believed to be the ancestral state in angiosperms. A recent analysis of extant, basal angiosperms revealed that diploid endosperms were also common. The female gametophyte in these species has four, not eight nuclei. At the moment, it is unclear whether diploid or triploid endosperms are the most primitive.

> **?** **Inquiry question** If the endosperm failed to develop in a seed, how do you think the fitness of that seed's embryo would be affected? Explain your answer.

> **Q** **Data analysis** Using the graph in figure 31.13, calculate the pollen tube growth rate, being sure to show your work. Compare your rate with the rate in the table in figure 31.13 and propose an explanation for any differences.

Learning Outcomes Review 31.3

Amborella, the closest living relative of the first angiosperms, provides clues to the origins of this very successful group. Although angiosperms are a clade, horizontal gene transfer has contributed genes from distantly related plant species. Angiosperms are characterized by ovules that at pollination are enclosed within an ovary at the base of a carpel, a structure unique to the phylum; a fruit develops from the ovary. Evolutionary innovations of angiosperms include flowers to attract pollinators, fruits to protect embryos and aid in their dispersal, and double fertilization, which provides endosperm to help nourish the embryo.

■ *How could genes from moss be transported into a flowering plant, other than via parasitic plants?*

31.4 Seeds

Learning Outcomes

1. *Describe four ways in which seeds help to ensure the survival of a plant's offspring.*
2. *List environmental conditions that can lead to seed germination in some plants.*

Following double fertilization in angiosperms and single fertilization in gymnosperms leading to embryonic development, a profoundly important event occurs: The embryo stops developing. In this section we focus on seed development in the angiosperms. In many plants, development of the embryo is arrested soon after the meristems and cotyledons differentiate. The integuments—the outer cell layers of the ovule—develop into a relatively impermeable **seed coat,** which encloses the seed with its dormant embryo and stored food (figure 31.14).

Seeds protect the embryo

The seed is a vehicle for dispersing the embryo to distant sites. Being encased in the protective layers of a seed allows a plant embryo to survive in environments that might kill a mature plant.

Seeds are an important adaptation in at least four ways:

1. Seeds maintain dormancy under unfavorable conditions and postpone development until better conditions arise. If conditions are marginal, a plant can "afford" to have some seeds germinate, because some of those that germinate may survive, while others remain dormant.
2. Seeds afford maximum protection to the young plant at its most vulnerable stage of development.
3. Seeds contain stored food that allows a young plant to grow and develop before photosynthetic activity begins.
4. Perhaps most important, seeds are adapted for dispersal, facilitating the migration of plant genotypes into new habitats.

A mature seed contains only about 5 to 20% water. Under these conditions, the seed and the young plant within it are very stable; its arrested growth is primarily due to the progressive and severe desiccation of the embryo and the associated reduction in metabolic activity. Germination

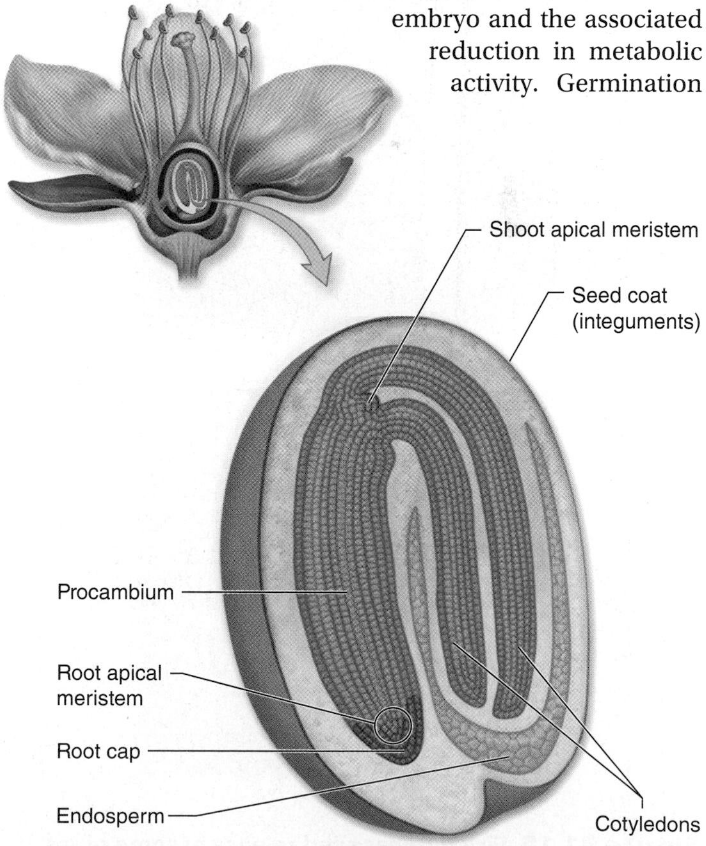

Figure 31.14 Seed development. The integuments of this mature angiosperm ovule are forming the seed coat. Note that the two cotyledons have grown into a bent shape to accommodate the tight confines of the seed. In some embryos, the shoot apical meristem will have already initiated a few leaf primordia as well.

cannot take place until water and oxygen reach the embryo. Seeds of some plants have been known to remain viable for hundreds and, in rare instances, thousands of years.

Specialized seed adaptations improve survival

Specific adaptations often help ensure that seeds will germinate only under appropriate conditions. Sometimes, seeds lie within tough cones that do not open until they are exposed to the heat of a fire (figure 31.15). This strategy causes the seed to germinate in an open, fire-cleared habitat where nutrients are relatively abundant, having been released from plants burned in the fire.

Seeds of other plants germinate only when inhibitory chemicals leach from their seed coats, thus guaranteeing their germination when sufficient water is available. Still other seeds germinate only after they pass through the intestines of birds or mammals or are regurgitated by them, which both weakens the seed coats and ensures dispersal. Sometimes seeds of plants thought to be extinct in a particular area may germinate under unique or improved environmental circumstances, and the plants may then reestablish themselves.

Figure 31.15 Fire induces seed release in some pines.
Fire can destroy adult jack pines, but stimulate growth of the next generation. *a.* The cones of a jack pine are tightly sealed and cannot release the seeds protected by the scales. *b.* High temperatures lead to the release of the seeds.

Learning Outcomes Review 31.4

The seed coat originates from the integuments and encloses the embryo and stored nutrients. The four advantages conferred by seeds are dormancy, protection of the embryo, nourishment, and a method of dispersal. Fire, heavy rains, or passage through an animal's digestive tract may be required for germination in some species.

■ *What type of seed dormancy would you expect to find in trees living in climates with cold winters?*

31.5 Fruits

Learning Outcomes

1. *Identify the structures from which fruits develop.*
2. *Distinguish among berries, legumes, drupes, and samaras.*

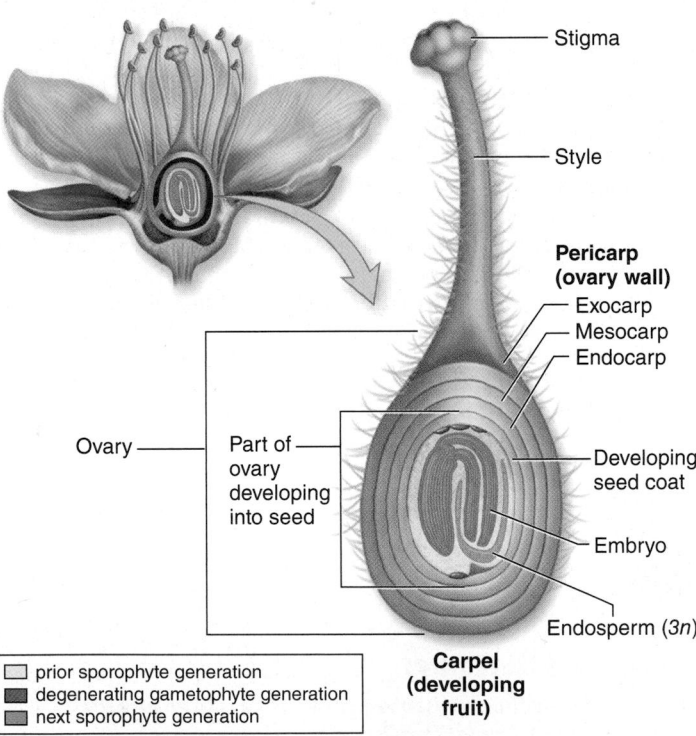

	prior sporophyte generation
	degenerating gametophyte generation
	next sporophyte generation

Figure 31.16 Fruit development. The carpel (specifically the ovary) wall is composed of three layers: the exocarp, mesocarp, and endocarp. One, some, or all of these layers develops to contribute to the recognized fruit in different species. The seed matures within this developing fruit.

? **Inquiry question** Three generations are represented in this diagram. Label the ploidy levels of the tissues of different generations shown here.

Survival of angiosperm embryos depends on fruit development as well as seed development. Fruits are most simply defined as mature ovaries (carpels). During seed formation, the flower ovary begins to develop into fruit (figure 31.16). In some cases, pollen landing on the stigma can initiate fruit development, but more frequently the coordination of fruit, seed coat, embryo, and endosperm development follow fertilization.

It is possible for fruits to develop without seed development. Commercial bananas, for example, have aborted seed development, but do produce mature, edible ovaries. Bananas are propagated asexually since no embryo develops.

Fruits are adapted for dispersal

Fruits form in many ways and exhibit a wide array of adaptations for dispersal. Three layers of ovary wall, also called the *pericarp*, can have distinct fates, which account for the diversity of fruit types from fleshy to dry and hard. The differences among some of the fruit types are shown in figure 31.17.

Developmentally, fruits are fascinating organs that contain three genotypes in one package. The fruit and seed coat are from the prior sporophyte generation. Remnants of the gametophyte generation that produced the egg are found in the developing seed, and the embryo represents the next sporophyte generation (see figure 31.16).

True Berries

The entire pericarp is fleshy, although there may be a thin skin. Berries have multiple seeds in either one or more ovaries. The tomato flower had four carpels that fused.
Each carpel contains multiple ovules that develop into seeds.

Outer pericarp
Fused carpels
Seed

Drupes

Single seed enclosed in a hard pit; peaches, plums, cherries. Each layer of the pericarp has a different structure and function, with the endocarp forming the pit.

Pericarp
Exocarp (skin)
Mesocarp
Endocarp (pit)
Seed

Aggregate Fruits

Derived from many ovaries of a single flower; strawberries, blackberries. Unlike tomato, these ovaries are not fused and covered by a continuous pericarp.

Sepals of a single flower
Ovary
Seed

Legumes

Split along two carpel edges (sutures) with seeds attached to edges; peas, beans. Unlike fleshy fruits, the three tissue layers of the ovary do not thicken extensively. The entire pericarp is dry at maturity.

Pericarp
Stigma
Seed
Style

Samaras

Not split and with a wing formed from the outer tissues; maples, elms, ashes.

Pericarp
Seed

Multiple Fruits

Individual flowers form fruits around a single stem. The fruits fuse as seen with pineapple.

Main stem
Pericarp of individual flower

Figure 31.17 Examples of some kinds of fruits. Legumes and samaras are examples of dry fruits. Legumes open to release their seeds, but samaras do not. Drupes and true berries are simple fleshy fruits; they develop from a flower with a single pistil composed of one or more carpels. Aggregate and multiple fruits are compound fleshy fruits; they develop from flowers with more than one pistil or from more than one flower.

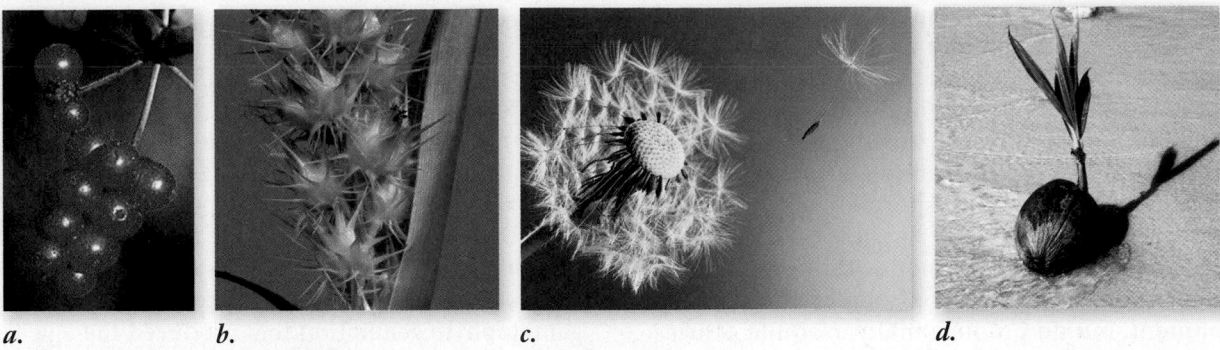

a.   *b.*   *c.*   *d.*

Figure 31.18 Animal-dispersed fruits. *a.* The bright red berries of this honeysuckle, *Lonicera hispidula,* are highly attractive to birds. After eating the fruits, birds may carry the seeds they contain for great distances either internally or, because of their sticky pulp, stuck to their feet or other body parts. *b.* You will know if you have ever stepped on the fruits of *Cenchrus incertus;* their spines adhere readily to any passing animal. *c.* False dandelion, *Pyrrhopappus carolinianus,* has "parachutes" that widely disperse the fruits in the wind, much to the gardener's despair. *d.* This fruit of the coconut palm, *Cocos nucifera,* is sprouting on a sandy beach. Coconuts, one of the most useful fruits for humans in the tropics, have become established on other islands by drifting there on the waves.

Fruits allow angiosperms to colonize large areas

Aside from the many ways fruits can form, they also exhibit a wide array of specialized dispersal methods. Fruits with fleshy coverings, often shiny black or bright blue or red, normally are dispersed by birds or other vertebrates (figure 31.18*a*). Like red flowers, red fruits signal an abundant food supply. By feeding on these fruits, birds and other animals may carry seeds from place to place and thus transfer plants from one suitable habitat to another. Such seeds require a hard seed coat to resist stomach acids and digestive enzymes.

Fruits with hooked spines, such as those of burrs (figure 31.18*b*), are typical of several genera of plants that occur in the northern deciduous forests. Such fruits are often disseminated by mammals, including humans, when they hitch a ride on fur or clothing. Squirrels and similar mammals disperse and bury fruits such as acorns and other nuts. Some of these sprout when conditions become favorable, such as after the spring thaw.

Other fruits, including those of maples, elms, and ashes, have wings that aid in their distribution by the wind. Orchids have minute, dustlike seeds, which are likewise blown away by the wind. The dandelion provides another familiar example of a fruit type that is wind-dispersed (figure 31.18*c*), and the dispersal of seeds from plants such as milkweeds, willows, and cottonwoods is similar. Water dispersal adaptations include air-filled chambers surrounded by impermeable membranes to prevent the entrance of H_2O.

Coconuts and other plants that characteristically occur on or near beaches are regularly spread throughout a region by floating in water (figure 31.18*d*). This sort of dispersal is especially important in the colonization of distant island groups, such as the Hawaiian Islands.

It has been calculated that the seeds of about 175 angiosperms, nearly one-third from North America, must have reached Hawaii to evolve into the roughly 970 species found there today. Some of these seeds blew through the air, others were transported on the feathers or in the guts of birds, and still others floated across the Pacific. Although the distances are rarely as great as the distance between Hawaii and the mainland, dispersal is just as important for mainland plant species that have discontinuous habitats, such as mountaintops, marshes, or north-facing cliffs.

Learning Outcomes Review 31.5

As a seed develops, the pericarp layers of the ovary wall develop into the fruit. A berry has a fleshy pericarp; a legume has a dry pericarp that opens to release seeds; the outer layers of a drupe pericarp are fleshy; and a samara is a dry structure with a wing. Animals often distribute the seeds of fleshy fruits and fruits with spines or hooks. Wind disperses lightweight seeds and samara forms.

■ *What features of fruits might encourage animals to eat them?*

31.1 The Evolution of Seed Plants

The seed protects the embryo.

An extra layer of sporophyte tissue surrounds the embryo, creating the ovule, which later hardens (figure 31.1). The seed protects the embryo, helps it resist drying out, and allows a dormant stage that pauses the life cycle until environmental conditions are favorable.

Water is not required to transport the male gametophyte to the female gametophyte.

The gametophytes of seed plants consist of only a few cells. Pollen grains are male gametophytes; each pollen grain contains sperm cells. Water is not required for fertilization. The female gametophyte develops within an ovule that forms the seed.

31.2 Gymnosperms: Plants with "Naked Seeds" (figure 31.3)

Gymnosperms have ovules that are not completely enclosed in sporophyte (diploid) tissue at the time of pollination.

The four groups of gymnosperms are coniferophytes, cycadophytes, gnetophytes, and ginkgophytes; all lack flowers and true fruits.

Conifers are the largest gymnosperm phylum.

Conifers include pines, spruces, firs, cedars, and many other groups. Both the tallest and the oldest vascular plants are conifers.

Pines are an exemplary conifer genus.

Pines have tough, needle-like leaves in clusters of two to five. They produce male and female cones.

In the male cones, microspore mother cells in the microsporangia give rise to microspores that then develop into four-celled pollen grains, the male gametophytes.

In the female cones, a megasporangium (nucellus) produces a single megaspore mother cell that becomes four megaspores; three of these break down, and the remaining one develops into a female gametophyte, which produces archegonia that carry eggs.

Upon fertilization, a pollen tube emerges from the pollen grain and grows through the nucellus. Eventually two sperm cells migrate through the tube, and one unites with the egg; the other disintegrates.

Cycads resemble palms, but are not flowering plants.

Most cycads can reach heights of 15 m or more. Cycads, unlike palms, produce cones; in some species the female cones are huge. Their life cycle is like that of conifers.

Gnetophytes have xylem vessels.

Xylem vessels, a common feature in angiosperms, are highly efficient conducting cells. Gnetophytes are the only gymnosperms that have them.

This group includes the strange *Welwitschia* genus of southwestern Africa and the numerous worldwide *Ephedra* species.

Only one species of the ginkgophytes remains extant.

Ginkgo biloba is a gymnosperm with broad leaves that it sheds in the fall. It is dioecious, and the fleshy seeds of the female tree have a foul odor.

31.3 Angiosperms: The Flowering Plants (figure 31.8)

Angiosperms are distinct from gymnosperms and other plants because their ovules are enclosed within diploid tissue called the ovary at the time of fertilization, and they form fruits.

Sorting out angiosperm origins is complicated.

No one is certain how the angiosperms arose, although the extinct family Archaefructaceae may have been a sister clade. The earliest extant angiosperm appears to be *Amborella trichopoda,* found on the island of New Caledonia.

Horizontal gene transfer occurred in land plants.

Moss genes have been found in the *Amborella* genome, possibly transferred by parasitic plants, although it is unknown if they have a function in *Amborella.*

Flowers house the gametophyte generation of angiosperms.

Flowers are considered to be modified stems that bear modified leaves. Flower parts are organized into four whorls: sepals, petals, androecium, and gynoecium (figure 31.11*a*).

The male androecium consists of the stamens where haploid pollen, the male gametophyte, is produced.

The female gynoecium consists of one or more carpels that contain the female gametophyte. The carpel has three major regions: the ovary, which later becomes the fruit; the stigma, which is the tip of the carpel; and the style, a stalk that connects the stigma and ovary.

Most species use flowers to attract pollinators and reproduce.

Many angiosperm flowers have nectaries near the base of the ovary that produce nectar containing nutrients and other molecules. Nectar and scent attracts animal pollinators, which carry pollen from one flower to another; some angiosperms are wind pollinated, however.

The angiosperm life cycle includes double fertilization (figure 31.12).

The megaspore produces eight haploid nuclei. The female gametophyte consists of a large embryo sac with the eight nuclei in seven cells. The egg and the two polar nuclei (in a single cell) are most important.

After landing on a receptive stigma, a pollen grain develops a pollen tube that grows toward the embryo sac. Eventually, two sperm pass through this tube. One fuses with the egg to form a zygote, and the other unites with the polar bodies to form a triploid endosperm nucleus that develops into endosperm to nourish the embryo.

31.4 Seeds (figure 31.14)

Seeds protect the embryo.

Seeds help to ensure the survival of the next generation by maintaining dormancy during unfavorable conditions, protecting the embryo, providing food for the embryo, and providing a means for dispersal.

Specialized seed adaptations improve survival.

Before a seed germinates, its seed coat must become permeable so that water and oxygen can reach the embryo. Adaptations have evolved to ensure germination under appropriate survival conditions. In certain gymnosperms, seeds may be released from cones after a fire. Alternatively, seeds may require passage through a digestive tract, freeze–thaw cycles, or abundant moisture.

31.5 Fruits (figure 31.16)

Fruits are adapted for dispersal.

In angiosperms, a fruit is a mature ovary. Fruit development is coordinated with embryo, endosperm, and seed coat development.

Angiosperms produce many types of fruit, which vary depending on the fate of the pericarp (carpel wall). Fruits can be dry or fleshy, and they can be simple (single carpel), aggregate (multiple carpels), or multiple (multiple flowers).

A fruit is genetically unique because it contains tissues from the parent sporophyte (the seed coat and fruit tissue), the gametophyte (remnants in the developing seed), and the offspring sporophyte (the embryo).

Fruits allow angiosperms to colonize large areas.
Fruits exhibit a wide array of dispersal mechanisms. They may be ingested and transported by animals, buried in caches by herbivores, carried away by birds and mammals, blown by the wind, or float away on water.

Review Questions

UNDERSTAND

1. The lack of seeds is a characteristic of all
   a. lycophytes.
   b. conifers.
   c. tracheophytes.
   d. gnetophytes.

2. Which of the following adaptations allows plants to pause their life cycle until environmental conditions are optimal?
   a. Stomata
   b. Phloem and xylem
   c. Seeds
   d. Flowers

3. Which of the following gymnosperms possesses a form of vascular tissue that is similar to that found in the angiosperms?
   a. Cycads
   b. Gnetophytes
   c. Ginkgophytes
   d. Conifers

4. In a pine tree, the microspores and megaspores are produced by the process of
   a. fertilization.
   b. mitosis.
   c. fusion.
   d. meiosis.

5. Which of the following terms is NOT associated with a male portion of a plant?
   a. Megaspore
   b. Antheridium
   c. Pollen grains
   d. Microspore

6. Which of the following potentially represents the oldest known living species of angiosperm?
   a. *Cooksonia*
   b. *Archaefructus*
   c. *Chlamydomonas*
   d. *Amborella*

7. The integuments of an ovule will develop into the
   a. embryo.
   b. endosperm.
   c. fruit.
   d. seed coat.

8. An example of a drupe is a
   a. strawberry.
   b. plum.
   c. bean.
   d. pineapple.

9. The pericarp is the
   a. ovary wall.
   b. developing seed coat.
   c. ovary.
   d. mature endosperm.

APPLY

1. Reproduction in angiosperms can occur more quickly than in gymnosperms because
   a. gymnosperm sperm requires water to swim to the egg.
   b. flowers always increase the rate of reproduction.
   c. angiosperm pollen tubes grow more quickly than gymnosperm pollen tubes.
   d. angiosperms have nectaries.

2. In double fertilization, one sperm produces a diploid ___, and the other produces a triploid ___.
   a. zygote; primary endosperm
   b. primary endosperm; microspore

   c. antipodal; zygote
   d. polar nuclei; zygote

3. Apply your understanding of angiosperms to identify which innovations likely contributed to the tremendous success of angiosperms.
   a. Homospory in angiosperms
   b. Fruits that attract animal dispersers
   c. Cones that protect the seed
   d. Dominant gametophyte generation

4. Comparing stems of two plant specimens under the microscope, you identify vessels in one sample and conclude the specimen
   a. with vessels must be an angiosperm or a gnetophyte.
   b. with vessels is either *Ephedra* or a cycad.
   c. without vessels is a pterophyte.
   d. without vessels must be a tracheophyte.

5. In a flower after fertilization, the following tissues are diploid:
   a. carpel, integuments, and megaspore mother cell.
   b. carpel, integuments, and megaspore.
   c. carpel, megaspore, and zygote.
   d. carpel, megaspore mother cell, and endosperm.

6. Fruits are complex organs that are specialized for dispersal of seeds. Which of the following plant tissues does NOT contribute to mature fruit?
   a. Sporophytic tissue from the previous generation
   b. Gametophytic tissue from the previous generation
   c. Sporophytic tissue from the next generation
   d. Gametophytic tissue from the next generation

SYNTHESIZE

1. You have been hired as a research assistant to investigate the origins of the angiosperms, specifically the boundary between a gymnosperm and an angiosperm. Which characteristics would you use to clearly define a new fossil as a gymnosperm? An angiosperm?

2. Assess the benefits and drawbacks of self-pollination for a flowering plant? Explain your answer.

3. The relationship between flowering plants and pollinators is often used as an example of coevolution. Many flowering plant species have flower structures that are adaptive to a single species of pollinator. Evaluate the benefits and drawbacks of using such a specialized relationship.

4. As you are eating an apple one day, you decide that you'd like to save the seeds and plant them. You do so, but they fail to germinate. Discuss all possible reasons that the seeds did not germinate and strategies you could try to improve your chances of success.

Chapter 32

Fungi

Chapter Contents

Introduction

The fungi, an often overlooked group of unicellular and multicellular organisms, have a profound influence on ecology and human health. Along with bacteria, they are important decomposers and disease-causing organisms. Fungi are found everywhere—from the tropics to the tundra and in both terrestrial and aquatic environments. Fungi made it possible for plants to colonize land by associating with rootless stems and aiding in the uptake of nutrients and water. Mushrooms and toadstools are the multicellular spore-producing part of fungi that grow rapidly under proper conditions. A single Armillaria fungus can cover 15 hectares underground and weigh 100 tons, making it the largest organism in the world based on area. Some puffball fungi are almost a meter in diameter and may contain 7 trillion spores—enough to circle the Earth's equator!

Yeasts are used to make bread and beer, but other fungi cause disease in plants and animals. These fungal killers are particularly problematic because fungi are animals' closest relatives. Drugs that can kill fungi often have toxic effects on animals, including humans. In this chapter we present the major groups of this intriguing life-form.

32.1 Defining Fungi

Learning Outcomes

1. *Identify characteristics that distinguish fungi from other eukaryotes.*
2. *Compare mitosis in fungi and animals.*
3. *Explain why fungi are useful for bioremediation.*

Mycologists, scientists who study fungi as well as fungi-like protists, believe there may be as many as 1.5 million fungal species. Fungi exist either as single-celled yeasts or in multicellular form with several different cell types. Their reproduction may be either sexual or asexual, and they exhibit an unusual form of mitosis. They are specialized to extract and absorb nutrients from their surroundings through external secretion of enzymes. Recent phylogenetic analysis of DNA sequences and protein sequences indicates that fungi are more closely related to animals than to plants. Fossils and molecular data indicate that animals and fungi last shared a common ancestor

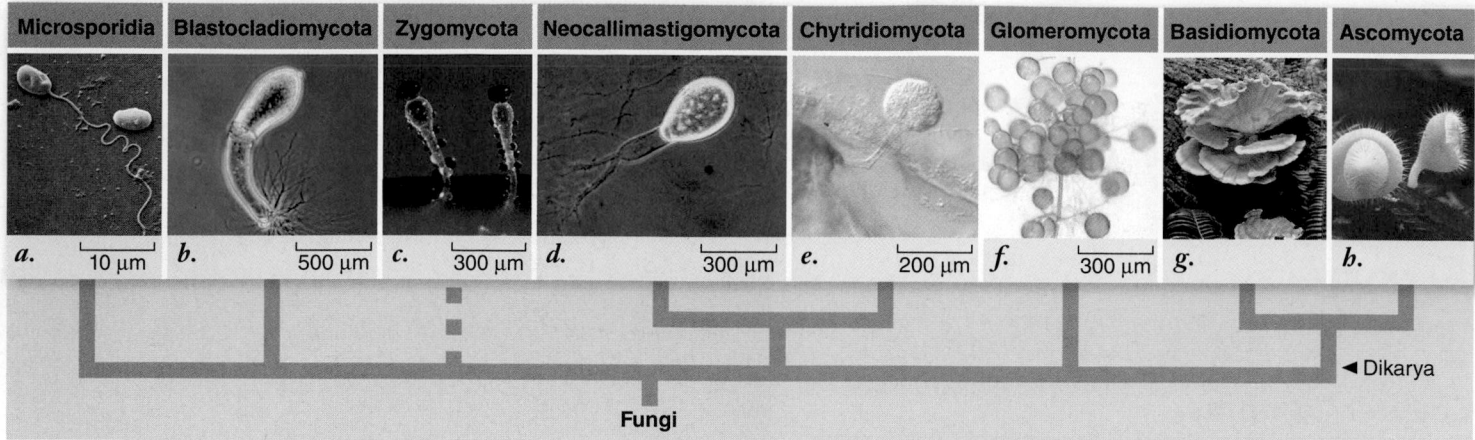

Figure 32.1 Seven major phyla of fungi. All phyla, except Zygomycota, are monophyletic. *a.* Microsporidia, including *Encephalitozoon cuniculi,* are animal parasites. *b. Allomyces arbuscula,* a water mold, is a blastocladiomycete. *c. Pilobolus,* a zygomycete, grows on animal dung and also on culture medium. Stalks about 10 mm long contain dark, spore-bearing sacs. *d.* Neocallimastigales, including *Piromyces communis,* decompose cellulose in the rumens of herbivores. *e.* Some chytrids, including members of the genus *Rhizophydium,* parasitize green algae. *f.* Spores of *Glomus intraradices,* a glomeromycete associated with roots. *g. Amanita muscaria,* the fly agaric, is a toxic basidiomycete. *h.* The cup fungus *Cookeina tricholoma* is an ascomycete from the rain forest of Costa Rica. In the cup fungi, the spore-producing structures line the cup; in basidiomycetes that form mushrooms such as *Amanita,* they line the gills beneath the cap of the mushroom. All visible structures of fungi, including the ones shown here, arise from an extensive network of filamentous hyphae that penetrates and is interwoven with the substrate on which they grow.

at least 460 MYA, but inconsistencies remain. The oldest fossil resembles extant members of the genus *Glomus* that arose within the Glomeromycota. One DNA analysis placed the animal/fungal divergence at 1.5500 BYA. This latter estimate is not generally accepted, but many researchers believe that the last common ancestor existed around close to 670 MYA based on an analysis of multiple genes. The last common ancestor with animals was a single cell, and unique solutions to multicellularity evolved both in fungi and in animals.

The phylogenetic relationships among fungi have been the cause of much debate. Traditionally, four fungal phyla were recognized, based primarily on characteristics of the cells undergoing meiosis: Chytridiomycota ("chytrids"), Zygomycota ("zygomycetes"), Ascomycota ("ascomycetes"), and Basidiomycota ("basidiomycetes"). The chytrids and zygomycetes are not monophyletic. The understanding of fungal phylogeny is going through rapid and exciting changes, aided by increasing molecular sequence data (figure 32.1 and table 32.1). In 2007, mycologists agreed on seven monophyletic phyla: Microsporidia, Blastocladiomycota, Neocallimastigomycota, Chytridiomycota, Glomeromycota, Basidiomycota, and Ascomycota (see figure 32.1). Blastomycetes and neocallimastigomycetes were formerly grouped with the chytrids. The Microsporidia are sister to all other fungi, but there is disagreement as to whether or not they are true fungi.

TABLE 32.1	Fungi		
Group	**Typical Examples**	**Key Characteristics**	**Approximate Number of Living Species**
Chytridiomycota	*Allomyces*	Aquatic, flagellated fungi that produce haploid gametes in sexual reproduction or diploid zoospores in asexual reproduction.	1000
Zygomycota	*Rhizopus, Pilobolus*	Multinucleate hyphae lack septa, except for reproductive structures; fusion of hyphae leads directly to formation of a zygote in zygosporangium, in which meiosis occurs just before it germinates; asexual reproduction is most common.	1050
Glomeromycota	*Glomus*	Form arbuscular mycorrhizae. Multinucleate hyphae lack septa. Reproduce asexually.	150
Ascomycota	Truffles, morels	In sexual reproduction, ascospores are formed inside a sac called an ascus; asexual reproduction is also common.	45,000
Basidiomycota	Mushrooms, toadstools, rusts	In sexual reproduction, basidiospores are borne on club-shaped structures called basidia; asexual reproduction occurs occasionally.	22,000

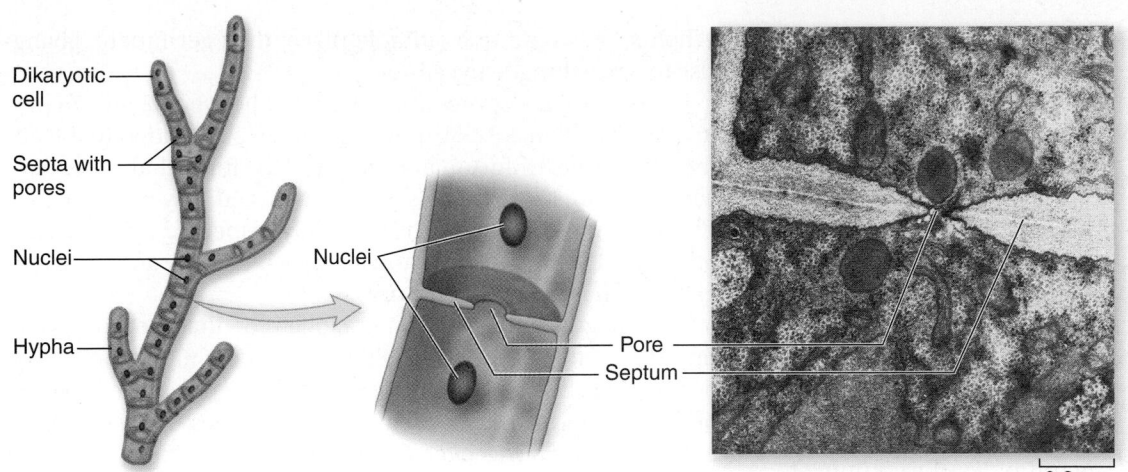

Figure 32.2 A septum. This transmission electron micrograph (and the accompanying schematic drawing) of a section through a hypha of an ascomycete shows a pore through which the cytoplasm streams.

0.2 μm

The body of a fungus is a mass of connected hyphae

Some hyphae are continuous or branching tubes filled with cytoplasm and multiple nuclei. Other hyphae are typically made up of long chains of cells joined end-to-end and divided by crosswalls called septa (singular, *septum*). The septa rarely form a complete barrier, except when they separate the reproductive cells. Even fungi with septa can be considered one long cell.

Cytoplasm characteristically flows or streams freely throughout the hyphae, passing through major pores in the septa (figure 32.2). Because of this streaming, proteins synthesized throughout the hyphae may be carried to their actively growing tips. As a result, fungal hyphae may grow very rapidly when food and water are abundant and the temperature is optimum. For example, you may have seen mushrooms suddenly appear in a lawn overnight after a rain in summer.

The mycelium

A mass of connected hyphae is called a **mycelium** (plural, mycelia). (This word and the term mycology, the study of fungi, are both derived from the Greek word *mykes,* meaning fungi.) The mycelium of a fungus (figure 32.3) constitutes a system that may, in the aggregate, be many meters long. This mycelium grows into the soil, wood, or other material and digestion of the material begins quickly. In two of the four major groups of fungi, reproductive structures formed of interwoven hyphae, such as mushrooms, puffballs, and morels, are produced at certain stages of the life cycle. These structures expand rapidly because of rapid inflation of the hyphae.

Cell walls with chitin

The cell walls of fungi are formed of polysaccharides, including chitin. In contrast, cell walls of plants and many protists contain cellulose, not chitin. Chitin is a modified cellulose consisting of linked glucose units to which nitrogen groups have been added; this polymer is then cross-linked with proteins. Chitin is the same material that makes up the major portion of the hard shells, or exoskeletons, of arthropods, a group of animals that includes insects and crustaceans (chapter 34). Chitin is one of the shared traits that has led scientists to believe that fungi and animals are more closely related than fungi and plants.

Fungal cells may have more than one nucleus

Fungi are different from most animals and plants in that each cell (or hypha) can house one, two, or more nuclei. A hypha that has only one nucleus is called **monokaryotic;** a cell with two nuclei is **dikaryotic.** In a dikaryotic cell, the two haploid nuclei exist independently. Dikaryotic hyphae have some of the genetic properties of diploids, because both genomes are transcribed.

Sometimes, many nuclei intermingle in the common cytoplasm of a fungal mycelium, which can lack distinct cells. If a dikaryotic or multinucleate hypha has nuclei that are derived from two genetically distinct individuals, the hypha is called **heterokaryotic.** Hyphae whose nuclei are genetically similar to one another are called **homokaryotic.**

Mitosis is not followed by cell division

Mitosis in multicellular fungi differs from that in most other organisms. Because of the linked nature of the cells, the cell itself is not the relevant unit of reproduction; instead, the nucleus is. The nuclear envelope does not break down and reform; instead, the spindle apparatus is formed *within* it.

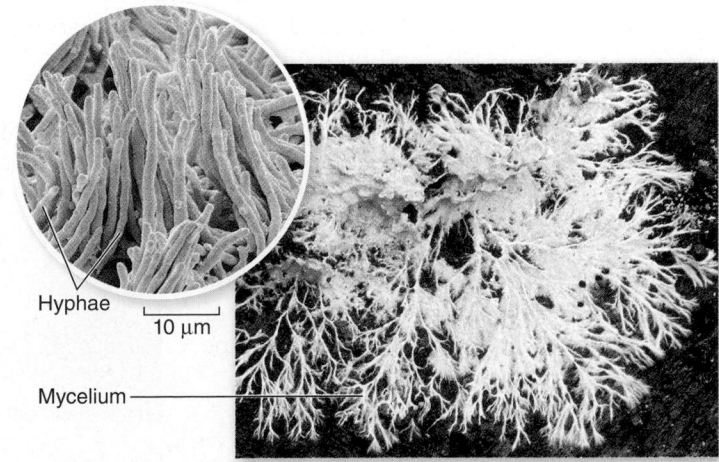

Hyphae
10 μm
Mycelium

Figure 32.3 Fungal mycelium. This mycelium, composed of hyphae, is growing through leaves on the forest floor in Maryland.

Figure 32.4
Fungal spores.
Scanning electron micrograph of fungal spores from *Aspergillus*.

10 μm

Centrioles are absent in all fungi except chytrids; instead, fungi regulate the formation of microtubules during mitosis with small, relatively amorphous structures called *spindle plaques*. This unique combination of features strongly suggests that fungi originated from some unknown group of single-celled eukaryotes with these characteristics.

Fungi can reproduce both sexually and asexually

Many fungi are capable of producing both sexual and asexual spores. When a fungus reproduces sexually, two haploid hyphae of compatible mating types may come together and fuse.

The dikaryon stage

In animals, plants, and some fungi, the fusion of two haploid cells during reproduction immediately results in a diploid cell ($2n$). But in other fungi, namely basidiomycetes and ascomycetes, an intervening dikaryotic stage ($1n + 1n$) occurs before the parental nuclei fuse and form a diploid nucleus. In ascomycetes, this dikaryon stage is brief, occurring in only a few cells of the sexual reproductive structure. In basidiomycetes, however, it can last for most of the life of the fungus, including both the feeding and sexual spore-producing structures.

Reproductive structures

Some fungal species produce specialized mycelial structures to house the production of spores. Examples are the mushrooms we see above ground, the "shelf" fungus that appears on the trunks of dead trees, and puffballs, which can house billions of spores.

As noted previously, the cytoplasm in fungal hyphae normally flows through perforated septa or moves freely in their absence. Reproductive structures are an important exception to this general pattern. When reproductive structures form, they are cut off by complete septa that lack perforations or that have perforations that soon become blocked.

Spores

Spores are the most common means of reproduction among fungi. They may form as a result of either asexual or sexual processes, and they are often dispersed by the wind.

When spores land in a suitable place, they germinate, giving rise to a new fungal mycelium.

Because the spores are very small, between 2 and 75 μm in diameter (figure 32.4), they can remain suspended in the air for a long time. Unfortunately, many of the fungi that cause diseases in plants and animals are spread rapidly by such means. The spores of other fungi are routinely dispersed by insects or other small animals. A few fungal phyla retain the ancestral flagella and have motile zoospores.

Biologists had believed for a long time that the worldwide presence of fungal species could be accounted for, on an evolutionary timescale, by the almost limitless, long-distance dispersal of fungal spores. Recent biogeographic studies, however, have examined the phylogenetic relationships among fungi in distant parts of the world and disproved this long-held assumption.

Fungi are heterotrophs that absorb nutrients

All fungi obtain their food by secreting digestive enzymes into their surroundings and then absorbing the organic molecules produced by this external digestion. The fungal body plan reflects this approach. Unicellular fungi have the greatest surface area-to-volume ratio of any fungus, maximizing the surface area for absorption. Extensive networks of hyphae also provide an enormous surface area for absorptive nutrition in a fungal mycelium.

Many fungi are able to break down the cellulose in wood, cleaving the linkages between glucose subunits and then absorbing the glucose molecules as food. Most fungi also digest lignin, an insoluble organic compound that strengthens plant cell walls. The specialized metabolic pathways of fungi allow them to obtain nutrients from dead trees and from an extraordinary range of organic compounds, including tiny roundworms called nematodes (figure 32.5*a*).

The mycelium of the edible oyster mushroom *Pleurotus ostreatus* (figure 32.5*b*) excretes a substance that paralyzes nematodes that feed on the fungus. When the worms become sluggish and inactive, the fungal hyphae envelop and penetrate their bodies. Then the fungus secretes digestive juices and absorbs the nematode's nutritious contents, just like it would from a plant source.

This fungus usually grows within living trees or on old stumps, obtaining the bulk of its glucose through the

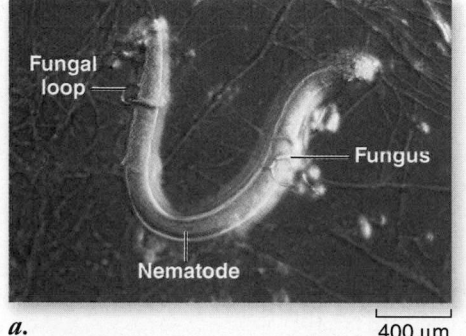

Fungal loop

Fungus

Nematode

a.

400 μm

b.

Figure 32.5 Carnivorous fungi. *a.* Fungus obtaining nutrients from a nematode. *b.* The oyster mushroom *Pleurotus ostreatus* not only decomposes wood but also immobilizes nematodes, which the fungus uses as a source of nitrogen.

enzymatic digestion of cellulose and lignin from plant cell walls. The nematodes it consumes apparently serve mainly as a source of nitrogen—a substance almost always in short supply in biological systems. Other fungi are even more active predators than *Pleurotus,* snaring, trapping, or firing projectiles into nematodes, rotifers, and other small animals on which they prey.

Because of their ability to break down almost any carbon-containing compound—even jet fuel—fungi are of interest for use in bioremediation, using organisms to clean up soil or water that is environmentally contaminated. As one example, some fungal species can remove selenium, an element that is toxic in high accumulations, from soils by combining it with other harmless volatile compounds.

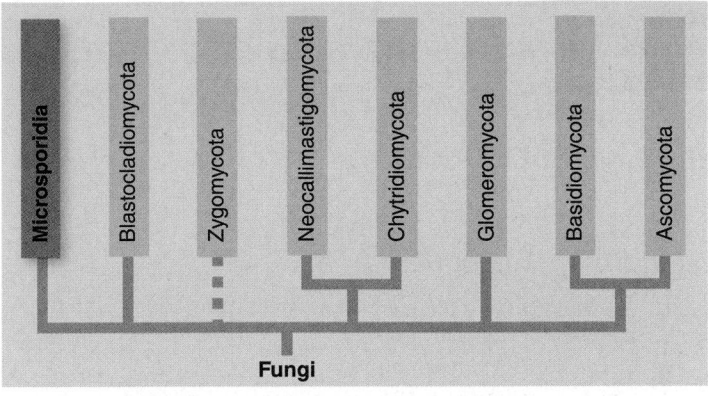

Learning Outcomes Review 32.1

Fungi are more closely related to animals than to plants. Many, but not all, fungi form seven monophyletic phyla. A fungus consists of a mass of hyphae (cells) termed a mycelium; cell walls contain the polysaccharide chitin. Mitosis in fungi divides the nucleus but not the hypha itself. Sexual reproduction may occur when hyphae of two different mating types fuse. Haploid nuclei from each type may persist separately in some groups, termed a dikaryon stage. Spores are produced sexually or asexually and are spread by wind or animals. Fungi secrete digestive enzymes externally and then absorb the products of the digestion. Fungi can break down almost any organic compound.

■ **What differentiates fungi from animals, since both are heterotrophs?**

■ **What protects hyphae from being digested by the fungus's own secreted enzymes?**

32.2 Microsporidia: Unicellular Parasites

Learning Outcomes

1. *Explain characteristics that led microsporidians to be classified as protists.*
2. *Describe evidence for placing microsporidians with fungi.*

Microsporidia are obligate, intracellular, animal parasites, long thought to be protists. The lack of mitochondria led biologists to believe that microsporidians were in a deep branch of protists that diverged before endosymbiosis led to mitochondria. Genome sequencing of the microsporidian *Encephalitozoon cuniculi* revealed genes related to mitochondrial functions within the tiny 2.9-Mb genome. Finding mitochondrial genes led to the hypothesis that microsporidia ancestors had mitochondria, and that greatly reduced, mitochondrion-derived organelles exist in microsporidia. Coupled with phylogenies derived from analyses of new sequence data, microsporidia have been tentatively moved from the protists to the fungi.

E. cuniculi and other microsporidia commonly cause disease in immunosuppressed patients, such as those with AIDS and people who have received organ transplants. Microsporidians infect hosts with their spores, which contain a polar tube (figure 32.6). The polar tube extrudes the contents of the spore into the cell and the parasite sets up housekeeping in a vacuole. *E. cuniculi* infects intestinal and neuronal cells, leading to diarrhea and neurodegenerative disease. Understanding the phylogenetic placement of the microsporidia is important in identifying effective disease treatments.

Learning Outcomes Review 32.2

Microsporidia lack mitochondria; however, the presence of mitochondrial genes indicates that at one time an ancestral form possessed them. As obligate parasites, microsporidia cause diseases in animals, including humans.

■ **How would you distinguish a microsporidian from the parasitic protistan Plasmodium?**

Figure 32.6 Polar tube of an *Encephalitozoon cuniculi* spore infects cells.

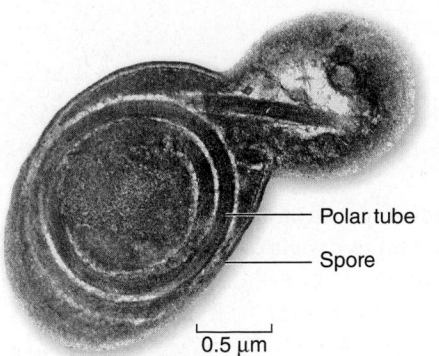

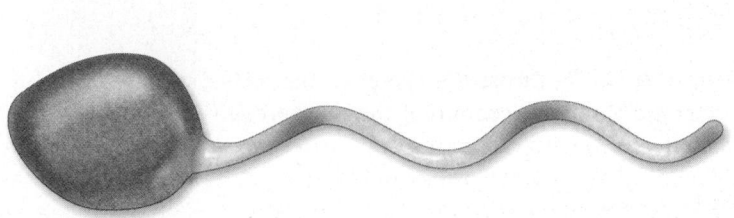

Polar tube
Spore
0.5 μm

Chytridiomycota and Relatives: Fungi with Flagellated Zoospores

Learning Outcomes

1. *Distinguish between blastocladiomycetes and microsporidians.*
2. *Explain the meaning of "chytrid."*
3. *Discuss possible uses of neocallimastigomycetes.*

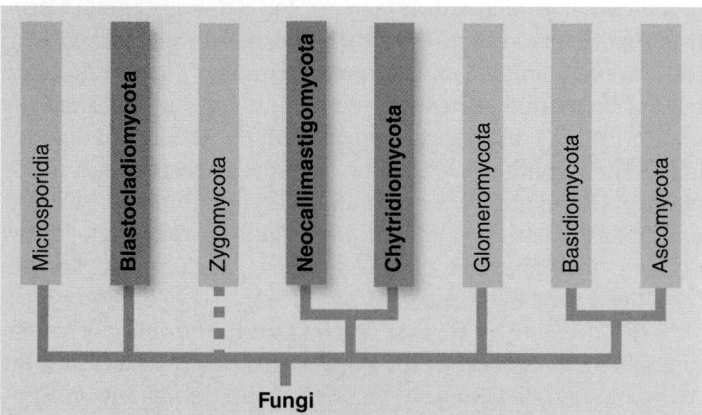

Members of phylum Chytridiomycota, the **chitridiomycetes** or **chytrids,** are aquatic, flagellated fungi that are closely related to ancestral fungi. Motile zoospores are a distinguishing character of this fungal group. Chytrid has its origins in the Greek word *chytridion,* meaning "little pot," referring to the structure that releases the flagellated zoospores (figure 32.7).

Chytrids include *Batrachochytrium dendrobatidis,* which has been implicated in the die-off of amphibians. Other chytrids have been identified as plant pathogens (figure 32.8). Traditionally the blastocladiomycetes and the neocallimastigomycetes have been grouped with the chytrids. Recent phylogenetic inferences have accorded the two groups their own phyla. Both proposed phyla are discussed here because of their many similarities with the chytrids.

Blastocladiomycota have single flagella

Blastocladiomycetes have uniflagellated zoospores. Blastocladiomycetes, neocallimastigomycetes, and chytridiomycetes were originally grouped as a single phylum because they all have flagella that have been lost in other groups, except the microsporidians. Inclusion of multiple genes in phylogenetic analyses has established that the three groups form three separate, monophyletic phyla.

Allomyces is a typical blastocladiomycete genus. A water mold, it exhibits a true haplodiplontic life cycle (figure 32.9). Reproduction in *Allomyces* species is enhanced by the secretion of a pheromone, a long-distance chemical signal, from the female gametes that attracts male gametes. Pheromones are similar to hormones, but work between organisms, rather than within organisms. The *Allomyces* pheromone is called sirenin (after the Sirens in Greek mythology) and was the first fungal sex hormone to be identified chemically.

Unlike microsporidia, which lack mitochondria, *A. macrogynus* has giant mitochondria in its zoospores. Each flagellated zoospore contains a single giant mitochondrion that is fragmented into several normal-sized organelles in vegetative cells.

Neocallimastigomycota digest cellulose in ruminant herbivores

Within the rumen (first digestive chamber) of mammalian herbivores, **neocallimastigomycetes** enzymatically digest the cellulose and lignin of the plant biomass in their grassy diet. Sheep, cows, kangaroos, and elephants all depend on these

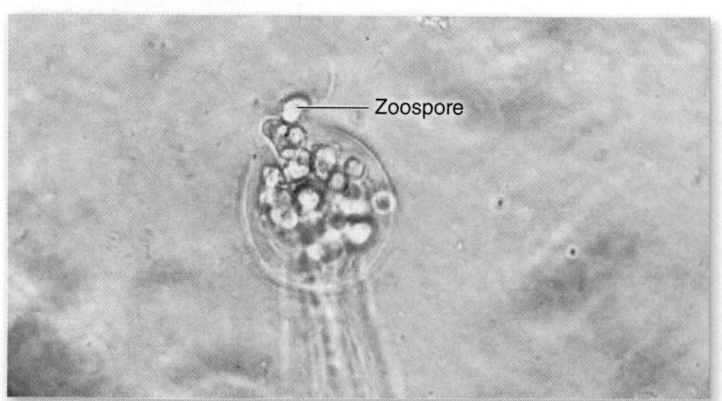

Figure 32.7 Zoospore release. The potlike structure (*chytridion* in Greek) containing the zoospores gives chytrids their name.

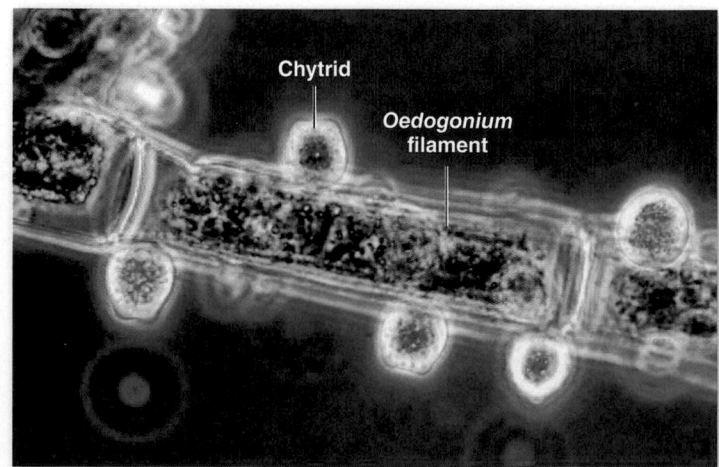

Figure 32.8 Chytridiomycota can be plant pathogens. *Rhizophydium granulosporum* sporangia on a filament of the green algae *Oedogonium.*

Figure 32.9
Allomyces,* a blastocladiomycete that grows in the soil. *a.* The** spherical sporangia can produce either diploid zoospores via mitosis or haploid zoospores via meiosis. ***b. Life cycle of *Allomyces,* which has both haploid and diploid multicellular stages (alternation of generations).

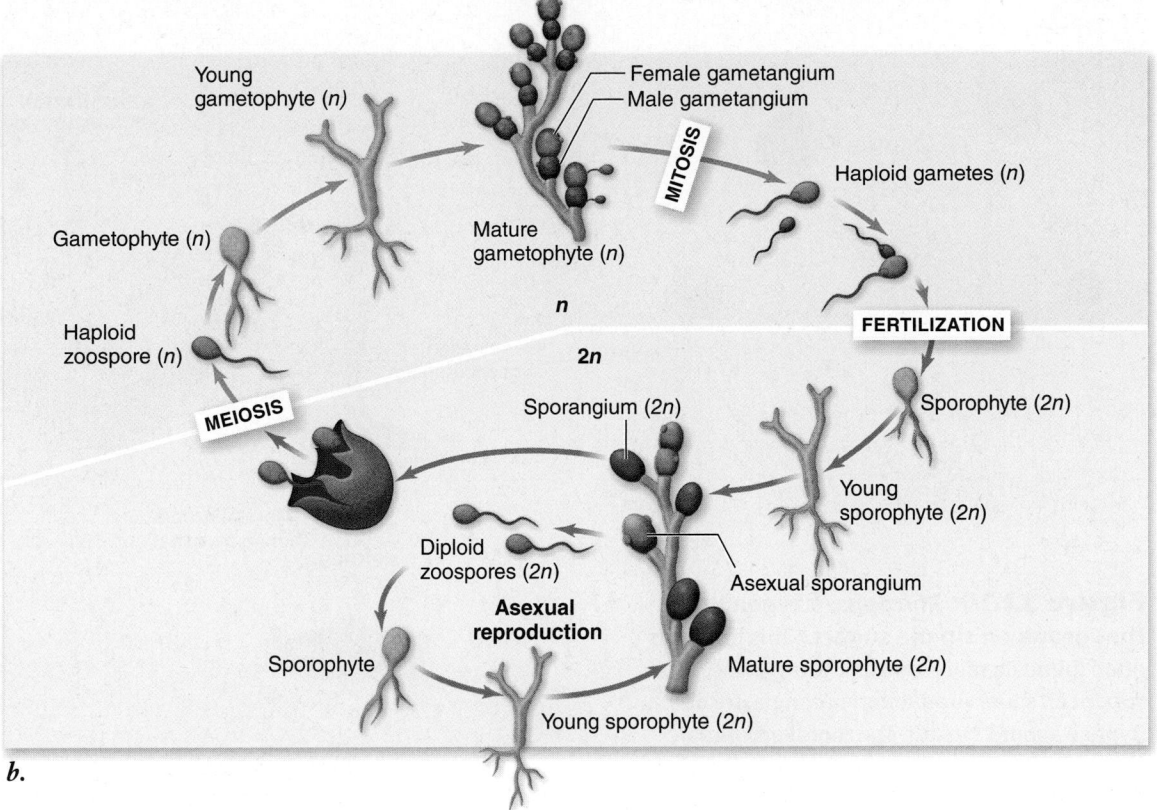

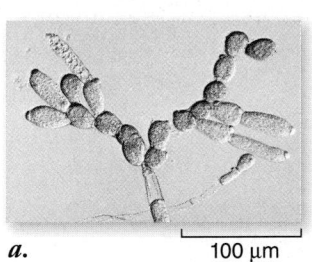

a.   100 μm   *b.*

fungi to obtain sufficient calories. These anaerobic fungi have greatly reduced mitochondria that lack cristae. Their zoospores have multiple flagella. "Mastig" in Neocalli*mastigo*mycota is Latin for "whips," referencing the multiple flagella.

The genus *Neocallimastix* can survive on cellulose alone. Genes encoding digestive enzymes such as cellulase made their way into the *Neocallimastix* genomes via horizontal gene transfer from bacteria. Horizontal gene transfer is covered in chapter 24.

The many enzymes that neocallimastigomycetes use to digest cellulose and lignin in plant cell walls may be useful in biofuel production from cellulose. Although it is possible to obtain ethanol from cellulose, breaking down the cellulose is a major technical hurdle. The use of neocallimastigomycetes fungi to produce cellulosic ethanol is a promising, cost-effective approach.

Learning Outcomes Review 32.3

Three closely related phyla, Chytridiomycota, Blastocladiomycota, and Neocallimastigomycota have flagellated zoospores. Chytrid refers to the potlike shape of the structure releasing the zoospores. *Allomyces,* a representative blastocladiomycete, has uniflagellated zoospores with giant mitochondria, and a haplodiplontic life cycle. Neocallimastigomycetes acquired cellulases from bacteria. They aid ruminant animals in digesting cellulose from plants and may be useful in production of biofuels.

■ **What two features differentiate blastocladiomycetes from microsporidians?**

32.4 ## *Zygomycota: Fungi That Produce Zygotes*

Learning Outcomes

1. Describe the defining feature of the zygomycetes.
2. Explain the advantage of zygospore formation.

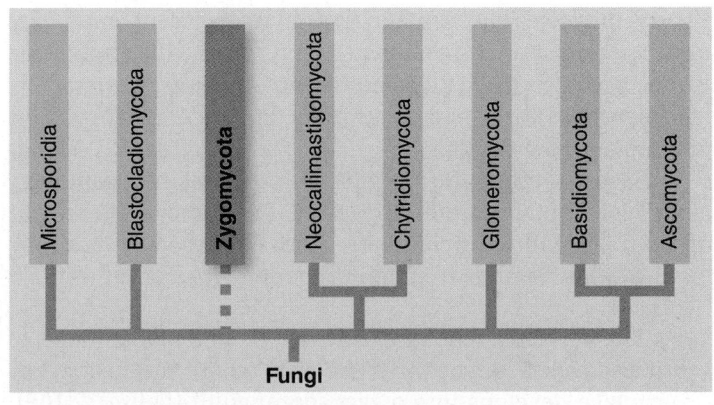

Zygomycetes (phylum Zygomycota) include only about 1050 named species, but they are incredibly diverse. The zygomycota are not monophyletic, but are included in this chapter as a group while research on their evolutionary history

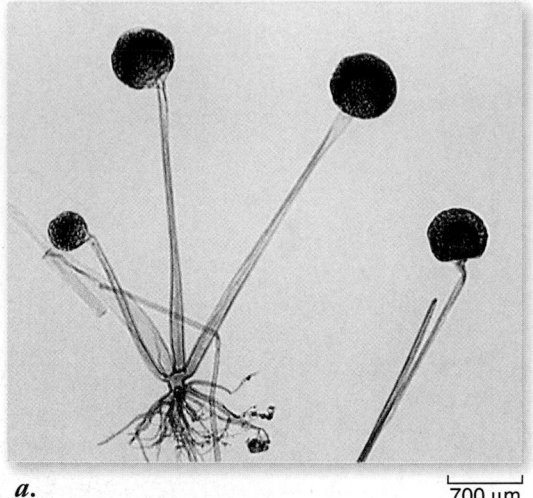

a.

700 μm

Figure 32.10 *Rhizopus,* **a zygomycete that grows on simple sugars.** This fungus is often found on moist bread or fruit. *a.* The dark, spherical, spore-producing sporangia are on hyphae about 1 cm tall. The rootlike hyphae (rhizoids) anchor the sporangia. *b.* Life cycle of *Rhizopus.* The Zygomycota group is named for the zygosporangia characteristic of *Rhizopus.* The (+) and (–) denote mating types.

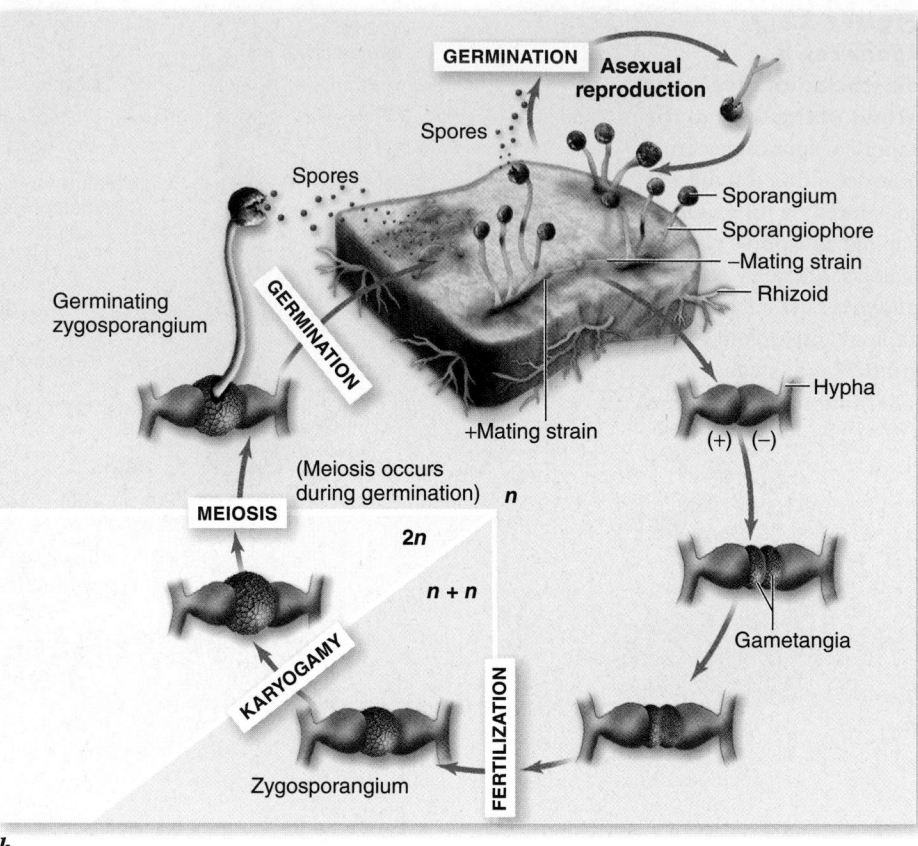

b.

continues. Among them are some of the more common bread molds, which have been assigned to the monophyletic subphylum Mucoromycotina (figure 32.10), as well as a variety of others found on decaying organic material, including strawberries and other fruits. A few human pathogens are in this group.

In sexual reproduction, zygotes form inside a zygosporangium

The zygomycetes lack septa in their hyphae except when they form sporangia or gametangia (structures in which spores or gametes are produced). The group is named after a characteristic feature of the sexual phase of the life cycle, the formation of diploid zygote nuclei.

Sexual reproduction begins with the fusion of gametangia, which contain numerous nuclei. The gametangia are cut off from the hyphae by complete septa. These gametangia may be formed on hyphae of different mating types or on a single hypha.

The haploid nuclei fuse to form a diploid zygote nucleus, a process called karyogamy. The area where the fusion has taken place develops into a zygosporangium (figure 32.10*b*), within which a **zygospore** develops. The zygospore, which may contain one or more diploid nuclei, acquires a thick coat that helps the fungus survive conditions not favorable for growth.

Meiosis, followed by mitosis, occurs during the germination of the zygospore, which releases haploid spores. Haploid

hyphae grow when these haploid spores germinate. Except for the zygote nuclei, all nuclei of the zygomycetes are haploid.

Asexual reproduction is more common

Asexual reproduction occurs much more frequently than sexual reproduction in the zygomycetes. During asexual reproduction, hyphae produce clumps of erect stalks, called **sporangiophores.** The tips of the sporangiophores form sporangia, which are separated by septa. Thin-walled haploid spores are produced within the sporangia. These spores are shed above the food substrate, in a position where they may be picked up by the wind and dispersed to a new food source.

Learning Outcomes Review 32.4

Zygomycetes are named for the production of diploid zygote nuclei by fusion of haploid nuclei, a process called karyogamy. The hyphae of zygomycetes are multinucleate, with septa only where gametangia or sporangia are separated. Many zygomycetes form characteristic resting structures called zygosporangia, which contain zygospores that are able to withstand harsh conditions.

■ *Under what conditions would you expect a zygomycete to produce zygospores rather than haploid spores?*

32.5 Glomeromycota: Asexual Plant Symbionts

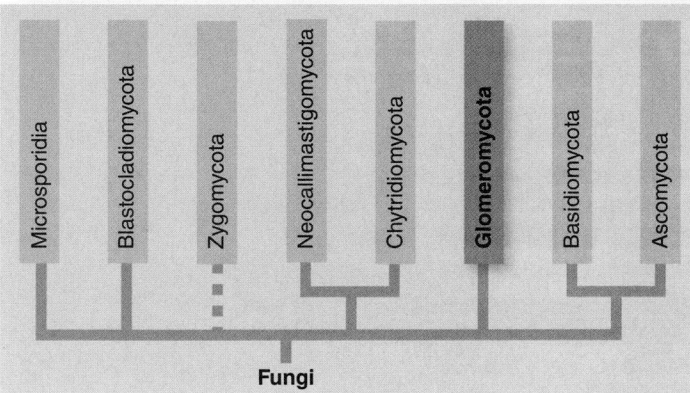

The glomeromycetes, a tiny group of fungi with approximately 150 described species, likely made the evolution of terrestrial plants possible. Tips of hyphae grow within the root cells of most trees and herbaceous plants, forming a branching structure that allows nutrient exchange. The intracellular associations with plant roots are called **arbuscular mycorrhizae.** The specifics of arbuscular mycorrhizal associations and other forms of mycorrhizal interactions are detailed in section 32.8.

Glomeromycetes cannot survive in the absence of a host plant, although the plant host is viable in the absence of the fungus. The symbiotic relationship is mutualistic, with the glomeromycetes providing essential minerals, especially phosphorous, and the plants providing carbohydrates.

The glomeromycetes are challenging to characterize, in part because there is no evidence of sexual reproduction. These fungi exemplify our emerging understanding of fungal phylogeny. Like zygomycetes, glomeromycetes lack septae in their hyphae and were once grouped with the zygomycetes. However, comparisons of DNA sequences of small-subunit rRNAs reveal that glomeromycetes are a monophyletic clade that is phylogenetically distinct from zygomycetes. Unlike zygomycetes, glomeromycetes lack zygospores. Glomeromycota originated at least 600 to 620 MYA, well before the split of the Ascomycota and Basidiomycota, which we will consider next.

32.6 Basidiomycota: The Club (Basidium) Fungi

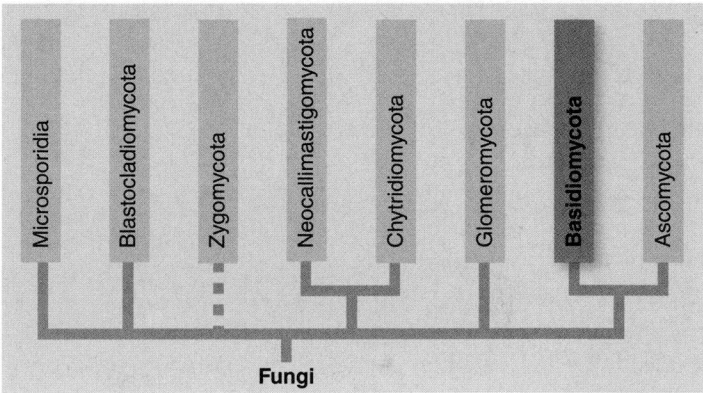

The basidiomycetes (phylum Basidiomycota) include some of the most familiar fungi. Among the basidiomycetes are not only the mushrooms, toadstools, puffballs, jelly fungi, and shelf fungi, but also many important plant pathogens, including rusts and smuts (figure 32.11a). Rust infections resemble rusting metals, and smut infections appear black and powdery due to the spores. Many mushrooms are used as food, but others are hallucinogenic or deadly poisonous.

Basidiomycetes sexually reproduce within basidia

Basidiomycetes are named for their characteristic sexual reproductive structure, the club-shaped **basidium** (plural, *basidia*). Karyogamy (fusion of two nuclei) occurs within the basidium, giving rise to the only diploid cell of the life cycle (figure 32.11b). Meiosis occurs immediately after karyogamy. In the basidiomycetes, the four haploid products of meiosis are incorporated into **basidiospores.** In most members of this phylum, the basidiospores are borne at the end of the basidia on slender projections (sterigmata).

The secondary mycelium of basidiomycetes is heterokaryotic

The life cycle of a basidiomycete continues with the production of monokaryotic hyphae after spore germination. These hyphae lack septa early in development. Eventually, septa form between the nuclei of the monokaryotic hyphae. A basidiomycete mycelium made up of monokaryotic hyphae is called a *primary mycelium.*

Different mating types of monokaryotic hyphae may fuse, forming a dikaryotic mycelium, or *secondary mycelium.* Such a

a.

Figure 32.11
Basidiomycetes.

a. The death cap mushroom, *Amanita phalloides.* These mushrooms are usually lethal to animals that eat them. *b.* Life cycle of a basidiomycete. The basidium is the reproductive structure.

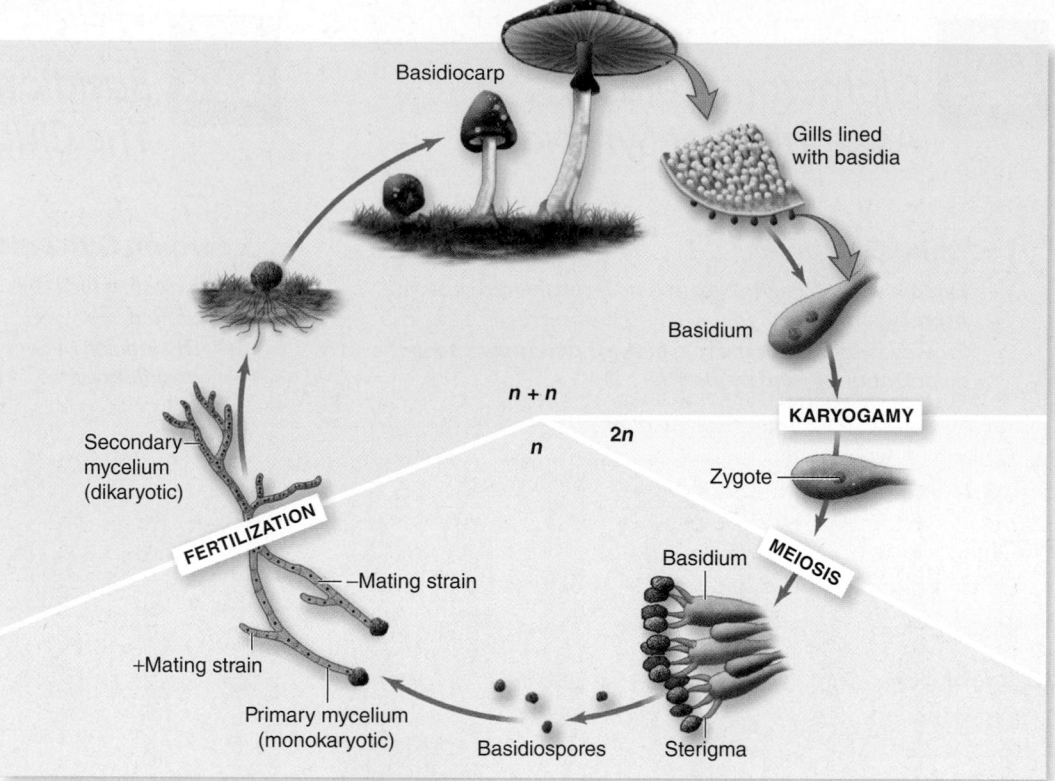

b.

mycelium is heterokaryotic, with two nuclei representing the two different mating types, between each pair of septa. This stage is the dikaryon stage described earlier as being a distinguishing feature of fungi. It is found in both the ascomycetes and the basidiomycetes. The two phyla are grouped as the subkingdom Dikarya because of this commonality. The maintenance of two genomes in the heterokaryon allows for more genetic plasticity than in a diploid cell with one nucleus. One genome may compensate for mutations in the other.

The **basidiocarps,** or mushrooms, are formed entirely of secondary (dikaryotic) mycelium. Gills, sheets of tissue on the undersurface of the cap of a mushroom, produce vast numbers of minute spores. It has been estimated that a mushroom with a cap measuring 7.5 cm in diameter produces as many as 40 million spores per hour!

 Data analysis You compared spore production of the basidiocarp *Cryptoporus volvatus* growing on fir trees by placing a stainless steel funnel under the fungus to collect and count spores produced between June 1st and September 1st. One of your samples had a surface area of 1.10 cm² and produced a total of 1.31×10^8 spores contrasted with a second with an 8.37 cm² surface area that produced 6.87×10^9 spores. Make a quantitative statement about the relative amount of spore production per square centimeter in the two fungi.

Learning Outcomes Review 32.6

Basidiomycetes undergo karyogamy, after which meiosis occurs within club-shaped basidia. The primary mycelium consists of monokaryotic hyphae resulting from spore germination. The secondary mycelium of basidiomycetes is the dikaryon stage, in which two nuclei exist within a single hyphal segment.

■ *What distinguishes a dikaryotic cell from a diploid cell?*

32.7 Ascomycota: The Sac (Ascus) Fungi

Learning Outcomes

1. *Compare the ascomycetes and the basidiomycetes.*
2. *List the ways ascomycetes affect humans.*

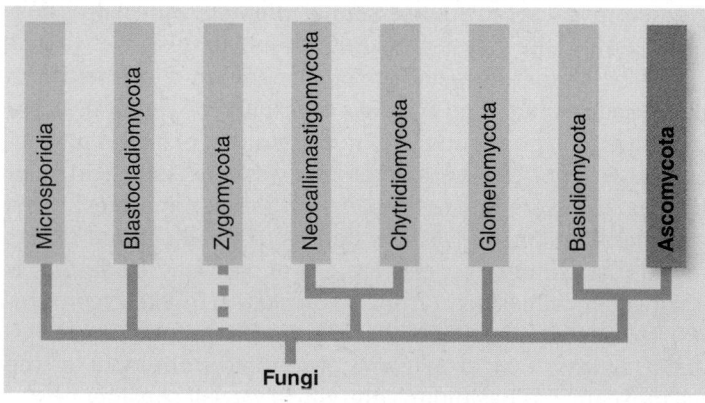

The phylum **Ascomycota** contains about 75% of the known fungi. Among the ascomycetes are such familiar and economically important fungi as bread yeasts, common molds, morels (figure 32.12*a*), cup fungi (figure 32.12*b*), and truffles. Also included in this phylum are many serious plant pathogens, including those that produce chestnut blight, *Cryphonectria parasitica,* and Dutch elm disease, *Ophiostoma ulmi.* Penicillin-producing ascomycetes are in the genus *Penicillium.*

Sexual reproduction occurs within the ascus

The ascomycetes are named for their characteristic reproductive structure, the microscopic, saclike **ascus** (plural, *asci*). Karyogamy, the production of the only diploid nucleus of the ascomycete life cycle (figure 32.12c), occurs within the ascus. The structure of an ascus differs from that of a basidium, although functionally the two are identical.

Asci are differentiated within a structure made up of densely interwoven hyphae, corresponding to the visible portions of a morel or cup fungus, called the **ascocarp.** Meiosis immediately follows karyogamy, forming four haploid daughter nuclei. These usually divide again by mitosis, producing eight haploid nuclei that become walled **ascospores.** The ascospores of the ascomycetes are borne internally in asci instead of externally as in basidiospores.

In many ascomycetes, the ascus becomes highly turgid at maturity and ultimately bursts, often at a preformed area. When this occurs, the ascospores may be thrown as far as 32 cm, an amazing distance considering that most ascospores are only about 10 μm long. This would be equivalent to throwing a baseball (diameter 7.5 cm) 1.25 km—about 10 times the length of a home run!

Asexual reproduction occurs within conidiophores

Asexual reproduction is very common in the ascomycetes. It takes place by means of **conidia** (singular, *conidium*), asexual spores cut off by septa at the ends of modified hyphae called conidiophores. Conidia allow for the rapid colonization of a new food source. Many conidia are multinucleate. The hyphae of ascomycetes are divided by septa, but the septa are perforated, and the cytoplasm flows along the length of each hypha. The septa that cut off the asci and conidia are initially perforated, but later become blocked.

Some ascomycetes have yeast morphology

Most yeasts are ascomycetes with a single-celled lifestyle. Most of yeasts' reproduction is asexual and takes place by cell fission or budding, when a smaller cell forms from a larger one (figure 32.13). Sometimes two yeast cells fuse, forming one cell containing two nuclei. This cell may then function as an ascus, with karyogamy followed immediately by meiosis. The resulting ascospores function directly as new yeast cells.

a.

b.

Figure 32.12
Ascomycetes. *a.* This morel, *Morchella esculenta,* is a delicious edible ascomycete that appears in early spring. *b.* A cup fungus. *c.* Life cycle of an ascomycete. Haploid ascospores form within the ascus.

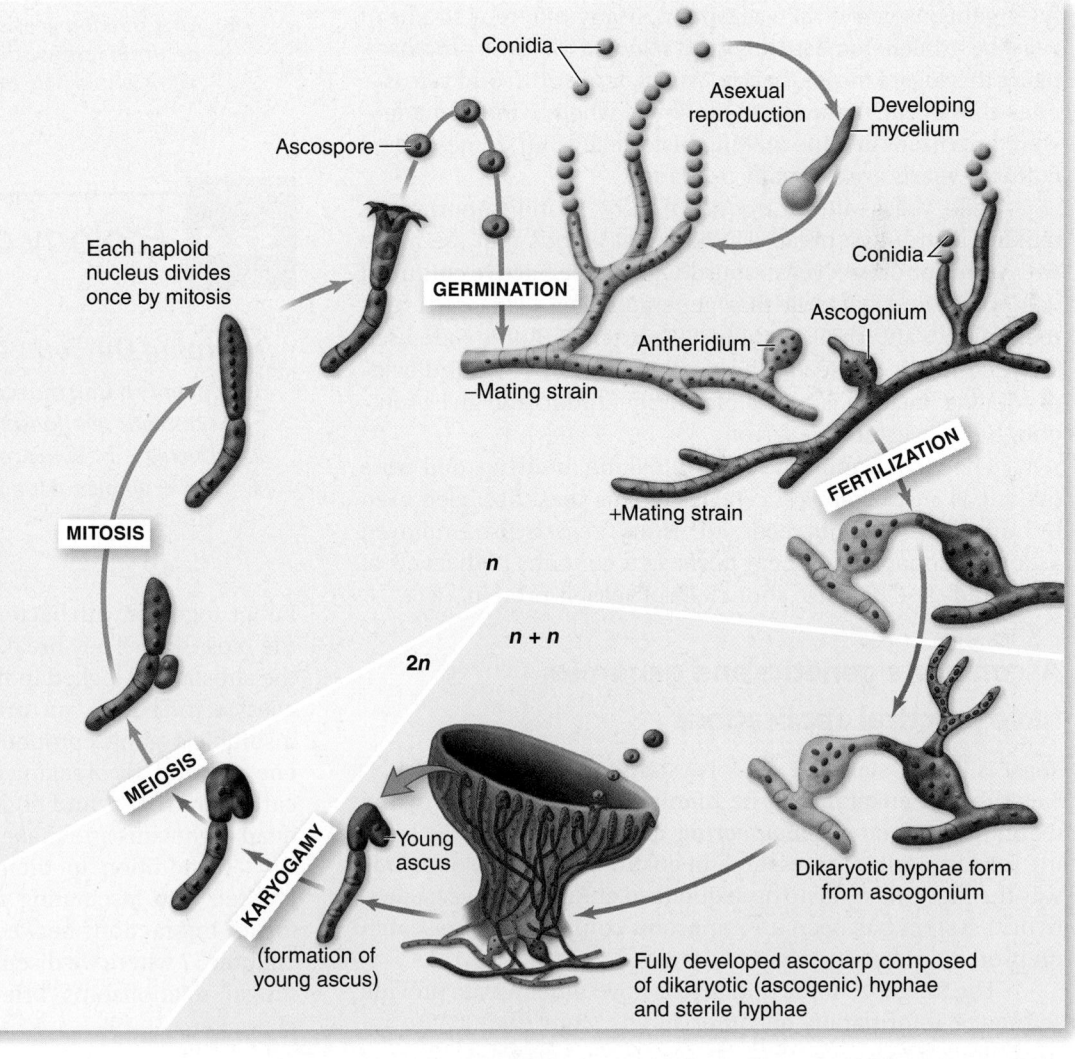

c.

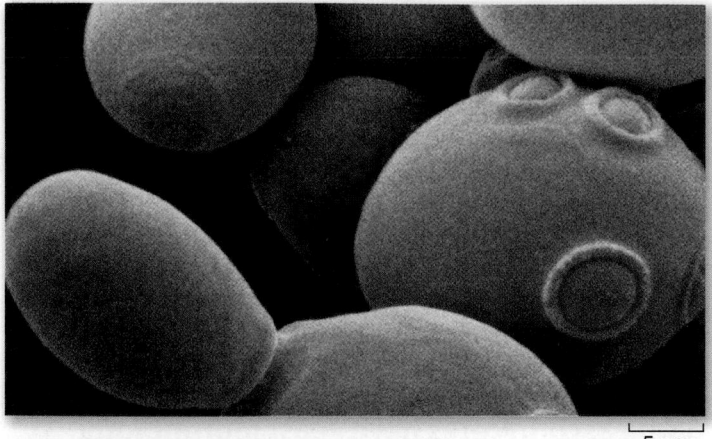

Figure 32.13 Budding in *Saccharomyces*. As shown in this scanning electron micrograph, the cells tend to hang together in chains, a feature that calls to mind the derivation of single-celled yeasts from multicellular ancestors.

The ability of yeasts to ferment carbohydrates, breaking down glucose to produce ethanol and carbon dioxide, is fundamental to the production of bread, beer, wine, and distilled spirits such as whiskey. About four billion of these tiny, powerful organisms can fit in a teaspoon. Many different strains of yeast have been domesticated and selected for these processes, using the sugars in rice, barley, wheat, and corn. Wild yeasts—ones that occur naturally in the areas where wine is made—were important in wine making historically, but domesticated cultured yeasts are normally used now.

Wild yeast, often *Candida milleri,* is still important in making sourdough bread. Unlike most breads that are made with pure cultures of yeast, sourdough uses an active culture of wild yeast and bacteria that generates acid. This culture is maintained, and small amounts—"starter" cultures—are used for each batch of bread. The combination of yeast and acid-producing bacteria is needed for fermentation and gives sourdough bread its unique flavor.

The most important yeast in baking, brewing, and wine making is *Saccharomyces cerevisiae.* This yeast has been used by humans throughout recorded history. Yeast is also employed as a nutritional supplement because it contains high levels of B vitamins and because about 50% of yeast is protein.

Ascomycete genetics and genomics have practical applications

Yeast is a long-standing model system for genetic research. It was the first eukaryote to be manipulated extensively by the techniques of genetic engineering, and they still play the leading role as models for research in eukaryotic cells. *S. cerevisiae* was the first eukaryote to be sequenced entirely. The yeast two-hybrid system has been an important component of research on protein interactions (see chapter 18).

The fungal genome initiative is now under way to provide sequence information on other fungi. More than 17 fungal genomes have been completed and 205 are being sequenced. These fungi were selected based on their effects on human health, including plant pathogens that threaten our food supply.

The ascomycetes *Coccidioides posadasii* and its relative *C. immitis* were included because they are endemic in soil in the southwestern portion of the United States, can cause a fatal infection called coccidioidomycosis ("valley fever"), and have been considered a possible bioterrorism threat. In the United States the annual infection rate is about 100,000 individuals, although only a small percentage of infected individuals die.

A second important criterion in selecting which fungi to sequence was the potential to provide information on fungal evolution. This new information will complement and expand our understanding of the diverse fungal kingdom.

Learning Outcomes Review 32.7

Ascomycetes undergo karyogamy within a characteristic saclike structure, the ascus. The function of the ascus is therefore identical to that of the basidium, although they are structurally different. Meiosis follows, resulting in the production of ascospores. Yeasts within this group generally reproduce asexually by budding. Ascomycetes include both beneficial forms used as foods, and in the production of foods, and harmful forms responsible for diseases and spoilage. Many ascomycetes are employed in scientific research.

■ *Coccidioidomycosis is caused by inhaling spores; it often occurs in farmworkers in the Southwest United States. What would help prevent this disease?*

32.8 *Ecology of Fungi*

Learning Outcomes

1. *Identify a trait that contributes to the value of fungi in symbiotic relationships.*
2. *Describe the living components of a lichen.*
3. *List examples of fungal associations with different organisms.*

Fungi, together with bacteria, are the principal decomposers in the biosphere. They break down organic materials and return the substances locked in those molecules to circulation in the ecosystem. Fungi can break down cellulose and lignin, an insoluble organic compound that is one of the major constituents of wood. By breaking down such substances, fungi release carbon, nitrogen, and phosphorus from the bodies of living or dead organisms and make them available to other organisms.

In addition to their role as decomposers, fungi have entered into fascinating relationships with a variety of lifeforms. Interactions between different species are described in chapter 57 where we discuss community ecology, but we cover fungal relationships briefly here because of their unique character.

Fungi have a range of symbioses

The interactions, or symbioses, between fungi and other living organisms fall into a broad range of categories. In some cases, the symbiosis is an **obligate symbiosis** (essential for survival), and in other cases it is a **facultative symbiosis** (the fungus can survive without the host). Within a group of closely related fungi, several different types of symbiosis can be found.

First, here is a summary of ways in which living things can interact. **Pathogens** and **parasites** gain resources from their host, but they have a negative effect on the host that can even lead to death. The difference between pathogens and parasites is that pathogens cause disease, but parasites do not, except in extreme cases.

Commensal relationships benefit one partner but do not harm the other. Fungi that are in a **mutualistic** relationship benefit both themselves and their hosts. Many of these relationships are described in the discussion that follows.

Endophytes live inside plants and may protect plants from parasites

Endophytic fungi live inside plants, actually in the intercellular spaces. Found throughout the plant kingdom, many of these relationships may be examples of parasitism or commensalism.

There is growing evidence that some of these fungi protect their hosts from herbivores by producing chemical toxins or deterrents. Most often, the fungus synthesizes alkaloids that protect the plant. As you will learn in chapter 39, plants also synthesize a wide range of alkaloids, many of which serve to defend the plant.

One way to assess whether an endophyte is enhancing the health of its host plant is to grow plots of plants with and without the same endophyte. An experiment with perennial ryegrass, *Lolium perenne,* demonstrated that it is more resistant to aphid feeding when an endophytic fungus, *Neotyphodium,* is present (figure 32.14).

 ## Lichens are an example of symbiosis between different kingdoms

Lichens (figure 32.15) are symbiotic associations between a fungus and a photosynthetic partner. Although many lichens are excellent examples of mutualism, some fungi are parasitic on their photosynthetic host.

Composition of a lichen

Ascomycetes are the fungal partners in all but about 20 of the approximately 15,000 species of lichens estimated to exist. Most of the visible body of a lichen consists of its fungus, but between the filaments of that fungus are cyanobacteria, green algae, or sometimes both (figure 32.16).

Specialized fungal hyphae penetrate or envelop the photosynthetic cell walls within them and transfer nutrients directly to the fungal partner. Note that although fungi penetrate the cell wall, they do not penetrate the plasma membrane.

SCIENTIFIC THINKING

Hypothesis: *Endophytic fungi can protect their host from herbivory.*
Prediction: *There will be fewer aphids (Rhopalosiphum padi, an herbivore) on perennial ryegrass (Lolium perenne) infected with endophytic fungi than on uninfected ryegrass.*
Test: *Place five adult aphids on each pot of 2-week-old grass plants with and without endophytic fungi. Place pots in perforated bags and grow for 36 days. Count the number of aphids in each pot.*

Fungal endophyte   No endophyte

Result: *Significantly more aphids were found on the uninfected grass plants.*

Conclusion: *Endophytic fungi protect host plants from herbivory.*
Further Experiments: *How do you think the fungi protect the plants from herbivory? If they secrete chemical toxins, could you use this basic experimental design to test specific fungal compounds?*

Figure 32.14 Effect of the fungal endophyte *Neotyphodium* on the aphid population living on perennial ryegrass (*Lolium perenne*).

Fruticose Lichen	Foliose Lichen	Crustose Lichen

a.   *b.*   *c.*

Figure 32.15 Lichens are found in a variety of habitats. *a.* A fruticose lichen, growing in the soil. *b.* A foliose ("leafy") lichen, growing on the bark of a tree in Oregon. *c.* A crustose lichen, growing on rocks leading to the breakdown of rock into soil.

Biochemical signals sent out by the fungus apparently direct its cyanobacterial or green algal component to produce metabolic substances that it does not produce when growing independently of the fungus.

The fungi in lichens are unable to grow normally without their photosynthetic partners, and the fungi protect their partners from strong light and desiccation. When fungal components of lichens have been experimentally isolated from their photosynthetic partner, they survive, but grow very slowly.

Ecology of lichens

The durable construction of the fungus combined with the photosynthetic properties of its partner have enabled lichens to invade the harshest habitats—the tops of mountains, the farthest northern and southern latitudes, and dry, bare rock faces in the desert. In harsh, exposed areas, lichens are often the first colonists, breaking down the rocks and setting the stage for the invasion of other organisms.

Lichens are often strikingly colored because of the presence of pigments that probably play a role in protecting the photosynthetic partner from the destructive action of the Sun's rays. These same pigments may be extracted from the lichens and used as natural dyes. The traditional method of manufacturing Scotland's famous Harris tweed used fungal dyes.

Lichens vary in sensitivity to pollutants in the atmosphere, and some species are used as bioindicators of air quality. Their sensitivity results from their ability to absorb substances dissolved in rain and dew. Lichens are generally absent in and around cities because of automobile traffic and industrial activity, but some are adapted to these conditions. As pollution decreases, lichen populations tend to increase.

 Mycorrhizae are fungi associated with roots of plants

The roots of about 90% of all plant families have species that are involved in mutualistic symbiotic relationships with certain kinds of fungi. It has been estimated that these fungi probably amount to 15% of the total weight of the world's plant roots. Associations of this kind are termed **mycorrhizae,** from the Greek words for fungus and root.

The fungi in mycorrhizal associations function as extensions of the root system. The fungal hyphae dramatically increase the amount of soil contact and total surface area for absorption. When mycorrhizae are present, they aid in the direct transfer of phosphorus, zinc, copper, and other mineral nutrients from the soil into the roots. The plant, on the other hand, supplies organic carbon to the fungus, so the system is an example of mutualism.

There are two principal types of mycorrhizae (figure 32.17). In arbuscular mycorrhizae, the fungal hyphae penetrate the outer cells of the plant root, forming coils, swellings, and minute branches; they also extend out into the surrounding soil. In **ectomycorrhizae,** the hyphae surround but do not penetrate the cell walls of the roots. In both kinds of

Algal cells

Fungal hyphae

40 μm

Figure 32.16 Stained section of a lichen. This section shows fungal hyphae *(purple)* more densely packed into a protective layer on the top and, especially, the bottom layer of the lichen. The blue cells near the upper surface of the lichen are those of a green alga. These cells supply carbohydrate to the fungus.

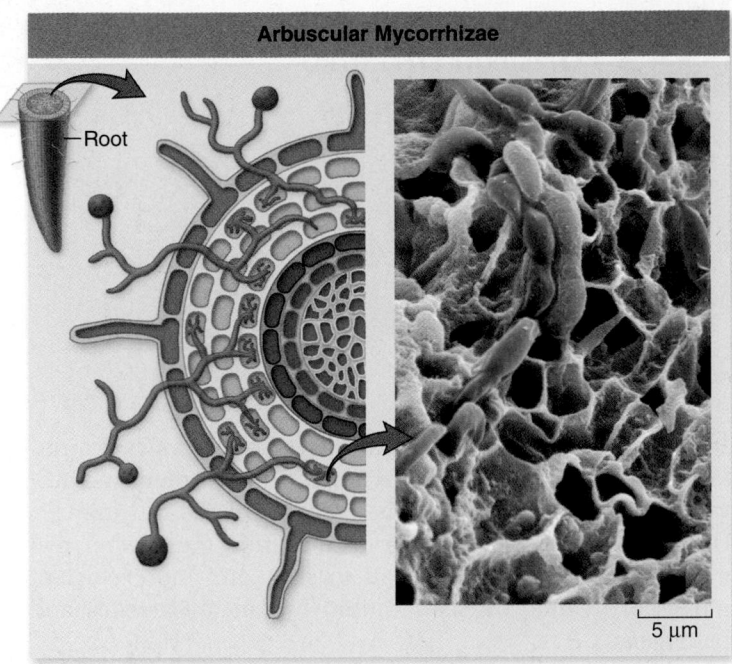

a.

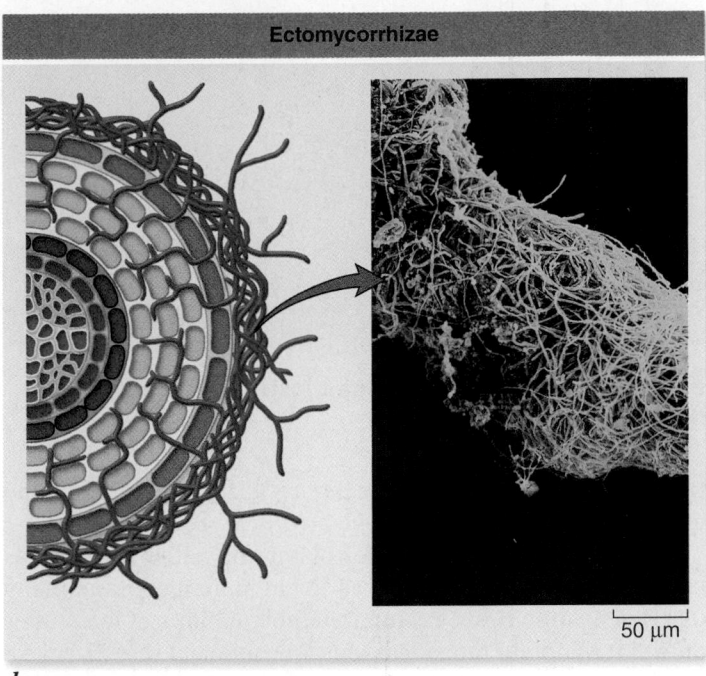

b.

Figure 32.17 **Arbuscular mycorrhizae and ectomycorrhizae.** *a.* In arbuscular mycorrhizae, fungal hyphae penetrate the root cell wall of plants but not the plant membranes. *b.* Ectomycorrhizae on the roots of a *Eucalyptus* tree do not penetrate root cells, but grow around and extend between the cells.

mycorrhizae, the mycelium extends far out into the soil. A single root may associate with many fungal species, dividing the root at a millimeter-by-millimeter level.

Arbuscular mycorrhizae

Arbuscular mycorrhizae are by far the more common of the two types, involving roughly 70 to 80% of all plant species (figure 32.17*a*). The fungal component in them are glomeromycetes, a monophyletic group that arose within one of the zygomycete lineages. The glomeromycetes are associated with more than 200,000 species of plants.

Unlike mushrooms, none of the glomeromycetes produce aboveground fruiting structures, and as a result, it is difficult to arrive at an accurate count of the number of extant species. Arbuscular mycorrhizal fungi are being studied intensively because they are potentially capable of increasing crop yields with lower phosphate and energy inputs.

The earliest fossil plants often show arbuscular mycorrhizal roots. Such associations may have played an important role in allowing plants to colonize land. The soils available at such times would have been sterile and lacking in organic matter. Plants that form mycorrhizal associations are particularly successful in infertile soils; considering the fossil evidence, it seems reasonable that mycorrhizal associations helped the earliest plants succeed on such soils. In addition, the closest living relatives of early vascular plants surviving today continue to depend strongly on mycorrhizae.

Some nonphotosynthetic plants also have mycorrhizal associations, but the symbiosis is one-way because the plant has no photosynthetic resources to offer. Instead of a two-partner symbiosis, a tripartite symbiosis is established. The fungal mycelium extends between a photosynthetic plant and a nonphotosynthetic, parasitic plant. This third, nonphotosynthetic member of the symbiosis is called an *epiparasite*. Not only does it obtain phosphate from the fungus, but it uses the fungus to channel carbohydrates from the photosynthetic plant to itself. Epiparasitism also occurs in ectomycorrhizal symbiosis.

Ectomycorrhizae

Ectomycorrhizae (see figure 32.17*b*) involve far fewer kinds of plants than do arbuscular mycorrhizae—perhaps a few thousand. Most ectomycorrhizal hosts are forest trees, such as pines, oaks, birches, willows, eucalyptus, and many others. The fungal components in most ectomycorrhizae are basidiomycetes, but some are ascomycetes.

Most ectomycorrhizal fungi are not restricted to a single species of plant, and most ectomycorrhizal plants form associations with many ectomycorrhizal fungi. Different combinations have different effects on the physiological characteristics of the plant and its ability to survive under different environmental conditions. At least 5000 species of fungi are involved in ectomycorrhizal relationships.

Fungi also form mutual symbioses with animals

A range of mutualistic fungi–animal symbioses has been identified. Ruminant animals host neocallimastigomycete fungi in their gut. The fungus gains a nutrient-rich environment in exchange for releasing nutrients from grasses with high cellulose and lignin content.

One tripartite symbiosis involves ants, plants, and fungi. Leaf-cutter ants are the dominant herbivore in the New World tropics. These ants, members of the phylogenetic tribe Attini, have an obligate symbiosis with specific fungi that they have domesticated and maintain in an underground garden. The ants provide fungi with leaves to eat and protection from

Figure 32.18 Ant–fungi symbiosis. Ants farming their fungal garden.

pathogens and other predators (figure 32.18). The fungi are the ants' food source.

Depending on the species of ant, the ant nest can be as small as a golf ball or as large as 50 cm in diameter and many feet deep. Some nests are inhabited by millions of leaf-cutter ants that maintain fungal gardens. These social insects have a caste system, and different ants have specific roles. Traveling on trails as long as 200 m, leaf-cutter ants search for foliage for their fungi. A colony of ants can defoliate an entire tree in a day. This ant farmer–fungi symbiosis has evolved multiple times and may have occurred as early as 50 MYA.

Learning Outcomes Review 32.8

Fungi are the primary decomposers in ecosystems. A range of symbiotic relationships have evolved between fungi and plants. Endophytes live inside tissues of a plant and may protect it from parasites. Lichens are a complex symbiosis between fungal species and cyanobacteria or green algae. Mycorrhizal associations between fungi and plant roots are mutually beneficial and in some cases are obligate symbioses. Fungi have also coevolved with animals in mutualistic relationships.

■ *How might the symbiosis between fungi and ants have evolved?*

32.9 Fungal Parasites and Pathogens

Learning Outcomes

1. Review the pathogenic effects of fungi and the targets they affect.
2. Explain why treating fungal disease in animals is particularly difficult.

Fungi can destroy a crop of plants and create significant problems for human health. A major problem in treatment and prevention is that fungi are eukaryotes, as are plants and animals. Understanding how fungi are distinct from these other two eukaryotic kingdoms may lead to safer and more efficient means of treating diseases caused by fungal parasites and pathogens.

Fungal infestation can harm plants and those who eat them

Fungal species cause many diseases in plants (figure 32.19), and they are responsible for billions of dollars in agricultural losses every year. Not only are fungi among the most harmful pests of living plants, but they also spoil food products that have been harvested and stored. In addition, fungi often secrete substances into the foods they are infesting that make these foods unpalatable, carcinogenic, or poisonous.

Pathogenic fungi–plant symbioses are numerous, and fungal pathogens of plants can also harm the animals that consume the plants. *Fusarium* species growing on spoiled food produce highly toxic substances, including vomitoxin, which has been implicated in brain damage in humans and animals in the southwestern United States.

Figure 32.19 World's largest organism?
a. *Armillaria*, a pathogenic fungus shown here afflicting three discrete regions of coniferous forest in Montana, grows out from a central focus as a single, circular clone. The large patch at the bottom of the picture is almost 8 hectares. **b.** Closeup of tree destroyed by *Armillaria*. **c.** *Armillaria* growing on a tree.

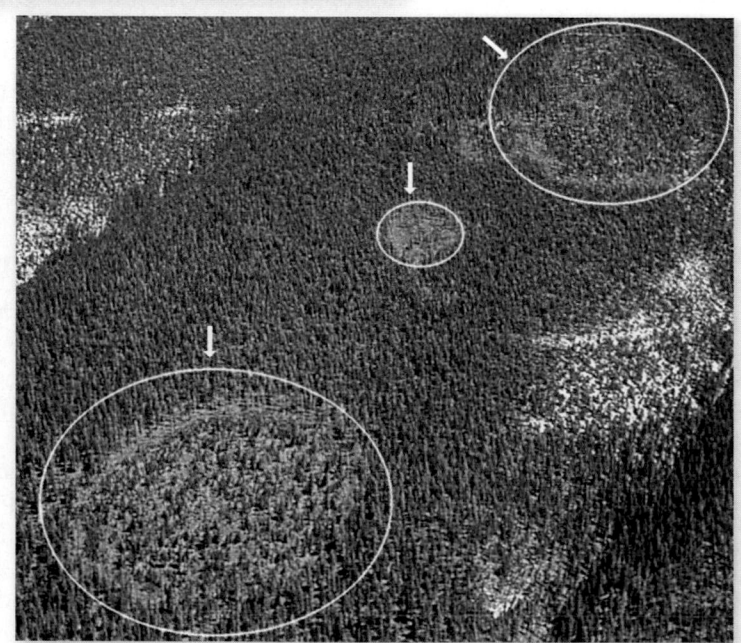

a.

b.

c.

a.   *b.*   5 µm

Figure 32.20 Maize (corn) fungal infections. *a. Ustilago maydis* infections of maize are a delicacy in Hispanic cuisine. *b.* A photomicrograph of *Aspergillus flavus* conidia. *Aspergillus flavus* infects maize and can produce aflatoxins that are harmful to animals.

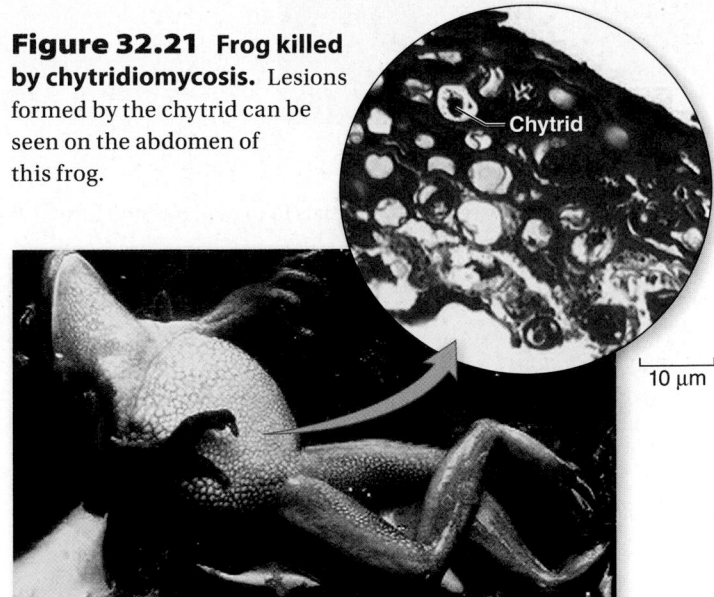

Figure 32.21 Frog killed by chytridiomycosis. Lesions formed by the chytrid can be seen on the abdomen of this frog.

Chytrid

10 µm

Aflatoxins, which are among the most carcinogenic compounds known, are produced by some *Aspergillus flavus* strains growing on corn, peanuts, and cotton seed (figure 32.20). Aflatoxins can also damage the kidneys and the nervous system of animals, including humans. Most developed countries have legal limits on the concentration of aflatoxin permitted in different foods. More recently, aflatoxins have been considered as possible bioterrorism agents.

In contrast, corn smut is a maize fungal disease that is harmful to the plant, but not animals that consume it (see figure 32.20). Corn smut is caused by the basidiomycete *Ustilago maydi* and is edible.

Fungal infections are difficult to treat in humans and other animals

Human and animal diseases can also be fungal in origin. Some common diseases, such as ringworm (which is not a worm but a fungus), athlete's foot, and nail fungus, can be treated with topical antifungal ointments and in some cases with oral medication.

Fungi can create devastating human diseases that are often difficult to treat because of the close phylogenetic relationship between fungi and animals. Yeast ascomycetes are important pathogens that cause diseases such as thrush, an infection of the mouth; the yeast *Candida* causes common oral or vaginal infections. *Pneumocystis jiroveci* (formerly *P. carinii*) invades the lungs, disrupting breathing, and can spread to other organs. In immune-suppressed AIDS patients, this infection can lead to death.

Mold allergies are common, and mold-infested "sick" buildings pose concerns for inhabitants. Individuals with suppressed immune systems and people undergoing steroid treatments for inflammatory disorders are particularly at risk for fungal disease.

An example of a parasitic fungi–animal symbiosis is **chytridiomycosis,** first identified in 1998 as an emerging infectious disease of amphibians. Amphibian populations have been declining worldwide for over three decades. The many possible causes for the decline in amphibian numbers are covered in chapter 60. In this chapter, a primary causative agent, fungal infection, is explored. The decline correlates with the presence of the chytrid *Batrachochytrium dendrobatidis* encased in the skin (figure 32.21), identified after extensive studies of frog carcasses. Sick and dead frogs were more likely than healthy frogs to have flasklike structures encased in their skin, which proved to be associated with chytrid spore production (see figure 32.21).

The connection with *B. dendrobatidis* has been supported by DNA sequence data, by isolating and culturing the chytrid, and by infecting healthy frogs with the organism and replicating disease symptoms. The *B. dendrobatidis* genome has now been sequenced. The infection reduces sodium and potassium transport across the skin, altering electrolyte balance which leads to cardiac arrest. Bathing frogs in antifungal drugs can halt the disease or even eliminate the chytrids.

How the disease emerged simultaneously on different continents is a yet unsolved mystery. Both environmental change and carriers are being considered.

Learning Outcomes Review 32.9

Fungi can severely harm or kill both plants and animals, either by direct infection or by secretion of toxins and carcinogens. Treatment of fungal disease and parasitism in animals is made difficult by the close relationship between fungi and animals; what is damaging to the fungus may also have ill effects on the host.

■   *What is likely to be the most common mechanism for the spread of fungal disease?*

Chapter Review

32.1 Defining Fungi

Fungi are more closely related to animals than to plants and form seven monophyletic phyla (figure 32.1).

Fungi are heterotrophic and have hyphal cells; their cell walls contain chitin. They may have a dikaryon stage and undergo nuclear mitosis.

The body of a fungus is a mass of connected hyphae.

A mass of connected hyphae is termed a mycelium. Hyphae can be continuous and multinucleate, or they may be divided into long chains of cells separated by cross-walls called septa.

The chitin found in fungi cell walls is the same material found in the exoskeletons of arthropods.

Fungal cells may have more than one nucleus.

A hypha with only one nucleus is monokaryotic; a hypha with two nuclei is dikaryotic. The two haploid nuclei exist independently, but both genomes are transcribed so that some properties of diploids may be observed.

Mitosis is not followed by cell division.

Because cells are linked, the cell is not the relevant unit of reproduction, but rather the nucleus is. The spindle forms inside the nuclear envelope, which does not break down and re-form.

Fungi can reproduce both sexually and asexually.

Fungi can reproduce sexually by fusion of hyphae from two compatible mating types or hyphae from the same fungus. Spores can form either by asexual or sexual reproduction and are usually dispersed by the wind.

Fungi are heterotrophs that absorb nutrients.

Fungi obtain their nutrients through excreting enzymes for external digestion and then absorbing the products.

32.2 Microsporidia: Unicellular Parasites

Microsporidia are obligate cellular parasites that lack mitochondria but may have had them at one time; they were previously classed with the protists.

32.3 Chytridiomycota and Relatives: Fungi with Flagellated Zoospores

Chytrids form symbiotic relationships; they have been implicated in the decline of amphibian species.

Blastocladiomycota have single flagella.

Allomyces, a blastocladiomycete, is an example of a fungus with a haplodiplontic life cycle (figure 32.9).

Neocallimastigomycota digest cellulose in ruminant herbivores.

Neocallimastigomycetes have enzymes that can digest cellulose and lignin; they may have uses in production of biofuels.

32.4 Zygomycota: Fungi That Produce Zygotes

In sexual reproduction, zygotes form inside a zygosporangium (figure 32.10).

Zygomycetes all produce a diploid zygote. In sexual reproduction, fusion (karyogamy) of the haploid nuclei of gametangia produces diploid zygote nuclei. These become zygospores.

Asexual reproduction is more common.

Sporangia produce airborne haploid spores; bread mold is a common example of a zygomycete.

32.5 Glomeromycota: Asexual Plant Symbionts

Glomeromycete hyphae form intracellular associations with plant roots and are called arbuscular mycorrhizae.

The glomeromycetes show no evidence of sexual reproduction.

32.6 Basidiomycota: The Club (Basidium) Fungi

The basidiocarp is the visible reproductive structure of this group, which includes mushrooms, toadstools, puffballs, and others.

Basidiomycetes sexually reproduce within basidia (figure 32.11).

Karyogamy occurs within the basidia, giving rise to a diploid cell. Meiosis then ultimately results in four haploid basidiospores.

The secondary mycelium of basidiomycetes is heterokaryotic.

Primary mycelium is monokaryotic, but different mating types may fuse to form the secondary mycelium. Maintenance of two haploid genomes allows greater genetic plasticity.

32.7 Ascomycota: The Sac (Ascus) Fungi

Sexual reproduction occurs within the ascus (figure 32.12).

Karyogamy occurs only in the ascus and results in a diploid nucleus. Meiosis and mitosis then result in eight haploid nuclei in walled ascospores.

Asexual reproduction occurs within conidiophores.

Asexual reproduction is very common and occurs by means of conidia formed at the end of modified hyphae called conidiophores.

Some ascomycetes have yeast morphology.

Yeasts usually reproduce by cell fission or budding.

Ascomycete genetics and genomics have practical applications.

32.8 Ecology of Fungi

Fungi are organisms capable of breaking down cellulose and lignin.

Fungi have a range of symbioses.

Fungi can be pathogenic or parasitic, commensal or mutualistic.

Endophytes live inside plants and may protect plants from parasites.

Lichens are an example of symbiosis between different kingdoms.

A lichen is composed of a fungus, usually an ascomycete, along with cyanobacteria, green algae, or both.

Mycorrhizae are fungi associated with roots of plants.

Arbuscular mycorrhizae are common and involve glomeromycetes; ectomycorrhizae are primarily found in forest trees and involve basidiomycetes and a few ascomycetes.

Fungi also form mutual symbioses with animals.

Some ants grow "farms" of fungi by providing plant material.

32.9 Fungal Parasites and Pathogens

Fungal infestation can harm plants and those who eat them.

Fungi spread via spores and can secrete chemicals that make food unpalatable, carcinogenic, or poisonous.

Fungal infections are difficult to treat in humans and other animals.

Treatment of fungal diseases in animals is difficult because of the similarities between the two kingdoms.

UNDERSTAND

1. Which of the following is NOT a characteristic of a fungus?

   a. Cell walls made of chitin
   b. A form of mitosis different from plants and animals
   c. Ability to conduct photosynthesis
   d. Filamentous structure

2. A fungal cell that contains two genetically different nuclei would be classified as

   a. monokaryotic.   c. homokaryotic.
   b. bikaryotic.   d. heterokaryotic.

3. Which of the following groups of fungi is NOT monophyletic?

   a. Zygomycota   c. Glomeromycota
   b. Basidiomycota   d. Ascomycota

4. Based on physical characteristics, the _____ represent the most ancient phylum of fungi.

   a. Basidiomycota   c. Ascomycota
   b. Zygomycota   d. Chytridiomycota

5. The early evolution of terrestrial plants was made possible by mycorrhizal relationships with the

   a. Zygomycetes.   c. Ascomycota.
   b. Glomeromycota.   d. Basidiomycota.

6. Symbiotic relationships occur between the fungi and

   a. plants.   c. animals.
   b. bacteria.   d. All of the choices are correct.

7. Which of the following species of fungi is NOT associated with diseases in humans?

   a. *Pneumocystis jiroveci*
   b. *Aspergillus flavus*
   c. *Candida albicans*
   d. *Batrachochytrium dendrobatidis*

APPLY

1. In a culture of hyphae of unknown origin you notice that the hyphae lack septa and that the fungi reproduce asexually by using clumps of erect stalks. However, at times sexual reproduction can be observed. To what group of fungi would you assign it?

   a. Chytridiomycota   c. Ascomycota
   b. Basidiomycota   d. Zygomycota

2. Examine the life cycle of a typical basidiomycetes and determine where you would expect to find a dikaryotic cell.

   a. Primary mycelium   c. In the basidiospores
   b. Secondary mycelium d. In the zygote

3. Determine which of the following is correct regarding the yeast *Saccharomyces cerevisiae*.

   a. It reproduces asexually by a process called budding.
   b. It produces an ascocarp during reproduction.
   c. It belongs in the group Zygomycota.
   d. All of the choices are correct.

4. Appraise the fungal relationship between a forest tree and a basidiomycetes and determine the most suitable classification for the symbiosis.

   a. Parasitism only
   b. An arbuscular mycorrhizae
   c. Ectomycorrhizae
   d. A lichen

5. Choose which of the following best reflects the symbiotic relationships between animals and fungi.

   a. Protection from bacteria
   b. Colonization of land
   c. Protection from desiccation
   d. Exchange of nutrients

SYNTHESIZE

1. Historically fungi have been classified as being more plantlike despite their lack of photosynthetic ability. Although we now know that fungi are more closely related to the animals than the plants, review characteristics that initially led scientists to place them closer to the plants.

2. The importance of fungi in the evolution of terrestrial life is typically understated. Evaluate the importance of fungi in the colonization of land.

3. Based on your understanding of fungi, hypothesize why antibiotics won't work in the treatment of a fungal infection.

4. You are looking through a seed catalog and discover you have a choice of pea seed that is treated or not treated with a fungicide. Identify one reason why you might choose the treated seed and one reason you might prefer the untreated seed.

ONLINE RESOURCE

www.ravenbiology.com

Understand, Apply, and Synthesize—enhance your study with animations that bring concepts to life and practice tests to assess your understanding. Your instructor may also recommend the interactive eBook, individualized learning tools, and more.

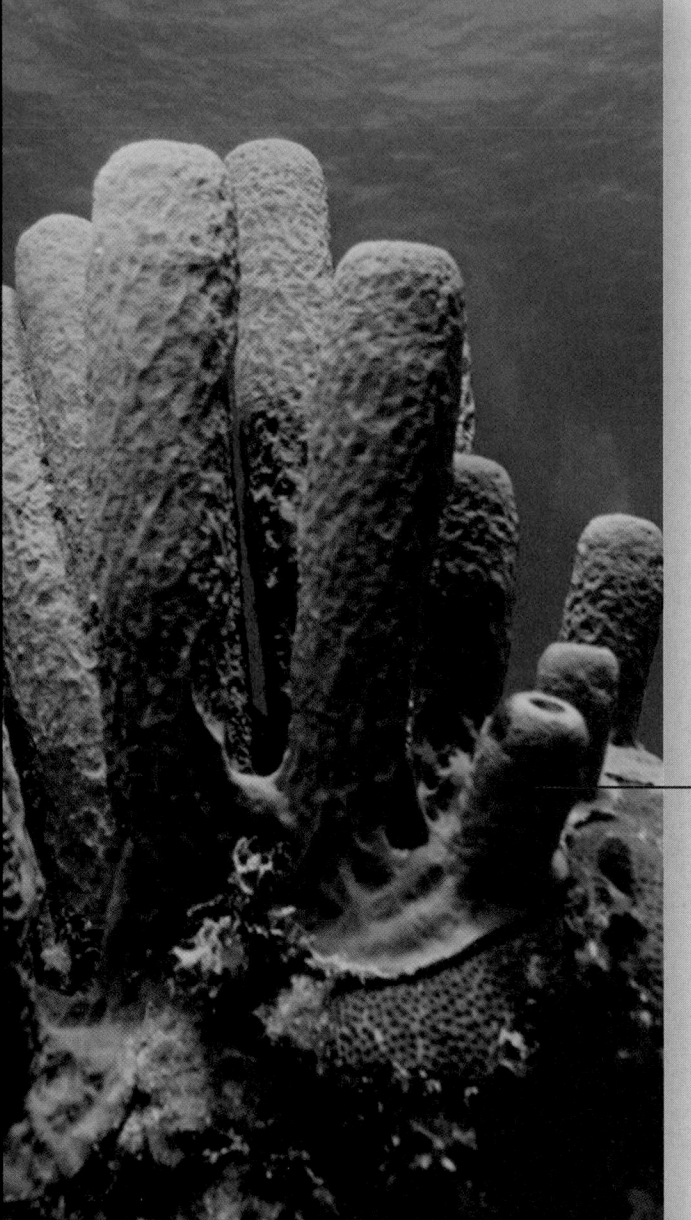

Chapter 33

Animal Diversity and the Evolution of Body Plans

Chapter Contents

Introduction

Animals are among the most abundant living organisms. Found in almost every habitat, they bewilder us with their diversity of form, habitat, behavior, and lifestyle. About a million and a half species have been described, and several million more are thought to await discovery. Despite their great diversity, animals have much in common. For example, locomotion is a distinctive characteristic, although not all animals can move about. Early naturalists thought that sponges and corals were plants because the adults are attached to the surface on which they live. In this and the next two chapters, we explore the great diversity of modern animals, the result of a long evolutionary history. We start in this chapter with a discussion of the features that characterize animals, and review the phyla at the base of the animal phylogeny. In the next two chapters, we discuss the two major groups of animals: the protostomes and deuterostomes.

33.1 Some General Features of Animals

Learning Outcome

1. *Identify three features that characterize all animals and three that characterize only some types of animals.*

Animals are so diverse that few criteria fit them all. But some, such as animals being eaters, or consumers, apply to all. Others, such as animals being mobile (they can move about), have exceptions. Taken together, the universal characteristics and other features of major importance exhibited by most species are convincing evidence that animals are monophyletic—that they descended from a common ancestor.

Animals share some general characteristics

Heterotrophy

All animals are heterotrophs—that is, they obtain energy and organic molecules by ingesting other organisms. Unlike autotrophic plants and algae, animals cannot construct organic molecules from inorganic chemicals. Some animals (herbivores) consume autotrophs; other animals (carnivores) consume heterotrophs; some animals (omnivores) consume both autotrophs and heterotrophs; and still others (detritivores) consume decomposing organisms.

Multicellularity

All animals are multicellular; many have complex bodies. Unicellular heterotrophic organisms, which were at one time regarded as simple animals, are now considered members of several different clades within the large and diverse group of protists, discussed in chapter 29.

No cell walls

Animal cells differ from those of other multicellular organisms: they lack rigid cell walls and are usually quite flexible. The many cells of animal bodies are held together by extracellular frames of structural proteins such as collagen. Other proteins form unique intercellular junctions between animal cells.

Active movement

Although other single-celled organisms are able to travel from place to place, animals move more rapidly and in more complex ways—this ability is perhaps their most striking characteristic, one directly related to the flexibility of their cells and the evolution of nerve and muscle tissues. A remarkable form of movement unique to animals is flying, a capability that is well developed among vertebrates and insects such as the butterfly (phylum Arthropoda) shown in figure 33.1a. Many animals cannot move from place to place (they are sessile) or do so rarely or slowly (they are sedentary) although they have muscles or muscle fibers that allow parts of their bodies to move. Sponges, however, have little capacity for movement.

a.

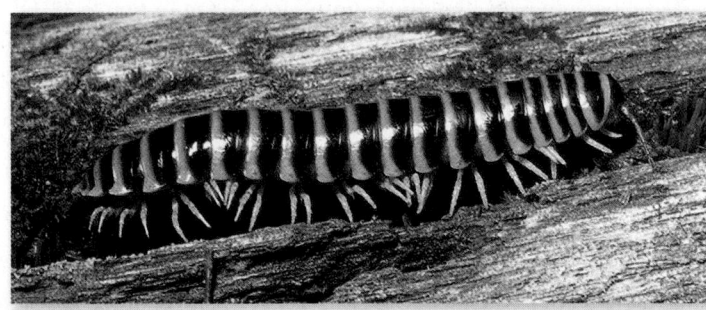
b.

c.

Figure 33.1 Some characteristics shared by animals. *a.* Many, but not all animals, are able to move from place to place. Other organisms are also able to move, but flying is unique to animals. *b.* Animals exhibit a wide array of different sizes and forms, but the vast majority of animals are invertebrates, lacking a backbone, such as this millipede. *c.* Animals undergo a process of development beginning with series of cell divisions called cleavage, producing a multicelled structure called a blastula.

Diversity of form

Animals vary greatly in form, ranging in size from organisms too small to see with the unaided eye to enormous whales and giant squids. Almost all animals lack a backbone—they are therefore called invertebrates, like the millipede (phylum Arthropoda) in figure 33.1*b*. Of the million known living animal species, fewer than 60,000 have a backbone—they are therefore referred to as vertebrates. Probably most of the many millions of animal species awaiting discovery are invertebrates.

Diversity of habitat

Animals are grouped into 35–40 phyla, most of which have members that occur only in the sea. Members of fewer phyla occur in fresh water, and members of fewer still occur on land. Members of three phyla that are successful in the marine environment—Arthropoda, Mollusca, and Chordata—also dominate animal life on land. Only one animal phylum, Onychophora (velvet worms), is entirely terrestrial.

Sexual reproduction

Most animals reproduce sexually. Animal eggs, which are non-mobile, are much larger than the small, usually flagellated sperm. In animals, cells formed in meiosis function as gametes. These haploid cells do not divide by mitosis first, as they do in plants and fungi, but rather fuse directly with each other to form the zygote. Consequently, there is no counterpart among animals to the alternation of haploid (gametophyte) and diploid (sporophyte) generations characteristic of plants. Some individuals of some species and all individuals of a very few others are incapable of sexual reproduction.

Embryonic development

An animal zygote first undergoes a series of mitotic divisions, called cleavage, and like the dividing frog's egg in figure 33.1*c*, cleavage produces a ball of cells, the blastula. In most animals, the blastula folds inward at one point to form a hollow sac with an opening at one end called the blastopore. An embryo at this stage is called a gastrula. The subsequent growth and movement of the cells of the gastrula differ from one group of animals to another, reflecting the evolutionary history of the group. Embryos of most kinds of animals develop into a larva, which looks unlike the adult of the species, lives in a different habitat, and eats different sorts of food; in most groups, it is very small. A larva undergoes metamorphosis, a radical reorganization, to transform into the adult body form.

Tissues

The cells of all animals except sponges are organized into structural and functional units called tissues—collections of cells that together are specialized to perform specific tasks.

Learning Outcome Review 33.1

All animals are multicellular and heterotrophic, and their cells lack cell walls. Most animals can move from place to place, can reproduce sexually, and possess unique tissues. Animals can be found in almost all habitats.

- ■ *Why could animals not have been the first type of life to have evolved?*

33.2 Evolution of the Animal Body Plan

Learning Outcomes

1. Differentiate between a pseudocoelom and a coelom.
2. Explain the difference between protostomes and deuterostomes.
3. Describe the advantages of segmentation.

The features described in the preceding section evolved over the course of millions of years. We can understand how the history of life has proceeded by examining the types of animal bodies and body plans present in fossils and in existence today. Five key innovations can be noted in animal evolution:

1. Symmetry
2. Tissues, allowing specialized structures and functions
3. A body cavity
4. Various patterns of embryonic development
5. Segmentation, or repeated body units

These innovations are explained in the sections that follow. Some innovations appear to have evolved only once, some twice or more. Innovations that have only evolved a single time can provide evidence of close evolutionary relationship, thus serving as a synapomorphy (shared derived character) for an evolutionary group composed of an ancestral species and all of its descendants, termed a clade (see chapter 23). On the other hand, some innovations evolve more than once in different clades. This is the phenomenon of convergent evolution (see chapter 23). Although not indicative of close evolutionary relationship, convergently evolved innovations may be evidence that distantly related species have adapted in similar ways to similar environmental conditions.

Most animals exhibit radial or bilateral symmetry

A typical sponge lacks definite symmetry, growing as an irregular mass. Virtually all other animals have a definite shape and symmetry that can be defined along an imaginary axis drawn through the animal's body. The two main types of symmetry are radial and bilateral (note that many animals are not perfectly symmetrical, but close enough that they are considered symmetrical).

Radial symmetry

The body of a member of phylum Cnidaria (jellyfish, sea anemones, and corals: the C of Cnidaria is silent; see chapter 34) exhibits **radial symmetry.** Its parts are arranged in such a way that any longitudinal plane passing through the central axis divides the organism into halves that are approximate mirror images. A pie, for example, is radially symmetrical, and so is a sea anemone (figure 33.2*a*).

Radial Symmetry

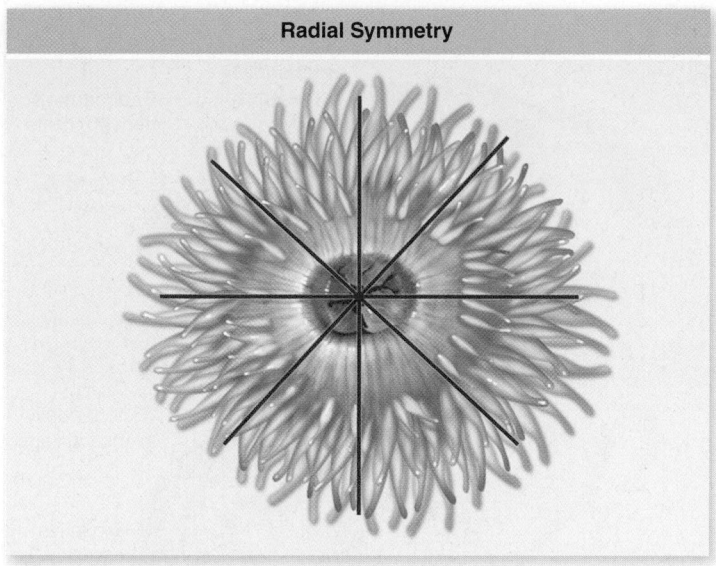

a.

Bilateral Symmetry

Dorsal

Sagittal plane

Posterior

Frontal plane

Anterior

Transverse plane

Ventral

b.

Figure 33.2 **A comparison of radial and bilateral symmetry.** *a.* Radially symmetrical animals, such as this sea anemone (phylum Cnidaria), can be bisected into equal halves by any longitudinal plane that passes through the central axis. *b.* Bilaterally symmetrical animals, such as this turtle (phylum Chordata), can only be bisected into equal halves in one plane (the sagittal plane).

Bilateral symmetry

The bodies of most animals other than sponges and cnidarians exhibit **bilateral symmetry,** in which the body has right and left halves that are mirror images of each other. Animals with this body plan are collectively termed the Bilateria. The sagittal plane defines these halves (figure 33.2*b*). A bilaterally symmetrical body has, in addition to left and right halves, dorsal and ventral portions, which are divided by the frontal plane, and anterior (front) and posterior (rear) ends, which are divided by the transverse plane (in an animal that walks on all fours, dorsal is the top side).

Sometimes, deciding whether an organism should be considered bilaterally or radially symmetrical is not easy. For example, in echinoderms (sea stars and their relatives), adults are radially symmetrical (actually pentaradially symmetrical, because the body has five clear sections), but the larvae are bilaterally symmetrical. In this case, examination of animal phylogeny indicates that the radial symmetry of the adult is an evolutionarily derived condition from a bilaterally symmetrical ancestral condition; combined with the form of the larvae, scientists consider echinoderms to be bilaterally symmetrical.

Bilateral symmetry constitutes a major evolutionary advance in the animal body plan. Bilaterally symmetrical animals have the ability to move through the environment in a consistent direction (typically with the anterior end leading)—a feat that is difficult for radially symmetrical animals. Associated with directional movement is the grouping of nerve cells into a brain, and sensory structures, such as eyes and ears, at the anterior end of the body. This concentration of nervous tissue at the anterior end, which appears to have occurred early in evolution, is called **cephalization.** Much of the layout of the nervous system in bilaterally symmetrical animals is centered on one or more major longitudinal nerve cords that transmit information from the anterior sense organs and brain to the rest of the body. Cephalization is often considered a consequence of the development of bilateral symmetry.

The evolution of tissues allowed for specialized structures and functions

The zygote (a fertilized egg), has the capability of giving rise to all the kinds of cells in an animal's body. That is, it is totipotent (all powerful). During embryonic development, cells specialize in carrying out particular functions. In all animals except sponges, the process is irreversible: once a cell differentiates to serve a function, it and its descendants can never serve any other.

A sponge cell that had specialized to serve one function (such as lining the cavity where feeding occurs) can lose the special attributes that serve that function and change to serve another function (such as being a gamete). Thus a sponge cell can dedifferentiate and redifferentiate. Cells of all other animals are organized into tissues, each of which is characterized by cells of particular morphology and capability. But their competence to dedifferentiate prevents sponge cells from forming clearly defined tissues (and therefore, of course, organs, which are composed of tissues).

Sponges are unique in being able to dedifferentiate and then redifferentiate. The fact that all other animals differentiate irreversibly suggests that organisms with bodies containing cells specialized to serve particular functions may have an advantage over those with cells that potentially have multiple functions. The evolution of specialized tissues may be a key innovation—a trait that fosters evolutionary diversification (see chapter 22).

A body cavity made the development of advanced organ systems possible

In the process of embryonic development, the cells of animals of most groups organize into three layers (called germ layers): an outer **ectoderm,** an inner **endoderm,** and an intermediate **mesoderm.** Animals with three embryonic cell layers are said

chapter **33** *Animal Diversity and the Evolution of Body Plans* **639** to be triploblastic. During maturation from the embryo, certain organs and organ systems develop from each germ layer. The ectoderm gives rise to the outer covering of the body and the nervous system; the endoderm gives rise to the digestive system, including the intestine; and the skeleton and muscles develop from the mesoderm. Cnidarians have only two layers (thus they are diploblastic)—the endoderm and the ectoderm—and lack organs. Sponges lack germ layers altogether; they, of course, have no tissues or organs. All triploblastic animals are members of the Bilateria.

A key innovation in the body plan of some bilaterians was a body cavity isolated from the exterior of the animal. This is different from the digestive cavity, which is open to the exterior at least through the mouth, and in most animals at the opposite end as well, via the anus. The evolution of efficient organ systems within the animal body was not possible until a body cavity evolved for accommodating and supporting organs (such as our heart and lungs), distributing materials, and fostering complex developmental interactions. The cavity is filled with fluid: in most animals, the fluid is liquid, but in vertebrates, it is gas—the body cavity of humans filling with liquid represents a life-threatening condition. A very few types of bilaterians have no body cavity, the space between tissues that develop from the mesoderm and those that develop from the endoderm being filled with cells and connective tissue. These are the so-called acoelomate animals (figure 33.3).

Body cavities

Body cavities appear to have evolved multiple times in the Bilateria. In some animals, a body cavity called the **pseudocoelom** develops embryologically between the mesoderm and endoderm and thus occurs in the adult between tissues derived from the mesoderm and those derived from endoderm; animals with this type of body cavity are termed pseudocoelomates (see figure 33.3). Although the word *pseudocoelom* means "false coelom," this is a true body space and characterizes many groups of animals. A **coelom** is a cavity that develops entirely within the mesoderm. The coelom is surrounded by a layer of epithelial cells derived from the mesoderm and termed the peritoneum.

The circulatory system

In many small animals, nutrients and oxygen are distributed and wastes are removed by fluid in the body cavity. Most larger animals, in contrast, have a **circulatory system,** a network of vessels that carry fluids to and from the parts of the body distant from the sites of digestion (gut) and gas exchange (gills or lungs). The circulating fluid carries nutrients and oxygen to the tissues and removes wastes, including carbon dioxide, by diffusion between the circulatory fluid and the other cells of the body.

In an **open circulatory system,** the blood passes from vessels into sinuses (open areas within the body), mixes with body fluid that bathes the cells of tissues, then reenters vessels in another location. In a **closed circulatory system,** the blood flows entirely within blood vessels, so it is physically separated from other body fluids. Blood moves through a closed circulatory system faster and more efficiently than it does through an open system; open systems are typical of animals that

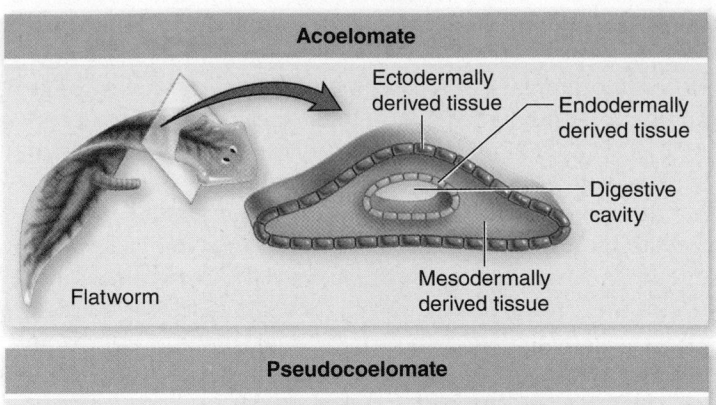
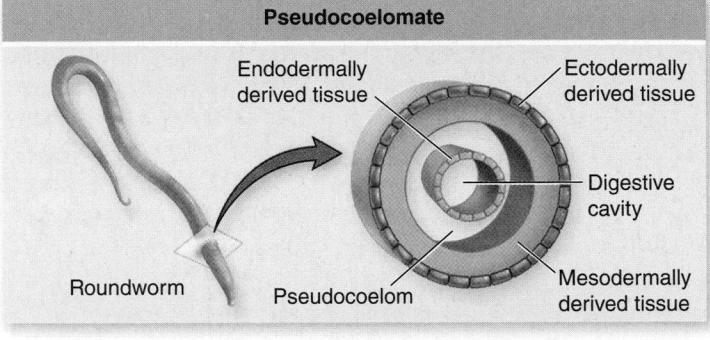
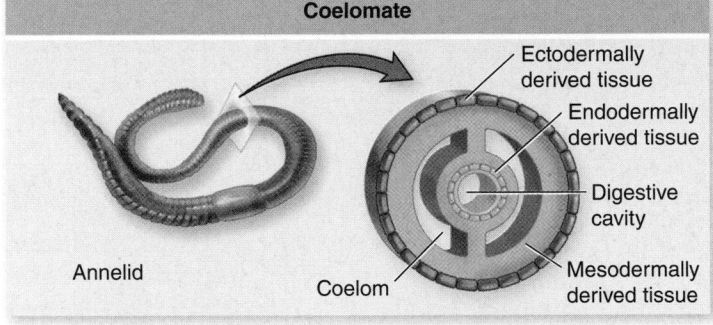

Figure 33.3 Three body plans for bilaterally symmetrical animals. Acoelomates, such as flatworms, have no body cavity between the digestive tract (derived from the endoderm) and the musculature layer (derived from the mesoderm). Pseudocoelomates have a body cavity, the pseudocoelom, between tissues derived from the endoderm and those derived from the mesoderm. Coelomates have a body cavity, the coelom, that develops entirely within tissues derived from the mesoderm, and so is lined on both sides by tissue derived from the mesoderm.

are relatively inactive and so do not have a high demand for oxygen. In small animals, blood can be pushed through a closed circulatory system by the animal's movement. In larger animals, the body musculature does not provide enough force, so the blood must be propelled by contraction of one or more hearts, which are specialized, muscular parts of the blood vessels.

Bilaterians have two main types of development

The processes of embryonic development in animals is discussed fully in chapter 53. Briefly, development of a bilaterally symmetrical animal begins with mitotic cell divisions (called cleavages) of the egg that lead to the formation of a hollow ball of cells, which subsequently indents to form a two-layered ball. The internal space that is created through such indentation (see figure 53.11) is the **archenteron** (literally the "primitive gut"); it communicates with the outside by a **blastopore.**

In a protostome, the mouth of the adult animal develops from the blastopore or from an opening near the blastopore (protostome means "first mouth"—the first opening becomes the mouth). **Protostomes** include most bilaterians, including flatworms, nematodes, mollusks, annelids, and arthropods. In some protostomes, both mouth and anus form from the embryonic blastopore; in other protostomes, the anus forms later in another region of the embryo. Two outwardly dissimilar groups—the echinoderms and the chordates—together with a few other small phyla, constitute the **deuterostomes,** in which the mouth of the adult animal does not develop from the blastopore. The deuterostome blastopore gives rise to the organism's anus, and the mouth develops from a second pore

that arises later in development (deuterostome means "second mouth"). Protostomes and deuterostomes differ in several other aspects of embryology too, as discussed later.

Cleavage patterns

The cleavage pattern relative to the embryo's polar axis determines how the resulting cells lie with respect to one another. In some protostomes, each new cell cleaves off at an angle oblique to the polar axis. As a result, a new cell nestles into the space between the older ones in a closely packed array. This pattern is called **spiral cleavage** because a line drawn through a sequence of dividing cells spirals outward from the polar axis (figure 33.4 top). Spiral cleavage is characteristic of annelids, mollusks, nemerteans, and related phyla; the clade of animals with this cleavage pattern is therefore known as the Spiralia.

In all deuterostomes, by contrast, the cells divide parallel to and at right angles to the polar axis. As a result, the pairs of cells from each division are positioned directly above and below one another, a process that gives rise to a loosely packed

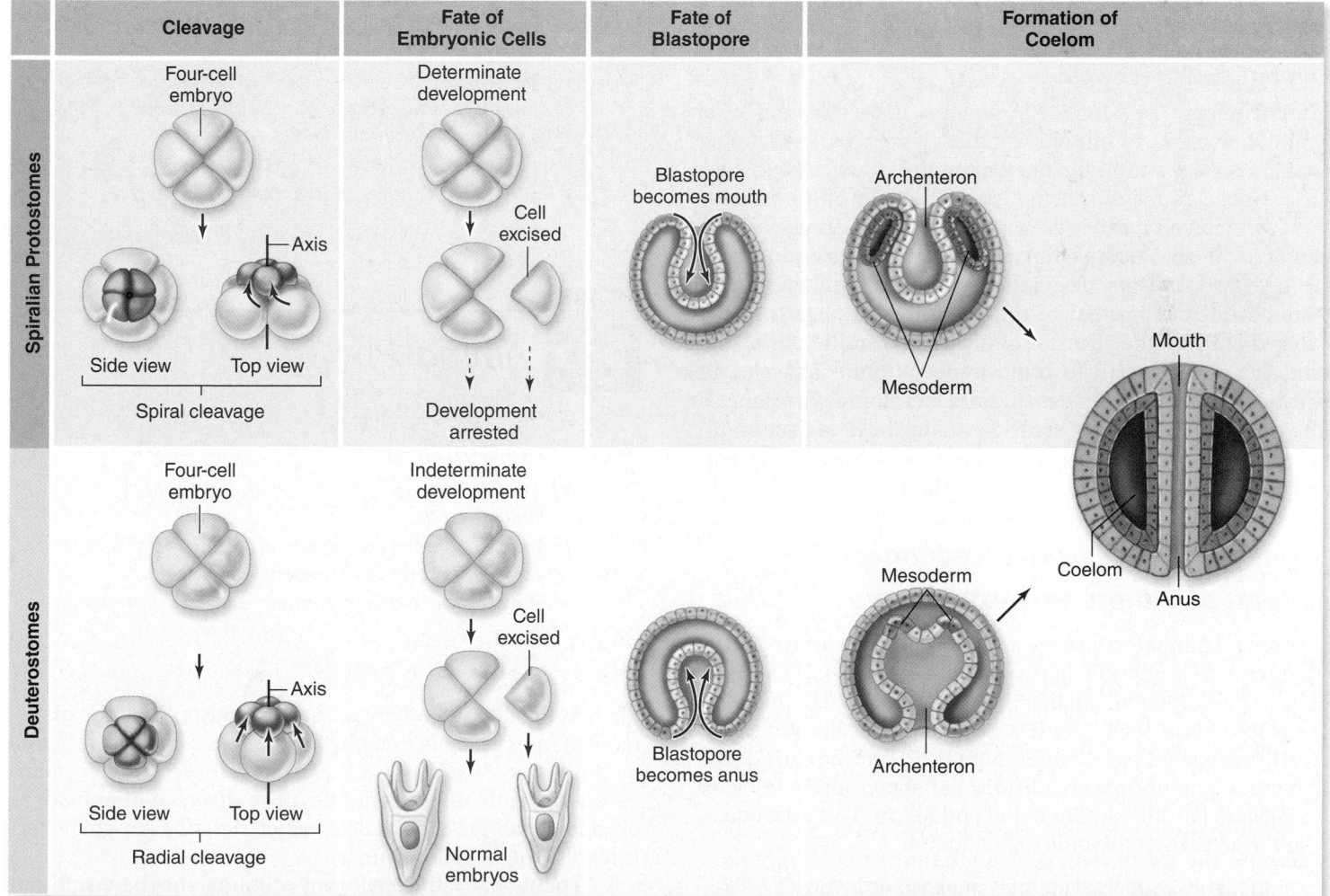

Figure 33.4 Embryonic development in protostomes and deuterostomes. In spiralian protostomes, embryonic cells cleave in a spiral pattern and exhibit determinate development; the blastopore becomes the animal's mouth, and the coelom originates from a split among endodermal cells. In deuterostomes, embryonic cells cleave radially and exhibit indeterminate development; the blastopore becomes the animal's anus, and the coelom originates from an invagination of the archenteron.

chapter **33** *Animal Diversity and the Evolution of Body Plans*

ball. This pattern is called **radial cleavage** because a line drawn through a sequence of dividing cells describes a radius outward from the polar axis (figure 33.4 bottom).

Determinate versus indeterminate development

Many protostomes exhibit **determinate development,** in which the type of tissue each embryonic cell will form in the adult is determined early, in many lineages even before cleavage begins, when the molecules that act as developmental signals are localized in different regions of the egg. Consequently, the cell divisions that occur after fertilization segregate molecular signals into different daughter cells, specifying the fate of even the very earliest embryonic cells. Each embryonic cell is destined to occur only in particular parts of the adult body, so if the cells are separated, development cannot proceed.

Deuterostomes, conversely, display **indeterminate development.** The first few cell divisions of the zygote produce identical daughter cells. If the cells are separated, any one can develop into a complete organism because the molecules that signal the embryonic cells to develop differently are not segregated into different cells until later in the embryo's development. (This is how identical twins are formed.) Thus, each cell remains totipotent and its fate is not determined for several cleavages.

Formation of the coelom

The coelom arises within the mesoderm. In protostomes, cells simply move apart from one another to create an expanding coelomic cavity within the mass of mesodermal cells. In deuterostomes, groups of cells pouch off the end of the archenteron, which you recall is the primitive gut—the hollow in the center of the developing embryo that is lined with endoderm.

The consistency of deuterostome development and its distinctiveness from that of the protostomes suggest that it evolved once, in the ancestor of the deuterostome phyla. The mode of development in protostomes is more diverse, but because of the distinctiveness of spiral development, scientists infer it also evolved once, in the common ancestor to all spiralian phyla. As we shall see, our modern understanding of animal phylogeny supports these conclusions.

Segmentation allowed for redundant systems and improved locomotion

Segmented animals consist of a series of linearly arrayed compartments that typically look alike (see figure 34.21), at least early in development, but that may have specialized functions. Development of segmentation is mediated at the molecular level by *Hox* genes (see chapters 19 and 25). During early development, segments first are obvious in the mesoderm but later are reflected in the ectoderm and endoderm. Two advantages result from early embryonic segmentation:

1. In highly segmental animals, such as earthworms (phylum Annelida), each segment may develop a more or less complete set of adult organ systems. Because these are redundant systems, damage to any one segment need not be fatal because other segments duplicate the damaged segment's functions.

2. Locomotion is more efficient when individual segments can move semi-independently. Because partitions isolate the segments, each can contract or expand autonomously. Therefore, a long body can move in ways that are often quite complex.

Segmentation underlies the organization of body plans of the most morphologically complex animals. In some adult arthropods, the segments are fused, but segmentation is usually apparent in embryological development. In vertebrates, the backbone and muscle blocks are segmented, although segmentation is often disguised in the adult form.

Previously, zoologists considered that true segmentation was found only in annelids, arthropods, and chordates, but segmentation is now recognized to be more widespread. Animals such as onychophorans (velvet worms), tardigrades (water bears), and kinorhynchs (mud dragons) are also segmented.

Learning Outcomes Review 33.2

Animals are distinguished on the basis of symmetry, tissues, type of body cavity, sequence of embryonic development, and segmentation. Bilateral animals are those that have bodies with a left and right side, which are mirror images. Within bilaterians, protostomes develop the mouth prior to the anus; deuterostomes develop the mouth after the anus has formed. Segmentation allows redundant systems and more efficient locomotion. A pseudocoelom is a space that develops between the mesoderm and endoderm; a coelom develops entirely within mesoderm.

■ *How is cephalization related to body symmetry?*

33.3 *Animal Phylogeny*

Learning Outcomes

1. *Identify the characters that distinguish the major animal phyla.*
2. *Identify patterns of convergent evolution in significant morphological and developmental characters.*
3. *Identify the placement of humans among the animal phyla.*

There is little disagreement among biologists about the placement of most animals into phyla, although zoologists disagree on the status of some phyla, particularly those with few members or recently discovered ones. The diversity of animals is obvious in tables 33.1 and 33.2, which describe key characteristics of 20 of the animal phyla.

Traditionally, the phylogeny of animals has been inferred using features of anatomy and aspects of embryological development, as discussed earlier, from which a broad consensus emerged over the last century concerning the main branches of the animal tree of life. However, in the past 30 years, gene sequence data have accumulated at an

TABLE 33.1	Animal Phyla with the Most Species			
Phylum	**Typical Examples**		**Key Characteristics**	**Approximate Number of Named Species**
Arthropoda (arthropods)	Beetles, other insects, crabs, spiders, krill, scorpions, centipedes, millipedes		Chitinous exoskeleton covers segmented, coelomate body. With paired, jointed appendages; many types of insects have wings. Occupy marine, terrestrial, and freshwater habitats. Most arthropods are insects (as are most animals!).	1,000,000
Mollusca (mollusks)	Snails, oysters, clams, octopuses, slugs		Coelomate body of many mollusks is covered by one or more shells secreted by a part of the body termed the mantle. Many kinds possess a unique rasping tongue, a radula. Members occupy marine, terrestrial, and freshwater habitats (35,000 species are terrestrial).	110,000
Chordata (chordates)	Mammals, fish, reptiles, amphibians		Each coelomate individual possesses a notochord, a dorsal nerve cord, pharyngeal slits, and a postanal tail at some stage of life. In vertebrates, the notochord is replaced during development by the spinal column. Members occupy marine, terrestrial, and freshwater habitats (20,000 species are terrestrial).	56,000
Platyhelminthes (flatworms)	Planarians, tapeworms, liver and blood flukes		Unsegmented, acoelomate, bilaterally symmetrical worms. Digestive cavity has only one opening; tapeworms lack a gut. Many species are parasites of medical and veterinary importance. Members occupy marine, terrestrial, and freshwater habitats (as well as the bodies of other animals).	20,000
Nematoda (roundworms)	*Ascaris*, pinworms, hookworms, filarial worms		Pseudocoelomate, unsegmented, bilaterally symmetrical worms; tubular digestive tract has mouth and anus. Members occupy marine, terrestrial, and freshwater habitats; some are important parasites of plants and animals, including humans.	25,000 (but it is thought by some that the number of nematode species may be much greater)
Annelida (segmented worms)	Earthworms, polychaetes, tube worms, leeches		Segmented, bilaterally symmetrical, coelomate worms with a complete digestive tract; most have bristles (chaetae) on each segment that anchor them in tubes or aid in crawling. Occupy marine, terrestrial, and freshwater habitats.	16,000
Cnidaria (cnidarians)	Jellyfish, *Hydra*, corals, sea anemones, sea fans		Radially symmetrical, acoelomate body has tissues but no organs. Mouth opens into a simple digestive sac and is surrounded by tentacles armed with stinging capsules (nematocysts). In some groups, individuals are joined into colonies; some can secrete a hard exoskeleton. The very few nonmarine species live in fresh water.	10,000
Echinodermata (echinoderms)	Sea stars, sea urchins, sand dollars, sea cucumbers		Adult body pentaradial (fivefold) in symmetry. Water-vascular system is a coelomic space; endoskeleton of calcium carbonate plates. Many can regenerate lost body parts. Fossils are more diverse in body plan than extant species. Exclusively marine.	7000
Porifera (sponges)	Barrel sponges, boring sponges, basket sponges, bath sponges		Bodies of most asymmetrical: defining "an individual" is difficult. Body lacks tissues or organs, being a meshwork of cells surrounding channels that open to the outside through pores, and that expand into internal cavities lined with food-filtering flagellated cells (choanocytes). Most species are marine (150 species live in fresh water).	7000
Bryozoa (moss animals) (also called Polyzoa and Ectoprocta)	Sea mats, sea moss		The only exclusively colonial phylum; each colony comprises numerous small, coelomate individuals (zooids) connected by an exoskeleton (calcareous in marine species, organic in most freshwater ones). A ring of ciliated tentacles (lophophore) surrounds the mouth of each zooid; the anus lies beyond the lophophore.	4500

TABLE 33.2

Some Important Animal Phyla with Fewer Species—and Three Recently Discovered Ones

Phylum	Typical Examples		Key Characteristics	Approximate Number of Named Species
Rotifera (wheel animals)	Rotifers		Small pseudocoelomates with a complete digestive tract including a set of complex jaws. Cilia at the anterior end beat so they resemble a revolving wheel. Some are very important in marine and freshwater habitats as food for predators such as fishes.	2000
Nemertea (ribbon worms) (also called Rhynchocoela)	*Lineus*		Protostome worms notable for their fragility—when disturbed, they fragment in pieces. Long, extensible proboscis occupies their coelom; that of some tipped by a spearlike stylet. Most marine, but some live in fresh water, and a few are terrestrial.	900
Tardigrada (water bears)	*Hypsibius*		Microscopic protostomes with five body segments and four pairs of clawed legs. An individual lives a week or less but can enter a state of suspended animation ("cryptobiosis") in which it can survive for many decades. Occupy marine, freshwater, and terrestrial habitats.	800
Brachiopoda (lamp shells)	*Lingula*		Protostomous animals encased in two shells that are oriented with respect to the body differently than in bivalved mollusks. A ring of ciliated tentacles (lophophore) surrounds the mouth. More than 30,000 fossil species are known.	300
Onychophora (velvet worms)	*Peripatus*		Segmented protostomous worms resembling tardigrades; with a chitinous soft exoskeleton and unsegmented appendages. Related to arthropods. The only exclusively terrestrial phylum, but what are interpreted as their Cambrian ancestors were marine.	110
Ctenophora (sea walnuts)	Comb jellies, sea walnuts		Gelatinous, almost transparent, often bioluminescent marine animals; eight bands of cilia; largest animals that use cilia for locomotion; complete digestive tract with anal pore.	100
Chaetognatha (arrow worms)	*Sagitta*		Small, bilaterally symmetrical, transparent marine worms with a fin along each side, powerful bristly jaws, and lateral nerve cords. Some inject toxin into prey and some have large eyes. It is uncertain if they are coelomates, and, if so, whether protostomes or deuterostomes.	100
Loricifera (loriciferans)	*Nanaloricus mysticus*		Tiny marine pseudocoelomates that live in spaces between grains of sand. The mouth is borne on the tip of a flexible tube. Discovered in 1983.	10
Cycliophora (cycliophorans)	*Symbion*		Microscopic animals that live on mouthparts of claw lobsters. Discovered in 1995.	3
Micrognathozoa (micrognathozoans)	*Limnognathia*		Microscopic animals with complicated jaws. Discovered in 2000 in Greenland.	1

accelerating pace for all animal groups. Phylogenies developed from different molecules sometimes suggest quite different evolutionary relationships among the same groups of animals. However, combining data from multiple genes has resolved the relationships of most phyla. Current studies are using sequences from hundreds of genes to try to fully resolve the animal tree of life.

Molecular data are helping to resolve some problems with the traditional phylogeny, such as puzzling groups that did not fit well into the widely accepted phylogeny. These data may be especially helpful in clarifying relationships that conventional data cannot, as, for example, in animals such as parasites. Through dependence on their host, the anatomy, physiology, and behavior of parasites tends to be greatly altered, so features that may reveal the phylogenetic affinities of free-living animals can be highly modified or lost.

Current understanding of phylogeny differs from traditional views

Although they differ from one another in some respects, phylogenies utilizing molecular data share some deep structure with the traditional animal tree of life. Figure 33.5 is a summary of animal phylogeny developed from morphological, molecular, life history, and other types of relevant data. Some parts of this phylogeny are not firmly established, and new studies are constantly appearing, often with somewhat different conclusions. It is an exciting time to be a systematist, but shifts in understanding of relationships among groups of animals can be frustrating to some! Like any scientific idea, a phylogeny is a hypothesis, open to challenge and to being revised in light of additional data.

One consistent result is that Porifera (sponges) constitutes a monophyletic group that shares a common ancestor with other animals. Some systematists had considered sponges to comprise two (or three) groups that are not particularly closely related, but molecular data support what had been the majority view, that phylum Porifera is monophyletic.

Among remaining animals, termed Eumetazoa, molecular data are in accord with the traditional view that cnidarians (hydras, sea jellies, and corals) branch off the tree before the origin of animals with bilateral symmetry, the Bilateria. Our understanding of the phylogeny of the deuterostome branch of Bilateria (discussed in chapter 35) has not changed much, but our understanding of the phylogeny of protostomes has been altered by molecular data.

The most revolutionary development is that annelids and arthropods, which had been considered closely related because both exhibit segmentation, are in fact not close relatives, but rather belong to separate clades. Now arthropods are grouped with protostomes that molt their cuticles at least once during their life. These are termed ecdysozoans, which means "molting animals" (see chapter 34). Molecular sequence data can help test our ideas of which morphological features reveal evolutionary relationships best; in this case, molecular data allowed us to see that, contrary to our hypothesis, segmentation seems to have evolved convergently, but molting did not.

But not all features are easy to diagnose, and molecular data do not resolve all uncertainties. The enigmatic phylum Ctenophora (comb jellies)—pronounced with a silent C—has been considered both diploblastic and triploblastic and has been thought to have both a complete gut and a blind gut. Likewise the enigmatic phylum Chaetognatha (arrow worms) has been considered both coelomate and pseudocoelomate, and if coelomate, both protostome and deuterostome. Their placement in phylogenies varies, seeming to depend on the features and methods used to construct the tree. Further research is needed to resolve these uncertainties.

Morphology-based phylogeny focused on the state of the coelom

Current understanding of animal phylogeny requires a rethinking of how some characters have evolved. Zoologists previously inferred that the first animals were acoelomate, that some of their descendants evolved a pseudocoelom, and that some pseudocoelomate descendants evolved the coelom. This view was so widespread that classification systems were based on the state of the coelom. However, as you saw in chapter 21, evolution rarely occurs in such a linear and directional way. Our understanding of animal phylogeny now makes clear that the state of the coelom has evolved many more times than previously realized, and consequently that it is not a reliable character to infer phylogenetic relationships.

In particular, a coelom appears to have evolved just once, in the ancestor of the clade comprising protostomes and deuterostomes (see figure 33.5). Subsequently, the pseudocoelomate condition arose several times from coelomate ancestors within the protostome clade. As a result, all deuterostomes have coeloms, but the condition in protostomes is mixed. In addition, although the acoelomate condition is the ancestral state for animals, some animals have lost the body space, becoming acoelomate secondarily.

Protostomes consist of spiralians and ecdysozoans

Two clades of protostomes are recognized as having evolved independently since ancient times: the spiralians and the ecdysozoans (see figure 33.5). Spiralian animals grow by gradual addition of mass to the body and undergo spiral cleavage (see figure 33.4). There are two main groups of spiralians: Lophotrochozoa and Platyzoa. Lophotrochozoans move by muscular contractions and include most protostomes with a coelom. Most platyzoans are acoelomates and move by ciliary action.

Ecdysozoans are animals that molt (figure 33.6), a phenomenon that seems to have evolved only once in the animal kingdom. Of the numerous phyla of protostomes assigned to the Ecdysozoa, Arthropoda contains the largest number of described species of any phylum.

Deuterostomes include chordates and echinoderms

Deuterostomes consist of fewer phyla and species than protostomes, and are more uniform in many ways, despite great differences in appearance. Echinoderms such as sea stars, and

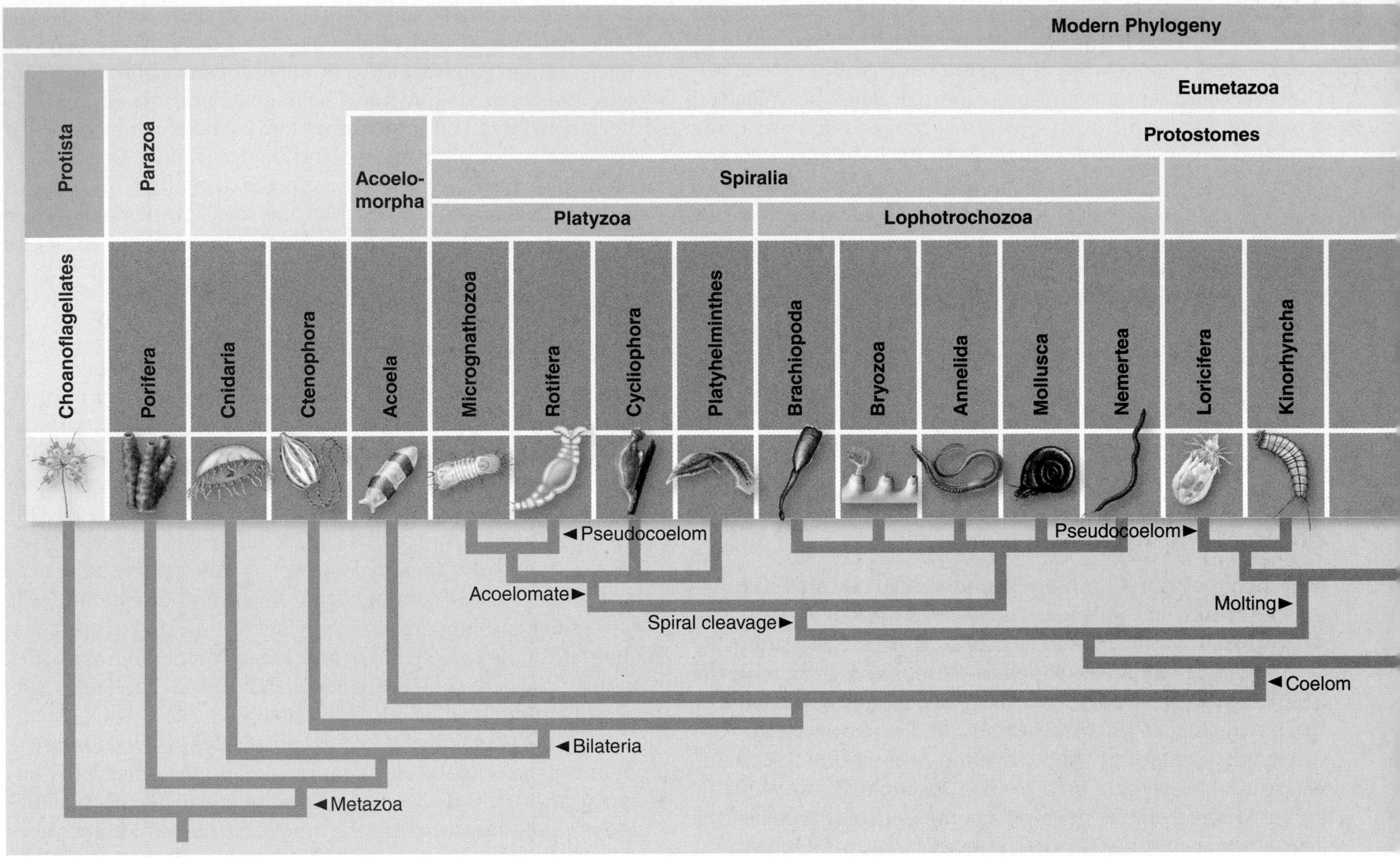

Modern Phylogeny

Eumetazoa

Protostomes

Spiralia

Acoelo-morpha

Platyzoa

Lophotrochozoa

Protista | Parazoa

Choanoflagellates | Porifera | Cnidaria | Ctenophora | Acoela | Micrognathozoa | Rotifera | Cycliophora | Platyhelminthes | Brachiopoda | Bryozoa | Annelida | Mollusca | Nemertea | Loricifera | Kinorhyncha

◀ Pseudocoelom

◀ Pseudocoelom ▶

Acoelomate ▶

Spiral cleavage ▶

Molting ▶

◀ Coelom

◀ Bilateria

◀ Metazoa

Figure 33.5 Proposed revision of the animal tree of life. A phylogeny of many of the 35–40 phyla reflects a consensus based on interpretation of anatomical and developmental data as well as results derived from molecular phylogenetic studies. Whether Chaetognatha is a protostome or a deuterostome is unclear.

? Inquiry question Which condition, acoelomate, pseudocoelomate, or coelomate, is a reliable indicator of phylogenetic relationships?

chordates such as humans, share a mode of development that is evidence of their evolution from a common ancestor, and separates them clearly from other animals.

Learning Outcomes Review 33.3

Scientists have defined phyla based on tissues, symmetry, characteristics such as presence or absence of a coelom or pseudocoelom, protostome versus deuterostome development, growth pattern and larval stages, and molecular data. Among protostomes, spiralian organisms have a growth pattern in which their body size simply increases; ecdysozoans must molt in order to grow larger. The condition of the coelom and the occurrence of segmentation have evolved convergently among animals, whereas other traits, such as molting, have arisen only once. Among the deuterostome phyla are echinoderms and chordates, the latter which include humans.

■ *Why do systematists attempt to characterize each group of animals by one or more features that have evolved only once?*

Figure 33.6 Blue crab undergoing ecdysis (molting). Members of Ecdysozoa grow step-wise because their external skeleton is rigid.

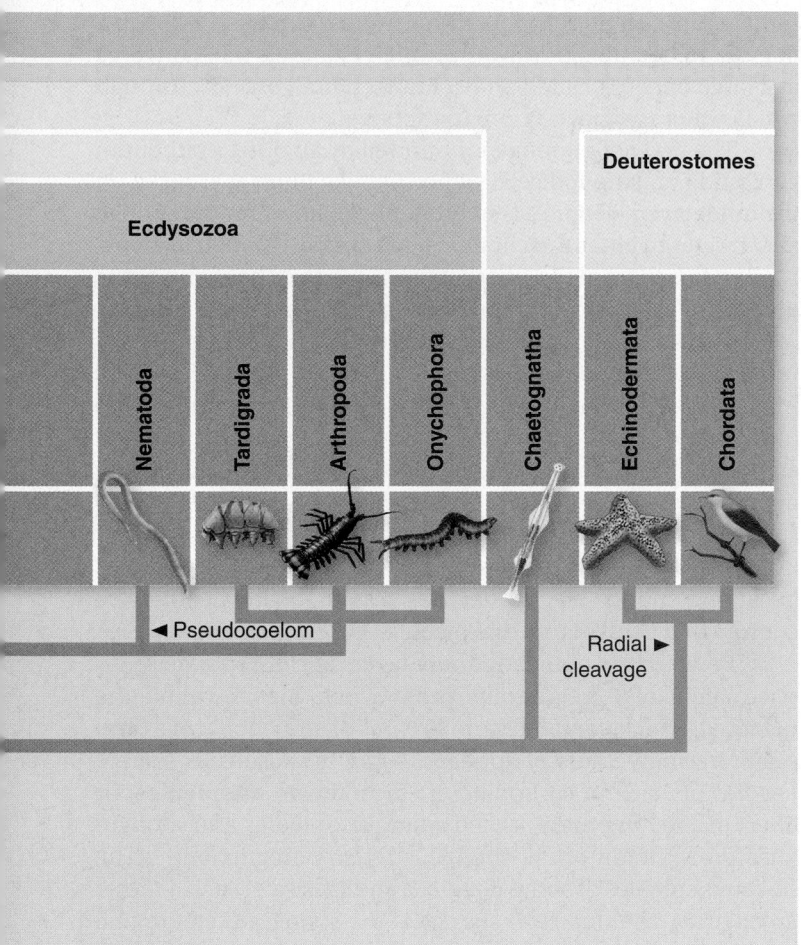

Ecdysozoa

Deuterostomes

Nematoda · Tardigrada · Arthropoda · Onychophora · Chaetognatha · Echinodermata · Chordata

◄ Pseudocoelom

Radial ► cleavage

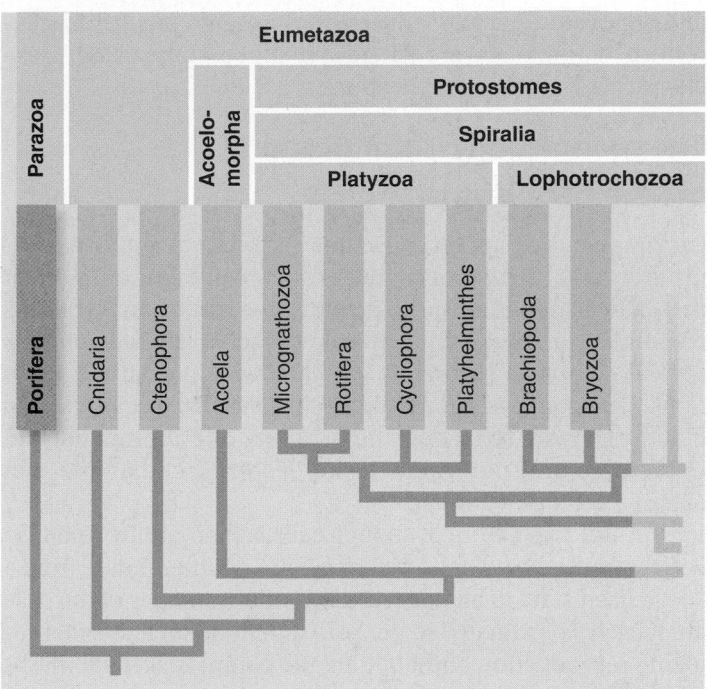

Eumetazoa

Parazoa

Acoelomorpha

Protostomes

Spiralia

Platyzoa · Lophotrochozoa

Porifera · Cnidaria · Ctenophora · Acoela · Micrognathozoa · Rotifera · Cycliophora · Platyhelminthes · Brachiopoda · Bryozoa

Parazoa: Animals That Lack Specialized Tissues

Learning Outcomes

1. *Describe the different types of cells in the sponge body.*
2. *Explain the function of choanocytes.*

In the remainder of this and the next two chapters, we will explore the great diversity of animals. We start with the morphologically simplest members of the animal kingdom—sponges, jellyfish, and some types of worms (fully a third of animal phyla are based on a "wormy" body plan!). Despite their simplicity, these animals can carry out all the essential functions of life, just as do more morphologically complex animals—eating, respiring, reproducing, and protecting themselves. The major organization of the animal body first evolved in these animals, the basic body plan from which all the rest of animals evolved.

Systematists traditionally divided the kingdom Animalia (also termed Metazoa) into two main branches. Parazoa ("near animals") comprises animals that, for the most part, lack

definite symmetry and do not possess tissues. These are the sponges, phylum Porifera. Because they are so different in so many ways from other animals, some scientists inferred that sponges were not closely related to other animals, which would mean that what we consider animals had two separate origins. Eumetazoa ("true animals") are animals that have a definite shape and symmetry. All have tissues, and most have organs and organ systems. Now most systematists agree that Parazoa and Eumetazoa are descended from a common ancestor, so animal life had a single origin. And although most phylogenies constructed with molecular data consider Parazoa to be at the base of the animal tree of life, some do not.

Sponges have a loose body organization

The major group of parazoans is phylum **Porifera,** the sponges. Despite their morphological simplicity, like all animals, sponges are truly multicellular.

Nearly 7000 species of sponges live in the sea, and perhaps 150 species live in fresh water. Marine sponges occur at all depths, and may be among the most abundant animals in the deepest part of the oceans. Although some sponges are small (no more than a few millimeters across), some may reach 2 m or more in diameter.

A few small sponges are radially symmetrical, but most members of this phylum lack symmetry. Some have a low and encrusting form and grow covering various sorts of surfaces; others are erect and lobed, some in complex patterns (figure 33.7a).

As is true of many marine invertebrate animals, larval sponges are free-swimming. After a sponge larva attaches to an appropriate surface, it metamorphoses into an adult and remains attached to that surface for the rest of its life. Thus adult sponges are sessile; that is, they are anchored, immobile, on rocks or other submerged objects. Sponges defend themselves by producing chemicals that repel potential predators and organisms that might overgrow them. As a result, pharmaceutical companies are interested in possibly using these chemicals for human applications.

The sponge body is composed of several cell types

Lacking head or appendages, mouth or anus, and the organized internal structure characteristic of all other animals, at first sight a sponge seems to be little more than a mass of cells embedded in a gelatinous matrix. In fact, a sponge contains several cell types (figure 33.7b), each with specialized functions. If a sponge is put through a fine sieve or coarse cloth so that the cells are separated, they will seek one another out and reassemble the entire sponge—a phenomenon that does not occur in any other animal.

A unique feature of sponge cells is their ability to differentiate from one type to another, and to dedifferentiate from a specialized state to an unspecialized one. Activities of the cells are loosely coordinated to perform functions such as synchronizing reproduction and building the complex skeletal meshwork. This distinguishes sponges as truly multicellular, by contrast with colonial protists, which may form aggregates of cells, but all are functionally identical (except for the reproductive cells).

A small, anatomically simple sponge has a vaselike shape. The walls of the "vase" have three functional layers. Facing the internal cavity are flagellated cells called **choanocytes,** or collar cells (see figure 33.7b). A larger and more complex sponge has many small chambers connected by channels

rather than a single chamber. Once water has passed through a flagellated chamber, it travels through channels that converge at a large opening called an **osculum** (plural, *oscula*), through which water is expelled from the sponge.

The body of a sponge is bounded by an outer epithelium consisting of flattened cells somewhat like those that make up the outer layers of animals in other phyla. Pores on the sponge allow water to enter the channels that course through its body, leading to and from the flagellated chambers. The name of the phylum, Porifera, refers to these pores, or ostia (singular, ostium); a large sponge has multiple oscula, but they are far, far fewer than the number of ostia.

Some epithelial cells are specialized to surround the ostia; they can contract when touched or exposed to appropriate stimuli, causing the ostia to close, thereby protecting the delicate inner cells from the entry of potentially harmful substances such as sand and noxious chemicals. Cells surrounding individual ostia operate independently of one another; because a sponge has no nervous system, actions cannot be coordinated across large distances.

Between the outer and inner layers of cells, sponges consist mainly of a gelatinous, protein-rich matrix called the **mesohyl,** which contains various types of amoeboid cells (and eggs). In many kinds of sponges, some of these cells secrete needles of calcium carbonate or silica known as **spicules,** or fibers of a tough protein called **spongin.** Spicules and spongin form the skeleton of the sponge, strengthening its body. The siliceous spicules of some deep-sea sponges can reach a meter in length! A genuine bath sponge is the spongin skeleton of a marine sponge; artificial sponges made of cellulose or plastic

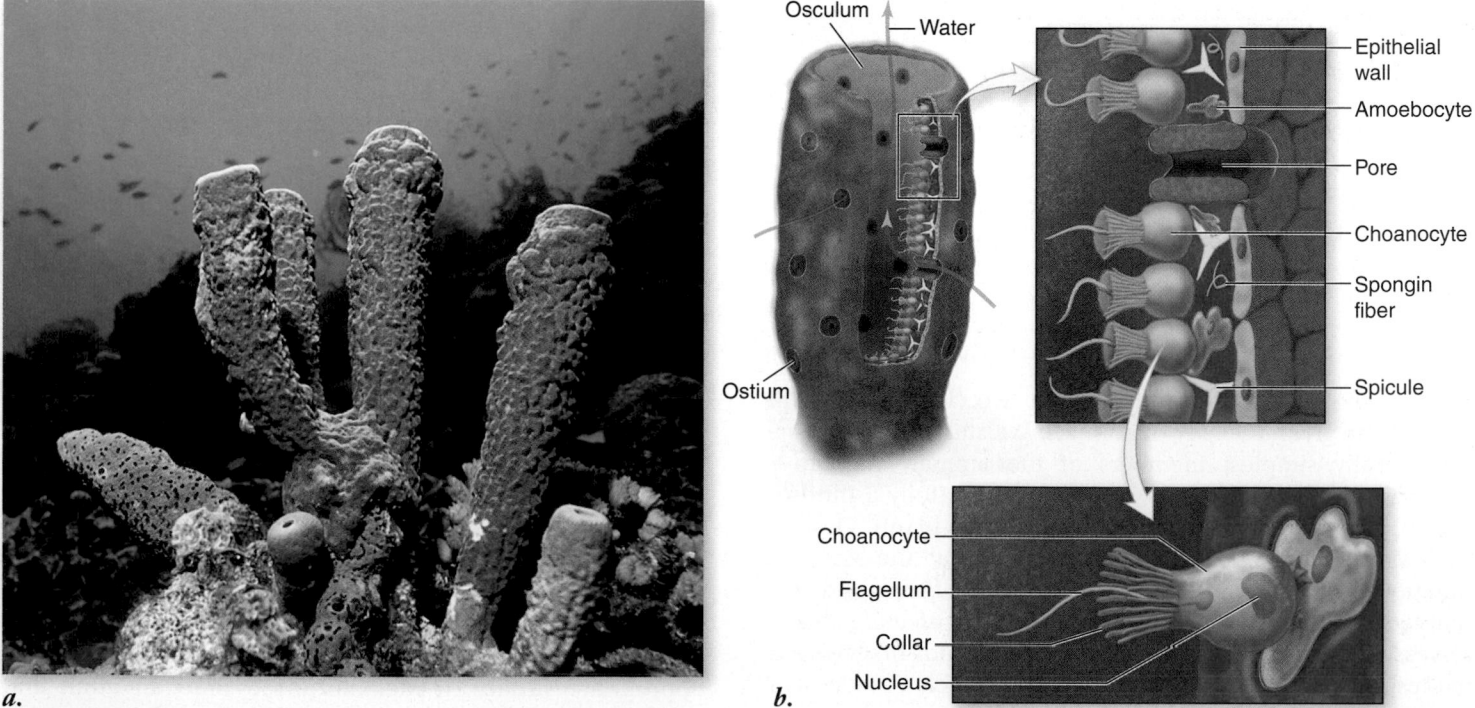

a.

b.

Figure 33.7 Phylum Porifera: Sponges. *a. Aplysina longissima.* This beautiful, bright orange and purple elongated sponge is found on deep coral reefs. *b.* Diagrammatic drawing of the simplest type of sponge. Sponges are composed of several distinct cell types, the activities of which are coordinated. The sponge body has no organized tissues, and most are not symmetrical.

are modeled on the body design of this animal, with a porous body adapted to contain large amounts of water. The three classes of sponges—Hexactinellida, Demospongiae, and Calcarea—are distinguished in part by the mineral form of their spicules.

Choanocytes help circulate water through the sponge

Each choanocyte resembles a protist with a single flagellum (plural, flagella) (see figure 33.7*b*)—a similarity that may reflect its evolutionary derivation. The pressure created by the beating flagella in the cavity contributes to circulating the water that brings in food and oxygen and carries out wastes. In large sponges, the inner wall of the body interior is convoluted, increasing the surface area and, therefore, the number of flagella. In such a sponge, 1 cm^3 of sponge can propel more than 20 L of water per day. Choanocytes also capture food particles from the passing water, engulfing and digesting them. Obviously, this arrangement restricts a sponge to feeding on particles considerably smaller than choanocytes—largely bacteria.

Sponges reproduce both asexually and sexually

Some sponges can reproduce asexually simply by breaking into fragments. Each fragment is able to continue growing as a new individual. Whether a sponge should be considered colonial is an illustration of the limitations of human language. A colony of invertebrate animals, such as coral, is generally defined as a group of individuals that are physically connected (and may be physiologically connected as well), all having been produced by asexual reproduction (such as budding or dividing) from a single progenitor that arose by sexual reproduction. Nearly all sponges grow by multiplying the number of flagellated chambers connected to a single osculum, but whether these units can be considered "individuals" is debatable.

Sponge sperm are created by the transformation of choanocytes, which are then released into the water where they may be carried into another sponge of the same species. When a sperm is captured by a choanocyte, it is carried to an egg cell, which is in the mesohyl (and in some sponges is also a transformed choanocyte). In many sponges, development of the externally ciliated larva occurs within the mother. In sponges of other species, the fertilized egg is released into the water, where development occurs. Whether a fertilized egg or a ciliated larva is released, after a short planktonic (drifting) stage, the larva settles on a suitable substrate where it transforms into an adult.

Learning Outcomes Review 33.4

Sponges possess multicellularity but have neither tissue-level development nor body symmetry. Cells that compose a sponge include a layer of choanocytes, a layer of epithelial cells, and amoeboid cells in the mesohyl between the two layers. Choanocytes have flagella that beat to circulate water through the sponge body, allowing food particles to be trapped.

■ *What features of a sponge make it seem to be a colony, and what features make it seem to be a single organism?*

33.5 Eumetazoa: Animals with True Tissues

Learning Outcomes

1. *Explain the defining feature of cnidarians.*
2. *Differentiate between cnidarians and ctenophores.*
3. *Discuss the question of symmetry of ctenophores.*

The Eumetazoa contains animals that evolved the first key transition in the animal body plan: distinct tissues. The embryonic cell layers differentiate into the tissues of the adult body, giving rise to the body plan characteristic of each group of animals.

Recall that the outer covering of the body (the epidermis) and the nervous system develop from the embryonic ectoderm, and the digestive tissue (called the **gastrodermis**) develops from the embryonic endoderm. In the Bilateria, the embryonic mesoderm, which lies between endoderm and ectoderm, forms the muscles. Eumetazoans also evolved body symmetry, the two main types of which are radial and bilateral symmetry.

All cnidarians are carnivores

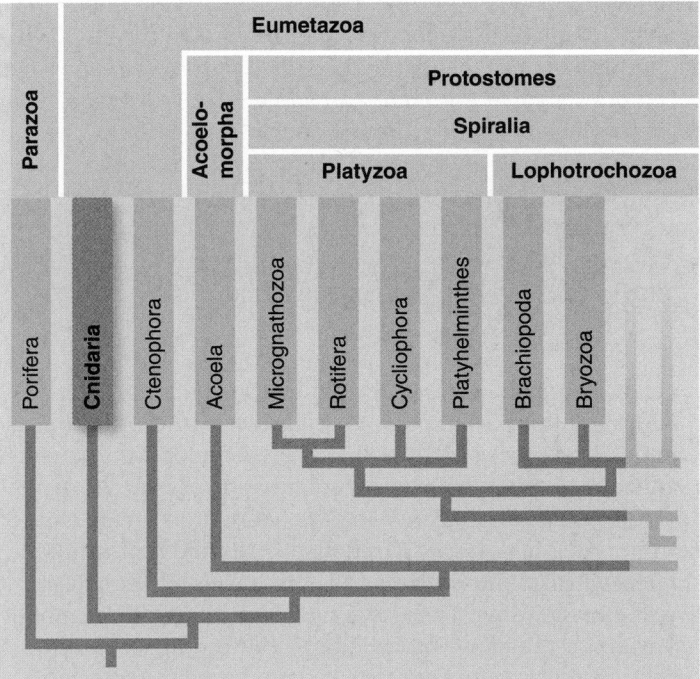

Most of the 10,000 species of cnidarians are marine; a very few live in fresh water. The bodies of these fascinating and simply constructed diploblastic animals are radially symmetrical and made of distinct tissues, although they do not have organs. Despite having no reproductive, circulatory, digestive, or excretory systems, cnidarians reproduce, exchange gas, capture and digest prey, and distribute the resulting organic molecules to all their cells. A cnidarian has no concentration of nervous tissue that could be considered a brain or even a

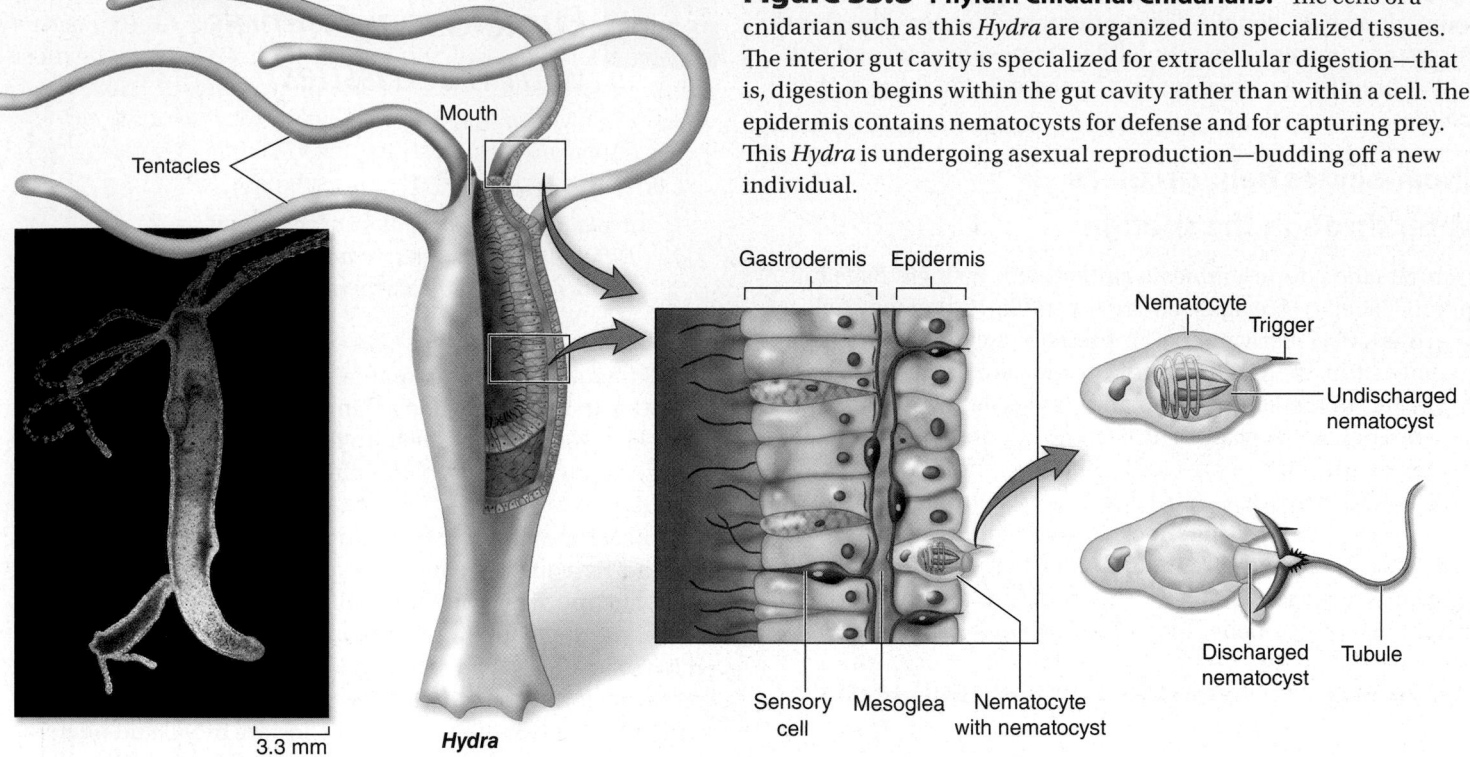

Figure 33.8 Phylum Cnidaria: Cnidarians. The cells of a cnidarian such as this *Hydra* are organized into specialized tissues. The interior gut cavity is specialized for extracellular digestion—that is, digestion begins within the gut cavity rather than within a cell. The epidermis contains nematocysts for defense and for capturing prey. This *Hydra* is undergoing asexual reproduction—budding off a new individual.

ganglion (a smaller group of nerve cells—see chapter 43). Rather, its nervous system is a latticework, the cells having junctions like those of bilaterians. All cnidarians have nervous receptors sensitive to touch, and some have gravity and light receptors, which include, in a few species, image-forming eyes.

Cnidarians capture their prey (which includes fishes, crustaceans, and many other kinds of animals) with nematocysts (figure 33.8), microscopic intracellular structures unique to the phylum. Food captured by a cnidarian is brought to the mouth by the tentacles that ring the mouth.

Basic body plans

A cnidarian exhibits one of two body forms: the polyp and the medusa (figure 33.9). A **polyp** is cylindrical, with a mouth surrounded by tentacles at the end of the cylinder opposite where it is attached. Attachment in most solitary polyps is to a firm substratum, but polyps that are part of a colony attach to the mass of common colonial tissue. A **medusa** is discoidal or umbrella-shaped, with a mouth surrounded by tentacles on one side; most live free in the water. These two seemingly quite different body forms share the same morphology—the mouth opens into a saclike gastrovascular cavity and is surrounded by tentacles.

The cnidarian body plan has a single opening leading to the gastrovascular space, which is the site of digestion, most gas exchange, waste discharge, and, in many cnidarians, formation of gametes. The body wall is composed of two layers: the epidermis, which covers the surfaces in contact with the outside environment, and the gastrodermis, which covers the gastrovascular cavity. Between these two layers is the **mesoglea,** which varies from acellular, being no more than a

glue holding gastrodermis to epidermis in *Hydra,* to thick and rubbery with many cells in large medusae called jellyfish (see figures 33.8, 33.9).

The gastrovascular space also serves as a hydrostatic skeleton (see chapter 46). A hydrostatic skeleton serves two of the roles that a skeleton made of bone or shells does: it provides a rigid structure against which muscles can operate, and it gives the animal shape. For muscles to operate against it, the fluid-filled space must be closed sufficiently tightly so the fluid is under pressure and does not escape when muscles contract around the space. Think of the gastrovascular space of a cnidarian full of water like an inflated, elongated, air-filled

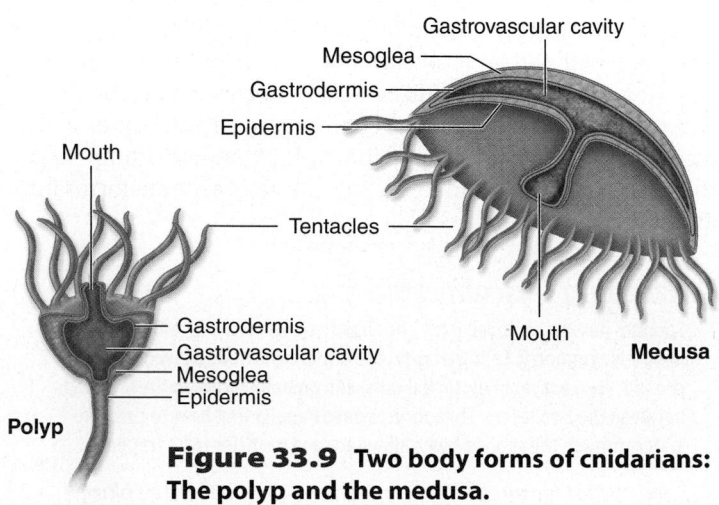

Figure 33.9 Two body forms of cnidarians: The polyp and the medusa.

balloon. Because it is firm, muscles that shorten this space cause it to broaden, and muscles that decrease its diameter lengthen it. However, if you inflate the balloon with air, but then do not hold the opening closed, the air will escape and the balloon will become flaccid. Similarly, when a cnidarian contracts greatly, it must open its mouth so water can escape. However, the mouth can close sufficiently tightly to retain water in the gastrovascular cavity, so the animal is turgid and can bend and extend its tentacles.

In addition to the hydrostatic skeleton of all polyps, those of many species build an exoskeleton of chitin or calcium carbonate around themselves; in many colonial species, the skeleton links the members of the colony. The polyps of a smaller number of species secrete internal skeletal elements. Polyps of some groups of cnidarians, such as sea anemones, have no skeleton at all. All medusae are solitary (that is, they do not form colonies) and form no skeleton.

The cnidarian life cycle

Cnidarians exhibit considerable variation in life history. Some cnidarians occur only as polyps, and others exist only as medusae, but many alternate between these two phases (see figure 33.9); both phases consist of diploid individuals. In general, in species having both polyp and medusa in the life cycle, the medusa forms gametes. The sexes are separate—a condition known as gonochorism—which means an individual is either male or female. An egg and a sperm unite to form a zygote, which develops into a planktonic ciliated **planula** larva that metamorphoses into a polyp. The polyp produces medusae asexually. Dispersal occurs in both medusa and larvae. In most species with such a life cycle, a polyp may also produce other polyps asexually; if they remain attached to one other, the resulting group is referred to as a colony.

In a small number of species, the planula produced by medusae may develop directly or indirectly into another medusa without passing through the polyp stage. More commonly, a polyp occurs but no medusa in the life cycle. In such cnidarians, the polyp can form gametes, and the resulting planula develops into another polyp. In some, but by no means all, of these species, the polyp can also produce other polyps asexually, by dividing, budding, or breaking off bits of itself that regenerate (grow the missing parts).

Digestion

A major evolutionary innovation in cnidarians is extracellular digestion of food inside the animal. Recall that in a sponge, digestion occurs within choanocytes, so food particles must be small enough to be engulfed by a choanocyte. In a cnidarian, digestion takes place partly in the gastrovascular cavity. Digestive enzymes, released from cells lining the cavity, partially break down food. Other cells lining the space engulf those food fragments by phagocytosis. This allows a cnidarian to feed on prey larger than a sponge can handle.

Nematocysts

Cnidarians capture food with the aid of **nematocysts,** microscopic stinging capsules unique to the phylum (see figure 33.8).

Although often referred to as "stinging cells," they are capsules, each secreted within a cell called a nematocyte. Cnidarians may be morphologically simple, but nematocysts are the most complex structures secreted by single animal cells. When appropriately stimulated, the closure at the end of the capsule springs open, and a tubule is emitted. Nematocyst discharge is one of the fastest cellular processes in nature. The mechanism of discharge of a nematocyst is unknown—several hypotheses for it have been proposed. Although the action of a nematocyst is often described as harpoon-like, the tubule actually everts (turns inside-out). It may penetrate or wrap around an object; the tubules of some nematocysts are barbed and some carry venom. In a very few species, the venoms are strong enough to kill a human.

Many thousand nematocytes typically are part of the epidermis of each tentacle. Each one can be used only once. In some types of cnidarians, nematocysts occur in parts of the body other than the tentacles, including in the gastrovascular cavity, where they may aid in digestion. In addition to being used offensively, nematocysts are the only defense of cnidarians—they are the reason jellyfish and fire coral sting. Some cnidarians have stinging capsules of types other than nematocysts, which are termed cnidae—hence the name of the phylum, Cnidaria.

Cnidarians are grouped into four—or five—classes

Traditionally, three classes of Cnidaria were recognized: Anthozoa (sea anemones, corals, sea fans), Hydrozoa (hydroids, *Hydra,* Portuguese man-of-war), and Scyphozoa (jellyfish). Now nearly all biologists accept that some scyphozoans are sufficiently distinct in life cycle and morphology that they constitute their own class, Cubozoa (box jellies), some of which have a sting sufficiently toxic to kill humans. Less widely accepted is the separation of some other scyphozoans into their own class, Staurozoa (star jellies).

Class Anthozoa: Sea anemones and corals

The largest class of Cnidaria is **Anthozoa,** consisting of approximately 6200 species of solitary and colonial polyps. They include soft-bodied sea anemones (figure 33.10), stony corals

Figure 33.10 Class Anthozoa. Crimson anemone, *Cribrinopsis fernaldi.*

Figure 33.11 Stony corals. *Tubastraea aurea,* a non–reef-building species from Malaysia.

(figure 33.11), and other groups known by such fanciful names as sea pens, sea pansies, sea fans, and sea whips. Many of these names reflect the plantlike appearance of the individual polyps or colonies.

An anthozoan polyp differs from a polyp of the other classes in that its gastrovascular cavity is compartmentalized by radially arrayed, longitudinal sheets of tissue called mesenteries. The gametes of an anthozoan develop in the mesenteries. The tentacles of an anthozoan are hollow, whereas those of many other cnidarians are solid.

Sea anemones (see figure 33.10) constitute a group of just over 1000 species of highly muscular and relatively complex soft-bodied anthozoans. Living in waters of all depths throughout the world, they range from a few millimeters to over a meter in diameter, and many are equally long.

Most corals are anthozoans, and most hard corals (about 1400 species, see figure 33.11) are closely related to sea anemones. The polyp of a coral secretes an exoskeleton of calcium carbonate around and under itself (this is the same material of which chalk is composed); as the individual or colony grows upward, dead skeleton accumulates below it. In shallow water of the tropics, these accumulations form coral reefs. Most waters in which coral reefs develop are nutrient-poor, but the corals are able to grow well because they contain within their cells symbiotic dinoflagellates (zooxanthellae) that photosynthesize, thereby providing energy for the animals. Obviously coral reefs are restricted to water sufficiently shallow for sunlight to reach the living animals (typically no more than 100 m), but corals that do not form reefs can grow deeper—to about 5000 m. Only about half the species of hard corals participate in forming reefs.

Coral reefs are economically important, serving as refuges for the young of many species of crustaceans and fishes that are eaten by humans, as well as protecting coasts of many

tropical islands. Unfortunately, they are threatened by global climate change. Water warmer than usual can cause the symbiosis with zooxanthellae to break down—a phenomenon known as "coral bleaching" because the underlying white skeleton of the animal becomes visible through its body in the absence of the symbionts. Bleaching does not always kill a coral, but it is stressful. As carbon dioxide in the atmosphere rises, it dissolves in water, forming carbonic acid, which lowers the pH of the water. This makes calcium carbonate less available; some species of calcifying marine plants and animals, including corals, have been shown to form less robust skeletons, and to form them more slowly, as pH drops.

Some "soft corals" secrete needles of calcium carbonate much like sponge spicules; these sclerites form a case around the polyp or are embedded in their tissues, presumably providing protection against predation. Some of these animals also secrete a horny rod that provides flexible support to the members of the colony growing around it (sea fans are such colonies).

Class Cubozoa: The box jellies

As their name implies, medusae of class **Cubozoa** are box-shaped, with a tentacle or group of tentacles hanging from each corner of the box (figure 33.12). Most of the 40 or so species are only a few centimeters in height, although some reach 25 cm. Box jellies are strong swimmers and voracious predators of fish in tropical and subtropical waters; both of these facts are related to some cubozoans having image-forming eyes! The stings of some species can be fatal to humans. The polyp stage is inconspicuous and in many cases unknown.

Class Hydrozoa: The hydroids

Most of the approximately 2700 species of class **Hydrozoa** have both polyp and medusa stages in their life cycle (see figure 33.9). The polyp stage of most species is colonial. The polyps of a colony may not be identical in structure or function: some may be specialized to feed but be unable to

**Figure 33.12
Class Cubozoa.**
Chironex fleckeri, a box jelly.

Figure 33.13
Portuguese man-of-war (*Physalia physalis*). This species has a very painful sting.

Figure 33.15
Star jelly. Stalked jellyfish (*Haliclystus auricula*).

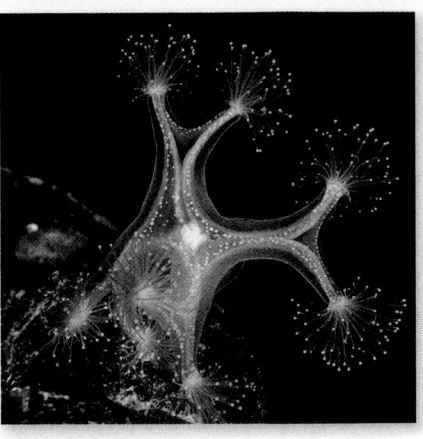

reproduce, whereas the opposite may be true of others (they are nourished by material transported from feeding polyps through the gastrovascular space connecting members of the colony). The Portuguese man-of-war (*Physalia physalis*) (figure 33.13) is not a jellyfish, but is a floating colony of highly integrated polypoid and medusoid individuals! Some marine hydroids and medusae are bioluminescent.

Hydrozoa is the only class of Cnidaria with freshwater members. A well-known hydroid is the freshwater *Hydra*, which is exceptional not only in its habitat, but in having no medusa stage and being solitary (see figure 33.8). Each polyp attaches by a basal disk, on which it can glide, aided by mucous secretions. It can also somersault—by bending over and attaching to the substrate by its tentacles, then looping over to a new location. If the polyp detaches from the substrate, it can float to the surface.

Class Scyphozoa: The jellyfish

In the approximately 200 species of class **Scyphozoa,** the medusa is much more conspicuous and complex than the polyp. Although many scyphomedusae are essentially colorless (figure 33.14), some are a striking orange, blue, or pink. The polyp is small, inconspicuous, simple in structure, and typically white in color, but is entirely lacking in a few scyphozoans that live in the open ocean.

The epithelium around the margin of a jellyfish contains a ring of muscle cells that can contract rhythmically to propel the animal through the water by jetting water from the

gastrovascular cavity or space beneath it; by contracting those on just one side, the animal can steer. A scyphomedusa (*Cyanea capilata*) was the killer in the Sherlock Holmes tale, *The Adventure of the Lion's Mane,* but, although some can inflict pain on humans, they are unlikely to be capable of killing.

Class Staurozoa: The star jellies

Among the reasons the 50 species of this group were separated from the Scyphozoa and placed in their own class (Staurozoa) is that the animal resembles a medusa in most ways but is attached to the substratum by a sort of stalk that emerges from the side opposite the mouth (figure 33.15). In addition, in all staurozoans known, the planula larva creeps rather than swims or drifts.

The comb jellies use cilia for movement

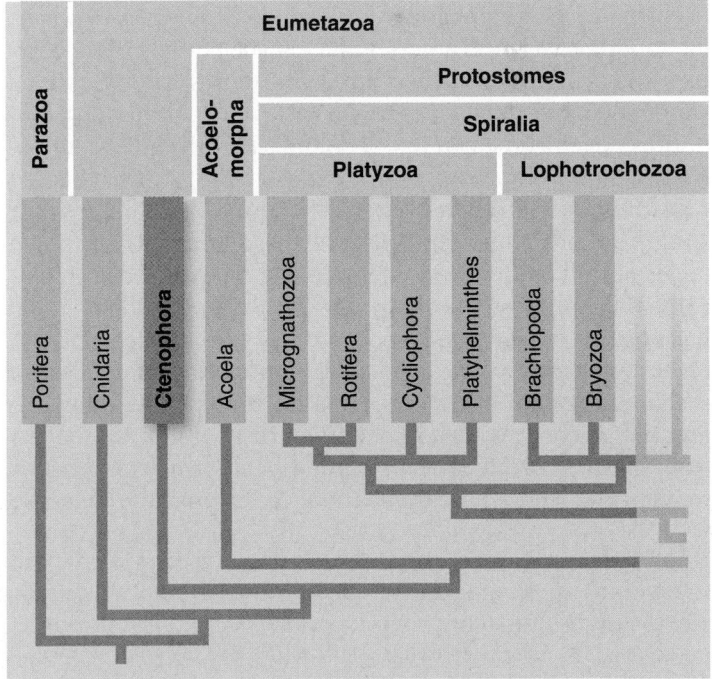

Pelagic members of the small marine phylum **Ctenophora** range from spherical to ribbon-like and are known as comb jellies, sea walnuts, or sea gooseberries. Abundant in the open ocean, most of these animals are transparent and only a few centimeters long, but rare ones are deeply pigmented and reach 1 m in length. They propel themselves through the water

Figure 33.14
Class Scyphozoa. *Aurelia aurita,* a jellyfish.

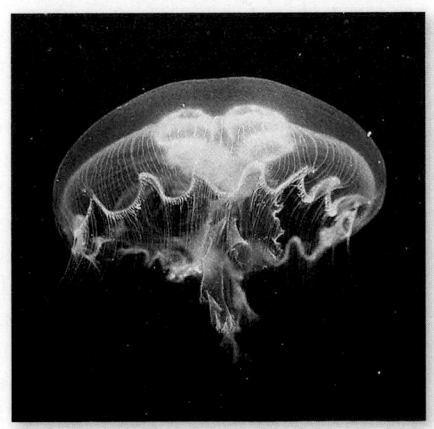

Figure 33.16

A comb jelly (phylum Ctenophora). Note the iridescent rows of comblike plates of fused cilia.

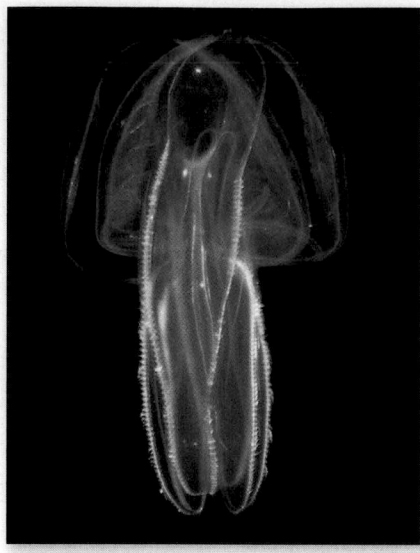

with eight rows of comblike plates of fused cilia that beat in a coordinated fashion, scattering light so they appear like rainbows (figure 33.16). Ctenophores are the largest animals that use cilia for locomotion. Many ctenophores are bioluminescent, giving off bright flashes of light that are particularly evident in the deep sea or on the surface of the ocean at night. Most ctenophores have two long, retractable tentacles used in prey capture. The epithelium of the tentacles contains **colloblasts,** a type of cell that bursts to discharge a strong adhesive material on contact with animal prey.

The phylogenetic position of Ctenophora is unclear. It was formerly considered closely related to Cnidaria because of the gelatinous, medusa-like form of most species. However, ctenophores lack nematocysts and are structurally more complex than cnidarians. They have anal pores through which water and other substances can exit the body. Although ctenophores have been considered diploblasts with radial symmetry, recent developmental studies have demonstrated them to have muscle cells derived from mesoderm, which means they should be considered triploblasts, like bilaterians. The two tentacles placed on opposite sides of a ctenophore's body impart some bilaterality to them, but whether their symmetry should be considered a modified type of radial symmetry is unclear. A recent molecular phylogeny places ctenophores at the base of the animal tree of life, as the sister group to all other metazoans. If this is borne out, it implies that the ancestral animal was triploblastic and bilaterally symmetrical. For the time being, we consider ctenophores to be the sister taxon of the Bilateria.

Learning Outcomes Review 33.5

Members of phylum Cnidaria have nematocysts, capsules used in defense and prey capture, and are radially symmetrical. Members of phylum Ctenophora lack nematocysts; they propel themselves by means of eight rows of comblike plates of fused cilia. Ctenophores have cells called colloblasts that assist in prey capture. The symmetry of ctenophores is a subject of debate.

■ *Why is the phylogenetic position of ctenophores uncertain?*

33.6 The Bilateria

Learning Outcomes

1. **List the distinguishing features of the Bilateria.**
2. **Understand the phylogeny of the major groups of Bilateria.**

The Bilateria is characterized by a key transition in the animal body plan—bilateral symmetry—which allowed animals to achieve high levels of specialization within parts of their bodies, such as the concentration of sensory structures at the anterior. (Even if ctenophores are truly bilateral as just mentioned, they lack these other specializations.) The Bilateria is divided into two clades. One comprises the protostomes and deuterostomes, which are discussed in the next two chapters. Here we briefly mention the other clade, the acoel flatworms.

Acoel flatworms appear to be distinct from other flatworms

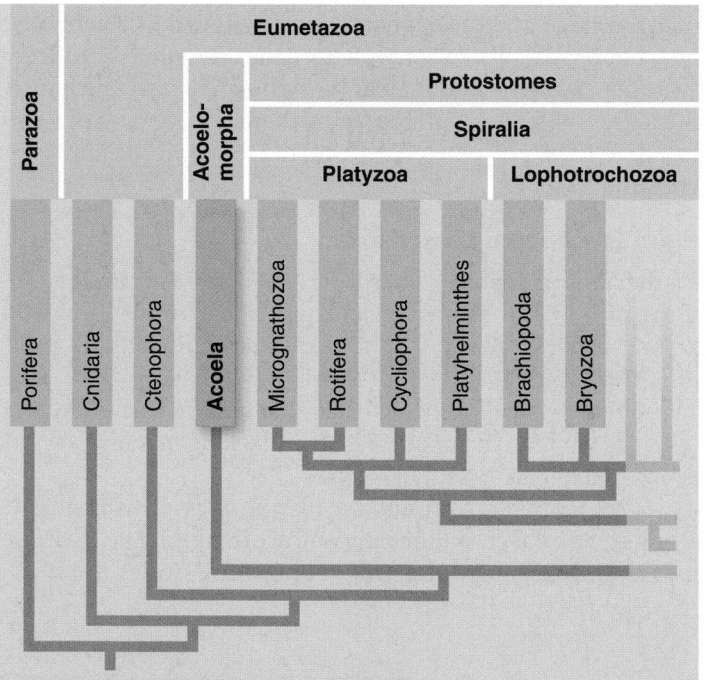

As its name implies, an acoel flatworm (figure 33.17) lacks a coelom. In addition, it has a nervous system that consists of a simple network of nerves with a minor concentration of neurons in the anterior end of the body. It lacks a permanent digestive cavity, so the mouth leads to a solid digestive syncytium (a mass of cells that have no cell membranes separating them).

These characteristics traditionally have been interpreted to classify acoel flatworms with flatworms in the phylum Platyhelminthes, which are discussed in the next chapter. However, based on molecular evidence, scientists have concluded that acoels are not closely related to members of phylum

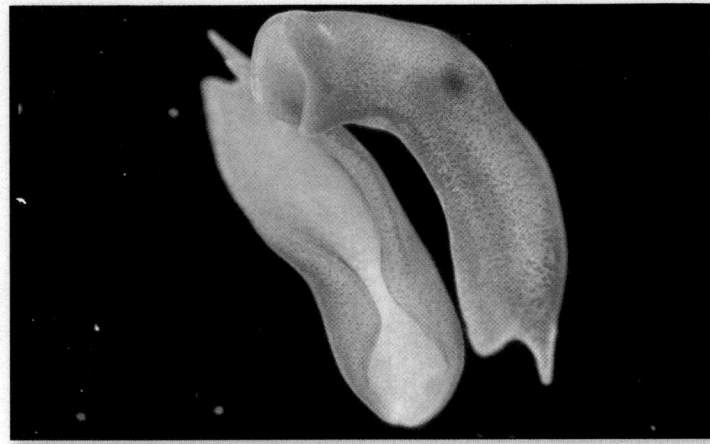

Figure 33.17 Phylum Acoela. An acoel flatworm of the genus *Waminoa*. These flatworms, long thought to be relatives of the Platyhelminthes, have a primitive nervous system and lack a permanent digestive cavity.

Platyhelminthes—similarities between the two groups are convergent. They belong in their own phylum, Acoela. Their precise position in the phylogenetic tree differs depending on the features used to construct it; one hypothesis, followed here, is that they are an early offshoot of the Bilateria, thus constituting the sister taxon to the clade comprising the protostomes and deuterostomes.

Learning Outcomes Review 33.6

The Bilateria is a clade characterized by bilateral symmetry and composed of two clades: the acoel flatworms and the clade comprising protostomes and deuterostomes. Traditionally, acoel flatworms were considered to be related to other types of flatworms, but recent molecular phylogenetic studies indicate that they are not closely related, and consequently that their morphological similarity is a result of convergent evolution.

■ *The acoel flatworms lack a coelom. Do any members of their sister taxon exhibit a similar condition?*

Chapter Review

33.1 Some General Features of Animals

Animals share some general characteristics.

Features common to all animals are multicellularity, heterotrophic lifestyle, and lack of a cell wall. Other features include specialized tissues, ability to move, and sexual reproduction.

33.2 Evolution of the Animal Body Plan

Most animals exhibit radial or bilateral symmetry.

Most sponges are asymmetrical, but other animals are bilaterally or radially symmetrical at some time during their life. The body parts of radially symmetrical animals are arranged around a central axis. The body of a bilaterally symmetrical animal has left and right halves. Most bilaterally symmetrical organisms are cephalized and can move directionally.

The evolution of tissues allowed for specialized structures and functions.

Each tissue consists of differentiated cells that have characteristic forms and functions.

A body cavity made the development of advanced organ systems possible.

Most bilaterian animals possess a body cavity other than the gut. A coelom is a cavity that lies within tissues derived from mesoderm. A pseudocoelom lies between tissues derived from mesoderm and the gut (which develops from endoderm). The acoelomate condition and the pseudocoelom appear to have evolved more than once, but the coelom evolved only once.

A circulatory system is an example of a specialized organ system that assists with distribution of nutrients and removal of wastes.

Bilaterians have two main types of development.

In a protostome, the mouth develops from or near the blastopore. A protostome has determinate development, and many have spiral cleavage.

In a deuterostome, the anus develops from the blastopore. A deuterostome has indeterminate development and radial cleavage.

Segmentation allowed for redundant systems and improved locomotion.

Segmentation, which evolved multiple times, allows for efficient and flexible movement because each segment can move somewhat independently. Another advantage to segmentation is redundant organ systems.

33.3 Animal Phylogeny

Animals are classified into 35 to 40 phyla based on shared characteristics.

Current understanding of phylogeny differs from traditional views.

By incorporating molecular data into phylogenetic analyses, some alterations have been made to the traditional view of how members of these phyla are related (figure 33.5).

Porifera (sponges) constitutes a monophyletic group that shares a common ancestor with other animals. Cnidarians (hydras, sea jellies, and corals) evolved before the origin of bilaterally symmetrical animals.

Annelids and arthropods had been considered closely related based on segmentation, but now arthropods are grouped with protostomes that molt their cuticles at least once during their life.

Morphology-based phylogeny focused on the state of the coelom.

The two groups of bilaterally symmetrical animals (the Bilateria)—protostomes and deuterostomes—differ in embryology. Possession of a coelom is an ancestral character of the clade comprising the protostome and deuterostomes. All deuterostomes have a coelom, but some protostomes have evolved a pseudocoelom and others have lost a coelom entirely.

Protostomes consist of spiralians and ecdysozoans.

Spiralia comprises the clades Lophotrochozoa and Platyzoa. Spiralian animals grow by gradual addition of mass to the body and undergo spiral cleavage.

Ecdysozoans grow by molting the external skeleton; Ecdysozoa includes many varied species, ranging from the pseudocoelomate, unsegmented Nematoda to the coelomate, segmented Arthropoda.

Deuterostomes include chordates and echinoderms.

The major groups of deuterostomes are the echinoderms, which include animals such as sea stars and sea urchins, and the chordates, which include vertebrates.

Deuterostome development indicates that echinoderms and chordates evolved from a common ancestor, distinguishing them clearly from other animals.

33.4 Parazoa: Animals That Lack Specialized Tissues

Parazoa ("near animals") comprises animals that, for the most part, lack definite symmetry, and that do not possess tissues.

Sponges have a loose body organization.

Sponges lack tissues and organs and a definite symmetry, but they do have a complex multicellularity. Larval sponges are free-swimming, and the adults are anchored onto submerged objects.

The sponge body is composed of several cell types.

Sponges have three layers: an external protective epithelial layer; a central protein-rich matrix called mesohyl with amoeboid cells; and an inner layer of choanocytes that circulate water and capture food particles (figure 33.7b).

Choanocytes help circulate water through the sponge.

The mesohyl may contain spicules and/or fibers of a tough protein called spongin that strengthen the body of the sponge.

Sponges reproduce both asexually and sexually.

Fragments of a sponge are able to grow into complete individuals. Sperm and eggs may be produced by mature individuals; these undergo fertilization to form zygotes that develop into free-swimming larvae that eventually become sessile adults.

33.5 Eumetazoa: Animals with True Tissues

The Eumetazoa contains animals with distinct tissues. The embryonic cell layers differentiate into the tissues of the adult body, giving rise to the body plan characteristic of each group of animals (figure 33.8).

All cnidarians are carnivores.

Members of the carnivorous and radially symmetrical Cnidaria have distinct tissues but no organs. They are diploblastic and have two body forms: a sessile, cylindrical polyp and a free-floating medusa.

Cnidarians are distinguished by capsules called nematocysts that are used in offense and defense.

Cnidarians have no circulatory, excretory, or respiratory systems. They have a latticework of nerve cells and are sensitive to touch; some have gravity receptors and light receptors.

Cnidarians are grouped into four—or five—classes.

The five classes of Cnidaria are the Anthozoa (sea anemones, corals, seafans); Cubozoa (box jellies); Hydrozoa (hydroids, Hydra), Scyphozoa (jellyfish); and Staurozoa (star jellies). The Staurozoa class is not accepted by all scientists.

The comb jellies use cilia for movement.

Ctenophora (comb jellies) is a small phylum of medusa-like animals that propel themselves with bands of fused cilia. They may be triploblastic. They capture prey with colloblasts, cells that release an adhesive.

33.6 The Bilateria

The Bilateria are characterized by bilateral symmetry, which allows for functional specialization such as having nerve receptors at the anterior end of the body.

Acoel flatworms appear to be distinct from other flatworms.

Acoel flatworms, phylum Acoela, were once considered basal to the phylum Platyhelminthes, but they appear to have evolved early in the history of the Bilateria and are the sister taxon to the clade comprised of protostomes and deuterostomes.

Review Questions

UNDERSTAND

1. Which of the following characteristics is unique to all animals?
   a. Sexual reproduction   c. Lack of cell walls
   b. Multicellularity   d. Heterotrophy

2. Animals are unique in the fact that they possess _____ for movement and _____ for conducting signals between cells.
   a. brains; muscles
   b. muscle tissue; nervous tissue
   c. limbs; spinal cords
   d. flagella; nerves

3. In animal sexual reproduction the gametes are formed by the process of
   a. meiosis.   c. fusion.
   b. mitosis.   d. binary fission.

4. The evolution of bilateral symmetry was a necessary precursor for the evolution of
   a. tissues.   c. a body cavity.
   b. segmentation.   d. cephalization.

5. A fluid-filled cavity that develops completely within mesodermal tissue is a characteristic of a
   a. coelomate.   c. acoelomate.
   b. pseudocoelomate. d. All of the choices are correct.

6. Which of the following statements is NOT true regarding segmentation?
   a. Segmentation allows the evolution of redundant systems.
   b. Segmentation is a requirement for a closed circulatory system.
   c. Segmentation enhances locomotion.
   d. Segmentation represents an example of convergent evolution.

7. Which of the following characteristics is used to distinguish between a parazoan and a eumetazoan?
   a. Presence of a true coelom
   b. Segmentation
   c. Cephalization
   d. Tissues

8. With regard to classification of animals, the study of which of the following is changing our understanding of how animals are organized?

   a. Molecular systematics
   b. Origin of tissues
   c. Patterns of segmentation
   d. Evolution of morphological characteristics

9. The coelom

   a. is a synapomorphy for the clade comprising protostomes and deuterostomes.
   b. evolved multiple times convergently.
   c. is always an indicator of close evolutionary relationship.
   d. forms within the endoderm.

10. A coelomate organism may have which of the following characteristics?

    a. Circulatory system
    b. Internal skeleton
    c. Larger size than a pseudocoelomate
    d. All of the choices are correct.

11. Which of the following cell types of a sponge possesses a flagellum?

    a. Choanocyte   c. Epithelial
    b. Amoebocyte   d. Spicules

12. The larval stage of a cnidarian is known as a

    a. medusa.   c. polyp.
    b. planula.   d. cnidocyte.

13. Acoel flatworms

    a. are closely related to other flatworms.
    b. have evolved an acoelom.
    c. are a member of the Parazoa.
    d. are the sister taxon to the clade comprised of protostomes and deuterostomes.

APPLY

1. The following diagram is of the blastopore stage of embryonic development. Based on the information in the diagram, which of the following statements is correct?

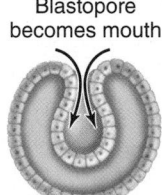
Blastopore becomes mouth

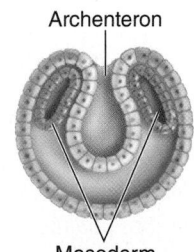
Archenteron

Mesoderm

a. Most animals exhibit this pattern of development.
b. It would have formed by spiralian cleavage.
c. It would exhibit determinate development.
d. All of the choices are correct.

2. Which of the following characteristics would not apply to a species in the Ecdysozoa?

   a. Bilateral
   b. Indeterminate cleavage
   c. Molt at least once in their life cycle
   d. Metazoan

3. In the rain forest you discover a new species that is terrestrial, has determinate development, molts during its lifetime, and possesses jointed appendages. To which phylum of animals should it be assigned?

4. Which of the following cell layers is not necessary to be considered a eumetazoan?

   a. Ectoderm
   b. Endoderm
   c. Mesoderm
   d. All of the above are found in all eumetazoans.

SYNTHESIZE

1. Worm evolution represents an excellent means of understanding the evolution of a body cavity. Using the phyla of worms Nematoda, Annelida, Platyhelminthes, and Nemertea, construct a phylogenetic tree based only on the form of body cavity (refer to figure 33.3 and table 33.1 for assistance). How does this differ from the phylogeny in figure 33.5? Should body cavity be used as the sole characteristic for classifying a worm?

2. Most students find it hard to believe that Echinodermata and Chordata are closely related phyla. If it were not for how they share a similar pattern of development, where would you place Echinodermata in the animal kingdom? Defend your answer.

ONLINE RESOURCE

www.ravenbiology.com

Understand, Apply, and Synthesize—enhance your study with animations that bring concepts to life and practice tests to assess your understanding. Your instructor may also recommend the interactive eBook, individualized learning tools, and more.

Chapter 34

Protostomes

Chapter Contents

Introduction

Almost all bilaterian animals belong to one of two clades, the protostomes and the deuterostomes. These clades differ in how they develop embryologically—in protostomes, the mouth of the adult animal develops from the blastopore or from an opening near it, whereas in deuterostomes, the blastopore gives rise to the anus and the mouth develops in other ways. The great majority of all animal species are protostomes, including insects, mollusks, worms, and many others.

The Clades of Protostomes

Learning Outcome

1. *List the major clades of protostomes and identify their distinguishing characteristics.*

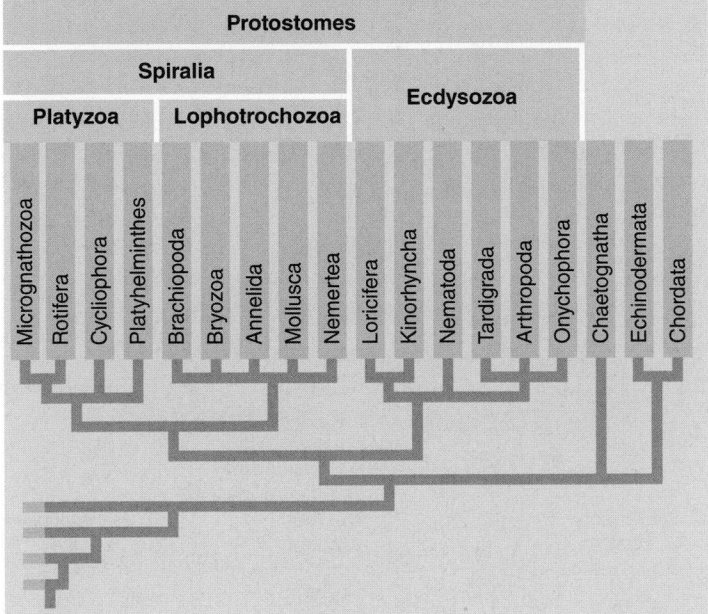

All protostomes belong to either the Spiralia or the Ecdysozoa (see figure 33.5). Before discussing the diversity of protostomes, we will briefly review the characteristics of these groups.

Spiralia

Spiralians develop as embryos using spiral cleavage (see figure 33.4). Most live in water and move through it using cilia or contractions of the body musculature. The two clades of spiralians are platyzoans and lophotrochozoans. The most prominent group of platyzoans is the flatworms (phylum Platyhelminthes), animals that have a simple body with no circulatory or respiratory system but a complex reproductive system. This group includes marine and freshwater planarians as well as the parasitic flukes and tapeworms.

Lophotrochozoa consists of two major phyla and several smaller ones. Many of the animals have a type of free-living larva known as a **trochophore** (figure 34.1a), and some have a feeding structure termed a **lophophore** (figure 34.1b), a horseshoe-shaped crown of ciliated tentacles around the mouth used in filter-feeding. The phyla characterized by a lophophore are Bryozoa and Brachiopoda. Lophophorate animals are sessile (anchored in place).

Among the lophotrochozoans with a trochophore are phyla Mollusca and Annelida. Mollusks are unsegmented, and their coelom is reduced to a hemocoel (open circulatory space) and some other small body spaces. This phylum includes animals as diverse as octopuses, snails, and clams. Annelids are segmented coelomate worms, the most familiar of which is the

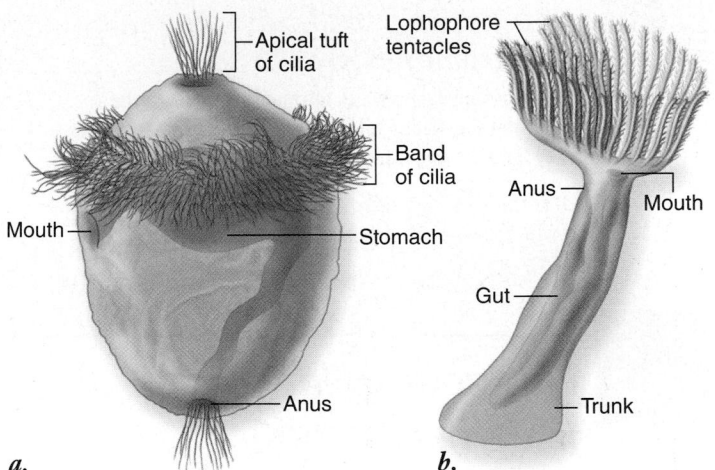

Figure 34.1 Trochophore *(a)* and lophophore *(b)*.

earthworm, but also includes leeches and the largely marine polychaetes.

Ecdysozoa

The other major clade of protostomes is the Ecdysozoa, which contains animals that molt. When an animal grows large enough that it completely fills its hard external skeleton, it must lose that skeleton by molting (a process also called ecdysis, hence the name Ecdysozoa). While the animal grows, it forms a new exoskeleton underneath the existing one. The first step in molting is for the body to swell until the existing exoskeleton cracks open and is shed (figure 34.2). Upon molting that skeleton, the animal inflates the soft, new one, expanding it using body fluids (and, in many insects and spiders, air as well). When the new one hardens, it is larger than the molted one had been and has room for growth. Thus, rather than being continuous, as in other animals, the growth of ecdysozoans is step-wise.

The Ecdysozoa contains the arthropods, which contains more species than any other phylum, and nematodes (roundworms), which are so numerous that, it is said, if everything in the world except nematodes were to disappear, an outline of what had been there would be visible in the remaining nematodes! Each phylum contains one of the model organisms used in laboratory studies that have informed much of our current understanding of genetics and development: the fruit fly *Drosophila melanogaster* and the roundworm *Caenorhabditis elegans.*

In this chapter, we review the diversity of the major phyla of protostomes, starting with the platyzoans, followed by the lophotrochozoans and then the ecdysozoans.

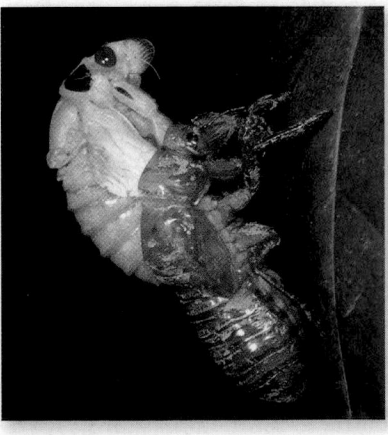

Figure 34.2 A cicada, a type of insect, molting.

34.2 Platyzoans: Flatworms (Platyhelminthes)

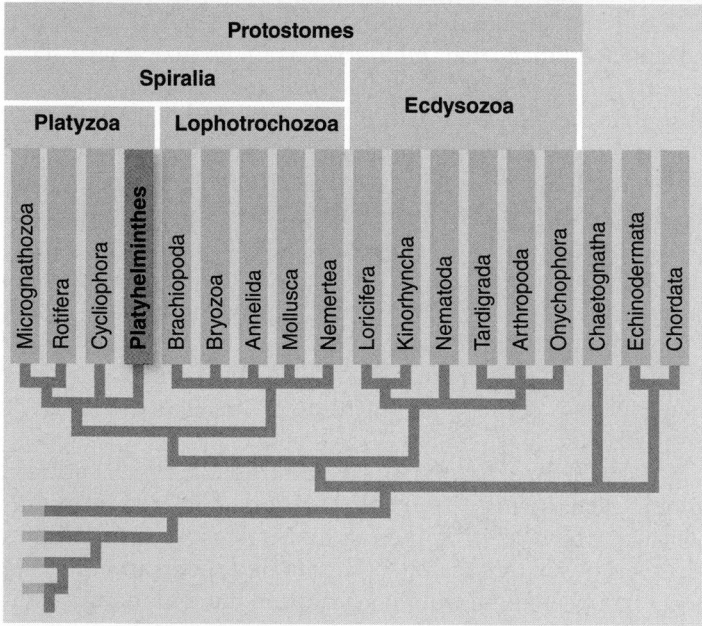

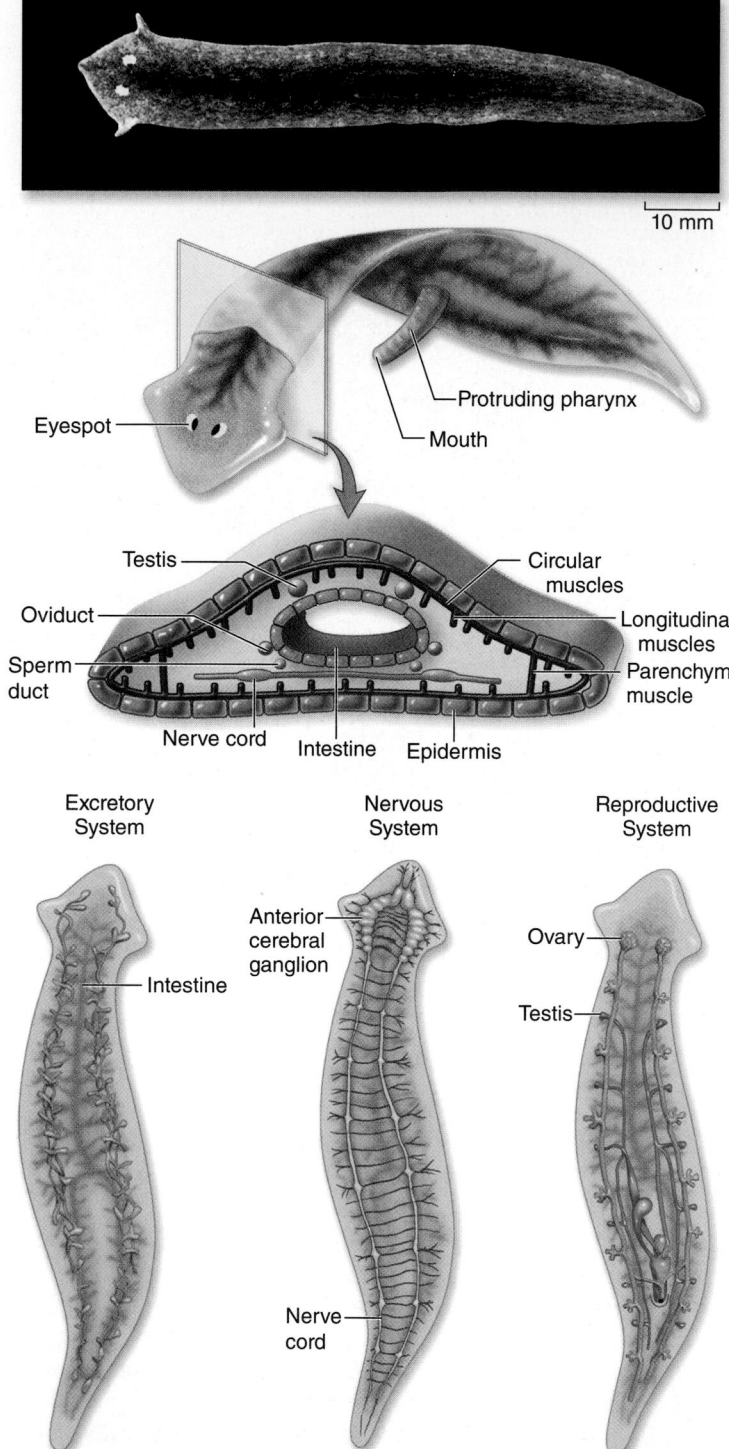

Figure 34.3 Architecture of a flatworm. A photo (*a*) and an idealized diagram (*b*) of the genus *Dugesia,* the familiar freshwater planarian of ponds and rivers. Upper schematic shows a whole animal and a transverse section through the anterior part of the body. Schematics below show the digestive, central nervous, and reproductive systems.

Flatworms have an incomplete gut

The phylum Platyhelminthes consists of some 20,000 species. These ciliated, soft-bodied animals are flattened dorsoventrally, the anatomy that gives them the name flatworms. Flatworm bodies are solid aside from an incomplete digestive cavity (figure 34.3). Although among the morphologically simplest of bilaterally symmetrical animals, they have some complex structures, like their reproductive apparatus, and they have the most complex life cycles among animals.

Free-living flatworms occur in a wide variety of marine, freshwater, and even moist terrestrial habitats. They are carnivores and scavengers, eating small animals and bits of organic debris. They move around by means of ciliated epithelial cells, which are particularly concentrated on their ventral surfaces, but they also have well-developed musculature. By contrast, many species of flatworms are parasitic, living inside the bodies of other animals. Flatworms range in length from 1 mm or less to many meters (some tapeworms).

Digestion in flatworms

Most flatworms have only a single opening for their digestive cavity, a mouth located on the bottom side of the animal at midbody. A flatworm ingests its food and tears it into small bits using muscular contractions in the upper end of the gut, the pharynx.

Like sponges, cnidarians, and ctenophores, a flatworm lacks a circulatory system for the transport of oxygen and food molecules. The thin body of a flatworm allows gas to diffuse between its cells and the surrounding environment (oxygen diffuses in and carbon dioxide diffuses out). Branches of the gut extend throughout the body, so the gut functions in both digestion and distribution of food. Cells that line the gut engulf most of the food particles by phagocytosis and digest them; but, as in cnidarians and most bilaterians, some particles are partly digested extracellularly. Tapeworms, which are parasitic, have a mouth at the front of their body and no digestive cavity at all; they absorb food directly through the body wall.

Excretion and osmoregulation

Unlike cnidarians and ctenophores, flatworms have an excretory system, which consists of a network of fine tubules that run throughout the body. Bulblike **flame cells,** so named because of the flickering movements of the flagella beating inside them, are located on the side branches of the tubules. Flagella in the flame cells move water and excretory substances into the tubules and then to pores located between the epidermal cells through which the liquid is expelled. Flame cells primarily regulate water balance of the organism; the excretory function appears to be secondary, and much of the metabolic waste of a flatworm diffuses into the gut and is eliminated through the mouth.

Nervous system and sensory organs

The flatworm nervous system is composed of an anterior cerebral ganglion and nerve cords that run down the body, with cross-connections that give it a ladder-like shape (see figure 34.3). Free-living flatworms are poorly cephalized, with eyespots on their heads (see figure 34.3). These inverted, pigmented cups, which contain light-sensitive cells connected to the nervous system, enable a worm to distinguish light from dark: most flatworms tend to move away from strong light.

Flatworm reproduction

The reproductive systems of flatworms are complex. Most are **hermaphroditic,** each individual containing both male and female sexual structures (see figure 34.3). In many members of Platyhelminthes, copulation is required between two individuals, and fertilization is internal, each partner depositing sperm in the copulatory sac of the other. The sperm travel along special tubes to reach the eggs.

In most freshwater flatworms, fertilized eggs are laid in cocoons strung in ribbons and hatch into miniature adults. In contrast, some marine species develop indirectly, the fertilized egg undergoing spiral cleavage, and the embryo giving rise to a larva that swims or drifts until metamorphosing, when it settles in an appropriate habitat.

Flatworms are known for their regenerative capacity: when a single individual of some species is divided into two or more parts, an entirely new flatworm can regrow what is missing from each bit.

Flatworms consist of two major groups

Most flatworms are parasitic; those that are not are referred to as free-living. Phylogenetic evidence indicates that the parasitic lifestyle evolved only once in platyhelminths, from free-living ancestors.

Turbellaria: Free-living flatworms

Flatworm phylogeny is in a state of flux. The group of free-living flatworms, Turbellaria, has been considered a class, but recent studies show it is not monophyletic so it is likely to be divided into several classes. One of the most familiar members of this group are freshwater members of the genus *Dugesia,* the common planarian used in biology laboratories.

Neodermata: Parasitic flatworms

All parasitic flatworms are placed in the subphylum Neodermata. That name means "new skin" and refers to the animal's outer surface, termed the neodermis. All neodermatans live as ectoparasites on or endoparasites in the bodies of other animals for some period of their lives. The neodermis is resistant to the digestive enzymes and immune defenses produced by the animals parasitized by these flatworms. These animals also lack other features of free-living flatworms such as eyespots, which are of no adaptive value to an organism living inside the body of another animal. Neodermata contains two subgroups: Trematoda, the flukes, and Cercomeromorpha, the tapeworms and their relatives.

Trematoda: The flukes

The more than 10,000 named species of flukes range in length from less than 1 mm to more than 8 cm. Flukes attach themselves within the bodies of their hosts by means of suckers, anchors, or hooks. A fluke takes in food (cells or fluids of the host) through its mouth, like its free-living relatives. The life cycle of some species involves only one host, usually a fish, but the life cycle of most flukes involves two or more hosts. The first intermediate host is almost always a snail, and the final host (in which the adult fluke lives and reproduces sexually) is almost always a vertebrate; in between there may be other intermediate hosts. Although the life of a parasite is secure within a host, which provides food and shelter, getting from one host to another is extremely risky, and most individuals die in the transition.

Flukes that cause disease in humans

The oriental liver fluke, *Clonorchis sinensis,* is an example of a flatworm that parasitizes humans (as well as cats, dogs, and pigs), living in the bile duct of the liver (figure 34.4). It is especially common in Asia. Each worm is 1 to 2 cm long and, like all flukes, has a complex life cycle. A fertilized egg containing a ciliated first-stage larva, the **miracidium,** is passed in the feces.

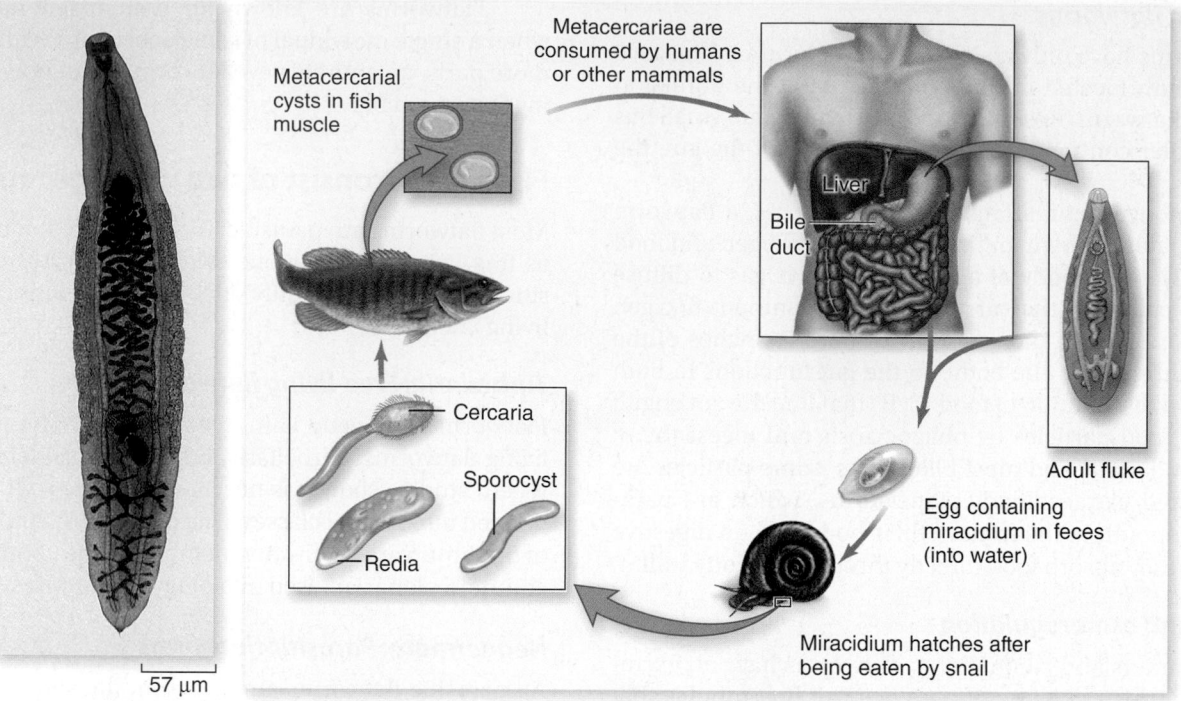

Figure 34.4 Life cycle of the oriental liver fluke, *Clonorchis sinensis*. *a.* Micrograph. *b.* Schematic of the life cycle.

If the larva reaches water, it may be ingested by an aquatic snail, but most do not reach water and most that do are not ingested. The prodigious number of eggs a parasitic flatworm produces is an adaptation to this life cycle full of risks. Within the snail, the ciliated larva transforms into a **sporocyst,** a baglike structure containing embryonic germ cells, each of which develops into a **redia** (plural, *rediae*), an elongated, nonciliated larva. Each of these larvae grows within the snail, then gives rise to several individuals of the next larval stage, the tadpole-like **cercaria** (plural, *cercariae*).

Cercariae escape into the water, where they swim about freely. When one encounters a fish of the family Cyprinidae—the family that includes carp and goldfish—it bores into the muscles, loses its tail, and encysts, transforming into a meta-cercaria. If a human or other mammal eats raw fish containing metacercariae, the cyst dissolves in the intestine, and the young fluke migrates to the bile duct, where it matures, thereby completing the cycle. Even if infected fish is cooked, the parasite can be transmitted if metacercariae stick to cutting boards or the hands of a person handling the raw fish flesh are then ingested. An individual fluke may live for 15 to 30 years in the liver; a heavy infection of liver flukes may cause cirrhosis of the liver and death in humans.

Perhaps the most important trematodes to human health are blood flukes of the genus *Schistosoma*. They afflict about 5% of the world's population, or more than 200 million people, in tropical Asia, Africa, Latin America, and the Middle East. About 800,000 people die each year from the disease called **schistosomiasis,** or bilharzia.

Schistosomes live in blood vessels associated with the intestine or the urinary bladder, depending on the species (figure 34.5). As a result, its fertilized eggs must first break through the wall of the blood vessel to get into the intestine or urinary bladder, from which it can exit the body. The damage done to the blood vessels and gut or bladder wall is considerable, especially considering that an individual worm may release 300 to 3000 eggs per day and live for many years.

Great effort is being made to control schistosomiasis. The worms protect themselves from the body's immune system in part by coating themselves with some of the host's own antigens that effectively render the worm immunologically invisible (see chapter 51). The search is on for a vaccine that would cause the host to develop antibodies to one of the antigens of young worms before they can protect themselves.

Figure 34.5 Schistosomes. The male lies within the groove on the female's ventral side, with its front and hind ends protruding.

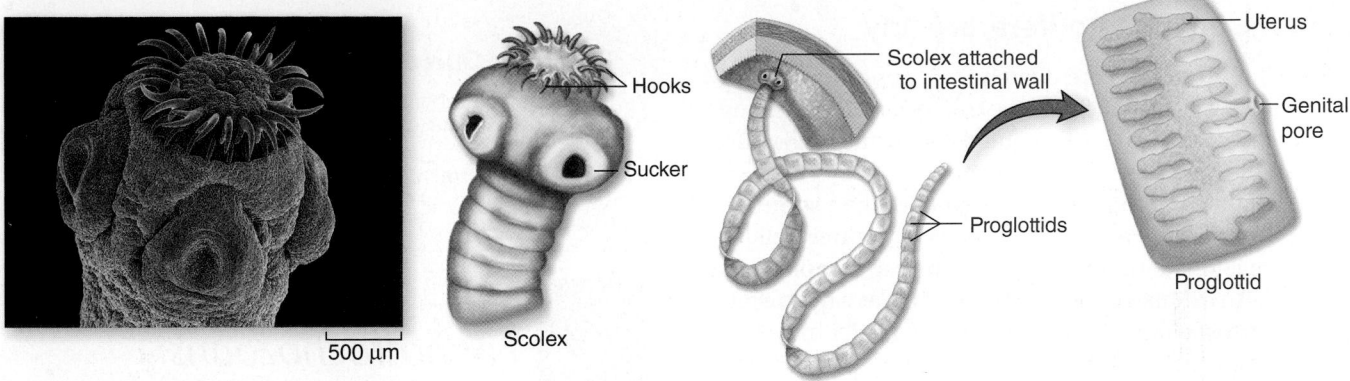

Figure 34.6 Tapeworms. *a.* Flatworms such as this beef tapeworm, *Taenia saginata,* live parasitically in the intestine of mammals. *b.* Schematic.

Cercomeromorpha: Tapeworms and their relatives

An adult tapeworm hangs onto the inner wall of its host's intestine by means of a terminal attachment structure. It lacks a digestive cavity as well as digestive enzymes, absorbing food from the host's gut through its outer surface. Most species of tapeworm occur in the intestines of vertebrates; about a dozen of them regularly occur in humans.

The long, flat body of a tapeworm is divided into three zones: the **scolex,** or attachment structure; the neck; and a series of repetitive sections, the **proglottids** (figure 34.6). The scolex of many species bears four suckers and may also have hooks. The scolex is not a head: it has neither concentrated nervous tissue nor a mouth. Each proglottid is a complete hermaphroditic unit, containing both male and female reproductive organs. Proglottids are formed continuously in a growth zone at the base of the neck, with maturing ones being pushed posteriorly as new ones are formed. The gonads in the proglottids progressively mature away from the neck, fertilization occurs, and the embryos form, the proglottids nearer the end of the body being more mature. The most terminal proglottid, filled with embryonated eggs, breaks off; either it ruptures, and the embryos, each surrounded by a shell, are carried out with the host's feces, or the entire proglottid is carried out, whereupon the embryos emerge from it through a pore or the ruptured body wall. Embryos are scattered in the environment, on leaves, in water, or in other places where they may be picked up by another animal.

The beef tapeworm, *Taenia saginata,* occurs as a juvenile in the intermuscular tissue of cattle; as an adult, it inhabits the intestines of human beings, where a mature worm may reach a length of 10 m or more. A worm attaches itself to the intestinal wall of the host by a scolex with four suckers. Shed proglottids pass from the human in the feces and may crawl onto vegetation, a favorable place to be picked up by cattle; they may remain viable for up to five months. If one is ingested by cattle, the larva burrows through the wall of the intestine and ultimately reaches muscle tissues through the blood or lymph vessels. About 1% of cattle in the United States are infected, and because some 20% of the beef consumed is not federally inspected, when humans eat infected beef that is cooked "rare," infection by these tapeworms can occur. As a result, the beef tapeworm is a frequent parasite of humans.

Learning Outcomes Review 34.2

Flatworms are compact, bilaterally symmetrical animals. Most have a blind digestive cavity with only one opening; they also have an excretory system consisting of flame cells within tubules that run throughout the body. Many are free-living, but some cause human diseases. The scolex of a tapeworm is not a head, but simply an anchoring device to hold the animal in the intestine where it acts as a parasite.

■ *How does the anatomy of a tapeworm relate to its way of life?*

34.3 Platyzoans: Rotifers (Rotifera)

Learning Outcome

1. *Explain why rotifers are referred to as "wheel animals."*

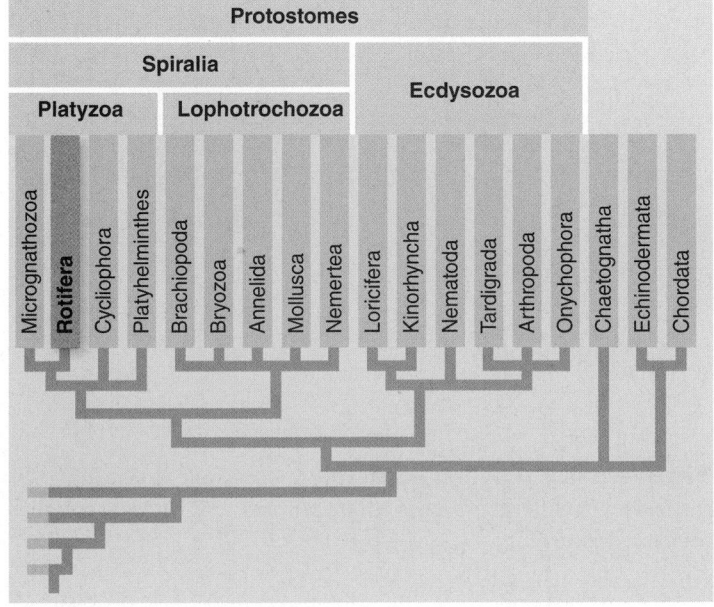

The rotifers, phylum Rotifera, are tiny

Rotifers (phylum **Rotifera**) are bilaterally symmetrical, unsegmented pseudocoelomates (figure 34.7*a*). Several features suggest their ancestors may have resembled flatworms, with which they are classified in the spiralian Platyzoa.

At 50 to 500 μm long, rotifers are smaller than some ciliate protists. But they have complex bodies with three cell layers, highly developed internal organs, and a complete gut (figure 34.7*b*). An extensive pseudocoelom acts as a hydrostatic skeleton; the cytoskeleton provides rigidity. A rotifer has a rigid external covering, but its body can lengthen and shorten greatly because the posterior part is tapered so it can fold up like a telescope. Many have adhesive toes used for clinging to vegetation and other such objects.

Diversity and distribution

About 1800 species are known, most of which occur in fresh water; a few rotifers live in soil, mosses, and the ocean. The lifespan of a rotifer is typically no longer than 1 or 2 weeks, but some species can survive in a desiccated, inactive state on the leaves of plants; when rain falls, the rotifers become active and feed in the film of water that temporarily covers the leaf.

Food gathering

The corona, a conspicuous ring of cilia at the anterior end (see figure 34.7*b*), is the source of the common name "wheel animals" for rotifers because its beating cilia make it appear that a wheel is rotating around the head of the animal. The corona is used for locomotion, but its cilia also sweep food into the rotifer's mouth. Once food is swallowed, it is crushed with a complex jaw in the pharynx.

Learning Outcome Review 34.3

Rotifers are extremely small but with a highly complex body structure; the name "wheel animals" comes from the apparent motion of their beating cilia.

■ What is the role of the cilia in rotifers?

34.4 Lophotrochozoans: Mollusks (Mollusca)

Learning Outcomes

1. List the defining features of phylum Mollusca.
2. Describe representatives of the four best-known groups of mollusks.
3. Explain the distinguishing features of cephalopods.

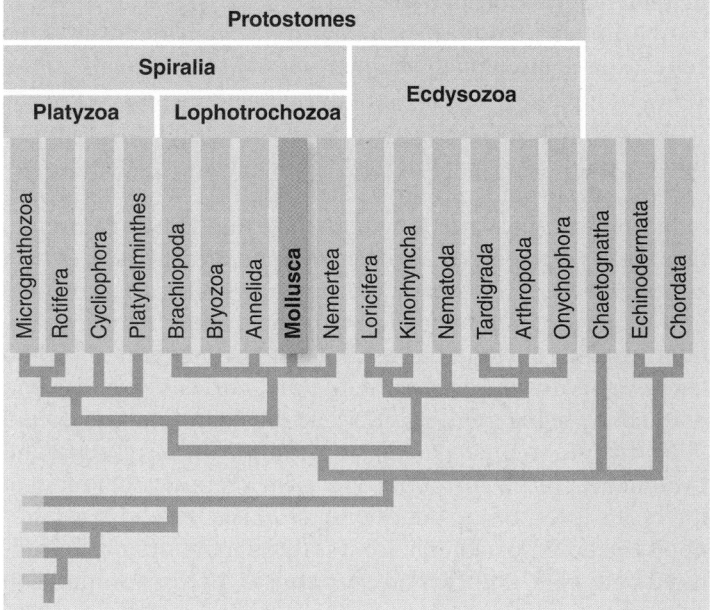

Mollusks (phylum Mollusca) are diverse morphologically and live in many types of environments. With more than 110,000 described species, the phylum is second in size only to arthropods. Mollusks include snails, slugs, clams, scallops, oysters, cuttlefish, octopuses, and many other familiar animals (figures 34.8 and 34.9). The shells of many mollusks are beautiful and elegant; they have long been collected, preserved, and studied by professional scientists and amateurs. Some mollusks, however, lack a shell.

Mollusks are extremely diverse—and important to humans

Mollusks range in size from almost microscopic to huge. Although most measure a few millimeters to centimeters in

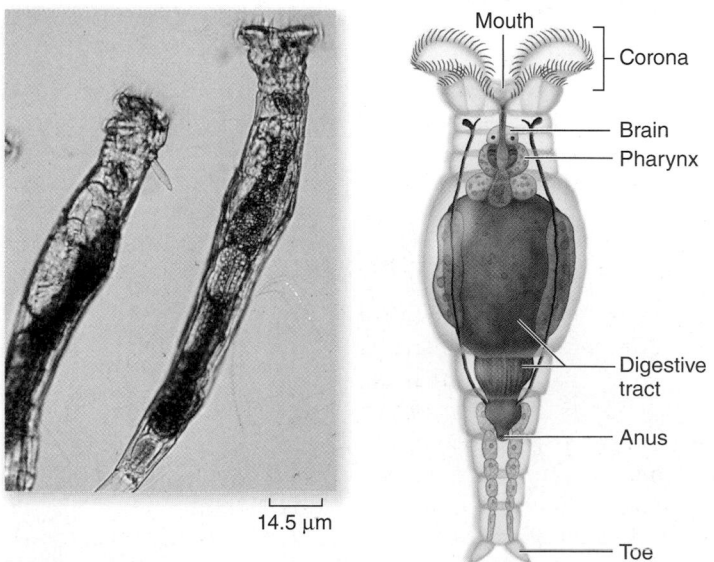

Figure 34.7 Phylum Rotifera. Microscopic in size (*a*), rotifers are smaller than some ciliate protists, and yet have complex internal organs (*b*).

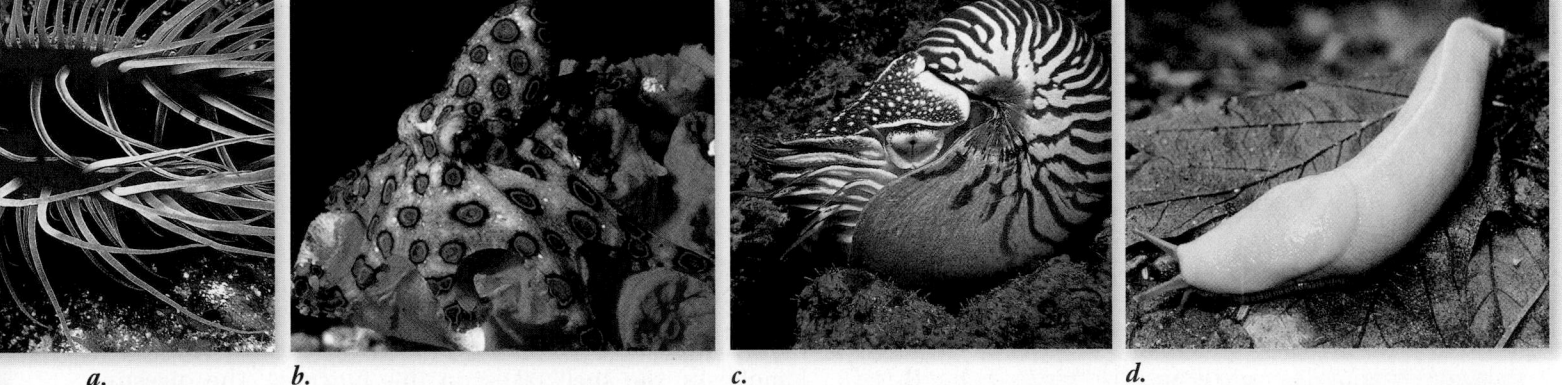

a.   *b.*   *c.*   *d.*

Figure 34.8 Mollusk diversity. Mollusks exhibit a broad range of variation. *a.* The flame scallop, *Lima scabra,* is a filter feeder. *b.* The blue-ringed octopus, *Hapalochlaena maculosa,* is one of the few mollusks dangerous to humans. Strikingly beautiful, it is equipped with a sharp beak that can deliver a poisonous bite! *c.* Nautiluses, such as this chambered nautilus, *Nautilus pompilius,* have been around since before the age of the dinosaurs. *d.* The banana slug, *Ariolimax columbianus,* native to the Pacific Northwest, is the second largest slug in the world, attaining a length of 25 cm.

their largest dimension, the giant squid may grow to more than 15 m long and weigh as much as 250 kg. It is therefore one of the heaviest invertebrates (although nemerteans can be longer, as discussed later). Other large mollusks are the giant clams of the genus *Tridacna,* which may be as long as 1.5 m and may weigh as much as 270 kg (see figure 34.9).

Like all major animal groups, mollusks evolved in the oceans, and most groups have remained there. Marine mollusks are widespread and many are abundant. Snails and slugs have invaded freshwater and terrestrial habitats, and freshwater mussels live in lakes and streams (the flat foot of a snail or slug allows it to crawl, but the foot of clams, mussels, and other bivalved mollusks is adapted to digging, so they cannot move about on land). Some places where terrestrial mollusks live, such as crevices of desert rocks, may appear dry, but if mollusks live there, the habitat has at least a temporary supply of water.

Mollusks—including oysters, clams, scallops, mussels, octopuses, and squids—are an important source of food for humans. They are also economically significant in other ways. For example, the material called mother-of-pearl (nacre), which is used for jewelry and other decorative objects, and formerly for buttons, comes from mollusk shells, most notably that of the abalone. Mollusks can also be pests. Bivalves called ship-worms burrow through wood exposed to the sea, damaging boats, docks, and pilings. The zebra mussel (*Dreissena polymorpha*) (see figure 59.17) has recently invaded many North American freshwater ecosystems. Many slugs and snails damage flowers, vegetable gardens, and crops. Other mollusks serve as hosts to the larval stages of many serious parasites, as discussed previously.

Figure 34.9 Giant clam. Second only to the arthropods in number of described species, members of the phylum Mollusca occupy almost every habitat on Earth. This giant clam, *Tridacna maxima,* contains symbiotic dinoflagellates (zooxanthellae) as do reef-forming corals. Through photosynthesis, the dinoflagellates probably contribute to the food of the clam, although it is a filter feeder like most bivalves.

The mollusk body plan is complex and varied

Some mollusk body plans are illustrated in figure 34.10. The **mantle,** a thick epidermal sheet, covers the dorsal side of the body, and bounds the mantle cavity. The mantle secretes the calcium carbonate of the shell in those mollusks with a shell. The muscular foot is the primary means of locomotion in mollusks other than cephalopods (octopuses, squid, and chambered nautilus). The head may be well developed or not. Mollusks are bilaterally symmetrical, but this symmetry is modified during development of many gastropods (snails and their relatives) which undergo torsion (a twisting of the body discussed in detail later in this chapter).

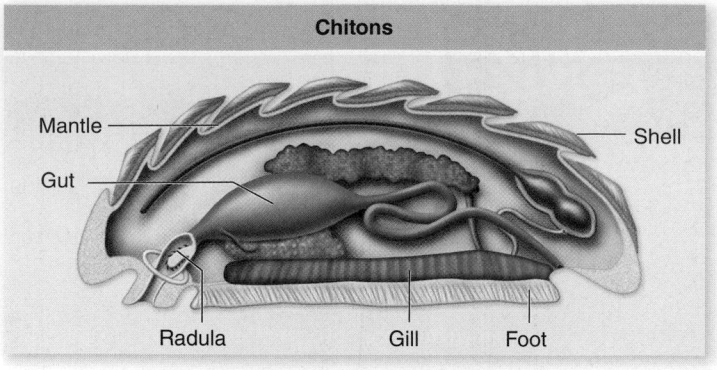

Chitons

Mantle

Gut

Shell

Radula   Gill   Foot

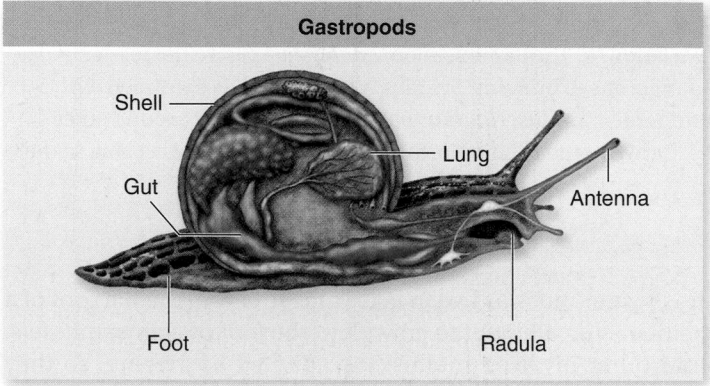

Gastropods

Shell

Lung

Gut

Antenna

Foot

Radula

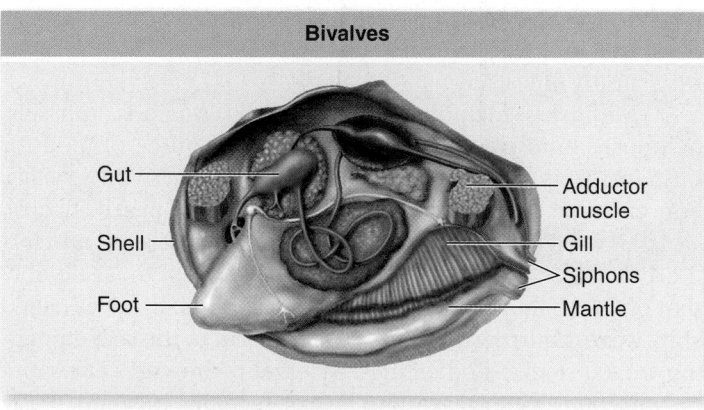

Bivalves

Gut

Shell

Foot

Adductor muscle

Gill

Siphons

Mantle

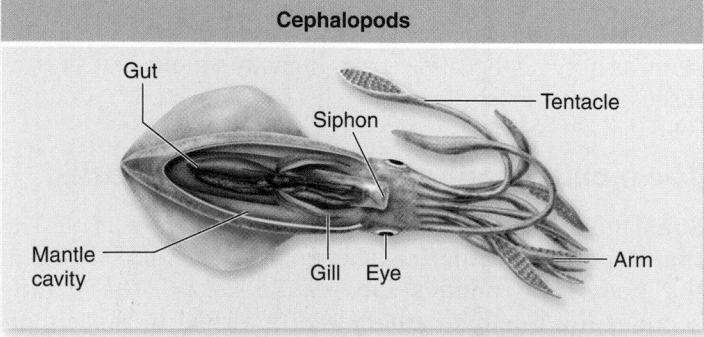

Cephalopods

Gut

Siphon

Tentacle

Mantle cavity

Gill   Eye

Arm

Figure 34.10 Body plans of some mollusks.

The muscular foot is variously adapted for locomotion, attachment, food capture, digging, or combinations of these functions. The foot of slugs and snails secretes mucus, forming a path that it glides along. In cephalopods, the foot is divided into tentacles and, in some cephalopods, arms. Clams burrow into mud and sand using their hatchet-shaped foot. In some mollusks that live in the open ocean, the foot is modified into winglike projections that can beat like wings and that have a large surface area that slows the rate of sinking.

Internal organs

In all mollusks, the coelom is highly reduced, being limited to small spaces around the excretory organs, heart, and part of the intestine. The coelom plays an important role by forming the hydrostatic skeleton in some other invertebrates, but in mollusks, the shell takes on this function. The digestive, excretory, and reproductive organs are concentrated in a visceral mass.

In aquatic mollusks, the gills (respiratory structures technically termed "ctenidia") project into the mantle cavity (see figure 34.10). They consist of filaments rich in blood vessels that greatly increase the surface area and capacity for gas exchange. The continuous stream of water that passes through the mantle cavity, propelled by cilia on the gills of all mollusks except cephalopods, carries oxygen in and carbon dioxide away. Mollusk ctenidia are so efficient that they can extract 50% or more of the dissolved oxygen from the water that passes through the mantle cavity. In addition to extracting oxygen from incoming water, the gills of most bivalves filter out food. Because the outlets from the excretory, reproductive, and digestive organs open into the mantle cavity, wastes and gametes are carried away from a mollusk's body with the exiting water stream.

Shells

One of the best-known characters of the phylum is the shell. A mollusk shell, which is secreted by the outer surface of the mantle, protects against predators and adverse environmental conditions. However, a shell is clearly not essential: reduction, internalization, and loss of the shell have evolved repeatedly. Examples of shell-less mollusks are cuttlefish, squids, and octopuses (cephalopods) as well as slugs (gastropods).

A typical mollusk shell consists of two layers of calcium carbonate, which is precipitated extracellularly. The outer layer consists of densely packed crystals. In some species, the inner layer is pearly in appearance, and is called mother-of-pearl or nacre. Pearls are formed when a foreign object, such as a grain of sand, becomes lodged between the mantle and the inner shell layer. The mantle coats the object with layer upon layer of nacre to reduce the irritation caused by the object. The most beautiful and highly valued pearls are produced by oysters depositing nacre.

Feeding and prey capture: Radula

A characteristic feature of most mollusks is the **radula** (plural, *radulae*), a rasping, tonguelike structure used in feeding. It consists of dozens to hundreds of microscopic, chitinous teeth arranged in rows on an underlying membrane and lies in a chamber at the anterior end of the gut (figure 34.11). The membrane wraps around a muscular support structure so

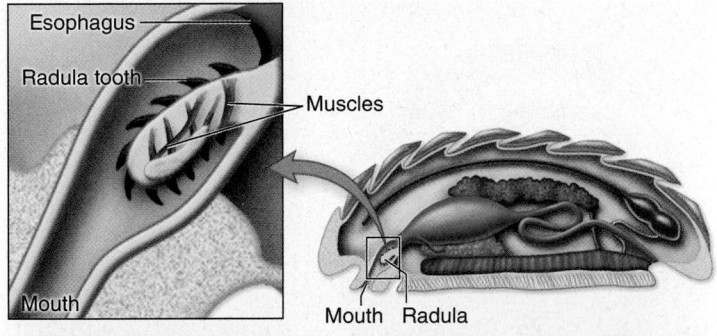

Figure 34.11 Structure of the radular apparatus in a chiton. The radula consists of rows of teeth made of chitin. As the animal feeds, its mouth opens, and the radula is thrust out to scrape food off surfaces. **a.** Schematic. **b.** Micrograph.

the radula can be protruded through the mouth and move something like a sanding belt over what is being rasped. Benthic mollusks use their radulae to scrape up algae and other food materials.

In some predatory gastropods (such as moon snails), the radula is modified to drill through clam shells so the snail can eat the clams. In snails of the genus *Conus,* the radula has been transformed into a harpoon that is associated with a venom gland; cone snails use this harpoon to capture prey such as fish, and some cone snails can harm—or even kill—humans.

Bivalves are the only mollusks that have no radula. The gills of most bivalves are adapted to filter food particles from the water, although primitive bivalves pick up bits of food from soft sediments (this is, they are deposit feeders) using appendages around the mouth.

Removal of wastes

Nitrogenous wastes are removed from the mollusk body by the **nephridium** (plural, *nephridia*), a sort of kidney. A typical nephridium has an open funnel, the nephrostome, which is lined with cilia. A coiled tubule runs from the **nephrostome** into a bladder, which in turn connects to an excretory pore. Wastes are gathered by the nephridia from the coelomic cavity and discharged into the mantle cavity. Sugars, salts, water, and other materials are reabsorbed by the walls of the nephridia and returned to the animal's body as needed to maintain osmotic balance.

The circulatory system

The main coelomic cavity of a mollusk is a hemocoel, which includes several sinuses and a network of vessels in the gills, where gas exchange takes place. Except for cephalopods, all mollusks have such an open circulatory system; blood (technically termed "hemolymph") is propelled by a heart through the aorta vessel that empties into the hemocoel. The blood moves through the hemocoel being recaptured by other venous vessels before re-entering the heart (see chapter 49). The heart of most, but not all mollusks has three chambers: two that collect aerated blood from the gills, and a third that pumps it into the hemocoel. The blood of the closed circulatory system of cephalopods is contained in a continuous system of vessels, so it does not contact other tissues directly.

Reproduction

Most mollusks have separate sexes although a few bivalves and many freshwater and terrestrial gastropods are hermaphroditic. Most hermaphroditic mollusks engage in cross-fertilization. Some oysters are able to change sex.

As is typical of animals living in the sea, many marine mollusks have external fertilization: gametes are released by males and females into the water where fertilization occurs. Most gastropods, however, have internal fertilization, with the male inserting sperm directly into the female's body (not, that is, into a preformed channel). Internal fertilization, a foot, and an efficient excretory system that prevents desiccation are some of the key adaptations that allowed gastropods to colonize the land.

A mollusk zygote undergoes spiral cleavage (thus Mollusca is part of the Spiralia; see chapter 33). The embryo develops into a free-swimming larva called a **trochophore** (figure 34.12a) that closely resembles the larval stage of many marine annelids and other lophotrochozoans. A trochophore swims by means of cilia that encircle the middle of its body. In most marine snails and in bivalves, the trochophore develops into a second free-swimming stage, the **veliger.** The veliger

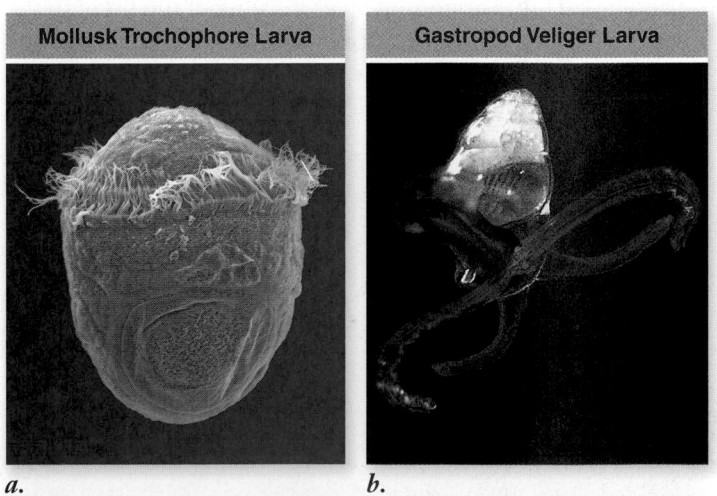

Figure 34.12 Stages in the molluskan life cycle. *a.* The trochophore larva. Similar larvae are characteristic of some annelid worms as well as a few other phyla. *b.* The veliger stage of a gastropod.

forms the beginnings of a foot, shell, and mantle (figure 34.12*b*). Trochophores and veligers drift widely in the ocean, dispersing these otherwise sedentary mollusks.

Four classes of mollusks show the diversity of the phylum

We examine four of the seven or eight recognized classes of mollusks: (1) Polyplacophora—chitons; (2) Gastropoda—limpets, snails, slugs, and their relatives; (3) Bivalvia—clams, oysters, scallops, and their relatives; and (4) Cephalopoda—squids, octopuses, cuttlefishes, and the chambered nautilus.

Chitons (Polyplacophora)

Chitons are exclusively marine and comprise perhaps 1000 species. The oval body is covered dorsally with eight overlapping dorsal calcareous plates (figure 34.13). The body is not segmented, but chitons do have eight sets of dorsoventral pedal retractor muscles and serially repeated gills. The broad, flat ventral foot, on which the animal creeps, is surrounded by a groove, the mantle cavity, in which the gills are suspended. Most chitons are grazing herbivores and live in shallow marine habitats, but chitons occur to depths of more than 7000 m.

Snails and slugs (Gastropoda)

The 40,000 or so species of limpets, snails, and slugs belong to this class, which is primarily marine. However, this group contains many freshwater species and the only terrestrial mollusks (figure 34.14). Most gastropods have a single shell, but some, such as slugs and nudibranchs (or sea slugs), have lost the shell through the course of evolution. Most gastropods creep on a foot, but in some it is modified for swimming.

The head of most gastropods has a pair of tentacles, which serve as chemo- or mechanoreceptors, with eyes at the base. A typical garden snail may have two sets of tentacles, one of them bearing eyes at the ends (see figure 34.14).

Uniquely among animals, gastropods undergo torsion during larval life. **Torsion** involves twisting of the body so the mantle cavity and anus are moved from a posterior location to the front of the body. Torsion may lead to the reduction or

Figure 34.14 A Gastropod mollusk. The terrestrial Oregon forest snail *Allogona townsendiana.*

disappearance of the nephridium, gonad, or other internal organs on one side. Most adult gastropods are therefore not bilaterally symmetrical. Torsion should not be confused with coiling, the spiral winding of the shell. Coiling also occurs in cephalopods. The fossil record suggests that the first gastropods were coiled but did not undergo torsion.

Like many gastropods, nudibranchs (sea slugs) are active predators. Nudibranchs get their name from their gills, which, instead of being enclosed within the mantle cavity, are exposed along the dorsal surface (figure 34.15). Nudibranchs would seem vulnerable to predation, but many secrete distasteful chemicals. They prey on animals that are avoided by other predators. Some are specialist feeders on sponges, most species of which form spicules and noxious chemicals. Other nudibranchs specialize in feeding on cnidarians, which are protected from most predators by their nematocysts. Some of these nudibranchs have the extraordinary ability to extract the nematocysts undischarged, transfer them through their digestive tract to the surface of their bodies, and use them for their own protection.

Figure 34.13 The noble chiton, *Eudoxochiton nobilis,* from New Zealand. The foot of the chiton can grip the substrate very strongly, making chitons hard to dislodge by waves or predators.

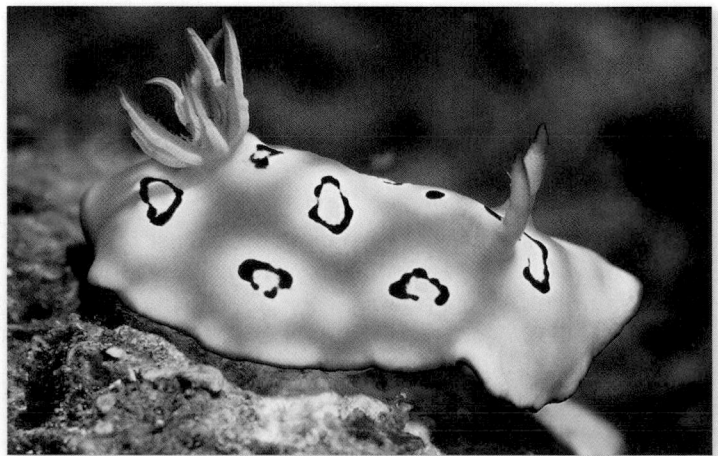

Figure 34.15 A nudibranch (or sea slug). The bright colors of many nudibranchs such as this alert predators to their repulsive taste.

In terrestrial gastropods, the mantle cavity, which is occupied by gills in aquatic snails, is extremely rich in blood vessels and serves as a lung. This lung absorbs oxygen from the air much more effectively than a gill could; however, a snail will drown if its lung fills with water; for this reason, terrestrial gastropods can close the opening of the lung to the outside.

Clams, mussels, and cockles (Bivalvia)

Most of the 10,000 species of bivalves are marine, but some live in fresh water. More than 500 species of pearly freshwater mussels, or naiads, occur in the rivers and lakes of North America.

Unlike other mollusks, a bivalve does not have a radula or distinct head (figure 34.16). The foot of most is wedge-shaped, adapted for burrowing or anchoring the animal in its burrow. Some species of clams can dig into sand or mud very rapidly by means of muscular contractions of their foot. Some species of scallops and file clams can move swiftly by clapping their shells rapidly together (although they cannot control the direction of their movement); the adductor muscle that allows this clapping is the part of a scallop eaten by humans. Projecting from the edge of a scallop's mantle are tentacle-like projections having complex eyes between them.

Bivalves, as their name implies, have two shells (valves) that are hinged dorsally so the shells are oriented laterally (left and right). A ligament lying along the hinge is structured so it causes the shells to gape open. One or two large adductor muscles link the shells internally (see figure 34.16), and when they contract, they counteract the hinge ligament to draw the shells together. The mantle covers the internal surface of the shells, enveloping the visceral mass on its inner side and secreting the shells on its outer side. As is typical of mollusks, the respiratory structures, a set of complexly folded gills on each side of the visceral mass, lie in the mantle cavity.

The edges of the mantle may be partly fused. In bivalves, in which this is the case, typically two areas are not fused and may be drawn out to form tubes called siphons (see figure 34.16). Water enters the mantle chamber through the **inhalant siphon** bringing oxygen and food, and water exits through the **exhalant siphon,** taking wastes and gametes with it. In bivalves that live buried deeply in mud or burrow into rock, the siphons allow the animals to eat and breathe, functioning essentially as snorkels.

Octopuses, squids, and nautiluses (Cephalopoda)

The more than 600 species of cephalopods are strictly marine. They are active predators that swim, often swiftly, and they are the only mollusks with a closed circulatory system. The foot has evolved into a series of arms equipped with suction cups, adhesive structures, or hooks that seize prey. Octopuses, as their name suggests, have eight arms; squids have eight arms and two tentacles; and the chambered nautilus has 80 to 90 tentacles (which lack suckers). After snaring prey with its arms, a cephalopod bites the prey with its strong, beaklike jaws, then pulls it into its mouth by the action of the radula. The salivary gland of many cephalopods secretes a toxin that can be injected into prey; the tiny blue-ringed octopus of Australia (see figure 34.8b) can kill a human with its deadly bite.

Cephalopods have the largest relative brain sizes among invertebrates and highly developed nervous systems. Many exhibit complex patterns of behavior and are highly intelligent (figure 34.17); octopuses can be easily trained to distinguish among classes of objects and have been known to climb out of their aquaria, move across a laboratory floor, enter another aquarium to capture and devour a crab, and then return to their own tank, leaving lab personnel to wonder why their crabs were disappearing. Cephalopod eyes are much like those of vertebrates, although they evolved independently (see chapter 44).

Aside from the chambered nautiluses, living cephalopods lack an external shell. Shelled cephalopods were formerly far more diverse, as evidenced by the many fossil cephalopods such as ammonites and belemnites. These cephalopods were extraordinarily successful because they could move in open water instead of on the sea bottom like other mollusks. However, once more maneuverable fishes evolved, shelled

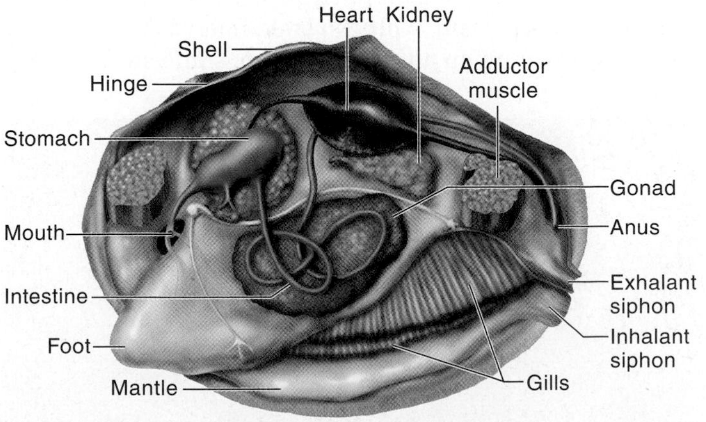

Figure 34.16 Diagram of a clam. The left shell and mantle are removed to show the internal organs and the foot. Bivalves such as this clam circulate water across their gills and filter out food particles.

Figure 34.17 Problem-solving by an octopus. This two-month-old common octopus (*Octopus vulgaris*) was presented with a crab in a jar. It tried to unscrew the lid to get to the crab. Although it failed in this attempt, in some cases it was successful.

Figure 34.18 Ink defense by a giant Pacific octopus (*Octopus dofleini*). When threatened, octopuses and squids expel a dark cloudy liquid.

mollusks declined, some dying out, others experiencing evolutionary reduction and eventual loss of their heavy shells. The cuttlebone of cuttlefish and the pen of squids are internal shells that support these animals and give them some buoyancy. Even the internal shell has disappeared in the lineage that gave rise to octopuses.

As in other mollusks, water passes through the mantle cavity. In a cephalopod, it is pumped in by muscles and exits through a siphon, which allows the animal to move by jet propulsion, and which can be directed to steer. The ink sac of cephalopods, which typically contains a purplish fluid, can eject its contents through the siphon as a cloud that may hide the cephalopod and confuse predators (figure 34.18).

Most octopuses and squids are capable of changing skin color and texture to match their background or to communicate with one another. They do so using chromatophores, epithelial cells that contain pigments. Some deep-sea squids harbor symbiotic luminescent bacteria. These may be emitted with the ink to produce a glimmering cloud (ink would not be seen at depths where sunlight cannot penetrate) or they may inhabit cells like chromatophores so they can light up the surface of the animal.

Another difference with many other mollusks is that cephalopods have direct development, that is, they lack a larval stage, hatching as miniature adults.

Learning Outcomes Review 34.4

Mollusks have a reduced coelom. Most have efficient excretory systems, gills for respiration, and a rasping structure—the radula—for gathering food. The mantle of mollusks not only secretes their protective shell but also forms structures essential to body functions. Chitons, gastropods, bivalves, and cephalopods are the four best-known groups.

■ *Why might the coelom of mollusks be reduced?*

34.5 Lophotrochozoans: Ribbon Worms (Nemertea)

Learning Outcome

1. *Describe the ways in which ribbon worms are similar and different from flatworms.*

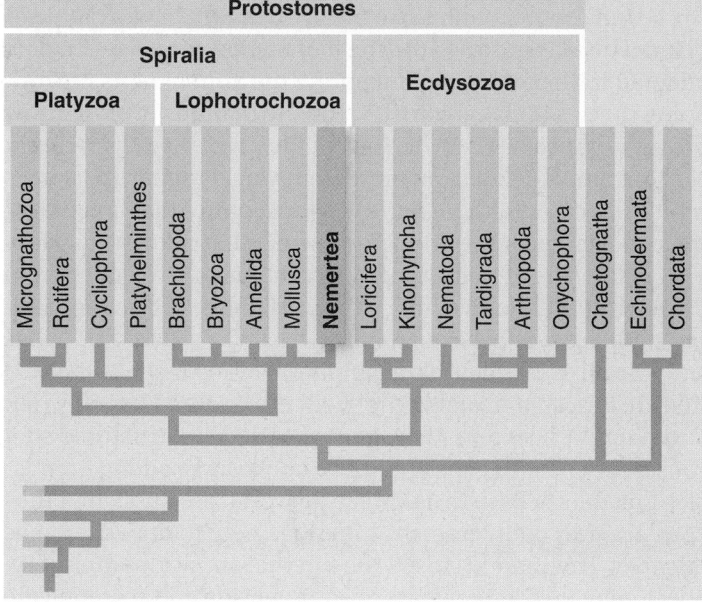

Nemerteans (phylum Nemertea) consist of about 900 species of cylindrical to flattened very long worms (figure 34.19). Most nemerteans are marine; a few species live in fresh water and humid terrestrial habitats. An individual may reach 10 to 20 cm in length, although the animals are difficult to measure because they can stretch, and many species break into pieces when disturbed or handled. The species *Lineus longissimus* has been reported to measure 60 m in length—the longest animal known!

The nemertean body plan resembles that of a flatworm, with networks of fine tubules constituting the excretory system, and with internal organs not lying in a body cavity. A bit of cephalization is present, with two lateral nerve cords extending posteriorly from an anterior ganglion; some animals have

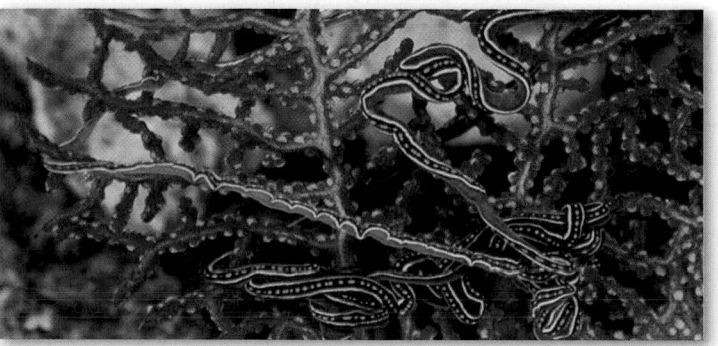

Figure 34.19 Phylum Nemertea: A ribbon worm of the genus *Lineus*. Some nemerteans can stretch to several meters in length.

eyespots on the head. But, by contrast with a platyhelminth, a nemertean has a complete gut, with both mouth and anus, connected by a straight tube. Nemerteans also possess a fluid-filled cavity called a rhynchocoel. This sac serves as a hydraulic power source for the proboscis, a long muscular tube that can be thrust out quickly from a sheath to capture animal prey.

Nemerteans all reproduce sexually. Some are capable of asexual reproduction by fragmentation. However, in most species, most fragments resulting from disturbance die, so nemertean regenerative powers may not be as great as is sometimes stated.

Blood of nemerteans flows entirely in vessels that are derived from the coelom. That and the rhynchocoel are good evidence that nemerteans are not related to flatworms, which they resemble superficially, but belong to the Lophotrochozoa, along with mollusks.

Learning Outcome Review 34.5

Nemerteans are very long worms that have a coelomic cavity and blood vessels derived from the coelom. They capture prey with a muscular proboscis. Unlike the acoelomate flatworms, nemerteans have a complete gut with mouth and anus.

■ *What would be some advantages of a flow-through digestive tract?*

34.6 Lophotrochozoans: Annelids (Annelida)

Learning Outcomes

1. Explain how circular and longitudinal muscles in a segmented body facilitate movement.
2. Distinguish between the classes Polychaeta and Clitellata.
3. Describe adaptations in leeches for feeding on the blood of animals.

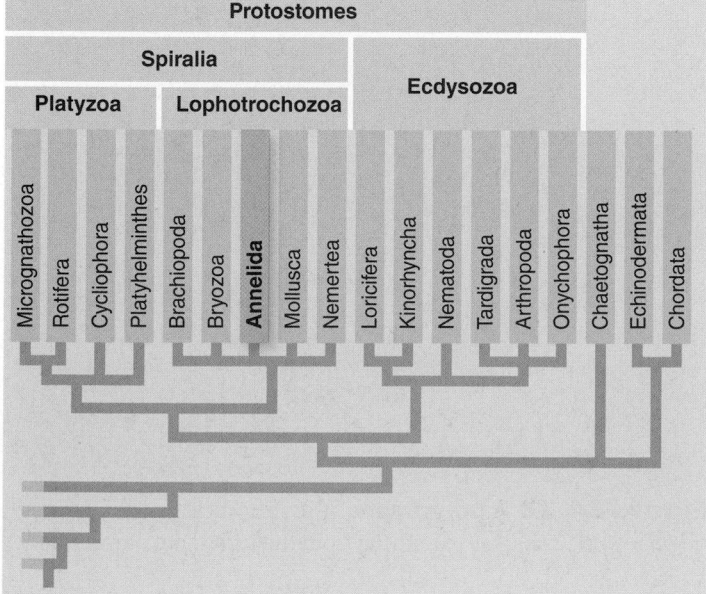

Figure 34.20 A polychaete annelid. *Nereis virens* is a wide-ranging, predatory, marine polychaete worm equipped with feathery parapodia for movement and respiration, as well as jaws for hunting. You may have purchased *Nereis* as fishing bait!

An important innovation in the animal body plan was segmentation, the building of a body from a series of repeated units (see chapter 33), which has evolved multiple times. One advantage of a segmented body is that the development and function of individual segments or groups of segments can differ. For example, some segments may be specialized for reproduction, whereas others are adapted for locomotion or excretion.

All animals that have been regarded as annelids are segmented (figure 34.20), so the animals were considered to constitute a natural group, but the monophyly of Annelida is being reconsidered because some unsegmented worms may belong to this clade.

The annelid body is composed of ringlike segments

The head, which contains a well-developed cerebral ganglion, or brain, and sensory organs occurs at the anterior end (front) of a series of ringlike segments that resemble a stack of coins (figure 34.21). Many species have eyes, which in some species

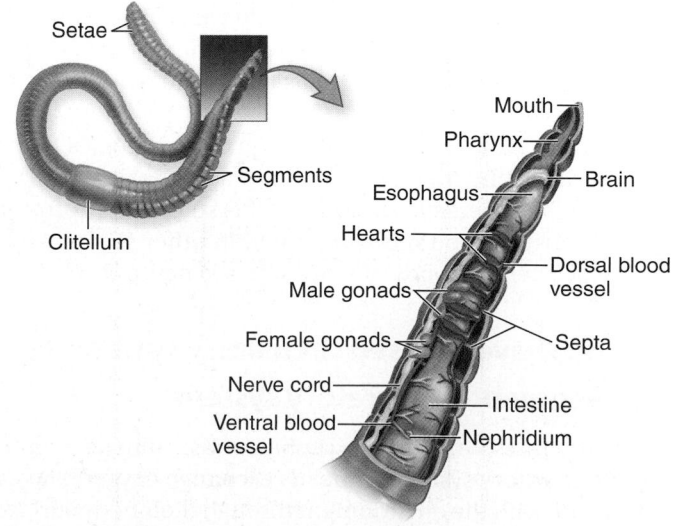

Figure 34.21 Phylum Annelida: An oligochaete. The earthworm body plan is based on repeated body segments. Segments are separated internally from each other by septa.

have lenses and retinas. Technically the head is not a segment, nor is the posterior end of the worm, the pygidium. In embryonic development, the head and tail form first, and then segments form between them; if a worm is cut in pieces, generally only those parts containing either head or tail can regenerate the missing parts, and those parts coming from the middle die.

Internally, the segments are divided from one another by partitions called septa, just as bulkheads separate the compartments of a submarine. Each segment has a pair of excretory organs, a ganglion, and locomotory structure; in most marine annelids, each also has a set of reproductive organs.

Although septa separate the segments, materials and biological signals do pass between segments. A closed circulatory system carries blood the length of the animal, anteriorly in the dorsal vessel and posteriorly in the ventral one. Connections from ventral to dorsal vessel in each segment bring the blood near enough to each cell so oxygen and food molecules diffuse from the blood into the cells of the body wall, and carbon dioxide and other wastes diffuse from the cells into the blood. A ventral nerve cord connects the ganglia in each segment with one another and with the brain. These neural connections allow the worm to function as a unified and coordinated organism.

Annelids move by contracting their segments

The basic annelid body plan is a tube within a tube, the digestive tract—extending from mouth to anus—passing through the septa, and suspended within the spacious coelom, which is surrounded by the body wall. Each portion of the digestive tract—pharynx, esophagus, crop, gizzard, and intestine—is specialized for a different function.

The coelomic fluid creates a hydrostatic skeleton that gives each segment rigidity, like an inflated balloon (see chapter 46). Annelids move by contraction of the circular and longitudinal muscles against the hydrostatic skeleton. When circular muscles are contracted around a segment, the segment decreases in diameter, so the coelomic fluid causes the segment to elongate. When longitudinal muscles contract, the segment shortens, so the coelomic fluid causes the segment to increase in diameter. Alternating these contractions and confining them to only some segments allows the worms to move in complex ways.

In most annelid groups, each segment possesses bristles of chitin called **chaetae** (or setae—singular, seta or chaeta). By extending the chaetae in some segments so that they protrude into the substrate and retracting them in other segments, the worm can extend its body, but not slip (see figure 46.1).

Annelids have a closed circulatory system but a segmented excretory system

Unlike mollusks, except for cephalopods, annelids have a closed circulatory system. Annelids exchange oxygen and carbon dioxide with the environment through their body surfaces, although some aquatic species have gills along the sides of the body or at the anterior end. Gases (and food molecules) are distributed throughout the body in blood vessels. Some of the vessels at the anterior end of the body are enlarged and heavily muscular, serving as hearts that pump the blood. An earthworm has five pulsating blood vessels on each side that help to move blood from the main dorsal vessel, the major pumping structure, to the ventral vessel.

The excretory system of annelids consists of ciliated, funnel-shaped nephridia like those of mollusks. Each segment has a pair of nephridia that collect wastes and transport them out of the body by way of excretory tubes. Some polychaete worms have protonephridia like the flame cells of planarians.

Annelida is composed of two— or three—classes

The roughly 12,000 described species of annelids occur in many habitats. They range in length from as little as 0.5 mm to giant Australian earthworms more than 3 m long. Although traditionally annelids have been classified into classes Polychaeta (mostly marine worms), Oligochaeta (mostly terrestrial worms, including earthworms), and Hirudinea (leeches), the monophyly of polychaetes is not well established. The classification of annelids may change in the near future, but we adopt the current two-class system of Polychaeta and Clitellata (which combines Oligochaeta and Hirudinea).

Polychaetes (Polychaeta)

Polychaetes include clamworms, scaleworms, lugworms, sea mice, tubeworms, and many others. Polychaetes are a crucial part of many marine food chains and are extremely abundant in particular habitats. Some of these worms are beautiful, with unusual forms and iridescent colors (figure 34.22).

On most segments, a polychaete has paired, fleshy, paddle-like lateral projections called **parapodia** (see figures 34.20 and 34.22). The parapodia bear chaetae; the word *polychaeta* means "many chaetae." Parapodia are used in swimming, burrowing, or crawling, and those of polychaetes that live in

Figure 34.22 A polychaete. The shiny bristleworm *Oenone fulgida*. Notice chaetae extending from the iridescent parapodia.

burrows or tubes may have chaetae with hooks that help anchor the worm. Parapodia can also play an important role in gas exchange because they greatly increase the surface area of the body, and in some species they bear or are even transformed into gill-like structures.

Polychaetes can swim or crawl, and some are active predators with powerful jaws. Other polychaetes live in tubes or burrows of hardened mud, sand, mucus-like secretions, or calcium carbonate. Sedentary polychaetes may project feathery tentacles that sweep the water for food, filter feeding; the tentacles may also serve as gills, exchanging gas. Some deep-sea tubeworms such as *Riftia* (figure 34.23) are gutless as adults. Projections from the body of these worms house sulfur-oxidizing bacteria that synthesize organic compounds used by the worm. These worms aggregate near hydrothermal vents where sulfur is plentiful and can grow to more than a meter in length.

Polychaete gametes typically are released into the water, where fertilization occurs externally. Many polychaetes have gonads in most segments, but in some groups, gonads are confined to certain segments. In palolo worms (*Palola vidiris*) and their relatives, these segments are at the end of the body; spawning involves the end of the worm breaking off and swimming to the surface of the sea, where it ruptures, releasing the gametes. The gamete-filled terminal parts of palolo worms are considered delicacies by some people in the South Pacific.

Fertilization results in spiral cleavage followed by the production of ciliated, mobile trochophore larvae similar to that of mollusks. Metamorphosis of a trochophore involves differentiation of a head and tail end, with development of segments in between, from a posterior growth zone.

Earthworms and leeches (Clitellata)

Some authorities still consider earthworms to belong to class Oligochaeta and leeches to class Hirudinea, but most now put earthworms and leeches into a single class, although, confusingly, it sometimes may be called Oligochaeta. More often it is called Clitellata, because of the feature that unites these seemingly quite different animals, the clitellum—a thickened band on the body, which is the familiar "saddle" of an earthworm (see figure 34.21).

Earthworms. The body of a typical earthworm consists of 100 to 175 similar segments. The head is not well differentiated. An earthworm has no parapodia, and its chaetae (which are fewer than in polychaetes—the word *oligochaeta* means "few chaetae") project directly from the body wall.

Earthworms eat their way through the soil, ingesting it by muscular action of their strong pharynx; organic material is ground in the gizzard. What passes through an earthworm is deposited outside the opening of its burrow as castings that form irregular mounds. In this way, earthworms contribute to loosening, aeration, and enrichment of the soil. In view of their underground lifestyle, it is not surprising that earthworms have no eyes. But they do have light-, chemo-, and touch-sensitive cells, most concentrated in segments near each end of the body—those regions most likely to encounter light or other stimuli.

Earthworms are hermaphroditic, another way in which they differ from most polychaetes, which have separate sexes, but they cross-fertilize through mating (figure 34.24). The clitellum secretes mucus that holds the worms together during copulation, their anterior ends pointing in opposite directions, their ventral surfaces touching. Sperm cells are released from pores in specialized segments of one partner into the sperm receptacles of the other, the process going in both directions simultaneously. Several days after the worms separate, the clitellum of each worm secretes a mucus cocoon, surrounded by a protective layer of chitin. As the worm moves, this sheath passes over the female pores of the body, receiving eggs and incorporating the deposited sperm so that fertilization takes place within the cocoon. When the cocoon passes over the end of the worm, its edges pinch together. Within the cocoon, the fertilized eggs develop directly into young worms similar to adults.

Leeches. Most leeches live in fresh water, although a few are marine and some tropical leeches occupy terrestrial habitats.

Figure 34.23 Giant tubeworms *(Riftia pachyptila)* living in deep-sea hydrothermal vents near the Galápagos. These tubeworms and associated animals (notice the small crab near the top right) are an example of a community dependent on hydrogen sulfide, rather than the Sun, as an energy source (see chapter 58).

Figure 34.24 Earthworms mating. The anterior ends are pointing in opposite directions.

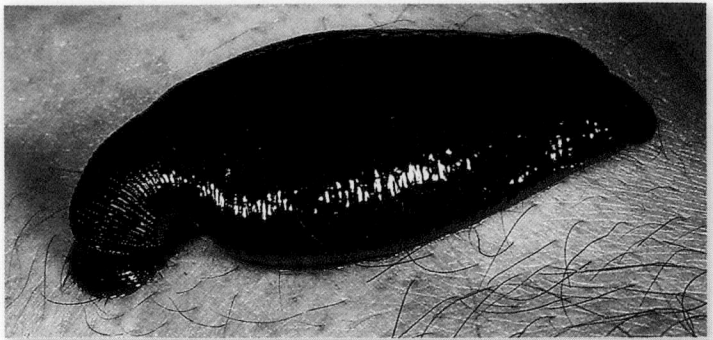

Figure 34.25 A leech. *Hirudo medicinalis,* the medicinal leech, feeding on a human arm. Leeches use chitinous, bladelike jaws to make an incision to access blood; they secrete an anticoagulant to keep the blood from clotting. Both the anticoagulant and the leech itself have made important contributions to modern medicine.

Dorsoventrally flattened, most leeches are 2 to 6 cm long, but one tropical species reaches 30 cm. Like earthworms, leeches are hermaphroditic, but the clitellum develops only during the breeding season. Leeches also cross-fertilize.

Unlike that of earthworms and polychaetes, the coelom of a leech is reduced and not divided into segments. The suckers at one or both ends of a leech's body are used for locomotion and to attach to prey. A leech with suckers at both ends moves by attaching first one and then the other end to the substrate, looping along. Many species are also capable of swimming. Except for one species, leeches have no chaetae.

About half the known species of leeches eat detritus or devour small animals. The others suck blood or other fluids from their hosts (figure 34.25). Blood-sucking leeches secrete an anticoagulant into the wound to prevent the blood from clotting, and vasodilators to keep the blood flowing; the leech's powerful pharynx pumps the blood out quickly once a hole has been opened. Anesthetics injected into the prey prevent the leech from being noticed while piercing the skin; they are usually detected only after detaching, when blood starts to flow from the wound. Freshwater parasitic leeches may remain on their hosts for long periods, sucking the host's blood from time to time. Leeches detect prey by sensing gradients of carbon dioxide in the environment. In some tropical forests, a few seconds after you stop on a trail, you can see dozens of leeches approaching you from all directions!

One of the best-known species, the medicinal leech *Hirudo medicinalis* (see figure 34.25), reaches 10 to 12 cm long, and has bladelike, chitinous jaws that can rasp through an animal's skin. Leeches were used in medicine for hundreds of years to treat patients whose diseases were mistakenly believed to be caused by an excess of blood. Today, however, leeches are used to remove excess blood after surgery or to keep blood from coagulating in severed appendages that have been reattached. Accumulations of blood can cause the tissue to die; when leeches remove such blood, new capillaries form in about a week, and the tissues remain healthy. The anticoagulant and anesthetic properties of the leech are also being investigated by pharmaceutical companies.

Learning Outcomes Review 34.6

Annelids generally exhibit segmentation. Each segment has its own excretory and locomotor elements; circular and longitudinal muscles in segments cause the body to extend and contract, respectively. Worms of class Polychaeta, which are mostly marine, have parapodia on their segments. Leeches and earthworms were formerly in separate classes, but a major morphological similarity, the clitellum, has been used to group them into a single class, Clitellata.

■ *What would be the advantages of having nervous and circulatory systems that serve the entire body instead of segmented systems?*

34.7 Lophophorates: Bryozoans (Bryozoa) and Brachiopods (Brachiopoda)

Learning Outcomes

1. Describe the lophophore and its function.
2. Distinguish between bryozoans and brachiopods.

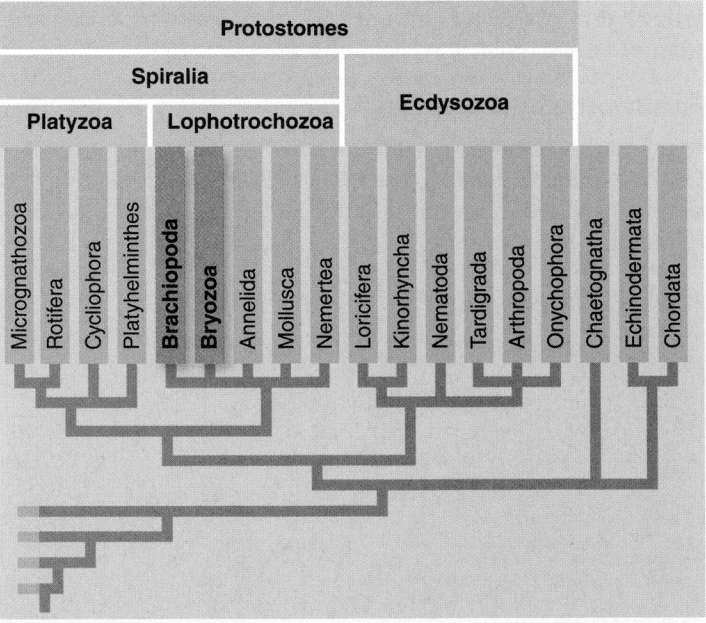

Two phyla of mostly marine animals—Bryozoa and Brachiopoda—are characterized by a lophophore, a circular or U-shaped ridge around the mouth bearing one or two rows of ciliated tentacles into which the coelom extends. The lophophore functions as a surface for gas exchange, and the cilia of the lophophore serve to guide the organic detritus and plankton on which the animal feeds to the mouth. Because of the lophophore, bryozoans and brachiopods have been considered related to one another, but some recent data indicate that the structures may have evolved convergently.

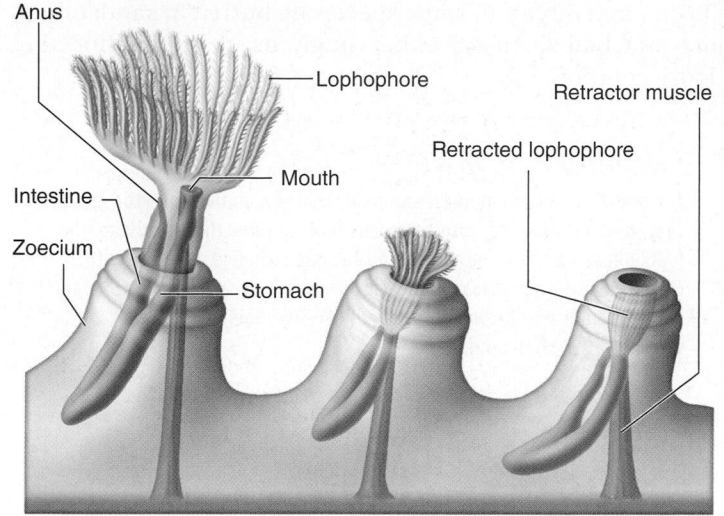

Anus

Lophophore

Mouth

Retractor muscle

Retracted lophophore

Intestine

Zoecium

Stomach

a.

Figure 34.26 Bryozoans (phylum Bryozoa). *a.* This drawing depicts a small portion of a colony of the freshwater bryozoan genus *Plumatella*, which grows on rocks. The individual at the left has a fully extended lophophore. The tiny individuals disappear into the zoecium when disturbed. *b. Plumatella repens,* another freshwater bryozoan.

b.

Brachiopods share some features with protostomes and others with deuterostomes. Cleavage in both brachiopods and bryozoans is mostly radial, as in deuterostomes. The formation of the coelom varies. In phoronids (once considered a phylum on their own but now considered part of Brachiopoda), the mouth forms from the blastopore (a feature of protostomes), whereas in the rest of brachiopods and in the bryozoans, it forms from the end of the embryo opposite the blastopore (a feature of deuterostomes). Molecular evidence allies lophophorates with protostomes. Because of the discrepancies between anatomical and developmental characters and molecular characters, the phylogeny of these animals continues to be a fascinating puzzle.

Bryozoans are the only exclusively colonial animals

Bryozoans are small—usually less than 0.5 mm long—and live in colonies that look like patches of moss on the surfaces of submerged objects (figure 34.26). Their common name, "moss-animals," is a direct translation of the Latin word *bryozoa.* The digestive system is U-shaped, with the anus opening near the mouth, as in many sessile animals.

The 4000 species of bryozoans include both marine and freshwater forms. Each individual bryozoan—a zooid—secretes a tiny chitinous chamber called a zoecium (plural, *zoecia*) that is attached to rocks or other substrates such as the leaves of marine plants and algae. Calcium carbonate is deposited in the wall of a zoecium in many marine bryozoans, and in early geological times, bryzoans formed reefs just as corals do today. A zooid can divide or bud to create asexually another zooid beside the existing one so one wall of the new zooid's zoecium is shared with that of the existing one; this expanding group of zoecia constitutes a colony. Individuals in the colony communicate chemically through pores between the zoecia. Not all zoecia of a colony may be identical; some are specialized for functions such as feeding, reproduction, or defense.

Brachiopods and phoronids are solitary lophophorates

Brachiopods, or lamp shells, superficially resemble clams because they have two calcified valves (figure 34.27). Recall that the shells of bivalves are lateral, but in brachiopods, the valves are dorsal and ventral. Many species attach to rocks or

Spiral portion of lophophore

Mouth

Mantle

Lateral arm of lophophore

Coelom

Ventral (pedicle) valve

Adductor muscle

Intestine

Nephridium

Stomach

Digestive gland

Pedicle

Dorsal (brachial) valve

Gonad

a.

b.

Figure 34.27 Brachiopods (phylum Brachiopoda). *a.* All the body structures except the pedicle lay within two calcified shells, or valves. *b.* The brachiopod *Terebratulina septentrionalis* is slightly opened so the lophophore is visible.

sand by the pedicle (a stalk) that protrudes through an opening in one shell, whereas in others one valve is cemented to the substrate and the animal lacks a pedicle. The lophophore lies on the body, between the shells. The gut in some brachiopods is U-shaped, as in bryozoans, whereas in others there is no anus at all.

Slightly more than 300 species of brachiopods exist today, but more than 30,000 fossil species are known. Because brachiopods were common in the Earth's oceans for millions of years and because their shells fossilize readily, many are used as index fossils, defining a particular geological period or sediment type.

Each **phoronid** (figure 34.28) secretes a chitinous tube around itself from which it can extend its lophophore to feed. The animal quickly withdraws into the tube when disturbed, as some polychaete worms do. Only about 10 phoronid species are known, ranging in length from a few millimeters to 30 cm. Individuals of some species lie buried in sand; others are attached to rocks, either singly or in groups, forming loose colonies.

Learning Outcomes Review 34.7

The two phyla of lophophorates probably share a common ancestor, and they show a mixture of protostome and deuterostome characteristics. The lophophore is a characteristic feeding and gas exchange structure with ciliated tentacles. Bryozoans are exclusively colonial, whereas brachiopods and phoronids are solitary. Brachiopods have two shells with dorsal and ventral valves, not lateral.

■ *How is a U-shaped digestive tract an advantage to bryozoans and brachiopods?*

34.8 Ecdysozoans: Roundworms (Nematoda)

Learning Outcomes

1. *Describe the musculature of a nematode that allows it to wriggle in a highly characteristic manner.*
2. *Explain the life cycle of a nematode and how it produces disease in humans.*

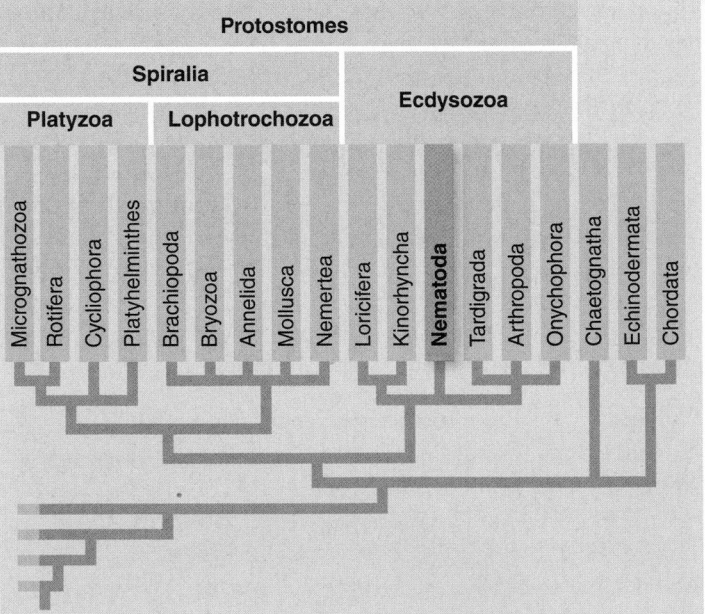

Vinegar eels, eelworms, and other roundworms constitute a large phylum, **Nematoda,** with some 20,000 recognized species; scientists estimate that the actual number might approach 100 times that many. Nematodes are abundant and diverse in marine and freshwater habitats, and many members of this phylum are parasites of plants and animals (figure 34.29). Many nematodes are microscopic and live in soil. A spadeful of fertile soil may contain, on the average, a million nematodes.

Figure 34.28 Phoronids. A phoronid lives in a chitinous tube that the animal secretes. The lophophore consists of two horseshoe-shaped ridges of tentacles and can be withdrawn into the tube when the animal is disturbed.

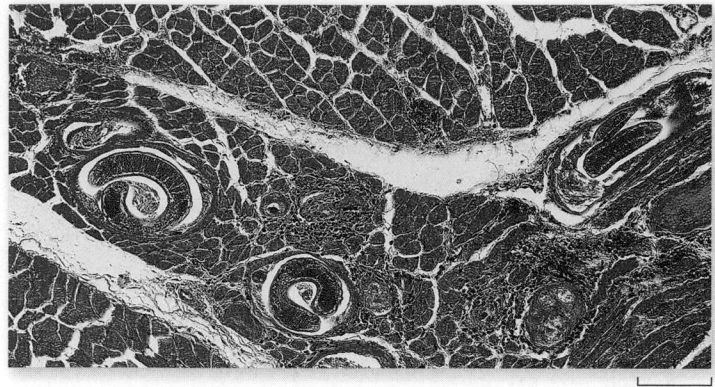

Figure 34.29 *Trichinella* **nematode encysted in pork.**
The serious disease trichinosis can result from eating undercooked pork or bear meat containing such cysts.

Nematode structure

Nematodes are bilaterally symmetrical, unsegmented worms covered by a flexible, thick cuticle that is molted as they grow—parasitic nematodes molt four times. Lacking specialized respiratory organs, nematodes exchange oxygen and carbon dioxide through their cuticles. Muscles beneath the epidermis, which underlies the cuticle, extend longitudinally, from anterior to posterior. Nematodes are unusual among worms in that they lack circular musculature, so they can shorten but not change diameter. The pulling of these longitudinal muscles against both the cuticle and the pseudocoelom produces the characteristic wriggling motion of nematodes.

Nematodes possess a well-developed digestive system and feed on a diversity of food sources. Near the mouth, at the anterior end, are hairlike sensory structures. The mouth may be equipped with piercing organs called **stylets.** Food passes into the mouth as a result of the sucking action produced by the rhythmic contraction of a muscular pharynx and continues through the intestine; waste is eliminated through the anus (figure 34.30).

Reproduction and development

Reproduction in nematodes is sexual. Nematode males and females differ in form, a state known as **sexual dimorphism** (meaning "two bodies"): the tail end of the smaller male is hooked (see figure 34.30), whereas that of a female is straight. Fertilization is internal, the male using its hooked end and associated structure to help inseminate the female. Development is indirect, meaning that an egg hatches into a larva, which does not grow directly into an adult, but must pass through several molts and, in parasitic species, transfer from one host to another.

The adults of some species consist of a fixed number of cells, a phenomenon known as eutely. For this reason, nematodes have become extremely important subjects for genetic and developmental studies (see chapter 19). The 1-mm-long *Caenorhabditis elegans* matures in three days; its body is transparent, and it has precisely 959 cells. It is the only animal whose complete developmental cellular anatomy is known.

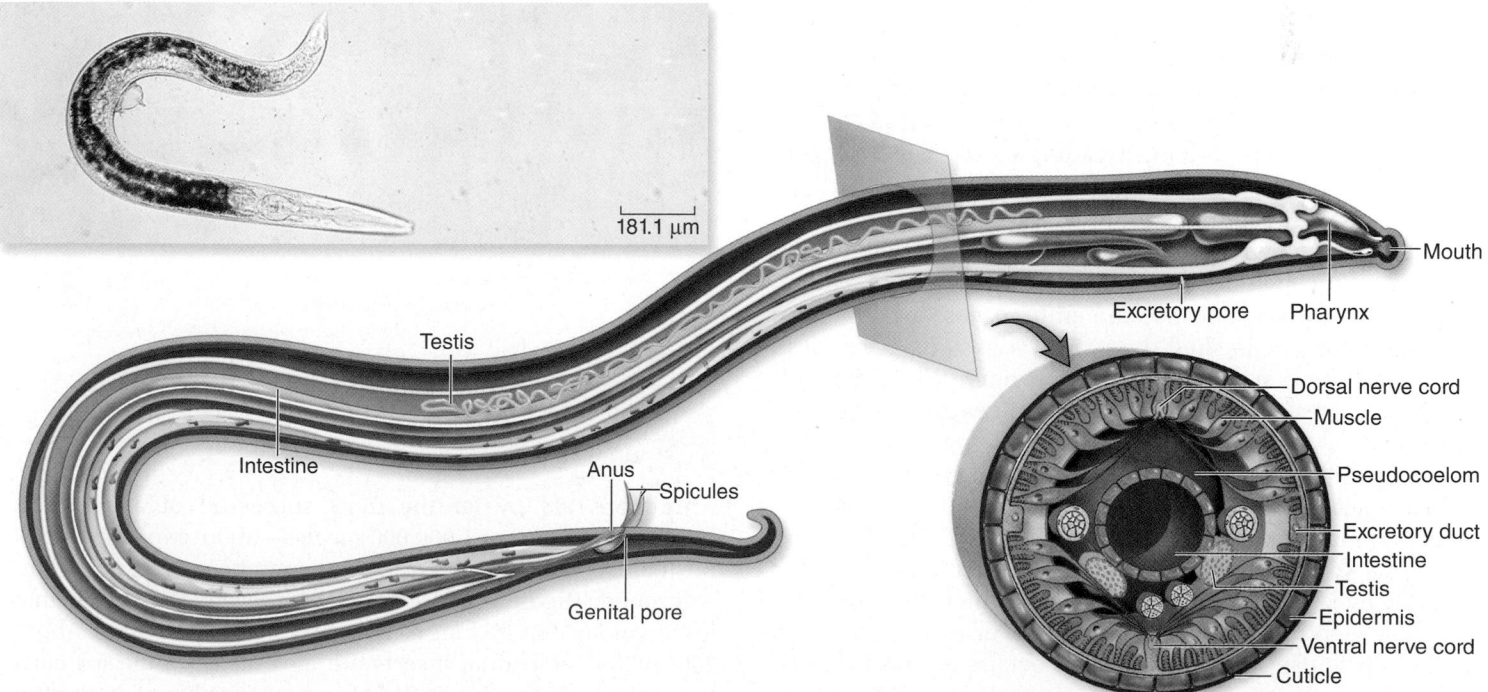

Figure 34.30 **Phylum Nematoda: Roundworms.** Roundworms such as this male nematode possess a body cavity between the gut and the body wall called the pseudocoelom. It allows nutrients to circulate throughout the body and prevents organs from being deformed by muscle movements.

Nematode lifestyles

Many nematodes are active hunters, preying on protists and other small animals. Many are parasites of plants or live within the bodies of larger animals. Almost every species of plant and animal that has been studied has been found to have at least one parasitic species of nematode. The largest known nematode, which can attain a length of 9 m, parasitizes the placenta of sperm whales.

Nematode-caused human diseases

About 50 species of nematodes, including several that are rather common in the United States, regularly parasitize human beings. Hookworms, most of the genus *Necator,* can be common in southern states. By sucking blood through the intestinal wall, they can produce anemia.

The most serious and common nematode-caused disease in temperate regions is trichinosis. Worms of the genus *Trichinella* (see figure 34.29) live in the small intestine of some mammals, especially pigs and bears, where fertilized females burrow through the intestinal wall and release live young (as many as 1500 per female). The young enter the lymph channels, which transport them to muscles throughout the body. There they mature and form highly resistant, calcified cysts. Eating undercooked or raw pork or bear in which cysts are present transmits the worm. Fatal infections, which can occur if the worms are abundant, are rare: in the United States, only about 20 deaths have been attributed to trichinosis during the past decade.

It is estimated that pinworms, *Enterobius vermicularis,* infect about 30% of children and 16% of adults in the United States. Adult pinworms live in the human rectum where they usually cause nothing more serious than itching of the anus; large numbers, however, can lead to prolapse of the rectum. The worms can easily be killed by drugs.

The intestinal roundworm *Ascaris lumbricoides* infects approximately one of six people worldwide, but is rare in areas with modern plumbing. An adult female, which can be as much as 30 cm long, can release as many as 20,000 fertilized eggs each day into the gut of its host. The eggs are carried from the body in the host's feces and can remain viable for years in the soil; dust may carry them onto food, eating implements, or lips. Because the embryo has developed in each egg before it is shed, once it is ingested, it hatches. The larva follows a circuitous path through the body, and metamorphoses into an adult, which lives in the human intestine.

Some nematode-caused diseases are extremely serious in the tropics. Filariasis is caused by several species of nematodes that infect at least 250 million people worldwide. Filarial worms of some species live in the circulatory system. Infection by *Wuchereria bancrofti* may produce the condition known as elephantiasis, in which the lower extremities may swell to disfiguring proportions. This occurs because worms clog the lymph nodes, causing severe inflammation and resulting in swelling by preventing the lymph from circulating. The larval filarial worms are transmitted by an intermediate host, typically a blood-sucking insect such as a mosquito.

34.9 Ecdysozoans: Arthropods (Arthropoda)

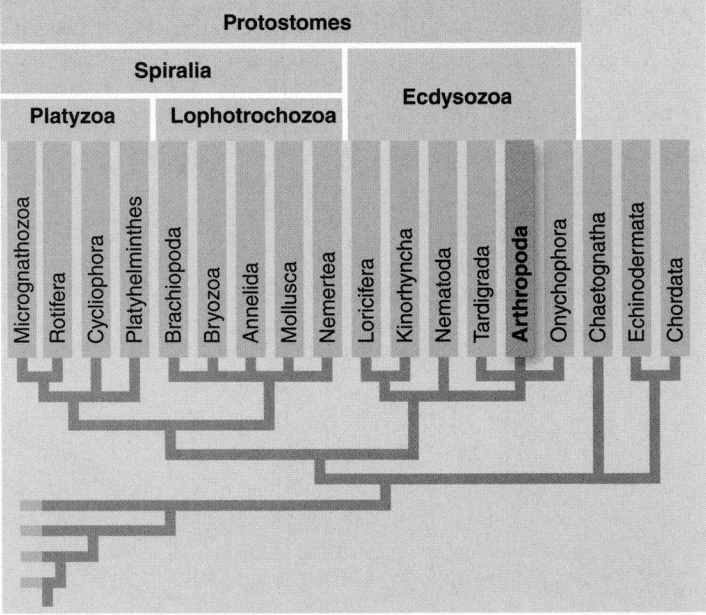

Arthropods are by far the most successful of all animals (table 34.1). Well over 1,000,000 species—about two-thirds of all the named species on Earth—are members of the phylum Arthropoda (figure 34.31). One scientist recently estimated that insects alone may include as many as 30 million species. About 200 million individual insects are alive at any time for each human! Insects (see figure 34.31) and other arthropods abound in every habitat on the planet, but there are few marine insects. Members of the phylum are small, generally a few millimeters in length, but adults range in size from about 80 μm long (some parasitic mites) to 3 m across (Japanese spider crabs).

TABLE 34.1		Major Groups of the Phylum Arthropoda	
Class		**Characteristics**	**Members**
Chelicerata		Mouthparts are chelicerae (pincers or fangs).	Spiders, mites, ticks, scorpions, daddy long-legs, horseshoe crabs
Crustacea		Mouthparts are mandibles; appendages are biramous ("two-branched"); the head has two pairs of antennae.	Lobsters, crabs, shrimps, isopods, barnacles
Hexapoda		Mouthparts are mandibles; the body consists of three regions: a head with one pair of antennae, a thorax, and an abdomen; appendages are uniramous ("single-branched").	Insects (beetles, bees, flies, fleas, true bugs, grasshoppers, butterflies, termites), springtails
Myriapoda		Mouthparts are mandibles; the body consists of a head with one pair of antennae, and numerous segments, each bearing paired uniramous appendages.	Centipedes, millipedes

Arthropods are of enormous economic importance, affecting all aspects of human life. They pollinate crops and are valuable as food for humans and other animals, but they also compete with humans for food and damage crops. Diseases spread by insects and ticks strike every kind of plant and animal, including human beings. Insects are by far the most important herbivores in terrestrial ecosystems: virtually every kind of plant is eaten by one or more species.

Although our understanding of the phylogenetic relationships among the groups of arthropods and their relationships to other animals may shift with new findings, taxonomists currently recognize four extant classes (a fifth, the trilobites, is extinct): chelicerates, crustaceans, hexapods, and myriapods. Mouthparts of chelicerates are chelicerae (pincers), whereas those of the other three classes are **mandibles** (biting jaws). Mandibles are inferred to have arisen (probably from a pair of limbs) in the common ancestor of crustaceans, hexapods, and myriapods, which means that these groups are more closely related to one another than any of them is to chelicerates.

Arthropods exhibit key features and organ systems

Part of arthropod success is explained by the modularity of the segmented body, the **exoskeleton,** and the jointed appendages. The advantages of segmentation were discussed in the section

on annelids. A hard exoskeleton confers protection against predators, but it acts something like a straight-jacket, restricting motion. Joints in the appendages maintain protection while providing some flexibility. With this system, arthropods have developed many efficient modes of locomotion, both in the oceans, where they originated, and on land, which they colonized early in the Devonian period, more than 400 MYA.

Segmentation

In members of some classes of arthropods, many body segments look alike. In others, the segments are specialized into functional groups, or **tagmata** (singular, *tagma*), such as the head, thorax, and abdomen of an insect (figure 34.32). The

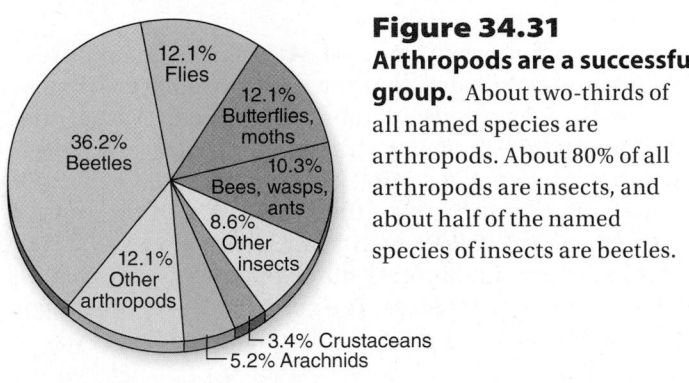

Figure 34.31 Arthropods are a successful group. About two-thirds of all named species are arthropods. About 80% of all arthropods are insects, and about half of the named species of insects are beetles.

12.1% Flies
12.1% Butterflies, moths
36.2% Beetles
10.3% Bees, wasps, ants
8.6% Other insects
12.1% Other arthropods
3.4% Crustaceans
5.2% Arachnids

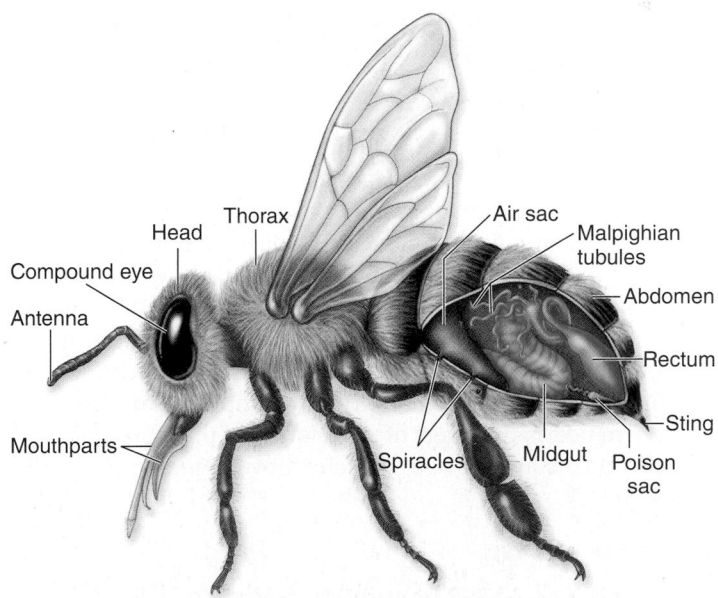

Figure 34.32 Phylum Arthropoda. This bee, like all insects and other arthropods, has a segmented body and jointed appendages. An insect body is composed of three tagmata: head, thorax, and abdomen. All arthropods have an exoskeleton made of chitin. Some insects evolved wings that permit them to fly.

fusion of segments, known as tagmatization, is of central importance in the evolution of arthropods. Typically, the segments can be distinguished during larval development, but fusion in development obliterates them. All arthropods have a distinct head; in many crustaceans and chelicerates, head and thorax fuse to form the cephalothorax, or **prosoma.**

An exoskeleton

The rigid external skeleton, or exoskeleton, is made of chitin and protein. In any animal, the skeleton provides antagonism for muscles (and in many animals, a surface for muscle attachment), support for the body, and protection against physical forces. The arthropod exoskeleton protects against water loss, which was a powerful advantage in insects colonizing land. As you learned in chapter 3, chitin is chemically similar to cellulose, the dominant structural component of plants, and shares with it properties of toughness and flexibility. The chitin and protein of an arthropod exoskeleton provide a covering that is very strong while being capable of flexing in response to the contraction of muscles attached to it.

An exoskeleton has inherent limitations. As arthropods increase in size, their exoskeletons must get disproportionately thick to bear the pull of the muscles. If beetles were as large as eagles, or crabs the size of cows, the exoskeleton would be so thick that the animal would be unable to move its great weight. Few terrestrial arthropods weigh more than a few grams, but aquatic ones can be heavier because water, being denser than air, provides more support. Another limitation is that, because the body is encased in a rigid skeleton, arthropods periodically must undergo **ecdysis,** or molting. Controlled by ecdysteroid hormones (see chapter 45), molting was explained in chapter 33. The anterior and posterior regions of the digestive tract as well as the **compound eyes** are covered with cuticle and therefore are also shed at ecdysis. The animal is especially vulnerable during molting while the exoskeleton is soft, and it may hide from predators until the new exoskeleton hardens.

Jointed appendages

The name *arthropod* means "jointed feet"; all arthropods have jointed appendages. Appendages may be modified into antennae, mouthparts of various kinds, or legs.

One advantage of jointed appendages is that they can be extended and retracted by bending. Imagine how difficult life would be if your arms and legs could not bend! In addition, joints serve as a fulcrum, or stable point, for appendage movement, so leverage is possible. A small muscle force on a lever can produce a large movement; for example, extending your lower arm takes advantage of the fulcrum of the elbow. A small contraction distance in your muscles moves your hand through a large arc.

Circulatory system

The circulatory system of arthropods is open. The principal component of an insect's circulatory system is a longitudinal muscular vessel called the heart, which is near the dorsal surface of the thorax and abdomen (figure 34.33). When the heart contracts, blood is pumped anteriorly. From there it gradually flows through the spaces between the tissues toward the posterior end. When the heart relaxes, blood

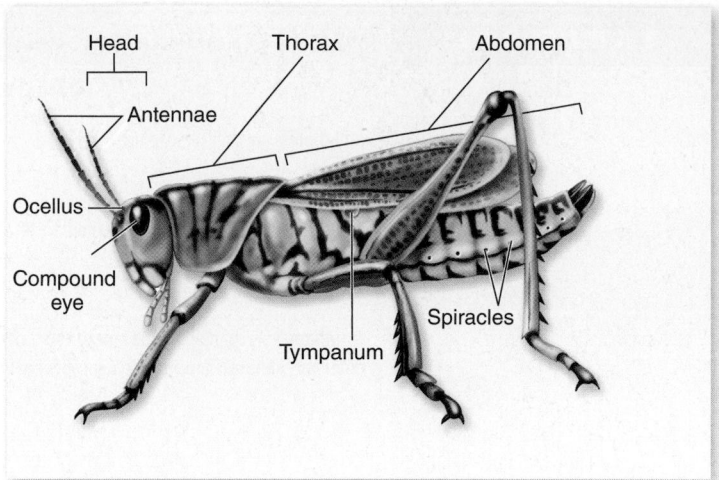

a.

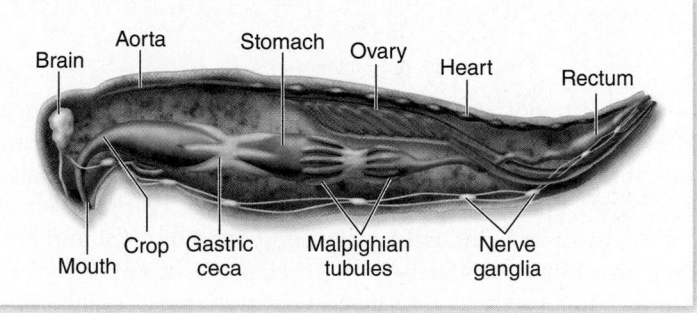

b.

Figure 34.33 A Grasshopper (order Orthoptera). This grasshopper illustrates the major structural features of the insects, the arthropod group with the greatest number of species. *a.* External anatomy. *b.* Internal anatomy.

returns to it from those spaces through one-way valves in the posterior region of the heart.

Nervous system

The central feature of the arthropod nervous system is a double chain of segmented ganglia along the animal's ventral surface (see figure 34.33). At the anterior end of the animal are three fused pairs of dorsal ganglia, which constitute the brain; however, ventral ganglia (generally a pair per segment) control much of the animal's activities. Therefore, an arthropod can carry out functions such as eating, moving, and copulating even if the brain has been removed. The brain seems to be a control point, or inhibitor, for various actions, rather than a stimulator, as it is in vertebrates.

Compound eyes (figure 34.34) occur in many insects, crustaceans, centipedes, and the extinct trilobites. They are composed of hundreds or more independent visual units called **ommatidia** (singular, *ommatidium*), each covered with a lens and including a complex of eight retinular cells and a light-sensitive central core, the rhabdom. Simple eyes, or **ocelli** (singular, *ocellus*) with single lenses, occur in some arthropods including those with compound eyes. Ocelli distinguish light from darkness. The ocelli of locusts and dragonflies function as horizon detectors to help the insect visually stabilize flight.

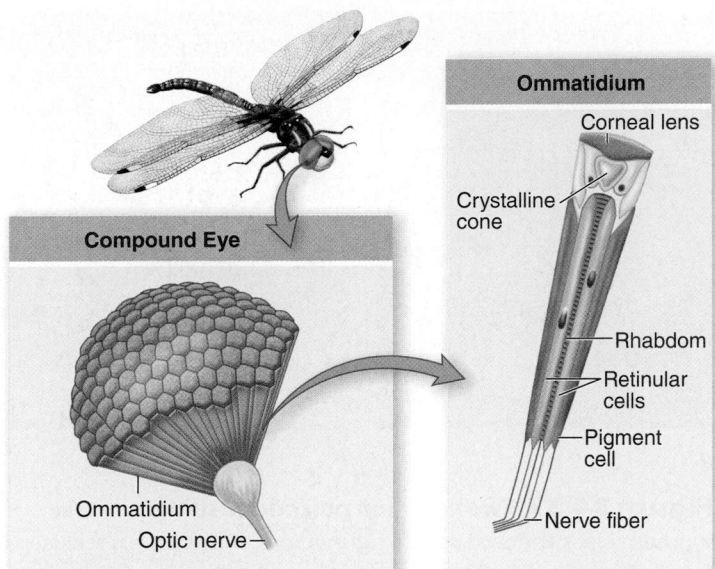

Compound Eye

Ommatidium

Optic nerve

Ommatidium

Corneal lens

Crystalline cone

Rhabdom

Retinular cells

Pigment cell

Nerve fiber

Figure 34.34 The compound eye. The compound eyes in insects are complex structures composed of many independent visual units called ommatidia.

Respiratory system

Marine arthropods such as crustaceans have gills, and marine chelicerates (such as horseshoe crabs) have book gills, flaps under the prosoma that appear to have evolved from legs. Some tiny arthropods lack any structures for exchanging oxygen, and their outer epithelium or gut have a respiratory function.

The respiratory system of most terrestrial arthropods consists of small, branched, cuticle-lined ducts called **tracheae** (singular, *trachea*) (figure 34.35) (the lining of which is shed at ecdysis). Tracheae ultimately branch into very small **tracheoles,** which are in direct contact with individual cells, allowing oxygen and carbon dioxide to diffuse across the plasma membranes. Because insects depend on the respiratory system rather than the circulatory system to carry oxygen to their tissues, all parts of the body must be near a respiratory

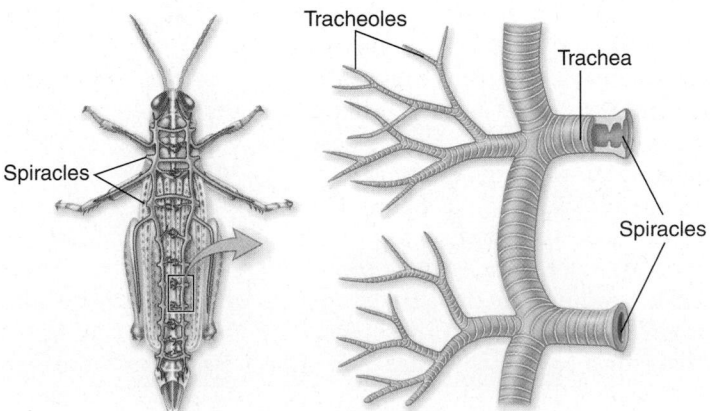

Tracheoles

Trachea

Spiracles

Spiracles

Figure 34.35 Tracheae and tracheoles. Tracheae and tracheoles are connected to the exterior by openings called spiracles and carry oxygen to all parts of a terrestrial insect's body.

passage. Along with the weight of the exoskeleton, this places severe limitations on arthropod size.

Air passes into the tracheae through openings in the exoskeleton called **spiracles** (see figures 34.32, 34.33, and 34.35), which, in most insects, can be opened and closed by valves. The ability to prevent water loss by closing the spiracles was a key adaptation that facilitated the arthropod invasion of land.

Many spiders have book lungs instead of or in addition to tracheae. A **book lung** is a series of leaflike plates within a chamber into which air is drawn and from which it is expelled by muscular contraction.

Excretory system

Various kinds of excretory systems occur in arthropods. In aquatic arthropods, much of the waste may diffuse from the blood in the gills.

Malpighian tubules, which occur in terrestrial insects, myriapods, and chelicerates, are slender projections from the digestive tract attached at the junction of the midgut and hindgut (see figures 34.32 and 34.33). Fluid passes through the walls of the Malpighian tubules to and from the blood in which the tubules are bathed. Nitrogenous wastes are precipitated as concentrated uric acid or guanine, emptied into the hindgut, and eliminated. Most of the water and salts in the fluid are reabsorbed by the hindgut and rectum to be returned to the arthropod's body. This efficient conservation of water by Malpighian tubules was another key adaptation facilitating invasion of the land by arthropods.

Spiders, mites, ticks, and horseshoe crabs (Chelicerata) have specialized anterior appendages

In the class Chelicerata, with some 57,000 species, the most anterior appendages, called **chelicerae** (singular, *chelicera*), may function as fangs or pincers. The body of a chelicerate is divided into two tagmata: the anterior prosoma, which bears all the appendages, and the posterior **opisthosoma,** which contains the reproductive organs. Chelicerates include familiar largely terrestrial arthropods such as spiders, ticks, mites, scorpions, and daddy longlegs. However, 4000 known species of mites and one species of spider live in freshwater habitats, and a few mites live in the sea. Exclusively marine groups of chelicerates are horseshoe crabs and sea spiders.

In addition to a pair of chelicerae, a chelicerate has a pair of **pedipalps,** and four pairs of walking legs on its prosoma. The pedipalps (often simply called palps) resemble legs but have one fewer segment and are not used for locomotion. In male spiders, the pedipalps are copulatory organs; in scorpions, they are large pincers; and in most other chelicerates, they are sensorial, acting like the antennae of other arthropods.

Most chelicerates are carnivorous, but mites are largely herbivorous. Aside from the daddy long-legs, which can ingest small particles, most cannot consume solid food. They subsist on liquids, including solid food that they liquefy by injecting with digestive enzymes and then suck up with the muscular pharynx.

Figure 34.36 Horseshoe crabs. Horseshoe crabs are aquatic chelicerates. This horseshoe crab, *Limulus polyphemus*, is found along the east coast of North America.

a.   *b.*

Figure 34.37 Two common poisonous spiders. *a.* The southern black widow, *Latrodectus mactans.* *b.* The brown recluse, *Loxosceles reclusa.* Both species are common throughout temperate and subtropical North America.

The four species of horseshoe crabs (figure 34.36) live off the North American Atlantic coast and in Southeast Asia. Another type of chelicerate, the sea spiders are such strange marine animals that some authorities exclude them from the Chelicerata. Most are small, but some can reach 150 mm or so across; many live in association with other marine animals, such as hydroids.

The 35,000 named species of spiders (order Araneae) play a major role in almost all terrestrial ecosystems. They are particularly important as predators of insects and other small animals. Spiders hunt their prey or catch it in silk webs of remarkable diversity. Silk is formed from a fluid protein that is forced out of **spinnerets** on the posterior portion of the spider's abdomen. Trap-door spiders construct silk-lined burrows with lids, seizing their prey as it passes by. Spiders such as the familiar wolf spiders and tarantulas hunt rather than spin webs.

All spiders have poison glands with channels through their chelicerae, which are pointed and are used to bite and paralyze prey. The bites of some, such as the western black widow *(Latrodectus hesperus)* and brown recluse *(Loxosceles reclusa)* (figure 34.37), can be fatal to humans and other large mammals.

The order Acari is the most diverse of the chelicerates. About 30,000 species of mites and ticks have been named, but scientists estimate that more than a million members of this order may exist. Acarines are found in nearly every habitat; they feed on a variety of organisms as predators and parasites.

Most mites are less than 1 mm long, but adults range from 100 nm to 2 cm. In most mites, the cephalothorax and abdomen are fused into an unsegmented, ovoid body. Respiration is by means of tracheae or directly through the body surface. Many mites pass through several stages during their life cycle. In most, an inactive eight-legged prelarva gives rise to an active six-legged larva, which in turn produces a succession of three eight-legged stages, and finally, the adult.

Ticks are parasites that attach to the surface of humans and other animals, causing discomfort by sucking blood. Ticks, which are larger than most other members of the order,

can carry disease-causing agents. For example, Rocky Mountain spotted fever is caused by bacteria; Lyme disease is caused by spirochaetes; and red-water fever, or Texas fever, is an important tick-borne protozoan disease of cattle, horses, sheep, and dogs.

Crabs, shrimps, lobsters, and pill bugs (Crustacea) are largely marine organisms

Crustaceans (class Crustacea) include 35,000 species of largely marine organisms, such as crabs, shrimps, lobsters, and barnacles. However, some groups, such as crayfish, occur in freshwater, and some crabs and copepods (figure 34.38) are among the most abundant multicellular organisms on Earth. A small number are terrestrial, including about half the estimated 4500 species of order Isopoda, the pillbugs; and some sand fleas or beach fleas (order Amphipoda) are semiterrestrial.

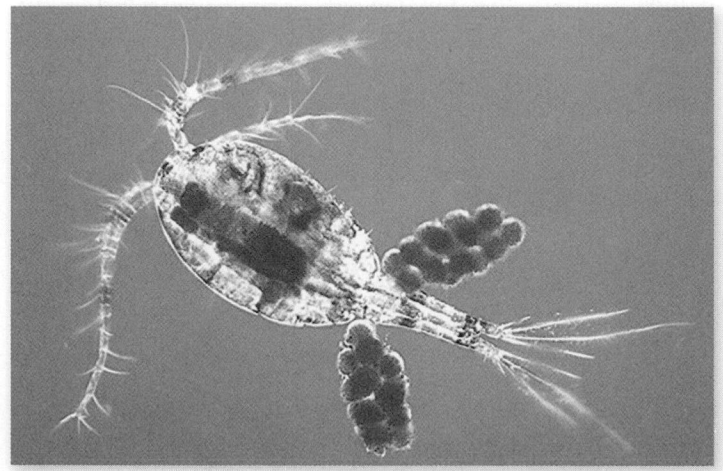

Figure 34.38 Freshwater crustacean. A copepod with attached eggs. Members of the marine and freshwater order Copepoda are important components of the plankton. Most are a few millimeters long.

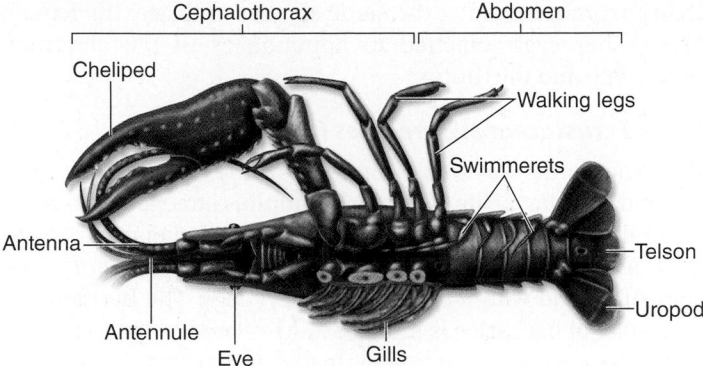

Figure 34.39 Decapod crustacean. Ventral view of a lobster, *Homarus americanus,* with some of its principal features labeled.

Some crustaceans (such as lobsters and crayfish) are valued as food for humans; planktonic crustaceans (such as krill) and larval crustaceans, which are abundant in the plankton, are the primary food of baleen whales and many smaller marine animals.

Crustacean body plans

A typical crustacean has three tagmata; the anteriormost two—the cephalon and thorax—may fuse to form the cephalothorax (figure 34.39). Most crustaceans have two pairs of antennae, three pairs of appendages for chewing and manipulating food, and various pairs of legs. Crustacean appendages, with the possible exception of the first pair of antennae, are biramous ("two-branched"). Crustaceans differ from hexapods, but resemble myriapods, in having appendages on their abdomen as well as their thorax. They are the only arthropods with two pairs of antennae.

Large crustaceans have feathery gills for respiration near the bases of their legs (see figure 34.39). Oxygen extracted from the gills is distributed through the circulatory system. In smaller crustaceans, gas exchange takes place directly through the thinner areas of the cuticle or the entire body.

Crustacean reproduction

In crustaceans, many kinds of copulation occur, and the members of some groups carry their eggs, either singly or in egg pouches, until they hatch. Crustaceans develop through a **nauplius** (plural, *nauplii*) stage (figure 34.40). The nauplius

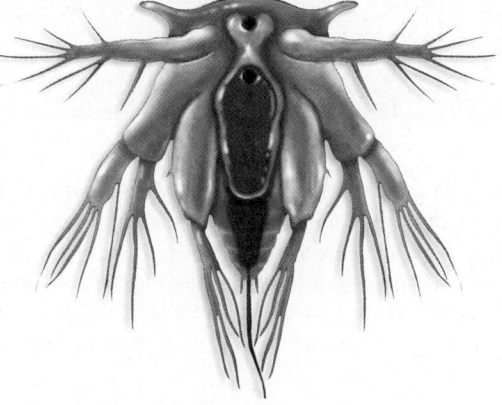

Figure 34.40 The nauplius larva. The nauplius of a crustacean is an important unifying feature found in members of this group.

hatches with three pairs of appendages and undergoes metamorphosis through several stages before reaching maturity. In many groups, this nauplius stage is passed in the egg, and the hatchling resembles a miniature adult.

The nauplius larva is characteristic of Crustacea, providing evidence that all members of this diverse group descended from a common ancestor that had a nauplius in its life cycle. The sessile barnacles, with their shell-like exoskeleton, had been thought to be related to mollusks until they were discovered to have a nauplius larva. More recently, the wormy pentastomids, which parasitize the respiratory tracts of vertebrates, were determined to be crustaceans in part through their nauplii (recall that parasites may become morphologically simplified and so can be difficult to place phylogenetically).

Shrimps, lobsters, crabs, and crayfish (Decapoda)

Large, primarily marine crustaceans such as shrimps, lobsters, and crabs, along with their freshwater relatives, the crayfish, belong to order Decapoda (see figure 34.39), which means "ten-footed" because of their five pairs of thoracic appendages. The exoskeleton of most is reinforced with calcium carbonate. The cephalothorax is covered by a dorsal shield, or carapace, which arises from the head. The pincers of many decapod crustaceans are used in obtaining food—for example, by crushing mollusk shells.

In lobsters and crayfish, appendages called **swimmerets** that occur along the ventral surface of the abdomen are used in reproduction and swimming. At the posterior end of the abdomen, paired flattened appendages known as **uropods** form a kind of paddle between which is a **telson,** a tail spine (see figure 34.39). When a lobster or crayfish contracts its abdominal muscle, the uropods and telson push water anteriorly, propelling the animal rapidly and forcefully backward through the water. It is this very large muscle that constitutes the "lobster tail" so valued by human diners!

One difference between crabs and lobsters is that the crab carapace is relatively much broader and the abdomen is just a small vestige tucked under the cephalothorax (figure 34.41). The abdomen of a male crab is much narrower

Figure 34.41 Cephalothorax. The cephalothorax of a Chesapeake Bay blue crab *(Callinectes sapidus).*

Figure 34.42 Gooseneck barnacles, *Lepas anatifera*.
These barnacles are filter feeding, which they do by sweeping their jointed legs through the water, gathering small particles of food. These are stalked barnacles; others lack a stalk.

than that of a female of the same species and size: the female carries her eggs attached to appendages of the abdomen between it and the thorax.

Sessile crustaceans: Barnacles (Cirripedia)

Barnacles (order Cirripedia; figure 34.42) are sessile as adults. At the end of its larval life, a barnacle nauplius attaches by its head to a piling, rock, or other submerged object, metamorphoses, grows calcareous plates around it, and spends the rest of its life capturing food with its jointed, feathery legs. The hermaphroditic state of barnacles is thought to be related to their sessility. Barnacles have the longest penis in the animal kingdom, relative to their size, which, considering that they can't move, is useful for mating with other barnacles located some distance away.

Insects (Hexapoda) are the most abundant animals on Earth

The insects, members of class Hexapoda, are by far the largest group of animals on Earth, in terms of number of species and number of individuals. Insects live in every habitat on land and

Order: Lepidoptera	Order: Homoptera	Order: Coleoptera
a.	*b.*	*c.*
Order: Diptera	Order: Orthoptera	Order: Isoptera
d.	*e.*	*f.*

Figure 34.43 Insect diversity. *a.* Luna moth, *Actias luna*. Luna moths and their relatives are among the most spectacular insects (order Lepidoptera). *b.* A thorn-shaped treehopper, *Umbonia crassicornis*. *c.* Boll weevil, *Anthonomus grandis*. Weevils are one of the largest groups of beetles. *d.* Soldier fly, *Ptecticus trivittatus*. *e.* Lubber grasshopper, *Romalea guttata*. *f.* Like ants, termites, such as these big-headed subterranean termites (Macrotermes sp.), have several castes with individuals specialized for different tasks (see chapter 54). The individual on the left is a soldier; its large jaws aid in defending the colony.

in fresh water, but very few have invaded the sea. More than half of named animal species are insects, and the actual proportion may be higher because millions of forms await detection, classification, and naming.

Approximately 90,000 described species occur in the United States and Canada; the actual number probably approaches 125,000. A hectare of lowland tropical forest is estimated to be inhabited by as many as 41,000 species of insects, and many suburban gardens may have 1500 or more species. It has been estimated that approximately a quintillion (10^{18}) individual insects are alive at any one time. A glimpse into the enormous diversity of insects is presented in figure 34.43 and table 34.2.

External features

Insects are primarily terrestrial, and aquatic insects probably had terrestrial ancestors. Most are small, ranging from 0.1 mm to about 30 cm in length or wingspan. Insect mouthparts all have the same basic structure; modifications reflect feeding habits (figure 34.44). Most insects have compound eyes, and many also have ocelli.

An insect body has three regions: the head, thorax, and abdomen (see figures 34.32, 34.33). The thorax consists of three segments, each with a pair of legs, which accounts for the name of the group, hexa (six) and poda (legs). Legs are absent in the larvae of certain groups—for example, most flies (order Diptera) and mosquitoes (figure 34.45). In addition, an insect may have one or two pairs of wings, which are not homologous to the other appendages, and which attach to the middle and posterior segments of the thorax (see figure 34.32). The wings, which consist of chitin and protein, arise as saclike outgrowths of the body wall; wings of moths and butterflies are covered with detachable scales that provide most of their bright colors

TABLE 34.2	Major Orders of Insects			
Order	**Typical Examples**		**Key Characteristics**	**Approximate Number of Named Species**
Coleoptera	Beetles		Two pairs of wings, the front one hard, protecting the rear one; heavily armored exoskeleton; biting and chewing mouthparts. Complete metamorphosis. The most diverse animal order.	350,000
Diptera	Flies		Front flying wings transparent; hindwings reduced to knobby balancing organs called halteres. Sucking, piercing, or lapping mouthparts; some bite people and other mammals. Complete metamorphosis.	120,000
Lepidoptera	Butterflies, moths		Two pairs of broad, scaly, flying wings, often brightly colored. Hairy body; tubelike, sucking mouthparts. Complete metamorphosis.	120,000
Hymenoptera	Bees, wasps, ants		Two pairs of transparent flying wings; mobile head and well-developed compound eyes; often possess stingers; chewing and sucking mouthparts. Many social. Complete metamorphosis.	100,000
Hemiptera and Homoptera	True bugs, bedbugs, leafhoppers, aphids, cicadas		Wingless or with two pairs of wings; piercing, sucking mouthparts, with which some draw blood, some feed on plants. Simple metamorphosis.	60,000
Orthoptera	Grasshoppers, crickets		Wingless or with two pairs of wings; among the largest insects; biting and chewing mouthparts in adults. Third pair of legs modified for jumping. Simple metamorphosis.	20,000
Odonata	Dragonflies		Two pairs of transparent flying wings that cannot fold back; large, long, and slender body; chewing mouthparts. Simple metamorphosis.	5000
Isoptera	Termites		Two pairs of wings, but some stages wingless; chewing mouthparts; simple metamorphosis. Social organization: labor divided among several body types. Some are among the few types of animals able to digest wood. Complete metamorphosis.	2000
Siphonaptera	Fleas		Wingless; flattened body with jumping legs; piercing and sucking mouthparts. Small; known for irritating bites. Complete metamorphosis.	1200

Figure 34.44
Mouthparts in three kinds of insects. Mouthparts are modified for **(a)** piercing in this mosquito of the genus *Culex;* **(b)** sucking nectar from flowers in the alfalfa butterfly of the genus *Colias;* and **(c)** sopping up liquids in the housefly, *Musca domestica.*

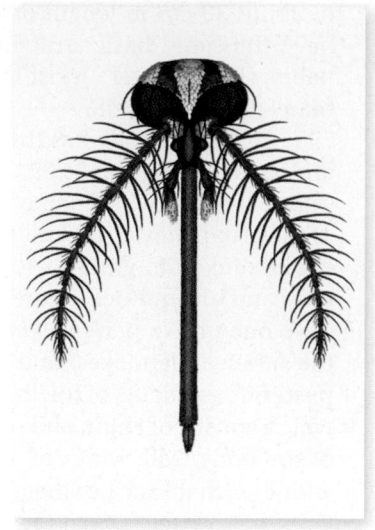 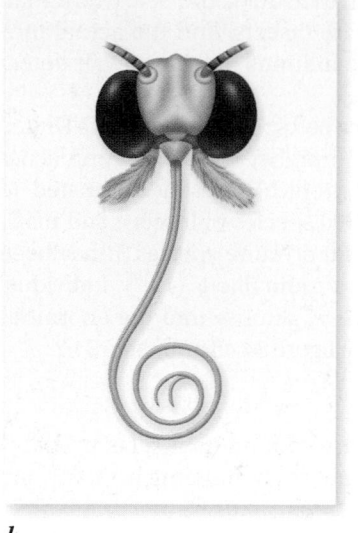 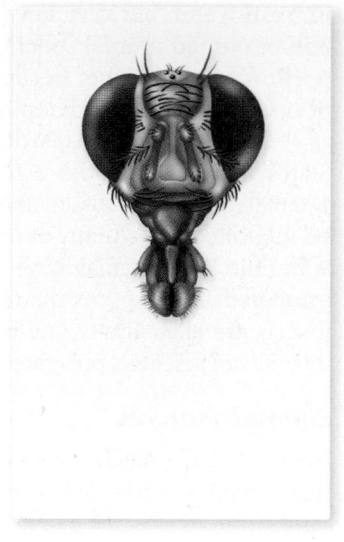

a.   *b.*   *c.*

(figure 34.46). Veins strengthen wings. An insect's thorax is almost entirely filled with muscles that operate the legs and wings. Fleas and lice are considered secondarily wingless, having descended from ancestors with wings. However, the ancestors of springtails and silverfish evolved before wings did, so those hexapods are considered primarily wingless.

Internal organization

The internal features of insects resemble those of other arthropods in many ways. The digestive tract is a tube about the same length as the body in some groups. However, in insects that feed on juices and so have sucking mouthparts, such as leafhoppers, cicadas, and many flies, the greatly coiled digestive tube may be several times longer than the body. Digestion takes place primarily in the stomach, or midgut, and excretion takes place through Malpighian tubules. Digestive enzymes are mainly secreted from the cells that line the midgut, although some are contributed by the salivary glands near the mouth.

In many winged insects, tracheae are dilated in various parts of the body, forming air sacs, which are surrounded by muscles to form a kind of bellows system that forces air deep into the body. The spiracles through which air enters the tracheal system are located on or between the segments along the sides of the thorax and abdomen. In most insects, the spiracles can be opened by muscular action. In some parasitic and aquatic groups of insects, the spiracles are permanently closed. In these groups, the tracheae are just below the surface of the insect, and gas exchange takes place by diffusion.

Sensory receptors

In addition to eyes, insects have several characteristic kinds of sense receptors. **Sensory setae** are hairlike structures that are widely distributed over the body. They are sensitive to mechanical and chemical stimulation and are linked to nerve cells. They are particularly abundant on the antennae and legs, the parts of the insect most likely to come into contact with other objects.

Sound, which may be of vital importance to insects such as grasshoppers, crickets, cicadas, and some moths, is detected

Figure 34.45 Larva of a mosquito, *Culex pipiens.*
The aquatic larvae of mosquitoes are quite active. They breathe through tubes at the surface of the water, as shown here. Covering the water with a thin film of oil suffocates them.

Figure 34.46 Scales on the wing of *Parnassius imperator,* a butterfly from China. Scales of this sort account for most of the colored patterns on the wings of butterflies and moths.

by a thin membrane, each called a **tympanum** (see figure 34.33), associated with the tracheal air sacs. In other groups of insects, sound waves are detected by sensory hairs. Male mosquitoes use thousands of sensory hairs on their antennae to detect sounds made by the vibrating wings of female mosquitoes.

In addition to sound, nearly all insects communicate by means of chemicals known as pheromones. These extremely diverse compounds are sent forth into the environment, where they convey a variety of messages, including mating signals and trail markers.

Insect life histories

During the course of their development, many insects undergo metamorphosis. For those such as grasshoppers, in which immature individuals are quite similar to adults, a series of molts results in an individual gradually getting bigger and more developed; this is termed simple metamorphosis. Those such as moths and butterflies have a life history involving a wormlike larval stage, a resting stage called a **pupa** or **chrysalis,** during which metamorphosis occurs, and then a final molt into the adult form or imago; this is termed complete metamorphosis.

Centipedes and millipedes (Myriapoda) have numerous legs

The body of a centipede and millipede consists of a head region posterior to which are numerous, more or less similar segments. Nearly all segments of a centipede have one pair of appendages (figure 34.47*a*), and nearly all segments of a millipede have two pairs of appendages (figure 34.47*b*). Each segment of a millipede is a simple tagma derived evolutionarily from two ancestral segments, which explains why millipedes have twice as many legs per segment as centipedes. Although the name *centipede* would imply an animal with 100 legs and the name *millipede* one with 1000, an adult centipede usually has fewer than 100 legs (most have 15, 21, or 23 pairs of legs); an adult millipede never reaches 1000 legs, most having 100 or fewer.

In both centipedes and millipedes, fertilization is internal, and all lay eggs. Young millipedes usually hatch with three pairs of legs; they add segments and legs as they pass through growth stages, but they do not change in general appearance. Centipedes have several types of development, young of some species hatching with their final number of legs and others adding legs after hatching. Centipedes that do not add legs as they grow tend to take care of their young, a behavior rather uncommon among invertebrates.

Centipedes, of which some 3000 species are known, are carnivorous, feeding mainly on insects. The appendages of the first trunk segment are modified into a pair of poison fangs. The poison may be toxic to humans, and although extremely painful, centipede bites are never fatal. In contrast, most millipedes are herbivores, feeding mainly on decaying vegetation such as leaf litter and rotting logs (which are typical habitats for the animals). Many millipedes can roll their bodies into a flat coil or sphere to defend themselves. More than 12,000 species of millipedes have been named, but this is estimated to be no more than one-sixth of the number of species that exists.

In each segment of their body, many millipedes have a pair of complex glands that produce a bad-smelling fluid, which they exude for defense through openings along the sides of the body. The chemistry of this material interests biologists because of the diversity of the compounds involved and their effectiveness in protecting millipedes from attack. Some species produce cyanide gas from segments near their head.

Learning Outcomes Review 34.9

Arthropods are segmented animals with exoskeletons and jointed appendages. The four living classes of arthropods are Chelicerata, with mouthparts (chelicerae) that function as fangs or pincers; Crustacea, in which all members have a nauplius developmental stage; Hexapoda, the insects, having three pairs of legs attached to fused segments of the thorax as adults; and Myriopoda, with one or two pairs of appendages on every body segment. An exoskeleton provides protection and muscle attachment, but growth requires molting. In some cases, larvae undergo metamorphosis into an adult with very different appearance.

■ **What would explain why the largest arthropods are found in marine environments?**

Centipede

Millipede

a.   *b.*

Figure 34.47 Myriapods. *a.* Centipedes, such as this member of the genus *Scolopendra,* are active predators. *b.* Millipedes, such as this member of the genus *Sigmoria,* are important herbivores and detritivores. Centipedes have one pair of legs per body segment, millipedes have two pairs per segment.

34.1 The Clades of Protosomes

Spiralians develop as embryos using spiral cleavage. Most live in water and move through it using cilia or contractions of the body musculature. The two clades of spiralians are platyzoans and lophotrochozoans.

Ecdysozoans are animals that can molt.

34.2 Platyzoans: Flatworms (Platyhelminthes)

Flatworms have an incomplete gut.

Free-living flatworms move by muscles and ciliated epithelial cells. They also exhibit a head and an incomplete gut.

Flatworms have an excretory system containing a fine network of tubules with flame cells. The primary function of this system is water balance.

Flatworms reproduce sexually and are hermaphroditic. They also have the capacity for asexual regeneration.

Flatworms consist of two major groups.

Free-living flatworms belong to the groups Turbellaria, which is likely not to be monophyletic.

Parasitic flatworms belong to the group Neodermata, of which there are two groups: the flukes (Trematoda), and the tapeworms and their relatives (Cercomeromorpha). Flukes and tapeworms can cause disease in humans.

34.3 Platyzoans: Rotifers (Rotifera)

Rotifers propel themselves and gather food with cilia and break down food with a complex jaw located in the pharynx. They are either free-swimming or sessile.

34.4 Lophotrochozoans: Mollusks (Mollusca)

Mollusks are extremely diverse—and important to humans.

Mollusks range from microscopic to huge and exhibit many different forms. They all have a coelom surrounding the heart.

The mollusk body plan is complex and varied.

Mollusks are generally bilaterally symmetrical, at least at some point in their lives (figure 34.10). They use a muscular foot (podium) for locomotion, attachment, food capture, or a combination.

The mantle is a thick epidermal sheet that forms the mantle cavity. It houses the respiratory structures (ctenidia or gills), and digestive, excretory, and reproductive products are discharged into it.

The outer mantle secretes a protective calcium carbonate shell. Some mollusks have internal or reduced shells, or none at all.

All mollusks except bivalves have a radula, a rasplike structure used in feeding (figure 34.10). Most have an open circulatory system, but cephalopods have a closed circulatory system.

In many mollusks, an embryo develops into a free-swimming trochophore larva (figure 34.12*a*). In some bivalves and gastropods, the embryo becomes a free-swimming veliger larva (figure 34.12*b*).

Four classes of mollusks show the diversity of the phylum.

The four best-known classes are Polyplacophora (chitons), Gastropoda (snails and slugs), Bivalvia (clams, mussels, and cockles), and Cephalopoda (octopuses and squids).

34.5 Lophotrochozoans: Ribbon Worms (Nemertea)

Nemerteans superficially resemble acoelomate flatworms, but they have a closed circulatory system and a complete digestive tract.

34.6 Lophotrochozoans: Annelids (Annelida)

The annelid body is composed of ringlike segments.

The segments of the annelid body are separated by septa. Each segment contains a pair of excretory organs, a ganglion, and, in most marine annelids, a set of reproductive organs. Anterior and posterior segments contain light-, chemo-, and touch receptors.

Segments are connected by a ventral nerve cord that includes an anterior brain region, and by a closed circulatory system.

Annelids move by contracting their segments.

The fluid-filled coelom acts as a hydrostatic skeleton. Each segment typically possesses chaetae, chitin bristles that help anchor the worm.

Annelids have a closed circulatory system but a segmented excretory system.

Annelids have a closed circulatory system; the dorsal vessel is connected to the ventral vessel by smaller vessels in the body wall.

Each segment contains a pair of nephridia that excrete wastes out of the body via the coelom and excretory tubes.

Annelida is composed of two—or three—classes.

Annelids have been grouped into three classes, but the currently accepted classification recognizes two: the Polychaeta, which exhibit parapodia and are mostly marine; and the Clitellata, which includes earthworms and leeches.

34.7 Lophophorates: Bryozoans (Bryozoa) and Brachiopods (Brachiopoda)

Bryozoa and Brachiopoda are characterized by a lophophore, a U-shaped ridge around the mouth bearing ciliated tentacles.

Bryozoans are the only exclusively colonial animals.

Each individual zooid produces a chitinous chamber called a zoecium that attaches to substrates and other colony members. They have deuterostome development.

Brachiopods and phoronids are solitary lophophorates.

The body of brachiopods is enclosed between two calcified shells that are dorsal and ventral, not lateral as in bivalves (figure 34.27). They exhibit some deuterostome characteristics and some protostome characteristics.

The phoronids, tube worms, are included with bryozoans because they are solitary lophophorates; however, they are protostomes.

34.8 Ecdysozoans: Roundworms (Nematoda)

Nematodes reproduce sexually and exhibit sexual dimorphism. More species of nematodes may exist than species of arthropods.

Some important human, veterinary, and plant diseases are caused by nematodes, including hookworm, pinworm, trichinosis, intestinal roundworm, and filariasis.

34.9 Ecdysozoans: Arthropods (Arthropoda)

Arthropods exhibit key features and organ systems.

Arthropods are segmented and exhibit an exoskeleton with muscles attached to the inside. The exoskeleton is molted during ecdysis, allowing the arthropod to grow. In some species, segments are fused into units called tagmata.

Arthropods' jointed appendages may be modified into mouth parts, antennae, or legs.

Many arthropods have compound eyes composed of ommatidia (figure 34.34); others have simple eyes (ocelli).

An open circulatory system includes a muscular heart. The respiratory system in terrestrial arthropods comprises spiracles, tracheae, and tracheoles (figure 34.35). The main components of the nervous system are an inhibitory brain and a ventral nerve cord.

The excretory system of terrestrial hexapods, myriapods, and chelicerates consists of Malpighian tubules that eliminate uric acid or guanine.

Spiders, mites, ticks, and horseshoe crabs (Chelicerata) have specialized anterior appendages.

Chelicerates have specialized anterior appendages, the chelicerae, that function as fangs or pincers.

Crabs, shrimps, lobsters, and pill bugs (Crustacea) are largely marine organisms.

Crustacea is characterized by a nauplius larva (figure 34.40). Crustaceans have three tagmata, and the two anterior ones fused to form a cephalothorax (figure 34.41).

Insects (Hexapoda) are the most abundant animals on Earth.

The hexapods are extraordinarily diverse. Insects have three pairs of legs attached to the thorax and may have wings. They have highly developed sensory systems.

Centipedes and millipedes (Myriapoda) have numerous legs.

Centipedes have one pair of appendages per segment, whereas millipedes have two pairs.

Review Questions

UNDERSTAND

1. In the flatworm, flame cells are involved in what metabolic process?
   a. Reproduction   c. Locomotion
   b. Digestion   d. Osmoregulation

2. The _____ of a mollusk is a highly efficient respiratory structure.
   a. nephridium   c. ctenidium
   b. radula   d. eliger

3. Torsion is a unique characteristic of the
   a. bivalves.   c. chitons.
   b. gastropods.   d. cephalopods.

4. Intelligence and complex behaviors are characteristics of the
   a. cephalopods.   c. brachiopods.
   b. polychaetes.   d. echinoderms.

5. Serial segmentation is a key characteristic of which of the following phyla?
   a. Mollusca   c. Bryozoa
   b. Brachiopoda   d. Annelida

6. The distinguishing feature of the Bryozoa and Brachiopoda is
   a. coelom.   c. chaetae.
   b. segmentation.   d. a lophophore.

7. Which of the following nematodes cause disease?
   a. Filarial worms   c. Trichina worms
   b. Pinworms   d. All of the choices are correct.

8. In terms of numbers of species, the most successful phylum on the planet is the
   a. Mollusca.   c. Echinodermata.
   b. Arthropoda.   d. Annelida.

9. Which of the following characteristics is NOT found in the arthropods?
   a. Jointed appendages   c. Closed circulatory system
   b. Segmentation   d. Segmented ganglia

10. Which of the following classes of arthropod possess chelicerae?
    a. Chilopoda   c. Hexapoda
    b. Crustacea   d. Chelicerata

11. Examples of decapods are
    a. centipedes and millipedes.
    b. barnacles.
    c. ticks and mites.
    d. lobsters and crayfish.

APPLY

1. To which of the following groups would a species that does not molt, possesses a coelom, and has a trochophore larva belong?
   a. Ecdysozoa   c. Platyzoa
   b. Parazoa   d. Lophotrochozoa

2. Which of the following is most closely related to lobsters and what characteristics are significant in determining this relationship?
   a. Bivalves   c. Earthworms
   b. Centipedes   d. Nematodes

3. Nematodes once were thought to be closely related to rotifers due to the presence of a pseudocoelom, but are now considered closer to the arthropods due to
   a. molting.   c. wings.
   b. jointed appendages.   d. a coelom.

SYNTHESIZE

1. Scientists studying the Chesapeake Bay have discovered that the decline in populations of clams and scallops has led to a drastic increase in the levels of pollution. What characteristic of this group could be attributed to this observation?

2. Chitin is present in a number of invertebrates, as well as the fungi. What does this tell you about the origin and importance of this substance?

3. What benefit would being a hermaphrodite confer on a parasitic species?

4. Does the lack of a digestive system in tapeworms indicate that it is a primitive, ancestral form of platyhelminthes? Explain your answer.

Chapter **35**

Deuterostomes

Chapter Contents

Introduction

At first glance, echinoderms such as sea stars and sea cucumbers and chordates such as hawks and elephants would seem to have little in common. Yet, both exhibit the same pattern of development, in which the anus is derived from the blastopore and the mouth originates from another part of the embryo. This synapomorphy unites the two clades as members of the Deuterostomia.

Echinoderms

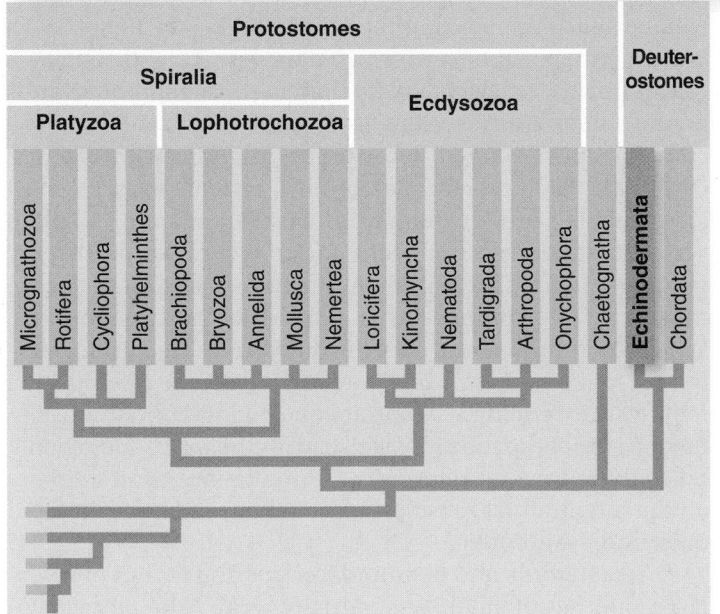

Members of the exclusively marine phylum Echinodermata are characterized by deuterostome development (see chapter 33) and possession of an endoskeleton. The endoskeleton is composed of hard, calcium carbonate plates that lie just beneath the delicate skin. The term *echinoderm,* which means "spiny skin," refers to the spines or bumps that occur on these plates in many species. Echinoderms, including sea stars (figure 35.1), brittle stars, sea urchins, sand dollars, and sea cucumbers, are some of the most familiar animals on the seashore. All have five axes of symmetry (that is, you could draw lines through their body in five different places that produced mirror images on each side) and thus are said to be **pentaradially symmetrical.**

Echinoderms are ancient and unmistakable

Echinodermata is an ancient group of marine animals that appeared nearly 600 MYA and that contains about 6000 living species. Echinoderms have an endoskeleton covered by living tissue. A unique feature of echinoderms is the hydraulic system that aids in movement or feeding. This fluid-filled system, called a **water-vascular system,** is a modification of one of several coelomic spaces and is composed of a central ring canal from which **radial canals** extend.

Although an excellent fossil record extends back into the Cambrian period, the origin of echinoderms remains unclear. Their phylogenetic placement (figure 33.5) suggests that they evolved from bilaterally symmetrical ancestors; this

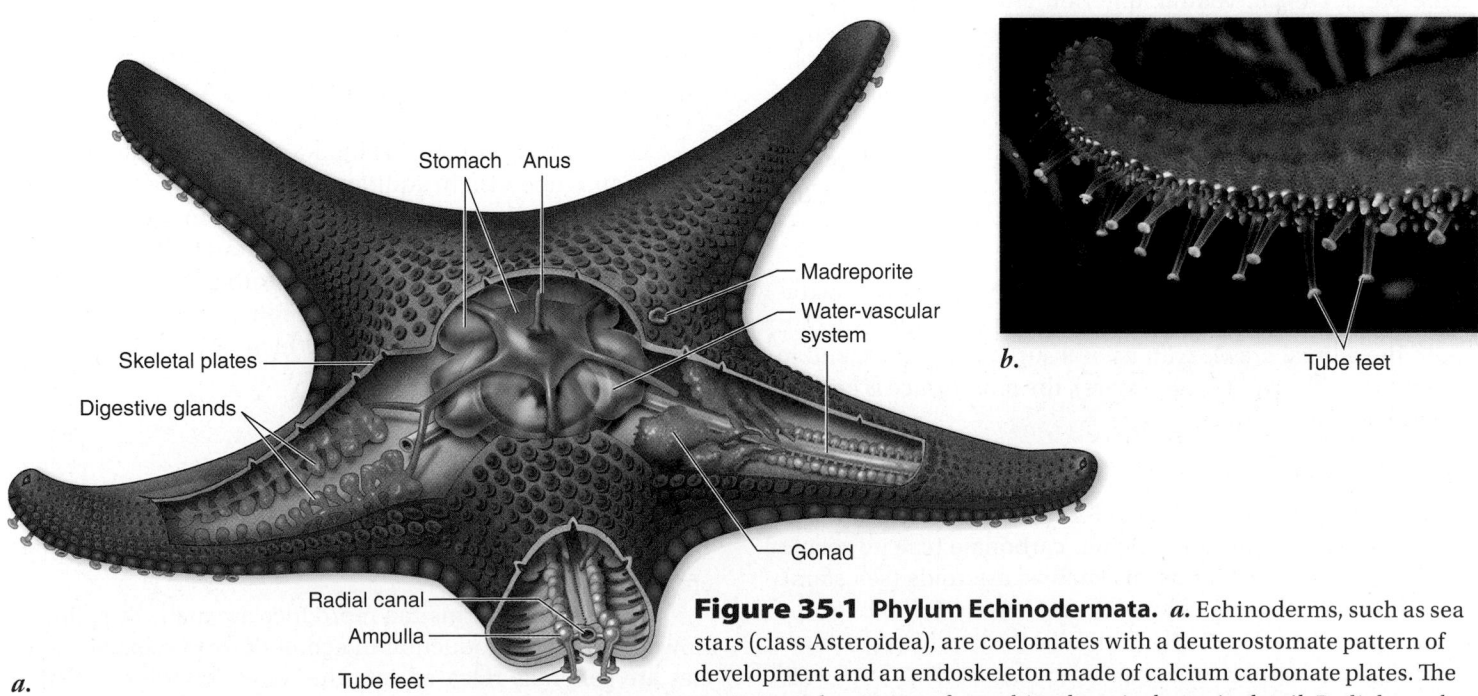

a.

b.

Labels: Stomach, Anus, Madreporite, Water-vascular system, Skeletal plates, Digestive glands, Radial canal, Ampulla, Tube feet, Gonad, Tube feet

Figure 35.1 Phylum Echinodermata. *a.* Echinoderms, such as sea stars (class Asteroidea), are coelomates with a deuterostomate pattern of development and an endoskeleton made of calcium carbonate plates. The water-vascular system of an echinoderm is shown in detail. Radial canals transport liquid to the tube feet. As the ampulla in each tube foot contracts, the tube foot extends and can attach to the substrate. When the muscles in the tube feet contract, the tube foot bends, pulling the animal forward. *b.* Extended nonsuckered tube feet of the sea star *Luidia magnifica.*

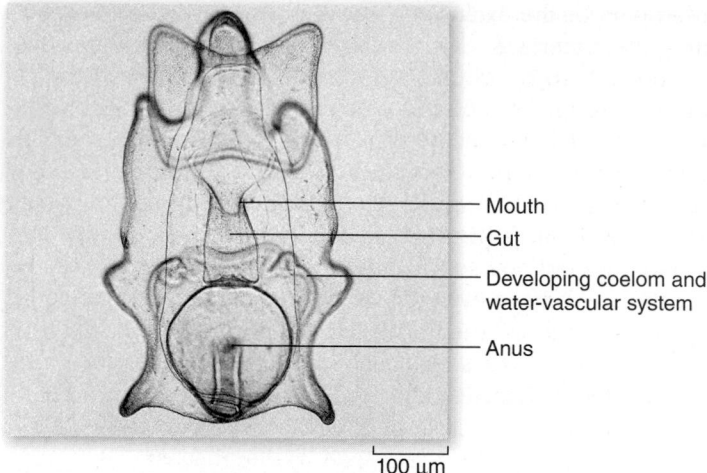

Mouth
Gut
Developing coelom and
water-vascular system
Anus

100 μm

Figure 35.2 **The free-swimming larva of the common sea star, *Asterias rubens*.** Such bilaterally symmetrical larvae suggest that the ancestors of the echinoderms may not have been radially symmetrical.

hypothesis is supported by the observation that echinoderm larvae are bilaterally symmetrical (figure 35.2).

Echinoderm symmetry is bilateral in larvae but pentaradial in adults

The body of echinoderms undergoes a fundamental shift during development, from bilaterally symmetrical larvae (see figure 35.2) to pentaradially symmetrical adults (see figure 35.1). Dorsal, ventral, anterior, and posterior have no meaning without a head or tail for orientation. Thus, orientation of the body of an adult echinoderm is described in reference to the mouth, which defines the **oral surface.** All systems of an adult echinoderm are organized with branches radiating from a center. For example, the nervous system consists of a central nerve ring from which branches arise; although the animals are capable of complex behavior patterns, there is no centralization of function.

In many echinoderms, the oral surface faces the substratum, although in sea cucumbers, the animal's axis is horizontal, so the animal crawls with its oral surface foremost, and in crinoids (sea lilies and feather stars), the oral surface is located opposite to the substrate.

The endoskeleton

Echinoderms have a delicate epidermis that stretches over an endoskeleton composed of calcium carbonate (calcite) plates called **ossicles.** In echinoderms such as asteroids (sea stars), the individual skeletal elements are loosely joined to one another. In others, especially echinoids (sea urchins and sand dollars), the ossicles abut one another tightly, forming a rigid shell (called a test). In sea cucumbers, by contrast, the ossicles are widely scattered, so the body wall is flexible. The ossicles in certain portions of the body of some echinoderms are perforated by pores. Tube feet, part of the water-vascular system discussed later, extend through these pores.

Echinoderms have collagenous tissue, which can change in texture from tough and rubbery to weak and fluid. This amazing tissue accounts for attributes of echinoderms such as the ability to autotomize (cast off) parts. This tissue is also responsible for a sea cucumber's being able to change from almost rigid to flaccid in a matter of seconds.

The water-vascular system

The water-vascular system is radially organized. From the ring canal, which encircles the animal's esophagus, a radial canal extends into each branch of the body (see figure 35.1). Water enters the water-vascular system through a **madreporite,** a sievelike plate that, in most echinoderms, is on the animal's surface, and flows to the ring canal through a stone canal, so named because it is reinforced by calcium carbonate. Each radial canal, in turn, extends through short side branches into the hollow tube feet (see figure 35.1*b*). In some echinoderms, each tube foot has a sucker at its end; in others, suckers are absent. At the base of each tube foot in most types of echinoderms is a muscular sac, the ampulla. When the ampulla contracts, the fluid, prevented from entering the radial canal by a one-way valve, is forced into the tube foot, thus extending it. Contraction of longitudinal muscles on one side of the tube foot wall causes the tube foot to bend; relaxation of the muscles in the ampulla and contraction of all the longitudinal muscles in the tube foot forces the fluid back into the ampulla.

In asteroids and echinoids, concerted action of a very large number of small, individually weak tube feet causes the animal to move across the sea floor. The tube feet around the mouth of sea cucumbers are used in feeding. In crinoids, tube feet that arise from the branches of the arms, which extend from the margins of an upward-directed cup, are used in capturing food from the surrounding water. The tube feet of brittle stars are pointed and specialized for feeding.

Gas exchange in most echinoderms is through the body surface and tube feet. In addition, a sea cucumber has paired respiratory trees, which branch off the hindgut. Water is drawn into them and exits from them through the anus. In an asteroid, one of the coelomic spaces other than the water-vascular system has branches into protrusions from the epidermis called papulae, through which gas exchange also occurs.

Regeneration and reproduction

Many echinoderms are able to regenerate lost parts. Some echinoderms can autotomize and then re-grow an arm. When some sea cucumber species are disturbed, they can cast out its entire digestive system.

Some echinoderms can reproduce asexually by splitting. However, most reproduction in echinoderms is sexual. Gametes are generally released into the water, where fertilization occurs, and free-swimming bilaterally symmetrical larvae develop (see figure 35.2). Each class of echinoderms has a characteristic type of larva. These larvae develop while floating in the ocean as plankton until they metamorphose into the sedentary adults.

Echinodermata contains five extant classes

There are five classes of living echinoderms, but more than 20 others have gone extinct. The living ones are (figure 35.3) (1) sea stars, or starfish, and sea daisies (Asteroidea); (2) sea cucumbers (Holothuroidea); (3) sea urchins and sand dollars (Echinoidea); (4) sea lilies and feather stars (Crinoidea); and (5) brittle stars (Ophiuroidea).

Adults of all classes exhibit a five-part body plan. Even sea cucumbers, which are shaped like their fruit namesake, have five longitudinal grooves along their bodies. Here we discuss three of the living classes that are most characteristic of the phylum.

Sea stars

Sea stars are perhaps the most familiar echinoderms. Important predators in many marine ecosystems, they range in size from a centimeter to a meter across. They are abundant in the intertidal zone, but also occur at the greatest depths of the ocean—10,000 m. Around 1500 species of sea stars are known.

A sea star consists of tapering arms that gradually merge into a central disk (see figure 35.1). Most sea stars have five arms, but others have many more, typically in multiples of five. The digestive space and gonads extend into the arms. The body is somewhat flattened, flexible, and covered with a pigmented epidermis. Asteroidea includes sea daisies, which were discovered in 1986 and were once considered to constitute their own class.

Brittle stars

Brittle stars constitute the largest class of echinoderms, with about 2000 species, and are probably the most abundant also. They resemble sea stars, but differ morphologically in several ways. The arms, which are of nearly equal diameter across their entire length, tapering only slightly from base to tip, merge abruptly into the central disk. They are nearly solid and can easily be autotomized (the source of the name "brittle"). The tube feet lack ampullae and suckers, being used for feeding, not locomotion. The animal has no anus; waste is eliminated through its mouth.

The most mobile of echinoderms, brittle stars move by pulling themselves along by "rowing" over the substrate by moving their slender arms, often in pairs or groups, from side to side. Some brittle stars use their arms to swim, an unusual habit among echinoderms. A brittle star has five arms, but in some larger ones (the basket stars), each arm may bifurcate several times. Brittle stars avoid light and are more active at night.

Sea urchins and sand dollars

Sand dollars and sea urchins lack arms. Five double rows of tube feet protrude through the plates of the calcareous skeleton. The protective, moveable spines that are attached to the skeleton by muscles and connective tissue are also arrayed pentamerally.

About 950 living species constitute the class Echinoidea. Their calcareous plates preserve well, so sea urchins and sand dollars are well represented in the fossil record, with more than 5000 extinct species described. Sand dollars are essentially flattened sea urchins; heart urchins are intermediate between the two.

Like holothurians, sea urchins are eaten by humans. In particular, the gonads, known as *uni* in Japan, are considered a delicacy.

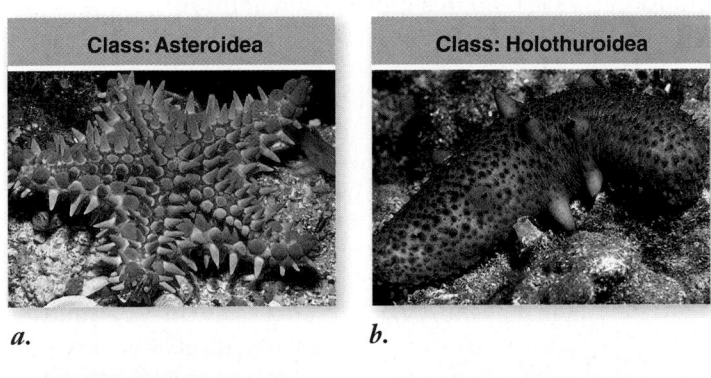

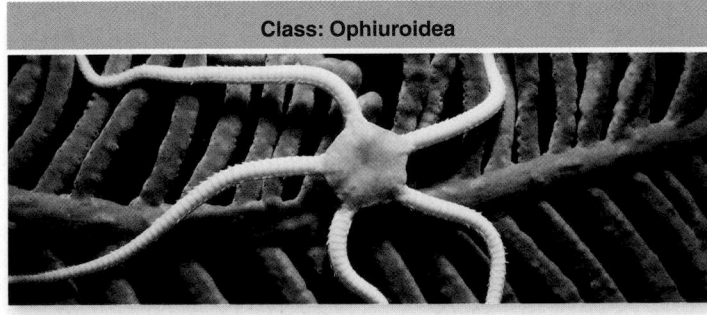

Figure 35.3 Diversity in echinoderms. *a.* Sea star, *Oreaster occidentalis*, in the Gulf of California, Mexico. *b.* Warty sea cucumber, *Parastichopus parvimensis*, Philippines. *c.* Sea urchin of the genus *Echinometra*, Carmel Bay, California. *d.* Feather star of the genus *Comatheria*, from Indonesia. *e.* Gaudy brittle star, *Ophioderma ensiferum*, Grand Turk Island, Caribbean Sea.

Learning Outcomes Review 35.1

Echinoderms are marine deuterostomes with endoskeletons. They are characterized by pentaradial symmetry in the adult, in which a line drawn in five directions produces mirror images. The water-vascular system and tube feet that act as suction cups aid in movement and feeding. The five living classes of echinoderms are Asteroidea (sea stars), Crinoidea (sea lilies), Echinoidea (sea urchins and sand dollars), Holothuroidea (sea cucumbers), and Ophiuroidea (brittle stars).

■ *How does phylogenetic information lead to the conclusion that echinoderms were bilaterally symmetrical?*

Learning Outcomes

1. List the defining characteristics of chordates.
2. Describe the evolutionary relationships of chordates to other taxa.

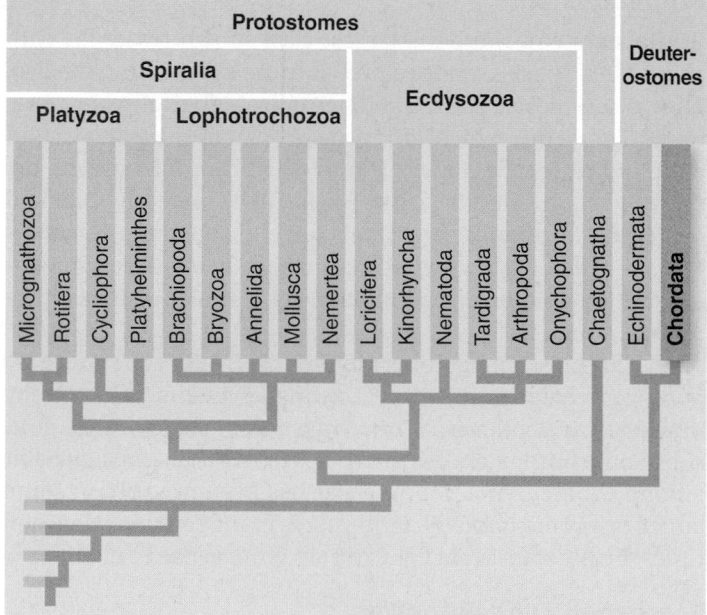

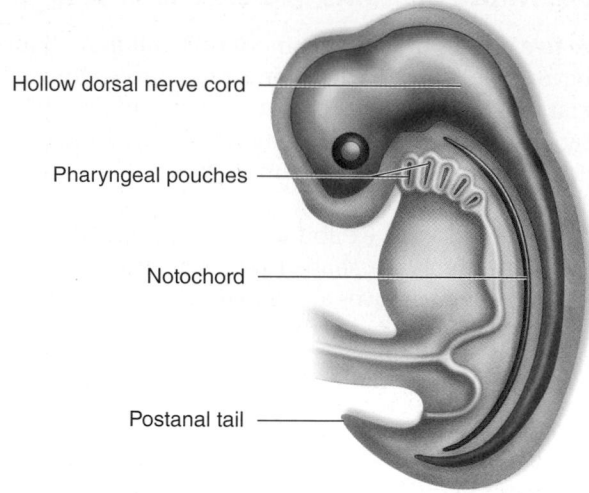

Figure 35.4 **The four principal features of the chordates, as shown in a generalized embryo.**

Members of the phylum Chordata exhibit great changes in the endoskeleton from that seen in echinoderms. The endoskeleton of echinoderms is functionally similar to the exoskeleton of arthropods—a hard shell with muscles attached to its inner surface. Chordates employ a very different kind of endoskeleton—one that is truly internal. Members of the phylum Chordata have a flexible rod that develops along the back of the embryo. Muscles attached to this rod allowed early chordates to swing their bodies from side to side, swimming through the water. This key evolutionary advance, attaching muscles to an internal element, started chordates along an evolutionary path that led to the vertebrates—and, for the first time, to truly large animals.

Four features characterize the chordates and have played an important role in the evolution of the phylum, which includes fish, amphibians, reptiles, birds, and mammals (figure 35.4):

1. A single, hollow **nerve cord** runs just beneath the dorsal surface of the animal. In vertebrates, the dorsal nerve cord differentiates into the brain and spinal cord.
2. A flexible rod, the **notochord,** forms on the dorsal side of the primitive gut in the early embryo and is present at some developmental stage in all chordates. The notochord is located just below the nerve cord. The notochord may persist in some chordates; in others it is replaced

during embryonic development by the vertebral column that forms around the nerve cord.

3. **Pharyngeal slits** connect the **pharynx,** a muscular tube that links the mouth cavity and the esophagus, with the external environment. In terrestrial vertebrates, the slits do not actually connect to the outside and are better termed **pharyngeal pouches.** Pharyngeal pouches are present in the embryos of all vertebrates. They become slits, open to the outside in animals with gills, but leave no external trace in those lacking gills. The presence of these structures in all vertebrate embryos provides evidence of their aquatic ancestry.
4. Chordates have a **postanal tail** that extends beyond the anus, at least during their embryonic development. Nearly all other animals have a terminal anus.

All chordates have all four of these characteristics at some time in their lives. For example, humans as embryos have

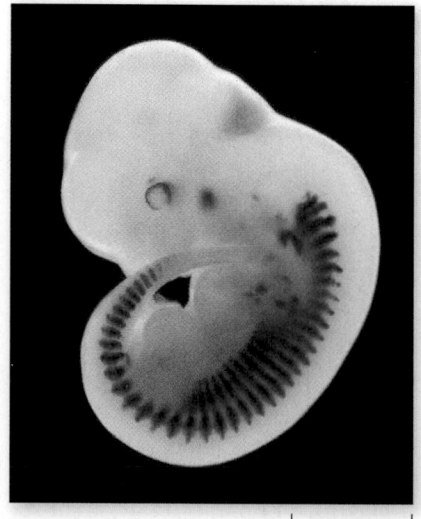

**Figure 35.5
A mouse embryo.**
At 11.5 days of development, the mesoderm is already divided into segments called somites (stained dark in this photo), reflecting the fundamentally segmented nature of all chordates.

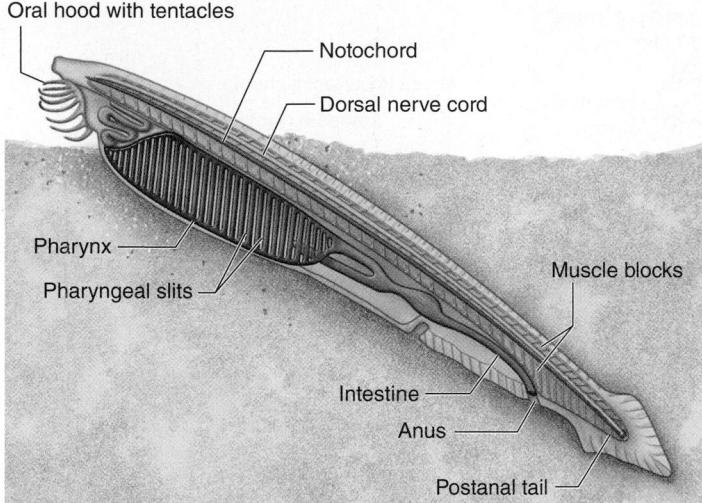

Oral hood with tentacles
Notochord
Dorsal nerve cord
Pharynx
Pharyngeal slits
Muscle blocks
Intestine
Anus
Postanal tail

Figure 35.6 Characteristics of chordates. Vertebrates, tunicates, and lancelets are chordates (phylum Chordata), coelomate animals with a flexible rod, the notochord, that provides resistance to muscle contraction and permits rapid lateral body movements. Chordates also possess pharyngeal pouches or slits (reflecting their aquatic ancestry and present habitat in some) and a hollow dorsal nerve cord. In nearly all vertebrates, the notochord is replaced during embryonic development by the vertebral column.

pharyngeal pouches, a dorsal nerve cord, a postanal tail, and a notochord. As adults, the nerve cord remains, and the notochord is replaced by the vertebral column. All but one pair of pharyngeal pouches is lost; this remaining pair forms the Eustachian tubes that connect the throat to the middle ear. The postanal tail regresses, forming the tail bone (coccyx). The presence of these traits in human embryos is an example of how patterns of development reflect evolutionary change, as discussed in chapter 21.

A number of other characteristics also fundamentally distinguish the chordates from other animals. Chordate muscles are arranged in segmented blocks, an arrangement which affects the basic organization of the chordate body and can often be clearly seen in embryos of this phylum (figure 35.5). Most chordates have an internal skeleton against which the muscles work. Either this internal skeleton or the notochord (figure 35.6) makes possible the extraordinary powers of locomotion characteristic of this group.

Learning Outcomes Review 35.2

Chordates are characterized by a hollow dorsal nerve cord, a notochord, pharyngeal pouches, and a postanal tail at some point in their development. The flexible notochord anchors internal muscles and allows rapid, versatile movement. Chordates are deuterostomes; their nearest relatives are the echinoderms.

■ *What distinguishes a chordate from an echinoderm?*

Learning Outcome

1. **Describe the nonvertebrate chordates and their characteristics.**

Phylum Chordata can be divided into three subphyla. Two of these, Urochordata and Cephalochordata, are nonvertebrate; the third subphylum is Vertebrata. The nonvertebrate chordates do not form vertebrae or other bones, and in the case of the urochordates, their adult form is greatly different from what we expect chordates to look like.

Tunicates have chordate larval forms

The tunicates and salps (subphylum Urochordata) are a group of about 1250 species of marine animals. Most of them are immobile as adults, with only the larvae having a notochord and nerve cord. As adults, they exhibit neither a major body cavity nor visible signs of segmentation (figure 35.7a, b). Most species occur in shallow waters, but some are found at great depths. In some tunicates, adults are colonial, living in masses on the ocean floor. The pharynx is lined with numerous cilia; beating of the cilia draws a stream of water into the pharynx, where microscopic food particles become trapped in a mucous sheet secreted from a structure called an *endostyle.*

The tadpole-like larvae of tunicates plainly exhibit all of the basic characteristics of chordates and mark the tunicates as having the most primitive combination of features found in any chordate (figure 35.7c). The larvae do not feed and have a poorly developed gut. They remain free-swimming for only a few days before settling to the bottom and attaching themselves to a suitable substrate by means of a sucker.

Tunicates change so much as they mature and adjust developmentally to an immobile, filter-feeding existence that it is difficult to discern their evolutionary relationships solely by examining an adult. Many adult tunicates secrete a tunic, a tough sac composed mainly of cellulose, a substance frequently found in the cell walls of plants and algae but rarely found in animals. The tunic surrounds the animal and gives the subphylum its name. Colonial tunicates may have a common sac and a common opening to the outside.

One group of Urochordates, the Larvacea, retains the tail and notochord into adulthood. One theory of vertebrate origins involves a larval form, perhaps that of a tunicate, which acquired the ability to reproduce.

Lancelets are small marine chordates

Lancelets (subphylum Cephalochordata) were given their English name because they resemble a lancet—a small, two-edged surgical knife. These scaleless chordates, a few centimeters long, occur widely in shallow water throughout the oceans

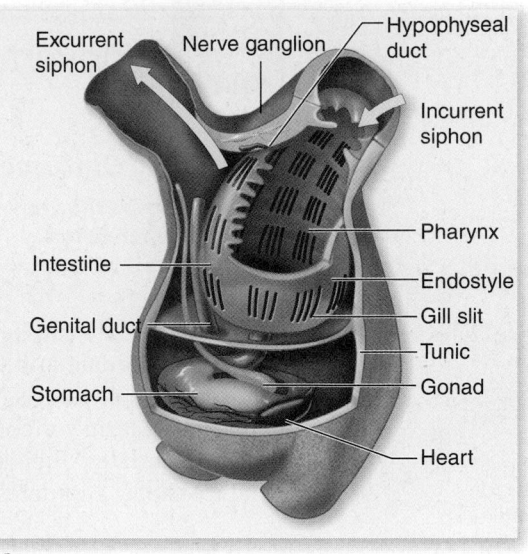

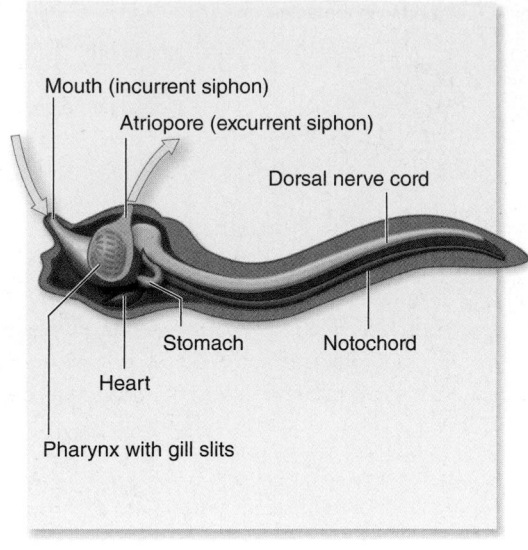

a. *b.* *c.*

Figure 35.7 Tunicates (phylum Chordata, subphylum Urochordata). *a.* The sea peach, *Halocynthia auranthium,* like other tunicates, does not move as an adult, but rather is firmly attached to the sea floor. *b.* Diagram of the structure of an adult tunicate. *c.* Diagram of the structure of a larval tunicate, showing the characteristic tadpole-like form. Larval tunicates resemble the postulated common ancestor of the chordates.

of the world. There are about 23 species of this subphylum. Most of them belong to the genus *Branchiostoma,* formerly called *Amphioxus,* a name still used widely. In lancelets, the notochord runs the entire length of the dorsal nerve cord and persists throughout the animal's life.

Lancelets spend most of their time partly buried in sandy or muddy substrates, with only their anterior ends protruding (figure 35.8). They can swim, although they rarely do so. Their muscles can easily be seen through their thin, transparent skin as a series of discrete blocks, called myomeres. Lancelets have many more pharyngeal gill slits than fishes do. Their skin lacks pigment and has only a single layer of cells, unlike the multilayered skin of vertebrates. The lancelet body is pointed at both ends. There is no distinguishable head or sensory structure other than pigmented light receptors.

Lancelets feed on microscopic plankton, using a current created by beating cilia that line the oral hood, pharynx, and

gill slits. The gill slits provide an exit for the water and are an adaptation for filter feeding. The oral hood projects beyond the mouth and bears sensory tentacles, which also ring the mouth.

The recent discovery of fossil forms similar to living lancelets in rocks 550 million years old argues for the antiquity of this group. Recent studies by molecular systematists further support the hypothesis that lancelets are the closest relatives of vertebrates.

Learning Outcome Review 35.3

Nonvertebrate chordates have notochords but no vertebrae or bones. Urochordates, such as tunicates, have obviously chordate larval forms but drastically different adult forms. Cephalochordates, such as lancelets, do not change body form as adults.

■ *How do lancelets and tunicates differ from each other, and from vertebrates?*

35.4 *Vertebrate Chordates*

Learning Outcomes

1. *Distinguish vertebrates from other chordates.*
2. *Explain how cartilage and bone contributed to increased size in vertebrates.*

Figure 35.8 Lancelets. Two lancelets, *Branchiostoma lanceolatum* (phylum Chordata, subphylum Cephalochordata), partly buried in shell gravel, with their anterior ends protruding. The muscle segments are clearly visible.

Vertebrates (subphylum Vertebrata) are chordates with a spinal column. The name *vertebrate* comes from the individual bony or cartilaginous segments called vertebrae that make up the spine.

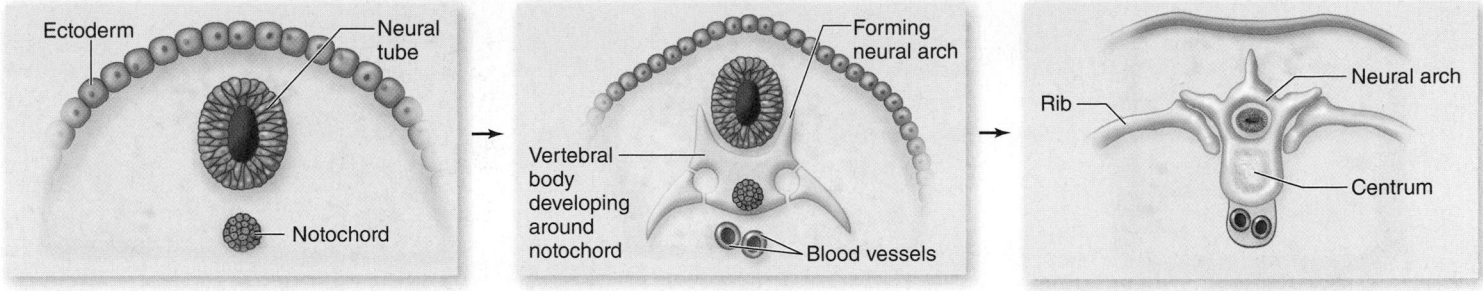

Figure 35.9 Embryonic development of a vertebra. During the evolution of animal development, the flexible notochord is surrounded and eventually replaced by a cartilaginous or bony covering, the centrum. The neural tube is protected by an arch above the centrum. The vertebral column functions as a strong, flexible rod that the muscles pull against when the animal swims or moves.

Vertebrates have vertebrae, a distinct head, and other features

Vertebrates differ from the tunicates and lancelets in two important respects:

Vertebral column. In all vertebrates except the earliest diverging fishes, the notochord is replaced during embryonic development by a vertebral column (figure 35.9). The column is a series of bony or cartilaginous vertebrae that enclose and protect the dorsal nerve cord like a sleeve.

Head. Vertebrates have a distinct and well-differentiated head with three pairs of well-developed sensory organs; the brain is encased within a protective box, the skull, or cranium, made of bone or cartilage.

In addition to these two key characteristics, vertebrates differ from other chordates in other important respects (figure 35.10):

Neural crest. A unique group of embryonic cells called the **neural crest** contributes to the development of many vertebrate structures. These cells develop on the crest of the neural tube as it forms by invagination and pinching together of the neural plate (see chapter 53 for a detailed account). Neural crest cells then migrate to various locations in the developing embryo, where they participate in the development of many different structures.

Internal organs. Internal organs characteristic of vertebrates include a liver, kidneys, and endocrine glands. The ductless endocrine glands secrete hormones that help regulate many of the body's functions. All vertebrates have a heart and a closed circulatory system. In both their circulatory and their excretory functions, vertebrates differ markedly from other animals.

Endoskeleton. The endoskeleton of most vertebrates is made of cartilage or bone. Cartilage and bone are specialized tissues containing fibers of the protein collagen compacted together (see chapter 46). Bone also contains crystals of a calcium phosphate salt. The great advantage of bone over chitin as a structural material is that bone is a dynamic, living tissue that is strong without being brittle. The vertebrate endoskeleton makes possible the great size and extraordinary powers of movement that characterize this group.

Vertebrates evolved half a billion years ago: An overview

The first vertebrates evolved in the oceans about 545 MYA, during the Cambrian period. Many of them looked like a flattened hot dog, with a mouth at one end and a fin at the other. The appearance of a hinged jaw was a major advance, opening up new food-gathering options, and jawed fishes became the dominant creatures in the sea. Their descendants, the amphibians, invaded the land. Amphibians, in turn, gave rise to the first reptiles about 300 MYA. Within 50 million years, reptiles, better suited to living out of water, replaced amphibians as the dominant land vertebrates.

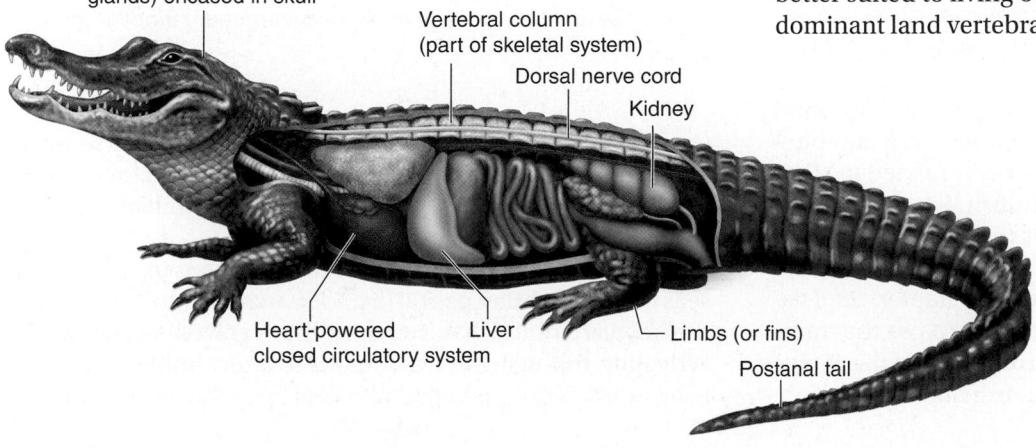

Figure 35.10 Major characteristics of vertebrates. Adult vertebrates are characterized by an internal skeleton of cartilage or bone, including a vertebral column and a skull. Several other internal and external features are characteristic of vertebrates.

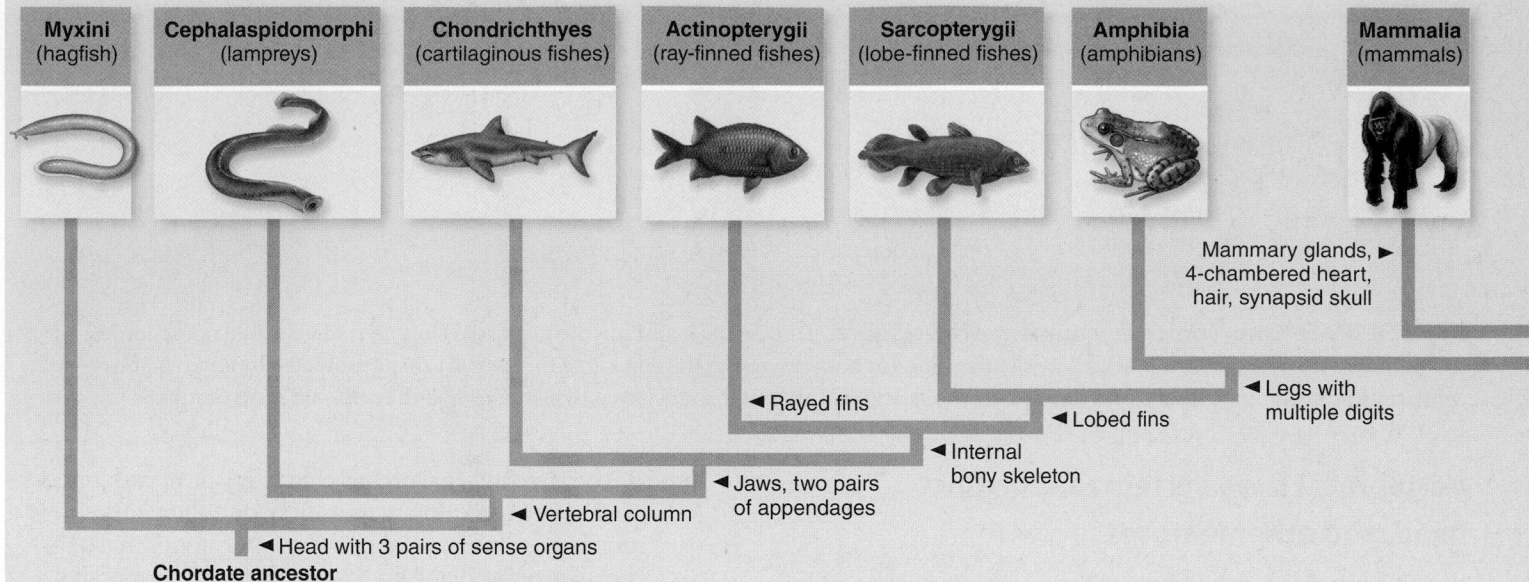

| Myxini (hagfish) | Cephalaspidomorphi (lampreys) | Chondrichthyes (cartilaginous fishes) | Actinopterygii (ray-finned fishes) | Sarcopterygii (lobe-finned fishes) | Amphibia (amphibians) | Mammalia (mammals) |

▶ Mammary glands, 4-chambered heart, hair, synapsid skull

◄ Legs with multiple digits

◄ Rayed fins

◄ Lobed fins

◄ Internal bony skeleton

◄ Jaws, two pairs of appendages

◄ Vertebral column

◄ Head with 3 pairs of sense organs

Chordate ancestor

Figure 35.11 Phylogeny of the living vertebrates. Some of the key characteristics that evolved among the vertebrate groups are shown in this phylogeny.

? Inquiry question Which vertebrate classes are paraphyletic and why?

With the success of reptiles, vertebrates truly came to dominate the surface of the Earth. Many kinds of reptiles evolved, ranging in size from smaller than a chicken to bigger than a bus, and including some that flew and others that swam. Among them evolved reptiles that gave rise to the two remaining great lines of terrestrial vertebrates: birds and mammals.

Dinosaurs and mammals appear at about the same time in the fossil record, 220 MYA. For over 150 million years, dinosaurs dominated the face of the Earth. Over all these million-and-a-half centuries, the largest mammal was no bigger than a medium-sized dog. Then, in the Cretaceous mass extinction, about 65 MYA, the dinosaurs and other types of reptiles abruptly disappeared. In their absence, mammals and birds quickly took their place, becoming abundant and diverse.

The history of vertebrates has been a series of evolutionary advances that have allowed vertebrates to first invade the land and then the air. In this chapter, we examine the key evolutionary advances that permitted vertebrates to invade the land successfully. As you will see, this invasion was a staggering evolutionary achievement, involving fundamental changes in many body systems.

While considering vertebrate evolution, keep in mind that two vertebrate groups—fish and reptiles—are paraphyletic (figure 35.11). Some fish are more closely related to other vertebrates than they are to some other fish. Similarly, some reptiles are more closely related to birds than they are to other reptiles. Biologists are currently divided about how to handle situations like this, as we discussed in chapter 23 (see figure 23.5). For the time being, we continue to use the traditional classification of vertebrates, bearing in mind the evolutionary implications of the paraphyly of fish and reptiles.

Learning Outcomes Review 35.4

Vertebrates are characterized by a vertebral column and a distinct head. Other distinguishing features are the development of a neural crest, a closed circulatory system, specialized organs, and a bony or cartilaginous endoskeleton that has the strength to support larger body size and powerful movements.

■ *In what ways would an exoskeleton limit the size of an organism?*

35.5 Fishes

Learning Outcomes

1. *Describe the major groups of fishes.*
2. *Discuss the significance of the evolutionary innovations of fishes.*

Over half of all vertebrates are fishes. The most diverse vertebrate group, fishes provided the evolutionary base for invasion of land by amphibians. In many ways, amphibians can be viewed as transitional—"fish out of water."

The story of vertebrate evolution started in the ancient seas of the Cambrian period (545–490 MYA). Figure 35.11 shows the key vertebrate characteristics that evolved subsequently. Wriggling through the water, jawless and toothless, the first fishes sucked up small food particles from the ocean floor like

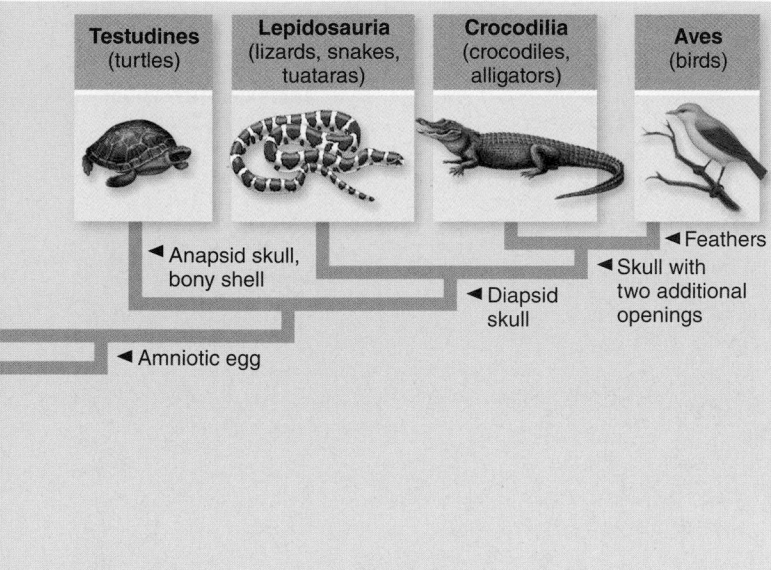

Figure 35.12 Fish. Fish are the most diverse vertebrates and include more species than all other kinds of vertebrates combined. *a.* Ribbon eel, *Rhinomuraena quaesita;* (*b*) Leafy sea-dragon, *Phycodurus eques;* (*c*) Yellowfin tuna, *Thunnus albacares.*

miniature vacuum cleaners. Most were less than a foot long, respired with gills, and had no paired fins or vertebrae (although some had rudimentary vertebrae); they did have a head and a primitive tail to push them through the water.

For 50 million years, during the Ordovician period (490–438 MYA), these simple fishes were the only vertebrates. By the end of this period, fish had developed primitive fins to help them swim and massive shields of bone for protection. Jawed fishes first appeared during the Silurian period (438–408 MYA), and along with them came a new mode of feeding.

Fishes exhibit five key characteristics

From 18-m-long whale sharks to tiny gobies no larger than your fingernail, fishes vary considerably in size, shape, color, and appearance (figure 35.12). Some live in freezing arctic seas, others in warm, freshwater lakes, and still others spend a lot of time entirely out of water. However varied, all fishes have important characteristics in common:

1. **Vertebral column.** Fish have an internal skeleton with a bony or cartilaginous spine surrounding the dorsal nerve cord, and a bony or cartilaginous skull encasing the brain. Exceptions are the jawless hagfish and lampreys. In hagfish, a cartilaginous skull is present, but vertebrae are not; the notochord persists and provides support. In lampreys, a cartilaginous skeleton and notochord are present, but rudimentary cartilaginous vertebrae also surround the notochord in places.
2. **Jaws and paired appendages.** Fishes other than lampreys and hagfish all have jaws and paired appendages, features that are also seen in tetrapods (see figure 35.11). Jaws allowed these fish to capture larger and more active prey. Most fishes have two pairs of fins: a pair of pectoral fins at the shoulder, and a pair of pelvic fins at the hip. In the lobe-finned fish, these pairs of fins became jointed.
3. **Internal gills.** Fishes are water-dwelling creatures and must extract oxygen dissolved in the water around them. They do this by directing a flow of water through their mouths and across their gills (see chapter 48). The gills are composed of fine filaments of tissue that are rich in blood vessels.
4. **Single-loop blood circulation.** Blood is pumped from the heart to the gills. From the gills, the oxygenated blood passes to the rest of the body, and then returns to the heart. The heart is a muscular tube-pump made of two chambers that contract in sequence.
5. **Nutritional deficiencies.** Fishes are unable to synthesize the aromatic amino acids (phenylalanine, tryptophan, and tyrosine; see chapter 3), so they must consume them in their foods. This inability has been inherited by all of their vertebrate descendants.

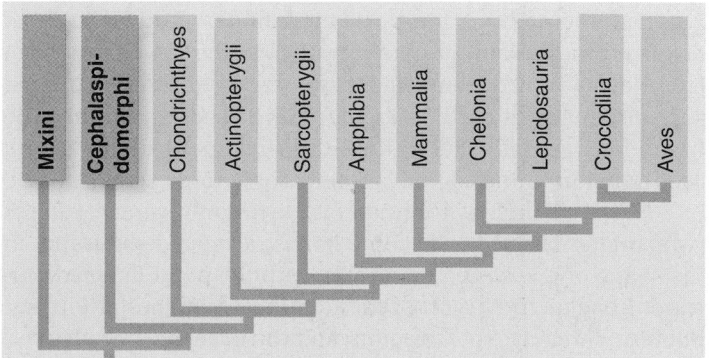

The first fishes

The first fishes did not have jaws, and instead had only a mouth at the front end of the body that could be opened to take in

TABLE 35.1 Major Classes of Fishes

Class	Typical Examples		Key Characteristics	Approximate Number of Living Species
Sarcopterygii	Lobe-finned fishes		Largely extinct group of bony fishes; ancestral to amphibians; paired lobed fins	8
Actinopterygii	Ray-finned fishes		Most diverse group of vertebrates; swim bladders and bony skeletons; paired fins supported by bony rays	30,000
Chondrichthyes	Sharks, skates, rays		Cartilaginous skeletons; no swim bladders; internal fertilization	750
Cephalaspidomorphi	Lampreys		Largely extinct group of jawless fishes with no paired appendages; parasitic and nonparasitic types; all breed in fresh water	35
Myxini	Hagfishes		Jawless fishes with no paired appendages; scavengers; mostly blind, but having a well-developed sense of smell	30
Placodermi	Armored fishes		Jawed fishes with heavily armored heads; many were quite large	Extinct
Acanthodii and Ostracoderms	Spiny fishes		Fishes with (acanthodians) or without (placoderms) jaws; paired fins supported by sharp spines; head shields made of bone; rest of skeleton cartilaginous	Extinct

food. Two groups survive today as hagfish (class Myxini; table 35.1) and lampreys (class Cephalaspidomorphi).

Another group were the *ostracoderms* (a word meaning "shell-skinned"). Only their head-shields were made of bone; their elaborate internal skeletons were constructed of cartilage. Many ostracoderms were bottom-dwellers, with a jawless mouth underneath a flat head, and eyes on the upper surface. Ostracoderms thrived in the Ordovician and Silurian periods (490–408 MYA), only to become almost extinct at the close of the Devonian period (408–360 MYA).

Evolution of the jaw

A fundamentally important evolutionary advance that occurred in the late Silurian period was the development of jaws. Jaws evolved from the most anterior of a series of arch supports made of cartilage, which were used to reinforce the tissue between gill slits to hold the slits open (figure 35.13). This transformation was not as radical as it might at first appear.

Each gill arch was formed by a series of several cartilages (which later evolved to become bones) arranged somewhat in the shape of a V turned on its side, with the point directed outward. Imagine the fusion of the front pair of arches at top and bottom, with hinges at the points, and you have the primitive vertebrate jaw. The top half of the jaw is not attached to the skull directly except at the rear. Teeth developed on the jaws from modified scales on the skin that lined the mouth (see figure 35.13).

Armored fishes called placoderms and spiny fishes called acanthodians both had jaws. Spiny fishes were very common during the early Devonian period, largely replacing

ostracoderms, but they became extinct themselves at the close of the Permian. Like ostracoderms, they had internal skeletons made of cartilage, but their scales contained small plates of bone, foreshadowing the much larger role bone would play in the future of vertebrates. Spiny fishes were jawed predators and far better swimmers than ostracoderms, with as many as seven fins to aid their swimming. All of these fins were reinforced with strong spines, giving these fishes their name.

By the mid-Devonian period, the heavily armored placoderms became common. A very diverse and successful group, seven orders of placoderms dominated the seas of the late Devonian, only to become extinct at the end of that period. The placoderm jaw was much improved over the primitive jaw of spiny fishes, with the upper jaw fused to the skull and the skull

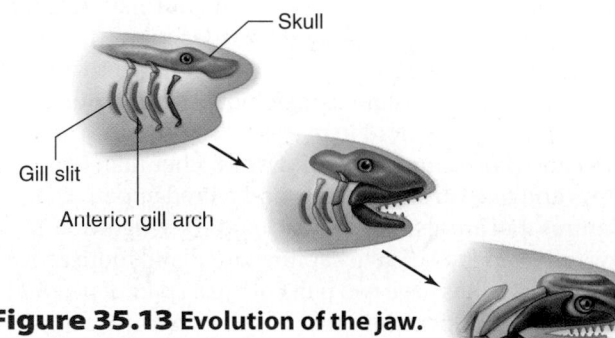

Figure 35.13 Evolution of the jaw.
Jaws evolved from the anterior gill arches of ancient, jawless fishes.

hinged on the shoulder. Many of the placoderms grew to enormous sizes, some over 30 feet long, with 2-foot skulls that had an enormous bite.

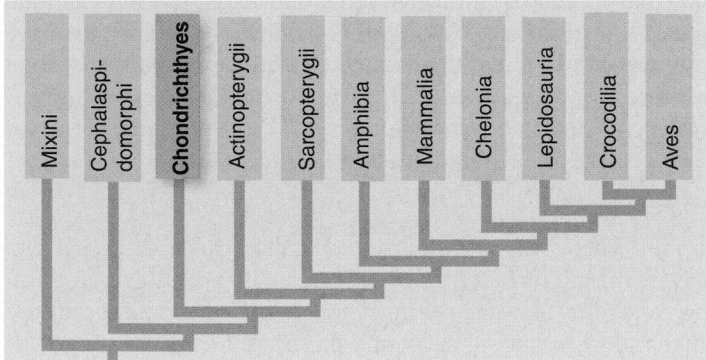

Sharks, with cartilaginous skeletons, became top predators

At the end of the Devonian period, essentially all of these pioneer vertebrates disappeared, replaced by sharks and bony fishes in one of several mass extinctions that occurred during Earth's history (see chapter 26). Sharks and bony fishes first evolved in the early Devonian, 400 MYA. In these fishes, the jaw was improved even further, with the upper part of the first gill arch behind the jaws being transformed into a supporting strut or prop, joining the rear of the lower jaw to the rear of the skull. This allowed the mouth to open much wider than was previously possible.

During the Carboniferous period (360–280 MYA), sharks became the dominant predators in the sea. Sharks, as well as skates and rays (all in the class Chondrichthyes), have a skeleton made of cartilage, like primitive fishes, but it is "calcified," strengthened by granules of calcium carbonate deposited in the outer layers of cartilage. The result is a very light and strong skeleton.

Streamlined, with paired fins and a light, flexible skeleton, sharks are superior swimmers (figure 35.14). Their pectoral fins are particularly large, jutting out stiffly like airplane wings—and that is how they function, adding lift to compensate for the downward thrust of the tail fin. Sharks are very aggressive predators, and some early sharks reached enormous size.

The evolution of teeth

Sharks were among the first vertebrates to develop teeth. These teeth evolved from rough scales on the skin and are not set into the jaw as human teeth are, but rather sit atop it. They are not

Figure 35.14 Chondrichthyes. Members of the class Chondrichthyes, such as this blue shark, *Prionace glauca,* are mainly predators or scavengers.

firmly anchored and are easily lost. In a shark's mouth, the teeth are arrayed in up to 20 rows; the teeth in front do the biting and cutting, and behind them other teeth grow and wait their turn. When a tooth breaks or is worn down, a replacement from the next row moves forward. A single shark may eventually use more than 20,000 teeth in its lifetime.

A shark's skin is covered with tiny, toothlike scales, giving it a rough "sandpaper" texture. Like the teeth, these scales are constantly replaced throughout the shark's life.

The lateral line system

Sharks, as well as bony fishes, possess a fully developed lateral line system. The **lateral line system** consists of a series of sensory organs that project into a canal beneath the surface of the skin. The canal runs the length of the fish's body and is open to the exterior through a series of sunken pits. Movement of water past the fish forces water through the canal. The pits are oriented so that some are stimulated no matter what direction the water moves. Details of the lateral line system's function are described in chapter 44. In a very real sense, the lateral line system is a fish's equivalent of hearing and therefore is a means of mechanoreception.

Reproduction in cartilaginous fishes

Reproduction in the class Chondrichthyes differs from that of most other fishes. Shark eggs are fertilized internally. During mating, the male grasps the female with modified fins called claspers, and sperm run from the male into the female through grooves in the claspers. Although a few species lay fertilized eggs, the eggs of most species develop within the female's body, and the pups are born alive (see figure 52.4).

This reproductive system can work to the detriment of sharks. Because of their long gestation periods and relatively few offspring, shark populations are not able to recover quickly from population declines. Unfortunately, in recent times sharks have been fished very heavily because shark fin soup has become popular in Asia and elsewhere. As a result, shark populations have declined greatly, and there is concern that many species may soon face extinction.

Shark evolution

Many of the early evolutionary lines of sharks died out during the great extinction at the end of the Permian period (248 MYA). The survivors thrived and underwent a burst of diversification during the Mesozoic era (248–65 MYA), when most of the modern groups of sharks appeared. Skates and rays, which are dorsoventrally flattened relatives of sharks, evolved at this time, some 200 million years after the sharks first appeared.

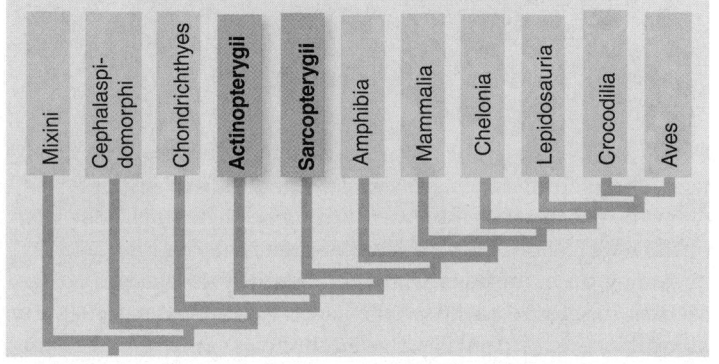

Bony fishes dominate the waters

Bony fishes evolved at the same time as sharks, some 400 MYA, but took quite a different evolutionary road. Instead of gaining speed through lightness, as sharks did, bony fishes adopted a heavy internal skeleton made completely of bone.

Bone is very strong, providing a base against which very strong muscles can pull. Not only is the internal skeleton ossified, but so is the outer covering of plates and scales. Most bony fishes have highly mobile fins, very thin scales, and completely symmetrical tails (which keep the fish on a straight course as it swims through the water). Bony fishes are the most species-rich group of fishes, indeed of all vertebrates. There are several dozen orders containing more than 30,000 living species.

The remarkable success of the bony fishes has resulted from a series of significant adaptations that have enabled them to dominate life in the water. These include the swim bladder and the gill cover (figure 35.15).

Swim bladder

Although bones are heavier than cartilaginous skeletons, most bony fishes are still buoyant because they possess a **swim bladder,** a gas-filled sac that allows them to regulate their buoyant density and so remain suspended at any depth in the water effortlessly. Sharks, by contrast, must move through the water or sink because, lacking a swim bladder, their bodies are denser than water.

In primitive bony fishes, the swim bladder is a dorsal outpocketing of the pharynx behind the throat, and these species fill the swim bladder by simply gulping air at the surface of the water. In most of today's bony fishes, the swim bladder is an independent organ that is filled and drained of gases, mostly nitrogen and oxygen, internally.

How do bony fishes manage this remarkable trick? It turns out that the gases are harvested from the blood by a unique gland that discharges the gases into the bladder when more buoyancy is required. To reduce buoyancy, gas is reabsorbed into the bloodstream through a structure called the oval body. A variety of physiological factors control the exchange of gases between the bloodstream and the swim bladder.

Gill cover

Most bony fishes have a hard plate called the **operculum** that covers the gills on each side of the head. Flexing the operculum permits bony fishes to pump water over their gills. The gills are suspended in the pharyngeal slits that form a passageway between the pharynx and the outside of the fish's body. When the operculum is closed, it seals off the exit.

When the mouth is open, closing the operculum increases the volume of the mouth cavity, so that water is drawn into the mouth. When the mouth is closed, opening the operculum decreases the volume of the mouth cavity, forcing water past the gills to the outside. Using this very efficient bellows, bony fishes can pass water over the gills while remaining stationary in the water. That is what a goldfish is doing when it seems to be gulping in a fish tank.

The evolutionary path to land ran through the lobe-finned fishes

Two major groups of bony fish are the ray-finned fishes (class Actinopterygii; figure 35.16*a*) and lobe-finned fishes (class Sarcopterygii). The groups differ in the structure of their fins (figure 35.16*b*). In ray-finned fishes, the internal skeleton of the fin is composed of parallel bony rays that support and stiffen each fin. There are no muscles within the fins; rather, the fins are moved by muscles within the body.

By contrast, lobe-finned fishes have paired fins that consist of a long fleshy muscular lobe (hence their name), supported by a central core of bones that form fully articulated joints with one another. There are bony rays only at the tips of each lobed fin. Muscles within each lobe can move the fin rays independently of one another, a feat no ray-finned fish could match.

Lobe-finned fishes evolved 390 MYA, shortly after the first bony fishes appeared. Only eight species survive today, two species of coelacanth (see figure 35.16*b*) and six species of lungfish. Although rare today, lobe-finned fishes played an important part in the evolutionary story of vertebrates. Amphibians almost certainly evolved from the lobe-finned fishes.

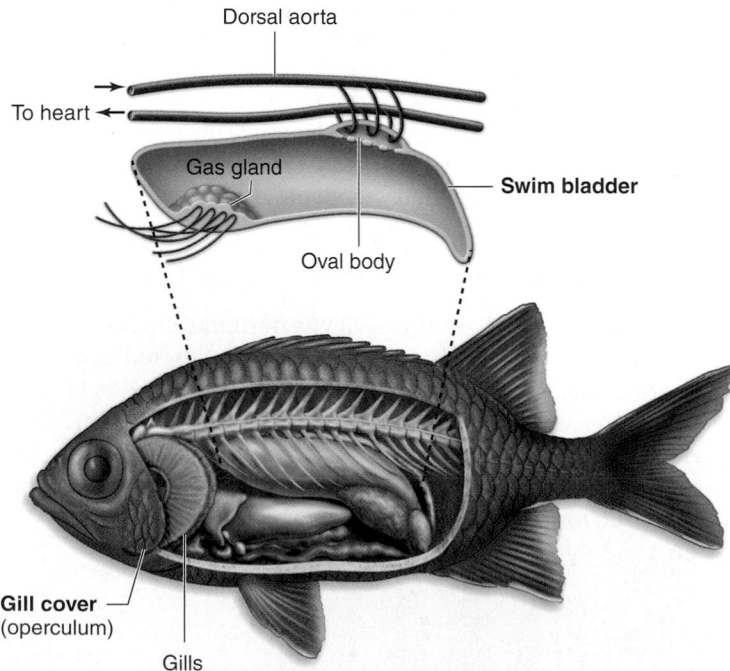

Figure 35.15 Diagram of a swim bladder. The bony fishes use this structure, which evolved as a dorsal outpocketing of the pharynx, to control their buoyancy in water. The swim bladder can be filled with or drained of gas to allow the fish to control buoyancy. Gases are taken from the blood, and the gas gland secretes the gases into the swim bladder; gas is released from the bladder by a muscular valve, the oval body.

Learning Outcomes Review 35.5

Fishes are generally characterized by the possession of a vertebral column, jaws, paired appendages, a lateral line system, internal gills, and single-loop circulation. Cartilaginous fishes have lightweight skeletons and were among the first vertebrates to develop teeth. The very successful bony fishes have unique characteristics such as swim bladders and gill covers, as well as ossified skeletons. One type of bony fish, the lobe-finned fish, gave rise to the ancestors of amphibians.

■   *What advantages do lobed fins have over ray fins?*

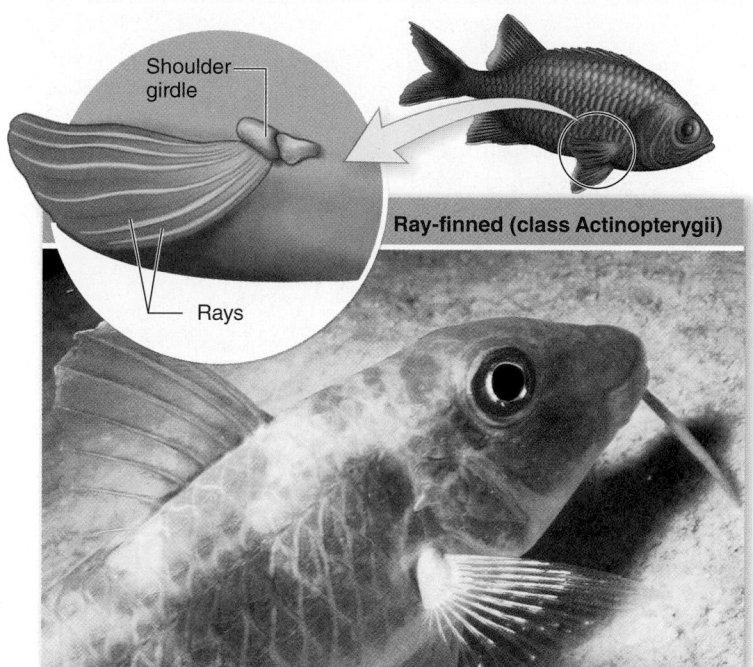

Ray-finned (class Actinopterygii)

Shoulder girdle

Rays

a.

Shoulder girdle

Rays

Lobe-finned (class Sarcopterygii)

Central core of bones in fleshy lobe

b.

Figure 35.16 Ray-finned and lobe-finned fishes. *a.* The ray-finned fishes, such as this Koi carp *(Cyprinus carpio),* are characterized by fins of only parallel bony rays. *b.* By contrast, the fins of lobe-finned fish have a central core of bones as well as rays. The coelacanth, *Latimeria chalumnae,* a lobe-finned fish (class Sarcopterygii), was discovered in the western Indian Ocean in 1938. This coelacanth represents a group of fishes thought to have been extinct for about 70 million years. Scientists who studied living individuals in their natural habitat at depths of 100–200 m observed them drifting in the current and hunting other fishes at night. Some individuals are nearly 3 m long; they have a slender, fat-filled swim bladder.

35.6 Amphibians

Learning Outcomes

1. **Describe the characteristics and major groups of amphibians.**
2. **Explain the challenges of moving from an aquatic to a terrestrial environment.**

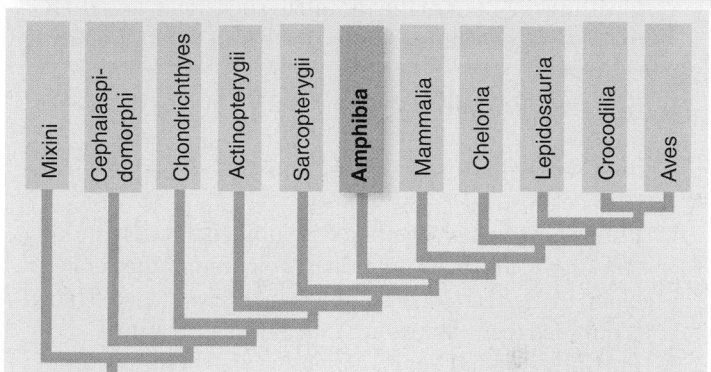

Frogs, salamanders, and caecilians, the damp-skinned vertebrates, are direct descendants of fishes. They are the sole survivors of a very successful group, the amphibians (class Amphibia)—the first vertebrates to walk on land. Most present-day amphibians are small and live largely unnoticed by humans, but they are among the most numerous of terrestrial vertebrates. Throughout the world, amphibians play key roles in terrestrial food chains.

Living amphibians have five distinguishing features

Biologists have classified living species of amphibians into three orders (table 35.2): 5000 species of frogs and toads in 22 families make up the order Anura ("without a tail"); 500 species of salamanders and newts in 9 families make up the order Caudata ("visible tail"); and 170 species (6 families) of

TABLE 35.2		Orders of Amphibians	
Order	**Typical Examples**	**Key Characteristics**	**Approximate Number of Living Species**
Anura	Frogs, toads	Compact, tailless body; large head fused to the trunk; rear limbs specialized for jumping	5000
Caudata	Salamanders, newts	Slender body; long tail and limbs set out at right angles to the body	500
Apoda	Caecilians	Tropical group with a snakelike body; no limbs; little or no tail	170

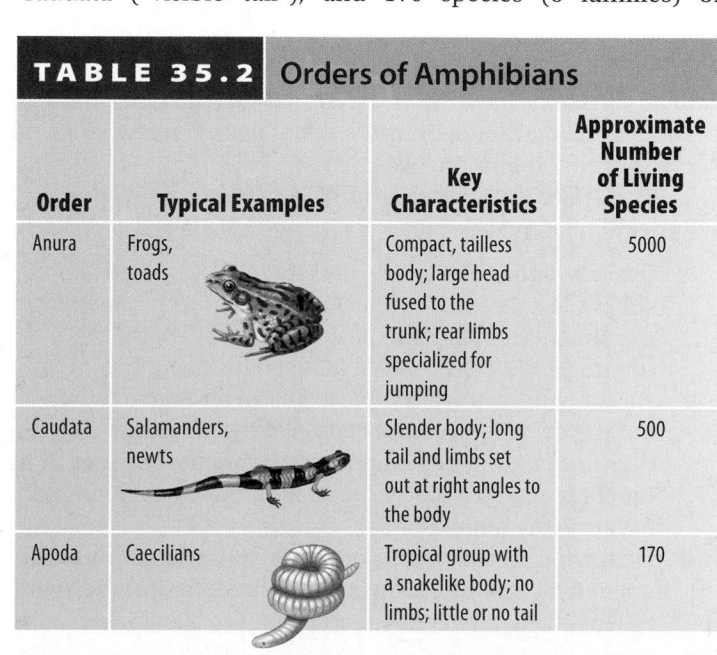

wormlike, nearly blind organisms called caecilians that live in the tropics make up the order Apoda ("without legs"). These amphibians have several key characteristics in common:

1. **Legs.** Frogs and most salamanders have four legs and can move about on land quite well. Legs were one of the key adaptations to life on land. Caecilians have lost their legs during the course of adapting to a burrowing existence.
2. **Lungs.** Most amphibians possess a pair of lungs, although the internal surfaces have much less surface area than do reptilian or mammalian lungs. Amphibians breathe by lowering the floor of the mouth to suck in air, and then raising it back to force the air down into the lungs (see chapter 48).
3. **Cutaneous respiration.** Frogs, salamanders, and caecilians all supplement the use of lungs by respiring through their skin, which is kept moist and provides an extensive surface area.
4. **Pulmonary veins.** After blood is pumped through the lungs, two large veins called pulmonary veins return the aerated blood to the heart for repumping. In this way aerated blood is pumped to the tissues at a much higher pressure.
5. **Partially divided heart.** A dividing wall helps prevent aerated blood from the lungs from mixing with nonaerated blood being returned to the heart from the rest of the body. The blood circulation is thus divided into two separate paths: pulmonary and systemic. The separation is imperfect, however, because no dividing wall exists in one chamber of the heart, the ventricle (see chapter 49).

Several other specialized characteristics are shared by all present-day amphibians. In all three orders, there is a zone of weakness between the base and the crown of the teeth. They also have a peculiar type of sensory rod cell in the retina of the eye called a "green rod." The function of this rod is unknown.

Amphibians overcame terrestrial challenges

The word *amphibia* means "double life," and it nicely describes the essential quality of modern-day amphibians, reflecting their ability to live in two worlds—the aquatic world of their fish ancestors and the terrestrial world they first invaded. Here, we review the checkered history of this group, almost all of whose members have been extinct for the last 200 million years. Then we examine in more detail what the few kinds of surviving amphibians are like.

The successful invasion of land by vertebrates posed a number of major challenges:

■ Because amphibian ancestors had relatively large bodies, supporting the body's weight on land as well as enabling movement from place to place was a challenge (figure 35.17). Legs evolved to meet this need.
■ Far more oxygen is available to gills in air than in water, but the delicate structure of fish gills requires the buoyancy of water to support them, and they cannot function in air. Therefore, other methods of obtaining oxygen were required.
■ Delivering great amounts of oxygen to the larger muscles needed for movement on land required modifications to the heart and circulatory system.

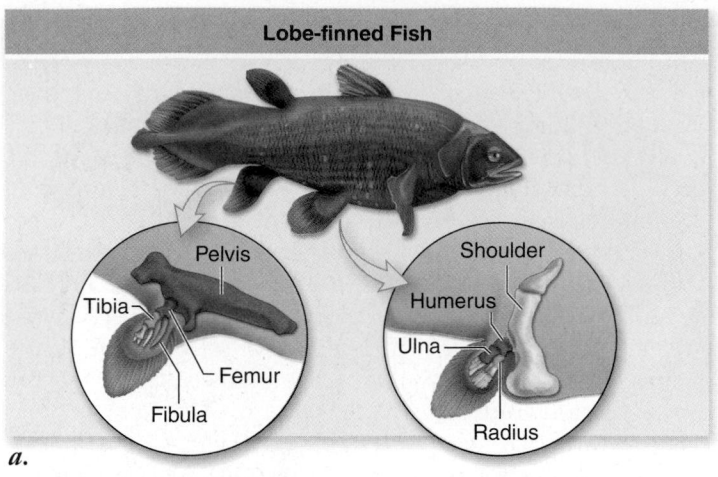

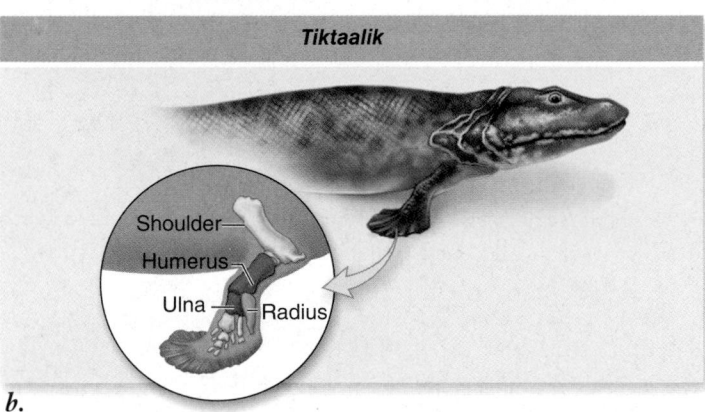

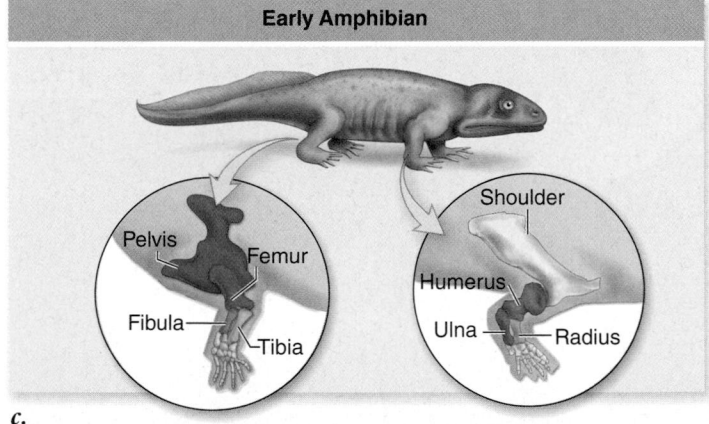

Figure 35.17 **A comparison between the limbs of a lobe-finned fish, *Tiktaalik,* and a primitive amphibian.** *a.* A lobe-finned fish. Some of these animals could probably move on land. *b. Tiktaalik.* The shoulder and limb bones are like those of an amphibian, but the fins are like those of a lobe-finned fish. The fossil of *Tiktaalik* did not contain the hindlimbs. *c.* A primitive amphibian. As illustrated by their skeletal structure, the legs of such an animal clearly could function better than those of its ancestors for movement on land.

■ Reproduction still had to be carried out in water so that eggs would not dry out.
■ Most importantly, the body itself had to be prevented from drying out.

The first amphibian

Amphibians solved these problems only partially, but their solutions worked well enough that amphibians have survived for 340 million years. Paleontologists agree that amphibians evolved from lobe-finned fish. *Ichthyostega*, one of the earliest amphibian fossils (figure 35.18), was found in a 370-million-year-old rock in Greenland. At that time, Greenland was part of what is now the North American continent and lay near the equator. All amphibian fossils from the next 100 million years are found in North America. Only when Asia and the southern continents merged with North America to form the supercontinent Pangea did amphibians spread throughout the world.

Ichthyostega was a strongly built animal, with sturdy forelegs well supported by shoulder bones. Unlike the skeleton of fish, the shoulder bones were no longer attached to the skull, so the limbs could support the animal's weight and the legs could move independently of the head. Because the hindlimbs were flipper-shaped, *Ichthyostega* probably moved like a seal, with the forelimbs providing the propulsive force for locomotion and the hindlimbs being dragged along with the rest of the body. To strengthen the backbone further, long, broad ribs evolved that overlap each other and formed a solid cage for the lungs and heart. The rib cage was so solid that it probably couldn't expand and contract for breathing. Instead, *Ichthyostega* probably obtained oxygen as many amphibians do today, by lowering the floor of the mouth to draw air in, and then raising it to push air down the windpipe into the lungs.

In 2006, an important transitional fossil between fish and *Ichthyostega* was discovered in northern Canada. *Tiktaalik*, which lived 375 MYA, had gills and scales like a fish, but a neck like an amphibian. Particularly significant, however, was the form of its forelimbs (see figure 35.17*b*): its shoulder, forearm, and wrist bones were like those of amphibians, but at the end of the limb was a lobed fin, rather than the toes of an amphibian. Ecologically, the 3-m long *Tiktaalik* was probably also intermediate between fish and amphibians, spending most of its time in the water, but capable of hauling itself out onto land to capture food or escape predators.

The rise and fall of amphibians

By moving onto land, amphibians were able to utilize many more resources and to access many habitats. Amphibians first became common during the Carboniferous period (360–280 MYA). Fourteen families of amphibians are known from the early Carboniferous, nearly all of them aquatic or semiaquatic, like *Ichthyostega* (see figure 35.18). By the late Carboniferous, much of North America was covered by low-lying tropical swamplands, and 34 families of amphibians thrived in this wet terrestrial environment, sharing it with pelycosaurs and other early reptiles.

In the early Permian period that followed (280–248 MYA), a remarkable change occurred among amphibians—they began to leave the marshes for dry uplands. Many of these terrestrial amphibians had bony plates and armor covering their bodies and grew to be very large, some as big as a pony. Both their large size and the complete covering of their bodies indicate that these amphibians did not use the cutaneous respiratory system of present-day amphibians, but rather had an impermeable leathery skin to prevent water loss. Consequently, they must have relied entirely on their lungs for respiration. By the mid-Permian period, there were 40 families of amphibians. Only 25% of them were still semiaquatic like *Ichthyostega*; 60% of the amphibians were fully terrestrial, and 15% were semiterrestrial. This was the peak of amphibian success, sometimes called the Age of Amphibians.

By the end of the Permian period, reptiles had evolved from amphibians. One group, therapsids, had become common and ousted the amphibians from their newly acquired niche on land. Following the mass extinction event at the end of the Permian, therapsids were the dominant land vertebrate, and most amphibians were aquatic. This trend continued in the following Triassic period (248–213 MYA), which saw the virtual extinction of amphibians from land.

Modern amphibians belong to three groups

All of today's amphibians descended from the three families of amphibians that survived the Age of the Dinosaurs. During the Tertiary period (65–2 MYA), these moist-skinned amphibians accomplished a highly successful invasion of wet habitats all over the world, and today there are over 5600 species of amphibians in 37 different families, constituting the orders Anura, Caudata, and Apoda.

Order Anura: Frogs and toads

Frogs and toads, amphibians without tails, live in a variety of environments, from deserts and mountains to ponds and puddles (figure 35.19*a*). Frogs have smooth, moist skin, a broad body, and long hind legs that make them excellent jumpers. Most frogs live in or near water and go through an aquatic tadpole stage before metamorphosing into frogs; however, some

Figure 35.18 Amphibians were the first vertebrates to walk on land. *Ichthyostega*, one of the first amphibians, had efficient limbs for crawling on land, an improved olfactory sense associated with a lengthened snout, and a relatively advanced ear structure for picking up airborne sounds. Despite these features, *Ichthyostega*, which lived about 350 MYA, was still quite fishlike in overall appearance and may have spent much of its life in water.

Order Anura	Order Caudata	Order Apoda

a.	*b.*	*c.*

Figure 35.19 Amphibians a. Red-eyed tree frog, *Agalychnis callidryas*. **b.** An adult tiger salamander, *Ambystoma tigrinum*. **c.** A caecilian, *Caecilia tentaculata*.

tropical species that don't live near water bypass this stage and hatch out as little froglets.

Unlike frogs, toads have a dry, bumpy skin and short legs, and are well adapted to dry environments. Toads do not form a monophyletic group; that is, all toads are not more closely related to each other than they are to some other frogs. Rather, the term *toad* is applied to those anurans that have adapted to dry environments by evolving a suite of adaptive characteristics; this convergent evolution has occurred many times among distantly related anurans.

Most frogs and toads return to water to reproduce, laying their eggs directly in water. Their eggs lack watertight external membranes and would dry out quickly on land. Eggs are fertilized externally and hatch into swimming larval forms called tadpoles. Tadpoles live in the water, where they generally feed on algae. After considerable growth, the body of the tadpole gradually undergoes metamorphosis into that of an adult frog.

Order Caudata: Salamanders

Salamanders have elongated bodies, long tails, and smooth, moist skin (figure 35.19*b*). They typically range in length from a few inches to a foot, although giant Asiatic salamanders of the genus *Andrias* are as much as 1.5 m long and weigh up to 33 kg. Most salamanders live in moist places, such as under stones or logs, or among the leaves of tropical plants. Some salamanders live entirely in water.

Salamanders lay their eggs in water or in moist places. Most species practice a type of internal fertilization in which the female picks up sperm packets deposited by the male. Like anurans, many salamanders go through a larval stage before metamorphosing into adults. However, unlike anurans, in which the tadpole is strikingly different from the adult frog, larval salamanders are quite similar to adults, although most live in water and have external gills and gill slits that disappear at metamorphosis.

Order Apoda: Caecilians

Caecilians, members of the order Apoda (also called Gymnophiona), are a highly specialized group of tropical burrowing amphibians (figure 35.19*c*). These legless, wormlike creatures average about 30 cm long, but can be up to 1.3 m long. They have very small eyes and are often blind. They

resemble worms but have jaws with teeth. They eat worms and other soil invertebrates. Fertilization is internal.

Learning Outcomes Review 35.6

Amphibians, which includes frogs and toads, salamanders, and caecilians, are generally characterized by legs, lungs, cutaneous respiration, and a more complex and a partially divided heart. All of these developed as adaptations to life on land. Most species rely on a water habitat for reproduction. Although some early forms reached the size of a pony, modern amphibians are generally quite small.

- **What challenges did amphibians overcome to make the transition to living on land?**
- **If reptiles evolved from amphibians, why aren't amphibians portrayed as being paraphyletic in figure 35.11?**

35.7 Reptiles

Learning Outcomes

1. *Describe the characteristics and major groups of reptiles.*
2. *Distinguish between synapsids and diapsids.*
3. *Explain the significance of the evolution of the amniotic egg.*

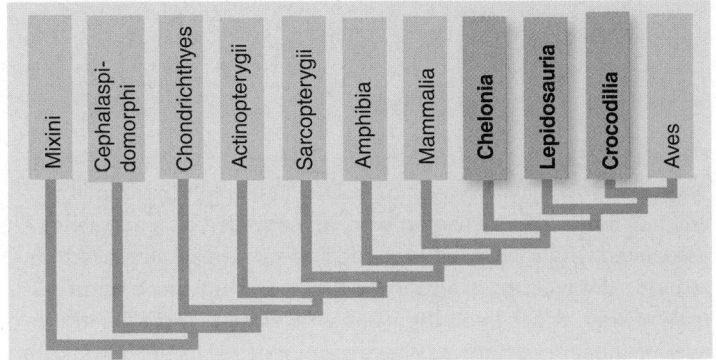

If we think of amphibians as a first draft of a manuscript about survival on land, then reptiles are the finished book. For each of the five key challenges of living on land, reptiles improved on the innovations of amphibians. The arrangement of legs evolved to support the body's weight more effectively, allowing reptile bodies to be bigger and to run. Lungs and heart became more efficient. The skin was covered with dry plates or scales to minimize water loss, and watertight coverings evolved for eggs.

Over 7000 species of reptiles (class Reptilia) now live on Earth (table 35.3). They are a highly successful group in today's world; there are more living species of snakes and lizards than there are of mammals.

TABLE 35.3	Major Orders of Reptiles		
Order	**Typical Examples**	**Key Characteristics**	**Approximate Number of Living Species**
Squamata, suborder Sauria	Lizards	Limbs set at right angles to body; anus is in transverse (sideways) slit; most are terrestrial	3800
Squamata, suborder Serpentes	Snakes	No legs; move by slithering; scaly skin is shed periodically; most are terrestrial	3000
Rhynchocephalia	Tuataras	Sole survivors of a once successful group that largely disappeared before dinosaurs; fused, wedgelike, socketless teeth; primitive third eye under skin of forehead	1
Chelonia	Turtles, tortoises, sea turtles	Armored reptiles with shell of bony plates to which vertebrae and ribs are fused; sharp, horny beak without teeth	250
Crocodylia	Crocodiles, alligators	Advanced reptiles with four-chambered heart and socketed teeth; anus is a longitudinal (lengthwise) slit; closest living relatives to birds	25
Ornithischia	Stegosaur	Dinosaurs with two pelvic bones facing backward, like a bird's pelvis; herbivores; legs under body	Extinct
Saurischia	Tyrannosaur	Dinosaurs with one pelvic bone facing forward, the other back, like a lizard's pelvis; both plant- and flesh-eaters; legs under body; birds evolved from Saurischian dinosaurs.	Extinct (except for bird descendants)
Pterosauria	Pterosaur	Flying reptiles; wings were made of skin stretched between fourth fingers and body; wingspans of early forms typically 60 cm, later forms nearly 8 m	Extinct
Plesiosaura	Plesiosaur	Barrel-shaped marine reptiles with sharp teeth and large, paddle-shaped fins; some had snakelike necks twice as long as their bodies	Extinct
Ichthyosauria	Ichthyosaur	Streamlined marine reptiles with many body similarities to sharks and modern fishes	Extinct

Reptiles exhibit three key characteristics

All living reptiles share certain fundamental characteristics, features they retain from the time when they replaced amphibians as the dominant terrestrial vertebrates. Among the most important are:

1. **Amniotic eggs.** Amphibians' eggs must be laid in water or a moist setting to avoid drying out. Most reptiles lay watertight eggs that contain a food source (the yolk) and a series of four membranes: the yolk sac, the amnion, the allantois, and the chorion (figure 35.20). Each membrane plays a role in making the egg an independent life-support system. All modern reptiles, as well as birds and mammals, show exactly this same pattern of membranes within the egg. These three classes are called **amniotes.**

   The outermost membrane of the egg is the **chorion,** which lies just beneath the porous shell. It allows exchange of respiratory gases but retains water. The **amnion** encases the developing embryo within a fluid-filled cavity. The **yolk sac** provides food from the yolk for the embryo via blood vessels connecting to the embryo's gut. The **allantois** surrounds a cavity into which waste products from the embryo are excreted.

2. **Dry skin.** Most living amphibians have moist skin and must remain in moist places to avoid drying out. Reptiles have dry, watertight skin. A layer of scales covers their bodies, greatly reducing water loss. These scales develop as surface cells fill with keratin, the same protein that forms claws, fingernails, hair, and bird feathers.

3. **Thoracic breathing.** Amphibians breathe by squeezing their throat to pump air into their lungs; this limits their breathing capacity to the volume of their mouths. Reptiles developed pulmonary breathing, expanding and contracting the rib cage and diaphragm to suck air into the lungs and then force it out. The capacity of this system is limited only by the volume of the lungs.

Reptiles dominated the Earth for 250 million years

During the 250 million years that reptiles were the dominant large terrestrial vertebrates, a series of different reptile groups appeared and then disappeared.

Synapsids

An important feature of reptile classification is the presence and number of openings behind the eyes (figure 35.21). Reptiles' jaw muscles were anchored to these holes, which allowed them to bite more powerfully. The first group to rise to dominance were the **synapsids,** whose skulls had a hole behind the openings for the eyes.

Pelycosaurs, an important group of early synapsids, were dominant for 50 million years and made up 70% of all land vertebrates; some species weighed as much as 200 kg. With long, sharp, "steak knife" teeth, these pelycosaurs were the first land vertebrates to kill beasts their own size (figure 35.22).

About 250 MYA, pelycosaurs were replaced by another type of synapsid, the therapsids (figure 35.23). Some evidence indicates that they may have been endotherms, able to produce heat internally, and perhaps even possessed hair. This would have permitted therapsids to be far more active than other vertebrates of that time, when winters were cold and long.

For 20 million years, therapsids (also called "mammal-like reptiles") were the dominant land vertebrate, until they were largely replaced 230 MYA by another group of reptiles, the diapsids. Most therapsids became extinct 170 MYA, but one group survived and has living descendants today—the mammals.

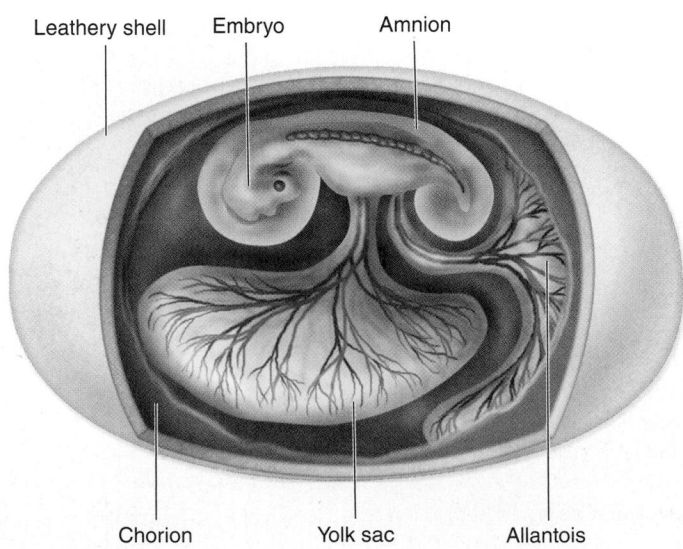

Figure 35.20 The watertight egg. The amniotic egg is perhaps the most important feature that allows reptiles to live in a wide variety of terrestrial habitats.

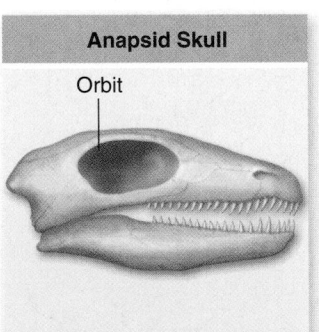

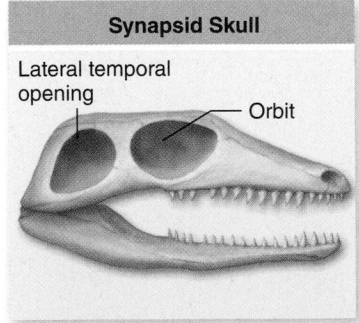

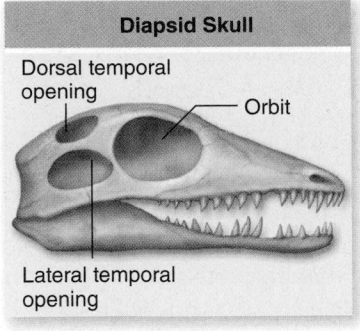

Figure 35.21 Skulls of reptile groups. Reptile groups are distinguished by the number of holes on the side of the skull behind the eye orbit: 0 (anapsids), 1 (synapsids), or 2 (diapsids). Turtles are the only living anapsids, although several extinct groups also had this condition.

Figure 35.22 A pelycosaur. *Dimetrodon,* a 4-m-long carnivorous pelycosaur, had a dorsal sail that is thought to have been used to regulate body temperature by dissipating body heat or gaining it by basking.

Figure 35.24 An early archosaur. *Euparkeria* had rows of bony plates along the sides of the backbone, as seen in modern crocodiles and alligators.

Archosaurs

Diapsids have skulls with two pairs of holes on each side of the head, and like amphibians and early reptiles, they were ectotherms. A variety of different diapsids occurred in the Triassic period (213–248 MYA), but one group, the archosaurs, were of particular evolutionary significance because they gave rise to crocodiles, pterosaurs, dinosaurs, and birds (figure 35.24).

Among the early archosaurs were the largest animals the world had seen, up to that point, and the first land vertebrates to be bipedal—to stand and walk on two feet. By the end of the Triassic period, however, one archosaur group rose to prominence: the dinosaurs.

Dinosaurs evolved about 220 MYA. Unlike previous bipedal diapsids, their legs were positioned directly underneath their bodies (figure 35.25). This design placed the weight of the body directly over the legs, which allowed dinosaurs to run with great speed and agility. Subsequently, a number of

types of dinosaur evolved to enormous size and reverted to a four-legged posture to support their massive weight. Other types—the theropods—became the most fearsome land predators the Earth has ever seen, and one theropod line evolved to become birds. Dinosaurs went on to become the most successful of all land vertebrates, dominating for more than 150 million years. All dinosaurs, except their bird descendants, became extinct rather abruptly 65 MYA, apparently as a result of an asteroid's impact.

Important characteristics of modern reptiles

As you might imagine from the structure of the amniotic egg, reptiles and other amniotes do not practice external fertilization as most amphibians do. Sperm would be unable to penetrate the membrane barriers protecting the egg. Instead, the male places sperm inside the female, where sperm fertilizes the egg before the protective membranes are formed. This is called internal fertilization.

Figure 35.23 A therapsid. This small, weasel-like cynodont therapsid, *Megazostrodon,* may have had fur. Living in the late Triassic period, this therapsid is so similar to modern mammals that some paleontologists consider it the first mammal.

Figure 35.25 Mounted skeleton of *Afrovenator*. This bipedal carnivore was about 30 feet long and lived in Africa about 130 MYA.

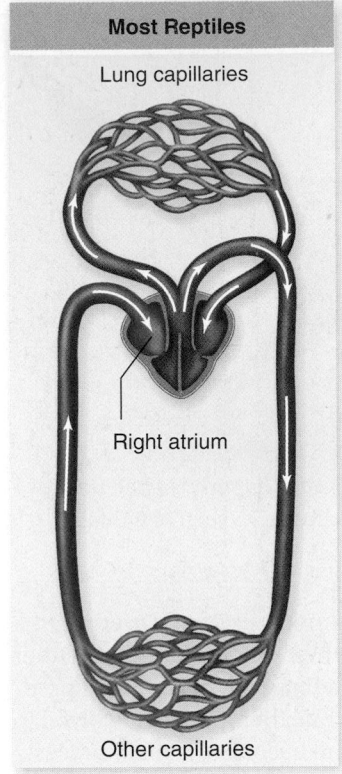

Most Reptiles

Lung capillaries

Right atrium

Other capillaries

a.

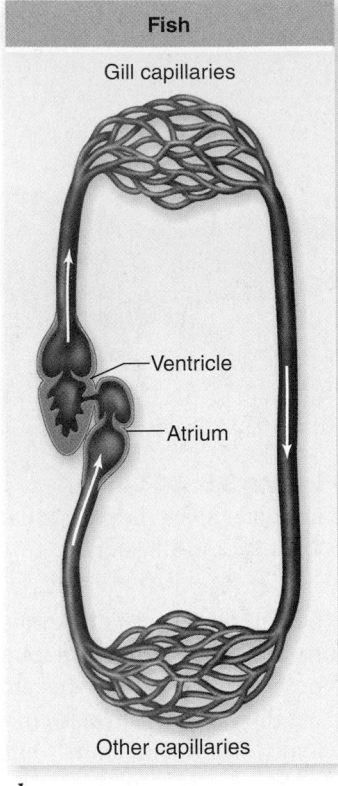

Fish

Gill capillaries

Ventricle

Atrium

Other capillaries

b.

Figure 35.26 A comparison of reptile and fish circulation. *a.* In most reptiles, oxygenated blood (red) is repumped after leaving the lungs, and circulation to the rest of the body remains vigorous. *b.* The blood in fishes flows from the gills directly to the rest of the body, resulting in slower circulation.

The circulatory system of reptiles is an improvement over that of fish and amphibians, providing oxygen to the body more efficiently (figure 35.26; see chapter 49). The improvement is achieved by extending the septum within the heart from the atrium partway across the ventricle. This septum creates a partial wall that tends to lessen mixing of oxygen-poor blood with oxygen-rich blood within the ventricle. In crocodiles, the septum completely divides the ventricle, creating a four-chambered heart, just as it does in birds and mammals (and probably did in dinosaurs).

All living reptiles are **ectothermic,** obtaining their heat from external sources. In contrast, **endothermic** animals are able to generate their heat internally (see chapter 42). Although they are ectothermic, that does not mean that reptiles cannot control their body temperature. Many species are able to precisely regulate their temperature by moving in and out of sunlight. In this way, some desert lizards can keep their bodies at a constant temperature throughout the course of an entire day. Of course, on cloudy days, or for species that live in shaded habitats, such thermoregulation is not possible, and in these cases, body temperature is the same as the temperature of the surrounding environment (see chapter 55).

Modern reptiles belong to four orders

The four surviving orders of reptiles contain about 7000 species. Reptiles occur worldwide except in the coldest regions, where it is impossible for ectotherms to survive. Reptiles are among the most numerous and diverse of terrestrial vertebrates.

Order Chelonia: Turtles and tortoises

The order Chelonia (figure 35.27*a*) consists of about 250 species of turtles (most of which are aquatic) and tortoises (which are

Figure 35.27 Living orders of reptiles. *a.* **Chelonia.** The red-bellied turtle, *Pseudemys rubriventris* (left) is shown basking, an effective means by which many ectotherms can precisely regulate their body temperature. The domed shells of tortoises, such as this Sri Lankan star tortoise, *Geochelone elegans,* provide protection against predators for these entirely terrestrial chelonians. *b.* **Tuatara (Sphenodon punctatus).** The sole living member of the ancient group Rhynchocephalia. Although they look like lizards, the common ancestor of rhynchocephalians and lizards diverged more than 250 MYA. *c.* **Squamata.** A collared lizard, *Crotaphytus collaris,* is shown left, and a smooth green snake, *Liochlorophis vernalis,* on the right. *d.* **Crocodylia.** Most crocodilians, such as the crocodile, *Crocodylus acutus,* and the gharial, *Gavialis gangeticus* (right), resemble birds and mammals in having four-chambered hearts; all other living reptiles have three-chambered hearts. Like birds, crocodiles are more closely related to dinosaurs than to any of the other living reptiles.

Order Chelonia

a.

Order Rhynchocephalia

b.

terrestrial). Turtles and tortoises lack teeth but have sharp beaks. They differ from all other reptiles because their bodies are encased within a protective shell. Many of them can pull their head and legs into the shell as well, for total protection from predators.

The shell consists of two basic parts. The carapace is the dorsal covering, and the plastron is the ventral portion. In a fundamental commitment to this shell architecture, the vertebrae and ribs of most turtle and tortoise species are fused to the inside of the carapace. All of the support for muscle attachment comes from the shell.

Whereas most tortoises have a dome-shaped shell into which they can retract their head and limbs, water-dwelling turtles have a streamlined, disc-shaped shell that permits rapid turning in water. Freshwater turtles have webbed toes; in marine turtles, the forelimbs have evolved into flippers.

Although marine turtles spend their lives at sea, they must return to land to lay their eggs. Many species migrate long distances to do this. Atlantic green turtles (*Chelonia mydas*) migrate from their feeding grounds off the coast of Brazil to Ascension Island in the middle of the South Atlantic—a distance of more than 2000 km—to lay their eggs on the same beaches where they themselves hatched.

Order Rhynchocephalia: Tuataras

Today, the order Rhynchocephalia contains a single species, the tuatara, a large, lizard-like animal about half a meter long (figure 35.27b). The only place in the world where this endangered species is found is a cluster of small islands off the coast of New Zealand. The limited diversity of modern rhynchocephalians belies their rich evolutionary past: In the Triassic period, rhynchocephalians experienced a great adaptive radiation, producing many species that differed greatly in size and habitat.

An unusual feature of the tuatara (and some lizards) is the inconspicuous "third eye" on the top of its head, called a parietal eye. Concealed under a thin layer of scales, the eye has a lens and a retina and is connected by nerves to the brain. Why have an eye, if it is covered up? The parietal eye may function to alert the tuatara when it has been exposed to too much sunlight, protecting it against overheating. Unlike most reptiles, tuataras are most active at low temperatures. They burrow during the day and feed at night on insects, worms, and other small animals.

Rhynchocephalians are the closest relatives of snakes and lizards, with whom they form the group **Lepidosauria**.

Order Squamata: Lizards and snakes

The order Squamata (figure 35.27c) includes 3800 species of lizards and about 3000 species of snakes. A distinguishing characteristic of this order is the presence of paired copulatory organs in the male. In addition, changes to the morphology of the head and jaws allow greater strength and mobility. Most lizards and snakes are carnivores, preying on insects and small animals, and these improvements in jaw design have made a major contribution to their evolutionary success.

Snakes, which evolved from a lizard ancestor, are characterized by the lack of limbs, movable eyelids, and external ears, as well as a great number of vertebrae (sometimes more than 300). Limblessness has actually evolved more than a dozen times in lizards; snakes are simply the most extreme case of this evolutionary trend.

Common lizards include iguanas, chameleons, geckos, and anoles. Most are small, measuring less than a foot in length. The largest lizards belong to the monitor family. The largest of all monitor lizards is the Komodo dragon of Indonesia, which reaches 3 m in length and can weigh more than 100 kg. Snakes also vary in length from only a few inches to more than 10 m.

Many lizards and snakes rely on agility and speed to catch prey and elude predators. Only two species of lizard are venomous, the Gila monster of the southwestern United States and the beaded lizard of western Mexico. Similarly, most species of snakes are nonvenomous. Of the 13 families of snakes, only 4 contain venomous species: the elapids (cobras, kraits, and coral snakes); the sea snakes; the vipers (adders, bushmasters, rattlesnakes, water moccasins, and copperheads); and some colubrids (African boomslang and twig snake).

Many lizards, including anoles, skinks, and geckos, have the ability to lose their tails and then regenerate a new one. This ability allows these lizards to escape from predators.

Order Crocodylia: Crocodiles and alligators

The order Crocodylia is composed of 25 species of large, primarily aquatic reptiles (figure 35.27d). In addition to crocodiles

Order Squamata

c.

Order Crocodylia

d.

and alligators, the order includes two less familiar animals: the caimans and the gavials, or gharials. Although all crocodilians are fairly similar in appearance today, much greater diversity existed in the past, including species that were entirely terrestrial and others that achieved a total length in excess of 50 feet.

Crocodiles are largely nocturnal animals that live in or near water in tropical or subtropical regions of Africa, Asia, and the Americas. The American crocodile (*Crocodylus acutus*) is found in southern Florida, Cuba, and throughout tropical Central America. Nile crocodiles (*Crocodylus niloticus*) and estuarine crocodiles (*Crocodylus porosus*) can grow to enormous size and are responsible for many human fatalities each year.

There are only two species of alligators: one living in the southern United States (*Alligator mississippiensis*) and the other a rare endangered species living in China (*Alligator sinensis*). Caimans, which resemble alligators, are native to Central America. Gharials are a group of fish-eating crocodilians with long, slender snouts that live only in India and Nepal.

All crocodilians are carnivores. They generally hunt by stealth, waiting in ambush for prey, and then attacking ferociously. Their bodies are well adapted for this form of hunting with eyes on top of their heads and their nostrils on top of their snouts, so they can see and breathe while lying quietly submerged in water. They have enormous mouths, studded with sharp teeth, and very strong necks. A valve in the back of the mouth prevents water from entering the air passage when a crocodilian feeds underwater.

In many ways, crocodiles resemble birds far more than they do other living reptiles. For example, crocodiles build nests and care for their young (traits they share with at least some dinosaurs), and they have a four-chambered heart, as birds do. Why are crocodiles more similar to birds than to other living reptiles? Most biologists agree that birds are in fact the direct descendants of dinosaurs. Both crocodiles and birds are more closely related to dinosaurs, and to each other, than they are to lizards and snakes.

Learning Outcomes Review 35.7

Reptiles have a hard, scaly skin that minimizes water loss, thoracic breathing, and an enclosed, amniotic egg that does not need to be laid in water. Synapsids had a single hole in the skull behind the eyes, and included ancestors of mammals; diapsids had two holes and are ancestors of modern reptiles and birds. The four living orders of reptiles include the turtles and tortoises, tuataras, lizards and snakes, and crocodiles.

■ *How do reptile eggs differ from those of amphibians?*

35.8 Birds

Learning Outcomes

1. *Name the key characteristics of birds.*
2. *Explain why some consider birds to be one type of reptile.*

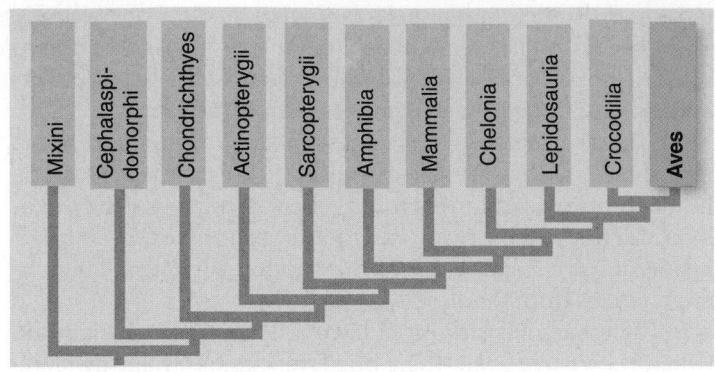

The success of birds lies in the development of a structure unique in the animal world—the feather. Developed from reptilian scales, feathers are the ideal adaptation for flight, serving as lightweight airfoils that are easily replaced if damaged (unlike the vulnerable skin wings of bats and the extinct pterosaurs). Today, birds (class **Aves**) are the most diverse of all terrestrial vertebrates, with 28 orders containing a total of 166 families and about 8600 species (table 35.4).

Key characteristics of birds are feathers and a lightweight skeleton

Modern birds lack teeth and have only vestigial tails, but they still retain many reptilian characteristics. For instance, birds lay amniotic eggs. Also, reptilian scales are present on the feet and lower legs of birds. Two primary characteristics distinguish birds from living reptiles:

1. **Feathers.** Feathers are modified reptilian scales made of keratin, just like hair and scales. Feathers serve two functions: providing lift for flight and conserving heat. The structure of feathers combines maximum flexibility and strength with minimum weight (figure 35.28).

   Feathers develop from tiny pits in the skin called follicles. In a typical flight feather, a shaft emerges from the follicle, and pairs of vanes develop from its opposite sides.

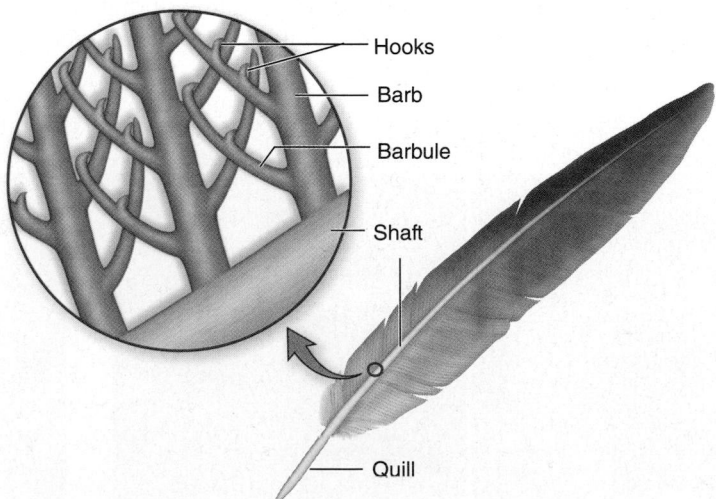

Figure 35.28 A feather. The enlargement shows how the secondary branches and barbs of the vanes are linked together by microscopic barbules.

TABLE 35.4

Major Orders of Birds

Order	Typical Examples	Key Characteristics	Approximate Number of Living Species
Passeriformes	Crows, mockingbirds, robins, sparrows, starlings, warblers	Well-developed vocal organs; perching feet; dependent young	5276
Apodiformes	Hummingbirds, swifts	Fast fliers; short legs; small bodies; rapid wing beat	428
Piciformes	Honeyguides, toucans, woodpeckers	Grasping feet; chisel-like, sharp bills can break down wood	383
Psittaciformes	Cockatoos, parrots	Large, powerful bills for crushing seeds; well-developed vocal organs	340
Charadriiformes	Auks, gulls, plovers, sandpipers, terns	Shorebirds; long, stiltlike legs; slender, probing bills	331
Columbiformes	Doves, pigeons	Perching feet; rounded, stout bodies	303
Falconiformes	Eagles, falcons, hawks	Carnivorous; keen vision; sharp, pointed beaks for tearing flesh; active during the day	288
Galliformes	Chickens, grouse, pheasants, quail	Often limited flying ability; rounded bodies	268
Gruiformes	Bitterns, coots, cranes, rails	Marsh birds; long, stiltlike legs; diverse body shapes; marsh-dwellers	209
Anseriformes	Ducks, geese, swans	Waterfowl; webbed toes; broad bill with filtering ridges	150
Strigiformes	Barn owls, screech owls	Nocturnal birds of prey; strong beaks; powerful feet	146
Ciconiiformes	Herons, ibises, storks	Waders; long-legged; large bodies	114
Procellariformes	Albatrosses, petrels	Seabirds; tube-shaped bills; capable of flying for long periods of time	104
Sphenisciformes	Emperor penguins, crested penguins	Marine; modified wings for swimming; flightless; found only in southern hemisphere; thick coats of insulating feathers	18
Dinornithiformes	Kiwis	Flightless; small; confined to New Zealand	2
Struthioniformes	Ostriches	Powerful running legs; flightless; only two toes; very large	1

At maturity, each vane has many branches called barbs. The barbs, in turn, have many projections called barbules that are equipped with microscopic hooks. These hooks link the barbs to one another, giving the feather a continuous surface and a sturdy but flexible shape.

Like scales, feathers can be replaced. Among living animals, feathers are unique to birds. Recent fossil finds suggest that some dinosaurs may have had feathers.

2. **Flight skeleton.** The bones of birds are thin and hollow. Many of the bones are fused, making the bird skeleton more rigid than a reptilian skeleton. The fused sections of backbone and of the shoulder and hip girdles form a sturdy frame that anchors muscles during flight. The power for active flight comes from large breast muscles that can make up 30% of a bird's total body weight. They stretch down from the wing and attach to the breastbone, which is greatly enlarged and bears a prominent keel for muscle attachment. Breast muscles also attach to the fused collarbones that form the so-called wishbone. No other living vertebrates have a fused collarbone or a keeled breastbone.

Birds arose about 150 MYA

A 150-million-year-old fossil of the first known bird, *Archaeopteryx* (see figures 21.9 and 35.29) was found in 1862 in a limestone quarry in Bavaria, the impression of its feathers stamped clearly into the rocks. The skeleton of *Archaeopteryx* shares many features with small theropod dinosaurs. About the size of a crow, *Archeopteryx* had a skull with teeth, and very few of its bones were fused to one another. Its bones are thought to have been solid, not hollow like a bird's. Also, it had a long, reptilian tail and no enlarged breastbone such as modern birds use to anchor flight muscles. Finally, the skeletal structure of the forelimbs was nearly identical to those of theropods.

Figure 35.29 *Archaeopteryx.* Closely related to its ancestors among the bipedal dinosaurs, the crow-sized *Archaeopteryx* lived in the forests of central Europe 150 MYA. The true feather colors of *Archaeopteryx* are not known.

Because of its many dinosaur features, several *Archaeopteryx* fossils in which the feathers had not been preserved were originally misclassified as *Compsognathus,* a small theropod dinosaur of similar size. What makes *Archaeopteryx* distinctly avian is the presence of feathers on its wings and tail.

The remarkable similarity of *Archaeopteryx* to *Compsognathus* has led almost all paleontologists to conclude that *Archaeopteryx* is the direct descendant of dinosaurs—indeed, that today's birds are "feathered dinosaurs." Some even speak flippantly of "carving the dinosaur" at Thanksgiving dinner. The recent discovery of fossils of a feathered dinosaur in China lends strong support to this inference. The dinosaur *Caudipteryx,* for example, is clearly intermediate between *Archaeopteryx* and dinosaurs, having large feathers on its tail and arms but also many features of dinosaurs like *Velociraptor* (figure 35.30 and see figure 23.11). Because the arms of

Figure 35.30 The evolutionary path to the birds. Almost all paleontologists now accept the theory that birds are the direct descendants of theropod dinosaurs.

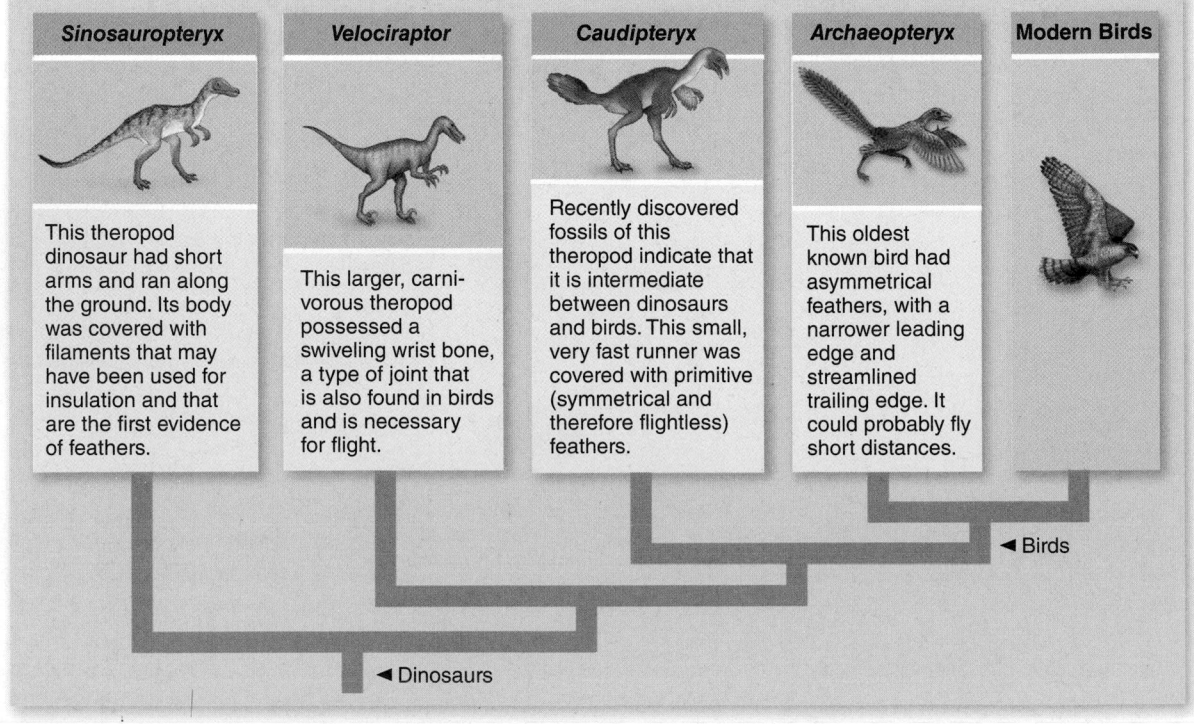

Sinosauropteryx
This theropod dinosaur had short arms and ran along the ground. Its body was covered with filaments that may have been used for insulation and that are the first evidence of feathers.

Velociraptor
This larger, carnivorous theropod possessed a swiveling wrist bone, a type of joint that is also found in birds and is necessary for flight.

Caudipteryx
Recently discovered fossils of this theropod indicate that it is intermediate between dinosaurs and birds. This small, very fast runner was covered with primitive (symmetrical and therefore flightless) feathers.

Archaeopteryx
This oldest known bird had asymmetrical feathers, with a narrower leading edge and streamlined trailing edge. It could probably fly short distances.

Modern Birds

◄ Birds

◄ Dinosaurs

Figure 35.31 A fossil bird from the early cretaceous. *Confuciornis* had long tail feathers. Some fossil specimens of this species lack the long tail feathers, suggesting that this trait was present in only one sex, as in some modern birds.

Caudipteryx were too short to use as wings, feathers probably didn't evolve for flight, but instead served as insulation, much as fur does for mammals, or perhaps as decorations used in sexual displays. This is an example of the early stages of a trait evolving for one reason, only subsequently to be further elaborated evolutionarily by selection for some other purpose, as described in Chapter 23.

Flight is an ability certain kinds of dinosaurs achieved as they evolved longer arms. We call these dinosaurs birds. Despite their close affinity to dinosaurs, birds exhibit three evolutionary novelties: feathers, hollow bones, and physiological mechanisms such as superefficient lungs that permit sustained, powered flight.

By the early Cretaceous period, only a few million years after *Archaeopteryx* lived, a diverse array of birds had evolved, with many of the features of modern birds. Fossils in Mongolia, Spain, and China discovered within the last few years reveal a diverse collection of toothed birds with the hollow bones and breastbones necessary for sustained flight (figure 35.31). Other fossils reveal highly specialized, flightless diving birds. These diverse birds shared the skies with pterosaurs for 70 million years until the flying reptiles went extinct at the end of the Cretaceous.

Because the impression of feathers is rarely fossilized and modern birds have hollow, delicate bones, the fossil record of birds is incomplete. Relationships among the 166 families of modern birds are mostly inferred from studies of the anatomy and degree of DNA similarity among living birds.

Modern birds are diverse but share several characteristics

The most ancient living birds appear to be the flightless birds, such as the ostrich. Ducks, geese, and other waterfowl evolved next, in the early Cretaceous, followed by a diverse group of woodpeckers, parrots, swifts, and owls. The largest of the bird orders, Passeriformes, evolved in the mid-Cretaceous and contains 60% of species alive today. Overall, there are 28 orders of birds, the largest consisting of over 5000 species (figure 35.32).

You can tell a great deal about the habits and food of a bird by examining its beak and feet. For instance, carnivorous birds such as owls have curved talons for seizing prey and sharp beaks for tearing apart the prey's flesh. The beaks of ducks are flat for shoveling through mud, whereas the beaks of finches are short, thick seed-crushers.

Order: Passeriformes

a.   b.   c.   d.

Figure 35.32 Diversity of *Passeriformes*, the largest order of birds. *a.* Summer tanager, *Piranga rubra*; *(b)* Indigo bunting, *Passerina cyanea*; *(c)* Stellar's jay, *Cyanositta stelleri*; *(d)* Bobolink, *Dolichonyx oryzivorus*.

Many adaptations enabled birds to cope with the heavy energy demands of flight, including respiratory and circulatory adaptations and endothermy.

Efficient respiration

Flight muscles consume an enormous amount of oxygen during active flight. The reptilian lung has a limited internal surface area, not nearly enough to absorb all the oxygen needed. Mammalian lungs have a greater surface area, but bird lungs satisfy this challenge with a radical redesign.

When a bird inhales, the air goes past the lungs to a series of air sacs located near and within the hollow bones of the back; from there, the air travels to the lungs and then to a set of anterior air sacs before being exhaled. Because air passes all the way through the lungs in a single direction, gas exchange is highly efficient. Respiration in birds is described in more detail in chapter 48.

Efficient circulation

The revved-up metabolism needed to power active flight also requires very efficient blood circulation, so that the oxygen captured by the lungs can be delivered to the flight muscles quickly. In the heart of most living reptiles, oxygen-rich blood coming from the lungs mixes with oxygen-poor blood returning from the body because the wall dividing the ventricle into two chambers is not complete. In birds, the wall dividing the ventricle is complete, and the two blood circulations do not mix—so flight muscles receive fully oxygenated blood (see chapter 49).

In comparison with reptiles and most other vertebrates, birds have a rapid heartbeat. A hummingbird's heart beats about 600 times a minute, and an active chickadee's heart beats 1000 times a minute. In contrast, the heart of the large, flightless ostrich averages 70 beats per minute—the same rate as the human heart.

Endothermy

Birds, like mammals, are endothermic (an example of convergent evolution). Many paleontologists believe the dinosaurs from which birds evolved were endothermic as well. Birds maintain body temperatures significantly higher than those of most mammals, ranging from 40° to 42°C (human body temperature is 37°C). Feathers provide excellent insulation, helping to conserve body heat.

The high temperatures maintained by endothermy permit metabolism in the bird's flight muscles to proceed at a rapid pace, providing the ATP necessary to drive rapid muscle contraction.

Learning Outcomes Review 35.8

Birds have the greatest diversity of species of all terrestrial vertebrates. *Archaeopteryx,* the oldest fossil bird, exhibited many traits shared with theropod dinosaurs. Key features of birds are feathers and a lightweight, hollow skeleton; additional features include auxiliary air sacs and a four-chambered heart.

■ *What traits do birds share with reptiles?*

35.9 Mammals

Learning Outcomes

1. *Describe the characteristics of mammals.*
2. *Compare the three groups of living mammals.*

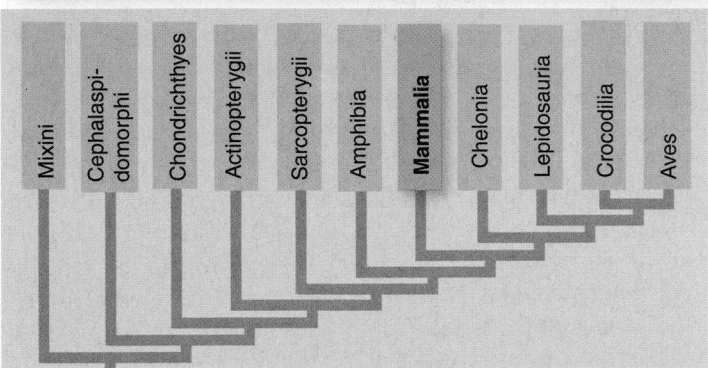

There are about 4500 living species of mammals (class Mammalia), fewer than the number of fishes, amphibians, reptiles, or birds. Most large, land-dwelling vertebrates are mammals. When we look out over an African plain, we see the big mammals—the lions, zebras, gazelles, and antelope. But the typical mammal is not that large. Of the 4500 species of mammals, 3200 are rodents, bats, shrews, or moles.

Mammals have hair, mammary glands, and other characteristics

Mammals are distinguished from all other classes of vertebrates by two fundamental characteristics—hair and mammary glands—and are marked by several other notable features:

1. **Hair.** All mammals have hair. Even apparently hairless whales and dolphins grow sensitive bristles on their snouts. The evolution of fur and the ability to regulate body temperature enabled mammals to invade colder climates that ectothermic reptiles do not inhabit. Mammals are endothermic animals and typically maintain body temperatures higher than the temperature of their surroundings. The dense undercoat of many mammals reduces the amount of body heat that escapes.

   Another function of hair is camouflage. The coloration and pattern of a mammal's coat usually matches its background. A little brown mouse is practically invisible against the brown leaf litter of a forest floor, and the orange and black stripes of a Bengal tiger disappear against the orange-brown color of the tall grass in which it hunts. Hairs also function as sensory structures. The whiskers of cats and dogs are stiff hairs that are very sensitive to touch. Mammals that are active at night or live underground often rely on their whiskers to locate prey or to avoid colliding with objects. Finally, hair can serve as a defensive weapon. Porcupines and hedgehogs protect themselves with long, sharp, stiff hairs called quills.

Unlike feathers, which evolved from modified reptilian scales, mammalian hair is a completely different form of skin structure. An individual mammalian hair is a long, protein-rich filament that extends like a stiff thread from a bulblike foundation beneath the skin known as a hair follicle. The filament is composed mainly of dead cells filled with the fibrous protein keratin.

2. **Mammary glands.** All female mammals possess mammary glands that can secrete milk. Newborn mammals, born without teeth, suckle this milk as their primary food. Even baby whales are nursed by their mother's milk. Milk is a very high-calorie food, important because of the high energy needs of a rapidly growing newborn mammal. About 50% of the energy in the milk comes from fat.

3. **Endothermy.** As stated previously, mammals are endothermic, a crucial adaptation that has allowed them to be active at any time of the day or night and to colonize extreme environments, from deserts to ice fields. Also, more efficient blood circulation provided by the four-chambered heart (see chapter 49) and more efficient respiration provided by the *diaphragm* (a special sheet of muscles below the rib cage that aids breathing; see chapter 48) make possible the higher metabolic rate on which endothermy depends.

4. **Placenta.** In most mammal species, females carry their developing young internally in a uterus, nourishing them through the placenta, and give birth to live young. The **placenta** is a specialized organ that brings the bloodstream of the fetus into close contact with the bloodstream of the mother (figure 35.33). Food, water, and oxygen can pass across from mother to child, and wastes can pass over to the mother's blood and be carried away.

In addition to these main characteristics, the mammalian lineage gave rise to several other adaptations in certain groups. These include specialized teeth, the ability of grazing animals to digest plants, hooves and horns made of keratin, and adaptations for flight in the bats.

Specialized teeth

Mammals have different types of teeth that are highly specialized to match particular eating habits (figure 35.34). It is usually possible to determine a mammal's diet simply by examining its teeth. A dog's long canine teeth, for example, are well suited for biting and holding prey, and some of its premolar and molar teeth are triangular and sharp for ripping off chunks of flesh.

In contrast, large herbivores such as deer lack canine teeth; instead, deer clip off mouthfuls of plants with flat, chisel-like incisors on its lower jaw. The deer's molars are large and covered with ridges to effectively shred and grind tough plant tissues.

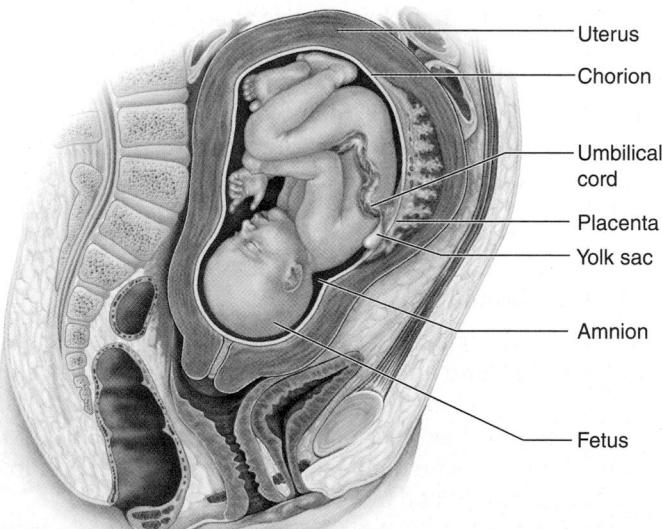

Figure 35.33 The placenta. The placenta is characteristic of the largest group of mammals, the placental mammals. It evolved from membranes in the amniotic egg. The umbilical cord evolved from the allantois. The chorion, or outermost part of the amniotic egg, forms most of the placenta itself. The placenta serves as the provisional lungs, intestine, and kidneys of the fetus, without ever mixing maternal and fetal blood.

> **?** **Inquiry question** How similar is the placenta to the egg of a reptile or bird? What can explain that similarity?

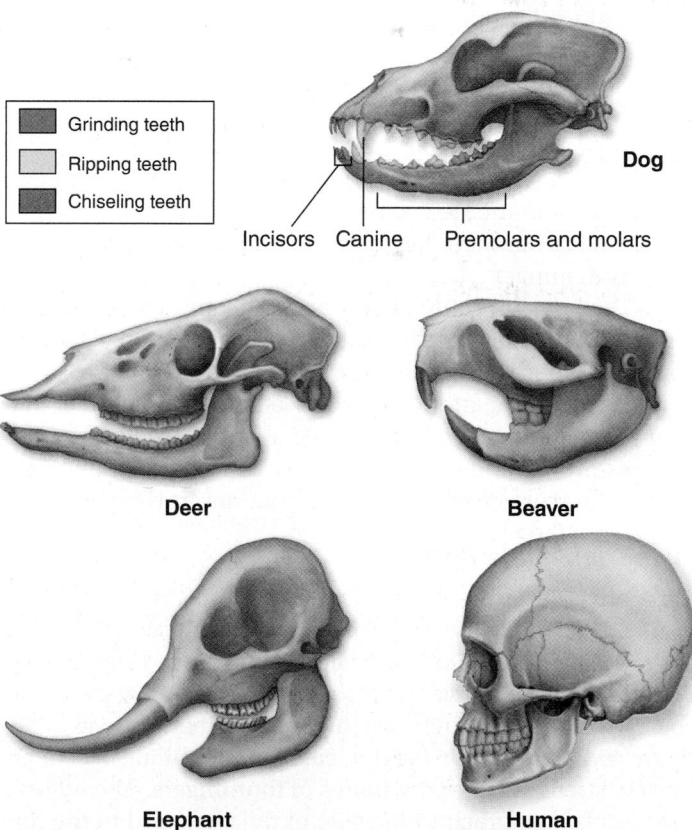

Figure 35.34 Mammals have different types of specialized teeth. Carnivores, such as dogs, have canine teeth that are able to rip food; some of the premolars and molars in dogs are also ripping teeth. Herbivores, such as deer, have incisors to chisel off vegetation and molars designed to grind up the plant material. In the beaver, the chiseling incisors dominate. In the elephant, the incisors have become specialized weapons, and molars grind up vegetation. Humans are omnivores; we have all three types: grinding, ripping, and chiseling teeth.

Digestion of plants

Most mammals are herbivores, eating mostly or only plants. Cellulose forms the bulk of a plant's body and is a major source of food for mammalian herbivores. Mammals do not have the necessary enzymes, however, for breaking the links between glucose molecules in cellulose. Herbivorous mammals rely on a mutualistic partnership with bacteria in their digestive tracts that have the necessary cellulose-splitting enzymes (see chapter 47).

Mammals such as cows, buffalo, antelopes, goats, deer, and giraffes have huge, four-chambered fermentation vats derived from the esophagus and stomach. The first chamber is the largest and holds a dense population of cellulose-digesting bacteria. Chewed plant material passes into this chamber, where the bacteria attack the cellulose. The material is then digested further in the other three chambers.

Rodents, horses, rabbits, and elephants, by contrast, have relatively small stomachs, and instead digest plant material in their large intestine, like a termite. The bacteria that actually carry out the digestion of the cellulose live in a pouch called the cecum that branches from the end of the small intestine.

Even with these complex adaptations for digesting cellulose, a mouthful of plant is less nutritious than a mouthful of meat. Herbivores must consume large amounts of plant material to gain sufficient nutrition. An elephant eats 135 to 150 kg (300–330 pounds) of plant foods each day.

Development of hooves and horns

Keratin, the protein of hair, is also the structural building material in claws, fingernails, and hooves. Hooves are specialized keratin pads on the toes of horses, cows, sheep, and antelopes. The pads are hard and horny, protecting the toe and cushioning it from impact.

The horns of cattle, sheep, and antelope are composed of a core of bone surrounded by a sheath of keratin. The bony core is attached to the skull, and the horn is not shed.

Deer antlers are made not of keratin, but of bone. Male deer grow and shed a set of antlers each year. As they grow during the summer, antlers are covered by a thin layer of skin known as velvet that provides a blood supply for the growing bone.

Flying mammals: Bats

Bats are the only mammals capable of powered flight (figure 35.35). Like the wings of birds and pterosaurs, bat wings are modified forelimbs, but even though these three types of vertebrates have convergently evolved wings, they have done so by modifying their forearms in different ways (see figure 46.22). The bat wing is a leathery membrane of skin and muscle stretched over the bones of four fingers. The edges of the membrane attach to the side of the body and to the hind leg. When resting, most bats prefer to hang upside down by their toe claws.

After rodents, bats are the second largest order of mammals. They have been a particularly successful group because many species have been able to utilize a food resource that most birds do not use—night-flying insects.

How do bats navigate in the dark? Late in the eighteenth century, the Italian biologist Lazzaro Spallanzani showed that a blinded bat could fly without crashing into things and still

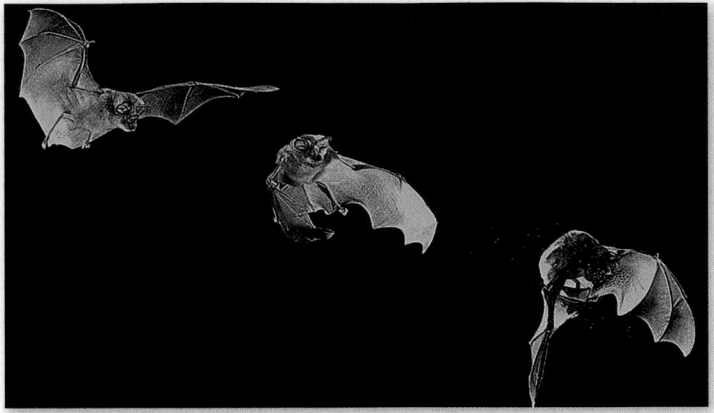

Figure 35.35 Greater horseshoe bat, *Rhinolophus ferrumequinum*. Bats are the only mammal capable of true flight.

capture insects. Clearly another sense other than vision was being used by bats to navigate in the dark. When Spallanzani plugged the ears of a bat, however, it was unable to navigate and collided with objects. Spallanzani concluded that bats "hear" their way through the night world.

Mammals diverged about 220 MYA

Mammals have been around since the time of the dinosaurs, about 220 MYA. Tiny, shrewlike creatures that lived in trees eating insects, the earliest mammals were only a minor element in a land that quickly came to be dominated by dinosaurs. Fossils reveal that these early mammals had large eye sockets, evidence that they may have been active at night. Early mammals also had a single lower jawbone. The fossil record shows a change in therapsids (the ancestors of mammals) from the reptile lower jaw, having several bones, to a jaw closer to the mammalian type. Two of the bones forming the therapsid jaw joint moved into the middle ear of mammals, linking with a bone already there to produce a three-bone structure that amplifies sound better than the reptilian ear (see figure 44.7).

The age of mammals

At the end of the Cretaceous period, 65 MYA, the dinosaurs and numerous other land and marine animals became extinct, but mammals survived, possibly because of the insulation their fur provided. In the Tertiary period (lasting from 65 MYA to 2 MYA), mammals rapidly diversified, taking over many of the ecological roles once dominated by dinosaurs.

Mammals reached their maximum diversity late in the Tertiary period, about 15 MYA. At that time, tropical conditions existed over much of the world. During the last 15 million years, world climates have changed, and the area covered by tropical habitats has decreased, causing a decline in the total number of mammalian species.

Modern mammals are placed into three groups

For 155 million years, while the dinosaurs flourished, mammals were a minor group of small insectivores and herbivores. The most primitive mammals were members of the subclass Prototheria. Most prototherians were small and resembled modern shrews. All prototherians laid eggs, as did their synapsid ancestors. The only prototherians surviving today are the monotremes.

The other major mammalian group is the subclass **Theria.** Therians are viviparous (that is, their young are born alive). The two living therian groups are marsupials, or pouched mammals (including kangaroos, opossums, and koalas), and the placental mammals (dogs, cats, humans, horses, and most other mammals).

Monotremes: Egg-laying mammals

The duck-billed platypus and two species of echidna are the only living **monotremes** (figure 35.36*a*). Among living mammals, only monotremes lay shelled eggs. The structure of their shoulder and pelvis is more similar to that of the early reptiles than to any other living mammal. Also like reptiles, monotremes have a cloaca, a single opening through which feces, urine, and reproductive products leave the body.

Despite the retention of some reptilian features, monotremes have the diagnostic mammalian characters: a single bone on each side of the lower jaw, fur, and mammary glands. Young monotremes drink their mother's milk after they hatch from eggs. Females lack well-developed nipples; instead, the milk oozes onto the mother's fur, and the babies lap it off with their tongues.

The platypus, found only in Australia, lives much of its life in the water and is a good swimmer. It uses its bill much as a duck does, rooting in the mud for worms and other soft-bodied animals. Echidnas of Australia (*Tachyglossus aculeatus*, the short-nosed echidna) and New Guinea (*Zaglossus bruijni*, the long-nosed echidna) have very strong, sharp claws, which they use for burrowing and digging. The echidna probes with its snout for insects, especially ants and termites.

Marsupials: Pouched mammals

The major difference between **marsupials** (figure 35.36*b*) and other mammals is their pattern of embryonic development. In marsupials, a fertilized egg is surrounded by chorion and amniotic membranes, but no shell forms around the egg as it does in monotremes. During most of its early development, the marsupial embryo is nourished by an abundant yolk within the egg. Shortly before birth, a short-lived placenta forms from the chorion membrane. Soon after, sometimes within eight days of fertilization, the embryonic marsupial is born. It emerges tiny and hairless, and crawls into the marsupial pouch, where it latches onto a mammary-gland nipple and continues its development.

Marsupials evolved shortly before placental mammals, about 125 MYA. Today, most species of marsupials live in Australia and South America, areas that have undergone long periods of geographic isolation. Marsupials in Australia and New Guinea have diversified to fill ecological positions occupied by placental mammals elsewhere in the world (see figure 21.18). The placental mammals in Australia and New Guinea today arrived relatively recently and include some introduced by humans. The only marsupial found in North America is the Virginia opossum, which has migrated north from Central America within the last three million years.

Figure 35.36 Today's mammals. *a.* Monotremes, the short-nosed echidna, *Tachyglossus aculeatus* (left), and the duck-billed platypus, *Ornithorhynchus anatinus* (right); *(b)* marsupials, the red kangaroo, *Macropus rufus* (left) and the opossum, *Didelphis virginiana,* (right); *(c)* placental mammals, the lion, *Panthera leo* (left) and the bottle-nosed dolphin, *Tursiops truncatus* (right).

Placental mammals

A placenta that nourishes the embryo throughout its entire development forms in the uterus of placental mammals (figure 35.36*c*). Most species of mammals living today, including humans, are in this group. Of the 19 orders of living mammals, 17 are placental mammals (although some scientists recognize four orders of marsupials, rather than one). Table 35.5 shows some of these orders. They are a very diverse group, ranging in size from 1.5-g pygmy shrews to 100,000-kg whales.

Early in the course of embryonic development, the placenta forms. Both fetal and maternal blood vessels are abundant in the placenta, and substances can be exchanged efficiently between the bloodstreams of mother and offspring (see figure 35.33). The fetal placenta is formed from the membranes of the chorion and allantois. In placental mammals, unlike in marsupials, the young undergo a considerable period of development before they are born.

TABLE 35.5	Major Orders of Placental Mammals		
Order	Typical Examples	Key Characteristics	Approximate Number of Living Species
Rodentia	Beavers, mice, porcupines, rats	Small plant-eaters; chisel-like incisor teeth	1814
Chiroptera	Bats	Flying mammals; primarily fruit- or insect-eaters; elongated fingers; thin wing membrane; mostly nocturnal; navigate by sonar	986
Insectivora	Moles, shrews	Small, burrowing mammals; insect-eaters; the most primitive placental mammals; spend most of their time underground	390
Carnivora	Bears, cats, raccoons, weasels, dogs	Carnivorous predators; teeth adapted for shearing flesh; no native families in Australia	274
Primates	Apes, humans, lemurs, monkeys	Tree-dwellers; large brain size; binocular vision; opposable thumb; group that evolved from a line that branched off early from other mammals	233
Artiodactyla	Cattle, deer, giraffes, pigs	Hoofed mammals with two or four toes Most species are herbivorous ruminants	211
Cetacea	Dolphins, porpoises, whales	Fully marine mammals; streamlined bodies; front limbs modified into flippers; no hindlimbs; blowholes on top of head; no hair except on muzzle	79
Lagomorpha	Rabbits, hares, pika	Four upper incisors (rather than the two seen in rodents); hind legs often longer than forelegs, an adaptation for jumping	69
Edentata	Anteaters, armadillos, sloths	Insectivorous; many are toothless, but some have degenerate, peglike teeth	30
Perissodactyla	Horses, rhinoceroses, tapirs	Hoofed mammals with odd number of toes; herbivorous teeth adapted for chewing	17
Proboscidea	Elephants	Long-trunked herbivores; two upper incisors elongated as tusks; largest living land animal	2

35.10 Evolution of the Primates

Learning Outcomes

1. *Describe the characteristics and major groups of primates.*
2. *List the distinguishing characteristics of hominids.*
3. *Explain the variations that form the basis for human races and why races do not represent evolutionarily distinct entities.*

Primates are the mammalian group that gave rise to our own species. Primates evolved two distinct features that allowed them to succeed as arboreal (tree-dwelling) insectivores.

1. **Grasping fingers and toes.** Unlike the clawed feet of tree shrews and squirrels, primates have grasping hands and feet that enable them to grip tree limbs, hang from branches, seize food, and in some primates, use tools. The first digit in many primates, the thumb, is opposable, and at least some, if not all, of the digits have nails.
2. **Binocular vision.** Unlike the eyes of shrews and squirrels, which sit on each side of the head, the eyes of primates are shifted forward to the front of the face. This produces overlapping binocular vision that lets the brain judge distance precisely—important to an animal moving through the trees and trying to grab or pick up food items.

Other mammals—for example, carnivorous predators—have binocular vision, but only primates have both binocular vision and grasping hands, making them particularly well adapted to their arboreal environment.

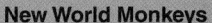

Figure 35.37 A prosimian. This tarsier, *Tarsius*, a prosimian native to tropical Asia, shows the characteristic features of primates: grasping fingers and toes and binocular vision.

The anthropoid lineage led to the earliest humans

About 40 MYA, the earliest primates split into two groups: the prosimians and the anthropoids. The **prosimians** ("before monkeys") looked something like a cross between a squirrel and a cat and were common in North America, Europe, Asia, and Africa. Only a few prosimians survive today—lemurs, lorises, and tarsiers (figure 35.37). In addition to grasping digits and binocular vision, prosimians have large eyes with increased visual acuity. Most prosimians are nocturnal, feeding on fruits, leaves, and flowers, and many lemurs have long tails for balancing.

Anthropoids

Anthropoids include monkeys, apes, and humans. Anthropoids are almost all diurnal—that is, active during the day—feeding mainly on fruits and leaves. Natural selection favored many changes in eye design, including color vision, that were adaptations to daytime foraging. An expanded brain governs the improved senses, with the braincase forming a larger portion of the head.

Anthropoids, like the relatively few diurnal prosimians, live in groups with complex social interactions. They tend to care for their young for prolonged periods, allowing for a long childhood of learning and brain development.

About 30 MYA, some anthropoids migrated to South America. Their descendants, known as the New World monkeys (figure 35.38*a*), are easy to identify: All are arboreal; they have flat, spreading noses; and many of them grasp objects with long, prehensile tails.

New World Monkeys	Old World Monkeys	Hominoids

a.   *b.*   *c.*

Figure 35.38 Anthropoids. *a.* New World Monkey, the squirrel monkey, *Saimiri oerstedii;* **(b)** Old World Monkey, the mandrill, *Mandrillus sphinx;* **(c)** hominoids, human, *Homo sapiens* (left) and gorilla, *Gorilla gorilla* (right).

Anthropoids that remained in Africa gave rise to two lineages: the Old World monkeys (figure 35.38*b*) and the hominoids (apes and humans, figure 35.38*c*). Old World monkeys include ground-dwelling as well as arboreal species. None of them have prehensile tails, their nostrils are close together, their noses point downward, and some have toughened pads of skin on their rumps for prolonged sitting.

Hominoids

The **hominoids** include the apes and the **hominids** (humans and their direct ancestors). The living apes consist of the gibbon (genus *Hylobates*), orangutan *(Pongo)*, gorilla *(Gorilla)*, and chimpanzee *(Pan)*. Apes have larger brains than monkeys, and they lack tails. With the exception of the gibbon, which is small, all living apes are larger than any monkey. Apes exhibit the most adaptable behavior of any mammal except human beings. Once widespread in Africa and Asia, apes are rare today, living in relatively small areas. No apes ever occurred in North or South America.

Studies of ape DNA have explained a great deal about how the living apes evolved. The Asian apes evolved first. The line of apes leading to gibbons diverged from other apes about 15 MYA, whereas orangutans split off about 10 MYA (figure 35.39). Neither group is closely related to humans.

The African apes evolved more recently, between 6 and 10 MYA. These apes are the closest living relatives to humans. The taxonomic group "apes" is a paraphyletic group; some apes are more closely related to hominids than they are to other apes. For this reason, some taxonomists have advocated placing humans and the African apes in the same zoological family, the Hominidae.

Fossils of the earliest hominids (humans and their direct ancestors), described later in this section, suggest that the common ancestor of the hominids was more like a chimpanzee than a gorilla. Based on genetic differences, scientists estimate that gorillas diverged from the line leading to chimpanzees and humans some 8 MYA.

Soon after the gorilla lineage diverged, the common ancestor of all hominids split off from the chimpanzee line to begin the evolutionary journey leading to humans. Because this split was so recent, few genetic differences between humans and chimpanzees have had time to evolve. For example, a human hemoglobin molecule differs from its chimpanzee counterpart in only a single amino acid. In general, humans and chimpanzees exhibit a level of genetic similarity normally found between closely related species of the same genus! In fact, when geneticists sequenced the genome of the chimpanzee, they found that it was 99% identical to the human genome (see chapter 24).

Comparing apes with hominids

The common ancestor of apes and hominids is thought to have been an arboreal climber. Much of the subsequent evolution of the hominoids reflected different approaches to locomotion. Hominids became bipedal, walking upright; in contrast, the apes evolved knuckle-walking, supporting their weight on the dorsal sides of their fingers. (Monkeys, by contrast, walk using the palms of their hands.)

Humans depart from apes in several areas of anatomy related to bipedal locomotion. Because humans walk on two

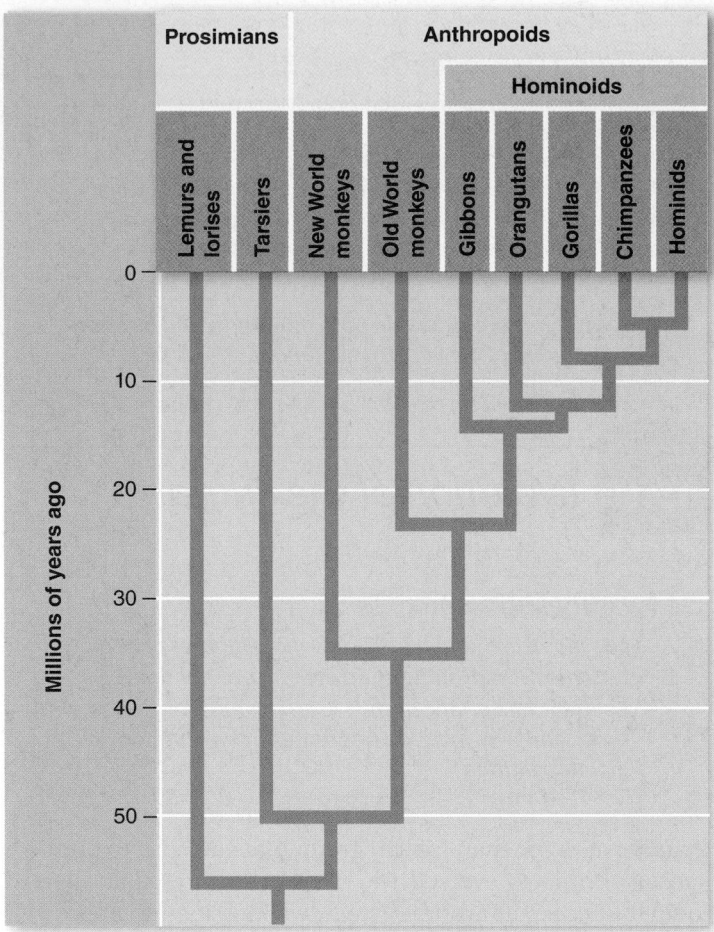

Figure 35.39 A primate evolutionary tree. Prosimians diverged early in primate evolution, whereas hominids diverged much more recently. Apes constitute a paraphyletic group because some apes are more closely related to nonape species (hominids) than they are to other apes.

legs, their vertebral column is more curved than an ape's, and the human spinal cord exits from the bottom rather than the back of the skull. The human pelvis has become broader and more bowl-shaped, with the bones curving forward to center the weight of the body over the legs. The hip, knee, and foot have all changed proportions.

Being bipedal, humans carry much of the body's weight on the lower limbs, which make up 32 to 38% of the body's weight and are longer than the upper limbs; human upper limbs do not bear the body's weight and make up only 7 to 9% of human body weight. African apes walk on all fours, with the upper and lower limbs both bearing the body's weight; in gorillas, the longer upper limbs account for 14 to 16% of body weight, the somewhat shorter lower limbs for about 18%.

Australopithecines were early hominids

Five to 10 MYA, the world's climate began to get cooler, and the great forests of Africa were largely replaced with savannas and open woodland. In response to these changes, a new kind of hominoid was evolving, one that was bipedal. These new hominoids are classified as hominids—that is, of the human line.

The major groups of hominids include three to seven species of the genus *Homo* (depending how you count them), seven species of the older, smaller-brained genus *Australopithecus,* and several even older lineages (figure 35.40). In every case where the fossils allow a determination to be made, the hominids are bipedal, the hallmark of hominid evolution.

In recent years, anthropologists have found a remarkable series of early hominid fossils extending as far back as 6 to 7 million years. Often displaying a mixture of primitive and modern traits, these fossils have thrown the study of early hominids into turmoil. Although the inclusion of these fossils among the hominids seems warranted, only a few specimens have been discovered, and they do not provide enough information to determine with certainty their relationships to australopithecines and humans. The search for additional early hominid fossils continues.

Early australopithecines

Our knowledge of australopithecines is based on hundreds of fossils, all found in South and East Africa (except for one specimen from Chad in West Africa). Australopithecines may have lived over a much broader area of Africa, but rocks of the proper age that might contain fossils are not exposed elsewhere. The evolution of hominids seems to have begun with an initial radiation of numerous species. The seven species identified so far provide ample evidence that australopithecines were a diverse group.

These early hominids weighed about 18 kg and were about 1 m tall. Their dentition was distinctly hominid, but their brains were no larger than those of apes, generally 500 cubic centimeters (cm^3) or less. *Homo* brains, by comparison, are usually larger than 600 cm^3; modern *H. sapiens* brains average 1350 cm^3.

The structure of australopithecine fossils clearly indicates that they walked upright. Evidence of bipedalism includes a set of some 69 hominid footprints found at Laetoli, East Africa. Two individuals, one larger than the other, walked upright side-by-side for 27 m, their footprints preserved in a layer of 3.7-million-year-old volcanic ash. Importantly, the big toe is not splayed out to the side as in a monkey or ape, indicating that these footprints were clearly made by hominids.

Bipedalism

The evolution of bipedalism marks the beginning of hominids. Bipedalism seems to have evolved as australopithecines left dense forests for grasslands and open woodland.

Whether larger brains or bipedalism evolved first was a matter of debate for some time. One school of thought proposed that hominid brains enlarged first, and then hominids became bipedal. Another school of thought saw bipedalism as a precursor to larger brains, arguing that bipedalism freed the forelimbs to manufacture and use tools, leading to the evolution of bigger brains. Recently, fossils unearthed in Africa have settled the debate. These fossils demonstrate that bipedalism extended back 4 million years; knee joint, pelvis, and leg bones

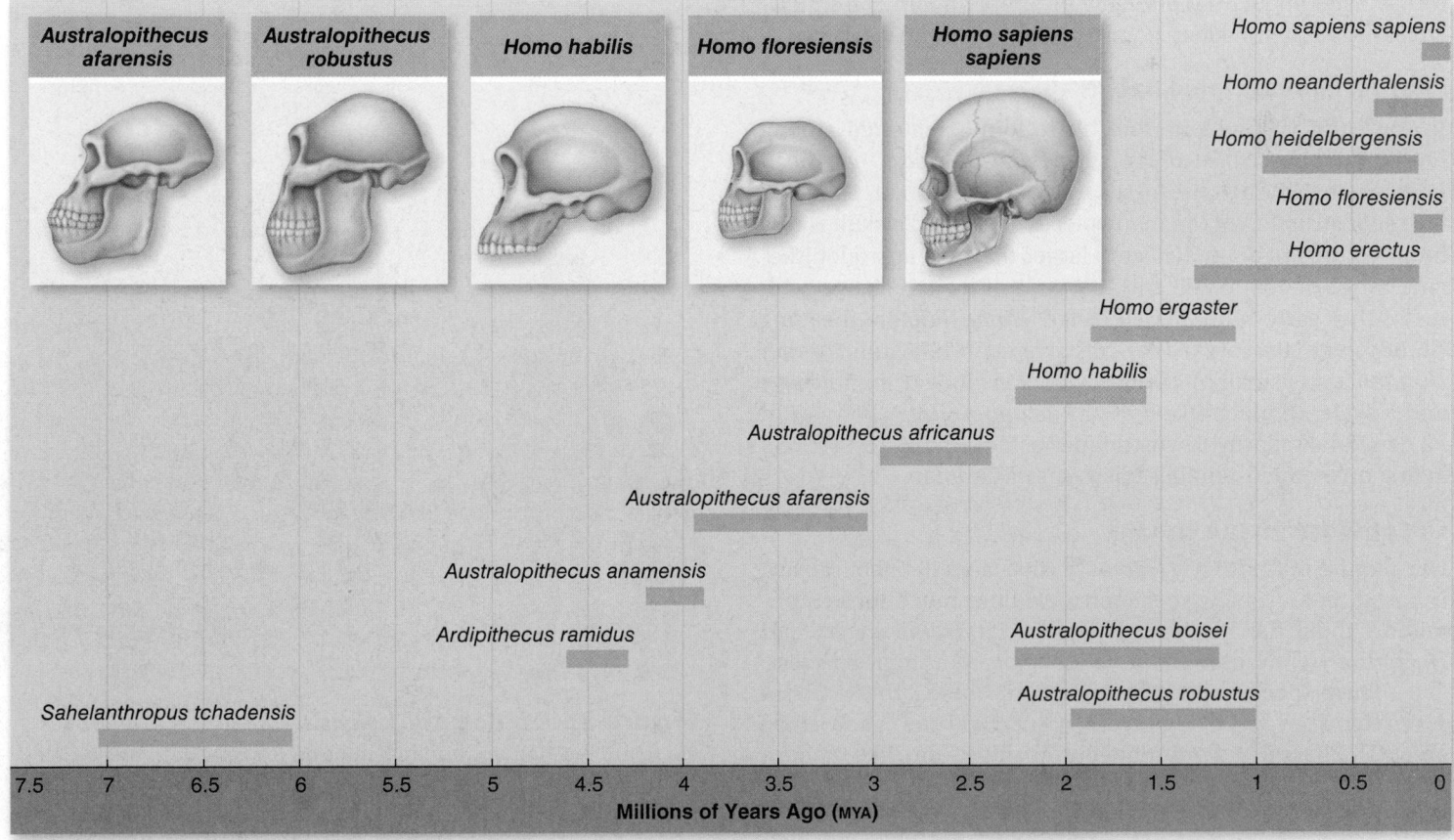

Figure 35.40 Hominid fossil history. Most, but not all, fossil hominoids are indicated here. New species are regularly being discovered, though great debate sometimes exists about whether a particular specimen represents a new species.

all exhibit the hallmarks of an upright stance. Substantial brain expansion, on the other hand, did not appear until roughly 2 MYA. In hominid evolution, upright walking clearly preceded large brains.

The reason bipedalism evolved in hominids remains a matter of controversy. No tools appeared until 2.5 MYA, so tool making seems an unlikely cause. Alternative ideas suggest that walking upright is faster and uses less energy than walking on four legs; that an upright posture permits hominids to pick fruit from trees and see over tall grass; that being upright reduces the body surface exposed to the Sun's rays; that an upright stance aided the wading of semiaquatic hominids; and that bipedalism frees the forelimbs of males to carry food back to females, encouraging pair-bonding. All of these suggestions have their proponents, and none is universally accepted. The origin of bipedalism, the key event in the evolution of hominids, remains a mystery.

The genus *Homo* arose roughly 2 MYA

The first humans (genus *Homo*) evolved from australopithecine ancestors about 2 MYA. The exact ancestor has not been clearly identified, but is commonly thought to be *Australopithecus afarensis* (the famous fossil "Lucy" was a member of this species). Only within the last 30 years have a significant number of fossils of early *Homo* been uncovered. An explosion of interest has fueled intensive field exploration, and new finds are announced regularly; every year, our picture of the base of the human evolutionary tree grows clearer. The following historical account will undoubtedly be supplanted by future discoveries, but it provides a good example of science at work.

The first human: Homo habilis

In the early 1960s, stone tools were found scattered among hominid bones close to the site where *A. boisei* had been unearthed. Although the fossils were badly crushed, painstaking reconstruction of the many pieces suggested a skull with a brain volume of about 680 cm³, larger than the australopithecine range of 400 to 550 cm³. Because of its association with tools, this early human was called *Homo habilis,* meaning "handy man." Partial skeletons discovered in 1986 indicate that *H. habilis* was small in stature, with arms longer than its legs and a skeleton much like that of *Australopithecus.* Because of its general similarity to australopithecines, many researchers at first questioned whether this fossil was human.

Out of Africa: Homo erectus

Our picture of what early *Homo* was like lacks detail because it is based on only a few specimens. We have much more information about the species that replaced it, *Homo erectus* and *H. ergaster,* which are sometimes considered a single species.

These species were a lot larger than *Homo habilis*—about 1.5 m tall. They had a large brain, about 1000 cm³, and walked erect. Their skull had prominent brow ridges and, like modern humans, a rounded jaw. Most interesting of all, the shape of the skull interior suggests that *H. erectus* was able to talk.

Far more successful than *H. habilis, H. erectus* quickly became widespread and abundant in Africa, and within 1 million years had migrated into Asia and Europe. A social species,

H. erectus lived in tribes of 20 to 50 people, often dwelling in caves. They successfully hunted large animals, butchered them using flint and bone tools, and cooked them over fires—a site in China contains the remains of horses, bears, elephants, and rhinoceroses.

Homo erectus survived for over a million years, longer than any other species of human. These very adaptable humans only disappeared in Africa about 500,000 years ago, as modern humans were emerging. Interestingly, they survived even longer in Asia, until 250,000 years ago.

A new addition to the human family: Homo floresiensis

The world was stunned in 2004 with the announcement of the discovery of fossils of a new human species from the tiny Indonesian island of Flores (figure 35.41). *Homo floresiensis* was notable for its diminutive stature; standing only a meter tall, and with a brain size of just 380 cm³, the species was quickly nicknamed "the Hobbit" after the characters in J. R. R. Tolkien's books. Just as surprising was the age of the fossils, the youngest of which was only 15,000 years old.

Despite its recency, a number of skeletal features suggest to most scientists that *H. floresiensis* is more closely related to *H. erectus* than to *H. sapiens.* If correct (and not all scientists agree), this result would indicate that the *H. erectus* lineage persisted much longer than previously thought—almost to the present day. It also would mean that until very recently, *H. sapiens* was not the only species of human on the planet. We

Figure 35.41 *Homo floresiensis.* This diminutive species (compare the modern human female on the right with the female *Homo floresiensis* on the left) occurred on the small island of Flores in what is now Indonesia. *H. floresiensis* preyed upon a dwarf species of elephant, *Stegodon sondaari,* which also occurred on Flores (compare with the larger African elephant, *Loxodonta africana,* in gray).

can only speculate about how *H. sapiens* and *H. floresiensis* may have interacted, and how these interactions may have been affected by the great difference in body size.

Why *H. floresiensis* evolved such small size is unknown, although a number of experts have pointed to the phenomenon of "island dwarfism," in which mammal species evolve to be much smaller on islands. Indeed, *H. floresiensis* coexisted with and preyed on a miniature species of elephant that also lived on Flores, but which also has gone extinct. These findings have rekindled interest in explaining why island dwarfism occurs.

Modern humans

The evolutionary journey entered its final phase when modern humans first appeared in Africa about 600,000 years ago. Investigators who focus on human diversity denote three species of modern humans: *Homo heidelbergensis, H. neanderthalensis,* and *H. sapiens* ("wise man").

The oldest modern human, *Homo heidelbergensis,* is known from a 600,000-year-old fossil from Ethiopia. Although it coexisted with *H. erectus* in Africa, *H. heidelbergensis* has more advanced anatomical features, including a bony keel running along the midline of the skull, a thick ridge over the eye sockets, and a large brain. Also, its forehead and nasal bones are very much like those of *H. sapiens.*

As *H. erectus* was becoming rarer, about 130,000 years ago, a new species of human arrived in Europe from Africa. *Homo neanderthalensis* likely branched off the ancestral line leading to modern humans as long as 500,000 years ago. Compared with modern humans, Neanderthals were short, stocky, and powerfully built; their skulls were massive, with protruding faces, heavy, bony ridges over the brows, and larger braincases.

Cro-Magnons and Neanderthals

The Neanderthals (classified by many paleontologists as a separate species, *Homo neanderthalensis*) were named after the Neander Valley of Germany where their fossils were first discovered in 1856. Rare at first outside of Africa, they became progressively more abundant in Europe and Asia, and by 70,000 years ago had become common.

The Neanderthals made diverse tools, including scrapers, spearheads, and hand axes. They lived in huts or caves. Neanderthals took care of their injured and sick and commonly buried their dead, often placing food, weapons, and even flowers with the bodies. Such attention to the dead strongly suggests that they believed in a life after death. This is the first evidence of the symbolic thinking characteristic of modern humans.

Fossils of *H. neanderthalensis* abruptly disappear from the fossil record about 34,000 years ago and are replaced by fossils of *H. sapiens* called the Cro-Magnons (named after the valley in France where their fossils were first discovered). Scientists have long debated whether Cro-Magnons outcompeted Neanderthals, or whether the two species interbred, merging their gene pools. Recent analyses of Neanderthal DNA reveal it to be quite distinct from Cro-Magnon DNA, indicating that the two species were quite distinct. Nonetheless, detailed comparison of the Neanderthal genome with that of humans from around the world reveals that some human populations contain bits of Neanderthal DNA in their genomes, evidence for at least some interbreeding in the distant past.

A variety of evidence indicates that Cro-Magnons came from Africa—fossils of essentially modern aspect but as much as 100,000 years old have been found there. Cro-Magnons seem to have replaced the Neanderthals completely in the Middle East by 40,000 years ago, and then spread across Europe, coexisting with the Neanderthals for several thousand years. The Cro-Magnons that replaced the Neanderthals had a complex social organization and are thought to have had full language capabilities. Elaborate and often beautiful cave paintings made by Cro-Magnons can be seen throughout Europe (figure 35.42).

Humans of modern appearance eventually spread across Siberia to North America, where they arrived at least 13,000 years ago, after the ice had begun to retreat and a land bridge still connected Siberia and Alaska. By 10,000 years ago, about 5 million people inhabited the entire world (compared with more than 6 billion today).

Our own species: Homo sapiens

Homo sapiens is the only surviving species of the genus *Homo* and indeed the only surviving hominid. Some of the best fossils of *H. sapiens* are 20 well-preserved skeletons with skulls found in a cave near Nazareth in Israel. Modern dating techniques estimate these humans to be between 90,000 and 100,000 years old. The skulls are modern in appearance and size, with high, short braincases, vertical foreheads with only slight brow

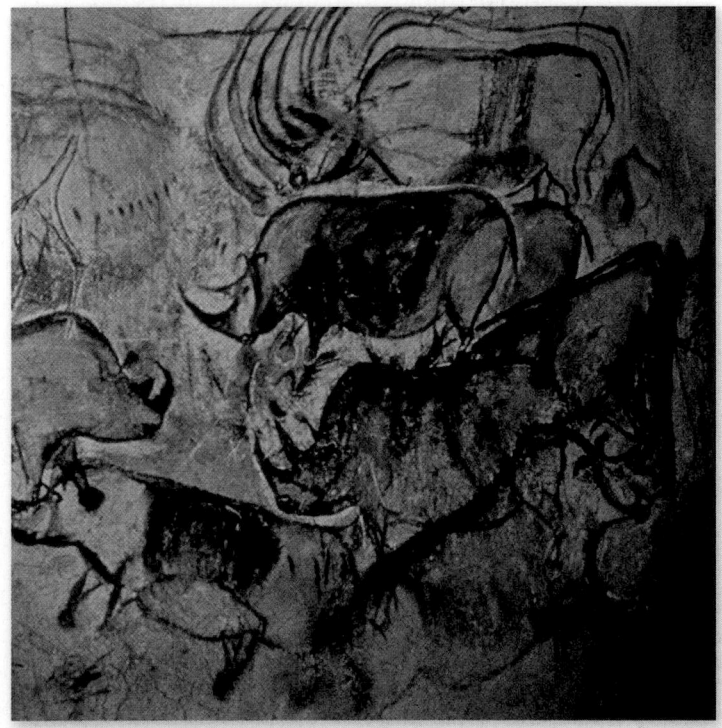

Figure 35.42 Cro-Magnon art. Rhinoceroses are among the animals depicted in this remarkable cave painting found in 1995 near Vallon-Pont d'Arc, France.

ridges, and a cranial capacity of roughly 1550 cm³. Our evolution has been marked by a progressive increase in brain size, distinguishing us from other animals in several ways. First, humans are able to make and use tools more effectively than any other animal—a capability that, more than any other factor, has been responsible for our dominant position in the world. Second, although not the only animal capable of conceptual thought, humans have refined and extended this ability until it has become the hallmark of our species. Finally, we use symbolic language and can, with words, shape concepts out of experience and transmit that accumulated experience from one generation to another.

Humans have undergone what no other animal ever has: extensive cultural evolution. Through culture, we have found ways to change and mold our environment, rather than changing evolutionarily in response to the environment's demands. We control our biological future in a way never before possible—an exciting potential and a frightening responsibility.

Human races

Human beings, like all other species, have differentiated in their characteristics as they have spread throughout the world. Local populations in one area often appear significantly different from those that live elsewhere. For example, northern Europeans often have blond hair, fair skin, and blue eyes, whereas Africans often have black hair, dark skin, and brown eyes. These traits may play a role in adapting the particular populations to their environments. Blood groups may be associated with immunity to diseases more common in certain geographical areas, and dark skin shields the body from the damaging effects of ultraviolet radiation, which is much stronger in the tropics than in temperate regions.

All human beings are capable of mating with one another and producing fertile offspring. The reasons that they do or do not choose to associate with one another are purely psychological and behavioral (cultural).

The number of groups into which the human species might logically be divided has long been a point of contention. Some contemporary anthropologists divide people into as many as 30 "races," others as few as three: Caucasoid, Negroid, and Oriental. American Indians, Bushmen, and Aborigines are examples of particularly distinctive subunits that are sometimes regarded as distinct groups.

The problem with classifying people or other organisms into races in this fashion is that the characteristics used to define the races are usually not well correlated with one another, and so the determination of race is always somewhat arbitrary. Humans are visually oriented; consequently, we have relied on visual cues—primarily skin color—to define races. However, when other types of characteristics, such as blood groups, are examined, patterns of variation correspond very poorly with visually determined racial classes. Indeed, if one were to break the human species into subunits based on overall genetic similarity, the groupings would be very different from those based on skin color or other visual features (figure 35.43).

In human beings, it is simply not possible to delimit clearly defined races that reflect biologically differentiated and well-defined groupings. The reason is simple: Different groups of people have constantly intermingled and interbred with one another during the entire course of history. This constant gene flow has prevented the human species from fragmenting into highly differentiated subspecies. Those characteristics that are differentiated among populations, such as skin color, represent classic examples of the antagonism between gene flow and natural selection. As you saw in chapter 20, when selection is strong enough, as it is for dark coloration in tropical regions, populations can differentiate even in the presence of gene flow. However, even in cases

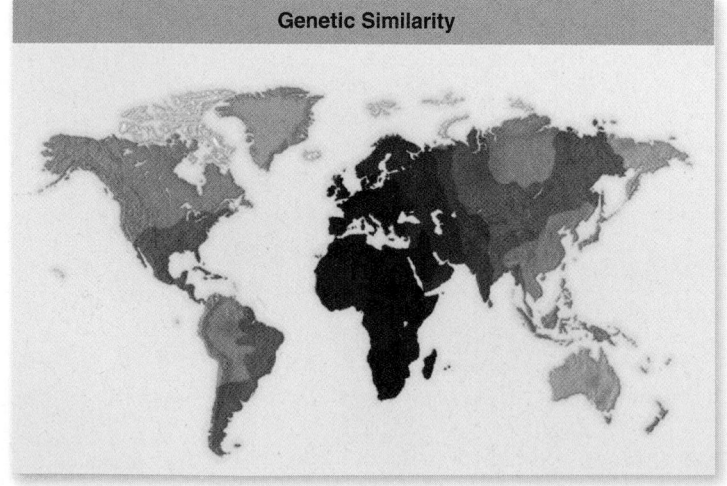

a.

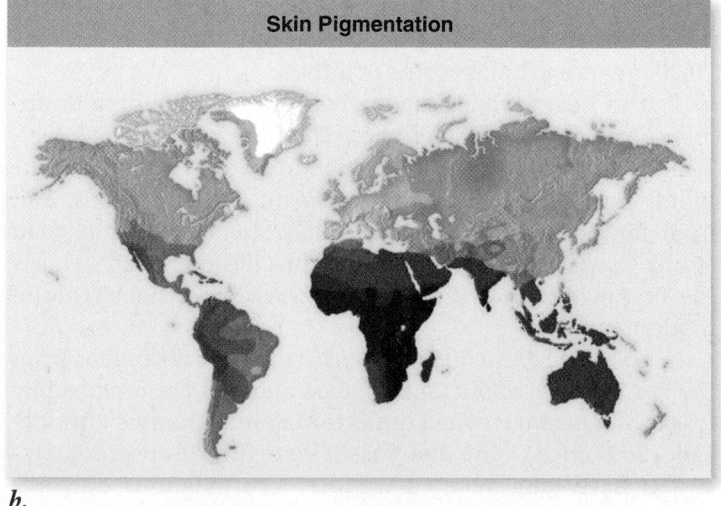

b.

Figure 35.43 **Patterns of genetic variation in human populations differ from patterns of skin color variation.** *a.* Genetic variation among *Homo sapiens*. The more similar areas are in color, the more similar they are genetically based on many enzyme and blood group genetic loci. *b.* Similarity among *Homo sapiens* based on skin color. The color of an area represents the skin pigmentation of the people native to that region.

such as this, gene flow will still ensure that populations are relatively homogeneous for genetic variation at other loci.

For this reason, relatively little of the variation in the human species represents differences between the described races. Indeed, one study calculated that only 8% of all genetic variation among humans could be accounted for as differences that exist among racial groups; in other words, the human racial categories do a very poor job in describing the vast majority of genetic variation that exists in humans. For this reason, most modern biologists reject human racial classifications as a means of reflecting patterns of biological differentiation in the human species. This is a sound biological basis for dealing with each human being on his or her own merits and not as a member of a particular "race."

Learning Outcomes Review 35.10

Primates include prosimians, monkeys, apes, and humans (hominids). Primates have grasping fingers and toes and binocular vision. Hominids diverged from other primates by developing bipedal locomotion that led to a number of additional adaptations, including modification of the spine, pelvis, and limbs. Several species of *Homo* evolved in Africa, and some migrated from there to Europe and Asia. Our own species, *Homo sapiens*, is proficient at conceptual thought and tool use and is the only animal that uses symbolic language. Considerable variation occurs among human populations, but the recognized races, which are based primarily on skin color, do not reflect significant biological differences.

■ *Which of these groups is monophyletic: prosimians, monkeys, apes, hominids?*

Chapter Review

35.1 Echinoderms

Echinoderms are ancient and unmistakable.
Echinoderms are characterized by deuterostome development, pentameral symmetry, an endoskeleton covered by a delicate epidermis, and a water-vascular system.

Echinoderm symmetry is bilateral in larvae but pentaradial in adults.
The larvae of echinoderms are bilateral, but the adult is pentaradial.

The water-vascular system, which is one of several coelomic compartments, aids in movement, feeding, circulation, respiration, and excretion.

Echinodermata contains five extant classes.
The five classes of echinoderms are Asteroidea (sea stars), Crinoidea (sea lilies), Echinoidea (sea urchins and sand dollars), Holothuroidea (sea cucumbers), and Ophiuroidea (brittle stars).

35.2 Chordates
Chordates share four features at some time during development: a single, hollow nerve cord; a flexible rod, the notochord; pharyngeal slits or pouches; and a postanal tail (figure 35.4).

35.3 Nonvertebrate Chordates

Tunicates have chordate larval forms.
Tunicates have a swimming larval form exhibiting all the features of a chordate, but their adult form is sessile and baglike.

Lancelets are small marine chordates.
Lancelets have chordate features throughout life, but as adults they lack bones and have no distinct head (figure 35.6).

35.4 Vertebrate Chordates

Vertebrates have vertebrae, a distinct head, and other features.
In vertebrates, a vertebral column encloses and protects the dorsal nerve cord. The distinct and well-differentiated head carries sensory organs. Vertebrates also have specialized internal organs and a bony or cartilaginous endoskeleton.

Vertebrates evolved half a billion years ago: An overview.
In fishes, evolution of a hinged jaw was a major advance. Other changes allowed a move into the terrestrial environment, giving rise to amphibians, reptiles, birds, and mammals (figure 35.11).

35.5 Fishes

Fishes exhibit five key characteristics.
The key characteristics of fishes are a vertebral column of bone or cartilage, jaws, paired appendages, internal gills, and a closed circulatory system.

Sharks, with cartilaginous skeletons, became top predators.
Sharks, rays, and skates are cartilaginous fishes. Sharks were streamlined for fast swimming, and they evolved teeth that enabled them to readily grab, kill, and devour prey.

The lateral line system of sharks and bony fishes is a sensory system that detects changes in pressure waves.

Bony fishes dominate the waters.
Bony fishes belong either to the ray-finned fishes (Actinopterygii), or the lobe-finned fishes (Sarcopterygii). Ray-finned fishes have fins stiffened with bony parallel rays.

The evolutionary path to land ran through the lobe-finned fishes.
The lobe-finned fishes have muscular lobes with bones connected by joints (figure 35.16). These structures could evolve into limbs capable of movement on land.

35.6 Amphibians

Living amphibians have five distinguishing features.
Amphibian adaptations include legs, lungs, cutaneous respiration, pulmonary veins, and a partially divided heart.

Amphibians overcame terrestrial challenges.
Terrestrial adaptations included being able to support large bodies against gravity, to breathe out of water, and to avoid desiccation.

Modern amphibians belong to three groups.
The Anura (frogs and toads) lack tails as adults; many have a larval tadpole stage. The Caudata (salamanders) have tails as adults and larvae similar to the adult form. The Apoda (caecilians) are legless.

35.7 Reptiles

Reptiles exhibit three key characteristics.

Reptiles posses a watertight amniotic egg; dry, watertight skin; and thoracic breathing (figure 35.20).

Modern reptiles practice internal fertilization and are ectothermic.

Reptiles dominated the Earth for 250 million years.

Synapsids gave rise to the therapsids that became the mammalian line. Diapsids gave rise to modern reptiles and the birds.

Modern reptiles belong to four orders.

The four orders of reptiles are Chelonia (turtles and tortoises); Rhynchocephalia (tuataras); Squamata (lizards and snakes); and Crocodylia (crocodiles and alligators).

35.8 Birds

Key characteristics of birds are feathers and a lightweight skeleton.

The feather is a modified reptilian scale. Feathers provide lift in gliding or flight and conserve heat (figure 35.28).

The lightweight skeleton of birds is an adaptation to flight.

Birds arose about 150 MYA.

Birds evolved from theropod dinosaurs. Feathers probably first arose to provide insulation, only later being modified for flight.

Modern birds are diverse but share several characteristics.

In addition to the key characteristics, birds have efficient respiration and circulation and are endothermic.

35.9 Mammals

Mammals have hair, mammary glands, and other characteristics.

Mammals are distinguished by fur and by mammary glands, which provide milk to feed the young. Mammals are also endothermic.

Mammals diverged about 220 MYA.

Mammals evolved from therapsids (synapsids) and reached maximum diversity about 15 MYA.

Modern mammals are placed into three groups.

The monotremes lay shelled eggs. In marsupials, an embryo completes development in a pouch. Placental mammals produce a placenta in the uterus to nourish the embryo.

35.10 Evolution of the Primates (figure 35.39)

Primates share two innovations: grasping fingers and toes, and binocular vision.

The anthropoid lineage led to the earliest humans.

The earliest primates were the prosimians; anthropoids, which include monkeys, apes, and humans, evolved later. Hominoids include the apes and the hominids, or humans.

Austrolopithecines were early hominids.

The distinguishing characteristics of hominids are upright posture and bipedal locomotion.

The genus Homo arose roughly 2 MYA.

Common features of early *Homo* species include a larger body and brain size. *Homo sapiens* is the only extant species. Humans exhibit conceptual thought, tool use, and symbolic language.

Review Questions

UNDERSTAND

1. Which of the following structures is not a component of the water-vascular system of an echinoderm?

    a. Ossicles   c. Radial canals
    b. Ampullae   d. Madreporites

2. Which of the following statements regarding all species of chordates is false?

    a. Chordates are deuterostomes.
    b. A notochord is present in the embryo.
    c. The notochord is surrounded by bone or cartilage.
    d. All possess a postanal tail during embryonic development.

3. In the figure, item A is the _____ and item B is the _____.

    a. complete digestive system; notochord
    b. spinal cord; nerve cord
    c. notochord; nerve cord
    d. pharyngeal slits; notochord

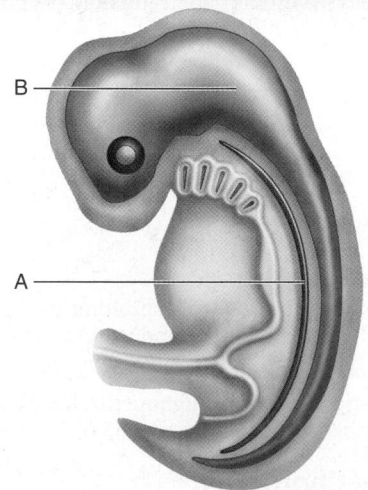

4. During embryonic development, a neural crest would be found in all of the following chordates, except

    a. cephalochordates.   c. birds.

    b. reptiles.   d. mammals.

5. The ___ of the bony fish evolved to counter the effects of increased bone density.

    a. gills   c. swim bladder

    b. jaws   d. teeth

6. Why was the evolution of the pulmonary veins important for amphibians?

    a. To move oxygen to and from the lungs

    b. To increase the metabolic rate

    c. For increased blood circulation to the brain

    d. None of the choices are correct.

7. Which of the following groups lacks a four-chambered heart?

    a. Birds   c. Mammals

    b. Crocodilians   d. Amphibians

8. All of the following are characteristics of reptiles, except

    a. cutaneous respiration.   c. thoracic breathing.

    b. an amniotic egg.   d. dry, watertight skin.

9. Which of the following evolutionary adaptations allows the birds to become efficient at flying?

    a. Structure of the feather

    b. High metabolic temperatures

    c. Increased respiratory efficiency

    d. All of the choices are correct.

APPLY

1. The reason that birds and crocodilians both build nests might be because they

    a. are both warm-blooded.

    b. both eat fish.

    c. both inherited the trait from a common ancestor.

    d. both are ectothermic.

2. Which of the following is the closest relative of lungfish?

    a. Hagfish   c. Ray-finned fish

    b. Sharks   d. Mammals

3. The fact that monotremes lay eggs

    a. indicates that they are more closely related to some reptiles than they are to some mammals.

    b. is a plesiomorphic trait.

    c. demonstrates that the amniotic egg evolved multiple times.

    d. is a result of ectothermy.

SYNTHESIZE

1. Some scientists believe the feathers did not evolve initially for flight, but rather for insulation. What benefits would this have had for early flightless birds?

2. Some people state that the dinosaurs have not "gone extinct," they are with us today. What evidence can be used to support this statement?

3. In what respect is the evolutionary diversification of hominids similar to that of horses discussed in chapter 21?

ONLINE RESOURCE

www.ravenbiology.com

Understand, Apply, and Synthesize—enhance your study with animations that bring concepts to life and practice tests to assess your understanding. Your instructor may also recommend the interactive eBook, individualized learning tools, and more.

Part **VI** Plant Form and Function

Chapter **36**

Plant Form

Chapter Contents

Introduction

Although the similarities among a cactus, an orchid, and a hardwood tree might not be obvious at first sight, most plants have a basic unity of structure. This unity is reflected in how the plants are constructed; in how they grow, manufacture, and transport their food; and in how their development is regulated. This chapter addresses the question of how a vascular plant is "built." We will focus on the cells, tissues, and organs that compose the adult plant body. The roots and shoots that give the adult plant its distinct above- and below-ground architecture are the final product of a basic body plan first established during embryogenesis, a process we will explore in detail in this chapter.

Organization of the Plant Body: An Overview

Learning Outcomes

1. *Describe the functions of three types of tissues in a plant.*
2. *Explain why meristems are vital after germination.*
3. *Compare the origins of primary growth and secondary growth.*

As you learned in chapters 30 and 31, the plant kingdom has great diversity, not only among its many phyla but even within species. The earliest vascular plants, many of which are extinct, did not have a clear differentiation of the plant body into specialized organs such as roots and leaves.

Among modern vascular plants, the presence of these organs reflects increasing specialization, particularly in relation to the demands of a terrestrial existence. Obtaining water, for example, is a major challenge, and roots are adapted for water absorption from the soil. Leaves, roots, branches, and flowers all exhibit variations in size and number from plant to plant. Development of the form and structure of these parts may be precisely controlled, but some aspects of leaf, stem, and root development are quite flexible. This chapter emphasizes the unifying aspects of plant form, using the flowering plants as a model.

Vascular plants have roots and shoots

A vascular plant consists of a root system and a shoot system (figure 36.1). Roots and shoots grow at their tips, which are called apices (singular, **apex**).

The **root system** anchors the plant and penetrates the soil, from which it absorbs water and ions crucial for the plant's nutrition. Root systems are often extensive, and growing roots can exert great force to move matter as they elongate and expand. Roots evolved later than the shoot system as an adaptation to living on land.

The **shoot system** consists of the stems and their leaves. Stems serve as a scaffold for positioning the leaves, the principal sites of photosynthesis. The arrangement, size, and other features of the leaves are critically important in the plant's production of food. Flowers, other reproductive organs, and ultimately, fruits and seeds are also formed on the shoot (flower morphology and plant reproduction is covered in chapter 41).

The repeating unit of the vegetative shoot consists of the internode, node, leaf, and axillary bud, but not reproductive structures. An axillary bud is a lateral shoot apex that allows the plant to branch or replace the main shoot if it is eaten by an herbivore. A vegetative axillary bud has the capacity to reiterate the development of the primary shoot. When the plant has shifted to the reproductive phase of development, these axillary buds may produce flowers or floral shoots.

Roots and shoots are composed of three types of tissues

Plant cell types can be distinguished by the size of their vacuoles, whether they are living or not at maturity, and by the thickness of secretions found in their cellulose cell walls, a distinguishing feature of plant cells (see chapter 4 to review cell structure). Some cells have only a primary cell wall of cellulose, synthesized by the protoplast near the cell membrane. Microtubules align within the cell and determine the orientation of

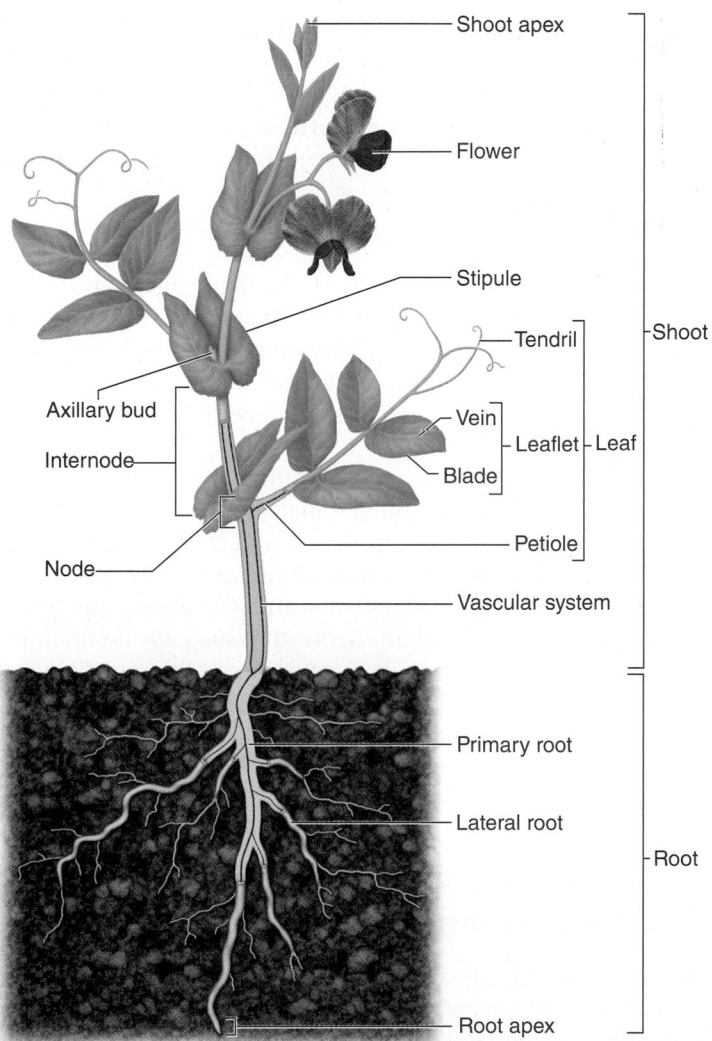

Figure 36.1 Diagram of a plant body. Branching root and shoot systems create the plant's architecture. Each root and shoot has an apex that extends growth. Leaves are initiated at the nodes of the shoot, which also contain axillary buds that can remain dormant, grow to form lateral branches, or make flowers. A leaf can be a simple blade or consist of multiple parts as shown here. Roots, shoots, and leaves are all connected with vascular (conducting) tissue.

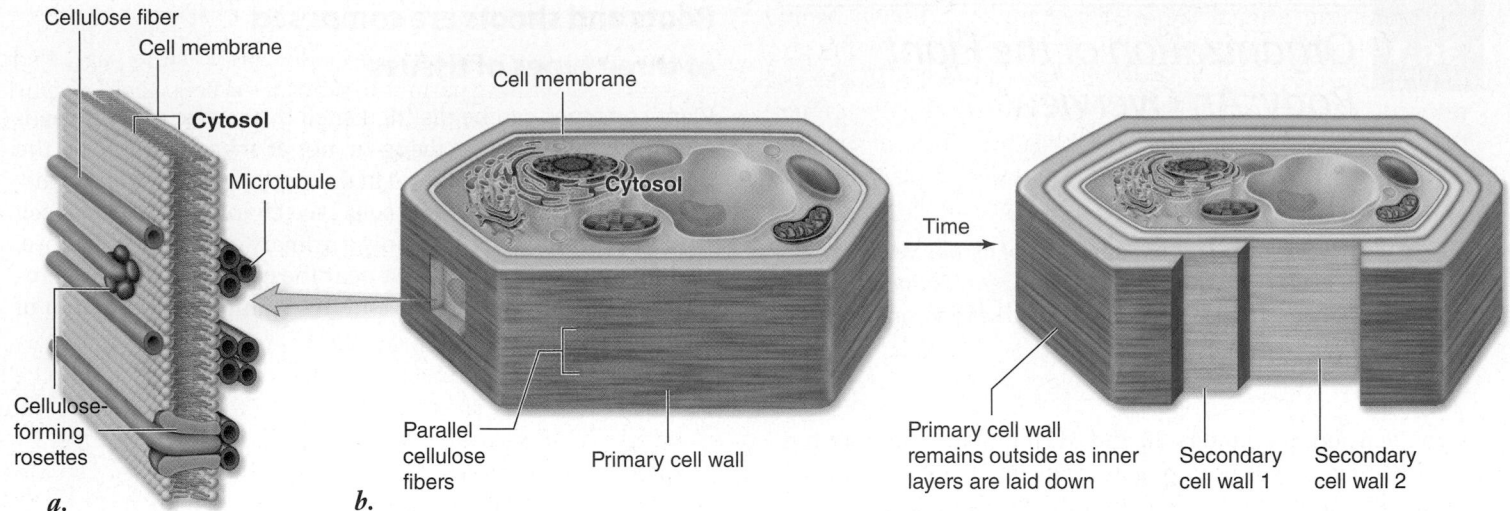

Figure 36.2 Synthesis of a plant cell wall. *a.* Cellulose is a glucose polymer that is produced at the cellulose-forming rosettes in the cell membrane to form the cell wall. Cellulose fibers are laid down parallel to microtubules inside the cell membrane. Additional substances that strengthen and waterproof the cell wall are added to the cell wall in some cell types. *b.* Some cells extrude additional layers of cellulose, increasing the mechanical strength of the wall. Because new cellulose is produced at the cell, the oldest layers of cellulose are on the outside of the cell wall. All cells have a primary cell wall. Additional layers of cellulose contribute to the secondary cell wall.

the cellulose fibers (figure 36.2*a*). Cells that support the plant body have more heavily reinforced cell walls with multiple layers of cellulose and other strengthening molecules, including lignin and pectin. Cellulose layers are laid down at angles to adjacent layers like plywood; this enhances the strength of the cell wall (figure 36.2*b*).

Roots, shoots, and leaves all contain three basic types of tissues: dermal, ground, and vascular tissue. Because each of these tissues extend through the root and shoot systems, they are called **tissue systems. Dermal tissue,** primarily *epidermis,* is one cell layer thick in most plants, and it forms an outer protective covering for the plant. **Ground tissue** cells function in storage, photosynthesis, and secretion, in addition to forming fibers that support and protect plants. **Vascular tissue** conducts fluids and dissolved substances throughout the plant body. Each of these tissues and their many functions are described in more detail in later sections.

Meristems elaborate the body plan throughout the plant's life

When a seed sprouts, only a tiny portion of the adult plant exists. Although embryo cells can undergo division and differentiation to form many cell types, the fate of most adult cells is more restricted. Further development of the plant body depends on the activities of **meristems** found in shoot and root apices, as well as other parts of the plant. Meristem cells are undifferentiated cells that can divide indefinitely and give rise to many types of differentiated cells.

Overview of meristems

Meristems are clusters of small cells with dense cytoplasm and proportionately large nuclei that act as stem cells do in animals. That is, one cell divides to give rise to two cells, of which

one remains meristematic, while the other undergoes differentiation and contributes to the plant body (figure 36.3). In this way, the population of meristem cells is continually renewed. Molecular genetic evidence supports the hypothesis that animal stem cells and plant meristem cells may also share some common pathways of gene expression. For example, both plant

Figure 36.3 Meristem cell division. Plant meristems consist of cells that divide to give rise to a differentiating daughter cell and a cell that persists as a meristem cell.

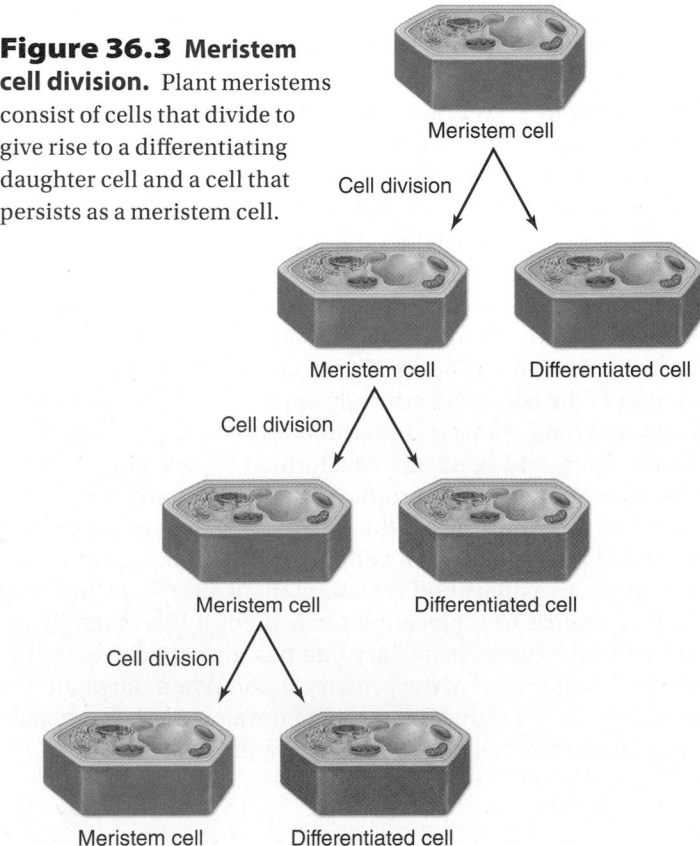

meristem and animal stem cells share the *Retinoblastoma* gene, which determines whether a cell continues dividing or differentiates. Extension of both root and shoot takes place as a result of repeated cell divisions and subsequent elongation of the cells produced by the **apical meristems.** In some vascular plants, including shrubs and most trees, **lateral meristems** produce an increase in root and shoot diameter.

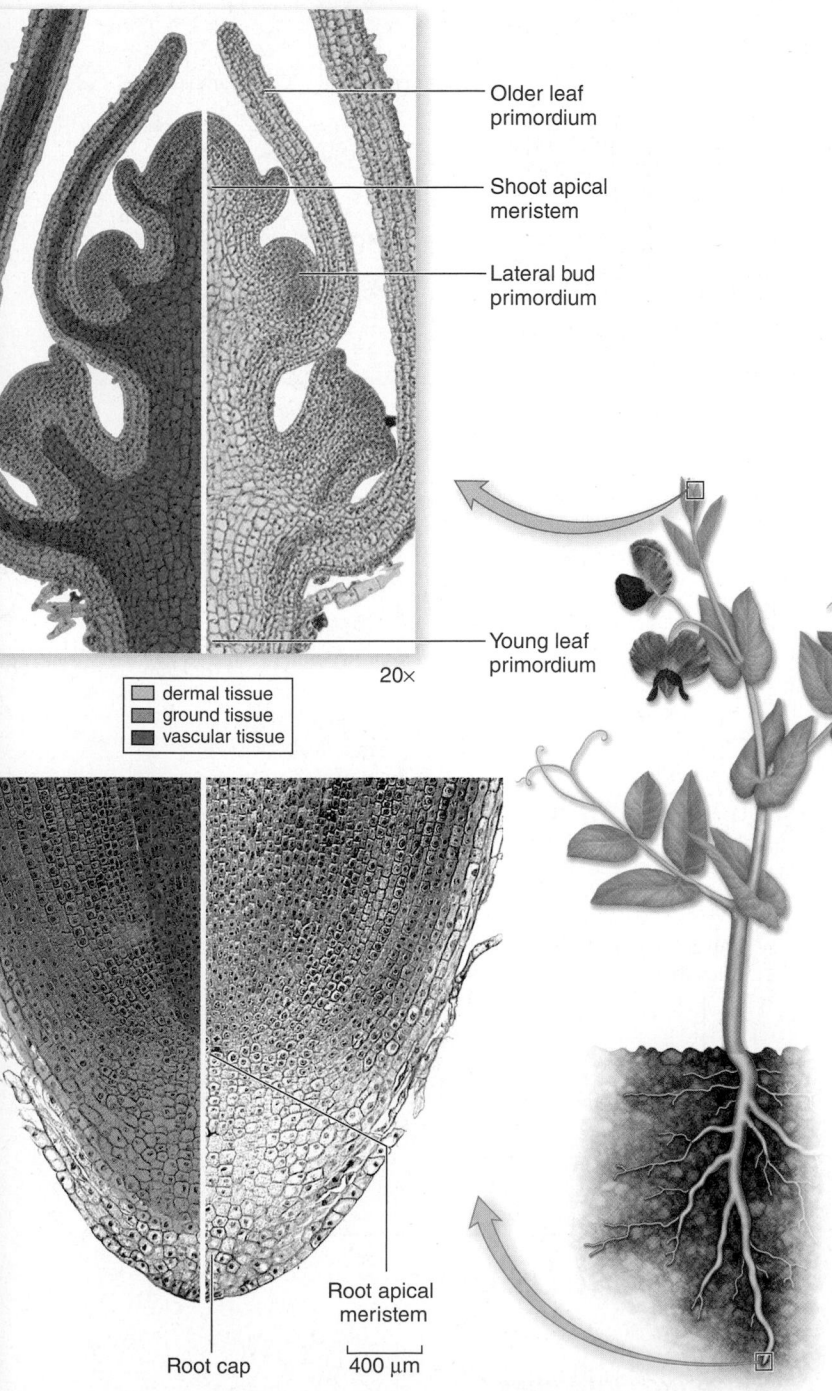

Older leaf primordium

Shoot apical meristem

Lateral bud primordium

Young leaf primordium

20×

- dermal tissue
- ground tissue
- vascular tissue

Root apical meristem

Root cap

400 μm

Figure 36.4 Apical meristems. Shoot and root apical meristems extend the plant body above and below ground. Leaf primordia protect the fragile shoot meristem, and the root meristem produces a protective root cap in addition to new root tissue.

Apical meristems

Apical meristems are located at the tips of stems and roots (figure 36.4). During periods of growth, the cells of apical meristems divide and continually add more cells at the tips. Tissues derived from apical meristems are called **primary tissues,** and the extension of the root and stem forms what is known as the **primary plant body.** The primary plant body comprises the young, soft shoots and roots of a tree or shrub, or the entire plant body in some plants.

Both root and shoot apical meristems are composed of delicate cells that need protection (see figure 36.4). The root apical meristem is protected by the root cap, the anatomy of which is described later on. Root cap cells are produced by the root meristem and are sloughed off and replaced as the root moves through the soil. In contrast, leaf primordia shelter the growing shoot apical meristem, which is particularly susceptible to desiccation because of its exposure to air and sun.

The apical meristem gives rise to the three tissue systems by first initiating **primary meristems.** The three primary meristems are the **protoderm,** which forms the epidermis; the **procambium,** which produces primary vascular tissues (primary xylem for water transport and primary phloem for nutrient transport); and the **ground meristem,** which differentiates further into ground tissue. In some plants, such as horsetails and corn, **intercalary meristems** arise in stem internodes (spaces between leaf attachments), adding to the internode lengths. If you walk through a cornfield on a quiet summer night when the corn is about knee high, you may hear a soft popping sound. This sound is caused by the rapid growth of the intercalary meristems. The amount of stem elongation that occurs in a very short time is quite surprising.

Lateral meristems

Many herbaceous plants (that is, plants with fleshy, not woody stems) exhibit only primary growth, but others also exhibit **secondary growth,** which may result in a substantial increase of diameter. Secondary growth is accomplished by the lateral meristems—peripheral cylinders of meristematic tissue within the stems and roots that increase the girth (diameter) of gymnosperms and most angiosperms. Lateral meristems form from ground tissue that is derived from apical meristems. Monocots are the major exception (figure 36.5).

Although secondary growth increases girth in many nonwoody plants, its effects are most dramatic in woody plants, which have two lateral meristems. Within the bark of a woody stem is the **cork cambium**—a lateral meristem that contributes to the outer bark of the tree. Just beneath the bark is the **vascular cambium**—a lateral meristem that produces secondary vascular tissue. The vascular cambium forms between the xylem and phloem in vascular bundles, adding secondary vascular tissue to both of its sides.

Secondary xylem is the main component of wood. Secondary phloem is very close to the outer surface of a woody stem. Removing the bark of a tree damages the phloem and may eventually kill the tree. Tissues formed from lateral meristems, which comprise most of the trunk, branches, and older roots of trees and shrubs, are known as **secondary tissues** and are collectively called the **secondary plant body.**

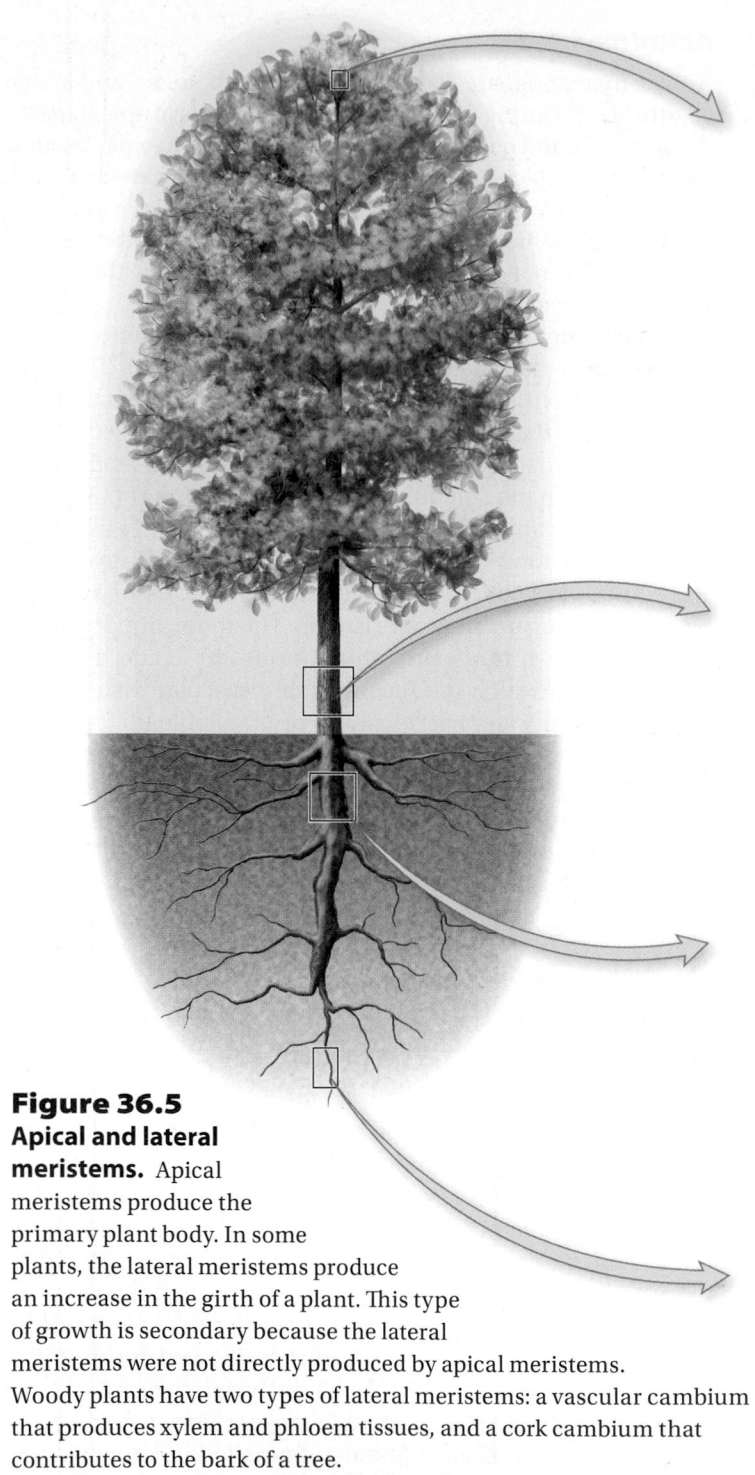

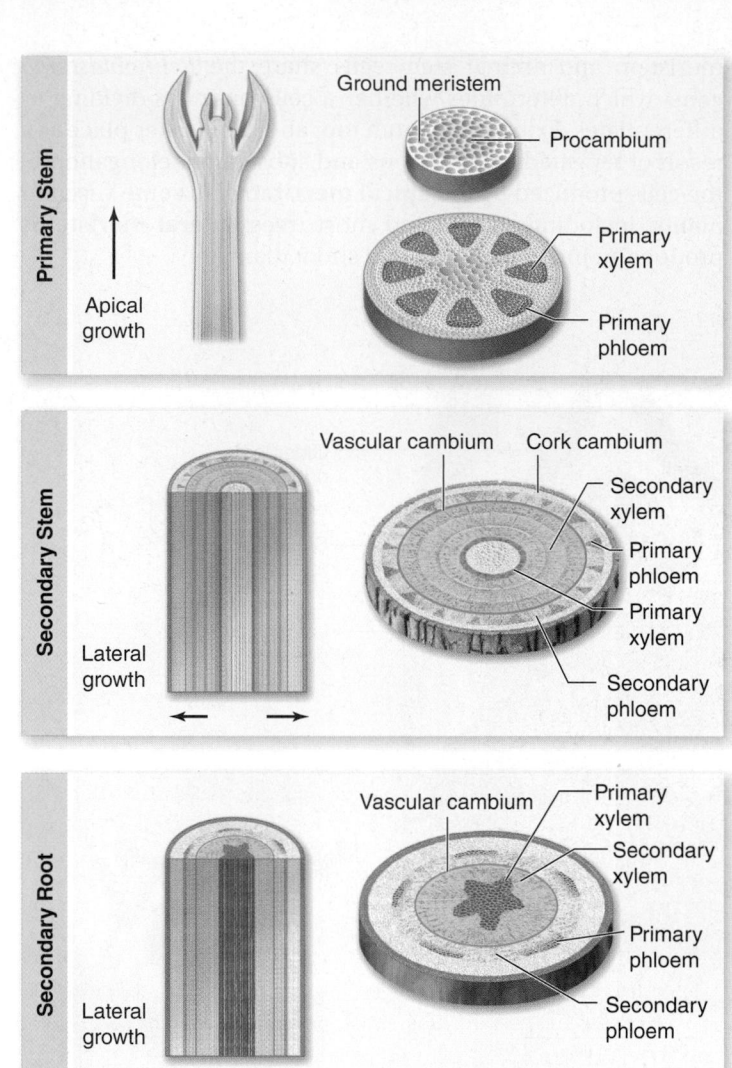

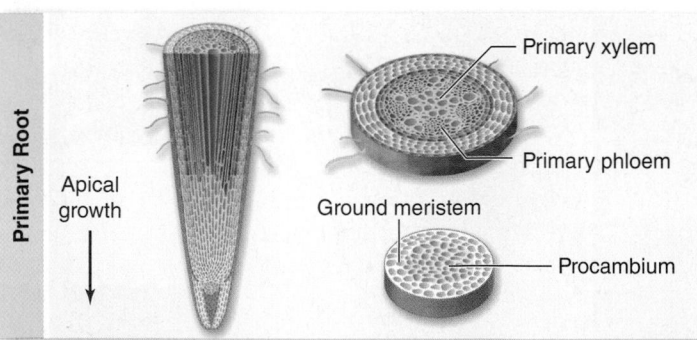

**Figure 36.5
Apical and lateral
meristems.** Apical
meristems produce the
primary plant body. In some
plants, the lateral meristems produce
an increase in the girth of a plant. This type
of growth is secondary because the lateral
meristems were not directly produced by apical meristems.
Woody plants have two types of lateral meristems: a vascular cambium
that produces xylem and phloem tissues, and a cork cambium that
contributes to the bark of a tree.

Learning Outcomes Review 36.1

The root system anchors plants and absorbs water and nutrients, whereas
the shoot system, consisting of stems, leaves, and flowers, carries out
photosynthesis and sexual reproduction. The three general types of tissue
in both roots and shoots are dermal, ground, and vascular tissue. Primary
growth is produced by apical meristems at the tips of roots and shoots;
secondary growth is produced by lateral meristems that are peripheral and
increase girth.

■ *Compare and contrast the locations and functions of
the different meristems in the shoot of a woody plant.*

36.2 Plant Tissues

Learning Outcomes

1. *Name the three cell types found in ground tissue and their
   functions.*
2. *Distinguish between xylem and phloem structure and
   function.*

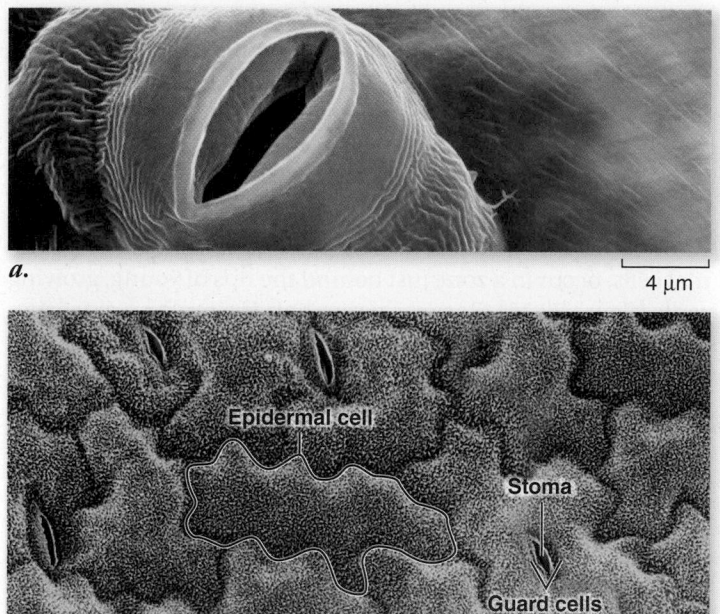

a.

4 μm

b.

Epidermal cell

Stoma

Guard cells

200 μm

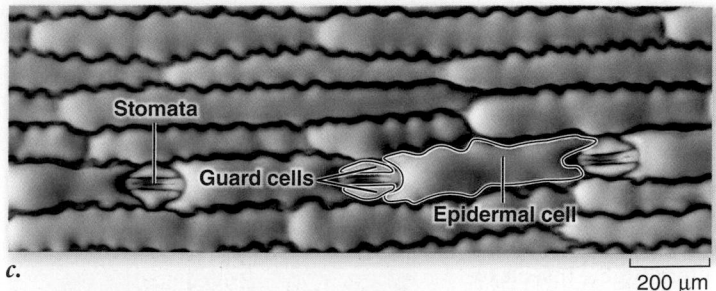

c.

Stomata

Guard cells

Epidermal cell

200 μm

Figure 36.6 **Stomata.** ***a.*** A stoma is the space between two guard cells that regulates the size of the opening. Stomata are evenly distributed within the epidermis of monocots and eudicots, but the patterning is quite different. ***b.*** A pea (eudicot) leaf with a random arrangement of stomata. ***c.*** A maize (corn, a monocot) leaf with stomata evenly spaced in rows. These photomicrographs also show the variety of cell shapes in plants. Some plant cells are boxlike, as seen in maize *(c),* and others are irregularly shaped, as seen in the jigsaw puzzle shapes of the pea epidermal cells, ***(b).***

Highly differentiated cell types distinguish the three tissue systems and correlate with the function of the tissue. Each tissue type is composed of several types of cells. In this section the structure and function of plant cells is considered.

Dermal tissue forms a protective interface with the environment

Dermal tissue derived from an embryo or apical meristem forms **epidermis.** This tissue is one cell layer thick in most plants and forms the outer protective covering of the plant. In young, exposed parts of the plant, the epidermis is covered with a fatty **cutin** layer constituting the **cuticle;** in plants such as desert succulents, several layers of wax may be added to the cuticle to limit water loss and protect against ultraviolet damage. In some cases, the dermal tissue forms the bark of trees.

Epidermal cells, which originate from the protoderm, cover all parts of the primary plant body. A number of types of specialized cells occur in the epidermis, including *guard cells, trichomes,* and *root hairs.*

Guard cells

Guard cells are paired, sausage-shaped cells flanking a stoma (plural, *stomata*), a mouth-shaped epidermal opening. Guard cells, unlike other epidermal cells, contain chloroplasts.

Stomata occur in the epidermis of leaves (figure 36.6*a*) and sometimes on other parts of the plant, such as stems or fruits. Stomata serve a dual function. Exchange of oxygen and carbon dioxide gases through the stomata is essential for photosynthesis. In addition, diffusion of water in vapor form takes place almost exclusively through the stomata, enabling transport of water and minerals from the roots to the leaves. There are from 1000 to more than 1 million stomata per square centimeter of leaf surface. In many plants, stomata are more numerous on the lower epidermis of the leaf than on the upper—a factor that helps minimize water loss. Some plants have

stomata only on the lower epidermis, and a few, such as water lilies, have them only on the upper epidermis to maximize gas exchange.

Guard cell formation is the result of an asymmetrical cell division producing a guard cell and a subsidiary cell that aids in the opening and closing of the stoma. The patterning of these asymmetrical divisions that results in stomatal distribution has intrigued developmental biologists (figure 36.6*b, c*).

Research on mutants that get "confused" about where to position stomata is providing information on the timing of stomatal initiation and the kind of intercellular communication that triggers guard cell formation. For example, the *too many mouths (tmm)* mutation in *Arabidopsis* disrupts the normal pattern of cell division that spatially separates stomata (figure 36.7). Investigations of this and other stomatal patterning genes revealed a coordinated network of cell–cell communication (see chapter 9) that informs cells of their position relative to other cells and determines cell fate. The *TMM* gene encodes a membrane-bound receptor that is part of a signaling pathway controlling asymmetrical cell division.

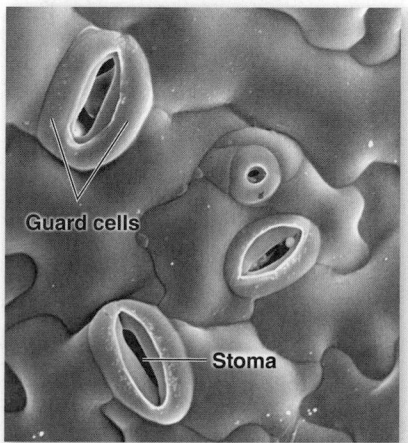

Guard cells

Stoma

270 μm

Figure 36.7
The too many mouths stomatal mutant. This *Arabidopsis* mutant plant lacks an essential signal for spacing stomata. Usually a differentiating guard cell pair inhibits differentiation of a nearby cell into a guard cell.

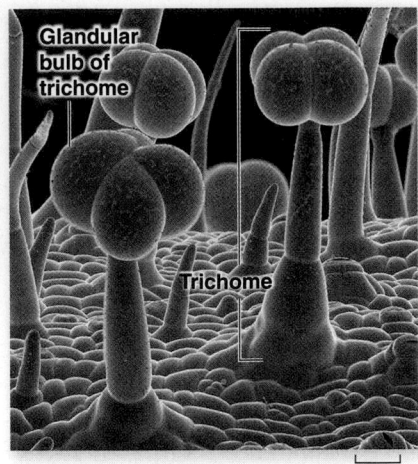

Figure 36.8 Trichomes. The trichomes with tan, bulbous tips on this tomato plant are glandular trichomes. These trichomes secrete substances that can literally glue insects to the trichome.

Glandular bulb of trichome

Trichome

35 μm

Trichomes

Trichomes are cellular or multicellular hairlike outgrowths of the epidermis (figure 36.8). They occur frequently on stems, leaves, and reproductive organs. A "fuzzy" or "woolly" leaf is covered with trichomes that can be seen clearly with a microscope under low magnification. Trichomes keep leaf surfaces cool and reduce evaporation by covering stomatal openings. They also protect leaves from high-intensity light and ultraviolet radiation and can buffer against temperature fluctuations. Trichomes can vary greatly in form; some consist of a single cell; others are multicellular. Some are glandular, secreting sticky or toxic substances to deter herbivory.

Genes that regulate trichome development have been identified, including *GLABROUS3 (GL3)* (figure 36.9). When trichome-initiating proteins, like GL3, reach a threshold level compared with trichome-inhibiting proteins, an epidermal cell becomes a trichome. Signals from this trichome cell now prevent neighbor cells from expressing trichome-promoting genes (see figure 36.9).

Root hairs

Root hairs, which are tubular extensions of individual epidermal cells, occur in a zone just behind the tips of young, growing roots (figure 36.10). Because a root hair is simply an extension of an epidermal cell and not a separate cell, no cross-wall isolates the hair from the rest of the cell. Root hairs keep the root in intimate contact with the surrounding soil particles and greatly increase the root's surface area and efficiency of absorption.

As a root grows, the extent of the root hair zone remains roughly constant as root hairs at the older end slough off while new ones are produced at the apex. Most of the absorption of water and minerals occurs through root hairs, especially in herbaceous plants. Root hairs should not be confused with lateral roots, which are multicellular structures and originate deep within the root.

The first land plants lacked roots, which later evolved from shoots. Some of the genes needed for trichome and stomatal differentiation in shoot epidermal cells were co-opted over evolutionary time for root hair development.

? **Inquiry question** Identify three dermal tissue traits that are adaptive for a terrestrial lifestyle and explain why these traits are advantageous.

Figure 36.9 Trichome patterning. Mutants have revealed genes involved in regulating the spacing and development of trichomes in *Arabidopsis.* *a.* Wild type. *b. glaborous3* mutant, which fails to initiate trichome development. *c.* In the presence of sufficient GL3 and when the levels of trichome-inhibiting proteins are sufficiently low, that cell develops a trichome. Once a cell begins trichome initiation, it signals neighboring cells and inhibits their ability to develop trichomes.

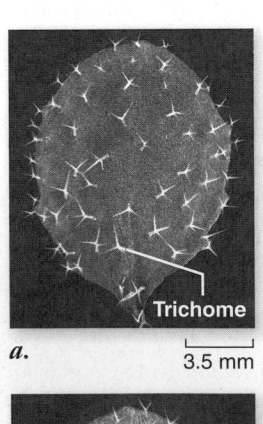

Trichome

a. 3.5 mm

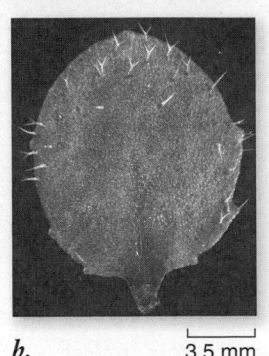

b. 3.5 mm

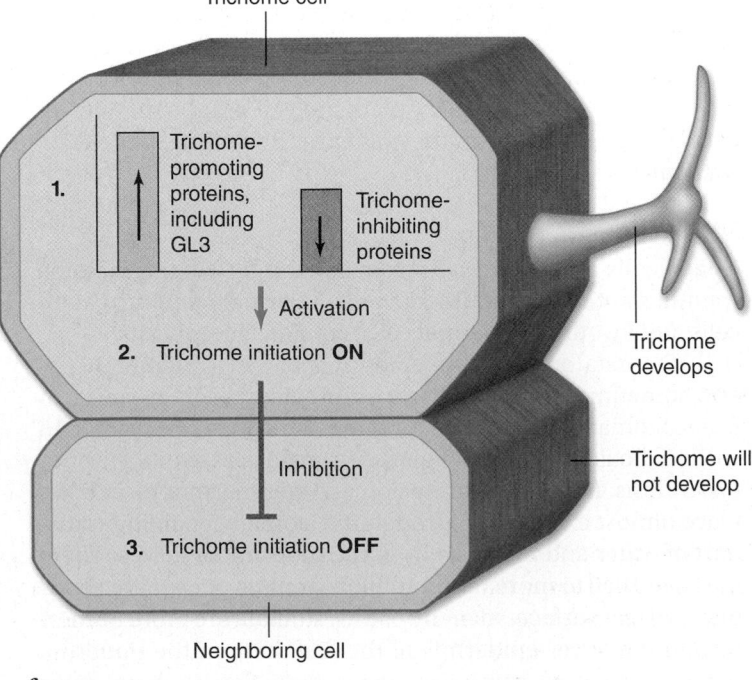

Trichome cell

1. Trichome-promoting proteins, including GL3

Trichome-inhibiting proteins

Activation

2. Trichome initiation **ON**

Inhibition

3. Trichome initiation **OFF**

Neighboring cell

Trichome develops

Trichome will not develop

c.

Figure 36.10 Root hairs. Root hair cells are a type of epidermal cell that increase the surface area of the root to enhance water and mineral uptake.

Ground tissue cells perform many functions, including storage, photosynthesis, and support

Ground tissue consists primarily of thin-walled *parenchyma cells* that function in storage, photosynthesis, and secretion. Other ground tissue, composed of *collenchyma cells* and *sclerenchyma cells,* provide support and protection.

Parenchyma

Parenchyma cells are the most common type of plant cell. They have large vacuoles, thin walls, and are initially (but briefly) more or less spherical. These cells, which have living protoplasts, push up against each other shortly after they are produced, however, and assume other shapes, often ending up with 11 to 17 sides.

Parenchyma cells may live for many years; in some plants (for example, cacti), they may live to be over 100 years old. They function in the storage of food and water, photosynthesis, and secretion. They are the most abundant cells of primary tissues and may also occur, to a much lesser extent, in secondary tissues (figure 36.11*a*).

Most parenchyma cells have only primary walls, which are walls laid down while the cells are still maturing. Parenchyma are less specialized than other plant cells, although many variations occur with special functions, such as nectar and resin secretion or storage of latex, proteins, and metabolic wastes. Some parenchyma contain chloroplasts, especially in leaves and in the outer parts of herbaceous stems. Such photosynthetic parenchyma tissue is called *chlorenchyma.*

Collenchyma

If celery "strings" have ever been caught between your teeth, you are familiar with tough, flexible **collenchyma cells.** Like parenchyma cells, collenchyma cells have living protoplasts and may live for many years. These cells, which are usually a little longer than wide, have walls that vary in thickness (figure 36.11*b*).

Flexible collenchyma cells provide support for plant organs, allowing them to bend without breaking. They often form strands or continuous cylinders beneath the epidermis of stems or leaf petioles (stalks) and along the veins in leaves.

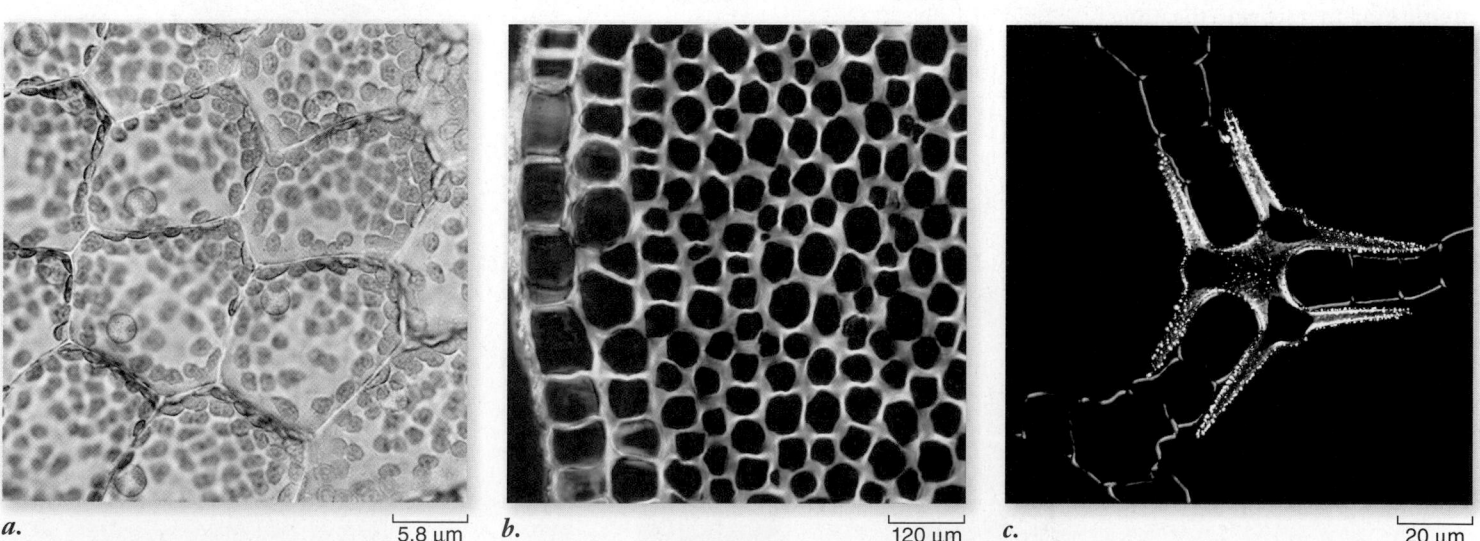

a.   5.8 μm   *b.*   120 μm   *c.*   20 μm

Figure 36.11 The three types of ground tissue. *a.* Parenchyma cells. Only primary cell walls are seen in this cross section of parenchyma cells from grass. *b.* Collenchyma cells. Thickened side walls are seen in this cross section of collenchyma cells from a young branch of elderberry *(Sambucus).* In other kinds of collenchyma cells, the thickened areas may occur at the corners of the cells or in other kinds of strips. *c.* Sclereids. Clusters of sclereids ("stone cells"), stained red in this preparation. The surrounding thin-walled cells, stained green, are parenchyma. Sclereids are one type of sclerenchyma tissue, which also contains fibers.

Strands of collenchyma provide much of the support for stems in the primary plant body.

Sclerenchyma

Sclerenchyma cells have tough, thick walls. Unlike collenchyma and parenchyma, they usually lack living protoplasts at maturity. Their secondary cell walls are often impregnated with **lignin,** a highly branched polymer that makes cell walls more rigid; for example, lignin is an important component in wood. Cell walls containing lignin are said to be *lignified.* Lignin is common in the walls of plant cells that have a structural or mechanical function. Some kinds of cells have lignin deposited in primary as well as secondary cell walls.

Sclerenchyma is present in two general types: fibers and sclereids. *Fibers* are long, slender cells that are usually grouped together in strands. Linen, for example, is woven from strands of sclerenchyma fibers that occur in the phloem of flax (*Linum* spp.) plants. *Sclereids* are variable in shape but often branched. They may occur singly or in groups; they are not elongated, but may have many different forms, including that of a star. The gritty texture of a pear is caused by groups of sclereids that occur throughout the soft flesh of the fruit (figure 36.11*c*). Sclereids are also found in hard seed coats. Both of these tough, thick-walled cell types serve to strengthen the tissues in which they occur.

 ## Vascular tissue conducts water and nutrients throughout the plant

Vascular tissue, as mentioned earlier, includes two kinds of conducting tissues: (1) *xylem,* which conducts water and dissolved minerals, and (2) *phloem,* which conducts a solution of carbohydrates—mainly sucrose—used by plants for food. The phloem also transports hormones, amino acids, and other substances that are necessary for plant growth. Xylem and phloem differ in structure as well as in function.

Xylem

Xylem, the principal water-conducting tissue of plants, usually contains a combination of *vessels,* which are continuous tubes formed from dead, hollow, cylindrical cells arranged end-to-end, and *tracheids,* which are dead cells that taper at the ends and overlap one another (figure 36.12). Primary xylem is derived from the procambium produced by the apical meristem. Secondary xylem is formed by the vascular cambium, a lateral meristem. Wood consists of accumulated secondary xylem.

In some plants (but not flowering plants), tracheids are the only water-conducting cells present; water passes in an unbroken stream through the xylem from the roots up through the shoot and into the leaves. When the water reaches the leaves, much of it diffuses in the form of water vapor into the intercellular spaces and out of the leaves into the surrounding air, mainly through the stomata. This diffusion of water vapor from a plant is known as **transpiration** (see chapter 37). In addition to conducting water, dissolved minerals, and inorganic ions such as nitrates and phosphates throughout the plant, xylem supplies support for the plant body.

Vessel members tend to be shorter and wider than tracheids. When viewed with a microscope, they resemble beverage cans with both ends removed. Both vessel members and tracheids have thick, lignified secondary walls and no living protoplasts at maturity. Lignin is produced by the cell and secreted to strengthen the cellulose cell walls before the protoplast dies, leaving only the cell wall.

Figure 36.12 Comparison between tracheids and vessel members. In tracheids, the water passes from cell to cell by means of pits. In vessel members, water moves by way of perforation plates. In gymnosperm wood, tracheids both conduct water and provide support; in most kinds of angiosperms, vessels are present in addition to tracheids. These two types of cells conduct water, and fibers provide additional support. The wood of red maple, *Acer rubrum,* contains both tracheids and vessels.

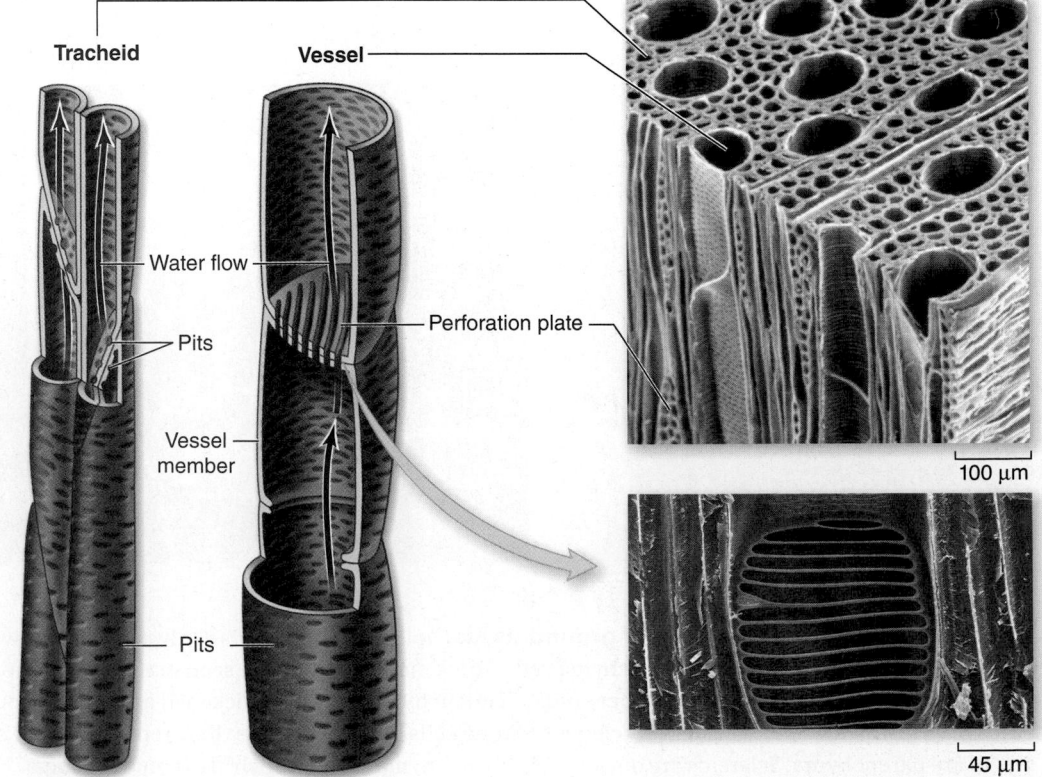

Tracheids contain *pits*, which are small, mostly rounded-to-elliptical areas where no secondary wall material has been deposited. The pits of adjacent cells occur opposite one another; the continuous stream of water flows through these pits from tracheid to tracheid. In contrast, vessel members, which are joined end to end, may be almost completely open or may have bars or strips of wall material across the open ends (see figure 36.12). Vessels appear to conduct water more efficiently than do the overlapping strands of tracheids. We know this partly because vessel members have evolved from tracheids independently in several groups of plants, suggesting that they are favored by natural selection.

In addition to conducting cells, xylem typically includes fibers and parenchyma cells (ground tissue cells). It is probable that some types of fibers have evolved from tracheids, becoming specialized for strengthening rather than conducting. The parenchyma cells, which are usually produced in horizontal rows called *rays* by special *ray initials* of the vascular cambium, function in lateral conduction and food storage. (An *initial* is another term for a meristematic cell. It divides to produce another initial and a cell that differentiates.)

In cross sections of woody stems and roots, the rays can be seen radiating out from the center of the xylem like the spokes of a wheel. Fibers are abundant in some kinds of wood, such as oak (*Quercus* spp.), and the wood is correspondingly dense and heavy. The arrangements of these and other kinds of cells in the xylem make it possible to identify most plant genera and many species from their wood alone.

Over 2000 years ago, paper was made in China by mashing herbaceous plants in water and separating out a thin layer of phloem fibers on a screen. Not until the third century of the common era did the secret of making paper make its way out of China. Today the ever-growing demand for paper is met by extracting xylem fibers from wood, including spruce, which is relatively soft because it has fewer ray fibers than oak. The lignin-rich cell walls yield brown paper that is often bleached. In addition, tissues from many other plants have been developed as sources of paper, including kenaf and hemp. U.S. paper currency is 75% cotton and 25% flax.

Phloem

Phloem, which is located toward the outer part of roots and stems, is the principal food-conducting tissue in vascular plants. If a plant is *girdled* (by removing a substantial strip of bark down to the vascular cambium around the entire circumference), the plant eventually dies from starvation of the roots.

Food conduction in phloem is carried out through two kinds of elongated cells: sieve cells and sieve-tube members. Gymnosperms, ferns, and horsetails have only sieve cells; most angiosperms have sieve-tube members. Both types of cells have clusters of pores known as sieve areas because the cell walls resemble sieves. Sieve areas are more abundant on the overlapping ends of the cells and connect the protoplasts of adjoining sieve cells and sieve-tube members. Both of these types of cells are living, but most sieve cells and all sieve-tube members lack a nucleus at maturity.

In sieve-tube members, some sieve areas have larger pores and are called sieve plates (figure 36.13). Sieve-tube members occur end to end, forming longitudinal series called sieve tubes. Sieve cells are less specialized than sieve-tube members, and the pores in all of their sieve areas are roughly of the same diameter. Sieve-tube members are more specialized, and presumably, more efficient than sieve cells.

Each sieve-tube member is associated with an adjacent, specialized parenchyma cell known as a *companion cell*. Companion cells apparently carry out some of the metabolic functions needed to maintain the associated sieve-tube member. In angiosperms, a common initial cell divides asymmetrically to produce a sieve-tube member cell and its companion cell. Companion cells have all the components of normal parenchyma cells, including nuclei, and numerous plasmodesmata (cytoplasmic connections between adjacent cells) connect their cytoplasm with that of the associated sieve-tube members.

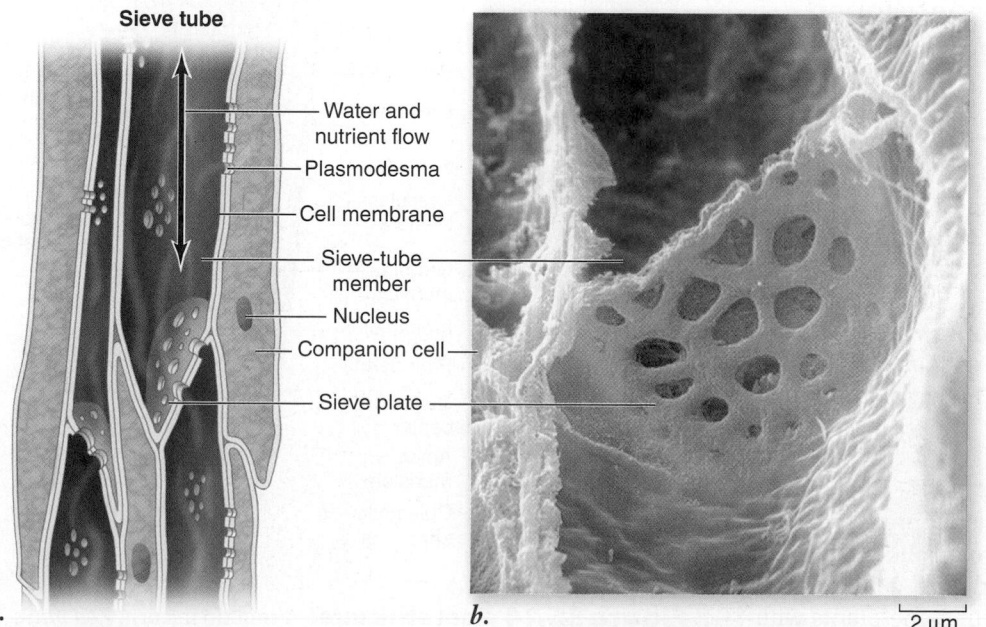

Sieve tube

- Water and nutrient flow
- Plasmodesma
- Cell membrane
- Sieve-tube member
- Nucleus
- Companion cell
- Sieve plate

a.

b.

2 μm

Figure 36.13 A sieve-tube member. *a.* Sieve-tube member cells are stacked, with sieve plates forming the connection. The narrow cell with the nucleus at the right of the sieve-tube member is a companion cell. This cell nourishes the sieve-tube members, which have plasma membranes, but no nuclei. *b.* Looking down into sieve plates in squash phloem reveals the perforations through which sucrose and hormones move.

© Dr. Richard Kessel & Dr. Gene Shih/Visuals Unlimited

Sieve cells in nonflowering plants have albuminous cells that function as companion cells. Unlike a companion cell, an albuminous cell is not necessarily derived from the same mother cell as its associated sieve cell. Fibers and parenchyma cells are often abundant in phloem.

Learning Outcomes Review 36.2

Dermal tissue protects a plant from its environment and contains specialized cells such as guard cells, trichomes, and root hairs. Ground tissue serves several functions, including storage (parenchyma cells), photosynthesis (specialized parenchyma called chlorenchyma), and structural support (collenchyma and sclerenchyma). Vascular tissue carries water through the xylem (primarily vessels) and nutrients through the phloem (primarily sieve-tube members).

■ **Contrast the structure and function of mature vessels and sieve-tube members.**

36.3 Roots: Anchoring and Absorption Structures

Learning Outcomes

1. *Describe key changes that occur in the four regions of a typical root.*
2. *Explain the function of root hairs.*
3. *Describe functions of modified roots.*

Roots have a simpler pattern of organization and development than stems, and we will consider them first. Keep in mind, however, that roots evolved after shoots and are a major innovation for terrestrial living.

Roots are adapted for growing underground and absorbing water and solutes

Four regions are commonly recognized in developing roots: the *root cap,* the *zone of cell division,* the *zone of elongation,* and the *zone of maturation* (figure 36.14). In these last three zones, the boundaries are not clearly defined.

When apical initials divide, daughter cells that end up on the tip end of the root become root cap cells. Cells that divide in the opposite direction pass through the three other zones before they finish differentiating. As you consider the different zones, visualize the tip of the root moving deeper into the soil, actively growing. This counters the static image of a root that diagrams and photos convey.

The root cap

The **root cap** has no equivalent in stems. It is composed of two types of cells: the inner *columella cells* (they look like columns), and the outer, lateral *root cap cells,* which are continuously replenished by the root apical meristem. In some plants with larger roots, the root cap is quite obvious. Its main function is to protect the delicate tissues behind it as growth extends the root through mostly abrasive soil particles.

Golgi bodies in the outer root cap cells secrete and release a slimy substance that passes through the cell walls to the outside. The root cap cells, which have an average life of less than a week, are constantly being replaced from the inside, forming a mucilaginous lubricant that eases the root through the soil. The slimy mass also provides a medium for the growth of beneficial nitrogen-fixing bacteria in the roots of plants such as legumes. A new root cap is produced when an existing one is artificially or accidentally removed from a root.

The root cap also functions in the perception of gravity. The columella cells are highly specialized, with the endoplasmic reticulum in the periphery and the nucleus located at either the middle or the top of the cell. They contain no large vacuoles. Columella cells contain *amyloplasts* (plastids with starch grains) that collect on the sides of cells facing the pull of gravity. When a potted plant is placed on its side, the amyloplasts drift or tumble down to the side nearest the source of gravity, and the root bends in that direction.

Lasers have been used to ablate (kill) individual columella cells in *Arabidopsis*. It turns out that only two columella cells are sufficient for gravity sensing! The precise nature of the gravitational response is unknown, but some evidence indicates that calcium ions in the amyloplasts influence the distribution of growth hormones (auxin in this case) in the cells. Multiple signaling mechanisms may exist, because bending has been observed in the absence of auxin. A current

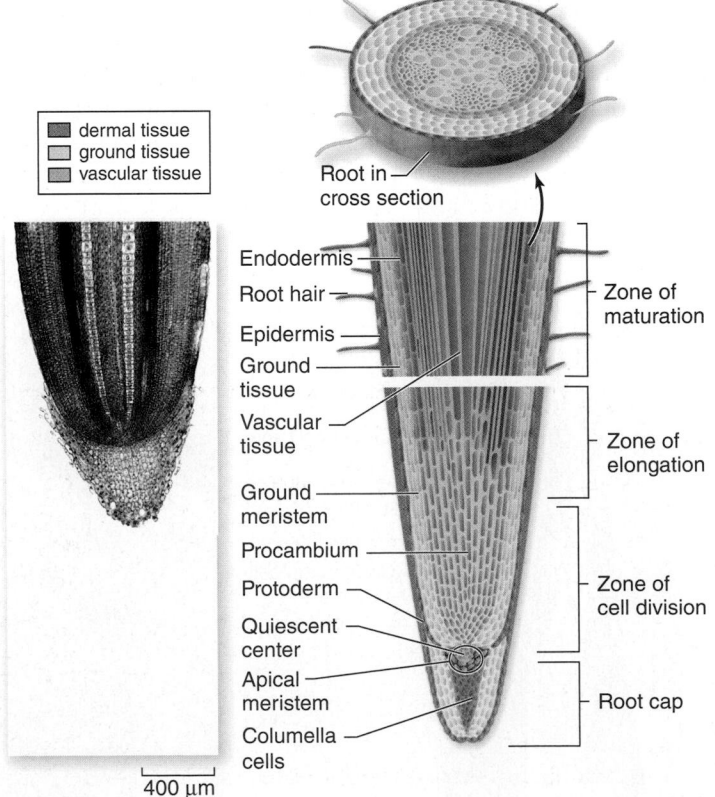

■ dermal tissue
■ ground tissue
■ vascular tissue

Root in cross section

Endodermis
Root hair
Epidermis
Ground tissue
Vascular tissue
Ground meristem
Procambium
Protoderm
Quiescent center
Apical meristem
Columella cells

Zone of maturation
Zone of elongation
Zone of cell division
Root cap

400 μm

Figure 36.14 Root structure. A root tip in corn, *Zea mays.*

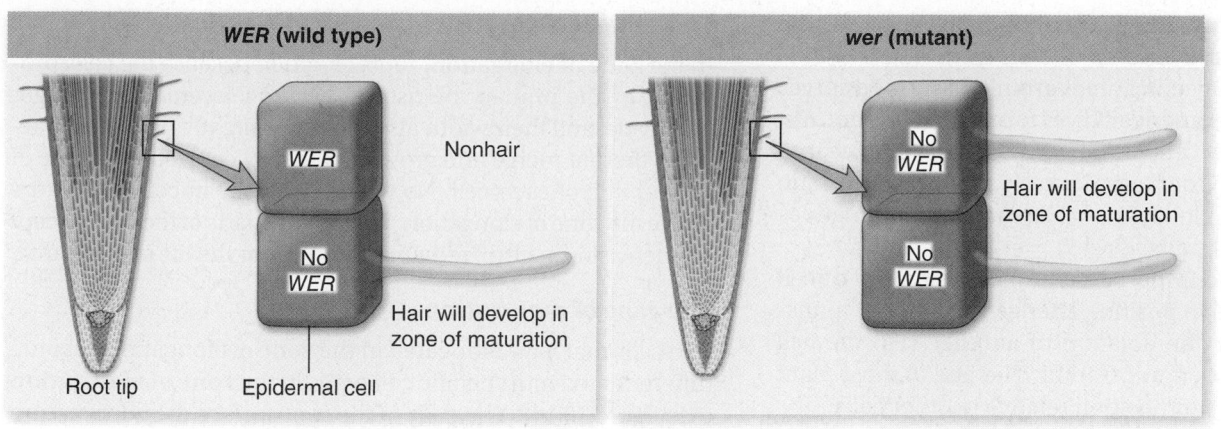
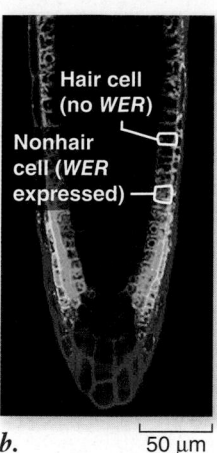

Figure 36.15 Tissue-specific gene expression. *a.* The *WEREWOLF* gene of *Arabidopsis* is expressed in some, but not all, epidermal cells and suppresses root hair development. The *wer* mutant is covered with root hairs. *b.* The *WER* promoter was attached to a gene coding for a green fluorescent protein and used to make a transgenic plant. The green fluorescence shows the nonhair epidermal cells where the gene is expressed. The red visually indicates cell boundaries because cell walls autofluoresce.

hypothesis is that an electrical signal moves from the columella cell to cells in the elongation zone (the region closest to the zone of cell division).

 Data analysis If 10 cells on the edges of a root apical meristem continually replenish a population of 100 root cap cells, how frequently must each of the cells divide?

The zone of cell division

The apical meristem is located in the center of the root tip in the area protected by the root cap. Most of the activity in this **zone of cell division** takes place toward the edges of the meristem, where the cells divide every 12 to 37 hours, often coordinately, reaching a peak of division once or twice a day.

Most of the cells are essentially cuboidal, with small vacuoles and proportionately large, centrally located nuclei. These rapidly dividing cells are daughter cells of the apical meristem. A group of cells in the center of the root apical meristem,

termed the *quiescent center,* divide only very infrequently. The presence of the quiescent center makes sense if you think about a solid ball expanding—the outer surface would have to increase far more rapidly than the very center.

The apical meristem daughter cells soon subdivide into the three primary tissues previously discussed: protoderm, procambium, and ground meristem. Genes have been identified in the relatively simple root of *Arabidopsis* that regulate the patterning of these tissue systems. The patterning of these cells begins in this zone, but the anatomical and morphological expression of this patterning is not fully revealed until the cells reach the zone of maturation.

For example, the *WEREWOLF (WER)* gene is required for the patterning of the two root epidermal cell types, those with and those without root hairs (figure 36.15). Plants with the *wer* mutation have an excess of root hairs because *WER* is needed to prevent root hair development in nonhair epidermal cells. Similarly, the *SCARECROW (SCR)* gene is necessary in ground cell differentiation (figure 36.16). A ground

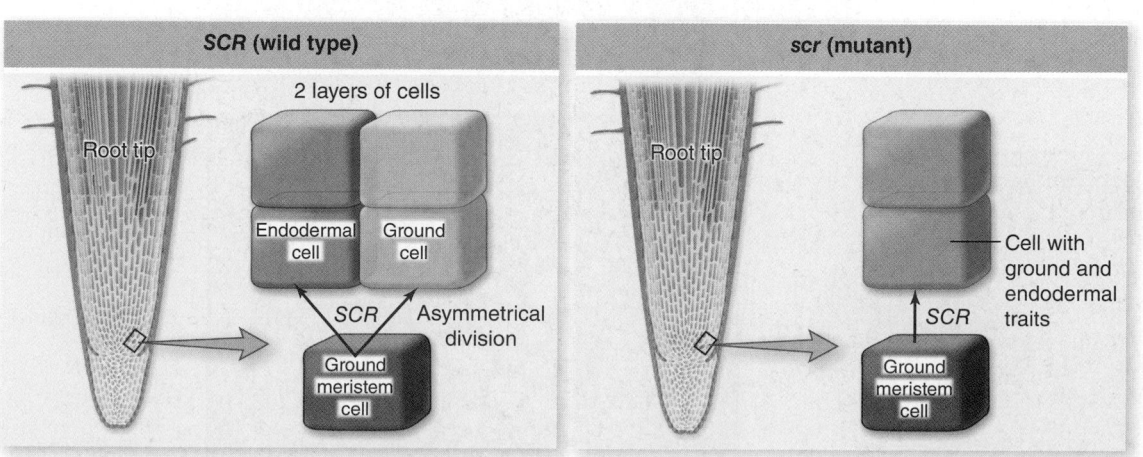
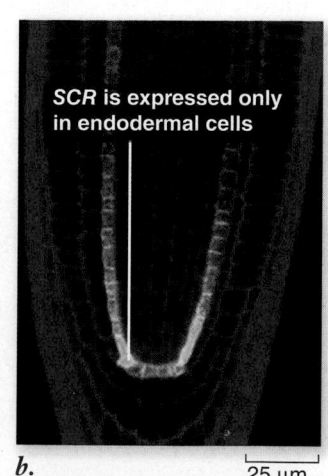

Figure 36.16 Scarecrow regulates asymmetrical cell division. *a.* *SCR* is needed for an asymmetrical cell division leading to the differentiation of daughter cells into endodermal and ground cells. *b.* The *SCR* promoter was attached to a gene coding for a green fluorescent protein to find out exactly where in the wild-type root *SCR* is expressed. *SCR* is only expressed in the endodermal cells, not the ground cells.

meristem cell undergoes an asymmetrical cell division that gives rise to two nested cylinders of cells from one if *SCR* is present. The outer cell layer becomes ground tissue and serves a storage function. The inner cell layer forms the endodermis, which regulates the intercellular flow of water and solutes into the vascular core of the root (see figure 36.5). The *scr* mutant, in contrast, forms a single layer of cells that have both endodermal and ground cell traits.

SCR illustrates the importance of the orientation of cell division. If a cell's relative position changes because of a mistake in cell division or the ablation of another cell, the cell develops according to its new position. The fate of most plant cells is determined by their position relative to other cells.

The zone of elongation

In the **zone of elongation,** roots lengthen because the cells produced by the primary meristems become several times longer than wide, and their width also increases slightly. The small vacuoles present merge and grow until they occupy 90% or more of the volume of each cell. No further increase in cell size occurs above the zone of elongation. The mature parts of the root, except for increasing in girth, remain stationary for the life of the plant.

The zone of maturation

The cells that have elongated in the zone of elongation become differentiated into specific cell types in the **zone of maturation** (see figure 36.14). The cells of the root surface cylinder mature into *epidermal cells,* which have a very thin cuticle, and include both root hair and nonhair cells. Although the root hairs are not visible until this stage of development, their fate was established much earlier, as you saw with the expression patterns of *WER* (see figure 36.15).

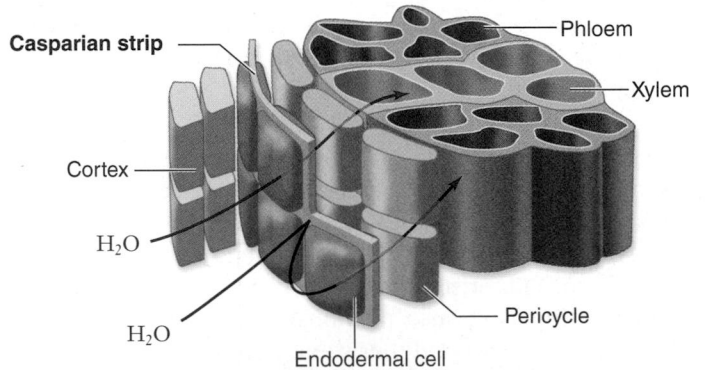

Figure 36.17 Cross sections of the zone of maturation of roots. Both monocot and eudicot roots have a Casparian strip as seen in the cross section of greenbriar *(Smilax),* a monocot, and buttercup *(Ranunculus),* a eudicot. The Casparian strip is a waterproofing band that forces water and minerals to pass through the plasma membranes of endodermal cells, rather than through the spaces in the cell walls.

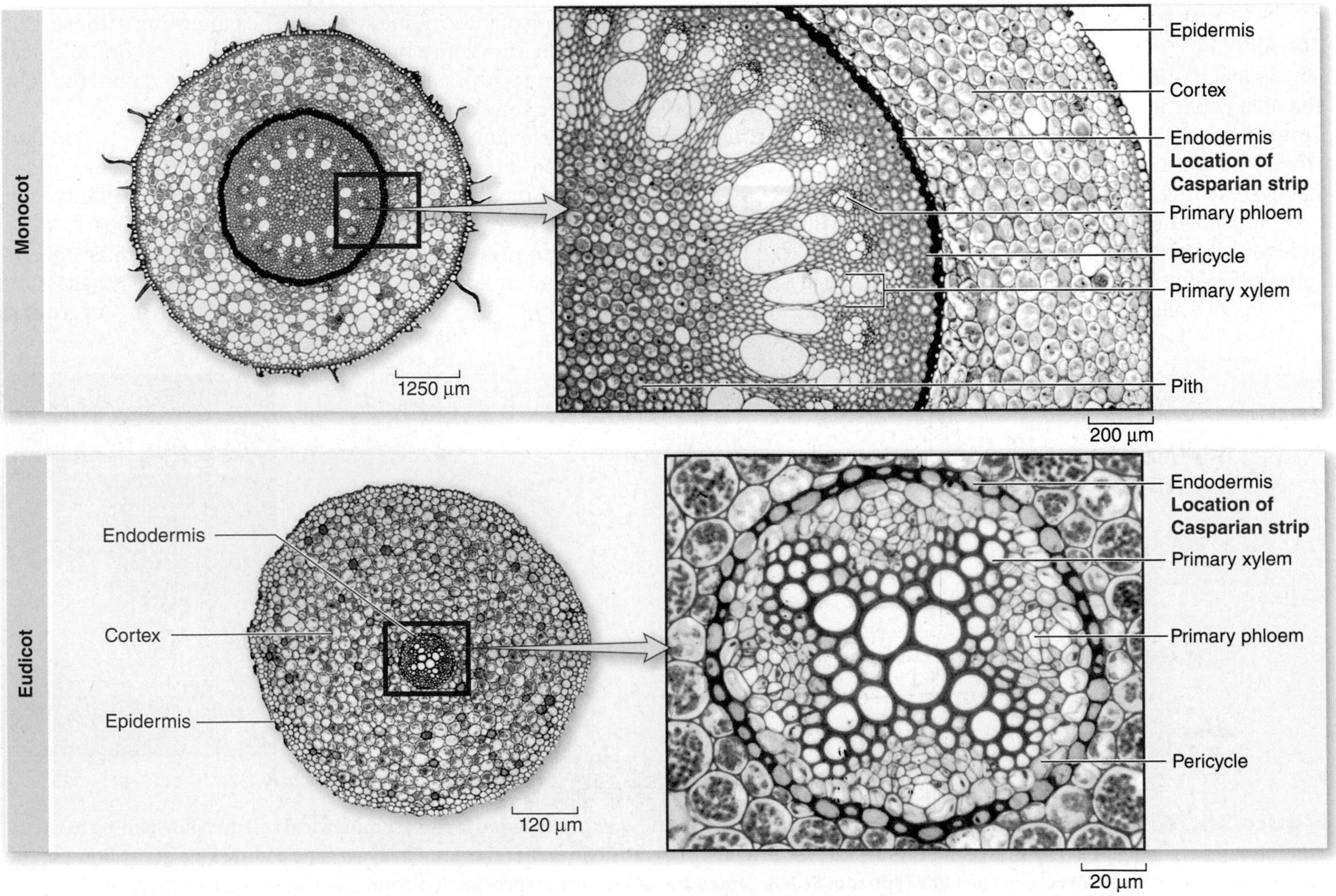

Root hairs can number many billions per plant and provide over 37,000 cm^2 of root surface; they greatly increase the surface area and therefore the absorptive capacity of the root. Symbiotic bacteria that fix atmospheric nitrogen into a form usable by legumes enter the plant via root hairs and "instruct" the plant to create a nitrogen-fixing nodule around it (see chapter 38).

Parenchyma cells are produced by the ground meristem immediately to the interior of the epidermis. This tissue, called the **cortex,** may be many cell layers wide and functions in food storage. As just described, the inner boundary of the cortex differentiates into a single-layered cylinder of **endodermis,** after an asymmetrical cell division regulated by *SCR* (see figures 36.16 and 36.17). Endodermal primary walls are impregnated with *suberin,* a fatty substance that is impervious to water. The suberin is produced in bands, called **Casparian strips,** that surround each adjacent endodermal cell wall perpendicular to the root's surface (see figure 36.17). These strips block transport between cells. The two surfaces that are parallel to the root surface are the only way into the vascular tissue of the root, and the plasma membranes control what passes through. Plants with a *scr* mutation lack this waterproof Casparian strip.

All the tissues interior to the endodermis are collectively referred to as the **stele.** Immediately adjacent and interior to the endodermis is a cylinder of parenchyma cells known as the **pericycle.** Pericycle cells divide, even after they mature. They can give rise to lateral (branch) roots or, in eudicots, to the two lateral meristems, the vascular cambium and the cork cambium.

The water-conducting cells of the primary xylem are differentiated as a solid core in the center of young eudicot roots. In a cross section of a eudicot root, the central core of primary xylem often is somewhat star-shaped, having from two to several radiating arms that point toward the pericycle (see figure 36.17). In monocot (and a few eudicot) roots, the primary xylem is in discrete vascular bundles arranged in a ring, which surrounds parenchyma cells, called *pith,* at the very center of the root (see figure 36.17). Primary phloem, composed of cells involved in food conduction, is differentiated in discrete groups of cells adjacent to the xylem in both eudicot and monocot roots.

In eudicots and other plants with secondary growth, part of the pericycle and the parenchyma cells between the phloem patches and the xylem become the root vascular cambium, which starts producing secondary xylem to the inside and secondary phloem to the outside. Eventually, the secondary tissues acquire the form of concentric cylinders. The primary phloem, cortex, and epidermis become crushed and are sloughed off as more secondary tissues are added.

In the pericycle of woody plants, the cork cambium contributes to the outer bark, which will be discussed in more detail when we look at stems. In the case of secondary growth in eudicot roots, everything outside the stele is lost and replaced with bark. Figure 36.18 summarizes the process of differentiation that occurs in plant tissue.

Modified roots accomplish specialized functions

Most plants produce either a taproot system, characterized by a single large root with smaller branch roots, or a fibrous root system, composed of many smaller roots of similar diameter. Some plants, however, have intriguing root modifications with specific functions in addition to those of anchorage and absorption.

Not all roots are produced by preexisting roots. Any root that arises along a stem or in some place other than the root of the plant is called an **adventitious root.** For example, climbing plants such as ivy produce roots from their stems; these can anchor the stems to tree trunks or to a brick wall. Adventitious root formation in ivy depends on the developmental stage of the shoot. When the shoot enters the adult phase of development, it is no longer capable of initiating these roots. Here we investigate functions of modified roots.

Prop roots. Some monocots, such as corn, produce thick adventitious roots from the lower parts of the stem. These

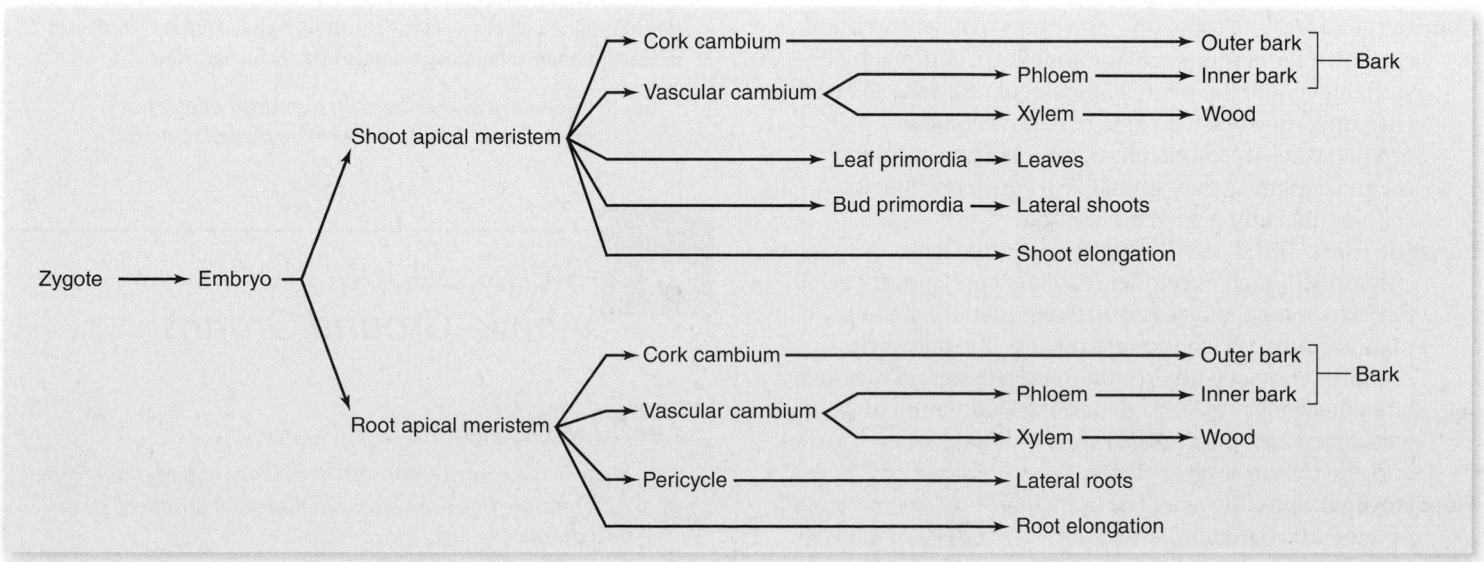

Figure 36.18 Stages in the differentiation of plant tissues.

Figure 36.19 Five types of modified roots. *a.* Maize (corn) prop roots originate from the stem and keep the plant upright. *b.* Epiphytic orchids attach to trees far above the tropical soil. Their roots are adapted to obtain water from the air rather than the soil. *c.* Pneumatophores *(foreground)* are spongy outgrowths from the roots below. *d.* A water storage root weighing over 25 kg (60 pounds). *e.* Buttress roots of a tropical fig tree.

a.

b.

so-called prop roots grow down to the ground and brace the plants against wind (figure 36.19*a*). Adventitious roots are common in wetland plants, allowing them to tolerate wet conditions.

Aerial roots. Plants such as epiphytic orchids, which are attached to tree branches and grow unconnected to the ground (but are not parasites), have roots that extend into the air (figure 36.19*b*). Some aerial roots have an epidermis that is several cell layers thick, an adaptation to reduce water loss. These aerial roots may also be green and photosynthetic, as in the vanilla orchid *(Vanilla planifolia)*.

Pneumatophores. Some plants that grow in swamps and other wet places may produce spongy outgrowths called *pneumatophores* from their underwater roots (figure 36.19*c*). The pneumatophores commonly extend several centimeters above water, facilitating oxygen uptake in the roots beneath (figure 36.19*c*).

Contractile roots. The roots from the bulbs of lilies and from several other plants, such as dandelions, contract by spiraling to pull the plant a little deeper into the soil each year, until they reach an area of relatively stable temperature. The roots may contract to one-third their original length as they spiral like a corkscrew due to cellular thickening and constricting.

Parasitic roots. The stems of certain plants that lack chlorophyll, such as dodder *(Cuscuta* spp.), produce peglike roots called *haustoria* that penetrate the host plants around which they are twined. The haustoria establish contact with the conducting tissues of the host and effectively parasitize their host. Dodder not only weakens plants but can also spread disease when it grows and attaches to several plants.

Food storage roots. The xylem of branch roots of sweet potatoes and similar plants produce at intervals many extra parenchyma cells that store large quantities of carbohydrates. Carrots, beets, parsnips, radishes, and

turnips have combinations of stem and root that also function in food storage. Cross sections of these roots reveal multiple rings of secondary growth.

Water storage roots. Some members of the pumpkin family (Cucurbitaceae), especially those that grow in arid regions, may produce water storage roots weighing 50 kg or more (figure 36.19*d*).

Buttress roots. Certain species of fig and other tropical trees produce huge buttress roots toward the base of the trunk, which provide considerable stability (figure 36.19*e*).

Learning Outcomes Review 36.3

The root cap protects the root apical meristem and helps to sense gravity. New cells formed in the zone of cell division grow in length in the zone of elongation. Cells differentiate in the zone of maturation, and root hairs appear here. Root hairs greatly increase the absorptive surface area of roots. Modified roots allow plants to carry out many additional functions, including bracing, aeration, and storage of nutrients and water.

■ *In what environment would a mutant allele that decreases root hair formation be likely to persist?*

36.4 Stems: Support for Above-Ground Organs

Learning Outcomes

1. *Identify the contributions of an axillary bud to plant form.*
2. *Differentiate between cross sections of a monocot stem and a eudicot stem.*
3. *Describe three functions of modified stems.*

c.

d.

e.

The supporting structure of a vascular plant's shoot system is the mass of stems that extend from the root system below ground into the air, often reaching great height. Stiff stems capable of rising upward against gravity are an ancient adaptation that allowed plants to move into terrestrial ecosystems.

Stems carry leaves and flowers and support the plant's weight

Like roots, stems contain the three types of plant tissue. Stems also undergo growth from cell division in apical and lateral meristems. The stem may be thought of as an axis from which other stems or organs grow. The shoot apical meristems are capable of producing these new stems and organs.

External stem structure

The shoot apical meristem initiates stem tissue and intermittently produces bulges (primordia) that are capable of developing into leaves, other shoots, or even flowers (figure 36.20). Leaves may be arranged in a spiral around the stem, or they may be in pairs opposite or alternate to one another; they also may occur in whorls (circles) of three or more (figure 36.21).

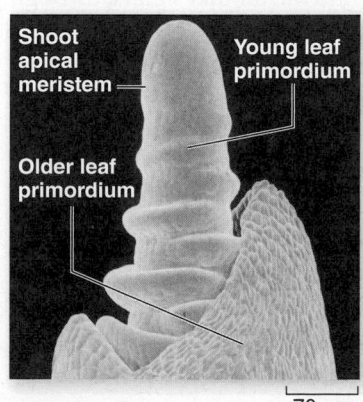

Figure 36.20 A shoot apex. Scanning electron micrograph of the apical meristem of wheat (*Triticum*).

The spiral arrangement is the most common, and for reasons still not understood, sequential leaves tend to be placed 137.5° apart. This angle relates to the golden mean, a mathematical ratio found in nature. The angle of coiling in shells of some gastropods is the same. The golden mean has been used in classical architecture (the Greek Parthenon wall dimensions), and even in modern art (for example, in paintings by Mondrian). In plants, the pattern of leaf arrangement, called **phyllotaxy,** may optimize the exposure of leaves to the Sun.

The region or area of leaf attachment to the stem is called a **node;** the portion of stem between two nodes is called an **internode.** A leaf usually has a flattened blade and sometimes a petiole (stalk). The angle between a leaf's petiole (or blade) and the stem is called an **axil.** An **axillary bud** is produced in each axil. This bud is a product of the primary shoot apical meristem, and it is itself a shoot apical meristem. Axillary buds frequently develop into branches with leaves or may form flowers.

Neither monocots nor herbaceous eudicot stems produce a cork cambium. The stems in these plants are usually

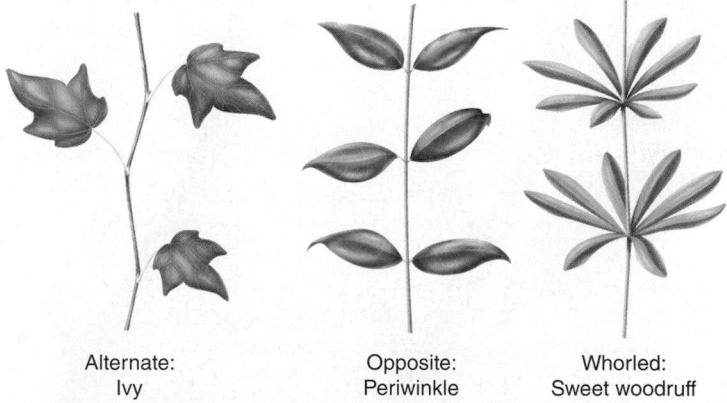

Alternate: Ivy

Opposite: Periwinkle

Whorled: Sweet woodruff

Figure 36.21 Types of leaf arrangements. The three common types of leaf arrangements are alternate, opposite, and whorled.

green and photosynthetic, with at least the outer cells of the cortex containing chloroplasts. Herbaceous stems commonly have stomata, and may have various types of trichomes (hairs).

Woody stems can persist over a number of years and develop distinctive markings, including scars from buds and leaves that are no longer attached (figure 36.22). Terminal buds usually extend the length of the shoot system during the growing season. Some buds, such as those of geraniums, are unprotected, but most buds of woody plants have protective winter bud scales that drop off, leaving tiny bud scale scars as the buds expand.

Some twigs have tiny scars of a different origin. A pair of butterfly-like appendages called *stipules* (part of the leaf) develop at the base of some leaves. The stipules can fall off and leave stipule *scars*. When the leaves of deciduous trees drop in the fall, they leave leaf scars with tiny bundle scars, marking where vascular connections were. The shapes, sizes, and other features of leaf scars can be distinctive enough to identify deciduous plants in winter, when they lack leaves (see figure 36.22).

Internal stem structure

A major distinguishing feature between monocot and eudicot stems is the organization of the vascular tissue system (figure 36.23). Most monocot vascular bundles are scattered throughout the ground tissue system, whereas eudicot vascular tissue is arranged in a ring with internal ground tissue *(pith)* and external ground tissues *(cortex)*. The arrangement of vascular tissue is directly related to the ability of the stem to undergo secondary growth. In eudicots, a vascular cambium may develop between the primary xylem and primary phloem (figure 36.24). In many ways, this is a connect-the-dots game in which the vascular cambium connects the ring of primary vascular bundles. There is no logical way to connect primary monocot vascular tissue that would allow a uniform increase in girth. Lacking a vascular cambium, therefore, monocots do not have secondary growth.

Rings in the stump of a tree reveal annual patterns of vascular cambium growth; cell size varies, depending on growth

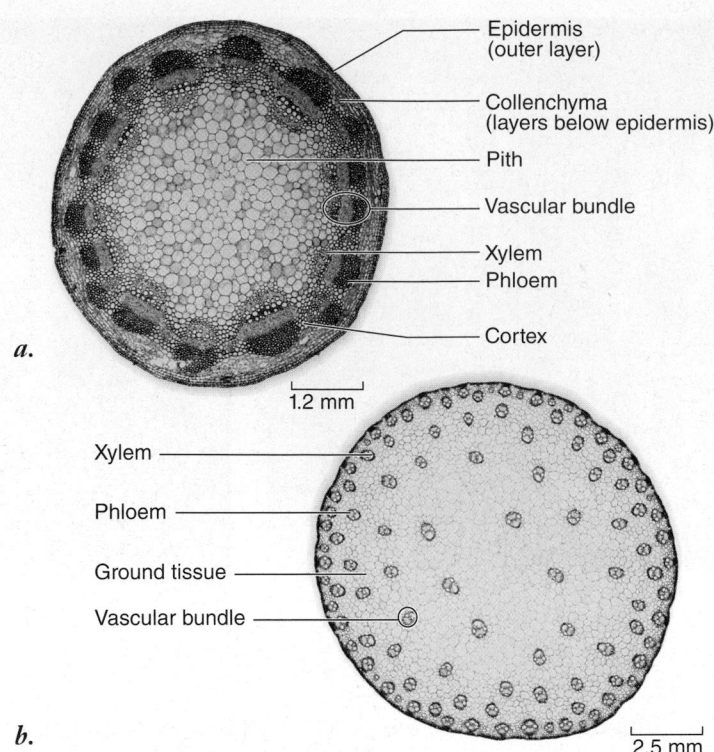

a.

1.2 mm

Xylem

Phloem

Ground tissue

Vascular bundle

b.

2.5 mm

Figure 36.23 Stems. Transverse sections of a young stem in *(a)* a eudicot, the common sunflower *(Helianthus annuus),* in which the vascular bundles are arranged around the outside of the stem; and *(b)* a monocot, corn *(Zea mays),* with characteristically scattered vascular bundles.

conditions (figure 36.25). Large cells form under favorable conditions such as abundant rainfalls. Rings of smaller cells mark the seasons where growth is limited. In woody eudicots and gymnosperms, a second cambium, the cork cambium, arises in the outer cortex (occasionally in the epidermis or phloem); it produces boxlike cork cells to the outside and also may produce parenchyma-like phelloderm cells to the inside (figure 36.26).

The cork cambium, cork, and phelloderm are collectively referred to as the *periderm* (see figure 36.26). Cork tissues, the cells of which become impregnated with water-repellent suberin shortly after they are formed and which then die, constitute the *outer bark*. The cork tissue cuts off water and food to the epidermis, which dies and sloughs off. In young stems, gas exchange between stem tissues and the air takes place through stomata, but as the cork cambium produces cork, it also produces patches of unsuberized cells beneath the stomata. These unsuberized cells, which permit gas exchange to continue, are called *lenticels* (figure 36.27).

Modified stems carry out vegetative propagation and store nutrients

Although most stems grow erect, some have modifications that serve special purposes, including natural vegetative propagation. In fact, the widespread artificial vegetative propagation of

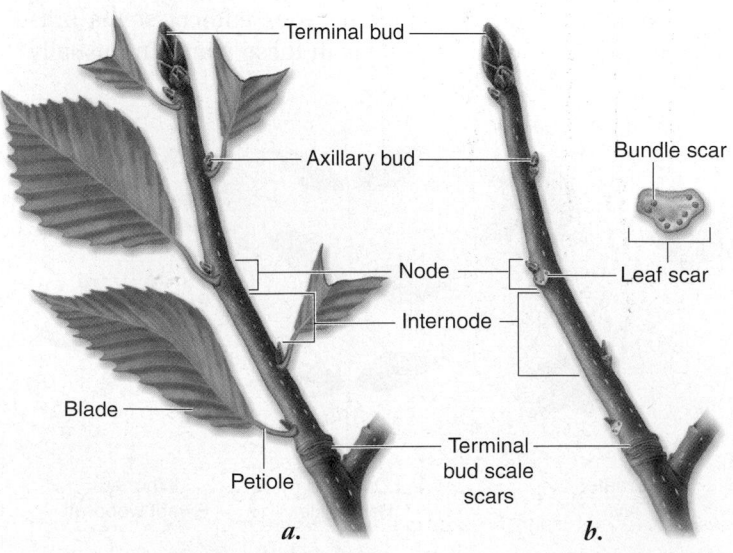

a.   *b.*

Figure 36.22 A woody twig. *a.* In summer. *b.* In winter.

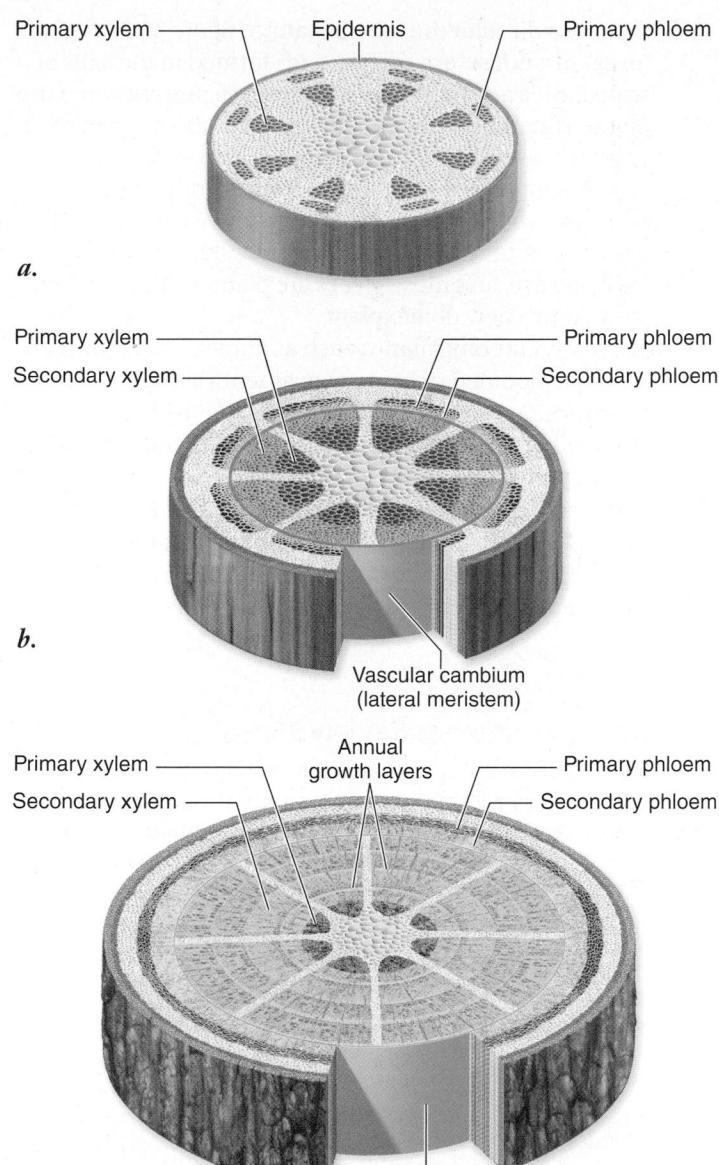

a.

Primary xylem — Epidermis — Primary phloem

b.

Primary xylem — Primary phloem
Secondary xylem — Secondary phloem

Vascular cambium
(lateral meristem)

c.

Primary xylem — Annual growth layers — Primary phloem
Secondary xylem — Secondary phloem

Vascular cambium
(lateral meristem)

Figure 36.24 Secondary growth. *a.* Before secondary growth begins in eudicot stems, primary tissues continue to elongate as the apical meristems produce primary growth. ***b.*** As secondary growth begins, the vascular cambium produces secondary tissues, and the stem's diameter increases. ***c.*** In this four-year-old stem, the secondary tissues continue to widen, and the trunk has become thick and woody. Note that the vascular cambium forms a cylinder that runs axially (up and down) in the roots and shoots that have them.

plants, both commercial and private, frequently involves cutting modified stems into segments, which are then planted, producing new plants. As you become acquainted with the following modified stems, keep in mind that stems have leaves at nodes, with internodes between the nodes, and buds in the axils of the leaves, whereas roots have no leaves, nodes, or axillary buds.

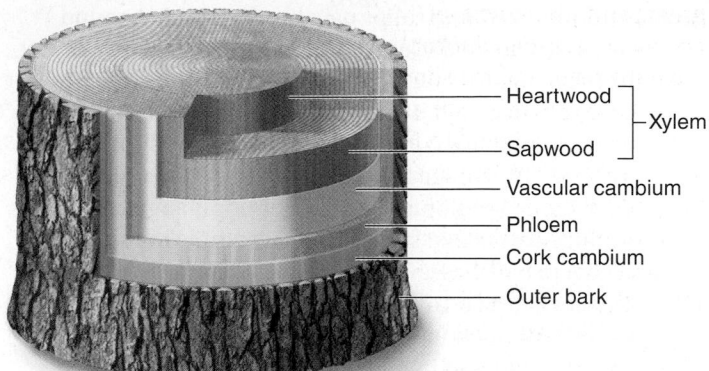

Heartwood ⎱ Xylem
Sapwood ⎰
Vascular cambium
Phloem
Cork cambium
Outer bark

Figure 36.25 Tree stump. The vascular cambium produces rings of xylem (sapwood and nonconducting heartwood) and phloem, and the cork cambium produces the cork.

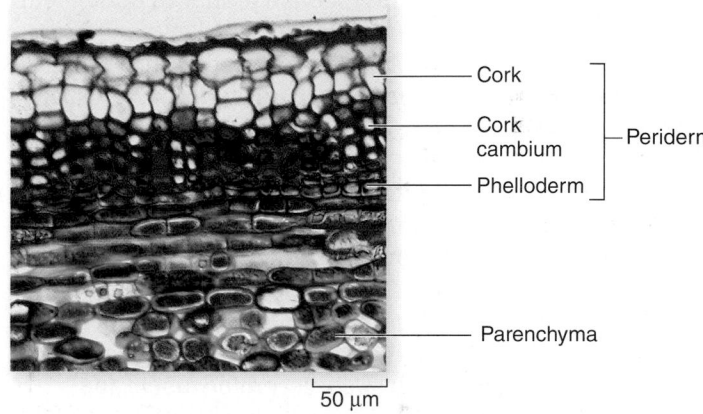

Cork
Cork cambium ⎱ Periderm
Phelloderm

Parenchyma

50 μm

Figure 36.26 Section of periderm. An early stage in the development of periderm in cottonwood, *Populus* sp.

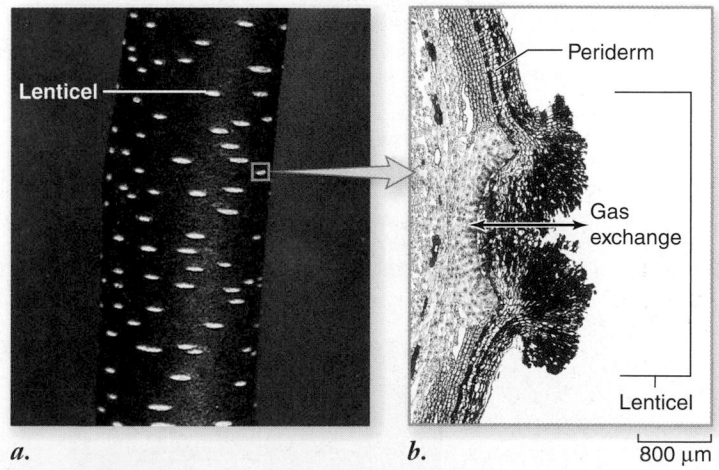

Lenticel

Periderm

Gas exchange

Lenticel

a. **b.**

800 μm

Figure 36.27 Lenticels. *a.* Lenticels, the numerous small, pale, raised areas shown here on cherry tree bark *(Prunus cerasifera),* allow gas exchange between the external atmosphere and the living tissues immediately beneath the bark of woody plants. ***b.*** Transverse section through a lenticel in a stem of elderberry, *Sambucus canadensis.*

Bulbs. Onions, lilies, and tulips have swollen underground stems that are really large buds with adventitious roots at the base (figure 36.28a). Most of a bulb consists of fleshy leaves attached to a small, knoblike stem. For most bulbs, next year's foliage comes from the tip of the shoot apex, protected by storage leaves from the previous year.

Corms. Crocuses, gladioluses, and other popular garden plants produce corms that superficially resemble bulbs. Cutting a corm in half, however, reveals no fleshy leaves. Instead, almost all of a corm consists of stem, with a few papery, brown nonfunctional leaves on the outside, and adventitious roots below.

Rhizomes. Perennial grasses, ferns, bearded iris, and many other plants produce rhizomes, which typically are horizontal stems that grow underground, often close to the surface (figure 36.28b). Each node has an inconspicuous scalelike leaf with an axillary bud; much larger photosynthetic leaves may be produced at the rhizome tip. Adventitious roots are produced throughout the length of the rhizome, mainly on the lower surface.

Runners and stolons. Strawberry plants produce horizontal stems with long internodes that unlike rhizomes, usually grow along the surface of the ground. Several runners may radiate out from a single plant (figure 36.28c). Some biologists use the term *stolon* synonymously with runner; others reserve the term *stolon* for a stem with long internodes (but no roots) that grows underground, as seen in potato plants (*Solanum* sp.). A potato itself, however, is another type of modified stem—a tuber.

Tubers. In potato plants, carbohydrates may accumulate at the tips of rhizomes, which swell, becoming tubers; the

rhizomes die after the tubers mature (figure 36.28d). The "eyes" of a potato are axillary buds formed in the axils of scalelike leaves. These leaves, which are present when the potato is starting to form, soon drop off; the tiny ridge adjacent to each "eye" of a mature potato is a leaf scar.

Crop potatoes are not grown from seeds produced by potato flowers, but propagated vegetatively from "seed potatoes." A tuber is cut up into pieces that contain at least one eye, and these pieces are planted. The eye then grows into a new potato plant.

Tendrils. Many climbing plants, such as grapes and English ivy, produce modified stems known as tendrils that twine around supports and aid in climbing (figure 36.28e). Some other tendrils, such as those of peas and pumpkins, are actually modified leaves or leaflets.

Cladophylls. Cacti and several other plants produce flattened, photosynthetic stems called cladophylls that resemble leaves (figure 36.28f). In cacti, the real leaves are modified as spines (see the following section).

Learning Outcomes Review 36.4

Shoots grow from apical and lateral meristems. Axillary buds may develop into branches, flowers, or leaves. In monocots, vascular tissue is evenly spaced throughout the stem ground tissue; in eudicots, vascular tissue is arranged in a ring with inner and outer ground tissues. Some plants produce modified stems for support, vegetative reproduction, or nutrient storage.

■ *Explain how a plant could flower and reproduce if an herbivore ate the shoot tip.*

Figure 36.28
Types of modified stems. *a.* Bulb. *b.* Adventitious roots. *c.* Runner. *d.* Stolon. *e.* Tendril. *f.* Cladophyll.

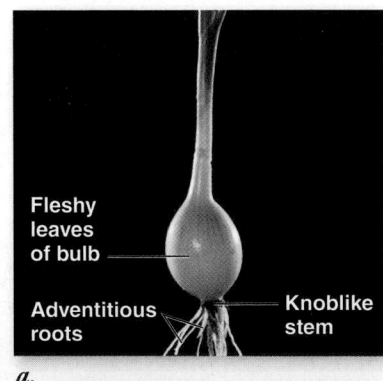

Fleshy leaves of bulb
Adventitious roots
Knoblike stem
a.

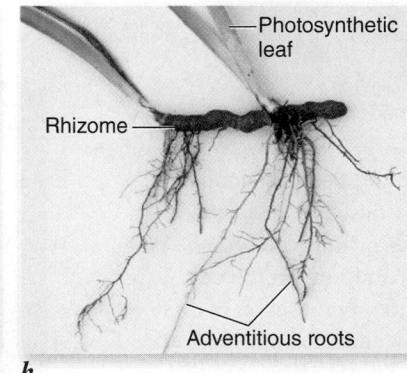

Photosynthetic leaf
Rhizome
Adventitious roots
b.

Runner
c.

Tuber (swollen tip of stolon)
Stolon
d.

Tendril
e.

Leaves
Cladophyll
f.

36.5 Leaves: Photosynthetic Organs

Learning Outcomes

1. Relate leaf structure to its role in photosynthesis.
2. Explain how leaf structure affects gas exchange and water loss.

Leaves, which are initiated as primordia by the apical meristems (see figure 36.20), are vital to life as we know it because they are the principal sites of photosynthesis on land, providing the base of the food chain. Leaves expand by cell enlargement and cell division. Like arms and legs in humans, they are determinate structures, which means their growth stops at maturity. Because leaves are crucial to a plant, features such as their arrangement, form, size, and internal structure are highly significant and can differ greatly. Different patterns have adaptive value in different environments.

Leaves are an extension of the shoot apical meristem and stem development. When they first emerge as primordia, they are not committed to being leaves. Experiments in which very young leaf primordia are isolated from fern and coleus plants and grown in culture have demonstrated this feature: If the primordia are young enough, they will form an entire shoot rather than a leaf. The positioning of leaf primordia and the initial cell divisions occur before those cells are committed to the leaf developmental pathway.

External leaf structure reflects vascular morphology

Leaves fall into two different morphological groups, which may reflect differences in evolutionary origin. A **microphyll** is a leaf with one vein branching from the vascular cylinder of the stem and not extending the full length of the leaf; microphylls are mostly small and are associated primarily with the phylum Lycophyta (see chapter 30). Most plants have leaves called **megaphylls,** which have several to many veins.

Most eudicot leaves have a flattened **blade** and a slender stalk, the **petiole,** while grasses and other monocots usually lack a petiole. The flattening of the leaf blade reflects a shift from radial symmetry to dorsal–ventral (top–bottom) symmetry. Leaf flattening increases the photosynthetic surface. Plant biologists are just beginning to understand how this shift occurs by analyzing mutants lacking distinct tops and bottoms (figure 36.29).

In addition, leaves may have a pair of **stipules,** which are outgrowths at the base of the petiole. The stipules, which may be leaflike or modified as spines (as in the black locust, *Robinia pseudo-acacia*) or glands (as in the purple-leaf plum tree *Prunus cerasifera*), vary considerably in size from microscopic to almost half the size of the leaf blade.

Veins (a term used for the vascular bundles in leaves) consist of both xylem and phloem and are distributed throughout the leaf blades. The main veins are parallel in most

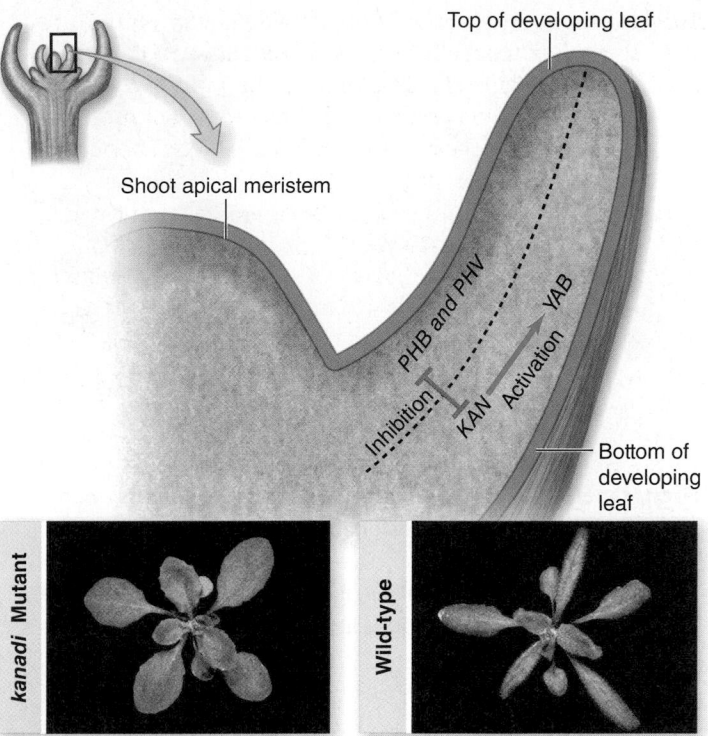

Figure 36.29 Establishing top and bottom in leaves. Several genes, including *PHABULOSA (PHB), PHAVOLUTA (PHV), KANADI (KAN),* and *YABBY (YAB)* make a flattened *Arabidopsis* leaf with a distinct upper and lower surface. *PHB* and *PHV* RNAs are restricted to the top; *KAN* and *YAB* are expressed in the bottom cells of a leaf. *PHB* and *KAN* have an antagonistic relationship, restricting expression of each to separate leaf regions. *KAN* leads to *YABBY* expression and lower leaf development. Without *KAN*, both sides of the leaf develop like the top portion.

monocot leaves; the veins of eudicots, on the other hand, form an often intricate network (figure 36.30).

Leaf blades come in a variety of forms, from oval to deeply lobed to having separate leaflets. In **simple leaves**

a.

b.

Figure 36.30 Eudicot and monocot leaves. *a.* The leaves of eudicots, such as this African violet relative from Sri Lanka, have netted, or reticulate, veins. *b.* Those of monocots, such as this cabbage palmetto, have parallel veins. The eudicot leaf has been cleared with chemicals and stained with a red dye to make the veins show more clearly.

a.   *b.*

Figure 36.31 Simple versus compound leaves. *a.* A simple leaf, its margin deeply lobed, from the English oak tree (*Quercus robur*). *b.* A pinnately compound leaf, from a black walnut (*Juglans nigra*). A compound leaf is associated with a single lateral bud, located where the petiole is attached to the stem.

(figure 36.31*a*), such as those of lilacs or birch trees, the blades are undivided, but simple leaves may have teeth, indentations, or lobes of various sizes, as in the leaves of maples and oaks.

In **compound leaves** (figure 36.31*b*), such as those of ashes, box elders, and walnuts, the blade is divided into *leaflets*. The relationship between the development of compound and simple leaves is an open question. Two explanations are being debated: (1) A compound leaf is a highly lobed simple leaf, or (2) a compound leaf utilizes a shoot development program, and each leaflet was once a leaf. To address this question, researchers are using single mutations that are known to convert compound leaves to simple leaves (figure 36.32).

Internal leaf structure regulates gas exchange and evaporation

The entire surface of a leaf is covered by a transparent epidermis, and most of these epidermal cells have no chloroplasts. As described earlier, the epidermis has a waxy cuticle, and different types of glands and trichomes may be present. Also, the lower epidermis (and occasionally the upper epidermis) of most leaves contains numerous slitlike or mouth-shaped stomata flanked by guard cells (figure 36.33).

The tissue between the upper and lower epidermis is called **mesophyll.** Mesophyll is interspersed with veins of various sizes.

Most eudicot leaves have two distinct types of mesophyll. Closest to the upper epidermis are one to several (usually two) rows of tightly packed, barrel-shaped to cylindrical chlorenchyma cells (parenchyma with chloroplasts) that constitute the palisade mesophyll (figure 36.34). Some plants, including species of *Eucalyptus,* have leaves that hang down, rather than extend horizontally. They have palisade mesophyll on both sides of the leaf.

Nearly all eudicot leaves have loosely arranged spongy mesophyll cells between the palisade mesophyll and the lower

Hypothesis: *The gene UNIFOLIATA (UNI) is necessary for compound leaf development in garden pea,* Pisum sativum.

Prediction: *Developing leaf primordia of a wild-type pea plant will express the* UNI *gene while* uni *mutant plants will not express the gene.*

Test: *Cut thin sections of wild-type and mutant pea shoots and place them on a microscope slide. Test for the presence of UNI RNA using a color-labeled, single-stranded DNA probe that will hybridize (bind) only the UNI RNA. View the labeled sections under a microscope.*

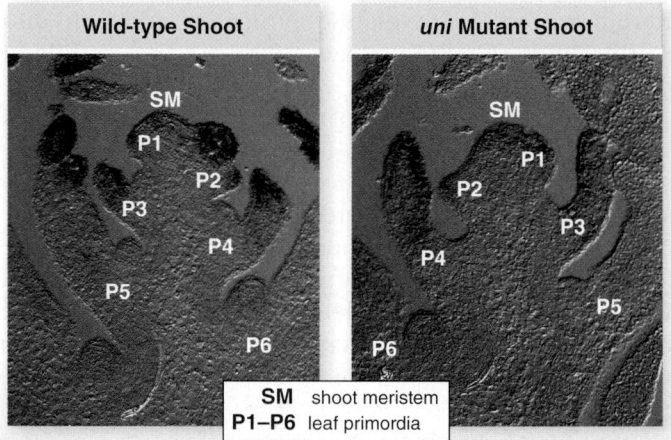

SM	shoot meristem
P1–P6	leaf primordia

Result: *UNI RNA was found in the wild-type and also in mutant leaves, but at lower levels.*

Conclusion: *The mutant* uni *gene is transcribed but at lower levels than the wild-type gene. Thus the prediction was incorrect.*

Further Experiments: *Although the* uni *gene is expressed, compound leaves do not develop. Refine the hypothesis and propose an experiment to test the revised hypothesis.*

Figure 36.32 Genetic regulation of leaf development.

epidermis, with many air spaces throughout the tissue. The interconnected intercellular spaces, along with the stomata, function in gas exchange and the passage of water vapor from the leaves.

The mesophyll of monocot leaves often is not differentiated into palisade and spongy layers, and there is often little distinction between the upper and lower epidermis. Instead, cells surrounding the vascular tissue are distinctive and are the site of carbon fixation. This anatomical difference often correlates with a modified photosynthetic pathway, C_4 *photosynthesis,* that maximizes the amount of CO_2 relative to O_2 to reduce energy loss through photorespiration (see chapter 8). The anatomy of a leaf directly relates to its juggling act of balancing

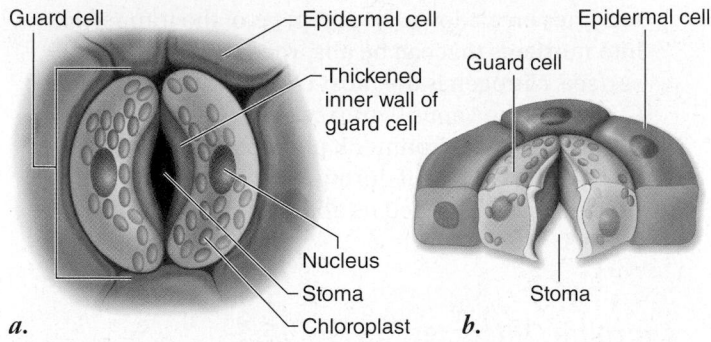

Figure 36.33 **A stoma.** *a.* Surface view. *b.* View in cross section.

water loss, gas exchange, and transport of photosynthetic products to the rest of the plant.

Modified leaves are highly versatile organs

As plants colonized a wide variety of environments, from deserts to lakes to tropical rain forests, plant organ modifications arose that would adapt the plants to their specific habitats. Leaves, in particular, have evolved some remarkable adaptations. A brief discussion of a few of these modifications follows:

Floral leaves (bracts). Poinsettias and dogwoods have relatively inconspicuous, small, greenish yellow flowers. However, both plants produce large modified leaves called *bracts* (mostly colored red in poinsettias and white or pink in dogwoods). These bracts surround the true flowers and perform the same function as showy petals.

In other plants, however, bracts can be quite small and inconspicuous.

Spines. The leaves of many cacti and other plants are modified as *spines* (see figure 36.28*f*). In cacti, sharp spines may deter predators. Spines should not be confused with *thorns*, such as those on the honey locust (*Gleditsia triacanthos*), which are modified stems, or with the prickles on raspberries, which are simply outgrowths from the epidermis or the cortex just beneath it.

Reproductive leaves. Several plants, notably *Kalanchoë*, produce tiny but complete plantlets along their margins. Each plantlet, when separated from the leaf, is capable of growing independently into a full-sized plant. The walking fern (*Asplenium rhizophyllum*) produces new plantlets at the tips of its fronds. Although many species can regenerate a whole plant from isolated leaf tissue, this in vivo regeneration is found among just a few species.

Window leaves. Several genera of plants growing in arid regions produce succulent, cone-shaped leaves with transparent tips. The leaves often become mostly buried in sand blown by the wind, but the transparent tips, which have a thick epidermis and cuticle, admit light to the hollow interiors. This strategy allows photosynthesis to take place beneath the surface of the ground.

Shade leaves. Leaves produced in the shade, where they receive little sunlight, tend to be larger in surface area, but thinner and with less mesophyll than leaves on the same tree receiving more direct light. This plasticity in development is remarkable. Environmental signals can have a major effect on development.

Insectivorous leaves. Almost 200 species of flowering plants are known to have leaves that trap insects; some plants digest the insects' soft parts. Plants with insectivorous leaves often grow in acid swamps that are deficient in

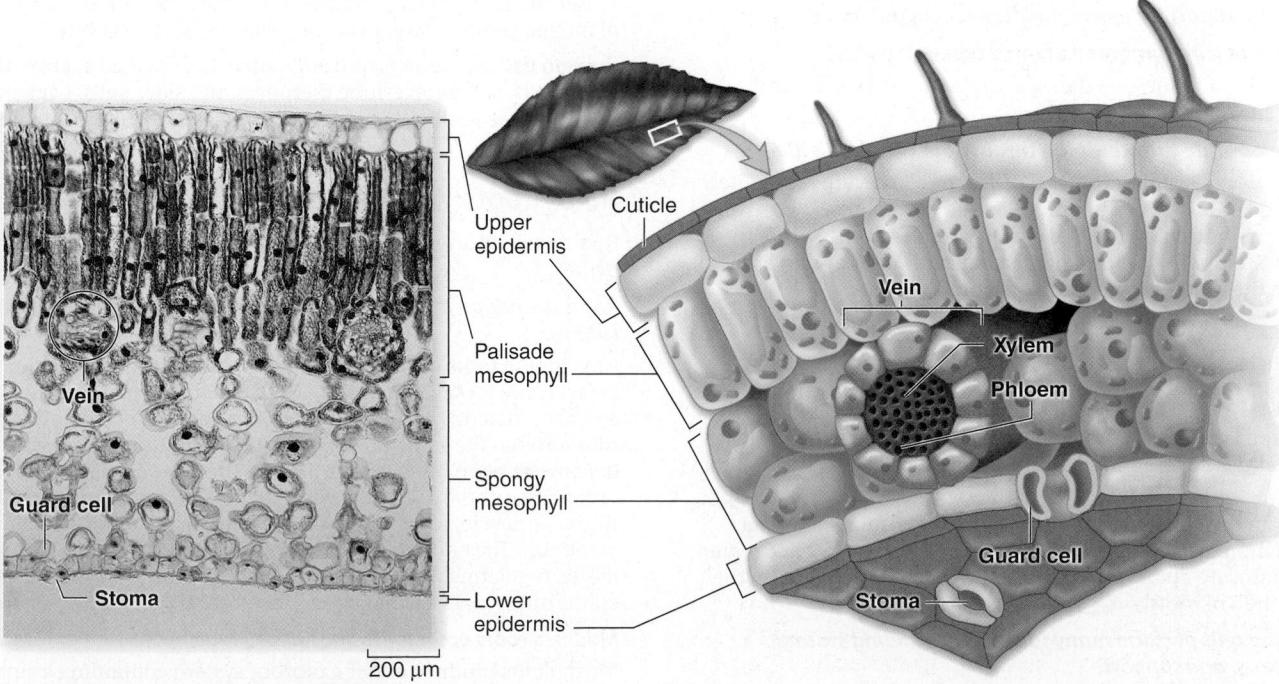

Figure 36.34 **A leaf in cross section.** Transection of a leaf showing the arrangement of palisade and spongy mesophyll, a vascular bundle or vein, and the epidermis with paired guard cells flanking the stoma.

needed elements or contain elements in forms not readily available to the plants; this inhibits the plants' capacities to maintain metabolic processes needed for their growth and reproduction. Their needs are met, however, by the supplementary absorption of nutrients from the animal kingdom.

Pitcher plants (for example, *Sarracenia, Darlingtonia,* or *Nepenthes* spp.) have cone-shaped leaves in which rainwater can accumulate. The insides of the leaves are very smooth, but stiff, downward-pointing hairs line the rim. An insect falling into such a leaf finds it very difficult to escape and eventually drowns. The leaf absorbs the nutrients released when bacteria, and in most species the plant's own digestive enzymes, decompose the insect bodies. Other plants, such as sundews *(Drosera),* have glands that secrete sticky mucilage that traps insects, which are then digested by enzymes.

The Venus flytrap *(Dionaea muscipula)* produces leaves that look hinged at the midrib. When tiny trigger hairs on the leaf blade are stimulated by a moving insect, the two halves of the leaf snap shut, and digestive enzymes break down the soft parts of the trapped insect into nutrients that can be absorbed through the leaf surface. Nitrogen is the most common nutrient needed. Curiously, the Venus flytrap cannot survive in a nitrogen-rich environment, perhaps as a result of a biochemical trade-off during the intricate evolutionary process that developed its ability to capture and digest insects.

Learning Outcomes Review 36.5

Leaves come in a range of forms. A simple leaf is undivided, whereas a compound leaf has a number of separate leaflets. Monocots typically produce leaves with parallel veins, whereas those of eudicots are netted. Mesophyll cells carry out photosynthesis; in monocots, mesophyll is undifferentiated, whereas in eudicots it is divided into palisade and spongy mesophyll. Leaves may be modified for reproduction, protection, water conservation, uptake of nutrients, and even as traps for insects.

■ *Contrast monocot and eudicot leaf anatomy in the context of photosynthesis.*

Chapter Review

36.1 Organization of the Plant Body: An Overview

Vascular plants have roots and shoots.

The root system is primarily below ground; roots anchor the plant and take up water and minerals. The shoot system is above ground and provides support for leaves and flowers (figure 36.1).

Roots and shoots are composed of three types of tissues.

The three types of tissues are dermal tissue, ground tissue, and vascular tissue.

Meristems elaborate the body plan throughout the plant's life.

Meristem cells are undifferentiated cells that can divide indefinitely and give rise to many types of differentiated cells (figure 36.3). Apical meristems are located on the tips of stems and near the tips of roots. Lateral meristems are found in plants that exhibit secondary growth. They add to the diameter of a stem or root.

36.2 Plant Tissues

Dermal tissue forms a protective interface with the environment.

Dermal tissue is primarily the epidermis, which is usually one cell layer thick and is covered with a fatty or waxy cuticle to retard water loss.

Guard cells in the epidermis control water loss through stomata, whereas trichomes protect the plant from both herbivory and water loss. Root hairs are epidermal cell structures that help increase the absorptive area of roots.

Ground tissue cells perform many functions, including storage, photosynthesis, and support.

Ground tissue is mainly composed of parenchyma cells, which function in storage, photosynthesis, and secretion. Collenchyma cells provide flexible support, and sclerenchyma cells provide rigid support.

Vascular tissue conducts water and nutrients throughout the plant.

Xylem tissue conducts water through dead cells called tracheids and vessel elements. Vessels evolved from tracheids and with a larger diameter, conduct larger volumes of water more rapidly.

Phloem tissue conducts nutrients such as dissolved sucrose through living cells called sieve-tube members and sieve cells. Lacking a nucleus, these cells rely on neighboring cells for some metabolic functions.

36.3 Roots: Anchoring and Absorption Structures

Roots evolved after shoots and are a major innovation for terrestrial living.

Roots are adapted for growing underground and absorbing water and solutes.

Developing roots exhibit four regions: (1) the root cap, which protects the root; (2) the zone of cell division, which contains the apical meristem; (3) the zone of elongation, which extends the root through the soil; and (4) the zone of maturation, in which cells become differentiated (figure 36.14). The zone of maturation has a radial organization with the vascular stele at the center, surrounded by the pericycle, which can produce lateral roots and vascular cambium. The endodermis is between the pericycle and the ground tissue, regulating what enters the vascular stele. At the surface, the epidermis regulates what enters the ground tissue.

Modified roots accomplish specialized functions.

Most plants produce either a taproot system containing a single large root with smaller branch roots, or a fibrous root system composed of many small roots.

Adventitious roots may be modified for support, stability, acquisition of oxygen, storage of water and food, or parasitism of a host plant.

36.4 Stems: Support for Above-Ground Organs

Stems carry leaves and flowers and support the plant's weight.

Leaves are attached to stems at nodes. The axil is the area between the leaf and stem, and an axillary bud develops in axils of eudicots.

The vascular bundles in stems of monocots are randomly scattered, whereas in eudicots the bundles are arranged in a ring.

Vascular cambium develops between the inner xylem and the outer phloem, allowing for secondary growth (figure 36.24).

Modified stems carry out vegetative propagation and store nutrients.

Bulbs, corms, rhizomes, runners and stolons, tubers, tendrils, and cladophylls are examples of modified stems. The tubers of potatoes are both a food source and a means of propagating new plants.

36.5 Leaves: Photosynthetic Organs

Leaves are the principle sites of photosynthesis. Leaf features such as their arrangement, form, size, and internal structure can be highly variable across environments.

External leaf structure reflects vascular morphology.

Vascular bundles are parallel in monocots, but form a network in eudicots. The leaves of most eudicots have a flattened blade and a slender petiole; monocots usually do not have a petiole.

Leaf blades may be simple or compound (divided into leaflets).

Internal leaf structure regulates gas exchange and evaporation.

The tissues of the leaf include the epidermis with guard cells, vascular tissue, and mesophyll in which photosynthesis takes place.

In eudicot leaves with a horizontal orientation, the mesophyll is partitioned into palisade cells near the upper surface and spongy cells near the lower surface.

The mesophyll of monocot leaves is often not differentiated.

Modified leaves are highly versatile organs.

Leaves vary greatly in form and are adapted to serve many different functions. Leaves may be modified for reproduction, protection, storage, mineral uptake, or even as insect traps to obtain nitrogen from insects.

Review Questions

UNDERSTAND

1. Which cells lack living protoplasts at maturity?
   a. Parenchyma
   b. Companion
   c. Collenchyma
   d. Sclerenchyma

2. The food-conducting cells in an oak tree are called
   a. tracheids.
   b. vessels.
   c. companion cells.
   d. sieve-tube members.

3. Root hairs form in the zone of
   a. cell division.
   b. elongation.
   c. maturation.
   d. More than one of the choices are correct.

4. Roots differ from stems because roots lack
   a. vessel elements.
   b. nodes.
   c. an epidermis.
   d. ground tissue.

5. Which statement about meristem cells in incorrect?
   a. One daughter cell of a dividing meristem cell remains meristematic and the other differentiates.
   b. Meristem cells have secondary cell walls for protection.
   c. Leaf primordia protect shoot apical meristems.
   d. Root apical meristems are protected by root caps.

6. Unlike eudicot stems, monocot stems lack
   a. vascular bundles.
   b. parenchyma.
   c. pith.
   d. epidermis.

7. The function of guard cells is to
   a. allow carbon dioxide uptake.
   b. repel insects and other herbivores.
   c. support leaf tissue.
   d. allow water uptake.

8. Palisade and spongy parenchyma are typically found in the mesophyll of
   a. monocots.
   b. eudicots.
   c. monocots and eudicots.
   d. neither monocots nor eudicots

9. In vascular plants, one difference between root and shoot systems is that
   a. root systems cannot undergo secondary growth.
   b. root systems undergo secondary growth, but do not form bark.
   c. root systems contain pronounced zones of cell elongation, whereas shoot systems do not.
   d. root systems can store food reserves, whereas stem structures do not.

10. Which of the following statements is NOT true of the stems of vascular plants?
    a. Stems are composed of repeating segments, including nodes and internodes.
    b. Primary growth only occurs at the shoot apical meristem.
    c. Vascular tissues may be arranged on the outside of the stem or scattered throughout the stem.
    d. Stems can contain stomata.

11. Which of the following plant cell type is mismatched to its function?
    a. Xylem—conducts mineral nutrients
    b. Phloem—serves as part of the bark
    c. Trichomes—reduces evaporation
    d. Collenchyma—performs photosynthesis

1. Fifteen years ago, your parents hung a swing from the lower branch of a large tree growing in your yard. When you go and sit in it today, you realize it is exactly the same height off the ground as it was when you first sat in it 15 years ago. The reason the swing is not higher off the ground as the tree has grown is that
   a. the tree trunk lacks secondary growth.
   b. the tree trunk is part of the primary growth system of the plant, but elongation is no longer occurring in that part of the tree.
   c. trees lack apical meristems and so do not get taller.
   d. you are hallucinating, because it is impossible for the swing not to have been raised off the ground as the tree grew.

2. A unique feature of plants is indeterminate growth. Indeterminate growth is possible because
   a. meristematic regions for primary growth occur throughout the entire plant body.
   b. all cell types in a plant often give rise to meristematic tissue.
   c. meristematic cells continually replace themselves.
   d. all cells in a plant continue to divide indefinitely.

3. If you were to relocate the pericycle of a plant root to the epidermal layer, how would it affect root growth?
   a. Secondary growth in the mature region of the root would not occur.
   b. The root apical meristem would produce vascular tissue in place of dermal tissue.
   c. Nothing would change because the pericycle is normally located near the epidermal layer of the root.
   d. Lateral roots would grow from the outer region of the root and fail to connect with the vascular tissue.

4. Many vegetables are grown today through hydroponics, in which the plant roots exist primarily in an aqueous solution. Which of the following root structures is no longer beneficial in hydroponics?
   a. Epidermis   c. Root cap
   b. Xylem   d. Bark

5. When you peel your potatoes for dinner, you are removing the majority of their
   a. dermal tissue.
   b. vascular tissue.
   c. ground tissue.
   d. Both a and b are correct.
   e. All of the choices are correct.

6. You can determine the age of an oak tree by counting the annual rings of _____ formed by the _____.
   a. primary xylem; apical meristem
   b. secondary phloem; vascular cambium
   c. dermal tissue; cork cambium
   d. secondary xylem; vascular cambium

7. Root hairs and lateral roots are similar in each respect except
   a. both increase the absorptive surface area of the root system.
   b. both are generally long-lived.
   c. both are multicellular.
   d. Both b and c are correct.

8. Plant organs form by
   a. cell division in meristematic tissue.
   b. cell migration into the appropriate position in the tissue.
   c. eliminating chromosomes in the precursor cells.
   d. Both a and b are corrrect.

9. Which is the correct sequence of cell types encountered in an oak tree, moving from the center of the tree out?
   a. Pith, secondary xylem, primary xylem, vascular cambium, primary phloem, secondary phloem, cork cambium, cork
   b. Pith, primary xylem, secondary xylem, vascular cambium, secondary phloem, primary phloem, cork cambium, cork
   c. Pith, primary xylem, secondary xylem, vascular cambium, secondary phloem, primary phloem, cork, cork cambium
   d. Pith, primary phloem, secondary phloem, vascular cambium, secondary xylem, primary xylem, cork cambium, cork

10. You've just bought a house with a great view of the mountains, but you have a neighbor who planted a bunch of trees that are now blocking your view. In an attempt to ultimately remove the trees and remain unlinked to the deed, you begin training several porcupines to enter the yard under the cover of night and perform a stealth operation. In order to most effectively kill the trees, you should train the porcupines to completely remove
    a. the vascular cambium.
    b. the cork.
    c. the cork cambium.
    d. the primary phloem.

SYNTHESIZE

1. If you were given an unfamiliar vegetable, how could you tell if it was a root or a stem, based on its external features and a microscopic examination of its cross section?

2. Potato tubers harvested from wet soil often have large lenticels. What is the adaptive significance of this?

3. Plant organs undergo many modifications to deal with environmental challenges. Design an imaginary, modified root, shoot, or leaf, and make a case for why it is the best example of a modified plant organ.

4. You have identified a mutant maize plant that cannot differentiate vessel cells. How would this affect the functioning of the plant?

5. Increasing human population on the planet is stretching our ability to produce sufficient food to support the world's population. If you could engineer the perfect crop plant, what features might it possess?

ONLINE RESOURCE

www.ravenbiology.com

Understand, Apply, and Synthesize—enhance your study with animations that bring concepts to life and practice tests to assess your understanding. Your instructor may also recommend the interactive eBook, individualized learning tools, and more.

Chapter 37

Transport in Plants

Chapter Contents

Introduction

Terrestrial plants face two major challenges: maintaining water and nutrient balance, and providing sufficient structural support for upright growth. The vascular system transports water, minerals, and organic molecules over great distances. Whereas the secondary growth of vascular tissue allows trees to achieve great heights, water balance alone keeps herbaceous plants upright. Think of a plant cell as a water balloon pressing against the insides of a soft-sided box, with many other balloon/box cells stacked on top. If the balloon springs a leak, the support is gone, and the box can collapse. How water, minerals, and organic molecules move between the roots and shoots of small and tall plants is the topic of this chapter.

How does water get from the roots to the top of a 10-story-high tree? Throughout human existence, curious people have wondered about this question. Plants lack muscle tissue and cannot pump fluid through a circulatory system. Nevertheless, water moves through the cell-wall spaces between the protoplasts of cells, through plasmodesmata (connections between cells), through plasma membranes, and through the interconnected, conducting elements extending throughout a plant (figure 37.1). Water first enters the roots and then moves to the xylem, the innermost vascular tissue of plants. Water rises through the xylem because of a combination of factors, and most of that water exits through the stomata in the leaves (figure 37.2).

Local changes result in long-distance movement of materials

The greatest distances traveled by water molecules and dissolved minerals are in the xylem. Water moves across cell membranes of root cells, into the vascular stele and the xylem. Once water enters the xylem of a redwood, for example, it can move upward as much as 100 m (see figure 37.2). Most of the force is "pulling" caused by **transpiration**—evaporation

from thin films of water in the stomata. This pulling occurs because water molecules stick to each other (cohesion) and to the walls of the tracheid or xylem vessel (adhesion). The result is an unusually stable column of liquid reaching great heights.

Long-distance transport also occurs in the phloem and is discussed later in this chapter. Unlike xylem transport, phloem transport is bidirectional (see figure 37.2).

Osmosis is enhanced by aquaporins

If a single plant cell is placed into water, the concentration of solutes inside the cell is greater than that of the external solution, and water moves into the cell by the process of **osmosis,** which you may recall from the discussion of membranes in chapter 5. The rate of osmosis across a membrane, however, is limited.

For a long time, scientists were puzzled when water was found to move more rapidly than predicted by osmosis alone. We now know that osmosis is enhanced by membrane water channels called aquaporins, which you first encountered in chapter 5 (figure 37.3). These transport channels occur in both plants and animals; in plants, they exist in vacuoles and plasma membranes and also allow for bulk flow across the membrane.

At least 30 different genes code for aquaporin-like proteins in *Arabidopsis.* Aquaporins speed up osmosis, but they do not change the direction of water movement. They are important in maintaining water balance within a cell and in moving water into the xylem.

Water potential and pressure gradients form a foundation for understanding local and long-distance transport in plants. The remaining sections of this chapter explore transport within and among different tissues and organs of the plant in more depth.

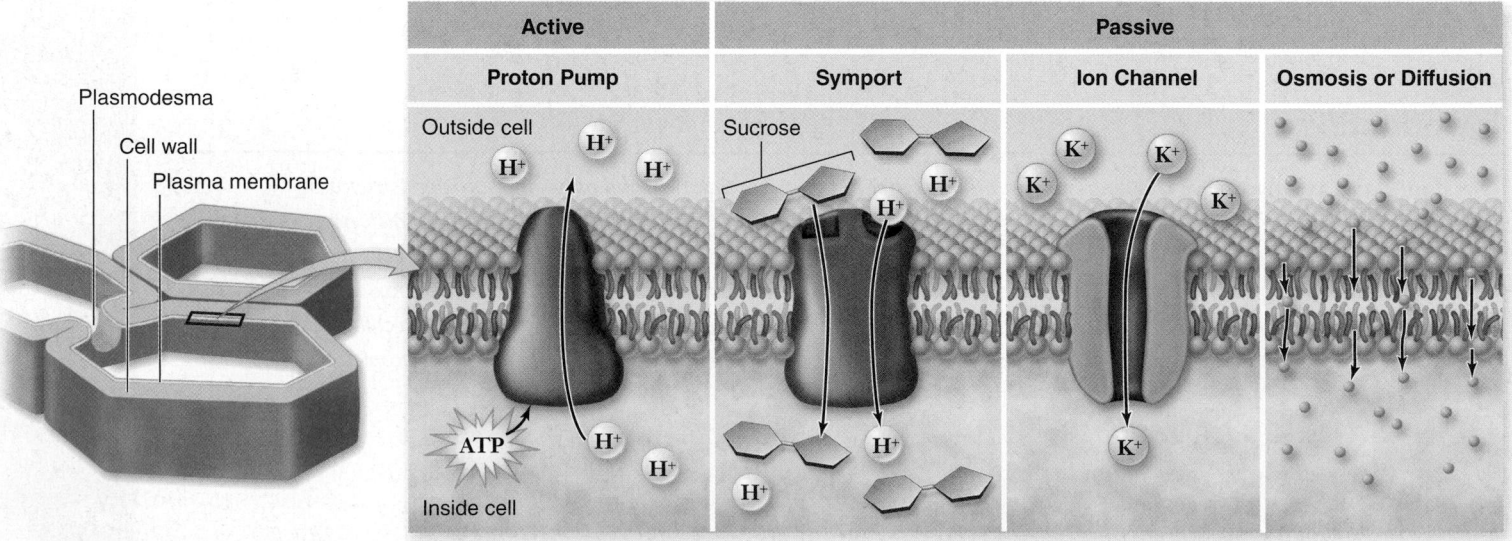

Figure 37.1 Transport between cells. Water, minerals, and organic molecules can diffuse across membranes, be actively or passively transported by membrane-bound transporters, or move through plasmodesmata. Details of membrane transport are found in chapter 5.

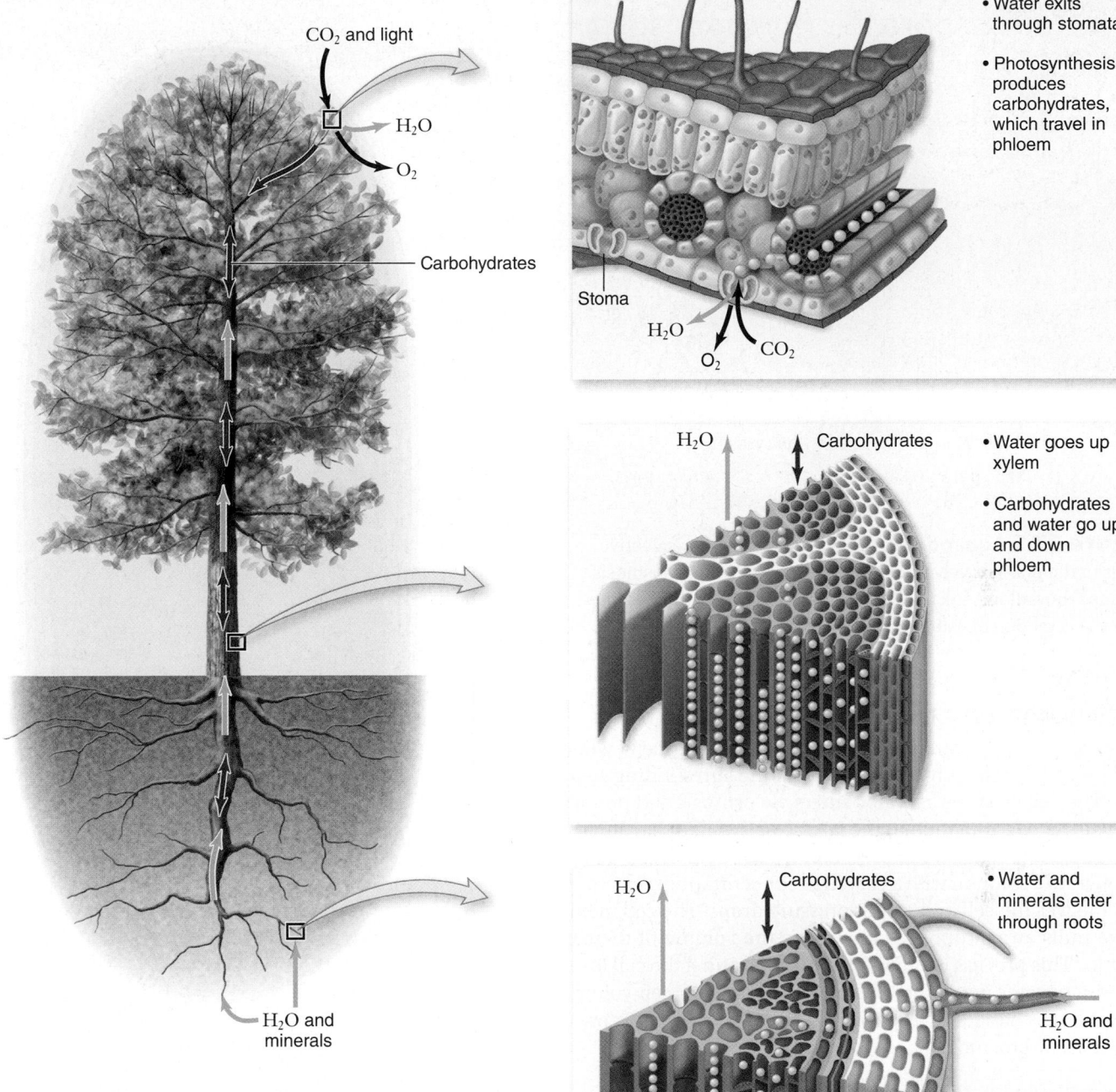

Figure 37.2 Water and mineral movement through a plant is unidirectional, whereas carbohydrates move bidirectionally.

Labels in figure:

CO₂ and light

H₂O

O₂

Carbohydrates

H₂O and minerals

- Water exits through stomata
- Photosynthesis produces carbohydrates, which travel in phloem

Stoma

H₂O

O₂

CO₂

- Water goes up xylem
- Carbohydrates and water go up and down phloem

H₂O

Carbohydrates

- Water and minerals enter through roots

H₂O

Carbohydrates

H₂O and minerals

Xylem   Phloem

Water potential regulates movement of water through the plant

Plant biologists explain the forces that act on water within a plant in terms of potentials. *Potentials* are a way of representing free energy (the potential to do work; see chapter 5). **Water potential,** abbreviated by the capital Greek letter psi (ψ) with a subscript W (ψ_w), is used to predict which way water will move. The key is to remember that water will move from a cell or solution with higher water potential to a cell or solution with lower water potential. Water potential is measured in units of pressure called **megapascals (MPa).** If you turn on your kitchen or bathroom faucet full blast, the water pressure should be between 0.2 and 0.3 MPa (30–45 psi).

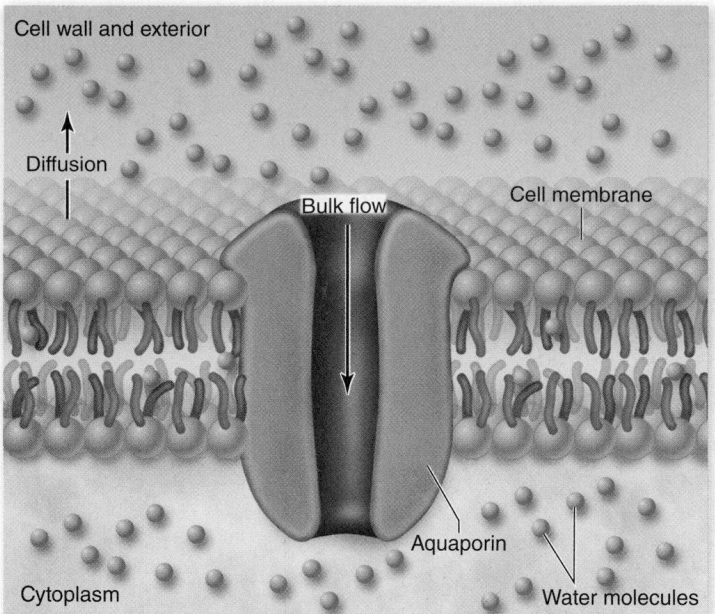

Figure 37.3 Aquaporins. Aquaporins are water-selective pores in the plasma membrane that increase the rate of osmosis because they allow bulk flow across the membrane. They do not alter the direction of water movement, however.

Water potential and turgor of cells

Plant cell walls constrain the expansion of plant cells when water enters a cell with a lower ψ_w than the surrounding solution. The cell expands as water enters via osmosis and presses against the cell wall, making it *turgid,* or swollen, because of the cell's increased turgor pressure. By contrast, if the cell is placed into a solution with a very high concentration of sucrose, water leaves the cell and turgor pressure drops. The cell membrane pulls away from the cell wall as the volume of the cell shrinks. This process is called **plasmolysis,** and if the cell loses too much water it will die. Even a tiny change in cell volume causes large changes in turgor pressure. When the turgor pressure falls to zero, most plants will wilt.

Calculation of water potential

A change in turgor pressure can be predicted more accurately by calculating the water potential of the cell and the surrounding solution. Water potential has two components: (1) physical forces, such as pressure on a plant cell wall or gravity, and (2) the concentration of solute in each solution (figure 37.4a). Water will always move, via osmosis, in the direction of lower water potential. For example, a waterfall moves downward because gravity is the major factor, and the water potential at the bottom of the waterfall is less than at the top because of the gravitational field.

In terms of physical forces, the contribution of gravity to water potential is so small at the level of a cell that it is generally not included in calculations unless you are considering a very tall tree. The turgor pressure, resulting from pressure against the cell wall, is referred to as **pressure potential (ψ_p).** As turgor pressure increases, ψ_p increases. A beaker of water containing dissolved sucrose, however, is not bounded by a cell

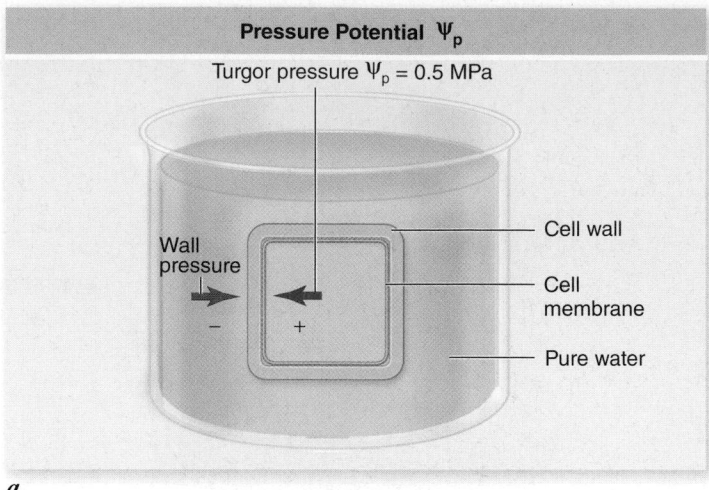

a.

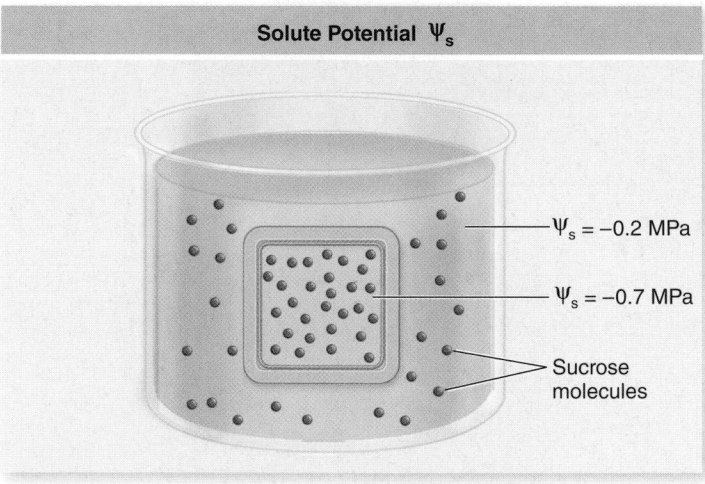

b.

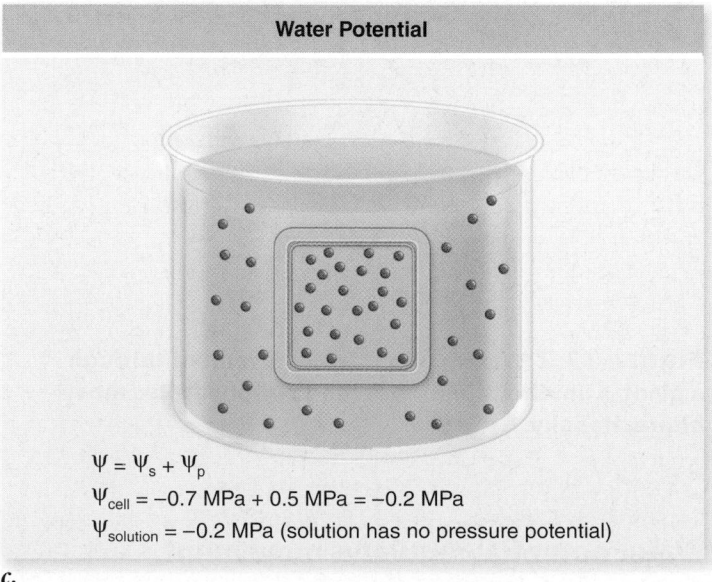

$$\psi = \psi_s + \psi_p$$
$$\psi_{cell} = -0.7 \text{ MPa} + 0.5 \text{ MPa} = -0.2 \text{ MPa}$$
$$\psi_{solution} = -0.2 \text{ MPa (solution has no pressure potential)}$$

c.

Figure 37.4 Determining water potential. *a.* Cell walls exert pressure in the opposite direction of cell turgor pressure. *b.* Using the given solute potentials, predict the direction of water movement based only on solute potential. *c.* Total water potential is the sum of ψ_s and ψ_p. Since the water potential inside the cell equals that of the solution, there is no net movement of water.

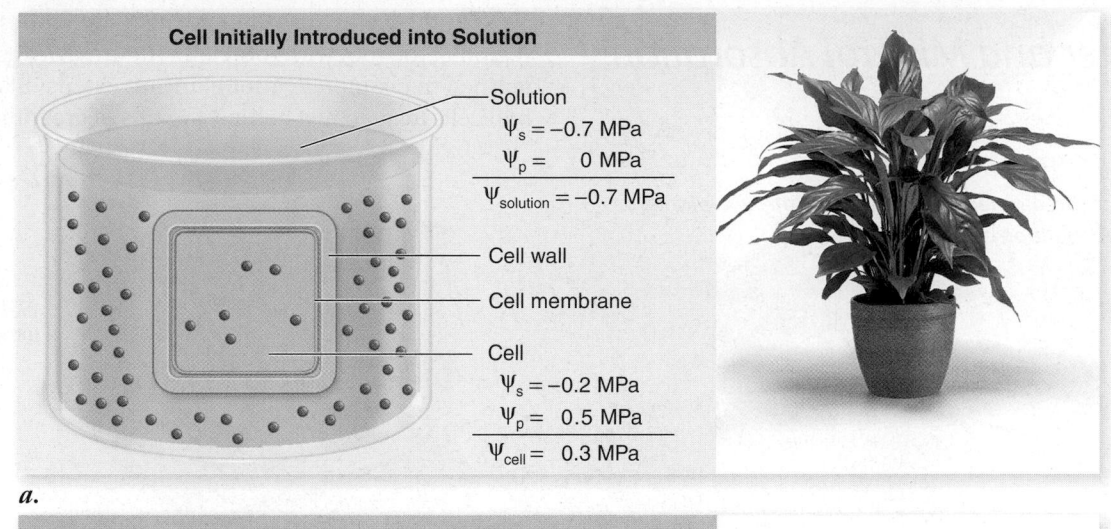

Cell Initially Introduced into Solution

Solution
$\psi_s = -0.7$ MPa
$\psi_p = 0$ MPa
$\overline{\psi_{\text{solution}} = -0.7 \text{ MPa}}$

Cell wall

Cell membrane

Cell
$\psi_s = -0.2$ MPa
$\psi_p = 0.5$ MPa
$\overline{\psi_{\text{cell}} = 0.3 \text{ MPa}}$

a.

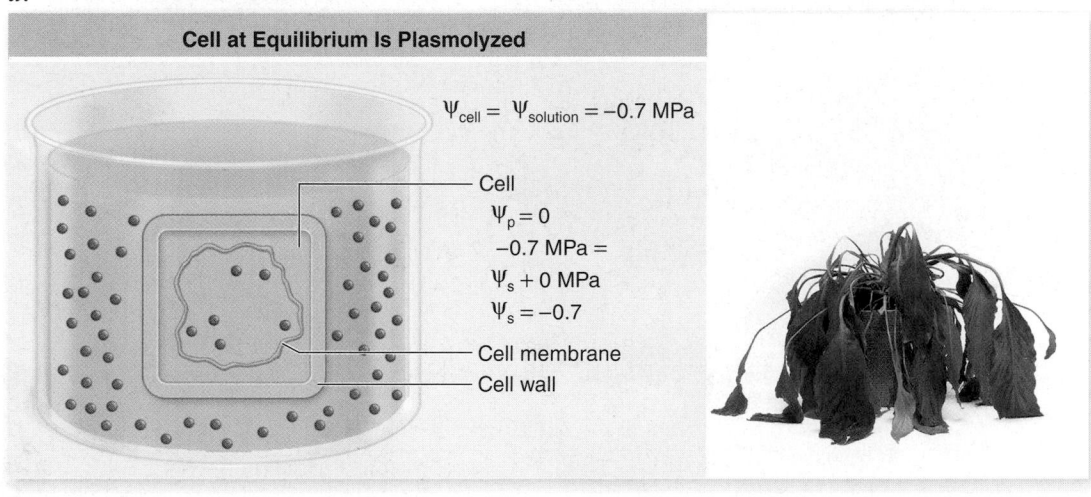

Cell at Equilibrium Is Plasmolyzed

$\psi_{\text{cell}} = \psi_{\text{solution}} = -0.7$ MPa

Cell
$\psi_p = 0$
-0.7 MPa $=$
$\psi_s + 0$ MPa
$\psi_s = -0.7$

Cell membrane
Cell wall

b.

Figure 37.5 Water potential at equilibrium. *a.* This cell initially had a larger ψ_w than the solution surrounding it. *b.* At osmotic equilibrium, the ψ_w of the cell and the solution should be the same. We assume that the cell is in a very large volume of solution of constant concentration. The final ψ_w of the cell should therefore equal the initial ψ_w of the solution. When a cell is plasmolyzed, $\psi_p = 0$. As the cell loses water, the cell's solution becomes concentrated.

membrane or a cell wall. Solutions that are not contained within a membrane cannot have turgor pressure, and they always have a ψ_p of 0 MPa.

The concentration of solutes also determines water potential and is referred to as **solute potential (ψ_s)**. Pure water has a solute potential of zero. As a solution increases in solute concentration, it decreases in ψ_s (<0 MPa, figure 37.4*b*). When solutes are added, water molecules form bonds with the solute molecules. Fewer free water molecules are available, which decreases the water potential. A solution with a higher solute concentration has a more negative ψ_s.

The total water potential (ψ_w) of a plant cell is the sum of its pressure potential (ψ_p) and solute potential (ψ_s); it represents the total potential energy of the water in the cell:

$$\psi_w = \psi_p + \psi_s$$

When the ψ_w inside the cell equals that of the solution, there is no net movement of water (figure 37.4*c*).

When a cell is placed into a solution with a different ψ_w, the tendency is for water to move in the direction that

eventually results in equilibrium—both the cell and the solution have the same ψ_w (figure 37.5). The ψ_p and ψ_s values may differ for cell and solution, but the sum ($= \psi_w$) should be the same.

 Data analysis What would ψ_w, ψ_s, and ψ_p of the cell in figure 37.5*a* be at equilibrium if it had been placed in a solution with a ψ_s of -0.5?

Learning Outcomes Review 37.1

The turgor of plant cells is determined by the direction of osmosis, which can be predicted based on the water potential of the cell and surrounding solution. Water potential is the sum of the pressure potential and the solute potential; water moves from an area of high water potential to an area of low water potential.

■ *Explain how physical pressure and solute concentration contribute to water potential.*

Learning Outcomes

1. *Compare the three water transport routes through plants.*
2. *Describe the function of Casparian strips.*

Most of the water absorbed by the plant comes in through the region of the root with root hairs (figures 37.6 and 37.7). As you learned in chapter 36, root hairs are extensions of root epidermal cells located just behind the tips of growing roots.

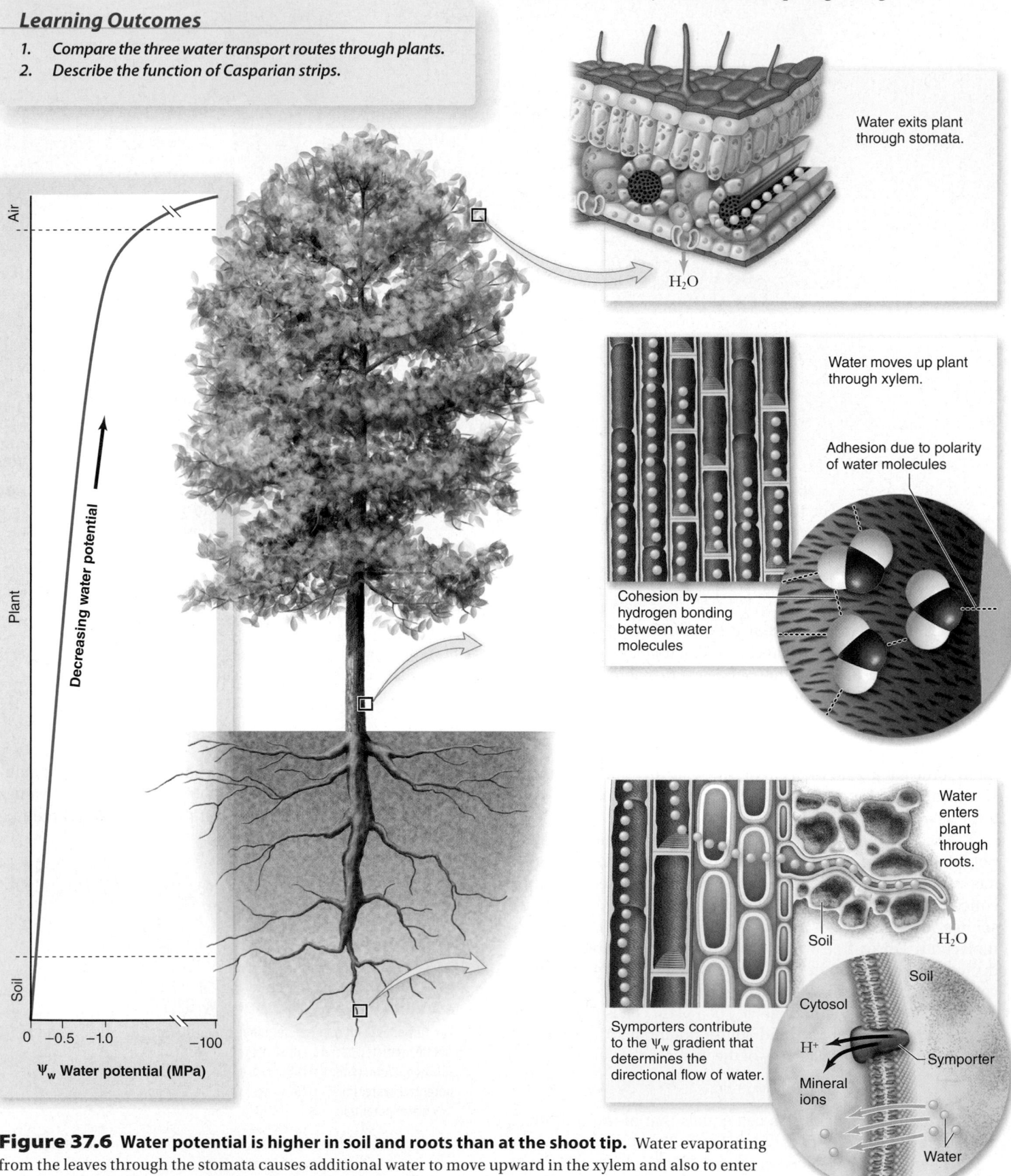

Water exits plant through stomata.

H_2O

Water moves up plant through xylem.

Adhesion due to polarity of water molecules

Cohesion by hydrogen bonding between water molecules

Water enters plant through roots.

Soil H_2O

Symporters contribute to the ψ_w gradient that determines the directional flow of water.

Soil

Cytosol

H^+ Symporter

Mineral ions

Water

Decreasing water potential

ψ_w **Water potential (MPa)**

Air

Plant

Soil

0 −0.5 −1.0 −100

Figure 37.6 Water potential is higher in soil and roots than at the shoot tip. Water evaporating from the leaves through the stomata causes additional water to move upward in the xylem and also to enter the plant through the roots. Water potential drops substantially in the leaves due to transpiration.

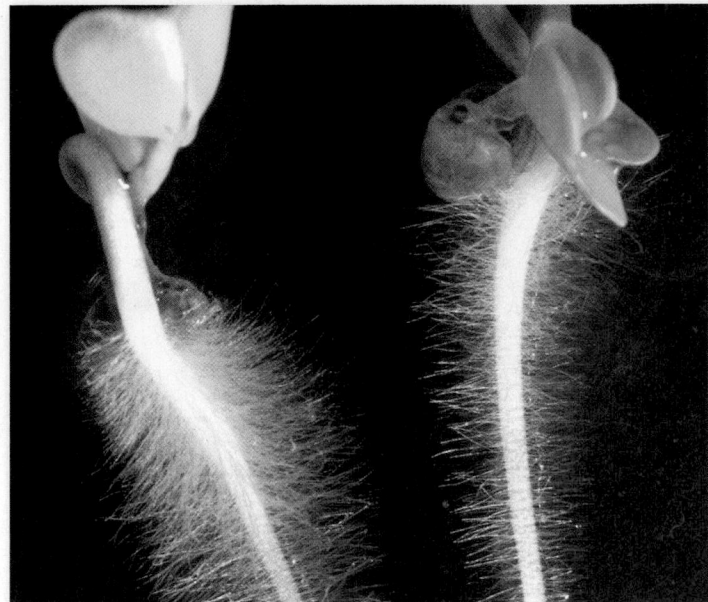

Figure 37.7 Water and minerals move into roots in regions rich with root hairs.

Surface area for the absorption of water and minerals is further increased in many species of plants by interacting with mycorrhizal fungi. These fungi extend the absorptive net far beyond that of root hairs and are particularly helpful in the uptake of phosphorous in the soil. Mycorrhizae are discussed in detail in chapter 32.

Once absorbed through root hairs, water and minerals must move across cell layers until they reach the vascular tissues; water and dissolved ions then enter the xylem and move throughout the plant.

Three transport routes exist through cells

Water and minerals can follow three pathways to the vascular tissue of the root (figure 37.8). The **apoplast route** includes movement through the cell walls and the space between cells.

Transport through the apoplast avoids membrane transport. The **symplast route** is the continuum of cytoplasm between cells connected by plasmodesmata. Once molecules are inside a cell, they can move between cells through plasmodesmata without crossing a plasma membrane. The **transmembrane route** involves membrane transport between cells and also across the membranes of vacuoles within cells. This route permits each cell the greatest amount of control over what substances enter and leave. These three routes are not exclusive, and molecules can change pathways at any time, until reaching the endodermis of the root.

 Inquiry question Which route would be the fastest for water movement? Would this always be the best way to move minerals into the plant?

Transport through the endodermis is selective

Eventually, on their journey inward, molecules reach the endodermis. Any further passage through the cell walls is blocked by the Casparian strips. As described in chapter 36, all cells in the cylinder of endodermis have connecting walls embedded with the waterproof material suberin (figure 37.9). Molecules must pass through the plasma membranes and protoplasts of the endodermal cells to reach the xylem. The endodermis, with its unique structure, along with the cortex and epidermis, controls water and nutrient flow to the xylem to regulate water potential and helps limit leakage of water out of the root.

Mineral ion concentration in the soil water is usually much lower than it is in the plant. Minerals are often taken up by root cells by active transport across the endodermis to increase the concentration in the stele. The plasma membranes of endodermal cells contain a variety of protein transport channels, through which proton pumps transport specific ions against even larger concentration gradients (refer to figure 37.1). Once inside the vascular stele, the ions, which are plant nutrients, are transported via the xylem throughout the plant.

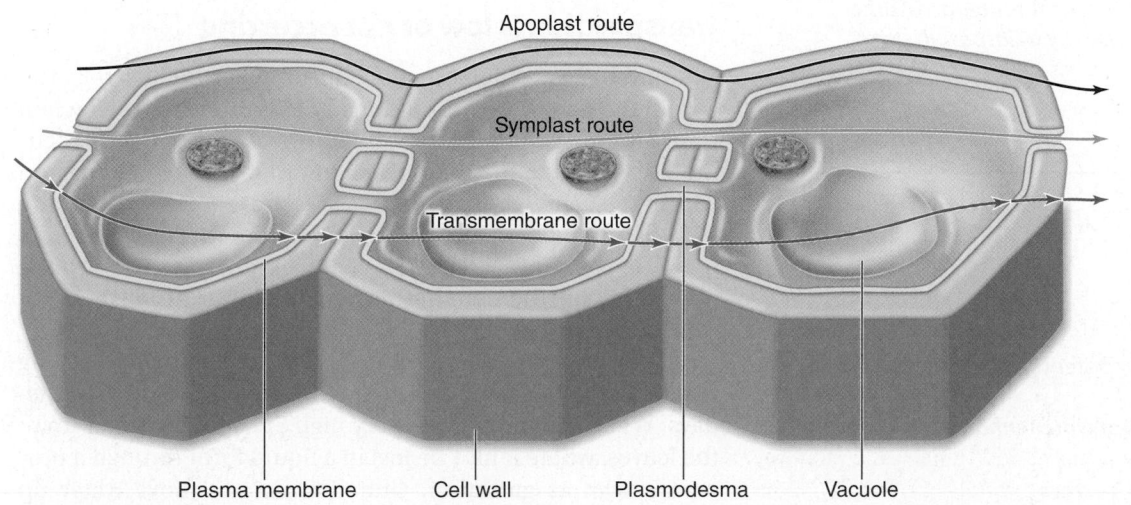

Apoplast route

Symplast route

Transmembrane route

Plasma membrane   Cell wall   Plasmodesma   Vacuole

Figure 37.8 Transport routes between cells.

Figure 37.9 The pathways of mineral transport in roots.

Minerals are absorbed at the surface of the root. In passing through the cortex, they must either follow the cell walls and the spaces between them or go directly through the plasma membranes and the protoplasts of the cells, passing from one cell to the next by way of the plasmodesmata. When they reach the endodermis, however, their further passage through the cell walls is blocked by the Casparian strips, and they must pass through the membrane and protoplast of an endodermal cell before they can reach the xylem.

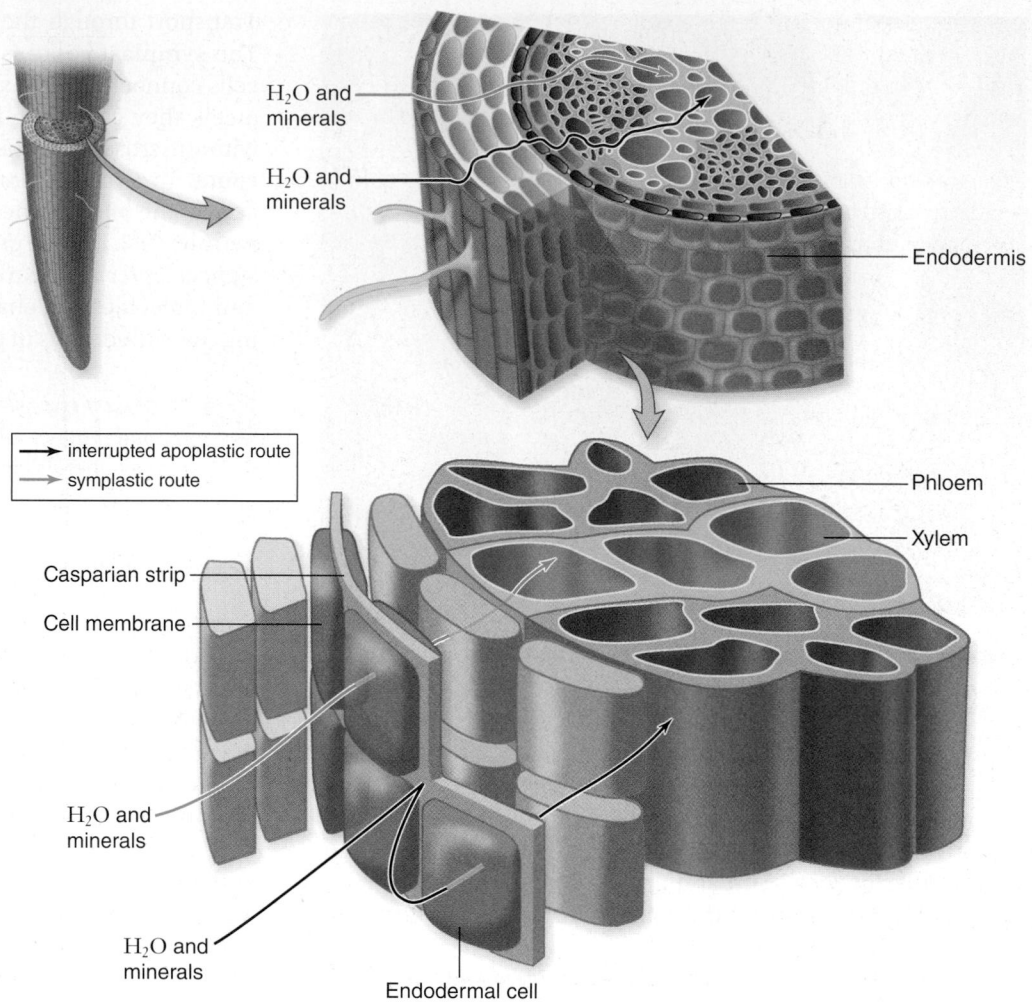

H₂O and minerals

H₂O and minerals

Endodermis

→ interrupted apoplastic route
→ symplastic route

Phloem

Xylem

Casparian strip

Cell membrane

H₂O and minerals

H₂O and minerals

Endodermal cell

Learning Outcomes Review 37.2

Water and minerals move into the plant from the soil, particularly in the region rich with root hairs. The three water transport routes are the apoplast route through the cells walls, the symplast route through plasmodesmata, and the transmembrane route across cell and vacuole membranes. Casparian strips force water and nutrients to move through the cell membranes of the endodermis, allowing selective control.

■ *Which of the three water transport routes across the endodermis would be blocked by a Casparian strip?*

37.3 Xylem Transport

Learning Outcomes

1. *Analyze the effects of water potential on water transport in xylem.*
2. *Explain how the properties of water enhance transpiration.*

The aqueous solution that passes through the membranes of endodermal cells enters the plant's vascular tissues and moves into the tracheids and vessel members of the xylem. As ions are actively pumped into the root or move via facilitated diffusion, their presence decreases the water potential and increases turgor pressure in the roots due to osmosis.

Root pressure is present even when transpiration is low or not occurring

Root pressure, which often occurs at night, is caused by the continued accumulation of ions in the roots at times when transpiration from the leaves is very low or absent. This accumulation results in an increasingly high ion concentration within the cells, which in turn causes more water to enter the root hair cells by osmosis. Ion transport further decreases the ψ_s of the roots. The result is movement of water into the plant and up the xylem columns, despite the absence of transpiration.

Under certain circumstances, root pressure is so strong that water will ooze out of a cut plant stem for hours or even days. When root pressure is very high, it can force water up to the leaves, where it may be lost in a liquid form through a process known as **guttation.** Guttation cannot move water up great heights or at rapid speeds. It does not take place through the stomata, but instead occurs through special groups of cells located near the ends of small veins that function only in this process. Guttation produces what is more commonly called dew on leaves.

Root pressure alone, however, is insufficient to explain xylem transport. Transpiration provides the main force for moving water and ionic solutes from roots to leaves.

A water potential gradient from roots to shoots enables transport

Water potential regulates the movement of water through a whole plant, as well as across cell membranes. Roots are the entry point. Water moves from the soil into the plant only if water potential of the soil is greater than in the root. Too much fertilizer or drought conditions lower the ψ_w of the soil and limit water flow into the plant. Water in a plant moves along a ψ_w gradient from the soil (where the ψ_w may be close to zero under wet conditions) to successively more negative water potentials in the roots, stems, leaves, and atmosphere (see figure 37.6).

Transpiration of water in a leaf creates negative pressure or tension in the xylem, which literally pulls water up the stem from the roots. The driving force for transpiration is the humidity gradient from 100% relative humidity inside the leaf to much less than 100% relative humidity outside the stomata. Molecules diffusing from the xylem replace evaporating water molecules.

Vessels and tracheids accommodate bulk flow

Water has an inherent **tensile strength** that arises from the cohesion of its molecules—their tendency to form hydrogen bonds with one another (see chapter 2). These two factors are the basis of the cohesion–tension theory of the bulk flow of water in the xylem. The tensile strength of a column of water varies inversely with the diameter of the column; that is, the smaller the diameter of the column, the greater the tensile strength. Because plant tracheids and vessels are tiny in diameter, the cohesive force of water is stronger than the pull of gravity. The water molecules also adhere to the sides of the tracheid or xylem vessels, further stabilizing the long column of water.

Given that a narrower column of water has greater tensile strength, it is intriguing that vessels, having diameters that are larger than tracheids, are found in so many plants. The difference in diameter has a larger effect on the mass of water in the column than on the tensile strength of the column. The volume of liquid moving in a column per second is proportional to r^4, where r is the radius of the column, at constant pressure. A twofold increase in radius would result in a 16-fold increase in the volume of liquid moving through the column. Given equal cross-sectional areas of xylem, a plant with larger-diameter vessels can move more water up its stems than a plant with narrower tracheids.

 Data analysis If a mutation increased the radius of a xylem vessel threefold, how would the movement of water through the plant be affected?

 Data analysis Comparative analyses of fossil and extant vascular plants reveal that the maximum diameter for a tracheid is 80 μm. Vessels, however, can reach diameters of up to 500 μm in diameter. How much more water could the largest vessels transport compared with the largest tracheids in the same amount of time? Your calculations are for an ideal situation. In the context of a growing plant, what other factors would affect the rate of transport through the tracheids and vessels?

The effect of cavitation

Tensile strength depends on the continuity of the water column; air bubbles introduced into the column when a vessel is broken or cut would cause the continuity and the cohesion to fail. A gas-filled bubble can expand because of tensile forces and block the tracheid or vessel, a process called **cavitation**. Cavitation stops water transport and can lead to dehydration and death of part or all of a plant (figure 37.10).

Anatomical adaptations can compensate for the problem of cavitation, including the presence of alternative pathways that can be used if one path is blocked. Individual tracheids and vessel members are connected to other tracheids or vessels by pits in their walls, and air bubbles are generally larger than these openings. In this way, bubbles cannot pass through the pits to further block transport. Freezing or deformation of cells can also cause small bubbles of air to form within xylem cells, especially with seasonal temperature changes. Cavitation is one reason older xylem often stops conducting water.

Mineral transport

Tracheids and vessels are essential for the bulk transport of minerals. Ultimately, the minerals that are actively transported into the roots are removed and relocated through the xylem to other metabolically active parts of the plant. Phosphorus, potassium, nitrogen, and sometimes iron may be abundant in the xylem during certain seasons. In many plants, this pattern of ionic concentration helps conserve these essential nutrients, which may move from mature deciduous parts such as leaves and twigs to areas of active growth, namely meristem regions.

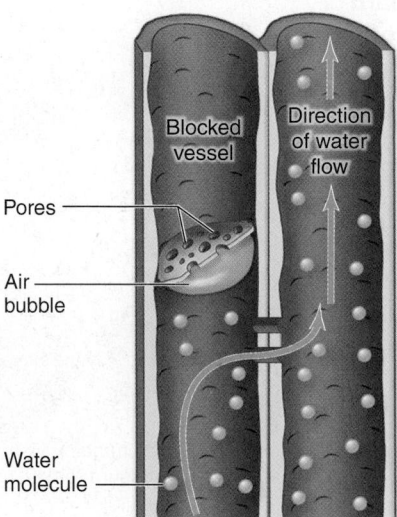

Figure 37.10 Cavitation. An air bubble can break the tensile strength of the water column. Bubbles are larger than pits and can block transport to the next tracheid or vessel. Water drains to surrounding tracheids or vessels.

Pores

Air bubble

Water molecule

Blocked vessel

Direction of water flow

Keep in mind that minerals that are relocated via the xylem must move with the generally upward flow through the xylem. Not all minerals can reenter the xylem conduit once they leave. Calcium, an essential nutrient, cannot be transported elsewhere once it has been deposited in a particular plant part. But some other nutrients can be transported in the phloem.

37.4 Rate of Transpiration

More than 90% of the water taken in by the roots of a plant is ultimately lost to the atmosphere. Water moves from the tips of veins into mesophyll cells, and from the surface of these cells it evaporates into pockets of air in the leaf. As discussed in chapter 36, these intercellular spaces are in contact with the air outside the leaf by way of the stomata.

Stomata open and close to balance H_2O and CO_2 needs

Water is essential for plant metabolism, but it is continuously being lost to the atmosphere. At the same time, photosynthesis requires a supply of CO_2 entering the chlorenchyma cells from the atmosphere. Plants therefore face two somewhat conflicting requirements: the need to minimize the loss of water to the atmosphere and the need to admit carbon dioxide. Structural features such as stomata and the cuticle have evolved in response to one or both of these requirements.

The rate of transpiration depends on weather conditions, including humidity and the time of day. As stated earlier, transpiration from the leaves decreases at night, when stomata are closed and the vapor pressure gradient between the leaf and the atmosphere is less. During the day, sunlight increases the temperature of the leaf, while transpiration cools the leaf through evaporative cooling.

On a short-term basis, closing the stomata can control water loss. This occurs in many plants when they are subjected to water stress. But the stomata must be open at least part of the time so that CO_2 can enter. As CO_2 enters the intercellular spaces, it dissolves in water before entering the plant's cells where it is used in photosynthesis. The gas dissolves mainly in water on the walls of the intercellular spaces below the stomata. The continuous stream of water that reaches the leaves from the roots keeps these walls moist.

Turgor pressure in guard cells causes stomata to open and close

The two sausage-shaped guard cells on each side of a stoma stand out from other epidermal cells not only because of their shape, but also because they are the only epidermal cells

Figure 37.11 Unequal cell-wall thickenings on guard cells result in the opening of stomata when the guard cells expand.

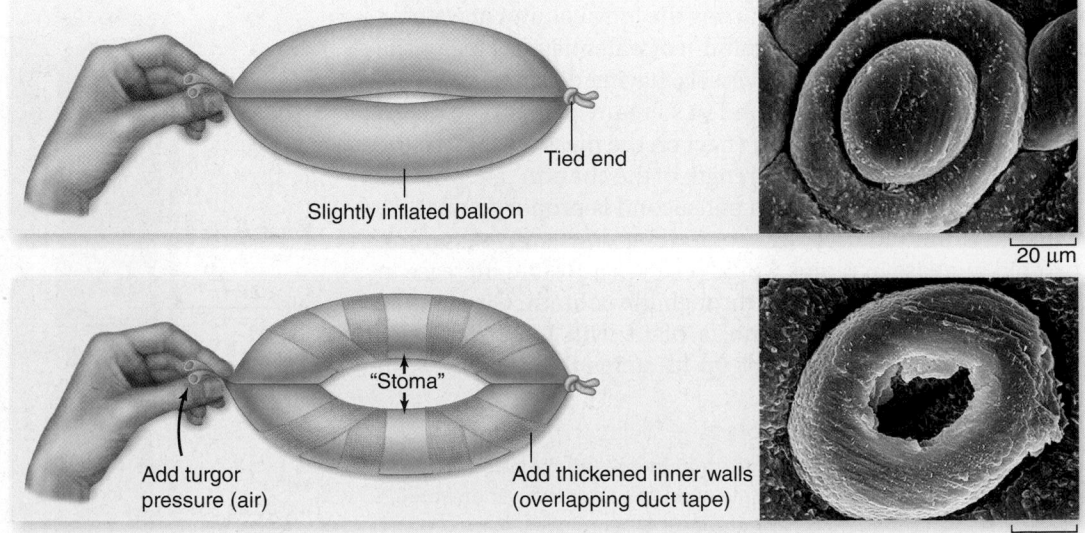

Tied end
Slightly inflated balloon
20 μm

"Stoma"
Add turgor pressure (air)
Add thickened inner walls (overlapping duct tape)
20 μm

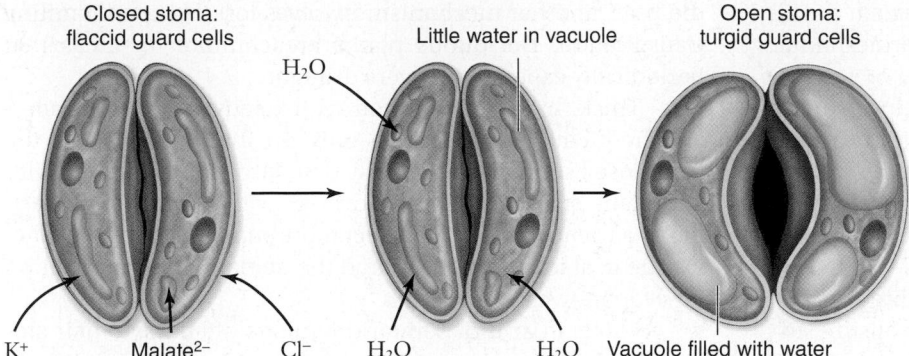

Figure 37.12 How a stoma opens. When H$^+$ ions are pumped from guard cells, K$^+$ and Cl$^-$ ions move in, and the guard cell turgor pressure increases as water enters by osmosis. The increased turgor pressure causes the guard cells to bulge, with the thick walls on the inner side causing each guard cell to bow outward, thereby opening the stoma.

containing chloroplasts. Their distinctive wall construction, which is thicker on the inside and thinner elsewhere, results in a bulging out and bowing when they become turgid.

You can make a model of this for yourself by taking two elongated balloons, tying the closed ends together, and inflating both balloons slightly. When you hold the two open ends together, there should be very little space between the two balloons. Now wrap duct tape around both balloons as shown in figure 37.11 (without releasing any air) and inflate each one a bit more. Hold the open ends together again. You should now be holding a roughly doughnut-shaped pair of "guard cells" with a "stoma" in the middle. Real guard cells rely on the influx and efflux of water, rather than air, to open and shut.

Turgor in guard cells results from the active uptake of potassium (K$^+$), chloride (Cl$^-$), and malate. As solute concentration increases, water potential decreases in the guard cells, and water enters osmotically. As a result, these cells accumulate water and become turgid, opening the stomata (figure 37.12). The energy required to move the ions across the guard cell membranes comes from the ATP-driven H$^+$ pump shown in figure 37.1.

The guard cells of many plant species regularly become turgid in the morning, when photosynthesis occurs, and lose turgor in the evening, regardless of the availability of water. During the course of a day, sucrose accumulates in the photosynthetic guard cells. The active pumping of sucrose out of guard cells in the evening may lead to loss of turgor and close the guard cell.

Environmental factors affect transpiration rates

Transpiration rates increase with temperature and wind velocity because water molecules evaporate more quickly. As humidity increases, the water potential difference between the leaf and the atmosphere decreases, but even at 95% relative humidity in the atmosphere, the vapor pressure gradient can sustain full transpiration. On a catastrophic level, when a whole plant wilts because insufficient water is available, the guard cells may lose turgor, and as a result, the stomata may close. Fluctuations in transpiration rate are tempered by opening or closing stomata.

Experimental evidence has indicated that several pathways regulate stomatal opening and closing. **Abscisic acid (ABA),** a plant hormone discussed in chapter 40, plays a primary role in allowing K$^+$ to pass rapidly out of guard cells, causing the stomata to close in response to drought. ABA binds to receptor sites in the plasma membranes of guard cells, triggering a signaling pathway that opens K$^+$, Cl$^-$, and malate ion channels. Turgor pressure decreases as water loss follows, and the guard cells close (figure 37.13).

CO$_2$ concentration, light, and temperature also affect stomatal opening. When CO$_2$ concentrations are high, the guard cells of many plant species close the stomata. Additional CO$_2$ is not needed at such times, and water is conserved when the

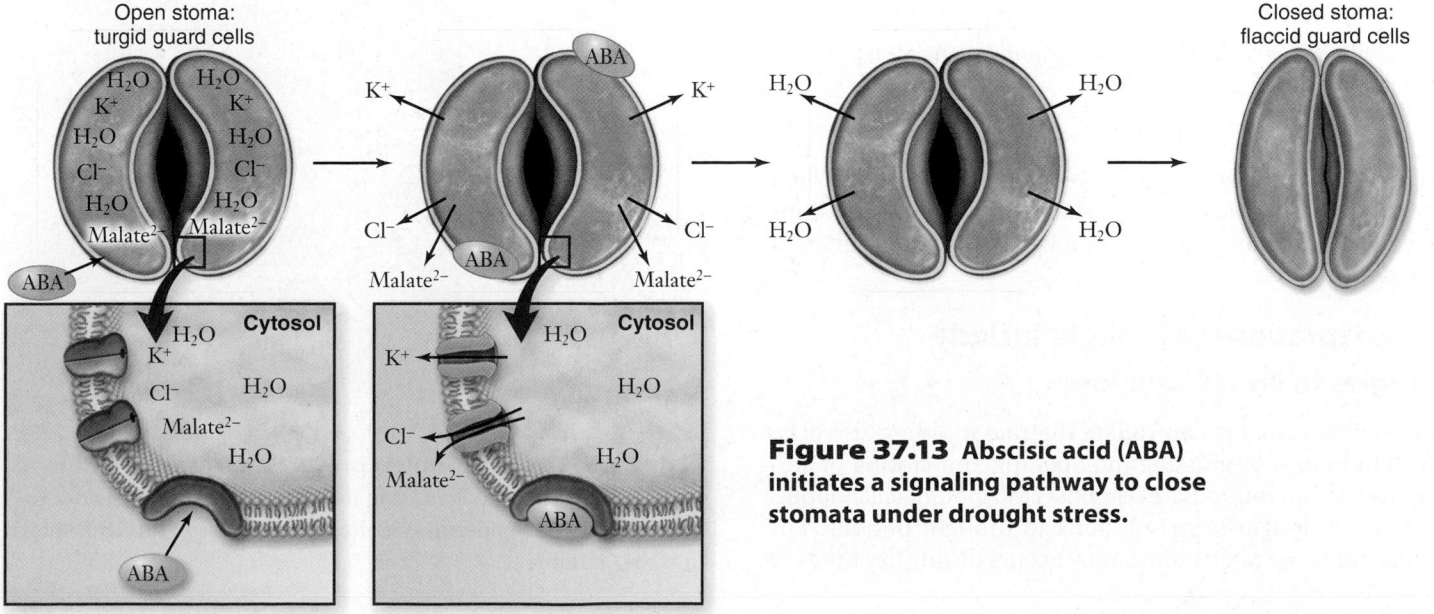

Figure 37.13 Abscisic acid (ABA) initiates a signaling pathway to close stomata under drought stress.

guard cells are closed. Although the exact mechanism is still being investigated, there appear to be some similarities in the signaling pathways used by ABA to close stomata.

Blue light regulates stomatal opening. This helps increase turgor to open the stomata when sunlight increases the evaporative cooling demands. K^+ transport against a concentration gradient is promoted by light. Blue light in particular triggers proton (H^+) transport, creating a proton gradient that drives the opening of K^+ channels.

The stomata may close when the temperature exceeds 30° to 34°C and water relations are unfavorable. To ensure sufficient gas exchange, these stomata open when it is dark and the temperature has dropped. Some plants are able to collect CO_2 at night in a modified form to be utilized in photosynthesis during daylight hours. In chapter 8, you learned about Crassulacean acid metabolism (CAM), which occurs in succulent plants such as cacti. In this process, stomata open and CO_2 is taken in at night and stored in organic compounds. These compounds are decarboxylated during the day, providing a source of CO_2 for fixation when stomata are closed. CAM plants are able to conserve water in dry environments.

Learning Outcomes Review 37.4

When guard cells of the stomata actively take up ions, their water potential decreases and they take up water by osmosis. When they become turgid they change shape, creating an opening in the stoma. Stomata close when a plant is under water stress, but they open when carbon dioxide is needed and transpiration does not cause excess water loss. Transpiration rates increase with high wind velocity, high temperatures, and low humidity.

■ *Draw a flowchart showing the sequence of events triggered in guard cells by blue light on a 28°C day. What other factors might alter the outcome?*

37.5 Water-Stress Responses

Learning Outcome

1. *Evaluate the adaptations for drought, flooding, and salt tolerance that enhance plant survival.*

Because plants cannot simply move to another location when water availability or salt concentrations change, adaptations have evolved to allow them to cope with such environmental fluctuations.

Plant adaptations to drought include strategies to limit water loss

Many mechanisms for controlling the rate of water loss have evolved in plants. Regulating the opening and closing of stomata provides an immediate response. Morphological adaptations provide longer-term solutions to drought periods. For example, for some plants dormancy occurs during dry times of

the year; another mechanism involves loss of leaves, limiting transpiration. Deciduous plants are common in areas that periodically experience severe drought.

Thick, hard leaves, often with relatively few stomata—and frequently with stomata only on the lower side of the leaf—lose water far more slowly than large, pliable leaves with abundant stomata. Leaves covered with masses of wooly-looking trichomes (hairs) reflect more sunlight, thereby reducing the heat load on the leaf and the demand for transpiration for evaporative cooling.

Plants in arid or semiarid habitats often have their stomata in crypts or pits in the leaf surface (figure 37.14). Within these depressions, the water surface tensions are altered, reducing the rate of water loss.

The search is on for genes responsible for drought tolerance. The *enhanced drought tolerance1* (*edt1*) gene in *Arabidopsis* increases root length, decreases stomatal density, and enhances ABA levels under drought conditions, which leads to stomatal closure. In experiments, plants with *edt1* recover from drought conditions that kill control plants.

Plant responses to flooding include short-term hormonal changes and long-term adaptations

Plants can also receive too much water, in which case they ultimately "drown." Flooding rapidly depletes available oxygen in the soil and interferes with the transport of minerals and carbohydrates in the roots. Abnormal growth often results.

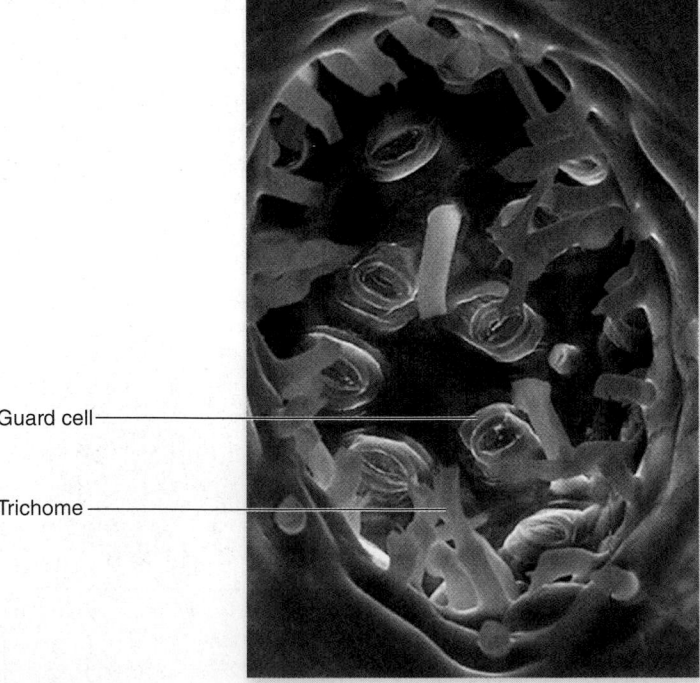

Guard cell

Trichome

35 μm

Figure 37.14 Anatomical protection from drought in leaves. Deeply embedded stomata, extensive trichomes, and multiple layers of epidermis minimize water loss in this leaf, shown in cross section.

Hormone levels change in flooded plants; ethylene, a hormone associated with suppression of root elongation, increases, while gibberellins and cytokinins, which enhance growth of new roots, usually decrease (see chapter 40). The hormonal changes contribute to the abnormal growth patterns.

Oxygen deprivation is among the most significant problems because it leads to decreased cellular respiration. Standing water has much less oxygen than moving water. Generally, standing-water flooding is more harmful to a plant (riptides excluded). Flooding that occurs when a plant is dormant is much less harmful than flooding when it is growing actively.

Physical changes that occur in the roots as a result of oxygen deprivation may halt the flow of water through the plant. Paradoxically, even though the roots of a plant may be standing in water, its leaves may be drying out. Plants can respond to flooded conditions by forming larger lenticels (which facilitate gas exchange—see figure 36.27) and adventitious roots that reach above flood level for gas exchange.

Whereas some plants survive occasional flooding, others have adapted to living in fresh water. One of the most frequent adaptations among plants to growing in water is the formation of **aerenchyma,** loose parenchymal tissue with large air spaces in it (figure 37.15). Aerenchyma is very prominent in water lilies and many other aquatic plants. Oxygen may be transported from the parts of the plant above water to those below by way of passages in the aerenchyma. This supply of oxygen allows oxidative respiration to take place even in the submerged portions of the plant.

Some plants normally form aerenchyma, whereas others, subject to periodic flooding, can form it when necessary. In corn, increased ethylene due to flooding induces aerenchyma formation.

Plant adaptations to high salt concentration include elimination methods

The algal ancestors of plants adapted to a freshwater environment from a saltwater environment before the "move" onto land. This adaptation involved a major change in controlling salt balance.

Growth in salt water

Plants such as mangroves that grow in areas normally flooded with salt water must not only provide a supply of oxygen to their submerged parts, but also control their salt balance. The salt must be excluded, actively secreted, or diluted as it enters. The black mangrove *(Avicennia germinans)* has long, spongy, air-filled roots that emerge above the mud. These roots, called *pneumatophores* (see chapter 36), have large lenticels on their above-water portions through which oxygen enters; it is then transported to the submerged roots (figure 37.16). In addition, the succulent leaves of some mangrove species contain large quantities of water, which dilute the salt that reaches them. Many plants that grow in such conditions also either secrete large quantities of salt or block salt uptake at the root level.

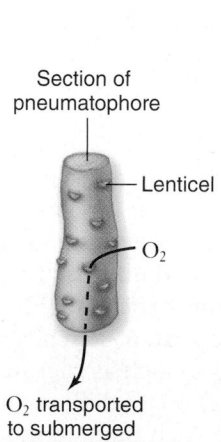

Figure 37.16 How mangroves get oxygen to their submerged parts. The black mangrove *(Avicennia germinans)* grows in areas that are commonly flooded, and much of each plant is usually submerged. However, modified roots called pneumatophores supply the submerged portions of the plant with oxygen because these roots emerge above the water and have large lenticels. Oxygen diffuses into the roots through the lenticels, passes into the abundant aerenchyma, and moves to the rest of the plant.

Figure 37.15 Aerenchyma. This tissue facilitates gas exchange in aquatic plants. *a.* Water lilies float on the surface of ponds, collecting oxygen and then transporting it to submerged portions of the plant. *b.* Large air spaces in the leaves of the water lily add buoyancy. The specialized parenchyma tissue that forms these open spaces is called aerenchyma. Gas exchange occurs through stomata found only on the upper surface of the leaf.

Growth in saline soil

Soil salinity is increasing, often caused by salt accumulation from irrigation. Currently 23% of the world's cultivated land has high levels of saline that reduce crop yield. The low water potential of saline soils results in water-stressed crops. Some plants, called **halophytes** (salt lovers), can tolerate soils with high salt concentrations. Mechanisms for salt tolerance are being studied with the goal of breeding more salt-tolerant plants. Some halophytes produce high concentrations of organic molecules within their roots to alter the water potential gradient between the soil and the root so that water flows into the root.

Learning Outcome Review 37.5

Adaptations to drought include dormancy, leaf loss, leaves that minimize water loss, and stomata that lie in depressions. When plants are exposed to flooding, oxygen deprivation leads to lower cellular respiration rates, impedance of mineral and carbohydrate transport, and changes in hormone levels. If a plant is exposed to a salty environment, it may exclude the salt from uptake, secrete it after it has been taken up, or dilute it.

- **Select one adaptation for drought, one for flooding, and one for salt tolerance and explain how each enhances plant survival.**

37.6 Phloem Transport

Learning Outcomes

1. **Explain the pressure–flow hypothesis.**
2. **Contrast phloem and xylem transport.**

Most carbohydrates manufactured in leaves and other green parts are distributed through the phloem to the rest of the plant. This process, known as **translocation,** provides suitable carbohydrate building blocks for the roots and other actively growing regions of the plant. Carbohydrates concentrated in storage organs such as tubers, often in the form of starch, are also converted into transportable molecules, such as sucrose, and moved through the phloem. In this section we discuss the ways by which carbohydrate- and nutrient-rich fluid, termed **sap,** is moved through the plant body.

Organic molecules are transported up and down the plant

The movement of sugars and other substances can be followed in phloem using radioactive labels (figure 37.17). Radioactive carbon dioxide ($^{14}CO_2$) can be incorporated into glucose as a result of photosynthesis. Glucose molecules are

---

Hypothesis: *As pea embryos develop in a pod, sugars will be transported through the phloem to the developing embryos.*

Prediction: *Radioactively labeled sugars will accumulate in developing pea embryos.*

Test: *Expose a healthy pea leaf to radioactive carbon dioxide ($^{14}CO_2$). Place photographic film over the entire plant at 1 and 12 hours after treatment and develop film.*

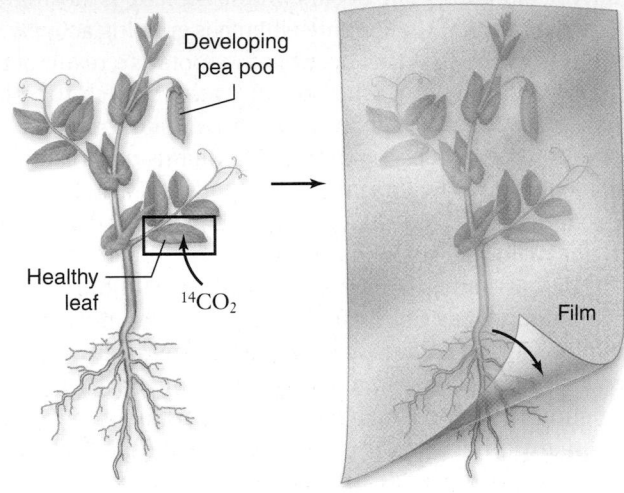

Result: *After 1 hour the radioactivity is concentrated near the application site. After 12 hours the radioactivity is concentrated in the developing embryo.*

At 1 Hour	At 12 Hours
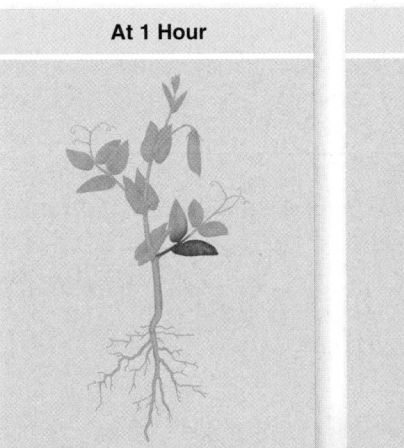	

Conclusion: *The $^{14}CO_2$ is incorporated into sugars during photosynthesis and transported to the developing embryo in the pod.*

Further Experiments: *Carrots take two years to flower. During the first season an underground root, the "carrot," develops and sugars are stored to be used for reproduction the next year. How could you test the hypothesis that sugars are transported to the developing storage root during the first season of growth for a carrot plant?*

Figure 37.17 Sucrose flow in phloem during fruit development.

used to make the disaccharide sucrose, which is transported in the phloem. Such studies have shown that sucrose moves both up and down in the phloem.

Aphids, a group of insects that extract plant sap for food, have been valuable tools in understanding translocation. Aphids thrust their stylets (piercing mouthparts) into phloem cells of leaves and stems to obtain the abundant sugars there. When a feeding aphid is removed by cutting its stylet, the liquid from the phloem continues to flow through the detached mouthpart and is thus available in pure form for analysis (figure 37.18). Sucrose is the primary solute in phloem. Using aphids to obtain the critical samples and radioactive tracers to mark them, plant biologists have demonstrated that substances in phloem can move remarkably fast, as much as 50 to 100 cm/h.

Phloem also transports plant hormones, and as will be explored in chapter 40, environmental signals can result in the rapid translocation of hormones in the plant. Recent evidence also indicates that mRNA can move through the phloem, providing a previously unknown mechanism for long-distance communication among cells. In addition, phloem carries other molecules, such as a variety of sugars, amino acids, organic acids, proteins, and ions.

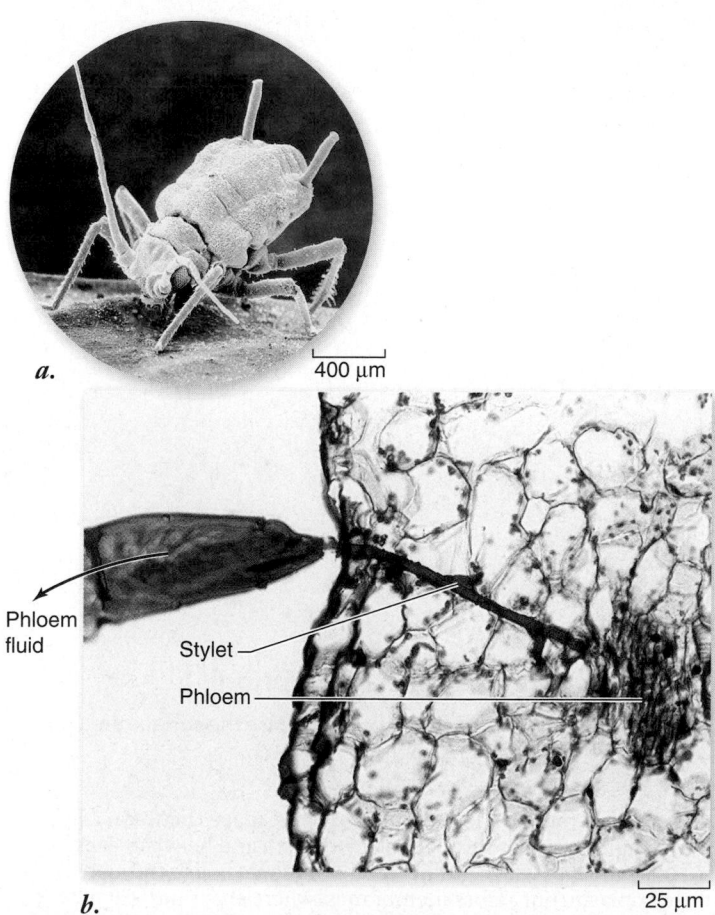

a.

400 μm

b.

25 μm

Phloem fluid

Stylet

Phloem

Figure 37.18 Feeding on phloem. *a.* Aphids, including this individual shown on the edge of a leaf, feed on the food-rich contents of the phloem, which they extract through *(b)* their piercing mouthparts, called stylets. When an aphid is separated from its stylet and the cut stylet is left in the plant, the phloem fluid oozes out of it and can then be collected and analyzed.

Turgor pressure differences drive phloem transport

The most widely accepted model of how carbohydrates in solution move through the phloem has been called the **pressure–flow hypothesis.** Dissolved carbohydrates flow from a *source* and are released at a *sink,* where they are utilized. Carbohydrate sources include photosynthetic tissues, such as the mesophyll of leaves. Food-storage tissues, such as the cortex of roots, can be either sources or sinks. Sinks also occur at the growing tips of roots and stems and in developing fruits. Also, because sources and sinks can change through time as needs change, the direction of phloem flow can change.

In a process known as **phloem loading,** carbohydrates (mostly sucrose) enter the sieve tubes in the smallest veins at the source. Some sucrose travels from mesophyll cells to the companion and sieve cells via the symplast (see figure 37.8). Much of the sucrose arrives at the sieve cell through apoplastic transport and is moved across the membrane via a sucrose and H^+ symporter (see chapter 5). This energy-requiring step is driven by a proton pump (see figure 37.1). Companion cells and parenchyma cells adjacent to the sieve tubes provide the ATP energy to drive this transport. Unlike vessels and tracheids, sieve cells must be alive to participate in active transport.

Bulk flow occurs in the sieve tubes without additional energy requirements. Because of the difference between the water potential in the sieve tubes and in the nearby xylem cells, water flows into the sieve tubes by osmosis. Turgor pressure in the sieve tubes thus increases, and this pressure drives the fluid throughout the plant's system of sieve tubes. At the sink, sucrose and hormones are actively removed from the sieve tubes, and water follows by osmosis. The turgor pressure at the sink drops, causing a mass flow from the stronger pressure at the source to the weaker pressure at the sink (figure 37.19). Most of the water at the sink then diffuses back into the xylem, where it may either be recirculated or lost through transpiration.

Transport of sucrose and other carbohydrates within sieve tubes does not require energy. But the pressure needed to drive the movement is created through energy-dependent loading and unloading of these substances from the sieve tubes.

Learning Outcomes Review 37.6

Translocation is the movement of dissolved carbohydrates and other substances from one part of the plant to another through the phloem. Sap in the phloem contains sucrose and other sugars, hormones, mRNA, amino acids, organic acids, proteins, and ions. According to the pressure–flow hypothesis, carbohydrates are loaded into sieve tubes, creating a difference in water potential. As a result, water enters the tubes and creates pressure to move fluid through the phloem.

- *Identify two differences in the mechanisms driving xylem and phloem transport.*

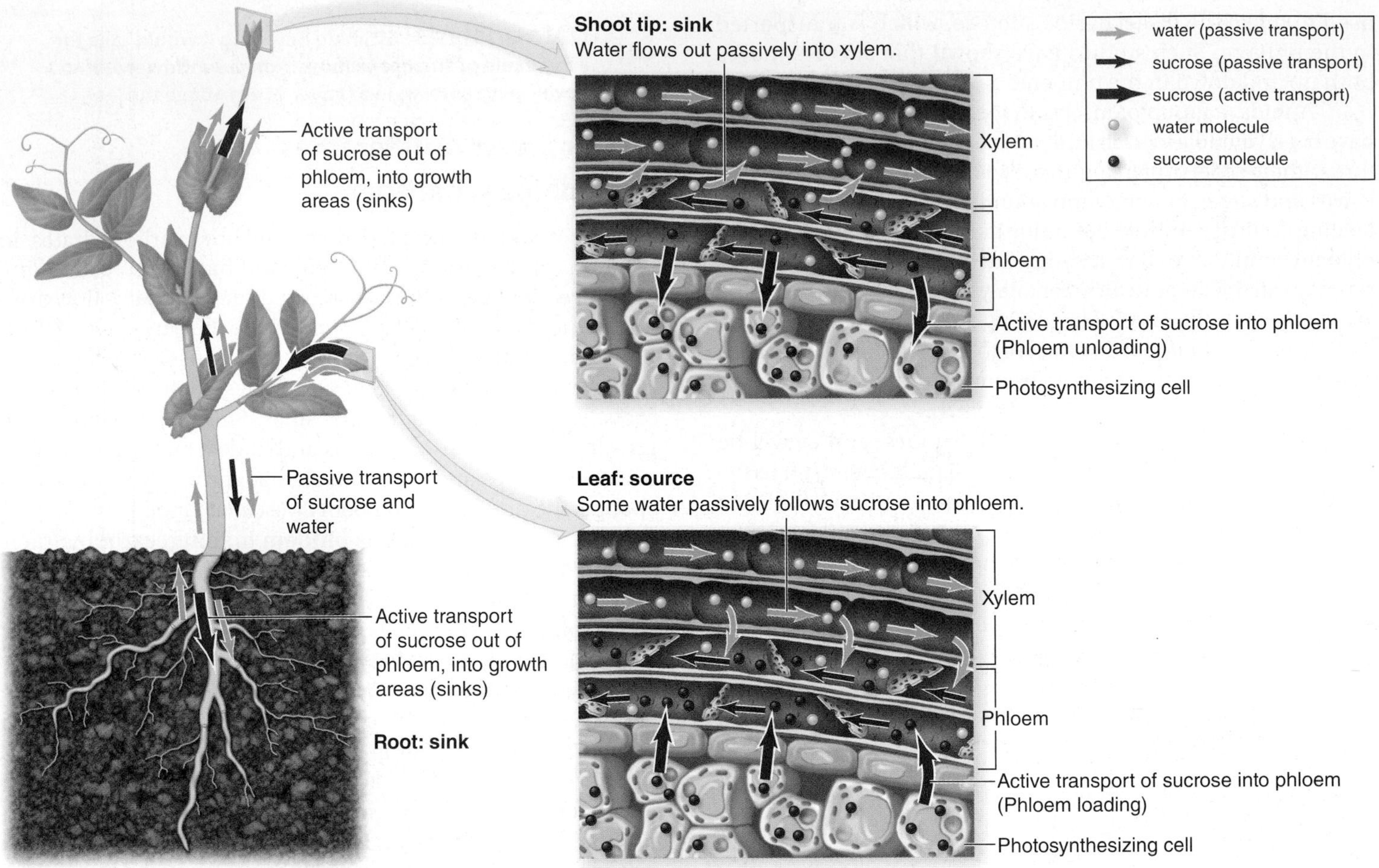

Shoot tip: sink
Water flows out passively into xylem.

Active transport
of sucrose out of
phloem, into growth
areas (sinks)

Xylem

Phloem

Active transport of sucrose into phloem
(Phloem unloading)

Photosynthesizing cell

Passive transport
of sucrose and
water

Active transport
of sucrose out of
phloem, into growth
areas (sinks)

Root: sink

Leaf: source
Some water passively follows sucrose into phloem.

Xylem

Phloem

Active transport of sucrose into phloem
(Phloem loading)

Photosynthesizing cell

→ water (passive transport)
→ sucrose (passive transport)
→ sucrose (active transport)
○ water molecule
● sucrose molecule

Figure 37.19 Diagram of mass flow. In this diagram, *red* dots represent sucrose molecules and *blue* dots symbolize water molecules. After moving from the mesophyll cells of a leaf or another part of the plant into the conducting cells of the phloem, the sucrose molecules are transported to other parts of the plant by mass flow and unloaded where they are required.

Chapter Review

37.1 Transport Mechanisms (figure 37.2)

Local changes result in long-distance movement of materials.
Properties of water, osmosis, and cellular activities predict the directions of water movement.

Osmosis is enhanced by aquaporins (figure 37.3).
Aquaporins are water channels in plasma membranes that allow water to move across the membrane more quickly.

Water potential regulates movement of water through the plant (figures 37.4 and 37.5).
The major force for water transport in a plant is the pulling of water by transpiration. Cohesion, adhesion, and osmosis all contribute to water movement.

Water potential is the sum of pressure potential and solute potential. Water moves from an area of high water potential to an area of low water potential.

37.2 Water and Mineral Absorption

Root hairs and mycorrhizal fungi can increase the surface area for absorption of water and minerals.

Three transport routes exist through cells.
The apoplast route is through cell walls and spaces between cells. The symplast route is through the cytoplasm and between cells via plasmodesmata. The transmembrane route is also through the cytoplasm, but across membranes, where entry and exit of substances can be controlled.

Transport through the endodermis is selective.
Casparian strips in the endoderm force water and nutrients to move across the cell membranes, allowing selective flow of water and nutrients to the xylem.

37.3 Xylem Transport

Root pressure is present even when transpiration is low or not occurring.

Root pressure results from the active transport of ions into the root cells, which causes water to move in through osmosis. Guttation occurs when water is forced out of a plant as a result of high root pressure.

Water has a high tensile strength due to its cohesive and adhesive properties, which are related to hydrogen bonding.

A water potential gradient from roots to shoots enables transport.

Water moves into plants when the soil water potential is greater than that of roots. Evaporation of water from leaves creates a negative water potential that pulls water upward through the xylem.

Vessels and tracheids accommodate bulk flow.

The volume of water that can be transported by a xylem vessel or tracheid is a function of its diameter. As diameter decreases, tensile strength increases; however, a larger volume of water can be transported through a tube with a larger radius.

Cavitation occurs when a gas bubble forms in a water column and water movement ceases.

37.4 Rate of Transpiration

Stomata open and close to balance H_2O and CO_2 needs.

More than 90% of the water absorbed by the roots is lost by evaporation through stomata. Stomata must open to take up carbon dioxide for photosynthesis and to allow evaporation for transpiration and cooling for the leaf (figure 37.12).

Turgor pressure in guard cells causes stomata to open and close.

Stomata open when the turgor pressure of guard cells increases due to the uptake of ions. The turgid guard cells change shape and create an opening between them. Stomata close when guard cells lose turgor pressure and become flaccid.

Environmental factors affect transpiration rates.

Transpiration rates increase as temperature and wind velocity increase and as humidity decreases. Stomata close at high temperatures or when carbon dioxide concentrations increase.

37.5 Water-Stress Responses

Plant adaptations to drought include strategies to limit water loss.

Plant adaptations to minimize water loss include closing stomata, becoming dormant, altering leaf characteristics to minimize water loss, and losing leaves.

Plant responses to flooding include short-term hormonal changes and long-term adaptations.

Flooding reduces oxygen availability for cellular respiration, results in abnormal growth, and reduces the efficiency of transport mechanisms.

Plants adapted to wet environments exhibit a variety of strategies, including lenticels, adventitious roots such as pneumatophores, and aerenchyma tissue to ensure oxygen for submerged parts.

Plant adaptations to high salt concentration include elimination methods.

Plants found in saline waters may exclude, secrete, or dilute salts that have been taken up.

Halophytes can take up water from saline soils by decreasing the water potential of their roots with high concentrations of organic molecules.

37.6 Phloem Transport

Organic molecules are transported up and down the plant.

Movement of organic nutrients from leaves to other parts of the plant through the phloem is called translocation.

The sap that moves through phloem contains sugars, plant hormones, mRNA, and other substances. Carbohydrates must be actively transported into the sieve tubes.

Turgor pressure differences drive phloem transport.

At the carbohydrate source, such as a photosynthetic leaf, active transport of sugars into the phloem causes a reduction in water potential.

As water moves into the phloem, turgor pressure drives the contents to a sink, such as a nonphotosynthetic tissue, where the sugar is unloaded.

Review Questions

UNDERSTAND

1. Which of the following is an active transport mechanism?
   a. Proton pump
   b. Ion channel
   c. Symport
   d. Osmosis

2. The water potential of a plant cell is the
   a. sum of the membrane potential and gravity.
   b. difference between membrane potential and gravity.
   c. sum of the pressure potential and solute potential.
   d. difference between pressure potential and solute potential.

3. Hydrogen bonding between water molecules results in
   a. submersion.
   b. adhesion.
   c. evaporation.
   d. cohesion.

4. Water movement through cell walls is
   a. apoplastic.
   b. symplastic.
   c. Both a and b are correct.
   d. Neither a nor b is correct.

5. Casparian strips are found in the root
   a. cortex.
   b. dermal tissue.
   c. endodermis.
   d. xylem.

6. The formation of an air bubble in the xylem is called
   a. agitation.
   b. cohesion.
   c. adhesion.
   d. cavitation.

7. Guttation is most likely to be observed on a
   a. cold winter day.
   b. cool summer night.
   c. warm sunny day.
   d. warm cloudy day.

8. Stomata open when guard cells
   a. take up potassium. c. take up ABA.
   b. lose potassium.   d. take up chloride.

9. Which of the following is NOT an adaptation to a high saline environment?
   a. Secretion of salts
   b. Lowering of root water potential
   c. Exclusion of salt
   d. Production of pneumatophores

10. A plant must expend energy to drive
   a. transpiration.
   b. translocation.
   c. both transpiration and translocation.
   d. neither transpiration nor translocation.

APPLY

1. Which of the following statements is inaccurate?
   a. Water moves to areas of low water potential.
   b. Xylem transports materials up the plant, and phloem transports materials down the plant.
   c. Water movement in the xylem is due, in part, to the cohesive and adhesive properties of water.
   d. Water movement across membranes is often due to differences in solute concentrations.

2. If you could override the control mechanisms that open stomata and force them to remain closed, what would you expect to happen to the plant?
   a. Sugar synthesis would likely slow down.
   b. Water transport would likely slow down.
   c. The plant could overheat.
   d. All would be the result of keeping stomata closed.

3. What will happen if a cell with a solute potential of –0.4 MPa and a pressure potential of 0.2 MPa is placed in a chamber filled with pure water that is pressurized with 0.5 MPa?
   a. Water will flow out of the cell.
   b. Water will flow into the cell.
   c. The cell will be crushed.
   d. The cell will explode.

4. If you were able to remove the aquaporins from cell membranes, which of the following would be the likely consequence?
   a. Water would no longer move across membranes.
   b. Plants would no longer be able to control the direction of water movement across membranes.
   c. The potassium symport would no longer function.
   d. Turgor pressor would increase.

5. What would be the consequence of removing the Casparian strip?
   a. Water and mineral nutrients would not be able to reach the xylem.
   b. There would be less selectivity as to what passed into the xylem.
   c. Water and mineral nutrients would be lost from the xylem back into the soil.
   d. Water and mineral nutrients would no longer be able to pass through the cell walls of the endodermis.

SYNTHESIZE

1. If you fertilize your houseplant too often, you may find that it looks wilted even when the soil is wet. Explain what has happened in terms of water potential.

2. How could you detect a plant with a mutation in a gene for an important aquaporin protein?

3. Contrast water transport mechanisms in plants with those in animals.

4. Measurements of tree trunk diameters indicate that the trunk shrinks during the day, with shrinkage occurring in the upper part of the trunk before it occurs in the lower part. Explain how these observations support the hypothesis that water is pulled through the trunk as a result of transpiration.

5. A carrot is a biennial plant. In the first year of growth, the seed germinates and produces a plant with a thick storage root. In the second year, a shoot emerges from the storage root and produces a flower stalk. Following fertilization, seeds are formed to start the life cycle again. Draw a carrot plant during the spring, summer, and fall of the two years of its life cycle and indicate the carbohydrate sources and sinks in each season.

ONLINE RESOURCE

www.ravenbiology.com

Understand, Apply, and Synthesize—enhance your study with animations that bring concepts to life and practice tests to assess your understanding. Your instructor may also recommend the interactive eBook, individualized learning tools, and more.

Chapter **38**

Plant Nutrition and Soils

Chapter Contents

Introduction

Vast energy inputs are required for the building and ongoing growth of a plant. In this chapter, you'll learn what inputs, besides energy from the Sun, a plant needs to survive. Plants, like animals, need various nutrients to remain healthy. Plants acquire these nutrients mainly through photosynthesis and from the soil. The lack of an important nutrient may slow a plant's growth or make it more susceptible to disease or even death. In addition to contributing nutrients, the soil hosts bacteria and fungi that aid plants in obtaining nutrients in a usable form. Getting sufficient nitrogen is particularly problematic because plants cannot directly convert atmospheric nitrogen into amino acids. A few plants are able to capture animals and secrete digestive juices to make nitrogen available for absorption.

38.1 Soils: The Substrates on Which Plants Depend

Learning Outcomes

1. *Explain how soil interacts with negatively and positively charged ions.*
2. *Explain how soil characteristics affect nutrient uptake by roots.*
3. *Describe cultivation approaches that can reduce soil erosion.*

Much of the activity that supports plant life is hidden within the soil. **Soil** is the highly weathered outer layer of the Earth's crust. It is composed of a mixture of ingredients, which may include sand, rocks of various sizes, clay, silt, **humus** (partially decomposed organic matter), and various other forms of mineral and organic matter. Pore spaces containing water and air occur between the particles of soil.

Soil is composed of minerals, organic matter, water, air, and organisms

The mineral fraction of soils varies according to the composition of the rocks. The Earth's crust includes about 92 naturally occurring elements (see chapter 2). Most elements are found in the form of inorganic compounds called minerals; most rocks consist of several different minerals.

The soil is also full of microorganisms that break down and recycle organic debris. For example, about 5 metric tons of carbon is tied up in the organisms present in the soil under a hectare of wheat in England—an amount that approximately equals the weight of 100 sheep!

Most roots are found in **topsoil** (figure 38.1), which is a mixture of mineral particles of varying size (most less than 2 mm in diameter), living organisms, and humus. Topsoils are characterized by their relative amounts of sand, silt, and clay. Soil composition determines the degree of water and nutrient binding to soil particles. Sand binds molecules minimally, but clay adsorbs water and nutrients quite tightly.

Water and mineral availability are determined by soil characteristics

Only minerals that are dissolved in water in the spaces or pores among soil particles are available for uptake by roots. Both mineral and organic soil particles tend to have negative charges, so they attract positively charged molecules and ions. The negatively charged anions stay in solution, creating a charge gradient between the soil solution and the root cells, so that positive ions would normally tend to move out of the cells. Proton pumps move H^+ out of the root to form a strong membrane potential (≈ -160 mV). The strong electrochemical gradient then causes K^+ and other ions to enter via ion channels. Some ions, especially anions, use cotransporters (figure 38.2). The membrane potential maintained by the root, as well as the water potential difference inside and outside the root, affects root transport of minerals. (Water potential is described in chapter 37.)

About half of the total soil volume is occupied by pores, which may be filled with air or water, depending on moisture conditions (figure 38.3). Some of the soil water is unavailable to plants. In sandy soil, for example, a substantial amount of water drains away immediately due to gravity. Another fraction of the water is held in small soil pores, which are generally less than about 50 μm in diameter. This water is readily available to plants. When this water is depleted through evaporation or root uptake, the plant wilts and will eventually die unless more water is added to the soil. However, as plants deplete water near the roots, the soil water potential decreases. This helps to move more water toward the roots since the soil water farther away has a higher water potential.

Soils have widely varying composition, and any particular soil may provide more or fewer plant nutrients. In addition, the soil's acidity and salinity, described shortly, can affect the availability of nutrients and water.

Cultivation can result in soil loss and nutrient depletion

When topsoil is lost because of erosion or poor landscaping, both the water-holding capacity and the nutrient relationships of the soil are adversely affected. Up to 50 billion tons of topsoil have been lost from fields in the United States in a single year.

Whenever the vegetative cover of soil is disrupted, such as by plowing and harvesting, erosion by water and wind increases—sometimes dramatically, as was the case in the 1930s in the southwestern Great Plains of the United States. This region became known as the "Dust Bowl" when a combination of poor

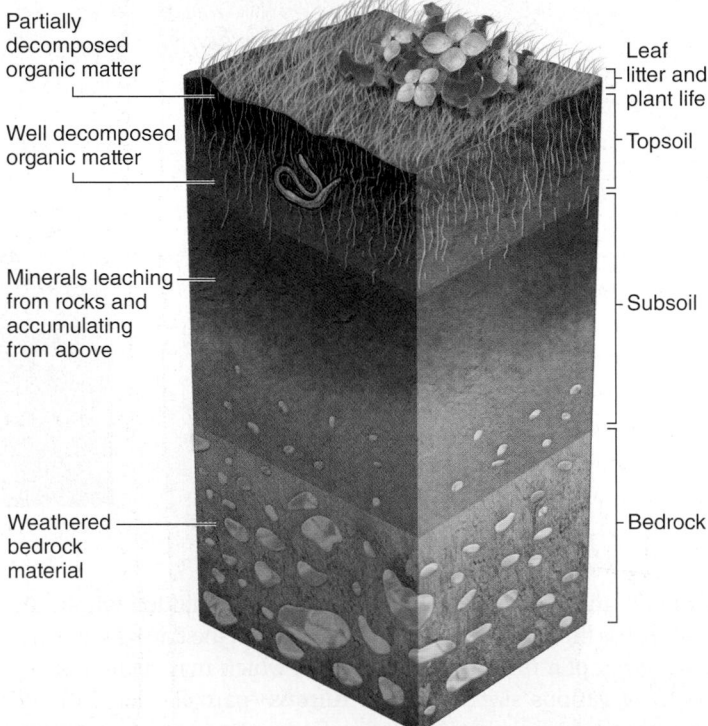

Figure 38.1 Most roots grow in the topsoil. Leaf litter and animal remains cover the uppermost layer in soil called topsoil. Topsoil contains organic matter, such as roots, small animals, and humus, and mineral particles of various sizes. Subsoil lies underneath the topsoil and contains larger mineral particles and relatively little organic matter. Beneath the subsoil are layers of bedrock, the raw material from which soil is formed over time and through weathering.

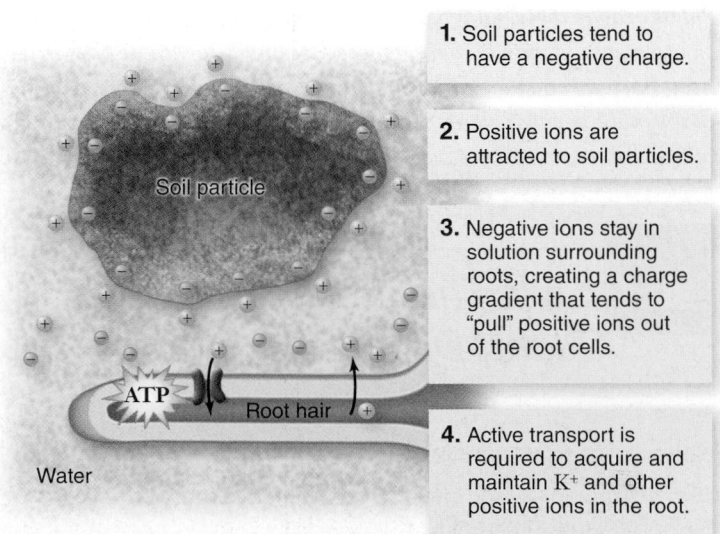

1. Soil particles tend to have a negative charge.

2. Positive ions are attracted to soil particles.

3. Negative ions stay in solution surrounding roots, creating a charge gradient that tends to "pull" positive ions out of the root cells.

4. Active transport is required to acquire and maintain K^+ and other positive ions in the root.

Figure 38.2 Role of soil charge in transport. Active transport is required to move positively charged ions into a root hair.

Figure 38.3 Water and air fill pores among soil particles. *a.* Without some space for air circulation in the soil, roots cannot respire. *b.* A balance of air and water in the soil is essential for root growth. *c.* Too little water decreases the soil water potential and prevents transpiration in plants.

farming practices and several years of drought made the soil particularly susceptible to wind erosion (figure 38.4*a*).

New approaches to cultivation are aimed at reducing soil loss. Intercropping (mixing crops in a field), conservation tillage, and not plowing under the fall crop detritus (no-till) are all erosion-prevention measures. Conservation tillage includes minimal till and even no-till approaches to farming.

Overuse of fertilizers in agriculture, lawns, and gardens can cause significant water pollution and its associated negative effects, such as overgrowth of algae in lakes (see chapter 57). Maintaining nutrient levels in the soil and preventing nutrient runoff into lakes, streams, and rivers improves crop growth and minimizes ecosystem damage.

One approach, site-specific farming, uses variable-rate fertilizer applicators guided by a computer and the global positioning system (GPS). Variable-rate application relies on information about local soil nutrient levels, based on analysis of soil samples. Another approach, integrated nutrient management, maximizes nutritional inputs using "green manure" (such as alfalfa tilled back into the soil), animal manure, and inorganic fertilizers. Green manures and animal manure have the advantage of releasing nutrients slowly as they are broken down by

decomposer organisms, so that nutrients may be utilized before leaching away. Sustainable agriculture integrates these conservation approaches.

pH and salinity affect water and mineral availability

Anything that alters water potential differences or ionic gradient balance between soil and roots can affect the ability of plants to absorb water and nutrients. Acid soils (having low pH) and saline soils (high in salts) can present problems for plant growth.

Acid soils

The pH of a soil affects the release of minerals from weathering rock. For example, at low pH aluminum, which is toxic to many plants, is released from rocks. Furthermore, aluminum can also combine with other nutrients and make them inaccessible to plants.

Most plants grow best at a neutral pH, but about 26% of the world's arable land is acidic. In the tropical Americas,

Figure 38.4 Soil degradation. *a.* Drought and poor farming practices led to wind erosion of farmland in the southwestern Great Plains of the United States in the 1930s. *b.* Draining marshland in Iraq resulted in a salty desert.

a.

b.

Background Information: *In acid soils, Al^{3+} is the common form of aluminum and is highly toxic to plants. Application of lime can increase soil pH and convert Al^{3+} to less toxic forms of aluminum, enhancing root growth.*

Hypothesis: *Aluminum-tolerant wheat plants are able to increase the pH around the root to enhance plant growth.*

Prediction: *When aluminum is added to culture solutions with wheat seedlings, the pH of the solution increases to a higher level in the tolerant plants than in the sensitive varieties.*

Test: *Grow sensitive (S) and tolerant (T) wheat varieties in culture solution starting at pH 4.5. Add 0, 25, 50, and 100 μm Al^{3+} in a dose experiment to both S and T plants and measure the pH after 24 hours.*

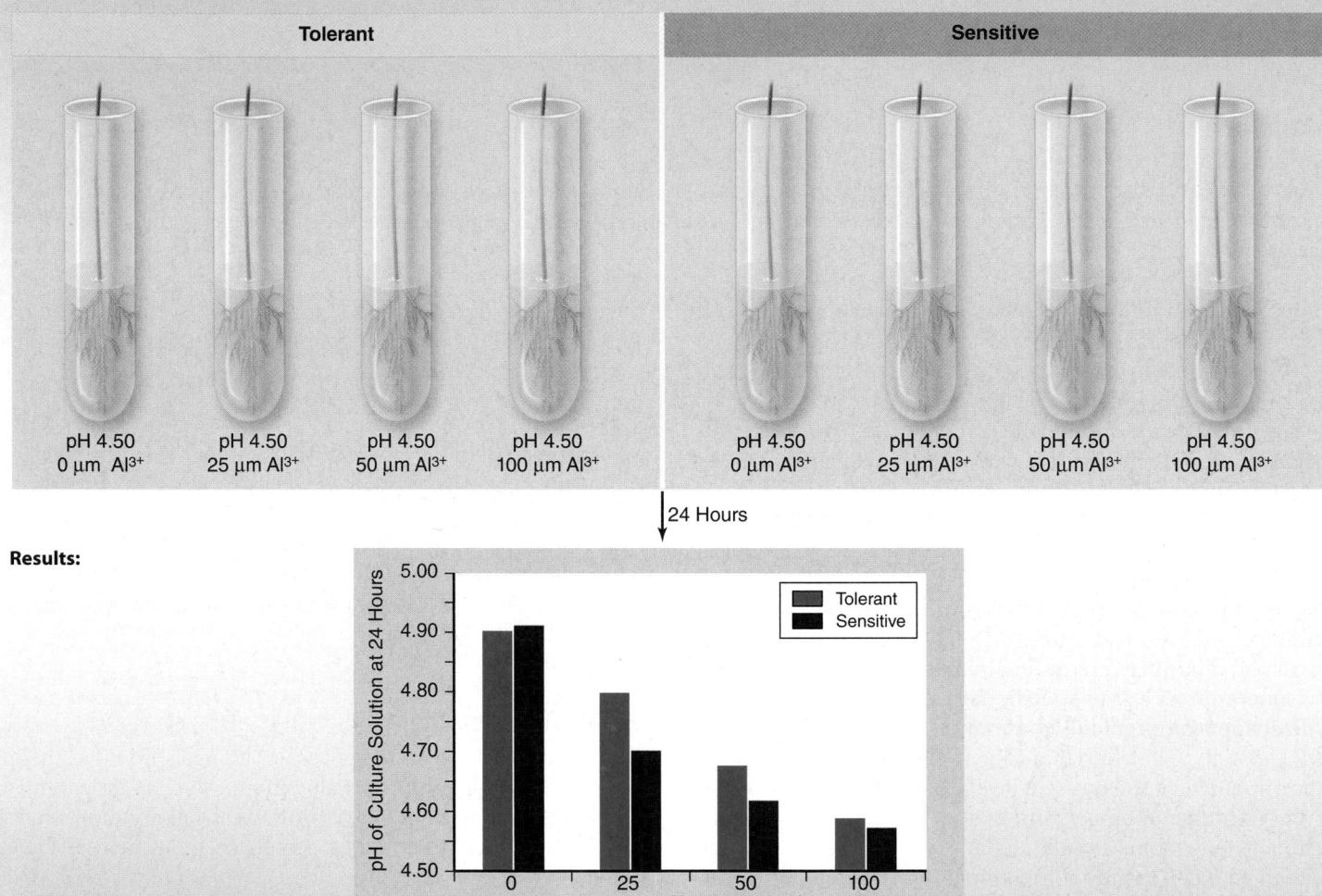

Results:

Conclusion: *Both sensitive and tolerant plants increase the pH around the roots in the absence of Al^{3+}. In the 25 and 50 μm range of Al^{3+}, the tolerant plants maintain a higher pH than the sensitive plants.*

Further Experiments: *To demonstrate the link between the pH of the culture solution and plant growth, you measure the rate of root growth in solutions of different pH. Assuming pH positively affects root growth, draw a possible graph of percent root elongation as a function of pH at 24 hours. Use the pH values found in the bar graph above.*

Figure 38.5 **Roots can modify local pH to decrease aluminum toxicity.**

68% of the soil is acidic. Aluminum toxicity in acid soils in Colombian fields can reduce maize (corn) yield fourfold (figure 38.5).

Breeding efforts in Colombia are producing aluminum-tolerant plants, and crop yields have increased 33%. In a few test fields, the yield increases have been as high as 70% compared with that for sensitive plants. The ability of plants to take up toxic metals can also be employed to clean up polluted soil, a topic explored later in this chapter.

Salinity

The accumulation of salt ions, usually Na^+ and Cl^-, in soil alters water potential, leading to the loss of turgor in plants. Approximately 23% of all arable land has salinity levels that limit plant growth. Saline soil is most common in dry areas where salts are introduced through irrigation. In such areas, precipitation is insufficient to remove the salts, which gradually accumulate in the soil.

One of the more dramatic examples of soil salinity occurs in the "cradle of civilization," Mesopotamia. The region once called the Fertile Crescent for its abundant agriculture is now largely a desert. Desertification was accelerated in southern Iraq. In the 1990s, most of 20,000 km² of marshlands was drained by redirecting water flow with dams, turning the marshes into a salty desert (see figure 38.4b). The dams were destroyed later, allowing water to enter the marshlands once again. Recovery of the marshlands is not guaranteed, but in areas where the entering water has lowered the salinity, there is hope.

Learning Outcomes Review 38.1

Topsoil is composed of mineral particles, living organisms, and humus. Roots use proton pumps to move protons (H^+) out of the root and into the soil. The result is an electrochemical gradient that causes positive mineral ions to enter the root through ion channels. The loss of topsoil by erosion can be reduced by intercropping, planting crop mixtures, conservation tillage, and no-till farming.

■ **Applying your knowledge of water potential, explain what follows after depletion of water close to the roots.**

38.2 Plant Nutrients

Learning Outcomes

1. Distinguish between macronutrients and micronutrients.
2. Explain how scientists determine the nutritional needs of plants.
3. Describe the goal of food fortification research.

The major source of plant nutrition is the fixation of atmospheric carbon dioxide (CO_2) into simple sugars using the energy of the Sun for photosynthesis. CO_2 enters through the stomata; oxygen (O_2) is a waste product of photosynthesis and an atmospheric component that also moves through the stomata. O_2 is used in cellular respiration to support growth and maintenance in the plant.

CO_2 and light energy are not sufficient, however, for the synthesis of all the molecules a plant needs. Plants require a number of inorganic nutrients as well. Some of these are **macronutrients,** which plants need in relatively large amounts, and others are **micronutrients,** required in trace amounts (table 38.1).

TABLE 38.1	Essential Nutrients in Plants		
Element	**Principal Form in Which Element Is Absorbed**	**Approximate Percent of Dry Weight**	**Examples of Important Functions**
MACRONUTRIENTS			
Carbon	CO_2	44	Major component of organic molecules
Oxygen	O_2, H_2O	44	Major component of organic molecules
Hydrogen	H_2O	6	Major component of organic molecules
Nitrogen	NO_3^-, NH_4^+	1–4	Component of amino acids, proteins, nucleotides, nucleic acids, chlorophyll, coenzymes, enzymes
Potassium	K^+	0.5–6	Protein synthesis, operation of stomata
Calcium	Ca^{2+}	0.2–3.5	Component of cell walls, maintenance of membrane structure and permeability; activates some enzymes
Magnesium	Mg^{2+}	0.1–0.8	Component of chlorophyll molecule, activates many enzymes
Phosphorus	$H_2PO_4^-$, HPO_4^-	0.1–0.8	Component of ADP and ATP, nucleic acids, phospholipids, several coenzymes
Sulfur	SO_4^{2-}	0.05–1	Components of some amino acids and proteins, coenzyme A
MICRONUTRIENTS (CONCENTRATIONS in ppm)			
Chlorine	Cl^-	100–10,000	Osmosis and ionic balance
Iron	Fe^{2+}, Fe^{3+}	25–300	Chlorophyll synthesis, cytochromes, nitrogenase
Manganese	Mn^{2+}	15–800	Activator of certain enzymes
Zinc	Zn^{2+}	15–100	Activator of many enzymes; active in formation of chlorophyll
Boron	BO_3^-, $B_4O_7^-$, or $H_2BO_3^-$	5–75	Possibly involved in carbohydrate transport, nucleic acid synthesis
Copper	Cu^2 or Cu^+	4–30	Activator or component of certain enzymes
Molybdenum	MoO_4^-	0.1–5	Nitrogen fixation, nitrate reduction

a.
b.

c.

d.

Figure 38.6 Mineral deficiencies in plants. *a.* Healthy tomato plant. *b.* Manganese-deficient leaves. *c.* Copper-deficient tomato leaf. *d.* Phosphorus deficiency in tomato.

? Inquiry question Identify the source of the majority of the mass in the trunk of a large oak tree.

Plants require nine macronutrients and seven micronutrients

The nine macronutrients are carbon, oxygen, and hydrogen—the three elements found in all organic compounds—plus nitrogen (essential for amino acids), potassium, calcium, magnesium (the center of the chlorophyll molecule), phosphorus, and sulfur. Each of these nutrients approaches or, in the case of carbon, may greatly exceed 1% of the dry weight of a healthy plant.

The seven micronutrient elements—chlorine, iron, manganese, zinc, boron, copper, and molybdenum—constitute from less than one to several hundred parts per million in most plants. A deficiency of any one can have severe effects on plant growth (figure 38.6). The macronutrients were generally discovered in the last century, but the micronutrients have been detected much more recently as technology developed to identify and work with such small quantities.

Nutritional requirements are assessed by growing plants in hydroponic cultures in which the plant roots are suspended in aerated water containing nutrients. For the purposes of testing, the solutions contain all the necessary nutrients in the right proportions, but with certain known or suspected nutrients left out. The plants are then allowed to grow and are studied for altered growth patterns and leaf coloration that might indicate a need for the missing element (figure 38.7). To give an idea of how small the needed quantities of micronutrients may be, the standard dose of molybdenum added to seriously deficient soils in Australia amounts to about 34 g (about one handful) per hectare (a square 100 m on a side—about 2.5 acres), once every 10 years!

Most plants grow satisfactorily in hydroponic cultures, if the roots are properly aerated. The method, although expensive, is occasionally practical for commercial purposes

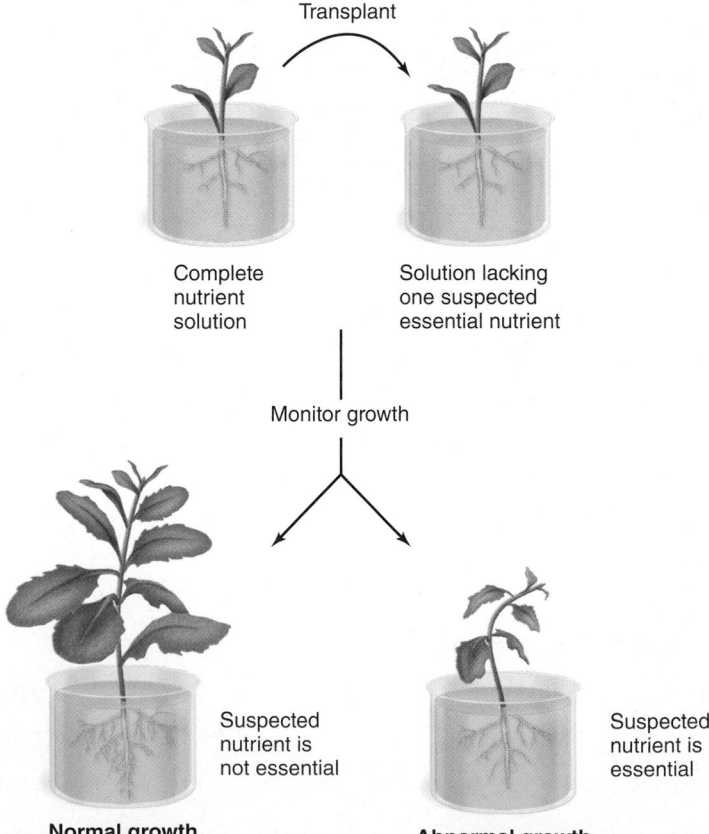

Figure 38.7 Identifying nutritional requirements of plants. A seedling is first grown in a complete nutrient solution. The seedling is then transplanted to a solution that lacks one nutrient thought to be essential. The growth of the seedling is studied for the presence of symptoms indicative of abnormal growth, such as discolored leaves or stunting. If the seedling's growth is normal, the nutrient that was left out may not be essential; if the seedling's growth is abnormal, the nutrient that is lacking is essential for growth.

Figure 38.8 Hydroponics. Soil provides nutrients and support, but both of these functions can be replaced in hydroponic systems to maximize growth. Here, tomato plants are suspended in the air, and the roots rotate through a nutrient bath.

(figure 38.8). Analytical chemistry has made it much easier to test plant material for levels of different molecules.

Food security is related to crop productivity and nutrient levels

Nutrient levels and crop productivity are a significant human concern. **Food security,** avoiding starvation, is a global issue. Increasing the nutritional value of crop species, especially in developing countries, could have tremendous benefits for human health.

Food fortification is an active area of research focused on ways to increase plants' uptake of minerals and the storage of minerals in roots and shoots for later human consumption. Phosphate uptake can be increased, for example, if it is more soluble in the soil. Some plants have been genetically modified to secrete citrate, an organic acid that solubilizes phosphate. As an added benefit, the citrate binds to aluminum, which can be toxic to plants and animals, and thus limits the uptake of aluminum into plants.

For other nutrients, such as iron, manganese, and zinc, plasma membrane transport is a limiting factor. Genes coding for these plasma membrane transporters have been cloned in other species and are being incorporated into crop plants. Eventually, breakfast cereals may be fortified with additional nutrients while the grains are growing in the field, as opposed to when they are processed in the factory.

Learning Outcomes Review 38.2

Plants require nine macronutrients in relatively large amounts and seven micronutrients in trace amounts. Plants are grown in controlled hydroponic solutions to determine which nutrients are required for growth. Scientists are studying ways to enhance the nutritional composition of food crops through enhancing nutrient uptake and storage. These methods of food fortification may enhance food security, the avoidance of human starvation.

■ *Why would a lack of magnesium in the soil limit food production?*

Learning Outcomes

1. *Explain the significance of nitrogen-fixing bacteria for plant nutrition.*
2. *Explain how mycorrhizal fungi benefit plants.*
3. *Describe the benefit gained by carnivorous plants when they capture insects.*

In some species, scarce nutrients have been obtained through the evolution of mutualistic associations with other organisms, parasitism, or even predation. One example is the requirement for nitrogen: Plants need ammonia (NH_3) or nitrate (NO_3^-) to build amino acids, but most of the nitrogen in the atmosphere is in the form of gaseous nitrogen (N_2). Plants lack the biochemical pathways (including the enzyme nitrogenase) necessary to convert N_2 to NH_3, but some bacteria have this capacity.

Bacteria living in close association with roots can provide nitrogen

Symbiotic relationships have evolved between some plant groups and bacteria that can convert gaseous nitrogen. Some of these bacteria live in close association with the roots of plants. Others end up being housed in tissues the plant grows especially for this purpose, called **nodules** (figure 38.9). Legumes and a few other plants can form root nodules. Hosting these bacteria costs the plant energy, but is well worth it when the soil lacks nitrogen compounds. To conserve energy, legume root hairs do not respond to bacterial signals when nitrogen levels are high.

Nitrogen fixation is the most energetically expensive reaction known to occur in any cell. Why should it be so difficult to add H_2 to N_2? The answer lies in the strength of the triple bond in N_2. Nitrogenase requires 16 ATP to make two molecules of NH_3. Making NH_3 without nitrogenase requires a contained system maintained at 450°C and 500 atm pressure—far beyond the maximums under which plants can survive.

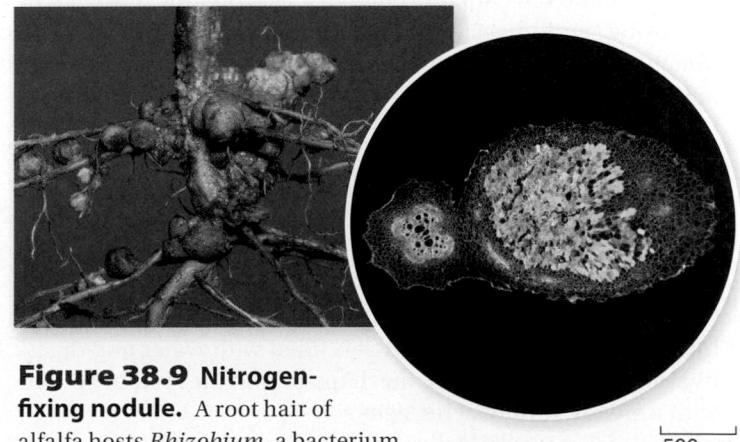

Figure 38.9 Nitrogen-fixing nodule. A root hair of alfalfa hosts *Rhizobium,* a bacterium that fixes nitrogen in exchange for carbohydrates.

500 μm

Rhizobium bacteria require oxygen and carbohydrates to support their energetically expensive lifestyle as nitrogen fixers. Carbohydrates are supplied through the vascular tissue of the plant, and leghemoglobin, which is structurally similar to animal hemoglobin, is produced by the plant to regulate oxygen availability to the bacteria. Without oxygen, the bacteria die; within the bacteria, however, nitrogenase has to be isolated from oxygen, which inhibits its activity. Leghemoglobin binds oxygen and controls its availability within the nodule to optimize both nitrogenase activity and cellular respiration.

Just how do legumes and nitrogen-fixing *Rhizobium* bacteria get together (figure 38.10)? Extensive signaling between the bacterium and the legume not only lets each organism know the other is present, but also checks whether the bacterium is the correct species for the specific legume. These highly evolved symbiotic relationships depend on exact species matches. Soybean and garden peas are both legumes, but each requires its own species of symbiotic *Rhizobium.*

Mycorrhizae aid a large portion of terrestrial plants

Nitrogen is not the only nutrient that is difficult for plants to obtain without assistance. Whereas symbiotic relationships with nitrogen-fixing bacteria are generally limited to some legume species, symbiotic associations with mycorrhizal fungi, described in chapter 32, are found in about 90% of vascular plants. Mycorrhizae play a significant role in enhancing phosphate transfer to the plant, and the uptake of some of the micronutrients is also facilitated. Functionally, the mycorrhizae substantially extend the surface area available for nutrient uptake.

Fungi most likely aided early rootless plants in colonizing land. Evidence now indicates that the signaling pathways that lead to plant symbiosis with some mycorrhizae may have been exploited to bring about the *Rhizobium*–legume symbiosis.

Carnivorous plants trap and digest animals to extract additional nutrients

Some plants are able to obtain nitrogen directly from other organisms, just as animals do. These carnivorous plants often grow in acidic soils, such as bogs, that lack organic nitrogen. By capturing and digesting small animals, primarily insects, directly, such plants obtain adequate nitrogen supplies and are able to grow in these seemingly unfavorable environments. Carnivorous plants have modified leaves adapted for luring and trapping prey. The plants often digest their prey with enzymes secreted from specialized types of glands.

Pitcher plants (*Nepenthes* spp.) attract insects by the bright, flower-like colors within their pitcher-shaped leaves, by scents, and perhaps also by sugar-rich secretions (figure 38.11*a*). Once inside the pitcher, insects slide down into the cavity of the leaf, which is filled with water and digestive enzymes. This passive mechanism provides pitcher plants with a steady supply of nitrogen.

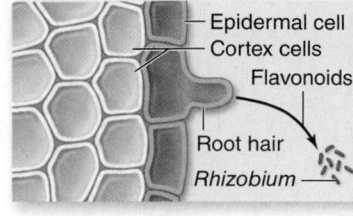

1. Pea roots produce flavonoids (a group of molecules used for plant defense and for making reddish pigment, among numerous other functions). The flavonoids are transported into the rhizobial cells.

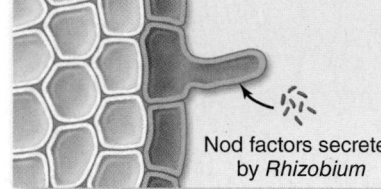

2. Flavonoids signal rhizobia to produce sugar-containing compounds called Nod (nodulation) factors.

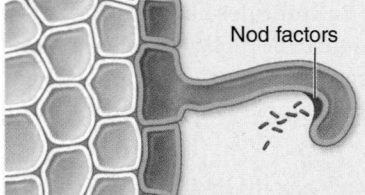

3. Nod factors are perceived by the surface of root hairs and signal the root hair to grow so that it curls around the rhizobia.

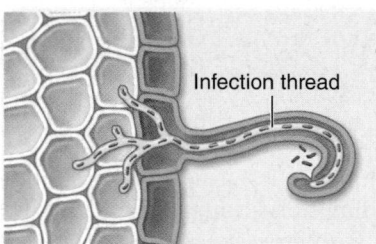

4. Rhizobia make an infection thread that grows in the root hair and moves into the cortex of the root. The rhizobia take control of cell division in the cortex and pericycle cells of the root (see chapter 36).

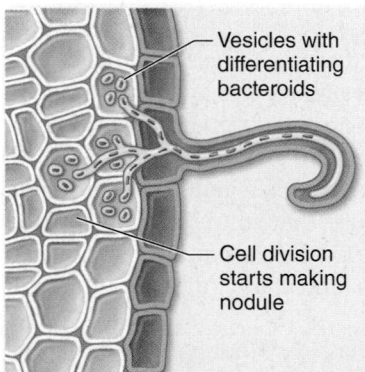

5. Rhizobia change shape and are now called bacteroids. Bacteroids produce an O_2-binding heme group that combines with a globin group from the pea to make leghemoglobin. Leghemoglobin gives a pinkish tinge to the nodule, and its function is much like that of hemoglobin, bringing O_2 to the rapidly respiring bacteroids, but isolating O_2 from nitrogenase.

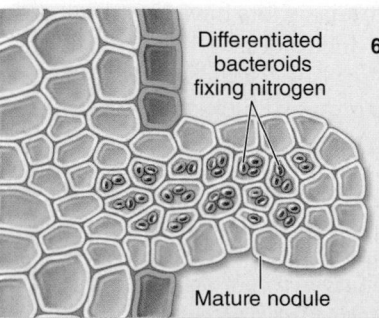

6. Bacteroids produce nitrogenase and begin fixing atmospheric nitrogen for the plant's use. In return, the plant provides organic compounds.

Figure 38.10 *Rhizobium*-**induced nodule formation.**

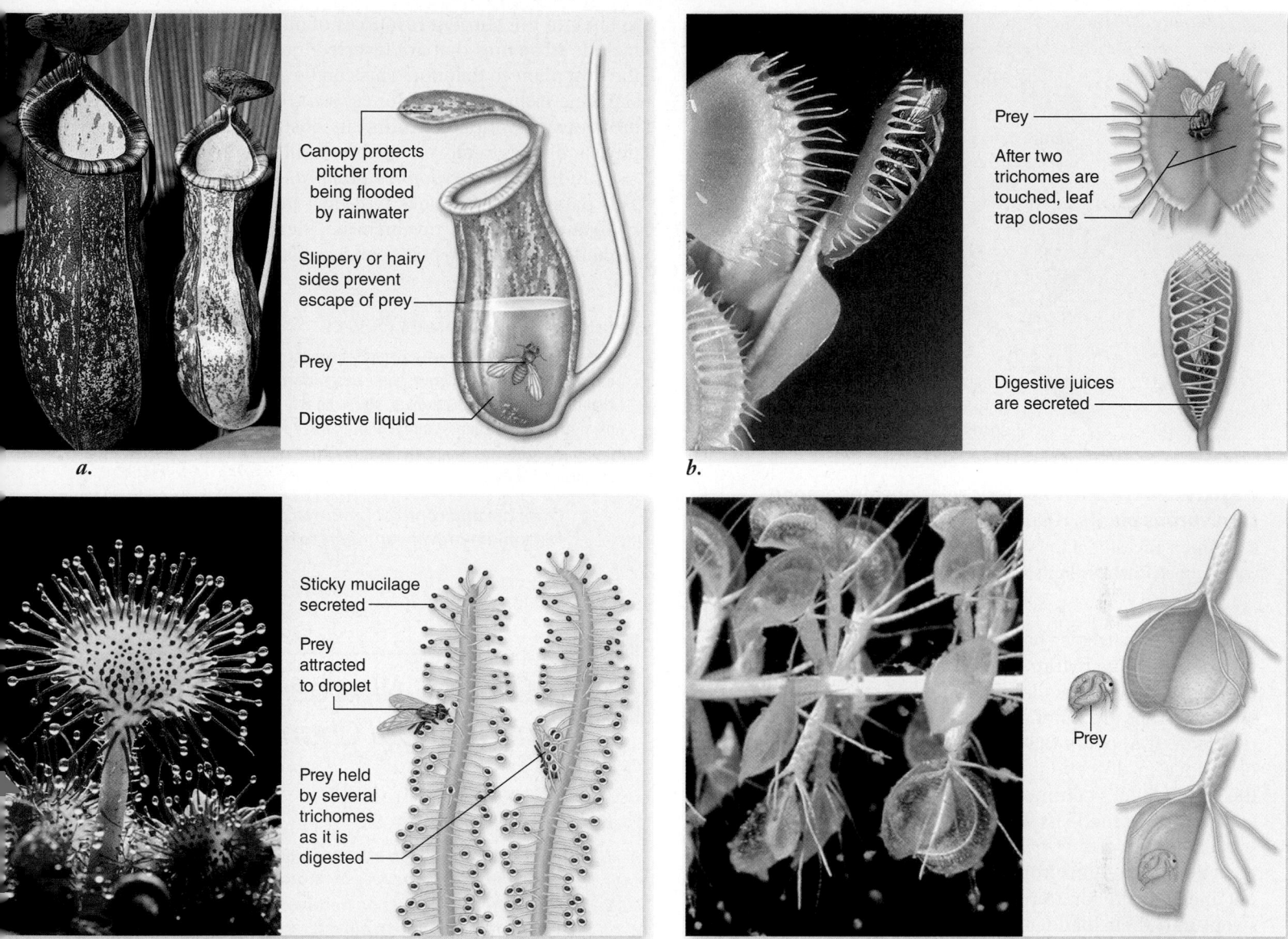

Figure 38.11 Nutritional adaptations. *a.* Asian pitcher plant, *Nepenthes.* Insects enter this carnivorous plant and are trapped and digested. Complex communities of invertebrate animals and protists inhabit the pitchers. *b.* Venus flytrap, *Dionaea.* If this fly touches two of the trichomes (hairs) on this modified leaf in a short time span, the trap will close. The plant secretes digestive enzymes that release nitrogen compounds from the fly, which will then be absorbed by the flytrap. *c.* Sundew, *Drosera,* traps insects with sticky secretions and then excretes digestive enzymes to obtain nutrients from the insect's body. *d.* Aquatic waterwheel, *Aldrovanda.* This close relative of the Venus flytrap snaps shut to capture and digest small aquatic animals. This aquatic plant's ancestor was a land dweller.

The Venus flytrap *(Dionaea muscipula)* grows in the bogs of coastal North and South Carolina. Three sensitive hairs on a leaf, when touched, trigger the two halves of the leaf to snap together in about 100 ms (figure 38.11*b*). The speed of trap closing has puzzled biologists as far back as Darwin. Turgor pressure changes can account for the movement; the speed, however, depends on the curved geometry of the leaf, which can snap between convex and concave shapes.

Once the Venus flytrap enfolds prey within a leaf, enzymes secreted from the leaf surfaces digest the prey. These flytraps use a growth mechanism to close, not just a decrease in

turgor pressure. The cells on the outer leaf surface irreversibly increase in size each time the trap shuts. As a result, they can only open and close a limited number of times.

In the sundews (*Drosera* spp.), another carnivorous group, glandular trichomes secrete both sticky mucilage, which traps small animals, and digestive enzymes; they do not close rapidly (figure 38.11*c*). Venus flytraps and the sundews share a common ancestor that lacked the snap-trap mechanism characteristic of the flytrap lineage (figure 38.12).

Aldrovanda vesicular, the aquatic waterwheel, is a closer relative of the flytraps. The waterwheel is a rootless

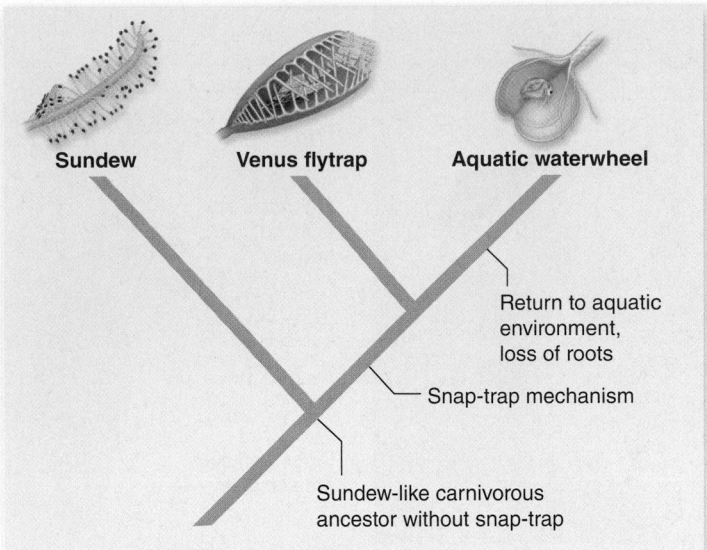

Figure 38.12 Phylogenetic relationships among carnivorous plants. The snap-trap mechanism was acquired by a common ancestor of the Venus flytrap and the aquatic waterwheel. Pitcher plants are not related to this clade.

plant that uses trigger hairs and a snap-trap mechanism like that of the Venus flytrap to capture and digest small animals (figure 38.11*d*). Molecular phylogenetic studies indicate that Venus flytraps are sister species with sundews, forming a sister clade. It appears that the snap-trap mechanism evolved only once in descendants of a sundew ancestor. Therefore, the waterwheel's common ancestor must have been a terrestrial plant that made its way back into the water.

Bladderworts (*Utricularia*) are aquatic, but appear to have different origins from the waterwheel, as well as a different mechanism for trapping organisms. Small animals are swept into their bladder-like leaves by the rapid action of a springlike trapdoor; then the leaves digest these animals.

Parasitic plants exploit resources of other plants

Parasitic plants come in photosynthetic and nonphotosynthetic varieties. In total, at least 3000 types of plants are known

Figure 38.13 Indian pipe, *Hypopitys uniflora.* This plant lacks chlorophyll and depends completely on nutrient transfer through the invasion of mycorrhizae and associated roots of other plants. Indian pipes are frequently found in northeastern U.S. forests.

to tap into the nutrient resources of other plants. Adaptations include structures that are inserted into the vascular tissue of the host plant so that nutrients can be siphoned into the parasite. One example is dodder (*Cuscuta* spp.), which looks like brown twine wrapped around its host. Dodder lacks chlorophyll and relies totally on its host for all its nutritional needs.

Indian pipe, *Hypopitys uniflora,* also lacks chlorophyll. This parasitic plant hooks into host trees through the fungal hyphae of the host's mycorrhizae (figure 38.13). The aboveground portion of the plant consists of flowering stems.

Learning Outcomes Review 38.3

Certain types of plants, such as legumes, produce root nodules in which nitrogen-fixing bacteria grow. These bacteria provide nitrogen compounds that the plant can use for growth. Mycorrhizal fungi live in association with plant roots and are important for phosphorus uptake. Carnivorous plants typically live in low-nitrogen soils and obtain nitrogen from the insects they capture and digest.

■ *Compare and contrast how myccorrhizal fungi and Rhizobium provide nutrients for their plant partners.*

38.4 Carbon–Nitrogen Balance and Global Change

Learning Outcomes

1. Describe the predicted effect of increased atmospheric carbon dioxide on the rate of photosynthesis in C_3 plants.
2. Explain the main effect on herbivores of a higher carbon:nitrogen ratio in plants.
3. Discuss why respiration rates increase with warmer temperatures.

The Intergovernmental Panel on Climate Change (IPCC), established by the United Nations and the World Meteorological Organization, has concluded that CO_2 is probably at its highest concentration in the atmosphere in at least 20 million years. In only the last 250 years, atmospheric CO_2 has increased 31%, which correlates with increases in many human activities, including the burning of fossil fuels.

The long-term effects of elevated CO_2 are complex and are not yet fully understood, but are associated with increased temperatures. The IPCC predicts the average global surface temperatures will continue to increase to between 1.4°C and 5.8°C above 1990 levels, by 2100. Chapter 58 explores the causal link between elevated CO_2 and global warming. Here, we consider how increased CO_2 may alter nutrient balance within plants, specifically the carbon and nitrogen balance.

The ratio of carbon to nitrogen in a plant is important for both plant health and the health of herbivores. Altering this ratio could alter plant–pest interactions as well as affect human nutrition.

Elevated CO₂ levels can alter photosynthesis and carbon levels in plants

First, we investigate the relationship between photosynthesis and the relative concentration of atmospheric CO_2. The two questions to be addressed in this section are (1) Does elevated CO_2 increase the rate of photosynthesis? and (2) Will elevated levels of CO_2 change the ratio of carbohydrates and proteins in plants?

The rate of photosynthesis

The Calvin cycle of photosynthesis fixes atmospheric CO_2 into sugar (see chapter 8). The first step of the Calvin cycle stars the most abundant protein on Earth, ribulose 1,5-bisphosphate carboxylase/oxygenase (rubisco). The active site of this enzyme can bind either CO_2 or O_2, and it catalyzes the addition of either molecule to a five-carbon molecule, ribulose 1,5-bisphosphate (RuBP) (figure 38.14). CO_2 is used to produce a three-carbon sugar that can in turn be used to synthesize glucose and sucrose; in contrast, O_2 is used in photorespiration, which results in neither nutrient nor energy storage. Photorespiration does not provide the plant with energy.

You may recall that C_4 plants have evolved a novel anatomical and biochemical strategy to reduce photorespiration (figure 38.15). CO_2 does not enter the Calvin cycle until it has been transported via another pathway to cells surrounding the vascular tissue. Here the level of CO_2 is increased relative to O_2 levels, and thus more CO_2 than O_2 is available to interact with rubisco's binding site.

In C_3 plants, as the relative amount of CO_2 increases, the Calvin cycle becomes more efficient. Thus, it is reasonable to hypothesize that the global increase in CO_2 should lead to increased photosynthesis and increased plant growth. Assuming that nutrient availability in the soil remains the same, the more rapidly growing plants should have lower levels of nitrogen-containing compounds, such as proteins, and also lower levels of minerals obtained from the soil. The ratio of carbon to nitrogen should increase. Long-term studies of plants grown under elevated CO_2 confirm this prediction.

Figure 38.14 Photorespiration.

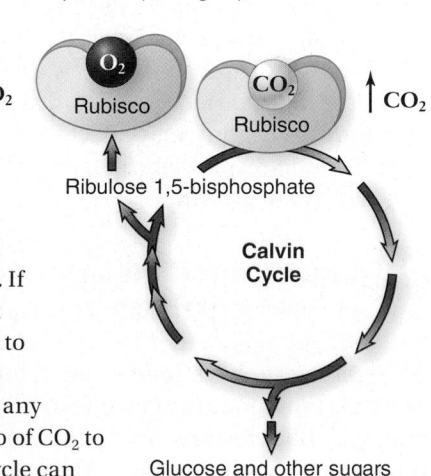

CO_2 and O_2 compete for the same site on the enzyme that catalyzes the first step in the Calvin cycle. If CO_2 binds, a three-carbon sugar is produced that can make glucose and sucrose. If O_2 binds, photorespiration occurs, and energy is used to break down a five-carbon molecule without yielding any useful product. As the ratio of CO_2 to O_2 increases, the Calvin cycle can produce more sugar.

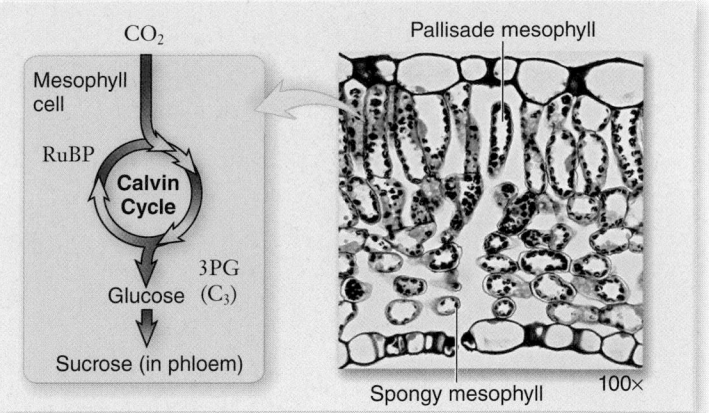

a. C_3 leaf

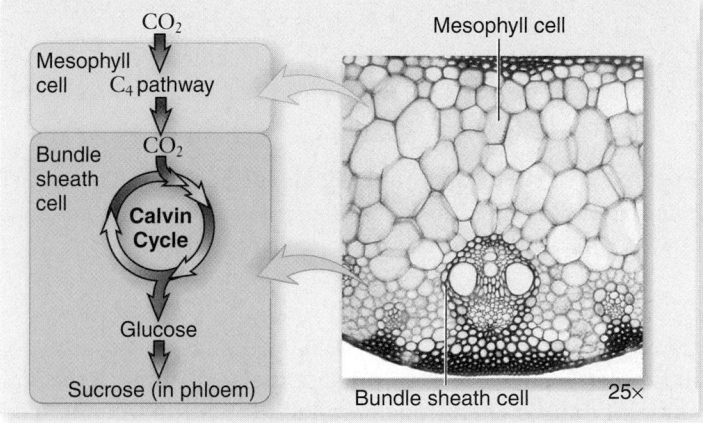

b. C_4 leaf (Kranz anatomy)

Figure 38.15 Reducing Photorespiration. C_4 plants reduce photorespiration by limiting the Calvin cycle to cells surrounding the vascular tissue, where O_2 levels are reduced. *a.* C_3 photosynthesis occurs in the mesophyll cells. *b.* C_4 photosynthesis uses an extra biochemical pathway to shuttle carbon deep within the leaf.

The optimal way to determine how CO_2 concentrations affect plant nutrition is to grow plants in an environment in which CO_2 levels can be precisely controlled. Experiments with potted plants in growth chambers are one approach, but far more information can be obtained in natural areas enriched with CO_2, called Free Air CO_2 Enrichment (FACE) studies. For example, the Duke Experimental Forest has rings of towers that release CO_2 toward the center of the ring (figure 38.16). These rings are 30 m in diameter and allow studies to be conducted at the ecosystem level. Such facilities allow for long-term studies of the effects of altered atmospheric conditions on ecosystems.

Extensive studies have yielded complex results. Potatoes grown in a European facility had a 40% higher photosynthetic rate when the concentration of CO_2 was approximately doubled. Potted plants often show an initial increase in photosynthesis, followed by a decrease over time that is associated with lower levels of rubisco production. Different species of plants in a Florida oak-shrub system showed different responses to elevated CO_2 levels, whereas over three years in the Duke Experimental Forest, plants achieved more biomass

Ring of towers emitting CO₂

Towers emitting CO₂

a.　　　　　　　　　　　　　　　　　　　　　　　　*b.*

Figure 38.16 Experimentally elevating CO₂. CO₂ rings in the Duke Experimental Forest FACE site provide ecosystem-level comparisons of plants grown in ambient and elevated CO₂ environments. *a.* Each ring is 30 m in diameter. *b.* Towers surrounding the rings blow CO₂ inward under closely monitored conditions.

in the CO_2 enclosures than outside the enclosures, if the soil contained sufficient nitrogen availability to support enhanced growth. C_3 plants show a greater increase in biomass than C_4 plants, and nitrogen fixing legumes, especially soybean, had larger increases in biomass than plants depending on nitrogen from the soil. In general, increased CO_2 corresponds to some increase in biomass, but also to an increase in the carbon:nitrogen ratio.

The ratio of proteins and carbohydrates

You learned earlier in this chapter that nitrogen availability limits plant growth. As CO_2 levels increase, relatively less nitrogen and other macronutrients are found in leaves. In that event, herbivores need to eat more biomass to obtain adequate nutrients, particularly protein. This situation would be of significant concern in agriculture, and it could affect human health. Insect infestations could be more devastating if each herbivore consumed more biomass. Protein deficiencies in human diets could result from decreased nitrogen in crops.

The relative decreases in nitrogen in some plants is greater than would be predicted by an increase in CO_2 fixation alone. The additional decrease in nitrogen incorporation into proteins has been accounted for by a decrease in photorespiration in plants using NO_3^- as their primary nitrogen source, but not in plants using ammonia. It is possible that energy-wasting photorespiration may actually be necessary for nitrogen to be incorporated into proteins in some plants.

This example illustrates how interdependent the biochemical pathways are that regulate carbon and nitrogen levels. Although global change is an ecosystem-level problem,

predictions about long-term effects hinge on understanding the physiological complexities of plant nutrition—an area of active research.

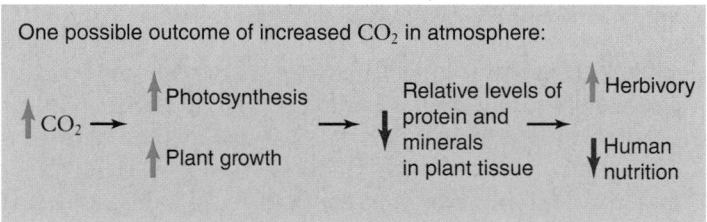

One possible outcome of increased CO₂ in atmosphere:

↑ CO₂ → ↑ Photosynthesis / ↑ Plant growth → ↓ Relative levels of protein and minerals in plant tissue → ↑ Herbivory / ↓ Human nutrition

Elevated temperature can affect respiration and carbon levels in plants

As much as half of all the carbohydrates produced from photosynthesis each day can be used in plant respiration that same day. The amount of carbohydrate available for respiration may be affected by atmospheric CO_2 and photosynthesis, as just discussed. Furthermore, the anticipated rise in temperature over the next century may affect the rate of respiration in other ways. Altered respiration rates can affect overall nutrient balance and plant growth.

? Inquiry question Why is plant respiration affected by temperature changes?

Biologists have known for a long time that respiration rates are temperature-sensitive in a broad range of plant species. Why does respiration rate change with temperature? One important factor is the effect of temperature on enzyme activity (see chapter 3). This effect is particularly important at very low temperatures and also very high temperatures that lead to protein denaturation.

Many responses of respiration rate to temperature change may be short-term, rather than long-term. Growing evidence indicates that respiration rate acclimates to a temperature increase over time, especially in leaves and roots that develop after the temperature shift. Over a long period at an elevated temperature, a plant could end up respiring at the same rate at which it had previously respired at a lower temperature.

Learning Outcomes Review 38.4

As atmospheric carbon dioxide increases, the rate of photosynthesis and the carbon:nitrogen ratio in plants are expected to increase as long as available soil nutrients do not change. A higher than normal carbon:nitrogen ratio would require herbivores to eat more biomass to meet their protein needs, which could, in turn, affect human health. As temperature increases, enzyme activity increases, accelerating the rate of respiration and breakdown of carbohydrates.

■ **How does the carbon:nitrogen ratio affect plants and herbivores (including humans)?**

■ **What strategies could help keep the carbon:nitrogen ratio in crop plants lower?**

38.5 Phytoremediation

Learning Outcomes

1. Define phytoremediation.
2. Explain how poplar trees have been used for phytoremediation.
3. Describe an advantage and a disadvantage of phytoremediation.

Some root cell membrane channels and transporters lack absolute specificity and can take up heavy metals like aluminum and other toxins. Although in most cases uptake of toxins is lethal or growth-limiting, some plants have evolved the ability to sequester or release these compounds into the atmosphere. These plants have potential for **phytoremediation,** the use of plants to concentrate or break down pollutants (figure 38.17).

Phytoremediation can work in a number of ways with both aquatic and soil pollutants. Plants may secrete a substance from their roots that breaks down the contaminant. More often, the harmful chemical enters the roots and is transported to the shoot system, making it easier to remove the chemical from the site. Some substances are simply stored by

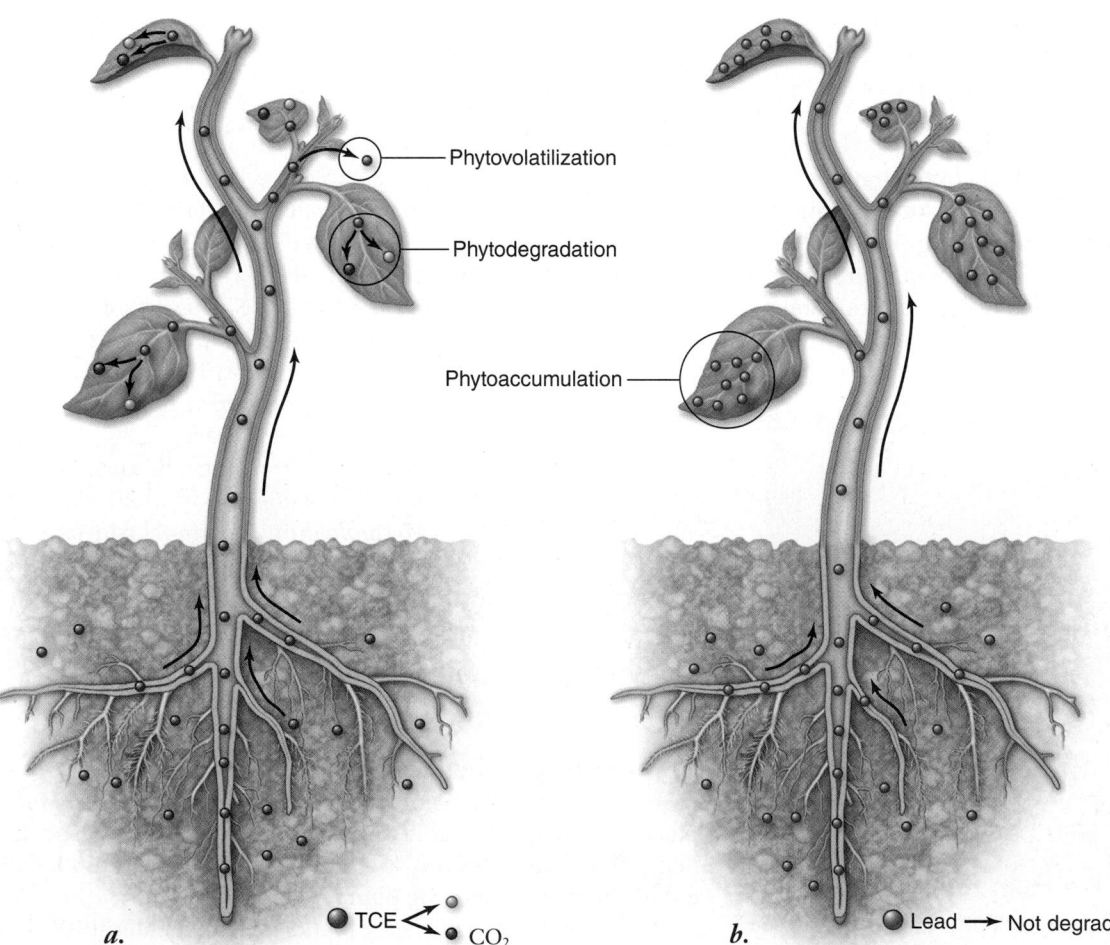

**Figure 38.17
Phytoremediation.**
Plants can use the same mechanisms to remove both nutrients and toxins from the soil. *a.* TCE (trichloroethylene) can be taken up by plants and degraded into CO_2 and chlorine before being released into the atmosphere. This process is called *phytodegradation.* Some of the TCE moves so rapidly through the xylem that it is not degraded before it is released through the stomata as a gas in a process called *phytovolatilization.* *b.* Other toxins, including heavy metals such as lead, can be taken up by plants, but not degraded. Such *phytoaccumulation* is particularly effective in removing toxins if they are stored in the shoot, where they can more easily be harvested.

Phytovolatilization

Phytodegradation

Phytoaccumulation

TCE < CO_2

a.

Lead → Not degraded

b.

the plant; later, the plant material is harvested, dried, and removed to a storage site.

For example, after the nuclear reactor disaster at Chernobyl in northern Ukraine, sunflowers effectively removed radioactive cesium from nearby lakes. The plants were floated in foam supports on the surface of the lakes and later collected. Because up to 85% of the weight of herbaceous plants can be water, drying down phytoremediators can restrict toxins like radioactive cesium to a small area.

In this section, we will explore several examples of soil phytoremediation.

Trichloroethylene may be removed by poplar trees

Trichloroethylene (TCE) is a volatile solvent that has been widely used as spot remover in the dry-cleaning industry, for degreasing engine turbines, as an ingredient in paints and cosmetics, and even as an anesthetic in human and veterinary medicine. Unfortunately, TCE is also a confirmed carcinogen, and chronic exposure can damage the liver.

In 1980, the Environmental Protection Agency (EPA) established a Superfund to clean up contamination in the United States. Forty percent of all sites funded by the Superfund include TCE contamination. How can we clean up 1900 hectares of soil in a Marine Corps Air Station in Orange County, California, that contain TCE once used to clean fighter jets? Landfills can isolate, but not eliminate, this volatile substance. Burning eliminates it from the site, but may release harmful substances into the atmosphere. A promising approach is to use plants to remove TCE from the soil.

As mentioned earlier, plants may take up a toxin from soil, allowing the toxin to be removed and concentrated elsewhere; but an even more successful strategy is for the plant to break down the contaminant into nontoxic by-products. Poplar trees (genus *Populus*) may provide just such a solution for TCE-contaminated sites (figure 38.18). Poplars naturally take up TCE from the soil and metabolize it into CO_2 and chlorine.

Figure 38.18 Phytoremediation for TCE. The U.S. Air Force is testing phytoremediation technology to clean up TCE at a former Air Force base in Fort Worth, Texas.

Other plant species can break down TCE as well, but poplars have the advantage of size and rapid transpiration. A five-year-old poplar can move between 100 and 200 L of water from its roots out through its leaves in a day. A plant that transpires less would not be able to remove as much TCE in a day.

Although removing TCE with poplar trees sounds like the perfect solution, this method has some limitations. Not all the TCE is metabolized, and given the rapid rate of transpiration in the poplar, some of the TCE enters the atmosphere via the leaves (phytovolatization). Once in the air, TCE has a half-life of 9 hours (half of it will break down into smaller molecules every 9 hours). Clearly, more risk assessment is needed before poplars are planted on every TCE-containing Superfund site.

The TCE that remains in the plant is metabolized quickly, and it is possible that the wood could be used after remediation is complete. It has been suggested that any remaining TCE would be eliminated if the wood were processed to make paper. Genetically modified poplars have been shown to metabolize about four times as much TCE as nonmodified poplars, so perhaps greater metabolic rates can be obtained.

As with any phytoremediation plan, it is critically necessary to estimate how much of a contaminant can be removed from a site by plants, and arriving at this estimate can be difficult. Possible risks, particularly when genetic modification is involved, must be weighed against the dangers posed by the contaminant.

Trinitrotoluene can be removed in limited amounts

In addition to volatile chemicals such as TCE, phytoremediation also holds promise for dealing with other environmental contaminants, including the explosive trinitrotoluene (TNT) and heavy metals. TNT is a solid, yellow material that was used widely in grenades and bombs until 1980. Contamination is found around factories that made TNT.

In some places, there is enough TNT in the soil to detonate, and thus incineration is not a viable option for removing TNT from most sites. Another issue is that TNT can seep into the groundwater; this is a matter of concern because TNT is carcinogenic and associated with liver disease.

TNT tends to stay near the top of the soil and to wash away quickly. Bean *(Phaseolus vulgaris)*, poplar, and the aquatic parrot feather *(Myriophyllum spicatum)* can take up and degrade low levels of TNT, but at higher concentrations, TNT is toxic to these plants.

Heavy metals can be successfully removed at lower cost

Heavy metals, including arsenic, cadmium, and lead, persist in soils and are toxic to animals in even small quantities. Many plants are also susceptible to heavy-metal toxicity, but species near mines have evolved strategies to partition certain heavy metals from the rest of the plant.

Four hundred species of plants have been identified that have the ability to hyperaccumulate toxic metals from the soil. For example, *Brassica juncea* (a relative of broccoli and

a.

b.

c.

Figure 38.19 Aznalcóllar mine spill. *a.* When the dike of a holding lagoon for mine waste broke, 5 million cubic meters of black sludge containing heavy metals were released into a national park and the Guadiamar River. *b.* Large amounts of sludge were removed mechanically. *c.* Phytoremediation appears to be a promising solution for treating the remaining heavy metals.

mustard plants) is especially effective at hyperaccumulating lead in the shoots of the plant. Unfortunately, *B. juncea* is a small, slow-growing plant, and eventually it becomes saturated with lead.

How would lead or cadmium travel from the soil into the leaves of a plant? There are some hints that root cell membranes may contain metal transporters that load the metal in the soil into the xylem. Citrate, mentioned earlier, can increase the rate of metal transport in xylem. The metals are sequestered inside vacuoles in the leaves. Trichomes, which are modified leaf epidermal cells, can sequester both lead and cadmium.

These hyperaccumulating plants are not a panacea for metal-contaminated soil because of concern that animals might move into the site and graze on lead- or cadmium-enriched plants. Harvesting and consolidating dried plant material is not a simple matter, but still phytoremediation is a promising technique. Estimated costs for phytoremediation are 50 to 80% lower than cleanup strategies that involve digging and dumping the contaminated soil elsewhere.

Phytoremediation may be a solution to the contamination resulting from a 1998 accident at the Aznalcóllar mine in Spain. A dam that contained the sludge from the mining operation broke, releasing 5 million cubic meters of sludge, composed of arsenic, cadmium, lead, and zinc, over 4300 hectares of land (figure 38.19). Much of the sludge was physically removed and dumped into an open mine pit. Phytoremediation solutions are being sought for the remaining contaminated soil.

Since the original spill, three plant species with the potential to hyperaccumulate some of the metals have begun growing in the area. These plants are fairly large and can accumulate a substantial amount of metal. They offer the advantage of being native species, thus reducing the dangers associated with introducing a nonnative, potentially invasive species to clean up the spill.

Learning Outcomes Review 38.5

Phytoremediation is the use of plants to concentrate or break down pollutants. Poplar trees can take up the soil contaminant TCE and break it down into nontoxic by-products. Compared with the alternative of removing contaminated soil, phytoremediation is less costly. A disadvantage is that animals could be harmed if they graze in an area where plants have taken up high levels of toxic compounds.

■ *Young poplar seedlings have been planted in an area contaminated with TCE. Identify potential safety concerns and strategies for minimizing any harmful effects.*

chapter **38** *Plant Nutrition and Soils*

38.1 Soils: The Substrates on Which Plants Depend

Soil is composed of minerals, organic matter, water, air, and organisms (figure 38.1).

Topsoil is a mixture of mineral particles, living organisms, and humus, which is partially decayed organic material. Microorganisms in the soil are important for nutrient recycling.

Water and mineral availability are determined by soil characteristics.

Minerals and organic soil particles are typically negatively charged so they draw positively charged ions away from the roots. Therefore, active transport of positively charged ions into the roots is required. Proton pumps in roots pump out H^+, creating an electrochemical gradient that draws mineral ions into the roots.

Approximately one-half the soil volume is made up of pores filled with air or water. Water added to the soil may drain through or be held in the pores, where it is available for root uptake.

Cultivation can result in soil loss and nutrient depletion.

Loss of topsoil through soil erosion results in reduced water-holding capacity and nutrient availability. Cultivation practices, including intercropping, conservation tillage, and no-till, have been developed to reduce soil erosion.

Overuse of fertilizers, pesticides, and herbicides causes water pollution.

pH and salinity affect water and mineral availability.

Acidic soils release minerals, such as aluminum, at levels that are toxic to plants.

Saline soils alter water potential, leading to a loss of water and turgor in plants. Saline soils are common where irrigation is practiced.

38.2 Plant Nutrients

Plants require nine macronutrients and seven micronutrients.

The nine macronutrients required by plants are carbon, oxygen, hydrogen, nitrogen, potassium, calcium, magnesium, phosphorus, and sulfur. The seven micronutrients are chlorine, iron, manganese, zinc, boron, copper, and molybdenum.

Food security is related to crop productivity and nutrient levels.

Plant breeding efforts to increase nutrient levels in food crops aim to provide health benefits and improve food security.

38.3 Special Nutritional Strategies

Bacteria living in close association with roots can provide nitrogen (figure 38.10).

Some plants, such as legumes, have a symbiotic relationship with nitrogen-fixing bacteria to obtain the nitrogen needed for protein synthesis. In exchange, the plants provide carbohydrates to the bacteria.

Mycorrhizae aid a large portion of terrestrial plants.

More than 90% of plants live in symbiotic association with mycorrhizal fungi. By extending the surface area of the root system, these fungi facilitate the uptake of phosphorus and micronutrients.

Carnivorous plants trap and digest animals to extract additional nutrients (figure 38.11).

Some plants that live in acidic, nitrogen-poor environments obtain mineral nutrients by capturing and digesting small animals such as insects.

Parasitic plants exploit resources of other plants.

Some parasitic plants produce chlorophyll, but others do not. They tap into host plants to obtain nutrients, including carbohydrates.

38.4 Carbon–Nitrogen Balance and Global Change

Elevated CO_2 levels can alter photosynthesis and carbon levels in plants (figure 38.14).

As CO_2 concentrations increase, the rate of photosynthesis increases and consequently biomass increases; however, the plant tissue that is produced is high in carbon relative to nitrogen, with a shift toward more carbohydrate and less protein.

As nutritional value decreases, more plant matter must be consumed to obtain the same amount of nutrients; the result is greater plant loss by herbivory.

Elevated temperature can affect respiration and carbon levels in plants.

The rate of enzyme reactions increases with ambient temperatures, increasing respiration. Because respiration breaks down carbohydrates, higher temperatures could cause additional changes in plant nutrient balance.

38.5 Phytoremediation

Phytoremediation utilizes plants to remove toxic contaminants from soil or water (figure 38.17).

Trichloroethylene may be removed by poplar trees.

Poplar trees have been used to remove TCE from the soil and convert it to nontoxic carbon dioxide and chlorine compounds.

Trinitrotoluene can be removed in limited amounts.

Some plants can take up low levels of TNT in the soil and degrade it. However, high levels are toxic to the plants.

Heavy metals can be successfully removed at lower cost.

Plants may accumulate high levels of contaminants in their shoots and store them there. The plants can be harvested and removed. If animals feed on these plants, however, they may be exposed to high concentrations of toxic compounds.

Phytoremediation is less expensive than removing contaminated soils.

UNDERSTAND

1. Which of the following is NOT found in topsoil?

   a. Humus
   b. Bedrock
   c. Bacteria
   d. Air

2. Mineral soil particles are typically

   a. negatively charged.
   b. positively charged.
   c. neutral.

3. What proportion of the soil volume is occupied by air and water?

   a. 10%
   b. 25%
   c. 50%
   d. 75%
   e. 90%

4. Which of the following is a micronutrient?

   a. Nitrogen
   b. Calcium
   c. Phosphorus
   d. Iron

5. The nodules of legume roots contain nitrogen-fixing

   a. bacteria.
   b. fungi.
   c. algae.
   d. plant cells.

6. Photorespiration occurs when

   a. glucose interacts with carbon dioxide.
   b. rubisco binds with oxygen.
   c. RuBP is converted to a sugar.
   d. the Sun provides energy for the breakdown of sugar.

7. In a C_4 plant, the Calvin cycle occurs in

   a. the epidermis.
   b. vascular tissue.
   c. bundle sheath cells.
   d. mesophyll cells.

8. One potential problem with using poplars to remove TCE from soils is that

   a. some TCE enters the atmosphere via the transpiration stream.
   b. poplar trees grow slowly.
   c. most TCE will wash away from the soil before it is removed.
   d. TCE interferes with chlorophyll production.

APPLY

1. You are performing an experiment to determine the nutrient requirements for a newly discovered plant and find that for some reason your plants die if you leave boron out of the growth medium but do fine with as low as 5 ppm in the solution. This suggests that boron is

   a. an essential macronutrient.
   b. a nonessential micronutrient.
   c. an essential micronutrient.
   d. a nonessential macronutrient.

2. If you wanted to conduct an experiment to determine the effects of varying levels of macronutrients on plant growth in your small greenhouse at home, which of the following macronutrients would be the most difficult to regulate?

   a. Carbon
   b. Nitrogen
   c. Potassium
   d. Phosphorus

3. Which of the following would decrease nitrogen availability for a pea plant?

   a. Inability of the plant to produce flavonoids
   b. Formation of Nod factors
   c. Presence of oxygen in the soil
   d. Production of leghemoglobin

4. Which of the following might you do to increase nutrient uptake by crop plants?

   a. Decrease the solubility of nutrients
   b. Create nutrients as positive ions
   c. Frequently plow the soil
   d. Genetically modify plants to increase the density of plasma membrane transporters in root cells

5. If you were to eat one ton (1000 kg) of potatoes, calculate approximately how much of the following minerals would you eat?

   a. Copper between 4 and 30 ppm
   b. Zinc between 15 and 100 ppm
   c. Potassium between 0.5 and 6%
   d. Iron between 25 and 300 ppm

SYNTHESIZE

1. A common farming practice involves fumigating the soil to kill harmful fungi. Fumigants may not be selective, though, so they may kill most microorganisms in the soil. What short-term and long-term effects might fumigation have on the soil?

2. Describe an experiment to determine the amount of boron needed for the normal growth of tomato seedlings.

3. You are setting up a business in your greenhouse to produce lettuce hydroponically in a system in which water and nutrients continually flows over the plant roots. In designing the instrumentation for your greenhouse, which factors in the air and water do you need to monitor and regulate and why?

ONLINE RESOURCE

www.ravenbiology.com

Understand, Apply, and Synthesize—enhance your study with animations that bring concepts to life and practice tests to assess your understanding. Your instructor may also recommend the interactive eBook, individualized learning tools, and more.

Chapter **39**

Plant Defense Responses

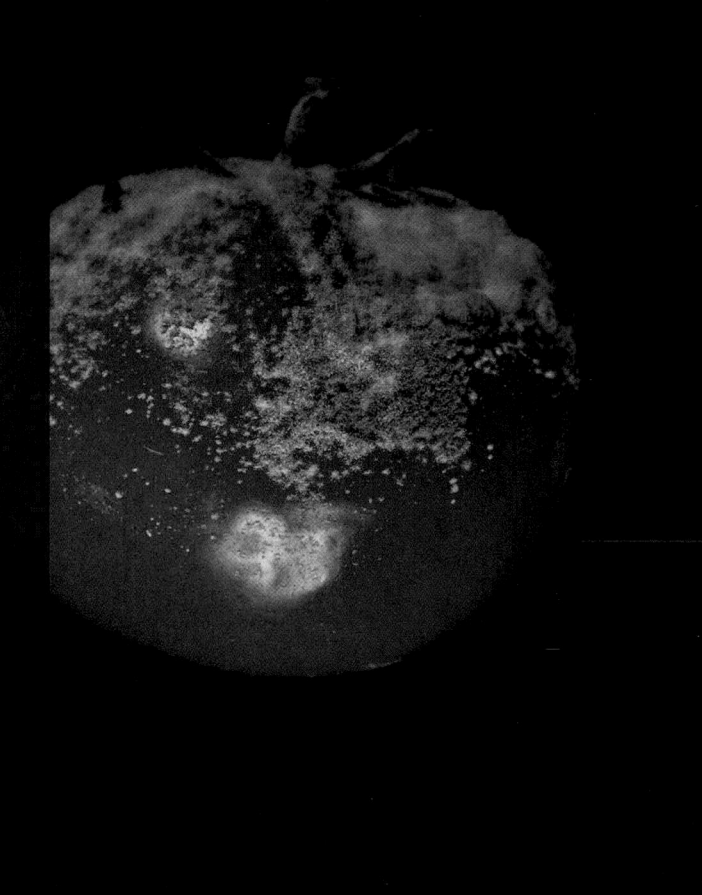

Chapter Contents

Introduction

Plants are constantly under attack by viruses, bacteria, fungi, animals, and even other plants. An amazing array of defense mechanisms has evolved to block or temper an invasion. Many plant–pest relationships undergo coevolution, with the plant winning sometimes and the pest winning with a new offensive adaptation at other times. The first line of plant defense is thick cell walls covered with a strong cuticle. Bark, thorns, and even trichomes can deter a hungry insect. When that first line of defense fails, a chemical arsenal of toxins is waiting. Many of these molecules have no effect on the plant. Some are modified by microbes in the intestine of an herbivore into a poisonous compound. Maintaining a toxin arsenal is energy-intensive; so, an alternative means of defense uses induced responses to protect and prevent future attacks.

Physical Defenses

Learning Outcomes

1. *Identify the compounds produced by the epidermis to protect against invasion.*
2. *Outline the steps taken by a fungus to invade a plant leaf.*
3. *Describe two beneficial associations between plants and microorganisms.*

Although abiotic factors such as weather and fire constitute genuine threats to an immobile plant, even greater daily threats exist in the form of viruses, bacteria, fungi, animals, and other plants. These enemies can tap into the nutrient resources of plants or use their DNA-replicating mechanisms to self-replicate. Some invaders kill the plant cells immediately, leading to necrosis (brown, dead tissue). Certain insects may tap into the phloem of a plant seeking carbohydrates, but leave behind a hitchhiking virus or bacterium.

The threat of these attackers is reduced when they have natural predators themselves. One of the greatest problems with nonnative invasive species, such as the emerald ash borer (figure 39.1), is the lack of natural predators in the new environment.

Dermal tissue provides first-line defense

The first defense all plants have is the dermal tissue system (see chapter 36). Epidermal cells throughout the plant secrete wax, which is a mixture of hydrophobic lipids. Layers of lipid material protect exposed plant surfaces from water loss and attack. Above-ground plant parts are also covered with cutin, a macromolecule consisting of long-chain fatty acids linked together. **Suberin,** another version of linked fatty acid chains, is found in cell walls of subterranean plant organs; suberin

Invaders can penetrate dermal defenses

Unfortunately, these exterior defenses can be penetrated in many ways. Mechanical wounds leave an open passageway through which microbial organisms can enter. Parasitic nematodes use their sharp mouthparts to get through the plant cell walls. Their actions either trigger the plant cells to divide, forming a tumorous growth, or, in species that attach to a single plant cell, cause the cell to enlarge and transfer carbohydrates from the plant to the hungry nematode (figure 39.2). In some cases, the wounding makes it easier for other pathogens, including fungi, to infect the plant (figure 39.3). In some cases simply having bacteria on the leaf surface can increase the risk of frost damage. The bacteria function as sites for ice nucleation; the resulting ice crystals severely damage the leaves.

Fungi strategically seek out the weak spot in the dermal system—the stomatal openings—to enter the plant. Some fungi have coevolved with a monocot that has evenly spaced stomata. These fungi appear to be able to measure distance to locate these evenly spaced stomatal openings and invade the plant. Figure 39.4 shows the phases of fungal invasion, which can include the following:

1. Windblown spores land on leaves. A germ tube emerges from the spore. Host recognition is necessary for spore germination.
2. The spore germinates and forms an adhesion pad, allowing it to stick to the leaf.
3. Hyphae grow through cell walls and press against the cell membrane.
4. Hyphae differentiate into specialized structures called haustoria (singular, *haustorium*). They expand, surrounded by cell membrane, and nutrient transfer begins.

Figure 39.1 Emerald ash borer (*Agrilus planipennus*). This invasive species arrived in Detroit in 2002, probably on wood packing materials, and has spread rapidly through parts of the United States. Emerald ash borers chew through the bark and into the phloem, damaging and killing tens of millions of ash trees.

a.   500×   *b.*

Figure 39.2 Nematodes attack the roots of crop plants. *a.* A nematode breaks through the epidermis of the root. *b.* Root-knot nematodes form tumors on roots.

Hypothesis: *Nematodes increase the severity of a potato wilt fungal disease by wounding roots and allowing the fungus to penetrate root tissue.*

Prediction: *Leaf wilt will be more severe when the root system is exposed to both the nematode and the fungus than when the root system is exposed to neither or either separately.*

Test: *Establish four treatments with four plants in each treatment group. Add the nematodes and the fungal pathogen to the soil of plants in group 1. Group 2 will only have fungus. Group 3 will have only nematodes, and group 4 will be untreated. Allow plants to grow for 42 days and record the extent of leaf wilt on each plant.*

Nematodes and fungus
(severe effect)

Fungus
(moderate effect)

Nematodes
(moderate effect)

Untreated (control)
(no effect)

Result: *Plants that are coinfected with the nematodes and fungus have more severe wilting than plants treated with nematodes or fungus alone. Control plants do not wilt.*

Conclusion: *Nematodes increase the severity of the wilting infection.*

Further Experiments: *Design an experiment to test the hypothesis that increased fungal wilt symptoms are the result of damage to the roots by the nematode, making it easier for the fungus to enter the plant. Could you mechanically wound the roots instead of exposing them to nematodes?*

Figure 39.3 Nematodes increase susceptibility of plant to fungal infection.

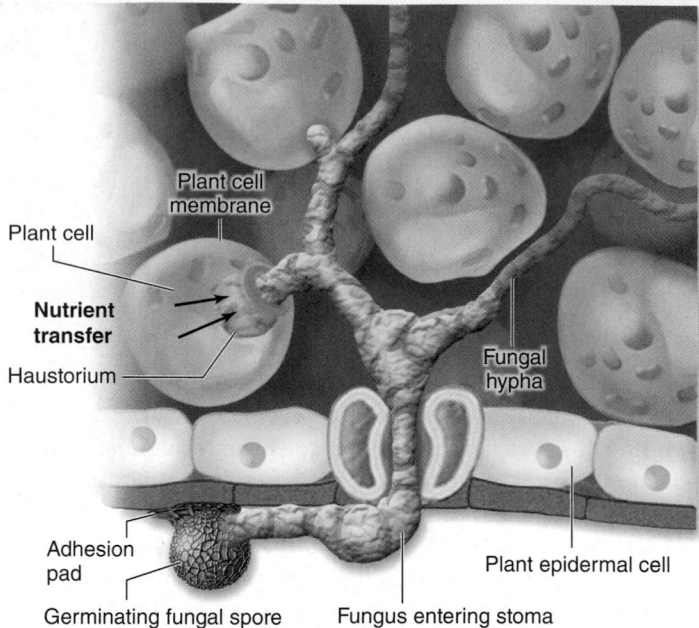

Figure 39.4 **Fungi sneak in through stomata.** Fungal hyphae penetrate cell walls, but not plasma membranes. The close contact between the fungal hyphae and the plant plasma membrane allows for ready transfer of plant nutrients to the fungus.

Bacteria and fungi can also be beneficial to plants

Mutualistic and parasitic relationships are often just opposite sides of the evolutionary coin. A parasitic relationship can evolve to become mutualistic, and a mutualistic relationship can transform into a parasitic one. In chapters 32 and 38, you saw how mycorrhizal fungi use a mechanism similar to the one just described to the mutual benefit of both the plant and the fungus. In the case of the relationship between legumes and nitrogen-fixing bacteria, the *Rhizobium* bacteria seeks out a root hair, infects it along with other tissues, and forms a root nodule. Other soil bacteria can also enhance plant growth, and are called plant growth-promoting rhizobacteria (PGPR). The term *rhizobacteria* refers to bacteria that live around the root system and often benefit from root exudates. In return they provide substances that support plant growth. *Azospirillum* spp., for example, provide gibberellins, which are growth hormones, for rice plants when the bacteria are living in close proximity to the root system. PGPR can also limit the growth of pathogenic soil bacteria.

Learning Outcomes Review 39.1

Epidermal cells secrete protective compounds, including wax, cutin, and suberin. Fungal spores may germinate and stick to plant leaves; hyphae enter the leaf through stomata, and produce haustoria to take up plant nutrients. Mutualistic partners with plants include mycorrhizal fungi that assist with nutrient absorption and nitrogen-fixing bacteria that provide nutrients.

■ *Why would protective substances on leaves include lipid-based compounds?*

39.2 Chemical Defenses

Many plants are filled with toxins that kill herbivores or, at the very least, make them quite ill. One example is the production of cyanide (HCN). Over 3000 species of plants produce cyanide-containing compounds called *cyanogenic glycosides* that break down into cyanide when cells are damaged. Cyanide stops electron transport, blocking cellular respiration.

Cassava (genus *Manihot*), a major food staple for many Africans, is filled with cyanogenic glycosides (specifically, manihotoxins) in the outer layers of the edible root. Unless these outer layers are scrubbed off, the cumulative effect of eating primarily cassava can be deadly.

Some toxins are unique to plants, but others are found in plants, vertebrates, and invertebrates and are called **defensins.** Defensins are small, cysteine-rich peptides with antimicrobial activity. The conservation of defensins in animals and plants reveals the ancient origins of innate immunity. Between 15 and 50 defensin genes have been identified in plant genomes, and over 317 defensin-like genes exist in the *Arabidopsis* genome. The exact mechanisms are being worked out, but in some cases plant defensins inhibit protein synthesis. When expression of defensin genes is suppressed in plants, they are more susceptible to bacterial and fungal infections. In addition to toxins that kill, plants can produce chemical compounds that make potential herbivores ill or that repel them with strong flavors or odors.

Plants maintain chemical arsenals

How did the biosynthetic pathways that produce these toxins evolve? Growing evidence indicates that the metabolic pathways needed to sustain life in plants have taken some evolutionary side trips, leading to the production of a stockpile of chemicals known as **secondary metabolites.** Many of these secondary metabolites affect herbivores, including humans (table 39.1).

Alkaloids, including caffeine, nicotine, cocaine, and morphine, can affect multiple cellular processes; if a plant cannot kill its attackers, it can overstimulate them with caffeine or sedate them with morphine. For example, the tobacco hornworm *(Manduca sexta)* can level a field of tobacco (figure 39.5); however, wild species of tobacco appear to have elevated levels of nicotine that are lethal to these worms.

Tannins bind to proteins and inactivate them. For example, some act by blocking enzymes that digest proteins, which reduces the nutritional value of the plant tissue. An insect that gets sick from a strong dose of tannins is likely to associate the

Figure 39.5 Herbivores can kill plants. Tobacco hornworms, *Manduca sexta,* consume huge amounts of tobacco leaf tissue, as well as tomato leaves.

flavor with illness and to avoid having that type of plant for lunch another time. Small doses of tannins and most other secondary metabolites are unlikely to cause any major digestive difficulties in larger animals, including humans. Animals can avoid many of the cumulative toxic effects of secondary metabolites by eating a varied diet.

Plant oils, particularly those found in plants of the mint family, which includes peppermint, sage, pennyroyal, and many others, repel insects with their strong odors. At high concentrations, some of these oils can also be toxic if ingested.

Why don't the toxins kill the plant? One strategy is for a plant to sequester a toxin in a membrane-bound structure, so that it does not come into contact with the cell's metabolic processes. The second solution is for the plant to produce a compound that is not toxic unless it is metabolized, often by microorganisms, in the intestine of an animal. Cyanogenic glycosides are a good example of the latter solution. The plant produces a sugar-bound cyanide compound that does not affect electron transport chains. Once an animal ingests cyanogenic glycoside, the compound is enzymatically broken down, releasing the toxic hydrogen cyanide.

Coevolution has led to defenses against some plant toxins. A tropical butterfly, *Helioconius sara,* can sequester the cyanogenic glycosides it ingests from its sole food source, the passion vine. Even more intriguing is a biochemical pathway that allows the butterfly to safely break down cyanogenic glycosides and use the released nitrogen in its own protein metabolism.

Plants can poison other plants

Some chemical toxins protect plants from other plants. **Allelopathy** occurs when a chemical compound secreted by the roots of one plant blocks the germination of nearby seeds or inhibits the growth of a neighboring plant. This strategy minimizes shading and competition for nutrients, while it maximizes the ability of a plant to use radiant sunlight for photosynthesis. Allelopathy works with both a plant's own species and different species. Black walnut trees *(Juglans nigra)* are a

TABLE 39.1 Secondary Metabolites

Compound	Source	Structure	Effect on Humans
Manihotoxin (cyanogenic glycoside)	Cassava, *Manihot esculenta*		Metabolized to release lethal cyanide
Genistein (phytoestrogen)	Soybean, *Glycine max*		Estrogen mimic
Taxol (terpenoid)	Pacific yew, *Taxus brevifolia*		Anticancer drug
Quinine (alkaloid)	Quinine bark, *Cinchona officinalis*		Antimalarial drug
Morphine (alkaloid)	Opium poppy, *Papaver somniferum*		Narcotic pain killer

good example. Very little vegetation will grow under a black walnut tree because of allelopathy (figure 39.6).

Humans are susceptible to plant toxins

Not only have humans been inadvertently poisoned by plants, but throughout much of human history, they have also been intentionally poisoned by other humans using plant products. Socrates, a Greek philosopher who lived 2390 years ago, was sentenced to death in Athens. He died after drinking a hemlock extract containing an alkaloid called coniine, which is similar in structure to nicotine and affects nicotine and nicotinic receptors. Coniine blocks the nicotinic receptor at neuromuscular junctions, thereby causing muscular paralysis beginning in the lower limbs and eventually leading to respiratory paralysis and death.

Ricin, an alkaloid found in castor beans (*Ricinus communis*), is six times more lethal than cyanide and twice as lethal as cobra venom. A single seed from the plant, which is still grown in flower gardens, can kill a young child if ingested. Ricin is found in the endosperm of the seed as a heterodimer composed of ricin A and ricin B, joined by a single disulfide bond. This heterodimer (proricin) is nontoxic, but when the disulfide bond is broken in humans or other animals, ricin A targets the 28S rRNA of the ribosome. Death occurs because ricin A cleaves the adenine from nucleotide 4324 in the 28S ribosome, resulting in a neutral ribose that changes the structure of rRNA and thus inhibits translation (figure 39.7). A single ricin A molecule can inactivate 1500 ribosomes per minute, blocking translation of proteins.

In 1978, Bulgarian expatriate and dissident Georgi Markov was about to board a bus in London on his way to work at the BBC when he felt a sharp stabbing pain in his thigh. A man near him picked up an umbrella from the ground and hurriedly left. Markov had been injected via a mechanism in the umbrella tip with a pinhead-sized metal sphere containing

**Figure 39.6
Black walnut trees are allelopaths**
Seedlings die when their roots come in contact with the root secretions of a black walnut.

0.2 mg of ricin. Markov died four days later. After the collapse of the Soviet Union, former KGB officers revealed that the assassination had been set up at the behest of the Bulgarian Communist Party leadership.

? Inquiry question Explain how ricin led to Markov's death.

Secondary metabolites may have medicinal value

Major research efforts on plant secondary metabolites are in progress because of their potential benefits, as well as dangers, to human health (see table 39.1).

Figure 39.7 Ricin from castor beans blocks translation. When the ricin A subunit is released from proricin, it binds to enzymatically inactivate rRNA in ribosomes and prevents mRNA from being translated into protein.

Soy and phytoestrogens

One example of the benefits and dangers is the presence of **phytoestrogens,** compounds very similar to the human hormone estrogen, in soybean products. In soybean plants, genistein is one of the major phytoestrogens.

Comparative studies between Asian populations that consume large amounts of soy foods and populations with lower dietary intake of soy products are raising intriguing questions and some conflicting results. For example, the lower rate of prostate cancer in Asian males might be accounted for by the down-regulation of androgen and estrogen receptors by a phytoestrogen. Soy is being marketed as a means for minimizing menopausal symptoms caused by declining estrogen levels in older women.

In humans, dietary phytoestrogens cross the placenta and can be found in the amniotic fluid during the second trimester of pregnancy. Questions have been raised about the effect of phytoestrogens on developing fetuses and even on babies who consume soy-based formula because of allergies to cow's milk formula. Because hormonal signaling is so complex, much more research is needed to fully understand how or even if phytoestrogens affect human physiology and development.

Taxol and breast cancer

Taxol, a secondary metabolite found in the Pacific yew *(Taxus brevifolia),* is effective in fighting cancer, especially breast cancer. The discovery of taxol's pharmaceutical value raised an environmental challenge. The very existence of the Pacific yew was being threatened as the shrubs were destroyed so that taxol could be extracted. Fortunately, organic chemists developed an effective and affordable strategy for synthesizing taxol in the laboratory.

Taxol is not an isolated case of drug discovery in plants. The hidden pharmaceutical value of many plants may lead to increased conservation efforts to protect plants that have the potential to make contributions toward human health. Although the plant pharmaceutical industry is growing, it is certainly not a new field. Until recent times, almost all medicines used by humans came from plants.

Quinine and malaria

In the 1600s, the Incas of Peru were treating malaria with a drink made from the bark of *Cinchona* trees. Malaria is caused by four types of human malaria parasites in the genus *Plasmodium,* which are carried by female *Anopheles* mosquitoes. *Plasmodium falciparum* is the most lethal of the four types. Symptoms include severe fevers and vomiting. The parasite feeds on red blood cells, and death can result from anemia or blocking of blood flow to the brain.

By 1820, the active ingredient in the bark of *Cinchona* trees, **quinine,** had been identified (see table 39.1). In the 19th century, British soldiers in India used quinine-containing "tonic water" to fight malaria. They masked the bitter taste of quinine with gin, creating the first gin and tonic drinks. In 1944, American chemists Robert Woodward and William Doering synthesized quinine. Now several other synthetic drugs are available to treat malaria.

Exactly how quinine and synthetic versions of this drug family work has puzzled researchers for a long time. Quinine can affect DNA replication, and also, when *P. falciparum* breaks down hemoglobin from red blood cells in its digestive vacuole, an intermediary toxic form of heme is released. Quinine may interfere with the subsequent polymerization of these hemes, leading to a buildup of toxic hemes that poison the parasite.

Unfortunately, even today malaria is a major threat to human health, causing over a million deaths per year. Ninety percent of these deaths occur in sub-Saharan Africa. An estimated 300,000,000 individuals are infected. *P. falciparum* strains have acquired resistance to synthetic drugs, and quinine is once again the drug of choice in some cases.

Herbal remedies have been used for centuries in most cultures. A resurgence of interest in plant-based remedies is resulting in a growing and unregulated industry. Although herbal remedies have great promise, we need to be aware that each plant contains many secondary metabolites, many of which have evolved to cause harm to herbivores, including humans.

Learning Outcomes Review 39.2

Plants accumulate secondary metabolites that can poison or otherwise harm herbivores. Plants also secrete chemicals that inhibit the growth of neighboring plants, a process termed allelopathy. Secondary metabolites may also have beneficial uses, such as phytoestrogens from soy, which may reduce menopausal symptoms in women; taxol from the Pacific yew, which acts as an anticancer agent; and quinine from *Cinchona* trees, which helps treat malaria.

■ *In what ways would a drug prepared from a whole plant differ from a drug prepared from an isolated chemical compound?*

39.3 Animals That Protect Plants

Learning Outcomes

1. *Describe the benefit Acacia trees receive from ants that live in them.*
2. *Explain how some plants use parasitoid wasps to destroy caterpillars.*

Not only do individual species and their traits evolve over time, but so do relationships between species. For example, evolution of chemicals to deter herbivores may often be accompanied over time by adaptation on the part of herbivores to withstand these chemicals. This evolutionary pattern is called coevolution. Here we consider two cases of mutualism that coevolved between animal and plant species.

***Acacia* trees and ants.** Several species of ants provide small armies to protect some species of *Acacia* trees from other herbivores. These stinging ants may inhabit an enlarged thorn of the tree; they attack other insects (figure 39.8) and sometimes small mammals and epiphytic plants. Some of the *Acacia* species provide their ants with sugar in nectaries located away from the flowers, and even with lipid food bodies at the tips of leaves.

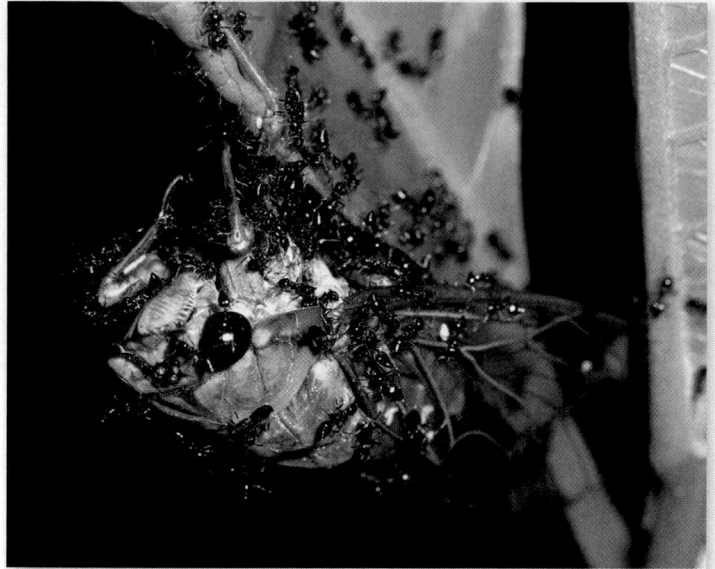

Figure 39.8 Ants attacking a katydid to protect "their" *Acacia* **tree.** Through coevolution, ants are sheltered by *Acacia* trees and attack otherwise harmful herbivores.

The only problem with ants chasing away other insects is that *Acacia* trees depend on bees to pollinate their flowers. What keeps the ants from swarming and stinging a bee that stops by to pollinate? Evidence indicates that when a flower opens on an *Acacia* tree, it produces some type of chemical ant deterrent that does not deter the bees. This chemical has not yet been identified.

Parasitoid wasps, caterpillars, and leaves. Caterpillars fill up on leaf tissue before they metamorphose into a moth or a butterfly. In some cases, proteinase inhibitors in leaves are sufficient to deter very hungry caterpillars. But some plants have developed another strategy. As the caterpillar chews away, a wound response in the plant leads to the release of a volatile compound. This compound wafts through the air, and if a female parasitoid wasp happens to be in the neighborhood, it is immediately attracted to the source. Parasitoid wasps are so named because they are parasitic on caterpillars. The wasp lays her fertilized eggs in the body of the caterpillar that is feeding on the leaf of the plant. These eggs hatch, and the emerging larvae kill and eat the caterpillar (figure 39.9).

Learning Outcomes Review 39.3

Mutualism is an interaction between species that is beneficial to both. Ants protect *Acacia* trees by attacking feeding herbivores. Parasitoid wasps are attracted by compounds released from plant tissues damaged by feeding caterpillars; they lay eggs in the caterpillars, which later are killed by the emerging larvae.

■ **Would you expect that wasps kill all the caterpillars? Explain.**

39.4 Systemic Responses to Invaders

Learning Outcomes

1. **Outline the sequence of events that leads to the production of a wound response.**
2. **Describe the gene-for-gene hypothesis.**
3. **Define systemic acquired resistance.**

So far, we have focused mainly on static plant responses to threats. Most of the deterrent chemicals such as toxins are maintained at steady-state levels. In addition, the morphological structures such as thorns or trichomes that help defend plants are part of the normal developmental program. Because these defenses are maintained whether an herbivore or other invader is present or not, they have an energetic downside. By

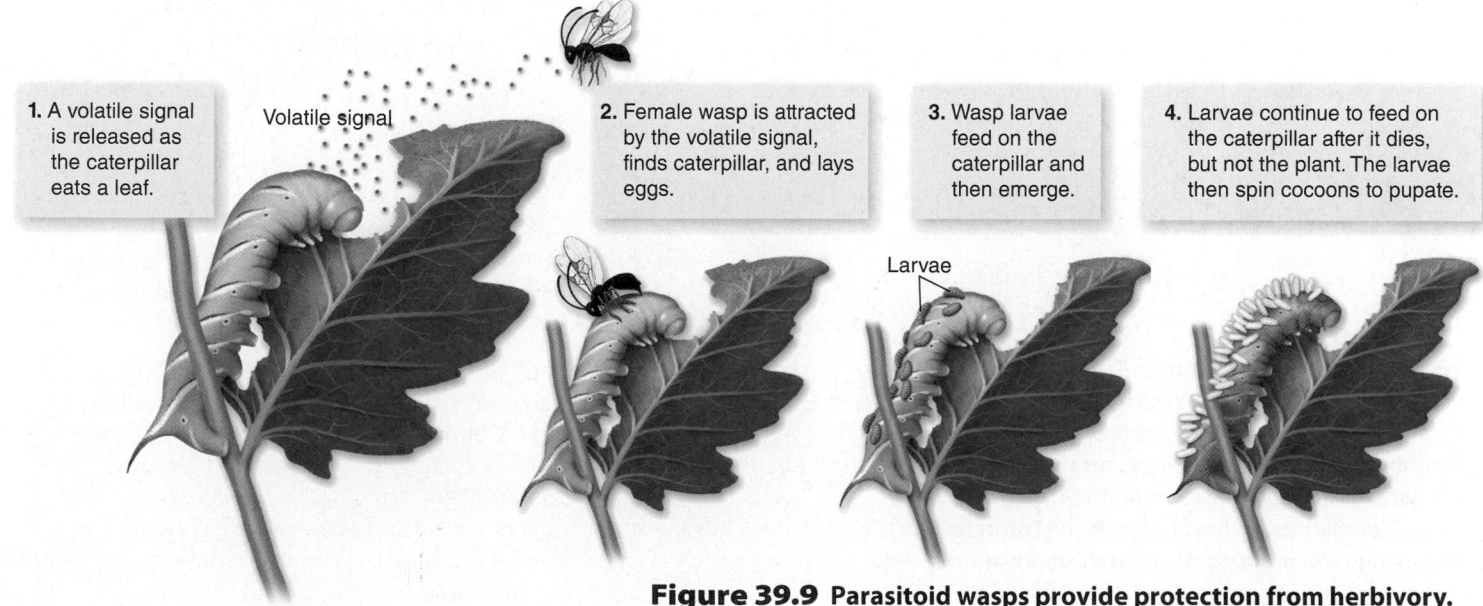

1. A volatile signal is released as the caterpillar eats a leaf.

Volatile signal

2. Female wasp is attracted by the volatile signal, finds caterpillar, and lays eggs.

3. Wasp larvae feed on the caterpillar and then emerge.

Larvae

4. Larvae continue to feed on the caterpillar after it dies, but not the plant. The larvae then spin cocoons to pupate.

Figure 39.9 Parasitoid wasps provide protection from herbivory.

contrast, resources could be conserved if the response to being under siege was inducible—that is, if the defense response could be launched only when a threat had been recognized. In this section, we explore these inducible defense mechanisms.

Wound responses protect plants from herbivores

As you just learned from the example of the parasitoid wasp, a **wound response** may occur when a leaf is chewed or injured. One induced outcome is the rapid production of proteinase inhibitors. These chemical toxins do not exist in the stockpile of defenses, but instead are produced in response to wounding.

Proteinase inhibitors bind to digestive enzymes in the gut of the herbivore. The proteinase inhibitors are produced throughout the plant, and not just at the wound site. How are cells in distant parts of the plant signaled to produce proteinase inhibitors? In tomato plants, the following sequence of events is responsible for this systemic response (figure 39.10):

1. Wounded leaves produce an 18-amino-acid peptide called **systemin** from a larger precursor protein.
2. Systemin moves through the apoplast (the space between cell walls) of the wounded tissue and into the nearby phloem. This small peptide-signaling molecule then moves throughout the plant in the phloem.
3. Cells with a systemin receptor bind the systemin, which leads to the production of **jasmonic acid.**
4. Jasmonic acid activates the transcription of defense genes, including the production of a proteinase inhibitor.

Although we know the most about the signaling pathway involving jasmonic acid, other molecules are involved in wound response as well. **Salicylic acid,** which is found in the bark of plants such as the white willow *(Salix alba),* is one example. Cell fragments also appear to be important signals for triggering an induced response, as is discussed shortly.

Mechanical damage separate from herbivore attack also elicits wound responses, which presents a challenge in designing plant experiments that involve cutting or otherwise mechanically damaging the tissue. Experimental controls, which should be cut or manipulated in the same way but without the test treatment, are especially important to ensuring that any changes observed are not due only to the wound response.

Defense responses can be pathogen-specific

Wound responses are independent of the type of herbivore or other agent causing the damage, but other responses are triggered by a specific pathogen.

Pathogen recognition

Both plants and animals are able to recognize common motifs on the surface of pathogens using pattern recognition receptors (PRRs). Recognition leads to a generalized defense mechanism, a type of innate immunity that is also discussed in chapter 51.

Another element of plant innate immunity is a more specific recognition of a pathogen but also followed by a generalized defense. More than half a century ago, the American geneticist H. H. Flor proposed the existence of a plant

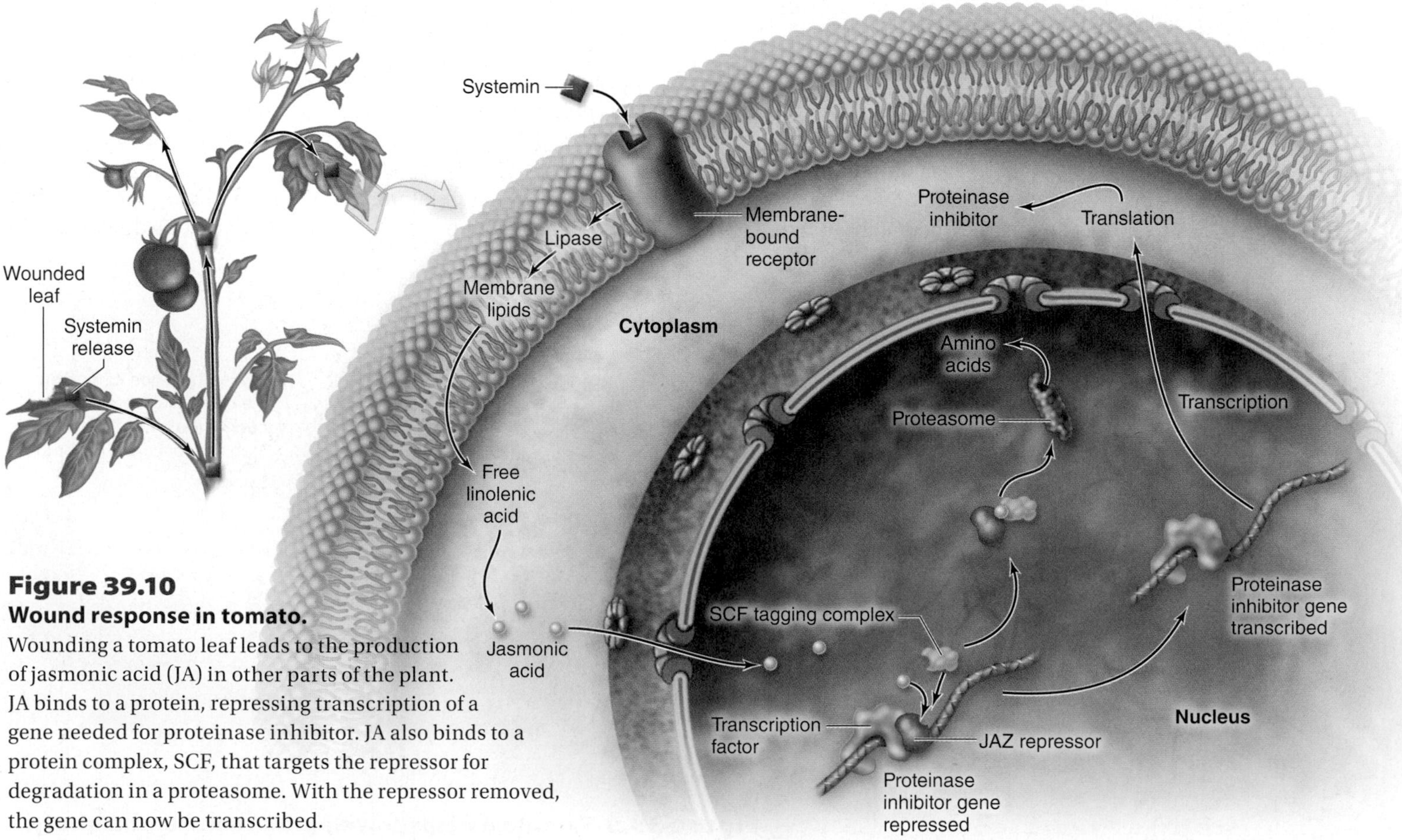

Figure 39.10
Wound response in tomato.
Wounding a tomato leaf leads to the production of jasmonic acid (JA) in other parts of the plant. JA binds to a protein, repressing transcription of a gene needed for proteinase inhibitor. JA also binds to a protein complex, SCF, that targets the repressor for degradation in a proteasome. With the repressor removed, the gene can now be transcribed.

Labels in figure: Systemin, Lipase, Membrane lipids, Membrane-bound receptor, Proteinase inhibitor, Translation, Cytoplasm, Amino acids, Proteasome, Transcription, Free linolenic acid, Wounded leaf, Systemin release, Jasmonic acid, SCF tagging complex, Transcription factor, JAZ repressor, Proteinase inhibitor gene repressed, Proteinase inhibitor gene transcribed, Nucleus resistance gene *(R)*, the product of which interacts with the product of an avirulence gene *(avr)* carried by a pathogen. *Avirulent* means not virulent (disease-causing). The pathogen's avr protein interacts with the plant's R protein to signal the pathogen's presence. In this way, the plant under attack can mount defenses, thus ensuring that the pathogen remains avirulent. If the pathogen's avr protein is not recognized by the plant, disease symptoms appear (figure 39.11). Pairs of *avr* and *R* genes have been cloned in different species pathogenized by microbes, fungi, and even insects in one case. Over 149 genes with putative R gene characteristics that might interact with avr proteins have been identified in *Arabidopsis,* but only a small number have been studied for actual function. This research has been motivated partially by the agronomic benefit of identifying genes that can be added via gene technology to crop plants to protect them from invaders.

The *avr/R* gene interaction is an example of ongoing coevolution. An avirulent invader can be detected and recognized. Mutations arising in the avirulent pathogen can result in a virulent pathogen that overcomes a plant's defenses and kills it—often leading to the pathogen's demise as well.

Specific defenses and the hypersensitive response

Much is now known about the signal transduction pathways that follow the recognition of the pathogen by *R* proteins. These pathways lead to the triggering of the **hypersensitive response (HR),** which leads to rapid cell death around the source of the invasion and also to a longer-term, whole-plant resistance (figures 39.11 and 39.12). A gene-for-gene response does not always occur, but plants still have defense responses to pathogens in general as well as to mechanical wounding. Some of the response pathways may be similar. Also, fragments of cell-wall carbohydrates may serve as recognition and signaling molecules.

When a plant is attacked and a gene-for-gene recognition occurs, the HR leads to very rapid cell death around the site of attack. This seals off the wounded tissue to prevent the pathogen or pest from moving into the rest of the plant. Hydrogen peroxide and nitric oxide are produced and may signal a cascade of biochemical events resulting in the localized death of host cells. These chemicals may also have negative effects on the pathogen, although protective mechanisms have coevolved in some pathogens.

Other antimicrobial agents produced include the **phytoalexins,** which are antimicrobial chemical defense agents. A variety of pathogenesis-related genes (*PR* genes) are also expressed, and their proteins can function as either antimicrobial agents or signals for other events that protect the plant.

In the case of virulent invaders for which there is no *R* recognition, changes in local cell walls at least partially block the pathogen or pest from moving further into the plant. In this case, an HR does not occur, and the local plant cells do not die.

Long-term protection

In addition to the HR or other local responses, plants are capable of a systemic response to a pathogen or pest attack, called a **systemic acquired resistance (SAR)** (figure 39.12). Several pathways lead to broad-ranging resistance that lasts for a period of days.

The long-distance signal that induces SAR is likely salicylic acid, rather than systemin, which is the long-distance signal in wound responses. At the cellular level, jasmonic acid (which was mentioned earlier in the context of the wound response pathways) is involved in SAR signaling. SAR allows the plant to respond more quickly if it is attacked again. This response, however, is not the same as the human or mammalian immune response, in which antibodies (proteins) that recognize specific antigens (foreign proteins) persist in the body. SAR is neither as specific nor as long-lasting.

Learning Outcomes Review 39.4

A wounded leaf initiates a signaling chain that stimulates production of proteinase inhibitors. When a plant has a resistance gene with a product that recognizes the product of an avirulence gene in the pathogen, the plant carries out a defense response; this recognition is called the gene-for-gene hypothesis. Systemic acquired resistance is a temporary broad form of resistance that may be induced by exposure to a pathogen.

■ **How does local cell death help preserve a plant under attack by a pathogen?**

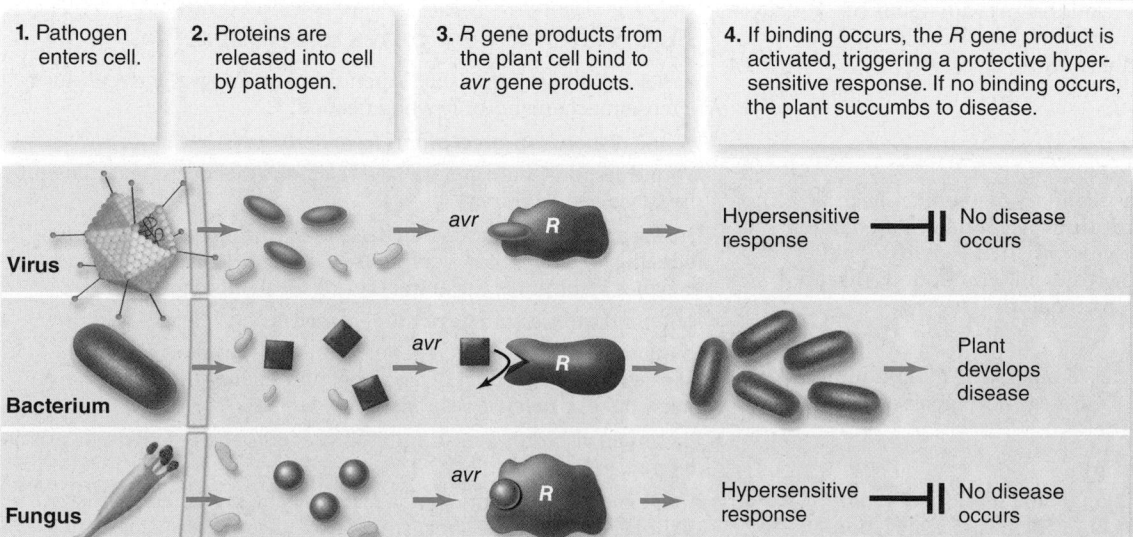

1. Pathogen enters cell.

2. Proteins are released into cell by pathogen.

3. *R* gene products from the plant cell bind to *avr* gene products.

4. If binding occurs, the *R* gene product is activated, triggering a protective hypersensitive response. If no binding occurs, the plant succumbs to disease.

Virus — avr — R — Hypersensitive response ⊣ No disease occurs

Bacterium — avr — R — Plant develops disease

Fungus — avr — R — Hypersensitive response ⊣ No disease occurs

Figure 39.11 Gene-for-gene hypothesis. H. H. Flor proposed that pathogens have an avirulence *(avr)* gene encoding a protein that is recognized by a plant resistance *(R) protein.* If the virus, bacterium, fungus, or insect has an *avr* protein that matches the *R* protein, a defense response will occur.

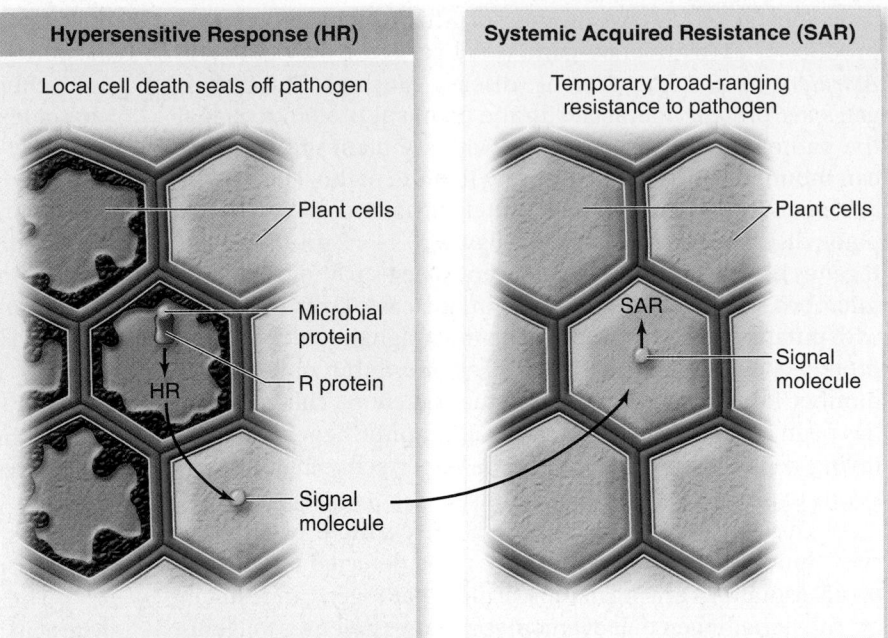

Figure 39.12 Plant defense responses.
The gene-for-gene response leads to local cell death (HR) and to the production of a mobile signal that provides longer-term resistance in the rest of the plant (SAR).

Chapter Review

39.1 Physical Defenses

Pathogens can harm plants in many ways, including exploiting nutrient resources and taking over DNA replication machinery.

Dermal tissue provides first-line defense.

Dermal tissues are covered with lipids such as cutin and suberin, which reduce water loss and prevent attack. Morphological features such as trichomes, bark, and thorns protect some plants.

Invaders can penetrate dermal defenses.

In spite of defense mechanisms, invaders can cause damage by piercing plants, eating plant parts, or entering the plant through the stomata.

Bacteria and fungi can also be beneficial to plants.

Mycorrhizal fungi form beneficial relationships with plants by enhancing uptake of water and minerals. Nitrogen-fixing bacteria provide nitrogen to plants in a usable form.

39.2 Chemical Defenses

Plants maintain chemical arsenals.

Plants may produce and stockpile secondary metabolites such as alkaloids, tannins, and oils that provide protection from predators (table 39.1). Plants protect themselves from their own toxins either by sequestering them in vesicles or producing compounds that are not toxic until they are ingested by a predator.

Plants can poison other plants.

Allelopathic plants secrete chemicals to block seed germination or inhibit growth of nearby plants. This strategy minimizes competition for resources such as light and nutrients.

Humans are susceptible to plant toxins.

Ricin is an example of a powerful plant toxin. It is found in the endosperm of castor beans (figure 39.7). Ricin is six times more lethal than cyanide.

Secondary metabolites may have medicinal value.

Plant secondary metabolites such as phytoestrogens, taxol, and quinine have pharmaceutical value for humans. Many other plant-based remedies have been used for centuries in human cultures.

39.3 Animals That Protect Plants

Mutualistic associations are beneficial to both the plant and animal partners. One example is the relationship between *Acacia* trees and ants, in which ants protect the trees from herbivores.

Another example is the association between certain plants, caterpillars, and parasitoid wasps. When chewed or damaged, the leaves release compounds that attract the wasps, which lay their eggs in the caterpillars. The wasps' larvae feed on the caterpillar, killing it (figure 39.9).

39.4 Systemic Responses to Invaders

Plants avoid an unnecessary expenditure of energy if they produce defense mechanisms only when needed.

Wound responses protect plants from herbivores.

Wound responses are generalized reactions that occur regardless of the cause of the injury.

During a wound response, a signal spreads throughout the phloem, inducing the production of proteinase inhibitors that bind to digestive enzymes in the gut of the animal eating the plant (figure 39.10).

Defense responses can be pathogen-specific.

In many plants, the plant *R* gene product may interact with an avirulence gene product of a pathogen in a gene-for-gene reaction that induces a defense response.

Plants can also produce antimicrobial agents such as phytoalexins as defense compounds.

After exposure to a pathogen, a plant may be protected against pathogen attack in the short-term future through a mechanism called systemic acquired resistance.

UNDERSTAND

1. Nonnative invasive species are often a threat to native species because they
   a. typically grow larger than other plants.
   b. are not susceptible to any diseases.
   c. are parasitic.
   d. do not have natural enemies in their new location.

2. Fungal pathogens transfer nutrients across a plant cell membrane using
   a. an adhesion pad.   c. guard cells.
   b. a haustorium.   d. tumors.

3. Casparian strips in roots contain _____, which helps to defend against invaders.
   a. wax   c. cutin
   b. suberin   d. cuticle

4. Which of the following is NOT a secondary metabolite?
   a. Caffeine   c. Taxol
   b. Morphine   d. Glucose

5. Parasitoid wasps protect plants from caterpillars by
   a. stinging them.   c. eating them.
   b. repelling them.   d. enclosing them in a capsule.

6. In response to wounding, a tomato plant first produces a peptide called
   a. systemin.   c. ricin.
   b. jasmonic acid.   d. salicylic acid.

7. When a cell undergoes a hypersensitive response, it
   a. builds cell walls quickly.
   b. releases defense response molecules from its vacuole.
   c. dies rapidly.
   d. destroys avirulence gene products.

8. The wound response products that bind to digestive enzymes in herbivores are
   a. proteinase inhibitors.   c. lipase inhibitors.
   b. proteinase promoters.   d. lipase promoters.

9. If a plant has been attacked by a pathogen, then it is likely to be able to respond more quickly to a subsequent attack due to a mechanism called
   a. basal defense.
   b. induced hypersensitive response.
   c. antimicrobial pathogen resistance.
   d. systemic acquired resistance.

APPLY

1. Some plants have developed a mutualistic relationship with parasitoid wasps. This mutualistic relationship would not occur if
   a. the plant quit producing nectar for the wasp.
   b. the wasp ceased to live on the plant.
   c. the plant quit producing volatile compounds that attract the wasp.
   d. the plant attracted too many caterpillars.

2. Both plant and animal immune systems can
   a. develop memory of past pathogens to more effectively deal with subsequent infections.
   b. initiate expression of proteins to help fight the infection.

   c. kill their own cells to prevent spread of the infection.
   d. All of the choices are correct.

3. Your friend informs you that it is highly likely all of the plants in your yard are "infected" with some kind of fungi or bacteria. The plants look perfectly healthy to you at this time. The most prudent thing for you to do would be:
   a. Remove all your plants because they are likely to die.
   b. Spray your plants with chemicals to remove all bacteria and fungi.
   c. Remove all your plants and replace the soil.
   d. Do nothing because many of these bacteria and fungi may be beneficial.

4. Some plants are recognized by fungal pathogens on the basis of their stomatal pores. Which of the following would provide these plants immunity from fungal infection?
   a. Removing all of the stomata from the plant
   b. Changing the spacing of stomatal pores in these plants
   c. Reinforcing the cell wall in the guard cells of stomatal pores
   d. Increasing the number of trichomes on the surfaces

5. You decide to plant a garden with a beautiful black walnut tree at one end and a majestic white oak at the other end. You are quite disappointed, however, when none of the seeds you plant around the walnut tree grow. What might explain this observation?
   a. The walnut tree filters out too much light, so the seeds fail to germinate.
   b. The roots of the walnut tree deplete all of the nutrients from the soil, so the new seedlings starve.
   c. The walnut tree produces chemical toxins that prevent seed germination.

6. If a pathogen contains an *avr* gene not recognized by a plant, the plant will most likely
   a. develop a disease.
   b. eliminate the pathogen because it is unrecognized.
   c. produce proteinase inhibitors.
   d. evolve a different *R* gene.

SYNTHESIZE

1. During the domestication of crops, humans have intentionally or inadvertently selected for lower levels of toxic compounds. Explain why each of these two types of selection would have occurred.

2. Parasitoid wasps seem like an effective method to control caterpillars. Discuss some limitations of this strategy. That is, outline some scenarios in which a plant might not be effectively protected by the wasps.

3. Systemin is transported through the phloem of a tomato plant to induce a wound response to herbivores. However, the direction of phloem movement is always from source to sink. Explain how the sites that receive the wound signal may vary with stages of the plant's life cycle.

ONLINE RESOURCE

www.ravenbiology.com

Understand, Apply, and Synthesize—enhance your study with animations that bring concepts to life and practice tests to assess your understanding. Your instructor may also recommend the interactive eBook, individualized learning tools, and more.

Chapter **40**

Sensory Systems in Plants

Chapter Contents

Introduction

All organisms sense and interact with their environments. This is particularly true of plants. Plant survival and growth are critically influenced by abiotic factors, including water, wind, and light. The effect of the local environment on plant growth also accounts for much of the variation in adult form within a species. In this chapter, we explore how a plant senses such factors and transduces these signals to elicit an optimal physiological, growth, or developmental response. Although responses can be observed on a macroscopic scale, the mechanism of response occurs at the level of the cell. Signals are perceived when they interact with a receptor molecule, causing a shape change and altering the receptor's ability to interact with signaling molecules. Hormones play an important role in the internal signaling that brings about environmental responses and are keyed in many ways to the environment.

40.1 *Responses to Light*

Learning Outcomes

1. *Compare the pigments phytochrome and chlorophyll.*
2. *List growth responses influenced by phytochrome.*
3. *Define phototropism.*

In chapter 8 we covered the details of photosynthesis, the process by which plants convert light energy into chemical bond energy. We described pigments, molecules that are capable of absorbing light energy; you learned that chlorophylls are the primary pigment molecules of photosynthesis. Plants contain other pigments as well, and one of the functions of these other pigments is to detect light and to mediate plants' response to light by passing on information.

Several environmental factors, including light, can initiate seed germination, flowering, and other critical developmental events in the life of a plant. **Photomorphogenesis** is the term used for nondirectional, light-triggered development. It can result in complex changes in form, including flowering.

Unlike photomorphogenesis, phototropisms are directional growth responses to light. Both photomorphogenesis and phototropisms compensate for the plant's inability to walk away from unfavorable environmental conditions and toward favorable ones.

P_{fr} facilitates expression of light-response genes

The pigment-containing protein **phytochrome (P)** is present in all groups of plants and in a few genera of green algae, but not in other protists, bacteria, or fungi. Phytochrome systems probably evolved among the green algae and were present in the common ancestor of the plants.

The phytochrome molecule exists in two interconvertible forms: The first form, **P_r**, absorbs red light at 660 nm wavelength; the second, **P_{fr}**, absorbs far-red light at 730 nm. Sunlight has more red than far-red light. P_r is biologically inactive; it is converted into P_{fr}, the active form, when red photons are present. P_{fr} is converted back into P_r when far-red photons are available. In other words, biological reactions that are affected by phytochrome occur when P_{fr} is present. When most of the P_{fr} has been replaced by P_r, the reaction cannot occur (figure 40.1).

The amount of P_{fr} is also regulated by degradation. Ubiquitin is a protein that tags P_{fr} for transport to the **proteasome**, a protein shredder composed of 28 proteins. The proteasome has a channel in the center, and as proteins pass through, they are clipped into amino acids that can be used to build other proteins as described in chapter 16. The process of tagging and recycling P_{fr} is precisely regulated to maintain needed amounts of phytochrome in the cell.

Phytochrome consists of two parts: a smaller part that is sensitive to light, called the *chromophore,* and a larger portion called the *apoprotein.* The apoprotein facilitates expression of light-response genes (figure 40.2). Over 2500 genes, 10% of the *Arabidopsis* genome, are involved in biological responses that begin with a conformational change in one of the phytochromes in response to red light. Phytochromes are involved in numerous signaling pathways that lead to gene expression. Some pathways also involve protein kinases or G proteins (described in chapter 9).

Phytochrome is found in the cytoplasm, but enters the nucleus to facilitate transcription of light-response genes. When P_r is converted to P_{fr}, it can move into the nucleus. Once in the nucleus, P_{fr} binds with other proteins that form a transcription complex, leading to the expression of light-regulated genes (figure 40.3). Phytochrome's protein-binding site (see figure 40.2) is essential for interactions with transcription factors.

Phytochrome also works through protein kinase-signaling pathways. When phytochrome converts to the P_{fr} form, the protein kinase domain of the apoprotein may phosphorylate a serine in the amino (N) terminus of the phytochrome itself (autophosphorylation), or it may phosphorylate

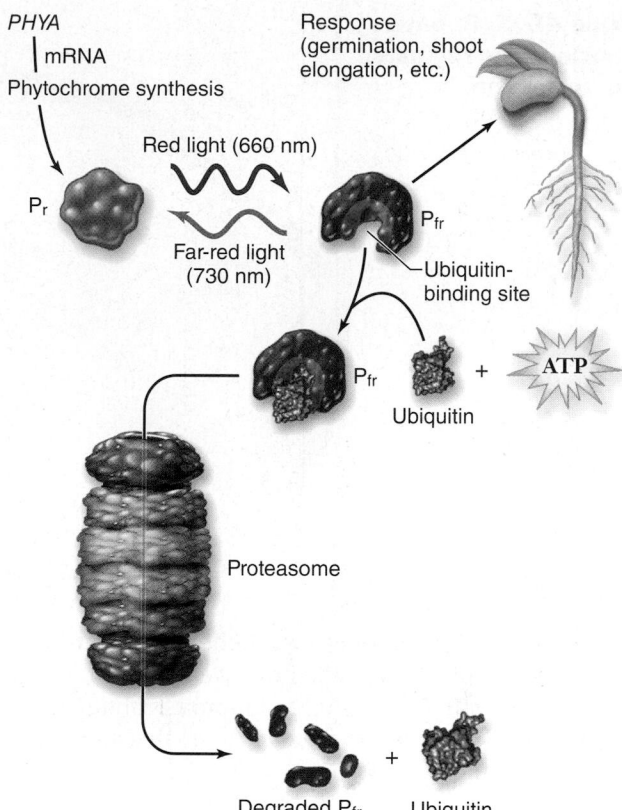

Figure 40.1 How phytochrome works. *PHYA* is one of the five *Arabidopsis* phytochrome genes. When exposed to red light, P_r changes to P_{fr}, the active form that elicits a response in plants. P_{fr} is converted to P_r when exposed to far-red light. The amount of P_{fr} is regulated by protein degradation. The protein ubiquitin tags P_{fr} for degradation in the proteasome.

a serine of another protein involved in light signaling (figure 40.4). Phosphorylation initiates a signaling cascade that can activate transcription factors and lead to the transcription of light-regulated genes.

Although phytochrome is involved in multiple signaling pathways, it does not directly initiate the expression of that 10%

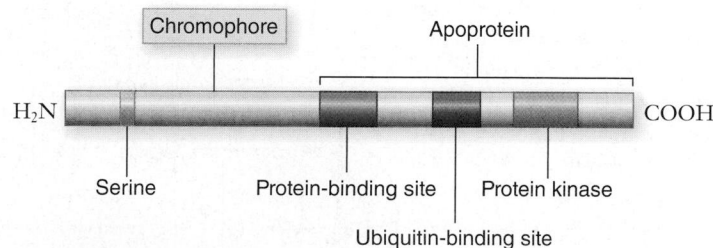

Figure 40.2 Phytochrome. Different parts of the phytochrome molecule have distinct roles in light regulation of growth and development. Phytochrome changes conformation when the chromophore responds to relative amounts of red and far-red light. The shape change affects the ability of phytochrome to bind to other proteins that participate in the signaling process. The ubiquitin-binding sites allow for degradation, and the protein kinase domain allows for further signaling via phosphorylation.

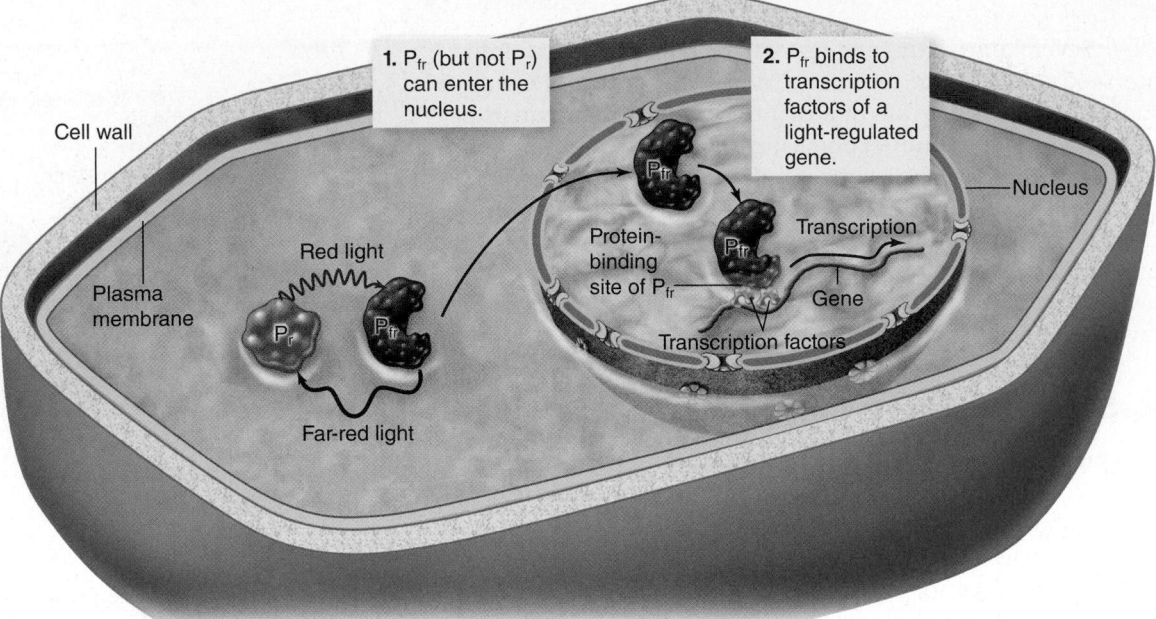

Figure 40.3 P_{fr} enters the nucleus and regulates gene expression.

1. P_{fr} (but not P_r) can enter the nucleus.

2. P_{fr} binds to transcription factors of a light-regulated gene.

Cell wall

Plasma membrane

Red light

Far-red light

P_r

P_{fr}

P_{fr}

P_{fr}

Protein-binding site of P_{fr}

Transcription factors

Transcription

Gene

Nucleus

of the *Arabidopsis* plant genome. Rather, phytochrome initiates expression of master regulatory genes that manage the complex interactions leading to photomorphogenesis and phototropisms. Gene expression is just the first step, with hormones playing important roles as well.

> **?** **Inquiry question** You are given the seed of a plant with a mutation in the protein kinase domain of phytochrome. Would you expect to see any red-light–mediated responses when you germinate the seed? Explain your answer.

Although we often refer to phytochrome as a single molecule here, several different phytochromes have been identified that appear to have specific functions. In *Arabidopsis* five forms of phytochrome, PHYA through PHYE, have been characterized, each playing overlapping but distinct roles in the light regulation of growth and development.

Both phytochrome and chlorophyll are pigments that have multiple forms and can absorb red light. Both sets of pigments absorb light in a very specific and narrow range of the spectrum. The key difference is whether the pigment uses the light as a source of information or a source of energy. In chapter 8, you learned that chlorophyll is the first step in an electron transport chain that is used to transduce energy. Phytochrome, however, is the first step in conveying information about the relative amounts of red and far red light in a signaling pathway that leads to gene expression (see figure 40.4). Phytochrome transduces information and chlorophyll transduces energy.

Many growth responses are linked to phytochrome action

Phytochrome is involved in a number of plant growth responses, including seed germination, shoot elongation, and detection of crowding by neighboring plants.

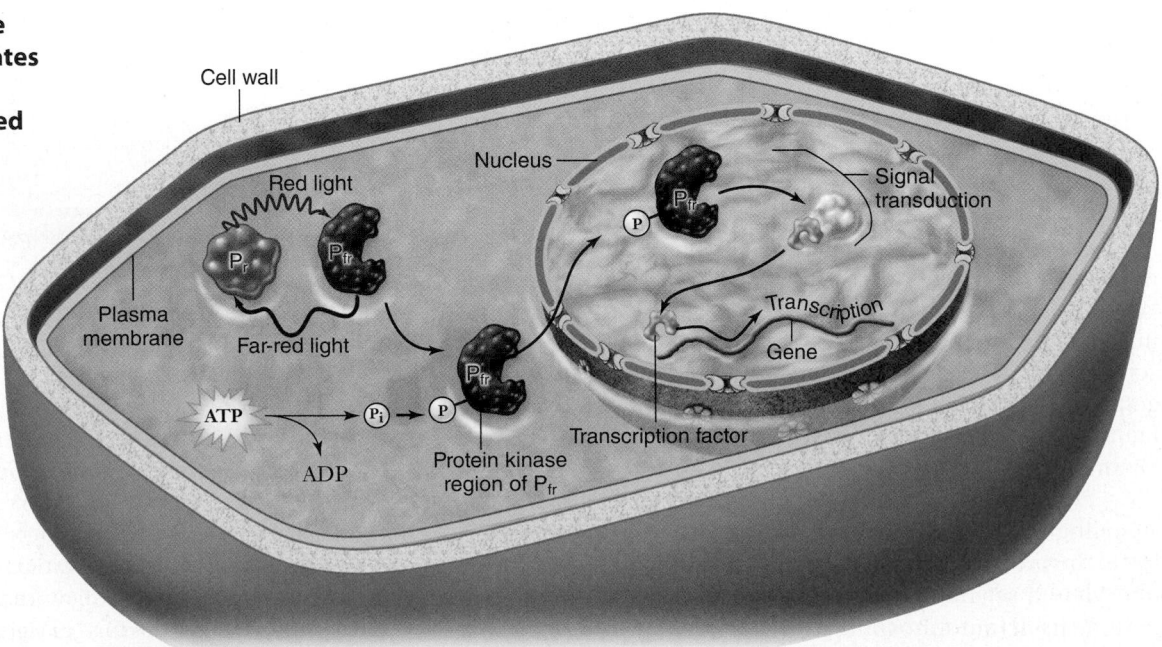

Figure 40.4 The kinase domain of P_{fr} phosphorylates P_{fr}, leading directly or indirectly to light-regulated gene expression. In this example, signaling leads to the release of a transcription factor from a protein complex.

Cell wall

Red light

Far-red light

Plasma membrane

P_r

P_{fr}

ATP

ADP

P_i

P

Protein kinase region of P_{fr}

P_{fr}

Nucleus

P

P_{fr}

Signal transduction

Transcription

Gene

Transcription factor

Seed germination

Seed germination is inhibited by far-red light and stimulated by red light in many plants. Because chlorophyll absorbs red light strongly but does not absorb far-red light, light filtered through the green leaves of canopy trees above a seed contains a reduced amount of red light. The far-red light inhibits seed germination by converting P_{fr} into the biologically inactive P_r form.

Consequently, seeds on the ground under deciduous plants, which lose their leaves in winter, are more apt to germinate in the spring after the leaves have decomposed and the seeds are exposed to direct sunlight and a greater amount of red light. This adaptation greatly improves the chances that seedlings will become established before leaves on taller plants shade the seedlings and reduce sunlight available for photosynthesis.

 Data analysis Sunlight has a red to far-red ratio of 1.2, whereas light under a canopy of leaves has a red to far-red ratio of 0.13. Explain both the reason for and the effects of the difference in relative amounts of red and far-red light.

Shoot elongation

Etiolation is a shoot elongation response in seedlings that begin growing in the dark. It is an energy conservation strategy to help plants reach light before they die. These seedlings are pale and slender because of a lack of red light. In the dark, they divert energy to internode elongation and green up when light becomes available. They assume a normal morphology when exposed to red light, increasing the amount of P_{fr}. This strategy is useful for seedlings when they have sprouted underground or under leaf cover.

The de-etiolated *(det2)* Arabidopsis mutant has a poor etiolation response, so its seedlings fail to elongate in the dark (figure 40.5). The *det2* mutants are defective in an enzyme necessary for biosynthesis of a brassinosteroid hormone, leading researchers to propose that brassinosteroids play a role in plant responses to light through phytochrome. (Brassinosteroids and other hormones are discussed later in this chapter.)

Detection of plant crowding

Red and far-red light also signal plant spacing. Leaf shading increases the amount of far-red light relative to red light. Plants somehow measure the amount of far-red light bounced back to them from neighboring plants. The closer together plants are, the more far-red relative to red light they perceive and the more likely they are to quickly grow tall, a strategy for outcompeting others for sunshine. If their perception is distorted by putting a light-blocking collar around the stem, the elongation response no longer occurs.

Light affects directional growth

Phototropisms, directional growth responses, contribute to the variety of overall plant shapes we see within a species as shoots grow toward light. Tropisms are particularly intriguing because they challenge us to connect environmental signals with cellular perception of the signal, transduction into biochemical pathways, and ultimately an altered growth response.

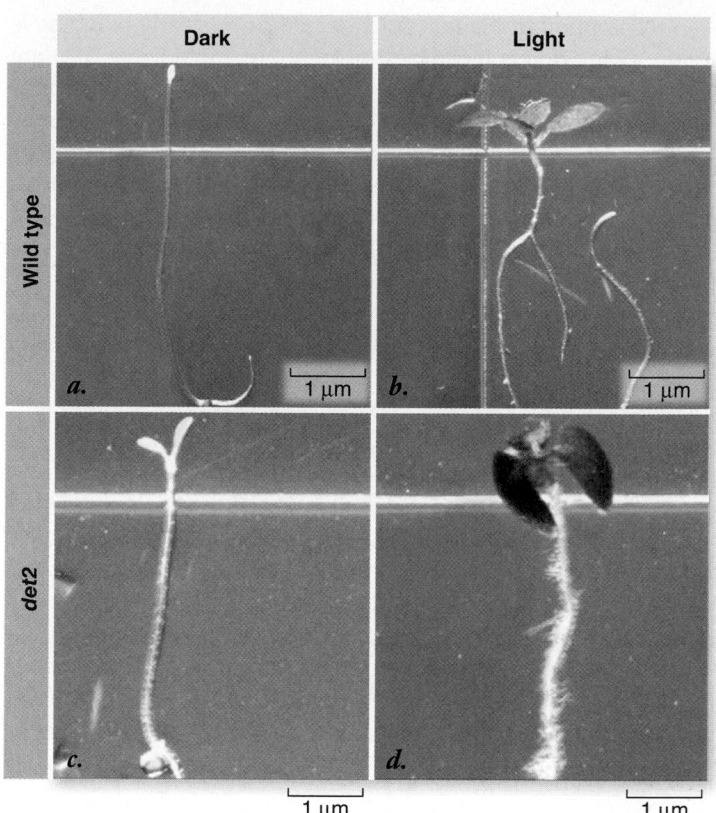

Figure 40.5 Etiolation is regulated by light and the *det2* gene in *Arabidopsis*. *det2* is needed for etiolation in dark-grown plants.

Positive phototropism in stems

Phototropic responses include the bending of growing stems and other plant parts toward sources of light with blue wavelengths (460-nm range) (figure 40.6). In general, stems are positively phototropic, growing toward a light source, but most roots do not respond to light, or in exceptional cases, exhibit only a weak negative phototropic response.

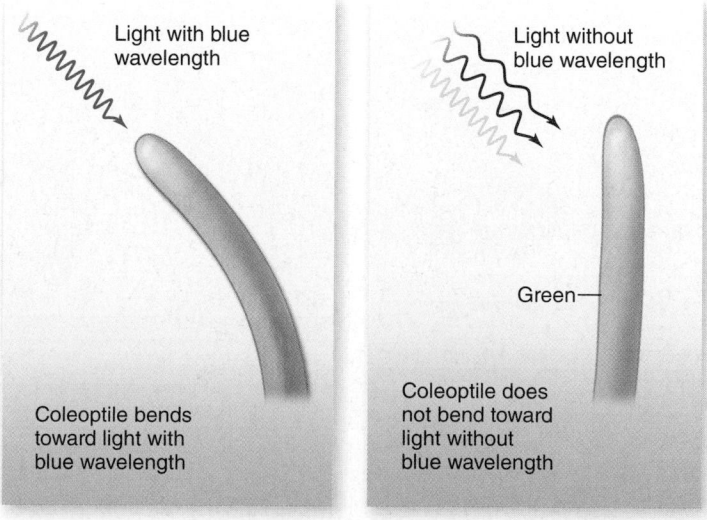

Figure 40.6 Phototropism. Oat coleoptiles growing toward light with blue wavelengths. Colors indicate the color of light shining on coleoptiles. Arrows indicate the direction of light.

The phototropic reactions of stems are clearly of adaptive value, giving plants greater exposure to available light. They are also important in determining the development of plant organs and, therefore, the appearance of the plant. Individual leaves may also display phototropic responses; the position of leaves is important to the photosynthetic efficiency of the plant. A plant hormone called *auxin,* discussed in a later section, is probably involved in most, if not all, of the phototropic growth responses of plants.

Blue-light receptors

The recent identification of blue-light receptors in plants is leading to exciting discoveries of how the light signal can ultimately be connected with a phototropic response. A blue-light receptor **phototropin 1 (PHOT1)** was identified through the characterization of a nonphototropic mutant.

The phot1 protein has two light-sensing regions, and they change conformation in response to blue light. This change activates another region of the protein that is a kinase. Both PHOT1 and a similar receptor, PHOT2, are receptor kinases unique to plants. A portion of PHOT1 is a kinase that autophosphorylates (figure 40.7). Currently, only the early steps in this signal transduction are understood. It will be intriguing to watch the story of the phot1 signal transduction pathway unfold, leading to an explanation of how plants grow toward the light.

 ## Circadian clocks are independent of light but are entrained by light

Although shorter and much longer naturally occurring rhythms also exist, **circadian rhythms** ("around the day") are particularly common and widespread among eukaryotic organisms. They relate the day–night cycle on Earth, although they are not exactly 24 hr in duration.

Jean de Mairan, a French astronomer, first identified circadian rhythms in 1729. He studied the sensitive plant *(Mimosa pudica),* which closes its leaflets and leaves at night. When de Mairan put the plants in total darkness, they continued "sleeping" and "waking" just as they had when exposed

to night and day. This is one of four characteristics of a circadian rhythm—it must continue to run in the absence of external inputs. Plants with a circadian rhythm do not actually have to be experiencing a pattern of daylight and darkness for their cycle to occur.

In addition, a circadian rhythm must be about 24 hr in duration, and the cycle can be reset or entrained. Although plants kept in darkness will continue the circadian cycle, the cycle's period may gradually move away from the actual day–night cycle, becoming desynchronized. In the natural environment, the cycle is entrained to a daily cycle through the action of phytochrome and blue-light photoreceptors.

Other eukaryotes, including humans, have circadian rhythms, and perhaps you have experienced a disruption of this rhythm as jet lag when you flew across a few time zones. Recovery from jet lag involves entrainment to the new time zone.

Another characteristic of a circadian cycle is that the "clock" can compensate for differences in temperature, so that the duration remains unchanged. This characteristic is unique, considering what we know about biochemical reactions, because most rates of reactions vary significantly based on temperature. Circadian clocks exist in many organisms, and they appear to have evolved independently multiple times.

The reversible circadian rhythm changes in leaf movements discussed earlier are typically brought about by alteration of cells' turgor pressure; we describe these changes in a later section.

Learning Outcomes Review 40.1

Plants grow and develop in response to environmental signals. Phytochrome, a red-light receptor, transduces information, whereas chlorophyll transduces energy. Phytochrome influences seed germination, shoot elongation, and other growth patterns. Phototropism is directional growth in response to light and is controlled by a blue-light receptor. Circadian rhythms are 24-hr cycles entrained to the day–night cycle.

■ *Why would it be an advantage to have both phytochromes and chlorophylls as pigments?*

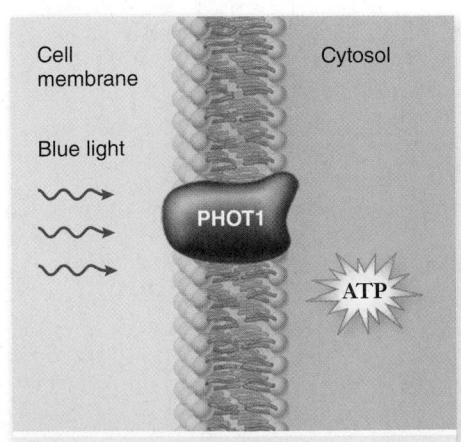
1. Light with blue wavelengths strikes plant cell membrane with phototropin 1 (PHOT1).

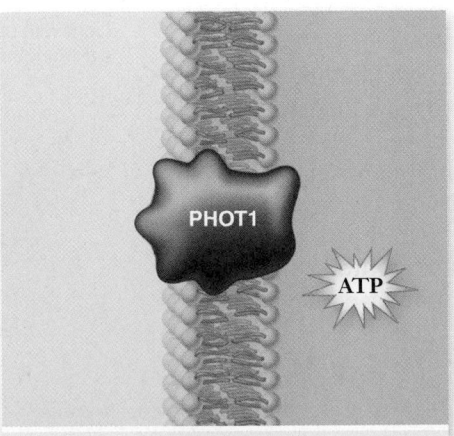
2. Blue light is absorbed by PHOT1, causing a change in conformation.

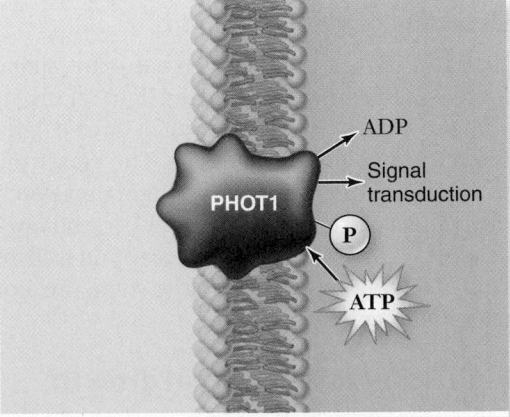
3. This conformational change results in auto-phosphorylation, triggering signal transduction.

Figure 40.7 Blue-light receptor. Blue light activates the light-sensing region of PHOT1, which in turn stimulates the kinase region of PHOT1 to autophosphorylate. This is just the first step in a signal transduction pathway that leads to phototropic growth.

When a potted plant is tipped over and left in place, the shoot bends and grows upward. The same thing happens when a storm pushes plants over in a field. These are examples of **gravitropism,** the response of a plant to the gravitational field of the Earth (figure 40.8). Because plants also grow in response to light, separating out phototropic effects is important in the study of gravitropism.

Plants align with the gravitational field: An overview

Gravitropic responses are present at germination, when the root grows down and the shoot grows up. Why does a shoot have a negative gravitropic response (growth away from gravity), but a root has a positive one? Auxins play a primary role in gravitropic responses, but they may not be the only way gravitational information is sent through the plant.

The opportunity to experiment in a gravity-free environment on the Space Shuttle and the International Space Station accelerated research in this area. Analysis of gravitropic mutants is also adding to our understanding of gravitropism. Investigators propose that four general steps lead to a gravitropic response:

1. Gravity is perceived by the cell.
2. A mechanical signal is transduced into a physiological signal in the cell that perceived gravity.
3. The physiological signal is transduced inside the cell and externally to other cells.
4. Differential cell elongation occurs, affecting cells in the "up" and "down" sides of the root or shoot.

Figure 40.8 Plant response to gravity. This plant was placed horizontally and allowed to grow for seven days. Note the negative gravitational response of the shoot.

Currently researchers are debating the steps involved in perception of gravity. In shoots, gravity is sensed along the length of the stem in the endodermal cells that surround the vascular tissue (figure 40.9*a*), and signaling occurs toward the outer epidermal cells. In roots, the cap is the site of gravity perception, and a signal must trigger differential cell elongation and division in the elongation zone (figure 40.9*b*).

In both shoots and roots, amyloplasts, plastids that contain starch, sink toward the center of the gravitational field and thus may be involved in sensing gravity. Amyloplasts interact with the cytoskeleton. Auxin evidently plays a role in transmitting a signal from the gravity-sensing cells that contain amyloplasts and the site where growth occurs. The link between amyloplasts and auxin is not fully understood.

? Inquiry question Where would you expect to find the highest concentration of auxin in a plant?

Stems bend away from a center of gravity

Increased auxin concentration on the lower side in stems causes the cells in that area to grow more than the cells on the

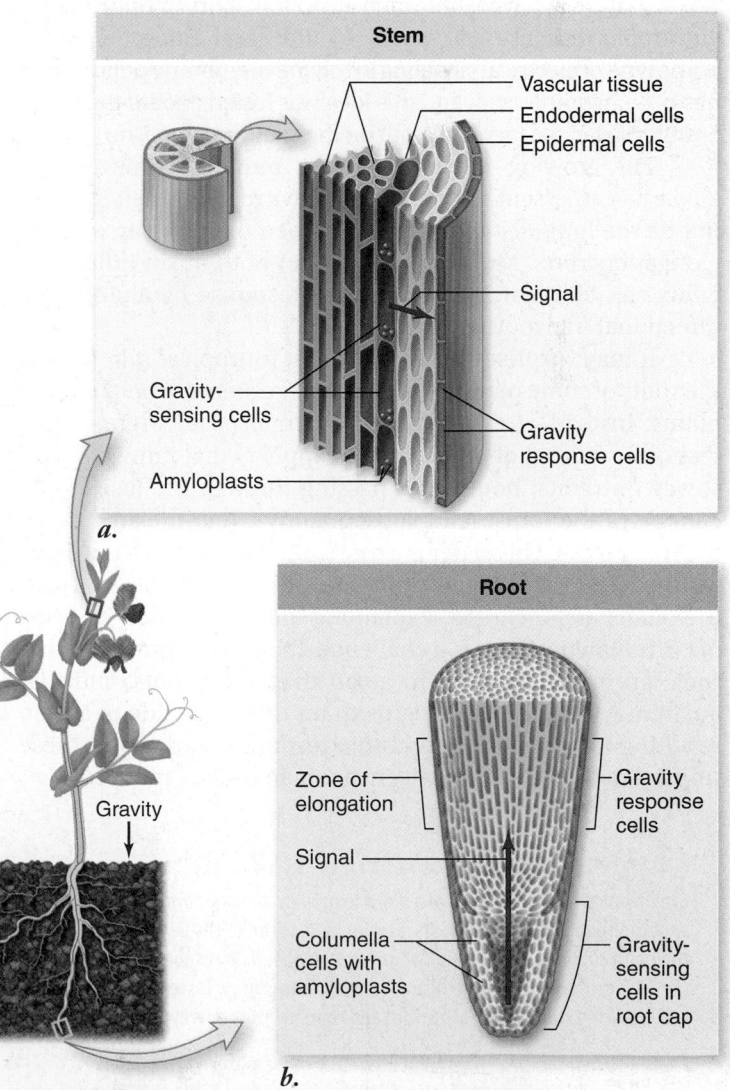

Figure 40.9 Sites of gravity sensing and response in roots and shoots.

upper side. The result is a bending upward of the stem against the force of gravity—in other words, a *negative gravitropic response*. Such differences in hormone concentration have not been as well documented in roots. Nevertheless, the upper sides of roots oriented horizontally grow more rapidly than the lower sides, causing the root ultimately to grow downward; this phenomenon is known as *positive gravitropic response*.

Two *Arabidopsis* mutants, *scarecrow (scr)* and *short root (shr)*, were initially identified by aberrant root phenotypes, but they also affect shoot gravitropism (figure 40.10). Both genes are needed for normal endodermal development (see figure 36.16). Without a fully functional endodermis, stems lack a normal gravitropic response. These endodermal cells carry amyloplasts in the stems, and in the mutants, stem endodermis fails to differentiate and produce gravity-sensing amyloplasts.

Roots bend toward a center of gravity

In roots, the gravity-sensing cells are located in the root cap, and the cells that actually undergo asymmetrical growth are in the distal elongation zone, which is closest to the root cap. How the information is transferred over this distance is an intriguing question. Auxin may be involved, but when auxin transport is suppressed, a gravitropic response still occurs in the distal elongation zone. Some type of electrical signaling involving membrane polarization has been hypothesized, and this idea was tested aboard the Space Shuttle. So far, the jury is still out on the exact mechanism.

The growing number of auxin mutants confirms that auxin has an essential role in root gravitropism, even if it may not be the long-distance signal between the root cap and the elongation zone. Mutations that affect both auxin influx and efflux can eliminate the gravitropic response by altering the directional transport of this hormone.

It may surprise you to learn that in tropical rain forests, the roots of some plants may grow up the stems of neighboring plants, instead of exhibiting the normal positive gravitropic responses typical of other roots. It appears that rainwater dissolves nutrients; both while passing through the lush upper canopy of the forest, and subsequently while trickling down the tree trunks. This water is a more reliable source of nutrients for the roots than the nutrient-poor rain forest soils in which the plants are anchored. Explaining this observation in terms of current hypotheses is a challenge. It has been proposed that roots are more sensitive to auxin than are shoots, and that auxin may actually inhibit growth on the lower side of a root, resulting in a positive gravitropic response. Perhaps in these tropical plants, the sensitivity to auxin in roots is reduced.

Learning Outcomes Review 40.2

Gravitropism is the response of a plant to gravity. In endodermis cells of shoots and root cap cells of roots, amyloplasts settle to the bottom, allowing the plant to sense the direction of gravitational pull. In response, cells on the lower side of stems and the upper side of roots grow faster than other cells, causing stems to grow upward and roots to grow downward.

■ **What would happen to the form of a plant growing under weightless conditions, such as in an orbiting spacecraft?**

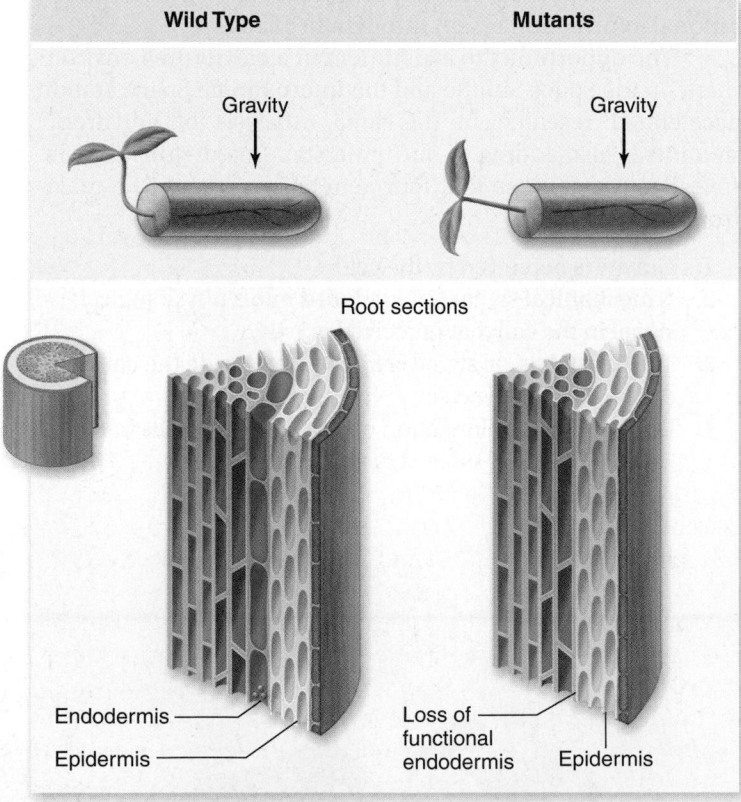

Figure 40.10 Amyloplasts in stem endoderm are needed for gravitropism. *a.* The *scr* and *shr* mutants of *Arabidopsis* have abnormal root development because they lack a fully differentiated endodermal layer. ***b.*** The endodermal defect extends into the stem, eliminating the positive gravitropic response of wild-type stems.

Responses to Mechanical Stimuli

Plants respond to touch and other mechanical stimuli in different ways, depending on the species and the type of stimulus. In some cases, plants permanently change form in response to mechanical stresses, a process termed **thigmomorphogenesis.** This change can be seen in trees growing where an almost constant wind blows from one direction (see the chapter-opening photo). Other responses are reversible and occur in the short term; for example, when mimosa leaves droop in response to touch. These responses are not tropisms, but rather turgor movements that come about due to changes in the internal water pressure of cells.

Touch can trigger irreversible growth responses

A **thigmotropism** is directional growth of a plant or plant part in response to contact with an object, animal, other plant, or even the wind. Thigmonastic responses are very similar to thigmotropisms, except that the direction of the growth response is the same regardless of the direction of the stimulus.

Tall, slender plants are more likely to snap during a wind or rainstorm than are plants with short, wide internodes. Environmental signals such as regularly occurring winds or the rubbing of one plant against another are sufficient to induce morphogenetic change leading to thicker, shorter internodes. In some cases, even repeated touching of a plant with a finger is enough to cause a change in plant growth.

Tendrils are modified stems that some species use to anchor themselves in the environment (figure 40.11). When a tendril makes contact with an object, specialized epidermal cells perceive the contact and promote uneven growth, causing the tendril to curl around the object, sometimes within only 3 to 10 min. Two hormones, auxin and ethylene, appear to be involved in tendril movements, and they can induce coiling even in the absence of any contact stimulus. Curiously, the tendrils of some species coil toward the site of the stimulus (thigmotropic growth), but those of other species may always coil clockwise, regardless of the side of the tendril that makes contact with an object. In some other plants, such as clematis, bindweed, and dodder, leaf petioles or unmodified stems twine around other stems or solid objects.

Perhaps the most dramatic touch response is the snapping of a Venus flytrap. As discussed in chapter 38, the modified leaves of the flytrap close in response to a touch stimulus, trapping insects or other potential sources of protein. A flytrap can shut in a mere 0.5 sec. The enlarged epidermal or mesophyll cells of the flytrap cause the trap to close. The speed of

Figure 40.11 Thigmotropism. The thigmotropic response of these twining stems causes them to coil around the object with which they have come in contact.

trap closure is enhanced by the shape of the leaf, which flips between a concave and convex form.

What is particularly amazing about this response is that the outer cells actually grow. The cell walls may soften in response to an electrical signal that moves through the leaf when the trigger hairs are touched, and the high pressure (turgor) of the water inside the cells pushes against the softened walls to enlarge the cell. This growth mechanism is distinct from other turgor movements (to be discussed shortly) because the water is already within the cell, not transferred into it in response to the electrical signal.

If digestible prey is caught, the trap will open about 24 hr later through the growth of inner cells of the flytrap. This growth response can only be triggered about four times before the leaf dies, presumably because so much energy is required for the individual flytrap to do this trick.

Arabidopsis is proving valuable as a model system to explore plant responses to touch. A gene has been identified that is expressed in 100-fold higher levels 10 to 30 min after touch. The gene codes for a calmodulin-like protein that binds Ca^{2+}, which is involved in a number of plant physiological processes. Given the value of a molecular genetics approach in dissecting the pathways leading from an environmental signal to a growth response, the touch gene provides a promising first step in understanding how plants respond to touch.

Reversible responses to touch and other stimuli involve turgor pressure

Unlike tropisms, some touch-induced plant movements are not based on growth responses, but instead result from reversible changes in the turgor pressure of specific cells. Turgor, as described in chapter 37, is pressure within a living cell resulting from diffusion of water into it. If water leaves turgid cells, the cells may collapse, causing plant movement; conversely,

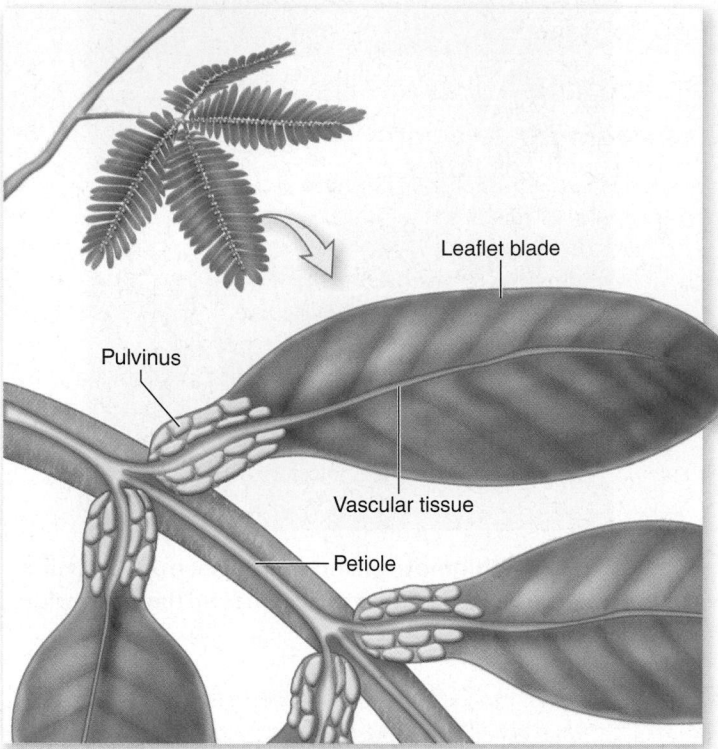

a.

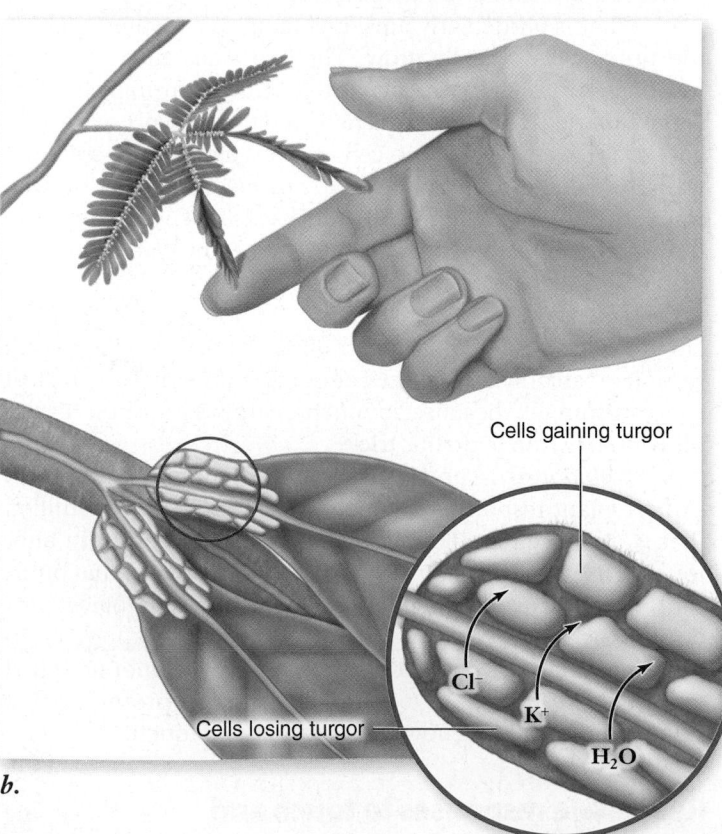

b.

Figure 40.12 Sensitive plant (*Mimosa pudica*).
a. The blades of *Mimosa* leaves are divided into numerous leaflets; at the base of each leaflet is a swollen structure called a pulvinus.
b. Changes in turgor cause leaflets to fold in response to a stimulus. When leaves are touched (center two leaves), ions move to the outer side of the pulvinus, water follows by osmosis, and the decreased interior turgor pressure leads to folding.

Figure 40.13 Tracking the Sun. These sunflowers follow the movement of the Sun every day.

water entering a limp cell may also cause movement as the cell once more becomes turgid.

Many plants, including those of the legume family (Fabaceae), exhibit leaf movements in response to touch or other stimuli. After exposure to a stimulus, the changes in leaf orientation are mostly associated with rapid turgor pressure changes in pulvini (singular, *pulvinus*), two-sided multicellular swellings located at the base of each leaf or leaflet. When leaves with pulvini, such as those of the sensitive plant (*Mimosa pudica*), are stimulated by wind, heat, touch, or in some instances, intense light, an electrical signal is generated. The electrical signal is translated into a chemical signal, with potassium ions being pumped from the cells in one-half of a pulvinus to the intercellular spaces in the other half, leading to the rapid osmosis of water to one side of the pulvinus.

The loss of turgor in half of the pulvinus causes the leaf to "fold." The movements of the leaves and leaflets of a sensitive plant are especially rapid; the folding occurs within a second or two after the leaves are touched (figure 40.12). Over a span of about 15 to 30 min after the leaves and leaflets have folded, water usually diffuses back into the same cells from which it left, and the leaf returns to its original position.

Some turgor movements are triggered by light. For example, the leaves of some plants may track the Sun, with their blades oriented at right angles to it; how their orientation is directed, however, is poorly understood. Such leaves can move quite rapidly (as much as 15° an hour). This movement maximizes photosynthesis and is analogous to solar panels designed to track the Sun (figure 40.13).

Some of the most familiar reversible changes due to turgor pressure are the circadian rhythms seen in leaves and flowers that open during the day and close at night, or vice versa. For example, the flowers of four o'clocks open in the afternoon, and evening primrose petals open at night. As described earlier, sensitive plant leaves also close at night. Bean leaves are horizontal during the day when their pulvini are turgid, but become more vertical at night as the pulvini lose turgor (figure 40.14). These sleep movements reduce water loss from transpiration during the night, but maximize photosynthetic surface area during the day.

Figure 40.14 Sleep movements in bean leaves. In the bean plant, leaf blades are oriented horizontally during the day and vertically at night.

Learning Outcomes Review 40.3

Thigmomorphogenesis is a change in growth form in response to a mechanical stress (physical contact or wind). Thigmotropism is directional growth, whereas a thigmonastic response has no directionality. A tropism is an irreversible growth response; a touch-induced plant movement, such as exhibited by *Mimosa pudica*, is reversible and is based on changes in turgor pressure.

■ *What would be some advantages of having leaves that fold when stimulated?*

40.4 Responses to Water and Temperature

Learning Outcomes

1. *List the environmental factors that can lead to dormancy.*
2. *Explain why seed dormancy is an important evolutionary innovation.*
3. *Identify the types of biological molecules that are most directly affected by low and high temperatures.*

Sometimes, modifying the direction of growth is not enough to protect a plant from harsh conditions. The ability to cease growth and go into a dormant stage when conditions become unfavorable, such as during seasonal changes in temperate climates, provides a survival advantage. The extreme example is seed dormancy, but there are intermediate approaches to waiting out the bad times as well.

Plants also have developed adaptations to more short-term fluctuations in temperature, such as might occur during a heat wave or cold snap. These strategies include changes in membrane composition and the production of heat shock proteins.

Dormancy is a response to water, temperature, and light

In temperate regions, we generally associate dormancy with winter, when freezing temperatures and the accompanying unavailability of water make it impossible for most plants to grow. During this season, buds of deciduous trees and shrubs remain dormant, and apical meristems remain well protected inside enfolding scales. Perennial herbs spend the winter underground, existing as stout stems or roots packed with stored food. Many other kinds of plants, including most annuals, pass the winter as seeds. Often dormancy begins with the dropping of leaves, which you have probably seen occur in deciduous trees in the autumn.

Organ abscission

Deciduous leaves are often shed as the plant enters dormancy. The process by which leaves or petals are shed is called **abscission.**

Abscission can be useful even before dormancy is established. For example, shaded leaves that are no longer photosynthetically productive can be shed. Petals, which are modified leaves, may senesce once pollination occurs. Orchid flowers remain fresh for long periods of time, even in a florist shop; however, once pollination occurs, a hormonal change is triggered that leads to petal senescence. This strategy makes sense in terms of allocation of energy resources because the petals are no longer necessary to attract a pollinator. One advantage of organ abscission, therefore, is that nutrient sinks can be discarded, conserving resources.

On a larger scale, deciduous plants in temperate areas produce new leaves in the spring and then lose them in the fall. In the tropics, however, the production and subsequent loss of leaves in some species is correlated with wet and dry seasons. Evergreen plants, such as most conifers, usually have a complete change of leaves every two to seven years, periodically losing some but not all of their leaves.

Abscission involves changes that take place in an *abscission zone* at the base of the petiole (figure 40.15). Young leaves produce hormones (especially cytokinins) that inhibit the development of specialized layers of cells in this zone. Hormonal changes take place as the leaf ages, however, and two layers of cells become differentiated. A *protective layer,* which may be several cells wide, develops on the stem side of the petiole base. These cells become impregnated with suberin, which you may recall is a fatty substance impervious to moisture. A *separation layer* develops on the leaf-blade side; the cells of the separation layer sometimes divide, swell, and become gelatinous.

When temperatures drop, when the duration and intensity of light diminishes, or when other environmental changes occur, enzymes break down the pectins in the middle lamellae of the separation cells. Wind and rain can then easily separate the leaf from the stem. Left behind is a sealed leaf scar that is protected from invasion by bacteria and other disease organisms.

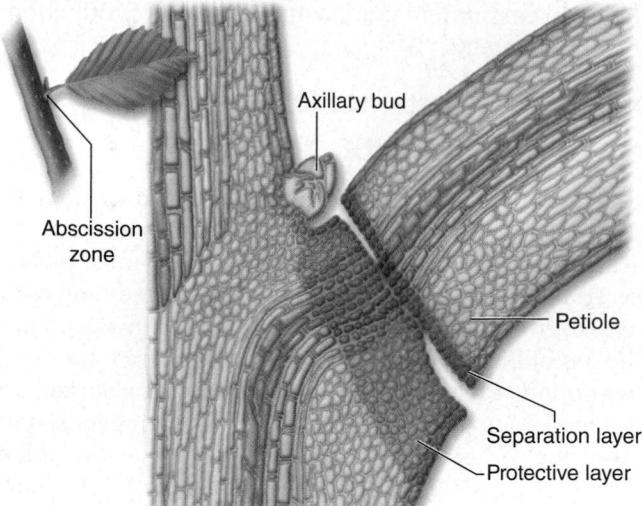

Figure 40.15 Leaf abscission. Hormonal changes in the leaf's abscission zone cause abscission. Two layers of cells in the abscission zone differentiate into a protective layer and a separation layer. As pectins in the separation layer break down, wind and rain can easily separate the leaf from the stem.

As the abscission zone develops, the green chlorophyll pigments present in the leaf break down, revealing the yellows and oranges of other pigments, such as carotenoids, that previously had been masked by the intense green colors. At the same time, water-soluble red or blue pigments called *anthocyanins* and *betacyanins* may also accumulate in the vacuoles of the leaf cells—all contributing to an array of fall colors in leaves (figure 40.16).

Seed dormancy

The extraordinary evolutionary innovation of the seed plants is the dormant seed that allows plant offspring to wait until conditions for germination are optimal. A seed moves into a dormant state during a complex process triggered by increased levels of the hormone abscisic acid (ABA) (figure 40.17). RNA and protein synthesis decreases as the seed dehydrates and the seed coat hardens to package the embryo into a time capsule.

Figure 40.16 Leaf color changes during abscission.

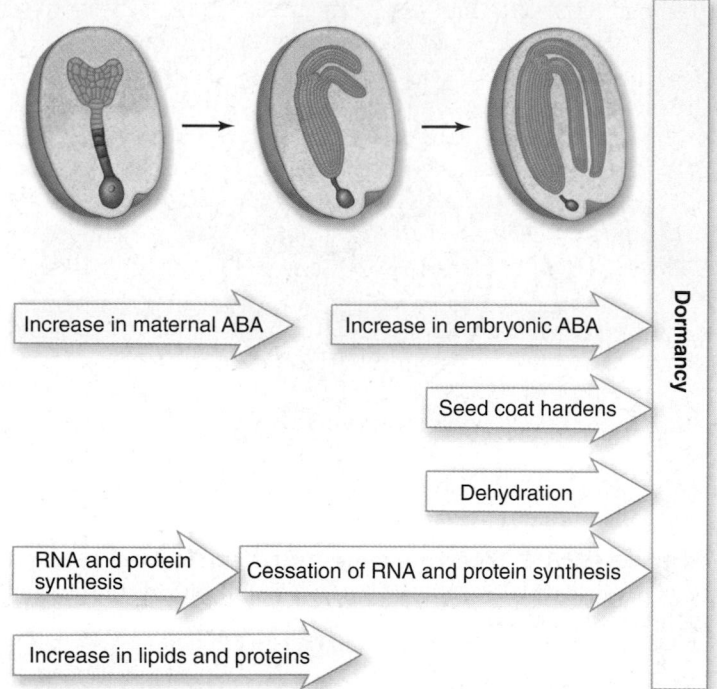

Figure 40.17 Seed dormancy. Accumulating food reserves, forming a protective seed coat, and dehydration are essential steps leading to dormancy. Abscisic acid (ABA) from both maternal and embryonic tissue is necessary for dormancy.

Sometimes the seeds can endure a wait of hundreds, even thousands, of years. In seasonally dry climates, seed dormancy occurs primarily during the dry season, often the summer. Rainfalls trigger germination when conditions for survival are more favorable.

Annual plants occur frequently in areas of seasonal drought. Seeds are ideal for allowing annual plants to bypass the dry season, when there is insufficient water for growth. When it rains, these seeds can germinate, and the plants can grow rapidly, having adapted to the relatively short periods when water is available.

Favorable temperatures, day length, and amounts of water can release buds, underground stems and roots, and seeds from a dormant state. Requirements vary among species. For example, some weed seeds germinate in cooler parts of the year and are inhibited from germinating by warmer temperatures. Day length differences can have dramatic effects on dormancy. For example, tree dormancy is common in temperate climates when the days are short, but is unusual in tropical trees growing near the equator, where day length remains about the same regardless of season.

Plants can survive temperature extremes

Sometimes temperatures change rapidly, and dormancy is not possible. How do plants survive temperature extremes? A number of adaptations, including some rapid response strategies, help plants overcome sudden chilling or extreme heat.

Chilling

Knowing the lipid composition of a plant's membranes can help predict whether the plant will be sensitive or resistant to chilling. Saturated lipids solidify at a higher temperature because they pack together more closely (see chapter 5), so the more unsaturated the membrane lipids are, the more resistant the plant is to chilling. *Arabidopsis* plants genetically modified to contain a higher percentage of saturated fatty acids have proved to be more sensitive to chilling.

When chilling occurs, the enzyme desaturase converts the single bonds in the saturated lipids to double bonds. This process lowers the temperature at which the membrane becomes rigid and cannot function properly.

Even highly unsaturated membranes are not enough to protect plants from freezing temperatures. At freezing, ice crystals form and the cells die from dehydration—not enough liquid water is available for metabolism. Some plants, however, have the ability to undergo deep supercooling and survive temperatures as low as –40°C. Supercooling occurs when ice crystal formation is limited, and the crystals occur in extracellular spaces where they cannot damage cell organelles. Furthermore, the cells of these plants must be able to withstand gradual dehydration.

Acquiring tolerance to chilling or freezing as the temperature drops can be explained by increased solute concentration. In addition, antifreeze proteins prevent ice crystals from forming. Ice crystals can also form (nucleate) around bacteria naturally found on the leaf surface. Some bacteria have been genetically engineered so that they do not nucleate ice crystals. Spraying leaves with these modified bacteria can provide frost tolerance in some crops.

High temperatures

High temperatures can be harmful because proteins denature and lose their function when heated. If temperatures suddenly rise 5° to 10°C, heat shock proteins (HSPs) are produced. These proteins can stabilize other proteins so that they don't unfold or misfold at higher temperatures. In some cases, HSPs induced by temperature increases can also protect plants from other stresses, including chilling.

Plants can survive otherwise lethal temperatures if they are gradually exposed to increasing temperature. These plants have *acquired thermotolerance*. More is being learned about temperature acclimation by isolating mutants that fail to acquire thermotolerance, including the aptly named *hot* mutants in *Arabidopsis*. One of the *HOT* genes codes for an HSP. Characterization of other *HOT* genes indicates that thermotolerance requires more than the synthesis of HSPs; some *HOT* genes stabilize membranes and are necessary for protein activity.

Learning Outcomes Review 40.4

Seasonal changes, such as reduction in temperature, light, and water availability, may lead to plant dormancy; in deciduous trees, leaf abscission is part of entering dormancy. Seed dormancy prevents germination until growth conditions are optimal. At low temperatures, lipids in membranes begin to solidify and ice crystals may form in tissues; at high temperatures, proteins denature.

■ *Why is it advantageous for broadleaf trees to drop leaves in autumn, when they must grow them again in spring?*

40.5 Hormones and Sensory Systems

Learning Outcomes

1. *Discuss properties of hormones.*
2. *Compare auxins with cytokinins.*
3. *Choose a hormone, and diagram how the hormone can affect gene transcription.*

Sensory responses that alter morphology rely on complex physiological networks. Many internal signaling pathways involve plant hormones, which are the focus of this section. Hormones are involved in responses to the environment, as well as in internally regulated development (see chapter 41).

The hormones that guide growth are keyed to the environment

Hormones are chemical substances produced in small, often minute quantities in one part of an organism and then transported to another part where they bring about physiological or developmental responses. How hormones act in a particular instance is influenced both by the hormone and the tissue that receives the message.

In animals, hormones are usually produced at definite sites, most commonly in organs such as glands. In plants, hormones are not produced in specialized tissues but, instead, in tissues that also carry out other, usually more obvious, functions. Eight major kinds of plant hormones have been identified: auxin, cytokinins, strigolactones, gibberellins, brassinosteroids, oligosaccharins, ethylene, and abscisic acid (table 40.1). Current research is focused on the biosynthesis of hormones and on characterizing the hormone receptors involved in signal transduction pathways. Much of the molecular basis of hormone function remains enigmatic.

Because hormones are involved in so many aspects of plant function and development, we have chosen to integrate examples of hormone activity with specific aspects of plant biology throughout the text. In this section, our goal is to give a brief overview of these hormones.

Auxin allows elongation and organizes the body plan

More than a century ago, an organic substance known as auxin was the first plant hormone to be discovered. **Auxin** increases the plasticity of plant cell walls and is involved in elongation of stems. Cells can enlarge in response to changes in turgor pressure, but cell walls must be fairly plastic for this expansion to occur. Auxin plays a role in softening cell walls. The discovery of auxin and its role in plant growth is an elegant example of thoughtful experimental design and is recounted here for that reason.

TABLE 40.1

TABLE 40.1 Functions of the Major Plant Hormones

Hormone		Major Functions	Where Produced or Found in Plant
Auxins		Promotion of stem elongation and growth; formation of adventitious roots; inhibition of leaf abscission; promotion of cell division (with cytokinins); inducement of ethylene production; promotion of lateral bud dormancy	Apical meristems; other immature parts of plants
Cytokinins		Stimulation of cell division, but only in the presence of auxin; promotion of chloroplast development; delay of leaf aging; promotion of bud formation	Root apical meristems; immature fruits
Strigolactones		Inhibition of axillary bud growth; positively regulate secondary growth; levels are regulated by auxin	Roots; cambium
Gibberellins		Promotion of stem elongation; stimulation of enzyme production in germinating seeds	Roots and shoot tips; young leaves; seeds
Brassinosteroids		Overlapping functions with auxins and gibberellins	Pollen, immature seeds, shoots, leaves
Oligosaccharins		Pathogen defense, possibly reproductive development	Cell walls
Ethylene		Control of leaf, flower, and fruit abscission; promotion of fruit ripening	Roots, shoot apical meristems; leaf nodes; aging flowers; ripening fruits

TABLE 40.1	Functions of the Major Plant Hormones *(continued)*		
Hormone		**Major Functions**	**Where Produced or Found in Plant**
Abscisic acid	$CH_3 \quad CH_3 \quad CH_3$ / OH / $O \quad CH_3 \quad COOH$	Inhibition of bud growth; control of stomatal closure; some control of seed dormancy; inhibition of effects of other hormones	Leaves, fruits, root caps, seeds

Discovery of auxin

Later in life, the great evolutionist Charles Darwin became increasingly devoted to the study of plants. In 1881, he and his son Francis published a book called *The Power of Movement of Plants*. In this book, the Darwins reported their systematic experiments on the response of growing plants to light—the responses that came to be known as phototropisms. They used germinating oat and canary grass seedlings in their experiments and made many observations in this field.

Charles and Francis Darwin knew that if light came primarily from one direction, seedlings would bend strongly toward it. If they covered the tip of a shoot with a thin glass tube, the shoot would bend as if it were not covered. However, if they used a metal foil cap to exclude light from the plant tip, the shoot would not bend (figure 40.18). They also found that using an opaque collar to exclude light from the stem below the tip did not keep the area above the collar from bending.

In explaining these unexpected findings, the Darwins hypothesized that when the shoots were illuminated from one side, they bent toward the light in response to an "influence" that was transmitted downward from its source at the tip of the shoot.

Some 30 years later, the Danish plant physiologist Peter Boysen-Jensen and the Hungarian plant physiologist Arpad Paal independently demonstrated that the substance causing the shoots to bend was a chemical. They showed that if the tip of a germinating grass seedling was cut off and then placed on top of a small block of agar separating it from the rest of the seedling, the seedling would still grow as if there had been no change. Something evidently was passing from the tip of the seedling through the agar into the region where the bending occurred.

? **Inquiry question** Propose a mechanism to explain how seedlings could bend in the light using what Paal discovered.

SCIENTIFIC THINKING

Hypothesis: *The shoot tip of a plant detects the direction of light.*

Prediction: *The shoot tip of a grass seedling will grow toward a unidirectional light source if it is not covered.*

Test: *Make four treatment groups, including (1) untreated seedling, (2) tip covered with lightproof cap, (3) tip covered with transparent cap, and (4) lightproof collar placed below tip.*

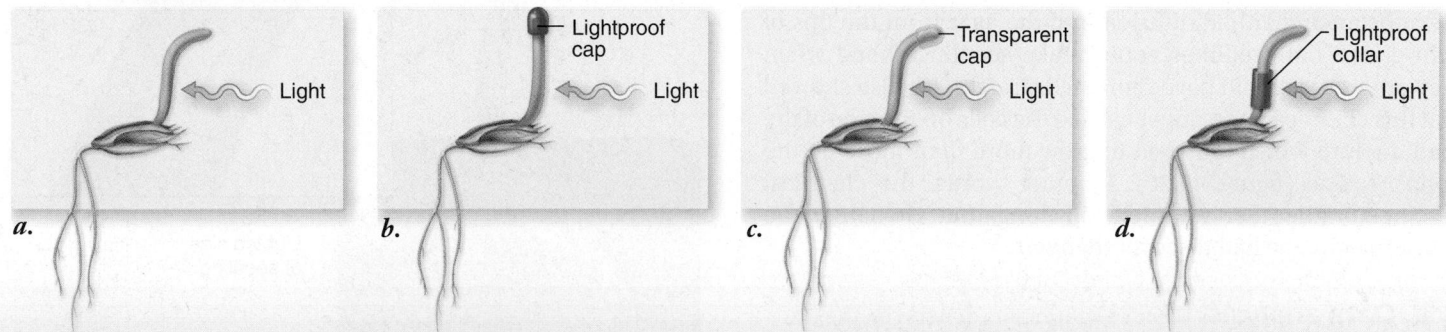

a.   b.   c.   d.

Result: *a. Young grass seedlings normally bend toward the light. b. The bending did not occur when the tip of a seedling was covered with a lightproof cap. c. Bending did occur when it was covered with a transparent one. d. When a collar was placed below the tip, the characteristic light response took place.*

Conclusion: *In response to light, an "influence" that caused bending was transmitted from the tip of the seedling to the area below, where bending normally occurs.*

Further Experiments: *How could you determine if the light response in a shoot tip requires the movement of a signal from one side of the shoot to the other? (Hint: See Went's experiment in figure 40.19).*

Figure 40.18 **Shoot tips perceive unidirectional light.**

Figure 40.19 Frits Went's experiment. Went concluded that a substance he named *auxin* promoted the elongation of the cells and that it accumulated on the side of an oat seedling away from the light.

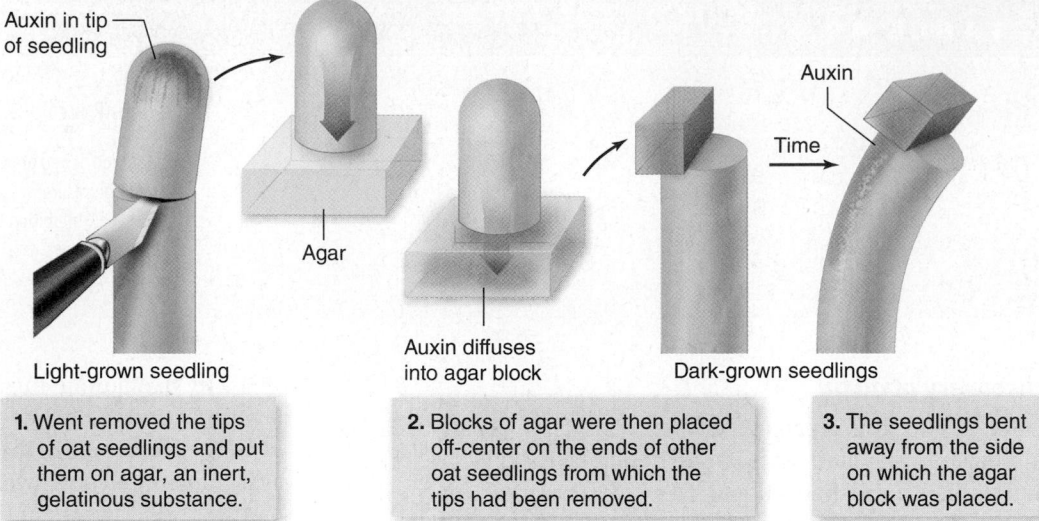

Auxin in tip of seedling

Auxin

Time

Agar

Light-grown seedling

Auxin diffuses into agar block

Dark-grown seedlings

1. Went removed the tips of oat seedlings and put them on agar, an inert, gelatinous substance.

2. Blocks of agar were then placed off-center on the ends of other oat seedlings from which the tips had been removed.

3. The seedlings bent away from the side on which the agar block was placed.

Then, in 1926, the Dutch plant physiologist Frits Went carried Paal's experiments a step further. Went cut off the tips of oat seedlings that had been illuminated normally and set these tips on agar. He then took oat seedlings that had been grown in the dark and cut off their tips in a similar way. Finally, Went cut tiny blocks from the agar on which the tips of the light-grown seedlings had been placed and placed them off-center on the tops of the decapitated dark-grown seedlings (figure 40.19). Even though these seedlings had not been exposed to the light themselves, they bent away from the side on which the agar blocks were placed.

As an experimental control, Went put blocks of pure agar on the decapitated stem tips and noted either no effect or a slight bending toward the side where the agar blocks were placed. Finally, Went cut sections out of the lower portions of the light-grown seedlings. He placed these sections on the tips of decapitated, dark-green oat seedlings and again observed no effect.

As a result of his experiments, Went was able to show that the substance that had diffused into the agar from the tips of light-grown oat seedlings could make seedlings bend when they otherwise would have remained straight. He also showed that this chemical messenger caused the cells on the side of the seedling into which it flowed to grow more than those on the opposite side (figure 40.20). In other words, the chemical enhanced rather than retarded cell elongation. He named the substance that he had discovered *auxin*.

 Data analysis Went's experiments depend on the experimenter's ability to distinguish between bent and unbent stems. Explain how you could quantify the extent of bending in a stem.

The effects of auxin

Auxin acts to adapt the plant to its environment in a highly advantageous way by promoting growth and elongation. Environmental signals directly influence the distribution of auxin in the plant. How does the environment—specifically, light—exert this influence? Theoretically, light might destroy the auxin, might decrease the cells' sensitivity to auxin, or might cause the auxin molecules to migrate away from the light into the shaded portion of the shoot. This last possibility has proved to be the case.

In a simple but effective experiment, American plant biologist Winslow Briggs inserted a thin sheet of transparent mica vertically between the half of the shoot oriented toward the light and the half of the shoot oriented away from it (figure 40.21). He found that light from one side does not cause a shoot with such a barrier to bend. When Briggs examined the illuminated plant, he found equal auxin levels on both the light and dark sides of the barrier. He concluded that a normal

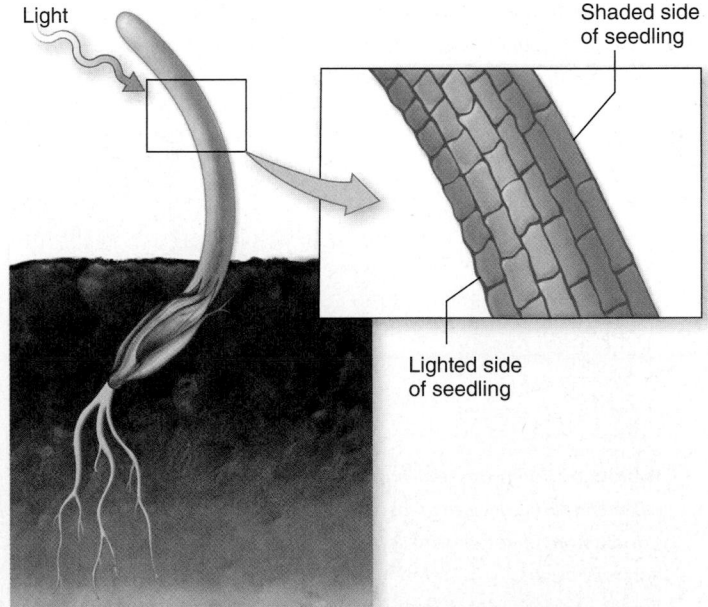

Light

Shaded side of seedling

Lighted side of seedling

Figure 40.20 Auxin causes cells on the dark side to elongate. Plant cells that are in the shade have more auxin and grow faster than cells on the lighted side, causing the plant to bend toward light. Further experiments showed exactly why there is more auxin on the shaded side of a plant.

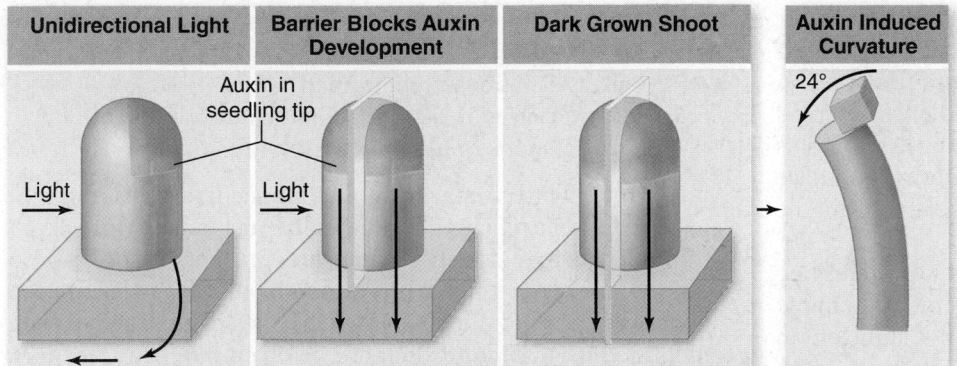

| Unidirectional Light | Barrier Blocks Auxin Development | Dark Grown Shoot | Auxin Induced Curvature |

Auxin in seedling tip

Light

Light

Light

24°

The same amount of total auxin is produced by a shoot tip grown with directional light, even when a barrier divides the shoot tip, and a shoot tip grown in the dark. All three blocks of agar cause the same amount of curvature in a tipless shoot.

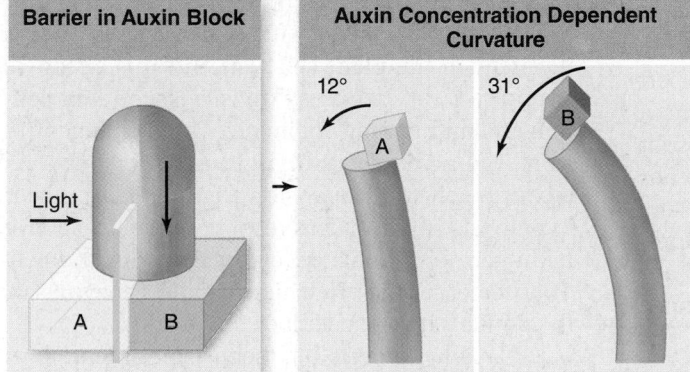

| Barrier in Auxin Block | Auxin Concentration Dependent Curvature |

Light

A B

12°

A

31°

B

Separating the base of the shoot tip and the agar block results in two agar blocks with different concentrations of auxin that produce different degrees of curvature in tipless shoots.

Figure 40.21 Phototropism and auxin: The Winslow Briggs experiments. Directional light causes the accumulation of auxin in the dark side of the shoot tip, which can move down the stem. Barriers inserted in the tip revealed that light affects auxin displacement rather than rate of auxin production.

plant's response to light from one direction involves auxin migrating from the light side to the dark side, and that the mica barrier prevented a response by blocking the migration of auxin.

The effects of auxin are numerous and varied. Auxin promotes the activity of the vascular cambium and the vascular tissues. Also, auxin is present in pollen in large quantities and plays a key role in the development of fruits. Synthetic auxins are used commercially for the same purpose. Fruits will normally not develop if fertilization has not occurred and seeds are not present, but frequently they will develop if auxin is applied. Pollination may trigger auxin release in some species, leading to fruit development even before fertilization has taken place.

How auxin works

In spite of this long history of research, auxin's molecular basis of action has been an enigma. The chemical structure of the most common auxin, **indoleacetic acid (IAA),** resembles that of the amino acid tryptophan, from which it is probably synthesized by plants (figure 40.22). Although other forms of auxin exist, IAA is the most common natural auxin.

An auxin-binding protein (ABP1) was identified two decades ago and has origins dating back to the algae. When auxin binds ABP1, the protein binds to a membrane receptor, initiating a signaling pathway that controls entry of the cell into the cell cycle. The ABP1 pathway is essential for auxin's role in cell division.

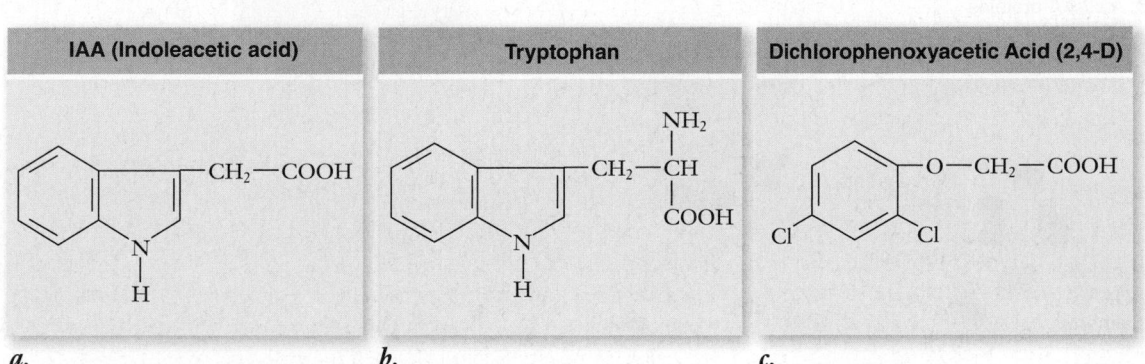

| IAA (Indoleacetic acid) | Tryptophan | Dichlorophenoxyacetic Acid (2,4-D) |

CH_2—COOH

NH_2

CH_2—CH

COOH

O—CH_2—COOH

Cl   Cl

a.

b.

c.

Figure 40.22 Auxins. *a.* Indoleacetic acid (IAA), the principal naturally occurring auxin. *b.* Tryptophan, the amino acid from which plants probably synthesize IAA. *c.* Dichlorophenoxyacetic acid (2,4-D), a synthetic auxin, is a widely used herbicide.

More recently, two families of proteins that mediate rapid, auxin-induced changes in gene expression have been identified: the auxin response factors (ARFs) and the Aux/IAA proteins. Transcription can be either enhanced or suppressed by ARFs, which are known to bind DNA. The Aux/IAA proteins function a bit earlier in the auxin response pathway and have been shown to bind to and repress proteins that activate the expression of *ARF* genes.

ARF genes are activated when Aux/IAA proteins are degraded by ubiquitin tagging and protein degradation in the proteasome. Auxin binding to ARF protein is not sufficient to initiate gene expression in response to auxin signaling because of Aux/IAA repression of ARF activity. How then does a plant sense auxin and degrade Aux/IAA proteins?

The identification of the elusive auxin receptor in 2005 hints at how plants sense and respond to auxin. Auxin binds directly to a protein called the transport inhibitor response protein 1 (TIR1). TIR1 is the enigmatic auxin receptor. It is part of a protein complex known as SCF, which is found throughout eukaryotes. SCF is shorthand for the three polypeptide subunits found in the complex: *Skp, Cullin*, and *F-box*. Auxin binds to TIR1 in the SCF complex if Aux/IAA proteins are present. Once auxin binds, the SCF complex degrades the Aux/IAA proteins through the ubiquitin pathway.

Five steps lead from auxin perception to auxin-induced gene expression (figure 40.23):

1. Auxin binds TIR1 in the SCF complex.
2. The activated SCF complex tags Aux/IAA proteins with ubiquitin.
3. Aux/IAA proteins are degraded in the proteasome.
4. Aux/IAA proteins are no longer available to bind and repress ARF transcriptional activators.
5. ARF transcription factors facilitate transcription of auxin-response genes.

Unlike with animal hormones, a specific signal is not sent to specific cells, eliciting a predictable response. Most likely, multiple auxin perception sites are present. Auxin is also unique among the plant hormones in that it is transported toward the base of the plant. Two families of genes have been identified in *Arabidopsis* that are involved in auxin transport. For example, one family of proteins (the *PIN*s) are involved in the top-to-bottom transport of auxin, and two other proteins function in the root tip to regulate the growth response to gravity, described earlier.

One of the direct effects of auxin is an increase in the plasticity of the plant cell wall, but this effect works only on young cell walls lacking extensive secondary cell-wall formation and may or may not involve rapid changes in gene expression. The **acid growth hypothesis** provides a model linking auxin to cell-wall expansion (figure 40.24). According to this hypothesis, auxin causes responsive cells to actively transport hydrogen ions from the cytoplasm into the cell-wall space. This decreases the pH, which activates enzymes that can break the bonds between cellulose fibers.

This hypothesis has been experimentally supported in several ways. Buffers that prevent cell-wall acidification block cell expansion. And, other compounds that release hydrogen ions from the cell can also cause cell expansion. Finally, the

Figure 40.23 Auxin regulation of gene expression. Auxin activates a ubiquitination pathway that releases ARF transcription factors from repression by Aux/IAA proteins. The result is auxin-induced gene expression.

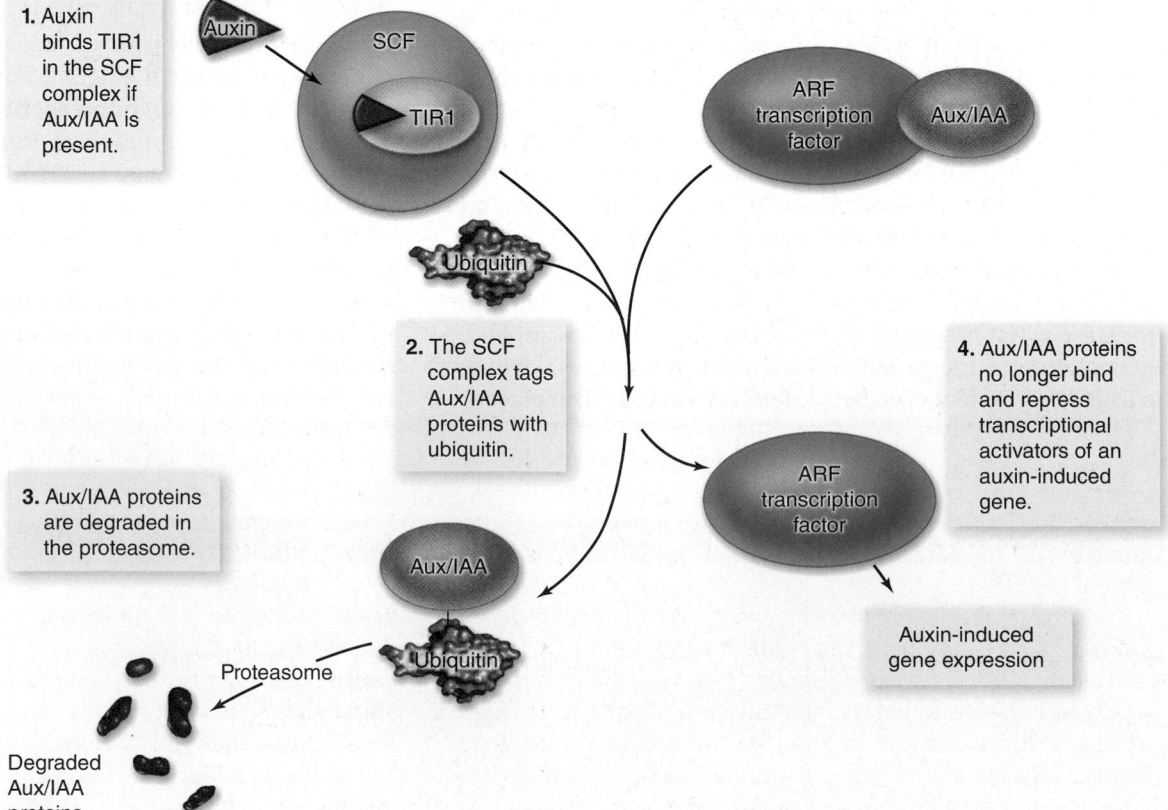

1. Auxin binds TIR1 in the SCF complex if Aux/IAA is present.

2. The SCF complex tags Aux/IAA proteins with ubiquitin.

3. Aux/IAA proteins are degraded in the proteasome.

4. Aux/IAA proteins no longer bind and repress transcriptional activators of an auxin-induced gene.

Auxin-induced gene expression

Degraded Aux/IAA proteins

Proteasome

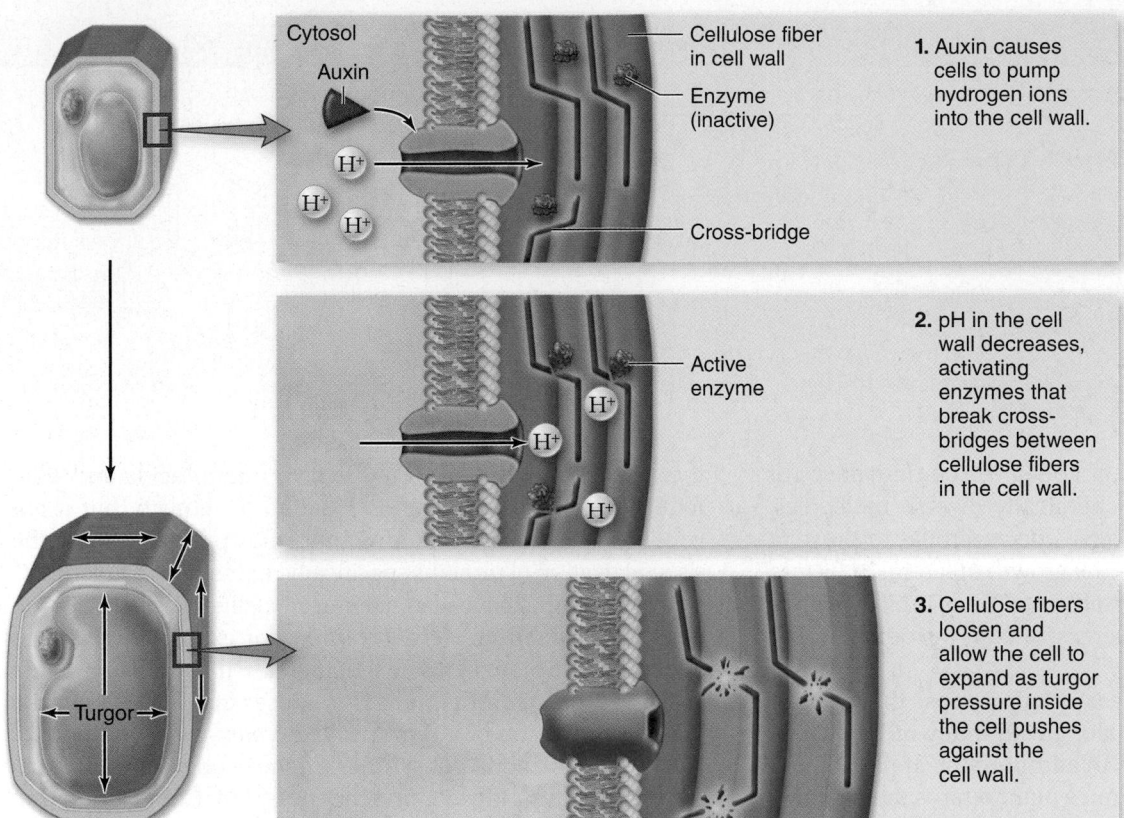

Figure 40.24 Acid growth hypothesis. Auxin stimulates the release of hydrogen ions (H^+) from the target cells, which alters the pH of the cell wall. This optimizes the activity of enzymes that break bonds in the cell wall, allowing the wall to expand.

1. Auxin causes cells to pump hydrogen ions into the cell wall.

2. pH in the cell wall decreases, activating enzymes that break cross-bridges between cellulose fibers in the cell wall.

3. Cellulose fibers loosen and allow the cell to expand as turgor pressure inside the cell pushes against the cell wall.

movement of hydrogen ions has been observed in response to auxin treatment. The snapping of the Venus flytrap is postulated to involve an acid growth response that allows cells to grow in just 0.5 sec and close the trap.

Synthetic auxins

Synthetic auxins, such as *naphthalene acetic acid (NAA)* and **indolebutyric acid (IBA),** have many uses in agriculture and horticulture, including the prevention of abscission. Synthetic auxins are used to prevent fruit drop in apples before they are ripe and to hold berries on holly that is being prepared for shipping during the winter season. Synthetic auxins are also used to promote flowering and fruiting in pineapples and to induce the formation of roots in cuttings.

Synthetic auxins are routinely used to control weeds. When used as herbicides, they are applied in higher concentrations than IAA would normally occur in plants. One of the most important synthetic auxin herbicides is *2,4-dichlorophenoxyacetic acid,* usually known as **2,4-D** (see figure 40.22c). It kills weeds in grass lawns by selectively eliminating broad-leaved dicots. The stems of the dicot weeds cease all axial growth.

The herbicide 2,4,5-trichlorophenoxyacetic acid, better known as 2,4,5-T, is closely related to 2,4-D. 2,4,5-T was widely used as a broad-spectrum herbicide to kill weeds and the seedlings of woody plants. It became notorious during the Vietnam War as a component of a jungle defoliant known as Agent Orange. When 2,4,5-T is manufactured, it is unavoidably contaminated with minute amounts of dioxin. Dioxin, in doses as low as a few parts per billion, has produced liver and lung diseases, leukemia, miscarriages, birth defects, and even death in laboratory animals. This chemical was banned in 1979 for most uses in the United States.

Cytokinins stimulate cell division and differentiation

Cytokinins constitute another group of naturally occurring growth hormones in plants. Studies by Austrian botanist Gottlieb Haberlandt around 1913 demonstrated the existence of an unknown chemical in various tissues of vascular plants that, when applied to cut potato tubers, would cause parenchyma cells to become meristematic and would induce the differentiation of a cork cambium. In other research, coconut milk, subsequently found to contain cytokinins, was used to promote the differentiation of organs in masses of plant tissue growing in culture. Subsequent studies have focused on the role cytokinins play in the differentiation of tissues from callus.

A *cytokinin* is a plant hormone that, in combination with auxin, stimulates cell division and differentiation. Most cytokinins are produced in the root apical meristems and transported throughout the plant. Developing fruits are also important sites of cytokinin synthesis. In mosses, cytokinins cause the formation of vegetative buds on the gametophyte. In all plants, cytokinins, working with other hormones, seem to regulate growth patterns.

Figure 40.25
Some cytokinins.
Two commonly used synthetic cytokinins: kinetin and 6-benzylamino purine. Note their resemblance to the purine base adenine.

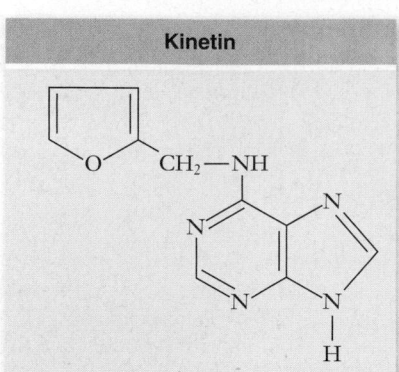

Kinetin

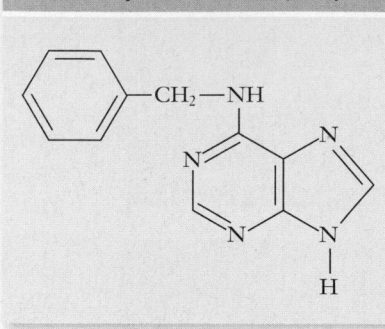

6-Benzylamino Purine (BAP)

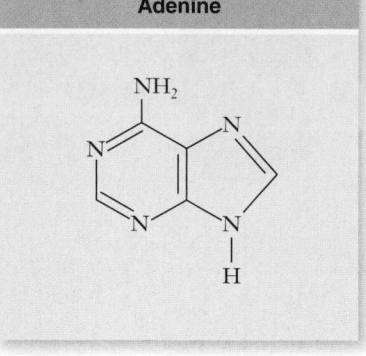

Adenine

Cytokinins are purines that appear to be derivatives of adenine (figure 40.25). Other chemically diverse molecules, not known to occur naturally, have effects similar to those of cytokinins. Cytokinins promote the growth of lateral buds into branches (figure 40.26). Conversely, cytokinins inhibit the formation of lateral roots, whereas auxins promote their formation.

As a consequence of these relationships, the balance between cytokinins and auxin, along with many other factors, determines the form of a plant. In addition, the application of cytokinins to leaves detached from a plant retards their yellowing. Therefore, they function as antiaging hormones.

The action of cytokinins, like that of other hormones, has been studied in terms of its effects on the growth and differentiation of masses of tissue growing in defined media. Plant tissue can form shoots, roots, or an undifferentiated mass, depending on the relative amounts of auxin and cytokinin (figure 40.27).

In the early cell-growth experiments in culture, coconut milk was an essential factor. Eventually, researchers discovered that coconut milk is not only rich in amino acids and other reduced nitrogen compounds required for growth, but it also contains cytokinins. Cytokinins apparently promote the synthesis or activation of proteins specifically required for cytokinesis.

Cytokinins have also been used against plants by pathogens. The bacterium *Agrobacterium,* for example, introduces genes into the plant genome that increase the rate of cytokinin, as well as auxin, production. This causes massive cell division and the formation of a tumor called *crown gall* (figure 40.28). How these hormone–biosynthesis genes ended up in a bacterium is an intriguing evolutionary question. Coevolution does not always work to a plant's advantage.

Strigolactones inhibit axillary bud growth

Plant form depends on the extent of axillary bud outgrowth, and for decades the extent of branching was attributed to the antagonistic actions of auxin and cytokinin. As branching mutants in garden pea (*Pisum sativum*), petunia (*Petunia hybrida*), and *Arabidopsis* were studied, a signal moving upward from the

Figure 40.26 **Cytokinins stimulate lateral bud growth.**
a. When the apical meristem of a plant is intact, auxin from the apical bud will inhibit the growth of lateral buds.
b. When the apical bud is removed, cytokinins are able to induce the growth of lateral buds into branches.
c. When the apical bud is removed and auxin is added to the cut surface, lateral bud outgrowth is suppressed.

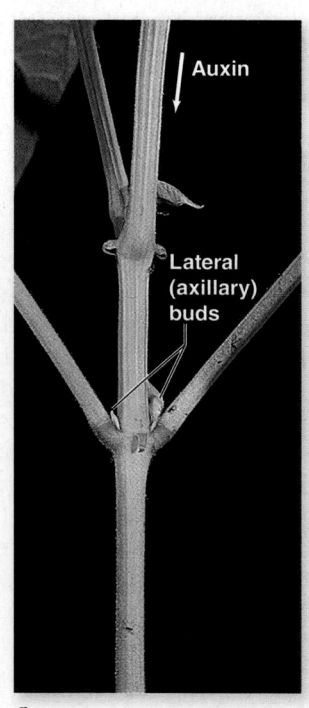

a.

b.

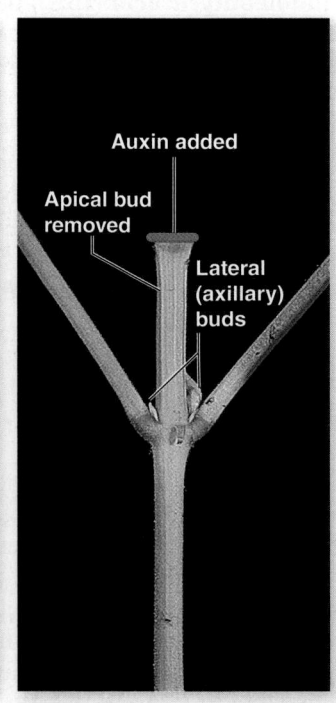

c.

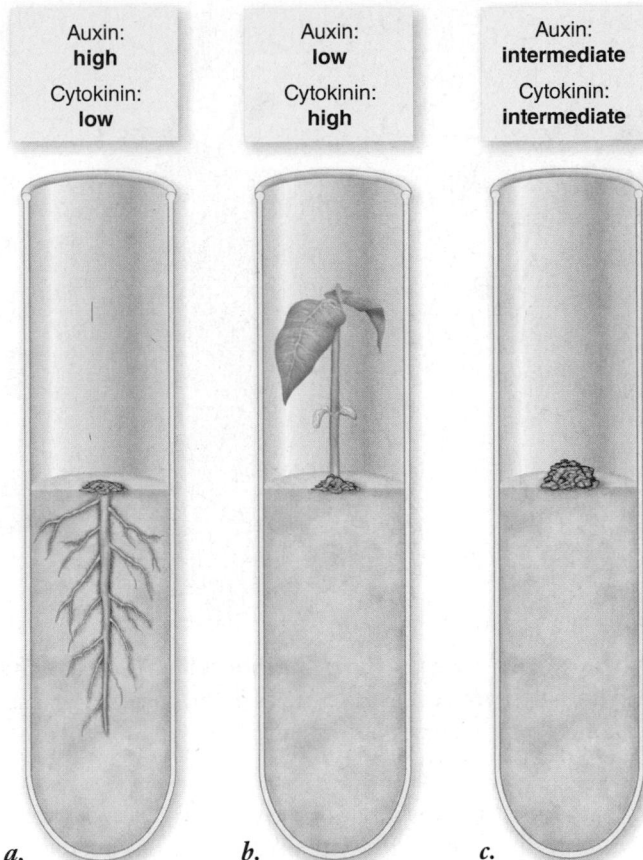

Figure 40.27 Relative amounts of cytokinins and auxin affect organ regeneration in culture. In tobacco, **a.** high auxin-to-cytokinin ratios favor root development; **b.** high cytokinin-to-auxin ratios favor shoot development; and **c.** intermediate concentrations result in the formation of undifferentiated cells. These developmental responses to cytokinin–auxin ratios in culture are species-specific.

Figure 40.28 Crown gall tumor. Sometimes cytokinins can be used against the plant by a pathogen. In this case, *Agrobacterium tumefaciens* (a bacterium) has incorporated a piece of its DNA into the plant genome. This DNA contains genes coding for enzymes necessary for cytokinin and auxin biosynthesis. The increased levels of these hormones in the plant cause massive cell division and the formation of a tumor.

roots was also found to inhibit axillary bud growth (auxin moves downward from the shoot tip). The signal was identified as a strigolactone, a new group of plant hormones. Strigolactones are derived from carotenoids, pigments that absorb blue light and play an important role in photosynthesis (figure 40.29).

Auxin can directly regulate the transcription of two strigolactone genes, *MAX3* and *MAX4*, involved in branching. Both genes contain auxin-response elements in their promoters. Treating plants with auxin results in increased expression of *MAX3* and *MAX4*.

New roles for strigolactone in plants continue to emerge. Most recently, strigolactones have been shown to regulate vascular cambium in coordination with auxin.

Gibberellins enhance plant growth and nutrient utilization

Gibberellins are named after the fungus *Gibberella fujikuroi*, which causes rice plants, on which it is parasitic, to grow abnormally tall. The Japanese plant pathologist Eiichi Kurosawa investigated bakanae ("foolish seedling") disease in the 1920s. He grew *Gibberella* in culture and obtained a substance

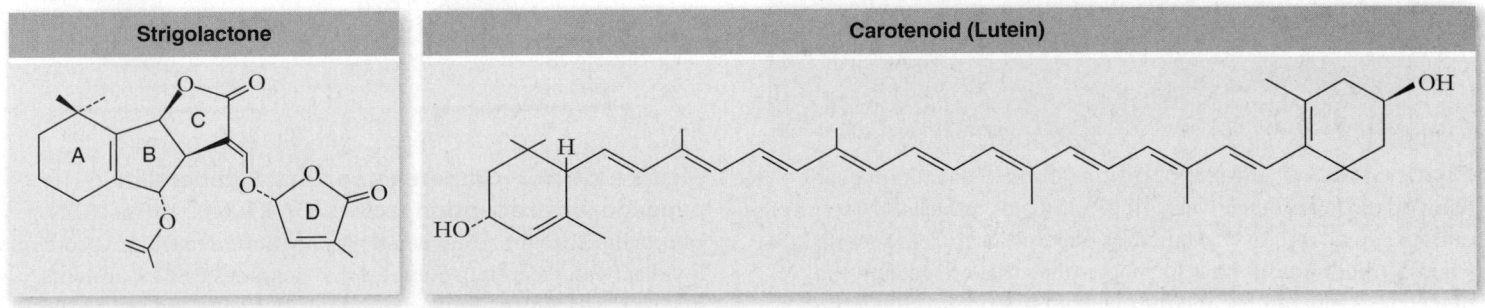

Figure 40.29 Strigolactones are plant hormones derived from carotenoids.

that, when applied to rice plants, produced bakanae. This substance was isolated and its structural formula identified by Japanese chemists in 1939. British chemists reconfirmed the formula in 1954.

Although such chemicals were first thought to be only a curiosity, they have since turned out to belong to a large class of more than 100 naturally occurring plant hormones. All are acidic and are usually abbreviated GA (for gibberellic acid), with a different subscript (GA_1, GA_2, and so forth) to distinguish each one.

Gibberellins, which are synthesized in the apical portions of stems and roots, have important effects on stem elongation. The elongation effect is enhanced if auxin is also present. The application of gibberellins to certain dwarf mutants is known to restore normal growth and development in many plants (figure 40.30). Some dwarf mutants produce insufficient amounts of gibberellin and respond to GA applications; others lack the ability to respond to gibberellin.

The large number of gibberellins are all part of a complex biosynthetic pathway that has been unraveled using gibberellin-deficient mutants in maize (corn). Although many of these gibberellins are intermediate forms in the production of GA_1, recent work shows that some forms may have specific biological roles.

Gibberellins also affect a number of other aspects of plant growth and development. In some cases, GAs hasten seed germination, apparently by substituting for the effects of cold or light requirements. Gibberellins are used commercially to increase space between grape flowers by extending internode length, so that the fruits have more room to grow. The result is a larger bunch of grapes containing larger individual fruits (figure 40.31).

Figure 40.31 Applications of gibberellins increase the space between grapes. Larger grapes (right) develop because there is more room between individual grapes.

Gibberllins elicit cell responses via a signaling pathway resulting in transcription of genes. The GA receptor has been identified. When GA binds to its receptor, it frees GA-dependent transcription factors from a repressor. These transcription factors can then directly affect gene expression (figure 40.32).

Although gibberellins function endogenously as hormones, they also function as pheromones in ferns. In ferns, gibberellin-like compounds released from one gametophyte can trigger the development of male reproductive structures on a neighboring gametophyte.

Figure 40.30 Effects of gibberellins. The mutant pea plant on the left is defective in gibberellin biosynthesis. This type of mutant can be rescued by applying gibberellins to the shoot tip (right). Other mutants have been identified that are defective in perceiving gibberellins, and they will not respond to gibberellin applications.

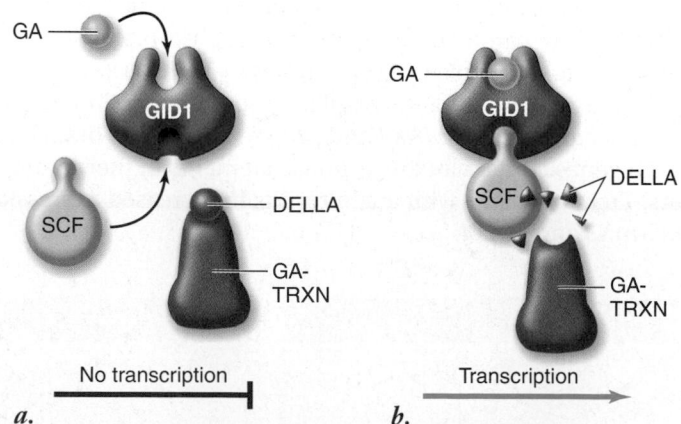

Figure 40.32 Gibberellins activate gibberellin-dependent transcription factors (GA-TRXN). *a.* GA-TRXN cannot bind to a promoter when they are bound to DELLA proteins. *b.* GA activates a protein complex that degrades DELLA proteins, freeing GA-TRXN to bind to a promoter, inducing gene transcription.

Brassinosteroids are structurally similar to animal hormones

Although plant biologists have known about *brassinosteroids* for 30 years, it is only recently that they have claimed their place as a class of plant hormones. They were first discovered in *Brassica* spp. pollen, hence the name. Their historical absence in discussions of hormones may be partially due to their functional overlap with other plant hormones, especially auxins and gibberellins. Additive effects among these three classes have been reported.

The application of molecular genetics to the study of brassinosteroids has advanced our understanding of how they are made and, to some extent, how they function in signal transduction pathways. What is particularly intriguing about brassinosteroids is their similarity to animal steroid hormones (figure 40.33). One of the genes coding for an enzyme in the brassinosteroid biosynthetic pathway has significant similarity to an enzyme used in the synthesis of testosterone and related steroids. Brassinosteroids have also been identified in algae, and they appear to be ubiquitous among the plants. It is plausible that their evolutionary origin predated the plant–animal split.

Brassinosteroids have a broad spectrum of physiological effects—elongation, cell division, bending of stems, vascular tissue development, delayed senescence, membrane polarization, and reproductive development. Environmental signals can trigger brassinosteroid actions. For example, reduced levels of blue-light increase brassinosteroid levels, resulting in elongated hypocotyls, yet another shade tolerance mechanism (figure 40.34).

Mutants have been identified that alter the response to a brassinosteroid, but signal transduction pathways remain to be uncovered. From an evolutionary perspective, it will be quite interesting to see how these pathways compare with animal steroid signal transduction pathways.

Oligosaccharins act as defense-signaling molecules

Plant cell walls are composed not only of cellulose but also of numerous complex carbohydrates called *oligosaccharides.*

Some evidence indicates that these cell-wall components (when degraded by pathogens) function as signaling molecules as well as structural wall components. Oligosaccharides that are proposed to have a hormone-like function are called *oligosaccharins.*

Oligosaccharins can be released from the cell wall by enzymes secreted by pathogens. These carbohydrates are believed to signal defense responses, such as the hypersensitive response (HR) discussed in chapter 39.

Another oligosaccharin has been shown to inhibit auxin-stimulated elongation of pea stems. These molecules are active at concentrations one to two orders of magnitude less than those of the traditional plant hormones; you have seen how auxin and cytokinin ratios can affect organogenesis in culture (see figure 40.27).

Oligosaccharins also affect the phenotype of regenerated tobacco tissue, inhibiting root formation and stimulating flower production in tissues that are competent to regenerate flowers. How the culture results translate to in vivo systems remains an open question.

Ethylene induces fruit ripening and aids plant defenses

Long before its role as a plant hormone was appreciated, the simple, gaseous hydrocarbon *ethylene* ($H_2C—CH_2$) was known to defoliate plants when it leaked from gaslights in old-fashioned streetlamps. Ethylene is, however, a natural product of plant metabolism that, in minute amounts, interacts with other plant hormones.

When auxin is transported down from the apical meristem of the stem, it stimulates the production of ethylene in the tissues around the lateral buds and thus retards their growth. Ethylene also suppresses stem and root elongation, probably in a similar way. An ethylene receptor has been identified and characterized, and it appears to have evolved early in the evolution of photosynthetic organisms, sharing features with environmental-sensing proteins identified in bacteria.

Ethylene plays a major role in fruit development. First, auxin, which is produced in significant amounts in pollinated flowers and developing fruits, stimulates ethylene production; this, in turn, hastens fruit ripening. Complex carbohydrates are broken down into simple sugars, chlorophylls are broken

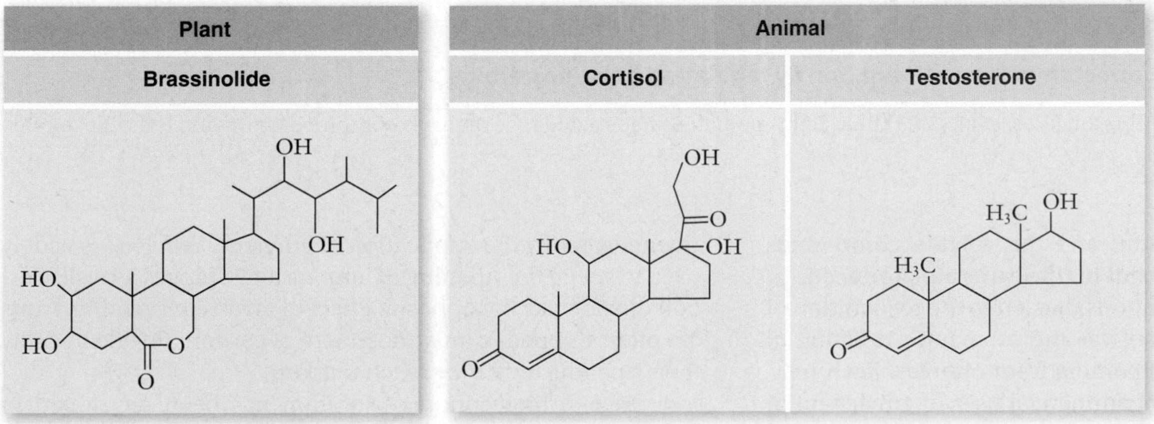

Plant	Animal	
Brassinolide	Cortisol	Testosterone

Figure 40.33 Brassinosteroids.
Brassinolide and other brassinosteroids have structural similarities to animal steroid hormones. Cortisol, testosterone, and estradiol (not shown) are animal steroid hormones.

Question: *Do brassinosteroids (BRs) control blue-light–mediated shade avoidance?*

Hypothesis: *Under low blue-light conditions, BR signaling is necessary for hypocotyl elongation.*

Prediction: *Plants with reduced levels of BR signaling will have shorter hypocotyls under low blue-light conditions than plants with typical levels of blue light.*

Experiment: *Arabidopsis mutants with impaired BR signaling (bri1-1) or biosynthesis (rot3-1), as well as wild-type (WT) plants treated with a BR inhibitor, brassinazole (Brz), are grown under full-spectrum light and light with low levels of blue. Hypocotyl length is measured 5 days after treatment.*

Result: *All three treatments reducing the BR signaling decreased the amount of hypocotyl elongation under low blue-light conditions, but had no effect on hypocotyl length under typical levels of blue light.*

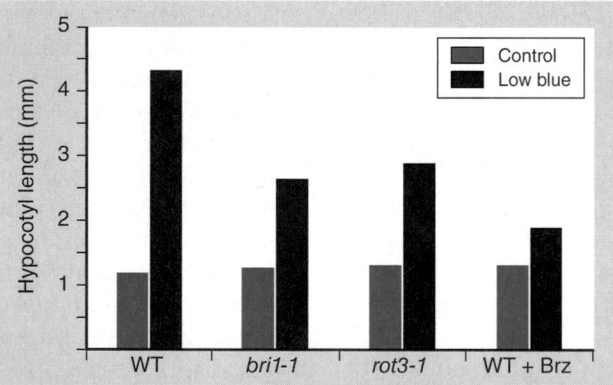

Further questions: *Compare the percent increase in hypocotyl length for WT and bri1-1. What factors might explain why inhibiting BR signaling does not fully suppress hypocotyl elongation under low blue light levels?*

Figure 40.34 Blue light affects hypocotyl elongation by regulating brassinosteroids.

Source: Keuskamp, DH, R Sasidharan, I Vos, et al. (2011) Blue-light-mediated shade avoidance requires combined auxin and brassinosteroid action in *Arabidopsis* seedlings.

down, cell walls become soft, and the volatile compounds associated with flavor and scent in ripe fruits are produced.

One of the first observations that led to the recognition of ethylene as a plant hormone was the premature ripening in bananas produced by gases coming from oranges. Such relationships have led to major commercial uses of ethylene. For example, tomatoes are often picked green and artificially ripened later by the application of ethylene. Ethylene is widely used to speed the ripening of lemons and oranges as well. Carbon dioxide has the opposite effect of arresting ripening; fruits are often shipped in an atmosphere of carbon dioxide to delay their ripening until they reach market.

Also, a biotechnology solution has been developed in which one of the genes necessary for ethylene biosynthesis has been cloned, and its antisense copy inserted into the tomato genome (figure 40.35). The antisense copy of the gene is a nucleotide sequence that is complementary to the sense copy of the gene. In this transgenic plant, both the sense and antisense sequences for the ethylene biosynthesis gene are transcribed. The sense and antisense mRNA sequences then pair with each other. This pairing blocks translation, which requires single-stranded RNA; as a result, ethylene is not synthesized, and the transgenic tomatoes do not ripen. In this way, the sturdy green tomatoes can be shipped without ripening and rotting. Exposing these tomatoes to ethylene later induces them to ripen.

Studies have shown that ethylene plays an important ecological role. Ethylene production increases rapidly when a plant is exposed to ozone and other toxic chemicals, temperature extremes, drought, attack by pathogens or herbivores, and other stresses. The increased production of ethylene that occurs can accelerate the loss of leaves or fruits that have been damaged by these stresses. Some of the damage associated with exposure to ozone is due to the ethylene produced by the plants.

The production of ethylene by plants attacked by herbivores or infected with pathogens may be a signal to activate the defense mechanisms of the plants and may include the production of molecules toxic to the pests.

Abscisic acid suppresses growth and induces dormancy

Abscisic acid appears to be synthesized mainly in mature green leaves, fruits, and root caps. The hormone earned its name because applications of it appear to stimulate fruit abscission in cotton, but there is little evidence that it plays an important role in this process. Ethylene is actually the chemical that promotes senescence and abscission.

Abscisic acid (ABA) probably induces the formation of winter buds—dormant buds that remain through the winter. The conversion of leaf primordia into bud scales follows (figure 40.36a). Like ethylene, ABA may also suppress growth of dormant lateral buds. It appears that ABA, by suppressing growth and elongation of buds, can counteract some of the effects of gibberellins; it also promotes senescence by counteracting auxin.

ABA plays a role in seed dormancy and is antagonistic to gibberellins during germination. As maize embryos develop in the kernels on the cob, increasing levels of ABA induce dormancy and prevent precocious germination, called vivipary

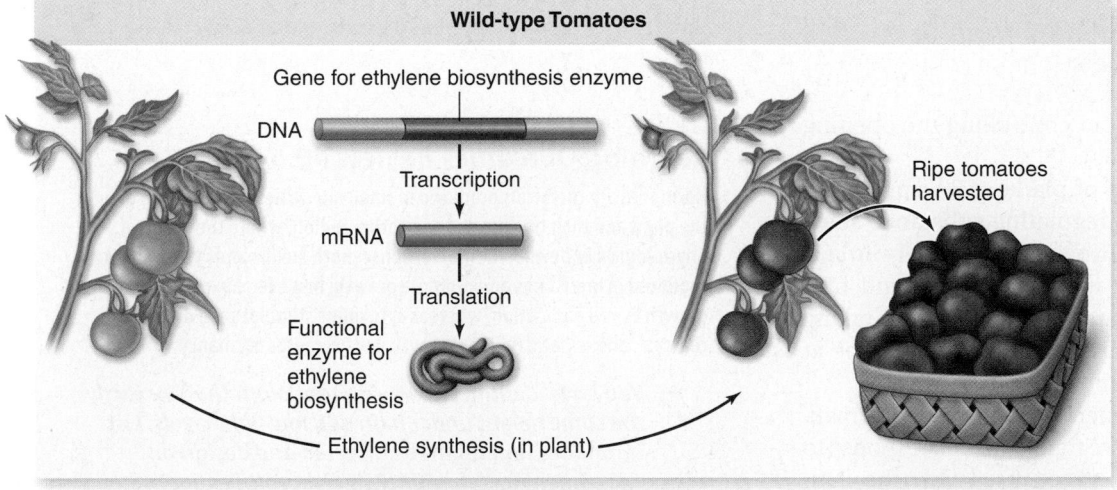

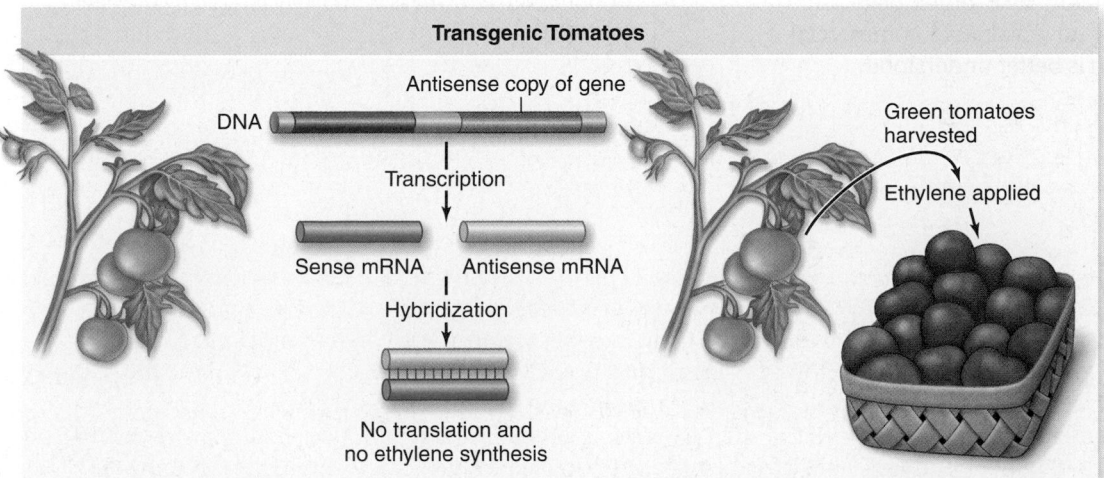

Figure 40.35 Genetic regulation of fruit ripening.
An antisense copy of the gene for ethylene biosynthesis prevents the formation of ethylene and subsequent ripening of transgenic fruit. The antisense strand is complementary to the sequence for the ethylene biosynthesis gene. After transcription, the antisense mRNA pairs with the sense mRNA, and the double-stranded mRNA cannot be translated into a functional protein. Ethylene is not produced, and the fruit does not ripen. The fruit is sturdier for shipping in its unripened form and can be ripened later with exposure to ethylene. Thus, while wild-type tomatoes may already be rotten and damaged by the time they reach stores, transgenic tomatoes stay fresh longer.

Figure 40.36 Effects of abscisic acid.
a. Abscisic acid (ABA) plays a role in the formation of these winter buds of an American basswood. These buds will remain dormant for the winter, and bud scales—modified leaves—will protect the buds from desiccation. ***b.*** In addition to bud dormancy, ABA is necessary for dormancy in seeds. This viviparous mutant in maize is deficient in ABA, and the embryos begin germinating on the developing cob. ***c.*** Abscisic acid also affects the closing of stomata by influencing the movement of potassium ions out of guard cells.

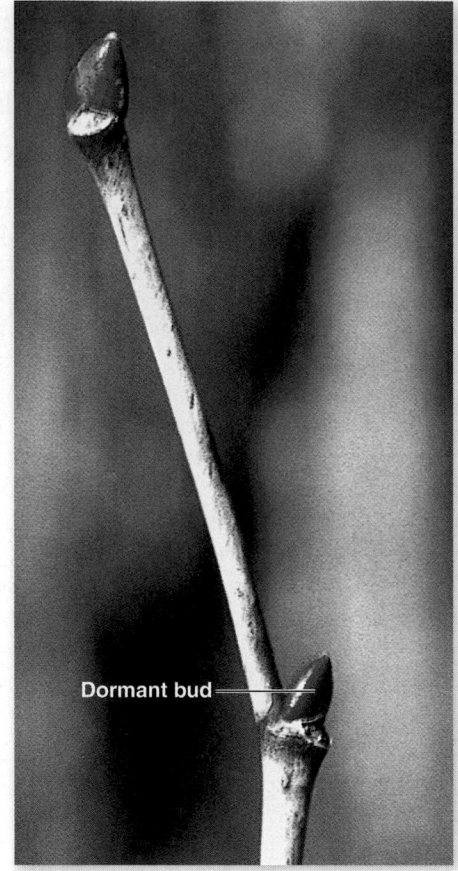

Seedling shoot

b.

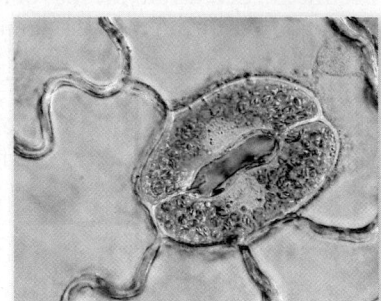

Dormant bud

a.

c.

20 μm

(figure 40.36*b*). It is also important in controlling the opening and closing of stomata (figure 40.36*c*).

Found to occur in all groups of plants, ABA apparently has been functioning as a growth-regulating substance since early in the evolution of the plant kingdom. Relatively little is known about the exact nature of its physiological and biochemical effects, but these effects are very rapid—often taking place within a minute or two—and therefore they must be at least partly independent of gene expression.

All of the genes have been sequenced in *Arabidopsis,* making it easier to identify which genes are transcribed in response to ABA. Abscisic acid levels become greatly elevated when the plant is subject to stress, especially drought. Like other plant hormones, ABA will probably prove to have valuable commercial applications when its mode of action is better understood.

Learning Outcomes Review 40.5

Hormones are chemicals produced in small quantities in one region of the plant and then transported to another region, where they cause a physiological or developmental response. Both auxins and cytokinins are produced in meristems and promote growth; however, auxins stimulate growth by cell elongation, whereas cytokinins stimulate cell division. In contrast, abscisic acid inhibits growth and promotes dormancy.

■ *You have identified a new mutant plant that is exactly the same height under high and low light levels. List hormones that might be affected and design an experiment to determine which hormone biosynthetic or signaling pathway is most likely altered by the new mutation.*

Chapter Review

40.1 Responses to Light

P_{fr} facilitates expression of light-response genes (figure 40.1).

Phytochrome exists as two interconvertible forms. The inactive form, P_r, absorbs red light and is converted to the active form, P_{fr}. P_{fr} absorbs far-red light and is converted to the inactive form, P_r. P_{fr} enters the nucleus and binds with other proteins to form a transcription complex, leading to expression of light-regulated genes. It can also activate a cascade of transcription factors.

Many growth responses are linked to phytochrome action.

P_{fr} is involved in seed germination, shoot elongation, and detection of plant crowding. Far-red light inhibits germination by inactivating P_{fr}, and red light stimulates it by activating P_r.

Crowded plants receive a greater proportion of far-red light, which is reflected from neighboring plants. The plants respond by growing taller to compete more effectively for sunlight.

Light affects directional growth.

Phototropisms are directional growth responses of stems toward blue light. Blue-light receptors such as phototropin 1 are a recent discovery.

Circadian clocks are independent of light but are entrained by light.

Circadian rhythms entrain to the daily cycle through the action of phytochrome and blue-light photoreceptors. In the absence of light, the cycle's period may become desynchronized, but it resets when light is available.

40.2 Responses to Gravity

Plants align with the gravitational field: An overview.

Gravitropism is the growth response to a gravitational field.

Certain cells in plants perceive gravity when amyloplasts are pulled downward. Following the detection of gravity, a physiological signal causes cell elongation in other cells. The hormone auxin is believed to transmit the signal.

Stems bend away from a center of gravity.

Shoots bend away from gravity, so they exhibit negative gravitropism. When auxin accumulates on the lower side of the stem, those cells elongate, causing the stem to bend upward.

Roots bend toward a center of gravity.

Roots bend toward gravity, so they exhibit positive gravitropism. If the root cap is horizontally oriented, the cells on the upper side of the root become elongated, causing the root to grow downward.

40.3 Responses to Mechanical Stimuli

Touch can trigger irreversible growth responses.

Thigmotropism is a permanent directional growth of a plant toward or away from a physical stimulus. It results in thigmomorphogenesis, a change in growth form.

Thigmonastic responses are independent of the direction of the stimulus and are usually produced by changes in turgor pressure.

Reversible responses to touch and other stimuli involve turgor pressure.

Touch-induced responses result from changes in turgor pressure. A stimulus causes an electrical signal, which results in a loss of potassium ions and water from cells of the pulvini. The loss of turgor causes the leaves to move.

Light can induce changes in turgor pressure, resulting in leaf tracking of sunlight, flower opening, and leaf sleep movements.

40.4 Responses to Water and Temperature

Dormancy is a response to water, temperature, and light.

Dormancy is the cessation of growth that occurs when a plant is exposed to environmental stress. Seasonal leaf abscission occurs in deciduous trees in the fall. Seed dormancy suspends germination until environmental conditions are optimal.

Plants can survive temperature extremes.

Plants respond to cold temperatures by increasing unsaturated lipids in membranes, limiting ice crystal formation to extracellular spaces, and producing antifreeze proteins.

When exposed to rapid increases in temperature, plants produce heat shock proteins, which help to stabilize other proteins.

40.5 Hormones and Sensory Systems

The hormones that guide growth are keyed to the environment.

Hormones are produced in small quantities in one part of a plant and then transported to another, where they bring about physiological or developmental responses.

Auxin allows elongation and organizes the body plan.

Auxins are produced in apical meristems and immature parts of a plant. They affect DNA transcription by binding to proteins. Auxins promote stem elongation, adventitious root formation, cell division, and lateral bud dormancy. They also inhibit leaf abscission and induce ethylene production.

Cytokinins stimulate cell division and differentiation.

Cytokinins are purines produced in root apical meristems and immature fruits. They promote mitosis, chloroplast development, and bud formation. Cytokinins also delay leaf aging.

Strigolactones inhibit axillary bud growth.

Strigolactones are produced in roots and move upward to regulate axillary bud growth and cambium development.

Gibberellins enhance plant growth and nutrient utilization.

Gibberellins are produced by root and shoot tips, young leaves, and seeds. They promote the elongation of stems and the production of enzymes in germinating seeds. In ferns, gibberellins function as pheromones.

Brassinosteroids are structurally similar to animal hormones.

Brassinosteroids are steroids produced in pollen, immature seeds, shoots, and leaves. They produce a broad spectrum of effects related to growth, senescence, and reproductive development.

Oligosaccharins act as defense-signaling molecules.

Pathogens secrete enzymes that release oligosaccharins from cell walls; these molecules induce pathogen defense responses. Oligosaccharins can also inhibit auxin-stimulated elongation, inhibit root formation, and stimulate flower production.

Ethylene induces fruit ripening and aids plant defenses.

Roots, shoot apical meristems, aging flowers, and ripening fruits produce ethylene, a gas that controls leaf, flower, and fruit abscission, promotes fruit ripening, and suppresses stem and root elongation. Ethylene may activate a defense response to attacks by pathogens and herbivores.

Abscisic acid suppresses growth and induces dormancy.

Mature green leaves, fruits, root caps, and seeds produce ABA. Abscisic acid inhibits bud growth and the effects of other hormones, induces seed dormancy, and controls stomatal closure.

UNDERSTAND

1. Which of the following is stimulated by blue light?

   a. Seed germination

   b. Detection of plant spacing

   c. Phototropism

   d. Shoot elongation

2. Stems and roots, respectively, exhibit

   a. a positive phototropic response and no phototropic response.

   b. a negative phototropic response and no phototropic response.

   c. no phototropic response and a positive phototropic response.

   d. no phototropic response and a negative phototropic response.

3. In stems, gravity is detected by cells of the

   a. epidermis.   c. periderm.

   b. cortex.   d. endodermis.

4. Chilling most directly affects

   a. nuclear proteins.   c. the cytoskeleton.

   b. vacuolar inclusions. d. membrane lipids.

5. Which of the following does NOT happen as a seed approaches a state of dormancy?

   a. The seed loses water.

   b. Abscisic acid levels in the embryo decrease.

   c. The seed coat hardens.

   d. Protein synthesis stops.

6. Dwarf mutants can sometimes be induced to grow normally by applying

   a. oligosaccharins   c. ethylene.

   b. abscisic acid.   d. gibberellin.

APPLY

1. If you exposed seeds to a series of red light versus far-red light treatments, which of the following exposure treatments would result in seed germination?

   a. Red; far-red

   b. Far-red; red

   c. Red; far-red; red; far-red; red; far-red; red; far-red

   d. Both a and c are correct.

2. If you were to plant a de-etiolated *(det2)* mutant *Arabidopsis* seed and keep it in a dark box, what would you expect to happen?

   a. The seed would germinate normally, but the plant would not become tall and spindly while it sought a light source.

   b. The seed would fail to germinate because it would not have light.

   c. The seed would germinate, and the plant would become tall and spindly while it sought a light source.

   d. The seed would germinate, and the plant would immediately die because it could not make sugar in the dark.

3. When Charles and Francis Darwin investigated phototropisms in plants, they discovered that

   a. auxin was responsible for light-dependent growth.

   b. light was detected at the shoot tip of a plant.

   c. light was detected below the shoot tip of a plant.

   d. only red light stimulated phototropism.

4. Which of the following defects would prevent auxin-induced gene expression?

   a. ARF transcription factor is unable to bind Aux/IAA proteins.

   b. TIR1 has a defect in its auxin binding site.

   c. Aux/IAA proteins have a defect that prevents them from binding ubiquitin, but they still can bind the ARF transcription factor.

   d. Both b and c are correct.

5. You have come up with a brilliant idea to stretch your grocery budget by buying green fruit in bulk and then storing it in a bag that you have blown up like a balloon. As you need fruit, you would take it out of the bag, and it would miraculously ripen. How would this work?

   a. The bag would block light from reaching the fruit, so it would not ripen.

   b. The bag would keep the fruit cool, so it would not ripen.

   c. The high CO_2 levels in the bag would prevent ripening.

   d. The high O_2 levels in the bag would prevent ripening.

6. You have identified a mutant *Arabidopsis* plant with a very short stem. You find that an important GA-induced gene is not being transcribed because no RNA for the gene is detected. Which of the following mutations would not result in this phenotype?

   a. A mutation in the GA transcription factor prevents DELLA proteins from binding.

   b. The enzyme needed for GA biosynthesis is defective.

   c. The GID1 protein cannot bind the SCF protein.

   d. There is a mutation in the promoter region of the GA-induced gene where the GA transcription factor would otherwise bind.

7. Which of the following might not be observed in a plant that is grown on the Space Shuttle in space?

   a. Phototropism   c. Circadian rhythms

   b. Photomorphogenesis   d. Gravitropism

SYNTHESIZE

1. If you buy a bag of potatoes and leave them in a dark cupboard for too long, they will begin to form long white sprouts with tiny leaves. Name this process and explain why the potatoes are behaving as they are.

2. Find the definition of taxis in the glossary. Compare and contrast tropism with taxis.

3. The current model for gravitropism suggests that the accumulation of amyloplasts on the bottom of a cell allows the cell to sense gravity. Suggest a plausible mechanism for the sensing of gravity that does not involve the settling out of particles.

4. Farmers who grow crops that are planted as seedlings may prepare them for their transition from the greenhouse to the field by brushing them gently every day for a few weeks. Why is this beneficial?

Chapter 41

Plant Reproduction

Chapter Contents

Introduction

The remarkable evolutionary success of flowering plants can be linked to their novel reproductive strategies. In this chapter, we explore the reproductive strategies of the angiosperms and how their unique features—flowers and fruits—have contributed to their success. This is, in part, a story of coevolution between plants and animals that ensures greater genetic diversity by dispersing plant gametes widely. Once fertilization occurs, a series of developmental events, coordinated with environmental inputs, leads to a waiting period and the germination and maturation of the next generation. In a stable environment, however, there are advantages to maintaining the status quo genetically; asexual reproduction, for example, is a strategy that produces cloned individuals. Plant life spans can vary from a few months to hundreds of years. An unusual twist to sexual reproduction in some flowering plants is that senescence (aging) and death of the parent plant immediately follow.

Reproductive Development

Learning Outcomes

1. *Describe the general life cycle of a flowering plant.*
2. *Define phase change.*
3. *Identify two* **Arabidopsis** *mutants that have been used to study phase change.*

In chapter 31, we noted that angiosperms represent an evolutionary innovation with their production of flowers and fruits. In this section, we begin with a vegetative plant and follow the developmental progression resulting in the elaborate structures associated with flowering (figure 41.1).

Plants go through developmental changes leading to reproductive maturity just as many animals do. This shift from juvenile to adult development is seen in the metamorphosis of a tadpole to an adult frog or a caterpillar to a butterfly that can then reproduce. Plants undergo a similar metamorphosis that leads to the production of a flower. Unlike the juvenile frog, which loses its tail, plants just keep adding structures to existing ones with their meristems.

Carefully regulated processes determine when and where flowers will form. Moreover, plants must often gain competence to respond to internal or external signals regulating flowering. Once plants are competent to reproduce, a

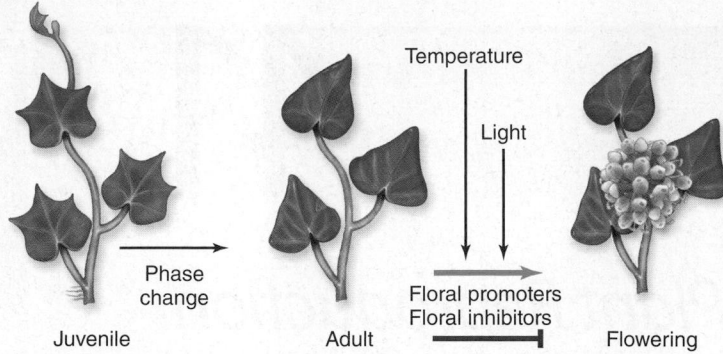

Figure 41.2 **Factors involved in initiating flowering.** This model depicts the environmentally cued and internally processed events that result in a shoot meristem initiating flowers. During phase change, the plant acquires competence to respond to flowering signals.

combination of factors—including light, temperature, and both promotive and inhibitory internal signals—determines when a flower is produced (figure 41.2). These signals turn on genes that specify formation of the floral organs—sepals, petals, stamens, and carpels. Once cells have instructions to become a specific floral organ, yet another developmental cascade leads to the three-dimensional construction of flower parts. We describe details of this process in the following sections.

The transition to flowering competence is termed phase change

At germination, most plants are incapable of producing a flower, even if all the environmental cues are optimal. Internal developmental changes allow plants to obtain competence to respond to external or internal signals (or both) that trigger flower formation. This transition is referred to as **phase change.**

Phase change can be morphologically obvious or very subtle. Take a look at an oak tree in the winter: Leaves will still be clinging to the lower branches until spring when the new buds push them off, but leaves on the upper branches will have fallen earlier (figure 41.3*a*). Those lower branches were initiated by a juvenile meristem. The fact that they did not respond to environmental cues and drop their leaves indicates that they are juvenile branches and have not made a phase change. Although the lower branches are older, their juvenile state was established when they were initiated and will not change.

Ivy also has distinct juvenile and adult phases of growth (figure 41.3*b*). Stem tissue produced by a juvenile meristem initiates adventitious roots that can cling to walls. If you look at very old brick buildings covered with ivy, you will notice that the uppermost branches are falling off because they have transitioned to the adult phase of growth and have lost the ability to produce adventitious roots.

It is important to note that even though a plant has reached the adult stage of development, it may or may not produce reproductive structures. Other factors may be necessary to trigger flowering.

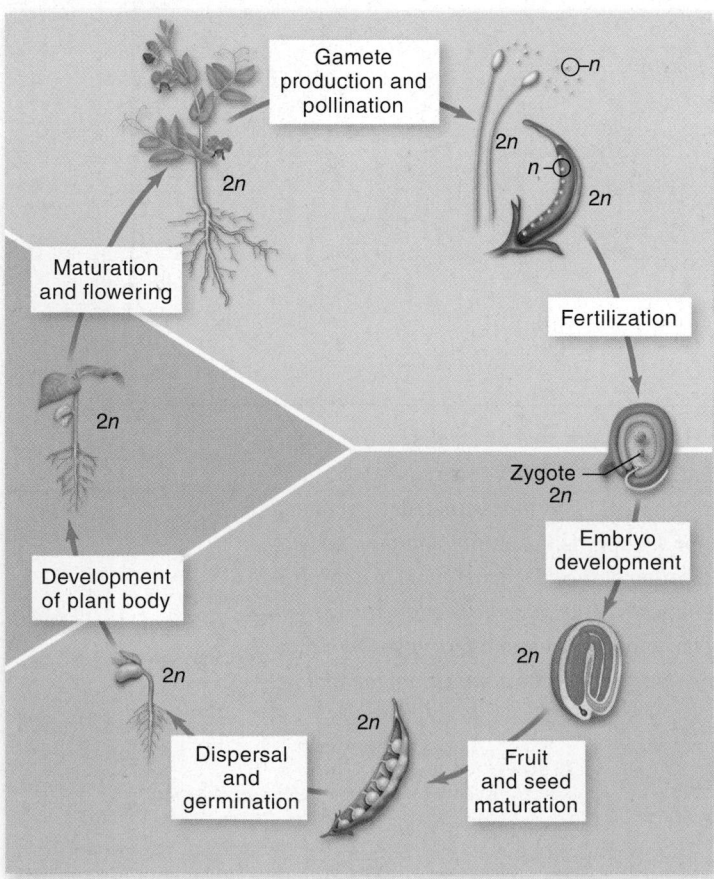

Figure 41.1 **Life cycle of a flowering plant (angiosperm).**

a.

b.

Figure 41.3 Phase change. *a.* The lower branches of this oak tree represent the juvenile phase of development; they cling to their leaves in the winter. The lower leaves are not able to form an abscission layer and break off the tree in the fall. Such visible changes are marks of phase change, but the real test is whether the plant is able to flower. *b.* Juvenile ivy (right) makes adventitious roots and has an alternating leaf phyllotaxy. Mature ivy (left) lacks adventitious roots, has spiral phyllotaxy, and can make flowers.

Mutations have clarified how phase change is controlled

Generally it is easier to get a plant to revert from an adult to juvenile state than to induce phase change experimentally. Applications of the plant hormone gibberellin and severe pruning can cause reversion. In the latter case, new vegetative growth occurs, as when certain shrubs (such as rhododendrons) are cut back and put out lush new growth in response.

The *embryonic flower (emf)* mutant of *Arabidopsis* flowers almost immediately (figure 41.4), which is consistent with the hypothesis that the wild-type allele suppresses flowering. As the wild-type plant matures, *EMF* expression decreases. This finding suggests that flowering is the default state, and that mechanisms have evolved to delay flowering. This delay presumably allows the plant to store more energy to be allocated for reproduction.

An example of inducing the juvenile-to-adult transition comes from overexpressing a gene necessary for flowering that is found in many species. This gene, *LEAFY (LFY)*, was cloned in *Arabidopsis,* and its promoter was replaced with a viral promoter that results in constant, high levels of *LFY* transcription. *LFY* with its viral promoter was then introduced into cultured aspen cells that were used to regenerate plants. When *LFY* is overexpressed in aspen, flowering occurs in weeks instead of years (figure 41.5).

Phase change requires both a sufficiently strong promotive signal and the ability to perceive the signal. Phase change can result in the production of receptors in the shoot to perceive a signal of a certain intensity. Alternatively an increase of promotive signal(s) or a decrease of inhibitory signal(s) can trigger phase change.

Phase change, as we said earlier, results in an adult plant, but not necessarily a flowering plant. The ability to reproduce is distinct from actual reproductive development. Flower production depends on a number of factors, which we explore next.

Learning Outcomes Review 41.1

In a flowering plant life cycle, fertilization produces an embryo in a seed. The embryo develops into a plant that eventually flowers, and the flowers once again produce gametes. Phase change is the transition from vegetative to reproductive growth. In *Arabidopsis,* expression of the embryonic flower mutant *(emf)* or overexpression of the *LEAFY* gene *(LFY)* result in early flowering.

■ *In evolutionary terms, why is flower production the default state in plants?*

Figure 41.4 Embryonic flower (EMF) gene prevents early flowering. Mutant plants that lack EMF protein flower as soon as they germinate. The flowers have malformed carpels and other defective floral structures close to the roots.

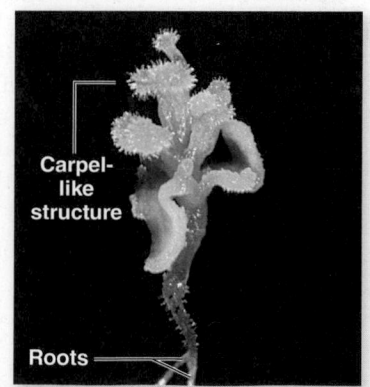

Carpel-like structure

Roots

Normal Flowering

a.

Accelerated Flowering

b.

Figure 41.5 Overexpression of a flowering gene can accelerate phase change. *a.* Normally, an aspen tree grows for several years before producing flowers (see inset). *b.* Overexpression of the *Arabidopsis* flowering gene, *LFY,* causes rapid flowering in a transgenic aspen (see inset).

Making Flowers

Four genetically regulated pathways to flowering have been identified: (1) the light-dependent pathway, (2) the temperature-dependent pathway, (3) the gibberellin-dependent pathway, and (4) the autonomous pathway. Plants can rely primarily on one pathway, but all four pathways can be present.

The environment can promote or repress flowering, and in some cases, it can be relatively neutral. For example, increasing light duration can be a signal that long summer days have arrived in a temperate climate and that conditions are favorable for reproduction. In other cases, plants depend on light to accumulate sufficient amounts of sucrose to fuel reproduction, but flower independently of day length.

Temperature can also be used as a signal. **Vernalization,** the requirement for a period of chilling of seeds or shoots for flowering, affects the temperature-dependent pathway. Assuming that regulation of reproduction first arose in more constant tropical environments, many of the day-length and temperature controls would have evolved as plants colonized more temperate climates. Plants with a vernalization requirement flower after, not during, a cold winter, enhancing reproductive success. The existence of redundant pathways to flowering helps ensure new generations.

The light-dependent pathway is geared to the photoperiod

Flowering requires much energy accumulated via photosynthesis. Thus, all plants require light for flowering, but this is distinct from the **photoperiodic,** or light-dependent, flowering pathway. Aspects of growth and development in most plants are keyed to changes in the proportion of light to dark in the daily 24-hr cycle (day length).

Sensitivity to the photoperiod provides a mechanism for organisms to respond to seasonal changes in the relative length of day and night. Day length changes with the seasons; the farther a region is from the equator, the greater the variation in day length.

Short-day and long-day plants

The flowering responses of plants to day length fall into several basic categories. In short-day plants, flowering is initiated when daylight becomes shorter than a critical length (figure 41.6). In **long-day plants,** flowering begins when daylight becomes longer. Other plants, such as snapdragons, roses,

Figure 41.6 How day length affects flowering. *a.* Clover (center panels) is a long-day plant that is stimulated by short nights to flower in the spring. Cocklebur (right-hand panels) is a short-day plant that, throughout its natural distribution in the northern hemisphere, is stimulated by long nights to flower in the fall. *b.* If the long night of late fall is artificially interrupted by a flash of light, the cocklebur will not flower, and the clover will. Although the terms *long-day* and *short-day* refer to the length of day, in each case, it is the duration of uninterrupted darkness that determines when flowering will occur.

and many native to the tropics, flower when mature, regardless of day length, as long as they have received enough light for normal growth. These are referred to as day-neutral plants. Still other plants, including ivy, have two critical photoperiods; they will not flower if the days are too long, and they also will not flower if the days are too short.

Although plants are referred to as long-day or short-day plants, it is actually the amount of darkness that determines whether a plant flowers. In obligate long- or short-day species, there is a sharp distinction between short and long nights, respectively. Flowering occurs in obligate long-day plants when the night length is less than the maximal amount of required darkness (critical night length) for that species. For obligate short-day plants, the amount of darkness must exceed the critical night length for the species.

In other long- or short-day plants, flowering occurs more rapidly or slowly depending on the length of day. These plants, which rely on other flowering pathways as well, are called **facultative long- or short-day** plants because the photoperiodic requirement is not absolute. The garden pea is an example of a facultative long-day plant.

Advantages of photoperiodic control of flowering

Using light as a cue permits plants to flower when abiotic environmental conditions are optimal, pollinators are available, and competition for resources with other plants may be less. For example, the spring herbaceous plants termed *ephemerals* flower in the woods before the tree canopy leafs out and blocks the sunlight necessary for photosynthesis. An example is the trailing arbutus *(Epigaea repens)* of the Northeast woods, which is also known as mayflower because of the time of year in which it blooms.

At middle latitudes, most long-day plants flower in the spring and early summer; examples of such plants include clover, irises, lettuce, spinach, and hollyhocks. Short-day plants usually flower in late summer and fall; these include chrysanthemums, goldenrods, poinsettias, soybeans, and many weeds, such as ragweed. Commercial plant growers use these responses to day length to bring plants into flower at specific times. For example, photoperiod is manipulated in greenhouses so that poinsettias flower just in time for the winter holidays (figure 41.7). The geographic distribution of certain plants may be determined by their flowering responses to day length.

The mechanics of light signaling

Photoperiod is perceived by several different forms of phytochrome and also by a blue-light–sensitive molecule (cryptochrome). Another type of blue-light–sensitive molecule (phototropin) was discussed in chapter 40. Phototropin affects photomorphogenesis, and cryptochrome affects photoperiodic responses.

The conformational change in a phytochrome or cryptochrome light-receptor molecule triggers a cascade of events that leads to the production of a flower. There is a link between light and the circadian rhythm regulated by an internal clock that facilitates or inhibits flowering. At a molecular level, the

Figure 41.7 Flowering time can be altered. Manipulation of photoperiod in greenhouses ensures that short-day poinsettias flower in time for the winter holidays. Even after flowering is induced, many developmental events must occur in order to produce species-specific flowers.

gaps in information about how light signaling and flower production are related are rapidly being filled in, and the control mechanisms have been found to be quite complex.

Photoperiodic Regulation of Transcription of the *CO* Gene. *Arabidopsis*, which as you know is commonly used in plant studies, is a facultative long-day plant that flowers in response to both far-red and blue light. Phytochrome and cryptochrome, the red- and blue-light receptors, respectively, regulate flowering via the gene *CONSTANS (CO)*. Precise levels of CO protein are maintained in accordance with the circadian clock, and phytochrome regulates the transcription of *CO*. Levels of *CO* mRNA are low at night and increase at daybreak. In addition, CO protein levels are modulated through the action of cryptochrome. *CO* is an important protein because it links the perception of day length with the production of a signal that moves from the leaves to the shoot where a change in gene transcription leads to the production of flowers.

? Inquiry question If levels of *CO* mRNA follow a circadian pattern, how could you determine whether protein levels are modulated by a mechanism other than transcription? Why would an additional level of control even be necessary?

The importance of posttranslational regulation of *CO* activity became apparent through studies of transgenic *Arabidopsis* plants. These plants contain a *CO* gene fused to a viral promoter that is always on and produces high levels of *CO* mRNA regardless of whether it is day or night. The regulation of *CO* gene expression by phytochrome A is therefore eliminated when this viral promoter is fused to the gene. Curiously, CO protein levels still follow a circadian pattern.

Although CO protein is produced day and night, levels of CO are lower at night because of targeted protein degradation. Ubiquitin tags the CO protein, and it is degraded by the proteasome as was described in chapter 40 for phytochrome degradation. Blue light acting via cryptochrome stabilizes CO during the day and protects it from ubiquitination and subsequent degradation.

CO and *LFY* Expression. CO is a transcription factor that turns on other genes, which results in the expression of *LFY*. As discussed in connection with phase change earlier in this section, *LFY* is one of the key genes that "tells" a meristem to switch over to flowering. We will see that other pathways also converge on this important gene. Genes that are regulated by *LFY* are discussed later in this chapter.

Florigen—The elusive flowering hormone

Long before any genes regulating flowering were cloned, a flowering hormone called florigen was postulated to trigger flower production. A considerable amount of evidence demonstrates the existence of substances that promote flowering and substances that inhibit it. Grafting experiments (binding the shoot system of one plant to the root system of another) have shown that these substances can move from leaves to shoots. The complexity of their interactions, as well as the fact that multiple chemical messengers are evidently involved, has made this scientifically and commercially interesting search very difficult.

Recent evidence indicates that the flowering locus T (FT) protein is the long elusive florigen. CO regulates the amount of FT in the leaf. FT then moves through the phloem to the shoot meristem and initiates additional gene expression, including *LFY*, leading to flower development.

The temperature-dependent pathway is linked to cold

Cold temperatures can accelerate or permit flowering in many species. As with light, this environmental connection ensures that plants flower at more optimal times.

Some plants require a period of chilling before flowering, called *vernalization*. This phenomenon was discovered in the 1930s by the Ukrainian scientist T. D. Lysenko while trying to solve the problem of winter wheat rotting in the fields. Because winter wheat would not flower without a period of chilling, Lysenko chilled the seeds and then planted them in the spring. The seeds successfully sprouted, grew, and produced grain.

Although this discovery was scientifically significant, Lysenko erroneously concluded that he had converted one species, winter wheat, to another, spring wheat, by simply altering the environment. Lysenko's point of view was supported by the communist philosophy of the time, which held that people could easily manipulate nature to increase production. Unfortunately, this philosophy led to a great many problems, including mistreatment of legitimate geneticists in the former Soviet Union. In addition, genetics and Darwinian evolution were suspect in the Soviet Union until the mid-1960s.

Vernalization is necessary for some seeds or plants in later stages of development. Analysis of mutants in *Arabidopsis* and pea plants indicate that vernalization is a separate flowering pathway.

The gibberellin-dependent pathway requires an increase in hormone level

In *Arabidopsis* and some other species, decreased levels of gibberellins delay flowering. Thus the gibberellin pathway is proposed to promote flowering. It is known that gibberellins enhance the expression of *LFY* in shoot meristems. Gibberellin actually binds the promoter of the *LFY* gene, so its effect on flowering is direct.

The autonomous pathway is independent of environmental cues

The autonomous pathway to flowering does not depend on external cues except for basic nutrition. Presumably, this delays flowering. A balance between floral promoting and inhibiting signals may regulate when flowering occurs in the autonomous pathway and the other pathways as well.

Determination for flowering is tested at the organ or whole-plant level by changing the environment and ascertaining whether the developmental fate has changed. In *Arabidopsis,* floral determination correlates with the increase of *LFY* gene expression, and it has already occurred by the time a second flowering gene, *APETALA1 (AP1),* is expressed. Because all four flowering pathways appear to converge with increased levels of *LFY,* this determination event should occur in species with a variety of balances among the pathways (figure 41.8).

 Inquiry question Why would it be advantageous for a plant to have four distinct pathways that all affect the expression of *LFY*?

Floral meristem identity genes activate floral organ identity genes

Arabidopsis and snapdragons are valuable model systems for identifying flowering genes and understanding their interactions. The four flowering pathways discussed earlier in this section lead to an adult meristem becoming a floral meristem

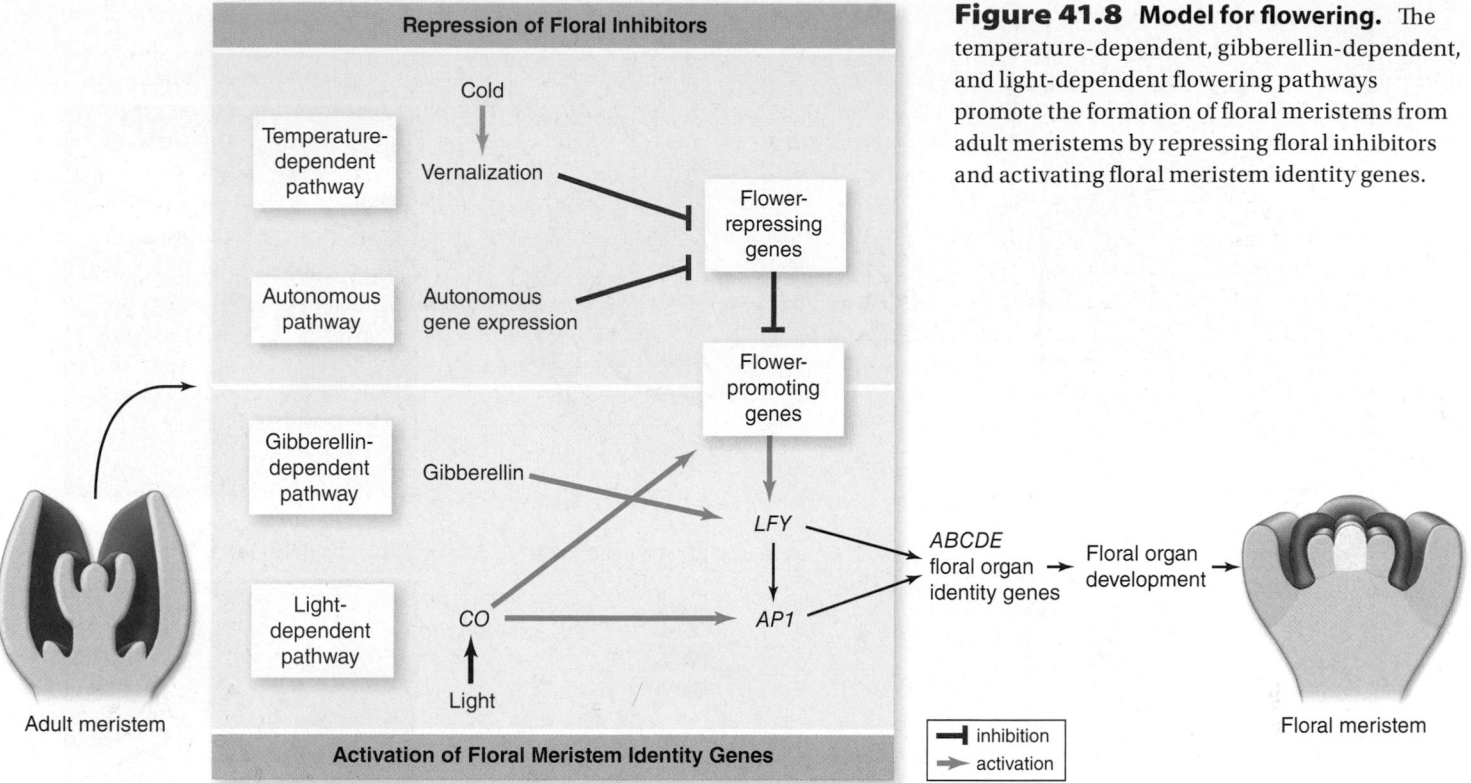

Figure 41.8 Model for flowering. The temperature-dependent, gibberellin-dependent, and light-dependent flowering pathways promote the formation of floral meristems from adult meristems by repressing floral inhibitors and activating floral meristem identity genes.

by either activating or repressing the inhibition of floral meristem identity genes (see figure 41.8). Two of the key floral meristem identity genes are *LFY* and *AP1*. These genes establish the meristem as a flower meristem. They then turn on floral organ identity genes. The floral organ identity genes define four concentric whorls, moving inward in the floral meristem, as sepal, petal, stamen, and carpel.

The ABC model

To explain how three classes of floral organ identity genes could specify four distinct organ types, the ABC model was developed (figure 41.9). The ABC model proposes that three classes of organ identity genes (*A, B,* and *C*) specify the floral organs in the four floral whorls. By studying mutants, the researchers have determined the following:

1. Class *A* genes alone specify the sepals.
2. Class *A* and class *B* genes together specify the petals.
3. Class *B* and class *C* genes together specify the stamens.
4. Class *C* genes alone specify the carpels.

The beauty of the ABC model is that it is entirely testable by making different combinations of floral organ identity mutants. Each class of genes is expressed in two whorls, yielding four different combinations of the gene products. When any one class is missing, aberrant floral organs occur in predictable positions.

Modifications to the ABC model

As compelling as the ABC model is, it cannot fully explain specification of floral meristem identity. Class *D* genes that are essential for carpel formation have been identified, but even this discovery did not explain why a plant lacking *A, B,* and *C* gene function produced four whorls of sepals rather than four whorls of leaves. Floral parts are thought to have evolved from leaves; therefore, if the floral organ identity genes are removed, whorls of leaves, rather than sepals, would be predicted.

The answer to this puzzle is found in the *E* genes, *SEPALATA1 (SEP1)* through *SEPALATA4 (SEP4)*. The triple mutant *sep1 sep2 sep3* and the *sep4* mutant both produce four whorls of leaves. The proteins encoded by the *SEP* genes can interact with class A, B, and C proteins and possibly affect transcription of genes needed for the development of floral organs. Identification of the *SEP* genes led to a new floral organ identity model that includes these class *E* genes (figure 41.10).

> **? Inquiry question** Would you expect plants to flower at a different time if there were no flower-repressing genes and the vernalization and autonomous pathway genes induced expression of flower-promoting genes?

It is important to recognize that the *ABCDE* genes are actually only the beginning of the making of a flower. These organ identity genes are transcription factors that turn on

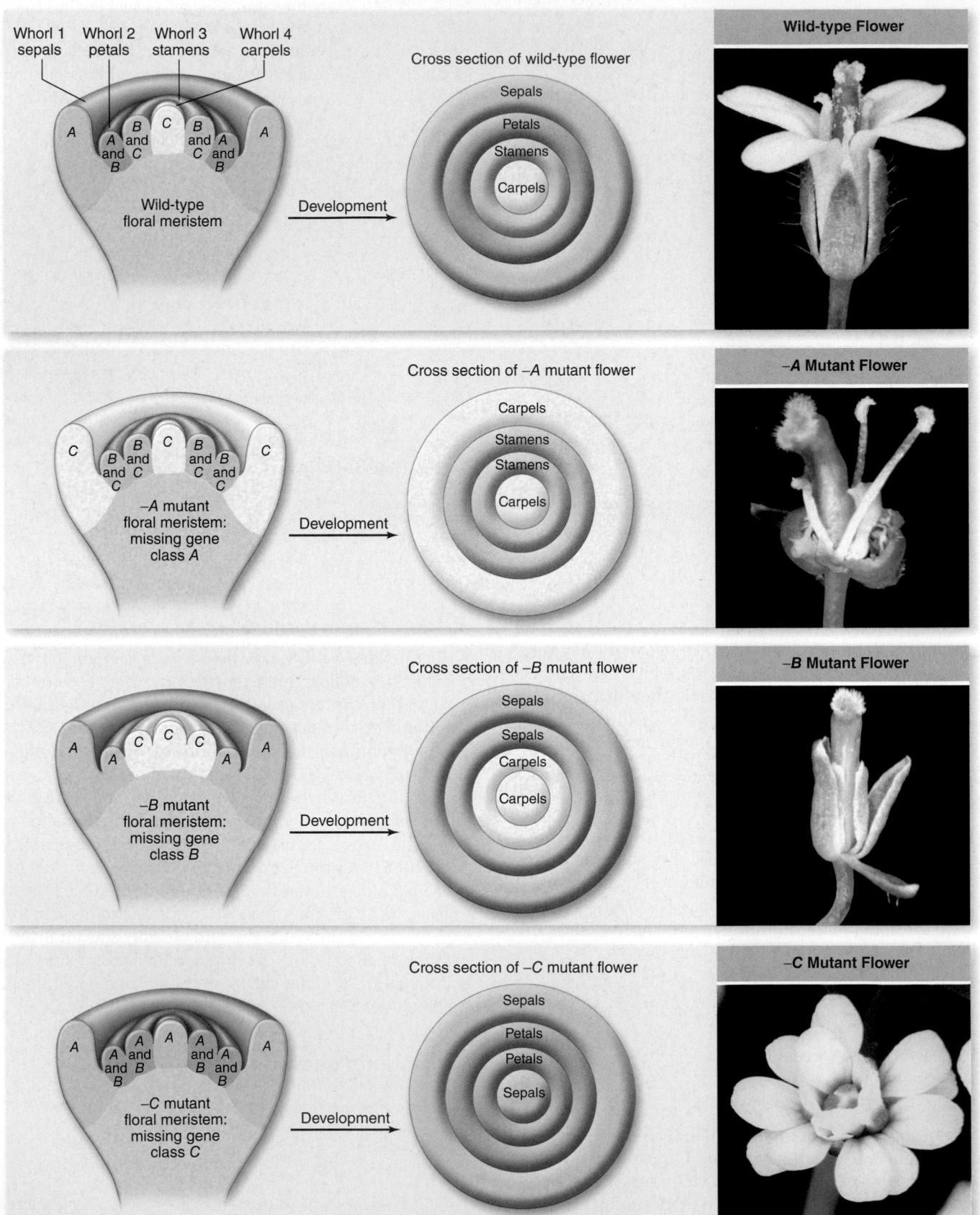

Figure 41.9 ABC model for floral organ specification. Letters labeling whorls indicate which gene classes are active. When *A* function is lost *(−A)*, *C* expands to the first and second whorls. When *B* function is lost *(−B)*, the outer two whorls have just *A* function, and both inner two whorls have just *C* function; none of the whorls have dual gene function. When *C* function is lost *(−C)*, *A* expands into the inner two whorls. These new combinations of gene expression patterns alter which floral structures form in each whorl.

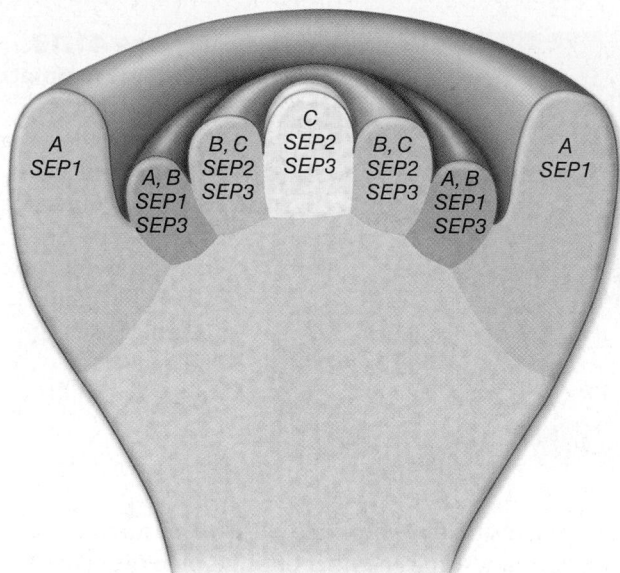

Figure 41.10 Class *E* genes are needed to specify floral organ identity. When all three *SEP* genes are mutated, four whorls of leaves are produced.

many more genes that actually give rise to the three-dimensional flower. Other genes "paint" the petals—that is, complex biochemical pathways lead to the accumulation of anthocyanin pigments in petal cell vacuoles. These pigments can be orange, red, or purple, and the actual color is influenced by pH as well.

41.3 Structure and Evolution of Flowers

Learning Outcomes

1. List the parts of a typical angiosperm flower.
2. Distinguish between bilateral symmetry and radial symmetry.
3. Differentiate between microgametophytes and megagametophytes.

The complex and elegant process that gives rise to the reproductive structure called the flower is often compared with metamorphosis in animals. It is indeed a metamorphosis, but the subtle shift from mitosis to meiosis in the megaspore mother cell that leads to the development of a haploid, gamete-producing gametophyte is perhaps even more critical. The same can be said for pollen formation in the anther of the stamen.

The flower not only houses the haploid generations that will produce gametes, but it also functions to increase the probability that male and female gametes from different (or sometimes the same) plants will unite.

Flowers evolved in the angiosperms

The evolution of the angiosperms is a focus of chapter 31. The diversity of angiosperms is partly due to the evolution of a great variety of floral phenotypes that may enhance the effectiveness of pollination. As mentioned previously, floral organs are thought to have evolved from leaves. In some early angiosperms, these organs maintain the spiral developmental pattern often found in leaves. The trend has been toward four distinct whorls of parts. A complete flower has four whorls (calyx, corolla, androecium, and gynoecium) (figure 41.11). An incomplete flower lacks one or more of the whorls.

Flower morphology

In both complete and incomplete flowers, the **calyx** usually constitutes the outermost whorl; it consists of flattened appendages, called sepals, which protect the flower in the bud. The petals collectively make up the **corolla** and may be fused. Many petals function to attract pollinators. Although these two outer whorls of floral organs are not involved directly in gamete production or fertilization, they can enhance reproductive success.

Male structures. Androecium is a collective term for all the **stamens** (male structures) of a flower. Stamens are specialized structures that bear the angiosperm microsporangia. Similar structures bear the microsporangia in the pollen cones of

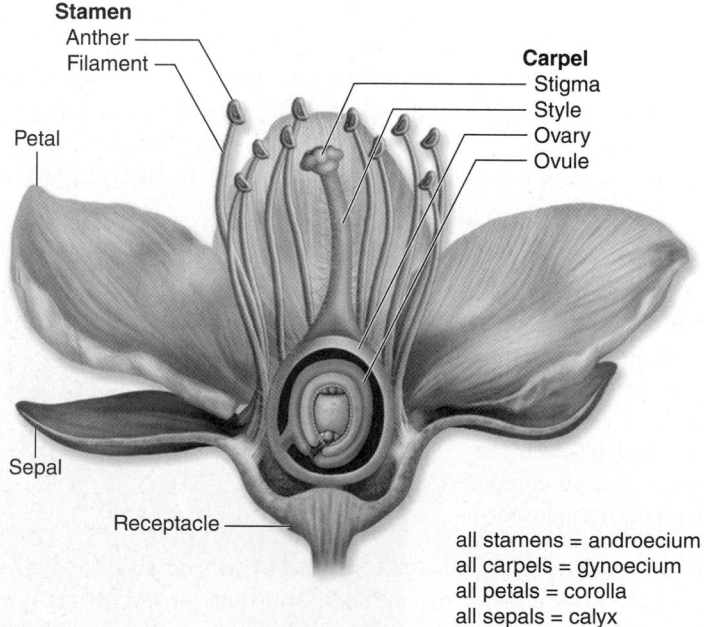

all stamens = androecium
all carpels = gynoecium
all petals = corolla
all sepals = calyx

Figure 41.11 A complete angiosperm flower.

gymnosperms. Most living angiosperms have stamens with filaments ("stalks") that are slender and often threadlike; four microsporangia are evident at the apex in a swollen portion, the **anther.** Some of the more primitive angiosperms have stamens that are flattened and leaflike, with the sporangia produced from the upper or lower surface.

Female structures. The gynoecium is a collective term for all the female parts of a flower. In most flowers, the gynoecium, which is unique to angiosperms, consists of a single carpel or two or more fused carpels. Single or fused carpels are often referred to as simple or compound pistils, respectively. Most flowers with which we are familiar—for example, those of tomatoes and oranges—have a compound pistil. Other, less specialized flowers—for example, buttercups and stonecups—may have several to many separate, simple pistils, each formed from a single carpel.

Ovules (which develop into seeds) are produced in the pistil's swollen lower portion, the **ovary,** which usually narrows at the top into a slender, necklike style with a pollen-receptive stigma at its apex. Sometimes the stigma is divided, with the number of stigma branches indicating how many carpels compose the particular pistil.

Carpels are essentially rolled floral leaves with ovules along the margins. It is possible that the first carpels were leaf blades that folded longitudinally; the leaf margins, which had hairs, did not actually fuse until the fruit developed, but the hairs interlocked and were receptive to pollen. In the course of evolution, evidence indicates that the hairs became localized into a stigma; a style was formed; and the fusing of the carpel margins ultimately resulted in a pistil. In many modern flowering plants, the carpels have become highly modified and are not visually distinguishable from one another unless the pistil is cut open.

Trends of floral specialization

Two major evolutionary trends led to the wide diversity of modern flowering plants: (1) Separate floral parts have grouped together, or fused, and (2) floral parts have been lost or reduced (figure 41.12).

In the more advanced angiosperms, the number of parts in each whorl has often been reduced from many to few. The spiral patterns of attachment of all floral parts in primitive angiosperms have, in the course of evolution, given way to a single whorl at each level. The central axis of many flowers

Figure 41.13
Bilateral symmetry in an orchid. Although more basal flowers are usually radially symmetrical, flowers of many derived groups, such as the orchid family (Orchidaceae), are bilaterally symmetrical.

has shortened, and the whorls are close to one another. In some evolutionary lines, the members of one or more whorls have fused with one another, sometimes joining into a tube. In other kinds of flowering plants, different whorls may be fused together.

Whole whorls may even be lost from the flower, which may lack sepals, petals, stamens, carpels, or various combinations of these structures. Modifications often relate to pollination mechanisms, and in plants such as the grasses, wind has replaced animals for pollen dispersal.

Trends in floral symmetry

Other trends in floral evolution have affected the symmetry of the flower. Primitive flowers such as those of buttercups are radially symmetrical; that is, one could draw a line anywhere through the center and have two roughly equal halves. Flowers of many advanced groups are bilaterally symmetrical; they are divisible into two equal parts along only a single plane. Examples of such flowers are snapdragons, mints, and orchids (figure 41.13). Bilaterally symmetrical flowers are also common among violets and peas. In these groups, they are often associated with advanced and highly precise pollination systems.

Bilateral symmetry has arisen independently many times. In snapdragons, the *CYCLOIDEA* gene regulates floral symmetry, and in its absence flowers are more radial (figure 41.14). Here the experimental alteration of a single gene

Figure 41.12
Trends in floral specialization. Wild geranium, *Geranium maculatum,* a typical eudicot. The petals are reduced to 5 each, the stamens to 10, compared with early angiosperms.

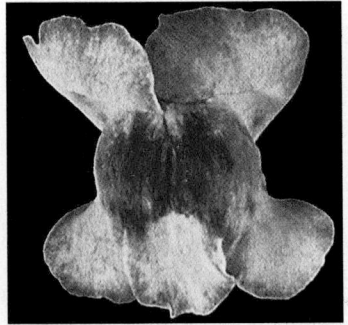 

a.   *b.*

Figure 41.14 Genetic regulation of asymmetry in flowers. *a.* Snapdragon flowers (*Antirrhinum* sp.) normally have bilateral symmetry. *b.* The *CYCLOIDEA* gene regulates floral symmetry, and *cycloidea* mutant snapdragons have radially symmetrical flowers.

is sufficient to cause a dramatic change in morphology. Whether the same gene or functionally similar genes arose naturally in parallel in other species is an open question.

The human influence on flower morphology

Although much floral diversity is the result of natural selection related to pollination, it is important to recognize the effect breeding (artificial selection) has on flower morphology. Humans have selected for practical or aesthetic traits that may have little adaptive value to species in the wild. For example, maize (corn) has been bred to satisfy the human palate. Human intervention ensures the reproductive success of each generation; however, in a natural setting, modern corn would not have the same protection from herbivores as its ancestors, and the fruit dispersal mechanism would be quite different.

Gametes are produced in the gametophytes of flowers

Reproductive success depends on uniting the gametes (egg and sperm) found in the embryo sacs and pollen grains of flowers. As you learned in chapters 30 and 31, plant sexual life cycles are characterized by an *alternation of generations,* in which a diploid sporophyte generation gives rise to a haploid gametophyte generation. In angiosperms, the gametophyte generation is very small and is completely enclosed within the tissues of the parent sporophyte. The male gametophytes, or microgametophytes, are **pollen grains.** The female gametophyte, or megagametophyte, is the **embryo sac.** Pollen grains and the embryo sac both are produced in separate, specialized structures of the angiosperm flower.

Like animals, angiosperms have separate structures for producing male and female gametes (figure 41.15), but the reproductive organs of angiosperms are different from those of animals in two ways. First, both male and female structures usually occur together in the same individual flower. Second, angiosperm reproductive structures are not permanent parts of the adult individual. Angiosperm flowers and reproductive organs develop seasonally, at times of the year most favorable for pollination. In some cases, reproductive structures are produced only once, and the parent plant dies. And, as you learned earlier in this chapter, the germ line in angiosperms is not set aside early on, but forms quite late during phase change.

Pollen formation

Anthers contain four microsporangia, which produce microspore mother cells *(2n).* Microspore mother cells produce microspores *(n)* through meiotic cell division. The microspores, through mitosis and wall differentiation, become pollen. Inside each pollen grain is a generative cell; this cell later divides to produce two sperm cells.

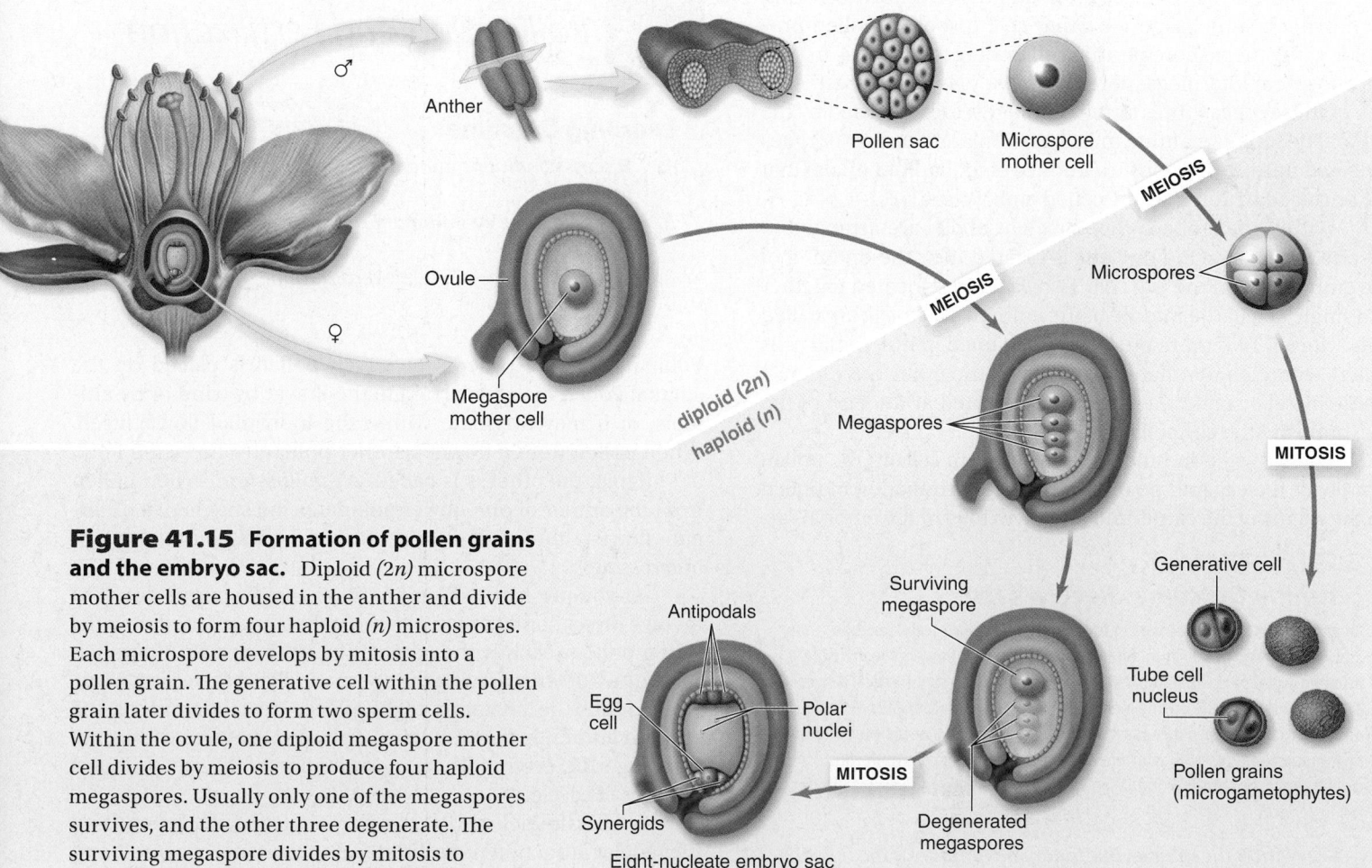

Figure 41.15 Formation of pollen grains and the embryo sac. Diploid *(2n)* microspore mother cells are housed in the anther and divide by meiosis to form four haploid *(n)* microspores. Each microspore develops by mitosis into a pollen grain. The generative cell within the pollen grain later divides to form two sperm cells. Within the ovule, one diploid megaspore mother cell divides by meiosis to produce four haploid megaspores. Usually only one of the megaspores survives, and the other three degenerate. The surviving megaspore divides by mitosis to produce an embryo sac with eight nuclei.

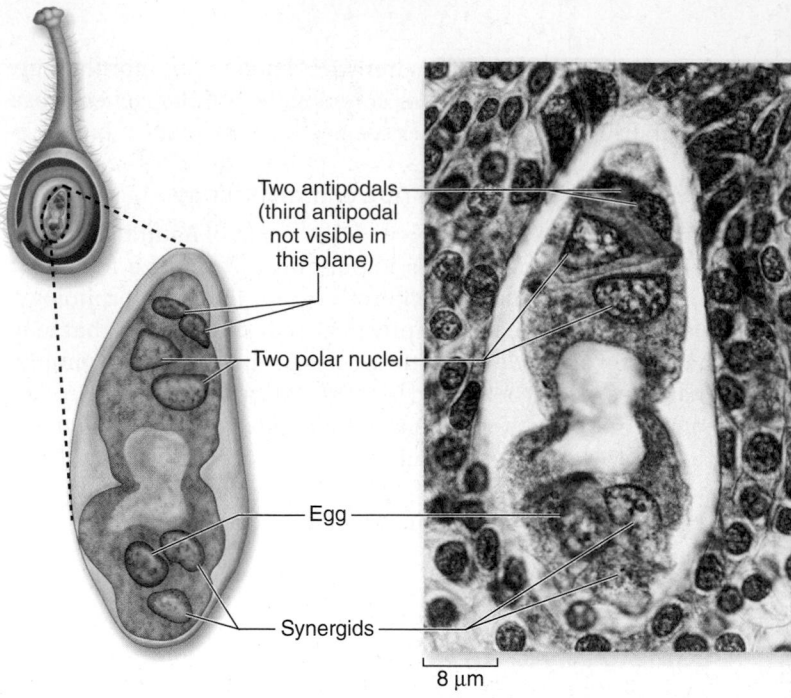

Figure 41.16 Pollen grains. _a._ In the Easter lily, _Lilium candidum,_ the pollen tube emerges from the pollen grain through the groove or furrow that occurs on one side of the grain. **_b._** In a plant of the sunflower family, _Hyoseris longiloba,_ three pores are hidden among the ornamentation of the pollen grain. The pollen tube may grow out through any one of them.

Pollen grain shapes are specialized for specific flower species. As discussed in more detail later in this section, fertilization requires that the pollen grain grow a tube that penetrates the style until it encounters the ovary. Most pollen grains have a furrow or pore from which this pollen tube emerges; some grains have three furrows (figure 41.16).

Embryo sac formation

Eggs develop in the ovules of the angiosperm flower. Within each ovule is a megaspore mother cell. Just as in pollen production, the megaspore mother cell undergoes meiosis to produce four haploid megaspores. In most plants, however, only one of these megaspores survives; the rest are absorbed by the ovule. The lone remaining megaspore enlarges and undergoes repeated mitotic divisions to produce eight haploid nuclei that are enclosed within a seven-celled embryo sac.

Within the embryo sac, the eight nuclei are arranged in precise positions. One nucleus is located near the opening of the embryo sac in the egg cell. Two others are located together in a single cell in the middle of the embryo sac; these are called _polar nuclei._ Two more nuclei are contained in individual cells called synergids that flank the egg cell; the other three nuclei reside in cells called the antipodals, located at the end of the sac, opposite the egg cell (figure 41.17).

The first step in uniting the two sperm cells in the pollen grain with the egg and polar nuclei is the germination of pollen on the stigma of the carpel and its growth toward the embryo sac.

Figure 41.17 A mature embryo sac of a lily. Eight nuclei are produced by mitotic divisions of the haploid megaspore. One is in the egg, two are polar nuclei, two occur in synergid cells, and three are in antipodal cells. The micrograph is falsely colored.

41.4 _Pollination and Fertilization_

Learning Outcomes

1. **Discuss conditions under which self-pollination may be favored.**
2. **Describe three evolutionary strategies that promote outcrossing.**
3. **List the products of double fertilization.**

Pollination is the process by which pollen is placed on the stigma. Pollen may be carried to the flower by wind or by animals, or it may originate within the individual flower itself. When pollen from a flower's anther pollinates the same flower's stigma, the process is called _self-pollination._ When pollen from the anther of one flower pollinates the stigma of a different flower, the process is termed _cross-pollination,_ or _outcrossing._

As you just learned, pollination in angiosperms does not involve direct contact between the pollen grain and the ovule. When pollen reaches the stigma, it germinates, and a pollen tube grows down, carrying the sperm nuclei to the embryo sac. After double fertilization takes place, development of the embryo and endosperm begins. The seed matures within the ripening fruit; eventually, the germination of the seed initiates another life cycle.

Successful pollination in many angiosperms depends on the regular attraction of **pollinators,** such as insects, birds, and other animals, which transfer pollen between plants of the

Learning Outcomes Review 41.3

An angiosperm flower consists of four concentric whorls: calyx, corolla, androecium, and gynoecium. Bilaterally symmetrical flowers are divisible into two equal parts along only a single plane; radially symmetrical flowers can be divided equally on any plane. The microspore mother cells in flowers undergo meiosis to produce microspores, which undergo mitosis to produce microgametophytes (pollen grains). Megaspore mother cells undergo a similar process to produce megaspores, which result in megagametophytes (embryo sacs).

■ **What is the main evolutionary advantage of the flower?**

same species. When animals disperse pollen, they perform the same function for flowering plants that they do for themselves when they actively search out mates.

The relationship between plant and pollinator can be quite intricate. Mutations in either partner can block reproduction. If a plant flowers at the "wrong" time, the pollinator may not be available. If the morphology of the flower or pollinator is altered, the result may be physical barriers to pollination. Clearly, floral morphology has coevolved with pollinators, and the result is a much more complex and diverse morphology, going beyond the simple initiation and development of four distinct whorls of organs.

Early seed plants were wind-pollinated

Early seed plants were pollinated passively, by the action of the wind. As in present-day conifers, great quantities of pollen were shed and blown about, occasionally reaching the vicinity of the ovules of the same species.

Individual plants of any wind-pollinated species must grow relatively close to one another for such a system to operate efficiently. Otherwise, the chance that any pollen will arrive at an appropriate destination is very small. The vast majority of windblown pollen travels less than 100 m. This short distance is significant compared with the long distances pollen is routinely carried by certain insects, birds, and other animals.

Flowers and animal pollinators have coevolved

The spreading of pollen from plant to plant by pollinators visiting flowers of an angiosperm species has played an important role in the evolutionary success of the group. It now seems clear that the earliest angiosperms, and perhaps their ancestors also, were insect-pollinated, and the coevolution of insects and plants has been important for both groups for over 100 million years. Such interactions have also been important in bringing about increased floral specialization. As flowers become increasingly specialized, so do their relationships with particular groups of insects and other animals.

Bees

Among insect-pollinated angiosperms, the most numerous groups are those pollinated by bees (figure 41.18). Like most insects, bees initially locate sources of food by odor and then orient themselves on the flower or group of flowers by its shape, color, and texture.

Flowers that bees characteristically visit are often blue or yellow. Many have stripes or lines of dots that indicate the location of the nectaries, which often occur within the throats of specialized flowers. Some bees collect nectar, which is used as a source of food for adult bees and occasionally for larvae. Most of the approximately 20,000 species of bees visit flowers to obtain pollen, which is used to provide food in cells where bee larvae complete their development.

Except for a few hundred species of social and semisocial bees and about 1000 species that are parasitic in the nests

Figure 41.18 Pollination by a bumblebee. As this bumblebee, *Bombus* sp., collects nectar, pollen sticks to its body. The pollen will be distributed to the next plant the bee visits.

of other bees, the great majority of bees—at least 18,000 species—are solitary. Solitary bees in temperate regions characteristically produce only a single generation in the course of a year. Often, they are active as adults for as little as a few weeks a year.

Solitary bees often use the flowers of a particular group of plants almost exclusively as sources of their larval food. The highly constant relationships of such bees with those flowers may lead to modifications, over time, in both the flowers and the bees. For example, the time of day when the flowers open may correlate with the time when the bees appear; the mouthparts of the bees may become elongated in relation to tubular flowers; or the bees' pollen-collecting apparatuses may be adapted to the anthers of the plants that they normally visit. When such relationships are established, they provide both an efficient mechanism of pollination for the flowers and a constant source of food for the bees that "specialize" on them.

Insects other than bees

Other than bees, among flower-visiting insects, a few groups are especially prominent. Flowers such as phlox, which are visited regularly by butterflies, often have flat "landing platforms" on which butterflies perch. They also tend to have long, slender floral tubes filled with nectar that is accessible to the long, coiled proboscis characteristic of Lepidoptera, the order of insects that includes butterflies and moths.

Flowers such as jimsonweed (*Datura stramonium*), evening primrose (*Oenothera biennis*), and others visited regularly by moths are often white, yellow, or some other pale color; they also tend to be heavily scented, making the flowers easy to locate at night (figure 41.19).

Birds

Several interesting groups of plants are regularly visited and pollinated by birds, especially the hummingbirds of North and

Hypothesis: *Moths are more effective than bumblebees at moving pollen long distances.*

Prediction: *The pollen donors of seeds of wild plants are more widely distributed if moths carried the pollen than if bees carried it.*

Test: *Locate a large natural patch of the wild plant. Make sure that both moths and bees are abundant and that the plants are variable for a genetically controlled trait. In this case, assume the population contains some purple-flowered plants (a dominant trait) and some with white flowers (a recessive trait). Remove all the flowers from the purple-flowered plants except those at the edge of the population. Find a white-flowered plant at the center of the population to use as the test plant. Cover some flowers during the day and uncover them in the evening so moths, but not bees, can pollinate them. With other flowers, cover in the evening but not during the day, so bees, but not moths, can pollinate them. Collect seeds from each set of flowers and grow the plants. For each treatment, count the number of plants that produce purple flowers. These will have pollen donors that were a long distance from the test plant.*

Cover some flowers during the day and others in the evening. Count the number of purple flowered plants obtained from each treatment.

Result: *Seeds produced by bee pollination produced the same number of plants with purple flowers as those produced by moth pollination.*

Conclusion: *The hypothesis is not supported. Bees carry pollen as far as moths do.*

Further Experiments: *Growing plants from seed to check for flower color is very time-consuming. Propose another way to determine the source of the pollen in this experiment.*

Figure 41.19 Effectiveness of moths as pollinators.

South America and the sunbirds of Africa (figure 41.20). Such plants must produce large amounts of nectar because birds will not continue to visit flowers if they do not find enough food to maintain themselves. But flowers producing large amounts of nectar have no advantage in being visited by insects, because an insect could obtain its energy requirements at a single flower and would not cross-pollinate the flower. How are these different selective forces balanced in flowers that are "specialized" for hummingbirds and sunbirds?

Figure 41.20 Hummingbirds and flowers. A long-tailed hermit hummingbird *(Phaethornis superciliosus)* extracts nectar from the flowers of *Heliconia imbricata* in the forests of Costa Rica. Note the pollen on the bird's beak. Hummingbirds of this group obtain nectar primarily from long, curved flowers that more or less match the length and shape of their beaks.

The answer involves the evolution of flower color. Ultraviolet light is highly visible to insects. Carotenoids, the yellow or orange pigments we described in chapter 8 in the context of photosynthesis, are responsible for the colors of many flowers, including sunflowers and mustard. Carotenoids reflect both in the yellow range and in the ultraviolet range, the mixture resulting in a distinctive color called "bee's purple." Such yellow flowers may also be marked in distinctive ways normally invisible to us, but highly visible to bees and other insects (figure 41.21). These markings can be in the form of a bull's-eye or a landing strip.

a.   *b.*

Figure 41.21 How a bee sees a flower. *a.* The yellow flower of *Ludwigia peruviana* (Peruvian primrose) photographed in normal light and *(b)* with a filter that selectively transmits ultraviolet light. The outer sections of the petals reflect both yellow and ultraviolet, a mixture of colors called "bee's purple"; the inner portions of the petals reflect yellow only and therefore appear dark in the photograph that emphasizes ultraviolet reflection. To a bee, this flower appears as if it has a conspicuous central bull's-eye.

In contrast, red does not stand out as a distinct color to most insects, but it is a very conspicuous color to birds. To most insects, the red upper leaves of poinsettias look just like the other leaves of the plant. Consequently, even though the flowers produce abundant supplies of nectar and attract hummingbirds, insects tend to bypass them. Thus, the red color both signals to birds the presence of abundant nectar and makes that nectar as inconspicuous as possible to insects. Red is also seen again in fruits that are dispersed by birds (see chapter 31).

Other animal pollinators

Other animals, including bats and small rodents, may aid in pollination. The signals here are also species-specific. As an example, the saguaro cactus (*Carnegeia gigantea*) of the Sonoran desert is pollinated by bats that feed on nectar at night, as well as by birds and insects.

These animals may also assist in dispersing the seeds and fruits that result from pollination. Monkeys are attracted to orange and yellow, and thus can be effective in dispersing fruits of this color in their habitats.

 ## Some flowering plants use wind pollination

A number of groups of angiosperms are wind-pollinated—a characteristic of early seed plants. Among these groups are oaks, birches, cottonwoods, grasses, sedges, and nettles. The flowers of these plants are small, greenish, and odorless; their corollas are reduced or absent (figures 41.22 and 41.23). Such flowers often are grouped together in fairly large numbers and may hang down in tassels that wave about in the wind and shed pollen freely.

Many wind-pollinated plants have stamen- and carpel-containing flowers separated between individuals or physically separated on a single individual. Maize is a good example, with pollen-producing tassels at the top of the plant and axillary shoots with female flowers lower down. Separation of pollen-producing and ovule-bearing flowers is a strategy that greatly promotes outcrossing, since pollen from one flower must land on a different flower for fertilization to have any chance of occurring. Some wind-pollinated plants, especially trees and shrubs, flower in the spring, before the development of their leaves can interfere with the wind-borne pollen. Wind-pollinated species do not depend on the presence of a pollinator for species survival, which may be another survival advantage.

Self-pollination is favored in stable environments

Thus far we have considered examples of pollination that tend to lead to outcrossing, which is as highly advantageous for plants and for eukaryotic organisms generally. Nevertheless, self-pollination also occurs among angiosperms, particularly in temperate regions. Most self-pollinating plants have small, relatively inconspicuous flowers that shed pollen directly onto the stigma, sometimes even before the bud opens.

Figure 41.22 Staminate and pistillate flowers of a birch, *Betula* sp. Birches are monoecious; their staminate flowers hang down in long, yellowish tassels, and their pistillate flowers mature into clusters of small, brownish, conelike structures.

Figure 41.23 Wind-pollinated flowers. The large yellow anthers, dangling on very slender filaments, are hanging out, about to shed their pollen to the wind. Later, these flowers will become pistillate, with long, feathery stigmas—well suited for trapping windblown pollen—sticking far out of them. Many grasses, like this one, are therefore dichogamous.

You might logically ask why many self-pollinated plant species have survived if outcrossing is as important genetically for plants as it is for animals. Biologists propose two basic reasons for the frequent occurrence of self-pollinated angiosperms:

1. Self-pollination is ecologically advantageous under certain circumstances because self-pollinators do not need to be visited by animals to produce seed. As a result, self-pollinated plants expend less energy in producing pollinator attractants and can grow in areas where the kinds of insects or other animals that might visit them are absent or very scarce—as in the Arctic or at high elevations.

2. In genetic terms, self-pollination produces progenies that are more uniform than those that result from outcrossing. Remember that because meiosis is involved, recombination still takes place, as described in chapter 11—and therefore the offspring will not be identical to the parent. However, such progenies may contain high proportions of individuals well-adapted to particular habitats.

Self-pollination in normally outcrossing species tends to produce large numbers of ill-adapted individuals because it brings together deleterious recessive alleles—but some of these combinations may be highly advantageous in particular habitats. In these habitats, it may be advantageous for the plant to continue self-pollinating indefinitely.

Several evolutionary strategies promote outcrossing

Outcrossing, as we have stressed, is critically important for the adaptation and evolution of all eukaryotic organisms, with a few exceptions. Often, flowers contain both stamens and pistils, which increases the likelihood of self-pollination. One general strategy to promote outcrossing, therefore, is to separate stamens and pistils. Another strategy involves self-incompatibility that prevents self-fertilization.

Separation of male and female structures in space or in time

In a number of species—for example, willows and some mulberries—staminate and pistillate flowers may occur on separate plants. Such plants, which produce only ovules or only pollen, are called dioecious, meaning "two houses." These plants clearly cannot self-pollinate and must rely exclusively on outcrossing. In other kinds of plants, such as oaks, birches, corn (maize), and pumpkins, separate male and female flowers may both be produced on the same plant. Such plants are called **monoecious,** meaning "one house" (see figure 41.22). In monoecious plants, the separation of pistillate and staminate flowers, which may mature at different times, greatly enhances the probability of outcrossing.

Even if, as usually is the case, functional stamens and pistils are both present in each flower of a particular plant species, these organs may reach maturity at different times. Plants in which this occurs are called **dichogamous.** If the stamens mature first, shedding their pollen before the stigmas are receptive, the flower is effectively staminate at that time. Once the stamens have finished shedding pollen, the stigma or stigmas may become receptive, and the flower may become essentially pistillate (see figures 41.23 and 41.24). This separation in time has the same effect as if

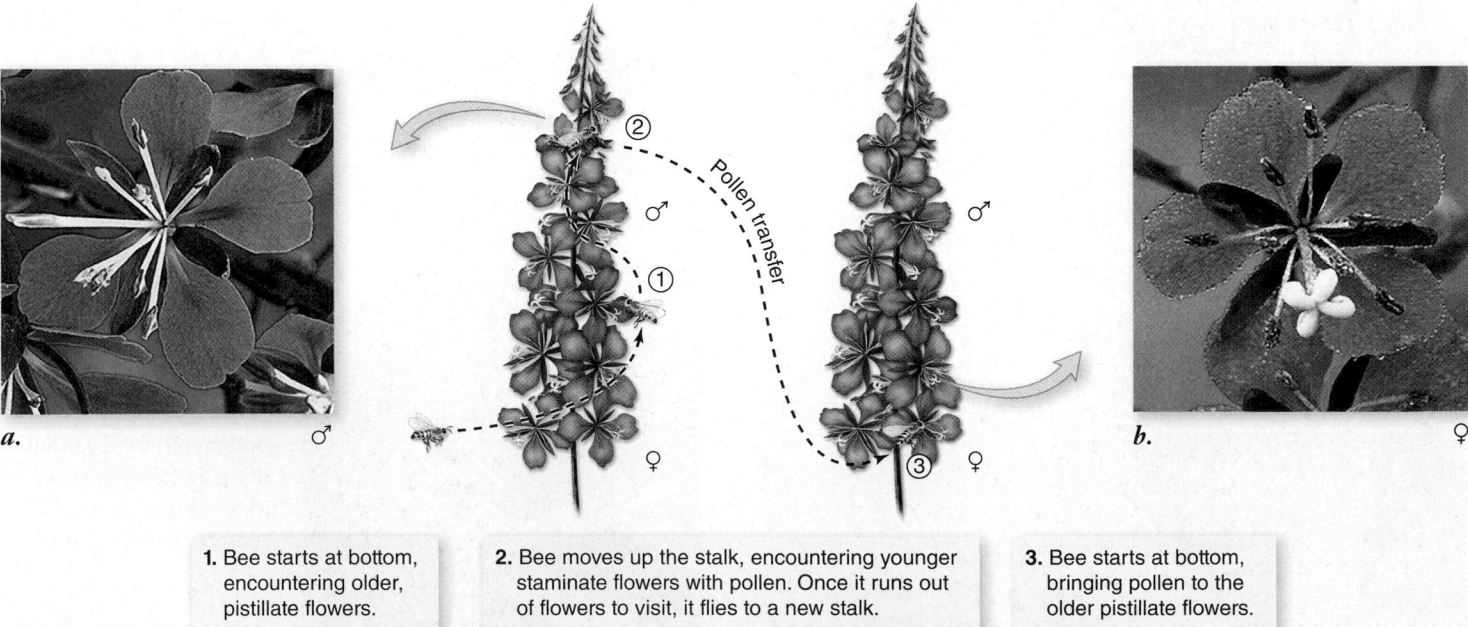

1. Bee starts at bottom, encountering older, pistillate flowers.

2. Bee moves up the stalk, encountering younger staminate flowers with pollen. Once it runs out of flowers to visit, it flies to a new stalk.

3. Bee starts at bottom, bringing pollen to the older pistillate flowers.

Figure 41.24 Dichogamy, as illustrated by the flowers of fireweed, *Epilobium angustifolium.* More than 200 years ago (in the 1790s) fireweed, which is outcrossing, was one of the first plant species to have its process of pollination described. First, the anthers shed pollen, and then the style elongates above the stamens while the four lobes of the stigma curl back and become receptive. Consequently, the flowers are functionally staminate at first, becoming pistillate about two days later. The flowers open progressively up the stem, so that the lowest are visited first, promoting outcrossing. Working up the stem, the bees encounter pollen-shedding, staminate-phase flowers and become covered with pollen, which they then carry to the lower, functionally pistillate flowers of another plant. Shown here are flowers in (*a*) the staminate phase and (*b*) the pistillate phase.

individuals were dioecious; the outcrossing rate is thereby significantly increased.

Many flowers are constructed such that the stamens and stigmas do not come in contact with each other. With this arrangement, the natural tendency is for the pollen to be transferred to the stigma of another flower, rather than to the stigma of its own flower, thereby promoting outcrossing.

Self-incompatibility

Even when a flower's stamens and stigma mature at the same time, genetic **self-incompatibility,** which is widespread in flowering plants, increases outcrossing. Self-incompatibility results when the pollen and stigma recognize each other as being genetically related, and pollen tube growth is blocked (figure 41.25).

The *S* (self-incompatibility) locus controls self-incompatibility. Many alleles at the *S* locus regulate recognition responses between pollen and stigma. Researchers have identified two types of self-incompatibility. *Gametophytic self-incompatibility* depends on the haploid *S* locus of the pollen and the diploid *S* locus of the stigma. If either of the *S* alleles in the stigma matches the pollen's *S* allele, pollen tube growth stops before it reaches the embryo sac. Petunias exhibit gametophytic self-incompatibility.

In *sporophytic self-incompatibility,* as occurs in broccoli, both *S* alleles of the pollen parent, not just the *S* allele of the pollen itself, are important. If the alleles in the stigma match either of the pollen parent's *S* alleles, the haploid pollen will not germinate.

Pollen-recognition mechanisms may have originated in a common ancestor of the gymnosperms. Fossils with pollen tubes from the Carboniferous period are consistent with the hypothesis that they had highly evolved pollen-recognition systems.

Angiosperms undergo double fertilization

Fertilization in angiosperms is a complex, somewhat unusual process in which two sperm cells are utilized in a unique process called double fertilization, as described in chapter 31. Double fertilization results in two key developments: (1) the fertilization of the egg, and (2) the formation of a nutrient substance called endosperm that nourishes the embryo.

Once a pollen grain has been spread by wind, by animals, or through self-pollination, it adheres to the sticky, sugary substance that covers the stigma and begins to grow a pollen tube that pierces the style (figure 41.26). The pollen tube, nourished by the sugary substance, grows until it reaches the ovule in the ovary. Meanwhile, the generative cell within the pollen grain tube cell divides to form two sperm cells.

The pollen tube eventually reaches the embryo sac in the ovule. At the entry to the embryo sac, one of the nuclei flanking the egg cell degenerates, and the pollen tube enters that cell. The tip of the pollen tube bursts and releases the two sperm cells. One of the sperm cells fertilizes the egg cell, forming a zygote. The other sperm cell fuses with the two polar nuclei located at the center of the embryo sac, forming the triploid ($3n$) primary endosperm nucleus. The primary endosperm nucleus eventually develops into the endosperm (food supply).

Once fertilization is complete, the embryo develops as its cells divide numerous times. Meanwhile, protective tissues enclose the embryo, resulting in the formation of the seed. The seed, in turn, is enclosed in another structure, called the fruit. These typical angiosperm structures evolved in response to the need for seeds to be dispersed over long distances to ensure genetic variability.

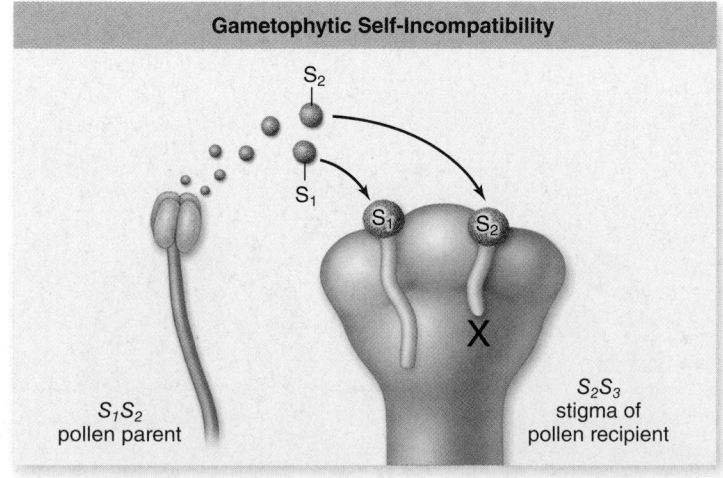

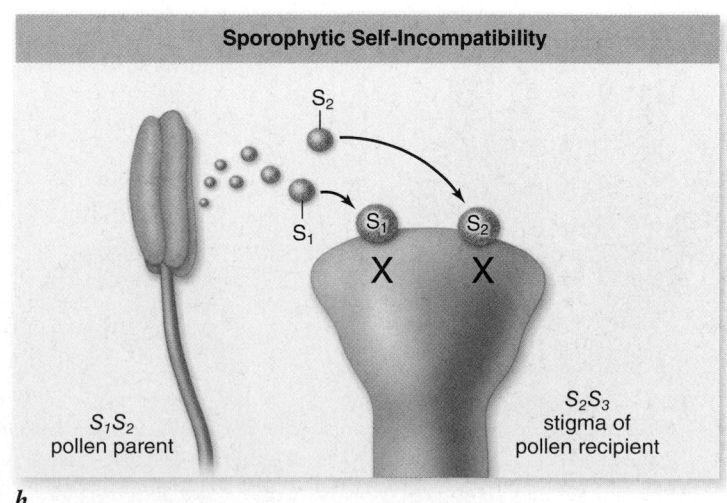

a. b.

Figure 41.25 Genetic control can block self-pollination. *a.* Gametophytic self-incompatibility is determined by the haploid pollen genotype. *b.* Sporophytic self-incompatibility recognizes the genotype of the diploid pollen parent, not just the haploid pollen genotype. The pollen contains proteins produced by the S_1S_2 parent. In both cases, the recognition is based on the *S* locus, which has many different alleles. The subscript numbers indicate the *S* allele genotype. In gametophytic self-incompatibility, the block comes after pollen tube germination. In sporophytic self-incompatibility, the pollen tube fails to germinate.

Figure 41.26 The formation of the pollen tube and double fertilization. When pollen lands on the stigma of a flower, the pollen tube cell grows toward the embryo sac, forming a pollen tube. While the pollen tube is growing, the generative cell divides to form two sperm cells. When the pollen tube reaches the embryo sac, it enters one of the synergids and releases the sperm cells. In a process called double fertilization, one sperm cell nucleus fuses with the egg cell to form the diploid (2n) zygote, and the other sperm cell nucleus fuses with the two polar nuclei to form the triploid (3n) endosperm nucleus.

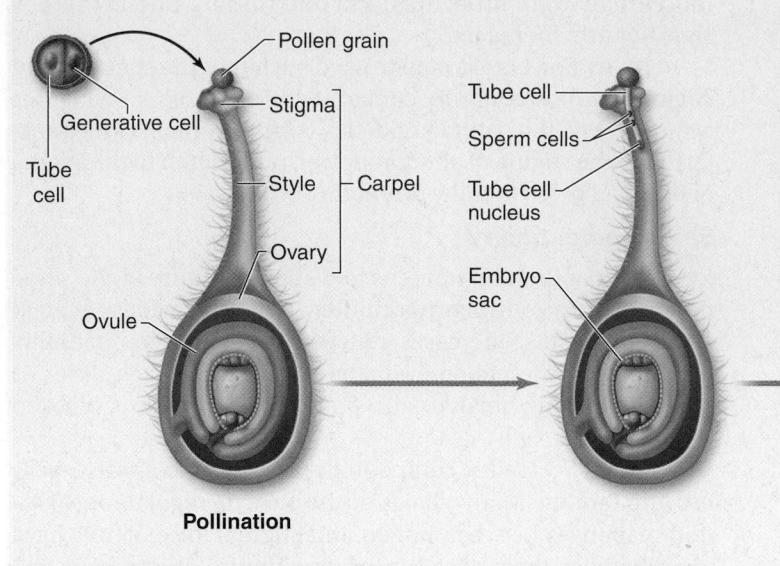

Pollination

Learning Outcomes Review 41.4

Self-pollination may be favored when pollinators are absent or when plants are adapted to a stable environment, and therefore uniform offspring are advantageous. Mechanisms to promote outcrossing include the production of separate male and female flowers, maturation of male flowers at a different time than female flowers, and genetically controlled self-incompatibility. Double fertilization produces a diploid embryo and triploid endosperm that provides nutrition.

■ *Are all offspring of a self-pollinating plant identical?*

41.5 *Embryo Development*

Learning Outcomes

1. *Name the two axes of the plant body that are established during embryonic development.*
2. *List the three tissue systems that develop in an embryo.*
3. *Describe the three critical events that must accompany seed development.*

Embryo development begins once the egg cell is fertilized. As described briefly in chapter 31, the growing pollen tube from a pollen grain enters the angiosperm embryo sac through one of the synergids, releasing two sperm cells (see figure 41.26) One sperm cell fertilizes the central cell with its polar nuclei, and the resulting cell division produces a nutrient source, the **endosperm,** for the embryo. The other sperm cell fertilizes the egg to produce a zygote, and cell division soon follows, creating the **embryo.**

A single cell divides to produce a three-dimensional body plan

The first division of the zygote (fertilized egg) in a flowering plant is asymmetrical and generates cells with two different fates (figure 41.27). One daughter cell is small, with dense

Figure 41.27 Stages of development in an angiosperm embryo. The very first cell division is asymmetrical. Differentiation begins almost immediately after fertilization.

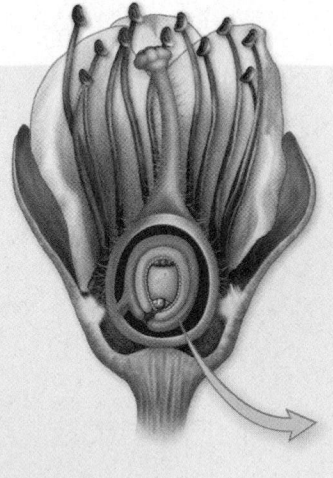
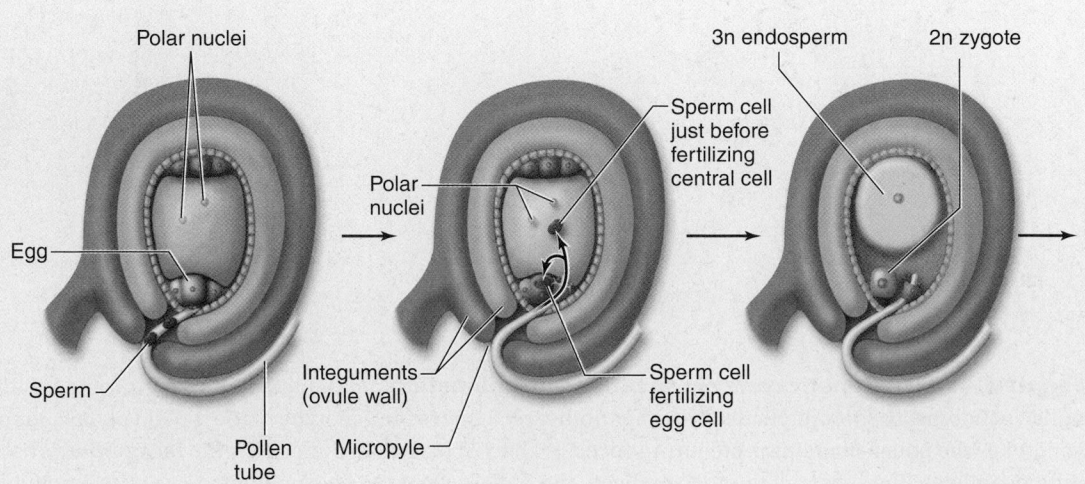

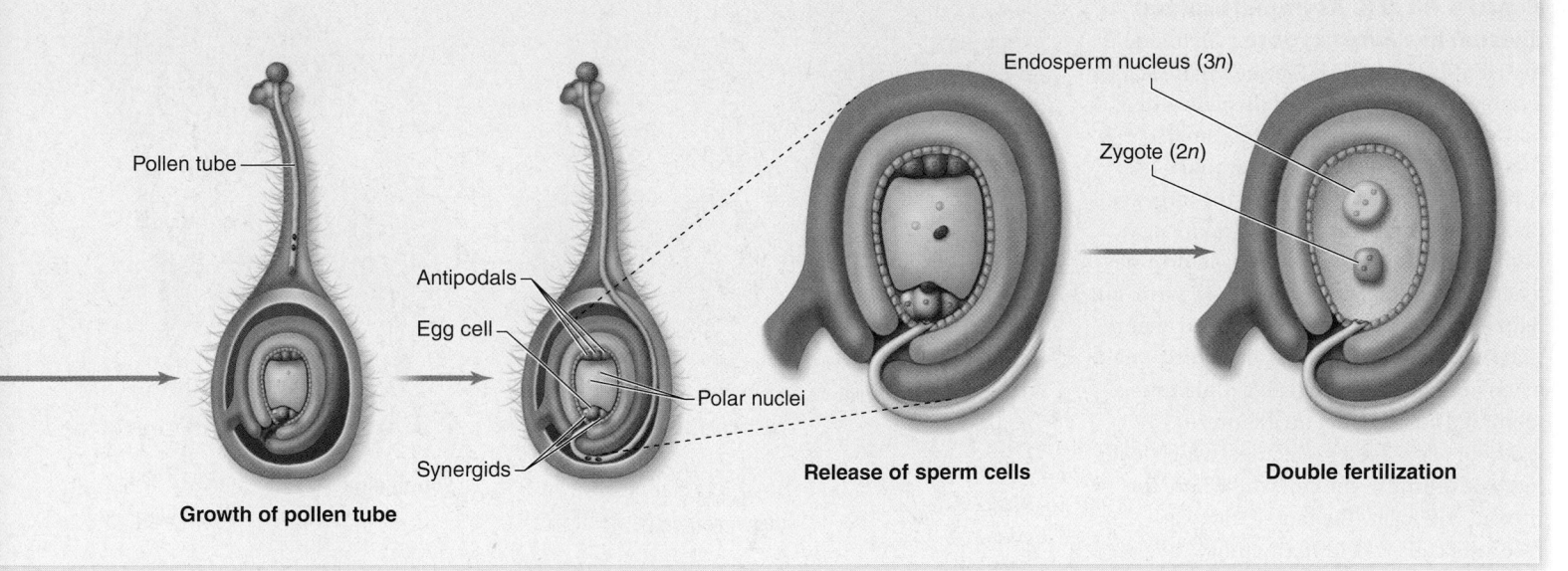

Pollen tube

Antipodals

Egg cell

Polar nuclei

Synergids

Growth of pollen tube

Endosperm nucleus (3*n*)

Zygote (2*n*)

Release of sperm cells

Double fertilization

cytoplasm. That cell, which is destined to become the embryo, begins to divide repeatedly in different planes, forming a ball of cells. The other, larger daughter cell divides repeatedly, forming an elongated structure called a **suspensor,** which links the embryo to the nutrient tissue of the seed. The suspensor also provides a route for nutrients to reach the developing embryo. The root–shoot axis also forms at this time; cells near the suspensor are destined to form a root, and those at the other end of the axis ultimately become a shoot.

Investigating mechanisms for establishing asymmetry in plant embryo development is difficult because the zygote is embedded within the female gametophyte, which is surrounded by sporophyte tissue (ovule and carpel tissue). To understand the cell biology of the first asymmetrical division of

zygotes, biologists have studied the brown alga *Fucus.* We must be cautious about inferring too much about angiosperm asymmetrical divisions from the brown algae because the last common ancestor of brown algae and the angiosperm line was a single-celled organism. Nevertheless, asymmetrical division extends far back in the tree of life. Even prokaryotes can divide asymmetrically

Zygote development in Fucus

In the brown alga *Fucus,* the egg is released prior to fertilization, so no extra tissues surround the zygote, making its development easier to observe. A bulge that develops on one side of the zygote establishes the vertical axis. Cell division occurs, and the original bulge becomes the smaller of the two daughter cells. The smaller cell develops into a rhizoid that anchors the

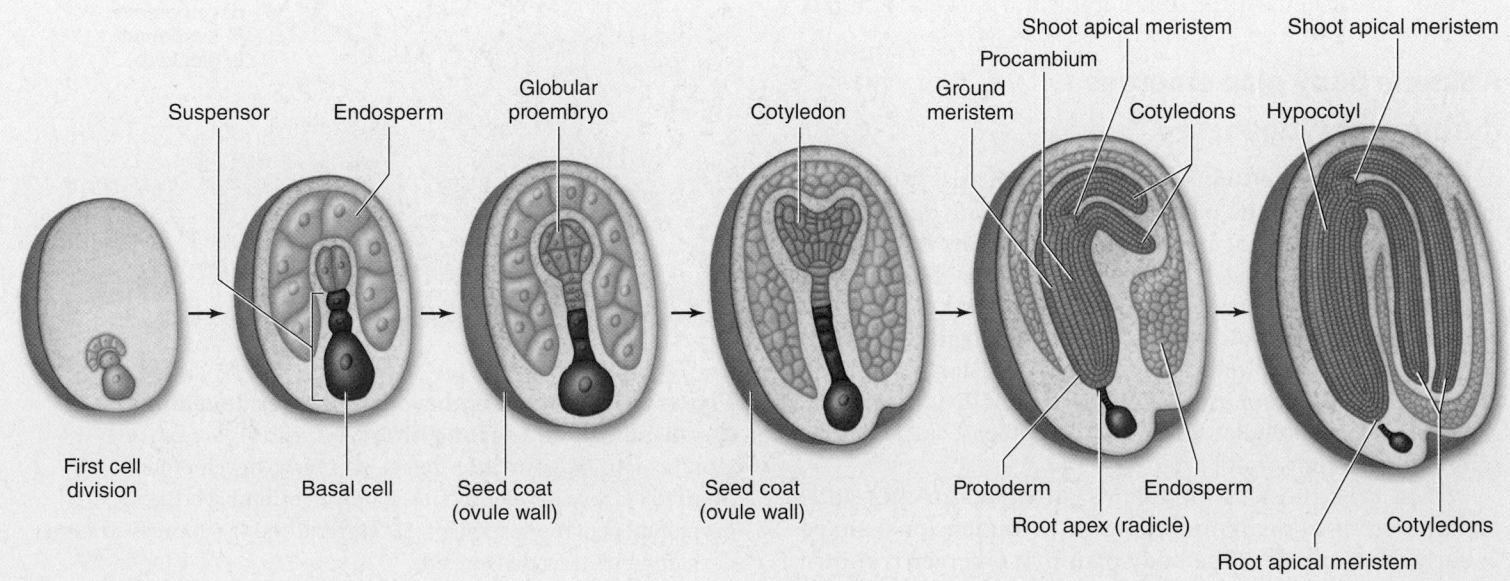

Suspensor   Endosperm   Globular proembryo   Cotyledon   Ground meristem   Procambium   Shoot apical meristem   Cotyledons   Shoot apical meristem   Hypocotyl

First cell division   Basal cell   Seed coat (ovule wall)   Seed coat (ovule wall)   Protoderm   Root apex (radicle)   Endosperm   Cotyledons   Root apical meristem

Figure 41.28 Asymmetrical cell division in a _Fucus_ zygote. An unequal distribution of material in the zygote leads to a bulge where the first cell division will occur. This division results in a smaller cell that will go on to divide and produce the rhizoid that anchors the alga; the larger cell divides to form the thallus, or main algal body. The point of sperm entry determines where the smaller rhizoid cell will form, but light and gravity can modify this to ensure that the rhizoid will point downward where it can anchor this brown alga. Calcium-mediated currents set up an internal gradient of charged molecules, which leads to a weakening of the cell wall where the rhizoid will form. The fate of the two resulting cells is held "in memory" by cell-wall components.

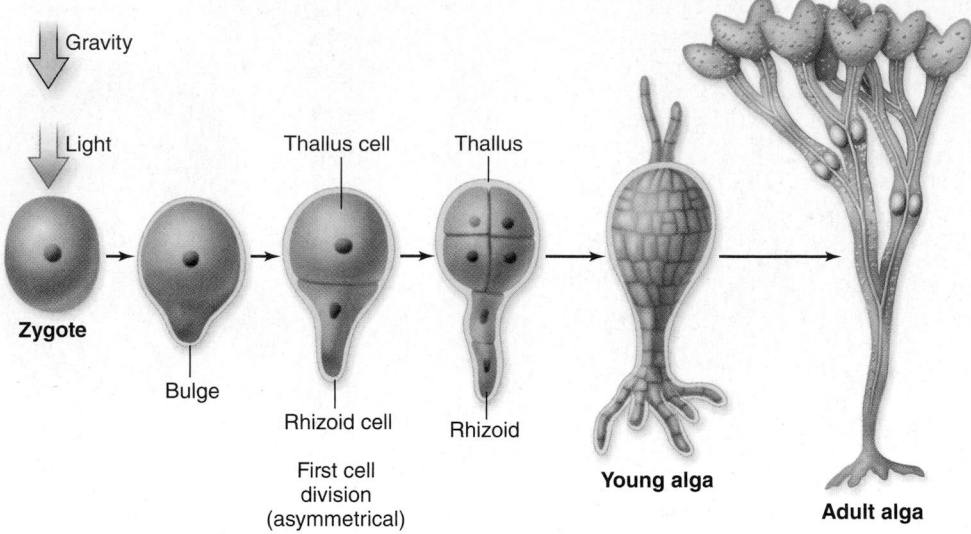

alga, and the larger cell develops into the main body, or thallus, of the sporophyte (figure 41.28).

This axis is first established by the point of sperm entry, but it can be changed by environmental signals, especially light and gravity, which ensure that the rhizoid attaches to a solid base and the thallus is oriented toward a light source. Internal gradients are established that specify where the rhizoid forms in response to environmental signals.

Establishing asymmetry in angiosperms

Genetic approaches make it possible to explore asymmetrical development in angiosperms. Studies of mutants reveal what can go wrong in development, which often makes it possible to infer normal developmental mechanisms. For example, the suspensor mutant in _Arabidopsis thaliana_ undergoes aberrant development in the embryo followed by embryo-like development of the suspensor (figure 41.29). Analysis of this mutant led to the conclusion that the presence of the embryo normally prevents the suspensor from developing into a second embryo.

A simple body plan emerges during embryogenesis

In plants, three-dimensional shape and form arise by regulating the amount and the pattern of cell division. We have just described how a vertical axis (root–shoot axis) becomes established at a very early stage; the same is true for establishment of a radial axis (inner–outer axis). Although the first cell division gives rise to a single row of cells, cells soon begin dividing in different directions, producing a three-dimensional solid ball of cells. The root–shoot axis lengthens as cells divide. New cell walls form perpendicular to the root–shoot axis, stacking new cells along the root–shoot axis.

The cells must divide in two directions in the radial plane in order to maintain proper three-dimensional shape in early development. The body plan that emerges is shown

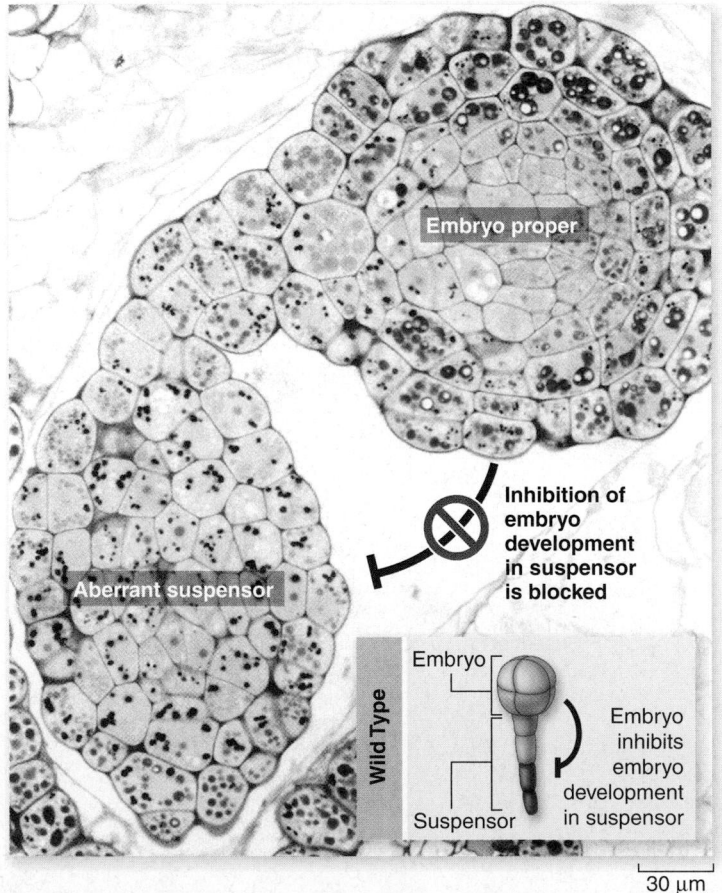

Figure 41.29 The embryo suppresses development of the suspensor as a second embryo. This _suspensor (sus)_ mutant of _Arabidopsis_ has a defect in embryo development. Aborted embryo development is followed by embryo-like development of the suspensor. _SUS_ is required to suppress embryo development in suspensor cells.

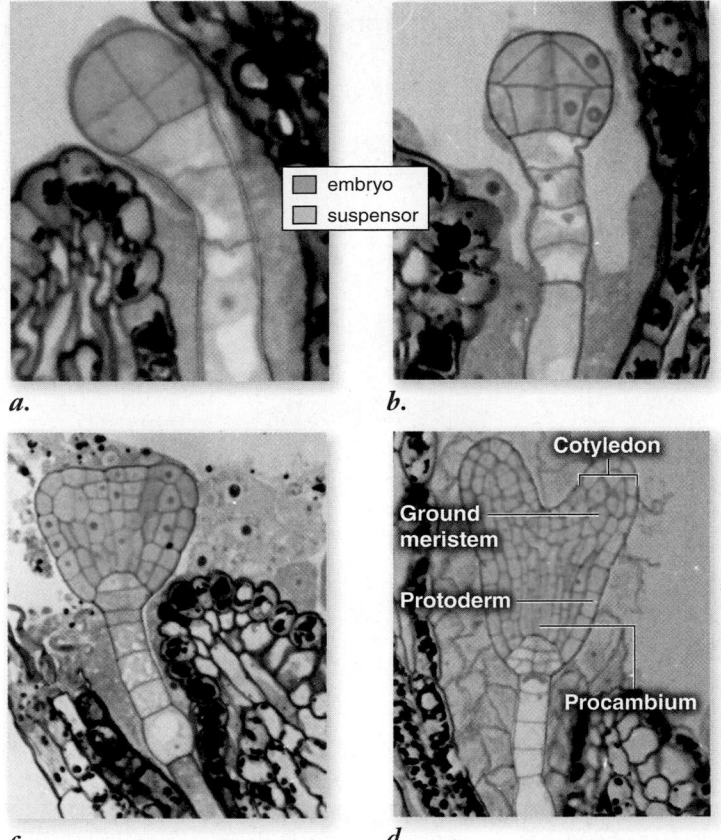

Figure 41.30 Early developmental stages of *Arabidopsis thaliana*. *a.* Early cell division has produced the embryo and suspensor. ***b.*** Globular stage results from cell divisions along both the root–shoot and radial axes. Cell differentiation, including establishment of the root and shoot apical meristems occurs at this stage. ***c, d.*** Heart-shaped stage. The cotyledons (seed leaves) are now visible, and the three tissue systems continue to differentiate.

in figure 41.30. Apical meristems, the actively dividing cell regions at the tips of roots and shoots, establish the root–shoot axis in the globular stage, from which the three basic tissue systems arise: *dermal, ground,* and *vascular* tissue (see chapter 36). These tissues are organized radially around the root–shoot axis.

Data analysis You sectioned an *Arabidopsis* embryo proper (not the suspensor) through several different planes and examined the sections under a light microscope. You estimate there are 128 cells. If the cells in the embryo divided at the same rate, how many synchronous sets of cell divisions occurred after the first cell division established the embryo and suspensor?

Root and shoot formation

Both the shoot and root meristems are apical meristems, but their formation is controlled independently. Shoot formation requires the *SHOOTMERISTEMLESS (STM)* gene in *Arabidopsis.* Plants that do not make STM protein fail to produce viable shoots, but do produce roots (figure 41.31).

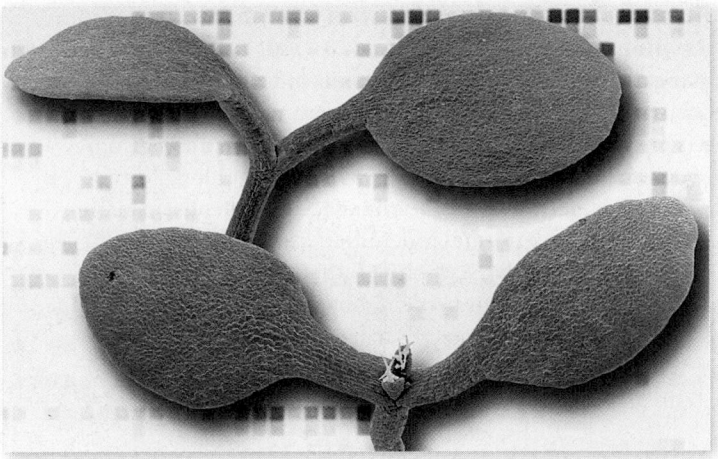

Figure 41.31 *SHOOTMERISTEMLESS* is needed for shoot formation. Shoot-specific genes specify formation of the shoot apical meristem, but are not necessary for root development. The *stm* mutant of *Arabidopsis* (shown on top) has a normal root meristem but fails to produce a shoot meristem between its two cotyledons. The *STM* wild type is shown below the *stm* mutant for comparison.

The *STM* gene codes for a transcription factor with a homeobox region, sharing a common evolutionary origin with the *Hox* genes that are important in establishing animal body plans (see chapters 19 and 25). Compared with animals, however, *Hox*-like genes have a more limited role in regulating plant body plans. Other gene families, encoding different transcription factors, also play key roles in patterning in plants.

Root formation requires the *HOBBIT* gene in *Arabidopsis* (figure 41.32). The *hobbit* mutants form shoot meristems, but no root meristems form. Cell divisions in *hobbit* roots occur in the wrong directions. Plants with a *hobbit* mutation accumulate a biochemical repressor of genes that are induced by auxin (a plant hormone). Based on the mutant phenotype, *HOBBIT* appears to repress the production of the repressor of auxin-induced genes. Or, more simply stated, HOBBIT protein allows auxin to induce the expression of a gene or genes needed for correct cell division to make a root meristem. Auxin is one of eight classes of hormones discussed in chapter 40 that regulate plant development and function.

? Inquiry question Referring to part *(e)* of figure 41.32, explain why this mutant fails to develop an embryonic root.

One way that auxin induces gene expression is by activating a transcription factor. *MONOPTEROS (MP)* is a gene that codes for an auxin-induced transcription factor (see figure 41.32), and like *HOBBIT,* it is necessary for root formation, but not shoot formation, in *Arabidopsis.* Once activated, MP protein binds to the promoter of another gene, leading to transcription of a gene or genes needed for root meristem formation.

? Inquiry question Predict the phenotype of a plant with a mutation in the *MP* gene that results in an MP protein that can no longer bind its repressor.

Figure 41.32 **Genetic control of embryonic root development in *Arabidopsis*.** *a.* HOBBIT inhibits the repression of the auxin response, allowing auxin-induced root development to occur. *b.* MONOPTEROS cannot act as a transcription factor when it is bound by a repressor. Auxin releases the repressor from MONOPTEROS, which then activates transcription of a root development gene. *c.* A wild-type seedling depends on auxin-induced genes for normal root initiation during embryogenesis. *d.* The *hobbit* seedling has a stub rather than a root because abnormal cell divisions prevent root meristem formation. *e.* The *monopteros* seedling also fails to develop a root.

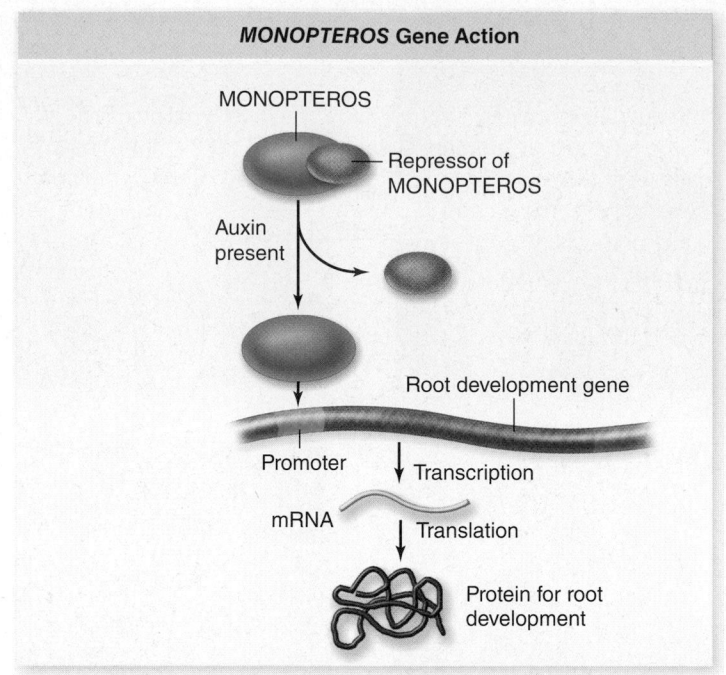

MONOPTEROS Gene Action

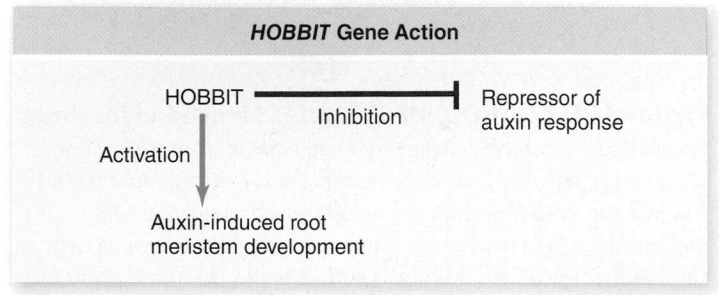

HOBBIT Gene Action

a.

b.

HOBBIT and MONOPTEROS
(both wild type)

c.   1,000 μm

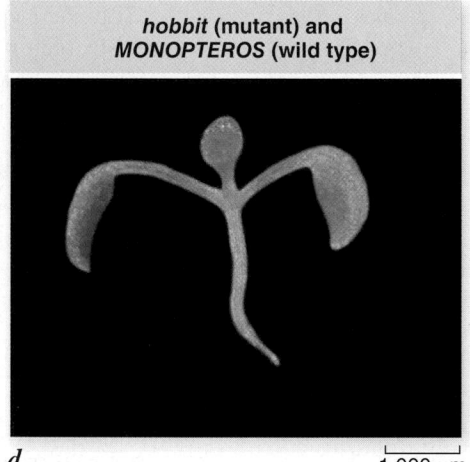

hobbit (mutant) and
MONOPTEROS (wild type)

d.   1,000 μm

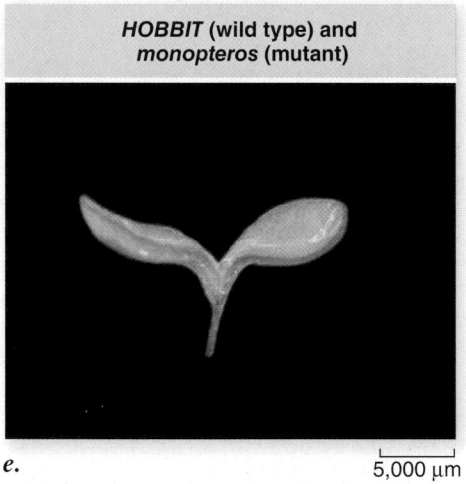

HOBBIT (wild type) and
monopteros (mutant)

e.   5,000 μm

Formation of the three tissue systems

Three basic tissues, called *primary meristems,* differentiate while the plant embryo is still a ball of cells (called the globular stage; see figure 41.30). No cell movements are involved in plant embryo development. The protoderm consists of the outermost cells in a plant embryo and will become *dermal tissue* (see chapter 36). These cells almost always divide with their cell plate perpendicular to the body surface, thus perpetuating a single outer layer of cells. Dermal tissue protects the plant from desiccation. Stomata that open and close to facilitate gas exchange and minimize water loss are derived dermal tissue.

A ground meristem gives rise to the bulk of the embryonic interior, consisting of *ground tissue* cells that eventually function in food and water storage.

Finally, procambium at the core of the embryo will form the future *vascular tissue,* which is responsible for water and nutrient transport.

Cell fates are generally more limited after embryogenesis, however, when embryo-specific genes are not expressed. For example, the *LEAFY COTYLEDON* gene in *Arabidopsis* is active in early and late embryo development, and it may be responsible for maintaining an embryonic environment. It is possible to turn this gene on later in development using recombinant DNA techniques described in chapter 17. When it is turned on, embryos can form on leaves!

Morphogenesis

The globular stage gives rise to a heart-shaped embryo with two bulges in one group of angiosperms (the eudicots, such as *A. thaliana* in figure 41.30*c, d*), and a ball with a bulge on a single side in another group (the monocots). These bulges are **cotyledons** ("first leaves") and are produced by the embryonic cells, and not by the shoot apical meristem that begins forming during the globular stage. This process, called morphogenesis (generation of form), results from changes in planes and rates of cell division (see figure 41.34).

Because plant cells cannot move, the form of a plant body is largely determined by the plane in which its cells divide. It is also controlled by changes in cell shape as cells expand osmotically after they form (figure 41.33). The position of the cell plate determines the direction of division, and both microtubules and actin play a role in establishing the cell plate's position. Plant hormones and other factors influence the orientation of bundles of microtubules on the interior of the plasma membrane. These microtubules also guide cellulose deposition as the cell wall forms around the outside of a new cell (see figure 36.2) where four of the six sides are reinforced more heavily with cellulose; the cell tends to expand and grow in the direction of the two sides having less reinforcement (figure 41.33b).

Much is being learned about morphogenesis at the cellular level from mutants that are able to divide but cannot control their plane of cell division or the direction of cell expansion. The lack of root meristem development in *hobbit* mutants is just one such example. As the procambium begins differentiating in the root, a critical division parallel to the root's surface is regulated by the gene *WOODEN LEG* (*WOL*, figure 41.34). Without that division, the cylinder of cells that would form

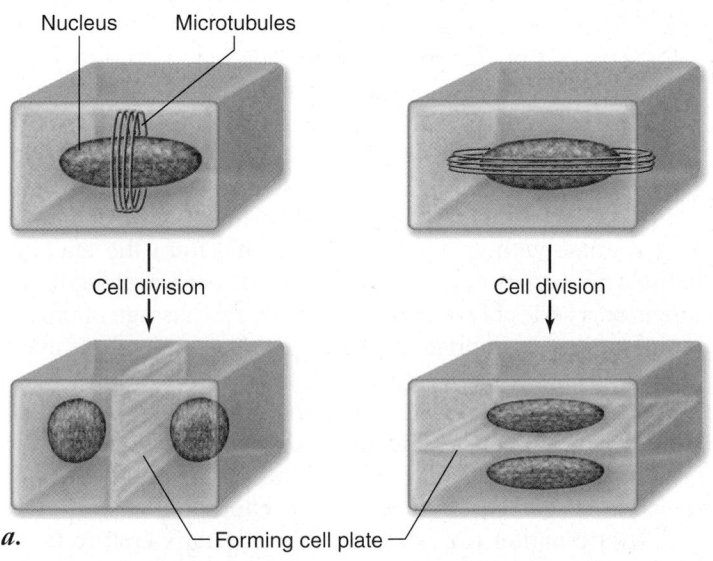

a.

b.

Figure 41.33 Cell division and expansion.
a. Orientation of microtubules determines the orientation of cell plate formation and thus the new cell wall. *b.* Not all sides of a plant cell have the same amount of cellulose reinforcement. With water uptake, cells expand in directions that have the least amount of cell-wall reinforcement.

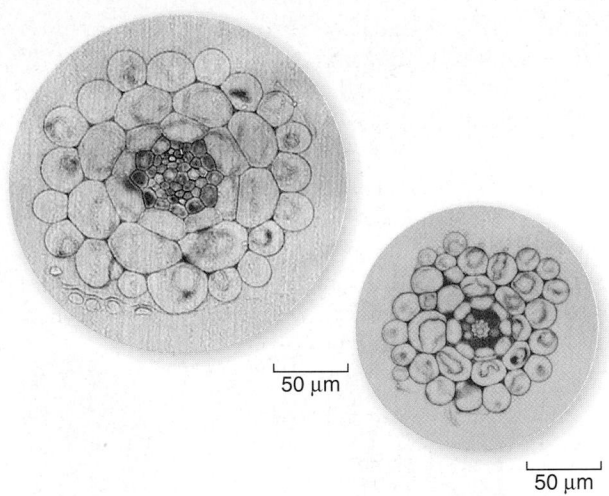

50 µm

50 µm

Figure 41.34 *WOODEN LEG* **is needed for phloem development.** The *wol* mutant (right) has less vascular tissue than wild-type *Arabidopsis* (left), but all of it is xylem.

phloem is missing. Only xylem forms in the vascular tissue system, giving the root a "wooden leg."

Early in embryonic development, most cells can give rise to a wide range of cell and organ types, including leaves. As development proceeds, the cells with multiple potentials are mainly restricted to the meristem regions. Many meristems are established by the time embryogenesis ends and the seed becomes dormant. After germination, apical meristems continue adding cells to the growing root and shoot tips. Apical meristem cells of corn, for example, divide every 12 hours, producing half a million cells per day in an actively growing corn plant. Lateral meristems can cause an increase in the girth of some plants, whereas intercalary meristems in the stems of grasses allow for elongation.

Food reserves form during embryogenesis

While the embryo is developing, three other critical events are occurring in angiosperms: (1) development of a food supply, (2) development of the seed coat, and (3) development of the fruit surrounding the seed. Nutritional reserves support the embryo during germination, while it gains photosynthetic capacity. In angiosperms, double fertilization produces endosperm for nutrition; in gymnosperms, the megagametophyte is the food source (see chapter 31). The seed coat is the result of the differentiation of ovule tissue (from the parental sporophyte) to form a hard, protective covering around the embryo. The seed then enters a dormant phase, signaling the end of embryogenesis. In angiosperms, the fruit develops from the carpel wall surrounding the ovule. Seed development and germination, as well as fruit development, are addressed later in this chapter. In this section, we focus on nutrient reserves.

Throughout embryogenesis, starch, lipids, and proteins are synthesized. The seed storage proteins are so abundant that the genes coding for them were the first cloning targets for plant molecular biologists. Providing nutritional resources is part of the evolutionary trend toward enhancing embryo survival.

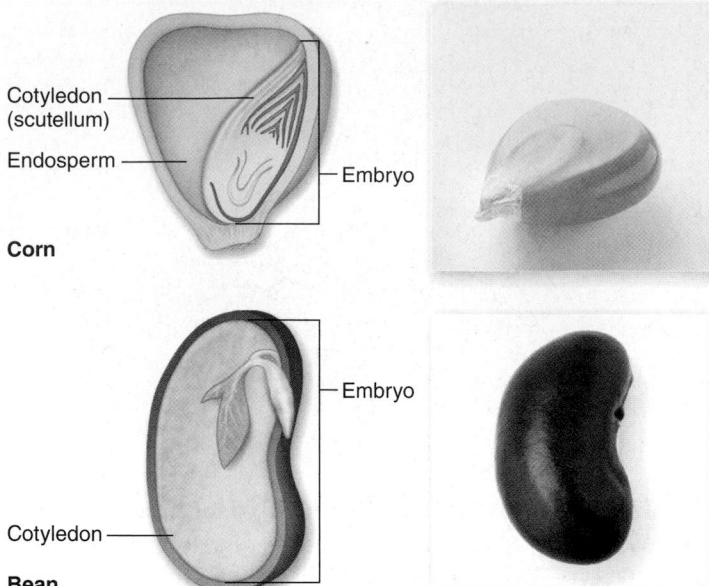

Cotyledon (scutellum)

Endosperm

Embryo

Corn

Embryo

Cotyledon

Bean

Figure 41.35 Endosperm in maize and bean. The maize kernel has endosperm that is still present at maturity, but the endosperm in the bean has disappeared. The bean embryo's cotyledons take over food storage functions.

The sporophyte transfers nutrients via the suspensor in angiosperms. (In gymnosperms, the suspensor serves only to push the embryo closer to the megagametophytic nutrient source.) This happens concurrently with the development of the endosperm, which is present only in angiosperms (although double fertilization has been observed in the gymnosperm *Ephedra*). Endosperm formation may be extensive or minimal.

Endosperm in coconut includes the "milk," a liquid. In corn, the endosperm is solid. In popping corn it expands with heat to form the white edible part of popped corn. In peas and beans, the endosperm is used up during embryo development, and nutrients are stored in thick, fleshy cotyledons (figure 41.35).

Because the photosynthetic machinery is built in response to light, it is critical that seeds have stored nutrients to aid in germination until the growing sporophyte can photosynthesize. A seed buried too deeply in the soil will use up all its reserves in cellular respiration before reaching the surface and sunlight.

Learning Outcomes Review 41.5

The root–shoot axis and the radial axis form during plant embryogenesis. The three tissues formed in an embryo are the protoderm, ground meristem, and procambium, which give rise to the three adult tissues. While the embryo is being formed, a food supply is being established for the embryo in the form of endosperm; a seed coat is forming from ovule tissues; and the fruit is developing from the carpel wall.

■ *How does the nutritive tissue of a gymnosperm seed differ from that of an angiosperm seed?*

41.6 Germination

Learning Outcomes

1. *Describe the events that occur during seed germination.*
2. *Contrast the pattern of shoot emergence in bean (eudicot) with that in maize (monocot).*

When conditions are satisfactory, the embryo emerges from its previously desiccated state, utilizes food reserves, and resumes growth. Although **germination** is a process characterized by several stages, it is often defined as the emergence of the **radicle** (first root) through the seed coat.

External signals and conditions trigger germination

Germination begins when a seed absorbs water and its metabolism resumes. The amount of water a seed can absorb is phenomenal, and osmotic pressure creates a force strong enough to break the seed coat. At this point, it is important that oxygen be available to the developing embryo because plants, like animals, require oxygen for cellular respiration. Few plants produce seeds that germinate successfully under water, although some, such as rice, have evolved a tolerance to anaerobic conditions.

Even though a dormant seed may have imbibed a full supply of water and may be respiring, synthesizing proteins and RNA, and apparently carrying on normal metabolism, it may fail to germinate without an additional signal from the environment. This signal may be light of the correct wavelength and intensity, a series of cold days, or simply the passage of time at temperatures appropriate for germination. The seeds of many plants will not germinate unless they have been **stratified**—held for periods of time at low temperatures. This phenomenon prevents the seeds of plants that grow in seasonally cold areas from germinating until they have passed the winter, thus protecting their tender seedlings from harsh, cold conditions.

Germination can occur over a wide temperature range (5°–30°C), although certain species may have relatively narrow optimum ranges. Some seeds will not germinate even under the best conditions. In some species, a significant fraction of a season's seeds remain dormant for an indeterminate length of time, providing a gene pool of great evolutionary significance to the future plant population. The presence of ungerminated seeds in the soil of an area is referred to as the **seed bank.**

Nutrient reserves sustain the growing seedling

Germination occurs when all internal and external requirements are met. Germination and early seedling growth require the utilization of metabolic reserves stored as starch in amyloplasts (colorless plastids) and protein bodies. Fats and oils, also stored, in some kinds of seeds, can readily be digested during germination to produce glycerol and fatty acids, which

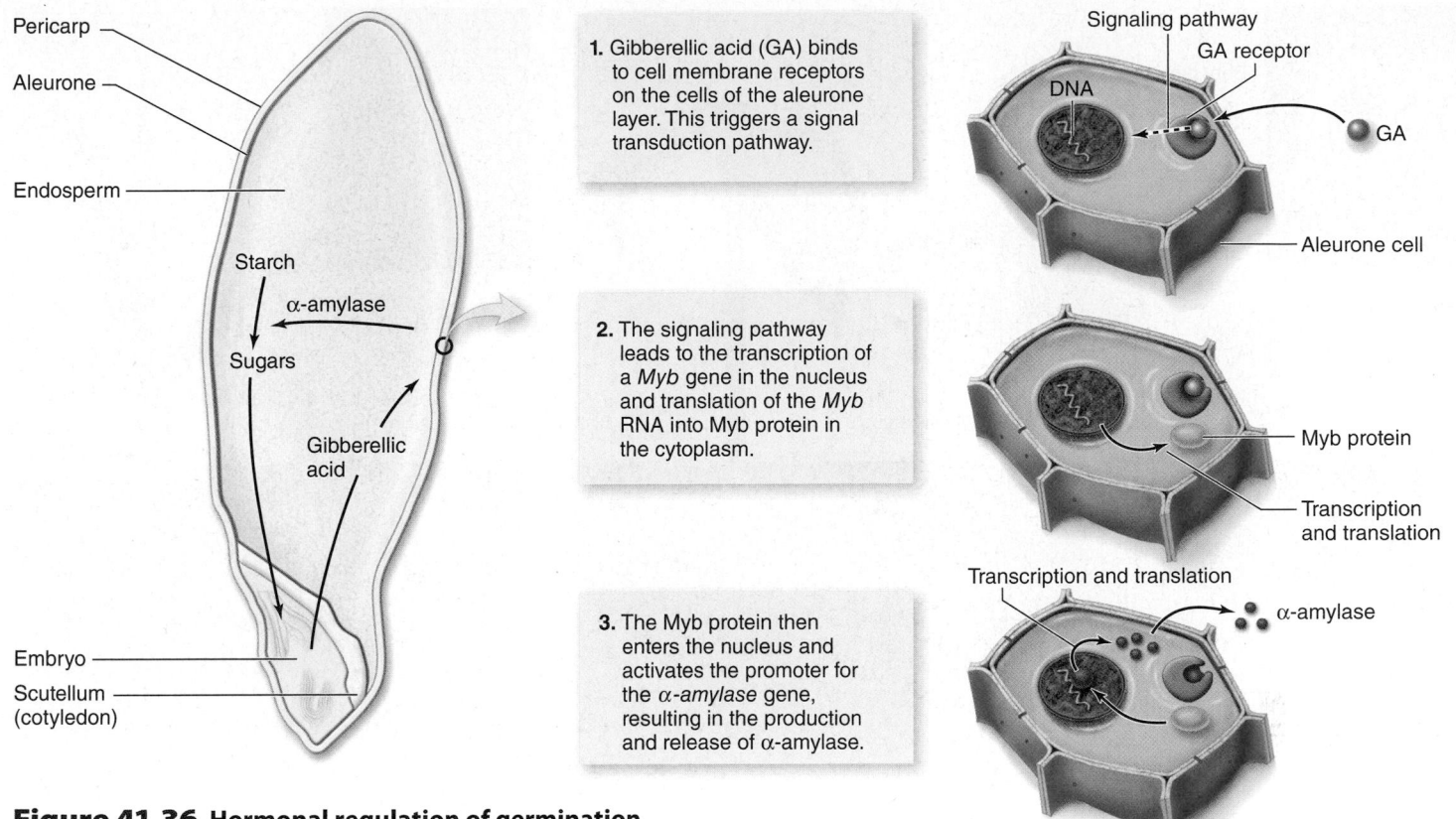

Pericarp

Aleurone

Endosperm

Starch

α-amylase

Sugars

Gibberellic
acid

Embryo

**Scutellum
(cotyledon)**

1. Gibberellic acid (GA) binds to cell membrane receptors on the cells of the aleurone layer. This triggers a signal transduction pathway.

Signaling pathway

GA receptor

DNA

GA

Aleurone cell

2. The signaling pathway leads to the transcription of a *Myb* gene in the nucleus and translation of the *Myb* RNA into Myb protein in the cytoplasm.

Myb protein

Transcription
and translation

Transcription and translation

α-amylase

3. The Myb protein then enters the nucleus and activates the promoter for the α-*amylase* gene, resulting in the production and release of α-amylase.

Figure 41.36 Hormonal regulation of germination.

yield energy through cellular respiration. They can also be converted to glucose. Depending on the kind of plant, any of these reserves may be stored in the embryo or in the endosperm.

In the kernels of cereal grains, the single cotyledon is modified into a relatively massive structure called the **scutellum** (figure 41.36). The abundant food stored in the scutellum is used up first during germination. Later, while the seedling is becoming established, the scutellum serves as a nutrient conduit from the endosperm to the rest of the embryo.

The utilization of stored starch by germinating plants is one of the best examples of how hormones modulate plant development (see figure 41.36). The embryo produces gibberellic acid, a hormone, which signals the outer layer of the endosperm, called the **aleurone,** to produce α-amylase. This enzyme is responsible for breaking down the endosperm's starch, primarily amylose, into sugars that are passed by the scutellum to the embryo. Abscisic acid, another plant hormone, which is important in establishing dormancy, can inhibit starch breakdown. Abscisic acid levels may be reduced when a seed beginning to germinate absorbs water. (The action of plant hormones is covered in chapter 40.)

The seedling becomes oriented in the environment, and photosynthesis begins

As the sporophyte pushes through the seed coat, it orients with the environment so that the root grows down and the shoot grows up. New growth comes from delicate meristems that are protected from environmental rigors. The shoot becomes photosynthetic, and the postembryonic phase of growth and development is under way. Figure 41.37 shows the process of germination and subsequent development of the plant body in eudicots and monocots.

The emerging shoot and root tips are protected by additional tissue layers in the monocots—the *coleoptile* surrounding the shoot, and the *coleorhiza* surrounding the radicle. Other protective strategies include having a bent shoot emerge so tissues with more rugged cell walls push through the soil.

The emergence of the embryonic root and shoot from the seed during germination varies widely from species to species. In most plants, the root emerges before the shoot appears and anchors the young seedling in the soil (see figure 41.37). In plants such as peas, the cotyledons may be held below ground; in other plants, such as beans, radishes, and onions, the cotyledons are held above ground. The cotyledons may become green and contribute to the nutrition of the seedling as it becomes established, or they may shrivel relatively quickly. The period from the germination of the seed to the establishment of the young plant is critical for the plant's survival; the seedling is unusually susceptible to disease, insect damage and drought during this period (figure 41.38). Soil composition and pH can also affect the survival of a newly germinated plant.

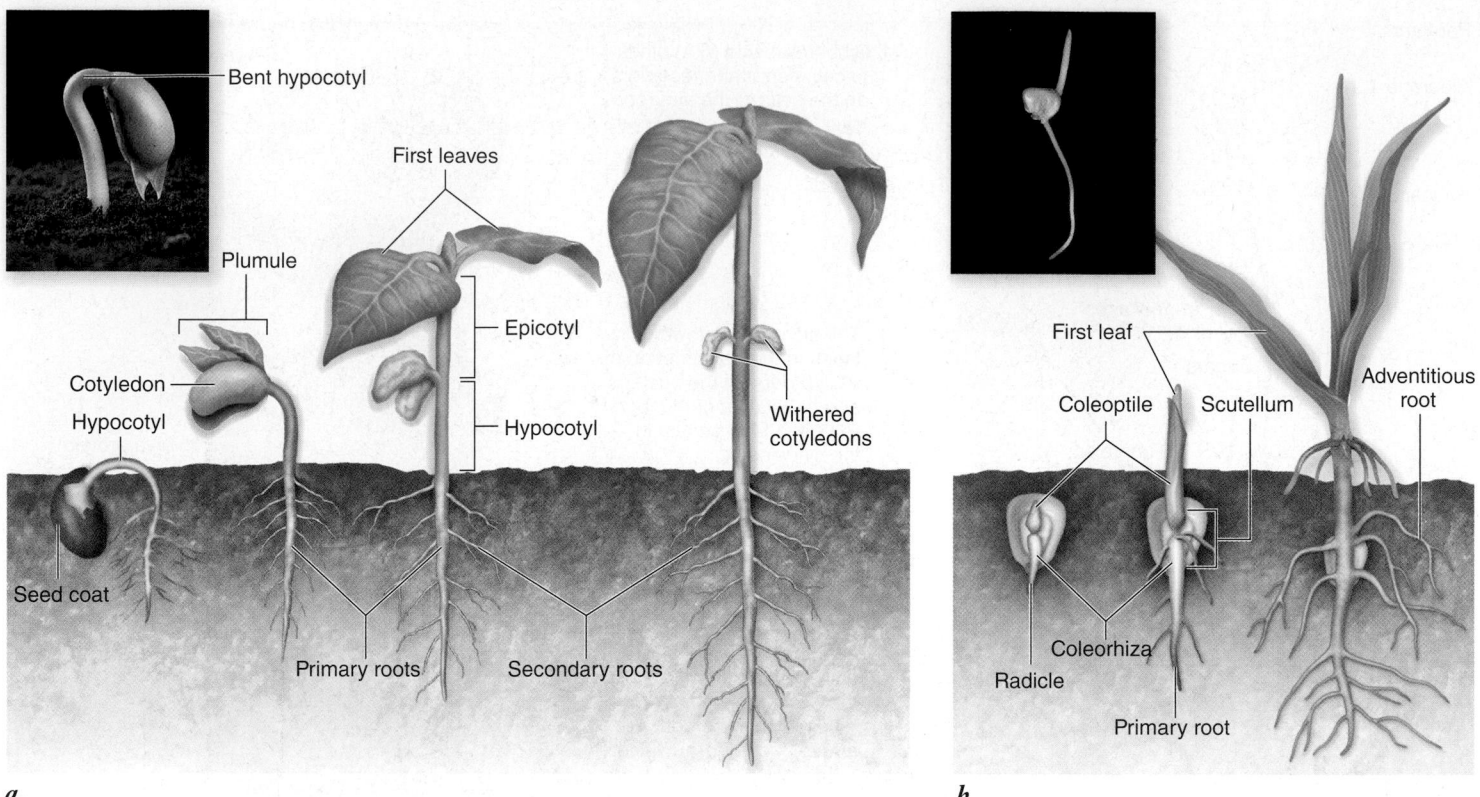

Figure 41.37 Germination. The stages shown are for *(a)* a eudicot, the common bean (*Phaseolus vulgaris*), and *(b)* a monocot, maize (*Zea mays*). Note that the bending of the hypocotyl (region below the cotyledons) protects the delicate bean shoot apex as it emerges through the soil. Maize radicles are protected by a protective layer of tissue called the coleorhiza, in addition to the root cap found in both bean and maize. A sheath of cells called the coleoptile, rather than a hypocotyl tissue, protects the emerging maize shoot tip.

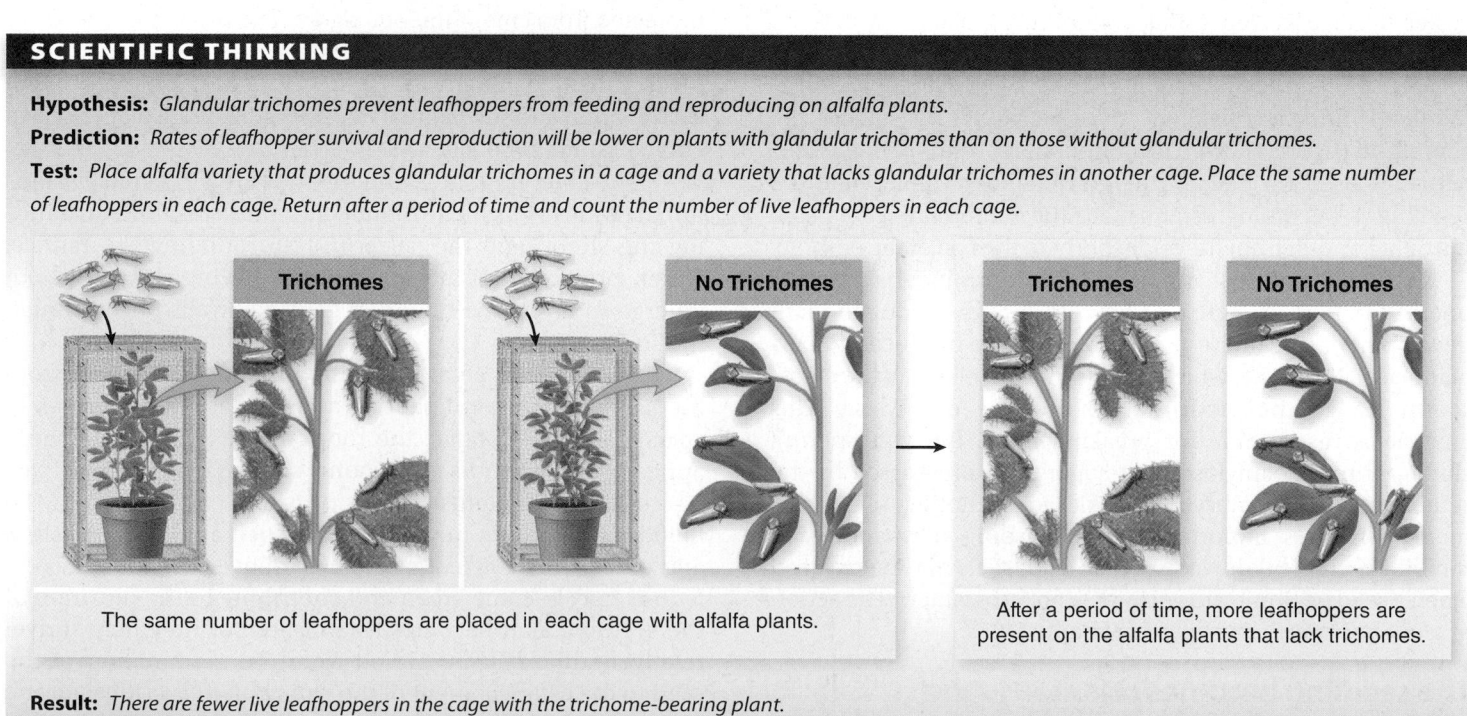

SCIENTIFIC THINKING

Hypothesis: *Glandular trichomes prevent leafhoppers from feeding and reproducing on alfalfa plants.*

Prediction: *Rates of leafhopper survival and reproduction will be lower on plants with glandular trichomes than on those without glandular trichomes.*

Test: *Place alfalfa variety that produces glandular trichomes in a cage and a variety that lacks glandular trichomes in another cage. Place the same number of leafhoppers in each cage. Return after a period of time and count the number of live leafhoppers in each cage.*

The same number of leafhoppers are placed in each cage with alfalfa plants.

After a period of time, more leafhoppers are present on the alfalfa plants that lack trichomes.

Result: *There are fewer live leafhoppers in the cage with the trichome-bearing plant.*

Conclusion: *The hypothesis is supported. The survival and reproduction rate of leafhoppers on plants with trichomes was lower than that on plants lacking trichomes.*

Further Experiments: *Design an experiment to determine if trichomes in general or just glandular trichomes can deter leafhoppers.*

Figure 41.38 Glandular trichomes can protect plants from insects.

41.7 Asexual Reproduction

Learning Outcomes

1. Define apomixis.
2. List examples of plant parts involved in vegetative reproduction.
3. Outline the steps involved in protoplast regeneration.

Self-pollination reduces genetic variability, but asexual reproduction results in genetically identical individuals because only mitotic cell divisions occur. In the absence of meiosis, individuals that are highly adapted to a relatively unchanging environment persist for the same reasons that self-pollination is favored. Should conditions change dramatically, there will be less variation in the population for natural selection to act on, and the species may be less likely to survive.

Asexual reproduction is also used in agriculture and horticulture to propagate a particularly desirable plant with traits that would be altered by sexual reproduction or even by self-pollination. Most roses and potatoes, for example, are vegetatively (asexually) propagated.

Apomixis involves development of diploid embryos

In certain plants, including some citruses, certain grasses (such as Kentucky bluegrass), and dandelions, the embryos in the seeds may be produced asexually from the parent plant. This kind of asexual reproduction is known as apomixis. Seeds produced in this way give rise to individuals that are genetically identical to their parents.

Although these plants reproduce by cloning diploid cells in the ovule, they also gain the advantage of seed dispersal, an adaptation usually associated with sexual reproduction. Asexual reproduction in plants is far more common in harsh or marginal environments, where there is little leeway for variation. For example, a greater proportion of asexual plants occurs in the Arctic than in temperate regions.

In vegetative reproduction, new plants arise from nonreproductive tissues

In a very common form of asexual reproduction called vegetative reproduction, new plant individuals are simply cloned from parts of adults (figure 41.39). The forms of vegetative reproduction in plants are many and varied.

Runners or stolons. Some plants reproduce by means of *runners* (also called stolons)—long, slender stems that grow along the surface of the soil. In the cultivated strawberry, for example, leaves, flowers, and roots are produced at every other node on the runner. Just beyond each second node, the tip of the runner turns up and becomes thickened. This thickened portion first produces adventitious roots and then a new shoot that continues the runner.

Rhizomes. Underground horizontal stems, or *rhizomes,* are also important reproductive structures, particularly in grasses and sedges. Rhizomes invade areas near the parent plant, and each node can give rise to a new flowering shoot. The noxious character of many weeds results from this type of growth pattern, and many garden plants, such as irises, are propagated almost entirely from rhizomes. Corms and bulbs are vertical underground stems. Tubers are also stems specialized for storage and reproduction. Tubers are the terminal storage portion of a rhizome. Potatoes (*Solanum* spp.) are propagated artificially from tuber segments, each with one or more "eyes." The eyes, or "seed pieces," of a potato give rise to the new plant.

Suckers. The roots of some plants—for example, cherry, apple, raspberry, and blackberry—produce *suckers,* or sprouts, which give rise to new plants. Commercial varieties of banana produce only sterile seeds and are propagated instead by suckers that develop from buds on underground stems. When the root of a dandelion is broken, as it may be if one attempts to pull it from the ground, each root fragment may give rise to a new plant.

Figure 41.39 Vegetative reproduction. Small plants arise from notches along the leaves of the house plant *Kalanchoë daigremontiana.* The plantlets can fall off and grow into new plants, an unusual form of vegetative reproduction.

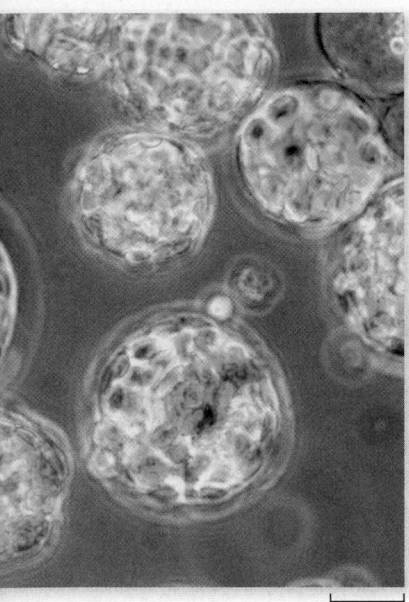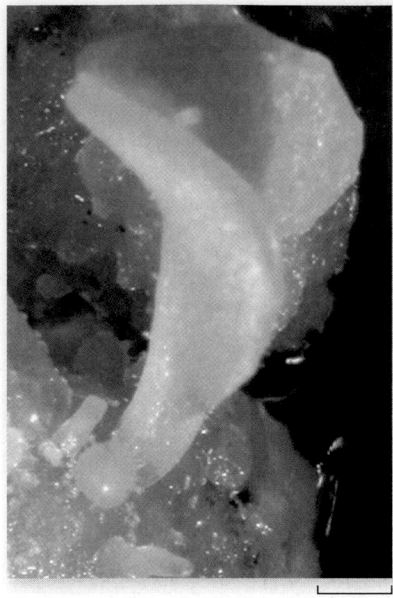

a.　　　　　　　100 μm　　b.　　　　　　　1 mm　　c.　　　　　　　1 mm　　d.

Figure 41.40 Protoplast regeneration. Different stages in the recovery of intact plants from single plant protoplasts of evening primrose. ***a.*** Individual plant protoplasts. ***b.*** Regeneration of the cell wall and the beginning of cell division. ***c.*** Production of somatic cell embryos from the callus. ***d.*** Recovery of a plantlet from the somatic cell embryo in culture. The plant can later be rooted in soil.

Adventitious plantlets. In a few plant species, even the leaves are reproductive. One example is the houseplant *Kalanchoë daigremontiana* (see figure 41.39), familiar to many people as the "maternity plant," or "mother of thousands." The common names of this plant are based on the fact that numerous plantlets arise from meristematic tissue located in notches along the leaves. The maternity plant is ordinarily propagated by means of these small plants, which, when they mature, drop to the soil and take root.

Plants can be cloned from isolated cells in the laboratory

Whole plants can be cloned by regenerating plant cells or tissues on nutrient medium with growth hormones. This is another form of asexual reproduction. Cultured leaf, stem, and root tissues can undergo organogenesis in culture and form roots and shoots. In some cases, individual cells can also give rise to whole plants in culture.

Individual cells can be isolated from tissues with enzymes that break down cell walls, leaving behind the *protoplast,* a plant cell enclosed only by a plasma membrane. Plant cells have greater developmental plasticity than most vertebrate animal cells, and many, but not all, cell types in plants maintain the ability to generate organs or an entire organism in culture. Consider the limited number of adult stem cells in vertebrates and the challenges associated with cloning discussed in chapter 19.

When single plant cells are cultured, wall regeneration takes place. Cell division follows to form a *callus,* an undifferentiated mass of cells (figure 41.40). Once a callus is formed, whole plants can be produced in culture. Whole-plant

development can go through an embryonic stage or can start with the formation of a shoot or root.

Tissue culture has many agricultural and horticultural applications. Virus-free raspberries and sugarcane can be propagated by culturing meristems, which are generally free of viruses, even in an infected plant. As with other forms of asexual reproduction, genetically identical individuals can be propagated.

Learning Outcomes Review 41.7

In apomixis, embryos are produced by mitosis rather than fertilization; in contrast, asexual vegetative reproduction occurs from vegetative plant parts. Examples include runners, stolons, rhizomes, suckers, and adventitious plant parts. In the laboratory, protoplasts are produced by isolating cells and removing the cell walls. Inducing mitosis results in a cluster of undifferentiated cells called a callus, which can then be stimulated to differentiate into a plant.

■ *Under what conditions would vegetative reproduction benefit survival?*

41.8 *Plant Life Spans*

Learning Outcomes

1. *Distinguish between herbaceous and woody perennials.*
2. *Define perennial and annual plants.*
3. *Describe the life cycle of a biennial plant.*

Once established, plants live for highly variable periods of time, depending on the species. Life span may or may not correlate with reproductive strategy. Woody plants, which have extensive secondary growth, nearly always live longer than herbaceous plants, which have limited or no secondary growth. Bristlecone pine, for example, can live upward of 4000 years.

Some herbaceous plants send new stems above the ground every year, producing them from woody underground structures. Others germinate and grow, flowering just once before they die. Shorter-lived plants rarely become very woody because there is not enough time for secondary tissues to accumulate. Depending on the length of their life cycles, herbaceous plants may be annual, biennial, or perennial, whereas woody plants are generally perennial (figure 41.41).

Determining life span is even more complicated for clonally reproducing organisms. Aspen trees (Populus tremuloides) form huge clones from asexual reproduction of their roots. Collectively, an aspen clone may form the largest "organism" on Earth. Other asexually reproducing plants may cover less territory but live for thousands of years. Creosote bushes (Larrea tridentata) in the Mojave Desert have been identified that are as much as 12,000 years old!

Perennial plants live for many years

Perennial plants continue to grow year after year and may be herbaceous (as are many woodland, wetland, and prairie wildflowers), or woody (as are trees and shrubs). The majority of vascular plant species are perennials. Perennial plants in general are able to flower and produce seeds and fruit for an indefinite number of growing seasons.

Herbaceous perennials rarely experience any secondary growth in their stems; the stems die each year after a period of relatively rapid growth and food accumulation. Food is often stored in the plants' roots or underground stems, which can become quite large in comparison with their less substantial aboveground counterparts.

Trees and shrubs generally flower repeatedly, but there are exceptions. Bamboo lives for many seasons as a nonreproducing plant, but senesces and dies after flowering. The same is true for at least one tropical tree (Tachigali versicolor), which achieves great heights before flowering and senescing. Considering the tremendous amount of energy that goes into the growth of a tree, this particular reproductive strategy is quite curious.

Trees and shrubs are either deciduous, with all the leaves falling at one particular time of year and the plants remaining bare for a period, or evergreen, with the leaves dropping throughout the year and the plants never appearing completely bare. In northern temperate regions, conifers are the most familiar evergreens, but in tropical and subtropical regions, most angiosperms are evergreen, except where severe seasonal drought occurs. In these areas, many angiosperms are deciduous, losing their leaves during the dry season and thus conserving water.

Annual plants grow, reproduce, and die in a single year

Annual plants grow, flower, and form fruits and seeds within one growing season and die when the process is complete. Many crop plants are annuals, including corn, wheat, and soybeans. Annuals generally grow rapidly under favorable conditions and in proportion to the availability of water or nutrients. The lateral meristems of some annuals, such as sunflowers or giant ragweed, do produce some secondary tissues for support, but most annuals are entirely herbaceous.

Annuals typically die after flowering once; the developing flowers or embryos use hormonal signaling to reallocate nutrients, so the parent plant literally starves to death. This can be demonstrated by comparing a population of bean plants in which the beans are continually picked with a population in which the beans are left on the plant. The frequently picked population will continue to grow and yield beans much longer than the untouched population. The process that leads to the death of a plant is called **senescence.**

a. *b.*

Figure 41.41 Annual and perennial plants. Plants live for very different lengths of time. *a.* Desert annuals complete their entire life span in a few weeks, flowering just once. *b.* Some trees, such as the giant redwood (Sequoiadendron giganteum), which occurs in scattered groves along the western slopes of the Sierra Nevada in California, live 2000 years or more, and flower year after year.

Biennial plants follow a two-year life cycle

Biennial plants, which are much less common than annuals, have life cycles that take two years to complete. During the first year, biennials store the products of photosynthesis in underground storage organs. During the second year of growth, flowering stems are produced using energy stored in the underground parts of the plant. Certain crop plants, including carrots, cabbage, and beets, are biennials, but these plants generally are harvested for food during their first season, before they flower. They are grown for their leaves or roots, not for their fruits or seeds.

Wild biennials include evening primroses, Queen Anne's lace *(Daucus carota)*, and mullein *(Verbascum thapsis)*. Many plants that are considered biennials actually do not flower until they are three or more years of age, but all biennial plants flower only once before they die.

Chapter Review

41.1 Reproductive Development

Plant life cycles are characterized by an alternation of generations (figure 41.1).

The transition to flowering competence is termed phase change.

Phase change prepares a plant to respond to external and internal signals to begin flowering. External factors include light and temperature, and internal factors include hormone production.

Mutations have clarified how phase change is controlled.

In experiments with *Arabidopsis,* plants that flower earlier than normal result from mutations in phase change genes. The implication is that mechanisms have evolved to delay flowering.

41.2 Making Flowers

Four genetically regulated pathways to flowering have been identified. The balance between floral-promoting and floral-inhibiting signals regulates flowering.

The light-dependent pathway is geared to the photoperiod (figure 41.6).

The light-dependent pathway induces flowering based on the length of the dark period a plant experiences during 24 hr. Plants may be short-day, long-day, or day-neutral, depending on their flowering response.

The temperature-dependent pathway is linked to cold.

Some plants require vernalization, or exposure of seeds or plants to chilling, in order to induce flowering.

The gibberellin-dependent pathway requires an increase in hormone level.

Decreased levels of gibberellins delay flowering in plants with this pathway. Gibberellins likely affect phase-change gene expression.

The autonomous pathway is independent of environmental cues.

The autonomous pathway is typical of day-neutral plants. A balance between floral-promoting and floral-inhibiting signals controls flower development.

Floral meristem identity genes activate floral organ identity genes.

Once floral organ identity genes are turned on, the four floral organs develop according to the ABC model. Class *A* genes alone specify sepals, classes *A* and *B* together specify petals, classes *B* and *C* specify stamens, and class *C* genes alone specify carpels.

41.3 Structure and Evolution of Flowers

Flowers evolved in the angiosperms.

Floral organs are believed to have evolved from leaves.

Complete flowers have four whorls corresponding to the four floral organs: the calyx, corolla, androecium, and gynoecium (figure 41.11). Incomplete flowers lack one or more of the whorls.

Angiosperms may have radially or bilaterally symmetrical flowers.

Gametes are produced in the gametophytes of flowers (figure 41.15).

Meiosis in the anthers produces microspores, which undergo mitosis to produce pollen grains, which are the male gametophytes or microgametophytes.

Each pollen grain contains the generative cell that later divides to produce two sperm cells and a tube cell.

Meiosis in the ovules produces megaspores, which undergo mitosis to produce embryo sacs, which are the female gametophytes or megagametophytes.

The embryo sac contains seven cells, one of which is the egg cell and one of which contains two polar nuclei. The latter cell develops into triploid endosperm after fertilization.

41.4 Pollination and Fertilization

Early seed plants were wind-pollinated.

Wind-pollination is a passive process and does not carry pollen over long distances. Consequently, plants must be relatively close together to ensure that pollination occurs.

Flowers and animal pollinators have coevolved.

Animal pollinators provide an efficient transfer of pollen that may cover long distances. Animal-pollinated flowers produce odors and visual cues to guide pollinators.

Some flowering plants use wind pollination.

Many wind-pollinated plant species have male and female flowers on separate individuals or on separate parts of each individual. The flowers are grouped in large numbers and exposed to the wind.

Self-pollination is favored in stable environments.

Plants adapted to a stable environment benefit from having uniform progeny that are likely to be more successful than those arising from cross-pollination. Offspring from self-pollination are not genetically identical, however.

Self pollination is also favored where animal pollinators are scarce.

Several evolutionary strategies promote outcrossing.

Outcrossing is promoted in plants in which male and female flowers are physically separated on the same plant or on different plants, or in which the two flowers mature on a different schedule.

Self-incompatibility prevents self-fertilization by preventing pollen tube growth (figure 41.25).

Angiosperms undergo double fertilization (figure 41.26).

Double fertilization produces a diploid zygote and triploid endosperm that provides nourishment to the zygote.

41.5 Embryo Development

A single cell divides to produce a three-dimensional body plan.

An angiosperm zygote divides to produce an embryo surrounded by endosperm (figure 41.27). In early divisions, the root-shoot axis and radial axis become established.

Developmental mutants in model plants reveal what can go wrong, allowing inferences about how development proceeds under normal conditions.

A simple body plan emerges during embryogenesis (figure 41.30).

Shoot and root apical meristems develop, and protoderm, ground meristem, and procambium differentiate; these will become the three types of tissue in an adult plant.

Morphogenesis creates a three-dimensional embryo that includes one or two cotyledons.

Food reserves form during embryogenesis.

While the embryo is being formed, a food supply is being established for the embryo. In angiosperms, this consists of the endosperm produced by double fertilization; in gymnosperms, the megagametophyte is the food source. In addition, a seed coat forms, and the fruit develops.

41.6 Germination

Seed germination is defined as the emergence of the radical through the seed coat.

External signals and conditions trigger germination.

A seed must imbibe water in order to germinate. Abundant oxygen is necessary to support the high metabolic rate of a germinating seed.

Environmental signals are often needed for germination. Examples include light of a certain wavelength, an appropriate temperature, and stratification (a period of chilling).

Nutrient reserves sustain the growing seedling.

Germination is a high-energy process, requiring stored nutrients such as starch, fats, and oils.

The endosperm acts as a starch reserve. Utilization of stored starch begins when the embryo produces the plant hormone gibberellic acid, which in turn stimulates production of an amylase to break down amylose (figure 41.36). Starch metabolism can be inhibited by abscisic acid, a plant hormone that has a role in dormancy.

The seedling becomes oriented in the environment, and photosynthesis begins.

In most plants, the root emerges before the shoot appears, anchoring the young seedling.

In many eudicots, the shoot is bent as it emerges from the soil, protecting the growing tip (figure 41.37). Monocots produce additional tissues to protect emerging shoots and roots.

During seedling emergence in eudicots such as beans, the cotyledons are often pulled up with the growing shoot. In monocots such as corn, the cotyledon remains underground.

A seedling enters the postembryonic phase of growth and development when the emerging shoot becomes photosynthetic.

41.7 Asexual Reproduction

Asexual reproduction results in genetically identical individuals because progeny are produced by mitosis.

Apomixis involves development of diploid embryos.

Apomixis is the production of embryos by mitosis rather than fertilization. These embryos develop in seeds.

In vegetative reproduction, new plants arise from nonreproductive tissues.

Vegetative parts such as runners, rhizomes, suckers, and adventitious plantlets may give rise to new individual clones (figure 41.39).

Plants can be cloned from isolated cells in the laboratory.

Stripping away the cell wall produces a protoplast, which can then be induced to undergo mitosis to produce a callus. With the proper treatments, the callus can differentiate into a complete plant.

41.8 Plant Life Spans

Perennial plants live for many years.

Perennials live for years, although they may undergo dormancy.

Annual plants grow, reproduce, and die in a single year.

Many crop plants are annuals and require replanting every year, such as corn, wheat, and soy beans.

Biennial plants follow a two-year life cycle.

During the first year, biennials grow and store nutrients. In the second year, they produce flowers and seeds. Biennial crop plants are often harvested during the first year, such as carrots.

Review Questions

UNDERSTAND

1. Morphogenesis is the development of
   a. growth form.
   c. a phase change.
   b. reproductive structures.
   d. meristems.

2. Vernalization induces flowering following exposure to
   a. water.
   c. cold.
   b. drought.
   d. heat.

3. Photoperiod is perceived by
   a. phytochrome and cryptochrome.
   b. phytochrome and chlorophyll.
   c. cryptochrome and chlorophyll.
   d. phytochrome, cryptochrome, and chlorophyll.

4. Which of the following is NOT a component of a flower?
   a. Sepal
   b. Stamen
   c. Carpel
   d. Bract

5. Megaspores are produced in
   a. anthers by mitosis.
   b. anthers by meiosis.
   c. ovules by mitosis.
   d. ovules by meiosis.

6. A stamen contains a
   a. style.
   b. stigma.
   c. filament.
   d. carpel.

7. Unlike bee-pollinated flowers, bird-pollinated flowers
   a. produce a strong fragrance.
   b. contain a landing pad.
   c. produce a bull's-eye pattern.
   d. are red.

8. After the first mitotic division of the zygote, the larger of the two cells becomes the
   a. embryo.
   b. endosperm.
   c. suspensor.
   d. micropyle.

9. Endosperm is produced by the union of
   a. a central cell with a sperm cell.
   b. a sperm cell with a synergid cell.
   c. an egg cell with a sperm cell.
   d. a suspensor with an egg cell.

10. During the globular stage of embryo development, apical meristems establish the
    a. embryo–suspensor axis.
    b. inner–outer axis.
    c. embryo–endosperm axis.
    d. root–shoot axis.

11. Which of the following is NOT a primary meristem?
    a. Cork cambium
    b. Ground meristem
    c. Procambium
    d. Protoderm

12. During seed germination, this hormone produces the signal for the aleurone to begin starch breakdown.
    a. Abscisic acid
    b. Ethylene
    c. Gibberellic acid
    d. Auxin

13. The shoot tip of an emerging maize seedling is protected by
    a. hypocotyl.
    b. epicotyl.
    c. coleoptile.
    d. plumule.

14. Asexual reproduction is likely to be most common in which ecosystem?
    a. Tropical rainforest
    b. Temperate grassland
    c. Arctic tundra
    d. Deciduous forest

15. Protoplasts are plant cells that lack
    a. nuclei.
    b. cell walls.
    c. plasma membranes.
    d. protoplasm.

16. Perennial plants are
    a. always herbaceous.
    b. always woody.
    c. either herbaceous or woody.
    d. neither herbaceous nor woody.

17. Senescence refers to
    a. plant aging.
    b. reproductive growth.
    c. pollination.
    d. the accumulation of storage reserves.

APPLY

1. Under which of the following conditions would pollen from an S_2S_5 plant successfully pollinate an S_1S_5 flower?
   a. Using pollen from a carpelate flower to fertilize a staminate flower would be successful.
   b. If the plants used gametophytic self-incompatibility, half of the pollen would be successful.
   c. If the plants used sporophytic self-incompatibility, half of the pollen would be successful.
   d. Pollen from an S_2S_5 plant can never pollinate an S_1S_5 flower.

2. Your roommate is taking biology with you this semester and thinks he understands short- and long-day plants. He purchases one plant of each type and decides to see the difference himself by first trying to cause the short-day plant to flower. He places both plants under the same conditions and exposes each to a regimen of 10-hr days, expecting that the short-day plant will flower, and the long-day plant will not. You play a trick on your roommate and reverse the outcome. Specifically, what did you have to do?
   a. Lengthen the time each is exposed to light
   b. Shorten the time each is exposed to light
   c. Quickly expose the plants to light during the middle of the night
   d. None of the choices is correct.

3. In Iowa, a company called Team Corn works to ensure that fields of seed corn outcross so that hybrid vigor can be maintained. They do this by removing the staminate (that is, pollen-producing) flowers from the corn plants. In an attempt to put Team Corn out of business, you would like to develop genetically engineered corn plants that
   a. contain Z genes to prevent germination of pollen on the stigmatic surface.
   b. contain S genes to stop pollen tube growth during self-fertilization.
   c. express B-type homeotic genes throughout developing flowers.
   d. express A-type homeotic genes throughout developing flowers.

4. Monoecious plants such as corn have either staminate or carpelate flowers. Knowing what you do about the molecular mechanisms of floral development, which of the following might explain the development of single-sex flowers?
   a. Expression of B-type genes in the presumptive carpel whorl will generate staminate flowers.
   b. Loss of A-type genes in the presumptive petal whorl will allow C-type and B-type genes to produce stamens instead of petals in that whorl.
   c. Restricting B-type gene expression to the presumptive petal whorl will generate carpelate flowers.
   d. All of the choices are correct.

5. One of the most notable differences between gamete formation in most animals and gamete formation in plants is that
   a. plants produce gametes in somatic tissue, whereas animals produce gametes in germ tissue.
   b. plants produce gametes by mitosis, whereas animals produce gametes by meiosis.
   c. plants produce only one of each gamete, but animals produce many gametes.
   d. plants produce gametes that are diploid, but animals produce gametes that are haploid.

6. A plant lacking the *WOODEN LEG* gene will likely
   a. be incapable of transporting water to its leaves.
   b. lack xylem and phloem.
   c. be incapable of transporting photosynthate.
   d. All of the choices are correct.

7. How would plant development change if the functions of the genes *SHOOTMERISTEMLESS (STM)* and *MONOPTEROUS (MP)* were reversed?
   a. The embryo–suspensor axis would be reversed.
   b. The embryo–suspensor axis would be duplicated.
   c. The root–shoot axis would be reversed.
   d. The root–shoot axis would be duplicated.

8. How would a loss-of-function mutation in the α-amylase gene affect seed germination?
   a. The seed could not imbibe water.
   b. The embryo would starve.
   c. The seed coat would not rupture.
   d. The seed would germinate prematurely.

9. Loss-of-function mutations in the *suspensor* gene in *Arabidopsis* lead to the development of two embryos in a seed. After analyzing the expression of this gene in early wild-type embryos, you find high levels of mRNA transcribed from the *suspensor* gene in the developing suspensor cells. What is the likely function of the suspensor protein?
   a. Suspensor protein likely stimulates development of the embryonic tissue.
   b. Suspensor protein likely stimulates development of the suspensor tissue.
   c. Suspensor protein likely inhibits embryonic development in the suspensor.
   d. Suspensor protein likely inhibits suspensor development in the embryo.

SYNTHESIZE

1. A commercial greenhouse in a remote location produces poinsettias. However, after a highway is built near the greenhouse, the poinsettias fail to flower. Explain what has happened.

2. If you live in a north temperate region, explain why it is advantageous to grow spinach for your salad in early spring rather than during the summer.

3. In wild columbine, flower morphology encourages cross-pollination. However, during the middle of the receptive period of the stigma, self-pollination can occur if the flower was not previously pollinated. If cross-pollination occurs after self-pollination, then that pollen reaches the base of the style before the self-pollen. Discuss the adaptive significance of this reproduction strategy.

4. In most parts of the world, commercial potato crops are produced asexually by planting tubers. However, in some regions of the world, such as Southeast Asia and the Andes, some potatoes are grown from true seeds. Discuss the advantages and disadvantages of growing potatoes from true seed.

5. Design an experiment to determine whether light or gravity is more important in determining the orientation of the rhizoid during zygote development in *Fucus*.

ONLINE RESOURCE

www.ravenbiology.com

Understand, Apply, and Synthesize—enhance your study with animations that bring concepts to life and practice tests to assess your understanding. Your instructor may also recommend the interactive eBook, individualized learning tools, and more.

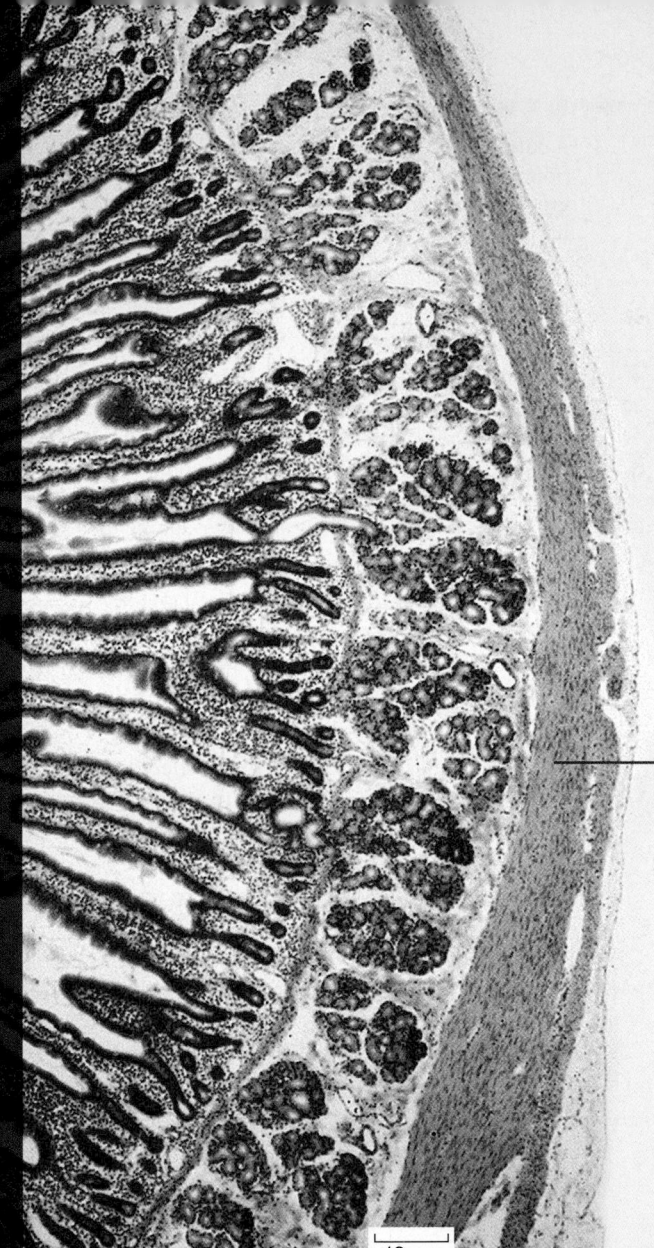

19 μm

Chapter **42**

The Animal Body and Principles of Regulation

Chapter Contents

Part **VII** Animal Form and Function

Introduction

When people think of animals, they may think of pet dogs and cats, the animals in a zoo, on a farm, in an aquarium, or wild animals living outdoors. When thinking about the diversity of animals, people may picture the differences between the predatory lions and tigers and the herbivorous deer and antelope, or between a dangerous shark and a playful dolphin. Despite the differences among these animals, they are all vertebrates. All vertebrates share the same basic body plan, with similar tissues and organs that operate in much the same way. The micrograph shows a portion of the duodenum, part of the digestive system, which is made up of multiple types of tissues. In this chapter, we begin a detailed consideration of the biology of the vertebrates and the fascinating structure and function of their bodies. We conclude this chapter by exploring the principles involved in regulation and control of complex functional systems.

Organization of the Vertebrate Body

Learning Outcomes

1. List the levels of organization in the vertebrate body.
2. Identify the tissue types found in vertebrates.
3. Describe how body cavities are organized.

The vertebrate body has four levels of organization: (1) cells, (2) tissues, (3) organs, and (4) organ systems. Like those of all animals, the bodies of vertebrates are composed of different cell types. Depending on the group, between 50 and several hundred different kinds of cells contribute to the adult vertebrate body. Humans have 210 different types of cells.

Tissues are groups of cells of a single type and function

Groups of cells that are similar in structure and function are organized into *tissues*. Early in development, the cells of the growing embryo differentiate into the three fundamental embryonic tissues, called **germ layers.** From the innermost to the outermost layers, these are the *endoderm, mesoderm,* and *ectoderm.* Each germ layer, in turn, differentiates into the scores of different cell types and tissues that are characteristic of the vertebrate body.

In adult vertebrates, there are four principal kinds of tissues, or **primary tissues:** (1) **epithelial,** (2) **connective,** (3) **muscle,** and (4) **nerve tissue.** Each type is discussed in separate sections of this chapter.

Organs and organ systems provide specialized functions

Organs are body structures composed of several different types of tissues that form a structural and functional unit (figure 42.1). One example is the heart, which contains cardiac muscle, connective tissue, and epithelial tissue. Nerve tissue connects the brain and spinal cord to the heart and helps regulate the heartbeat.

An **organ system** is a group of organs that cooperate to perform the major activities of the body. For example, the circulatory system is composed of the heart and blood vessels (arteries, capillaries, and veins) (see chapter 49). These organs cooperate in the transport of blood and help distribute substances about the body. The vertebrate body contains 11 principal organ systems.

The general body plan of vertebrates is a tube within a tube, with internal support

The bodies of all vertebrates have the same general architecture. The body plan is essentially a tube suspended within a tube. The inner tube is the digestive tract, a long tube that travels from the mouth to the anus. An internal skeleton made of jointed bones or cartilage that grows as the body grows supports the outer tube, which forms the main vertebrate body. The outermost layer of the vertebrate body is the integument, or skin, and its many accessory organs and parts—hair, feathers, scales, and sweat glands.

Vertebrates have both dorsal and ventral body cavities

Inside the main vertebrate body are two identifiable cavities. The *dorsal body cavity* forms within a bony skull and a column of bones, the vertebrae. The skull surrounds the brain, and within the stacked vertebrae is a channel that contains the spinal cord.

The *ventral body cavity* is much larger and extends anteriorly from the area bounded by the rib cage and vertebral column posteriorly to the area contained within the ventral body muscles (the abdominals) and the pelvic girdle. In mammals, a sheet of muscle, the diaphragm, breaks the ventral body cavity anteriorly into the *thoracic cavity,* which contains the heart and lungs, and posteriorly into the *abdominopelvic cavity,* which contains many organs, including the stomach, intestines, liver, kidneys, and urinary bladder (figure 42.2a).

Recall from the discussion of the animal body plan in chapter 33 that a coelom is a fluid-filled body cavity completely formed within the embryonic mesoderm layer of some animals (vertebrates included). The coelom is present in

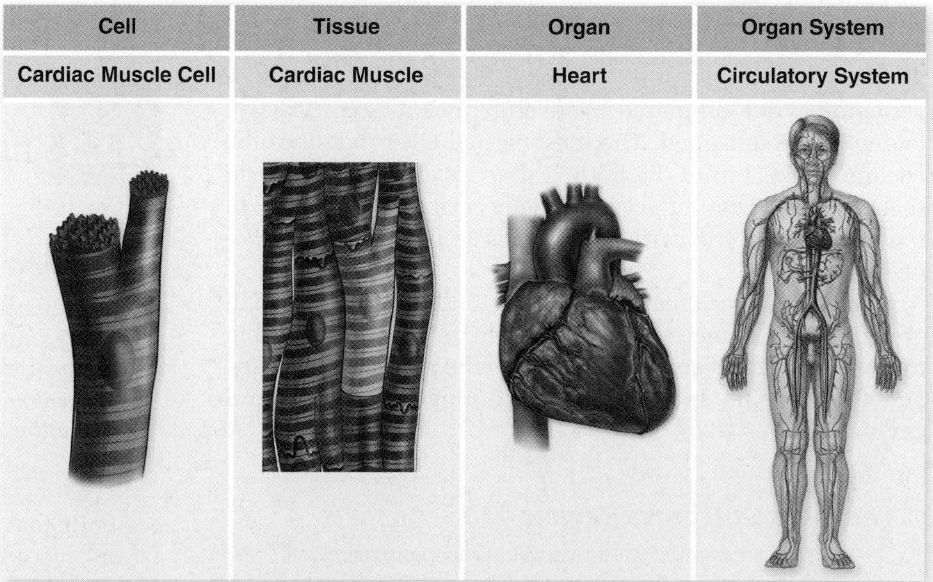

Cell	Tissue	Organ	Organ System
Cardiac Muscle Cell	Cardiac Muscle	Heart	Circulatory System

Figure 42.1 Levels of organization within the body. Similar cell types operate together and form tissues. Tissues functioning together form organs such as the heart, which is composed primarily of cardiac muscle with a lining of epithelial tissue. An organ system consists of several organs working together to carry out a function for the body. An example of an organ system is the circulatory system, which consists of the heart, blood vessels, and blood.

Figure 42.2 Architecture of the vertebrate body. *a.* All vertebrates have dorsal and ventral body cavities. The dorsal cavity divides into the cranial (contains the brain) and vertebral (contains the spinal cord) cavities. In mammals, a muscular diaphragm divides the ventral cavity into the thoracic and abdominopelvic cavities. *b.* Cross sections through three body regions show the relationships between body cavities, major organs, and coeloms (pericardial, pleural, and peritoneal cavities).

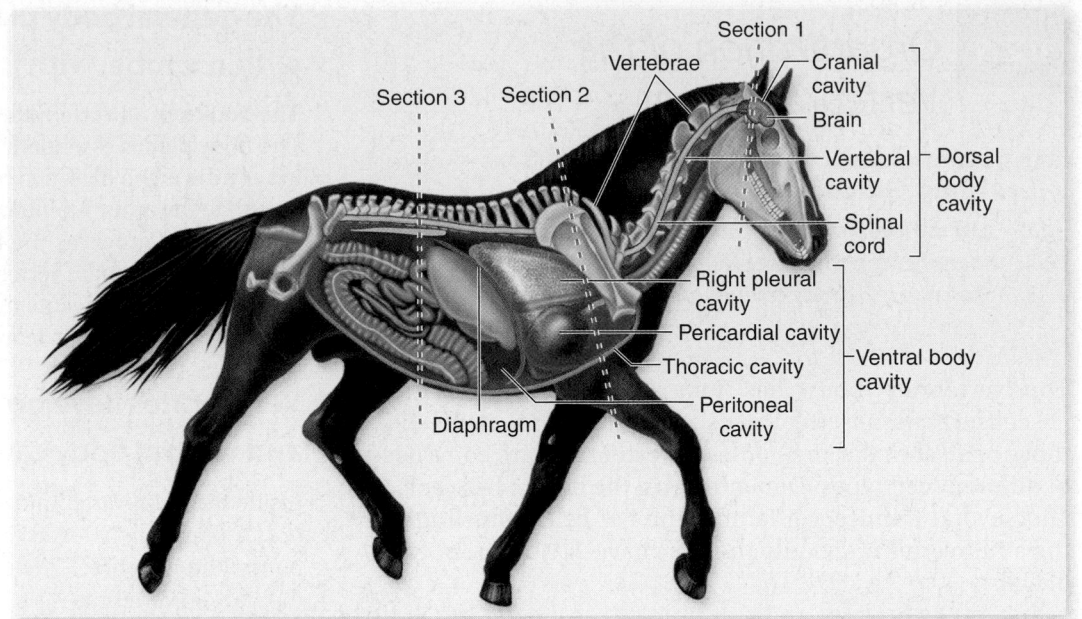

a.

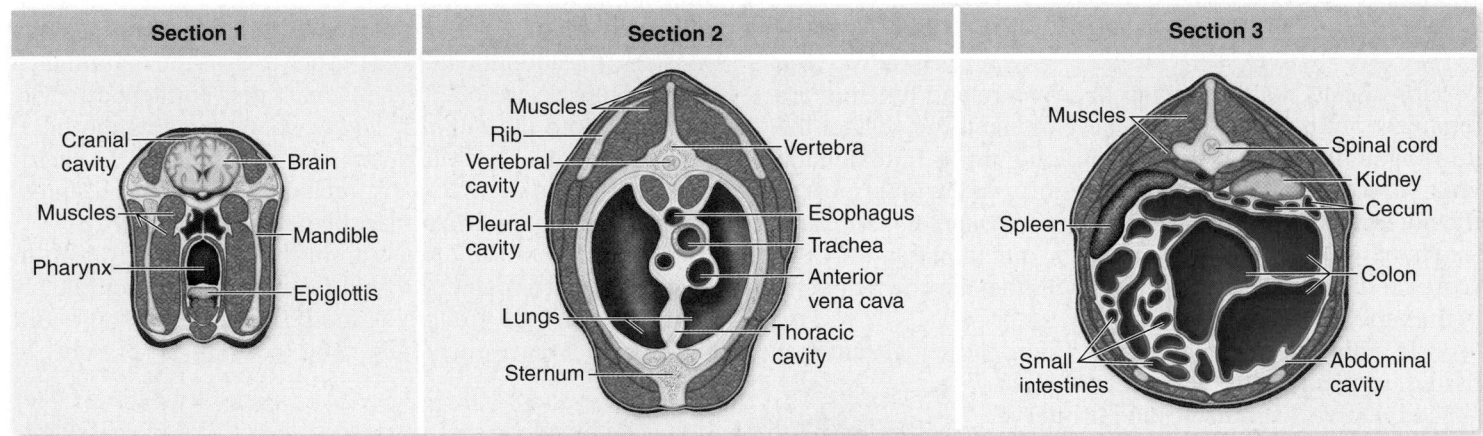

b.

vertebrates, but compared to invertebrates it is constricted, folded, and subdivided. The mesodermal layer that lines the coelom extends from the body wall to envelop and suspend several organs within the ventral body cavity (figure 42.2*b*). In the abdominopelvic cavity, the coelomic space is the *peritoneal cavity.*

In the thoracic cavity, the heart and lungs invade and greatly constrict the coelomic space. The thin space within mesodermal layers around the heart is the **pericardial cavity,** and the two thin spaces around the lungs are the **pleural cavities** (figure 42.2*b*).

Learning Outcomes Review 42.1

The body's cells are organized into tissues, which in turn are organized into organs and organ systems. The main types of tissues in vertebrates are epithelial, connective, muscle, and nerve tissue. The bodies of humans and other mammals contain dorsal and ventral cavities. The ventral cavity is divided by the diaphragm into thoracic and abdominopelvic cavities. The adult coelom subdivides into the peritoneal, pericardial, and pleural cavities.

■ *Can an organ be made of more than one tissue?*

42.2 Epithelial Tissue

Learning Outcomes

1. *Describe the structure and function of an epithelium.*
2. *Illustrate the cell types found in an epithelial membrane.*
3. *Explain the structure and function of different epithelia.*

An epithelial membrane, or **epithelium** (plural, *epithelia*), covers every surface of the vertebrate body. Epithelial membranes can come from any of the three germ layers. For example, the epidermis, derived from ectoderm, constitutes the outer portion of the skin. An epithelium derived from endoderm lines the inner surface of the digestive tract, and the inner surfaces of blood vessels derive from mesoderm. Some epithelia change in the course of embryonic development into glands, which are specialized for secretion.

Epithelium forms a barrier

Because epithelial membranes cover all body surfaces, a substance must pass through an epithelium in order to enter or leave the body. Epithelial membranes thus provide a barrier that can impede the passage of some substances while facilitating the passage of others. For land-dwelling vertebrates, the relative impermeability of the surface epithelium (the epidermis) to water offers essential protection from dehydration and from airborne pathogens. The epithelial lining of the digestive tract, in contrast, must allow selective entry of the products of digestion while providing a barrier to toxic substances. The epithelium of the lungs must allow for the rapid diffusion of gases into and out of the blood.

A characteristic of all epithelia is that the cells are tightly bound together, with very little space between them. Nutrients and oxygen must diffuse to the epithelial cells from blood vessels supplying underlying connective tissues. This places a limit on the thickness of epithelial membranes; most are only one or a few cell layers thick.

Epithelial regeneration

Epithelium possesses remarkable regenerative powers, constantly replacing its cells throughout the life of the animal. For example, the liver, a gland formed from epithelial tissue, can readily regenerate, even after surgical removal of substantial portions. The epidermis renews every two weeks, and the epithelium inside the stomach is completely replaced every two to three days. This ability to regenerate is useful in a surface tissue because it constantly renews the surface and also allows quick replacement of the protective layer should damage or injury occur.

Structure of epithelial tissues

Epithelial tissues attach to underlying connective tissues by a fibrous membrane. The secured side of the epithelium is called the *basal surface,* and the free side is the *apical surface.* This difference gives epithelial tissues an inherent polarity, which is often important in the function of the tissue. For example, proteins stud the basal surfaces of some epithelial tissues in the kidney tubules; these proteins actively transport Na$^+$ into the intercellular spaces, creating an osmotic gradient that helps return water to the blood (see chapter 50).

Epithelial types reflect their function

The two general classes of epithelial membranes are termed *simple* (single layer of cells) and *stratified* (multiple layers of cells). These classes are further subdivided into squamous, cuboidal, and columnar, based on the shape of the cells (table 42.1). *Squamous cells* are flat, *cuboidal cells* are about as wide as they are tall, and *columnar cells* are taller than they are wide.

Simple epithelium

As mentioned, *simple epithelial membranes* are one cell thick. A simple squamous epithelium is composed of squamous epithelial cells that have a flattened shape when viewed in cross section. Examples of such membranes are those that line the lungs and blood capillaries, where the thin, delicate nature of these membranes permits the rapid movement of molecules (such as the diffusion of gases).

A simple cuboidal epithelium lines kidney tubules and several glands. In the case of glands, these cells are specialized for secretion.

A simple columnar epithelium lines the airways of the respiratory tract and the inside of most of the gastrointestinal tract, among other locations. Interspersed among the columnar epithelial cells of mucous membranes are numerous *goblet cells,* which are specialized to secrete mucus. The columnar epithelial cells of the respiratory airways contain cilia on their apical surface (the surface facing the lumen, or cavity), which move mucus and dust particles toward the throat. In the small intestine, the apical surface of the columnar epithelial cells forms fingerlike projections called *microvilli,* which increase the surface area for the absorption of food.

The expanded size of both cuboidal and columnar cells accommodates the added intracellular machinery needed for production of glandular secretions, active absorption of materials, or both. The glands of vertebrates form from invaginated epithelia. In **exocrine glands,** the connection between the gland and the epithelial membrane remains as a duct. The duct channels the product of the gland to the surface of the epithelial membrane, and thus to the external environment (or to an interior compartment that opens to the exterior, such as the digestive tract). A few examples of exocrine glands include sweat and sebaceous (oil) glands as well as the salivary glands. **Endocrine glands** are ductless glands; their connections with the epithelium from which they are derived has been lost during development. Therefore, their secretions (hormones) do not channel onto an epithelial membrane. Instead, hormones enter blood capillaries and circulate through the body. Endocrine glands are covered in more detail in chapter 45.

Stratified epithelium

Stratified epithelial membranes are two to several cell layers thick and are named according to the features of their apical cell layers. For example, the epidermis is a *stratified squamous epithelium;* its properties are discussed in chapter 51. In terrestrial vertebrates, the epidermis is further characterized as a *keratinized epithelium* because its upper layer consists of dead squamous cells and is filled with a water-resistant protein called *keratin.*

The deposition of keratin in the skin increases in response to repeated abrasion, producing calluses. The water-resistant property of keratin is evident when comparing the skin of the face to the red portion of the lips, which can easily become dried and chapped. Lips are covered by a nonkeratinized, stratified squamous epithelium.

Learning Outcomes Review 42.2

Epithelial tissues generally form barriers and include membranes that cover all body surfaces and glands. An epidermis has a basal surface that attaches to an underlying connective tissue and an apical surface that is free. Some epithelia are specialized for protection, whereas those that cover the surfaces of hollow organs may be specialized for transport and secretion. Simple epithelium has a single cell layer and may be classified as squamous, cuboidal, columnar, or pseudostratified; stratified epithelium is primarily squamous.

■ *How does the epithelium in a gland function differently from that in the lining of your gut?*

TABLE 42.1   Epithelial Tissue

SIMPLE EPITHELIUM

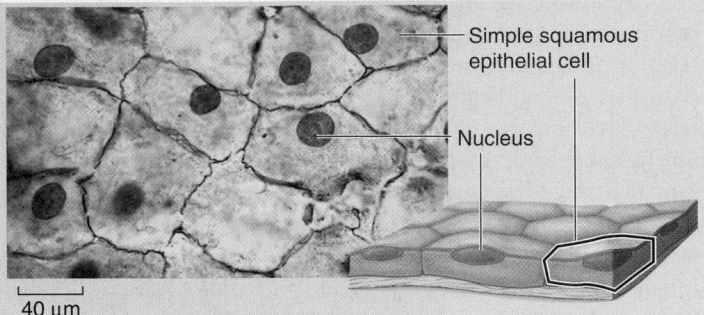

Simple squamous epithelial cell

Nucleus

40 μm

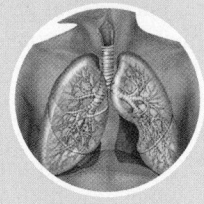

Squamous

Typical Location

Lining of lungs, capillary walls, and blood vessels

Function

Cells form thin layer across which diffusion can readily occur

Characteristic Cell Types

Epithelial cells

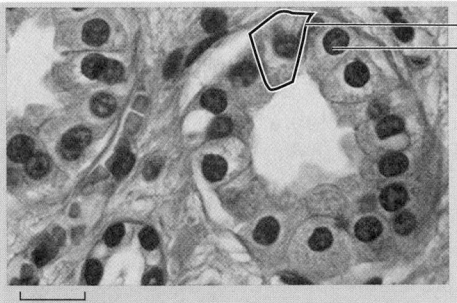

Cuboidal epithelial cell
Nucleus

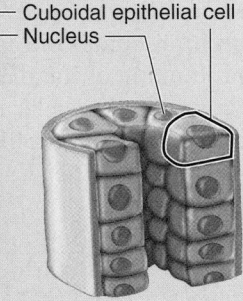

50 μm

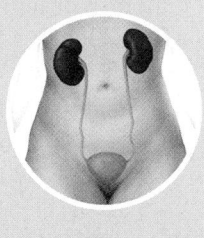

Cuboidal

Typical Location

Lining of some glands and kidney tubules; covering of ovaries

Function

Cells rich in specific transport channels; functions in secretion and absorption

Characteristic Cell Types

Gland cells

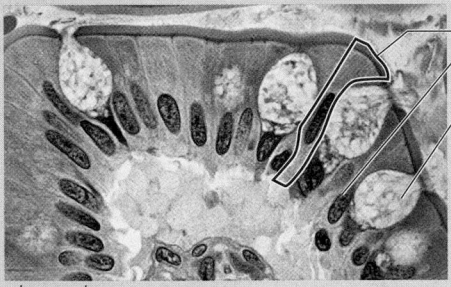

Columnar epithelial cell

Nucleus

Goblet cell

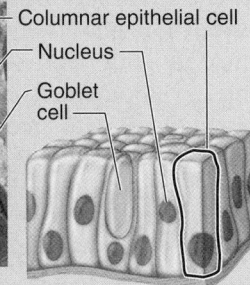

40 μm

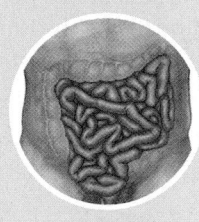

Columnar

Typical Location

Surface lining of stomach, intestines, and parts of respiratory tract

Function

Thicker cell layer; provides protection and functions in secretion and absorption

Characteristic Cell Types

Epithelial cells

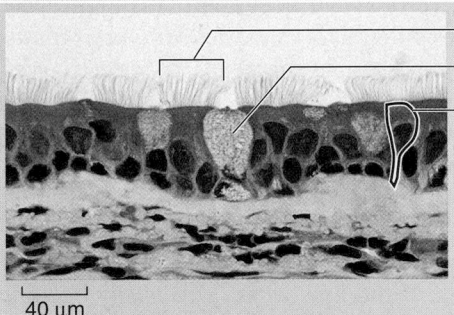

Cilia

Goblet cell

Pseudostratified columnar cell

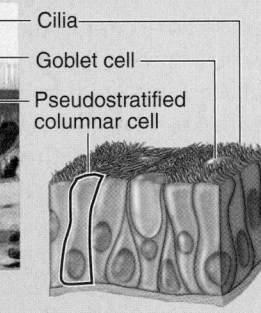

40 μm

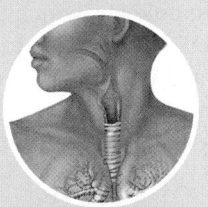

Pseudostratified Columnar

Typical Location

Lining of parts of the respiratory tract

Function

Secretes mucus; dense with cilia that aid in movement of mucus; provides protection

Characteristic Cell Types

Gland cells; ciliated epithelial cells

STRATIFIED EPITHELIUM

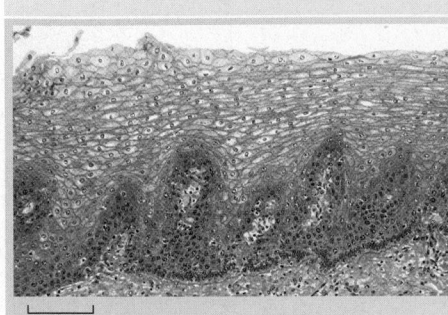

50 μm

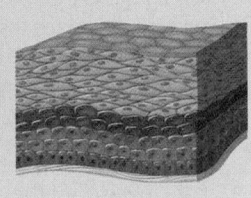

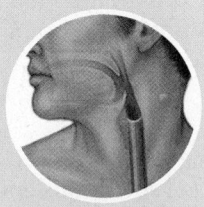

Squamous

Typical Location

Outer layer of skin; lining of mouth

Function

Tough layer of cells; provides protection

Characteristic Cell Types

Epithelial cells

42.3 Connective Tissue

Learning Outcomes

1. Describe the structure and function of connective tissue.
2. Differentiate among forms of connective tissue.
3. List the cells that make connective tissue.

Connective tissues derive from embryonic mesoderm and occur in many different forms (table 42.2). We divide these various forms into two major classes: *connective tissue proper*, which further divides into loose and dense connective tissues, and **special connective tissues,** which include cartilage, bone, and blood.

At first glance, it may seem odd that such diverse tissues are in the same category. Yet all connective tissues share a common structural feature: They all have abundant extracellular material because their cells are spaced widely apart. This extracellular material is called the **matrix** of the tissue. In bone, the matrix contains crystals that make the bones hard; in blood, the matrix is plasma, the fluid portion of the blood. The matrix itself consists of protein fibers and **ground substance,** the fluid material between cells and fibers containing a diverse array of proteins and polysaccharides.

Connective tissue proper may be either loose or dense

During the development of both loose and dense connective tissues, cells called fibroblasts produce and secrete the extracellular matrix. Loose connective tissue contains other cells as well, including mast cells and macrophages—cells of the immune system.

Loose connective tissue

Loose connective tissue consists of cells scattered within a matrix that contains a large amount of ground substance. This gelatinous material is strengthened by a loose scattering of protein fibers such as collagen, which supports the tissue by forming a meshwork (figure 42.3), elastin, which makes the tissue elastic, and reticulin, which helps support the network of collagen. The flavored gelatin of certain desserts consists primarily of extracellular material extracted from the loose connective tissues of animals.

Adipose cells, more commonly termed fat cells, are important for nutrient storage, and they also occur in loose connective tissue. In certain areas of the body, including under the skin, in bone marrow, and around the kidneys, these cells can develop in large groups, forming **adipose tissue** (figure 42.4).

Each adipose cell contains a droplet of triglycerides within a storage vesicle. When needed for energy, the adipose cell hydrolyzes its stored triglyceride and secretes fatty acids into the blood for oxidation by the cells of the muscles, liver, and other organs. Adipose cells cannot divide; the number of adipose cells in an adult is generally fixed. When a person

Figure 42.3 Collagen fibers. These fibers, shown under an electron microscope, are composed of many individual collagen strands and can be very strong under tension.

gains weight, the cells become larger, and when weight is lost, the cells shrink.

Dense connective tissue

Dense connective tissue, with less ground substance, contains tightly packed collagen fibers, making it stronger than loose connective tissue. It consists of two types: regular and irregular. The collagen fibers of *dense regular connective tissue* line up in parallel, like the strands of a rope. This is the structure of tendons, which bind muscle to bone, and ligaments, which bind bone to bone.

In contrast, the collagen fibers of *dense irregular connective tissue* have many different orientations. This type of connective tissue produces the tough coverings that package organs, such as the capsules of the kidneys and adrenal glands. It also covers muscle, nerves, and bones.

Special connective tissues have unique characteristics

The special connective tissues—cartilage, bone, and blood—each have unique cells and matrices that allow them to perform their specialized functions.

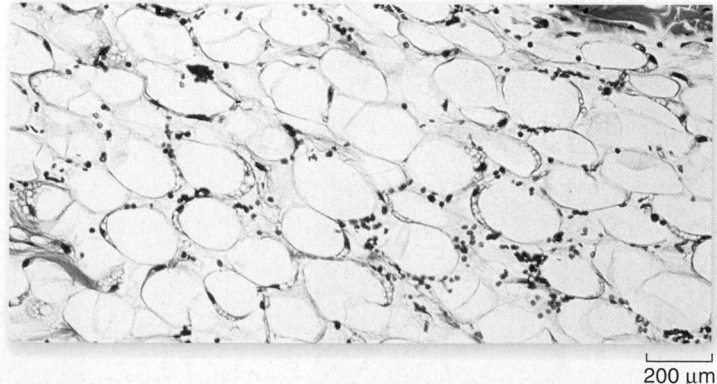

Figure 42.4 Adipose tissue. Fat is stored in globules of adipose tissue, a type of loose connective tissue. As a person gains or loses weight, the size of the fat globules increases or decreases. A person cannot decrease the number of fat cells by losing weight.

TABLE 42.2 Connective Tissue

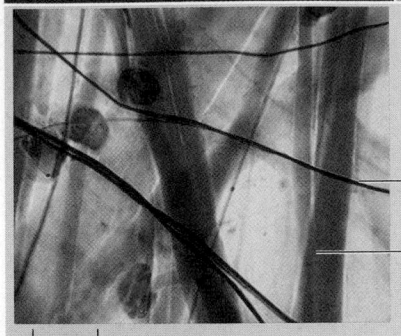

Elastin

Collagen

58 μm

Loose Connective Tissue
Typical Location
Beneath skin; between organs
Function
Provides support, insulation, food storage, and nourishment for epithelium
Characteristic Cell Types
Fibroblasts, macrophages, mast cells, fat cells

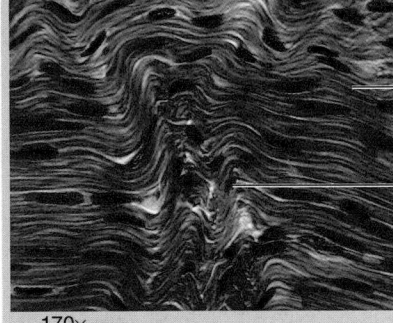
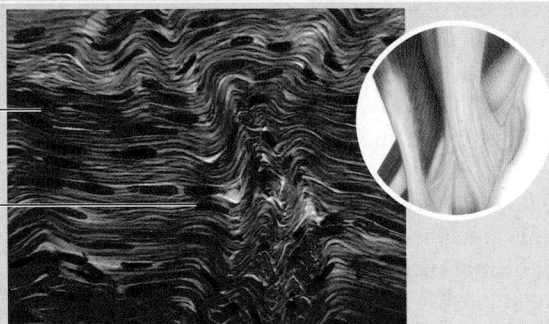

Collagen fibers

Nuclei of fibroblasts

170×

Dense Connective Tissue
Typical Location
Tendons; sheath around muscles; kidney; liver; dermis of skin
Function
Provides flexible, strong connections
Characteristic Cell Types
Fibroblasts

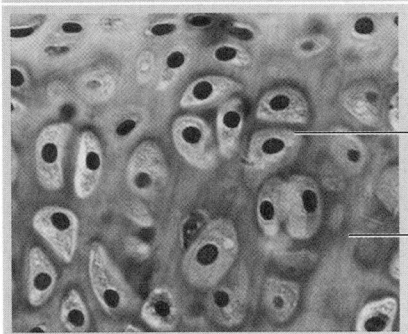
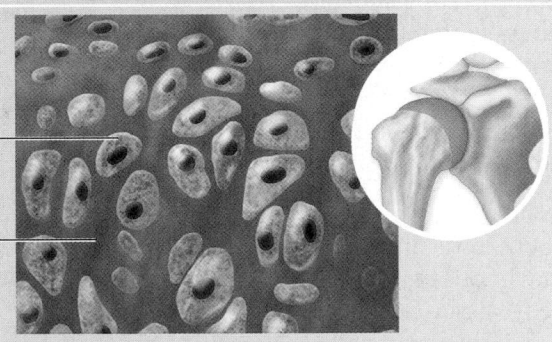

Chondrocyte

Ground substance

100 μm

Cartilage
Typical Location
Spinal disks; knees and other joints; ear; nose; tracheal rings
Function
Provides flexible support, shock absorption, and reduction of friction on load-bearing surfaces
Characteristic Cell Types
Chondrocytes

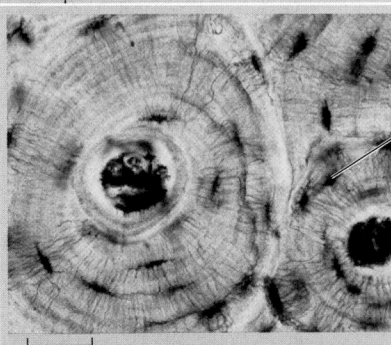
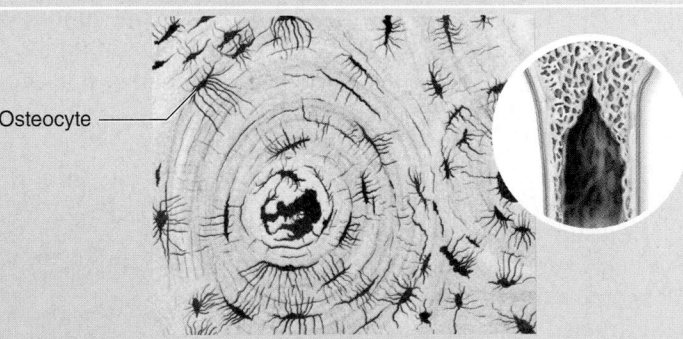

Osteocyte

100 μm

Bone
Typical Location
Most of skeleton
Function
Protects internal organs; provides rigid support for muscle attachment
Characteristic Cell Types
Osteocytes

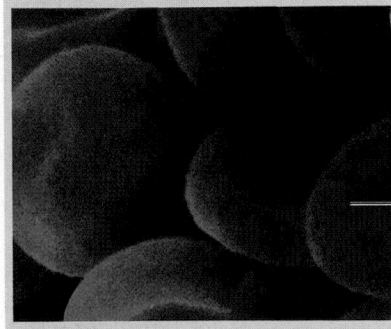
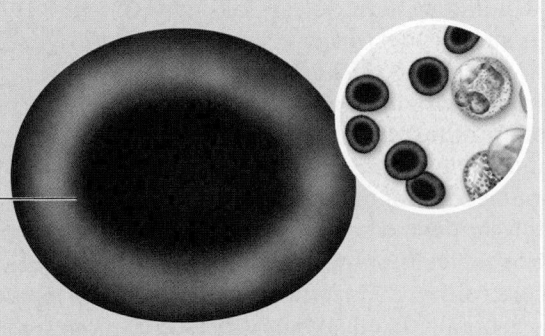

Red blood cell

6 μm

Blood
Typical Location
Circulatory system
Function
Functions as highway of immune system; carries nutrients and waste; and is the primary means of communication between organs
Characteristic Cell Types
Erythrocytes, leukocytes

Cartilage

Cartilage (see table 42.2) is a specialized connective tissue in which the ground substance forms from a characteristic type of glycoprotein, called *chondroitin,* and collagen fibers laid down along lines of stress in long, parallel arrays. The result is a firm and flexible tissue that does not stretch, is far tougher than loose or dense connective tissue, and has great tensile strength.

Cartilage makes up the entire skeletal system of the modern agnathans and cartilaginous fishes (see chapter 35). In most adult vertebrates, however, cartilage is restricted to the joint surfaces of bones that form freely movable joints and certain other locations. In humans, for example, the tip of the nose, the outer ear, the intervertebral disks of the backbone, the larynx, and a few other structures are composed of cartilage.

Chondrocytes, the cells of cartilage, live within spaces called **lacunae** within the cartilage ground substance. These cells remain alive even though there are no blood vessels within the cartilage matrix; they receive oxygen and nutrients by diffusion through the cartilage ground substance from surrounding blood vessels. This diffusion can only occur because the cartilage matrix is well hydrated and not calcified, as is bone.

Bone

Bone cells, or **osteocytes,** remain alive even though the extracellular matrix becomes hardened with crystals of calcium phosphate. Blood vessels travel through central canals into the bone, providing nutrients and removing wastes. Osteocytes extend cytoplasmic processes toward neighboring osteocytes through tiny canals, or *canaliculi.* Osteocytes communicate with the blood vessels in the central canal through this cytoplasmic network. Bone is described in more detail in chapter 46 along with muscle.

In the course of fetal development, the bones of vertebrate fins, arms, and legs, among other appendages, are first "modeled" in cartilage. The cartilage matrix then calcifies at particular locations, so that the chondrocytes are no longer able to obtain oxygen and nutrients by diffusion through the matrix. Living bone replaces the dying and degenerating cartilage.

Blood

We classify *blood* as a connective tissue because it contains abundant extracellular material, the fluid plasma. The cells of blood are *erythrocytes,* or red blood cells, and *leukocytes,* or white blood cells. Blood also contains platelets, or *thrombocytes,* which are fragments of a type of bone marrow cell. We discuss blood more fully in chapter 49.

All connective tissues have similarities

Although the descriptions of the types of connective tissue suggest numerous different functions for these tissues, they have some similarities. As mentioned, connective tissues originate as embryonic mesoderm, and they all contain abundant extracellular material called matrix; however, the extracellular matrix material is different in different types of connective tissue. Embedded within the extracellular matrix of each tissue type are varieties of cells, each with specialized functions.

42.4 Muscle Tissue

Muscles are the motors of the vertebrate body. The characteristic that makes muscle cells unique is the relative abundance and organization of actin and myosin filaments within them. Although these filaments form a fine network in all eukaryotic cells, where they contribute to movement of materials within the cell, they are far more abundant and organized in muscle cells, which are specialized for contraction.

Vertebrates possess three kinds of muscle: *smooth, skeletal,* and *cardiac* (table 42.3). Skeletal and cardiac muscles are also known as *striated muscles* because their cells appear to have transverse stripes when viewed in longitudinal section under the microscope. The contraction of each skeletal muscle is under voluntary control, whereas the contraction of cardiac and smooth muscles is generally involuntary.

Smooth muscle is found in most organs

Smooth muscle was the earliest form of muscle to evolve, and it is found throughout most of the animal kingdom. In vertebrates, smooth muscle occurs in the organs of the internal environment, or *viscera,* and is also called *visceral muscle.* Smooth muscle tissue is arranged into sheets of long, spindle-shaped cells, each cell containing a single nucleus. In some tissues, the cells contract only when a nerve stimulates them—and then all of the cells in the sheet contract as a unit.

In vertebrates, muscles of this type line the walls of many blood vessels and make up the iris of the eye, which contracts in bright light. In other smooth muscle tissues, such as those in the wall of the digestive tract, the muscle cells themselves may spontaneously initiate electrical impulses, leading to a slow, steady contraction of the tissue. Here nerves regulate, rather than cause, the activity.

Skeletal muscle moves the body

Skeletal muscles are usually attached to bones by tendons, so that their contraction causes the bones to move at their joints. A skeletal muscle is made up of numerous, very long muscle

TABLE 42.3   Muscle Tissue

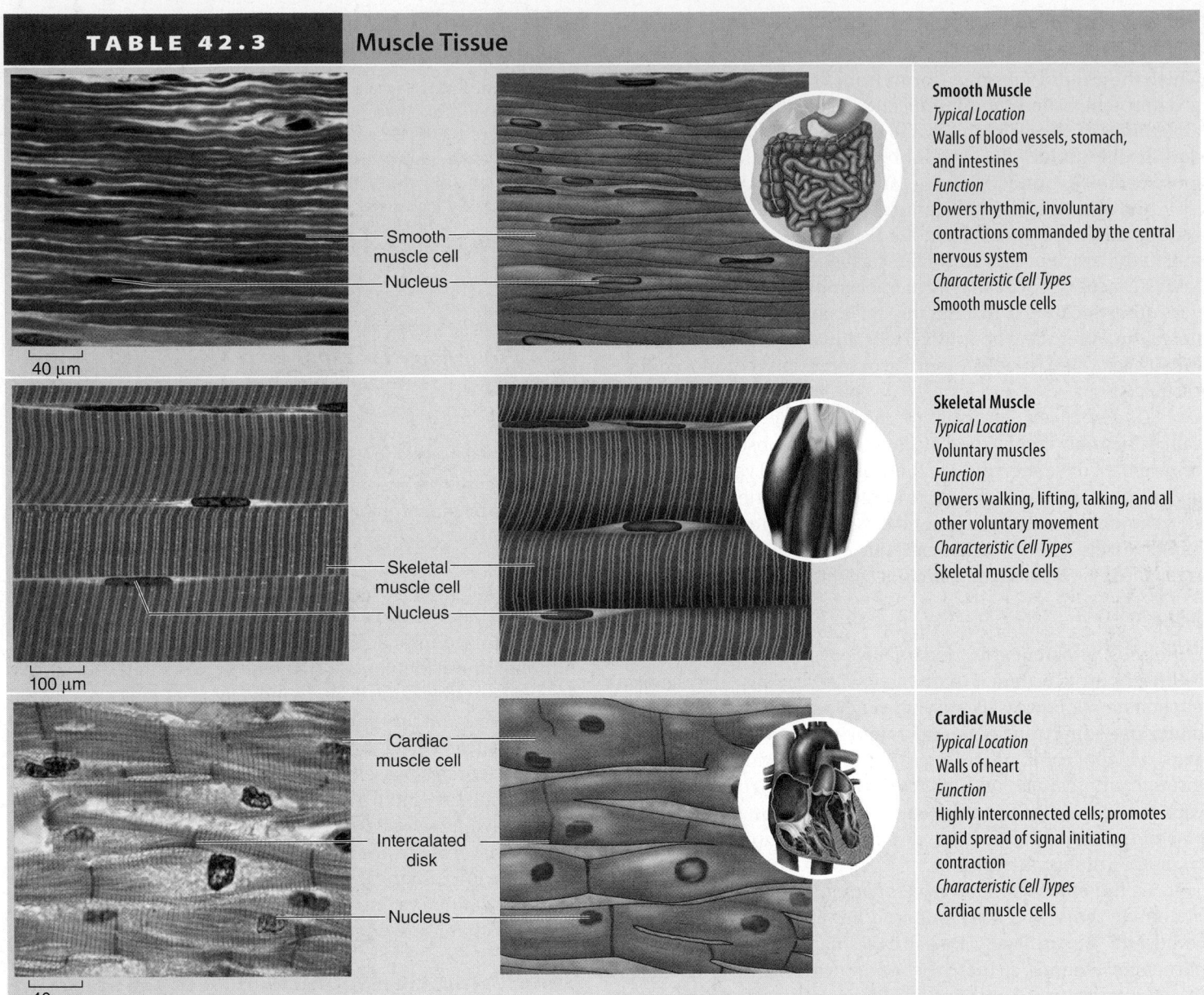

Smooth Muscle

Typical Location
Walls of blood vessels, stomach, and intestines

Function
Powers rhythmic, involuntary contractions commanded by the central nervous system

Characteristic Cell Types
Smooth muscle cells

Smooth muscle cell
Nucleus
40 μm

Skeletal Muscle

Typical Location
Voluntary muscles

Function
Powers walking, lifting, talking, and all other voluntary movement

Characteristic Cell Types
Skeletal muscle cells

Skeletal muscle cell
Nucleus
100 μm

Cardiac Muscle

Typical Location
Walls of heart

Function
Highly interconnected cells; promotes rapid spread of signal initiating contraction

Characteristic Cell Types
Cardiac muscle cells

Cardiac muscle cell
Intercalated disk
Nucleus
40 μm

cells called **muscle fibers,** which have multiple nuclei. The fibers lie parallel to each other within the muscle and are connected to the tendons on the ends of the muscle. Each skeletal muscle fiber is stimulated to contract by a motor neuron.

The nervous system controls the overall strength of a skeletal muscle contraction by controlling the number of motor neurons that fire, and therefore the number of muscle fibers stimulated to contract. Each muscle fiber contracts by means of substructures called **myofibrils** containing highly ordered arrays of actin and myosin myofilaments. These filaments give the muscle fiber its striated appearance.

Skeletal muscle fibers are produced during development by the fusion of several cells, end to end. This embryological development explains why a mature muscle fiber contains many nuclei. The structure and function of skeletal muscle is explained in more detail in chapter 46.

The heart is composed of cardiac muscle

The hearts of vertebrates are made up of striated muscle cells arranged very differently from the fibers of skeletal muscle. Instead of having very long, multinucleate cells running the length of the muscle, **cardiac muscle** consists of smaller, interconnected cells, each with a single nucleus. The interconnections between adjacent cells appear under the microscope as dark lines called **intercalated disks.** In reality, these lines are regions where gap junctions link adjacent cells. As noted in chapter 4, gap junctions have openings that permit the movement of small substances and ions from one cell to another. These interconnections enable the cardiac muscle cells to form a single functioning unit.

Certain specialized cardiac muscle cells can generate electrical impulses spontaneously, but the nervous system usually

Question: *Is the heartbeat a function of the nervous system, or does it originate in the heart itself?*

Hypothesis: *Cells in the heart are capable of generating an action potential without stimulation by the nervous system.*

Prediction: *If the heartbeat is produced by cells in the heart, then an isolated heart should continue to beat.*

Test: *Remove a frog's heart and keep in a bath of nutrient solution and oxygen.*

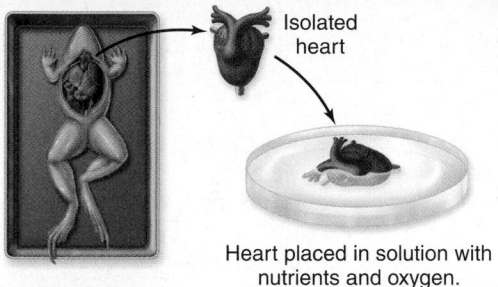

Isolated heart

Heart placed in solution with nutrients and oxygen.

Result: *The heart continues to contract with no connection to the nervous system.*

Conclusion: *The heartbeat is intrinsic to the heart.*

Further Experiments: *How would you integrate the conclusions from this experiment with the one described in figure 43.12?*

Figure 42.5 **The source of the heartbeat.**

regulates the rate of impulse activity (figure 42.5). The impulses generated by the specialized cell groups spread across the gap junctions from cell to cell, synchronizing the heart's contraction. Chapter 49 describes this process more fully.

Learning Outcomes Review 42.4

Muscles are the motors of the body; they are able to contract to change their length. Muscle tissue is of three types: smooth, skeletal, and cardiac. Smooth muscles provide a variety of visceral functions. Skeletal muscles enable the vertebrate body to move. Cardiac muscle forms a muscular pump, the heart.

■ *Why is it important that cardiac muscle cells have gap junctions?*

42.5 Nerve Tissue

Learning Outcomes

1. Describe the basic structure of neurons.
2. Identify the two divisions of the nervous system.

The fourth major class of vertebrate tissue is nerve tissue (table 42.4). Its cells include neurons and their supporting cells, called neuroglia. Neurons are specialized to produce and conduct electrochemical events, or impulses.

Neurons sometimes extend long distances

Most **neurons** consist of three parts: a cell body, dendrites, and an axon. The *cell body* of a neuron contains the nucleus. *Dendrites* are thin, highly branched extensions that receive incoming stimulation and conduct electrical impulses to the cell body. The *axon* is a single extension of cytoplasm that conducts impulses away from the cell body. Axons and dendrites can be quite long. For example, the cell bodies of neurons that control the muscles in your feet lie in the spinal cord, and their axons may extend over a meter to your feet.

Neuroglia provide support for neurons

Neuroglia do not conduct electrical impulses, but instead support and insulate neurons and eliminate foreign materials in and around neurons. In many neurons, neuroglia cells associate with the axons and form an insulating covering, a *myelin sheath,* produced by successive wrapping of the membrane around the axon. Gaps in the myelin sheath, known as *nodes of Ranvier,* serve as sites for accelerating an impulse (see chapter 43).

Two divisions of the nervous system coordinate activity

The nervous system is divided into the **central nervous system (CNS),** which includes the brain and spinal cord, and the **peripheral nervous system (PNS),** which includes *nerves* and *ganglia.* Nerves consist of axons in the PNS that are bundled together in much the same way as wires are bundled together in a cable. Ganglia are collections of neuron cell bodies. The CNS generally has the role of integration and interpretation of input, such as input from the senses; the PNS communicates signals to and from the CNS to the rest of the body, such as to muscle cells or endocrine glands.

Learning Outcomes Review 42.5

Nerve tissue is composed of neurons and neuroglia. Neurons are specialized to receive and conduct electrical signals; they generally have a cell body with a nucleus, dendrites that receive incoming signals, and axons that conduct impulses away from the cell body. Neuroglia have support functions, including providing insulation to axons. The central nervous system (CNS) and peripheral nervous system (PNS) both contain neurons and neuroglia.

■ *In chapter 4 you read that the surface-area-to-volume ratio limits cell size. How do neurons reach up to a meter in length in spite of this?*

TABLE 42.4   Nerve Tissue

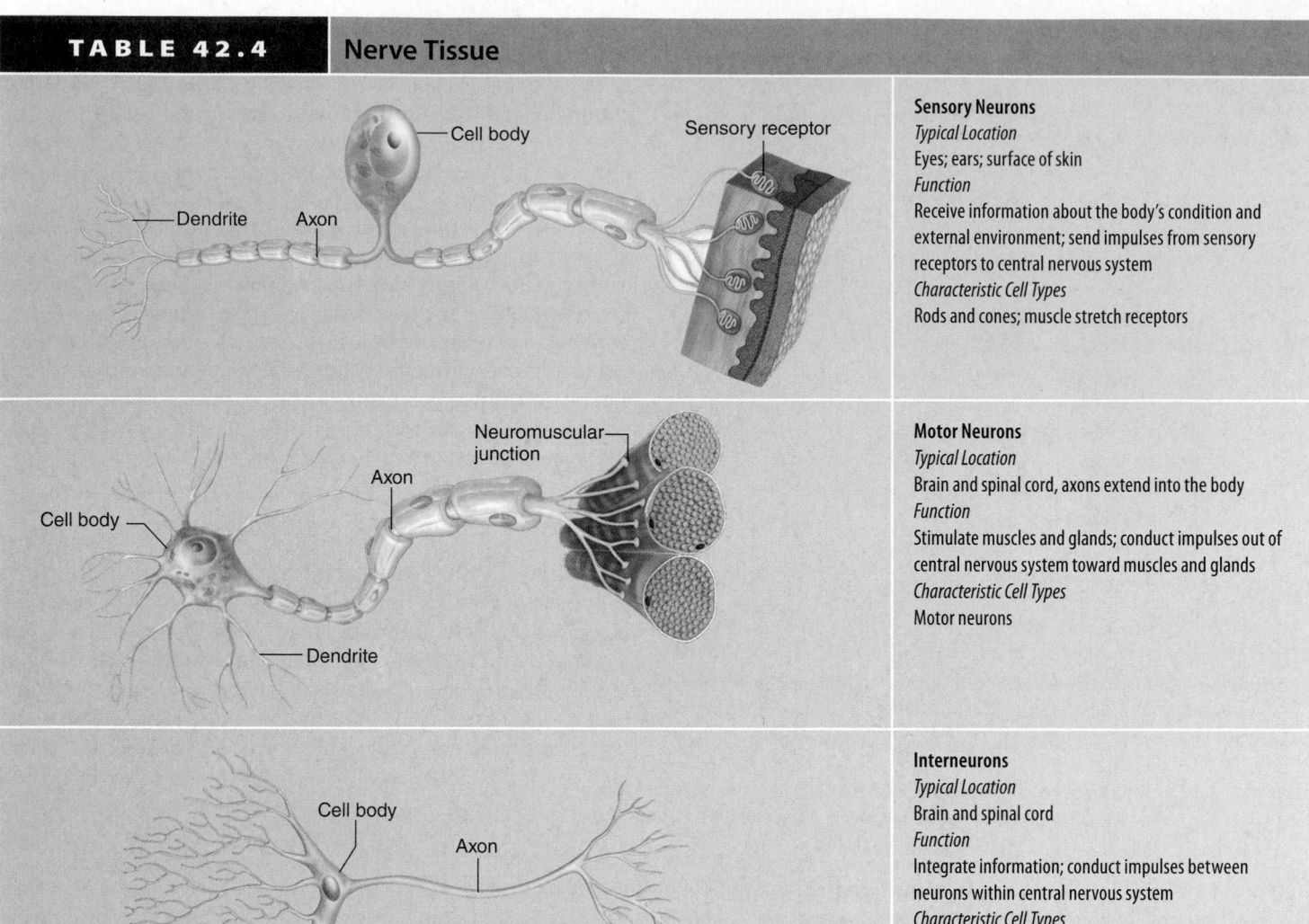

Sensory Neurons
Typical Location
Eyes; ears; surface of skin
Function
Receive information about the body's condition and external environment; send impulses from sensory receptors to central nervous system
Characteristic Cell Types
Rods and cones; muscle stretch receptors

Motor Neurons
Typical Location
Brain and spinal cord, axons extend into the body
Function
Stimulate muscles and glands; conduct impulses out of central nervous system toward muscles and glands
Characteristic Cell Types
Motor neurons

Interneurons
Typical Location
Brain and spinal cord
Function
Integrate information; conduct impulses between neurons within central nervous system
Characteristic Cell Types
Interneurons

42.6 Overview of Vertebrate Organ Systems

Learning Outcomes

1. *Identify the different organ systems in vertebrates.*
2. *Explain the functional organization of these systems.*

In the chapters that follow, we look closely at the major organ systems of vertebrates (figure 42.6). In each chapter, you will be able to see the intimate relationship of structure and function. We approach the organ systems by placing them in the following functional groupings:

- Communication and integration
- Support and movement
- Regulation and maintenance
- Defense
- Reproduction and development

Communication and integration sense and respond to the environment

Two organ systems detect external and internal stimuli and coordinate the body's responses. The **nervous system,** which consists of the brain, spinal cord, nerves, and sensory organs, detects internal sensory feedback and external stimuli such as light, sound, and touch. This information is collected and integrated, and then the appropriate response is made.

The **sensory systems** are a subset of the nervous system we consider in chapter 44. These include the organs and tissues that sense external stimuli, such as vision, hearing, smell, and so on.

Nervous System

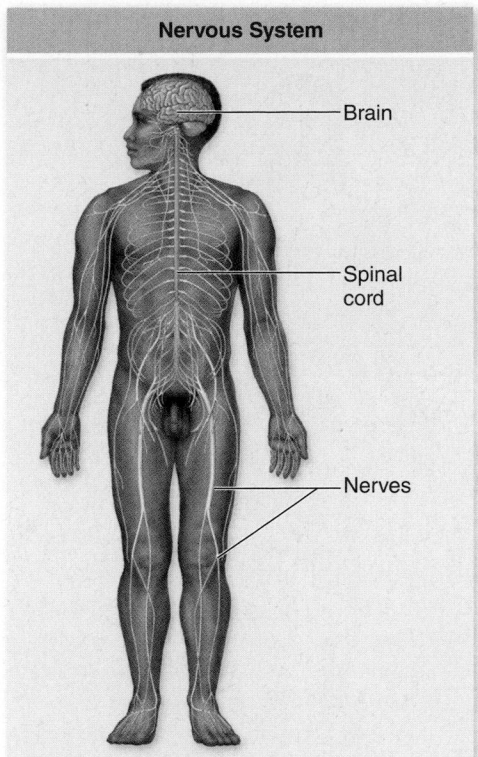

Brain

Spinal cord

Nerves

Endocrine System

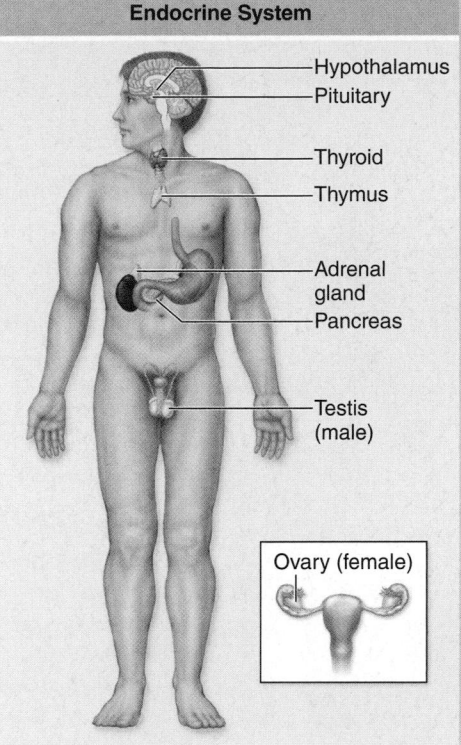

Hypothalamus

Pituitary

Thyroid

Thymus

Adrenal gland

Pancreas

Testis (male)

Ovary (female)

Skeletal System

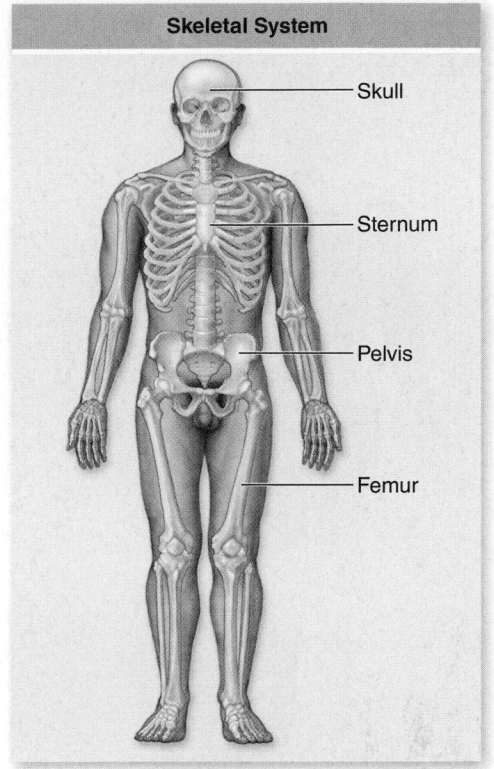

Skull

Sternum

Pelvis

Femur

Muscular System

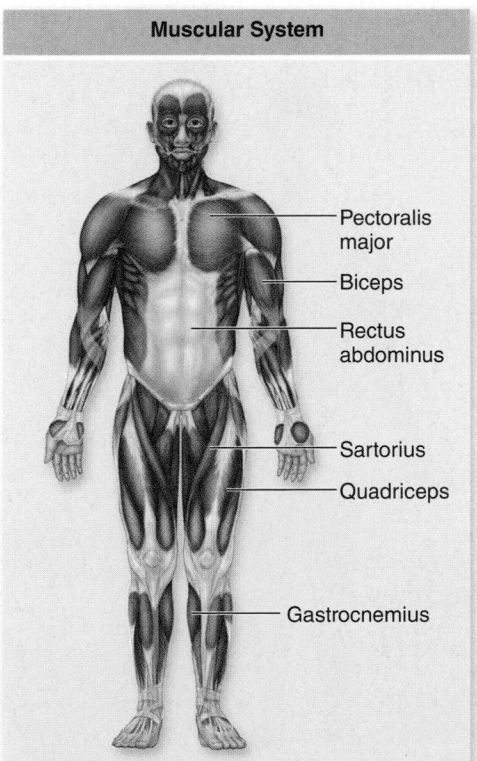

Pectoralis major

Biceps

Rectus abdominus

Sartorius

Quadriceps

Gastrocnemius

Digestive System

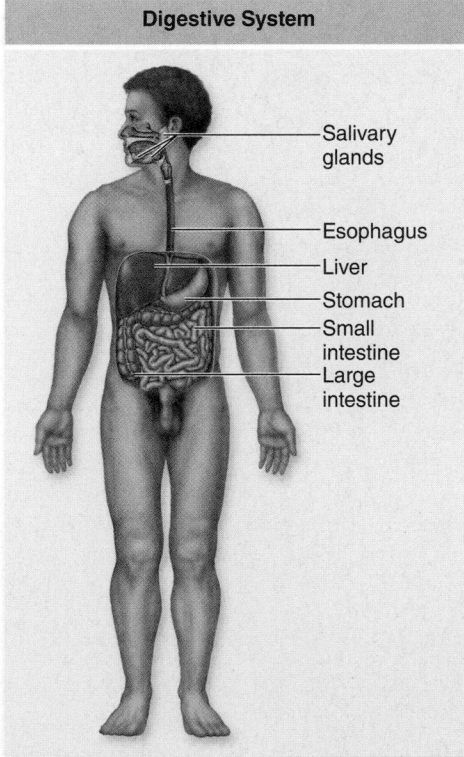

Salivary glands

Esophagus

Liver

Stomach

Small intestine

Large intestine

Circulatory System

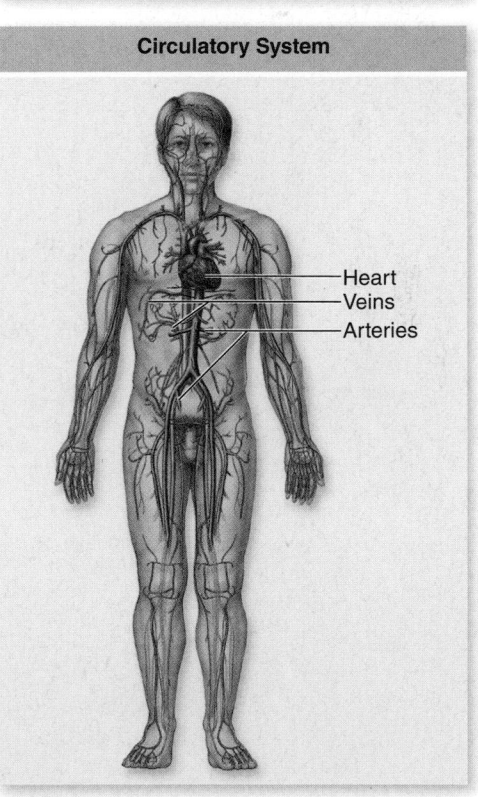

Heart

Veins

Arteries

Figure 42.6 Vertebrate organ systems. Shown are the 11 principal organ systems of the human body, including both male and female reproductive systems.

(continued)

Working in parallel with the nervous system, the **endocrine system** (see chapter 45) issues chemical signals that regulate and fine-tune the myriad chemical processes taking place in all other organ systems.

Skeletal support and movement are vital to all animals

The **musculoskeletal system** consists of two interrelated organ systems. Muscles are most obviously responsible for

Respiratory System

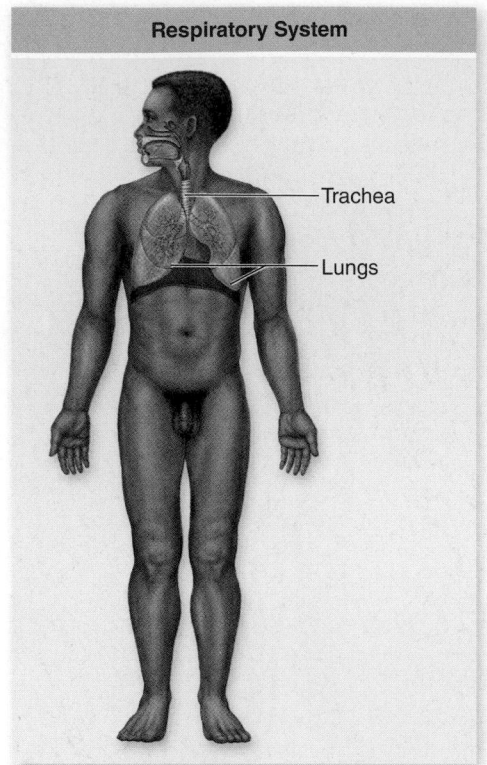

- Trachea
- Lungs

Urinary System

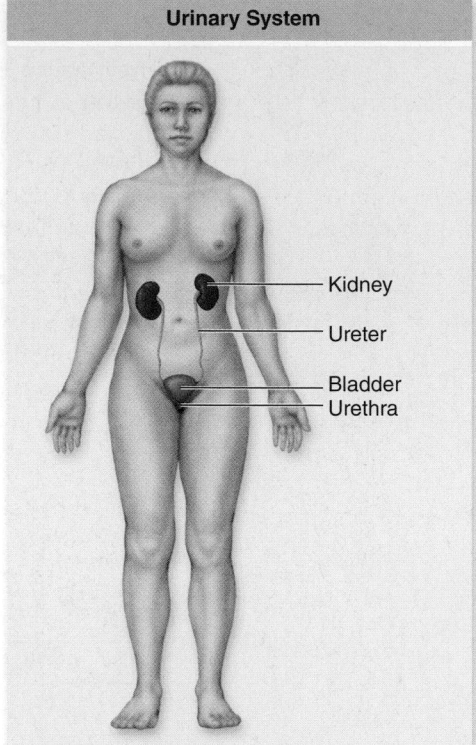

- Kidney
- Ureter
- Bladder
- Urethra

Integumentary System

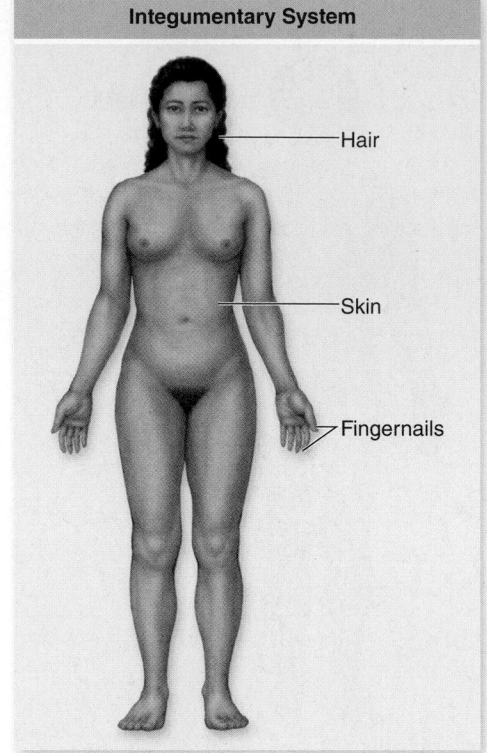

- Hair
- Skin
- Fingernails

Lymphatic/Immune System

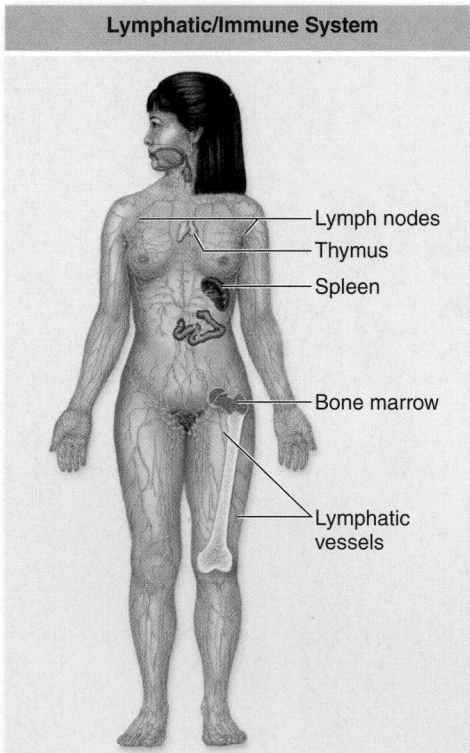

- Lymph nodes
- Thymus
- Spleen
- Bone marrow
- Lymphatic vessels

Reproductive System (male)

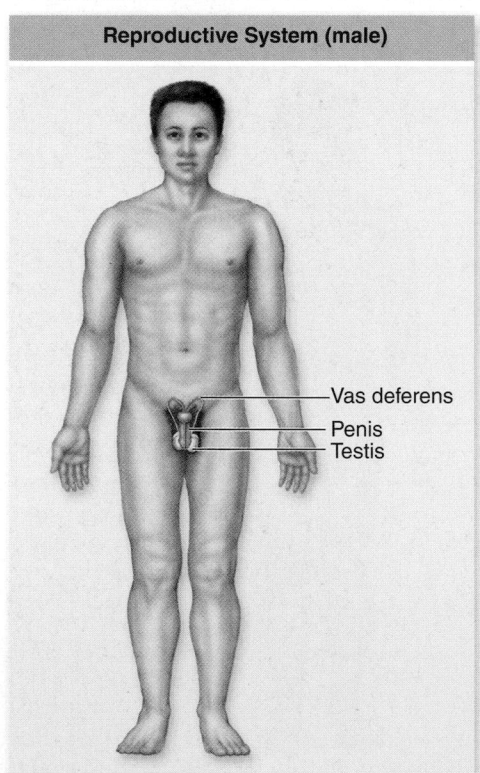

- Vas deferens
- Penis
- Testis

Reproductive System (female)

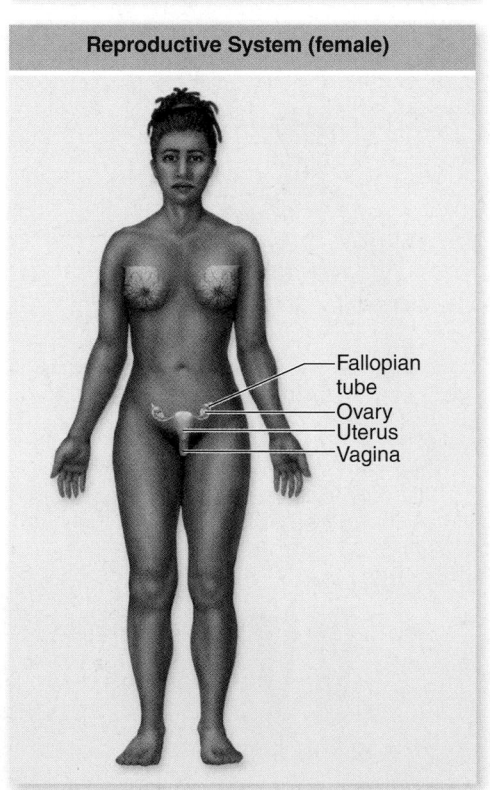

- Fallopian tube
- Ovary
- Uterus
- Vagina

Figure 42.6 Vertebrate organ systems, *continued*

movement, but without something to pull on, a muscle is useless. The skeletal system is the rigid framework against which most muscles pull. Vertebrates have internal skeletons, but many other animals exhibit external skeletons (such as insects) or hydrostatic skeletons (earthworms). Together, these two organ systems enable animals to exhibit a wide array of finely controlled movements.

Regulation and maintenance of the body's chemistry ensures continued life

The organ systems grouped under regulation and maintenance participate in nutrient acquisition, waste disposal, material distribution, and maintenance of the internal environment. Chapter 47 on the **digestive system** describes how we eat, absorb nutrients, and eliminate solid wastes. The heart and vessels of the **circulatory system** pump and distribute blood, carrying nutrients and other substances throughout the body. The body acquires oxygen and expels carbon dioxide via the **respiratory system.**

Finally, vertebrates tightly regulate the concentration of their body fluids. We explore these processes in Chapter 50 on osmoregulation, which is largely carried out by the **urinary system.**

The body can defend itself from attackers and invaders

Every animal faces assault by bacteria, viruses, fungi, protists, and even other animals. The body's first line of defense is the **integumentary system**—intact skin. Disease-causing agents that penetrate the first defense encounter a host of other protective **immune system** responses, including the production of antibodies and specialized cells that attack invading organisms.

Reproduction and development ensure continuity of species

The biological continuity of vertebrates is the province of the **reproductive system.** Male and female reproductive systems consist of organs where male and female gametes develop, as well as glands and tubes that nurture gametes and allow gametes of complementary sexes to come into contact with one another. The female reproductive system in many vertebrates also has systems for nurturing the developing embryo and fetus.

After gametes have fused to form a *zygote,* an elaborate process of cell division and development takes place to change this beginning diploid cell into a multicellular adult. This process is explored in chapter 53 on animal development.

Learning Outcomes Review 42.6

Vertebrate organ systems include the nervous, endocrine, skeletal, muscular, digestive, circulatory, respiratory, urinary, integumentary, lymphatic/immune, and reproductive systems. These may be grouped functionally based on their roles in communication and integration, support and movement, regulation and maintenance, defense, and reproduction and development.

■ *Is there any overlap between the different organ systems?*

42.7 *Homeostasis*

Learning Outcomes

1. *Explain homeostasis.*
2. *Illustrate how negative feedback can limit a response.*
3. *Illustrate how antagonistic effectors can maintain a system at a set point.*

As animals have evolved, specialization of body structures has increased. Each cell is a sophisticated machine, finely tuned to carry out a precise role within the body. Such specialization of cell function is possible only when extracellular conditions stay within narrow limits. Temperature, pH, the concentrations of glucose and oxygen, and many other factors must remain relatively constant for cells to function efficiently and interact properly with one another. The dynamic constancy of the internal environment is called *homeostasis.* The term *dynamic* is used because conditions are never constant, but fluctuate continuously within narrow limits. Homeostasis is essential for life, and most of the regulatory mechanisms of the vertebrate body are involved with maintaining homeostasis.

Negative feedback mechanisms keep values within a range

To maintain internal constancy, the vertebrate body uses a type of control system known as a **negative feedback.** In negative feedback, conditions within the body as well as outside it are detected by specialized sensors, which may be cells or membrane receptors. If conditions deviate too far from a set point, biochemical reactions are initiated to change conditions back toward the set point.

This *set point* is analogous to the temperature setting on a space heater. When room temperature drops, the change is detected by a temperature-sensing device inside the heater controls—the **sensor.** The thermostat on which you have indicated the set point for the heater contains a **comparator;** when the sensor information drops below the set point, the comparator closes an electrical circuit. The flow of electricity through the heater then produces more heat. Conversely, when the room temperature increases, the change causes the circuit to open, and heat is no longer produced. Figure 42.7 summarizes the negative feedback loop.

In a similar manner, the human body has set points for body temperature, blood glucose concentration, electrolyte (ion) concentration, the tension on a tendon, and so on. The integrating center is often a particular region of the brain or spinal cord, but in some cases, it can also be cells of endocrine glands. When a deviation in a condition occurs, a message is sent to increase or decrease the activity of particular target organs, termed *effectors.* Effectors are generally muscles or glands, and their actions can change the value of the condition in question back toward the set point value.

Mammals and birds are *endothermic;* they can maintain relatively constant body temperatures independent of the environmental temperature. In humans, when the blood temperature exceeds 37°C (98.6°F), neurons in a part of the brain called the **hypothalamus** detect the temperature change. Acting through the control of motor neurons, the hypothalamus responds by promoting the dissipation of heat through sweating, dilation of blood vessels in the skin, and other mechanisms. These responses tend to counteract the rise in body temperature.

Figure 42.7 Generalized diagram of a negative feedback loop. Negative feedback loops maintain a state of homeostasis, or dynamic constancy of the internal environment. Changing conditions are detected by sensors, which feed information to an integrating center that compares conditions to a set point. Deviations from the set point lead to a response to bring internal conditions back to the set point. Negative feedback to the sensor terminates the response.

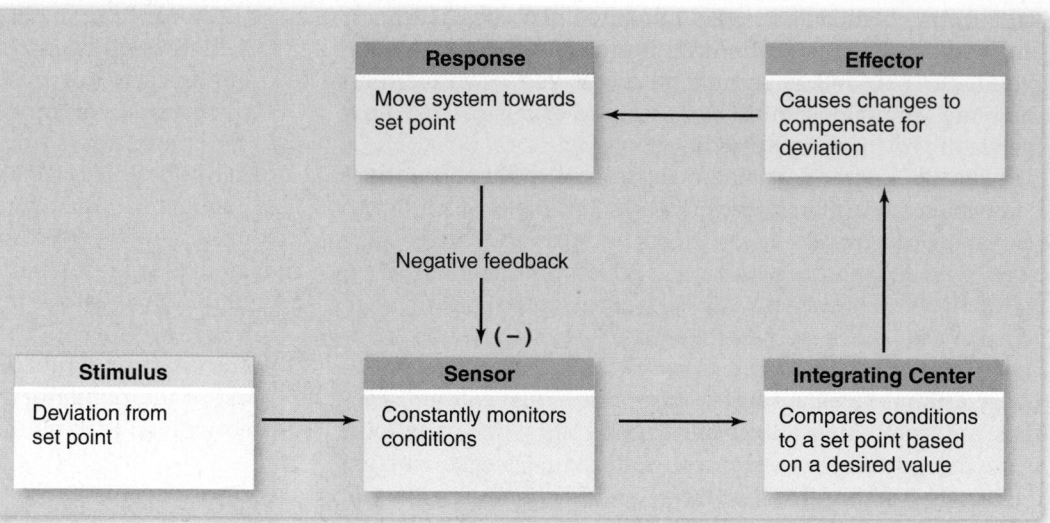

Antagonistic effectors act in opposite directions

The negative feedback mechanisms that maintain homeostasis often oppose each other to produce a finer degree of control. Most factors in the internal environment are controlled by several effectors, which often have antagonistic (opposing) actions. Control by antagonistic effectors is sometimes described as "push–pull," in which the increasing activity of one effector is accompanied by decreasing activity of an antagonistic effector. This affords a finer degree of control than could be achieved by simply switching one effector on and off.

To return to our earlier example, room temperature can be maintained by just turning the heater on and off, or turning an air conditioner on and off. A much more stable temperature is possible, however, if a thermostat controls both the air conditioner and heater. Then the heater turns on when the air conditioner shuts off, and vice versa (figure 42.8a).

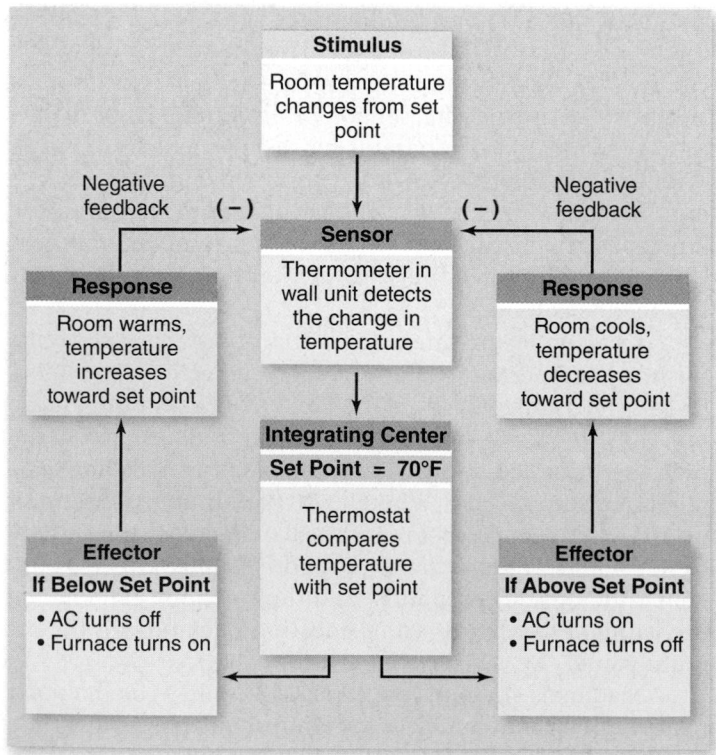

a.

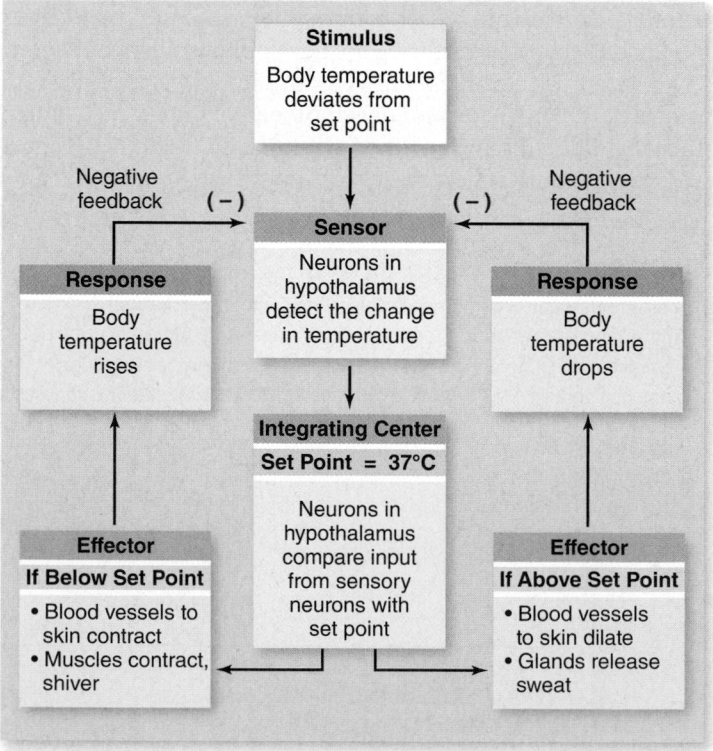

b.

Figure 42.8 Room and body temperature are maintained by negative feedback and antagonistic effectors. *a.* If a thermostat senses a low temperature (as compared with a set point), the furnace turns on and the air conditioner turns off. If the temperature is too high, the air conditioner turns on and the furnace turns off. *b.* The hypothalamus of the brain detects an increase or decrease in body temperature. The comparator (also in the hypothalamus) then processes the information and activates effectors, such as surface blood vessels, sweat glands, and skeletal muscles. Negative feedback results in a reduction in the difference of the body's temperature compared with the set point. Consequently, the stimulation of the effectors by the comparator is also reduced.

Antagonistic effectors are similarly involved in the control of body temperature. When body temperature falls, the hypothalamus coordinates a different set of responses, such as constriction of blood vessels in the skin and initiation of shivering, muscle contractions that help produce heat. These responses raise body temperature and correct the initial challenge to homeostasis (figure 42.8b).

Positive feedback mechanisms enhance a change

In a few cases, the body uses *positive feedback* mechanisms, which push or accentuate a change further in the same direction. In a positive feedback loop, the effector drives the value of the controlled variable even farther from the set point. As a result, systems in which there is positive feedback are highly unstable, analogous to a spark that ignites an explosion. They do not help to maintain homeostasis.

Nevertheless, such systems are important components of some physiological mechanisms. For example, positive feedback occurs in blood clotting, in which one clotting factor activates another in a cascade that leads quickly to the formation of a clot. Positive feedback also plays a role in the contractions of the uterus during childbirth (figure 42.9). In this case, stretching of the uterus by the fetus stimulates contraction, and contraction causes further stretching; the cycle continues until the uterus expels the fetus.

In the body, most positive feedback systems act as part of some larger mechanism that maintains homeostasis. In the examples we have described, formation of a blood clot stops bleeding and therefore tends to keep blood volume constant, and expulsion of the fetus reduces the contractions of the uterus, stopping the cycle.

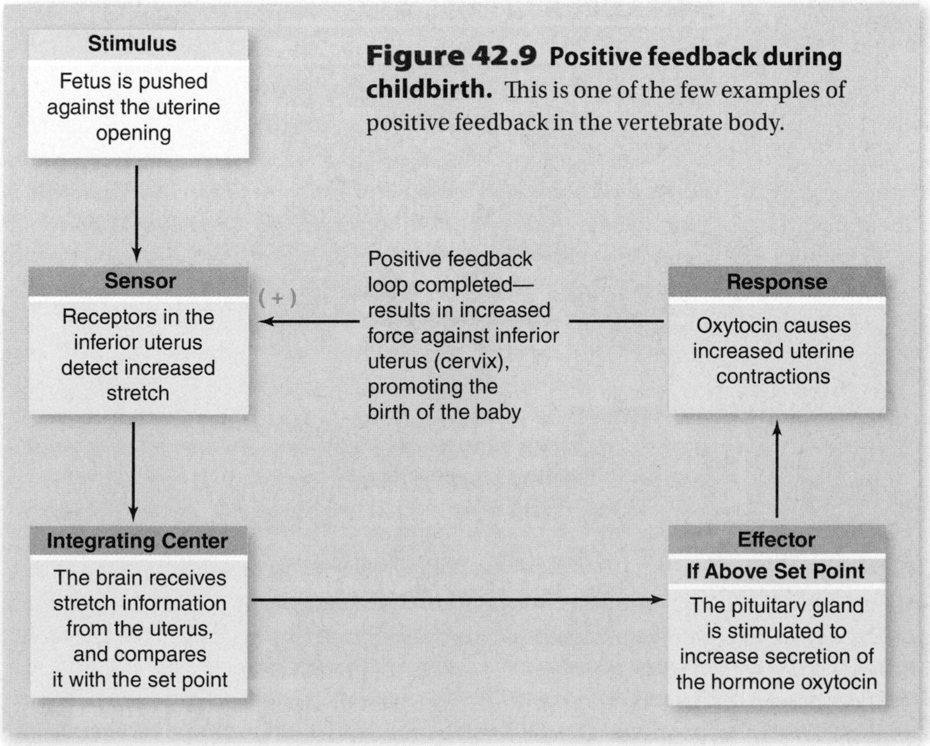

42.8 Regulating Body Temperature

Temperature is one of the most important aspects of the environment that all organisms must contend with. This provides a good example to apply the principles of homeostatic regulation from the last section. As we will see, some organisms have a body temperature that conforms to the environment and others regulate their body temperature. First, let's consider why temperature is so important.

Figure 42.9 Positive feedback during childbirth.
This is one of the few examples of positive feedback in the vertebrate body.

Stimulus
Fetus is pushed against the uterine opening

Sensor
Receptors in the inferior uterus detect increased stretch

(+)

Positive feedback loop completed—results in increased force against inferior uterus (cervix), promoting the birth of the baby

Response
Oxytocin causes increased uterine contractions

Integrating Center
The brain receives stretch information from the uterus, and compares it with the set point

Effector
If Above Set Point
The pituitary gland is stimulated to increase secretion of the hormone oxytocin

Q_{10} is a measure of temperature sensitivity

The rate of any chemical reaction is affected by temperature: The rate increases with increasing temperature, and it decreases with decreasing temperature. Reactions catalyzed by enzymes show the same kinetic effects, but the enzyme itself is also affected by temperature.

We can make this temperature dependence quantitative by examining the rate of a reaction at two different temperatures. The ratio between the rates of a reaction at two temperatures that differ by 10°C is called the Q_{10} for the enzyme:

$$Q_{10} = R_{T+10}/R_T$$

For most enzymes the Q_{10} value is around 2, which means for every 10°C increase in temperature, the rate of the reaction doubles. Obviously, this cannot continue forever since at high temperatures the enzyme's structure is affected and it can no longer be active.

The Q_{10} concept can also be applied to overall metabolism. The equation remains the same, but instead of the rate of a single reaction, the overall metabolic rate is used. When this has been measured, most organisms have a Q_{10} for metabolic rate around 2 to 3. This observation implies that the effect of temperature is mainly on the enzymes that make up metabolism.

In a few rare cases—for example, in some intertidal invertebrates—the Q_{10} is close to 1. Notice that this value means no change in metabolic rate with temperature. In the case of these intertidal invertebrates, they are exposed to large temperature fluctuations as they are alternately flooded with relatively cold water and exposed to direct sunlight and much higher air temperatures. These organisms have adapted to deal with these large temperature swings, probably through the evolution of different enzymes in a single metabolic pathway that have large differences in optimal temperature. This allows one enzyme to "make up" for others with decreased activity at a particular temperature.

Temperature is determined by internal and external factors

Body temperature appears simple, yet a large number of variables influence it. These variables include both internal and external factors, as well as behavior. As you may recall from chapter 6, the second law of thermodynamics indicates that no energy transaction is 100% efficient. Thus the reactions that make up metabolism are constantly producing heat as a result of this inefficiency. This heat must either be dissipated or can be used to raise body temperature.

Overall metabolic rate and body temperature are interrelated. Lower body temperatures do not allow high metabolic rates because of the temperature dependence of enzymes discussed earlier. Conversely, high metabolic rates may cause unacceptable heating of the body, requiring cooling.

Organisms therefore must deal with external and internal factors that relate body heat, metabolism, and the environment. The simplest model for temperature, more accurately body heat, is:

$$\text{body heat} = \text{heat produced} + (\text{heat gained} - \text{heat lost})$$

we can simplify this further to:

$$\text{body heat} = \text{heat produced} + \text{heat transferred}$$

Notice that the heat transferred can be either positive or negative, that is, it can be used for both heating and cooling.

We recognize four mechanisms of heat transfer that are relevant to biological systems: radiation, conduction, convection, and evaporation (figure 42.10).

- **Radiation.** The transfer of heat by electromagnetic radiation, such as from the Sun, does not require direct contact. Heat is transferred from hotter bodies to colder bodies by radiation.

Figure 42.10 Methods of heat transfer. Heat can be gained or lost by conduction, convection, and radiation. Heat can also be lost by evaporation of water on the surface of an animal.

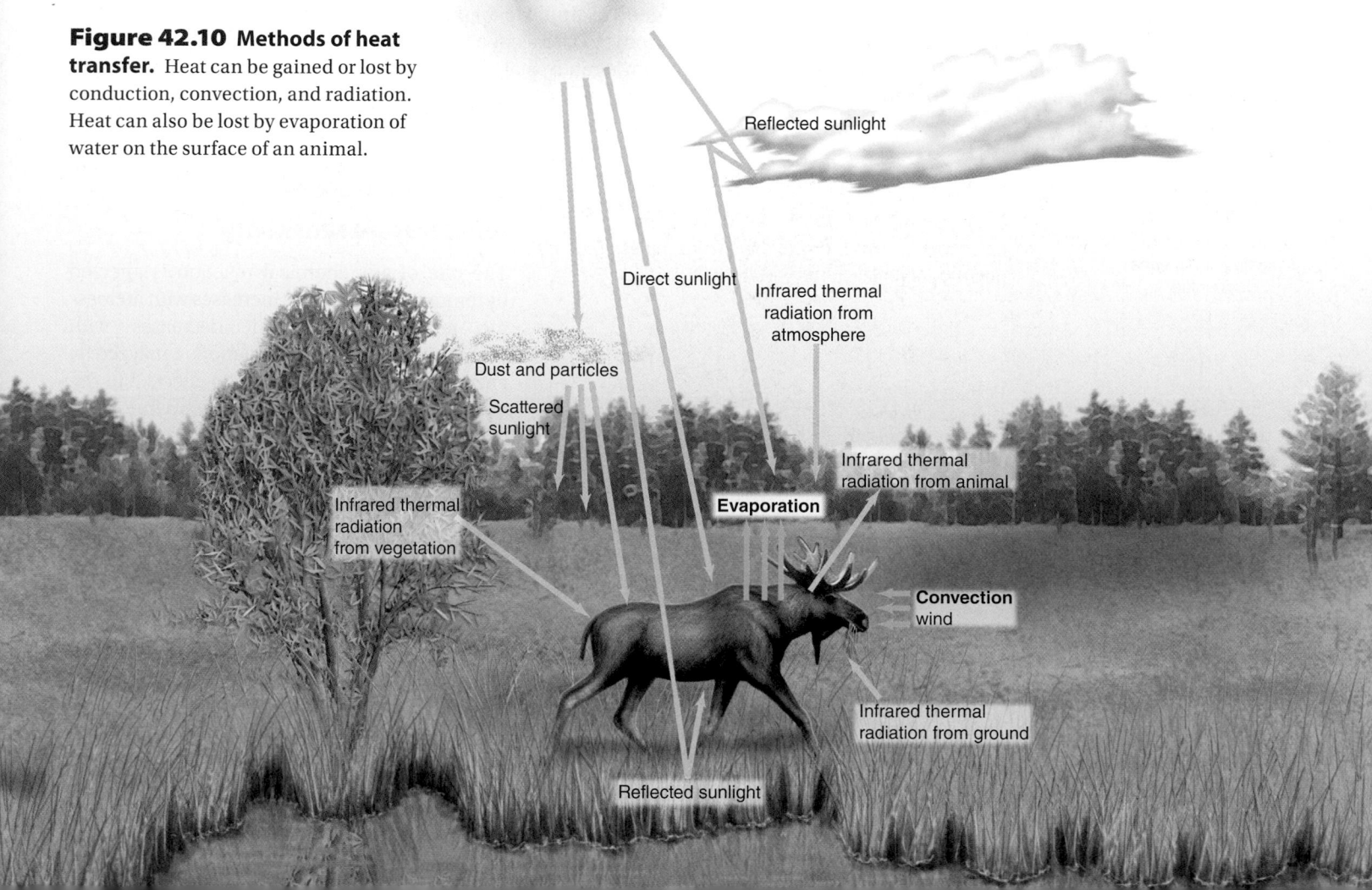

- **Conduction.** The direct transfer of heat between two objects is called conduction. It is literally a direct transfer of kinetic energy between the molecules of the two objects in contact. Energy is transferred from hotter objects to colder ones.
- **Convection.** Convection is the transfer of heat brought about by the movement of a gas or liquid. This movement may be externally caused (wind) or may be due to density differences related to heating and cooling—for example, heated air is less dense and rises; the same is true for water.
- **Evaporation.** All substances have a heat of vaporization, that is, the amount of energy needed to change them from a liquid to a gas phase. Water, as you saw in chapter 2, has a high heat of vaporization, and many animals use this attribute of water as a source of cooling.

Other factors

The overall rate of heat transfer by the methods just listed depends on a number of factors that influence these physical processes. These factors include surface area, temperature difference, and specific heat conduction. Taking these in order, the larger the surface area relative to overall mass, the greater the conduction of heat. Thus, small organisms have a relatively larger surface area for their mass, and they gain or lose heat more readily to the surroundings. This can be affected to a small extent by changing posture, and by extending or pulling in the limbs.

Temperature difference is also important; the greater the difference between ambient temperature and body temperature, the greater the heat transfer. The closer an animal's temperature is to the ambient temperature, the less heat is gained or lost.

Finally, an animal with high heat conductance tends to have a body temperature close to the ambient temperature. For animals that regulate temperature, surrounding the body with a substance with lower heat conductance has an advantage: It acts as insulation. Insulating substances include such features as feathers, fur, and blubber. For animals that regulate body temperature through behavior, a high heat conductance can maximize heat transfer.

Organisms are classified based on heat source

For many years, physiologists classified animals according to whether they maintained a constant body temperature, or their body temperature fluctuated with environment. Animals that regulated their body temperature about a set point were called *homeotherms,* and those that allow their body temperature to conform to the environment were called *poikilotherms.*

Because homeotherms tended to maintain their body temperature above the ambient temperature, they were also colloquially called "warm-blooded"; poikilotherms were termed "cold-blooded." The problem with this terminology is that a poikilotherm in an environment with a stable temperature (for example, many deep-sea fish species) has a more constant body temperature than some homeotherms.

These limitations to the dichotomy based on temperature regulation led to another view based on how body heat is generated. Animals that use metabolism to generate body heat and maintain their temperatures above the ambient temperature are called *endotherms.* Animals with a relatively low metabolic rate that do not use metabolism to produce heat are called *ectotherms.* Endotherms tend to have a lower thermal conductivity due to insulating mechanisms, and ectotherms tend to have high thermal conductivity and lack insulation.

These two terms represent ideal end points of a spectrum of physiology and adaptations. Many animals fall in between these extremes and can be considered *heterotherms.* It is a matter of judgment how a particular animal is classified if it exhibits characteristics of each group.

Ectotherms regulate temperature using behavior

Despite having low metabolic rates, ectotherms can regulate their temperature using behavior. Most invertebrates use behavior to adjust their temperature. Many butterflies, for example, must reach a certain body temperature before they can fly. In the cool of the morning, they orient their bodies so as to maximize their absorption of sunlight. Moths and many other insects use a shivering reflex to warm their thoracic flight muscles so that they may take flight (figure 42.11).

Vertebrates other than mammals and birds are also ectothermic, and their body temperatures are more or less

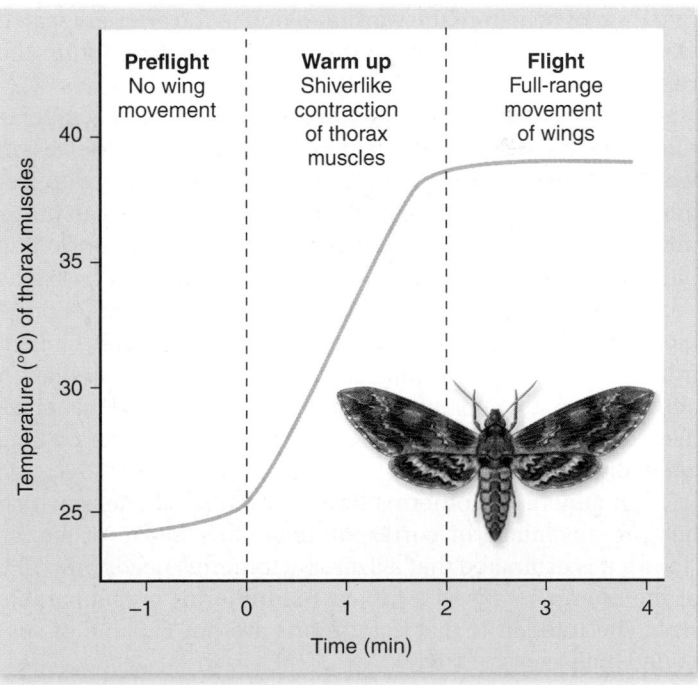

Figure 42.11 Thermoregulation in insects. Some insects, such as the sphinx moth, contract their thoracic muscles to warm up for flight.

? Inquiry question Why does muscle temperature stop warming after 2 min?

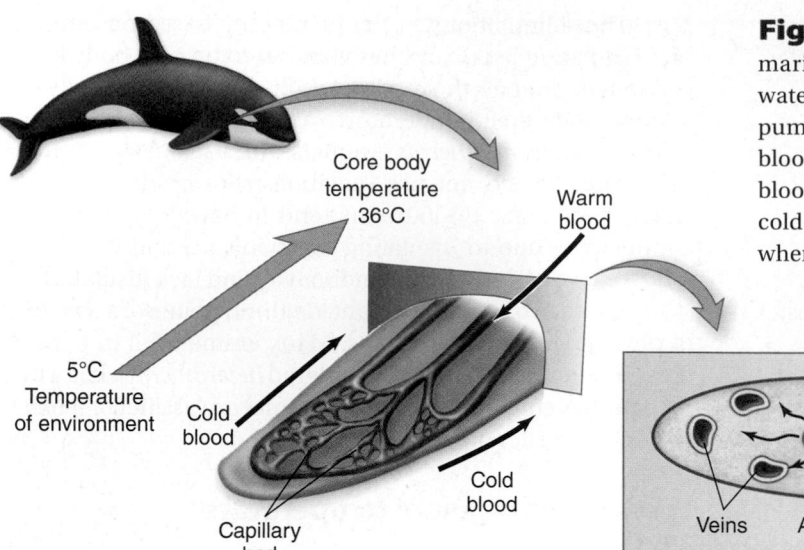

dependent on the environmental temperature. This does not mean that these animals cannot maintain high and relatively constant body temperatures, but they must use behavior to do this. Many ectothermic vertebrates are able to maintain temperature homeostasis, that is, are homeothermic ectotherms.

For example, certain large fish, including tuna, swordfish, and some sharks, can maintain parts of their body at a significantly higher temperature than that of the water. They do so using *countercurrent heat exchange.* This circulatory adaptation allows the cooler blood in the veins to be warmed through radiation of heat from the warmer blood in the arteries located close to the veins. The arteries carry warmer blood from the center of the body (figure 42.12).

Reptiles attempt to maintain a constant body temperature through behavioral means—by placing themselves in varying locations of sunlight and shade. That's why you frequently see lizards basking in the sun. Some reptiles can maximize the effect of behavioral regulation by also controlling blood flow. The marine iguana can increase and decrease its heart rate and control the extent of dilation or contraction of blood vessels to regulate the amount of blood available for heat transfer by conduction. Increased heart rate and vasodilation allows them to maximize heating when on land, whereas decreased heart rate and vasoconstriction minimize cooling when diving for food.

In general, ectotherms have low metabolic rates, which has the advantage of correspondingly low intake of energy (food). It is estimated that a lizard (ectotherm) needs only 10% of the energy intake of a mouse (endotherm) of comparable size. The tradeoff is that ectotherms are not capable of sustained high-energy activity.

Endotherms create internal metabolic heat for conservation or dissipation

For endotherms, the generation of internal heat via high metabolic rate can be used to warm the organism if it is cold, but also represents a source of heat that must be dissipated at higher temperatures.

The simplest response that affects heat transfer is to control the amount of blood flow to the surface of the animal. Dilating blood vessels increases the amount of blood flowing to the surface, which in turn increases thermal heat exchange and dissipation of heat. In contrast, constriction of blood vessels decreases the amount of blood flowing to the surface and decreases thermal heat exchange, limiting the amount of heat lost due to conduction.

When ambient temperatures rise, many endotherms take advantage of evaporative cooling in the form of sweating or panting. Sweating is found in some mammals, including humans, and involves the active extrusion of water from sweat glands onto the surface of the body. As the water evaporates, it cools the skin, and this cooling can be transferred internally by capillaries near the surface of the skin. Panting is a similar adaptation used by some mammals and birds that takes advantage of respiratory surfaces for evaporative cooling. For evaporative cooling to be effective, the animal must be able to tolerate the loss of water.

The advantage of endothermy is that it allows sustained high-energy activity. The tradeoff for endotherms is that the high metabolic rate has a corresponding cost in requiring relatively constant and high rates of energy intake (food).

Body size and insulation

Size is one important characteristic affecting animal physiology. Changes in body mass have a large effect on metabolic rate. Smaller animals consume much more energy per unit body mass than larger animals. This relationship is summarized in the "mouse to elephant" curve that shows the non-proportionality of metabolic rate versus size of mammals (figure 42.13).

For small animals with a high metabolic rate, surface area is also large relative to their volume. In a cold environment, this can be disastrous as they cannot produce enough internal heat to balance conductive loss through their large surface area. Thus, small endotherms in cold environments require significant insulation to maintain their body temperature. The amount of insulation can also vary seasonally

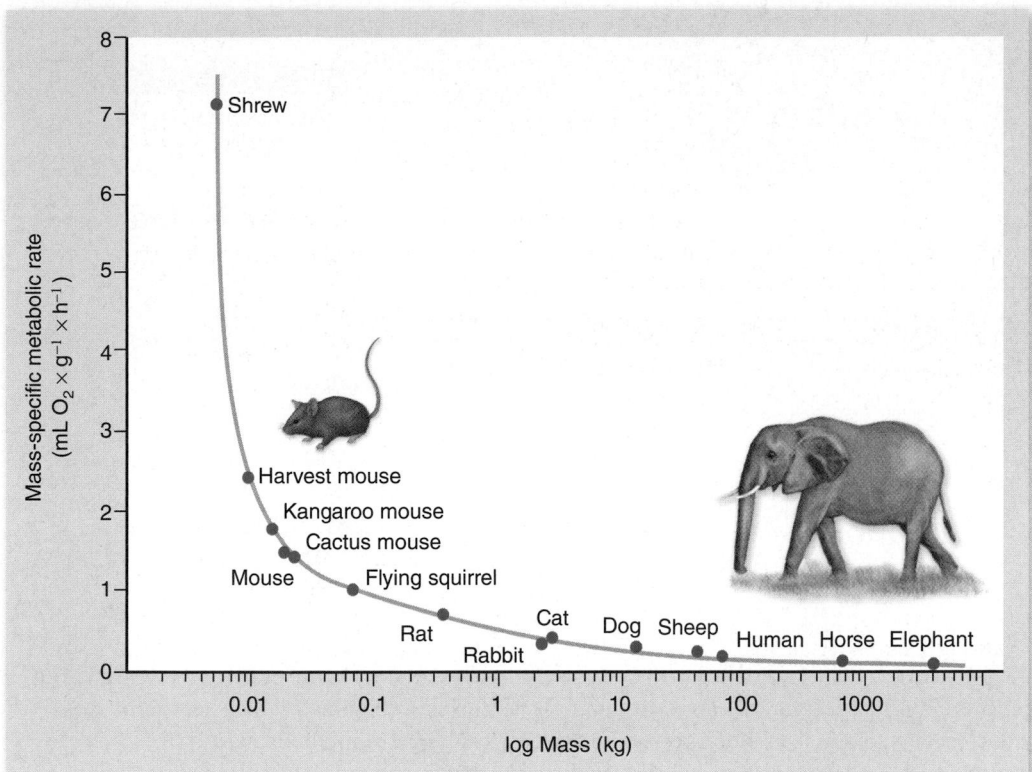

Figure 42.13 Relationship between body mass and metabolic rate in mammals. Smaller animals have a much higher metabolic rate per unit body mass relative to larger animals. In the figure, mass-specific metabolic rate (expressed as O_2 consumption per unit mass) is plotted against body mass. Note that the body mass axis is a logarithmic scale.

🔍 **Data analysis** What does this graph predict about the different challenges faced by smaller versus larger mammals in hot and cold environments?

and geographically with thicker coats in the north and in winter.

Conversely, large animals in hot environments have the opposite problem: Although their metabolic rate is relatively low, they still produce a large amount of heat with much less relative surface area to dissipate this heat by conduction. Thus large endotherms in hot environments usually have little insulation and will use behavior to lose heat, such as elephants flapping their ears to increase convective heat loss.

Thermogenesis

When temperatures fall below a critical lower threshold, normal endothermic responses are not sufficient to warm an animal. In this case, the animal resorts to **thermogenesis,** or the use of normal energy metabolism to produce heat. Thermogenesis takes two forms: shivering and nonshivering thermogenesis.

In nonshivering thermogenesis, fat metabolism is altered to produce heat instead of ATP. Nonshivering thermogenesis takes place throughout the body, but in some mammals, special stores of fat called brown fat are utilized specifically for this purpose. This brown fat is stored in small deposits in the neck and between the shoulders. This fat is highly vascularized, allowing efficient transfer of heat away from the site of production.

Shivering thermogenesis uses muscles to generate heat without producing useful work. It occurs in some insects, such as the earlier example of a butterfly warming its flight muscles, and in endothermic vertebrates. Shivering involves the use of antagonistic muscles to produce little net generation of movement, but hydrolysis of ATP, generating heat.

Mammalian thermoregulation is controlled by the hypothalamus

Mammals that maintain a relatively constant core temperature need an overall control system (summarized in figure 42.14). The system functions much like the heating/cooling system in your house that has a thermostat connected to a furnace to produce heat and an air conditioner to remove heat. Such a system maintains the temperature of your house about a set point by alternately heating or cooling as necessary.

When the temperature of your blood exceeds 37°C (98.6°F), neurons in the hypothalamus detect the temperature change (see chapters 43 and 44). This leads to stimulation of the *heat-losing center* in the hypothalamus. Nerves from this area cause a dilation of peripheral blood vessels, bringing more blood to the surface to dissipate heat. Other nerves stimulate the production of sweat, leading to evaporative cooling. Production of hormones that stimulate metabolism is also inhibited.

When your temperature falls below 37°C, an antagonistic set of effects are produced by the hypothalamus. This is under control of the *heat-promoting center,* which has nerves that constrict blood vessels to reduce heat transfer, and inhibit sweating to prevent evaporative cooling. The adrenal medulla is stimulated to produce epinephrine, and the anterior pituitary to produce TSH, both of which stimulate metabolism. In the case of TSH, this is indirect as it stimulates the thyroid to produce thyroxin, which stimulates metabolism (see chapter 45). A combination of epinephrine and autonomic nerve stimulation of fat tissue can induce thermogenesis to produce more internal heat. Again, as temperature rises, negative feedback to the hypothalamus reduces the heat-producing response.

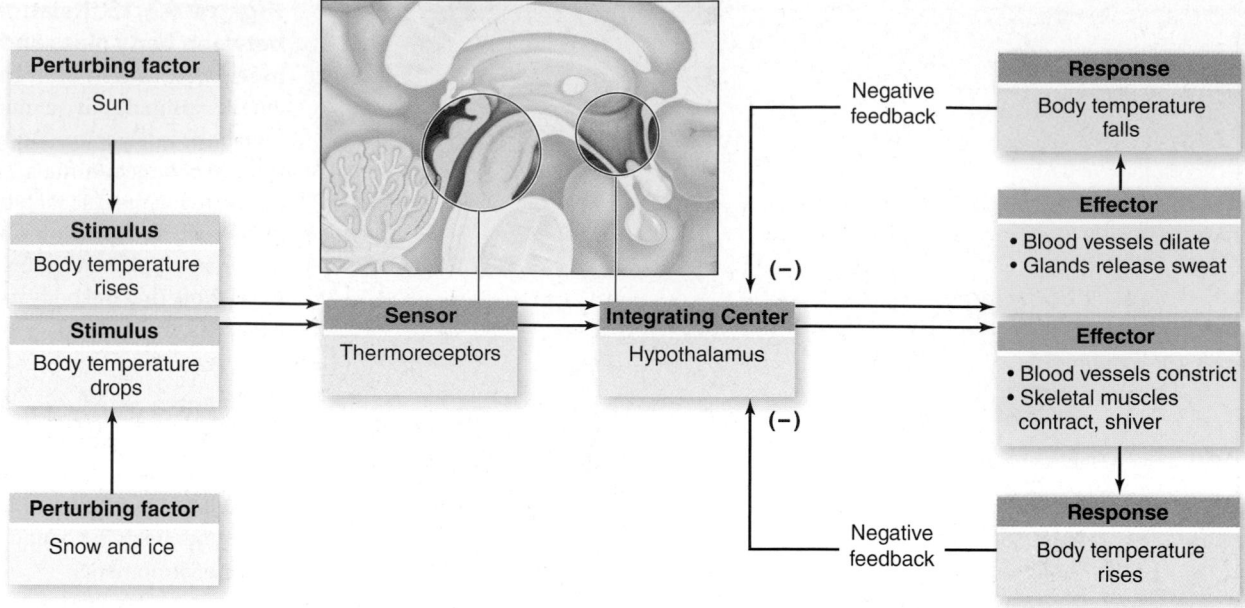

Figure 42.14 The control of body temperature by the hypothalamus. Central thermoreceptors in the brain and interior of the abdomen sense changes in core temperature. These thermoreceptors synapse with neurons on the hypothalamus, which acts as an integrating center. The hypothalamus then controls effectors such as blood vessels and sweat glands via sympathetic nerves. The hypothalamus also causes the release of hormones that stimulate the thyroid to produce thyroxin, which modulates metabolism.

 Data analysis Pyrogens are compounds that lead to increased temperature, or fever. What part of this system would be the target of pyrogens?

Fever

Substances that cause a rise in temperature are called **pyrogens,** and they produce the state we call **fever.** Fever is a result of resetting the body's normal set point to a higher temperature. A number of gram-negative bacteria have components in their cell walls called endotoxins that act as pyrogens. Substances produced by circulating white blood cells act as pyrogens as well. Pyrogens act on the hypothalamus to increase the set point.

The adaptive value of fever seems to be that increased temperature can inhibit the growth of bacteria. Evidence for this comes from the observation that some ectotherms respond to pyrogens as well. When desert iguanas were injected with pyrogen-producing bacteria, they spent more time in the sun, producing an elevated temperature: They induced fever behaviorally!

These observations have led to a reevaluation of fever as a state that should be treated medically. Fever is a normal response to infection, and treatment to reduce fever may be working against this natural defense system. Extremely high fevers, however, can be dangerous, inducing symptoms ranging from seizures to delirium.

Torpor

Endotherms can also reduce both metabolic rate and body temperature to produce a state of dormancy called *torpor.* Torpor allows an animal to reduce the need for food intake by reducing metabolism. Some birds, such as the hummingbird, allow their body temperature to drop as much as 25°C at night. This strategy is found in smaller

endotherms; larger mammals have too large a mass to allow rapid cooling.

Hibernation is an extreme state in which deep torpor lasts for several weeks or even several months. In this case, the animal's temperature may drop as much as 20°C below its normal set point for an extended period of time. The animals that practice hibernation seem to be in the midrange of size; smaller endotherms quickly consume more energy than they can easily store, even by reducing their metabolic rate.

Very large mammals do not appear to hibernate. It was long thought that bears hibernate, but in reality their temperature is reduced only a few degrees. They instead undergo a prolonged winter sleep. With their large thermal mass and low rate of heat loss, they do not seem to require the additional energy savings of hibernation.

Learning Outcomes Review 42.8

The Q_{10} value of an enzyme indicates how its activity changes with a 10°C rise in temperature. The Q_{10} can also be applied to an organism's overall metabolism. Body heat is equal to heat produced plus heat transferred. Heat is transferred by conduction, convection, radiation, and evaporation. Organisms that generate heat and can maintain a temperature above ambient levels are called endotherms. Organisms that conform to their surroundings are called ectotherms. Both types can regulate temperature, but ectotherms mainly do so with behavior. Mammals maintain a consistent body temperature through regulation of metabolic rate by the hypothalamus. Two negative feedback loops act to raise or lower temperature as needed.

■ *Why are the terms "cold-blooded" and "warm-blooded" outmoded and inaccurate?*

42.1 Organization of the Vertebrate Body (figure 42.1)

Tissues are groups of cells of a single type and function.

Adult vertebrate primary tissues are epithelial, connective, muscle, and nerve tissues.

Organs and organ systems provide specialized function.

Organs consist of a group of different tissues that form a structural and functional unit. An organ system is a group of organs that collectively perform a function.

The general body plan of vertebrates is a tube within a tube, with internal support (figure 42.2).

The tube of the digestive tract is surrounded by the skeleton and accessory organs and is enclosed in the integument.

Vertebrates have both dorsal and ventral body cavities.

The dorsal body cavity lies within the skull and vertebrae. The ventral body cavity, bounded by the rib cage and abdominal muscles, comprises the thoracic cavity and the abdominopelvic cavity.

The coelomic space of the abdominopelvic cavity is the peritoneal cavity; that surrounding the heart is the pericardial cavity; and those around the lungs are the pleural cavities.

42.2 Epithelial Tissue

Epithelium forms a barrier.

Epithelial cells are tightly bound together, forming a selective barrier. Epithelial cells are replaced constantly and can regenerate in wound healing. Epithelium has a basal surface attached to underlying connective tissues, and a free apical surface.

Epithelial types reflect their function.

Epithelium is divided into two general classes: simple (one cell layer) and stratified (multiple cells thick). These are further divided into squamous, cuboidal, and columnar based on the shape of cells (table 42.1). Vertebrate glands form from invaginated epithelia.

42.3 Connective Tissue

Connective tissue proper may be either loose or dense.

Connective tissues contain various cells in an extracellular matrix of proteins and ground substance. Connective tissue proper is divided into loose connective tissue and dense connective tissue.

Special connective tissues have unique characteristics.

Special connective tissues, such as cartilage, rigid bone, and blood, have unique cells and matrices (table 42.2). Cartilage is formed by chondrocytes and bone by osteocytes.

All connective tissues have similarities.

All connective tissues originate from mesoderm and contain a variety of cells within an extracellular matrix.

42.4 Muscle Tissue (table 42.3)

Smooth muscle is found in most organs.

Involuntary smooth muscle occurs in the viscera and is composed of long, spindle-shaped cells with a single nucleus.

Skeletal muscle moves the body.

Voluntary skeletal or striated muscle is usually attached by tendons to bones, and the cells (fibers) have multiple nuclei and contain contractile myofibrils.

The heart is composed of cardiac muscle.

Cardiac muscle consists of striated muscle cells connected to each other by gap junctions that allow coordination.

42.5 Nerve Tissue (table 42.4)

Neurons sometimes extend long distances.

Neurons have a cell body with a nucleus; dendrites, which receive impulses; and an axon, which transmits impulses away.

Neuroglia provide support for neurons.

Neuroglia help regulate the neuronal environment. Some types form the myelin sheaths that surround some axons.

Two divisions of the nervous system coordinate activity.

The central nervous system is the brain and spinal cord, and the peripheral nervous system contains nerves and ganglia.

42.6 Overview of Vertebrate Organ Systems (figure 42.6)

Communication and integration sense and respond to the environment.

The three organ systems involved in communication and integration are the nervous, sensory, and endocrine systems.

Skeletal support and movement are vital to all animals.

The musculoskeletal system consists of muscles and the skeleton they act upon.

Regulation and maintenance of the body's chemistry ensures continued life.

The digestive, circulatory, respiratory, and urinary systems accomplish ingestion of nutrients and elimination of wastes.

The body can defend itself from attackers and invaders.

The integumentary system forms a barrier against attack; the immune system mounts a counterattack to foreign pathogens.

Reproduction and development ensure continuity of species.

All vertebrate species are capable of sexual reproduction.

42.7 Homeostasis

Homeostasis refers to the dynamic constancy of the internal environment and is essential for life.

Negative feedback mechanisms keep values within a range.

Negative feedback loops include a sensor, an integration center, and effectors that respond to deviations from a set point.

Antagonistic effectors act in opposite directions.

Negative feedback mechanisms often occur in antagonistic pairs that push and pull against each other.

Positive feedback mechanisms enhance a change.

In a positive feedback loop changes in one direction bring about further changes in the same direction.

42.8 Regulating Body Temperature

Q_{10} is a measure of temperature sensitivity.

Q_{10} is the ratio of reaction rates at two temperatures 10°C apart. For chemical reactions Q_{10} is about 2. Most organisms have a Q_{10} around 2 to 3, indicating temperature affects mainly enzymatic reactions.

Temperature is determined by internal and external factors.

Internal factors include metabolic rate; external factors affect heat transfer. Heat is transferred through radiation, conduction, convection, and evaporation (figure 42.10).

Organisms are classified based on heat source.

Endotherms have high metabolic rates and generate heat internally. Ectotherms have low metabolic rates and conform to ambient temperature.

Ectotherms regulate temperature using behavior.

Ectotherms move around in an environment to alter their temperature (figures 42.11 and 42.12).

Endotherms create internal metabolic heat for conservation or dissipation.

Endotherms regulate temperature by changes in metabolic rate, blood flow, and sweating or panting. Thermogenesis occurs when temperature falls below a critical level.

Mammalian thermoregulation is controlled by the hypothalamus (figure 42.14).

The hypothalamus acts through a heat-losing and heat-promoting center to keep the blood temperature near a set point. Fever is an increase in body temperature; torpor is a lowered metabolic state associated with dormancy.

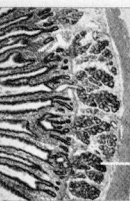

Review Questions

UNDERSTAND

1. Which of the following cavities would contain your stomach?

   a. Peritoneal   c. Pleural
   b. Pericardial   d. Thoracic

2. Epithelial tissues do all of the following except

   a. form barriers or boundaries.
   b. absorb nutrients in the digestive tract.
   c. transmit information in the central nervous system.
   d. allow exchange of gases in the lung.

3. Ectotherms

   a. cannot regulate their body temperatures.
   b. regulate their internal temperature using metabolic energy.
   c. can regulate temperature using behavior.
   d. regulate temperature by dissipating but not generating heat.

4. Connective tissues include a diverse group of cells, yet they all share

   a. cuboidal shape.
   b. the ability to produce hormones.
   c. the ability to contract.
   d. the presence of an extracellular matrix.

5. Skeletal muscle cells differ from the "typical" mammalian cell in that they

   a. contain multiple nuclei.
   b. have mitochondria.
   c. have no plasma membrane.
   d. are not derived from embryonic tissue.

6. Examples of smooth muscle sites include

   a. the lining of blood vessels.
   b. the iris of the eye.
   c. the wall of the digestive tract.
   d. All of the choices are correct.

7. The function of neuroglia is to

   a. carry messages from the PNS to the CNS.
   b. support and protect neurons.
   c. stimulate muscle contraction.
   d. store memories.

8. Skeletal muscle cells are

   a. large multinucleate cells that arise by growth.
   b. large multinucleate cells that arise by fusion of smaller cells.
   c. small cells connected by many gap junctions.
   d. large cells with a single nucleus.

9. Connective tissues, although quite diverse in structure and location, do share a common theme; the connection between other types of tissues. Although all of the following seem to fit that criterion, one of the tissues listed is not a type of connective tissue. Which one?

   a. Blood   c. Adipose
   b. Muscle   d. Cartilage

APPLY

1. What do all the organs of the body have in common?

   a. Each contains the same kinds of cells.
   b. Each is composed of several different kinds of tissue.
   c. Each is derived from ectoderm.
   d. Each can be considered part of the circulatory system.

2. Rheumatoid arthritis is an autoimmune disease that attacks the linings of joints within the body. The cells that line these joints, and whose destruction causes the symptoms of arthritis, are known as

   a. osteocytes.   c. chondrocytes.
   b. erythrocytes.   d. thrombocytes.

3. Suppose that an alien virus arrives on Earth. This virus causes damage to the nervous system by attacking the structures of neurons. Which of the following structures would be immune from attack?

   a. Axon   c. Neuroglia
   b. Dendrite   d. All of these would be attacked by the virus.

4. Homeostasis

   a. is a dynamic process.
   b. describes the maintenance of the internal environment of the body.
   c. is essential to life.
   d. All of the choices are correct.

5. Which of the following scenarios correctly describes positive feedback?

   a. If the temperature increases in your room, your furnace increases its output of warm air.
   b. If you drink too much water, you produce more urine.
   c. If the price of gasoline increases, drivers decrease the length of their trips.
   d. If you feel cold, you start to shiver.

6. The three types of muscle all share

   a. a structure that includes striations.
   b. a membrane that is electrically excitable.
   c. the ability to contract.
   d. the characteristic of self-excitation.

SYNTHESIZE

1. Suppose that you discover a new disease that affects nutrient absorption in the gut as well as causes problems with the skin. Is it possible that one disease could involve the same tissues? How could this occur?

2. Which organ systems are involved in regulation and maintenance? Why do you think they are linked in this way?

3. We have all experienced hunger pangs. Is hunger a positive or negative feedback stimulus? Describe the steps involved in the response to this stimulus.

4. Why is homeostasis described as a dynamic process? How does negative feedback function in this process? How can antagonistic effectors result in a constant level, and how does this relate to the idea of a dynamic process?

ONLINE RESOURCE

www.ravenbiology.com

Understand, Apply, and Synthesize—enhance your study with animations that bring concepts to life and practice tests to assess your understanding. Your instructor may also recommend the interactive eBook, individualized learning tools, and more.

Chapter **43**

The Nervous System

Chapter Contents

20 μm

Introduction

All animals except sponges use a network of nerve cells to gather information about the body's condition and the external environment, to process and integrate that information, and to issue commands to the body's muscles and glands. As we saw in chapter 42, homeostasis of the animal's body is accomplished by negative feedback loops that maintain conditions within narrow limits. Negative feedback implies not only detection of appropriate stimuli but also communication of information to begin a response. The nervous system, composed of neurons, such as the one pictured here, is a fast communication system and a part of many feedback systems in the body.

An animal must be able to respond to environmental stimuli. A fly escapes a flyswatter; the antennae of a crayfish detect food and the crayfish moves toward it. To accomplish these actions, animals must have *sensory receptors* that can detect the stimulus and *motor effectors* that can respond to it. In most invertebrate phyla and in all vertebrate classes, sensory receptors and motor effectors are linked by way of the nervous system.

The central nervous system is the "command center"

As described in chapter 42, the nervous system consists of neurons and supporting cells. Figure 43.1 shows the three types of neurons. In vertebrates, **sensory neurons** (or afferent neurons) carry impulses from sensory receptors to the *central nervous system (CNS),* which is composed of the brain and spinal cord. **Motor neurons** (or efferent neurons) carry impulses from the CNS to effectors—muscles and glands. A third type of neuron is present in the nervous systems of most invertebrates and all vertebrates: **interneurons** (or association neurons). Interneurons are located in the brain and spinal cord of vertebrates, where they help provide more complex reflexes and higher associative functions, including learning and memory.

The peripheral nervous system collects information and carries out responses

Together, sensory and motor neurons constitute the *peripheral nervous system (PNS)* in vertebrates. Motor neurons that stimulate skeletal muscles to contract make up the **somatic nervous system;** those that regulate the activity of the smooth muscles, cardiac muscle, and glands compose the **autonomic nervous system.**

The autonomic nervous system is further broken down into the *sympathetic* and *parasympathetic* divisions. These divisions counterbalance each other in the regulation of many organ systems. Figure 43.2 illustrates the relationships among the different parts of the vertebrate nervous system.

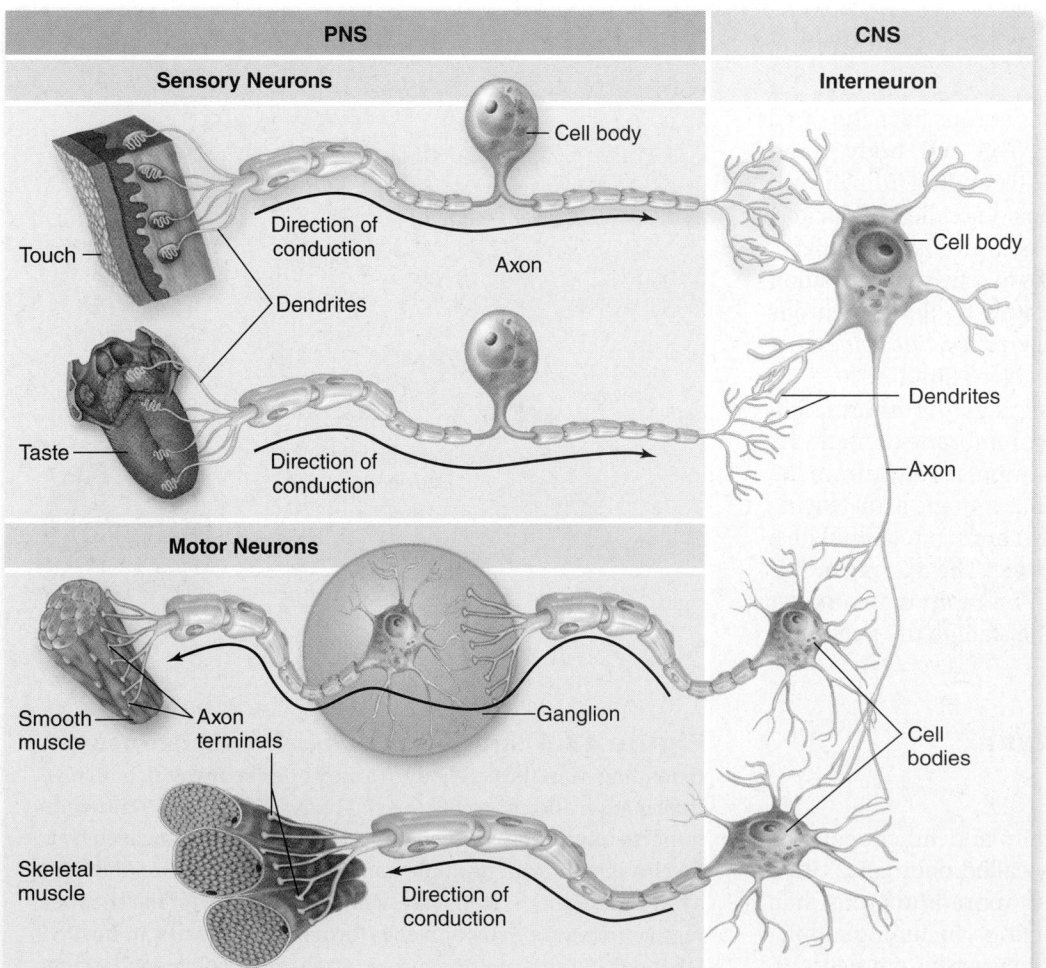

Figure 43.1 Three types of neurons. The brain and spinal cord form the central nervous system (CNS) of vertebrates, and sensory and motor neurons form the peripheral nervous system (PNS). Sensory neurons of the peripheral nervous system carry information about the environment to the CNS. Interneurons in the CNS provide links between sensory and motor neurons. Motor neurons of the PNS system carry impulses or "commands" to muscles and glands (effectors).

Figure 43.2 Divisions of the vertebrate nervous system. The major divisions are the central and peripheral nervous systems. The brain and spinal cord make up the central nervous system (CNS). The peripheral nervous system (PNS) includes everything outside the CNS and is divided into sensory and motor pathways. Sensory pathways can detect either external or internal stimuli. Motor pathways are divided into the somatic nervous system that activates voluntary muscles and the autonomic nervous system that activates involuntary muscles. The sympathetic and parasympathetic nervous systems are subsets of the autonomic nervous system that trigger opposing actions.

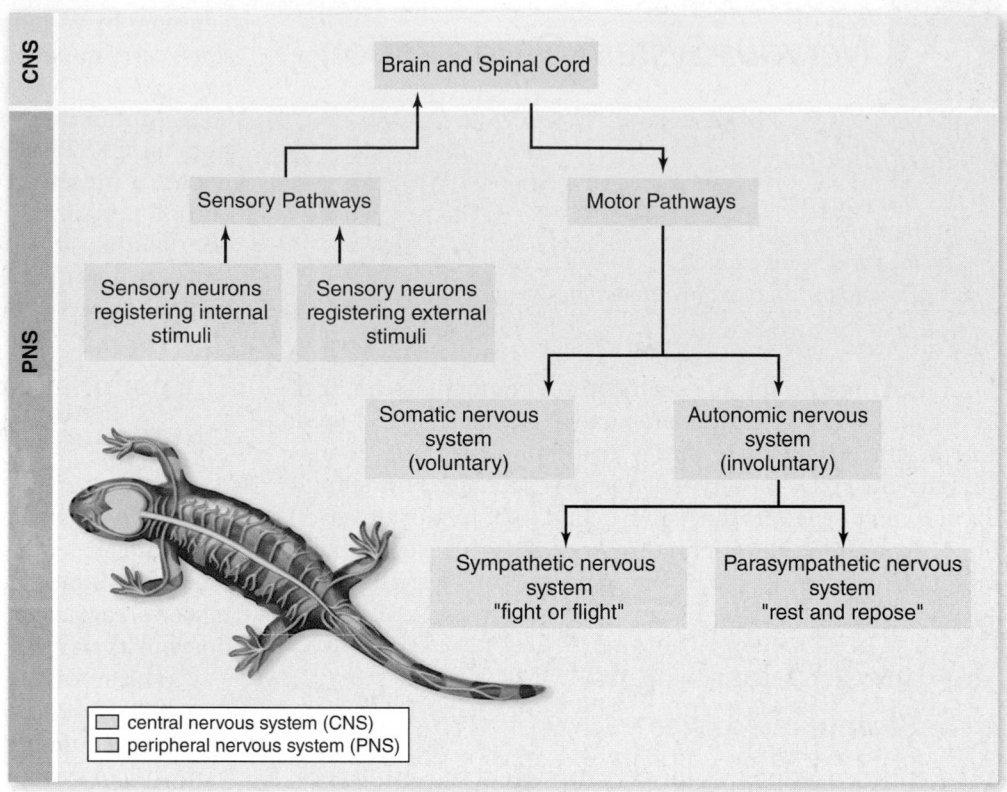

CNS

PNS

Brain and Spinal Cord

Sensory Pathways

Motor Pathways

Sensory neurons registering internal stimuli

Sensory neurons registering external stimuli

Somatic nervous system (voluntary)

Autonomic nervous system (involuntary)

Sympathetic nervous system "fight or flight"

Parasympathetic nervous system "rest and repose"

☐ central nervous system (CNS)
☐ peripheral nervous system (PNS)

The structure of neurons supports their function

Despite their varied appearances, most neurons have the same functional architecture (figure 43.3). The **cell body** is an enlarged region containing the nucleus. Extending from the cell body are one or more cytoplasmic extensions called **dendrites.** Motor and association neurons possess a profusion of highly branched dendrites, enabling those cells to receive information from many different sources simultaneously. Some neurons have extensions from the dendrites called *dendritic spines* that increase the surface area available to receive stimuli.

The surface of the cell body integrates the information arriving at its dendrites. If the resulting membrane excitation is sufficient, it triggers the conduction of impulses away from the cell body along an **axon.** Each neuron has a single axon leaving its cell body, although an axon may also branch to stimulate a number of cells. An axon can be quite long: The axons controlling the muscles in a person's feet can be more than a meter long, and the axons that extend from the skull to the pelvis in a giraffe are about 3 m long.

Supporting cells include Schwann cells and oligodendrocytes

Neurons are supported both structurally and functionally by supporting cells, which are collectively called neuroglia. These cells are one-tenth as big and 10 times more numerous than neurons, and they serve a variety of functions, including supplying the neurons with nutrients, removing wastes from neurons, guiding axon migration, and providing immune functions.

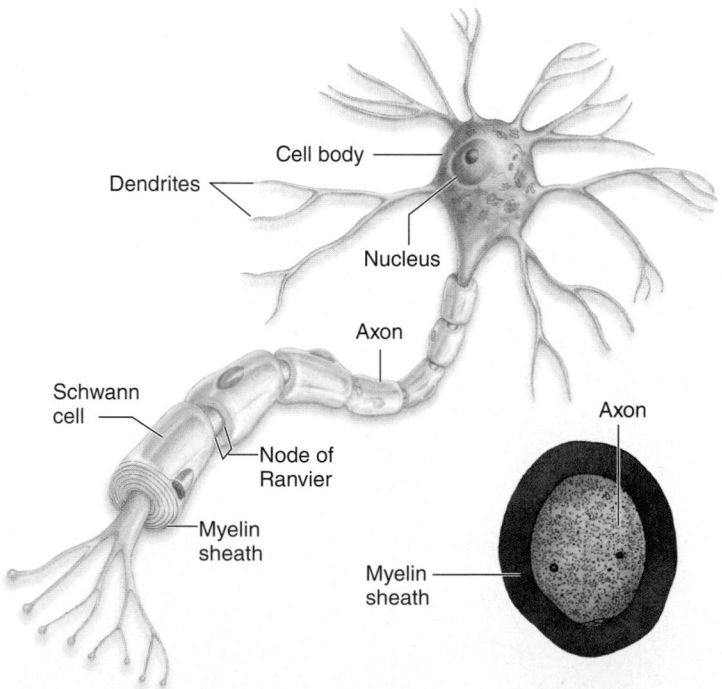

Cell body

Dendrites

Nucleus

Axon

Schwann cell

Axon

Node of Ranvier

Myelin sheath

Myelin sheath

Figure 43.3 Structure of a typical vertebrate neuron. Extending from the cell body are many dendrites, which receive information and carry it to the cell body. A single axon transmits impulses away from the cell body. Many axons are encased by a myelin sheath, with multiple membrane layers that insulate the axon. Small gaps, called nodes of Ranvier, interrupt the sheath at regular intervals. Schwann cells form myelin sheaths in the PNS (as shown for this motor neuron); extensions of oligodendrocytes form myelin sheaths in the CNS.

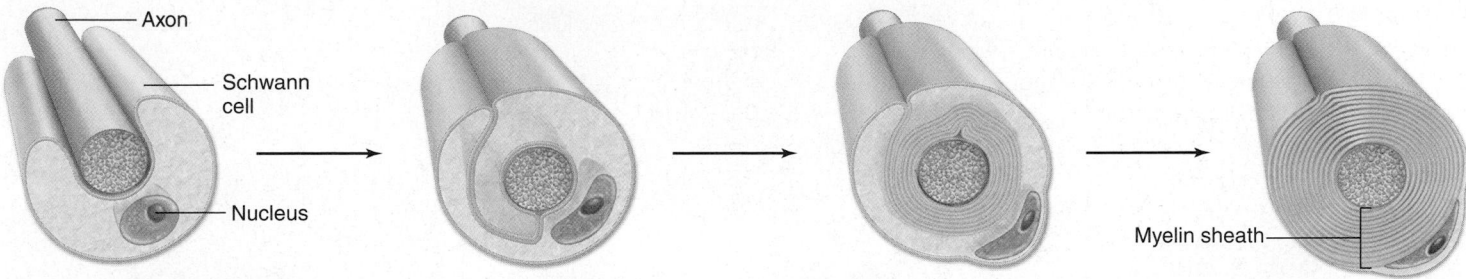

Figure 43.4 The formation of the myelin sheath around a peripheral axon. The myelin sheath forms by successive wrappings of Schwann cell membranes around the axon.

Two of the most important kinds of neuroglia in vertebrates are **Schwann cells** and **oligodendrocytes,** which produce **myelin sheaths** that surround the axons of many neurons. Schwann cells produce myelin in the PNS, and oligodendrocytes produce myelin in the CNS. During development, these cells wrap themselves around each axon several times to form the myelin sheath—an insulating covering consisting of multiple layers of compacted membrane (figure 43.4).

Axons that have myelin sheaths are said to be myelinated, and those that don't are unmyelinated. In the CNS, myelinated axons form the *white matter,* and the unmyelinated dendrites and cell bodies form the *gray matter.* In the PNS, myelinated axons are bundled together, much like wires in a cable, to form nerves.

Small gaps, known as **nodes of Ranvier** (see figure 43.3), interrupt the myelin sheath at intervals of 1 to 2 μm. We discuss the role of the myelin sheath in impulse conduction in the next section.

Learning Outcomes Review 43.1

The vertebrate nervous system consists of the central nervous system (CNS) and peripheral nervous system (PNS). The PNS comprises the somatic nervous system and autonomic nervous system; the latter has sympathetic and parasympathetic divisions. A neuron consists of a cell body, dendrites that receive information, and a single axon that sends signals. Neurons carry out nervous system functions; they are supported by a variety of neuroglia.

■ *Which division of the PNS is under conscious control?*

43.2 The Mechanism of Nerve Impulse Transmission

Learning Outcomes

1. *Compare the relative concentrations of important ions inside and outside the cell.*
2. *Describe the production of the resting potential.*
3. *Explain how the action of voltage-gated channels produces an action potential.*

Neuronal function depends on the ability to create an electric potential across the plasma membrane and then alter this potential to propagate signals. Since cells are aqueous solutions, electric charge is carried by ions, and cells manipulate the concentrations of a number of important ions across the plasma membrane.

When a neuron is stimulated, electrical changes in the plasma membrane spread or propagate from one part of the cell to another. This signaling depends on the properties of a variety of specialized membrane transport proteins. First, we examine some of the basic electrical properties common to the membrane of most animal cells that produce a membrane potential, then we see how neurons send signals (action potentials) through changes in this potential along an axon.

An electrical difference exists across the plasma membrane

In chapter 5 you learned that ions can cross the cell membrane through specialized membrane proteins called ion channels. These ion channels are specific for different ions, such as K^+ or Na^+, and they can either be leakage channels (open all the time) or gated channels (open in response to a stimulus).

Any separation of electric charges of opposite sign represents an electric potential that is capable of doing work; we encounter this when we use a flashlight, which draws current from such a potential in a battery. Cells maintain an electric potential across the plasma membrane, in this case the interior of the membrane is the negative pole, and the exterior is the positive pole. Because cells are very small, their membrane potential is also very small. The resting membrane potential of many vertebrate neurons ranges from –40 to –90 millivolts (mV), or 0.04 to 0.09 volts (V). For the examples and figures in this chapter, we use an average resting membrane potential value of –70 mV. The minus sign indicates that the inside of the cell is negative with respect to the outside.

Contributors to membrane potential

The inside of the cell is more negatively charged in relation to the outside because of two factors:

1. The sodium–potassium pump, described in chapter 5, brings two potassium ions (K^+) into the cell for every three sodium ions (Na^+) it pumps out (figure 43.5). This helps establish and maintain concentration differences that result in high K^+ and low Na^+ concentrations inside the cell, and high Na^+ and low K^+ concentrations outside the cell.

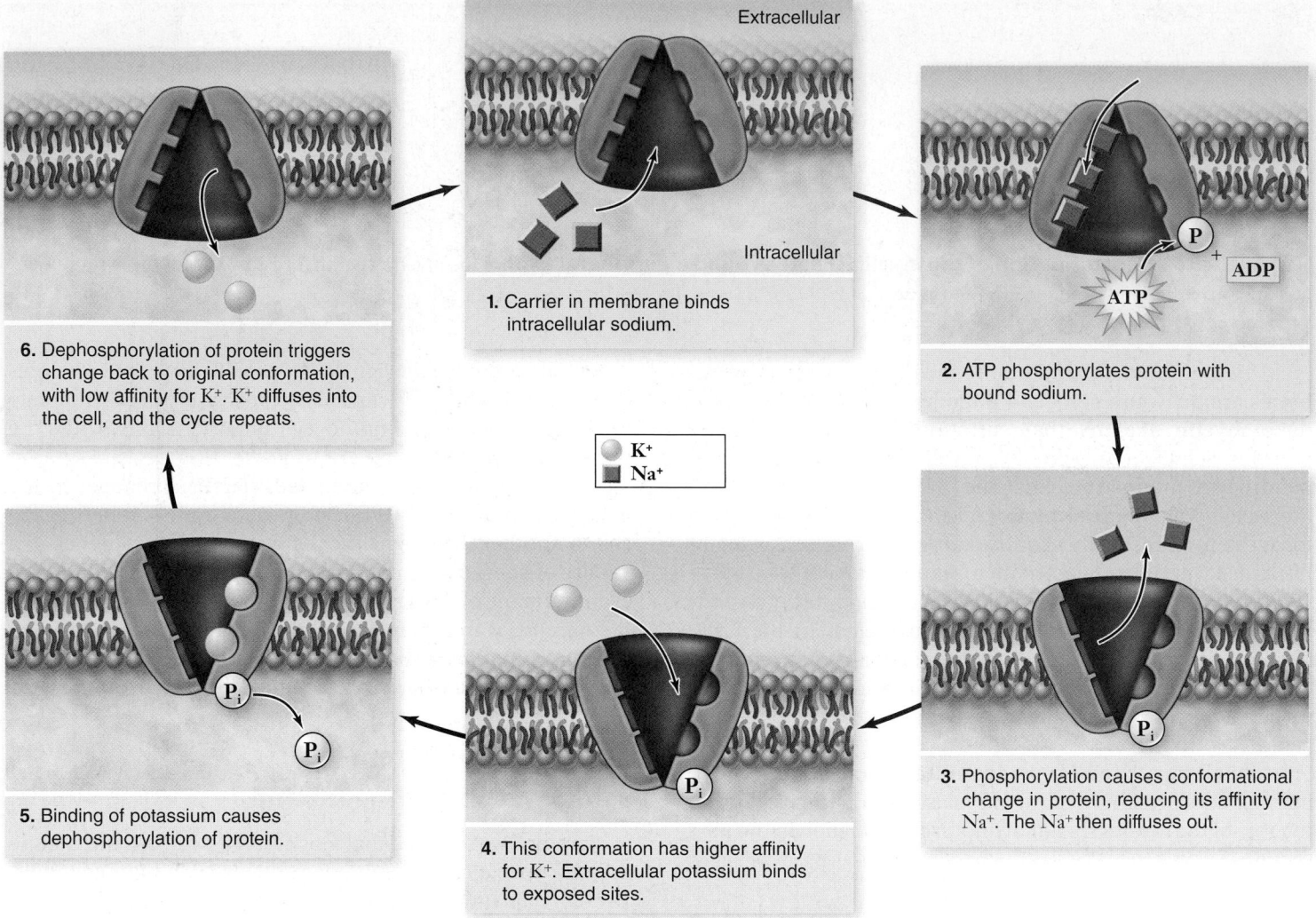

Extracellular

Intracellular

1. Carrier in membrane binds intracellular sodium.

2. ATP phosphorylates protein with bound sodium.

K⁺
Na⁺

6. Dephosphorylation of protein triggers change back to original conformation, with low affinity for K⁺. K⁺ diffuses into the cell, and the cycle repeats.

3. Phosphorylation causes conformational change in protein, reducing its affinity for Na⁺. The Na⁺ then diffuses out.

5. Binding of potassium causes dephosphorylation of protein.

4. This conformation has higher affinity for K⁺. Extracellular potassium binds to exposed sites.

Figure 43.5 **The sodium–potassium pump.** This pump transports three Na⁺ to the outside of the cell and simultaneously transports two K⁺ to the inside of the cell. This is an active transport carrier requiring the (phosphorylating) energy of ATP.

2. Ion channels in the cell membrane are more numerous for K⁺ than for Na⁺, making the membrane more permeable for K⁺. **Ion channels** are membrane proteins that form pores through the membrane, allowing diffusion of specific ions across the membrane. Because there are more ion channels for K⁺, the membrane is more permeable to K⁺ and it will diffuse out of the cell.

The resting potential: Balance between two forces

The **resting potential** arises due to the action of the sodium–potassium pump and the greater permeability of the membrane to K⁺. The pump moves three Na⁺ outside for every two K⁺ inside, which creates a small imbalance in cations outside the cell. This has only a minor effect; however, the concentration gradients created by the pump are significant. The concentration of K⁺ is much higher inside the cell than outside, leading to diffusion of K⁺ through K⁺ leakage channels that are always open. Since the membrane is not permeable to the negative ions that could counterbalance this (mainly organic phosphates, amino acids, and proteins), positive charge builds up outside the membrane and negative charge inside the

membrane. This electrical potential then is an attractive force pulling K⁺ ions back inside the cell. The balance between the diffusional force and the electrical force produces an **equilibrium potential** (table 43.1). By relating the work done by each type of force, we can derive a quantitative expression for this equilibrium potential called the Nernst equation. This assumes the action of a single ion, and for a positive ion with charge equal to +1, the Nernst equation for K⁺ ions is:

$$E_K = 58 \text{ mV} \ \log([K^+]_{out}/[K^+]m_{in})$$

The calculated equilibrium potential for K⁺ is –90 mV (see table 43.1), close to the measured value of –70 mV. The calculated value for Na⁺ is +60 mV, clearly not at all close to the measured value, but the leakage of a small amount of Na⁺ back into the cell is responsible for lowering the equilibrium potential of K⁺ to the –70 mV value observed. The resting membrane potential of a neuron can be measured and viewed or graphed using a voltmeter and a pair of electrodes, one outside and one inside the cell (figure 43.6).

Neurons are unique not because of the production and maintenance of the resting membrane potential, but rather due

	Concentration in ECF		Ratio	Equilibrium Potential
TABLE 43.1 The Ionic Composition of Cytoplasm and Extracellular Fluid (ECF)				
Ion	(mM)	(mM)	(ECF:cytoplasm)	(mV)
Na^+	150	15	10:1	+60
K^+	5	150	1:30	−90
Cl^-	110	7	15:1	−70

to the sudden temporary disruptions to the resting membrane potential that occur in response to stimuli. Two types of changes can be observed: **graded potentials** and **action potentials.** Graded potentials are small *continuous* changes to membrane potential, and action potentials are *transient* disruptions of the potential triggered by a threshold change in potential. Action potentials are the actual signals that propagate along an axon.

Changes in membrane potential are due to changes in membrane permeability

The resting potential arose because of the permeability of the membrane to K^+ through leakage channels. Changes in potential are due to the action of a different class of channels called gated channels. These gated channels act like a door that can be either open or closed in response to a stimulus. There are two types of gated channels that we will see: chemically gated, or **ligand-gated,** channels that respond to a chemical signal, and **voltage-gated channels** that respond to changes in membrane potential. Ligand-gated channels lead to graded

potentials that determine whether an axon will fire, and voltage-gated channels produce action potentials.

Graded potentials: Depolarizing or hyperpolarizing

In most neurons, gated ion channels in dendrites respond to the binding of signaling molecules (figure 43.7; see also figure 9.4*a*). Hormones and neurotransmitters act as ligands, inducing opening of ligand-gated channels, and causing changes in plasma membrane permeability that lead to changes in membrane voltage.

Permeability changes are measurable as depolarizations or hyperpolarizations of the membrane potential. A **depolarization** makes the membrane potential less negative (more positive), whereas a **hyperpolarization** makes the membrane potential more negative. For example, a change in potential from –70 mV to –65 mV is a depolarization; a change from –70 mV to –75 mV is a hyperpolarization.

These small changes in membrane potential result in graded potentials because their size depends on either the strength of the stimulus or the amount of ligand available to bind

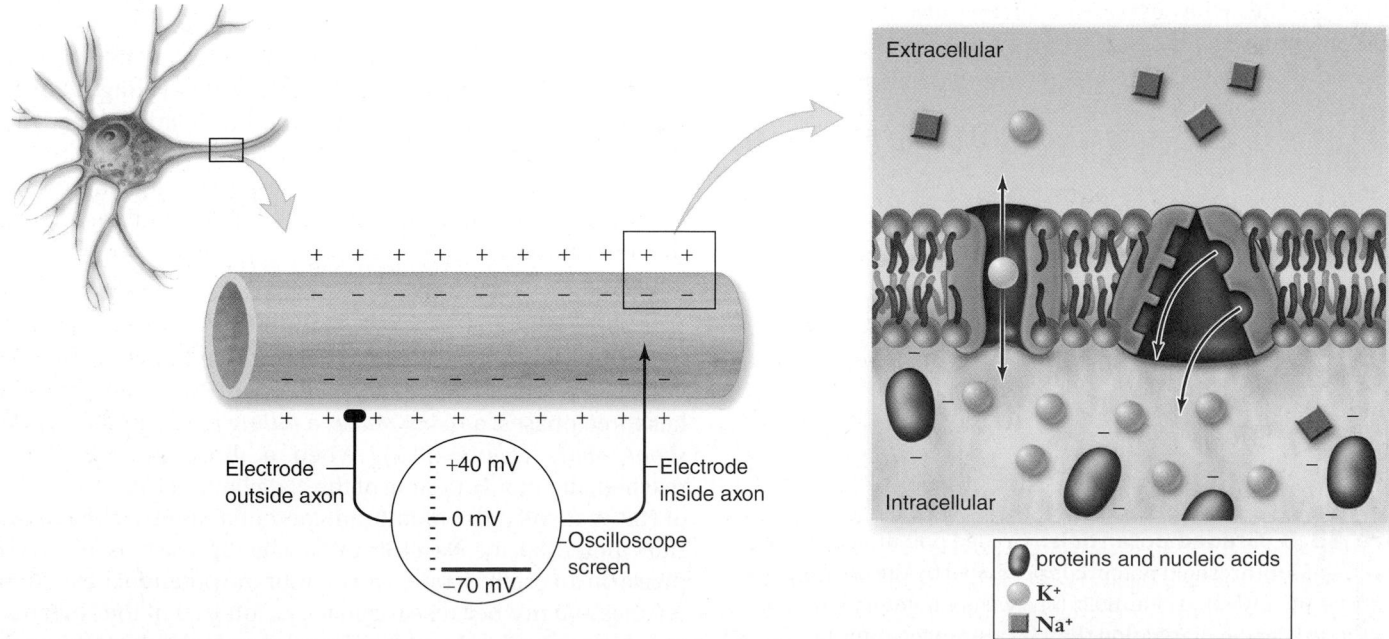

Figure 43.6 Establishment of the resting membrane potential. A voltmeter placed with one electrode inside an axon and the other outside the membrane. The electric potential inside is –70 mV relative to the outside of the membrane. K^+ diffuses out of the cell through ion channels because its concentration is higher inside than outside. Negatively charged proteins and nucleic acids inside the cell cannot leave the cell and attract cations from outside the cell, such as K^+. This balance of electrical and diffusional forces produces the resting potential. The sodium–potassium pump maintains cell equilibrium by counteracting the effects of Na^+ leakage into the cell and contributes to the resting potential by moving 3 Na^+ outside for every 2 K^+ moved inside.

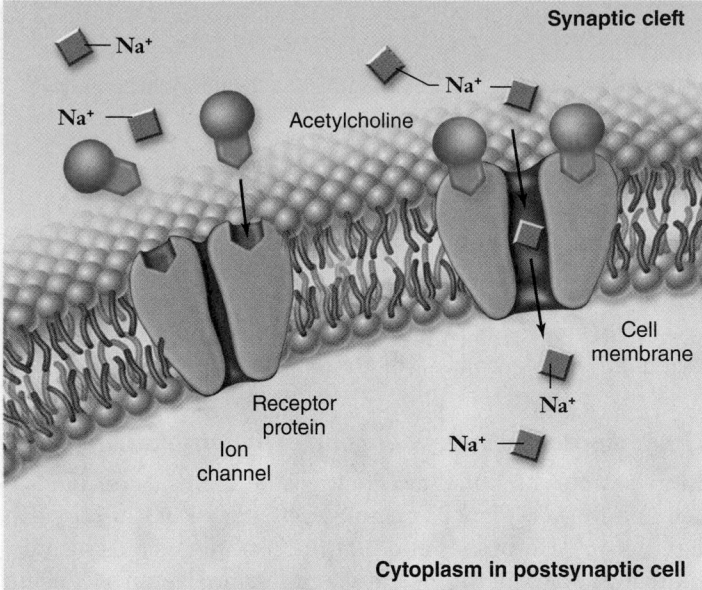

Figure 43.7 A chemically gated ion channel. The acetylcholine (ACh) receptor is a chemically gated channel that can bind the neurotransmitter ACh. Binding of ACh causes the channel to open allowing Na^+ ions to flow into the cell by diffusion.

with their receptors. These potentials diminish in amplitude as they spread from their point of origin. Depolarizing or hyperpolarizing potentials can add together to amplify or reduce their effects, just as two waves can combine to make a bigger one when they meet in synchronization or can cancel each other out when a trough meets with a crest. The ability of graded potentials to combine is called **summation** (figure 43.8). We will return to this topic in the next section after we discuss the nature of action potentials.

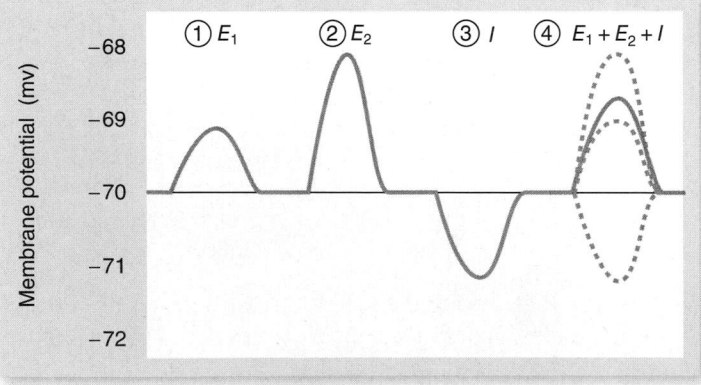

Figure 43.8 Graded potentials. Graded potentials are the summation of subthreshold potentials produced by the opening of different chemically gated channels. (1) A weak excitatory stimulus, E_1, elicits a smaller depolarization than (2) a stronger stimulus, E_2. (3) An inhibitory stimulus, I, produces a hyperpolarization. (4) If all three stimuli occur very close together, the resulting polarity change will be the sum of the three individual changes.

 Data analysis Draw the resulting potentials for all possible combinations of E_1, E_2, and I.

Action potentials result when depolarization reaches a threshold

When a particular level of depolarization is reached (about −55 mV in some mammalian axons), a nerve impulse, or action potential, is produced in the region where the axon arises from the cell body. The level of depolarization needed to produce an action potential is called the **threshold potential.** Depolarizations bring a neuron closer to the threshold, and hyperpolarizations move the neuron farther from the threshold.

The action potential is caused by another class of ion channels: **voltage-gated ion channels.** These channels open and close in response to changes in membrane potential; the flow of ions controlled by these channels creates the action potential. Voltage-gated channels are found in neurons and in muscle cells. Two different channels are used to create an action potential in neurons: **voltage-gated Na^+ channels** and **voltage-gated K^+ channels.**

Sodium and potassium voltage-gated channels

The behavior of the voltage-gated Na^+ channel is more complex than that of the K^+ channel, so we will consider it first. The channel has two gates: an activation gate and an inactivation gate. In its resting state the activation gate is closed and the inactivation gate is open. When the threshold voltage is reached, the activation gate opens rapidly, leading to an influx of Na^+ ions due to both concentration and voltage gradients. After a short period the inactivation gate closes, stopping the influx of Na^+ ions and leaving the channel in a temporarily inactivated state. The channel is returned to its resting state by the activation gate closing and the inactivation gate opening. The result of this is a transient influx of Na^+ that depolarizes the membrane in response to a threshold voltage.

The K^+ channel has a single activation gate that is closed in the resting state. In response to a threshold voltage, it opens slowly. With the high concentration of K^+ inside the cell, and the membrane now far from the equilibrium potential, an efflux of K^+ begins. The positive charge now leaving the cell counteracts the effect of the Na^+ channel and repolarizes the membrane.

Tracing an action potential's changes

Let us now put all of this together and see how the changing flux of ions leads to an action potential. The action potential has three phases: a *rising phase,* a *falling phase,* and an *undershoot phase* (figure 43.9). When a threshold potential is reached, the rapid opening of the Na^+ channel causes an influx of Na^+ that shifts the membrane potential toward the equilibrium potential for Na (+60 mV). This appears as the rising phase on an oscilloscope. The membrane potential never quite reaches +60 mV because the inactivation gate of the Na^+ channel closes, terminating the rising phase. At the same time, the opening of the K^+ channel leads to K^+ diffusing out of the cell, repolarizing the membrane in the falling phase. The K^+ channels remain open longer than necessary to restore the resting potential, resulting in a slight undershoot. This entire sequence of events for a single action potential takes about a millisecond.

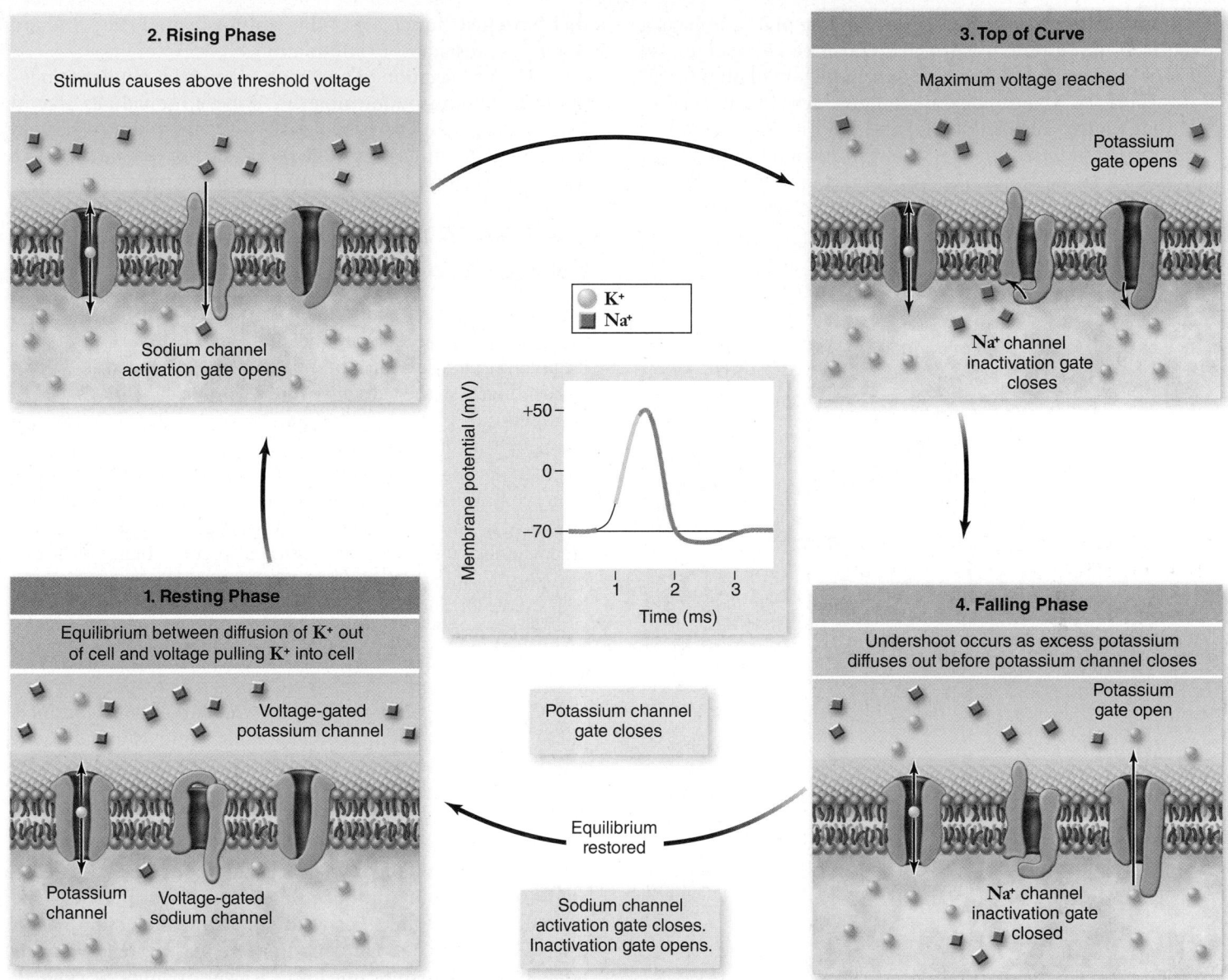

Figure 43.9 **The action potential.** (1) At resting membrane potential, voltage-gated ion channels are closed, but there is some diffusion of K+. In response to a stimulus, the cell begins to depolarize, and once the threshold level is reached, an action potential is produced. (2) Rapid depolarization occurs (the rising portion of the spike) because voltage-gated sodium channel activation gates open, allowing Na+ to diffuse into the axon. (3) At the top of the spike, Na+ channel inactivation gates close, and voltage-gated potassium channels that were previously closed begin to open. (4) With the K+ channels open, repolarization occurs because of the diffusion of K+ out of the axon. An undershoot occurs before the membrane returns to its original resting potential.

The nature of action potentials

Action potentials are separate, all-or-none events. An action potential occurs if the threshold voltage is reached, but not while the membrane remains below threshold. Action potentials do not add together or interfere with one another, as graded potentials can. After Na+ channels "fire" they remain in an inactivated state until the inactivation gate reopens, preventing any summing of effects. This is called the absolute refractory period when the membrane cannot be stimulated. There is also a relative refractory period during which stimulation produces action potentials of reduced amplitude.

The production of an action potential results entirely from the passive diffusion of ions. However, at the end of each action potential, the cytoplasm contains a little more Na+ and a little less K+ than it did at rest. Although the number of ions moved by a single action potential is tiny relative to the concentration gradients of Na+ and K+, eventually this would have an effect. The constant activity of the sodium–potassium pump compensates for these changes. Thus, although active transport is not required to produce action potentials, it is needed to maintain the ion gradients.

Action potentials are propagated along axons

The movement of an action potential through an axon is not generated by ions flowing from the base of the axon to the end. Instead an action potential originates at the base of the axon, and is then recreated in adjacent stretches of membrane along the axon.

Each action potential, during its rising phase, reflects a reversal in membrane polarity. The positive charges due to influx of Na⁺ can depolarize the adjacent region of membrane to threshold, so that the next region produces its own action potential (figure 43.10). Meanwhile, the previous region of membrane repolarizes back to the resting membrane potential. The signal does not back up because the Na⁺ channels that have just "fired" are still in an inactivated state and are refractory (resistant) to stimulation.

The propagation of an action potential is similar to people in a stadium performing the "wave": Individuals stay in place as they stand up (depolarize), raise their hands (peak of the action potential), and sit down again (repolarize). The wave travels around the stadium, but the people stay in place.

There are two ways to increase the velocity of nerve impulses

Action potentials are conducted without decreasing in amplitude, so the last action potential at the end of an axon is just as large as the first action potential. Animals have evolved two ways to increase the velocity of nerve impulses. The velocity of conduction is greater if the diameter of the axon is large or if the axon is myelinated (table 43.2).

Increasing the diameter of an axon increases the velocity of nerve impulses due to the electrical property of resistance. Electrical resistance is inversely proportional to cross-sectional area, which is a function of diameter, so larger diameter axons have less resistance to current flow. The positive charges carried by Na⁺ flow farther in a larger diameter axon, leading to a higher than threshold voltage farther from the origin of Na⁺ influx.

Larger diameter axons are found primarily in invertebrates. For example, in the squid, the escape response is controlled by a so-called giant axon. This large axon conducts nerve impulses faster than other smaller squid axons, allowing a rapid escape response. The squid giant axon was used by Alan Lloyd Hodgkin and Andrew Huxley in their pioneering studies of nerve transmission.

Myelinated axons conduct impulses more rapidly than unmyelinated axons because the action potentials in myelinated axons are only produced at the nodes of Ranvier. One action potential still serves as the depolarization stimulus for the next, but the depolarization at one node spreads quickly beneath the insulating myelin to trigger opening of voltage-gated channels at the next node. The impulses therefore seem to jump from node to node (figure 43.11) in a process called **saltatory conduction** (Latin *saltare,* "to jump").

To see how saltatory conduction speeds impulse transmission, let's return for a moment to the stadium wave analogy

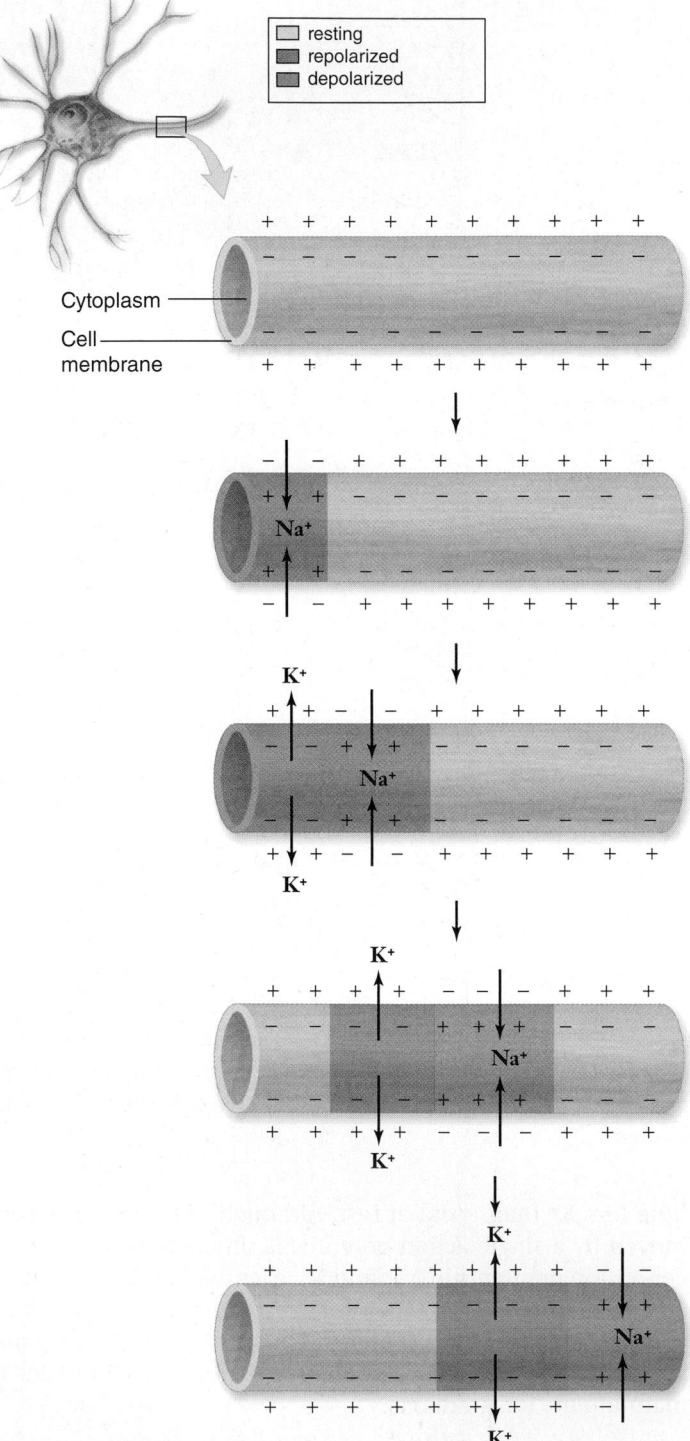

Figure 43.10 Propagation of an action potential in an unmyelinated axon. When one region produces an action potential and undergoes a reversal of polarity, it serves as a depolarization stimulus for the next region of the axon. In this way, action potentials regenerate along each small region of the unmyelinated axon membrane.

TABLE 43.2	Conduction Velocities of Some Axons		
	Axon Diameter (μm)	Myelin	Conduction Velocity (m/s)
Squid giant axon	500	No	25
Large motor axon to human leg muscle	20	Yes	120
Axon from human skin pressure receptor	10	Yes	50
Axon from human skin temperature receptor	5	Yes	20
Motor axon to human internal organ	1	No	2

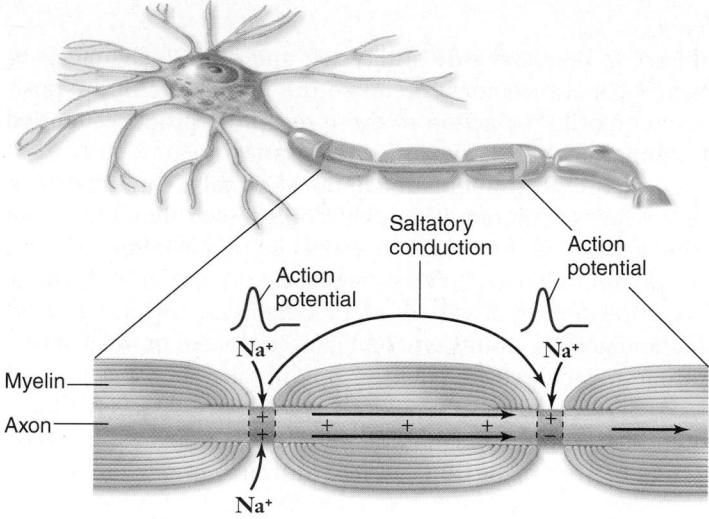

Figure 43.11 Saltatory conduction in a myelinated axon. Action potentials are only produced at the nodes of Ranvier in a myelinated axon. One node depolarizes the next node so that the action potentials can skip between nodes. As a result, saltatory ("jumping") conduction in a myelinated axon is more rapid than conduction in an unmyelinated axon.

to describe propagation of an action potential. The wave moves across the seats of a crowded stadium as fans seeing the people in the adjacent section stand up are triggered to stand up in turn. Because the wave skips sections of empty bleachers, it actually progresses around the stadium even faster with more empty sections. The wave doesn't have to "wait" for the missing people to stand, so it simply moves to the next populated section—just as the action potential jumps the nonconducting regions of myelin between exposed nodes.

Learning Outcomes Review 43.2

Neurons maintain high K$^+$ levels inside the cell, and high Na$^+$ levels outside the cell. Diffusion of K$^+$ to the outside leads to a resting potential of about −70 mV. Opening of ligand-gated channels can depolarize or hyperpolarize the membrane, causing a graded potential. Action potentials are triggered when membrane potential exceeds a threshold value. Voltage-gated Na$^+$ channels open, and depolarization occurs; subsequent opening of K$^+$ channels leads to repolarization.

■ *How can only positive ions result in depolarization and repolarization of the membrane during an action potential?*

43.3 Synapses: Where Neurons Communicate with Other Cells

Learning Outcomes

1. Distinguish between electrical and chemical synapses.
2. List the different chemical neurotransmitters.
3. Explain the effects of addictive drugs on the nervous system.

An action potential passing down an axon eventually reaches the end of the axon and all of its branches. Neurons communicate with other neurons, with muscle cells, or with gland cells through specialized intercellular junctions called **synapses** that are found at the end of an axon. The neuron whose axon transmits action potentials to the synapse is termed the *presynaptic cell,* and the cell receiving the signal on the other side of the synapse is the *postsynaptic cell.*

The two types of synapses are electrical and chemical

The nervous systems of animals have two basic types of synapses: electrical and chemical. **Electrical synapses** involve direct cytoplasmic connections formed by gap junctions between the pre- and postsynaptic neurons (see chapter 4; figure 4.28). Membrane potential changes, including action potentials, pass directly and rapidly from one cell to the other through the gap junctions. Electrical synapses are common in invertebrate nervous systems, but are rare in vertebrates.

The vast majority of vertebrate synapses are *chemical synapses* (figure 43.12). When synapses are viewed under a light microscope, the presynaptic and postsynaptic cells

---

SCIENTIFIC THINKING

Question: *Is communication between neurons, and between neurons and muscle, chemical or electrical?*

Hypothesis: *Signaling between a neuron and heart muscle is chemical.*

Prediction: *Application of chemical solutes from one heart will affect the activity of another heart.*

Test: *Two frog hearts are placed in saline, one with vagus nerve attached, the other without. The vagus heart is stimulated, then fluid from around the vagus nerve is removed and applied to the other heart.*

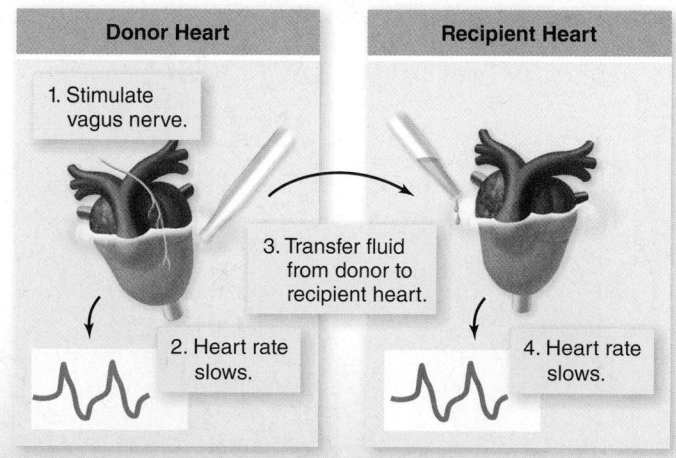

Result: *Heart that was not stimulated by the vagus nerve slows as though it was stimulated.*

Conclusion: *The nerve released a chemical signal that slowed heart rate.*

Further Experiments: *How does this conclusion extend the experiment described in Fig 42.5?*

Figure 43.12 Synaptic signaling.

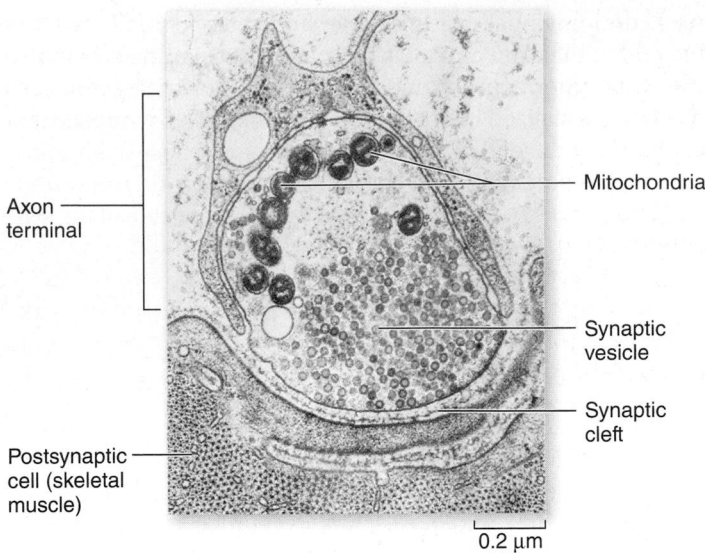

Figure 43.13 **A synaptic cleft.** An electron micrograph showing a neuromuscular synapse. Synaptic vesicles have been colored green.

Labels on figure: Axon terminal; Mitochondria; Synaptic vesicle; Synaptic cleft; Postsynaptic cell (skeletal muscle); 0.2 μm

appear to touch, but when viewed with an electron microscope most have a **synaptic cleft,** a narrow space that separates these two cells (figure 43.13).

The end of the presynaptic axon is swollen and contains numerous **synaptic vesicles,** each packed with chemicals called *neurotransmitters.* When action potentials arrive at the end of the axon, they stimulate the opening of voltage-gated calcium (Ca^{2+}) channels, causing a rapid inward diffusion of Ca^{2+}. This influx of Ca^{2+} triggers a complex series of events that leads to the fusion of synaptic vesicles with the plasma membrane and the release of neurotransmitter by exocytosis (see chapter 5; figure 43.14).

The higher the frequency of action potentials in the presynaptic axon, the greater the number of vesicles that release their contents of neurotransmitters. The neurotransmitters diffuse to the other side of the cleft and bind to chemical- or ligand-gated receptor proteins in the membrane of the postsynaptic cell. The action of these receptors produces graded potentials in the postsynaptic membrane.

Neurotransmitters are chemical signals in an otherwise electrical system, requiring tight control over the duration of their action. Neurotransmitters must be rapidly removed from the synaptic cleft to allow new signals to be transmitted. This is accomplished by a variety of mechanisms, including enzymatic digestion in the synaptic cleft, reuptake of neurotransmitter molecules by the neuron, and uptake by glial cells.

Several different types of neurotransmitters have been identified, and they act in different ways. We next consider the action of a few of the important neurotransmitter chemicals.

Many different chemical compounds serve as neurotransmitters

No single chemical characteristic defines a neurotransmitter, although we can group certain types according to chemical similarities. Some, such as acetylcholine, have wide use in the nervous system, particularly where nerves connect with muscles. Other neurotransmitters are found only in very specific types of junctions, such as in the CNS.

Acetylcholine

Acetylcholine (ACh) is the neurotransmitter that crosses the synapse between a motor neuron and a muscle fiber. This synapse is called a **neuromuscular junction** (figures 43.14, 43.15). Acetylcholine binds to its receptor proteins in the postsynaptic membrane and causes ligand-gated ion channels within these proteins to open (see figure 43.7). As a result, that site on the postsynaptic membrane produces a depolarization (figure 43.16a) called an *excitatory postsynaptic potential (EPSP).* The EPSP, if large enough, can open the voltage-gated channels for Na^+ and K^+ that are responsible for action potentials. Because the postsynaptic cell in this case is a skeletal

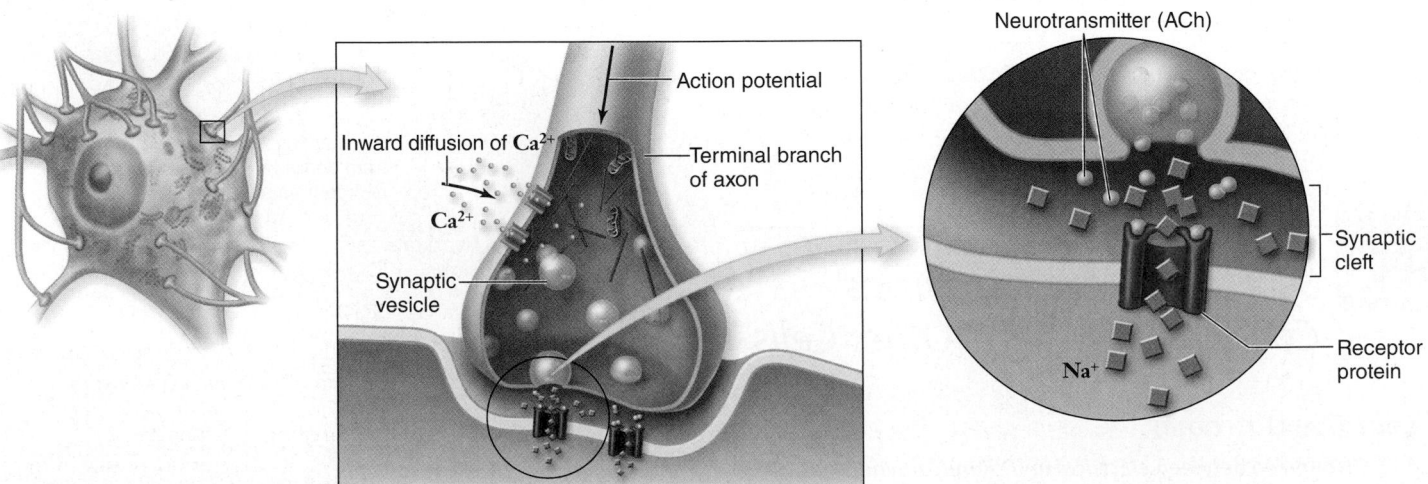

Labels on figure: Action potential; Inward diffusion of Ca^{2+}; Terminal branch of axon; Ca^{2+}; Synaptic vesicle; Neurotransmitter (ACh); Synaptic cleft; Receptor protein; Na^+

Figure 43.14 **The release of neurotransmitter.** Action potentials arriving at the end of an axon trigger inward diffusion of Ca^{2+}, which causes synaptic vesicles to fuse with the plasma membrane and release their neurotransmitters (acetylcholine [ACh] in this case). Neurotransmitter molecules diffuse across the synaptic gap and bind to ligand-gated receptors in the postsynaptic membrane.

Figure 43.15 Neuromuscular junctions. A light micrograph shows axons branching to make contact with several individual muscle fibers.

15 μm

muscle fiber, the action potentials it produces stimulate muscle contraction through mechanisms discussed in chapter 46.

For the muscle to relax, ACh must be eliminated from the synaptic cleft. *Acetylcholinesterase (AChE),* an enzyme in the postsynaptic membrane, eliminates ACh. This enzyme, one of the fastest known, cleaves ACh into inactive fragments. Nerve gas and the agricultural insecticide parathion are potent inhibitors of AChE; in humans, they can produce severe spastic paralysis and even death if paralysis affects the respiratory muscles. Although ACh acts as a neurotransmitter between motor neurons and skeletal muscle cells, many neurons also use ACh as a neurotransmitter at their synapses with the dendrites or cell bodies of other neurons.

Amino acids

Glutamate is the major excitatory neurotransmitter in the vertebrate CNS. Excitatory neurotransmitters act to stimulate action potentials by producing EPSPs. Some neurons in the brains of people suffering from Huntington disease undergo changes that render them hypersensitive to glutamate, leading to neurodegeneration.

Glycine and γ-aminobutyric acid (GABA) are inhibitory neurotransmitters. These neurotransmitters cause the opening of ligand-gated channels for the chloride ion (Cl^-), which has a concentration gradient favoring its diffusion into the neuron. Because Cl^- is negatively charged, it makes the inside of the membrane even more negative than it is at rest—for example, from –70 mV to –85 mV (see figure 43.16*b*). This hyperpolarization is called an *inhibitory postsynaptic potential (IPSP),* and it is very important for neural control of body movements and other

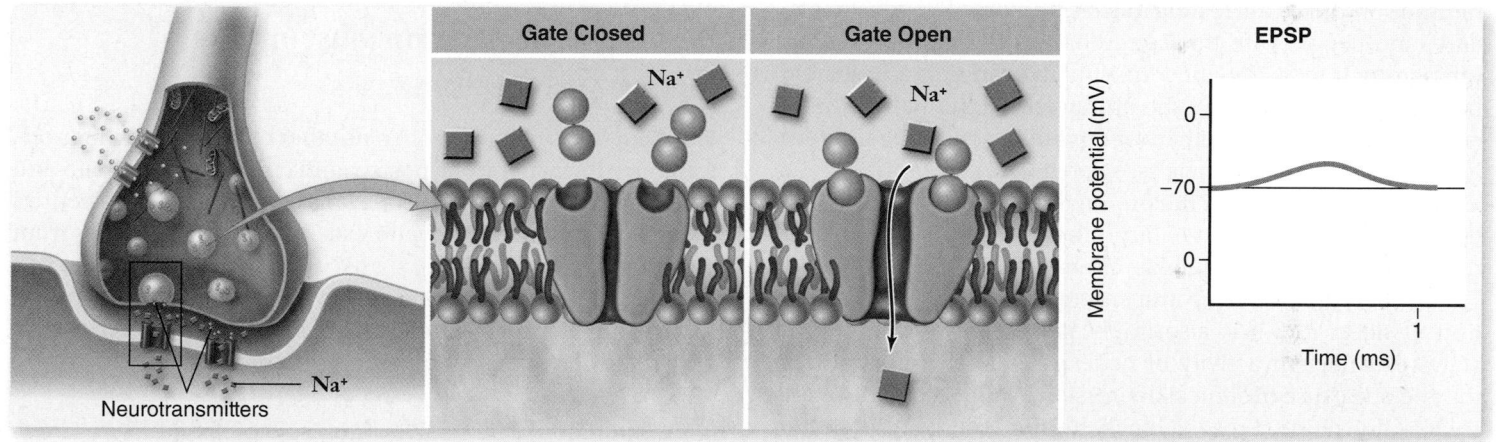

a.

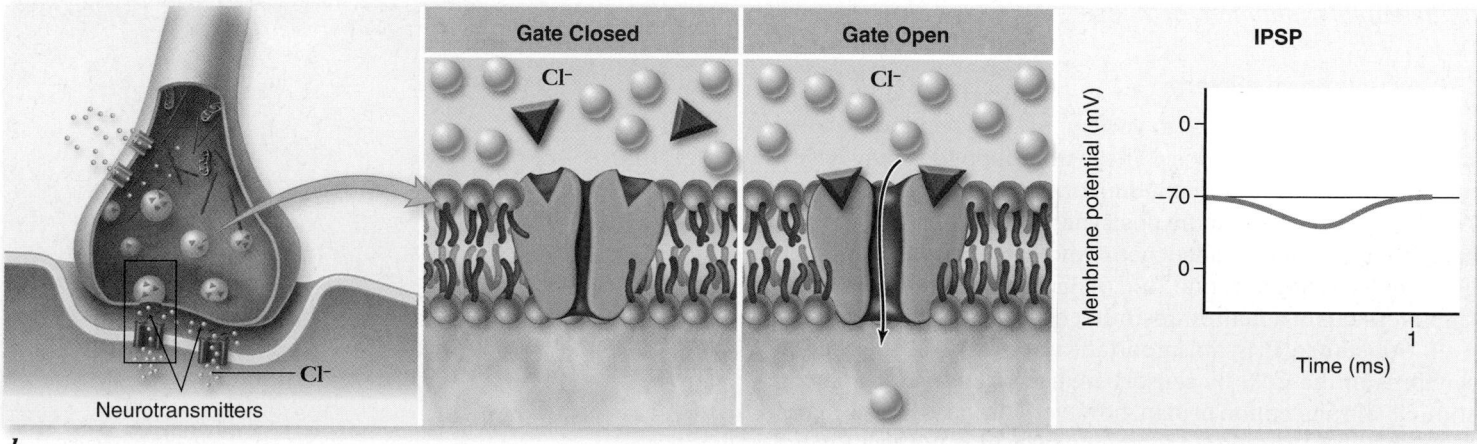

b.

Figure 43.16 Different neurotransmitters can have different effects. *a.* An excitatory neurotransmitter promotes a depolarization, or excitatory postsynaptic potential (EPSP). *b.* An inhibitory neurotransmitter promotes a hyperpolarization, or inhibitory postsynaptic potential (IPSP).

Q Data analysis If a cell body received input from two neurons at the same time that generated the two potentials shown on the right of figure 43.16, what would be the overall effect on membrane potential?

brain functions. The drug diazepam (Valium) causes its sedative and other effects by enhancing the binding of GABA to its receptors, thereby increasing the effectiveness of GABA at the synapse.

Biogenic amines

The **biogenic amines** include the hormone epinephrine (adrenaline), together with the neurotransmitters dopamine, norepinephrine, and serotonin. Epinephrine, norepinephrine, and dopamine are derived from the amino acid tyrosine and are included in the subcategory of *catecholamines*. Serotonin is a biogenic amine derived from a different amino acid, tryptophan.

Epinephrine is released into the blood as a hormonal secretion, while **norepinephrine** is released at synapses of neurons in the sympathetic nervous system (discussed in detail later on). The effects of these neurotransmitters on target receptors are responsible for the "fight or flight" response—faster and stronger heartbeat, increased blood glucose concentration, and diversion of blood flow into the muscles and heart.

Dopamine is a very important neurotransmitter used in some areas of the brain controlling body movements and other functions. Degeneration of particular dopamine-releasing neurons produces the resting muscle tremors of Parkinson disease, and people with this condition are treated with L-dopa (an acronym for L–3,4–dihydroxyphenylalanine), a precursor from which dopamine can be produced. Additionally, studies suggest that excessive activity of dopamine-releasing neurons in other areas of the brain is associated with schizophrenia. As a result, drugs that block the production of dopamine, such as the dopamine antagonist chlorpromazine (Thorazine), sometimes help patients with schizophrenia.

Serotonin is a neurotransmitter involved in the regulation of sleep, and it is also implicated in various emotional states. Insufficient activity of neurons that release serotonin may be one cause of clinical depression. Antidepressant drugs, such as fluoxetine (Prozac), block the elimination of serotonin from the synaptic cleft; these drugs are termed *selective serotonin reuptake inhibitors,* or SSRIs.

Other neurotransmitters

Axons can also release a variety of polypeptides, called **neuropeptides,** at synapses. These neuropeptides may have a typical neurotransmitter function, or they may have more subtle, long-term action on the postsynaptic neurons. In the latter case, they are often called **neuromodulators.** A given axon generally releases only one kind of neurotransmitter, but many can release both a neurotransmitter and a neuromodulator.

Substance P is an important neuropeptide released at synapses in the CNS by sensory neurons activated by painful stimuli. The perception of pain, however, can vary depending on circumstances. An injured football player may not feel the full extent of his trauma, for example, until he is out of the game.

The intensity with which pain is perceived partly depends on the effects of neuropeptides called *enkephalins* and *endorphins*. **Enkephalins,** released by axons descending from the brain into the spinal cord, inhibit the passage of pain information back up to the brain. **Endorphins,** released by neurons in the brain stem, also block the perception of pain. Opium and

its derivatives, morphine and heroin, have an analgesic (pain-reducing) effect because they are similar enough in chemical structure to bind to the receptors normally used by enkephalins and endorphins. For this reason, the enkephalins and the endorphins are referred to as *endogenous opiates.*

Nitric oxide (NO) is the first gas known to act as a regulatory molecule in the body. Because NO is a gas, it diffuses through membranes, so it cannot be stored in vesicles. It is produced as needed from the amino acid arginine. Nitric oxide diffuses out of the presynaptic axon and into neighboring cells by simply passing through the lipid portions of the plasma membranes.

In the PNS, nitric oxide is released by some neurons that innervate the gastrointestinal tract, penis, respiratory passages, and cerebral blood vessels. These autonomic neurons cause smooth-muscle relaxation in their target organs. This relaxation can produce the engorgement of the spongy tissue of the penis with blood, causing an erection. The drug sildenafil (Viagra) increases the release of NO in the penis, thus enabling and prolonging an erection. The brain releases nitric oxide as a neurotransmitter, where it appears to participate in the processes of learning and memory.

A postsynaptic neuron must integrate input from many synapses

Different types of input from a number of presynaptic neurons influence the activity of a postsynaptic neuron in the brain and spinal cord of vertebrates. For example, a single motor neuron in the spinal cord can have in excess of 50,000 synapses from presynaptic axons.

Each postsynaptic neuron may receive both excitatory and inhibitory synapses (figure 43.17). The EPSPs (depolarizations) and IPSPs (hyperpolarizations) from these synapses interact

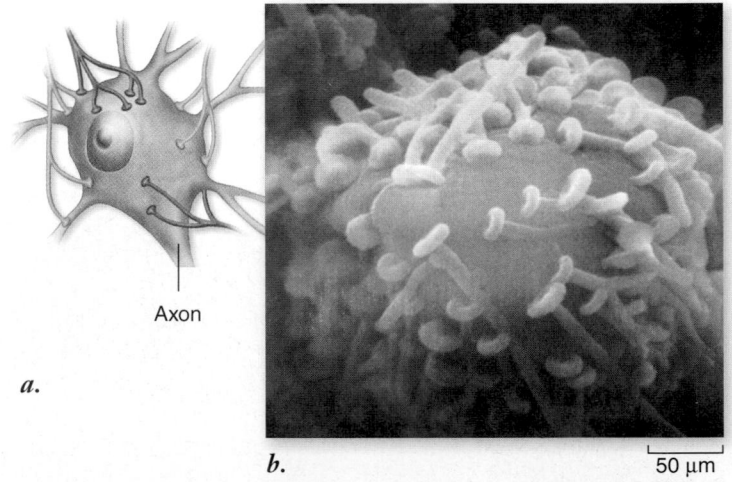

Axon

a.

b.   50 μm

Figure 43.17 Integration of EPSPs and IPSPs takes place on the neuronal cell body. *a.* The synapses made by some axons are excitatory *(green);* the synapses made by other axons are inhibitory *(red).* The summed influence of all of these inputs determines whether the axonal membrane of the postsynaptic cell will be sufficiently depolarized to produce an action potential. *b.* Micrograph of a neuronal cell body with numerous synapses.

with each other when they reach the cell body of the neuron. Small EPSPs add together to bring the membrane potential closer to the threshold, and IPSPs subtract from the depolarizing effect of the EPSPs, deterring the membrane potential from reaching threshold. This process is called *synaptic integration.*

Because of the all-or-none characteristic of an action potential, a postsynaptic neuron is like a switch that is either turned on or remains off. Information may be encoded in the pattern of firing over time, but each neuron can only fire or not fire when it receives a signal.

The events that determine whether a neuron fires may be extremely complex and involve many presynaptic neurons. There are two ways the membrane can reach the threshold voltage: by many different dendrites producing EPSPs that sum to the threshold voltage, or by one dendrite producing repeated EPSPs that sum to the threshold voltage. We call the first **spatial summation** and the second **temporal summation.**

In spatial summation, graded potentials due to dendrites from different presynaptic neurons that occur at the same time add together to produce an above-threshold voltage. All of this input does not need to be in the form of EPSPs, just so the potential produced by summing all of the EPSPs and IPSPs is greater than the threshold voltage. When the membrane at the base of the axon is depolarized above the threshold, it produces an action potential and a nerve impulse is sent down the axon.

In temporal summation, a single dendrite can produce sufficient depolarization to produce an action potential if it produces EPSPs that are close enough in time to sum to a depolarization that is greater than threshold. A typical EPSP can last for 15 ms, so for temporal summation to occur, the next impulse must arrive in less time. If enough EPSPs are produced to raise the membrane at the base of the axon above threshold, then an impulse will be sent.

The distinction between these two methods of summation is like filling a hole in the ground with soil: you can have many shovels that add soil to the hole until it is filled, or a single shovel that adds soil at a faster rate to fill the hole. When the hole is filled, the axon will fire.

Neurotransmitters play a role in drug addiction

When certain cells of the nervous system are exposed to a constant stimulus that produces a chemically mediated signal for a prolonged period, the cells may lose their ability to respond to that stimulus, a process called **habituation.** You are familiar with this loss of sensitivity—when you sit in a chair, for example, your awareness of the chair diminishes after a certain length of time.

Some nerve cells are particularly prone to this loss of sensitivity. If receptor proteins within synapses are exposed to high levels of neurotransmitter molecules for prolonged periods, the postsynaptic cell often responds by decreasing the number of receptor proteins in its membrane. This feedback is a normal function in all neurons, one of several mechanisms that have evolved to make the cell more efficient. In this case, the cell adjusts the number of receptors downward because plenty of stimulating neurotransmitter is available. In the case of artificial

neurotransmitter effects produced by drugs, long-term drug use means that more of the drug is needed to obtain the same effect.

Cocaine

The drug cocaine causes abnormally large amounts of neurotransmitter to remain in the synapses for long periods. Cocaine affects neurons in the brain's "pleasure pathways" (the *limbic system,* described later). These cells use the neurotransmitter dopamine. Cocaine binds tightly to the transporter proteins on presynaptic membranes that normally remove dopamine from the synaptic cleft. Eventually the dopamine stays in the cleft, firing the receptors repeatedly. New signals add more and more dopamine, firing the pleasure pathway more and more often (figure 43.18).

Nicotine

Nicotine has been found to have no affinity for proteins on the presynaptic membrane, as cocaine does; instead, it binds directly to a specific receptor on postsynaptic neurons of the brain. Because nicotine does not normally occur in the brain, why should it have a receptor there?

Researchers have found that "nicotine receptors" are a class of receptors that normally bind the neurotransmitter acetylcholine. Nicotine evolved in tobacco plants as a secondary compound—it affects the CNS of herbivorous insects, and therefore helps to protect the plant. It is an "accident of nature" that nicotine is also able to bind to some human ACh

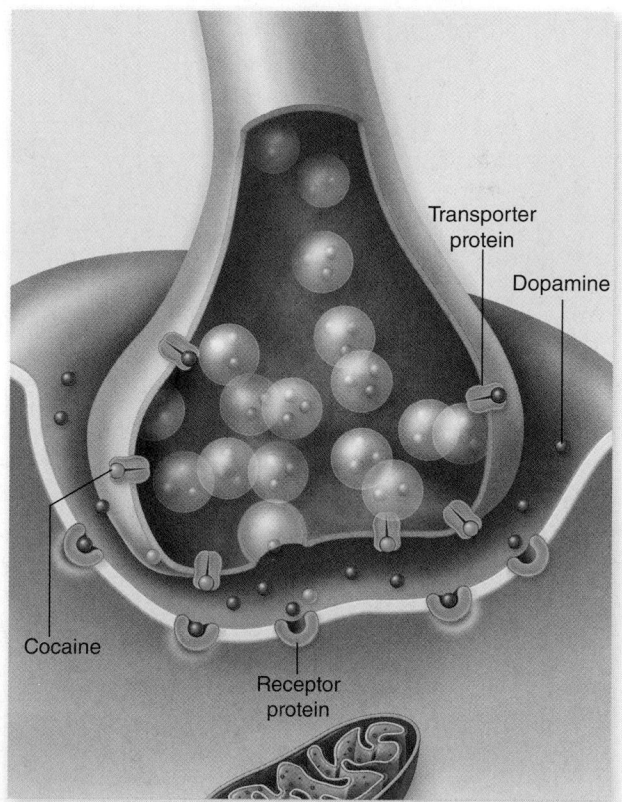

Figure 43.18 How cocaine alters events at the synapse. When cocaine binds to the dopamine transporters, it prevents reuptake of dopamine so the neurotransmitter survives longer in the synapse and continues to stimulate the postsynaptic cell. Cocaine thus acts to intensify pleasurable sensations.

receptors. When neurobiologists compare the nerve cells in the brains of smokers with those of nonsmokers, they find changes in both the number of nicotine receptors and the levels of RNA used to make the receptors. The brain adjusts to prolonged, chronic exposure to nicotine by "turning down the volume" in two ways: (1) by making fewer receptor proteins to which nicotine can bind; and (2) by altering the pattern of activation of the nicotine receptors—that is, their sensitivity to stimulation by neurotransmitters.

Having summarized the physiology and chemistry of neurons and synapses, we turn now to the structure of the vertebrate nervous system, beginning with the CNS and then the PNS.

Learning Outcomes Review 43.3

Electrical synapses involve direct cytoplasmic connections between two neurons; chemical synapses involve chemicals that cross the synaptic cleft, which separates neurons. Neurotransmitters include acetylcholine, epinephrine, glycine, GABA, biogenic amines, substance P, and nitric oxide. Many addictive drugs bind to sites that normally bind neurotransmitters or to membrane transport proteins in synapses.

■ *Why is tobacco use such a difficult habit to overcome?*

Learning Outcomes

1. **Describe the organization of the brain in vertebrates.**
2. **Describe characteristics of the human cerebrum.**
3. **Explain how a simple reflex works.**

The complex nervous system of vertebrate animals has a long evolutionary history. In this section we describe the structures making up the CNS, namely the brain and the spinal cord. First, it is helpful to review the origin and development of the vertebrate nervous system.

As animals became more complex, so did their nervous systems

Among the noncoelomate invertebrates (see chapter 33), sponges are the only major phylum that lack nerves. The simplest nervous systems occur among cnidarians (figure 43.19),

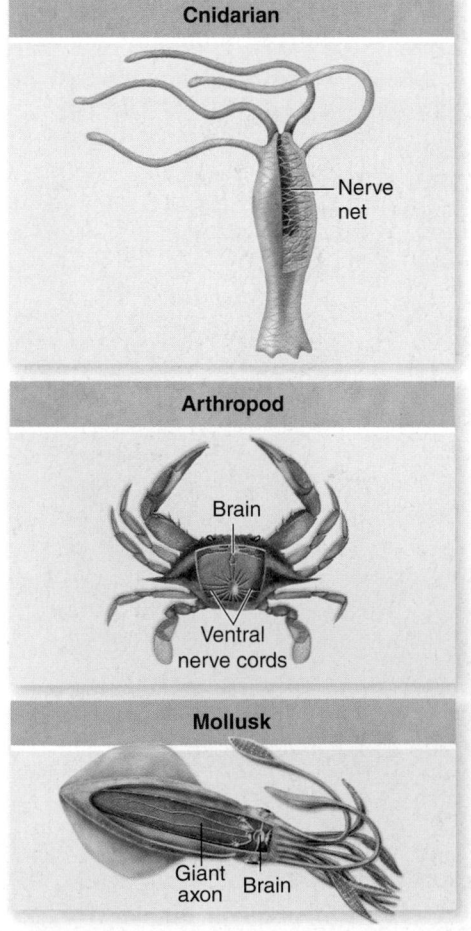
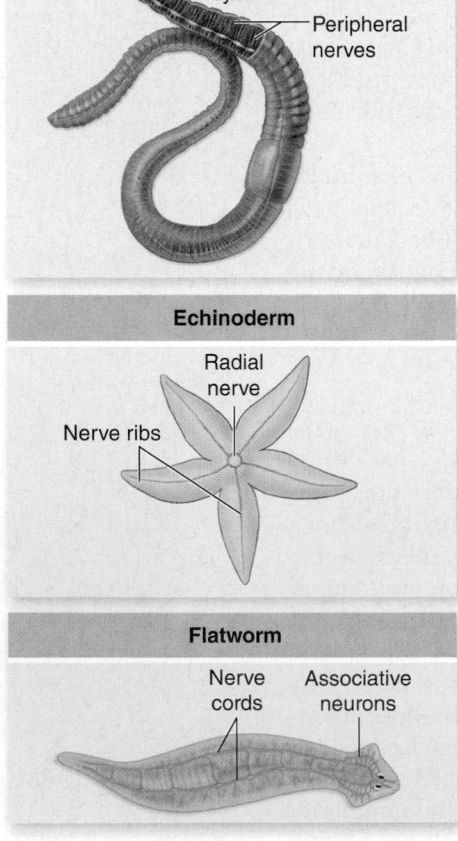
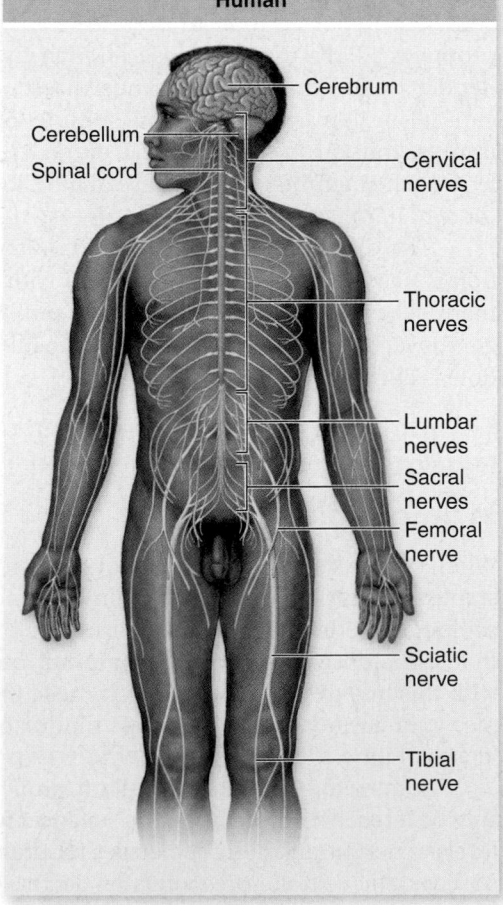

Figure 43.19 Diversity of nervous systems. Nervous systems in animals range from simple nerve nets to paired nerve cords with primitive brains to elaborate brains and sensory systems. Bilateral symmetry is correlated with the concentration of nervous tissue and sensory structures in the front end of the nerve cord. This evolutionary process is referred to as cephalization.

in which all neurons are similar and linked to one another in a web, or **nerve net.** There is no associative activity, no control of complex actions, and little coordination.

The simplest animals with associative activity in the nervous system are the free-living flatworms, phylum Platyhelminthes. Running down the bodies of these flatworms are two nerve cords, from which peripheral nerves extend outward to the muscles of the body. The two nerve cords converge at the front end of the body, forming an enlarged mass of nervous tissue that also contains interneurons with synapses connecting neurons to one another. This primitive "brain" is a rudimentary central nervous system and permits a far more complex control of muscular responses than is possible in cnidarians.

All of the subsequent evolutionary changes in nervous systems can be viewed as a series of elaborations on the characteristics already present in flatworms. For example, among coelomate invertebrates (see chapter 34), earthworms exhibit a central nervous system that is connected to all other parts of the body by peripheral nerves. And in arthropods, the central coordination of complex responses is increasingly localized in the front end of the nerve cord. As this region evolved, it came to contain a progressively larger number of interneurons and to develop tracts, which are major information highways within the brain.

Vertebrate brains have three basic divisions

Casts of the interior braincases of fossil agnathans, fishes that swam 500 MYA (see chapter 35), have revealed much about the early evolutionary stages of the vertebrate brain. Although small, these brains already had the three divisions that characterize the brains of all contemporary vertebrates:

1. the *hindbrain,* or rhombencephalon;
2. the *midbrain,* or mesencephalon; and
3. the *forebrain,* or prosencephalon (figure 43.20 and table 43.3).

The hindbrain in fishes

The hindbrain was the major component of these early brains, as it still is in fishes today. Composed of the **cerebellum, pons,** and **medulla oblongata,** the hindbrain may be considered an extension of the spinal cord devoted primarily to coordinating motor reflexes. Tracts containing large numbers of axons run like cables up and down the spinal cord to the hindbrain. The hindbrain, in turn, integrates the many sensory signals coming from the muscles and coordinates the pattern of motor responses.

Much of this coordination is carried on within a small extension of the hindbrain called the cerebellum ("little cerebrum"). In more advanced vertebrates, the cerebellum plays an increasingly important role as a coordinating center for movement and it is correspondingly larger than it is in the fishes. In all vertebrates, the cerebellum processes data on the current position and movement of each limb, the state of relaxation or contraction of the muscles involved, and the general position of the body and its relation to the outside world.

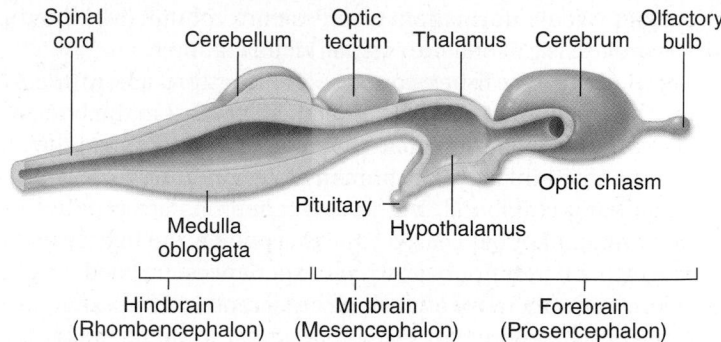

Figure 43.20 The basic organization of the vertebrate brain can be seen in the brains of primitive fishes. The brain is divided into three regions that are found in differing proportions in all vertebrates: the hindbrain, which is the largest portion of the brain in fishes; the midbrain, which in fishes is devoted primarily to processing visual information; and the forebrain, which is concerned mainly with olfaction (the sense of smell) in fishes. In terrestrial vertebrates, the forebrain plays a far more dominant role in neural processing than it does in fishes.

The midbrain and forebrain of fishes

In fishes, the remainder of the brain is devoted to the reception and processing of sensory information. The midbrain is composed primarily of the *optic tectum,* which receives and

TABLE 43.3	Subdivisions of the Central Nervous System
Major Subdivision	**Function**
SPINAL CORD	Spinal reflexes; relays sensory and motor information
BRAIN	
Hindbrain (Rhombencephalon)	
Medulla oblongata	Sensory nuclei; reticular-activating system; autonomic functions
Pons	Reticular-activating system; autonomic functions
Cerebellum	Coordination of movements; balance
Midbrain (Mesencephalon)	Reflexes involving eyes and ears
Forebrain (Prosencephalon)	
Diencephalon	
Thalamus	Relay station for ascending sensory and descending motor tracts; autonomic functions
Hypothalamus	Autonomic functions; neuroendocrine control
Telencephalon (cerebrum)	
Basal ganglia	Motor control
Corpus callosum	Connects and relays information between the two hemispheres
Hippocampus (limbic system)	Memory; emotion
Cerebral cortex	Higher cognitive functions; integrates and interprets sensory information; organizes motor output

processes visual information, whereas the forebrain is devoted to the processing of olfactory (smell) information.

The brains of fishes continue growing throughout their lives. This continued growth is in marked contrast to the brains of other classes of vertebrates, which generally complete their development by infancy. The human brain continues to develop through early childhood, but few new neurons are produced once development has ceased. One exception is the hippocampus, which has control over which experiences are filed away into long-term memory and which are forgotten. The extent of neurogenesis (production of new neurons) in adult brains is controversial, and one area of active current research.

The dominant forebrain in more recent vertebrates

Starting with the amphibians and continuing more prominently in the reptiles, processing of sensory information is increasingly centered in the forebrain. This pattern was the dominant evolutionary trend in the further development of the vertebrate brain (figure 43.21).

The forebrain in reptiles, amphibians, birds, and mammals is composed of two elements that have distinct functions. The *diencephalon* consists of the thalamus and hypothalamus. The **thalamus** is an integration and relay center between incoming sensory information and the cerebrum. The hypothalamus participates in basic drives and emotions and controls the secretions of the pituitary gland. The **telencephalon,** or "end brain," is located at the front of the forebrain and is devoted largely to associative activity. In mammals, the telencephalon is called the cerebrum. The telencephalon also includes structures we discuss later on when describing the human brain.

The expansion of the cerebrum

In examining the relationship between brain mass and body mass among the vertebrates, a remarkable difference is observed between fishes and reptiles on the one hand, and birds and mammals on the other. Mammals have brains that are particularly large relative to their body mass. This is especially true of porpoises and humans.

The increase in brain size in mammals largely reflects the great enlargement of the cerebrum, the dominant part of the mammalian brain. The **cerebrum** is the center for correlation, association, and learning in the mammalian brain. It receives sensory data from the thalamus and issues motor commands to the spinal cord via descending tracts of axons.

In vertebrates, the central nervous system is composed of the brain and the spinal cord (see table 43.3). These two structures are responsible for most of the information processing within the nervous system and they consist primarily of interneurons and neuroglia. Ascending tracts carry sensory information to the brain. Descending tracts carry impulses from the brain to the motor neurons and interneurons in the spinal cord that control the muscles of the body.

The human forebrain exhibits exceptional information-processing ability

The human cerebrum is so large that it appears to envelop the rest of the brain. It is split into right and left **cerebral hemispheres,** which are connected by a tract called the **corpus callosum** (figure 43.22). The hemispheres are further divided into the *frontal, parietal, temporal,* and *occipital lobes.*

Each hemisphere primarily receives sensory input from the opposite, or contralateral, side of the body and exerts motor control primarily over that side. Therefore, a touch on the right hand is relayed primarily to the left hemisphere, which may then initiate movement of the right hand in response to the touch. Damage to one hemisphere due to a stroke often results in a loss of sensation and paralysis on the contralateral side of the body.

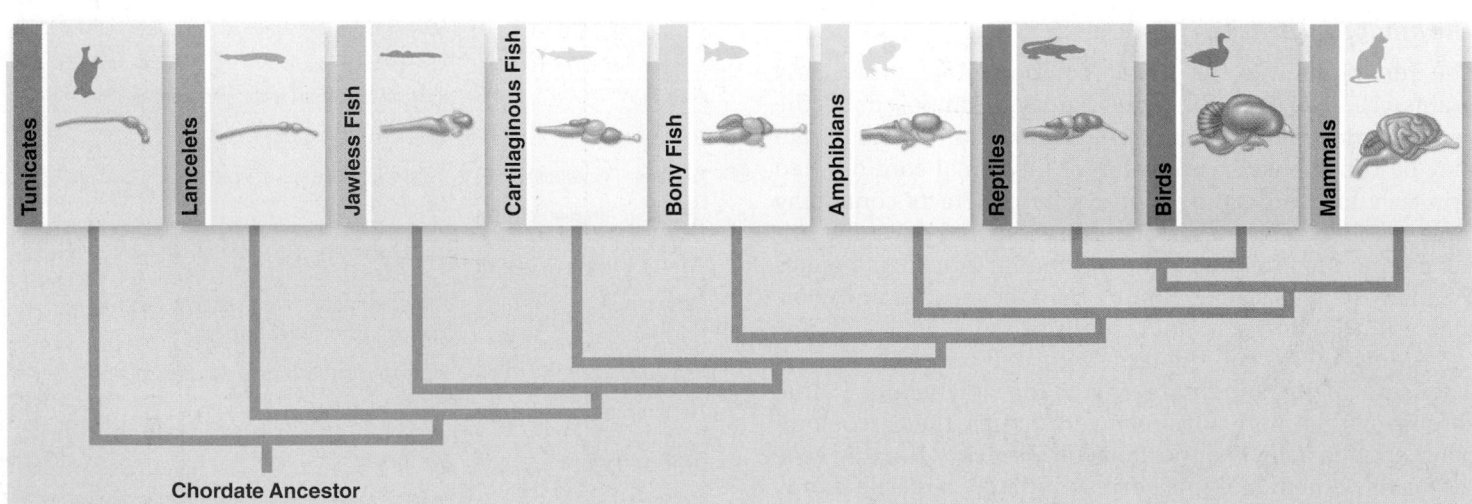

Figure 43.21 Evolution of the vertebrate brain. The relative sizes of different brain regions have changed as vertebrates have evolved. In sharks and other fishes, the hindbrain is predominant, and the rest of the brain serves primarily to process sensory information. In amphibians and reptiles, the forebrain is far larger, and it contains a larger cerebrum devoted to associative activity. In birds, which evolved from reptiles, the cerebrum is even more pronounced. In mammals, the cerebrum covers the optic tectum and is the largest portion of the brain. The dominance of the cerebrum is greatest in humans, in whom it envelops much of the rest of the brain.

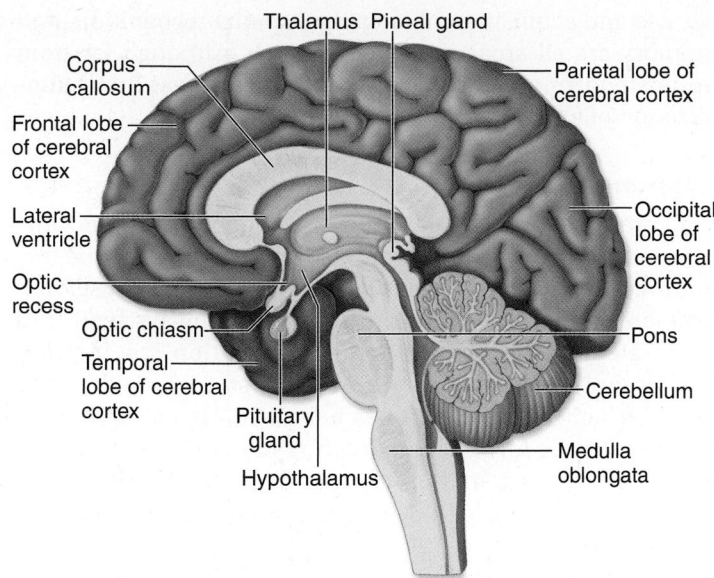

Figure 43.22 A section through the human brain. In this sagittal section showing one cerebral hemisphere, the corpus callosum, a fiber tract connecting the two cerebral hemispheres, can be clearly seen.

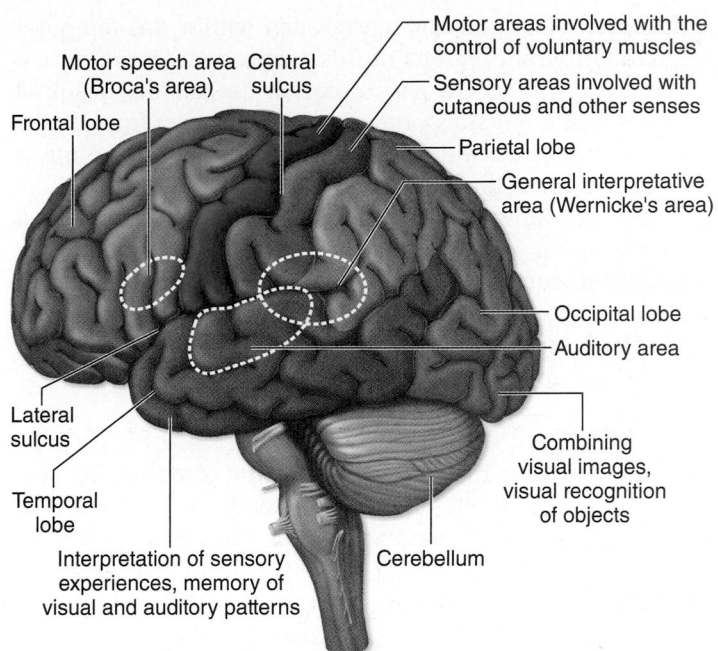

Figure 43.23 The cerebrum. This diagram shows the lobes of the cerebrum and indicates some of the known regions of specialization.

The cerebral cortex

Much of the neural activity of the cerebrum occurs within a layer of gray matter only a few millimeters thick on its outer surface. This layer, called the **cerebral cortex,** is densely packed with nerve cells. In humans, it contains over 10 billion nerve cells, amounting to roughly 10% of all the neurons in the brain. The surface of the cerebral cortex is highly convoluted; this is particularly true in the human brain, where the convolutions increase the surface area of the cortex threefold.

The activities of the cerebral cortex fall into one of three general categories: motor, sensory, and associative. Each of its regions correlates with a specific function (figure 43.23). The **primary motor cortex** lies along the *gyrus* (convolution) on the posterior border of the frontal lobe, just in front of the central *sulcus* (crease). Each point on the surface of the motor cortex is associated with the movement of a different part of the body (figure 43.24, right).

Just behind the central sulcus, on the anterior edge of the parietal lobe, lies the **primary somatosensory cortex.** Each point in this area receives input from sensory neurons serving skin and muscle senses in a particular part of the body (figure 43.24, left). Large areas of the primary motor cortex and primary somatosensory cortex are devoted to the fingers, lips, and tongue because of the need for manual dexterity

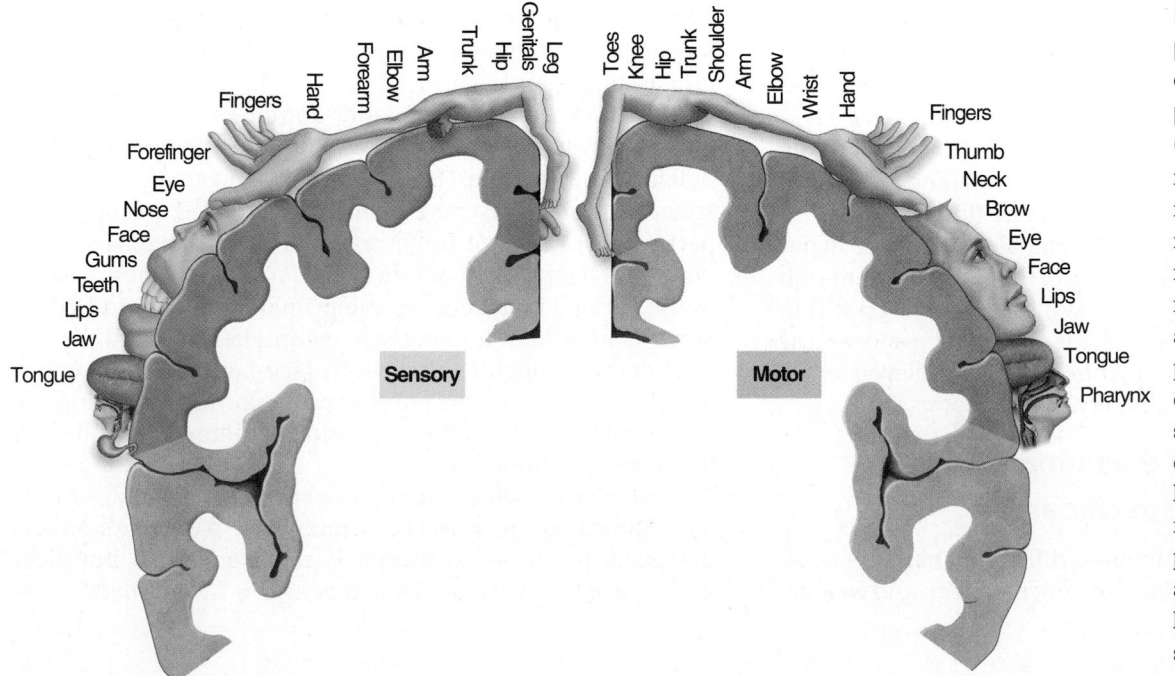

Figure 43.24 The primary somatosensory cortex (left) and the primary motor cortex (right). Each of these regions of the cerebral cortex is associated with a different region of the body, as indicated in this stylized map. The areas of the body are drawn in relative proportion to the amount of cortex dedicated to their sensation or control. For example, the hands have large areas of sensory and motor control, and the pharynx has a considerable area of motor control but little area devoted to the sensations of the pharynx.

and speech. The auditory cortex lies within the temporal lobe, and different regions of this cortex deal with different sound frequencies. The visual cortex lies on the occipital lobe, with different sites processing information from different positions on the retina, equivalent to particular points in the visual fields of the eyes.

The portion of the cerebral cortex that is not occupied by these motor and sensory cortices is referred to as the **association cortex.** The site of higher mental activities, the association cortex reaches its greatest extent in primates, especially humans, where it makes up 95% of the surface of the cerebral cortex.

Basal ganglia

Buried deep within the white matter of the cerebrum are several collections of cell bodies and dendrites that produce islands of gray matter. These aggregates of neuron cell bodies, which are collectively termed the *basal ganglia,* receive sensory information from ascending nerve tracts and motor commands from the cerebral cortex and cerebellum.

Outputs from the basal ganglia are sent down the spinal cord, where they participate in the control of body movements. Damage to specific regions of the basal ganglia can produce the resting tremor of muscles that is characteristic of Parkinson disease.

Thalamus and hypothalamus

The thalamus is a primary site of sensory integration in the brain. Visual, auditory, and somatosensory information is sent to the thalamus, where the sensory tracts synapse with association neurons. The sensory information is then relayed via the thalamus to the occipital, temporal, and parietal lobes of the cerebral cortex, respectively. The transfer of each of these types of sensory information is handled by specific aggregations of neuron cell bodies within the thalamus.

The hypothalamus integrates the visceral activities. It helps regulate body temperature, hunger and satiety, thirst, and—along with the limbic system—various emotional states. The hypothalamus also controls the pituitary gland, which in turn regulates many of the other endocrine glands of the body. By means of its interconnections with the cerebral cortex and with control centers in the *brainstem* (a term used to refer collectively to the midbrain, pons, and medulla oblongata), the hypothalamus helps coordinate the neural and hormonal responses to many internal stimuli and emotions.

The *hippocampus* and **amygdala,** along with the hypothalamus, are the major components of the **limbic system**—an evolutionarily ancient group of linked structures deep within the cerebrum that are responsible for emotional responses, as described earlier. The hippocampus is also believed to be important in the formation and recall of memories.

Complex functions of the human brain may be controlled in specific areas

Although studying brain function is difficult, it has long fascinated researchers. The distinction between sleep and waking, the use and acquisition of language, spatial recognition, and memory are all areas of active research. Although far from understood, one generalization that emerged was the regionalization of function.

Sleep and arousal

The brainstem contains a diffuse collection of neurons referred to as the *reticular formation.* One part of this formation, the *reticular-activating system,* controls consciousness and alertness. All of the sensory pathways feed into this system, which monitors the information coming into the brain and identifies important stimuli. When the reticular-activating system has been stimulated to arousal, it increases the level of activity in many parts of the brain. Neural pathways from the reticular formation to the cortex and other brain regions are depressed by anesthetics and barbiturates.

The reticular-activating system controls both sleep and the waking state. It is easier to sleep in a dark room than in a lighted one because there are fewer visual stimuli to stimulate the reticular-activating system. In addition, activity in this system is reduced by serotonin, a neurotransmitter discussed earlier. Serotonin causes the level of brain activity to fall, bringing on sleep.

Brain state can be monitored by means of an electro-encephalogram (EEG), a recording of electrical activity. Awake but relaxed individuals with eyes closed exhibit a brain pattern of large, slow waves termed *alpha waves.* In an alert individual with eyes open, the waves are more rapid *(beta waves)* and more desynchronized as sensory input is being received. *Theta waves* and *delta waves* are very slow waves seen during sleep. When an individual is in REM sleep—characterized by rapid eye movements with the eyes closed—the EEG is more like that of an awake, relaxed individual.

Language

Although the two cerebral hemispheres seem structurally similar, they are responsible for different activities. The most thoroughly investigated example of this lateralization of function is language.

The left hemisphere is the "dominant" hemisphere for language in 90% of right-handed people and nearly two-thirds of left-handed people. (By *dominant,* we mean it is the hemisphere in which most neural processing related to language is performed.) Different brain regions control language in the dominant hemisphere (figure 43.25). Wernicke's area, located in the parietal lobe between the primary auditory and visual areas, is important for language comprehension and the formulation of thoughts into speech (see figure 43.23). Broca's area, found near the part of the motor cortex controlling the face, is responsible for the generation of motor output needed for language communication.

Damage to these brain areas can cause language disorders known as *aphasias.* For example, if Wernicke's area is damaged, the person's speech is rapid and fluid but lacks meaning; words are tossed together as in a "word salad."

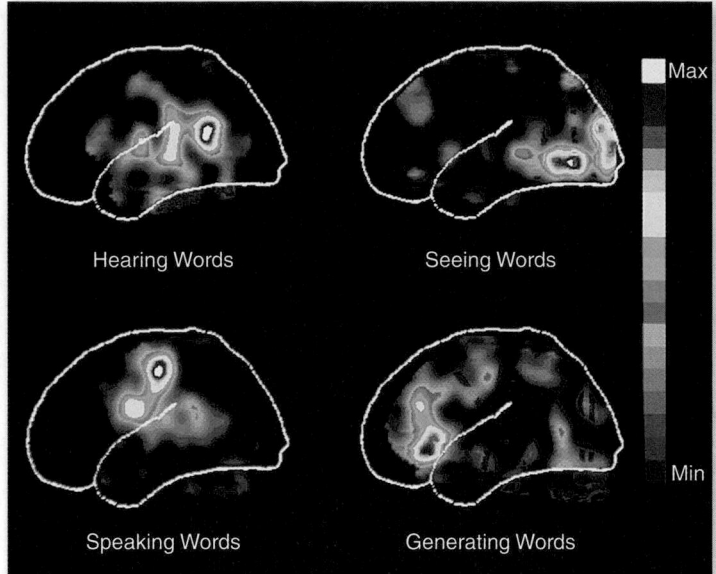

Figure 43.25 Different brain regions control various language activities. This illustration shows how the brain reacts in human subjects asked to listen to a spoken word, to read that same word silently, to repeat the word out loud, and then to speak a word related to the first. Regions of white, red, and yellow show the greatest activity. Compare this with figure 43.23 to see how regions of the brain are mapped.

Spatial recognition

Whereas the dominant hemisphere for language is adept at sequential reasoning, like that needed to formulate a sentence, the nondominant hemisphere (the right hemisphere in most people) is adept at spatial reasoning, the type of reasoning needed to assemble a puzzle or draw a picture. It is also the hemisphere primarily involved in musical ability—a person with damage to Broca's speech area in the left hemisphere may not be able to speak but may retain the ability to sing.

Damage to the nondominant hemisphere may lead to an inability to appreciate spatial relationships and may impair musical activities such as singing. Even more specifically, damage to the inferior temporal cortex in that hemisphere eliminates the capacity to recall faces, a condition known as prosopagnosia. Reading, writing, and oral comprehension remain normal, and patients with this disability can still recognize acquaintances by their voices. The nondominant hemisphere is also important for the consolidation of memories of nonverbal experiences.

Memory and learning

One of the great mysteries of the brain is the basis of memory and learning. Memory appears dispersed across the brain. Specific cortical sites cannot be identified for particular memories because relatively extensive cortical damage does not selectively remove memories. Although memory is impaired if portions of the brain, particularly the temporal lobes, are removed, it is not lost entirely. Many memories persist in spite of the damage, and the ability to access them is gradually recovered with time.

Fundamental differences appear to exist between short-term and long-term memory. Short-term memory is transient, lasting only a few moments. Such memories can readily be erased by the application of an electrical shock, leaving previously stored long-term memories intact. This result suggests that short-term memories are stored in the form of a transient neural excitation. Long-term memory, in contrast, appears to involve structural changes in certain neural connections within the brain.

Two parts of the temporal lobes, the hippocampus and the amygdala, are involved in both short-term memory and its consolidation into long-term memory. Damage to these structures impairs the ability to process recent events into long-term memories.

Synaptic plasticity

Part of the cellular basis of learning and memory seems to be long-term changes in the strength of synaptic connections. Two examples of this synaptic plasticity are long-term potentiation (LTP), and long-term depression (LTD). The mechanism of LTP is complex and not completely understood. One well-studied form involves synapses that release the neurotransmitter glutamate, and have N-methyl-D-aspartic acid (NMDA) type of receptors. When either the same synapse is stimulated repeatedly, or neighboring synapses are stimulated, the postsynaptic membrane becomes significantly depolarized. This releases a block of the NMDA receptor by Mg^{2+} such that glutamate binding causes an influx of Ca^{2+} that stimulates a signal transduction pathway involving calcium/calmodulin-dependent protein kinase II. This pathway leads to the insertion of another receptor type, the α-amino-3-hydroxyl-5-methyl-4-isoxazole-propionate (AMPA) receptor, into the postsynaptic membrane, making the synapse more sensitive to future stimulation (figure 43.26).

If the stimulation of an NMDA receptor is less, and the postsynaptic membrane is less depolarized, LTD can result. In this case, a different Ca^{2+}-dependent signaling pathway results in the loss of AMPA receptors from the membrane. Taken together, these two mechanisms can make a synapse more or less sensitive to future stimulation.

Alzheimer disease: Degeneration of brain neurons

In the past, little was known about *Alzheimer disease,* a condition in which the memory and thought processes of the brain become dysfunctional. Scientists disagree about the biological nature of the disease and its cause. Two hypotheses have been proposed: One suggests that nerve cells in the brain are killed from the outside in, and the other that the cells are killed from the inside out.

In the first hypothesis, external proteins called β-amyloid exist in an abnormal form, which then forms aggregates, or plaques. The plaques begin to fill in the brain and then damage and kill nerve cells. However, these amyloid plaques have been found in autopsies of people who did not exhibit Alzheimer disease.

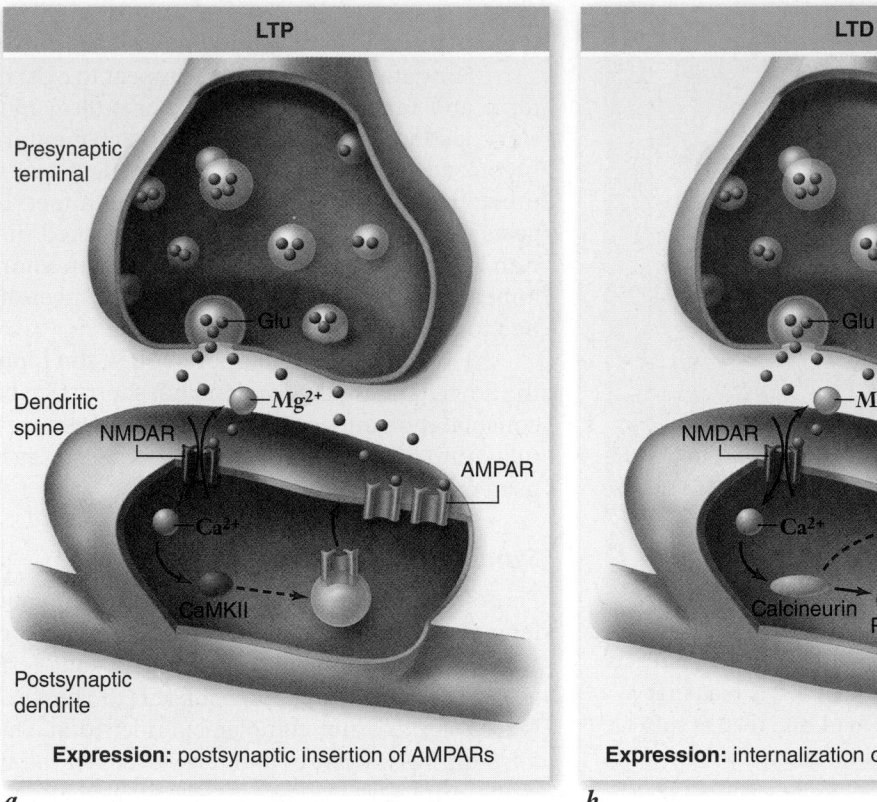

Expression: postsynaptic insertion of AMPARs

a.

Expression: internalization of postsynaptic AMPARs

b.

Figure 43.26 LTP and LTD modulate synaptic function. *a.* When the postsynaptic membrane is significantly depolarized, when GABA binds to the *N*-methyl-D-aspartic acid receptor (NMDAR), the influx of Ca^{2+} leads to the insertion of α-amino-3-hydroxyl-5-methyl-4-isoxazole-propionate receptors (AMPAR). This potentiates the synapse for future stimulation. *b.* When the postsynaptic membrane does not have as large a depolarization, or there is less GABA, then GABA binding to NMDA receptor triggers a different pathway that results in removal of AMPA receptors. This depresses the synapse for future stimulation. CaMKII, calmodulin-dependent protein kinase 2.

The second hypothesis maintains that the nerve cells are killed by an abnormal form of an internal protein called tau (τ), which normally functions to maintain protein transport microtubules. Abnormal forms of τ-protein assemble into helical segments that form tangles, which interfere with the normal functioning of the nerve cells. At this point, the association of tangles with actual neuronal death is stronger.

The spinal cord conveys messages and controls some responses directly

The spinal cord is a cable of neurons extending from the brain down through the backbone (figure 43.27). It is enclosed and protected by the vertebral column and layers of membranes called *meninges,* which also cover the brain. Inside the spinal cord are two zones.

The inner zone is gray matter and primarily consists of the cell bodies of interneurons, motor neurons, and neuroglia. The outer zone is white matter and contains cables of sensory axons in the dorsal columns and motor axons in the ventral columns. These nerve tracts may also contain the dendrites of other nerve cells. Messages from the body and the brain run up and down the spinal cord, the body's "information highway."

In addition to relaying messages, the spinal cord also functions in **reflexes,** the sudden, involuntary movement of

muscles. A reflex produces a rapid motor response to a stimulus because the sensory neuron passes its information to a motor neuron in the spinal cord, without higher level processing. One of the most frequently used reflexes in your body is blinking, a reflex that protects your eyes. If an object such as an insect or a cloud of dust approaches your eye, the eyelid blinks before you realize what has happened. The reflex occurs before the cerebrum is aware the eye is in danger.

Figure 43.27 A view down the human spinal cord. Pairs of spinal nerves can be seen extending from the spinal cord. Along these nerves, as well as the cranial nerves that arise from the brain, the central nervous system communicates with the rest of the body.

Figure 43.28
The knee-jerk reflex.
This is the simplest reflex, involving only sensory and motor neurons.

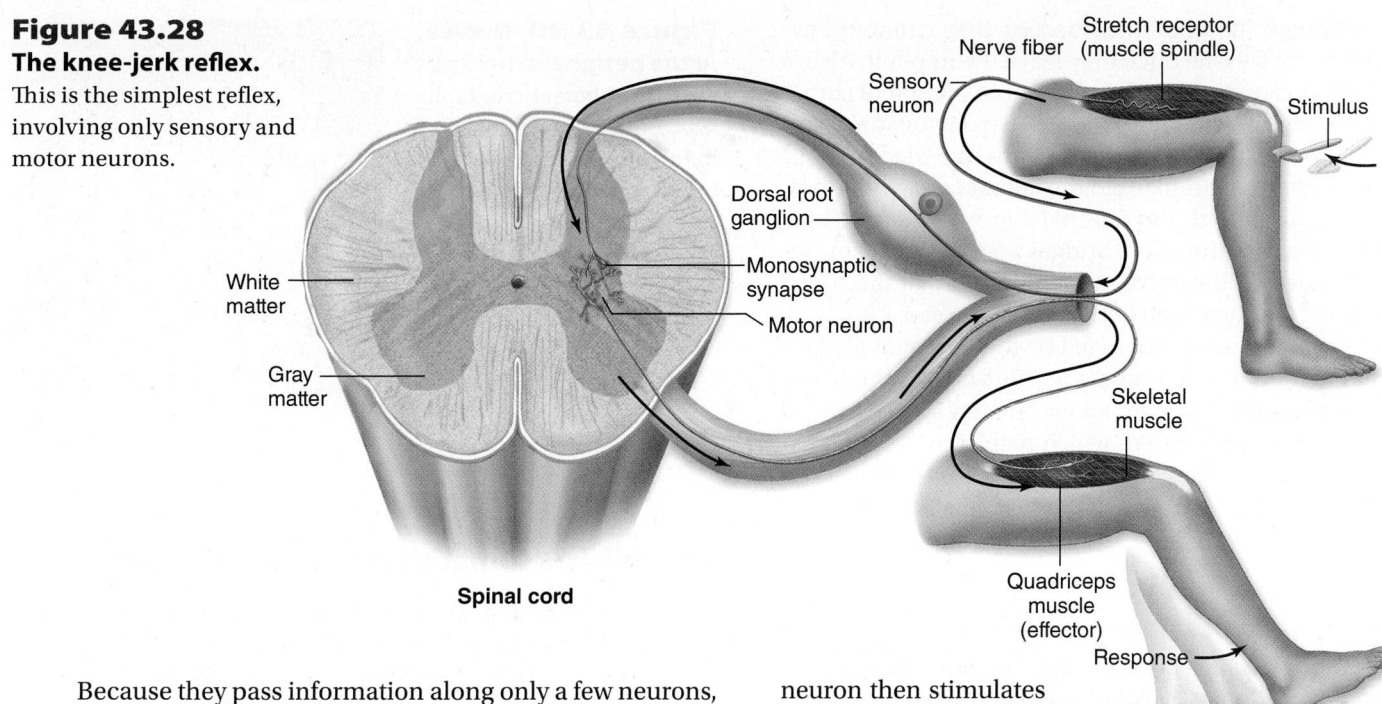

Spinal cord

Because they pass information along only a few neurons, reflexes are very fast. A few reflexes, such as the knee-jerk reflex (figure 43.28), are monosynaptic reflex arcs. In these, the sensory nerve cell makes synaptic contact directly with a motor neuron in the spinal cord whose axon travels directly back to the muscle.

Most reflexes in vertebrates, however, involve a single connecting interneuron between the sensory neuron and the motor neuron (figure 43.29). The withdrawal of a hand from a hot stove or the blinking of an eye in response to a puff of air involves a relay of information from a sensory neuron through one or more interneurons to a motor neuron. The motor neuron then stimulates the appropriate muscle to contract. Notice that the sensory neuron may also connect to other interneurons to send signals to the brain. Although you jerked your hand away from the stove, you will still feel pain.

Spinal cord regeneration

In the past, scientists tried to repair severed spinal cords by installing nerves from another part of the body to bridge the gap and act as guides for the spinal cord to regenerate. But most of these experiments failed. Although axons may

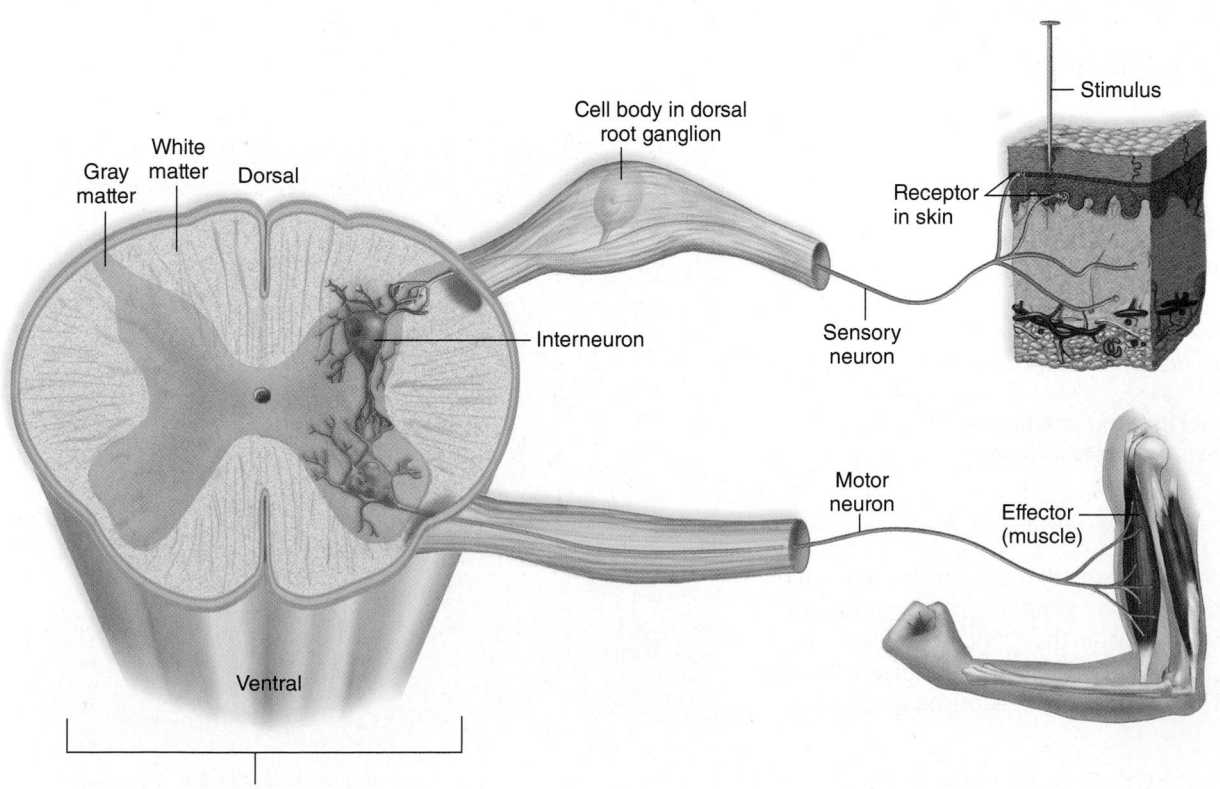

Figure 43.29
A cutaneous spinal reflex. This reflex is more complex than a knee-jerk reflex because it requires interneurons as well as sensory and motor neurons. Interneurons connect a sensory neuron with a motor neuron to cause muscle contraction as shown. Other interneurons inhibit motor neurons, allowing antagonistic muscles to relax.

regenerate through the implanted nerves, they cannot penetrate the spinal cord tissue once they leave the implant. Also, a factor that inhibits nerve growth is present in the spinal cord.

After discovering that fibroblast growth factor stimulates nerve growth, neurobiologists working with rats tried "gluing" the nerves on, from the implant to the spinal cord, with fibrin that had been mixed with the fibroblast growth factor. Three months later, rats with the nerve bridges began to show movement in their lower bodies. Dye tests indicated that the spinal cord nerves had regrown from both sides of the gap.

Many scientists are encouraged by the potential to use a similar treatment in human medicine. But most spinal cord injuries in humans do not involve a completely severed spinal cord; often, nerves are crushed, which results in different tissue damage. Also, even though the rats with nerve bridges did regain some ability to move, tests indicated that they were barely able to walk or stand.

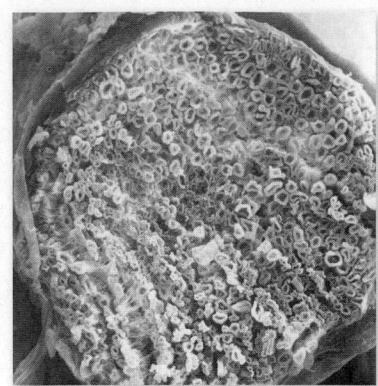

Figure 43.30 Nerves in the peripheral nervous system. Photomicrograph showing a cross section of a bullfrog nerve. The nerve is a bundle of axons bound together by connective tissue. Many myelinated axons are visible, each looking somewhat like a doughnut.

5 μm

Learning Outcomes Review 43.4

The vertebrate brain has three primary regions: the hindbrain, midbrain, and forebrain. The cerebrum, part of the forebrain, is composed of two cerebral hemispheres in which gray matter of the cerebral cortex overlays white matter and islands of gray matter (nuclei) called the basal ganglia. The spinal cord relays messages to and from the brain; a reflex occurs when the spinal cord processes sensory information directly and initiates a motor response.

■ *What is the advantage of having reflexes?*

43.5 The Peripheral Nervous System: Sensory and Motor Neurons

Learning Outcomes

1. *Describe the organization of the peripheral nervous system.*
2. *Explain the actions of sensory and somatic neurons.*
3. *Distinguish between the somatic and autonomic nervous systems.*
4. *Differentiate between the sympathetic and parasympathetic divisions of the autonomic nervous system.*

The PNS consists of nerves, the cablelike collections of axons (figure 43.30), and **ganglia** (singular, *ganglion*), aggregations of neuron cell bodies located outside the CNS. To review, the function of the PNS is to receive information from the environment, convey it to the CNS, and to carry responses to effectors such as muscle cells.

The PNS has somatic and autonomic systems

At the spinal cord, a spinal nerve separates into sensory and motor components. The axons of sensory neurons enter the dorsal surface of the spinal cord and form the **dorsal root** of the spinal nerve, whereas motor axons leave from the ventral surface of the spinal cord and form the **ventral root** of the spinal nerve. The cell bodies of sensory neurons are grouped together outside each level of the spinal cord in the **dorsal root ganglia.** The cell bodies of somatic motor neurons, on the other hand, are located within the spinal cord and so are not located in ganglia.

As mentioned earlier, somatic motor neurons stimulate skeletal muscles to contract, and autonomic motor neurons innervate involuntary effectors—smooth muscles, cardiac muscle, and glands. A comparison of the somatic and autonomic nervous systems is provided in table 43.4; we discuss each system in turn.

TABLE 43.4	Comparison of the Somatic and Autonomic Nervous Systems	
Characteristic	**Somatic**	**Autonomic**
Effectors	Skeletal muscle	Cardiac muscle

		Smooth muscle
		    Gastrointestinal tract
		    Blood vessels
		    Airways

		Exocrine glands
Effect on motor nerves	Excitation	Excitation or inhibition
Innervation of effector cells	Always single	Typically dual
Number of sequential neurons in path to effector	One	Two
Neurotransmitter	Acetylcholine	Acetylcholine, norepinephrine

The somatic nervous system controls movements

Somatic motor neurons stimulate the skeletal muscles of the body to contract in response to conscious commands and as part of reflexes that do not require conscious control. Voluntary control of skeletal muscles is achieved by activation of tracts of axons that descend from the cerebrum to the appropriate level of the spinal cord. Some of these descending axons stimulate spinal cord motor neurons directly, and others activate interneurons that in turn stimulate the spinal motor neurons.

When a particular muscle is stimulated to contract, however, its antagonist must be inhibited. In order to flex the arm, for example, the flexor muscles must be stimulated while the antagonistic extensor muscle is inhibited (see chapter 46). Descending motor axons produce this necessary inhibition by causing hyperpolarizations (IPSPs) of the spinal motor neurons that innervate the antagonistic muscles.

The autonomic nervous system controls involuntary functions through two divisions

The autonomic nervous system is composed of the *sympathetic* and *parasympathetic* divisions plus the medulla oblongata of the hindbrain, which coordinates this system. Although they differ, the sympathetic and parasympathetic divisions share several features. In both, the efferent motor pathway involves two neurons: The first has its cell body in the CNS and sends an axon to an autonomic ganglion; it is called *preganglionic neuron*. These neurons release acetylcholine at their synapses.

The second neuron has its cell body in the autonomic ganglion and sends its axon to synapse with a smooth muscle, cardiac muscle, or gland cell (figure 43.31). This second neuron is termed the *postganglionic neuron*. Those in the parasympathetic division release ACh, and those in the sympathetic division release norepinephrine.

The sympathetic division

In the sympathetic division, the preganglionic neurons originate in the thoracic and lumbar regions of the spinal cord (figure 43.32, left). Most of the axons from these neurons synapse in two parallel chains of ganglia immediately outside the spinal cord. These structures are usually called the *sympathetic chain* of ganglia. The sympathetic chain contains the cell bodies of postganglionic neurons, and it is the axons from these neurons that innervate the different visceral organs.

There are some exceptions to this general pattern, however. The axons of some preganglionic sympathetic neurons pass through the sympathetic chain without synapsing and, instead, terminate within the medulla of the adrenal gland (see chapter 45). In response to action potentials, the adrenal medulla cells secrete the hormone epinephrine (adrenaline). At the same time, norepinephrine is released at the synapses of the postganglionic neurons. As described earlier, both of these neurotransmitters prepare the body for action by heightening metabolism and blood flow.

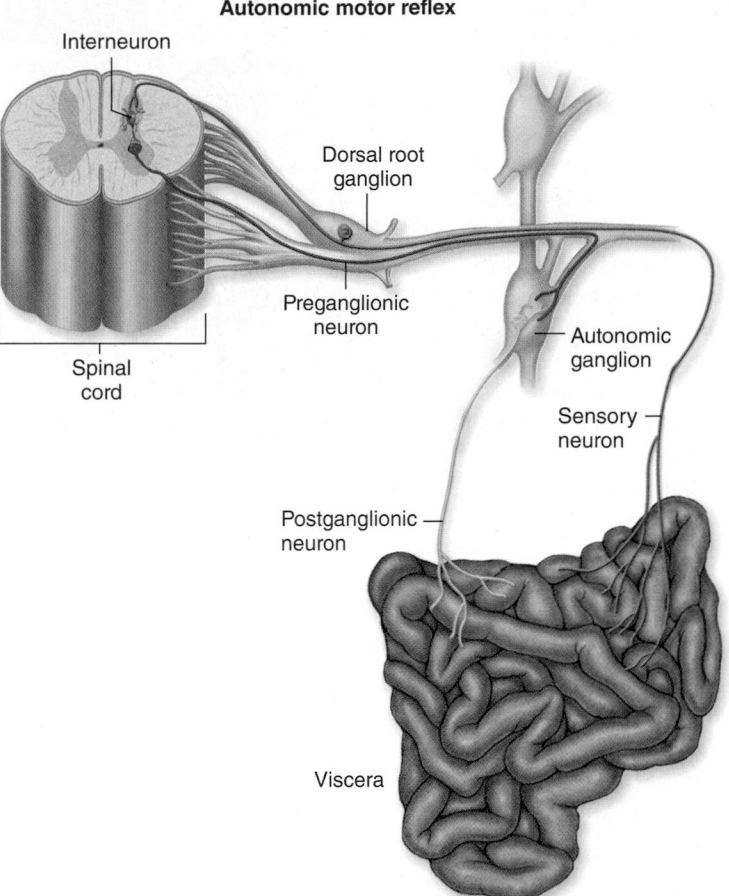

Autonomic motor reflex

Figure 43.31 An autonomic neural path. There are two motor neurons in the efferent pathway. The first, or preganglionic neuron, exits the CNS and synapses at an autonomic ganglion. The second, or postganglionic neuron, exits the ganglion and regulates the visceral effectors (smooth muscle, cardiac muscle, or glands).

The parasympathetic division

The actions of the sympathetic division are antagonized by the parasympathetic division. Preganglionic parasympathetic neurons originate in the brain and sacral regions of the spinal cord (see figure 43.32, right). Because of this origin, there cannot be a chain of parasympathetic ganglia analogous to the sympathetic chain. Instead, the preganglionic axons, many of which travel in the vagus (tenth cranial) nerve, terminate in ganglia located near or even within the internal organs. The postganglionic neurons then regulate the internal organs by releasing ACh at their synapses. Parasympathetic nerve effects include a slowing of the heart, increased secretions and activities of digestive organs, and so on. Table 43.5 compares the actions of the sympathetic and parasympathetic divisions.

G proteins mediate cell responses to autonomic signals

You might wonder how release of ACh can slow the heart rate—an inhibitory effect—when it has excitatory effects elsewhere.

Figure 43.32
The sympathetic and parasympathetic divisions of the autonomic nervous system. The preganglionic neurons of the sympathetic division exit the thoracic and lumbar regions of the spinal cord, and those of the parasympathetic division exit the brain and sacral region of the spinal cord. The ganglia of the sympathetic division are located near the spinal cord; and those of the parasympathetic division are located near the organs they innervate. Most of the internal organs are innervated by both divisions.

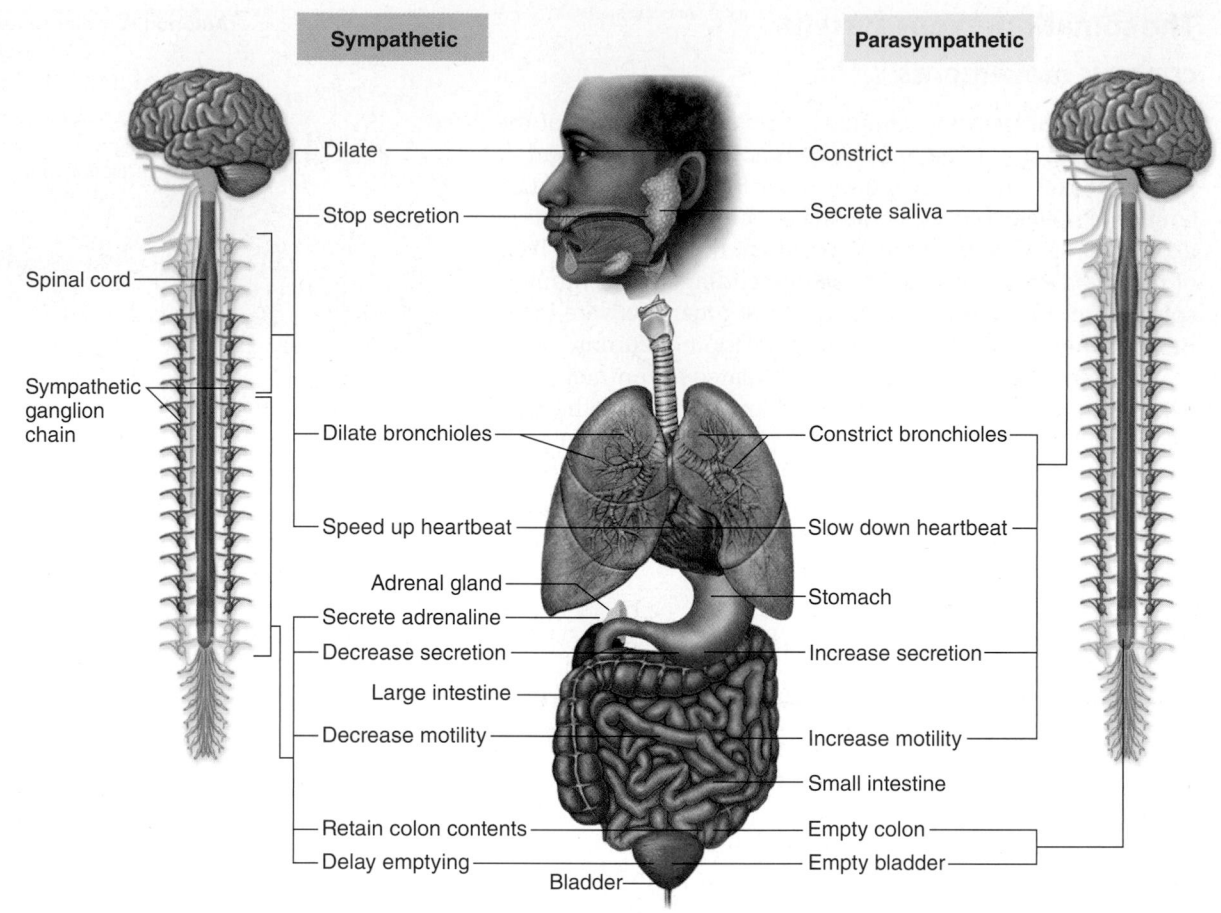

TABLE 43.5	Autonomic Innervation of Target Tissues	
Target Tissue	**Sympathetic Stimulation**	**Parasympathetic Stimulation**
Pupil of eye	Dilation	Constriction
Glands		
Salivary	Vasoconstriction; slight secretion	Vasodilation; copious secretion
Gastric	Inhibition of secretion	Stimulation of gastric activity
Liver	Stimulation of glucose secretion	Inhibition of glucose secretion
Sweat	Sweating	None
Gastrointestinal tract		
Sphincters	Increased tone	Decreased tone
Wall	Decreased tone	Increased motility
Gallbladder	Relaxation	Contraction
Urinary bladder		
Muscle	Relaxation	Contraction
Sphincter	Contraction	Relaxation
Heart muscle	Increased rate and strength	Decreased rate
Lungs	Dilation of bronchioles	Constriction of bronchioles
Blood vessels		
In muscles	Dilation	None
In skin	Constriction	None
In viscera	Constriction	Dilation

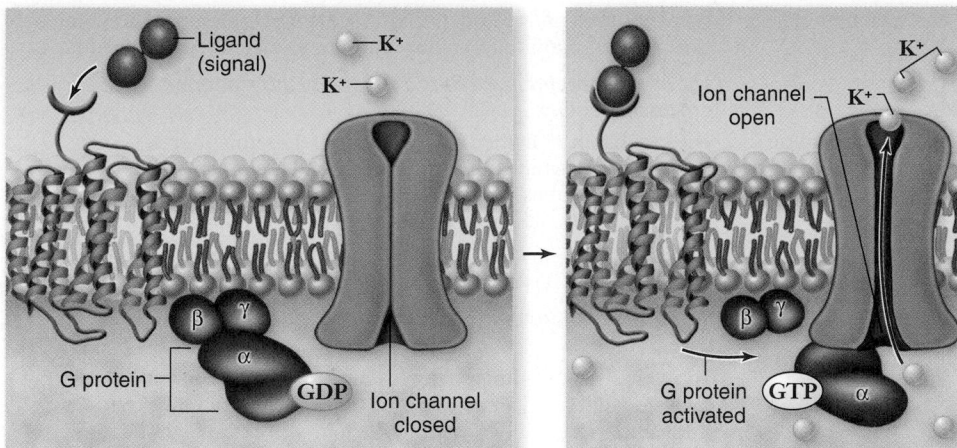

Figure 43.33 The parasympathetic effects of ACh require the action of G proteins. The binding of ACh to its receptor causes dissociation of a G protein complex, releasing some components of this complex to move within the membrane and bind to other proteins that form ion channels. Shown here are the effects of ACh on the heart, where the G protein components cause the opening of K⁺ channels. This leads to outward diffusion of potassium and hyperpolarization, slowing the heart rate.

The answer is simple: the cells involved in each case have different receptors for ACh that produce different effects. In the neuromuscular junction, the receptor for ACh is a ligand-gated Na⁺ channel that when open allows an influx of Na⁺ that depolarizes the membrane. In the case of the heart, the inhibitory effect on the pacemaker cells is produced because binding of ACh to a different receptor leads to K⁺ channels opening, resulting in the outward diffusion of K⁺, hyperpolarizing the membrane. The ACh receptor in the heart is a member of the class of receptors called G protein-coupled receptors.

In chapter 9, you learned that G protein-coupled receptors consist of a membrane receptor and effector protein that are coupled by the action of a G protein. The receptor is activated by binding to its ligand, in this case ACh, and the receptor activates a G protein that in turn activates an effector protein, in this case a K⁺ channel (figure 43.33).

This kind of system can also lead to excitation in other organs if the G protein acts on different effector proteins. For example, the parasympathetic nerves that innervate the stomach can cause increased gastric secretions and contractions.

The sympathetic nerve effects also are mediated by the action of G protein-coupled receptors. Stimulation by norepinephrine from sympathetic nerve endings and epinephrine from the adrenal medulla requires G proteins to activate the target cells. We describe these interactions in more detail, together with hormone action, in chapter 45.

Learning Outcomes Review 43.5

The PNS comprises the somatic (voluntary) and autonomic (involuntary) nervous systems. A spinal nerve contains sensory neurons, which carry information from sense organs to the CNS, and motor neurons, which carry directives from the CNS to targets such as muscle cells. The sympathetic division of the autonomic nervous system activates the body for fight-or-flight responses; the parasympathetic division generally promotes relaxation and digestion.

- *Why would having the sympathetic and parasympathetic divisions be more advantageous than having a single system?*

Chapter Review

43.1 Nervous System Organization

The three types of neurons in vertebrates are sensory neurons, motor neurons, and interneurons (figure 43.1).

The central nervous system is the "command center."
The CNS consists of the brain and the spinal cord, where sensory input is integrated and responses originate.

The peripheral nervous system collects information and carries out responses.
The PNS comprises sensory neurons that carry impulses to the CNS and motor neurons that carry impulses from the CNS to effectors.

The somatic nervous system primarily acts on skeletal muscles; the autonomic nervous system is involuntary and consists of the antagonistic sympathetic and parasympathetic divisions.

The structure of neurons supports their function.
Neurons have a cell body, dendrites that receive information, and a long axon that conducts impulses away from the cell.

Supporting cells include Schwann cells and oligodendrocytes.
Neuroglia are supporting cells of the nervous system. Schwann cells (PNS) and oligodendrocytes (CNS) produce myelin sheaths that surround and insulate axons (figure 43.3).

43.2 The Mechanism of Nerve Impulse Transmission

An electrical difference exists across the plasma membrane.
The sodium–potassium pump moves Na⁺ outside the cell and K⁺ into the cell. Leakage of K⁺ also moves positive charge outside the cell.

The resting potential is a balance between diffusion of K^+ out of the cell and attraction back in by negative charge.

Changes in membrane potential are due to changes in membrane permeability.

Membrane potential can be changed by the action of gated channels. Gated channels respond to chemical or electrical stimulus. Graded potentials combine in an additive way (summation).

Action potentials result when depolarization reaches a threshold.

Action potentials are all-or-nothing events resulting from the rapid and sequential opening of votage-gated ion channels (figure 43.9).

Action potentials are propagated along axons.

Influx of Na^+ during an action potential causes the adjacent region to depolarize, producing its own action potential (figure 43.10).

There are two ways to increase the velocity of nerve impulses.

The speed of nerve impulses increases as the diameter of the axon increases. Saltatory conduction, in which impulses jump from node to node, also increases speed (figure 43.11).

43.3 Synapses: Where Neurons Communicate with Other Cells

An action potential terminates at the end of the axon at the synapse—a gap between the axon and another cell.

The two types of synapses are electrical and chemical.

Electrical synapses consist of gap junctions; chemical synapses release neurotransmitters to cross the synapse (figure 43.14).

Many different chemical compounds serve as neurotransmitters.

Neurotransmitter molecules include acetylcholine, amino acids, biogenic amines, neuropeptides, and a gas—nitric oxide.

A postsynaptic neuron must integrate input from many synapses.

Excitatory postsynaptic potentials (EPSPs) depolarize the membrane; inhibitory postsynaptic potentials (IPSPs) hyperpolarize it. The additive effect may or may not produce an action potential.

Neurotransmitters play a role in drug addiction.

Addictive drugs often act by mimicking a neurotransmitter or by interfering with neurotransmitter reuptake.

43.4 The Central Nervous System: Brain and Spinal Cord

As animals became more complex, so did their nervous systems.

The nervous system has evolved from a nerve net composed of linked nerves, to nerve cords with association nerves, and to the development of coordination centers (figure 43.19).

Vertebrate brains have three basic divisions.

The vertebrate brain is divided into hindbrain, midbrain, and forebrain (see figure 43.20). The forebrain is divided further into the diencephalon and telencephalon. The telencephalon, called the cerebrum in mammals, is the center for association and learning.

The human forebrain exhibits exceptional information-processing ability.

The cerebrum is divided into right and left hemispheres (figure 43.22), which are subdivided into frontal, parietal, temporal, and occipital lobes (figure 43.23). The cerebrum contains the primary motor and somatosensory cortexes as well as the basal ganglia.

The limbic system consists of the hypothalamus, hippocampus, and amygdala, and it is responsible for emotional states.

Complex functions of the human brain may be controlled in specific areas.

The reticular activating system in the brainstem controls consciousness and alertness. Short-term memory may be stored as transient neural excitation; long-term memory involves changes in neural connections.

The spinal cord conveys messages and controls some responses directly.

Reflexes are the sudden, involuntary movement of muscles in response to a stimulus (figures 43.28 and 43.29).

43.5 The Peripheral Nervous System: Sensory and Motor Neurons

The PNS has somatic and autonomic systems.

Sensory axons (inbound) form the dorsal root of the spinal nerve. The cell bodies are in the dorsal root ganglia.

Motor axons (outbound) form the ventral root of the spinal nerve. Cell bodies are located in the spinal cord.

The somatic nervous system controls movements.

Somatic motor neurons stimulate skeletal muscles in response to conscious commands and involuntary reflexes.

The autonomic nervous system controls involuntary functions through two divisions.

Sympathetic neurons originate in the thoracic and lumbar regions of the spinal cord and synapse at an autonomic ganglion outside the spinal cord (figure 43.31). Parasympathetic neurons originate in the brain and in sacral regions of the spinal cord and terminate in ganglia near or within internal organs (figure 43.32 and table 43.5).

G proteins mediate cell responses to autonomic signals.

The binding of ACh activates a G protein that in turn activates a K^+ channel, allowing outflow of K^+ and hyperpolarization of the membrane and slowing heart rate.

Review Questions

UNDERSTAND

1. Which of the following best describes the electrical state of a neuron at rest?
   a. The inside of a neuron is more negatively charged than the outside.
   b. The outside of a neuron is more negatively charged than the inside.
   c. The inside and the outside of a neuron have the same electrical charge.
   d. Potassium ions leak into a neuron at rest.

2. The ____ cannot be controlled by conscious thought.
   a. motor neurons
   b. somatic nervous system
   c. autonomic nervous system
   d. skeletal muscles

3. A fight-or-flight response in the body is controlled by the

   a. sympathetic division of the nervous system.
   b. parasympathetic division of the nervous system.
   c. release of acetylcholine from postganglionic neurons.
   d. somatic nervous system.

4. Inhibitory neurotransmitters

   a. hyperpolarize postsynaptic membranes.
   b. hyperpolarize presynaptic membranes.
   c. depolarize postsynaptic membranes.
   d. depolarize presynaptic membranes.

5. White matter is____, and gray matter is____.

   a. comprised of axons; comprised of cell bodies and dendrites
   b. myelinated; unmyelinated
   c. found in the CNS; also found in the CNS
   d. All of the choices are correct.

6. During an action potential

   a. the rising phase is due to an influx of Na^+.
   b. the falling phase is due to an influx of K^+.
   c. the falling phase is due to an efflux of K^+.
   d. Both a and c are correct.

7. A functional reflex requires

   a. only a sensory neuron and a motor neuron.
   b. a sensory neuron, the thalamus, and a motor neuron.
   c. the cerebral cortex and a motor neuron.
   d. only the cerebral cortex and the thalamus.

APPLY

1. Imagine that you are doing an experiment on the movement of ions across neural membranes. Which of the following plays a role in determining the equilibrium concentration of ions across these membranes?

   a. Ion concentration gradients
   b. Ion pH gradients
   c. Ion electrical gradients
   d. Both a and c are correct.

2. The Na^+/K^+ ATPase pump is

   a. not required for action potential firing.
   b. important for long-term maintenance of resting potential.
   c. important only at the synapse.
   d. used to stimulate graded potentials.

3. Botox, a derivative of the botulinum toxin that causes food poisoning, inhibits the release of acetylcholine at the neuromuscular junction. How could this strange-sounding treatment produce desired cosmetic effects?

   a. By inhibiting the parasympathetic branch of the autonomic nervous system
   b. By inhibiting the sympathetic branch of the autonomic nervous system
   c. By causing paralysis of facial muscles, which decreases wrinkles in the face
   d. By causing facial muscles to contract, whereby the skin is stretched tighter, thereby reducing wrinkles

4. The following is a list of the components of a chemical synapse. A mutation in the structure of which of these would affect only the reception of the message, not its release or the response?

   a. Membrane proteins in the postsynaptic cell
   b. Proteins in the presynaptic cell
   c. Cytoplasmic proteins in the postsynaptic cell
   d. Both a and b are correct.

5. Suppose that you stick your finger with a sharp pin. The area affected is very small and only one pain receptor fires. However, it fires repeatedly at a rapid rate (it hurts!). This is an example of

   a. temporal summation.   c.   habituation.
   b. spatial summation.   d.   repolarization.

6. As you sit quietly reading this sentence, the part of the nervous system that is most active is the

   a. somatic nervous system.
   b. sympathetic nervous system.
   c. parasympathetic nervous system.
   d. None of the choices is correct.

7. G protein–coupled receptors are involved in the nervous system by

   a. controlling the release of neurotransmitters.
   b. controlling the opening and closing of Na^+ channels during an action potential.
   c. controlling the opening and closing of K^+ channels during an action potential.
   d. acting as receptors for neurotransmitters on postsynaptic cells.

SYNTHESIZE

1. Tetraethylammonium (TEA) is a drug that blocks voltage-gated K^+ channels. What effect would TEA have on the action potentials produced by a neuron? If TEA could be applied selectively to a presynaptic neuron that releases an excitatory neurotransmitter, how would it alter the synaptic effect of that neurotransmitter on the postsynaptic cell?

2. Describe the status of the Na^+ and K^+ channels at each of the following stages: rising, falling, and undershoot.

3. Describe the steps required to produce an excitatory postsynaptic potential (EPSP). How would these differ at an inhibitory synapse?

4. Your friend Karen loves caffeine. However, lately she has been complaining that she needs to drink more caffeinated beverages in order to get the same effect she used to. Excellent student of biology that you are, you tell her that this is to be expected. Why?

ONLINE RESOURCE

www.ravenbiology.com

Understand, Apply, and Synthesize—enhance your study with animations that bring concepts to life and practice tests to assess your understanding. Your instructor may also recommend the interactive eBook, individualized learning tools, and more.

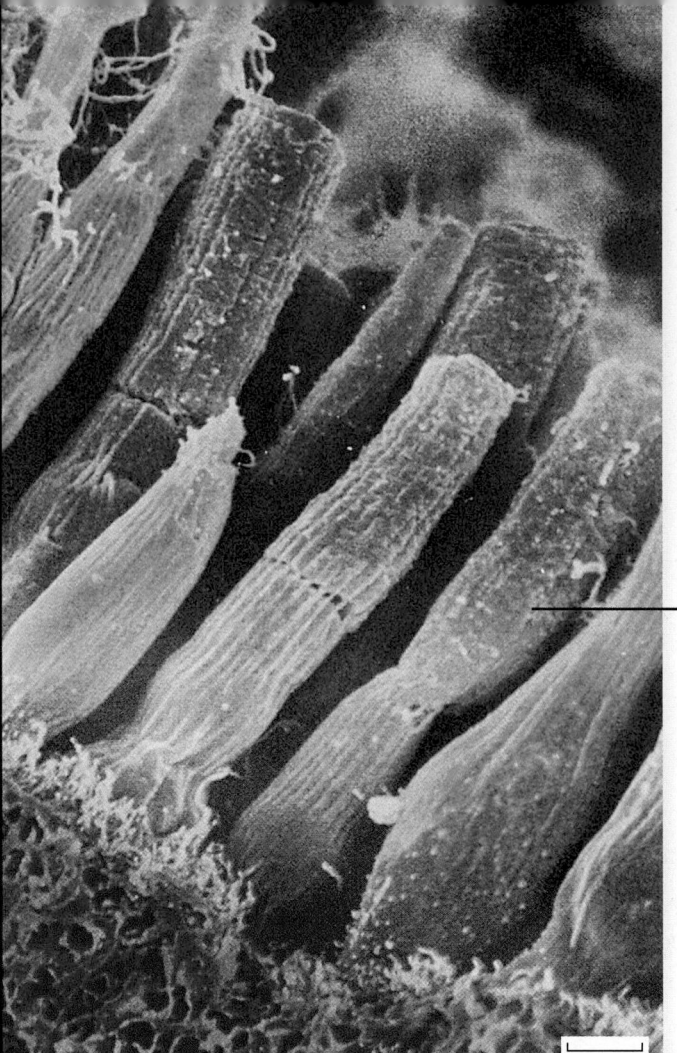

5 µm

Chapter **44**

Sensory Systems

Chapter Concepts

Introduction

All input from sensory neurons to the central nervous system (CNS) arrives in the same form, as electrical signals. Sensory neurons receive input from a variety of different kinds of sense receptor cells, such as the rod and cone cells found in the vertebrate eye shown in the micrograph. Different sensory neurons lead to different brain regions and so are associated with the different senses. The intensity of the sensation depends on the frequency of action potentials conducted by the sensory neuron. The brain distinguishes a sunset, a symphony, and searing pain only in terms of the identity of the sensory neuron carrying the action potentials and the frequency of these impulses. Thus, if the auditory nerve is artificially stimulated, the brain perceives the stimulation as sound. But if the optic nerve is artificially stimulated in exactly the same manner and degree, the brain perceives a flash of light.

In this chapter, we examine sensory systems, primarily in vertebrates. We also compare some of these systems with their counterparts in invertebrates.

Overview of Sensory Receptors

When we think of sensory receptors, the senses of vision, hearing, taste, smell, and touch come to mind—the senses that provide information about our environment. Certainly this external information is crucial to the survival and success of animals, but sensory receptors also provide information about internal states, such as stretching of muscles, position of the body, and blood pressure. In this section, we take a general look at types of receptors and how they work.

Sensory receptors detect both external and internal stimuli

Exteroceptors are receptors that sense stimuli that arise in the external environment. Almost all of a vertebrate's exterior senses evolved in water before the invasion of land. Consequently, many senses of terrestrial vertebrates emphasize stimuli that travel well in water, using receptors that have been retained in the transition from sea to land. Mammalian hearing, for example, converts an airborne stimulus into a waterborne one, using receptors similar to those that originally evolved in the water.

A few vertebrate sensory systems that function well in the water, such as the electrical organs of fish, cannot function in the air and are not found among terrestrial vertebrates. In contrast, some land-dwellers have sensory systems that could not function in water, such as infrared heat detectors.

Interoceptors sense stimuli that arise from within the body. These internal receptors detect stimuli related to muscle length and tension, limb position, pain, blood chemistry, blood volume and pressure, and body temperature. Many of these receptors are simpler than those that monitor the external environment and are believed to bear a closer resemblance to primitive sensory receptors. In the rest of this chapter, we consider the different types of exteroceptors and interoceptors according to the kind of stimulus each is specialized to detect.

Receptors can be grouped into three categories

Sensory receptors differ with respect to the nature of the environmental stimulus that best activates their sensory dendrites. Broadly speaking, we can recognize three classes of receptors:

1. **Mechanoreceptors** are stimulated by mechanical forces such as pressure. These include receptors for touch, hearing, and balance.

2. **Chemoreceptors** detect chemicals or chemical changes. The senses of smell and taste rely on chemoreceptors.
3. **Electromagnetic receptors** react to heat and light energy. The photoreceptors of the eyes that detect light are an example, as are the thermal receptors found in some reptiles.

The simplest sensory receptors are free nerve endings that react to bending or stretching of the sensory neuron's membrane in response to changes in temperature or to chemicals such as oxygen in the extracellular fluid. Other sensory receptors are more complex, involving the association of the sensory neurons with specialized epithelial cells.

Sensory information is conveyed in a four-step process

Sensory information picked up by sensory neurons is conveyed to the CNS, where the impulses are perceived in a four-step process (figure 44.1):

1. *Stimulation.* A physical stimulus impinges on a sensory neuron or an associated, but separate, sensory receptor.
2. *Transduction.* The stimulus energy is transformed into graded potentials in the dendrites of the sensory neuron.
3. *Transmission.* Action potentials develop in the axon of the sensory neuron and are conducted to the CNS along an afferent nerve pathway.
4. *Interpretation.* The brain creates a sensory perception from the electrochemical events produced by afferent stimulation. We actually perceive the five senses with our brains, not with our sense organs.

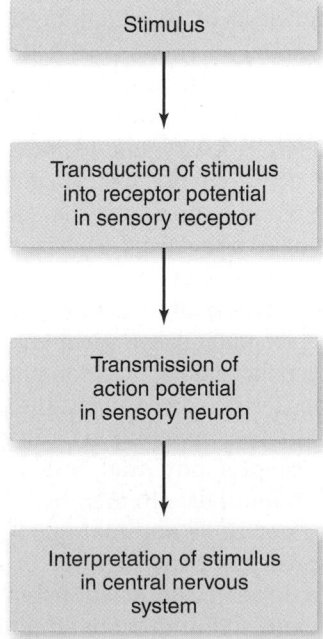

Figure 44.1 The path of sensory information. Sensory stimuli are transduced into receptor potentials, which can trigger sensory neuron action potentials that are conducted to the brain for interpretation.

Figure 44.2 Events in sensory transduction.

a. Depolarization of a free nerve ending leads to a receptor potential that spreads by local current flow to the axon. *b.* Action potentials are produced in the axon in response to a sufficiently large receptor potential.

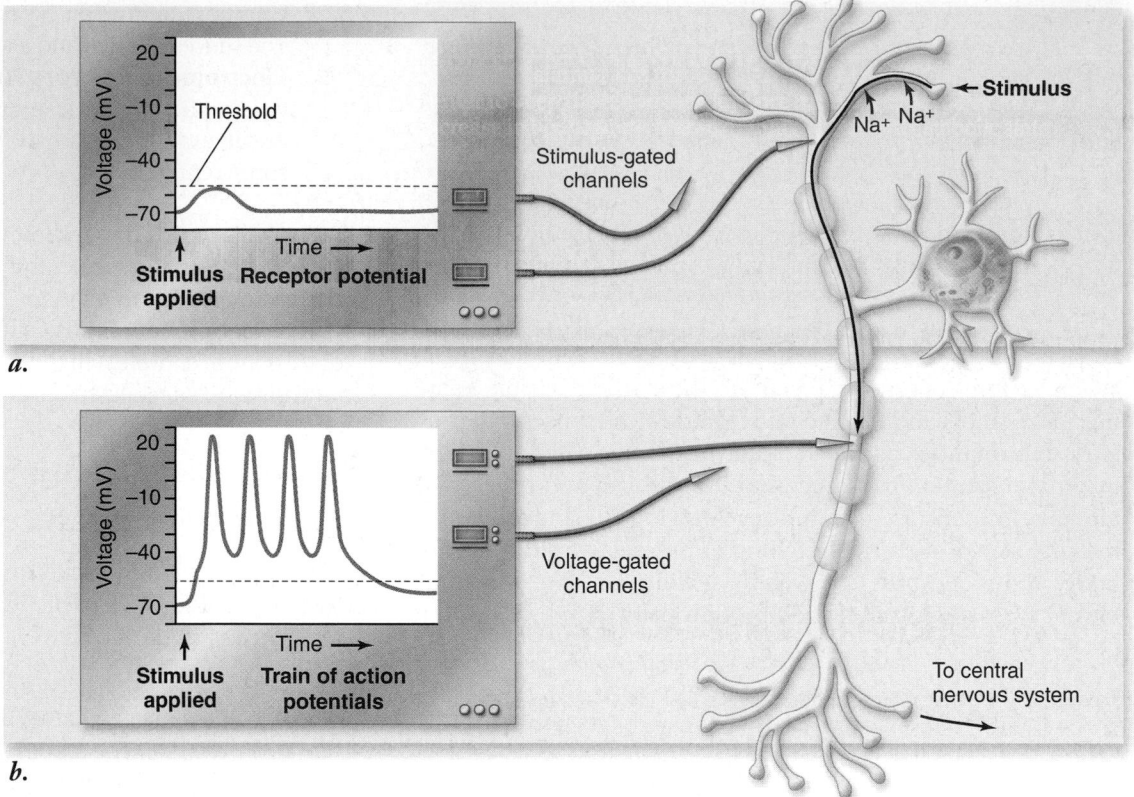

Sensory transduction involves gated ion channels

Sensory cells respond to stimuli because they possess **stimulus-gated ion channels** in their membranes. The sensory stimulus causes these ion channels to open or close, depending on the sensory system involved. In most cases, the sensory stimulus produces a depolarization of the receptor cell, analogous to the excitatory postsynaptic potential (EPSP, described in chapter 43) produced in a postsynaptic cell in response to a neurotransmitter. A depolarization that occurs in a sensory receptor on stimulation is referred to as a *receptor potential* (figure 44.2*a*).

Like an EPSP, a receptor potential is a graded potential: The larger the sensory stimulus, the greater the degree of depolarization. Receptor potentials also decrease in size with distance from their source. This prevents small, irrelevant stimuli from reaching the cell body of the sensory neuron. If the receptor potential or the summation of receptor potentials is great enough to generate a threshold level of depolarization, an action potential is produced that propagates along the sensory axon into the CNS (figure 44.2*b*).

The greater the sensory stimulus, the greater the depolarization of the receptor potential and the higher the frequency of action potentials. (Remember that frequency of action potentials, not their summation, is responsible for conveying the intensity of the stimulus.)

Generally, a logarithmic relationship exists between stimulus intensity and action potential frequency—for example, a particular sensory stimulus that is 10 times greater than another sensory stimulus produces action potentials at twice the frequency of the other stimulus. This relationship allows the CNS to interpret the strength of a sensory stimulus based on the frequency of incoming signals.

Learning Outcomes Review 44.1

Sensory receptors include mechanoreceptors, chemoreceptors, and electromagnetic energy-detecting receptors. The four steps by which information is conveyed to the CNS are stimulation, transduction, transmission, and interpretation in the CNS. Gated ion channels open or close in response to stimuli, altering membrane potential; if this change exceeds a threshold, an action potential is generated.

■ *Why is the relationship between intensity of stimulus and frequency of action potentials said to be logarithmic?*

44.2 Mechanoreceptors: Touch and Pressure

Learning Outcomes

1. *Explain how mechanoreceptors detect touch.*
2. *Distinguish among nociceptors, thermoreceptors, proprioceptors, and baroreceptors.*

Although the receptors of the skin, called the cutaneous receptors, are classified as interoceptors, they in fact respond to stimuli at the border between the external and internal environments. These receptors serve as good examples of the specialization of receptor structure and function, responding to pain, heat, cold, touch, and pressure.

Pain receptors alert the body to damage or potential damage

A stimulus that causes or is about to cause tissue damage is perceived as pain. The receptors that transmit impulses perceived as pain are called **nociceptors,** so named because they can be sensitive to noxious substances as well as tissue damage. Although specific nociceptors exist, many hyperstimulated sensory receptors can also produce the perception of pain in the brain.

Most nociceptors consist of free nerve endings located throughout the body, especially near surfaces where damage is most likely to occur. Different nociceptors may respond to extremes in temperature, very intense mechanical stimulation such as a hard impact, or specific chemicals in the extracellular fluid, including some that are released by injured cells. The thresholds of these sensory cells vary; some nociceptors are sensitive only to actual tissue damage, but others respond before damage has occurred.

Transient receptor potential ion channels

One kind of tissue damage can be due to extremes of temperature, and in this case the molecular details of how a noxious stimulus can result in the sensation of pain are becoming clear. A class of ion channel protein found in nociceptors, the transient receptor potential (TRP) ion channel, can be stimulated by temperature to produce an inward flow of cations, primarily Na^+ and Ca^{2+}. This depolarizing current causes the sensory neuron to fire, leading to the release of glutamate and an EPSP in neurons in the spinal cord, ultimately producing the pain response.

TRP channels that respond to both hot and cold have been found. Differences have also been found in the sensitivity of TRP channels to the degree of temperature change, with some responding only to temperature changes that damage tissues and others that respond to milder changes. Thus, we can respond to the feelings of hot and cold as well as feel pain associated with extremes of hot and cold.

The first such TRP channel identified responds to the chemical capsaicin, found in chili peppers, as well as to heat. This explains the sensation of heat we feel when we eat chili peppers, as well as the associated pain! A cold-responsive TRP receptor also responds to the chemical menthol, explaining how this substance is perceived as "cold." Chemical stimulation of TRP channels can reduce the body's pain response by desensitizing the sensory neuron. This analgesic response is why menthol is found in cough drops.

Thermoreceptors detect changes in heat energy

The skin contains two populations of **thermoreceptors,** which are naked dendritic endings of sensory neurons that are sensitive to changes in temperature. (Nociceptors are similar in that they consist of free nerve endings.) These thermoreceptors contain TRP ion channels that are responsive to hot and cold.

Cold receptors are stimulated by a fall in temperature and are inhibited by warming, whereas warm receptors are stimulated by a rise in temperature and inhibited by cooling. Cold receptors are located immediately below the epidermis; they are three to four times more numerous than are warm receptors. Warm receptors are typically located slightly deeper, in the dermis.

Thermoreceptors are also found within the hypothalamus of the brain, where they monitor the temperature of the circulating blood and thus provide the CNS with information on the body's internal (core) temperature. Information from the hypothalamic thermoreceptors alter metabolism and stimulate other responses to increase or decrease core temperature as needed.

Different receptors detect touch, depending on intensity

Several types of mechanoreceptors are present in the skin, some in the dermis and others in the underlying subcutaneous tissue (figure 44.3).

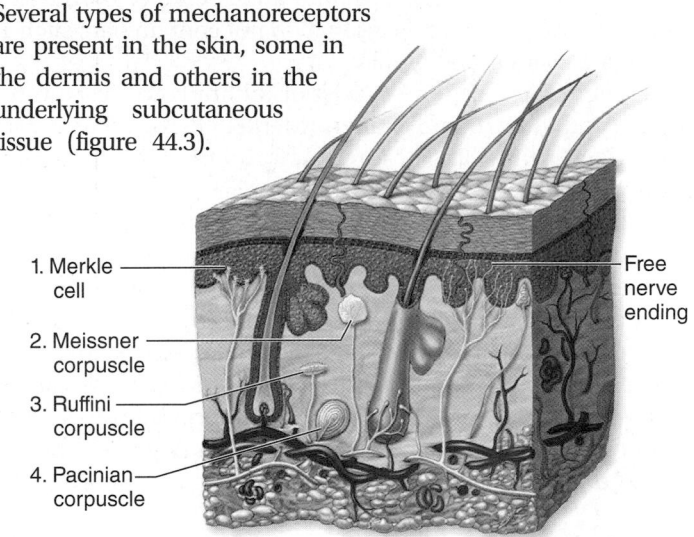

1. Merkle cell
2. Meissner corpuscle
3. Ruffini corpuscle
4. Pacinian corpuscle
Free nerve ending

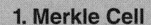

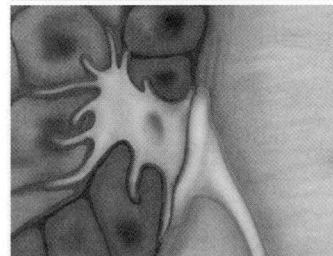

1. Merkle Cell

Tonic receptors located near the surface of the skin that are sensitive to touch pressure and duration.

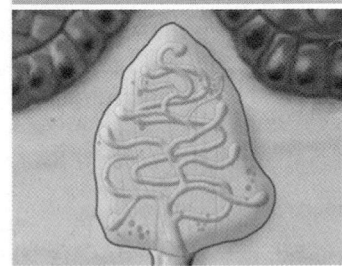

2. Meissner Corpuscle

Phasic receptors sensitive to fine touch, concentrated in hairless skin.

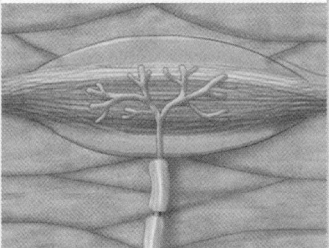

3. Ruffini Corpuscle

Tonic receptors located near the surface of the skin that are sensitive to touch pressure and duration.

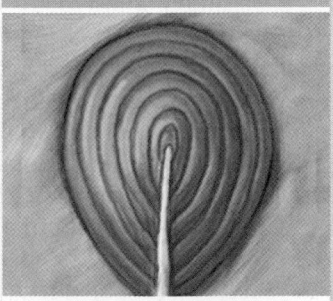

4. Pacinian Corpuscle

Pressure-sensitive phasic receptors deep below the skin in the subcutaneous tissue.

Figure 44.3 Sensory receptors in human skin.
Cutaneous receptors may be free nerve endings or sensory dendrites in association with other supporting structures.

These receptors contain sensory cells with ion channels that open in response to mechanical distortion of the membrane. They detect various forms of physical contact, known as the sense of touch.

Morphologically specialized receptors that respond to fine touch are most concentrated on areas such as the fingertips and face. They are used to localize cutaneous stimuli very precisely. These receptors can be either phasic (intermittently activated) or tonic (continuously activated). The phasic receptors include hair follicle receptors and Meissner corpuscles, which are present on surfaces that do not contain hair, such as the fingers, palms, and nipples.

The tonic receptors consist of Ruffini corpuscles in the dermis and touch dome endings (Merkel's disks) located near the surface of the skin. These receptors monitor the duration of a touch and the extent to which it is applied.

Deep below the skin in the subcutaneous tissue lie phasic, pressure-sensitive receptors called Pacinian corpuscles. Each of these receptors consists of the end of an afferent axon surrounded by a capsule of alternating layers of connective tissue cells and extracellular fluid. When sustained pressure is applied to the corpuscle, the elastic capsule absorbs much of the pressure, and the axon ceases to produce impulses. Pacinian corpuscles thus monitor only the onset and removal of pressure, as may occur repeatedly when something that vibrates is placed against the skin.

Muscle length and tension are monitored by proprioceptors

Buried within the skeletal muscles of all vertebrates except the bony fishes are **muscle spindles,** sensory stretch receptors that lie in parallel with the rest of the fibers in the muscle (figure 44.4). Each spindle consists of several thin muscle fibers wrapped together and innervated by a sensory neuron, which becomes activated when the muscle, and therefore the spindle, is stretched.

Muscle spindles, together with other receptors in tendons and joints, are known as **proprioceptors.** These sensory receptors provide information about the relative position or movement of the animal's body parts. The sensory neurons conduct action potentials into the spinal cord, where they synapse with somatic motor neurons that innervate the muscle. This pathway constitutes the muscle stretch reflex, including the knee-jerk reflex mentioned in chapter 43. When the muscle is briefly stretched by tapping the patellar ligament with a rubber mallet, the muscle spindle apparatus is also stretched. The spindle apparatus is embedded within the muscle, and, like the muscle fibers outside the spindle, is stretched along with the muscle. The result is the action potential that activates the somatic motor neurons and causes the leg to jerk.

When a muscle contracts, it exerts tension on the tendons attached to it. The **Golgi tendon organs,** another type of proprioceptor, monitor this tension. If it becomes too high, they elicit a reflex that inhibits the motor neurons innervating the muscle. This reflex helps ensure that muscles do not contract so strongly that they damage the tendons to which they are attached.

Baroreceptors detect blood pressure

Blood pressure is monitored at two main sites in the body. One is the carotid sinus, an enlargement of the left and right internal carotid arteries that supply blood to the brain. The other is

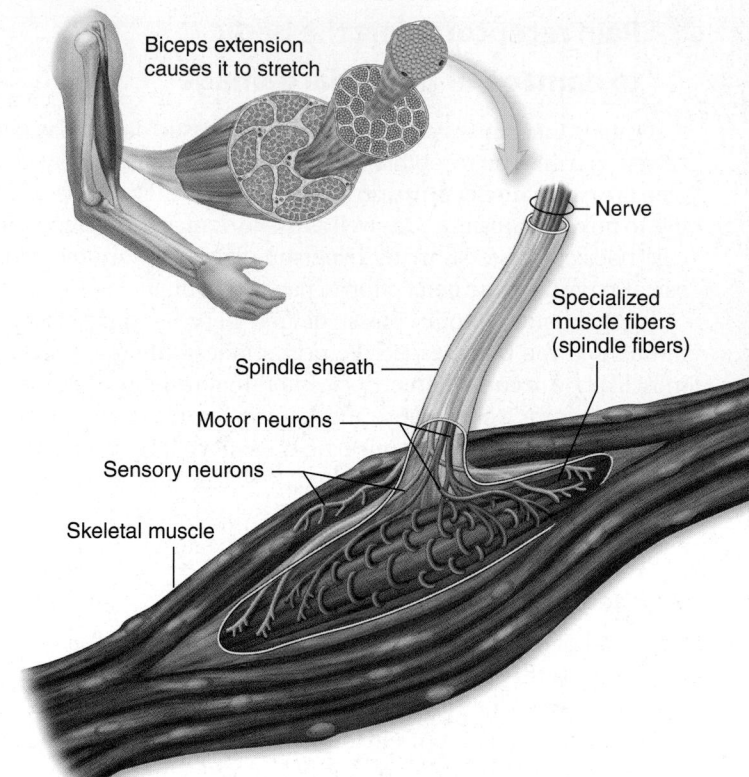

Figure 44.4 How a muscle spindle works. A muscle spindle is a stretch receptor embedded within skeletal muscle. Stretching of the muscle elongates the spindle fibers and stimulates the sensory dendritic endings wrapped around them. This causes the sensory neurons to send impulses to the CNS, where they synapse with interneurons and, in some cases, motor neurons.

the aortic arch, the portion of the aorta very close to its emergence from the heart. The walls of the blood vessels at both sites contain a highly branched network of afferent neurons called baroreceptors, which detect tension or stretch in the walls.

When the blood pressure decreases, the frequency of impulses produced by the baroreceptors decreases. The CNS responds to this reduced input by stimulating the sympathetic division of the autonomic nervous system, causing an increase in heart rate and vasoconstriction. Both effects help raise the blood pressure, thus maintaining homeostasis. A rise in blood pressure increases baroreceptor impulses, which conversely reduces sympathetic activity and stimulates the parasympathetic division, slowing the heart and lowering the blood pressure.

Learning Outcomes Review 44.2

Mechanical distortion of the plasma membrane of mechanoreceptors produces nerve impulses. Nociceptors detect damage or potential damage to tissues and cause pain; thermoreceptors sense changes in heat energy; proprioceptors monitor muscle length; and baroreceptors monitor blood pressure within arteries.

■ *Why is it important to detect stretching of muscles?*

Hearing, Vibration, and Detection of Body Position

Learning Outcomes

1. *Explain how sound waves in the environment lead to production of action potentials in the inner ear.*
2. *Describe how hearing differs between aquatic and terrestrial animals.*
3. *Describe how body position and movement are detected by hearing-associated structures.*

Hearing, the detection of sound waves, actually works better in water than in air because water transmits pressure waves more efficiently. Despite this limitation, hearing is widely used by terrestrial vertebrates to monitor their environments, communicate with other members of their species, and detect possible sources of danger.

Sound is a result of vibration, or waves, traveling through a medium, such as water or air. Detection of sound waves is possible through the action of specialized mechanoreceptors that first evolved in aquatic organisms. The cells that are involved in the detection of sound are also evolutionarily related to the gravity-sensing systems discussed in the end of this section.

The lateral line system in fish detects low-frequency vibrations

In addition to hearing, the lateral line system in fish provides a sense of "distant touch," enabling them to sense objects that reflect pressure waves and low-frequency vibrations. This enables a fish to detect prey, for example, and to swim in synchrony with the rest of its school. It also enables a blind cave fish to sense its environment by monitoring changes in the patterns of water flow past the lateral line receptors.

The lateral line system is found in amphibian larvae, but is lost at metamorphosis and is not present in any terrestrial vertebrate. The sense provided by the lateral line system supplements the fish's sense of hearing, which is performed by the sensory structures in their ears.

The lateral line system consists of hair cells within a longitudinal canal in the fish's skin that extends along each side of the body and within several canals in the head (figure 44.5a). The hair cells' surface processes project into a gelatinous membrane called a cupula. The hair cells are innervated by sensory neurons that transmit impulses to the brain.

Hair cells have several hairlike processes, called stereocilia, and one longer process called a **kinocilium** (figure 44.5b). The stereocilia are actually microvilli containing actin fibers, and the kinocilium is a true cilium that contains microtubules. Vibrations carried through the fish's environment produce movements of the cupula, which cause the processes to bend.

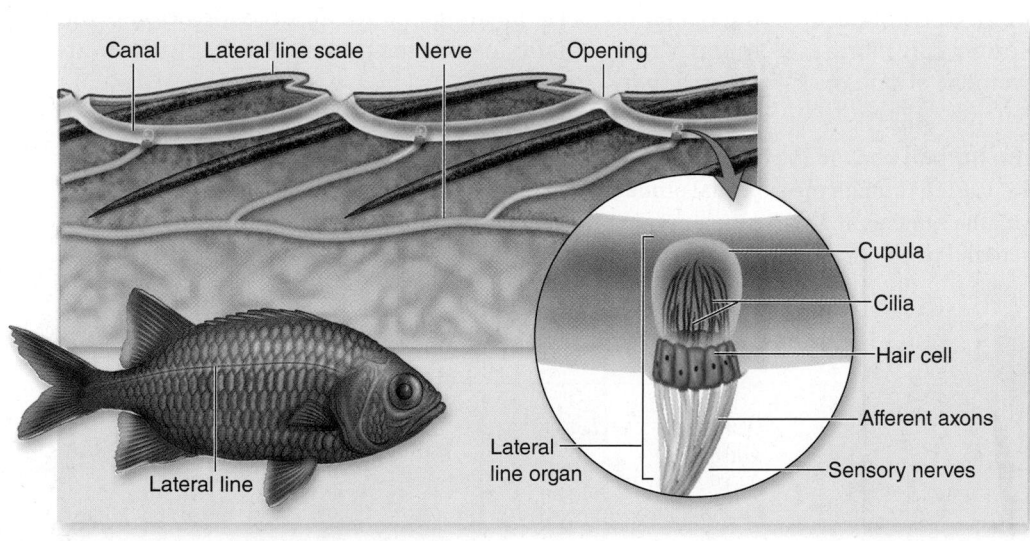

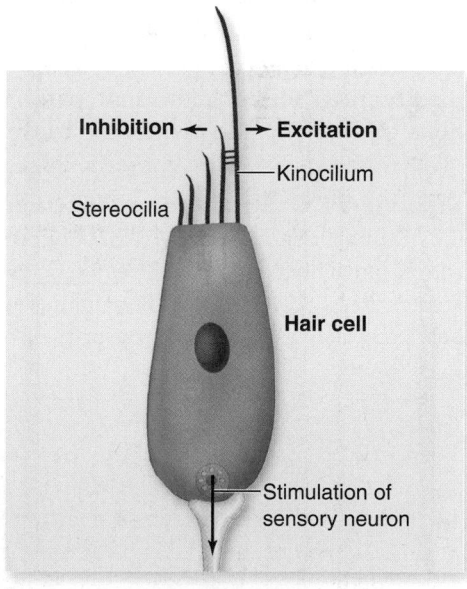

a. b.

Figure 44.5 The lateral line system. *a.* This system consists of canals running the length of the fish's body beneath the surface of the skin. Within these canals are sensory structures containing hair cells with cilia that project into a gelatinous cupula. Pressure waves traveling through the water in the canals deflect the cilia and depolarize the sensory neurons associated with the hair cells. *b.* Hair cells are mechanoreceptors with hairlike cilia that project into a gelatinous membrane. The hair cells of the lateral line system (and the membranous labyrinth of the vertebrate inner ear) have a number of smaller cilia called stereocilia and one larger kinocilium. When the cilia bend in the direction of the kinocilium, the hair cell releases a chemical transmitter that depolarizes the associated sensory neuron. Bending of the cilia in the opposite direction has an inhibitory effect.

? Inquiry question How would the lateral line system of a shark detect an injured and thrashing fish?

When the stereocilia bend in the direction of the kinocilium, the associated sensory neurons are stimulated and generate a receptor potential. As a result, the frequency of action potentials produced by the sensory neuron is increased. In contrast, if the stereocilia are bent in the opposite direction, then the activity of the sensory neuron is inhibited.

Ear structure is specialized to detect vibration

The structure of the ear allows pressure waves to be transduced into nerve impulses based on mechanosensory cells like those in the lateral line system. We will first consider the structure of the ear in fish, which is related to the lateral line system that senses pressure waves in water. Then we will consider how the structure of the ear of terrestrial vertebrates allows the sensing of pressure waves in air.

Hearing structures in fish

Sound waves travel through the body of a fish as easily as through the surrounding water because the fish's body is composed primarily of water. For sound to be detected, therefore, an object of different density is needed. In many fish, this function is served by the **otoliths,** literally "ear rocks," composed of calcium carbonate crystals. Otoliths are contained in the otolith organs of the membranous **labyrinth,** a system of fluid-filled chambers and tubes also present in other vertebrates. When otoliths in fish vibrate against hair cells in the otolith organ, action potentials are produced. Hair cells are so-called because of the stereocilia that project from their surface.

Hearing structures of terrestrial vertebrates

In the ears of terrestrial vertebrates, vibrations in air may be channeled through an ear canal to the eardrum, or tympanic membrane. These structures are part of the **outer ear.** Vibrations of the tympanic membrane cause movement of one or more small bones that are located in a bony cavity known as the **middle ear.**

Amphibians and reptiles have a single middle ear bone, the **stapes** (stirrup), but mammals have two others: the **malleus** (hammer) and **incus** (anvil) (figure 44.6a, b). Where did these two additional bones come from?

The fossil record makes clear that the malleus and incus of modern mammals is derived from the two bones in the lower jaws of synapsid reptiles (figure 44.7). Through evolutionary time, these bones became progressively smaller and came to lie closer to the stapes. Eventually, in modern mammals, they became completely disconnected from the jawbone and moved within the middle ear itself.

The middle ear is connected to the throat by the Eustachian tube, also known as the auditory tube, which equalizes the air pressure between the middle ear and the external environment. The "ear popping" you may have experienced when flying in an airplane or driving on a mountain is caused by pressure equalization between the two sides of the eardrum.

The stapes vibrates against a flexible membrane, the oval window, which leads into the **inner ear.** Because the oval window is smaller in diameter than the tympanic membrane, vibrations against it produce more force per unit area, transmitted into the inner ear. The inner ear consists of the **cochlea,** a bony structure containing part of the membranous labyrinth called the cochlear duct. The cochlear duct is located in the center of the cochlea; the area above the cochlear duct is the vestibular canal, and the area below is the tympanic canal (figure 44.6c). All three chambers are filled with fluid. The oval window opens to the upper vestibular canal, so that when the stapes causes it to vibrate, it produces pressure waves of fluid. These pressure waves travel down to the tympanic canal, pushing another flexible membrane, the round window, that transmits the pressure back into the middle ear cavity.

Figure 44.6 **Structure and function of the human ear.** The structure of the human ear is shown in successive enlargements illustrating functional parts (*a* to *d*). Sound waves passing through the ear canal produce vibrations of the tympanic membrane, which causes movement of the middle ear ossicles (the malleus, incus, and stapes) against an inner membrane (the oval window). This vibration creates pressure waves in the fluid in the vestibular and tympanic canals of the cochlea. These pressure waves cause cilia in hair cells to bend, producing signals from sensory neurons.

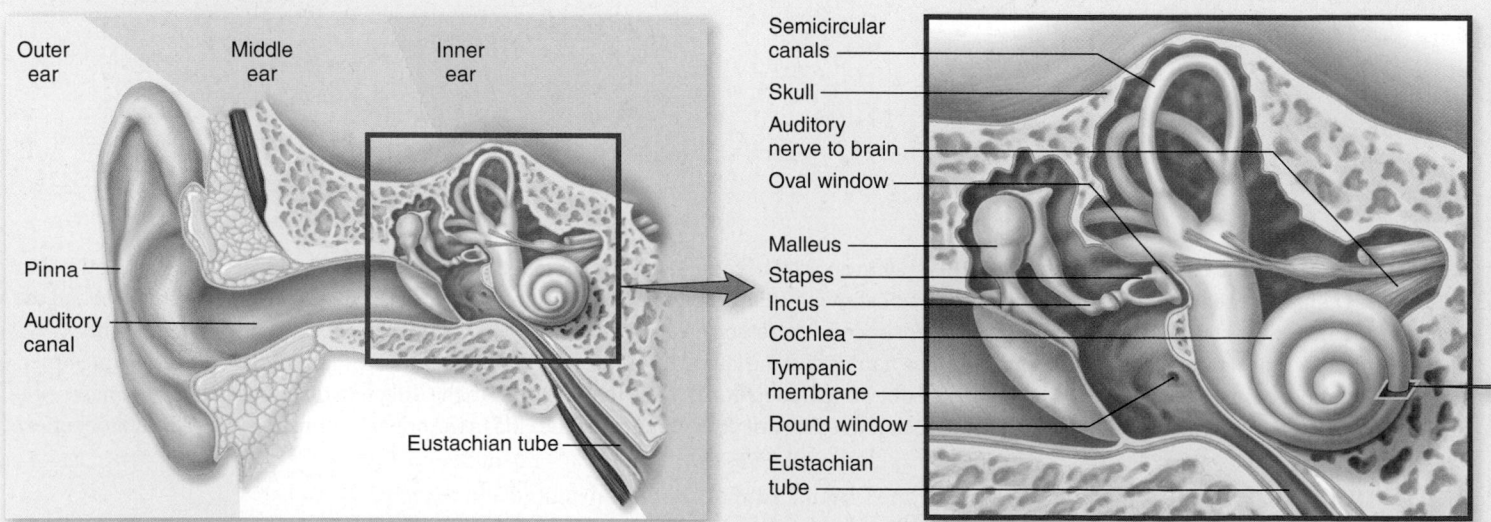

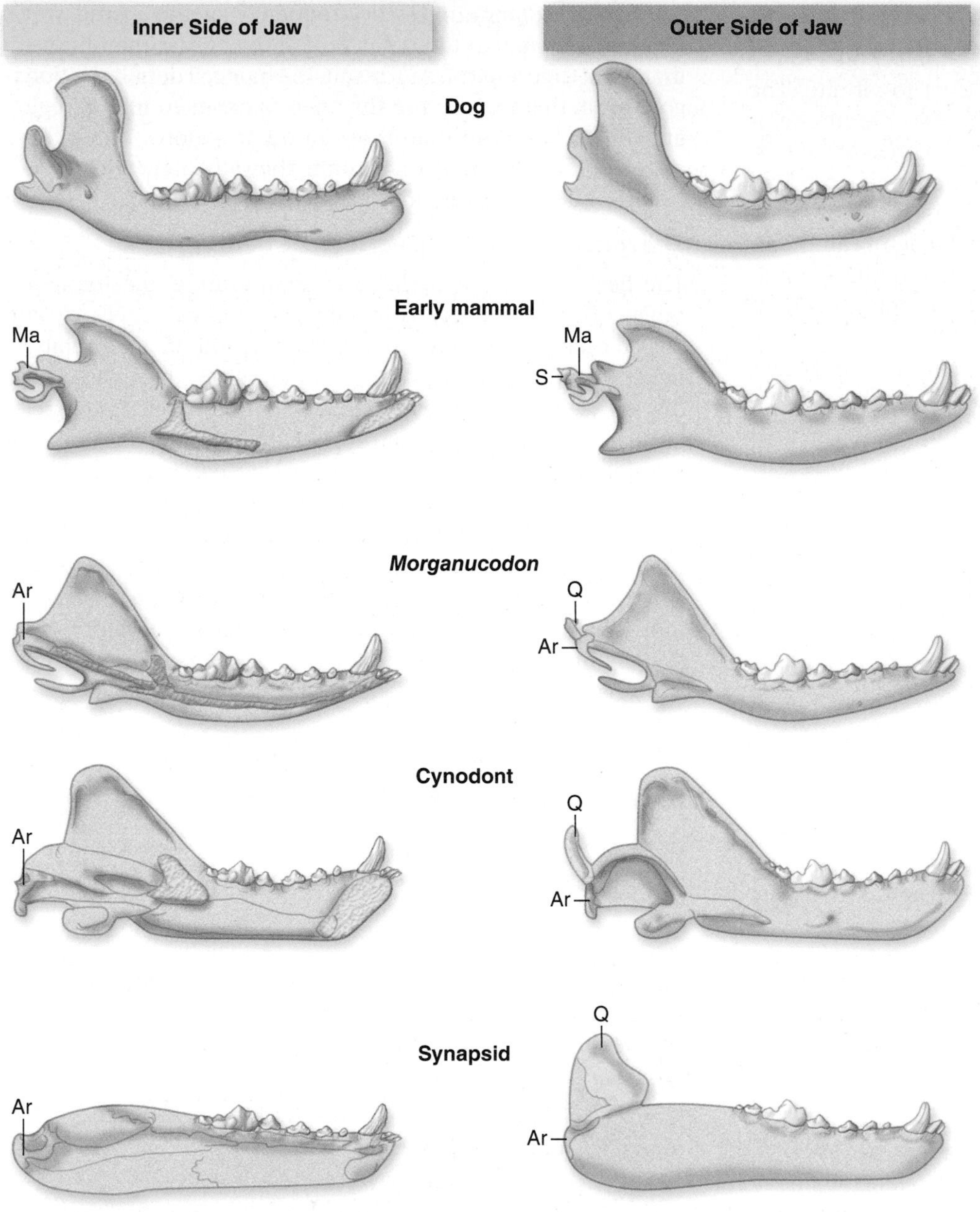

Inner Side of Jaw

Dog

Early mammal

Ma

Morganucodon

Ar

Cynodont

Ar

Synapsid

Ar

Outer Side of Jaw

Ma
S

Q
Ar

Q

Ar

Q

Ar

Figure 44.7 **Evolution of the mammalian inner ear.** Two of the bones in the inner ear of modern mammals, the stapes (S) and malleus (Ma), are derived from the quadrate (Q) and articular (Ar) bones, respectively, of their reptilian ancestors. The transition from an early ancestor of mammals, a synapsid (which lived more than 250 million years ago), through several transition forms, to a modern dog is illustrated (see chapter 35 for more information on the evolution of mammals from reptilian ancestors). Note how the bones become smaller and change position, ultimately disappearing from the lower jaw entirely in modern mammals (represented by the dog) and becoming parts of the inner ear. During embryology in modern mammals, these bones develop in association with the lower jawbone before moving inward to the inner ear, providing further evidence of their evolutionary origin.

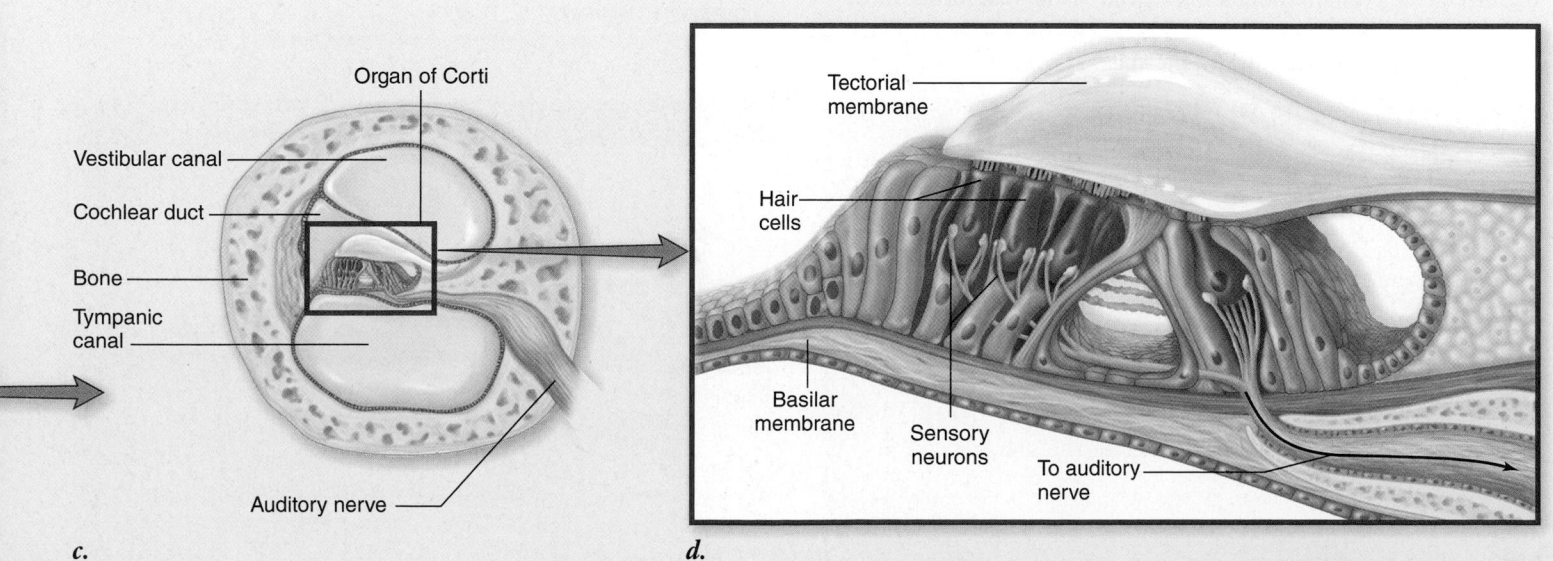

c.

Organ of Corti

Vestibular canal

Cochlear duct

Bone

Tympanic canal

Auditory nerve

d.

Tectorial membrane

Hair cells

Basilar membrane

Sensory neurons

To auditory nerve

Transduction occurs in the cochlea

As pressure waves are transmitted through the cochlea to the round window, they cause the cochlear duct to vibrate. The bottom of the cochlear duct, called the basilar membrane, is quite flexible and vibrates in response to these pressure waves. The surface of the basilar membrane contains sensory hair cells. The stereocilia from the hair cells project into an overhanging gelatinous membrane, the tectorial membrane. This sensory apparatus, consisting of the basilar membrane, hair cells with associated sensory neurons, and tectorial membrane, is known as the organ of Corti (figure 44.6d).

As the basilar membrane vibrates, the cilia of the hair cells bend in response to the movement of the basilar membrane relative to the tectorial membrane. The bending of these stereocilia in one direction depolarizes the hair cells. Bending in the opposite direction repolarizes or even hyperpolarizes the membrane. The hair cells, in turn, stimulate the production of action potentials in sensory neurons that project to the brain, where they are interpreted as sound.

Frequency localization in the cochlea

The basilar membrane of the cochlea consists of elastic fibers of varying length and stiffness, like the strings of a musical instrument, embedded in a gelatinous material. At the base of the cochlea (near the oval window), the fibers of the basilar membrane are short and stiff. At the far end of the cochlea (the apex), the fibers are 5 times longer and 100 times more flexible. Therefore, the resonant frequency of the basilar membrane is higher at the base than at the apex; the base responds to higher pitches, the apex to lower pitches.

When a wave of sound energy enters the cochlea from the oval window, it initiates an up-and-down motion that travels the length of the basilar membrane. However, this wave imparts most of its energy to that part of the basilar membrane with a resonant frequency near the frequency of the sound wave, resulting in a maximum deflection of the basilar membrane at that point (figure 44.8). As a result, the hair cell depolarization is greatest in that region, and the afferent axons from that region are stimulated more than those of other regions. When these action potentials arrive in the brain, they are interpreted as representing a sound of a particular frequency, or pitch.

The range of terrestrial vertebrate hearing

The flexibility of the basilar membrane limits the frequency range of human hearing to between approximately 20 and 20,000 cycles per second (hertz, Hz) in children. Our ability to hear high-pitched sounds decays progressively throughout middle age. Other vertebrates can detect sounds at frequencies lower than 20 Hz and much higher than 20,000 Hz. Dogs, for example, can detect sounds at 40,000 Hz, enabling them to hear high-pitched dog whistles that seem silent to a human listener.

Hair cells are also innervated by efferent axons from the brain, and impulses in those axons can make hair cells less sensitive. This central control of receptor sensitivity can increase an individual's ability to concentrate on a particular auditory signal (for example, a single voice) in the midst of background noise, which is effectively "tuned out" by the efferent axons.

Some vertebrates have the ability to navigate by sound

Because terrestrial vertebrates have two ears located on opposite sides of the head, the information provided by

Figure 44.8 Frequency localization in the cochlea.

The cochlea is shown unwound, so that the length of the basilar membrane can be seen. The fibers within the basilar membrane vibrate in response to different frequencies of sound, related to the pitch of the sound. Thus, regions of the basilar membrane show maximum vibrations in response to different sound frequencies. *a.* Notice that high-frequency (pitch) sounds vibrate the basilar membrane more toward the base whereas medium frequencies *(b)* and low frequencies *(c)* cause vibrations more toward the apex.

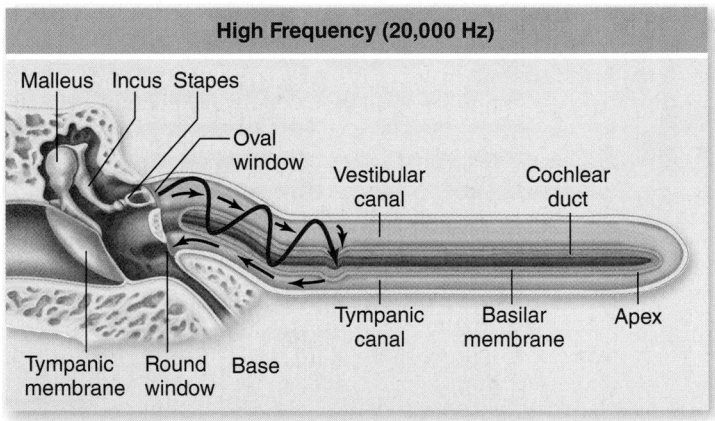

a.

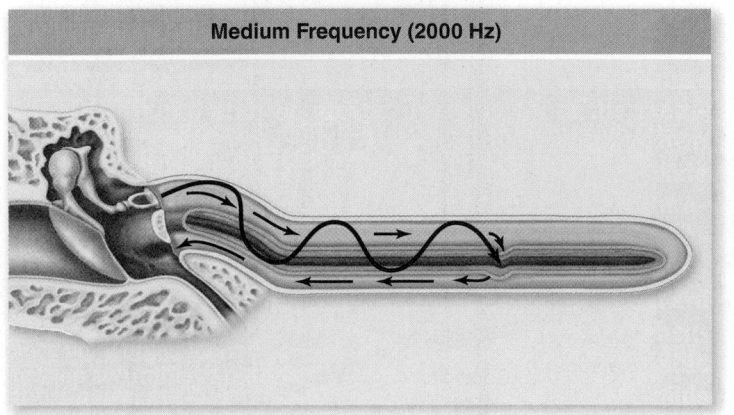

b.

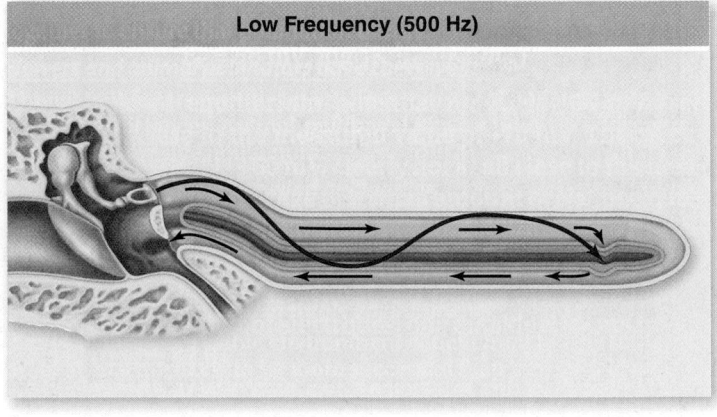

c.

hearing can be used to determine the direction of a sound source with some precision. Sound sources vary in strength, however, and sounds are weakened and reflected to varying degrees by the presence of objects in the environment. For these reasons, auditory sensors do not provide a reliable measure of distance.

A few groups of mammals that live and obtain their food in dark environments have circumvented the limitations of darkness. A bat flying in a completely dark room easily avoids objects placed in its path—even a wire less than a millimeter in diameter. Shrews use a similar form of "lightless vision" beneath the ground, as do whales and dolphins beneath the sea. All of these mammals are able to perceive presence and distance of objects by sound.

These mammals emit sounds and then determine the time it takes these sounds to reach an object and return to the animal. This process is called **echolocation.** A bat, for example, produces clicks that last 2 to 3 ms and are repeated several hundred times per second. By calculating the time each click takes to hit an object and return, bats can calculate the location, direction of movement, and speed of objects in their environment. The human inventions sonar and radar are based on the same principles of echolocation.

The three-dimensional imaging achieved with such an auditory sonar system is quite sophisticated. Bats can track and intercept rapidly maneuvering aerial prey and can distinguish one type of insect from another.

Body position and movement are detected by systems associated with hearing systems

The evolutionary strategy of using internal calcium carbonate crystals as a way to detect vibration has also allowed the development of sensory organs that detect body position in space and movements such as acceleration.

Most invertebrates can orient themselves with respect to gravity due to a sensory structure called a **statocyst.** Statocysts generally consist of ciliated hair cells with the cilia embedded in a gelatinous membrane containing crystals of calcium carbonate. These stones, or statoliths, increase the mass of the gelatinous membrane so that it can bend the cilia when the animal's position changes. If the animal tilts to the right, for example, the statolith membrane bends the cilia on the right side and activates associated sensory neurons.

A similar structure is found in the membranous labyrinth of the inner ear of vertebrates. This labyrinth is surrounded by bone and perilymph, which is similar in ionic content to interstitial fluid. Inside, the chambers and tubes are filled with endolymph fluid, which is similar in ionic content to intracellular fluid. Though intricate, the entire structure is very small; in a human, it is about the size of a pea.

Structure of the labyrinth and semicircular canals

The receptors for gravity in most vertebrates consist of two chambers of the membranous labyrinth called the **utricle** and **saccule** (figure 44.9). Within these structures are hair cells with

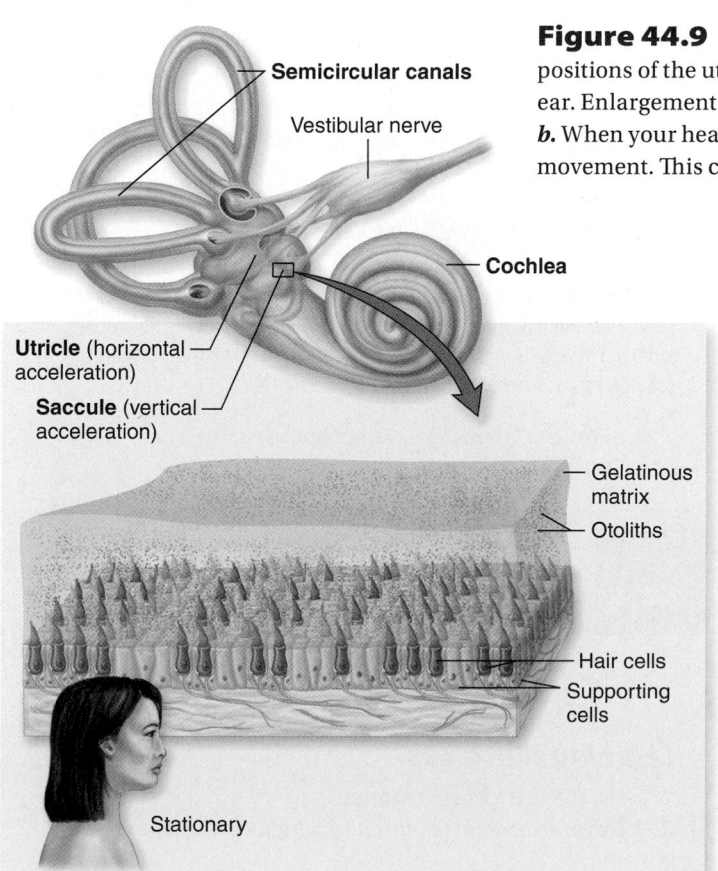

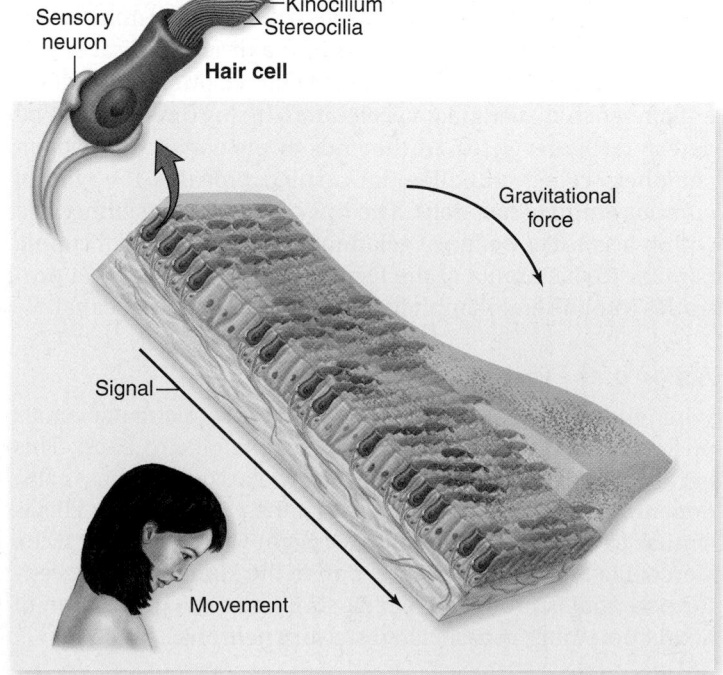

Figure 44.9 Structure and function of the utricle and saccule. *a.* The relative positions of the utricle and saccule within the membranous labyrinth of the human inner ear. Enlargement shows the gelatinous matrix containing otoliths covering hair cells. *b.* When your head bends forward, gravity distorts the matrix in the direction of movement. This causes the stereocilia in hair cells to bend, stimulating sensory neurons.

a.

b.

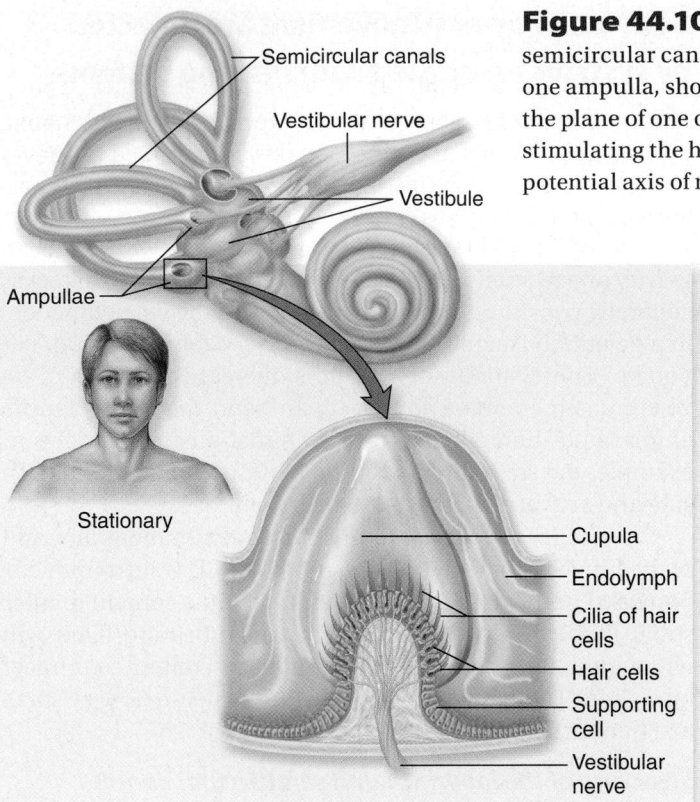

Semicircular canals

Vestibular nerve

Vestibule

Ampullae

Stationary

Cupula

Endolymph

Cilia of hair cells

Hair cells

Supporting cell

Vestibular nerve

a.

Figure 44.10 **The structure of the semicircular canals.** The position of the semicircular canals in relation to the rest of the inner ear. *a.* Enlargement of a section of one ampulla, showing how hair cell cilia insert into the cupula. *b.* Angular acceleration in the plane of one of the semicircular canals causes bending of the cupula, thereby stimulating the hair cells. Each inner ear contains three semicircular canals, one for each potential axis of rotation.

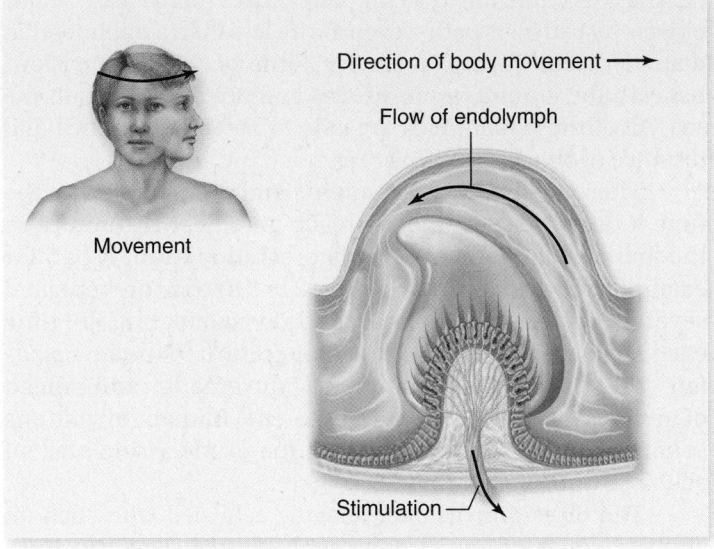

Direction of body movement →

Flow of endolymph

Movement

Stimulation

b.

stereocilia and a kinocilium, similar to those in the lateral line system of fish. The hairlike processes are embedded within a gelatinous membrane, the otolith membrane, containing calcium carbonate crystals. Because the otolith organ is oriented differently in the utricle and saccule, the utricle is more sensitive to horizontal acceleration (as in a moving car) and the saccule to vertical acceleration (as in an elevator). In both cases, the acceleration causes the stereocilia to bend, and consequently produces action potentials in an associated sensory neuron.

The membranous labyrinth of the utricle and saccule is continuous with three **semicircular canals,** oriented in different planes so that angular acceleration in any direction can be detected (figure 44.10). At the ends of the canals are swollen chambers called ampullae, into which protrude the cilia of another group of hair cells. The tips of the cilia are embedded within a sail-like wedge of gelatinous material called a cupula (similar to the cupula of the fish lateral line system) that protrudes into the endolymph fluid of each semicircular canal.

Action of the vestibular apparatus

When the head rotates, the fluid inside the semicircular canals pushes against the cupula and causes the cilia to bend. This bending either depolarizes or hyperpolarizes the hair cells, depending on the direction in which the cilia are bent. This is similar to the way the lateral line system works in a fish: If the stereocilia are bent in the direction of the kinocilium, a receptor potential is produced, which stimulates the production of action potentials in associated sensory neurons.

The saccule, utricle, and semicircular canals are collectively referred to as the **vestibular apparatus.** The saccule and utricle provide a sense of linear acceleration, and the semicircular canals provide a sense of angular acceleration. The brain uses information that comes from the vestibular apparatus about the body's position to maintain balance and equilibrium.

Learning Outcomes Review 44.3

Sound waves cause middle ear ossicles to vibrate; fluid in the inner ear is vibrated in turn, bending hair cells and causing action potentials. In terrestrial animals, sound waves in air must transition to the fluid in the inner ear. Hair cells in the vestibular apparatus of terrestrial vertebrates provide a sense of acceleration and balance.

■ *Why is a lateral line system not useful to adult amphibians?*

44.4 Chemoreceptors: Taste, Smell, and pH

Learning Outcomes

1. *List the five taste categories.*
2. *Describe how taste buds and olfactory neurons function.*

Some sensory cells, called chemoreceptors, contain membrane proteins that can bind to particular chemicals or ligands in the extracellular fluid. In response to this chemical interaction, the membrane of the sensory neuron becomes depolarized and produces action potentials. Chemoreceptors are used in the senses of taste and smell and are also important in monitoring the chemical composition of the blood and cerebrospinal fluid.

Taste detects and analyzes potential food

The perception of taste (gustation), like the perception of color, is a combination of physical and psychological factors. This is commonly broken down into five categories: sweet, sour, salty, bitter, and umami (perception of glutamate and other amino acids that give a hearty taste to many protein-rich foods such as meat, cheese, and broths). Taste buds—collections of chemosensitive epithelial cells associated with afferent neurons—mediate the sense of taste in vertebrates. In a fish, the taste buds are scattered over the surface of the body. These are the most sensitive vertebrate chemoreceptors known. They are particularly sensitive to amino acids; a catfish, for example, can distinguish between two different amino acids at a concentration of less than 100 parts per billion (1 g in 10,000 L of water)! The ability to taste the surrounding water is very impor-

tant to bottom-feeding fish, enabling them to sense the presence of food in an often murky environment.

The taste buds of all terrestrial vertebrates occur in the epithelium of the tongue and oral cavity, within raised areas called papillae (figure 44.11). Taste buds are onion-shaped structures of between 50 and 100 taste cells; each cell has fingerlike projections called microvilli that poke through the top of the taste bud, called the taste pore (figure 44.11c). Chemicals from food dissolve in saliva and contact the taste cells through the taste pore.

Within a taste bud, the chemicals that produce salty and sour tastes act directly through ion channels. The prototypical salty taste is due to Na^+ ions, which diffuse through Na^+ channels into cells in receptor cells in the taste bud. This Na^+ influx depolarizes the membrane, causing the receptor cell to release neurotransmitter and activate a sensory neuron that sends an impulse to the brain. The cells that detect sour taste act in a similar fashion except that the ion detected is H^+. Sour tastes are associated with increased concentration of protons that can also depolarize the membrane when they diffuse through ion channels.

Sweet, bitter, and umami substances bind to G protein–coupled receptors (see chapter 9) specific for each category. The nature and distribution of these receptors is an area of

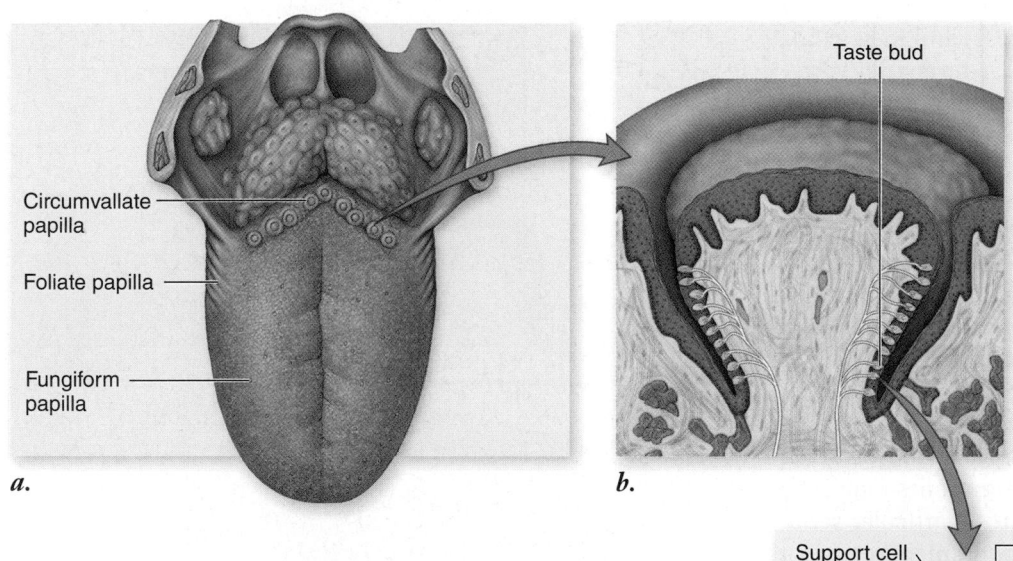

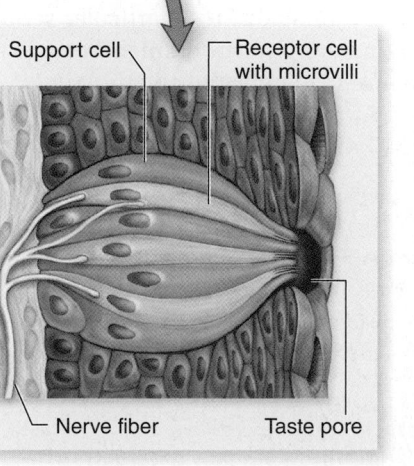

Figure 44.11 Taste. *a.* Human tongues have projections called papillae that bear taste buds. Different sorts of taste buds are located on different regions of the tongue. *b.* Groups of taste buds are embedded within a papilla. *c.* Individual taste buds are bulb-shaped collections of chemosensitive receptors that open out into the mouth through a pore. *d.* Photomicrograph of taste buds in papillae.

active investigation, but recent data indicate that individual receptor cells in the taste bud express only one type of receptor. This leads to cells that have receptors for sweet, for bitter, or for umami tastes. Activation of any of these G protein–coupled receptors then stimulates a single signaling pathway that leads the release of neurotransmitter from receptor cells to activate a sensory neuron and send an impulse to the brain. There they interact with other sensory neurons carrying information related to smell, described next. In this model, the different tastes are encoded to the brain based on which receptor cells are activated.

Like vertebrates, many arthropods also have taste chemoreceptors. For example, flies, because of their mode of searching for food, have taste receptors in sensory hairs located on their feet. The sensory hairs contain a variety of chemoreceptors that are able to detect sugars, salts, and other tastes by the integration of stimuli from these chemoreceptors (figure 44.12). If they step on potential food, their proboscis (the tubular feeding apparatus) extends to feed.

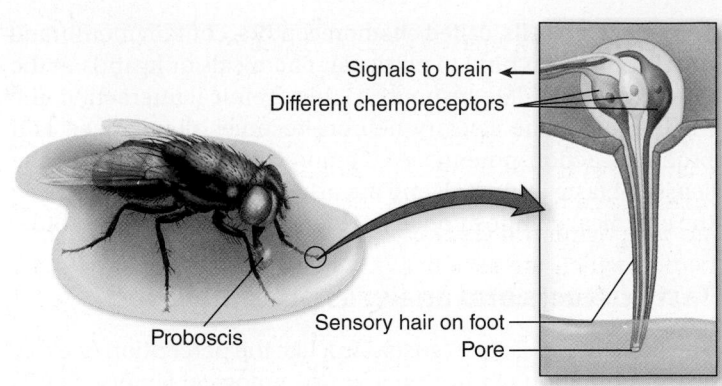

Figure 44.12 Many insects taste with their feet. In the blowfly shown here, chemoreceptors extend into the sensory hairs on the foot. Different chemoreceptors detect different types of food molecules. When the fly steps in a food substance, it can taste the different food molecules and extend its proboscis for feeding.

Smell can identify a vast number of complex molecules

In terrestrial vertebrates, the sense of smell (olfaction) involves chemoreceptors located in the upper portion of the nasal passages (figure 44.13). These receptors, whose dendrites end in tassels of cilia, project into the nasal mucosa, and their axons project directly into the cerebral cortex. A terrestrial vertebrate uses its sense of smell in much the same way that a fish uses its sense of taste—to sample the chemical environment around it.

Because terrestrial vertebrates are surrounded by air, their sense of smell has become specialized to detect airborne particles—but these particles must first dissolve in extracellular fluid before they can activate the olfactory receptors. The sense of smell can be extremely acute in many mammals, so much so that a single odorant molecule may be all that is needed to excite a given receptor.

Although humans can detect only five modalities of taste, they can discern thousands of different smells. New research suggests that as many as a thousand different genes may code for different receptor proteins for smell. The particular set of olfactory neurons that respond to a given odor might serve as a "fingerprint" the brain can use to identify the odor.

Figure 44.13 Smell. Humans detect smells by means of olfactory neurons (receptor cells) located in the lining of the nasal passages. The axons of these neurons transmit impulses directly to the brain via the olfactory nerve. Basal cells regenerate new olfactory neurons to replace dead or damaged cells. Olfactory neurons typically live about a month.

? Inquiry question In what ways do the senses of taste and smell share similarities? How are they different?

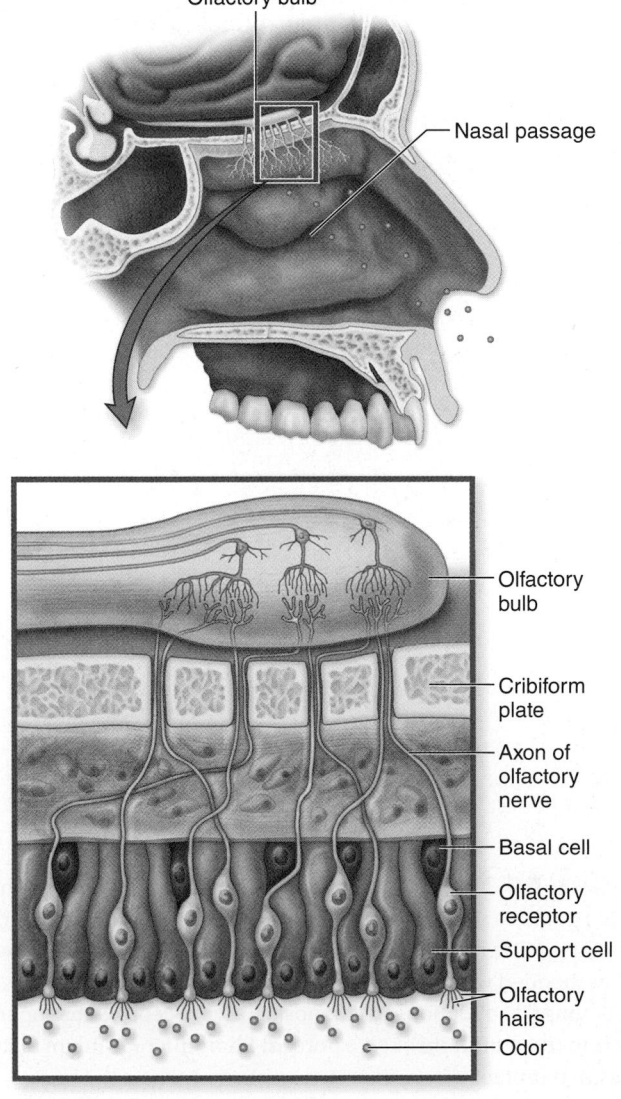

Internal chemoreceptors detect pH and other characteristics

Sensory receptors within the body detect a variety of chemical characteristics of the blood or fluids derived from the blood, including cerebrospinal fluid. Included among these receptors are the **peripheral chemoreceptors** of the aortic and carotid bodies, which are sensitive primarily to plasma pH, and the **central chemoreceptors** in the medulla oblongata of the brain, which are sensitive to the pH of cerebrospinal fluid. When the breathing rate is too low, the concentration of plasma CO_2 increases, producing more carbonic acid and causing a fall in the blood pH. The carbon dioxide can also enter the cerebrospinal fluid and lower the pH, thereby stimulating the central chemoreceptors. This stimulation indirectly affects the respiratory control center of the brainstem, which increases the breathing rate. The aortic bodies can also respond to a lowering of blood oxygen concentrations, but this effect is normally not significant unless a person goes to a high altitude where the partial pressure of oxygen is lower.

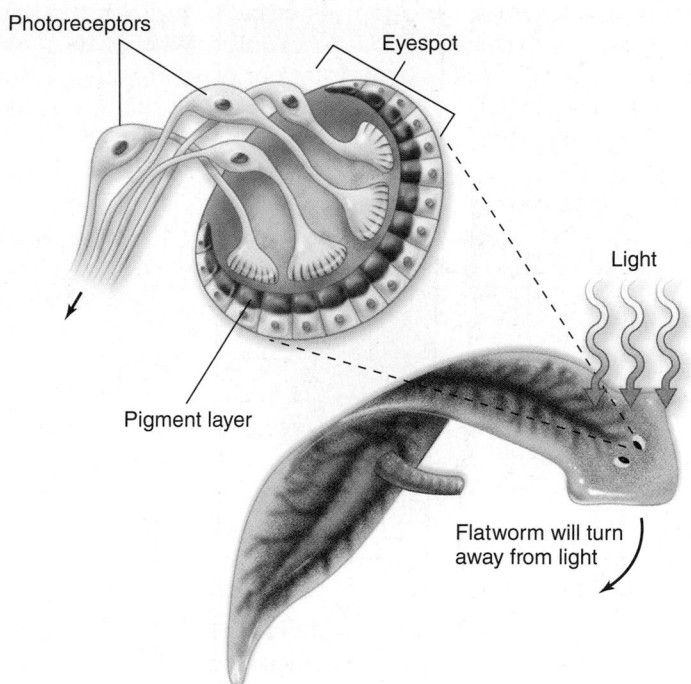

Figure 44.14 Simple eyespots in the flatworm. Eyespots can detect the direction of light because a pigmented layer on one side of the eyespot screens out light coming from the back of the animal. Light is thus detected more readily coming from the front of the animal; flatworms respond by turning away from the light.

> ### Learning Outcomes Review 44.4
>
> The five tastes humans perceive are sweet, sour, salty, bitter, and umami (amino acids). Taste and smell chemoreceptors detect chemicals from outside the body; olfactory receptors can identify thousands of different odors. Internal chemoreceptors monitor acid–base balance within the body and help regulate breathing.
>
> ■ **What are the advantages of insects having taste receptors on their feet?**

44.5 Vision

> ### Learning Outcomes
>
> 1. Compare invertebrate and vertebrate eyes.
> 2. Explain how a vertebrate eye focuses an image.
> 3. Describe how photoreceptors function.

The ability to perceive objects at a distance is important to most animals. Predators locate their prey, and prey avoid their predators, based on the three long-distance senses of hearing, smell, and vision. Of these, vision can act most distantly; with the naked eye, humans can see stars thousands of light years away—and a single photon is sufficient to stimulate a cell of the retina to send an action potential.

Vision senses light and light changes at a distance

Vision begins with the capture of light energy by **photoreceptors.** Because light travels in a straight line and arrives virtually instantaneously regardless of distance, visual information can be used to determine both the direction and the distance of an object. Other stimuli, which spread out as they travel and move more slowly, provide much less precise information.

Invertebrate eyes

Many invertebrates have simple visual systems with photoreceptors clustered in an eyespot. Simple eyespots can be made sensitive to the direction of a light source by the addition of a pigment layer that shades one side of the eye. Flatworms have a screening pigmented layer on the inner and back sides of both eyespots, allowing stimulation of the photoreceptor cells only by light from the front of the animal (figure 44.14). The flatworm will turn and swim in the direction in which the photoreceptor cells are the least stimulated. Although an eyespot can perceive the direction of light, it cannot be used to construct a visual image.

The members of four phyla—annelids, mollusks, arthropods, and chordates—have evolved well-developed, image-forming eyes. True image-forming eyes in these phyla, although strikingly similar in structure, are believed to have evolved independently, an example of convergent evolution (figure 44.15). Interestingly, the photoreceptors in all of these image-forming eyes use the same light-capturing molecule, suggesting that not many alternative molecules are able to play this role.

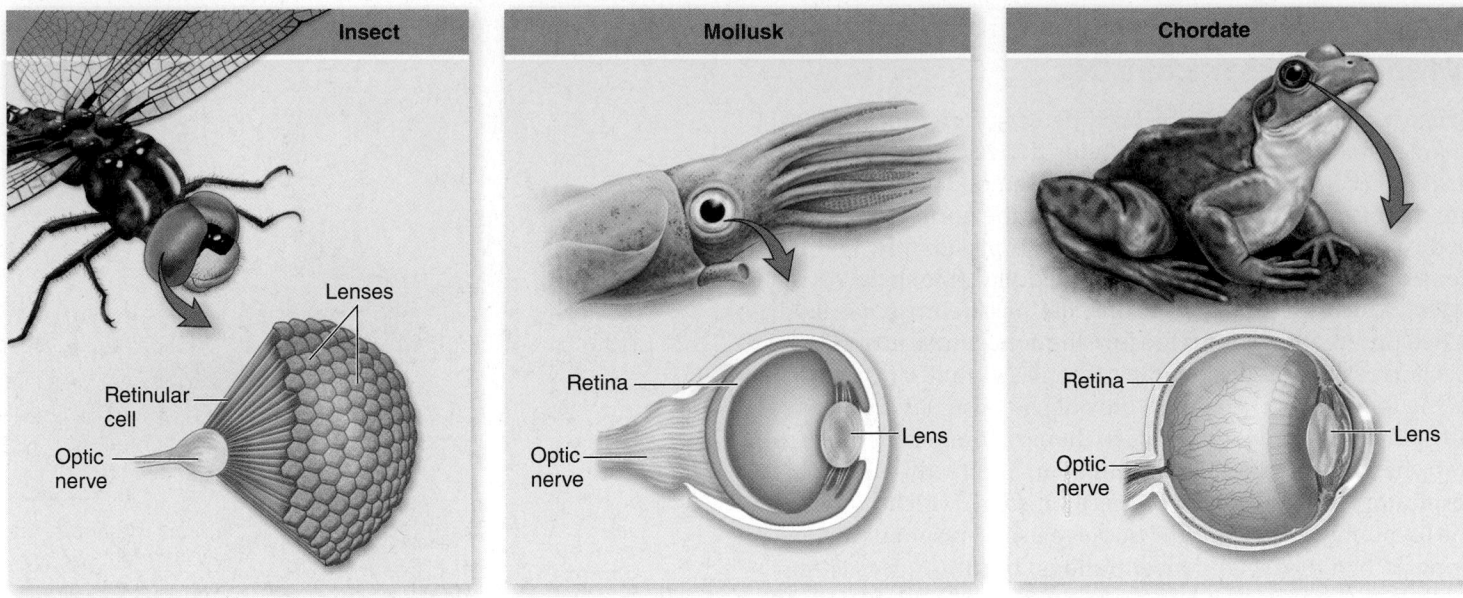

Figure 44.15 Eyes in three phyla of animals. Although they are superficially similar, these eyes differ greatly in structure from one another (see also figure 21.14 for a detailed comparison of mollusk and chordate eye structure). Each has evolved separately and, despite the apparent structural complexity, has done so from simpler structures.

Structure of the vertebrate eye

The human eye is typical of the vertebrate eye (figure 44.16). The "white of the eye" is the **sclera,** formed of tough connective tissue. Light enters the eye through a transparent **cornea,** which begins to focus the light. Focusing occurs because light is refracted (bent) when it travels into a medium of different density. The colored portion of the eye is the **iris;** contraction of the iris muscles in bright light decreases the size of its opening, the pupil. Light passes through the pupil to the **lens,** a transparent structure that completes the focusing of the light onto the retina at the back of the eye. The lens is attached by the suspensory ligament to the ciliary muscles.

The shape of the lens is influenced by the amount of tension in the suspensory ligament, which surrounds the lens and attaches it to the circular ciliary muscle. When the ciliary muscle contracts, it puts slack in the suspensory ligament, and the lens becomes more rounded and bends light more strongly. This rounding is required for close vision. In distance vision, the ciliary muscles relax, moving away from the lens and tightening the suspensory ligament. The lens thus becomes more flattened and bends light less, keeping the image focused on the retina. People who are nearsighted or farsighted do not properly focus the image on the retina (figure 44.17). Interestingly, the lens of an amphibian or a fish does not change shape; these animals instead focus images by moving their lens in and out, just as you would do to focus a camera.

Vertebrate photoreceptors are rod cells and cone cells

The vertebrate retina contains two kinds of photoreceptor cells, called rods and cones (figure 44.18). **Rods,** which get

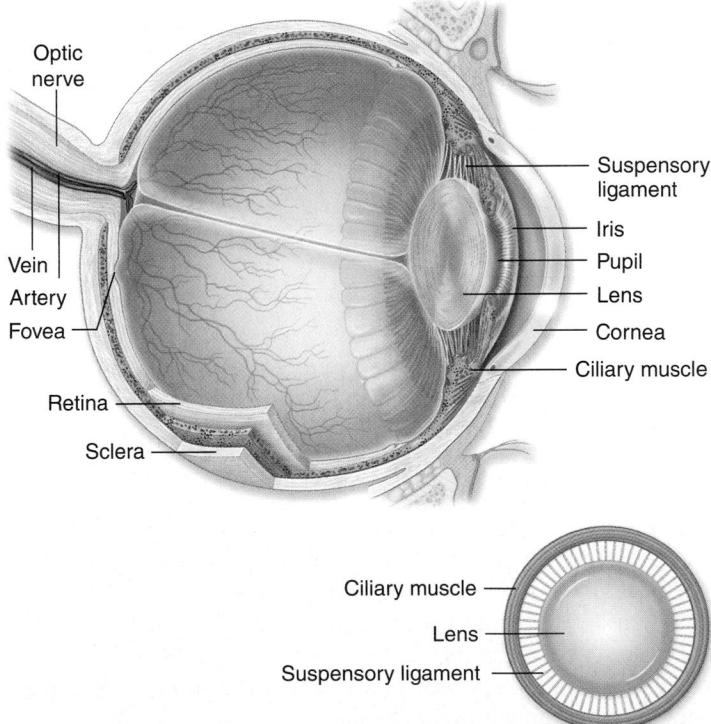

Figure 44.16 Structure of the human eye. The transparent cornea and lens focus light onto the retina at the back of the eye, which contains the photoreceptors (rods and cones). The center of each eye's visual field is focused on the fovea. Focusing is accomplished by contraction and relaxation of the ciliary muscle, which adjusts the curvature of the lens.

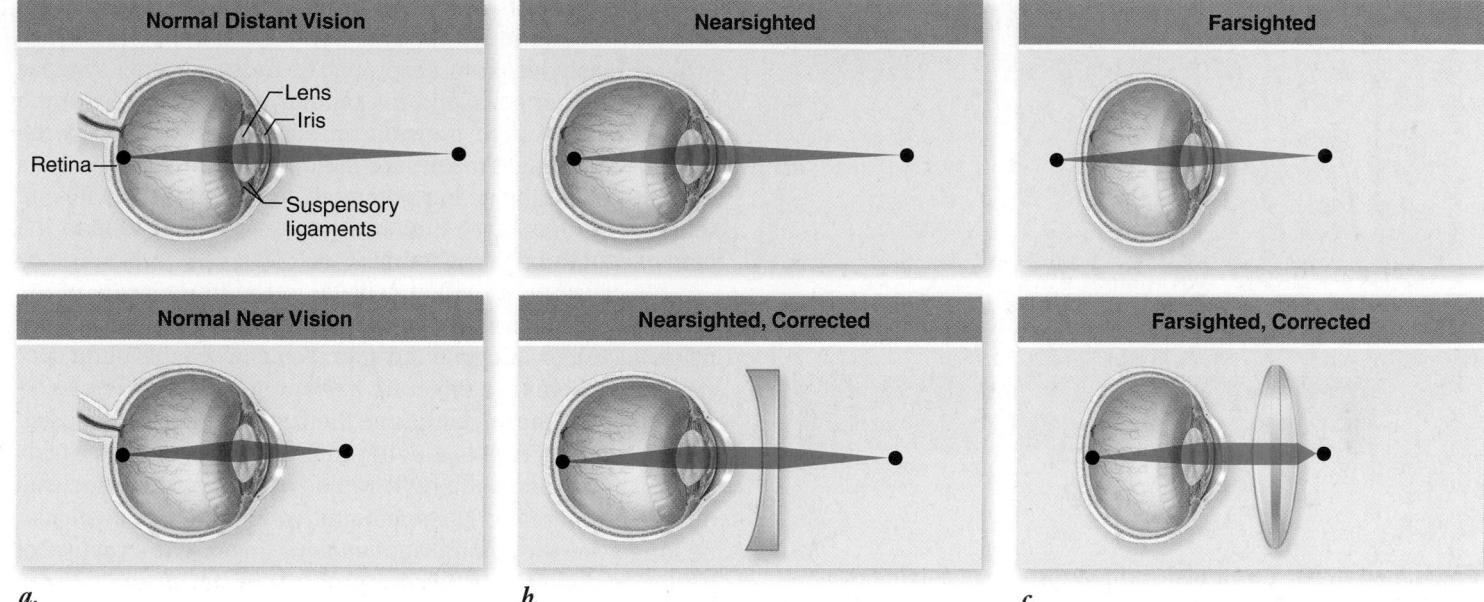

Figure 44.17 Focusing the human eye. *a.* In people with normal vision, the image remains focused on the retina in both near and far vision because of changes produced in the curvature of the lens. When a person with normal vision stands 20 feet or more from an object, the lens is in its least convex form, and the image is focused on the retina. *b.* In nearsighted people, the image comes to a focus in front of the retina, and the image thus appears blurred. *c.* In farsighted people, the focus of the image would be behind the retina because the distance from the lens to the retina is too short. Corrective lenses adjust the angle of the light as it enters the eye, focusing it on the retina.

their name from the shape of their outer segment, are responsible for black-and-white vision when the illumination is dim. In contrast, **cones** are responsible for high visual acuity (sharpness) and color vision; cones have a cone-shaped outer segment. Humans have about 100 million rods and 3 million cones in each retina. Most of the cones are located in the central region of the retina known as the **fovea,** where the eye forms its sharpest image. Rods are almost completely absent from the fovea.

Structure of rods and cones

Rods and cones have the same basic cellular structure. An inner segment rich in mitochondria contains numerous vesicles filled with neurotransmitter molecules. It is connected by a narrow stalk to the outer segment, which is packed with hundreds of flattened disks stacked on top of one another. The light-capturing molecules, or photopigments, are located on the membranes of these disks (see figure 44.18).

In rods, the photopigment is called **rhodopsin.** It consists of the protein opsin bound to a molecule of *cis*-retinal, which is produced from vitamin A. Vitamin A is derived from carotene, a photosynthetic pigment in plants.

The photopigments of cones, called **photopsins,** are structurally very similar to rhodopsin. Humans have three kinds of cones, each of which possesses a photopsin consisting of *cis*-retinal bound to a protein with a slightly different amino acid sequence. These differences shift the absorption maximum, the region of the electromagnetic spectrum that is

best absorbed by the pigment (figure 44.19). The absorption maximum of the *cis*-retinal in rhodopsin is 500 nanometers (nm); in contrast, the absorption maxima of the three kinds of cone photopsins are 420 nm (blue-absorbing), 530 nm (green-absorbing), and 560 nm (red-absorbing). These differences in the light-absorbing properties of the photopsins

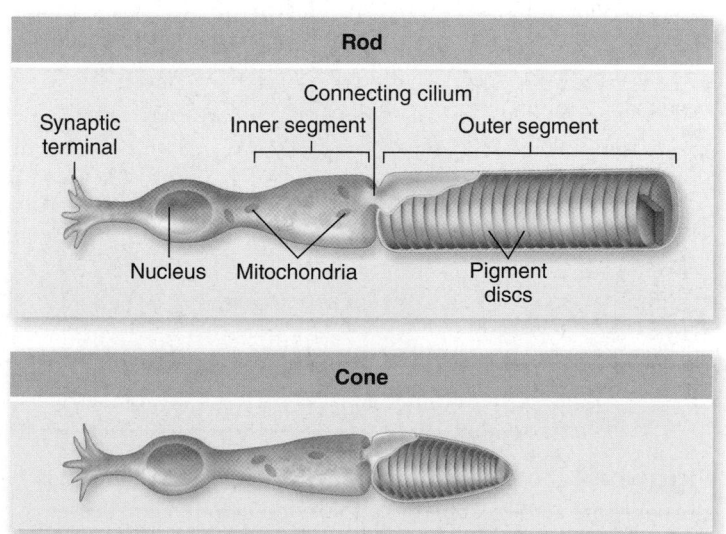

Figure 44.18 Rods and cones. The pigment-containing outer segment in each of these cells is separated from the rest of the cell by a partition through which there is only a narrow passage, the connecting cilium.

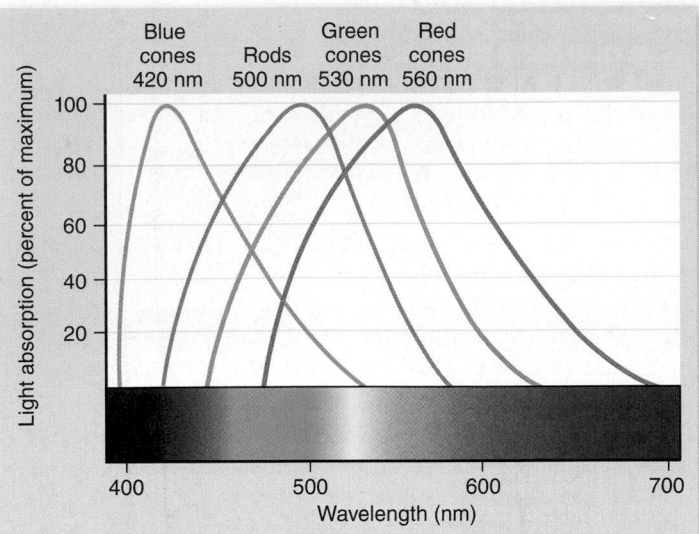

Figure 44.19 Color vision. The absorption maximum of *cis*-retinal in the rhodopsin of rods is 500 nm. However, the "blue cones" have their maximum light absorption at 420 nm; the "green cones" at 530 nm; and the "red cones" at 560 nm. The brain perceives all other colors from the combined activities of these three cones' systems.

are responsible for the different color sensitivities of the three kinds of cones, which are often referred to as simply blue, green, and red cones.

The **retina,** the inside surface of the eye, is made up of three layers of cells (figure 44.20): The layer closest to the external surface of the eyeball consists of the rods and cones; the next layer contains **bipolar cells;** and the layer closest to the

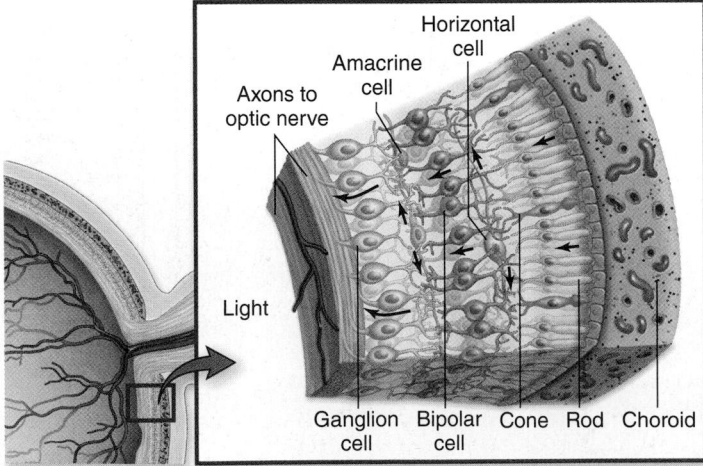

Figure 44.20 Structure of the retina. Note that the rods and cones are at the rear of the retina, not the front. Light passes through four other types of cells (ganglion, amacrine, bipolar, and horizontal) in the retina before it reaches the rods and cones. Once the photoreceptors are activated, they stimulate bipolar cells, which in turn stimulate ganglion cells. The flow of sensory information in the retina is thus opposite to the direction of light.

cavity of the eye is composed of **ganglion cells.** Thus, light must first pass through the ganglion cells and bipolar cells in order to reach the photoreceptors. The rods and cones synapse with the bipolar cells, and the bipolar cells synapse with the ganglion cells, which transmit impulses to the brain via the optic nerve. Ganglion cells are the only neurons of the retina capable of sending action potentials to the brain. The flow of sensory information in the retina is therefore opposite to the path of light through the retina.

Because the ganglion cells lie in the inner cavity of the eye, the optic nerve must intrude through the retina (see figure 44.16), creating a blind spot. You can see this blind spot yourself by holding a finger up in front of your face. Put a colored object on the finger tip, and then, with your left eye closed, focus on a point next to, but beyond, the fingertip. Now slowly move your finger to the right while keeping your eye focused on the distant point. At some point, you'll notice that you can no longer see the colored spot on your finger. The structure of the eye of mollusks avoids this problem by having the sensory neurons attach behind, rather than in front of, the retina (see figure 21.14).

The retina contains two additional types of neurons called horizontal cells and amacrine cells. Stimulation of horizontal cells by photoreceptors at the center of a spot of light on the retina can inhibit the response of photoreceptors peripheral to the center. This lateral inhibition enhances contrast and sharpens the image.

Most vertebrates, particularly those that are diurnal (active during the day), have color vision, as do many insects and some other invertebrates. Indeed, honeybees—as well as some birds, lizards, and other vertebrates (figure 44.21)—can see light in the near-ultraviolet range, which is invisible to the human eye. Color vision requires the presence of more than one photopigment in different receptor cells, but not all animals with color vision have the three-cone system characteristic of humans and other primates. Fish, turtles, and birds, for example, have four or five kinds of cones; the "extra" cones enable these animals to see near-ultraviolet light and to distinguish shades of colors that we cannot detect. On the other hand, many mammals, for example, squirrels and dogs, have only two types of cones and thus have more limited ability to distinguish different colors.

? Inquiry question How do additional kinds of cones affect color vision?

Sensory transduction in photoreceptors

The transduction of light energy into nerve impulses follows a sequence that is the opposite of the usual way that sensory stimuli are detected. In the dark, the photoreceptor cells release an inhibitory neurotransmitter that hyperpolarizes the bipolar neurons. This prevents the bipolar neurons from releasing excitatory neurotransmitter to the ganglion cells that signal to the brain.

The production of inhibitory neurotransmitter by photoreceptor cells is due to the presence of ligand-gated Na^+

Hypothesis: *Birds can see light in the ultraviolet range.*

Prediction: *Birds will respond to individuals differently depending on how much ultraviolet is detected in their feathers.*

Test: *Zebra finch feathers reflect a moderate amount of ultraviolet light. Female zebra finches were exposed to different males, some of which were behind a filter that screened out UV light, whereas others were behind a control filter that let the UV pass through.*

Result and Conclusion: *Females preferred to spend time near the UV-positive males. Not only can female zebra finches see light in the UV range, but they prefer males with UV in the feathers.*

Further Experiments: *What are two hypotheses about why females prefer UV-positive males? How would you test these hypotheses?*

Figure 44.21 Ultraviolet vision in birds. Humans cannot distinguish colors in the near ultraviolet range, whereas many animals can. This photograph was taken with a special film that shows ultraviolet patterns on a zebra finch (*Taeniopygia guttata*) that are not detectable by humans.

 Data analysis For animals that can detect near-ultraviolet light, where would the maximum light absorption peak for this receptor be in relation to the other photoreceptors in figure 44.19? (*Hint:* Refer to figure 8.4 to see the entire electromagnetic spectrum.)

channels. In the dark, many of these channels are open, allowing an influx of Na^+. This flow of Na^+ in the absence of light, called the dark current, depolarizes the membrane of photoreceptor cells. In this state, the cells produce inhibitory neurotransmitter that hyperpolarizes the membrane of bipolar cells.

The control of the dark current depends on the ligand for the Na^+ channels in the photoreceptor cells: the nucleotide cyclic guanosine monophosphate (cGMP). In the dark, the level of cGMP is high, and the channels are open. The system is made sensitive to light by the nature and structure of the photopigments. Photopigments in the eye are actually G protein–coupled receptor proteins that are activated by absorbing light.

In the light, the process works in the opposite way. When a photopigment absorbs light, *cis*-retinal isomerizes and dissociates from the receptor protein, opsin, in what is known as the bleaching reaction. As a result of this dissociation, the opsin receptor protein changes shape, activating its associated G protein. The activated G protein then activates its effector protein, the enzyme phosphodiesterase, which cleaves cGMP to GMP. The loss of cGMP causes the cGMP-gated Na^+ channels to close, reducing the dark current (figure 44.22). Each opsin is associated with over 100 regulatory G proteins, which, when activated, release subunits that activate hundreds of molecules of the phosphodiesterase enzyme. Each enzyme molecule can convert thousands of cGMP to GMP, closing the Na^+ channels at a rate of about 1000 per second and inhibiting the dark current.

The absorption of a single photon of light can block the entry of more than a million Na^+, without changing K^+ permeability—the photoreceptor becomes hyperpolarized and releases less inhibitory neurotransmitter. Freed from inhibition, the bipolar cells release excitatory neurotransmitter to the ganglion cells, which send impulses to the brain (figure 44.23).

Visual processing takes place in the cerebral cortex

Action potentials propagated along the axons of ganglion cells are relayed through structures called the **lateral geniculate nuclei** of the thalamus and projected to the occipital lobe of the cerebral cortex (see figure 44.23). There the brain interprets this information as light in a specific region of the eye's receptive field. The pattern of activity among the ganglion cells across the retina encodes a point-to-point map of the receptive field, allowing the retina and brain to image objects in visual space.

The frequency of impulses in each ganglion cell provides information about the light intensity at each point. At the same time, the relative activity of ganglion cells connected (through bipolar cells) with the three types of cones provides color information.

 Visual acuity

The relationship between receptors, bipolar cells, and ganglion cells varies in different parts of the retina. In the fovea, each cone makes a one-to-one connection with a bipolar cell, and each bipolar cell synapses with one ganglion cell. This point-to-point relationship is responsible for the high acuity of foveal vision.

Outside the fovea, many rods can converge on a single bipolar cell, and many bipolar cells can converge on a single ganglion cell. This convergence permits the summation of neural activity, making the area of the retina outside the fovea more sensitive to dim light than the fovea, but at the expense of acuity and color vision. This is why dim objects, such as faint stars at night, are best seen when you don't look directly at them. It has been said that we use the periphery of the eye as a detector, and the fovea as an inspector.

Figure 44.22 Signal transduction in the vertebrate eye. In the absence of light, cGMP keeps Na⁺ channels open causing a Na⁺ influx that leads to the release of inhibitory neurotransmitter. Light is absorbed by the retinal in rhodopsin, changing its structure. This causes rhodopsin to associate with a G protein. The activated G protein stimulates phosphodiesterase, which converts cGMP to GMP. Loss of cGMP closes Na⁺ channels and prevents release of inhibitory neurotransmitter, which causes bipolar cells to stimulate ganglion cells.

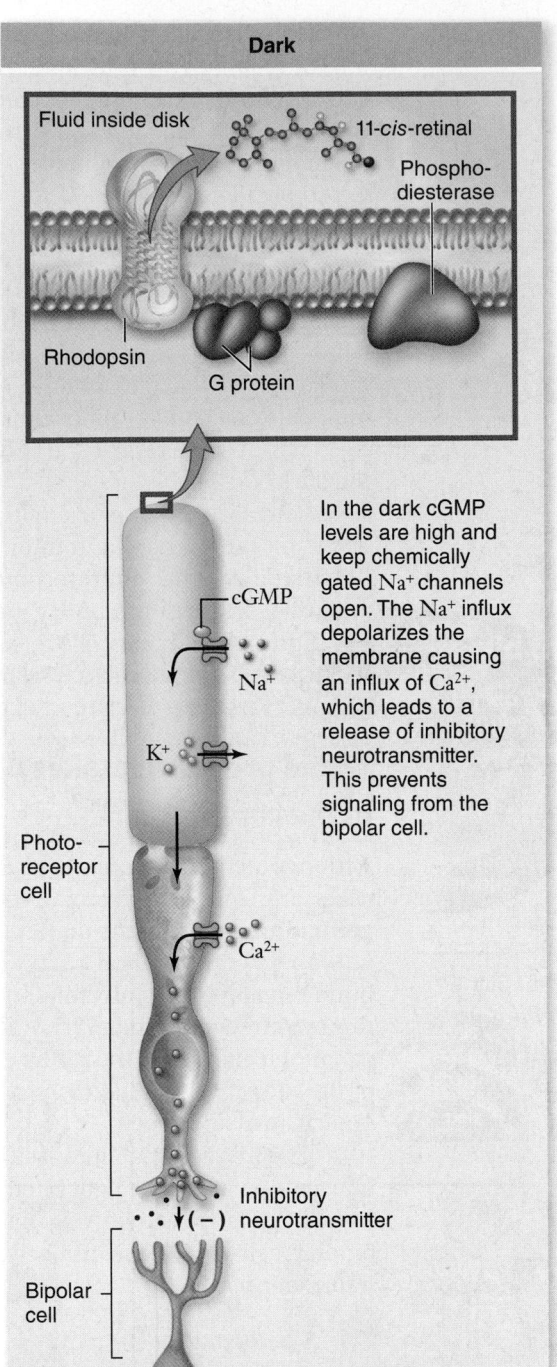

Dark

Fluid inside disk   11-*cis*-retinal   Phospho-diesterase

Rhodopsin   G protein

cGMP

Na⁺

K⁺

Ca²⁺

Photo-receptor cell

In the dark cGMP levels are high and keep chemically gated Na⁺ channels open. The Na⁺ influx depolarizes the membrane causing an influx of Ca²⁺, which leads to a release of inhibitory neurotransmitter. This prevents signaling from the bipolar cell.

Inhibitory neurotransmitter   (−)

Bipolar cell

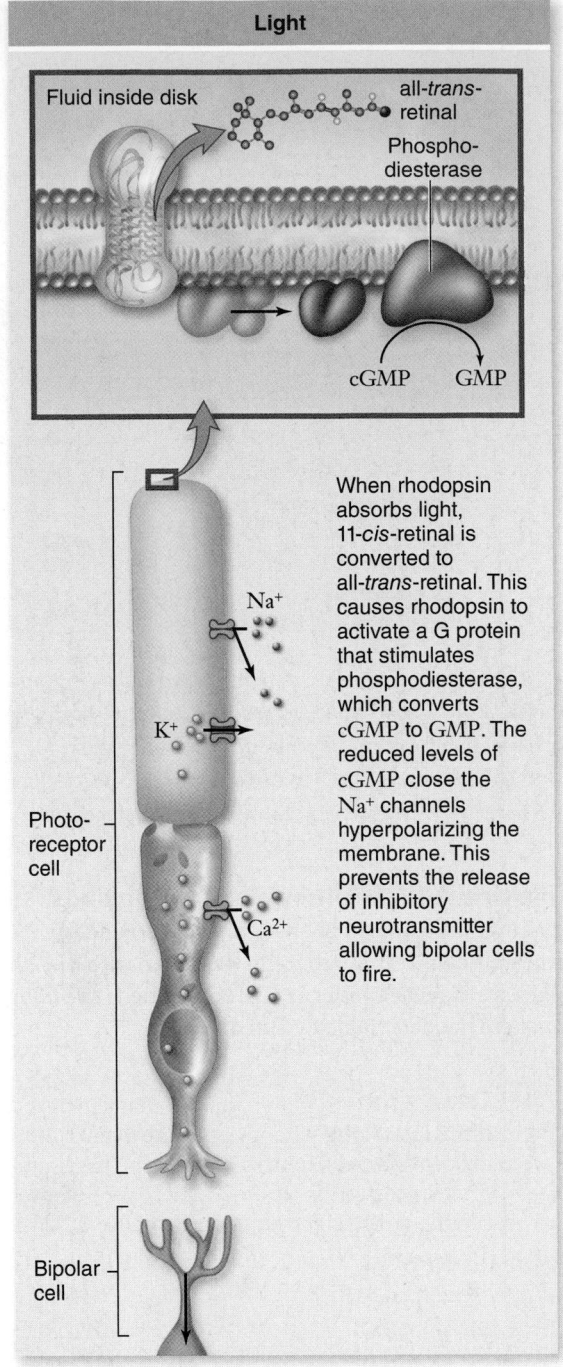

Light

Fluid inside disk   all-*trans*-retinal   Phospho-diesterase

cGMP   GMP

Na⁺

K⁺

Ca²⁺

Photo-receptor cell

When rhodopsin absorbs light, 11-*cis*-retinal is converted to all-*trans*-retinal. This causes rhodopsin to activate a G protein that stimulates phosphodiesterase, which converts cGMP to GMP. The reduced levels of cGMP close the Na⁺ channels hyperpolarizing the membrane. This prevents the release of inhibitory neurotransmitter allowing bipolar cells to fire.

Bipolar cell

Color blindness can result from an inherited lack of one or more types of cones. People with normal color vision are trichromats; that is they have all three cones. Those with only two types of cones are dichromats. For example, people with red-green color blindness may lack red cones and have difficulty distinguishing red from green. Color blindness resulting from absence of one type of cone is a sex-linked recessive trait (see chapter 13), and therefore it is most often exhibited in males. Red-green color blindness can also result from a shift in the sensitivity curve of the absorption spectrum for one type of cone, resulting in the different cone types being stimulated by the same electromagnetic wavelengths and causing the individual to be unable to distinguish between red and green.

Binocular vision

Primates (including humans) and most predators have two eyes, one located on each side of the face. When both eyes are trained on the same object, the image that each eye sees is slightly different because the views have a slightly different angle. This slight displacement of the images (an effect called parallax) permits **binocular vision,** the ability to perceive three-dimensional images and to sense depth. Having eyes facing forward maximizes the field of overlap in which this stereoscopic vision occurs.

In contrast, prey animals generally have eyes located to the sides of the head, preventing binocular vision but

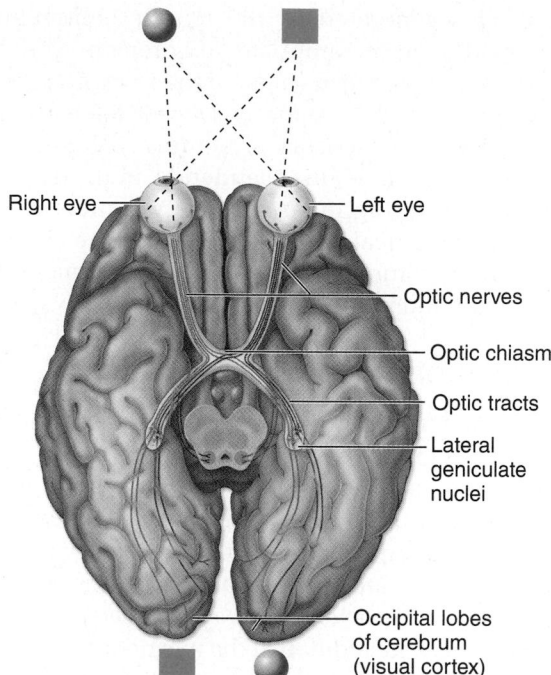

Figure 44.23 The pathway of visual information. Action potentials in the optic nerves are relayed from the retina to the lateral geniculate nuclei, and from there to the visual cortex of the occipital lobes. Note that half the optic nerves (the medial fibers arising from the inner portion of the retinas) cross to the other side at the optic chiasm, so that each hemisphere of the cerebrum receives input from both eyes.

enlarging the overall receptive field. It seems that natural selection has favored the detection of potential predators over depth perception in many prey species. The eyes of the American woodcock *(Scolopax minor)*, for example, are located at exactly opposite sides of the bird's skull so that it has a 360° field of view without turning its head.

Most birds have laterally placed eyes and, as an adaptation, have two foveas in each retina. One fovea provides sharp frontal vision, like the single fovea in the retina of mammals, and the other fovea provides sharper lateral vision.

Learning Outcomes Review 44.5

Many invertebrate groups have eyespots that detect light without forming images. Annelids, mollusks, arthropods, and chordates have independently evolved image-forming eyes. The vertebrate eye admits light through a pupil and then focuses it with an adjustable lens onto the retina, which contains photoreceptors. Photoreceptor rods and cones contain the photopigment *cis*-retinal, which indirectly activates bipolar neurons and then ganglion cells. The latter then transmit action potentials that ultimately reach the occipital lobe of the brain.

- *Can an individual with red-green color blindness learn to distinguish these two colors? Why or why not?*

Learning Outcomes

1. List examples of uncommon special senses.
2. Explain how ampullae of Lorenzini work.

Vision is the primary sense used by all vertebrates that live in a light-filled environment, but visible light is by no means the only part of the electromagnetic spectrum that vertebrates use to sense their environment.

Some snakes have receptors capable of sensing infrared radiation

Electromagnetic radiation with wavelengths longer than those of visible light is too low in energy to be detected by photoreceptors. Radiation from this infrared portion of the spectrum is what we normally think of as radiant heat.

Heat is an extremely poor environmental stimulus in water because water readily absorbs heat. Air, in contrast, has a low thermal capacity, so heat in air is a potentially useful stimulus. The only vertebrates known to have the ability to sense infrared radiation, however, are several types of snakes.

One type, the pit vipers, possess a pair of heat-detecting **pit organs** located on either side of the head between the eye and the nostril (figure 44.24). Each pit organ is composed of

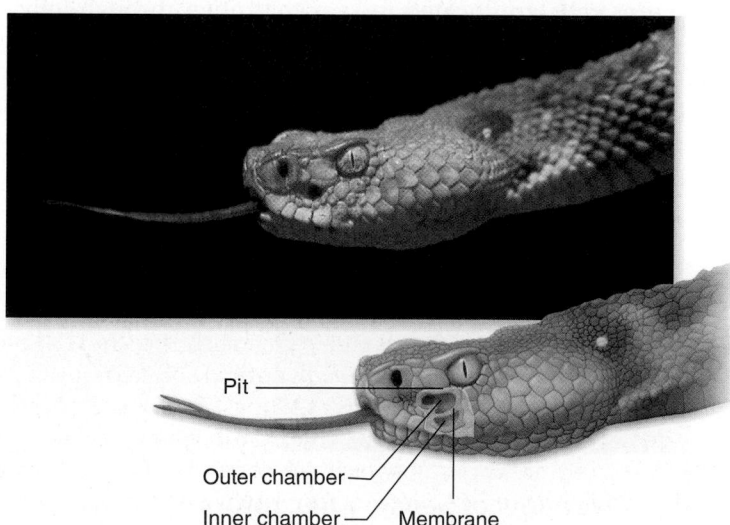

Figure 44.24 "Seeing" heat. The depression between the nostril and the eye of this rattlesnake opens into the pit organ. In the cutaway portion of the diagram, you can see that the organ is composed of two chambers separated by a membrane. Snakes known as pit vipers have the ability to sense infrared radiation (heat).

two chambers separated by a membrane. The infrared radiation falls on the membrane and warms it. Thermal receptors on the membrane are stimulated. The nature of these receptors is not known; they probably consist of temperature-sensitive neurons innervating the two chambers.

The paired pit organs appear to provide stereoscopic information, in much the same way that two eyes do. In fact, the nerves from the pits are connected to the optic tectum, the same part of the brain that controls vision; recent research suggests that information from the pits and from the eyes are overlain on each other, allowing snakes to combine visual and infrared thermal data. In fact, the pits are designed like a pinhole camera, and to some extent can focus a thermal image!

As a result, these exteroceptors are extraordinarily sensitive. Blind pit vipers can strike as accurately as a normal snake, and snakes deprived of their senses of sight and smell can accurately strike a target only 0.2° warmer than the background. Many pit vipers hunt endothermic prey at night, so the value of these capabilities is obvious.

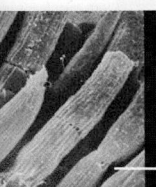

Some vertebrates can sense electrical currents

Although air does not readily conduct an electrical current, water is a good conductor. All aquatic animals generate electrical currents from contractions of their muscles. A number of different groups of fishes can detect these electrical currents. The so-called electrical fish even have the ability to produce electrical discharges from specialized electrical organs. Electrical fish use these weak discharges to locate their prey and mates and to construct a three-dimensional image of their environment, even in murky water.

The elasmobranchs (sharks, rays, and skates) have electroreceptors called the **ampullae of Lorenzini.** The receptor cells are located in sacs that open through jelly-filled canals to pores on the body surface. The jelly is a very good conductor, so a negative charge in the opening of the canal can depolarize the receptor at the base, causing the release of neurotransmitter and increased activity of sensory neurons. This allows sharks, for example, to detect the

electrical fields generated by the muscle contractions of their prey. Although the ampullae of Lorenzini were lost in the evolution of teleost fish (most of the bony fish), electroreception reappeared in some groups of teleost fish that developed analogous sensory structures. Electroreceptors evolved yet another time, independently, in the duck-billed platypus, an egg-laying mammal. The receptors in its bill can detect the electrical currents created by the contracting muscles of shrimp and fish, enabling the mammal to detect its prey at night and in muddy water.

Some organisms detect magnetic fields

Eels, sharks, bees, and many birds appear to navigate along the magnetic field lines of the Earth. Even some bacteria use such forces to orient themselves.

Birds kept in dark cages, with no visual cues to guide them, peck and attempt to move in the direction in which they would normally migrate at the appropriate time of the year. They do not do so, however, if the cage is shielded from magnetic fields by steel. In addition, if the magnetic field of a blind cage is deflected 120° clockwise by an artificial magnet, a bird that normally orients to the north will orient toward the east-southeast. The nature of magnetic receptors in these vertebrates is the subject of much current research; it appears that birds have two systems, one using specialized pigments in the visual system and the other involving elements in the nervous system, both of which are magnetically sensitive and which work together to allow navigation by magnetic fields.

Learning Outcomes Review 44.6

Pit vipers can detect infrared radiation (heat). Many aquatic vertebrates can locate prey and perceive environmental contours by means of electroreceptors. The ampullae of Lorenzini, electroreceptors found in sharks and their relatives, contain a highly conductive jelly that triggers sensory neurons. Magnetic receptors may aid in bird migration.

- **Would a heat-sensing organ be useful for hunting ectothermic prey?**

Chapter Review

44.1 Overview of Sensory Receptors

Sensory receptors detect both external and internal stimuli.

Exteroceptors sense stimuli from the external environment, whereas interoceptors sense stimuli from the internal environment.

Receptors can be grouped into three categories.

Receptors differ with respect to the environmental stimulus to which they respond: mechanoreceptors, chemoreceptors, and electromagnetic receptors.

Sensory information is conveyed in a four-step process.

Once detected, sensory information is conveyed in four steps: stimulation, transduction, transmission, and interpretation.

Sensory transduction involves gated ion channels.

Sensory transduction produces a graded receptor potential. A single potential or a sum of potentials may exceed a threshold to produce an action potential (figure 44.2). A logarithmic relationship exists between stimulus intensity and action potential frequency.

44.2 Mechanoreceptors: Touch and Pressure

Pain receptors alert the body to damage or potential damage.

Nociceptors are free nerve endings located in the skin that respond to damaging stimuli, which is perceived as pain. Extreme temperatures can affect transient receptor potential (TRP) ion channels and cause depolarization by inflow of Na^+ and Ca^{2+}.

Thermoreceptors detect changes in heat energy.

Thermoreceptors are naked dendritic endings of sensory neurons that also contain TRP ion channels and respond to cold or heat.

Different receptors detect touch, depending on intensity.

Various receptors in the skin respond to mechanical distortion of the membrane to convey touch (figure 44.3).

Muscle length and tension are monitored by proprioceptors.

Proprioceptors provide information about the relative position or movement of body parts and the degree of muscle stretching.

Baroreceptors detect blood pressure.

44.3 Hearing, Vibration, and Detection of Body Position

Hearing, the detection of sound or pressure waves, works best in water and provides directional information.

The lateral line system in fish detects low-frequency vibrations (figure 44.5).

Ear structure is specialized to detect vibration.

The outer ear of terrestrial vertebrates channels sound to the eardrum (tympanic membrane) (figure 44.6). Vibrations are transferred through middle ear bones to the oval window and into the cochlea, where the organ of Corti transduces them.

Transduction occurs in the cochlea.

The basilar membrane of the cochlea consists of fibers that respond to different frequencies of sound (figure 44.8).

Some vertebrates have the ability to navigate by sound.

Echolocation allows bats, whales, and other species to navigate by sound.

Body position and movement are detected by systems associated with hearing systems.

Body position is detected by statocysts, ciliated hair cells embedded in a gelatinous matrix containing statoliths (figure 44.9). Body movement is detected by hair cells located in the saccule and utricle (figure 44.10).

44.4 Chemoreceptors: Taste, Smell, and pH

Taste detects and analyzes potential food.

Taste buds are collections of chemosensitive epithelial cells located on papillae (figure 44.11). Tastes are broken down into five categories: sweet, sour, salty, bitter, and umami.

Smell can identify a vast number of complex molecules.

Smell, or olfaction, involves chemoreceptors located in the upper portion of the nasal passages (figure 44.13). Their axons connect directly to the cerebral cortex.

Internal chemoreceptors detect pH and other characteristics.

Internal chemoreceptors of the aorta detect changes in blood pH, and central chemoreceptors in the medulla oblongata are sensitive to the pH of the cerebrospinal fluid.

44.5 Vision

Vision senses light and light changes at a distance.

Four phyla—annelids, mollusks, arthropods, and chordates—have independently evolved image-forming eyes (figure 44.15).

In the vertebrate eye, light enters through the pupil, with intensity controlled by the iris. The lens, controlled by the ciliary muscle, focuses the light on the retina (figure 44.16).

Vertebrate photoreceptors are rod cells and cone cells.

Rods detect black and white; cones are necessary for visual acuity and color vision (figure 44.18).

In the retina, photoreceptors synapse with bipolar cells, which in turn synapse with ganglion cells; the ganglion cells send action potentials to the brain (figure 44.20).

Visual processing takes place in the cerebral cortex (figure 44.23).

In the fovea, a region of the retina responsible for high acuity, each cone cell is connected to a single bipolar cell/ganglion cell, unlike in areas outside the fovea.

Primates and most predators have binocular vision—images from each eye overlap to produce a three-dimensional image.

44.6 The Diversity of Sensory Experiences

Some snakes have receptors capable of sensing infrared radiation.

The pit organ of pit vipers detects heat.

Some vertebrates can sense electrical currents.

Electroreceptors in elasmobranchs and the duck-billed platypuses can detect electrical currents.

Some organisms detect magnetic fields.

Many organisms appear to navigate along magnetic field lines, most likely using magnetically-sensitive elements in the visual and nervous systems.

Review Questions

UNDERSTAND

1. Which of these is NOT a method by which sensory receptors receive information about the internal or external environment?
   a. Changes in pressure
   b. Light or heat changes
   c. Changes in molecular concentration
   d. All of these are used by sensory receptors.

2. Which of the following correctly lists the steps of perception?
   a. Interpretation, stimulation, transduction, transmission
   b. Stimulation, transduction, transmission, interpretation
   c. Interpretation, transduction, stimulation, transmission
   d. Transduction, interpretation, stimulation, transmission

3. All sensory receptors are able to initiate nerve impulses by opening or closing
   a. voltage-gated ion channels.
   b. exteroceptors.
   c. interoceptors.
   d. stimulus-gated ion channels.

4. In the fairy tale, Sleeping Beauty fell asleep after pricking her finger. What kind of receptor responds to that kind of painful stimulus?
   a. Mechanoreceptor   c. Thermoreceptor
   b. Nociceptor   d. Touch receptor

5. The ear detects sound by the movement of
   a. the basilar membrane.
   b. the tectorial membrane.
   c. the Eustachian tube.
   d. fluid in the semicircular canals.

6. Hair cells in the vestibular apparatus of terrestrial vertebrates
   a. measure temperature changes within the body.
   b. sense sound in very low range of hearing.
   c. provide a sense of acceleration and balance.
   d. measure changes in blood pressure.

7. _____ is the photopigment contained within both rods and cones of the eye.
   a. Carotene   c. Photochrome
   b. *Cis*-retinal   d. Chlorophyll

8. Which of the following is NOT a method used by vertebrates to gather information about their environment?
   a. Infrared radiation
   b. Magnetic fields
   c. Electrical currents
   d. All of these are methods used for sensory reception.

9. The lobe of the brain that recognizes and interprets visual information is the
   a. occipital lobe.   c. parietal lobe.
   b. frontal lobe.   d. temporal lobe.

APPLY

1. What do the sensory systems of annelids, mollusks, arthropods, and chordates have in common?
   a. They all use the same stimuli for taste.
   b. They all use neurons to detect vibration.
   c. They all have image-forming eyes that evolved independently.
   d. They all use chemoreceptors in their skin to detect food.

2. Animals can more easily tell the direction of a visual signal than an auditory signal because
   a. light travels in straight lines.
   b. the wind provides too much background noise.
   c. sound travels faster underwater.
   d. eyes are more sensitive than ears.

3. Some birds have broader color perception than humans, who are trichromatic. How does a bird that is pentachromatic differ from a human?
   a. They have different photoreceptor cells.
   b. They can see four different colors.
   c. They have photoreceptors that detect five different wavelengths of light.
   d. They can't see in the dark, but have the same daylight vision as humans.

4. The ability of some insects, birds, and lizards to see ultraviolet light is
   a. a result of a common diet eaten by those species.
   b. an example of convergent evolution in cone cell sensitivity.
   c. the ancestral state inherited from flatworms.
   d. an adaptation for nocturnal activity.

SYNTHESIZE

1. When blood pH falls too low, a potentially fatal condition known as acidosis results. Among the variety of responses to this condition, the body changes the breathing rate. How does the body sense this change? How does the breathing rate change? How does this increase pH?

2. The function of the vertebrate eye is unusual compared with other processes found within the body. For example, the direction in which sensory information flows is actually opposite to the path that light takes through the retina. Explain the sequence of events involved in the movement of light and information through the structures of the eye, and explain why they move in opposite directions. Consider, also, how this sequence of events compares to the functioning of the mollusk eye.

3. How would the otolith organs of an astronaut respond to zero gravity? Would the astronaut still have a subjective impression of motion? Would the semicircular canals detect angular acceleration equally well at zero gravity?

ONLINE RESOURCE

www.ravenbiology.com

Understand, Apply, and Synthesize—enhance your study with animations that bring concepts to life and practice tests to assess your understanding. Your instructor may also recommend the interactive eBook, individualized learning tools, and more.

Chapter **45**

The Endocrine System

Chapter Contents

Introduction

Diabetes is a disease in which well-fed people appear to starve to death. The disease was known to Roman and Greek physicians, who described a "melting away of flesh" coupled with excessive urine production "like the opening of aqueducts." Until 1922, the diagnosis of diabetes in children was effectively a death sentence. In that year, Frederick Banting and Charles Best extracted the molecule insulin from the pancreas. Injections of insulin into the bloodstream dramatically reversed the symptoms of the disease. This served as an impressive confirmation of a new concept: that certain internal organs produced powerful regulatory chemicals that were distributed via the blood.

We now know that the tissues and organs of the vertebrate body cooperate to maintain homeostasis through the actions of many regulatory mechanisms. Two systems, however, are devoted exclusively to the regulation of the body organs: the nervous system and the endocrine system. Both release regulatory molecules that control the body organs by binding to receptor proteins on or in the cells of those organs. In this chapter, we examine the regulatory molecules of the endocrine system, the cells and glands that produce them, and how they function to regulate the body's activities.

Regulation of Body Processes by Chemical Messengers

There are four mechanisms of cell communication: direct contact, synaptic signaling, endocrine signaling, and paracrine signaling. In chapter 19 we considered the cellular and molecular details of cell signaling. Here we will see how this form of communication is used to control processes at the level of the whole animal.

As discussed in chapter 43, the axons of neurons secrete chemical messengers called neurotransmitters into the synaptic cleft. These chemicals diffuse only a short distance to the postsynaptic membrane, where they bind to their receptor proteins and stimulate the postsynaptic cell. Synaptic transmission generally affects only the postsynaptic cell that receives the neurotransmitter.

A *hormone,* in contrast, is a regulatory chemical that is secreted into extracellular fluid and carried by the blood, and can therefore act at a distance from its source. Organs that are specialized to secrete hormones are called *endocrine glands,* but some organs, such as the liver and the kidney, can produce hormones in addition to performing other functions. The organs and tissues that produce hormones are collectively called the **endocrine system.**

The blood carries hormones to every cell in the body, but only target cells with the appropriate receptor for a given hormone can respond to it. Hormone receptor proteins function in a similar manner to neurotransmitter receptors. The receptor proteins specifically bind the hormone and activate signal transduction pathways that produce a response to the hormone. The highly specific interaction between hormones and their receptors enable hormones to be active at remarkably small concentrations. It is not unusual to find hormones circulating in the blood at concentrations of 10^{-8} to 10^{-10} M. In addition to the chemical messengers released as neurotransmitters and as hormones, other molecules are released and act within an organ on nearby cells as local regulators. These chemicals are termed **paracrine regulators.** They act in a way similar to endocrine hormones, but they do not travel through the blood to reach their target. This allows cells of an organ to regulate one another.

Cells can also release signaling molecules that affect their own behavior, or *autocrine signaling*. This is common in the immune system, and is also seen in cancer cells that may release growth factors that stimulate their own growth.

Chemical communication is not limited to cells within an organism. *Pheromones* are chemicals released into the environment to communicate among individuals of a single species. These aid in communication between animals and may alter the behavior or physiology of the receiver, but are not involved in the normal metabolic regulation of an animal.

Figure 45.1 compares the different types of chemical messengers used for internal regulation.

Some molecules act as both circulating hormones and neurotransmitters

Blood delivery of hormones enables endocrine glands to coordinate the activity of large numbers of target cells distributed throughout the body, but that may not be the only role for these molecules. A molecule produced by an endocrine gland and used as a hormone may also be produced and used as a neurotransmitter by neurons. The hormone norepinephrine, for example, is secreted into the blood by the adrenal glands, but it is also released as a neurotransmitter by sympathetic nerve endings. Norepinephrine acts as a hormone to coordinate the activity of the heart, liver, and blood vessels during response to stress.

Neurons can also secrete a class of hormones called **neurohormones** that are carried by blood. The neurohormone antidiuretic hormone, for example, is secreted by neurons in the brain. Some specialized regions of the brain contain not only neurotransmitting neurons, but also clusters of

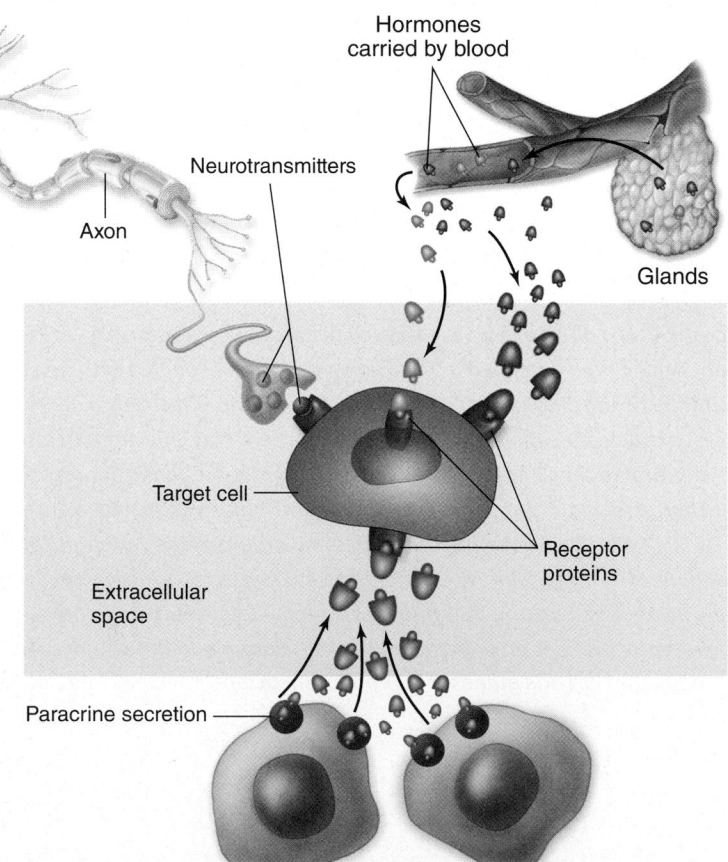

Figure 45.1 Different types of chemical messengers. The functions of organs are influenced by neural, paracrine, and endocrine regulators. Each type of chemical regulator binds to specific receptor proteins on the surface or within the cells of target organs.

neurons producing neurohormones. In this way, neurons can deliver chemical messages beyond the nervous system itself.

The secretory activity of many endocrine glands is controlled by the nervous system. As you will see, the hypothalamus controls the hormonal secretions of the anterior-pituitary gland, and produces the hormones of the posterior pituitary.

The secretion of a number of hormones, however, can be independent of neural control. For example, the release of insulin by the pancreas and aldosterone by the adrenal cortex is stimulated by increases in the blood concentrations of glucose and potassium (K^+), respectively.

Endocrine glands produce three chemical classes of hormones

The endocrine system (figure 45.2) includes all of the organs that secrete hormones—the thyroid gland, pituitary gland, adrenal glands, and so on (table 45.1). Cells in these organs secrete hormones into extracellular fluid, where it diffuses into

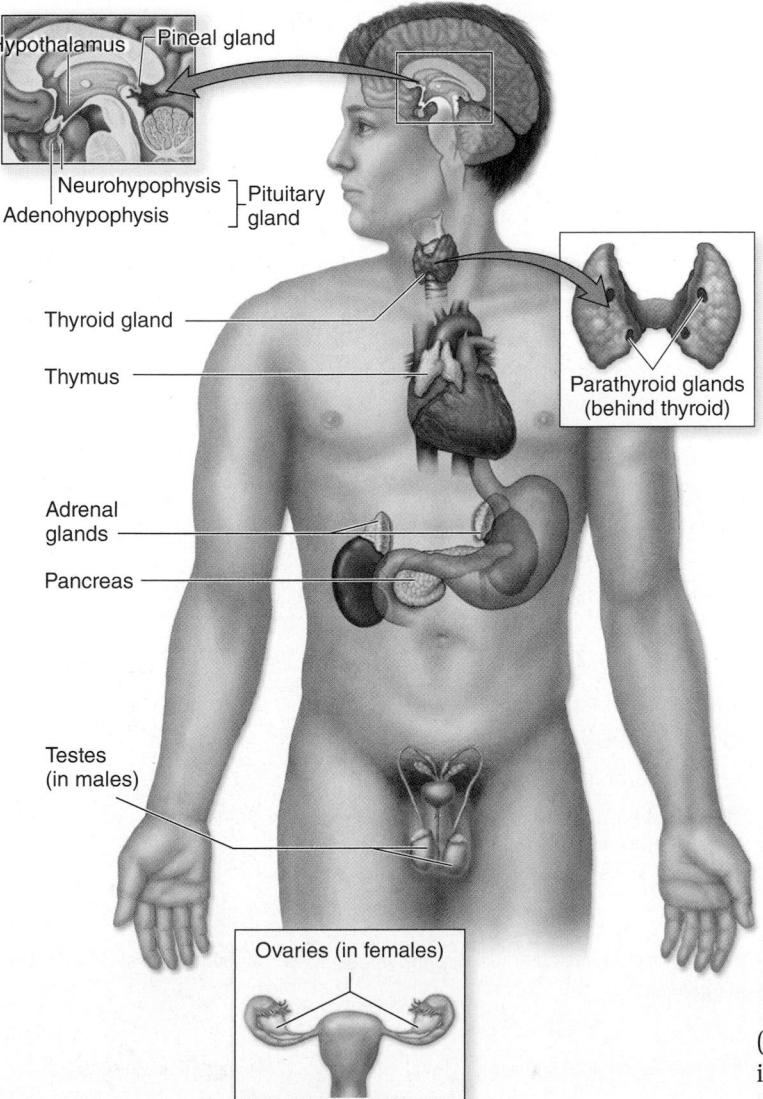

Figure 45.2 The human endocrine system. The major endocrine glands are shown, but many other organs secrete hormones in addition to their primary functions.

surrounding blood capillaries. For this reason, hormones are referred to as endocrine secretions. In contrast, cells of some glands excrete their products into a duct to outside the body, or into the gut. For example, the pancreas excretes hydrolytic enzymes into the lumen of the small intestine. These glands are termed exocrine glands.

Molecules that function as hormones must exhibit two basic characteristics. First, they must be sufficiently complex to convey regulatory information to their targets. Simple molecules such as carbon dioxide, or ions such as Ca^{2+}, do not function as hormones. Second, hormones must be adequately stable to resist destruction prior to reaching their target cells. Three primary chemical categories of molecules meet these requirements.

1. **Peptides and proteins** are composed of chains of amino acids. Some important examples of peptide hormones include antidiuretic hormone (9 amino acids), insulin (51 amino acids), and growth hormone (191 amino acids). These hormones are encoded in DNA and produced by the same cellular machinery responsible for transcription and translation of other peptide molecules. The most complex are glycoproteins composed of two peptide chains with attached carbohydrates. Examples include thyroid-stimulating hormone and luteinizing hormone.

2. **Amino acid derivatives** are hormones manufactured by enzymatic modification of specific amino acids; this group comprises the biogenic amines discussed in chapter 43. They include hormones secreted by the adrenal medulla (the inner portion of the adrenal gland), thyroid, and pineal glands. Those secreted by the adrenal medulla are derived from tyrosine. Known as **catecholamines,** they include epinephrine (adrenaline) and norepinephrine (noradrenaline). Other hormones derived from tyrosine are the **thyroid hormones,** secreted by the thyroid gland. The pineal gland secretes a different amine hormone, **melatonin,** derived from tryptophan.

3. **Steroids** are lipids manufactured by enzymatic modifications of cholesterol. They include the hormones testosterone, estradiol, progesterone, aldosterone, and cortisol. Steroid hormones can be subdivided into sex steroids, secreted by the testes, ovaries, placenta, and adrenal cortex, and corticosteroids (mineralocorticoids and cortisol), secreted only by the adrenal cortex.

Hormones can be categorized as lipophilic or hydrophilic

The manner in which hormones are transported and interact with their targets differs depending on their chemical nature. Hormones may be categorized as lipophilic (nonpolar), which are fat-soluble, or hydrophilic (polar), which are water-soluble. The lipophilic hormones include the steroid hormones and thyroid hormones. Most other hormones are hydrophilic.

This distinction is important in understanding how these hormones regulate their target cells. Hydrophilic hormones are freely soluble in blood, but cannot pass through the membrane of

TABLE 45.1

TABLE 45.1 — Principal Mammalian Endocrine Glands and Their Hormones*

Endocrine Gland and Hormone	Target Tissue	Principal Actions	Chemical Nature
Hypothalamus			
Releasing hormones	Adenohypophysis	Activate release of adenohypophyseal hormones	Peptides
Inhibiting hormones	Adenohypophysis	Inhibit release of adenohypophyseal hormones	Peptides (except prolactin-inhibiting factor, which is dopamine)
Neurohypophysis (Posterior-pituitary gland)			
Antidiuretic hormone (ADH)	Kidneys	Conserves water by stimulating its reabsorption from urine	Peptide (9 amino acids)
Oxytocin (OT)	Uterus	Stimulates contraction	Peptide (9 amino acids)
	Mammary glands	Stimulates milk ejection	
Adenohypophysis (Anterior-pituitary gland)			
Adrenocorticotropic hormone (ACTH)	Adrenal cortex	Stimulates secretion of adrenal cortical hormones such as cortisol	Peptide (39 amino acids)
Melanocyte-stimulating hormone (MSH)	Skin	Stimulates color change in reptiles and amphibians; various functions in mammals	Peptide (two forms; 13 and 22 amino acids)
Growth hormone (GH)	Many organs	Stimulates growth by promoting bone growth, protein synthesis, and fat breakdown	Protein
Prolactin (PRL)	Mammary glands	Stimulates milk production	Protein
Thyroid-stimulating hormone (TSH)	Thyroid gland	Stimulates thyroxine secretion	Glycoprotein
Luteinizing hormone (LH)	Gonads	Stimulates ovulation and corpus luteum formation in females; stimulates secretion of testosterone in males	Glycoprotein
Follicle-stimulating hormone (FSH)	Gonads	Stimulates spermatogenesis in males; stimulates development of ovarian follicles in females	Glycoprotein
Thyroid Gland			
Thyroid hormones (thyroxine and triiodothyronine)	Most cells	Stimulates metabolic rate; essential to normal growth and development	Amino acid derivative (iodinated)
Calcitonin	Bone	Inhibits loss of calcium from bone	Peptide (32 amino acids)

*These are hormones released from endocrine glands. Hormones are released from organs that have additional, nonendocrine functions, such as the liver, kidney, and intestine.

Endocrine Gland and Hormone	Target Tissue		Principal Actions	Chemical Nature
Parathyroid Glands				
Parathyroid hormone (PTH)	Bone, kidneys, digestive tract		Raises blood calcium level by stimulating bone breakdown; stimulates calcium reabsorption in kidneys; activates vitamin D	Peptide (34 amino acids)
Adrenal Medulla				
Epinephrine (adrenaline) and norepinephrine (noradrenaline)	Smooth muscle, cardiac muscle, blood vessels		Initiates stress responses; raises heart rate, blood pressure, metabolic rate; dilates blood vessels; mobilizes fat; raises blood glucose level	Amino acid derivatives
Adrenal Cortex				
Glucocorticoids (e.g., cortisol)	Many organs		Adaptation to long-term stress; raises blood glucose level; mobilizes fat	Steroid
Mineralocorticoids (e.g., aldosterone)	Kidney tubules		Maintains proper balance of Na^+ and K^+ in blood	Steroid
Pancreas				
Insulin	Liver, skeletal muscles, adipose tissue		Lowers blood glucose level; stimulates glycogen, fat, protein synthesis	Peptide (51 amino acids)
Glucagon	Liver, adipose tissue		Raises blood glucose level; stimulates breakdown of glycogen in liver	Peptide (29 amino acids)
Ovary				
Estradiol	General		Stimulates development of female secondary sex characteristics	Steroid
	Female reproductive structures		Stimulates growth of sex organs at puberty and monthly preparation of uterus for pregnancy	
Progesterone	Uterus		Completes preparation for pregnancy	Steroid
	Mammary glands		Stimulates development	
Testis				
Testosterone	Many organs		Stimulates development of secondary sex characteristics in males and growth spurt at puberty	Steroid
	Male reproductive structures		Stimulates development of sex organs; stimulates spermatogenesis	
Pineal Gland				
Melatonin	Gonads, brain, pigment cells		Regulates biological rhythms	Amino acid derivative

Figure 45.3
The life of hormones.
Endocrine glands produce both hydrophilic and lipophilic hormones, which are transported to targets through the blood. Lipophilic hormones bind to transport proteins that make them soluble in blood. Target cells have membrane receptors for hydrophilic hormones, and intracellular receptors for lipophilic hormones. Hormones are eventually destroyed by their target cells or cleared from the blood by the liver or the kidney.

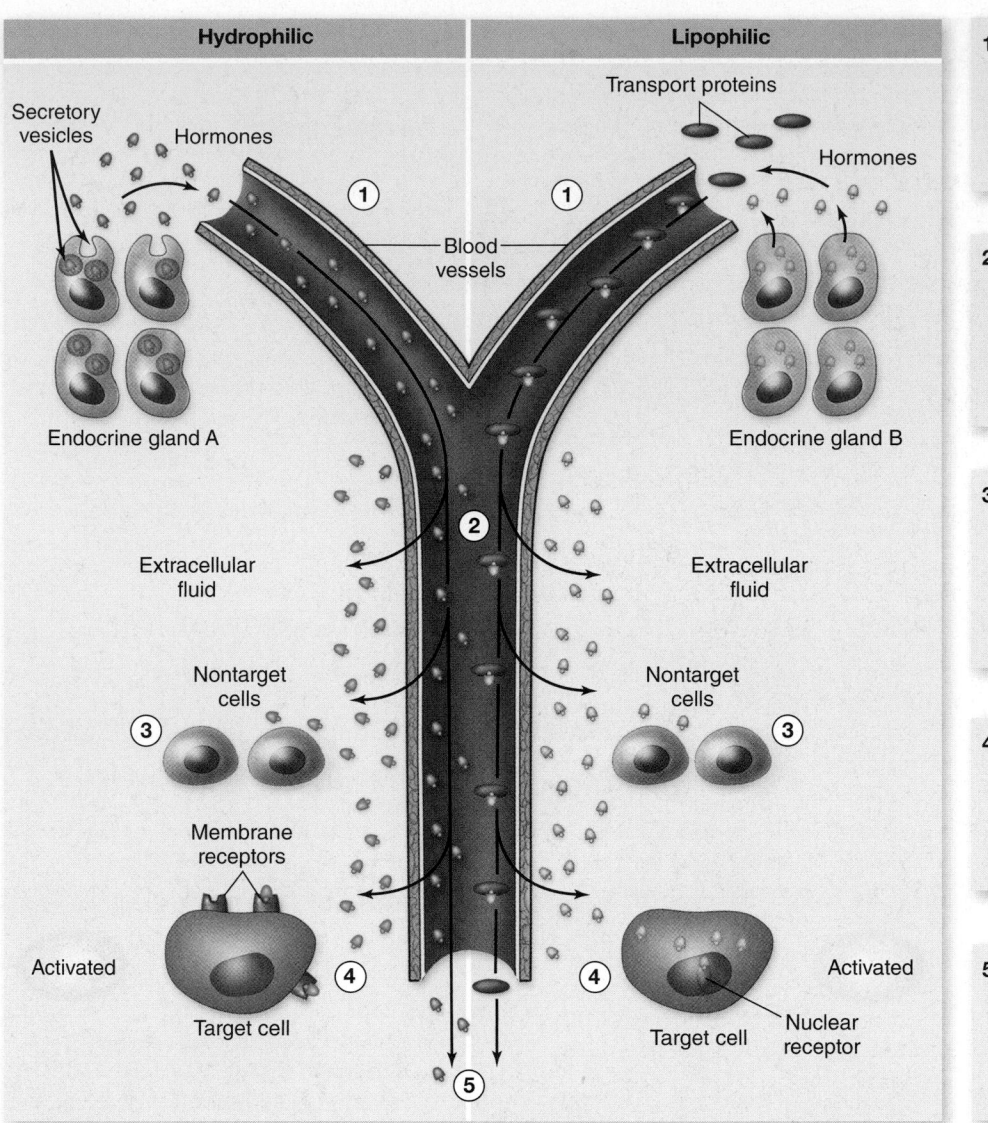

1. Hormones secreted into extracellular fluid and diffuse into bloodstream.

2. Hormones distributed by blood to all cells. Diffuse from blood to extracellular fluid.

3. Nontarget cells lack receptors, and cell stimulation does not occur.

4. Target cells possess receptors, and are activated by hormones.

5. Unused, deactivated hormones are removed by the liver and kidney.

target cells. They must therefore activate their receptors from outside the cell membrane. In contrast, lipophilic hormones travel in the blood attached to transport proteins (figure 45.3). Their lipid solubility enables them to cross cell membranes and bind to intracellular receptors.

Both types of hormones are eventually destroyed or otherwise deactivated after their use, eventually being excreted in bile or urine. However, hydrophilic hormones are deactivated more rapidly than lipophilic hormones. Hydrophilic hormones tend to act over relatively brief periods of time (minutes to hours), whereas lipophilic hormones generally are active over prolonged periods, such as days to weeks.

Paracrine regulators exert powerful effects within tissues

Paracrine regulation occurs in most organs and among the cells of the immune system. **Growth factors,** proteins that promote growth and cell division in specific organs, are among the most important paracrine regulators. Growth factors play a critical role in regulating mitosis throughout life (see chapter 10). For example, *epidermal growth factor* activates

mitosis of skin and development of connective tissue cells, whereas *nerve growth factor* stimulates the growth and survival of neurons. *Insulin-like growth factor* stimulates cell division in developing bone as well as protein synthesis in many other tissues. **Cytokines** (described in chapter 51) are growth factors specialized to control cell division and differentiation in the immune system, whereas **neurotropins** are growth factors that regulate the nervous system.

The importance of growth factor function is underscored by the observation that damage to the genes coding for growth factors or their receptors can lead to the unregulated cell division and development of tumors.

Paracrine regulation of blood vessels

The gas nitric oxide (NO), which can function as a neurotransmitter (see chapter 43), is also produced by the endothelium of blood vessels. In this context, it is a paracrine regulator because it diffuses to the smooth muscle layer of the blood vessel and promotes vasodilation. One of its major roles involves the control of blood pressure by dilating arteries. The endothelium of blood vessels is a rich source of paracrine regulators, including *endothelin,* which stimulates vasoconstriction, and *bradykinin,* which promotes vasodilation. Paracrine regulation supplements the regulation of blood vessels by autonomic nerves, enabling vessels to respond to local conditions, such as increased pressure or reduced oxygen.

Prostaglandins

A particularly diverse group of paracrine regulators are the **prostaglandins.** A prostaglandin is a 20-carbon-long fatty acid that contains a five-membered carbon ring. This molecule is derived from the precursor molecule *arachidonic acid,* released from phospholipids in the cell membrane under hormonal or other stimulation. Prostaglandins are produced in almost every organ and participate in a variety of regulatory functions. Some prostaglandins are active in promoting smooth muscle contraction. Through this action, they regulate reproductive functions such as gamete transport, labor, and possibly ovulation. Excessive prostaglandin production may be involved in premature labor, endometriosis, or dysmenorrhea (painful menstrual cramps). They also participate in lung and kidney regulation through effects on smooth muscle.

In fish, prostaglandins have been found to function as both a hormone and a paracrine regulator. Prostaglandins produced in the fish's ovary during ovulation can travel to the brain to synchronize associated spawning behavior.

Prostaglandins are produced at locations of tissue damage, where they promote many aspects of inflammation, including swelling, pain, and fever. This effect of prostaglandins has been well studied. Drugs that inhibit prostaglandin synthesis, such as aspirin, help alleviate these symptoms.

Aspirin is the most widely used of the *nonsteroidal anti-inflammatory drugs (NSAIDs),* a class of drugs that also includes indomethacin and ibuprofen. These drugs act to inhibit two related enzymes: cyclooxygenase-1 and 2 (COX-1 and COX-2). The anti-inflammatory effects are due to the inhibition of COX-2, which is necessary for the production of prostaglandins from arachidonic acid. This reduces inflammation and associated pain from the action of prostaglandins. Unfortunately, the inhibition of COX-1 produces unwanted side effects, including gastric bleeding and prolonged clotting time.

More recently developed pain relievers, called *COX-2 inhibitors,* selectively inhibit COX-2 but not COX-1. COX-2 inhibitors may be of potentially great benefit to arthritis sufferers and others who must use pain relievers regularly, but concerns have been raised that they may also affect other aspects of prostaglandin function in the cardiovascular system. Some COX-2 inhibitors were removed from the market when a greater risk of heart attack and stroke was detected. Some have remained in use, however, and others may be reintroduced upon FDA approval. Aside from the possibly lessened gastrointestinal side effects, COX-2 inhibitors are not more effective for pain than the older NSAIDs.

Learning Outcomes Review 45.1

Hormones coordinate the activity of specific target cells. The three chemical classes of endocrine hormones are peptides and proteins, amino acid derivatives, and steroids. Lipophilic hormones such as steroids can cross membranes, but need carriers in the blood; hydrophilic hormones move readily in the blood, but cannot cross membranes. Paracrine regulators act within the organ in which they are produced.

■ **How do hormones and neurotransmitters differ?**

45.2 Actions of Lipophilic Versus Hydrophilic Hormones

Learning Outcomes

1. **Explain how steroid hormone receptors function.**
2. **Explain how the signal carried by peptide hormones crosses the membrane.**
3. **Describe the different types of membrane receptors.**

As mentioned previously, hormones can be divided into the lipophilic (lipid-soluble) and the hydrophilic (water-soluble). The receptors and actions of these two broad categories have notable differences, which we explore in this section.

Lipophilic hormones activate intracellular receptors

The lipophilic hormones include all of the steroid hormones and thyroid hormones (figure 45.4) as well as other lipophilic

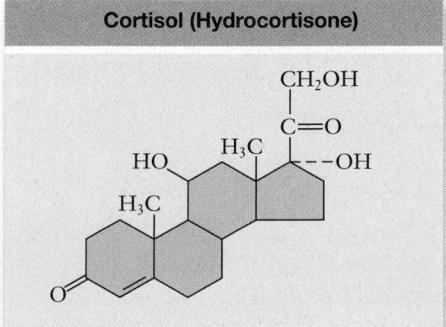
Cortisol (Hydrocortisone)

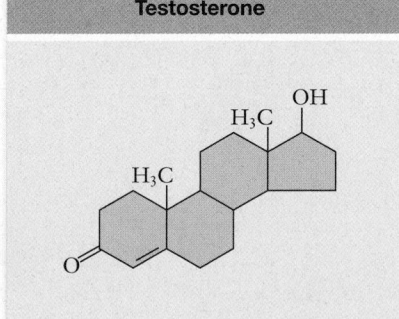
Testosterone

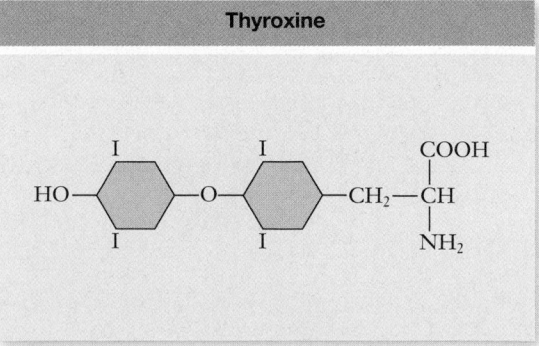
Thyroxine

Figure 45.4 Chemical structures of lipophilic hormones. Steroid hormones are derived from cholesterol. The two steroid hormones shown, cortisol and testosterone, differ slightly in chemical structure yet have widely different effects on the body. The thyroid hormone, thyroxine, is formed by coupling iodine to the amino acid tyrosine.

regulatory molecules including the retinoids, or vitamin A. Lipophilic hormones can enter cells because the lipid portion of the plasma membrane does not present a barrier. Once inside the cell the lipophilic regulatory molecules all have a similar mechanism of action.

Transport and receptor binding

These hormones circulate bound to transport proteins (see figure 45.3), which make them soluble and prolong their survival in the blood. When the hormones arrive at their target cells, they dissociate from their transport proteins and pass through the plasma membrane of the cell (figure 45.5). The hormone then binds to an intracellular receptor protein.

Some steroid hormones bind to their receptors in the cytoplasm, and then move as a hormone-receptor complex into the nucleus. Other steroids and the thyroid hormones travel directly into the nucleus before encountering their receptor proteins. Whether the hormone finds its receptor in the nucleus or translocates with its receptor into the nucleus from the cytoplasm, the rest of the story is similar.

Activation of transcription in the nucleus

The hormone receptor, activated by binding to the hormone, is now also able to bind to specific regions of the DNA. These DNA regions, located in the promoters of specific genes, are known as **hormone response elements.** The binding of the hormone-receptor complex has a direct effect on the level of transcription at that site by activating, or in some cases deactivating, gene transcription. Receptors therefore function as *hormone-activated transcription factors* (see chapters 9 and 16).

The proteins that result from activation of these transcription factors often have activity that changes the metabolism of the target cell in a specific fashion; this change constitutes the cell's response to hormone stimulation. When estrogen binds to its receptor in liver cells of chickens, for example, it activates the cell to produce the protein vitellogenin, which is then transported to the ovary to form the yolk of eggs. In contrast, when thyroid hormone binds to its receptor in the anterior pituitary of humans, it inhibits the expression of the gene for thyrotropin, a mechanism of negative feedback (described later).

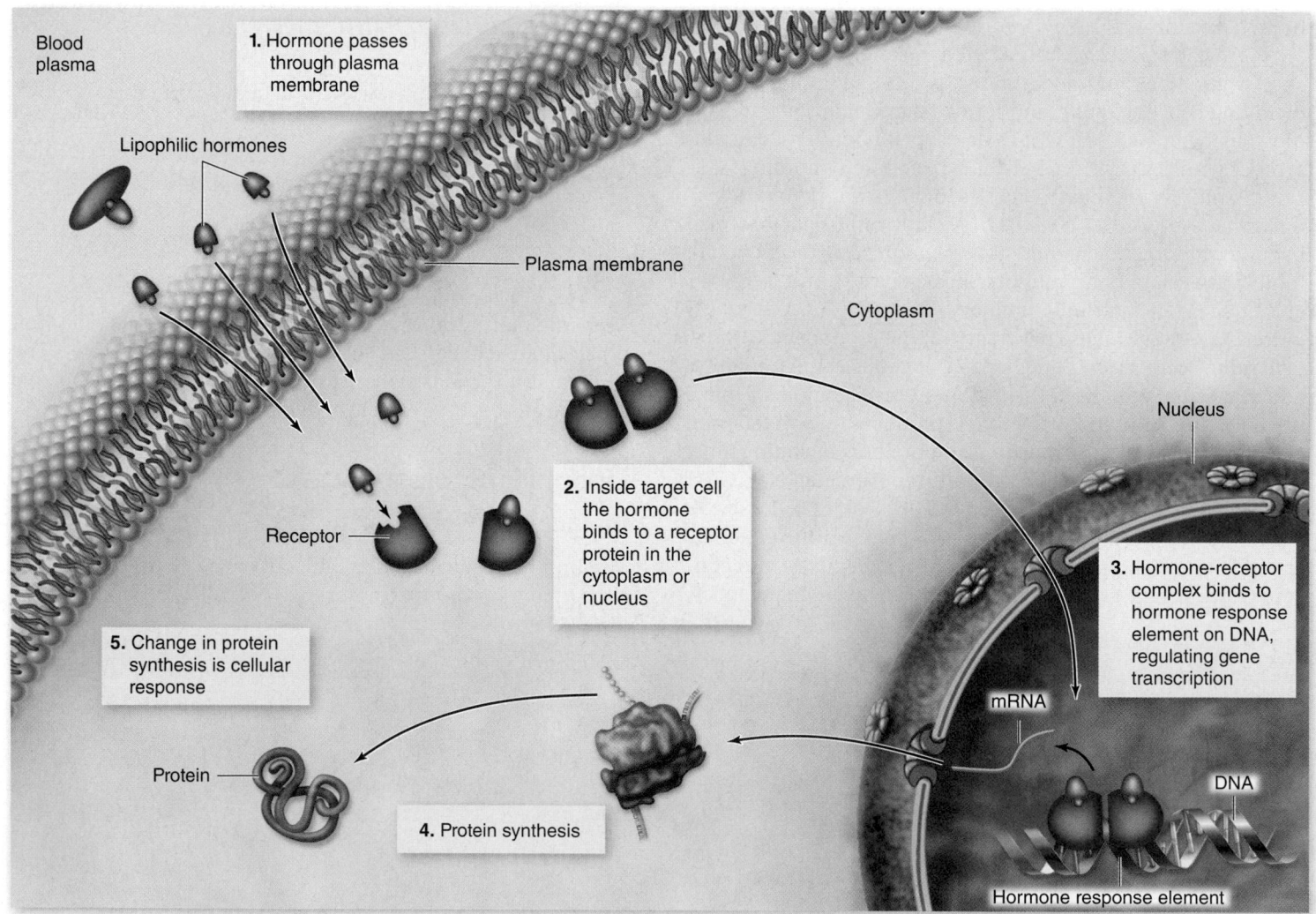

Figure 45.5 The mechanism of lipophilic hormone action. Lipophilic hormones diffuse through the plasma membrane of cells and bind to intracellular receptor proteins. The hormone-receptor complex then binds to specific regions of the DNA (hormone response elements), regulating the production of messenger RNA (mRNA). Most receptors for these hormones reside in the nucleus; if the hormone is one that binds to a receptor in the cytoplasm, the hormone-receptor complex moves together into the nucleus.

Because this activation and transcription process requires alterations in gene expression, it often takes several hours before the response to lipophilic hormone stimulation is apparent in target cells.

Hydrophilic hormones activate receptors on target cell membranes

Hormones that are too large or too polar to cross the plasma membranes of their target cells include all of the peptide, protein, and glycoprotein hormones, as well as the catecholamine hormones. These hormones bind to receptor proteins located on the outer surface of the plasma membrane. This binding must then activate the hormone response inside the cell, initiating the process of signal transduction. The cellular response is most often achieved through receptor-dependent activation of the powerful intracellular enzymes called *protein kinases*. As described in chapter 9, protein kinases are critical regulatory enzymes that activate or deactivate intracellular proteins by phosphorylation. By regulating protein kinases, hydrophilic hormone receptors exert a powerful influence over the broad range of intracellular functions.

Receptor kinases

For some hormones, such as insulin, the receptor itself is a kinase (figure 45.6), and it can directly phosphorylate intracellular proteins that alter cellular activity. In the case of insulin, this action results in the placement in the plasma membrane of glucose transport proteins that enable glucose to enter cells. Other peptide hormones, such as growth hormone, work through similar mechanisms, although the receptor itself is not a kinase. Instead, the hormone-bound receptor recruits and activates intracellular kinases, which then initiate the cellular response.

Second-messenger systems

Many hydrophilic hormones, such as epinephrine, work through second-messenger systems. A number of different molecules in the cell can serve as second messengers, as you saw in chapter 9. The interaction between the hormone and its receptor activates mechanisms in the plasma membrane that increase the concentration of the second messengers within the target cell cytoplasm.

In the early 1960s, Earl Sutherland showed that activation of the epinephrine receptor on liver cells increases intracellular

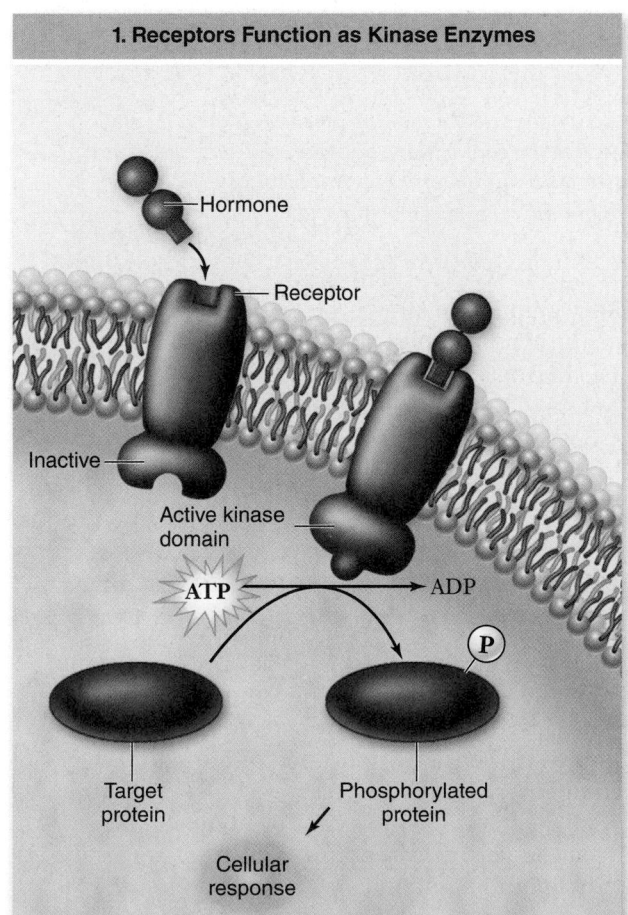

1. Receptors Function as Kinase Enzymes

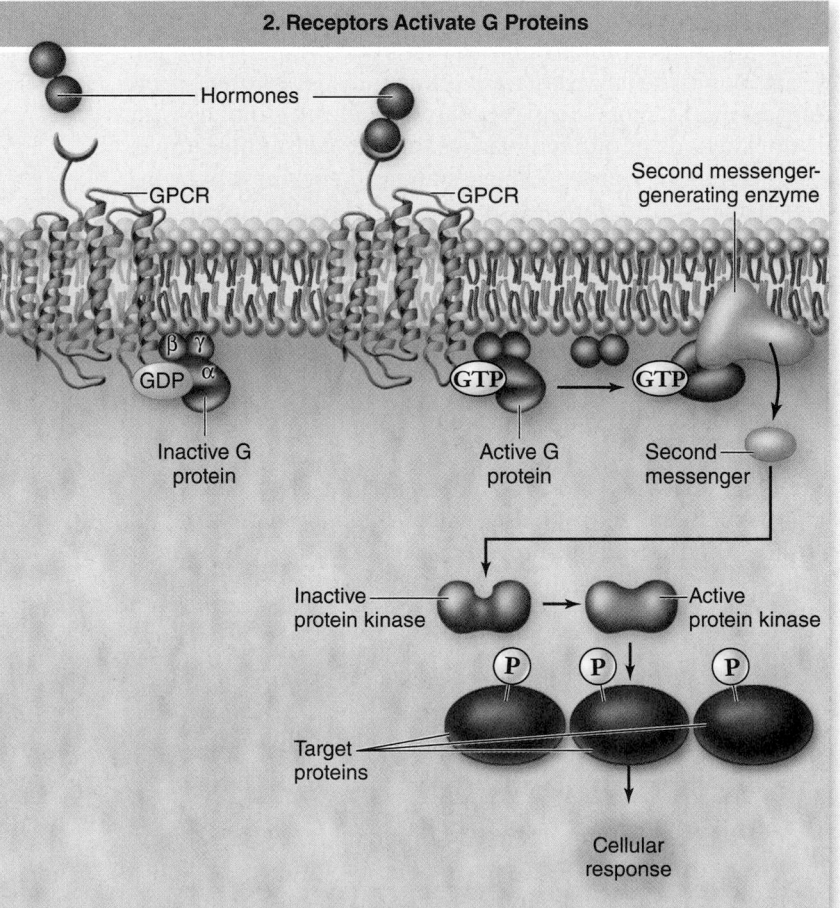

2. Receptors Activate G Proteins

Figure 45.6 The action of hydrophilic hormones. Hydrophilic hormones cannot enter cells and must therefore work extracellularly via activation of transmembrane receptor proteins. (1) These receptors can function as kinase enzymes, activating phosphorylation of other proteins inside cells. (2) Alternatively, acting through intermediary G proteins, the hormone-bound receptor activates production of a second messenger. The second messenger activates protein kinases that phosphorylate and thereby activate other proteins. GPCR, G protein–coupled receptor.

cyclic adenosine monophosphate, or cyclic AMP (cAMP), which then serves as an intracellular second messenger. The cAMP second-messenger system was the first such system to be described. Since that time, another hormonally regulated second-messenger system has been described that generates two lipid messengers: **inositol triphosphate (IP₃)** and diacyl-glycerol (DAG). These systems were described in chapter 9.

The action of G proteins

Receptors that activate second messengers do not manufacture the second messenger themselves. Rather, they are linked to a second-messenger-generating enzyme via membrane proteins called *G proteins* [that is, they are G protein–coupled receptors (GPCR); see chapter 9]. The binding of the hormone to its receptor causes the G protein to shuttle within the plasma membrane from the receptor to the second-messenger-generating enzyme (see figure 45.6). When the G protein activates the enzyme, the result is an increase in second-messenger molecules inside the cell.

In the case of epinephrine, the G protein activates an enzyme called *adenylyl cyclase,* which catalyzes the formation of the second messenger cAMP from ATP. The second messenger formed at the inner surface of the plasma membrane then diffuses within the cytoplasm, where it binds to and activates protein kinases.

The identities of the proteins that are subsequently phosphorylated by the protein kinases vary from one cell type to the next and include enzymes, membrane transport proteins, and transcription factors. This diversity provides hormones with distinct actions in different tissues. In liver cells, for example, cAMP-dependent protein kinases activate enzymes that convert glycogen into glucose. In contrast, cardiac muscle cells express a different set of cellular proteins such that a cAMP increase activates an increase in the rate and force of cardiac muscle contraction.

Activation versus inhibition

The cellular response to a hormone depends on the type of G protein activated by the hormone's receptor. Some receptors are linked to G proteins that activate second-messenger-producing enzymes, whereas other receptors are linked to G proteins that inhibit their second-messenger-generating enzyme. As a result, some hormones stimulate protein kinases in their target cells, and others inhibit their targets. Furthermore, a single hormone can have distinct actions in two different cell types if the receptors in those cells are linked to different G proteins.

Epinephrine receptors in the liver, for example, produce cAMP through the enzyme adenylyl cyclase, mentioned earlier. The cAMP they generate activates protein kinases that promote the production of glucose from glycogen. In smooth muscle, by contrast, epinephrine receptors can be linked through a different stimulatory G protein to the IP₃-generating enzyme phospholipase C. As a result, epinephrine stimulation of smooth muscle results in IP₃-regulated release of intracellular calcium, causing muscle contraction.

Duration of hydrophilic hormone effects

The binding of a hydrophilic hormone to its receptor is reversible and usually very brief; hormones soon dissociate from receptors or are rapidly deactivated by their target cells after binding.

Additionally, target cells contain specific enzymes that rapidly deactivate second messengers and protein kinases. As a result, hydrophilic hormones are capable of stimulating immediate responses within cells, but often have a brief duration of action (minutes to hours).

Learning Outcomes Review 45.2

Steroids, which are lipophilic, pass through a target cell's membrane and bind to intracellular receptor proteins. The hormone-receptor complex then binds to the hormone response element of the promotor region of the target gene. Hydrophilic hormones such as peptides bind externally to membrane receptors that activate protein kinases directly or that operate through second-messenger systems such as cAMP or IP₃/DAG.

■ *How can a single hormone, such as epinephrine, have different effects in different tissues?*

45.3 The Pituitary and Hypothalamus: The Body's Control Centers

Learning Outcomes

1. *Describe the structure of the pituitary gland.*
2. *Describe the connections between the hypothalamus, posterior pituitary, and anterior pituitary.*
3. *Explain the roles of releasing and inhibiting hormones.*

The **pituitary gland,** also known as the **hypophysis,** hangs by a stalk from the hypothalamus at the base of the brain posterior to the optic chiasm. The hypothalamus is a part of the central nervous system (CNS) that has a major role in regulating body processes. Both these structures were described in chapter 43; here we discuss in detail how they work together to bring about homeostasis and changes in body processes.

The pituitary is a compound endocrine gland

A microscopic view reveals that the gland consists of two parts, one of which appears glandular and is called the **anterior pituitary,** or **adenohypophysis.** The other portion appears fibrous and is called the **posterior pituitary,** or **neurohypophysis.** These two portions of the pituitary gland have different embryonic origins, secrete different hormones, and are regulated by different control systems. These two regions are conserved in all vertebrate animals, suggesting an ancient and important function of each.

The posterior pituitary stores and releases two neurohormones

The posterior pituitary appears fibrous because it contains axons that originate in cell bodies within the hypothalamus and that extend along the stalk of the pituitary as a tract of fibers. This anatomical relationship results from the way the posterior pituitary is formed in embryonic development. As the floor of the third ventricle of the brain forms the hypothalamus, part of this neural tissue grows downward to produce the posterior pituitary. The hypothalamus and posterior pituitary thus remain directly interconnected by a tract of axons.

Antidiuretic hormone

The endocrine role of the posterior pituitary first became evident in 1912, when a remarkable medical case was reported: A man who had been shot in the head developed the need to urinate every 30 minutes or so, 24 hours a day. The bullet had lodged in his posterior pituitary. Subsequent research demonstrated that removal of this portion of the pituitary produces the same symptoms.

In the early 1950s investigators isolated a peptide from the posterior pituitary, **antidiuretic hormone (ADH).** ADH stimulates water reabsorption by the kidneys (figure 45.7), and in doing so inhibits diuresis (urine production). When ADH is missing, as it was in the shooting victim, the kidneys do not reabsorb as much water, and excessive quantities of urine are produced. This is why the consumption of alcohol, which inhibits ADH secretion, leads to frequent urination. The role of ADH in kidney function is covered in chapter 50.

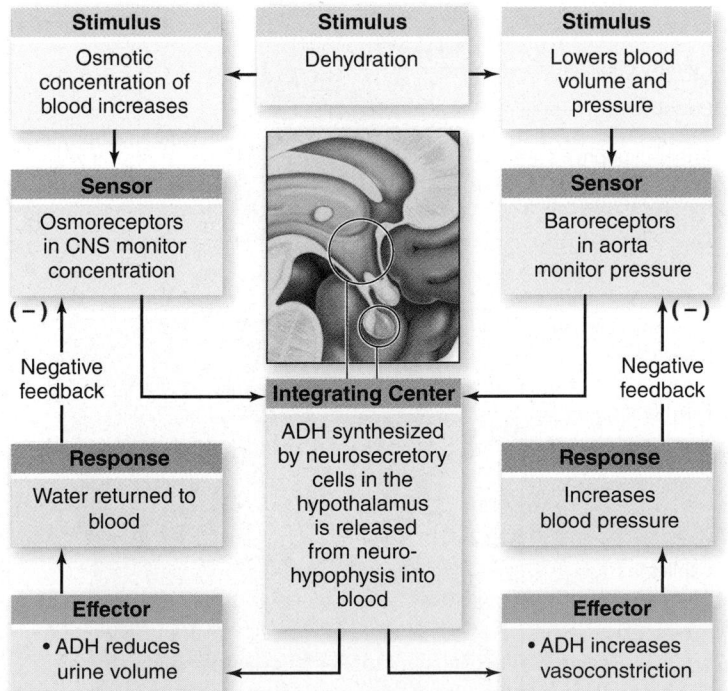

Figure 45.7 **The effects of antidiuretic hormone (ADH).** Dehydration increases the osmotic concentration of the blood and lowers blood pressure, stimulating the neurohypophysis to secrete ADH. ADH increases reabsorption of water by the kidneys and causes vasoconstriction, increasing blood pressure. Decreased blood osmolarity and increased blood pressure complete negative feedback loops to maintain homeostasis.

Oxytocin

The posterior pituitary also secretes **oxytocin,** a second peptide neurohormone that, like ADH, is composed of nine amino acids. In mammals, oxytocin stimulates the milk ejection reflex. During suckling, sensory receptors in the nipples send impulses to the hypothalamus, which triggers the release of oxytocin. Oxytocin is also needed to stimulate uterine contractions in women during childbirth.

Oxytocin secretion continues after childbirth in a woman who is breast-feeding; as a result, the uterus of a nursing mother contracts and returns to its normal size after pregnancy more quickly than the uterus of a mother who does not breast-feed.

A related posterior-pituitary neurohormone, *arginine vasotocin,* exerts similar effects in nonmammalian species. For example, in chickens and sea turtles, arginine vasotocin activates oviduct contraction during egg laying.

More recently, oxytocin has been identified as an important regulator of reproductive behavior. In both men and women, it is thought to be involved in promoting pair bonding (leading to its being called the "cuddle hormone") as well as regulating sexual responses, including arousal and orgasm. For these effects, it most likely functions in a paracrine fashion inside the CNS, much like a neurotransmitter.

Hypothalamic production of the neurohormones

ADH and oxytocin are actually produced by neuron cell bodies located in the hypothalamus. These two neurohormones are transported along the axon tract that runs from the hypothalamus to the posterior pituitary, where they are stored. In response to the appropriate stimulation—increased blood plasma osmolality in the case of ADH, the suckling of a baby in the case of oxytocin—the neurohormones are released by the posterior pituitary into the blood.

Because this reflex control involves both the nervous and the endocrine systems, ADH and oxytocin are said to be secreted by a **neuroendocrine reflex.**

The anterior pituitary produces seven hormones

The anterior pituitary, unlike the posterior pituitary, does not develop from growth of the brain; instead, it develops from a pouch of epithelial tissue that pinches off from the roof of the embryo's mouth. In spite of its proximity to the brain, it is not part of the nervous system.

Because it forms from epithelial tissue, the anterior pituitary is an independent endocrine gland. It produces at least seven essential hormones, many of which stimulate growth of their target organs, as well as production and secretion of other hormones from additional endocrine glands. Therefore, several hormones of the anterior pituitary are collectively termed *tropic hormones,* or *tropins.* Tropic hormones act on other endocrine glands to stimulate secretion of hormones produced by the target gland.

The hormones produced and secreted by different cell types in the anterior pituitary can be categorized into three structurally similar families: the *peptide hormones,* the *protein hormones,* and the *glycoprotein hormones.*

Peptide hormones

The **peptide hormones** of the anterior pituitary are cleaved from a single precursor protein, and therefore they share some common sequence. They are fewer than 40 amino acids in size.

1. **Adrenocorticotropic hormone** (**ACTH,** or *corticotropin*) stimulates the adrenal cortex to produce corticosteroid hormones, including cortisol (in humans) and corticosterone (in many other vertebrates). These hormones regulate glucose homeostasis and are important in the response to stress.
2. **Melanocyte-stimulating hormone** (**MSH**) stimulates the synthesis and dispersion of melanin pigment, which darkens the epidermis of some fish, amphibians, and reptiles, and can control hair pigment color in mammals.

Protein hormones

The **protein hormones** each comprise a single chain of approximately 200 amino acids, and they share significant structural similarities.

1. **Growth hormone** (**GH,** or *somatotropin*) stimulates the growth of muscle, bone (indirectly), and other tissues, and it is also essential for proper metabolic regulation.
2. **Prolactin** (**PRL**) is best known for stimulating the mammary glands to produce milk in mammals; however, it has diverse effects on many other targets, including regulation of ion and water transport across epithelia, stimulation of a variety of organs that nourish young, and activation of parental behaviors.

Glycoprotein hormones

The largest and most complex hormones known, the *glycoprotein hormones* are dimers, containing alpha (α) and beta (β) subunits, each around 100 amino acids in size, with covalently linked sugar residues. The α subunit is common to all three hormones. The β subunit differs, endowing each hormone with a different target specificity.

1. **Thyroid-stimulating hormone** (**TSH,** or *thyrotropin*) stimulates the thyroid gland to produce the hormone thyroxine, which in turn regulates development and metabolism by acting on nuclear receptors.
2. **Luteinizing hormone** (**LH**) stimulates the production of estrogen and progesterone by the ovaries and is needed for ovulation in female reproductive cycles (see chapter 52). In males, it stimulates the testes to produce testosterone, which is needed for sperm production and for the development of male secondary sexual characteristics.
3. **Follicle-stimulating hormone** (**FSH**) is required for the development of ovarian follicles in females. In males, it is required for the development of sperm. FSH stimulates the conversion of testosterone into estrogen in females, and into dihydroxytestosterone in males. FSH and LH are collectively referred to as *gonadotropins*.

Hypothalamic neurohormones regulate the anterior pituitary

The anterior pituitary, unlike the posterior pituitary, is not derived from the brain and does not receive an axon tract from the hypothalamus. Nevertheless, the hypothalamus controls the production and secretion of its hormones. This control is itself exerted hormonally rather than by means of nerve axons.

Neurons in the hypothalamus secrete two types of neurohormones, **releasing hormones** and **inhibiting hormones,** that diffuse into blood capillaries at the base of the hypothalamus (figure 45.8). These capillaries drain into small veins that run within the stalk of the pituitary to a second bed of capillaries in the anterior pituitary. This unusual system of vessels is known as the *hypothalamohypophyseal portal system.* In a portal system, two capillary beds are linked by veins. In this case, the hormone enters the first capillary bed, and the vein delivers this to the second capillary bed where the hormone exits and enters the anterior pituitary.

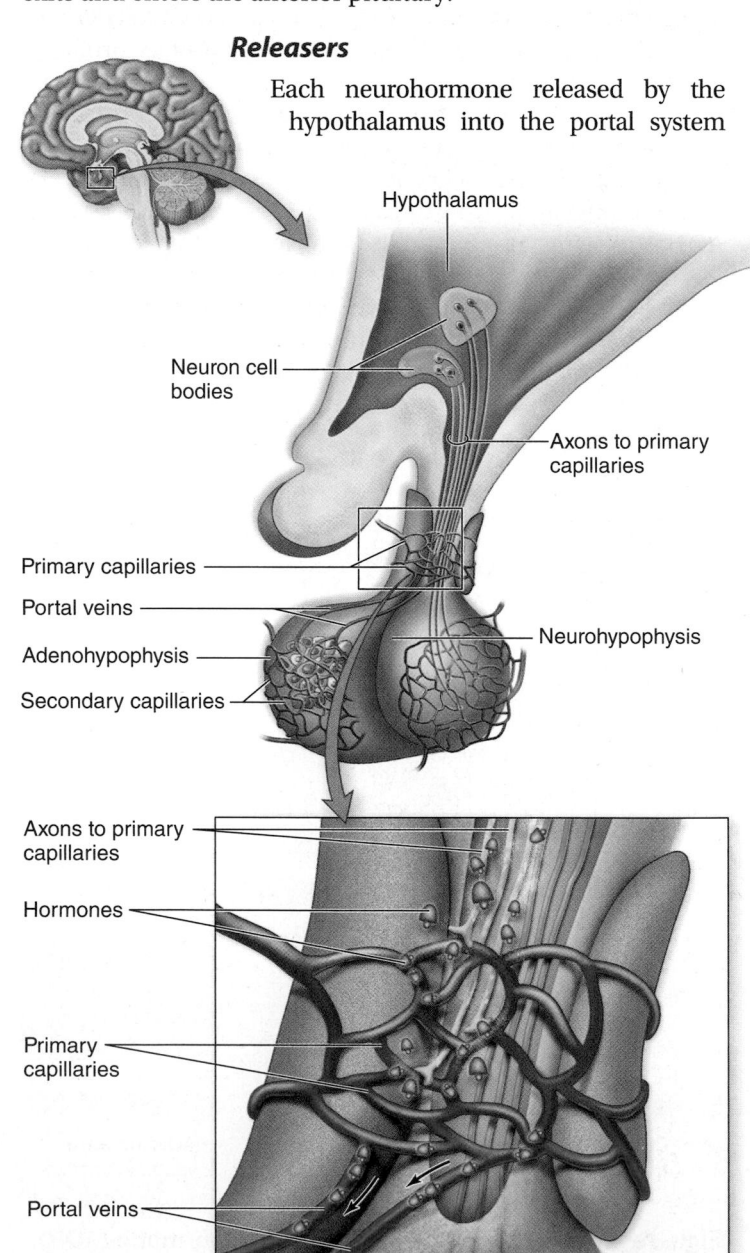

Figure 45.8 Hormonal control of the adenohypophysis by the hypothalamus. Neurons in the hypothalamus secrete hormones that are carried by portal blood vessels directly to the adenohypophysis, where they either stimulate or inhibit the secretion of hormones from the adenohypophysis.

regulates the secretion of a specific hormone in the anterior pituitary. Releasing hormones are peptide neurohormones that stimulate release of other hormones; specifically, *thyrotropin-releasing hormone* (TRH) stimulates the release of TSH; *corticotropin-releasing hormone* (CRH) stimulates the release of ACTH; and *gonadotropin-releasing hormone* (GnRH) stimulates the release of FSH and LH. A releasing hormone for growth hormone, called *growth hormone-releasing hormone* (GHRH), has also been discovered, and TRH, oxytocin and vasoactive intestinal peptide all appear to act as releasing hormones for prolactin.

Inhibitors

The hypothalamus also secretes neurohormones that inhibit the release of certain anterior-pituitary hormones. To date, three such neurohormones have been discovered: *Somatostatin,* or *growth hormone-inhibiting hormone* (GHIH), which inhibits the secretion of GH; *prolactin-inhibiting factor* (PIF), which inhibits the secretion of prolactin and has been found to be the neurotransmitter dopamine; and *MSH-inhibiting hormone* (MIH), which inhibits the secretion of MSH.

Feedback from peripheral endocrine glands regulates anterior-pituitary hormones

Because hypothalamic hormones control the secretions of the anterior pituitary, and because the hormones of the anterior pituitary in turn control the secretions of other endocrine glands, it may seem that the hypothalamus is in charge of hormonal secretion for the whole body. This however, ignores a crucial aspect of endocrine control: The hypothalamus and the anterior pituitary are themselves partially controlled by the very hormones whose secretion they stimulate. In most cases,

this control is inhibitory (figure 45.9). This type of control system is called *negative feedback,* and it acts to maintain relatively constant levels of the target cell hormone.

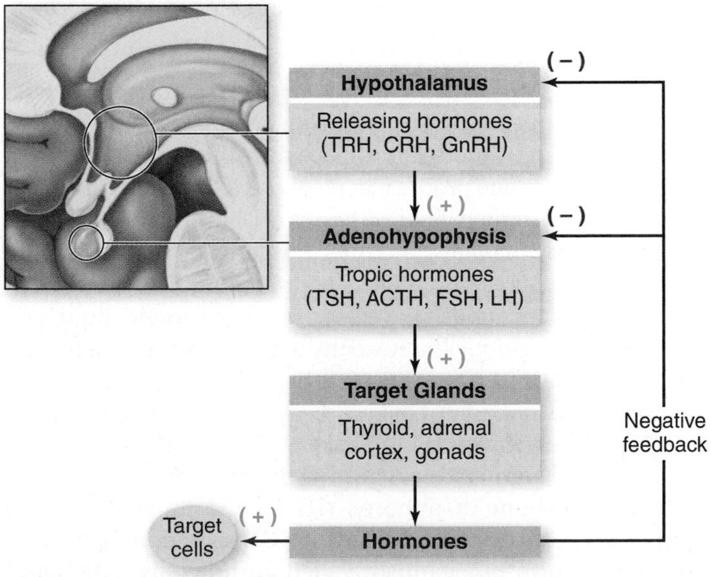

An example of negative feedback: Thyroid gland control

To illustrate how important the negative feedback mechanism is, let's consider the hormonal control of the thyroid gland. The hypothalamus secretes TRH into the hypothalamohypophyseal portal system, which stimulates the anterior pituitary to secrete TSH. TSH in turn causes the thyroid gland to release **thyroxine.** Thyroxine and other thyroid hormones affect metabolic rate, as described in the following section.

Among thyroxine's many target organs are the hypothalamus and the anterior pituitary themselves. Thyroxine acts on these organs to inhibit their secretion of TRH and TSH, respectively. This negative feedback inhibition is essential for homeostasis because it keeps the thyroxine levels fairly constant.

The hormone thyroxine contains the element iodine; without iodine, the thyroid gland cannot produce thyroxine. Individuals living in iodine-poor areas (such as central prairies distant from seacoasts and the fish that are the natural source of iodine) lack sufficient iodine to manufacture thyroxine, so the hypothalamus and anterior pituitary receive far less negative feedback inhibition than is normal. This reduced inhibition results in elevated secretion of TRH and TSH.

High levels of TSH stimulate the thyroid gland, whose cells enlarge in a futile attempt to manufacture more thyroxine. Because they cannot without iodine, the thyroid gland keeps getting bigger and bigger—a condition known as a goiter (figure 45.10). Goiter size can be reduced by providing iodine in the diet. In most countries, goiter is prevented through the addition of iodine to table salt.

Figure 45.9 Negative feedback inhibition. The hormones secreted by some endocrine glands feed back to inhibit the secretion of hypothalamic releasing hormones and adenohypophysis tropic hormones. ACTH, adrenocorticotropic hormone; CRH, corticotropin-releasing hormone; FSH, follicle-stimulating hormone; GnRH, gonadotropin-releasing hormone; LH, luteinizing hormone; TRH, thyroid-releasing hormone; TSH, thyroid-stimulating hormone

Figure 45.10 A woman with a goiter. This condition is caused by a lack of iodine in the diet. As a result, thyroxine secretion is low, so there is less negative feedback inhibition of TSH. The elevated TSH secretion, in turn, stimulates the thyroid to enlarge in an effort to produce additional thyroxine.

An example of positive feedback: Ovulation

Positive feedback in the control of the hypothalamus and anterior pituitary by the target glands is uncommon because positive feedback causes deviations from homeostasis. Positive feedback accentuates change, driving the change in the same direction. One example is the control of **ovulation,** the explosive release of a mature egg (an oocyte) from the ovary.

As the oocyte grows, follicle cells surrounding it produce increasing levels of the steroid hormone estrogen, resulting in a progressive rise in estrogen in the blood. Peak estrogen levels signal the hypothalamus that the oocyte is ready to be ovulated. Estrogen then exerts positive feedback on the hypothalamus and pituitary, resulting in a surge of LH from the anterior pituitary. This LH surge causes the follicle cells to rupture and release the oocyte to the oviduct, where it can potentially be fertilized. The positive feedback cycle is then terminated because the tissue remaining of the ovarian follicle forms the corpus luteum, which secretes progesterone and estrogen that feed back to inhibit secretion of FSH and LH. This process is discussed in more detail in chapter 52.

Hormones of the anterior pituitary work directly and indirectly

Early in the 20th century, experimental techniques were developed for surgical removal of the pituitary gland (a procedure called *hypophysectomy*). Hypophysectomized animals exhibited a number of deficits, including reduced growth and development, diminished metabolism, and failure of reproduction. These powerful and diverse effects earned the pituitary a reputation as the "master gland." Indeed, many of these are *direct effects,* resulting from anterior-pituitary hormones activating receptors in nonendocrine targets, such as liver, muscle, and bone. The tropic hormones produced by the anterior pituitary have *indirect effects,* however, through their ability to activate other endocrine glands, such as the thyroid, adrenal glands, and gonads. Of the seven anterior-pituitary hormones, growth hormone, prolactin, and MSH work primarily through direct effects, whereas the tropic hormones ACTH, TSH, LH, and FSH have endocrine glands as their exclusive targets.

Effects of growth hormone

The importance of the anterior pituitary is illustrated by a condition known as *gigantism,* characterized by excessive growth of the entire body or any of its parts. The tallest human being ever recorded, Robert Wadlow, had gigantism (figure 45.11). Born in 1928, he stood 8 feet 11 inches tall, weighed 485 pounds, and was still growing before he died from an infection at the age of 22.

We now know that gigantism is caused by the excessive secretion of GH in a growing child. By contrast, a deficiency in GH secretion during childhood results in **pituitary dwarfism**—a failure to achieve normal stature.

GH stimulates protein synthesis and growth of muscles and connective tissues; it also indirectly promotes the elongation of bones by stimulating cell division in the cartilaginous epiphyseal growth plates of bones (see chapter 46). Researchers found that this stimulation does not occur in the absence of

Figure 45.11 The Alton giant. This photograph of Robert Wadlow of Alton, Illinois, taken on his 21st birthday, shows him at home with his father and mother and four siblings. Born normal size, he developed a growth-hormone-secreting pituitary tumor as a young child and never stopped growing during his 22 years of life, reaching a height of 8 ft 11 in.

blood plasma, suggesting that GH must work in concert with another hormone to exert its effects on bone. We now know that GH stimulates the production of **insulin-like growth factors,** which liver and bone produce in response to stimulation by GH. The insulin-like growth factors then stimulate cell division in the epiphyseal growth plates, and thus the elongation of the bones.

Although GH exhibits its most dramatic effects on juvenile growth, it also functions in adults to regulate protein, lipid, and carbohydrate metabolism. Recently a peptide hormone named **ghrelin,** produced by the stomach between meals, was identified as a potent stimulator of GH release, establishing an important linkage between nutrient intake and GH production.

Because human skeletal growth plates transform from cartilage into bone at puberty, GH can no longer cause an increase in height in adults. Excessive GH secretion in an adult results in a form of gigantism called **acromegaly,** characterized by bone and soft tissue deformities such as a protruding jaw, elongated fingers, and thickening of skin and facial features. Our knowledge of the regulation of GH has led to the development of drugs that can control its secretion, for example through activation of somatostatin, or by mimicking ghrelin. As a result, gigantism is much less common today.

Animals that have been genetically engineered to express additional copies of the GH gene grow to larger than normal size (see figure 17.16), making agricultural applications of GH manipulation an active area of investigation. Among other actions, GH has been found to increase milk yield in cows, promote weight gain in pigs, and increase the length of fish. The growth-promoting actions of GH thus appear to have been conserved throughout the vertebrates.

Other hormones of the anterior pituitary

Like growth hormone, prolactin acts on organs that are not endocrine glands. In contrast to GH, however, the actions of prolactin appear to be very diverse. In addition to stimulating production of milk in mammals, prolactin has been implicated in the regulation of tissues important in birds for the nourishment and incubation of young, such as the crop (which produces "crop milk," a nutritional fluid fed to chicks by regurgitation) and the brood patch (a vascular area on the abdomen of birds used to warm eggs).

In amphibians, prolactin promotes transformation of salamanders from terrestrial forms to aquatic breeding adults. Associated with these reproductive actions is an ability of prolactin to activate associated behaviors, such as parental care in mammals, broodiness in birds, and "water drive" in amphibians.

Prolactin also has varied effects on electrolyte balance through actions on the kidneys of mammals, the gills of fish, and the salt glands of marine birds. This variation suggests that although prolactin may have an ancient function in the regulation of salt and water movement across membranes, its actions have diversified with the appearance of new vertebrate species. The field of comparative endocrinology studies questions about hormone action across diverse species, with the objective of understanding the mechanisms of hormone evolution.

Unlike growth hormone and prolactin, the other adenohypophyseal hormones act on relatively few targets. TSH stimulates the thyroid gland, and ACTH stimulates the adrenal cortex. The gonadotropins, FSH and LH, act on the gonads. Although both FSH and LH act on the gonads, they each target different cells in the gonads of both females and males (see chapter 52). These hormones all share the common characteristic of activating target endocrine glands.

The final pituitary hormone, MSH regulates the activity of cells called melanophores, which contain the black pigment **melanin.** In response to MSH, melanin is dispersed throughout these cells, darkening the skin of reptiles, amphibians, or fish. In mammals, which lack melanophores but have similar cells called melanocytes, MSH can darken hair by increasing melanin deposition in the developing hair shaft.

Learning Outcomes Review 45.3

The posterior pituitary develops from neural tissue; the anterior pituitary develops from epithelial tissue. Axons from the hypothalamus extend into the posterior pituitary and produce neurohormones; these neurons also secrete factors that release or inhibit hormones of the anterior pituitary. Releasers stimulate secretion of hormones; TRH causes TSH release. Inhibitors suppress secretion; GHIH inhibits GH release.

■ *Could someone with a pituitary tumor causing gigantism be treated with GHIH? What outcome would you predict?*

45.4 The Major Peripheral Endocrine Glands

Learning Outcomes

1. Identify the major peripheral endocrine glands.
2. Describe the components of Ca^{2+} homeostasis.
3. Explain the action of pancreatic hormones on blood glucose.

Although the pituitary produces an impressive array of hormones, many endocrine glands are found in other locations. Some of these may be controlled by tropic hormones of the pituitary, but others, such as the adrenal medulla and the pancreas, are independent of pituitary control. Several endocrine glands develop from derivatives of the primitive pharynx, which is the most anterior segment of the digestive tract (see chapter 47). These glands, which include the *thyroid* and *parathyroid* glands, produce hormones that regulate processes associated with nutrient uptake, such as carbohydrate, lipid, protein, and mineral metabolism.

The thyroid gland regulates basal metabolism and development

The thyroid gland varies in shape in different vertebrate species, but is always found in the neck area, anterior to the heart. In humans it is shaped like a bow tie and lies just below the Adam's apple in the front of the neck.

The thyroid gland secretes three hormones: primarily thyroxine, smaller amounts of triiodothyronine (collectively referred to as thyroid hormones), and calcitonin. As described earlier, thyroid hormones are unique in being the only molecules in the body containing iodine (thyroxine contains four iodine atoms, triiodothyronine contains three).

Thyroid-related disorders

Thyroid hormones work by binding to nuclear receptors located in most cells in the body, influencing the production and activity of a large number of cellular proteins. The importance of thyroid hormones first became apparent from studies of human thyroid disorders. Adults with hypothyroidism have low metabolism due to underproduction of thyroxine, including a reduced ability to utilize carbohydrates and fats. As a result, they are often fatigued, overweight, and feel cold. Hypothyroidism is particularly concerning in infants and children, where it impairs growth, brain development, and reproductive maturity. Fortunately, because thyroid hormones are small, simple molecules, people with hypothyroidism can take thyroxine orally as a pill.

People with hyperthyroidism, by contrast, often exhibit opposite symptoms: weight loss, nervousness, high metabolism, and overheating because of overproduction of thyroxine. Drugs are available that block thyroid hormone synthesis in the thyroid gland, but in some cases portions of the thyroid gland must be removed surgically or by radiation treatment.

Actions of thyroid hormones

Thyroid hormones regulate enzymes controlling carbohydrate and lipid metabolism in most cells, promoting the appropriate use of these fuels for maintaining the body's basal metabolic rate. Thyroid hormones often function cooperatively, or *synergistically*, with other hormones, promoting the activity of growth hormone, epinephrine, and reproductive steroids. Through these actions, thyroid hormones function to ensure that adequate cellular energy is available to support metabolically demanding activities.

In humans, which exhibit a relatively high metabolic rate at all times, thyroid hormones are maintained in the blood at constantly elevated levels. In contrast, in reptiles, amphibians, and fish, which undergo seasonal cycles of activity, thyroid hormone levels in the blood increase during periods of metabolic activation (such as growth, reproductive development, migration, or breeding) and diminish during periods of inactivity in cold months.

Some of the most dramatic effects of thyroid hormones are observed in their regulation of growth and development. In developing humans, for example, thyroid hormones promote growth of neurons and stimulate maturation of the CNS. Children born with hypothyroidism are stunted in their growth and suffer severe mental retardation, a condition called *cretinism*. Early detection through measurement of thyroid hormone levels allows this condition to be treated with thyroid hormone administration.

The most impressive demonstration of the importance of thyroid hormones in development is displayed in amphibians. Thyroid hormones direct the metamorphosis of tadpoles into frogs, a process that requires the transformation of an aquatic, herbivorous larva into a terrestrial, carnivorous juvenile (figure 45.12). If the thyroid gland is removed from a tadpole, it will not change into a frog. Conversely, if an immature tadpole is fed pieces of a thyroid gland, it will undergo premature metamorphosis and become a miniature frog. This illustrates the powerful actions thyroid hormones can elicit by regulating the expression of multiple genes.

Calcium homeostasis is regulated by several hormones

Calcium is a vital component of the vertebrate body both because of its being a structural component of bones and because of its role in ion-mediated processes such as muscle contraction. The thyroid and parathyroid glands act with vitamin D to regulate calcium homeostasis.

Calcitonin secretion by the thyroid

In addition to the thyroid hormones, the thyroid gland also secretes **calcitonin**, a peptide hormone that plays a role in maintaining proper levels of calcium (Ca^{2+}) in the blood. When the blood Ca^{2+} concentration rises too high, calcitonin stimulates the uptake of calcium into bones, thus lowering its

Figure 45.12 Thyroxine triggers metamorphosis in amphibians. In tadpoles at the premetamorphic stage, the hypothalamus stimulates the adenohypophysis to secrete TSH (thyroid-stimulating hormone). TSH then stimulates the thyroid gland to secrete thyroxine. Thyroxine binds to its receptor and initiates the changes in gene expression necessary for metamorphosis. As metamorphosis proceeds, thyroxine reaches its maximal level, after which the forelimbs begin to form and the tail is reabsorbed.

Data analysis

Hypophysectomy is the surgical removal of the pituitary. If this is performed on an early tadpole, how would this affect thyroxine levels and metamorphosis? What if the surgery was performed on a late tadpole?

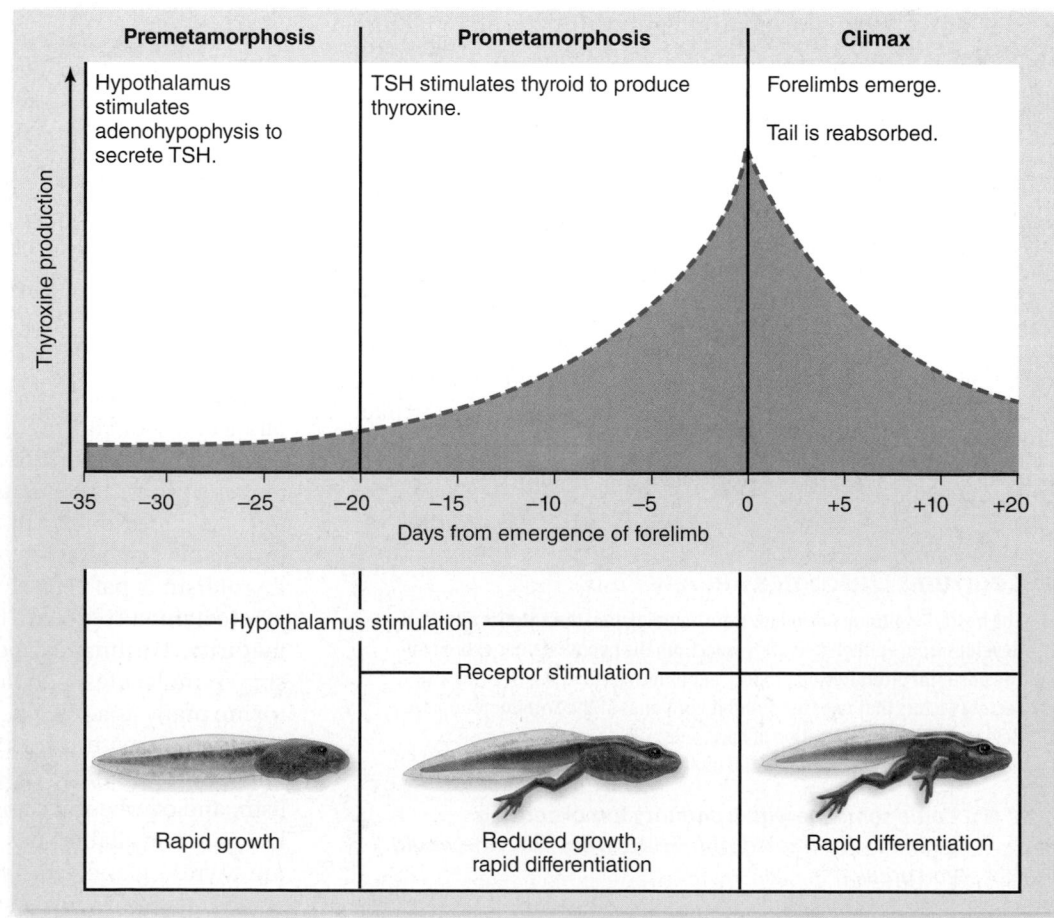

level in the blood. Although calcitonin may be important in the physiology of some vertebrates, it appears less important in the day-to-day regulation of Ca^{2+} levels in adult humans. It may, however, play an important role in bone remodeling in rapidly growing children.

Parathyroid hormone (PTH)

The parathyroid glands are four small glands attached to the thyroid. Because of their size, researchers ignored them until well into the 20th century. The first suggestion that these organs have an endocrine function came from experiments on dogs: If their parathyroid glands were removed, the Ca^{2+} concentration in the dogs' blood plummeted to less than half the normal value. The Ca^{2+} concentration returned to normal when an extract of parathyroid gland was administered. However, if too much of the extract was administered, the dogs' Ca^{2+} levels rose far above normal as the calcium phosphate crystals in their bones were dissolved. It was clear that the parathyroid glands produce a hormone that stimulates the release of calcium from bone.

The hormone produced by the parathyroid glands is a peptide called **parathyroid hormone (PTH).** PTH is synthesized and released in response to falling levels of Ca^{2+} in the blood. This decline cannot be allowed to continue uncorrected because a significant fall in the blood Ca^{2+} level can cause severe muscle spasms. A normal blood Ca^{2+} level is important for the functioning of muscles, including the heart, and for proper functioning of the nervous and endocrine systems.

PTH stimulates the osteoclasts (bone cells) in bone to dissolve the calcium phosphate crystals of the bone matrix and release Ca^{2+} into the blood (figure 45.13). PTH also stimulates the kidneys to reabsorb Ca^{2+} from the urine and leads to the activation of vitamin D, needed for the absorption of calcium from food in the intestine.

Vitamin D

Vitamin D is produced in the skin from a cholesterol derivative in response to ultraviolet light. It is called an essential vitamin because in temperate regions of the world a dietary source is needed to supplement the amount produced by the skin. (In the tropics, people generally receive enough exposure to sunlight to produce adequate vitamin D.) Diffusing into the blood from the skin, vitamin D is actually an inactive form of a hormone. In order to become activated, the molecule must gain two hydroxyl groups (—OH); one of these is added by an enzyme in the liver, the other by an enzyme in the kidneys.

The enzyme needed for this final step is stimulated by PTH, thereby producing the active form of vitamin D known as 1,25-dihydroxyvitamin D. This hormone stimulates the intestinal absorption of Ca^{2+} and thereby helps raise blood Ca^{2+} levels so that bone can become properly mineralized. A diet deficient in vitamin D thus leads to poor bone formation, a condition called rickets.

To ensure adequate amounts of this essential hormone, vitamin D is now added to commercially produced milk in the United States and some other countries. This is certainly a preferable alternative to the prior method of vitamin D administration, the dreaded dose of cod liver oil.

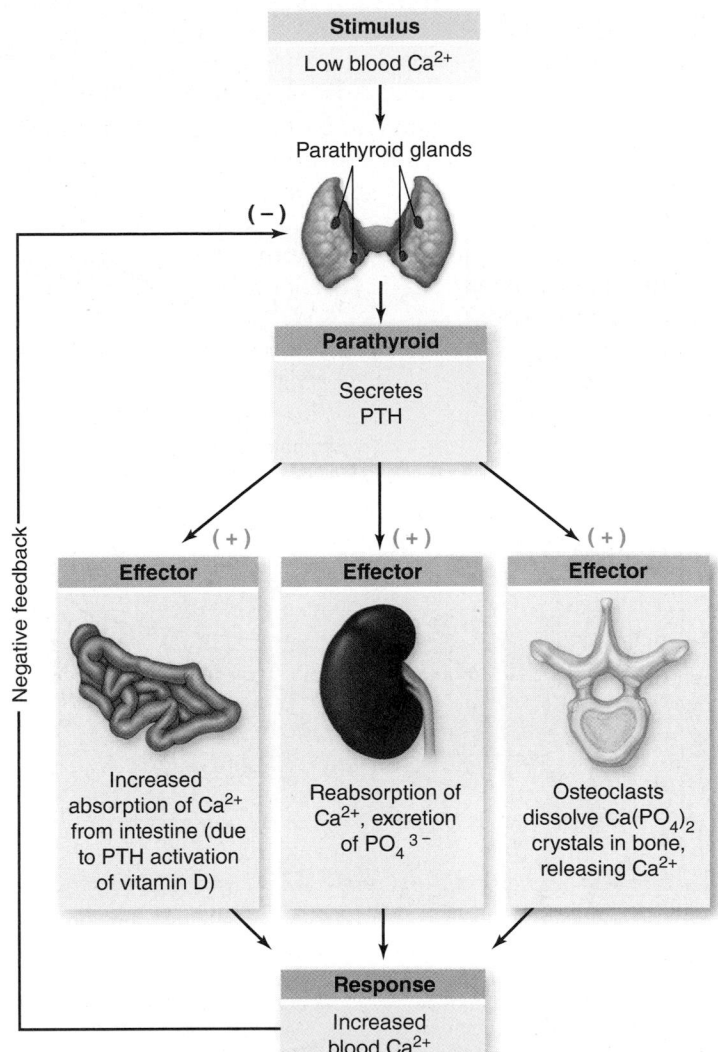

Figure 45.13 Regulation of blood Ca^{2+} levels by parathyroid hormone (PTH). When blood Ca^{2+} levels are low, PTH is released by the parathyroid glands. PTH directly stimulates the dissolution of bone and the reabsorption of Ca^{2+} by the kidneys. PTH indirectly promotes the intestinal absorption of Ca^{2+} by stimulating the production of the active form of vitamin D.

The adrenal gland releases both catecholamine and steroid hormones

The **adrenal glands** are located just above each kidney (figure 45.14). Each gland is composed of an inner portion, the *adrenal medulla,* and an outer layer, the *adrenal cortex.*

The adrenal medulla

The adrenal medulla receives neural input from axons of the sympathetic division of the autonomic nervous system, and it secretes the catecholamines epinephrine and norepinephrine in response to stimulation by these axons. The actions of these hormones trigger "alarm" responses similar to those elicited by the sympathetic division, helping to prepare the body for extreme efforts. Among the effects of these hormones are an increased heart rate, increased blood pressure, dilation of the bronchioles, elevation in blood glucose, reduced blood flow to the skin and digestive organs, and

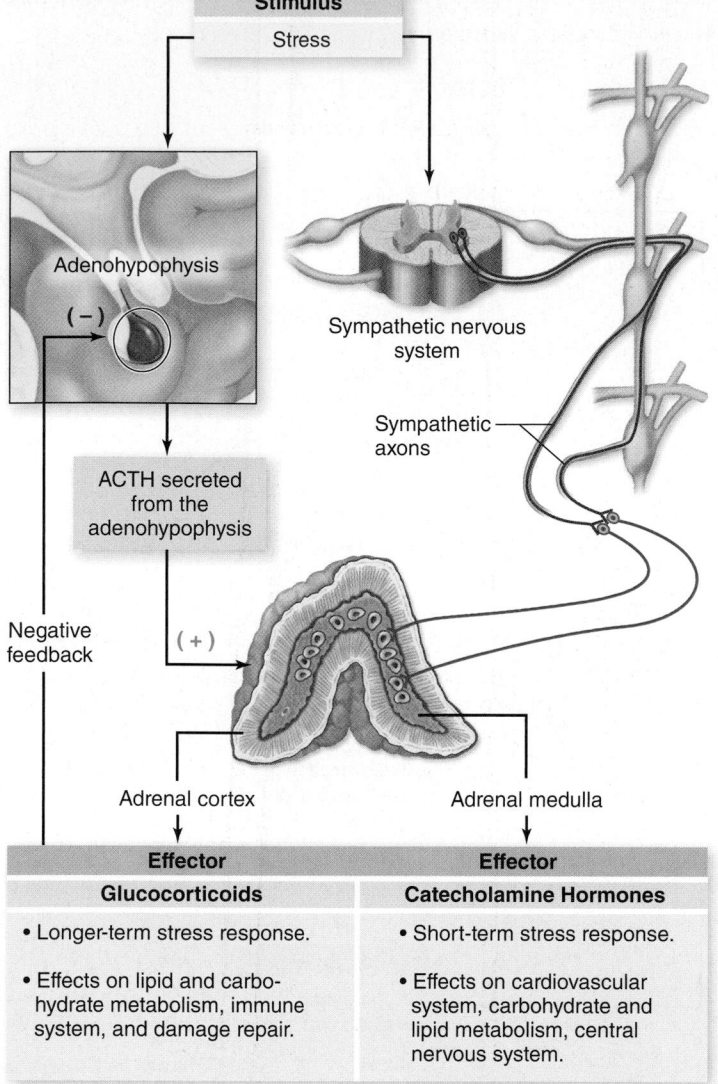

Stimulus

Stress

Adenohypophysis

(−)

ACTH secreted from the adenohypophysis

Negative feedback

(+)

Sympathetic nervous system

Sympathetic axons

Adrenal cortex

Adrenal medulla

Effector	Effector
Glucocorticoids	**Catecholamine Hormones**
• Longer-term stress response.	• Short-term stress response.
• Effects on lipid and carbohydrate metabolism, immune system, and damage repair.	• Effects on cardiovascular system, carbohydrate and lipid metabolism, central nervous system.

Figure 45.14 The adrenal glands. The adrenal medulla produces the catecholamines epinephrine and norepinephrine, which initiate a response to acute stress. The adrenal cortex produces steroid hormones, including the glucocorticoid cortisol. In response to stress, cortisol secretion increases glucose production and stimulates the immune response.

increased blood flow to the heart and muscles. The actions of epinephrine, released as a hormone, supplement those of neurotransmitters released by the sympathetic nervous system.

The adrenal cortex

The hormones from the adrenal cortex are all steroids and are referred to collectively as *corticosteroids*. *Cortisol* (also called hydrocortisone) and related steroids secreted by the adrenal cortex act on various cells in the body to maintain glucose homeostasis. In mammals, these hormones are referred to as glucocorticoids, and their secretion is primarily regulated by ACTH from the anterior pituitary.

The glucocorticoids stimulate the breakdown of muscle protein into amino acids, which are carried by the blood to the liver. They also stimulate the liver to produce the enzymes needed for gluconeogenesis, which can convert amino acids

into glucose. Glucose synthesis from protein is particularly important during very long periods of fasting or exercise, when blood glucose levels might otherwise become dangerously low.

Whereas glucocorticoids are important in the daily regulation of glucose and protein, they, like the adrenal medulla hormones, are also secreted in large amounts in response to stress. It has been suggested that during stress they activate the production of glucose at the expense of protein and fat synthesis.

In addition to regulating glucose metabolism, the glucocorticoids modulate some aspects of the immune response. The physiological significance of this action is still unclear, and it may be apparent only when glucocorticoids are maintained at elevated levels for long periods of time (such as long-term stress). Glucocorticoids are used to suppress the immune system in persons with immune disorders (such as rheumatoid arthritis) and to prevent the immune system from rejecting organ and tissue transplants. Derivatives of cortisol, such as prednisone, have widespread medical use as anti-inflammatory agents.

Aldosterone, the other major corticosteroid, is classified as a mineralocorticoid because it helps regulate mineral balance. The secretion of aldosterone from the adrenal cortex is activated by angiotensin II, a product of the renin–angiotensin system described in chapter 50, as well as high blood K^+. Angiotensin II activates aldosterone secretion when blood pressure falls.

A primary action of aldosterone is to stimulate the kidneys to reabsorb Na^+ from the urine. (Blood levels of Na^+ decrease if Na^+ is not reabsorbed from the urine.) Sodium is the major extracellular solute; it is needed for the maintenance of normal blood volume and pressure, as well as for the generation of action potentials in neurons and muscles. Without aldosterone, the kidneys would lose excessive amounts of blood Na^+ in the urine.

Aldosterone-stimulated reabsorption of Na^+ also results in kidney excretion of K^+ in the urine. Aldosterone thus prevents K^+ from accumulating in the blood, which would lead to malfunctions in electrical signaling in nerves and muscles. Because of these essential functions performed by aldosterone, removal of the adrenal glands, or diseases that prevent aldosterone secretion, are invariably fatal without hormone therapy.

Pancreatic hormones are primary regulators of carbohydrate metabolism

The pancreas is located adjacent to the stomach and is connected to the duodenum of the small intestine by the pancreatic duct. It secretes bicarbonate ions and a variety of digestive enzymes into the small intestine through this duct (see chapter 47), and for a long time the pancreas was thought to be solely an exocrine gland.

Insulin

In 1869, however, a German medical student named Paul Langerhans described some unusual clusters of cells scattered throughout the pancreas; these clusters came to be called *islets of Langerhans.* They are now more commonly called pancreatic islets. Laboratory workers later observed that the surgical removal of the pancreas caused glucose to appear in the urine, the hallmark of the disease diabetes mellitus. This led to the discovery that the pancreas, specifically the islets of Langerhans, produced a hormone that prevents this disease.

That hormone is **insulin,** secreted by the beta (β) cells of the islets. Insulin was not isolated until 1922 when Banting and Best succeeded where many others had not. On January 11, 1922, they injected an extract purified from beef pancreas into a 13-year-old diabetic boy, whose weight had fallen to 65 pounds and who was not expected to survive. With that single injection, the glucose level in the boy's blood fell 25%. A more potent extract soon brought the level down to near normal. The doctors had achieved the first instance of successful insulin therapy.

Glucagon

The islets of Langerhans produce another hormone; the alpha (α) cells of the islets secrete **glucagon,** which acts antagonistically to insulin (figure 45.15). When a person eats carbohydrates, the blood glucose concentration rises. Blood glucose directly activates the secretion of insulin by the β cells and inhibits the secretion of glucagon by the α cells. Insulin promotes the cellular uptake of glucose into the liver, muscle, and fat cells. It also activates the storage of glucose as glycogen in liver and muscle or as fat in fat cells. Between meals, when the concentration of blood glucose falls, insulin secretion decreases, and glucagon secretion increases. Glucagon promotes the hydrolysis of stored glycogen in the liver and fat in

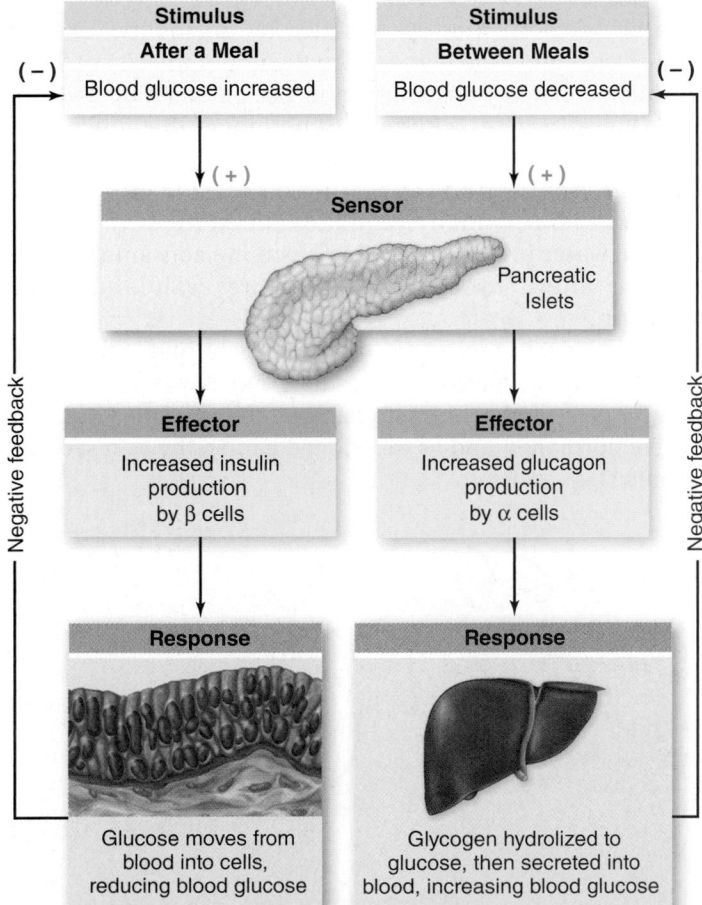

Figure 45.15 The antagonistic actions of insulin and glucagon on blood glucose. Insulin stimulates the cellular uptake of blood glucose into skeletal muscles, adipose cells, and the liver after a meal. Glucagon stimulates the hydrolysis of liver glycogen between meals, so that the liver can secrete glucose into the blood. These antagonistic effects help to maintain homeostasis of the blood glucose concentration.

adipose tissue. As a result, glucose and fatty acids are released into the blood and can be taken up by cells and used for energy.

Treatment of diabetes

Although many hormones favor the movement of glucose into cells, insulin is the only hormone that promotes movement of glucose from blood into cells. For this reason, disruptions in insulin signaling can have serious consequences. People with *type I,* or insulin-dependent, diabetes mellitus, lack the insulin-secreting β cells and consequently produce no insulin. Treatment for these patients consists of insulin injections. (Because insulin is a peptide hormone, it would be digested if taken orally and must instead be injected subcutaneously.)

In the past, only insulin extracted from the pancreas of pigs or cattle was available, but today people with insulin-dependent diabetes can inject themselves with human insulin produced by genetically engineered bacteria. Active research on the possibility of transplanting islets of Langerhans holds much promise of a lasting treatment for these patients.

Most diabetic patients, however, have *type II,* or noninsulin-dependent, diabetes mellitus. They generally have normal or even above-normal levels of insulin in their blood, but their cells have a reduced sensitivity to insulin. These people may not require insulin injections and can often control their diabetes through diet and exercise. It is estimated that over 90% of the cases of diabetes in North America are type II. Worldwide at least 346 million suffer from diabetes, and it is expected that this number will grow. Type II diabetes is especially common in developed countries, and it has been suggested that there is a linkage between type II diabetes and obesity.

Learning Outcomes Review 45.4

The major peripheral endocrine glands are the thyroid and parathyroid glands, the adrenal glands, and the pancreas. Calcium homeostasis results from the action of calcitonin, parathyroid hormone, and vitamin D. The adrenal glands produce stress hormones. Insulin and glucagon, antagonists from the pancreas, help maintain blood glucose at a normal level.

■ *Why does your body need two hormones to maintain blood sugar at a constant level?*

45.5 Other Hormones and Their Effects

Learning Outcomes

1. *Characterize the role of sex steroids in development.*
2. *List nonendocrine sources of hormones.*
3. *Identify the insect hormones involved in molting and metamorphosis.*

A variety of vertebrate and invertebrate processes are regulated by hormones and other chemical messengers, and in this section we review the most important ones.

Sex steroids regulate reproductive development

The ovaries and testes in vertebrates are important endocrine glands, producing the sex steroid hormones, including estrogens, progesterone, and testosterone (to be described in detail in chapter 52). Estrogen and progesterone are the primary "female" sex steroids, and testosterone and its immediate derivatives are the primary "male" sex steroids, or androgens. Both types of hormone can be found in both sexes, however.

During embryonic development, testosterone production in the male embryo is critical for the development of male sex organs. In mammals, sex steroids are responsible for the development of secondary sexual characteristics at puberty. These characteristics include breasts in females, body hair, and increased muscle mass in males. Because of this latter effect, some athletes have misused androgens to increase muscle mass. Use of steroids for this purpose has been condemned by virtually all major sports organizations, and it can cause liver disorders as well as a number of other serious side effects.

In females, sex steroids are especially important in maintaining the sexual cycle. Estrogen and progesterone produced in the ovaries are critical regulators of the menstrual and ovarian cycles. During pregnancy, estrogen production in the placenta maintains the uterine lining, which protects and nourishes the developing embryo.

Melatonin is crucial to circadian cycles

Another major endocrine gland is the pineal gland, located in the roof of the third ventricle of the brain in most vertebrates (see figure 43.22). It is about the size of a pea and is shaped like a pinecone, which gives it its name.

The pineal gland evolved from a medial light-sensitive eye (sometimes called a "third eye," although it could not form images) at the top of the skull in primitive vertebrates. This pineal eye is still present in primitive fish (cyclostomes) and some modern reptiles. In other vertebrates, however, the pineal gland is buried deep in the brain, and it functions as an endocrine gland by secreting the hormone melatonin.

Melatonin was named for its ability to cause blanching of the skin of lower vertebrates by reducing the dispersal of melanin granules. We now know, however, that it serves as an important timing signal delivered through the blood. Melatonin levels in the blood increase in darkness and fall during the daytime.

The secretion of melatonin is regulated by activity of the *suprachiasmatic nucleus* (*SCN*) of the hypothalamus. The SCN is known to function as the major biological clock in vertebrates, entraining (synchronizing) various body processes to a circadian rhythm—one that repeats every 24 hr. Through regulation by the SCN, the secretion of melatonin by the pineal gland is activated in the dark.

This daily cycling of melatonin release regulates wake-sleep and temperature cycles. Disruptions of these cycles, as occurs with jet lag or night shift work, can sometimes be minimized by melatonin administration. Melatonin also helps regulate reproductive cycles in some vertebrate species that have distinct breeding seasons.

Some hormones are not produced by endocrine glands

A variety of hormones are secreted by organs that are not exclusively endocrine glands. The thymus is the site of T cell production in many vertebrates and T cell maturation in mammals. It also secretes a number of hormones that function in the regulation of the immune system.

The right atrium of the heart secretes *atrial natriuretic hormone,* which stimulates the kidneys to excrete salt and water in the urine. This hormone acts antagonistically to aldosterone, which promotes salt and water retention.

The kidneys secrete *erythropoietin,* a hormone that stimulates the bone marrow to produce red blood cells. Other organs, such as the liver, stomach, and small intestine, also secrete hormones, and as mentioned earlier, the skin secretes vitamin D.

Figure 45.16 Hormonal control of metamorphosis in the silkworm moth, *Bombyx mori.* Molting hormone, ecdysone, controls when molting occurs. Brain hormone stimulates the prothoracic gland to produce ecdysone. Juvenile hormone determines the result of a particular molt. Juvenile hormone is produced by bodies near the brain called the corpora allata. High levels of juvenile hormone inhibit the formation of the pupa. Low levels of juvenile hormone are necessary for the pupal molt and metamorphosis.

Data analysis Juvenile hormone has been used as an insecticide. How would this affect the life cycle of an insect? Would it be lethal?

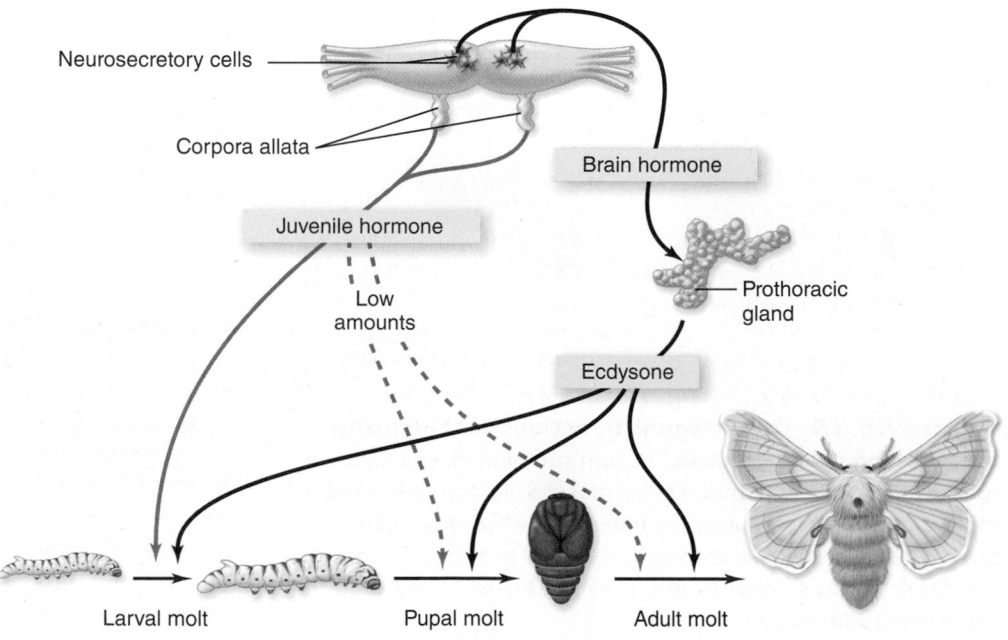

Neurosecretory cells

Corpora allata

Brain hormone

Juvenile hormone

Low amounts

Prothoracic gland

Ecdysone

Larval molt   Pupal molt   Adult molt

Insect hormones control molting and metamorphosis

Most invertebrate groups produce hormones as well; these control reproduction, growth, and color change. A dramatic action of hormones in insects is similar to the role of thyroid hormones in amphibian metamorphosis.

As insects grow during postembryonic development, their hardened exoskeletons do not expand. To overcome this problem, insects undergo a series of molts wherein they shed their old exoskeleton (figure 45.16) and secrete a new, larger one. In some insects, a juvenile insect, or larva, undergoes a radical transformation to the adult form during a single molt. This process is called metamorphosis.

Hormonal secretions influence both molting and metamorphosis in insects. Prior to molting, neurosecretory cells on the surface of the brain secrete a small peptide, **prothoracicotropic hormone (PTTH),** which in turn stimulates a gland in the thorax called the prothoracic gland to produce **molting hormone,** or **ecdysone** (see figure 45.16). High levels of ecdysone bring about the biochemical and behavioral changes that cause molting to occur.

Another pair of endocrine glands near the brain, called the *corpora allata,* produce a hormone called **juvenile hormone.** High levels of juvenile hormone prevent the transformation to the adult and result in a larval-to-larval molt. If the level of juvenile hormone is low, however, the molt will result in metamorphosis (figure 45.17).

Cancer cells may alter hormone production or have altered hormonal responses

Hormones and paracrine secretions actively regulate growth and cell division. Normally, hormone production is kept under precise control, but malfunctions in signaling systems can sometimes occur. Unregulated hormone stimulation can then lead to serious physical consequences.

Tumors that develop in endocrine glands, such as the anterior pituitary or the thyroid, can produce excessive amounts of hormones, causing conditions such as gigantism or hyperthyroidism. Spontaneous mutations can damage receptors or intracellular signaling proteins, with the result that target cell responses are activated even in the absence of hormone stimulation. Mutations in growth factor receptors, for example, can activate excessive cell division, resulting in tumor formation. Some tumors that develop in steroid-responsive tissues, such as the breast and prostate, remain sensitive to hormone stimulation. Blocking steroid hormone production can therefore diminish tumor growth.

The important effects of hormones on development and differentiation are illustrated by the case of diethylstilbestrol (DES). DES is a synthetic estrogen that was given to pregnant women from 1940 to 1970 to prevent miscarriage. It was subsequently discovered that daughters who had been exposed to DES as fetuses had an elevated probability of developing a rare form of cervical cancer later in life. Developmental alterations elicited by hormone treatment may thus take many years to become apparent.

www.ravenbiology.com

SCIENTIFIC THINKING

Hypothesis: *Juvenile hormone blocks or inhibits the stimulation of gene expression by ecdysone.*

Prediction: *Treatment of isolated imaginal discs with ecdysone plus increasing amounts of JH should show a decrease in ecdysone stimulated transcription.*

Test: *Discs dissected from late third instar* Drosophila *larvae are incubated in the presence of ecdysone, with and without JH. Incorporation of ³H-UTP into RNA was used as a measure of gene expression.*

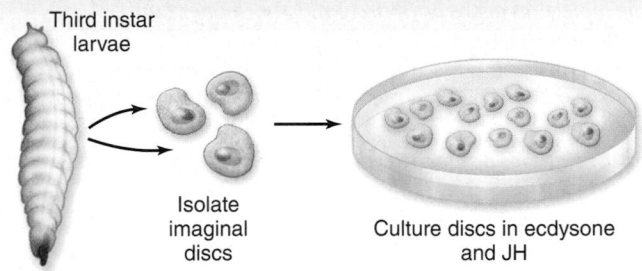

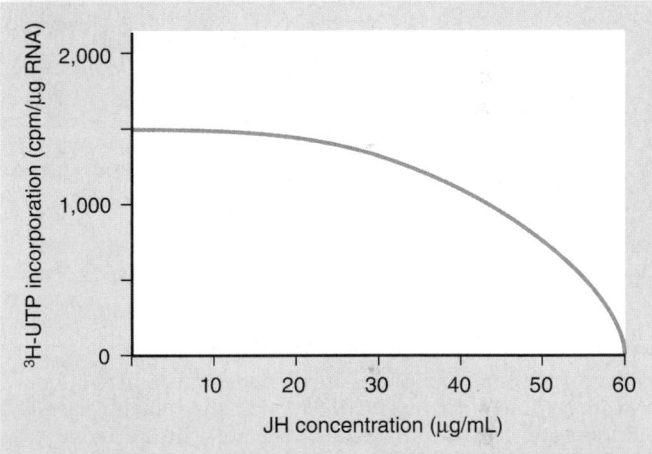

Result: *The graph shows a relatively high incorporation of ³H-UTP in the presence of ecdysone alone. The addition of JH causes dose-dependent reduction of RNA synthesis.*

Conclusion: *JH inhibits the ecdysone-stimulated synthesis of RNA in imaginal discs.*

Further Experiments: *How else can this system with isolated imaginal discs be used to analyze metamorphosis?*

Figure 45.17 **Effect of ecdysone and juvenile hormone on RNA synthesis in isolated *Drosophila* imaginal discs.**

Learning Outcomes Review 45.5

Testosterone causes an embryo to develop as a male; testosterone and estrogen produced at puberty are responsible for secondary sex characteristics. The female menstrual cycle is regulated by sex hormone balance. The thymus, the right atrium of the heart, and the kidneys secrete hormones although it is not their main function. In insects, molting hormone elicits molting, and low levels of juvenile hormone cause metamorphosis.

- ■ *Atrial natriuretic hormone reduces blood volume; would this affect blood pressure?*

45.1 Regulation of Body Processes by Chemical Messengers

Hormones are signaling molecules carried by the blood and may have distant targets. Paracrine regulators act locally, and pheromones released into the environment communicate between individuals of the same species.

Some molecules act as both circulating hormones and neurotransmitters.

Norepinephrine is a neurotransmitter in the sympathetic nervous system and also is a hormone that is released into the blood by the adrenal glands.

Endocrine glands produce three chemical classes of hormones.

The three classes of endocrine hormones are peptides and proteins, such as TSH; amino acid derivatives, such as thyroxine; and steroids, such as estrogen and testosterone (table 45.1).

Hormones can be categorized as lipophilic or hydrophilic.

Lipophilic hormones are fat-soluble and can cross the cell membrane; hydrophilic hormones are water-soluble and cannot cross membranes.

Paracrine regulators exert powerful effects within tissues.

Paracrine regulation occurs in most organs and among immune-system cells. Prostaglandins are involved in inflammation, and they are the target of NSAIDs.

45.2 Actions of Lipophilic Versus Hydrophilic Hormones

Lipophilic hormones activate intracellular receptors.

Circulating lipophilic hormones are carried in the blood bound to transport proteins (figure 45.3). They pass through the plasma membrane and activate intracellular receptors. The hormone-receptor complex can bind to specific gene promoter regions termed hormone response elements to activate transcription.

Hydrophilic hormones activate receptors on target cell membranes.

Hydrophilic hormones bind to a membrane receptor to initiate a signal transduction pathway (figure 45.6). Many receptors are kinases that phosphorylate proteins directly. Others are G protein–coupled receptors that activate a second-messenger system. Hydrophilic hormones tend to be short-lived, but lipophilic hormones tend to have effects of longer duration.

45.3 The Pituitary and Hypothalamus: The Body's Control Centers

The pituitary is a compound endocrine gland.

The anterior pituitary (adenohypophysis) is composed of glandular tissue derived from epithelial tissue; the posterior pituitary (neurohypophysis) is fibrous and is derived from neural tissue.

The posterior pituitary stores and releases two neurohormones.

The posterior pituitary contains axons from the hypothalamus that release neurohormones. One of these is ADH, involved in water reabsorption; the other is oxytocin.

The anterior pituitary produces seven hormones.

The hormones produced by the anterior pituitary include peptide, protein and glycoprotein hormones. These hormones tend to stimulate growth, and many are tropic hormones that stimulate other endocrine glands (table 45.1).

Hypothalamic neurohormones regulate the anterior pituitary.

Releasing and inhibiting hormones produced in the hypothalamus pass to the anterior pituitary through a portal system and regulate the anterior-pituitary's hormone production (figure 45.8).

Feedback from peripheral endocrine glands regulates anterior-pituitary hormones.

The activity of the anterior pituitary is also regulated by negative feedback; for example, thyroxine, produced by the thyroid in response to TSH, inhibits further secretion of TSH (figure 45.9).

Hormones of the anterior pituitary work directly and indirectly.

Three of the seven hormones, GH, prolactin, and MSH, work directly on nonendocrine tissues; the other four, ACTH, TSH, LH, and FSH, are tropic hormones that have endocrine glands as their targets. Defects in GH production can lead to either pituitary dwarfism (low), or gigantism (high).

45.4 The Major Peripheral Endocrine Glands

Some endocrine glands are controlled by tropic hormones of the pituitary, others are independent of pituitary control.

The thyroid gland regulates basal metabolism and development.

The thyroid hormones thyroxine and triiodothyronine regulate basal metabolism in vertebrates and trigger metamorphosis in amphibians (figure 45.12).

Calcium homeostasis is regulated by several hormones.

Blood calcium is regulated by calcitonin, which lowers blood calcium levels, and parathyroid hormone, which raises blood calcium levels (figure 45.13).

The adrenal gland releases both catecholamine and steroid hormones.

Catecholamines, epinephrine and norepinephrine, trigger "alarm" responses (figure 45.14). Corticosteroids maintain glucose homeostasis and modulate some aspects of the immune response.

Pancreatic hormones are primary regulators of carbohydrate metabolism.

Blood glucose is controlled by antagonistic hormones. The pancreas secretes insulin, which reduces blood glucose, and glucagon, which raises blood glucose (figure 45.15). Type I diabetes arises from loss of insulin-producing cells, and type II is a result of insulin insensitivity.

45.5 Other Hormones and Their Effects

Sex steroids regulate reproductive development.

Sex steroids regulate sexual development and reproduction. The ovaries primarily produce estrogen and progesterone, which are responsible for the menstrual cycle. The testes produce testosterone.

Melatonin is crucial to circadian cycles.

The pineal gland produces melatonin, which can control the dispersion of pigment granules and the daily wake–sleep cycles.

Some hormones are not produced by endocrine glands.

The thymus secretes hormones that regulate the immune system. The right atrium of the heart secretes atrial natriuretic hormone, which acts antagonistically to aldosterone. The skin manufactures and secretes vitamin D.

Insect hormones control molting and metamorphosis.

In insects the hormone ecdysone stimulates molting, and juvenile hormone levels control the nature of the molt. Metamorphosis requires high ecdysone and low juvenile hormone.

Cancer cells may alter hormone production or have altered hormonal responses.

Cancer developing from cells targeted by hormones, such as in the breast and prostate, may still be stimulated by those hormones.

Review Questions

UNDERSTAND

1. Which of the following best describes hormones?
   a. Hormones are relatively unstable and work only in the area adjacent to the gland that produced them.
   b. Hormones are long-lasting chemicals released from glands.
   c. All hormones are lipid-soluble.
   d. Hormones are chemical messengers that are released into the environment.

2. Steroid hormones
   a. can diffuse through the membrane without a carrier.
   b. have a direct effect on gene expression.
   c. bind to membrane receptors.
   d. Both a and b are correct.

3. Second messengers are activated in response to
   a. steroid hormones.   c. hydrophilic hormones.
   b. thyroxine.   d. All of the choices are correct.

4. Which of the following is true about lipophilic hormones?
   a. They are freely soluble in the blood.
   b. They require a transport protein in the bloodstream.
   c. They cannot enter their target cells.
   d. They are rapidly deactivated after binding to their receptors.

5. An organ is classified as part of the endocrine system if it
   a. produces cholesterol.
   b. is capable of converting amino acids into hormones.
   c. has intracellular receptors for hormones.
   d. secretes hormones into the circulatory system.

6. Hormones released from the pituitary gland have two different sources. Those that are produced by the neurons of the hypothalamus are released through the _____, and those produced within the pituitary are released through the _____.
   a. thalamus; hippocampus
   b. neurohypophysis; adenohypophysis
   c. right pituitary; left pituitary
   d. cortex; medulla

7. Which of the following conditions is unrelated to the production of growth hormone?
   a. Control of blood calcium
   b. Pituitary dwarfism
   c. Increased milk production in cows
   d. Acromegaly

APPLY

1. You think one of your teammates is using anabolic steroids to build muscle. You know that continued use of steroids can cause profound changes in cell function. This is due in part to the fact that these hormones act
   a. to regulate gene expression.
   b. by activating second messengers.
   c. as protein kinases.
   d. via G protein–coupled receptors.

2. Your Uncle Sal likes to party. When he goes out drinking, he complains that he needs to urinate more often. You explain to him that this is because alcohol suppresses the release of the hormone
   a. thyroxine, which increases water reabsorption from the kidney.
   b. thyroxine, which decreases water reabsorption from the kidney.
   c. ADH, which decreases water reabsorption from the kidney.
   d. ADH, which increases water reabsorption from the kidney.

3. Your new research project is to design a pesticide that will disrupt the endocrine systems of arthropods without harming humans and other mammals. Which of the following substances should be the target of your investigations?
   a. Insulin   c. Juvenile hormone
   b. ADH   d. Cortisol

4. Coat color in mammals is controlled by a hormone receptor called the melanocortin receptor. When this receptor is bound by the hormone MSH, pigment cells produce dark eumelanin. When the receptor is bound by an MSH antagonist that prevents MSH binding, pigment cells make yellow/red pheomelanin. In the Irish Setter, the overall red coat color could be due to a mutation in the
   a. receptor that prevents the antagonist from binding.
   b. receptor that prevents MSH from binding.
   c. MSH protein such that it binds the receptor more efficiently.
   d. antagonist such that it no longer binds to the receptor.

5. Tumors that affect the pituitary can lead to decreases in some, but not all, hormones released by the pituitary. A patient with such a tumor exhibits fatigue, weight loss, and low blood sugar. This is probably due to lack of production of
   a. GH, which leads to loss of muscle mass.
   b. ACTH, which leads to loss of production of glucocorticoids.
   c. TSH, which leads to loss of production of thyroxin.
   d. ADH, which leads to excess urine production.

6. You experience a longer period than normal between meals. Your body's response to this will be to produce
    a. insulin to raise your blood sugar.
    b. glucagon to raise your blood sugar.
    c. insulin to lower your blood sugar.
    d. glucagon to lower your blood sugar.

7. Mild vitamin D deficiency can lead to osteoporosis, or reduced bone mineral density. This is thought to be due to an association with increased levels of
    a. calcitonin, which leads to an increase in serum Ca^{2+} and bone loss.
    b. PTH, which leads to an increase in serum Ca^{2+} and bone loss.
    c. ADH, which reduces blood pressure and leads to bone loss.
    d. insulin, which leads to a decrease in blood glucose and bone loss.

SYNTHESIZE

1. How can blocking hormone production decrease cancerous tumor growth?

2. Suppose that two different organs, such as the liver and heart, are sensitive to a particular hormone (such as epinephrine). The cells in both organs have identical receptors for the hormone, and hormone-receptor binding produces the same intracellular second messenger in both organs. However, the hormone produces different effects in the two organs. Explain how this can happen.

3. Many physiological parameters, such as blood Ca^{2+} concentration and blood glucose levels, are controlled by two hormones that have opposite effects. What is the advantage of achieving regulation in this manner instead of by using a single hormone that changes the parameters in one direction only?

ONLINE RESOURCE

www.ravenbiology.com

Understand, Apply, and Synthesize—enhance your study with animations that bring concepts to life and practice tests to assess your understanding. Your instructor may also recommend the interactive eBook, individualized learning tools, and more.

Chapter **46**

The Musculoskeletal System

Chapter Contents

Introduction

The ability to move is so much a part of our daily lives that we tend to take it for granted. It is made possible by the combination of a semirigid skeletal system, joints that act as hinges, and a muscular system that can pull on this skeleton. Animal locomotion can be thought of as muscular action that produces a change in body shape, which places a force on the outside environment. When a racehorse runs down the track, its legs move forward and backward. As its feet contact the ground, the force they exert move its body forward at a considerable speed. In a similar way, when a bird takes off into flight, its wings exert force on the air; a swimming fish's movements push against the water. In this chapter, we will examine the nature of the muscular and skeletal systems that allow animal movement.

Types of Skeletal Systems

Muscles have to pull against something to produce the changes that cause movement. This necessary form of supporting structure is called a skeletal system. Zoologists commonly recognize three types of skeletal systems in animals: **hydrostatic skeletons, exoskeletons,** and **endoskeletons.**

Hydrostatic skeletons use water pressure inside a body wall

Hydrostatic skeletons are found primarily in soft-bodied terrestrial invertebrates, such as earthworms and slugs, and soft-bodied aquatic invertebrates, such as jellyfish, and squids.

Musculoskeletal action in earthworms

In these animals a fluid-filled central cavity is encompassed by two sets of muscles in the body wall: circular muscles that are repeated in segments and run the length of the body, and longitudinal muscles that oppose the action of the circular muscles.

Muscles act on the fluid in the body's central space, which represents the hydrostatic skeleton. As locomotion begins (figure 46.1), the anterior circular muscles contract, pressing

on the inner fluid and forcing the front of the body to become thin as the body wall in this region extends forward.

On the underside of a worm's body are short, bristle-like structures called chaetae. When circular muscles act, the chaetae of that region are pulled up close to the body and lose contact with the ground. Circular-muscle activity is passed backward, segment by segment, to create a backward wave of contraction.

As this wave continues, the anterior circular muscles now relax, and the longitudinal muscles take over, thickening the front end of the worm and allowing the chaetae to protrude and regain contact with the ground. The chaetae now prevent that body section from slipping backward. This locomotion process proceeds as waves of circular muscle contraction are followed by waves of longitudinal muscle effects.

Exoskeletons consist of a rigid outer covering

Exoskeletons are a rigid, hard case that surrounds the body. Arthropods, such as crustaceans and insects, have exoskeletons made of the polysaccharide *chitin* (figure 46.2a). As you learned in earlier chapters, chitin is found in the cell walls of fungi and some protists as well as in the exoskeletons of arthropods.

A chitinous exoskeleton resists bending and thus acts as the skeletal framework of the body; it also protects the internal organs and provides attachment sites for the muscles, which lie inside the exoskeletal casing. But in order to grow, the animal must periodically molt, shedding the exoskeleton (see chapter 34). The animal is vulnerable to predation until the new (slightly larger) exoskeleton forms. Molting crabs and lobsters often hide until the process is completed.

Exoskeletons have other limitations. The chitinous framework is not as strong as a bony, internal one. This fact by itself would set a limit for insect size, but there is a more important factor: Insects breathe through openings in their body that lead into tiny tubes, and as insect size increases beyond a certain limit, the ratio between the inside surface area of the tubes and the volume of the body overwhelms this sort of

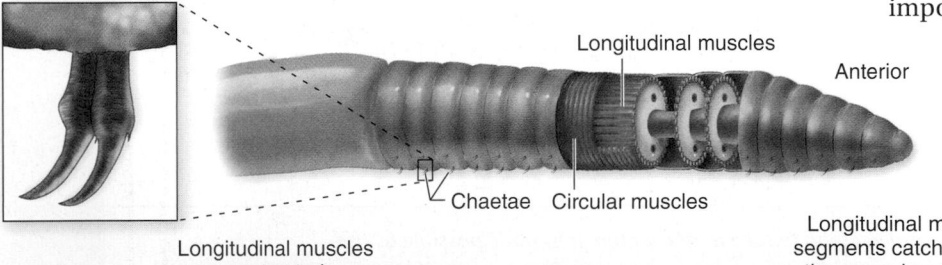

Longitudinal muscles

Anterior

Chaetae   Circular muscles

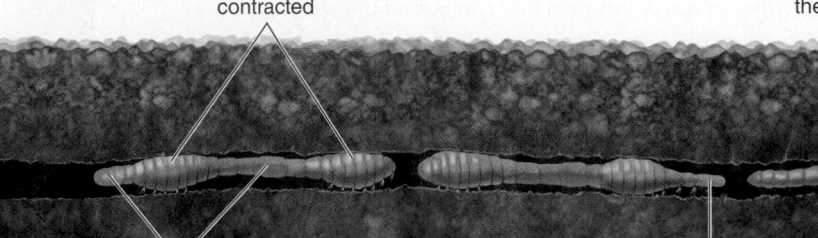

Longitudinal muscles contracted

Longitudinal muscles contract, and segments catch up. Chaetae attach to the ground and prevent backsliding.

Circular muscles contracted

Circular muscles contract, and anterior end moves forward. Chaetae lose attachment to ground.

Circular muscles contract, and anterior end moves forward.

Figure 46.1 Locomotion in earthworms. The hydrostatic skeleton of the earthworm uses muscles to move fluid within the segmented body cavity, changing the shape of the animal. When circular muscles contract, the pressure in the fluid rises. At the same time the longitudinal muscles relax, and the body becomes longer and thinner. When the longitudinal muscles contract and the circular muscles relax, the chaetae of the worm's lower surface extend to prevent backsliding. A wave of circular followed by longitudinal muscle contractions down the body produces forward movement.

Exoskeleton

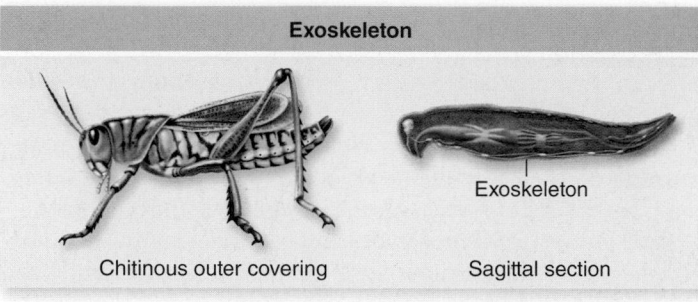

Chitinous outer covering   Sagittal section   Exoskeleton

a.

Endoskeleton

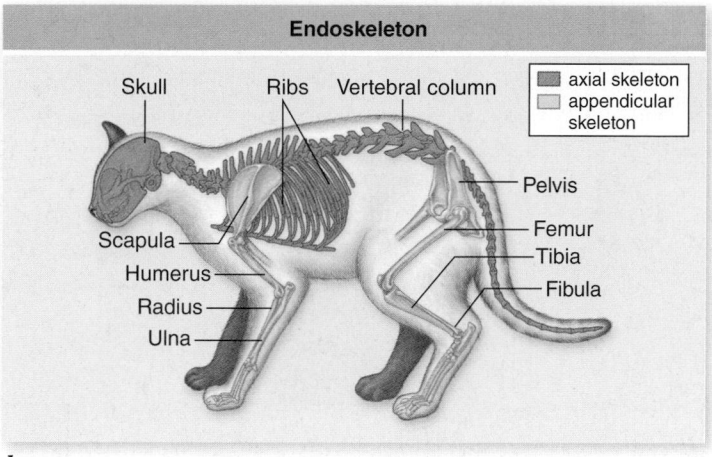

Skull   Ribs   Vertebral column

☐ axial skeleton
☐ appendicular skeleton

Scapula
Humerus
Radius
Ulna

Pelvis
Femur
Tibia
Fibula

b.

Figure 46.2 Exoskeleton and endoskeleton.
a. The hard, tough outer covering of an arthropod, such as this grasshopper, is its exoskeleton and is composed of chitin.
b. Vertebrates, such as this cat, have endoskeletons formed of bone and cartilage. Some of the major bony features are labeled.

respiratory system. Finally, when muscles are confined within an exoskeleton, they cannot enlarge in size and power with increased use, as they can in animals with endoskeletons.

Endoskeletons are composed of hard, internal structures

Endoskeletons, found in vertebrates and echinoderms, are rigid internal skeletons that form the body's framework and offer surfaces for muscle attachment. Echinoderms, such as sea urchins and sand dollars, have skeletons made of calcite, a crystalline form of calcium carbonate. This calcium compound is different from that in bone, which is based on calcium phosphate.

Vertebrate skeletal tissues

The vertebrate endoskeleton (figure 46.2*b*) includes fibrous dense connective tissue along with the more rigid special connective tissues, cartilage or bone (see chapter 42). Cartilage is strong and slightly flexible, a characteristic important in such functions as padding the ends of bones where they come together in a joint. Although some large, active animals such as sharks have totally cartilaginous skeletons, bone is the main component in vertebrate skeletons. Bone is much stronger than cartilage and much less flexible.

Unlike chitin, both cartilage and bone are living tissues. Bone, particularly, can have high metabolic activity

especially if bone cells are present throughout the matrix, a common condition. Bone, and to some extent cartilage, can change and remodel itself in response to injury or to physical stresses.

Learning Outcomes Review 46.1

With a hydrostatic skeleton, muscle contraction puts pressure on the fluid inside the body, forcing the body to extend. Opposing muscles then shorten the body to draw the animal forward. Invertebrate exoskeletons consist of hard chitin; they must be shed and renewed (molting) for the animal to grow. Endoskeletons are composed of fibrous dense connective tissue along with cartilage or mineralized bone.

■ **What limitations does an exoskeleton impose on terrestrial invertebrates?**

46.2 A Closer Look at Bone

Learning Outcomes

1. Compare intramembranous and endochondral development.
2. Describe how growth occurs in epiphyses.
3. Explain how bone remodeling occurs.

Bone is a hard but resilient tissue that is unique to vertebrate animals. This connective tissue first appeared over 520 MYA and is now found in all vertebrates except cartilaginous fishes (see chapter 35).

Bones can be classified by two modes of development

Bone tissue itself can be of several types classified in a few different ways. The most common system is based on the way in which bone develops.

Intramembranous development

In intramembranous development, bones form within a layer of connective tissue. Many of the flat bones that make up the exterior of the skull and jaw are intramembranous.

Typically, the site of the intramembranous bone-to-be begins in a designated region in the dermis of the skin. During embryonic development, the dermis is formed largely of **mesenchyme**—a loose tissue consisting of undifferentiated mesenchyme cells and other cells that have arisen from them—along with collagen fibers. Some of the undifferentiated mesenchyme cells differentiate to become specialized cells called **osteoblasts** (figure 46.3). These osteoblasts arrange themselves along the collagenous fibers and begin to secrete the enzyme alkaline phosphatase, which causes calcium phosphate salts to form in a crystalline configuration called *hydroxyapatite*. The crystals merge along the fibers to encase them.

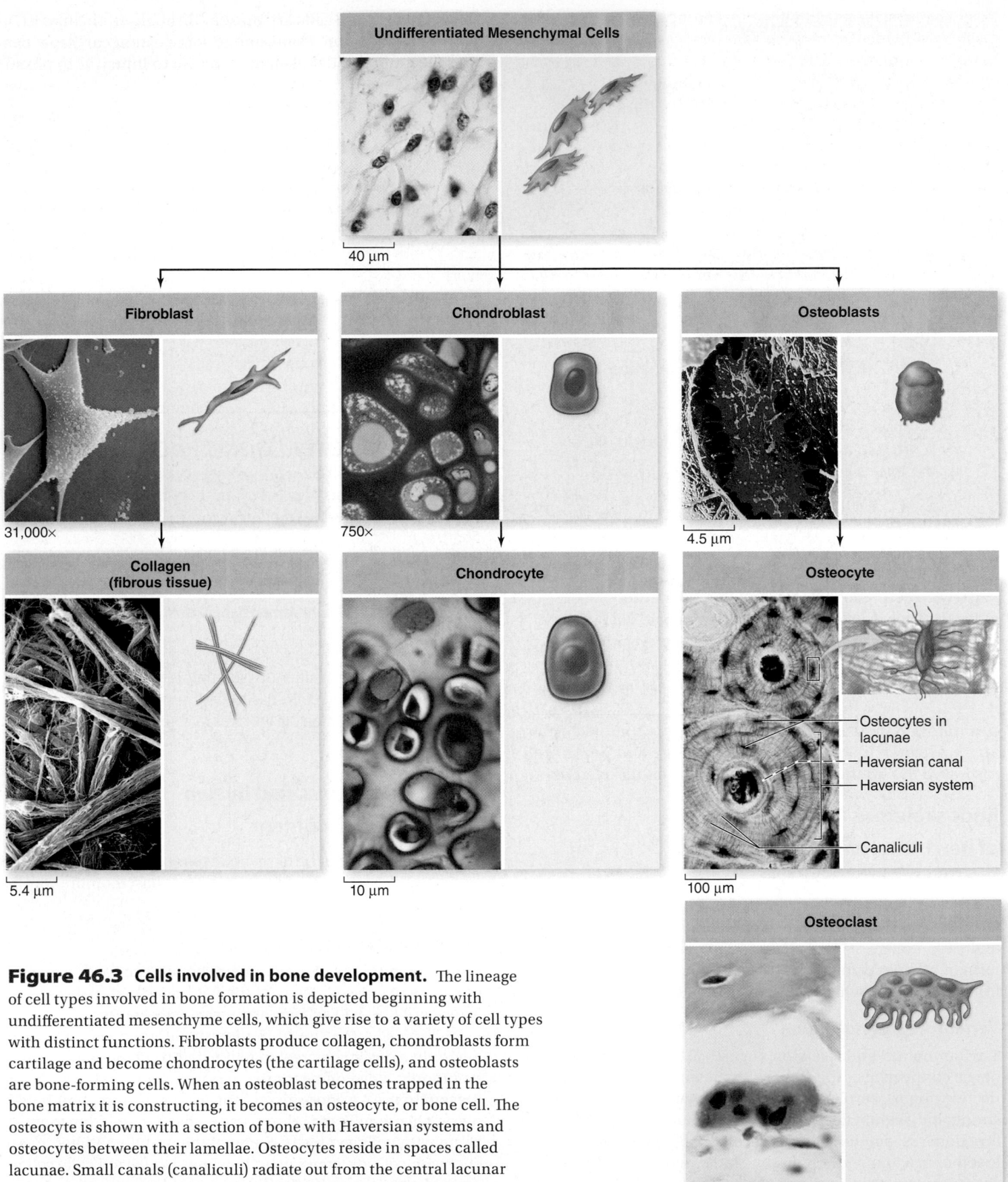

Figure 46.3 **Cells involved in bone development.** The lineage of cell types involved in bone formation is depicted beginning with undifferentiated mesenchyme cells, which give rise to a variety of cell types with distinct functions. Fibroblasts produce collagen, chondroblasts form cartilage and become chondrocytes (the cartilage cells), and osteoblasts are bone-forming cells. When an osteoblast becomes trapped in the bone matrix it is constructing, it becomes an osteocyte, or bone cell. The osteocyte is shown with a section of bone with Haversian systems and osteocytes between their lamellae. Osteocytes reside in spaces called lacunae. Small canals (canaliculi) radiate out from the central lacunar space, which contains the arms of the osteocyte. Osteoclasts, bone-removing cells, are not derived from mesenchyme cells but are formed by fusion of monocytes, a type of white blood cell.

The crystals give the bone its hardness, but without the resilience afforded by collagen's stretching ability, bone would be rigid but dangerously brittle. Typical bones have roughly equal volumes of collagen and hydroxyapatite, but hydroxyapatite contributes about 65% to the bone's weight.

As the osteoblasts continue to make bone crystals, some become trapped in the bone matrix and undergo dramatic changes in shape and function, now becoming cells called osteocytes (see figure 46.3). They lie in tight spaces within the bone matrix called lacunae. Little canals extending from the lacunae, called **canaliculi,** permit contact of the starburst-like extensions of each osteocyte with those of its neighbors (see figure 46.3). In this way, many cells within bones can participate in intercellular communication.

As an intramembranous bone grows, it requires alterations of shape. Imagine that you were modeling with clay, and you wanted to take a tiny clay bowl and make it larger. Simply putting more clay on the outside would not work; you would need to remove clay from the inside to increase the bowl's capacity as well. As bone grows, it must also undergo a remodeling process, with matrix being added in some regions and removed in others. This is where osteoclasts come in. These unusual cells are formed from the fusion of monocytes, a type of white blood cell, to form large multinucleate cells. Their function is to break down the bone matrix.

Endochondral development

Bones that form through endochondral development are typically those that are deeper in the body and form its architectural framework. Examples include vertebrae, ribs, bones of the shoulder and pelvis, long bones of the limbs, and the most

internal of the skull bones. Endochondral bones begin as tiny, cartilaginous models that have the rough shape of the bones that eventually will be formed. Bone development of this kind consists of adding bone to the outside of the cartilaginous model, while replacing the interior cartilage with bone.

Bone added to the outside of the model is produced in the fibrous sheath that envelopes the cartilage. This sheath is tough and made of collagen fibers, but it also contains undifferentiated mesenchyme cells. Osteoblasts arise and sort themselves out along the fibers in the deepest part of the sheath. Bone is then formed between the sheath and the cartilaginous matrix. This process is somewhat similar to what occurs in the dermis in the production of intramembranous bone.

As the outer bone is formed, the interior cartilage begins to calcify. The calcium source for this process seems to be the cartilage cells themselves. As calcification continues, the inner cartilaginous tissue breaks down into pieces of debris. Blood vessels from the sheath, now called the periosteum, force their way through the outer bony jacket, thus entering the interior of the cartilaginous model, and cart off the debris. Again, trapped osteoblasts transform into osteocytes, and osteoclasts for bone remodeling arise from cell fusions in the same manner as occurs in intramembranous bone. Growth in bone thickness occurs by adding additional bone layers just beneath the periosteum.

Endochondral bones increase in length in a different way, unlike growth in intramembranous development. As an example, consider a long bone such as a mammalian humerus (in humans, the upper arm bone). Like many limb bones, it is formed of a slender shaft with widened ends, called **epiphyses** (figure 46.4). Within the epiphyses are the *epiphyseal growth*

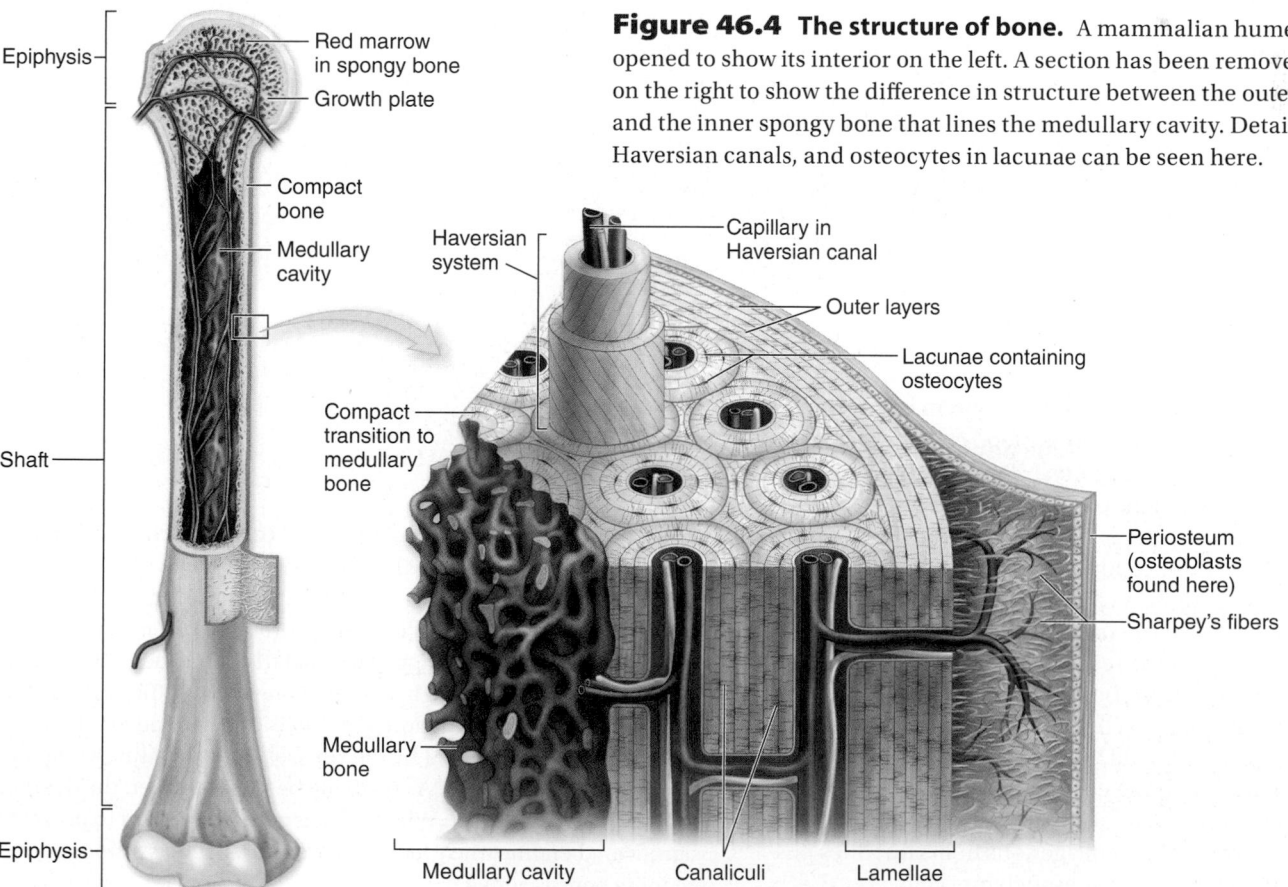

Figure 46.4 The structure of bone. A mammalian humerus is partly opened to show its interior on the left. A section has been removed and magnified on the right to show the difference in structure between the outer compact bone and the inner spongy bone that lines the medullary cavity. Details of basic layers, Haversian canals, and osteocytes in lacunae can be seen here.

plates that separate the epiphyses from the shaft itself. As long as the bone is growing in length, these growth plates are composed of cartilage (see figure 46.4). The actual events taking place in the plates are not simple, but they can be simply summarized.

1. During growth of a long bone, the cartilage of the growth plates is actively growing in the lengthwise direction to thicken the plate.
2. This growth pushes the epiphysis farther away from the slender shaft portion, which effectively increases the length of the bone.
3. At the same time, from the shaft's side, a process of cartilage calcification encroaches on the cartilaginous growth plate, so that the bony portion of the shaft elongates.

As long as the rate of new cartilage thickening stays ahead of the creeping calcification, the bone continues to grow in length. Eventually the cartilaginous expansion slows and is overtaken by the calcification, which obliterates this region of growth.

Growth in length usually ceases in humans by late adolescence. Although growth of the bone length is curtailed at this time, growth in width is not. The diameter of the shaft can be enhanced by bone addition just beneath the periosteum throughout an individual's life.

Bone structure may include blood vessels and nerves

Developing bone often has an internal blood supply, which is especially evident in endochondral bones. The internal blood routes, however, do not necessarily remain after the bones have completed development. In most mammals the endochondral bones retain internal blood vessels and are called **vascular bones.** Vascular bone is also found in many reptiles and a few amphibians. *Cellular bones* contain osteocytes, and many such bones are also vascular. This bone remains metabolically active (see figure 46.4).

In fishes and birds, bones are **avascular.** Typically avascular bone does not contain osteocytes and is termed *acellular bone.* This type of bone is fairly inert except for its surface, where the periosteum with its mesenchyme cells is capable of repairing the bone.

Many bones, particularly the endochondral long bones, contain a central cavity termed the *medullary cavity.* In many vertebrates, the medullary cavity houses the bone marrow, important in the manufacture of red and white blood cells. In such cases this cavity is termed the **marrow cavity.** Not all medullary cavities contain marrow, however. Light-boned birds, for example, have huge interior cavities, but they are empty of marrow. Birds depend on stem cells in other body locations to produce red blood cells.

Bone lining the medullary cavities differs from the smooth, dense bone found closer to the outer surface. Based on density and texture, bone falls into three categories: the outer dense **compact bone,** the **medullary bone** that lines the internal cavity, and **spongy bone** that has a honeycomb structure and typically forms the epiphyses inside a thick shell of compact bone. Both compact and spongy bone contribute to a bone's strength. Medullary cavities are lined with thin tissues called the **endosteum,** which contains no collagenous fibers but does possess other constituents including mesenchyme cells.

Vascular bone usually has a special internal organization called the **Haversian system.** Beneath the outer basic layers, endochondral bone is constructed of concentric layers called *Haversian lamellae.* These concentric tubes are laid down around narrow channels called *Haversian canals* that run parallel to the length of the bone. Haversian canals may contain nerve fibers but always contain blood vessels that keep the osteocytes alive even though they are entombed in the bony matrix.

The small vessels within the canals include both arterioles and venules or capillaries, and they connect to larger vessels that extend internally from both the periosteum and endosteum and that run in canals perpendicular to the Haversian canals.

Bone remodeling allows bone to respond to use or disuse

It is easy to think of bones as being inert, especially since we rarely encounter them except as the skeletons of dead animals. But just as muscles, skin, and other body tissues may change depending on the stresses of the environment, bone also is a dynamic tissue that can change with demands made on it.

Mechanical stresses such as compression at joints, the forces of muscles on certain portions and features of a bone, and similar effects may all be remodeling factors that not only shape the bone during its embryonic development, but after birth as well. Depending on the directions and magnitudes of forces impinging on a bone, it may thicken; the size and shape of surface features to which muscles, tendons, or ligaments attach may change in size and shape; even the direction of the tiny bony struts that make up spongy bone may be altered. Exercise and frequent use of muscles for a particular task change more than just the muscles; blood vessels and fibrous connective tissue increase, and the skeletal frame becomes more robust through bone thickening and enhancement.

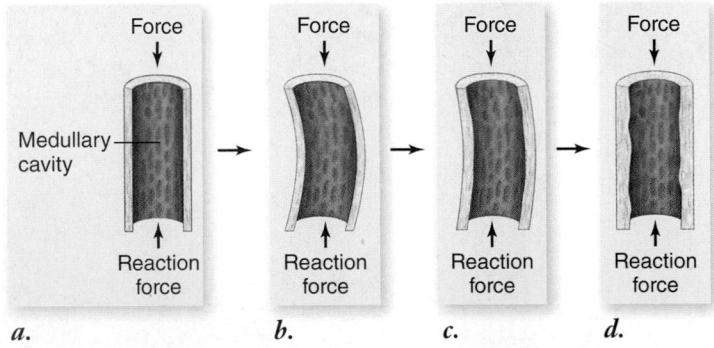

Figure 46.5 **Model of stress and remodeling in a long bone.** This figure shows a diagrammatic section of a long bone, such as a leg bone. The section is placed under a load or force, which causes a reaction force from the ground the leg is standing upon. *a.* Under a mild compressive load the bone does not bend. *b.* If the load is large enough, and the bone is not sufficiently thick, the bone will bend (the bending shown is exaggerated for clarity). *c.* Osteoblasts are signaled by the stresses in the bending section to produce additional bone. As the bone becomes thicker, the degree of bending is reduced. *d.* When sufficient bone is added to prevent significant bending, the production of new osteoblasts stops and no more bone is added.

The phenomenon of remodeling is known for all bones, but it is easiest to demonstrate in a long bone. Small forces may not have much of an effect on the bone, but larger ones—if frequent enough—can initiate remodeling (figure 46.5). In the example shown, larger compressive forces may tend to bend a bone, even if the bend is imperceptible to the eye. This bending stress promotes bone formation that thickens the bone. As the bone becomes thicker the amount of bending is reduced (figure 46.5c). Further bone addition produces sufficient bone thickness to entirely prevent significant bending (figure 46.5d). Once this point has been attained, the bone addition stops. This is another example of a negative-feedback system.

The effect of remodeling can be seen by examining bone thickness in rodents forced to exercise. The continual stresses placed on the limb bones cause additional bone to be deposited, leading to thicker and stronger bone (figure 46.6).

The same phenomenon is seen in human weight lifters, who develop stronger and thicker arm bones, and equestrians, whose femurs become much thicker on the inside because of the forces they exert to remain mounted.

This phenomenon also has important medical implications. Osteoporosis, which is characterized by a loss of bone mineral density, is a debilitating and potentially life-threatening ailment that afflicts more than 10 million people with another 34 million at increased risk in the United States, affecting primarily postmenopausal women, but also those suffering from malnutrition and a number of diseases. One treatment is a regimen of weight-lifting to stimulate bone deposition and thus counter the effects of osteoporosis.

SCIENTIFIC THINKING

Hypothesis: *Bone remodeling strengthens bones in response to external pressures.*

Prediction: *Bones that are used in more strenuous activities will deposit more bone and become stronger.*

Test: *Provide laboratory mice with an exercise wheel and make sure they run for several hours a day; keep a control group without a wheel.*

Mouse with exercise wheel   Mouse without exercise wheel

Result: *After 10 weeks, the running mice developed thicker limb bones.*

Further Experiments: *Modern microelectronics allow the development of stress sensors small enough to implant on the limb bone of a mouse. With such sensors, experiments can quantify how much stress different activities place on a bone and can more accurately investigate the relationship between the direction and magnitude of forces placed on a bone and the extent to which the bone remodels.*

Figure 46.6 The effect of exercise on bone remodeling.

46.3 Joints and Skeletal Movement

Learning Outcomes

1. *Define the different types of joints.*
2. *Explain how muscles produce movement at joints.*
3. *Describe how antagonistic muscles work at a joint.*

Movements of the endoskeleton are powered by the skeletal musculature. The skeletal movements that respond to muscle action occur at **joints,** or articulations, where one bone meets another.

Moveable joints have different ranges of motion, depending on type

Each movable joint within the skeleton has a characteristic range of motion. Four basic joint movement patterns can be distinguished: *ball-and-socket, hinge, gliding,* and *combination.*

Ball-and-socket joints are like those of the hip, where the upper leg bone forms a ball fitting into a socket in the pelvis. This type of joint can perform universal movement in all directions, plus twisting of the ball (figure 46.7a).

The simplest type of joint is the **hinge joint,** such as the knee, where movement of the lower leg is restricted to rotate forward or backward, but not side to side (figure 46.7b).

Gliding joints can be found in the skulls of a number of nonmammalian vertebrates, and are also present between the lateral vertebral projections in many vertebrates, including mammals (figure 46.7c). The vertebral projections are paired and extend from the front and back of each vertebra. The projections in front are a little lower, and each can slip along the undersurface of the posterior projection from the vertebra just ahead of it. This sliding joint gives stability to the vertebral column while allowing some flexibility of movement between vertebrae.

Combination joints are, as you might suppose, those that have movement characteristics of two or more joint types. The typical mammalian jaw joint is a good example.

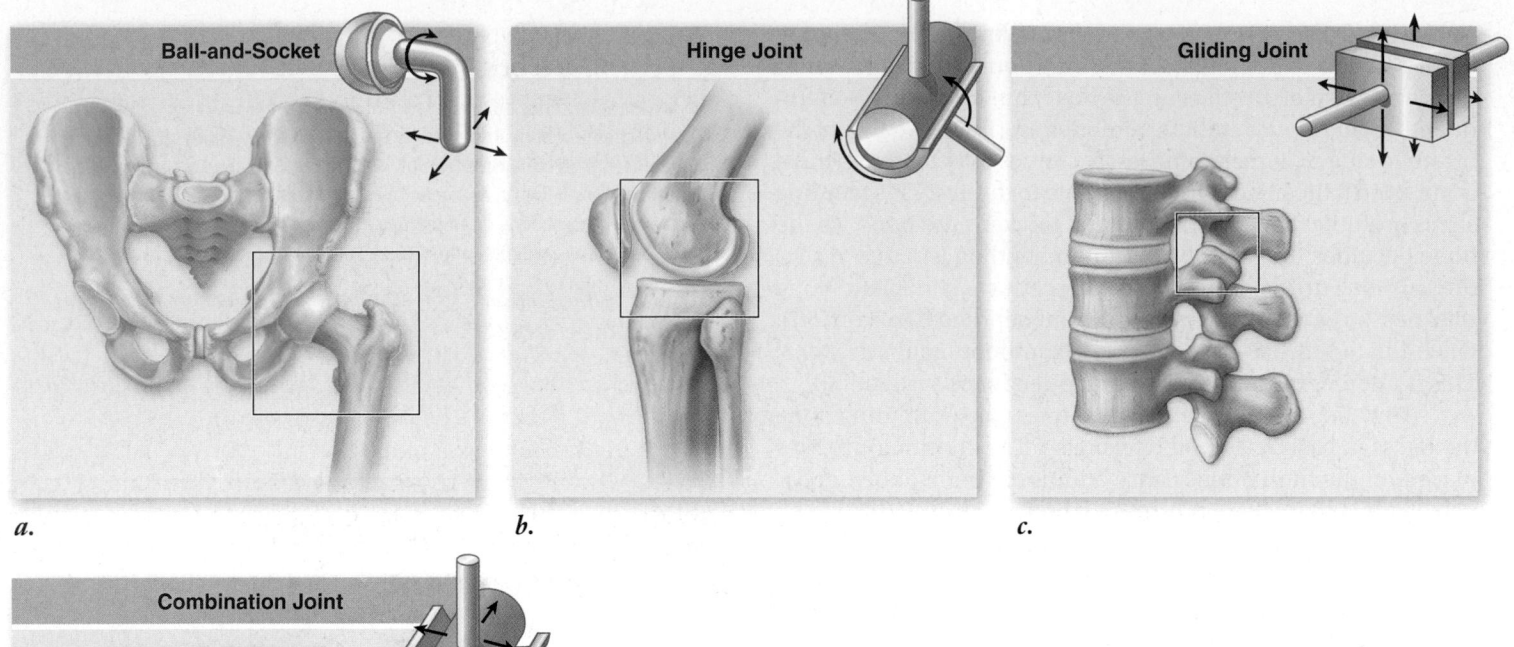

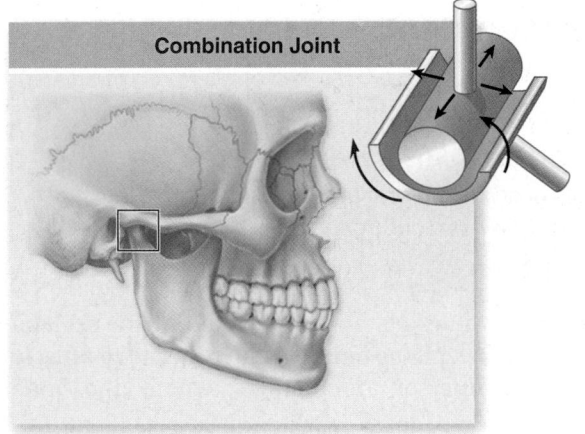

Figure 46.7 Patterns of joint movement. *a.* Ball-and-socket joints, such as the hip joint, permit movement and twisting of the leg within the hip socket. *b.* A hinge joint, as the term implies, allows movement in only one plane. *c.* Gliding joints are well represented by the lateral vertebral joints (not the central ones) that permit sliding of one surface on another. *d.* Combination joints have features of more than one type of joint, such as the mammalian jaw joint that allows both rotation and side-to-side sliding.

Most mammals chew food into small pieces. To chew food well, the lower jaw needs to move from side to side to get the best contact between upper and lower teeth. The lower jaw can also slip forward and backward to some extent. At the same time, the jaw joint must be shaped to allow the hinge-like opening and closing of the mouth. The mammalian joint conformation thus combines features from hinge and gliding joints (figure 46.7d).

Skeletal muscles pull on bones to produce movement at joints

Skeletal muscles produce movement of the skeleton when they contract. Usually, the two ends of a skeletal muscle are attached to different bones, although some may be attached to other structures, such as skin. There are two means of bone attachment: Muscle fibers may connect directly to the periosteum, the bone's fibrous covering, or sheets of muscle may be connected to bone by a dense connective tissue strap or cord, called a *tendon* that attaches to the periosteum (figure 46.8).

One attachment of the muscle, the origin, remains relatively stationary during a contraction. The other end, the insertion, is attached to a bone that moves when the muscle

contracts. For example, contraction of the quadriceps muscles of the leg causes the lower leg to rotate forward relative to the upper leg section.

Typically, muscles are arranged so that any movement produced by one muscle can be reversed by another. The leg flexor muscles, called hamstrings (see figure 46.8), draw the lower leg back and upward, bending the knee. Its movement is countered by the quadriceps muscles. The important concept is that two muscles or muscle groups can be mutually antagonistic, with the action of one countered by the action of the other.

Learning Outcomes Review 46.3

Types of joints include ball-and-socket, hinge, gliding, and combination joints. Muscles, positioned across joints, cause movement of bones relative to each other by contracting and exerting pulling force. Antagonistic muscles oppose each other, a key feature since muscles can only contract and cannot push.

■ *In what ways does a bony endoskeleton overcome the limitations of an exoskeleton for terrestrial life-forms?*

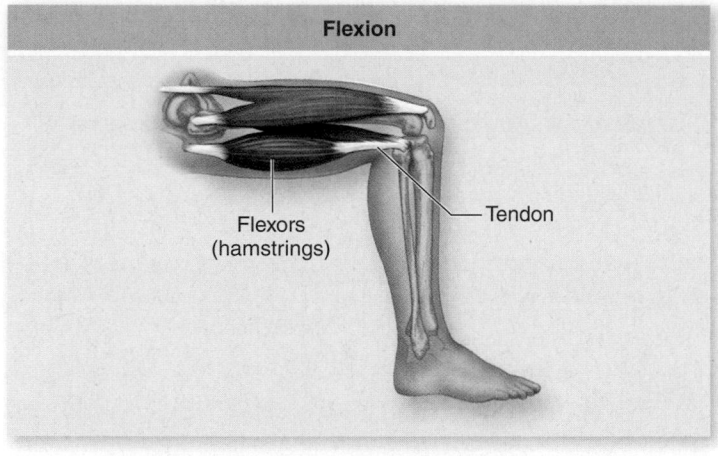

Flexion

Flexors
(hamstrings)

Tendon

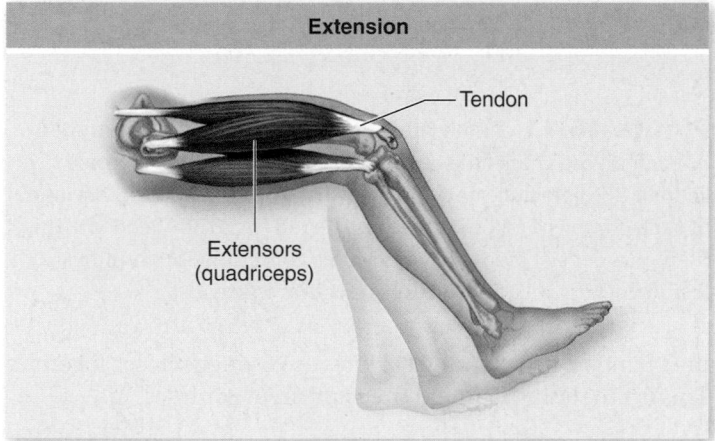

Extension

Tendon

Extensors
(quadriceps)

Figure 46.8 Flexor and extensor muscles of the leg.
Antagonistic muscles act in opposite ways. In humans, the
hamstrings, a group of three muscles, cause the lower leg to move
backward relative to the upper leg, whereas the quadriceps, a
group of four muscles, pull the lower leg forward.

> **?** **Inquiry question** Would the antagonistic muscles
> work in the same way in the legs of an animal with an
> exoskeleton, such as the grasshopper in figure 46.2?

46.4 *Muscle Contraction*

Learning Outcomes

1. *Explain the sliding filament mechanism of muscle
   contraction.*
2. *Describe the role of calcium in muscle contraction.*
3. *Differentiate between slow-twitch and fast-twitch
   muscle fibers.*

This section concentrates on the skeletal muscle of vertebrates. Vertebrate muscle has enjoyed the most attention
and is thus the best understood of animal muscular

function. Each skeletal muscle contains numerous muscle
fibers, as described in chapter 42. Each muscle fiber
encloses a bundle of 4 to 20 elongated structures called
myofibrils. Each myofibril, in turn, is composed of thick
and thin **myofilaments** (figure 46.9).

Under a microscope, the myofibrils have alternating
dark and light bands, which give skeletal muscle fiber its
striped appearance. The thick myofilaments are stacked
together to produce the dark bands, called *A bands;* the
thin filaments alone are found in the light bands, or
I bands.

Each I band in a myofibril is divided in half by a disk of
protein called a *Z line* because of its appearance in electron
micrographs. The thin filaments are anchored to these disks. In
an electron micrograph of a myofibril (figure 46.10), the structure of the myofibril can be seen to repeat from Z line to Z line.
This repeating structure, called a **sarcomere,** is the smallest
subunit of muscle contraction.

Muscle fibers contract as overlapping filaments slide together

The thin filaments overlap with thick filaments on each side of
an A band, but in a resting muscle, they do not project all the
way to the center of the A band. As a result, the center of an
A band (called an *H band*) is lighter than the areas on each
side, which have interdigitating thick and thin filaments. This
appearance of the sarcomeres changes when the muscle
contracts.

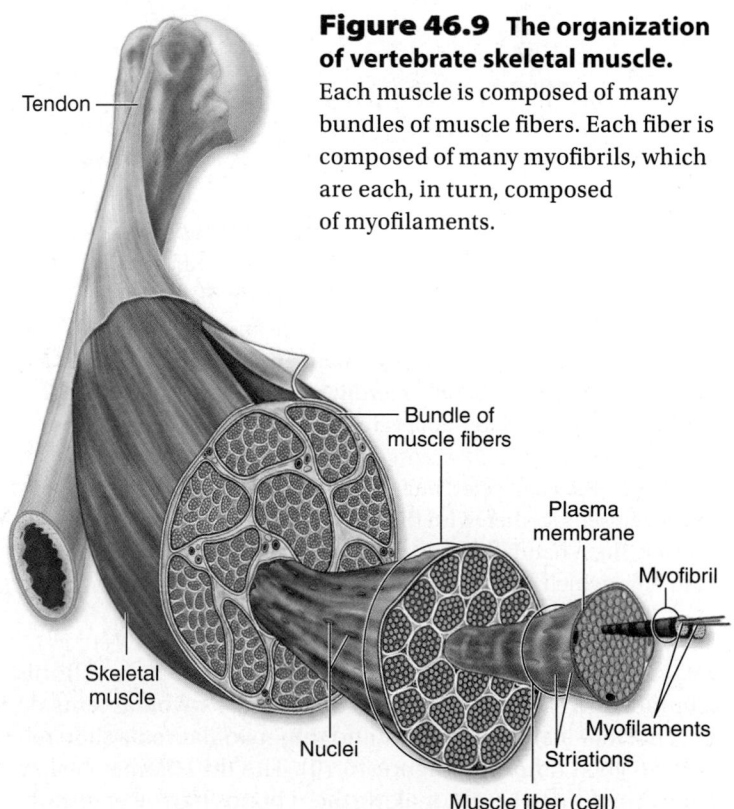

**Figure 46.9 The organization
of vertebrate skeletal muscle.**
Each muscle is composed of many
bundles of muscle fibers. Each fiber is
composed of many myofibrils, which
are each, in turn, composed
of myofilaments.

Tendon

Bundle of
muscle fibers

Plasma
membrane

Myofibril

Myofilaments

Striations

Nuclei

Skeletal
muscle

Muscle fiber (cell)

Relaxed Muscle

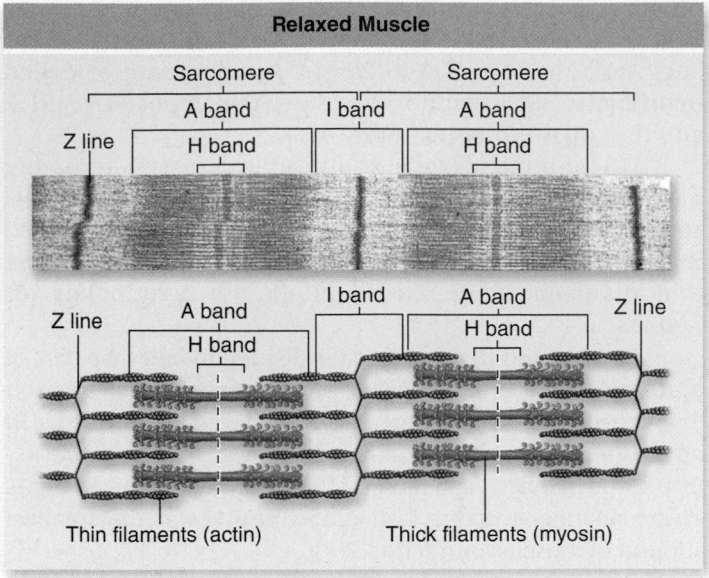

0.49 μm

Contracted Muscle

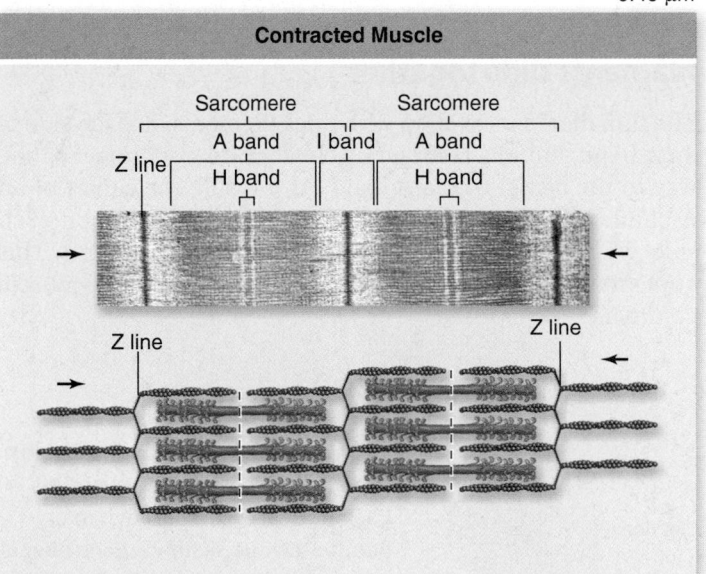

0.45 μm

Figure 46.10 **The structure of sarcomeres in relaxed and contracted muscles.** Two sarcomeres are shown in micrographs and as drawings of thick and thin filaments. The Z lines form the borders of each sarcomere and the A bands represent thick filaments. The thin filaments are within the I bands and extend into the A bands interdigitated with thick filaments. The H band is the lighter-appearing central region of the A band containing only thick filaments. The muscle on the top is shown relaxed. In the contracted muscle in the bottom, the Z lines have moved closer together, with the I bands and H bands becoming shorter. The A band does not change in size as it contains the thick filaments, which do not change in length.

A muscle contracts and shortens because its myofibrils contract and shorten. When this occurs, the myofilaments do *not* shorten; instead, the thick and thin myofilaments slide relative to each other (see figure 46.10). The thin filaments slide deeper into the A bands, making the H bands narrower until, at maximal shortening, they disappear entirely. This also makes

Myosin Molecule

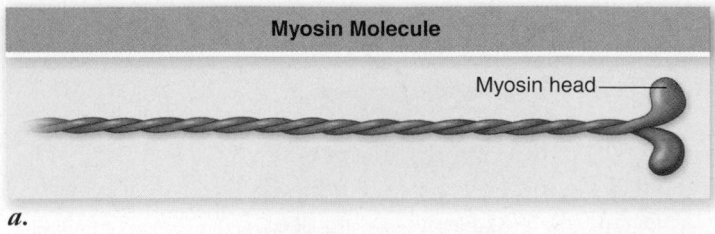

a.

Thick Filament

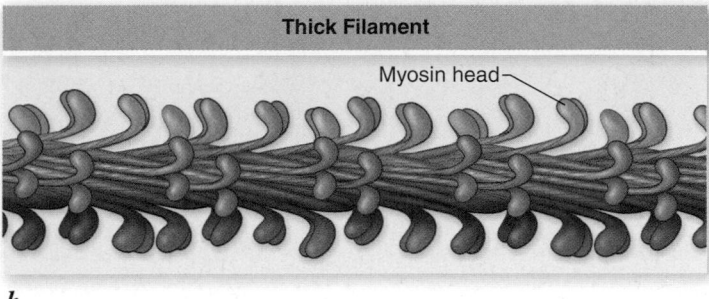

b.

Figure 46.11 **Thick filaments are composed of myosin.** *a.* Each myosin molecule consists of two polypeptide chains shaped like golf clubs and wrapped around each other; at the end of each chain is a globular region referred to as the "head." *b.* Thick filaments consist of myosin molecules combined into bundles from which the heads protrude at regular intervals.

the I bands narrower, as the Z lines are brought closer together. This is the sliding filament mechanism of contraction.

The sliding filament mechanism

Electron micrographs reveal cross-bridges that extend from the thick to the thin filaments, suggesting a mechanism that might cause the filaments to slide. To understand how this is accomplished requires examining the thick and thin filaments at a molecular level. Biochemical studies show that each thick filament is composed of many subunits of the protein myosin packed together. The myosin protein consists of two subunits, each shaped like a golf club with a head region that protrudes from a long filament, with the filaments twisted together. Thick filaments are composed of many copies of myosin arranged with heads protruding from along the length of the fiber (figure 46.11). The myosin heads form the cross-bridges seen in electron micrographs.

Each thin filament consists primarily of many globular actin proteins arranged into two fibers twisted into a double helix (figure 46.12). If we were able to see a sarcomere at the molecular level, it would have the structure depicted in figure 46.13.

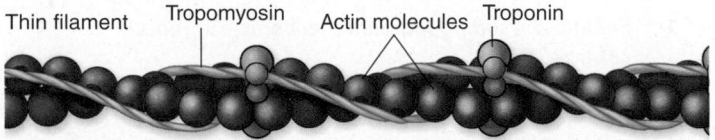

Figure 46.12 **Thin filaments are composed of globular actin proteins.** Two rows of actin proteins are twisted together in a helix to produce the thin filaments. Other proteins, tropomyosin and troponin, associate with the strands of actin and are involved in muscle contraction. These other proteins are discussed later in the chapter.

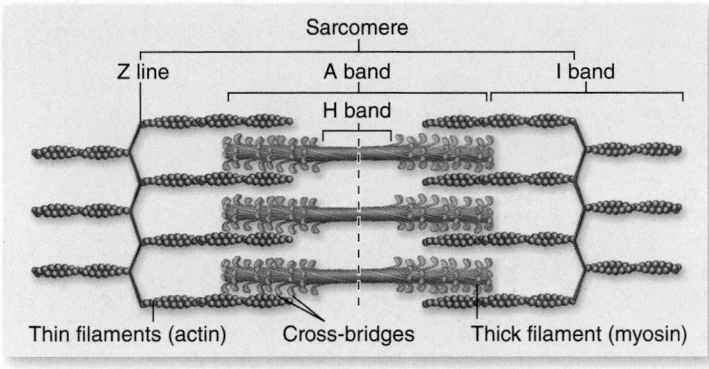

Sarcomere

Z line | A band | I band

H band

Thin filaments (actin) | Cross-bridges | Thick filament (myosin)

a.

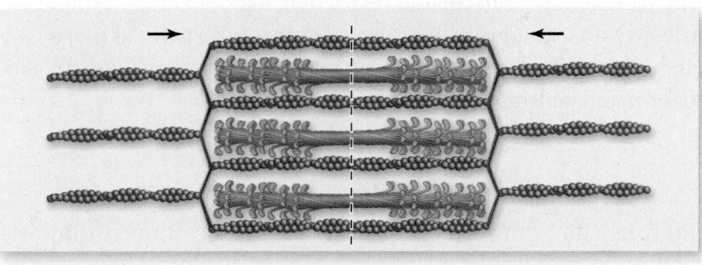

b.

Figure 46.13 **The interaction of thick and thin filaments in striated muscle sarcomeres.** *a.* The heads on the two ends of the thick filaments are oriented in opposite directions so that the cross-bridges pull the thin filaments and the Z lines on each side of the sarcomere toward the center. *b.* This sliding of the filaments produces muscle contraction.

Myosin is a member of the class of protein called *motor proteins* that are able to convert the chemical energy in ATP into mechanical energy (see chapter 4). This occurs by a series of events called the cross-bridge cycle (figure 46.14). When the myosin heads hydrolyze ATP into ADP and P_i, the conformation of myosin is changed, activating it for the later power stroke. The ADP and P_i both remain attached to the myosin head, keeping it in this activated conformation. The analogy to a mousetrap, set and ready to spring, is often made to describe this action. In this set position, the myosin head can bind to actin, forming cross-bridges. When a myosin head binds to actin, it releases the P_i and undergoes another conformational change, pulling the thin filament toward the center of the sarcomere in the *power stroke,* at which point it loses the ADP (see figures 46.13*b,* 46.14). At the end of the power stroke, the myosin head binds to a new molecule of ATP, which displaces it from actin. This cross-bridge cycle repeats as long as the muscle is stimulated to contract. This sequence of events can be thought of like pulling a rope hand-over-hand. The myosin heads are the hands and the actin fibers the rope.

In death, the cell can no longer produce ATP, and therefore the cross-bridge cycle cannot be broken—causing the muscle stiffness of death called *rigor mortis.* A living cell, however, always has enough ATP to allow the myosin heads to detach from actin. How, then, is the cross-bridge cycle arrested so that the muscle can relax? We discuss the regulation of contraction and relaxation next.

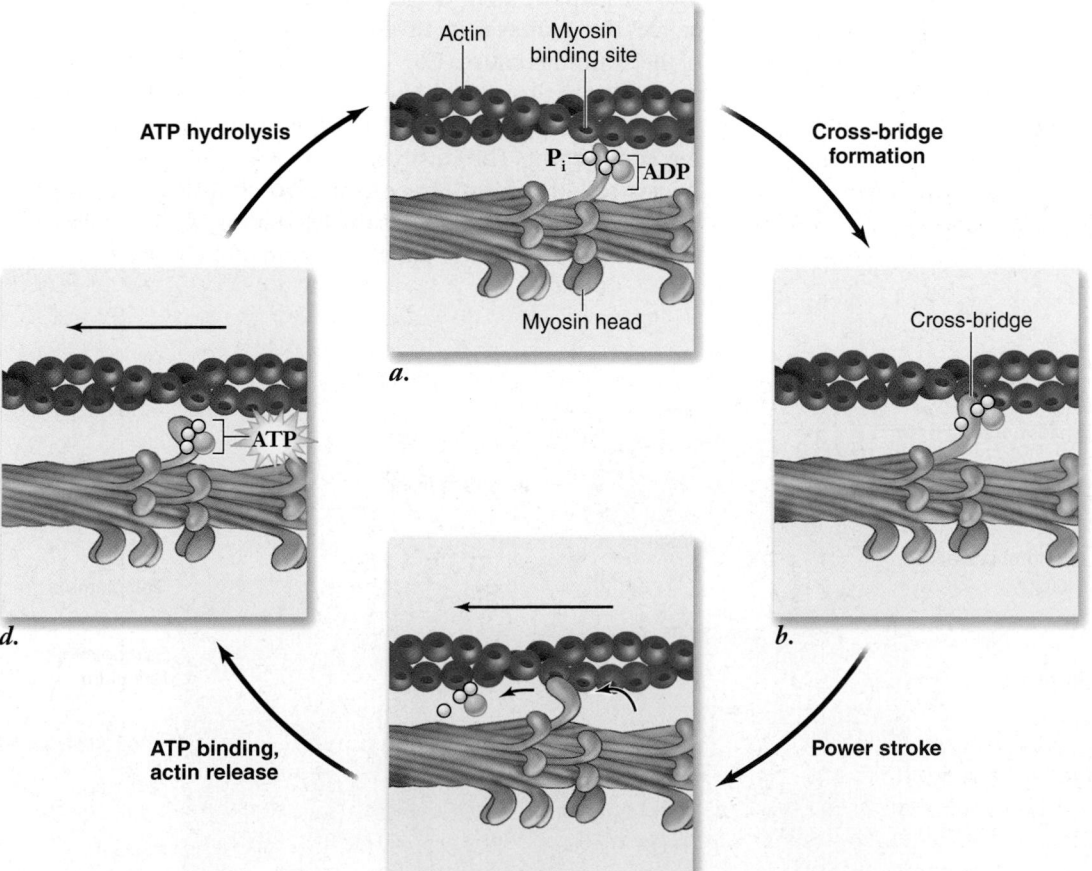

ATP hydrolysis

Actin | Myosin binding site

P_i | ADP

Myosin head

a.

Cross-bridge formation

Cross-bridge

b.

ATP

d.

ATP binding, actin release

Power stroke

c.

Figure 46.14 **The cross-bridge cycle in muscle contraction.** *a.* Hydrolysis of ATP by myosin causes a conformational change that moves the head into an energized state. The ADP and P_i remain bound to the myosin head, which can bind to actin. *b.* Myosin binds to actin forming a cross-bridge. *c.* During the power stroke, myosin returns to its original conformation, releasing ADP and P_i. *d.* ATP binds to the myosin head breaking the cross-bridge. ATP hydrolysis returns the myosin head to its energized conformation, allowing the cycle to begin again.

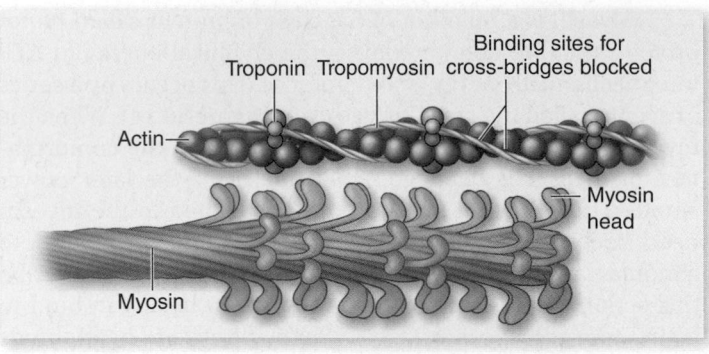
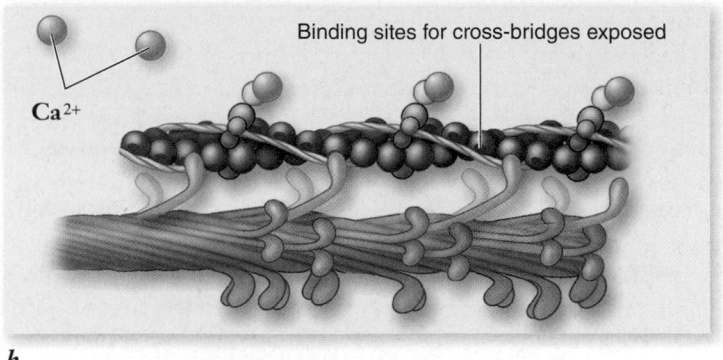

a. *b.*

Figure 46.15 How calcium controls striated muscle contraction. *a.* When the muscle is at rest, a long filament of the protein tropomyosin blocks the myosin-binding sites on the actin molecule. Because myosin is unable to form cross-bridges with actin at these sites, muscle contraction cannot occur. *b.* When Ca^{2+} binds to another protein, troponin, the Ca^{2+}–troponin complex displaces tropomyosin and exposes the myosin-binding sites on actin, permitting cross-bridges to form and contraction to occur.

Contraction depends on calcium ion release following a nerve impulse

When a muscle is relaxed, its myosin heads are in the activated conformation bound to ADP and P_i, but they are unable to bind to actin. In the relaxed state, the attachment sites for the myosin heads on the actin are physically blocked by another protein, known as **tropomyosin,** in the thin filaments. Cross-bridges therefore cannot form and the filaments cannot slide.

For contraction to occur, the tropomyosin must be moved out of the way so that the myosin heads can bind to the uncovered actin-binding sites. This requires the action of **troponin,** a regulatory protein complex that holds tropomyosin and actin together. The regulatory interactions between troponin and tropomyosin are controlled by the calcium ion (Ca^{2+}) concentration of the muscle fiber cytoplasm.

When the Ca^{2+} concentration of the cytoplasm is low, tropomyosin inhibits cross-bridge formation (figure 46.15*a*). When the Ca^{2+} concentration is raised, Ca^{2+} binds to troponin, altering its conformation and shifting the troponin–tropomyosin complex. This shift in conformation exposes the myosin-binding sites on the actin. Cross-bridges can thus form, undergo power strokes, and produce muscle contraction (figure 46.15*b*).

Muscles need a reliable supply of Ca^{2+}. Muscle fibers store Ca^{2+} in a modified endoplasmic reticulum called a **sarcoplasmic reticulum (SR)** (figure 46.16). When a muscle fiber is stimulated to contract, the membrane of the muscle fiber becomes depolarized. This is transmitted deep into the muscle fiber by invaginations of the cell membrane called the **transverse tubules (T tubules).** Depolarization of the T tubules causes Ca^{2+} channels in the SR to open, releasing Ca^{2+} into the cytosol. Ca^{2+} then diffuses into the myofibrils, where it binds to troponin, altering its conformation and allowing contraction. The involvement of Ca^{2+} in muscle contraction is called **excitation–contraction coupling** because it is the release of Ca^{2+} that links the excitation of the muscle fiber by the motor neuron to the contraction of the muscle.

Figure 46.16 Relationship between the myofibrils, transverse tubules, and sarcoplasmic reticulum.
Neurotransmitter released at a neuromuscular junction binds chemically gated Na^+ channels, causing the muscle cell membrane to depolarize. This depolarization is conducted along the muscle cell membrane and down the transverse tubules to stimulate the release of Ca^{2+} from the sarcoplasmic reticulum. Ca^{2+} diffuses through the cytoplasm to myofibrils, causing contraction.

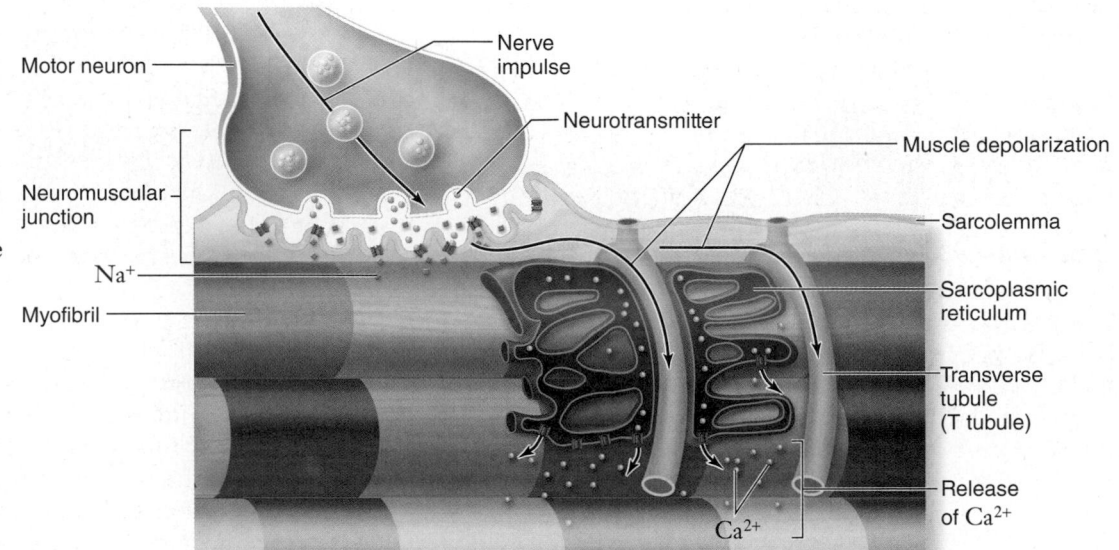

Nerve impulses from motor neurons

Muscles are stimulated to contract by motor neurons. The motor neurons that stimulate skeletal muscles are called *somatic motor neurons.* The axon of a somatic motor neuron extends from the neuron cell body and branches to make synapses with a number of muscle fibers. These synapses between neurons and muscle cells are called *neuromuscular junctions* (see figure 46.16). One axon can stimulate many muscle fibers, and in some animals, a muscle fiber may be innervated by more than one motor neuron. However, in humans, each muscle fiber has only a single synapse with a branch of one axon.

When a somatic motor neuron delivers electrochemical impulses, it stimulates contraction of the muscle fibers it innervates (makes synapses with) through the following events:

1. The motor neuron, at the neuromuscular junction, releases the neurotransmitter acetylcholine (ACh). ACh binds to receptors in the muscle cell membrane to open Na^+ channels. The influx of Na^+ ions depolarizes the muscle cell membrane.
2. The impulses spread along the membrane of the muscle fiber and are carried into the muscle fibers through the T tubules.
3. The T tubules conduct the impulses toward the sarcoplasmic reticulum, opening Ca^{2+} channels and releasing Ca^{2+}. The Ca^{2+} binds to troponin, exposing the myosin-binding sites on the actin myofilaments and stimulating muscle contraction.

When impulses from the motor neuron cease, it stops releasing ACh, in turn stopping the production of impulses in the muscle fiber. Another membrane protein in the SR then uses energy from ATP hydrolysis to pump Ca^{2+} back into the SR by active transport. Troponin is no longer bound to Ca^{2+}, so tropomyosin returns to its inhibitory position, allowing the muscle to relax.

Motor units and recruitment

A single muscle fiber can produce variable tension depending on the frequency of stimulation. The response of an entire muscle depends on the number of individual fibers involved and their degree of tension. The set of muscle fibers innervated by all the axonal branches of a motor neuron, plus the motor neuron itself, is defined as a **motor unit** (figure 46.17).

Every time the motor neuron produces impulses, all muscle fibers in that motor unit contract together. The division of the muscle into motor units allows the muscle's strength of contraction to be finely graded, a requirement for coordinated movements. Muscles that require a finer degree of control, such as those that move the eyes, have smaller motor units (fewer muscle fibers per neuron). Muscles that require less precise control but must exert more force, such as the large muscles of the legs, have more fibers per motor neuron.

Most muscles contain motor units in a variety of sizes, and these can be selectively activated by the nervous system. The weakest contractions of a muscle involve activation of a few small motor units. If a slightly stronger contraction is necessary, additional small motor units are also activated. The initial increments of increased force are therefore relatively small. As ever greater forces are required, more units and larger units are

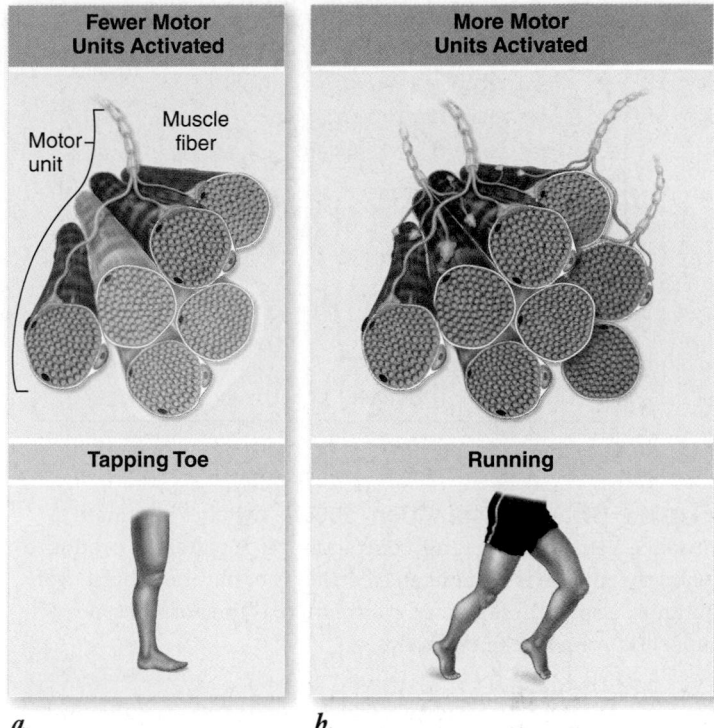

Fewer Motor Units Activated	More Motor Units Activated

a. *b.*

Figure 46.17 The number and size of motor units. A motor unit consists of a motor neuron and all of the muscle fibers it innervates. *a.* Precise muscle contractions require smaller motor units. *b.* Large muscle movements require larger motor units. The more motor units activated, the stronger the contraction.

brought into action, and the force increments become larger. This cumulative increase of numbers and sizes of motor units to produce a stronger contraction is termed **recruitment.**

The two main types of muscle fibers are slow-twitch and fast-twitch

An isolated skeletal muscle can be studied by stimulating it artificially with electric shocks. A muscle stimulated with a single electric shock quickly contracts and relaxes in a response called a twitch. Increasing the stimulus voltage increases the strength of the twitch up to a maximum. If a second electric shock is delivered immediately after the first, it produces a second twitch that may partially "ride piggyback" on the first. This cumulative response is called summation (figure 46.18).

An increasing frequency of electric shocks shortens the relaxation time between successive twitches as the strength of contraction increases. Finally, at a particular frequency of stimulation, no visible relaxation occurs between successive twitches. Contraction is smooth and sustained, as it is during normal muscle contraction in the body. This sustained contraction is called **tetanus.** (The disease known as tetanus gets its name because the muscles of its victims go into an agonizing state of contraction.)

Skeletal muscle fibers can be divided on the basis of their contraction speed into *slow-twitch* and *fast-twitch fibers.* The muscles that move the eyes, for example, have a high proportion of fast-twitch fibers and reach maximum tension in about 7.3 milliseconds (msec); the soleus muscle

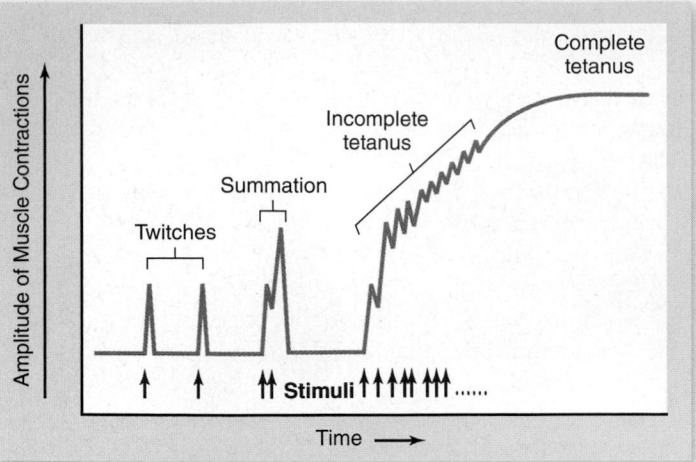

Figure 46.18 Summation. Muscle twitches summate to produce a sustained, tetanic contraction. This pattern is produced when the muscle is stimulated electrically or naturally by neurons. Tetanus, a smooth, sustained contraction, is the normal type of muscle contraction in the body.

? Inquiry question What determines the maximum amplitude of a summated muscle contraction?

in the leg, by contrast, has a high proportion of slow-twitch fibers and requires about 100 msec to reach maximum tension (figure 46.19).

Slow-twitch fibers

Slow-twitch fibers have a rich capillary supply, numerous mitochondria and aerobic respiratory enzymes, and a high concentration of **myoglobin** pigment. Myoglobin is a red pigment similar to the hemoglobin in red blood cells, but its higher affinity for oxygen improves the delivery of oxygen to the

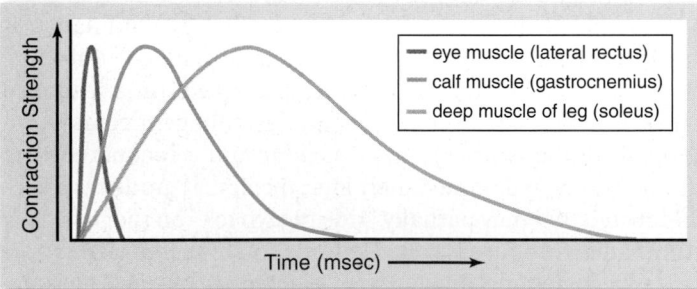

Figure 46.19 Skeletal muscles have different proportions of fast-twitch and slow-twitch fibers. The muscles that move the eye contain mostly fast-twitch fibers, whereas the deep muscle of the leg (the soleus) contains mostly slow-twitch fibers. The calf muscle (gastrocnemius) is intermediate in its composition.

? Inquiry question How would you determine if the calf muscle contains a mix of fast-twitch and slow-twitch fibers, or instead is composed of an intermediate form of fiber?

slow-twitch fibers. Because of their high myoglobin content, slow-twitch fibers are also called *red fibers.* These fibers can sustain action for a long period of time without fatigue.

Fast-twitch fibers

The thicker **fast-twitch fibers** have fewer capillaries and mitochondria than slow-twitch fibers and not as much myoglobin; hence, these fibers are also called *white fibers.* Fast-twitch fibers can respire anaerobically by using a large store of glycogen and high concentrations of glycolytic enzymes. The "dark meat" and "white meat" found in chicken and turkey consists of muscles with primarily red and white fibers, respectively. Fast-twitch fibers are adapted for the rapid generation of power and can grow thicker and stronger in response to weight training; however, they lack the endurance characteristics of slow-twitch fibers.

In addition to the type I and type II fibers, human muscles have an intermediate form of fibers that are fast-twitch, but they also have a high oxidative capacity and so are more resistant to fatigue. Endurance training increases the proportion of these fibers in muscles.

In general, human sprinters tend to have more fast-twitch fibers, whereas long-distance runners have more slow-twitch fibers. These differences are paralleled in the animal world. Comparisons of closely related species that differ in their lifestyles show that species that rely on short, high-speed movements to capture prey or evade predators tend to have more fast-twitch fibers, whereas closely related species that move more slowly, but for longer periods of time, have more slow-twitch fibers.

Muscle metabolism changes with the demands made on it

Skeletal muscles at rest obtain most of their energy from the aerobic respiration of fatty acids (see chapter 7). During use of the muscle, such as during exercise, muscle stores of glycogen and glucose delivered by the blood are also used as energy sources. The energy obtained by cellular respiration is used to make ATP, which is needed for the movement of the cross-bridges during muscle contraction and the pumping of Ca^{2+} back into the sarcoplasmic reticulum during muscle relaxation.

Skeletal muscles respire anaerobically for the first 45 to 90 sec of moderate-to-heavy exercise because the cardiopulmonary system requires this amount of time to increase the oxygen supply to the muscles. If exercise is not overly strenuous, aerobic respiration then contributes the major portion of the skeletal muscle energy requirements following the first 2 min of exercise. However, more vigorous exercise may require more ATP than can be provided by aerobic respiration, in which case anaerobic respiration continues to provide ATP as well.

Whether exercise is light, moderate, or intense for a particular individual depends on that person's maximal capacity for aerobic exercise. The maximum rate of oxygen consumption in the body is called the *aerobic capacity.* In general, individuals in better condition have greater aerobic capacity and thus can sustain higher levels of aerobic exercise for longer periods without having to also use anaerobic respiration.

Physical training increases aerobic capacity and muscle strength

Muscle fatigue refers to the use-dependent decrease in the ability of a muscle to generate force. Fatigue is highly variable and can arise from a number of causes. The intensity of contraction as well as duration of contraction are involved. In addition, fatigue is affected by cellular metabolism: aerobic or anaerobic. In the case of short-duration maximal exertion, fatigue was long thought to be caused by a buildup of lactic acid (from anaerobic metabolism). More recent data also implicate a buildup in inorganic phosphate (P_i) from the breakdown of creatine phosphate, which also occurs during anaerobic metabolism. In longer term, lower intensity exertion, fatigue appears to result from depletion of glycogen.

Because the depletion of muscle glycogen places a limit on exercise, any adaptation that spares muscle glycogen will improve physical endurance. Trained athletes have an increased proportion of energy derived from the aerobic respiration of fatty acids, resulting in a slower depletion of their muscle glycogen reserve. Athletes also have greater muscle vascularization, which facilitates both oxygen delivery and lactic acid removal. Because the aerobic capacity of endurance-trained athletes is higher than that of untrained people, athletes can perform for longer and put forth more effort before muscle fatigue occurs.

Endurance training does not increase muscle size. Muscle enlargement is produced only by frequent periods of high-intensity exercise in which muscles work against high resistance, as in weight lifting. Resistance training increases the thickness of type II (fast-twitch) muscle fibers, causing skeletal muscles to grow by hypertrophy (increased cell size) rather than by cell division and an increased number of cells.

Learning Outcomes Review 46.4

Sliding of myofilaments within muscle myofibrils is responsible for contraction; it involves the motor protein myosin, which forms cross-bridges on actin fibers. The process of shortening is controlled by Ca^{2+} ions released from the sarcoplasmic reticulum. The Ca^{2+} binds to troponin, making myosin-binding sites in actin available. Slow-twitch fibers can sustain activity for a longer period of time; fast-twitch fibers use glycogen for rapid generation of power.

■ *What advantages do increased myoglobin and mitochondria confer on slow-twitch fibers?*

46.5 Modes of Animal Locomotion

Learning Outcomes

1. Describe how friction and gravity affect locomotion.
2. Discuss how lift is created by wings.
3. Explain how evolution has shaped structures used for locomotion.

Animals are unique among multicellular organisms in their ability to move actively from one place to another. Locomotion requires both a propulsive mechanism and a control mechanism. There are a wide variety of propulsive mechanisms, most involving contracting muscles to generate the necessary force. Ultimately, it is the nervous system that activates and coordinates the muscles used in locomotion. In large animals, active locomotion is almost always produced by appendages that oscillate—*appendicular locomotion*—or by bodies that undulate, pulse, or undergo peristaltic waves—*axial locomotion*.

Although animal locomotion occurs in many different forms, the general principles remain much the same in all groups. The physical constraints to movement—gravity and friction—are the same in every environment, differing only in degree.

Swimmers must contend with friction when moving through water

For swimming animals, the buoyancy of water reduces the effect of gravity. As a result, the primary force retarding forward movement is frictional drag, so body shape is important in reducing the force needed to push through the water.

Some marine invertebrates move about using hydraulic propulsion. For example, scallops clap the two sides of their shells together forcefully, and squids and octopuses squirt water like a marine jet, as described in chapter 34.

In contrast, many invertebrates and all aquatic vertebrates swim. Swimming involves pushing against the water with some part of the body. At one extreme, eels and sea snakes swim by sinuous undulations of the entire body (figure 46.20a). The undulating body waves of eel-like swimming are created by waves of muscle contraction alternating between the left and right axial musculature. As each body segment in turn pushes against the water, the moving wave forces the eel forward.

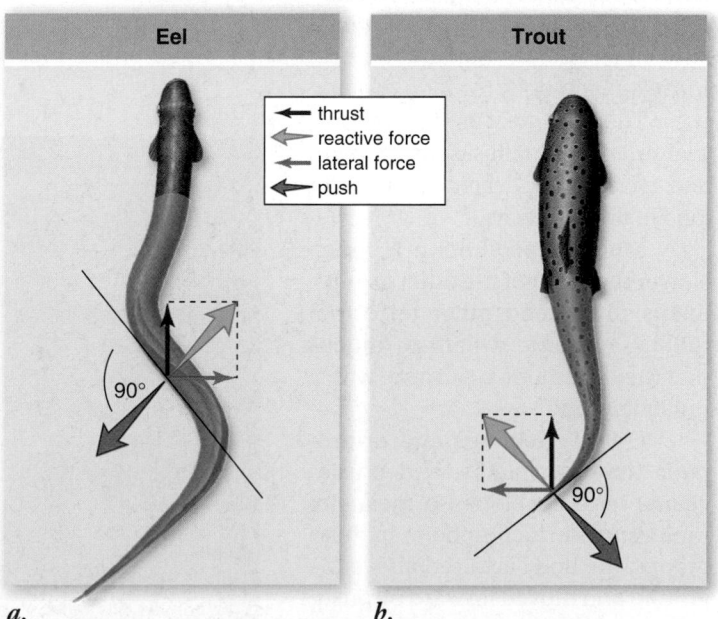

Figure 46.20 **Movements of swimming fishes.** *a.* An eel pushes against the water with its whole body, whereas (*b*) a trout pushes only with its posterior half.

Other types of fish use similar mechanics as the eel but generate most of their propulsion from the posterior part of the body using the caudal (rear) fin (figure 46.20b). This also allows considerable specialization in the front end of the body without sacrificing propulsive force. Reptiles, such as alligators, swim in the same manner using undulations of the tail.

Whales and other marine mammals such as sea lions have evolutionarily returned to an aquatic lifestyle (see figure 21.10) and have convergently evolved a similar form of locomotion. Like fish, marine mammals also swim using undulating body waves. However, unlike any of the fishes, the waves pass from top to bottom and not from side to side. This difference illustrates how past evolutionary history can shape subsequent evolutionary change. The mammalian vertebral column is structured differently from that of fish in a way that stiffens the spine and allows little side-to-side flexibility. For this reason, when the ancestor of whales reentered aquatic habitats, they evolved adaptations for swimming that used dorsoventral (top-to-bottom) flexing.

Many terrestrial tetrapod vertebrates are able to swim, usually through movement of their limbs. Most birds that swim, such as ducks and geese, propel themselves through the water by pushing against it with their hind legs, which typically have webbed feet. Frogs and most aquatic mammals also swim with their hind legs and have webbed feet. Tetrapod vertebrates that swim with their forelegs usually have these limbs modified as flippers and "fly" through the water using motions very similar to those used by aerial fliers; examples include sea turtles, penguins, and fur seals.

Terrestrial locomotion must deal primarily with gravity

Air is a much less dense medium than water, and thus the frictional forces countering movement on land are much less than those in water. Instead, countering the force of gravity is the biggest challenge for nonaquatic organisms, which either must move on land or fly through the air.

The three great groups of terrestrial animals—mollusks, arthropods, and vertebrates—each move over land in different ways.

Mollusk locomotion is much slower than that of the other groups. Snails, slugs, and other terrestrial mollusks secrete a path of mucus that they glide along, pushing with a muscular foot.

Only vertebrates and arthropods (insects, spiders, and crustaceans) have developed a means of rapid surface locomotion. In both groups, the body is raised above the ground and moved forward by pushing against the ground with a series of jointed appendages, the legs.

Although animals may walk on only two legs or more than 100,

the same general principles guide terrestrial locomotion. Because legs must provide support as well as propulsion, it is important that the sequence of their movements not shove the body's center of gravity outside the legs' zone of support, unless the duration of such imbalance is short. Otherwise, the animal will fall. The need to maintain stability determines the sequence of leg movements, which are similar in vertebrates and arthropods.

The apparent differences in the walking gaits of these two groups reflect the differences in leg number. Vertebrates walk on two or four legs; all arthropods have six or more limbs. Although the many legs of arthropods increase stability during locomotion, they also appear to reduce the maximum speed that can be attained.

The basic walking pattern of quadrupeds, from salamanders to most mammals, is left hind leg, right foreleg, right hind leg, left foreleg. The highest running speeds of quadruped mammals, such as the gallop of a horse, may involve the animal being supported by only one leg, or even none at all. This is because mammals have evolved changes in the structure of both their axial and appendicular skeleton that permit running by a series of leaps.

Vertebrates such as kangaroos, rabbits, and frogs are effective leapers (figure 46.21). However, insects are the true Olympians of the leaping world. Many insects, such as grasshoppers, have enormous leg muscles, and some small insects can jump to heights more than 100 times the length of their body!

Flying uses air for support

The evolution of flight is a classic example of convergent evolution, having occurred independently four times, once in insects and three times among vertebrates (figure 46.22a). All three vertebrate fliers modified the forelimb into a wing structure, but they did so in different ways, illustrating how natural selection can sometimes build similar structures through different evolutionary pathways (figure 46.22b). In both birds and pterosaurs (an extinct group of reptiles that flourished alongside the

Figure 46.21 **Animals that hop or leap use their rear legs to propel themselves through the air.** The powerful leg muscles of this frog allow it to explode from a crouched position to a takeoff in about 100 msec.

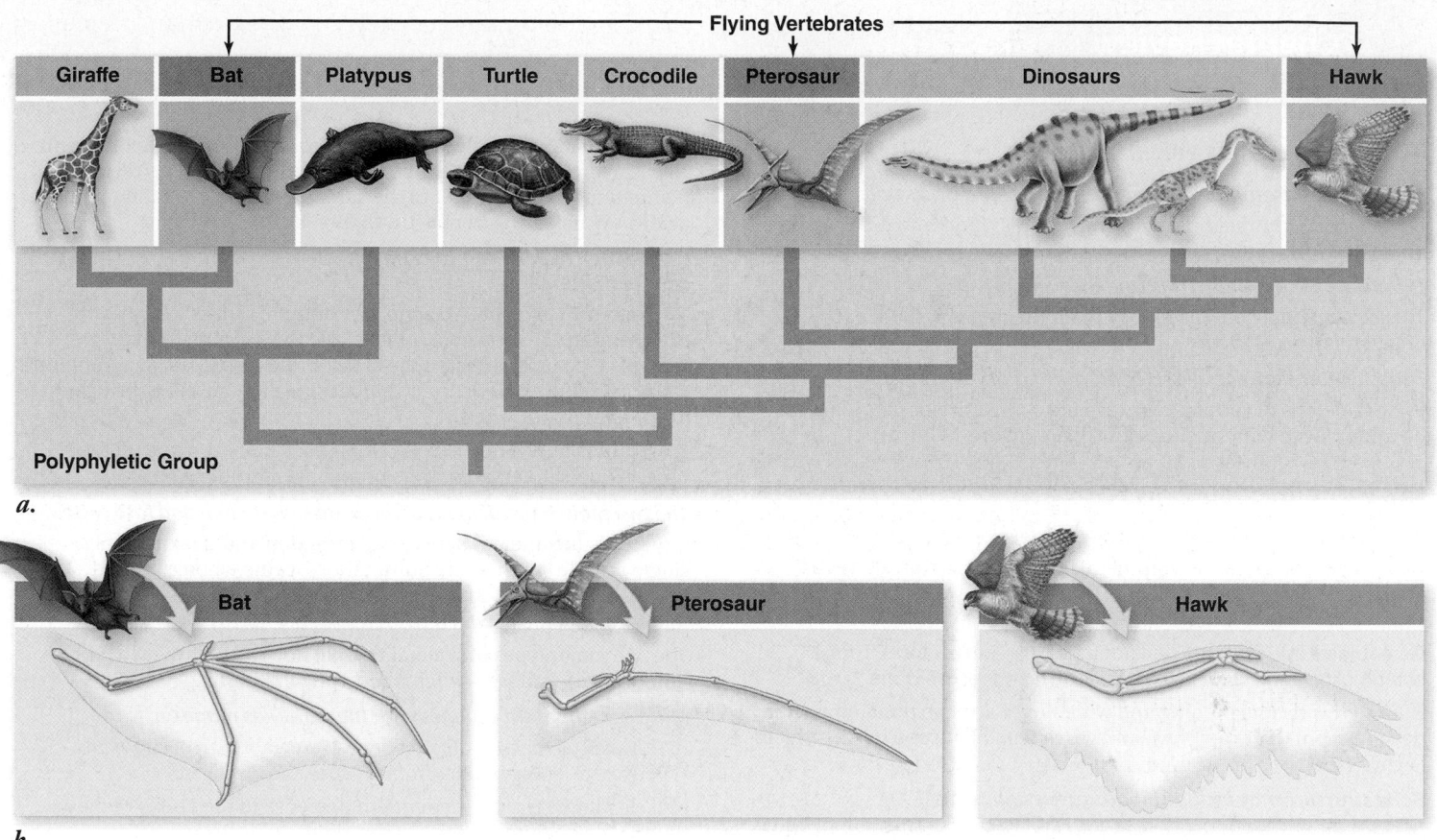

Figure 46.22 Convergent evolution of wings in vertebrates. Wings evolved independently in birds, bats, and pterosaurs, in each case by elongation of different elements of the forelimb.

dinosaurs), the wing is built on a single support, but in birds it is elongation of the radius, ulna, and wrist bones, whereas in pterosaurs it is an elongation of the fourth finger bone. By contrast, in bats the wing is supported by multiple bones, each of which is an elongated finger bone. A second difference is that the wings of pterosaurs and bats are composed of a membrane formed from skin, whereas birds use feathers, which are modified from reptile scales.

In all groups, active flying takes place in much the same way. Propulsion is achieved by pushing down against the air with wings. This alone provides enough lift to keep insects in the air. Vertebrates, being larger, need greater lift, obtaining it with wings whose upper surface is more convex (in cross section) than the lower. Because air travels farther over the top surface, it moves faster. A fluid, like air, decreases its internal pressure the faster it moves. Thus, there is a lower pressure on top of the wing and higher pressure on the bottom of the wing. This is the same principle used by airplane wings.

In birds and most insects, the raising and lowering of the wings is achieved by the alternate contraction of extensor muscles (elevators) and flexor muscles (depressors). Four insect orders (including those containing flies, mosquitoes, wasps, bees, and beetles) beat their wings at frequencies ranging from 100 to more than 1000 times per second, faster than nerves can carry successive impulses!

In these insects, the flight muscles are not attached to the wings at all, but rather to the stiff wall of the thorax, which is distorted in and out by their contraction. The reason these muscles can contract so fast is that the contraction of one muscle set stretches the other set, triggering its contraction in turn without waiting for the arrival of a nerve impulse.

In addition to active flight, many species have evolved adaptations—primarily flaps of skin that increase surface area and thus slow down the rate of descent—to enhance their ability to glide long distances. Gliders have done this in many ways, including flaps of skin along the body in flying squirrels, snakes, and lizards; webbing between the toes in frogs; and the evolution in some lizards of ribs that extend beyond the body wall and that are connected by skin that can be spread out to form a large gliding surface.

Learning Outcomes Review 46.5

Locomotion involves friction and pressure created by body parts, often appendages, against water, air, or ground. Walking, running, and flying require supporting the body against gravity's pull. Flight is achieved when a pressure difference between air flowing over the top and bottom of a wing creates lift. Solutions to locomotion have evolved convergently many times, in both homologous and nonhomologous structures.

■ *In what ways would locomotion by a series of leaps be more advantageous than by alternation of legs?*

46.1 Types of Skeletal Systems

Hydrostatic skeletons use water pressure inside a body wall.

By muscular contractions, earthworms press fluid into different parts of the body, causing them to move (figure 46.1).

Exoskeletons consist of a rigid outer covering.

The exoskeleton, composed of hard chitin, must be shed for the organism to grow (figure 46.2a).

Endoskeletons are composed of hard, internal structures.

Endoskeletons of vertebrates are living connective tissues that may be mineralized with calcium phosphate (figure 46.2b).

46.2 A Closer Look at Bone

Bones can be classified by two modes of development.

In intramembranous development, bone forms within a layer of connective tissue (figure 46.3). In endochondral development, bone fills in a cartilaginous model.

Osteoblasts initiate bone development; osteocytes form from osteoblasts; and osteoclasts break down and resorb bone.

Bones grow by lengthening and widening. Cartilage remaining after development of the epiphyses serves as a pad between bone surfaces (figure 46.4).

Bone structure may include blood vessels and nerves.

In birds and fishes, bone is avascular and basically acellular. In other vertebrates, bone contains bone cells, blood capillaries, and nerves collected in Haversian systems.

Bone remodeling allows bone to respond to use or disuse.

Bone structure may thicken or thin depending on use and on forces impinging on the bone (figure 46.5).

46.3 Joints and Skeletal Movement

Moveable joints have different ranges of motion, depending on type.

Ball-and-socket joints can perform movement in all directions; hinge joints have restricted movement; gliding joints slide, providing stability and flexibility; and combination joints allow rotation and sliding (figure 46.7).

Skeletal muscles pull on bones to produce movement at joints.

Muscles attach to the periosteum directly or through a tendon. Skeletal muscles occur in antagonistic pairs that oppose each other's movement (figure 46.8).

46.4 Muscle Contraction

Muscle fibers contract as overlapping filaments slide together.

The different myofibril bands seen microscopically result from the degree of overlap of actin and myosin filaments (figure 46.10). Muscle contraction occurs when actin and myosin filaments form cross-bridges and slide relative to each other.

The globular head of myosin forms a cross-bridge with actin when ATP is hydrolyzed to ADP and P_i. Upon bridging, it pulls the thin filament toward the center of the sarcomere. The head then binds to a new ATP, releasing from actin (figure 46.14).

Contraction depends on calcium ion release following a nerve impulse.

Tropomyosin, attached to actin by troponin, blocks formation of a cross-bridge. Nerve stimulation releases calcium from the sarcoplasmic reticulum into the cytosol, and formation of a troponin–calcium complex displaces tropomyosin, allowing cross-bridges to form (figures 46.15, 46.16).

Motor units are composed of a single motor neuron and all the muscle fibers innervated by its branches (figure 46.17).

The two main types of muscle fibers are slow-twitch and fast-twitch.

A twitch is the interval between contraction and relaxation of a single muscle stimulation. Summation occurs when a second twitch "piggybacks" on the first twitch. Tetanus is the state when no relaxation occurs between twitches (figure 46.18).

The two major types of skeletal muscle fibers are slow-twitch fibers (endurance), and fast-twitch fibers (power bursts).

Muscle metabolism changes with the demands made on it.

At rest skeletal muscles obtain energy by metabolism of fatty acids. When active, energy comes from glucose and glycogen.

Muscle fatigue is a use-dependent decrease in the ability of the muscle to generate force.

Endurance training does not increase muscle size; high-intensity exercise with resistance increases the size of the muscle (hypertrophy).

46.5 Modes of Animal Locomotion

Swimmers must contend with friction when moving through water.

Among vertebrates, aquatic locomotion occurs by pushing some or all of the body against the water. Many vertebrates undulate the body or tail for propulsion, but others use their limbs (figure 46.20).

Terrestrial locomotion must deal primarily with gravity.

Most terrestrial animals move by lifting their bodies off the ground and pushing against the ground with appendages. Terrestrial animals that walk or run use fundamentally the same mechanisms during locomotion.

Flying uses air for support.

Propulsion is accomplished as wings push down against the air. Lift in larger organisms is created by a pressure difference as air flows above and below a convex wing.

In both flying and gliding, convergent evolution has produced the same outcome through different evolutionary pathways.

UNDERSTAND

1. Exoskeletons and endoskeletons differ in that
   a. an exoskeleton is rigid, and an endoskeleton is flexible.
   b. endoskeletons are found only in vertebrates.
   c. exoskeletons are composed of calcium, and endoskeletons are built from chitin.
   d. exoskeletons are external to the soft tissues, and endoskeletons are internal.

2. Worms use a hydrostatic skeleton to generate movement. How do they do this?
   a. Their bones are filled with water, which provides the weight of the skeleton.
   b. The change in body structure is caused by contraction of muscles compressing the watery body fluid.
   c. The muscles contain water vacuoles, which, when filled, provide a rigid internal structure.
   d. The term *hydrostatic* simply refers to moist environment. They generate movement just as arthropods do.

3. Bone develops by one of two mechanisms depending on the underlying scaffold. Which pairing correctly describes these mechanisms?
   a. Intramembranous and extramembranous
   b. Endochondral and exochondral
   c. Extramembranous and exochondral
   d. Endochondral and intramembranous

4. Which of the following statements best describes the sliding filament mechanism of muscle contraction?
   a. Actin and myosin filaments do not shorten, but rather, slide past each other.
   b. Actin and myosin filaments shorten and slide past each other.
   c. As they slide past each other, actin filaments shorten, but myosin filaments do not shorten.
   d. As they slide past each other, myosin filaments shorten, but actin filaments do not shorten.

5. Motor neurons stimulate muscle contraction via the release of
   a. Ca^{2+}.
   b. ATP.
   c. acetylcholine.
   d. hormones.

6. Which of the following statements about muscle metabolism is false?
   a. Skeletal muscles at rest obtain most of their energy from muscle glycogen and blood glucose.
   b. ATP can be quickly obtained by combining ADP with phosphate derived from creatine phosphate.
   c. Exercise intensity is related to the maximum rate of oxygen consumption.
   d. ATP is required for the pumping of the Ca^{2+} back into the sarcoplasmic reticulum.

7. If you wanted to study the use of ATP during a single contraction cycle within a muscle cell, which of the following processes would you use?
   a. Summation
   b. Twitch
   c. Treppe
   d. Tetanus

8. Place the following events in the correct order.
   1. Sarcoplasmic reticulum releases Ca^{2+}.
   2. Myosin binds to actin.
   3. Action potential arrives from neuron.
   4. Ca^{2+} binds to troponin.

   a. 1, 2, 3, 4
   b. 3, 1, 2, 4
   c. 2, 4, 3, 1
   d. 3, 1, 4, 2

APPLY

1. You take X-rays of two individuals. Ray has been a weight lifter and body builder for 30 years; Ben has led a mostly sedentary life. What differences would you expect in their X-rays?
   a. No difference, they would both have thicker bones than a younger person due to natural thickening with age.
   b. No difference, lifestyle does not affect bone density.
   c. Ray would have thicker bones due to reshaping as a result of physical stress.
   d. Ben would have thicker bones because bone accumulates like fat tissue from a sedentary lifestyle.

2. You have identified a calcium storage disease in rats. How would this inability to store Ca^{2+} affect muscle contraction?
   a. Ca^{2+} would be unable to bind to tropomyosin, which enables troponin to move and reveal binding sites for cross-bridges.
   b. Ca^{2+} would be unable to bind to troponin, which enables tropomyosin to move and reveal binding sites for cross-bridges.
   c. Ca^{2+} would be unable to bind to tropomyosin, which enables troponin to release ATP.
   d. Ca^{2+} would be unable to bind to troponin, which enables tropomyosin to release ATP.

3. How do the muscles move your hand through space?
   a. By contraction
   b. By attaching to two bones across a joint
   c. By lengthening
   d. Both a and b are correct.

4. How can osteocytes remain alive within bone?
   a. Bones are composed of only dead or dormant cells.
   b. Haversian canals are bone structures that contain blood vessels that provide materials for the osteocytes.
   c. Osteocytes have membrane extensions that protrude from bone and allow them to exchange materials with the surrounding fluids.
   d. Bones are hollow in the middle and the low pressure there draws fluid from the blood that nourishes the osteocytes.

5. Swimming underwater using forelimbs for propulsion is similar to flying through the air because
   a. birds are the only class of vertebrates that have species that do both.
   b. both involve coordinating movements of the forelimbs and hindlimbs.
   c. both must counter strong forces caused by friction.
   d. both involve generating lift by pushing down on the air or water to counter gravity.

6. If a drug inhibits the release of ACh, what will happen?
   a. Somatic motor neurons will fail to activate.
   b. Somatic motor neuron impulses will not lead to muscle fiber contraction.
   c. Myosin molecules will fail to release ADP.
   d. An influx of sodium ions will lead to muscle cell membrane depolarization.

SYNTHESIZE

1. You are designing a space-exploration vehicle to use on a planet with a gravity greater than Earth. Given a choice between a hydrostatic or an exoskeleton, which would you choose? Why?

2. You start running as fast as you can. Then, you settle into a jog that you can easily maintain. How do energy sources utilized by your skeletal muscles change during the switch? Why?

3. The nerve gas methylphosphonofluoridic acid (sarin) inhibits the enzyme acetylcholinesterase, required to break down acetylcholine. Based on this information, what are the likely effects of this nerve gas on muscle function?

4. If natural selection favors the evolution of wings in different types of vertebrates, why didn't it produce structures that were built in the same way?

ONLINE RESOURCE

www.ravenbiology.com

Understand, Apply, and Synthesize—enhance your study with animations that bring concepts to life and practice tests to assess your understanding. Your instructor may also recommend the interactive eBook, individualized learning tools, and more.

Chapter **47**

The Digestive System

Chapter Contents

Introduction

Plants and other photosynthetic organisms can produce the organic molecules they need from inorganic components. Therefore, they are autotrophs, or self-sustaining. Animals, such as the chipmunk shown here, are heterotrophs: They must consume organic molecules present in other organisms. The molecules heterotrophs eat must be digested into smaller molecules in order to be absorbed into the animal's body. Once these products of digestion enter the body, the animal can use them for energy in cellular respiration or for the construction of the larger molecules that make up its tissues. The process of animal digestion is the focus of this chapter.

Types of Digestive Systems

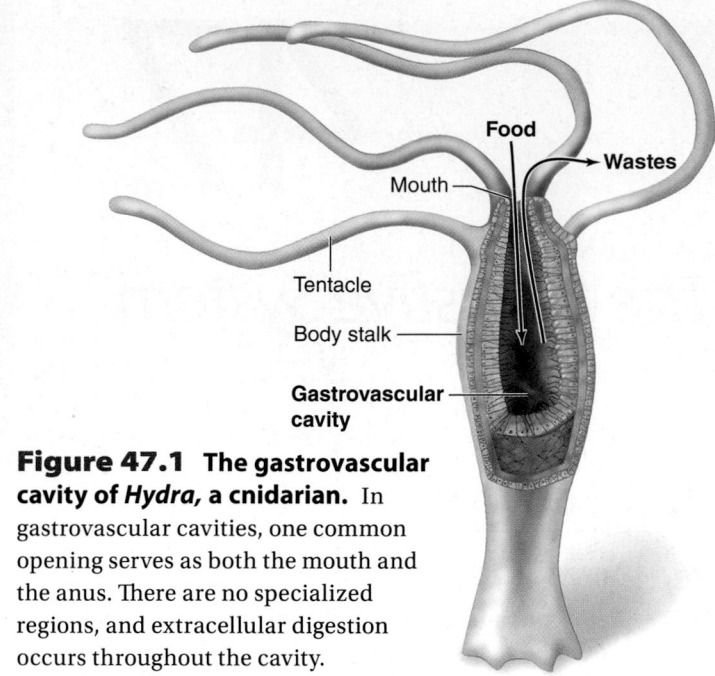

Figure 47.1 The gastrovascular cavity of *Hydra*, a cnidarian. In gastrovascular cavities, one common opening serves as both the mouth and the anus. There are no specialized regions, and extracellular digestion occurs throughout the cavity.

Heterotrophs are divided into three groups on the basis of their food sources. Animals that eat plants exclusively are classified as **herbivores;** common examples include algae-eating snails, sapsucking insects, and vertebrates such as cows, horses, rabbits, and sparrows. Animals that eat other animals, such as crabs, squid, many insects, cats, eagles, trout, and frogs, are **carnivores.** Animals that eat both plants and other animals are **omnivores.** Humans are omnivores, as are pigs, bears, and crows.

Invertebrate digestive systems are bags or tubes

Single-celled organisms as well as sponges digest their food intracellularly. Other multicellular animals digest their food extracellularly, within a digestive cavity. In this case, the digestive enzymes are released into a cavity that is continuous with the animal's external environment. In cnidarians and in flatworms such as planarians, the digestive cavity has only one opening that serves as both mouth and anus (see chapters 33 and 34). There is no specialization within this type of digestive system, called a *gastrovascular cavity,* because every cell is exposed to all stages of food digestion (figure 47.1).

Specialization occurs when the digestive tract, or alimentary canal, has a separate mouth and anus, so that transport of food is one-way. The most primitive digestive tract is seen in nematodes (phylum Nematoda), where it is simply a tubular *gut* lined by an epithelial membrane. Earthworms (phylum Annelida) have a digestive tract specialized in different regions for the ingestion, storage, fragmentation, digestion, and absorption of food. All more complex animal groups, including all vertebrates, show similar specializations (figure 47.2).

The ingested food may be stored in a specialized region of the digestive tract or it may first be subjected to physical fragmentation. This fragmentation may occur through the chewing action of teeth (in the mouth of many vertebrates) or the grinding action of pebbles (in the gizzard of earthworms and birds). Chemical digestion then occurs, breaking down the larger food molecules of polysaccharides and disaccharides, fats, and proteins into their smallest subunits.

Chemical digestion involves hydrolysis reactions that liberate the subunit molecules—primarily monosaccharides, amino acids, and fatty acids—from the food. These products of chemical digestion pass through the epithelial lining of the gut into the blood, in a process known as *absorption.* Any molecules in the food that are not absorbed cannot be used by the animal. These waste products are excreted, or defecated, from the anus.

Vertebrate digestive systems include highly specialized structures molded by diet

In humans and other vertebrates, the digestive system consists of a tubular gastrointestinal tract and accessory digestive organs (figure 47.3).

Nematode	Earthworm	Salamander
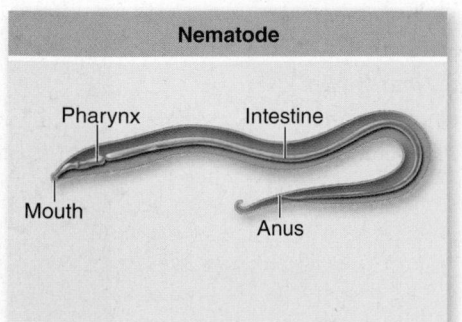	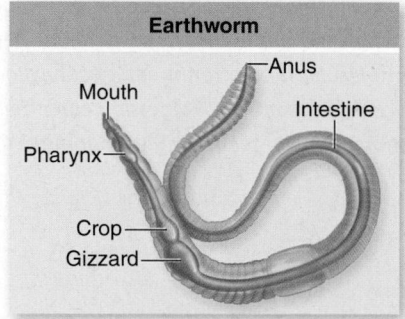	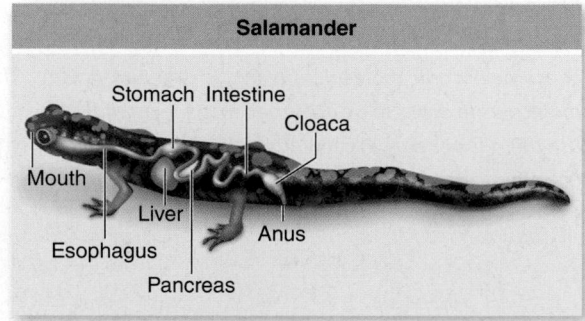

Figure 47.2 The one-way digestive tract of nematodes, earthworms, and vertebrates. One-way movement through the digestive tract allows different regions of the digestive system to become specialized for different functions.

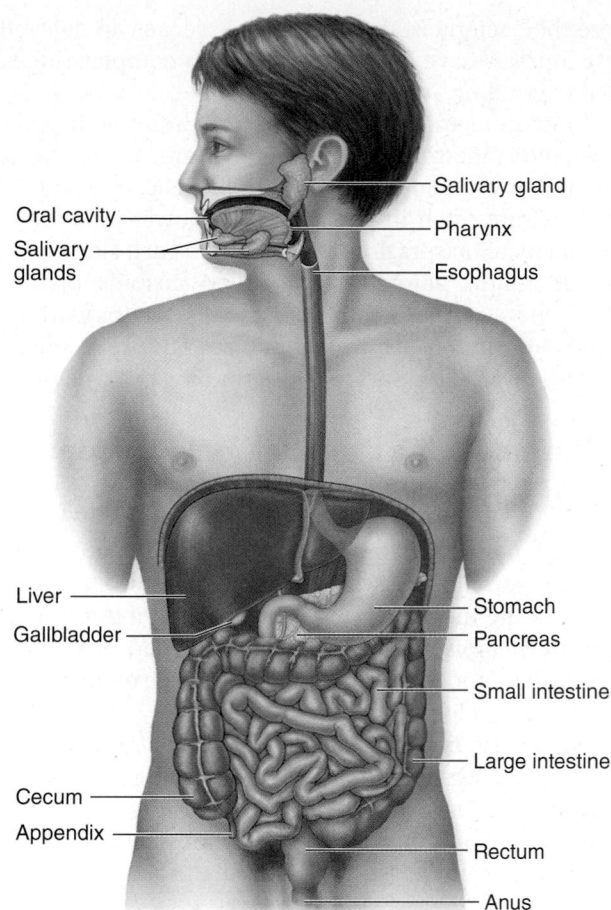

Figure 47.3 The human digestive system. The human digestive system consists of the oral cavity, esophagus, stomach, small intestine, large intestine, rectum, and anus; and is aided by accessory organs.

Overview of the digestive tract

The initial components of the gastrointestinal tract are the mouth and the pharynx, which is the common passage of the oral and nasal cavities. The pharynx leads to the esophagus, a muscular tube that delivers food to the stomach, where some preliminary digestion occurs.

From the stomach, food passes to the small intestine, where a battery of digestive enzymes continues the digestive process. The products of digestion, together with minerals and water, are absorbed across the wall of the small intestine into the bloodstream. What remains is emptied into the large intestine, where some of the remaining water and minerals are absorbed.

In most vertebrates other than mammals, the waste products emerge from the large intestine into a cavity called the cloaca (see figure 47.2), which also receives the products of the urinary and reproductive systems. In mammals, the urogenital products are separated from the fecal material in the large intestine; the fecal material enters the rectum and is expelled through the anus.

The accessory digestive organs include the liver, which produces *bile* (a green solution that emulsifies fat), the gallbladder, which stores and concentrates the bile, and the pancreas. The pancreas produces *pancreatic juice,* which contains digestive enzymes and bicarbonate buffer. Both bile and

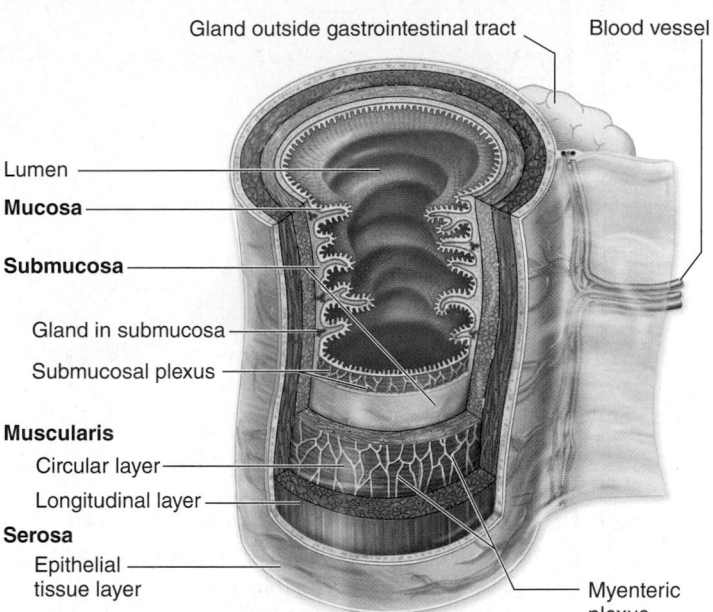

Figure 47.4 The layers of the gastrointestinal tract. The mucosa contains an epithelial lining; the submucosa is composed of connective tissue; and the muscularis consists of smooth muscles. Glands secrete substances via ducts into specific regions of the tract.

pancreatic juice are secreted into the first region of the small intestine, the duodenum, where they aid digestion.

Tissues of the digestive tract

The tubular gastrointestinal tract of a vertebrate has a characteristic layered structure (figure 47.4). The innermost layer is the **mucosa,** an epithelium that lines the interior, or lumen, of the tract. The next major tissue layer, made of connective tissue, is called the **submucosa.**

Just outside the submucosa is the **muscularis,** which consists of a double layer of smooth muscles. The muscles in the inner layer have a circular orientation and serve to constrict the gut, whereas those in the outer layer are arranged longitudinally and work to shorten it. Another epithelial tissue layer, the **serosa,** covers the external surface of the tract. Nerve networks, intertwined in *plexuses* between muscle layers, are located in the submucosa and help regulate the gastrointestinal activities.

In the rest of this chapter, we focus on the details of the vertebrate digestive system's structure and function. We close the chapter with discussion of nutrients that are essential to vertebrates.

> ### Learning Outcomes Review 47.1
>
> As digestive systems become more specialized, they become one way—food entering at one end and exiting at the other—and different regions become specialized for different roles. The digestive system of vertebrates includes mouth and pharynx, esophagus, stomach, small and large intestines, cloaca or rectum, anus, and accessory organs. The layers of tissue that compose the tubular tract are the mucosa, the submucosa, the muscularis, and the serosa.
>
> ■ *What might be the advantages of a one-way digestive system?*

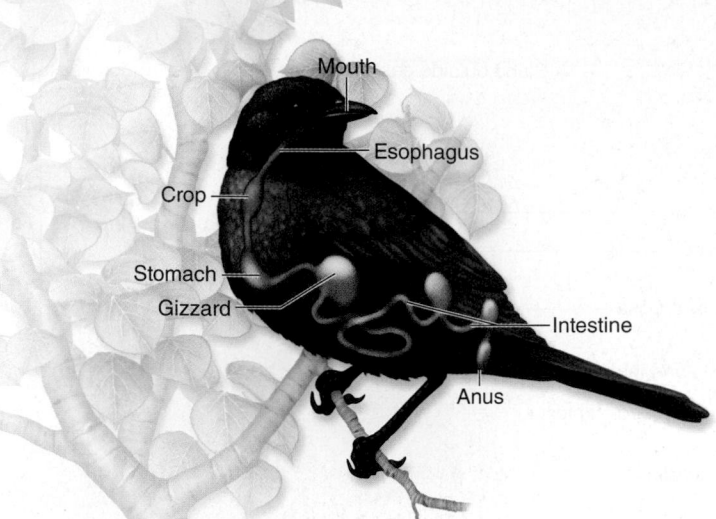

Figure 47.5 The digestive tract of birds. Birds lack teeth but have a muscular chamber called the gizzard that works to break down food. Birds swallow gritty objects or pebbles that lodge in the gizzard and pulverize food before it passes into the intestine. Food is stored in the crop.

47.2 The Mouth and Teeth: Food Capture and Bulk Processing

Learning Outcomes

1. Identify adaptive variation in vertebrate tooth shape.
2. Understand the role of the mouth in the digestive process.

Specializations of the digestive systems in different kinds of vertebrates reflect the way these animals live. Birds, which lack teeth, break up food in their two-chambered stomachs (figure 47.5). In one of these chambers, called the *gizzard,* small pebbles ingested by the bird are churned together with the food by muscular action. This churning grinds up the seeds and other hard plant material into smaller chunks that can be digested more easily.

 Vertebrate teeth are adapted to different types of food items

Many vertebrates have teeth (figure 47.6), used for chewing, or *mastication,* that break up food into small particles and mix it with fluid secretions. Carnivorous mammals have pointed teeth that lack flat grinding surfaces. Such teeth are adapted for cutting and shearing. Carnivores often tear off pieces of their prey but have little need to chew them, because digestive enzymes can act directly on animal cells. By contrast, grass-eating herbivores must pulverize the cellulose cell walls of plant tissue

before the bacteria in their rumens or cecae can digest them. These animals have large, flat teeth with complex ridges well suited to grinding.

Human teeth are specialized for eating both plant and animal food. Viewed simply, humans are carnivores in the front of the mouth and herbivores in the back (see figure 47.6). The four front teeth in the upper and lower jaws are sharp, chisel-shaped incisors used for biting. On each side of the incisors are sharp, pointed teeth called cuspids (sometimes referred to as "canine" teeth), which are used for tearing food. Behind the canines are two premolars and three molars, all with flattened, ridged surfaces for grinding and crushing food.

 The mouth is a chamber for ingestion and initial processing

Inside the mouth, the tongue mixes food with a mucous solution, saliva. In humans, three pairs of salivary glands secrete saliva into the mouth through ducts in the mouth's mucosal lining. Saliva moistens and lubricates the food so that it is easier to swallow and does not abrade the tissue of the esophagus as it passes through.

Saliva also contains the hydrolytic enzyme salivary amylase, which initiates the breakdown of the polysaccharide starch into the disaccharide maltose. This digestion is usually minimal in humans, however, because most people don't chew their food very long.

Stimulation of salivation

The secretions of the salivary glands are controlled by the nervous system, which in humans maintains a constant flow of about half a milliliter per minute when the mouth is empty of food. This continuous secretion keeps the mouth moist.

The presence of food in the mouth triggers an increased rate of secretion. Taste buds as well as olfactory (smell) neurons send impulses to the brain, which responds by stimulating the salivary glands (see chapter 45). The most potent stimuli are acidic solutions; lemon juice, for example, can

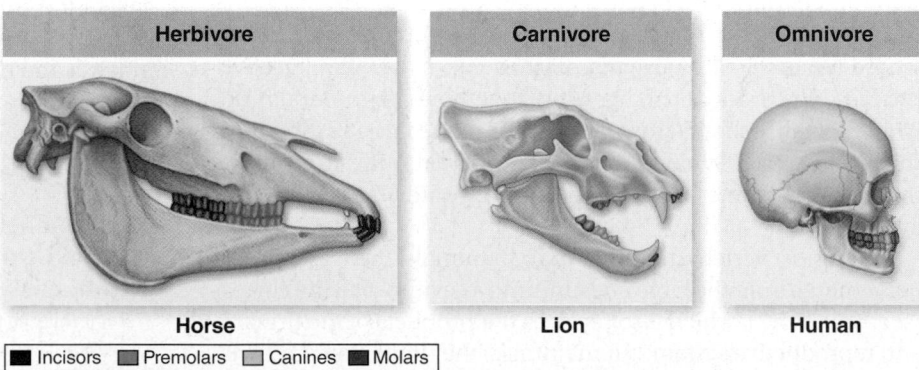

Figure 47.6 Patterns of dentition depend on diet. Different mammals (herbivore, carnivore, or omnivore) have evolved specific variations from a generalized pattern of dentition depending on their diets.

? Inquiry question Why don't all animals have teeth like omnivores, capable of eating anything?

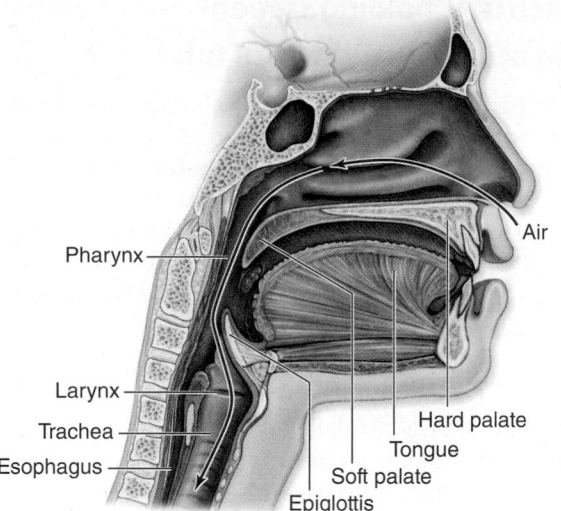
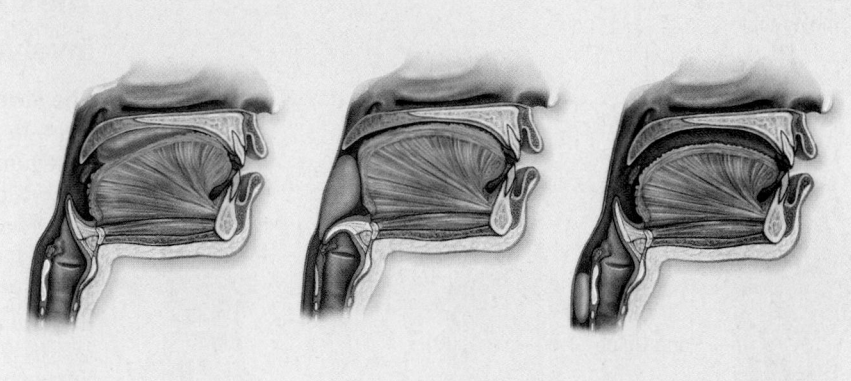

Pharynx

Air

Larynx

Trachea

Esophagus

Hard palate

Tongue

Soft palate

Epiglottis

1. As food moves to the back of the mouth, the soft palate seals off the nasal cavity.

2. During swallowing, the larynx rises and is sealed off by the epiglottis. This forces the bolus into the esophagus and prevents entry into the trachea. As the bolus moves into the esophagus the larynx relaxes.

Figure 47.7 The mechanics of swallowing. Cross section through head and throat showing relevant structures (left). During swallowing (right) the tongue pushes the palate upward, and the soft palate seals off the nasal cavity. Elevation of the larynx causes the epiglottis to seal off the trachea, thus preventing food from entering the airway.

? **Inquiry question** What goes wrong to cause someone to choke?

increase the rate of salivation eightfold. The sight, sound, or smell of food can stimulate salivation markedly in many animals; in humans, thinking or talking about food can also have this effect.

Swallowing

Swallowing is initiated by voluntary action, then is continued under involuntary control. When food is ready to be swallowed, the tongue moves it to the back of the mouth. In mammals, the process of swallowing begins when the soft palate elevates, pushing against the back wall of the pharynx (figure 47.7). Elevation of the soft palate seals off the nasal cavity and prevents food from entering it. Pressure against the pharynx triggers an automatic, involuntary response, the swallowing reflex. Because it is a reflex, swallowing cannot be stopped once it is initiated.

Neurons within the walls of the pharynx send impulses to the swallowing center in the brain. In response, electrical impulses in motor neurons stimulate muscles to contract and raise the **larynx** (voice box). This pushes the glottis, the opening from the larynx into the trachea (windpipe), against a flap of tissue called the **epiglottis.** These actions keep food out of the respiratory tract, directing it instead into the esophagus.

Learning Outcomes Review 47.2

In vertebrates with teeth, tooth shape exhibits adaptations to diet: herbivores have large, flat teeth for grinding, whereas carnivores have pointed teeth for tearing. The mouth serves as an initial processing center, tasting ingested food, breaking it down, and beginning digestion with saliva secretion prior to swallowing.

■ *Explain how the role of the mouth in digestion depends on the type of teeth.*

47.3 ### The Esophagus and the Stomach: The Early Stages of Digestion

Learning Outcomes

1. *Describe how food moves through the esophagus.*
2. *Explain what digestive processes take place in the stomach.*

Swallowed food enters a muscular tube called the esophagus, which connects the pharynx to the stomach. The esophagus actively moves a processed lump of food, called a **bolus,** through the action of muscles. Food from a meal is stored in the stomach where it undergoes early stages of digestion.

Muscular contractions of the esophagus move food to the stomach

In adult humans, the esophagus is about 25 cm long; the upper third is enveloped in skeletal muscle for voluntary control of swallowing, whereas the lower two-thirds is surrounded by involuntary smooth muscle. The swallowing center stimulates successive one-directional waves of contraction in these muscles that move food along the esophagus to the stomach. These rhythmic waves of muscular contraction are called **peristalsis** (figure 47.8); they enable humans and other vertebrates to swallow even if they are upside down.

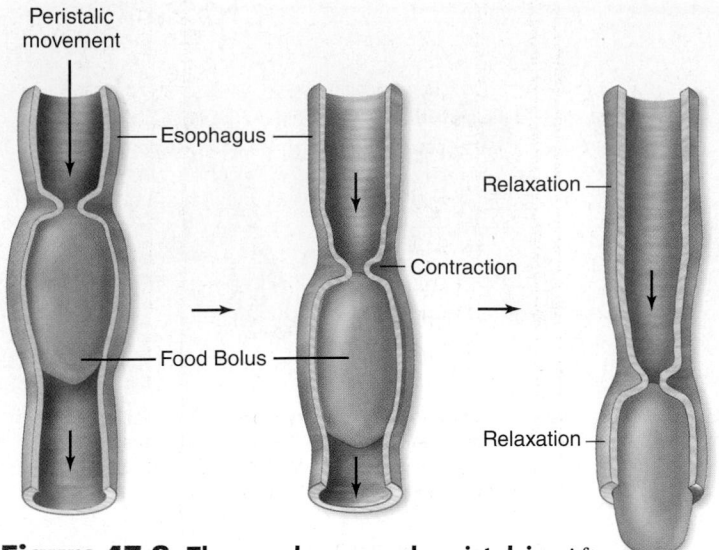

Figure 47.8 The esophagus and peristalsis. After food has entered the esophagus, rhythmic waves of muscular contraction, called peristalsis, move the food down to the stomach.

In many vertebrates, the movement of food from the esophagus into the stomach is controlled by a ring of circular smooth muscle, or a *sphincter,* that opens in response to the pressure exerted by the food. Contraction of this sphincter prevents food in the stomach from moving back into the esophagus. Rodents and horses have a true sphincter at this site, and as a result they cannot regurgitate; humans lack a true sphincter. Normally, the esophagus is closed off except during swallowing.

The stomach is a "holding station" involved in acidic breakdown of food

The **stomach** (figure 47.9) is a saclike portion of the digestive tract. Its inner surface is highly convoluted, enabling it to fold up when empty and open out like an expanding balloon as it fills with food. For example, the human stomach has a volume of only about 50 mL when empty, but it may expand to contain 2 to 4 L of food when full. Carnivores that engage in sporadic gorging as an important survival strategy possess stomachs that are able to distend even more.

Secretory systems

The stomach contains a third layer of smooth muscle for churning food and mixing it with **gastric juice,** an acidic secretion of the tubular gastric glands of the mucosa (see figure 47.9). These exocrine glands contain three kinds of secretory cells: *mucus-secreting cells, parietal cells,* which secrete hydrochloric acid (HCl), and *chief cells,* which secrete **pepsinogen,** the inactive form of the protease (protein-digesting enzyme) **pepsin.**

Pepsinogen has 44 additional amino acids that block its active site. HCl causes pepsinogen to unfold, exposing the active site, which then acts to remove the 44 amino acids. This yields the active protease, pepsin. This process of secreting an inactive form that is then converted into an active enzyme outside the cell prevents the chief cells from digesting themselves. In the stomach, mucus produced by mucus-secreting cells serves the same purpose, covering the interior walls and preventing them from being digested.

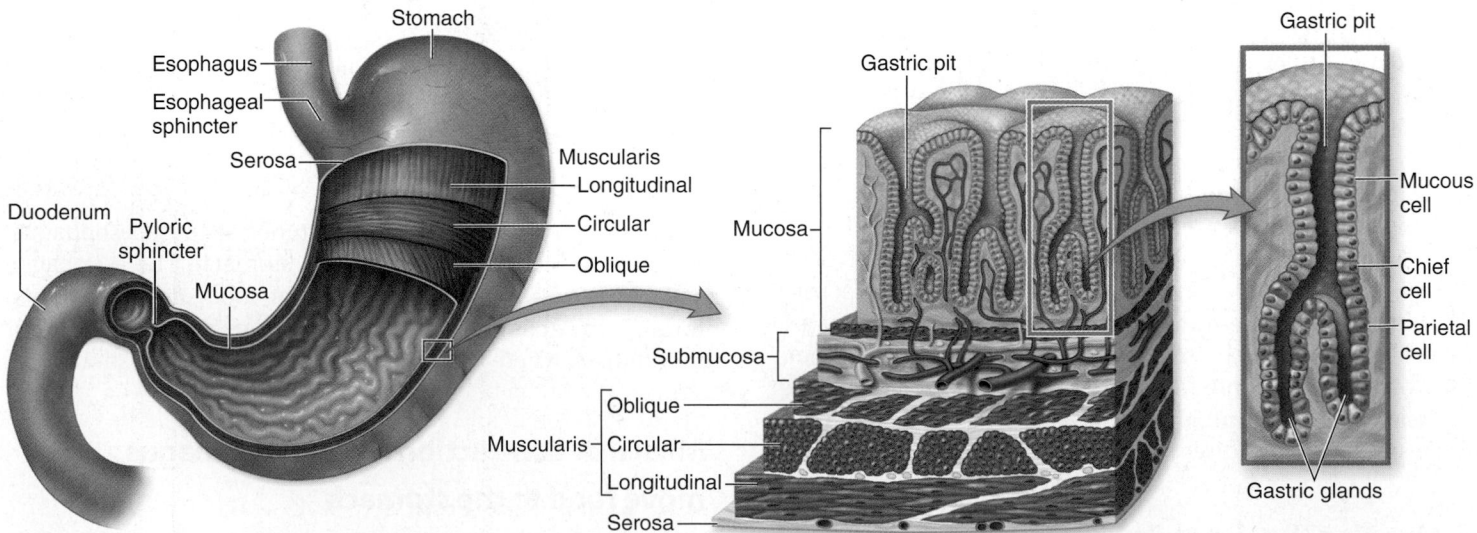

Figure 47.9 The stomach and duodenum. Food enters the stomach from the esophagus. A ring of smooth muscle called the pyloric sphincter controls the entrance to the duodenum, the upper part of the small intestine. The epithelial walls of the stomach are dotted with deep infoldings called gastric pits that contain gastric glands. The gastric glands consist of mucous cells, chief cells that secrete pepsinogen, and parietal cells that secrete HCl. Gastric pits are the openings of the gastric glands.

? Inquiry question How does the digestive system keep from being digested by the gastric secretions it produces?

In addition to producing HCl, the parietal cells of the stomach also secrete **intrinsic factor,** a polypeptide needed for the intestinal absorption of vitamin B$_{12}$. Because this vitamin is required for the production of red blood cells, people who lack sufficient intrinsic factor develop a type of anemia (low red blood cell count) called *pernicious anemia.*

Action of acid

The human stomach produces about 2 L of HCl and other gastric secretions every day, creating a very acidic solution. The concentration of HCl in this solution is about 10 milli-molar (mM), equal to a pH of 2. Thus, gastric juice is about 250,000 times more acidic than blood, whose normal pH is 7.4.

The low pH in the stomach helps denature food proteins, making them easier to digest, and keeps pepsin maximally active. Active pepsin hydrolyzes food proteins into shorter chains of polypeptides that are not fully digested until the mixture enters the small intestine. The mixture of partially digested food and gastric juice is called **chyme.** In adult humans, only proteins are partially digested in the stomach—no significant digestion of carbohydrates or fats occurs there.

The acidic solution within the stomach also kills most of the bacteria that are ingested with the food. The few bacteria that survive the stomach and enter the intestine intact are able to grow and multiply there, particularly in the large intestine. In fact, vertebrates harbor thriving colonies of bacteria within their intestines, and bacteria are a major component of feces. As we discuss later, bacteria that live within the digestive tracts of ruminants play a key role in the ability of these mammals to digest cellulose.

Ulcers

Overproduction of gastric acid can occasionally eat a hole through the wall of the stomach or the duodenum, causing a peptic ulcer. Although we once blamed consumption of spicy food, the most common cause of peptic ulcers is now thought to be infection with the bacterium *Heliocobacter pylori.*

H. pylori can grow on the lining of the human stomach, surviving the acid pH by secreting substances that buffer the pH of its immediate surroundings. Although infection with *H. pylori* is common in the United States (about 20% of people younger than 40 and 50% older than 60), most people are asymptomatic. However, in some cases, infection by *H. pylori* can reduce or weaken the mucosal layer in the stomach or duodenum, allowing acidic secretions to attack the underlying epithelium. Antibiotic treatment of the infection can reduce symptoms and often even cure the ulcer.

Leaving the stomach

Chyme leaves the stomach through the *pyloric sphincter* (see figure 47.9) to enter the small intestine. This is where all terminal digestion of carbohydrates, lipids, and proteins occurs and where the products of digestion—amino acids, glucose, and so on—are absorbed into the blood. Only some of the water in chyme and a few substances, such as aspirin and alcohol, are absorbed through the wall of the stomach.

47.4 The Intestines: Breakdown, Absorption, and Elimination

The capacity of the small intestine is limited, and its digestive processes take time. Consequently, efficient digestion requires that only relatively small amounts of chyme be introduced from the stomach into the small intestine at any one time. Coordination between gastric and intestinal activities is regulated by neural and hormonal signals, which we will describe in section 47.6.

The structure of the small intestine is specialized for digestion and nutrient uptake

The small intestine is approximately 4.5 m long in a living person, but 6 m long at autopsy when all the muscles have relaxed. The first 25 cm is the **duodenum;** the remainder of the small intestine is divided into the **jejunum** and the **ileum.**

The duodenum receives acidic chyme from the stomach, digestive enzymes and bicarbonate from the pancreas, and bile from the liver and gallbladder. Enzymes in the pancreatic juice digest larger food molecules into smaller fragments. This digestion occurs primarily in the duodenum and jejunum.

The epithelial wall of the small intestine is covered with tiny, fingerlike projections called **villi** (singular, *villus;* figure 47.10). In turn, each epithelial cell lining the villi is covered on its apical surface (the side facing the lumen) by many foldings of the plasma membrane that form cytoplasmic extensions called **microvilli.** These are quite tiny and can be seen clearly only with an electron microscope. Under a light micrograph, the microvilli resemble the bristles of a brush, and for that reason the epithelial wall of the small intestine is also called a *brush border.*

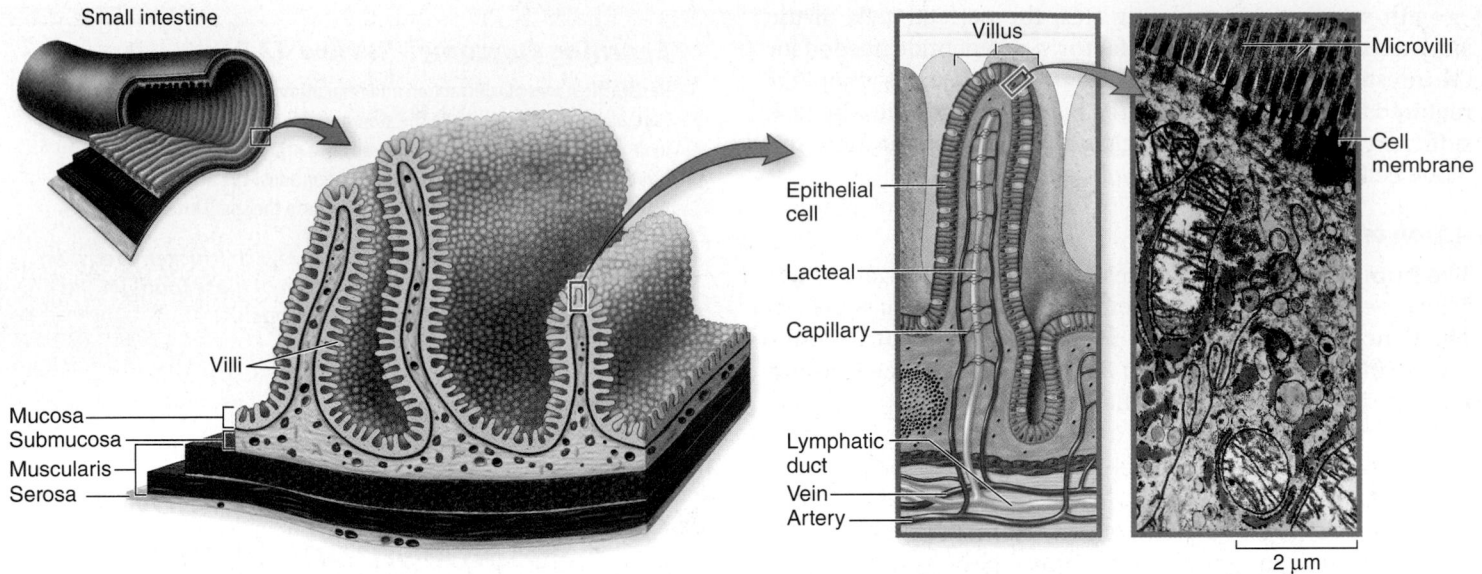

Figure 47.10 The small intestine. Successive enlargements show folded epithelium studded with villi that increase the surface area. The micrograph shows an epithelial cell with numerous microvilli.

The villi and microvilli greatly increase the surface area of the small intestine; in humans, this surface area is 300 m²— about 3200 square feet, larger than a tennis court! It is over this vast surface that the products of digestion are absorbed.

Data analysis Explain how to convert square meters to square feet and calculate the exact area of the small intestine in square feet.

The microvilli also participate in digestion because a number of digestive enzymes are embedded within the epithelial cells' plasma membranes, with their active sites exposed to the chyme. These brush border enzymes include those that hydrolyze the disaccharides lactose and sucrose, among others. Many adult humans lose the ability to produce the brush border enzyme lactase and therefore cannot digest lactose (milk sugar), a rather common condition called *lactose intolerance.* The brush border enzymes complete the digestive process that started with the action of salivary amylase in the mouth.

Accessory organs secrete enzymes into the small intestine

The main organs that aid digestion are the pancreas, liver, and gallbladder. They empty their secretions, primarily enzymes, through ducts directly into the small intestine.

Secretions of the pancreas

The pancreas (figure 47.11), a large gland situated near the junction of the stomach and the small intestine, secretes pancreatic fluid into the duodenum through the *pancreatic duct;* thus, the pancreas functions as an exocrine gland. This fluid contains a host of enzymes, including **trypsin** and **chymotrypsin,** which digest proteins; **pancreatic amylase,** which digests starch; and **lipase,** which digests fat. Like pepsin in the stomach, these enzymes are released into the duodenum primarily as inactive

enzymes and are then activated by trypsin, which is first activated by a brush border enzyme of the intestine.

Pancreatic enzymes digest proteins into smaller polypeptides, polysaccharides into shorter chains of sugars, and fats into free fatty acids and monoglycerides. Digestion of

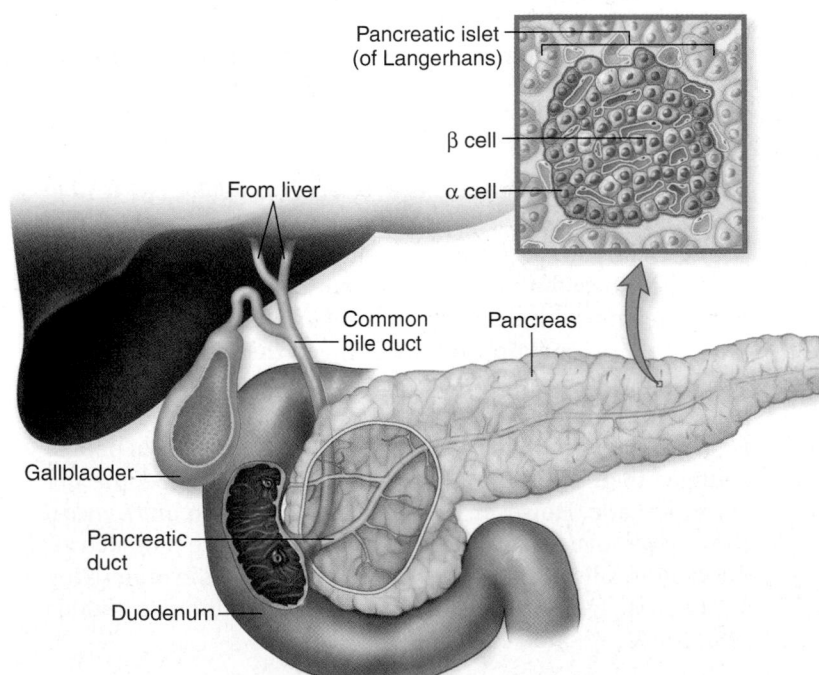

Figure 47.11 The pancreas. The pancreatic and bile ducts empty into the duodenum. The pancreas secretes pancreatic juice into the pancreatic duct. The pancreatic islets of Langerhans secrete hormones into the blood; α cells secrete glucagon, and β cells secrete insulin. The liver secretes bile, which consists of bile pigments (waste products from the liver) and bile salts. Bile salts play a role in the digestion of fats. Bile is concentrated and stored in the gallbladder until it is needed in the duodenum on the arrival of fatty food.

proteins and carbohydrates is then completed by the brush border enzymes. Pancreatic fluid also contains bicarbonate, which neutralizes the HCl from the stomach and gives the chyme in the duodenum a slightly alkaline pH. The digestive enzymes and bicarbonate are produced by clusters of secretory cells known as **acini.**

In addition to its exocrine role in digestion, the pancreas also functions as an endocrine gland, secreting several hormones into the blood that control the blood levels of glucose and other nutrients. These hormones are produced in the **islets of Langerhans,** clusters of endocrine cells scattered throughout the pancreas. The two most important pancreatic hormones, insulin and glucagon, were described in chapter 45; their actions are also discussed later on.

Liver and gallbladder

The **liver** is the largest internal organ of the body (see figure 47.3), weighing about 1.5 kg, the size of a football. The liver plays many roles, including detoxification of harmful substances, protein synthesis, and glycogen storage, among others. Chronic abuse of alcohol can overtax the liver's detoxification capabilities, leading to cirrhosis and eventually to liver failure.

The main exocrine secretion of the liver is bile, a fluid mixture consisting of *bile pigments* and *bile salts* that is delivered into the duodenum during the digestion of a meal. The bile pigments do not participate in digestion; they are waste products resulting from the liver's destruction of old red blood cells and are ultimately eliminated with the feces. If the excretion of bile pigments by the liver is blocked, the pigments can accumulate in the blood and cause a yellow staining of the tissues known as *jaundice.*

In contrast, the bile salts play a very important role in preparing fats for subsequent enzymatic digestion. Because fats are insoluble in water, they enter the intestine as drops within the watery chyme. The bile salts, which are partly lipid-soluble and partly water-soluble, work like detergents, dispersing the large drops of fat into a fine suspension of smaller droplets. This emulsification action produces a greater surface area of fat for the action of lipase enzymes, and thus allows the digestion of fat to proceed more rapidly.

After bile is produced in the liver, it is stored and concentrated in the gallbladder. The arrival of fatty food in the duodenum triggers a neural and endocrine reflex that stimulates the gallbladder to contract, causing bile to be transported through the common bile duct and injected into the duodenum (these reflexes are the topic of a later section). Gallstones are hardened precipitates of cholesterol that form in some individuals. If these stones block the bile duct, contraction of the gallbladder causes intense pain, often felt in the back. In severe cases of blockage, surgical removal of the gallbladder may be performed.

Absorbed nutrients move into blood or lymph capillaries

After their enzymatic breakdown, proteins and carbohydrates are absorbed as amino acids and monosaccharides, respectively. They are transported across the brush border into the epithelial cells that line the intestine by a combination of active transport and facilitated diffusion (figure 47.12*a*). Glucose is transported by coupled transport with Na^+ ions (also called secondary active transport). Fructose, found in most fruit, is transported by facilitated diffusion. Most amino acids are transported by active transport using a variety of different transporters. Some of these carrier proteins use cotransport with Na^+ ions; others transport only amino acids. Once they

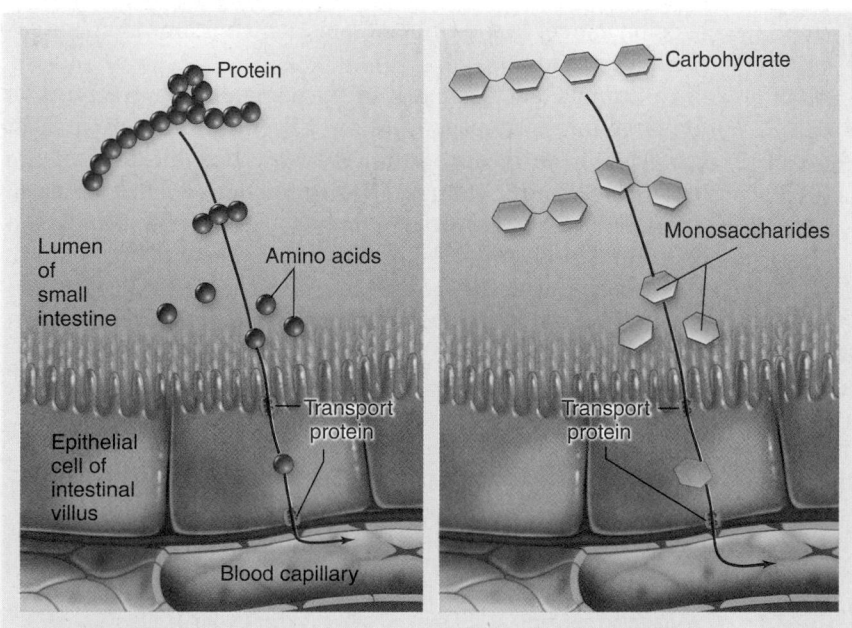

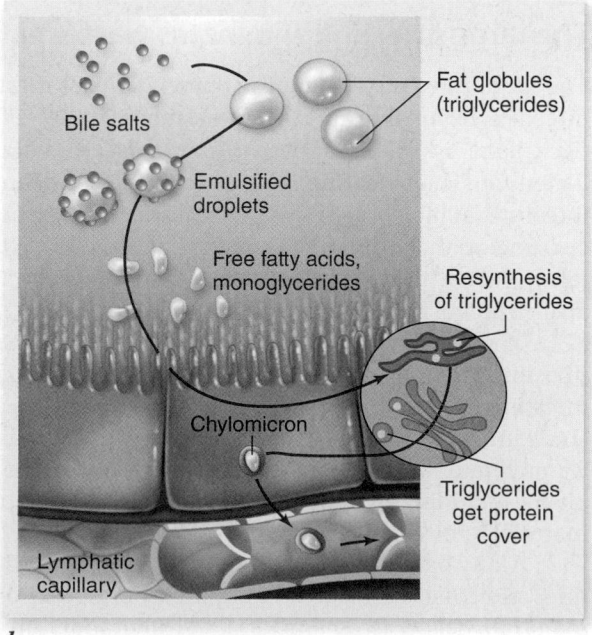

a.

b.

Figure 47.12 Absorption of the products of digestion. *a.* Monosaccharides and amino acids are transported into blood capillaries. *b.* Fatty acids and monoglycerides within the intestinal lumen are absorbed and converted within the intestinal epithelial cells into triglycerides. These are then coated with proteins to form structures called chylomicrons, which enter lymphatic capillaries.

have entered epithelial cells across the apical membrane, these monosaccharides and amino acids move through the cytoplasm and are transported across the basolateral membrane and into the blood capillaries within the villi.

The blood carries these products of digestion from the intestine to the liver via the hepatic portal vein. A portal vein connects two beds of capillaries instead of returning to the heart. In this case, the intestine is connected to the liver by the hepatic portal vein, thus the liver receives blood-borne molecules from the intestine. Because of the hepatic portal vein, the liver is the first organ to receive most of the products of digestion, except for fat.

The products of fat digestion are absorbed by a different mechanism (figure 47.12b). Fats (triglycerides) are hydrolyzed into fatty acids and monoglycerides by digestion. These fatty acids and monoglycerides are nonpolar and can thus enter epithelial cells by simple diffusion. Once inside the intestinal epithelial cells they are reassembled into triglycerides. The triglycerides then combine with proteins to form small particles called **chylomicrons,** which are too bulky to enter blood capillaries in the intestine. Instead of entering the hepatic portal circulation, the chylomicrons are absorbed into lymphatic capillaries (see chapter 49), which empty their contents into the blood in veins near the neck. Chylomicrons can make the blood plasma appear cloudy if a sample of blood is drawn after a fatty meal.

The amount of fluid passing through the small intestine in a day is startlingly large: approximately 9 L. However, almost all of this fluid is absorbed into the body rather than eliminated in the feces: About 8.5 L is absorbed in the small intestine and an additional 350 mL in the large intestine. Only about 50 g of solid and 100 mL of liquid leaves the body as feces. The normal fluid absorption efficiency of the human digestive tract approaches 99%, which is very high indeed.

The large intestine eliminates waste material

The large intestine, or **colon,** is much shorter than the small intestine, occupying approximately the last meter of the digestive tract; it is called "large" because of its larger diameter, not its length. The small intestine empties directly into the large intestine at a junction where two vestigial structures, the **cecum** and the **appendix,** remain (figure 47.13). No digestion takes place within the large intestine, and only about 4% of the absorption of fluids by the intestine occurs there.

The large intestine is not as convoluted as the small intestine, and its inner surface has no villi. Consequently, the large intestine has less than 1/30 the absorptive surface area of the small intestine. The function of the large intestine is to absorb water, remaining electrolytes, and products of bacterial metabolism (including vitamin K). The large intestine prepares waste material to be expelled from the body.

Many bacteria live and reproduce within the large intestine, and the excess bacteria are incorporated into the refuse material, called *feces.* Bacterial fermentation produces gas within the colon at a rate of about 500 mL per day. This rate increases greatly after the consumption of beans or other types of vegetables because the passage of undigested plant material

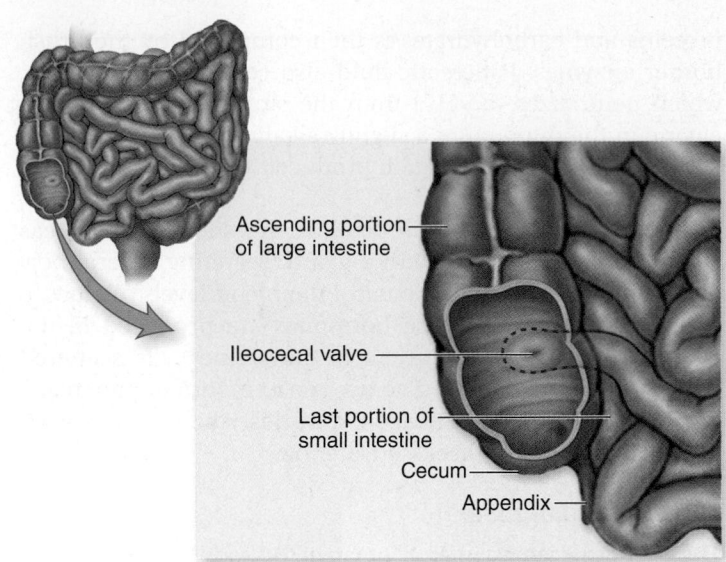

Figure 47.13 The junction of the small and large intestines in humans. The large intestine, or colon, starts with the cecum, which is relatively small in humans compared with that in other mammals. A vestigial structure called the appendix extends from the cecum. When the appendix becomes inflamed, which usually results from objects filling its interior, appendicitis results. Without surgery, appendicitis is often fatal.

(fiber) into the large intestine provides substrates for bacterial fermentation.

The human colon has evolved to process food with a relatively high fiber content. Diets that are low in fiber, which are common in the United States and other developed countries, result in a slower passage of food through the colon. Low dietary fiber content is thought to be associated with the level of colon cancer in the United States, which is among the highest in the world.

Compacted feces, driven by peristaltic contractions of the large intestine, pass from the large intestine into a short tube called the rectum. From the rectum, the feces exit the body through the anus. Two sphincters control passage through the anus. The first is composed of smooth muscle and opens involuntarily in response to pressure inside the rectum. The second, composed of striated muscle, can be controlled voluntarily by the brain, thus permitting a conscious decision to delay defecation.

Learning Outcomes Review 47.4

The small intestine is where most digestion takes place; its inner surface is covered with villi that increase its absorptive surface area. The large intestine absorbs water, electrolytes, and bacterial metabolites. Digestion is accomplished by a combination of enzymes from the pancreas and by bile salts released from the liver. Glucose and amino acids are absorbed by active transport and facilitated diffusion. Fat is absorbed by simple diffusion.

■ *Why does fat not require transport to cross the intestinal epithelium?*

Learning Outcomes

1. Explain how vertebrates digest cellulose.
2. Describe how rumination works.
3. Discuss convergent evolution at the molecular level in herbivores.

Animals lack the enzymes necessary to digest cellulose, but the digestive tracts of some animals contain bacteria and protists that convert cellulose into substances the host can absorb. Although digestion by gastrointestinal microorganisms plays a relatively small role in human nutrition, it is an essential element in the nutrition of many other kinds of animals, including insects such as termites and cockroaches, and a few groups of herbivorous mammals. The relationships between these microorganisms and their animal hosts are mutually beneficial and provide an excellent example of symbiosis (see chapter 56).

Plant cellulose is particularly resistant to digestion. As a result, herbivores tend to have much longer digestive tracts than carnivores, allowing greater time for digestion to occur (figure 47.14). In addition, many herbivores have modified their digestive tracts to enhance digestion of plant material.

Ruminants rechew regurgitated food

Ruminants have a four-chambered stomach (figure 47.15). The first three portions include the reticulum, the rumen, and the omasum. These are followed by the true stomach, the abomasum.

The rumen, which may hold up to 50 gallons, serves as a fermentation vat where bacteria and protists convert cellulose and other molecules into a variety of simpler compounds. The location of the rumen at the front of the four chambers allows the animal to regurgitate and rechew the contents of the rumen, an activity called *rumination,* or "chewing the cud." This breaks tougher fiber in the diet into smaller particles, increasing the surface area for microbial attachment.

After chewing, the cud is swallowed for further microbial digestion in the rumen, then passes to the omasum, and then to the abomasum, where it is finally mixed with gastric juice. This process leads to far more efficient digestion of cellulose in ruminants than in mammalian herbivores that lack a rumen, such as horses.

Nonruminant Herbivore	**Ruminant Herbivore**
Simple stomach, large cecum	Four-chambered stomach with large rumen; long small and large intestine

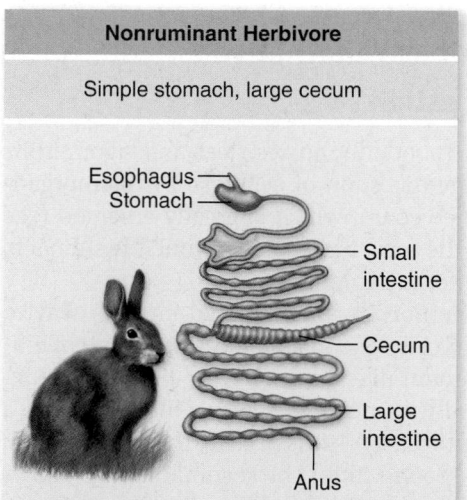

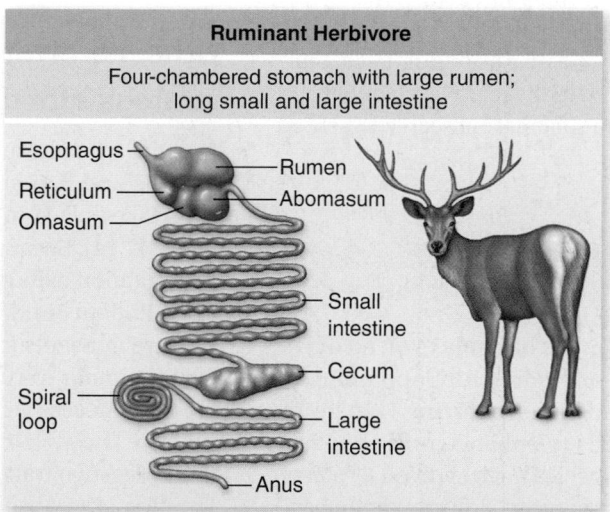

Insectivore	**Carnivore**
Short intestine, no cecum	Short intestine and colon, small cecum

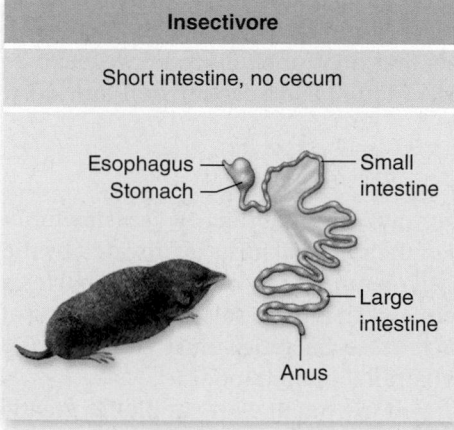

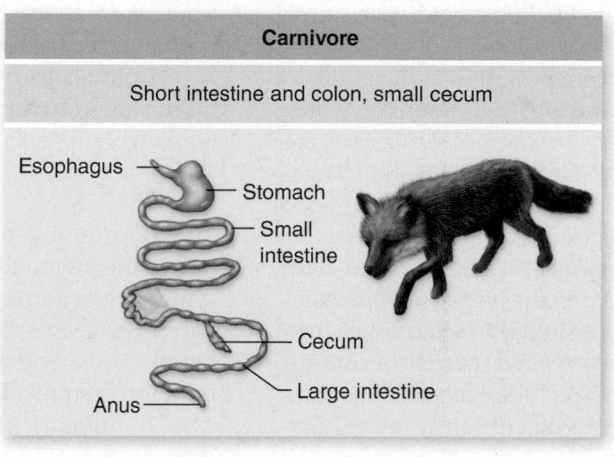

Figure 47.14 The digestive systems of different mammals reflect their diets. Herbivores, such as rabbits and deer, require long digestive tracts with specialized compartments for the breakdown of plant matter. Diets composed of animal matter, thus lacking cellulose, are more easily digested; insectivorous and carnivorous mammals, such as voles and foxes, respectively, have short digestive tracts with few specialized pouches.

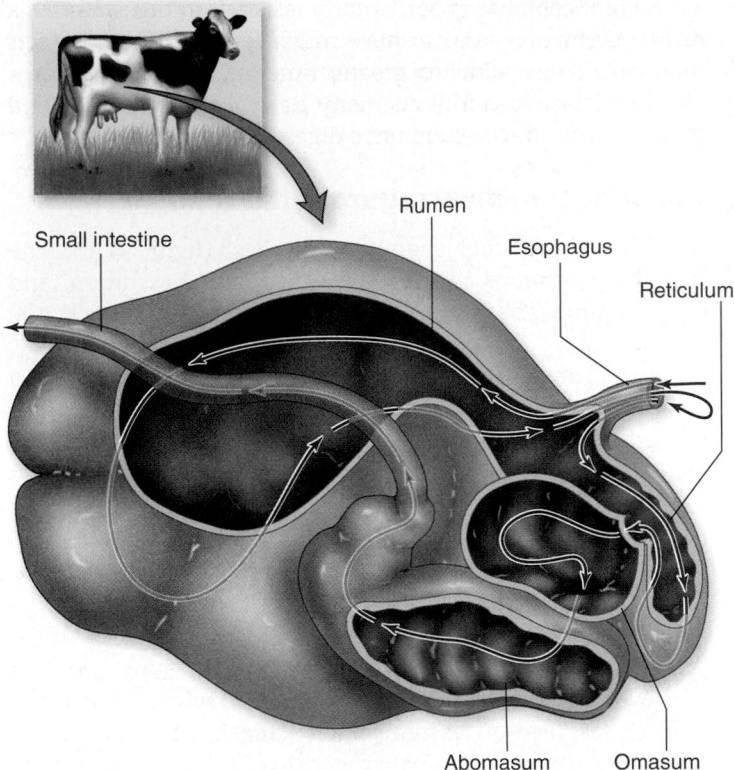

Figure 47.15 Four-chambered stomach of a ruminant.
Grass and other plants eaten by ruminants enter the rumen, where they are partially digested. The rumen contains bacteria that break down cellulose from the plant cell walls. Before moving into a second chamber, the reticulum, the food may be regurgitated and rechewed. The food is then transferred to the rear two chambers: the omasum and abomasum. Only the abomasum secretes gastric juice as in the human stomach.

Foregut fermentation has evolved convergently many times

Although the four-chambered stomach has only evolved once, many other types of herbivores—including hippopotamuses, langur monkeys, sloths, kangaroos, and hoatzins (a type of bird)—have evolved large stomachs to enhance microbial fermentation. In many cases, these species have evolved a variety of other anatomical structures that serve to slow down the passage of food through the stomach, leading to increased time for fermentation.

A remarkable case of convergent evolution at the molecular level is exhibited by ruminants and the langur monkey, which subsists primarily on leaves. In most mammals, lysozymes are enzymes found in saliva and tears, which attack invading bacteria. However, in ruminants and langurs, lysozymes have been modified to take on a new role, digesting bacteria in the stomach. In both cases, five identical amino acid changes have evolved (figure 47.16); the result is that the lysozyme molecules of ruminants and langurs are more similar to each other than they are to lysozymes in more closely related species. In contrast to many cases of convergent evolution, this example illustrates that convergent evolution has occurred in distantly related species by the exact same evolutionary changes.

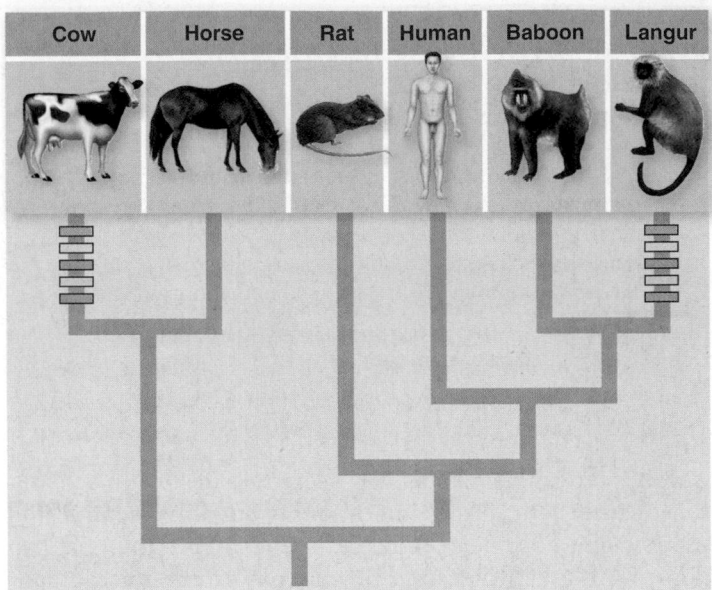

Figure 47.16 Convergent evolution of lysozyme structure in ruminants (represented by the cow) and leaf-eating hanuman langur *(Presbytis entellus).* The same five amino acid changes evolved independently in both groups.

 Data analysis If you constructed a phylogeny using molecular data from lysozyme, what would it look like?

Other herbivores have alternative strategies for digestion

In some animals, such as rodents, horses, deer, and lagomorphs (rabbits and hares), the digestion of cellulose by microorganisms takes place in the cecum, which is greatly enlarged (see figure 47.14). Because the cecum is located beyond the stomach, regurgitation of its contents is impossible.

Rodents and lagomorphs have evolved another way to capture nutrients from cellulose that achieves a degree of efficiency similar to ruminant digestion. They do this by producing some feces packed with nutrients, which they then eat, a practice known as coprophagy—thus passing the food through their digestive tract a second time. The second passage allows the animal to absorb the nutrients produced by the microorganisms in its cecum. Coprophagic animals cannot remain healthy if they are prevented from eating their feces.

Animals with diets that don't include cellulose, such as insectivores or carnivores, don't have a cecum, or if they do, it is greatly reduced.

Vitamin K

Another example of the way intestinal microorganisms function in the metabolism of their animal hosts is provided by the synthesis of vitamin K. All mammals rely on intestinal bacteria to synthesize this vitamin, which is necessary for the clotting of blood. Birds, which lack these bacteria, must consume the required quantities of vitamin K in their food.

In humans, prolonged treatment with antibiotics greatly reduces the populations of bacteria in the intestine; under such circumstances, it may be necessary to provide supplementary vitamin K. Restoring the normal flora of the digestive tract with beneficial bacteria may also help replace vitamin K.

Learning Outcomes Review 47.5

The digestive tracts of many herbivores harbor colonies of cellulose-digesting microorganisms. Complex fermentation chambers have also evolved in the digestive tract. In rumination, partially digested food is regurgitated from the rumen for additional processing by the mouth. In distantly related herbivorous species, similar digestive enzymes have evolved by identical but independent changes.

■ *Would you expect identical mutations to be successful in different species? Why or why not?*

47.6 Neural and Hormonal Regulation of the Digestive Tract

Learning Outcomes

1. Explain how the nervous system stimulates the digestive process.
2. Identify the major enterogastrones.

The activities of the gastrointestinal tract are coordinated by the nervous system and the endocrine system. The nervous system, for example, stimulates salivary and gastric secretions in response to the sight, smell, and consumption of food. When food arrives in the stomach, proteins in the food stimulate the secretion of a stomach hormone called **gastrin,** which in turn stimulates the secretion of pepsinogen and HCl from the gastric glands (figure 47.17). The secreted HCl then lowers the pH of the gastric juice, which acts to inhibit further secretion of gastrin in a negative feedback loop. In this way, the secretion of gastric acid is kept under tight control.

The passage of chyme from the stomach into the duodenum of the small intestine inhibits the contractions of the stomach, so that no additional chyme can enter the duodenum until the previous amount can be processed. This stomach or gastric inhibition is mediated by a neural reflex and by duodenal hormones secreted into the blood. These hormones are collectively known as the **enterogastrones.**

The major enterogastrones include **cholecystokinin (CCK), secretin,** and **gastric inhibitory peptide (GIP).** Chyme with high fat content is the strongest stimulus for CCK and GIP secretions, whereas increasing chyme acidity primarily influences the release of secretin. All three of these enterogastrones inhibit gastric motility (churning action) and gastric juice secretions; the result is that fatty meals remain in the stomach longer than nonfatty meals, allowing more time for digestion of complex fat molecules.

In addition to gastric inhibition, CCK and secretin have other important regulatory functions in digestion. CCK also stimulates increased pancreatic secretions of digestive enzymes and gallbladder contractions. Gallbladder contractions inject more bile into the duodenum, which enhances the emulsification and efficient digestion of fats. The other major function of secretin is to stimulate the pancreas to release more bicarbonate, which neutralizes the acidity of the chyme. Secretin has the distinction of being the first hormone ever discovered. Table 47.1 summarizes the actions of the digestive hormones and enzymes.

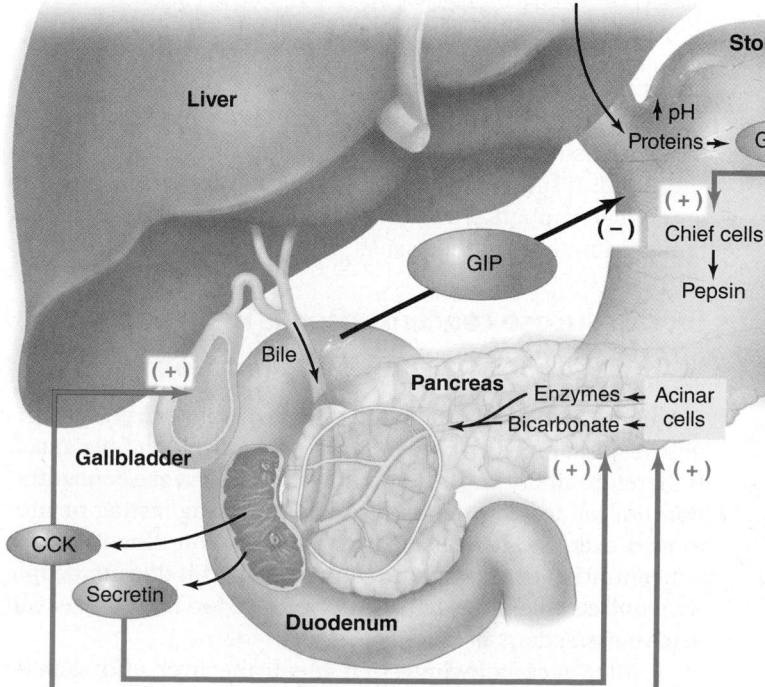

Learning Outcomes Review 47.6

Sensory input such as sight, smell, and taste stimulate salivary and gastric activity, as does the arrival of food in the stomach. The major enterogastrones are cholecystokinin (CCK), secretin, and gastric inhibitory peptide (GIP); these regulate passage of chyme into the duodenum and also release of pancreatic enzymes and bile.

■ *Would you expect anosmia, an inability to perceive scents, to affect digestion?*

Figure 47.17 Hormonal control of the gastrointestinal tract. Gastrin, secreted by the mucosa of the stomach, stimulates the secretion of hydrochloric acid (HCl) and pepsinogen (which is converted into pepsin). The duodenum secretes three hormones: cholecystokinin (CCK), which stimulates contraction of the gallbladder and secretion of pancreatic enzymes; secretin, which stimulates secretion of pancreatic bicarbonate; and gastric inhibitory peptide (GIP), which inhibits stomach emptying.

TABLE 47.1 Hormones and Enzymes of Digestion

HORMONES

Hormone	Class	Source	Stimulus	Action	Note
Gastrin	Polypeptide	Pyloric portion of stomach	Entry of food into stomach	Stimulates secretion of HCl and pepsinogen by stomach	Acts on same organ that secretes it
Cholecystokinin (CCK)	Polypeptide	Duodenum	Fatty chyme in duodenum	Stimulates gallbladder contraction and secretion of digestive enzymes by pancreas	Structurally similar to gastrin
Gastric inhibitory peptide (GIP)	Polypeptide	Duodenum	Fatty chyme in duodenum	Inhibits stomach emptying	Also stimulates insulin secretion
Secretin	Polypeptide	Duodenum	Acidic chyme in duodenum	Stimulates secretion of bicarbonate by pancreas	The first hormone to be discovered (1902)

ENZYMES

Location	Enzymes	Substrates	Digestion Products
Salivary glands	Amylase	Starch, glycogen	Disaccharides
Stomach	Pepsin	Proteins	Short peptides
Pancreas	Lipase	Triglycerides	Fatty acids, monoglycerides
	Trypsin, chymotrypsin	Proteins	Peptides
	DNase	DNA	Nucleotides
	RNase	RNA	Nucleotides
Small intestine (brush border)	Peptidases	Short peptides	Amino acids
	Nucleases	DNA, RNA	Sugars, nucleic acid bases
	Lactase, maltase, sucrase	Disaccharides	Monosaccharides

47.7 Accessory Organ Function

Learning Outcomes

1. Describe the liver's role in maintaining homeostasis.
2. Explain how the pancreas acts to control blood glucose concentration.

The liver and pancreas both have critical roles beyond the production of digestive enzymes. The liver is a key organ in the breakdown of toxins, and the pancreas secretes hormones that regulate the blood glucose level, in part through actions on liver cells.

The liver modifies chemicals to maintain homeostasis

Because the hepatic portal vein carries blood from the stomach and intestine directly to the liver, the liver is in a position to chemically modify the substances absorbed in the gastrointestinal tract before they reach the rest of the body. For example, ingested alcohol and other drugs are taken into liver cells and metabolized; this is one reason that the liver is often damaged as a result of alcohol and drug abuse.

The liver also removes toxins, pesticides, carcinogens, and other poisons, converting them into less toxic forms. For example, the liver converts the toxic ammonia produced by intestinal bacteria into urea, a compound that can be contained safely and carried by the blood at higher concentrations.

Similarly, the liver regulates the levels of many compounds produced within the body. Steroid hormones, for instance, are converted into less active and more water-soluble forms by the liver. These molecules are then included in the bile and eliminated from the body in the feces or are carried by the blood to the kidneys and excreted in the urine.

The liver also produces most of the proteins found in blood plasma. The total concentration of plasma proteins is significant because it must be kept within certain limits to maintain osmotic balance between blood and interstitial (tissue) fluid. If the concentration of plasma proteins drops too low, as can happen as a result of liver disease such as cirrhosis, fluid accumulates in the tissues, a condition called *edema*.

Blood glucose concentration is maintained by the actions of insulin and glucagon

The neurons in the brain obtain energy primarily from the aerobic respiration of glucose obtained from the blood plasma. It is therefore vitally important that the blood glucose concentration not fall too low, as might happen during fasting or prolonged exercise. It is also important that the blood glucose concentration not stay at too high a level, as it does in people with untreated diabetes mellitus, because too high a level can lead to tissue damage.

After a carbohydrate-rich meal, the liver and skeletal muscles remove excess glucose from the blood and store it as

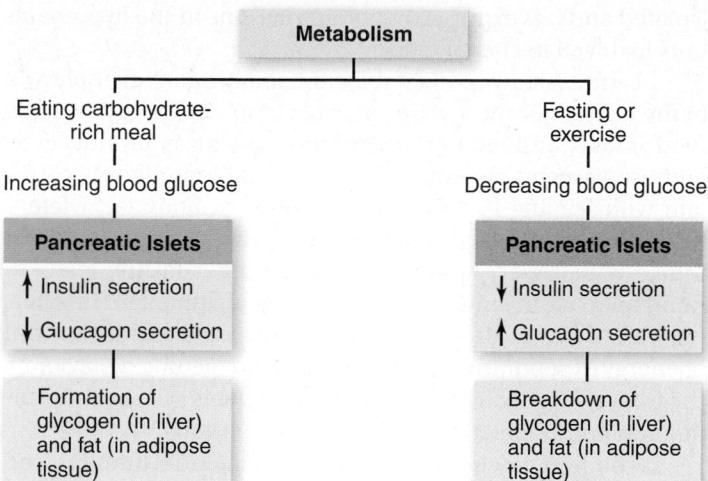

Figure 47.18 The actions of insulin and glucagon.
After a meal, an increased secretion of insulin by the β cells of the pancreatic islets promotes the deposition of glycogen and fat. During fasting or exercising, increased glucagon secretion by the α cells of the pancreatic islets and decreased insulin secretion promote the breakdown (through hydrolysis reactions) of glycogen and fat.

the polysaccharide glycogen. This process is stimulated by the hormone insulin, secreted by the β (beta) cells in the pancreatic islets of Langerhans (figure 47.18).

When blood glucose levels decrease, as they do between meals, during periods of fasting, and during exercise, the liver secretes glucose into the blood. This glucose is obtained in part from the breakdown of liver glycogen to glucose-6-phosphate, a process called **glycogenolysis.** The phosphate group is then removed, and free glucose is secreted into the blood. Skeletal muscles lack the enzyme needed to remove the phosphate group, and so, even though they have glycogen stores, they cannot secrete glucose into the blood. However, muscle cells can use this glucose directly for energy metabolism because glucose-6-phosphate is actually the product of the first reaction in glycolysis. The breakdown of liver glycogen is stimulated by another hormone, glucagon, which is secreted by the α (alpha) cells of the islets of Langerhans in the pancreas (see figure 47.18).

If fasting or exercise continues, the liver begins to convert other molecules, such as amino acids and lactic acid, into glucose. This process is called **gluconeogenesis** ("new formation of glucose"). The amino acids used for gluconeogenesis are obtained from muscle protein, which explains the severe muscle wasting that occurs during prolonged fasting.

Learning Outcomes Review 47.7

The liver is responsible for neutralizing potentially harmful toxins and also for modification of steroid hormones. The liver also produces vital plasma proteins. Pancreatic hormones and the liver regulate blood glucose concentrations. Insulin stimulates the formation of glycogen and fat in the liver. Glucagon stimulates the breakdown of glycogen in the liver, which releases glucose into the blood.

■ *What is one important advantage of the hepatic portal system?*

47.8 Food Energy, Energy Expenditure, and Essential Nutrients

Learning Outcomes

1. Explain the basal metabolic rate and the effect of exercise.
2. List hormones involved in regulating appetite and body weight.
3. Name the essential nutrients.

The ingestion of food serves two primary functions: It provides a source of energy and it provides raw materials the animal is unable to manufacture for itself.

Even an animal completely at rest requires energy to support its metabolism; the minimum rate of energy consumption under defined resting conditions is called the **basal metabolic rate (BMR).** The BMR is relatively constant for a given individual, depending primarily on the person's age, sex, and body size.

Exertion increases metabolic rate

Physical exertion raises the metabolic rate above the basal levels, so the amount of energy the body consumes per day is determined not only by the BMR but also by the level of physical activity. If food energy taken in is greater than the energy consumed per day, the excess energy will be stored in glycogen and fat (figure 47.18). Because glycogen reserves are limited, however, continued ingestion of excess food energy results primarily in the accumulation of fat.

The intake of food energy is measured in **kilocalories** (1 kilocalorie = 1000 calories; nutritionists use Calorie with a capital C instead of kilocalorie). The measurement of kilocalories in food is determined by the amount of heat generated when the food is "burned," either literally, in a testing device called a calorimeter, or in the body, when the food is digested and later oxidized during cellular respiration. Caloric intake can be altered by the choice of foods, and the amount of energy expended can be changed by the choice of lifestyle.

The daily energy expenditures (metabolic rates) of humans vary between 1300 and 5000 kilocalories per day, depending on the person's BMR and level of physical activity. When the total kilocalories ingested exceeds the metabolic rate for a sustained period, a person accumulates an amount of fat that is deleterious to health, a condition called obesity. In the United States, about 34% of all adults between 40 and 59 are classified as obese. If 20- to 40-year-olds are added to this group, the percentage of obese individuals drops some but is still fully 30%.

Food intake is under neuroendocrine control

For many years the neuronal and hormonal basis of appetite was a mystery. Experiments with fasting and overfeeding in rats showed an increase in food intake when fasting ends. This increase restores lost body weight to baseline values and food intake then drops. These experiments indicated the existence of control mechanisms to link food intake to energy balance. The presence of a hormonal *satiety factor* produced by adipose tissue was hypothesized to explain these observations. It has also been shown that regions of the hypothalamus are involved in feeding behavior. Other studies in rodent models had identified a number of genes that can lead to obesity. As modern molecular genetics has allowed the cloning of many of these genes, the outlines of a model to link dietary intake to energy balance have emerged. This model involves afferent signaling from adipose tissue and feeding behavior into the central nervous system (CNS), and efferent signaling outward from the CNS tied to energy expenditure, storage, reproduction and feeding behavior. We will discuss the relevant hormones first, then show how they fit into an overall control circuit.

Leptin

One of the rodent models for obesity, the obese mouse, is caused by a mutation in a single gene named *ob* (for obese). Animals homozygous for the recessive mutant allele become obese compared with wild-type mice (figure 47.19). When the gene responsible for this dramatic phenotype was isolated, it proved to encode a peptide hormone named leptin, leading to the hypothesis that the lack of leptin production in mutant individuals is responsible for obesity. Sure enough, when *ob/ob* animals are injected with leptin, they stop overeating and lose weight (see figure 47.19). These experiments identified leptin as the main satiety factor, and the key to the control of appetite. The gene for the leptin receptor *(db)* has also been

isolated and it is expressed in brain neurons in the hypothalamus involved in energy intake.

Leptin is now thought to be the main signaling molecule in the afferent portion of the control circuit for energy sensing, food intake, and energy expenditure. Leptin is produced by adipose tissue in response to feeding, and leptin levels correlate with feeding behavior and amount of body fat. Dietary restriction reduces leptin levels, signaling the brain that food intake is necessary, whereas refeeding after fasting leads to rapid increase in leptin levels and a loss of appetite. The efferent part of this control circuit is complex and includes control of energy expenditure, energy storage, and feeding behavior by the CNS. Reproduction is even affected by this system as reproduction is inhibited under starvation conditions.

The leptin gene has also been isolated in humans and leptin appears to function in humans much as it does in mice. However, recent studies in humans show that the activity of the *ob* gene and the blood concentrations of leptin are actually higher in obese than in lean people, and that the leptin produced by obese people appears to be normal. It has been suggested that, in contrast with the mutant mice, most cases of human obesity may result from a reduced sensitivity to the actions of leptin in the brain, rather than from reduced leptin production by adipose cells. Research on leptin in humans is ongoing and is of great interest to both academic scientists and to the pharmaceutical industry.

Insulin

Although the extreme obesity associated with the loss-of-function mutations in the *ob* gene indicate that other hormonal signals cannot substitute for leptin signaling, other hormones are also involved. Insulin has been implicated in signaling satiety as well, and insulin levels also fall with fasting and rise with obesity. As insulin's primary role is homeostasis of blood glucose, as described earlier, its role in the control circuit of energy balance is complex.

Gut hormones (enterogastrones)

The gut produces a number of hormones that control the physiology of digestion described earlier. Several of these have also been implicated in the regulation of food intake. They are produced directly in response to feeding, necessary for their role in digestion.

The hormones GIP and CCK have receptors in the hypothalamus and seem to send the same kind of inhibitory signals to the brain as leptin and insulin. The levels of these gut hormones also vary with feeding behavior in a pattern similar to leptin and insulin.

The gut hormone ghrelin has the opposite effect of these appetite-suppressing hormones. Ghrelin also has receptors in the hypothalamus, but ghrelin appears to stimulate food intake. This role is supported by studies in rats showing that chronic administration of ghrelin leads to obesity. Ghrelin levels appear to rise before feeding and may be involved in initiating feeding behavior. One of the treatments for severe obesity, gastric bypass surgery, leads to reduced levels of ghrelin. It has been suggested that this is one of the reasons for the suppression of appetite seen after this surgery.

SCIENTIFIC THINKING

Question: *Why are ob/ob mice obese?*

Hypothesis: *Lack of leptin production stimulates appetite and leads to overeating.*

Experiment: *Inject mutant mice with leptin.*

Results: *In just two weeks, injected mice lost 30% of their body weight with no apparent side effects.*

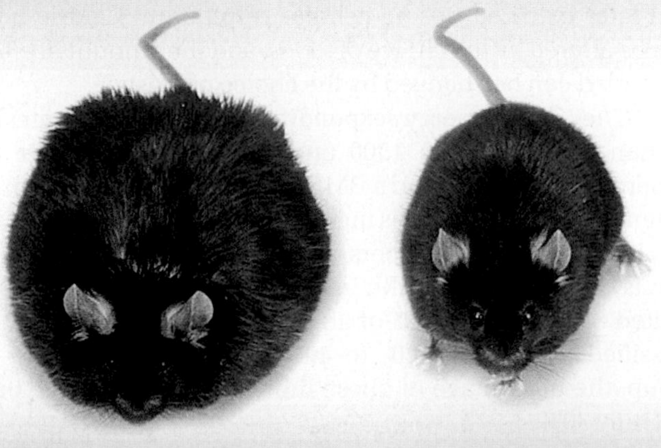

Further questions: *What is the molecular mechanism by which leptin has its effect? Could mutations in other genes have the same effect?*

Figure 47.19 Effects of the hormone leptin.

Neuropeptides

The efferent control over feeding and energy balance is less clear than the afferent control detailed earlier. The central regulator is the hypothalamus and two brain neuropeptides have been implicated: neuropeptide Y (NPY) and alpha-melanocyte-stimulating hormone (α-MSH). These peptides are antagonistic, with NPY inducing feeding activity and α-MSH suppressing it.

Evidence for this comes from experiments that show that production and release of α-MSH is stimulated by leptin and that administration of α-MSH suppresses feeding. Loss of function for the α-MSH receptor also leads to obesity. In contrast, the expression of NPY is negatively regulated by leptin and administration of NPY stimulates feeding behavior.

Model for energy balance

The current model for energy balance and feeding behavior is summarized in figure 47.20. The role of both leptin and insulin is long-term regulation of the afferent portion of this signaling network. Leptin and insulin are produced by adipose tissue and the pancreas, respectively, in response to the effects of feeding behavior, not as a direct response to feeding itself. This leads to circulating levels of leptin that correlate with the amount of adipose tissue. The extreme example of this is the very high level of leptin seen in obese individuals. High levels of leptin and insulin then act on the hypothalamus to increase levels of α-MSH and reduce levels of NPY. This causes a reduction in appetite and increased energy expenditure and allows reproduction and growth. Low levels of these hormones act on the hypothalamus to reduce α-MSH levels and increase NPY levels. This leads to increased appetite and decreased energy expenditure. If very low levels of leptin persist, this can inhibit reproduction and growth.

The gut hormones CCK and GIP are produced in response to feeding and represent short-term regulators of the afferent portion of the energy balance control circuit. Their action is the same as that of leptin and insulin. The gut hormone ghrelin is also a short-term regulator that stimulates feeding.

Essential nutrients are those that the body cannot manufacture

Over the course of evolution, many animals have lost the ability to synthesize certain substances that nevertheless continue to play critical roles in their metabolism. Substances that an animal cannot manufacture for itself but that are necessary for its health and must be obtained in the diet are referred to as essential nutrients.

Included among the essential nutrients are **vitamins,** certain organic substances required in trace amounts. For

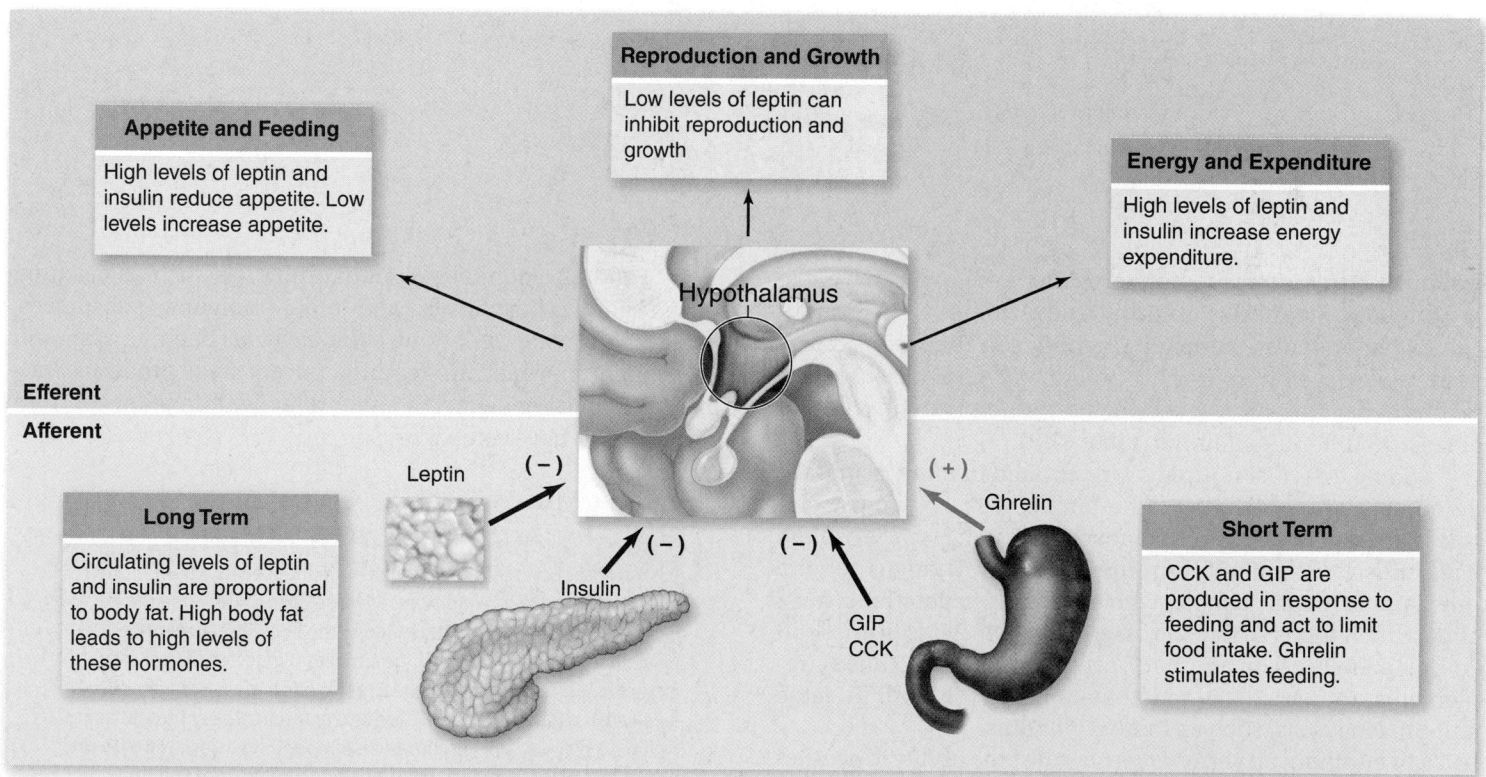

Figure 47.20 Hormonal control of feeding behavior. The control of feeding behavior is under both long-term control related to the amount of adipose tissue, and short-term control related to the act of feeding. This control is mediated by the CNS. The major brain region involved is the hypothalamus.

? Inquiry question Suppose the GIP and CCK sensors in the hypothalamus didn't work. How would this affect levels of leptin production?

TABLE 47.2	Major Vitamins Required by Humans		
Vitamin	**Function**	**Source**	**Deficiency Symptoms**
Vitamin A (retinol)	Used in making visual pigments, maintaining epithelial tissues	Green vegetables, milk products, liver	Night blindness, flaky skin
B-complex vitamins			
B_1	Coenzyme in CO_2 removal during cellular respiration	Meat, grains, legumes	Beriberi, weakening of heart, edema
B_2 (riboflavin)	Part of coenzymes FAD and FMN, which play metabolic roles	Many different kinds of foods	Inflammation and breakdown of skin, eye irritation
B_3 (niacin)	Part of coenzymes NAD^+ and $NADP^+$	Liver, lean meats, grains	Pellagra, inflammation of nerves, mental disorders
B_5 (pantothenic acid)	Part of coenzyme-A, a key connection between carbohydrate and fat metabolism	Many different kinds of foods	Rare: fatigue, loss of coordination
B_6 (pyridoxine)	Coenzyme in many phases of amino acid metabolism	Cereals, vegetables, meats	Anemia, convulsions, irritability
B_{12} (cyanocobalamin)	Coenzyme in the production of nucleic acids	Red meats, dairy products	Pernicious anemia
Biotin	Coenzyme in fat synthesis and amino acid metabolism	Meat, vegetables	Rare: depression, nausea
Folic acid	Coenzyme in amino acid and nucleic acid metabolism	Green vegetables	Anemia, diarrhea
Vitamin C	Important in forming collagen, cementum of bone, teeth, connective tissue of blood vessels; may help maintain resistance to infection	Fruit, green leafy vegetables	Scurvy, breakdown of skin, blood vessels
Vitamin D (calciferol)	Increases absorption of calcium and promotes bone formation	Dairy products, cod liver oil	Rickets, bone deformities
Vitamin E (tocopherol)	Protects fatty acids and cell membranes from oxidation	Margarine, seeds, green leafy vegetables	Rare
Vitamin K	Essential to blood clotting	Green leafy vegetables	Severe bleeding

example, humans, apes, monkeys, and guinea pigs have lost the ability to synthesize ascorbic acid (vitamin C). If vitamin C is not supplied in sufficient quantities in their diets, these mammals develop scurvy, a potentially fatal disease that results in degeneration of connective tissues. Humans require at least 13 different vitamins (table 47.2).

Some essential nutrients are required in more than trace amounts. Many vertebrates, for example, are unable to synthesize one or more of the 20 amino acids. These *essential amino acids* must be obtained from food they eat. Humans require nine amino acids. People who are strict vegetarians must choose their foods so that the essential amino acids in one food complement those in another. Vegetarians may also need supplements to provide certain vitamins not found in large amounts in plants, such as some B vitamins.

In addition, all vertebrates have lost the ability to synthesize certain long-chain unsaturated fatty acids and therefore must obtain them in food. In contrast, some essential nutrients that vertebrates can synthesize cannot be manufactured by the members of other animal groups. For example, vertebrates can synthesize cholesterol, a key component of steroid hormones, but some carnivorous insects cannot.

Food also supplies essential minerals such as calcium, magnesium, phosphorus, and other inorganic substances, including a wide variety of *trace elements* such as zinc and molybdenum, which are required in very small amounts. Animals obtain trace elements either directly from plants or from animals that have eaten plants.

Learning Outcomes Review 47.8

Basal metabolic rate is the minimum amount of energy consumption under defined resting conditions. Exercise does not increase the basal metabolic rate, but it does add to the body's total energy expenditure. Obesity results if the amount of ingested food energy exceeds energy expenditure over a prolonged period. Hormones involved in regulating appetite are leptin; insulin; the enterogastrones including CCK, GIP, and ghrelin; and neuropeptides associated with the hypothalamus. The essential nutrients for humans are 13 vitamins, essential minerals, and essential amino acids and fatty acids that the body cannot synthesize.

■ *What might explain obesity in a person with normal leptin levels?*

47.1 Types of Digestive Systems

Invertebrate digestive systems are bags or tubes.

In cnidarians and flatworms, the incomplete digestive system is a gastrovascular cavity with only one opening (figure 47.1). In contrast, a complete digestive system, with a one-way tube from mouth to anus, allows specialization of digestive organs.

Vertebrate digestive systems include highly specialized structures molded by diet.

The gastrointestinal tract includes the mouth and pharynx, esophagus, stomach, small and large intestines, cloaca or rectum, and anus (figure 47.3). The four tissue layers of the tract are the mucosa, submucosa, muscularis, and serosa (figure 47.4).

47.2 The Mouth and Teeth: Food Capture and Bulk Processing

Vertebrate teeth are adapted to different types of food items.

Birds lack teeth but have a gizzard where small pebbles grind food. The teeth of mammals are adapted to reflect their feeding habits (figure 47.6).

The mouth is a chamber for ingestion and initial processing.

Salivary glands secrete saliva containing amylase that moistens food and begins digestion as the food is chewed. Swallowing, once begun, is involuntary (figure 47.7).

47.3 The Esophagus and the Stomach: The Early Stages of Digestion

Muscular contractions of the esophagus move food to the stomach.

Rhythmic muscular contractions and relaxation, called peristalsis, propel a bolus of food to the stomach.

The stomach is a "holding station" involved in acidic breakdown of food.

In the stomach, hydrochloric acid breaks down food and converts pepsinogen into pepsin, an active protease. The mixture of food and gastric juice, termed chyme, moves through the pyloric sphincter to the small intestine.

47.4 The Intestines: Breakdown, Absorption, and Elimination

The structure of the small intestine is specialized for digestion and nutrient uptake.

The surface area of the small intestine is increased by fingerlike projections called villi (figure 47.10). The duodenum receives digestive secretions from the pancreas and liver.

Accessory organs secrete enzymes into the small intestine.

Accessory organs include the salivary glands, pancreas, liver, and gallbladder (figure 47.11). The pancreas secretes digestive enzymes and bicarbonate. The liver secretes bile, which is stored in the gallbladder. Bile disperses fats into small droplets.

Absorbed nutrients move into blood or lymph capillaries.

Amino acids and monosaccharides move into epithelial cells by active transport and facilitated diffusion (figure 47.12) and then pass into the bloodstream. Fatty acids and monoglycerides simply diffuse into epithelial cells. They are reassembled into chylomicrons that enter the lymphatic system.

Absorbed molecules that pass into the bloodstream are transported to the liver through the hepatic portal vein.

The large intestine eliminates waste material.

The large intestine absorbs water and concentrates waste material, which is stored in the rectum until it can be eliminated.

47.5 Variations in Vertebrate Digestive Systems

Ruminants rechew regurgitated food.

The four-chambered stomach of ruminants consists of the rumen, reticulum, omasum, and abomasum. Food initially processed in the rumen is regurgitated for further chewing.

Foregut fermentation has evolved convergently many times.

Enlarged foreguts have evolved in many species to provide a chamber for microbial fermentation. In some unrelated herbivores, identical changes in lysozyme have evolved.

Other herbivores have alternative strategies for digestion.

In some herbivores, digestion of cellulose by microorganisms takes place in the cecum, located beyond the stomach.

47.6 Neural and Hormonal Regulation of the Digestive Tract

The activities of the gastrointestinal tract are coordinated by the nervous and endocrine systems.

Duodenal hormones regulate passage of chyme into the duodenum. High fat content in the chyme stimulates the release of CCK and GIP; low chyme pH stimulates the release of secretin. In turn, CCK stimulates release of pancreatic enzymes and bile. Secretin stimulates release of bicarbonate.

47.7 Accessory Organ Function

The liver modifies chemicals to maintain homeostasis.

The liver is involved in detoxification, regulation of steroid hormone levels, and production of proteins found in the blood plasma.

Blood glucose concentration is maintained by the actions of insulin and glucagon.

Insulin lowers blood glucose and increases glycogen storage; glucagon increases blood glucose and utilization of glycogen.

47.8 Food Energy, Energy Expenditure, and Essential Nutrients

Exertion increases metabolic rate.

The basal metabolic rate is the minimum rate of energy consumption under resting conditions. Activity leads to an increase in the metabolic rate.

Food intake is under neuroendocrine control.

Food intake is regulated by the hormones leptin and insulin, by enterogastrones, and by neuropeptides (figure 47.20).

Essential nutrients are those that the body cannot manufacture.

Essential nutrients are those that cannot be synthesized by animals. For humans, they are 13 vitamins (table 47.2), the essential amino acids, essential minerals, and certain fatty acids.

Review Questions

UNDERSTAND

1. How is the digestion of fats different from that of proteins and carbohydrates?

   a. Fat digestion occurs in the small intestine, and the digestion of proteins and carbohydrates occurs in the stomach.

   b. Fats are absorbed into cells as fatty acids and monoglycerides but are then modified for absorption; amino acids and glucose are not modified further.

   c. Fats enter the hepatic portal circulation, but digested proteins and carbohydrates enter the lymphatic system.

   d. Digested fats are absorbed in the large intestine, and digested proteins and carbohydrates are absorbed in the small intestine.

2. Although the stomach is normally thought of as the major player in the digestive process, the bulk of chemical digestion actually occurs in the

   a. mouth.   c. duodenum.
   b. appendix.   d. large intestine.

3. After being absorbed through the intestinal mucosa, glucose and amino acids are

   a. absorbed directly into the systemic circulation.

   b. used to build glycogen and peptides before being released to the body cells.

   c. transported directly to the liver by the hepatic portal vein.

   d. further digested by bile before release into the circulation.

4. Which of these pairings is incorrect?

   a. Fat transport/lymphatic system
   b. Glucose transport/lymphatic system
   c. Amino acid transport/circulatory system
   d. All of these pairings are correct.

5. Intestinal microorganisms aid digestion and absorption by

   a. digesting cellulose.   c. synthesizing vitamin K.
   b. producing glucose.   d. Both a and c are correct.

6. The _____ and _____ play important roles in the digestive process by producing chemicals that are required to digest proteins, lipids, and carbohydrates.

   a. liver; pancreas   c. kidneys; appendix
   b. liver; gallbladder   d. pancreas; gallbladder

7. Which of the following represents the action of insulin?

   a. Increases blood glucose levels by the hydrolysis of glycogen

   b. Increases blood glucose levels by stimulating glucagon production

   c. Decreases blood glucose levels by forming glycogen

   d. Increases blood glucose levels by promoting cellular uptake of glucose

APPLY

1. The small intestine is specialized for absorption because it

   a. is the last section of the digestive tract and retains food the longest.

   b. has saclike extensions along its length that collect food.

   c. has no outlet so food remains within it for longer periods of time.

   d. has an extremely large surface area that allows extended exposure to food.

2. The primary function of the large intestine is to concentrate wastes into solid form (feces) for release from the body. How does it accomplish this?

   a. By adding additional cells from the mucosal layer
   b. By absorbing water
   c. By releasing salt
   d. All of these are methods used by the large intestine.

3. Inactive forms of some molecules are secreted

   a. because they take less material and energy.

   b. because they must combine with water to be activated.

   c. so their activity can be regulated.

   d. to prevent them from clogging the gland from which they are secreted.

4. Obese humans probably have high levels of leptin because

   a. leptin stimulates eating.

   b. something is wrong with the leptin receptors in their brain, leading to increased leptin production to make up for the apparent shortage.

   c. weight gain leads to the production of leptin.

   d. leptin responds to mechanical stimulation in the adrenal cortex.

SYNTHESIZE

1. Many birds possess crops, although few mammals do. Suggest a reason for this difference between birds and mammals.

2. Suppose that you wanted to develop a new treatment for obesity based on the hormone leptin. What structures in the body produce leptin? What does it do? Should your treatment cause an increase in blood levels of leptin or a decrease? Could this treatment affect any other systems in the body?

3. How could a drop in plasma proteins and a decrease in bile production be related to alcohol and drug abuse?

4. Unlike many cases of convergence (see section 47.5), ruminants and langur monkeys have modified the enzyme lysozyme in the same way to achieve the same end result. Why might this case be different?

5. Birds do not have teeth. Do you think they have adaptations to processing different types of food, comparable to the diversity seen in mammals? If so, what might these adaptive differences be?

ONLINE RESOURCE

www.ravenbiology.com

Understand, Apply, and Synthesize—enhance your study with animations that bring concepts to life and practice tests to assess your understanding. Your instructor may also recommend the interactive eBook, individualized learning tools, and more.

Chapter 48

The Respiratory System

Chapter Contents

Introduction

Every cell in the animal body must exchange materials with its surrounding environment. In single-celled organisms, this exchange occurs directly across the cell membrane to and from the external environment. In multicellular organisms, however, most cells are not in contact with the external environment and must rely on specialized systems for transport and exchange. Although these systems aid in bulk transport, the properties of transport across the plasma membrane do not change. Many structural adaptations throughout the animal kingdom increase surface areas where transport occurs, so that the needs of every cell are met. The interface between air from the environment and blood in the mammalian lungs provides an excellent example of the efficiency associated with increased surface area. In the time it takes you to breathe in, trillions of oxygen molecules have been transported across 80 m² of alveolar membrane into blood capillaries. In this and the next chapter, we describe respiration and circulation, the two systems that directly support the other organ systems and tissues of the body.

48.1 Gas Exchange Across Respiratory Surfaces

Learning Outcomes

1. *Describe gas exchange across membranes.*
2. *Explain Fick's Law of Diffusion.*
3. *Compare evolutionary strategies for maximizing gas diffusion.*

One of the major physiological challenges facing all multicellular animals is obtaining sufficient oxygen (O_2) and disposing of excess carbon dioxide (CO_2) (figure 48.1). Oxygen is used in mitochondria for cellular respiration—a process that also produces CO_2 as waste (see chapter 7). Respiration at the body system level involves a host of processes not found at the cellular level, ranging from the mechanics of breathing to the exchange of O_2 and CO_2 in respiratory organs.

Invertebrates display a wide variety of respiratory organs, including the epithelium, tracheae, and gills. Some vertebrates, such as fish and larval amphibians, also use gills; adult amphibians use their skin or other epithelia either as a supplemental or primary external respiratory organ.

Many adult amphibians, reptiles, birds, and mammals have lungs. In both aquatic and terrestrial animals, these highly vascularized respiratory organs are the site at which oxygen diffuses into the blood, and carbon dioxide diffuses out. In the body tissues, the direction of gas diffusion is the reverse of that in the respiratory organs.

Figure 48.1 Elephant seals are respiratory champions. Diving to great depths, elephant seals can hold their breath for over 2 hr, descend and ascend rapidly in the water, and endure repeated dives without suffering any apparent respiratory distress.

The mechanics, structure, and evolution of respiratory systems, along with the principles of gas diffusion between the blood and tissues, are the subjects of this chapter.

Gas exchange involves diffusion across membranes

Because plasma membranes must be surrounded by water to be stable, the external environment in gas exchange is always aqueous. This is true even in terrestrial vertebrates; in these cases, oxygen from air dissolves in a thin layer of fluid that covers the respiratory surfaces.

In vertebrates, the gases diffuse into the aqueous layer covering the epithelial cells that line the respiratory organs. The diffusion process is passive, driven only by the difference in O_2 and CO_2 concentrations on the two sides of the membranes and their relative solubilities in the plasma membrane. For dissolved gases, concentration is usually expressed as pressure; we explain this more fully a little later.

In general, the rate of diffusion between two regions is governed by a relationship known as **Fick's Law of Diffusion.** Fick's Law states that for a dissolved gas, the rate of diffusion (R) is directly proportional to the pressure difference (Δp) between the two sides of the membrane and the area (A) over which the diffusion occurs. Furthermore, R is inversely proportional to the distance (d) across which the diffusion must occur. A molecule-specific diffusion constant, D, accounts for the size of molecule, membrane permeability, and temperature. Shown as a formula, Fick's Law is stated as:

$$R = \frac{DA\,\Delta p}{d}$$

Major evolutionary changes in the mechanism of respiration have occurred to optimize the rate of diffusion (see figure 48.1). R can be optimized by changes that (1) increase the surface area, A; (2) decrease the distance, d; or (3) increase the concentration difference, as indicated by Δp. The evolution of respiratory systems has involved changes in all of these factors.

 Data analysis Which would increase the rate of diffusion more, increasing the area of the membrane or increasing the difference in pressure on the two sides of the membrane?

Evolutionary strategies have maximized gas diffusion

The levels of oxygen needed for cellular respiration cannot be obtained by diffusion alone over distances greater than about 0.5 mm. This restriction severely limits the size and structure of organisms that obtain oxygen entirely by diffusion from the environment. Bacteria, archaea, and protists are small enough that such diffusion can be adequate, even in some colonial forms (figure 48.2a), but most multicellular animals require structural adaptations to enhance gas exchange.

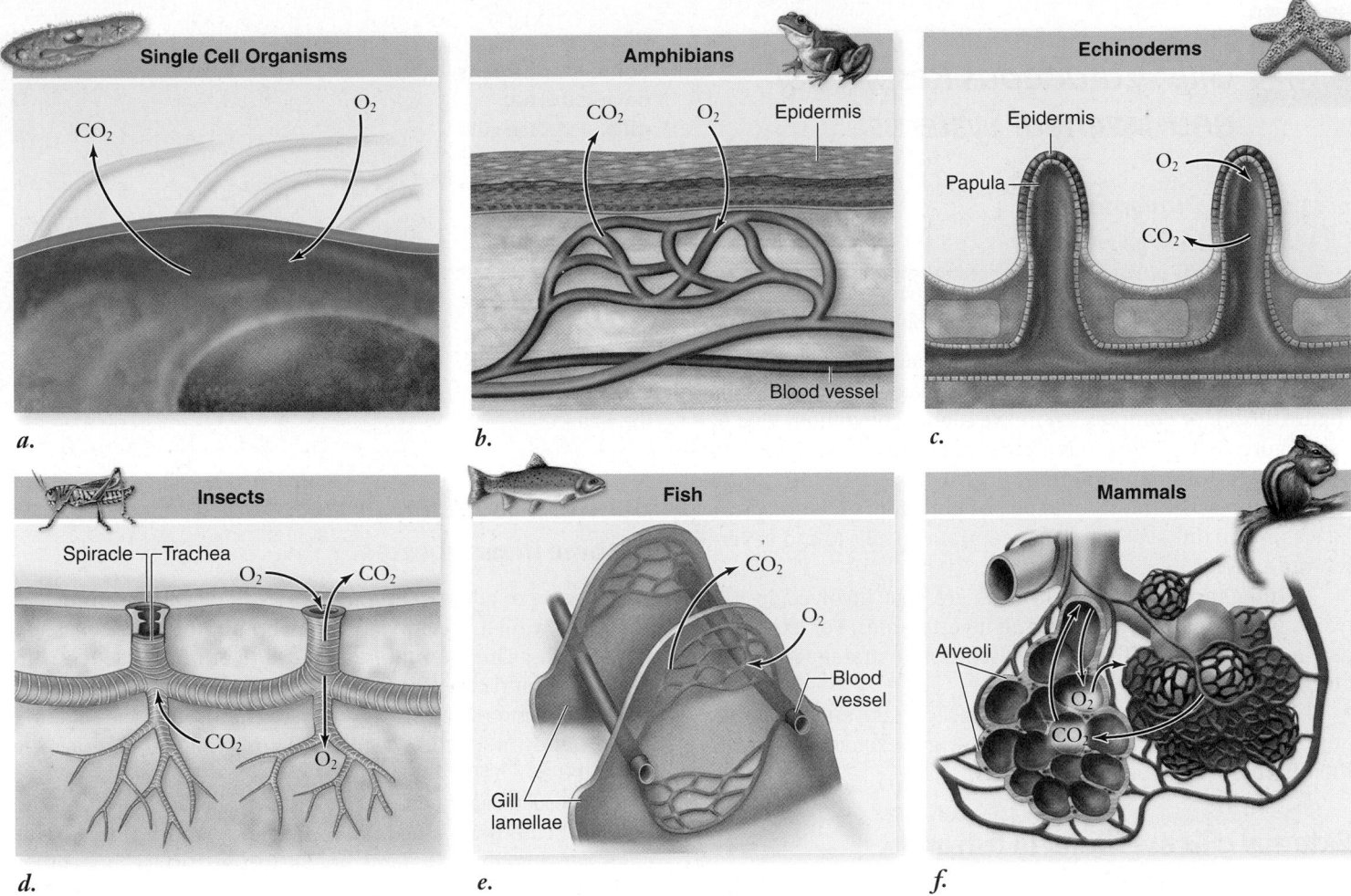

Figure 48.2 Different gas exchange systems in animals. *a.* Gases diffuse directly into single-celled organisms. *b.* Most amphibians and many other animals respire across their skin. Amphibians also exchange gases via lungs. *c.* Echinoderms have protruding papulae, which provide an increased respiratory surface area. *d.* Insects respire through an extensive tracheal system. *e.* The gills of fishes provide a very large respiratory surface area and countercurrent exchange. *f.* The alveoli in mammalian lungs provide a large respiratory surface area but do not permit countercurrent exchange. Inhaled fresh air contains some CO_2, but levels are higher in the lungs, so more CO_2 is exhaled than inhaled; similarly, O_2 levels are higher in fresh air, leading to an influx of O_2.

Increasing oxygen concentration difference

Most phyla of invertebrates lack specialized respiratory organs, but they have developed means of improving diffusion. Many organisms create a water current that continuously replaces the water over the respiratory surfaces; often, beating cilia produce this current. Because of this continuous replenishment of water, the external oxygen concentration does not decrease along the diffusion pathway. Although some of the oxygen molecules that pass into the organism have been removed from the surrounding water, new water continuously replaces the oxygen-depleted water. This maximizes the concentration difference—the Δp of the Fick equation.

Increasing area and decreasing distance

Other invertebrates (mollusks, arthropods, echinoderms) and vertebrates possess respiratory organs—such as gills, tracheae, and lungs—that increase the surface area available for diffusion (see figure 48.2). These adaptations also bring the external environment (either water or air) close to the internal fluid, which is usually circulated throughout the body—such as blood or hemolymph. The respiratory organs thus increase the rate of diffusion by maximizing surface area (A) and decreasing the distance (d) the diffusing gases must travel.

Learning Outcomes Review 48.1

Gases must be dissolved to diffuse across living membranes. Direction of diffusion is driven by a concentration difference (gradient) between the two sides. Fick's Law states that the rate of diffusion is increased by a greater pressure difference and membrane area, and decreased by greater distance. Evolutionary strategies have therefore aimed to increase gradient and area and to lessen the distance gases must travel.

■ *Which factor is affected by continuously beating cilia?*

Gills, Cutaneous Respiration, and Tracheal Systems

Learning Outcomes

1. Describe how gills work.
2. Explain the advantage of countercurrent flow.

Gills are specialized extensions of tissue that project into water. Gills can be simple, as in the papulae of echinoderms (figure 48.2c), or complex, as in the highly convoluted gills of fish (figure 48.2e). The great increase in diffusion surface area that gills provide enables aquatic organisms to extract far more oxygen from water than would be possible from their body surface alone. In this section we concentrate on gills found in vertebrate animals.

Other moist external surfaces are also involved in gas exchange in some vertebrates and invertebrates. For example, gas exchange across the skin is a common strategy in many amphibian groups.

Terrestrial arthropods such as insects take an alternative approach; their tracheal systems allow gas exchange through their hard exoskeletons (figure 48.2d).

External gills are found in fish and amphibian larvae

External gills are not enclosed within body structures. Examples of vertebrates with external gills are the larvae of many fish and amphibians, as well as amphibians such as the axolotl, which retains larval features throughout life (figure 48.3).

One of the disadvantages of external gills is that they must constantly be moved to ensure contact with fresh water having high oxygen content. The highly branched gills, however, offer significant resistance to movement, making this form of respiration ineffective except in smaller animals. Another disadvantage is that external gills, with their thin epithelium for gas exchange, are easily damaged.

Figure 48.3 Some amphibians have external gills. External gills are used by aquatic amphibians, both larvae and some species that live their entire lives in water such as this axolotl (*Ambystoma mexicanum*), to extract oxygen from the water.

Branchial chambers protect gills of some invertebrates

Other types of aquatic animals evolved specialized *branchial chambers,* which provide a means of pumping water past stationary gills. The internal *mantle cavity* of mollusks opens to the outside and contains the gills. Contraction of the muscular walls of the mantle cavity draws water in through the inhalant siphon and then expels it through the exhalant siphon (see chapter 34).

In crustaceans, the branchial chamber lies between the bulk of the body and the hard exoskeleton of the animal. This chamber contains gills and opens to the surface beneath a limb. Movement of the limb draws water through the branchial chamber, thus creating currents over the gills.

Gills of bony fishes are covered by the operculum

The gills of bony fishes are located between the oral cavity, sometimes called the buccal (mouth) cavity, and the *opercular cavities* where the gills are housed (figure 48.4). The two sets of cavities function as pumps that expand alternately to move water into the mouth, through the gills, and out of the fish through the open operculum, or gill cover.

Figure 48.4 How most bony fishes respire. The gills are suspended between the buccal (mouth) cavity and the opercular cavity. Respiration occurs in two stages. The oral valve in the mouth is opened and the jaw is depressed, drawing water into the buccal cavity while the opercular cavity is closed. The oral valve is closed and the operculum is opened, drawing water through the gills to the outside.

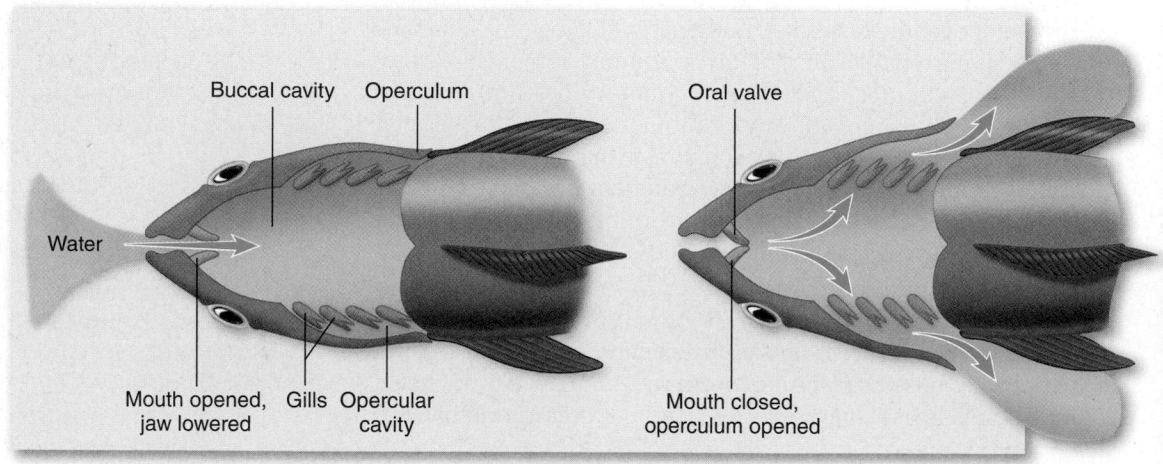

Buccal cavity   Operculum   Oral valve

Water

Mouth opened, jaw lowered   Gills   Opercular cavity

Mouth closed, operculum opened

Some bony fishes that swim continuously, such as tuna, have practically immobile opercula. These fishes swim with their mouths partly open, constantly forcing water over the gills in what is known as *ram ventilation*. Most bony fishes, however, have flexible gill covers. For example, the remora, a fish that rides "piggyback" on sharks, uses ram ventilation while the shark is swimming, but employs the pumping action of its opercula when the shark stops swimming.

There are between three and seven gill arches on each side of the fish's head. Each gill arch is composed of two rows of *gill filaments,* and each gill filament contains thin membranous plates, or *lamellae,* that project out into the flow of water (figure 48.5). Water flows past the lamellae in one direction only.

Within each lamella, blood flows opposite to the direction of water movement. This arrangement is called **countercurrent flow,** and it acts to maximize the oxygenation

of the blood by maintaining a positive oxygen gradient along the entire pathway for diffusion, increasing Δp in Fick's Law of Diffusion. The advantages of a countercurrent flow system are illustrated in figure 48.6a. Countercurrent flow ensures that an oxygen concentration gradient remains between blood and water throughout the length of the gill lamellae. This permits oxygen to continue to diffuse all along the lamellae, so that the blood leaving the gills has nearly as high an oxygen concentration as the water entering the gills.

If blood and water flowed in the same direction, the flow would be *concurrent* (figure 48.6b). In this case, the concentration difference across the gill lamellae would fall rapidly as the water lost oxygen to the blood, and net diffusion of oxygen

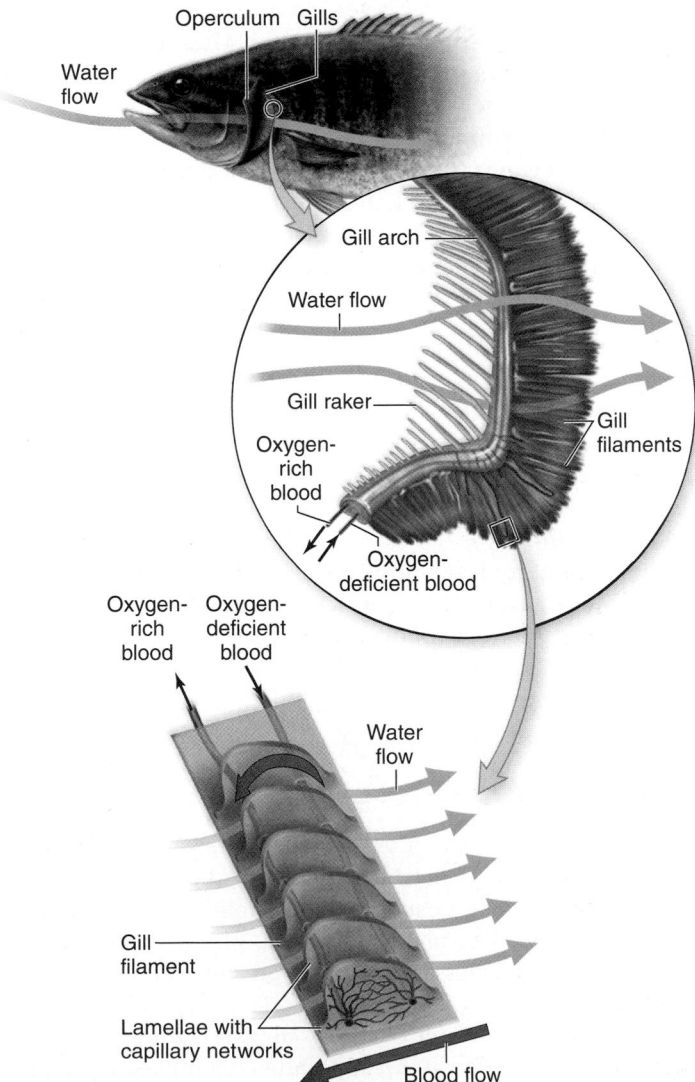

Figure 48.5 Structure of a fish gill. Water passes from the gill arch over the filaments (from left to right in the diagram). Water always passes the lamellae in a direction opposite to the direction of blood flow through the lamellae. The success of the gill's operation critically depends on this countercurrent flow of water and blood.

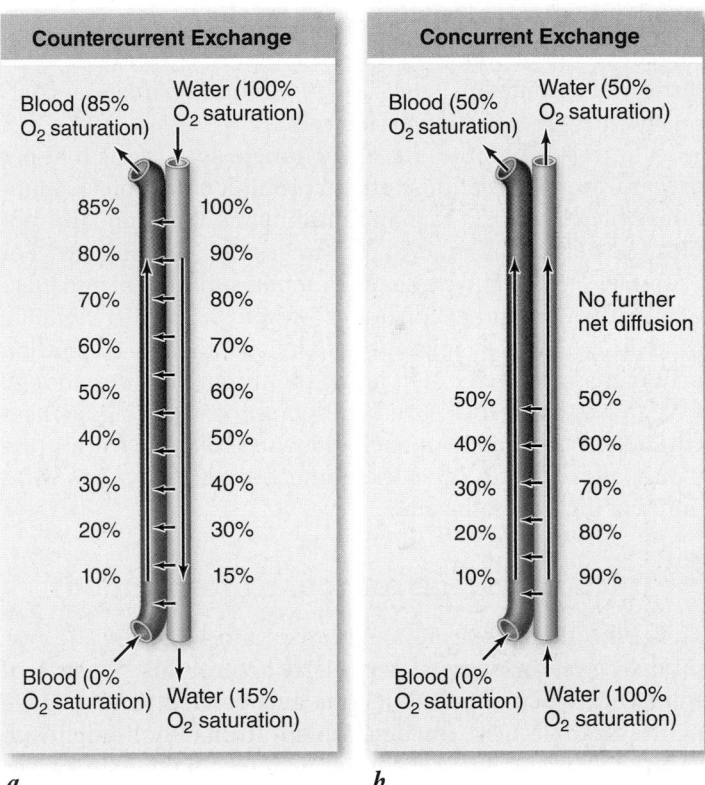

Figure 48.6 Countercurrent exchange. This process allows for the most efficient blood oxygenation. When blood and water flow in opposite directions (**a**), the initial oxygen (O_2) concentration difference between water and blood is small, but is sufficient for O_2 to diffuse from water to blood. As more O_2 diffuses into the blood, raising the blood's O_2 concentration, the blood encounters water with ever-higher O_2 concentrations. At every point, the O_2 concentration is higher in the water, so that diffusion continues. In this example, blood attains an O_2 concentration of 85%. When blood and water flow in the same direction (**b**), O_2 can diffuse from the water into the blood rapidly at first, but the diffusion rate slows as more O_2 diffuses from the water into the blood, until finally the concentrations of O_2 in water and blood are equal. In this example, blood's O_2 concentration cannot exceed 50%.

Data analysis In the countercurrent exchange system, where along the blood vessel would the most oxygen diffuse from the water into the blood? Where would this occur in the concurrent exchange system?

would cease when the level of oxygen became the same in the water and in the blood.

Because of the countercurrent exchange of gases, fish gills are the most efficient of all respiratory organs.

 ## Cutaneous respiration requires constant moisture

Oxygen and carbon dioxide can diffuse across cutaneous (skin) surfaces in some vertebrates (see figure 48.2*b*). Most commonly, these vertebrates are aquatic, such as amphibians and some turtles, and they have highly vascularized areas of thin epidermis. The process of exchanging oxygen and carbon dioxide across the skin is called **cutaneous respiration.** In amphibians, cutaneous respiration supplements—and sometimes replaces—the action of lungs. Although not common, some terrestrial amphibians, such as plethodontid salamanders, rely on cutaneous respiration exclusively.

Terrestrial reptiles have dry, tough, scaly skins that not only prevent desiccation, but also prohibit cutaneous respiration, which is utilized by many amphibians. Some aquatic reptiles, however, have the ability to respire cutaneously. For example, soft-shelled turtles can remain submerged and inactive in river sediment for hours without having to ventilate their lungs. At that level of activity, cutaneous respiration occurring through the skin lining the throat provides enough oxygen to the tissues. Even the common pond slider uses cutaneous respiration to help stay submerged. During the winter, these turtles can stay submerged for many days without needing to breathe air.

 ## Tracheal systems are found in arthropods

The arthropods have no single respiratory organ. The respiratory system of most terrestrial arthropods consists of small, branched cuticle-lined air ducts called *tracheae* (see figure 48.2*d*). These trachea, which ultimately branch into very small tracheoles, are a series of tubes that transmit gases throughout the body. Tracheoles are in direct contact with individual cells, and oxygen diffuses directly across the plasma membranes.

Air passes into the trachea by way of specialized openings in the exoskeleton called *spiracles,* which, in most terrestrial arthropods, can be opened and closed by valves. The ability to prevent water loss by closing the spiracles was a key adaptation that facilitated the invasion of land by arthropods.

Learning Outcomes Review 48.2

Gills are highly subdivided structures providing a large surface area for exchange. In countercurrent flow, blood in the gills flows opposite to the direction of water to maintain a gradient difference and maximize gas exchange. Some amphibians rely on cutaneous respiration. Highly subdivided tracheal systems have evolved in arthropods, and these have been modified with valves as an adaptation to terrestrial life.

■ *What are the anatomical requirements for a countercurrent flow system?*

48.3 Lungs

Despite the high efficiency of gills as respiratory organs in aquatic environments, gills were replaced in terrestrial animals for two principal reasons:

1. **Air is less supportive than water.** The fine membranous lamellae of gills lack structural strength and rely on water for their support. A fish out of water, although awash in oxygen, soon suffocates because its gills collapse into a mass of tissue. Unlike gills, internal air passages such as tracheae and lungs can remain open because the body itself provides the necessary structural support.
2. **Water evaporates.** Air is rarely saturated with water vapor, except immediately after a rainstorm. Consequently, terrestrial organisms constantly lose water to the atmosphere. Gills would provide an enormous surface area for water loss.

The lung minimizes evaporation by moving air through a branched tubular passage. The tracheal system of arthropods also uses internal tubes to minimize evaporation.

The air drawn into the respiratory passages becomes saturated with water vapor before reaching the inner regions of the lung. In these areas, a thin, wet membrane permits gas exchange. Unlike the one-way flow of water that is so effective in the respiratory function of gills, gases move in and out of lungs by way of the same airway passages, a two-way flow system. Birds have an exceptional respiratory system, as described later on.

Breathing of air takes advantage of partial pressures of gases

Dry air contains 78.09% nitrogen, 20.95% oxygen, 0.93% argon and other inert gases, and 0.03% carbon dioxide. Convection currents cause the atmosphere to maintain a constant composition to altitudes of at least 100 km, although the *amount* (number of molecules) of air that is present decreases as altitude increases.

Because of the force of gravity, air exerts a pressure downward on objects below it. An apparatus that measures air pressure is called a *barometer,* and 760 mm Hg is the barometric pressure of the air at sea level. A pressure of 760 mm Hg is also defined as one atmosphere (1.0 atm) of pressure.

Each type of gas contributes to the total atmospheric pressure according to its fraction of the total molecules present. The pressure contributed by a gas is called its **partial pressure,** and it is indicated by P_{N_2}, P_{O_2}, P_{CO_2}, and so on. At sea level, the partial pressures of N_2, O_2, and CO_2 are as follows:

$$P_{N_2} = 760 \times 79.02\% = 600.6 \text{ mm Hg}$$

$$P_{O_2} = 760 \times 20.95\% = 159.2 \text{ mm Hg}$$

$$P_{CO_2} = 760 \times 0.03\% = 0.2 \text{ mm Hg}$$

Humans do not survive for long at altitudes above 6000 m. Although the air at these altitudes still contains 20.95% oxygen, the atmospheric pressure is only about 380 mm Hg, so the P_{O_2} is only 80 mm Hg ($380 \times 20.95\%$), half the amount of oxygen available at sea level.

In the following sections, we describe respiration in vertebrates with lungs, beginning with reptiles and amphibians. We then summarize mammalian lungs and the highly adapted and specialized lungs of birds.

Amphibians and reptiles breathe in different ways

The lungs of amphibians are formed as saclike outpouchings of the gut (figure 48.7). Although the internal surface area of these sacs is increased by folds, much less surface area is available for gas exchange in amphibian lungs than in the lungs of other terrestrial vertebrates. Each amphibian lung is connected to the rear of the oral cavity, or pharynx, and the opening to each lung is controlled by a valve, the glottis.

Amphibians do not breathe the same way as other terrestrial vertebrates. Amphibians force air into their lungs; they fill their oral cavity with air (figure 48.7a), close their mouth and nostrils, and then elevate the floor of their oral cavity. This pushes air into their lungs in the same way that a pressurized tank of air is used to fill balloons (figure 48.7b). This is called **positive pressure breathing;** in humans, it would be analogous to forcing air into a person's lungs by performing mouth-to-mouth resuscitation.

Most reptiles breathe in a different way, by expanding their rib cages by muscular contraction. This action creates a lower pressure inside the lungs compared with the atmosphere, and the greater atmospheric pressure moves air into the lungs. This type of ventilation is termed **negative pressure breathing** because of the air being "pulled in" by the animal, like sucking water through a straw, rather than being "pushed in."

Mammalian lungs have greatly increased surface area

Endothermic animals, such as birds and mammals, have consistently higher metabolic rates and thus require more oxygen (see chapter 7). Both these vertebrate groups exhibit more complex and efficient respiratory systems than ectothermic animals. The evolution of more efficient respiratory systems accommodates the increased demands on cellular respiration of endothermy.

The lungs of mammals are packed with millions of **alveoli,** tiny sacs clustered like grapes (figure 48.8). This provides each lung with an enormous surface area for gas exchange. Each alveolus is composed of an epithelium only

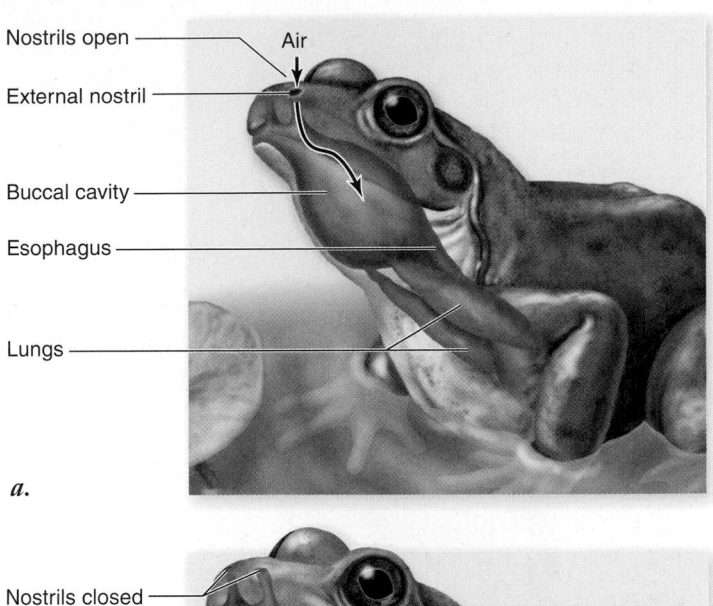

a.

b.

Figure 48.7 Amphibian lungs. Each lung of this frog is an outpouching of the gut and is filled with air by the creation of a positive pressure in the buccal cavity. *a.* The buccal cavity is expanded and air flows through the open nostrils. *b.* The nostrils are closed and the buccal cavity is compressed, thus creating the positive pressure that fills the lungs. The amphibian lung lacks the structures present in the lungs of other terrestrial vertebrates that provide an enormous surface area for gas exchange, and so are not as efficient as the lungs of other vertebrates.

one cell thick, and is surrounded by blood capillaries with walls that are also only one cell layer thick. Thus, the distance d across which gas must diffuse is very small—only 0.5 to 1.5 μm.

Inhaled air is taken in through the mouth and nose past the pharynx to the larynx (voice box), where it passes through an opening in the vocal cords, the *glottis,* into a tube supported by C-shaped rings of cartilage, the trachea (windpipe). The term *trachea* is used both for the vertebrate windpipe and the respiratory tubes of arthropods, although the structures are obviously not homologous. The mammalian trachea bifurcates into right and left bronchi (singular, *bronchus*), which enter each lung and further subdivide into bronchioles that deliver the air into the alveoli.

The alveoli are surrounded by an extensive capillary network. All gas exchange between the air and blood takes place

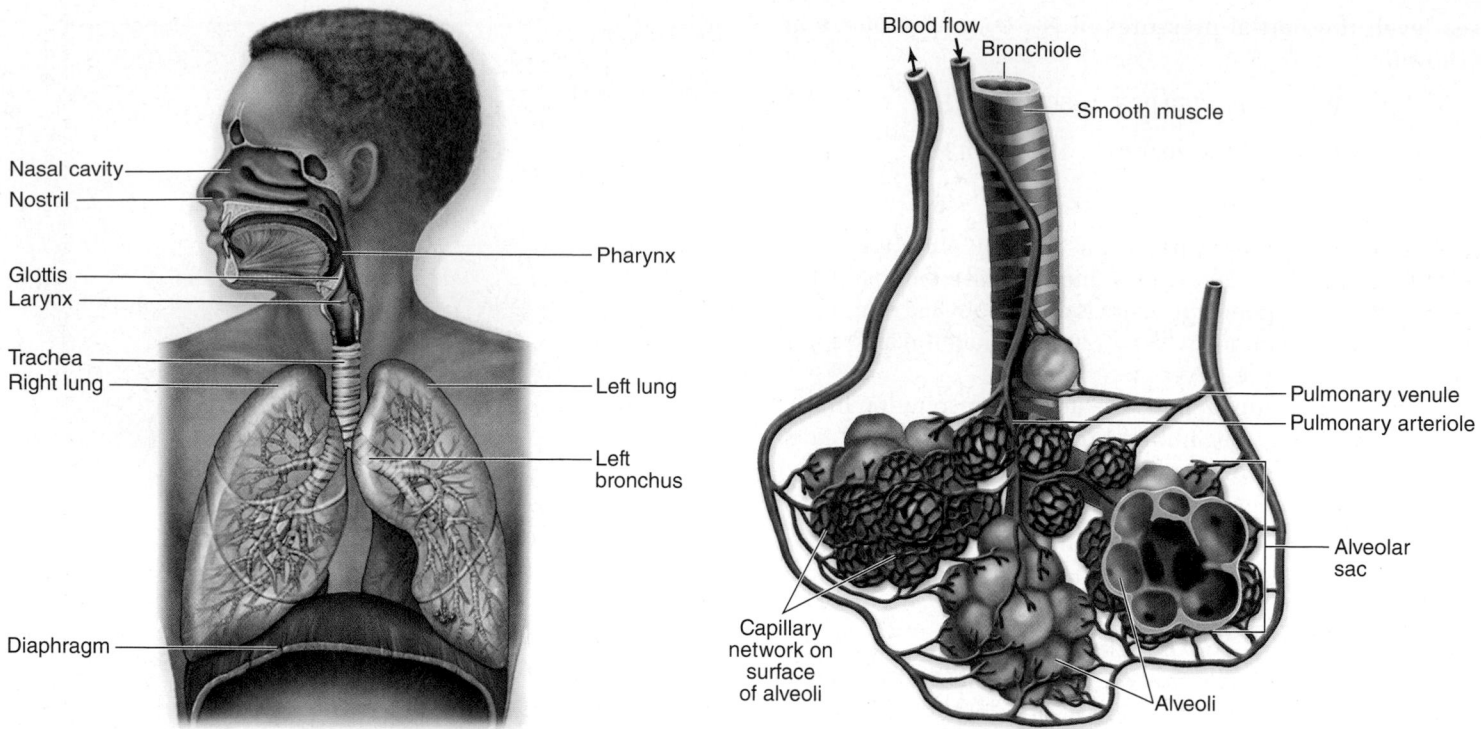

Figure 48.8 The human respiratory system and the structure of the mammalian lung. The lungs of mammals have an enormous surface area because of the millions of alveoli that cluster at the ends of the bronchioles. This provides for efficient gas exchange with the blood.

across the walls of the alveoli. The branching of bronchioles and the vast number of alveoli combine to increase the respiratory surface area far above that of amphibians or reptiles. In humans, each lung has about 300 million alveoli, and the total surface area available for diffusion can be as much as 80 m², or about 42 times the surface area of the body. Details of gas exchange at the alveolar interface with blood capillaries is described in sections that follow.

The respiratory system of birds is a highly efficient flow-through system

The avian respiratory system is a unique structure that affords birds the most efficient respiration of all terrestrial vertebrates. Unlike the mammalian lung, which ends in blind alveoli, the bird lung channels air through tiny air vessels called parabronchi, where gas exchange occurs. Air flows through the parabronchi in one direction only. This flow is similar to the unidirectional flow of water through a fish gill.

In other terrestrial vertebrates, inhaled fresh air is mixed with "old," oxygen-depleted air left from the previous breathing cycle. The lungs of amphibians, reptiles, and mammals are never completely empty of the gases within them. In birds, only fresh air enters the parabronchi of the lung, and the "old" air exits the lung by a different route. The unidirectional flow of air is achieved through the action of anterior and posterior air sacs unique to birds (figure 48.9a). When these sacs are expanded during inhalation, they take in air, and when they are compressed during exhalation, they push air into and through the lungs.

Respiration in birds occurs in two cycles (figure 48.9b). Each cycle has an inhalation and exhalation phase—but the air inhaled in one cycle is not exhaled until the second cycle.

Upon inhalation, both anterior and posterior air sacs expand. The inhaled air, however, only enters the posterior air sacs; the anterior air sacs fill with air pulled from the lungs. Upon exhalation, the air forced out of the anterior air sacs is released outside the body, but the air forced out of the posterior air sacs now enters the lungs. This process is repeated in the second cycle.

The unidirectional flow of air also permits further respiratory efficiency: The flow of blood through the avian lung runs at a 90° angle to the air flow. This crosscurrent flow is not as efficient as the 180° countercurrent flow in fishes' gills, but it has a greater capacity to extract oxygen from the air than does a mammalian lung.

Because of these respiratory adaptations, a sparrow can be active at an altitude of 6000 m, whereas a mouse, which has a similar body mass and metabolic rate, would die from lack of oxygen in a fairly short time.

Learning Outcomes Review 48.3

Lungs provide a large surface area for gas exchange while minimizing evaporation; unlike gills, they contain structural support that prevents their collapse. Amphibians push air into their lungs; most reptiles and all birds and mammals pull air into their lungs by expanding the thoracic cavity. The respiratory system of birds has efficient, one-way air flow and crosscurrent blood flow through the lungs.

■ *What selection pressure would bring about the evolution of birds' highly efficient lungs?*

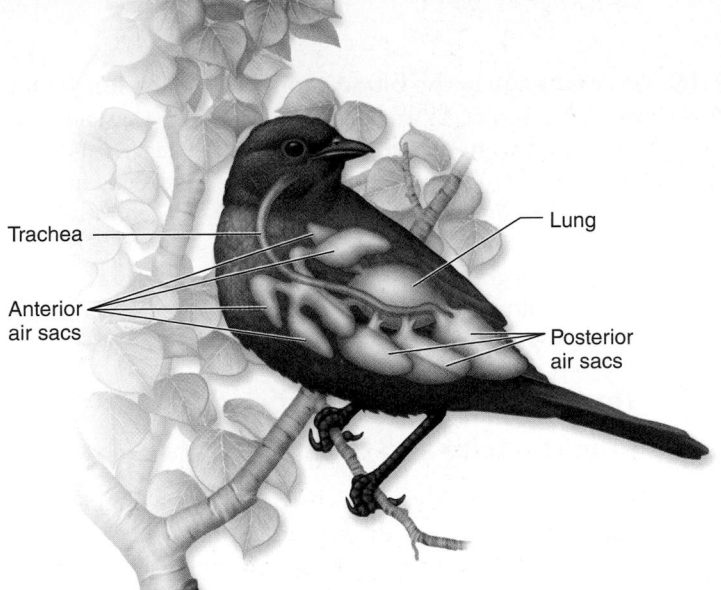

Trachea

Anterior
air sacs

Lung

Posterior
air sacs

a.

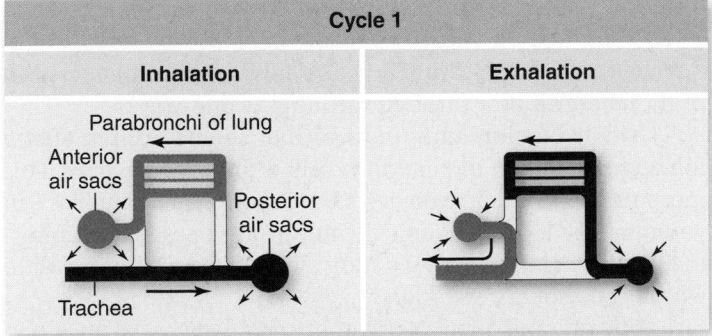

Cycle 1

| **Inhalation** | **Exhalation** |

Parabronchi of lung

Anterior
air sacs

Posterior
air sacs

Trachea

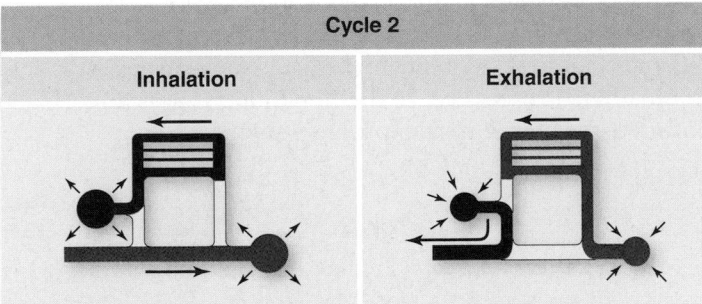

Cycle 2

| **Inhalation** | **Exhalation** |

b.

Figure 48.9 How a bird breathes. *a.* Birds have a system of air sacs, divided into an anterior group and posterior group, that extend between the internal organs and into the bones.
b. Breathing occurs in two cycles. Cycle 1: Inhaled air (shown in red) is drawn from the trachea into the posterior air sacs (shown expanding as it fills with air) and then is exhaled into the lungs (posterior air sacs deflate). Cycle 2: Air is drawn from the lungs into the anterior air sacs, which expand, and then is exhaled from these air sacs through the trachea. Passage of air through the lungs is always in the same direction, from posterior to anterior (right to left in this diagram). These cycles are always going on simultaneously; during inhalation, fresh air enters the posterior air sacs at the same time that air from the previous breath that was in the lungs moves into the anterior air sacs. In exhalation, the newer air moves from the posterior air sacs to the lung at the same time that air in the anterior air sacs is exhaled from the body. At the same time, another breath of inhaled air (purple) is moving through cycle 1.

Structures, Mechanisms, and Control of Ventilation in Mammals

Learning Outcomes

1. *Explain how contraction and relaxation of the diaphragm and intercostal muscles result in breathing.*
2. *Describe how the nervous system regulates breathing.*
3. *List and characterize the major respiratory diseases.*

About 30 billion capillaries can be found in each lung, roughly 100 capillaries per alveolus. Thus, an alveolus can be visualized as a microscopic air bubble whose entire surface is bathed by blood. Gas exchange occurs very rapidly at this interface.

Blood returning from the systemic circulation, depleted in oxygen, has a partial oxygen pressure (P_{O_2}) of about 40 mm Hg. By contrast, the P_{O_2} in the alveoli is about 105 mm Hg. The difference in pressures, namely the Δp of Fick's Law, is 65 mm Hg, leading to oxygen moving into the blood. The blood leaving the lungs, as a result of this gas exchange, normally contains a P_{O_2} of about 100 mm Hg. As you can see, the lungs do a very effective, but not perfect, job of oxygenating the blood. These changes in the P_{O_2} of the blood, as well as the changes in plasma carbon dioxide (indicated as the P_{CO_2}), are shown in figure 48.10.

 Lung structure and function supports the respiratory cycle

In humans and other mammals, the outside of each lung is covered by a thin membrane called the **visceral pleural membrane.** A second membrane, the **parietal pleural membrane,** lines the inner wall of the thoracic cavity. The space between these two membrane sheets, the **pleural cavity,** is normally very small and filled with fluid. This fluid causes the two membranes to adhere, effectively coupling the lungs to the thoracic cavity. The pleural membranes package each lung separately—if one lung collapses due to a perforation of the membranes, the other lung can still function.

During inhalation, the thoracic volume is increased through contraction of two sets of muscles: the *external intercostal muscles* and the *diaphragm.* Contraction of the external intercostal muscles between the ribs raises the ribs and expands the rib cage. Contraction of the **diaphragm,** a convex sheet of striated muscle separating the thoracic cavity from the abdominal cavity, causes the diaphragm to lower and assume a more flattened shape. This expands the volume of the thorax and lungs, bringing about negative pressure ventilation (figure 48.11*a*).

The thorax and lungs have a degree of elasticity; expansion during inhalation places these structures under elastic tension. The relaxation of the external intercostal muscles and diaphragm produces unforced exhalation because the elastic tension is released, allowing the thorax and lungs to recoil. You

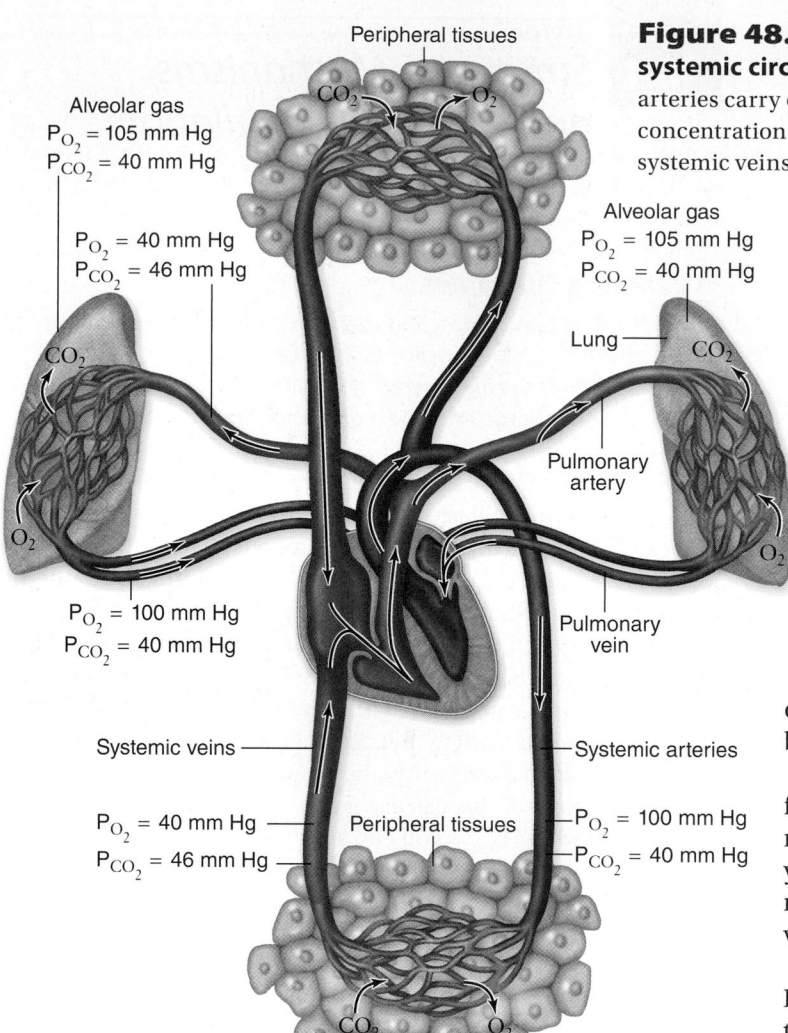

Peripheral tissues

Alveolar gas
P_{O_2} = 105 mm Hg
P_{CO_2} = 40 mm Hg

P_{O_2} = 40 mm Hg
P_{CO_2} = 46 mm Hg

CO_2

O_2

P_{O_2} = 100 mm Hg
P_{CO_2} = 40 mm Hg

Systemic veins

P_{O_2} = 40 mm Hg
P_{CO_2} = 46 mm Hg

Peripheral tissues

CO_2 O_2

Alveolar gas
P_{O_2} = 105 mm Hg
P_{CO_2} = 40 mm Hg

Lung CO_2

Pulmonary
artery

O_2

Pulmonary
vein

Systemic arteries

P_{O_2} = 100 mm Hg
P_{CO_2} = 40 mm Hg

CO_2 O_2

Figure 48.10 **Gas exchange in the blood capillaries of the lungs and systemic circulation.** As a result of gas exchange in the lungs, the systemic arteries carry oxygenated blood with a relatively low carbon dioxide (CO_2) concentration. After the oxygen (O_2) is unloaded to the tissues, the blood in the systemic veins has a lowered O_2 content and an increased CO_2 concentration.

can produce a greater exhalation force by actively contracting your abdominal muscles—such as when blowing up a balloon (figure 48.11*b*).

Ventilation efficiency depends on lung capacity and breathing rate

A variety of terms are used to describe the volume changes of the lung during breathing. In a person at rest, each breath moves a tidal volume of about 500 mL of air into and out of the lungs. About 150 mL of the tidal volume is contained in the tubular passages (trachea, bronchi, and bronchioles), where no gas exchange occurs—termed the *anatomical dead space.* The gases in this space mix with fresh air during inhalation. This mixing is one reason that respiration in mammals is not as efficient as in birds, where air flow through the lungs is one-way.

The maximum amount of air that can be expired after a forceful, maximum inhalation is called the vital capacity. This measurement, which averages 4.6 L in young men and 3.1 L in young women, can be clinically important because an abnormally low vital capacity may indicate damage to the alveoli in various pulmonary disorders.

The rate and depth of breathing normally keeps the blood P_{O_2} and P_{CO_2} within a normal range. If breathing is insufficient to maintain normal blood gas measurements (a rise in the blood P_{CO_2} is the best indicator), the person is hypoventilating. If breathing is excessive, so that the blood P_{CO_2} is abnormally lowered, the person is said to be **hyperventilating.**

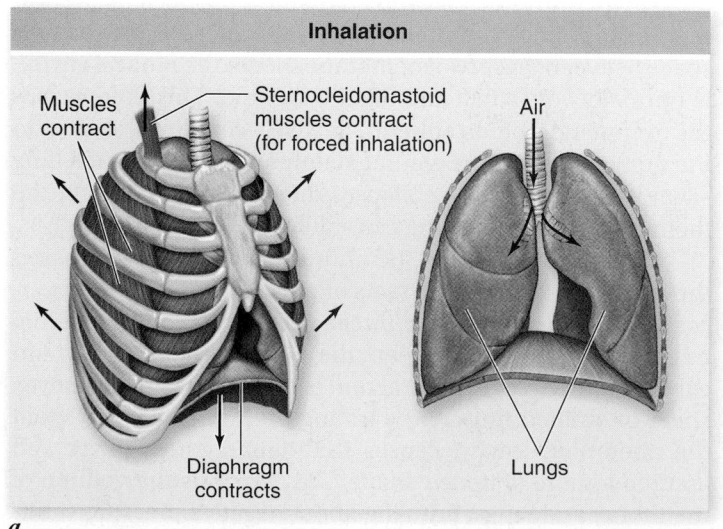

Inhalation

Muscles contract

Sternocleidomastoid muscles contract (for forced inhalation)

Air

Diaphragm contracts

Lungs

a.

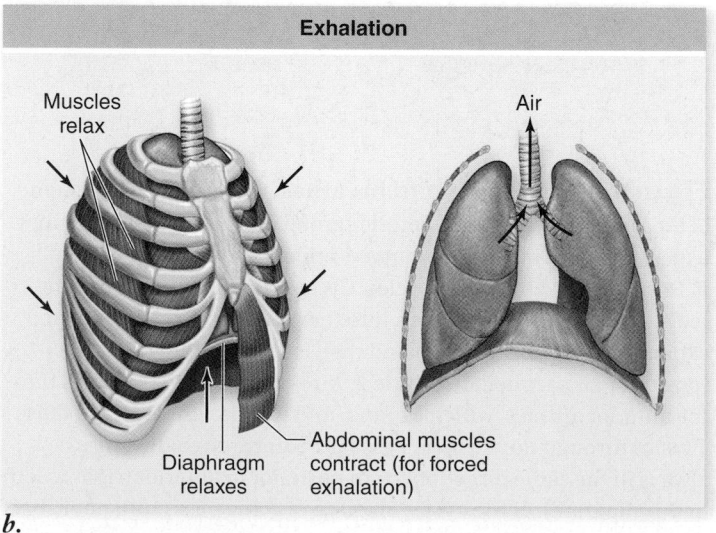

Exhalation

Muscles relax

Air

Diaphragm relaxes

Abdominal muscles contract (for forced exhalation)

b.

Figure 48.11 **How a human breathes.** *a.* Inhalation. The diaphragm contracts and the walls of the chest cavity expand, increasing the volume of the chest cavity and lungs. As a result of the larger volume, air is drawn into the lungs. *b.* Exhalation. The diaphragm and chest walls return to their normal positions as a result of elastic recoil, reducing the volume of the chest cavity and forcing air out of the lungs through the trachea. Note that inhalation can be forced by contracting accessory respiratory muscles (such as the sternocleidomastoid), and exhalation can be forced by contracting abdominal muscles.

The increased breathing that occurs during moderate exertion is not necessarily hyperventilation because the faster and more forceful breathing is matched to the higher metabolic rate, and blood gas measurements remain normal. Next, we describe how breathing is regulated to keep pace with metabolism.

Ventilation is under nervous system control

Each breath is initiated by neurons in a *respiratory control center* located in the medulla oblongata. These neurons stimulate the diaphragm and external intercostal muscles to contract, causing inhalation. When these neurons stop producing impulses, the inspiratory muscles relax and exhalation occurs. Although the muscles of breathing are skeletal muscles, they are usually controlled automatically. This control can be voluntarily overridden, however, as in hypoventilation (breath holding) or hyperventilation.

Neurons of the medulla oblongata must be responsive to changes in blood P_{O_2} and P_{CO_2} in order to maintain homeostasis. You can demonstrate this mechanism by simply holding your breath. Your blood carbon dioxide level immediately rises, and your blood oxygen level falls. After a short time, the urge to breathe induced by the changes in blood gases becomes overpowering. The rise in blood carbon dioxide, as indicated by a rise in P_{CO_2}, is the primary initiator, rather than the fall in oxygen levels.

A rise in P_{CO_2} causes an increased production of carbonic acid (H_2CO_3), which lowers the blood pH. A fall in blood pH stimulates chemosensitive neurons in the **aortic** and **carotid bodies,** in the aorta and the carotid artery (figure 48.12b).

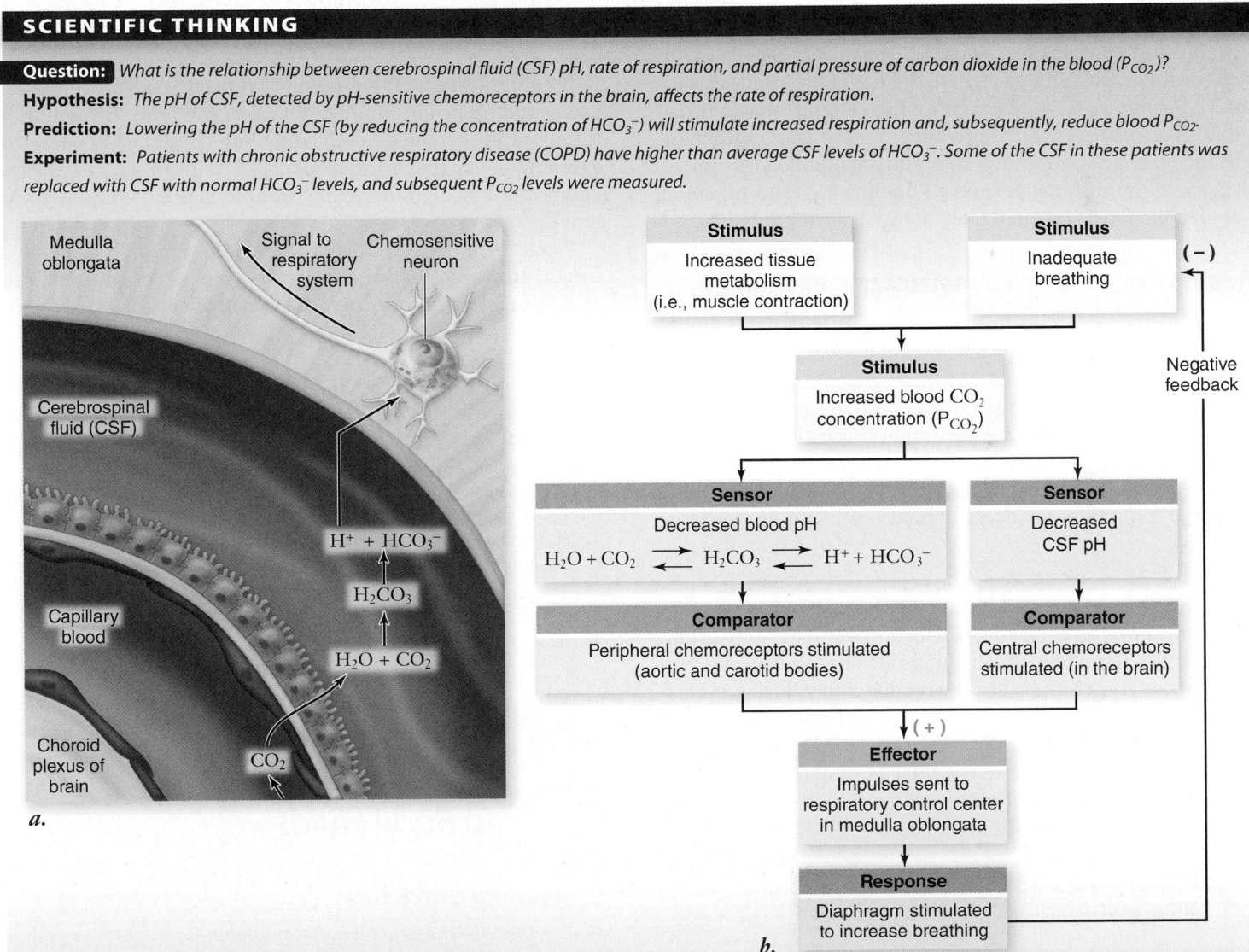

SCIENTIFIC THINKING

Question: *What is the relationship between cerebrospinal fluid (CSF) pH, rate of respiration, and partial pressure of carbon dioxide in the blood (P_{CO2})?*

Hypothesis: *The pH of CSF, detected by pH-sensitive chemoreceptors in the brain, affects the rate of respiration.*

Prediction: *Lowering the pH of the CSF (by reducing the concentration of HCO_3^-) will stimulate increased respiration and, subsequently, reduce blood P_{CO2}.*

Experiment: *Patients with chronic obstructive respiratory disease (COPD) have higher than average CSF levels of HCO_3^-. Some of the CSF in these patients was replaced with CSF with normal HCO_3^- levels, and subsequent P_{CO2} levels were measured.*

Result: *The reduced HCO_3^- levels (and corresponding drop in CSF pH) resulted in increased respiration, which subsequently resulted in lower arterial P_{CO2}.*

Conclusion: *The drop in pH (caused by the change in HCO_3^- levels) is detected by H^+ ion chemoreceptors in the brain. The brain sends impulses to the respiratory control center in the medulla oblongata, which directs an increase in breathing. Likewise, when the concentration of CO_2 in the blood rises as a result of inadequate breathing or increased tissue metabolism, the pH of the blood and the CSF decreases, stimulating the central chemoreceptors in the brain and the peripheral chemoreceptors in the aortic and carotid bodies. The receptors signal the brain stem control center, increasing ventilation and reducing the P_{CO2} to normal levels.*

Figure 48.12 **Regulation of breathing by pH-sensitive chemoreceptors.**

These peripheral receptors send impulses to the respiratory control center, which then stimulates increased breathing. The brain also contains central chemoreceptors that are stimulated by a drop in the pH of cerebrospinal fluid (CSF) (figure 48.12a).

A person cannot voluntarily hyperventilate for too long. The decrease in plasma P_{CO_2} and increase in pH of plasma and CSF caused by hyperventilation suppress the reflex drive to breathe. Deliberate hyperventilation allows people to hold their breath longer—not because it increases oxygen in the blood, but because the carbon dioxide level is lowered and takes longer to build back up, postponing the need to breathe.

In people with normal lungs, P_{O_2} becomes a significant stimulus for increased breathing rates only at high altitudes, where the P_{O_2} of the atmosphere is low. The symptoms of low oxygen at high altitude are known as mountain sickness, which may include feelings of weakness, headache, nausea, vomiting, and reduced mental function. All of these symptoms are related to the low P_{O_2}, and breathing supplemental oxygen often may remove all symptoms.

Respiratory diseases restrict gas exchange

Chronic obstructive pulmonary disease (COPD) refers to any disorder that obstructs air flow on a long-term basis. The major COPDs are asthma, chronic bronchitis, and emphysema. In **asthma,** an allergen triggers the release of histamine and other inflammatory chemicals that cause intense constriction of the bronchi and sometimes suffocation. Other COPDs are commonly caused by cigarette smoking but can also result from air pollution or occupational exposure to airborne irritants.

Emphysema

In **emphysema,** alveolar walls break down and the lung exhibits larger but fewer alveoli. The lungs also become fibrotic and less elastic. The air passages open adequately during inhalation but they tend to collapse and obstruct the outflow of air. People with emphysema become exhausted because they expend three to four times the normal amount of energy just to breathe. Eighty to 90% of emphysema deaths are caused by cigarette smoking.

? Inquiry question How does emphysema affect the diffusion of gases in and out of the lung, based on Fick's Law?

Lung cancer

Lung cancer accounts for more deaths than any other form of cancer. The most important cause of lung cancer is cigarette smoking, distantly followed by air pollution (figure 48.13). Lung cancer follows or accompanies COPD.

Over 90% of lung tumors originate in the mucous membranes of the large bronchi. As a tumor invades the bronchial

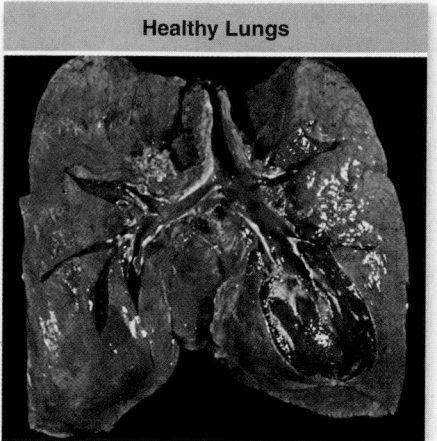 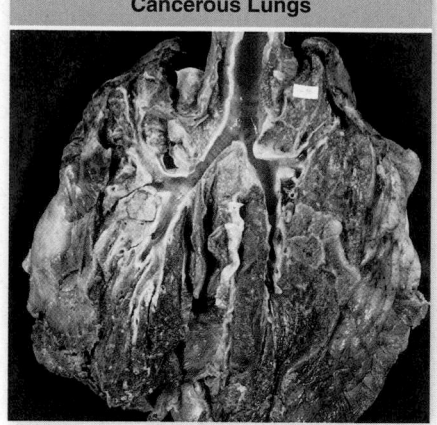

Healthy Lungs **Cancerous Lungs**

Figure 48.13 Comparison of healthy lung and a lung with cancer.

wall and grows around it, it compresses the airway and may cause collapse of more distal parts of the lung. Growth of a tumor often produces coughing, but coughing is such an everyday occurrence for smokers, it seldom causes alarm. Often, the first sign of serious trouble is the coughing up of blood.

Lung cancer metastasizes (spreads) so rapidly that it has usually invaded other organs by the time it is diagnosed. The chance of recovery from metastasized lung cancer is poor, with only 3% of patients surviving for 5 years after diagnosis.

Learning Outcomes Review 48.4

In humans, each breath moves a tidal volume of about 500 mL in and out of the lungs; 150 mL remains in the tubular passages where no gases are exchanged (anatomical dead space). Depth and rate of ventilation is regulated primarily by neurons in the medulla oblongata that detect CO_2 concentration. Diseases such as COPD limit gas exchange by obstructing air flow. Lung cancer, associated with tobacco use, has a low survival rate.

■ *How do mammals breathe differently from birds?*

48.5 Transport of Gases in Body Fluids

Learning Outcomes

1. *Describe the structure of hemoglobin.*
2. *Predict hemoglobin's oxygen affinity in various environmental conditions.*
3. *Explain how carbon dioxide is transported by the blood.*

The amount of oxygen that can be dissolved in the blood plasma depends directly on the P_{O_2} of the air in the alveoli, as explained earlier. When mammalian lungs are functioning normally, the blood plasma leaving the lungs has almost as much dissolved oxygen as is theoretically possible, given the P_{O_2} of the air. Because of oxygen's low solubility, however, blood plasma can contain a maximum of only about 3 mL of O_2 per liter. But whole blood normally carries almost 200 mL of O_2 per liter. Most of the oxygen in the blood is bound to molecules of hemoglobin inside red blood cells.

Respiratory pigments bind oxygen for transport

Hemoglobin is a protein composed of four polypeptide chains and four organic compounds called *heme groups*. At the center of each heme group is an atom of iron, which can bind to a molecule of oxygen (figure 48.14). Thus, each hemoglobin molecule can carry up to four molecules of oxygen.

Hemoglobin loads up with oxygen in the alveolar capillaries of the pulmonary circulation, forming oxyhemoglobin. This molecule has a bright red color. As blood passes through capillaries in the systemic circulation, some of the oxyhemoglobin releases oxygen, becoming **deoxyhemoglobin.** Deoxyhemoglobin has a darker red color; but it imparts a bluish tinge to tissues. Illustrations of the cardiovascular system show vessels carrying oxygenated blood with a red color and vessels that carry oxygen-depleted blood with a blue color.

Hemoglobin is an ancient protein; it is not only the oxygen-carrying molecule in all vertebrates, but is also used as an oxygen carrier by many invertebrates, including annelids, mollusks, echinoderms, flatworms, and even some protists. Many other invertebrates, however, employ different oxygen carriers, such as **hemocyanin.** In hemocyanin, the oxygen-binding atom is copper instead of iron. Hemocyanin is not found associated with blood cells, but is instead one of the free proteins in the circulating fluid (hemolymph) of arthropods and some mollusks.

Hemoglobin and myoglobin provide an oxygen reserve

At a blood P_{O_2} of 100 mm Hg, the level found in blood leaving the alveoli, approximately 97% of the hemoglobin within red

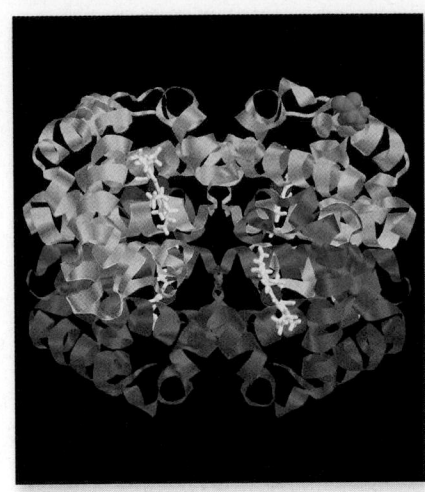

Figure 48.14 The structure of the adult hemoglobin protein. Hemoglobin consists of four polypeptide chains: two α chains (blue) and two β chains (gold). Each chain is associated with a heme group (in white), and each heme group has a central iron atom (red ball), which can bind to a molecule of O_2.

blood cells is in the form of oxyhemoglobin—indicated as a percent oxyhemoglobin saturation of 97%.

In a person at rest, blood that returns to the heart in the systemic veins has a P_{O_2} that is decreased to about 40 mm Hg. At this lower P_{O_2}, the percent saturation of hemoglobin is only 75%. In a person at rest, therefore, 22% (97% minus 75%) of the oxyhemoglobin has released its oxygen to the tissues. Put another way, roughly one-fifth of the oxygen is unloaded in the tissues, leaving four-fifths of the oxygen in the blood as a reserve. A graphic representation of these changes is called an oxyhemoglobin dissociation curve (figure 48.15).

This large reserve of oxygen serves an important function. It enables the blood to supply the body's oxygen needs during exertion as well as at rest. During exercise, for example, the muscles' accelerated metabolism uses more oxygen and decreases the venous blood P_{O_2}. The P_{O_2} of the venous blood could drop to 20 mm Hg; in this case, the percent saturation of hemoglobin would be only 35% (see figure 48.15). Because arterial blood would still contain 97% oxyhemoglobin, the amount of oxygen unloaded would now be 62% (97% minus 35%), instead of the 22% at rest.

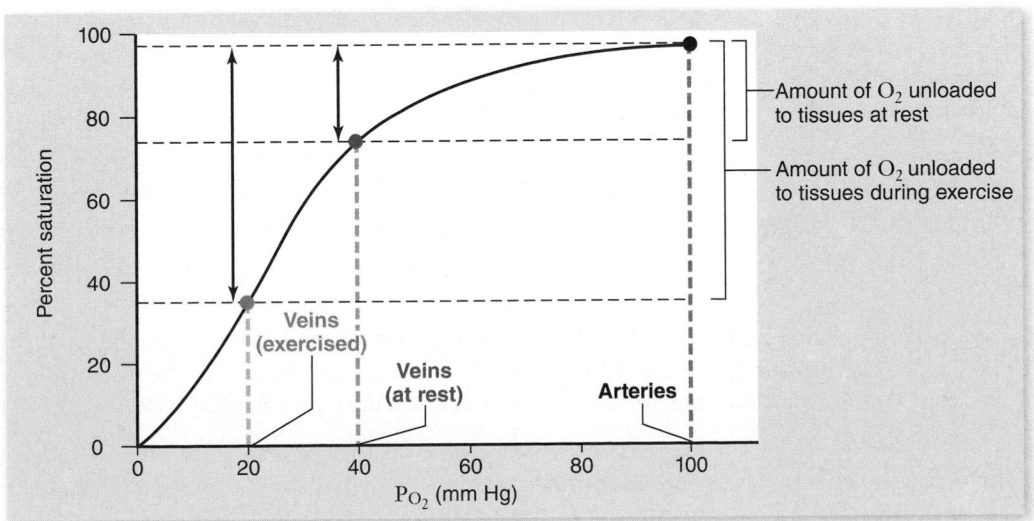

Figure 48.15 The oxyhemoglobin dissociation curve. Hemoglobin combines with O_2 in the lungs, and this oxygenated blood is carried by arteries to the body cells. After O_2 is removed from the blood to support cellular respiration, the blood entering the veins contains less O_2.

 Data analysis How would you determine how much oxygen was unloaded to the tissues?

In addition to this function, the oxygen reserve also ensures that the blood contains enough oxygen to maintain life for 4 to 5 min if breathing is interrupted or if the heart stops pumping.

A second oxygen reserve is available in myoglobin, an oxygen-binding molecule found in muscle cells. Myoglobin is composed of a single polypeptide chain with an iron atom that can bind to an O_2 molecule. Myoglobin has a higher affinity for oxygen than hemoglobin, which means that when oxygen levels fall in muscle cells, myoglobin will contain oxygen after the hemoglobin supplies have been exhausted. Deep sea-diving mammals, such as the elephant seal in figure 48.1, are able to stay under water for long periods in part because of the high levels of oxygen stored in the myoglobin in their muscles.

 Data analysis Based on the preceding information, would an otherwise healthy person benefit significantly from breathing 100% oxygen following a bout of intense exercise such as a 400-m sprint?

Hemoglobin's affinity for oxygen is affected by pH and temperature

Oxygen transport in the blood is affected by other conditions including temperature and pH. The CO_2 produced by metabolizing tissues combines with H_2O to form carbonic acid (H_2CO_3). H_2CO_3 dissociates into bicarbonate (HCO_3^-) and H^+, thereby lowering blood pH. This reaction occurs primarily inside red blood cells, where the lowered pH reduces

hemoglobin's affinity for oxygen, causing it to release oxygen more readily.

The effect of pH on hemoglobin's affinity for oxygen, known as the **Bohr effect** or **Bohr shift,** is the result of H^+ binding to hemoglobin. It is shown graphically by a shift of the oxyhemoglobin dissociation curve to the right (figure 48.16a).

Increasing temperature has a similar effect on hemoglobin's affinity for oxygen (figure 48.16b). Because skeletal muscles produce carbon dioxide more rapidly during exercise, and because active muscles produce heat, the blood unloads a higher percentage of the oxygen it carries during exercise.

Carbon dioxide is primarily transported as bicarbonate ion

About 8% of the CO_2 in blood is simply dissolved in plasma; another 20% is bound to hemoglobin. Because CO_2 binds to the protein portion of hemoglobin, and not to the iron atoms of the heme groups, it does not compete with oxygen; however, it does cause hemoglobin's shape to change, lowering its affinity for oxygen.

The remaining 72% of the CO_2 diffuses into the red blood cells, where the enzyme carbonic anhydrase catalyzes the combining of CO_2 with water to form H_2CO_3. H_2CO_3 dissociates into HCO_3^- and H^+ ions. The H^+ binds to deoxyhemoglobin, and the HCO_3^- moves out of the erythrocyte into the plasma via a transporter that exchanges one Cl^- for a HCO_3^- (this is called the "chloride shift").

This reaction removes large amounts of CO_2 from the plasma, maintaining a diffusion gradient that allows additional

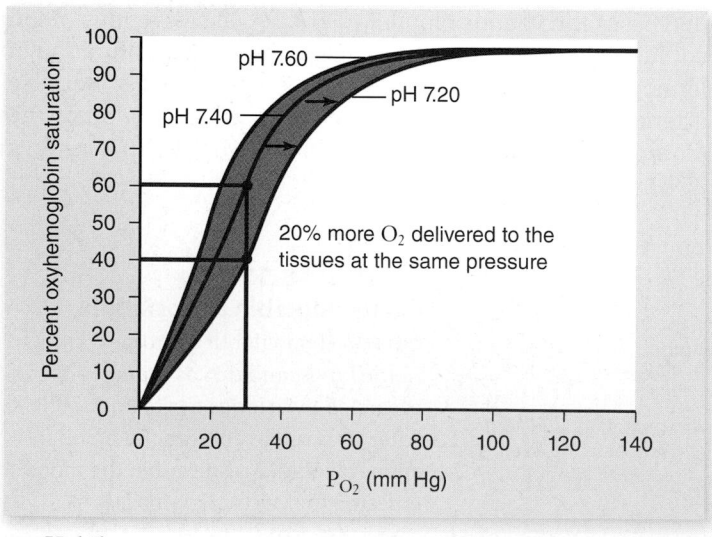

a. pH shift

b. Temperature shift

Figure 48.16 **The effect of pH and temperature on the oxyhemoglobin dissociation curve.** *a.* Lower blood pH and (*b*) higher blood temperatures shift the oxyhemoglobin dissociation curve to the right, facilitating O_2 unloading. In this example, this can be seen as a lowering of the oxyhemoglobin percent saturation from 60% to 40%, indicating that the difference of 20% more O_2 is unloaded to the tissues.

Data analysis For a given P_{O_2}, how does oxyhemoglobin saturation percentage vary in response to changes in temperature?

Inquiry question What effect does high blood pressure have on oxygen unloading to the tissues during exercise?

CO_2 to move into the plasma from the surrounding tissues (figure 48.17a). The formation of H_2CO_3 is also important in maintaining the acid–base balance of the blood; HCO_3^- serves as the major buffer of the blood plasma.

In the lungs, the lower P_{CO_2} of the gas mixture inside the alveoli causes the carbonic anhydrase reaction to proceed in the reverse direction, converting H_2CO_3 into H_2O and CO_2 (figure 48.17b). The CO_2 diffuses out of the red blood cells and into the alveoli, so that it can leave the body in the next exhalation.

Other dissolved gases are also transported by hemoglobin, most notably nitric oxide (NO), which plays an important role in vessel dilation. Carbon monoxide (CO) binds more strongly to hemoglobin than does oxygen, which is why carbon monoxide poisoning can be deadly. Victims of carbon monoxide poisoning often have bright red skin due to hemoglobin's binding with CO.

Learning Outcomes Review 48.5

Hemoglobin consists of four polypeptide chains, each associated with an iron-containing heme group that can bind O_2. Hemoglobin's affinity for oxygen is affected by pH and temperature; more O_2 is released into tissues at lower pH and at higher temperature. Carbon dioxide is transported in the blood in three ways: dissolved in the plasma, bound to hemoglobin, and as bicarbonate in the plasma following a reaction with carbonic anhydrase in the red blood cells.

■ *What are the differences in the way that oxygen and carbon dioxide are transported in blood?*

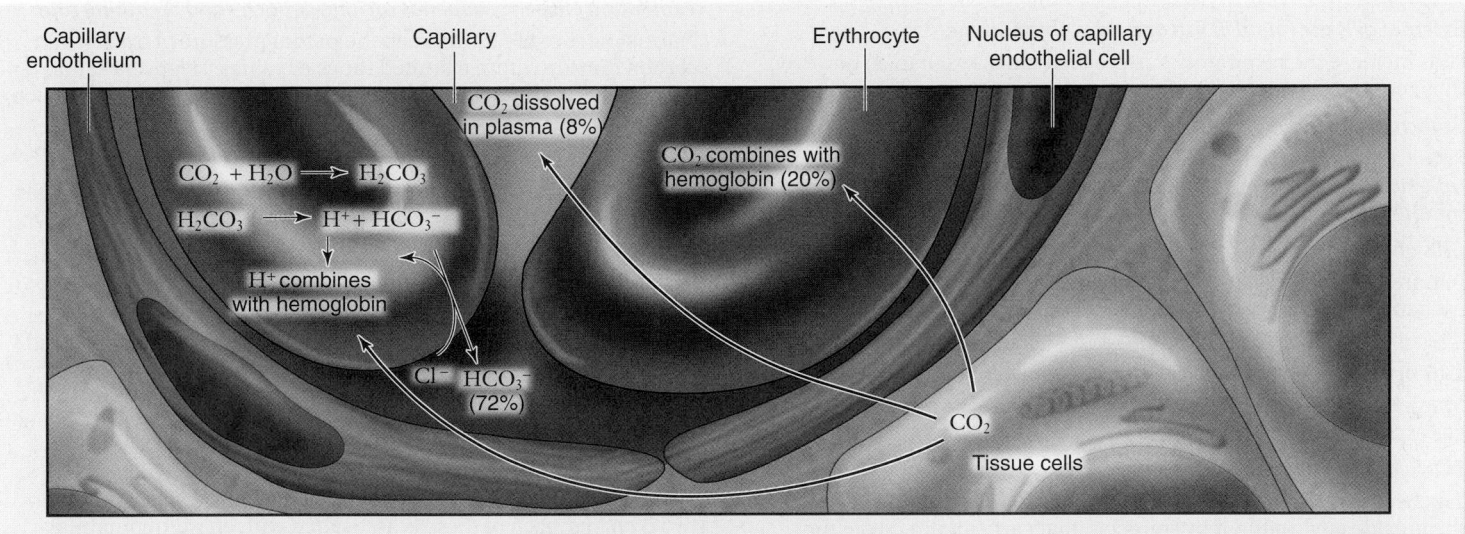

a.

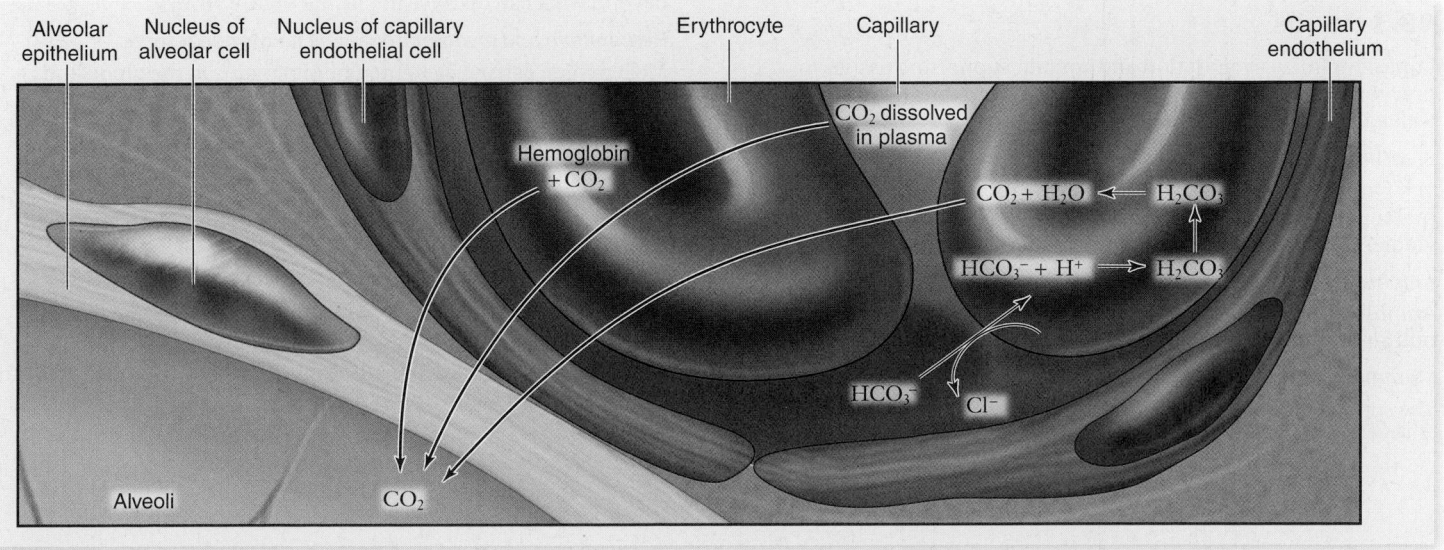

b.

Figure 48.17 The transport of carbon dioxide by the blood. *a.* Passage into bloodstream. CO_2 is transported in three ways: dissolved in plasma, bound to the protein portion of hemoglobin, and as bicarbonate (HCO_3^-), which forms in red blood cells. The reaction of CO_2 with H_2O to form H_2CO_3 (carbonic acid) is catalyzed by the enzyme carbonic anhydrase in red blood cells. *b.* Removal from bloodstream. When the blood passes through the pulmonary capillaries, these reactions are reversed so that CO_2 gas is formed, which is exhaled.

48.1 Gas Exchange Across Respiratory Surfaces

Gas exchange involves diffusion across membranes.

Diffusion is a passive process; the rate of diffusion (R) increases with a higher concentration gradient and greater surface area, but decreases with distance (Fick's Law).

Evolutionary strategies have maximized gas diffusion.

Most invertebrate phyla lack specialized respiratory organs, but have evolved ways to increase oxygen concentration differences. Most other animals possess respiratory organs.

48.2 Gills, Cutaneous Respiration, and Tracheal Systems

External gills are found in fish and amphibian larvae.

Gills increase the respiratory surface area for gas exchange; however, they require an aqueous environment.

Branchial chambers protect gills of some invertebrates.

Some aquatic invertebrates have branchial chambers in which oxygenated water is pumped past stationary gills. Mollusks possess a mantle in which water is drawn in and expelled.

Gills of bony fishes are covered by the operculum.

In bony fishes, diffusion of gases is maximized by countercurrent exchange, in which blood in gills flows in a direction opposite the flow of water over the gills (figures 48.4 and 48.5).

Cutaneous respiration requires constant moisture.

Many amphibians and a few reptiles use cutaneous respiration for gas exchange.

Tracheal systems are found in arthropods.

Tracheae and tracheoles are a series of small tubes, connected with the outside environment by spiracles, that carry air directly to the cells. The ability to open and close the spiracles allowed arthropods to invade the land.

48.3 Lungs

Lungs minimize evaporation and contain supporting tissues to prevent collapse of exchange membranes, and thus have become well adapted to terrestrial living.

Breathing of air takes advantage of partial pressures of gases.

The partial pressure of gases refers to the proportion of atmospheric pressure attributed to each gas. It is responsible for the pressure gradient that brings about gas exchange.

Amphibians and reptiles breathe in different ways.

Amphibians force air into their lungs by positive pressure; reptiles pull air in using negative pressure (figure 48.7).

Mammalian lungs have greatly increased surface area.

The surface area of mammalian lungs is enormous due to numerous alveoli, encased by an extensive capillary network (figure 48.8).

The respiratory system of birds is a highly efficient flow-through system.

The respiratory system of birds involves one-way direction of air flow. Air moves through the respiratory system in a two-cycle process so that fresh and used air never mix (figure 48.9).

48.4 Structures, Mechanisms, and Control of Ventilation in Mammals

Lung structure and function supports the respiratory cycle.

Gas exchange is driven by differences in partial pressures. Lungs are filled by contraction of the diaphragm and external intercostal muscles, creating negative pressure (figure 48.11).

Ventilation efficiency depends on lung capacity and breathing rate.

Normal rates of breathing keep the partial pressure of oxygen and carbon dioxide within a limited range of values. Hypoventilation occurs when carbon dioxide levels are too high, and hyperventilation when they are too low.

Ventilation is under nervous system control.

Each breath is initiated by neurons in the respiratory control center, primarily those that detect CO_2 levels. Humans can voluntarily hypo- or hyperventilate, but only for a limited time.

Respiratory diseases restrict gas exchange.

Emphysema occurs when alveolar walls break down, which makes breathing very energetically expensive. Lung cancer is highly deadly and caused primarily by smoking.

48.5 Transport of Gases in Body Fluids

Respiratory pigments bind oxygen for transport.

Hemoglobin consists of four polypeptide chains, two α chains and two β chains; each of these is associated with an iron-containing heme group that can bind to O_2 (figure 48.14).

Hemoglobin increases the ability of the blood to transport oxygen beyond what can dissolve in plasma (figure 48.15).

Hemoglobin and myoglobin provide an oxygen reserve.

Most oxygen carried by hemoglobin remains in the blood and is available when needed. In addition, myoglobin molecules in muscle cells retain oxygen at lower partial pressures than hemoglobin and thus serve as an additional oxygen reserve.

Hemoglobin's affinity for oxygen is affected by pH and temperature.

The affinity of hemoglobin for oxygen decreases as pH decreases and as temperature increases (figure 48.16). Therefore at lower pH and higher temperature, more oxygen is released.

Carbon dioxide is primarily transported as bicarbonate ion.

Most carbon dioxide diffuses into red blood cells and combines with water to form bicarbonate atoms in a reaction catalyzed by the enzyme carbonic anhydrase.

UNDERSTAND

1. If you hold your breath for a long time, body CO_2 levels are likely to ____, and the pH of body fluids is likely to ____.

   a. increase; increase
   b. decrease; increase
   c. increase; decrease
   d. decrease; decrease

2. Increased efficiency of gas exchange in vertebrates has been brought about by all of the following mechanisms except

   a. cutaneous respiration.
   b. unidirectional air flow.
   c. crosscurrent blood flow.
   d. cartilaginous rings in the trachea.

3. Which of the following is the primary method by which carbon dioxide is transported to the lungs?

   a. Dissolved in plasma
   b. Bound to hemoglobin
   c. As carbon monoxide
   d. As bicarbonate

4. Gills are found in

   a. fish.
   b. amphibians.
   c. aquatic invertebrates.
   d. All of the choices are correct.

5. Fick's Law of Diffusion states the rate of diffusion is directly proportional to

   a. the area differences between the cross section of the blood vessel and the tissue.
   b. the pressure differences between the two sides of the membrane and area over which the diffusion occurs.
   c. the pressure differences between the inside of the organism and the outside.
   d. the temperature of the gas molecule.

6. Cutaneous respiration requires

   a. moist and highly vascularized skin.
   b. the absence of gills and lungs.
   c. an environment rich in oxygen.
   d. low temperatures.

7. Hyperventilation occurs

   a. as a result of breathing rapidly.
   b. when oxygen levels become low.
   c. when tidal volumes are unusually low.
   d. when the partial pressure of carbon dioxide is low.

8. Most carbon dioxide is

   a. dissolved in the plasma.
   b. bound to hemoglobin.
   c. combined with water in red blood cells to form carbonic acid.
   d. stored in the lungs prior to exhalation.

APPLY

1. When you take a deep breath, your stomach moves out because

   a. swallowing air increases the volume of the thoracic cavity.
   b. your stomach shouldn't move out when you take a deep breath because you want the volume of your chest cavity to increase, not your abdominal cavity.
   c. contracting your abdominal muscles pushes your stomach out, generating negative pressure in your lungs.
   d. when your diaphragm contracts, it moves down, pressing your abdominal cavity out.

2. Marine mammals are able to hold their breath for extended periods underwater because

   a. unlike humans, they don't hypoventilate.
   b. partial pressure of carbon dioxide does not increase underwater.
   c. myoglobin in muscle tissue provides an oxygen reserve.
   d. the brains of marine mammals do not have receptors that respond to impulses initiated in the aortic and carotid bodies.

3. Countercurrent flow systems do not occur in lungs because they

   a. require oxygen suspended in flowing water.
   b. are limited to fish.
   c. only work in moving organisms.
   d. cannot operate in the presence of carbon dioxide.

4. Respiratory organs of invertebrates and vertebrates are similar in that

   a. they use negative pressure breathing.
   b. they take advantage of countercurrent flow systems.
   c. they increase the surface area available for diffusion.
   d. the air flows through the organ in one direction.

5. Mountain climbers may have difficulty at high elevations because

   a. the partial pressure of oxygen is lower at higher elevations.
   b. more CO_2 occurs at higher altitudes.
   c. the concentration of all elements of the air is lower at higher elevations.
   d. cooler temperatures restrict the metabolic activity of oxygen at high elevations.

6. During exercise more oxygen is delivered to the muscles because

   a. active muscles produce more CO_2, lowering the pH of the blood.
   b. active muscles produce heat.
   c. Both a and b are correct.
   d. Neither a nor b is correct.

SYNTHESIZE

1. Compare the operation and efficiency of fish gills with amphibian, bird, and mammal lungs.

2. What happens when, during exercise, the oxygen needs of the peripheral tissues increase greatly?

3. Explain how bacteria, archaea, protists, and many phyla of invertebrates can survive without respiratory organs.

ONLINE RESOURCE

www.ravenbiology.com

Understand, Apply, and Synthesize—enhance your study with animations that bring concepts to life and practice tests to assess your understanding. Your instructor may also recommend the interactive eBook, individualized learning tools, and more.

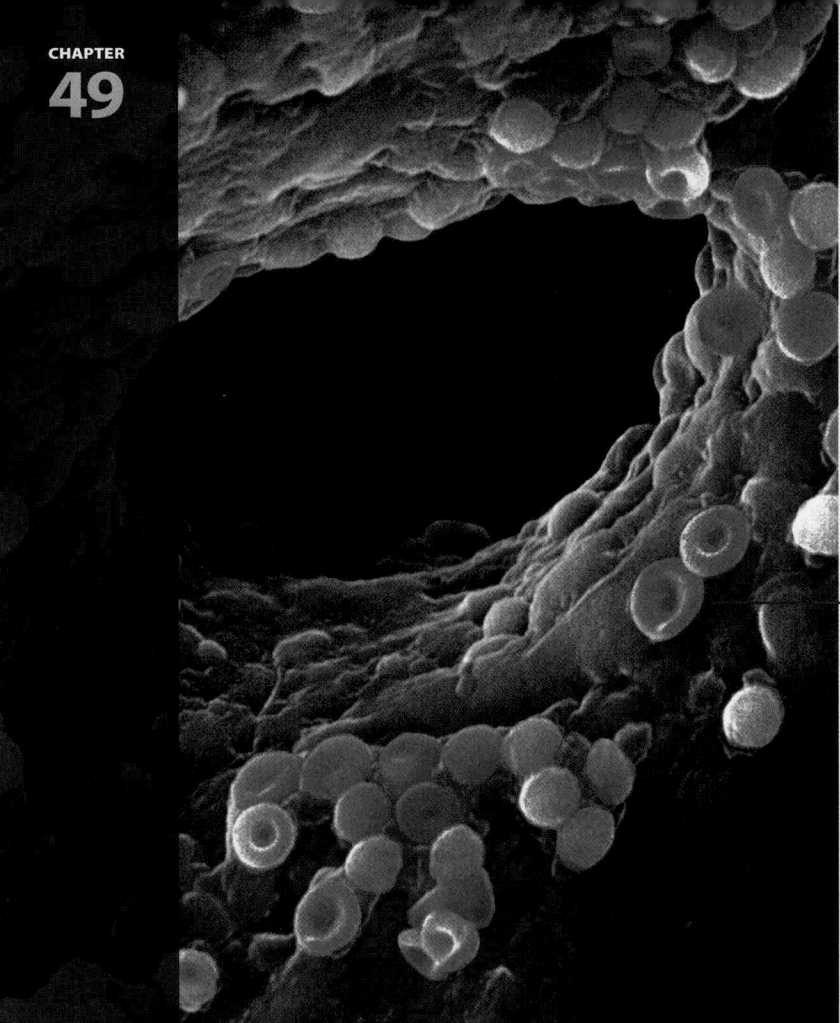

Chapter **49**

The Circulatory System

Chapter Contents

Introduction

In multicellular organisms, oxygen obtained by the respiratory system and nutrients processed by the digestive system must be transported to cells throughout the body. Conversely, carbon dioxide and other waste products produced within the cells must be returned to the respiratory, digestive, and urinary systems for elimination from the body. These tasks are the responsibility of the circulatory system. All multicellular organisms have a heart that pumps fluids through the body. Many invertebrates have an open system in which fluids move through the body cavity. Vertebrates also have a system like this that transports lymph through the body; however, the primary circulatory fluid is blood, which moves through a closed system of blood vessels.

49.1 The Components of Blood

Learning Outcomes

1.   *Describe the functions of circulating blood.*
2.   *Distinguish between the types of formed elements.*
3.   *Delineate the process of blood clotting.*

Blood is a connective tissue composed of a fluid matrix, called **plasma,** and several different kinds of cells and other **formed elements** that circulate within that fluid (figure 49.1). Blood

platelets, although included in figure 49.1, are not complete cells; rather, they are fragments of cells that are produced in the bone marrow. (We describe the action of platelets in blood clotting later in this section.)

Circulating blood has many functions:

1.   **Transportation.** All of the substances essential for cellular metabolism are transported by blood. Red blood cells transport oxygen attached to hemoglobin; nutrient molecules are carried in the plasma, sometimes bound to carriers; and metabolic wastes are eliminated as blood passes through the liver and kidneys.

2.   **Regulation.** The cardiovascular system transports regulatory hormones from the endocrine glands and

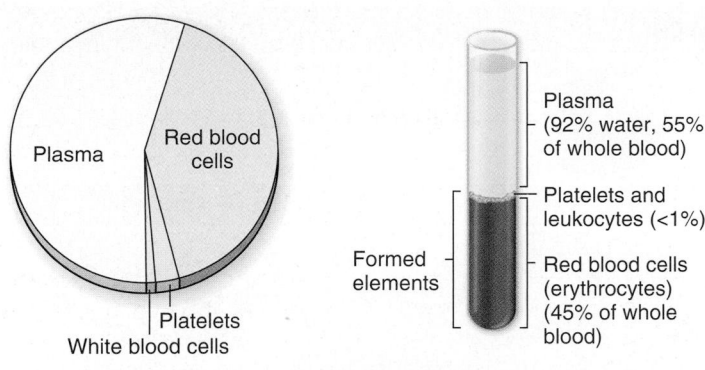

Blood Plasma	Red Blood Cells	Platelets
Plasma proteins (7%) Albumin (54%) Globulins (38%) Fibrinogen (7%) All others (1%) **Water** (91.5%) **Other solutes** (1.5%) Electrolytes Nutrients Gases Regulatory substances Waste products	4 million–6 million/ mm^3 blood	150,000–300,000/ mm^3 blood
	Neutrophils	**Eosinophils**
	60–70%	2–4%
Monocytes	**Basophils**	**Lymphocytes**
3–8%	0.5–1%	20–25%

Figure 49.1 Composition of blood.

also participates in temperature regulation. Contraction and dilation of blood vessels near the surface of the body (just beneath the epidermis) helps to conserve or to dissipate heat as needed.

3. **Protection.** The circulatory system protects against injury and foreign microbes or toxins introduced into the body. Blood clotting helps to prevent blood loss when vessels are damaged. White blood cells, or leukocytes, help to disarm or disable invaders such as viruses and bacteria (see chapter 51).

Blood plasma is a fluid matrix

Blood plasma is the matrix in which blood cells and platelets are suspended. Interstitial (extracellular) fluids originate from the fluid present in plasma.

Although plasma is 92% water, it also contains the following solutes:

1. **Nutrients, wastes, and hormones.** Dissolved within the plasma are all of the nutrients resulting from digestive breakdown that can be used by cells, including glucose,

amino acids, and vitamins. Also dissolved in the plasma are wastes such as nitrogen compounds and CO_2 produced by metabolizing cells. Endocrine hormones released from glands are also carried through the blood to their target cells.

2. **Ions.** Blood plasma is a dilute salt solution. The predominant plasma ions are Na^+, Cl^-, and bicarbonate ions (HCO_3^-). In addition, plasma contains trace amounts of other ions such as Ca^{2+}, Mg^{2+}, Cu^{2+}, K^+, and Zn^{2+}.

3. **Proteins.** The liver produces most of the plasma proteins, including **albumin,** which constitutes most of the plasma protein; the alpha (α) and beta (β) **globulins,** which serve as carriers of lipids and steroid hormones; and **fibrinogen,** which is required for blood clotting. Blood plasma with the fibrinogen removed is called **serum.**

Formed elements include circulating cells and platelets

The formed elements of blood cells and cell fragments include red blood cells, white blood cells, and platelets. Each element has a specific function in maintaining the body's health and homeostasis.

Erythrocytes

Each microliter of blood contains about 5 million **red blood cells,** or **erythrocytes.** The fraction of the total blood volume that is occupied by erythrocytes is called the blood's *hematocrit;* in humans, the hematocrit is typically around 45%.

Each erythrocyte resembles a doughnut-shaped disk with a central depression that does not go all the way through. Mature mammalian erythrocytes lack nuclei. The erythrocytes of vertebrates contain hemoglobin, a pigment that binds and transports oxygen. (Hemoglobin was described more fully in the previous chapter when we discussed respiration.) In vertebrates, hemoglobin is found only in erythrocytes. In invertebrates, the oxygen-binding pigment (not always hemoglobin) is also present in plasma.

Leukocytes

Less than 1% of the cells in human blood are **white blood cells,** or **leukocytes;** there are only 1 or 2 leukocytes for every 1000 erythrocytes. Leukocytes are larger than erythrocytes and have nuclei. Furthermore, leukocytes are not confined to the blood as erythrocytes are, but can migrate out of capillaries through the intercellular spaces into the surrounding interstitial (tissue) fluid.

Leukocytes come in several varieties, each of which plays a specific role in defending against invading microorganisms and other foreign substances, as described in chapter 51. **Granular leukocytes** include neutrophils, eosinophils, and basophils, which are named according to the staining properties of granules in their cytoplasm. **Nongranular leukocytes** include monocytes and lymphocytes. In humans, neutrophils are the most numerous of the leukocytes, followed in order by lymphocytes, monocytes, eosinophils, and basophils.

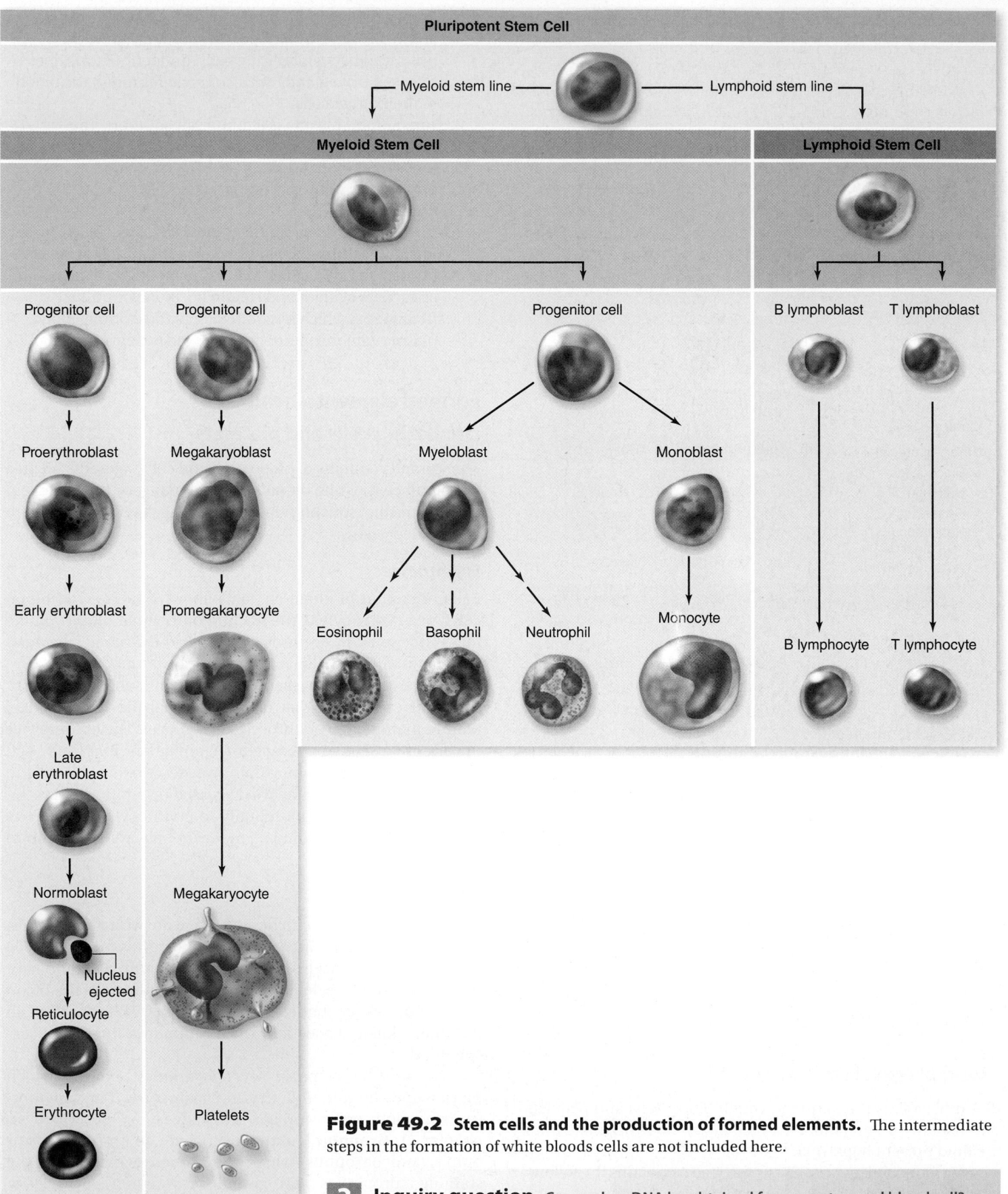

Pluripotent Stem Cell

Myeloid stem line — Lymphoid stem line

Myeloid Stem Cell

Lymphoid Stem Cell

Progenitor cell

Progenitor cell

Progenitor cell

B lymphoblast   T lymphoblast

Proerythroblast

Megakaryoblast

Myeloblast

Monoblast

Early erythroblast

Promegakaryocyte

Eosinophil   Basophil   Neutrophil

Monocyte

B lymphocyte   T lymphocyte

Late
erythroblast

Normoblast

Nucleus
ejected

Megakaryocyte

Reticulocyte

Erythrocyte

Platelets

Figure 49.2 Stem cells and the production of formed elements. The intermediate steps in the formation of white bloods cells are not included here.

? Inquiry question Can nuclear DNA be obtained from a mature red blood cell?

Platelets

Platelets are cell fragments that pinch off from larger cells in the bone marrow. They are approximately 3 μm in diameter, and following an injury to a blood vessel, the liver releases *prothrombin* into the blood. In the presence of this clotting factor, fibrinogen is converted into insoluble threads of **fibrin.** Fibrin then aggregates to form the clot.

Formed elements arise from stem cells

The formed elements of blood each have a finite life span and therefore must be constantly replaced. Many of the old cell fragments are digested by phagocytic cells of the spleen; however, many products from the old cells, such as iron and amino acids, are incorporated into new formed elements. The creation of new formed elements begins in the bone marrow (see chapter 46).

All of the formed elements develop from **pluripotent stem cells** (see chapter 19). The production of blood cells occurs in the bone marrow and is called **hematopoiesis.** This process generates two types of stem cells with a more restricted fate: a lymphoid stem cell that gives rise to lymphocytes and a myeloid stem cell that gives rise to the rest of the blood cells (figure 49.2).

When the oxygen available in the blood decreases, the kidney converts a plasma protein into the hormone **erythropoietin.** Erythropoietin then stimulates the production of erythrocytes from the myeloid stem cells through a process called **erythropoiesis.**

In mammals, maturing erythrocytes lose their nuclei prior to release into circulation. In contrast, the mature erythrocytes of all other vertebrates remain nucleated. *Megakaryocytes* are examples of committed cells formed in bone marrow from stem cells. Pieces of cytoplasm are pinched off the megakaryocytes to form the platelets.

 Inquiry question Why do you think the use of erythropoietin as a drug is banned in the Olympics and in some other sports?

Blood clotting is an example of an enzyme cascade

When a blood vessel is broken or cut, smooth muscle in the vessel walls contracts, causing the vessel to constrict. Platelets then accumulate at the injured site and form a plug by sticking to one another and to the surrounding tissues (figure 49.3). A cascade of enzymatic reactions is triggered by the platelets, plasma factors, and molecules released from the damaged tissue.

One of the results of this cascade is that fibrinogen, normally dissolved in the plasma, comes out of solution in a reaction that forms fibrin. The platelet plug is then reinforced by fibrin threads, which contract to form a tighter mass. The tightening plug of platelets, fibrin, and often trapped erythrocytes constitutes a blood clot.

Once the tissue damage is healed, the careful process of dissolving the blood clot begins. This process is significant because if a clot breaks loose and travels in the circulatory system, it may end up blocking a blood vessel in the brain, causing a stroke, or in the heart, causing a heart attack.

Learning Outcomes Review 49.1

The circulatory system functions in transport of materials, regulation of temperature and body processes, and protection of the body. Formed elements in blood include red blood cells, white blood cells, and platelets. Blood clotting involves a cascade of enzymatic reactions triggered by platelets and plasma factors to produce insoluble fibrin from fibrinogen.

■ *How does a blood clot form?*

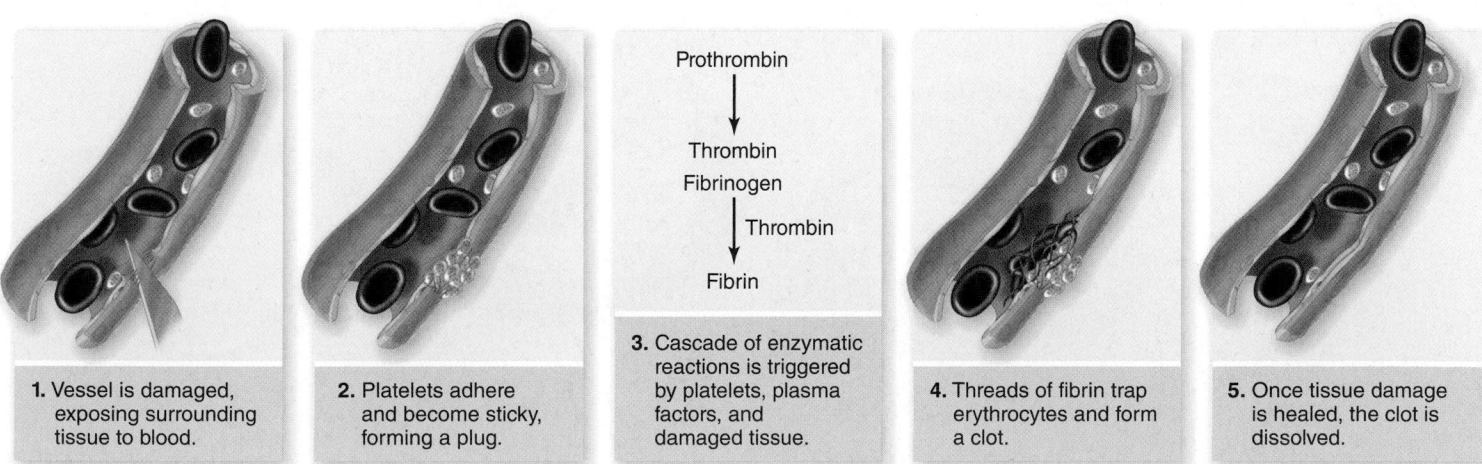

1. Vessel is damaged, exposing surrounding tissue to blood.

2. Platelets adhere and become sticky, forming a plug.

3. Cascade of enzymatic reactions is triggered by platelets, plasma factors, and damaged tissue.

Prothrombin
↓
Thrombin
Fibrinogen
↓ Thrombin
Fibrin

4. Threads of fibrin trap erythrocytes and form a clot.

5. Once tissue damage is healed, the clot is dissolved.

Figure 49.3 Blood clotting. Fibrin is formed from a soluble protein, fibrinogen, in the plasma. This reaction is catalyzed by the enzyme thrombin, which is converted from an inactive enzyme called prothrombin in the presence of platelets, plasma factors, and molecules released from the damaged tissue. The activation of thrombin is the last step in a cascade of enzymatic reactions that produces a blood clot when a blood vessel is damaged.

49.2 Invertebrate Circulatory Systems

Learning Outcomes

1. Distinguish between open and closed circulatory systems.
2. Define hemolymph.

The nature of the circulatory system in multicellular invertebrates is directly related to the size, complexity, and lifestyle of the organism in question. Sponges and most cnidarians utilize water from the environment as a circulatory fluid. Sponges pass water through a series of channels in their bodies, and *Hydra* and other cnidarians circulate water through a **gastrovascular cavity** (figure 49.4*a*). Because the body wall in *Hydra* species is only two cell layers thick, each cell layer is in direct contact with either the external environment or the gastrovascular cavity.

Invertebrates with a pseudocoelom, such as roundworms and rotifers (see chapter 34) use the fluids of the body cavity for circulation. Most of these invertebrates are quite small or are long and thin, and therefore adequate circulation is accomplished by movements of the body against the body fluids, which are in direct contact with the internal tissues and organs. Larger animals, however, have tissues that are several

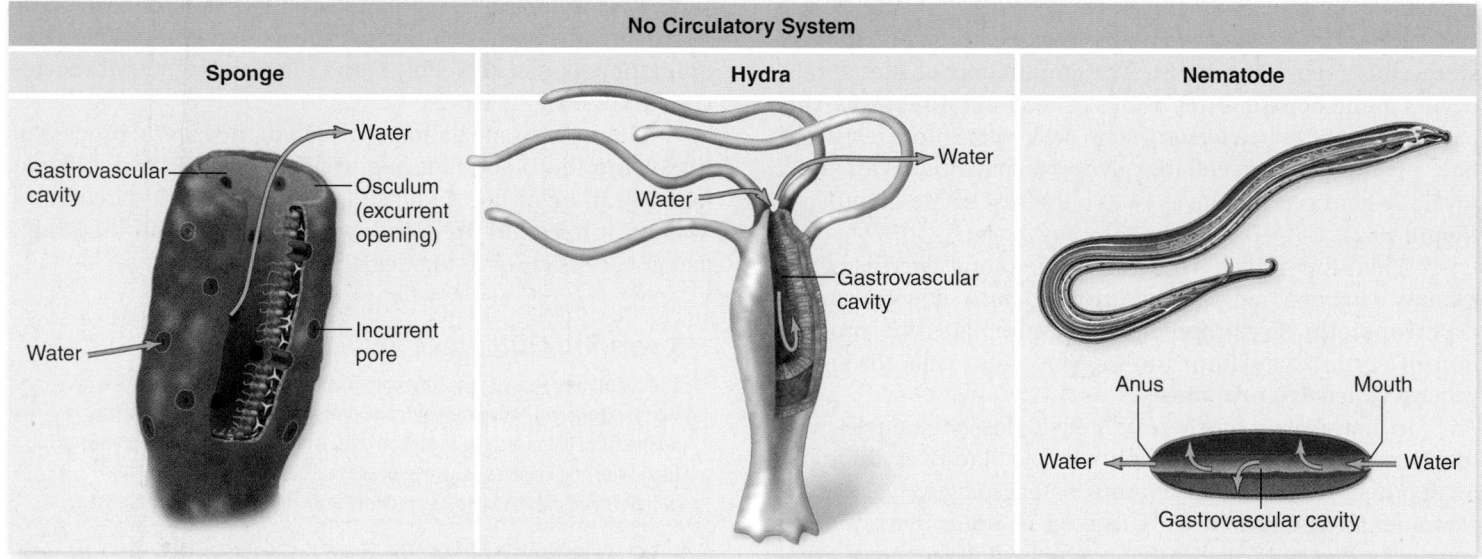

a.

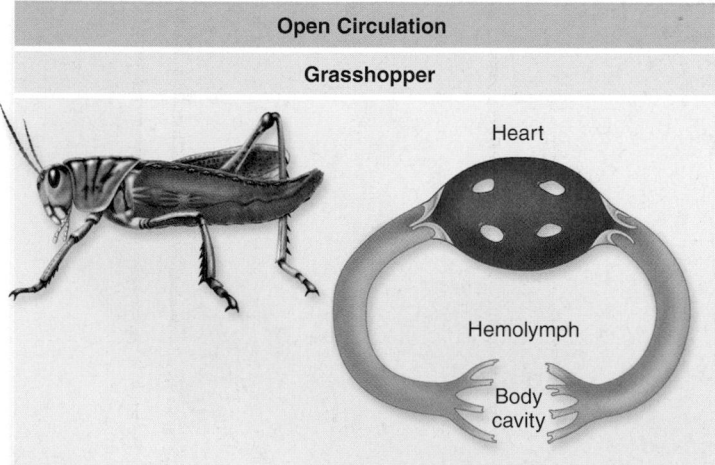

b.

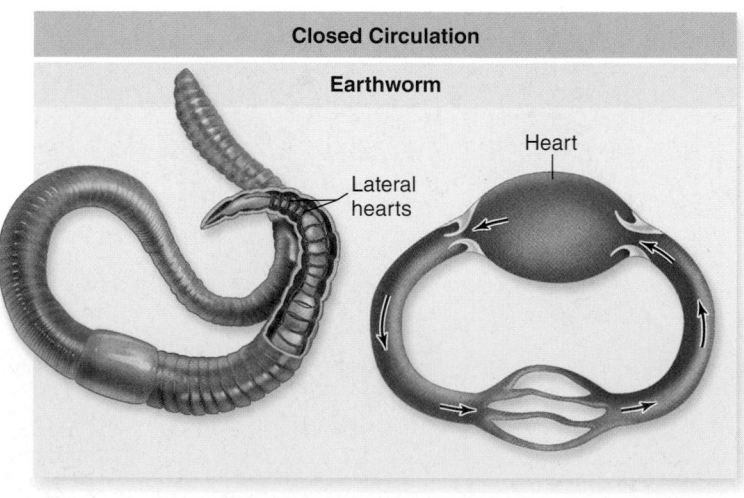

c.

Figure 49.4 Circulatory systems of the animal kingdom. *a.* Sponges (left panel) do not have a separate circulatory system. They circulate water using many incurrent pores and one excurrent pore. The gastrovascular cavity of a hydra (middle panel) serves as both a digestive and a circulatory system, delivering nutrients directly to the tissue cells by diffusion from the digestive cavity. The nematode (right panel) is thin enough that the digestive tract can also be used as a circulatory system. Larger animals require a separate circulatory system to carry nutrients to and wastes away from tissues. *b.* In the open circulation of an insect, hemolymph is pumped from a tubular heart into cavities in the insect's body; the hemolymph then returns to the blood vessels so that it can be recirculated. *c.* In the closed circulation of the earthworm, blood pumped from the hearts remains within a system of vessels that returns it to the hearts. All vertebrates also have closed circulatory systems.

cell layers thick, so that many cells are too far away from the body surface or digestive cavity to directly exchange materials with the environment. Instead, oxygen and nutrients are transported from the environment and digestive cavity to the body cells by an internal fluid within a circulatory system.

Open circulatory systems move fluids in a one-way path

The two main types of circulatory systems are *open* and *closed.* In an open circulatory system, such as that found in most mollusks and in arthropods (figure 49.4*b*), there is no distinction between the circulating fluid and the extracellular fluid of the body tissues. This fluid is thus called **hemolymph.**

In insects, a muscular tube, or **heart,** pumps hemolymph through a network of channels and cavities in the body. The fluid then drains back into the central cavity.

Closed circulatory systems move fluids in a loop

In a closed circulatory system, the circulating fluid, blood, is always enclosed within blood vessels that transport it away from and back to the heart (figure 49.4*c*). Some invertebrates, such as cephalopod mollusks and annelids (see chapter 34), and all vertebrates have a closed circulatory system.

In annelids such as earthworms, a dorsal vessel contracts rhythmically to function as a pump. Blood is pushed through five small connecting arteries, which also function as pumps, to a ventral vessel, which transports the blood posteriorly until it eventually reenters the dorsal vessel. Smaller vessels branch from each artery to supply the tissues of the earthworm with oxygen and nutrients and to remove waste products.

Learning Outcomes Review 49.2

In invertebrates, open circulatory systems pump hemolymph into tissues, from which it then drains into a central cavity. Closed circulatory systems move fluid in a loop to and from a muscular pumping region such as a heart. Hemolymph (invertebrates) is identical to the extracellular fluid in the tissues.

■ *In the open circulatory system of insects, how does hemolymph get back to the heart?*

49.3 Vertebrate Circulatory Systems

Learning Outcomes

1. Trace the evolution of the chambered heart from lancelets to birds and mammals.
2. Delineate the flow of blood through the circulatory system in birds and mammals.

The evolution of large and complex hearts and closed circulatory systems put a premium on efficient circulation. In response, vertebrates have evolved a remarkable set of adaptations inextricably linking circulation and respiration, which has facilitated diversification throughout aquatic and terrestrial habitats and permitted the evolution of large body size.

In fishes, more efficient circulation developed concurrently with gills

Chordates ancestral to the vertebrates are thought to have had simple tubular hearts, similar to those now seen in lancelets (see chapter 35). The heart was little more than a specialized zone of the ventral artery that was more heavily muscled than the rest of the arteries; it contracted in simple peristaltic waves.

The development of gills by fishes required a more efficient pump, and in fishes we see the evolution of a true chamber-pump heart. The fish heart is, in essence, a tube with four structures arrayed one after the other to form two pumping chambers (figure 49.5). The first two structures—the **sinus venosus** and **atrium**—form the first chamber; the second two, the **ventricle** and **conus arteriosus,** form the second chamber. The sinus venosus is the first to contract, followed by the atrium, the ventricle, and finally the conus arteriosus.

Despite shifts in the relative positions of these structures, this heartbeat sequence is maintained in all vertebrates. In fish, the electrical impulse that produces the contraction is initiated in the sinus venosus; in other vertebrates, the electrical impulse is initiated by a structure homologous to the sinus venosus—the **sinoatrial (SA) node.**

After blood leaves the conus arteriosus, it moves through the gills, becoming oxygenated. Blood leaving the gills then flows through a network of arteries to the rest of the body, finally returning to the sinus venosus. This simple loop has one serious limitation: in passing through the capillaries in the gills, blood pressure drops significantly. This slows circulation from the gills to the rest of the body and can limit oxygen delivery to tissues.

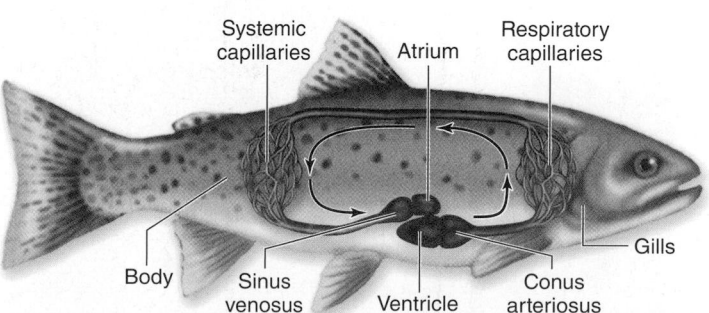

Figure 49.5 The heart and circulation of a fish. Diagram of a fish heart, showing the structures in series with each other (sinus venosus; atrium; ventricles; conus arteriosus) that form two pumping chambers. Blood is pumped by the ventricle through the gills and then to the body. Blood rich in oxygen (oxygenated) is shown in red; blood low in oxygen (deoxygenated) is shown in blue.

In amphibians and most reptiles, lungs required a separate circulation

The advent of lungs in amphibians (see chapter 48) involved a major change in the pattern of circulation, a second pumping circuit. After blood is pumped by the heart through the *pulmonary arteries* to the lungs, it does not go directly to the tissues of the body. Instead, it is returned via the *pulmonary veins* to the heart. Blood leaves the heart a second time to be circulated through other tissues. This system is termed **double circulation:** One system, the **pulmonary circulation,** moves blood between heart and lungs, and another, the **systemic circulation,** moves blood between the heart and the rest of the body.

Amphibian circulation

Optimally, oxygenated blood from lungs would go directly to tissues, rather than being mixed in the heart with deoxygenated blood returning from the body. The amphibian heart has two structural features that significantly reduce this mixing (figure 49.6). First, the atrium is divided into two chambers: The right atrium receives deoxygenated blood from the systemic circulation, and the left atrium receives oxygenated blood from the lungs. These two types of blood, therefore, do not mix in the atria.

Because an amphibian heart has a single ventricle, the separation of the pulmonary and systemic circulations is incomplete. The extent of mixing when the contents of each atrium enter the ventricle is reduced by internal channels created by recesses in the ventricular wall. The conus arteriosus is partially separated by a dividing wall, which directs deoxygenated blood into the pulmonary arteries and oxygenated blood into the *aorta,* the major artery of the systemic circulation.

Amphibians living in water can obtain additional oxygen by diffusion through their skin. Thus, amphibians have a *pulmocutaneous circuit* that sends blood to both the lungs and the skin. Cutaneous respiration is also seen in turtles.

Reptilian circulation

Among reptiles, additional modifications have further reduced the mixing of blood in the heart. In addition to having two separate atria, reptiles have a septum that partially subdivides the ventricle. This separation is complete in one order of reptiles, the crocodilians, which have two separate ventricles divided by a complete septum (see the following section). Another change in the circulation of reptiles is that the conus arteriosus has become incorporated into the trunks of the large arteries leaving the heart.

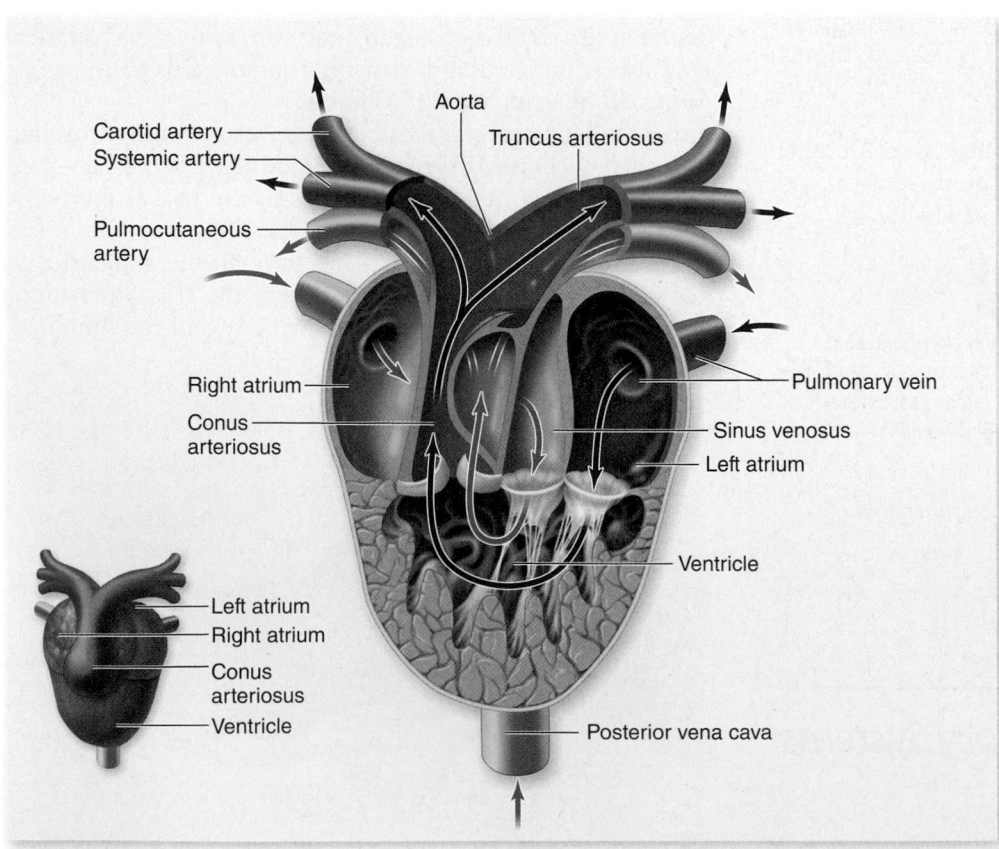

a.

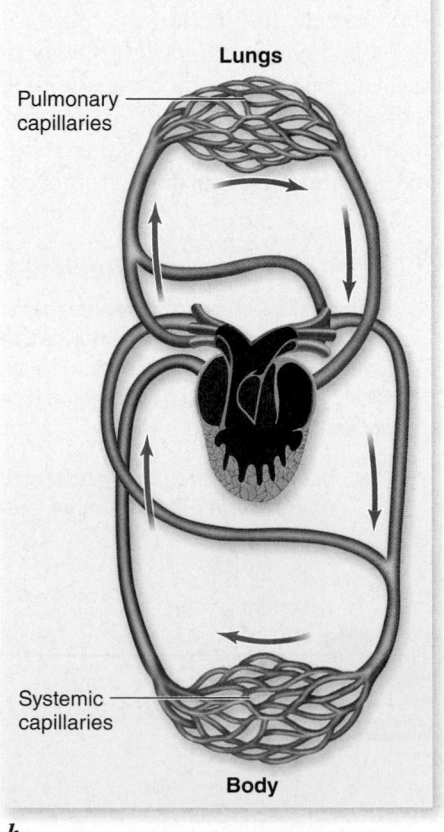

b.

Figure 49.6 The heart and circulation of an amphibian. *a.* The frog has a three-chambered heart with two atria but only one ventricle, which pumps blood both to the lungs and to the body. *b.* Despite the potential for mixing, the oxygenated and deoxygenated bloods (red and blue lines, respectively) mix little as they are pumped to the body and lungs. Oxygenation of blood also occurs by gas exchange through the skin.

Mammals, birds, and crocodilians have two completely separated circulatory systems

Mammals, birds, and crocodilians have a four-chambered heart with two separate atria and two separate ventricles (figure 49.7). The hearts of birds and crocodiles exhibit some differences, but overall are quite similar, which is not surprising given their close evolutionary relationship (figure 49.8). However, the extreme similarity of the hearts of birds and mammals—so alike that a single illustration can suffice for both (see figure 49.7)—is a remarkable case of convergent evolution (see figure 49.8).

In a four-chambered heart, the right atrium receives deoxygenated blood from the body and delivers it to the right ventricle, which pumps the blood to the lungs. The left atrium receives oxygenated blood from the lungs and delivers it to the left ventricle, which pumps the oxygenated blood to the rest of the body (see figure 49.7).

The heart in these vertebrates is a two-cycle pump. Both atria fill with blood and simultaneously contract, emptying their blood into the ventricles. Both ventricles also contract at the same time, pushing blood simultaneously into the pulmonary and systemic circulations.

The increased efficiency of the double circulatory system in mammals and birds is thought to have been important in the evolution of endothermy. More efficient circulation is necessary to support the high metabolic rate required for maintenance of internal body temperature about a set point.

Throughout the evolutionary history of the vertebrate heart, the sinus venosus has served as a pacemaker, the site where the impulses that initiate the heartbeat originate. Although the sinus venosus constitutes a major chamber in the fish heart, it is reduced in size in amphibians and is further reduced in reptiles. In mammals and birds, the sinus venosus is no longer present as a separate chamber, although some of its tissue remains in the wall of the right atrium. This tissue, the sinoatrial (SA) node, is still the site where each heartbeat originates as detailed later in the chapter.

Learning Outcomes Review 49.3

The chordate heart has evolved from a muscular region of a vessel, to the two-chambered heart of fish, the three-chambered heart of amphibians and most reptiles, and the four-chambered heart of crocodilians, birds, and mammals. Deoxygenated blood travels in the pulmonary circuit from the right atrium into the right ventricle and then to the lungs; it returns to the left atrium. Oxygenated blood travels in the systemic circuit from the left atrium into the left ventricle and then to the body; it returns to the right atrium.

■ *What is the physiological advantage of having separated ventricles?*

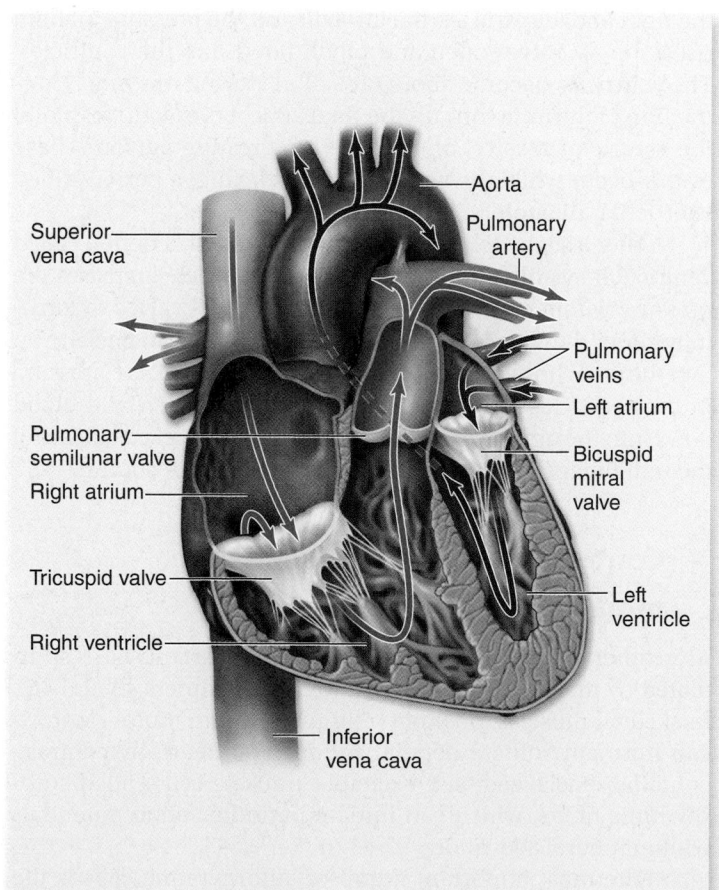

a.

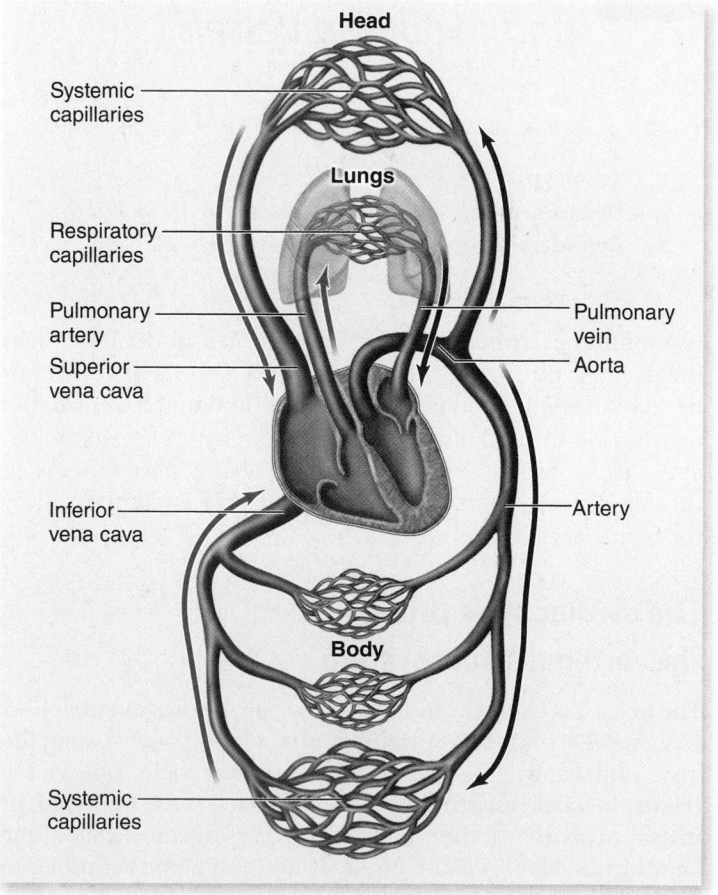

b.

Figure 49.7 The heart and circulation of mammals and birds. *a.* The path of blood through the four-chambered heart. *b.* The right side of the heart receives deoxygenated blood and pumps it to the lungs; the left side of the heart receives oxygenated blood and pumps it to the body. In this way, the pulmonary and systemic circulations are kept completely separate.

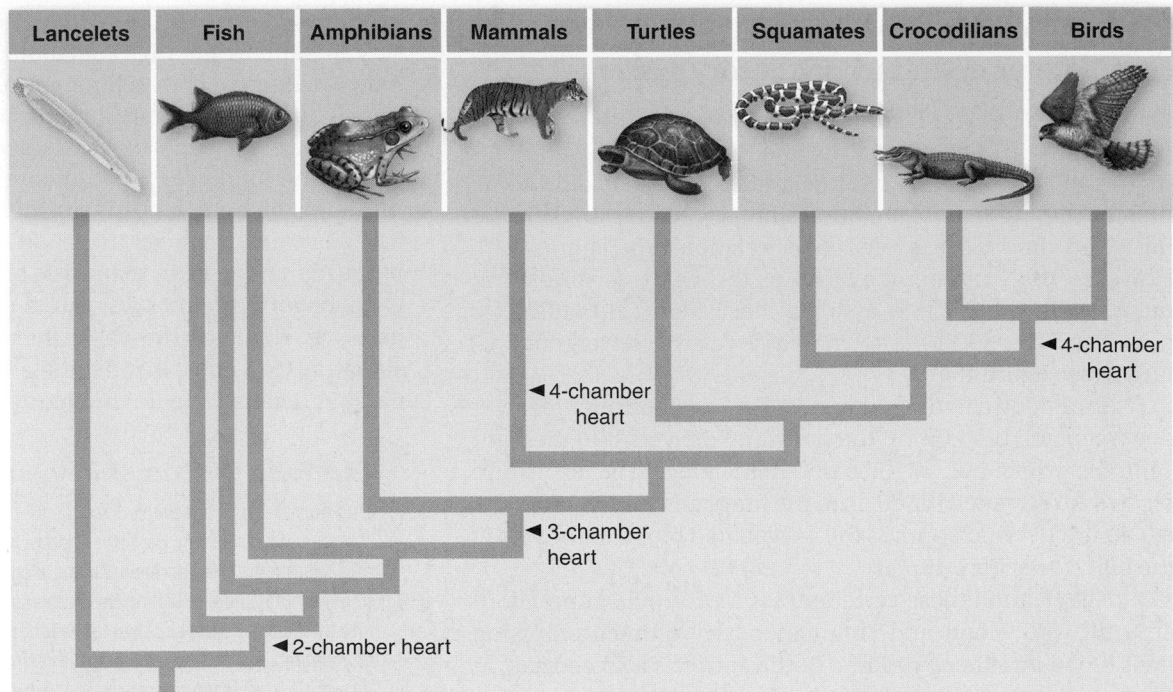

Figure 49.8
Evolution of the heart in vertebrates. Despite their similarity, the four-chambered hearts of mammals and birds evolved convergently.

Lancelets | Fish | Amphibians | Mammals | Turtles | Squamates | Crocodilians | Birds

◄4-chamber heart

◄4-chamber heart

◄3-chamber heart

◄2-chamber heart

49.4 The Four-Chambered Heart and the Blood Vessels

Learning Outcomes

1. Explain the cardiac cycle.
2. Describe the role of autorhythmic cells of the SA node.
3. Define blood pressure and how it is measured.

As mentioned earlier, the heart of mammals, birds, and crocodilians goes through two contraction cycles, one of atrial contraction to send blood to the ventricles, and one of ventricular contraction to send blood to the pulmonary and systemic circuits. These two contractions plus the resting period between these make up the complete **cardiac cycle** encompassed by the heartbeat.

The cardiac cycle drives the cardiovascular system

The heart has two pairs of valves. One pair, the **atrioventricular (AV) valves,** maintains unidirectional blood flow between the atria and ventricles. The AV valve on the right side is the **tricuspid valve,** and the AV valve on the left is the **bicuspid,** or **mitral, valve.** Another pair of valves, together called the **semilunar valves,** ensure one-way flow out of the ventricles to the arterial systems. The **pulmonary valve** is located at the exit of the right ventricle, and the **aortic valve** is located at the exit of the left ventricle. These valves open and close as the heart goes through its cycle. The closing of these valves produces the "lub-dub" sounds heard with a stethoscope.

The cardiac cycle is portrayed in figure 49.9. It begins as blood returns to the resting heart through veins that empty into the right and left atria. As the atria fill and the pressure in them rises, the AV valves open and blood flows into the ventricles. The ventricles become about 80% filled during this time. Contraction of the atria tops up the final 20% of the 80 mL of blood the ventricles receive, on average, in a resting person. These events occur while the ventricles are relaxing, a period called ventricular **diastole.**

After a slight delay, the ventricles contract, a period called ventricular **systole.** Contraction of each ventricle increases the pressure within each chamber, causing the AV valves to forcefully close (the "lub" sound), preventing blood from backing up into the atria. Immediately after the AV valves close, the pressure in the ventricles forces the semilunar valves open and blood flows into the arterial systems. As the ventricles relax, closing of the semilunar valves prevents backflow (the "dub" sound).

Contraction of heart muscle is initiated by autorhythmic cells

As in other types of muscle, contraction of heart muscle is stimulated by membrane depolarization (see chapters 43 and 46). In skeletal muscles, only nerve impulses from motor neurons can normally initiate depolarization. The heart, by contrast, contains specialized "self-excitable" muscle cells called autorhythmic fibers, which can initiate periodic action potentials without neural activation.

The most important group of autorhythmic cells is the sinoatrial (SA) node, described earlier (figure 49.10). Located in the wall of the right atrium, the SA node acts as a pacemaker for the rest of the heart by producing spontaneous action potentials at a faster rate than other autorhythmic cells. These spontaneous action potentials are due to a constant leakage of

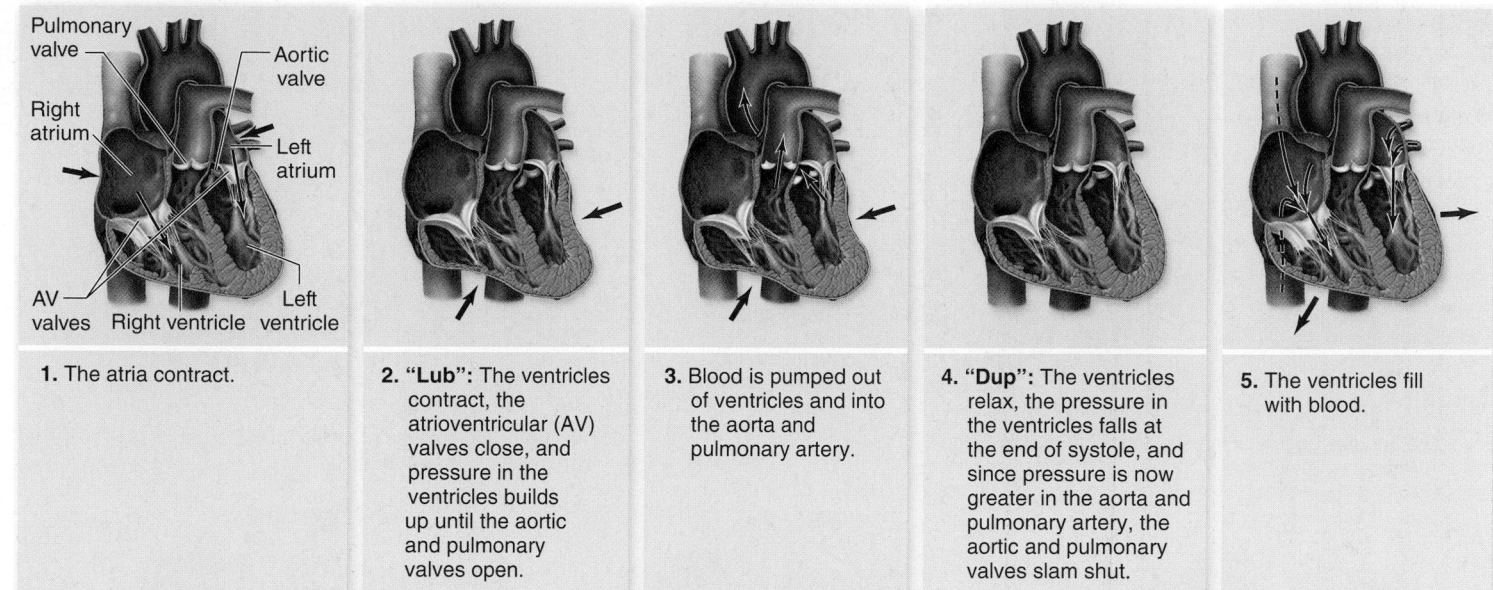

1. The atria contract.

2. "Lub": The ventricles contract, the atrioventricular (AV) valves close, and pressure in the ventricles builds up until the aortic and pulmonary valves open.

3. Blood is pumped out of ventricles and into the aorta and pulmonary artery.

4. "Dup": The ventricles relax, the pressure in the ventricles falls at the end of systole, and since pressure is now greater in the aorta and pulmonary artery, the aortic and pulmonary valves slam shut.

5. The ventricles fill with blood.

a.

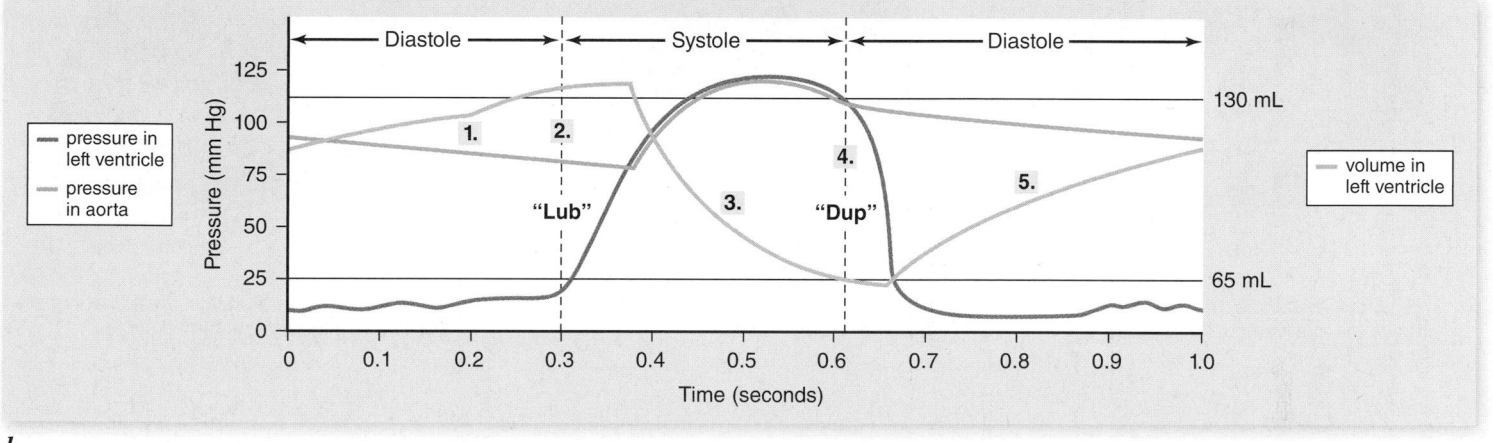

b.

Figure 49.9 The cardiac cycle. *a.* Contraction and relaxation of the atria and ventricles moves blood through the heart. *b.* Blood pressure and volume changes through the cardiac cycle, shown here for the left ventricle.

Na^+ ions into the cell that depolarize the membrane. When the threshold is reached, an action potential occurs. At the end of the action potential, the membrane is again below threshold and the process begins again. The cells of the SA node generate an action potential every 0.6 sec, equivalent to about 100 a minute. As we will see later in the chapter, the autonomic nervous system can modulate this rate.

Each depolarization initiated by this pacemaker is transmitted through two pathways: one to the cardiac muscle fibers of the left atrium, and the other to the right atrium and the atrioventricular (AV) node. Once initiated, depolarizations spread quickly from one muscle fiber to another in a wave that envelops the right and left atria nearly simultaneously. The rapid spread of depolarization is made possible because special conducting fibers are present and because the cardiac muscle cells are coupled by groups of gap junctions located within intercalated disks (see chapter 43).

A sheet of connective tissue separating the atria from the ventricles blocks the spread of excitation through muscle fibers from one chamber to the other. The AV node provides the only pathway for conduction of the depolarization from the atria to the ventricles. The fibers of the AV node slow down the conduction of the depolarizing signals, delaying the contraction of the ventricle by about 0.1 sec. This delay permits the atria to finish contracting and emptying their blood into the ventricles before the ventricles contract.

From the AV node, the wave of depolarization is conducted rapidly over both ventricles by a network of fibers called the atrioventricular bundle, or bundle of His. These fibers relay the depolarization to Purkinje fibers, which directly stimulate the myocardial cells of the left and right ventricles, causing their almost simultaneous contraction.

The stimulation of myocardial cells produces an action potential that leads to contraction. Contraction is controlled by Ca^{2+} and the troponin/tropomyosin system similar to skeletal muscle (see chapter 46), but the shape of the action potential is different. The initial rising phase due to an influx of Na^+ from voltage-gated Na^+ channels is followed by a plateau phase that leads to more sustained contraction. The plateau phase is due to the opening of voltage-gated Ca^{2+} channels. The resulting

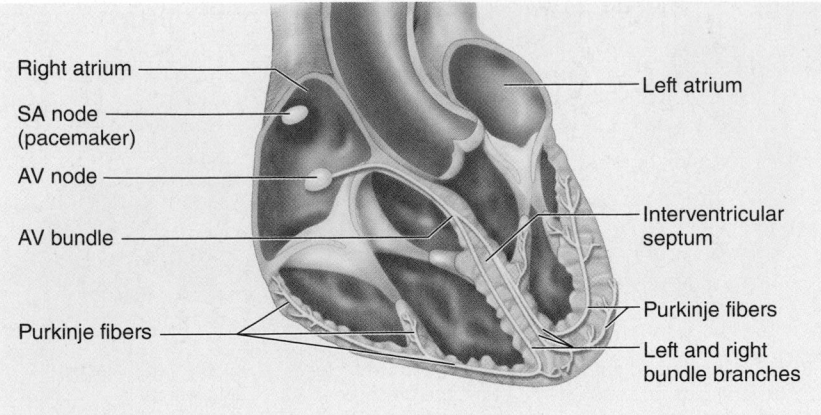

Right atrium

SA node (pacemaker)

AV node

AV bundle

Purkinje fibers

Left atrium

Interventricular septum

Purkinje fibers

Left and right bundle branches

1. The impulse begins at the SA node and travels to the AV node.

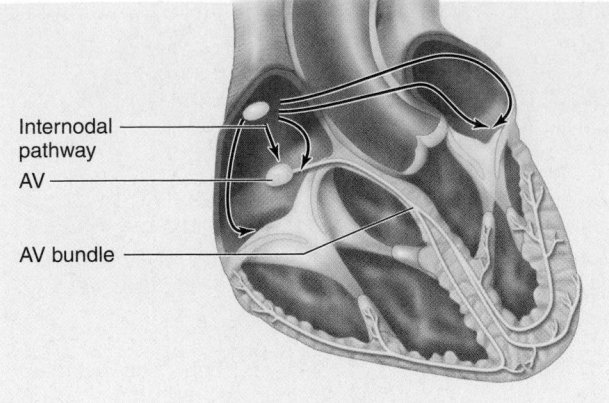

Internodal pathway

AV

AV bundle

2. The impulse is delayed at the AV node. It then travels to the AV bundle.

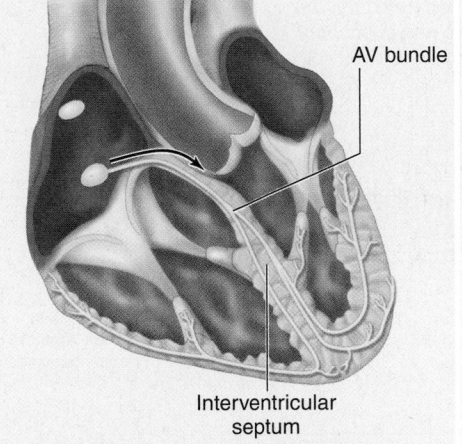

AV bundle

Interventricular septum

3. From the AV bundle, the impulse travels down the interventricular septum.

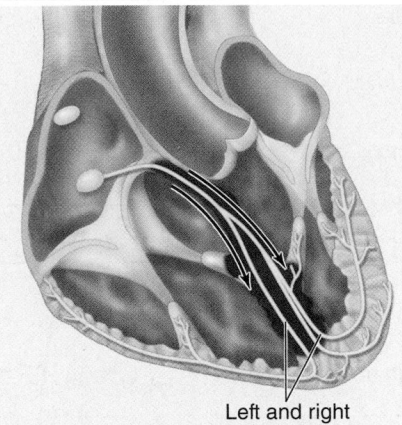

Left and right bundle branches

4. The impulse spreads to branches from the interventricular septum.

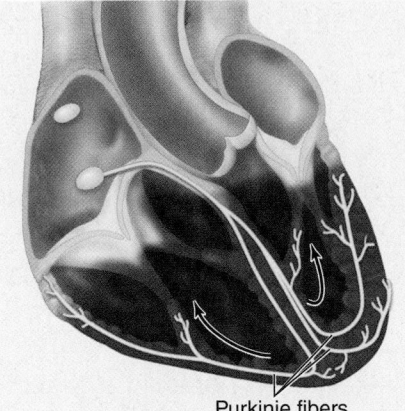

Purkinje fibers

5. Finally reaching the Purkinje fibers, the impulse is distributed throughout the ventricles.

Figure 49.10 The path of electrical excitation in the heart. The events occurring during contraction of the heart are correlated with the measurement of electrical activity by an electrocardiogram (ECG, also called EKG). The depolarization/ contraction of the atrium is shown in green above and corresponds to the P wave of the ECG (also in green). The depolarization/ contraction of the ventricle is shown in red above and corresponds to the QRS wave of the ECG (also in red). The T wave on the ECG corresponds to the repolarization of the ventricles. The atrial repolarization is masked by the QRS wave.

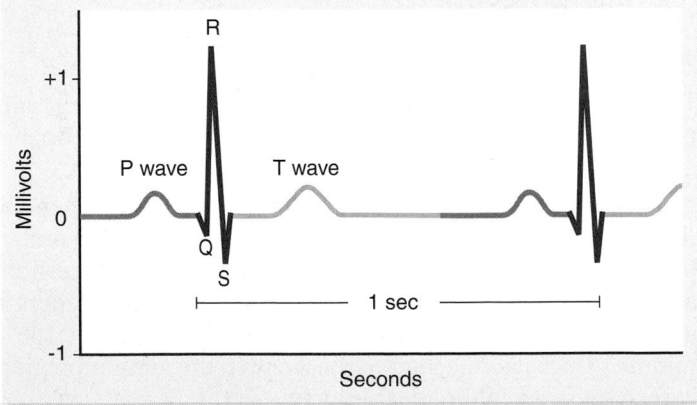

influx of Ca^{2+} keeps the membrane depolarized when the Na^+ channels inactivate. This, in turn, leads to more voltage-gated Ca^{2+} channels in the sarcoplasmic reticulum opening. The additional Ca^{2+} in the cytoplasm produces a more sustained contraction. The Ca^{2+} is removed from the cytoplasm by a pump in the sarcoplasmic reticulum similar to skeletal muscle, and an additional carrier in the plasma membrane pumps Ca^{2+} into the interstitial space.

The electrical activity of the heart can be recorded from the surface of the body with electrodes placed on the limbs and chest. The recording, called an electrocardiogram (ECG or EKG), shows how the cells of the heart depolarize and repolarize during the cardiac cycle (see figure 49.10). Depolarization causes contraction of the heart, and repolarization causes relaxation.

The first peak in the recording, P, is produced by the depolarization of the atria, and is associated with atrial systole. The second, larger peak, QRS, is produced by ventricular depolarization; during this time, the ventricles contract (ventricular systole). The last peak, T, is produced by ventricular repolarization; at this time, the ventricles begin diastole.

Arteries and veins branch to and from all parts of the body

The right and left **pulmonary arteries** deliver oxygen-depleted blood from the right ventricle to the right and left lungs. As previously mentioned, the **pulmonary veins** return oxygenated blood from the lungs to the left atrium of the heart.

The **aorta** and all its branches are systemic arteries, carrying oxygen-rich blood from the left ventricle to all parts of the body. The **coronary arteries** are the first branches off the aorta; these supply oxygenated blood to the heart muscle itself (see figure 49.7b). Other systemic arteries branch from the aorta as it makes an arch above the heart and as it descends and traverses the thoracic and abdominal cavities.

The blood from the body's organs, now lower in oxygen, returns to the heart in the systemic veins. These eventually empty into two major veins: the **superior vena cava,** which drains the upper body, and the **inferior vena cava,** which drains the lower body. These veins empty into the right atrium, completing the systemic circulation.

The flow of blood through the arteries, capillaries, and veins is driven by the pressure generated by ventricular contraction. The ventricles must contract forcefully enough to move the blood through the entire circulatory system.

Arterial blood pressure can be measured

As the ventricles contract, great pressure is generated within them and transferred through the arteries once the aortic valve opens. The pulse that you can detect in your wrist or neck results from changes in pressure as elastic arteries expand and contract with the periodic blood flow. Doctors use blood pressure as a general indicator of cardiovascular health because a variety of conditions can cause increases or decreases in pressure.

A *sphygmomanometer* measures the blood pressure in the brachial artery found on the inside part of the arm, above the elbow (figure 49.11). A cuff wrapped around the upper part of the arm is tightened enough to stop the flow of blood to the lower part of the arm. As the cuff is slowly loosened, eventually the blood pressure produced by the heart is greater than the constricting pressure of the cuff and blood begins pulsating through the artery, producing a sound that can be detected using a stethoscope. The point at which this pulsing sound begins marks the peak pressure, or **systolic pressure,** at which ventricles are contracting. As the cuff is loosened further, the point is reached where the pressure of the cuff is lower than the blood pressure throughout the cardiac cycle, at which time the blood vessel is no longer distorted and the pulsing sound stops. This point marks the minimum pressure between heartbeats or **diastolic pressure,** at which the ventricles are relaxed.

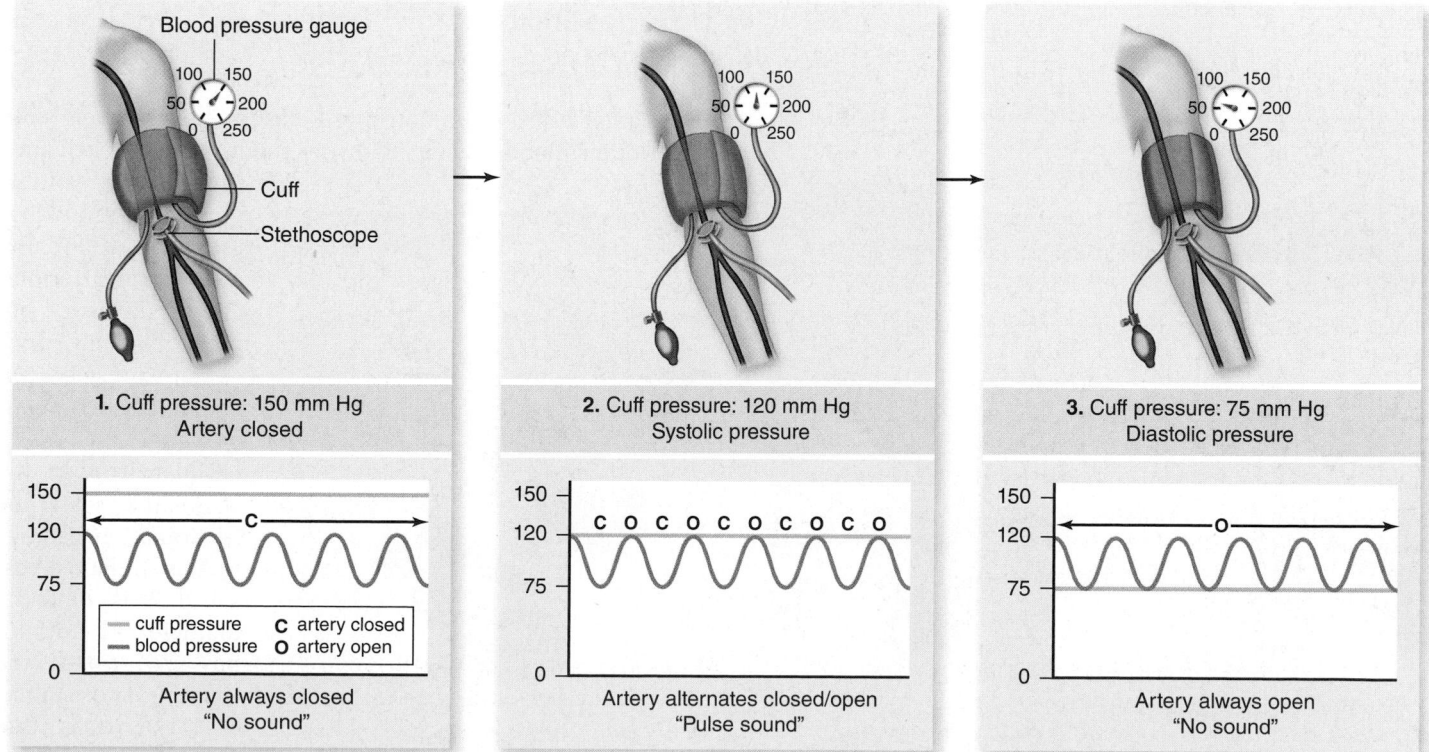

Figure 49.11 Measurement of blood pressure. The blood pressure cuff is tightened to stop the blood flow through the brachial artery. As the cuff is loosened, the maximal (systolic) pressure becomes greater than the cuff pressure and blood can momentarily pass through, producing a pulse that can be heard with a stethoscope. The pressure at this point is recorded as the systolic pressure. As the cuff pressure continues to drop, blood pressure is greater than cuff pressure for larger portions of the cardiac cycle. Eventually, even the minimum pressure during the cycle is greater than the cuff pressure, at which time the blood vessel is no longer distorted and silent laminar flow returns, replacing the pulsing sound. The diastolic pressure is recorded as the pressure at which a sound is no longer heard.

The blood pressure is written as a ratio of systolic over diastolic pressure, and for a healthy person in his or her twenties, a typical blood pressure is 120/75 (measured in millimeters of mercury, or mm Hg). The medical condition called **hypertension** (high blood pressure) is defined as either a systolic pressure greater than 150 mm Hg or a diastolic pressure greater than 90 mm Hg.

Learning Outcomes Review 49.4

The cardiac cycle consists of systole and diastole; the ventricles contract at systole and relax at diastole. The SA node in the right atrium initiates waves of depolarization that stimulate first the atria and then travel to the AV node, which stimulates the ventricles. Blood pressure is expressed as the ratio of systolic pressure over diastolic pressure and is measured with a device called a sphygmomanometer.

■ *What would happen without a delay between auricular and ventricular contraction?*

49.5 *Characteristics of Blood Vessels*

Learning Outcomes

1. *Describe the four tissue layers in blood vessels.*
2. *Explain the distinctions among arteries, capillaries, and veins.*
3. *Describe how the lymphatic system operates.*

You already know that blood leaves the heart through vessels known as **arteries.** These continually branch, forming a hollow "tree" that enters each organ of the body. The finest, microscopic branches of the arterial tree are the **arterioles.** Blood from the arterioles enters the **capillaries,** an elaborate latticework of very narrow, thin-walled tubes. After traversing the capillaries, the blood is collected into microscopic **venules,** which lead to larger vessels called **veins,** and these carry blood back to the heart.

Larger vessels are composed of four tissue layers

Arteries, arterioles, veins, and venules all have the same basic structure (figure 49.12). The innermost layer is an epithelial sheet called the *endothelium.* Covering the endothelium is a thin layer of elastic fibers, a smooth muscle layer, and a connective tissue layer. The walls of these vessels, therefore, are thick enough to significantly reduce exchange of materials between the blood and the tissues outside the vessels.

The walls of capillaries, in contrast, are composed only of endothelium, so molecules and ions can leave the blood plasma by diffusion, by filtration through pores between the cells of the capillary walls, and by transport through the endothelial cells. Therefore, exchange of gases and metabolites between the blood and the interstitial fluids and cells of the body takes place through the capillaries.

Arteries and arterioles have evolved to withstand pressure

The larger arteries contain more elastic fibers in their walls than other blood vessels, allowing them to recoil each time they receive a volume of blood pumped by the heart. Smaller arteries and arterioles are less elastic, but their relatively thick smooth muscle layer enables them to resist bursting.

The narrower the vessel, the greater the frictional resistance to flow. In fact, a vessel that is half the diameter of another has *16 times* the frictional resistance. The reason is that resistance to blood flow increases to the fourth power in inverse proportion to the amount that the radius of the vessel decreases. Therefore,

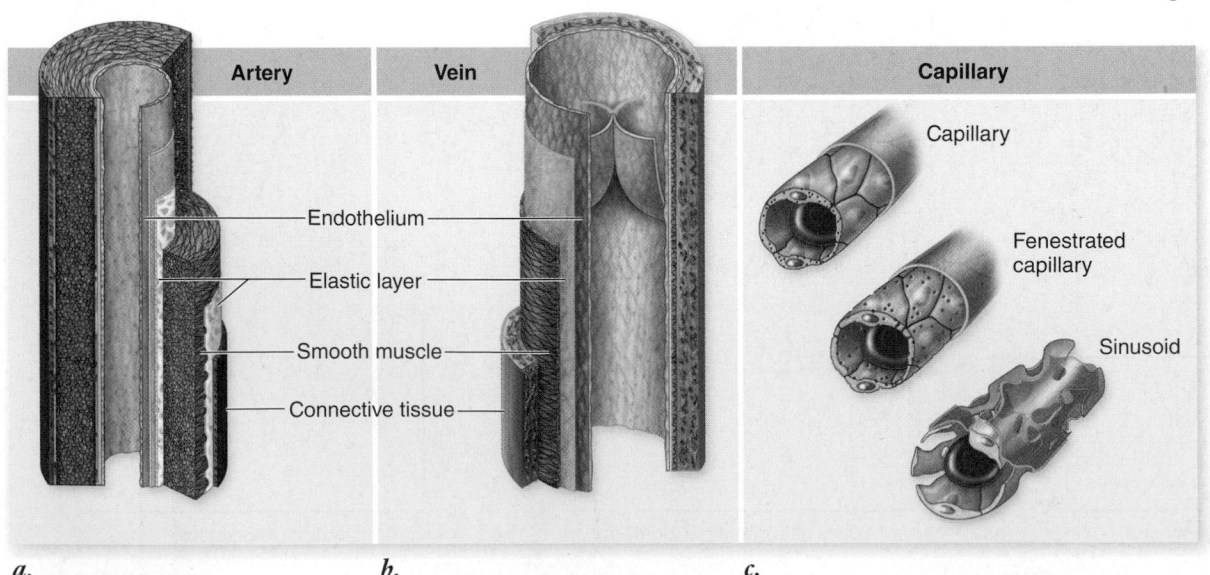

a.   *b.*   *c.*

Figure 49.12 The structure of blood vessels. Arteries (*a*) and veins (*b*) have the same tissue layers, but the smooth muscle layer in arteries is much thicker and there are two elastic layers. *c.* Capillaries are composed of only a single layer of endothelial cells. (Not to scale.)

🔍 **Data analysis** If a blood vessel has a radius of 60 μm and a capillary in the same system has a radius of 4 μm, what is the difference in resistance to blood flow in the two?

within the arterial tree, the small arteries and arterioles provide the greatest resistance to blood flow.

Contraction of the smooth muscle layer of the arterioles results in **vasoconstriction,** which greatly increases resistance and decreases flow. Relaxation of the smooth muscle layer results in **vasodilation,** decreasing resistance and increasing blood flow to an organ. Chronic vasoconstriction of the arterioles can result in hypertension, or high blood pressure.

Vasoconstriction and vasodilation are important means of regulating body heat in both ectotherms and endotherms (figure 49.13). By increasing blood flow to the skin, an animal can increase the rate of heat exchange, which is beneficial for gaining or losing heat. Conversely, shunting blood away from the skin is effective when an animal needs to minimize heat exchange, as might happen in cold weather.

Capillaries form a vast network for exchange of materials

The huge number and extensive branching of the capillaries ensure that every cell in the body is within 100 micrometers (μm)

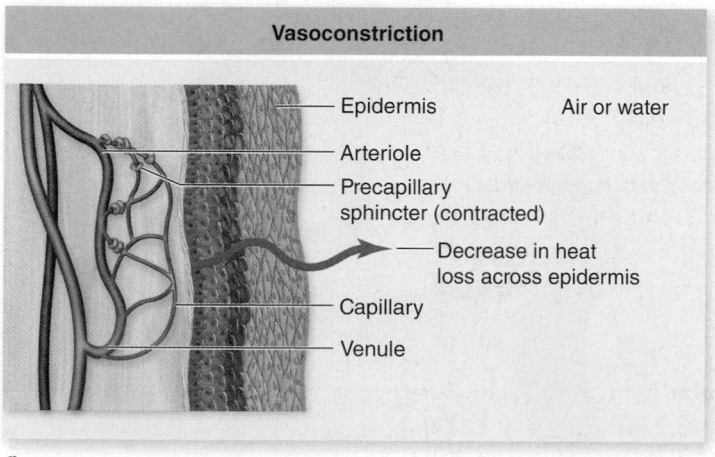

a.

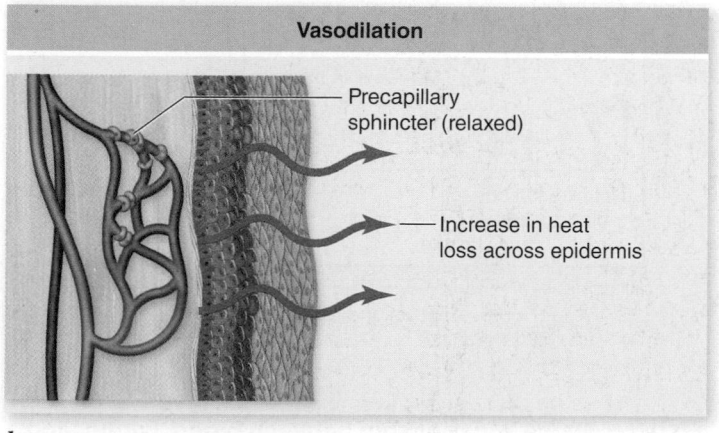

b.

Figure 49.13 Regulation of heat exchange. The amount of heat gained or lost at the body's surface can be regulated by controlling the flow of blood to the surface. *a.* Constriction of surface blood vessels limits flow and heat loss when the animal is warmer than the surrounding air; when the animal is cooler than the surrounding air (not shown here), constriction minimizes heat gain; (*b*) dilation of these vessels increases flow and heat exchange.

of a capillary. On the average, capillaries are about 1 mm long and 8 μm in diameter; this diameter is only slightly larger than a red blood cell (5 to 7 μm in diameter). Despite the close fit, normal red blood cells are flexible enough to squeeze through capillaries without difficulty.

Although each capillary is very narrow, so many of them exist that the capillaries have the greatest *total* cross-sectional area of any other type of vessel. Consequently, blood moving through capillaries goes more slowly and has more time to exchange materials with the surrounding extracellular fluid. By the time the blood reaches the end of a capillary, it has released some of its oxygen and nutrients and picked up carbon dioxide and other waste products. Blood loses pressure and velocity as it moves through the arterioles and capillaries, but as cross-sectional area decreases in the venous side, velocity increases.

Venules and veins have less muscle in their walls

Venules and veins have the same tissue layers as arteries, but they have a thinner layer of smooth muscle. Less muscle is needed because the pressure in the veins is only about one-tenth that in the arteries. Most of the blood in the cardiovascular system is contained within veins, which can expand to hold additional amounts of blood. You can see the expanded veins in your feet when you stand for a long time.

The venous pressure alone is not sufficient to return blood to the heart from the feet and legs, but several other sources of pressure provide help. Most significantly, skeletal muscles surrounding the veins can contract to move blood by squeezing the veins, a mechanism called the **venous pump.** Blood moves in one direction through the veins back to the heart with the help of **venous valves** (figure 49.14). When a

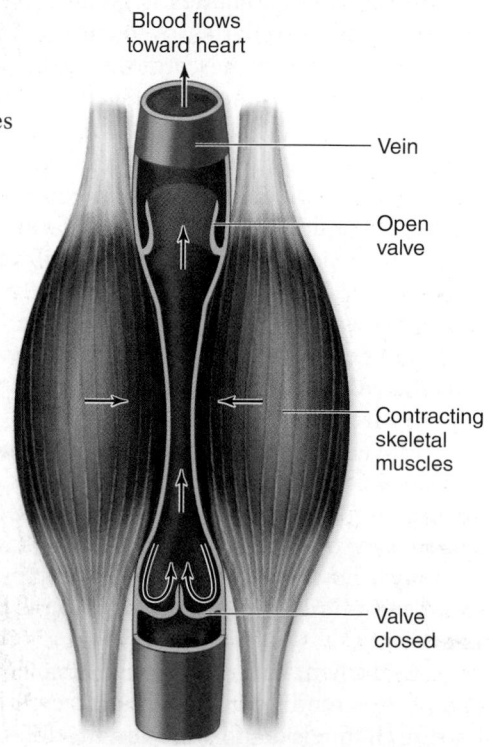

Figure 49.14 One-way flow of blood through veins. Venous valves ensure that blood moves through the veins in only one direction, back to the heart.

person's veins expand too much with blood, the venous valves may no longer work and the blood may pool in the veins. Veins in this condition are known as varicose veins.

The lymphatic system handles fluids that leave the cardiovascular system

The cardiovascular system is considered a closed system because all its vessels are connected with one another—none are simply open-ended. But a significant amount of water and solutes in the blood plasma filter through the walls of the capillaries to form the interstitial (tissue) fluid. Most of the fluid leaves the capillaries near their arteriolar ends, where the blood pressure is higher; it is returned to the capillaries near their venular ends (figure 49.15).

Fluid returns by osmosis (see chapter 5). Most of the plasma proteins cannot escape through the capillary pores because of their large size, and so the concentration of proteins in the plasma is greater than the protein concentration in the interstitial fluid. The difference in protein concentration produces an osmotic pressure gradient that causes water to move into the capillaries from the interstitial space.

High capillary blood pressure can cause too much interstitial fluid to accumulate. In pregnant women, for example, the enlarged uterus, carrying the fetus, compresses veins in the abdominal cavity, thereby adding to the capillary blood pressure in the woman's lower limbs. The increased interstitial fluid can cause swelling of the tissues, or **edema,** of the feet.

Edema may also result if the plasma protein concentration is too low. Fluids do not return to the capillaries, but remain as interstitial fluid. Low protein concentration in the plasma may be caused either by liver disease, because the liver produces most of the plasma proteins, or by insufficient dietary protein such as occurs in starvation.

Even under normal conditions, the amount of fluid filtered out of the capillaries is greater than the amount that returns to the capillaries by osmosis. The remainder does eventually return to the cardiovascular system by way of an open circulatory system called the **lymphatic system.**

The lymphatic system consists of lymphatic capillaries, lymphatic vessels, lymph nodes, and lymphatic organs, including the spleen and thymus. Excess fluid in the tissues drains into blind-ended lymph capillaries with highly permeable walls. This fluid, now called **lymph,** passes into progressively larger lymphatic vessels, which resemble veins and have one-way valves (similar to figure 49.14). The lymph eventually enters two major lymphatic vessels, which drain into the left and right subclavian veins located under the collarbones.

Movement of lymph in mammals results as skeletal muscles squeeze against the lymphatic vessels, a mechanism similar to the venous pump that moves blood through veins. In some cases, the lymphatic vessels also contract rhythmically. In many fishes, all amphibians and reptiles, bird embryos, and some adult birds, movement of lymph is propelled by **lymph hearts.**

As the lymph moves through lymph nodes and lymphatic organs, it is modified by phagocytic cells (see chapter 4) that line the channels of those organs. In addition, the lymph nodes

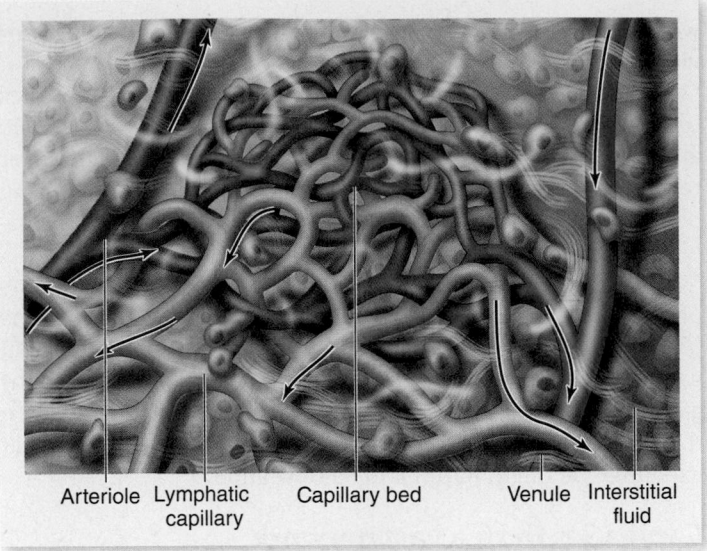

a.

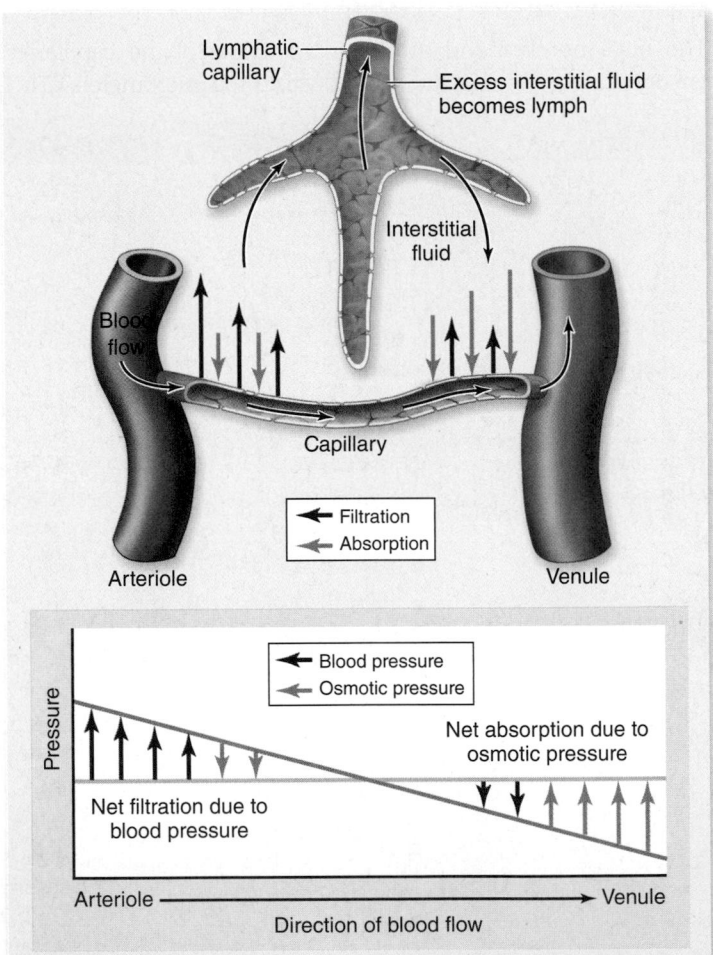

b.

Figure 49.15 Relationship between blood, lymph, and interstitial fluid. *a.* Vessels of the circulatory and lymphatic systems with arrows indicating the direction of flow of fluid in the vessels. *b.* Plasma fluid, minus proteins, is filtered out of capillaries, forming interstitial fluid that bathes tissues. Much of this fluid is returned to the capillaries by osmosis due to the higher protein concentration in plasma. Excess interstitial fluid drains into open-ended lymphatic capillaries, which ultimately return the fluid to the cardiovascular system.

and lymphatic organs contain *germinal centers,* where the activation and proliferation of lymphocytes occurs.

Cardiovascular diseases affect the delivery system

Cardiovascular diseases are the leading cause of death in the United States; more than 80 million people have some form of cardiovascular disease. Many disease conditions result from problems in arteries, such as blockage or rupture.

Atherosclerosis, or **hardening of the arteries,** is an accumulation within the arteries of fatty materials, abnormal amounts of smooth muscle, deposits of cholesterol or fibrin, or various kinds of cellular debris. These accumulations cause an increase in vascular resistance, which impedes blood flow (figure 49.16). The lumen (interior) of the artery may be further narrowed by a clot that forms as a result of the atherosclerosis. In the severest cases, the artery becomes completely blocked.

The accumulation of cholesterol in vessels is affected by a number of factors including total serum cholesterol and the levels of different cholesterol carrier proteins. Because cholesterol is not very water-soluble, it is carried in blood in the form of lipoprotein complexes. Two main forms are observed that differ in density: low-density lipoproteins (LDL) and high-density lipoproteins (HDL)—often called "bad cholesterol" and "good cholesterol," respectively. The reason for this is that HDLs tend to take cholesterol out of circulation, transporting it to the liver for elimination, and LDL is the carrier that brings cholesterol to all cells in the body. The problem arises when cells have enough cholesterol. This causes a reduction in the amount of LDL receptors, leading to high levels of circulating LDLs, which can end up being deposited in blood vessels.

Atherosclerosis is promoted by genetic factors, smoking, hypertension (high blood pressure), and the effects of cholesterol just discussed. Stopping smoking is the single most effective action a smoker can take to reduce the risk of atherosclerosis.

Arteriosclerosis occurs when calcium is deposited in arterial walls. It tends to occur when atherosclerosis is severe. Not only do such arteries have restricted blood flow, but they also lack the ability to expand as normal arteries do. This decrease in flexibility forces the heart to work harder because blood pressure increases to maintain flow.

Heart attacks (myocardial infarctions) are the main cause of cardiovascular deaths in the United States, accounting for about one-fifth of all deaths. Heart attacks result from an insufficient supply of blood to one or more parts of the heart muscle, which causes myocardial cells in those parts to die. Heart attacks may be caused by a blood clot forming somewhere in the coronary arteries and may also result if an artery is blocked by atherosclerosis. Recovery from a heart attack is possible if the portion of the heart that was damaged is small enough that the heart can still contract as a functional unit.

Angina pectoris, which literally means "chest pain," occurs for reasons similar to those that cause heart attacks, but it is not as severe. The pain may occur in the heart and often also in the left arm and shoulder. Angina pectoris is a warning sign that the blood supply to the heart is inadequate but is still sufficient to avoid myocardial cell death.

Strokes are caused by an interference with the blood supply to the brain. They may occur when a blood vessel bursts in the brain (hemorrhagic stroke), when blood flow in a cerebral artery is blocked by a blood clot or by atherosclerosis (ischemic stroke). The effects of a stroke depend on the severity of the damage and where in the brain the stroke occurs.

Learning Outcomes Review 49.5

The four layers of blood vessels are (1) endothelium, (2) an elastic layer, (3) smooth muscle, and (4) connective tissue. In contrast, capillaries have only endothelium. Arteries have more muscle in their walls than do veins to help withstand greater pressure; large arteries also have more elastic fibers for recoil. Excess interstitial fluid, called lymph, is returned to the cardiovascular system via the lymphatic system, a one-way system.

■ *What is the connection between the lymphatic and circulatory systems?*

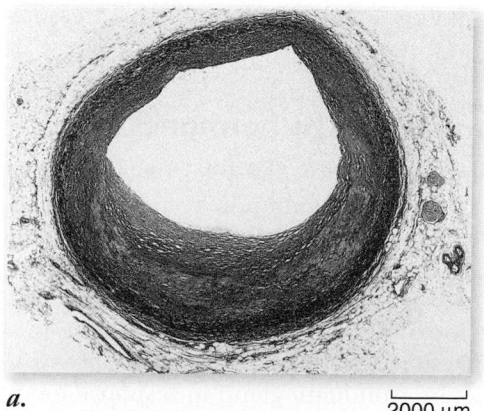
a.
2000 μm

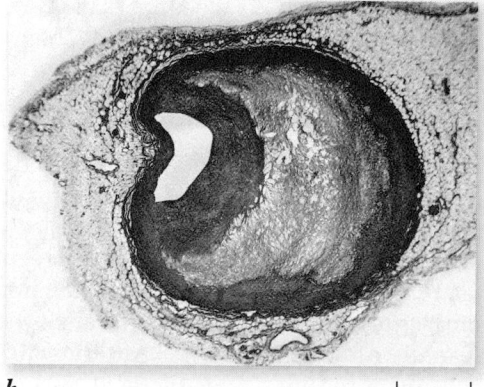
b.
2500 μm

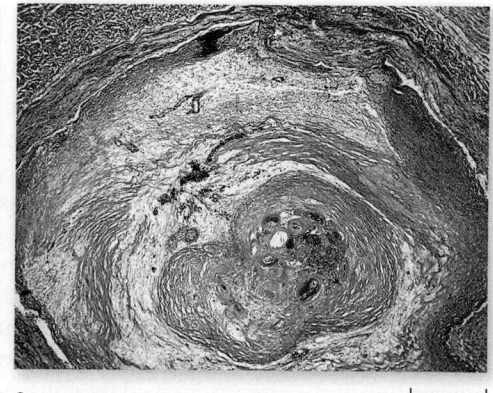
c.
1000 μm

Figure 49.16 Atherosclerosis. *a.* The coronary artery shows only minor blockage. *b.* The artery exhibits severe atherosclerosis—much of the passage is blocked by buildup on the interior walls of the artery. *c.* The coronary artery is essentially completely blocked.

49.6 Regulation of Blood Flow and Blood Pressure

Learning Outcomes

1. *Describe how exertion affects cardiac output.*
2. *Explain how hormones regulate blood volume.*

Although the autonomic nervous system does not initiate the heartbeat, it does modulate its rhythm and force of contraction. In addition, several mechanisms regulate characteristics of the cardiovascular system, including cardiac output, blood pressure, and blood volume.

The nervous system may speed up or slow down heart rate

Heart rate is under the control of the autonomic nervous system. The cardiac center of the medulla oblongata (a part of the hindbrain; see chapter 43) consists of two neuronal centers that modulate heart rate. The **cardioacceleratory center** sends signals by way of the sympathetic cardiac accelerator nerves to the SA node, AV node, and myocardium. These nerves secrete norepinephrine, which increases the heart rate. Sympathetic nervous system stimulation can also increase contractility of the heart muscle itself, thus ejecting more blood per contraction (stroke volume).

The **cardioinhibitory center** sends signals via the parasympathetic fibers in the vagus nerve to the SA and AV nodes. The vagus nerve secretes acetylcholine, which inhibits the development of action potentials and so slows the heart down.

Cardiac output increases with exertion

Cardiac output is the volume of blood pumped by each ventricle per minute. It is calculated by multiplying the heart rate by the *stroke volume,* which is the volume of blood ejected by each ventricle per beat. For example, if the heart rate is 72 beats per minute and the stroke volume is 70 mL, the cardiac output is approximately 5 L/min, which is about average in a resting human.

Cardiac output increases during exertion because of an increase in both heart rate and stroke volume. When exertion begins, such as running, the heart rate increases up to about 100 beats per minute to provide more oxygen to cells in the body. As movement becomes more intense, skeletal muscles squeeze on veins more vigorously, returning blood to the heart more rapidly. In addition, the ventricles contract more strongly, so they empty more completely with each beat.

During exercise, the cardiac output increases to a maximum of about 25 L/min in an average young adult. Although the cardiac output has increased fivefold, not all organs receive five times the blood flow; some receive more, others less. Arterioles in some organs, such as in the digestive system, constrict, while the arterioles in the working muscles and heart dilate.

 Data analysis If a person's heart at rest typically beats at 72 beats per minute with a stroke volume of 70 mL/beat, how much increase in beat number and volume is needed to increase cardiac output to 25 L/min?

The baroreceptor reflex maintains homeostasis in blood pressure

The arterial blood pressure (BP) depends on two factors: the cardiac output (CO) and the resistance (R) to blood flow in the vascular system. This relationship can be expressed as:

$$BP = CO \times R$$

An increased blood pressure, therefore, could be produced by an increase in either heart rate or blood volume (because both increase the cardiac output), or by vasoconstriction, which increases the resistance to blood flow. Conversely, blood pressure falls if the heart rate slows or if the blood volume is reduced—for example, by dehydration or excessive bleeding (hemorrhage).

Changes in arterial blood pressure are detected by **baroreceptors** located in the arch of the aorta and in the carotid arteries (see chapter 44). These sensors are stretch receptors sensitive to expansion and contraction of arteries. When the baroreceptors detect a fall in blood pressure, the number of impulses to the cardiac center is decreased, resulting in increased sympathetic stimulation and decreased parasympathetic stimulation of the heart and other targets. This increases heart rate and stroke volume to amplify cardiac output. This also causes vasoconstriction of blood vessels in the skin and viscera, raising resistance. These combine to increase blood pressure, closing the feedback loop in this direction (figure 49.17, top).

When baroreceptors detect a rise in blood pressure, the number of impulses to the cardiac center is increased. This has the opposite effect of decreasing sympathetic stimulation and increasing parasympathetic simulation of the heart. This lowers heart rate and stroke volume to reduce cardiac output. The cardiac center also sends signals causing vasodilation of blood vessels in the skin and viscera, lowering resistance. These combine to decrease blood pressure, closing the feedback loop in this direction. Thus, the baroreceptor reflex forms a negative feedback loop responding to changes in blood pressure (figure 49.17, bottom).

Blood volume is regulated by hormones

Blood pressure depends in part on the total blood volume because this can affect the cardiac output. A decrease in blood volume decreases blood pressure, if all else remains equal. Blood volume regulation involves the effects of four hormones: (1) antidiuretic hormone, (2) aldosterone, (3) atrial natriuretic hormone, and (4) nitric oxide.

Antidiuretic hormone (ADH), also called **vasopressin,** is secreted by the posterior-pituitary gland in response to an increase in the osmolarity of the blood plasma (see chapter 45). Dehydration, for example, causes the blood volume to decrease. Osmoreceptors in the hypothalamus promote thirst

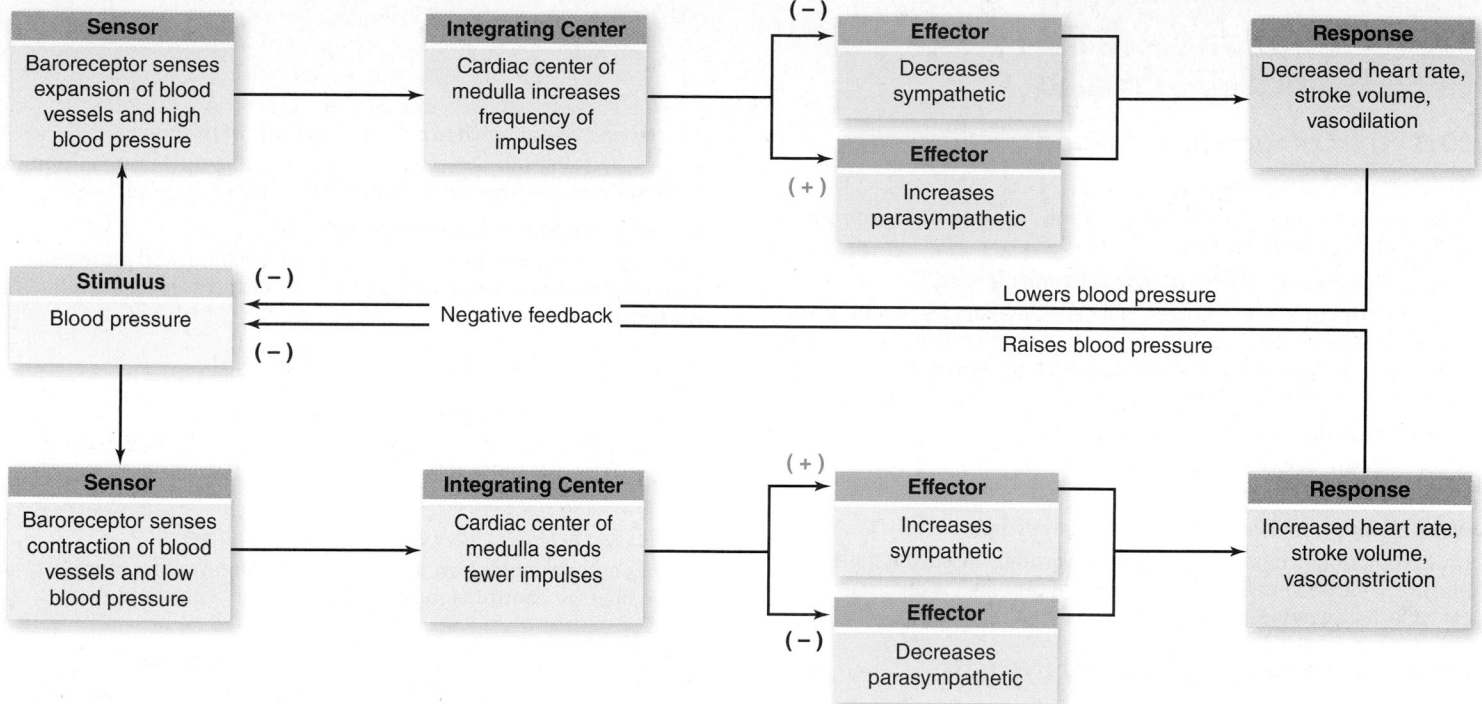

Figure 49.17 Baroreceptor negative feedback loops control blood pressure. Baroreceptors form the afferent portion of a feedback loop controlling blood pressure. The frequency of nerve impulses from these stretch receptors correlates with blood pressure. This information is processed in the cardiac center of the medulla. The efferent portion of the loop involves sympathetic and parasympathetic nerves that innervate the heart. This control can raise or lower heart rate and stroke volume to raise and lower blood pressure in response to baroreceptors signaling.

and stimulate ADH secretion from the posterior-pituitary gland. ADH, in turn, stimulates the kidneys to retain more water in the blood, excreting less in the urine. A dehydrated person thus drinks more and urinates less, helping to raise the blood volume and restore homeostasis.

Whenever the kidneys experience a decreased blood flow, a group of kidney cells initiate the release of an enzyme known as renin into the blood. Renin activates a blood protein, angiotensin, which stimulates vasoconstriction throughout the body while stimulating the adrenal cortex to secrete **aldosterone.** This steroid hormone acts on the kidneys to promote the retention of Na^+ and water in the blood (see chapter 45).

When excess Na^+ is present, less aldosterone is secreted by the adrenals, so that less Na^+ is retained by the kidneys. Na^+ excretion in the urine is promoted by another hormone, **atrial natriuretic hormone.** This hormone is secreted by the right atrium of the heart in response to stretching caused by an increased blood volume. The action of atrial natriuretic hormone completes a negative feedback loop, lowering the blood volume and pressure.

Nitric oxide (NO) is a gas produced by endothelial cells of blood vessels. As described in chapter 45, it is one of a number of paracrine regulators of blood vessels. In solution, NO passes outward through the cell layers of the vessel, causing the smooth muscles that encase it to relax and the blood vessels to dilate (become wider). For over a century, heart patients have been prescribed nitroglycerin to relieve chest pain, but only now has it become clear that nitroglycerin acts by releasing nitric oxide.

Learning Outcomes Review 49.6

Cardiac output is the heart rate times the heart's stroke volume. As exertion increases, cardiac output increases to meet the body's demands. Blood pressure depends on cardiac output and the resistance to blood flow due to constriction of the arteries. The blood volume is regulated by antidiuretic hormone, aldosterone, and atrial natriuretic hormone; nitric oxide causes vasodilation that lessens resistance.

■ *What are the connections between regulation of heart rate and breathing rate?*

49.1 The Components of Blood

Blood plasma is a fluid matrix.

Plasma is 92% water plus nutrients, hormones, ions, plasma proteins, and wastes (figure 49.1).

Formed elements include circulating cells and platelets.

Blood cells include erythrocytes (red cells), leukocytes (white cells), and platelets. Erythrocytes contain hemoglobin for oxygen transport, and leukocytes are part of the immune system. Platelets help initiate blood clotting (figure 49.3).

Formed elements arise from stem cells.

Blood cells are derived from pluripotent stem cells in bone marrow by hematopoiesis (figure 49.2).

Blood clotting is an example of an enzyme cascade.

Upon initiation of clotting, fibrinogen, normally dissolved in the plasma, is turned into fibrin, an insoluble protein, via an enzyme cascade. As a wound heals, the clot must be dissolved.

49.2 Invertebrate Circulatory Systems

Open circulatory systems move fluids in a one-way path.

Sponges pass water through channels, and cnidarians circulate water through a gastrovascular cavity. Small animals can use body cavity fluids for circulation.

Closed circulatory systems move fluids in a loop.

Closed systems have a distinct circulatory fluid, such as blood, enclosed in vessels and transported in a loop.

49.3 Vertebrate Circulatory Systems

In fishes, more efficient circulation developed concurrently with gills.

Fishes have a linear heart with two pumping chambers to increase efficiency of blood flow through the gills; from the gills, the blood moves into the rest of the body (figure 49.5).

In amphibians and most reptiles, lungs required a separate circulation.

Pulmonary circulation pumps blood to the lungs, and systemic circulation pumps blood to the body.

Amphibian hearts have two atria that separate blood flow to the lungs and body, and a single ventricle (figure 49.6). The heart of most reptiles has a septum that partially divides the ventricle, reducing mixing of blood from the atria.

Mammals, birds, and crocodilians have two completely separated circulatory systems.

The four-chambered heart has two ventricles (figure 49.7). The extreme similarity between the heart of mammals and birds is an example of convergent evolution.

49.4 The Four-Chambered Heart and the Blood Vessels

The cardiac cycle drives the cardiovascular system.

The unidirectional flow of blood through the heart is maintained by two atrioventricular valves (figure 49.9). During diastole ventricles relax and atria contract; during systole ventricles contract.

Contraction of heart muscle is initiated by autorhythmic cells.

Contraction is initiated by the SA node, a natural pacemaker, and impulses then travel to the AV node (figure 49.10).

Arteries and veins branch to and from all parts of the body.

Arteries and arterioles carry oxygenated blood to the body; veins and venules return deoxygenated blood to the heart (figure 49.7).

Arterial blood pressure can be measured.

A sphygmomanometer measures the peak (systolic) and minimum (diastolic) blood pressure. Blood pressure is expressed as the ratio of systolic to diastolic.

49.5 Characteristics of Blood Vessels

Larger vessels are composed of four tissue layers.

Arteries and veins consist of endothelium, elastic fibers, smooth muscle, and connective tissues (figure 49.12). Capillaries have only one layer of endothelium.

Arteries and arterioles have evolved to withstand pressure.

Arteries and arterioles have a thicker muscular layer and more elastic fibers to control blood flow and to recoil with changes in blood pressure.

Capillaries form a vast network for exchange of materials.

Capillaries are the region of the circulatory system where exchange takes place with the body's tissues (figure 49.13).

Venules and veins have less muscle in their walls.

The return of blood to the heart through veins is facilitated by skeletal muscle contractions and one-way valves (figure 49.14).

The lymphatic system handles fluids that leave the cardiovascular system.

Fluid from plasma filters out of capillaries, then returns via the separate, one-way lymphatic system (figure 49.15). The lymphatic system connects with the blood circulation at the subclavian veins.

Cardiovascular diseases affect the delivery system.

Atherosclerosis is an accumulation of fatty materials in arteries; it is one cause of a heart attack, which results from an insufficient supply of blood to heart muscle. Strokes are caused by blockage of the blood supply to the brain.

49.6 Regulation of Blood Flow and Blood Pressure

The nervous system may speed up or slow down heart rate.

Norepinephrine from sympathetic neurons increases heart rate; acetylcholine from parasympathetic neurons decreases the rate.

Cardiac output increases with exertion.

Both heart rate and stroke volume increase with exertion.

The baroreceptor reflex maintains homeostasis in blood pressure.

Arterial blood pressure is monitored by baroreceptors in the aortic arch and carotid arteries, which relay impulses to the cardiac center (figure 49.17).

Blood volume is regulated by hormones.

Blood volume regulation and arterial resistance involves the effects of four hormones: (1) antidiuretic hormone, (2) aldosterone, (3) atrial natriuretic hormone, and (4) nitric oxide.

UNDERSTAND

1. An ECG measures
   a. changes in electrical potential during the cardiac cycle.
   b. Ca^{2+} concentration of the ventricles in diastole.
   c. the force of contraction of the atria during systole.
   d. the volume of blood being pumped during the contraction cycle.

2. Systole is vitally important to heart function and begins in the heart with the
   a. activation of the AV node.
   b. activation of the SA node.
   c. opening of the voltage-gated potassium gates.
   d. opening of the semilunar valves.

3. Which of the following is the correct sequence of events in the circulation of blood?
   a. Heart ⟶ arteries ⟶ arterioles ⟶ capillaries ⟶ venules ⟶ lymph ⟶ heart
   b. Heart ⟶ arteries ⟶ arterioles ⟶ capillaries ⟶ veins ⟶ venules ⟶ heart
   c. Heart ⟶ arteries ⟶ arterioles ⟶ capillaries ⟶ venules ⟶ veins ⟶ heart
   d. Heart ⟶ arterioles ⟶ arteries ⟶ capillaries ⟶ venules ⟶ veins ⟶ heart

4. Which of the following statements is not true?
   a. Only arteries carry oxygenated blood.
   b. Both arteries and veins have a layer of smooth muscle.
   c. Both arteries and veins branch out into capillary beds.
   d. Precapillary sphincters regulate blood flow through capillaries.

5. The lymphatic system is like the circulatory system in that they both
   a. have nodes that filter out pathogens.
   b. have a network of arteries.
   c. have capillaries.
   d. are closed systems.

6. Which pairing of structure and function is incorrect?
   a. Erythrocytes: oxygen transport
   b. Platelets: blood clotting
   c. Plasma: waste transport
   d. All of the choices are correct.

7. When a sphygmomanometer is used,
   a. blood pulses through the vein when systolic pressure is greater than the pressure caused by the cuff.
   b. pulsing ceases when blood pressure falls below the systolic pressure.
   c. blood does not move through the vein when cuff pressure is greater than maximal blood pressure.
   d. cuff pressure stops decreasing when it equals systolic pressure.

8. Blood clots are made of
   a. fibrin.
   b. fibrinogen.
   c. prothrombin.
   d. All of the choices are correct.

APPLY

1. In vertebrate hearts, atria contract from the top, and ventricles contract from the bottom. How is this accomplished?
   a. Depolarization from the SA node proceeds across the atria from the top; depolarization from the AV node is carried to the bottom of the ventricles before it emanates over ventricular tissue.
   b. The depolarization from the SA node is initiated from motor neurons coming down from our brain; depolarization from the AV node is initiated from motor neurons coming up from our spinal cord.
   c. Gravity carries the depolarization from the SA node down from the top of the heart; contraction of the diaphragm forces depolarization from the AV node from the bottom up.
   d. This statement is false; both contract from the bottom.

2. A molecule of CO_2 that is generated in the cardiac muscle of the left ventricle would *not* pass through which of the following structures before leaving the body?
   a. Right atrium   c. Right ventricle
   b. Left atrium   d. Left ventricle

3. The difference between the amphibian and mammal hearts is that
   a. in the amphibian heart, oxygenated and deoxygenated blood mix completely in the single ventricle.
   b. in the amphibian heart, there are two SA nodes so that contractions occur simultaneously throughout the heart.
   c. in the ventricle in the amphibian heart, internal channels reduce mixing of blood.
   d. in the amphibian heart, only the left aorta pumps oxygen obtained by diffusion through the skin.

4. Contraction of the smooth muscle layers of the arterioles
   a. increases the frictional resistance to blood flow.
   b. may be a way of increasing heat exchange through the skin.
   c. can increase blood flow to an organ.
   d. All of the choices are correct.

SYNTHESIZE

1. Humans have a number of mechanisms that help to maintain blood pressure, particularly when it falls too low. Explain how the kidney and the endocrine systems help to maintain blood pressure.

2. What is the difference among blood, lymph, and hemolymph?

3. Is the evolution of the four-chambered heart related to the evolution of endothermy?

4. What do you think are the clinical symptoms indicating that a person requires the surgical implantation of a mechanical pacemaker?

ONLINE RESOURCE

www.ravenbiology.com

Understand, Apply, and Synthesize—enhance your study with animations that bring concepts to life and practice tests to assess your understanding. Your instructor may also recommend the interactive eBook, individualized learning tools, and more.

Chapter 50

Osmotic Regulation and the Urinary System

Chapter Contents

Introduction

The majority of your body weight is actually water, but you exist in a very dehydrating environment. The kangaroo rat in the picture lives in a desert environment that is even more dehydrating and yet is so parsimonious with water that it never needs to drink; it generates sufficient water as a by-product of oxidizing its food. Fish can exist in both freshwater and marine environments, facing the challenge of either gaining or losing water, respectively. Life in these different environments is possible because elaborate mechanisms enable organisms to control the osmotic strength of their blood and extracellular fluids. The regulation of internal fluid and its composition is an example of homeostasis—the ability of living organisms to maintain internal conditions within an optimal range. In this chapter, we describe the osmoregulatory systems of a number of animals, including the mammalian urinary system. These organ systems maintain the water and ionic balance of fluids in the body.

50.1 Osmolarity and Osmotic Balance

Learning Outcomes

1. Explain the importance of osmotic balance.
2. Classify organisms based on their osmotic regulation.

Water in a multicellular animal's body is distributed between the intracellular and extracellular compartments (figure 50.1). To maintain osmotic balance, the extracellular compartment of an animal's body (including its blood plasma) must be able to take water from the environment and to excrete excess water into the environment. Inorganic ions must also be exchanged between the extracellular body fluids and the external environment to maintain homeostasis. Exchanges of water and electrolytes between the body and the external environment occur across specialized epithelial cells and, in most vertebrates, through a filtration process in the kidneys.

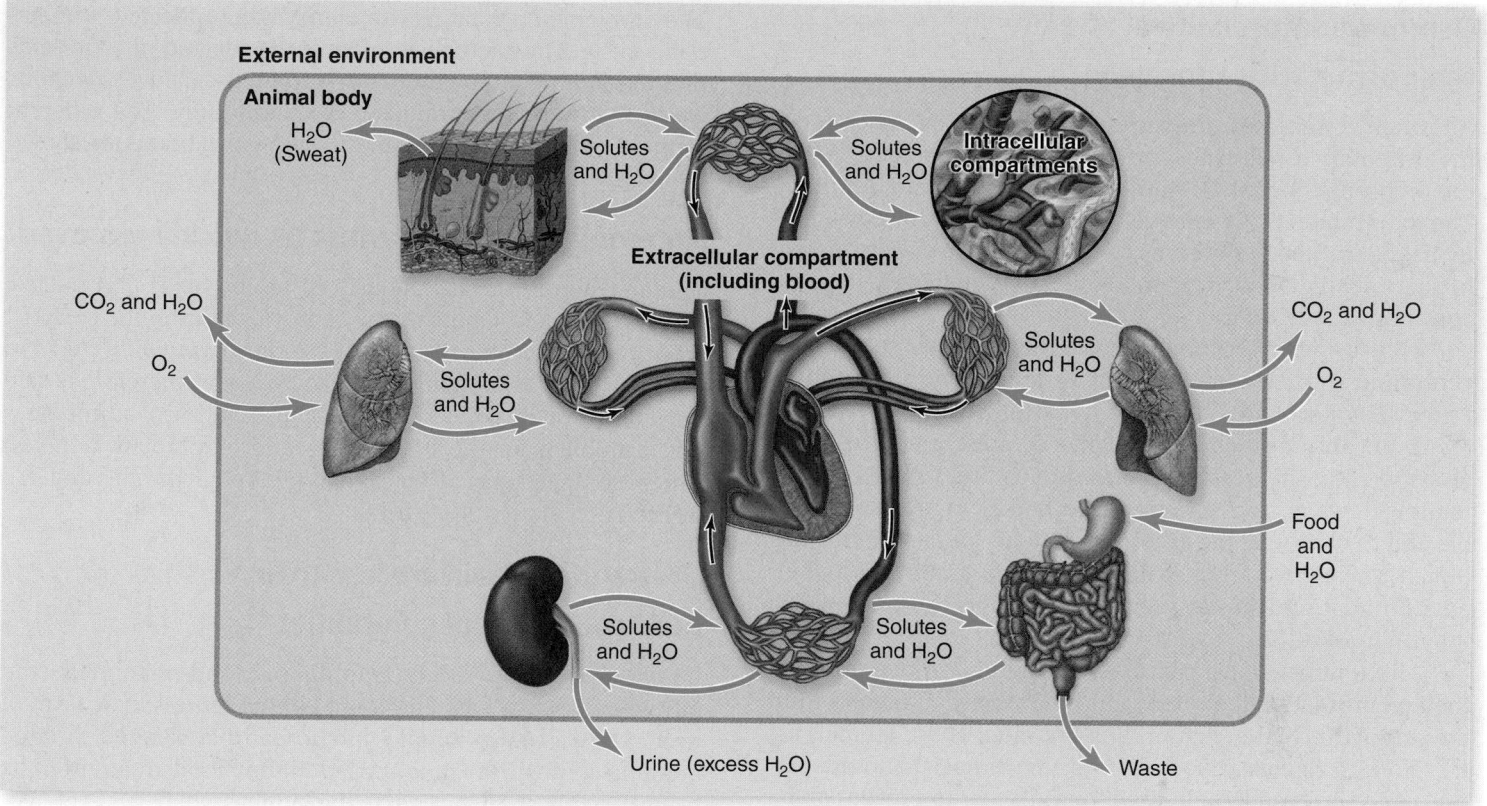

Figure 50.1 **The interaction between intracellular and extracellular compartments of the body and the external environment.** Water can be taken in from the environment or lost to the environment. Exchanges of water and solutes between the extracellular fluids of the body and the environment occur across transport epithelia, and water and solutes can be filtered out of the blood by the kidneys. Overall, the amount of water and solutes that enters and leaves the body must be balanced in order to maintain homeostasis.

Most vertebrates maintain homeostasis for both the total solute concentration of their extracellular fluids and the concentration of specific inorganic ions. Sodium (Na^+) is the major cation in extracellular fluids, and chloride (Cl^-) is the major anion. The divalent cations, calcium (Ca^{2+}) and magnesium (Mg^{2+}), the monovalent cation potassium (K^+), as well as other ions, also have important functions and are maintained at constant levels.

Osmotic pressure is a measure of concentration difference

You learned in chapter 5 that osmosis is the diffusion of water across a semipermeable membrane. Osmotic movement of water always occurs from a more dilute solution (with a lower solute concentration) to a less dilute solution (with a higher solute concentration). The osmotic pressure of a solution is a measure of its tendency to take in water by osmosis. This is the amount of pressure needed to balance the pressure created by the movement of water.

A solution with a higher concentration of solute exerts more osmotic pressure. This is measured as the **osmolarity** of a solution, the number of osmotically active moles of solute per liter of solution. Notice that osmolarity can differ from molar concentration if a substance dissociates in solution into more than one osmotically active particle. For example, a 1 molar (M) solution of sucrose is also 1 osmolar (Osm), but a

1 M solution of NaCl is 2 Osm as it dissociates into two osmotically active ions.

The **tonicity** of a solution is a measure of the ability of the solution to change the volume of a cell by osmosis. An animal cell placed in a *hypertonic* solution loses water to the surrounding solution and shrinks. In contrast, an animal cell placed in a *hypotonic* solution gains water and expands. A cell in an *isotonic* solution shows no net water movement. In medical care, isotonic solutions such as normal saline and 5% dextrose are used to bathe exposed tissues and are given as intravenous fluids.

Osmoconformers live in marine environments

The osmolarity of body fluids in most marine invertebrates is the same as that of seawater (although the concentrations of particular solutes, such as Mg^{2+}, are not equal). Because the extracellular fluids are isotonic to seawater, no osmotic gradient exists, and there is no tendency for water to leave or enter the body. Such organisms are termed **osmoconformers**—they are in osmotic equilibrium with their environment.

Among the vertebrates, only the primitive hagfish are strict osmoconformers. The sharks and their relatives in the class Chondrichthyes (cartilaginous fish) are also isotonic to seawater, even though their blood level of NaCl is lower than that of seawater; the difference in total osmolarity is made up by retaining urea, as described later on.

Osmoregulators control their osmolarity internally

All other vertebrates are **osmoregulators**—that is, animals that maintain a relatively constant blood osmolarity despite the different concentration in the surrounding environment. The maintenance of a relatively constant body fluid osmolarity has permitted vertebrates to exploit a wide variety of ecological niches. Achieving this constancy, however, requires continuous regulation.

Freshwater vertebrates have a much higher solute concentration in their body fluids than that of the surrounding water. In other words, they are hypertonic to their environment. Because of their cells' higher osmotic pressure, water tends to enter their bodies. Consequently, they have adapted to prevent water from entering their bodies as much as possible and to eliminate the excess water that does enter. In addition, freshwater vertebrates tend to lose inorganic ions to their environment and so must actively transport these ions back into their bodies.

In contrast, most marine vertebrates are hypotonic to their environment; their body fluids have only about one-third the osmolarity of the surrounding seawater. These animals are therefore in danger of losing water by osmosis, and adaptations have evolved to help them retain water to prevent dehydration. They do this by drinking seawater and eliminating the excess ions through kidneys and gills.

The body fluids of terrestrial vertebrates clearly have a higher concentration of water than does the air surrounding them. Therefore, they tend to lose water to the air by evaporation from the skin and lungs. All reptiles, birds, and mammals, as well as amphibians during the time when they live on land, face this problem. Urinary/osmoregulatory systems have evolved in these vertebrates that help them retain water.

Learning Outcomes Review 50.1

Osmotic balance must be maintained so that tissues can carry out metabolic functions. Physiological mechanisms help most vertebrates keep blood osmolarity and ion concentrations relatively constant. Marine invertebrates are osmoconformers; their body fluids are isotonic to their environment. Most vertebrates are osmoregulators; their body fluids are either hypertonic or hypotonic compared to their environment.

- **During osmosis, does water move toward regions of higher or lower osmolarity?**

50.2 Nitrogenous Wastes: Ammonia, Urea, and Uric Acid

Learning Outcome

1. *Identify the relative toxicity and solubility of different forms of nitrogenous waste.*

The problem of regulating osmolarity is complicated by metabolic waste. The catabolism of amino acids and nucleic acids produces nitrogen-containing by-products called nitrogenous waste that must be eliminated from the body. The different forms of nitrogenous waste produced by vertebrates are shown in figure 50.2.

Ammonia is toxic and must be quickly removed

The first step in the metabolism of amino acids and nucleic acids is the deamination, that is, removal of the amino ($-NH_2$) group, and its combination with H^+ to form *ammonia* (NH_3) in the liver. Ammonia is quite toxic to cells and therefore is safe only in very dilute concentrations. The excretion of ammonia is not a problem for the bony fishes and amphibian tadpoles, which eliminate most of it by diffusion through the gills and less by excretion in very dilute urine.

Urea and uric acid are less toxic but have different solubilities

In elasmobranchs, adult amphibians, and mammals, the nitrogenous wastes are eliminated in the far less toxic form of **urea.** Urea is water-soluble and so can be excreted in large amounts in the urine. It is carried in the bloodstream from its place of synthesis in the liver to the kidneys where it is excreted.

Reptiles, birds, and insects excrete nitrogenous wastes in the form of **uric acid,** which is only slightly soluble in water. As a result of its low solubility, uric acid precipitates and thus can be excreted using very little water. Uric acid forms the pasty white material in bird droppings called *guano.* It costs the animal energy to synthesize uric acid, but this is offset by the conservation of water.

The ability to synthesize uric acid in these groups of animals is also important because their eggs are encased within shells, and nitrogenous wastes build up as the embryo grows within the egg. The formation of uric acid, although a lengthy process that requires considerable energy, produces a compound that crystallizes and precipitates. As a solid precipitate, uric acid is unable to affect the embryo's development even though it is still inside the egg.

Mammals also produce some uric acid, but it is a waste product of the degradation of purine nucleotides, not of amino acids. Most mammals have an enzyme called *uricase,* which converts uric acid into a more soluble derivative, **allantoin.** Only humans, apes, and the Dalmatian dog lack this enzyme, and they must excrete the uric acid. In humans, excessive accumulation of uric acid in the joints produces a condition known as *gout.*

Learning Outcome Review 50.2

Metabolic breakdown of amino acids and nucleic acids produces ammonia as a by-product. Bony fishes and gilled amphibians excrete ammonia; other vertebrates convert nitrogenous wastes into urea (reptiles and birds) and uric acid (mammals and adult amphibians), which are less toxic. Most mammals produce a small amount of uric acid that is broken down by uricase except in humans, apes, and the Dalmatian dog.

- **Why is nitrogenous waste problematic?**

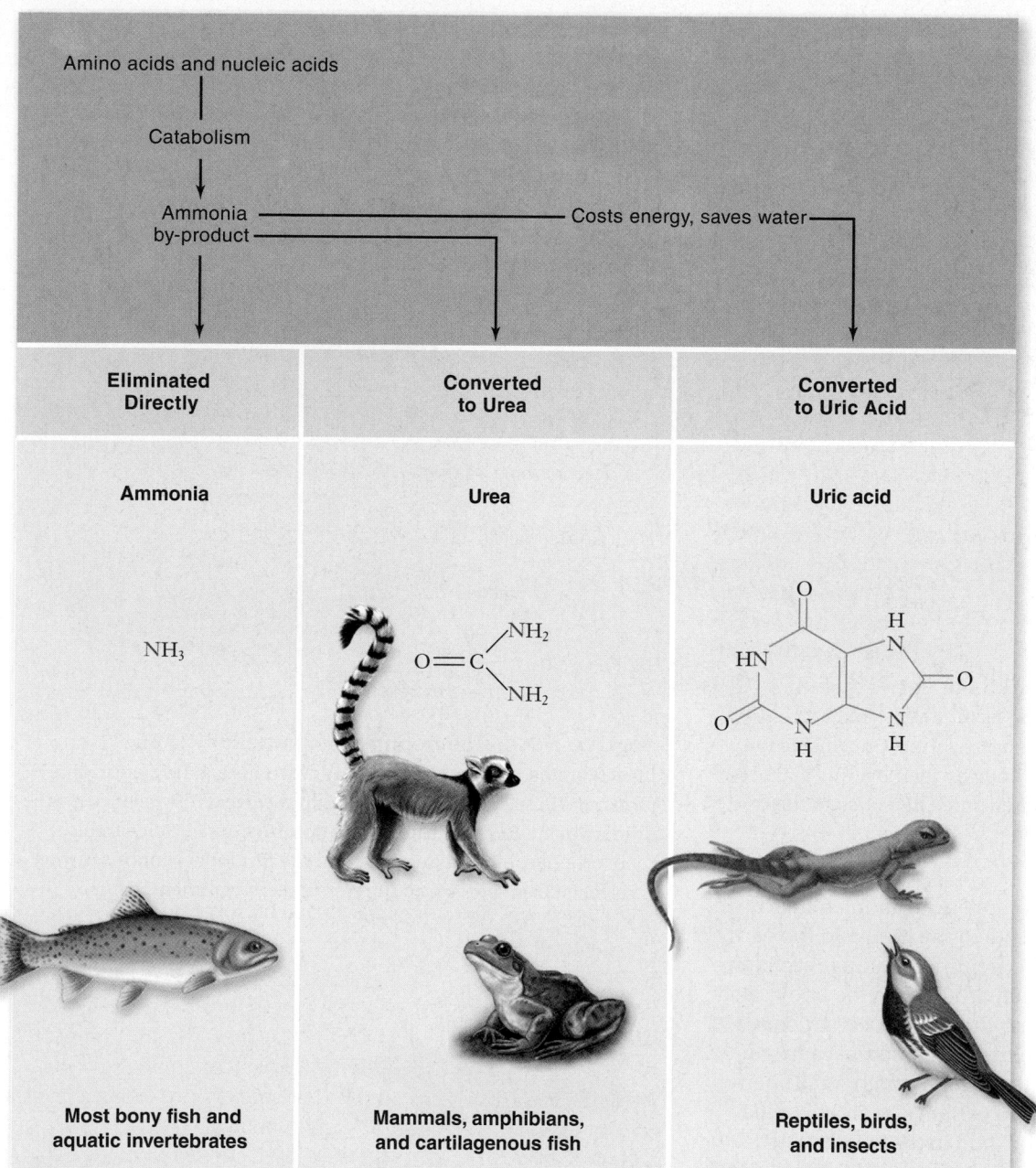

Figure 50.2
Nitrogenous wastes.
When amino acids and nucleic acids are metabolized, the immediate by-product is ammonia, which is quite toxic but can be eliminated through the gills of teleost fish. Mammals convert ammonia into urea, which is less toxic. Birds and terrestrial reptiles convert it instead into uric acid, which is insoluble in water. Production of uric acid is the most energetically expensive of the three but also saves the most water.

Amino acids and nucleic acids

↓

Catabolism

↓

Ammonia by-product ———— Costs energy, saves water ————

Eliminated Directly	Converted to Urea	Converted to Uric Acid
Ammonia	Urea	Uric acid
NH_3		
Most bony fish and aquatic invertebrates	Mammals, amphibians, and cartilagenous fish	Reptiles, birds, and insects

---

50.3 Osmoregulatory Organs

Learning Outcomes

1. Describe invertebrate osmoregulatory organs.
2. Define reabsorption and secretion.

A variety of mechanisms have evolved in animals to cope with problems of nitrogenous waste and water balance. In many animals, the removal of water or salts from the body is coupled with the removal of metabolic wastes through the excretory system. Single-celled protists employ contractile vacuoles for this purpose, as do sponges. Other multicellular animals have a system of excretory tubules (little tubes) that expel fluid and wastes from the body. In addition, more elaborate systems can be found in invertebrates. In vertebrates, the urinary system is highly complex.

Invertebrates make use of specialized cells and tubules

In flatworms, tubules called **protonephridia** branch throughout the body into bulblike flame cells (figure 50.3). Although these simple excretory structures open to the outside of the body, they do not open to the inside; rather, the movement of cilia within the flame cells must draw in fluid from the body. Water and metabolites are then reabsorbed, and the substances to be excreted are expelled through excretory pores.

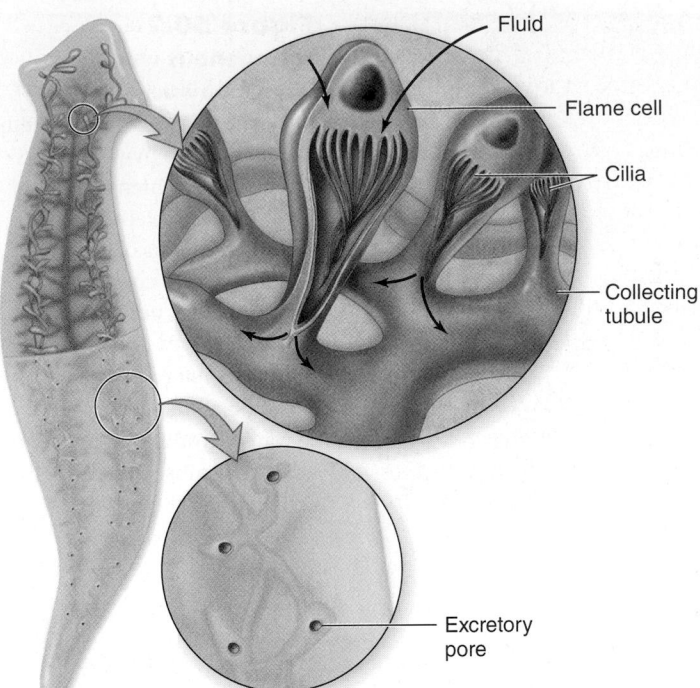

Figure 50.3 The protonephridia of flatworms. A branching system of tubules, bulblike flame cells, and excretory pores make up the protonephridia of flatworms. Cilia inside the flame cells draw in fluids from the body by their beating action. Substances are then expelled through pores that open to the outside of the body.

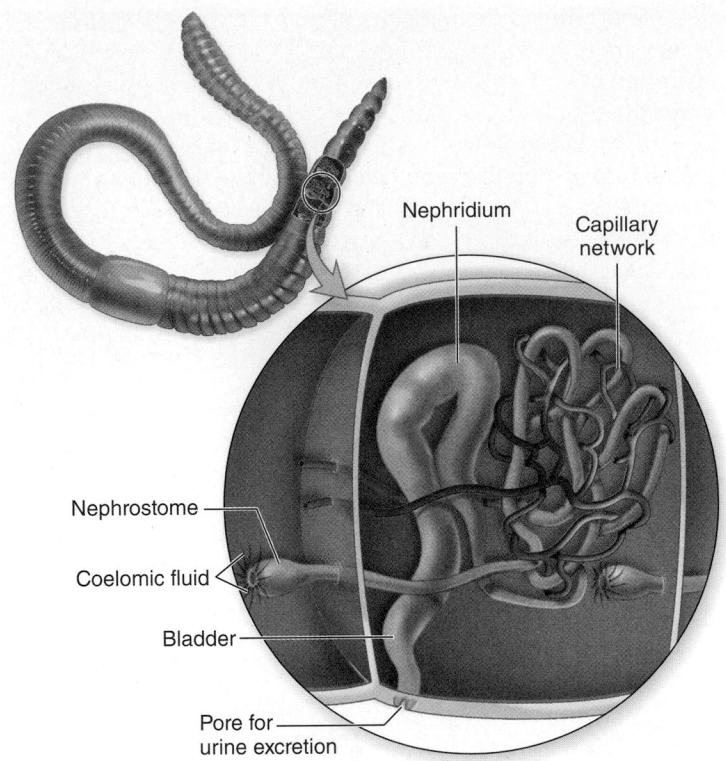

Figure 50.4 The nephridia of annelids. Most invertebrates, such as the annelid shown here, have nephridia (*orange*). These consist of tubules that receive a filtrate of coelomic fluid, which enters the funnel-like nephrostomes. Salt can be reabsorbed from these tubules, and the fluid that remains, urine, is released from pores into the external environment.

Other invertebrates have a system of tubules that open both to the inside and to the outside of the body. In the earthworm, these tubules are known as nephridia (orange structures in figure 50.4). The nephridia obtain fluid from the body cavity through a process of filtration into funnel-shaped structures called *nephrostomes*. The term *filtration* is used because the fluid is formed under pressure and passes through small openings, so that molecules larger than a certain size are excluded. This filtered fluid is isotonic to the fluid in the coelom, but as it passes through the tubules of the nephridia, NaCl is removed by active transport processes.

The general term for transport out of the tubule and into the surrounding body fluids is **reabsorption.** Because salt is reabsorbed from the filtrate, the urine excreted is more dilute than the body fluids—that is, the urine is hypotonic. The kidneys of mollusks and the excretory organs of crustaceans (called *antennal glands*) also produce urine by filtration and reclaim certain ions by reabsorption.

Insects have a unique osmoregulatory system

The excretory organs in insects are the Malpighian tubules (figure 50.5), extensions of the digestive tract that branch off anterior to the hindgut. Urine is not formed by filtration in these tubules because there is no pressure difference between the blood in the body cavity and the tubule. Instead, waste molecules and potassium (K^+) ions are secreted into the tubules by active transport.

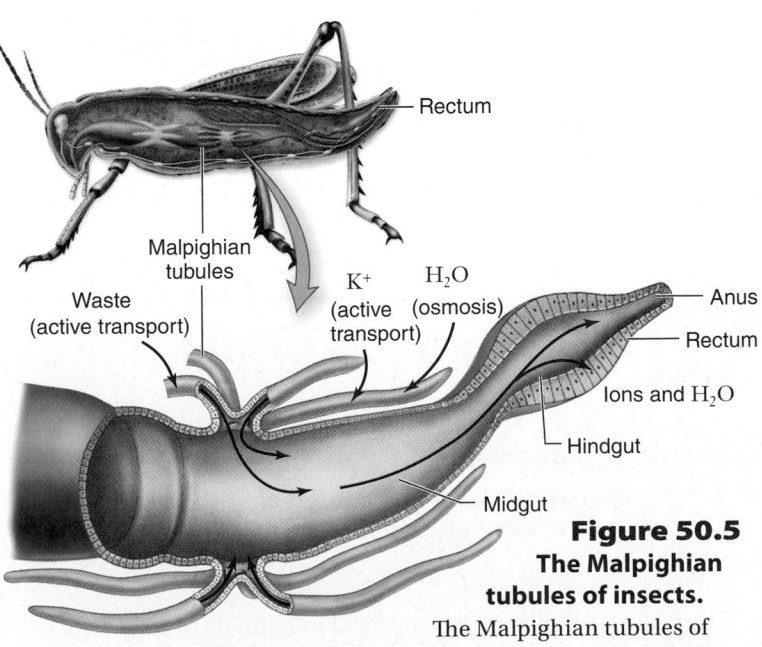

Figure 50.5 The Malpighian tubules of insects. The Malpighian tubules of insects are extensions of the digestive tract that collect water and wastes from the body's circulatory system. K^+ is secreted into these tubules, drawing water with it osmotically. Much of this water (*arrows*) is reabsorbed across the wall of the hindgut.

Secretion is the opposite of reabsorption—ions or molecules are transported from the body fluid into the tubule. The secretion of K⁺ creates an osmotic gradient that causes water to enter the tubules by osmosis from the body's open circulatory system. Most of the water and K⁺ is then reabsorbed into the circulatory system through the epithelium of the hindgut, leaving only small molecules and waste products to be excreted from the rectum along with feces. Malpighian tubules thus provide a very efficient means of water conservation.

The vertebrate kidney filters and then reabsorbs

The **kidneys** of vertebrates, unlike the Malpighian tubules of insects, create a tubular fluid by filtering the blood under pressure. In addition to waste products and water, the filtrate contains many small molecules, including glucose, amino acids, and vitamins, that are of value to the animal. These molecules and most of the water are reabsorbed from the tubules into the blood, while wastes remain in the filtrate. Additional wastes may be secreted by the tubules and added to the filtrate, and the final waste product, urine, is eliminated from the body.

It may seem odd that the vertebrate kidney should filter out almost everything from blood plasma (except proteins, which are too large to be filtered) and then spend energy to take back or reabsorb what the body needs. But selective reabsorption provides great flexibility. Various vertebrate groups have evolved the ability to reabsorb molecules that are especially valuable in particular habitats. This flexibility is a key factor underlying the successful colonization of many diverse environments by the vertebrates. In the rest of this chapter, we focus on the vertebrate kidney and its elimination of waste materials, notably nitrogen compounds.

Learning Outcomes Review 50.3

Many invertebrates filter fluid into a system of tubules and then reabsorb ions and water, leaving waste products for excretion. Insects create an excretory fluid by secreting K⁺ and waste products into tubules, which draw water osmotically. The vertebrate kidney produces a filtrate that enters tubules and is modified to become urine.

- **How are the function of Malpighian tubules and kidneys similar?**

50.4 Evolution of the Vertebrate Kidney

Learning Outcomes

1. Differentiate between osmotic adaptations in freshwater fish versus those in marine fish.
2. Explain the role of the loop of Henle in the evolution of vertebrate kidneys.

The kidney is a complex organ made up of thousands of repeating units called **nephrons,** each with the structure of a loop that penetrates deep into the medulla of the kidney (shown schematically in figure 50.6). Blood pressure forces the fluid in blood out of a ball of capillaries called the *glomerulus* into *Bowman's capsule,* the beginning of the tubule system. This process filters the blood forming the tubular filtrate that can then be modified by the rest of the nephron. The glomerulus retains blood cells, proteins, and other useful large molecules in the blood, but it allows the water, and the small molecules and wastes dissolved in it, to pass through and into the tubule system of the nephron. As the filtered fluid passes through the nephron tube, useful nutrients and ions are reabsorbed from it by both active and passive transport mechanisms, leaving the water and metabolic wastes behind in a fluid urine. (The details of this process are described in the following section.)

Although the same basic design has been retained in all vertebrate kidneys, a few modifications have occurred. Because

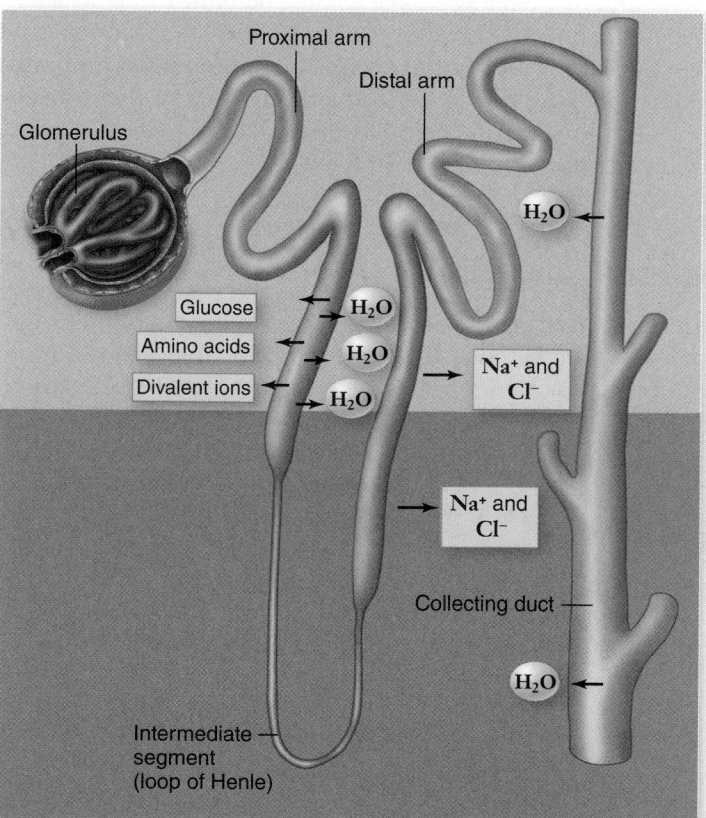

Figure 50.6 **Organization of the vertebrate nephron.**
The nephron tubule is a basic design that has been retained in the kidneys of vertebrates. Sugars, amino acids, water, important monovalent ions, and divalent ions are reabsorbed in the proximal arm; water and monovalent ions such as Na⁺ and Cl⁻ are reabsorbed in the loop of Henle; varying amounts of water and monovalent ions (Na⁺ and Cl⁻) can be reabsorbed in the distal arm and the collecting duct, depending on hormonal influences.

 Data analysis The disease diabetes mellitus results in a higher concentration of glucose in the blood than can be reabsorbed in the kidney. Would this lead to gain or loss of water?

the original glomerular filtrate is isotonic to blood, all vertebrates can produce a urine that is isotonic to blood by reabsorbing ions and water in equal proportions. Or, they can produce a urine that is hypotonic to blood—more dilute than the blood—by reabsorbing relatively less water. Only birds and mammals can reabsorb enough water from their glomerular filtrate to produce a urine that is hypertonic to blood—more concentrated than the blood—by reabsorbing relatively more water.

Freshwater fishes must retain electrolytes and keep water out

Kidneys are thought to have evolved among the freshwater teleosts, or bony fishes. Because the body fluids of a freshwater fish are hypertonic with respect to the surrounding water, these animals face two serious problems: (1) Water tends to enter the body from the environment; and (2) solutes tend to leave the body and enter the environment.

Freshwater fish address the first problem by *not* drinking water and by excreting a large volume of dilute urine, which is hypotonic to their body fluids. They address the second problem by reabsorbing ions across the nephron tubules, from the glomerular filtrate back into the blood. In addition, they actively transport ions across their gill surfaces from the surrounding water into the blood (figure 50.7, left).

Marine bony fishes must excrete electrolytes and keep water in

Although most groups of animals seem to have evolved first in the sea, marine bony fish (teleosts) probably evolved from freshwater ancestors. They faced significant new problems in making the transition to the sea because their body fluids were, and are, hypotonic to seawater. Consequently, water tends to leave their bodies by osmosis across their gills, and they also lose water in their urine. To compensate for this continuous water loss, marine fish drink large amounts of seawater (figure 50.7, right).

Many of the divalent cations (principally, Ca^{2+} and Mg^{2+}) in the seawater that a marine fish drinks remain in the digestive tract and are eliminated through the anus. Some, however, are absorbed into the blood, as are the monovalent ions K^+, Na^+, and Cl^-. Most of the monovalent ions are actively transported out of the blood across the gill surfaces, whereas the divalent ions that enter the blood are secreted into the nephron tubules and excreted in the urine. In these two ways, marine bony fish eliminate the ions they get from the seawater they drink. The urine they excrete is isotonic to their body fluids. It is more concentrated than the urine of freshwater fish, but not as concentrated as that of birds and mammals.

Cartilaginous fishes pump out electrolytes and retain urea

The elasmobranchs, including sharks and rays, are by far the most common subclass in the class Chondrichthyes (cartilaginous fish). Elasmobranchs have solved the osmotic problem posed by their seawater environment in a different way. Instead of having body fluids that are hypotonic to seawater, so that they have to continuously drink seawater and actively pump out ions, the elasmobranchs reabsorb urea from the nephron tubules and maintain a blood urea concentration that is 100 times higher than that of mammals.

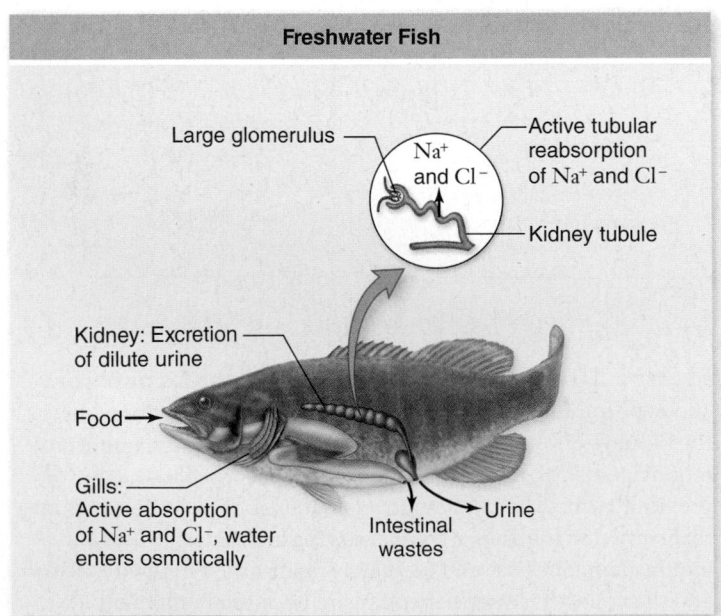

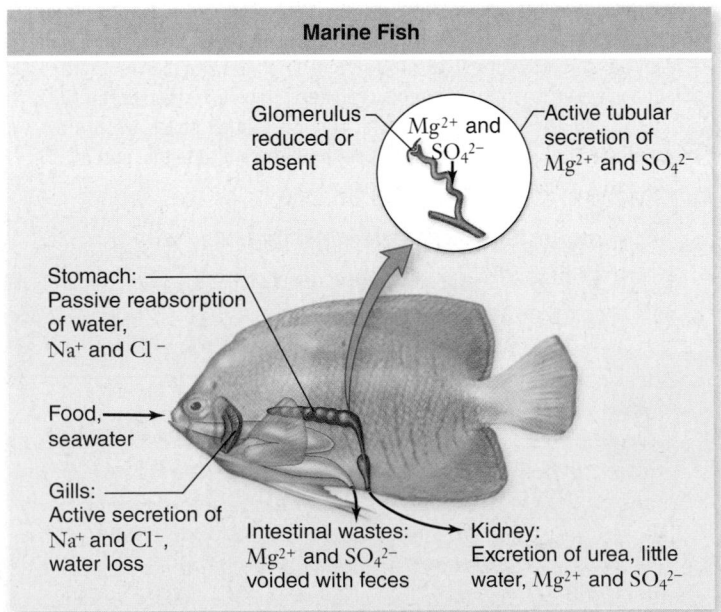

Figure 50.7 Freshwater and marine teleosts face different osmotic problems. Whereas the freshwater teleost is hypertonic to its environment, the marine teleost is hypotonic to seawater. To compensate for its tendency to take in water and lose ions, a freshwater fish excretes dilute urine, avoids drinking water, and reabsorbs ions across the nephron tubules. To compensate for its osmotic loss of water, the marine teleost drinks seawater and eliminates the excess ions through active transport across epithelia in the gills and kidneys.

The added urea makes elasmobranchs' blood approximately isotonic to the surrounding sea. Because no net water movement occurs between isotonic solutions, water loss is therefore prevented. As a result, these fishes do not need to drink seawater for osmotic balance, and their kidneys and gills do not have to remove large amounts of ions from their bodies. The enzymes and tissues of the cartilaginous fish have evolved to tolerate the high urea concentrations.

Amphibians and reptiles have osmotic adaptations to their environments

The first terrestrial vertebrates were the amphibians, and the amphibian kidney is identical to that of freshwater fish. This is not surprising because amphibians spend a significant portion of their time in fresh water, and when on land, they generally stay in wet places. Amphibians produce a very dilute urine and compensate for their loss of Na^+ by actively transporting Na^+ across their skin from the surrounding water.

Reptiles, on the other hand, live in diverse habitats. Those living mainly in fresh water occupy a habitat similar to that of the freshwater fish and amphibians, and thus have similar kidneys. Marine reptiles, including some crocodilians, sea turtles, sea snakes, and one lizard, possess kidneys similar to those of their freshwater relatives, but they face opposite problems—they tend to lose water and take in salts. Like marine bony fish, they drink seawater and excrete an isotonic urine. Marine reptiles eliminate excess salt through salt glands located near the nose or the eye.

The kidneys of terrestrial reptiles also reabsorb much of the salt and water in their nephron tubules, helping somewhat to conserve blood volume in dry environments. Like fish and amphibians, they cannot produce urine that is more concentrated than the blood plasma; however, they don't really excrete urine, but instead empty it into a cloaca (the common exit of the digestive and urinary tracts), where additional water can be reabsorbed and the wastes excreted with the feces.

Mammals and birds are able to excrete concentrated urine and retain water

Mammals and birds are the only vertebrates able to produce urine with a higher osmotic concentration than their body fluids. These vertebrates are therefore able to excrete their waste products in a small volume of water, so that more water can be retained in the body.

Human kidneys can produce urine that is as much as 4.2 times as concentrated as blood plasma, but the kidneys of some other mammals are even more efficient at conserving water. For example, camels, gerbils, and pocket mice of the genus *Perognathus* can excrete urine that is 8, 14, and 22 times as concentrated as their blood plasma, respectively. The kidneys of kangaroo rats (genus *Dipodomys*) are so efficient that they never have to drink water; they can obtain all the water they need from their food and from water produced in aerobic cellular respiration.

The production of hypertonic urine is accomplished by the *loop of Henle* portion of the nephron (see figures 50.6 and 50.11), found only in mammals and birds. The degree of

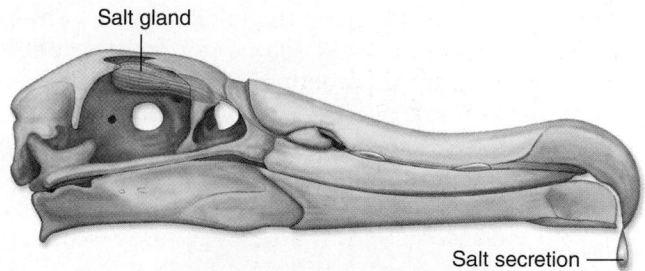

Figure 50.8 How marine birds cope with excess salt. Marine birds drink seawater and then excrete the salt through salt glands. The extremely salty fluid excreted by these glands can then dribble down the beak.

concentration depends on the length of the loop; most mammals have some nephrons with short loops and other nephrons with much longer loops. Birds, however, have relatively few or no nephrons with long loops, so they cannot produce urine that is as concentrated as that of mammals. At most, they can only reabsorb enough water to produce a urine that is about twice the concentration of their blood. Marine birds solve the problem of water loss by drinking salt water and then excreting the excess salt from salt glands near the eyes (figure 50.8).

The moderately hypertonic urine of a bird is delivered to its cloaca, along with the fecal material from its digestive tract. If needed, additional water can be absorbed across the wall of the cloaca to produce a semisolid white paste or pellet, which is excreted.

Learning Outcomes Review 50.4

The kidneys of freshwater fishes must excrete copious amounts of very dilute urine. Marine bony fishes drink seawater and excrete an isotonic urine; cartilaginous fishes retain urea, which prevents water loss. In mammals and birds, the loop of Henle allows reabsorption of water from the urine, making it hypertonic to their body fluids.

■ *Mammals and birds have nephrons with a loop of Henle, but reptiles do not. What are the possible evolutionary explanations for this?*

50.5 The Mammalian Kidney

Learning Outcomes

1. *Describe the actions of filtration, reabsorption, and secretion.*
2. *Name the primary components of the kidney.*
3. *Describe the main parts of a nephron.*

In humans, the kidneys are fist-sized organs located in the lower back. Each kidney receives blood from a renal artery, and from this blood, urine is produced. Urine drains from each kidney through a **ureter,** which carries the urine to a

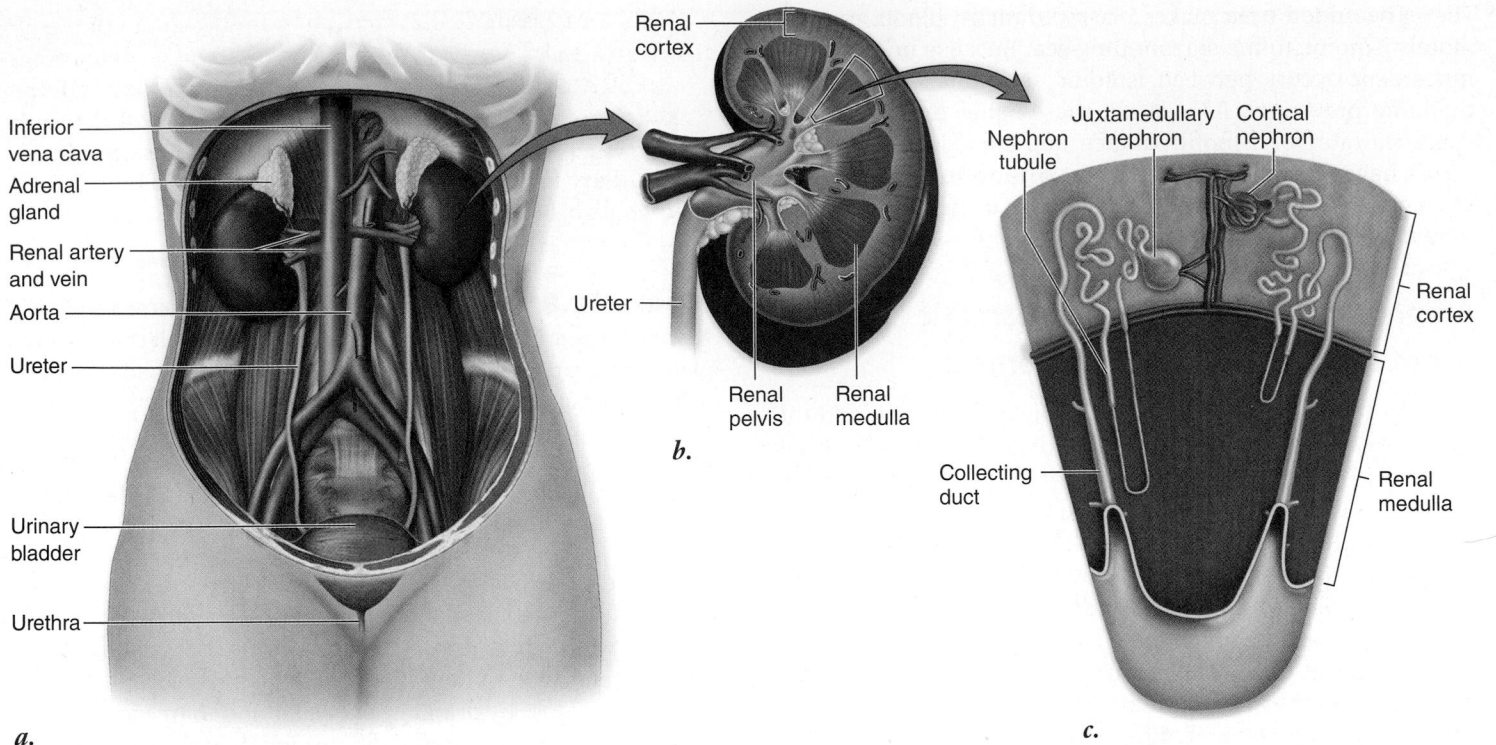

Figure 50.9 **The human renal system.** *a.* The positions of the organs of the urinary system. *b.* A sectioned kidney, revealing the internal structure. *c.* The position of nephrons in the mammalian kidney. Cortical nephrons are located predominantly in the renal cortex; juxtamedullary nephrons have long loops that extend deep into the renal medulla.

urinary bladder. From the bladder, urine is passed out of the body through the **urethra** (figure 50.9).

Within the kidney, the mouth of the ureter flares open to form a funnel-like structure, the *renal pelvis*. The renal pelvis, in turn, has cup-shaped extensions that receive urine from the renal tissue. The renal tissue is divided into an outer **renal cortex** and an inner **renal medulla.**

The kidney has three basic functions summarized in figure 50.10:

1. *Filtration:* Fluid in the blood is filtered into the tubule system, leaving cells and large protein in the blood and a filtrate composed of water and all of the blood solutes. This filtrate is modified by the rest of the kidney to produce urine for excretion.
2. *Reabsorption:* Reabsorption is the selective movement of important solutes such as glucose, amino acids, and a variety of inorganic ions, out of the filtrate in the tubule system to the extracellular fluid, then back into the bloodstream via peritubular capillaries. The process of reabsorption can utilize active or passive processes depending on the solute. Water is also reabsorbed, and this can be controlled to regulate the amount of water loss.
3. *Secretion:* Secretion is the movement of substances from the blood into the extracellular fluid, then into the filtrate in the tubule system. Unlike reabsorption, which preserves substances in the body, this adds to what will be expelled from the body and can be used to remove toxic substances.

The nephron is the filtering unit of the kidney

On a microscopic level, each kidney contains about a million functioning *nephrons.* Mammalian kidneys contain a mixture of **juxtamedullary nephrons,** which have long loops that dip

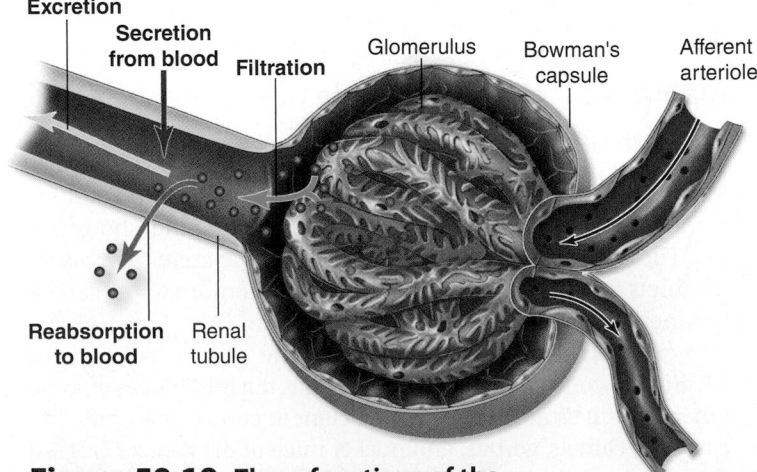

Figure 50.10 **Three functions of the kidney.** Molecules enter the urine by filtration out of the glomerulus and by secretion into the tubules from surrounding peritubular capillaries. Molecules that entered the filtrate can be returned to the blood by reabsorption from the tubules into surrounding peritubular capillaries. The fluid exiting the kidney is eliminated from the body by excretion through the tubule to a ureter and then to the bladder.

deeply into the medulla, and **cortical nephrons** with shorter loops. The significance of the length of the loops will be explained a little later.

The production of filtrate

Each nephron consists of a long tubule and associated small blood vessels (figure 50.11). First, blood is carried by an *afferent arteriole* to a tuft of capillaries in the renal cortex—the **glomerulus.** Here the blood is filtered as the blood pressure forces fluid through the porous capillary walls. Blood cells and plasma proteins are too large to enter this glomerular filtrate, but large amounts of the plasma, consisting of water and dissolved molecules, leave the vascular system at this step. The filtrate immediately enters the first region of the nephron tubules. This region, **Bowman's capsule,** envelops the glomerulus much as a large, soft balloon surrounds your hand if you press your fist into it. The capsule has slit openings so that the glomerular filtrate can enter the system of nephron tubules.

Blood components that were not filtered out of the glomerulus drain into an *efferent arteriole,* which then empties into a second bed of capillaries called **peritubular capillaries** that surround the tubules. This is only one of several locations in the body where two capillary beds occur in series. In juxtamedullary nephrons, efferent arteriole and peritubular capillaries also feed the **vasa recta** capillaries that surround the loop of Henle. As described later, the peritubular capillaries are needed for the processes of reabsorption and secretion.

After the filtrate enters Bowman's capsule, it goes into a portion of the nephron called the **proximal convoluted tubule,** located in the cortex. In a cortical nephron, the fluid then flows through the **loop of Henle** that dips only minimally into the medulla before ascending back into the cortex. In juxtamedullary nephrons, the loop of Henle extends much deeper into the medulla before ascending back up into the cortex. More water can be reabsorbed from juxtamedullary nephrons than from cortical nephrons. The fluid then moves deeper into the medulla and back up again into the cortex in a loop of

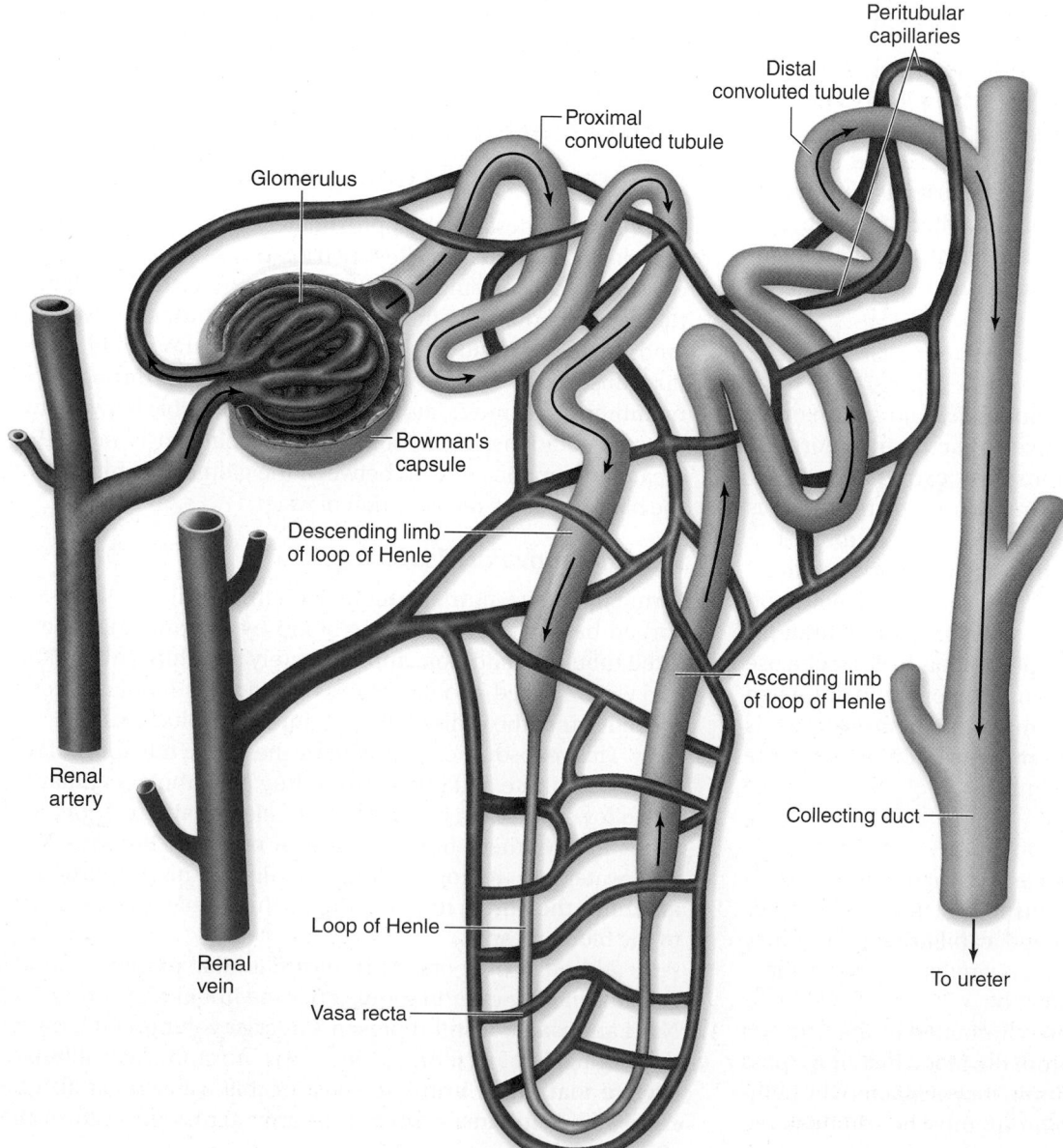

Figure 50.11
A nephron in a mammalian kidney.
The nephron tubule is surrounded by peritubular capillaries in the cortex, and their vasa recta extensions surround the loop of Henle in the medulla. This capillary bed carries away molecules and ions that are reabsorbed from the filtrate.

Peritubular capillaries

Distal convoluted tubule

Proximal convoluted tubule

Glomerulus

Bowman's capsule

Descending limb of loop of Henle

Renal artery

Renal vein

Ascending limb of loop of Henle

Loop of Henle

Collecting duct

Vasa recta

To ureter

Henle. As mentioned earlier, only the kidneys of mammals and birds have loops of Henle, and this is why only birds and mammals have the ability to concentrate their urine.

Collection of urine

After leaving the loop, the fluid is delivered to a **distal convoluted tubule** in the cortex that next drains into a **collecting duct.** The collecting duct again descends into the medulla, where it merges with other collecting ducts to empty its contents, now called urine, into the renal pelvis.

Water, some nutrients, and some ions are reabsorbed; other molecules are secreted

Most of the water and dissolved solutes that enter the glomerular filtrate must be returned to the blood by reabsorption, or the animal would literally urinate to death. In a human, for example, approximately 2000 L of blood passes through the kidneys each day, and 180 L of water leaves the blood and enters the glomerular filtrate.

Water

Because humans have a total blood volume of only about 5 L and produce only 1 to 2 L of urine per day, it is obvious that each liter of blood is filtered many times per day, and most of the filtered water is reabsorbed. Water is reabsorbed from the filtrate by the proximal convoluted tubule, as it passes through the descending loop of Henle and the collecting duct. The selective reabsorption in the collecting duct is driven by an osmotic gradient produced by the loop of Henle, as is described shortly.

Glucose and other nutrients

The reabsorption of glucose, amino acids, and many other molecules needed by the body is driven by active transport and secondary active transport (cotransport) carriers. As in all carrier-mediated transport, a maximum rate of transport is reached whenever the carriers are saturated (see chapter 5).

In the case of the renal glucose carriers in the proximal convoluted tubule, saturation occurs when the concentration of glucose in the blood (and thus in the glomerular filtrate) is about 180 mg/100 mL of blood. If a person has a blood glucose concentration in excess of this amount, as happens in untreated diabetes mellitus, the glucose remaining in the filtrate is expelled in the urine. Indeed, the presence of glucose in the urine is diagnostic of diabetes mellitus.

Secretion of wastes

The secretion of foreign molecules and particular waste products of the body involves the transport of these molecules across the membranes of the blood capillaries and kidney tubules into the filtrate. This process is similar to reabsorption, but it proceeds in the opposite direction.

Some secreted molecules are eliminated in the urine so rapidly that they may be cleared from the blood in a single pass through the kidneys. This rapid elimination explains why penicillin, which is secreted by the nephrons, must be administered in very high doses and several times per day.

Excretion of toxins and excess ions maintains homeostasis

A major function of the kidney is the elimination of a variety of potentially harmful substances that animals eat and drink. In addition, urine contains nitrogenous wastes, described earlier, that are products of the catabolism of amino acids and nucleic acids. Urine may also contain excess K^+, H^+, and other ions that are removed from the blood.

Urine's generally high H^+ concentration (pH 5 to 7) helps maintain the acid–base balance of the blood within a narrow range (pH 7.35 to 7.45). Moreover, the excretion of water in urine contributes to the maintenance of blood volume and pressure (see chapter 49); the larger the volume of urine excreted, the lower the blood volume.

The purpose of kidney function is therefore homeostasis; the kidneys are critically involved in maintaining the constancy of the internal environment. When disease interferes with kidney function, it causes a rise in the blood concentration of nitrogenous waste products, disturbances in electrolyte and acid–base balance, and a failure in blood pressure regulation. Such potentially fatal changes highlight the central importance of the kidneys in normal body physiology.

Each part of the mammalian nephron performs a specific transport function

As previously described, approximately 180 L of isotonic glomerular filtrate enters the Bowman's capsules of human kidneys each day. After passing through the remainder of the nephron tubules, this volume of fluid would be lost as urine if it were not reabsorbed back into the blood. It is clearly impossible to produce this much urine, yet water is only able to pass through a cell membrane by osmosis, and osmosis is not possible between two isotonic solutions. Therefore, some mechanism is needed to create an osmotic gradient between the glomerular filtrate and the blood, to allow reabsorption of water.

Proximal convoluted tubule

Virtually all the nutrient molecules in the filtrate are reabsorbed back into the systemic blood by the proximal convoluted tubule. In addition, approximately two-thirds of the NaCl and water filtered into Bowman's capsule is immediately reabsorbed across the walls of the proximal convoluted tubule.

This reabsorption is driven by the active transport of Na^+ out of the filtrate and into surrounding peritubular capillaries. Cl^- follows Na^+ passively because of electrical attraction, and water follows them both because of osmosis. Because NaCl and water are removed from the filtrate in proportionate amounts, the filtrate that remains in the tubule is still isotonic to the blood plasma.

Although only one-third of the initial volume of filtrate remains in the nephron tubule after the initial reabsorption of NaCl and water, it still represents a large volume (60 L out of the original 180 L of filtrate). Obviously, no animal can afford to excrete that much urine, so most of this water must also be reabsorbed. It is reabsorbed primarily across the wall of the collecting duct.

Loop of Henle

The function of the loop of Henle is to create a gradient of increasing osmolarity from the cortex to the medulla. This allows water to be reabsorbed by osmosis in the collecting duct as it runs down into the medulla past the loop of Henle. The descending and ascending limbs of the loop of Henle differ structurally and in their permeability to ions and water. This produces a gradient of increasing osmolarity from cortex to medulla (figure 50.12). The structure of the loop also forms another example of a countercurrent system, this time acting to increase the osmolarity of interstitial fluid. To understand the functioning of the loop of Henle, it is easiest to start in the ascending limb:

1. The entire ascending limb is impermeable to water. The thick portion of the ascending limb actively transports Na^+ out of the tubule, with Cl^- passively following. The thin ascending limb is permeable to both Na^+ and Cl^-, which move out by diffusion.

2. The descending limb is thin and permeable to water but not to NaCl. Because of the Na^+ and Cl^- lost by the ascending limb, the osmolarity of the interstitial fluid is higher than in the descending limb, and water moves out of the descending limb by osmosis. This also increases the osmolarity of the fluid in the tubule such that as it turns at the bottom, it will lose NaCl by diffusion in the thin ascending loop as described earlier.

3. The loss of water from the descending limb multiplies the concentration that can be achieved at each level of the loop through the active extrusion of Na^+ (with Cl^- following passively) by the ascending limb. The longer the loop of Henle, the longer the region of interaction between the descending and ascending limbs, and the greater the total concentration that can be achieved. In a human kidney, the concentration of filtrate entering the loop is 300 milliosmolar (mOsm), and this concentration is multiplied to more than 1200 mOsm at the bottom of the longest loops of Henle in the renal medulla.

4. The NaCl pumped out of the ascending limb of the loop is reabsorbed from the surrounding interstitial fluid into the loops of the *vasa recta* capillaries, so that NaCl can diffuse from the blood leaving the medulla to the blood entering the medulla. Thus, the vasa recta also functions in a countercurrent exchange, similar to that described for the countercurrent flow of blood in the fins of large aquatic vertebrates for heat exchange (see chapter 42), and of water and blood through gills to enhance oxygen exchange (see chapter 48). In the case of the vasa recta, this exchange prevents the flow of blood through the capillaries from destroying the osmotic gradient established by the loop of Henle. Thus, blood can be supplied to this region of the kidney without affecting the ability of the collecting duct to selectively reabsorb water.

Because fluid flows in opposite directions in the two limbs of the loop, the action of the loop of Henle in creating a hypertonic renal medulla is known as the countercurrent multiplier system. The osmotic gradient that is established is greater than what would be produced by just active transport of salts out of the tubule system.

The high solute concentration of the renal medulla is primarily the result of NaCl accumulation by the countercurrent multiplier system, but urea also contributes to the total osmolarity of the medulla. The descending limb of the loop of Henle and the collecting duct are both permeable to urea, which leaves these regions of the nephron by diffusion.

Distal convoluted tubule and collecting duct

Because NaCl was pumped out of the ascending limb, the filtrate that arrives at the distal convoluted tubule and enters the collecting duct in the renal cortex is hypotonic (with a concentration of only 100 mOsm). The collecting duct carrying this dilute fluid now plunges into the medulla. As a result of the hypertonic interstitial fluid of the renal medulla, a strong osmotic gradient pulls water out of the collecting duct and into the surrounding blood vessels.

The osmotic gradient is normally constant, but the permeability of the distal convoluted tubule and the collecting duct to water is adjusted by a hormone, *antidiuretic hormone* (ADH), mentioned in chapters 45 and 49. When an animal needs to conserve water, the posterior-pituitary gland secretes more ADH, and this hormone increases the number of water channels in the plasma membranes of the collecting duct cells. This increases the permeability of the collecting ducts to water

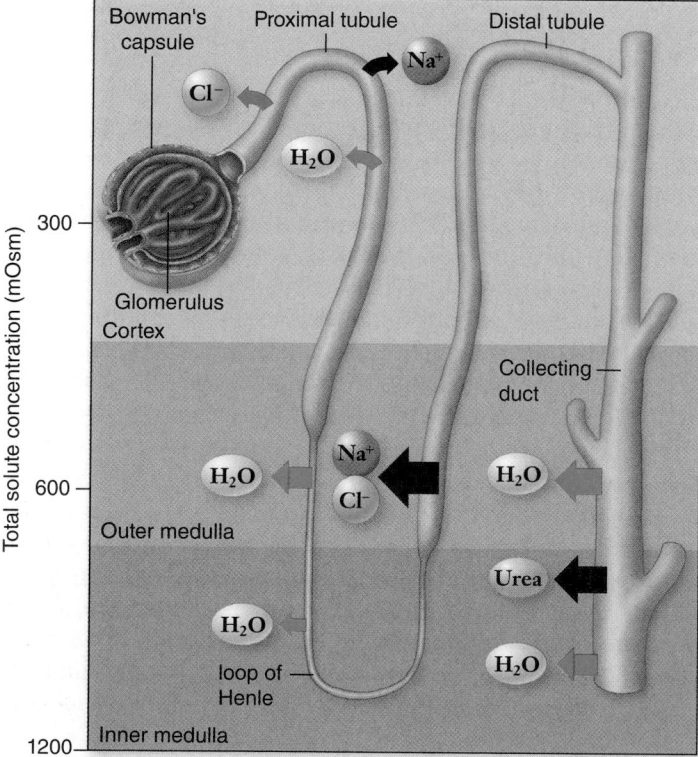

Figure 50.12 The reabsorption of salt and water in the mammalian kidney. Active transport of Na^+ out of the proximal tubules is followed by the passive movement of Cl^- and water. Active extrusion of NaCl from the ascending limb of the loop of Henle creates the osmotic gradient required for the reabsorption of water from the descending limb of the loop of Henle and the collecting duct. The two limbs of the loop form a countercurrent multiplier system that increases the osmotic gradient. The changes in osmolarity from the cortex to the medulla are indicated to the left of the figure.

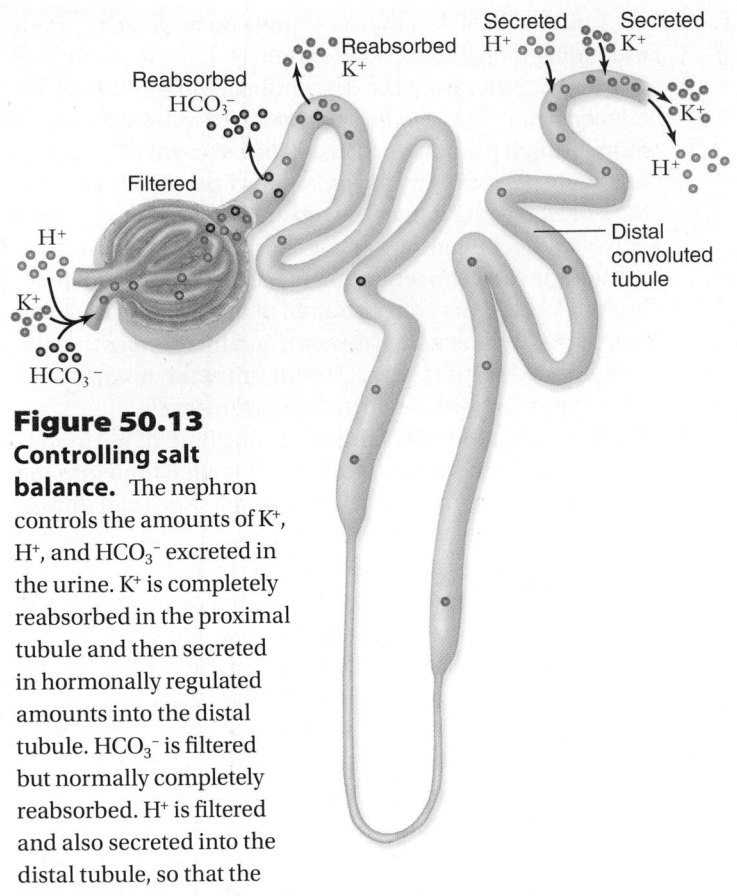

Figure 50.13
Controlling salt balance. The nephron controls the amounts of K⁺, H⁺, and HCO₃⁻ excreted in the urine. K⁺ is completely reabsorbed in the proximal tubule and then secreted in hormonally regulated amounts into the distal tubule. HCO₃⁻ is filtered but normally completely reabsorbed. H⁺ is filtered and also secreted into the distal tubule, so that the final urine has an acidic pH.

so that more water is reabsorbed and less is excreted in the urine. The animal thus excretes a hypertonic urine.

In addition to regulating water balance, the kidneys regulate the balance of electrolytes in the blood by reabsorption and secretion. For example, the kidneys reabsorb K⁺ in the proximal tubule and then secrete an amount of K⁺ needed to maintain homeostasis into the distal convoluted tubule (figure 50.13). The kidneys also maintain acid–base balance by excreting H⁺ into the urine and reabsorbing HCO₃⁻.

The reabsorption of NaCl in the distal convoluted tubule and collecting duct depends on the needs of the body and is under the control of the hormone *aldosterone*. Both ADH and aldosterone influence the distal convoluted tubule and collecting duct, although aldosterone is more significant in terms of NaCl. We present more about hormonal control of excretion in the next section.

Learning Outcomes Review 50.5

Fluid and certain solutes move out of the blood and into the tubular systems of the kidneys through filtration (passive) and secretion (transported); important solutes and water are returned to the blood through reabsorption. The mammalian kidney is divided into a cortex and a medulla and contains about a million nephrons. The parts of a nephron include the glomerulus and Bowman's capsule, proximal convoluted tubule, loop of Henle, distal convoluted tubule, and collecting duct.

■ *The compound mannitol is filtered but cannot be reabsorbed. How would this affect the volume of urine produced?*

50.6 Hormonal Control of Osmoregulatory Functions

Learning Outcomes

1. *Explain the actions of ADH, aldosterone, and ANH.*
2. *Analyze the relationship between blood osmolarity and blood pressure.*

In mammals and birds, the amount of water excreted in the urine, and thus the concentration of the urine, varies according to the changing needs of the body. Acting through the mechanisms described next, the kidneys excrete a hypertonic urine when the body needs to conserve water. If an animal drinks excess water, the kidneys excrete a hypotonic urine.

As a result, the volume of blood, the blood pressure, and the osmolarity of blood plasma are maintained relatively constant by the kidneys, no matter how much water you drink. The kidneys also regulate the plasma K⁺ and Na⁺ concentrations and blood pH within very narrow limits. These homeostatic functions of the kidneys are coordinated primarily by hormones.

Antidiuretic hormone causes water to be conserved

Antidiuretic hormone (ADH) is produced by the hypothalamus and secreted by the posterior-pituitary gland. The primary stimulus for ADH secretion is an increase in the osmolarity of the blood plasma. When a person is dehydrated or eats salty food, the osmolarity of plasma increases. Osmoreceptors in the hypothalamus respond to the elevated blood osmolarity by sending increasing action potentials to the integration center (also in the hypothalamus). This, in turn, triggers a sensation of thirst and an increase in the secretion of ADH (figure 50.14).

ADH causes the walls of the distal convoluted tubules and collecting ducts in the kidney to become more permeable to water. Water channels called aquaporins (see chapter 5) are contained within the membranes of intracellular vesicles in the epithelium of the distal convoluted tubules and collecting ducts; ADH stimulates the fusion of the vesicle membrane with the plasma membrane, similar to the process of exocytosis. The aquaporins are now in place and allow water to flow out of the tubules and ducts in response to the hypertonic condition of the renal medulla. This water is reabsorbed into the bloodstream.

When secretion of ADH is reduced, the plasma membrane pinches in to form new vesicles that contain aquaporins. This removes the aquaporins from the plasma membrane of the distal convoluted tubule and collecting duct, making them less permeable to water. Thus more water is excreted in urine.

Under conditions of maximal ADH secretion, a person excretes only 600 mL of highly concentrated urine per day. A person who lacks ADH due to pituitary damage has the disorder known as *diabetes insipidus* and constantly excretes a large volume of dilute urine. Such a person is in danger of becoming

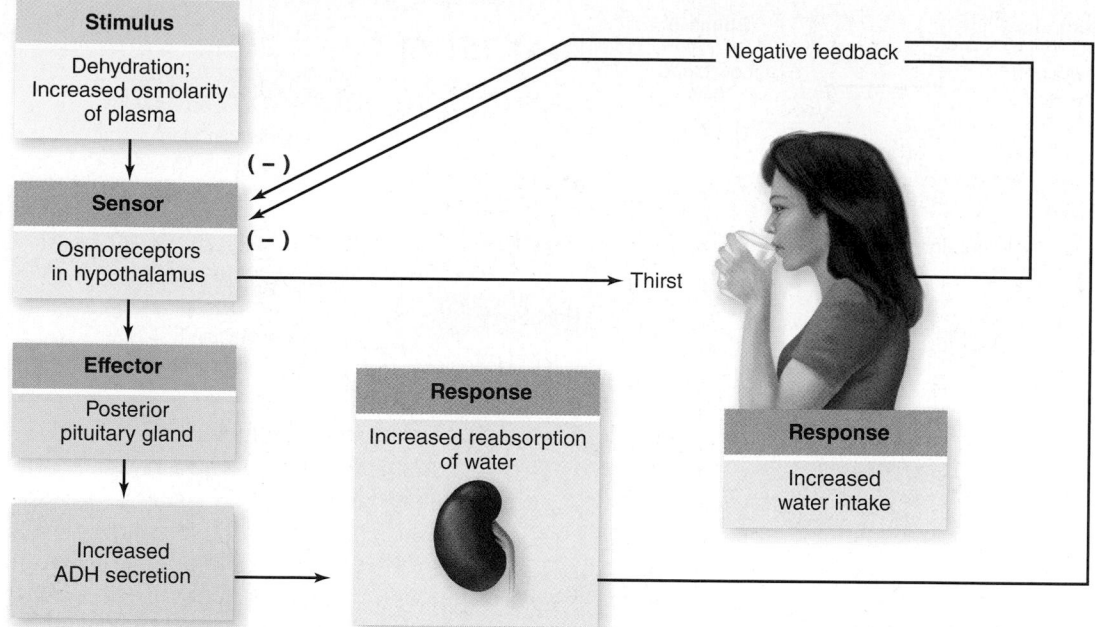

Figure 50.14 **Antidiuretic hormone stimulates the reabsorption of water by the kidneys.** This action completes a negative feedback loop and helps to maintain homeostasis of blood volume and osmolarity.

severely dehydrated and succumbing to dangerously low blood pressure.

Homeostasis via ADH action is also affected by the common drugs ethanol and caffeine, both of which inhibit secretion of ADH. This is the basis for the dehydration that is the after effect of drinking too much alcohol.

Aldosterone and atrial natriuretic hormone control sodium ion concentration

Sodium ions are the major solute in the blood plasma. When the blood concentration of Na$^+$ falls, therefore, the blood osmolarity also falls. This drop in osmolarity inhibits ADH secretion, causing more water to remain in the collecting duct for excretion in the urine. As a result, the blood volume and blood pressure decrease.

A decrease in extracellular Na$^+$ also causes more water to be drawn into cells by osmosis, partially offsetting the drop in plasma osmolarity, but further decreasing blood volume and blood pressure. If Na$^+$ deprivation is severe, the blood volume may fall so low that blood pressure is insufficient to sustain life. For this reason, salt is necessary for life. Many animals have a "salt hunger" and actively seek salt, such as when deer gather at "salt licks."

A drop in blood Na$^+$ concentration is normally compensated for by the kidneys under the influence of the hormone *aldosterone*, which is secreted by the adrenal cortex. Aldosterone stimulates the distal convoluted tubules and collecting ducts to reabsorb Na$^+$, decreasing the excretion of Na$^+$ in the urine. Indeed, under conditions of maximal aldosterone secretion, Na$^+$ may be completely absent from the urine. The reabsorption of Na$^+$ is followed by reabsorption of Cl$^-$ and water, so aldosterone has the net effect of promoting the retention of both salt and water. It thereby helps to maintain blood volume, osmolarity, and pressure.

The secretion of aldosterone in response to a decreased blood level of Na$^+$ is indirect. Because a fall in blood Na$^+$ is accompanied by decreased blood volume, the flow of blood past a group of cells called the juxtaglomerular apparatus is reduced. The juxtaglomerular apparatus is located in the region of the kidney between the distal convoluted tubule and the afferent arteriole (figure 50.15).

When blood flow is reduced, the juxtaglomerular apparatus responds by secreting the enzyme *renin* into the blood (figure 50.16). Renin catalyzes the production of the polypeptide angiotensin I from the protein angiotensinogen. Angiotensin I is then converted by another enzyme into angiotensin II, which stimulates blood vessels to constrict and the adrenal cortex to secrete aldosterone. Thus, homeostasis of blood volume and pressure can be maintained by the activation of this renin–angiotensin–aldosterone system. In addition to stimulating Na$^+$ reabsorption, aldosterone also promotes the secretion of K$^+$ into the distal convoluted tubules and collecting ducts. Consequently, aldosterone lowers the blood K$^+$ concentration, helping to maintain constant blood K$^+$ levels in the face of changing amounts of K$^+$ in the diet. People who lack the ability to produce aldosterone will die if untreated because of the excessive loss of salt and water in the urine and the buildup of K$^+$ in the blood.

The action of aldosterone in promoting salt and water retention is opposed by another hormone, *atrial natriuretic hormone (ANH)*, mentioned in chapter 49. This hormone is secreted by the right atrium of the heart in response to an increased blood volume, which stretches the atrium. Under these conditions, aldosterone secretion from the adrenal cortex decreases, and ANH secretion increases, thus promoting the excretion of salt and water in the urine and lowering the blood volume.

Learning Outcomes Review 50.6

ADH stimulates the insertion of water channels into the cells of the distal convoluted tubule and collecting duct, making them more permeable to water and increasing its reabsorption. Aldosterone promotes reabsorption of Na$^+$, Cl$^-$, and water across the distal convoluted tubule and collecting duct, as well as the secretion of K$^+$ into the tubules. ANH decreases Na$^+$ and Cl$^-$ reabsorption. When blood osmolarity increases, more water is retained by the release of ADH, and blood pressure increases; if blood osmolarity falls, ADH release is inhibited, less water is retained, and blood pressure drops.

■ *What would be the effect of a compound that blocks aquaporin water channels?*

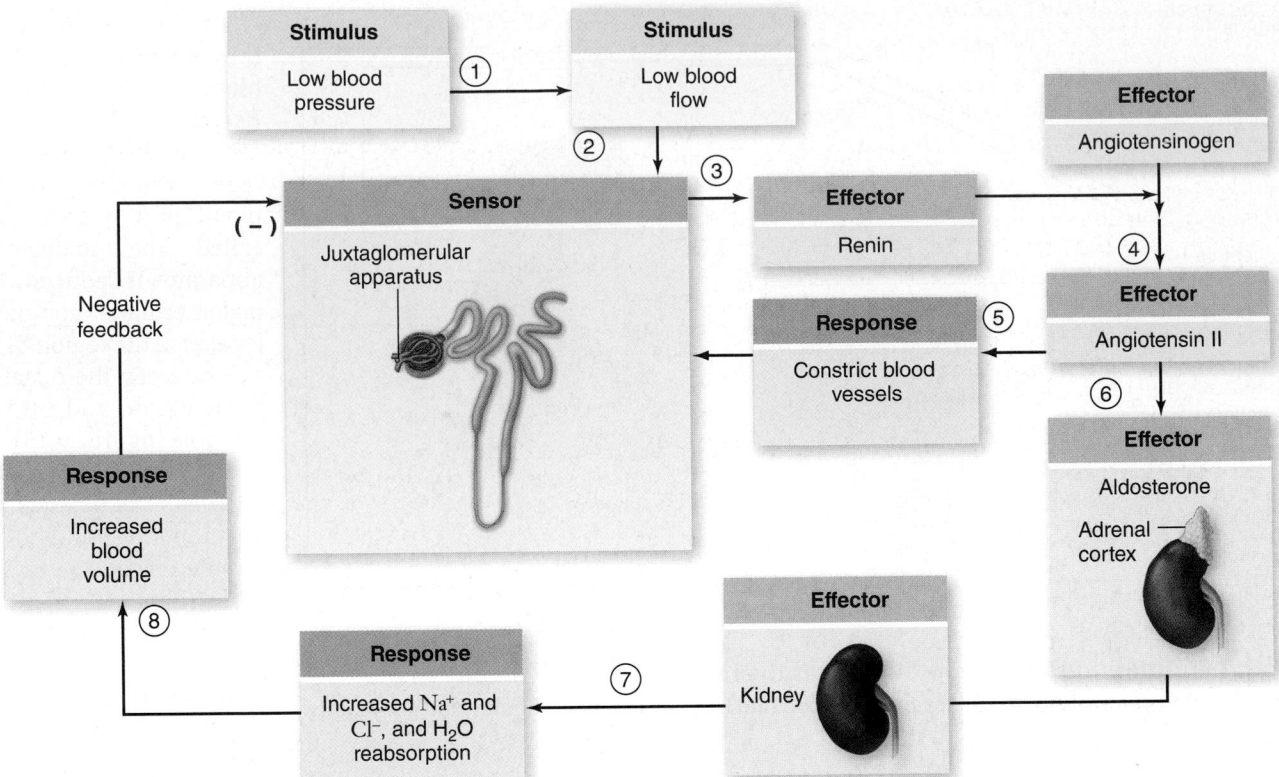

Figure 50.15 **A lowering of blood volume activates the renin–angiotensin–aldosterone system.** (1) Low blood volume and a decrease in blood Na$^+$ levels reduce blood pressure. (2) Reduced blood flow past the juxtaglomerular apparatus triggers (3) the release of renin into the blood, which catalyzes the production of angiotensin I from angiotensinogen. (4) Angiotensin I converts into a more active form, angiotensin II. (5) Angiotensin II stimulates blood vessel constriction and (6) the release of aldosterone from the adrenal cortex. (7) Aldosterone stimulates the reabsorption of Na$^+$ in the distal convoluted tubules. Increased Na$^+$ reabsorption is followed by the reabsorption of Cl$^-$ and water. (8) This increases blood volume. An increase in blood volume may also trigger the release of an atrial natriuretic hormone, which inhibits the release of aldosterone. These two systems work together to maintain homeostasis.

Figure 50.16 Reduced blood flow to the kidney can cause hypertension.

SCIENTIFIC THINKING

Question: *What is the relationship between atherosclerosis and elevated blood pressure?*

Hypothesis: *Atherosclerotic plaques in the renal artery will reduce blood flow to the kidney and consequently filtration pressure. The system will respond by raising blood pressure.*

Prediction: *If blood flow to the renal artery is reduced, this will mimic the effect of plaque deposition and should result in increased blood pressure.*

Test: *Clamp a renal artery to restrict blood flow, and measure blood pressure before, while clamped, and with clamp removed.*

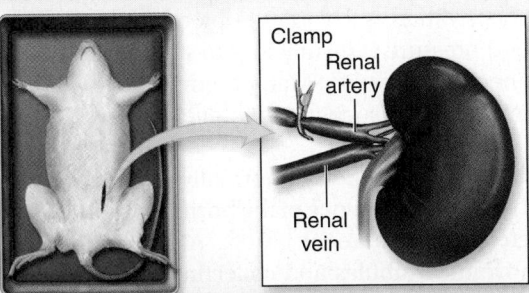

Result: *Clamping the renal artery results in increased blood pressure. This increase is relieved by removing the clamp.*

Conclusion: *Restricting blood flow to the kidney causes a homeostatic response to increase blood pressure.*

Further Experiments: *If this increase in blood pressure involved the renin–angiotensis system, what changes would you expect in the level and function of these proteins?*

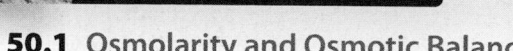
50.1 Osmolarity and Osmotic Balance

Osmotic pressure is a measure of concentration difference.

Osmotic pressure is a solution's propensity to take in water by osmosis. Osmolarity is defined as moles of solute per liter of solution. Cells in a hypertonic solution lose water, and cells in a hypotonic solution gain water. In an isotonic solution, cells are at equilibrium.

Osmoconformers live in marine environments.

Osmoconformers are in osmotic equilibrium with their environment. Most marine invertebrates are osmoconformers.

Osmoregulators control their osmolarity internally.

Tissues of freshwater vertebrates are hypertonic to their environment, and those of marine vertebrates are hypotonic; both must maintain their osmolarity. Terrestrial vertebrates are all osmoregulators and have adaptations to retain water.

50.2 Nitrogenous Wastes: Ammonia, Urea, and Uric Acid

Ammonia is toxic and must be quickly removed.

When animals catabolize amino acids and nucleic acids they produce toxic nitrogenous wastes (figure 50.2).

In fishes and gilled amphibians, abundant water is used to flush ammonia quickly from the body, or it is eliminated through gills.

Urea and uric acid are less toxic but have different solubilities.

Mammals convert ammonia to urea, which is less toxic. Urea requires less water, but requires energy to manufacture. Birds and terrestrial reptiles convert ammonia to uric acid. Uric acid is the least toxic and requires the least water to eliminate, but is the most energetically expensive.

50.3 Osmoregulatory Organs

Invertebrates make use of specialized cells and tubules.

Flatworms use tubular protonephridia connected to flame cells that draw fluid from the body (figure 50.3). Other invertebrates have nephridia open to both the inside and outside of the body for filtering and reabsorption (figure 50.4).

Insects have a unique osmoregulatory system.

Insects have Malpighian tubules through which uric acid and wastes are secreted into the excretory organ, and water and salts are reabsorbed before excretion (figure 50.5).

The vertebrate kidney filters and then reabsorbs.

Kidneys of vertebrates produce urine by filtration, secretion, and selective reabsorption of water and important solutes.

50.4 Evolution of the Vertebrate Kidney

The kidney is made up of a cortex, a medulla, and thousands of units called nephrons that regulate body fluids (figure 50.6).

Freshwater fishes must retain electrolytes and keep water out.

Freshwater fishes maintain hypertonicity by excreting large quantities of dilute, hypotonic urine and reabsorbing ions (figure 50.7).

Marine bony fishes must excrete electrolytes and keep water in.

Saltwater fish maintain hypotonicity by drinking large amounts of seawater and excreting or actively transporting ions out through the gills. Their urine is isotonic to the blood.

Cartilaginous fishes pump out electrolytes and retain urea.

Cartilaginous fishes are isotonic to their environment because they retain urea, and actively pump out electrolytes. They also produce isotonic urine.

Amphibians and reptiles have osmotic adaptations to their environments.

Freshwater amphibians and reptiles act like freshwater fishes, and the kidneys of marine reptiles function like marine bony fishes. Terrestrial reptiles can't produce hypertonic urine.

Mammals and birds are able to excrete concentrated urine and retain water.

Mammals and birds can excrete hypertonic urine and retain water. The degree of concentration of urine depends on the length of the loop of Henle.

50.5 The Mammalian Kidney (figure 50.10)

The nephron is the filtering unit of the kidney.

Blood is filtered through the capillaries of the glomerulus driven by blood pressure. The filtrate passes through the Bowman's capsule, proximal convoluted tubule, loop of Henle, distal convoluted tubule, and collecting duct (figure 50.11). Blood passes from the afferent arteriole to the glomerulus, the efferent arteriole, the peritubular capillaries, and the vasa recta.

Water, some nutrients, and some ions are reabsorbed; other molecules are secreted.

Glucose, amino acids, and many other molecules are reabsorbed by active transport. Water is reabsorbed osmotically in the proximal convoluted tubule and the collecting duct. Foreign molecules and some wastes are secreted from the blood capillaries and into the tubules by active transport.

Excretion of toxins and excess ions maintains homeostasis.

The kidney eliminates many potentially harmful substances including nitrogenous wastes, excess ions, and toxins. The kidneys therefore are critical to homeostasis.

Each part of the mammalian nephron performs a specific transport function.

Reabsorption of nutrients and NaCl occurs in the proximal tubule. The loop of Henle creates a gradient of increasing osmolarity from the cortex to the medulla. The gradient allows selective reabsorption of water from the collecting duct. A longer loop of Henle produces more concentrated urine (figure 50.12).

50.6 Hormonal Control of Osmoregulatory Functions

Antidiuretic hormone causes water to be conserved.

ADH, produced by the hypothalamus, increases the permeability of the collecting duct (figure 50.14), allowing greater reabsorption of water.

Aldosterone and atrial natriuretic hormone control sodium ion concentration.

Low Na^+ levels inhibit ADH secretion, and aldosterone stimulates Na^+ uptake by the distal convoluted tubule. ANH antagonizes the action of aldosterone.

UNDERSTAND

1. Which of the following is NOT an ion homeostatically maintained in vertebrates?

   a. Cl^-
   b. Na^+
   c. Ca^{2+}
   d. Fl^-

2. Suppose that your research mentor has decided to do a project on the filtering capabilities of Malpighian tubules. Which of the following creatures will you be spending your summer studying?

   a. Ants
   b. Birds
   c. Mammals
   d. Earthworms

3. A shark's blood is isotonic to the surrounding seawater because of the reabsorption of ____ in its blood.

   a. ammonia
   b. uric acid
   c. urea
   d. NaCl

4. An important function of the excretory system is to eliminate excess nitrogen produced by metabolic processes. Which of the following organisms is most efficient at packaging nitrogen for excretion?

   a. Frog
   b. Freshwater fish
   c. Iguana
   d. Camel

5. Which of the following is a function of the kidneys?

   a. The kidneys remove harmful substances from the body.
   b. The kidneys recapture water for use by the body.
   c. The kidneys regulate the levels of salt in the blood.
   d. All of the choices are correct.

6. Humans excrete their excess nitrogenous wastes as

   a. uric acid crystals.
   b. compounds containing protein.
   c. ammonia.
   d. urea.

7. An osmoregulator would maintain its internal fluids at a concentration that is ____ relative to its surroundings.

   a. isotonic
   b. hypertonic
   c. hypotonic
   d. hypertonic, hypotonic, or isotonic

APPLY

1. In comparing invertebrate and vertebrate excretory systems, you conclude that

   a. both filter body fluids and then reabsorb water and solutes.
   b. both use a tubule system to process the filtrate.
   c. both reabsorb ions and water to control osmotic balance.
   d. only vertebrates filter fluids and reabsorb water and ions.

2. A viral infection that specifically interferes with the reabsorption of ions from the glomerular filtrate would attack cells located in the

   a. Bowman's capsule.
   b. glomerulus.
   c. renal tubules.
   d. collecting duct.

3. Diuretics are drugs that can be used to treat high blood pressure by increasing urinary output. Possible mechanisms of action in the kidney include

   a. increasing ADH secretion.
   b. inhibition of NaCl reabsorption from the loop of Henle or the proximal tubule.
   c. increasing permeability of the collecting duct.
   d. increasing NaCl reabsorption in the proximal tubule.

4. Caffeine inhibits the secretion of ADH. Prior to an exam, you have a large coffee. During the exam, you can expect

   a. greater water reabsorption from the collecting duct.
   b. less water reabsorption from the collecting duct.
   c. an increase in reabsorption of glucose from the proximal convoluted tubule.
   d. a decrease in reabsorption of glucose from the proximal convoluted tubule.

5. You and your study partner want to draw the pathway that controls the reabsorption of sodium ion when blood pressure falls. Which of the following is the correct sequence of events?

   1. Aldosterone is released.
   2. Kidney tubules reabsorb Na^+.
   3. Renin is released.
   4. Juxtaglomerular apparatus recognizes a drop in blood pressure.
   5. Angiotensin II is produced.

   a. 1, 3, 5, 2, 4
   b. 4, 2, 3, 1, 5
   c. 4, 3, 5, 1, 2
   d. 2, 4, 3, 1, 5

6. You are studying renal function in different species of mammals that are found in very different environments. You look at species from a desert environment and compare them with ones from a tropical environment. The desert species would be expected to have

   a. shorter loops of Henle than the tropical species.
   b. longer loops of Henle than the tropical species.
   c. shorter proximal convoluted tubule than the tropical species.
   d. longer distal convoluted tubules than the tropical species.

SYNTHESIZE

1. Indicate the areas of the nephron that the following hormones target, and describe when and how the hormones elicit their actions.

   a. Antidiuretic hormone
   b. Aldosterone
   c. Atrial natriuretic hormone

2. John's doctor is concerned that John's kidneys may not be functioning properly due to a circulatory condition. The doctor wants to determine if the blood volume flowing through the kidneys (called renal blood flow rate) is within normal range. Calculate what would be a "normal" renal blood flow rate based on the following information:

   John weighs 90 kg. Assume a normal total blood volume is 80 mL/kg of body weight, and a normal heart pumps the total blood volume through the heart once per minute (cardiac output). Also assume that the normal renal blood flow rate is 21% of cardiac output.

ONLINE RESOURCE

www.ravenbiology.com

Understand, Apply, and Synthesize—enhance your study with animations that bring concepts to life and practice tests to assess your understanding. Your instructor may also recommend the interactive eBook, individualized learning tools, and more.

Chapter

51

The Immune System

Chapter Contents

Introduction

When you consider how animals defend themselves, it is natural to think of turtles and armadillos with their obvious external armor. However, armor offers little protection against the greatest dangers vertebrates face—microorganisms and viruses. We live in a world awash with organisms too tiny to see with the naked eye, and no vertebrate could long withstand their onslaught unprotected. We survive because we have evolved a variety of very effective defenses. However, our defenses are far from perfect. Some 40 million people died from influenza in 1918–1919, and more than a million people will die of malaria this year. Attempts to improve our defenses against infectious diseases are being actively researched.

51.1 Innate Immunity

Learning Outcomes

1. *Distinguish between innate and adaptive immunity.*
2. *Explain how the innate system recognizes pathogens.*
3. *Describe the inflammatory response.*

For many years, the response of vertebrates to microbial invasion was divided into specific and nonspecific forms of defense. It has now become clear that this is not only an oversimplification, it is not an accurate view of the immune system. We now view the response as much more integrated and consisting of two parts: innate and adaptive immunity. The key to the function of the immune system is the ability to distinguish self from nonself cells, and the two branches of immunity do this in very different ways.

The innate system recognizes specific molecules that are conserved in particular pathogens, such as lipopolysaccharide in gram-negative bacteria. The receptor proteins that bind to these conserved proteins do not result from the kind of genomic rearrangements found in adaptive immunity, but do involve recognition of invading pathogens. The characteristic of this system is a rapid response that brings cells to the site of infection and uses soluble antimicrobial proteins to fight the pathogen.

Innate immunity is evolutionarily ancient, with some of the proteins involved recently being identified in cnidarians.

Parts of the complement system (described later) have also been identified in horseshoe crabs, indicating that this system is also more ancient than previously thought. This also implies that the lack of complement in other protostomes is probably due to loss. Together, this implies that the ancestor to all bilaterians had some form of innate immunity.

Adaptive immunity is characterized by the genetic rearrangements that generate a diverse set of molecules that can recognize virtually any invading pathogen. This is the basis for a slower, but highly specific response to invading pathogens, and for the more rapid response to a second attack that is the basis for vaccines. In this chapter, we will discuss innate and adaptive immunity and how they are interrelated. We will begin with a brief description of the barrier that a pathogen must cross to gain access to the interior of the body.

The skin is a barrier to infection

The skin is the largest organ of the body, accounting for 15% of an adult human's total weight. The integument not only defends the body by providing a nearly impenetrable barrier, but also reinforces this defense with chemical weapons on the surface. Oil and sweat glands give the skin's surface a pH of 3 to 5, which is acidic enough to inhibit the growth of many pathogenic microorganisms. Sweat also contains the enzyme lysozyme, which digests bacterial cell walls. Epithelial cells also produce a variety of small antimicrobial peptides.

The skin is also home to many normal flora, nonpathogenic bacteria or fungi that are well adapted to the skin conditions in different regions of the body. Pathogenic bacteria that might attempt to colonize the skin generally are unable to compete with the normal flora. The epidermis of skin is approximately 10 to 30 cells thick, about as thick as this page. The outer layer contains cells that are continuously abraded, injured, and worn by friction and stress during the body's many activities. Cells are shed continuously and are replaced by new cells produced in the innermost layer of the epidermis.

Mucosal epithelial surfaces also prevent entry of pathogens

In addition to the skin, three other potential routes of entry by microorganisms and viruses must be guarded: the digestive tract, the respiratory tract, and the urogenital tract. Recall that each of these tracts opens to the external environment. Each of these tracts is lined by epithelial cells, which are continuously replaced, as are those of the skin.

A layer of mucus, secreted by specialized cells scattered between the epithelial cells, covers all these epithelial surfaces. Pathogens are frequently trapped within this mucus layer and are eliminated by mechanisms specific to the particular tract.

Microbes are present in food, but many are killed by saliva (which contains lysozyme), by the very acidic environment of the stomach, and by digestive enzymes in the intestine. Additionally, the gastrointestinal tract is home to a vast array of nonpathogenic normal flora, whose presence inhibits the growth of pathogenic competitors. These nonpathogenic organisms not only outcompete pathogens, but they also may secrete substances that kill harmful agents.

Microorganisms present in inhaled air are trapped by the mucus within the smaller bronchi and bronchioles before they can reach the warm, moist lungs, which would provide ideal breeding grounds for them. The epithelial cells lining these passages have cilia that continually sweep the mucus toward the glottis. There the mucus can be swallowed, carrying potential invaders out of the lungs and into the digestive tract. One of the pitfalls of smoking is that nicotine paralyzes the cilia of the respiratory system so that this natural cleaning of the air passages does not take place.

Vaginal secretions are sticky and acidic, and they also promote the growth of normal flora; all of these characteristics help prevent foreign invasion. In both males and females, acidic urine continually washes potential pathogens from the urinary tract. In addition to these physical and chemical barriers to pathogen invasion, the body also uses defense mechanisms such as vomiting, diarrhea, coughing, and sneezing to expel potential pathogens.

Innate immunity recognizes characteristic pathogen molecules

Innate immunity is a response to invading pathogens that involves both soluble factors and a variety of different types of blood cells. This innate response to invading pathogens is based on the recognition of molecules that are characteristic of a type of pathogen. Examples include the lipopolysaccharide (LPS) found in gram-negative bacterial cell walls; peptidoglycan, which is found in all bacterial cell walls; and viral DNA and RNA. Different receptor proteins exist that bind to each of these classes of molecules. These receptors can be soluble proteins or membrane proteins on the surface of blood cells.

Toll-like receptors

The best studied of these innate receptor proteins is the Toll receptor in *Drosophila* and the Toll-like receptors (TLR) found in many species. In *Drosophila*, Toll was originally discovered as a part of the dorsal–ventral patterning pathway. Later, the same membrane receptor was found to mediate a response to fungal infection.

In vertebrates, 11 TLRs have been found in humans and 13 in the mouse. These bind to a variety of specific targets important to pathogen survival, which therefore do not vary greatly. These include gram-negative LPS, bacterial lipoproteins, bacterial peptidoglycan fragments, yeast cell-wall components, unmethylated CpG motifs in bacterial DNA, and viral RNA. This represents a wide range of possible invading pathogens that vertebrates have been host to over a long period of evolutionary time.

TLRs contain repeated leucine-rich regions that fold to form binding pockets that can accommodate a variety of shapes. In different TLRs, these pockets recognize different classes of molecules. Because the molecules binding to these pockets are critical to the pathogen, a single receptor type can recognize a range of pathogens that share a feature such as LPS or peptidoglycan.

Activation of innate receptors turns on signal transduction pathways that enhance the response of both innate and adaptive immune responses. This provides a connection

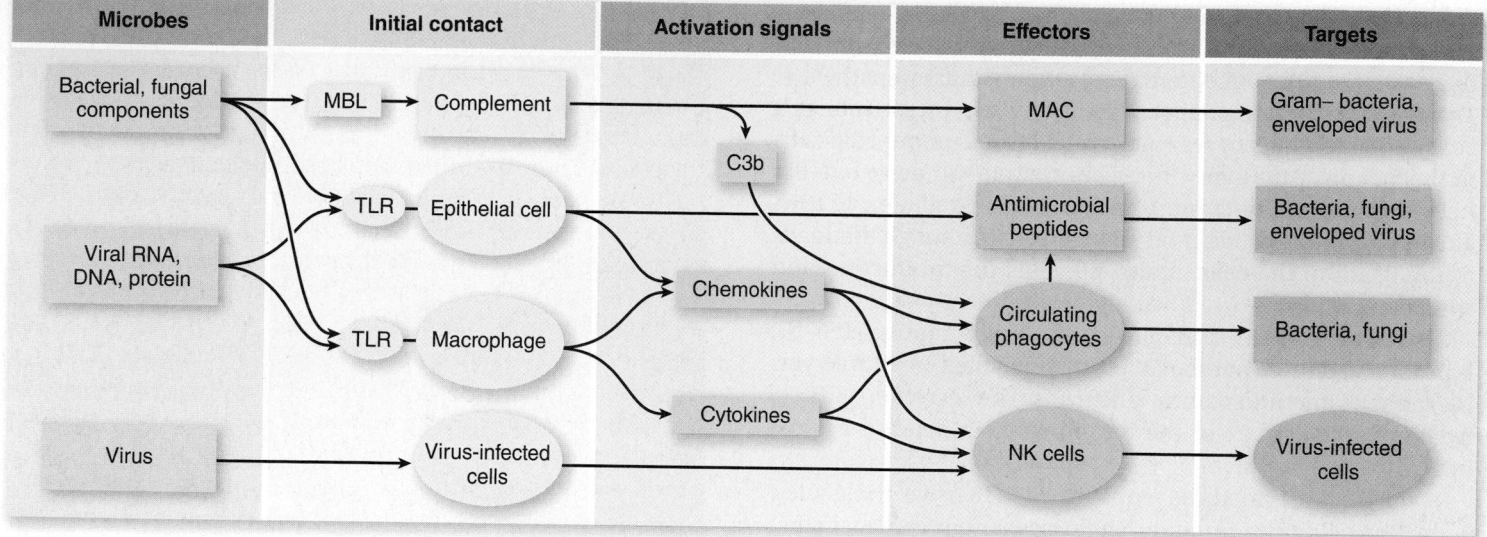

Figure 51.1 Overview of innate immunity. Pathogens have critical molecules that adhere to either membrane-bound (TLR), or soluble receptors (MBL). This results in the production of cytokines and chemokines that attract phagocytes, of antimicrobial peptides, the membrane attack complex (MAC) of the complement cascade, and the activation of natural killer cells (NK cells).

between innate and adaptive immunity. This leads to induction of the inflammatory response (described later on); to the production of antimicrobial peptides; and to the production of cytokines that attract phagocytic cells as well as B and T cells. All of these together make up the innate response to infection.

Cytoplasmic receptors

Since the discovery of Toll and TLR proteins, a second class of receptors was discovered localized to the cytoplasm. These internal receptors also bind to characteristic pathogen molecules and can recognize invading pathogens in the cytoplasm of cells after phagocytosis. These receptors also are part of the response to viral RNA.

Soluble receptors

Soluble receptors also circulate in serum and can respond to specific pathogen molecules, including some of the lectin family that bind to mannose, which is found in bacterial cell walls. These lectin proteins are important in activating the complement system described later.

Innate immunity leads to diverse responses to a pathogen

Recognition of an invading pathogen by innate receptors such as the TLR family leads to a signal transduction pathway (see chapter 19) to produce a coordinated response. This includes the production of secreted signaling molecules, production of antimicrobial peptides, and activation of complement. This response is shown in overview in figure 51.1. The antimicrobial peptides are similar to those normally found in the integument. There are several classes of these peptides with different modes of action. One well-studied example is the defensins. The cysteines in defensins interact with positively charged amino acids on the surface of a pathogen. By binding to the outer membrane of gram-negative bacteria, they can both

disrupt the membrane and enhance phagocytosis. This action is also effective against enveloped viruses (figure 51.2). The antimicrobial enzyme lysozyme that can degrade bacterial cell walls can also be induced by this response.

Another class of proteins induced by innate defenses that plays a key role in body defense are interferons. Interferons are

---

SCIENTIFIC THINKING

Hypothesis: *Defensin peptides have activity against viruses.*

Prediction: *Viruses incubated with the human neutrophil defensin one (HNP-1) in vitro will have reduced infectivity.*

Test: *Direct test by incubating different viruses with HNP-1, then using standard infectivity assay expressed as plaque-forming units (PFU)/mL.*

Virus	Mean log$_{10}$ Reduction in PFU/mL		
	25 µg/mL	50 µg/mL	100 µg/mL
HSV-1	2	2.9	3
HSV-2	0.8	1.2	2
Vesicular stomatitus virus	0.4	0.7	0.9
Influenza virus	0.4	0.5	0.7
Cytomegalovirus	-0.02	0.09	0.3

Result: *HNP-1 has activity against a variety of viruses as shown in the table.*

Conclusion: *HNP-1 has activity against different enveloped viruses, although this activity is not equally effective against the different viruses.*

Further Experiments: *How could you determine the mechanisms of action?*

Figure 51.2 Activity of defensin against different viruses.

important secreted signaling molecules with diverse functions. The two classes of interferons, type I and type II, have different receptors and activate different signal transduction pathways. The type I interferons are synthesized when a virus infects a cell and act as messengers that protect normal uninfected cells in the vicinity. Although viruses are still able to penetrate the neighboring cells, interferons induce the degradation of RNA and block protein production in these cells. Although this leads to the death of the cells, it also prevents the production and spread of the virus.

Type II interferon, in humans interferon gamma (IFN-γ), is produced only by particular leukocytes called T lymphocytes (described later) and natural killer cells. The secretion of IFN-γ by these cells is part of the immunological defense against infection and cancer.

In addition to these nonspecific defensive molecules, activation of innate immunity leads to two other responses that are important enough to be discussed separately later: activation of the inflammatory response and activation of the complement pathway. Signaling from both TLR and internal receptors can lead to the secretion of a variety of cytokines, or regulatory signaling molecules. These attract other nonspecific phagocytic cells, cause inflammation, and even signal to the adaptive immune system.

Phagocytic cells are associated with innate immunity

Among the most important innate defenses are cells that can nonspecifically kill invading pathogens. These are a type of leukocyte, or white blood cell, that circulates through the body and attacks pathogens within tissues (see chapter 49 for an overview of blood and blood cells). Three basic kinds of defending leukocytes have been identified, and each kills invading microorganisms differently.

Macrophages

Macrophages ("big eaters") are large, irregularly shaped cells that kill microorganisms by ingesting them through phagocytosis (figure 51.3). Once within the macrophage, the membrane-bound phagosome fuses with a lysosome. Fusion activates lysosomal enzymes that kill and digest the microorganism. Additionally, large quantities of oxygen-containing free radicals are frequently produced within the phagosome; these free radicals are very reactive and degrade the pathogen.

In addition to bacteria, macrophages also engulf viruses, cellular debris, and dust particles in the lungs. Macrophages roam continuously in the extracellular fluid that bathes tissues. In response to an infection, monocytes, which are undifferentiated macrophages found in the blood, squeeze through the endothelial cells of capillaries to enter the connective tissues. There, at the site of the infection, the monocytes mature into active, phagocytic macrophages.

Neutrophils

Neutrophils are the most abundant circulating leukocytes, accounting for 50 to 70% of the peripheral blood leukocytes. They are the first type of cell to appear at the site of tissue

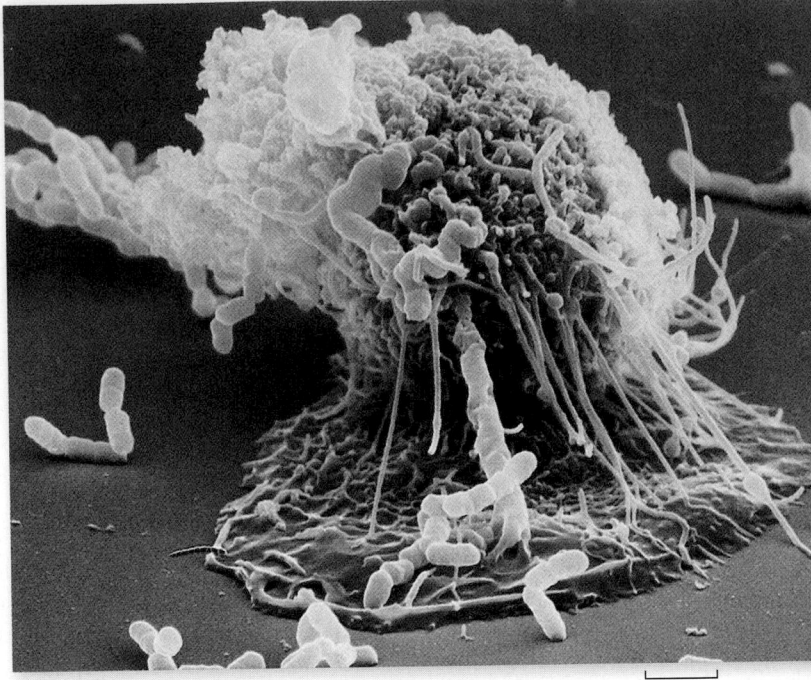

Figure 51.3 A macrophage in action. In this scanning electron micrograph, a macrophage is "fishing" with long, sticky cytoplasmic extensions. Bacterial cells that come in contact with the extensions are drawn toward the macrophage and engulfed.

damage or infection. Like macrophages, they squeeze between capillary endothelial cells to enter infected tissues, where they ingest a variety of pathogens by phagocytosis. Their mechanism of pathogen destruction is similar to that of macrophages except that they produce an even greater range of reactive oxygen radicals. Neutrophils also produce defensin peptides.

Natural killer cells

Natural killer (NK) cells do not attack invading microbes, instead they kill cells of the body that have been infected with viruses. They kill not by phagocytosis, but rather by inducing apoptosis (programmed cell death) of the target cell (see chapter 19). Proteins called perforins, released from the NK cells, insert into the membrane of the target cell creating a pore in the membrane. Other NK-produced proteins, called granzymes, enter these pores and activate proteins in the target cells that induce apoptosis (figure 51.4). Macrophages ingest the resulting membrane-bounded vesicular cell debris.

NK cells also attack tumor cells, often before the tumor cells have had a chance to divide sufficiently to be detectable as a tumor. The vigilant surveillance by NK cells is one of the body's most potent defenses against cancer. Thus, these cells are often said to play a role in immune surveillance.

The inflammatory response is a nonspecific response to infection or tissue injury

The inflammatory response involves several systems of the body, and it may be either localized or systemic. An acute

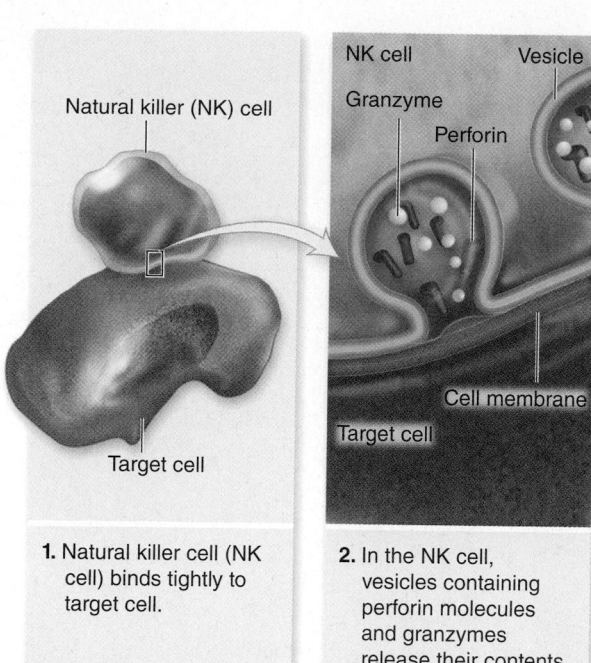
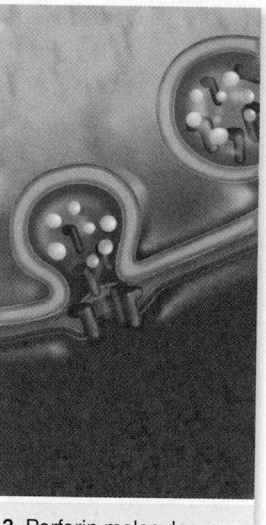
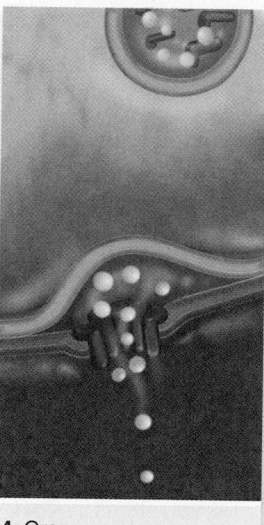
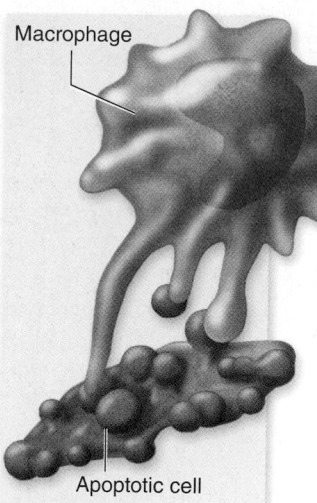

Natural killer (NK) cell

Target cell

NK cell   Vesicle

Granzyme

Perforin

Cell membrane

Target cell

Macrophage

Apoptotic cell

1. Natural killer cell (NK cell) binds tightly to target cell.

2. In the NK cell, vesicles containing perforin molecules and granzymes release their contents by exocytosis.

3. Perforin molecules polymerize in the plasma membrane of the target cell, forming pores.

4. Granzymes pass through the pores and activate caspase enzymes that induce apoptosis in the target cell.

5. The apoptotic cell is broken down into vesicles. Macrophages then phagocytose these vesicles.

Figure 51.4 How natural killer cells eliminate target cells. Natural killer cells kill virally infected cells by programmed cell death, or apoptosis. This is accomplished by secreting proteins that form pores in the cell to be killed, along with proteins that diffuse through these pores and induce apoptosis.

? Inquiry question What would happen if an NK cell killed a virally infected target cell by simply causing the cell to burst, releasing all the cell contents into the tissues?

response is one that generally starts rapidly but lasts for only a relatively short while.

Certain infected or injured cells release chemical alarm signals—most notably histamine, along with prostaglandins and bradykinin (see chapter 45). These chemicals cause vasodilation of local blood vessels. This increases the flow of blood to the site and causes the area to become red and warm, two of the hallmark signs of inflammation. These chemicals also increase the permeability of capillaries in the area, producing the third hallmark sign of inflammation, the edema (tissue swelling) often associated with infection. Swelling puts pressure on nerve endings in the region, and this, in combination with the release of other mediators, leads to pain and potential loss of function, the final two hallmark signs of inflammation.

Increased capillary permeability initially promotes the migration of phagocytic neutrophils from the blood to the extracellular fluid bathing the tissues, where the neutrophils can ingest and degrade pathogens; the pus associated with some infections is a mixture of dead or dying pathogens, tissue cells, and neutrophils. The neutrophils also secrete signaling molecules that attract monocytes several hours later; as the monocytes differentiate into macrophages, they too engulf pathogens and the remains of the dead cells (figure 51.5). The inflammatory response is accompanied by an acute-phase response. One manifestation of this response is an elevation of body temperature, or fever (see chapter 42). When a macrophage with a TLR on its surface binds to an invading pathogen,

the cytokine called **interleukin-1 (IL-1)** is released and is carried by the blood to the brain. IL-1 causes neurons in the hypothalamus to raise the body's temperature several degrees above the normal value of 37°C (98.6°F). This increase in body temperature promotes the activity of phagocytic cells and impedes the growth of some microorganisms.

Fever contributes to the body's defense by stimulating phagocytosis and causing the liver and spleen to store iron. This storage reduces blood levels of iron, which bacteria need in large amounts to grow. Very high fevers are hazardous, however, because excessive heat may denature critical enzymes. In general, temperatures greater than 39.4°C (103°F) are considered dangerous for humans, and those greater than 40.6°C (105°F) are often fatal.

A group of proteins collectively referred to as acute-phase proteins are also released from cells of the liver during an inflammatory response, sometimes at levels 1000-fold above the normal serum concentration. These proteins bind to a variety of microorganisms and promote their ingestion by phagocytic cells—the neutrophils and macrophages.

Complement can form a membrane attack complex

The cellular defenses of vertebrates are enhanced by a very effective chemical defense called the **complement system**—a group of approximately 30 different proteins that circulate freely in the blood plasma. Usually they occur in an inactive form, which can enter the tissues during an inflammatory response. Complement can be activated by mannose-binding

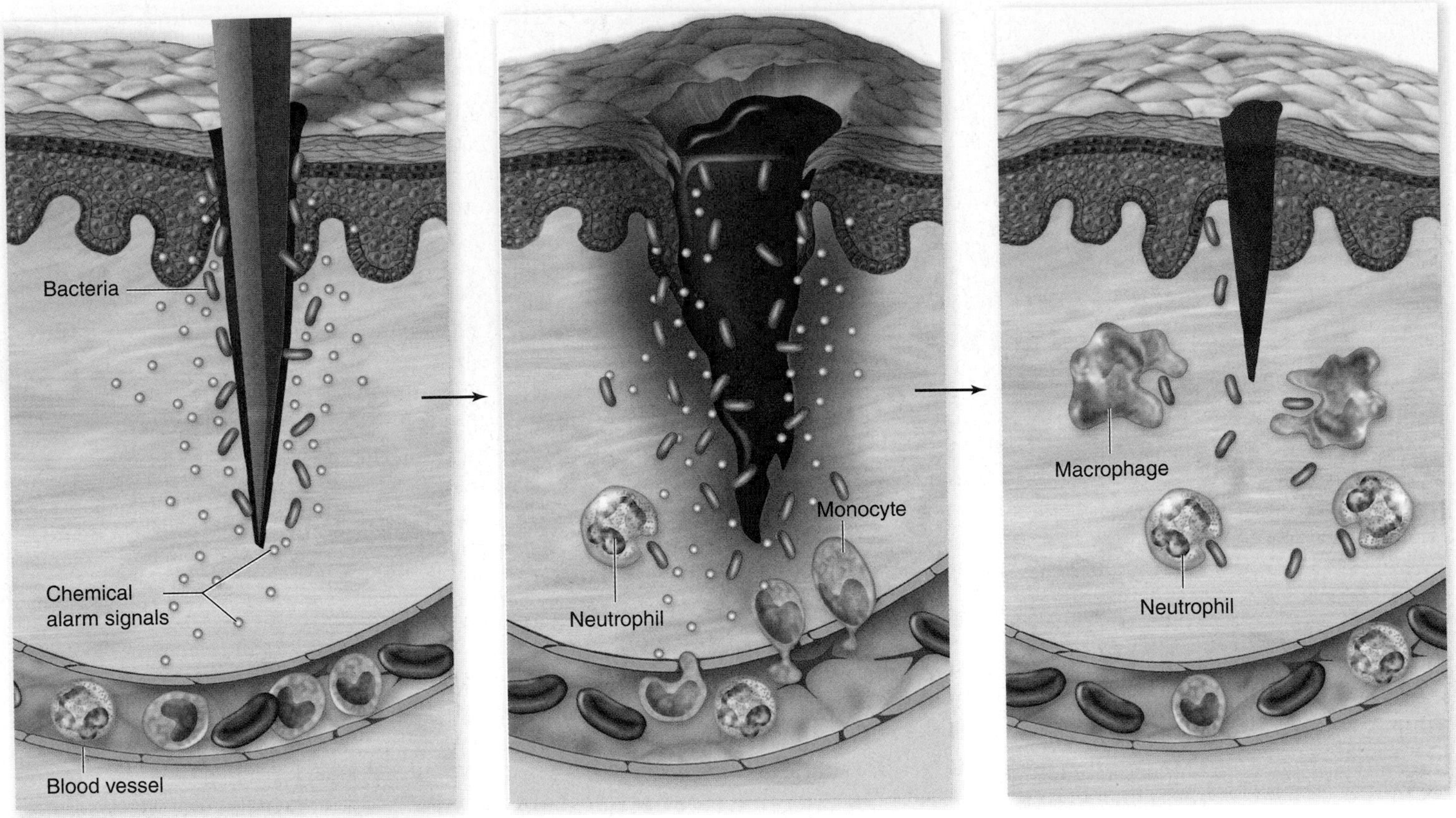

Figure 51.5 The events in local inflammation. When invading pathogens have penetrated an epithelial surface, chemical alarm signals, such as histamine and prostaglandins, released from damaged cells, cause nearby blood vessels to dilate and increase in permeability. Increased blood flow causes swelling and promotes the accumulation of phagocytic cells, specifically neutrophils followed by macrophages, which attack and engulf invading the pathogens.

lectin protein (MBL), one of the soluble sensors of innate immunity, or can be activated by a complex series of reactions involving charged species on the surface of pathogens.

When the complement system is activated, complement proteins aggregate to form a **membrane attack complex (MAC)** that inserts itself into the pathogen's plasma membrane (or the lipid membrane around an enveloped virus), forming a pore. Extracellular fluid enters the pathogen through this pore, causing the pathogen to swell and burst. Activation of the complement proteins is also triggered in a specific fashion when antibodies (which are secreted by B lymphocytes) are bound to invading pathogens, as we describe in a later section.

Other complement proteins, particularly one known as *C3b,* may coat the surface of invading pathogens. Phagocytic neutrophils and macrophages, which have receptors for C3b, may thus be "directed" to bind to the pathogens, promoting their phagocytosis and destruction. This is a particularly efficient way of eliminating those pathogens that do not have an outer lipid membrane into which a MAC may insert itself. Some complement proteins stimulate the release of histamine and other mediators by cells known as mast cells and basophils, promoting the dilation and increased permeability of capillaries; other complement proteins attract more phagocytes, especially neutrophils, to the area of infection through the more-permeable blood vessels.

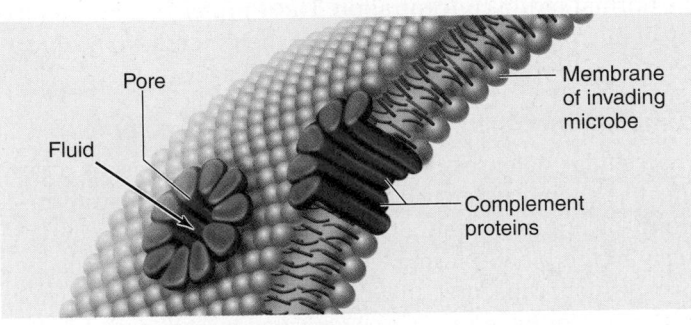

Learning Outcomes Review 52.1

Innate immunity is ancient and recognizes molecular patterns; adaptive immunity involves genetic rearrangements to attack specific pathogens. The molecular patterns that innate immunity recognizes include bacterial lipopolysaccharide and peptidoglycan, as well as viral RNA and DNA. The inflammatory response begins with histamine release and involves a variety of molecules and signals that attract neutrophils, increase permeability, activate the complement system, and trigger fever.

■ *Is innate immunity nonspecific?*

51.2 Adaptive Immunity

Few of us pass through childhood without contracting a variety of infectious illnesses. Prior to the advent of an effective vaccine in about 1991, most children contracted chicken pox before reaching their teens. Chicken pox and some other such diseases were considered diseases of childhood because most people, once recovered, never experienced them again. They developed immunity to the chicken pox–causing *varicella-zoster* virus and maintained this immunity as long as their immune systems remained intact. Similarly, immunization today with a nonpathogenic form of *varicella* virus can also confer protection. This immunity is produced by adaptive immune defense mechanisms, also called acquired immunity.

Immunity had long been observed, but the mechanisms have only recently been understood

Societies have known for over 2000 years that an individual who experiences an infectious disease is often protected against a subsequent occurrence of the same disease. The scientific study of immunity, however, did not begin until 1796, when an English country doctor, Edward Jenner, carried out an experiment to protect people against smallpox.

Jenner and the smallpox virus

Smallpox, caused by the *variola* virus, was a common and deadly disease in the 1700s and earlier centuries. As with chicken pox, those who survived smallpox rarely caught the disease again, and people had been known to deliberately infect themselves through inoculation hoping to survive a mild case and become immune. Jenner observed, however, that milkmaids who had caught a much milder form of "the pox" called cowpox (presumably from cows) rarely experienced smallpox.

Jenner set out to test the idea that cowpox could confer protection against smallpox. He inoculated a healthy child with fluid from a cowpox vesicle and later deliberately infected him with fluid from a smallpox vesicle; as he had predicted, the child did not become ill. (Jenner's experiment would be considered unethical today.) Subsequently, many people were protected from smallpox by immunization with fluid from cowpox vesicles, a much less risky proposition (figure 51.6).

We now know that smallpox and cowpox are caused by two different viruses that have similar surfaces. Jenner's patients who were injected with the cowpox virus mounted a defense that was also effective against a later infection of the smallpox virus.

Jenner's procedure of injecting a harmless agent to confer resistance to a dangerous one is called *vaccination*. Modern attempts to develop resistance to malaria, herpes, and other diseases often involve delivering antigens via a harmless *vaccinia* virus related to the cowpox virus (see chapter 17).

Pasteur and avian cholera

Many years passed before anyone learned how exposure to an infectious agent could confer resistance to a disease. A key step toward answering this question was taken more than a half-century later by the famous French scientist Louis Pasteur. Pasteur was studying avian cholera, a form that infects birds. He isolated a culture of bacteria from diseased chickens that would produce the disease if injected into healthy birds.

It is reported that before departing on a two-week vacation, he accidentally left his bacterial culture out on a shelf. When he returned, he injected this old culture into healthy birds and found that it had apparently been weakened; the injected birds became only slightly ill and then recovered. Surprisingly, however, those birds did not get sick when subsequently infected with fresh cholera bacteria that did produce the disease in control chickens. Clearly, something about the bacteria could elicit immunity, as long as the bacteria did not kill the animals first. We now know that molecules protruding from the surfaces of the bacteria evoked active immunity in the chickens.

Antigens stimulate specific immune responses

An **antigen** is a molecule that provokes a specific immune response. The most effective antigens are large, complex molecules such as proteins. The greater their "foreignness," or put another way, their phylogenetic distance from the host, the greater will be the immune response they elicit.

Figure 51.6
The birth of immunology.
This painting shows Edward Jenner inoculating patients with cowpox in the 1790s and thus protecting them from smallpox.

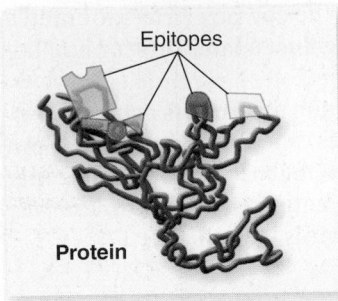

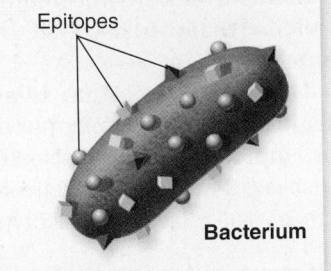

Figure 51.7 **Many different epitopes are exhibited by any one antigen.** *a.* A single protein, with associated carbohydrate, may have many different antigenic determinants called epitopes, each of which can stimulate a distinct immune response. *b.* A pathogen such as a bacterium has many proteins on its surface, and there are likely to be multiple copies of each. Note that the protein and bacterium are not drawn to scale with respect to each other.

Antigens may be components of a microorganism or a virus, but they may also be proteins or glycoproteins on the surface of transfused red blood cells or on transplanted tissue. They may also be components of foods or pollens. A large antigen is likely to have many different parts, known as *antigenic determinants,* or *epitopes* (figure 51.7), each of which can stimulate a distinct immune response.

Hematopoiesis gives rise to the cells of the immune system

All the cells that are found in the blood are derived from the division and differentiation of hematopoietic stem cells, a process called hematopoiesis (see chapter 49). Embryologically, these stem cells are initially found in the yolk sac, then migrate to the fetal liver and spleen and finally to the bone marrow. Stem cells give rise to lymphoid progenitors and myeloid progenitors. A lymphoid progenitor, in turn, gives rise to both the B and T lymphocytes as well as to natural killer cells. A myeloid progenitor gives rise to all the other cells of the immune system as well as to erythrocytes and platelets (see figure 49.2).

Although the lymphocytes are responsible for adaptive immunity, all the other leukocytes illustrated in figure 49.2 play supporting roles in this specific response or are part of innate immunity. **Monocytes** give rise to the macrophages, and these along with the **neutrophils** are phagocytic cells. **Eosinophils** are important in the elimination of helminths (flatworms; see chapter 34), either via secretion of digestive enzymes through perforin pores inserted in the plasma membrane of the helminths or occasionally by phagocytosis. They also play a role in exacerbating chronic inflammatory diseases such as asthma or inflammatory bowel disease.

Basophils and **mast cells** are not phagocytic but rather secrete inflammatory mediators such as histamine and prostaglandin in response to the binding of complement proteins during the elimination of pathogens. These cells, and mast cells in particular, are also activated during an allergic response,

TABLE 51.1	Cells of the Immune System	
Cell Type		**Function**
Helper T cell		Specifically recognizes foreign peptides on antigen-presenting cells, inducing the release of cytokines that activate B cells or macrophages
Cytotoxic T cell		Specifically recognizes and kills "altered-self" cells: virally infected or tumor cells
B cell		Binds specific soluble antigens with its membrane-bound antibody; serves as an antigen-presenting cell to T$_H$ cells; on activation differentiates into plasma and memory B cells
Plasma cell		Derived from activated B cell; is a biochemical factory devoted to the secretion of antibodies directed against specific antigens
Natural killer cell		Rapidly recognizes and kills virally infected or tumor cells
Monocyte		Precursor of macrophage; located in blood
Macrophage		Phagocytic tissue cell that is a component of the body's first cellular line of defense; also serves as an antigen-presenting cell to T$_H$ cells
Neutrophil		A phagocytic cell that is a component of the body's first cellular line of defense; found in the blood in large numbers until attracted to tissues during inflammation
Eosinophil		Important to the elimination of parasites and involved in chronic inflammatory diseases
Basophil		Circulating cell that releases mediators such as histamine that promote inflammation
Mast cell		Located primarily under mucosal surfaces and releases mediators such as histamine that promote inflammation; triggered both during inflammatory and allergic responses
Dendritic cell		Important antigen-presenting cell to naive T$_H$ cells; also helps in the activation of naive T$_C$ cells

and the inflammatory mediators they release cause the symptoms of allergy.

Dendritic cells are important in the activation of T cells, as will be described further. Dendritic cells also form a link between innate and adaptive immunity. Dendritic cells have a variety of TLRs that recognize pathogens and stimulate secretion of cytokines and the inflammatory response. Thus they can present antigens and recognize pathogen patterns via innate receptors as well. The roles of these cells are summarized in table 51.1.

Lymphocytes carry out the adaptive immune responses

The adaptive immune system is characterized by

1. Specificity of recognition of antigen
2. Wide diversity of antigens can be specifically recognized
3. Memory, whereby the immune system responds more quickly and more intensely to an antigen it encountered previously than to one it is meeting for the first time
4. Ability to distinguish self-antigens from nonself

The cells in the blood involved in the adaptive immune response are leukocytes derived from a stem cell line called lymphoid progenitor cells (see figure 49.2). These **lymphocytes** have receptor proteins on their surfaces that recognize specific epitopes on an antigen and direct an immune response against either the antigen in solution or on the cell surface (figure 51.8). This response is also affected by signals derived from the innate system described earlier. The innate system dominates early in infection by a new pathogen, and the adaptive response dominates in later stages of infection.

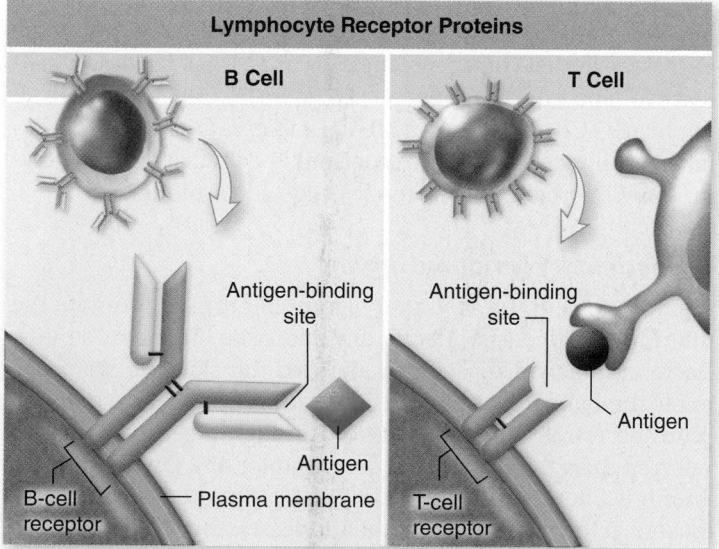

Figure 51.8 **B- and T-cell receptors bind antigens.**
B-cell receptors are immunoglobulin (Ig) molecules with a characteristic Y-shaped structure. Every B cell has a single kind of Ig on its surface that binds to a single antigenic determinant. T-cell receptors are simpler than Ig molecules, but also bind to specific antigenic determinants. T cells only bind to antigens bound to another cell.

Lymphocytes and antigen recognition

Although all the receptor proteins on any one lymphocyte have the same epitope specificity, it is rare that any two lymphocytes have identical specificities. This feature produces the diversity of immune responses that ensures that at least some epitopes of any antigen that might be encountered are recognized.

A lymphocyte that has never before encountered antigen is referred to as a *naive lymphocyte.* When a naive lymphocyte binds to a foreign antigen, the lymphocyte is activated, causing it to divide producing a clone of cells with identical antigen specificity, a process called **clonal selection.** Some of these cells respond immediately to the antigen, and others become memory cells, which can remain in our bodies for years and perhaps for the remainder of our lives. Memory cells are easily and rapidly activated on subsequent encounters with the same antigen.

B cells

Lymphocytes called **B lymphocytes,** or **B cells,** respond to antigens by secreting proteins called **antibodies,** or **immunoglobulins (Ig).** Antigen recognition occurs when an antigen binds to immunoglobulins on the B cell's membrane. Binding to antigen, in conjunction with other signals to be described later, initiates a signaling pathway that leads to the production of plasma cells that secrete antibodies specific for the epitope recognized by the antibody in the B-cell membrane. This B-cell–mediated response producing secreted antibodies is called **humoral immunity.**

T cells

Other lymphocytes, called **T lymphocytes,** or **T cells,** do not secrete antibodies but instead regulate the immune responses of other cells or directly attack the cells that carry the specific antigens. These cells participate in the other arm of adaptive immunity called **cell-mediated immunity.** Both cell-mediated and humoral immunity processes are described in detail in later sections.

? Inquiry question Jenner used cowpox virus to elicit an immune response against smallpox. What does this tell us about the antigenic properties of the two viruses?

Adaptive immunity can be active or passive

Immunity can be acquired in different ways. First, an individual can gain immunity when infected by a pathogen and perhaps developing the disease it causes. Alternatively, an individual can be immunized with portions of a pathogen or with a less virulent form of the pathogen. Both of these situations result in *active immunity,* associated with the activation of specific lymphocytes and the generation of memory cells by the individual. Second, an individual can gain immunity by obtaining antibodies from another individual. This happened to you before you were born, as some antibodies made by your mother were transferred to your body across the placenta. Immunity gained in this way is called *passive immunity,* and it does not result in the generation of memory cells. The immunity is only effective as long as the antibodies remain in your body. Like any other proteins, they will degrade in time.

The immune system is supported by two classes of organs

The organs of the immune system consist of the **primary lymphoid organs**—the bone marrow and the thymus—as well as the **secondary lymphoid organs**—the lymph nodes, spleen, and mucosa-associated lymphoid tissue, or MALT (figure 51.9).

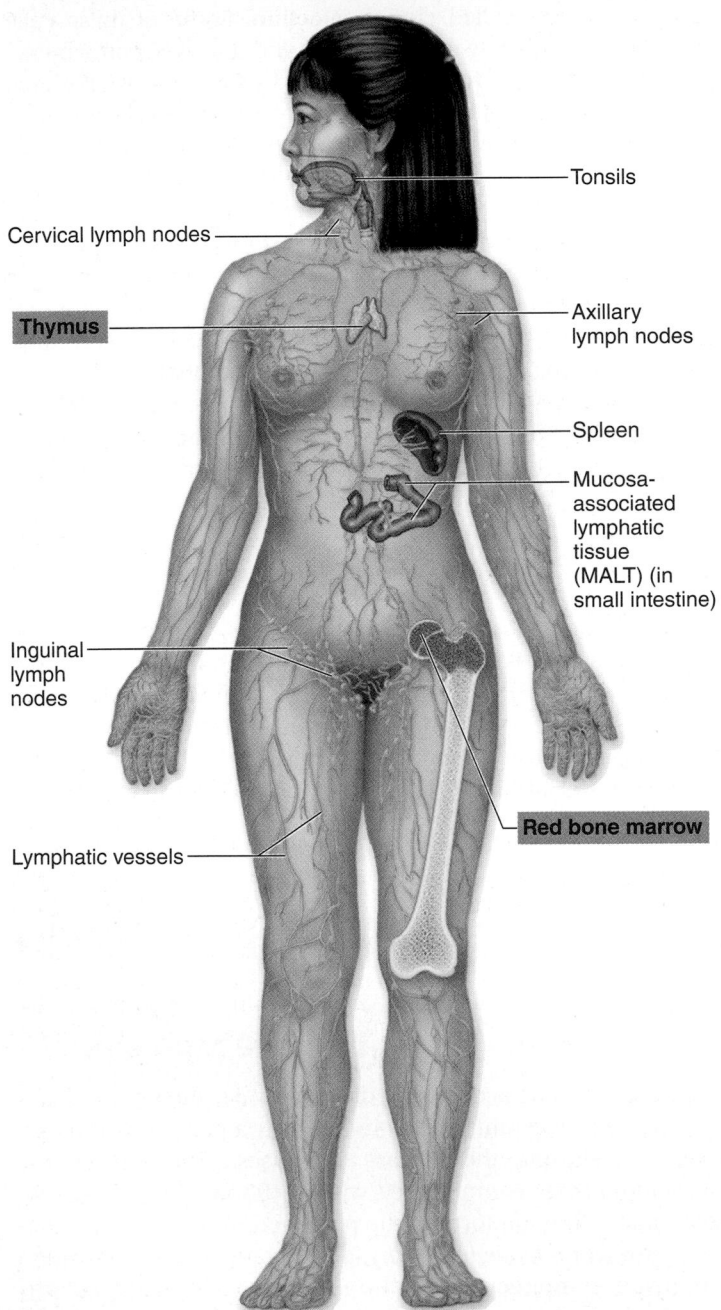

Figure 51.9 Organs of the specific immune system.
There are two types of immune system organs: primary lymphoid organs (red boxes), in which B and T lymphocytes mature and acquire their specific receptors, and secondary lymphoid organs (labeled in black) in which antigen is collected and through which the mature naive lymphocytes circulate in order to meet and be stimulated by antigen.

The primary lymphoid organs

The **bone marrow** is not only the source of stem cells, it is where B cells mature. After hematopoiesis gives rise to the most immature B cells, progenitor B cells, these cells complete their maturation in the bone marrow. It is here that DNA rearrangements of the immunoglobulin genes, to be discussed later, dictate the specificity of each B cell. Every B cell has about 10^5 Ig molecules on its surface, all with identical specificity of epitope binding and all different from cell to cell.

Any lymphocytes that are likely to bind to self-antigens undergo apoptosis (figure 51.10a). The remainder are released to circulate in the blood and lymph and pass through the secondary lymphoid organs, where they may encounter antigen.

After their origin in the bone marrow, progenitor T cells migrate to the **thymus,** a primary lymphoid organ located just above the heart. The thymus is very large in infants; it starts to shrink in the teenage years, which it continues to do throughout life.

The antigen receptor on T cells is designated the **T-cell receptor,** or **TCR.** The TCR is produced by gene rearrangements as T cells mature in the thymus, similar to those that occur for Ig genes of progenitor B cells. Thus, T cells may express about 10^5 identical TCRs per T cell, all likely to be different from one T cell to the next.

B cells recognize an epitope of an intact antigen that may or may not be a protein. In contrast, T cells recognize only a peptide fragment of a protein antigen, and this peptide fragment must be bound to one of a series of self-proteins that are present on the surface of almost all of the body's cells. These proteins are encoded by genes in the **major histocompatibility complex,** or **MHC.** The MHC is discussed in detail in subsequent sections.

During selection in the thymus, T cells are exposed to many thymic cells, all expressing self-MHC proteins with bound self-peptides on their surfaces. If a T cell's TCRs bind too strongly to these self-MHC protein complexes, that T cell becomes "self-reactive" and undergoes apoptosis (figure 51.10b). Conversely, if the T cell's TCR does not bind MHC complexes at all, it is also eliminated. Only about 5% of the progenitor T cells that enter the thymus pass this rigorous two-step selection and avoid apoptosis.

The secondary lymphoid organs

The locations of the secondary lymphoid organs promote the filtering of antigens that enter any part of an individual's body. Bacteria attached to a thorn stuck in the skin, for example, enter the lymph that bathes the tissues. Lymph is eventually returned to the blood circulation through a series of vessels referred to as the lymphatics (see chapter 49). On its way, the lymph is filtered in the thousands of lymph nodes, which are located at the junction of lymphatic vessels (see figure 51.9).

The many mature but naive B and T lymphocytes that have entered a lymph node on exiting the primary lymphoid organs, or the memory cells that are located here, become activated on meeting with antigen. Antibodies secreted on activation of B cells in the lymph nodes, as well as the clonal progeny of the activated B and T cells, then leave the lymph node and enter the blood circulation when the lymph is returned to blood vessels near the heart.

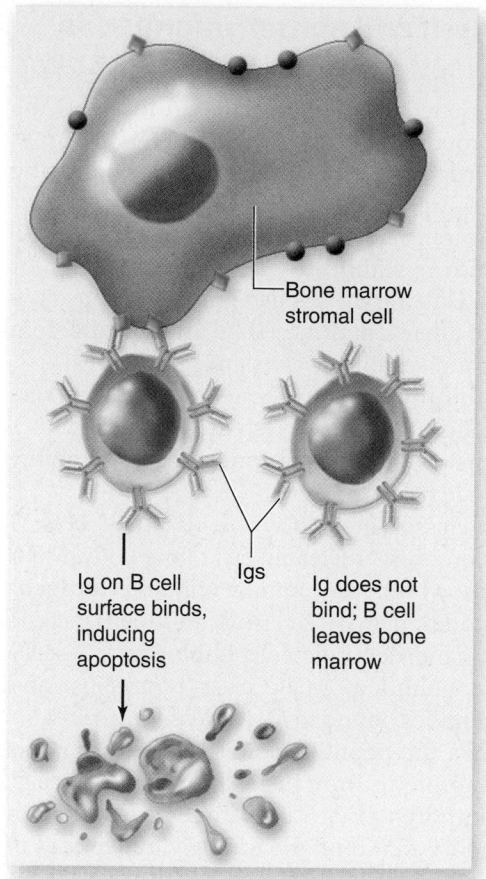

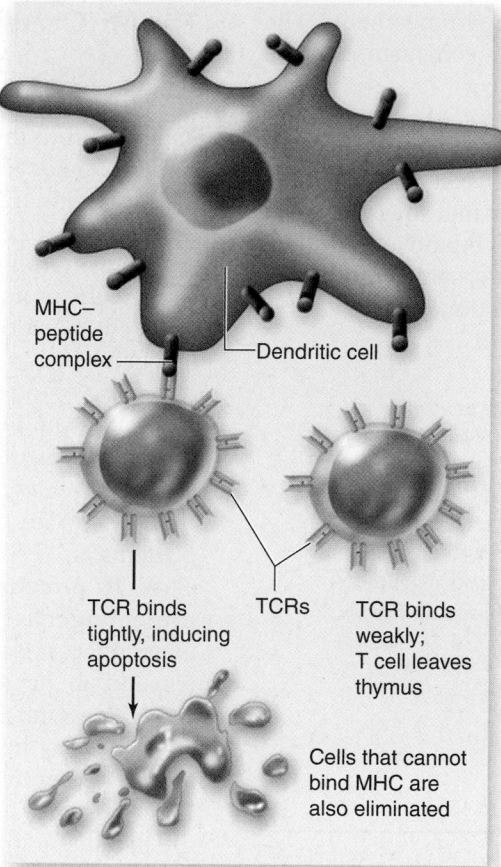

Figure 51.10 Selection against self-reactive lymphocytes in primary lymphoid organs. After B and T lymphocytes acquire their specific receptors, self-reactive cells are eliminated by apoptosis. *a.* If Igs on the surface of a maturing B cell bind to an epitope on a bone marrow stromal cell, that cell will undergo apoptosis. The small percentage (10%) of B cells whose Igs do not recognize stromal cell epitopes will be released from the bone marrow. *b.* If TCRs on a maturing T cell bind too tightly to self-MHC/self-peptide complexes on dendritic cells in the thymus, that T cell will undergo apoptosis. Cells that don't bind MHC complexes at all are also eliminated. The very small percentage (2–5%) of maturing T cells that bind MHC peptide complexes with intermediate affinity are released from the thymus. These cells bind self-MHC/foreign peptide complexes with high affinity.

Lymphocytes responding to antigens in a lymph node may pass out of capillaries supplying blood to the lymph node and enter the node's tissues. This is the cause of the "swollen glands" that sometimes accompany infection. The local lymph nodes enlarge due to the vast influx of lymphocytes.

Some antigens are found primarily in the blood, or in the blood as well as in the tissues. One example is the bacterium *Neisseria meningitidis,* a cause of a potentially fatal meningitis (infection of the meninges, layers of membranes covering the brain). Immune responses to such antigens occur in the spleen.

The splenic artery carries blood to the spleen where it then subdivides into arterioles. Antigens released into the ground tissue of the spleen are recognized by B and T cells present in the white pulp, regions of the spleen immediately surrounding the arterioles. Lymphocytes in the white pulp may be activated, as in the lymph node. Antibodies along with some of the activated lymphocytes exit via the splenic vein.

The final important secondary lymphoid organ is the **mucosa-associated lymphoid tissue (MALT),** which includes the tonsils, the appendix, and a large number of follicles located in the connective tissue under mucosal surfaces. These follicles are composed of lymphocytes, primarily B cells but also some T cells, and some macrophages. Any antigens that pass through the mucosa immediately encounter lymphocytes in these follicles and their entry farther into the body may be stopped at this point.

If invading organisms manage to escape or evade innate defenses of mucosal surfaces as well as the specific responses of the lymphocytes in the MALT, then they still face a further chance of being stopped by responses in the other secondary lymphoid organs.

Two forms of adaptive immunity have evolved

Adaptive immunity, involving the ability to distinguish between self and nonself, was long thought to have evolved once in vertebrates. The type of adaptive immunity described in this chapter first arose in the cartilaginous fish that evolved some 450 MYA (see chapter 35).

Sharks and rays possess a thymus and a spleen, as well as a rather diffuse MALT. These animals mount cell-mediated responses with T cells bearing TCRs, and humoral responses with B cells that secrete Ig. Bone marrow in which hematopoiesis occurs appeared first in amphibians, although its exact role appears to vary in different species. Lymph nodes appeared first in birds, and their immune system differs little from that of mammals.

Recently, a second form of adaptive immunity has been described in jawless fish. This system does not involve B and T cells with their characteristic receptors. Instead, lymphocytes have receptor proteins composed of variable repeats rich in the amino acid leucine. These proteins appear to function much like Ig, but with a completely different protein architecture. The number of different receptor proteins produced by this system appears similar to the number of potential Ig. The generation of diversity in the two systems appears to have some similarity as the different lymphocyte receptors in jawless fish are also assembled by DNA rearrangements. The makeup of the genes involved and the mechanism of these rearrangements is currently unknown.

It is unclear whether this newly described form of adaptive immunity was present in the ancestor to all chordates, or if it evolved in the lineage that gave rise to jawless fish. Given the differences in the two systems, it is likely that they represent independent events. If this other form of adaptive immunity was present in the ancestor to all chordates, some vestige may remain in modern vertebrates, including humans.

Learning Outcomes Review 51.2

Adaptive immunity is able to recognize individual pathogens and mount a specific response. Lymphocytes, produced in bone marrow, must acquire their specific receptors and undergo selection for self-reactivity in primary lymphoid organs. These mature but naive lymphocytes circulate to secondary lymphoid organs, where they may encounter foreign antigens. B cells produce circulating antibodies (humoral immunity); T cells kill pathogens or help other cells respond to them (cell-mediated immunity).

■ **What type of adult stem cells are found in the immune system?**

51.3   Cell-Mediated Immunity

Learning Outcomes

1. *Describe the function of cytotoxic T cells.*
2. *Explain the role of helper T cells.*

T cells may be characterized as either **cytotoxic T cells (T_C)** or *helper T cells (T_H)*. These cells can also be identified based on cell surface markers. T_C cells have CD8 protein on their cell surface, making them CD8$^+$ cells. T_H cells have CD4 protein on their cell surface, making them CD4$^+$ cells.

To be activated, both of these T cell types must recognize peptide fragments bound to MHC proteins, but the two cell types may be distinguished by (1) recognition of different classes of MHC proteins, which have distinct cell distributions, and (2) differing roles of the T cells after they are activated.

The MHC carries self and nonself information

As discussed earlier, the surfaces of most vertebrate cells exhibit glycoproteins encoded by the MHC. In humans, the name given to the proteins encoded by the MHC complex is **human leukocyte antigens (HLAs).** The genes encoding the MHC proteins are highly polymorphic (have many alleles). For example, the HLA proteins are specified by genes that are the most polymorphic known, with nearly 500 alleles detected for some of the proteins. Only rarely will two individuals have the same combination of alleles, and the HLAs are thus different for each individual, much as fingerprints are.

MHC proteins on the tissue cells serve as self markers that enable an individual's immune system, specifically its T cells, to distinguish its own cells from foreign cells, an ability called *self versus nonself recognition.*

There are two classes of MHC proteins. **MHC class I proteins** are present on every nucleated cell of the body. **MHC class II proteins,** however, are found only on **antigen-presenting cells** (in addition to MHC class I); these cells include macrophages, B cells, and dendritic cells (table 51.2). T_C cells respond to peptides bound to MHC class I proteins, and T_H cells respond to peptides bound to MHC class II proteins.

Most of the time, the peptides bound to MHC proteins are derived from self-proteins from the individual's own cells. For this reason, it is important that T cells undergo selection in the thymus so that those that bind too strongly to peptides of self-proteins on self-MHC are eliminated. In this way, T cells normally are activated only outside the primary lymphoid organs in which they mature, when they encounter peptides of foreign proteins on self-MHC—for example, in the case of viral infection or cancer.

Cytotoxic T cells eliminate virally infected cells and tumor cells

Activated cytotoxic T cells recognize "altered-self" cells, particularly those that are virally infected or tumor cells. The TCRs of cytotoxic T lymphocytes recognize peptides of endogenous antigens bound to MHC class I proteins. Peptides of endogenous antigens are generated in a cell's cytosol and then are pumped by special transport proteins into the rough endoplasmic reticulum where they become bound to MHC class I proteins. These proteins continue on their way through the endomembrane system to the cell surface.

TABLE 51.2	Lymphocyte Recognition of Antigen			
	Recognize epitopes of soluble or particulate antigen	**Recognize peptides bound to self-MHC proteins**	**Class of MHC proteins recognized**	**Cell types on which recognized MHC is expressed**
B cells	Yes	No	None	NA
T_H (CD4$^+$) cells	No	Yes	Class II	Antigen-presenting cells: dendritic cells, B cells, and macrophages
T_C (CD8$^+$) cells	No	Yes	Class I	All nucleated cells

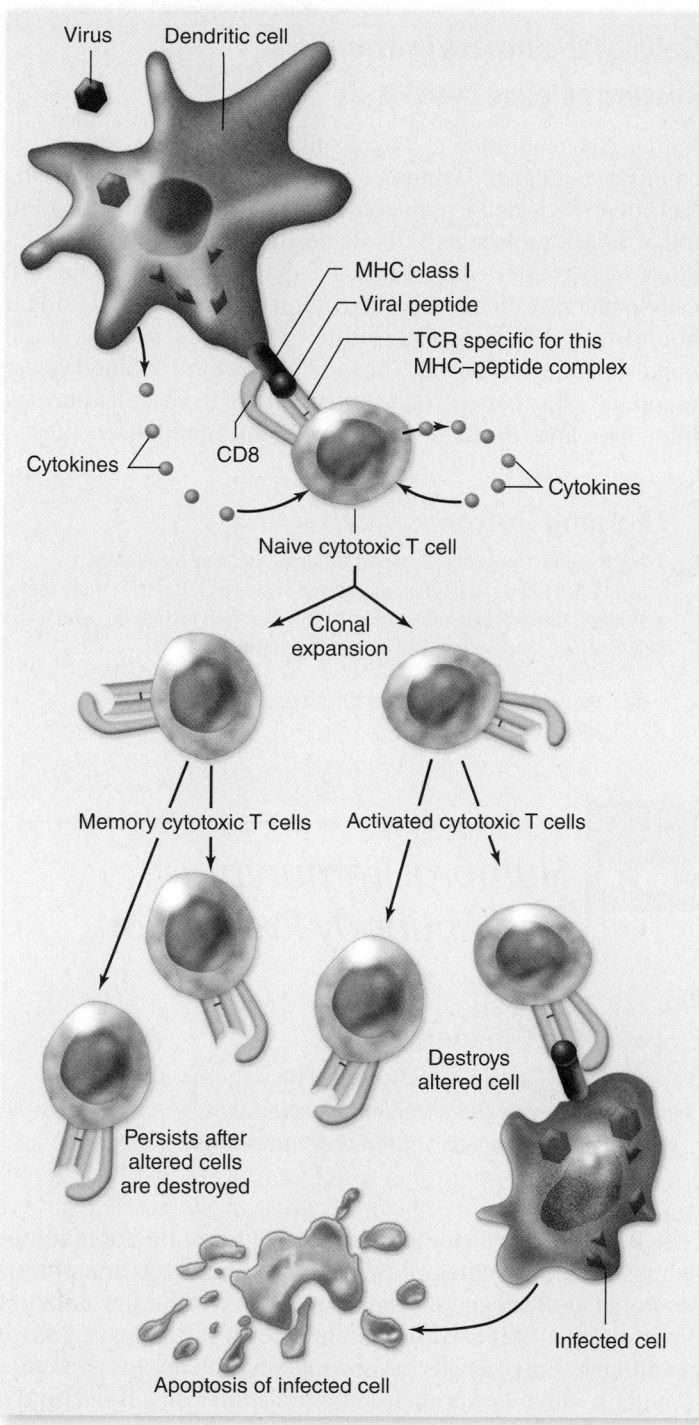

An endogenous antigen may be a self-protein, or it may be a viral protein produced within a virally infected cell or an unusual protein produced by a cancerous cell. T_C cells respond only to the peptides of these unusual proteins bound to self-MHC class I. T-cell activation occurs in a secondary lymphoid organ, as described earlier. In a lymph node, for example, T cells encounter antigen-presenting cells. Dendritic cells in particular often present antigens that activate T_C cells.

Because not all viruses can infect dendritic cells, the dendritic cells must ingest viruses or tumor cells and then, through a mechanism referred to as cross-presentation, place the viral or tumor peptides on MHC class I proteins. Binding of the T_C cell through its TCR and its CD8 site to the dendritic cell induces clonal expansion of the T_C cell, generating many activated T_C cells as well as memory T_C cells (figure 51.11). The activated T_C cells then circulate around the body where they bind to "target" host cells that express the same combination of foreign peptide on self-MHC class I (figure 51.12).

Apoptosis of the target cell is induced in a very similar fashion to that used by NK cells; a T_C cell secretes perforin monomers that create pores in the target's membrane; granzymes enter and activate caspases, which in turn cause apoptosis of the target.

Helper T cells secrete proteins that direct immune responses

Activated helper T cells, T_H cells, secrete low-molecular-weight proteins known as **cytokines.** A vast array of cytokines is known, many but not all of which are secreted by T_H cells. These cytokines bind to specific receptors on the membranes of many other cells, particularly but not exclusively those of the immune system. On binding, they initiate signaling cascades in these cells that promote their activation or differentiation.

Figure 51.11 Cytotoxic T cells induce apoptosis of "altered-self" cells. Naive cytotoxic T cells are initially activated on TCR recognition of foreign peptide displayed on self-MHC class I proteins on dendritic cells in a secondary lymphoid organ. Activation results in clonal expansion and differentiation into memory cells and activated cells. Activated progeny of the T_C cell can induce apoptosis of any cell in the periphery (outside the secondary lymphoid organ) that displays the same self-MHC class I–peptide combination on its surface. This will most likely be a virally infected cell or a tumor cell.

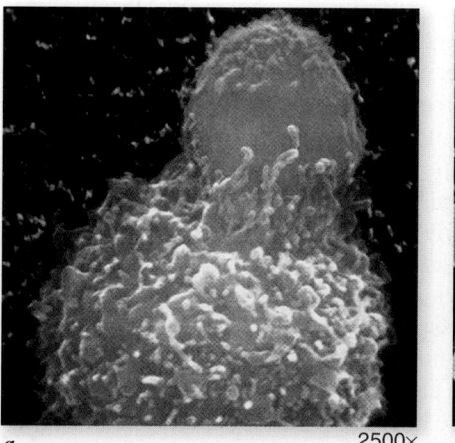
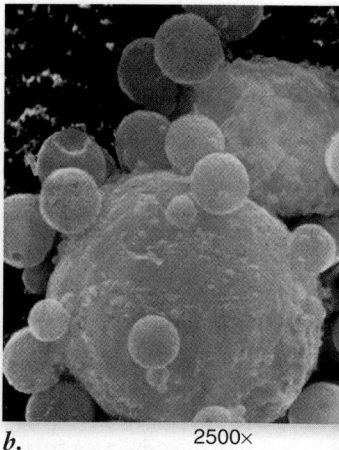

a.   2500×   b.   2500×

Figure 51.12 Cytotoxic T cells destroy tumor cells.
a. The cytotoxic T cell (orange) comes into contact with a tumor cell (purple). *b.* The T cell recognizes that the tumor cell is "altered-self" and induces the apoptosis of the tumor cell.

Because cytokines are quite potent, they are generally secreted at very low concentrations so that, with a few exceptions, they bind only to nearby cells. IL-1 is an exception in that it travels to the hypothalamus to induce the fever response. Different subsets of T_H cells secrete cytokines specific for different cell receptors, so it is largely the T_H cells and the cytokines they secrete that determine whether an immune response will be humoral or cell-mediated in nature.

T_H cells respond to exogenous antigen that has been brought into an antigen-presenting cell. Macrophages or dendritic cells acquire these antigens by phagocytosis or endocytosis, and B cells gain them through receptor-mediated endocytosis. Once inside these cells, the antigen is gradually degraded in increasingly acidic endosomes or lysosomes. Peptides of the antigen join with MHC class II proteins in certain of these endosomes, and the MHC class II–peptide complexes are then transported to and displayed on the cell surface of the antigen-presenting cell. T_H cells encounter these cells within the secondary lymphoid organs and bind to the complexes. The CD4 protein of the T_H cells additionally bind to conserved regions of MHC class II.

A naive T_H cell expresses a protein called CD28 that must bind to a protein called B7 if that T cell is to be activated. B7 is found only on antigen-presenting cells and is at highest levels on dendritic cells. This requirement ensures that T_H cells are activated only when needed; this careful regulation is necessary due to the potency of the cytokines these cells release.

As with T_C cells, an activated T_H cell gives rise to a clone of T_H cells including both effector T_H cells and memory T_H cells, with identical TCR specificity. Most of the effector cells will leave the lymphoid organ and circulate around the body.

T cells are the primary cells that mediate transplant rejection

When T cells encounter the nonself MHC–peptide complexes present on transplanted tissue, such as a kidney, the TCRs on many of the T cells can weakly bind to these complexes. This is simply a case of cross-reactivity: The structure of a nonself MHC–peptide complex sufficiently resembles that of the self-MHC–foreign-peptide complex. The result is that the T cell binds to the foreign tissue cell.

Although the interactions between TCRs and nonself MHC–peptide complexes are relatively weak, many interactions occur between any one T cell and any one transplanted cell because a high density of MHC proteins is present on the surface of all cells. This activates the T cells and initiates the attack on the foreign tissues.

Because of the genetic basis of MHC proteins, the more closely two individuals are related, the less their MHC proteins vary, and thus the more likely they will be to tolerate each other's tissues. As a result, relatives are often sought as donors for patients in need of an organ transplant, and HLA typing is done to find matching alleles.

A variety of drugs are used to suppress immune system rejection of a transplant; most individuals with a non-MHC-matched transplant continue to take some of these drugs for the remainder of their lives. One very effective drug is cyclosporin, which blocks the activation of lymphocytes.

Cells of the innate immune system release cytokines

Many cells in addition to T_H cells release cytokines, always in a carefully regulated fashion. For example, macrophages that have been activated by phagocytosis of antigen, or by the binding of PAMP molecules to TLRs on their surface, release cytokines such as interleukin-12 (IL-12) that can, in turn, bind to T_H cells to increase their level of activation. Macrophages with TLRs bound to PAMP also release other cytokines, such as tumor necrosis factor-α (TNF-α). These cytokines bind to blood vessels to induce a local or even systemic increase in vascular permeability. This links the innate response to the adaptive response.

Learning Outcomes Review 51.3

T cells respond to peptides of foreign antigens displayed on self-MHC proteins. Activated T_C cells induce apoptosis of altered self cells—those that are virally infected or are tumor cells. T_H cells secrete cytokines that promote either cell-mediated or humoral immune responses.

■ How are T-cell receptors different from Toll-like receptors?

51.4 Humoral Immunity and Antibody Production

Learning Outcomes

1. Explain how antibody diversity is generated.
2. List the five classes of immunoglobulins.
3. Explain how vaccination prevents disease.

The B-cell receptors for antigen are the immunoglobulin molecules present as integral proteins in the plasma membrane. As noted earlier, each B cell exhibits about 10^5 immunoglobulin molecules of identical specificity for a particular epitope of an antigen. Naive B cells in secondary lymph organs encounter antigens. When immunoglobulin molecules on a B cell bind to a specific epitope on an antigen, and the B cell receives additional required signals, particularly cytokines secreted by T_H cells, then that B cell becomes activated, proliferating into plasma cells and memory cells (figure 51.13).

Each plasma B cell is a miniature factory producing soluble antibodies of the same specificity as the membrane-bound antibodies of the parent B cell. These antibodies enter the lymph and blood circulation as well as the extracellular fluid, and they bind to the appropriate epitopes of antigen encountered anywhere in the body. Any one antigen may present a variety of epitopes, so that different B cells might recognize different epitopes of a single antigen.

Once immunoglobulins coat an antigen, many other cells and processes may be activated to eliminate the antigen. The immunity to avian cholera that Pasteur observed in his chickens resulted from such antibodies and from the continued presence of the progeny of the B cells that produced them.

Immunoglobulin structure reveals variable and constant regions

Each immunoglobulin molecule consists of two identical short polypeptides called *light chains* and two identical longer polypeptides called **heavy chains** (figure 51.14). The four chains in an immunoglobulin molecule are held together by disulfide bonds, forming a Y-shaped molecule (figure 51.14*a*). Each "arm" of the molecule is referred to as an Fab region, and the "stem" is the Fc region (figure 51.14*b*).

Antibody specificity: The variable region

Comparison of the amino acid sequences of many different immunoglobulin molecules has demonstrated that the specificity of immunoglobulins for antigen epitopes resides in the amino-terminal half of each Fab region. This half of the Fab has an amino acid sequence that varies from one immunoglobulin to the next and is thus designated the *variable region*. Both the light chain and the heavy chain have a variable region.

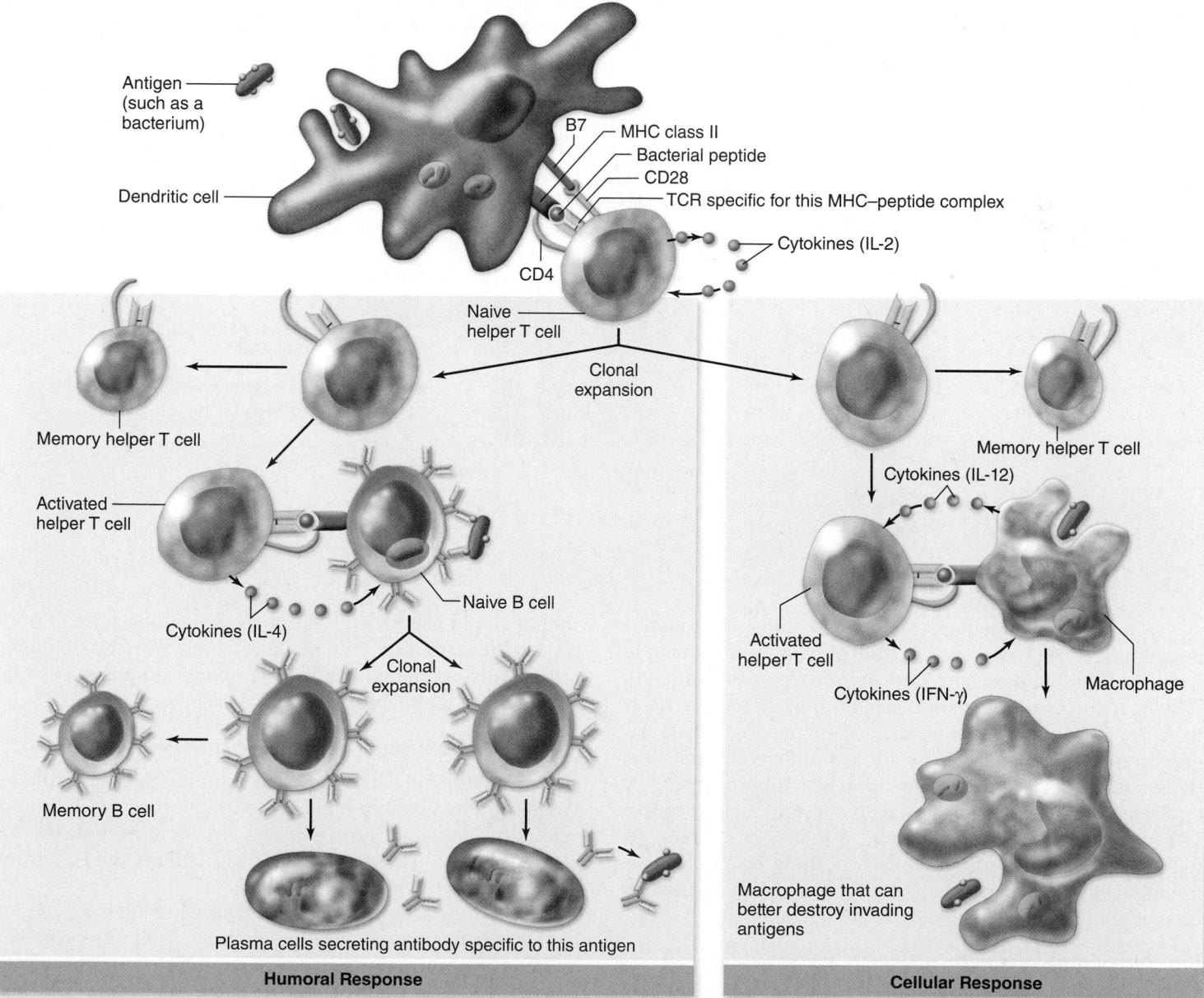

Figure 51.13 **Helper T cells secrete cytokines promoting either cell-mediated or humoral immune responses.** Naive helper T cells are initially activated by TCR bound to a foreign peptide displayed on self-MHC class II proteins on dendritic cells. Activation results in clonal expansion and differentiation into memory cells and activated cells. T_H cells promote the humoral response when they recognize the same antigen displayed by a B cell. Cytokines such as interleukin-4 (IL-4) released from the T_H cell then activate the B cell, producing memory cells and plasma that secrete antibodies against the antigen. T_H cells also secrete interferon-γ (IFN-γ), which stimulates cells involved in the cellular response such as the macrophage shown here. Macrophages secrete other cytokines that stimulate T_H cells.

Figure 51.14 **The structure of an immunoglobulin molecule.** *a.* In this model of an immunoglobulin (Ig) molecule, amino acids in the peptide chains are represented by small spheres. The molecule consists of two heavy chains *(brown)* and two light chains *(yellow)*. The four chains form a Y shape, with two identical antigen-binding sites at the arms of the Y, the Fab regions, and a stem, or Fc region. The two Fab regions are joined to the Fc region by a flexible hinge. *b.* A more schematic depiction showing heavy chains *(brown)* and light chains *(yellow)* as rods. The two identical halves of the molecule are joined by disulfide bonds *(red)* as are the heavy and light chains of each half. *c.* Ig molecule shown as a membrane protein. This depiction highlights the domain structure of heavy and light chains. Each chain contains a series of domains, each about 110 amino acids, which include an immunoglobulin fold motif. These are represented as loops with globular structure maintained by disulfide bonds *(red)*. The amino-terminal half of each Fab is a variable region *(blue)* that binds to an epitope and the remainder of the molecule is the constant region.

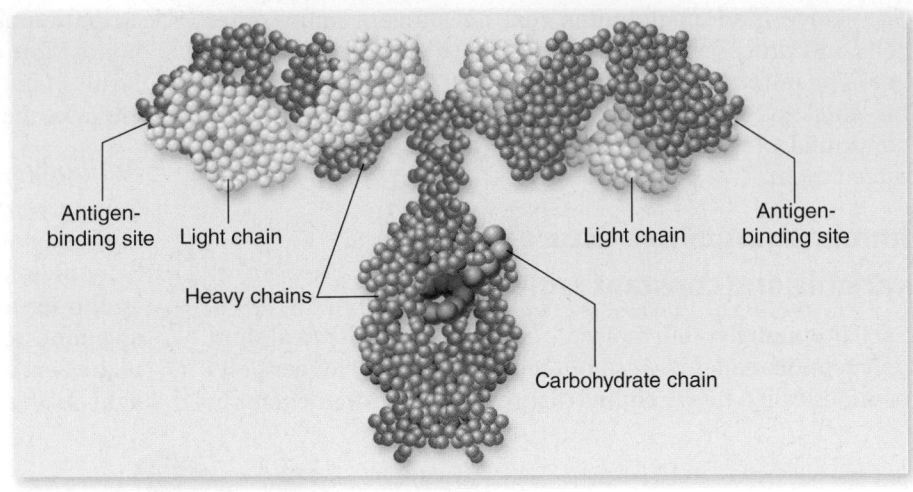

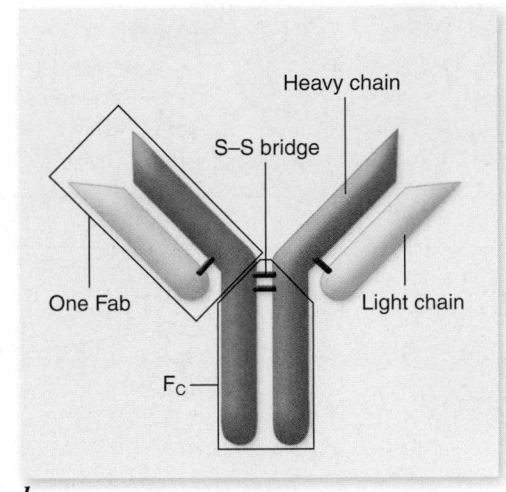

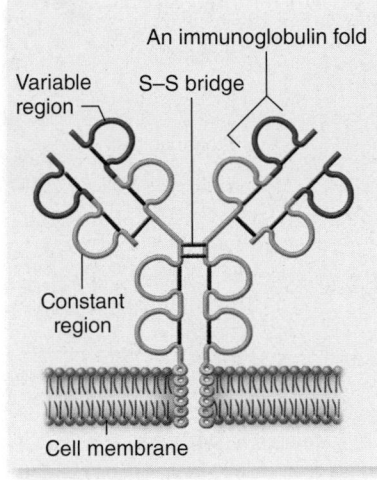

The amino acid sequence of the remainder of the immunoglobulin is relatively constant from one immunoglobulin to the next and is thus designated the *constant region* (figure 51.14). Both light and heavy chains also exhibit constant regions. Careful analysis shows that light-chain constant regions of mammalian immunoglobulins consist of two different sequences, designated κ (kappa) and λ (lambda), which have apparently equivalent function. The heavy-chain constant regions consist of five different sequences: μ (mu), δ (delta), γ (gamma), α (alpha), and ε (epsilon). When each of these heavy chains is bound to either type of light chain, they give rise to a particular class of immunoglobulin: IgM, IgD, IgG, IgA, and IgE.

Binding of antibody with antigen

The variable regions of the heavy and light chains fold together to form a sort of cleft, the *antigen-binding site* (see figure 51.14). The size and shape of the antigen-binding site, as well as which amino acids line its surface, determine the specificity of each immunoglobulin for an antigen epitope.

Because each immunoglobulin is composed of two identical halves, they can each bind with two identical epitopes, although not generally on the same antigen because of steric (shape) constraints. This ability to bind with two epitopes allows the formation of antigen–antibody complexes containing multiple antibody and antigen molecules (figure 51.15*a*).

Function of antibody classes: The constant region

Although the specificity of each immunoglobulin is determined by its variable region, the function of the immunoglobulin depends on its class, as determined by the heavy-chain constant region, and particularly the Fc portion of the constant region.

Many cells have Fc receptors that can bind the Fc region of a particular class of immunoglobulin. Therefore, when an immunoglobulin binds to an antigen through its antigen-binding site, another cell, such as a phagocytic cell, may be brought close to the antigen by binding to the Fc region of the immunoglobulin (figure 51.15*c*). This binding of antigen–antibody complex to Fc receptors can also activate these cells. In this way, specific immunoglobulins can promote the interaction of nonspecific cells with the antigen, generally resulting in the elimination of the antigen.

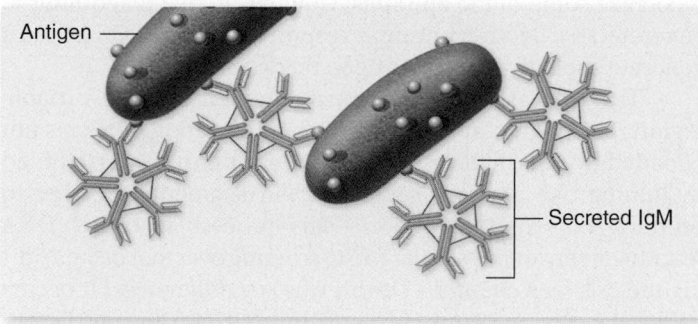

a.

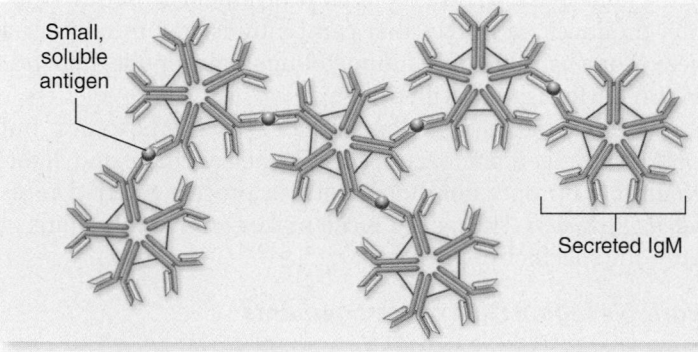

b.

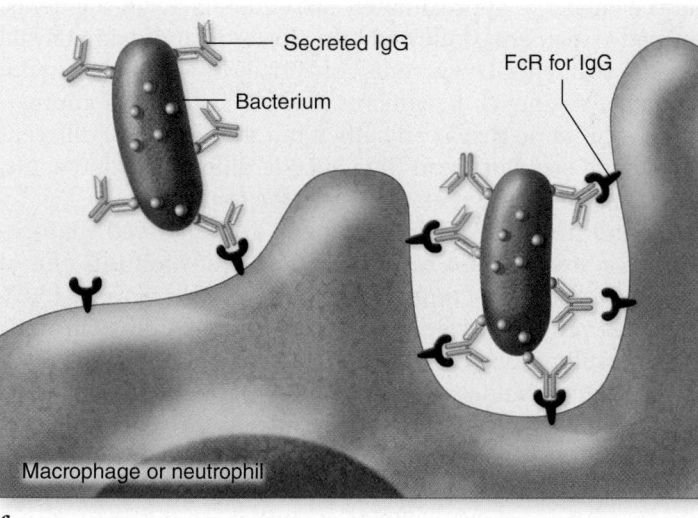

c.

Figure 51.15 Binding of antibody to antigens can cause agglutination, precipitation, or neutralization of the antigens. *a.* Binding of secreted IgM to larger particulate antigens leads to the clumping, or agglutination, of the antigens. *b.* Binding of secreted IgM to small soluble antigens can lead to their precipitation. Secreted IgG, due to its high concentration (75% of plasma Ig), can also agglutinate and precipitate antigens. IgG does not precipitate as efficiently as IgM because of the pentameric nature of secreted IgM. *c.* Secreted IgG can coat or neutralize an antigen by blocking its ability to bind to a host. Macrophages and neutrophils that have Fc receptors for IgG can thus attach to an antigen–antibody complex, which they will then phagocytose and destroy.

The five classes of immunoglobulins have different functions

The five classes of antibodies are based on the sequence and structure of the constant regions of their heavy chains. These five classes have different functions in the protection of an individual. Characteristics of the different classes are summarized in table 51.3 and are described in the following sections.

Keep in mind that antibodies don't kill invading pathogens directly; rather, they cause destruction of pathogens by targeting them for attack by other, nonspecific cells or by activating the complement system.

IgM is a receptor on the surface of all mature, naive B cells and is the first type of antibody to be secreted during an immune response. Although IgM in the membrane of a B cell is monomeric in form, it is secreted as a pentamer (five units) of about 900,000 kDa. Its large size restricts it to the circulation, but its pentameric form means that it very efficiently promotes agglutination of larger antigens (figure 51.15*a*) and precipitation of

TABLE 51.3	Five Classes of Immunoglobulins
Class	**Function**
IgM Pentamer	First antibody secreted during the primary immune response; promotes agglutination and precipitation reactions and activates complement
IgD Monomer	Present only on surfaces of B cells; serves as antigen receptor
IgG Monomer	Major antibody secreted during the secondary response; neutralizes antigens and promotes their phagocytosis and activates complement
IgA Dimer	Most abundant form of antibody in body secretions; high density of IgA-secreting plasma cells in the MALT
IgE Monomer	Fc binds to mast cells and basophils; allergen binding to V regions promotes the release of mediators, which triggers allergic reactions

soluble antigens (figure 51.15*b*). IgM bound to an antigen also activates a complement protein cascade, triggered by the binding of certain complement proteins to the exposed Fc ends.

IgD is also present, along with IgM, on mature naive B cells. The B cells can be activated by cross-linking of two IgD molecules, although under normal circumstances this class of immunoglobulin is not secreted by the cells. On B-cell activation, IgD is no longer displayed on the cell surface. Other roles for IgD remain elusive.

IgG is the major form of antibody in the blood plasma and in most tissues, making up about 75% of plasma antibodies. It is the most common form of antibody produced in a secondary immune response (any response triggered on a subsequent exposure to an antigen). IgG can bind to an antigen in such quantity that the antigen—a virus, bacterium, or bacterially derived toxin—is said to be neutralized, meaning that it can no longer bind to the host. Macrophages and neutrophils have Fc receptors that bind to IgGs bound to antigens, and in this way IgG binding or coating of antigens facilitates their elimination by phagocytosis (figure 51.15*c*). IgG is also important in providing passive immunity to a fetus; it readily crosses the placenta from the mother. Finally, IgG can also activate complement, although not as efficiently as IgM, leading to pathogen elimination.

IgA is the major form of antibody in external secretions, such as saliva, tears, and the mucus that coats the gastrointestinal tract, bronchi, and genitourinary tract. IgA plays a major role in protection of these surfaces; it is usually secreted as a dimer. The many plasma cells present in the MALT, under the mucosal surfaces, secrete IgA that crosses the epithelial cells to the lumen of these tracts; here it can bind and neutralize antigens. Additionally, any pathogen that passes through a mucosal surface becomes bound to IgA because it is secreted by cells in follicles under that surface. The bound IgA crosses the epithelial cells into the lumen, taking the pathogen with it. The pathogen can then be eliminated by innate defenses. IgA also provides passive immunity to a nursing infant since it is present in mother's milk.

IgE is present at very low concentration in the plasma. On secretion, most becomes bound to mast cells and basophils that recognize the Fc portion of IgE. As described later, binding of certain normally harmless antigens to IgE molecules bound to mast cells and basophils produces the symptoms of allergy, such as the runny nose and itchy eyes of hay fever. IgE is also often secreted in response to an infection by helminth worms. In this instance, secreted IgE binds to epitopes on the worms and is then recognized by Fc receptors on eosinophils. The eosinophils generally kill the worms by secreting digestive enzymes through perforin pores into the worms.

Immunoglobulin diversity is generated through DNA rearrangement

The vertebrate immune system is capable of recognizing as foreign virtually any nonself molecule presented to it. It is estimated that human or mouse B cells can generate antibodies with over 10^{10} different antigen-binding sites. Although an individual probably does not have antibodies specific to all epitopes of an antigen, it is fairly certain that antibodies will recognize some of the epitopes, which is all that is required to generate an effective immune response. How do vertebrates generate such diversity of antigen recognition?

The answer lies in the unusual genetics of the variable region. This region in each chain of an immunoglobulin is not encoded by one single stretch of DNA but rather is assembled by joining two or three separate DNA segments together to produce the variable region. This process is called DNA rearrangement and is similar to the crossing over that occurs during meiosis (see chapter 11) with two key differences: It occurs between loci on the same chromosome and it is site-specific.

DNA rearrangement occurs as a progenitor B cell matures in the bone marrow. After DNA rearrangement, RNA transcription produces an mRNA that can be translated into either a heavy- or a light-chain immunoglobulin polypeptide, depending on the locus transcribed.

Cells contain homologous pairs of chromosomes, but DNA rearrangement occurs for the heavy-chain and light-chain loci on only one homologue, a process referred to as *allelic exclusion*. Thus, each B cell makes immunoglobulins of only one specificity.

Variable region DNA rearrangements

Sequencing of human immunoglobulin heavy-chain gene loci from several different individuals shows that the locus contains a cluster of approximately 50 sequential DNA segments, termed V segments, followed by a cluster of approximately 30 smaller segments, D segments, and finally by another cluster of 6 smaller segments, J segments. Each V segment is approximately the same size as any other, but they are all of different nucleotide sequence and thus encode different amino acids; the situation is similar for the D and the J segments.

The first DNA rearrangement during B-cell maturation is a site-specific recombination event joining one of the D segments to one of the J segments (figure 51.16). Recombination between two sites on the same chromosome results in the deletion of the intervening DNA, which is subsequently degraded.

This is followed by another site-specific recombination joining a V segment to the rearranged DJ, with the deletion of all the intervening DNA. Which V, which D, and which J are chosen by any cell appears to be completely random.

Because of the many combinations of V, D, and J that can be formed, one can calculate the generation of about 9000 different heavy-chain variable-region sequences. A similar situation occurs for light-chain variable region formation, except that each light-chain variable region is encoded by only a V segment and a J segment.

Other processes contribute even further to the diversity of variable region sequence. As the DNA segments are joined to each other, a few nucleotides may be added to or deleted from the ends of each segment, and this is generally followed by somewhat imprecise joining of the segments to each other, resulting in a shift of the reading frame. B cells may end up expressing any heavy-chain variable region with any light-chain variable region during its maturation. Lastly, these genes show an elevated mutation rate, termed somatic hyper-mutation. Taking all these processes into account has allowed the estimate of more than 10^{10} possible variable regions.

Transcription and translation

After the DNA rearrangements that encode the variable region are complete, pre-mRNA transcripts are formed with 5′ ends that begin at the rearranged variable region-encoding segments and continue through constant region exons. More specifically, heavy-chain-encoding pre-mRNA transcripts start at the rearranged VDJ and continue through exons encoding μ and δ constant regions (see figure 51.16).

Alternative splicing of these RNA transcripts removes any extra J segments that remain 3′ to the rearranged VDJ, as well as either δ or μ sequences, resulting in transcripts that all encode the same variable region but either μ or δ constant region exons, respectively. Translation results in a μ or δ heavy-chain polypeptide, which associates with a light-chain polypeptide in the rough endoplasmic reticulum. Thus, the mature naive B cell expresses both IgM and IgD on its surface, both having the same antigen-binding specificity (see figure 51.16).

T-cell receptors

At this point, we must briefly return to T-cell receptors and examine their similarity to the immunoglobulins of B cells. The structure of a TCR is essentially like a single Fab region of an immunoglobulin molecule (figure 51.17).

The TCR is a dimeric (two-chain) protein, with about 95% of TCRs composed of an α chain and a β chain. The amino terminal halves of the two polypeptides are the variable domains that bind to self-MHC plus peptide, and the membrane-proximal halves are the constant domains of each polypeptide. The TCR variable-region gene loci also contain multiple DNA segments—V, D, and J, or only V and J—that are

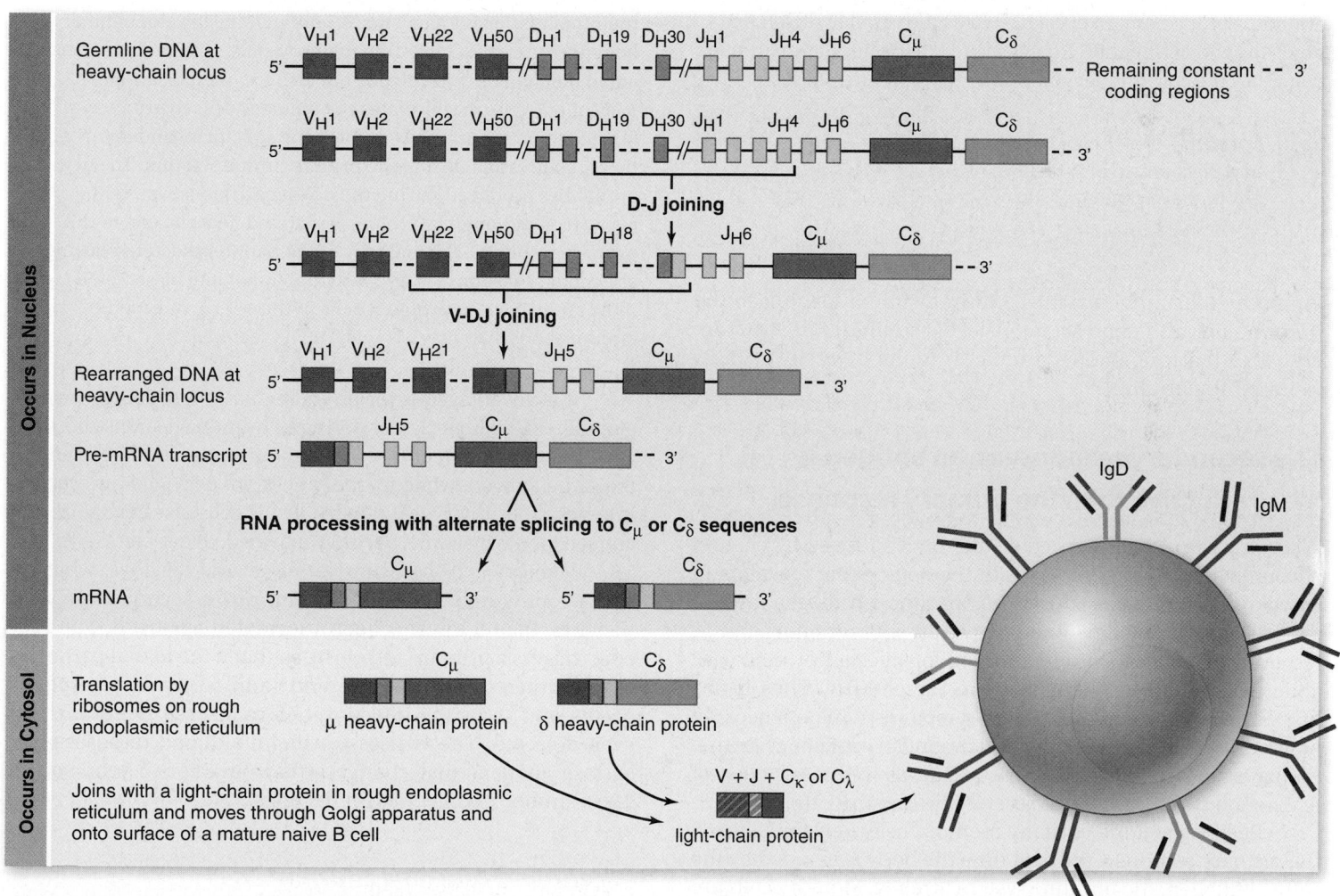

Figure 51.16 **Immunoglobulin diversity is generated by rearrangement of segments of DNA.** An immunoglobulin (Ig) protein is encoded by different segments of DNA: a V (variable), a D (diversity), and a J (joining) segment, plus a constant region. These are joined by a precise sequence of DNA rearrangements during maturation in the bone marrow. This first joins a D to a J segment, then this combined DJ joins to a V segment. Other cells will select other V, D, and J segments, contributing to Ig diversity. Transcription starts at the rearranged VDJ and continues through constant-region exons. PreRNA splicing joins the variable region to either a μ or a δ constant region. These transcripts are translated by ribosomes on the RER to produce heavy-chain polypeptides that join with light chains (encoded by a V, a J, and a C). These proteins are transported to the cell surface, resulting in a mature naive B cell that expresses both IgM (μ constant region) and IgD (δ constant region), with the same variable region and thus the same antigen specificity.

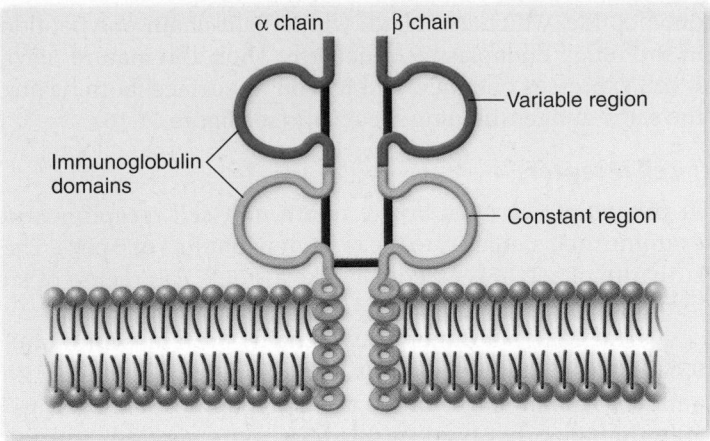

Figure 51.17 The structure of a TCR is similar to an immunoglobulin Fab. TCRs are composed of two chains—generally α and β—joined by a disulfide bond (red). Each also includes two immunoglobulin domains as in an Ig Fab. The amino-terminal domain of each chain is the variable region that binds to an MHC–peptide complex, and the membrane-proximal domain is the constant region. Unlike Igs, TCRs are not secreted.

> **? Inquiry question** What does the common structure and mechanism of formation of Igs and TCRs suggest about the evolution of B and T lymphocytes and these proteins?

joined by the same enzymes and in a similar fashion to the immunoglobulin gene segments. This similarity in structure and DNA rearrangements produces similar diversity of TCRs as immunoglobulins.

The secondary response to an antigen is more effective than the primary response

When a particular antigen enters the body, it must, by chance, encounter naive lymphocytes with the appropriate receptors to provoke an immune response. The first time a pathogen invades the body only a few B or T cells may exist with receptors able to recognize the pathogen's epitopes or, for infected or otherwise abnormal cells, foreign peptides bound to self-MHC. Thus, in this first encounter, a person develops symptoms of illness because only a few cells are present that can mount an immune response. Clonal expansion of T and B cells occurs, as well as secretion of IgM antibodies, but this takes several days (figure 51.18).

Because a clone of many memory cells develops during the primary response, the next time the body is invaded by the same pathogen, the immune system is ready. Memory cells are more rapidly activated than are naive lymphocytes, so a secondary immune response both initiates and peaks much more rapidly than a primary response. Further clonal expansion takes place, along with the secretion of large amounts of antibodies that are generally of the IgG class, although IgA and IgE are also possible (see figure 51.18). The class of immunoglobulin produced is dictated by the identity of the cytokines, derived from activated T_H memory cells, that bind to the B cells during their secondary response.

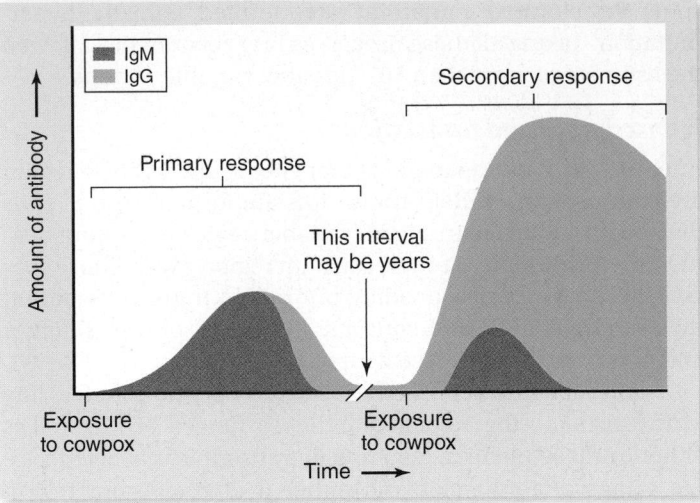

Figure 51.18 The development of active immunity. Immunity to smallpox in Jenner's patients occurred because their inoculation with cowpox stimulated the development of lymphocyte clones, including memory cells, with receptors that could bind not only to cowpox but also to smallpox antigens. A second exposure stimulates the memory cells to produce large amounts of antibody of the same specificity and much more rapidly than during the primary immunization. The first antibodies produced during the primary response are IgM in class (red), although IgG (blue) is secreted near the end of the primary response. The majority of the antibody secreted during a secondary response is IgG, although IgA could be secreted if the antigen has activated B cells in the MALT, or in some circumstances, such as allergies, IgE is secreted.

It is advantageous for an individual to produce immuno-globulins of different classes during an immune response because each class has a different function. During a second exposure to the same antigen, while memory cells are activated and secrete isotypes other than IgM, other naive B cells also recognize the antigen for the first time, become activated, and secrete IgM.

Memory cells can survive for several decades, which is why people rarely contract chicken pox a second time after they have had it once or been vaccinated against it. The vaccine triggers a primary response, so that if the actual pathogen is encountered later, a large and rapid secondary response occurs and stops the infection before disease symptoms are even detected. The viruses causing childhood diseases have surface antigens that change little from year to year, so the same antibody is effective for decades.

> ### Learning Outcomes Review 51.4
> Antibodies have variable regions by which they recognize and bind to an antigen. Variable regions are encoded by joining distinct DNA segments, providing recognition diversity. Each antibody also has one of five kinds of constant region that determines its function; these five classes are IgA, IgD, IgE, IgG, and IgM. Vaccination artificially presents an antigen to elicit the primary response; when encountered later, a pathogen with this antigen is eliminated quickly by the secondary response.
>
> ■ *How do Ig receptors differ from TLR innate receptors?*

51.5 Autoimmunity and Hypersensitivity

Learning Outcomes

1. Define autoimmune diseases.
2. Explain the cellular basis of the allergic reaction.

Sometimes the immune system is the cause of disease rather than the cure. Inappropriate responses to self-antigens may occur, as well as inappropriate or greatly heightened responses to foreign antigens, which, in turn, causes tissue damage.

A mature animal's immune system normally does not respond to that animal's own tissue. This acceptance of self cells is known as **immunological tolerance.** The immune system of a fetus undergoes the process of tolerance to lose the ability to respond to self-molecules as its development proceeds.

We now know that not all self-reactive T and B lymphocytes undergo apoptosis during selection in primary lymphoid organs. Normal healthy individuals are known to possess mature, potentially self-reactive lymphocytes. The activity of these cells, however, is regulated or suppressed so that they do not respond to the self-antigens they encounter. When this regulation or suppression breaks down, then humoral or cell-mediated responses can occur against self-antigens, causing serious and sometimes fatal disease.

Additionally, an immune response against a foreign antigen may be a greater one than is actually required to eliminate the antigen, or the response may be seemingly inappropriate to the antigen. Thus, instead of eliminating the antigen with only a localized inflammatory response, extensive tissue damage and occasionally death occurs.

Autoimmune diseases result from immune system attack on the body's own tissues

Autoimmune diseases are produced by the failure of immunological tolerance. Autoreactive T cells become activated, and autoreactive B cells produce autoantibodies, causing inflammation and organ damage. More than 40 known or suspected autoimmune diseases exist, affecting 5 to 7% of the population. For reasons that are not understood, two-thirds of the people with autoimmune diseases are women.

Autoimmune diseases can result from a variety of mechanisms. For example, the self-antigen may normally be hidden from the immune system; if later it is exposed, the immune system may treat it as foreign. This happens, for example, when a protein normally trapped in the thyroid follicles triggers autoimmune destruction of the thyroid (Hashimoto thyroiditis). It also occurs in sympathetic ophthalmia in which antigens are released from the eye.

Because the immune attack triggers inflammation, and inflammation causes tissue and organ damage, the immune system must be suppressed to alleviate the symptoms of auto-immune diseases. Immune suppression is generally accomplished by administering corticosteroids and nonsteroidal anti-inflammatory drugs, including aspirin.

Allergies are caused by IgE secretion in response to antigens

The most common form of allergy is known as immediate hypersensitivity. It is the result of excessive IgE production in response to antigens, generally referred to as allergens in this context. Allergens that provoke immediate hypersensitivity include various foods, the venom in insect stings, molds, animal danders, and pollen grains. The most common allergy of this type is seasonal hay fever, which may be provoked by the pollen from ragweed (*Ambrosia* spp.) or other plants. Allergies have earned the designation "immediate" because a response to an allergen occurs within seconds or minutes.

The first time or even the first few times that one encounters an allergen, the allergen binds to and activates B cells, which start to secrete allergen-specific immunoglobulins. Activated T_H cells release cytokines such as IL-4, which bind to the B cells and dictate that the antibodies secreted should be IgE. The B cells rapidly switch from the more common IgG secretion to IgE secretion.

Unlike IgG, IgE rapidly binds to mast cells and basophils. When the individual is again exposed to the same allergen, the allergen now specifically binds to the exposed variable regions of identical IgE molecules attached to mast cells and basophils. This binding triggers these cells to secrete histamine, prostaglandins, and other chemical mediators, which produce the symptoms of allergy (figure 51.19).

In *systemic anaphylaxis,* the allergic reaction is severe and potentially life-threatening because of the rapid inflammatory response and release of chemical mediators. The individual experiences a tremendous drop in blood pressure; swelling of the epiglottis can block the trachea, and bronchial constriction can prevent the exit of air from the lungs. This combination of effects is referred to as *anaphylactic shock.* Death can result within 20 to 30 min without prompt medical treatment.

Fortunately, most people with allergies only experience *local anaphylaxis,* such as the itchy welts from hives, or the respiratory constriction of mild asthma. Diarrhea from response to food allergens is another form of local anaphylaxis.

Allergies have traditionally been treated with antihistamines that prevent histamine released by mast cells from binding to its receptor. More recently, a variety of drugs have been developed that block the activation of mast cells and basophils, so that they do not release their mediators. Hyposensitization treatment is another alternative; it consists of the injection, over several months, of an increasing concentration of the allergens to which one is allergic. In some individuals, particularly those with allergic rhinitis (runny nose and eyes) or asthma, this treatment seems to cause a preferential secretion of IgG rather than IgE, and allergy symptoms diminish over time.

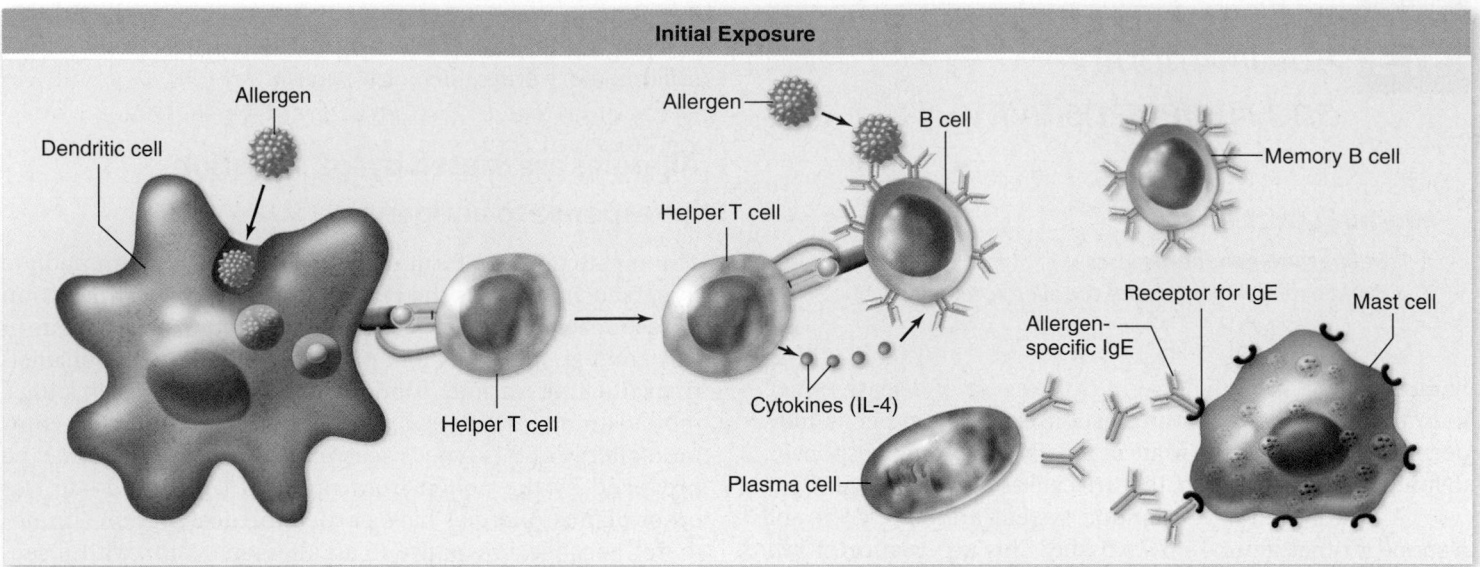

Initial Exposure

Allergen

Dendritic cell

Helper T cell

Helper T cell

Allergen

B cell

Memory B cell

Cytokines (IL-4)

Plasma cell

Allergen-specific IgE

Receptor for IgE

Mast cell

Figure 51.19 An allergic response. On initial exposure to an allergen, B cells are activated to secrete antibodies of the IgE class. These antibodies are in very low levels in the plasma but rapidly bind to Fc receptors on mast cells or basophils. On subsequent exposure to allergen, the allergen cross-links variable regions of two neighboring IgEs with the same epitope specificity on mast cells and basophils. This induces the cells to release histamine and other mediators of inflammation that will cause the symptoms of allergy.

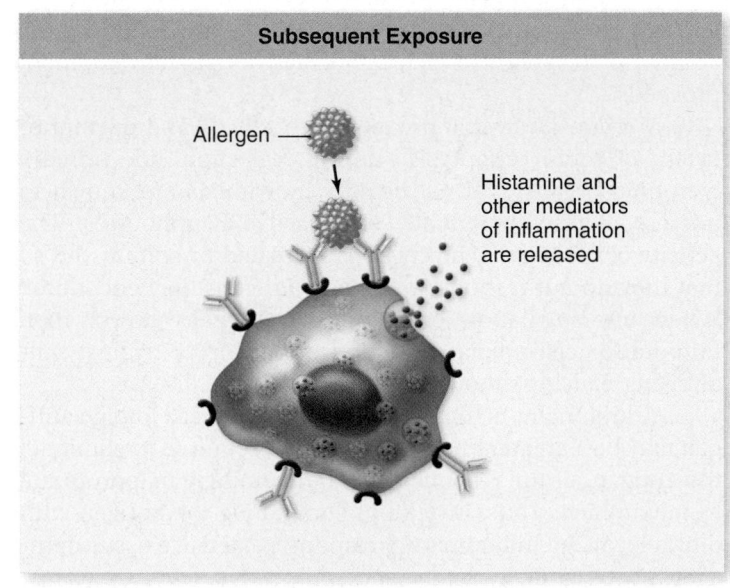

Subsequent Exposure

Allergen

Histamine and other mediators of inflammation are released

Delayed-type hypersensitivity is mediated by T$_H$ cells and macrophages

Delayed-type hypersensitivity, which is mediated by T$_H$ cells and macrophages, produces symptoms within about 48 hr after a second exposure to an antigen. (A first exposure causes a much slower primary response, such as described earlier, frequently without the manifestation of any symptoms.) A form of delayed-type hypersensitivity is contact dermatitis, caused by such varied materials as poison ivy, nickel in jewelry, and some cosmetics. After contact with poison ivy, oils that enter the skin complex with skin proteins, causing the proteins to appear foreign. A delayed-type hypersensitivity response requires that antigen entering the body travel to a secondary lymphoid organ, generally lymph nodes, where T$_H$ cells can be activated. These activated T$_H$ cells then recirculate around the body, and on encountering macrophages that

have ingested the antigen, they release cytokines that activate the macrophages. This induces the macrophages to release other cytokines, and in the case of poison ivy, itchy welts on the skin erupt. The time required for the activation of T$_H$ cells and then of macrophages is the reason for the "delayed" response to the antigens.

Learning Outcomes Review 51.5

Autoimmunity, allergies (immediate hypersensitivity), and delayed hypersensitivity are all examples of inappropriate or heightened immune responses. Autoimmunity results from a loss of self-tolerance. Allergies are associated with a rapid response from mast cells when an allergen binds to IgE on these cells' membrane. Delayed hypersensitivity to pathogens or irritants such as poison ivy is mediated by T$_H$ cells and macrophages.

■ *How are allergies different from autoimmune diseases?*

51.6 Antibodies in Medical Treatment and Diagnosis

Learning Outcomes

1. Explain antigen–antibody reactions in the ABO blood group system.
2. Define monoclonal antibodies.
3. Describe the use of monoclonal antibodies in diagnosis.

The vertebrate immune system can have a range of effects on medical treatment of disease. As two examples, we discuss blood type and its effect on transfusion, as well as the use of monoclonal antibodies for diagnosis and treatment.

Blood type indicates the antigens present on an individual's red blood cells

A person's blood type is determined by certain antigens found on the red blood cell surface. These antigens are clinically important because they must be matched between donor and recipient during a blood transfusion.

ABO groups

In chapter 12, you learned about the genetic basis of the human *ABO blood groups*. The protein–sugar complex on the surface of the red blood cells acts as an antigen, and these antigens differ with regard to the sugar present (or absent, in the case of type O). The immune system is tolerant to its own red blood cell antigens but makes antibodies that bind to those that differ, causing agglutination (clumping) and lysis of foreign red blood cells. Apparently, IgM antibodies made in response to carbohydrates on bacteria that are part of our normal flora also recognize the monosaccharide differences on red blood cells. Such antibodies are not made against carbohydrate patterns that are also present on our own cells.

Rh factor

Another important blood-borne antigen is the Rh antigen or *Rh factor*. This protein is either present (Rh positive) or absent (Rh negative) on the surface of red blood cells. An Rh-negative person who receives an Rh-positive blood transfusion produces antibodies to the foreign Rh protein on the transfused cells.

An additional complication occurs when Rh-negative mothers carry Rh-positive fetuses, which may result in the infant exhibiting a condition called hemolytic disease of newborns (HDN). A first child is usually not harmed; however, at the time of the first birth, the Rh-negative mother's immune system may be exposed to fetal blood. As a result, the woman may become sensitized and produce antibodies and memory B cells against the Rh antigen. If any exposure to fetal blood occurs during a subsequent pregnancy, IgG antibodies, secreted on activation of these memory cells, can cross the placenta and cause destruction of the red blood cells of the fetus.

Blood typing is done by taking advantage of the circulating IgM antibodies, which are produced against foreign blood antigens but not against self. If type A blood is mixed with serum from a person with type B or type O blood, the anti-A antibodies in the serum cause the type A red blood cells to agglutinate. This does not happen if type A blood is mixed with serum from another type A individual or from a type AB individual.

Similarly, if serum from an Rh-negative individual is added to red blood cells, agglutination of the red blood cells indicates that they came from an Rh-positive individual. This individual's blood would not be an appropriate match for transfusion.

Typing of blood prior to transfusions prevents destruction of mismatched cells by a transfusion recipient, as described next. Over 20 blood groups, including the ABO and Rh groups, have been identified; most variants of these other blood groups are rare, but individuals at risk for mismatch may need to "stockpile" their own blood before elective surgery—a practice termed *autologous blood donation.*

Transfusion reactions result from mismatched blood transfusions

Prior to the advent of blood typing in the early 20th century, transfusion of blood was a last resort because of the danger of death from transfusion reaction. An immediate transfusion reaction occurs when an individual receives blood that is not correctly ABO matched. Typically, within 5 to 8 hr of the start of the transfusion, tremendous intravascular hemolysis (rupture) of the transfused red blood cells is detected. This rupture is a result of IgM binding to foreign antigens and activating the complement system. The result is the formation of MACs in the red blood cell membranes and rapid osmotic lysis of the cells.

The hemoglobin released from the red blood cells is converted to a molecule called *bilirubin,* which is particularly toxic to cells and can cause severe organ damage, especially to the kidneys. The major treatment in such situations is to stop the transfusion immediately and to administer large amounts of intravenous fluids to "wash" the bilirubin from the body.

Monoclonal antibodies are a valuable tool for diagnosis and treatment

Antibodies to a known antigen may be obtained by chemically purifying an antigen and then injecting it into a laboratory animal (vertebrate). Periodic bleeding of the animal after a few immunizations allows the isolation of serum antibodies against the antigen. But because an antigen typically has many different epitopes, the antibodies obtained by this method are *polyclonal,* that is, they are secreted by B-cell clones with many different specificities. Their polyclonal nature decreases their sensitivity to any one particular epitope, and it may result in some degree of cross-reactivity with closely related epitopes of different antigens.

Monoclonal antibodies, by contrast, exhibit specificity for one epitope only. In the preparation of monoclonal antibodies, an animal, generally a mouse, is immunized several times with an antigen and is subsequently killed. B lymphocytes, many of

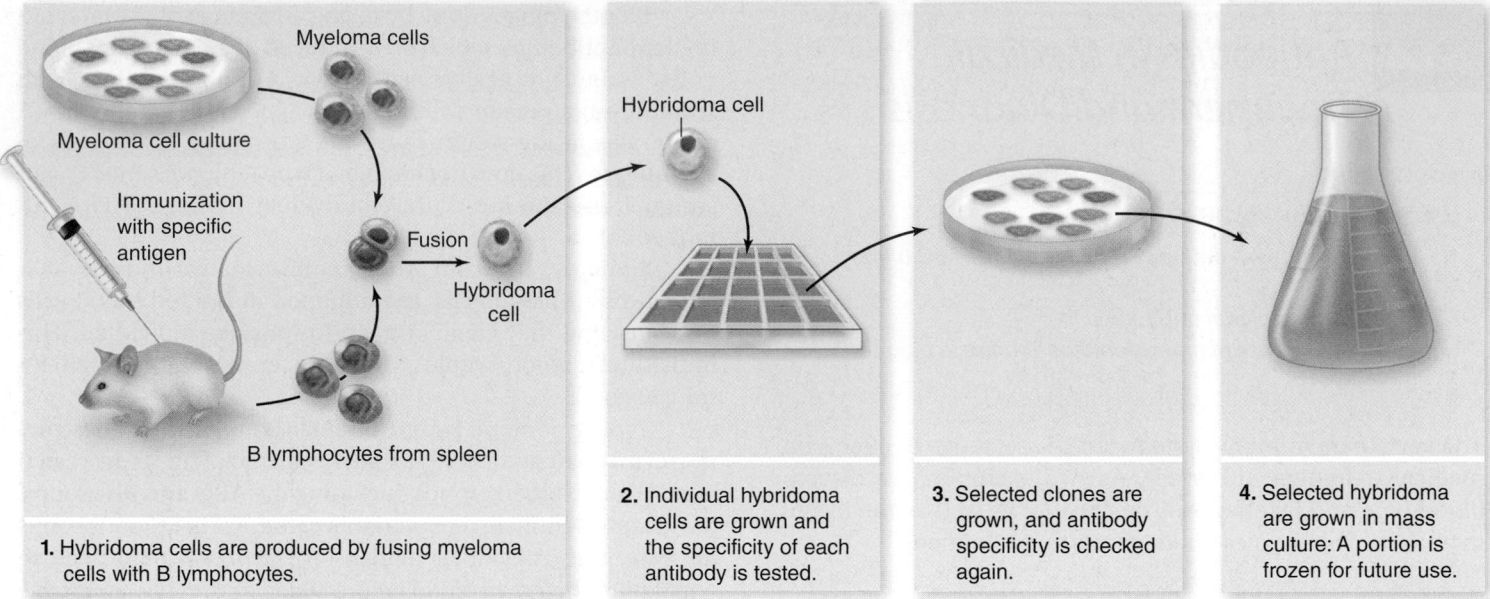

Figure 51.20 The production of monoclonal antibodies. These antibodies are of a single specificity and are produced by "hybridoma" cells. These result from the fusion of B cells, specific for a particular antigenic determinant, with myeloma cells, a B-cell tumor that no longer secretes Ig but provides immortality to the fusion. After hybridoma production, the antibody produced by each hybridoma is tested to see whether it produces specific antibodies against the desired antigen. Selected hybridomas are grown in mass culture for antibody production and are frozen for future use.

which should now be specific for epitopes of the antigen, are collected from the animal's spleen. These B cells would soon die in culture, but utilizing a technique first described in 1975, they are fused with cancerous multiple myeloma cells. These myeloma cells have all the characteristics of plasma cells, except for the secretion of immunoglobulins—but more importantly, they are immortal, meaning they will divide indefinitely. The outcome of a B-cell/myeloma cell fusion is a *hybridoma cell,* that can divide indefinitely and that continues to secrete a large quantity of identical, monoclonal antibodies of the specificity produced by a single B cell (figure 51.20).

Monoclonal antibodies and diagnostic testing

The availability of large quantities of pure monoclonal antibodies has allowed the development of much more sensitive clinical laboratory tests. Some pregnancy tests, for example, use a monoclonal antibody produced against the hormone human chorionic gonadotrophin (hCG, secreted early in pregnancy). The test uses hCG-coated latex particles that are exposed to a urine sample and anti-hCG antibody. If the urine contains hCG, it will block binding of the antibody to the hCG-coated particles and prevent their agglutination, indicating pregnancy based on the presence of hCG (figure 51.21).

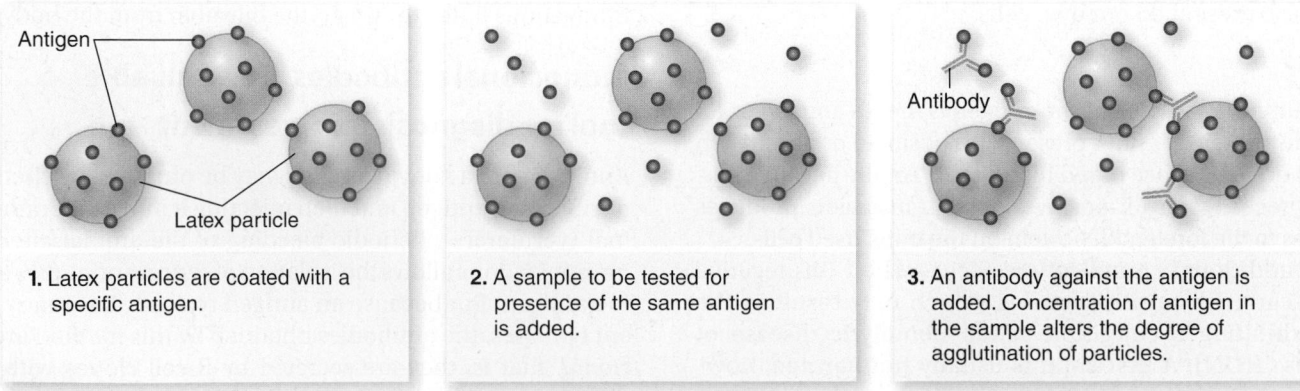

Figure 51.21 Using monoclonal antibodies to detect an antigen. Many clinical tests, such as pregnancy testing, use monoclonal antibodies. A specific antigen is attached to latex beads that are mixed with the test sample and a monoclonal antibody specific for the antigen. If no antigen is present in the sample, the antibody will cause agglutination of the beads. If the sample contains the antigen, it will bind to the antibodies and prevent agglutination of the beads by the antibody.

Data analysis How would a high level of HCG present in a urine sample be indicated in this agglutination test?

Acquired immunodeficiency syndrome (AIDS) is characterized in part by destruction of T_H cells. The progression of this disease can be monitored by examining the reactivity of a patient's leukocytes with a monoclonal antibody against CD4, a marker of T_H cells, to track a decrease in the number of these cells.

Learning Outcomes Review 51.6

Blood group antibodies in plasma made blood transfusion risky and often fatal in the past. Type O RBCs have no surface antigen, but serum from a type O person has both anti-A and anti-B antibodies; someone with type A produces anti-B, and someone with B produces anti-A. Type AB serum lacks A or B antibodies, but RBCs have both surface antigens. Monoclonal antibodies are specific for only a single epitope (antigenic determinant). Hybridoma technology has allowed production of monoclonal antibodies for use in diagnostic tests and elimination of tumors.

■ *Why do diagnostic kits use monoclonal rather than polyclonal antibodies?*

51.7 Pathogens That Evade the Immune System

Learning Outcomes

1. *Explain how pathogens can change antigenic specificity.*
2. *Describe how the immune system has affected the evolution of pathogens.*

For any pathogen to establish itself in a host and to cause a productive infection in which the pathogen successfully reproduces, the pathogen must evade both the nonspecific and specific immune systems. In response to the selective pressure caused by the development of previous immunity against specific epitopes on the pathogen, many pathogens can alter the structure of their surface antigens so that they are no longer recognized. This is a form of natural selection that allows pathogens with altered surface antigens to survive and continue to cause infection. Other pathogens have simply evolved ways to evade destruction. Infection by still other pathogens can actually cause the death of cells of the immune system.

Many pathogens change surface antigens to avoid immune system detection

Influenza virus is perhaps the most universally known example of an organism or virus altering its surface antigens and thus avoiding immune system recognition and destruction. Because of this tendency of the virus to change, yearly immunizations against influenza virus are recommended.

The two viral proteins expressed on the influenza virus' envelope, as well as on the surface of cells infected by influenza virus, are hemagglutinin (HA) and neuraminidase (NA). Because this virus has an RNA genome, it is replicated by a viral RNA polymerase that lacks proofreading ability. As a result, mutations are likely to accumulate over time, including point mutations to the HA and NA genes. This is referred to as **antigen drift.**

Even more dramatically but less frequently, the HA and NA proteins may also undergo **antigen shift,** referring to the sudden appearance of a new subtype of influenza virus in which the expressed HA or NA proteins (or both) are completely different. Such a change makes the population particularly susceptible to infection. Immunization with a new vaccine created every year using the most common strains of the virus attempts to establish immunity in the population prior to infection by the strains circulating.

Antigen shifting and the resulting lack of immunity is the reason for the recent interest in "bird flu." The subtype of influenza that causes bird flu is characterized as H5N1, a primarily avian form of influenza to which people have no immunity. There is no evidence as of this writing, however, that the H5N1 virus can infect people except through contact with infected birds. This strain has been a source of concern due to the high mortality rates seen in human infections. Even more recently, a flu strain with the subtype H1N1 has jumped from pigs to humans and is proving to be very infectious. This virus, with genetic contributions from bird, pig and human strains, was the virus responsible for the pandemic of 2009–2010, declared by the WHO in May 2009 (see chapter 27).

Many other pathogens can alter or shift their surface antigens in order to avoid immune system destruction. As another example, every year, more than 1 million people, the vast majority of whom are African children under the age of 5, die from malaria. This disease, as described in chapter 29, is caused by the protozoan parasite *Plasmodium* and contracted when humans are bitten by an infected *Anopheles* mosquito. These protozoans have several life-cycle stages and are hidden from the immune system alternately within host hepatocytes or red blood cells. In addition, they can alter some of the proteins expressed during certain life cycle stages. Continued use of certain anti-*Plasmodium* drugs has also selected for the emergence of multidrug-resistant organisms. Work is ongoing to develop a vaccine that would induce effective immunity to specific life cycle stages and that would thus promote immune system elimination of the *Plasmodium*.

> **? Inquiry question** Why were we able to eliminate smallpox virus using a vaccine but cannot eliminate influenza?

Many mechanisms have evolved in bacteria to evade immune system attack

Salmonella typhimurium, a common cause of food poisoning, can alternate between expression of two different flagellar proteins, so that antibodies made to one protein do not recognize the other protein and therefore cannot be used to promote phagocytosis of the bacteria.

Mycobacterium tuberculosis bacteria, once phagocytosed into macrophages, inhibit fusion of the phagosome with lysosomes. These organisms can then multiply quite successfully within the macrophages.

Other bacteria that invade mucosal surfaces, such as *Neisseria meningitidis* or *Neisseria gonorrhoeae,* secrete proteases that degrade the IgA antibodies that protect the mucosal surface. External capsules on many particularly pathogenic strains of bacteria block binding of the phagocytosis-inducing complement protein C3b, slowing the phagocytosis response. Because bacteria utilizing any of these mechanisms are better able to survive, the immune response acts as selective pressure favoring the evolution of such mechanisms.

HIV infection kills T$_H$ cells and causes immunosuppression

One mechanism for defeating vertebrate defenses is to attack the adaptive immune system itself. CD4$^+$ T$_H$ cells play a central role in the activity of the immune system: The cytokines they secrete directly or indirectly affect the activity of all other cells of the immune system.

HIV, human immunodeficiency virus, mounts a direct attack on T$_H$ cells (see chapter 27). It binds to the CD4 proteins present on these cells and utilizes these proteins to promote endocytosis into the cells. (The virus infects monocytes as well because they too express CD4.) HIV-infected cells die only after releasing replicated viruses that infect other CD4$^+$ cells (figure 51.22). Over time, the number of T$_H$ cells in an infected individual decreases.

An individual is considered to have AIDS when his or her T$_H$ cell levels have dropped dramatically, leading to an increase in infections due to opportunistic organisms as well as other diseases.

The progression of HIV infection

The immune system initially controls an HIV infection by the production of antibodies to the virus and by the elimination of virally infected cells by cytotoxic T cells. For a time, the level of HIV in the serum does not increase beyond a steady state, and the number of T$_H$ cells does not significantly decrease.

As the virus reproduces in the T$_H$ cells, it rapidly kills some of them, but many others continue to divide on antigen stimulation. Eventually, however, HIV kills T$_H$ cells more rapidly than they can proliferate. HIV-encoded proteins also cause a decrease in the expression of MHC class I on the infected cells, so that these cells are less likely to be recognized and killed by T$_C$ cells.

Finally, because HIV is a retrovirus (see chapter 27), it can integrate itself into the genome of infected cells and hide in a "latent" form. When any of these cells divides, the HIV genome is present in the progeny cells and thus these progeny can start to produce HIV at any point in time.

The combined effect of these responses to HIV infection is to ravage the human immune system. With little defense against infection, any of a variety of otherwise commonplace infections may prove fatal. Death by cancer also becomes far more likely. In fact, AIDS was first recognized as a disease when a cluster of previously healthy young men all died of *Pneumocystis jiroveci*–induced pneumonia, a disease that

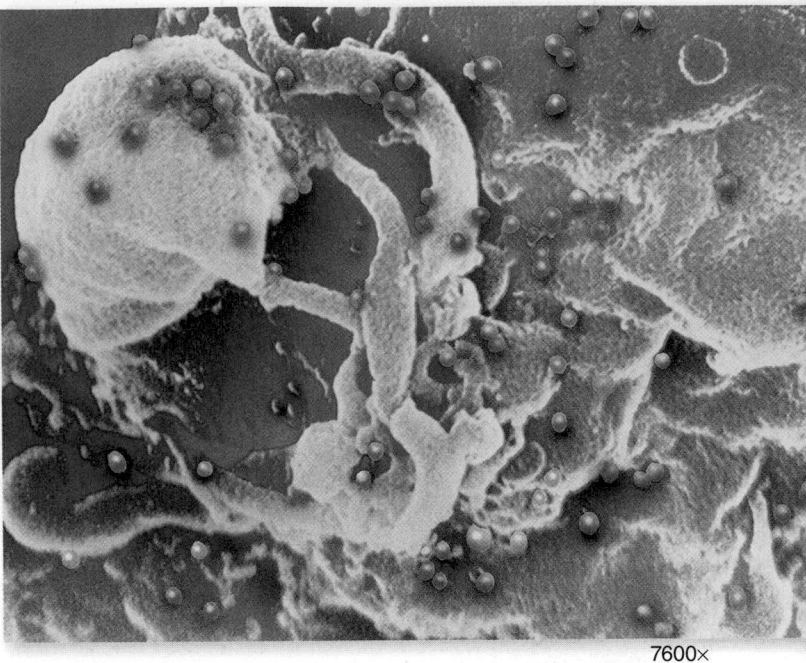

7600×

Figure 51.22 HIV, the virus that causes AIDS. Viruses released from infected CD4$^+$ T$_H$ cells soon spread to neighboring T$_H$ cells, infecting them in turn. The individual viruses, colored red in this scanning electron micrograph, are extremely small; over 200 million would fit on the period at the end of this sentence.

generally affects the immunosuppressed, or of Kaposi's sarcoma, a rare form of cancer.

The human effect of HIV

Although HIV became a human disease only recently, AIDS is already clearly one of the most serious diseases in human history. The WHO estimates that for the year 2010 about 34 million people are living with HIV, with the greatest number in sub-Saharan Africa (about 23 million), followed by south and southeast Asia (3.5 million). For 2010, the WHO estimates there were 2.7 million people newly infected, and an estimated 1.8 million AIDS deaths worldwide.

The fatality rate of untreated AIDS is close to 100%. No patient exhibiting the symptoms of AIDS has been known to survive more than a few years without treatment. The disease is not highly contagious, however; it is transmitted from one individual to another through the transfer of internal body fluids, typically semen and blood.

Learning Outcomes Review 51.7

Pathogens such as influenza virus frequently alter epitopes on their surfaces and thus evade specific immune system recognition. Other pathogens, such as HIV, infect and destroy T$_H$ cells, simply disabling the immune system. The diseases associated with AIDS often result from destruction of the immune system.

■ *Polio is a viral disease against which vaccines have been very successful. How would you say this virus differs from influenza virus?*

Chapter Review

51.1 Innate Immunity

The skin is a barrier to infection.

In addition to being a physical barrier, skin has a low surface pH, lysozyme secreted in sweat, and a population of nonpathogenic organisms, all of which deter pathogens.

Mucosal epithelial surfaces also prevent entry of pathogens.

The epithelia of the digestive, respiratory, and urogenital tracts produce mucus to trap microorganisms.

Innate immunity recognizes characteristic pathogen molecules.

Toll-like receptors have leucine-rich regions that recognize molecules such as LPS and peptidoglycan.

Innate immunity leads to diverse responses to a pathogen.

Binding of molecules specific to a pathogen leads to production of antimicrobial peptides and cytokines, and activation of complement, among other actions.

Phagocytic cells are associated with innate immunity.

Macrophages and neutrophils are associated with phagocytosis (figure 51.4). Natural killer (NK) cells induce apoptosis.

The inflammatory response is a nonspecific response to infection or tissue injury.

Histamines increase blood flow and permeability of capillaries (figure 51.5). Acute-phase fever promotes phagocytic activity and impedes growth of microbes.

Complement can form a membrane attack complex.

Complement forms pores in invading cells, coats pathogens with C3b proteins, and targets cells for destruction.

51.2 Adaptive Immunity

Immunity had long been observed, but the mechanisms have only recently been understood.

Jenner's and Pasteur's work on cowpox and avian cholera indicated that prior exposure to a disease prevented subsequent infections

Antigens stimulate specific immune responses.

Surface receptors on lymphocytes recognize antigens and direct a specific immune response (figure 51.8).

Hematopoiesis gives rise to the cells of the immune system (table 51.1).

Lymphocytes carry out the adaptive immune responses.

A naive lymphocyte binds to a foreign antigen and divides, producing a clone of activated cells and memory cells.

Humoral immunity is the production of Ig by B cells. Cell-mediated immunity involves T cells that regulate the immune responses of other cells or directly attack cells.

Adaptive immunity can be active or passive.

The immune system is supported by two classes of organs.

Immune system organs include the primary lymphoid organs and secondary lymphoid organs (figure 51.9).

Two forms of adaptive immunity have evolved.

An immune system based on proteins with variable repeats has been found in jawless fishes, in contrast to the immunoglobulin system of other vertebrates.

51.3 Cell-Mediated Immunity

The MHC carries self and nonself information.

Most cells exhibit glycoproteins encoded by the MHC. In humans these proteins are called human leukocyte antigens (HLAs). MHC class I proteins, found on every nucleated cell, present antigen brought into the cell by phagocytosis; MHC class II proteins are found only on antigen-presenting cells.

Cytotoxic T cells eliminate virally infected cells and tumor cells.

T_C cells recognize virally infected cells and tumor cells. They destroy cells in a fashion similar to NK cells (figure 51.11).

Helper T cells secrete proteins that direct immune responses.

T_H cells secrete cytokines in response to foreign antigens and promote both cell-mediated and humoral immune responses. Activated T_C and T_H produce both effector and memory cells.

T cells are the primary cells that mediate transplant rejection.

Cells of the innate immune system release cytokines.

Macrophages release interleukin-12 and tumor necrosis factor-α.

51.4 Humoral Immunity and Antibody Production

Immunoglobulin structure reveals variable and constant regions.

B cells are activated by membrane Ig molecules binding to a specific epitope on an antigen. Activated B cells produce antibody secreting plasma cells and memory cells.

Immunoglobulins consist of two light chain and two longer heavy-chain polypeptides, with the binding site in the Fab region (figure 51.14). Antibodies can agglutinate, precipitate, or neutralize antigens (figure 51.15).

The five classes of immunoglobulins have different functions (table 51.3).

Immunoglobulin diversity is generated through DNA rearrangement.

Ig diversity is generated by DNA rearrangements (figure 51.16). T cell receptors (TCRs) are similar to a single Fab region of an Ig. Their diversity also results from DNA rearrangements.

The secondary response to an antigen is more effective than the primary response.

On second exposure to a pathogen, a rapid secondary immune response is launched due to memory cells (figure 51.18).

51.5 Autoimmunity and Hypersensitivity

Autoimmune diseases result from immune system attack on the body's own tissues.

The acceptance of self cells is called immunological tolerance. Autoimmunity is a failure of immunological tolerance.

Allergies are caused by IgE secretion in response to antigens.

Immediate hypersensitivity is caused by allergens' binding to IgE, triggering the release of histamine (figure 51.19). Anaphylaxis is a severe reaction with rapid inflammation and release of chemical mediators.

Delayed-type hypersensitivity is mediated by T_H cells and macrophages.

Symptoms appear about 48 hr after second exposure.

51.6 Antibodies in Medical Treatment and Diagnosis

Blood type indicates the antigens present on an individual's red blood cells.

The ABO blood group antigens, Rh factor antigens, and many others constitute the blood type.

HLA antigens can be used to match tissues between donors and recipients in transplants.

Transfusion reactions result from mismatched blood transfusions.
Before blood type was understood, many died from mismatched blood transfusions.

Monoclonal antibodies are a valuable tool for diagnosis and treatment.
Monoclonal antibodies are specific to a single epitope and can be used in testing and manufacture of immunotoxins.

51.7 Pathogens That Evade the Immune System

Many pathogens change surface antigens to avoid immune system detection.
Antigen drift and antigen shift, such as exhibited in influenza viruses, allow alteration of surface antigens to avoid immune detection.

Many mechanisms have evolved in bacteria to evade immune system attack.
Some bacteria have evolved mechanisms to inhibit normal immune system processes, such as slowing phagocytosis.

HIV infection kills T_H cells and causes immunosuppression.
Destruction of T_H cells lowers the body's ability to fend off infections, leading to the characteristic illnesses of AIDS.

Review Questions

UNDERSTAND

1. Cells that target and kill body cells infected by viruses are
   a. macrophages.
   b. natural killer cells.
   c. monocytes.
   d. neutrophils.

2. Structures on invading cells recognized by the adaptive immune system are known as
   a. antigens.
   b. interleukins.
   c. antibodies.
   d. lymphocytes.

3. Which one of the following acts as the "alarm signal" to activate the body's adaptive immune system by stimulating helper T cells?
   a. B cells
   b. Interleukin-1
   c. Complement
   d. Histamines

4. Cytotoxic T cells are called into action by the
   a. presence of histamine.
   b. presence of interleukin-1.
   c. presence of interleukin-2.
   d. interferon.

5. Receptors that trigger innate immune responses
   a. are antibodies recognizing specific antigens.
   b. are T-cell receptors recognizing specific antigens.
   c. recognize pathogen-associated molecular patterns.
   d. are not specific at all.

6. Diseases in which the person's immune system no longer recognizes its own MHC proteins are called
   a. allergies.
   b. autoimmune diseases.
   c. immediate hypersensitivity.
   d. delayed hypersensitivity.

7. Suppose that a new disease is discovered that suppresses the immune system. Which of the following would indicate that the disease specifically affects the B cells rather than the helper or cytotoxic T cells?
   a. A decrease in the production of interleukin-2
   b. A decrease in interferon production
   c. A decrease in the number of plasma cells
   d. A decrease in the production of interleukin-1

APPLY

1. You start a new job in a research lab. The lab protocols state that you should check your hands for any breaks in the skin before handling infectious agents. This is because the epidermis fights microbial infections by
   a. making the surface of the skin acidic.
   b. excreting lysozyme to attack bacteria.
   c. producing mucus to trap microorganisms.
   d. All of the choices are correct.

2. In comparing T-cell receptors and immunoglobulins
   a. the proteins have unrelated structures, but diversity is generated by a similar mechanism.
   b. the proteins have related structures, but diversity is generated by different mechanisms.
   c. the proteins have related structures, and diversity is generated by a similar mechanism.
   d. the proteins have unrelated structures, and diversity is generated by different mechanisms.

3. If you have type AB blood, which of the following results would be expected?
   a. Your blood agglutinates with anti-A antibodies only.
   b. Your blood agglutinates with anti-B antibodies only.
   c. Your blood agglutinates with both anti-A and anti-B antibodies.
   d. Your blood would not agglutinate with either anti-A or anti-B antibodies.

4. Suppose that you get a paper cut while studying. Arrange the following into the correct order in time.
   a. Injured epidermal cells release histamine.
   b. Bacteria enter the cut.
   c. Helper T cells are activated.
   d. Macrophages engulf bacteria.

5. If you wanted to cure allergies by bioengineering an antibody that would bind and disable the antibody responsible for allergic reactions, which of the following would you target?
   a. IgG
   b. IgA
   c. IgE
   d. IgD

6. Why do we need to be repeatedly vaccinated for influenza viruses?
   a. They attack only the helper T cells, thereby suppressing the immune system.
   b. They alter their surface proteins and thus avoid immune recognition.
   c. They don't actually generate an immune response; the "flu" is actually an inflammatory response.
   d. They are too small to serve as good antigens.

7. If you wanted to design an artificial cell that could safely carry drugs inside the body, which of the following molecules would you need to mimic to deter the immune system?
   a. MHC-1   c. Antigen
   b. Interleukin-1   d. Complement

SYNTHESIZE

1. Suppose you take a job in the marketing department of a cosmetics company. Always seeking a competitive advantage, the vice president of marketing has decided to advertise the new skin lotion as having immune-enhancing effects. The lotion is produced from secretions of a plant that releases extremely watery, alkaline fluid. Explain how you will market this product as an immune-enhancer.

2. Your new kitten scratches your roommate. Her skin is reddened and feels warm and sore to the touch; she thinks she has contracted some kind of fatal infection. In order to deflect her anger (she is definitely not a cat person!), you try telling her about the activities of the innate defense system. Explain what is actually happening to her skin.

3. Some people claim that they never catch colds. How could you show that this is due to a difference in receptors on their respective cell surfaces?

4. Toll-like receptors have been found in a wide variety of organisms, including both protostomes and deuterostomes, and now in cnidarians. In addition, parts of the signaling system have been found in a wide variety of organisms as have parts of the complement system. What does this say about the evolution of innate immunity?

ONLINE RESOURCE

www.ravenbiology.com

Understand, Apply, and Synthesize—enhance your study with animations that bring concepts to life and practice tests to assess your understanding. Your instructor may also recommend the interactive eBook, individualized learning tools, and more.

Chapter **52**

The Reproductive System

Chapter Contents

Introduction

Bird song in the spring, insects chirping outside the window, frogs croaking in swamps, and wolves howling in a frozen northern forest are all sounds of evolution's essential act, reproduction. These distinctive noises, as well as the bright coloration of some animals, function to attract mates. Few subjects pervade our everyday thinking more than sex, and few urges are more insistent. This chapter deals with sex and reproduction among the vertebrates, including humans.

52.1 Animal Reproductive Strategies

Learning Outcomes

1. *Distinguish between sexual and asexual methods of reproduction.*
2. *Describe the different types of hermaphroditism.*
3. *Explain factors that influence sex determination.*

Most animals, including humans, reproduce sexually. As described in chapter 11, sexual reproduction requires a specialized form of cell division, meiosis, to produce haploid gametes, each of which has a single complete set of chromosomes. These gametes, including sperm and eggs (or *ova;* singular *ovum*), are united by fertilization to restore the diploid complement of chromosomes. The diploid fertilized egg, or zygote, develops by mitotic division into a new multicellular organism.

Reproduction can occur asexually

Bacteria, archaea, protists, and multicellular animals including cnidarians and tunicates, as well as many other types of animals, reproduce asexually. In asexual reproduction, genetically identical cells are produced from a single parent cell through mitosis. In single-celled organisms, an individual organism divides, a process called fission, and then each part becomes a separate but identical organism. Cnidarians commonly reproduce by budding, whereby a part of the parent's body becomes separated from the rest and differentiates into a new individual (figure 52.1). The new individual may become an independent animal or may remain attached to the parent, forming a colony.

Another form of asexual reproduction, **parthenogenesis,** is common in many species of arthropods. In parthenogenesis, females produce offspring from unfertilized eggs. Some species are exclusively parthenogenic (and all female), whereas others switch between sexual reproduction and parthenogenesis, producing progeny that are both diploid and haploid, respectively. In honeybees, for example, a queen bee mates only once and stores the sperm. She then can control the release of the sperm. If no sperm are released, the eggs develop parthenogenetically into haploid drones, which are males. If sperm are allowed to fertilize the eggs, the fertilized eggs develop into diploid worker bees, which are female. However, when fertilized eggs are exposed to the appropriate hormone, they will develop into queens.

In 1958, the Russian biologist Ilya Darevsky reported one of the first cases of unusual modes of reproduction among vertebrates. He observed that some populations of small lizards of the genus *Lacerta* were exclusively female, and he suggested that these lizards could lay eggs that were viable even if they were not fertilized. In other words, they were capable of asexual reproduction in the absence of sperm, a type of parthenogenesis. Further work has shown that parthenogenesis has evolved a number of times in lizards, as well as in fish and salamanders.

In some species, individuals can be both male and female

Another variation in reproductive strategies is hermaphroditism, in which one individual has both testes and ovaries, and so can produce both sperm and eggs. A tapeworm is

Figure 52.2 Hermaphroditism and protogyny. The bluehead wrasse, *Thalassoma bifasciatium,* is protogynous—females sometimes turn into males. Here, a large male, which had been a female before changing sex, is seen among females, which are typically much smaller.

> **?** **Inquiry question** Why might it be adaptive for an individual to be female at small size and become male when very large? Under what circumstances might the opposite condition, protandry, evolve?

hermaphroditic and can fertilize itself, a useful strategy because it is unlikely to encounter another tapeworm. Most hermaphroditic animals, however, require another individual in order to reproduce. Two earthworms, for example, are required for reproduction—each functions as both male and female during copulation, and each leaves the encounter with fertilized eggs.

Numerous fish genera include species whose individuals can change their sex, a process called *sequential hermaphroditism.* Among coral reef fish, for example, both protogyny ("first female," a change from female to male) and protandry ("first male," a change from male to female) occur. In fish that practice protogyny (figure 52.2), the sex change appears to be under social control. These fish commonly live in large groups, or schools, where successful reproduction is typically limited to one or a few large, dominant males. If those males are removed, the largest female rapidly changes sex and becomes a dominant male.

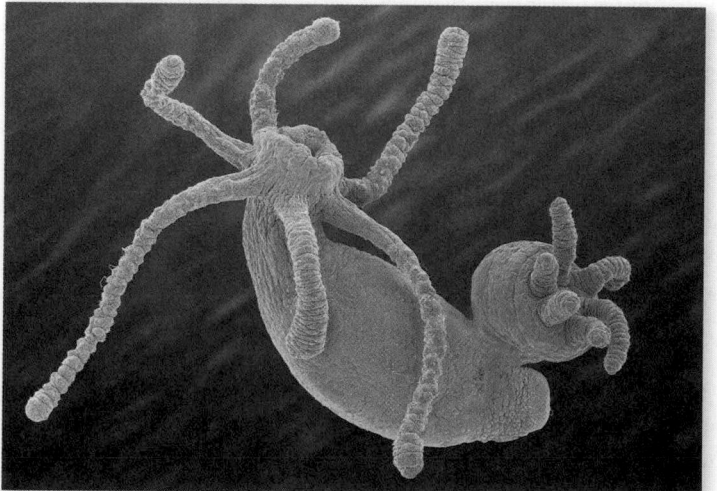

14×

Figure 52.1 Cnidarian budding. This cnidarian reproduces asexually by budding. The new individual can be seen in the lower right of the micrograph.

Sex can be determined genetically or by environmental conditions

An individual's sex is determined during development as an embryo, but the way in which it occurs varies among species.

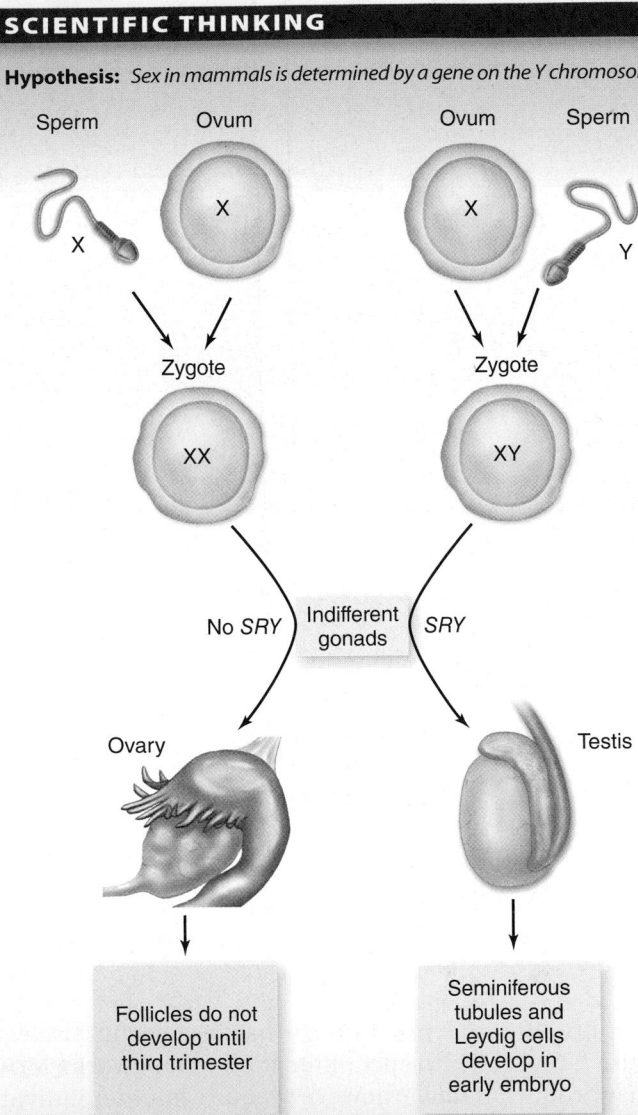

Hypothesis: *Sex in mammals is determined by a gene on the Y chromosome.*

Sperm   Ovum   Ovum   Sperm

X   X   X   Y

Zygote   Zygote

XX   XY

No *SRY*   Indifferent gonads   *SRY*

Ovary   Testis

Follicles do not develop until third trimester

Seminiferous tubules and Leydig cells develop in early embryo

Prediction 1: *Individuals lacking a Y chromosome will develop as females.*

Test: *Compare individuals with unusual sets of sex chromosomes.*

Results: *Individuals with Turner's Syndrome are XO (they only have a single X chromosome) and develop as females; individuals with Klinefelter's Syndrome are XXY and develop as males.*

Prediction 2: *The SRY gene on the Y chromosome is responsible for the development of male sexual characteristics.*

Test: *Breed transgenic mice that are XY but have a mutation of the SRY gene.*

Results: *These mice develop as females.*

Figure 52.3 Sex determination in mammals. The sex-determining region of the mammalian Y chromosome is designated *SRY*. Testes are formed when the Y chromosome and *SRY* are present; ovaries are formed when they are absent.

 Temperature-sensitive sex determination

In many fish and reptiles, an individual's sex is determined by the temperature it experiences during development. In some species, cold temperatures produce males and warm temperatures produce females, but in others, the opposite occurs. In still others, males are produced at both high and low temperatures, and females at temperatures in-between.

Phylogenetic analyses indicate that temperature-sensitive sex determination has evolved many times from ancestors in which sex was genetically determined. The molecular mechanisms that determine sex in this species are now being discovered, but why this trait has evolved so often is not well understood. One possibility is that sometimes it may be advantageous for a female to be able to determine the sex of her offspring by laying eggs in appropriate locations. Currently, the jury is still out on this hypothesis.

Genetic sex determination

In all birds and mammals and many other vertebrates, sex is determined by an individual's genes. In mammals and some other animals, individuals with an X and a Y chromosome are males, whereas individuals with two X chromosomes are females. However, in other animals, such as birds, it is females that are the heterozygous sex.

Sexual differentiation in humans

The reproductive systems of human males and females appear similar for the first 40 days after conception. During this time, the cells that will give rise to ova or sperm migrate from the yolk sac to the embryonic gonads, which have the potential to become either ovaries in females or testes in males (figure 52.3). For this reason, the embryonic gonads are said to be "indifferent."

If the embryo is a male, a gene on the Y chromosome converts the indifferent gonads into testes. In females, which lack a Y chromosome, this gene and the protein it encodes are absent, and the gonads become ovaries. An important gene involved in sex determination is known as *SRY* (for "sex-determining region of the Y chromosome").

Once testes have formed in the embryo, they secrete testosterone and other hormones that promote the development of the male external genitalia and accessory reproductive organs.

If the embryo lacks the *SRY* gene, the embryo develops female external genitalia and accessory organs. In other words, all mammalian embryos will develop into females unless a functional *SRY* gene is present.

Learning Outcomes Review 52.1

Sexual reproduction involves fusion of gametes derived from different individuals of a species. Asexual reproduction in some species may be accomplished by fission, budding, or parthenogenesis. In hermaphroditic species, an individual may have both testes and ovaries; in sequential hermaphroditism, an individual may change from one sex to the other. Sex can be determined either by genes or by environmental conditions such as temperature experienced while an embryo.

■ **Why might natural selection favor genetic sex determination?**

52.2 Vertebrate Fertilization and Development

Learning Outcomes

1. Distinguish among viviparity, oviparity, and ovoviviparity.
2. Describe the advantages of internal fertilization.

Vertebrate sexual reproduction evolved in the ocean before vertebrates colonized the land. The females of most species of marine bony fish produce eggs in batches and release them into the water. The males generally release their sperm into the water containing the eggs, where the union of the free gametes occurs. This process is known as external fertilization.

Although seawater is not a hostile environment for gametes, it does cause the gametes to disperse rapidly, so their release by females and males must be almost simultaneous. Thus, most marine fish restrict the release of their eggs and sperm to a few brief and well-defined periods. Some reproduce just once a year, but others do so more frequently. The ocean has few seasonal clues that organisms can use as signals for synchronizing reproduction, but one all-pervasive signal is the cycle of the Moon. Once each month, the Moon approaches closer to the Earth than usual, and when it does, its increased gravitational attraction causes somewhat higher tides. Many marine organisms sense the tidal changes and link the production and release of their gametes to the lunar cycle.

Once vertebrates began living on land, they encountered a new danger—desiccation, a problem that can be especially severe for the small and vulnerable gametes. On land, the gametes could not simply be released near each other because they would soon dry up and perish. Consequently, intense selective pressure resulted in the evolution of internal fertilization in terrestrial vertebrates (as well as some groups of fish)—that is, the introduction of male gametes directly into the female reproductive tract. By this means, fertilization still occurs in a nondesiccating environment, even when the adult animals are fully terrestrial.

Internal fertilization has led to three strategies for development of offspring

The vertebrates that practice internal fertilization exhibit three strategies for embryonic and fetal development, namely *oviparity*, *ovoviviparity*, and *viviparity*.

1. **Oviparity** is found in some bony fish, most reptiles, some cartilaginous fish, some amphibians, a few mammals, and all birds. The eggs, after being fertilized internally, are deposited outside the mother's body to complete their development.
2. **Ovoviviparity** is found in some bony fish (including mollies, guppies, and mosquito fish), some cartilaginous fish, and many reptiles. The fertilized eggs are retained within the mother to complete their development, but the

Figure 52.4 Viviparous fish carry live, mobile young within their bodies. The young complete their development within the body of the mother and are then released as small but competent adults. Here, a lemon shark *(Negaprion brevirostris)* has just given birth to a young shark, which is still attached by the umbilical cord.

embryos still obtain all of their nourishment from the egg yolk. The young are fully developed when they are hatched and released from the mother.

3. **Viviparity** is found in most cartilaginous fish, some amphibians, a few reptiles, and almost all mammals (figure 52.4). The young develop within the mother and obtain nourishment directly from their mother's blood, rather than from the egg yolk. A placenta, the structure through which blood and gas exchange occurs, has not evolved only in mammals (see chapter 53), but also several times in fishes and lizards.

Evolution of reproductive systems

Live birth—either viviparity or ovoviviparity—has evolved many times in vertebrates: once in mammals, but many times independently in fishes, amphibians, and reptiles (figure 52.5). The evolution of live birth appears to be a one-way evolutionary street: once it evolves, it is almost never lost. Live birth requires internal fertilization so that the eggs can develop within the body of the female; internal fertilization has evolved only once in amniotes (the clade composed of reptiles, birds, and mammals), but multiple times within fishes and amphibians.

Internal fertilization requires some means of transferring the sperm from the male to the female. Salamanders evolved one way: males deposit their sperm on top of a mass of eggs and then the female positions her cloaca above them and then lowers her body, picking up the fertilized eggs. All other vertebrates have taken a different approach, evolving an intromittent organ that the male uses to transfer the sperm directly into the female's body. The variety of structures that has evolved to accomplish this includes a modified pelvic fin in cartilaginous fishes; a modification of the cloaca (the common opening through which waste and reproductive products exit the body) in some frogs, caecilians, and birds; penises derived from different embryological structures in turtles, crocodiles, and

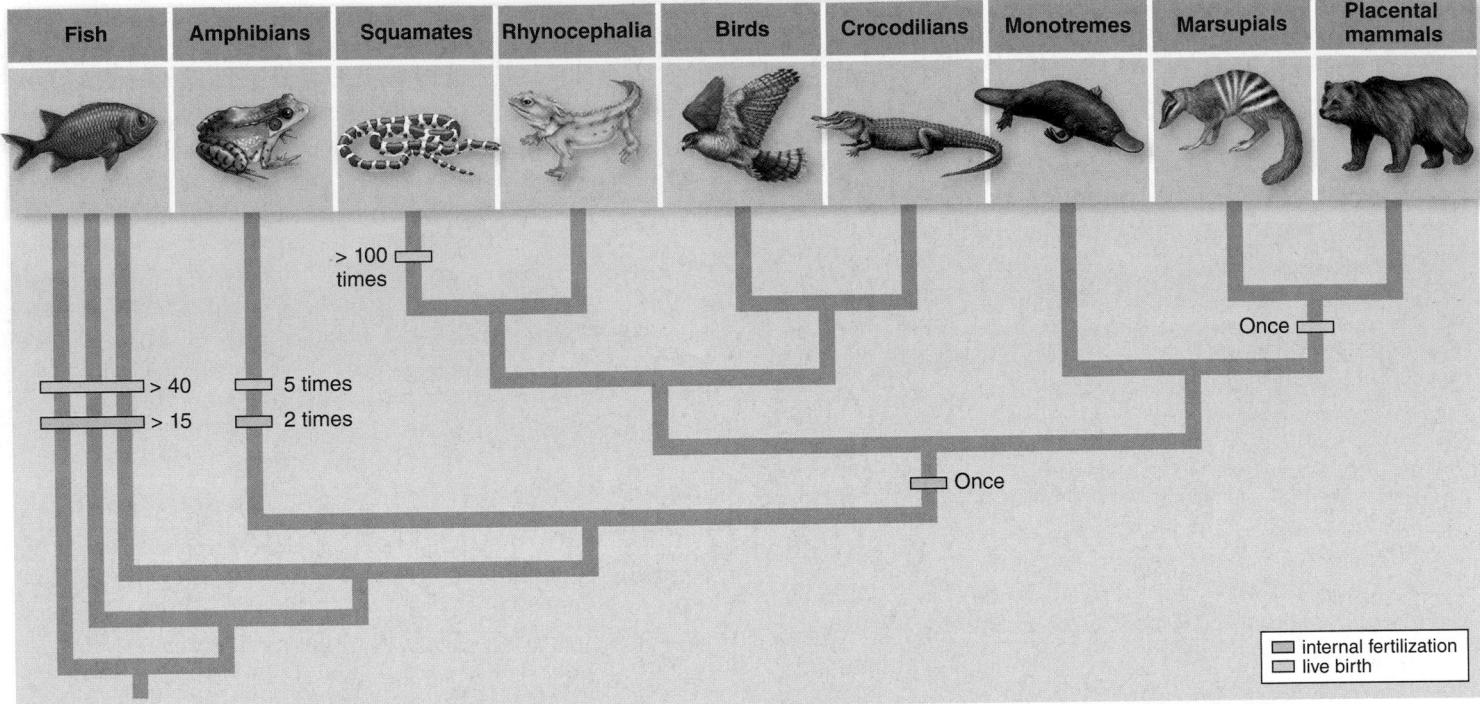

| Fish | Amphibians | Squamates | Rhynocephalia | Birds | Crocodilians | Monotremes | Marsupials | Placental mammals |

Figure 52.5 Evolution of internal fertilization and live birth in vertebrates. Although live birth has evolved many times in fishes and squamate reptiles, most species in both groups lay eggs. Evolutionary reversal from live birth to egg-laying has occurred very rarely, if at all. Estimates of the number of origins in fishes and squamates is based on detailed phylogenetic analyses within each group; uncertainty in numbers is a result of incomplete information in some groups.

? Inquiry question Why do you think that egg-laying rarely evolves from live-bearing?

mammals; and a pair of hemipenises—one on each side—in snakes and lizards. Moreover, intromittent organs have been lost entirely in birds and rhynchocephalians; to achieve internal fertilization, males and females of these species simply align their cloacae and pass the sperm from male to female.

Most fishes and amphibians have external fertilization

Most fishes and amphibians, unlike other vertebrates, reproduce by means of external fertilization, although internal fertilization has arisen many times.

Fishes

Fertilization in most species of bony fish (teleosts) is external, and the eggs contain only enough yolk to sustain the developing embryo for a short time. After the initial supply of yolk has been exhausted, the young fish must seek its food from the waters around it. Development is speedy, and the young that survive mature rapidly. Although thousands or even millions of eggs are fertilized in a single mating, many of the resulting individuals succumb to microbial infection or predation, and few grow to maturity.

In marked contrast to the bony fish, fertilization in most cartilaginous fish is internal. Development of the young is generally viviparous, and the female usually gives birth to few, well-developed offspring.

Amphibians

The life cycle of amphibians is still tied to the water. Fertilization is external in most amphibians. Gametes from both males and females are released through the cloaca. Among the frogs and toads, the male grasps the female and discharges fluid containing the sperm onto the eggs as the female releases them into the water (figure 52.6).

Although the eggs of most amphibians develop in the water, there are some interesting exceptions (figure 52.7). In some frogs, for example, the eggs develop in the back of the parents; in others, males carry around the tadpoles in their vocal sacs, and the young frogs leave through their parents' mouths.

Reptiles and birds have internal fertilization

All birds and about 80% of reptile species are oviparous. After the eggs are fertilized internally, they are deposited outside the mother's body to complete their development.

 ### Reptiles

Most oviparous reptiles lay eggs and then abandon them. These eggs are surrounded by a leathery shell that is deposited as the egg passes through the oviduct, the part of the female reproductive tract leading from the ovary. Other species of reptiles are ovoviviparous, forming eggs that develop into embryos within the body of the mother, and some species are viviparous.

Figure 52.6 The eggs of frogs are fertilized externally.
When frogs mate, the clasp of the male induces the female
to release a large mass of mature eggs, over which the male
discharges his sperm.

Birds

All birds practice internal fertilization, though most male
birds lack a penis (some, including swans, geese, and
ostriches, have modified the wall of the cloaca to serve as an
intromittent organ).

As the egg passes along the oviduct, glands secrete albumin proteins (the egg white) and the hard, calcareous shell
that distinguishes bird eggs from reptilian eggs. Although
modern reptiles are ectotherms, birds are endotherms (see
chapter 42); therefore, most birds incubate their eggs after

Figure 52.8 Crested penguins incubating their egg.
This nesting pair is changing the parental guard in a
stylized ritual.

laying them to keep them warm (figure 52.8). The young that
hatch from the eggs of most bird species are unable to survive
unaided because their development is still incomplete. These
young birds are fed and nurtured by their parents, and they
grow to maturity gradually.

The shelled eggs of reptiles and birds constitute one of
the most important adaptations of these vertebrates to life on
land. As described in chapter 35, these eggs are known as
amniotic eggs because the embryo develops within a fluid-
filled cavity surrounded by a membrane called the *amnion.*
Other extraembryonic membranes in amniotic eggs include
the *chorion,* which lines the inside of the eggshell, the *yolk sac,*
and the *allantois.* Together, these extraembryonic membranes
in combination with the external calcareous shell help form a
desiccation-resistant egg that can be laid in dry places. In contrast, the eggs of fish and amphibians contain only one extraembryonic membrane, the yolk sac, and must be deposited in
an aquatic habitat to keep from drying out.

The viviparous mammals, including humans, also have
extraembryonic membranes, as described in chapter 53.

a.

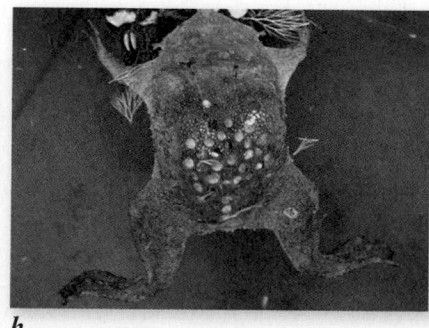
b.

c.

d.

Figure 52.7 Different ways young develop in frogs. *a.* In poison arrow frogs (family Dendrobatidae), the male carries the
tadpoles on his back. *b.* In the female Surinam toad *(Pipa pipa),* froglets develop from eggs in special brooding pouches on the back.
c. In the South American pygmy marsupial frog *(Flectonotus pygmaeus),* the female carries the developing larvae in a pouch on her back.
d. Tadpoles of the Darwin's frog *(Rhinoderma darwinii)* develop into froglets in the vocal pouch of the male and emerge from the mouth.

Mammals generally do not lay eggs, but give birth to their young

Some mammals are seasonal breeders, reproducing only once a year, while others have more frequent reproductive cycles. Among the latter, the females generally undergo reproductive cycles, and the males are more constant in their reproductive capability.

Female reproductive cycles

Cycling in females involves the periodic release of a mature ovum from the ovary in a process known as *ovulation*. Most female mammals are "in heat," or sexually receptive to males, only around the time of ovulation. This period of sexual receptivity is called **estrus,** and the reproductive cycle is therefore called an estrous cycle. Reproductive cycles continue in females until they become pregnant.

In the estrous cycle of most mammals, changes in the secretion by the anterior-pituitary gland of follicle-stimulating hormone (FSH) and luteinizing hormone (LH) cause changes in egg cell development and hormone secretion in the ovaries (see chapter 45). Humans and apes have menstrual cycles that are similar to the estrous cycles of other mammals in their pattern of hormone secretion and ovulation. Unlike mammals with estrous cycles, however, human and ape females bleed when they shed the inner lining of their uterus, a process called **menstruation,** and they may engage in copulation at any time during the cycle.

Rabbits and cats differ from most other mammals in that they are induced ovulators. Instead of ovulating in a cyclic fashion regardless of sexual activity, the females ovulate only after copulation, as a result of a reflex stimulation of LH secretion.

Monotremes, marsupials, and placental mammals

The most primitive mammals, the **monotremes** (consisting solely of the duck-billed platypus and the echidna), are oviparous, like the reptiles from which they evolved. They incubate their eggs in a nest (figure 52.9*a*) or specialized pouch, and the young hatchlings obtain milk from their mother's mammary glands by licking her skin (because monotremes lack nipples).

All other mammals are viviparous and are divided into two subcategories based on how they nourish their young. The **marsupials,** a group that includes opossums and kangaroos,

give birth to small, fetuslike offspring that are incompletely developed. The young complete their development in a pouch of their mother's skin, where they can obtain nourishment from nipples of the mammary glands (figure 52.9*b*).

The placental mammals (figure 52.9*c*) retain their young for a much longer period of development within the mother's uterus. The fetuses are nourished by a structure known as the placenta, which is derived from both an extraembryonic membrane (the chorion) and the mother's uterine lining. Because the fetal and maternal blood vessels are in very close proximity in the placenta, the fetus can obtain nutrients by diffusion from the mother's blood. The functioning of the placenta is discussed in more detail in chapter 53.

Learning Outcomes Review 52.2

Oviparous species lay eggs; the amniotic eggs of reptiles and birds protect the embryo from desiccation. Females of ovoviviparous species retain fertilized eggs inside their bodies and release fully developed young when eggs hatch. Most mammals are viviparous, giving birth to young that have been nourished by the mother's body. Internal fertilization allows embryos to develop inside the female's body and thus receive protection from external environmental conditions.

■ *Under what circumstances would an estrous cycle be advantageous?*

52.3 Structure and Function of the Human Male Reproductive System

Learning Outcomes

1. Describe the sequence of events in spermatogenesis.
2. Describe semen and explain how it is released during mating.
3. Explain how hormones regulate male reproductive function.

Figure 52.9 Reproduction in mammals. *a.* Monotremes lay eggs in a nest, such as the duck-billed platypus *(Ornithorhynchus anatinus)* shown here with newly hatched offspring. *b.* Marsupials, such as this red kangaroo *(Macropus rufus),* give birth to small offspring that complete their development in a pouch. *c.* In placental mammals, such as this spotted deer doe *(Axis axis)* nursing her fawn, the young remain inside the mother's uterus for a longer period of time and are born relatively more developed.

a.

b.

c.

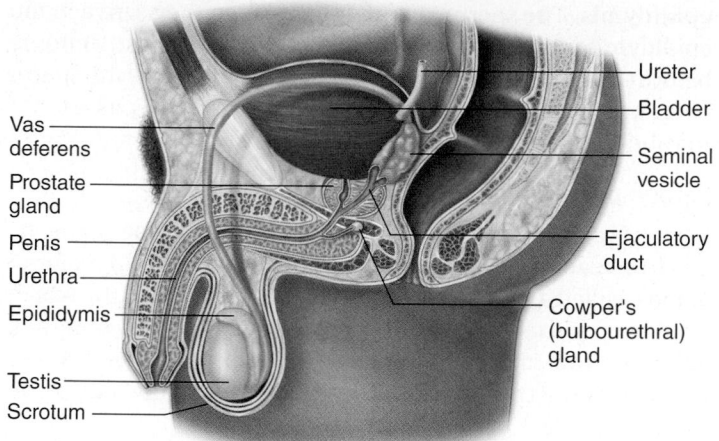

Figure 52.10 Organization of the human male reproductive system. The penis and scrotum are the external genitalia, the testes are the gonads, and the other organs are accessory sex organs, aiding the production and ejaculation of semen.

The structures of the human male reproductive system, typical of male mammals, are illustrated in figure 52.10. When testes form in the human embryo, they develop seminiferous tubules, the sites of sperm production, beginning around 43 to 50 days after conception. At about 9 to 10 weeks, the Leydig cells, located in the interstitial tissue between the seminiferous tubules, begin to secrete testosterone (the major male sex hormone, or androgen). Testosterone secretion during embryonic development converts indifferent structures into the male external genitalia, the *penis* and the *scrotum,* the latter being a sac that contains the testes. In the absence of testosterone,

these structures develop into the female external genitalia. Testosterone is also responsible at puberty for male secondary sex characteristics, such as development of the beard, a deeper voice, and body hair.

In an adult, each testis is composed primarily of the highly convoluted seminiferous tubules (figure 52.11, left). Although the testes are actually formed within the abdominal cavity, shortly before birth they descend through an opening called the inguinal canal into the scrotum, which suspends them outside the abdominal cavity. The scrotum maintains the testes at around 34°C, slightly lower than the core body temperature (37°C). This lower temperature is required for normal sperm development in humans.

Sperm cells are produced by the millions

The wall of the seminiferous tubule consists of spermatogonia, or *germ cells,* and supporting Sertoli cells. The germ cells near the outer surface of the seminiferous tubule are diploid and are the only cells that will undergo meiosis to produce gametes (see chapter 11). The developing gamete cells, located closer to the lumen of the tubule, are haploid.

Cell divisions leading to sperm

A spermatogonium cell divides by mitosis to produce two diploid cells. One of these two cells then undergoes meiotic division to produce four haploid cells that will become sperm while the other remains as a spermatogonium. In that way, the male never runs out of spermatogonia to produce sperm. Adult males produce an average of 100 to 200 million sperm each day and can continue to do so throughout most of the rest of their lives.

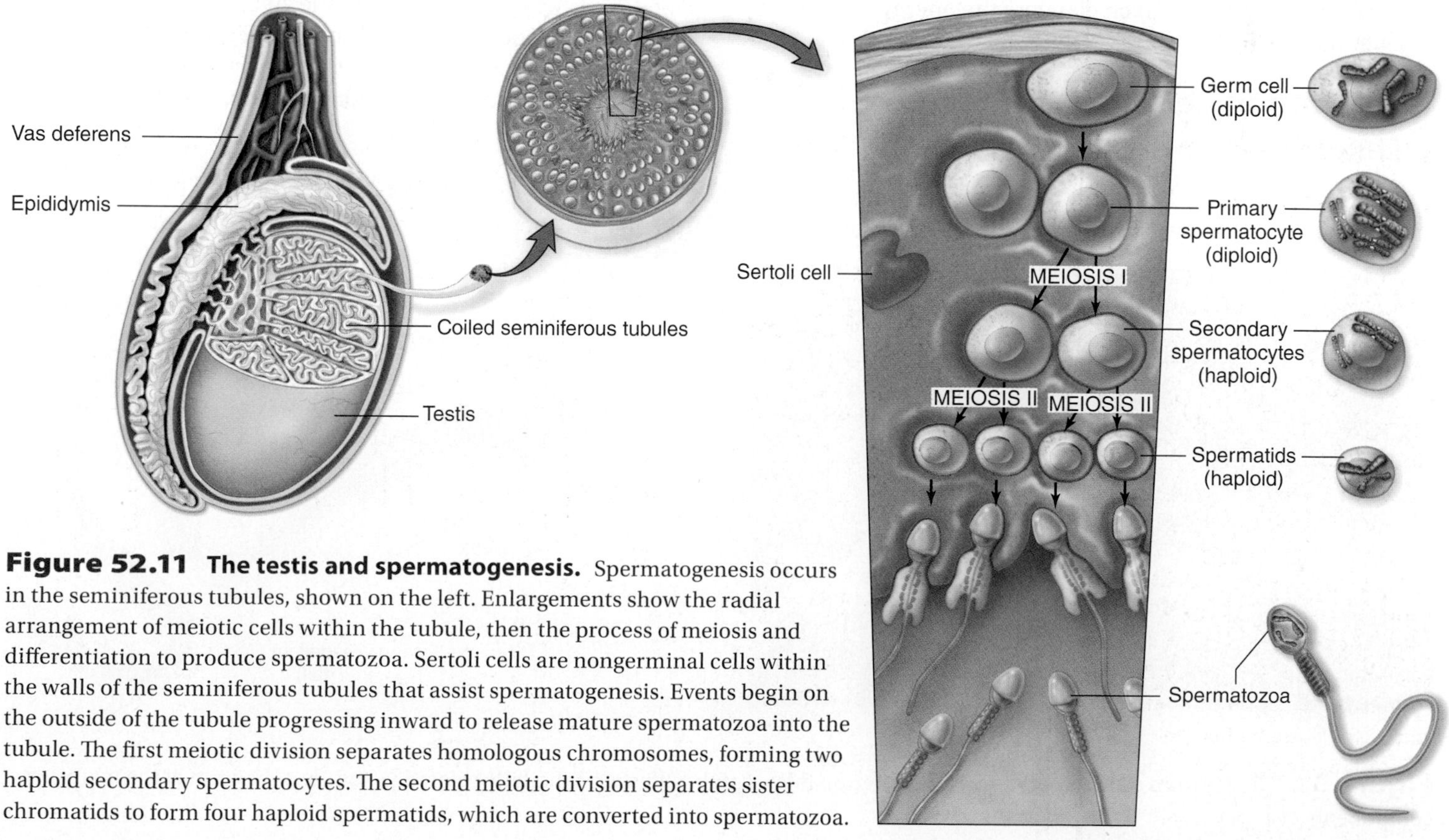

Figure 52.11 The testis and spermatogenesis. Spermatogenesis occurs in the seminiferous tubules, shown on the left. Enlargements show the radial arrangement of meiotic cells within the tubule, then the process of meiosis and differentiation to produce spermatozoa. Sertoli cells are nongerminal cells within the walls of the seminiferous tubules that assist spermatogenesis. Events begin on the outside of the tubule progressing inward to release mature spermatozoa into the tubule. The first meiotic division separates homologous chromosomes, forming two haploid secondary spermatocytes. The second meiotic division separates sister chromatids to form four haploid spermatids, which are converted into spermatozoa.

The diploid daughter cell that begins meiosis is called a primary spermatocyte. In humans it has 23 pairs of chromosomes (46 chromosomes total), and each chromosome is duplicated, with two chromatids. The first meiotic division separates the homologous chromosome pairs, producing two haploid secondary spermatocytes. However, each chromosome still consists of two duplicate chromatids.

Each of these cells then undergoes the second meiotic division to separate the chromatids and produce two haploid cells, the **spermatids.** Therefore, a total of four haploid spermatids are produced from each primary spermatocyte (see figure 52.11, right). All of these cells constitute the germinal epithelium of the seminiferous tubules because they "germinate" the gametes.

Supporting tissues

In addition to the germinal epithelium, the walls of the seminiferous tubules contain nongerminal cells such as the Sertoli cells mentioned earlier. These cells nurse the developing sperm and secrete products required for spermatogenesis. They also help convert the spermatids into **spermatozoa (sperm)** by engulfing their extra cytoplasm.

Sperm structure

Spermatozoa are relatively simple cells, consisting of a head, body, and flagellum (tail) (figure 52.12). The head encloses a compact nucleus and is capped by a vesicle called an acrosome, which is derived from the Golgi complex. The acrosome contains enzymes that aid in the penetration of the protective layers surrounding the egg. The body and tail provide a propulsive mechanism: Within the tail is a flagellum, and inside the body are a centriole, which acts as a basal body for the flagellum, and mitochondria, which generate the energy needed for flagellar movement.

Male accessory sex organs aid in sperm delivery

After the sperm are produced within the seminiferous tubules, they are delivered into a long, coiled tube called the **epididymis.** The sperm are not motile when they arrive in the epididymis, and they must remain there for at least 18 hours before their motility develops. From the epididymis, the sperm enter another long tube, the **vas deferens,** which passes into the abdominal cavity via the inguinal canal.

Semen production

Semen is a complex mixture of fluids and sperm. The vas deferens from each testis joins with one of the ducts from a pair of glands called the seminal vesicles (see figure 52.10), which produce a fructose-rich fluid constituting about 60% of semen volume. From this point, the vas deferens continues as the ejaculatory duct and enters the prostate gland at the base of the urinary bladder.

In humans, the **prostate gland** is about the size of a golf ball and is spongy in texture. It contributes up to 30% of the bulk of the semen. Within the prostate gland, the ejaculatory duct merges with the urethra from the urinary bladder. The urethra carries the semen out of the body through the tip of the penis. A pair of pea-sized bulbourethral glands add secretions to make up the last 10% of semen, also secreting a fluid that lines the urethra and lubricates the tip of the penis prior to coitus (sexual intercourse).

Structure of the penis and erection

In addition to the urethra, the penis has two columns of erectile tissue, the corpora cavernosa, along its dorsal side and one column, the corpus spongiosum, along the ventral side (figure 52.13). Penile erection is produced by neurons in the parasympathetic division of the autonomic nervous system, which release nitric oxide (NO), causing arterioles in the penis to dilate. The erectile tissue becomes turgid as it engorges with blood. This increased pressure in the erectile tissue compresses the veins, so blood flows into the penis but cannot flow out.

Most mammals have a bone in the penis, called a baculum, that contributes to its stiffness during erection, but humans do not.

Ejaculation

The result of erection and continued sexual stimulation is ejaculation, the ejection from the penis of about 2 to 5 mL of

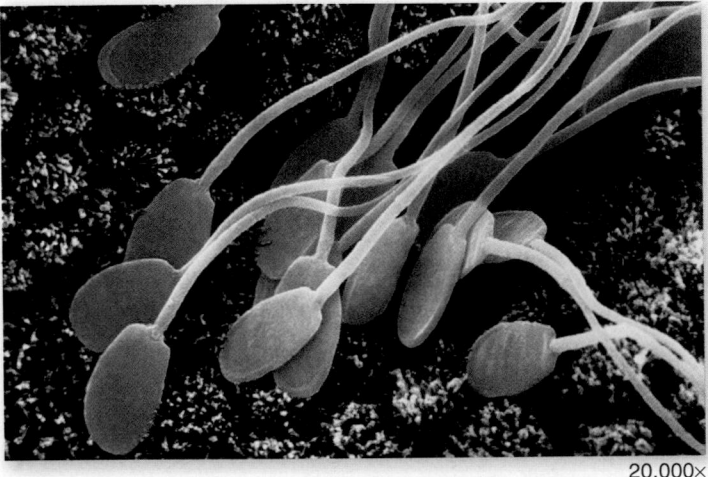

20,000×

a.

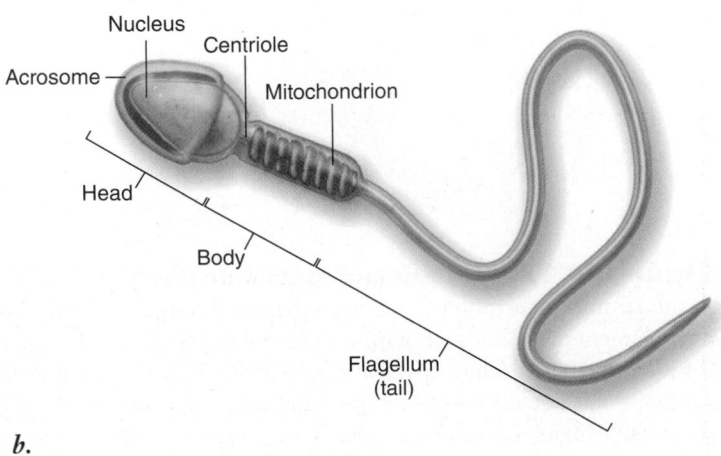

b.

Figure 52.12 Human sperm. *a.* A scanning electron micrograph with sperm digitally colored yellow. *b.* A diagram of the main components of a sperm cell.

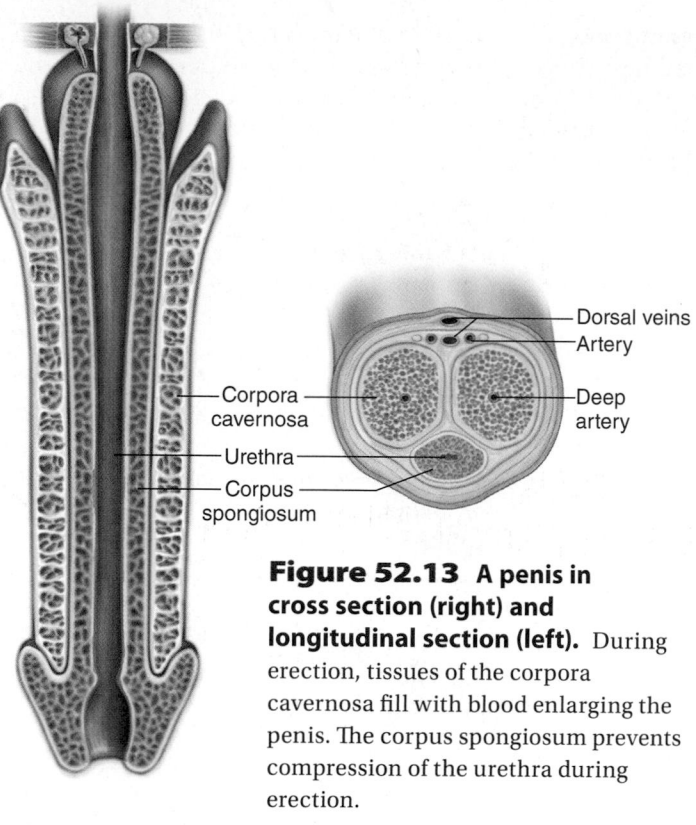

Corpora cavernosa
Urethra
Corpus spongiosum

Dorsal veins
Artery
Deep artery

Figure 52.13 A penis in cross section (right) and longitudinal section (left). During erection, tissues of the corpora cavernosa fill with blood enlarging the penis. The corpus spongiosum prevents compression of the urethra during erection.

Hormones regulate male reproductive function

As you saw in chapter 45, the anterior-pituitary gland secretes two gonadotropic hormones: follicle-stimulating hormone (FSH) and luteinizing hormone (LH). Although these hormones are named for their actions in the female, they are also involved in regulating male reproductive function (table 52.1). In males, FSH stimulates the Sertoli cells to facilitate sperm development, and LH stimulates the Leydig cells to secrete testosterone.

The principle of negative feedback inhibition applies to the control of FSH and LH secretion (figure 52.14). The hypothalamic hormone gonadotropin-releasing hormone (GnRH) stimulates the anterior-pituitary gland to secrete both FSH and LH. FSH causes the Sertoli cells to release a peptide hormone called inhibin, which specifically inhibits FSH secretion. Similarly, LH stimulates testosterone secretion, and testosterone feeds back to inhibit the release of LH, both directly at the anterior-pituitary gland and indirectly by reducing GnRH release from the hypothalamus.

The importance of negative feedback inhibition can be demonstrated by removing the testes; in the absence of testosterone and inhibin, the secretion of FSH and LH from the anterior pituitary is greatly increased.

semen containing an average of 300 million sperm. Successful fertilization requires such a high sperm count because the odds against any one sperm cell completing the journey to the egg and fertilizing it are extraordinarily high, and the acrosomes of many sperm need to interact with the egg before a single sperm can penetrate the egg (fertilization is described in chapter 53). Males with fewer than 20 million sperm per milliliter are generally considered sterile. Despite their large numbers, sperm constitute only about 1% of the volume of the semen ejaculated.

Learning Outcomes Review 52.3

Each of the spermatogonia lining the seminiferous tubules of the testes undergoes mitosis; one of the two daughter cells then undergoes meiosis to produce four haploid sperm cells. Semen consists of sperm from the testes and fluid from the seminal vesicles and prostate gland. Sexual stimulation causes erection of the penis, and continued stimulation leads to ejaculation of semen. Production of sperm and secretion of testosterone from the testes are controlled by FSH and LH from the anterior pituitary.

■ *Would natural selection favor those males that produce more sperm? Explain your answer.*

TABLE 52.1	Mammalian Reproductive Hormones
MALE	
Follicle-stimulating hormone (FSH)	Stimulates spermatogenesis via Sertoli cells
Luteinizing hormone (LH)	Stimulates secretion of testosterone by Leydig cells
Testosterone	Stimulates development and maintenance of male secondary sexual characteristics, accessory sex organs, and spermatogenesis
FEMALE	
Follicle-stimulating hormone (FSH)	Stimulates growth of ovarian follicles and secretion of estradiol
Luteinizing hormone (LH)	Stimulates ovulation, conversion of ovarian follicles into corpus luteum, and secretion of estradiol and progesterone by corpus luteum
Estradiol (estrogen)	Stimulates development and maintenance of female secondary sexual characteristics; prompts monthly preparation of uterus for pregnancy
Progesterone	Completes preparation of uterus for pregnancy; helps maintain female secondary sexual characteristics
Oxytocin	Stimulates contraction of uterus and milk-ejection reflex
Prolactin	Stimulates milk production

Figure 52.14 Hormonal interactions between the testes and anterior pituitary.
The hypothalamus secretes GnRH, which stimulates the anterior pituitary to produce LH and FSH. LH stimulates the Leydig cells to secrete testosterone, which is involved in development and maintenance of secondary sexual characteristics, and stimulates spermatogenesis. FSH stimulates the Sertoli cells of the seminiferous tubules, which facilitate spermatogenesis. FSH also stimulates Sertoli cells to secrete inhibin. Testosterone and inhibin exert negative feedback inhibition on the secretion of LH and FSH, respectively.

? Inquiry question Why do you think the brain is affected when the testes are surgically removed (termed *castration*)?

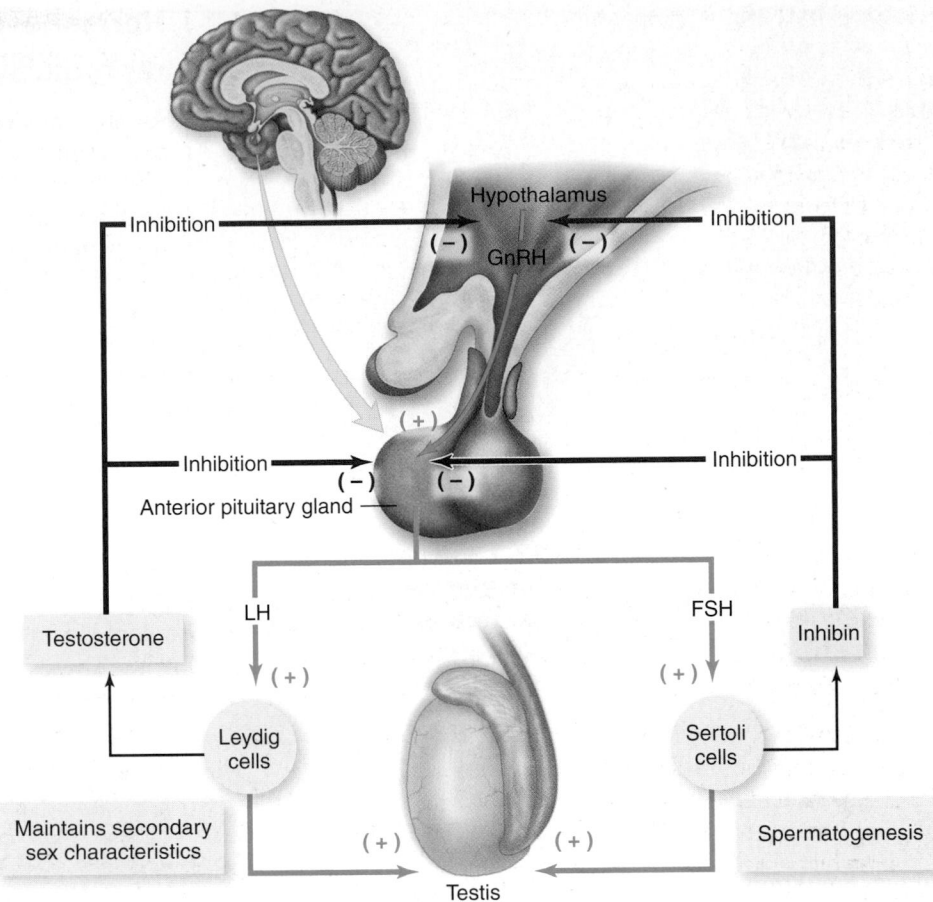

52.4 Structure and Function of the Human Female Reproductive System

Learning Outcomes

1. *Describe the sequence of events in production of an oocyte.*
2. *Explain ovulation and the female reproductive cycle.*
3. *Explain how hormones regulate female reproductive function.*

The structures of the reproductive system in a human female are shown in figure 52.15. In contrast to the testes, the ovaries develop much more slowly. In the absence of testosterone, the female embryo develops a **clitoris** and labia majora from the same embryonic structures that produce a penis and a scrotum in males. Thus, the clitoris and penis, and the labia majora and scrotum, are homologous structures. The clitoris, like the penis, contains corpora cavernosa and is therefore erectile.

The ovaries contain microscopic structures called ovarian follicles, which each contains a potential egg cell called a primary oocyte and smaller **granulosa cells.**

At puberty, the granulosa cells begin to secrete the major female sex hormone, estradiol (also called estrogen), triggering

menarche, the onset of menstrual cycling. Estradiol also stimulates the formation of the female secondary sexual characteristics, including breast development and the production of pubic hair. In addition, estradiol and another steroid hormone, progesterone, help maintain the female accessory sex organs: the fallopian tubes, uterus, and vagina.

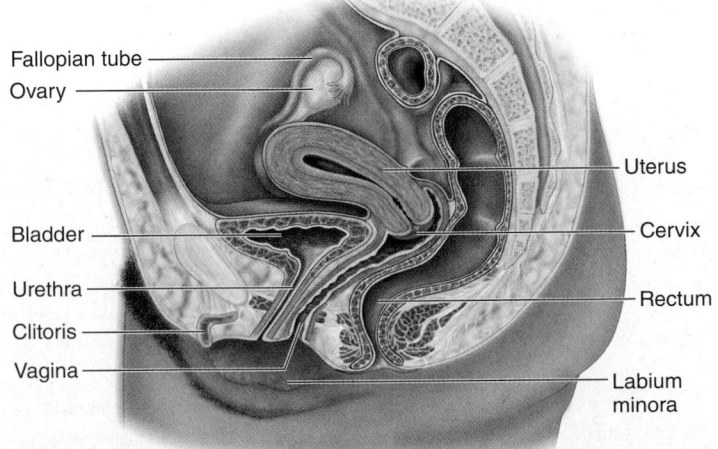

Figure 52.15 Organization of the human female reproductive system. The ovaries are the gonads, the Fallopian tubes receive the ovulated ova, and the uterus is the womb, the site of development of an embryo if the egg cell becomes fertilized.

Usually only one egg is produced per menstrual cycle

At birth, a female's ovaries contain about 1 million follicles, each containing a **primary oocyte** that has begun meiosis but is arrested in prophase of the first meiotic division. During each menstrual cycle a group of follicles is recruited to reinitiate development, but the process of follicle maturation takes many months, such that at any given time there are follicles at many different stages of maturation in the ovaries. The human menstrual cycle lasts approximately one month (28 days on average) and can be divided in terms of ovarian activity into a follicular phase and luteal phase, with the two phases separated by the event of ovulation (figure 52.16).

At the beginning of each menstrual cycle a single dominant follicle emerges from the group that was recruited many months earlier and continues its development while the other follicles in that group enter a pathway for destruction.

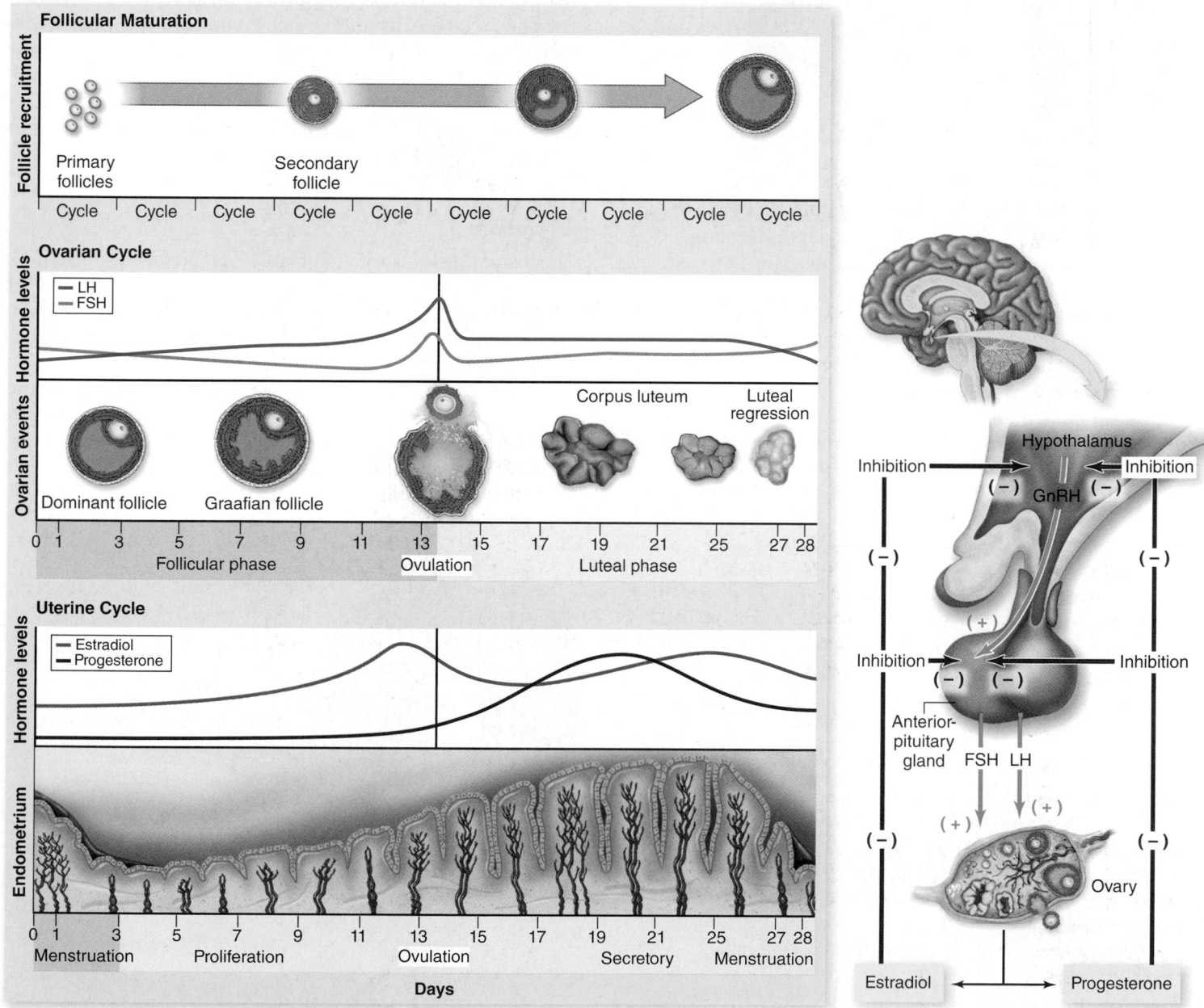

Figure 52.16 The human menstrual cycle. Left: Hormone levels during the cycle are correlated with ovulation (middle) and the growth of the endometrial lining of the uterus (bottom). Growth and thickening of the endometrium is stimulated by estradiol during the proliferative phase. Estradiol and progesterone maintain and regulate the endometrium during the secretory phase. Decline in the levels of these two hormones triggers menstruation. Follicles take many months to develop before being ready for ovulation (top). Right: Production of estradiol and progesterone by the anterior pituitary is controlled by negative feedback.

Data analysis At what points in the menstrual cycle are levels of follicle-stimulating hormone greater than those of luteinizing hormone?

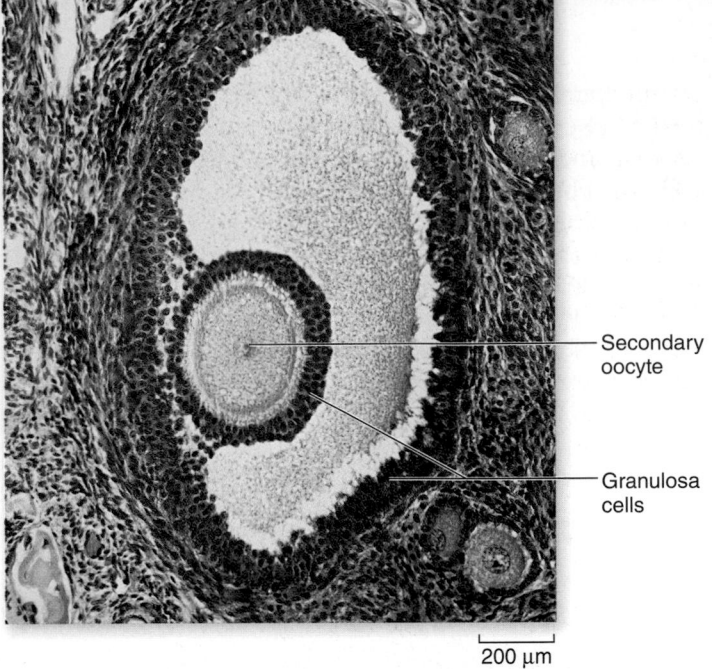

Figure 52.17 A mature Graafian follicle in a cat ovary.
Note the ring of granulosa cells that surrounds the secondary oocyte. This ring will remain around the egg cell when it is ovulated, and sperm must tunnel through the ring in order to reach the plasma membrane of the secondary oocyte.

Follicular phase

During the *follicular phase,* the dominant follicle achieves full maturity as a **late tertiary,** or **Graafian, follicle** under FSH stimulation. This follicle forms a thin-walled blister on the surface of the ovary. The uterus is lined with a simple columnar epithelial membrane called the endometrium; during the follicular phase estradiol causes growth of the endometrium. This phase is therefore also known as the **proliferative phase** of the endometrium (see figure 52.16).

The primary oocyte within the Graafian follicle completes the first meiotic division during the follicular phase. Instead of forming two equally large daughter cells, however, it produces one large daughter cell, the secondary oocyte (figure 52.17), and one tiny daughter cell, called a **polar body.** Thus, the secondary oocyte acquires almost all of the cytoplasm from the primary oocyte (unequal cytokinesis), increasing its chances of sustaining the early embryo should the oocyte be fertilized. The polar body, on the other hand, disintegrates.

The secondary oocyte then begins the second meiotic division, but its progress is arrested at metaphase II. It is in this form that the potential egg cell is discharged from the ovary at ovulation, and it does not complete the second meiotic division unless it becomes fertilized in the Fallopian tube.

Ovulation

The increasing level of estradiol in the blood during the follicular phase stimulates the anterior-pituitary gland to secrete LH about midcycle. This sudden secretion of LH causes the fully developed Graafian follicle to burst in the process of ovulation, releasing its secondary oocyte.

The released oocyte enters the abdominal cavity near the fimbriae, the feathery projections surrounding the opening to the Fallopian tube. The ciliated epithelial cells lining the Fallopian tube draw in the oocyte and propel it through the Fallopian tube toward the uterus.

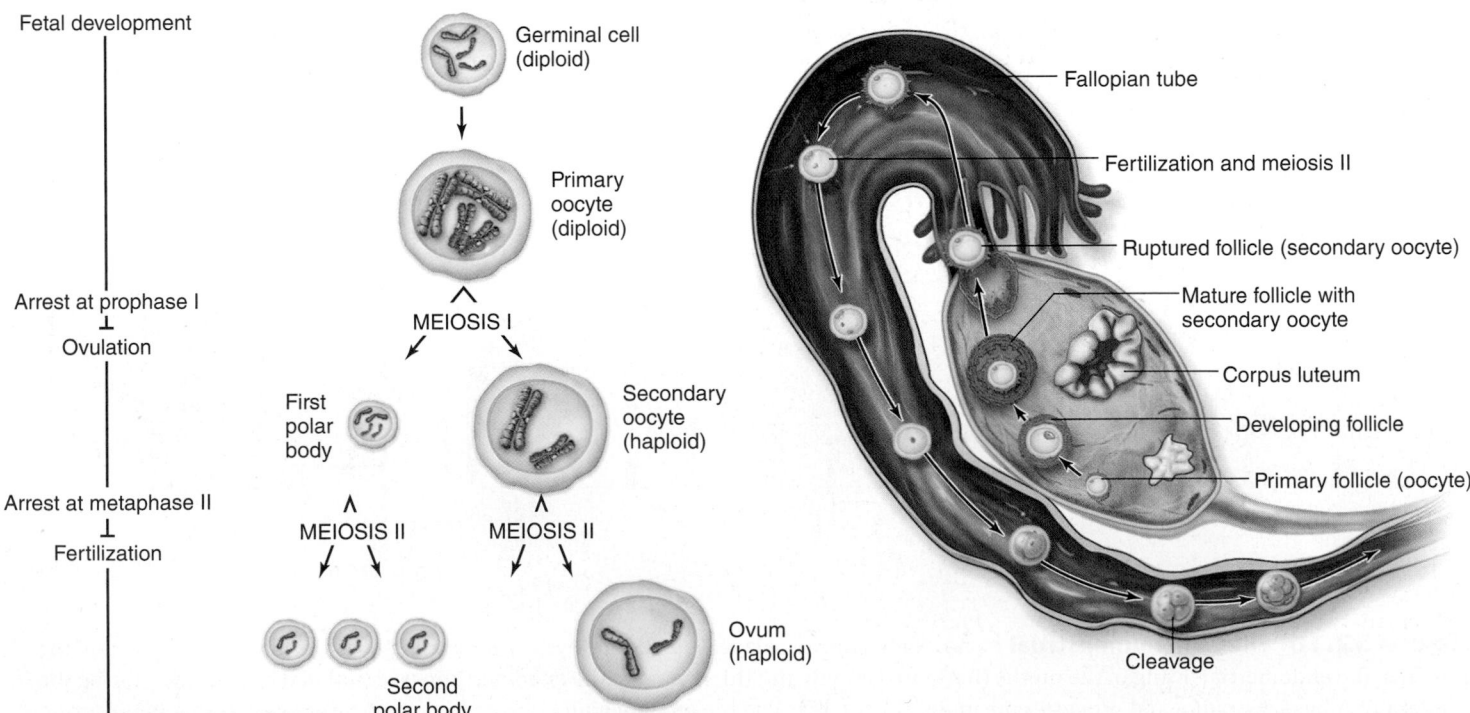

Figure 52.18 The meiotic events of oogenesis in humans. A primary oocyte is diploid. At the completion of the first meiotic division, one division product is eliminated as a polar body, and the other, the secondary oocyte, is released during ovulation. The secondary oocyte does not complete the second meiotic division until after fertilization; that division yields a second polar body and a single haploid egg, or ovum. Fusion of the haploid egg nucleus with a haploid sperm nucleus produces a diploid zygote.

If it is not fertilized, the oocyte disintegrates within a day following ovulation. If it is fertilized, the stimulus of fertilization allows it to complete the second meiotic division, forming a fully mature ovum and a second polar body (figure 52.18). Fusion of the nuclei from the ovum and the sperm produces a diploid zygote. Fertilization normally occurs in the upper one-third of the Fallopian tube, and in humans the zygote takes approximately 3 days to reach the uterus and then another 2 to 3 days to implant in the endometrium (figure 52.19).

Luteal phase

After ovulation, LH stimulation completes the development of the Graafian follicle into a structure called the **corpus luteum.** For this reason, the second half of the menstrual cycle is referred to as the **luteal phase.** The corpus luteum secretes both estradiol and another steroid hormone, progesterone. The high blood levels of estradiol and progesterone during the luteal phase now exert negative feedback inhibition of FSH and LH secretion by the anterior-pituitary gland (see figure 52.16). This inhibition during the luteal phase is in contrast to the stimulation exerted by estradiol on LH secretion at midcycle, which caused ovulation. The inhibitory effect of estradiol and progesterone after ovulation acts as a natural contraceptive mechanism, preventing both the development of additional follicles and continued ovulation.

During the luteal phase of the cycle, the combination of estradiol and progesterone cause the endometrium to become more vascular, glandular, and enriched with glycogen deposits. Because of the endometrium's glandular appearance and function, this portion of the cycle is known as the **secretory phase** of the endometrium. These changes prepare the uterine lining for embryo implantation.

In the absence of fertilization, the corpus luteum degenerates due to the decreasing levels of LH and FSH near the end of the luteal phase. Estradiol and progesterone, which the corpus luteum produces, inhibit the secretion of LH, the hormone needed for its survival. The disappearance of the corpus luteum results in an abrupt decline in the blood concentration of estradiol and progesterone at the end of the luteal phase, causing the built-up endometrium to be sloughed off with accompanying bleeding. This is menstruation; the portion of the cycle in which it occurs is known as the *menstrual phase* of the endometrium.

If the ovulated oocyte is fertilized, however, the tiny embryo prevents regression of the corpus luteum and subsequent menstruation by secreting *human chorionic gonadotropin (hCG)*, an LH-like hormone produced by the chorionic membrane of the embryo. By maintaining the corpus luteum, hCG keeps the levels of estradiol and progesterone high and thereby prevents menstruation, which would terminate the pregnancy. Because hCG comes from the embryonic chorion and not from the mother, it is the hormone tested for in all pregnancy tests.

Mammals with estrous cycles

Menstruation is absent in mammals with an estrous cycle. Although such mammals do cyclically shed cells from the endometrium, they don't bleed in the process. The estrous cycle is divided into four phases: proestrus, estrus, metestrus, and diestrus, which correspond to the proliferative, midcycle, secretory, and menstrual phases of the endometrium in the menstrual cycle.

Female accessory sex organs receive sperm and provide nourishment and protection to the embryo

The Fallopian tubes (also called uterine tubes or oviducts) transport ova from the ovaries to the uterus. In humans,

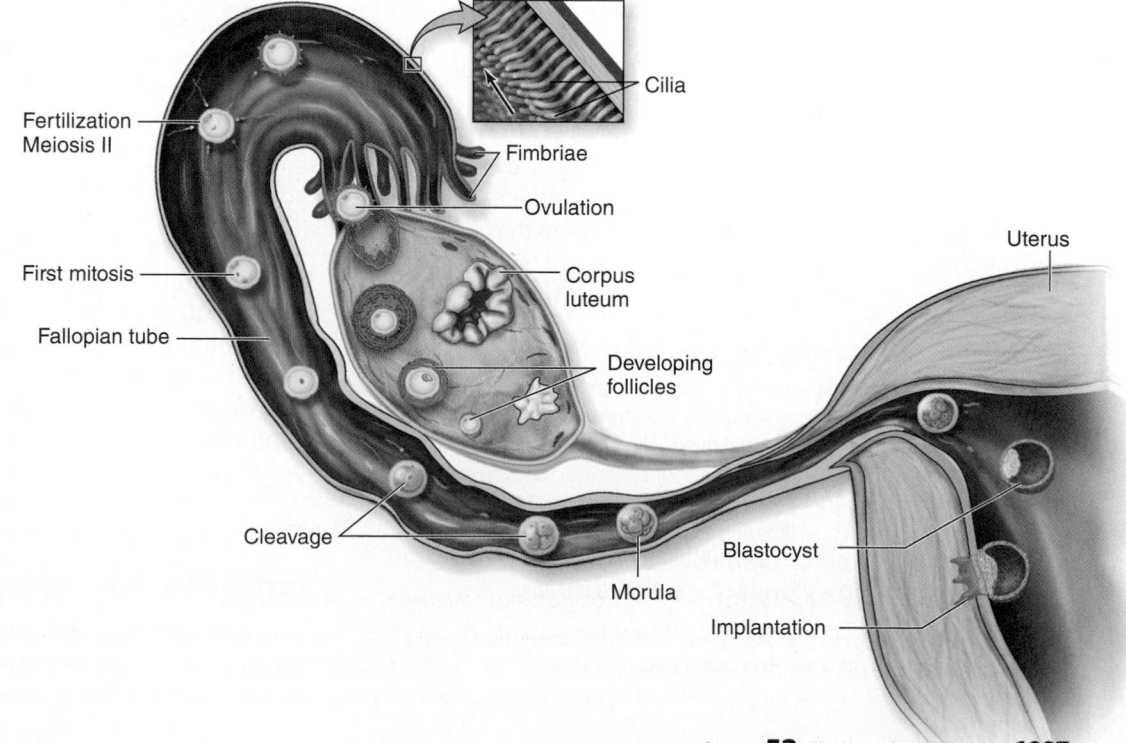

Figure 52.19 The journey of an egg. Produced within a follicle and released at ovulation, the secondary oocyte is swept into a Fallopian tube and carried along by waves of ciliary motion in the tube walls. Sperm journeying upward from the vagina penetrate the secondary oocyte, meiosis is completed and fertilization of the resulting ovum occurs within the Fallopian tube. The resulting zygote undergoes several mitotic divisions while still in the tube. By the time it enters the uterus, it is a hollow sphere of cells called a blastocyst. The blastocyst implants within the wall of the uterus, where it continues its development. (The egg and its subsequent stages have been enlarged for clarification.)

Cilia
Fimbriae
Ovulation
Corpus luteum
Developing follicles
Uterus
Blastocyst
Implantation
Morula
Cleavage
Fallopian tube
First mitosis
Fertilization Meiosis II

Figure 52.20
A comparison of mammalian uteruses.
a. Humans and other primates; (*b*) cats, dogs, and cows; and (*c*) marsupials.

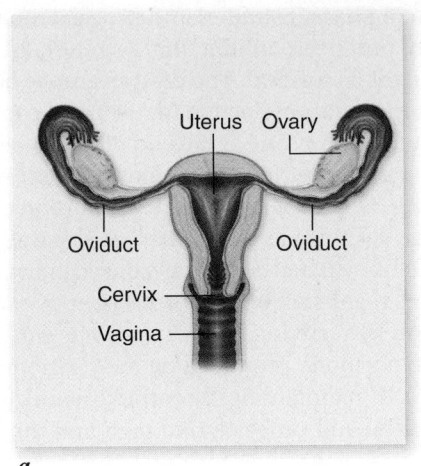

a.

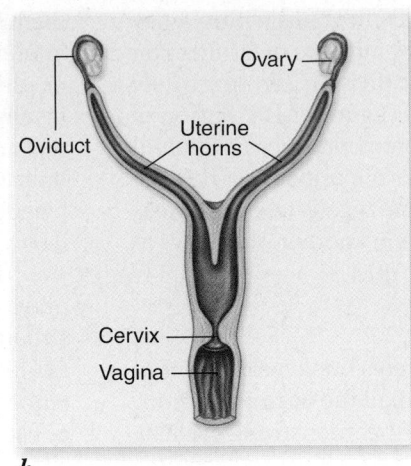

b.

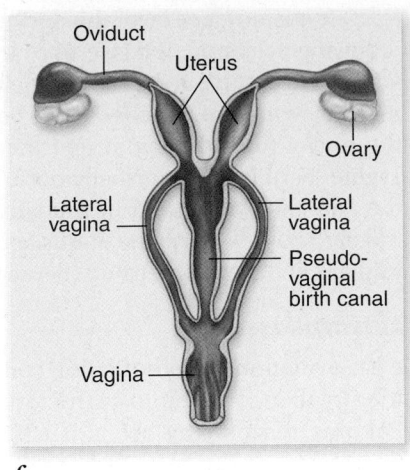

c.

the **uterus** is a muscular, pear-shaped organ that narrows to form a neck, the cervix, which leads to the vagina (figure 52.20*a*).

The entrance to the vagina is initially covered by a membrane called the *hymen*. This will eventually be disrupted by vigorous activity or actual sexual intercourse. In the latter case, this can make the first experience painful when the hymen is ruptured.

During sexual arousal, the labia minora, clitoris, and vagina all become engorged with blood, much like the male erectile tissues. The clitoris has many sensory nerve endings and is one of the most sensitive and responsive areas for female arousal. During sexual arousal, glands located near the vaginal opening, called Bartholin's glands, secrete a lubricating fluid that facilitates penetration by the penis. Ejaculation by the male introduces sperm cells that must then make the long swim out of the vagina and up the Fallopian tubes to encounter a secondary oocyte for fertilization to occur.

Mammals other than primates have more complex female reproductive tracts, in which part of the uterus divides to form uterine "horns," each of which leads to an oviduct (figure 52.20*b, c*). Cats, dogs, and cows, for example, have one cervix but two uterine horns separated by a septum, or wall. Marsupials, such as opossums, carry the split even further, with two unconnected uterine horns, two cervices, and two vaginas. A male marsupial has a forked penis that can enter both vaginas simultaneously.

Learning Outcomes Review 52.4

Primary oocytes reside in follicles in the ovaries. At puberty, some oocytes are triggered by FSH to develop with every menstrual cycle. Unequal cytokinesis produces a single egg and three polar bodies from each primary oocyte. During the follicular phase, a dominant follicle matures; ovulation is the release of this follicle's secondary oocyte triggered by LH. This oocyte completes division only if fertilization occurs. During the luteal phase, development of additional oocytes is inhibited. If fertilization does not occur, the endometrium is sloughed off as menstrual bleeding.

■ *Would more than one offspring per pregnancy be favored by natural selection? Under what conditions?*

52.5 Contraception and Infertility Treatments

Learning Outcomes

1. Compare the different types of birth control.
2. Describe causes of infertility.

In most vertebrates, copulation is associated solely with reproduction. Reflexive behavior that is deeply ingrained in the female limits sexual receptivity to those periods of the sexual cycle when she is fertile. In humans and a few species of apes, the female can be sexually receptive throughout her reproductive cycle, and this extended receptivity to sexual intercourse serves a second important function—it reinforces pair-bonding, the emotional relationship between two individuals.

Sexual intercourse may be a necessary and important part of humans' emotional lives—and yet not all couples desire to initiate a pregnancy every time they engage in sex. Throughout history, people and cultures have attempted to control reproduction while still being able to engage in sexual intercourse. The prevention of pregnancy or giving birth is known as birth control. Physiologically, pregnancy begins not at fertilization but approximately a week later with successful implantation. Methods of birth control that act prior to implantation are usually termed contraception.

In contrast, some couples desire to have children, but find for a variety of reasons that pregnancy is not occurring—a condition termed infertility. Technologies have also been developed to assist these couples in having children.

Contraception is aimed at preventing fertilization or implantation

A variety of approaches, differing in effectiveness and in their acceptability to different couples, religions, and cultures, are commonly taken to prevent pregnancy (figure 52.21 and table 52.2).

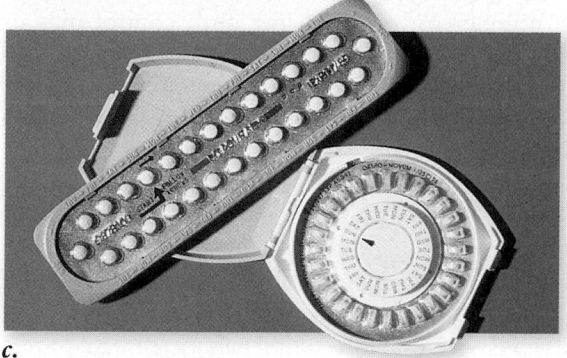
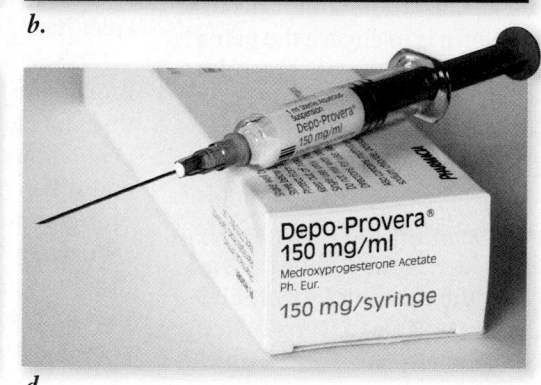

a.

b.

c.

d.

Figure 52.21 Four common methods of birth control. *a.* Condom; (*b*) diaphragm and spermicidal jelly; (*c*) oral contraceptives; (*d*) medroxyprogesterone acetate (Depo-Provera).

TABLE 52.2	**Methods of Birth Control**			
Device	Action	Failure Rate*	Advantages	Disadvantages
Oral contraceptive	Hormones (progesterone analogue alone or in combination with other hormones) primarily prevent ovulation	1–5, depending on type	Convenient; highly effective; provides significant noncontraceptive health benefits such as protection against ovarian and endometrial cancers	Must be taken regularly; possible minor side effects, which new formulations have reduced; not for women with cardiovascular risks (mostly smokers over age 35)
Condom	Thin sheath for penis collects semen; "female condoms" sheath vaginal walls	3–15	Easy to use, effective, inexpensive, protects against some sexually transmitted diseases	Requires male cooperation, may diminish spontaneity, may deteriorate on the shelf
Diaphragm	Soft rubber cup covers entrance to uterus; prevents sperm from reaching egg, holds spermicide	4–25	No dangerous side effects; reliable if used properly; provides some protection against sexually transmitted diseases and cervical cancer	Requires careful fitting, some inconvenience associated with insertion and removal; may be dislodged during intercourse
Intrauterine device (IUD)	Small plastic or metal device placed in the uterus, prevents implantation; some contain copper, others release hormones	1–5	Convenient, highly effective; infrequent replacement	Can cause excess menstrual bleeding and pain; risk of perforation, infection, expulsion, pelvic inflammatory disease, and infertility; not recommended for those who eventually intend to conceive or are not monogamous; dangerous in pregnancy
Cervical cap	Miniature diaphragm covers cervix closely, prevents sperm from reaching egg, holds spermicide	Probably similar to that of diaphragm	No dangerous side effects; fairly effective; can remain in place longer than diaphragm	Problems with fitting and insertion; comes in limited number of sizes
Foams, creams, jellies, vaginal suppositories	Chemical spermicides inserted in vagina before intercourse prevent sperm from entering uterus	10–25	Can be used by anyone who is not allergic; protect against some sexually transmitted diseases; no known side effects	Relatively unreliable; sometimes messy; must be used 5–10 minutes before each act of intercourse
Implant (levonorgestrel; Norplant)	Capsules surgically implanted under skin slowly release hormone that blocks ovulation	0.03	Very safe, convenient, and effective; very long-lasting (5 years); may have nonreproductive health benefits like those of oral contraceptives	Irregular or absent periods; minor surgical procedure needed for insertion and removal; some scarring may occur
Injectable contraceptive (medroxyprogesterone; Depo-Provera)	Injection every 3 months of a hormone that is slowly released and prevents ovulation	1	Convenient and highly effective; no serious side effects other than occasional heavy menstrual bleeding	Animal studies suggest it may cause cancer, though new studies in humans are mostly encouraging; occasional heavy menstrual bleeding

*Failure rate is expressed as pregnancies per 100 actual users per year.

Source: Data from American College of Obstetricians and Gynecologists: Contraception, Patient Education Pamphlet No. AP005. ACOG, Washington, D.C., 1990.

Abstinence

The most reliable way to avoid pregnancy is to not have sexual intercourse at all, which is called *abstinence*. Of all the methods of contraception, this is the most certain. It is also the most limiting and the most difficult method to sustain. The drive to engage in sexual intercourse is compelling, and many unwanted pregnancies result when a couple who desire each other and are attempting to adhere to abstinence fail in the attempt.

Sperm blockage

If sperm cannot reach the uterus, fertilization cannot occur. One way to prevent the delivery of sperm is to encase the penis within a thin sheath, or condom. Some males do not favor the use of condoms, which tend to decrease males' sensory pleasure during intercourse. In principle, this method is easy to apply and foolproof, but in practice it has a failure rate of 3 to 15% per year because of incorrect or inconsistent use or condom failure. Nevertheless, condom use is the most commonly employed form of contraception in the United States. Condoms are also widely used to prevent the transmission of AIDS and other sexually transmitted diseases (STDs). Over a billion condoms are sold in the United States each year.

A second way to prevent the entry of sperm into the uterus is to place a cover over the cervix. The cover may be a relatively tight-fitting cervical cap, which is worn for days at a time, or a rubber dome called a diaphragm, which is inserted before intercourse. Because the dimensions of individual cervices vary, a cervical cap or diaphragm must be initially fitted by a physician. Pregnancy rates average 4 to 25% per year for women using diaphragms. Failure rates for cervical caps are somewhat lower.

Sperm destruction

A third general approach to pregnancy prevention is to eliminate the sperm after ejaculation. This can be achieved in principle by washing out the vagina immediately after intercourse, before the sperm have a chance to enter the uterus. Such a procedure is called a douche. The douche method is difficult to apply well, because it involves a rapid dash to the bathroom immediately after ejaculation and a very thorough washing. Douching can, in fact, increase the possibility of conception by forcing sperm farther up into the vagina and uterus, thereby accounting for its high failure rate (40%).

Alternatively, sperm delivered to the vagina can be destroyed there with spermicidal agents, jellies, or foams. These treatments generally require application immediately before intercourse. Their failure rates vary from 10 to 25%. The use of a spermicide with a condom or diaphragm increases the effectiveness over each method used independently.

Prevention of ovulation

Since about 1960, a widespread form of contraception in the United States has been the daily ingestion of birth control pills, or oral contraceptives, by women. These pills contain analogues of progesterone, sometimes in combination with estrogens. As described earlier, progesterone and estradiol act by negative feedback to inhibit the secretion of FSH and LH during the luteal phase of the ovarian cycle, thereby preventing follicle development and ovulation. They also cause a buildup of the endometrium. The hormones in birth control pills have the same effects. Because the pills block ovulation, no ovum is available to be fertilized.

A woman generally takes the hormone-containing pills for 3 weeks; during the fourth week, she takes pills without hormones, allowing the levels of those hormones in her blood to fall, which causes menstruation.

Oral contraceptives provide a very effective means of birth control, with a failure rate of only 1 to 5% per year. In a variation of the oral contraceptive, hormone-containing capsules are implanted beneath the skin. These implanted capsules have failure rates below 1%.

A small number of women using birth control pills or implants experience undesirable side effects, such as blood clotting and nausea. These side effects have been reduced in newer generations of birth control pills, which contain less estrogen and different analogues of progesterone. Moreover, these new oral contraceptives provide a number of benefits, including reduced risks of endometrial and ovarian cancer, cardiovascular disease, and osteoporosis (for older women). However, they may increase the risk of developing breast cancer and cervical cancer, though the evidence to date is very uncertain.

The risks involved with birth control pills increase in women who smoke and increase greatly in women over 35 who smoke. The current consensus is that, for many women, the health benefits of oral contraceptives outweigh their risks, although a physician must help each woman determine the relative risks and benefits.

Prevention of embryo implantation

The insertion of an intrauterine device (IUD), such as a coil or other irregularly shaped object, is an effective means of contraception because the irritation it produces prevents the implantation of an embryo. IUDs have a failure rate of only 1 to 5%. Their high degree of effectiveness probably reflects their convenience; once they are inserted, they can be forgotten. The great disadvantage of this method is that almost a third of the women who attempt to use IUDs experience cramps, pain, and sometimes bleeding and therefore must discontinue using them. There is also a risk of uterine infection with insertion of the IUD.

Another method of preventing embryo implantation is the "morning-after pill," or Plan B, which contains 50 times the dose of estrogen present in birth control pills. The pill works by temporarily stopping ovum development, by preventing fertilization, or by stopping the implantation of a fertilized ovum. Its failure rate is 1 to 10% per use.

Many women are uneasy about taking such high hormone doses because side effects can be severe. This pill is not designed as a regular method of pregnancy prevention, but rather as a method of emergency contraception.

Sterilization

Sterilization is usually accomplished by the surgical removal of portions of the tubes that transport the gametes from the gonads (figure 52.22). It is an almost 100% effective means of contraception. Sterilization may be performed on either males or females, preventing sperm from entering the semen in males and preventing an ovulated oocyte from reaching the uterus in females.

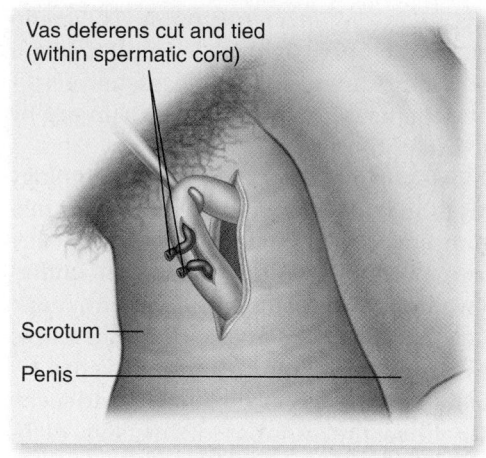

Vas deferens cut and tied
(within spermatic cord)

Scrotum

Penis

a.

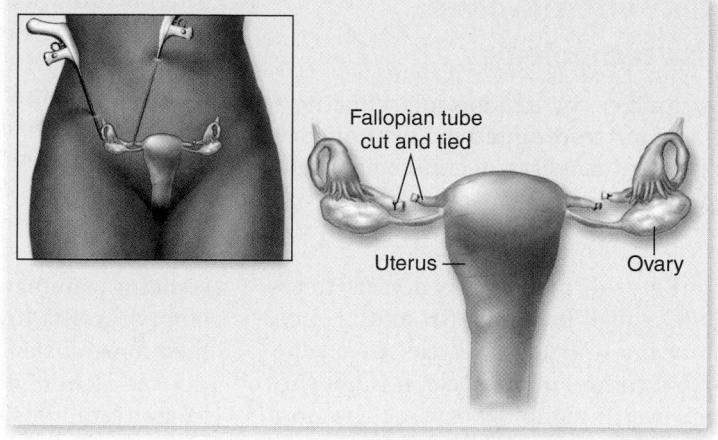

Fallopian tube
cut and tied

Uterus

Ovary

b.

**Figure 52.22
Birth control
through
sterilization.**
a. Vasectomy;
(b) tubal ligation.

In males, sterilization involves a vasectomy, the removal and tying off of a portion of the vas deferens from each testis. In females, the comparable operation, called tubal ligation, involves the removal of a section of each Fallopian tube and tying off the tube. In very rare cases, it is possible for the tubes to grow back together, restoring fertility. This is more common in vasectomy but does occur in both at a very low level. This accounts for the less than 100% effectiveness statistically. Both methods can also be reversed surgically, though for vasectomy the surgery is both expensive and frequently unsuccessful.

Infertility occurs in both males and females

Infertility is defined as the inability to conceive after 12 months of contraception-free sexual intercourse. In about 40% of cases, the failure to conceive is due to problems on the male side with about 45% due to problems on the female side, leaving another 15% unexplained (idiopathic infertility). Given these background statistics, it is clear that we still have a lot to learn about human fertility, despite a significant amount of study.

Female infertility

Infertility in females can occur due to a failure at any stage from the production of an oocyte to the implantation of the zygote. The most common problems arise from failure to ovulate, and from some kind of mechanical blockage preventing either fertilization or implantation.

The leading cause of infertility worldwide is pelvic inflammatory disease (PID). This can be caused by infection with a number of different bacteria that all lead to blockage of the Fallopian tubes. This blockage then causes problems in sperm passage, and of transfer of fertilized eggs to the uterus.

Endometriosis, the presence of ectopic endometrial tissue, can lead to infertility by a mechanism similar to PID. The body responds to the ectopic tissue by trying to wall it off with scar tissue. The buildup of scar tissue can then prevent the transfer of eggs to the uterus.

Another common cause of infertility in females is age, or premature ovarian failure (POF). Fertility declines significantly in females with age, and the incidence of some genetic abnormalities caused by nondisjunction of chromosomes increases (see chapter 13). If a women younger than 40 has a diminished supply of eggs, this is considered diagnostic of POF.

Disruption of the normal hormonal control of ovulation discussed earlier is also a common cause of infertility in females. Decreased levels of GnRH will disrupt ovulation, a condition referred to as hypogonadotropic hypogonadism. This can arise from damage to the hypothalamus or pituitary, or by any disorder that affects normal levels of hypothalamic hormones. For example, diabetes, thyroid disease, and excessive adrenal androgen production all affect hormonal feedback to the hypothalamus and can disrupt its normal function, leading to decreased levels of GnRH and infertility. Excessive exercise and anorexia can also lead to reduced GnRH levels and produce infertility.

Hormonal imbalances can occur during the luteal phase as well. Inadequate levels of progesterone during the luteal phase reduce the thickening of the uterine wall. If the uterine wall is inadequately prepared, implantation may not occur or can lead to an increased likelihood of spontaneous abortion.

Male infertility

Infertility in males can be due to a reduced number, viability, or motility of sperm in the ejaculate. These can be due to a variety of factors from infection to hormonal imbalances. Analysis on the male side is easier since sperm collection is noninvasive. Sperm can be easily analyzed for number, viability, morphology, and motility.

Infertility can arise from autoimmunity to sperm, leading to sperm loss, as well as due to abnormalities of all of the glands that contribute to the production of semen. Damage to the vas deferens or to the seminiferous tubules can also result in infertility. Anything that disrupts the maturation process of sperm can result in possible infertility.

After all possible causes have been ruled out, up to 5% of infertile men suffer from unexplained infertility. This may be due to genetic causes as the numbers seem to be similar worldwide despite different environments. It has been estimated in studies of *Drosophila* that up to 1500 recessive genes contribute to male fertility. Work is ongoing to examine the human genome for evidence of similar genes.

Treatment of infertility often involves assisted reproductive technologies

There are two basic possibilities for treating infertility: hormonal treatment and **assisted reproductive technologies (ART)**. The number and variety of assisted technologies available today is large and growing.

Hormone treatment

In the case of female infertility due to ovulatory defects, treatment is designed to produce high levels of FSH and LH at a single point during the normal menstrual cycle. Given the complexity of the hormonal control of the cycle, it is not surprising that this can be achieved in a number of ways. The most common drug currently used is clomiphene (Clomid), which is a competitive inhibitor of the estrogen receptor. This interferes with the negative feedback loop controlling estradiol production by the ovaries and consequently increases FSH and LH levels. If this is not successful, gonadotropins can be injected to stimulate ovulation.

Assistive reproductive technology

The simplest method to assist reproduction is to use artificial insemination, a process by which sperm are introduced into the female reproductive tract artificially. This is widely used in reproduction of domestic animals and is also used in humans. This has also been extended in cases of infertility in which both sperm and egg are introduced artificially by a technique called *gametic intrafallopian transfer,* or *GIFT.*

The birth of the first "test tube baby" in 1978 was heralded as the beginning of a new age of reproductive technology. Even the early pioneers may not have envisioned how far this technology would proceed. The basic technique of external fertilization is called *in vitro fertilization* (*IVF*), and transfer of the developing embryo is called simply *embryo transfer* (*ET*). When the sperm are unable to successfully fertilize an egg in vitro, they can be directly injected into an egg by *intracytoplasmic sperm injection* (*ICSI*).

One of the downsides of much of this assisted technology is multiple births. This is due to the common practice of transferring more than one embryo to ensure that at least one implants and develops normally. With advances in understanding of human development, it is possible to monitor early embryo growth to select the "best" embryos for transfer and to therefore transfer fewer embryos to reduce multiple births.

It is also possible to freeze sperm, eggs, and even human embryos to reduce the number of invasive techniques such as harvesting oocytes. Live births have been achieved using all combinations of frozen eggs, sperm, and embryos. This allows the transfer of a single embryo while freezing others produced by in vitro fertilization. If the first embryo transferred does not implant, then the others can be thawed and transferred later.

Learning Outcomes Review 52.5

Pregnancy can be prevented by a variety of contraception methods, including abstinence, barrier contraceptives, hormonal inhibition, and sterilization surgery. Some methods are more susceptible to human error than others and thus have a lower success rate. Infertility can be treated by hormonal manipulation to induce ovulation or by the use of assisted reproductive technologies. These assisted technologies include in vitro fertilization and intracytoplasmic sperm injection.

■ *Why isn't there a male birth control pill?*

Chapter Review

52.1 Animal Reproductive Strategies

Reproduction can occur asexually.

Sexual reproduction involves production by meiosis of haploid gametes (eggs and sperm). These join at fertilization to produce a diploid zygote.

Asexual reproduction produces offspring with the same genes as the parent organism.

In budding, a part of an individual becomes separated and develops into a new, identical individual. In parthenogenesis, females produce offspring from unfertilized eggs.

In some species, individuals can be both male and female.

In hermaphroditism, an individual has both testes and ovaries (simultaneous) or may change sex (sequential).

Sex can be determined genetically or by environmental conditions.

In some animals, the temperature an individual experiences as an embryo determines its sex. In mammals, sex is genetically determined by the presence of a Y chromosome (figure 52.3).

52.2 Vertebrate Fertilization and Development

Internal fertilization has led to three strategies for development of offspring.

Vertebrates with internal fertilization exhibit three strategies for development: oviparity, ovoviviparity, and viviparity. Both internal fertilization and live birth have evolved many times.

Most fishes and amphibians have external fertilization.

Most fish and amphibians release eggs and sperm into the water, where the gametes unite by chance. Few fertilized eggs grow to maturity.

Reptiles and birds have internal fertilization.

The embryos of reptiles and birds develop in a fluid-filled cavity surrounded by the amnion and extraembryonic membranes and a shell to help prevent desiccation.

Mammals generally do not lay eggs, but give birth to their young.

Mammals are also amniotic, but most species are viviparous. Most mammals have an estrus cycle, but primates have a menstrual cycle.

52.3 Structure and Function of the Human Male Reproductive System

Sperm cells are produced by the millions.

Haploid sperm are produced by meiosis of spermatogonia with the aid of Sertoli cells (figure 52.11). Each spermatogonium produces four sperm cells. A sperm cell has three parts: a head with an acrosome, a body containing mitochondria, and a flagellar tail.

Male accessory sex organs aid in sperm delivery.

Semen is a complex mixture of sperm and fluids from the seminal vesicles, prostate gland, and bulbourethral glands.

The urethra of the penis transports both sperm and urine and contains two columns of erectile tissue, blood vessels, and nerves (figure 52.13). Ejaculation is the ejection of semen from the penis by smooth muscle contraction.

Hormones regulate male reproductive function.

Male reproductive function is controlled by the hormones FSH and LH and negative feedback loops (figure 52.14, table 52.1).

52.4 Structure and Function of the Human Female Reproductive System

Usually only one egg is produced per menstrual cycle.

The female clitoris and labial lips have the same embryonic origin as the penis and scrotum. They develop in the absence of testosterone.

In adult females, FSH stimulates follicular development, which in turn produces estrogen. LH stimulates ovulation and corpus luteum development, which produces progesterone and more estrogen. Estrogen and progesterone are necessary to develop and maintain the uterine lining (figure 52.16).

The ovarian cycle has three phases: follicular phase, ovulation, and luteal phase. The uterine cycle has three stages that mirror the ovarian cycle: menstruation, proliferation, and secretion.

At birth, all primary oocytes are arrested in the first meiotic division. Each oocyte is capable of producing one ovum and three polar bodies. Each month, one oocyte completes meiosis I. This secondary oocyte begins the second meiotic division and arrests until the egg is fertilized (figure 52.18).

A fertilized egg, or zygote, develops into a blastocyst and implants in the wall of the uterus. Here it produces hCG, which maintains the corpus luteum and prevents menstruation.

If fertilization and implantation do not occur, the production of hormones declines, causing the built-up endometrium in the uterus to be sloughed off during menstruation.

Female accessory sex organs receive sperm and provide nourishment and protection to the embryo.

The Fallopian tubes transport ova from the ovaries to the uterus. The vagina receives sperm, which enters the uterus via the cervix (figure 52.20). Other female organs are involved in sexual response.

52.5 Contraception and Infertility Treatments

Contraception is aimed at preventing fertilization or implantation.

Pregnancy can be avoided by abstinence, by blocking sperm from reaching the ovum, by destroying sperm after ejaculation, by preventing ovulation or embryo implantation, or by sterilization. Some methods are more successful in practice than others.

Infertility occurs in both males and females.

Female infertility ranges from failure of oocyte production to failure of zygote implantation. Male infertility is usually due to reduction in sperm number, viability, or motility; hormonal imbalance; or damage to the sperm delivery system.

Treatment of infertility often involves assisted reproductive technologies.

Hormonal treatment may be used to correct ovulatory defects or sperm production defects. Assistive reproduction technologies involve artificial insemination, in vitro fertilization and embryo transfer, or intracytoplasmic sperm injection.

Review Questions

UNDERSTAND

1. You have discovered a new organism living in tide pools at your favorite beach. Every so often, one of the creature's appendages will break off and gradually grow into a whole new organism, identical to the first. This is an example of
   a. sexual reproduction.   c. budding.
   b. fission.   d. parthenogenesis.

2. If you decided that the organism you discovered in question 1 used parthenogenesis, what would you also know about this species?
   a. It is asexual.
   b. All the individuals are female.
   c. Each individual develops from an unfertilized egg.
   d. All of the choices are correct.

3. Which of the following terms describes your first stage as a diploid organism?
   a. Sperm   c. Gamete
   b. Egg   d. Zygote

4. Which of the following structures is the site of spermatogenesis?
   a. Prostate   c. Urethra
   b. Bulbourethral gland   d. Seminiferous tubule

5. FSH and LH are produced by the
   a. ovaries.   c. anterior pituitary.
   b. testes.   d. adrenal glands.

6. Gametogenesis requires the conclusion of meiosis II. When does this occur in females?
   a. During fetal development
   b. At the onset of puberty
   c. After fertilization
   d. After implantation

7. Mutations that affect proteins in the acrosome would impede which of the following functions?
   a. Fertilization   c. Meiosis
   b. Locomotion   d. Semen production

8. In humans, fertilization occurs in the_____, and implantation of the zygote occurs in the_____.
   a. seminiferous tubules; uterus
   b. vagina; oviduct
   c. oviduct; uterus
   d. urethra; uterus

9. The testicles of male mammals are suspended in the scrotum because
   a. the optimum temperature for sperm production is less than the normal core body temperature of the organism.
   b. the optimum temperature for sperm production is higher than the normal core body temperature of the organism.
   c. there is not enough room in the pelvic area for the testicles to be housed internally.
   d. it is easier for the body to expel sperm during ejaculation.

APPLY

1. The major difference between an estrous cycle and a menstrual cycle is that
   a. sexual receptivity occurs only around ovulation in the estrous cycle, but it can occur during any time of the menstrual cycle.
   b. estrous cycles occur in reptiles, but menstrual cycles occur in mammals.
   c. estrous cycles are determined by FSH, but menstrual cycles are determined by LH.
   d. estrous cycles occur monthly, but menstrual cycles occur sporadically.

2. Which of the following is a major difference between spermatogenesis and oogenesis?
   a. Spermatogenesis involves meiosis, and oogenesis involves mitosis.
   b. Spermatogenesis is continuous, but oogenesis is variable.
   c. Spermatogenesis produces fewer gametes per precursor cell than oogenesis.
   d. All of these are significant differences between oogenesis and spermatogenesis.

3. In species with environmental sex determination,
   a. sex is determined during development as an embryo.
   b. environmental conditions determine the sex of an individual.
   c. hermaphroditism always occurs.
   d. Both a and b are correct.

4. Internal and external fertilization differ in that all species that
   a. produce an amniotic egg have internal fertilization.
   b. do not produce an amniotic egg have external fertilization.
   c. produce live young have a penis or other intromittent organ.
   d. lay eggs are external fertilizers.

SYNTHESIZE

1. Suppose that the *SRY* gene mutated such that a male embryo could not produce functional protein. What kinds of changes would you expect to see in the embryo?

2. Why do you think that amphibians and many fish have external fertilization, whereas lizards, birds, and mammals rely on internal fertilization?

3. How are the functions of FSH and LH similar in male and female mammals? How do they differ?

4. You are interested in developing a contraceptive that blocks hCG receptors. Will it work? Why or why not?

5. Why are all parthenogenic parents female?

ONLINE RESOURCE

www.ravenbiology.com

Understand, Apply, and Synthesize—enhance your study with animations that bring concepts to life and practice tests to assess your understanding. Your instructor may also recommend the interactive eBook, individualized learning tools, and more.

Chapter

53

Animal Development

Chapter Contents

Introduction

Sexual reproduction in all but a few animals unites two haploid gametes to form a single diploid cell called a zygote. The zygote develops by a process of cell division and differentiation into a complex multicellular organism, composed of many different tissues and organs, as the picture illustrates. At the same time, a group of cells that constitute the germ line *are set aside to enable the developing organism to engage in sexual reproduction as an adult. In this chapter, we focus on the stages that all coelomate animals pass through during embryogenesis: fertilization, cleavage, gastrulation, and organogenesis (table 53.1). Development is a dynamic process, and so the boundaries between these stages are somewhat artificial. Although differences can be found in the details, developmental genes and cellular pathways have been greatly conserved, and they create similar structures in different organisms.*

TABLE 53.1	Stages of Animal Development (Using a Mammal as an Example)	
Fertilization	The haploid male and female gametes fuse to form a diploid zygote.	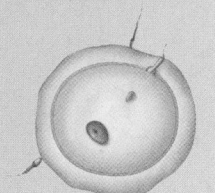
Cleavage	The zygote rapidly divides into many cells, with no overall increase in size. In many animals, these divisions affect future development because different cells receive different portions of the egg cytoplasm and, hence, different cytoplasmic determinants. Cleavage ends with formation of a blastula (called a blastocyst in mammals), which varies in structure among animal embryos.	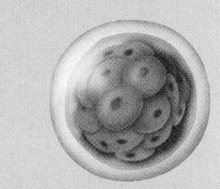 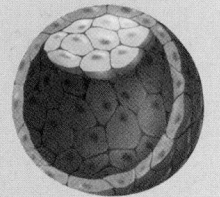 Blastocyst
Gastrulation	The cells of the embryo move, forming the three primary germ layers: ectoderm, mesoderm, and endoderm.	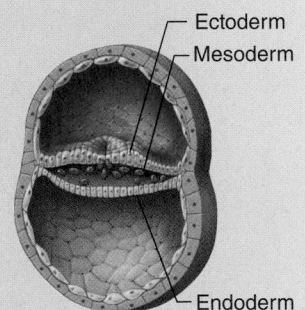 Ectoderm — Mesoderm — Endoderm
Organogenesis	Cells from the three primary germ layers interact in various ways to produce the organs of the body. In chordates, organogenesis begins with formation of the notochord and the hollow dorsal nerve cord in the process of neurulation.	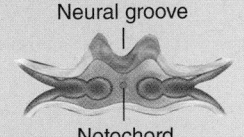 Neural groove — Notochord 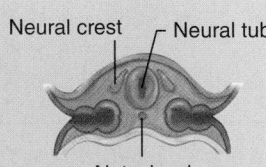 Neural crest — Neural tube — Notochord 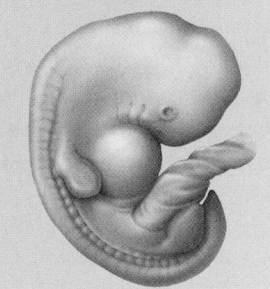

53.1 Fertilization

Learning Outcomes

1. **Describe the events necessary for fertilization to occur.**
2. **List different ways that polyspermy is blocked.**

In all sexually reproducing animals, the first step in development is the union of male and female gametes, a process called *fertilization*. As you learned in the preceding chapter, fertilization is typically external in aquatic animals. In contrast, internal fertilization is used by most terrestrial animals to provide a nondesiccating environment for the gametes.

One physical challenge of sexual reproduction is for gametes to get together. Many elaborate strategies have evolved to enhance the likelihood of such encounters. For example, most marine invertebrates release hundreds of millions of eggs and sperm into the surrounding seawater on spawning; others use lunar cycles to time gamete release. Elaborate courtship behaviors are typical of many animals that utilize internal fertilization (see chapter 52). Fertilization itself consists of three events: sperm penetration and membrane fusion, egg activation, and fusion of nuclei.

A sperm must penetrate to the plasma membrane of the egg for membrane fusion to occur

Embryonic development begins with the fusion of the sperm and egg plasma membranes. But the unfertilized egg presents a challenge to this process, since it is enveloped by one or more protective coats. These protective coats include the *chorion* of insect eggs, the *jelly layer* and *vitelline envelope* of sea urchin and frog eggs, and the *zona pellucida* of mammalian eggs. Mammalian oocytes are also surrounded by a layer of supporting granulosa cells (figure 53.1). Thus, the first challenge of fertilization is that sperm have to penetrate these external layers to reach the plasma membrane of the egg.

A saclike organelle named the **acrosome** is positioned between the plasma membrane and the nucleus of the sperm head. The acrosome contains digestive enzymes, which are released by the process of exocytosis when a sperm reaches the outer layers of the egg. These enzymes create a hole in the protective layers, enabling the sperm to tunnel its way through to the egg's plasma membrane.

In sea urchin sperm, actin monomers assemble into cytoskeletal filaments just under the plasma membrane to create a long narrow offshoot—the *acrosomal process*. The acrosomal process extends through the vitelline envelope to the egg's plasma membrane, and the sperm nucleus then passes through the acrosomal process to enter the egg.

In mice, an acrosomal process is not formed, and the entire sperm head burrows through the zona pellucida to the egg. Membrane fusion of the sperm and egg then allows the sperm nucleus to pass directly into the egg cytoplasm. In many species, egg cytoplasm bulges out at membrane fusion to engulf the head of the sperm (figure 53.2).

Sperm
Jelly layer
Plasma membrane
Vitelline envelope
Nucleus of egg
Cytoplasm
Cortical granules
a.
Sperm
Granulosa cell
Zona pellucida
Plasma membrane
First polar body
Cortical granules
Cytoplasm
b.
Oocyte
Granulosa cell
c.
25 μm
d.
20 μm

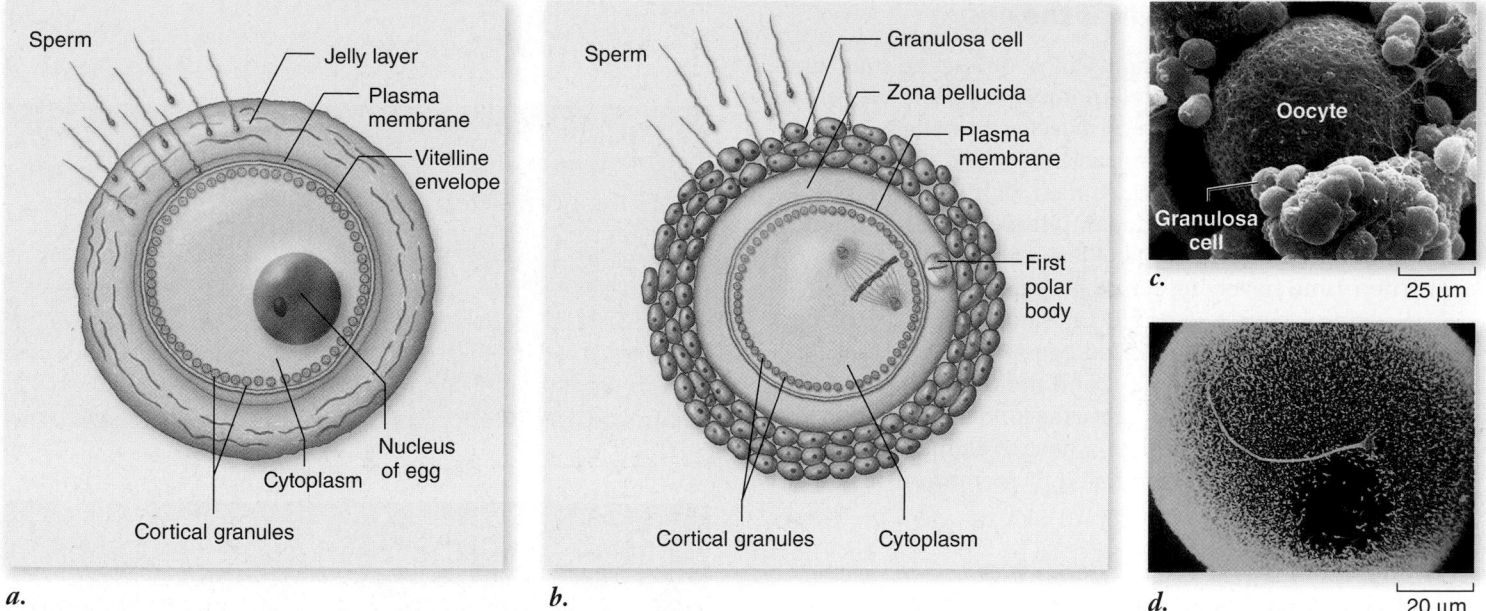

Figure 53.1 Animal reproductive cells. ***a.*** The structure of a sea urchin egg at fertilization. This diagram also shows the relative sizes of the sperm and egg. ***b.*** A mammalian sperm must penetrate a layer of granulosa cells and then a glycoprotein layer called the zona pellucida before it reaches the oocyte membrane. The scanning electron micrographs show ***(c)*** a human oocyte surrounded by numerous granulosa cells and ***(d)*** a human sperm on an egg.

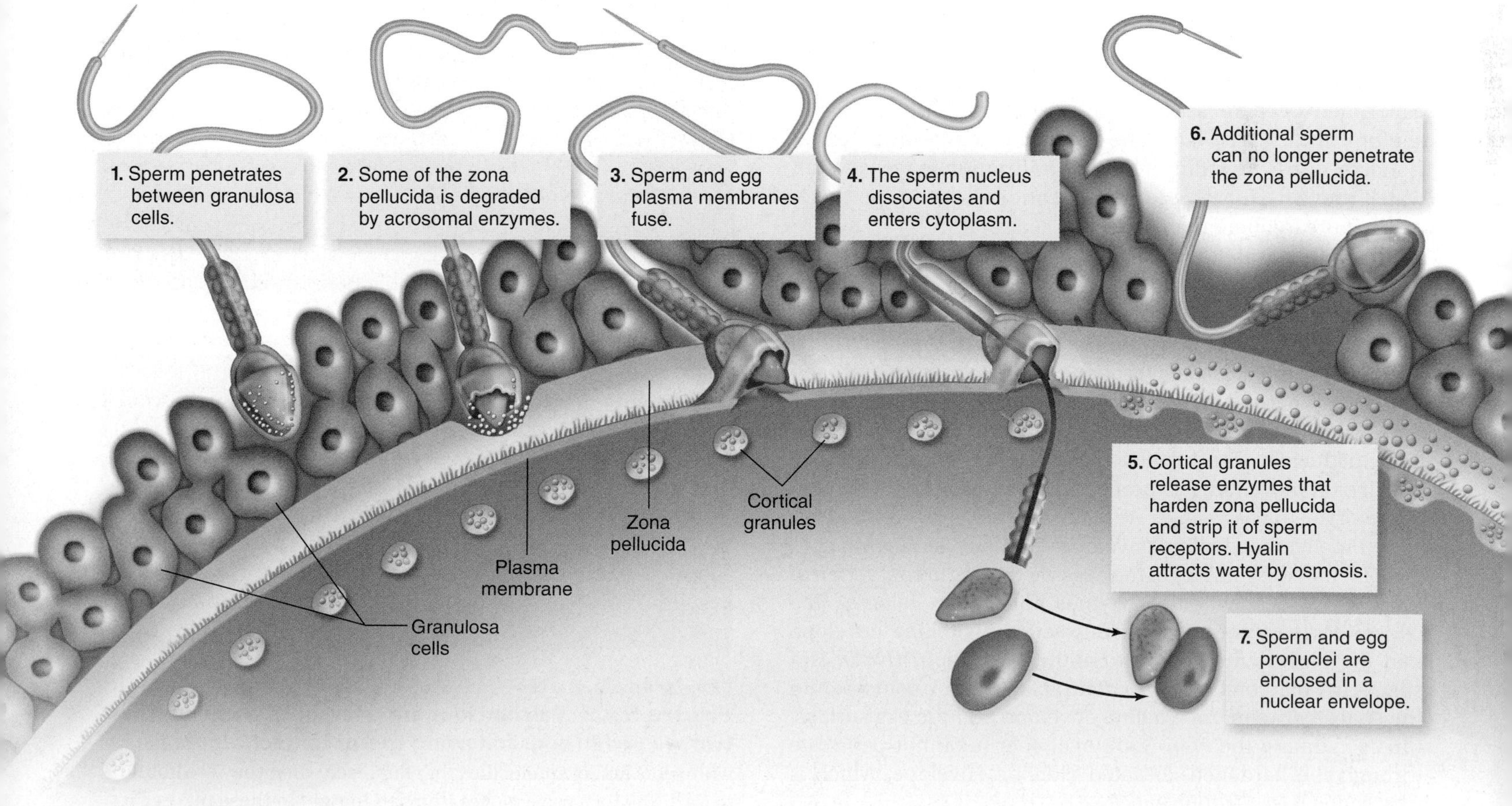

1. Sperm penetrates between granulosa cells.
2. Some of the zona pellucida is degraded by acrosomal enzymes.
3. Sperm and egg plasma membranes fuse.
4. The sperm nucleus dissociates and enters cytoplasm.
6. Additional sperm can no longer penetrate the zona pellucida.
5. Cortical granules release enzymes that harden zona pellucida and strip it of sperm receptors. Hyalin attracts water by osmosis.
7. Sperm and egg pronuclei are enclosed in a nuclear envelope.
Zona pellucida
Cortical granules
Plasma membrane
Granulosa cells

Figure 53.2 Sperm penetration and fusion. The sperm must penetrate the outer layers around the egg before fusion of sperm and egg plasma membranes can occur. Fusion activates the egg and leads to a series of events that prevent polyspermy.

Membrane fusion activates the egg

After ovulation, the egg remains in a quiescent state until fusion of the sperm and egg membranes triggers reactivation of the egg's metabolism. In most species, there is a dramatic increase in the levels of free intracellular Ca^{2+} ions in the egg shortly after the sperm makes contact with the egg's plasma membrane. This increase is due to release of Ca^{2+} from internal, membrane-bounded organelles, starting at the point of sperm entry and traversing across the egg.

Scientists have been able to watch this wave of Ca^{2+} release by preloading unfertilized eggs with a dye that fluoresces when bound to free Ca^{2+}, and then fertilizing the eggs (figure 53.3). The released Ca^{2+} act as second messengers in the cytoplasm of the egg, to initiate a host of changes in protein activity. These many events initiated by membrane fusion are collectively called *egg activation.*

Blocking of additional fertilization events

Because large numbers of sperm are released during spawning or ejaculation, many more than one sperm is likely to reach, and try to fertilize, a single egg. Multiple fertilization would result in a zygote that has three or more sets of chromosomes, a condition known as *polyploidy.* Polyploidy is incompatible with animal development, although it is frequently found in plants. As a result, an early response to sperm fusion in many animal eggs is to prevent fusion of additional sperm—in other words, to initiate a block to *polyspermy.*

In sea urchins, membrane contact by the first sperm results in a rapid, transient change in membrane potential of the egg, which prevents other sperm from fusing to the egg's plasma membrane. The importance of this event was shown by experiments where sea urchin eggs are fertilized in low-sodium, artificial seawater. The change in membrane potential is mostly due to an influx of Na^+, so fertilization in low-sodium water prevents this. Under these conditions polyspermy is much more frequent than in normal seawater.

Many animals use additional mechanisms to permanently alter the composition of the exterior egg coats, preventing any further sperm from penetrating through these layers. In sea urchins and mammals, specialized vesicles called **cortical granules,** located just beneath the plasma membrane of the egg, release their contents by exocytosis into the space between the plasma membrane and the vitelline envelope or zona pellucida, respectively. In each case, cortical granule enzymes remove critical sperm receptors from the outer coat of the egg.

Finally, the vitelline envelopes in many sea urchin species "lift off" the surfaces of the eggs via the combined action of different cortical granule enzymes and hyalin release. The enzymes digest connections between the vitelline envelope and the plasma membrane to allow separation. *Hyalin* is a sugar-rich macromolecule that attracts water by osmosis into the space between the vitelline envelope and the egg surface, thus separating the two. Additional sperm cannot penetrate through the hardened, elevated vitelline envelope, which is now called a *fertilization envelope.*

Many animals do not utilize any specific mechanisms to prevent multiple sperm from entering an egg. In these species,

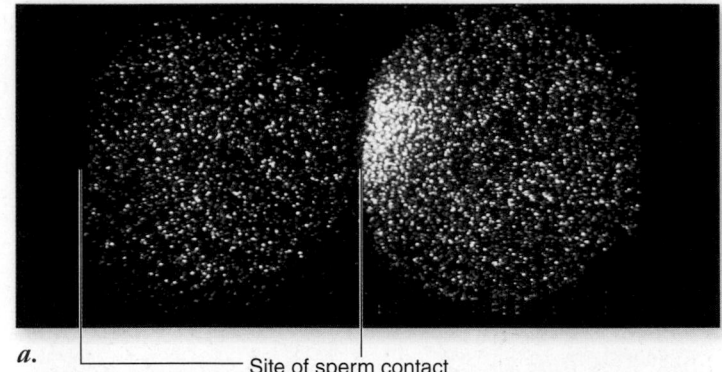

a. ———— Site of sperm contact

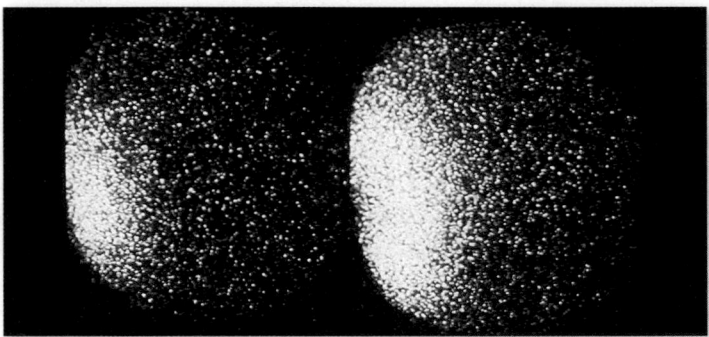

b.

c.

d.

Figure 53.3 Calcium ions are released in a wave across two sea urchin eggs following sperm contact. The bright white dots are dye molecules that fluoresce when they are bound to Ca^{2+}. The Ca^{2+} wave moves from left to right in these two eggs *(a–d).* The egg on the right was fertilized a few seconds before the egg on the left. The wave takes about 30 sec to cross the entire egg.

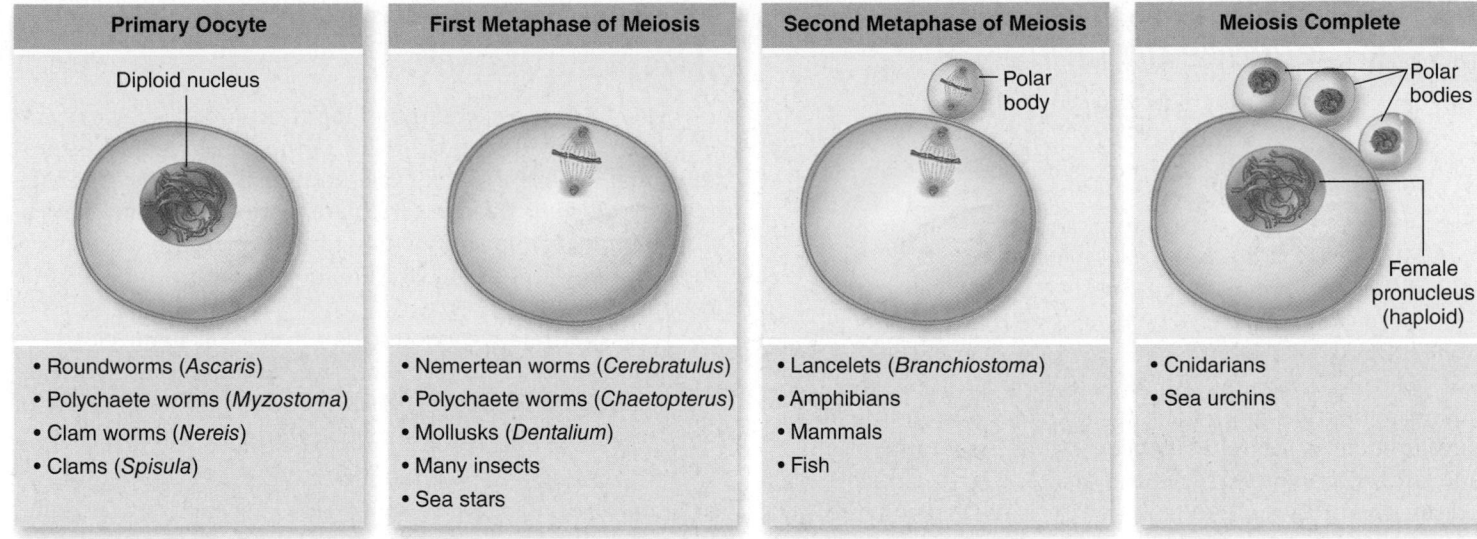

Primary Oocyte	First Metaphase of Meiosis	Second Metaphase of Meiosis	Meiosis Complete
Diploid nucleus		Polar body	Polar bodies / Female pronucleus (haploid)
• Roundworms (*Ascaris*) • Polychaete worms (*Myzostoma*) • Clam worms (*Nereis*) • Clams (*Spisula*)	• Nemertean worms (*Cerebratulus*) • Polychaete worms (*Chaetopterus*) • Mollusks (*Dentalium*) • Many insects • Sea stars	• Lancelets (*Branchiostoma*) • Amphibians • Mammals • Fish	• Cnidarians • Sea urchins

Figure 53.4 Stage of egg maturation at time of sperm binding in representative animals.

all but one of the sperm nuclei is degraded or subsequently extruded from the egg to prevent polyploidy.

Other effects of sperm penetration

In addition to the previously mentioned surface changes, sperm penetration can have three other effects on the egg. First, in many animals, the nucleus of the unfertilized egg is not yet haploid because it had not entered or completed meiosis prior to ovulation (figure 53.4). Fusion of the sperm plasma membrane then triggers the eggs of these animals to complete meiosis. In mammals, a single large egg with a haploid nucleus and one or more small polar bodies, which contain the other nuclei, are produced (see chapter 52).

Second, sperm penetration in many animals triggers movements of the egg cytoplasm. In chapter 19, we discussed the cytoplasmic rearrangements of newly fertilized tunicate eggs, which result in the asymmetrical localization of pigment granules that determine muscle development. In amphibian embryos, the point of sperm entry is the focal point of cytoplasmic movements in the egg, and these movements ultimately establish the bilateral symmetry of the developing animal.

In some frogs, for example, sperm penetration causes an outer pigmented cap of egg cytoplasm to rotate toward the point of entry, uncovering a gray crescent of interior cytoplasm opposite the point of penetration (figure 53.5). The position of this gray crescent determines the orientation of the first cell division. A line drawn between the point of sperm entry and the gray crescent would bisect the right and left halves of the future adult.

Third, activation is characterized by a sharp increase in protein synthesis and an increase in metabolic activity in general. Experiments demonstrate that the burst of protein synthesis in an activated egg uses mRNAs that were deposited into the cytoplasm of the egg during oogenesis.

In some animals, it is possible to artificially activate an egg without the entry of a sperm, simply by pricking the egg membrane. An egg that is activated in this way may go on to develop parthenogenetically. A few kinds of amphibians, fish, and reptiles rely entirely on parthenogenetic reproduction in nature, as we mentioned in chapter 52.

The fusion of nuclei restores the diploid state

In the third and final stage of fertilization, the haploid sperm nucleus fuses with the haploid egg nucleus to form the diploid nucleus of the zygote. The process involves migration of the two nuclei toward each other along a microtubule-based aster. A centriole that enters the egg cell with the sperm nucleus organizes the microtubule array, which is made from stored tubulin proteins in the egg's cytoplasm.

In mammals, including humans, the nuclei do not actually fuse. Instead sperm and egg nuclear membranes each break down prior to the formation of a new diploid nucleus. A new nuclear membrane forms around the two sets of chromosomes.

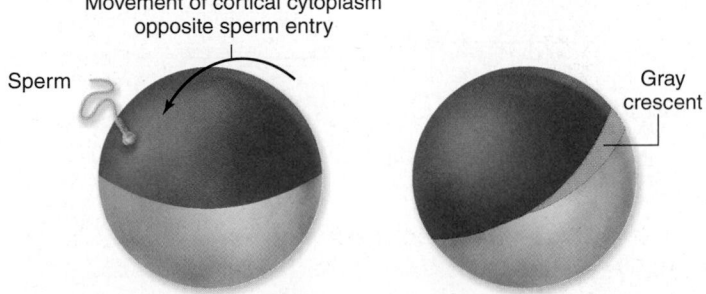

Movement of cortical cytoplasm opposite sperm entry

Sperm

Gray crescent

Figure 53.5 Gray crescent formation in frog eggs.
The gray crescent forms on the side of the egg opposite the point of penetration by the sperm.

Learning Outcomes Review 53.1

Following penetration, fusion of sperm with the egg membrane initiates a series of events including egg activation, blocks to polyspermy, and major rearrangements of cytoplasm. Polyspermy is blocked by changes in membrane polarity, release of enzymes that remove sperm receptors, and release of hyalin that lifts the vitelline envelope from the cell membrane. Egg and sperm nuclei then fuse to create a diploid zygote.

■ *What is the role of Ca^{2+} in egg activation?*

53.2 *Cleavage and the Blastula Stage*

Learning Outcomes

1. *Define the terms cleavage and blastula.*
2. *Describe the different patterns of cleavage.*
3. *Explain what is meant by regulative development.*

Following fertilization, the second major event in animal development is the rapid division of the zygote into a larger and larger number of smaller and smaller cells (see table 53.1). This period of division, called *cleavage,* is not accompanied by an increase in the overall size of the embryo. Each individual cell in the resulting tightly packed mass of cells is referred to as a *blastomere.* In many animals, the two ends of the egg and subsequent embryo are traditionally referred to as the **animal pole** and the **vegetal pole.** In general, the blastomeres of the animal pole go on to form the external tissues of the body, and those of the vegetal pole form the internal tissues.

The blastula is a hollow mass of cells

In many animal embryos, the outermost blastomeres in the ball of cells produced during cleavage become joined to one another by tight junctions, belts of protein that encircle a cell and weld it to its neighbors (see chapter 4). These tight junctions create a seal that isolates the interior of the cell mass from the surrounding medium.

Subsequently, cells in the interior of the mass begin to pump Na^+ from their cytoplasm into the spaces between cells. The resulting osmotic gradient causes water to be drawn into the center of the embryo, enlarging the intercellular spaces. Eventually, the spaces coalesce to form a single large cavity within the embryo. The resulting hollow ball of cells is called a *blastula* (or *blastocyst* in mammals), and the fluid-filled cavity within the blastula is known as the **blastocoel** (see table 53.1).

Cleavage patterns are highly diverse and distinctive

Cleavage divisions are quite rapid in most species, and chapter 19 provides an overview of the conserved set of proteins that control the cell cycle in animal embryos. Cleavage patterns are quite diverse, and there are about as many ways to divide up the cytoplasm of an animal egg during cleavage as there are phyla of animals! Nonetheless, we can make some generalizations.

First, the relative amount of nutritive yolk in the egg is the characteristic that most affects the cleavage pattern of an animal embryo (figure 53.6). Vertebrates exhibit a variety of developmental strategies involving different patterns of yolk utilization.

Cleavage in insects

Insects have yolk-rich eggs, and in chapter 19 we discussed the *syncytial blastoderm* of insects, in which multiple mitotic divisions of the nucleus occur in the absence of cytokinesis. Because there are no membranes separating the early embryonic nuclei of insects, gradients of diffusible proteins termed *morphogens* within the egg's cytoplasm can directly and differentially affect the activity of these embryonic nuclei, and thus the pattern of the early embryo. The nuclei eventually migrate to the periphery of the egg, where cell membranes form around each nucleus. The resulting *cellular blastoderm* of an insect has a single layer of cells surrounding a central mass of yolk (see figure 19.12 and table 53.2).

Cleavage of eggs with moderate or little yolk

In eggs that contain moderate to little yolk, cleavage occurs throughout the whole egg, a pattern called **holoblastic cleavage** (figure 53.7). This pattern of cleavage is characteristic of invertebrates such as mollusks, annelids, echinoderms, and tunicates, and also of amphibians and mammals (described shortly).

In sea urchins, holoblastic cleavage results in the formation of a symmetrical blastula composed of a single layer of cells of approximately equal size surrounding a spherical

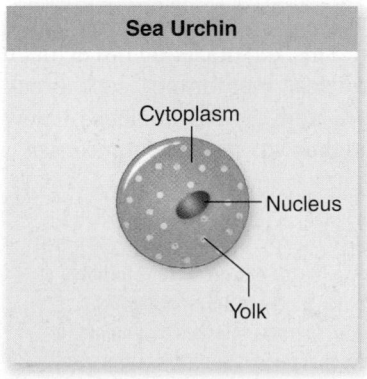

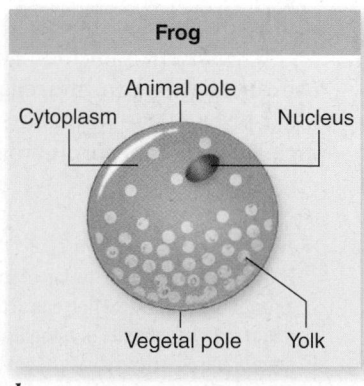

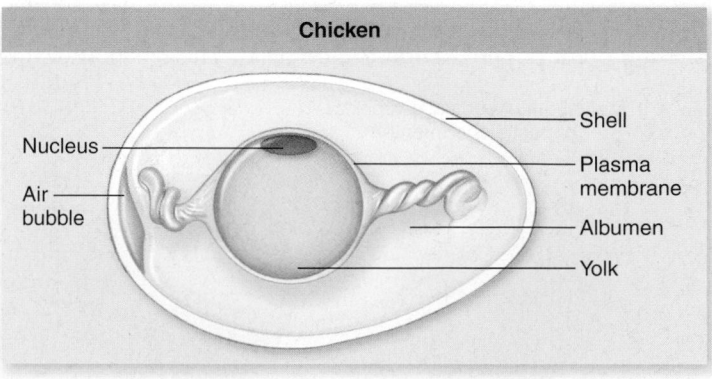

a. *b.* *c.*

Figure 53.6 Yolk distribution in three kinds of eggs. *a.* In a sea urchin egg, the cytoplasm contains a small amount of evenly distributed yolk and a centrally located nucleus. *b.* In a frog egg, there is much more yolk, and the nucleus is displaced toward one pole. *c.* Bird eggs are complex, with the nucleus contained in a small disc of cytoplasm that sits on top of a large, central yolk mass.

TABLE 53.2	The Major Cleavage Patterns of Animal Embryos

HOLOBLASTIC (COMPLETE) CLEAVAGE

Isolecithal (Sparse, evenly distributed yolk)

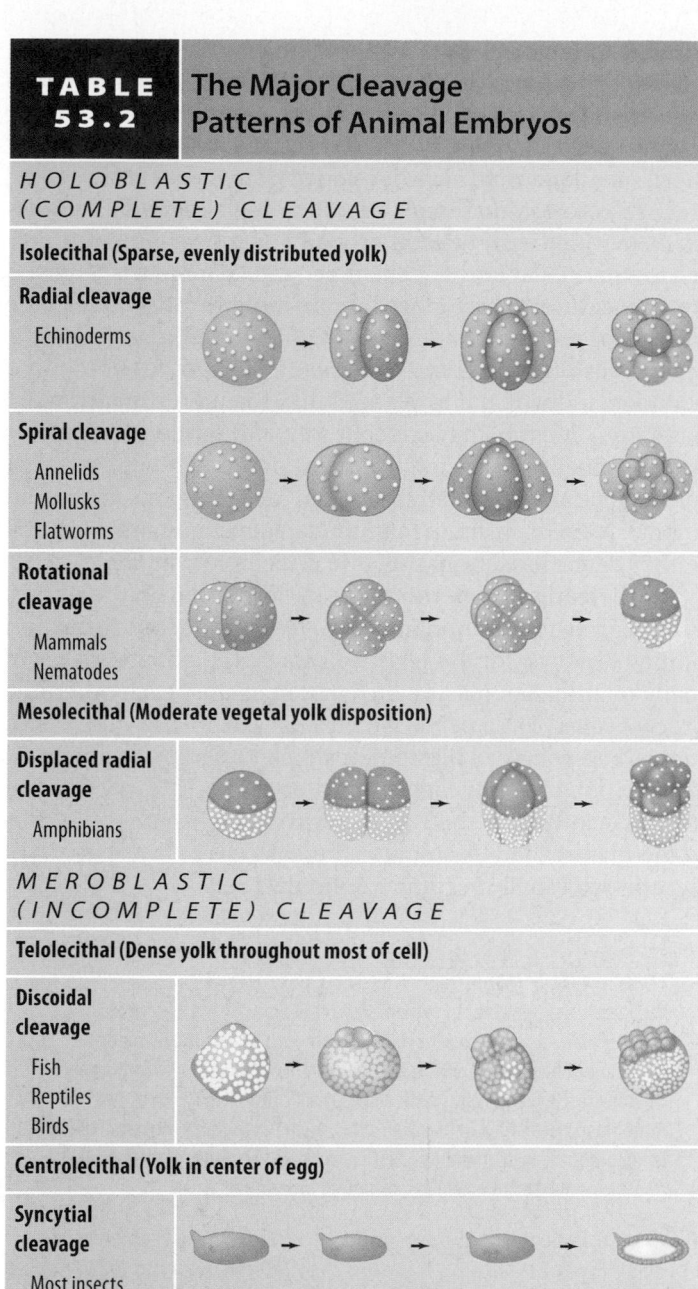

Radial cleavage Echinoderms	
Spiral cleavage Annelids Mollusks Flatworms	
Rotational cleavage Mammals Nematodes	

Mesolecithal (Moderate vegetal yolk disposition)

Displaced radial cleavage Amphibians	

MEROBLASTIC (INCOMPLETE) CLEAVAGE

Telolecithal (Dense yolk throughout most of cell)

Discoidal cleavage Fish Reptiles Birds	

Centrolecithal (Yolk in center of egg)

Syncytial cleavage Most insects	

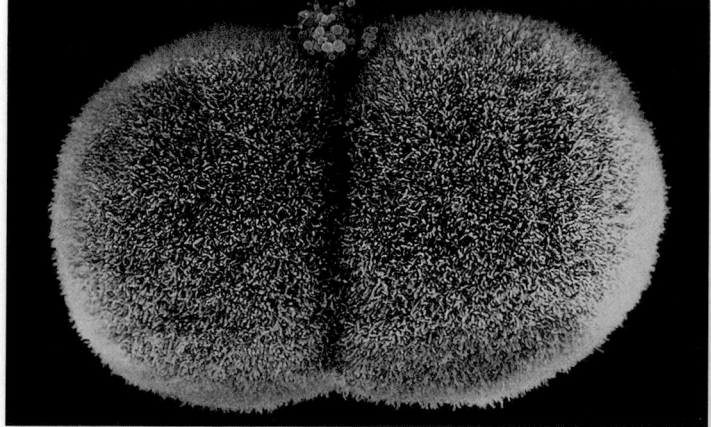

3.3 μm

Figure 53.7 Holoblastic cleavage. In this type of cleavage, which is characteristic of eggs with relatively small amounts of yolk, cell division occurs throughout the entire egg.

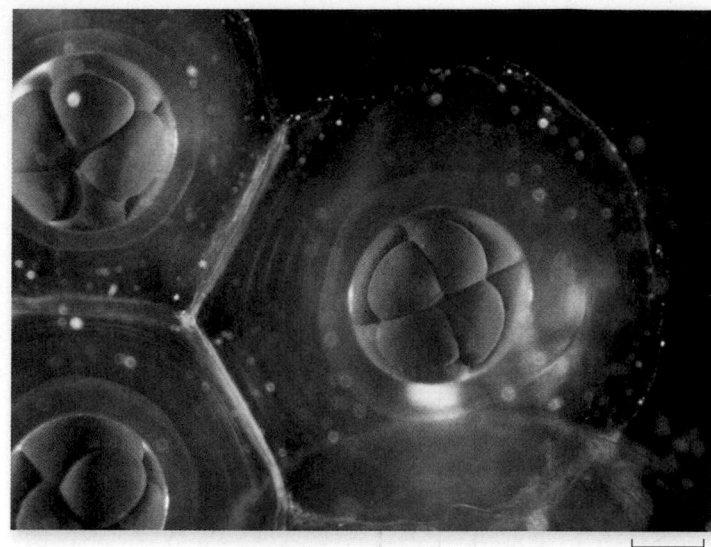

a.   350 μm

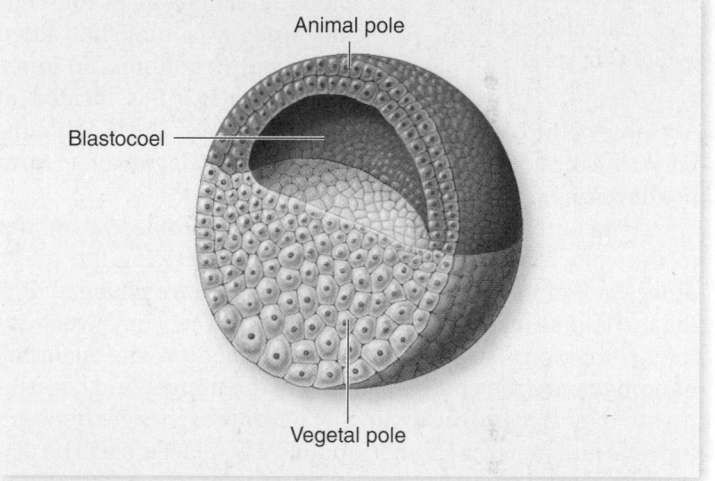

b.

Figure 53.8 Frog cleavage and blastula formation.
a. The closest cells in this photo (those near the animal pole) divide faster and are smaller than those near the vegetal pole (below cells of the animal pole). *b.* A cross section of a frog blastula, showing an eccentric blastocoel, larger yolk-filled cells at the vegetal pole, and smaller cells with little yolk at the animal pole.

blastocoel. In contrast, amphibian eggs contain much more cytoplasmic yolk in the vegetal hemisphere than in the animal hemisphere. Because yolk-rich regions divide much more slowly than areas with little yolk, horizontal cleavage furrows are displaced toward the animal pole (figure 53.8*a*). Thus, holoblastic cleavage in frog eggs results in an asymmetrical blastula, with a displaced blastocoel. The blastula consists of large cells containing a lot of yolk at the vegetal pole, and smaller, more numerous cells containing little yolk at the animal pole (figure 53.8*b*).

Cleavage of eggs with large amounts of yolk

The eggs of reptiles, birds, and some fishes are composed almost entirely of yolk, with a small amount of clear cytoplasm concentrated at one pole called the **blastodisc.** Cleavage in

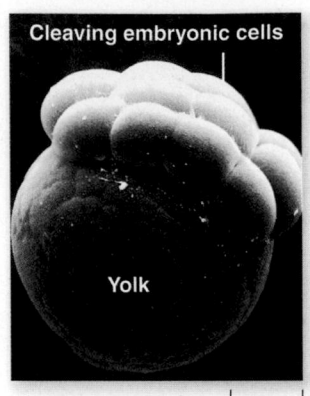

Cleaving embryonic cells

Yolk

25 µm

Figure 53.9
Meroblastic cleavage.
Only a portion of the egg
actively divides to form a
mass of cells in this type of
cleavage, which occurs in
eggs with relatively large
amounts of yolk.

these eggs is restricted to the
blastodisc. The yolk is essentially
an inert mass. This type of cleavage pattern is called **meroblastic cleavage** (figure 53.9). The resulting embryo is not spherical, but
rather has the form of a thin cap
perched on the yolk.

Cleavage in mammals

Mammalian eggs contain very
little yolk; however, mammalian
embryogenesis has many similarities to development of their
reptilian and avian relatives.

Because cleavage is not
impeded by yolk in mammalian
eggs, it is holoblastic, forming a
structure called a *blastocyst,* in
which a single layer of cells
surrounds a central fluid-filled
blastocoel. In addition, an **inner
cell mass (ICM)** is located at
one pole of the blastocoel cavity (figure 53.10). The ICM is similar to the blastodisc of reptiles and birds, and it goes on to form
the developing embryo.

The outer layer of cells, called the **trophoblast,** is similar
to the cells that form the membranes underlying the tough
outer shell of the reptilian egg. These cells have changed during the course of mammalian evolution to carry out a very different function: Part of the trophoblast enters the maternal
endometrium (the epithelial lining of the uterus) and contributes to the *placenta,* the organ that permits exchanges between
the fetal and maternal blood supplies. The placenta will be discussed in more detail in a later section.

The major cleavage patterns of animal embryos are summarized in table 53.2.

Blastomeres may or may not be committed to developmental paths

Viewed from the outside, cleavage-stage embryos often look like
a simple ball or disc of similar cells. In many animals, this
appearance is misleading; for example, the unequal segregation

ICM

Blastocoel

Trophoblast

Blastodisc

Yolk

**Figure 53.10 The embryos of mammals and birds
are more similar than they seem.** A mammalian blastula
(left), called a blastocyst, is composed of a sphere of cells, the
trophoblast, surrounding a cavity, the blastocoel, and an inner cell
mass (ICM). An avian (bird) blastula consists of a cap of cells, the
blastodisc, resting atop a large yolk mass *(right).* The blastodisc
will form an upper and a lower layer with a compressed blastocoel
in between.

of cytoplasmic determinants into specific blastomeres of tunicate embryos (described in chapter 19) commits those cells to
different developmental paths. The experimental destruction or
removal of these committed cells results in embryos deficient in
the tissues that would have developed from those cells.

In contrast, mammals exhibit highly *regulative development,* in which early blastomeres do not appear to be committed to a particular fate. For example, if a blastomere is removed
from an early eight-cell stage human embryo (as is done in the
process of preimplantation genetic diagnosis), the remaining
seven cells of the embryo will "regulate" and develop into a
complete individual if implanted into the uterus of a woman.
Similarly, embryos that are split into two (either naturally or
experimentally) form identical twins. It therefore appears that
inheritance of maternally encoded determinants is not an
important mechanism in mammalian development, and body
form is determined primarily by cell–cell interactions.

The earliest patterning events in mammalian embryos
occur during the preimplantation stages that lead to formation
of the blastocyst. At the eight-cell stage, the outer surfaces of
many mammalian blastomeres flatten against each other in a
process called *compaction,* which serves to polarize the blastomeres. The polarized blastomeres then undergo asymmetrical
cell divisions. Cell lineage studies have shown that cells that
are in the interior of the embryo most often become ICM cells
of the mammalian blastocyst, whereas cells on the exterior of
the embryo usually become trophoblast cells.

Learning Outcomes Review 53.2

Cleavage is a series of rapid cell divisions that transforms the zygote into
the blastula—a hollow ball of cells. The amount of yolk is the major
determinant of cleavage pattern. Eggs with little yolk cleave completely
(holoblastic cleavage); eggs with a large yolk cannot cleave completely
(meroblastic cleavage). In many animals, each blastomere is committed to a
developmental path; in mammals, blastomeres are not committed but can
regulate as needed to produce a complete individual.

■ *If the cells of a mammalian embryo were separated at
the four-cell stage, would they develop normally? What
about a frog embryo at the four-cell stage?*

53.3 Gastrulation

Learning Outcomes

1. *Describe the outcome of gastrulation.*
2. *Compare gastrulation in different animals.*
3. *Name the extraembryonic membranes in amniotes.*

In a complex series of cell shape changes and cell movements,
the cells of the blastula rearrange themselves to form the basic
body plan of the embryo. This process, called *gastrulation,*
forms the three primary germ layers and converts the blastula
into a bilaterally symmetrical embryo with a central progenitor
gut and visible anterior–posterior and dorsal–ventral axes.

TABLE 53.3	Developmental Fates of the Primary Germ Layers in Vertebrates
Ectoderm	Epidermis of skin, nervous system, sense organs
Mesoderm	Skeleton, muscles, blood vessels, heart, blood, gonads, kidneys, dermis of skin
Endoderm	Lining of digestive and respiratory tracts, liver, pancreas, thymus, thyroid

Gastrulation produces the three germ layers

Gastrulation creates the three primary *germ layers:* endoderm, ectoderm, and mesoderm. The cells in each germ layer have very different developmental fates. The cells that move into the embryo to form the tube of the primitive gut are *endoderm;* they give rise to the lining of the gut and its derivatives (pancreas, lungs, liver, etc.). The cells that remain on the exterior are *ectoderm,* and their derivatives include the epidermis on the outside of the body and the nervous system. The cells that move into the space between the endoderm and ectoderm are *mesoderm;* they eventually form the notochord, bones, blood vessels, connective tissues, muscles and internal organs such as the kidneys and gonads (table 53.3).

Cells move during gastrulation using a variety of cell shape changes. Some cells use broad, actin-filled extensions called *lamellipodia* to crawl over neighboring cells. Other cells send out narrow extensions called *filopodia,* which are used to "feel out" the surfaces of other cells or the extracellular matrix. Once a satisfactory attachment is made, the filopodia retract to pull the cell forward. Contractions of actin filament bundles are responsible for many of these cell shape changes. Cells that are tightly attached to one another via desmosomes or adherens junctions will move as cell sheets.

In embryos with little yolk and a hollow blastula, the cell sheet at the vegetal pole of the blastula **invaginates** (dents inward) to form the primitive gut tube. In embryos with large yolky cells that are hard to move, sheets of smaller cells **involute** (roll inward) from the surface of the blastula and move over the basal surfaces of the outer cells. Other cells break away from cell sheets and migrate as individual cells during **ingression.**

Avian and mammalian gastrulation begins with **delamination,** in which one sheet of cells splits into two sheets. Each migrating cell possesses particular cell-surface glycoproteins, which adhere to specific molecules on the surfaces of other cells or in the extracellular matrix. Changes in cell adhesiveness, as described in chapter 19, are key events in gastrulation. The extracellular matrix protein fibronectin and the corresponding integrin receptors of cells are essential molecules of gastrulation in many animals.

Gastrulation patterns also vary according to the amount of yolk

Just as in cleavage patterns, yolk quantity also affects the types of cell movements that occur during gastrulation. Here, we examine gastrulation in four representative classes of embryos with differing quantities of yolk.

Gastrulation in sea urchins

Echinoderms such as sea urchins develop from relatively yolk-poor eggs and form hollow, symmetrical blastulas. Gastrulation begins when cells at the vegetal surface of the blastula change their shape to form a flattened **vegetal plate.** In an example of ingression, a subset of cells in the vegetal plate breaks away from the blastula wall and moves into the blastocoel cavity. These **primary mesenchyme cells** are future mesoderm cells, and they use *filopodia* to migrate through the blastocoel cavity (figure 53.11). Eventually, they become

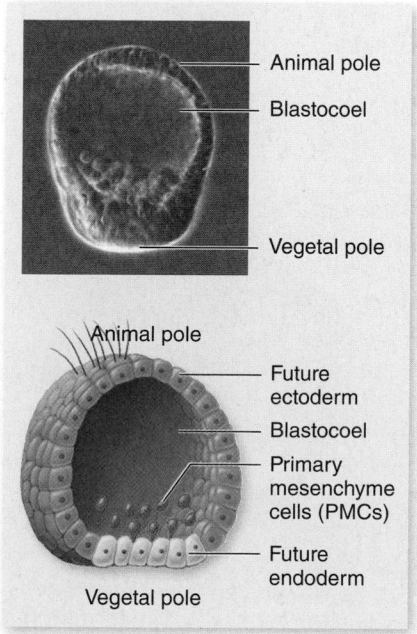

a.

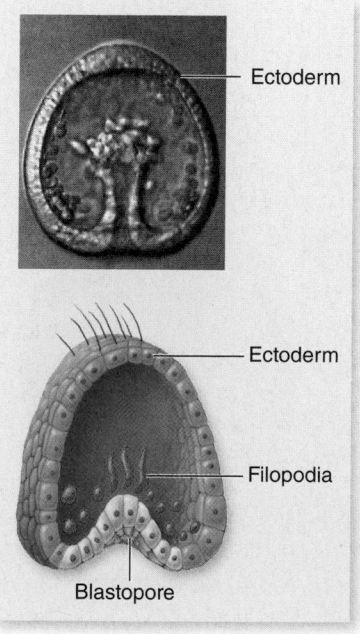

b.

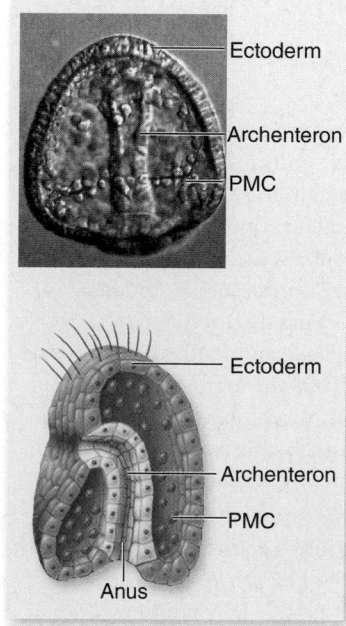

c.

Figure 53.11
Gastrulation in a sea urchin. *a.* Gastrulation begins with formation of the vegetal plate and ingression of primary mesenchyme cells (prospective mesoderm cells) into the blastocoel cavity. *b.* The endoderm is then formed by invagination of the remaining vegetal plate cells and extension of a cellular tube to produce the primitive gut, or archenteron. *c.* Cells that remain on the surface form the ectoderm.

localized in the ventrolateral corners of the blastocoel, where they form the larval skeleton.

The remaining cells of the vegetal plate then invaginate into the blastocoel to form the endoderm layer, creating a structure that looks something like an indented tennis ball. Eventually, the inward-moving tube of cells contacts the opposite side of the gastrula and stops moving. The hollow structure resulting from the invagination is called the *archenteron,* and it is the progenitor of the digestive tube. The opening of the archenteron, the future anus, is known as the *blastopore.* A secondary opening develops at the point where the archenteron contacts the opposite side of the gastrula, forming the mouth (see figure 53.11). Animals in which the anus develops first and the mouth second are termed *deuterostomes,* as was discussed in chapter 33.

Gastrulation in frogs

The blastula of an amphibian has an asymmetrical yolk distribution, and the yolk-laden cells of the vegetal pole are less numerous but much larger than the yolk-free cells of the animal pole. Consequently, gastrulation is more complex than it is in sea urchins. In frogs, a layer of surface cells first invaginates to form a small, crescent-shaped slit, which initiates formation

of the blastopore. Next, cells from the animal pole involute over the dorsal lip of the blastopore (see figure 53.12*a*), which forms at the same location as the gray crescent of the fertilized egg (see figure 53.5).

The involuting cell layer eventually presses against the inner surface of the opposite side of the embryo, eliminating the blastocoel and producing an archenteron with a blastopore. In this case, however, the blastopore is filled with yolk-rich cells, forming the **yolk plug** (figure 53.12*b, c*). The outer layer of cells resulting from these movements is the ectoderm, and the inner layer is the endoderm. Other cells that involute over the dorsal lip and ventral lip (the two lips of the blastopore that are separated by the yolk plug) migrate between the ectoderm and endoderm to form the third germ layer—the mesoderm (figure 53.12*c–e*).

Gastrulation in birds

At the end of cleavage in a bird or reptile, the developing embryo is a small cap of cells called the **blastoderm,** which sits on top of the large ball of yolk (figure 53.13*a*). As a result, gastrulation proceeds somewhat differently.

In birds, the blastoderm first separates into two layers, and a blastocoel cavity forms between them (figure 53.13*b*).

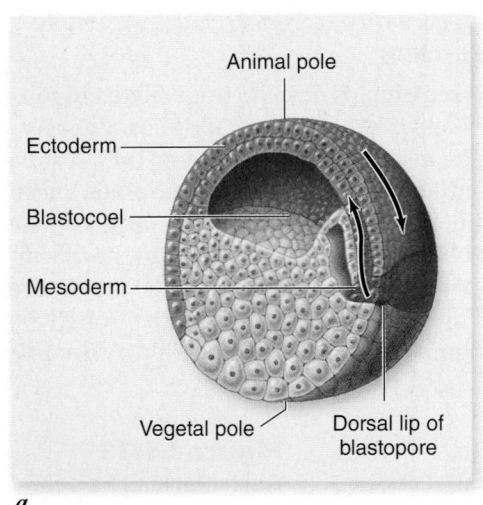

a.

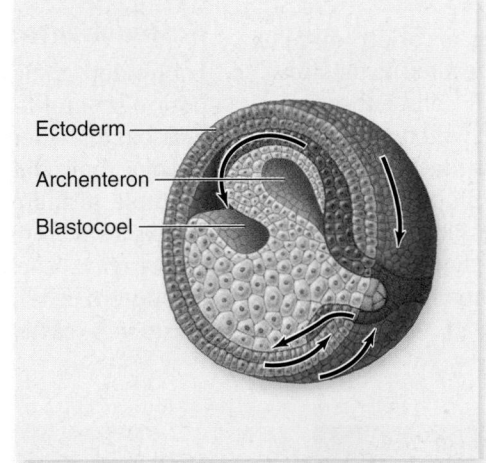

b.

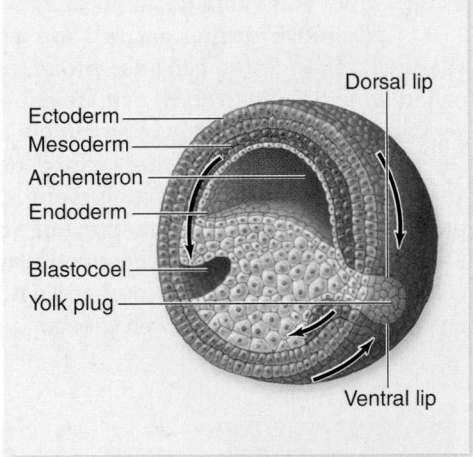

c.

Figure 53.12 Frog gastrulation. *a.* A layer of cells from the animal pole moves toward the vegetal pole, ultimately involuting through the dorsal lip of the blastopore. *b.* Cells in the dorsal lip zone then involute into the hollow interior, or blastocoel, eventually pressing against the far wall. The three primary germ tissues (ectoderm, mesoderm, and endoderm) become distinguished. Ectoderm is shown in blue, mesoderm in red, and endoderm in yellow. *c.* The movement of cells through the blastopore creates a new internal cavity, the archenteron, which displaces the blastocoel. *d.* Organogenesis begins when the neural plate forms from dorsal ectoderm to begin the process of neurulation. *e.* The neural plate next forms a neural groove and then a neural tube. The cells of the neural ectoderm are shown in purple.

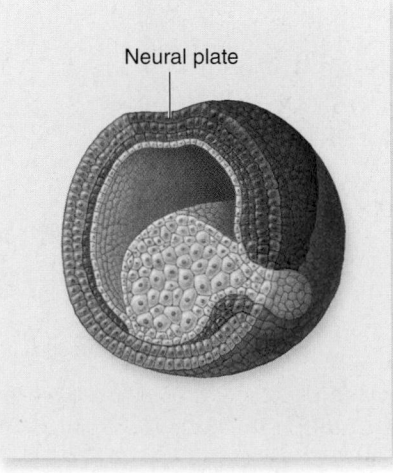

d.

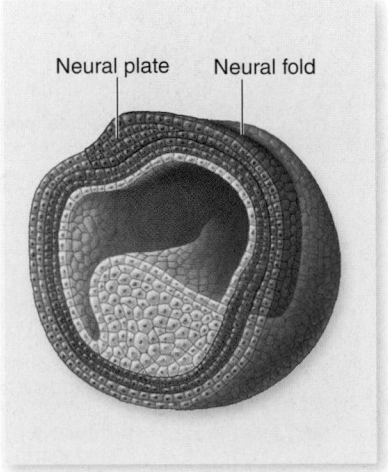

e.

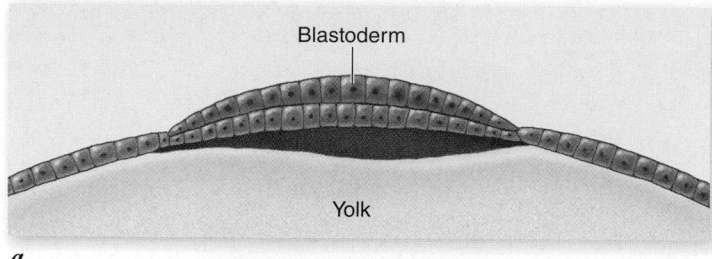

a.

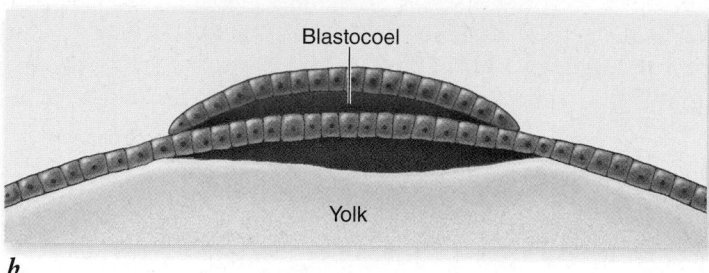

b.

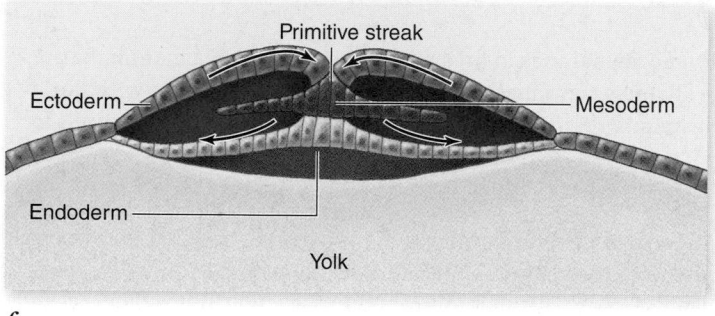

c.

Figure 53.13 Gastrulation in birds. *a.* The avian blastula is made up of a disc of cells sitting atop the large yolk mass. *b.* Gastrulation commences with the delamination of the blastoderm into two layers. All three germ layers are derived from the upper layer of the blastoderm. *c.* Cells that migrate through the primitive streak into the interior of the embryo are future endoderm or mesoderm cells. Cells that remain in the upper layer form the ectoderm.

The deep, internal layer of the bilayered blastoderm gives rise to extraembryonic tissues only (described later on), whereas all cells of the embryo proper are derived from the upper layer of cells. Thus, the upper layer of the blastoderm gives rise to all three germ layers.

Some of the surface cells begin moving to the midline, where they break away from the surface sheet of cells and ingress into the blastocoel cavity. A furrow along the longitudinal midline marks the site of this ingression (figure 53.13*c*). This furrow, analogous to an elongated blastopore, is called the **primitive streak.** Some cells migrate through the primitive streak and across the blastocoel cavity to displace cells in the lower layer. These deep-migrating cells form the endoderm. Other cells that move through the primitive streak migrate laterally into intermediate regions and form a new layer—the mesoderm. Cells that remain on the surface and do not enter the primitive streak form the ectoderm.

Gastrulation in mammals

Mammalian gastrulation proceeds much the same as it does in birds. In both types of animals, the embryo develops from a flattened collection of cells—the blastoderm in birds or the inner cell mass in mammals. Although the blastoderm of a bird is flattened because it is pressed against a mass of yolk, the inner cell mass of a mammal is flat despite the absence of a yolk mass.

In mammals, the placenta has made yolk dispensable; the embryo obtains nutrients from its mother following implantation into the uterine wall. However, the embryo still gastrulates as though it were sitting on top of a ball of yolk.

In mammals, a primitive streak forms, and cell movements through the primitive streak give rise to the three primary germ layers, much the same as in birds (figure 53.14). Similarly, mammalian embryos envelop their "missing" yolk by forming a yolk sac from extraembryonic cells that migrate away from the lower layer of the blastoderm and line the blastocoel cavity.

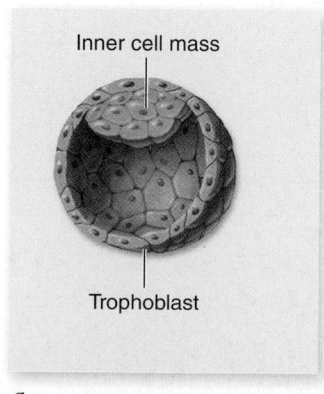

a.

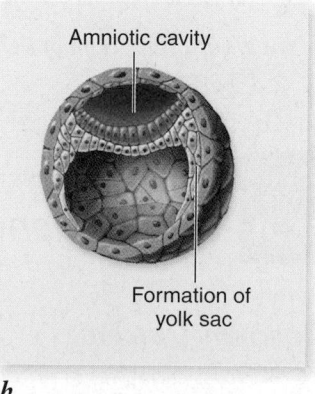

b.

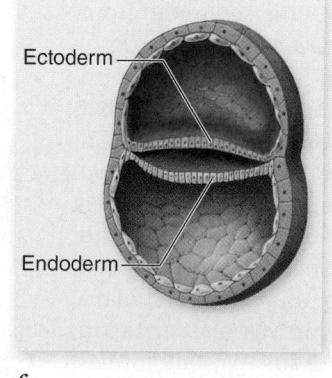

c.

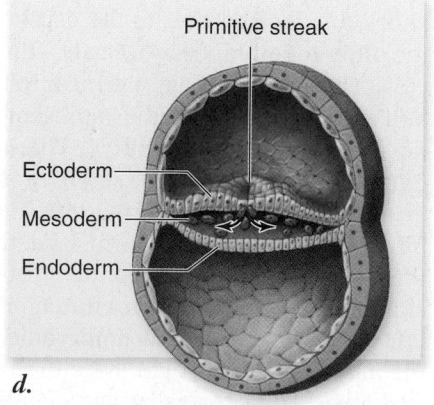

d.

Figure 53.14 Mammalian gastrulation. *a.* Cross section of the mammalian blastocyst at the end of cleavage. *b.* The amniotic cavity forms between the inner cell mass (ICM) and the pole of the embryo. Meanwhile, the ICM flattens and delaminates into two layers that will become ectoderm and endoderm. *b.* and *c.* Cells of the lower layer migrate out to line the blastocoel cavity to form the yolk sac. *d.* A primitive streak forms the ectoderm layer, and cells destined to become mesoderm migrate into the interior, similar to gastrulation in birds.

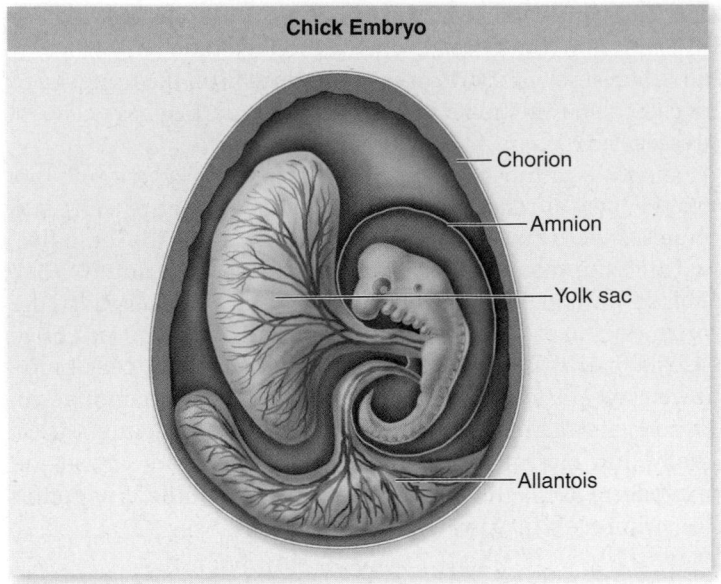

Chick Embryo

- Chorion
- Amnion
- Yolk sac
- Allantois

a.

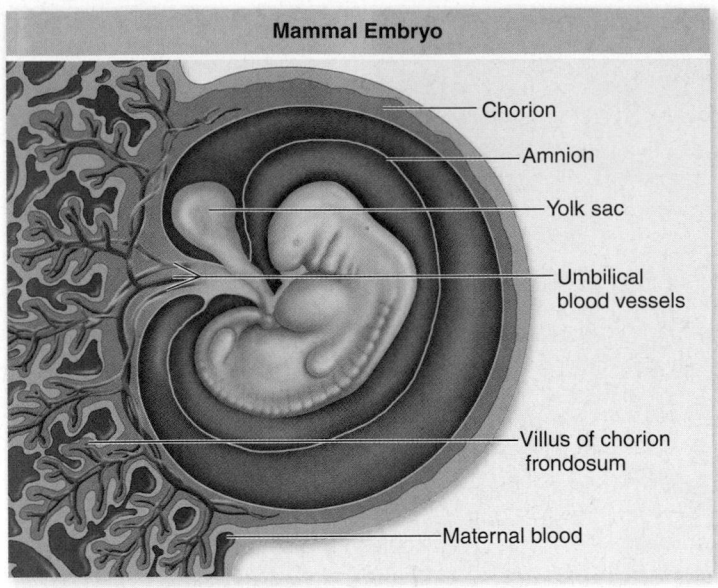

Mammal Embryo

- Chorion
- Amnion
- Yolk sac
- Umbilical blood vessels
- Villus of chorion frondosum
- Maternal blood

b.

Figure 53.15 The extraembryonic membranes. The extraembryonic membranes in (*a*) a chick embryo and (*b*) a mammalian embryo share some of the same characteristics. However, in the chick, the allantois continues to grow until it eventually unites with the chorion just under the eggshell, where it is involved in gas exchange. In the mammalian embryo, the allantois contributes blood vessels to the developing umbilical cord.

Extraembryonic membranes are an adaptation to life on dry land

As an adaptation to terrestrial life, the embryos of reptiles, birds, and mammals develop within a fluid-filled *amniotic membrane,* or *amnion* (see chapter 35). The amniotic membrane and several other membranes form from embryonic cells, but they are located outside of the body of the embryo. For this reason, they are known as **extraembryonic membranes.** The extraembryonic membranes include the amnion, chorion, yolk sac, and allantois.

In birds, the amnion and chorion arise from two folds that grow to completely surround the embryo (figure 53.15*a*). The amnion is the inner membrane that surrounds the embryo and suspends it in *amniotic fluid,* thereby mimicking the aquatic environments of fish and amphibian embryos. The chorion is located next to the eggshell and is separated from the other membranes by a cavity—the *extraembryonic coelom.*

The *yolk sac* plays a critical role in the nutrition of bird and reptile embryos; it is also present in mammals, although it does not nourish the embryo. The *allantois* is derived as an outpouching of the gut and serves to store the uric acid excreted in the urine of birds. During development, the allantois of a bird embryo expands to form a sac that eventually fuses with the overlying chorion, just under the eggshell. The fusion of the allantois and chorion form a functioning unit, the chorioallantoic membrane, in which embryonic blood vessels, carried in the allantois, are brought close to the porous eggshell for gas exchange. The chorioallantoic membrane is thus the respiratory membrane of a bird embryo.

In mammals, the trophoblast cells of the blastocyst implant into the endometrial lining of the mother's uterus and become the chorionic membrane (figure 53.15*b*). The part of the chorion in contact with endometrial tissue contributes to the placenta. The other part of the placenta is composed of modified endometrial tissue of the mother's uterus, as is described in more detail in a later section. The allantois in mammals contributes blood vessels to the structure that will become the umbilical cord, so that fetal blood can be delivered to the placenta for gas exchange.

Learning Outcomes Review 53.3

Gastrulation involves cell rearrangement and migration to produce ectoderm, mesoderm, and endoderm. In sea urchins, endoderm forms by invagination of the blastula; mesodermal cells form from other surface cells. In vertebrates with moderate to extensive amounts of yolk, surface cells move through a blastopore or a primitive streak, respectively. Mammalian gastrulation is similar to gastrulation in birds. Extraembryonic membranes of amniote species form from embryonic cells outside the embryo's body and include the yolk sac, amnion, chorion, and allantois.

■ *What kind of cellular behaviors are necessary for gastrulation?*

53.4 *Organogenesis*

Learning Outcomes

1. *Discuss examples of organogenesis.*
2. *Describe neurulation and somitogenesis.*
3. *Explain the migration and role of neural crest cells.*

Gastrulation establishes the basic body plan and creates the three primary germ layers of animal embryos. The stage is now set for *organogenesis*—the formation of the organs in their proper locations—which occurs by interactions of cells within and between the three germ layers. Thus, organogenesis follows rapidly on the heels of gastrulation, and in many animals begins before gastrulation is complete. Over the course of subsequent development, tissues develop into organs and animal embryos assume their unique body form (see table 53.1).

Changes in gene expression lead to cell determination

All of the cells in an animal's body, with the exception of a few specialized ones that have lost their nuclei, have the same complement of genetic information. Despite the fact that all of its cells are genetically identical, an adult animal contains dozens to hundreds of cell types, each expressing some unique aspect of the total genetic information for that individual. The information for other cell types is not lost, but most cells within a developing organism progressively lose the capacity to express ever-larger portions of their genomes. What factors determine which genes are to be expressed in a particular cell?

To a large degree, a cell's location in the developing embryo determines its fate. By changing a cell's location, an experimenter can often alter its developmental destiny, as mentioned in chapter 19. But this is only true up to a certain point in the cell's development. At some stage, every cell's ultimate fate becomes fixed, a process referred to as *cell determination*.

A cell's fate can be established by inheritance of cytoplasmic determinants or by interactions with neighboring cells. The process by which a cell or group of cells instructs neighboring cells to adopt a particular fate is called *induction*. If a nonporous barrier, such as a layer of cellophane, is imposed between the inducer and the target tissue, no induction takes place. In contrast, a porous filter, through which proteins can pass, does permit induction to occur.

In these experiments, researchers concluded that the inducing cells secrete a paracrine signal molecule that binds to the cells of the target tissue. Such signal molecules are capable of producing changes in the patterns of gene transcription in the target cells. You will learn more about the origin of embryonic induction a little later in this chapter.

Development of selected systems in *Drosophila* illustrates organogenesis

In chapter 19, you saw how the creation of morphogen gradients in a fruit fly embryo leads to hierarchies of gene expression that direct cell fate decisions along both the anterior–posterior and dorsal–ventral axes. These two axes form a coordinate system to specify the position of tissues and organs within the *Drosophila* embryo. In this section we look at development of three different organs: salivary glands, the heart, and the tracheae of the respiratory system.

Salivary gland development

The fruit fly larva is a mobile eating machine, and thus it has very active salivary glands. The primordia of the salivary glands

develop as simple tubular invaginations of ectodermal cells on the ventral surface of the third head segment.

Salivary glands develop only from an anterior strip of cells that express the *sex combs reduced (scr)* gene. No salivary glands form in *scr*-deficient embryos, whereas experimental expansion of *scr* expression along the anterior–posterior axis results in the formation of additional salivary gland primordia along the length of the embryo.

The *scr* gene is one of the homeotic genes in the Antennapedia complex, which encode transcription factors that bind to DNA via their homeodomains to regulate gene expression (see chapter 19). One downstream target of the *scr* gene is the *fork head (fkh)* gene, which has Scr-binding sites in its enhancer. The *fkh* gene is required for secretory cell development in salivary gland rudiments, and it encodes a transcription factor that directly activates expression of salivary gland-specific genes. Thus, action of the *scr* gene activates *fkh* expression at the proper anterior location for salivary gland formation.

The inhibitory action of a dorsally expressed protein, Decapentaplegic (Dpp), determines the ventral position of the salivary glands. Activation of the Dpp-signaling pathway represses salivary gland specification in neighboring cells. This restricts development of salivary gland rudiments to their specific ventral patch of ectoderm cells (figure 53.16). In mutant

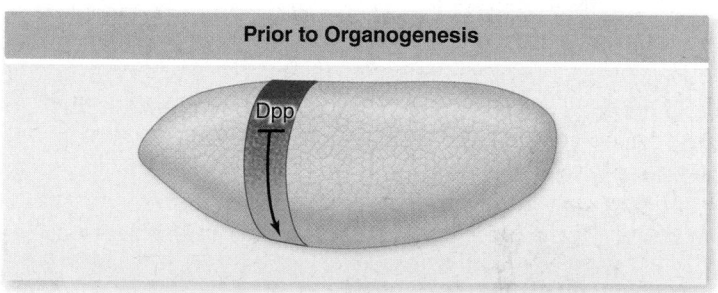

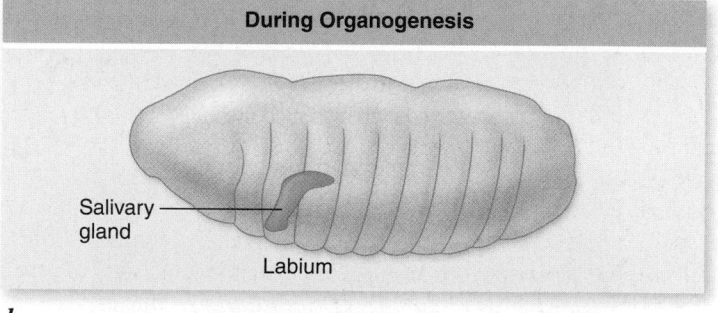

Figure 53.16 **Salivary gland formation in *Drosophila*.** Prospective salivary gland cells are determined by the intersection of the anterior–posterior and dorsal–ventral axes. *a.* Prior to organogenesis, the sex combs reduced *(scr)* gene is expressed in an anterior band of cells (shaded blue). At the same time, Decapentaplegic protein (Dpp) is released by cells on the dorsal side of the embryo, forming a gradient in the dorsal–ventral direction. Dpp specifies dorsal cell fates and inhibits formation of salivary gland rudiments. *b.* During organogenesis, the salivary glands develop in areas where Scr is expressed but Dpp is absent. Each salivary gland rudiment forms as a ventral invagination of the surface ectoderm on either side of the third head segment (the labium).

embryos deficient for Dpp or any of the downstream Dpp-signaling proteins, salivary gland rudiments are not restricted to this ventral patch, and they form from the entire ectoderm of the third segment.

Heart development

The heart is a mesoderm-derived structure in all animals, and it is the first organ to become functional during embryonic development. The dorsal vessel is the heart-equivalent structure in *Drosophila melanogaster*. The homeobox-containing gene *tinman* is expressed in the prospective heart mesoderm and in the developing dorsal vessel, and its activity is required for dorsal vessel development in *Drosophila* (figure 53.17).

Dorsal vessel development in *Drosophila* is also dependent on two other types of transcription factors (known as GATA and T-box factors). In an illuminating case of evolutionary conservation, scientists have discovered gene families similar to each of these three *Drosophila* genes in vertebrates. Moreover, members of these gene families play important roles in vertebrate heart specification.

This evolutionary conservation includes not just the structure of these genes, but their function as well. Researchers have discovered that specification of cardiac mesoderm is subject to inductive signals from adjoining germ layers in both *Drosophila* and vertebrates. In vertebrates, the heart develops in an internal location, and the inductive signals come from the underlying anterior endoderm. In *Drosophila*, the dorsal vessel forms in a more superficial location, and the signals come from the overlying ectoderm.

Despite the different sources, the signals that regulate the expression of these three key types of transcription factors are themselves conserved between *Drosophila* and vertebrates. Given the critical and conserved circulatory function of the heart, it is perhaps not surprising that similar gene families mediate the specification of heart mesoderm in both *Drosophila* and vertebrates.

Tracheae: Branching morphogenesis

As you learned in chapters 34 and 48, insects exchange gases via a branching system of finer and finer tubes called *tracheae*. The repeated branching of simple epithelial tubes that leads to formation of the tracheal system is an example of **branching morphogenesis.**

Mutations in the *branchless* gene in *Drosophila* result in embryos with greatly reduced tracheal systems. The *branchless* gene encodes a member of the large family of **fibroblast growth factors (FGF),** which bind to receptor tyrosine kinase proteins (see chapter 9) to stimulate proliferation of target cells. In another interesting case of evolutionary conservation, the mammalian FGF homologue of the *branchless* gene is required for branching morphogenesis that creates the alveolar passageways in the mammalian lung.

In both animals, loose clusters of mesenchymal cells adjacent to distal regions of the epithelial tube secrete FGF. The FGF binds to a specific FGF receptor in the membrane of the epithelial cells, stimulating them to proliferate and to grow out into a new tube bud.

In vertebrates, organogenesis begins with neurulation and somitogenesis

The process of organogenesis in vertebrates begins with the formation of two morphological features found only in chordates: the *notochord* and the hollow **dorsal nerve cord** (see chapter 35). The development of the dorsal nerve cord is called *neurulation.*

Development of the neural tube

The notochord forms from mesoderm and is first visible soon after gastrulation is complete. It is a flexible rod located along the dorsal midline in the embryos of all chordates, although its function as a supporting structure is supplanted by the subsequent development of the vertebral column in the vertebrates. After the notochord has been laid down, the region of dorsal ectodermal cells situated above the notochord begins to thicken to form the *neural plate.*

The thickening is produced by the elongation of the dorsal ectoderm cells. Those cells then assume a wedge shape because of contracting bundles of actin filaments at their apical end. This change in shape causes the neural tissue to roll up into a **neural groove** running down the long axis of the embryo. The edges of the neural groove then move toward each other

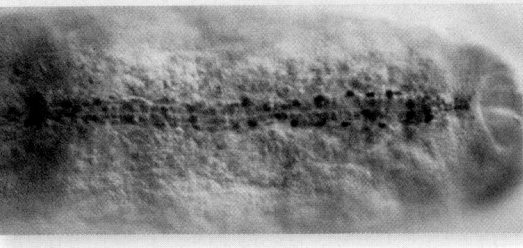
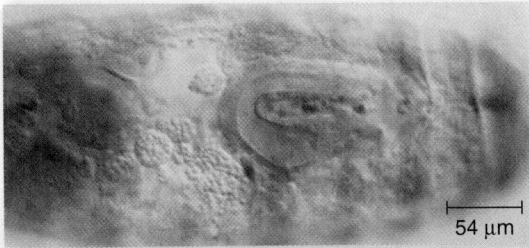
Figure 53.17 A gene necessary for heart formation in *Drosophila*.

and fuse, creating a long hollow cylinder, the **neural tube** (figure 53.18). The neural tube eventually pinches off from the surface ectoderm to end up beneath the surface of the embryo's back. Regional changes, which are under control of the *Hox* gene complexes (see chapter 19), then occur in the neural tube as it differentiates into the spinal cord and brain.

Generation of somites

While the neural tube is forming from dorsal ectoderm, the rest of the basic architecture of the body is being rapidly established by changes in the mesoderm. The sheets of mesoderm on either side of the developing notochord separate into a series of rounded regions called **somitomeres.** The somitomeres then separate into segmented blocks called **somites** (see figure 53.18). The mesoderm in the head region does not separate into discrete somites but remains connected as somitomeres, which form the skeletal muscles of the face, jaws, and throat.

Somites form in an anterior–posterior wave with a regular periodicity that can be easily timed—for example, by using

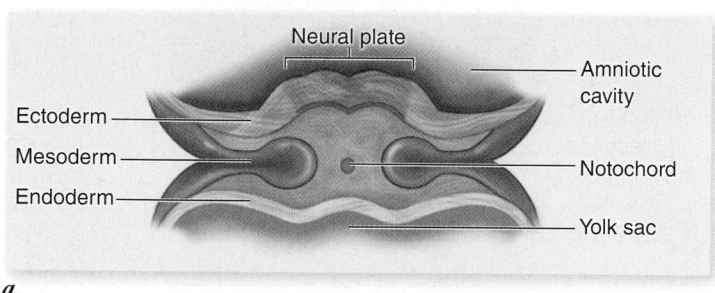

a.

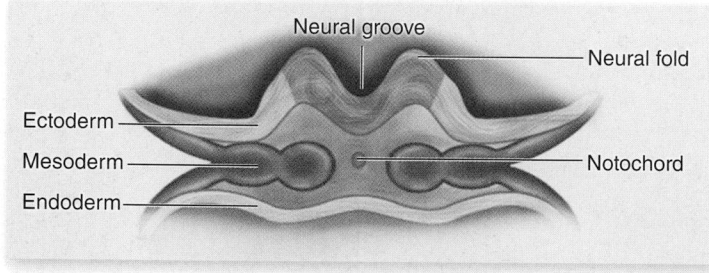

b.

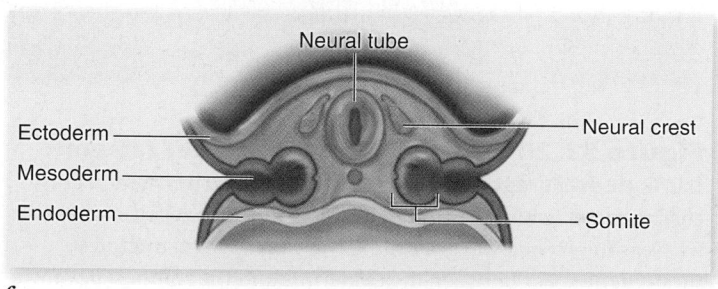

c.

Figure 53.18 Mammalian neural tube formation.
a. The neural plate forms from ectoderm above the notochord.
b. The cells of the neural plate fold together to form the neural groove. *c.* The neural groove eventually closes to form a hollow tube called the neural tube, which will become the brain and spinal cord. As the tube closes, some of the cells from the dorsal margin of the neural tube differentiate into the neural crest, migratory cells that form a variety of structures and are characteristic of vertebrates.

a vital dye, which marks cells without killing them, to mark each somite as it forms in a chick embryo. Cells at the presumptive boundary regions in the presomitic mesoderm instruct cells anterior to them to condense and separate into somites at specific times (for example, every 90 min in a chick embryo). This "clock" appears to be regulated by contact-mediated cell signaling between neighboring cells.

Somites themselves are transient embryonic structures, and soon after their formation, cells disperse and start differentiating along different pathways to ultimately form the skeleton, skeletal musculature, and associated connective tissues. The total number of somites formed is species-specific; for example, chickens form 50 somites, whereas some species of snakes form as many as 400 somites.

Some body organs, including the kidneys, adrenal glands, and gonads, develop within a strip of mesoderm that runs lateral to each row of somites. The remainder of the mesoderm, which is most ventrally located, moves out and around the endoderm and eventually surrounds it completely. As a result of this movement, the mesoderm becomes separated into two layers. The outer layer is associated with the inner body wall, and the inner layer is associated with the outer lining of the gut tube. Between these two layers of mesoderm is the *coelom* (see chapter 33), which becomes the body cavity of the adult. Figure 53.19 shows the major mesoderm lineages of amniote embryos.

Migratory neural crest cells differentiate into many cell types

Neurulation occurs in all chordates, and the process in the simple lancelet, a nonvertebrate chordate, is much the same as it is in a human. However, neurulation is accompanied by an additional step in vertebrates. Just before the neural groove closes to form the neural tube, its edges pinch off, forming a

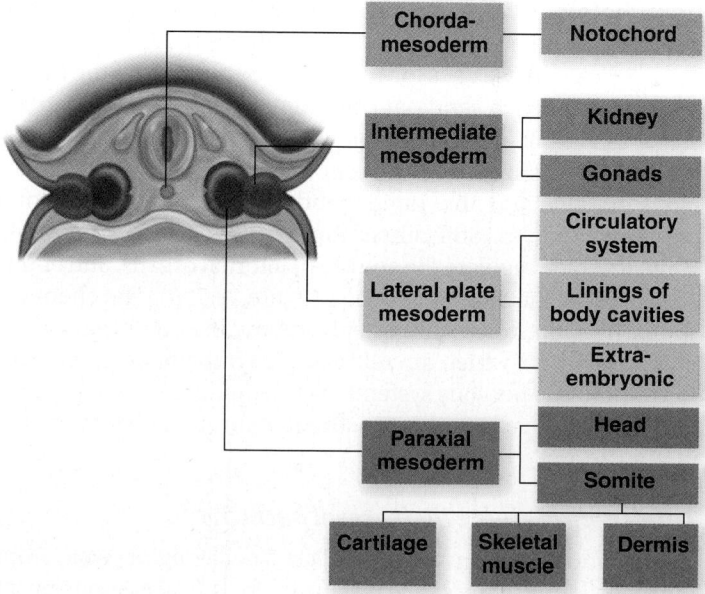

Figure 53.19 Mesoderm-derived structures of birds and mammals.

small cluster of cells—the *neural crest*—between the roof of the neural tube and the surface ectoderm (figure 53.18c).

In another example of extensive cell movements during animal development, the neural crest cells then migrate away from the neural tube to colonize many different regions of the developing embryo. The appearance of the neural crest was a key event in the evolution of the vertebrates because neural crest cells, after reaching their final destinations, ultimately develop into many structures characteristic of the vertebrate body.

The differentiation of neural crest cells depends on their migration pathway and final location. Neural crest cells migrate along one of three pathways in the embryo. Cranial neural crest cells are anterior cells that migrate into the head and neck; trunk neural crest cells migrate along one of two different pathways (to be described shortly). Each population of neural crest cells develops into a variety of cell types.

Cranial neural crest cells' migration

Cranial neural crest cells contribute significantly to development of the skeletal and connective tissues of the face and skull, as well as differentiating into nerve and glial cells of the nervous system, and melanocyte pigment cells. Changes in the placement of cranial neural crest cells during development have led to the evolution of the great complexity and variety of vertebrate heads.

There are two waves of cranial neural crest cell migration. The first produces both dorsal and ventral structures, and the second produces only dorsal structures and makes much less cartilage and bone. Transplantation experiments indicate that the developmental potential of the cells in these two waves is identical. The differences in cell fate are due to the environment the migrating cells encounter and not due to prior determination of cell fate.

Trunk neural crest cells: Ventral pathway

Neural crest cells located in more posterior positions have very different developmental fates depending on their migration pathway. The first trunk neural crest cells that migrate away from the neural tube pass through the anterior half of each adjoining somite to ventral locations (figure 53.20a).

Some of these cells form the sensory neurons of the dorsal root ganglia, which send out projections to connect the periphery of the animal with the spinal cord (see chapter 43). Others become specialized as Schwann cells, which insulate nerve fibers to facilitate the rapid conduction of impulses along peripheral nerves. Still others form nerves of the autonomic ganglia, which regulate the activity of internal organs, and endocrine cells of the adrenal medulla (figure 53.20b). The chemical similarity of the hormone epinephrine and the neurotransmitter norepinephrine, which are released by sympathetic neurons of the autonomic nervous systems, may result because both adrenal medullary cells and sympathetic neurons derive from the neural crest.

Trunk neural crest cells: Lateral pathway

The second group of trunk neural crest cells migrate away from the neural tube in the space just under the surface ectoderm, to occupy this space around the entire body of the embryo. There, they will differentiate into the pigment cells of the skin

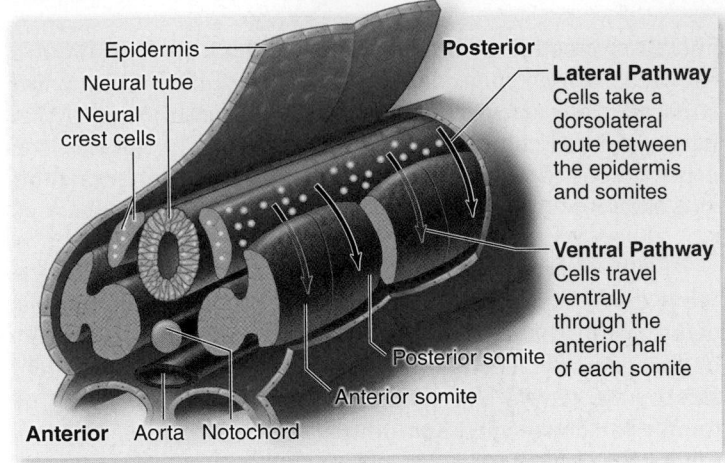

a.

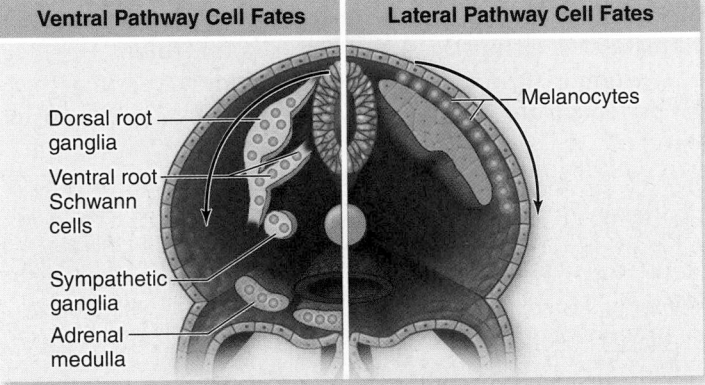

b.

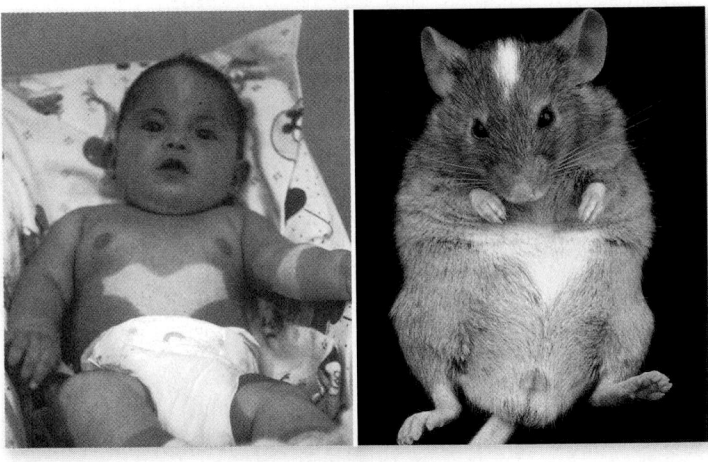

c.

Figure 53.20 Migration pathways and cell fates of trunk neural crest cells. *a.* The first wave of trunk neural crest cells migrates ventrally through the anterior half of each somite, whereas the second wave of cells leaves dorsally and migrates through the space between the epidermis and the somites. *b.* Ventral pathway neural crest cells differentiate into a variety of specialized cell types, but lateral pathway cells develop into the melanocytes (pigment cells) of the skin. *c.* A mutation in a gene that promotes survival of neural crest cells in all mammals leads to white spotting on the bellies and foreheads of both human babies and mice!

 Data analysis Two different genes have mutant alleles that cause this spotting phenotype. One gene encodes a signaling molecule. What is the likely function of the other gene product?

(figure 53.20*a, b*). Mutations in genes that affect the survival and migration of neural crest cells lead to white spotting in the skin on ventral surfaces, as well as internal problems in other neural crest-derived tissues (figure 53.20*c*).

Because the fate of a neural crest cell is dictated by its migration pathway, many studies have been done to identify the molecules that control the migration pathways of neural crest cells. Cell adhesion molecules on cell surfaces and in the extracellular matrix are expected to play prominent roles. For example, prospective neural crest cells down-regulate the expression of *N*-cadherin on their surfaces, which enables them to break away from the neural tube. Then, soon after leaving the neural tube, integrin receptors appear on the surfaces of neural crest cells, allowing them to interact with proteins in the extracellular matrix pathways along which they will migrate.

Neural crest derivatives are important in vertebrate evolution

Primitive chordates such as lancelets are filter feeders, using the rapid beating of cilia to draw water into their mouths, which then exits through slits in their pharynx. These pharyngeal slits evolved into the vertebrate gill chamber, a structure that provides a greatly improved means of gas exchange. Thus, evolution of the gill chamber was certainly a key event in the transition from filter feeding to active predation, which requires a much higher metabolic rate.

In the development of the gill chamber, some of the cranial neural crest cells form cartilaginous bars between the embryonic pharyngeal slits. Other cranial neural crest cells induce portions of the mesoderm to form muscles along the cartilage, and still others to form neurons that carry impulses between the central nervous system and these muscles.

Many of the unique vertebrate adaptations that contribute to their varied ecological roles involve structures that arise from neural crest cells. The vertebrates became fast-swimming predators with much higher metabolic rates. This accelerated metabolism permitted a greater level of activity than was possible among the more primitive chordates. Other evolutionary changes associated with the derivatives of the neural crest provided better detection of prey, a greatly improved ability to orient spatially during prey capture, and the means to respond quickly to sensory information. The evolution of the neural crest and of the structures derived from it were thus crucial steps in the evolution of the vertebrates (figure 53.21).

Learning Outcomes Review 53.4

Genetic control of organogenesis relies on conserved families of cell-signaling molecules and transcription factors. The control of heart development in *Drosophila* and mammals uses some of the same proteins. The process of neurulation forms the basic nervous system in vertebrates. Somitogenesis is the division of mesoderm into somites. Neural crest cells arise from the neural tube and migrate to many sites to form a variety of cell types. The evolution of the neural crest led to the appearance of many vertebrate-specific adaptations.

■ *Are neural crest cells determined prior to migration?*

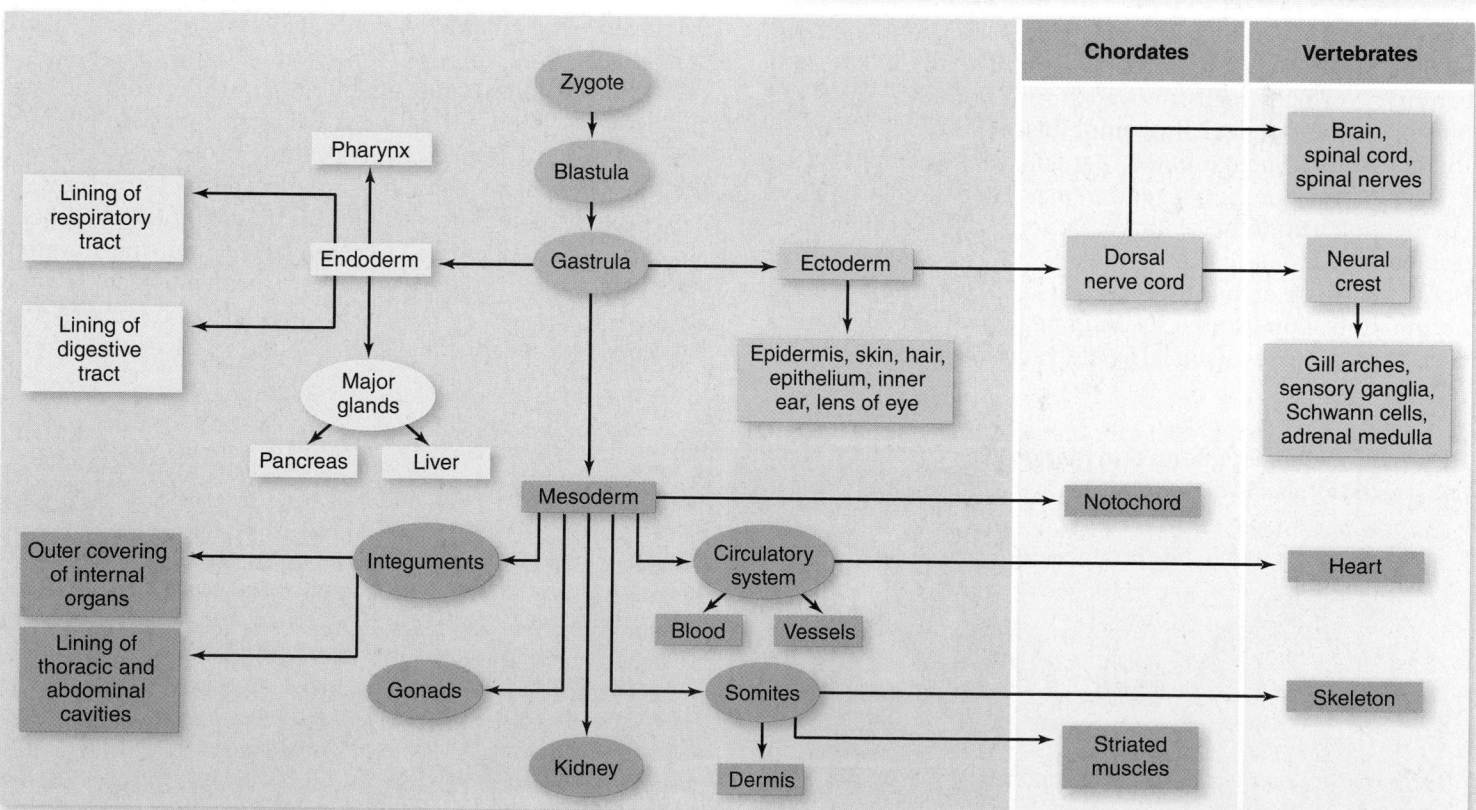

Figure 53.21 Germ-layer derivation of the major tissue types in animals. The three germ layers that form during gastrulation give rise to all the organs and tissues in the body, but the neural crest cells that form from ectodermal tissue give rise to structures that are prevalent in vertebrates, such as gill arches and bones of the face and skull.

Vertebrate Axis Formation

Learning Outcomes

1. **Describe the Spemann-Mangold experiment.**
2. **Explain the function of the organizer.**
3. **Distinguish between primary and secondary inductive events.**

In animal development, the relative position of cells in particular germ layers determines, to a large extent, the organs that develop from them. In *Drosophila,* you have seen that formation of morphogen gradients in the syncytial blastoderm establishes the anterior–posterior and dorsal–ventral axes of the embryo. The *Hox* gene complexes in vertebrates function similarly to the homeotic genes of *Drosophila* to specify the position of organs along the anterior–posterior axis. But how is cell fate selection along the dorsal–ventral axis accomplished in vertebrate embryos? Put another way, how do cells of the dorsal ectoderm "know" they are above the mesoderm-derived notochord, and thus fated to develop into the neural tube? The solution to this puzzle is one of the outstanding accomplishments of experimental embryology.

The Spemann organizer determines dorsal–ventral axis

The renowned German biologist Hans Spemann and his student Hilde Mangold solved this puzzle early in the 20th century. Normally, cells derived from the dorsal lip of the blastopore of a gastrulating amphibian embryo give rise to the notochord. Spemann and Mangold removed cells of the dorsal lip from one embryo and transplanted them to a different location on another embryo (figure 53.22). The new location corresponded to that of the animal's future belly. They found that some of the embryos developed two notochords: a normal dorsal one, and a second one along the belly. Moreover, a complete set of dorsal axial structures (e.g., notochord,

neural tube, and somites) formed at the ventral transplantation site in most of these embryos.

By using genetically different donor and host blastulas, Spemann and Mangold were able to show that the second notochord produced by transplanting dorsal lip cells contained host cells as well as transplanted ones. The transplanted dorsal lip cells had thus acted as *organizers,* stimulating cells that would normally form skin and belly structures to develop into dorsal axial structures. The belly cells must clearly contain the genetic information for dorsal axial developmental program, but they do not express it in the normal course of their development. Signals from the transplanted dorsal lip cells, however, must have caused them to do so.

How the organizer works

An organizer is a cluster of cells that release diffusible signal molecules, which then convey positional information to other cells. As seen earlier, organizers can have a profound influence on the development of surrounding tissues. Working as signal beacons, they inform surrounding cells of their distance from the organizer. The closer a particular cell is to an organizer, the higher the concentration of the signal molecule *(morphogen)* it experiences. Organizers and the diffusible morphogens that they release are thought to be part of a widespread mechanism for determining relative position and cell fates during vertebrate development.

The action of morphogens

The action of morphogens can be studied by using isolated portions of the blastula. The blastula can be bisected into an animal half (the animal cap) and vegetal half (the vegetal cap). If animal caps are removed from a frog blastula and cultured alone, they will form only ectoderm-derived epidermal cells. Similarly, cultured vegetal caps will form only endodermal cells. However, if animal caps are cultured combined with vegetal caps, the animal caps will form mesodermal structures.

The molecules involved in this induction have not been unambiguously identified. Members of the transforming growth factor beta (TGF-β) family have been implicated. These include activin, and *Xenopus* nodal-related proteins (Xnrs). Evidence for the inducing activity of these molecules includes

Figure 53.22 Spemann and Mangold's dorsal lip transplant experiment. Tissue from the dorsal lip of a donor embryo induced the formation of a second axis in the future belly region of a second, recipient embryo.

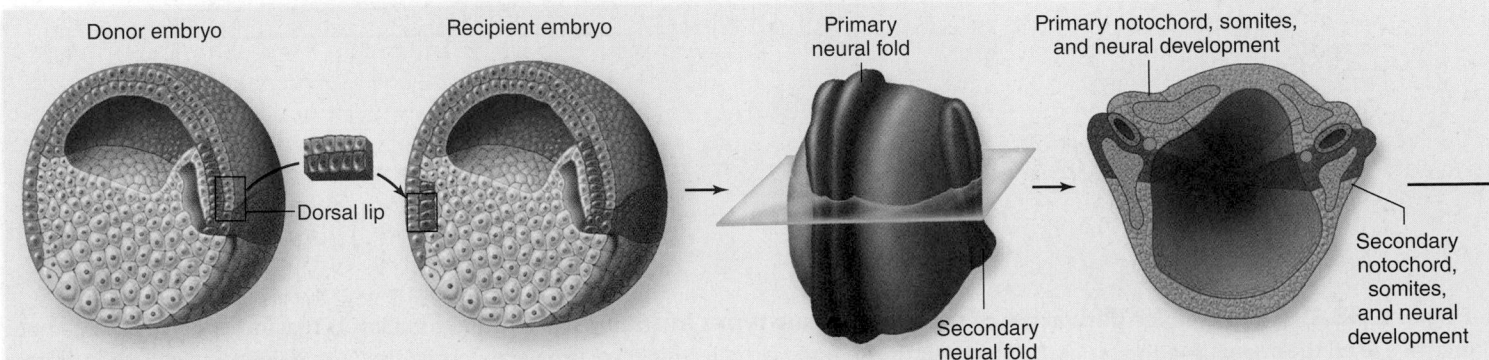

the timing and pattern of expression of these proteins correlates with inducing tissue, and depleting developing embryos of these proteins prevents induction.

The origin of the organizer

How do cells of the frog blastopore's dorsal lip become the Spemann organizer and how do they acquire their ability to specify cell fate along the dorsal–ventral axis? In frogs, as in fruit flies, this process starts during oogenesis in the mother. At that time, maternally encoded dorsal determinants are put into the developing oocyte, one of which accumulates at the vegetal pole of the unfertilized egg. At fertilization, cytoplasmic rearrangements cause this determinant to shift to the future dorsal side of the egg.

First, a signal from the point of sperm entry initiates the assembly of a microtubule array, which enables the egg's plasma membrane and the underlying cortical cytoplasm to rotate over the surface of the deeper cytoplasm. This physical rotation shifts this maternally encoded dorsal determinant to the opposite side of the egg from the point of sperm entry (figure 53.23a, b). In some frogs, a gray crescent forms opposite the sperm entry point, as mentioned earlier, and this crescent marks the future site of the dorsal lip.

Cells that form in this area during cleavage (called the Nieuwkoop center for Pieter Nieuwkoop who did the previously mentioned animal cap studies) receive the dorsal determinants that moved during cortical rotation. The dorsal determinants cause a change in gene expression in these cells, producing a signaling molecule that induces the cells above them to develop into the dorsal lip of the blastopore (figure 53.23c).

Maternally encoded dorsal determinants activate Wnt signaling

Experiments carried out over the last 15 years suggest that the maternally encoded dorsal determinants in *Xenopus* are mRNAs for proteins that function in the intracellular **Wnt** signaling pathway. *Wnt* genes encode a large family of cell-signaling proteins that affect the development of a number of structures in both vertebrates and invertebrates. Turning on the Wnt pathway in the dorsal vegetal cells of the Nieuwkoop center leads ultimately to activation of a transcription factor, which moves into the nucleus to activate the expression of genes necessary for organizer specification.

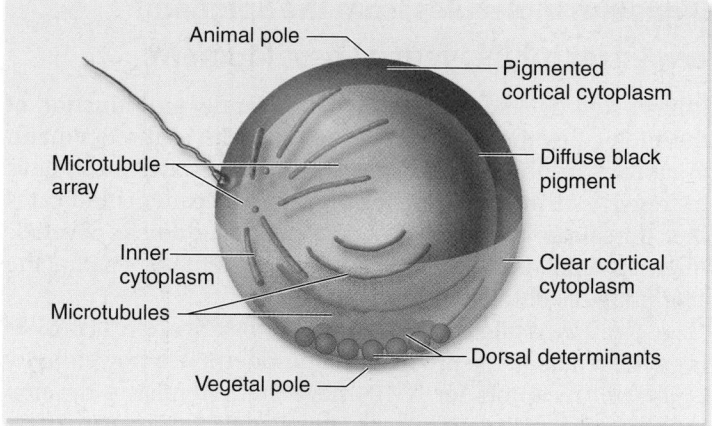

a.

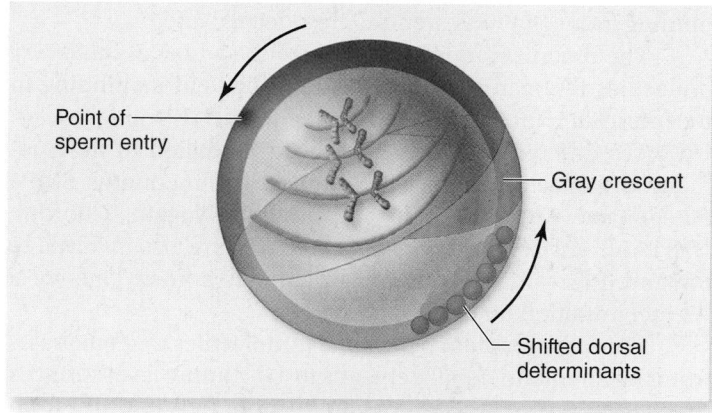

b.

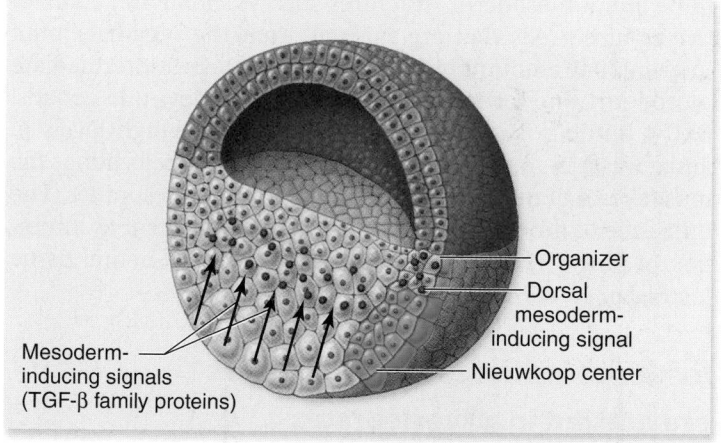

c.

Figure 53.23 **Creation of the Spemann organizer.** *a.* Dorsal determinants are localized at the vegetal pole of the unfertilized frog egg. At fertilization, a microtubule array forms at the site of sperm entry. These microtubules organize parallel microtubules to line the vegetal half of the egg between the cortex and cytoplasm. *b.* The cortical cytoplasm and dorsal determinants ride on this parallel array of microtubules, shifting to a site opposite sperm entry. *c.* Cells that inherit these shifted dorsal determinants form the Nieuwkoop center, which releases diffusible signaling molecules that specify the overlying cells to become the organizer. The organizer forms at the area of the gray crescent, visible following the cytoplasmic rearrangements at fertilization.

Signaling molecules from the Spemann organizer inhibit ventral development

It has taken decades to establish the identity and function of the molecules that are synthesized by cells of the Spemann organizer to subsequently specify dorsal mesoderm cell fates in frogs. A surprising finding of recent experiments indicates that dorsal lip cells do not directly *activate* dorsal development. Instead, dorsal mesoderm development is a result of the *inhibition* of ventral development.

A protein called **bone morphogenetic protein 4 (BMP4)** is expressed in all the prospective mesoderm of a frog embryo. Cells with receptors for BMP4 have the potential to develop into mesodermal derivatives. The specific mesodermal fate depends on how many receptors bind BMP4: More BMP4 binding induces a more ventral mesodermal fate.

The organizer functions by secreting a host of *inhibitory* molecules that can bind to BMP4 and prevent its binding to receptor. Such molecules are referred to as BMP4 antagonists. Up to 13 different proteins have been identified in the Spemann organizer, most of which appear to function as BMP4 antagonists. These include the proteins Noggin, Chordin, Dickkopf, and Cerebrus. Noggin and BMP4 are also involved in toe and finger joint formation, so humans homozygous for a *Noggin* mutation have fused joints.

Thus, the gradient of *inhibitory* molecules that emanates from the Spemann organizer leads to a declining level of BMP4 *function* in the ventral-to-dorsal direction. Cells farthest from the organizer bind the highest levels of BMP4 and differentiate into ventral mesoderm structures such as blood and connective tissues. Cells that are midway from the organizer bind intermediate amounts of BMP4, differentiate into intermediate mesoderm, and form organs such as the kidneys and gonads. BMP4 binding is completely inhibited by the high levels of antagonists in the organizer itself. Thus, these cells adopt the most dorsal of mesoderm fates and develop into somites. The influence of the organizer also extends to ectoderm as inhibition of BMP4 in ectoderm leads to formation of neural tissue instead of epidermis (figure 53.24).

Evidence indicates that organizers are present in all vertebrates

In chicks, a group of cells at the anterior limit of the primitive streak called *Hensen's node* functions similarly to the dorsal lip of the blastopore: Hensen's node induces a second axis when transplanted to another area of a chick embryo. Recent studies have shown that cells of Hensen's node act like the Spemann organizer, secreting molecules that inhibit ventral development. These molecules are the same as those found in frog embryos. Therefore, these experiments once again illustrate the evolutionary conservation of particular genes in animal development.

In addition, notochord signaling acts to pattern the neural tube. The notochord produces the signaling molecule sonic hedgehog (Shh), which is related to a signaling molecule in *Drosophila* called hedgehog. Signaling by Shh specifies ventral

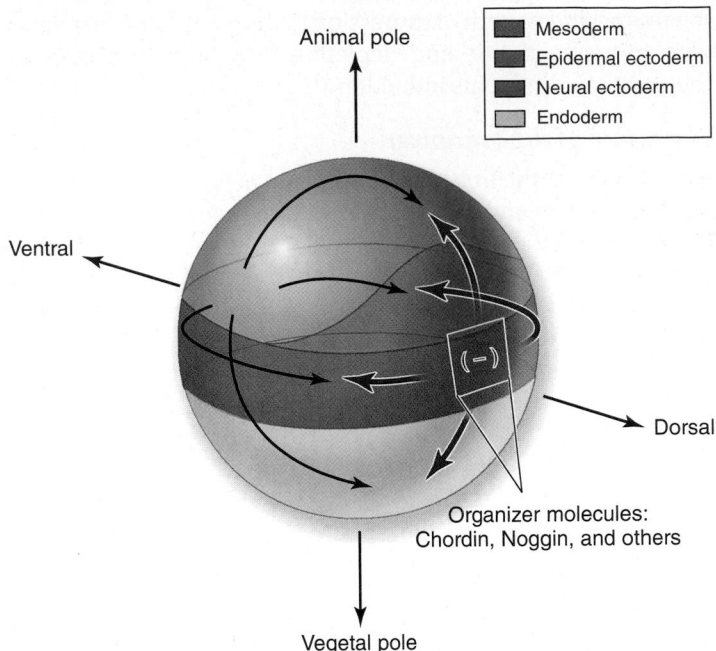

Figure 53.24 Function of the Spemann organizer. The organizer is a hotbed of secreted molecules that bind to and antagonize the action of BMP4, a morphogen that at high levels specifies ventral mesoderm cell fates.

cell fate with dose-related effects similar to those described for the TGF-β family proteins discussed earlier. In this way, induction by the notochord causes somites to form vertebrae, ribs, muscle, and skin, depending on the levels of Shh cells are exposed to.

Induction can be primary or secondary

The process of induction that Spemann initially discovered appears to be a fundamental mode of development in vertebrates. Inductions between the three primary germ layers—ectoderm, mesoderm, and endoderm—are referred to as **primary inductions.** The differentiation of the central nervous system during neurulation by the interaction of dorsal ectoderm and dorsal mesoderm to form the neural tube is an example of primary induction.

Inductions between tissues that have already been specified to develop along a particular developmental pathway are called **secondary inductions.** An example of secondary induction is the development of the lens of the vertebrate eye. The eye develops as an extension of the forebrain, a stalk that grows outward until it comes into contact with the surface ectoderm (figure 53.25). At a point directly above the growing stalk, a layer of the surface ectoderm pinches off, forming a transparent lens. The formation of lens from the surface ectoderm requires induction by the underlying neural ectoderm.

This was shown by transplantation experiments performed by Spemann. When the optic stalks of the two eyes have just started to project from the brain prior to lens formation, one of the budding stalks can be removed and

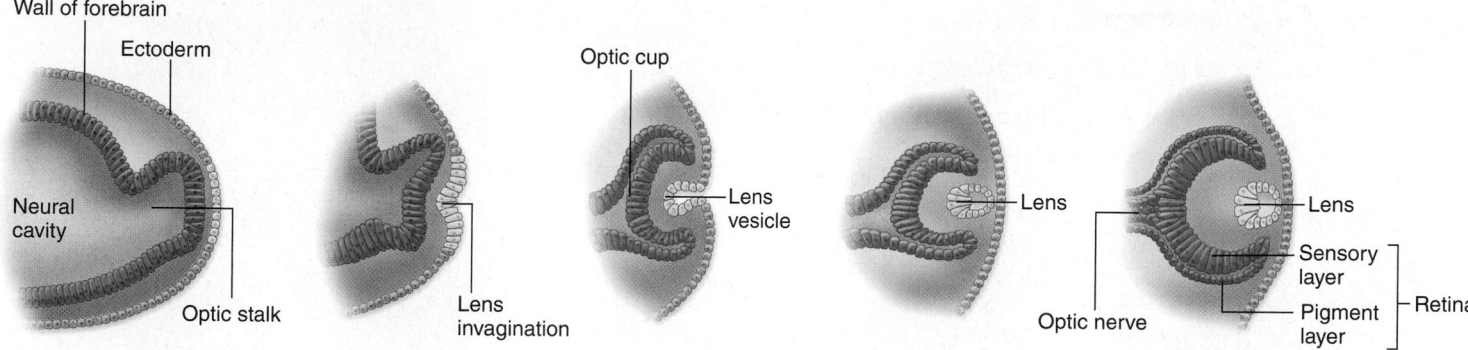

Figure 53.25 Development of the vertebrate eye by induction. An extension of the optic stalk grows until it contacts the surface ectoderm, where it induces a section of the ectoderm to pinch off and form the lens. Other structures of the eye develop from the optic stalk, with lens cells reciprocally inducing the formation of photoreceptors in the optic cup.

transplanted underneath surface ectoderm in a region that would normally develop into the epidermis of the skin (such as that of the belly). When this is done, a lens forms from belly ectoderm cells in the region above where the budding stalk was transplanted. This lens forms due to inductive signals from the underlying optic stalk.

Learning Outcomes Review 53.5

The Spemann-Mangold experiment showed that transplanted cells of the dorsal lip of the blastopore act as organizers stimulating development of a notochord. Hensen's node plays an equivalent role in vertebrates. By inhibiting BMP4, the organizer induces ectoderm to form neural tissue and mesoderm to form dorsal mesoderm. Primary inductions between germ layers lead to development of the vertebrate nervous system, whereas secondary inductions result in formation of structures such as the lens of the eye.

■ **How can the organizer function by inhibiting the action of other molecules?**

53.6 Human Development

Learning Outcomes

1. Describe the major developmental events in the first trimester.
2. Explain the role of the placenta.
3. Describe the hormonal control of the birth process.

Human development from fertilization to birth takes an average of 266 days, or about 9 months. This time is commonly divided into three periods called *trimesters*. We describe here the development of the embryo as it takes place during these trimesters. Later, we summarize the process of birth, nursing of the infant, and postnatal development.

During the first trimester, the zygote undergoes rapid development and differentiation

About 30 hr after fertilization, the zygote undergoes its first cleavage; the second cleavage occurs about 30 hr after that. By the time the embryo reaches the uterus, 6 to 7 days after fertilization, it has differentiated into a blastocyst. As mentioned earlier, the blastocyst consists of an inner cell mass, which will become the body of the embryo, and a surrounding layer of trophoblast cells (see figure 53.10).

The trophoblast cells of the blastocyst digest their way into the endometrial lining of the uterus in the process known as **implantation.** The blastocyst begins to grow rapidly and initiates the formation of the amnion and the chorion.

Development in the first month

During the second week after fertilization, the developing chorion and the endometrial tissues of the mother engage to form the placenta (figure 53.26). Within the placenta, the mother's blood and the blood of the embryo come into close proximity but do not mix. Gases are exchanged, however, and the placenta provides nourishment for the embryo, detoxifies certain molecules that may pass into the embryonic circulation, and secretes hormones. Certain substances, such as alcohol, drugs, and antibiotics, are not stopped by the placenta and pass from the mother's bloodstream into the embryo.

One of the hormones released by the placenta is human chorionic gonadotropin (hCG), which was discussed in chapter 52. This hormone is secreted by the trophoblast cells even before they become the chorion, and it is the hormone assayed in pregnancy tests. Human chorionic gonadotropin maintains the mother's corpus luteum. The corpus luteum, in turn, continues to secrete estradiol and progesterone, thereby preventing menstruation and further ovulations.

Gastrulation also takes place in the second week after fertilization, and the three germ layers are formed. Neurulation occurs in the third week. The first somites appear, which give rise to the muscles, vertebrae, and connective tissues. By the end of the third week, over a dozen somites are evident, and the blood vessels and gut have begun to develop. At this point, the embryo is about 2 mm long.

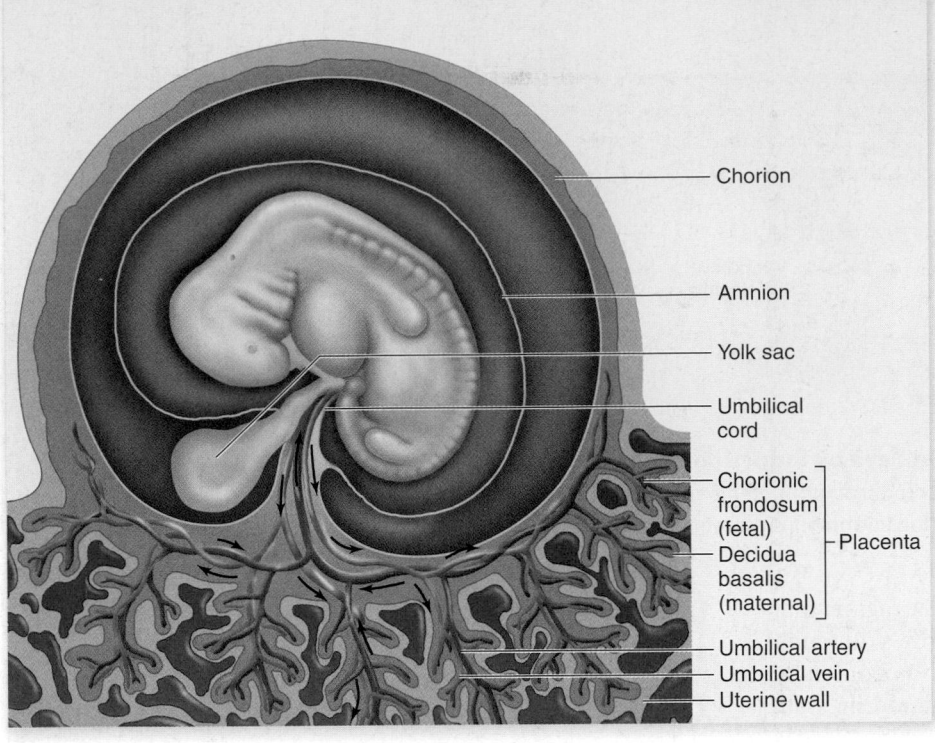
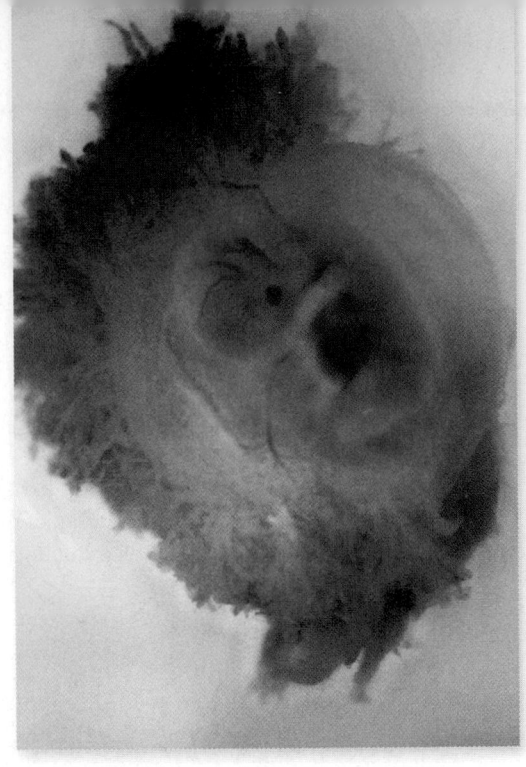

a.

b.

Figure 53.26 Structure of the placenta. *a.* The placenta contains a fetal component, the chorionic frondosum, and a maternal component, the decidua basalis. Deoxygenated fetal blood from the umbilical arteries (shown in blue) enters the placenta, where it picks up oxygen and nutrients from the mother's blood. Oxygenated fetal blood returns in the umbilical vein (shown in red) to the fetus. ***b.*** Note that the 7-week embryo is surrounded by a fluid-filled amniotic sac.

Organogenesis begins during the fourth week (figure 53.27*a*). The eyes form. The tubular heart develops its four chambers and starts to pulsate rhythmically, as it will for the rest of the individual's life. At 70 beats per minute, the heart is destined to beat more than 2.5 billion times during a lifetime of 70 years. Over 30 pairs of somites are visible by the end of the fourth week, and the arm and leg buds have begun to form. The embryo has increased in length to about 5 mm. Although the developmental scenario is now far advanced, many women are still unaware they are pregnant at this stage. Most spontaneous abortions (miscarriages), which frequently occur in the case of a defective embryo, occur during this period.

The second month

Organogenesis continues during the second month (figure 53.27*b*). The miniature limbs of the embryo assume their adult shapes. The arms, legs, knees, elbows, fingers, and toes can all be seen—as well as a short bony tail. The bones of the embryonic tail, an evolutionary reminder of our past, later fuse to form the coccyx.

Within the abdominal cavity, the major organs, including the liver, pancreas, and gallbladder, become evident. By the end of the second month, the embryo has grown to about 25 mm in length, weighs about 1 g, and begins to look distinctly human. The ninth week marks the transition from embryo to fetus. At this time, all of the major organs of the body have been established in their proper locations.

The third month

The nervous system develops during the third month, and the arms and legs start to move (figure 53.27*c*). The embryo begins to show facial expressions and carries out primitive reflexes such as the startle reflex and sucking.

At around 10 weeks, the secretion of hCG by the placenta declines, and the corpus luteum regresses as a result. However, menstruation does not occur because the placenta itself secretes estradiol and progesterone (figure 53.28).

The high levels of estradiol and progesterone in the blood during pregnancy continue to inhibit the release of FSH and LH, thereby preventing ovulation. They also help maintain the uterus and eventually prepare it for labor and delivery, and they stimulate the development of the mammary glands in preparation for lactation after delivery.

During the second trimester, the basic body plan develops further

Bones actively enlarge during the fourth month (figure 53.27*d*), and by the end of the month, the mother can feel the baby kicking. By the end of the fifth month, the rapid heartbeat of the fetus can be heard with a stethoscope, although it can also be detected as early as 10 weeks with a fetal monitor.

Growth begins in earnest in the sixth month; by the end of that month, the fetus weighs 600 g (1.3 lb) and is over 300 mm (1 ft) long. Most of its prebirth growth is still to come, however. The fetus cannot yet survive outside the uterus without special medical intervention.

During the third trimester, organs mature to the point at which the baby can survive outside the womb

The third trimester is predominantly a period of growth and maturation of organs. The weight of the fetus doubles several

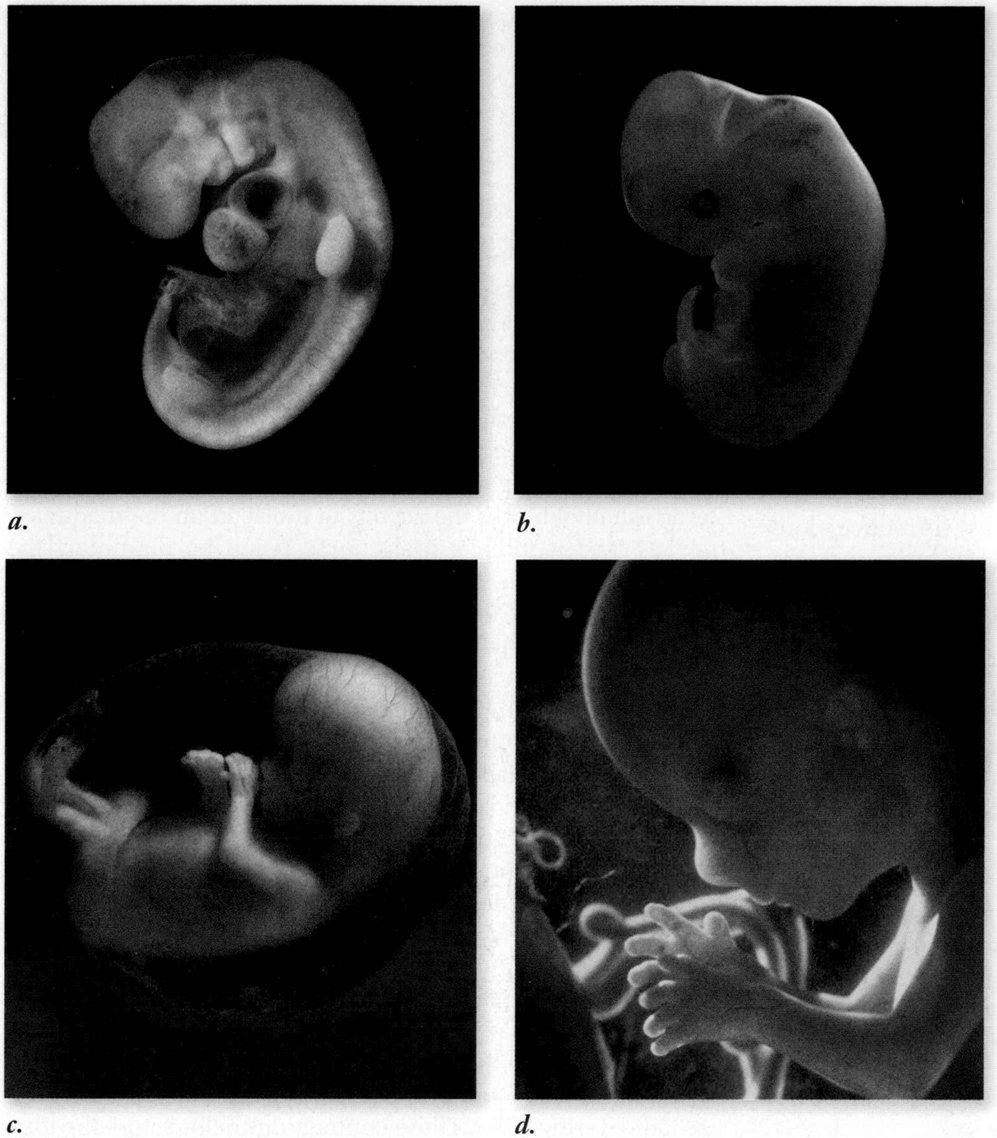

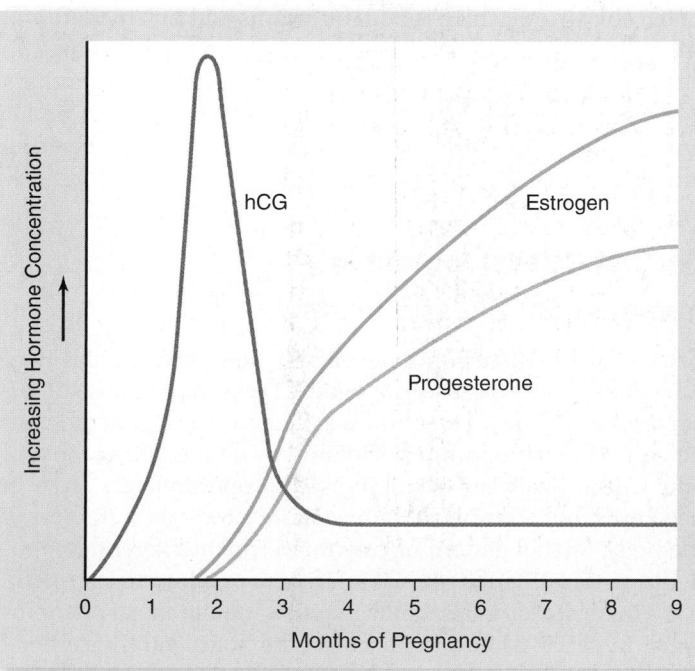

Figure 53.27 The developing human. *(a)* 4 weeks, *(b)* end of 5th week, *(c)* 3 months, and *(d)* 4 months.

a.

b.

c.

d.

times, but this increase in bulk is not the only kind of growth that occurs. Most of the major nerve tracts in the brain, as well as many new neurons (nerve cells), are formed during this period. Neurological growth is far from complete when birth takes place, however. If the fetus remained in the uterus until its neurological development was complete, it would grow too large for safe delivery through the pelvis. Instead, the infant is

Figure 53.28 Hormonal secretion by the placenta. The placenta secretes human chorionic gonadotropin (hCG), which peaks in the second month and then declines. After 5 weeks, it secretes increasing amounts of estrogen and progesterone.

? **Inquiry question** The high levels of estradiol and progesterone secreted by the placenta prevent ovulation and thus formation of any additional embryos during pregnancy. What would be the expected effect of these high hormone levels in the absence of pregnancy?

born as soon as the probability of its survival is high, and its brain continues to develop and produce new neurons for months after birth.

Critical changes in hormones bring on birth

In some mammals, changing hormone levels in the developing fetus initiate the process of birth. The fetuses of these mammals have an extra layer of cells in their adrenal cortex, which secrete corticosteroids that induce the uterus of the mother to manufacture prostaglandins. Prostaglandins trigger powerful contractions of the uterine smooth muscles.

In humans, fetal secretion of cortisol increases during late pregnancy, which appears to stimulate estradiol secretion by the placenta. The mother's uterus releases prostaglandins, possibly as a result of the high levels of estradiol secreted by the placenta. Estradiol also stimulates the uterus to produce more oxytocin receptors, and as a result, the uterus becomes increasingly sensitive to oxytocin.

Prostaglandins begin the uterine contractions, but then sensory feedback from the uterus stimulates the release of oxytocin from the mother's posterior-pituitary gland. Working together, oxytocin and prostaglandins further stimulate uterine contractions, forcing the fetus downward (figure 53.29). This positive feedback mechanism accelerates during labor. Initially, only a few contractions occur each hour, but the rate eventually increases to one contraction every 2 to 3 min. Finally, strong contractions, aided by the mother's voluntary pushing, expel the fetus, which is now a newborn baby, or *neonate.*

After birth, continuing uterine contractions expel the placenta and associated membranes, collectively called the *afterbirth.* The umbilical cord is still attached to the baby, and

to free the newborn, a doctor or midwife clamps and cuts the cord. Blood clotting and contraction of muscles in the cord prevent excessive bleeding.

Nursing of young is a distinguishing feature of mammals

Milk production, or *lactation,* occurs in the alveoli of mammary glands when they are stimulated by the anterior-pituitary hormone prolactin. Milk from the alveoli is secreted into a series of alveolar ducts, which are surrounded by smooth muscle and lead to the nipple.

During pregnancy, high levels of progesterone stimulate the development of the mammary alveoli, and high levels of estradiol stimulate the development of the alveolar ducts. However, estradiol blocks the actions of prolactin on the mammary glands, and it inhibits prolactin secretion by promoting the release of prolactin-inhibiting hormone from the hypothalamus. During pregnancy, therefore, the mammary glands are prepared for, but prevented from, lactating. The growth of mammary glands is also stimulated by the placental hormones human chorionic somatomammotropin, a prolactin-like hormone, and human somatotropin, a growth hormone-like hormone.

When the placenta is discharged after birth, the concentrations of estradiol and progesterone in the mother's blood decline rapidly. This decline allows the anterior-pituitary gland to secrete prolactin, which stimulates the mammary alveoli to produce milk. Sensory impulses associated with the baby's suckling trigger the posterior-pituitary gland to release oxytocin. Oxytocin stimulates contraction of the smooth muscle surrounding the alveolar ducts, thus causing milk to be ejected by the breast. This pathway is known as the *milk let-down reflex,* and it is found in other mammals as well. The secretion of oxytocin during lactation also causes some uterine contractions, as it did during labor. These contractions help restore the tone of uterine muscles in mothers who are breast-feeding.

The first milk produced after birth is a yellowish fluid called colostrum, which is both nutritious and rich in maternal antibodies. Milk synthesis begins about 3 days following the birth and is referred to as the milk "coming in." Many mothers nurse for a year or longer. When nursing stops, the accumulation of milk in the breasts signals the brain to stop secreting prolactin, and milk production ceases.

Postnatal development in humans continues for years

Growth of the infant continues rapidly after birth. Babies typically double their birth weight within 2 months. Because different organs grow at different rates and cease growing at different times, the body proportions of infants are different from those of adults. The head, for example, is disproportionately large in newborns, but after birth it grows more slowly than the rest of the body. Such a pattern of growth, in which different components grow at different rates, is referred to as **allometric growth.**

In most mammals, brain growth is mainly a fetal phenomenon. In chimpanzees, for instance, the brain and the cerebral

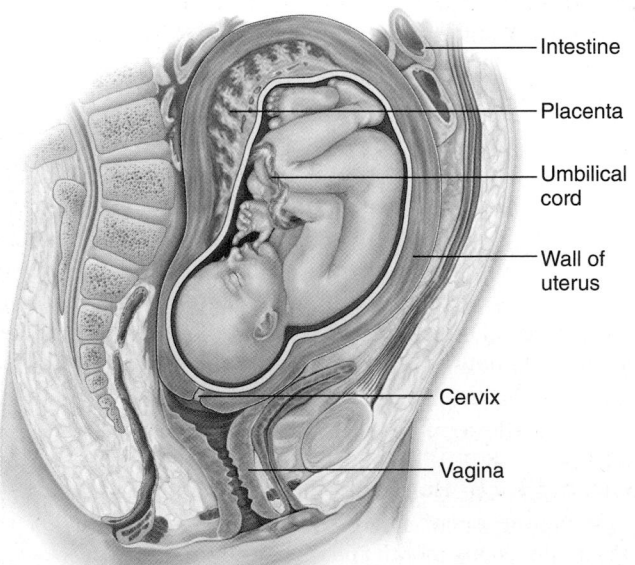

Figure 53.29 Position of the fetus just before birth.
A developing fetus causes major changes in a woman's anatomy. The stomach and intestines are pushed far up, and considerable discomfort often results from pressure on the lower back. In a normal vaginal delivery, the fetus exits through the cervix, which must dilate (expand) considerably to permit passage.

Labels: Intestine, Placenta, Umbilical cord, Wall of uterus, Cervix, Vagina

portion of the skull grow very little after birth, whereas the bones of the jaw continue to grow. As a result, the head of an adult chimpanzee looks very different from that of a fetal or infant chimpanzee. In human infants, by contrast, the brain and cerebral skull grow at the same rate as the jaw. Therefore, the jaw–skull proportions do not change after birth, and the head of a human adult looks very similar to that of a human fetus or infant.

The fact that the human brain continues to grow significantly for the first few years of postnatal life means that adequate nutrition and a safe environment are particularly crucial during this period for the full development of a person's intellectual potential.

Learning Outcomes Review 53.6

The critical stages of human development occur in the first trimester of gestation; the subsequent 6 months involve growth and maturation. Growth of the brain is not complete at birth and must be completed postnatally. Hormones in the mother's blood maintain the nutritive uterine environment for the developing fetus; changes in hormone secretion and levels stimulate birth (prostaglandins and oxytocin) and lactation (oxytocin and prolactin).

■ **Why are teratogens (agents that cause birth defects) most potent in the first trimester?**

Chapter Review

53.1 Fertilization

A sperm must penetrate to the plasma membrane of the egg for membrane fusion to occur.

The sperm's acrosome releases digestive enzymes to penetrate the egg's external layers (figure 53.1). Fusion with the egg's membrane allows the sperm nucleus to pass into the egg's cytoplasm.

Membrane fusion activates the egg.

Fusion of membranes triggers egg activation by the release of calcium (figure 53.2). Blocks to polyspermy include changes in membrane potential and alterations to the external coat of the egg. Upon egg activation, meiosis is completed (figures 53.4 and 53.5).

The fusion of nuclei restores the diploid state.

Fertilization is complete when the haploid sperm nucleus fuses with the haploid egg nucleus.

53.2 Cleavage and the Blastula Stage

The blastula is a hollow mass of cells.

Cleavage is a rapid series of cell divisions that produces blastomeres, which form a hollow ball of cells called a blastula.

Cleavage patterns are highly diverse and distinctive.

Cleavage patterns are primarily influenced by the amount of yolk (table 53.2). With little or no yolk, cleavage is holoblastic (involving the whole egg); where more yolk is present, cleavage is meroblastic (involving the blastodisc only). Cleavage in mammals is holoblastic.

Blastomeres may or may not be committed to developmental paths.

In many animals, unequal segregation of cytoplasmic determinants commits each blastomere to a different path. Mammals exhibit regulative development in which the fate of early blastomeres is not predetermined.

53.3 Gastrulation

Gastrulation produces the three germ layers.

During gastrulation the three germ layers differentiate: endoderm, ectoderm, and mesoderm (table 53.3). Cells move during gastrulation using a variety of cell shape changes.

Gastrulation patterns also vary according to the amount of yolk.

The amount of yolk also influences cell movement. In frogs, a layer of cells involutes through the dorsal lip of the blastopore. In birds, surface cells migrate through the primitive streak. Mammalian gastrulation is similar to that of birds.

Extraembryonic membranes are an adaptation to life on dry land.

The yolk sac, amnion, chorion, and allantois prevent desiccation and nourish and protect the developing embryo (figure 53.15).

53.4 Organogenesis

Changes in gene expression lead to cell determination.

A cell's location in the developing embryo often determines its fate. Differentiation can be established by inheritance of cytoplasmic determinants and by interactions with other cells (induction).

Development of selected systems in **Drosophila** *illustrates organogenesis.*

The development of salivary glands, the dorsal vessel, and tracheae all demonstrate the action of gene expression on development.

In vertebrates, organogenesis begins with neurulation and somitogenesis (figures 55.18–55.20).

Neurulation is the formation of the neural tube from ectoderm near the notochord; somitogenesis is the establishment of mesoderm into units called somites.

Migratory neural crest cells differentiate into many cell types.

Neural crest cells migrate widely to become connective tissue, nerve and glial cells, melanocytes, sensory neurons, and other cells.

Neural crest derivatives are important in vertebrate evolution.

Many of the unique adaptations of vertebrates have arisen from neural crest cells (figure 53.21).

53.5 Vertebrate Axis Formation

The Spemann organizer determines dorsal–ventral axis.

Organizers are a cluster of cells that produce gradients of diffusible signal molecules, conveying positional information to other cells.

Maternally encoded dorsal determinants activate Wnt signaling.

Turning on the Wnt pathway activates organizer specification.

Signaling molecules from the Spemann organizer inhibit ventral development.

Morphogens can either activate or inhibit development along a certain path. The Spemann organizer induces formation of the dorsum by inhibiting ventral development (figure 53.24).

Evidence indicates that organizers are present in all vertebrates.

Cells at the anterior edge of the primitive streak, termed Hensen's node, function similarly to the Spemann organizer.

Induction can be primary or secondary.

Primary induction occurs between the three germ layers; secondary induction occurs between already determined tissues.

53.6 Human Development

During the first trimester, the zygote undergoes rapid development and differentiation.

Implantation of the blastocyst occurs at the end of the first week of pregnancy. During the second week, the embryonic chorion and the mother's endometrial tissues form the placenta, and gastrulation occurs. Organogenesis begins during the fourth week. The eighth week marks the transition from embryo to fetus.

During the second trimester, the basic body plan develops further.

During the third trimester, organs mature to the point at which the baby can survive outside the womb.

Critical changes in hormones bring on birth.

Birth is initiated by secretions of steroids from the fetal adrenal cortex that induce prostaglandins, which cause contractions.

Nursing of young is a distinguishing feature of mammals.

Nursing involves a neuroendocrine reflex, causing the release of oxytocin and the milk let-down response.

Postnatal development in humans continues for years.

Postnatal development continues with different organs growing at different rates—called allometric growth.

Review Questions

UNDERSTAND

1. Which of the following events occur immediately after fertilization?
   a. Egg activation
   b. Polyspermy defense
   c. Cytoplasm changes
   d. All of these occur after fertilization.

2. Which of the following plays the greatest role in determining how cytoplasmic division occurs during cleavage?
   a. Number of chromosomes
   b. Amount of yolk
   c. Orientation of the vegetal pole
   d. Sex of the zygote

3. Gastrulation is a critical event during development. Why?
   a. Gastrulation converts a hollow ball of cells into a bilaterally symmetrical structure.
   b. Gastrulation causes the formation of a primitive digestive tract.
   c. Gastrulation causes the blastula to develop a dorsal–ventral axis.
   d. All of these are significant events that occur during gastrulation.

4. Gastrulation in a mammal would be most similar to gastrulation in
   a. a gecko.
   b. a tuna.
   c. an eagle.
   d. no other species; mammalian gastrulation is unique.

5. Somites
   a. begin forming at the tail end of the embryo and then move forward in a wavelike fashion.
   b. are derived from endoderm.
   c. develop into only one type of tissue per somite.
   d. may vary in number from one species to the next.

6. Of the following processes, which occurs last?
   a. Cleavage   c. Gastrulation
   b. Neurulation   d. Fertilization

7. Which of the following would qualify as a secondary induction?
   a. The formation of the lens of the eye due to induction by the neural ectoderm
   b. Differentiation during neurulation by the dorsal ectoderm and mesoderm
   c. Both of the choices are correct.
   d. Neither of the choices is correct.

APPLY

1. Your cousin just had twins. She tells you that twinning occurs when two sperm fertilize the same egg. You reply that
   a. yes, she is right, that is the most common source of twinning.
   b. no, only one sperm survives passage through the uterine cervix, so two sperm are never present at fertilization.
   c. no, cortical granules are used to prevent additional sperm penetration.
   d. no, twinning occurs when unfertilized eggs divide spontaneously and thus is parthenogenic in nature.

2. In the Spemann experiment, when the dorsal lip is transplanted, the recipient embryo then has a second source of molecules that
    a. specifies ventral fate.
    b. inhibits the molecules that specify ventral fate.
    c. specifies dorsal fate.
    d. inhibits the molecules that specify dorsal fate.

3. Suppose that a burst of electromagnetic radiation were to strike the blastomeres of only the animal pole of a frog embryo. Which of the following would be most likely to occur?
    a. A change or mutation relevant to the epidermis or skin
    b. A switching of the internal organs so that reverse orientation (left/right) occurs along the midline of the body
    c. The migration of the nervous system to form outside of the body
    d. Failure of the reproductive system to develop

4. Your Aunt Ida thinks that babies can stimulate the onset of their own labor. You tell her that
    a. among mammals the onset of labor has been most closely linked to a change in the phases of the Moon.
    b. it is the mother's circadian clock that determines the onset of labor.
    c. body weight determines the onset of labor.
    d. changes in fetal hormone levels can affect the onset of labor.

5. Drug or alcohol exposure during which of the following stages is most likely to have a profound effect on the neural development of the fetus?
    a. Preimplantation   c. Second trimester
    b. First trimester   d. Third trimester

6. Axis formation in amniotic embryos could be affected by
    a. mutations in cells in the dorsal lip of the blastopore.
    b. mutations in cells in the primitive streak.
    c. Both of the choices are correct.
    d. Neither of the choices is correct.

SYNTHESIZE

1. Suppose you discover a new species whose development mechanisms have not been documented before. How could you determine at what stage the cell fate is determined?

2. You look up from your studying to see your dog, Fifi, acting silly again. Using this as a teachable moment, compare and contrast the homeoboxes in your dog and the fruit fly she just ate.

3. Why doesn't a woman menstruate while she is pregnant?

4. Spemann and Mangold were able to demonstrate that some cells act as "organizers" during development. What types of cells did they use? How did they determine that these cells were organizers?

ONLINE RESOURCE

www.ravenbiology.com

Understand, Apply, and Synthesize—enhance your study with animations that bring concepts to life and practice tests to assess your understanding. Your instructor may also recommend the interactive eBook, individualized learning tools, and more.

Chapter 54

Behavioral Biology

Chapter Contents

Introduction

The study of behavior is at the center of many disciplines of biology. Observing behavior provides important insights into the workings of the brain and nervous system, the influences of genes and the environment, when and how animals reproduce, and how they adapt to their environment. Behavior is shaped by natural selection and is controlled by internal mechanisms involving genes, hormones, neurotransmitters, and neural circuits. In this chapter, we explore how behavioral biology integrates approaches from several branches of biological science to provide a detailed understanding of the mechanisms that underscore behavior and its evolution.

Observing animal behavior and making inferences about what one sees is at once simple and profound. Behavior is what an animal does. It is the most immediate way an animal responds to its environment by tracking environmental cues and signals such as odors, sounds, or visual signals associated with food, predators, or mates. Behavior also concerns thinking and cognition, monitoring one's social environment, and making decisions as to whether or not to cooperate or act altruistically. Behavior allows animals to survive and reproduce and is thus critical to the evolutionary process. The work of behavioral biologists has provided important insights into animal behavior, including the very meaning of human behavior.

Behavior can be analyzed in terms of mechanisms and evolutionary origin

Why does an animal behave in a particular way? Consider hearing a bird sing. We could ask how it vocalizes or determine the time of the year it sings most frequently. We could also ask about the function of the song, that is, ask why it sings. Answers to questions about how birds sing consider the role of internal factors such as hormones and nerve cells and other physiological processes. Such questions concern proximate causation: the mechanisms that produce the behavior. To analyze the proximate cause of birdsong, we could measure hormone levels or study the development of brain regions and neural circuits associated with singing. For example, a male songbird may sing during the breeding season because of an increased level of the steroid sex hormone testosterone, which binds to receptors in the brain and triggers the production of song. Additionally, neural connections between the brain and the syrinx (the bird's vocal organ) must develop to allow songs to be produced. These explanations describe the proximate cause of birdsong.

Asking about the function of a behavior (once again, birdsong) is to ask why it evolved. To answer this question, we would determine how it influenced survival or reproductive success. A male bird sings to defend a territory from other males and to attract a female with which to reproduce. This is the ultimate, or evolutionary, explanation for the male's vocalization. Now we can understand its ultimate causation, or adaptive value. Researchers often study behavior from both perspectives to fully appreciate its mechanisms and ecological function, and thus its role in evolution. Behavior can be analyzed at four levels: (1) physiology (how it is influenced by hormones, nerve cells, and other internal factors); (2) ontogeny (how it develops in an individual); (3) phylogeny (its origin in groups of related species); and (4) adaptive significance (its role in survival and fitness). We'll begin by tracing the history of the study of mechanisms of behavior by focusing on the work of ethologists—biologists who first began to study behavior at the turn of the 20th century.

Ethology emphasizes the study of instinct and its origins

Ethology is the study of the natural history of behavior, with an emphasis on behaviors that form an animal's instincts, or programmed behaviors. Ethologists observed that individuals of a given species behaved in stereotyped ways, showing the same pattern of behavior in response to a particular stimulus. Because their behavior seemed reflexive, they considered it to be instinctive, or *innate.* Behaviors were thought to be programmed by the nervous system, which in turn was designed by genes, and responses would occur without experience. Ethologists based their instinct model on observations and experiments of simple behaviors such as egg retrieval by geese. Geese incubate their eggs in a nest. If an egg falls out of the nest, the goose will roll the egg back into the nest with a side-to-side motion of its neck while the egg is tucked beneath its bill (figure 54.1). Even if the egg is removed during retrieval, the goose will still complete the egg-retrieval sequence, as if driven by a program activated by the initial sight of the egg outside the nest.

This example is one paradigm of instinct theory and illustrates the way ethologists conceptualized the mechanisms of behavior. Egg retrieval behavior is triggered by a *key stimulus* (sometimes called a *sign stimulus*); this is the egg out of the nest. Early ethologists thought the nervous system regulated behavior via the *innate releasing mechanism,* a neural circuit involved in the perception of the key stimulus and the triggering of a motor program, the *fixed action pattern,* in this case the act of guiding the egg back to the nest. Ethologists generalized

Sign stimulus

Fixed action pattern

Figure 54.1 Innate egg-rolling response in geese. The series of movements used by a goose to retrieve an egg is a fixed action pattern. Once it detects the sign stimulus (in this case, an egg outside the nest), the goose goes through the entire set of movements: It will extend its neck toward the egg, get up, and roll the egg back into the nest with a side-to-side motion of its neck while the egg is tucked beneath its bill.

that the key stimulus is a cue or signal in the environment that initiates neural events that cause behavior. The innate releasing mechanism involves the sensory apparatus that detects the signal and the neural circuit controlling muscles to generate the fixed action pattern.

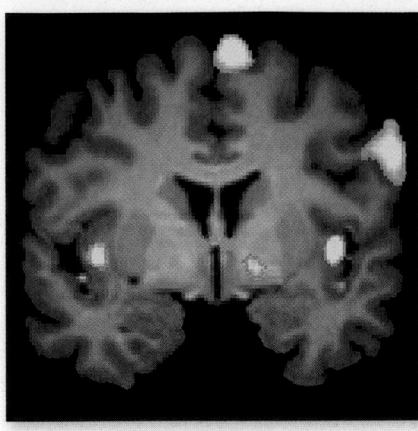

Figure 54.2
Functional magnetic resonance imaging (fMRI). MRIs reveal neural activity in specific regions of the brain. In this case, activity in part of the brain called the nucleus accumbens is associated with viewing images of food.

Learning Outcomes Review 54.1

Proximate causation of behavior involves the immediate mechanisms that bring about an action; ultimate causation refers to the adaptive value of a behavior. Ethology is the study of the nature of behavior, emphasizing instinct and the regulation of behavior by internal factors such as genes, nerve cells, and hormones. Instinctual behavior is thought to occur when key stimuli affect innate releasing mechanisms, which trigger fixed motor programs.

■ *Why is it important to understand the phylogeny (evolutionary origins) of behavior?*

54.2 *Nerve Cells, Neurotransmitters, Hormones, and Behavior*

Learning Outcomes

1. *Relate the structure of neural circuits to their function.*
2. *Describe the role of hormones and neurotransmitters in behavior.*

Although early ethologists had little understanding of neurobiology, they hypothesized elements of the nervous system (the innate releasing mechanism) controlled behavior. Today, neuroethologists—researchers who examine the neurobiology of behavior—can describe in detail how information in the environment is processed by sensory cells and how nerve impulses are transmitted to other neurons and muscles to form neural circuits that regulate behaviors important to survival. Behavior reflects the organization of the peripheral and central nervous system, and studying behavior can help us understand how neurons function individually and in combination with other neurons in circuits (see chapters 43 and 44).

Behaviors that must occur rapidly, like those used to capture prey or flee predators, involve neural mechanisms that enable such functions. Some moths have an earlike sensory organ equipped with sensory neurons designed to detect the ultrasonic cries of bats; the neuron fires so rapidly that the moth can initiate evasive action before the bat is even aware of the moth's presence. Specialized cells in a frog's retina detect moving objects like insects and release the tongue in fractions of a second once suitable prey is sighted. Likewise, the jaws of a predatory ant snap shut when prey trigger sensory hairs between the mandibles. Rapid responses to predators or prey often involve large nerve cell axons that can quickly transmit impulses to muscles. In the example of "trap jaw" ants, large axons of the mandibular motor neuron—the fastest neuron yet identified—fire nerve impulses that close the jaws in only 33 msec. Neural

circuits that enable quick responses often are made up of few sensory and motor neurons, and their connecting nerve cells.

Behavioral biologists examine the relationship of hormones to behavior to understand the endocrine mechanisms that are the foundation of reproduction, parental care, aggression, and stress (see chapter 45). In this way, the effects of the steroid sex hormones estrogen and testosterone on the behavior of vertebrates have been determined. Testosterone in the male, for example, regulates territorial behavior and courtship, whereas estrogen in the female controls her mating behavior. Glucocorticoid hormones are involved in stress.

Neuroscientists may measure levels of neurotransmitters such as serotonin and dopamine in the nervous system or blood and associate these chemicals with behavior (see chapter 43). These chemicals are released by nerve cells and can affect activity in different brain regions. Serotonin has been shown to influence aggression in an incredibly wide range of animals including lobsters, mice, and humans. Researchers may inject a neurotransmitter or pharmacologically change its level in the brain to examine how it affects behavior.

The techniques of neuroethology include identifying and mapping individual neurons, their dendrites and connections to other neurons, and studying how their impulses and neurochemicals regulate behavior. Today, techniques such as functional magnetic resonance imaging (fMRI) are generating exciting data on the specialized functions of different regions of the human brain. One striking example concerns how the brain responds to images of food (figure 54.2). In contrast to expectation, the brain's response does not occur in the visual cortex, the region associated with object recognition, but in a circuit in the nucleus accumbens in the forebrain, normally involved in reward and pleasure.

Learning Outcomes Review 54.2

Instinctive behaviors appear to involve programmed circuits in the nervous system that are likely to be genetically controlled. Research in neuroethology supports the instinct concept of behavior by describing the organization of neural circuits governing behavior. Chemical signals provided by hormones and by neurotransmitters such as serotonin and dopamine cause behaviors to occur.

■ *If a male songbird is injected with testosterone two weeks earlier than when these birds normally start to sing in the spring, what would you expect to happen?*

Learning Outcomes

1. *Discuss the types of studies that have provided evidence to link genes and behavior.*
2. *Explain how single genes can influence behavior.*
3. *Describe the role of genes in complex behaviors such as aggression, parental care, and pair bonding.*

Instinct theory assumed that genes play a role in behavior, but ethologists did not conclusively demonstrate the role genes can play. The study of genes and behavior has often been highly controversial, as ethologists and social scientists engaged in a seemingly endless debate over whether behavior is determined more by an individual's genes (nature) or by its learning and experience (nurture). One problem with this nature/nurture controversy is that the question is framed as an "either/or" proposition, which fails to consider that both instinct and experience can have significant roles, often interacting in complex ways to shape behavior.

Behavioral genetics deals with the contribution that heredity makes to behavior. It is obvious that genes, the units of heredity, are passed from one generation to the next and guide the development of the nervous system and potentially the behavioral responses it regulates. But animals may also develop in a rich social environment and have experiences that guide behavior. The importance of "nature" and "nurture" to behavior can be seen by first reviewing the history of studies in behavioral genetics and next examining the importance of experience and development. We'll then consider their interaction.

Artificial selection, hybrid, and twin studies link genes and behavior

Pioneering research indicated that behavioral differences among individuals result from genetic differences. Research on a variety of animals demonstrated that hybrids showed behaviors involved in nest building and courtship that were intermediate between those of parents. These early efforts to define the role of genes in behavior demonstrated that behavior can have a heritable component, but fell short of identifying the genes involved. With the development of molecular biology, far greater precision was added to the analysis of the genetics of behavior.

Learning itself can be influenced by genes. In one classic study, rats had to find their way through a maze of blind alleys and only one exit, where a reward of food awaited them. Some rats quickly learned to zip through the maze to the food, making few mistakes, but other rats made more errors in learning the correct path. Researchers bred rats that made few errors with one another to establish a "maze-bright" group, and error-prone rats were interbred, forming a "maze-dull" group. Offspring in each group were then tested for their maze-learning ability. The offspring of maze-bright rats learned to negotiate the maze with fewer errors than their parents, while the offspring of maze-dull parents performed more poorly. Repeating this artificial selection method for several generations led to two behaviorally distinct types of rat with very different maze-learning abilities (figure 54.3). This type of study suggests the ways in which natural selection could shape behavior over time, making genes for certain abilities more prevalent.

The interaction of genes and the environment can be seen in humans by comparing the behavior of identical twins (who are genetically the same), raised in the same environment or separated at birth and raised apart in different environments. Data on human twins raised together or raised apart allows researchers to determine whether similarities in behavior result from their genetic similarity or from shared environmental experiences. Twins studies indicate many similarities in a wide range of personality traits even though twins were raised in very different environments. On the other hand, these studies show that some traits, such as antisocial behavior, result from a combination of genes and experience during childhood. These studies indicate that genetics plays a role in behavior even in humans, although the relative importance of genetics versus environment is still debated.

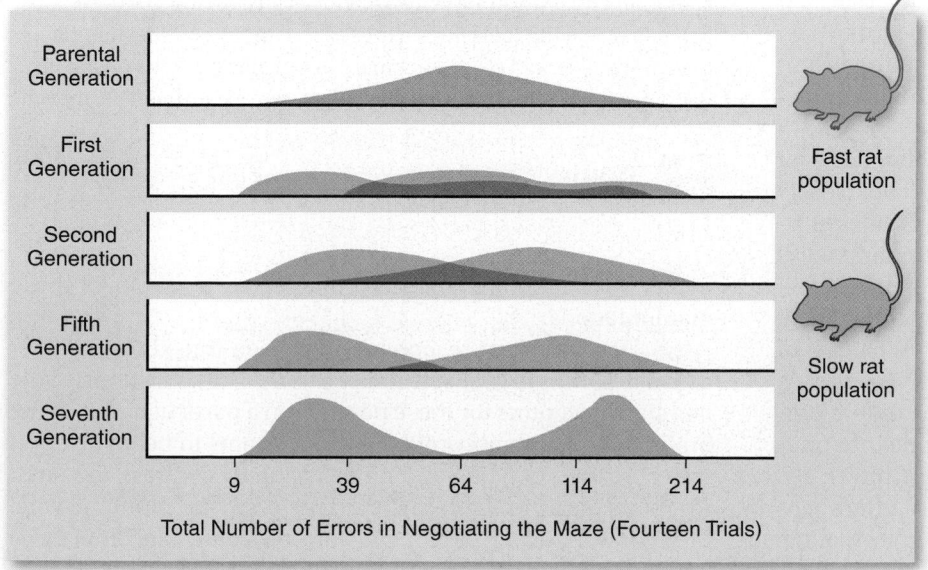

Figure 54.3 **The genetics of learning.** Rats that made the fewest errors in the parental population were interbred to select for rats that had improved maze-learning ability (green), and rats that made the most errors were interbred to select for rats that were error-prone (red).

? Inquiry question What would happen if, after the seventh generation, rats were randomly assigned mates regardless of their ability to learn the maze?

🔍 Data analysis By approximately how much has the mean number of errors changed from the parental population to the seventh generation of the fast rat population? What type of selection does this illustrate?

Some behaviors appear to be controlled by a single gene

Artificial selection, hybrid, and twin studies indicated a role for genes in behavior, but more recent work has taken advantage of advances in molecular biology and identified the actual genes involved. Fruit flies, *Drosophila*, have traditionally provided a useful model system in which the effect of single genes have been identified. Single genes have also been shown to influence behavior in animals ranging from mice to humans.

In fruit flies, individuals that possess alternative alleles for a particular gene differ greatly in their feeding behavior as larvae: Larvae with one allele move around a great deal as they eat, whereas individuals with the alternative allele move hardly at all. A wide variety of experimentally induced mutations at other genes affect courtship behavior in males and females. For example, *fru* is a regulatory gene whose transcription products govern the design of the courtship center of the fruit fly brain. This gene turns on other genes involved in the neural circuitry of courtship.

Single genes in mice are associated with spatial memory and parenting. For example, some mice with a particular mutation have trouble remembering recently learned information about where objects are located. This is apparently because they lack the ability to produce the enzyme α-calcium-calmodulin-dependent kinase II, which plays an important role in the functioning of the hippocampus, a part of the brain important for spatial learning.

Single genes can play a role even in behaviors as complex as maternal care. For example, the *fosB* gene determines whether female mice nurture their young effectively. Females with both *fosB* alleles disabled initially investigate their newborn babies, but then ignore them, in stark contrast to the caring and protective maternal behavior displayed by normal females (figure 54.4). The cause of this inattentiveness appears to result from a chain reaction. When mothers of new babies initially inspect them, information from their auditory, olfactory, and tactile senses is transmitted to the hypothalamus, where *fosB* alleles are activated. The *fosB* alleles produce a protein, which in turn activates other enzymes and genes that affect the neural circuitry of the hypothalamus. These modifications in the brain cause the female to behave maternally. If mothers lack the *fosB* alleles, this process is stopped midway. No protein is activated, the brain's neural circuitry is not rewired, and maternal behavior does not result. The "maternal instincts" of mice can thus be defined genetically!

Another fascinating example of the genetic basis of behavior concerns prairie and montane voles, two closely related species of North American rodents that differ profoundly in their social behavior. Male and female prairie voles form monogamous pair bonds and share parental care, whereas montane voles are promiscuous (meaning they mate with multiple partners and go their separate ways). The act of mating leads to the release of the neuropeptides vasopressin and oxytocin, and the response to these peptides differs dramatically in each species. Injection of either peptide into prairie voles leads to pair bonding even without mating. Conversely, injecting a chemical that blocks the action of these neuropeptides causes prairie voles not to form pair bonds after mating.

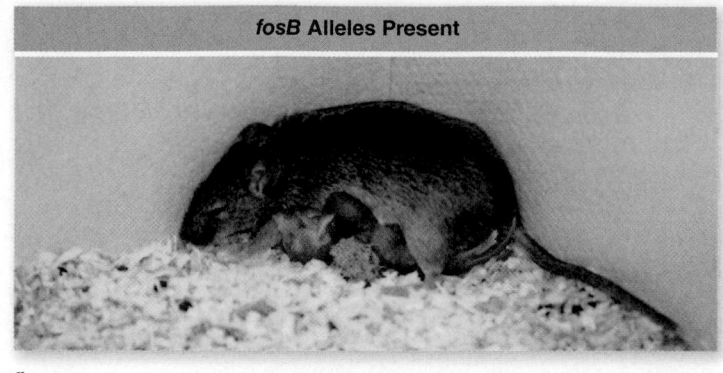

fosB Alleles Present

a.

fosB Alleles Inactivated

b.

c.   *d.*

Figure 54.4 Genetically caused defect in maternal care.
a. In mice, normal mothers take very good care of their offspring, retrieving them if they move away and crouching over them.
b. Mothers with the mutant *fosB* allele perform neither of these behaviors, leaving their pups exposed. *c.* Amount of time female mice were observed crouching in a nursing posture over offspring. *d.* Proportion of pups retrieved when they were experimentally moved.

? Inquiry question Why does the lack of *fosB* alleles lead to maternal inattentiveness?

By contrast, montane voles are unaffected by either of these manipulations.

These different responses have been traced to interspecific differences in brain structure (figure 54.5). The prairie vole has many receptors for these peptides in a particular part of the brain, the nucleus accumbens, which seems to be involved in the expression of pair-bonding behavior. By contrast, few such receptors occur in the same brain region in the montane vole. In laboratory experiments with prairie voles, blocking these

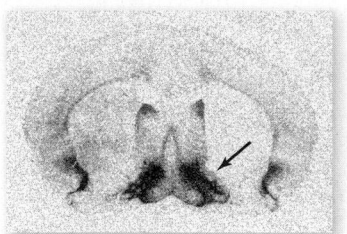

a. Prairie vole   *b.* Montane vole

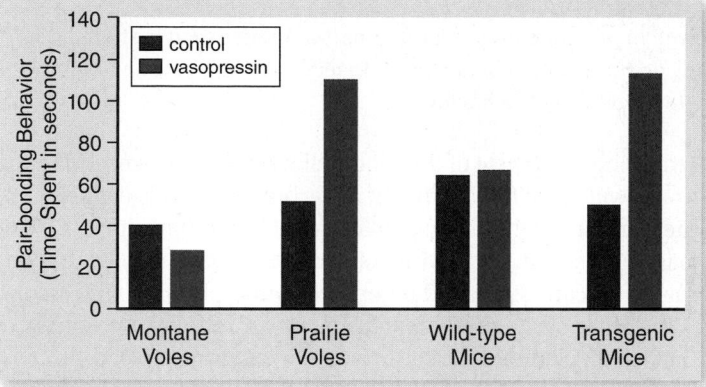

c.

Figure 54.5 Genetic basis of differences in pair-bonding behavior in two rodent species. *a.* and *b.* The prairie *(Microtus ochrogaster)* and montane *(M. montanus)* voles differ in the distribution of one type of vasopressin receptor in the brain. *c.* Transgenic laboratory mice created with the prairie-vole version of the receptor genes respond to injections of vasopressin by exhibiting heightened levels of pair-bonding behavior in 5-min trials compared with their response to a control injection. By contrast, normal wild-type mice show no increase in such behaviors.

 Data analysis Which group of mice shows the highest level of pair-bonding behavior under normal conditions? Which shows the least?

receptors tends to prevent pair-bonding, whereas stimulating them leads to pair-bonding behavior. The gene that codes for the peptide receptors has also been identified, and a difference in the DNA structure between the species has been discovered. To test the hypothesis that this genetic difference was responsible for the differences in behavior, scientists created transgenic laboratory mice with the prairie vole version of the gene, and sure enough, when injected with vasopressin, the transgenic mice exhibited pair-bonding behavior very similar to that of prairie voles, whereas normal mice showed no response (see figure 54.5).

The vasopressin receptor gene also varies in structure among primate species that vary in degree of pair bonding. In human males, the gene has recently been found to be associated with the strength of marital bonds and satisfaction in marriage.

Another example of single-gene effects involves the production of monoamine oxidases (MAOs), enzymes that degrade neurotransmitters such as serotonin and dopamine. Transgenic mice that lack MAOA (monoamine oxidase-A) are

highly aggressive. In humans, a single point mutation results in the lack of the ability to produce MAOA, resulting in antisocial behavior and violence. MAO abnormalities are also associated with mood disorders in humans.

Learning Outcomes Review 54.3

A relationship between genes and behavior has been demonstrated in many ways, including artificial selection experiments and studies on the effects of single genes. Genes can regulate behavior by producing molecular factors that influence the function of the nervous system; mutations altering these factors have been found to affect behavior.

■ *What would you infer about the role of genes in pair-bonding in prairie voles if you learned that males sometimes seek to copulate with females other than their own mate?*

54.4 Learning

Learning Outcomes

1. *Describe the mechanisms of learning.*
2. *Define learning preparedness.*
3. *Explain how instinct influences learning preparedness.*

Instincts can guide an animal's actions, but behavior can also develop from previous experiences, a process termed learning. Traditionally, psychologists studied the mechanisms of learning using laboratory rodents, but today both proximate and ultimate causes of learning are understood by integrating learning into an ecological and evolutionary framework.

Learning mechanisms include habituation and association

Habituation is a simple form of learning defined as a decrease in response to a repeated stimulus that has no positive or negative consequences. Initially, the stimulus may evoke a strong response, but the response declines with repeated exposure. For example, young birds see many types of objects moving overhead. At first, they may respond by crouching and remaining still. But frequently seen objects, such as falling leaves or members of their own species flying overhead, have no positive or negative consequence to the nestlings. Over time, the young birds may habituate to such stimuli and stop responding. Thus, habituation can be thought of as learning not to respond to a stimulus.

One ecological context in which habituation has adaptive value is prey defense. Birds that feed on insects search for suitable prey in a visually complex environment. Insects that have camouflaged bodies appear to be twigs or leaves, which are commonly encountered as birds search for prey. Because birds see these objects very frequently, they habituate to their

a.   *b.*   *c.*

Figure 54.6 **Learning what is edible.** Associative learning is involved in predator–prey interactions. *a.* A naive toad is offered a bumblebee as food. *b.* The toad is stung, and *(c)* subsequently avoids feeding on bumblebees or any other insects having black-and-yellow coloration. The toad has associated the appearance of the insect with pain and modifies its behavior.

appearance. Insects that look like twigs or leaves are therefore protected because they do not trigger an attack, and they survive to reproduce.

More complex forms of learning concern changes in behavior through an association between two stimuli or between a stimulus and a response. In associative learning, for example, (figure 54.6) a behavior is modified, or conditioned, through the association. The two major types of associative learning—classical conditioning and operant conditioning—differ in the way the associations are established. In **classical conditioning,** the paired presentation of two different kinds of stimuli causes the animal to form an association between the stimuli. Classical conditioning is also called **Pavlovian conditioning,** after the Russian psychologist Ivan Pavlov, who first described it.

Pavlov presented meat powder, an unconditioned stimulus, to a dog and noted that the dog responded by salivating, an unconditioned response. If an unrelated stimulus, such as the ringing of a bell, was repeatedly presented at the same time as the meat powder, the dog would soon salivate in response to the sound of the bell alone. The dog had learned to associate the unrelated sound stimulus with the meat powder stimulus. Its response to the sound stimulus was, therefore, conditioned, and the sound of the bell is referred to as a conditioned stimulus.

In **operant conditioning,** an animal learns to associate its behavioral response with a reward or punishment. American psychologist B. F. Skinner studied operant conditioning in rats by placing them in an apparatus that came to be called a "Skinner box." As the rat explored the box, it would occasionally press a lever by accident, causing a pellet of food to appear. Soon it learned to associate pressing the lever (the behavioral response) with obtaining food (the reward). This sort of trial-and-error learning is of major importance to most vertebrates. Learning provides flexibility that allows behavior to be fine-tuned to an environment.

Instinct governs learning preparedness

Psychologists once believed that any two stimuli could be linked through learning and that animals could be conditioned to perform any learnable behavior. This view has changed. Today, researchers believe that instinct guides learning by determining what type of information can be learned. Animals may have innate predispositions toward forming certain associations. For example, if a rat is offered a food pellet at the same time it is exposed to X-rays (which later produce nausea), the rat

remembers the taste of the food pellet but not its size, and in the future will avoid food with that taste, but will readily eat pellets of the same size if they have a different taste. Similarly, pigeons can learn to associate food with colors, but not with sounds. In contrast, they can associate danger with sounds, but not with colors.

These examples of learning preparedness demonstrate that what an animal can learn is biologically influenced—that is, learning is possible only within the boundaries set by evolution. Innate programs for learning have evolved because they lead to adaptive responses. In nature, food that is toxic to a rat is likely to have a particular taste; thus, it is adaptive to be able to associate a taste with a feeling of sickness that may develop hours later. The seed a pigeon eats may have a distinctive color that the pigeon can see, but it makes no sound the pigeon can hear.

An animal's ecology is key to understanding its learning capabilities. Some species of birds, such as Clark's nutcracker, feed on seeds. When seeds are abundant, these birds store them in buried caches so they will have food during the winter. Seed caches (up to 2000!) may be buried and then recovered as long as nine months later. One would expect these birds to have an extraordinary spatial memory, and this is indeed what has been found (figure 54.7). Clark's nutcracker, and other seed-

Figure 54.7 **The Clark's nutcracker has an extraordinary memory.** A Clark's nutcracker (*Nucifraga columbiana*) can remember the locations of up to 2000 seed caches months after hiding them. After conducting experiments, scientists have concluded that the birds use features of the landscape and other surrounding objects as spatial references to memorize the locations of the caches.

hoarding birds, have an unusually large hippocampus, the center for memory storage in the brain. This illustrates how feeding ecology (caching seeds to survive the winter) affects the evolution of brain anatomy (an enlarged hippocampus).

Figure 54.8 An unlikely parent. The eager goslings follow Konrad Lorenz as if he were their mother. He is the first object they saw when they hatched, and they have used him as a model for imprinting. Lorenz won the 1973 Nobel Prize in Physiology or Medicine for this work.

Learning Outcomes Review 54.4

Habituation is a diminishing response to a repeated stimulus that is neither positive nor negative. Association may occur as either classical conditioning or operant conditioning. Animals can change their behavior through learning in a variety of ways. Although learning mechanisms may be similar across species, animals also differ in their learning abilities according to their ecology.

■ *In some rodents, males travel far while females remain close to the nest. Do males or females have greater spatial memory? What experiment could you conduct to test your hypothesis?*

54.5 The Development of Behavior

Learning Outcomes

1. *Discuss the role of the critical period in imprinting.*
2. *Explain how social contact can influence growth and development.*
3. *Explain how the study of song learning in white-crowned sparrows illustrates the interaction of instinct and learning.*

Behavioral biologists recognize that behavior has both genetic and learned components. Thus far in this chapter, we have discussed the influence of genes and learning separately. But as you will see, these factors interact during development to shape behavior.

Parent–offspring interactions influence how behavior develops

As an animal matures, it may form social attachments to other individuals or develop preferences that will influence behavior later in life. This process of behavioral development is called **imprinting.** The success of imprinting is highest during a critical period (roughly 13 to 16 hours after hatching in geese). During this time, information required for normal development must be acquired. In **filial imprinting,** social attachments form between parents and offspring. For example, young birds like ducks and geese begin to follow their mother within a few hours after hatching, and their following response results in a social bond between mother and young. The young birds' initial experience, through imprinting, can determine how social behavior develops later in life. The ethologist Konrad Lorenz showed that geese will follow the first object they see after hatching and direct their social behavior toward that object, even if it is not their mother! Lorenz raised geese from

eggs, and when he offered himself as a model for imprinting, the goslings treated him as if he were their parent, following him dutifully (figure 54.8).

Interactions between parents and offspring are key to the normal development of social behavior. The psychologist Harry Harlow gave orphaned rhesus monkey infants the opportunity to form social attachments with two surrogate "mothers," one made of soft cloth covering a wire frame and the other made only of wire (figure 54.9). The infants chose to

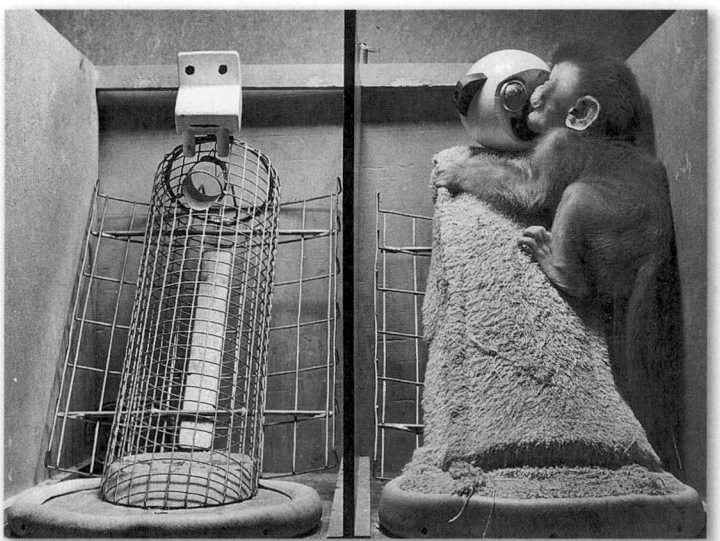

Figure 54.9 Choice trial on infant monkeys. Given a choice between a wire frame that provided food and a similar frame covered with cloth and given a monkey-like head, orphaned rhesus monkeys (*Macaca mulatta*) chose the monkeylike figure over the food.

spend time with the cloth mother, even if only the wire mother provided food, indicating that texture and tactile contact, rather than provision of food, may be among the key qualities in a mother that promote infant social attachment. If infant monkeys are deprived of normal social contact, their development is abnormal. Greater degrees of deprivation lead to greater abnormalities in social behavior during childhood and adulthood. Studies of orphaned human infants similarly suggest that a constant "mother figure" is required for normal growth and psychological development.

Recent research has revealed a biological need for the stimulation that occurs during parent–offspring interactions early in life. Female rats lick their pups after birth, and this stimulation inhibits the release of a brain peptide that can block normal growth. Pups that receive normal tactile stimulation also have more brain receptors for glucocorticoid hormones, thus a greater tolerance for stress, and longer-lived brain cells. Premature human infants who are massaged gain weight rapidly. These studies indicate that the need for normal social interaction is based in the brain, and that touch and other aspects of contact between parents and offspring are important for physical as well as behavioral development.

Instinct and learning may interact as behavior develops

We began this chapter by considering the proximate and ultimate causation of birdsong. Let's continue with the classic studies by neurobiologist Peter Marler on song learning in white-crowned sparrows to examine how innate programs and experience each contribute to the development of behavior.

Mature male white-crowned sparrows sing a species-specific courtship song during the mating season. Through a series of elegant experiments, Marler asked if the song was the result of an instinctive program, learning, or both. Marler reared male birds in soundproof incubators equipped with speakers and microphones to control what a bird heard as it matured, and then recorded the song it produced as an adult. Males that heard no song at all during development sang a poorly developed song as adults (figure 54.10), indicating that instinct alone did not guide song production. In a second study, males were played only the song of a different species, the song sparrow. These males sang a poorly structured song as well. This experiment showed that males would not imitate any song they heard to learn to sing. But birds that heard the song of their own species, or that heard the songs of both the white-crowned sparrow and the song sparrow, sang a fully developed, white-crowned sparrow song as adults.

These results suggest males have a selective **genetic template,** or innate program, that guides them to learn the appropriate song. During a critical period in development, the template will accept the white-crowned sparrow song as a model. Thus, song acquisition depends on learning, but only the song of the correct species can be learned; the genetic template limits what can be learned.

But learning plays a prominent role as well. If a young male becomes deaf after it hears its species' song during the critical period, it will sing a poorly developed song as an adult.

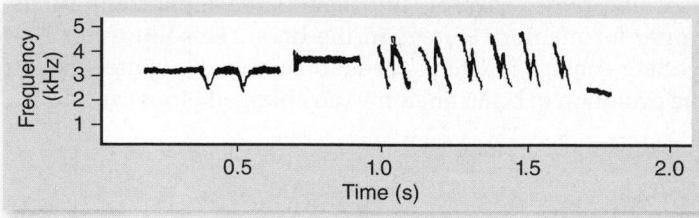

a.

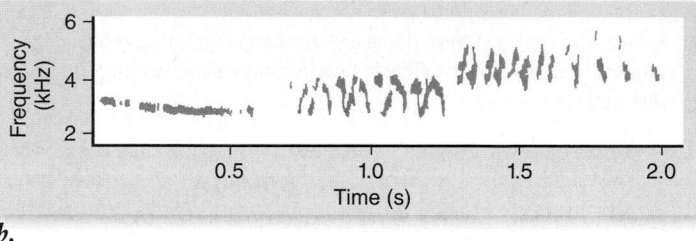

b.

Figure 54.10 Song development in birds. *a.* The sonograms of songs produced by male white-crowned sparrows (*Zonotrichia leucophrys*) that had been exposed to their own species' song during development are different from (*b*) those of male sparrows that heard no song during rearing. This difference indicates that the genetic program itself is insufficient to produce a normal song.

Therefore, the bird must hear the correct song at the right time, and then "practice" listening to himself sing, matching what he hears to the model his genetic template has accepted.

Although this explanation of song development stood unchallenged for many years, white-crowned sparrow males can learn another species' song under certain conditions. If a live male strawberry finch *(Amandava amandava)* is placed in a cage next to a young male sparrow, the young sparrow will learn to sing the strawberry finch's song. This finding indicates that social stimuli—in this case, being able to see, hear, and interact with another bird—is more effective than a tape-recorded song in altering the innate template that guides song development.

The males of some bird species may have no opportunity to hear the song of their own species. In such cases, it appears that the males instinctively "know" their own species' song. For example, cuckoos are **brood parasites;** females lay their eggs in the nest of another species of bird, and the young that hatch are reared by the foster parents (figure 54.11). When the cuckoos become adults, they sing the song of their own species rather than that of their foster parents. Because male brood parasites would most likely hear the song of their host species during development, it is adaptive for them to ignore such "incorrect" stimuli. They hear no adult males of their own species singing, so no correct song models are available. In these species, natural selection has produced a completely genetically guided song. Other birds can also sing a correct species-typical song, even if reared in isolation.

? Inquiry question Imagine there is only one bird species on an island. Do you think instinct, learning, or both will guide song development?

Figure 54.11 Brood parasite. Cuckoos lay their eggs in the nests of other species of birds. Because the young cuckoos (large bird to the right) are raised by a different species (such as this meadow pipit, smaller bird to the left), they have no opportunity to learn the cuckoo song; the cuckoo song they later sing is innate.

a.   *b.*

Figure 54.12 Animal thinking? *a.* This chimpanzee is stripping the leaves from a twig, which it will then use to probe a termite nest. This behavior strongly suggests that the chimpanzee is consciously planning ahead, with full knowledge of what it intends to do. *b.* This sea otter is using a rock as an "anvil," against which it bashes a clam to break it open. A sea otter will often keep a favorite rock for a long time, as though it has a clear idea of its future use of the rock. Behaviors such as these suggest that animals have cognitive abilities.

Learning Outcomes Review 54.5

During the critical period, offspring must engage in certain social interactions for normal behavioral development. Parent–offspring contact stimulates the release of physiological factors, such as hormones and brain receptors, crucial to growth and brain development. In white-crowned sparrows, young males must hear their species' song to sing it correctly, indicating that both instinct and learning affect song development.

■ *Some researchers have tried to link IQ and genes in humans. Why would this research be seen as controversial?*

54.6 Animal Cognition

Learning Outcome

1. *Explain why behavioral biologists today are more open to considering that animals can think.*

For many decades, students of animal behavior flatly rejected the notion that nonhuman animals can think. Now serious attention is given to animal awareness. The central question of whether animals show **cognitive behavior**—that is, they process information and respond in a manner that suggests thinking—is widely supported (figure 54.12).

In a series of classic experiments conducted in the 1920s, a chimpanzee was left in a room with bananas hanging from the ceiling out of reach. Also in the room were several boxes lying on the floor. After some unsuccessful attempts to jump up and grab the bananas, the chimp stacked the boxes beneath the suspended bananas, and climbed up to claim its prize (figure 54.13). Field researchers have observed that Japanese macaques learned to wash sand off potatoes and to float grain to separate it from sand. Chimpanzees pull leaves off a tree branch and then stick the branch into the entrance of a termite nest to "fish" for food. Chimps also crack open nuts using pieces of wood in a "hammer and anvil" technique. Even more remarkable is that parents appear to teach nut cracking to their offspring!

Recent studies have found that chimpanzees and other primates show behaviors that provide strong evidence of cognition. For example, chimps cooperate with other chimps in ways that suggest an understanding of past success. Cognitive ability is not limited to primates: Ravens and other corvid birds also show extraordinary insight and problem-solving ability (figure 54.14) as do octopuses (see figure 34.17).

Figure 54.13 Problem solving by a chimpanzee. Unable to get the bananas by jumping, the chimpanzee devises a solution.

Figure 54.14 Problem solving by a raven. Confronted with a problem it has never previously faced, the raven figures out how to get the meat at the end of the string by repeatedly pulling up a bit of string and stepping on it.

Learning Outcome Review 54.6

Research has provided compelling evidence that some nonhuman animals are able to solve problems and use reasoning, cognitive abilities once thought uniquely human.

- *How could you determine whether a chimpanzee has the ability to count objects?*

54.7 Orientation and Migratory Behavior

Learning Outcomes

1. Define migration.
2. Distinguish between orientation and navigation.
3. Describe different systems for navigation.

Monarch butterflies and many birds travel thousands of miles over continents to overwintering sites in the tropics. Many animals travel away from a nest and then return. To do so, they track cues in the environment, often showing exceptional skill at orientation. Animals with a homing instinct, such as pigeons, recognize complex features of the environment to return to their home. Despite decades of study, our understanding of animal orientation is far from complete.

Migration often involves populations moving large distances

Long-range, two-way movements are known as migrations. Each fall, ducks, geese, and many other birds migrate south along flyways from Canada across the United States, heading as far as South America, and then returning each spring.

Monarch butterflies also migrate each fall from central and eastern North America to their overwintering sites in several small, geographically isolated areas of coniferous forest in the mountains of central Mexico (figure 54.15). Each August, the butterflies begin a flight southward and at the end of winter, the monarchs begin the return flight to their summer breeding ranges. Two to five generations may be produced as the butterflies fly north: butterflies that migrate in the autumn to the precise locations in Mexico have never been there before!

Recent geographic range expansions by some migrating birds have revealed how migratory patterns change. When colonies of bobolinks became established in the western United States, far from their normal range in the Midwest and East, they did not migrate directly to their winter range in South America. Instead, they migrated east to their ancestral range, and then south along the original flyway (figure 54.16). Rather than changing the original migration pattern, they simply added a new segment. Scientists continue to study the western bobolinks to learn whether, in time, a more efficient migration path will evolve or

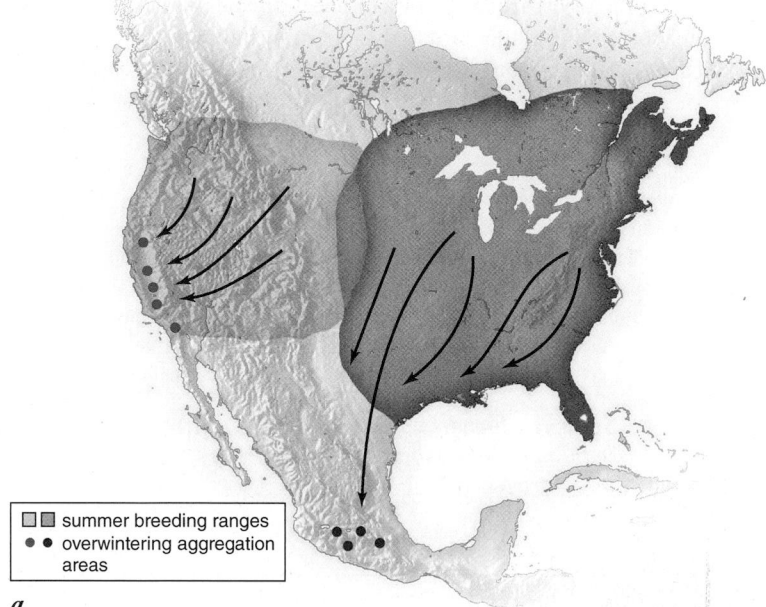

☐☐ summer breeding ranges
● ● overwintering aggregation areas

a.

b.

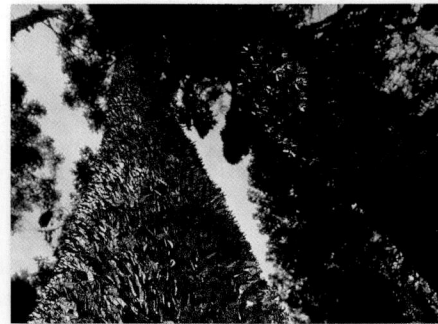

c.

Figure 54.15 Migration of monarch butterflies (Danaus plexippus). *a.* Monarchs from western North America overwinter in areas of mild climate along the Pacific coast. Those from eastern North America migrate over 3000 kilometers to Mexico. *b.* Monarch butterflies arrive at the remote forests of the overwintering grounds in Mexico, where they *(c)* form aggregations on the tree trunks.

whether the birds will always follow their ancestral course. The behavior of butterflies and birds accentuate the mysteries of the mechanism employed during migration.

Migrating animals must be capable of orientation and navigation

To get from one place to another, animals must have a "map" (that is, know where to go) and a "compass" (use environmental cues to guide their journey). Orientation requires following a bearing such as a source of light, but navigation is the ability to set or adjust a bearing, and then follow it. The former is analogous to using a compass, while the latter is like using a compass in conjunction with a map. The nature of the "map" animals use is unclear.

Birds and other animals navigate by looking at the sun during the day and the stars at night. The indigo bunting is a short-distance nocturnal migrant bird. It flies during the day using the Sun as a guide, and compensates for the movement of the Sun in the sky as the day progresses. These birds use the positions of constellations around the North Star in the night sky as a compass.

Many migrating birds also have the ability to detect Earth's magnetic field and to orient themselves with respect

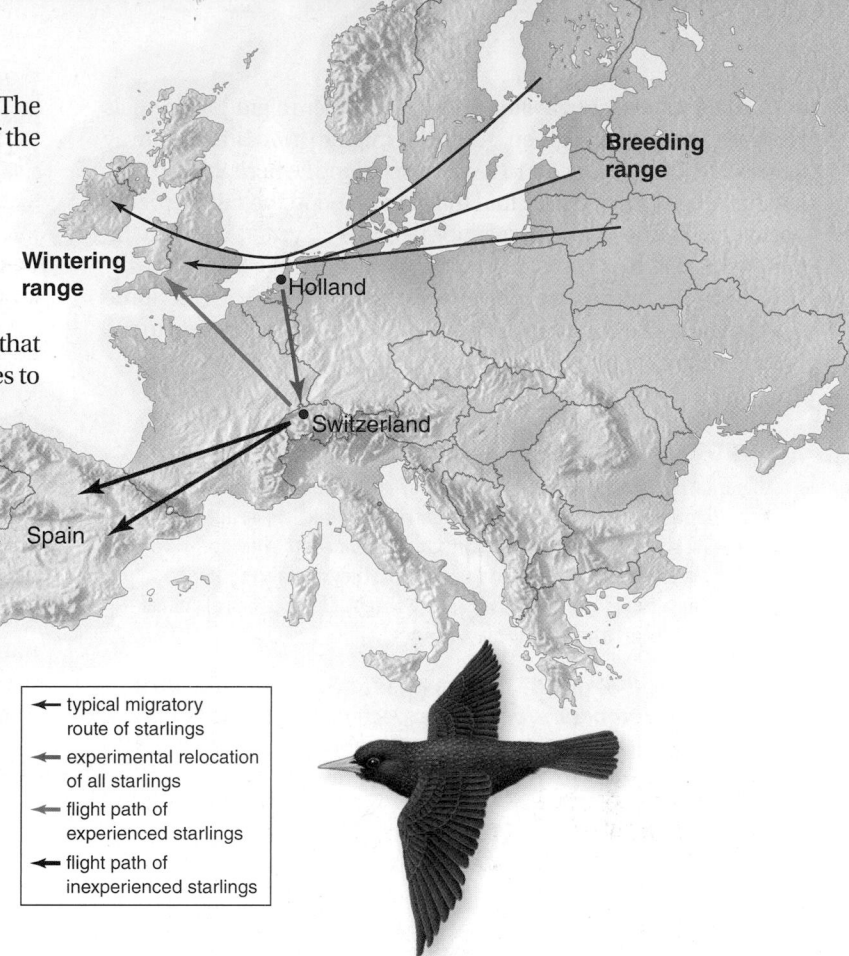

Legend:
- typical migratory route of starlings
- experimental relocation of all starlings
- flight path of experienced starlings
- flight path of inexperienced starlings

Figure 54.17 Migratory behavior of starlings (Sturnus vulgaris). The navigational abilities of inexperienced birds differ from those of adults that have made the migratory journey before. Starlings were captured in Holland, halfway along their full migratory route from Baltic breeding grounds to wintering grounds in the British Isles; these birds were transported to Switzerland and released. Experienced older birds compensated for the displacement and flew toward the normal wintering grounds (blue arrow). Inexperienced young birds kept flying in the same direction, on a course that took them toward Spain (red arrows). These observations imply that inexperienced birds fly by orientation, but experienced birds learn true navigation.

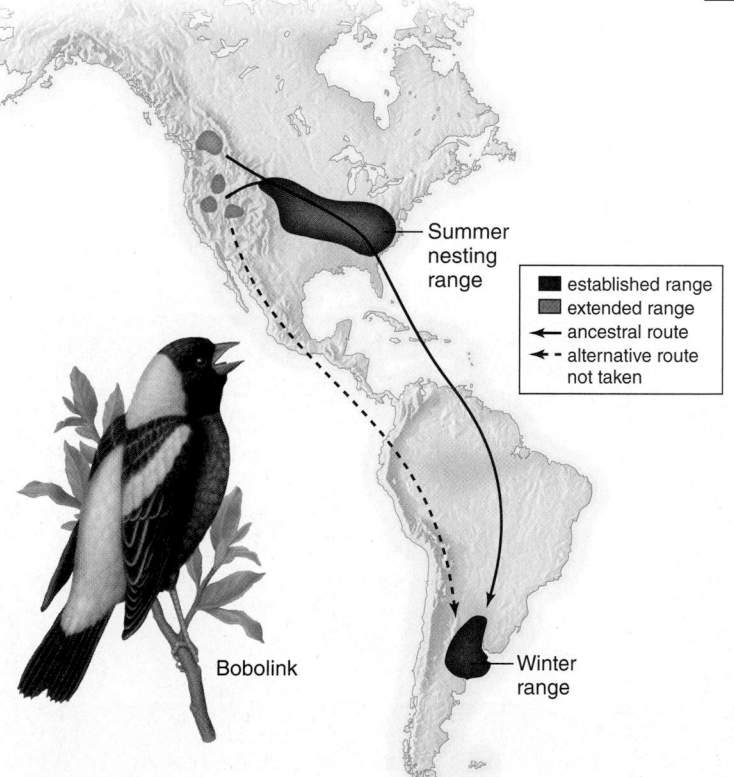

Legend:
- established range
- extended range
- ancestral route
- alternative route not taken

Figure 54.16 Birds on the move. The summer range of bobolinks (*Dolichonyx oryzivorus*) recently extended to the far western United States from their established range in the Midwest. When birds in these newly established populations migrate to South America in the winter, they do not fly directly to the winter range; instead, they first fly to the Midwest and then use the ancestral flyway, going much farther than if they flew directly to their winter range.

to it when cues from the Sun or stars are not available. In an indoor cage, they will attempt to move in the correct geographic direction, even though there are no visible external cues. However, the placement of a magnet near the cage can alter the direction in which the birds attempt to move. Researchers have found magnetite, a magnetized iron ore, in the eyes and upper beaks of some birds, but how these sensory organs function is not known.

The first migration of a bird appears to be innately guided by both celestial cues (the birds fly mainly at night) and Earth's magnetic field. When the two cues are experimentally manipulated to give conflicting directions, the information provided by the stars seems to override the magnetic information. Recent studies, however, indicate that celestial cues indicate the general direction for migration, whereas magnetic cues indicate the specific migratory path (perhaps a turn the bird must make mid-route). Experiments on starlings indicate that inexperienced birds migrate by orientation, but older birds that have migrated previously use true navigation (figure 54.17).

We know relatively little about how other migrating animals navigate. For instance, green sea turtles migrate from Brazil halfway across the Atlantic Ocean to Ascension Island, where the females lay their eggs. How do they find this tiny island in the middle of the ocean, which they haven't seen for perhaps 30 years? How do the young that hatch on the island know how to find their way to Brazil? Newly hatched turtles use wave action as a cue to head to sea. Some sea turtles use the Earth's magnetic field to maintain position in the North Atlantic, but turtle migration is still largely a mystery.

Learning Outcomes Review 54.7

Migration is the long-distance movement of a population, often in a cyclic way. Orientation refers to following a bearing or a direction; navigation involves setting a bearing or direction based on some sort of map or memory. Many species use celestial navigation; they may also be able to detect magnetic fields when those cues are absent. The precision of animal migration remains a mystery in many species.

■ *Animals as diverse as butterflies and birds migrate over long distances. Would you expect them to use different navigation systems? Why or why not?*

54.8 Animal Communication

Learning Outcomes

1. *Explain the nature of signals used in mate attraction.*
2. *Explain the role of courtship signals in reproductive isolation.*
3. *Describe how honeybees communicate information about the location of new food sources.*

Communication is central to species recognition and reproductive isolation, and to the interactions that are essential to social behavior. Much research in behavior analyzes the nature of communication signals, determining how they are produced and received, and identifies their ecological roles and evolutionary origins. Communication involves several signal modalities, including visual, acoustic, chemical, electric, and vibrational signals.

Figure 54.18 A stimulus–response chain. Stickleback courtship involves a sequence of behaviors leading to the fertilization of eggs.

1. Female gives head-up display to male

2. Male swims zigzag to female and then leads her to nest

3. Male shows female entrance to nest

4. Female enters nest and spawns while male stimulates tail

5. Male enters nest and fertilizes eggs

Successful reproduction depends on appropriate signals and responses

During courtship, animals produce signals to communicate with potential mates. A stimulus–response chain sometimes occurs, in which the behavior of the male in turn releases a behavior in the female, resulting in mating (figure 54.18). These signals are usually highly species-specific. Many studies on communication involve designing experiments to determine which key stimuli associated with an animal's visual appearance, sounds, or odors convey information about the nature of the signals produced by the sender. One classical study analyzed territorial defense and courtship communication in stickleback fish (figure 54.19).

Finding a mate: Communicating information about species identity

Courtship signals often restrict communication to members of the same species and in doing so serve a key function in reproductive isolation (see chapter 22). The flashes of fireflies (which are actually beetles) are species-specific signals: females recognize conspecific males by their flash pattern (figure 54.20), and males recognize conspecific females by their flash

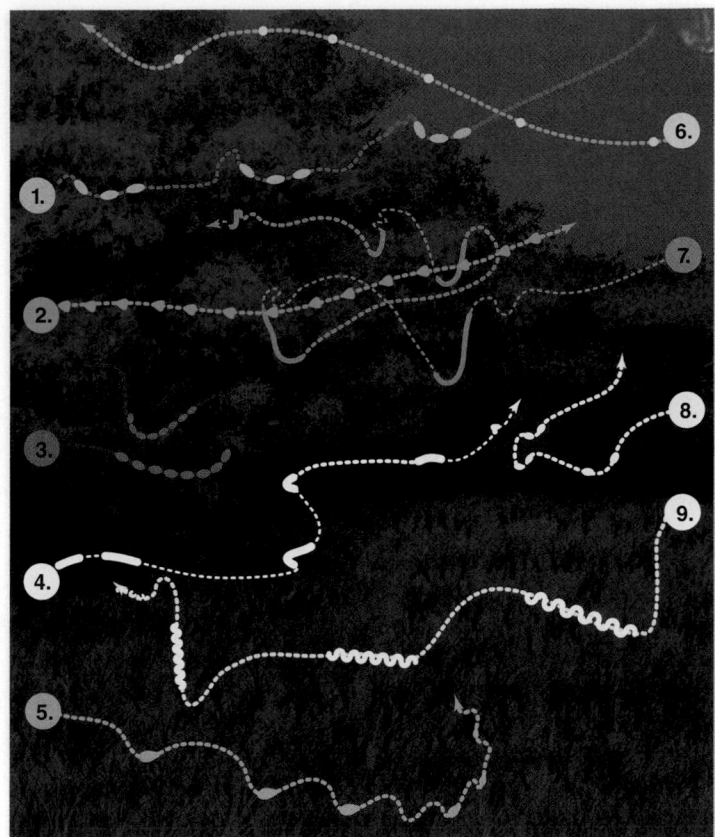

Figure 54.20 Firefly fireworks. The bioluminescent displays of these lampyrid beetles are species-specific and serve as behavioral mechanisms of reproductive isolation. Each number represents the flash pattern of a male of a different species.

response. This series of reciprocal responses provides a continuous "check" on the species identity of potential mates.

Pheromones, chemical messengers used for communication between individuals of the same species, serve as sex attractants in many animals. Female silk moths *(Bombyx mori)* produce a sex pheromone called bombykol in a gland associated with the reproductive system. The male's antennae contain numerous highly sensitive sensory receptors, and neurophysiological studies show they specifically detect bombykol. In some moth species, males can detect extremely low concentrations of sex pheromone and locate females from as far as 7 km away!

Many insects, amphibians, and birds produce species-specific acoustic signals to attract mates. Bullfrog males call by inflating and discharging air from their vocal sacs, located beneath the lower jaw. Females can distinguish a conspecific male's call from those of other frogs that may be in the same habitat and calling at the same time. As mentioned earlier, male birds sing to advertise their presence and to attract females. In many species, variations in the males' songs identify individual males in a population. In these species, the song is individually specific as well as species-specific.

Vibrations, like sound signals, are a form of mechanical communication used by insects, amphibians, and other animals. For example, as we saw in chapter 22, lacewings distinguish one species from another by the pattern of

---

SCIENTIFIC THINKING

Hypothesis: *The red underside of male stickleback is the key stimulus that releases an aggressive response by a territory-holding male.*

Prediction: *Models with red coloration will trigger an attack by a resident male.*

Test: *Construct plastic models, some of which accurately resemble a stickleback male, but lack the red underside. Construct other models that vary in their fishlike appearance, but have red-colored undersides. Expose a territorial male to the models one at a time, and record the number of attacks.*

Result: *Realistic models lacking red elicit no response. Odd-shape models trigger an attack if they have red undersides, even if they poorly resemble fish.*

Conclusion: *The red underside of a male stickleback is the key stimulus that triggers aggressive behavior.*

Further Experiments: *How would you determine if the color of a male stickleback was a releaser of aggressive behavior? How would you know if sound was important? Could you determine if stimuli have additive effects? How? (See also figure 54.18.)*

Figure 54.19 Key stimulus in stickleback fish.

Figure 54.21 **Alarm calling by a prairie dog (Cynomys ludovicianus).** When a prairie dog sees a predator, it stands on its hind legs and gives an alarm call, which causes other prairie dogs to rapidly return to their burrows.

vibrations they make on branches with their abdomens (see figure 22.4).

Courtship behaviors play a major role in sexual selection, which we discuss later in this chapter.

Communication enables information exchange among group members

Many insects, fish, birds, and mammals live in social groups in which information is communicated between group members. For example, some individuals in mammalian societies serve as sentinels, vigilantly on the lookout for danger. When a predator appears, they give an alarm call, and group members respond by seeking shelter (figure 54.21). Social insects such as ants and honeybees produce alarm pheromones that trigger attack behavior. Ants also deposit trail pheromones between the nest and a food source to lead other colony members to food. Honeybees have an extremely complex dance language that directs hivemates to nectar sources.

The dance language of the honeybee

The European honeybee lives in colonies of tens of thousands of individuals whose behaviors are integrated into a complex, cooperative society. Worker bees may forage miles from the hive, collecting nectar and pollen from a variety of plants and switching between plant species depending on their energetic rewards. Food sources used by bees tend to occur in patches, and each patch offers much more food than a single bee can transport to the hive. A colony is able to exploit the resources of a patch because of the behavior of scout bees, which locate patches and communicate their location to hivemates through a dance language. Over many years, Nobel laureate Karl von Frisch (who shared the 1973 prize with Tinbergen and Lorenz), together with generations of students and colleagues, was able to unravel the details of dance language communication.

When a scout bee returns after finding a distant food source, she performs a remarkable behavior pattern called a waggle dance on a vertical comb in the darkness of the hive. The path of the bee during the dance resembles a figure-eight. On the straight part of the path (indicated with dashes in figure 54.22), the bee vibrates ("waggles") her abdomen while producing bursts of sound. The bee may stop periodically to give her hivemates a sample of the nectar carried in her crop. As she dances, she is followed closely by other bees, which soon appear at the new food source to assist in collecting food.

Von Frisch and his colleagues performed experiments to show that hivemates use information in the waggle dance to locate new food sources. The scout bee indicates the direction of the food source by representing the angle between the food source, the hive, and the Sun as the deviation from vertical of the straight run of the dance performed on the hive comb. Thus if the bee danced with the straight run pointing directly up, then the food source would be in the direction of the Sun. If the food is at a 30° angle to the right of the Sun's position, then the straight run would be oriented upward at a 30° angle to the right of vertical) (figure 54.22a). The distance to the food source is indicated by the duration of the straight run. One ingenious experiment used computer-controlled robot bees to give hivemates incorrect information, demonstrating that bees use the directions coded in the dance!

Figure 54.22 **The waggle dance of honeybees (Apis mellifera).** **a.** The angle between the food source, the nest, and the Sun is represented by a dancing bee as the angle between the straight part of the dance and vertical. The food is 30° to the right of the Sun, and the straight part of the bee's dance on the hive is 30° to the right of vertical. **b.** A scout bee dances on a comb in the hive.

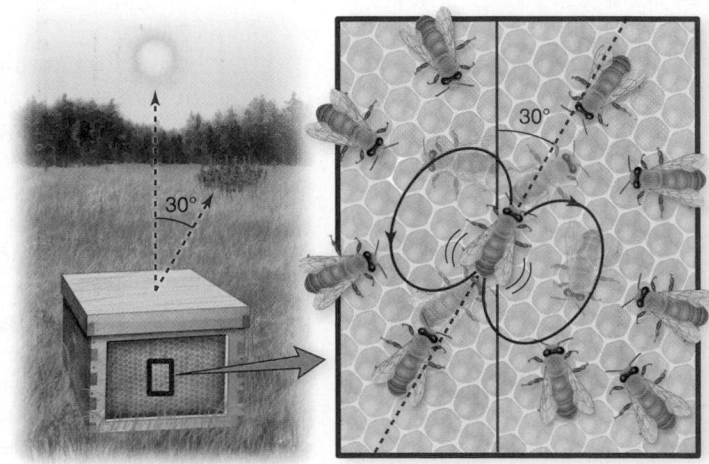

a.

b.

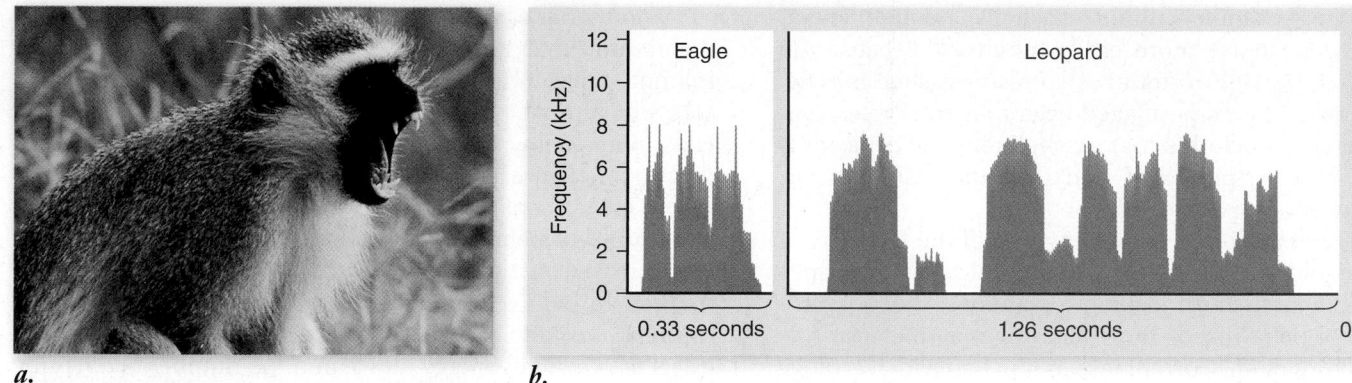

a.

b.

Figure 54.23 Primate semantics. Vervet monkeys (*Cercopithecus aethiops)* give different alarm calls *(a)* when troop members sight an eagle, leopard, or snake. *b.* Each distinctive call elicits a different and adaptive escape behavior.

Language in nonhuman primates and humans

Evolutionary biologists have sought the origins of human language in the communication systems of monkeys and apes. Some nonhuman primates have a "vocabulary" that allows individuals to signal the identity of specific predators. Different vocalizations of African vervet monkeys, for example, indicate eagles, leopards, or snakes, among other threats (figure 54.23).

The complexity of human language would at first appear to defy biological explanation, but closer examination suggests that the differences are in fact superficial—all languages share many basic similarities. All of the roughly 3000 languages draw from the same set of 40 consonant and vowel sounds (English uses two dozen of them), and humans of all cultures can acquire and learn them. Researchers believe these similarities reflect the way our brains handle abstract information.

Learning Outcomes Review 54.8

Animal communication involves production and reception of signals, in the form of sounds, chemicals, or movements, that primarily have an ecological function. Courtship signals are highly species-specific and serve as a mechanism of reproductive isolation. Animals living in social groups, such as honeybees, may use complex systems of communication to exchange information about food and predators.

■ *Two species of moth use the same sex pheromone to locate mates. Explain how these species could nevertheless be reproductively isolated.*

54.9 Behavioral Ecology

Learning Outcomes

1. *Describe behavioral ecology.*
2. *Discuss the economic analysis of behaviors.*

Niko Tinbergen pioneered the study of the adaptive function of behavior. Stated simply, this is how behavior allows an animal to stay alive and keep its offspring alive. For example,

Tinbergen observed that after gull nestlings hatch, the parents remove the eggshells from the nest. To understand why (ultimate causation), he painted chicken eggs to resemble gull eggs (figure 54.24), which had camouflage coloration to allow them to be inconspicuous against the natural background. He distributed them throughout the area in which the gulls were nesting, placing broken eggshells with their prominent white interiors next to some of the eggs. As a control, he left other

Figure 54.24 The adaptive value of egg coloration. Niko Tinbergen, a winner of the 1973 Nobel Prize in Physiology or Medicine, painted chicken eggs to resemble the mottled brown camouflage of gull eggs. The eggs were used to test the hypothesis that camouflaged eggs are more difficult for predators to find and thus increase the young's chances of survival.

camouflaged eggs alone without eggshells. He then noted which eggs were found more easily by crows. Because the crows could use the white interior of a broken eggshell as a cue, they ate more of the camouflaged eggs that were near eggshells. Tinbergen concluded that eggshell removal behavior is adaptive: it reduces predation and thus increases the offspring's chances of survival.

Tinbergen is credited with being one of the founders of **behavioral ecology,** the study of how natural selection shapes behavior. This branch of ecology examines the adaptive significance of behavior, or how behavior may increase survival and reproduction. Current research in behavioral ecology focuses on how behavior contributes to an animal's reproductive success, or fitness. As we saw in section 54.3, differences in behavior among individuals often result from genetic differences. Therefore, natural selection operating on behavior has the potential to produce evolutionary change.

Consequently, the field of behavioral ecology is concerned with two questions. First, is behavior adaptive? Although it is tempting to assume that behavior must in some way represent an adaptive response to the environment, this need not be the case. As you saw in chapter 20, traits can appear for many reasons other than natural selection, such as genetic drift, gene flow, or the correlated consequences of selection on other traits. Moreover, traits may be present in a population because they evolved as adaptations in the past, but are no longer useful. These possibilities hold true for behavioral traits as much as for any other kind of trait.

If behavior is adaptive, the next question is: How is it adaptive? Although the ultimate criterion is reproductive success, behavioral ecologists are interested in how behavior can lead to greater reproductive success. Does a behavior enhance energy intake, thus increasing the number of offspring produced? Does it increase mating success? Does it decrease the chance of predation? The job of a behavioral ecologist is to determine the effect of a behavioral trait—for example, foraging efficiency—on each of these activities and then to discover whether increases translate into increased fitness. Benefits and costs of these effects, estimated in terms of energy or offspring, are often used to analyze the adaptive nature of behavior.

Foraging behavior can directly influence energy intake and individual fitness

A useful way to understand the approach of behavioral ecology is by focusing on foraging behavior. For many animals, food comes in a variety of sizes. Larger foods may contain more energy but may be harder to capture and less abundant. In addition, animals may forage for some types of food that are farther away than other types. For these animals foraging involves a trade-off between a food's energy content and the cost of obtaining it. The net energy (estimated in calories or joules) gained by feeding on prey of each size is simply the energy content of the prey minus the energy costs of pursuing and handling it. According to **optimal foraging theory,** natural selection favors individuals whose foraging behavior is as energetically efficient as possible. In other words, animals tend to feed on prey that maximize their net energy intake per unit of foraging time.

A number of studies have demonstrated that foragers do prefer prey that maximize energy return. Shore crabs, for example, tend to feed primarily on intermediate-sized mussels, which provide the greatest energy return; larger mussels yield more energy, but also take considerably more energy to crack open (figure 54.25).

This optimal foraging approach assumes natural selection will favor behavior that maximizes energy acquisition if the increased energy reserves lead to increases in reproductive success. In both Columbian ground squirrels (*Spermophilus columbianus*) and captive zebra finches, a direct relationship exists between net energy intake and the number of offspring raised; similarly, the reproductive success of orb-weaving spiders is related to how much food they can capture.

Animals have other needs beside energy, however, and sometimes these needs conflict. One obvious need is the avoidance of predators: Often, the behavior that maximizes energy intake is not the one that minimizes predation risk. In this case, the behavior that maximizes fitness often may reflect a trade-off between obtaining the most energy at the least risk of being eaten. Not surprisingly, many studies have shown that a wide variety of animal species alter their foraging behavior—becoming less active, spending more time watching

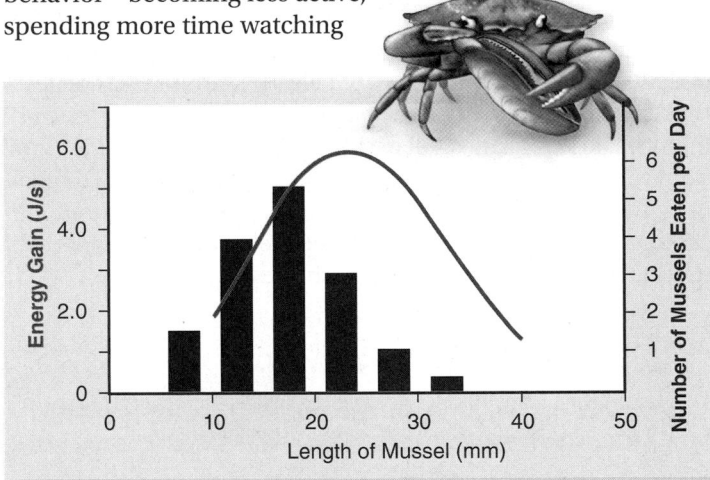

Figure 54.25 Optimal diet. The shore crab selects a diet of energetically profitable prey. The curve describes the net energy gain (equal to energy gained minus energy expended) derived from feeding on different sizes of mussels. The bar graph shows the numbers of mussels of each size in the diet. Shore crabs tend to feed on those mussels that provide the most energy.

? Inquiry question What factors might be responsible for the slight difference in peak prey length relative to the length optimal for maximum energy gain?

🔍 Data analysis Suppose a raccoon forages for 10 hours a day. It can either look for frogs in a stream or insects in the dirt. It takes on average 30 minutes to find a frog, but only 12 minutes to find a cricket. The raccoon obtains 10 joules of energy eating a frog and 2 joules for each cricket. Where should the raccoon forage, the stream or the dirt?

In reality, we must also include the time that it takes to capture and eat prey (called "handling time" as well as the time it takes to find them). Suppose it takes 15 minutes to subdue and ingest every frog, as opposed to 3 minutes per cricket. Does that change your answer?

for predators, or staying nearer to cover—when predators are present. Compromises, in this case a trade-off between vigilance and feeding, may thus be made during foraging.

Optimal foraging theory assumes that energy-maximizing behavior has evolved by natural selection. Therefore, it must have a genetic basis. For example, female zebra finches particularly successful in maximizing net energy intake tend to have similarly successful offspring. In this study, young birds were removed from their mothers before they were able to leave the nest, so this similarity indicates that foraging behavior probably has a genetic component. Studies on other animals show that age, experience, and learning are also important to the development of efficient foraging.

Territorial behavior evolves if the benefits of holding a territory exceed the costs

Animals often move over a large area, or home range, during their course of activity. In many species, the home range of several individuals overlaps in time or in space, but each individual defends a portion of its home range and uses it and its resources exclusively. This behavior is called **territoriality** (figure 54.26).

The defining characteristic of territorial behavior is defense against intrusion and resource use by other individuals. Territories are defended by displays advertising that territories are occupied, and by overt aggression. A bird sings from its perch within a territory to prevent take-over by a neighboring bird. If a potential usurper is not deterred by the song, the territory owner may attack and try to drive it away. But territorial defense has its costs. Singing is energetically expensive, and attacks can lead to injury. Using a signal (a song or visual display) to advertise occupancy can reveal a bird's position to a predator.

Why does an animal bear the costs of territorial defense? Energetic benefits of territoriality may take the form of increased food intake due to exclusive use of resources, access to mates, or access to refuges from predators. Studies of nectar-feeding birds such as hummingbirds and sunbirds provide an example (figure 54.27). A bird benefits from having the exclusive use of a patch of flowers because it can efficiently harvest the nectar the

Figure 54.27 The benefit of territoriality. Sunbirds (on the left), found in Africa and ecologically similar to New World hummingbirds (on the right), protect their food source by attacking other sunbirds that approach flowers in their territory.

flowers produce. To maintain exclusive use, the bird must actively defend the patch. The benefits of exclusive use outweigh the costs of defense only under certain conditions.

Sunbirds, for example, expend 3000 calories per hour chasing intruders from a territory. Whether the benefit of defending a territory will exceed this cost depends on the amount of nectar in the flowers and how efficiently the bird can collect it. When flowers are very scarce or nectar levels are very low, a nectar-feeding bird may not gain enough energy to balance the energy used in defense. Under these conditions, it is not energetically advantageous to be territorial. Similarly, when flowers are very abundant, a bird can efficiently meet its daily energy requirements without behaving territorially and adding the costs of defense. Again, from an energetic standpoint, defending abundant resources isn't worth the cost, either. Territoriality therefore only occurs at intermediate levels of flower availability and nectar production, when the benefits of defense outweigh the costs.

In many species, access to females is a more important determinant of territory size for males than is food availability. In some lizards, for example, males maintain enormous territories during the breeding season. These territories, which encompass the territories of several females, are much larger than would be required to supply enough food, and they are defended vigorously. In the nonbreeding season, by contrast, male territory size decreases dramatically, as does aggressive territorial behavior.

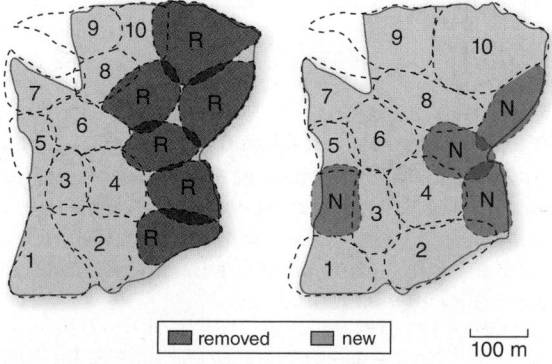

Figure 54.26 Competition for space. Territory size in birds is adjusted according to the number of competitors. When six pairs of great tits *(Parus major)* were removed from their territories (indicated by R in the left figure), their territories were taken over by other birds in the area and by four new pairs (indicated by N in the right figure). Numbers correspond to the birds present before and after.

removed   new   100 m

Learning Outcomes Review 54.9

Behavioral ecology is the study of the adaptive significance of behavior—that is, how it affects survival and reproductive success. An economic approach estimates the energy benefits and costs of a behavior and assumes that animals gain more from a behavior than they expend, obtaining a fitness advantage. Foraging behavior and defense of a territory can be analyzed in this way. Apart from energy gains, considerations such as avoiding predators are also important to fitness.

■ *The Hawaiian honeycreeper, a nectar-feeding bird, fails to defend flowers that are either infrequently encountered or very abundant. Why?*

54.10 Reproductive Strategies and Sexual Selection

Learning Outcomes

1. *Explain parental investment and the prediction it makes about mate choice.*
2. *Describe how sexual selection leads to the evolution of secondary sexual characteristics.*
3. *Explain why some species are generally monogamous and others are polygynous.*

During the breeding season, animals make several important life-history "decisions" concerning their choice of mates, how many mates to have, and how much time and energy to devote to rearing offspring. These decisions are all aspects of an animal's **reproductive strategy,** a set of behaviors that presumably have evolved to maximize reproductive success. Energetic costs of reproduction appear to have been critically important to behavioral differences between females and males. Ecological factors such as the way food resources, nest sites, and members of the opposite sex are spatially distributed in the environment, as well as disease, are important in the evolution of reproductive decisions.

The sexes often have different reproductive strategies

Males and females have the common goal of improving the quantity and quality of offspring they produce, but usually differ in the way they attempt to maximize fitness. Such a difference in reproductive behavior is clearly seen in mate choice. Darwin was the first to observe that females often do not mate with the first male they encounter, but instead seem to evaluate a male's quality and then decide whether to mate. Peahens prefer to mate with peacocks that have more eyespots on their elaborate tail feathers (figure 54.28b, c). Similarly, female frogs prefer to mate with males having more acoustically complex, and thus attractive, calls. This behavior, called mate choice, is well known in many invertebrate and vertebrate species.

Males are selective in choosing a mate much less frequently than females. Why should this be? Many of the differences in reproductive strategies between the sexes can be understood by comparing the parental investment made by males and females. **Parental investment** refers to the energy and time each sex "invests" in producing and rearing offspring; it is, in effect, an estimate of the energy expended by males and females in each reproductive event.

Numerous studies have shown that females generally have a higher parental investment. One reason is that eggs are much larger than sperm—195,000 times larger in humans! Eggs contain proteins and lipids in the yolk and other nutrients for the developing embryo, but sperm are little more than mobile DNA packages. In some groups of animals (mammals, for example), females are responsible for gestation and lactation, costly reproductive functions only they can carry out.

The consequence of such inequalities in reproductive investment is that the sexes face very different selective pressures. Because any single reproductive event is relatively inexpensive for males, they can best increase their fitness by mating with as many females as possible. This is because male fitness is likely not limited by the amount of sperm they can produce. By contrast, each reproductive event for females is much more

a.

b.

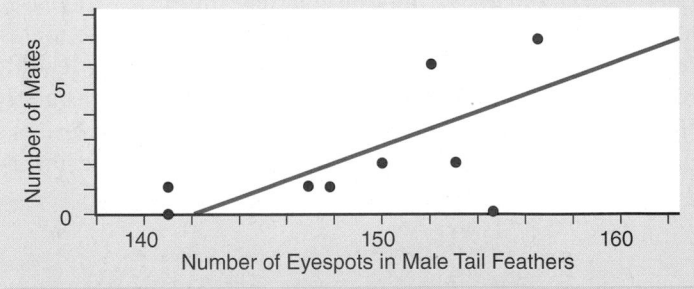

c.

Figure 54.28 Products of sexual selection. In bird species such as (*a*) the raggiana bird of paradise (*Paradisaea raggiana*) and (*b*) the peacock (*Pavo cristatus*), males use their much longer tail feathers in courtship displays. *c.* Female peahens prefer to mate with males having greater numbers of eyespots in their tail feathers.

? Inquiry question Why do females prefer males with more spots?

🔍 Data analysis Suppose a male had 155 eyespots. How many females would you expect him to mate with? If you had multiple males with 155 eyespots, do you think they all would mate with the same number of females?

costly, and the number of eggs that can be produced often limits reproductive success. For this reason, a female should be choosy, trying to pick the male that can provide the greatest benefit to her offspring and thus improve her fitness.

These conclusions hold only when female reproductive investment is much greater than that of males. In species with biparental care, males may contribute equally to the cost of raising young; in this case, the degree of mate choice should be more equal between the sexes.

In some cases, male investment exceeds that of females. For example, male Mormon crickets transfer a protein-containing packet (a spermatophore) to females during mating. Almost 30% of a male's body weight is made up by the spermatophore, which provides nutrition for the female and helps her develop her eggs. As we might expect from our model of mate choice, in this case it is the females that compete with one another for access to males, which are the choosy sex. Indeed, males are quite selective, favoring heavier females. Heavier females have more eggs; thus, males that choose larger females leave more offspring (figure 54.29).

Males care for eggs and developing young in many species, including seahorses and a number of birds and insects. As with Mormon crickets, these males are often choosy, and females compete for mates.

Sexual selection occurs through mate competition and mate choice

As discussed in chapter 20, the reproductive success of an individual is determined by how long the individual lives, how frequently it mates, and how many offspring it produces per mating. The second of these factors, competition for mates, is termed **sexual selection.** Some people consider sexual selection to be distinctive from natural selection, but others see it as a subset of natural selection, just one of the many factors affecting an organism's fitness.

Sexual selection involves both **intrasexual selection,** or competitive interactions between members of one sex ("the power to conquer other males in battle," as Darwin put it), and **intersexual selection,** which is another name for mate choice ("the power to charm"). Sexual selection leads to the evolution of structures used in combat with other males, such as a deer's antlers and a ram's horns, as well as ornaments used to "persuade" members of the opposite sex to mate, such as long tail feathers and bright plumage (see figure 54.28a, b). These traits are called **secondary sexual characteristics.**

Selection strongly favors any trait that confers greater ability in mate competition. Larger body size is a great advantage if dominance is important, as it is in territorial species. Males may thus be considerably larger than females. Such differences between the sexes are referred to as **sexual dimorphism.** In other species, structures used for fighting, such as horns, antlers, and large canine teeth, have evolved to be larger in males because of the advantage they give in intrasexual competition.

Sometimes **sperm competition** occurs between the sperm of different males if females mate with multiple males. This type of competition, which occurs after mating, has selected for sperm-transfer organs designed to remove the sperm of a prior mating, large testes to produce more sperm per mating, and sperm that hook themselves together to swim more rapidly. These traits enhance the likelihood of fertilizing an egg.

Intrasexual selection

In many species, individuals of one sex—usually males—compete with one another for the opportunity to mate. Competition can occur for a territory in which females feed or bear young. Males may also directly compete for the females themselves. A few successful males may engage in an inordinate number of matings, while most males do not mate at all. For example, elephant seal males control territories on breeding beaches and a few dominant males do most of the breeding (figure 54.30). On one beach, for example, eight males impregnated 348 females, while the remaining males mated rarely, if at all.

Intersexual selection

Intersexual selection concerns the active choice of a mate. Mate choice has both direct and indirect benefits.

Direct benefits of mate choice. In some cases, the benefits of mate choice are obvious. If males help raise offspring, females benefit by choosing the male that can provide the best care—the better the parent, the more offspring she is likely to rear. In other species, males provide no care, but maintain territories that provide food, nesting sites, and predator refuges. In red deer, males that hold territories with the highest quality grasses mate with the most females. In this case, there is a direct benefit of a female mating with such a territory owner: She feeds with little disturbance on quality food.

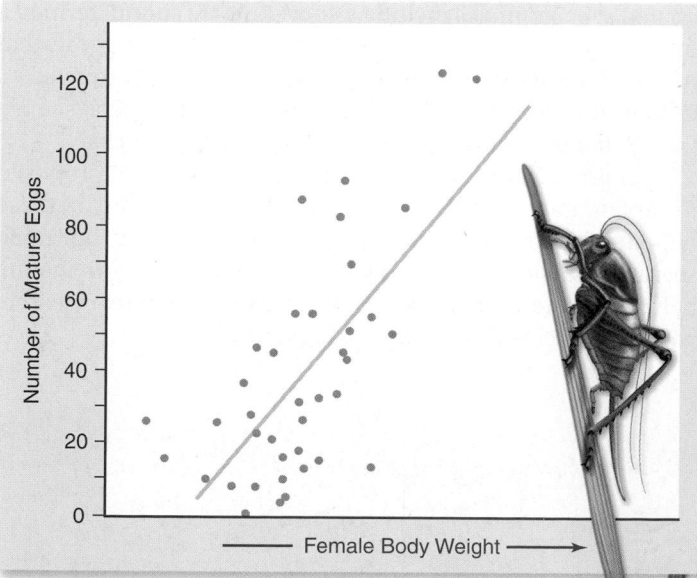

Figure 54.29 The advantage of male mate choice. Male Mormon crickets (*Anabrus simplex*) choose heavier females as mates, and larger females have more eggs. Thus, male mate selection increases fitness.

? Inquiry question Is there a benefit to females for mating with large males?

Figure 54.30 Male elephant seals *(Mirounga angustirostris)* fight with one another for possession of territories. Only the largest males can hold territories, which contain many females.

Indirect benefits of mate choice. In many species, however, males provide no direct benefits of any kind to females. In such cases, it is not intuitively obvious what females have to gain by being "choosy." Moreover, what could be the possible benefit of choosing a male with an extremely long tail or a complex song?

A number of theories have been proposed to explain the evolution of such preferences. One idea is that females choose the male that is the healthiest or oldest. Large males, for example, have probably been successful at living long, acquiring a lot of food, and resisting parasites and disease. In other species, features other than size may indicate a male's condition. In guppies and some birds, the brightness of a male's color reflects the quality of his diet and overall health. Females may gain two benefits from mating with the healthiest males. First, healthy males are less likely to be carrying diseases, which might be transmitted to the female during mating. Second, to the extent that the males' success in living long and prospering is the result of his genetic makeup, the female will be ensuring that her offspring receive good genes from their father.

Several experimental studies in fish and moths have examined whether female mate choice leads to greater reproductive success. In these experiments, females in one group were allowed to choose the males with which they would mate, whereas males were randomly mated to a different group of females. Offspring of females that chose their mates were more vigorous and survived better than offspring from females given no choice, which suggests that females preferred males with a better genetic makeup.

A variant of this theory goes one step further. In some cases, females prefer mates with traits that appear to be detrimental to survival (see figure 54.28c). The long tail of the peacock is a hindrance in flying and makes males more vulnerable to predators. Why should females prefer males with such traits? The **handicap hypothesis** states that only genetically superior mates can survive with such a handicap. By choosing a male with the largest handicap, the female is ensuring that her offspring will receive these quality genes. Of course, the male offspring will also inherit the genes for the handicap. For this reason, evolutionary biologists are still debating the merit of this hypothesis.

Alternative theories about the evolution of mate choice. Some courtship displays appear to have evolved from a predisposition in the female's sensory system to respond to certain stimuli. For example, females may be better able to detect particular colors or sounds at a certain frequency, and thus be attracted to such signals. **Sensory exploitation** involves the evolution in males of a signal that "exploits" these preexisting biases. For example, if females are particularly adept at detecting red objects, then red coloration may evolve in males as part of a courtship display.

To understand the evolution of courtship calls, consider the vocalizations of the Túngara frog (figure 54.31). Unlike related species, males include a short burst of sound, termed a "chuck," at the end of their calls. Recent research suggests that not only are females of this species particularly attracted to calls of this sort, but so are females of related species, even though males of these species do not produce "chucks."

A great variety of other hypotheses have been proposed to explain the evolution of mating preferences. Many of these hypotheses may be correct in some circumstances, but none seems capable of explaining all of the variation in mating behavior in the animal world. This is an area of vibrant research, with new discoveries appearing regularly.

a.

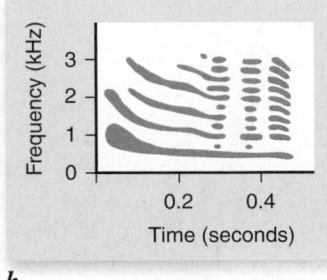
b.

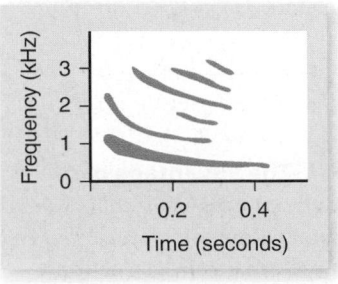
c.

Figure 54.31 **Male Túngara frog** *(Physalaemus pustulosus)* **calling.** Female frogs of several species in the genus *Physalaemus* prefer males that include a "chuck" in their call. However, only males of the Túngaru frog *(a)* produce such calls *(b)*; males of other species do not *(c).*

Mating systems reflect the ability of parents to care for offspring and are influenced by ecology

The number of individuals with which an animal mates during the breeding season varies among species. Mating systems include monogamy (one male mates with one female), polygyny (one male mates with more than one female; see figure 54.30), and polyandry (one female mates with more than one male). Only monogamous mating includes a pair bond (like prairie voles). Like mate choice, mating systems have evolved to allow females and males to maximize fitness.

The option of having more than one mate may be constrained by the need for offspring care. If females and males are able to care for young, then the presence of both parents may be necessary for young to be reared successfully. Monogamy may thus be favored. Generally this is the case for birds, in which over 90% of all species appear to be monogamous. A male may either remain with his mate and provide care for the offspring or desert that mate to search for others; both strategies may increase his fitness. The strategy that natural selection will favor depends on the requirement for male assistance in feeding or defending the offspring. In some species (like humans), offspring are **altricial**—they require prolonged and extensive care. In these species, the need for care by two parents reduces the tendency for the male to desert his mate and seek other matings. In species in which the young are **precocial** (requiring little parental care), males may be more likely to be polygynous because the need for their parenting is lower. In mammals, only females lactate, freeing males from feeding offspring. It follows that most mammals are polygynous.

Mating systems are strongly influenced by ecology. A male may defend a territory that holds nest sites or food sources sufficient for more than one female. If territories vary in quality or quantity of resources, a female's fitness is maximized if she mates with a male holding a high-quality territory, even if he has mated. Although a male may already have a mate, it is still more advantageous for the female to breed with a mated male holding a high-quality territory than with an unmated male holding a low-quality territory. This favors the evolution of polygyny.

Polyandry is relatively rare, but the evolution of multiple mating by females is becoming better understood. It is best known in birds like spotted sandpipers and jacanas living in highly productive environments such as marshes and wetlands. Here, females take advantage of the increased resources available to rear offspring by laying clutches of eggs with more than one male. Males provide all incubation and parenting, and females mate and leave eggs with two or more males.

Females may also mate with several males to genetically diversify their offspring, which in turn increases disease resistance. This appears to be the case in honey bees, for example, in which a queen may mate with many males.

Extra-pair copulations

The "monogamy" of many bird species has been reevaluated as DNA fingerprinting (see chapter 17) has become commonly used to determine paternity and precisely quantify the reproductive success of individual males (figure 54.32a). In red-winged blackbirds (figure 54.32b), researchers established that half of all nests contained at least one hatchling fertilized by a male other than the territory owner; overall, 20% of the offspring were the result of such **extra-pair copulations (EPCs).**

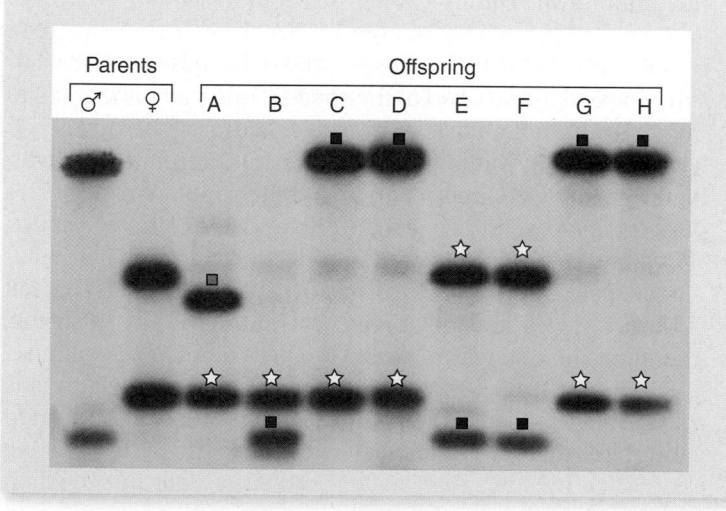

a.

b.

Source: a: (unpubl, Douglas Ross)

Figure 54.32 The study of paternity. *a.* DNA fingerprinting gel from parents and offspring in a nest of pied flycatchers (*Ficedula hypoleuca*). The bands represent fragments of different DNA lengths. The female who lays the egg and her mate are both heterozygous and do not share bands, allowing assignment of parentage. All offspring have one of the female's bands, indicated by a yellow star, as would be expected given that the eggs are in her nest. Seven of the offspring have an allele from the father, indicated by a red symbol, but the eighth, A, has an allele not present in either parent, indicated by a blue symbol, demonstrating that this bird was the result of an extra-pair copulation with another male in the population. *b.* Results of a DNA fingerprinting study in red-winged blackbirds (*Agelaius phoeniceus*). Fractions indicate the proportion of offspring fathered by the male in whose territory the nest occurred. Arrows indicate how many offspring were fathered by particular males outside of each territory. Nests on some territories were not sampled.

What is the evolutionary advantage of EPCs? For males, the answer is obvious: increased reproductive success. Females, on the other hand, may mate with genetically superior individuals even if already paired with a male, thus enhancing the genes passed on to their offspring. The female doesn't produce more offspring, but offspring of better genetic quality. In some birds and other animals, EPCs may help females increase the amount of care they get from males to raise their offspring. This is exactly what happens in a common English bird. Females mate not only with the territory owner, but also with subordinate males that hang around the edge of the territory. If these subordinates mate a sufficient number of times with a female, they will help raise her young, presumably because they may have fathered some of these young.

 Alternative mating strategies

Natural selection has led to the evolution of many ways of increasing reproductive success. For example, in many species of fish, there are two genetic classes of males. One group is large and defends territories to obtain matings. The other group is small and adopts a completely different strategy. These males do not maintain territories, but loiter at the edge of the territories of large males. Just at the end of a male's courtship, when the female is laying her eggs and the territorial male is depositing sperm, the smaller male darts in and releases its own sperm into the water, thus fertilizing some of the eggs. If this strategy is successful, natural selection will favor the evolution of these two different male reproductive strategies.

Similar patterns are seen in other organisms. In some dung beetles, territorial males have large horns that they use to guard the chambers in which females reside, whereas genetically small males don't have horns. The smaller males can't possibly hope to outcompete the larger ones in one-on-one battles. However, they have a different strategy. Because of the size of their horns, the large males cannot enter the females' chambers. Taking advantage of this, the smaller males dig side tunnels and attempt to intercept and mate with the female inside her chamber. Isopods take this one step further by having three genetic size classes. The medium-sized males pass for females and enter a large male's territory in this way; males in the smallest class are so tiny, they are able to sneak in completely undetected.

This is just a glimpse of the rich diversity in mating systems and mating tactics that have evolved. The bottom line is: If there is a way of increasing reproductive success, natural selection will favor its evolution.

Learning Outcomes Review 54.10

The sex that invests more in reproduction (parental investment) tends to exhibit mate choice. Females or males can be selective, depending on the energy and time they devote to parental care. Sexual selection governs evolution of secondary sex characteristics in that mates are chosen on the basis of phenotype and competitive success. Reproductive success influences whether males and females mate monogamously or with multiple partners.

■ *Pipefish males incubate young in a brood pouch. Which sex would you expect to show mate choice? Why?*

54.11 Altruism

Learning Outcomes

1. *Explain altruism and its benefits.*
2. *Explain kin selection and inclusive fitness.*
3. *Discuss how haplodiploidy influences kin selection in eusocial insects.*

Understanding the evolution of altruism has been a particular challenge to evolutionary biologists, including Darwin himself. Why should an individual decrease his or her own fitness to help another? How could genes for altruism be favored by natural selection, given that the frequency of such genes should decrease in populations through time?

In fact, there can be great benefits to being an altruist, even if the altruism leads an individual to forego reproduction or even sacrifice its own life. Let's examine how this can work.

Altruism is behavior that benefits another individual at a cost to the actor. Humans sacrificing themselves in times of war or placing themselves in jeopardy to help their children are examples, but altruism also has been described in an extraordinary variety of organisms. In many bird species, for example, there are "helpers at the nest"—birds other than parents who assist in raising their young. In both mammals and birds, individuals that spy a predator may give an alarm call, alerting other members of their group to allow them to escape, even though such an act might call the predator's attention to the caller. And in social insects like ants, workers are sterile offspring that help their mother, the colony's queen, to reproduce.

A number of explanations have been put forward to explain the evolution of altruism. Once it was thought that altruism evolved for the "good of the species." Individuals that fail to mate, for example, have been called "altruists" because their lack of success in competition has been misinterpreted as a willingness to forego reproduction so that the population or species does not increase in size, exhaust its resources, and go extinct. This group selection explanation (selection acting on a population or species) is simply incorrect because individuals that fail to secure mates and not breed will not leave any offspring. Therefore, their "altruism" would not be favored by selection.

Current studies of altruism note that seemingly altruistic acts are in fact selfish. For example, helpers at the nest are often young birds that gain valuable parenting experience by assisting established breeders; this may give them an advantage when they breed. Moreover, they may have limited opportunities to reproduce on their own, and by hanging around breeding pairs, may inherit the territory when established breeders die.

Reciprocity theory explains altruism between unrelated individuals

One explanation of altruism proposes that genetically unrelated individuals may form "partnerships" in which mutual exchanges of altruistic acts occur because they benefit both participants. Partners are willing to give aid at one time and delay "repayment" for the good deed to a time in the future when they themselves are in need. In **reciprocal altruism,** the partnerships are stable because "cheaters" (nonreciprocators) are discriminated against and do not receive future aid. According to this hypothesis, if the altruistic act is relatively inexpensive, the small benefit a cheater receives by not reciprocating is far outweighed by the potential cost of not receiving future aid. Under these conditions, cheating behavior should be eliminated by selection.

Vampire bats roost in hollow trees, caves, and mines in groups of 8 to 12 individuals (figure 54.33). Because bats have a high metabolic rate, individuals that have not fed recently may die. Bats that have found a host imbibe a great deal of blood, so giving up a small amount to keep a roostmate from starvation presents no great energy cost to the donor. Vampire bats tend to share blood with past reciprocators that are not necessarily relatives. If an individual fails to give blood to a bat from which it received blood in the past, it will be excluded from future bloodsharing. Reciprocity routinely occurs in many primates, including humans.

Kin selection theory proposes a direct genetic advantage to altruism

The great population geneticist J. B. S. Haldane once passionately said in a pub that he would willingly lay down his life for two brothers or eight first cousins.

Evolutionarily speaking, this sacrifice makes sense, because for each allele Haldane received from his parents, his brothers each had a 50% chance of receiving the same allele (figure 54.34). Statistically, it is expected that two of his brothers would pass on as many of Haldane's particular combination of alleles to the next generation as Haldane himself would.

Figure 54.33 Truth is stranger than fiction: Reciprocal altruism in vampire bats (Desmodus rotundus). Vampire bats do feed on the blood of large mammals, but they don't transform into people and sleep in coffins. Vampires live in groups and share blood meals. They remember which bats have provided them with blood in the past and are more likely to share with those bats that have shared with them previously. The bats here are feeding on cattle in Brazil.

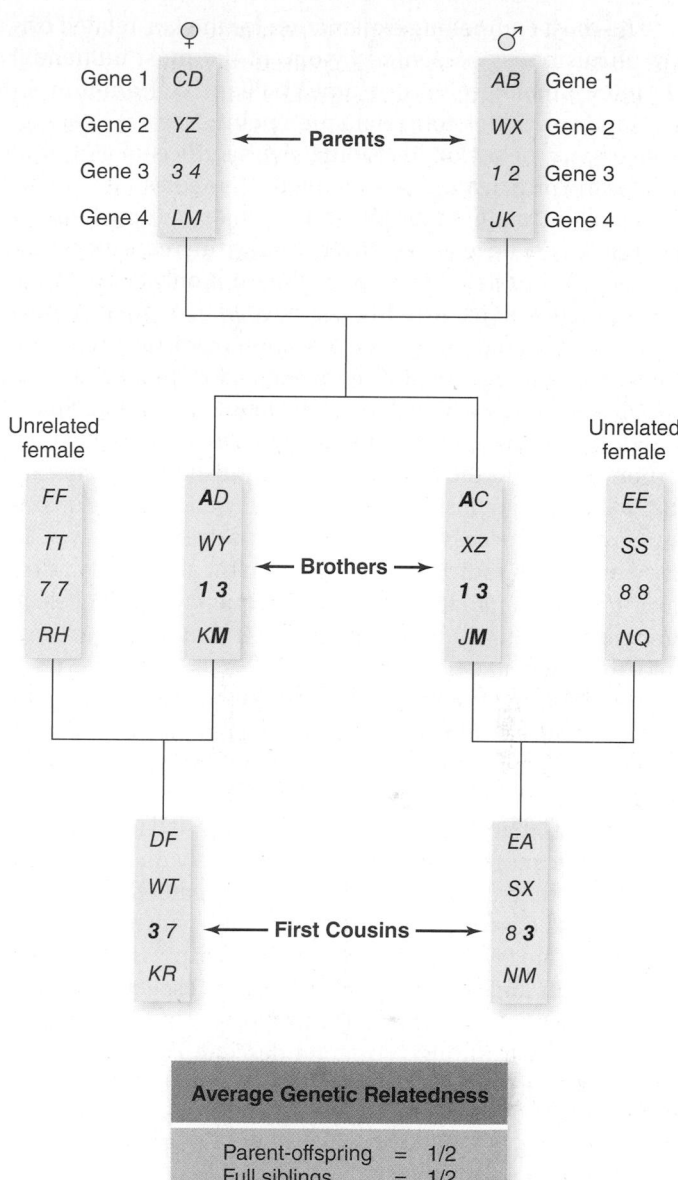

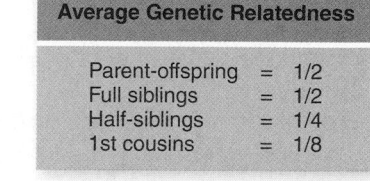

Average Genetic Relatedness	
Parent-offspring	= 1/2
Full siblings	= 1/2
Half-siblings	= 1/4
1st cousins	= 1/8

Figure 54.34 Hypothetical example of genetic relationships. On average, full siblings share half of their alleles (for illustrative purposes, the parents are shown as each having two different alleles for each gene; for example, the father has alleles *A* and *B* for Gene *1*, whereas the mother has alleles *C* and *D*). By contrast, cousins only share one-eighth of their alleles on average. Each letter and number represents a different allele.

 Data analysis On average, how genetically similar are you to your aunt?

Similarly, Haldane and a first cousin would share an eighth of their alleles (see figure 54.34). Their parents, who are siblings, would each share half their alleles, and each of their children would receive half of these, of which half on the average would be in common: $1/2 \times 1/2 \times 1/2 = 1/8$. Eight first cousins would therefore pass on as many of those alleles to the next generation as Haldane himself would.

The most compelling explanation for the kin-related origin of altruism was presented by one of the most influential evolutionary biologists of our time, William D. Hamilton, in 1964. Hamilton understood Haldane's point: Natural selection will favor any behavior, including the sacrifice of life, that increases the propagation of an individual's alleles.

Hamilton mathematically showed that by directing aid toward close genetic relatives, an altruist may increase the reproductive success of its relatives enough to not only compensate for the reduction in its own fitness, but even increase its fitness beyond what would be possible without assisting relatives. Because the altruist's behavior increases the propagation of alleles in relatives, it will be favored by natural selection. Selection that favors altruism directed toward relatives is called kin selection. Although the behaviors are altruistic, the genes are actually "behaving selfishly," because they encourage the organism to favor the success of copies of themselves in relatives. In other words, if an individual has a dominant allele that causes altruism, any action that increases the frequency of this allele in future generations will be favored, even if that action is detrimental to the actor.

Hamilton then defined reproductive success with a new concept—inclusive fitness. Inclusive fitness considers gene propagation through both direct (personal fitness) and indirect (the fitness of relatives) reproduction. Hamilton's kin selection model predicts that altruism is likely to be directed toward close relatives. The more closely related two individuals are, the greater the potential genetic payoff, and the greater the inclusive fitness. This is described by Hamilton's rule, which states that altruistic acts are favored when $rb > c$. In this expression, b and c are the benefits and costs of the altruistic act, respectively, and r is the coefficient of relatedness, the proportion of alleles shared by two individuals through common descent. For example, an individual should be willing to have one less child ($c = 1$) if such actions allow a half-sibling, which shares one-quarter of its genes ($r = 0.25$), to have five or more additional offspring ($b = 5$).

Haplodiploidy and altruism in ants, bees, and wasps

The relationship between genetic relatedness, kin selection, and altruism can be best understood using social insects as an example. A hive of honeybees consists of a single queen, who is the sole egg-layer, and tens of thousands of her offspring, female workers with nonfunctional ovaries (figure 54.35). Honeybees are eusocial ("truly" social): their societies are defined by reproductive division of labor (only the queen reproduces), cooperative care of the brood (workers nurse, clean, and forage), and overlap of generations (the queen lives with several generations of her offspring).

Darwin was perplexed by eusociality. How could natural selection favor the evolution of sterile workers that left no offspring? It remained for Hamilton to explain the origin of eusociality in hymenopterans (bees, wasps, and ants) using his kin selection model. In these insects, males are haploid (produced from unfertilized eggs) and females are diploid. This system of sex determination, called haplodiploidy, leads to unusual genetic relatedness among colony members. If the queen is fertilized by a single male, then all female offspring will inherit exactly the same alleles from their father (because he is haploid and has only one copy of each allele). Female offspring

Figure 54.35 Reproductive division of labor in honeybees. The queen (center) is the sole egg-layer. Her daughters are sterile workers.

(workers and future queens) will also share among themselves, on average, half of the alleles they get from their mother, the queen. Consequently, they will share, on average, 75% of their alleles with each sister (to verify this, rework figure 54.34, but allow the father to only have one allele for each gene).

Now recall Haldane's statement of commitment to family while you read this section. If a worker should have offspring of her own, she would share only half of her alleles with her young (the other half would come from their father). Thus, because of this close genetic relatedness due to haplodiploidy, workers would propagate more of their own alleles by giving up their own reproduction to assist their mother in rearing their sisters, some of whom will be new queens, start new colonies, and reproduce.

In this way, the unusual haplodiploid system may have set the "genetic stage" for the evolution of eusociality. Indeed, eusociality has evolved at least 12 separate times in the Hymenoptera. One wrinkle in this theory, however, is that eusocial systems have evolved in other insects (thrips, some weevils, and termites), and mammals (naked mole rats). Although the thrips and weevils are also haplodiploid, termites and naked mole rats are not. Thus, although haplodiploidy may have facilitated the evolution of eusociality, other factors can influence social evolution.

 Data analysis Suppose that males are haploid (only have one chromosome, and thus only one allele for each gene), as in hymenopterans. How closely related would female first cousins be if their moms were sisters? How closely related would they be if, instead, their fathers were brothers (remember that first cousins are the offspring of siblings (either two brothers, two sisters, or a brother and a sister)? You may want to refer to Figure 54.34 to help you organize your approach to this question.

Other examples of kin selection

Kin selection may explain altruism in other animals. Belding's ground squirrels (*Spermophilus beldingi*) give alarm calls when they spot a predator such as a coyote or a badger. Such predators may attack a calling squirrel, so giving the signal places the caller at risk. A ground squirrel colony consists of a female and her daughters, sisters, aunts, and nieces. When males mature, they disperse long distances from where they

Figure 54.36 Kin selection in the white-fronted bee-eater (*Merops bullockoides*). Bee-eaters are small insectivorous birds that live in Africa in large colonies. Bee-eaters often help others raise their young; helpers usually choose to help close relatives.

are born, so adult males in the colony are not genetically related to the females. By marking all squirrels in a colony with an individual dye pattern on their fur and by recording which individuals gave calls and the social circumstances of their calling, researchers found that females who have relatives living nearby are more likely to give alarm calls than females with no kin nearby. Males tend to call much less frequently, as would be expected because they are not related to most colony members.

Another example of kin selection is provided by the white-fronted bee-eater, a bird which lives along river banks in Africa in colonies of 100 to 200 individuals (figure 54.36). In contrast to ground squirrels, the male bee-eaters usually remain in the colony in which they were born, and the females disperse to join new colonies. Many bee-eaters do not raise their own offspring, but instead help others. Most helpers are young birds, but older birds whose nesting attempts have failed may also be helpers. The presence of a single helper, on average, doubles the number of offspring that survive. Two lines of evidence support the idea that kin selection is important in determining helping behavior in this species. First, helpers are normally males, which are usually related to other birds in the colony, and not females, which are not related. Second, when birds have the choice of helping different parents, they almost invariably choose the parents to which they are most closely related.

Learning Outcomes Review 54.11

Genetic and ecological factors have contributed to evolution of altruism, a behavior that benefits another individual at a cost to the actor. Individuals may benefit directly if cooperative acts are reciprocated among unrelated interactants. Kin selection explains how altruistic acts directed toward relatives, which share alleles, increase an individual's inclusive fitness. Haplodiploidy has resulted in eusociality among some insects by increasing genetic relatedness; it is not found in vertebrates.

■ *Imagine that you witness older group members rescuing infants in a troupe of monkeys when a predator appears. How would you test whether the altruistic act you see is reciprocity or kin selection?*

54.12 The Evolution of Group Living and Animal Societies

Learning Outcomes

1. *Explain the possible advantages of group living.*
2. *Contrast the nature of insect and vertebrate societies.*
3. *Discuss social organization in African weaver birds and how it is influenced by ecology.*

Organisms from cnidarians and insects to fish, whales, chimpanzees, and humans live in social groups. To encompass the wide variety of social phenomena, we can broadly define a society as a group of organisms of the same species that are organized in a cooperative manner.

Why have individuals in some species given up a solitary existence to become members of a group? One hypothesis is that individuals in groups benefit directly from social living. For example, a bird in a flock may be better protected from predators. As flock size increases, the risk of predation decreases because there are more individuals to scan the environment for predators.

A member of a flock may also increase its feeding success if it can acquire information from other flock members about the location of new, rich food sources. In some predators, hunting in groups can increase success and allow the group to tackle prey too large for any one individual.

Insect societies form efficient colonies containing specialized castes

We've already discussed the origin of eusociality in the insect order Hymenoptera (ants, bees, and wasps). Additionally, all termites (order Isoptera) are also eusocial, and a few other insect and arthropod species are eusocial. Social insect colonies are composed of different *castes,* groups of individuals that differ in reproductive ability (queens vs. workers), size, and morphology and perform different tasks. Workers nurse, maintain the nest, and forage; soldiers are large and have powerful jaws specialized for defense.

The structure of an insect society is illustrated by leaf-cutters, which form colonies of as many as several million individuals. These ants cut leaves and use it to grow crops of fungi beneath the ground. Workers divide the tasks of leaf cutting, defense, mulching the fungus garden, and implanting fungal hyphae according to their body size (figure 54.37).

The structure of a vertebrate society is related to ecology

In contrast to the highly structured and integrated insect societies and their remarkable forms of altruism, vertebrate social groups are usually less rigidly organized and less cohesive. It seems paradoxical that vertebrates, which have larger brains and are capable of more complex behaviors, are generally less altruistic than insects (the exception, of course, is humans). Reciprocity and kin-selected altruism are common in vertebrate societies, although there is often more conflict and aggression among group members. Conflicts generally center on access to food and mates and occur because a vertebrate society is made up of individuals striving to improve their own fitness.

Social groups of vertebrates have a size, stability of members, number of breeding males and females, and type of mating system characteristic of a given species. Diet and predation are important factors in shaping social groups. For example, meerkats take turns watching for predators while other group members forage for food (figure 54.38).

African weaver birds, which construct nests from vegetation, provide an excellent example of the relationship between

Figure 54.38 Foraging and predator avoidance. A meerkat sentinel on duty. Meerkats (*Suricata suricata*) are a species of highly social mongoose living in the semiarid sands of the Kalahari Desert in southern Africa. This meerkat is taking its turn to act as a lookout for predators. Under the security of its vigilance, the other group members can focus their attention on foraging.

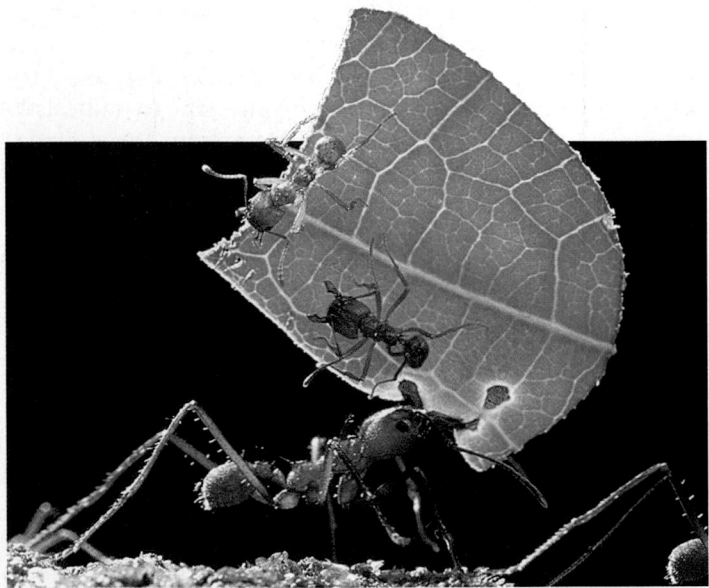

Figure 54.37 Castes of ants. These leaf-cutter ants are members of different castes. The large ant is a worker carrying leaves to the nest, whereas the smaller ants are protecting the worker from attack by small parasitic flies.

ecology and social organization. Their roughly 90 species can be divided according to the type of social group they form. One group of species lives in the forest and builds camouflaged, solitary nests. Males and females are monogamous; they forage for insects to feed their young. The second group of species nests in colonies in trees on the savanna. They are polygynous and feed in flocks on seeds.

The feeding and nesting habits of these two groups of species are correlated with their mating systems. In the forest, insects are hard to find, and both parents must cooperate in feeding the young. The camouflaged nests do not call the attention of predators to their brood. On the open savanna, building a hidden nest is not an option. Rather, savanna-dwelling weaver birds protect their young from predators by nesting in trees, which are not very abundant. This shortage of safe nest sites means that birds must nest together in colonies. Because seeds occur abundantly, a female can acquire all the food needed to rear young without a male's help. The male, free from the duties of parenting, spends his time courting many females—a polygynous mating system.

One exception to the general rule that vertebrate societies are not organized like those of insects is the naked mole rat (*Heterocephalus glaber*), a small, hairless rodent that lives in and near East Africa. Unlike other kinds of mole rats, which live alone or in small family groups, naked mole rats form large underground colonies with a far-ranging system of tunnels and a central nesting area. It is not unusual for a colony to contain 80 individuals.

Naked mole rats feed on bulbs, roots, and tubers, which they locate by constant tunneling. As in insect societies, there is a division of labor among the colony members, with some individuals working as tunnelers while others perform different tasks, depending on the size of their bodies. Large mole rats defend the colony and dig tunnels.

Naked mole rat colonies have a reproductive division of labor similar to the one normally associated with the eusocial insects. All of the breeding is done by a single female, or "queen," who has one or two male consorts. The workers, consisting of both sexes, keep the tunnels clear and forage for food.

Learning Outcomes Review 54.12

Advantages of group living include protection from predators and increased feeding success. Eusocial insects form complex, highly altruistic societies that increase the fitness of the colony. The members of vertebrate societies exhibit more conflict and competition, but also cooperate and behave altruistically, especially toward kin. African weaver birds have developed different types of societies depending on the ecology of their habitat, particularly the safety of nesting sites.

- ■ *What are the benefits and costs associated with living in social groups?*
- ■ *Why is altruism directed toward kin considered to be selfish behavior?*
- ■ *Is a human army more like an insect society or a vertebrate society? Explain your answer.*

Chapter Review

54.1 The Natural History of Behavior

Behavior can be analyzed in terms of mechanisms and evolutionary origin.

Proximate causation refers to the mechanisms of behavior. Ultimate causation examines a behavior's evolutionary significance.

Ethology emphasizes the study of instinct and its origins.

Innate, or instinctive, behavior is a response to an environmental stimulus or trigger that does not require learning (figure 54.1).

54.2 Nerve Cells, Neurotransmitters, Hormones, and Behavior

Instinctive behaviors are accomplished by neural circuits, which develop under genetic control. Hormones and neurotransmitters can act to regulate behavior.

54.3 Behavioral Genetics

Artificial selection, hybrid and twin studies link genes and behavior.

Breeding fast-learning and slow-learning rats for several generations produced two distinct behavioral populations (figure 54.3).

Some behaviors appear to be controlled by a single gene.

54.4 Learning

Learning mechanisms include habituation and association.

Habituation, a form of nonassociative learning, is a decrease in response to repeated nonessential stimuli. Associative learning is a change in behavior by association of two stimuli or of a behavior and a response (conditioning).

Classical (Pavlovian) conditioning occurs when two stimuli are associated with each other. Operant conditioning occurs when an animal associates a behavior with reward or punishment.

Instinct governs learning preparedness.

What an animal can learn is biologically influenced—that is, learning is possible only within the boundaries set by evolution.

54.5 The Development of Behavior

Parent–offspring interactions influence how behavior develops.

In imprinting, a young animal forms an attachment to other individuals or develops preferences that influence later behavior.

Instinct and learning may interact as behavior develops.

Animals may have an innate genetic template that guides their learning as behavior develops, such as song development in birds. Studies on twins reveal a role for both genes and environment in human behavior.

54.6 Animal Cognition

Some animals exhibit cognitive behavior and can devise solutions to novel situations (figures 54.12, 54.13).

54.7 Orientation and Migratory Behavior

Migration often involves populations moving large distances.

Migrating animals must be capable of orientation and navigation (figure 54.16).

Orientation is the mechanism by which animals move by tracking environmental stimuli such as celestial clues or Earth's magnetic field. Navigation is following a route based on orientation and some sort of "map." The nature of the map in animals is not known.

54.8 Animal Communication

Successful reproduction depends on appropriate signals and responses.

Courtship signals are usually species-specific and help to ensure reproductive isolation (figure 54.19).

Communication enables information exchange among group members (figures 54.20, 54.21).

54.9 Behavioral Ecology

Foraging behavior can directly influence energy intake and individual fitness.

Natural selection favors optimal foraging strategies in which energy acquisition (cost) is minimized and reproductive success (benefit) is maximized.

Territorial behavior evolves if the benefits of holding a territory exceed the costs.

54.10 Reproductive Strategies and Sexual Selection

The sexes often have different reproductive strategies.

One sex may be choosier than the other, and which one often depends on the degree of parental investment.

Sexual selection occurs through mate competition and mate choice.

Intrasexual selection involves competition among members of the same sex for the chance to mate. Intersexual selection is one sex choosing a mate.

Mate choice may provide direct benefits (increased resource availability or parental care) or indirect benefits (genetic quality of the mate).

Mating systems reflect the ability of parents to care for offspring and are influenced by ecology.

Mating systems include monogamy, polygyny, and polyandry; they are influenced by ecology and constrained by needs of offspring.

54.11 Altruism

Reciprocity theory explains altruism between unrelated individuals.

Mutual exchanges benefit both participants; a participant that does not reciprocate would not receive future aid.

Kin selection theory proposes a direct genetic advantage to altruism.

Kin selection increases the reproductive success of relatives and increased frequency of alleles shared by kin, and thus increases an individual's inclusive fitness.

Ants, bees, and wasps have haplodiploid reproduction, and therefore a high degree of gene sharing.

54.12 The Evolution of Group Living and Animal Societies

A social system is a group organized in a cooperative manner.

Insect societies form efficient colonies containing specialized castes (figure 54.37).

Social insect societies are composed of different castes that are specialized to reproduce or to perform certain colony maintenance tasks.

The structure of a vertebrate society is related to ecology.

Vertebrate social systems are less rigidly organized and cohesive and are influenced by food availability and predation.

Review Questions

UNDERSTAND

1. A key stimulus, innate releasing mechanism, and fixed action pattern
   a. are mechanisms associated with behaviors that are learned.
   b. are components of behaviors that are innate.
   c. involve behaviors that cannot be explained in terms of ultimate causation.
   d. involve behaviors that are not subject to natural selection.

2. In operant conditioning
   a. an animal learns that a particular behavior leads to a reward or punishment.
   b. an animal associates an unconditioned stimulus with a conditioned response.
   c. learning is unnecessary.
   d. habituation is required for an appropriate response.

3. The study of song development in sparrows showed that
   a. the acquisition of a species-specific song is innate.
   b. there are two components to this behavior: a genetic template and learning.
   c. song acquisition is an example of associative learning.
   d. All of the choices are correct.

4. The difference between following a set of driving directions given to you by somebody on the street (for example ". . . take a right at the next light, go four blocks and turn left . . .") and using a map to find your destination is
   a. the difference between navigation and orientation, respectively.
   b. the difference between learning and migration, respectively.
   c. the difference between orientation and navigation, respectively.
   d. why birds are not capable of orientation.

5. In courtship communication
   a. the signal itself is always species-specific.
   b. the sign communicates species identity.
   c. a stimulus–response chain is involved.
   d. courtship signals are produced only by males.

6. Behavioral ecology assumes
   a. that all behavioral traits are innate.
   b. learning is the dominant determinant of behavior.
   c. behavioral traits are subject to natural selection.
   d. behavioral traits do not affect fitness.

7. According to optimal foraging theory

 a. individuals minimize energy intake per unit of time.

 b. energy content of a food item is the only determinant of a forager's food choice.

 c. time taken to capture a food item is the only determinant of a forager's food choice.

 d. a higher energy item might be less valuable than a lower energy item if it takes too much time to capture the larger item.

8. The elaborate tail feathers of a male peacock evolved because they

 a. improve reproductive success of males and females.

 b. improve male survival.

 c. reduce survival.

 d. None of the choices is correct.

9. From the perspective of females, extra-pair copulations (EPCs)

 a. are always disadvantageous to females.

 b. can be associated with receiving male aid.

 c. are unimportant to fitness because they rarely lead to fertilization of eggs.

 d. can only be of benefit if the EPC male has elaborate secondary sexual traits.

10. In the haplodiploidy system of sex determination, males are

 a. haploid.

 b. diploid.

 c. sterile.

 d. not present because bees exist as single-sex populations.

11. According to kin selection, saving the life of your _____ would do the least for increasing your inclusive fitness.

 a. mother c. sister-in-law

 b. brother d. niece

12. Altruism

 a. is only possible through reciprocity.

 b. is only possible through kin selection.

 c. can only be explained by group selection.

 d. will only occur when the fitness benefit of a given act is greater than the fitness cost.

APPLY

1. Refer to figure 54.25. Data on size of mussels eaten by shore crabs suggest they eat sizes smaller than expected by an optimal foraging model. Suggest a hypothesis for why and describe an experiment to test your hypothesis.

2. Refer to figure 54.26. Six pairs of birds were removed but only four pairs moved in. Where did the new pairs come from? Additionally, it appears that many of the birds that were not removed expanded their territories and that the new residents ended up with smaller territories than the pairs they replaced. Explain.

3. Refer to figure 54.28. Peahens prefer to mate with peacocks that have more eyespots in their tail feathers (that is, longer tail feathers). It has also been suggested that the longer the tail feathers, the more impaired the flight of the males. One possible hypothesis to explain such a preference by females is that the males with the longest tail feathers experience the most severe handicap, and if they can nevertheless survive, it reflects their "vigor." Suggest some studies that would allow you to test this idea. Your description should include the kinds of traits that you would measure and why.

4. An altruistic act is defined as one that benefits another individual at a cost to the actor. There are two theories to explain how such behavior evolves: reciprocity and kin selection. How would you distinguish between the two in a field study? In the context of natural selection, is an altruistic act "costly" to an individual who performs it?

SYNTHESIZE

1. Insects that sting or contain toxic chemicals often have black and yellow coloration and consequentially are not eaten by predators. How could you determine if a predator has an innate avoidance of insects that are colored this way, or if the avoidance is learned? If avoidance is learned, how would you determine the learning mechanism involved? How would you measure the adaptive significance of the black and yellow coloration to the prey insect?

2. Behavioral genetics has made great advances from detailed studies of a single animal such as the fruit fly as a model system to develop general principles of how genes regulate behavior. What are advantages and disadvantages of this "model system" approach? How would you determine how broadly applicable the results of such studies are to other animals?

3. If a female bird chooses to live in the territory of a particular male, why might she mate with a male other than the territory owner?

ONLINE RESOURCE

www.ravenbiology.com

Understand, Apply, and Synthesize—enhance your study with animations that bring concepts to life and practice tests to assess your understanding. Your instructor may also recommend the interactive eBook, individualized learning tools, and more.

Chapter 55

Ecology of Individuals and Populations

Chapter Contents

Introduction

Ecology, the study of how organisms relate to one another and to their environments, is a complex and fascinating area of biology that has important implications for each of us. In our exploration of ecological principles, we first consider how organisms respond to the abiotic environment in which they exist and how these responses affect the properties of populations, emphasizing population dynamics. In chapter 56, we discuss communities of coexisting species and the interactions that occur among them. In subsequent chapters, we discuss the functioning of entire ecosystems and of the biosphere, concluding with a consideration of the problems facing our planet and our fellow species.

55.1 The Environmental Challenges

Learning Outcomes

1. List some challenges that organisms face in their environments.
2. Describe ways in which individuals respond to environmental changes.
3. Explain how species adapt to environmental conditions.

The nature of the physical environment in large measure determines which organisms live in a particular climate or region. Key elements of the environment include:

Temperature. Most organisms are adapted to live within a relatively narrow range of temperatures and will not thrive if temperatures are colder or warmer. The growing season of plants, for example, is importantly influenced by temperature.

Water. All organisms require water. On land, water is often scarce, so patterns of rainfall have a major influence on life.

Figure 55.1 Meeting the challenge of obtaining moisture. On the dry sand dunes of the Namib Desert in southwestern Africa, the fog-basking beetle (*Onymacris unguicularis)* collects moisture from the fog by holding its abdomen up at the crest of a dune to gather condensed water; water condenses as droplets and trickles down to the beetle's mouth.

Sunlight. Almost all ecosystems rely on energy captured by photosynthesis; the availability of sunlight influences the amount of life an ecosystem can support, particularly below the surface in marine communities.

Soil. The physical consistency, pH, and mineral composition of the soil often severely limit terrestrial plant growth, particularly the availability of nitrogen and phosphorus.

An individual encountering environmental variation may maintain a "steady-state" internal environment, a condition known as *homeostasis*. Many animals and plants actively employ physiological, morphological, or behavioral mechanisms to maintain homeostasis. The beetle in figure 55.1 is using a behavioral mechanism to cope with drastic changes in water availability. Other animals and plants are known as conformers because they conform to the environment in which they find themselves, their bodies adopting the temperature, salinity, and other physical aspects of their surroundings.

Responses to environmental variation can be seen over both the short and the long term. In the short term, spanning periods of a few minutes to an individual's lifetime, organisms have a variety of ways of coping with environmental change. Over longer periods, natural selection can operate to make a population better adapted to its environment.

Organisms are capable of responding to environmental changes that occur during their lifetime

During the course of a day, a season, or a lifetime, an individual organism must cope with a range of living conditions. They do so through the physiological, morphological, and behavioral abilities they possess. These abilities are a product of natural selection acting in a particular environmental setting over time, which explains why an individual organism that is moved to a different environment may not survive.

TABLE 55.1	Physiological Changes at High Elevation
Increased rate of breathing	
Increased erythrocyte production, raising the amount of hemoglobin in the blood	
Decreased binding capacity of hemoglobin, increasing the rate at which oxygen is unloaded in body tissues	
Increased density of mitochondria, capillaries, and muscle myoglobin	

 Physiological responses

Many organisms are able to respond to environmental change by making physiological adjustments. For example, you sweat when it is hot, increasing evaporative heat loss and thus preventing overheating. Similarly, people who visit high altitudes may initially experience altitude sickness—the symptoms of which include heart palpitations, nausea, fatigue, headache, mental impairment, and in serious cases, pulmonary edema— because of the lower atmospheric pressure and consequent lower oxygen availability in the air. After several days, however, the same people usually feel fine, because a number of physiological changes have increased the delivery of oxygen to their body tissues (table 55.1).

Some insects avoid freezing in the winter by adding glycerol "antifreeze" to their blood; others tolerate freezing by converting much of their glycogen reserves into alcohols that protect their cell membranes from freeze damage.

 Morphological capabilities

Animals that maintain a constant internal temperature (endotherms) in a cold environment have adaptations that tend to minimize energy expenditure. For example, many mammals grow thicker coats during the winter, their fur acting as insulation to retain body heat. In general, the thicker the fur, the greater the insulation (figure 55.2). Thus, a wolf's fur is about three times thicker in winter than in summer and insulates more than twice as well.

Figure 55.2 Morphological adaptation. Fur thickness in North American mammals has a major effect on the degree of insulation the fur provides.

Behavioral responses

Many animals deal with variation in the environment by moving from one patch of habitat to another, avoiding areas that are unsuitable. The tropical lizard in figure 55.3 manages to maintain a fairly uniform body temperature in an open habitat by basking in patches of sunlight and then retreating to the shade when it becomes too hot. By contrast, in shaded forests, the same lizard does not have the opportunity to regulate its body temperature through behavioral means. Thus, it becomes a conformer and adopts the temperature of its surroundings.

Behavioral adaptations can be extreme. Spadefoot toads (genus *Scaphiophus*), which are widely distributed in North America, can burrow nearly a meter below the surface and remain there for as long as nine months of each year, their metabolic rates greatly reduced as they live on fat reserves. When moist, cool conditions return, the toads emerge and breed. The young toads mature rapidly and burrow underground.

Natural selection leads to evolutionary adaptation to environmental conditions

The ability of an individual to alter its physiology, morphology, or behavior is itself an evolutionary adaptation, the result of natural selection. The results of natural selection can also be detected by comparing closely related species that live in different environments. In such cases, species often have evolved striking adaptations to the particular environment in which they live.

For example, animals that live in different climates show many differences. Mammals from colder climates tend to have shorter ears and limbs—a phenomenon termed *Allen's rule*—which reduces the surface area across which animals lose heat. Lizards that live in different climates exhibit physiological adaptations for coping with life at different temperatures. Desert lizards are unaffected by high temperatures that would kill a lizard from northern Europe, but the northern lizards are capable of running, capturing prey, and digesting food at cooler temperatures at which desert lizards would be completely immobilized.

Many species also exhibit adaptations to living in areas where water is scarce. Everyone knows of the camel and other desert animals that can go extended periods without drinking water. Another example of desert adaptation is seen in frogs. Most frogs have moist skins through which water permeates readily. Such animals could not survive in arid climates because they would rapidly dehydrate and die. However, some frogs have solved this problem by evolving a greatly reduced rate of water loss through the skin. One species, for example, secretes a waxy substance from specialized glands that water-proofs its skin and reduces rates of water loss by 95%.

Adaptation to different environments can also be studied experimentally. For example, when strains of *E. coli* were grown at high temperatures (42°C), the speed at which the bacteria utilized

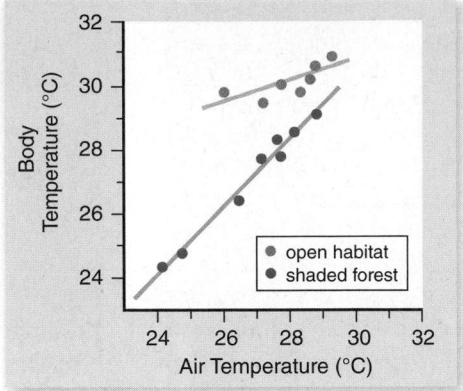

Figure 55.3 Behavioral adaptation. In open habitats, the Puerto Rican crested lizard *(Anolis cristatellus)* maintains a relatively constant temperature by seeking out and basking in patches of sunlight; as a result, it can maintain a relatively high temperature even when the air is cool. In contrast, in shaded forests, this behavior is not possible, and the lizard's body temperature conforms to that of its surroundings.

? Inquiry question When given the opportunity, lizards regulate their body temperature to maintain a temperature optimal for physiological functioning. Would lizards in open habitats exhibit different escape behaviors from lizards in shaded forest?

Data analysis Can the slope of the line tell us something about the behavior of the lizard?

resources improved through time. After 2000 generations, this ability increased 30% over what it had been when the experiment started. The mechanism by which efficiency of resource use increased is unknown and is the focus of current research.

Learning Outcomes Review 55.1

Environmental conditions include temperature, water and light availability, and soil characteristics. When the environment changes, individual organisms use a variety of physiological, morphological, and behavioral mechanisms to adjust. Over time, adaptations to different environments may evolve in populations.

- **How might a species respond if its environment grew steadily warmer over time?**

55.2 Populations: Groups of a Single Species in One Place

Learning Outcomes

1. Distinguish between a population and a metapopulation.
2. Understand what causes a species' geographic range to change through time.

Organisms live as members of a **population,** groups of individuals that occur together at one place and time. In the rest of this chapter, we consider the properties of populations, focusing on factors that influence whether a population grows or shrinks, and at what rate. The explosive growth of the world's human population in the last few centuries provides a focus for our inquiry.

The term population can be defined narrowly or broadly. This flexibility allows us to speak in similar terms of the world's human population, the population of protists in the gut of a termite, or the population of deer that inhabit a forest. Sometimes the boundaries defining a population are sharp, such as the edge of an isolated mountain lake for trout, and sometimes they are fuzzier, as when deer readily move back and forth between two forests separated by a cornfield.

Three characteristics of population ecology are particularly important: (1) population range, the area throughout which a population occurs; (2) the pattern of spacing of individuals within that range; and (3) how the population changes in size through time.

A population's geographic distribution is termed its range

No population, not even one composed of humans, occurs in all habitats throughout the world. Most species, in fact, have relatively limited geographic ranges, and the range of some species is miniscule. For example, the Devil's Hole pupfish lives in a single spring in southern Nevada (figure 55.4), and the Socorro isopod *(Thermosphaeroma thermophilus)* is known from a single spring system in New Mexico. At the other extreme, some species are widely distributed. The common dolphin *(Delphinus delphis),* for example, is found throughout all the world's oceans.

As discussed earlier, organisms must be adapted to the environment in which they occur. Polar bears are exquisitely adapted to survive the cold of the Arctic, but you won't find them in the tropical rain forest. Certain prokaryotes can live in the near-boiling waters of Yellowstone's geysers, but they do not occur in cooler streams nearby. Each population has its own requirements—temperature, humidity, certain types of food, and a host of other factors—that determine where it can live and reproduce and where it can't. In addition, in places that are otherwise suitable, the presence of predators, competitors, or parasites may prevent a population from occupying an area, a topic we will take up in chapter 56.

Ranges undergo expansion and contraction

Population ranges are not static but change through time. These changes occur for two reasons. In some cases, the environment changes. As the glaciers retreated at the end of the last ice age, approximately 10,000 years ago, many North American plant and animal populations expanded northward. At the same time, as climates warmed, species experienced shifts in the elevation at which they could live (figure 55.5).

In addition, populations can expand their ranges when they are able to circumvent inhospitable habitat to colonize suitable, previously unoccupied areas. For example, the cattle egret is native to Africa. Some time in the late 1800s, these birds appeared in northern South America, having made the nearly 3500-km transatlantic crossing, perhaps aided by

Figure 55.4 The Devil's Hole pupfish (Cyprinodon diabolis). This fish has the smallest range of any vertebrate species in the world.

strong winds. Since then, they have steadily expanded their range and now can be found throughout most of the United States (figure 55.6).

The human effect

By altering the environment, humans have allowed some species, such as coyotes, to expand their ranges and move into areas they previously did not occupy. Moreover, humans have

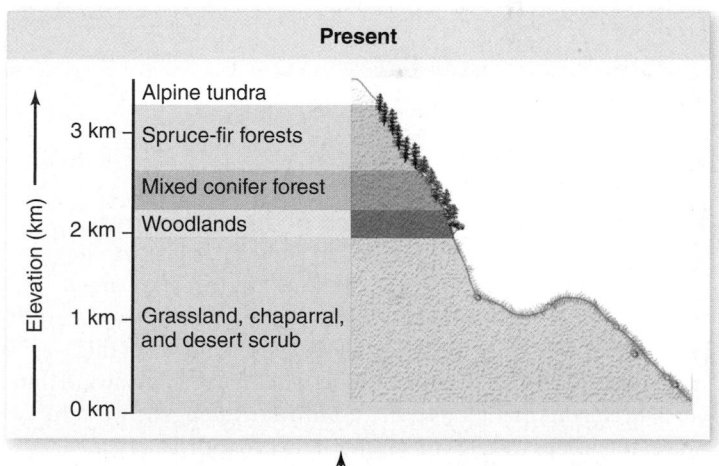

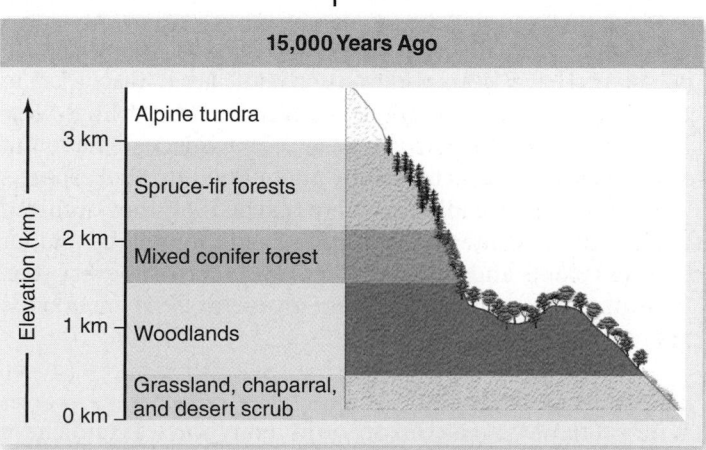

Figure 55.5 Altitude shifts in altitudinal distributions of trees in the mountains of southwestern North America. During the glacial period 15,000 years ago, conditions were cooler than they are now. As the climate warmed, tree species that require colder temperatures shifted their range upward in altitude so that they continue to live in the climatic conditions to which they are adapted.

Figure 55.6 Range expansion of the cattle egret (Bubulcus ibis). The cattle egret—so named because it follows cattle and other hoofed animals, catching any insects or small vertebrates it disturbs—first arrived in South America from Africa in the late 1800s. Since the 1930s, the range expansion of this species has been well documented, as it has moved northward into much of North America, as well as southward along the western side of the Andes to near the southern tip of South America.

served as an agent of dispersal for many species. Some of these transplants have been widely successful, as is discussed in greater detail in chapter 59. For example, 100 starlings were introduced into New York City in 1896 in a misguided attempt to establish every species of bird mentioned by Shakespeare. Their population steadily spread so that by 1980, they occurred throughout the United States. Similar stories could be told for countless plants and animals, and the list increases every year. Unfortunately, the success of these invaders often comes at the expense of native species.

Dispersal mechanisms

Dispersal to new areas can occur in many ways. Lizards have colonized many distant islands, as one example, probably due to individuals or their eggs floating or drifting on vegetation. Bats are the only mammals on many distant islands because they can fly to them.

Seeds of plants are designed to disperse in many ways (figure 55.7). Some seeds are aerodynamically designed to be blown long distances by the wind. Others have structures that stick to the fur or feathers of animals, so that they are carried

long distances before falling to the ground. Still others are enclosed in fleshy fruits. These seeds can pass through the digestive systems of mammals or birds and then germinate where they are defecated. Finally, seeds of mistletoes *(Arceuthobium)* are violently propelled from the base of the fruit in an explosive discharge. Although the probability of long-distance dispersal events leading to successful establishment of new populations is low, over millions of years, many such dispersals have occurred.

Individuals in populations exhibit different spacing patterns

Another key characteristic of population structure is the way in which individuals of a population are distributed. They may be randomly spaced, uniformly spaced, or clumped.

Random spacing

Random spacing of individuals within populations occurs when they do not interact strongly with one another and when they are not affected by nonuniform aspects of their environment. Random distributions are not common in nature. Some species of trees, however, appear to exhibit random distributions in Panamanian rain forests.

Uniform spacing

Uniform spacing within a population may often, but not always, result from competition for resources. This spacing is accomplished, however, in many different ways.

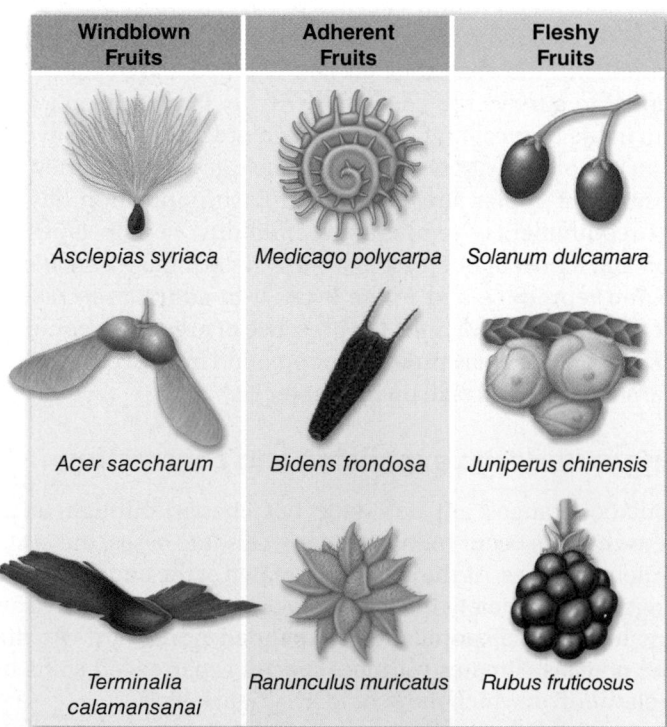

Windblown Fruits	Adherent Fruits	Fleshy Fruits
Asclepias syriaca	*Medicago polycarpa*	*Solanum dulcamara*
Acer saccharum	*Bidens frondosa*	*Juniperus chinensis*
Terminalia calamansanai	*Ranunculus muricatus*	*Rubus fruticosus*

Figure 55.7 Some of the many adaptations for dispersal of seeds. Seeds have evolved a number of different means of facilitating dispersal from their maternal plant. Some seeds can be transported great distances by the wind, whereas seeds enclosed in adherent or fleshy fruits can be transported by animals.

In animals, uniform spacing often results from behavioral interactions, as described in chapter 54. In many species, individuals of one or both sexes defend a territory from which other individuals are excluded. These territories provide the owner with exclusive access to resources, such as food, water, hiding refuges, or mates, and tend to space individuals evenly across the habitat. Even in nonterritorial species, individuals often maintain a defended space into which other animals are not allowed to intrude.

Among plants, uniform spacing is also a common result of competition for resources. Closely spaced individual plants compete for available sunlight, nutrients, or water. These contests can be direct, as when one plant casts a shadow over another, or indirect, as when two plants compete by extracting nutrients or water from a shared area. In addition, some plants, such as the creosote bush, produce chemicals in the surrounding soil that are toxic to other members of their species. In all of these cases, only plants that are spaced an adequate distance from each other will be able to coexist, leading to uniform spacing.

Clumped spacing

Individuals clump into groups or clusters in response to uneven distribution of resources in their immediate environments. Clumped distributions are common in nature because individual animals, plants, and microorganisms tend to occur in habitats defined by soil type, moisture, or other aspects of the environment to which they are best adapted.

Social interactions also can lead to clumped distributions. Many species live and move around in large groups, which go by a variety of names (for example, flock, herd, pride). These groupings can provide many advantages, including increased awareness of and defense against predators, decreased energy cost of moving through air and water, and access to the knowledge of all group members.

On a broader scale, populations are often most densely populated in the interior of their range and less densely distributed toward the edges. Such patterns usually result from the manner in which the environment changes in different areas.

Populations are often best adapted to the conditions in the interior of their distribution. As environmental conditions change, individuals are less well adapted, and thus densities decrease. Ultimately, the point is reached at which individuals cannot persist at all; this marks the edge of a population's range.

A metapopulation comprises distinct populations that may exchange members

Species often exist as a network of distinct populations that interact with one another by exchanging individuals. Such networks, termed **metapopulations,** usually occur in areas in which suitable habitat is patchily distributed and is separated by intervening stretches of unsuitable habitat.

Dispersal and habitat occupancy

The degree to which populations within a metapopulation interact depends on the amount of dispersal; this interaction is often not symmetrical: Populations increasing in size tend to send out many dispersers, whereas populations at low levels tend to receive more immigrants than they send off. In addition, relatively isolated populations tend to receive relatively few arrivals.

Not all suitable habitats within a metapopulation's area may be occupied at any one time. For a number of reasons, some individual populations may become extinct, perhaps as a result of an epidemic disease, a catastrophic fire, or the loss of genetic variation following a population bottleneck (see chapter 59). Dispersal from other populations, however, may eventually recolonize such areas. In some cases, the number of habitats occupied in a metapopulation may represent an equilibrium in which the rate of extinction of existing populations is balanced by the rate of colonization of empty habitats.

Source–sink metapopulations

A species may also exhibit a metapopulation structure in areas in which some habitats are suitable for long-term population maintenance, but others are not. In these situations, termed **source–sink metapopulations,** the populations in the better areas (the sources) continually send out dispersers that bolster the populations in the poorer habitats (the sinks). In the absence of such continual replenishment, sink populations would have a negative growth rate and would eventually become extinct.

Metapopulations of butterflies have been studied particularly intensively. In one study, researchers sampled populations of the Glanville fritillary butterfly at 1600 meadows in southwestern Finland (figure 55.8). On average, every year,

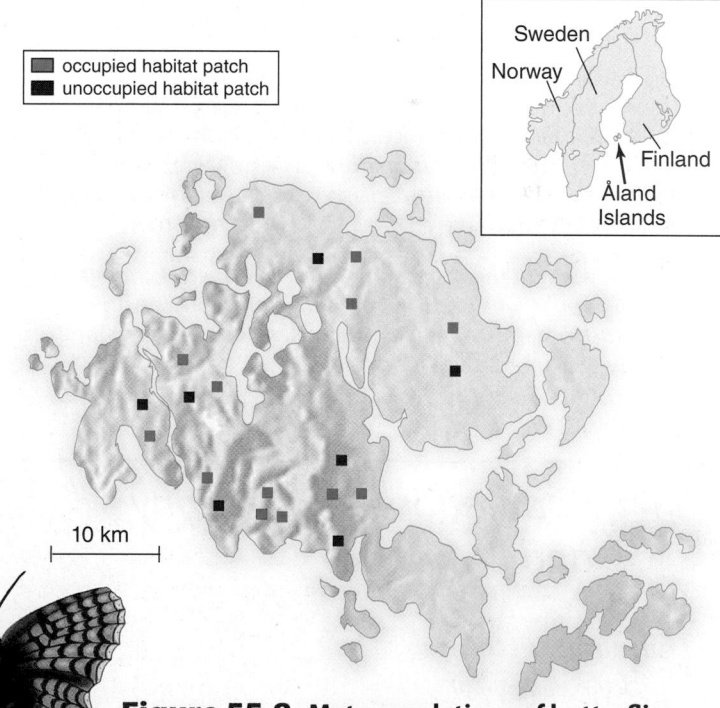

Figure 55.8 Metapopulations of butterflies. The Glanville fritillary butterfly *(Melitaea cinxia)* occurs in metapopulations in southwestern Finland on the Åland Islands. None of the populations is large enough to survive for long on its own, but continual immigration of individuals from other populations allows some populations to survive. In addition, continual establishment of new populations tends to offset extinction of established populations, although in recent years, extinctions have outnumbered colonizations.

chapter **55** *Ecology of Individuals and Populations* **1167**

200 populations became extinct, but 114 empty meadows were colonized. A variety of factors seemed to increase the likelihood of a population's extinction, including small population size, isolation from sources of immigrants, low resource availability (as indicated by the number of flowers on a meadow), and lack of genetic variation within the population.

The researchers attribute the greater number of extinctions than colonizations to a string of very dry summers. Because none of the populations is large enough to survive on its own, continued survival of the species in southwestern Finland would appear to require the continued existence of a metapopulation network in which new populations are continually created and existing populations are supplemented by immigrants. Continued bad weather thus may doom the species, at least in this part of its range.

Metapopulations, where they occur, can have two important implications for the range of a species. First, through continuous colonization of empty patches, metapopulations prevent long-term extinction. If no such dispersal existed, then each population might eventually perish, leading to disappearance of the species from the entire area. Moreover, in source-sink metapopulations, the species occupies a larger area than it otherwise might, including marginal areas that could not support a population without a continual influx of immigrants. For these reasons, the study of metapopulations has become very important in conservation biology as natural habitats become increasingly fragmented.

Learning Outcomes Review 55.2

A population is a group of individuals of a single species existing together in an area. A population's range, the area it occupies, changes over time. Populations, in turn, may form a network, or metapopulation, connected by individuals that move from one group to another. Within a population, the distribution of individuals can be random, uniform, or clumped, and the distribution is determined in part by the availability of resources.

■ *How might the geographic range of a species change if populations could not exchange individuals with each other? How would your answer depend on what type of metapopulation existed?*

55.3 Population Demography and Dynamics

Learning Outcomes

1. *Define demography.*
2. *Describe the factors that influence a species' demography.*
3. *Explain the significance of survivorship curves.*

The dynamics of a population—how it changes through time—are affected by many factors. One important factor is the age distribution of individuals—that is, what proportion of individuals are adults, juveniles, and young.

Demography is the quantitative study of populations. How the size of a population changes through time can be studied at two levels: as a whole or broken down into parts. At the most inclusive level, we can study the whole population to determine whether it is increasing, decreasing, or remaining constant. Put simply, populations grow if births outnumber deaths and shrink if deaths outnumber births (ignoring, for the moment, immigration and extinction). Understanding these trends is often easier, however, if we break the population into smaller units composed of individuals of the same age (for example, 1-year-olds) and study the factors affecting birth and death rates for each unit separately.

Sex ratio and generation time affect population growth rates

Population growth can be influenced by the population's sex ratio. The number of births in a population is usually directly related to the number of females; births may not be as closely related to the number of males in species in which a single male can mate with several females. In many species, males compete for the opportunity to mate with females, as you learned in chapter 54; consequently, a few males have many matings, and many males do not mate at all. In such species, the sex ratio is female-biased and does not affect population growth rates; reduction in the number of males simply changes the identities of the reproductive males without reducing the number of births. By contrast, among monogamous species, pairs may form long-lasting reproductive relationships, and a reduction in the number of males can then directly reduce the number of births.

Generation time is the average interval between the birth of an individual and the birth of its offspring. This factor can also affect population growth rates. Species differ greatly in generation time. Differences in body size can explain much of this variation—mice go through approximately 100 generations during the course of one elephant generation (figure 55.9). But small size does not always mean short generation time. Newts, for example, are smaller than mice, but have considerably longer generation times.

In general, populations with short generations can increase in size more quickly than populations with long generations. Conversely, because generation time and life span are usually closely correlated, populations with short generation times may also diminish in size more rapidly if birth rates suddenly decrease.

Age structure is determined by the numbers of individuals in different age groups

A group of individuals of the same age is referred to as a cohort. In most species, the probability that an individual will reproduce or die varies through its life span. As a result, within a population, every cohort has a characteristic birth rate, or **fecundity,** defined as the number of offspring produced in a standard time (for example, per year), and death rate, or **mortality,** the number of individuals that die in that period.

The relative number of individuals in each cohort defines a population's **age structure.** Because different cohorts have different fecundity and death rates, age structure has a critical

Figure 55.9 The relationship between body size and generation time.

In general, larger organisms have longer generation times, although there are exceptions.

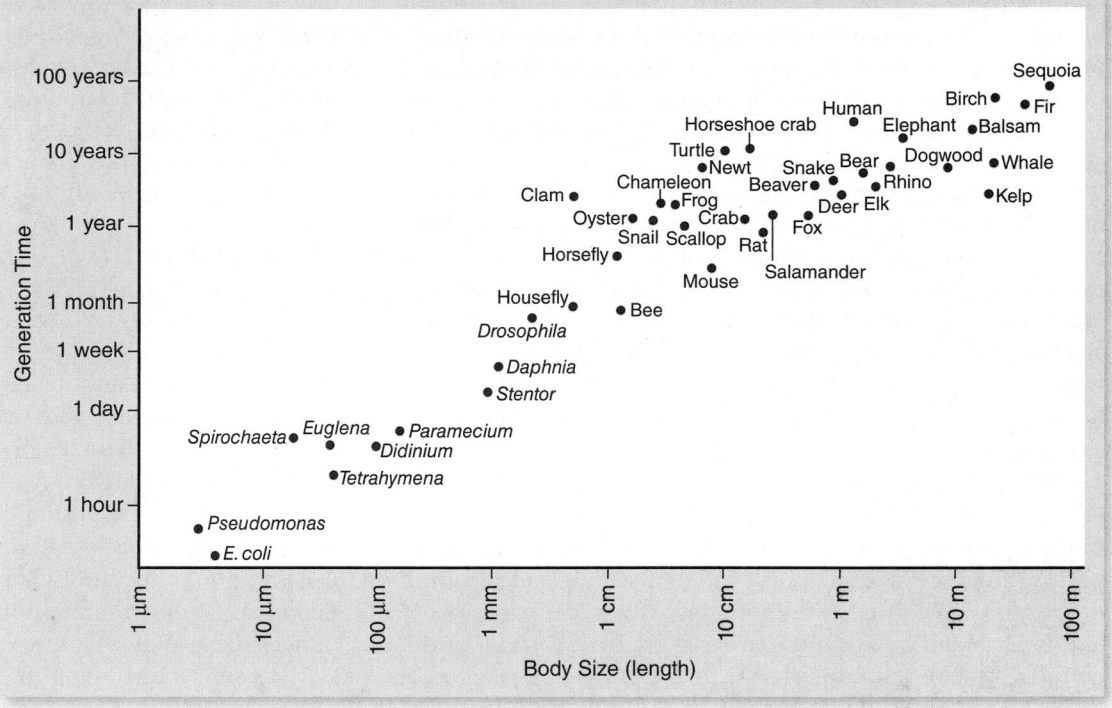

influence on a population's growth rate. Populations with a large proportion of young individuals, for example, tend to grow rapidly because an increasing proportion of their individuals are reproductive. Human populations in many developing countries are an example, as will be discussed later in this chapter. Conversely, if a large proportion of a population is relatively old, populations may decline. This phenomenon now characterizes Japan and some countries in Europe.

Life tables show probability of survival and reproduction through a cohort's life span

To assess how populations in nature are changing, ecologists use a **life table,** which tabulates the fate of a cohort from birth until death, showing the number of offspring produced and the number of individuals that die each time period. Table 55.2 is

TABLE 55.2	Life Table of the Annual Bluegrass (*Poa annua*) for a Cohort Containing 843 Seedlings						
Age (in 3-month intervals)	Number Alive at Beginning of Time Interval	Proportion of Cohort Alive at Beginning of Time Interval (survivorship)	Deaths During Time Interval	Mortality Rate During Time Interval	Seeds Produced During Time Interval	Seeds Produced per Surviving Individual (fecundity)	Seeds Produced per Member of Cohort (fecundity × survivorship)
0	843	1.000	121	0.144	0	0.00	0.00
1	722	0.856	195	0.270	303	0.42	0.36
2	527	0.625	211	0.400	622	1.18	0.74
3	316	0.375	172	0.544	430	1.36	0.51
4	144	0.171	90	0.626	210	1.46	0.25
5	54	0.064	39	0.722	60	1.11	0.07
6	15	0.018	12	0.800	30	2.00	0.04
7	3	0.004	3	1.000	10	3.33	0.01
8	0	0.000	—		Total = 1665		Total = 1.98

an example of a life table analysis from a study of the annual bluegrass. This study follows the fate of 843 individuals through time, charting how many survive in each interval and how many offspring each survivor produces.

In table 55.2, the first column indicates the age of the cohort (that is, the number of 3-month intervals from the start of the study). The second and third columns indicate the number of survivors and the proportion of the original cohort still alive at the beginning of that interval. The fifth column presents the **mortality rate,** the proportion of individuals that started that interval alive but died by the end of it. The seventh column indicates the average number of seeds produced by each surviving individual in that interval, and the last column shows the number of seeds produced relative to the size of the original cohort.

Much can be learned by examining life tables. In the case of the annual bluegrass, we see that both the probability of dying and the number of offspring produced per surviving individual steadily increases with age. By adding up the numbers in the last column, we get the total number of offspring produced per individual in the initial cohort. This number is almost 2, which means that for every original member of the cohort, on average two new individuals have been produced. A value of 1.0 would be the break-even number, the point at which the population was neither growing nor shrinking. In this case, the population appears to be growing rapidly.

Data analysis If annual bluegrass were a weed and you wanted to target one age group to remove, would you focus on the age class that produces the most seeds per individual? Can you imagine any circumstance in which your answer would be different?

In most cases, life table analysis is more complicated than this. First, except for organisms with short life spans, it is difficult to track the fate of a cohort until the death of the last individual. An alternative approach is to construct a cross-sectional study, examining the fate of cohorts of different ages in a single period. In addition, many factors—such as offspring reproducing before all members of their parents' cohort have died—complicate the interpretation of whether populations are growing or shrinking.

Survivorship curves demonstrate how survival probability changes with age

The percentage of an original population that survives to a given age is called its **survivorship.** One way to express some aspects of the age distribution of populations is through a *survivorship curve.* Examples of different survivorship curves are shown in figure 55.10. Oysters produce vast numbers of offspring, only a few of which live to reproduce. However, once they become established and grow into reproductive individuals, their mortality rate is extremely low (type III survivorship curve). In hydra, animals related to jellyfish, individuals are equally likely to die at any age. The result is a straight survivorship curve (type II). Finally, mortality rates in humans, as in many other animals and in protists, rise steeply later in life (type I survivorship curve).

Of course, these descriptions are just generalizations, and many organisms show more complicated patterns. Examination of the data for annual bluegrass, for example, reveals that it is most similar to a type II survivorship curve (figure 55.11).

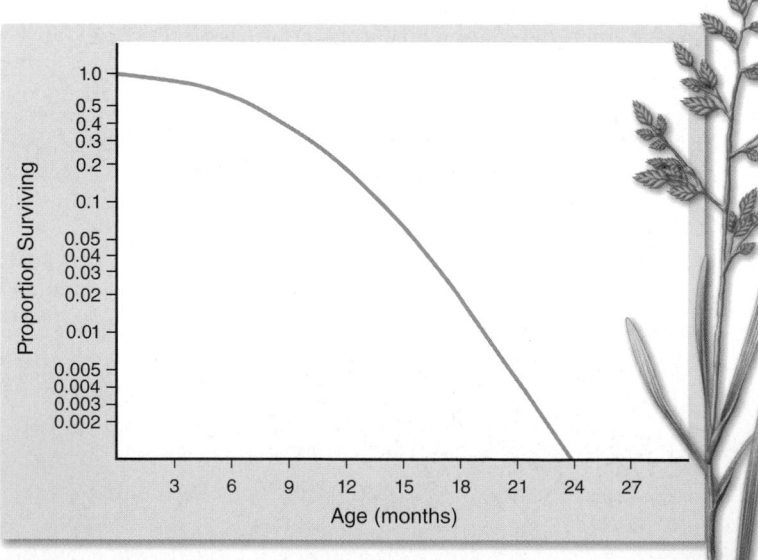

Figure 55.11 Survivorship curve for a cohort of annual bluegrass. After several months of age, mortality increases at a constant rate through time.

Inquiry question Suppose you wanted to keep annual bluegrass in your room as a houseplant. Suppose, too, that you wanted to buy an individual plant that was likely to live as long as possible. What age plant would you buy? How might the shape of the survivorship curve affect your answer?

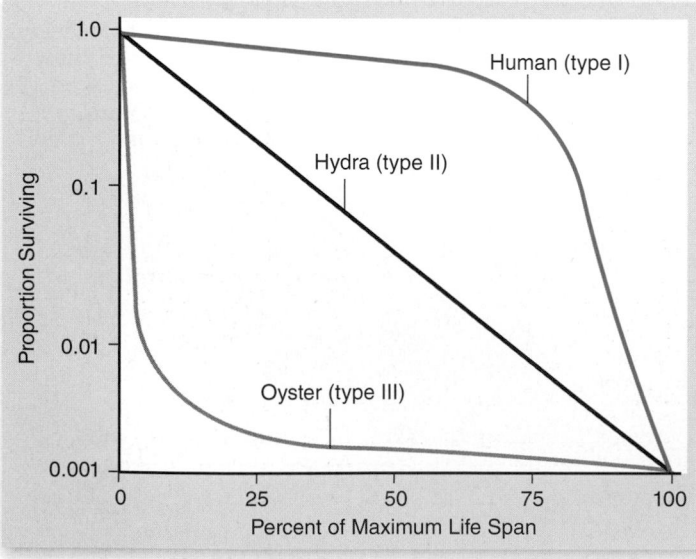

Figure 55.10 Survivorship curves. By convention, survival (the vertical axis) is plotted on a log scale. Humans have a type I life cycle, hydra (an animal related to jellyfish) type II, and oysters type III.

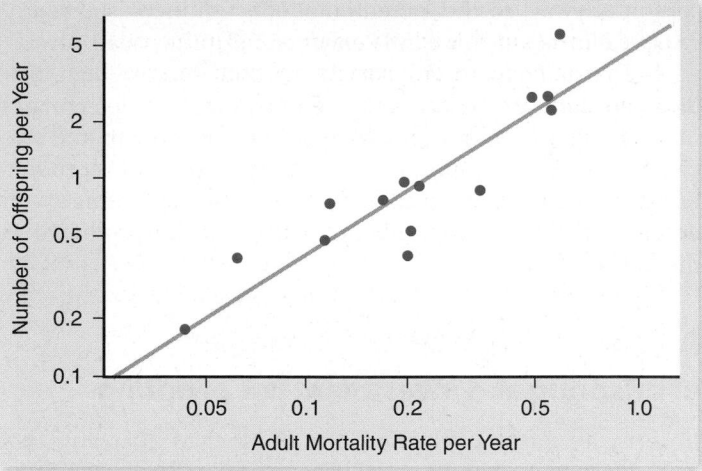

Figure 55.12 Reproduction has a price. Data from many bird species indicate that increased fecundity in birds correlates with higher mortality, ranging from the albatross (lowest) to the sparrow (highest). Species that raise more offspring per year have a higher probability of dying during that year.

55.4 Life History and the Cost of Reproduction

Natural selection favors traits that maximize the number of surviving offspring left in the next generation by an individual organism. Two factors affect this quantity: how long an individual lives, and how many young it produces each year.

Why doesn't every organism reproduce immediately after its own birth, produce large families of offspring, care for them intensively, and perform these functions repeatedly throughout a long life, while outcompeting others, escaping predators, and capturing food with ease? The answer is that no one organism can do all of this, simply because not enough resources are available. Consequently, organisms allocate resources either to current reproduction or to increasing their prospects of surviving and reproducing at later life stages.

The complete life cycle of an organism constitutes its life history. All life histories involve significant trade-offs. Because resources are limited, a change that increases reproduction may decrease survival and reduce future reproduction. As one example, a Douglas fir tree that produces more cones increases its current reproductive success—but it also grows more slowly. Because the number of cones produced is a function of how large a tree is, this diminished growth will decrease the number of cones it can produce in the future. Similarly, birds that have more offspring each year have a higher probability of dying during that year or of producing smaller clutches the following year (figure 55.12). Conversely, individuals that delay reproduction may grow faster and larger, enhancing future reproduction.

In one elegant experiment, researchers changed the number of eggs in the nests of a bird, the collared flycatcher (figure 55.13). Birds whose clutch size (the number of eggs produced in one breeding event) was decreased expended less energy raising their young and thus were able to lay more eggs the next year, whereas those given more eggs worked harder and consequently produced fewer eggs the following

year. Ecologists refer to the reduction in future reproductive potential resulting from current reproductive efforts as the **cost of reproduction.**

Natural selection favors the life history that maximizes lifetime reproductive success. When the cost of reproduction is low, individuals should produce as many offspring as possible because there is little cost. Low costs of reproduction may occur when resources are abundant and may also be relatively low when overall mortality rates are high. In the latter case, individuals may be unlikely to survive to the next breeding

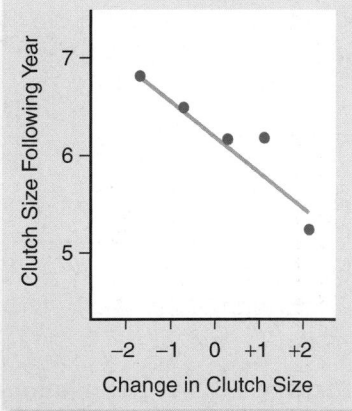

Figure 55.13 Reproductive events per lifetime. Adding eggs to nests of collared flycatchers (*Ficedula albicollis*), which increases the reproductive efforts of the female rearing the young, decreases clutch size the following year; removing eggs from the nest increases the next year's clutch size. This experiment demonstrates the trade-off between current reproductive effort and future reproductive success.

Data analysis Which has a greater effect, removing or adding eggs to a nest?

season anyway, so the incremental effect of increased reproductive efforts may have little effect on future survival.

Alternatively, when costs of reproduction are high, lifetime reproductive success may be maximized by deferring or minimizing current reproduction to enhance growth and survival rates. This situation may occur when costs of reproduction significantly affect the ability of an individual to survive or decrease the number of offspring that can be produced in the future.

A trade-off exists between number of offspring and investment per offspring

In terms of natural selection, the number of offspring produced is not as important as how many of those offspring themselves survive to reproduce. Assuming that the amount of energy to be invested in offspring is limited, a balance must be reached between the number of offspring produced and the size of each offspring (figure 55.14). This trade-off has been experimentally demonstrated in the side-blotched lizard, which normally lays between four and five eggs at a time. When some of the egg follicles are removed surgically early in the reproductive cycle, the female lizard produces only one to three eggs, but supplies each of these eggs with greater amounts of yolk, producing eggs and, subsequently, hatchlings that are much larger than normal (figure 55.15).

In the side-blotched lizard and many other species, the size of offspring is critical—larger offspring have a

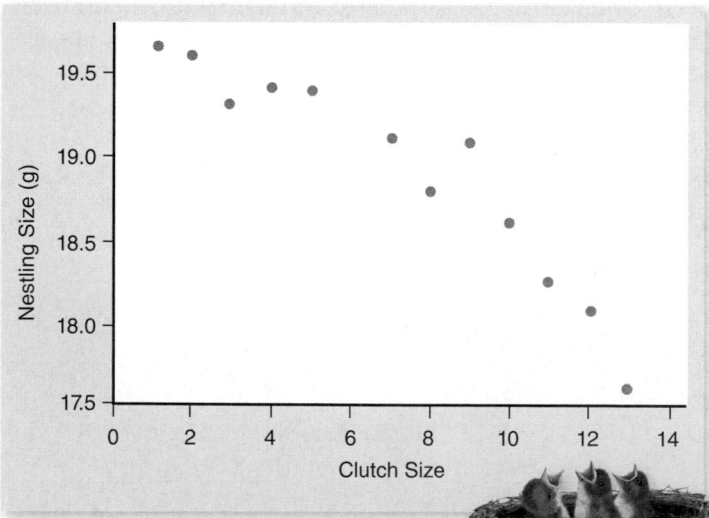

Figure 55.14 The relationship between clutch size and offspring size. In great tits *(Parus major),* the size of the nestlings is inversely related to the number of eggs laid. The more mouths they have to feed, the less the parents can provide to any one nestling.

🔍 **Data analysis** Overall, what clutch size probably requires the greatest energy expenditure by the parents?

❓ **Inquiry question** Would natural selection favor producing many small young or a few large ones?

Figure 55.15 Variation in the size of baby side-blotched lizards *(Uta stansburiana)* produced by experimental manipulations. In clutches in which some developing eggs were surgically removed, the remaining offspring were larger (center) than lizards produced in control clutches in which all the eggs were allowed to develop (right). The causal relationship between yolk provided by the mother and size of the offspring was demonstrated in experiments in which some of the yolk was removed from the eggs and smaller lizards resulted (left).

greater chance of survival. Producing many offspring with little chance of survival might not be the best strategy, but producing only a single, extraordinarily robust offspring also would not maximize the number of surviving offspring. Rather, an intermediate situation, in which several fairly large offspring are produced, should lead to the maximum number of surviving offspring.

Reproductive events per lifetime represent an additional trade-off

The trade-off between age and fecundity plays a key role in many life histories. Annual plants and most insects focus all their reproductive resources on a single large event and then die. This life history adaptation is called **semelparity.** Organisms that produce offspring several times over many seasons exhibit a life history adaptation called **iteroparity.**

Species that reproduce yearly must avoid overtaxing themselves in any one reproductive episode so that they will be able to survive and reproduce in the future. Semelparity, or "big bang" reproduction, is usually found in short-lived species that have a low probability of staying alive between broods, such as plants growing in harsh climates. Semelparity is also favored when fecundity entails large reproductive cost, exemplified by Pacific salmon migrating upriver to their spawning grounds. In these species, rather than investing some resources in an unlikely bid to survive until the next breeding season, individuals put all their resources into one reproductive event.

Age at first reproduction correlates with life span

Among mammals and many other animals, longer-lived species put off reproduction longer than short-lived species, relative to expected life span. The advantage of delayed reproduction is that juveniles gain experience before expending the high costs of reproduction. In long-lived animals, this advantage outweighs the energy that is invested in survival and growth rather than reproduction.

In shorter-lived animals, on the other hand, time is of the essence; thus, quick reproduction is more critical than juvenile training, and reproduction tends to occur earlier.

Learning Outcomes Review 55.4

Life history adaptations involve many trade-offs between reproductive cost and investment in survival. These trade-offs take a variety of forms, from laying fewer than the maximum possible number of eggs to putting all energy into a single bout of reproduction. Natural selection favors maximizing reproductive success, but number of offspring produced must be tempered by available resources.

■ *How might the life histories of two species differ if one was subject to high levels of predation and the other had few predators?*

55.5 Environmental Limits to Population Growth

Learning Outcomes

1. *Explain exponential growth.*
2. *Discuss why populations cannot grow exponentially forever.*
3. *Define carrying capacity and explain what might cause it to change.*

Populations often remain at a relatively constant size, regardless of how many offspring are born. As you saw in chapter 1, Darwin based his theory of natural selection partly on this seeming contradiction. Natural selection occurs because of checks on reproduction, with some individuals producing fewer surviving offspring than others. To understand populations, we must consider how they grow and what factors in nature limit population growth.

The exponential growth model applies to populations with no growth limits

The rate of population increase, *r*, is defined as the difference between the birth rate, *b*, and the death rate, *d*, corrected for movement of individuals in or out of the population (*e*, rate of movement out of the area; *i*, rate of movement into the area). Thus,

$$r = (b - d) + (i - e)$$

Movements of individuals can have a major influence on population growth rates. For example, the increase in human population in the United States during the closing decades of the 20th century was mostly due to immigration.

The simplest model of population growth assumes that a population grows without limits at its maximal rate and also that rates of immigration and emigration are equal. This rate, called the **biotic potential,** is the rate at which a population of a given species increases when no limits are placed on its rate of growth. In mathematical terms, this is defined by the following formula:

$$\frac{dN}{dt} = r_i N$$

where *N* is the number of individuals in the population, *dN/dt* is the rate of change in its numbers over time, and r_i is the intrinsic rate of natural increase for that population—its innate capacity for growth.

The biotic potential of any population is exponential (red line in figure 55.16). Even when the *rate* of increase remains constant, the actual *number* of individuals accelerates rapidly as the size of the population grows. The result of unchecked exponential growth is a population explosion.

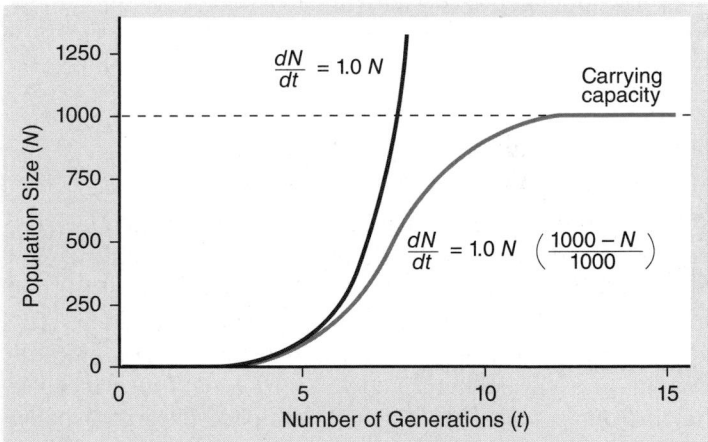

Figure 55.16 Two models of population growth.
The red line illustrates the exponential growth model for a population with an *r* of 1.0. The blue line illustrates the logistic growth model in a population with *r* = 1.0 and *K* = 1000 individuals. At first, logistic growth accelerates exponentially; then, as resources become limited, the death rate increases and growth slows. Growth ceases when the death rate equals the birth rate. The carrying capacity *(K)* ultimately depends on the resources available in the environment.

 Data analysis Suppose in one species of mouse, each pair produced four surviving offspring, whereas in another population, each pair produced five offspring. If this rate of population increase remained unchanged, after 10 generations, how much larger would the population of the second species be?

A single pair of houseflies, laying 120 eggs per generation, could produce more than 5 trillion descendants in a year. In 10 years, their descendants would form a swarm more than 2 m thick over the entire surface of the Earth! In practice, such patterns of unrestrained growth prevail only for short periods, usually when an organism reaches a new habitat with abundant resources. Natural examples of such short period of unrestrained growth include dandelions arriving in the fields, lawns, and meadows of North America from Europe for the first time; algae colonizing a newly formed pond; or cats introduced to an island with many birds, but previously lacking predators.

Carrying capacity

No matter how rapidly populations grow, they eventually reach a limit imposed by shortages of important environmental factors, such as space, light, water, or nutrients. A population ultimately may stabilize at a certain size, called the **carrying capacity** of the particular place where it lives. The carrying capacity, symbolized by K, is the maximum number of individuals that the environment can support.

The logistic growth model applies to populations that approach their carrying capacity

As a population approaches its carrying capacity, its rate of growth slows greatly, because fewer resources remain for each new individual to use. The growth curve of such a population, which is always limited by one or more factors in the environment, can be approximated by the following logistic growth equation:

$$\frac{dN}{dt} = rN\left(\frac{K-N}{K}\right)$$

In this model of population growth, the growth rate of the population *(dN/dt)* is equal to its intrinsic rate of natural increase (r multiplied by N, the number of individuals present at any one time), adjusted for the amount of resources available. The adjustment is made by multiplying rN by the fraction of K, the carrying capacity, still unused $[(K - N)/K]$. As N increases, the fraction of resources by which r is multiplied becomes smaller and smaller, and the rate of increase of the population declines.

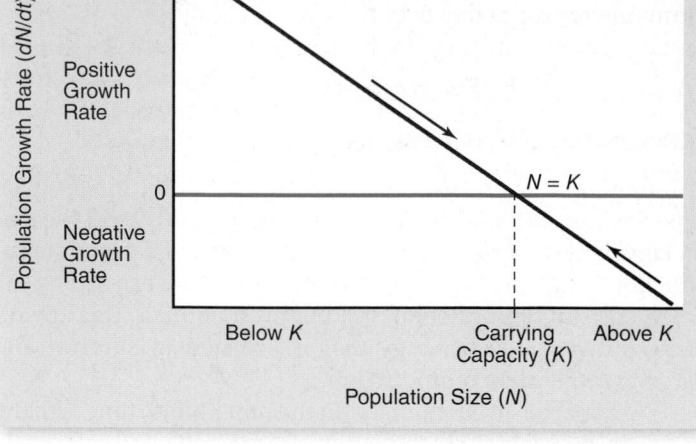

Figure 55.17 Relationship between population growth rate and population size. Populations far from the carrying capacity (K) have high growth rates—positive if the population is below K, and negative if it is above K. As the population approaches K, growth rates approach zero.

> **? Inquiry question** Why does the growth rate converge on zero?

Graphically, if you plot N versus t (time), you obtain a **sigmoidal growth curve** characteristic of many biological populations. The curve is called "sigmoidal" because its shape has a double curve like the letter S. As the size of a population stabilizes at the carrying capacity, its rate of growth slows, eventually coming to a halt (blue line in figure 55.16).

In mathematical terms, as N approaches K, the *rate of population growth (dN/dt)* begins to slow, reaching 0 when $N = K$ (figure 55.17). Conversely, if the population size exceeds the carrying capacity, then $K - N$ will be negative, and the population will experience a negative growth rate. As the population size then declines toward the carrying capacity, the magnitude of this negative growth rate will decrease until it reaches 0 when $N = K$.

Notice that the population tends to move toward the carrying capacity regardless of whether it is initially above or

Figure 55.18 Many populations exhibit logistic growth. *a.* A fur seal *(Callorhinus ursinus)* population on St. Paul Island, Alaska. *b.* Two laboratory populations of the cladoceran *Bosmina longirostris*. Note that the populations first exceeded the carrying capacity, before decreasing to what appears to be a size that was then maintained.

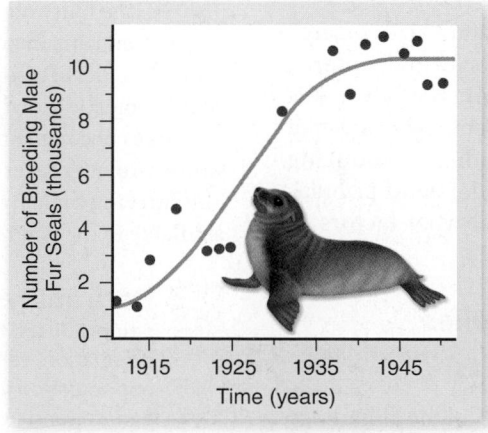

a.

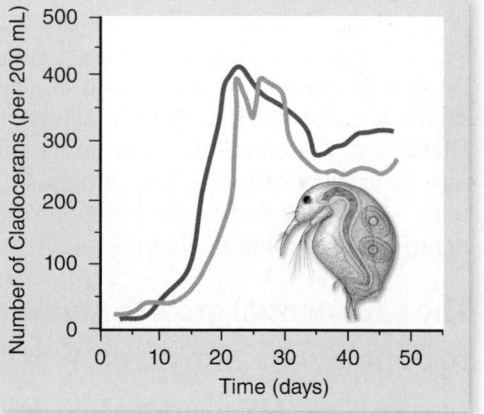

b.

below it. For this reason, logistic growth tends to return a population to the same size. In this sense, such populations are considered to be in equilibrium because they would be expected to be at or near the carrying capacity at most times.

In many cases, real populations display trends corresponding to a logistic growth curve. This is true not only in the laboratory, but also in natural populations (figure 55.18*a*). In some cases, however, the fit is not perfect (figure 55.18*b*), and as we shall see shortly, many populations exhibit other patterns.

Learning Outcomes Review 55.5

Exponential growth refers to population growth in which the number of individuals accelerates even when the rate of increase remains constant; it results in a population explosion. Exponential growth is eventually limited by resource availability. The size at which a population in a particular location stabilizes is defined as the carrying capacity of that location for that species. Populations often grow to the carrying capacity of their environment.

■ *What might cause a population's carrying capacity to change, and how would the population respond?*

55.6 Factors That Regulate Populations

Learning Outcomes

1. *Compare density-dependent and density-independent factors.*
2. *Evaluate why the size of some populations cycles.*
3. *Consider how the life history adaptations of species may differ depending on how often populations are at their carrying capacity.*

A number of factors may affect population size through time. Some of these factors depend on population size and are therefore termed *density-dependent*. Other factors, such as

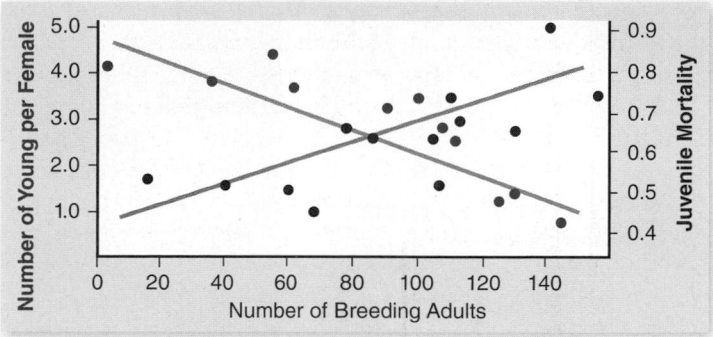

Figure 55.20 Density dependence in the song sparrow (*Melospiza melodia*) on Mandarte Island. Reproductive success decreases and mortality rates increase as population size increases.

? Inquiry question What would happen if researchers supplemented the food available to the birds?

natural disasters, affect populations regardless of size; these factors are termed *density-independent*. Many populations exhibit cyclic fluctuations in size that may result from complex interactions of factors.

Density-dependent effects occur when reproduction and survival are affected by population size

The reason population growth rates are affected by population size is that many important processes have **density-dependent effects.** That is, as population size increases, either reproductive rates decline or mortality rates increase, or both, a phenomenon termed *negative feedback* (figure 55.19).

Populations can be regulated in many different ways. When populations approach their carrying capacity, competition for resources can be severe, leading both to a decreased birth rate and an increased risk of death (figure 55.20). In addition, predators often focus their attention on a particularly common prey species, which also results in increasing rates of mortality as populations increase. High population densities can also lead to an accumulation of toxic wastes in the environment.

Behavioral changes may also affect population growth rates. Some species of rodents, for example, become antisocial, fighting more, breeding less, and generally acting stressed-out. These behavioral changes result

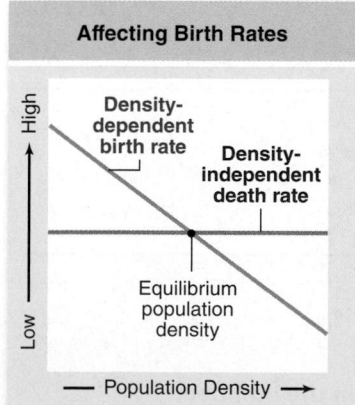

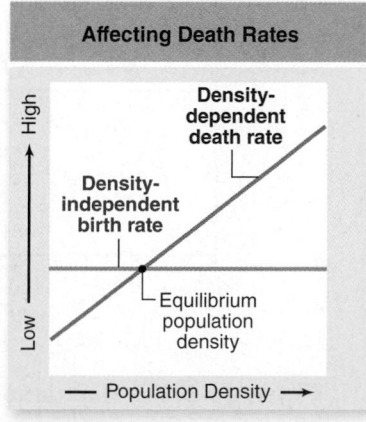

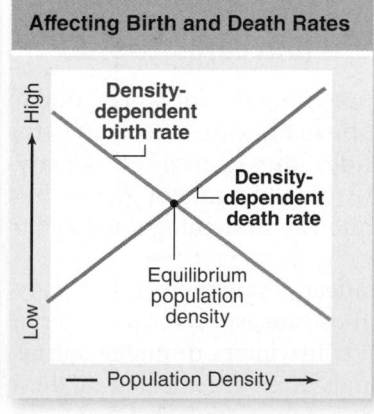

Figure 55.19 Density-dependent population regulation. Density-dependent factors can affect birth rates, death rates, or both.

? Inquiry question Why might birth rates be density-dependent?

chapter **55** *Ecology of Individuals and Populations*

Figure 55.21 Density-dependent effects. Migratory locusts (*Locusta migratoria*) are a legendary plague of large areas of Africa and Eurasia. At high population densities, the locusts have different hormonal and physical characteristics and take off as a swarm.

from hormonal actions, but their ultimate cause is not yet clear; most likely, they have evolved as adaptive responses to situations in which resources are scarce. In addition, in crowded populations, the population growth rate may decrease because of an increased rate of emigration of individuals attempting to find better conditions elsewhere (figure 55.21).

However, not all density-dependent factors are negatively related to population size. In some cases, growth rates increase with population size. This phenomenon is referred to as the **Allee effect** (after American zoologist Warder Allee, who first described it), and is an example of *positive feedback*. The Allee effect can take several forms. Most obviously, in populations that are too sparsely distributed, individuals may have difficulty finding mates. Moreover, some species may rely on large groups to deter predators or to provide the necessary stimulation for breeding activities. The Allee effect is a major threat for many endangered species, which may never recover from decreased population sizes caused by habitat destruction, overexploitation, or other causes (see chapter 59).

Density-independent effects include environmental disruptions and catastrophes

Growth rates in populations sometimes do not correspond to the logistic growth equation. In many cases, such patterns result because growth is under the control of **density-independent effects.** In other words, the rate of growth of a population at any instant is limited by something unrelated to the size of the population.

A variety of factors may affect populations in a density-independent manner. Most of these are aspects of the external environment, such as extremely cold winters, droughts, storms, or volcanic eruptions. Individuals often are affected by these occurrences regardless of the size of the population.

Populations in areas where such events occur relatively frequently display erratic growth patterns in which the populations increase rapidly when conditions are benign, but exhibit large reductions whenever the environment turns

hostile (figure 55.22). Needless to say, such populations do not produce the sigmoidal growth curves characteristic of the logistic equation.

Population cycles may reflect complex interactions

In some populations, density-dependent effects lead not to an equilibrium population size but to cyclic patterns of increase and decrease. For example, ecologists have studied cycles in hare populations since the 1820s. They have found that the North American snowshoe hare (*Lepus americanus*) follows a "10-year cycle" (in reality, the cycle varies from 8 to 11 years). Hare population numbers fall 10-fold to 30-fold in a typical cycle, and 100-fold changes can occur (figure 55.23). Two factors appear to be generating the cycle: food plants and predators.

Food plants. The preferred foods of snowshoe hares are willow and birch twigs. As hare density increases, the quantity of these twigs decreases, forcing the hares to feed on high-fiber (low-quality) food. Lower birth rates, low juvenile survivorship, and low growth rates follow. The hares also spend more time searching for food, an activity that increases their exposure to predation. The result is a precipitous decline in willow and birch twig abundance, and a corresponding fall in hare abundance. It takes 2 to 3 years for the quantity of mature twigs to recover.

Predators. A key predator of the snowshoe hare is the Canada lynx. The Canada lynx shows a "10-year" cycle of abundance that seems remarkably entrained to the hare

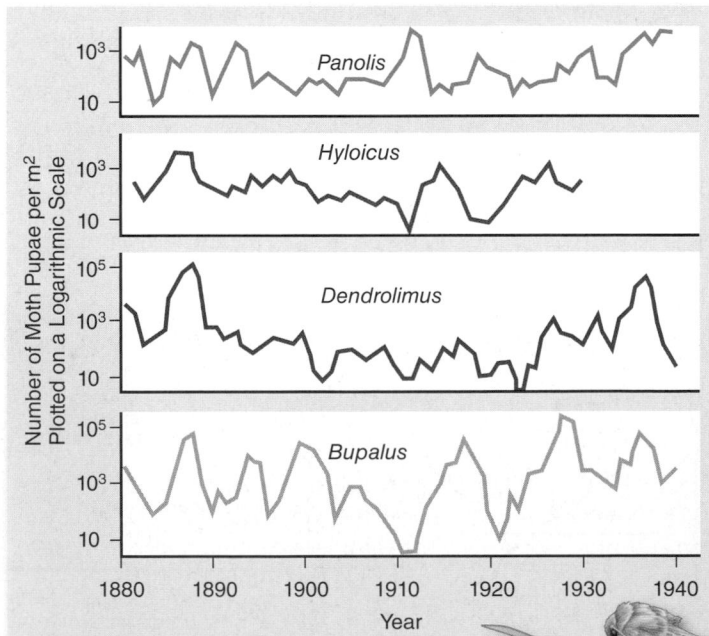

Figure 55.22 Fluctuations in the number of pupae of four moth species in Germany. The population fluctuations suggest that density-independent factors are regulating population size. The concordance in trends through time suggests that the same factors are regulating population size in all four species.

abundance cycle (see figure 55.23). As hare numbers increase, lynx numbers do too, rising in response to the increased availability of the lynx's food. When hare numbers fall, so do lynx numbers, their food supply depleted.

Which factor is responsible for the predator–prey oscillations? Do increasing numbers of hares lead to overharvesting of plants (a hare–plant cycle), or do increasing numbers of lynx lead to overharvesting of hares (a hare–lynx cycle)? Field experiments carried out by Charles Krebs and coworkers in 1992 provide an answer.

In Canada's Yukon, Krebs set up experimental plots that contained hare populations. If food is added (no food shortage effect) and predators are excluded (no predator effect) in an experimental area, hare numbers increase 10-fold and stay there—the cycle is lost. However, the cycle is retained if either of the factors is allowed to operate alone: exclude predators but don't add food (food shortage effect alone), or add food in the presence of predators (predator effect alone). Thus, both factors can affect the cycle, which in practice seems to be generated by the interaction between the two.

Population cycles are not as rare as traditionally thought; one review of nearly 700 long-term (25 years or more) studies of trends within populations found that cycles were not uncommon; nearly 30% of the studies—including birds, mammals, fish, and crustaceans—provided evidence of some cyclic pattern in population size through time, although most of these cycles are nowhere near as dramatic in amplitude as the hare–lynx cycles. In some cases, such as that of the snowshoe hare and lynx, density-dependent factors may be involved, whereas in other cases, density-independent factors, such as cyclic climatic patterns, may be responsible.

Resource availability affects life history adaptations

As you have seen, some species usually maintain stable population sizes near the carrying capacity, whereas in other species population sizes fluctuate markedly and are often far below carrying capacity. The selective factors affecting such species differ markedly. Individuals in populations near their carrying capacity may face stiff competition for limited resources; by contrast, individuals in populations far below carrying capacity have access to abundant resources.

We have already described the consequences of such differences. When resources are limited, the cost of reproduction often will be very high. Consequently, selection will favor individuals that can compete effectively and utilize resources efficiently. Such adaptations often come at the cost of lowered reproductive rates. Such populations are termed **K-selected** because they are adapted to thrive when the population is near its carrying capacity (K). Table 55.3 lists some of the typical features of K-selected populations. Examples of K-selected species include coconut palms, whooping cranes, whales, and humans.

By contrast, in populations far below the carrying capacity, resources may be abundant. Costs of reproduction are low, and selection favors those individuals that can produce the maximum number of offspring. Selection here favors individuals with the highest reproductive rates; such populations are termed **r-selected.** Examples of organisms displaying r-selected life history adaptations include dandelions, aphids, mice, and cockroaches.

Most natural populations show life history adaptations that exist along a continuum ranging from completely r-selected traits to completely K-selected traits. Although these

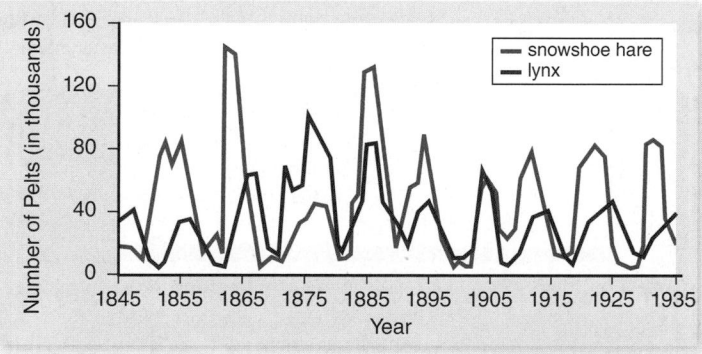

Figure 55.23 Linked population cycles of the snowshoe hare (Lepus americanus) and the northern lynx (Lynx canadensis). These data are based on records of fur returns from trappers in the Hudson Bay region of Canada. The lynx population carefully tracks that of the snowshoe hare, but lags behind it slightly.

? Inquiry question Suppose experimenters artificially kept the hare population at a high and constant level; what would happen to the lynx population? Conversely, if experimenters artificially kept the lynx population at a high and constant level, what would happen to the hare population?

TABLE 55.3	r-Selected and K-Selected Life History Adaptations	
Adaptation	**r-Selected Populations**	**K-Selected Populations**
Age at first reproduction	Early	Late
Life span	Short	Long
Maturation time	Short	Long
Mortality rate	Often high	Usually low
Number of offspring produced per reproductive episode	Many	Few
Number of reproductions per lifetime	Few	Many
Parental care	None	Often extensive
Size of offspring or eggs	Small	Large

tendencies hold true as generalities, few populations are purely *r*- or *K*-selected and show all of the traits listed in table 55.3. These attributes should be treated as generalities, with the recognition that many exceptions exist.

Learning Outcomes Review 55.6

Density-dependent factors such as resource availability come into play particularly when population size is larger; density-independent factors such as natural disasters operate regardless of population size. Population density may be cyclic due to complex interactions such as resource cycles and predator effects. Populations with density-dependent regulation often are near their carrying capacity; in species with populations well below carrying capacity, natural selection may favor high rates of reproduction when resources are abundant.

■ **Can a population experience both positive and negative density-dependent effects?**

55.7 Human Population Growth

Learning Outcomes

1. **Explain how the rate of human population growth has changed through time.**
2. **Describe the effects of age distribution on future growth.**
3. **Evaluate the relative importance of rapid population growth and resource consumption as threats to the biosphere and human welfare.**

Humans exhibit many *K*-selected life history traits, including small brood size, late reproduction, and a high degree of parental care. These life history traits evolved during the early history of hominids, when the limited resources available from the environment controlled population size. Throughout most of human history, our populations have been regulated by food availability, disease, and predators. Although unusual disturbances, including floods, plagues, and droughts, no doubt affected the pattern of human population growth, the overall size of the human population grew slowly during our early history.

Two thousand years ago, perhaps 130 million people populated the Earth. It took a thousand years for that number to double, and it was 1650 before it had doubled again, to about 500 million. In other words, for over 16 centuries, the human population was characterized by very slow growth. In this respect, human populations resembled many other species with predominantly *K*-selected life history adaptations.

Human populations have grown exponentially

Starting in the early 1700s, changes in technology gave humans more control over their food supply, enabled them to develop superior weapons to ward off predators, and led to the development of cures for many diseases. At the same time, improvements in shelter and storage capabilities made humans less vulnerable to climatic uncertainties. These changes allowed humans to expand the carrying capacity of the habitats in which they lived and thus to escape the confines of logistic growth and re-enter the exponential phase of the sigmoidal growth curve.

Responding to the lack of environmental constraints, the human population has grown explosively over the last 300 years. Although the birth rate has decreased over time, the death rate has decreased to a much greater extent. The difference between birth and death rates meant that the population grew as much as 2% per year, although the rate has now declined to 1.2% per year.

A 1.2% annual growth rate may not seem large, but it has produced a current human population that has recently surpassed 7 billion people (figure 55.24). At this growth rate, 78 million people would be added to the world population in the next year, and the human population would double in 58 years. Both the current human population level and the projected growth rate have potentially grave consequences for our future.

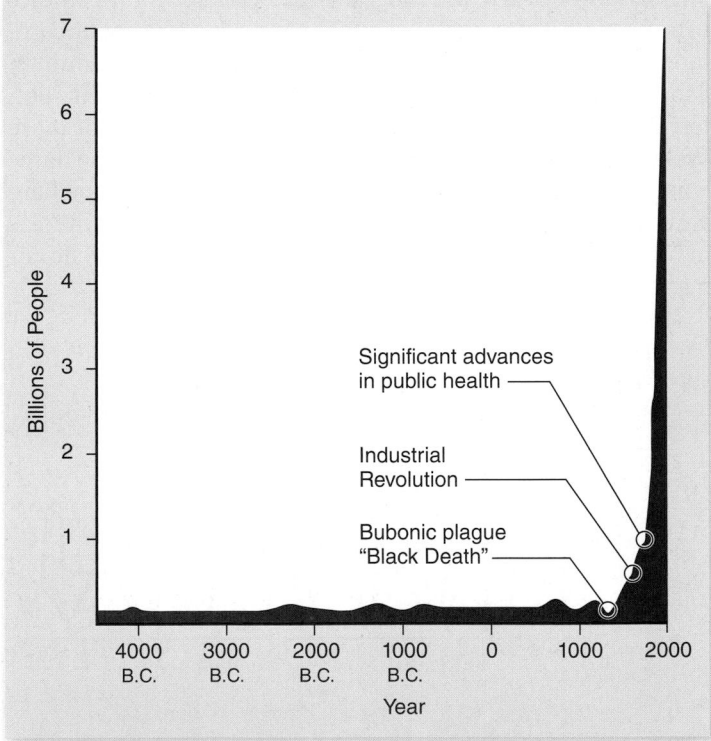

Figure 55.24 History of human population size. Temporary increases in death rate, even a severe one such as that occurring during the Black Death of the 1300s, have little lasting effect. Explosive growth began with the Industrial Revolution in the 1800s, which produced a significant, long-term lowering of the death rate. The current world population has recently exceeded 7 billion, and if it continued to grow at the present rate, would double in 58 years.

? Inquiry question Based on what we have learned about population growth, what do you predict will happen to human population size?

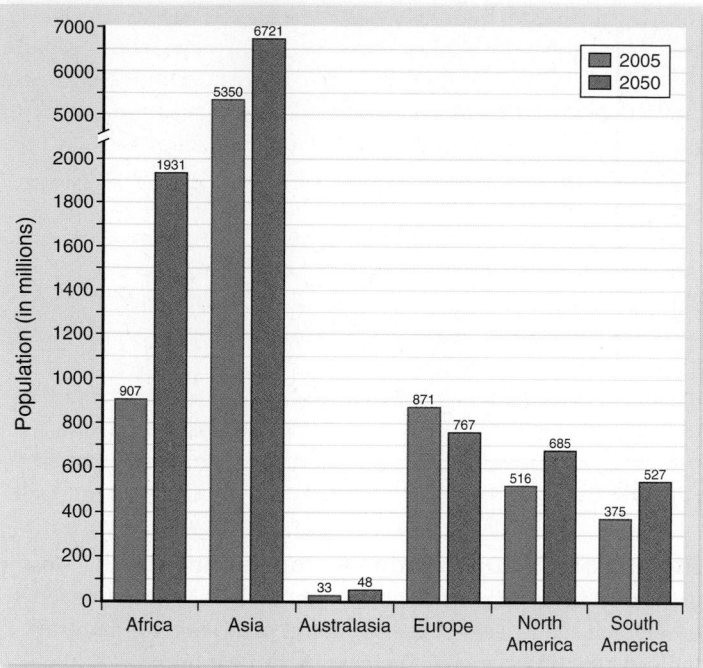

Figure 55.25 Projected population growth in 2050.
Developed countries are predicted to grow little; almost all of the population increase will occur in less-developed countries.

Population pyramids show birth and death trends

Although the human population as a whole continues to grow rapidly at the beginning of the 21st century, this growth is not occurring uniformly over the planet. Rather, most of the population growth is occurring in Africa and Asia (figure 55.25). By contrast, populations are actually decreasing in some countries in Europe.

The rate at which a population can be expected to grow in the future can be assessed graphically by means of a **population pyramid,** a bar graph displaying the numbers of people in each age category (figure 55.26). Males are conventionally shown to the left of the vertical age axis, females to the right. A human population pyramid thus displays the age composition of a population by sex. In most human population pyramids, the number of older females is disproportionately large compared with the number of older males, because females in most regions have a longer life expectancy than males.

Viewing such a pyramid, we can predict demographic trends in births and deaths. In general, a rectangular pyramid is characteristic of countries whose populations are stable, neither growing nor shrinking. A triangular pyramid is characteristic of a country that will exhibit rapid future growth because most of its population has not yet entered the childbearing years. Inverted triangles are characteristic of populations that are shrinking, usually as a result of sharply declining birth rates.

Examples of population pyramids for Sweden and Kenya in 2010 are shown in figure 55.26. The two countries exhibit very different age distributions. The nearly rectangular population pyramid for Sweden indicates that its population is not expanding because birth rates have decreased and average life span has increased. The very triangular pyramid of Kenya, by contrast, results from relatively high birth rates and shorter average life spans, which can lead to explosive future growth. The difference is most apparent when we consider that only

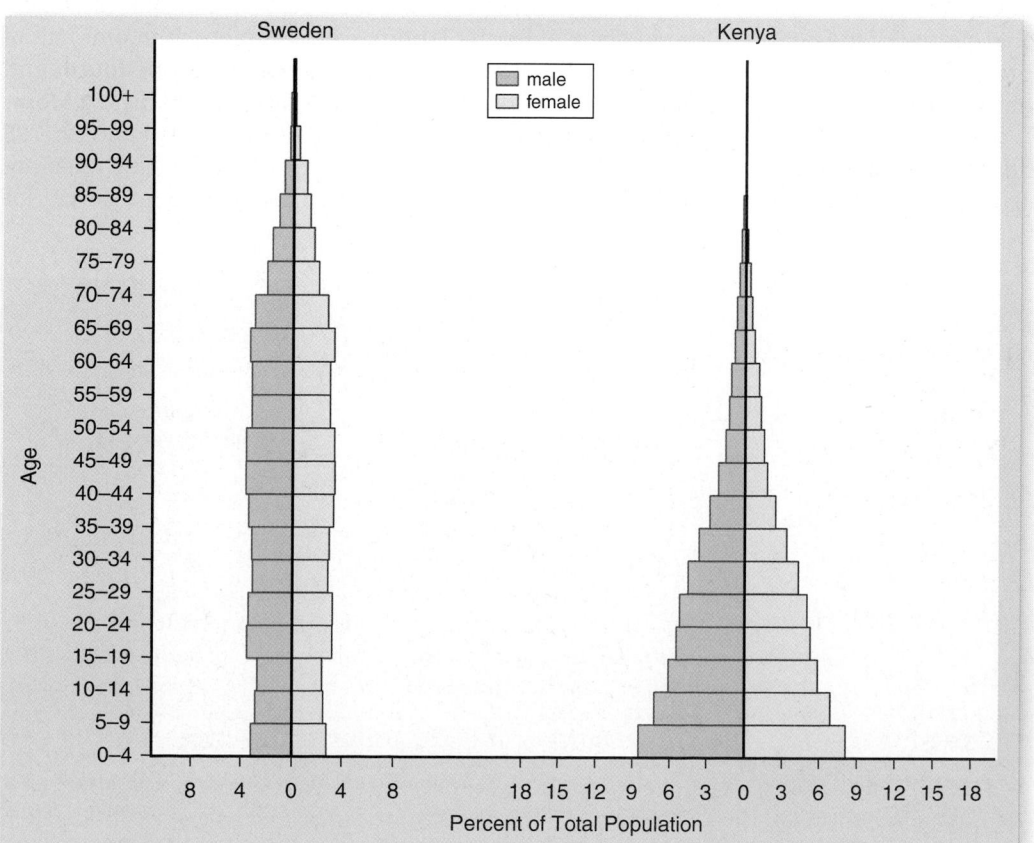

Figure 55.26 Population pyramids from 2010. Population pyramids are graphed according to a population's age distribution. Kenya's pyramid has a broad base because of the great number of individuals below childbearing age. When the young people begin to bear children, the population will experience rapid growth. The Swedish pyramid exhibits a slight bulge among middle-aged Swedes, the result of the "baby boom" that occurred in the middle of the 20th century, and many postreproductive individuals resulting from Sweden's long average life span.

? Inquiry question
What will the population distributions look like in 20 years?

16% of Sweden's population is less than 15 years old, compared with nearly half of all Kenyans. Moreover, the fertility rate (offspring per woman) in Sweden is 1.7; in Kenya, it is 4.7. As a result, Kenya's population could double in less than 35 years, whereas Sweden's will remain stable.

Humanity's future growth is uncertain

Earth's rapidly growing human population constitutes perhaps the greatest challenge to the future of the biosphere, the world's interacting community of living things. Humanity is adding 78 million people a year to its population—over a million every 5 days, 150 every minute! In more rapidly growing countries, the resulting population increase is staggering (table 55.4). India, for example, had a population of 1.05 billion in 2002; by 2050, its population likely will exceed 1.6 billion.

A key element in the world's population growth is its uneven distribution among countries. Of the billion people added to the world's population in the 1990s, 90% live in developing countries (figure 55.27). The fraction of the world's population that lives in industrialized countries is therefore diminishing. In 1950, fully one-third of the world's population lived in industrialized countries; by 1996, that proportion had fallen to one-quarter; and in 2020, the proportion will have fallen to one-sixth. In the future, the world's population growth will be centered in the parts of the world least equipped to deal with the pressures of rapid growth.

Rapid population growth in developing countries has had the harsh consequence of increasing the gap between rich and poor. Today, the 19% of the world's population that lives in the industrialized world have a per capita income of $22,060, but 81% of the world's population lives in developing countries and has a per capita income of only $3,580. Furthermore, of the people in the developing world, about one-quarter of the population gets by on $1 per day. Eighty percent of all the energy used today is consumed by the industrialized world, but only 20% is used by developing countries.

No one knows whether the world can sustain today's population of 7 billion people, much less the far greater

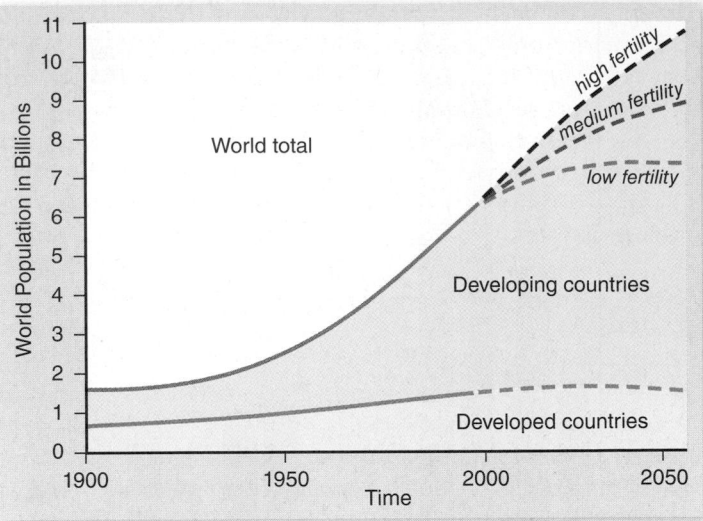

Figure 55.27 Distribution of population growth. Most of the worldwide increase in population since 1950 has occurred in developing countries. The age structures of developing countries indicate that this trend will increase in the near future. World population in 2050 likely will be between 7.3 and 10.7 billion, according to a recent United Nations study. Depending on fertility rates, the population at that time will either be increasing rapidly or slightly, or in the best case, declining slightly.

numbers expected in the future. As chapter 57 outlines, the world ecosystem is already under considerable stress. We cannot reasonably expect to expand its carrying capacity indefinitely, and indeed we already seem to be stretching the limits.

Despite using an estimated 45% of the total biological productivity of Earth's landmasses and more than one-half of all renewable sources of fresh water, between one-fourth and one-eighth of all people in the world are malnourished. Moreover, as anticipated by Thomas Malthus in his famous 1798 work, *Essay on the Principle of Population,* death rates are beginning to rise in some areas. In sub-Saharan Africa, for

TABLE 55.4	A Comparison of 2010 Population Data in Developed and Developing Countries		
	United States (highly developed)	**Brazil (moderately developed)**	**Ethiopia (poorly developed)**
Fertility rate	2.1	2.2	6.0
Doubling time at current rate (years)	80	55	28
Infant mortality rate (per 1000 births)	7	28.3	86.9
Life expectancy at birth (years)	78	74	53
Per capita GDP (U.S. $)*	$48,387	$11,769	$1093
Population < 15 years old (%)	13	25	44

*GDP, gross domestic product.

example, population projections for the year 2025 have been scaled back from 1.33 billion to 1.05 billion (a 21% decrease) because of the effect of AIDS. Similar decreases are projected for Russia as a result of higher death rates due to disease.

If we are to avoid catastrophic increases in the death rate, birth rates must fall dramatically. Faced with this grim dichotomy, significant efforts are underway worldwide to lower birth rates.

The population growth rate has declined

The world population growth rate is declining, from a high of 2.0% in the period 1965–1970 to 1.2% in 2008. Nonetheless, because of the larger population, this amounts to an increase of 78 million people per year to the world population, compared with 53 million per year in the 1960s.

Most countries are devoting considerable attention to slowing the growth rate of their populations, and there are genuine signs of progress. For example, from 1984 to 2008, family planning programs in Kenya succeeded in reducing the fertility rate from 8.0 to 4.7 children per couple, thus lowering the population growth rate from 4.0% per year to 2.8% per year. Because of these efforts, the global population may stabilize at about 8.9 billion people by the middle of the current century. How many people the planet can support sustainably depends on the quality of life that we want to achieve; there are already more people than can be sustainably supported with current technologies.

Consumption in the developed world further depletes resources

Population size is not the only factor that determines resource use; per capita consumption is also important. In this respect, we in the industrialized world need to pay more attention to lessening the impact each of us makes because, even though the vast majority of the world's population is in developing countries, the overwhelming percentage of consumption of resources occurs in the industrialized countries. Indeed, the wealthiest 20% of the world's population accounts for 86% of the world's consumption of resources and produces 53% of the world's carbon dioxide emissions, whereas the poorest 20% of the world is responsible for only 1.3% of consumption and 3% of carbon dioxide emissions. Looked at another way, in terms of resource use, a child born today in the industrialized world will consume many more resources over the course of his or her life than a child born in the developing world.

One way of quantifying this disparity is by calculating what has been termed the **ecological footprint,** which is the amount of productive land required to support an individual at the standard of living of a particular population through the course of his or her life. This figure estimates the acreage used for the production of food (both plant and animal), forest products, and housing, as well as the area of forest required to absorb carbon dioxide produced by the combustion of fossil fuels. As figure 55.28 illustrates, the ecological footprint of an individual in the United States is more than 10 times greater than that of someone in India.

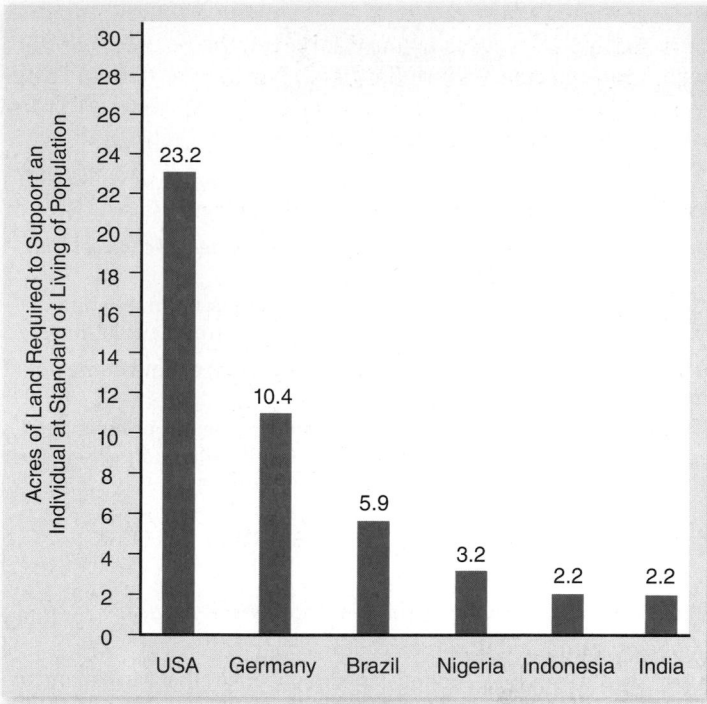

Figure 55.28 Ecological footprints of individuals in different countries. An ecological footprint calculates how much land is required to support a person through his or her life, including the acreage used for production of food, forest products, and housing, in addition to the forest required to absorb the carbon dioxide produced by the combustion of fossil fuels.

? Inquiry question Which is a more important cause of resource depletion, overpopulation or overconsumption? Explain.

Based on these measurements, researchers have calculated that resource use by humans is now one-third greater than the amount that nature can sustainably replace. Moreover, consumption is increasing rapidly in parts of the developing world; if all humans lived at the standard of living in the industrialized world, two additional planet Earths would be needed.

Building a sustainable world is the most important task facing humanity's future. The quality of life available to our children will depend to a large extent on our success in limiting both population growth and the amount of per capita resource consumption.

Learning Outcomes Review 55.7

For most of its history, the K-selected human population increased gradually. In the last 400 years, with resource control, the human population has grown exponentially; at the current rate, it would double in 58 years. A population pyramid shows the number of individuals in different age categories. Pyramids with a wide base are undergoing faster growth than those that are uniform from top to bottom. Growth rates overall are declining, but consumption per capita in the developed world is still a significant drain on resources.

- *Which is more important, reducing global population growth or reducing resource consumption levels in developed countries?*

55.1 The Environmental Challenges

Key environmental factors include temperature, water, sunlight, and soil type. Individuals seek to maintain internal homeostasis.

Organisms are capable of responding to environmental changes that occur during their lifetime.

Most individuals can cope with variations in their natural habitat, such as short-term changes in temperature and water availability.

Natural selection leads to evolutionary adaptation to environmental conditions.

Over evolutionary time, physiological, morphological, or behavioral adaptations evolve that make organisms better suited to the environment in which they live.

55.2 Populations: Groups of a Single Species in One Place

A population's geographic distribution is termed its range.

Ranges undergo expansion and contraction.

Most populations have limited geographic ranges that can expand or contract through time as the environment changes.

Dispersal mechanisms may allow some species to cross a barrier and expand their range. Human actions have led to range expansion of some species, often with detrimental effects.

Individuals in populations exhibit different spacing patterns.

Within a population, individuals are distributed randomly, uniformly, or are clumped. Nonrandom distributions may reflect resource distributions or competition for resources.

A metapopulation comprises distinct populations that may exchange members.

The degree of exchange between populations in a metapopulation is highest when populations are large and more connected.

Metapopulations may act as a buffer against extinction by permitting recolonization of vacant areas or marginal areas.

55.3 Population Demography and Dynamics

Sex ratio and generation time affect population growth rates.

Abundant females, a short generation time, or both can be responsible for more rapid population growth.

Age structure is determined by the numbers of individuals in different age groups.

Every age cohort has a characteristic fecundity and death rate, and so the age structure of a population affects growth.

Life tables show probability of survival and reproduction through a cohort's life span.

Survivorship curves demonstrate how survival probability changes with age (figures 55.10, 55.11).

In some populations, survivorship is high until old age, whereas in others, survivorship is lowest among the youngest individuals.

55.4 Life History and the Cost of Reproduction

Because resources are limited, reproduction has a cost. Resources allocated toward current reproduction cannot be used to enhance survival and future reproduction (figure 55.13).

A trade-off exists between number of offspring and investment per offspring.

When reproductive cost is high, fitness can be maximized by deferring reproduction, or by producing a few large-sized young that have a greater chance of survival.

Reproductive events per lifetime represent an additional trade-off.

Semelparity is reproduction once in a single large event. Iteroparity is production of offspring several times over many seasons.

Age at first reproduction correlates with life span.

Longer-lived species delay first reproduction longer compared with short-lived species, in which time is of the essence.

55.5 Environmental Limits to Population Growth

The exponential growth model applies to populations with no growth limits.

The rate of population increase, r, is defined as the difference between birth rate, b, and death rate, d.

Exponential growth occurs when a population is not limited by resources or by other species (figure 55.16).

The logistic growth model applies to populations that approach their carrying capacity.

Logistic growth is observed as a population reaches its carrying capacity. Usually, a population's growth rate slows to a plateau. In some cases the population overshoots and then drops back to the carrying capacity.

55.6 Factors That Regulate Populations

Density-dependent effects occur when reproduction and survival are affected by population size.

Density-dependent factors include increased competition and disease. To stabilize a population size, birth rates must decline, death rates must increase, or both.

Density-independent effects include environmental disruptions and catastrophes.

Density-independent factors are not related to population size and include environmental events that result in mortality.

Population cycles may reflect complex interactions.

In some cases, population size is cyclic because of the interaction of factors such as food supply and predation (figure 55.23).

Resource availability affects life history adaptations.

Populations at carrying capacity have adaptations to compete for limited resources; populations well below carrying capacity exhibit a high reproductive rate to use abundant resources.

55.7 Human Population Growth

Human populations have grown exponentially.

Technology and other innovations have simultaneously increased the carrying capacity and decreased mortality in the past 300 years.

Population pyramids show birth and death trends.

Populations with many young individuals are likely to experience high growth rates as these individuals reach reproductive age.

Humanity's future growth is uncertain.

The human population is unevenly distributed. Rapid growth in developing countries has resulted in poverty, whereas most resources are utilized by the industrialized world.

The population growth rate has declined.

Even at lower growth rates, the number of individuals on the planet is likely to plateau at 7 to 10 billion.

Consumption in the developed world further depletes resources.

Resource consumption rates in the developed world are very high; a sustainable future requires limits both to population growth and to per capita resource consumption.

UNDERSTAND

1. Source–sink metapopulations are distinct from other types of metapopulations because
   a. exchange of individuals only occurs in the former.
   b. populations with negative growth rates are a part of the former.
   c. populations never go extinct in the former.
   d. all populations eventually go extinct in the former.

2. The potential for social interactions among individuals should be maximized when individuals
   a. are randomly distributed in their environment.
   b. are uniformly distributed in their environment.
   c. have a clumped distribution in their environment.
   d. None of the choices is correct.

3. When ecologists talk about the cost of reproduction they mean
   a. the reduction in future reproductive output as a consequence of current reproduction.
   b. the amount of calories it takes for all the activity used in successful reproduction.
   c. the amount of calories contained in eggs or offspring.
   d. None of the choices is correct.

4. A life history trade-off between clutch size and offspring size
   a. means that as clutch size increases, offspring size increases.
   b. means that as clutch size increases, offspring size decreases.
   c. means that as clutch size increases, adult size increases.
   d. means that as clutch size increases, adult size decreases.

5. The difference between exponential and logistic growth rates is
   a. exponential growth depends on birth and death rates and logistic does not.
   b. in logistic growth, emigration and immigration are unimportant.
   c. that both are affected by density, but logistic growth is slower.
   d. that only logistic growth reflects density-dependent effects on births or deaths.

6. The logistic population growth model, $dN/dt = rN[(K - N)/K]$, describes a population's growth when an upper limit to growth is assumed. As N approaches (numerically) the value of K
   a. dN/dt increases rapidly.
   b. dN/dt approaches 0.
   c. dN/dt increases slowly.
   d. the population becomes threatened by extinction.

7. Which of the following is an example of a density-dependent effect on population growth?
   a. An extremely cold winter
   b. A tornado
   c. An extremely hot summer in which cool burrow retreats are fewer than number of individuals in the population
   d. A drought

APPLY

1. If the size of a population is reduced due to a natural disaster such as a flood
   a. population growth rates may increase because the population is no longer near its carrying capacity.
   b. population growth rates may decrease because individuals have trouble finding mates.
   c. population rates may remain unchanged if the population was already well below the carrying capacity.
   d. All of the choices are correct.

2. In populations subjected to high levels of predation
   a. individuals should invest little in reproduction so as to maximize their survival.
   b. individuals should produce few offspring and invest little in any of them.
   c. individuals should invest greatly in reproduction because their chance of surviving to another breeding season is low.
   d. individuals should stop reproducing altogether.

3. In a population in which individuals are uniformly distributed
   a. the population is probably well below its carrying capacity.
   b. natural selection should favor traits that maximize the ability to compete for resources.
   c. immigration from other populations is probably keeping the population from going extinct.
   d. None of the choices is correct.

4. The elimination of predators by humans
   a. will cause its prey to experience exponential growth until new predators arrive or evolve.
   b. will lead to an increase in the carrying capacity of the environment.
   c. may increase the population size of a prey species if that prey's population was being regulated by predation from the predator.
   d. will lead to an Allee effect.

SYNTHESIZE

1. Refer to figure 55.8. What are the implications for evolutionary divergence among populations that are part of a metapopulation versus populations that are independent of other populations?

2. Refer to figure 55.13. Given a trade-off between current reproductive effort and future reproductive success (the so-called cost of reproduction), would you expect old individuals to have the same "optimal" reproductive effort as young individuals?

3. Refer to figure 55.14. Because the number of offspring that a parent can produce is often a trade-off with the size of individual offspring, many circumstances lead to an

intermediate number and size of offspring being favored. If the size of an offspring was completely unrelated to the quality of that offspring (its chances of surviving until it reaches reproductive age), would you expect parents to fall on the left or right side of the *x*-axis (clutch size)? Explain.

4. Refer to figure 55.26. Would increasing the mean generation time have the same kind of effect on population growth rate as reducing the number of children that an individual female has over her lifetime? Which effect would have a bigger influence on population growth rate? Explain.

Chapter 56

Community Ecology

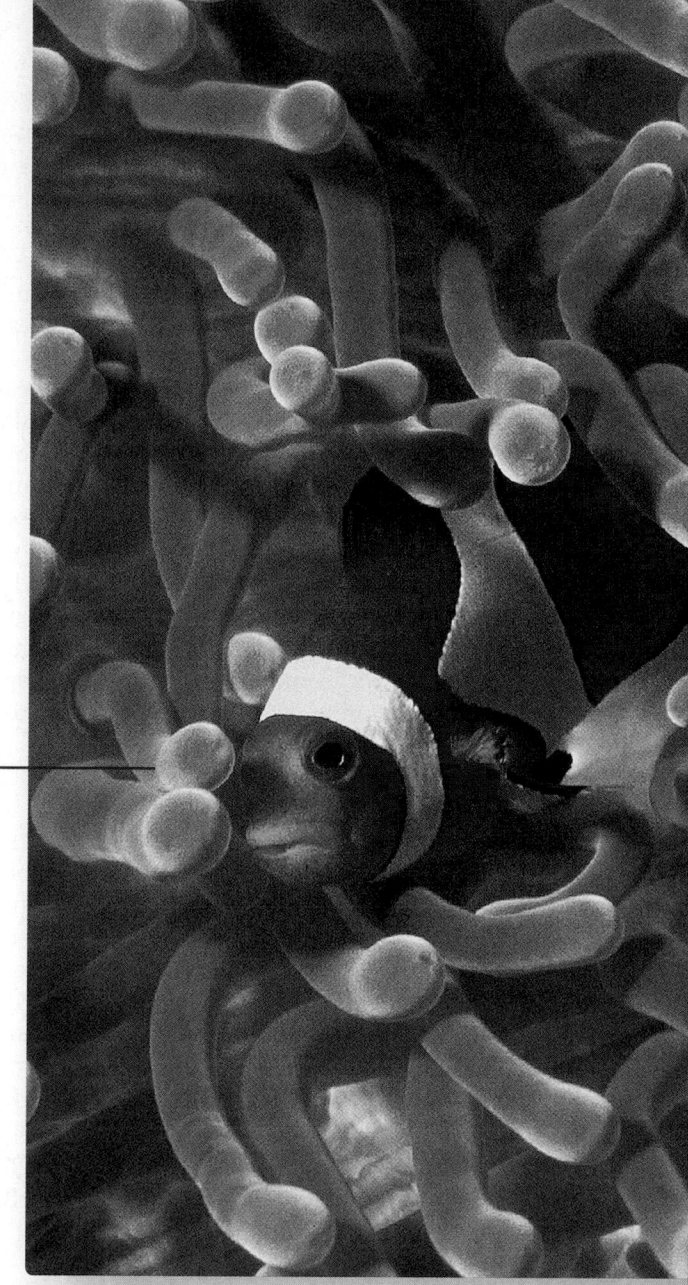

Chapter Contents

Introduction

All the organisms that live together in a place are members of a community. The myriad of species that inhabit a tropical rain forest are a community. Indeed, every inhabited place on Earth supports its own particular array of organisms. Over time, the different species that live together have made many complex adjustments to community living, evolving together and forging relationships that give the community its character and stability. Both competition and cooperation have played key roles; in this chapter, we look at these and other factors in community ecology.

56.1 Biological Communities: Species Living Together

Learning Outcomes

1. *Define community.*
2. *Describe how community composition may change across a geographic landscape.*

Almost any place on Earth is occupied by species, sometimes by many of them, as in the rain forests of the Amazon, and sometimes by only a few, as in the near-boiling waters of Yellowstone's geysers (where a number of microbial species live). The term **community** refers to the species that occur at any particular locality (figure 56.1). Communities can be characterized either by their constituent species or by their properties, such as **species richness** (the number of species present) or **primary productivity** (the amount of energy produced).

Interactions among community members govern many ecological and evolutionary processes. These interactions, such as predation and mutualism, affect the population biology of particular species—whether a population increases or decreases in abundance, for example—as well as the ways in which energy and nutrients cycle through the ecosystem. Moreover, the community context affects the patterns of natural selection faced by a species, and thus the evolutionary course it takes.

Scientists study biological communities in many ways, ranging from detailed observations to elaborate, large-scale experiments. In some cases, studies focus on the entire community, whereas in other cases only a subset of species that are likely to interact with one another are studied.

Communities change over space and time

For the most part, species seem to respond independently to changing environmental conditions. As a result, community composition changes gradually across landscapes as some species appear and become more abundant, while others decrease in abundance and eventually disappear.

A famous example of this pattern is the abundance of tree species in the Santa Catalina Mountains of Arizona along a geographic gradient running from very dry to very moist. Figure 56.2 shows that species can change abundance in patterns that are for the most part independent of one another. As a result, tree communities at different localities in these mountains fall on a continuum, one merging into the next, rather than representing discretely different sets of species.

Similar patterns through time are seen in paleontological studies. For example, a very good fossil record exists for the trees and small mammals that occurred in North America over the past 20,000 years. Examination of prehistoric communities shows little similarity to those that occur today. Many species that occur together today were never found together in the past. Conversely, species that used to occur in the same communities often do not overlap in their geographic ranges today. These findings suggest that as climate has changed during the waxing and waning of the Ice Ages, species have responded independently, rather than shifting their distributions together.

Nonetheless, in some cases the abundance of species in a community does change geographically in a synchronous pattern. Often, this occurs at **ecotones,** places where the environment changes abruptly. For example, in the western United States, certain patches of habitat have serpentine soils. This soil differs from normal soil in many ways—for example, high concentrations of nickel, chromium, and iron; low concentrations of copper and calcium. Comparison of the plant species that occur on different soils shows that distinct communities exist on each type, with an abrupt transition from one to the other over a short distance (figure 56.3). Similar transitions are seen wherever greatly different habitats come into contact, such as at the interface between terrestrial and aquatic habitats or where grassland and forest meet.

Figure 56.1 An African savanna community. A community consists of all the species—plants, animals, fungi, protists, and prokaryotes—that occur at a locality, in this case Etosha National Park in Namibia.

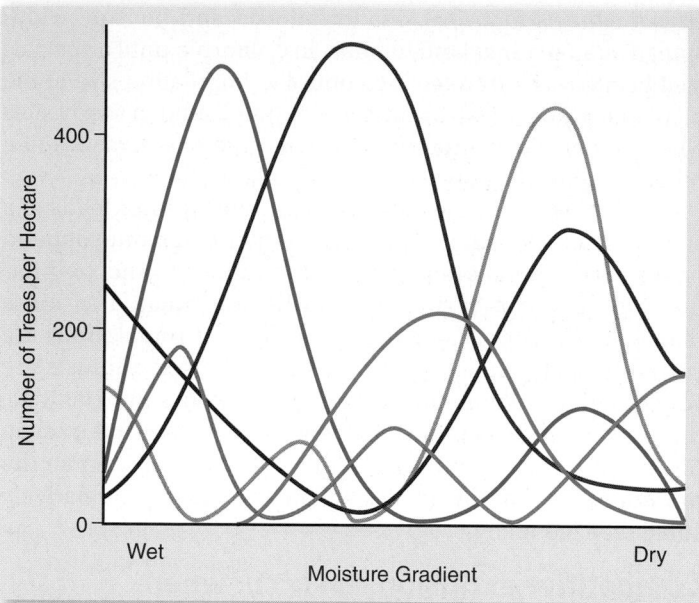

Figure 56.2 Abundance of tree species along a moisture gradient in the Santa Catalina Mountains of southeastern Arizona. Each line represents the abundance of a different tree species. The species' patterns of abundance are independent of one another. Thus, community composition changes continually along the gradient.

> **❓ Inquiry question** Why do species exhibit different patterns of response to change in moisture?

Learning Outcomes Review 56.1

A community comprises all species that occur at one site. In most cases, the abundance of community members appears to vary independently across space and through time. Community composition also changes gradually depending on environmental factors when moving from one location to another, such as from a very dry area to a very moist area.

- ■ *In a community, would you expect greater variation over time in abundance of animal life or plant life? Why?*

56.2 *The Ecological Niche Concept*

Learning Outcomes

1. *Define niche and resource partitioning.*
2. *Differentiate between fundamental and realized niches.*
3. *Explain how the presence of other species can affect a species' realized niche.*

Each organism in a community confronts the challenge of survival in a different way. The **niche** an organism occupies is the

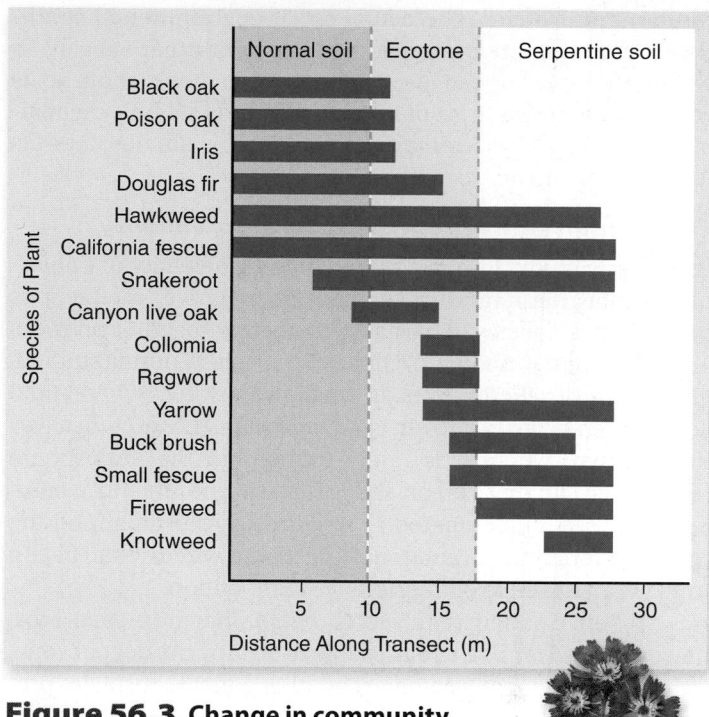

Figure 56.3 Change in community composition across an ecotone. The plant communities on normal and serpentine soils are greatly different, and the transition from one community to another occurs over a short distance.

> **🔍 Data analysis** In this field study, how close is the serpentine soil to the normal soil?

> **❓ Inquiry question** Why is there a sharp transition between the two community types?

total of all the ways it uses the resources of its environment. A niche may be described in terms of space utilization, food consumption, temperature range, appropriate conditions for mating, requirements for moisture, and other factors.

Sometimes species are not able to occupy their entire niche because of the presence or absence of other species. Species can interact with one another in a number of ways, and these interactions can either have positive or negative effects. One type of interaction, **interspecific competition,** occurs when two species use the same resource and there is not enough to satisfy both. Physical interactions over access to resources—such as fighting to defend a territory or displacing an individual from a particular location—are referred to as **interference competition;** consuming the same resources is called **exploitative competition.**

Fundamental niches are potential; realized niches are actual

The entire niche that a species is capable of using, based on its physiological tolerance limits and resource needs, is called the

fundamental niche. The actual set of environmental conditions, including the presence or absence of other species, in which the species can establish a stable population is its **realized niche.** Because of interspecific interactions, the realized niche of a species may be considerably smaller than its fundamental niche.

Competition between species for niche occupancy

In a classic study, Joseph Connell of the University of California, Santa Barbara, investigated competitive interactions between two species of barnacles that grow together on rocks along the coast of Scotland. Of the two species Connell studied, *Chthamalus stellatus* lives in shallower water, where tidal action often exposes it to air, and *Semibalanus balanoides* lives lower down, where it is rarely exposed to the atmosphere (figure 56.4). In these areas, space is at a premium and *S. balanoides* always outcompeted *C. stellatus* by crowding it off the rocks, undercutting it, and replacing it even where it had begun to grow, an example of interference competition.

When Connell removed *S. balanoides* from this area, however, *C. stellatus* was easily able to occupy the deeper zone, indicating that no physiological or other general obstacles prevented it from becoming established there. In contrast, *S. balanoides* could not survive in the shallow-water habitats where *C. stellatus* normally occurs; it does not have the physiological adaptations to warmer temperatures that allow *C. stellatus* to occupy this zone. Thus, the fundamental niche of *C. stellatus* includes both shallow and deeper zones, but its realized niche is much narrower because *C. stellatus* can be outcompeted by *S. balanoides* in parts of its fundamental niche. By contrast, the realized and fundamental niches of *S. balanoides* appear to be identical.

Other causes of niche restriction

Processes other than competition can also restrict the realized niche of a species. For example, the plant St. John's wort (*Hypericum perforatum*) was introduced and became widespread in open rangeland habitats in California until a specialized beetle was introduced to control it. Population size of the plant quickly decreased, and it is now only found in shady sites where the beetle cannot thrive. In this case, the presence of a predator limits the realized niche of a plant.

In some cases, the absence of another species leads to a smaller realized niche. Many North American plants depend on insects for pollination; indeed, the value of insect pollination for American agriculture has been estimated as more than $2 billion per year. However, pollinator populations are currently declining for several reasons. Conservationists are concerned that if these insects disappear from some habitats, the realized niche of many plant species will decrease or even disappear entirely. In this case, the absence—rather than the presence—of another species will be the cause of a relatively small realized niche.

Competitive exclusion can occur when species compete for limited resources

In classic experiments carried out in 1934 and 1935, Russian ecologist Georgii Gause studied competition among three species of *Paramecium*, a tiny protist. Each of the three species grew well in culture tubes by themselves, preying on bacteria and yeasts that fed on oatmeal suspended in the culture fluid (figure 56.5*a*). However, when Gause grew *P. aurelia* together with *P. caudatum* in the same culture tube, the numbers of *P. caudatum* always declined to extinction, leaving *P. aurelia* the only survivor (figure 56.5*b*). Why did this happen? Gause found that *P. aurelia* could grow six times faster than its competitor *P. caudatum* because it was able to better utilize the limited available resources, an example of exploitative competition.

From experiments such as this, Gause formulated what is now called the principle of **competitive exclusion.** This

Figure 56.4
Competition among two species of barnacles.
The fundamental niche of *Chthamalus stellatus* includes both deep and shallow zones. *C. stellatus* is forced out of the part of its fundamental niche that overlaps the realized niche of *Semibalanus balanoides*.

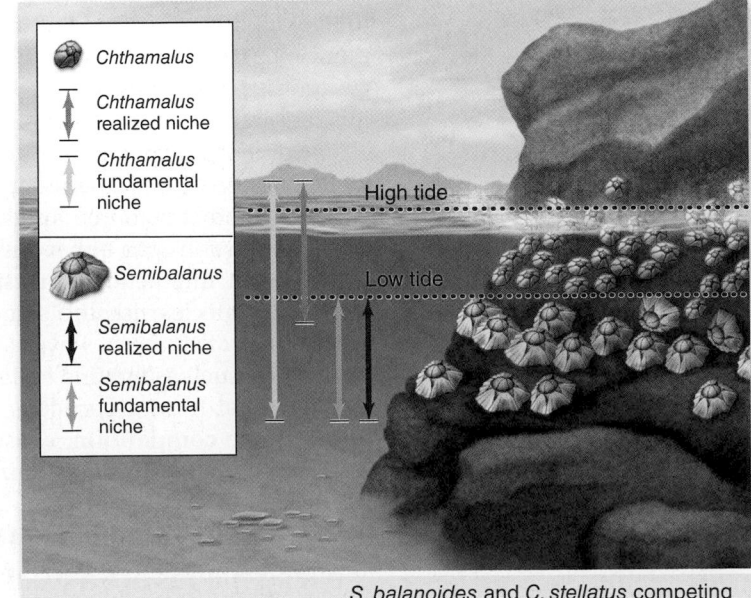

Chthamalus

↕ *Chthamalus* realized niche

↕ *Chthamalus* fundamental niche

Semibalanus

↕ *Semibalanus* realized niche

↕ *Semibalanus* fundamental niche

High tide

Low tide

S. balanoides and *C. stellatus* competing

C. stellatus fundamental and realized niches are identical when *S. balanoides* is removed.

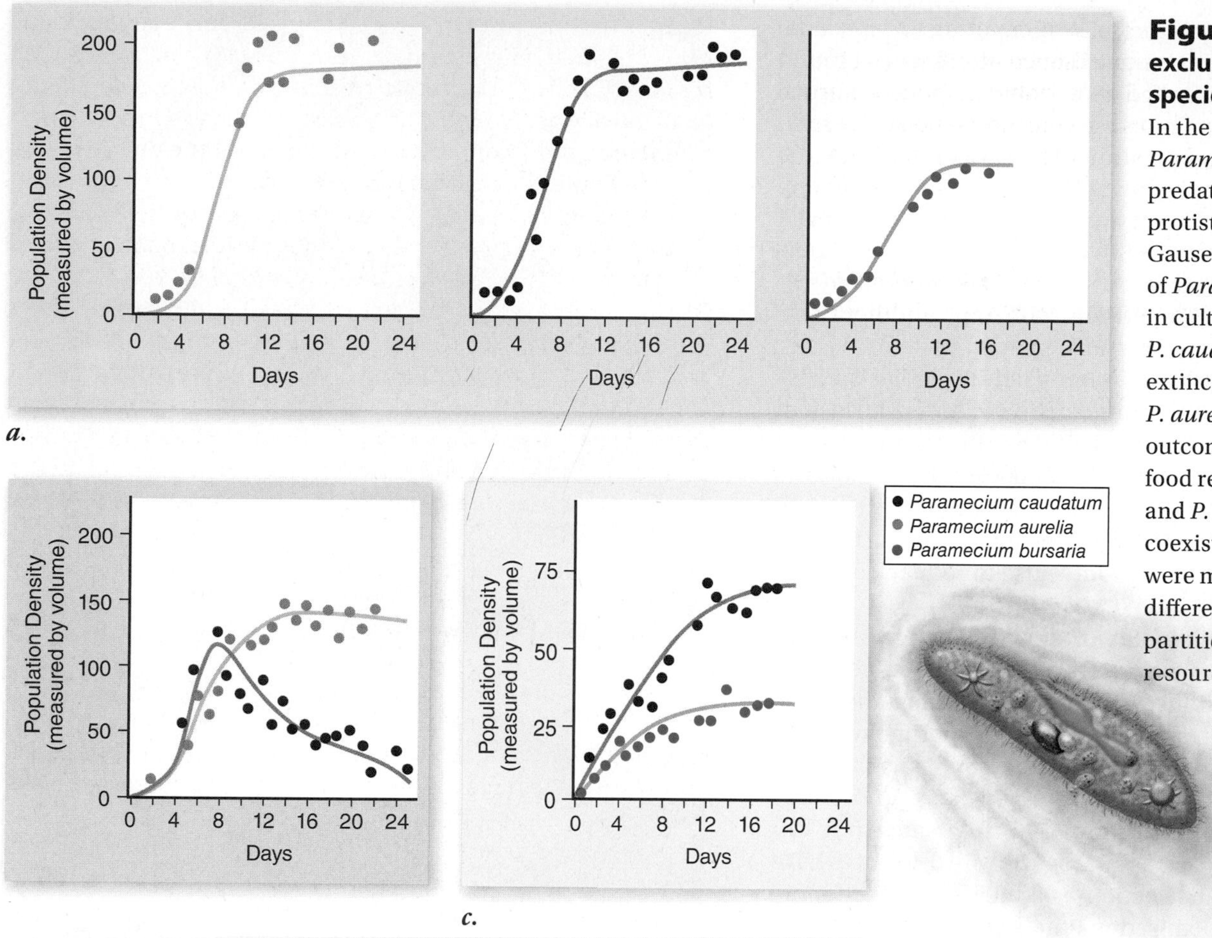

Figure 56.5 Competitive exclusion among three species of *Paramecium*. In the microscopic world, *Paramecium* is a ferocious predator that preys on smaller protists. ***a.*** In his experiments, Gause found that three species of *Paramecium* grew well alone in culture tubes. ***b.*** However, *P. caudatum* declined to extinction when grown with *P. aurelia*. *P. aurelia* outcompeted *P. caudatum* for food resources. ***c.*** *P. caudatum* and *P. bursaria* were able to coexist because the two species were more proficient at using different resources and thus partitioned the available resources.

● *Paramecium caudatum*
● *Paramecium aurelia*
● *Paramecium bursaria*

? Inquiry question What evidence, if any, do these graphs provide that *P. caudatum* and *P. bursaria* compete for resources?

principle states that if two species are competing for a limited resource such as food or water, the species that uses the resource more efficiently will eventually eliminate the other locally. In other words, no two species with the same niche can coexist when resources are limiting.

Niche overlap and coexistence

In a revealing experiment, Gause challenged *Paramecium caudatum*—the defeated species in his earlier experiments— with a third species, *P. bursaria*. Because he expected these two species to also compete for the limited bacterial food supply, Gause thought one would win out, as had happened in his previous experiments. But that's not what happened. Instead, both species survived in the culture tubes, dividing the food resources.

The explanation for the species' coexistence is simple. In the upper part of the culture tubes, where the oxygen concentration and bacterial density were high, *P. caudatum* dominated because it was better able to feed on bacteria. In the lower part of the tubes, however, the lower oxygen concentration favored the growth of a different potential food, yeast, and *P. bursaria* was better able to eat this food. The fundamental niche of each species was the whole culture tube, but the realized niches of the species were different portions of the tube, allowing them to coexist. However, competition did have a

negative effect on the participants (figure 56.5c). When grown without a competitor, both species reached densities three times greater than when they were grown with a competitor.

Competitive exclusion refined

Gause's principle of competitive exclusion can be restated as: No two species can occupy identical niches *indefinitely* when resources are limiting. Certainly species can and do coexist while competing for some of the same resources, but Gause's hypothesis predicts that when two species coexist on a long-term basis, either resources must not be limited or their niches will always differ in one or more features; otherwise, one species will outcompete the other, and the extinction of the second species will inevitably result.

Competition may lead to resource partitioning

Gause's competitive exclusion principle has a very important consequence: If competition for a limited resource is intense, then either one species will drive the other to extinction, or the species will evolve differences that reduce the competition between them.

When the ecologist Robert MacArthur studied five species of warblers, small insect-eating forest songbirds, he discovered that they appeared to be competing for the same resources. But when he investigated in greater detail he found that each species actually fed in a different part of spruce trees and so ate different subsets of insects. One species fed on insects near the tips of branches, a second within the dense foliage, a third on the lower branches, a fourth high on the trees, and a fifth at the very apex of the trees. Thus, each species of warbler had evolved so as to utilize a different portion of the spruce tree resource. They had *subdivided the niche* to avoid direct competition with one another. This niche subdivision is termed **resource partitioning.**

Resource partitioning is often seen in similar species that occupy the same geographic area. Such sympatric species often avoid competition by living in different portions of the habitat or by using different food or other resources (figure 56.6). This pattern of resource partitioning is thought to result from the process of natural selection causing initially similar species to diverge in resource use to reduce competitive pressures.

Whether such evolutionary divergence has occurred can be investigated by comparing species whose ranges only partially overlap. Where the two species occur together, they often tend to exhibit greater differences in morphology (the form and structure of an organism) and resource use than do allopatric populations of the same species that do not occur with the other species. Called *character displacement,* the differences evident between sympatric species are thought to have been favored by natural selection as a means of partitioning resources and thus reducing competition (see chapter 22).

As an example, the two Darwin's finches in figure 56.7 have bills of similar size where the finches are allopatric (that is, each living on an island where the other does not occur). On islands where they are sympatric (that is, occur together), the two species have evolved beaks of different sizes, one adapted to larger seeds and the other to smaller ones. Character displacement such as this may play an important role in adaptive radiation, leading new species to adapt to different parts of the environment, as discussed in chapter 22.

Detecting interspecific competition can be difficult

It is not simple to determine when two species are competing. The fact that two species use the same resources need not imply competition if that resource is not in limited supply. Even if the population sizes of two species are negatively correlated, such that where one species has a large population, the other species has a small population and vice versa, the two species may not be competing for the same limiting resource. Instead, the two species might be independently responding to the same feature of the environment—perhaps one species thrives best in warm conditions and the other where it's cool.

Experimental studies of competition

Some of the best evidence for the existence of competition comes from experimental field studies. By setting up experiments in which two species occur either alone or together,

a.

b.

c.

d.

Figure 56.6 Resource partitioning among sympatric lizard species. Species of *Anolis* lizards on Caribbean islands partition their habitats in a variety of ways. Some species *(a)* occupy leaves and branches in the canopy of trees, *(b)* others use twigs on the periphery, and *(c)* still others are found at the base of the trunk. In addition, *(d)* some use grassy areas in the open. When two species occupy the same part of the tree, they either utilize different-sized insects as food or partition the thermal microhabitat; for example, one might only be found in the shade, whereas the other would only be found in the open, basking in the light.

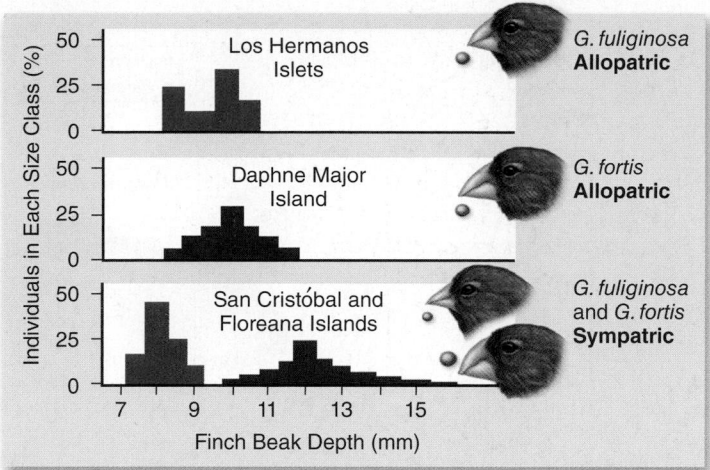

Figure 56.7 Character displacement in Darwin's finches. These two species of finches (genus *Geospiza*) have beaks of similar size when allopatric, but different size when sympatric.

? **Inquiry question** Why do the two species have bills of the same size when they are allopatric?

scientists can determine whether the presence of one species has a negative effect on a population of the second species.

For example, a variety of seed-eating rodents occur in North American deserts. In 1988, researchers set up a series of 50-m × 50-m enclosures to investigate the effect of kangaroo rats on smaller, seed-eating rodents. Kangaroo rats were removed from half of the enclosures, but not from the others. The walls of all of the enclosures had holes that allowed rodents to come and go, but in the plots in which the kangaroo rats had been removed, the holes were too small to allow the kangaroo rats to reenter.

Over the course of the next 3 years, the researchers monitored the number of the smaller rodents present in the plots. As figure 56.8 illustrates, the number of other rodents was substantially higher in the absence of kangaroo rats, indicating that kangaroo rats compete with the other rodents and limit their population sizes.

A great number of similar experiments have indicated that interspecific competition occurs between many species of plants and animals. The effects of competition can be seen in aspects of population biology other than population size, such as behavior and individual growth rates. For example, two species of *Anolis* lizards occur on the Caribbean island of St. Maarten. When one of the species, *A. gingivinus,* is placed in 12-m × 12-m enclosures without the other species, individual lizards grow faster and perch lower than do lizards of the same species when placed in enclosures in which *A. pogus,* a species normally found near the ground, is also present.

Limitations of experimental studies

Experimental studies are a powerful means of understanding interactions between coexisting species and are now commonly conducted by ecologists. Nonetheless, they have their limitations.

First, care is necessary in interpreting the results of field experiments. Negative effects of one species on another do not automatically indicate the existence of competition. For example, many similarly sized fish have a negative effect on one another, but it results not from competition, but from the fact that adults of each species prey on juveniles of the other species.

In addition, the presence of one species may attract predators or parasites, which then also prey on the second species. In this case, even if the two species are not competing, the second species may have a lower population size in the presence of the first species due to predators or parasites.

SCIENTIFIC THINKING

Question: *Does interspecific interaction occur between rodent species?*
Hypothesis: *The larger kangaroo rat will have a negative effect on other species.*
Experiment: *Build large cages in desert areas. Remove kangaroo rats from some cages, leaving them present in others.*
Result: *In the absence of kangaroo rats, the number of other rodents increases quickly and remains higher than in the control cages throughout the course of the experiment.*

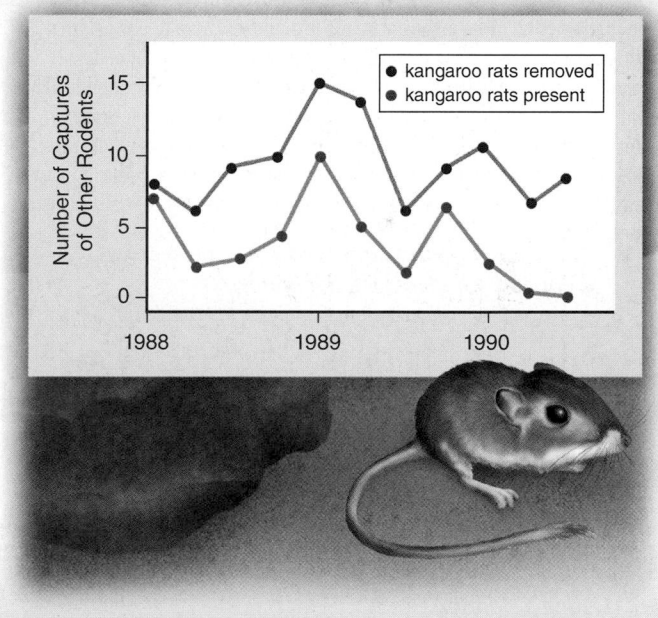

Interpretation: *Why do you think population sizes rise and fall in synchrony in the two cages?*

Figure 56.8 Detecting interspecific competition.
This experiment tested how removal of kangaroo rats affected the population size of other rodents. Immediately after kangaroo rats were removed, the number of other rodents increased relative to the enclosures that still contained kangaroo rats. Notice that population sizes (as estimated by number of captures) changed in synchrony in the two treatments, probably reflecting changes in the weather.

? **Inquiry question** Why are there more individuals of other rodent species when kangaroo rats are excluded?

Indeed, we can't rule out this possibility with the results of the kangaroo rat exclusion study just mentioned, although the close proximity of the enclosures (they were adjacent) would suggest that the same predators and parasites were present in all of them. Thus, experimental studies are most effective when combined with detailed examination of the ecological mechanisms causing the observed effect of one species on another.

Second, experimental studies are not always feasible. For example, the coyote population has increased in the United States in recent years concurrently with the decline of the grey wolf. Is this trend an indication that the species compete? Because of the size of the animals and the large geographic areas occupied by each individual, manipulative experiments involving fenced areas with only one or both species—with each experimental treatment replicated several times for statistical analysis—are not practical. Similarly, studies of slow-growing trees might require many centuries to detect competition between adult trees. In such cases, detailed studies of the ecological requirements of each species are our best bet for understanding interspecific interactions.

56.3 Predator–Prey Relationships

Learning Outcomes

1. *Define predation.*
2. *Describe the effects predation can have on a population.*

Predation is the consuming of one organism by another. In this sense, predation includes everything from a leopard capturing and eating an antelope, to a deer grazing on spring grass.

When experimental populations are set up under simple laboratory conditions, as illustrated in figure 56.9 with the predatory protist *Didinium* and its prey *Paramecium,* the predator often exterminates its prey and then becomes extinct itself, having nothing left to eat. If refuges are provided for the *Paramecium,* however, its population drops to low levels but not to extinction. Low prey population levels then provide inadequate food for the *Didinium,* causing the predator population to decrease. When this occurs, the prey population can recover.

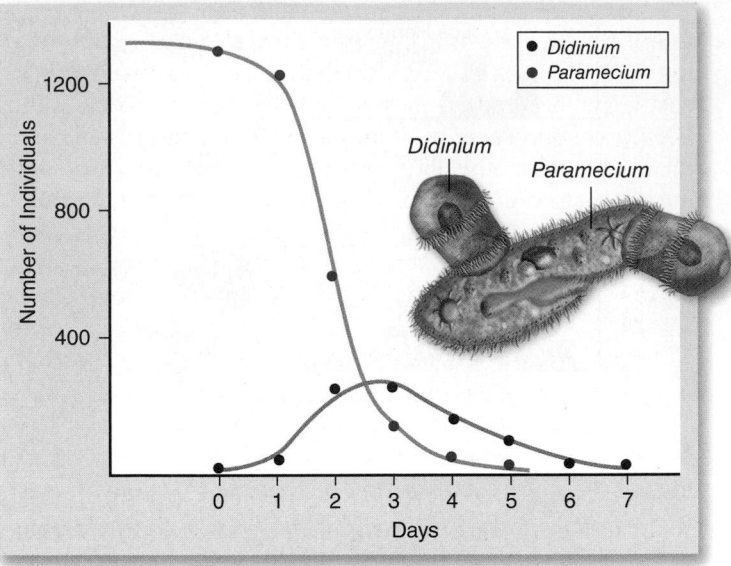

Figure 56.9 Predator–prey in the microscopic world.
When the predatory *Didinium* is added to a *Paramecium* population, the numbers of *Didinium* initially rise, and the numbers of *Paramecium* steadily fall. When the *Paramecium* population is depleted, however, the *Didinium* individuals also die.

 Inquiry question Can you think of any ways this experiment could be changed so that *Paramecium* might not go extinct?

Predation strongly influences prey populations

In nature, predators often have large effects on prey populations. As the previous example indicates, however, the interaction is a two-way street: prey can also affect the dynamics of predator populations. The outcomes of such interactions are complex and depend on a variety of factors.

Prey population explosions and crashes

Some of the most dramatic examples of the interconnection between predators and their prey involve situations in which humans have either added or eliminated predators from an area. For example, the elimination of large carnivores from much of the eastern United States has led to population explosions of white-tailed deer, which strip the habitat of all edible plant life within their reach. Similarly, when sea otters were hunted to near extinction on the western coast of the United States, populations of sea urchins, a principal prey item of the otters, exploded.

Conversely, the introduction of rats, dogs, and cats to many islands around the world has led to the decimation of native fauna. Populations of Galápagos tortoises on several islands are endangered by introduced rats, pigs, dogs, and cats, which eat the eggs and the young tortoises. Similarly, in New Zealand, several species of birds and reptiles have been eradicated by rat predation and now only occur on a few offshore islands that the rats have not reached. For example, every individual of the now-extinct Stephens Island wren was killed by a single lighthouse keeper's cat.

A classic example of the role predation can play in a community involves the introduction of prickly pear cactus to Australia in the 19th century. In the absence of predators, the cactus spread rapidly, so that by 1925 it occupied 12 million hectares of rangeland in an impenetrable morass of spines that made cattle ranching difficult. To control the cactus, a predator from its natural habitat in Argentina, the moth *Cactoblastis cactorum,* was introduced, beginning in 1926. By 1940, cactus populations had been greatly reduced and it now usually occurs in small populations.

Predation and coevolution

Predation provides strong selective pressures on prey populations. Any feature that would decrease the probability of capture should be strongly favored. In turn, the evolution of such features causes natural selection to favor counteradaptations in predator populations. The process by which these adaptations are selected in lockstep fashion in two or more interacting species is termed **coevolution.** A coevolutionary "arms race" may ensue in which predators and prey are constantly evolving better defenses and better means of circumventing these defenses. In the sections that follow, you'll learn more about these defenses and responses.

Plant adaptations defend against herbivores

Plants have evolved many mechanisms to defend themselves from herbivores. The most obvious are morphological defenses: Thorns, spines, and prickles play an important role in discouraging large plant eaters, and plant hairs, especially those that have a glandular, sticky tip, deter insect herbivores. Some plants, such as grasses, deposit silica in their leaves, both strengthening and protecting themselves. If enough silica is present, these plants are simply too tough to eat.

Chemical defenses

As significant as morphological adaptations are, the chemical defenses that occur so widely in plants are even more widespread. Plants exhibit some amazing chemical adaptations to combat herbivores. For example, recent work demonstrates that when attacked by caterpillars, wild tobacco plants emit a chemical into the air that attracts a species of bug that feeds on that caterpillar (discussed in greater detail in chapter 39).

The best-known and perhaps most important of the chemical defenses of plants against herbivores are *secondary chemical compounds.* These chemicals are distinguished from primary compounds, which are the components of a major metabolic pathway, such as respiration. Many plants, and apparently many algae as well, contain structurally diverse secondary compounds that are either toxic to most herbivores or disturb their metabolism greatly, preventing, for example, the normal development of larval insects. Consequently, most herbivores tend to avoid the plants that possess these compounds.

The mustard family (Brassicaceae) produces a group of chemicals known as mustard oils. These substances give the pungent aromas and tastes to plants such as mustard, cabbage, watercress, radish, and horseradish. Although we enjoy these flavors, the chemicals are toxic to many groups of insects.

Similarly, plants of the milkweed family (Asclepiadaceae) and the related dogbane family (Apocynaceae) produce a milky sap that deters herbivores from eating them. In addition, these plants usually contain cardiac glycosides, molecules that can produce drastic deleterious effects on the heart function of vertebrates.

The coevolutionary response of herbivores

Some herbivores have evolved ways of circumventing plant defenses, allowing them to feed on these plants without harm, often as their exclusive food source.

For example, cabbage butterfly caterpillars (subfamily Pierinae) feed almost exclusively on plants of the mustard and caper families, as well as on a few other small families of plants that also contain mustard oils (figure 56.10). Similarly, caterpillars of monarch butterflies and their relatives (subfamily Danainae) feed on plants of the milkweed and dogbane families. How do these animals manage to avoid the chemical defenses of the plants, and what are the evolutionary precursors and ecological consequences of such patterns of specialization?

We can offer a potential explanation for the evolution of these particular patterns. Once the ability to manufacture mustard oils evolved in the ancestors of the caper and mustard families, the plants were protected for a time against most or all herbivores that were feeding on other plants in their area. At some point, certain groups of insects—for example, the cabbage butterflies—evolved the ability to break down mustard oils and thus feed on these plants without harming themselves. Having developed this new capability, the butterflies were able to use a new resource without competing with other herbivores for it. As we saw in chapter 22, exposure to an underutilized resource often leads to evolutionary diversification and adaptive radiation.

**Figure 56.10
Insect herbivores well suited to their plant hosts.** *a.* The green caterpillars of the cabbage white butterfly (*Pieris rapae*) are camouflaged on the leaves of cabbage and other plants on which they feed. Although mustard oils protect these plants against most herbivores, the cabbage white butterfly caterpillars are able to break down the mustard oil compounds. *b.* An adult cabbage white butterfly.

Animal adaptations defend against predators

Some animals that feed on plants rich in secondary compounds receive an extra benefit. For example, when the caterpillars of monarch butterflies feed on plants of the milkweed family, they do not break down the cardiac glycosides that protect these plants from herbivores. Instead, the caterpillars concentrate and store these compounds in fat bodies; they then pass them through the chrysalis stage to the adult and even to the eggs of the next generation.

The incorporation of cardiac glycosides protects all stages of the monarch life cycle from predators. A bird that eats a monarch butterfly quickly regurgitates it (figure 56.11) and in the future avoids the conspicuous orange-and-black pattern that characterizes the adult monarch. Some bird species have evolved the ability to tolerate the protective chemicals; these birds eat the monarchs.

Chemical defenses

Animals also manufacture and use a startling array of defensive substances. Bees, wasps, predatory bugs, scorpions, spiders, and many other arthropods use chemicals to defend themselves and to kill their own prey. In addition, various chemical defenses have evolved among many marine invertebrates, as well as a variety of vertebrates, including frogs, snakes, lizards, fishes, and some birds.

The poison-dart frogs of the family Dendrobatidae produce toxic alkaloids in the mucus that covers their brightly colored skin; these alkaloids are distasteful and sometimes deadly to animals that try to eat the frogs (figure 56.12). Some of these toxins are so powerful that a few micrograms will kill a person if injected into the bloodstream. More than 200 different alkaloids have been isolated from these frogs, and some are playing important roles in neuromuscular research. Similarly intensive investigations of marine animals, venomous reptiles, algae,

Figure 56.12 Vertebrate chemical defenses. Frogs of the family Dendrobatidae, abundant in the forests of Central and South America, are extremely poisonous to vertebrates; 80 different toxic alkaloids have been identified from different species in this genus. Dendrobatids advertise their toxicity with bright coloration. As a result of either instinct or learning, predators avoid such brightly colored species that might otherwise be suitable prey.

and flowering plants are underway in search of new drugs to fight cancer and other diseases, or to use as sources of antibiotics.

Defensive coloration

Many insects that feed on milkweed plants are brightly colored; they advertise their poisonous nature using an ecological strategy known as warning coloration.

Showy coloration is characteristic of animals that use poisons and stings to repel predators; organisms that lack specific chemical defenses are seldom brightly colored. In fact, many have cryptic coloration—color that blends with the surroundings and thus hides the individual from predators (figures 56.10 and 56.13). Camouflaged animals usually do not live together in groups because a predator that discovers one individual gains a valuable clue to the presence of others.

Mimicry allows one species to capitalize on defensive strategies of another

During the course of their evolution, many species have come to resemble distasteful ones that exhibit warning coloration. The mimic gains an advantage by looking like the distasteful model. Two types of mimicry have been identified: Batesian mimicry and Müllerian mimicry.

Batesian mimicry

Batesian mimicry is named for Henry Bates, the British naturalist who first brought this type of mimicry to general attention in 1857. In his journeys to the Amazon region of South America, Bates discovered many instances of palatable insects that resembled brightly colored, distasteful species. He reasoned that the mimics would be avoided by predators, who would be fooled by the disguise into thinking the mimic was the distasteful species.

Many of the best-known examples of Batesian mimicry occur among butterflies and moths. Predators of these insects

a.   *b.*

Figure 56.11 A blue jay learns not to eat monarch butterflies. *a.* This cage-reared jay had never seen a monarch butterfly before it tried eating one. *b.* The same jay regurgitated the butterfly a few minutes later. After having such a bad experience, birds usually do not try to capture orange-and-black insects of any kind.

Figure 56.13 Cryptic coloration and form. An inchworm caterpillar *(Nacophora quernaria)* closely resembles the twig on which it is hanging.

must use visual cues to hunt for their prey; otherwise, similar color patterns would not matter to potential predators. Increasing evidence indicates that Batesian mimicry can involve nonvisual cues, such as olfaction, although such examples are less obvious to humans.

The kinds of butterflies that provide the models in Batesian mimicry are, not surprisingly, members of groups whose caterpillars feed on only one or a few closely related plant families. The plant families on which they feed are strongly protected by toxic chemicals. The model butterflies incorporate the poisonous molecules from these plants into their bodies. The mimic butterflies, in contrast, belong to groups in which the feeding habits of the caterpillars are not so restricted. As caterpillars, these butterflies feed on a number of different plant families that are unprotected by toxic chemicals.

One often-studied mimic among North American butterflies is the tiger swallowtail, whose range occurs throughout the eastern United States and into Canada (figure 56.14*a*). In areas in which the poisonous pipevine swallowtail occurs, female tiger swallowtails are polymorphic and one color form is extremely similar in appearance to the pipevine swallowtail.

The caterpillars of the tiger swallowtail feed on a variety of trees, including tulip, aspen, and cherry, and neither caterpillars nor adults are distasteful to birds. Interestingly, the Batesian mimicry seen in the adult tiger swallowtail butterfly does not extend to the caterpillars: Tiger swallowtail caterpillars are camouflaged on leaves, resembling bird droppings, but the pipevine swallowtail's distasteful caterpillars are very conspicuous.

Müllerian mimicry

Another kind of mimicry, **Müllerian mimicry,** was named for the German biologist Fritz Müller, who first described it in 1878. In Müllerian mimicry, several unrelated but protected animal species come to resemble one another (figure 56.14*b*). If animals that resemble one another are all poisonous or dangerous, they gain an advantage because a predator will learn more quickly to avoid them. In some cases, predator populations even evolve an innate avoidance of species; such evolution may occur more quickly when multiple dangerous prey look alike.

In both Batesian and Müllerian mimicry, mimic and model must not only look alike but also act alike. For example, the members of several families of insects that closely resemble wasps behave surprisingly like the wasps they mimic, flying often and actively from place to place.

Learning Outcomes Review 56.3

Predation is the consuming of one organism by another. High predation can drive prey populations to extinction; conversely, in the absence of predators, prey populations often explode and exhaust their resources. Defensive adaptations may evolve in prey species, such as becoming distasteful or poisonous, or having defensive structures, appearance, or capabilities.

■ *A nonpoisonous scarlet king snake has red, black, and yellow bands of color similar to that of the poisonous eastern coral snake. What type of mimicry is being exhibited?*

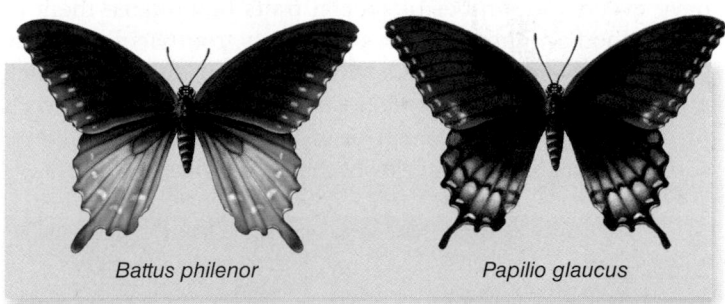

Battus philenor   Papilio glaucus

a. **Batesian mimicry:** Pipevine swallowtail butterfly (*Battus philenor*) is poisonous; Tiger swallowtail (*Papilio glaucus*) is a palatable mimic.

Heliconius erato   Heliconius melpomene

Heliconius sapho   Heliconius cydno

b. **Müllerian mimicry:** Two pairs of mimics; all are distasteful.

Figure 56.14 Mimicry. *a.* Batesian mimicry. Pipevine swallowtail butterflies (*Battus philenor*) are protected from birds and other predators by the poisonous compounds they derive from the food they eat as caterpillars and store in their bodies. Adult pipevine swallowtails advertise their poisonous nature with warning coloration. Tiger swallowtails (*Papilio glaucus*) are Batesian mimics of the poisonous pipevine swallowtail and are not chemically protected. *b.* Pairs of Müllerian mimics. *Heliconius erato* and *H. melpomene* are sympatric, and *H. sapho* and *H. cydno* are sympatric. All of these butterflies are distasteful. They have evolved similar coloration patterns in sympatry to minimize predation; predators need only learn one pattern to avoid.

56.4 The Many Types of Species Interactions

Learning Outcomes

1. *Explain the different forms of symbiosis.*
2. *Describe how coevolution occurs between mutualistic partners.*
3. *Explain how the occurrence of one ecological process may affect the outcome of another occurring at the same time.*

The plants, animals, protists, fungi, and prokaryotes that live together in communities have changed and adjusted to one another continually over millions of years. We have already discussed competition and predation, but other types of ecological interactions commonly occur. For example, many features of flowering plants have evolved in relation to plant pollination by animals (figure 56.15). These animals, in turn, have evolved a number of special traits that enable them to obtain food or other resources efficiently from the plants they visit, often from their flowers. While doing so, the animals pick up pollen, which they may deposit on the next plant they visit, or seeds, which may be left elsewhere in the environment, sometimes a great distance from the parent plant.

Figure 56.15 Pollination by a bat. Many flowers have coevolved with other species to facilitate pollen transfer. Insects are widely known as pollinators, but they're not the only ones: birds, bats, and even small marsupials and lizards serve as pollinators for some species. Notice the cargo of pollen on the bat's snout.

Symbiosis involves long-term interactions

In symbiosis, two or more kinds of organisms interact in often elaborate and more-or-less permanent relationships. All symbiotic relationships carry the potential for coevolution between the organisms involved, and in many instances the results of this coevolution are fascinatingly complex.

Examples of symbiosis include lichens, which are associations of certain fungi with green algae or cyanobacteria. Another important example are mycorrhizae, associations between fungi and the roots of most kinds of plants. The fungi expedite the plant's absorption of certain nutrients, and the plants in turn provide the fungi with carbohydrates (both mycorrhizae and lichens are discussed in greater detail in chapter 32). Similarly, root nodules that occur in legumes and certain other kinds of plants contain bacteria that fix atmospheric nitrogen and make it available to their host plants.

In the tropics, leaf-cutter ants are often so abundant that they can remove a quarter or more of the total leaf surface of the plants in a given area in a single year (see figure 32.18). They do not eat these leaves directly; rather, they take them to underground nests, where they chew them up and inoculate them with the spores of particular fungi. These fungi are cultivated by the ants and brought from one specially prepared bed to another, where they grow and reproduce. In turn, the fungi constitute the primary food of the ants and their larvae. The relationship between leaf-cutter ants and these fungi is an excellent example of symbiosis. Recent phylogenetic studies using DNA and assuming a molecular clock (see chapter 23) suggest that these symbioses are ancient, perhaps originating more than 50 MYA.

The major kinds of symbiotic relationships include (1) mutualism, in which both participating species benefit; (2) **parasitism,** in which one species benefits but the other is harmed; and (3) commensalism, in which one species benefits and the other neither benefits nor is harmed. Parasitism can also be viewed as a form of predation, although the organism that is preyed on does not necessarily die.

Mutualism benefits both species

Mutualism is a symbiotic relationship between organisms in which both species benefit. Mutualistic relationships are of fundamental importance in determining the structure of biological communities.

Mutualism and coevolution

Some of the most spectacular examples of mutualism occur among flowering plants and their animal visitors, including insects, birds, and bats. During the course of flowering-plant evolution, the characteristics of flowers evolved in relation to the characteristics of the animals that visit them for food and, in the process, spread their pollen from individual to individual. At the same time, characteristics of the animals have changed, increasing their specialization for obtaining food or other substances from particular kinds of flowers.

Another example of mutualism involves ants and aphids. Aphids are small insects that suck fluids from the phloem of living plants with their piercing mouthparts. They extract a certain amount of the sucrose and other nutrients from this fluid, but they excrete much of it in an altered form through their anus. Certain ants have taken advantage of this—in effect, domesticating the aphids. Like ranchers taking cattle to fresh fields to graze, the ants carry the aphids to new plants and then consume as food the "honeydew" that the aphids excrete.

Ants and acacias: A prime example of mutualism

A particularly striking example of mutualism involves ants and certain Latin American tree species of the genus *Acacia.* In these species, certain leaf parts, called stipules, are modified as paired, hollow thorns. The thorns are inhabited by stinging ants of the genus *Pseudomyrmex,* which do not nest anywhere else (figure 56.16). Like all thorns that occur on plants, the acacia thorns serve to deter herbivores.

At the tip of the leaflets of these acacias are unique, protein-rich bodies called Beltian bodies, named after the 19th-century British naturalist Thomas Belt. Beltian bodies do not occur in species of *Acacia* that are not inhabited by ants, and their role is clear: they serve as a primary food for the ants. In addition, the plants secrete nectar from glands near the bases of their leaves. The ants consume this nectar as well, feeding it and the Beltian bodies to their larvae.

Obviously, this association is beneficial to the ants, and one can readily see why they inhabit acacias of this group. The ants and their larvae are protected within the swollen thorns, and the trees provide a balanced diet, including the sugar-rich nectar and the protein-rich Beltian bodies. What, if anything, do the ants do for the plants?

Whenever any herbivore lands on the branches or leaves of an acacia inhabited by ants, the ants, which continually patrol the acacia's branches, immediately attack and devour the herbivore. The ants that live in the acacias also help their hosts compete with other plants by cutting away any encroaching branches that touch the acacia in which they are living. They create, in effect, a tunnel of light through which the acacia can grow, even in the lush tropical rain forests of lowland Central America. In fact, when an ant colony is experimentally removed

from a tree, the acacia is unable to compete successfully in this habitat. Finally, the ants bring organic material into their nests. The parts they do not consume, together with their excretions, provide the acacias with an abundant source of nitrogen.

When mutualism may not be mutualism

Things are not always as they seem. Ant–acacia associations also occur in Africa; in Kenya, several species of acacia ants occur, but only a single species is found on any one tree. One species, *Crematogaster nigriceps,* is competitively inferior to two of the other species. To prevent invasion by these other ant species, *C. nigriceps* prunes the branches of the acacia, preventing it from coming into contact with branches of other trees, which would serve as a bridge for invaders.

Although this behavior is beneficial to the ant, it is detrimental to the tree because it destroys the tissue from which flowers are produced, essentially sterilizing the tree. In this case, what initially evolved as a mutualistic interaction has instead become a parasitic one.

Parasitism benefits one species at the expense of another

Parasitism is harmful to the prey organism and beneficial to the parasite. In many cases, the parasite kills its host, and thus the ecological effects of parasitism can be similar to those of predation. In the past parasitism was studied mostly in terms of its effects on individuals and the populations in which they live, but in recent years researchers have realized that parasitism can be an important factor affecting community structure.

External parasites

Parasites that feed on the exterior surface of an organism are external parasites, or ectoparasites (figure 56.17). Many

Figure 56.16 Mutualism: Ants and acacias. Ants of the genus *Pseudomyrmex* live within the hollow thorns of certain species of acacia trees in Latin America. The nectaries at the bases of the leaves and the Beltian bodies at the ends of the leaflets provide food for the ants. The ants, in turn, supply the acacias with organic nutrients and protect the acacias from herbivores and shading from other plants.

Figure 56.17 An external parasite. The yellow vines are the flowering plant dodder *(Cuscuta),* a parasite that has lost its chlorophyll and its leaves in the course of its evolution. Because it is heterotrophic (unable to manufacture its own food), dodder obtains its food from the host plants it grows on.

instances of external parasitism are known in both plants and animals. **Parasitoids** are insects that lay eggs in or on living hosts. This behavior is common among wasps, whose larvae feed on the body of the unfortunate host, often killing it.

Internal parasites

Parasites that live within the body of their hosts, termed **endoparasites,** occur in many different phyla of animals and protists. Internal parasitism is generally marked by much more extreme specialization than external parasitism, as shown by the many protist and invertebrate parasites that infect humans.

The more closely the life of the parasite is linked with that of its host, the more its morphology and behavior are likely to have been modified during the course of its evolution (the same is true of symbiotic relationships of all sorts). Conditions within the body of an organism are different from those encountered outside and are apt to be much more constant. Consequently, the structure of an internal parasite is often simplified, and unnecessary armaments and structures are lost as it evolves (for example, see descriptions of tapeworms in chapter 34).

Parasites and host behaviors

Many parasites have complex life cycles that require several different hosts for growth to adulthood and reproduction. Recent research has revealed the remarkable adaptations of certain parasites that alter the behavior of the host and thus facilitate transmission from one host to the next. For example, many parasites cause their hosts to behave in ways that make them more vulnerable to their predators; when the host is ingested, the parasite is able to infect the predator.

One of the most famous examples involves a parasitic flatworm, *Dicrocoelium dendriticum,* which lives in ants as an intermediate host, but reaches adulthood in large herbivorous mammals such as cattle and deer. Transmission from an ant to a deer might seem difficult because deer do not normally eat insects. The flatworm, however, has evolved a remarkable adaptation. When an ant is infected, one of the flatworms migrates to the brain and causes the ant to climb to the top of vegetation and lock its mandibles onto a grass blade at the end of the day, just when herbivores are grazing (figure 56.18). The result is that the ant is eaten along with the grass, leading to infection of the grazer.

Commensalism benefits one species and is neutral to the other

In commensalism, one species benefits and the other is neither hurt nor helped by the interaction. In nature, individuals of one species are often physically attached to members of another. For example, epiphytes are plants that grow on the branches of other plants. In general, the host plant is unharmed, and the epiphyte that grows on it benefits. An example is Spanish moss, which hangs on trees in the southern United States. This plant and other members of its genus, which is in the

Figure 56.18 Parasitic manipulation of host behavior. Due to a parasite in its brain, an ant climbs to the top of a grass blade, where it may be eaten by a grazing herbivore, thus passing the parasite from insect to mammal.

pineapple family, grow on trees to gain access to sunlight; they generally do not harm the trees (figure 56.19).

Similarly, various marine animals, such as barnacles, grow on other, often actively moving sea animals, such as whales, and thus are carried passively from place to place.

Figure 56.19 An example of commensalism. Spanish moss *(Tillandsia usneoides)* benefits from using trees as a substrate, but the trees generally are not affected positively or negatively.

These "passengers" presumably gain more protection from predation than they would if they were fixed in one place, and they also reach new sources of food. The increased water circulation that these animals receive as their host moves around may also be of great importance, particularly if the passengers are filter feeders. Unless the number of these passengers gets too large, the host species is usually unaffected.

When commensalism may not be commensalism

One of the best known examples of symbiosis involves the relationships between certain small tropical fishes (clownfish) and sea anemones, shown in the opening figure of this chapter. The fish have evolved the ability to live among the stinging tentacles of sea anemones, even though these tentacles would quickly paralyze other fishes that touched them. The clownfish feed on food particles left from the meals of the host anemone, remaining uninjured under remarkable circumstances.

On land, an analogous relationship exists between birds called oxpeckers and grazing animals such as cattle or antelopes (figure 56.20). The birds spend most of their time clinging to the animals, picking off parasites and other insects, carrying out their entire life cycles in close association with the host animals.

No clear-cut boundary exists between commensalism and mutualism; in each of these cases, it is difficult to be certain whether the second partner receives a benefit or not. A sea anemone may benefit by having particles of food removed from its tentacles because it may then be better able to catch other prey. Similarly, although often thought of as commensalism, the association of grazing mammals and gleaning birds is actually an example of mutualism. The mammal benefits by having parasites and other insects removed from its body, but the birds also benefit by gaining a dependable source of food.

On the other hand, commensalism can easily transform itself into parasitism. Oxpeckers are also known to pick not only parasites, but also scabs off their grazing hosts. Once the scab is picked, the birds drink the blood that flows from the wound. Occasionally, the cumulative effect of persistent attacks can greatly weaken the herbivore, particularly when conditions are not favorable, such as during droughts.

Ecological processes have interactive effects

We have seen the different ways in which species can interact with one another. In nature, however, more than one type of interaction often occurs at the same time. In many cases, the outcome of one type of interaction is modified or even reversed when another type of interaction is also occurring.

Predation reduces competition

When resources are limiting, a superior competitor can eliminate other species from a community through competitive

Figure 56.20 Commensalism, mutualism, or parasitism? In this symbiotic relationship, oxpeckers definitely receive a benefit in the form of nutrition from the ticks and other parasites they pick off their host (in this case, an impala, *Aepyceros melampus*). But the effect on the host is not always clear. If the ticks are harmful, their removal benefits the host, and the relationship is mutually beneficial. If the oxpeckers also pick at scabs, causing blood loss and possible infection, the relationship may be parasitic. If the hosts are unharmed by either the ticks or the oxpeckers, the relationship may be an example of commensalism.

exclusion. However, predators can prevent or greatly reduce exclusion by lowering the numbers of individuals of competing species.

A given predator may often feed on two, three, or more kinds of plants or animals in a given community. The predator's choice depends partly on the relative abundance of the prey options. In other words, a predator may feed on species A when it is abundant and then switch to species B when A is rare. Similarly, a given prey species may become a primary source of food for increasing numbers of species as it becomes more abundant. In this way, superior competitors may be prevented from competitively excluding other species.

Such patterns are often characteristic of communities in marine intertidal habitats. For example, by preying selectively on bivalves, sea stars prevent bivalves from monopolizing a habitat, opening up space for many other organisms

Question: *Does predation affect the outcome of interspecific competitive interactions?*

Hypothesis: *In the absence of predators, prey populations will increase until resources are limiting, and some species will be competitively excluded.*

Experiment: *Remove predatory sea stars (Pisaster ochraceus) from some areas of rocky intertidal shoreline and monitor populations of species the sea stars prey upon. In control areas, pick up sea stars, but replace them where they were found to control for the effects of people walking through the study area.*

a.   *b.*

Result: *In the absence of sea stars, the population of the mussel* Mytilus californianus *exploded, occupying all available space and eliminating many other species from the community.*

Interpretation: *What would happen if sea stars were returned to the experimental plots?*

Figure 56.21 Predation reduces competition. *a.* In a controlled experiment in a coastal ecosystem, Robert Paine of the University of Washington removed a key predator, sea stars *(Pisaster)*. *b.* In response, fiercely competitive mussels, a type of bivalve mollusk, exploded in population growth, effectively crowding out seven other indigenous species.

(figure 56.21). When sea stars are removed from a habitat, species diversity falls precipitously, and the seafloor community comes to be dominated by a few species of bivalves.

Predation tends to reduce competition in natural communities, so it is usually a mistake to attempt to eliminate a major predator, such as wolves or mountain lions, from a community. The result may be a decrease in biological diversity.

Parasitism may counter competition

Parasites may affect sympatric species differently and thus influence the outcome of interspecific interactions. One classic experiment investigated interactions between two sympatric flour beetles, *Tribolium castaneum* and *T. confusum,* with and without an intracellular parasite. In the absence of the parasite, *T. castaneum* is dominant, and *T. confusum* normally becomes extinct. When the parasite is present, however, the outcome is reversed, and *T. castaneum* perishes.

Similar effects of parasites in natural systems have been observed in many species. For example, in the *Anolis* lizards of St. Maarten mentioned previously, the competitively inferior species is resistant to lizard malaria (a disease related to human malaria), whereas the other species is highly susceptible. In places where the parasite occurs, the competitively inferior species can hold its own and the two species coexist;

elsewhere, the competitively dominant species outcompetes and eliminates it.

Indirect effects

In some cases, species may not directly interact, yet the presence of one species may affect a second by way of interactions with a third. Such effects are termed indirect effects.

Many desert rodents eat seeds, and so do the ants in their community; thus, we might expect them to compete with each other. But when all rodents were removed from experimental enclosures and not allowed back in (unlike the previous experiment, no holes were placed in the enclosure walls), ant populations first increased but then declined (figure 56.22).

The initial increase was the expected result of removing a competitor. Why did it then reverse? The answer reveals the intricacies of natural ecosystems. Rodents prefer large seeds, whereas ants prefer smaller ones. Furthermore, in this system, plants with large seeds are competitively superior to plants with small seeds. The removal of rodents therefore led to an increase in the number of plants with large seeds, which reduced the number of small seeds available to ants, which in turn led to a decline in ant populations. In summary, the effect of rodents on ants is complicated: a direct, negative effect of resource competition and an indirect, positive effect mediated by plant competition.

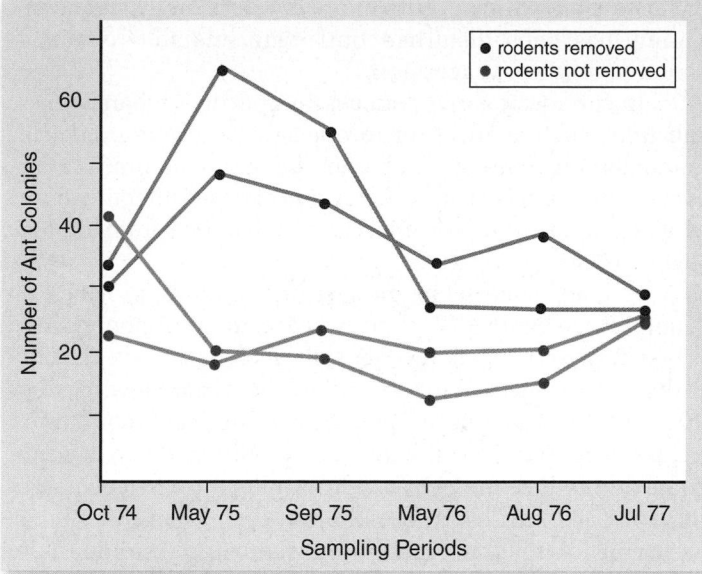

a.

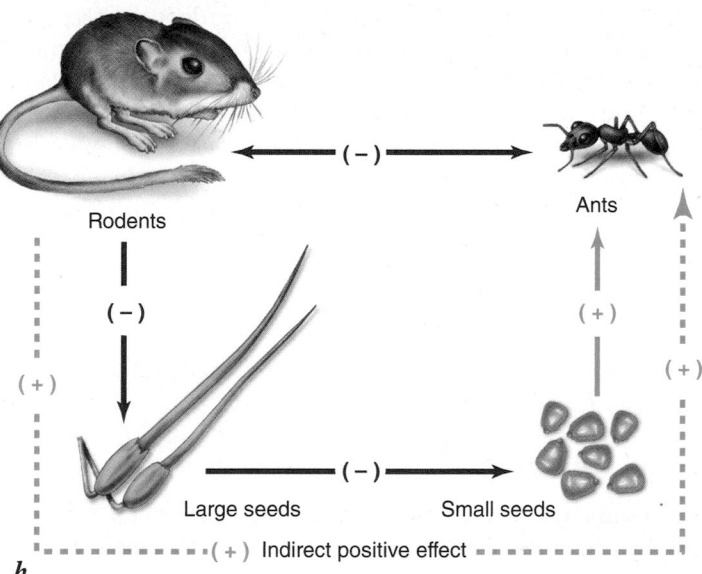

b.

Figure 56.22 **Direct and indirect effects in an ecological community.** *a.* In the enclosures in which rodents had been removed, ants initially increased in population size relative to the ants in the control enclosures, but then these ant populations declined. *b.* Rodents and ants both eat seeds, so the presence of rodents has a direct negative effect on ants, and vice versa. However, the presence of rodents has a negative effect on large seeds. In turn, the number of plants with large seeds has a negative effect on plants that produce small seeds, which the ants eat. Hence, the presence of rodents should increase the number of small seeds. In turn, the number of small seeds has a positive effect on ant populations. Thus, indirectly, the presence of rodents has a positive effect on ant population size.

🔍 **Data analysis** Based on the flow chart in *(b)*, explain why the number of ant colonies first increases and then declines when rodents are removed.

❓ **Inquiry question** How would you test the hypothesis that plant competition mediates the positive effect of rodents on ants?

Figure 56.23 **Example of a keystone species.** Beavers, by constructing dams and transforming flowing streams into ponds, create new habitats for many plant and animal species.

Keystone species have major effects on communities

Species whose effects on the composition of communities are greater than one might expect based on their abundance are termed **keystone species.** Predators, such as the sea star described earlier, can often serve as keystone species by preventing one species from outcompeting others, thus maintaining high levels of species richness in a community.

A wide variety of other types of keystone species also exist. Some species manipulate the environment in ways that create new habitats for others. Beavers, for example, change running streams into small impoundments, altering the flow of water and flooding areas (figure 56.23). Similarly, alligators excavate deep holes at the bottoms of lakes. In times of drought, these holes are the only areas where water remains, thus allowing aquatic species that otherwise would perish to persist until the drought ends and the lake refills.

Learning Outcomes Review 56.4

The types of symbiosis include mutualism, in which both participants benefit; commensalism, in which one benefits and the other is neutrally affected; and parasitism, in which one benefits at the expense of the other. Mutualistic species often undergo coevolution, such as the shape of flowers and the features of animals that feed on and pollinate them. Ecological interactions can affect many processes in a community; for example, predation and parasitism may lessen resource competition.

■ *How could the presence of a predator positively affect populations of a species on which it preys?*

56.5 Ecological Succession, Disturbance, and Species Richness

Learning Outcomes

1. Define succession and distinguish primary versus secondary.
2. Describe how early colonizers may affect subsequent occurrence of other species.
3. Explain how disturbance can either positively or negatively affect species richness.

Even when the climate of an area remains stable year after year, communities have a tendency to change from simple to complex in a process known as **succession.** This process is familiar to anyone who has seen a vacant lot or cleared woods slowly become occupied by an increasing number of species.

Succession produces a change in species composition

If a wooded area is cleared or burned and left alone, plants will slowly reclaim the area. Eventually, all traces of the clearing will disappear, and the area will again be woods. This kind of succession, which occurs in areas where an existing community has been disturbed but organisms still remain, is called **secondary succession.**

In contrast, **primary succession** occurs on bare, lifeless substrate, such as rocks, or in open water, where organisms gradually move into an area and change its nature. Primary succession occurs in lakes and on land exposed after the retreat of glaciers, and on volcanic islands that rise from the sea (figure 56.24).

Primary succession on glacial moraines provides an example (see figure 56.24). On the bare, mineral-poor ground exposed when glaciers recede, soil pH is basic as a result of carbonates in the rocks, and nitrogen levels are low. Lichens are the first vegetation able to grow under such conditions. Acidic secretions from the lichens help break down the substrate and reduce the pH, as well as adding to the accumulation of soil. Mosses then colonize these pockets of soil, eventually building up enough nutrients in the soil for alder shrubs to take hold. Over a hundred years, the alders, which have symbiotic bacteria that fix atmospheric nitrogen (described in chapter 28), increase soil nitrogen levels, and their acidic leaves further lower soil pH. Eventually, spruce trees grow above the alders and shade them, crowding them out entirely and forming a dense spruce forest.

In a similar example, an *oligotrophic* lake—one poor in nutrients—may gradually, by the accumulation of organic matter, become *eutrophic*—rich in nutrients. As this occurs, the composition of communities will change, first increasing in species richness and then declining.

Why succession happens

Succession happens because species alter the habitat and the resources available in it in ways that favor other species. Three dynamic concepts are of critical importance in the process: establishment, facilitation, and inhibition.

1. **Establishment.** Early successional stages are characterized by weedy, *r*-selected species that are tolerant of the harsh,

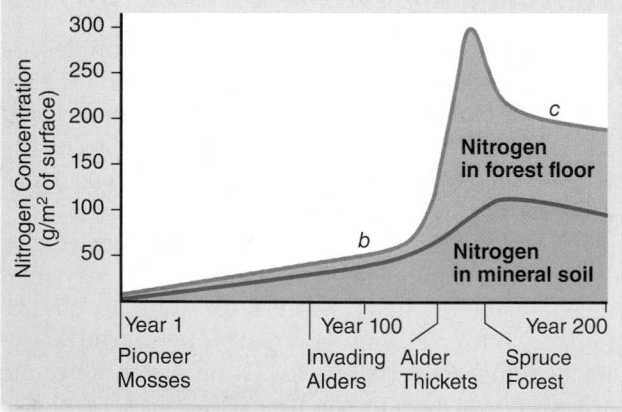

a.

Figure 56.24 Primary succession at Alaska's Glacier Bay.
a. Initially, the glacial moraine at Glacier Bay, Alaska, had little soil nitrogen *b.* The first invaders of these exposed sites are pioneer moss species with nitrogen-fixing, mutualistic microbes. *c.* Within 20 years, young alder shrubs take hold. Rapidly fixing nitrogen, they soon form dense thickets. *d.* Eventually spruce overgrow the mature alders, forming a forest.

b.

c.

d.

abiotic conditions in barren areas (chapter 55 discussed *r*-selected and *K*-selected species).

2. **Facilitation.** The weedy early successional stages introduce local changes in the habitat that favor other, less weedy species. Thus, the mosses in the Glacier Bay succession convert nitrogen to a form that allows alders to invade (see figure 56.24). Similarly, the nitrogen build-up produced by the alders, though not necessary for spruce establishment, leads to more robust forests of spruce better able to resist attack by insects.

3. **Inhibition.** Sometimes the changes in the habitat caused by one species, while favoring other species, also inhibit the growth of the original species that caused the changes. Alders, for example, do not grow as well in acidic soil as the spruce and hemlock that replace them.

Over the course of succession, the number of species typically increases as the environment becomes more hospitable. In some cases, however, as ecosystems mature, more *K*-selected species replace *r*-selected ones, and superior competitors force out other species, leading ultimately to a decline in species richness.

Succession in animal communities

The species of animals present in a community also change through time in a successional pattern. As the vegetation changes during succession, habitat disappears for some species and appears for others.

A particularly striking example occurred on the Krakatau islands, which were devastated by an enormous volcanic eruption in 1883. Initially composed of nothing but barren ash-fields, the three islands of the group experienced rapid successional change as vegetation became reestablished. A few blades of grass appeared the next year, and within 15 years the coastal vegetation was well established and the interior was covered with dense grasslands. By 1930, the islands were almost entirely forested (figure 56.25).

The fauna of Krakatau changed in synchrony with the vegetation. Nine months after the eruption, the only animal found was a single spider, but by 1908, 200 animal species were found in a 3-day exploration. For the most part, the first animals were grassland inhabitants, but as trees became established, some of these early colonists, such as the zebra dove and the long-tailed shrike (a type of predatory bird), disappeared and were replaced by forest-inhabiting species, such as fruit bats and fruit-eating birds.

Although patterns of succession of animal species have typically been caused by vegetational succession, changes in the composition of the animal community in turn have affected plant occurrences. In particular, many plant species that are animal-dispersed or pollinated could not colonize Krakatau until their dispersers or pollinators had become established. For example, fruit bats were slow to colonize Krakatau, and until they appeared, few bat-dispersed plant species were present.

Disturbances can play an important role in structuring communities

Traditionally, many ecologists considered biological communities to be in a state of equilibrium, a stable condition that resisted change and fairly quickly returned to its original state if disturbed by humans or natural events. Such stability was usually attributed to the process of interspecific competition.

In recent years, this viewpoint has been reevaluated. Increasingly, scientists are recognizing that communities are constantly changing as a result of climatic changes, species invasions, and disturbance events. As a result, many ecologists now invoke nonequilibrium models that emphasize change, rather than stability. A particular focus of ecological research concerns the role that disturbances play in determining the structure of communities.

Disturbances can be widespread or local. Severe disturbances, such as forest fires, drought, and floods, may affect large areas. Animals may also cause severe disruptions. Gypsy moths can devastate a forest by consuming all of the leaves on its trees. Unregulated deer populations may grow explosively, the deer overgrazing and so destroying the forest in which they

a. *b.*

Figure 56.25 Succession after a volcanic eruption. A major volcanic explosion in 1883 on the island of Krakatau destroyed all life on the island. *a.* This photo shows a later, much less destructive eruption of the volcano. *b.* Krakatau, forested and populated by animals.

live. On the other hand, local disturbances may affect only a small area, as when a tree falls in a forest or an animal digs a hole and uproots vegetation.

Intermediate disturbance hypothesis

In some cases, disturbance may act to increase the species richness of an area. According to the *intermediate disturbance hypothesis,* communities experiencing moderate amounts of disturbance will have higher levels of species richness than communities experiencing either little or great amounts of disturbance.

Two factors could account for this pattern. First, in communities where moderate amounts of disturbance occur, patches of habitat exist at different successional stages. Within the area as a whole, then, species diversity is greatest because the full range of species—those characteristic of all stages of succession—are present. For example, a pattern of intermittent episodic disturbance that produces gaps in the rain forest (as when a tree falls) allows invasion of the gap by other species (figure 56.26). Eventually, the species inhabiting the gap will go through a successional sequence, one tree replacing another, until a canopy tree species comes again to occupy the gap. But if there are many gaps of different ages in the forest, many different species will be coexisting, some in young gaps and others in older ones.

Second, moderate levels of disturbance may prevent communities from reaching the final stages of succession, in which a few dominant competitors eliminate most of the other species. In contrast, too much disturbance might leave the community continually in the earliest stages of succession, when species richness is relatively low.

Ecologists are increasingly realizing that disturbance is common, rather than exceptional, in many communities. As a result, the idea that communities inexorably move along a successional trajectory culminating in the development of a predictable end-state, or "climax," community is no longer widely accepted. Rather, predicting the state of a community in the future may be difficult because the unpredictable occurrence of disturbances will often counter successional changes. Understanding the role that disturbances play in structuring communities is currently an important area of investigation in ecology.

Figure 56.26 Intermediate disturbance. A single fallen tree created a small light gap in the tropical rain forest of Panama. Such gaps play a key role in maintaining the high species diversity of the rain forest. In this case, a sunlight-loving plant is able to sprout up among the dense foliage of trees in the forest.

Learning Outcomes Review 56.5

Communities change through time by a process termed succession. Primary succession occurs on bare, lifeless substrate; secondary succession occurs where an existing community has been disturbed. Early-arriving species alter the environment in ways that allow other species to colonize, and new colonizers may have negative effects on species already present. Sometimes, moderate levels of disturbance can lead to increased species richness because species characteristic of all levels of succession may be present.

■ *From a community point of view, would clear-cutting a forest be better than selective harvest of individual trees? Why or why not?*

Chapter Review

56.1 Biological Communities: Species Living Together

A community is a group of different species that occupy a given location.

Communities change over space and time.

In accordance with the individualistic view, species generally respond independently to environmental conditions, and community composition gradually changes over space and time.

However, in locations where conditions rapidly change, species composition may change greatly over short distances.

56.2 The Ecological Niche Concept

Fundamental niches are potential; realized niches are actual.

A niche is the total of all the ways a species uses environmental resources. The fundamental niche is the entire niche a species is capable of using if there are no intervening factors. The realized niche is the set of actual environmental conditions that allow establishment of a stable population.

Realized niches are usually smaller than fundamental niches because interspecific interactions limit a species' use of some resources.

Competitive exclusion can occur when species compete for limited resources.

The principle of competitive exclusion states that if resources are limiting, two species cannot simultaneously occupy the same niche; rather, one species will be eliminated.

Competition may lead to resource partitioning.

By using different resources (partitioning), sympatric species can avoid competing with each other and can coexist with reduced realized niches.

Detecting interspecific competition can be difficult.

Although experimentation is a powerful means of testing the hypothesis that species compete, practical limitations exist. Detailed knowledge of the ecology of species is important to evaluate the results of experiments and possible interactions.

56.3 Predator–Prey Relationships

Predation strongly influences prey populations.

Predation is the consuming of one organism by another, and includes not only one animal eating another, but also an animal eating a plant.

Natural selection strongly favors adaptations of prey species to prevent predation. In turn, sometimes predators evolve counter-adaptations, leading to an evolutionary "arms race."

Plant adaptations defend against herbivores.

Plants produce secondary chemical compounds that deter herbivores. Sometimes the herbivores evolve an ability to ingest the compounds and use them for their own defense.

Animal adaptations defend against predators.

Animal adaptations include chemical defenses and defensive coloration such as warning coloration or camouflage.

Mimicry allows one species to capitalize on defensive strategies of another.

In Batesian mimicry, a species that is edible or nontoxic evolves warning coloration similar to that of an inedible or poisonous species. In Müllerian mimicry, two species that are both toxic evolve similar warning coloration.

56.4 The Many Types of Species Interactions

Symbiosis involves long-term interactions.

Many symbiotic species have coevolved and have permanent relationships.

Mutualism benefits both species.

One example is the case of ants and acacias, in which *Acacia* plants provide a home and food for a species of stinging ants that protect them from herbivores.

Parasitism benefits one species at the expense of another.

Many organisms have parasitic lifestyles, living on or inside one or more host species and causing damage or disease as a result.

Commensalism benefits one species and is neutral to the other.

Examples of commensal relationships include epiphytes growing on large plants and barnacles growing on sea animals.

Ecological processes have interactive effects.

Because many processes may occur simultaneously, species may affect one another not only through direct interactions but also through their effects on other species in the community.

Keystone species have major effects on communities.

Keystone species are those that maintain a more diverse community by reducing competition between species or by altering the environment to create new habitats.

56.5 Ecological Succession, Disturbance, and Species Richness

Succession produces a change in species composition.

Primary succession begins with a barren, lifeless substrate, whereas secondary succession occurs after an existing community is disrupted by fire, clearing, or other events.

Disturbances can play an important role in structuring communities.

Community composition changes as a result of local and global disturbances that "reset" succession.

Intermediate levels of such disturbance may maximize species richness in two ways: by creating a patchwork of different habitats harboring different species, and by preventing communities from reaching the final stage of succession, which may be dominated by only a few, competitively superior species.

Review Questions

UNDERSTAND

1. Studies that demonstrate that species living in an ecological community change independently of one another in space and time
   a. indicate that the species arrived in the community at different times.
   b. indicate that the species' realized niches are regulated by different aspects of the environment.
   c. suggest species interactions are the sole determinant of which species coexist in a community.
   d. None of the choices is correct.

2. If two species have very similar realized niches and are forced to coexist and share a limiting resource indefinitely,
   a. both species would be expected to coexist.
   b. both species would be expected to go extinct.
   c. the species that uses the limiting resource most efficiently should drive the other species extinct.
   d. both species would be expected to become more similar to one another.

3. According to the idea of coevolution between predator and prey, when a prey species evolves a novel defense against a predator

   a. the predator is expected to always go extinct.
   b. the prey population should increase irreversibly out of control of the predator.
   c. the predator population should increase.
   d. evolution of a predator response should be favored by natural selection.

4. In order for mimicry to be effective in protecting a species from predation, it must

   a. occur in a palatable species that looks like a distasteful species.
   b. have cryptic coloration.
   c. occur such that mimics look and act like models.
   d. occur in only poisonous or dangerous species.

5. Which of the following is an example of commensalism?

   a. A tapeworm living in the gut of its host
   b. A clownfish living among the tentacles of a sea anemone
   c. An acacia tree and acacia ants
   d. Bees feeding on nectar from a flower

6. A species whose effect on the composition of a community is greater than expected based on its abundance can be called a

   a. predator.
   b. primary succession species.
   c. secondary succession species.
   d. keystone species.

7. When a predator preferentially eats the superior competitor in a pair of competing species

   a. the inferior competitor is more likely to go extinct.
   b. the superior competitor is more likely to persist.
   c. coexistence of the competing species is more likely.
   d. None of the choices is correct.

8. Species that are the first colonists in a habitat undergoing primary succession

   a. are usually the fiercest competitors.
   b. help maintain their habitat constant so their persistence is ensured.
   c. may change their habitat in a way that favors the invasion of other species.
   d. must first be successful secondary succession specialists.

APPLY

1. Which of the following can cause the realized niche of a species to be smaller than its fundamental niche?

   a. Predation   c. Parasitism
   b. Competition   d. All of the choices are correct.

2. The presence of a predatory species

   a. always drives a prey species to extinction.
   b. can positively affect a prey species by having a detrimental effect on competing species.
   c. indicates that the climax stage of succession has been reached.
   d. None of the choices is correct.

3. Resource partitioning by sympatric species

   a. always occurs when species have identical niches.
   b. may not occur in the presence of a predator, which reduces prey population sizes.

c. results in the fundamental and realized niches being the same.
d. is more common in herbivores than carnivores.

4. Parasitism differs from predation because

   a. the presence of parasitism doesn't lead to selection for defensive adaptations in parasitized species.
   b. parasites and the species they parasitize never engage in an evolutionary "arms race."
   c. parasites don't have strong effects on the populations of the species they parasitize.
   d. None of the choices is correct.

5. The presence of one species (A) in a community may benefit another species (B) if

   a. a commensalistic relationship exists between the two.
   b. The first species (A) preys on a predator of the second species (B).
   c. The first species (A) preys on a species that competes with a species that is eaten by the second species (B).
   d. All of the choices are correct.

SYNTHESIZE

1. Competition is traditionally indicated by documenting the effect of one species on the population of another. Are there alternative ways to study the potential effects of competition on organisms that are impractical to study with experimental manipulations because they are too big or live too long?

2. Refer to figure 56.9. If the single prey species of *Paramecium* was replaced by several different potential prey species that varied in their palatability or ease of subduing by the predator (leading to different levels of preference by the predator) what would you expect the dynamics of the system to look like; that is, would the system be more or less likely to go to extinction?

3. Refer to figure 56.22. Are there alternative hypotheses that might explain the increase followed by the decrease in ant colony numbers subsequent to rodent removal in the experiment described in figure 56.22? If so, how would you test the mechanism hypothesized in the figure?

4. Refer to figure 56.7. Examine the pattern of beak size distributions of two species of finches on the Galápagos Islands. One hypothesis that can be drawn from this pattern is that character displacement has taken place. Are there other hypotheses? If so, how would you test them?

5. Is it possible that some species function together as an integrated, holistic community, whereas other species at the same locality behave more individualistically? If so, what factors might determine which species function in which way?

ONLINE RESOURCE

www.ravenbiology.com

Understand, Apply, and Synthesize—enhance your study with animations that bring concepts to life and practice tests to assess your understanding. Your instructor may also recommend the interactive eBook, individualized learning tools, and more.

Chapter **57**

Dynamics of Ecosystems

Chapter Contents

Introduction

The Earth is a relatively closed system with respect to chemicals. It is an open system in terms of energy, however, because it receives energy at visible and near-visible wavelengths from the Sun and steadily emits thermal energy to outer space in the form of infrared radiation. The organisms in ecosystems interact in complex ways as they participate in the cycling of chemicals and as they capture and expend energy. All organisms, including humans, depend on the specialized abilities of other organisms—plants, algae, animals, fungi, and prokaryotes—to acquire the essentials of life. Interactions between species determine both how chemicals and energy move through ecosystems and how communities are structured.

Biogeochemical Cycles

Learning Outcomes

1. *Define ecosystem.*
2. *List four chemicals whose cyclic interactions are critical to organisms.*
3. *Describe how human activities disrupt these cycles.*

An ecosystem includes all the organisms that live in a particular place, plus the abiotic (nonliving) environment at that location. Ecosystems are intrinsically dynamic in a number of ways, including their processing of matter and energy. We start with matter.

The atomic constituents of matter cycle within ecosystems

During the biological processing of matter, the atoms of which it is composed, such as the atoms of carbon or oxygen, maintain their integrity even as they are assembled into new compounds and the compounds are later broken down. The Earth has an essentially fixed number of each of the types of atoms of biological importance, and the atoms are recycled.

Each organism assembles its body from atoms that previously were in the soil, the atmosphere, other parts of the abiotic environment, or other organisms. When the organism dies, its atoms are released unaltered to be used by other organisms or returned to the abiotic environment. Because of the cycling of the atomic constituents of matter, your body is

likely during your life to contain a carbon or oxygen atom that once was part of Julius Caesar's body or Cleopatra's.

The atoms of the various chemical elements are said to move through ecosystems in **biogeochemical cycles,** a term emphasizing that the cycles of chemical elements involve not only biological organisms and processes, but also geological (abiotic) systems and processes. Biogeochemical cycles include processes that occur on many spatial scales, from cellular to planetary, and they also include processes that occur on multiple time-scales, from seconds (biochemical reactions) to millennia (weathering of rocks).

Biogeochemical cycles usually cross the boundaries of ecosystems to some extent, rather than being self-contained within individual ecosystems. For example, one ecosystem might import or export carbon to others.

In this section, we consider the cycles of some major elements along with the compound water. We also present an example of biogeochemical cycles in a forest ecosystem.

Carbon, the basis of organic compounds, cycles through most ecosystems

Carbon is a major constituent of the bodies of organisms because carbon atoms help form the framework of all organic compounds (see chapter 3); almost 20% of the weight of the human body is carbon. From the viewpoint of the day-to-day dynamics of ecosystems, carbon dioxide (CO_2) is the most significant carbon-containing compound in the abiotic environments of organisms. It makes up 0.03% of the volume of the atmosphere, meaning the atmosphere contains about 750 billion metric tons of carbon. In aquatic ecosystems, CO_2 reacts spontaneously with the water to form bicarbonate ions (HCO_3^-).

Figure 57.1 The carbon cycle. Photosynthesis by plants and algae captures carbon in the form of organic chemical compounds. Aerobic respiration by organisms and fuel combustion by humans return carbon to the form of carbon dioxide (CO_2) or bicarbonate (HCO_3^-). Microbial methanogens living in oxygen-free microhabitats, such as the mud at the bottom of the pond, might produce methane (CH_4), a gas that would enter the atmosphere and then gradually be oxidized abiotically to carbon dioxide (shown in green circled inset).

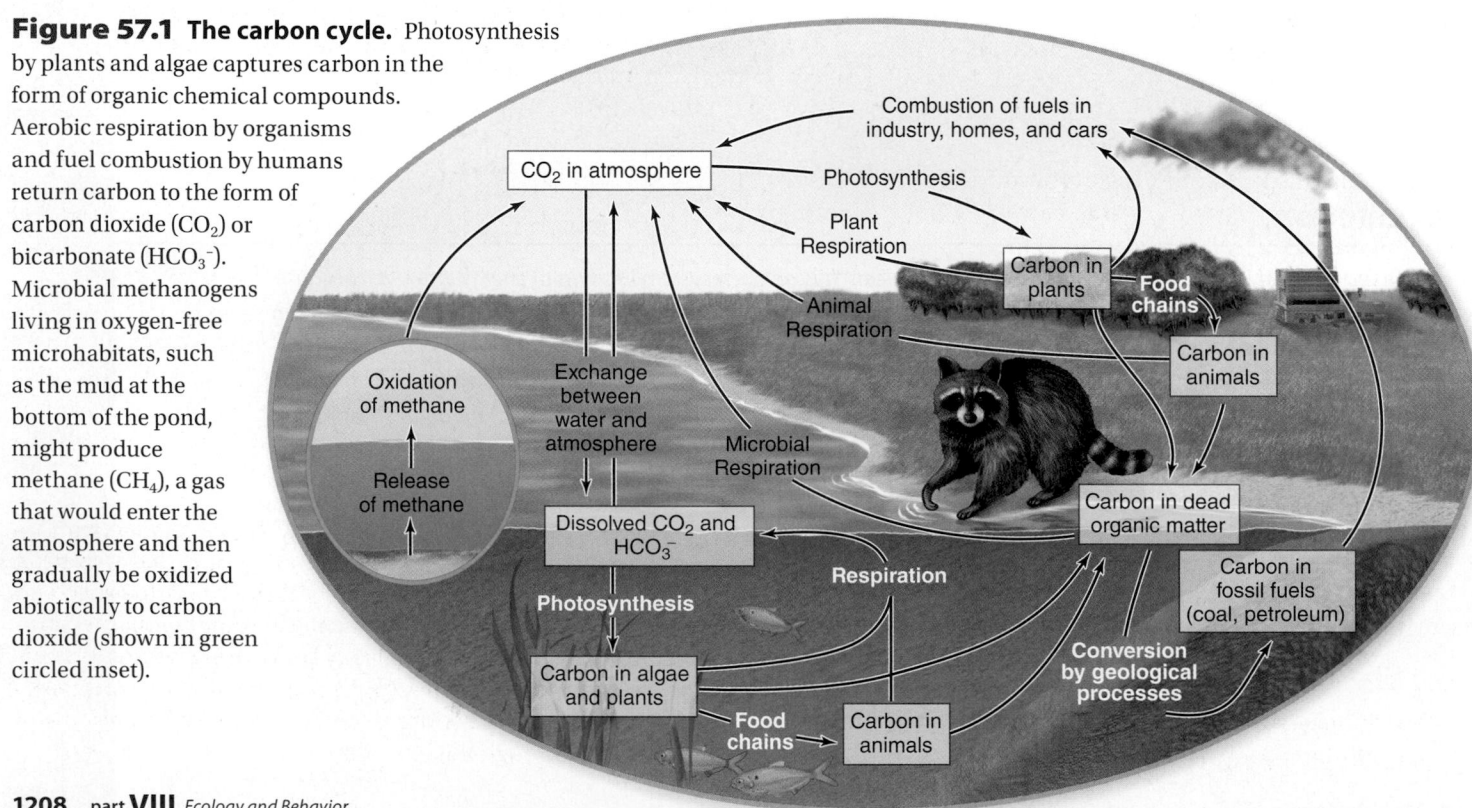

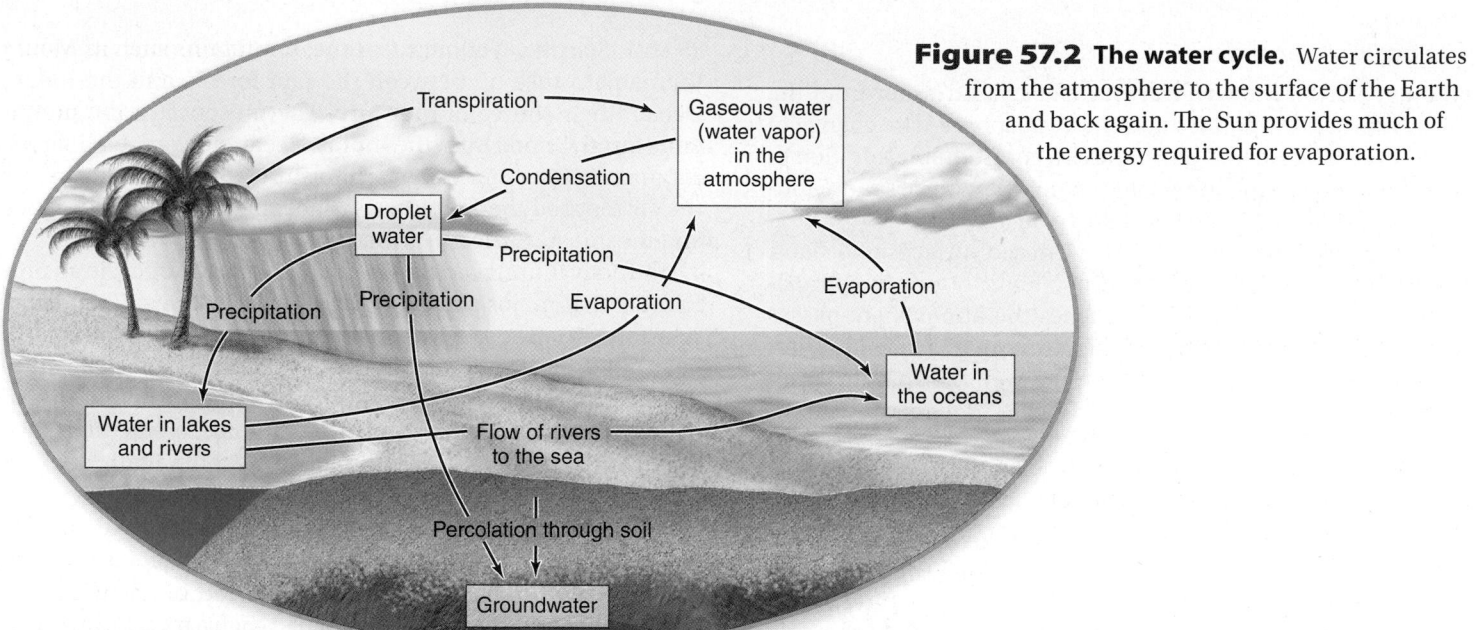

Figure 57.2 The water cycle. Water circulates from the atmosphere to the surface of the Earth and back again. The Sun provides much of the energy required for evaporation.

The basic carbon cycle

The carbon cycle is straightforward, as shown in figure 57.1. In terrestrial ecosystems, plants and other photosynthetic organisms take in CO_2 from the atmosphere and use it in photosynthesis to synthesize the carbon-containing organic compounds of which they are composed (see chapter 8). The process is sometimes called carbon fixation; fixation refers to metabolic reactions that make nongaseous compounds from gaseous ones.

Animals eat the photosynthetic organisms and build their own tissues by making use of the carbon atoms in the organic compounds they ingest. Both the photosynthetic organisms and the animals obtain energy during their lives by breaking down some of the organic compounds available to them, through aerobic cellular respiration (see chapter 7). When they do this, they produce CO_2. Decaying organisms also produce CO_2. Carbon atoms returned to the form of CO_2 are available once more to be used in photosynthesis to synthesize new organic compounds (carbon molecules occurring within an organism are referred to as "organic," whereas those outside of an organism are called "inorganic").

In aquatic ecosystems, the carbon cycle is fundamentally similar, except that inorganic carbon is present in the water not only as dissolved CO_2, but also as HCO_3^- ions, both of which act as sources of carbon for photosynthesis by algae and aquatic plants.

Methane producers

Microbes that break down organic compounds by anaerobic cellular respiration (see chapter 7) provide an additional dimension to the global carbon cycle. Methanogens, for example, are microbes that produce methane (CH_4) instead of CO_2. One major source of CH_4 is wetland ecosystems, where methanogens live in the oxygen-free sediments. Methane that enters the atmosphere is oxidized abiotically to CO_2, but CH_4 that remains isolated from oxygen can persist for great lengths of time.

The rise of atmospheric carbon dioxide

Another dimension of the global carbon cycle is that over long stretches of time, some parts of the cycle may proceed more rapidly than others. These differences in rate have ordinarily been relatively minor on a year-to-year basis; in any one year, the amount of CO_2 made by breakdown of organic compounds almost matches the amount of CO_2 used to synthesize new organic compounds.

Small mismatches, however, can have large consequences if continued for many years. The Earth's present reserves of coal were built up over geologic time. Organic compounds such as cellulose accumulated by being synthesized faster than they were broken down, and then they were transformed by geological processes into the fossil fuels. Most scientists believe that the world's petroleum and natural gas reserves were created in the same way.

Human burning of fossil fuels today is creating large contemporary imbalances in the carbon cycle. Carbon that took millions of years to accumulate in the reserves of fossil fuels is being rapidly returned to the atmosphere, driving the concentration of CO_2 in the atmosphere upward year by year and helping to spur fears of global warming (see chapter 58).

The availability of water is fundamental to terrestrial ecosystems

The water cycle, seen in figure 57.2, is probably the most familiar of all biogeochemical cycles. All life depends on the presence of water. The bodies of most organisms consist mainly of water. The adult human body, for example, is about 60% water by weight. The amount of water available in an ecosystem often determines the nature and abundance of the organisms present, as illustrated by the difference between forests and deserts (see chapter 58).

Each type of biogeochemical cycle has distinctive features. A distinctive feature of the water cycle is that water is a compound, not an element, and thus it can be synthesized and broken down. It is synthesized during aerobic cellular respiration (see chapter 7) and chemically split during photosynthesis (see chapter 8). The rates of these processes are ordinarily about equal, and therefore a relatively constant amount of water cycles through the biosphere.

The basic water cycle

One key part of the water cycle is that liquid water from the Earth's surface evaporates into the atmosphere. The change of water from a liquid to a gas requires a considerable addition of thermal energy, explaining why evaporation occurs more rapidly when solar radiation beats down on a surface.

Evaporation occurs directly from the surfaces of oceans, lakes, and rivers. In terrestrial ecosystems, however, approximately 90% of the water that reaches the atmosphere passes through plants. Trees, grasses, and other plants take up water from soil via their roots, and then the water evaporates from their leaves and other surfaces through a process called transpiration (see chapter 37).

Evaporated water exists in the atmosphere as a gas, just like any other atmospheric gas. The water can condense back into liquid form, however, mostly because of cooling of the air. Condensation of gaseous water (water vapor) into droplets or crystals causes the formation of clouds, and if the droplets or crystals are large enough, they fall to the surface of the Earth as precipitation (rain or snow).

Groundwater

Less obvious than surface water, which we see in rivers and lakes, is water under ground—termed groundwater. Groundwater occurs in **aquifers,** which are permeable, underground layers of rock, sand, and gravel that are often saturated with water. Groundwater is the most important reservoir of water on land in many parts of the world, representing over 95% of all fresh water in the United States, for example.

Groundwater consists of two subparts. The upper layers of the groundwater constitute the water table, which is unconfined in the sense that it flows into streams and is partly accessible to the roots of plants. The lower, confined layers of the groundwater are generally out of reach to streams and plants, but can be tapped by wells. Groundwater is recharged by water that percolates downward from above, such as from precipitation. Water in an aquifer flows much more slowly than surface water, anywhere from a few millimeters to a meter or so per day.

In the United States, groundwater provides about 25% of the water used by humans for all purposes, and it supplies about 50% of the population with drinking water. In the Great Plains states, the deep Ogallala Aquifer is tapped extensively as a water source for agricultural and domestic needs. The aquifer is being depleted faster than it is recharged—a local imbalance in the water cycle—posing an ominous threat to the agricultural production of the area. Similar threats exist in many of the drier portions of the globe.

Changes in ecosystems brought about by changes in the water cycle

Water is so crucial for life that changes in its supply in an ecosystem can radically alter the nature of the ecosystem. Such changes have occurred often during the Earth's geological history.

Consider, for example, the ecosystem of the Serengeti Plain in Tanzania, famous for its seemingly endless grasslands occupied by vast herds of antelopes and other grazing animals. The semiarid grasslands of today's Serengeti were rain forests 25 MYA. Starting at about that time, mountains such as Mount Kilimanjaro rose up between the rain forests and the Indian Ocean, their source of moisture. The presence of the mountains forced winds from the Indian Ocean upward, cooling the air and causing much of its moisture to precipitate before the air reached the rain forests. The land became much drier, and the forests turned to grasslands.

Today, human activities can alter the water cycle so profoundly that major changes occur in ecosystems. Changes in rain forests caused by deforestation provide an example. In healthy tropical rain forests, more than 90% of the moisture that falls as rain is taken up by plants and returned to the air by transpiration. Plants, in a very real sense, create their own rain: The moisture returned to the atmosphere falls back on the forests.

When human populations cut down or burn the rain forests in an area, the local water cycle is broken. Water that falls as rain thereafter drains away in rivers instead of rising to form clouds and fall again on the forests. Just such a transformation is occurring today in many tropical rain forests (figure 57.3). Large areas in Brazil, for example, were transformed in the 20th century from lush tropical forest to semiarid desert, depriving many unique plant and animal species of their native habitat.

The nitrogen cycle depends on nitrogen fixation by microbes

Nitrogen is a component of all proteins and nucleic acids and is required in substantial amounts by all organisms; proteins are 16% nitrogen by weight. In many ecosystems, nitrogen is the chemical element in shortest supply relative to the needs of organisms. A paradox is that the atmosphere is 78% nitrogen by volume.

Figure 57.3 Deforestation disrupts the local water cycle. Tropical deforestation can have severe consequences, such as the extensive erosion in this area in the Amazon region of Brazil.

Nitrogen availability

How can nitrogen be in short supply if the atmosphere is so rich with it? The answer is that the nitrogen in the atmosphere is in its elemental form—molecules of nitrogen gas (N_2)—and the vast majority of organisms, including all plants and animals, have no way to use nitrogen in this chemical form.

For animals, the ultimate source of nitrogen is nitrogen-containing organic compounds synthesized by plants or by algae or other microbes. Herbivorous animals, for example, eat plant or algal proteins and use the nitrogen-containing amino acids in them to synthesize their own proteins.

Plants and algae use a number of simple nitrogen-containing compounds as their sources of nitrogen to synthesize proteins and other nitrogen-containing organic compounds in their tissues. Two commonly used nitrogen sources are ammonia (NH_3) and nitrate ions (NO_3^-). As described in chapter 38, certain prokaryotic microbes can synthesize ammonia and nitrate from N_2 in the atmosphere, thereby constituting a part of the nitrogen cycle that makes atmospheric nitrogen accessible to plants and algae (figure 57.4). Other prokaryotes turn NH_3 and NO_3^- into N_2, making the nitrogen inaccessible. The balance of the activities of these two sets of microbes determines the accessibility of nitrogen to plants and algae.

Microbial nitrogen fixation, nitrification, and denitrification

The synthesis of nitrogen-containing compounds from N_2 is known as **nitrogen fixation.** The first step in this process is the synthesis of NH_3 from N_2, and biochemists sometimes use the term *nitrogen fixation* to refer specifically to this step. After NH_3 has been synthesized, other prokaryotic microbes oxidize part of it to form NO_3^-, a process called **nitrification.**

Certain genera of prokaryotes have the ability to accomplish nitrogen fixation using a system of enzymes known as the nitrogenase complex (the *nif* gene complex; see chapter 28). Most of the microbes are free-living, but on land some are found in symbiotic relationships with the roots of legumes (plants of the pea family, Fabaceae), alders, myrtles, and other plants.

Additional prokaryotic microbes (including both bacteria and archaea) are able to convert the nitrogen in NO_3^- into N_2 (or other nitrogen gases such as N_2O), a process termed **denitrification.** Ammonia can be subjected to denitrification indirectly by being converted first to NO_3^- and then to N_2.

Nitrogenous wastes and fertilizer use

Most animals, when they break down proteins in their metabolism, excrete the nitrogen from the proteins as NH_3. Humans and other mammals excrete nitrogen as urea in their urine (see chapter 50); a number of types of microbes convert the urea to NH_3. The NH_3 from animal excretion can be picked up by plants and algae as a source of nitrogen.

Human populations are radically altering the global nitrogen cycle by the use of fertilizers on lawns and agricultural fields. The fertilizers contain forms of fixed nitrogen that crops can use, such as ammonium (NH_4) salts manufactured industrially from atmospheric N_2. Partly because of the production of fertilizers, humans have already doubled the rate of transfer of N_2 in usable forms into soils and waters.

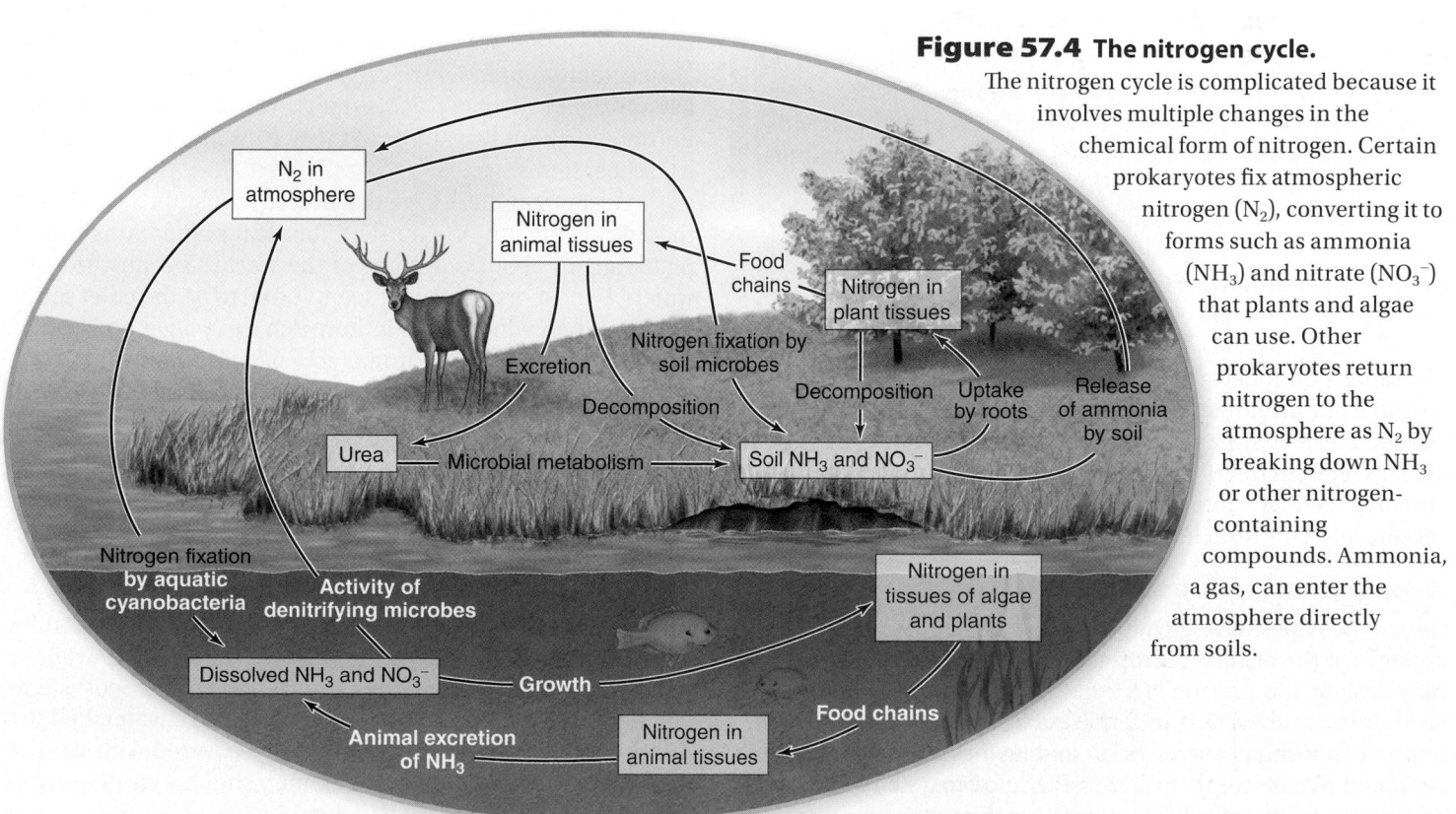

Figure 57.4 The nitrogen cycle.
The nitrogen cycle is complicated because it involves multiple changes in the chemical form of nitrogen. Certain prokaryotes fix atmospheric nitrogen (N_2), converting it to forms such as ammonia (NH_3) and nitrate (NO_3^-) that plants and algae can use. Other prokaryotes return nitrogen to the atmosphere as N_2 by breaking down NH_3 or other nitrogen-containing compounds. Ammonia, a gas, can enter the atmosphere directly from soils.

Figure 57.5 The phosphorus cycle. In contrast to carbon, water, and nitrogen, phosphorus occurs only in the liquid and solid states and thus does not enter the atmosphere.

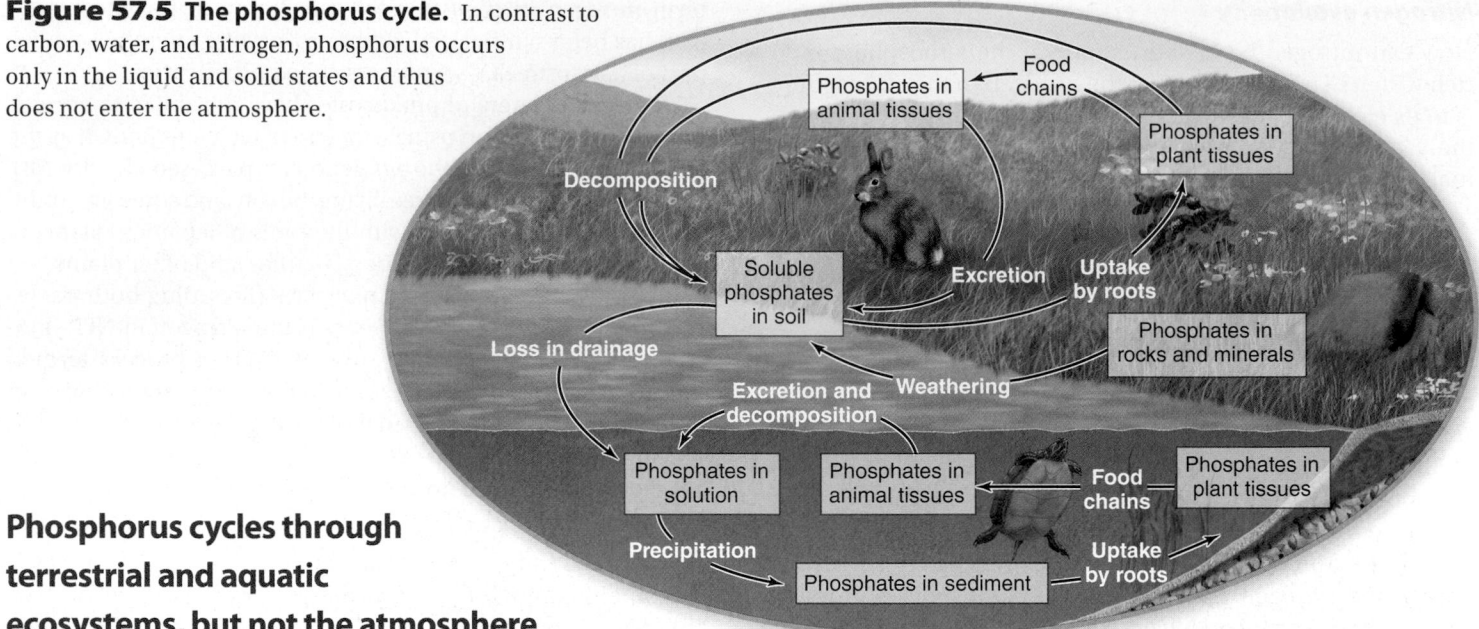

Phosphorus cycles through terrestrial and aquatic ecosystems, but not the atmosphere

Phosphorus is required in substantial quantities by all organisms; it occurs in nucleic acids, membrane phospholipids, and other essential compounds, such as adenosine triphosphate (ATP).

Unlike carbon, water, and nitrogen, phosphorus (P) has no significant gaseous form and does not cycle through the atmosphere (figure 57.5). In this respect, the phosphorus cycle exemplifies the sorts of cycles also exhibited by calcium, silicon, and many other mineral elements. Another feature that greatly simplifies the phosphorus cycle compared with the nitrogen cycle is that phosphorus exists in ecosystems in just a single oxidation state, phosphate (PO_4^{3-}).

Phosphate availability

Plants and algae use free inorganic PO_4^{3-} in the soil or water for synthesizing their phosphorus-containing organic compounds. Animals then tap the phosphorus in plant or algal tissue compounds to build their own phosphorus compounds. When organisms die, decay microbes—in a process called phosphate remineralization—break up the organic compounds in their bodies, releasing phosphorus as inorganic PO_4^{3-} that plants and algae again can use.

The phosphorus cycle includes critical abiotic chemical and physical processes. Free PO_4^{3-} exists in soil in only low concentrations both because it combines with other soil constituents to form insoluble compounds and because it tends to be washed away by streams and rivers. Weathering of many sorts of rocks releases new PO_4^{3-} into terrestrial systems, but then rivers carry the PO_4^{3-} into the ocean basins. There is a large one-way flux of PO_4^{3-} from terrestrial rocks to deep-sea sediments.

Phosphates as fertilizers

Human activities have greatly modified the global phosphorus cycle since the advent of crop fertilization. Fertilizers are typically designed to provide PO_4^{3-} because crops might otherwise be short of it; the PO_4^{3-} in fertilizers is typically derived from crushed phosphate-rich rocks and bones. Detergents are another potential culprit in adding PO_4^{3-} to ecosystems, but laws now mandate low-phosphate detergents in much of the world.

Limiting nutrients in ecosystems are those in short supply relative to need

A chain is only as strong as its weakest link. For the plants and algae in an ecosystem to grow—and to thereby provide food for animals—they need many different chemical elements. The simplest theory is that in any particular ecosystem, one element will be in shortest supply relative to the needs for it by the plants and algae. That element is the limiting nutrient—the weak link—in the ecosystem.

The cycle of a limiting nutrient is particularly important because it determines the rate at which the nutrient is made available for use. We gave the nitrogen and phosphorus cycles close attention precisely because those elements are the limiting nutrients in many ecosystems. Nitrogen is the limiting nutrient in about two-thirds of the oceans and in many terrestrial ecosystems.

Oceanographers have discovered in just the last 15 years that iron is the limiting nutrient for algal populations (phytoplankton) in about one-third of the world's oceans. In these waters, wind-borne soil dust seems often to be the chief source of iron. When wind brings in iron-rich dust, algal populations proliferate, provided the iron is in a usable chemical form. In this way, sand storms in the Sahara Desert, by increasing the dust in global winds, can increase algal productivity in Pacific waters (figure 57.6).

Biogeochemical cycling in a forest ecosystem has been studied experimentally

An ongoing series of studies at the Hubbard Brook Experimental Forest in New Hampshire has yielded much of the available information about the cycling of nutrients in forest ecosystems. Hubbard Brook is the central stream of a large watershed that drains the hillsides of a mountain range covered with temperate deciduous forest. Multiple tributary streams carry water off the hillsides into Hubbard Brook.

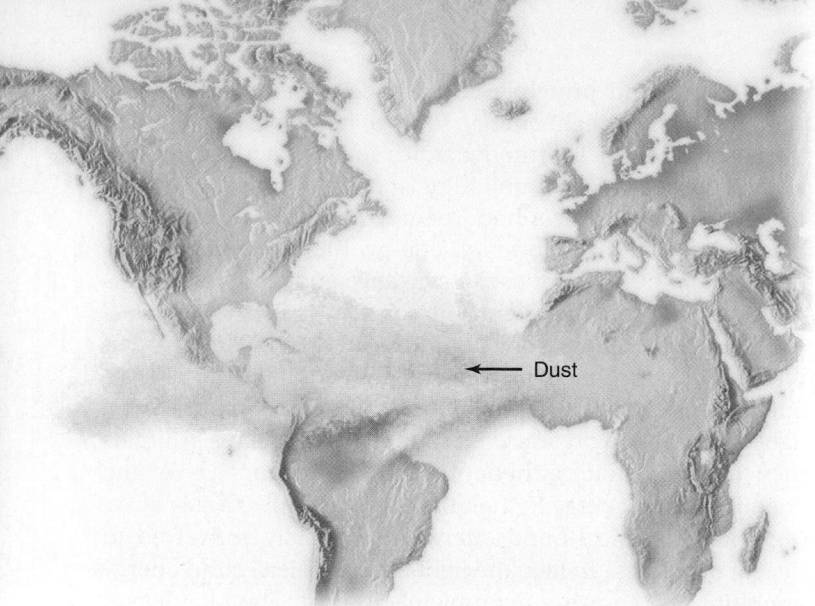

Dust

Figure 57.6 One world. Every year, millions of metric tons of iron-rich dust is carried westward by the trade winds from the Sahara Desert and neighboring Sahel area. A working hypothesis of many oceanographers is that this dust fertilizes parts of the ocean, including parts of the Pacific Ocean, where iron is the limiting nutrient. Land use practices in Africa, which are increasing the size of the north African desert, can thus affect ecosystems on the other side of the globe.

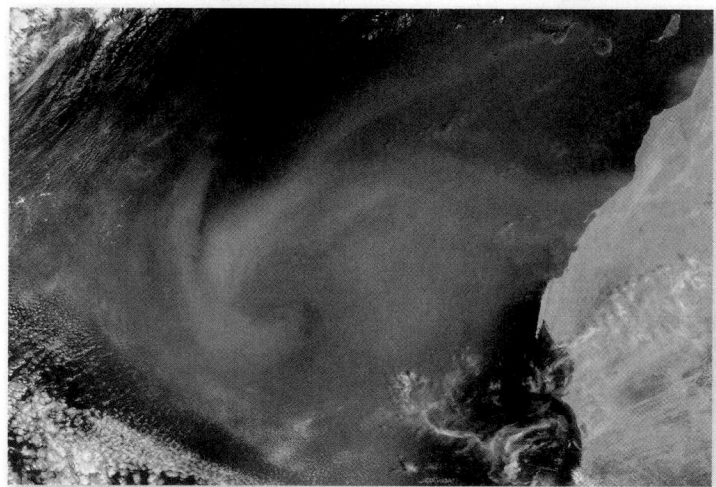

Six tributary streams, each draining a particular valley, were equipped with measurement devices when the study was started. All of the water that flowed out of each valley had to pass through the measurement system, where the flow of water and concentrations of nutrients was quantified.

The undisturbed forests around Hubbard Brook are efficient at retaining nutrients. In a year, only small quantities of nutrients enter a valley from outside, doing so mostly as a result of precipitation. The quantities carried out in stream waters are small also. When we say "small," we mean the influxes and outfluxes represent just minor fractions of the total amounts of nutrients in the system—about 1% in the case of calcium, for example.

In 1965 and 1966, the investigators felled all the trees and cleared all shrubs in one of the six valleys and prevented regrowth (figure 57.7a). The effects were dramatic. The amount of water running out of that valley increased by 40%, indicating that water previously taken up by vegetation and evaporated into the atmosphere was now running off. The amounts of a number of nutrients running out of the system also greatly increased. For example, the rate of loss of calcium increased ninefold. Phosphorus, on the other hand, did not increase in the stream water; it apparently was locked up in insoluble compounds in the soil.

The change in the status of nitrogen in the disturbed valley was especially striking (figure 57.7b). The undisturbed forest in this valley had been accumulating NO_3^- at a rate of about 5 kg per hectare per year, but the deforested ecosystem lost NO_3^- at a rate of about 53 kg per hectare per year. The NO_3^- concentration in the stream water rapidly increased. The fertility of the valley decreased dramatically, while the run-off of nitrate generated massive algal blooms downstream.

This experiment is particularly instructive at the start of the 21st century because forested land continues to be cleared worldwide (see chapter 58).

a.

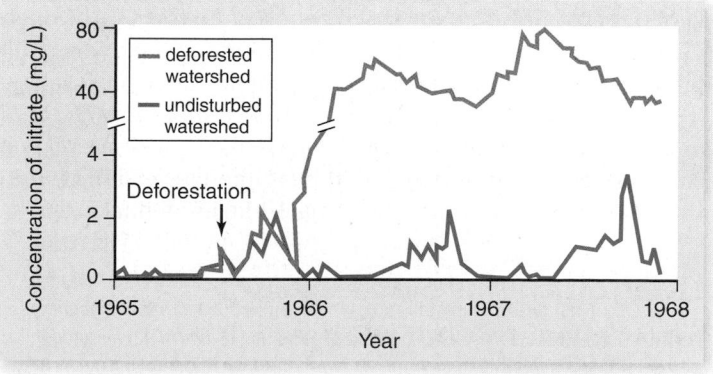

b.

Figure 57.7 The Hubbard Brook experiment. *a.* A 38-acre watershed was completely deforested, and the runoff monitored for several years. *b.* Deforestation greatly increased the loss of nutrients in runoff water from the ecosystem. (Note: The values on the *y*-axis scale are interrupted [double line] to save space.) The orange curve shows the nitrate concentration in the runoff water from the deforested watershed; the green curve shows the nitrate concentration in runoff water from an undisturbed neighboring watershed.

🔍 **Data analysis** About how many times greater was the nitrate runoff in the deforested area than in the neighboring undisturbed area?

57.2 The Flow of Energy in Ecosystems

Learning Outcomes

1. *Describe the different trophic levels.*
2. *Distinguish between energy and heat.*
3. *Explain how energy moves through trophic levels.*

The dynamic nature of ecosystems includes the processing of energy as well as that of matter. Energy, however, follows very different principles than does matter. Energy is never recycled. Instead, radiant energy from the Sun that reaches the Earth makes a one-way pass through our planet's ecosystems before being converted to heat and radiated back into space, signifying that the Earth is an open system for energy.

Energy can neither be created nor destroyed, but changes form

Why is energy so different from matter? A key part of the answer is that energy exists in several different forms, such as light, chemical-bond energy, motion, and heat. Although energy is neither created nor destroyed in the biosphere (the First Law of Thermodynamics), it frequently changes form.

A second key point is that organisms cannot convert heat to any of the other forms of energy. Thus, if organisms convert some chemical-bond or light energy to heat, the conversion is one-way; they cannot cycle that energy back into its original form.

Living organisms can use many forms of energy, but not heat

To understand why the Earth must function as an open system with regard to energy, two additional principles need to be recognized. The first is that organisms can use only certain forms of energy. For animals to live, they must have energy specifically as chemical-bond energy, which they acquire from their foods. Plants must have energy as light. Neither animals nor plants (nor any other organisms) can use heat as a source of energy.

The second principle is that whenever organisms use chemical-bond or light energy, some of it is converted to heat; the Second Law of Thermodynamics states that a partial conversion to heat is inevitable. Put another way, animals and plants require chemical-bond energy and light to stay alive, but as they use these forms of energy, they convert them to heat, which they cannot use to stay alive and which they cannot cycle back into the original forms.

Fortunately for organisms, the Earth functions as an open system for energy. Light arrives every day from the Sun. Plants and other photosynthetic organisms use the newly arrived light to synthesize organic compounds and stay alive. Animals then eat the photosynthetic organisms, making use of the chemical-bond energy in their organic molecules to stay alive. Light and chemical-bond energy are partially converted to heat at every step. In fact, the light and chemical-bond energy are ultimately converted completely to heat. The heat leaves the Earth by being radiated into outer space at invisible, infrared wavelengths of the electromagnetic spectrum. For life to continue, new light energy is always required.

The Earth's incoming and outgoing flows of radiant energy must be equal for global temperature to stay constant. One concern is that human activities are changing the composition of the atmosphere in ways that impede the outgoing flow—the so-called *greenhouse effect,* which is described in chapter 58. Heat may be accumulating on Earth, causing global warming (see chapter 58).

Energy flows through trophic levels of ecosystems

In chapter 7, we introduced the concepts of autotrophs ("self-feeders") and heterotrophs ("fed by others"). **Autotrophs** synthesize the organic compounds of their bodies from inorganic precursors such as CO_2, water, and NO_3^- using energy from an abiotic source. Some autotrophs use light as their source of energy and therefore are **photoautotrophs;** they are the photosynthetic organisms, including plants, algae, and cyanobacteria. Other autotrophs are **chemoautotrophs** and obtain energy by means of inorganic oxidation reactions, such as the microbes that use hydrogen sulfide available at deep water vents (see chapter 58). All chemoautotrophs are prokaryotic. The photoautotrophs are of greatest importance in most ecosystems, and we focus on them in the remainder of this chapter.

Heterotrophs are organisms that cannot synthesize organic compounds from inorganic precursors, but instead live by taking in organic compounds that other organisms have made. They obtain the energy they need to live by breaking up some of the organic compounds available to them, thereby liberating chemical-bond energy for metabolic use (see chapter 7). Animals, fungi, and many microbes are heterotrophs.

When living in their native environments, species are often organized into chains that eat each other sequentially. For example, a species of insect might eat plants, and then a species of shrew might eat the insect, and a species of hawk might eat the shrew. Food passes through the four species in the sequence: plants $\longrightarrow$ insect $\longrightarrow$ shrew $\longrightarrow$ hawk. A sequence of species like this is termed a food chain.

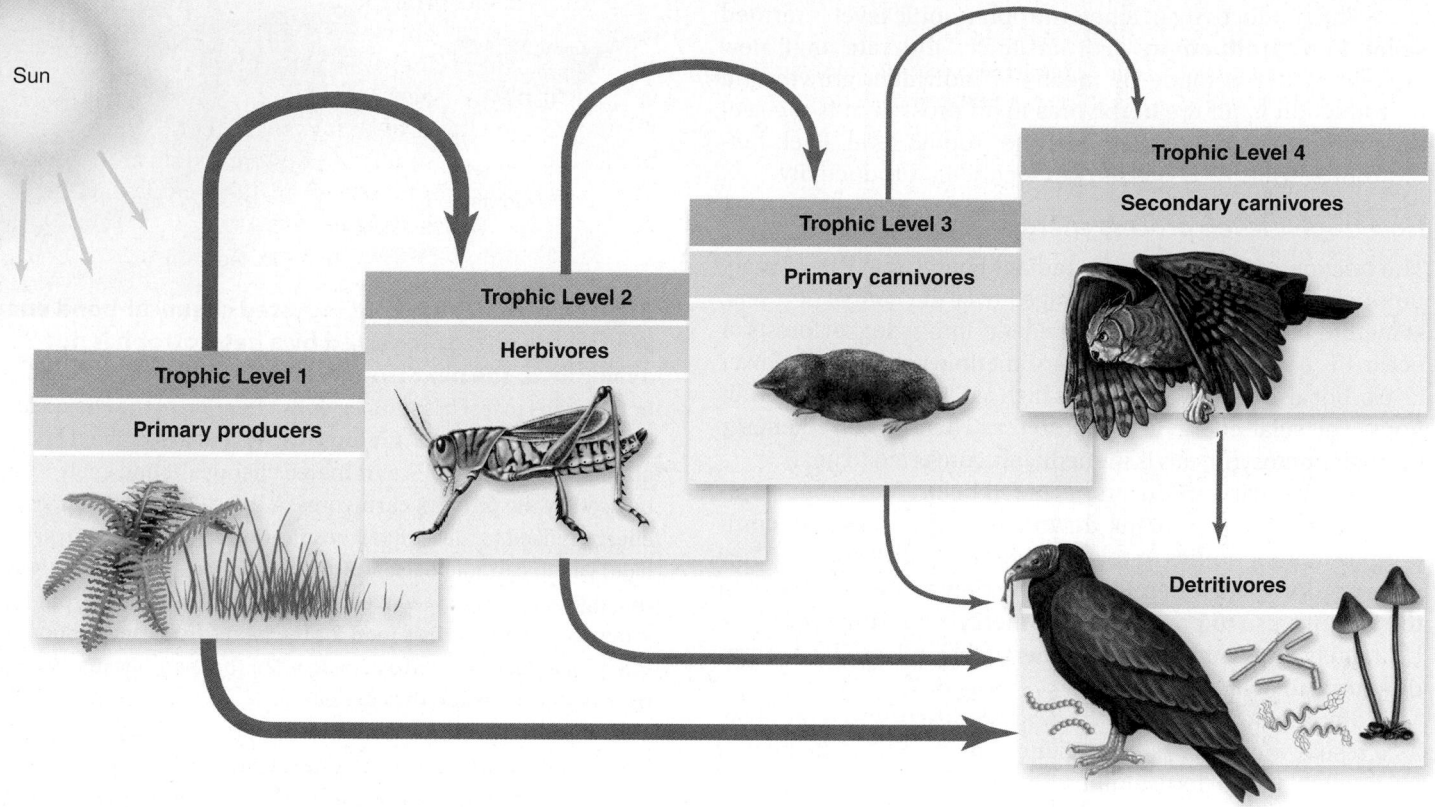

Figure 57.8 Trophic levels within an ecosystem. Primary producers such as plants obtain their energy directly from the Sun, placing them in trophic level 1. Animals that eat plants, such as plant-eating insects, are herbivores and are in trophic level 2. Animals that eat the herbivores, such as shrews, are primary carnivores and are in trophic level 3. Animals that eat the primary carnivores, such as owls, are secondary carnivores in trophic level 4. Each trophic level, although illustrated here by a particular species, consists of all the species in the ecosystem that function in a similar way in terms of what they eat. The organisms in the detritivore trophic level consume dead organic matter they obtain from all the other trophic levels.

In a whole ecosystem, many species play similar roles; there is typically not just a single species in each role. For example, the animals that eat plants might include not just a single insect species, but perhaps 30 species of insects, plus perhaps 10 species of mammals. To organize this complexity, ecologists recognize a limited number of feeding, or **trophic, levels** (figure 57.8).

Definitions of trophic levels

The first trophic level in an ecosystem, called the **primary producers,** consists of all the autotrophs in the system. The other trophic levels consist of the heterotrophs—the **consumers.** All the heterotrophs that feed directly on the primary producers are placed together in a trophic level called the **herbivores.** In turn, the heterotrophs that feed on the herbivores (eating them or being parasitic on them) are collectively termed **primary carnivores,** and those that feed on the primary carnivores are called **secondary carnivores.**

Advanced studies of ecosystems need to take into account that organisms often do not line up in simple linear sequences in terms of what they eat; some animals, for example, eat both primary producers and other animals. Nonetheless, a linear sequence of trophic levels is a useful organizing principle for many purposes.

An additional consumer level is the **detritivore** trophic level. Detritivores differ from the organisms in the other trophic

levels in that they feed on the remains of already-dead organisms; detritus is dead organic matter. A subcategory of detritivores is the **decomposers,** which are mostly microbes and other minute organisms that live on and break up dead organic matter.

Concepts to describe trophic levels

Trophic levels consist of whole populations of organisms. For example, the primary-producer trophic level consists of the whole populations of all the autotrophic species in an ecosystem. Ecologists have developed a special set of terms to refer to the properties of populations and trophic levels.

The **productivity** of a trophic level is the rate at which the organisms in the trophic level collectively synthesize new organic matter (new tissue substance). **Primary productivity** is the productivity of the primary producers. An important complexity in analyzing the primary producers is that not only do they synthesize new organic matter by photosynthesis, but they also break down some of the organic matter to release energy by means of aerobic cellular respiration (see chapter 7). The **respiration** of the primary producers, in this context, is the rate at which they break down organic compounds. **Gross primary productivity (GPP)** is simply the raw rate at which the primary producers synthesize new organic matter; **net primary productivity (NPP)** is the GPP minus the respiration of the primary producers. The NPP represents the organic matter available for herbivores to use as food.

The productivity of a heterotroph trophic level is termed **secondary productivity.** For instance, the rate that new organic matter is made by means of individual growth and reproduction in all the herbivores in an ecosystem is the secondary productivity of the herbivore trophic level. Each heterotroph trophic level has its own secondary productivity.

How trophic levels process energy

The fraction of incoming solar radiant energy that the primary producers capture is small. Averaged over the course of a year, something around 1% of the solar energy impinging on forests or oceans is captured. Investigators sometimes observe far lower levels, but also see percentages as high as 5% under some conditions. The solar energy not captured as chemical-bond energy through photosynthesis is immediately converted to heat.

The primary producers, as noted before, carry out respiration in which they break down some of the organic compounds in their bodies to release chemical-bond energy. They use a portion of this chemical-bond energy to make ATP, which they in turn use to power various energy-requiring processes. Ultimately, the chemical-bond energy they release by respiration turns to heat.

Remember that organisms cannot use heat to stay alive. As a result, whenever energy changes form to become heat, it loses much or all of its usefulness for organisms as a fuel source. What we have seen so far is that about 99% of the solar energy impinging on an ecosystem turns to heat because it fails to be used by photosynthesis. Then some of the energy captured by photosynthesis also becomes heat because of respiration by the primary producers. All the heterotrophs in an ecosystem must live on the chemical-bond energy that is left.

An example of energy loss between trophic levels

As chemical-bond energy is passed from one heterotroph trophic level to the next, a great deal of the energy is diverted all along the way. This principle has dramatic consequences. It means that, over any particular period of time, the amount of chemical-bond energy available to primary carnivores is far less than that available to herbivores, and the amount available to secondary carnivores is far less than that available to primary carnivores.

Why does the amount of chemical-bond energy decrease as energy is passed from one trophic level to the next? Consider the use of energy by the herbivore trophic level as an example (figure 57.9). After an herbivore such as a leaf-eating insect ingests some food, it produces feces. The chemical-bond energy in the compounds in the feces is not passed along to the primary carnivore trophic level. The remaining chemical-bond energy of the food is assimilated by the herbivore and is used for a number of functions. Part of the assimilated energy is liberated by cellular respiration to be used for tissue repair, body movements, and other such functions. The energy used in these ways turns to heat and is not passed along to the carnivore trophic level. The remainder of the chemical-bond energy is built into the tissues of the herbivore and can serve as food for a carnivore. However, some herbivore individuals die of disease or accident rather than being eaten by predators.

In the end, of course, some of the initial chemical-bond energy acquired from the leaf is built into the tissues of herbivore individuals that are eaten by primary carnivores. Much of

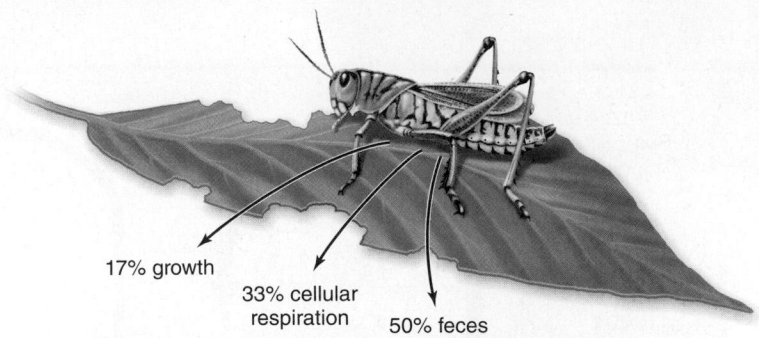

Figure 57.9 The fate of ingested chemical-bond energy: Why all the energy ingested by a heterotroph is not available to the next trophic level. A heterotroph such as this herbivorous insect assimilates only a fraction of the chemical-bond energy it ingests. In this example, 50% is not assimilated and is eliminated in feces; this eliminated chemical-bond energy cannot be used by the primary carnivores. A third (33%) of the ingested energy is used to fuel cellular respiration and thus is converted to heat, which cannot be used by the primary carnivores. Only 17% of the ingested energy is converted into insect biomass through growth and can serve as food for the next trophic level, but not even that percentage is certain to be used in that way because some of the insects die before they are eaten.

the initial chemical-bond energy, however, is diverted into heat, feces, and the bodies of herbivore individuals that carnivores do not get to eat. The same scenario is repeated at each step in a series of trophic levels (figure 57.10).

Ecologists figure as a rule of thumb that the amount of chemical-bond energy available to a trophic level over time is about 10% of that available to the preceding level over the same period of time. In some instances the percentage is higher, even as high as 30%.

Heat as the final energy product

Essentially all of the chemical-bond energy captured by photosynthesis in an ecosystem eventually becomes heat as the chemical-bond energy is used by various trophic levels. To see this important point, recognize that when the detritivores in the ecosystem metabolize all the dead bodies, feces, and other materials made available to them, they produce heat just like the other trophic levels do.

Productive ecosystems

Ecosystems vary considerably in their NPP. Wetlands and tropical rain forests are examples of particularly productive ecosystems (figure 57.11); in them, the NPP, measured as dry weight of new organic matter produced, is often around 2000 g/m^2/year. By contrast, the corresponding figures for some other types of ecosystems are 1200 to 1300 for temperate forests, 900 for savanna, and 90 for deserts. (These general ecosystem types, termed *biomes,* are described in chapter 58.)

The number of trophic levels is limited by energy availability

The rate at which chemical-bond energy is made available to organisms in different trophic levels decreases exponentially as energy makes its way from primary producers to herbivores and

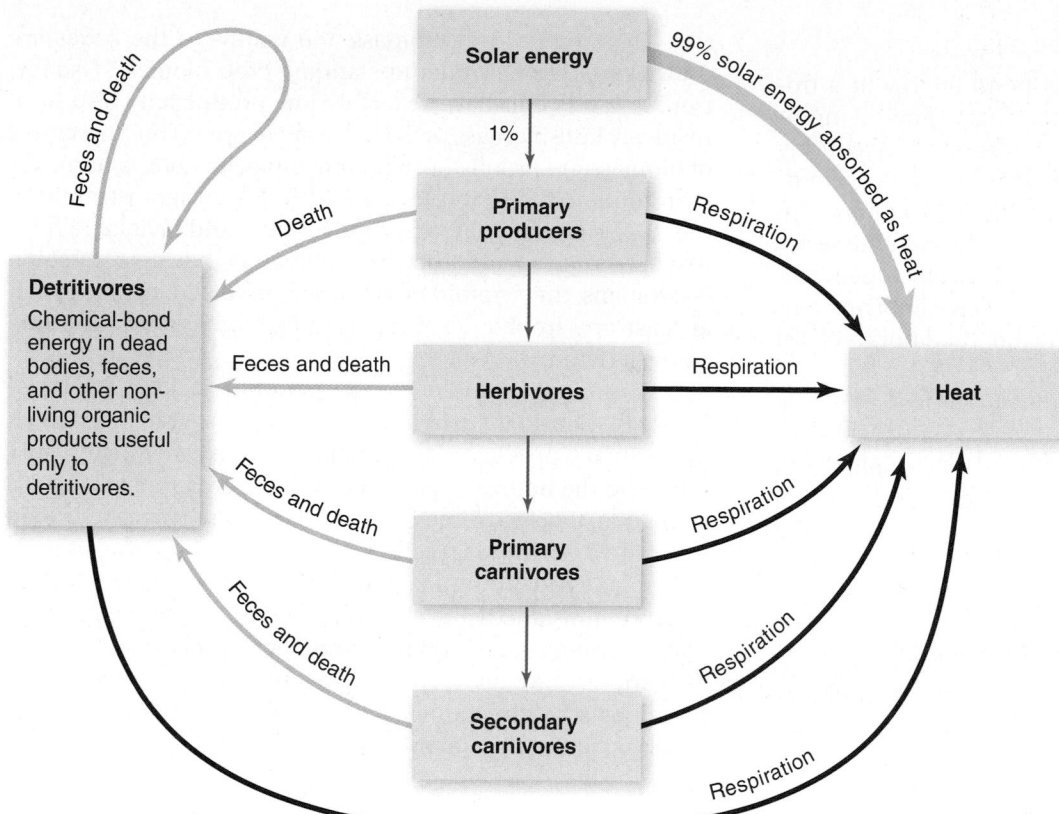

Figure 57.10 The flow of energy through an ecosystem. Blue arrows represent the flow of energy that enters the ecosystem as light and is then passed along as chemical-bond energy to successive trophic levels. At each step energy is diverted, meaning that the chemical-bond energy available to each trophic level is less than that available to the preceding trophic level. Red arrows represent diversions of energy into heat. Tan arrows represent diversions of energy into feces and other organic materials useful only to the detritivores. Detritivores may be eaten by carnivores, so some of the chemical-bond energy returns to higher trophic levels.

then to various levels of carnivores. To envision this critical point, assume for simplicity that the primary producers in an ecosystem gain 1000 units of chemical-bond energy over a period of time. If the energy input to each trophic level is 10% of the input to the preceding level, then the input of chemical-bond energy to the herbivore trophic level is 100 units, to the primary carnivores, 10 units, and to the secondary carnivores, 1 unit over the same period of time.

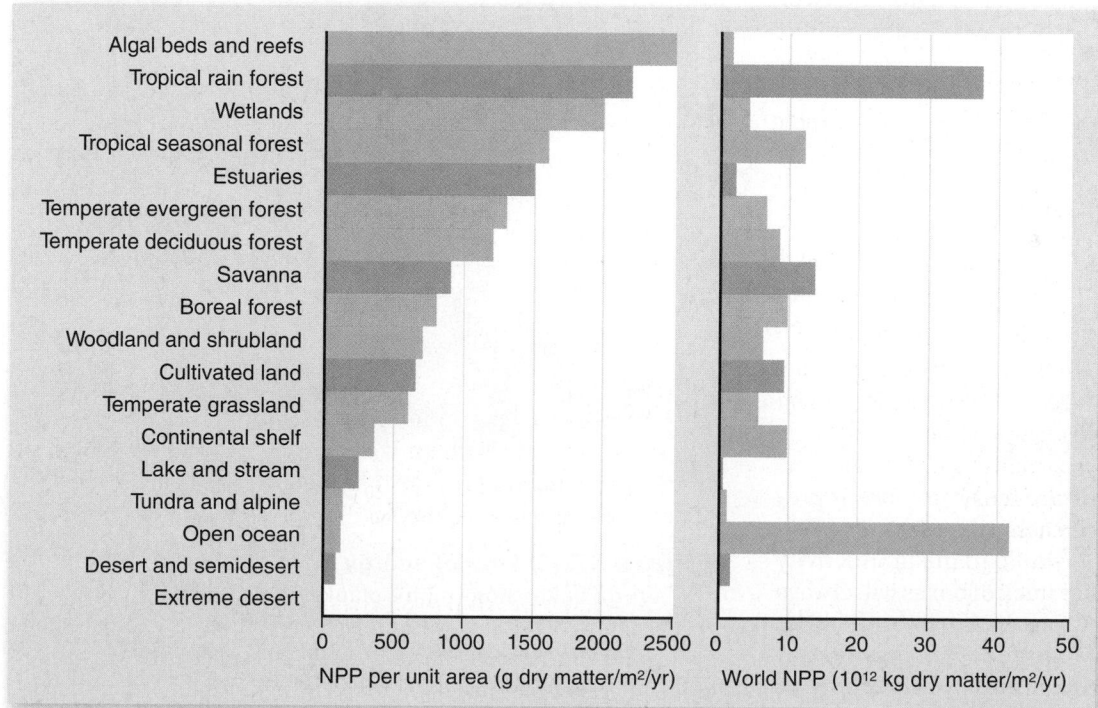

Figure 57.11 Ecosystem productivity per year. The first column of data shows the average net primary productivity (NPP) per square meter per year. The second column of data factors in the area covered by the ecosystem type; it is the product of the productivity per square meter per year multiplied by the number of square meters occupied by the ecosystem type worldwide. Note that an ecosystem type that is very productive on a square-meter basis may not contribute much to global productivity if it is an uncommon type, such as wetlands. On the other hand, a very widespread ecosystem type, such as the open ocean, can contribute greatly to global productivity even if its productivity per square meter is low.

Source: Data in: Begon, M., J.L. Harper, and C. R. Townsend, *Ecology* 3/e, Blackwell Science, 1996, page 715. Original source: Whittaker, R. H. *Communities and Ecosystems*, 2/e, Macmillan, London, 1975.

 Data analysis Is there a relationship among the habitat types between the NPP per unit area and the World NPP?

Limits on top carnivores

The exponential decline of chemical-bond energy in a trophic chain limits the lengths of trophic chains and the numbers of top carnivores an ecosystem can support. According to our model calculations, if an ecosystem includes secondary carnivores, only about one-thousandth of the energy captured by photosynthesis passes all the way through the series of trophic levels to reach these animals as usable chemical-bond energy. Tertiary carnivores would receive only one ten-thousandth. This helps explain why no predators subsist solely on eagles or lions.

The decline of available chemical-bond energy also helps explain why the numbers of individual top-level carnivores in an ecosystem tend to be low. The whole trophic level of top carnivores receives relatively little energy, and yet such carnivores tend to be big: They have relatively large individual body sizes and great individual energy needs. Because of these two factors, the population numbers of top predators tend to be small.

The longest trophic chains probably occur in the oceans. Some tunas and other top-level ocean predators probably function as third- and fourth-level carnivores at times. The challenge of explaining such long trophic chains is obvious, but the solutions are not well understood presently.

Humans as consumers: A case study

The flow of energy in Cayuga Lake in upstate New York (figure 57.12) helps illustrate how the energetics of trophic levels can affect the human food supply. Researchers calculated from the actual properties of this ecosystem that about 150 of each 1000 calories of chemical-bond energy captured by primary producers in the lake were transferred into the bodies of herbivores. Of these calories, about 30 were transferred into the bodies of smelt, small fish that were the principal primary carnivores in the system.

If humans ate the smelt, they gained about 6 of the 1000 calories that originally entered the system. If trout ate the smelt and humans ate the trout, the humans gained only about 1.2 calories. For human populations in general, more energy is available if plants or other primary producers are eaten than if animals are eaten—and more energy is available if herbivores rather than carnivores are consumed.

Ecological pyramids illustrate the relationship of trophic levels

Imagine that the trophic levels of an ecosystem are represented as boxes stacked on top of each other. Imagine also that the width of each box is proportional to the productivity of the trophic level it represents. The stack of boxes will always have the shape of a pyramid; each box is narrower than the one under it because of the inviolable rules of energy flow. A diagram of this sort is called a pyramid of energy flow or pyramid of productivity (figure 57.13a). It is an example of an ecological pyramid.

There are several types of ecological pyramids. Pyramid diagrams can be used to represent standing crop biomass or numbers of individuals, as well as productivity.

In a **pyramid of** biomass, the widths of the boxes are drawn to be proportional to standing crop biomass. Usually, trophic levels that have relatively low productivity also have relatively little biomass present at a given time. Thus, pyramids of biomass are usually upright, meaning each box is narrower than the one below it (figure 57.13b). An upright pyramid of biomass is not mandated by fundamental and inviolable rules like an upright pyramid of productivity is, however. In some ecosystems, the pyramid of biomass is **inverted,** meaning that at least one trophic level has greater biomass than the one below it (figure 57.13c).

How is it possible for the pyramid of biomass to be inverted? Consider a common sort of aquatic system in which the primary producers are single-celled algae (phytoplankton), and the herbivores are rice grain-sized animals (such as copepods) that feed directly on the algal cells. In such a system, the turnover of the algal cells is often very rapid: The cells multiply rapidly, but the animals consume them equally rapidly. In these circumstances, the algal cells never develop a large population size or large biomass. Nonetheless, because the algal cells are very productive, the ecosystem can support a substantial biomass of the animals, a biomass larger than that ever observed in the algal population.

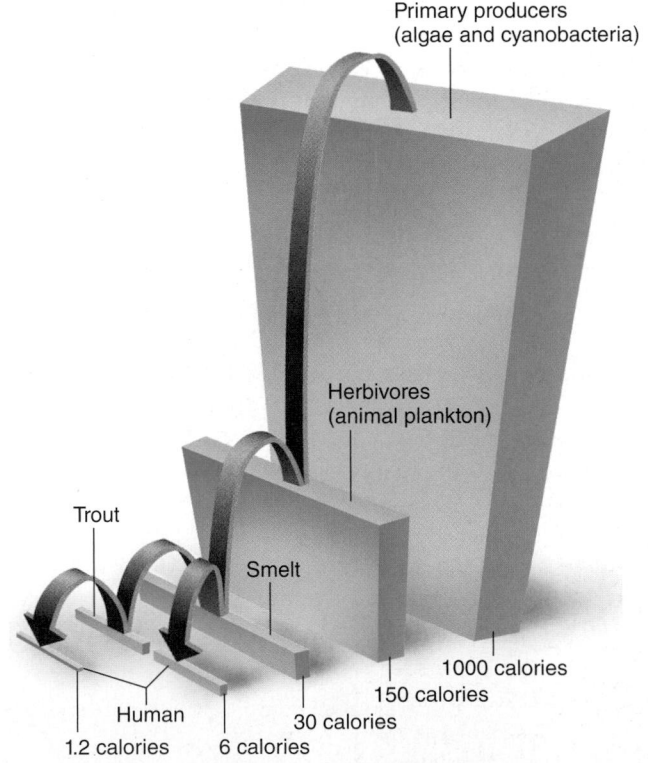

Primary producers (algae and cyanobacteria)

Herbivores (animal plankton)

Trout

Smelt

1000 calories

150 calories

Human

30 calories

1.2 calories

6 calories

Figure 57.12 Flow of energy through the trophic levels of Cayuga Lake. Autotrophic plankton (algae and cyanobacteria) fix the energy of the Sun, the herbivores (animal plankton) feed on them, and both are consumed by smelt. The smelt are eaten by trout. The amount of fish flesh produced per unit time for human consumption is at least five times greater if people eat smelt rather than trout, but people typically prefer to eat trout.

? Inquiry question Why does it take so many calories of algae to support so few calories of humans?

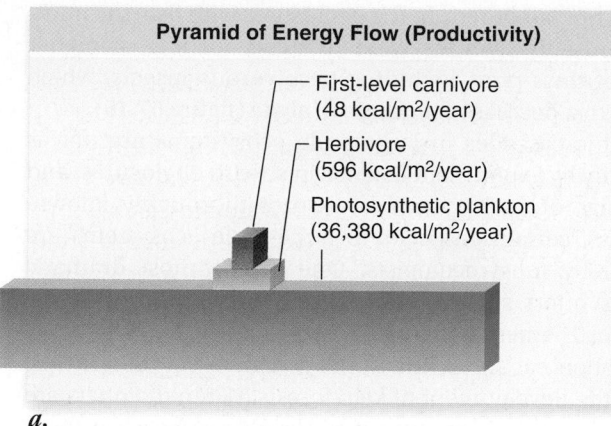

Pyramid of Energy Flow (Productivity)

First-level carnivore
(48 kcal/m²/year)

Herbivore
(596 kcal/m²/year)

Photosynthetic plankton
(36,380 kcal/m²/year)

a.

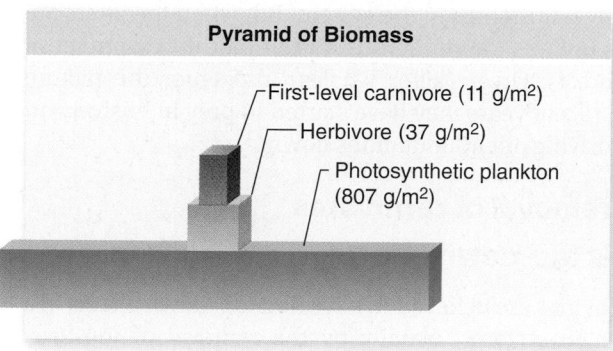

Pyramid of Biomass

First-level carnivore (11 g/m²)

Herbivore (37 g/m²)

Photosynthetic plankton
(807 g/m²)

b.

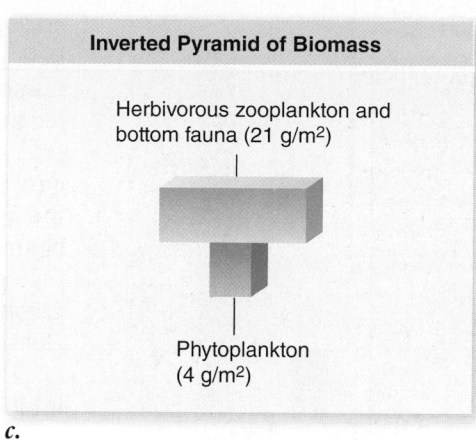

Inverted Pyramid of Biomass

Herbivorous zooplankton and
bottom fauna (21 g/m²)

Phytoplankton
(4 g/m²)

c.

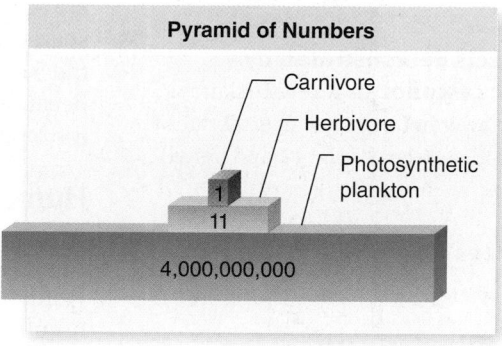

Pyramid of Numbers

Carnivore

Herbivore

Photosynthetic
plankton

1

11

4,000,000,000

d.

Figure 57.13 Ecological pyramids. In an ecological pyramid, successive trophic levels in an ecosystem are represented as stacked boxes, and the widths of the boxes represent the magnitude of an ecological property in the various trophic levels. Ecological pyramids can represent several different properties. *a.* Pyramid of energy flow (productivity). *b.* Pyramid of biomass of the ordinary type. *c.* Inverted pyramid of biomass. *d.* Pyramid of numbers.

Inquiry question
How can the existence of inverted pyramids of biomass be explained?

In a pyramid of numbers, the widths of the boxes are proportional to the numbers of individuals present in the various trophic levels (figure 57.13*d*). Such pyramids are usually, but not always, upright.

Learning Outcomes Review 57.2

Trophic levels in an ecosystem include primary producers, herbivores, primary carnivores, and secondary carnivores. Detritivores consume dead or waste matter from all levels. As energy passes from one level to another, some is inevitably lost as heat, which cannot be reclaimed. Photosynthetic primary producers capture about 1% of solar energy as chemical-bond energy. As this energy is passed through the other trophic levels, some is diverted at each step into heat, feces, and dead matter; only about 10% is available to the next level.

■ *Describe the different ways that matter, such as carbon atoms, and energy move through ecosystems?*

57.3 Trophic-Level Interactions

Learning Outcomes

1. *Explain the meaning of trophic cascade.*
2. *Distinguish between top-down and bottom-up effects.*

The existence of food chains creates the possibility that species in any one trophic level may have effects on more than one trophic level. Primary carnivores, for example, may have effects not only on the animals they eat, but also, indirectly, on the plants or algae eaten by their prey. Conversely, increases in primary productivity may provide more food not just to herbivores, but also, indirectly, to carnivores.

The process by which effects exerted at an upper trophic level flow down to influence two or more lower levels is termed a **trophic cascade.** The effects themselves are called **top-down effects.** When an effect flows up through a trophic chain, such as from primary producers to higher trophic levels, it is termed a **bottom-up effect.**

Top-down effects occur when changes in the top trophic level affect primary producers

The existence of top-down effects has been confirmed by controlled experiments in some types of ecosystems, particularly freshwater ones. For example, in one study, sections of a stream were enclosed with a mesh that prevented fish from entering. Brown trout—predators on invertebrates—were added to some enclosures but not others. After 10 days, the numbers of invertebrates in the enclosures with trout were only two-thirds as great as the numbers in the no-fish enclosures (figure 57.14). In turn, the biomass of algae, which the invertebrates ate, was five times greater in the trout enclosures than the no-fish ones.

The logic of the trophic cascade just described leads to the expectation that if secondary carnivores are added to enclosures, they would also cause cascading effects. The secondary carnivores would be predicted to keep populations of primary carnivores in check, which would lead to a profusion of herbivores and a scarcity of primary producers.

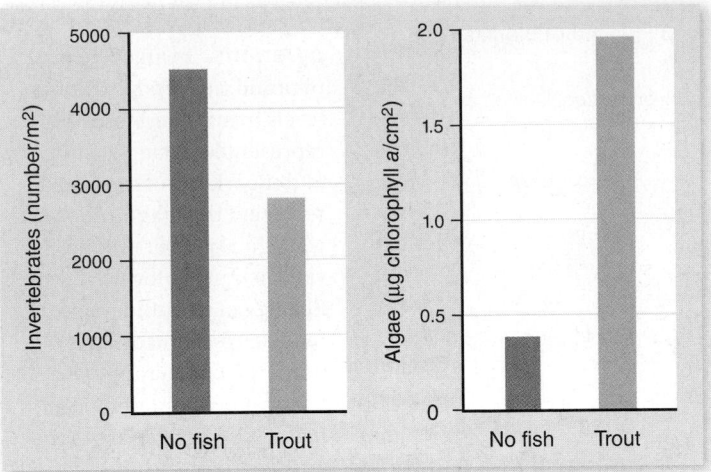

Figure 57.14 Top-down effects demonstrated by experiment in a simple trophic cascade. In a New Zealand stream, enclosures with trout had fewer herbivorous invertebrates (see the left-hand panel) and more algae (see the right-hand panel) than ones without trout.

? Inquiry question Why do streams with trout have more algae?

In an experiment similar to the one just described, enclosures were created in free-flowing streams in northern California. In these streams, the principal primary carnivores were damselfly larvae (termed nymphs). Fish that preyed on the nymphs and on other primary carnivores were added to some enclosures but not others. In the enclosures with fish, the numbers of damselfly nymphs were reduced, leading to higher numbers of their prey, including herbivorous insects, which led in turn to a decreased biomass of algae (figure 57.15).

Trophic cascades in large-scale ecosystems are not as easy to verify by experiment as ones in stream enclosures, and the workings of such cascades are not thoroughly known. Nonetheless, certain cascades in large-scale ecosystems are recognized by most ecologists. One of the most dramatic involves sea otters, sea urchins, and kelp forests along the West Coast of North America (figure 57.16).

The otters eat the urchins, and the urchins eat young kelps, inhibiting the development of kelp forests. When the otters are abundant, the kelp forests are well developed because there are relatively few urchins in the system. But when the otters are sparse, the urchins are numerous and impair development of the kelp forests. Orcas (killer whales) also enter the picture because in recent years they have started to prey intensively on the otters, driving otter populations down.

Human removal of carnivores produces top-down effects

Human activities are believed to have had top-down effects in a number of ecosystems, usually by the removal of top-level carnivores. The great naturalist Aldo Leopold posited such effects long before the trophic cascade hypothesis had been scientifically articulated when he wrote in *Sand County Almanac:*

"I have lived to see state after state extirpate its wolves. I have watched the face of many a new wolfless mountain, and seen the south-facing slopes wrinkle with a maze of new deer

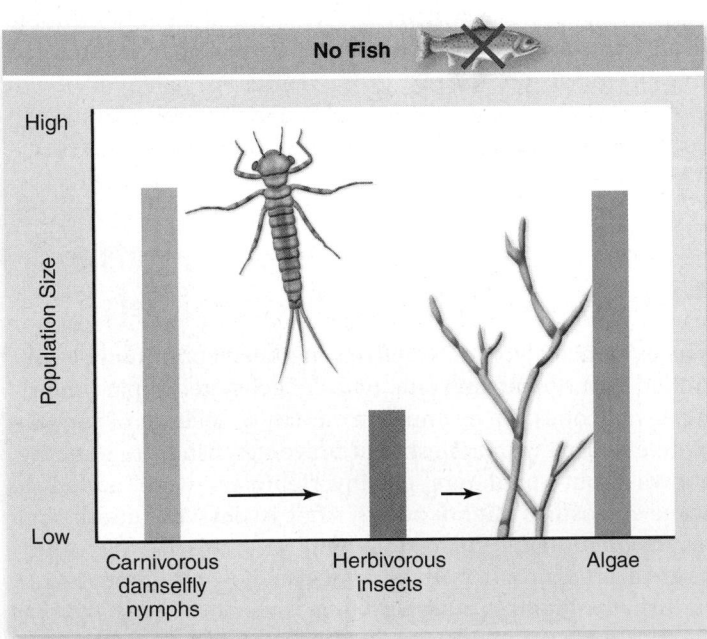
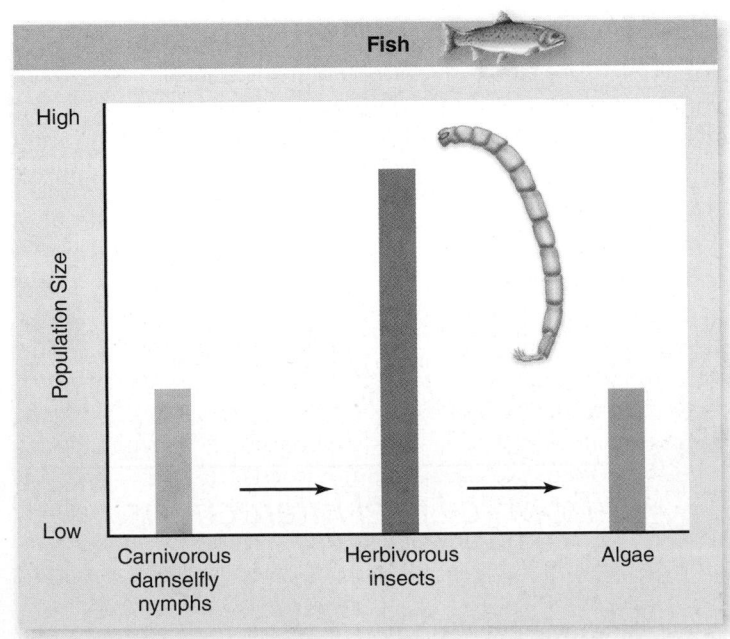

Figure 57.15 Top-down effects demonstrated by an experiment in a four-level trophic cascade. Stream enclosures with large, carnivorous fish *(on right)* have fewer primary carnivores, such as damselfly nymphs, more herbivorous insects (exemplified here by the number of chironomids, a type of aquatic insect), and lower levels of algae.

? Inquiry question Explain why the presence of fish leads to less algae in this experiment, and more algae in the experiment in figure 57.14. What might be the effect if snakes that prey on fish were added to the enclosures?

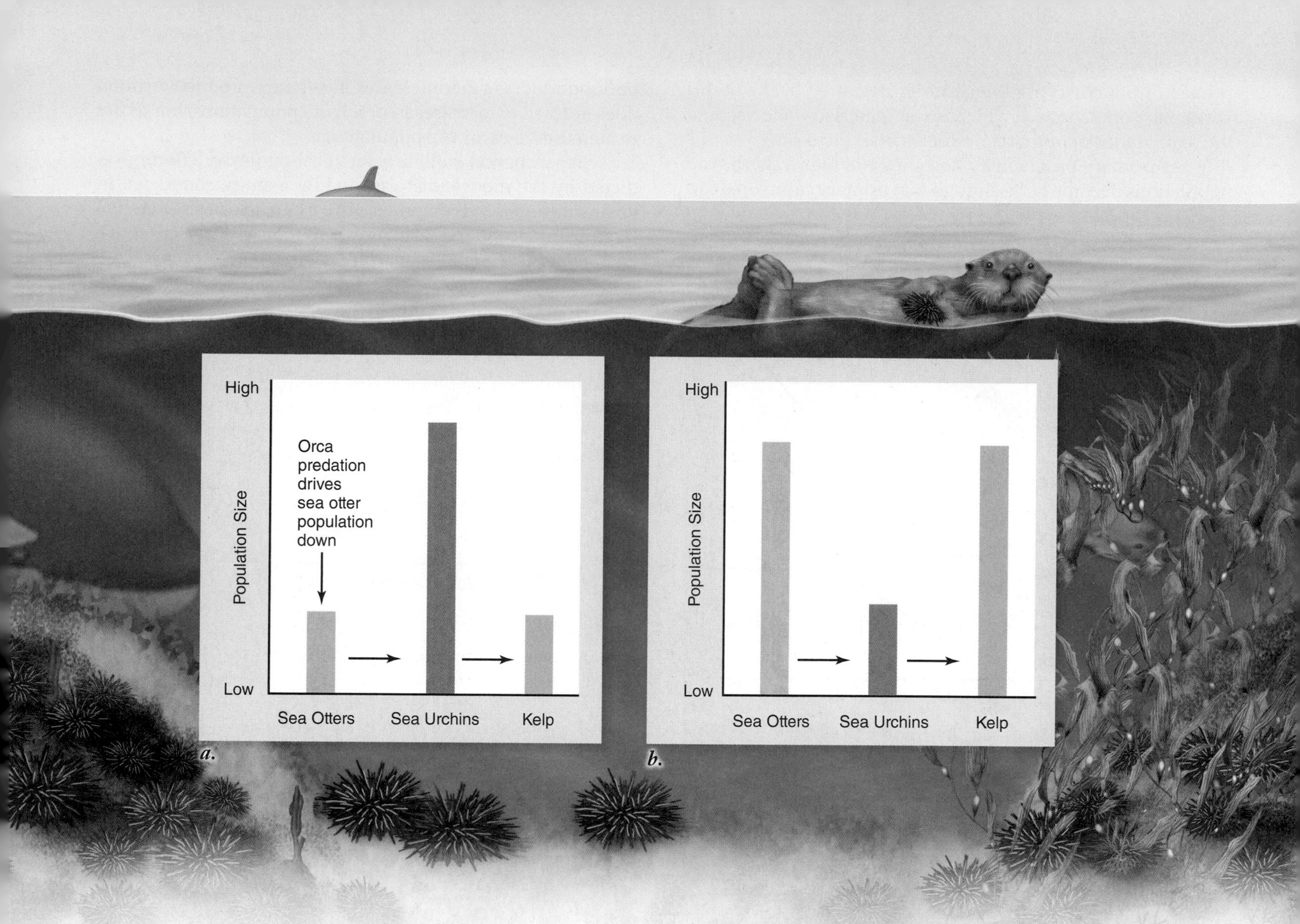

Figure 57.16 A trophic cascade in a large-scale ecosystem. Along the West Coast of North America, the sea otter/sea urchin/kelp system exists in two states: In the state shown in panel *a,* low populations of sea otters permit high populations of urchins, which suppress kelp populations; in the state shown in panel *b,* high populations of otters keep urchins in check, permitting profuse kelp growth. According to a recent hypothesis, a switch of orcas to preying on otters rather than other mammals is leading the ecosystem today to be mostly in the state represented on the left.

trails. I have seen every edible bush and seedling browsed, first to anemic desuetude, and then to death. I have seen every edible tree defoliated to the height of a saddle horn."

Many similar examples exist in which the removal of predators has led to cascading effects on lower trophic levels. Large predators such as jaguars and mountain lions are absent on Barro Colorado Island, a hilltop turned into an island by the construction of the Panama Canal at the beginning of the last century. As a result, smaller predators whose populations are normally held in check—including monkeys, peccaries (a relative of the pig), coatimundis, and armadillos—have become extraordinarily abundant. These animals eat almost anything they find. Ground-nesting birds are particularly vulnerable, and many species have declined; at least 15 bird species have vanished from the island entirely.

Similarly, in the world's oceans, large predatory fish such as billfish and cod have been reduced by overfishing to an average of 10% of their previous numbers in virtually all parts of the world's oceans. In some regions, the prey of cod—such as certain shrimp and crabs—have become many times more abundant than they were before, and further cascading effects are evident at still lower trophic levels.

Bottom-up effects occur when changes to primary producers affect higher trophic levels

In predicting bottom-up effects, ecologists must take account of the life histories of the organisms present. A model of bottom-up effects thought to apply to a number of types of ecosystems is diagrammed in figure 57.17.

According to the model, when primary productivity is low, producer populations cannot support significant herbivore populations. As primary productivity increases, herbivore populations become a feature of the ecosystem. Increases in primary productivity are then entirely devoured by the herbivores, the populations of which increase in size while keeping the populations of primary producers from increasing.

As primary productivity becomes still higher, herbivore populations become large enough that primary carnivores can

be supported. Further increases in primary productivity then does not lead to increases in herbivore populations, but rather to increases in carnivore populations.

Experimental evidence for the bottom-up effects predicted by the model was provided by a study conducted in enclosures on a river (figure 57.18). The enclosures excluded large fish (secondary carnivores). A roof was placed above each enclosure. Some roofs were clear, whereas others were tinted to various degrees, so that the enclosures differed in the amount of sunlight entering them.

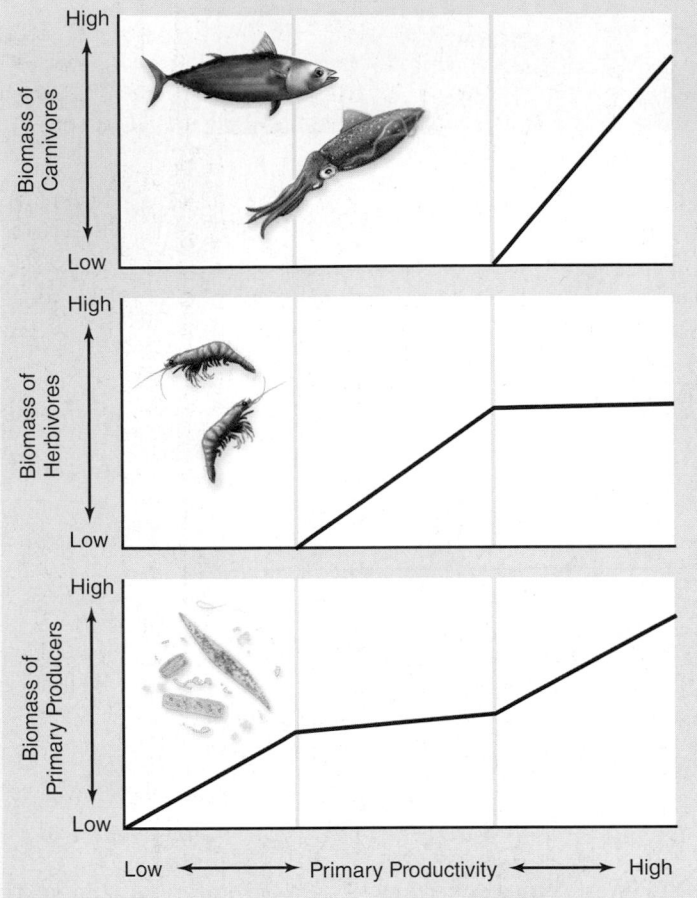

Figure 57.17 A model of bottom-up effects. At low levels of primary productivity, herbivore populations cannot obtain enough food to be maintained; without herbivory, the standing crop biomass of the primary producers such as these diatoms increases as their productivity increases. Above some threshold, increases in primary productivity lead to increases in herbivore populations and herbivore biomass; the biomass of the primary producers then does not increase as primary productivity increases because the increasing productivity is cropped by the herbivores. Above another threshold, populations of primary carnivores can be sustained. As primary productivity increases above this threshold, the carnivores consume the increasing productivity of the herbivores, so the biomass of the herbivore populations remains relatively constant while the biomass of the carnivore populations increases. The biomass of the primary producers is no longer constrained by increases in the herbivore populations and thus also increases with increasing primary productivity. A key to understanding the model is to maintain a distinction between the concepts of productivity and standing crop biomass.

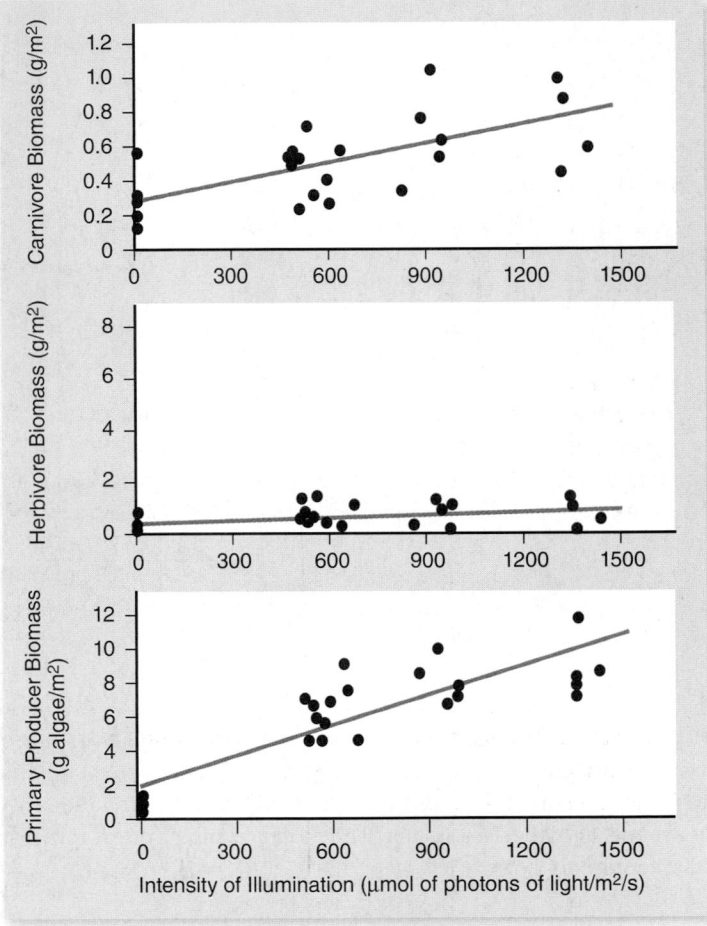

Figure 57.18 An experimental study of bottom-up effects in a river ecosystem. This system, studied on the Eel River in northern California, exhibited the patterns modeled by the red graphs of figure 57.17. Increases in the intensity of illumination led to increases in primary productivity and in the biomass of the primary producers. The biomass of the carnivore populations also increased. However, herbivore biomass did not increase much with increasing primary productivity because increases in herbivore productivity were consumed by the carnivores.

Data analysis What is the significance of the angle of the regression line in the three graphs?

Inquiry question Suppose secondary carnivores existed at all light levels in this experimental ecosystem. How would you expect the biomass of primary producers, herbivores, and primary carnivores to change with increasing light levels?

Inquiry question How is it possible for the biomass of the primary producers to stay relatively constant as the primary productivity increases?

The primary productivity was highest in the enclosures with clear roofs and lowest in the ones with darkly tinted roofs. As primary productivity increased in parallel with illumination, the biomass of the primary producers increased, as did the biomass of the carnivores. However, the biomass of the trophic level sandwiched in between, the herbivores, did not increase much, as predicted by the model in figure 57.17 (see red graph lines).

Learning Outcomes Review 57.3

Populations of species at different trophic levels affect one another, and these effects can propagate through the levels. Top-down effects, termed trophic cascades, are observed when changes in carnivore populations affect lower trophic levels. Bottom-up effects are observed when changes in primary productivity affect the higher trophic levels.

■ **Could top-down and bottom-up effects occur simultaneously?**

57.4 Biodiversity and Ecosystem Stability

Learning Outcomes

1. Define ecosystem stability.
2. Describe the effects of species richness on ecosystem function.
3. Name possible factors that contribute to species richness in the tropics.

In chapter 56, we discussed *species richness*—the number of species present in a community. Ecologists have long debated the consequences of differences in species richness between communities. One theory is that species-rich communities are more stable—that is, more constant in composition and better able to resist disturbance. This hypothesis has been elegantly studied by David Tilman and colleagues at the University of Minnesota's Cedar Creek Natural History Area.

Species richness may increase stability: The Cedar Creek studies

Workers monitored 207 small rectangular plots of land (8–16 m²) for 11 years (figure 57.19a). In each plot, they counted the number of prairie plant species and measured the total amount of plant biomass (that is, the mass of all plants on the plot). Over the course of the study, plant species richness was related to community stability—plots with more species showed less year-to-year variation in biomass. Moreover, in two drought years, the decline in biomass was negatively related to species richness—that is, plots with more species were less affected by drought.

These findings were subsequently confirmed by an experiment in which plots were seeded with different numbers

Question: *Does species richness affect the invasibility of a community?*
Hypothesis: *The rate of successful invasion will be lower in communities with greater richness.*
Experiment: *Add seeds from the same number of non-native plants to experimental plots that differ in the number of plant species.*

a.

b.

Result: *Although the number of successful invasive species is highly variable, more species-rich plots on average are invaded by fewer species.*
Interpretation: *What might explain why so much variation exists in the number of successful invading species in communities with the same species richness?*

Figure 57.19 Effect of species richness on ecosystem stability. *a.* One of the Cedar Creek experimental plots. *b.* Community stability can be assessed by looking at the effect of species richness on community invasibility. Each dot represents data from one experimental plot in the Cedar Creek experimental fields. Plots with more species are harder to invade by nonnative species.

 Data analysis How well can you predict the number of invading species from knowing the number of species on a plot?

of species. Again, more species-rich plots had greater year-to-year stability in biomass over a 10-year period.

In a related experiment, when seeds of other plant species were added to different plots, the ability of these species to become established was negatively related to species richness (figure 57.19b). More diverse communities, in other words, are more resistant to invasion by new species, which is another measure of community stability.

Species richness may also affect other ecosystem processes. Tilman and colleagues monitored 147 experimental plots that varied in number of species to estimate how much growth was occurring and how much nitrogen the growing plants were taking up from the soil. They found that the more species a plot had, the greater the nitrogen uptake and total amount of biomass produced. In his study, increased biodiversity clearly appeared to lead to greater productivity.

Laboratory studies on artificial ecosystems have provided similar results. In one elaborate study, ecosystems covering 1 m^2 were constructed in laboratory growth chambers that controlled temperature, light levels, air currents, and atmospheric gas concentrations. A variety of plants, insects, and other animals were introduced to construct ecosystems composed of 9, 15, or 31 species, with the lower diversity treatments containing a subset of the species in the higher diversity enclosures. As with Tilman's experiments, the amount of biomass produced was related to species richness, as was the amount of carbon dioxide consumed, another measure of the productivity of the ecosystem.

Tilman's conclusion that healthy ecosystems depend on diversity is not accepted by all ecologists, however. Critics question the validity and relevance of these biodiversity studies, arguing that the more species are added to a plot, the greater the probability that one species will be highly productive. To show that high productivity results from high species richness per se, rather than from the presence of particular highly productive species, experimental plots have to exhibit "overyielding"; in other words, plot productivity has to be greater than that of the single most productive species grown in isolation.

Although this point is still debated, recent work at Cedar Creek and elsewhere has provided evidence of overyielding, supporting the claim that species richness of communities enhances community productivity and stability.

Species richness is influenced by ecosystem characteristics

A number of factors are known or hypothesized to affect species richness in a community. We discussed some in chapter 56, such as loss of keystone species and moderate physical disturbance. Here we discuss three more: primary productivity, habitat heterogeneity, and climatic factors.

Primary productivity

Ecosystems differ substantially in primary productivity (see figure 57.11). Some evidence indicates that species richness is related to primary productivity, but the relationship between them is not linear. In a number of cases, for example, ecosystems with intermediate levels of productivity tend to have the greatest number of species (figure 57.20*a*).

Why this is so is debated. One possibility is that levels of productivity are linked with numbers of consumers. Applying this concept to plant species richness, the argument is that at low productivity, there are few herbivores, and superior competitors among the plants are able to eliminate most other plant species. In contrast, at high productivity so many herbivores are present that only the plant species most resistant to grazing survive, reducing species diversity. As a result, the greatest numbers of plant species coexist at intermediate levels of productivity and herbivory.

Habitat heterogeneity

Spatially heterogeneous abiotic environments are those that consist of many habitat types—such as soil types, for example. These heterogeneous environments can be expected to accommodate more species of plants than spatially homogeneous environments. What's more, the species richness of animals can be expected to reflect the species richness of plants present. An example of this latter effect is seen in figure 57.20*b*: The

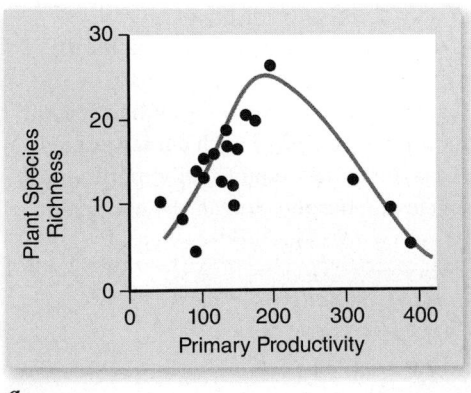

a.

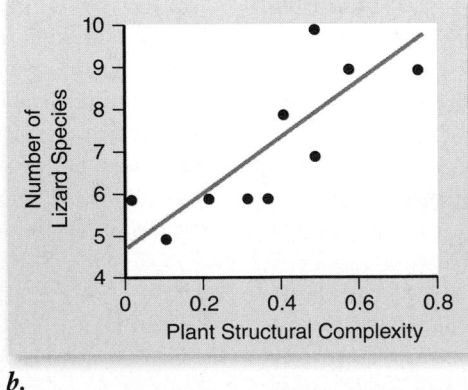

b.

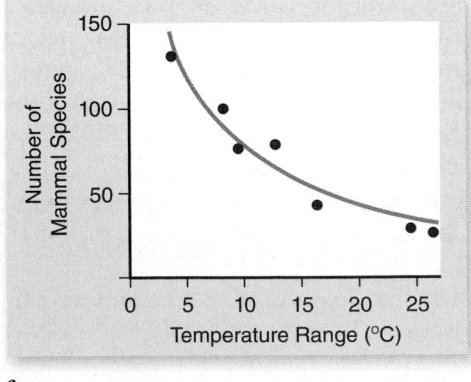

c.

Figure 57.20 Factors that affect species richness. *a. Productivity:* In plant communities of mountainous areas of South Africa, species richness of plants peaks at intermediate levels of productivity (biomass). *b. Spatial heterogeneity:* The species richness of desert lizards is positively correlated with the structural complexity of the plant cover in desert sites in the American Southwest. *c. Climate:* The species richness of mammals is inversely correlated with monthly mean temperature range along the West Coast of North America.

? Inquiry question (a) Why is species richness greatest at intermediate levels of productivity? (b) Why do more structurally complex areas have more species? (c) Why do areas with less variation in temperature have more species?

number of lizard species at various sites in the American Southwest mirrors the local structural diversity of the plants.

Climatic factors

The role of climatic factors is more difficult to predict. On the one hand, more species might be expected to coexist in a seasonal environment than in a constant one because a changing climate may favor different species at different times of the year. On the other hand, stable environments are able to support specialized species that would be unable to survive where conditions fluctuate. The number of mammal species at locations along the West Coast of North America is inversely correlated with the amount of local temperature variation—the wider the variation, the fewer mammalian species—supporting the latter line of argument (figure 57.20c).

Tropical regions have the highest diversity, although the reasons are unclear

Biologists have long recognized that more different kinds of animals and plants inhabit the tropics than the temperate regions. For many types of organisms, there is a steady increase in species richness from the arctic to the tropics. This species diversity gradient in numbers of species correlated with latitude has been reported for plants and animals, including birds (figure 57.21), mammals, and reptiles.

For the better part of a century, ecologists have puzzled over the species diversity gradient from the arctic to the tropics. The difficulty has not been in forming a reasonable hypothesis of why more species exist in the tropics, but rather in sorting through these many reasonable hypotheses. Here, we consider five of the most commonly discussed suggestions.

Evolutionary age of tropical regions

Scientists have frequently proposed that the tropics have more species than temperate regions because the tropics have existed over long, uninterrupted periods of evolutionary time, whereas temperate regions have been subject to repeated glaciations. The greater age of tropical communities would have allowed complex population interactions to coevolve within them, fostering a greater variety of plants and animals.

Recent work suggests that the long-term stability of tropical communities has been greatly exaggerated, however. An examination of pollen within undisturbed soil cores reveals that during glaciations, the tropical forests contracted to a few small refuges surrounded by grassland. This suggests that the tropics have not had a continuous record of species richness over long periods of evolutionary time.

Increased productivity

A second often-advanced hypothesis is that the tropics contain more species because this part of the Earth receives more solar radiation than do temperate regions. The argument is that more solar energy, coupled to a year-round growing season, greatly increases the overall photosynthetic activity of tropical plants.

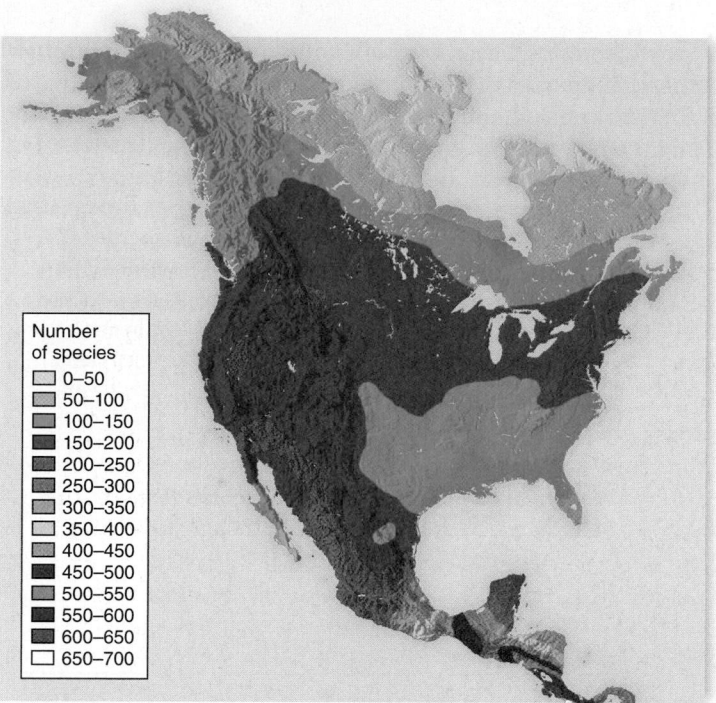

Figure 57.21 A latitudinal gradient in species richness. Among North and Central American birds, a marked increase in the number of species occurs moving toward the tropics. Fewer than 100 species are found at arctic latitudes, but more than 600 species live in southern Central America.

Legend: Number of species
0–50
50–100
100–150
150–200
200–250
250–300
300–350
350–400
400–450
450–500
500–550
550–600
600–650
650–700

If we visualize the tropical forest's total resources as a pie, and its species niches as slices of the pie, we can see that a larger pie accommodates more slices. But as noted earlier, many field studies have indicated that species richness is highest at intermediate levels of productivity. Accordingly, increasing productivity would be expected to lead to lower, not higher, species richness.

Stability/constancy of conditions

Seasonal variation, though it does exist in the tropics, is generally substantially less than in temperate areas. This reduced seasonality might encourage specialization, with niches subdivided to partition resources and so avoid competition. The expected result would be a larger number of more specialized species in the tropics, which is what we see. Many field tests of this hypothesis have been carried out, and almost all support it, reporting larger numbers of narrower niches in tropical communities than in temperate areas.

Predation

Many reports indicate that predation may be more intense in the tropics. In theory, more intense predation could reduce the importance of competition, permitting greater niche overlap and thus promoting greater species richness.

Spatial heterogeneity

As noted earlier, spatial heterogeneity promotes species richness. Tropical forests, by virtue of their complexity, create a variety of microhabitats and so may foster larger numbers of species. Perhaps the long vertical column of vegetation through which light passes in a tropical forest produces a wide range of light frequencies and intensities, creating a greater variety of light environments and so promoting species diversity.

Learning Outcomes Review 57.4

An ecosystem is stable if it remains relatively constant in composition and is able to resist disturbance. Experimental field studies support the conclusion that species-rich communities are better able to resist invasion by new species, as well as have increased biomass production at the primary level, although not all ecologists agree with these conclusions. Species richness is greatest in the tropics, and the reasons may include habitat variation, increased sunlight, and long-term climate and seasonal stability.

■ *What might be the effects on primary productivity if air pollution decreased the amount of sunlight reaching Earth's surface?*

57.5 Island Biogeography

Learning Outcomes

1. *Describe the species–area relationship.*
2. *Explain how area and isolation affect rates of colonization and extinction.*

One of the most reliable patterns in ecology is the observation that larger islands contain more species than do smaller islands. In 1967, Robert MacArthur of Princeton University and Edward O. Wilson of Harvard University proposed that this species–area relationship was a result of the effect of geographic area and isolation on the likelihood of species extinction and colonization.

The equilibrium model proposes that extinction and colonization reach a balance point

MacArthur and Wilson reasoned that species are constantly being dispersed to islands, so islands have a tendency to accumulate more and more species. At the same time that new species are added, however, other species are lost by extinction. As the number of species on an initially empty island increases, the rate of colonization must decrease as the pool of potential colonizing species not already present on the island becomes depleted. At the same time, the rate of extinction should increase—the more species on an island, the greater the likelihood that any given species will perish.

As a result, at some point the number of extinctions and colonizations should be equal, and the number of species should then remain constant. Every island of a given size, then, has a characteristic equilibrium number of species that tends to persist through time (the intersection point in figure 57.22*a*)—though the species composition will change as some species become extinct and new species colonize.

MacArthur and Wilson's equilibrium model proposes that island species richness is a dynamic equilibrium between colonization and extinction. Both island size and distance from the mainland would affect colonization and extinction. We would expect smaller islands to have higher rates of extinction because their population sizes would, on average, be smaller. Also, we would expect fewer colonizers to reach islands that lie farther from the mainland. Thus, small islands far from the mainland would have the fewest species; large islands near the mainland would, have the most (figure 57.22*b*).

The predictions of this simple model bear out well in field data. Asian Pacific bird species (figure 57.22*c*) exhibit a positive correlation of species richness with island size, but a negative correlation of species richness with distance from the source of colonists.

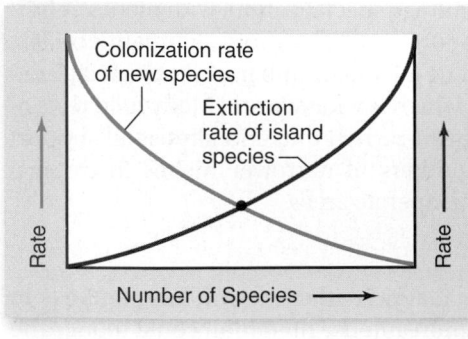

a.

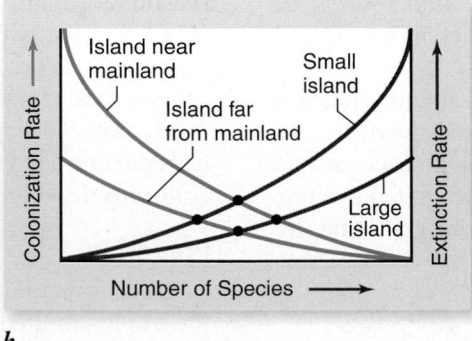

b.

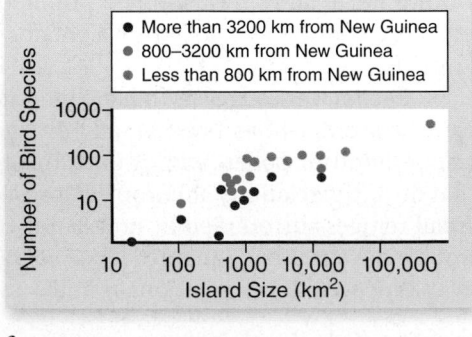

c.

Figure 57.22 The equilibrium model of island biogeography. *a.* Island species richness reaches an equilibrium (black dot) when the colonization rate of new species equals the extinction rate of species on the island. *b.* The equilibrium shifts depending on the rate of colonization, the size of an island, and its distance to sources of colonists. Species richness is positively correlated with island size and inversely correlated with distance from the mainland. Smaller islands have higher extinction rates, shifting the equilibrium point to the left. Similarly, more distant islands have lower colonization rates, again shifting the equilibrium point leftward. *c.* The effect of distance from a larger island, which can be the source of colonizing species, is readily apparent. More distant islands have fewer bird species than do nearer islands of the same size.

The equilibrium model is still being tested

Wilson and Dan Simberloff, then a graduate student, performed initial studies in the mid-1960s on small mangrove islands in the Florida keys. These islands were censused, cleared of animal life by fumigation, and then allowed to recolonize, with censuses being performed at regular intervals. These and other such field studies have tended to support the equilibrium model.

Long-term experimental field studies, however, are suggesting that the situation is more complicated than MacArthur and Wilson envisioned. Their model predicts a high level of species turnover as some species perish and others arrive. But studies of island birds and spiders indicate that very little turnover occurs from year to year. Those species that do come and go, moreover, comprise a subset of species that never attain high populations. A substantial proportion of the species appear to maintain high populations and rarely go extinct.

These studies have been going on for a relatively short period of time. It is possible that over periods of centuries, the equilibrium model is a good description of what determines island species richness.

Learning Outcomes Review 57.5

The species–area relationship is an observation that an island of larger area contains more species. Species richness on islands appears to be a dynamic equilibrium between colonization and extinction. Distance from a mainland also affects the rates of colonization and extinction, and therefore fewer species would be found on small, isolated islands far from a mainland.

■ *Under what circumstances would a smaller island be expected to have more species than a larger island?*

 ## Chapter Review

57.1 Biogeochemical Cycles

The atomic constituents of matter cycle within ecosystems.

The atoms of chemical elements move through ecosystems in biogeochemical cycles.

Carbon, the basis of organic compounds, cycles through most ecosystems.

The carbon cycle usually involves carbon dioxide, which is fixed through photosynthesis and released by respiration. Carbon is also present as bicarbonate ions and as methane. Burning of fossil fuels has created an imbalance in the carbon cycle (figure 57.1).

The availability of water is fundamental to terrestrial ecosystems.

Water enters the atmosphere via evaporation and transpiration and returns to the Earth's surface as precipitation. It is broken down during photosynthesis and also produced during cellular respiration. Much of the Earth's water, including the groundwater in aquifers, is polluted, and human activities alter the water supply of ecosystems (figure 57.2).

The nitrogen cycle depends on nitrogen fixation by microbes.

Nitrogen is usually the element in shortest supply even though N_2 makes up 78% of the atmosphere. Nitrogen must be converted into usable forms by nitrogen-fixing microorganisms. Human use of nitrates in fertilizers has doubled the available nitrogen (figure 57.4).

Phosphorus cycles through terrestrial and aquatic ecosystems, but not the atmosphere.

Phosphorus, another limiting nutrient, is released by weathering of rocks; it flows into the oceans where it is deposited in deep-sea sediments. Humans also use phosphates as fertilizers (figure 57.5).

Limiting nutrients in ecosystems are those in short supply relative to need.

The cycle of a limiting nutrient, such as nitrogen, determines the rate at which the nutrient is made available for use.

Biogeochemical cycling in a forest ecosystem has been studied experimentally.

Ongoing experiments indicate that severe disturbance of an ecosystem results in mineral depletion and runoff of water.

57.2 The Flow of Energy in Ecosystems

Energy can neither be created nor destroyed, but changes form.

Energy exists in forms such as light, stored chemical-bond energy, motion, and heat. In any conversion, some energy is lost.

Living organisms can use many forms of energy, but not heat.

The Second Law of Thermodynamics states that whenever organisms use chemical-bond or light energy, some of it is inevitably converted to heat and cannot be retrieved.

Energy flows through trophic levels of ecosystems.

Organic compounds are synthesized by autotrophs and are utilized by both autotrophs and heterotrophs. As energy passes from organism to organism, each level is termed a trophic level, and the sequence through progressive trophic levels is called a food chain (figure 57.8).

The base trophic level includes the primary producers; herbivores that consume primary producers are the next level. They in turn are eaten by primary carnivores, which may be consumed by secondary carnivores. Detritivores feed on waste and the remains of dead organisms.

Only about 1% of the solar energy that impinges on the Earth is captured by photosynthesis. As energy moves through each trophic level, very little (approximately 10%) remains from the preceding trophic level (figure 57.10).

The number of trophic levels is limited by energy availability.

The exponential decline of energy between trophic levels limits the length of food chains and the numbers of top carnivores that can be supported.

Ecological pyramids illustrate the relationship of trophic levels.

Ecological pyramids based on energy flow, biomass, or numbers of organisms are usually upright. Inverted pyramids of biomass or numbers are possible if at least one trophic level has a greater biomass or more organisms than the level below it (figure 57.13).

57.3 Trophic-Level Interactions

Top-down effects occur when changes in the top trophic level affect primary producers.

A trophic cascade, or top-down effect, occurs when a change exerted at an upper trophic level affects a lower level (figure 57.15).

Human removal of carnivores produces top-down effects.

Removal of carnivores causes an increase in the abundance of species in lower trophic levels, such as an increase in deer populations when wolves or other predators are destroyed.

Bottom-up effects occur when changes to primary producers affect higher trophic levels.

An increase of producers may lead to the appearance or increase of herbivores; however, further increase in producers may then lead to increase in carnivores, without a comparable increase in herbivores (figure 57.17).

57.4 Biodiversity and Ecosystem Stability

Species richness may increase stability: The Cedar Creek studies.

The Cedar Creek studies indicate that higher species richness results in less year-to-year variation in biomass and in greater resistance to drought and invasion by nonnative species.

Species richness is influenced by ecosystem characteristics.

Primary production, habitat heterogeneity, and climatic factors all affect the number of species in an ecosystem (figure 57.20).

Tropical regions have the highest diversity, although the reasons are unclear.

The higher diversity of tropical regions may reflect long evolutionary time, higher productivity from increased sunlight, less seasonal variation, greater predation that reduces competition, or spatial heterogeneity (figure 57.21).

57.5 Island Biogeography

The species–area relationship reflects that larger islands contain more species than do smaller ones.

The equilibrium model proposes that extinction and colonization reach a balance point (figure 57.22).

Smaller islands have fewer species because of higher rates of extinction. Islands near a mainland have more species than distant islands because of higher rates of colonization. An equilibrium is reached when the extinction rate balances the colonization rate.

The equilibrium model is still being tested.

Long-term studies are needed to clarify all the factors involved.

Review Questions

UNDERSTAND

1. Which of the statements about groundwater is NOT accurate?
   a. In the United States, groundwater provides 50% of the population with drinking water.
   b. Groundwaters are being depleted faster than they can be recharged.
   c. Groundwaters are becoming increasingly polluted.
   d. Removal of pollutants from groundwaters is easily achieved.

2. Photosynthetic organisms
   a. fix carbon dioxide.
   b. release carbon dioxide.
   c. fix oxygen.
   d. Both a and b are correct.
   e. Both a and c are correct.

3. Some bacteria have the ability to "fix" nitrogen. This means
   a. they convert ammonia into nitrites and nitrates.
   b. they convert atmospheric nitrogen gas into biologically useful forms of nitrogen.
   c. they break down nitrogen-rich compounds and release ammonium ions.
   d. they convert nitrate into nitrogen gas.

4. Nitrogen is often a limiting nutrient in many ecosystems because
   a. there is much less nitrogen in the atmosphere than carbon.
   b. elemental nitrogen is very rapidly used by most organisms.
   c. nitrogen availability is being reduced by pollution due to fertilizer use.
   d. most organisms cannot use nitrogen in its elemental form.

5. Which of the following statements about the phosphorus cycle is correct?
   a. Phosphorus is fixed by plants and algae.
   b. Most phosphorus released from rocks is carried to the oceans by rivers.
   c. Animals cannot get their phosphorus from eating plants and algae.
   d. Fertilizer use has not affected the global phosphorus budget.

6. As a general rule, how much energy is lost in the transmission of energy from one trophic level to the one immediately above it?
   a. 1%   c. 90%
   b. 10%   d. 50%

7. Inverted ecological pyramids of real systems usually involve
   a. energy flow.
   b. biomass.
   c. energy flow and biomass.
   d. None of the choices is correct.

8. Bottom-up effects on trophic structure result from
   a. a limitation of energy flowing to the next higher trophic level.
   b. actions of top predators on lower trophic levels.
   c. climatic disruptions on top consumers.
   d. stability of detritivores in ecosystems.

9. Species diversity
   a. increases with latitude as you move away from the equator to the arctic.
   b. decreases with latitude as you move away from the equator to the arctic.
   c. stays the same as you move away from the equator to the arctic.
   d. increases with latitude as you move north of the equator and decreases with latitude as you move south of the equator.

10. The equilibrium model of island biogeography suggests all of the following *except*
    a. larger islands have more species than smaller islands.
    b. the species richness of an island is determined by colonization and extinction.
    c. smaller islands have lower rates of extinction.
    d. islands closer to the mainland will have higher colonization rates.

APPLY

1. Based on results from studies at Hubbard Brook Experimental Forest, what would be the predicted effect of clearing trees from a watershed?
   a. Increased loss of water and nutrients from a watershed
   b. Decreased loss of water and nutrients from a watershed
   c. Increased availability of phosphorus
   d. Increased availability of nitrate

2. According to the trophic cascade hypothesis, the removal of carnivores from an ecosystem may result in
   a. a decline in the number of herbivores and a decline in the amount of vegetation.
   b. a decline in the number of herbivores and an increase in the amount of vegetation.
   c. an increase in the number of herbivores and an increase in the amount of vegetation.
   d. an increase in the number of herbivores and a decrease in the amount of vegetation.

3. At Cedar Creek Natural History Area, experimental plots showed reduced numbers of invaders as species diversity of plots increased
   a. suggesting that low species diversity increases stability of ecosystems.
   b. suggesting that ecosystem stability is a function of primary productivity only.
   c. consistent with the theory that intermediate disturbance results in the highest stability.
   d. None of the choices is correct.

SYNTHESIZE

1. Given that ectotherms do not utilize a large fraction of ingested food energy to maintain a high and constant body temperature (generate heat), how would you expect the food chains of systems dominated by ectothermic herbivores and carnivores to compare with systems dominated by endothermic herbivores and carnivores?

2. Given that, in general, energy input is greatest at the bottom trophic level (primary producers) and decreases with increasing transfers across trophic levels, how is it possible for many lakes to show much greater standing biomass in herbivorous zooplankton than in the phytoplankton they consume?

3. Ecologists often worry about the potential effects of the loss of species (e.g., due to pollution, habitat degradation, or other human-induced factors) on an ecosystem for reasons other than just the direct loss of the species. Using figure 57.17 explain why.

4. Explain several detailed ways in which increasing plant structural complexity could lead to greater species richness of lizards (figure 57.20*b*). Could any of these ideas be tested? How?

ONLINE RESOURCE

www.ravenbiology.com

Understand, Apply, and Synthesize—enhance your study with animations that bring concepts to life and practice tests to assess your understanding. Your instructor may also recommend the interactive eBook, individualized learning tools, and more.

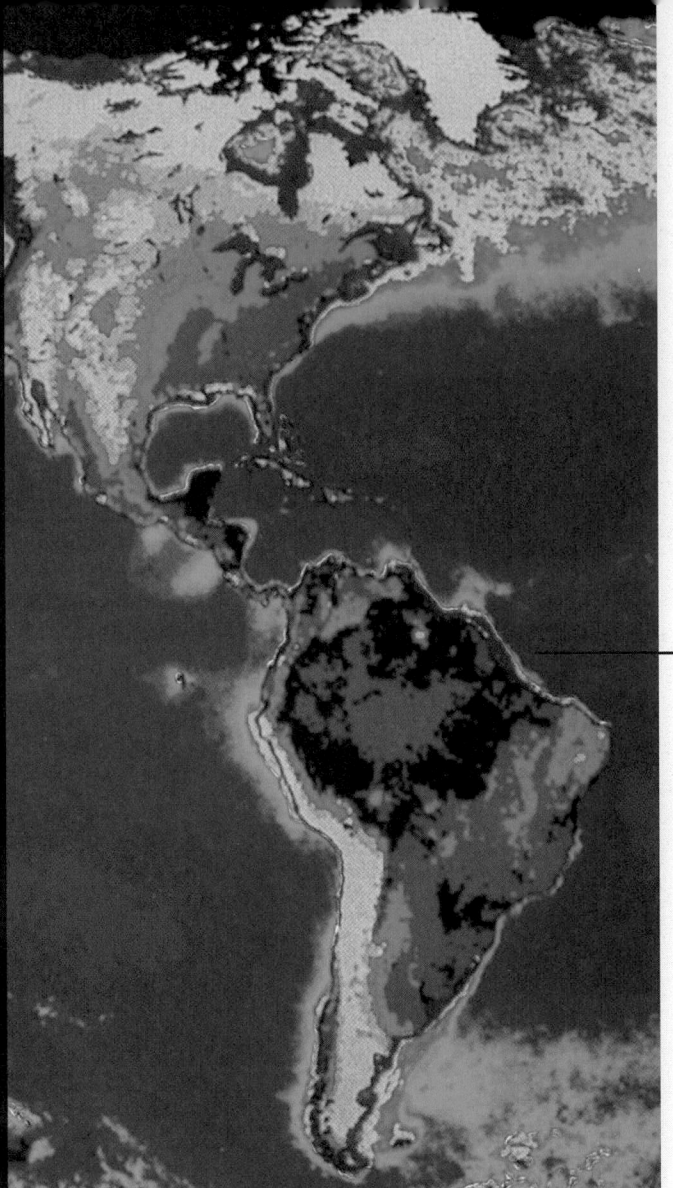

Chapter 58

The Biosphere

Chapter Contents

Introduction

The biosphere includes all living communities on Earth, from the profusion of life in the tropical rainforests to the planktonic communities in the world's oceans. In a very general sense, the distribution of life on Earth reflects variations in the world's abiotic environments, such as the variations in temperature and availability of water from one terrestrial environment to another. The figure on this page is a satellite image of the Americas. The colors are keyed to the relative abundance of chlorophyll, an indicator of the richness of biological communities. Green and dark green areas on land are areas with high primary productivity (such as thriving forests), whereas yellow areas include the deserts of the Americas and the tundra of the far north, which have lower productivity.

58.1 Ecosystem Effects of Sun, Wind, and Water

Learning Outcomes

1. *Describe changes in wind and current direction with latitude.*
2. *Explain the Coriolis effect.*
3. *Describe how temperature changes with altitude and latitude.*

The great global patterns of life on Earth are heavily influenced by (1) the amount of solar radiation that reaches different parts of the Earth and seasonal variations in that radiation; and (2) the patterns of global atmospheric circulation and the resulting patterns of oceanic circulation. Local characteristics, such as soil types and the altitude of the land, interact with the global patterns in sunlight, winds, and water currents to determine the conditions under which life exists and thus the distributions of ecosystems.

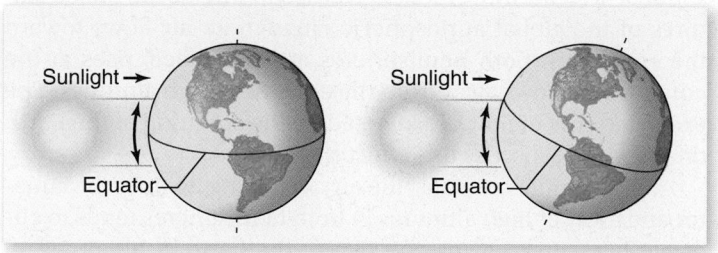

a.

Figure 58.1 **Relationships between the Earth and the Sun are critical in determining the nature and distribution of life on Earth.** *a.* A beam of solar energy striking the Earth in the middle latitudes of the northern hemisphere (or the southern) spreads over a wider area of the Earth's surface than an equivalent beam striking the Earth at the equator. *b.* The fact that the Earth orbits the Sun each year has a profound effect on climate. In the northern and southern hemispheres, temperature changes in an annual cycle because the Earth's axis is not perpendicular to its orbital plane and, consequently, each hemisphere tilts toward the Sun in some months but away from the Sun in others.

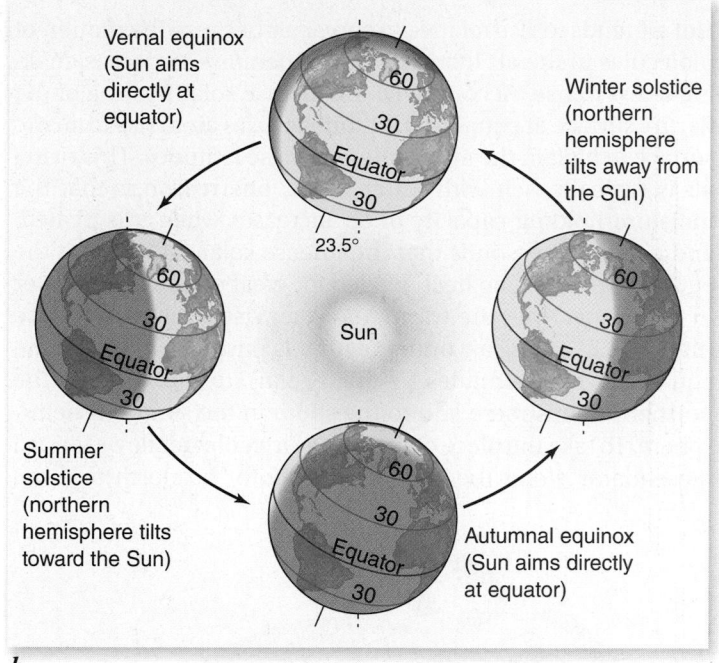

b.

Solar energy and the Earth's rotation affect atmospheric circulation

The Earth receives energy from the Sun at a high rate in the form of electromagnetic radiation at visible and near-visible wavelengths. Each square meter of the upper atmosphere receives about 1400 joules per second (J/sec), which is equivalent to the output of fourteen 100-watt (W) lightbulbs.

As the solar radiant energy passes through the atmosphere, its intensity and wavelength composition are modified. About half of the energy is absorbed within the atmosphere, and half reaches the Earth's surface. The gases in the atmosphere absorb some wavelengths strongly while allowing other wavelengths to pass freely through. As a result, the wavelength composition of the solar energy that reaches the Earth's surface is different from that emitted by the Sun. For example, the band of ultraviolet wavelengths known as UV-B is strongly absorbed by ozone (O_3) in the atmosphere, and thus this wavelength is greatly reduced in the solar energy that reaches the Earth's surface.

How solar radiation affects climate

Some parts of the Earth's surface receive more energy from the Sun than others. These differences have a great effect on climate.

A major reason for differences in solar radiation from place to place is the fact that Earth is a sphere, or nearly so (figure 58.1*a*). The tropics are particularly warm because the Sun's rays arrive almost perpendicular to the surface of the Earth in regions near the equator. Closer to the poles, the angle at which the Sun's rays strike, called the *angle of incidence,* spreads the solar energy out over more of the Earth's surface, providing less energy per unit of surface area. As figure 58.2 shows, the highest annual mean temperatures occur near the equator (0° latitude).

The Earth's annual orbit around the Sun and its daily rotation on its own axis are also important in determining patterns of solar radiation and their effects on climate (figure 58.1*b*). The axis of rotation of the Earth is not perpendicular to the plane in which the earth orbits the Sun. Because the axis is tilted by approximately 23.5°, a progression of seasons occurs on all parts of the Earth, especially at latitudes far from the equator. The northern hemisphere, for example, tilts toward the Sun during some months but away during others, giving rise to summer and winter; the further away from the equator, the greater the difference between summer and winter, as figure 58.2 demonstrates.

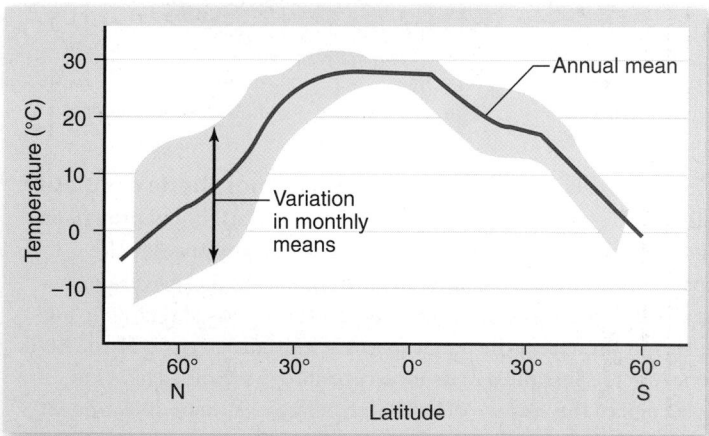

Figure 58.2 **Annual mean temperature varies with latitude.** The red line represents the annual mean temperature at various latitudes, ranging from near the North Pole at the left to near Antarctica at the right; the equator is at 0° latitude. At each latitude, the upper edge of the blue zone is the highest mean monthly temperature observed in all the months of the year, and the lower edge is the lowest mean monthly temperature.

 Data analysis How is seasonal variation in temperature related to latitude and why?

chapter **58** *The Biosphere*

Global circulation patterns in the atmosphere

Hot air tends to rise relative to cooler air because the motion of molecules in the air increases as temperature increases, making it less dense. Accordingly, the intense solar heating of the Earth's surface at equatorial latitudes causes air to rise from the surface to high in the atmosphere at these latitudes. This rising air is typically rich with water vapor; one reason is that the moisture-holding capacity of air increases when it is heated, and a second reason is that the intense solar radiation at the equator provides the heat needed for great quantities of water to evaporate. After the warm, moist air rises from the surface (figure 58.3), rising air underneath it is pushed away from the equator at high altitudes (above 10 km), to the north in the northern hemisphere and to the south in the southern hemisphere. To take the place of the rising air, cooler air flows toward the equator along the surface from both the north and the south. These air movements give rise to one of the major features of the global atmospheric circulation: air flows toward the equator in both hemispheres at the surface, rises at the equator, and flows away from the equator at high altitudes. The exact patterns of flow are affected by the spinning of the Earth on its axis; we discuss this effect shortly.

For complex reasons, the air circulating up from the equator and away at high altitudes in both hemispheres tends to circulate back down to the surface of the Earth at about 30° of latitude, both north and south (see figure 58.3). During the course of this movement, the moisture content of the air changes radically because of the changes in temperature the air undergoes. Cooling dramatically decreases air's ability to hold water vapor. Consequently, much of the water vapor in the air rising from the equator condenses to form clouds and rain as the air moves upward. This rain falls in the latitudes near the equator, latitudes that experience the greatest precipitation on Earth.

By the time the air starts to descend back to the Earth's surface at latitudes near 30°, it is cold and thus has lost most of its water vapor. Although the air rewarms as it descends, it does not gain much water vapor on the way down. Many of the greatest deserts occur at latitudes near 30° because of the steady descent of dry air to the surface at those latitudes. The Sahara Desert is the most dramatic example.

The air that descends at latitudes near 30° flows only partly toward the equator after reaching the surface of the Earth. Some of it flows toward the poles, helping to give rise in each hemisphere to winds that blow over the Earth's surface from 30° toward 60° latitude. At latitudes near 60° air tends to rise from the surface toward high altitudes.

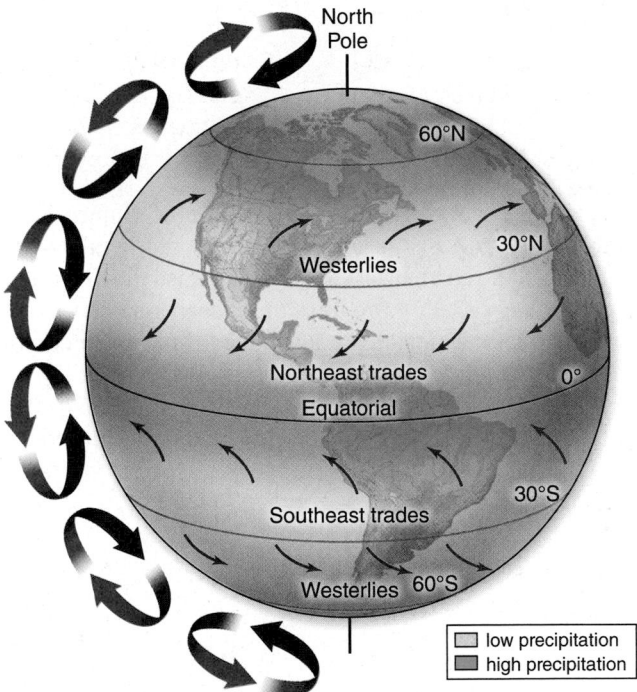

> **?** **Inquiry question** Why is it hotter at latitudes near 0°?

The Coriolis effect

If Earth did not rotate on its axis, global air movements would follow the simple patterns already described. Air currents—the winds—move across a rotating surface, however. Because the solid Earth rotates under the winds, the winds move in curved paths across the surface, rather than straight paths. The curvature of the paths of the winds due to Earth's rotation is termed the **Coriolis effect,** after the 19th-century French mathematician, Gaspard-Gustave Coriolis, who described it.

If you were standing on the North Pole, the Earth would appear to be rotating counterclockwise on its axis, but if you were at the South Pole, the Earth would appear to be rotating clockwise. This property of a rotating sphere, that its direction of rotation is opposite when viewed from its two poles, explains why the direction of the Coriolis effect is opposite in the two hemispheres. In the northern hemisphere, winds always curve to the right of their direction of motion; in the southern hemisphere, they always curve to the left.

The reason for these wind patterns is that the circumference of a sphere, the Earth, changes with latitude. It is zero at the poles and 38,000 km at the equator. Thus, land surface speed changes from about 0 to 1500 km per hour going from the poles to the equator. Air descending at 30° north latitude may be going roughly the same speed as the land surface below it. As it moves toward the equator, however, it is moving more

Figure 58.3 Global patterns of atmospheric circulation. The diagram shows the patterns of air circulation that prevail on average over weeks and months of time (on any one day the patterns might be dramatically different from these average patterns). Rising air that is cooled creates bands of relatively high precipitation near the equator and at latitudes near 60°N and 60°S. Air that has lost most of its moisture at high altitudes tends to descend to the surface of the Earth at latitudes near 30°N and 30°S, creating bands of relatively low precipitation. The red arrows show the winds blowing at the surface of the Earth; the blue arrows show the direction the winds blow at high altitude. The winds travel in curved paths relative to the Earth's surface because the Earth is rotating on its axis under them (the Coriolis effect). A terminological problem to recognize is that the formal names given to winds refer to the directions from which they come, rather than the directions toward which they go; thus, the winds between 30° and 60° are called Westerlies because they come out of the west. Unfortunately, oceanographers use the opposite approach, naming water currents for the directions in which they go.

slowly than the surface below it, so it is deflected to its right in the northern hemisphere and to its left in the southern hemisphere. In other words, in both the northern and southern hemispheres, the winds blow westward as well as toward the equator. The result (see figure 58.3) is that winds on both sides of the equator—called the Trade Winds—blow out of the east and toward the west.

Conversely, air masses moving north from 30° are moving more rapidly than underlying land surfaces and thus are deflected again to their right, which in this case is eastward. Similarly, in the southern hemisphere, air masses between 30° and 60° are deflected eastward, to the left. In both hemispheres, therefore, winds between 30° and 60° blow out of the west and toward the east; these winds are called Westerlies.

Global currents are largely driven by winds

The major ocean currents are driven by the winds at the surface of the Earth, which means that indirectly the currents are driven by solar energy. The radiant input of heat from the Sun sets the atmosphere in motion as already described, and then the winds set the ocean in motion.

In the north Atlantic Ocean (figure 58.4), the global winds follow this pattern: Surface winds tend to blow out of the east and toward the west near the equator, but out of the west and toward the east at midlatitudes (between 30° and

60°). Consequently, surface waters of the north Atlantic Ocean tend to move in a giant closed curve—called a **gyre**—flowing from North America toward Europe at midlatitudes, then returning from Europe and Africa to North America at latitudes near the equator.

Water currents are affected by the Coriolis effect. Thus, the Coriolis effect contributes to this clockwise closed-curve motion. Water flowing across the Atlantic toward Europe at midlatitudes tends to curve to the right and enters the flow from east to west near the equator. This latter flow also tends to curve to its right and enters the flow from west to east at midlatitudes. In the south Atlantic Ocean, the same processes occur in a sort of mirror image, and similar clockwise and counterclockwise gyres occur in the north and south Pacific Ocean as well.

Regional and local differences affect terrestrial ecosystems

The environmental conditions at a particular place are affected by regional and local effects of solar radiation, air circulation, and water circulation, not just the global patterns of these processes. In this section we look at just a few examples of regional and local effects, focusing on terrestrial systems. These include rain shadows, monsoon winds, elevation, and presence of microclimate factors.

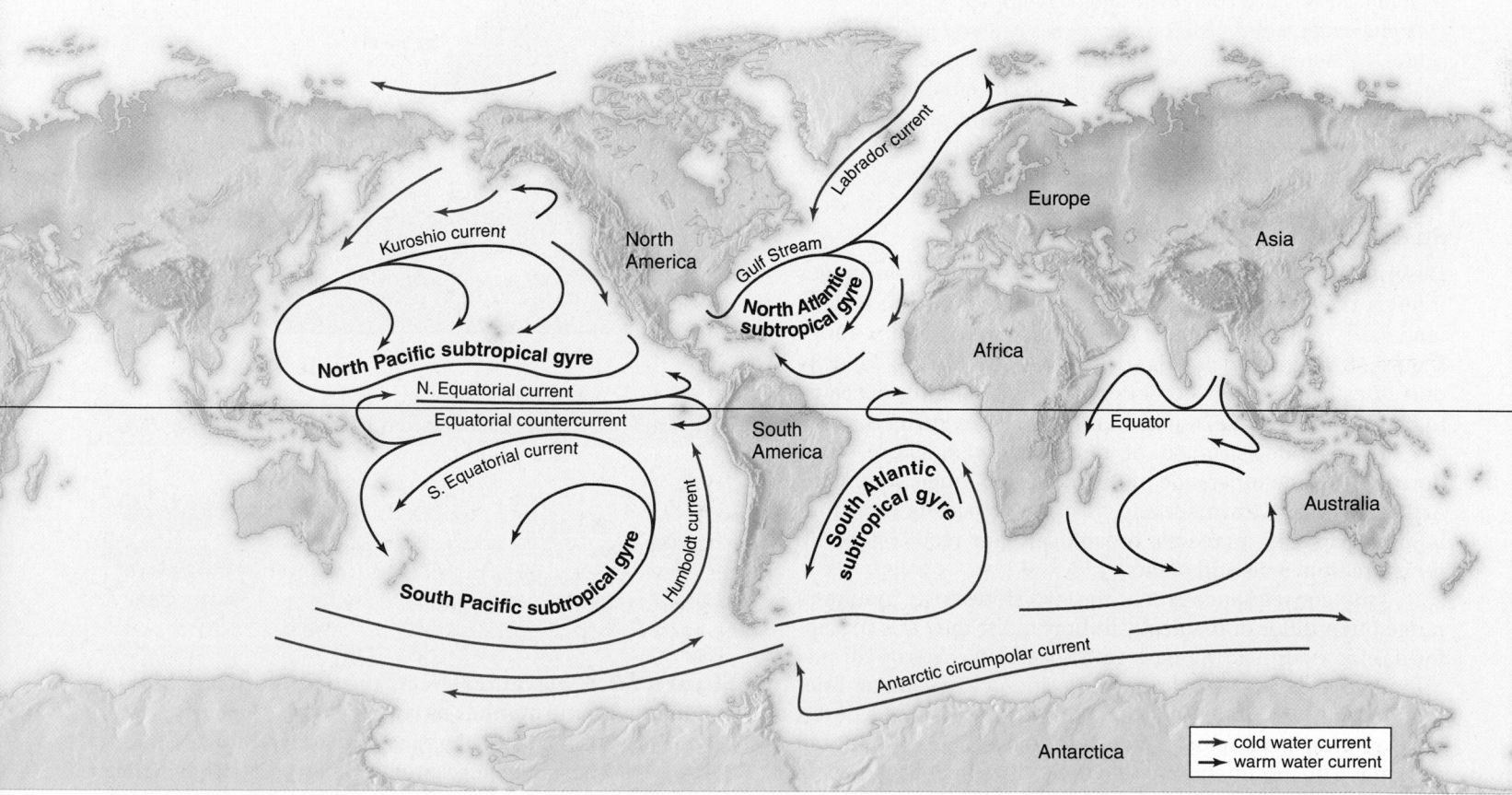

Figure 58.4 Ocean circulation. In the centers of several of the great ocean basins, surface water moves in great closed-curve patterns called gyres. These water movements affect biological productivity in the oceans and sometimes profoundly affect the climate on adjacent landmasses, as when the Gulf Stream brings warm water to the region of the British Isles.

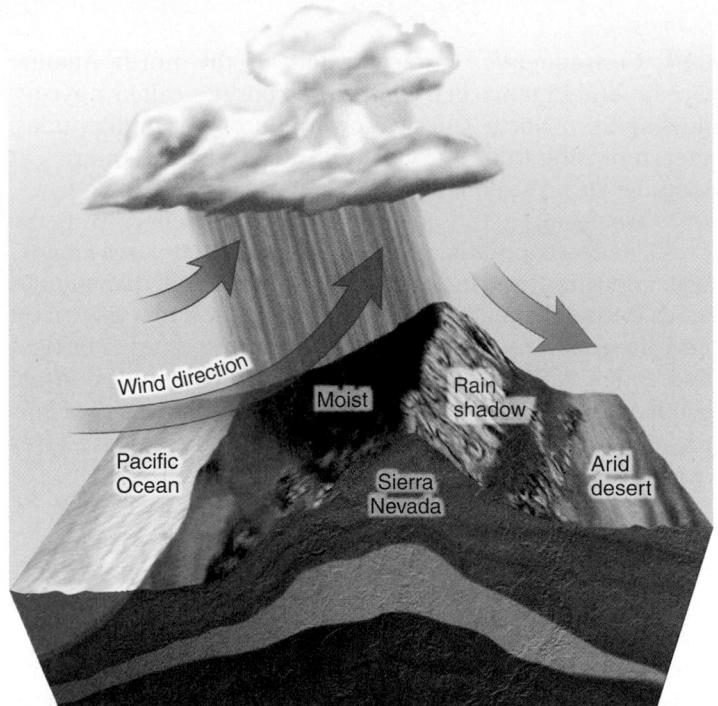

Figure 58.5 The rain shadow effect exemplified in California. Moisture-laden winds from the Pacific Ocean rise and are cooled when they encounter the Sierra Nevada Mountains. As the moisture-holding capacity of the air decreases at colder, higher altitudes, precipitation occurs, making the seaward-facing slopes of the mountains moist; tall forests occur on those slopes, including forests that contain the famous giant sequoias *(Sequoiadendron giganteum)*. As the air descends on the eastern side of the mountain range, its moisture-holding capacity increases again, and the air picks up moisture from its surroundings. As a result, the eastern slopes of the mountains are arid, and rain shadow deserts sometimes occur.

Rain shadows

Deserts on land sometimes occur because mountain ranges intercept moisture-laden winds from the sea. When air flowing landward from the oceans encounters a mountain range (figure 58.5), the air rises, and its moisture-holding capacity decreases because it becomes cooler at higher altitude, causing precipitation to fall on the mountain slopes facing the sea.

As the air—stripped of much of its moisture—then descends on the other side of the mountain range, it remains dry even as it is warmed, and as it is warmed its moisture-holding capacity increases, meaning it can readily take up moisture from soils and plants.

One consequence is that the two slopes of a mountain range often differ dramatically in how moist they are; in California, for example, the eastern slopes of the Sierra Nevada Mountains—facing away from the Pacific Ocean—are far drier than the western slopes. Another consequence is that a desert may develop on the dry side, the Mojave Desert being an example. The mountains are said to produce a rain shadow.

Monsoons

The continent of Asia is so huge that heating and cooling of its surface during the passage of the seasons causes massive regional shifts in wind patterns. During summer, the surface of the Asian landmass heats up more than the surrounding oceans, but during winter the landmass cools more than the oceans. The consequence is that winds tend to blow off the water into the interior of the Asian continent in summer, particularly in the region of the Indian Ocean and western tropical Pacific Ocean. These winds reverse to flow off the continent out over the oceans in winter. These seasonally shifting winds are called the monsoons. They affect rainfall patterns, and their duration and strength can spell the difference between food sufficiency and starvation for hundreds of millions of people in the region each year.

Elevation

Another significant regional pattern is that in mountainous regions, temperature and other conditions change with elevation. At any given latitude, air temperature falls about 6°C for every 1000-m increase in elevation. The ecological consequences of the change of temperature with elevation are similar to those of the change of temperature with latitude (figure 58.6).

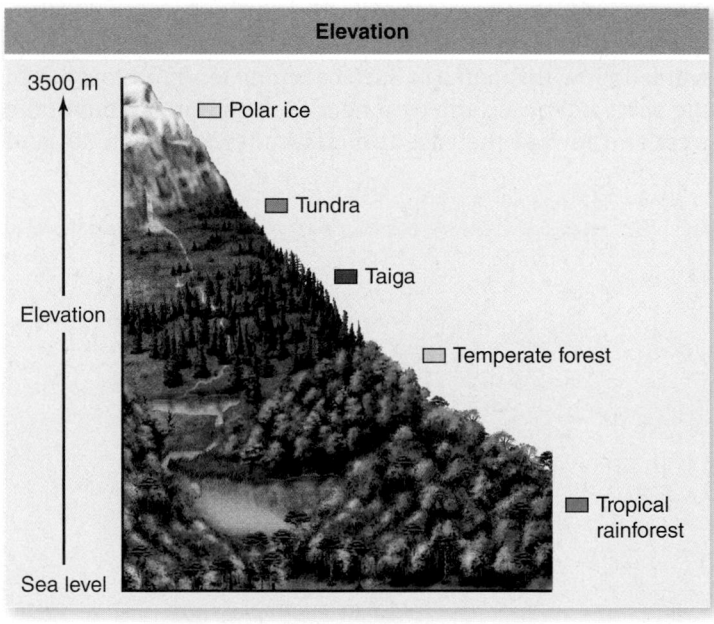

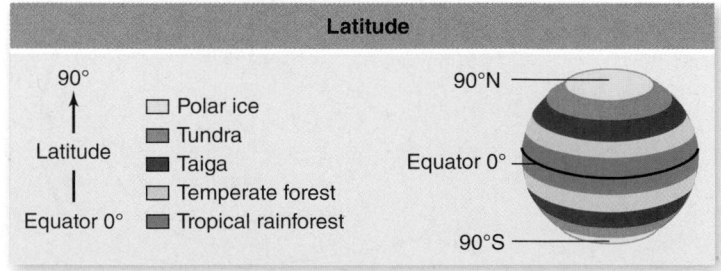

Figure 58.6 Elevation affects the distribution of biomes in much the same manner as latitude does. Biomes that normally occur far north of the equator at sea level also occur in the tropics at high mountain elevations. Thus, on a tall mountain in the tropics, one might see a sequence of biomes like the one illustrated above. In North America, a 1000-m increase in elevation results in a temperature drop equal to that of an 880-km increase in latitude.

Microclimates

Conditions also vary in significant ways on very small spatial scales. For example, in a forest, a bird sitting in an open patch may experience intense solar radiation, a high air temperature, and a low humidity, even while a mouse hiding under a log 10 feet away may experience shade, a cool temperature, and air saturated with water vapor. Such highly localized sets of climatic conditions are called microclimates.

In some cases, species avoid competing by adapting to use different microclimates. Sympatric salamanders, for example, may be specialized for the different levels of moisture found in different parts of the habitat.

Learning Outcomes Review 58.1

More intense solar heating of some global regions relative to others sets up global patterns of atmospheric circulation, which in turn cause global patterns of water circulation in the oceans. The Coriolis effect is caused by the Earth spinning beneath the moving air masses of the atmosphere. These patterns—plus seasonal changes—strongly affect the conditions that exist for living organisms in different parts of the world. In general, temperature declines as altitude or latitude increases.

■ *How would global air movement patterns be different if the Earth turned in the opposite direction?*

58.2 Earth's Biomes

Learning Outcomes

1. Define biome.
2. Explain the primary factors that determine which type of biome is found in a particular place.

Biomes are major types of ecosystems on land. Each biome has a characteristic appearance and is distributed over wide areas of land defined largely by sets of regional climatic conditions. Biomes are named according to their vegetational structures, but they also include the animals that are present.

As you might imagine from the broad definition given for biomes, there are a number of ways to classify terrestrial ecosystems into biomes. Here we recognize eight principal biomes: (1) tropical rainforest, (2) savanna, (3) desert, (4) temperate grassland, (5) temperate deciduous forest, (6) temperate evergreen forest, (7) taiga, and (8) tundra.

Six additional biomes recognized by some ecologists are: polar ice, mountain zone, chaparral, warm moist evergreen forest, tropical monsoon forest, and semidesert. Other ecologists lump these six with the eight major ones. Figure 58.7 shows the distributions of all 14 biomes.

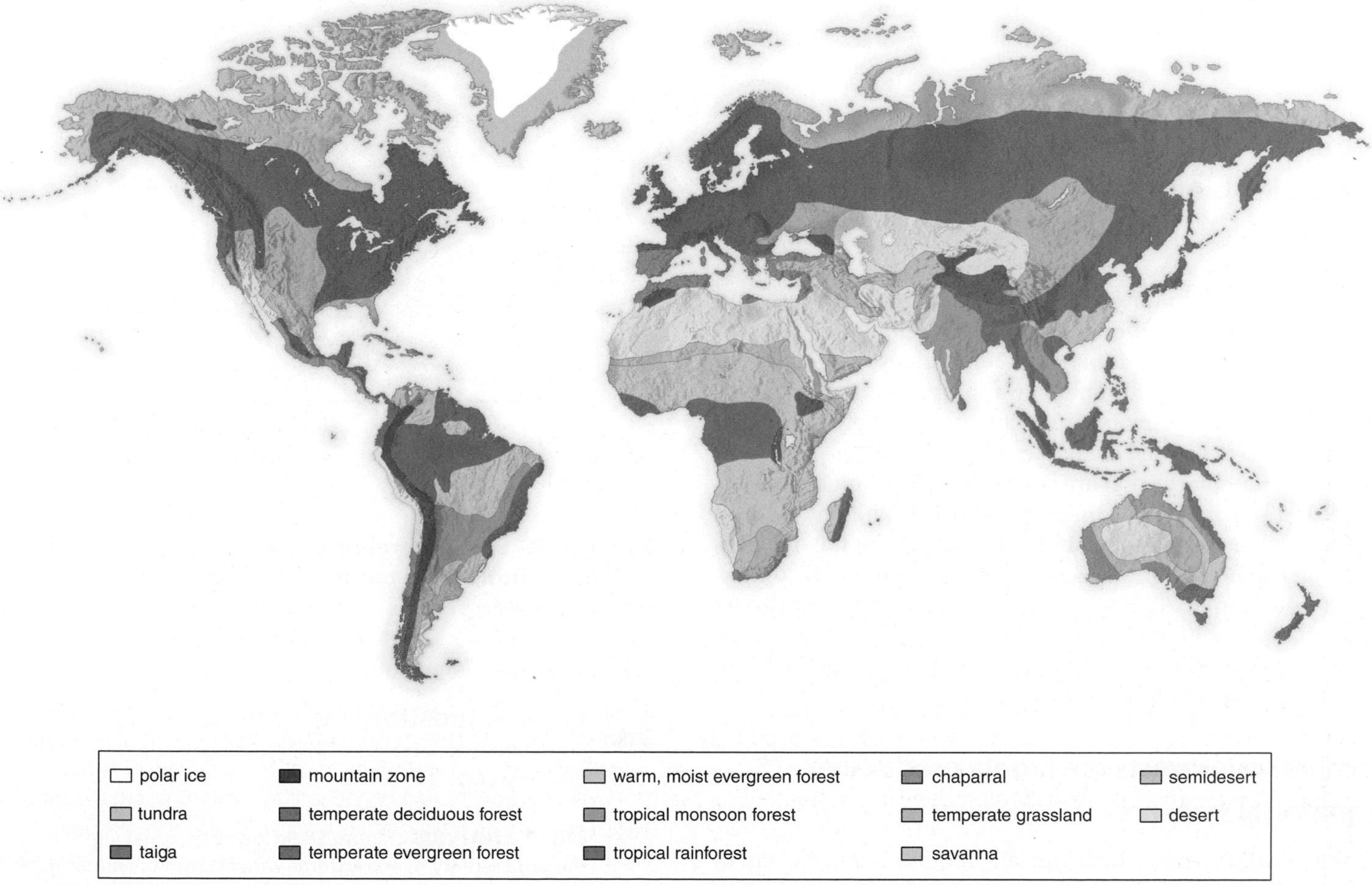

☐ polar ice	■ mountain zone	☐ warm, moist evergreen forest	▨ chaparral	▨ semidesert
▨ tundra	■ temperate deciduous forest	☐ tropical monsoon forest	■ temperate grassland	☐ desert
▨ taiga	■ temperate evergreen forest	■ tropical rainforest	▨ savanna	

Figure 58.7 The distributions of biomes. Each biome is similar in vegetational structure and appearance wherever it occurs.

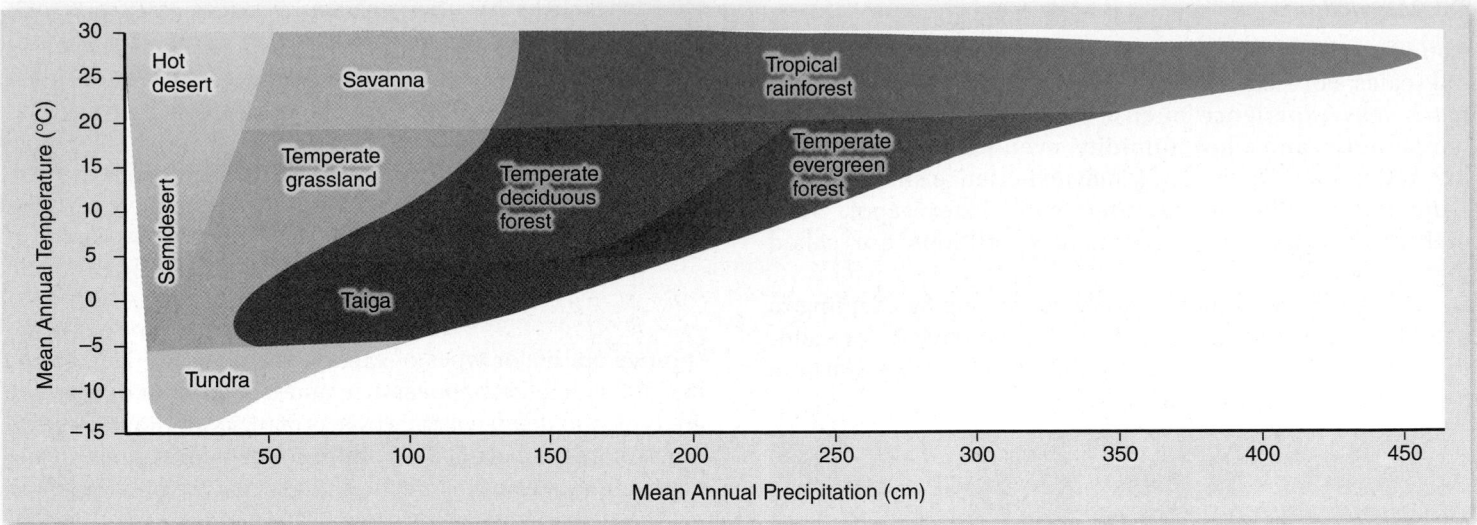

Figure 58.8 Predictors of biome distribution. Temperature and precipitation are quite useful predictors of biome distribution, although other factors sometimes also play critical roles.

Biomes are defined by their characteristic vegetational structures and associated climatic conditions, rather than by the presence of particular plant species. Two regions assigned to the same biome thus may differ in the species that dominate the landscape. Tropical rainforests around the world, for example, are all composed of tall, lushly vegetated trees, but the tree species that dominate a South American tropical rainforest are different from those in an Indonesian one. The similarity between such forests results from convergent evolution (see chapter 21).

Temperature and moisture often determine biomes

In determining which biomes are found where, two key environmental factors are temperature and moisture. As seen in figure 58.8, if you know the mean annual temperature and mean annual precipitation in a terrestrial region, you often can predict the biome that dominates. Temperature and moisture affect ecosystems in a number of ways. One reason they are so influential is that primary productivity is strongly correlated with them, as described in chapter 57 (figure 58.9).

Different places that are similar in mean annual temperature and precipitation sometimes support different biomes, indicating that temperature and moisture are not the only factors that can be important. Soil structure and mineral composition (see chapter 38) are among the other factors that can be influential. The biome that is present may also depend on whether the conditions of temperature and precipitation are strongly seasonal or relatively constant.

Tropical rainforests are highly productive equatorial systems

Tropical rainforests, which typically require 140 to 450 cm of rain per year, are the richest ecosystems on land (figure 58.10). They are very productive because they enjoy the advantages of

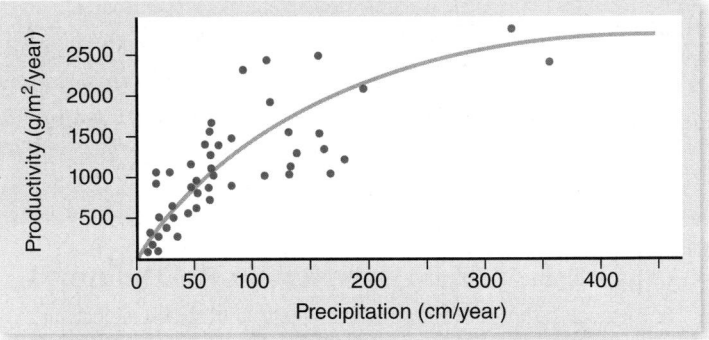

a.

b.

Figure 58.9 The correlations of primary productivity with precipitation and temperature. The net primary productivity of ecosystems at 52 locations around the globe correlates significantly with (**a**) mean annual precipitation and (**b**) mean annual temperature.

? Inquiry question Why might you expect primary productivity to increase with increasing precipitation and temperature?

Data analysis Figure 58.9*a* indicates great variation in the productivity of ecosystems with intermediate amounts of precipitation (~125 cm/year). How can information in figures 58.9*b* and 58.8 explain this?

Figure 58.10 Tropical rainforest.

both high temperature and high precipitation (see figure 58.9). They also exhibit very high biodiversity, being home to at least half of all the species of terrestrial plants and animals in the world—over 2 million species! In a single square mile of Brazilian rainforest, there can be 1200 species of butterflies—twice the number found in all of North America. Tropical rainforests recycle nutrients rapidly, so their soils often lack great reservoirs of nutrients.

Savannas are tropical grasslands with seasonal rainfall

Savannas are tropical or subtropical grasslands, often dotted with widely spaced trees or shrubs. On a global scale, savannas often occur as transitional ecosystems between tropical rainforests and deserts; they are characteristic of warm places where annual rainfall (50–125 cm) is too little to support rainforest, but not so little as to produce desert conditions.

Rainfall is often highly seasonal in savannas. The Serengeti ecosystem in East Africa is probably the world's most famous example of the savanna biome. In most of the Serengeti, no rain falls for many months of the year, but during other months rain is abundant. The huge herds of grazing animals in the ecosystem respond to the seasonality of the rain; a number of species migrate away from permanently flowing rivers only during the months when rain falls.

Deserts are regions with little rainfall

Deserts are dry places where rain is both sparse (annual rainfall often less than 25–40 cm) and unpredictable. The unpredictability means that plants and animals cannot depend on experiencing rain even once each year. As mentioned earlier, many of the largest deserts occur at latitudes near 30°N and 30°S because of global air circulation patterns (see figure 58.3). Other deserts result from rain shadows (see figure 58.5).

Vegetation is sparse in deserts, and survival of both plants and animals depends on water conservation. Many desert organisms enter inactive stages during rainless periods. To avoid extreme temperatures, small desert vertebrates often live in deep, cool, and sometimes even somewhat moist burrows. Some emerge only at night. Among large desert animals, camels drink large quantities of water when it is available and then conserve it so well that they can survive for weeks without drinking. Oryxes (large, desert-dwelling antelopes) survive opportunistically on moisture in leaves or roots that they dig up, as well as drinking water when possible.

Temperate grasslands have rich soils

Halfway between the equator and the poles are temperate regions where rich temperate grasslands grow. These grasslands, also called **prairies,** once covered much of the interior of North America, and they were widespread in Eurasia and South America as well.

The roots of perennial grasses characteristically penetrate far into the soil, and grassland soils tend to be deep and fertile. Temperate grasslands are often highly productive when converted to agricultural use, and vast areas have been transformed in this way. In North America prior to this change in land use, huge herds of bison and pronghorn antelope inhabited the temperate grasslands, migrating seasonally as resources changed over the course of the year. Natural temperate grasslands are one of the biomes adapted to periodic fire and therefore need fires to prosper.

Temperate deciduous forests are adapted to seasonal change

Mild but seasonal climates (warm summers and cold winters), plus plentiful rains, promote the growth of temperate deciduous forests in the eastern United States, eastern Canada, and Eurasia (figure 58.11). A deciduous tree is one that drops its leaves in the winter. Deer, bears, beavers, and raccoons are familiar animals of these forests.

Temperate evergreen forests are coastal

Temperate evergreen forests occur along coastlines with temperate climates, such as in the northwest of the United States. The dominant vegetation includes trees, such as spruces, pines, and redwoods, that do not drop their leaves (thus, they are *ever green*).

Taiga is the northern forest where winters are harsh

Taiga and tundra (described next) differ from other biomes in that both stretch in great unbroken circles around the entire

Figure 58.11 Temperate deciduous forest.

globe (see figure 58.7). The taiga consists of a great band of northern forest dominated by coniferous trees (spruce, hemlock, and fir) that retain their needle-like leaves all year long.

The taiga is one of the largest biomes on Earth. Winters in taiga regions are severely long and cold, and most of the limited precipitation falls in the summer. Many large herbivores, including elk, moose, and deer, plus carnivores such as wolves, bears, lynx, and wolverines, are characteristic of the taiga.

Tundra is a largely frozen treeless area with a short growing season

In the far north, at latitudes above the taiga but south of the polar ice, few trees grow. The landscape that occurs in this band, called tundra, is open, windswept, and often boggy. This enormous biome covers one-fifth of the Earth's land surface. Little rain or snow falls. **Permafrost**—soil ice that persists throughout all seasons—usually exists within a meter of the ground surface.

What trees can be found are small and mostly confined to the margins of streams and lakes. Large grazing mammals, including musk-oxen and reindeer (caribou), and carnivores such as wolves, foxes, and lynx, live in the tundra. Populations of lemmings (a small rodent native to the Arctic) rise and fall dramatically, with important consequences for the animals that prey on them.

Learning Outcomes Review 58.2

Major types of ecosystems called biomes can be distinguished in different climatic regions on land. These biomes are much the same wherever they are found on the Earth. Annual mean temperature and precipitation are effective predictors of biome type; however, the range of seasonal variation and the soil characteristics of a region also come into play.

■ *Why do different biomes occur at different latitudes?*

58.3 Freshwater Habitats

Learning Outcomes

1. Define photic zone.
2. Explain what causes spring and fall overturns in lakes.
3. Distinguish between eutrophic and oligotrophic lakes.

Of the major habitats, fresh water covers by far the smallest percentage of the Earth's surface: Only 2%, compared with 27% for land and 71% for ocean. The formation of fresh water starts with the evaporation of water into the atmosphere, which removes most dissolved constituents, much like distillation does. When water falls back to the Earth's surface as rain or snow, it arrives in an almost pure state, although it may have picked up biologically significant dissolved or particulate matter from the atmosphere.

Freshwater wetlands—marshes, swamps, and bogs—represent intermediate habitats between the freshwater and terrestrial realms. Wetlands are highly productive (see figure 57.11). They also play key additional roles, such as acting as water storage basins that moderate flooding.

Primary production in freshwater bodies is carried out by single-celled algae (phytoplankton) floating in the water, by algae growing as films on the bottom, and by rooted plants such as water lilies. In addition, a considerable amount of organic matter—such as dead leaves—enters some bodies of fresh water from plant communities growing on the land nearby.

Life in freshwater habitats depends on oxygen availability

The concentration of dissolved oxygen (O_2) is a major determinant of the properties of freshwater communities. Oxygen dissolves in water just like sugar or salt does. Fish and other aquatic organisms obtain the oxygen they need by taking it up from solution. The solubility of oxygen is therefore critically important.

In reality, oxygen is not very soluble in water. Consequently, even when fresh water is fully aerated and at equilibrium with the atmosphere, the amount of oxygen it contains per liter is only 5%, or less, of that in air. This means that, in terms of acquiring the oxygen they need, freshwater organisms have a far smaller margin of safety than air-breathing ones.

Oxygen is constantly added to and removed from any body of fresh water. Oxygen is added by photosynthesis and by aeration from the atmosphere, and it is removed by animals and other heterotrophs. If a lot of decaying organic matter is present in a body of water, the oxygen demand of the decay microbes can be high and affect other life-forms. Under conditions in which the rate of oxygen removal from water exceeds the rate of addition, the concentration of dissolved oxygen can fall so low that many aquatic animals cannot survive in it.

Lake and pond habitats change with water depth

Bodies of relatively still fresh water are called lakes if large and ponds if small. Water absorbs light passing through it, and the intensity of sunlight available for photosynthesis decreases sharply with increasing depth. In deep lakes, only water relatively near the surface receives enough light for phytoplankton to exhibit a positive net primary productivity (figure 58.12). Those waters are described as the **photic zone.**

The photic zone

The thickness of the photic zone depends on how much particulate matter is in the water. Water that is relatively free of particulate matter and clear allows light to penetrate to a depth

of 10 m at sufficient intensity to support phytoplankton. Water that is thick with surface algal cells or soil from erosion may not allow light to penetrate very far before its intensity becomes too diminished for algal growth.

The supply of dissolved oxygen to the deep waters of a lake can be a problem because all oxygen enters any aquatic system near its surface. In the still waters of a lake, mixing between the surface and deeper layers may not occur except occasionally. When photosynthesis produces oxygen, it adds it to the photic zone of the lake near the surface. Thermal stratification commonly affects how readily oxygen enters the deep waters from the surface waters.

Thermal stratification

Thermal stratification is characteristic of many lakes and large ponds. In summer, as shown at the bottom of figure 58.13, water warmed by the Sun forms a layer known as the *epilimnion* at the surface—because warm water is less dense than

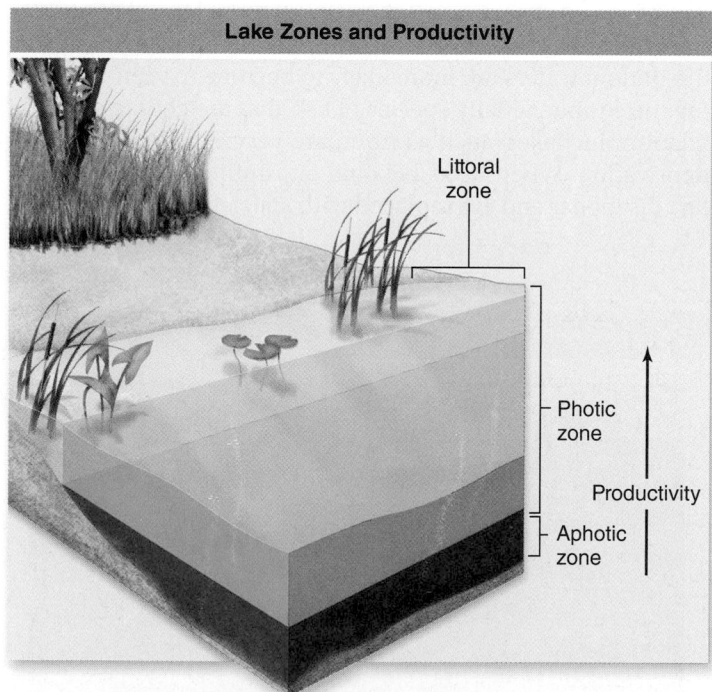

Figure 58.12 Light in a lake. The intensity of the sunlight available for photosynthesis decreases with depth in a lake. Consequently, only some of the upper waters—termed the photic zone—receive sufficient light for the net primary productivity of phytoplankton to be positive. The depth of the photic zone depends on how cloudy the water is. The shallows at the edge of a lake are called the littoral zone. They are well-illuminated to the bottom, so rooted plants and bottom algae can thrive there.

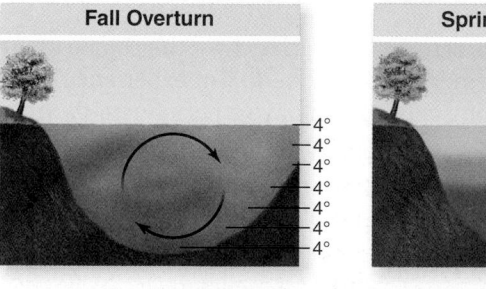

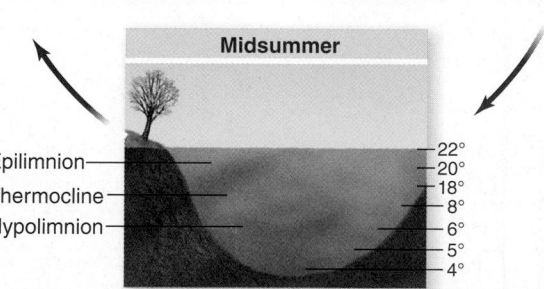

Figure 58.13 The annual cycle of thermal stratification in a temperate-zone lake. During the summer (lower diagram), water warmed by the Sun (the epilimnion) floats on top of colder, denser water (the hypolimnion). The lake is also thermally stratified in winter (upper diagram) when water that is near freezing or frozen floats on top of water that is at 4°C (the temperature of greatest density for fresh water). Stratification is disrupted in the spring and fall overturns, when the lake is at an approximately uniform temperature and winds mix it from top to bottom.

cold water and tends to float on top. Colder, denser water, called the *hypolimnion,* lies below. Between the warm and cold layers is a transitional layer, the thermocline. Although here we are focusing on fresh water, a similar thermal structuring of the water column occurs also in many parts of the ocean.

In a lake, thermal stratification tends to cut off the oxygen supply to the bottom waters; a consequence of the stratification is that the upper waters that receive oxygen do not mix with the bottom waters. The concentration of oxygen at the bottom may then gradually decline over time as the organisms living there use oxygen faster than it is replaced. If the rate of oxygen use is high, the bottom waters may run out of oxygen and become oxygen-free before summer is over. Oxygen-free conditions, if they occur, kill most animals in a lake.

In autumn, the temperature of the upper waters in a stratified lake drops until it is about the same as the temperature of the deep waters. The densities of the two water layers become similar, and the tendency for them to stay apart is weakened. Winds can then force the layers to mix, a phenomenon called the fall overturn (see figure 58.13). High oxygen concentrations are then restored in the bottom waters.

Chapter 2 discussed the unique properties of water. Fresh water is densest when its temperature is 4°C, and ice, at 0°, floats on top of this dense water. As a lake is cooled toward the freezing point with the onset of winter, the whole lake first reaches 4°C. Then, some water cools to an even lower temperature, and when it does, it becomes less dense and rises to the top. Further cooling of this surface water causes it to freeze into a layer of ice covering the lake. In spring, the ice melts, the surface water warms up, and again winds are able to mix the whole lake—the spring overturn.

Because temperature changes less over the course of the year in the tropics, many lakes there do not experience turnover. As a result, tropical lakes can have a permanent thermocline with depletion of oxygen near the bottom.

Lakes differ in oxygen and nutrient content

Bodies of fresh water that are low in algal nutrients (such as nitrate or phosphate) and low in the amount of algal material per unit of volume are termed *oligotrophic.* Such waters are often crystal clear. Oligotrophic streams and rivers tend to be high in dissolved oxygen because the movement of the flowing water aerates them; the small amount of organic matter in the water means that oxygen is used at a relatively low rate. Similarly, oligotrophic lakes and ponds tend to be high in dissolved oxygen at all depths all year because they also have a low rate of oxygen use. Because the water is relatively clear, light can penetrate the waters readily, allowing photosynthesis to occur through much of the water column, from top to bottom (figure 58.14).

Eutrophic bodies of water are high in algal nutrients and often populated densely with algae. They are more likely to be low in dissolved oxygen, especially in summer. In a eutrophic body of water, decay microbes often place high demands on the oxygen available because when thick populations of algae die, large amounts of organic matter are made available for decomposition. Moreover, light does not penetrate eutrophic waters well because of all the organic matter in the water; photosynthetic oxygen addition is therefore limited to just a relatively thin layer of water at the top.

Human activities have often transformed oligotrophic lakes into eutrophic ones. For example, when people overfertilize their lawns or fields, nitrate and phosphate from the fertilizers wash off into local water systems. Lakes that receive these nutrients become more eutrophic. A consequence is that the bottom waters are more likely to become oxygen-free during the summer. Many species of fish that are characteristic of oligotrophic lakes, such as trout, are very sensitive to oxygen deprivation. When lakes become eutrophic, these species of fish disappear and are replaced with species like carp that can

Oligotrophic Lake

a.

Eutrophic Lake

b.

Figure 58.14 Oligotrophic and eutrophic lakes. *a.* Oligotrophic lakes are low in algal nutrients, have high levels of dissolved oxygen, and are clear. *b.* Eutrophic lakes have high levels of algal nutrients and low levels of dissolved oxygen. Light does not penetrate deeply in such lakes.

better tolerate low oxygen concentrations. Lakes can return toward an oligotrophic state over time if steps are taken to eliminate the addition of excess nitrates, phosphates, and foreign organic matter such as sewage.

Learning Outcomes Review 58.3

The photic zone is the layer near the surface into which light penetrates. Photosynthesis can occur only in the photic zone. Thermal stratification is a major determinant of oxygen levels. In temperate lakes, mixing of different layers occurs when the layers reach the same temperature in spring and fall, and winds can cause the layers to mix. This overturn prevents oxygen depletion near the lake bottom. Eutrophic lakes are high in nutrients for algae but are low in dissolved oxygen; oligotrophic lakes are low in nutrients but high in dissolved oxygen at all depths.

■ *Why do tropical lakes often not experience seasonal turnover, and what effect is this likely to have on the ecosystems of these lakes?*

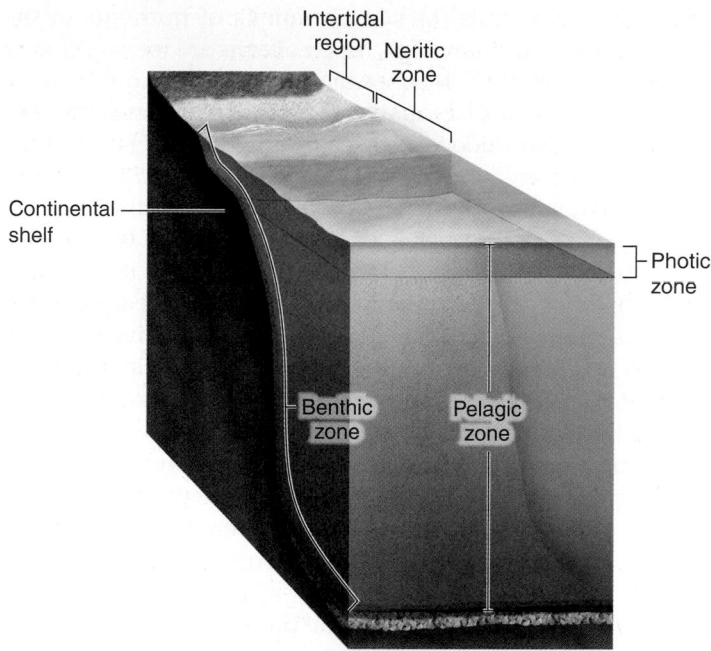

Figure 58.15 Basic concepts and terminology used in describing marine ecosystems. The continental shelf is the submerged edge of the continent. The waters over it are termed neritic and, on a worldwide average basis, are only 130 m deep at their deepest. The region where the tides rise and fall along the shoreline is called the intertidal region. The bottom is called the benthic zone, whereas the water column in the open ocean is called the pelagic zone. The photic zone is the part of the pelagic zone in which enough light penetrates for the phytoplankton to have a positive net primary productivity. The vertical scale of this drawing is highly compressed; whereas the outer edge of the continental shelf is 130 m deep, the open ocean in fact averages 35 times deeper (4000–5000 m deep).

58.4 Marine Habitats

Learning Outcomes

1. Know the different marine habitats.
2. Explain why El Niño events occur.

About 71% of the Earth's surface is covered by ocean. Near the coastlines of the continents are the continental shelves, where the water is not especially deep (figure 58.15); the shelves, in essence, represent the submerged edges of the continents. Worldwide, the shelves average about 80 km wide, and the depth of the water over them increases from 1 m to about 130 m as one travels from the coast toward the open ocean.

Beyond the continental shelves, the depth suddenly becomes much greater. The average depth of the open ocean is 4000 to 5000 m, and some parts—called trenches—are far deeper, reaching 11,000 m in the Marianas Trench in the western Pacific Ocean.

In most of the ocean, the principal primary producers are phytoplankton floating in the well-lit surface waters. A revolution is currently underway in scientific understanding of the limiting nutrients for ocean phytoplankton (see chapter 57). Primary production by the phytoplankton is presently understood to be nitrogen-limited in about two-thirds of the world's ocean, but iron-limited in about one-third. The principal known iron-limited areas are the great Southern Ocean surrounding Antarctica, parts of the equatorial Pacific Ocean, and parts of the subarctic, northeast Pacific Ocean. Where the water is shallow along coastlines, primary production is carried out not just by phytoplankton but also by rooted plants such as seagrasses and by bottom-dwelling algae, including seaweeds.

The world's ocean is so vast that it includes many different types of ecosystems. Some, such as coral reefs and estuaries, are high in their net primary productivity per unit of area (see figure 57.11), but others are low in productivity per unit area. Ocean ecosystems are of four major types: open oceans, continental shelf ecosystems, upwelling regions, and deep sea.

Open oceans have low primary productivity

In speaking of the open oceans, we mean the waters far from land (beyond the continental shelves) that are near enough to the surface to receive sunlight or to interact on a daily or weekly basis with those waters. We will discuss the deep sea separately later on.

The intensity of solar illumination in the open oceans drops from being high at the surface to being essentially zero at 200 m of depth; photosynthesis is limited to this level of the ocean. However, nutrients for phytoplankton, such as nitrate, tend to be present at low concentrations in the photic zone because over eons of time in the past, ecological processes have exported nitrate and other nutrients from the upper waters to the deep waters, and no vigorous forces exist in the open ocean to return the nutrients to the sunlit waters.

Because of the low concentrations of nutrients in the photic zone, large parts of the open oceans are low in primary productivity per unit area (see figure 57.11) and aptly called a "biological desert." These parts—which correspond to the centers of the great midocean gyres (see figure 58.4)—are often collectively termed the *oligotrophic ocean* (figure 58.16) in reference to their low nutrient levels and low productivity.

People fish the open oceans today for only a few species, such as tunas and some species of squids and whales. Fishing in the open oceans is limited to relatively few species for two reasons. First, because of the low primary productivity per unit of area, animals tend to be thinly distributed in the open oceans. The only ones that are commercially profitable to catch are those that are individually large or that tend to gather together in tight schools. Second, costs for travelling far from land are high. All authorities agree that as we turn to the sea to help feed the burgeoning human population, we cannot expect the open ocean regions to supply great quantities of food.

Continental shelf ecosystems provide abundant resources

Many of the ecosystems on the continental shelves are relatively high in productivity per unit area. An important reason is that the waters over the shelves—termed the **neritic waters** (see figure 58.15)—tend to have relatively high concentrations of nitrate and other nutrients, averaged over the year.

Because the waters over the shelves are shallow, they have not been subject, over the eons of time, to the loss of nutrients into the deep sea, as the open oceans have. Over the shelves, nutrient-rich materials that sink hit the shallow bottom, and the nutrients they contain are stirred back into the water column by stormy weather. In addition, nutrients are continually replenished by run-off from nearby land.

Around 99% of the food people harvest from the ocean comes from continental shelf ecosystems or nearby upwelling regions. The shelf ecosystems are also particularly important to humankind in other ways. Mineral resources taken from the ocean, such as petroleum, come almost exclusively from the shelves. In addition, almost all recreational uses of the ocean, from sailing to scuba diving, take place on the shelves. The shelves feature prominently in these ways because they are close to coastlines and relatively shallow.

Estuaries

Estuaries are one of the types of shelf ecosystems. An estuary is a place along a coastline, such as a bay, that is partially surrounded by land and in which fresh water from streams or rivers mixes with ocean water, creating intermediate (brackish) salinities.

Estuaries, besides being bodies of water, include intertidal marshes or swamps. An **intertidal** habitat is an area that is exposed to air at low tide but under water at high tide. The marshes of the intertidal zone are called **salt marshes.** Intertidal swamps called **mangrove swamps** (dominated by trees and bushes) occur in tropical and subtropical parts of the world.

Estuaries are a vital and highly productive ecosystem—they provide shelter and food for many aquatic animals, especially the larvae and young, that people harvest for food.

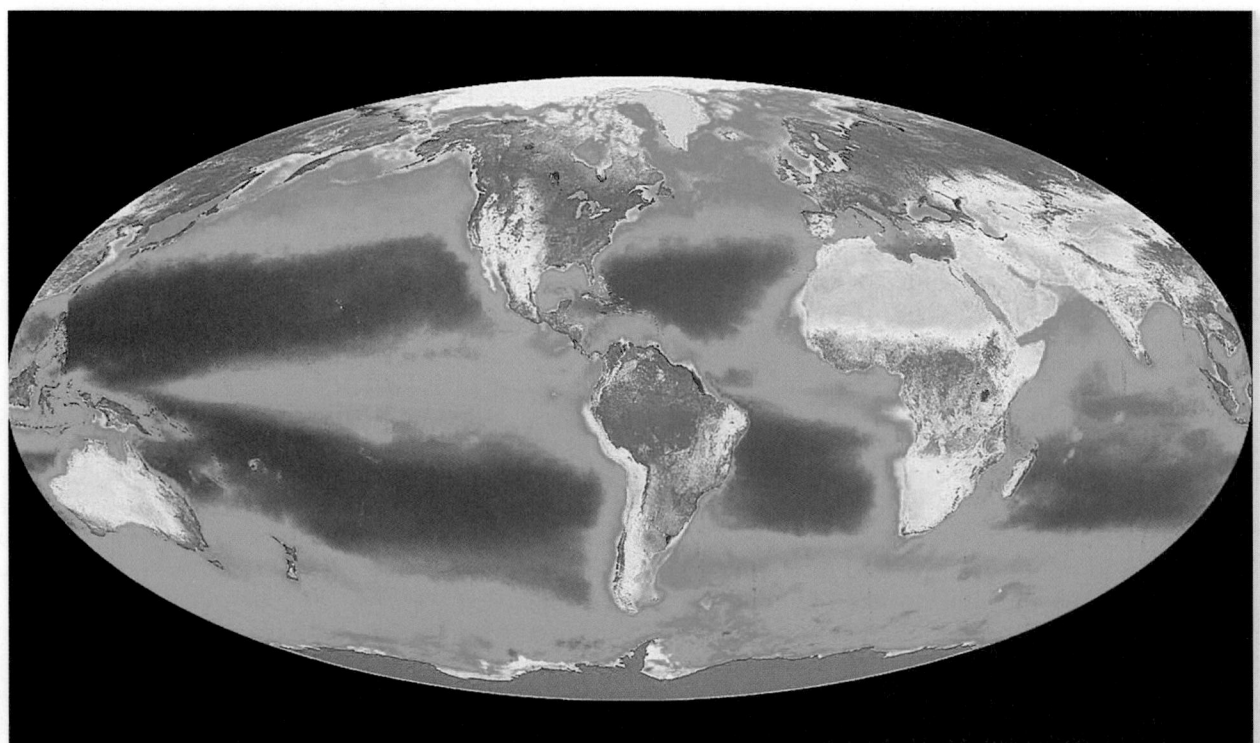

Figure 58.16 Major functional regions of the ocean. The regions classed as oligotrophic ocean (*colored dark blue*) are "biological deserts" with low productivity per unit area. Continental shelf ecosystems (*green at the edge of continents*) are typically medium to high in productivity. Upwelling regions (*yellow at the edge of continents*) such as along the western coasts of North and South America and southern Africa are the highest in productivity per unit area and rank with the most productive of all ecosystems on Earth.

Estuaries are also important to a very large number of other animal species, such as migrating birds.

Banks and coral reefs

Other types of shelf ecosystems include banks and coral reefs. **Banks** are local shallow areas on the shelves, often extremely important as fishing grounds. Georges Bank, 100 km off the shore of Massachusetts, was formerly one of the most productive and famous; much of this area has been closed to fishing since the mid-1990s because of overexploitation.

Coral reef ecosystems occur in subtropical and tropical latitudes. Their defining feature is that in them, stony corals—corals that secrete a solid, calcified type of skeleton—build three-dimensional frameworks that form a unique habitat in which many other distinctive organisms live, including reef fish and soft corals (figure 58.17).

All the 700 or so species of reef-building corals are animal–algal symbioses; the animals are cnidarians, and dinoflagellate symbionts live within the cells of their inner cell layer (the gastrodermis); see chapter 33. These corals depend on photosynthesis by the algal symbionts, and thus require clear waters through which sunlight can readily penetrate. Reef-building corals are threatened worldwide, as described later in this chapter.

Upwelling regions experience mixing of nutrients and oxygen

The upwelling regions of the ocean are localized places where deep water is drawn consistently to the surface because of the action of local forces such as local winds. The deep water is often rich in nitrate and other nutrients. Upwelling therefore steadily brings nutrients into the well-lit surface layers. Phytoplankton respond to the abundance of nutrients and light with

Figure 58.17 A coral reef ecosystem. Reef-building corals, which consist of symbioses between cnidarians and algae, construct the three-dimensional structure of the reef and carry out considerable primary production. Fish and many other kinds of animals find food and shelter, making these ecosystems among the most diverse. About 20% of all fish species occur specifically in coral reef ecosystems.

prolific growth and reproduction. Upwelling regions have the highest primary productivity per unit area in the world's ocean.

The most famous upwelling region (see figure 58.16) is found along the coast of Peru and Ecuador, where upwelling occurs year-round. Another important upwelling region is the coastline of California, along which upwelling occurs during about half the year in the summer, explaining why swimmers find cold water at the beaches even in July and August.

Upwelling regions support prolific but vulnerable fisheries. Sardine fishing in the California upwelling region crashed a few decades ago, but previously was enormously important to the region, as Nobel Prize–winning author John Steinbeck chronicled in a number of his books, most notably *Cannery Row.*

El Niño Southern Oscillation (ENSO)

The phenomenon named El Niño first came to the attention of science in studies of the Peru–Ecuador upwelling region. In that region, every 2 to 7 years on an irregular and relatively unpredictable basis, the water along the coastline becomes profoundly warm, and simultaneously the primary productivity becomes unusually low.

Because of the low primary productivity, the ordinarily prolific fish populations weaken, and populations of seabirds and sea mammals that depend on the fish are stressed and plummet. The local people had named a mild annual warming event, which occurred around Christmas each year, "El Niño" (literally, "the child," after the Christ Child). Scientists adopted the term El Niño Southern Oscillation (ENSO) to refer to those dramatic warming events.

The immediate cause of El Niño took several decades to figure out, but research ultimately showed that the cause is a weakening of the east-to-west Trade Winds in the region. The Trade Winds ordinarily blow warm surface water to the west, away from the Peru–Ecuador coast. This thins the warm surface layer of water along the coast, so that deep water—cold but highly rich in nutrients—is drawn to the surface, leading to high primary production.

Weakening of the Trade Winds allows the warm surface layer to become thicker. Upwelling continues, but under such circumstances it merely recirculates the thick warm surface layer, which is nutrient-depleted.

After these fundamentals had been discovered, researchers in the 1980s realized that the weakening of the Trade Winds is actually part of a change in wind circulation patterns that recurs irregularly. One reason the Trade Winds blow east-to-west in ordinary times is that the surface waters in the western equatorial Pacific are warmer than those in the eastern equatorial Pacific; air rises from the warm western areas, creating low pressure at the surface there, and air blows out of the east into the low pressure. During an El Niño, the warmer the eastern ocean gets, the more similar it becomes to the western ocean, reducing the difference in pressure across the ocean. Thus, once the Trade Winds weaken a bit, the pressure difference that makes them blow is lessened, weakening the Trade Winds further. Warm water ordinarily kept in the west by the Trade Winds creeps progressively eastward at equatorial latitudes because of this self-reinforcing series of events. Ultimately, effects of El Niño occur across large parts of the world's weather

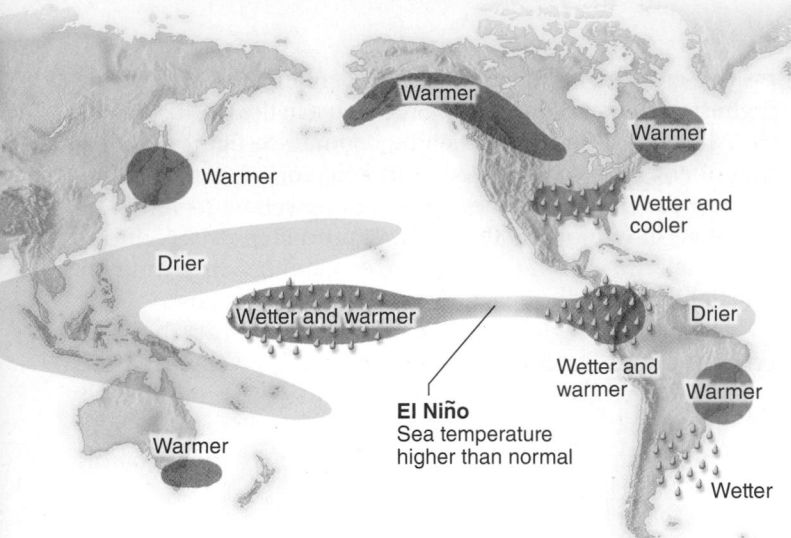

Figure 58.18 An El Niño winter. This diagram shows just some of the worldwide alterations of weather that are often associated with the El Niño phenomenon.

systems, affecting sea temperatures in California, rainfall in the southwestern United States, and even systems as far distant as Africa.

One specific result is to shift the weather systems of the western Pacific Ocean 6000 km eastward. The tropical rainstorms that usually drench Indonesia and the Philippines occur when warm seawater abutting these islands causes the air above it to rise, cool, and condense its moisture into clouds. When the warm water moves east, so do the clouds, leaving the previously rainy areas in drought. Conversely, the western edge of Peru and Ecuador, which usually receives little precipitation, gets a soaking.

El Niño can wreak havoc on ecosystems. During an El Niño event, plankton can drop to 1/20 of their normal abundance in the waters of Peru and Ecuador, and because of the drop in plankton productivity, commercial fish stocks virtually disappear (figure 58.18). In the Galápagos Islands, for example, seabird and sea lion populations crash as animals starve due to the lack of fish. By contrast, on land, the heavy rains produce a bumper crop of seeds, and land birds flourish. In Chile, similar effects on seed abundance propagate up the food chain, leading first to increased rodent populations and then to increased predator populations, a nice example of a bottom-up trophic cascade, as was discussed in chapter 57.

The deep sea is a cold, dark place with some fascinating communities

The deep sea is by far the single largest habitat on Earth, in the sense that it is a huge region characterized by relatively uniform conditions throughout the globe. The deep sea is seasonless, cold (2–5°C), totally dark, and under high pressure (400–500 atmospheres where the bottom is 4000–5000 m deep).

In most regions of the deep sea, food originates from photosynthesis in the sunlit waters far above. Such food—in the form of carcasses, fecal pellets, and mucus—can take as much as a month to drift down from the surface to the bottom, and along the way about 99% of it is eaten by animals living in

the water column. Thus, the bottom communities receive only about 1% of the primary production and are food-poor. Nonetheless, a great many species of animals—most of them small-bodied and thinly distributed—are now known to live in the deep sea. Some of the animals are bioluminescent (figure 58.19a) and thereby able to communicate or attract prey by use of light.

Hydrothermal vent communities

The most astounding communities in the deep sea are the hydrothermal vent communities. Unlike most parts of the deep sea, these communities are thick with life (figure 58.19b), including large-bodied animals such as worms the size of baseball bats. The reason such a profusion of life can be supported is that these communities live on vigorous, local primary production rather than depending on the photic zone far above.

The hydrothermal vent communities occur at places where tectonic plates are moving apart, and seawater—circulating through porous rock—is able to come into contact with very hot rock under the seafloor. This water is heated to temperatures in excess of 350°C and, in the process, becomes rich in hydrogen sulfide.

As the water rises up out of the porous rock, free-living and symbiotic bacteria oxidize the sulfide, and from this reaction they obtain energy, which, in a manner analogous to photosynthesis, they use to synthesize their own cellular substance, grow, and reproduce. These sulfur-oxidizing bacteria are chemoautotrophs (see chapter 57). Animals in the

Figure 58.19 Life in the deep sea. a. The luminous spot below the eye of this deep-sea fish results from the presence of a symbiotic colony of bioluminescent bacteria. Bioluminescence is a fairly common feature of mobile animals in the parts of the ocean that are so deep as to be dark. It is more common among species living partway down to the bottom than in ones living at the bottom. **b.** These large worms (shown in the larger photograph) live along vents where hot water containing hydrogen sulfide rises through cracks in the seafloor crust (inset photo). Much of the body of each worm is devoted to a colony of symbiotic sulfur-oxidizing bacteria. The worms transport sulfide and oxygen to the bacteria, which oxidize the sulfur and use the energy thereby obtained for primary production of new organic compounds, which they share with their worm hosts.

a.   *b.*

communities either survive on the bacteria or eat other animals that do. The hydrothermal vent communities are among the few communities on Earth that do not depend on the Sun's energy for primary production.

58.5 Human Impacts on the Biosphere: Pollution and Resource Depletion

We all know that human activities can cause adverse changes in ecosystems. In discussing these, it is important to recognize that creative people can often come up with rational solutions to such problems.

An outstanding example is provided by the history of DDT in the United States. DDT is a highly effective insecticide that was sprayed widely in the decades following World War II, often on wetlands to control mosquitoes. During the years of heavy DDT use, populations of ospreys, bald eagles, and brown pelicans—all birds that catch large fish—plummeted. Ultimately, the use of DDT was connected with the demise of these birds.

Scientists established that DDT and its metabolic products became more and more concentrated in the tissues of animals as the compounds were passed along food chains (figure 58.20). Animals at the bottom of food chains accumulated relatively low concentrations in their fatty tissues. But the primary carnivores that preyed on them accumulated higher concentrations from eating great numbers, and the secondary carnivores accumulated higher concentrations yet. Top-level carnivores, such as the birds that eat large fish, were dramatically affected by the DDT. In these birds, scientists found that metabolic products of DDT disrupted the formation of eggshells. The birds laid eggs with such thin shells that they often cracked before the young could hatch.

Researchers concluded that the demise of the fish-eating birds could be reversed by a rational plan to clean ecosystems of DDT, and laws were passed banning its use. Now, three decades later, populations of ospreys, eagles, and pelicans are rebounding dramatically. For some people, a major reason to study science is the opportunity to be part of success stories of this sort.

Freshwater habitats are threatened by pollution and resource depletion

Fresh water is not just the smallest of the major habitats, but also the most threatened. One of the simplest yet most ominous threats to fresh water is that burgeoning human populations often extract excessive amounts of water from rivers, lakes, or streams. The Colorado River, for example, is one of the greatest rivers in North America, originating with snow melt in the Rocky Mountains and flowing through Utah, Arizona, Nevada, California, and northern Mexico before emptying into the ocean. Today, water is pumped out of the river all along its way to meet the water needs of cities (even ones as distant as Los Angeles) and to irrigate crops. The river now frequently runs out of water and dries up in the desert, never reaching the sea. Worldwide, many crises in the supply of fresh water loom on the horizon.

Pollution: Point source versus diffuse

Pollution of fresh water is a global problem. Point-source pollution comes from an identifiable location—such as easily

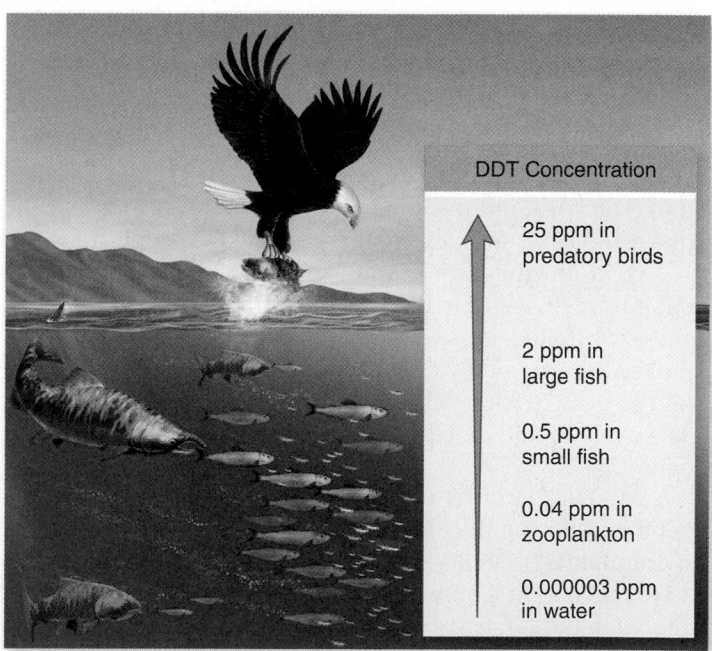

Figure 58.20 Biological magnification of DDT concentration. Because all the DDT an animal eats in its food tends to accumulate in its fatty tissues, DDT becomes increasingly concentrated in animals at higher levels of the food chain. The concentrations at the right are in parts per million (ppm). Before DDT was banned in the United States, bird species that eat large fish underwent drastic population declines because metabolic products of DDT made their eggshells so thin that the shells broke during incubation.

identified factories or other facilities that add pollutants at defined locations, such as an outfall pipe. Examples include sewage-treatment plants, which discharge treated effluents at specific spots on rivers, and factories that sometimes discharge water contaminated with heavy metals or chemicals. Laws and technologies can readily be brought to bear to moderate point-source pollution because the exact locations and types of pollution are well defined. In many countries, great progress has been made, but in other countries, often in the developing world, water pollution is still a major problem.

Diffuse pollution is exemplified by eutrophication caused by excessive run-off of nitrates and phosphates from lawn and agricultural field fertilization. When excessive nitrates and phosphates enter rivers and lakes, the character of the bodies of water is changed for the worse; the concentration of dissolved oxygen declines, and fish species such as carp take the place of more desirable species. The problem is exacerbated when rivers empty into the ocean. The eutrophication caused by the accumulation of chemicals can lead to enormous areas of water with no oxygen, causing massive die-offs of fish and other animals. The most famous such area, covering approximately 17,500 km² in 2011, occurs where the Mississippi River empties into the Gulf of Mexico, but other "dead zones" occur in places around the world.

The nitrates and phosphates that cause these problems originate on thousands of farms and lawns spread over whole watersheds, and they often enter fresh waters at virtually countless locations. The diffuseness of this sort of pollution renders it difficult to modify by simple technical fixes. Instead, solutions often depend on public education and political action.

Pollution from coal burning: Acid precipitation

A type of pollution that has properties intermediate between the point-source and diffuse types is the pollution that can arise from burning of coal for power generation. Although each smokestack is a point source, there are many stacks, and the smoke and gases from these stacks spread over wide areas.

Acid precipitation is one aspect of this problem. When coal is burned, sulfur in the coal is oxidized. The sulfur oxides, unless controlled, are spewed into the atmosphere in the stack smoke, and there they combine with water vapor to produce sulfuric acid. Falling rain or snow picks up the acid and is excessively acidic when it reaches the surface of the Earth (figure 58.21).

Mercury emitted in stack smoke is a second potential problem. Burning of coal can be one of the major sources of environmental mercury, a serious public health issue because just small amounts of mercury can interfere with brain development in human fetuses and infants.

Acid precipitation and mercury pollution affect freshwater ecosystems. At pH levels below 5.0, many fish species and other aquatic animals die, unable to reproduce. Thousands of lakes and ponds around the world no longer support fish because of pH shifts induced by acid precipitation. Mercury that falls from atmospheric emissions into lakes and ponds accumulates in the tissues of food fish. In the Great Lakes region of the United States, people—especially pregnant women—are advised to eat little or no locally caught fish because of its mercury content.

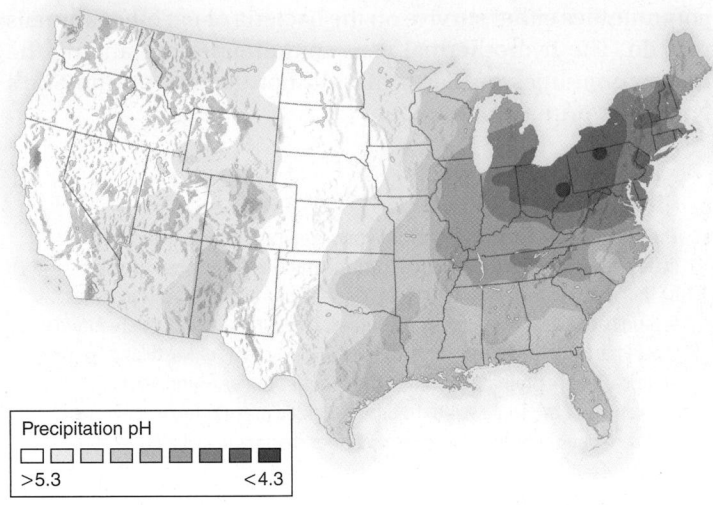

Precipitation pH
>5.3   <4.3

Figure 58.21 pH values of rainwater in the United States. pH values of less than 7 represent acid conditions; the lower the values, the greater the acidity. Precipitation in parts of the United States, especially in the Northeast, is commonly more acidic than natural rainwater, which has a pH of 5.6 or higher.

Forest ecosystems are threatened in tropical and temperate regions

Probably the single greatest problem for terrestrial habitats worldwide is deforestation by cutting or burning. There are many reasons for deforestation. In poverty-stricken countries, deforestation is often carried out diffusely by the general population; people burn wood to cook or stay warm, and they collect it from the local forests.

At the other extreme, corporations still cut large tracts of virgin forests in an industrialized fashion, often shipping the wood halfway around the world to buyers. Tropical hardwoods, such as mahogany, from Southeast Asian rainforests are shipped to the United States for use in furniture, and softwood logs are shipped from Alaska to East Asia for pulping and paper production. Forests are sometimes simply burned to open up land for farming or ranching (figure 58.22a).

Loss of habitat

The loss of forest habitat can have dire consequences. Particularly diverse sets of species depend on tropical rainforests for their habitat, for example. Thus, when rainforests are cleared, the loss of biodiversity can be extreme. Many tropical forest regions have been severely degraded, and recent estimates suggest that less than half of the world's tropical rainforests remain in pristine condition. All of the world's tropical rainforests will be degraded or gone in about 30 years at present rates of destruction.

Besides loss of habitat, deforestation can have numerous secondary consequences, depending on local contexts. In the Sahel region, south of the Sahara Desert in Africa, deforestation has been a major contributing factor in increased desertification. In the forests of the northeastern United States, as the Hubbard Brook experiment shows (see figure 57.7), deforestation can lead to both a loss of nutrients from forest soils and a simultaneous nutrient enrichment of bodies of water downstream.

a.   *b.*

Figure 58.22   **Destroying the tropical rainforests.**
a. These fires are destroying a tropical rainforest in Brazil to clear it for cattle pasture. *b.* The consequences of deforestation can be seen on these middle-elevation slopes in Madagascar, which once supported tropical rainforest, but now support only low-grade pastures and permit topsoil to erode into the rivers (note the color of the water, stained brown by high levels of soil erosion). This sort of picture is seen in a number of places around the world, including Ecuador and Haiti as well as Madagascar.

Disruption of the water cycle

As discussed in chapter 57, cutting of a tropical rainforest often interrupts the local water cycle in ways that permanently alter the landscape. After an area of tropical rainforest is cleared, rainwater often runs off the land to distant places, rather than being returned to the atmosphere immediately above by transpiration. This change may render conditions unsuitable for the rainforest trees that originally lived there. Then the poorly vegetated land—exposed and no longer stabilized by thick root systems—may be ravaged by erosion (figure 58.22b).

Acid rain

Deforestation can be a problem in temperate regions, as well as in the tropics. In addition, acid rain affects forests as well as lakes and streams; large tracts of trees in temperate regions have been adversely affected by acid rain. By changing the acidity of the soil, acid rain can lead to widespread tree mortality (figure 58.23).

Marine habitats are being depleted of fish and other species

Overfishing of the ocean has risen to crisis proportions in recent decades and probably represents the single greatest current problem in the ocean realm. The ocean is so huge that it has tended to be more immune than freshwater or terrestrial ecosystems to global human alteration. Nonetheless, the total world fish catch has been pushed to its maximum for over two decades, even as demand for fish has continued to rise. Fishing pressure is so excessive that 25% to 30% of the world's ocean fish stocks are presently officially rated as being overexploited,

Figure 58.23 **Damage to trees by acid precipitation at Clingman's Dome, Tennessee.** Acid precipitation weakens trees and makes them more susceptible to pests and predators.

depleted, or in recovery; another 40% to 50% are rated as being maximally exploited.

Major cod fisheries in waters off of Nova Scotia, Massachusetts, and Great Britain have been closed to fishing in the past 15 years because of collapse (figure 58.24). Overfishing can have disturbing indirect effects. In impoverished parts of Africa, poaching on primates and other wild mammals in national parks increases when fish catches decline.

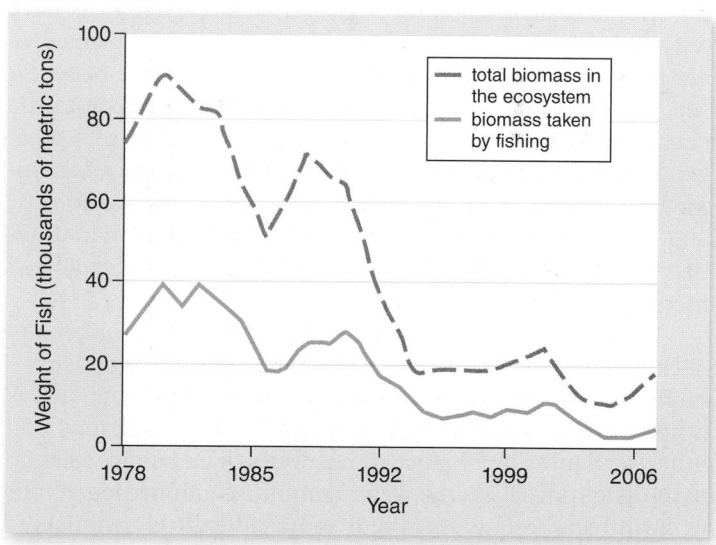

Figure 58.24 **The collapse of a fishery.** The red line shows the biomass of cod (*Gadus morhua*) in the Georges Bank ecosystem as estimated by the U.S. National Marine Fisheries Service based on data collected by scientific sampling. The biomass declined steeply between the 1970s and 1990s because of fishing pressure. As the years passed, commercial landings of cod (*blue line*) remained fairly constant, in part because ships worked harder and harder to catch cod, until catches fell precipitously toward zero and the fishery collapsed in the mid-1990s. Regulatory agencies closed the fishery in the mid-1990s to permit the cod to recover, but even in 2009 recovery of cod was weak, and production from the fishery was far below historical norms.

Aquaculture: At present only a quick fix

Production of fish by aquaculture has grown steadily in the last two decades, and it is often viewed as a straightforward solution to the fisheries problem. But the dietary protein needs of many aquacultured fish, such as salmon, are met largely with wild-caught fish. In this case, exploitation has simply shifted to different species.

In addition, current aquaculture practices often damage natural ocean ecosystems. One example is the clearing of mangrove swamps along coasts to create shrimp and fish ponds, which are abandoned when their productivity declines. Research is needed to ameliorate these problems.

Pollution effects

As large as the ocean is, enough pollutants are being added that at the start of the 21st century, polluting materials are easily detectable on a global basis. An expedition to some of the most remote, uninhabited islands in the vast Pacific Ocean recently reported, for example, that considerable amounts of plastic could be found washed up on the beaches. Similarly, even the waters of the Arctic are laced with toxic chemicals; biopsy samples of tissue from Arctic killer whales (*Orcinus orca*) revealed extremely high levels of many chemicals, including pesticides and a flame-retardant chemical often used in carpets. Nonetheless, because of the ocean's vastness, concentrations of pollutants are not at crisis levels in the ocean at large.

Just as in freshwater habitats, pollution from coal burning and other sources is causing increased acidity of the ocean, but in this case, the culprit is carbon dioxide. Scientists have only recently realized that one consequence of the massive amounts of carbon dioxide entering the atmosphere (see next section) is that much of it ends up being dissolved into the ocean. And when it does, it combines with water to form hydrogen ions (H^+) and bicarbonate (HCO_3^-) ions. By some measurements, the result has been an increase in the ocean's acidity by 30% since the 1950s, making the ocean's pH lower than it has been any time in the last 20 million years, with a projected increase of 150% by the end of this century.

The biological consequences of this acidification are not clear, but a primary concern is that many marine organisms—including corals, echinoderms, oyster larvae and many others—use calcium carbonate to build their skeletons and other structures. As the ocean becomes more acidic, calcium carbonate ($CaCO_3$) in the water combines with hydrogen ions to form bicarbonate, making it more difficult for animals to build the structures they need to survive and grow. In fact, the opposite can occur, and calcium carbonate already incorporated into skeletons or other structures may actually dissolve. The magnitude of these effects is not yet clear, but many scientists fear that coral reefs and many other elements of ocean ecosystems may be in peril.

Destruction of coastal ecosystems

Deterioration of coastal ecosystems is also a grave problem. Estuaries along coastlines are often subject to severe eutrophication; since about 1970, for example, the bottom waters of the Chesapeake Bay near Washington, DC, have become oxygen-free each summer because of the decay of excessive amounts of organic matter.

Another coastal problem is destruction of salt marshes, which (like freshwater wetlands) are often perceived as disposable. Most authorities believe that the loss of salt marshes in the 20th century was a major contributing factor to the destruction of New Orleans by Hurricane Katrina in 2005; had the salt marshes and cypress swamps been present at their full extent, they would have absorbed a great deal of the flooding water and buffered the city from some of the storm's violence.

Stratospheric ozone depletion has led to an ozone "hole"

The colors of the satellite photo in figure 58.25*a* represent different concentrations of ozone (O_3) located 20 to 25 km above the Earth's surface in the stratosphere. Stratospheric ozone is depleted over Antarctica (purple region in the figure) to between one-half and one-third of its historically normal concentration, a phenomenon called the ozone hole.

Although depletion of stratospheric ozone is most dramatic over Antarctica, it is a worldwide phenomenon. Over the United States, the ozone concentration has been reduced by about 4%, according to the U.S. Environmental Protection Agency.

Stratospheric ozone and UV-B

Stratospheric ozone is important because it absorbs ultraviolet (UV) radiation—specifically the wavelengths called **UV-B**—from incoming solar radiation. UV-B is damaging to living organisms in a number of ways; for instance, it increases risks of cataracts and skin cancer in people. Depletion of stratospheric ozone permits more UV-B to reach the Earth's surface and therefore increases the risks of UV-B damage. Every 1% drop in stratospheric ozone is estimated to lead to a 6% increase in the incidence of skin cancer, for example. In the southern hemisphere, where ozone depletion is the greatest, rates of skin cancer are much higher than elsewhere in the world; in Australia, for example, two out of every three people will develop skin cancer by age 70. UV exposure also may be detrimental to many types of animals, such as amphibians (figure 58.26)

Ozone depletion and CFCs

The major cause of the depletion of stratospheric ozone is the addition of industrially produced chlorine- and bromine-containing compounds to the atmosphere. Of particular concern are chlorofluorocarbons (CFCs), used until recently as refrigerants in air conditioners and refrigerators, and in manufacturing. CFCs released into the atmosphere can ultimately liberate free chlorine atoms, which in the stratosphere catalyze the breakdown of ozone molecules (O_3) to form ordinary oxygen (O_2). Ozone is continually being made and broken down, and free chlorine atoms tilt the balance toward a faster rate of breakdown.

The extreme depletion of ozone seen in the ozone hole is a consequence of the unique weather conditions that exist

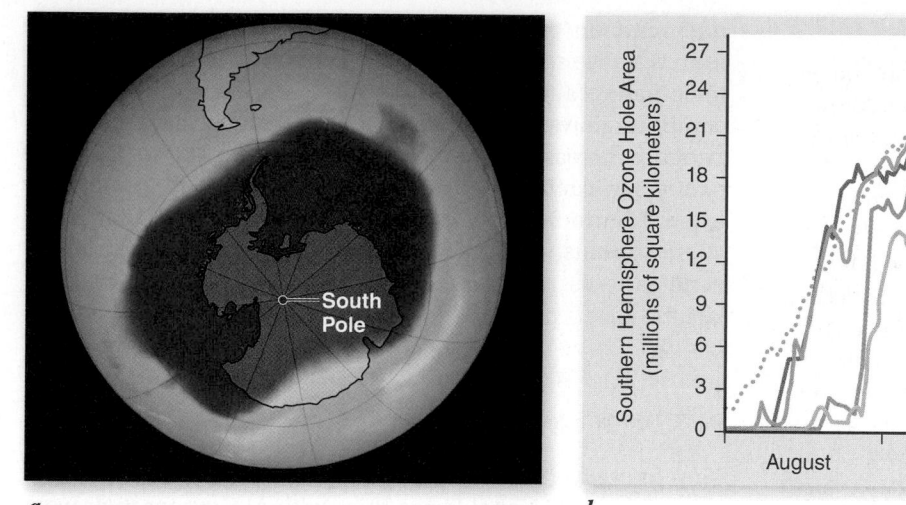

a.

b.

Figure 58.25 **The ozone hole over Antarctica.** NASA satellites currently track the extent of ozone depletion in the stratosphere over Antarctica each year. Every year since about 1980, an area of profound ozone depletion, called the ozone hole, has appeared in August (early spring in the southern hemisphere) when sunlight triggers chemical reactions in cold air trapped over the South Pole during the Antarctic winter. The hole intensifies during September before tailing off as temperature rises in November–December. *a.* In September 2006, the 11.4 million-square-mile hole (*purple* in the satellite image shown) covered an area larger than the United States, Canada, and Mexico combined, the largest hole ever recorded. *b.* Concentrations of ozone-depleting chemical compounds in the atmosphere have probably peaked in the last few years and are expected to decline slowly over the decades ahead.

over Antarctica. During the continuous dark of the Antarctic winter, a strong stratospheric wind, the polar-night jet, develops and, blowing around the full circumference of the Earth, isolates the stratosphere over Antarctica from the rest of the atmosphere. As a result, the Antarctic stratosphere stays extremely cold (–80°C or lower) for many weeks, permitting

unique types of ice clouds to form. Reactions associated with the particles in these clouds lead to accumulation of diatomic chlorine, Cl_2. When sunlight returns in the early Antarctic spring, the diatomic chlorine is photochemically broken up to form free chlorine atoms in great abundance, and the ozone-depleting reactions ensue.

SCIENTIFIC THINKING

Question: *Does exposure to UV radiation affect the survival of amphibian eggs?*

Hypothesis: *Direct UV exposure is detrimental to eggs.*

Experiment: *Fertilized eggs from several frog species are placed into enclosures in full sunlight. All enclosures have screens, some of which filter out UV radiation, whereas others do not affect UV transmission. Eggs are monitored to see whether they survive to hatching or whether they die.*

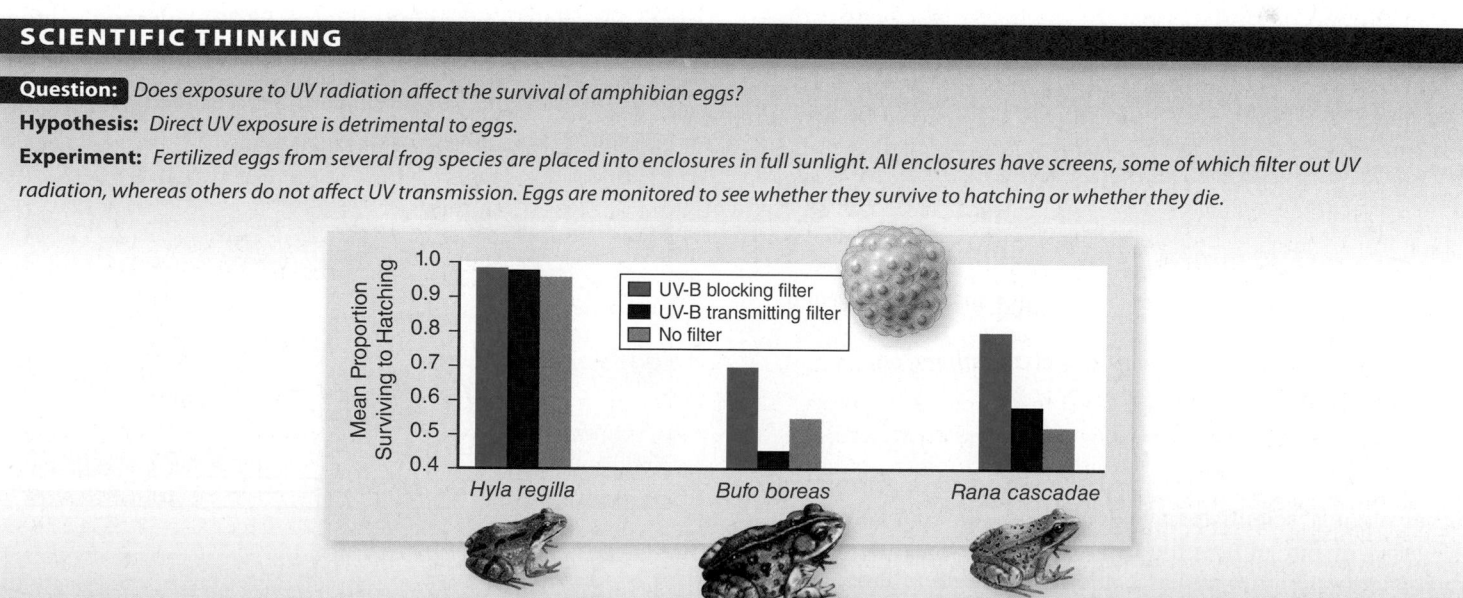

Result: *Egg survival was greatly decreased in two of three species in the enclosures where UV radiation was not filtered out, as compared to survival in the filtered enclosures. Therefore, the hypothesis is confirmed: UV exposure is detrimental to amphibian eggs.*

Further Questions: *What factors might explain why some species are affected by UV exposure and others are not? How could your hypotheses be tested?*

Figure 58.26 **The effect of UV radiation on amphibian eggs.**

Phase-out of CFCs

After research revealed the causes of stratospheric ozone depletion, worldwide agreements were reached to phase out the production of CFCs and other compounds that lead to ozone depletion. Manufacture of such compounds ceased in the United States in 1996, and there is now general awareness about the importance of using "ozone-safe" alternative chemicals. The atmosphere will cleanse itself of ozone-depleting compounds only slowly because the substances are chemically stable. Nonetheless, the problem of ozone depletion is diminishing and is expected to be substantially corrected by the second half of the 21st century.

The CFC story is an excellent example of how environmental problems arise and can be solved. Initially, CFCs were heralded as an efficient and cost-effective way to provide cooling, a clear improvement over previous technologies. At that time, their harmful consequences were unknown. Once the problems were identified, international agreements led to an effective solution, and creative technological advances led to replacements that solved the problem at little cost.

Learning Outcomes Review 58.5

Pollution and resource depletion are the major human effects on the environment, with freshwater habitats being most threatened. Point-source pollution comes from identifiable locations, such as factories, whereas diffuse pollution comes from numerous sources, such as fertilized lawns. Deforestation is a major problem in that it destroys habitat, disrupts communities, depletes resources, and changes the local water cycle and weather patterns. Overfishing is the greatest problem in the oceans.

■ Were CFCs an example of point-source or diffuse pollution? In general, how do efforts to combat pollution depend on their source?

58.6 Human Impacts on the Biosphere: Climate Change

Learning Outcomes

1. Explain the link between atmospheric carbon dioxide and global warming.
2. Describe the consequences of global warming on ecosystems and human health.

By studying the Earth's history and making comparisons with other planets, scientists have determined that concentrations of gases in our atmosphere, particularly CO_2, maintain the average temperature on Earth about 25°C higher than it would be if these gases were absent. This fact emphasizes that the composition of our atmosphere is a key consideration for life on Earth. Unfortunately, human activities are now changing the composition of the atmosphere in ways that most authorities conclude will be damaging or, in the long run, disastrous.

Because of changes in atmospheric composition, the average temperature of the Earth's surface is increasing, a

phenomenon called global warming. As you might imagine from what we said at the beginning of this chapter, changes in temperature alter global wind and water-current patterns in complicated ways. This means that as the average global temperature increases, some particular regions of the world warm to a lesser extent, whereas other regions heat up to a greater extent (figure 58.27). It also means that rainfall patterns are altered because global precipitation patterns depend on global wind patterns. Enormous computer models are used to calculate the effects predicted in all parts of the world.

Independent computer models predict global changes

The Intergovernmental Panel on Climate Change, which shared the 2007 Nobel Peace Prize with Al Gore for their work on global climate change, recently released its fourth assessment report. Based on a variety of different scenarios, computer models predicted that global temperatures would increase 1.1°C to 6.4°C (2.0–11.5°F) by the end of this century.

More ominous perhaps than temperature are some of the predictions for precipitation. For example, although northern Europe is expected to receive more precipitation than today, another recent study predicted that parts of southern Europe will receive about 20% less, disrupting natural ecosystems, agriculture, and human water supplies. Some European countries may come out ahead economically, but others will come out behind, and political relationships among countries will likely change as some shift from being food exporters to the more tenuous role of requiring food imports.

Carbon dioxide is a major greenhouse gas

Carbon dioxide is the gas usually emphasized in discussing the cause of global warming (figure 58.28), although other

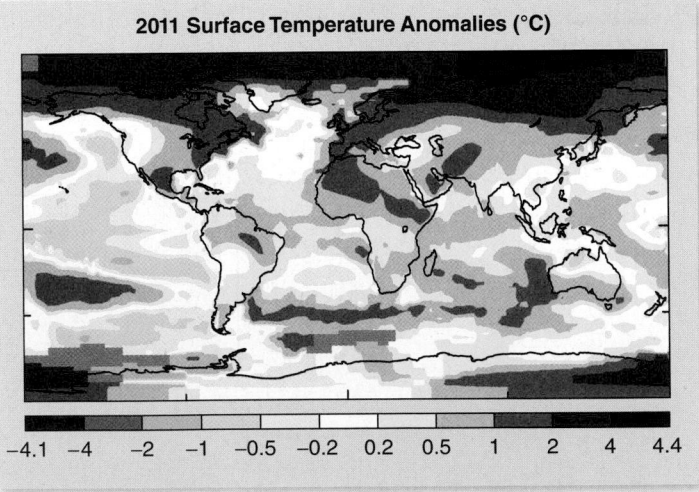

Figure 58.27 Geographic variation in global warming.
The 10 warmest years since record keeping began in 1880 all occur within the 12-year period 1998–2011, but some areas of the globe heated up more than others. Colors indicate how much warming occurred in 2011 relative to the mean temperature during a reference period (1951–1980) prior to full onset of the modern greenhouse effect.

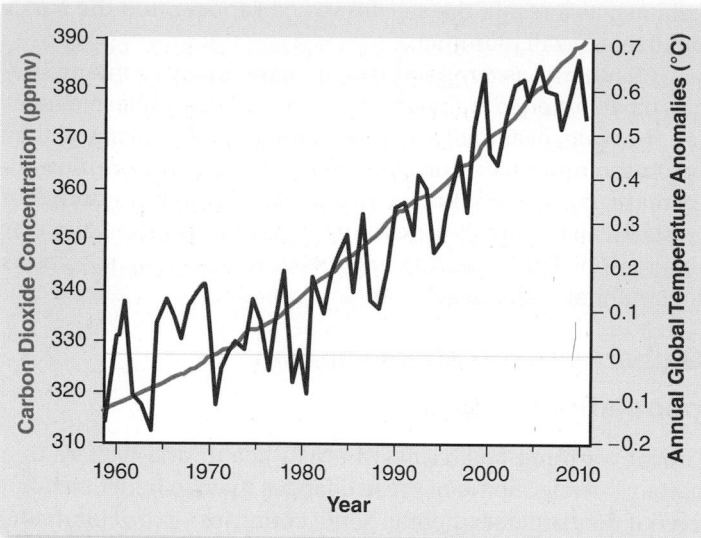

Figure 58.28 The greenhouse effect. The concentration of carbon dioxide in the atmosphere has increased steadily since the 1950s, as shown by the blue line. The red line shows the change in average global temperatures for the same period. The temperatures are recorded as anomalies relative to the mean temperature during a reference period (1951–1980).

atmospheric gases are also involved. A monitoring station on the top of the Mauna Loa volcano on the island of Hawaii has monitored the concentration of atmospheric CO_2 since the 1950s. This station is particularly important because it is in the middle of the Pacific Ocean, far from the great continental landmasses where most people live, and it is therefore able to monitor the state of the global atmosphere without confounding influences of local events.

In 1958, the atmosphere was 0.031% CO_2. By 2009, the concentration had risen to 0.039%. All authorities agree that the cause of this steady rise in atmospheric CO_2 is the burning of coal and petroleum products by the increasing (and increasingly energy-demanding) human population.

How carbon dioxide affects temperature

The atmospheric concentration of CO_2 affects global temperature because carbon dioxide strongly absorbs electromagnetic radiant energy at some of the wavelengths that are critical for the global heat budget. As stressed in chapter 57, the Earth not only receives radiant energy from the Sun, but also emits radiant energy into outer space. The Earth's temperature will be constant only if the rates of these two processes are equal.

The incoming solar energy is at relatively short wavelengths of the electromagnetic spectrum: wavelengths that are visible or near-visible. The outgoing energy from the Earth is at different, longer wavelengths. Carbon dioxide absorbs energy at certain of the important long-wave infrared wavelengths. This means that although carbon dioxide does not interfere with the arrival of radiant energy at short wavelengths, it retards the rate at which energy travels away from the Earth at long wavelengths into outer space.

Carbon dioxide is often called a greenhouse gas because its effects are analogous to those of a greenhouse. The reason that a glass greenhouse gets warm inside is that window glass is

transparent to light but only slightly transparent to long-wave infrared radiation. Energy that strikes a greenhouse as light enters the greenhouse freely. Once inside, the energy is absorbed as heat and then re-radiated as long-wave infrared radiation. The infrared radiation cannot easily get out through the glass, and therefore energy accumulates inside.

Other greenhouse gases

Carbon dioxide is not the only greenhouse gas. Others include methane and nitrous oxide. The effect of any particular greenhouse gas depends on its molecular properties and concentration. For example, molecule-for-molecule, methane has about 20 times the heat-trapping effect of carbon dioxide; on the other hand, methane is less concentrated and less long-lived in the atmosphere than carbon dioxide.

Methane is produced in globally significant quantities in anaerobic soils and in the fermentation reactions of ruminant mammals, such as cows. Huge amounts of methane are presently locked up in Arctic permafrost. Melting of the permafrost could cause a sudden and large perturbation in global temperature by releasing methane rapidly.

Agricultural use of fertilizers is the largest source of nitrous oxide emissions, with energy consumption second and industrial use third.

Global temperature change has affected ecosystems in the past and is doing so now

Evidence for warming can be seen in many ways. For example, on a worldwide statistical basis, ice on lakes and rivers forms later and melts sooner than it used to; on average, ice-free seasons are now 2.5 weeks longer than they were a century ago. Also, the extent of ice at the North Pole has decreased substantially, and glaciers are retreating around the world (figure 58.29).

Figure 58.29 Disappearing glaciers. Mount Kilimanjaro in Tanzania in 1970 (*top*) and 2000 (*bottom*). Note the decrease in glacier coverage over three decades.

Global warming—and cooling—have occurred in the past, most recently during the ice ages and intervening warm periods. Species often responded by shifting their geographic ranges, tracking their environments. For example, a number of cold-adapted North American tree species that are now found only in the far north, or at high elevations, lived much farther south or at substantially lower elevations 10,000–20,000 years ago, when conditions were colder. Present-day global warming is having similar effects. For example, many butterfly and bird species have shifted northward in recent decades (figure 58.30).

Many migratory birds arrive earlier at their summer breeding grounds than they did decades ago. Many insects and amphibians breed earlier in the year, and many plants flower earlier. In Australia, recent research shows that wild fruit fly populations have undergone changes in gene frequencies in the past 20 years, such that populations in cool parts of the country now genetically resemble ones in warm parts.

Reef-building corals seem to have narrow margins of safety between the sea temperatures to which they are accustomed and the maximum temperatures they can survive. Global warming seems already to be threatening some corals by inducing mass "bleaching," a disruption of the normal and necessary symbiosis between the cnidarians and algal cells.

There are reasons to think that the effects of global warming on natural ecosystems today may, overall, be more severe than those of warming events in the distant past. One concern is that the rate of warming today is rapid, and therefore evolutionary adaptations would need to occur over relatively few generations to aid the survival of species. Another concern is that natural areas no longer cover the whole landscape but often take the form of parks that are completely surrounded by cities or farms. The parks are at fixed geographic locations and in general cannot be moved. If climatic conditions in a park become unsuitable for its inhabitants, the species cannot shift their range because the park is surrounded by inhospitable

Figure 58.30 **Butterfly range shift.** The distribution of the speckled wood butterfly, *Pararge aegeria*, in Great Britain in 1970–1997 (green) included areas far to the north of the distribution in 1915–1939 (black).

habitats; as a result, the species may disappear and the park's biodiversity will plummet.

Similarly, as temperatures increase, many montane species have shifted to higher altitudes to find their preferred habitat. However, eventually they can shift no higher because they reach the mountain's peak. As the temperature continues to increase, the species' habitat disappears entirely. A number of Costa Rican frog species are thought to have become extinct for this reason. The same fate may befall many Arctic species as their habitat melts away.

Global warming affects human populations as well

Global warming could affect human health and welfare in a variety of ways. Some of these changes may be beneficial, but even if they are detrimental, some countries—particularly the wealthier ones—will be able to adjust. But poorer countries may not be able to transform as quickly, and some changes will require extremely costly countermeasures that even wealthy countries will be hard-pressed to afford.

Rising sea levels

During the second half of the 20th century, sea level rose at 2 to 3 cm per decade. The U.S. Environmental Protection Agency predicts that sea level is likely to rise two or three times faster in the 21st century because of two effects of global warming: (1) the melting of polar ice and glaciers, adding water to the ocean, and (2) the increase of average ocean temperature, causing an increase in volume because water expands as it warms. Such an increase would cause increased erosion and inundation of low-lying land and coastal marshes, and other habitats would also be imperiled. As many as 200 million people would be affected by increased flooding. Should sea levels continue to rise, coastal cities and some entire islands, such as the Maldives in the Indian Ocean, would be in danger of becoming submerged.

Other climatic effects

Global warming is predicted to have a variety of effects besides increased temperatures. In particular, the frequency or severity of extreme meteorological events—such as heat waves, droughts, severe storms, and hurricanes—is expected to increase, and El Niño events, with their attendant climatic effects, may become more common.

In addition, rainfall patterns are likely to shift, and those geographic areas that are already water-stressed, which are currently home to nearly 2 billion people, will likely face even graver water shortage problems in the years to come. Some evidence suggests that these effects are already evident in the increase in powerful storms, hurricanes, and the frequency of El Niño events over the past few years.

Effects on agriculture

Global warming may have both positive and negative effects on agriculture. On the positive side, warmer temperatures and increased atmospheric carbon dioxide tend to increase growth of some crops and thus may increase agricultural yields. Other crops, however, may be negatively affected. Furthermore, most crops will be affected by increased frequencies of droughts. Moreover, although crops in north temperate regions may flourish with higher temperatures, many tropical crops are already growing at their maximal temperatures, so increased temperatures may lead to reduced crop yields.

Also on the negative side, changes in rainfall patterns, temperature, pest distributions, and various other factors will require many adjustments. Such changes may come relatively easily for farmers in the developed world, but the associated costs may be devastating for those in the developing countries.

Effects on human health

Increasingly frequent storms, flooding, and drought will have adverse consequences on human health. Aside from their direct effect, such events often disrupt the fragile infrastructure of developing countries, leading to the loss of safe drinking water and other problems. As a result, epidemics of cholera and other diseases may be expected to occur more often.

In addition, as temperatures rise, areas suitable for tropical organisms will expand northward. Of particular concern are those organisms that cause human diseases. Many diseases currently limited to tropical areas may expand their range and become problematic in nontropical countries. Diseases transmitted by mosquitoes, such as malaria (see chapter 29), dengue fever, and several types of encephalitis, are examples. The distribution of mosquitoes is limited by cold; winter freezes kill many mosquitoes and their eggs. As a result, malaria only occurs in areas where temperatures are usually above 16°C, and yellow fever and dengue fever, transmitted by a different mosquito species from malaria, occur in areas where temperatures are normally above 10°C. Moreover, at higher temperatures, the malaria pathogen matures more rapidly.

Malaria already kills 1 million people every year; some projections suggest that the percentage of the human population at risk for malaria may increase by 33% by the end of the 21st century. Moreover, as predicted, malaria already appears to be on the move. By 1980, malaria had been eradicated from all of the United States except California, but in recent years it has appeared in a variety of southern, and even a few northern, states.

Dengue fever (sometimes called "breakbone fever" because of the pain it causes) is also spreading. Previously a disease restricted to the tropics and subtropics, where it infects 50 to 100 million people a year, it now occurs in the United States, southern South America, and northern Australia.

One of the most alarming aspects of these diseases is that no vaccines are available. Drug treatment is available (for malaria), but the parasites are rapidly evolving resistance and rendering the drugs ineffective. There is no drug treatment for dengue fever.

Solving the problem

The release of the IPCC's fourth assessment in 2007 may come to be seen as a turning point in humanity's response to climate change. Global warming is now recognized, even by former skeptics, as an ongoing phenomenon caused in large part by human actions. Even formerly recalcitrant governments now seem poised to take action, and corporations are recognizing the opportunities provided by the need to reverse human impacts. The resulting "green" technologies and practices are becoming increasingly common. With concerted efforts from citizens, corporations, and governments, the more serious consequences of global climate change hopefully can be averted, just as ozone depletion was reversed in the last century.

Learning Outcomes Review 58.6

Carbon dioxide is a significant greenhouse gas, meaning that it prevents heat from escaping the Earth so that temperatures rise. Global warming caused by changes in atmospheric composition—most notably CO_2 accumulation—may increase desertification and cause some habitats and species to disappear. Global warming may also melt ice caps and glaciers, altering coastlines as water levels rise. Violent weather events, disruption of water availability, and flooding of low-lying areas, as well as increased incidence of tropical diseases, may also occur.

■ **In what ways does global climate change pose different questions from those posed by ozone depletion?**

Chapter Review

58.1 Ecosystem Effects of Sun, Wind, and Water

Solar energy and the Earth's rotation affect atmospheric circulation.
The amount of solar radiation reaching the Earth's surface has a great effect on climate. The seasons result from changes in the Earth's position relative to the Sun (figure 58.1). Hot air with its increased water content rises at the equator, then cools and loses its moisture, creating the equatorial rainforests (figure 58.3).

As the drier cool air of the upper atmosphere moves away from the equator and then descends to Earth, it removes moisture from the Earth's surface and creates deserts on its way back to the equator.

Winds travel in curved paths relative to the Earth's surface because the Earth rotates on its axis (the Coriolis effect; see figure 58.3).

Global currents are largely driven by winds (figure 58.4).
Four large circular gyres in ocean currents can be found, driven by wind direction. These also are influenced by the Coriolis effect.

Regional and local differences affect terrestrial ecosystems.
A rain shadow occurs when a range of mountains removes moisture from air moving over it from the windward side, creating a drier environment on the opposite side (figure 58.5).

For every 1000-m increase in elevation, temperature drops approximately 6°C (figure 58.6).

Microclimates are small-scale differences in conditions.

58.2 Earth's Biomes

Temperature and moisture often determine biomes.

Average annual temperature and rainfall, as well as the range of seasonal variation, determine different biomes. Eight major types of biomes are recognized.

Tropical rainforests are highly productive equatorial systems.

Savannas are tropical grasslands with seasonal rainfall.

Deserts are regions with little rainfall.

Temperate grasslands have rich soils.

Temperate deciduous forests are adapted to seasonal change.

Temperate evergreen forests are coastal.

Taiga is the northern forest where winters are harsh.

Tundra is a largely frozen treeless area with a short growing season.

58.3 Freshwater Habitats

Life in freshwater habitats depends on oxygen availability.

Oxygen is not very soluble in water. Oxygen is constantly added by photosynthesis of aquatic plants and removed by heterotrophs.

Lake and pond habitats change with water depth.

The photic zone, near the surface, is the zone of primary productivity; its depth varies with water clarity (figure 58.12).

In the summer, the warmer water (epilimnion) floats on top of the colder water (hypolimnion). Freshwater lakes turn over twice a year as the temperature at the surface and at depth become the same, and the layers are set in motion by wind (figure 58.13).

Lakes differ in oxygen and nutrient content.

Oligotrophic lakes have high oxygen and low nutrients, whereas eutrophic lakes are the opposite.

58.4 Marine Habitats

The ocean is divided into several zones: intertidal, neritic, photic, benthic, and pelagic zones (figure 58.15).

Open oceans have low primary productivity.

Phytoplankton is the primary producer in open waters, and primary production is low due to low nutrient levels.

Continental shelf ecosystems provide abundant resources.

Neritic waters are found over continental shelves and have higher nutrient levels (figure 58.15). Estuaries frequently contain rich intertidal zones. Other ecosystems include productive banks on continental shelves and symbiotic coral reef ecosystems.

Upwelling regions experience mixing of nutrients and oxygen.

In upwelling regions, local winds bring up nutrient-rich deep waters, creating the highest rates of primary production. El Niño events occur when Trade Winds weaken, restricting upwelling to surface waters rather than to the deeper nutrient-rich waters.

The deep sea is a cold, dark place with some fascinating communities.

The deep sea is the single largest habitat. Hydrothermal vent communities occur where tectonic plates are moving apart; chemoautotrophs living there obtain energy from oxidation of sulfur.

58.5 Human Impacts on the Biosphere: Pollution and Resource Depletion

Dangerous chemicals like DDT are biomagnified as energy moves up the food chain (figure 58.20).

Freshwater habitats are threatened by pollution and resource depletion.

Point-source and diffuse pollution, acid precipitation, and overuse threaten freshwater habitats (figure 58.21).

Forest ecosystems are threatened in tropical and temperate regions.

Deforestation leads to loss of habitat, disruption of the water cycle, and loss of nutrients. Acid rain has a major detrimental effect on forests as well as on lakes and streams (figure 58.23).

Marine habitats are being depleted of fish and other species.

Many fisheries, such as the Georges Bank ecosystem, have collapsed and have not recovered.

Stratospheric ozone depletion has led to an ozone "hole."

Increased transmission of UV-B radiation is harmful to life. Global regulation of CFCs seems to be reversing ozone depletion.

58.6 Human Impacts on the Biosphere: Climate Change

Independent computer models predict global changes.

Carbon dioxide is a major greenhouse gas.

Carbon dioxide allows solar radiation to pass through the atmosphere but prevents heat from leaving the Earth, creating warmer conditions.

Global temperature change has affected ecosystems in the past and is doing so now.

If temperatures change rapidly, natural selection cannot occur rapidly enough to prevent many species from becoming extinct.

Global warming affects human populations as well.

Changing sea levels, increased frequency of extreme climatic events, direct and indirect effects on agriculture, and the expansion of tropical diseases can all affect human life.

UNDERSTAND

1. The Coriolis effect

   a. drives the rotation of the Earth.
   b. is responsible for the relative lack of seasonality at the equator.
   c. drives global wind circulation patterns.
   d. drives global wind and ocean circulation patterns.

2. What two factors are most important in biome distribution?

   a. Temperature and latitude
   b. Rainfall and temperature
   c. Latitude and rainfall
   d. Temperature and soil type

3. In a rain shadow, air is cooled as it rises and heated as it descends, often producing a wet and dry side because the water-holding capacity of the air

   a. is directly related to air temperature.
   b. is inversely related to air temperature.
   c. is unaffected by air temperature.
   d. produces changes in air temperature.

4. Thermal stratification in a lake

   a. is not modified by fall and spring overturn.
   b. leads to higher oxygen in deep versus surface waters.
   c. leads to higher oxygen in surface versus deep waters.
   d. is reduced when ice forms on the surface of the lake.

5. Oligotrophic lakes have

   a. low oxygen, and high nutrient availability.
   b. high oxygen, and high nutrient availability.
   c. high oxygen, and low nutrient availability.
   d. low oxygen, and low nutrient availability.

6. Deep-sea hydrothermal vent communities

   a. get their energy from photosynthesis in the photic zone near the surface.
   b. use bioluminescence to generate food.
   c. are built on the energy produced by the activity of chemoautotrophs that oxidize sulfur.
   d. contain only bacteria and other microorganisms.

7. Biological magnification occurs when

   a. pollutants increase in concentration in tissues at higher trophic levels.
   b. the effect of a pollutant is magnified by chemical interactions within organisms.
   c. an organism is placed under a dissecting scope.
   d. a pollutant has a greater than expected effect once ingested by an organism.

8. Which of the following is a point source of pollution?

   a. Lawns
   b. Smokestacks of coal-fired power plants
   c. Factory effluent pipe draining into a river
   d. Acid rain

APPLY

1. If the Earth were not tilted on its axis of rotation, the annual cycle of seasons in the northern and southern hemispheres

   a. would be reversed.   c. would be reduced.
   b. would stay the same.   d. would not exist.

2. Oligotrophic lakes can be turned into eutrophic lakes as a result of human activities such as

   a. overfishing of sensitive species, which disrupts fish communities.
   b. introducing nutrients into the water, which stimulates plant and algal growth.
   c. disrupting terrestrial vegetation near the shore, which causes soil to run into the lake.
   d. spraying pesticides into the water to control aquatic insect populations.

3. If a pesticide is harmless at low concentrations (such as DDT) and used properly, how can it become a threat to nontarget organisms?

   a. Because after exposure to DDT, some species develop allergic reactions even at low levels of exposure
   b. Because DDT molecules can combine so that their concentration increases through time
   c. Because the concentration of chemicals such as DDT is increasingly concentrated at higher trophic levels
   d. Because global warming and exposure to UV-B radiation renders molecules such as DDT increasingly potent

4. If there are many greenhouse gases, why is only carbon dioxide considered a cause of global warming?

   a. The other gases do not cause global warming.
   b. Scientists are concerned about other causes; for example, release of methane from melting permafrost could have significant effects on global warming.
   c. Other gases occur in such low quantities that they have little effect on the climate.
   d. Carbon dioxide is the only gas that absorbs long-wavelength infrared radiation.

SYNTHESIZE

1. Discuss how figure 58.1 explains the pattern observed in figure 58.2.

2. Why are most of the Earth's deserts found at approximately 30° latitude?

3. If the world has experienced global warming many times in the past, why should we be concerned about it happening again now?

ONLINE RESOURCE

www.ravenbiology.com

Understand, Apply, and Synthesize—enhance your study with animations that bring concepts to life and practice tests to assess your understanding. Your instructor may also recommend the interactive eBook, individualized learning tools, and more.

Chapter *59*

Conservation Biology

Chapter Contents

Introduction

Among the greatest challenges facing the biosphere is the accelerating pace of species extinctions. Not since the end of the Cretaceous period 65 MYA have so many species become extinct in so short a time span. This challenge has led to the emergence of the discipline of conservation biology. Conservation biology is an applied science that seeks to learn how to preserve species, communities, and ecosystems. It studies the causes of declines in species richness and attempts to develop methods for preventing such declines. In this chapter, we first examine the biodiversity crisis and its importance. Then, using case histories, we identify and study factors that have played key roles in many extinctions. We finish with a review of recovery efforts at the species and community levels.

59.1 Overview of the Biodiversity Crisis

Learning Outcomes

1. *Describe the history of extinction through time.*
2. *Explain the importance of hotspots to biodiversity conservation.*

Extinction is a fact of life. Most species—probably all—become extinct eventually. More than 99% of species known to science

(most from the fossil record) are now extinct. Current rates of extinction are alarmingly high, however. Taking into account the current rapid and accelerating loss of habitat, especially in the tropics, it has been calculated that as much as 20% of the world's biodiversity may disappear by the middle of this century. In addition, many of these species may disappear before we are even aware of their existence. Scientists estimate that no more than 15% of the world's eukaryotic organisms have been discovered and given scientific names, and this proportion is probably much lower for tropical species.

These losses will affect more than poorly known groups. As many as 50,000 species of the world's total of 250,000 species of plants, 4000 of the world's 20,000 species of butterflies,

Figure 59.1 North America before human inhabitants. Animals found in North America prior to the arrival of humans included birds and large mammals, such as the saber-toothed cat, giant ground sloth, and teratorn vulture. This painting by Charles R. Knight is titled, *Rancho La Brea Tar Pit.* It's displayed in the George C. Page Museum in Los Angeles.

and nearly 2000 of the world's 8600 species of birds could be lost during this period. Considering that the human species has been in existence for less than 200,000 years of the world's 4.5-billion-year history, and that our ancestors developed agriculture only about 10,000 years ago, this is an astonishing—and dubious—accomplishment.

Prehistoric humans were responsible for local extinctions

A great deal can be learned about current rates of extinction by studying the past. In prehistoric times, members of *Homo sapiens* wreaked havoc whenever they entered a new area. For example, at the end of the last Ice Age, approximately 12,000 years ago, the fauna of North America was composed of a diversity of large mammals similar to those living in Africa today: mammoths and mastodons, horses, camels, giant ground sloths, saber-toothed cats, and lions, among others (figure 59.1).

Shortly after humans arrived, 74% to 86% of the megafauna (that is, animals weighing more than 100 lb) became extinct. These extinctions are thought to have been caused by hunting and, indirectly, by burning and clearing of forests. (Some scientists attribute these extinctions to climate change, but that hypothesis doesn't explain why the ends of earlier Ice Ages were not associated with mass extinctions, nor does it explain why extinctions occurred primarily among larger animals, with smaller species being relatively unaffected.)

Around the globe, similar results have followed the arrival of humans. Forty thousand years ago, Australia was occupied by a wide variety of large animals, including marsupials similar in size and ecology to hippos and leopards, a kangaroo 9 ft tall, and a 20-ft-long monitor lizard. These all disappeared at approximately the same time as humans arrived.

Smaller islands have also been devastated. Madagascar has seen the extinction of at least 15 species of lemurs, including one the size of a gorilla; a pygmy hippopotamus; and the flightless elephant bird, *Aepyornis,* the largest bird to ever live

(more than 3 m tall and weighing 450 kg). On New Zealand, 30 species of birds went extinct, including all 13 species of moas, another group of large, flightless birds. Interestingly, one continent that seems to have been spared these megafaunal extinctions is Africa. Scientists speculate that this lack of extinction in prehistoric Africa may have resulted because much of human evolution occurred in Africa. Consequently, African species had been coevolving with humans for several million years and thus had evolved counteradaptations to human predation.

Extinctions have continued in historical time

Historical extinction rates are best known for birds and mammals because these species are conspicuous—that is, relatively large and well studied. Estimates of extinction rates for other species are much rougher. The data presented in table 59.1, based on the best available evidence, show recorded extinctions from 1600 to the present. These estimates indicate that about 85 species of mammals and 113 species of birds have become extinct since the year 1600. That is about 2.1% of known mammal species and 1.3% of known birds.

The majority of extinctions have occurred in the last 150 years: one species every year during the period from 1850 to 1950, and four species per year between 1986 and 1990. This increase in the rate of extinction is the heart of the biodiversity crisis.

Unfortunately, the situation is worsening. For example, the number of bird species recognized as "endangered" increased 13% from 1999 to 2009, and a recent report suggested that as many as half of Earth's plant species may be threatened with extinction. Some researchers predict that two-thirds of all vertebrate species could perish by the end of this century.

The majority of historic extinctions—though by no means all of them—have occurred on islands. For example, of the 85 species of mammals that have gone extinct in the last 400 years, 60% lived on islands. The particular vulnerability of island species probably results from a number of factors: Such species

TABLE 59.1

Taxon	RECORDED EXTINCTIONS				Approximate Number of Species	Percent of Taxon Extinct
	Mainland	Island	Ocean	Total		
Mammals	30	51	4	85	4,000	2.1
Birds	21	92	0	113	8,600	1.3
Reptiles	1	20	0	21	6,300	0.3
Fish	22	1	0	23	24,000	0.1
Invertebrates*	49	48	1	98	1,000,000+	0.01
Flowering plants	245	139	0	384	250,000	0.2

Recorded Extinctions Since 1600

*Number of extinct invertebrates is probably greatly underestimated due to lack of knowledge for many species (other groups are probably underestimated to a lesser extent for the same reason).

have often evolved in the absence of predators, and so have lost their ability to escape both humans and introduced predators such as rats and cats. In addition, humans have introduced competitors and diseases; malaria, for example, has devastated the bird fauna of the Hawaiian Islands. Finally, island populations are often relatively small, and thus particularly vulnerable to extinction, as we shall see later in this chapter.

In recent years, the extinction crisis has moved from islands to continents. Most species now threatened with extinction occur on continents, and these areas will bear the brunt of the extinction crisis in this century.

Some people have argued that we should not be concerned, because extinctions are a natural event and mass extinctions have occurred in the past. Indeed, mass extinctions have taken place several times over the past half-billion years (figure 59.2). However, the current mass extinction event is notable in several respects. First, it is the only such event triggered by a single species (us!). Moreover, although species diversity usually recovers after a few million years, this is a long time to deny our descendants the benefits and joys of biodiversity.

In addition, it is not clear that biodiversity will rebound this time. After previous mass extinctions, new species have evolved to utilize resources newly available due to extinctions of the species that previously used them. Today, however, such resources are unlikely to be available, because humans are destroying the habitats and taking the resources for their own use.

Endemic species hotspots are especially threatened

A species found naturally in only one geographic area and no place else is said to be endemic to that area. The area over which an endemic species is found may be very large. For example, the black cherry tree *(Prunus serotina)* is endemic to all of temperate North America. More typically, however, endemic species

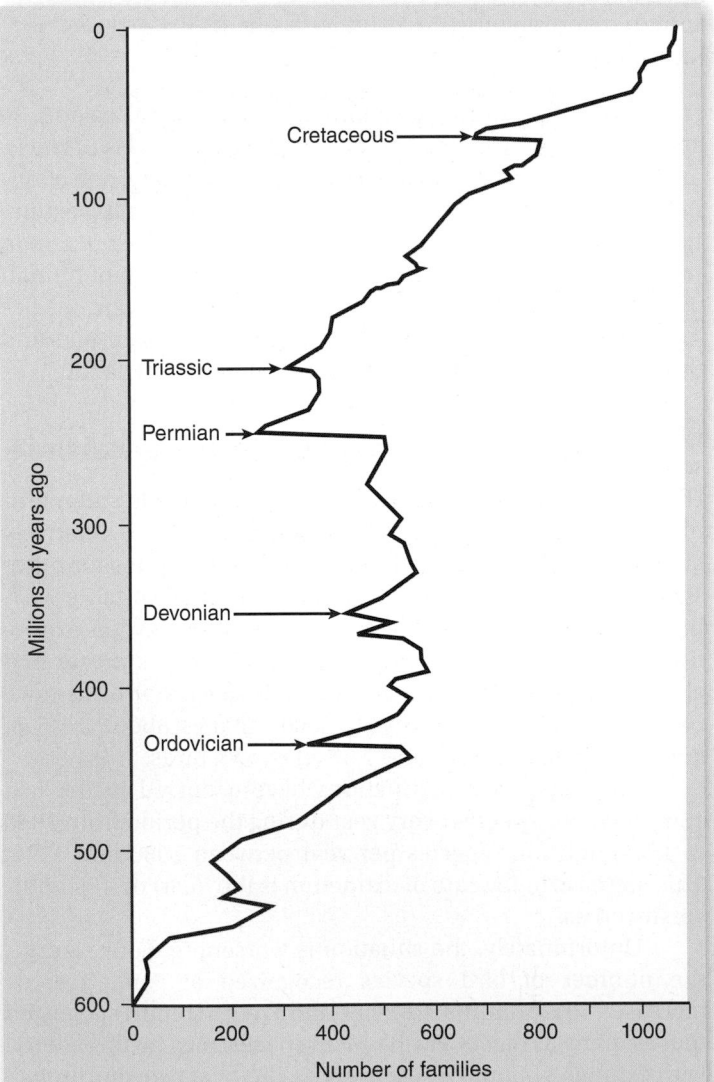

Figure 59.2 Mass extinctions measured by number of families of marine animals. Five major mass extinctions have been identified, with the most severe one occurring at the end of the Permian period when more than half of all plant and animal families and as much as 96% of all species may have perished. The most famous and well-studied extinction occurred at the end of the Cretaceous period (65 MYA), when a large asteroid slammed into Earth, causing fires and throwing particles into the air that obscured the Sun for months. This led to the extinction of the dinosaurs and a variety of other organisms.

Figure 59.3 Haleakala silversword (*Argyroxiphium sandwicense*). Many species of silverswords are endemic to very small areas. This photo illustrates two stages in the plant's life cycle.

occupy restricted ranges. The Komodo dragon (*Varanus komodoensis*) lives only on a few small islands in the Indonesian archipelago, and the Mauna Kea (*Argyroxiphium sandwicense*) and Haleakala silverswords (*A. s. macrocephalum*) each lives in a single volcano crater on the island of Hawaii (figure 59.3). Isolated geographic areas, such as oceanic islands, lakes, and mountain peaks, often have high percentages of endemic species, many in significant danger of extinction.

The number of endemic plant species can vary greatly from one place to another. In the United States, for example,

379 plant species are found in Texas and nowhere else, whereas New York has only one endemic plant species. California, with its varied array of habitats, including deserts, mountains, seacoast, old-growth forests, and grasslands, is home to more endemic plant species than any other state.

Species hotspots

Worldwide, notable concentrations of endemic species occur in particular regions. Conservationists have recently identified areas, termed hotspots, that have high endemism and are disappearing at a rapid rate. Such hotspots include Madagascar, a variety of tropical rainforests, the eastern Himalayas, areas with Mediterranean climates such as California, South Africa, and Australia, and several other areas (figure 59.4 and table 59.2). Overall, 25 such hotspots have been identified, which in total contain nearly half of all the terrestrial species in the world.

Why these areas contain so many endemic species is a topic of active scientific research. Some of these hotspots occur in areas of high species diversity; for these hotspots, the explanations for high species diversity in general, such as high productivity, probably apply (see chapter 57). In addition, some hotspots occur on isolated islands, such as New Zealand, New Caledonia, and the Hawaiian Islands, where evolutionary diversification over long periods has resulted in rich biotas composed of plant and animal species found nowhere else in the world.

Figure 59.4 Hotspots of high endemism. These areas are rich in endemic species under threat of imminent extinction.

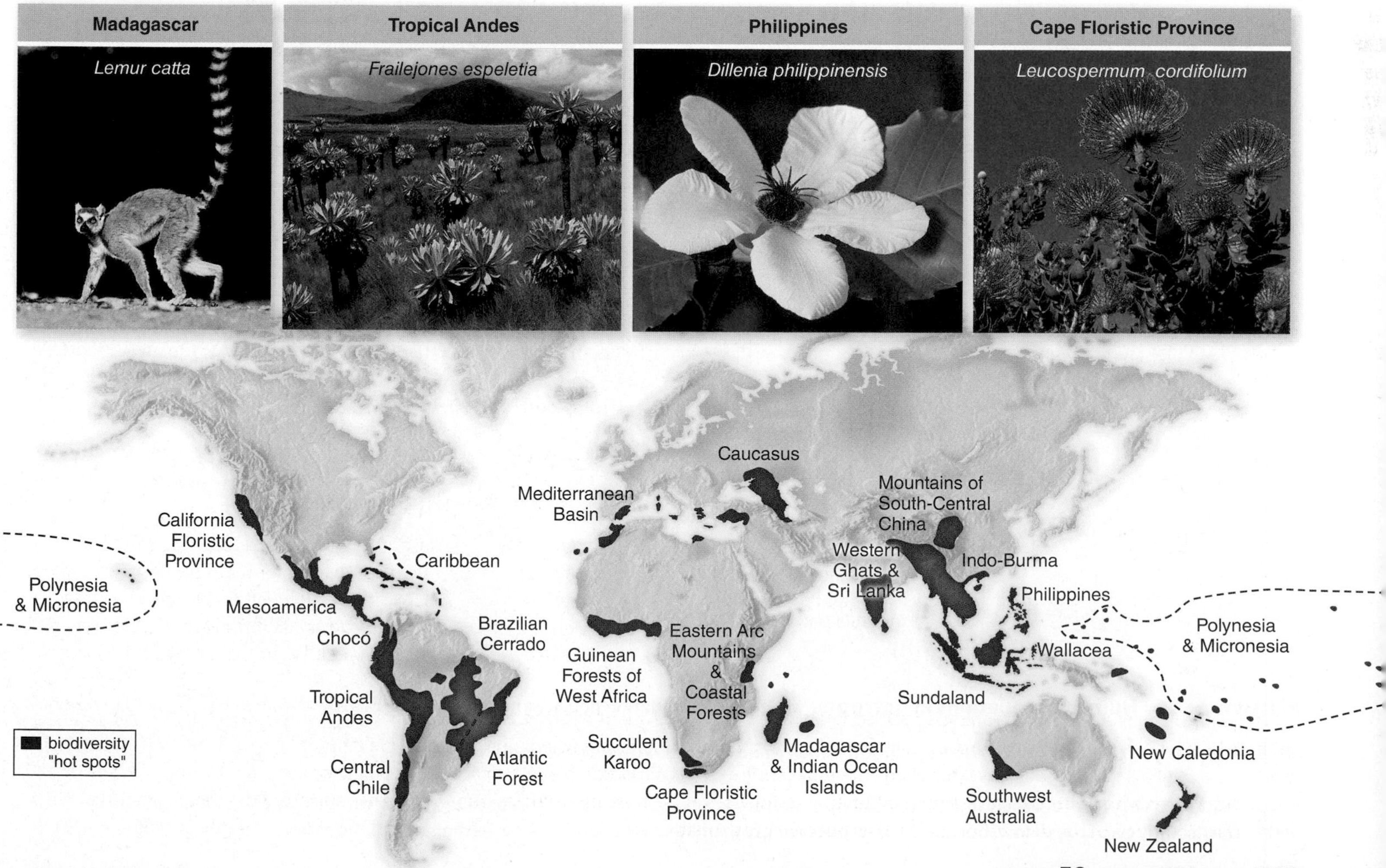

Madagascar	Tropical Andes	Philippines	Cape Floristic Province
Lemur catta	*Frailejones espeletia*	*Dillenia philippinensis*	*Leucospermum cordifolium*

■ biodiversity "hot spots"

Polynesia & Micronesia

California Floristic Province

Mesoamerica

Chocó

Tropical Andes

Central Chile

Caribbean

Brazilian Cerrado

Atlantic Forest

Mediterranean Basin

Guinean Forests of West Africa

Succulent Karoo

Cape Floristic Province

Caucasus

Eastern Arc Mountains & Coastal Forests

Madagascar & Indian Ocean Islands

Mountains of South-Central China

Western Ghats & Sri Lanka

Indo-Burma

Philippines

Wallacea

Sundaland

Southwest Australia

New Caledonia

New Zealand

Polynesia & Micronesia

TABLE 59.2	Numbers of Endemic Species in Some Hotspot Areas			
Region	Mammals	Reptiles	Amphibians	Plants
Atlantic coastal forest (Brazil)	160	60	253	6,000
South American Chocó	60	63	210	2,250
Philippines	115	159	65	5,832
Tropical Andes	68	218	604	20,000
Southwestern Australia	7	50	24	4,331
Madagascar	84	301	187	9,704
Cape region (South Africa)	9	19	19	5,682
California Floristic Province	30	16	17	2,125
New Caledonia	6	56	0	2,551
South-Central China	75	16	51	3,500

Human population growth in hotspots

Because of the great number of endemic species that hotspots contain, conserving their biological diversity must be an important component of efforts to safeguard the world's biological heritage. Or, to look at it another way, by protecting just 1.4% of the world's land surface, 44% of the world's vascular plants and 35% of its terrestrial vertebrates can be preserved.

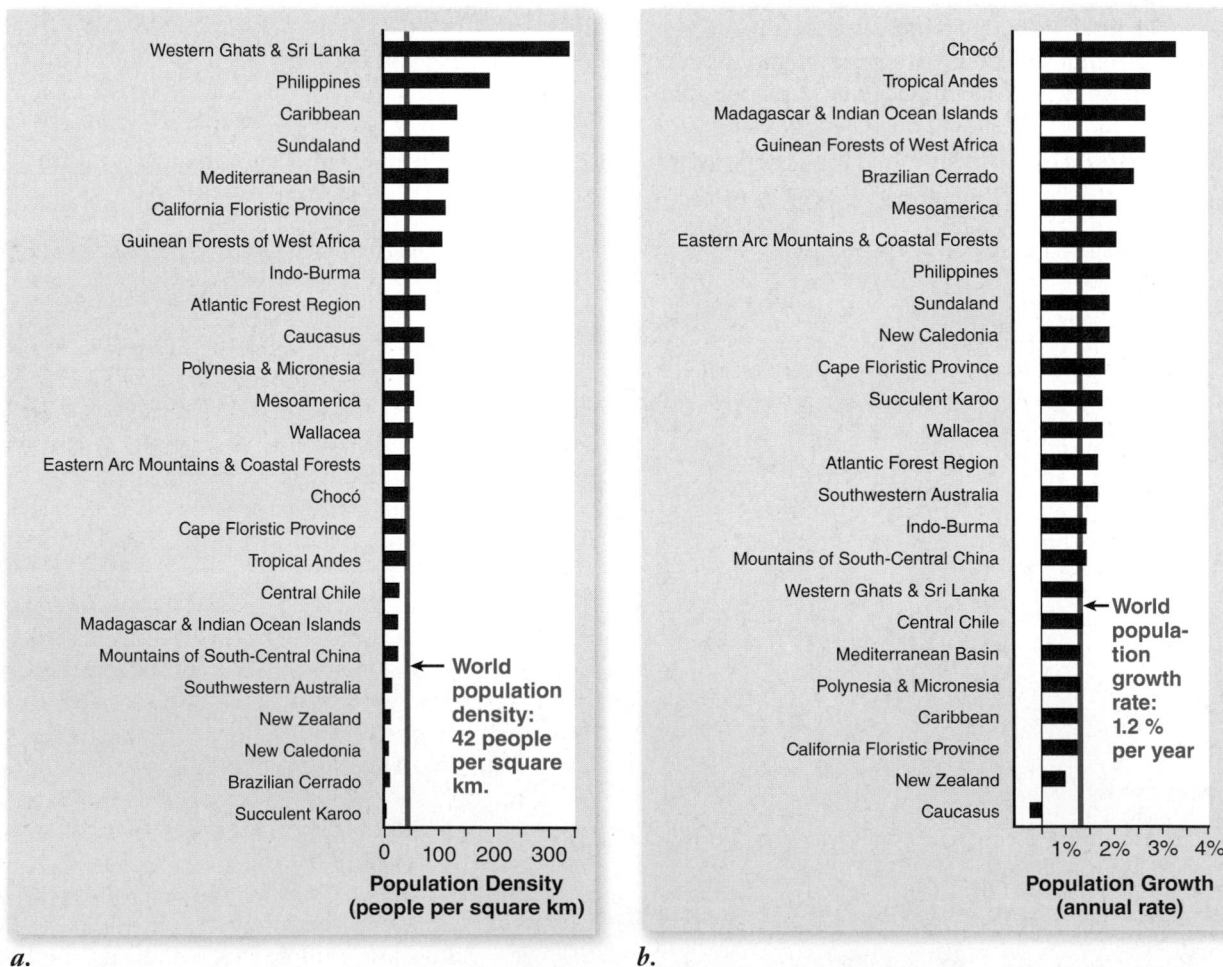

Figure 59.5 Human populations in hotspots. *a.* Human population density and *(b)* population growth rate in biodiversity hotspots.

? Inquiry question Why do population density and growth rates differ among hotspots?

🔍 Data analysis If we assume that population density is a good indicator of threat to the environment, which hotspots are not in particularly bad shape now, but are likely to become problematic in the future because their populations are currently growing quickly?

Unfortunately, hotspots contain not only many endemic species, but also growing human populations. In 1995, these areas contained 1.1 billion people—20% of the world's population—sometimes at high densities (figure 59.5a). More important, human populations were growing in all but one of these hotspots both because birth rates are much higher than death rates and because rates of immigration into these areas are often high. Overall, the rate of growth exceeded the global average in 19 hotspots (figure 59.5b). In some hotspots, the rate of growth is nearly twice that of the rest of the world.

Not surprisingly, many of these areas are experiencing high rates of habitat destruction as land is cleared for agriculture, housing, and economic development. More than 70% of the original area of each hotspot has already disappeared, and in 14 hotspots, 15% or less of the original habitat remains. In Madagascar, it is estimated that 90% of the original forest has already been lost—this on an island where 85% of the species are found nowhere else in the world. In the forests of the Atlantic coast of Brazil, the extent of deforestation is even higher: 95% of the original forest is gone.

Population pressure is not the only cause of habitat destruction in hotspots. Commercial exploitation to meet the demands of more affluent people in the developed world also plays an important role. For example, large-scale logging of tropical rainforests occurs in countries around the world to provide lumber, most of which ends up in the United States, western Europe, and Japan. Similarly, many forests in Central and South America are cleared to make way for cattle ranches that produce cheap meat for fast-food restaurants. Hotspots in more affluent countries are often at risk because they occur in areas where land has great value for real estate and commercial purposes, such as in Florida and California in the United States.

Learning Outcomes Review 59.1

The most recent losses of biodiversity have resulted from human activities, in both prehistoric and historical times. Endemic species are found in only a single region on Earth; regions with a high number of endemic species, known as hotspots, are particularly threatened by human encroachment and their preservation is critical.

■ *Why does so much resource exploitation occur in areas that are biodiversity hotspots?*

59.2 The Value of Biodiversity

Learning Outcomes

1. Distinguish between the direct and indirect economic value of biodiversity.
2. Explain what is meant by the aesthetic value of biodiversity.

Why should we worry about loss of biodiversity? The reason is that biodiversity is valuable to us in a number of ways:

■ *Direct economic value* of products we obtain from species of plants, animals, and other groups

a.   *b.*

Figure 59.6 **Plants of pharmaceutical importance.**
a. Two drugs extracted from the rosy periwinkle *(Catharanthus roseus),* vinblastine and vincristine, effectively treat common forms of childhood leukemia, increasing chances of survival from 20% to over 95%. *b.* Cancer-fighting drugs, such as taxol, have been developed from the bark of the Pacific yew *(Taxus brevifolia).*

■ *Indirect economic value* of benefits produced by species without our consuming them
■ *Ethical and aesthetic values*

The direct economic value of biodiversity includes resources for our survival

Many species have direct value as sources of food, medicine, clothing, biomass (for energy and other purposes), and shelter. Most of the world's food crops, for example, are derived from a small number of plants that were originally domesticated from wild plants in tropical and semiarid regions. As a result, many of our most important crops, such as corn, wheat, and rice, contain relatively little genetic variation (equivalent to a founder effect; see chapter 20), whereas their wild relatives have great diversity.

In the future, genetic variation from wild strains of these species may be needed if we are to improve yields or find a way to breed resistance to new pests. In fact, recent agricultural breeding experiments have illustrated the value of conserving wild relatives of common crops. For example, by breeding commercial varieties of tomato with a small, oddly colored wild tomato species from the mountains of Peru, scientists were able to increase crop yields by 50%, while increasing both nutritional content and color.

About 70% of the world's population depends directly on wild plants as their source of medicine. In addition, about 40% of the prescription and nonprescription drugs used today have active ingredients extracted from plants or animals. Aspirin, the world's most widely used drug, was first extracted from the leaves of the tropical willow, *Salix alba*. The rosy periwinkle from Madagascar has yielded potent drugs for combating childhood leukemia (figure 59.6), and drugs effective in treating several forms of cancer and other diseases have been produced from the Pacific yew. Overall, 62% of cancer drugs were developed from products derived from plants and animals.

Only recently have biologists perfected the techniques that make possible the transfer of genes from one species to another. We are just beginning to use genes obtained from other species to our advantage (see chapter 17). So-called "gene prospecting" of the genomes of plants and animals for useful genes has only begun. We have been able to examine only a minute proportion of the world's organisms to see whether any of their genes have useful properties for humans.

By conserving biodiversity, we maintain the option of finding useful benefits in the future. Unfortunately, many of the most promising species occur in habitats, such as tropical rainforests, that are being destroyed at an alarming rate.

Indirect economic value is derived from ecosystem services

Diverse biological communities are of vital importance to healthy ecosystems. They help maintain the chemical quality of natural water, buffer ecosystems against storms and drought, preserve soils and prevent loss of minerals and nutrients, moderate local and regional climate, absorb pollution, and promote the breakdown of organic wastes and the cycling of minerals.

In chapter 57, we discussed the evidence that the stability and productivity of ecosystems is related to species richness. By destroying biodiversity, we are creating conditions of instability and lessened productivity and promoting desertification, waterlogging, mineralization, and many other undesirable outcomes throughout the world.

The value of intact habitats

Economists have recently been able to compare the societal value, in monetary terms, of intact habitats compared with the value of destroying those habitats. Surprisingly, in most studies conducted so far, intact ecosystems are more valuable than the products derived by destroying them. In Thailand, as one example, coastal mangrove habitats are commonly cleared so that shrimp farms can be established. Although the shrimp produced are valuable, their value is vastly outweighed by the benefits in timber, charcoal production, offshore fisheries, and storm protection provided by the mangroves (figure 59.7a).

Similarly, intact tropical rainforest in Cameroon, West Africa, provides fruit and other forest materials. Clearing the forest for agriculture or palm plantations leads to stream-polluting erosion as well as increased flooding. Combining all the costs and benefits of the three options, maintaining intact forests has the highest economic value (figure 59.7b).

Case study: New York City watersheds

Probably the most famous example of the value of intact ecosystems is provided by the watersheds of New York City. Ninety percent of the water for the New York area's 8 million residents comes from the Catskill Mountains and the nearby headwaters of the Delaware River (figure 59.8). Water that runs off from over 4000 km² of rural, mountainous areas is collected into reservoirs and then transported by aqueduct more than 130 km to New York City at a rate of 4.9 billion liters per day.

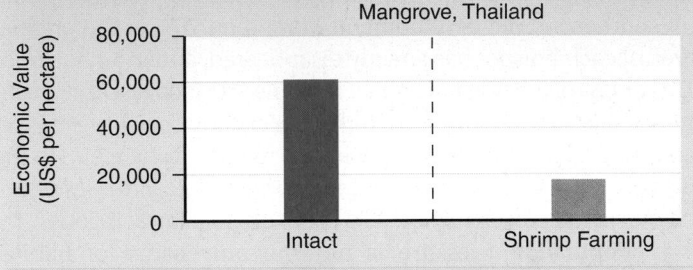

a.

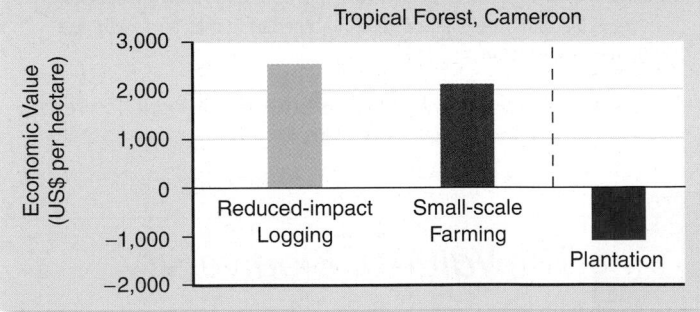

b.

Figure 59.7 The economic value of maintaining habitats. *a.* Mangroves in Thailand are more valuable than shrimp farms. *b.* Rainforests in Cameroon provide more economic benefits if they are left standing than if they are destroyed and the land used for other purposes.

 Data analysis If shrimp farms established on cleared mangrove habitats make money, how can clearing mangroves not be an economic plus?

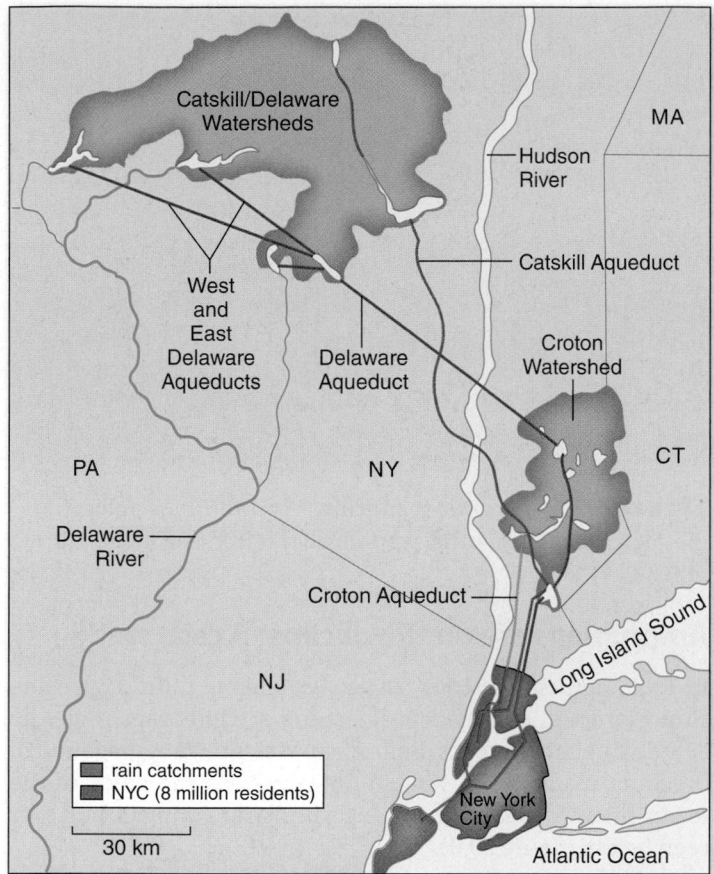

Figure 59.8 New York City's water source. New York gets its water from distant rain catchments. Preserving the ecological integrity of these areas is cheaper than building new water treatment plants.

In the 1990s, New York City faced a dilemma. New federal water regulations were requiring ever cleaner water, even as development and pollution in the source areas of the water were threatening to compromise water quality. The city had two choices: either work to protect the functioning ecosystem so that it could produce clean water, or construct filtration plants to clean it on arrival. Economic analysis made the choice clear: Building the plants would cost $6 billion, with annual operating costs of $300 million, whereas spending a billion dollars over 10 years could preserve the ecosystem and maintain water purity. The decision was easy.

Economic trade-offs

These examples provide some idea of the value of the services that ecosystems provide. But maintaining ecosystems is not always more valuable than converting them to other uses. Certainly, when the United States was being settled and land was plentiful, ecosystem conversion was beneficial. Even today, habitat destruction sometimes makes economic sense. Nonetheless, we still have only a rudimentary knowledge of the many ways intact ecosystems provide services. Often, it is not until they are lost that the value becomes clear, as unexpected negative effects, such as increased flooding and pollution, decreased rainfall, or vulnerability to hurricanes become apparent.

The same argument can be made for preserving particular species within ecosystems. Given how little we know about the biology of most species, particularly in the tropics, it is impossible to predict all the consequences of removing a species.

Imagine taking a parts list for an airplane and randomly changing a digit in one of the part numbers. You might change a seat cushion into a roll of toilet paper—or you might just as easily change a key bolt holding up a wing into a pencil. By removing biodiversity, we are gambling with the future of the ecosystems on which we depend and whose functioning we understand very little.

In recent years, the field of ecological economics has developed to study how the societal benefits provided by species and ecosystems can be appropriately valued. The problem is twofold. First, until recently, we have not had a good estimate of the monetary value of services provided by ecosystems, a situation which, as you've just seen, is now changing.

The second problem, however, is that the people who gain the benefits of environmental degradation are often not the same as the people who pay the costs. For instance, in the Thai mangrove example, the shrimp farmers reap the financial rewards, while the local people bear the costs. The same is true of factories that produce air or water pollution. Environmental economists are devising ways to appropriately value and regulate the use of the environment in ways that maximize the benefits relative to the costs to society as a whole.

Ethical and aesthetic values are based on our conscience and our consciousness

Many people believe that preserving biodiversity is an ethical issue because every species is of value in its own right, even if humans are not able to exploit or benefit from it. These people feel that along with the power to exploit and destroy other species comes responsibility: As the only organisms capable of eliminating large numbers of species and entire ecosystems, and as the only organisms capable of reflecting on what we are doing, humans should act as guardians or stewards for the diversity of life around us.

Almost no one would deny the aesthetic value of biodiversity—a wild mountain range, a beautiful flower, or a noble elephant—but how do we place a value on beauty or on the renewal many of us feel when we are in natural surroundings? Perhaps the best we can do is to consider the deep sense of loss we would feel if it no longer existed.

Learning Outcomes Review 59.2

The direct value of biodiversity includes resources for our survival, such as natural products and medicines that enhance our lives and can be used in a sustainable way. Indirect value includes economic benefits provided by healthy ecosystems, such as availability of clean water and recreational benefits. The aesthetic value of biodiversity refers to our sense of beauty and peace when experiencing a natural environment.

■ *What arguments could you use to convince shrimp farmers to stop operations and remediate the area they are using?*

59.3 Factors Responsible for Extinction

Learning Outcomes

1. List the major causes of species extinction.
2. Explain how these causes can interact to bring about extinction.

A variety of causes, independently or in concert, are responsible for extinctions (table 59.3). Historically, overexploitation was the major cause of extinction; although it is still a factor, habitat loss is the major problem for most groups today, and introduced species rank second. Many other factors can contribute to species extinctions as well, including disruption of ecosystem interactions, pollution, loss of genetic variation, and catastrophic disturbances, either natural or human-caused.

More than one of these factors may affect a species. In fact, a chain reaction is possible in which the action of one factor predisposes a species to be more severely affected by another factor. For example, habitat destruction may lead to decreased birth rates and increased mortality rates. As a result, populations become smaller and more fragmented, making them more vulnerable to disasters such as floods or forest fires, which may eliminate populations. Also, as the habitat becomes more fragmented, populations become isolated, so that genetic interchange ceases and areas devastated by disasters are not recolonized. Finally, as populations become very small, inbreeding increases, and genetic variation is lost through genetic drift, further decreasing population fitness. Which factor acts as the final coup de grace may be irrelevant; many factors, and the interactions between them, may have contributed to a species' eventual extinction.

Figure 59.9 An extinct species. A breeding assemblage of the golden toad (*Bufo periglenes*) which was last seen in the wild in 1989.

Amphibians are on the decline: A case study

In 1963, herpetologist Jay Savage was hiking through pristine cloud forest in Costa Rica. Reaching a windswept ridge, he couldn't believe his eyes. Before him was a huge aggregation of breeding toads. What was so amazing was the color of the toads: bright, eye-dazzling orange, unlike anything he had ever seen before (figure 59.9).

The color of the toads was so amazing and unexpected that Savage briefly considered the possibility that his colleagues had played a practical joke, getting to the clearing before him and somehow coloring normal toads orange. Realizing that this could not be, he went on to study the toads, eventually describing a species new to science, the golden toad, *Bufo periglenes.*

For the next 24 years, large numbers of toads were seen during the breeding season each spring. Their home was legally recognized as the Monteverde Cloud Forest Reserve,

TABLE 59.3	Causes of Extinctions				
	PERCENTAGE OF SPECIES INFLUENCED BY A GIVEN FACTOR*				
Group	**Habitat Loss**	**Overexploitation**	**Species Introduction**	**Other**	**Unknown**
E X T I N C T I O N S					
Mammals	19	23	20	2	36
Birds	20	11	22	2	45
Reptiles	5	32	42	0	21
Fish	35	4	30	4	48
T H R E A T E N E D E X T I N C T I O N S					
Mammals	68	54	6	20	—
Birds	58	30	28	2	—
Reptiles	53	63	17	9	—
Fish	78	12	28	2	—

*Some species may be influenced by more than one factor; thus, some rows may exceed 100%.

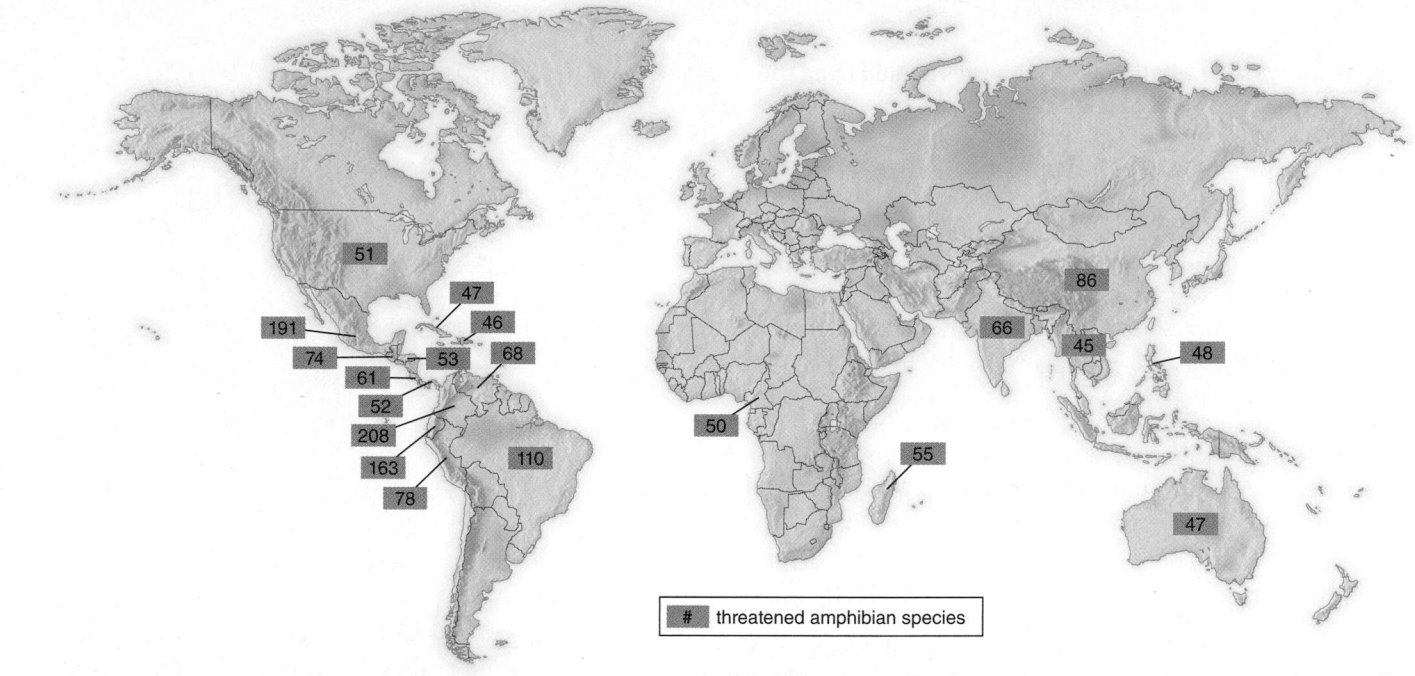

Venezuela	Panama	Madagascar	Australia
Dendrobates leucomelas	*Atelopus zeteki*	*Mantella aurantiaca*	*Litoria caerulea*

Figure 59.10 Amphibian extinction crisis. Boxes indicate the number of threatened species around the world in 2004. These numbers will undoubtedly be revised upward as scientists focus their attention on little-known species, many of which turn out to be in grave danger.

a well-protected, intact, and functioning ecosystem, seemingly a successful model of conservation. Then, in 1988, few toads were seen, and in 1989, only a single male was observed. Since then, despite exhaustive efforts, no more golden toads have been found.

Despite living in a well-protected ecosystem, with no obvious threats from pollution, introduced species, overexploitation, or any other factor, the species appears to have gone extinct, right under the eyes of watchful scientists and conservationists. How could this happen?

Frogs in trouble

At the first World Herpetological Congress in 1989 in Canterbury, England, frog experts from around the world met to discuss conservation issues relating to frogs and toads. At this meeting, it became clear that the golden toad story was not unique. Experts reported case after case of similar losses: Frog populations that had once been abundant were now decreasing or entirely gone.

Since then, scientists have devoted a great deal of time and effort to determining whether frogs and other amphibian species truly are in trouble and, if so, why. Unfortunately, the situation appears to be even worse than originally suspected. Amphibian experts recently reported that 43% of all amphibian

species have experienced decreases in population size, and one-third of all amphibian species are threatened with extinction in countries as different as Ecuador, Venezuela, Australia, and the United States (figure 59.10).

Moreover, these numbers are probably underestimates; little information exists from many areas of the world, such as Southeast Asia and central Africa. Indeed researchers think that as many as 100 species from the island nation of Sri Lanka have recently gone extinct, perhaps not surprisingly because 95% of that nation's rainforests have also disappeared in recent times.

Cause for concern

Amphibian declines are worrisome for several reasons. First, many of the species—including the golden toad—have declined in pristine, well-protected habitats. If species are becoming extinct in such areas, it brings into question our ability to preserve global biological diversity.

Second, many amphibian species are particularly sensitive to the state of the environment because of their moist skin, which allows chemicals from the environment to pass into the body, and their use of aquatic habitats for larval stages, which requires unpolluted water. In other words, amphibians may be analogous to the canaries formerly used in coal mines to detect problems with air quality: If the canaries keeled over, the miners knew they had to get out.

Third, no single cause for amphibian declines is apparent. Although a single cause would be of concern, it would also suggest that a coordinated global effort could reverse the trend, as happened with chlorofluorocarbons and decreasing ozone levels (see chapter 58). However, different species are afflicted by different problems, including habitat destruction, the effects of global warming, pollution, decreased stratospheric ozone levels, disease epidemics, and introduced species.

The implication is that the global environment is deteriorating in many different ways. Could amphibians be global "canaries," serving as indicators that the world's environment is in serious trouble?

 ## Habitat loss devastates species richness

As table 59.3 indicates, habitat loss is the most important cause of modern-day extinction. Given the tremendous amounts of ongoing destruction of all types of habitat, from rainforest to ocean floor, this should come as no surprise. Natural habitats may be adversely affected by humans in four ways:

1. **destruction,**
2. **pollution,**
3. **disruption,** and
4. **habitat fragmentation.**

In addition to these causes, global climate change, discussed in chapter 58, is an insidious threat that combines many of these factors. As climate changes, habitats will change—or disappear entirely, as is the case for polar bears (*Thalarctos maritimus),* which require ice floes on which to hunt their seal prey. Some studies estimate that as many as 30% of all species may be imperiled by global warming.

Destruction of habitat

A proportion of the habitat available to a particular species may simply be destroyed. This destruction is a common occurrence in the "clear-cut" harvesting of timber, in the burning of tropical forest to produce grazing land, and in urban and industrial development. Deforestation has been, and continues to be, by far the most pervasive form of habitat disruption (figure 59.11). Many tropical forests are being cut or burned at a rate of 1% or more per year.

To estimate the effect of reductions in habitat available to a species, biologists often use the well-established observation that larger areas support more species (see figure 57.22). Although this relationship varies according to geographic area, type of organism, and type of area, in general a 10-fold increase in area leads to approximately a doubling in the number of species. This relationship suggests, conversely, that if the area of a habitat is reduced by 90%, so that only 10% remains, then half of all species will be lost. Evidence for this hypothesis comes from a study in Finland of extinction rates of birds on habitat islands (that is, islands of a particular type of habitat surrounded by unsuitable habitat) where the population extinction rate was found to be inversely proportional to island size (figure 59.12).

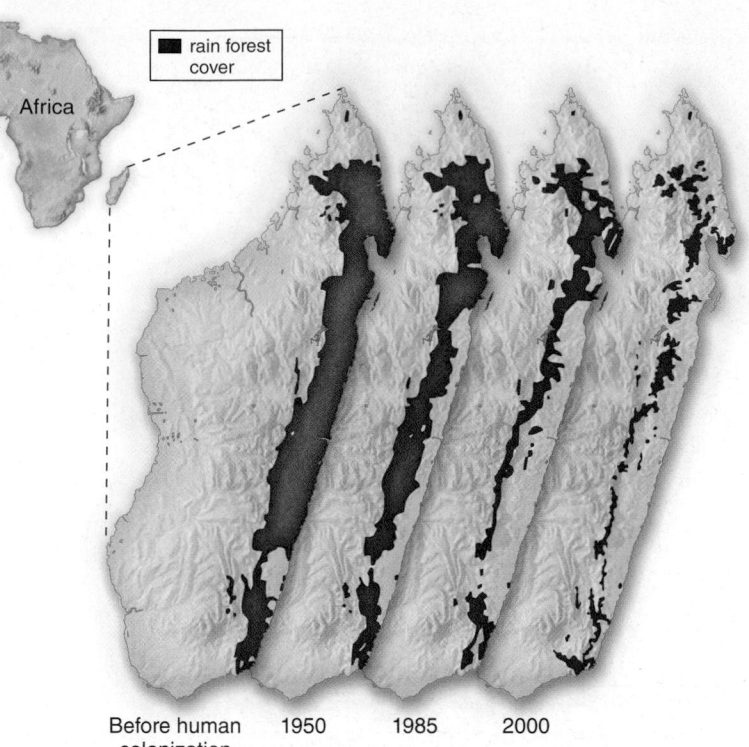

Figure 59.11 Extinction and habitat destruction. The rainforest covering the eastern coast of Madagascar, an island off the coast of East Africa, has been progressively destroyed and fragmented as the island's human population has grown. Ninety percent of the eastern coast's original forest cover is now gone. Many species have become extinct, and many others are threatened, including 16 of Madagascar's 31 primate species.

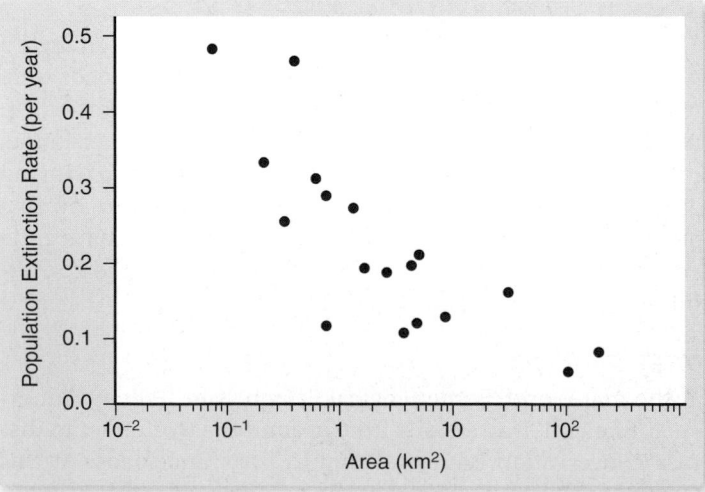

Figure 59.12 Extinction and island area. The data present percent extinction rates for populations as a function of habitat area for birds on a series of Finnish habitat islands. Smaller islands experience far greater extinction rates.

 Data analysis Would the extinction rate increase if an area were decreased in size by 90%? If so, by how much?

 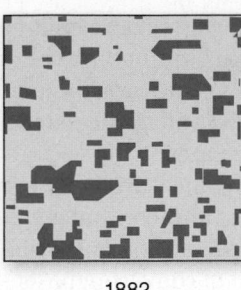 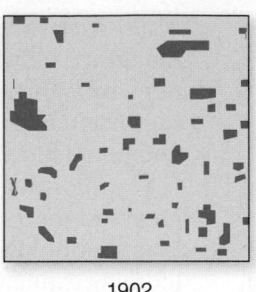 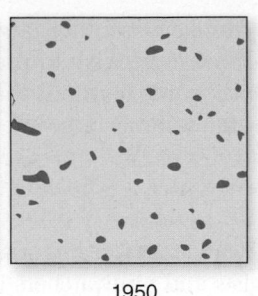

1831   1882   1902   1950

Figure 59.13 **Fragmentation of woodland habitat.** From the time of settlement of Cadiz Township, Wisconsin, the forest has been progressively reduced from a nearly continuous cover to isolated woodlots covering less than 1% of the original area.

Pollution

Habitat may be degraded by pollution to the extent that some species can no longer survive there. Degradation occurs as a result of many forms of pollution, from acid rain to pesticides. Aquatic environments are particularly vulnerable; for example, many northern lakes in both Europe and North America have been essentially sterilized by acid rain (see chapter 58).

Disruption

Human activities may disrupt a habitat enough to make it untenable for some species. For example, visitors to caves in Alabama and Tennessee caused significant population declines in bats over an 8-year period, some as great as 100%. When visits were fewer than one per month, less than 20% of bats were lost, but caves having more than four visits per month suffered population declines of 86% to 95%.

More generally, humans often alter the interactions that occur among species, such as the predator–prey or symbiotic relationships discussed in chapter 56. These disruptions can have far-ranging effects throughout an ecosystem. For example, when pollinating insects are killed off by insecticides, many plants do not reproduce, thus affecting all the animals that depend on the plants and their seeds for food.

Habitat fragmentation

Loss of habitat by a species frequently results not only in lowered population numbers, but also in fragmentation of the population into unconnected patches (figure 59.13). A habitat also may become fragmented in nonobvious ways, as when roads and habitation intrude into forest. The effect is to carve the populations living in the habitat into a series of smaller populations, often with disastrous consequences because of the relationship between range size and extinction rate. Although detailed data are not available, fragmentation of wildlife habitat in developed temperate areas is thought to be very substantial.

As habitats become fragmented and shrink in size, the relative proportion of the habitat that occurs on the boundary, or edge, increases. **Edge effects** can significantly degrade a population's chances of survival. Changes in microclimate (such as temperature, wind, humidity) near the edge may reduce appropriate habitat for many species more than the physical fragmentation suggests. In isolated fragments of rainforest, for example, trees on the edge are exposed to direct sunlight. As a result, these trees experience hotter and drier conditions than those normally encountered in the cool, moist forest interior, leading to negative effects on their survival and growth. In one study, the biomass of trees within 100 m of the forest edge decreased by 36% in the first 17 years after fragment isolation.

Also, increasing habitat edges opens up opportunities for some parasite and predator species that are more effective at edges. As fragments decrease in size, the proportion of habitat that is distant from any edge decreases, and consequently, more and more of the habitat is within the range of these species. Habitat fragmentation is blamed for local extinctions in a wide range of species.

The impact of habitat fragmentation can be seen clearly in a study conducted in Manaus, Brazil, where the rainforest was commercially logged. Landowners agreed to preserve patches of rainforest of various sizes, and censuses of these patches were taken before the logging started, while they were still part of a continuous forest. After logging, species began to disappear from the now-isolated patches (figure 59.14). First to go were the monkeys, which have large home ranges. Birds that prey on the insects flushed out by marching army ants followed, disappearing from patches too small to maintain enough army ant colonies to support them. As expected, the extinction rate was negatively related to patch size, but even the largest patches (100 hectares) lost half of their bird species in less than 15 years.

Figure 59.14 **A study of habitat fragmentation.** Landowners in Manaus, Brazil, agreed to preserve patches of rainforest of different sizes to examine the effect of patch size on species extinction. Biodiversity was monitored in the isolated patches before and after logging. Fragmentation led to significant species loss within patches. Army ants were one of the species that disappeared from smaller patches.

Because some species, such as monkeys, require large patches, large fragments are indispensable if we wish to preserve high levels of biodiversity. The take-home lesson is that preservation programs will need to provide suitably large habitat fragments to avoid this effect.

Case study: Songbird declines

Every year since 1966, the U.S. Fish and Wildlife Service has organized thousands of amateur ornithologists and birdwatchers in an annual bird count called the Breeding Bird Survey. In recent years, a shocking trend has emerged. Although year-round residents that prosper around humans, such as robins, starlings, and blackbirds, have increased their numbers and distribution over the last 45 years, forest songbirds have declined severely. The decline has been greatest among long-distance migrants such as thrushes, orioles, tanagers, vireos, buntings, and warblers. These birds nest in northern forests in the summer, but spend their winters in South or Central America or the Caribbean Islands.

In many areas of the eastern United States, more than three-quarters of the tropical migrant bird species have declined significantly. Rock Creek Park in Washington, D.C., for example, has lost 90% of its long-distance migrants in the past 20 years. Nationwide, American redstarts declined about 50% in the single decade of the 1970s. Studies of radar images from National Weather Service stations in Texas and Louisiana indicate that only about half as many birds fly over the Gulf of Mexico each spring as did in the 1960s. This suggests a total loss of about half a billion birds.

The culprit responsible for this widespread decline appears to be habitat fragmentation and loss. Fragmentation of breeding habitat and nesting failures in the summer nesting grounds of the United States and Canada have had a major negative effect on the breeding of woodland songbirds. Many of the most threatened species are adapted to deep woods and need an area of 25 acres or more per pair to breed and raise their young. As woodlands are broken up by roads and developments, it is becoming increasingly difficult for them to find enough contiguous woods to nest successfully.

A second and perhaps even more important factor is the availability of critical winter habitat in Central and South America. Studies of the American redstart clearly indicate that birds with better winter habitat have a superior chance of successfully migrating back to their breeding grounds in the spring. In a recent study, scientists were able to determine the quality of the habitat that particular birds used during the winter by examining levels of the stable carbon isotope ^{13}C in their blood. Plants growing in the best habitats in Jamaica and Honduras (mangroves and wetland forests) have low levels of ^{13}C, and so do the redstarts that feed on the insects that live in them. Of these wet-forest birds, 65% maintained or gained weight over the winter.

By contrast, plants growing in substandard dry scrub have high levels of ^{13}C, and so do the redstarts that feed in those habitats. Scrub-dwelling birds lost up to 11% of their body mass over the winter. Now here's the key: Birds that winter in the substandard scrub leave later in the spring on the long flight to northern breeding grounds, arrive later at their summer homes, and have fewer young (figure 59.15).

The proportion of ^{13}C in birds arriving in New Hampshire breeding grounds increases as spring wears on and scrub-overwintering stragglers belatedly arrive. Thus, loss of mangrove habitat in the neotropics is having a quantifiable negative influence. As the best habitat disappears, overwintering birds fare poorly, and this leads to decreased reproduction and population declines.

Unfortunately, the Caribbean lost about 10% of its mangroves in the 1980s, and continues to lose about 1% per year. This loss of key habitat appears to be a driving force in the looming extinction of some songbirds.

Overexploitation wipes out species quickly

Species that are hunted or harvested by humans have historically been at grave risk of extinction, even when the species is initially very abundant. A century ago, the skies of North America were darkened by huge flocks of passenger pigeons, but after being hunted as free and tasty food, they were driven to extinction. The bison that used to migrate in enormous herds across the central plains of North America only narrowly escaped the same fate.

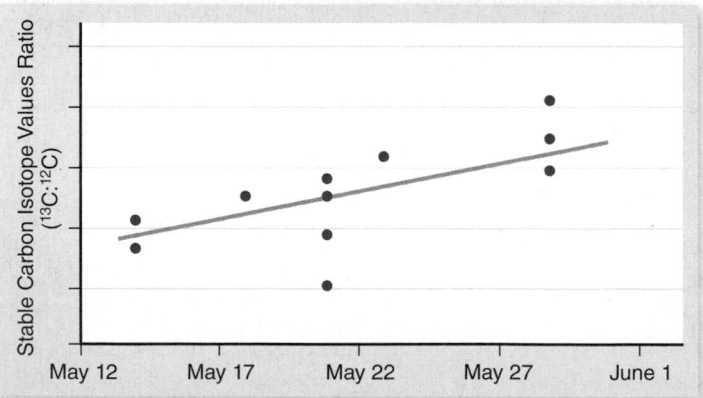

Figure 59.15 The American redstart (*Setophaga ruticilla*) a migratory songbird. The numbers of this species are in serious decline. The graph presents data on the ratio of ^{13}C to ^{12}C in male redstarts arriving at summer breeding grounds. Early arrivals, which have higher reproductive success, have lower proportions of ^{13}C to ^{12}C, indicating they wintered in more favorable mangrove–wetland forest habitats.

 Data analysis Based on the information in the text, what would the relationship between arrival date and reproductive success look like?

Commercial motivation for exploitation

The existence of a commercial market often leads to overexploitation of a species. The international trade in furs, for example, has severely reduced the numbers of chinchilla, vicuña, otter, and many cat species. The harvesting of commercially valuable trees provides another example: Almost all West Indies mahogany trees *(Swietenia mahogani)* have been logged, and the extensive cedar forests of Lebanon, once widespread at high elevations, now survive in only a few isolated groves.

A particularly telling example of overexploitation is the commercial harvesting of fish in the North Atlantic. During the 1980s, fishing fleets continued to harvest large amounts of cod off the coast of Newfoundland, even as the population numbers declined precipitously. By 1992, the cod population had dropped to less than 1% of its original numbers. The American and Canadian governments have closed the fishery, but no one can predict whether the fish populations will recover. The Atlantic bluefin tuna has experienced a 90% population decline in the past 10 years and swordfish has declined even further. In both cases, the drop has led to even more intense fishing of the remaining populations.

Case study: Whales

Whales, the largest living animals that ever evolved, are rare in the world's oceans today, their numbers driven down by commercial whaling. Before the advent of cheap, high-grade oils manufactured from petroleum in the early 20th century, oil made from whale blubber was an important commercial product in the worldwide marketplace. In addition, the fine, lattice-like structure termed "baleen" used by baleen whales to filter-feed plankton from seawater was used in women's undergarments. Because a whale is such a large animal, each individual captured is of significant commercial value.

In the 18th century, right whales were the first to bear the brunt of commercial whaling. They were called "right" whales because they were slow, easy to capture, did not sink when killed and provided up to 150 barrels of blubber oil and abundant baleen, making them the right whale for a commercial whaler to hunt.

As the right whale declined, whalers turned to the gray, humpback, and bowhead whales. As their numbers declined, whalers turned to the blue, the largest of all whales, and when those were decimated, to the fin, then the Sei, and then the sperm whales. As each species of whale became the focus of commercial whaling, its numbers began a steep decline (figure 59.16).

Hunting of right whales was made illegal in 1935. By then, they had been driven to the brink of extinction, their numbers less than 5% of what they had been. Although protected ever since, their numbers have not recovered in either the North Atlantic or the North Pacific. By 1946, several other whale species faced imminent extinction, and whaling nations formed the International Whaling Commission (IWC) to regulate commercial whale hunting. Like a fox guarding the henhouse, the IWC for decades did little to limit whale harvests, and whale numbers continued to decline steeply.

Finally, in 1974, when the numbers of all but the small minke whales had been driven down, the IWC banned hunting of blue, gray, and humpback whales, and instituted partial bans on other species. The rule was violated so often, however, that the IWC in 1986 instituted a worldwide moratorium on all commercial killing of whales. Although some commercial whaling continues, often under the guise of harvesting for scientific studies, annual whale harvests have dropped dramatically in the last 20 years.

Some species appear to be recovering, but others are not. Humpback numbers have more than doubled since the early 1960s, increasing nearly 10% annually, and Pacific gray whales have fully recovered to their previous numbers of about 20,000 animals, after having been hunted to fewer than 1000. Right, sperm, fin, and blue whales have not recovered, and no one knows whether they ever will.

Introduced species threaten native species and habitats

Colonization, a natural process by which a species expands its geographic range, occurs in many ways: A flock of birds gets blown off course, a bird eats a fruit and defecates its seed miles away, or lowered sea levels connect two previously isolated landmasses, allowing species to freely move back and forth. Such events—particularly those leading to successful establishment of a new population—probably occur rarely, but when they do, the resulting change to natural communities can be large. The reason is that colonization brings together species with no history of interaction. Consequently, ecological interactions may be particularly strong because the species have not evolved ways of adjusting to the presence of one

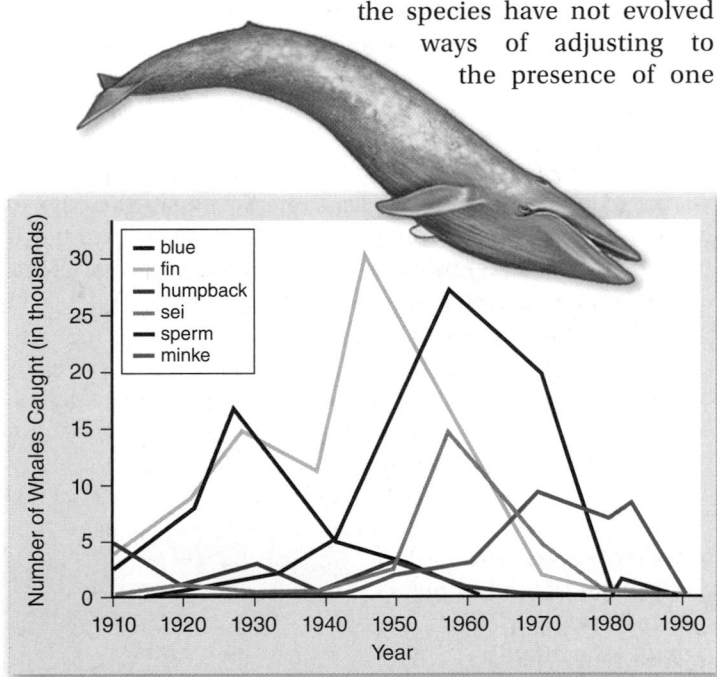

Figure 59.16 World catch of some whale species in the 20th century. Each species is hunted in turn until its numbers fall so low that hunting it becomes commercially unprofitable.

? Inquiry question Why might whale populations fail to recover once hunting is stopped?

Q Data analysis Only two species of whales were heavily hunted at the start of the 20th century. Does this graph provide an explanation for the timing of increased hunting of the other species?

Figure 59.17 Zebra mussels (*Dreissena polymorpha*) clogging a pipe. These mussels were introduced from Europe, and are now a major problem in North American rivers.

another, such as adaptations to avoid predation or to minimize competitive effects.

The paleontological record documents many cases in which geologic changes brought previously isolated species together, such as when the Isthmus of Panama emerged above the sea approximately 3 MYA, connecting the previously isolated fauna and flora of North and South America. In some cases, the result has been an increase in species diversity, but in other cases, invading species have led to the extinction of natives.

Human influence on colonization

Unfortunately, what was naturally a rare process has become all too common in recent years, thanks to the actions of humans. Species introductions due to human activities occur in many ways, sometimes intentionally, but usually not. Plants and animals can be transported in the ballast of large ocean vessels; in nursery plants; as stowaways in boats, cars, and planes; as beetle larvae within wood products—even as seeds or spores in the mud stuck to the bottom of a shoe. Overall, some researchers estimate that as many as 50,000 species have been introduced into the United States.

The effects of introductions on humans have been enormous. In the United States alone, nonnative species cost the economy an estimated $140 billion per year. For example, dozens of foreign weeds in Colorado have covered more than a million acres. Just three of these species cost wheat farmers tens of millions of dollars. At the same time, leafy spurge, a plant from Europe, outcompetes native grasses, ruining rangeland for cattle at a price tag of $144 million per year.

The zebra mussel, a mollusk native to the Black Sea region, is a huge problem throughout much of the eastern and central United States, where it can attain densities as high as 700,000/m², clogging pipes, including those for water and power plants, and causing an estimated $3 to $5 billion damage a year (figure 59.17).

Introduced species can also affect human health. For example, West Nile fever was probably introduced from Africa or the Middle East to the United States in the late 1990s.

The effect of species introductions on native ecosystems is equally dramatic. Islands have been particularly affected. For example, as mentioned in chapter 56, a single lighthouse keeper's cat wiped out an entire species, the Stephens Island wren. Rats had a devastating effect throughout the South Pacific where bird species nested on the ground and had no defense against the voracious predators to which they were evolutionarily naive. More recently, the brown tree snake, introduced to the island of Guam, essentially eliminated all species of forest birds.

In Hawaii, the problem has been slightly different: Introduced mosquitoes brought with them malaria, to which the native species had evolved no resistance. The result is that more than 100 species (>70% of the native fauna) either became extinct or are now restricted to higher and cooler elevations where the mosquitoes don't occur (figure 59.18).

The effects of introduced species may reverberate throughout an ecosystem. For example, the Argentine ant has spread through much of the southern United States, greatly reducing populations of most native ant species with which it comes in contact. The extinction of these ant species has had a dramatic negative effect on the coast horned lizard *(Phrynosoma coronatum)*, which feeds on the larger native species. In their absence, the lizards have shifted to less-preferred prey species. In addition, the native ant species consume seeds, and

Figure 59.18 The akiapolaau (*Hemignathus munroi*) and the palila (*Loxioides bailleui*), endangered Hawaiian birds. More than two-thirds of Hawaii's native bird species are now extinct or have been greatly reduced in population size. Bird faunas on islands around the world have experienced similar declines after human arrival.

in the process, play an important role in seed dispersal. Argentine ants, by contrast, do not eat seeds. In South Africa, where the Argentine ant has also appeared, at least one plant species has experienced decreased reproductive success due to the loss of its dispersal agent.

The most dramatic effects of introduced species, however, occur when entire ecosystems are transformed. Some plant species can completely overrun a habitat, displacing all native species and turning the area into a monoculture (that is, an area occupied by a single species). In California, the yellow star thistle (*Centaurea solstitialis*) now covers 4 million hectares of what was once highly productive grassland. In Hawaii, a small tree native to the Canary Islands, *Myrica faya*, has spread widely. Because it is able to fix nitrogen at high rates, it has caused a 90-fold increase in the nitrogen content of the soil, thus allowing other, nitrogen-requiring species to invade.

Efforts to combat introduced species

Once an introduced species becomes established, eradicating it is often extremely difficult, expensive, and time-consuming. Some efforts—such as the removal of goats and rabbits from certain small islands—have been successful, but many other efforts have failed. The best hope for stopping the ravages of introduced species is to prevent them from being introduced in the first place. Although easier said than done, government agencies are now working strenuously to put into place procedures that can intercept species in transit, before they have the opportunity to become established.

Case study: Lake Victoria cichlids

Lake Victoria, an immense, shallow, freshwater sea about the size of Switzerland in the heart of equatorial East Africa, used to be home to an incredibly diverse collection of over 450 species of cichlid fishes (see figure 22.16). These small, perchlike fish range from 5 to 13 cm in length, with males having endless varieties of color. Today, most of these cichlid species are threatened, endangered, or extinct.

What happened to bring about the abrupt loss of so many endemic cichlid species? In 1954, the Nile perch, a commercial fish with a voracious appetite, was purposely introduced on the Ugandan shore of Lake Victoria. Nile perch, which grow to over a meter in length, were to form the basis of a new fishing industry (figure 59.19). For decades, these perch did not seem to have a significant effect; over 30 years later, in 1978, Nile perch still made up less than 2% of the fish harvested from the lake.

Then something happened to cause the Nile perch population to explode and to spread rapidly through the lake, eating their way through the cichlids. By 1986, Nile perch constituted nearly 80% of the total catch of fish from the lake and the endemic cichlid species were virtually gone. Over 70% of cichlid species disappeared, including all open-water species.

So what happened to kick-start the mass extinction of the cichlids? The trigger seems to have been eutrophication. Before 1978, Lake Victoria had high oxygen levels at all depths, down to the bottom layers more than 60 m deep. However, by 1989 high inputs of nutrients from agricultural runoff and sewage from towns and villages had led to algal blooms that severely depleted oxygen levels in deeper parts of the lake. Cichlids feed on algae,

Figure 59.19 Nile perch (Lates niloticus). This predatory fish, which can reach a length of 2 m and a mass of 200 kg, was introduced into Lake Victoria as a potential food source. It is responsible for the virtual extinction of hundreds of species of cichlid fishes.

and initially their population numbers are thought to have risen in response to this increase in their food supply, but unlike the conditions during similar algal blooms of the past, the Nile perch was present to take advantage of the situation. With a sudden increase in its food supply (cichlids), the numbers of Nile perch exploded, and they simply ate all available cichlids.

Since 1990, the situation has been compounded by the introduction into Lake Victoria of a floating water weed from South America, the water hyacinth *Eichhornia crassipes*. Reproducing quickly under eutrophic conditions, thick mats of water hyacinth soon covered entire bays and inlets, choking off the coastal habitats of non-open-water cichlids.

Disruption of ecosystems can cause an extinction cascade

Species often become vulnerable to extinction when their web of ecological interactions becomes seriously disrupted. Because of the many relationships linking species in an ecosystem (see chapter 57), human activities that affect one species can have ramifications throughout an ecosystem, ultimately affecting many other species.

A recent case in point involves the sea otters that live in the cold waters off Alaska and the Aleutian Islands. Otter populations have declined sharply in recent years. In a 500-mile stretch of coastline, otter numbers have dropped from 53,000 in the 1970s to an estimated 6000, a plunge of nearly 90%. Investigating this catastrophic decline, marine ecologists uncovered a chain of interactions among the species of the ocean and kelp forest ecosystems, a falling-domino series of lethal effects that illustrates the concepts of both top-down and bottom-up trophic cascades discussed in chapter 57.

Case study: Alaskan near-shore habitat

The first in a series of events leading to the sea otter's decline seems to have been the heavy commercial harvesting of whales, described earlier in this chapter. Without whales to keep their numbers in check, ocean zooplankton thrived, leading in turn to proliferation of a species of fish called pollock that feeds on the abundant zooplankton. Given this ample food supply, the pollock proved to compete very successfully with other northern Pacific fish, such as herring and ocean perch, so that levels of these other fish fell steeply in the 1970s.

Then the falling chain of dominoes began to accelerate. The decline in the nutritious forage fish led to an ensuing crash in Alaskan populations of sea lions and harbor seals, for which pollock did not provide sufficient nourishment. This decline may also have been hastened by orcas (also called killer whales) switching from feeding on the less-available whales to feeding on seals and sea lions; the numbers of these pinniped species have fallen precipitously since the 1970s.

When pinniped numbers crashed, some orcas, faced with a food shortage, turned to the next best thing: sea otters. In one bay where the entrance from the sea was too narrow and shallow for orcas to enter, only 12% of the sea otters have disappeared, while in a similar bay that orcas could enter easily, two-thirds of the otters disappeared in a year's time.

Without otters to eat them, the population of sea urchins exploded, eating the kelp and thus "deforesting" the kelp forests and denuding the ecosystem (figure 59.20). As a result, fish species that live in the kelp forest, such as sculpins and greenlings, are declining.

Loss of keystone species

As discussed in chapter 56, a keystone species is a species that exerts a greater influence on the structure and functioning of an ecosystem than might be expected solely on the basis of its abundance. The sea otters of figure 59.20 are a keystone species of the kelp forest ecosystem, and their removal can have disastrous consequences.

No hard-and-fast line allows us to clearly identify keystone species. Rather, it is a qualitative concept, a statement that indicates a species plays a particularly important role in its community. Keystone species are usually characterized by the strength of their effect on their community.

Case study: Flying foxes

The severe decline of many species of "flying foxes," a type of bat (figure 59.21), in the Old World tropics is an example of how the loss of a keystone species can dramatically affect the other species living within an ecosystem, sometimes even leading to a cascade of further extinctions.

These bats have very close relationships with important plant species on the islands of the Pacific and Indian Oceans. The family Pteropodidae contains nearly 200 species, approximately one-quarter of them in the genus *Pteropus,* and is widespread on the islands of the South Pacific, where they are the most important—and often the only—pollinators and seed dispersers.

Figure 59.20 Disruption of the kelp forest ecosystem. Overharvesting by commercial whalers altered the balance of fish in the ocean ecosystem, inducing killer whales to feed on sea otters, a keystone species of the kelp forest ecosystem.

1. Whales
Overharvesting of plankton-eating whales may have caused an increase in plankton-eating pollock populations.

2. Nutritious fish
Populations of nutritious fish like ocean perch and herring declined, likely due to competition with pollock.

3. Sea lions and harbor seals
Sea lion and harbor seal populations drastically declined in Alaska, probably because the less-nutritious pollock could not sustain them.

4. Killer whales
With the decline in their prey populations of sea lions and seals, killer whales turned to a new source of food: sea otters.

5. Sea otters
Sea otter populations declined so dramatically that they disappeared in some areas.

6. Sea urchins
Usually the preferred food of sea otters, sea urchin populations now exploded and fed on kelp.

7. Kelp forests
Severely thinned by the sea urchins, the kelp beds no longer support a diversity of fish species, which may lead to a decline in populations of eagles that feed on the fish.

Figure 59.21 **The importance of keystone species.**
Flying foxes, a type of fruit-eating bat, are keystone species on many Old World tropical islands. It pollinates many plants and is a key disperser of seeds. Its elimination due to hunting and habitat loss is having a devastating effect on the ecosystems of many South Pacific Islands.

A study in Samoa found that 80% to 100% of the seeds landing on the ground during the dry season were deposited by flying foxes, which eat the fruits and defecate the seeds, often moving them great distances in the process. Many species are entirely dependent on these bats for pollination. Some have evolved features such as night-blooming flowers that prevent any other potential pollinators from taking over the role of the fruit bats.

In Guam, the two local species of flying fox have recently been driven extinct or nearly so, with a substantial impact on the ecosystem. Botanists have found that some plant species are not fruiting or are doing so only marginally, producing fewer fruits than normal. Fruits are not being dispersed away from parent plants, so seedlings are forced to compete, usually unsuccessfully, with adult trees.

Flying foxes are being driven to extinction by human hunters who kill them for food and for sport, and by orchard farmers who consider them pests. Flying foxes are particularly vulnerable because they live in large and obvious groups of up to a million individuals. Because they move in regular and predictable patterns and can be easily tracked to their home roost, hunters can easily kill thousands at a time.

Programs aimed at preserving particular species of flying foxes are only just beginning. One particularly successful example is the program to save the Rodrigues fruit bat, *Pteropus rodricensis,* which occurs only on Rodrigues Island in the Indian Ocean near Madagascar. The population dropped from about 1000 individuals in 1955 to fewer than 100 by 1974, largely due to the loss of the fruit bat's forest habitat to farming. Since 1974, the species has been legally protected, and the forest area of the island is being increased through a tree-planting program. Eleven captive-breeding colonies have been established, and the bat population is now increasing rapidly. The combination of legal protection, habitat restoration, and captive breeding has in this instance produced a very effective preservation program.

Small populations are particularly vulnerable

Because of the factors just discussed, populations of many species are fragmented and reduced in size. Such populations are particularly prone to extinction.

Demographic factors

Small populations are vulnerable to events that decrease survival or reproduction. For example, by nature of their size, small populations are ill-equipped to withstand catastrophes, such as a flood, forest fire, or disease epidemic. One example is provided by the history of the heath hen. Although the species was once common throughout the eastern United States, hunting pressure in the 18th and 19th centuries eventually eliminated all but one population, on the island of Martha's Vineyard near Cape Cod, Massachusetts. Protected in a nature preserve, the population was increasing in number until a fire destroyed most of the preserve's habitat. The small surviving population was then ravaged the next year by an unusual congregation of predatory birds, followed shortly thereafter by a disease epidemic. The last sighting of a heath hen, a male, was in 1932 (figure 59.22*a*).

When populations become extremely small, bad luck can spell the end. For example, the dusky seaside sparrow (figure 59.22*b*), a now-extinct subspecies that was found on the east coast of Florida, dwindled to a population of five individuals, all of which happened to be males. In a large population, the probability that all individuals will be of one sex is infinitesimal. But in small populations, just by the luck of the draw, it is possible that 5 or 10 or even 20 consecutive births will all be individuals of one sex, and that can be enough to send a species to extinction. In addition, when populations are small, individuals may have trouble finding each other (the Allee effect discussed in chapter 55), thus leading the population into a downward spiral toward extinction.

a.

b.

Figure 59.22 **Alive no more.** *a.* A museum specimen of the heath hen (*Tympanuchus cupido cupido*) which went extinct in 1932. *b.* This male was one of the last dusky seaside sparrows (*Ammodramus maritimus nigrescens*).

Lack of genetic variability

Small populations face a second dilemma. Because of their low numbers, such populations are prone to the loss of genetic variation as a result of genetic drift (figure 59.23). Indeed, many small populations contain little or no genetic variability. The result of such genetic homogeneity can be catastrophic. Genetic variation is beneficial to a population both because of heterozygote advantage (see chapter 20) and because genetically variable individuals tend not to have two copies of deleterious recessive alleles. Populations lacking variation are often composed of sickly, unfit, or sterile individuals. Laboratory groups of rodents and fruit flies that are maintained at small population sizes often perish after a few generations as each generation becomes less robust and fertile than the preceding one.

Although it is difficult to demonstrate that a species has gone extinct because of lack of genetic variation, studies of both zoo and natural populations clearly reveal that more genetically variable individuals have greater fitness. Furthermore, in the longer term, populations with limited genetic variation have diminished ability to adapt to changing environments, a particular concern given the way humans are changing the environment in so many ways (see chapter 58).

Interaction of demographic and genetic factors

As populations decrease in size, demographic and genetic factors combine to cause what has been termed an "extinction vortex." That is, as a population gets smaller, it becomes more vulnerable to demographic catastrophes. In turn, genetic variation starts to be lost, causing reproductive rates to decline and population numbers to decline even further, and so on. Eventually, the population disappears entirely, but attributing its demise to one particular factor would be misleading.

Case study: Prairie chickens

The greater prairie chicken, a close relative of the now-extinct heath hen, is a showy, 2-lb bird renowned for its flamboyant mating rituals (figure 59.24). Abundant in many midwestern states, the prairie chickens in Illinois have in the past six decades undergone a population collapse.

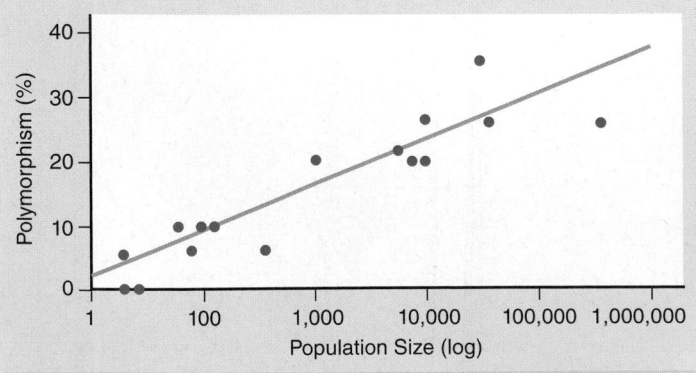

Figure 59.23 Loss of genetic variability in small populations. The percentage of genes that are polymorphic in isolated populations of the tree *Halocarpus bidwillii* in the mountains of New Zealand is a sensitive function of population size.

? **Inquiry question** Why do small populations lose genetic variation?

Figure 59.24 A mating ritual. The male greater prairie chicken (*Tympanuchus cupido pinnatus*) inflates bright orange air sacs, part of his esophagus, into balloons on each side of his head. As air is drawn into the sacs, it creates a three-syllable low frequency "boom-boom-boom" that can be heard for kilometers.

Once, enormous numbers of birds occurred throughout the state, but with the 1837 introduction of the steel plow, the first that could slice through the deep, dense root systems of prairie grasses, the Illinois prairie began to be replaced by farmland. By the turn of the 20th century, the prairie had all but vanished, and by 1931, the heath hen had become locally extinct in Illinois. The greater prairie chicken fared little better, its numbers falling to 25,000 statewide in 1933 and then to 2000 by 1962. In surrounding states with less intensive agriculture, it continued to prosper.

In 1962 and 1967, sanctuaries were established in Illinois to attempt to preserve the greater prairie chicken. But privately owned grasslands kept disappearing, along with their prairie chickens, and by the 1980s the birds were extinct in Illinois except for two preserves, and even there, their numbers kept falling. By 1990, the egg hatching rate, which at one time had averaged between 91% and 100%, had dropped to an extremely low 38%. By the mid-1990s, the count of males had dropped to as low as six in each sanctuary.

What was wrong with the sanctuary populations? One suggestion was that because of very small population sizes and a mating ritual whereby one male may dominate a flock, the Illinois prairie chickens had lost so much genetic variability as to create serious genetic problems. To test this idea, biologists at the University of Illinois compared DNA from frozen tissue samples of birds that had died in Illinois between 1974 and 1993, and found that in recent years Illinois birds had indeed become genetically less diverse.

The researchers then extracted DNA from tissue in the roots of feathers from stuffed birds collected in the 1930s from the same population. They found that Illinois birds had lost fully one-third of the genetic diversity of birds living in the same place before the population collapse of the 1970s. By contrast, prairie chicken populations in other states still contained much of the genetic variation that had disappeared from Illinois populations.

Now the stage was set to halt the Illinois prairie chicken's race toward extinction in Illinois. Wildlife managers began to transplant birds from genetically diverse populations of Minnesota, Kansas, and Nebraska to Illinois. Between 1992 and 1996, a total of 518 out-of-state prairie chickens were brought in to interbreed with the Illinois birds, and hatching rates were back up to 94% by 1998. It looks as though the prairie chickens have been saved from extinction in Illinois.

The key lesson here is the importance of not allowing things to go too far—not to drop down to a single isolated population. Without the outlying genetically different populations, the prairie chickens in Illinois could not have been saved. When the last population of the dusky seaside sparrow lost its last female, there was no other source of females, and the subspecies went extinct.

Learning Outcomes Review 59.3

Factors responsible for extinction include habitat destruction, pollution, disruption, and fragmentation. Overexploitation can reduce populations to low levels or eliminate them entirely. Introduced species can wreak havoc on native communities. Finally, small populations have less ability to rebound from catastrophes and are vulnerable to loss of genetic variation. Interaction of all these factors can hasten species' decline into extinction.

- **Does it make sense to take endangered species out of the wild to preserve them if their habitat is allowed to disappear? Explain.**

59.4 Approaches for Preserving Endangered Species and Ecosystems

Learning Outcomes

1. Distinguish between restoration of species and restoration of ecosystem functioning.
2. List the strategies for habitat restoration.
3. Explain the rationale for captive breeding programs.

Once the cause of a species' endangerment is known, it becomes possible to design a recovery plan. If the cause is commercial overharvesting, regulations can be issued to restrict harvesting and protect the threatened species. If the cause is habitat loss, plans can be instituted to restore the habitat. Loss of genetic variability in isolated subpopulations can be countered by transplanting individuals from genetically different populations. Populations in immediate danger of extinction can be captured, introduced into a captive-breeding program, and later reintroduced to other suitable habitat.

All of these solutions are extremely expensive. But as Bruce Babbitt, Secretary of the Interior in the Clinton administration, noted, it is much more economical to prevent "environmental trainwrecks" from occurring than to clean them up afterward. Preserving ecosystems and monitoring species before they are threatened is the most effective means of protecting the environment and preventing extinctions.

Destroyed habitats can sometimes be restored

Conservation biology typically concerns itself with preserving populations and species in danger of decline or extinction. Conservation, however, requires that there be something left to preserve; in many situations, conservation is no longer an option. Species, and in some cases whole communities, have disappeared or been irretrievably modified. The clear-cutting of the temperate forests of Washington State leaves little behind to conserve, as does converting a piece of land into a wheat field or an asphalt parking lot. Redeeming these situations requires restoration rather than conservation.

Three quite different sorts of habitat restoration programs might be undertaken, depending on the cause of the habitat loss.

Pristine restoration

In ecosystems where all species have been effectively removed, conservationists might attempt to restore the plants and animals that are the natural inhabitants of the area, if these species can be identified. When abandoned farmland is to be restored to prairie, as in figure 59.25, how would conservationists know what to plant?

a.

b.

Figure 59.25 Habitat restoration. The University of Wisconsin–Madison Arboretum has pioneered restoration ecology. *a.* The restoration of the prairie was at an early stage in November 1935. *b.* The prairie as it looks today. This picture was taken at approximately the same location as the 1935 photograph.

Although it is in principle possible to reestablish each of the original species in their original proportions, rebuilding a community requires knowing the identities of all the original inhabitants and the ecologies of each of the species. We rarely have this much information, so no restoration is ever truly pristine.

Increasingly, restoration biologists are working on restoring the functioning of an ecosystem, rather than trying to recreate the same community composition. This approach shifts the focus from restoring species to reconstructing the processes that operated in the natural habitat.

Removing introduced species

Sometimes the habitat has been destroyed by a single introduced species. In such a case, habitat restoration involves removing the introduced species. Restoration of the once-diverse cichlid fishes to Lake Victoria will require more than breeding and restocking the endangered species. The introduced water hyacinth and Nile perch populations will have to be brought under control or removed, and eutrophication will have to be reversed.

It is important to act quickly if an introduced species is to be removed. When aggressive African bees (the so-called "killer bees") were inadvertently released in Brazil, they remained confined to the local area for only one season. Now they occupy much of the western hemisphere.

Cleanup and rehabilitation

Habitats seriously degraded by chemical pollution cannot be restored until the pollution is cleaned up. The successful restoration of the Nashua River in New England is one example of how a concerted effort can succeed in restoring a heavily polluted habitat to a relatively pristine condition.

Once so heavily polluted by chemicals from dye manufacturing plants that it was different colors in different places, the river is now clean and used for many recreational activities.

Captive breeding programs have saved some species

Recovery programs, particularly those focused on one or a few species, must sometimes involve direct intervention in natural populations to avoid an immediate threat of extinction.

Case study: The peregrine falcon

American populations of birds of prey, such as the peregrine falcon, began an abrupt decline shortly after World War II. Of the approximately 350 breeding pairs east of the Mississippi River in 1942, all had disappeared by 1960. The culprit proved to be the chemical pesticide DDT (see chapter 58).

The use of DDT was banned by federal law in 1972, causing levels in the eastern United States to fall quickly. However, no peregrine falcons were left in the eastern United States to reestablish a natural population. Falcons from other parts of the country were used to establish a captive-breeding program

at Cornell University in 1970, with the intent of reestablishing the peregrine falcon in the eastern United States by releasing offspring of these birds. By the end of 1986, over 850 birds had been released in 13 eastern states, producing an astonishingly strong recovery (figure 59.26).

Case study: The California condor

The number of California condors (*Gymnogyps californianus*), a large, vulture-like bird with a wingspan of nearly 3 m, has been declining gradually for the past 200 years. By 1985, condor numbers had dropped so low that the bird was on the verge of extinction. Six of the remaining 15 wild birds disappeared in that year alone. The entire breeding population of the species consisted of the birds remaining in the wild and an additional 21 birds in captivity.

In a last-ditch attempt to save the condor from extinction, the remaining birds were captured and placed in a captive-breeding population. The breeding program was set up in zoos, with the goal of releasing offspring on a large, 5300-hectare ranch in prime condor habitat. Birds were isolated from human contact as much as possible, and closely related individuals were prevented from breeding.

By early 2009, the captive population of California condors had reached over 160 individuals. After extensive pre-release training to avoid power poles and people, captive-reared condors have been released successfully in California at two sites in the mountains north of Los Angeles, as well as at the Grand Canyon. Many of the released birds are doing well, and the wild population now numbers over 200 birds. Biologists are particularly excited by breeding activities that resulted in the first-ever

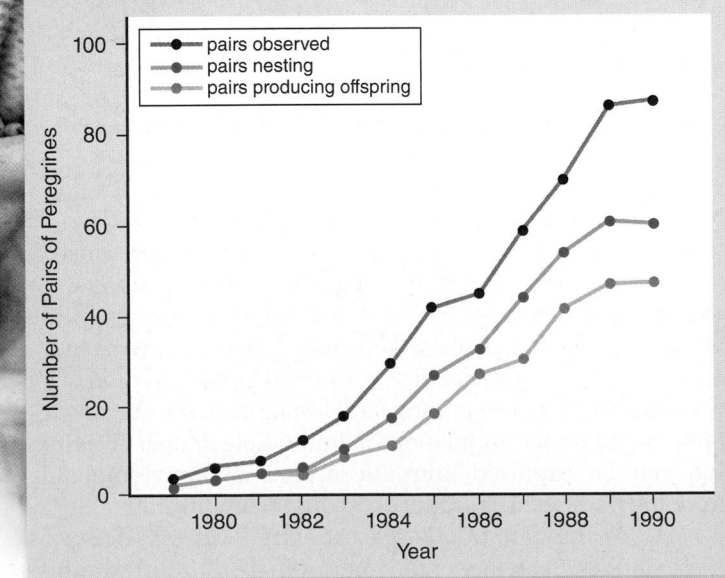

Figure 59.26 Success of captive breeding. The peregrine falcon *(Falco peregrinus)* has been reestablished in the eastern United States by releasing captive-bred birds over a period of 10 years.

offspring produced in the wild by captive-reared parents in both California and Arizona.

Case study: Yellowstone wolves

The ultimate goal of captive-breeding programs is not simply to preserve interesting species, but rather to restore ecosystems to a balanced, functional state. Yellowstone Park has been an ecosystem out of balance, due in large part to the systematic extermination of the gray wolf *(Canis lupus)* in the park early in the 20th century. Without these predators to keep their numbers in check, herds of elk and deer expanded rapidly, damaging vegetation so that the elk themselves starve in time of scarcity.

In an attempt to restore the park's natural balance, two complete wolf packs from Canada were released into the park in 1995 and 1996. The wolves adapted well, breeding so successfully that by 2002 the park contained 16 free-ranging packs and by 2005 the greater Yellowstone area contained over 300 wolves.

Although ranchers near the park have been unhappy about the return of the wolves, little damage to livestock has been noted, and the ecological equilibrium of Yellowstone Park seems well on the way to being regained. Elk are congregating in larger herds and are avoiding areas near rivers where they are vulnerable. As a result, riverside trees such as willows are increasing in number, in turn providing food for beavers, whose dams lead to the creation of ponds, a habitat type that had become rare in Yellowstone. This newly restored habitat, in turn, has led to increases in some species of birds such as the redstart that had been in decline for decades or disappeared entirely.

Current conservation approaches are multidimensional

Historically, conservationists strived to solve the problems of habitat fragmentation by focusing solely on preserving as much land as possible in a pristine state in national parks and reserves. Increasingly, however, it has become apparent that the amount of land that can be preserved in such a state is limited; moreover, many areas that are not completely protected nonetheless provide suitable habitat for many species.

As a result, conservation plans are becoming multidimensional, including not only pristine areas, but also surrounding areas in which some level of human disturbance is permitted. As discussed previously, isolated patches of habitat lose species far more rapidly than large preserves do. By including these other, less pristine areas, the total amount of area available for many species is increased.

The key to managing such large tracts of land successfully over a long time is to operate them in a way compatible with local land use. For example, although no economic activity is allowed in the core pristine area, the remainder of the land may be used for nondestructive harvesting of resources. Even areas in which hunting of some species is allowed provide protection for many other species.

Corridors of dispersal are also being provided that link the pristine areas, thus effectively increasing population sizes and allowing recolonization if a population disappears in one area due to a catastrophe. Corridors can also provide protection to species that move over great distances during the course of a year. Corridors in East Africa have protected the migration routes of ungulates. In Costa Rica, a corridor linking the lowland rainforest at the La Selva Biological Station to the montane rainforest in Braulio Carrillo National Park permits the altitudinal migration of many species of birds, mammals, and butterflies (figure 59.27).

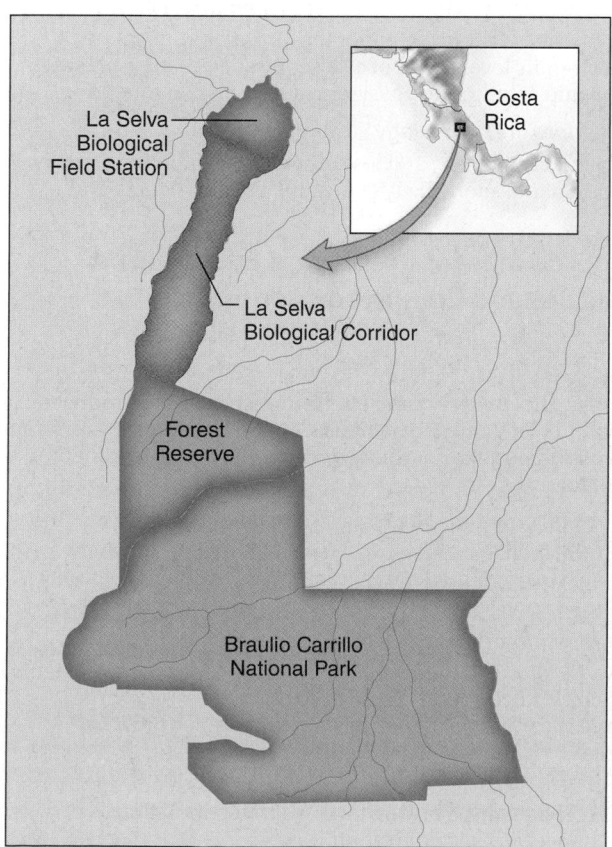

Figure 59.27 Corridor connecting two reserves. *a.* The Organization of Tropical Studies' La Selva Biological Station in Costa Rica is connected to Braulio Carrillo National Park. *b.* The corridor allows migration of birds, mammals, butterflies, and other animals from La Selva at 35 m above sea level to mountainous habitats up to 2900 m elevation.

In addition to this focus on maintaining large enough reserves, in recent years conservation biologists also have recognized that the best way to preserve biodiversity is to focus on preserving intact ecosystems, rather than particular species. For this reason, attention in many cases is turning to identifying those ecosystems most in need of preservation and devising the means to protect not only the species within the ecosystem, but the functioning of the ecosystem itself. This entails making sure that reserves are not only large enough, but also that they protect elements such as watersheds so that activities outside the reserve won't threaten the ecosystem within it.

Chapter Review

59.1 Overview of the Biodiversity Crisis

Prehistoric humans were responsible for local extinctions.

Shortly after humans arrived in North America after the last Ice Age, at least 75% of large mammals became extinct. The same pattern has been observed in other parts of the world.

Extinctions have continued in historical time.

The majority of historical extinctions have occurred within the last 150 years and on islands. The current mass extinction is the only such event triggered by one species, *Homo sapiens,* and the only one in which resources will not be widely available for evolutionary recovery afterward.

Endemic species hotspots are especially threatened.

Endemic species are found in one restricted range and are thus vulnerable to extinction. Hotspots are areas with many endemic species; many hotspots are the site of large human population growth and high rates of extinction.

59.2 The Value of Biodiversity

The direct economic value of biodiversity includes resources for our survival.

Many products are obtained from different species and ecosystems, including food, materials for clothing and shelter, and medicines.

Indirect economic value is derived from ecosystem services.

Intact ecosystems provide services such as maintaining water quality, preserving soils and nutrients, moderating local climates, and recycling nutrients. The value of intact ecosystems is often not apparent until they are lost.

Ethical and aesthetic values are based on our conscience and our consciousness.

Humans can and should make ethical decisions to protect the esthetic, ecological, and economic values of ecosystems.

59.3 Factors Responsible for Extinction

Amphibians are on the decline: A case study.

Almost half of all amphibian species have experienced decreases in population size. No single cause has been identified, which implies that global environmental changes may be responsible.

Habitat loss devastates species richness.

Habitat may be destroyed, polluted, disrupted, or fragmented. As habitats become more fragmented, the relative proportion of the remaining habitat that occurs on the boundary or edge increases rapidly, exposing species to parasites, nonnative invasive species, and predators (figure 59.13).

Overexploitation wipes out species quickly.

Hunting and harvesting of wild species pose a risk of extinction. The collapse of the cod fisheries of the North Atlantic and the decline of whale species are only two of many examples.

Introduced species threaten native species and habitats.

Natural or accidental introductions of new species results in large and often negative changes to a community because of lack of checks and balances on introduced species' growth in the form of species interactions.

Disruption of ecosystems can cause an extinction cascade.

Extinction cascades may occur either top-down or bottom-up through the trophic levels. Loss of a keystone species may increase competition and greatly alter ecosystem structure and function.

Small populations are particularly vulnerable.

Catastrophes, lack of mates, and loss of genetic variability all make reduced populations more likely to become extinct (figure 59.23).

59.4 Approaches for Preserving Endangered Species and Ecosystems

Destroyed habitats can sometimes be restored.

Restoration by removal of introduced species is very difficult and is most successful if done very soon after a new species is introduced. Severely polluted or damaged habitats sometimes cannot be restored to original conditions, but they may be restored to provide different environmental services.

Captive breeding programs have saved some species.

Species may be bred in captivity and returned to the wild when the factors that caused their endangerment are no longer a threat. Preservation of habitat may be a key in successful reintroduction.

Current conservation approaches are multidimensional.

The best way to preserve biodiversity is to preserve intact ecosystems rather than individual species. The key to management of large tracts of land is to operate them in a way compatible with local human needs.

Corridors of dispersal can link habitat fragments with one another and with larger habitats, allowing for increased population size, genetic exchange, and recolonization.

UNDERSTAND

1. Conservation hotspots are best described as
   a. areas with large numbers of endemic species, in many of which species are disappearing rapidly.
   b. areas where people are particularly active supporters of biological diversity.
   c. islands that are experiencing high rates of extinction.
   d. areas where native species are being replaced with introduced species.

2. The economic value of indirect ecosystem services
   a. is unlikely to exceed the economic value derived from uses after ecosystem conversion.
   b. has never been carefully determined.
   c. can greatly exceed the value derived after ecosystem conversion.
   d. is entirely aesthetic.

3. The amphibian decline is best described as
   a. global disappearance of amphibian populations due to the pervasiveness of local habitat destruction.
   b. global shrinkage of amphibian populations due to global climate change.
   c. the unexplained disappearance of golden toads in Costa Rica.
   d. None of the choices is correct.

4. Habitat fragmentation can negatively affect populations by
   a. restricting gene flow among areas that were previously continuous.
   b. increasing the relative amount of edge in suitable habitat patches.
   c. creating patches that are too small to support a breeding population.
   d. All of the choices are correct.

5. When populations are drastically reduced in size, genetic diversity and heterozygosity
   a. are likely to increase, enhancing the probability of extinction.
   b. are likely to decrease, enhancing the probability of extinction.
   c. are usually not factors that influence the probability of extinction.
   d. automatically respond in a way that protects populations from future changes.

6. A captive-breeding program followed by release to the wild
   a. is very likely, all by itself, to save a species threatened by extinction.
   b. is only likely to succeed when genetic variation of wild populations is very low.
   c. may be successful when combined with proper regulations and habitat restoration.
   d. None of the choices is correct.

APPLY

1. Historically, island species have tended to become extinct faster than species living on a mainland. Which of the following reasons can be used to explain this phenomenon?
   a. Island species have often evolved in the absence of predators and have no natural avoidance strategies.
   b. Humans have introduced diseases and competitors to islands, which negatively affect island populations.
   c. Island populations are usually smaller than mainland populations.
   d. All of the choices are correct.

2. Ninety-nine percent of all the species that ever existed have gone extinct,
   a. serving as evidence that current extinction rates are not higher than normal.
   b. but most of these losses have occurred in the last 400 years.
   c. which argues that the world just had too many species.
   d. None of the choices is correct.

3. To effectively address the biodiversity crisis, the protection of individual species
   a. must be used in concert with a principle of ecosystem management and restoration.
   b. is a sufficient management approach that merely needs to be expanded to more species.
   c. has no role to play in addressing the biodiversity crisis.
   d. usually conflicts with the principle of ecosystem management.

4. The introduction of a nonnative predator to an ecosystem could cause extinction by
   a. causing a top-down trophic cascade (see chapter 57).
   b. outcompeting a native carnivore (see chapter 56).
   c. transmitting parasites to which the native species are not adapted.
   d. All of the choices are correct.

SYNTHESIZE

1. If 99% of the species that ever existed are now extinct, why is there such concern over the extinction rates over the last several centuries?

2. Ecosystem conversion always has a cost and a benefit. Usually the benefit flows to a segment of society (a business or one group of people, for instance), but the costs are borne by all of society. That is what makes decisions about how and when to convert ecosystems difficult. However, is that a problem unique to conversion of ecosystems in the way we understand it today (for example, the conversion of the mangrove to a shrimp farm)? Are there other examples we can look to for guidance in how to make these decisions?

3. There is concern and evidence that amphibian populations are declining worldwide as a consequence of factors acting globally. Given that we know that species extinction is a natural process, how do we determine if there is a global decline that is different from normal species extinction?

4. Given what you learned in chapter 56 about interactions between species and in chapter 57 about interactions among trophic levels, how can the extinction of one species have far-ranging effects on an ecosystem? Is it possible to predict which species would be particularly likely to affect many other species if they were to go extinct?

5. All populations become small before going extinct. Is small population size really a cause of extinction, or just something that happens as a result of other factors that cause extinction?

Answer Key

CHAPTER 1

LEARNING OUTCOME QUESTIONS

1.1 No. The study of biology encompasses information/tools from chemistry, physics, and geology—in fact all of the "natural sciences."

1.2 A scientific theory has been tested by experimentation. A hypothesis is a starting point for explaining a body of observations. When predictions generated using the hypothesis have been tested, it gains the confidence associated with a theory. A theory still cannot be "proved," however, as new data can always force us to re-evaluate a theory.

1.3 No. Natural selection explains the patterns of living organisms we see at present and allows us to work back in time, but it is not intended to explain how life arose. This does not mean that we can never explain this, but merely that natural selection does not do this.

1.4 Viruses do not fit well into our definition of living systems. It is a matter of controversy whether viruses should be considered "alive." They lack the basic cellular machinery, but they do have genetic information. Some theories for the origin of cells view viruses as being a step from organic molecules to cell, but looking at current organisms, they do not fulfill our definition of life.

INQUIRY AND DATA ANALYSIS QUESTIONS

Page 10 Data analysis: Reducing the factor by which the geometric progression increases (lowering the value of the exponent) reduces the difference between numbers of people and amount of food production.

Page 10 Inquiry question: It can be achieved by lowering family size or delaying childbearing.

Page 11 Inquiry question: A snake would fall somewhere near the bird, as birds and snakes are closely related.

UNDERSTAND

1. b   2. c   3. a   4. b   5. d   6. b   7. c   8. c

APPLY

1. a   2. d   3. c   4. d   5. d   6. d   7. a

SYNTHESIZE

1. For something to be considered living it would demonstrate organization, possibly including a cellular structure. The organism would gain and use energy to maintain homeostasis, respond to its environment, and grow and reproduce. These latter properties would be difficult to determine if the evidence of life from other planets comes from fossils. Similarly, the ability of an alien organism to evolve could be difficult to establish.

2. a. The variables that were held the same between the two experiments include the broth, the flask, and the sterilization step.
   b. The shape of the flask influences the experiment because any cells present in the air can enter the flask with the broken neck, but they are trapped in the neck of the other flask.
   c. If cells can arise spontaneously, then cell growth will occur in both flasks. If cells can only arise from preexisting cells (cells in the air), then only the flask with the broken neck will grow cells. Breaking the neck exposes the broth to a source of cells.
   d. If the sterilization step did not actually remove all cells, then growth would have occurred in both flasks. This result would seem to support the hypothesis that life can arise spontaneously.

CHAPTER 2

LEARNING OUTCOME QUESTIONS

2.1 If the number of protons exceeds neutrons, there is no effect on charge; if the number of protons exceeds electrons, then the charge is (+).

2.2 Atoms are reactive when their outer electron shell is not filled with electrons. The noble gases have filled outer electron shells and are thus unreactive.

2.3 An ionic bond results when there is a transfer of electrons, resulting in positive and negative ions that are attracted to each other. A covalent bond is the result of two atoms sharing electrons. Polar covalent bonds involve unequal sharing of electrons. This produces regions of partial charge, but not ions.

2.4 C and H have about the same electronegativity, and thus form nonpolar covalent bonds. This would not result in a cohesive or adhesive fluid.

2.5 Since ice floats, a lake will freeze from the top down, not the bottom up. This means that water remains fluid on the bottom of the lake, allowing living things to overwinter.

2.6 Since pH is a log scale, this would be a change of 100-fold in $[H^+]$.

INQUIRY AND DATA ANALYSIS QUESTION

Page 30 Data analysis: From the graph, about four volumes of base must be added to change the pH from 4 to 6.

UNDERSTAND

1. b   2. d   3. b   4. a   5. c   6. d   7. b

APPLY

1. c   2. b   3. a   4. c   5. d   6. Chemical reactions involve changes in the electronic configuration of atoms. Radioactive decay involves the actual decay of the nucleus itself, producing another atom and emitting radiation.

SYNTHESIZE

1. A cation is an element that tends to lose an electron from its outer energy level, leaving behind a net positive charge due to the presence of the protons in the atomic nucleus. Electrons are only lost from the outer energy level if that loss is energetically favorable, that is, if it makes the atom more stable by virtue of obtaining a filled outer energy level (the octet rule). You can predict which elements are likely to function as cations by calculating which of the elements will possess one (or two) electrons in their outer energy level. Recall that each orbital surrounding an atomic nucleus can only hold two electrons. Energy level K is a single *s* orbital and can hold two electrons. Energy level L consists of another *s* orbital plus three *p* orbitals—holding a total of eight electrons. Use the atomic number of each element to predict the total number of electrons present. Examples of other cations would include hydrogen (H), lithium (Li), magnesium (Mg), and beryllium (Be).

2. Silicon has an atomic number of 14. This means that there are four unpaired electrons in its outer energy level (comparable to carbon). Based on this fact, you can conclude that silicon, like carbon, could form four covalent bonds. Silicon also falls within the group of elements with atomic masses less than 21, a property of the elements known to participate in the formation of biologically important molecules. Interestingly, silicon is much more prevalent than carbon on Earth. Although silicon dioxide is found in the cell walls of plants and single-celled organisms called diatoms, silicon-based life has not been identified on this planet. Given the abundance of silicon on Earth you can conclude that some other aspect of the chemistry of this atom makes it incompatible with the formation of molecules that make up living organisms.

3. Water is considered to be a critical molecule for the evolution of life on Earth. It is reasonable to assume that water on other planets could play a similar role. The key properties of water that would support its role in the evolution of life are:
   - The ability of water to act as a solvent. Molecules dissolved in water could move and interact in ways that would allow for the formation of larger, more complex molecules such as those found in living organisms.
   - The high specific heat of water. Water can modulate and maintain its temperature, thereby protecting the molecules or organisms within it from temperature extremes—an important feature on other planets.

- The difference in density between ice and liquid water. The fact that ice floats is a simple, but important feature of water environments since it allows living organisms to remain in a liquid environment protected under a surface of ice. This possibility is especially intriguing, given recent evidence of ice-covered oceans on Europa, a moon of the planet Jupiter.

CHAPTER 3

LEARNING OUTCOME QUESTIONS

3.1 Hydrolysis is the reverse reaction of dehydration. Dehydration is a synthetic reaction involving the loss of water and hydrolysis is cleavage by addition of water.

3.2 Starch and glycogen are both energy-storage molecules. Their highly branched nature allows the formation of droplets, and the similarity in the bonds holding adjacent glucose molecules together mean that the enzymes we have to break down glycogen allow us to break down starch. The same enzymes do not allow us to break down cellulose. The structure of cellulose leads to the formation of tough fibers.

3.3 The sequence of bases in an RNA would be identical to one strand and complementary to the other strand of the DNA with the exception that U would be in place of T (complementary to A).

3.4 If an unknown protein has sequence similarity to a known protein, we can infer its function is also similar. If an unknown protein has known functional domains or motifs, we can also use these to help predict function.

3.5 Phospholipids have a charged group replacing one of the fatty acids in a triglyceride. This leads to an amphipathic molecule that has both hydrophobic and hydrophilic regions. This will spontaneously form bilayer membranes in water.

UNDERSTAND

1. b   2. a   3. d   4. c   5. b   6. b   7. c   8. b

APPLY

1. c   2. d   3. b   4. d   5. b   6. b   7. d

SYNTHESIZE

1. The four biological macromolecules all have different structure and function. In comparing carbohydrates, nucleic acids, and proteins, we can think of these as being polymers with different monomers. In the case of carbohydrates, the polymers are all polymers of the simple sugar glucose. These are energy-storage molecules (with many C—H bonds) and structural molecules such as cellulose that make tough fibers.

    Nucleic acids are formed of nucleotide monomers, each of which consists of ribose, phosphate, and a nitrogenous base. These molecules are informational molecules that encode information in the sequence of bases. The bases interact in specific ways: A base-pairs with T, and G base-pairs with C. This is the basis for their informational storage.

    Proteins are formed of amino acid polymers. There are 20 different amino acids, and thus an incredible number of different proteins. These can have an almost unlimited number of functions. These functions arise from the amazing flexibility in structure of protein chains.

2. *Nucleic Acids*—Hydrogen bonds are important for complementary base-pairing between the two strands of nucleic acid that make up a molecule of DNA. Complementary base-pairing can also occur within the single nucleic acid strand of an RNA molecule.

    Proteins—Hydrogen bonds are involved in both the secondary and tertiary levels of protein structure. The α helices and β-pleated sheets of secondary structure are stabilized by hydrogen bond formation between the amino and carboxyl groups of the amino acid backbone. Hydrogen bond formation between R-groups helps stabilize the three-dimensional folding of the protein at the tertiary level of structure.

    Carbohydrates—Hydrogen bonds are less important for carbohydrates; however, these bonds are responsible for the formation of the fibers of cellulose that make up the cell walls of plants.

    Lipids—Hydrogen bonds are not involved in the structure of lipid molecules. The inability of fatty acids to form hydrogen bonds with water is key to their hydrophobic nature.

3. We have enzymes that can break down glycogen. Glycogen is formed from α-glucose subunits. Starch is also formed from α-glucose units, but cellulose is formed from β-glucose units. The enzymes that break the α-glycosidic linkages cannot break the β-glycosidic linkages. Thus we can degrade glycogen and starch but not cellulose.

CHAPTER 4

LEARNING OUTCOME QUESTIONS

4.1 The statement about all cells coming from preexisting cells might need to be modified. It would really depend on whether these Martian life-forms had a similar molecular/cellular basis as that of terrestrial life.

4.2 Bacteria and archaea both tend to be single cells that lack a membrane-bounded nucleus, and have extensive internal endomembrane systems. They both have a cell wall, although the composition is different. They do not undergo mitosis, although the proteins involved in DNA replication and cell division are not similar.

4.3 Part of what gives different organs their unique identities are the specialized cell types found in each. That does not mean that there will not be some cell types common to all (epidermal cells for example) but organs tend to have specialized cell types.

4.4 They don't!

4.5 The nuclear genes that encode organellar proteins moved from the organelle to the nucleus. There is evidence for a lot of "horizontal gene transfer" across domains; this is an example of how that can occur.

4.6 It provides structure and support for larger cells, especially in animal cells that lack a cell wall.

4.7 Microtubules and microfilaments are both involved in cell motility and in movement of substance around cells. Intermediate filaments do not have this dynamic role, but are more structural.

4.8 Cell junctions help to put together cells into higher level structures that are organized and joined in different ways. Different kinds of junctions can be used for different functional purposes.

INQUIRY AND DATA ANALYSIS QUESTIONS

Page 63 Inquiry question: Stretch, dent, convolute, fold, add more than one nucleus, anything which would increase the amount of diffusion between the cytoplasm and the external environment.

Page 75 Inquiry question: Both the cristae of mitochondria and the thylakoids of chloroplasts, where many of the reactions take place leading to the production of ATP, are highly folded. This allows for a large surface area, increasing the efficiency of the mechanisms of oxidative phosphorylation.

Page 80 Inquiry question: Ciliated cells in the trachea help to remove particulate matter from the respiratory tract so that it can be expelled or swallowed and processed in the digestive tract.

UNDERSTAND

1. d   2. d   3. c   4. a   5. c   6. d   7. b

APPLY

1. c   2. b   3. c   4. b   5. c   6. b   7. a

SYNTHESIZE

1. Your diagram should start at the SER and then move to the RER, Golgi apparatus, and finally to the plasma membrane. Small transport vesicles are the mechanism that would carry a phospholipid molecule between two membrane compartments. Transport vesicles are small "membrane bubbles" composed of a phospholipid bilayer.

2. If these organelles were free-living bacteria, they would have the features found in bacteria. Mitochondria and chloroplasts both have DNA but no nucleus, and they lack the complex organelles found in eukaryotes. At first glance, the cristae may seem to be an internal membrane system, but they are actually infoldings of the inner membrane. If endosymbiosis occurred, this would be the plasma membrane of the endosymbiont, and the outer membrane would be the plasma membrane of the engulfing cell. Another test would be to compare DNA in these organelles with current bacteria. This has actually shown similarities that make us confident of the identity of the endosymbionts.

3. The prokaryotic and eukaryotic flagella are examples of an analogous trait. Both flagella function to propel the cell through its environment by converting chemical energy into mechanical force. The key difference is in the structure of the flagella. The bacterial flagellum is composed of a single protein emerging from a basal body anchored within the cell's plasma membrane and using the potential energy of a proton gradient to cause a rotary movement. In contrast, the flagellum of the eukaryote is composed of many different proteins assembled into a complex axoneme structure that uses ATP energy to cause an undulating motion.

4. Eukaryotic cells are distinguished from prokaryotic cells by the presence of a system of internal membrane compartments and membrane-bounded organelles such as mitochondria and chloroplasts. As outlined in figure 4.19, the first step in the evolution of the eukaryotic cell was the infolding of the plasma membrane to create separate internal membranes such as the nuclear envelope and the endoplasmic reticulum. The origins of mitochondria and chloroplasts are hypothesized to be the result of a bit of cellular "indigestion" in which aerobic or photosynthetic prokaryotes were engulfed but not digested by the larger ancestor eukaryote. Given this information, there are two possible scenarios for the origin of *Giardia*. In the first scenario, the ancestor of *Giardia* split off from the eukaryotic lineage after the evolution of the nucleus but before the acquisition of mitochondria. In the second scenario, the ancestor of *Giardia* split off after the acquisition of mitochondria and subsequently lost the mitochondria. At present, neither of these two scenarios can be rejected. The first case was long thought to be the best explanation, but recently it has been challenged by evidence for the second case.

CHAPTER 5

LEARNING OUTCOME QUESTIONS

5.1 Cells would not be able to control their contents. Nonpolar molecules would be able to cross the membrane by diffusion, as would small polar molecules, but without proteins to control the passage of specific molecules, it would not function as a semipermeable membrane.

5.2 No. The nonpolar interior of the bilayer would not be soluble in the solvent. The molecules would organize with their nonpolar tails in the solvent, but the negative charge on the phosphates would repel other phosphates.

5.3 Transmembrane domains anchor protein in the membrane. They associate with the hydrophobic interior, thus they must be hydrophobic as well. If they slide out of the interior, they are repelled by water.

5.4 The concentration of the IV will be isotonic with your blood cells. If it were hypotonic, your blood cells would take on water and burst; if it were hypertonic, your blood cells would lose water and shrink.

5.5 Channel proteins are aqueous pores that allow facilitated diffusion. They cannot actively transport ions. Carrier proteins bind to their substrates and couple transport to some form of energy for active transport.

5.6 In all cases, there is recognition and specific binding of a molecule by a protein. In each case this binding is necessary for biological function.

INQUIRY AND DATA ANALYSIS QUESTIONS

Page 94 Inquiry question: As the name suggests for the fluid mosaic model, cell membranes have some degree of fluidity. The degree of fluidity varies with the composition of the membrane, but in all membranes, phospholipids are able to move about within the membrane. Also, due to the hydrophobic and hydrophilic opposite ends of phospholipid molecules, phospholipid bilayers form spontaneously. Therefore, if stressing forces happen to damage a membrane, adjacent phospholipids automatically move to fill in the opening.

Page 95 Inquiry question: Integral membrane proteins are those that are embedded within the membrane structure and provide passageways across the membrane. Because integral membrane proteins must pass through both polar and nonpolar regions of the phospholipid bilayer, the protein portion held within the nonpolar fatty acid interior of the membrane must also be nonpolar. The amino acid sequence of an integral protein would have polar amino acids at both ends, with nonpolar amino acids making up the middle portion of the protein.

UNDERSTAND

1. d   2. a   3. d   4. d   5. b   6. d   7. a

APPLY

1. c   2. b   3. d   4. c   5. d

SYNTHESIZE

1. Since the membrane proteins become intermixed in the absence of the energy molecule, ATP, one can conclude that chemical energy is not required for their movement. Since the proteins do not move and intermix when the temperature is cold, one can also conclude that the movement is temperature-sensitive. The passive diffusion of molecules also depends on temperature and does not require chemical energy; therefore, it is possible to conclude that membrane fluidity occurs as a consequence of passive diffusion.

2. The inner half of the bilayer of the various endomembranes becomes the outer half of the bilayer of the plasma membrane.

3. Lipids can be inserted into one leaflet to produce asymmetry. When lipids are synthesized in the SER, they can be assembled into asymmetric membranes. There are also enzymes that can flip lipids from one leaflet to the other.

CHAPTER 6

LEARNING OUTCOME QUESTIONS

6.1 At the bottom of the ocean, light is not an option as it does not penetrate that deep. However, there is a large source of energy in the form of reduced minerals, such as sulfur compounds, that can be oxidized. These are abundant at hydrothermal vents found at the junctions of tectonic plates. This supports whole ecosystems dependent on bacteria that oxidize reduced minerals available at the hydrothermal vents.

6.2 In a word, no. Enzymes only alter the rate of a reaction; they do not change the thermodynamics of the reaction. The action of an enzyme does not change the ΔG for the reaction.

6.3 In the text, it stated that the average person turns over approximately their body weight in ATP per day. This gives us enough information to determine approximately the amount of energy released:

$$100 \text{ kg} = 1.0 \times 10^5 \text{ g}$$
$$(1.0 \times 10^5 \text{ g})/(507.18 \text{ g/mol}) = 197.2 \text{ mol}$$
$$(197.2 \text{ mol})(7.3 \text{ kcal/mol}) = 1439 \text{ kcal}$$

6.4 This is a question that cannot be definitely answered, but we can give some reasonable conjectures. First, DNA's location is in the nucleus and not the cytoplasm, where most enzymes are found. Second, the double-stranded structure of DNA works well for information storage, but would not necessarily function well as an enzyme. Each base interacts with a base on the opposite strand, which makes for a very stable linear molecule, but does not encourage folding into the kind of complex 3-D shape found in enzymes.

6.5 Feedback inhibition is common in pathways that synthesize metabolites. In these anabolic pathways, when the end-product builds up, it feeds back to inhibit its own production. Catabolic pathways are involved in the degradation of compounds. Feedback inhibition makes less biochemical sense in a pathway that degrades compounds as these are usually involved in energy metabolism or in recycling or removal of compounds. Thus the end-product is destroyed or removed and cannot feed back.

INQUIRY AND DATA ANALYSIS QUESTION

Page 113 Data analysis: The overall ΔG would be the sum of the individual ΔGs, or 3.9 kcal/mol. This makes the overall process exergonic.

UNDERSTAND

1. b   2. a   3. b   4. a   5. d   6. b   7. d

APPLY

1. b   2. c   3. a   4. c   5. c   6. c

SYNTHESIZE

1. a. At 40°C the enzyme is at its optimum. The rate of the reaction is at its highest level.
   b. Temperature is a factor that influences enzyme function. This enzyme does not appear to function at either very cold or very hot temperatures. The shape of the enzyme is affected by temperature, and the enzyme's structure is altered enough at extreme temperatures that it no longer binds substrate. Alternatively, the enzyme may be denatured—that is a complete loss of normal three-dimensional shape at extreme temperatures. Think about frying an egg: What happens to the proteins in the egg?
   c. Everyone's body is slightly different. If the temperature optimum was very narrow, then the cells that make up a body would be vulnerable. Having a broad range of temperature optimums keeps the enzyme functioning.

2. a. The reaction rate would be slow because of the low concentration of the substrate ATP. The rate of reaction depends on substrate concentration.
   b. ATP acts like a noncompetitive, allosteric inhibitor when ATP levels are very high. If ATP binds to the allosteric site, then the reaction should slow down.
   c. When ATP levels are high, the excess ATP molecules bind to the allosteric site and inhibit the enzyme. The allosteric inhibitor functions by causing a change in the shape of the active site in the enzyme. This reaction is an

example of feedback regulation because ATP is a final product of the overall series of reactions associated with glycolysis. The cell regulates glycolysis by regulating this early step catalyzed by phosphofructokinase; the allosteric inhibitor is the "product" of glycolysis (and later stages) which is ATP.

CHAPTER 7

LEARNING OUTCOME QUESTIONS

7.1 Cells require energy for a wide variety of functions. The reactions involved in the oxidation of glucose are complex, and linking these to the different metabolic functions that require energy would be inefficient. Thus cells make and use ATP as a reusable source of energy.

7.2 The location of glycolysis does not argue for or against the endosymbiotic origin of mitochondria. It could have been located in the mitochondria previously and moved to the cytoplasm or could have always been located in the cytoplasm in eukaryotes.

7.3 For an enzyme like pyruvate decarboxylase the complex reduces the distance for the diffusion of substrates for the different stages of the reaction. Any possible unwanted side reactions are prevented. Finally the reactions occur within a single unit and thus can be controlled in a coordinated fashion. The main disadvantage is that since the enzymes are all part of a complex their evolution is more constrained than if they were independent.

7.4 At the end of the Krebs cycle, the electrons removed from glucose are all carried by soluble electron carriers. Most of these are in NADH, and a few are in $FADH_2$. All of these are all fed into the electron transport chain under aerobic conditions where they are used to produce a proton gradient.

7.5 A hole in the outer membrane would allow protons in the intermembrane space to leak out. This would destroy the proton gradient across the inner membrane, stopping the phosphorylation of ADP by ATP synthase.

7.6 Chemiosmosis means that the ATP/NADH ratio is not dependent on the number of "pumping" stations. Instead, it is dependent on the number of protons needed for one turn of the enzyme, and on the number of binding sites on the enzyme for ADP/ATP.

7.7 Glycolysis, which is the starting point for respiration from sugars, is regulated at the enzyme phosphofructokinase. This enzyme is just before the 6-C skeleton is split into two 3-C molecules. The allosteric effectors for this enzyme include ATP and citrate. Thus the "end-product" ATP, and an intermediate from the Krebs cycle, both feed back to inhibit the first part of this process.

7.8 The first obvious point is that the most likely type of ecosystem would be one where oxygen is nonexistent or limiting. This includes marine, aquatic, and soil environments. Any place where oxygen is in short supply is expected to be dominated by anaerobic organisms, and respiration produces more energy than fermentation.

7.9 The short answer is no. The reason is twofold. First the oxidation of fatty acids feeds acetyl units into the Krebs cycle. The primary output of the Krebs cycle is electrons that are fed into the electron transport chain to eventually produce ATP by chemiosmosis. The second reason is that the process of β oxidation that produces the acetyl units is oxygen dependent as well. This is because β oxidation uses FAD as a cofactor for oxidation, and the $FADH_2$ is oxidized by the electron transport chain.

7.10 The evidence for the origins of metabolism is indirect. The presence of O_2 in the atmosphere is the result of photosynthesis, so the record of when we went from a reducing to an oxidizing atmosphere chronicles the rise of oxygenic photosynthesis. Glycolysis is a universal pathway that is found in virtually all types of cells. This indicates that it is an ancient pathway that likely evolved prior to other types of energy metabolism. Nitrogen fixation probably evolved in the reducing atmosphere that preceded oxygenic photosynthesis as it is poisoned by oxygen, and aided by the reducing atmosphere.

INQUIRY AND DATA ANALYSIS QUESTION

Page 142 Data analysis: During the catabolism of fats, each round of β oxidation uses one molecule of ATP and generates one molecule each of $FADH_2$ and NADH. For a 16-carbon fatty acid, seven rounds of β oxidation would convert the fatty acid into eight molecules of acetyl-CoA. The oxidation of each acetyl-CoA in the Krebs cycle produces 10 molecules of ATP. The overall ATP yield from a 16-carbon fatty acid would be a net gain of 21 ATP from seven rounds of β oxidation [gain of 4 ATP per round – 1 per round to prime reactions] + 80 ATP from the oxidation of 8 acetyl-CoAs = 101 molecules of ATP.

UNDERSTAND

1. d 2. d 3. c 4. c 5. a 6. d 7. c

APPLY

1. b 2. b 3. c 4. c 5. c 6. b 7. b

SYNTHESIZE

1.

Molecules	Glycolysis	Cellular Respiration
Glucose	*Is the starting material for the reaction*	*Does not directly use glucose; however, does use pyruvate derived from glucose*
Pyruvate	*The end product of glycolysis*	*The starting material for cellular respiration*
Oxygen	*Not required*	*Required for aerobic respiration, but not for anaerobic respiration*
ATP	*Produced through substrate-level phosphorylation*	*Produced through oxidative phosphorylation. More produced than in glycolysis*
CO_2	*Not produced*	*Produced during pyruvate oxidation and Krebs cycle*

2. The electron transport chain of the inner membrane of the mitochondria functions to create a hydrogen ion concentration gradient by pumping protons into the intermembrane space. In a typical mitochondrion, the protons can only diffuse back down their concentration gradient by moving through the ATP synthase and generating ATP. If protons can move through another transport protein, then the potential energy of the hydrogen ion concentration gradient would be "lost" as heat.

3. If brown fat persists in adults, then the uncoupling mechanism to generate heat described above could result in weight loss under cold conditions. There is now some evidence to indicate that this may be the case.

CHAPTER 8

LEARNING OUTCOME QUESTIONS

8.1 Both chloroplasts and mitochondria have an outer membrane and an inner membrane. The inner membrane in both forms an elaborate structure. These inner membrane systems have electron transport chains that move protons across the membrane to allow for the synthesis of ATP by chemiosmosis. They also both have a soluble compartment in which a variety of enzymes carry out reactions.

8.2 All of the carbon in your body comes from carbon fixation by autotrophs. Thus, all of the carbon in your body was once CO_2 in the atmosphere, before it was fixed by plants.

8.3 The action spectrum for photosynthesis refers to the most effective wavelengths. The absorption spectrum for an individual pigment shows how much light is absorbed at different wavelengths.

8.4 Before the discovery of photosystems, we assumed that each chlorophyll molecule absorbed photons resulting in excited electrons.

8.5 Without a proton gradient, synthesis of ATP by chemiosmosis would be impossible. However, NADPH could still be synthesized because electron transport would still occur as long as photons were still being absorbed to begin the process.

8.6 A portion of the Calvin cycle is the reverse of glycolysis (the reduction of 3-phosphoglycerate to glyceraldehyde-3-phosphate).

8.7 Both C_4 plants and CAM plants fix carbon by incorporating CO_2 into the 4-carbon malate and then use this to produce high local levels of CO_2 for the Calvin cycle. The main difference is that in C_4 plants, this occurs in different cells, and in CAM plants this occurs at different times.

INQUIRY AND DATA ANALYSIS QUESTIONS

Page 150 Data analysis: Light energy is used in light-dependent reactions to reduce NADP$^+$ and to produce ATP. Molecules of chlorophyll absorb photons of light energy, but only within narrow energy ranges (specific wavelengths of light). When all chlorophyll molecules are in use, no additional increase in light intensity will increase the rate at which they can absorb light energy.

Page 154 Data analysis: The curves would be similar to the observed curve shown, but would plateau at a higher and lower level, respectively. Even the higher level would still be below the expected curve for all chlorophyll molecules being used.

Page 157 Data analysis: You could conclude that the two photosystems function together and not in series.

UNDERSTAND

1. c 2. a 3. a 4. b 5. c 6. c 7. a 8. b

APPLY

1. d 2. b 3. d 4. c 5. d 6. d 7. a 8. a

SYNTHESIZE

1. In C_3 plants CO_2 reacts with ribulose 1,5-bisphosphate (RuBP) to yield 2 molecules of PGA. This reaction is catalyzed by the enzyme rubisco. Rubisco also catalyzes the oxidation of RuBP. Which reaction predominates depends on the relative concentrations of reactants. The reactions of the Calvin cycle reduce the PGA to G3P, which can be used to make a variety of sugars including RuBP. In C_4 and CAM plants, an initial fixation reaction incorporates CO_2 into malate. The malate then can be decarboxylated to pyruvate and CO_2 to produce locally high levels of CO_2. The high levels of CO_2 get around the oxidation of RuBP by rubisco. In C_4 plants malate is produced in one cell, then shunted into an adjacent cell that lacks stomata to produce high levels of CO_2. CAM plants fix carbon into malate at night when their stomata are open, then use this during the day to fuel the Calvin cycle. Both are evolutionary innovations that have arisen in hot dry climates that allow plants to more efficiently fix carbon and prevent desiccation.

2. Figure 8.19 diagrams this relationship. The oxygen produced by photosynthesis is used as a final electron acceptor for electron transport in respiration. The CO_2 that results from the oxidation of glucose (or fatty acids) is incorporated into organic compounds via the Calvin cycle. Respiration also produces water, while photosynthesis consumes water.

3. Yes. Plants use their chloroplasts to convert light energy into chemical energy. During light reactions ATP and NADPH are created, but these molecules are consumed during the Calvin cycle and are not available for the cell's general use. The G3P produced by the Calvin cycle stores the chemical energy from the light reactions within its chemical bonds. Ultimately, this energy is stored in glucose and retrieved by the cell through the process of glycolysis and cellular respiration.

CHAPTER 9

LEARNING OUTCOME QUESTIONS

9.1 Ligands bind to receptors based on complementary shapes. This interaction based on molecular recognition is similar to the way enzymes interact with their ligands.

9.2 Hydrophobic molecules can cross the membrane and are thus more likely to have an internal receptor.

9.3 Intracellular receptors have direct effects on gene expression. This generally leads to effects with longer duration.

9.4 Ras protein occupies a central role in signaling pathways involving growth factors. A number of different kinds of growth factors act through Ras. So it is not surprising mutated Ras protein is a factor in a number of different cancers.

9.5 GPCRs are a very ancient and flexible receptor/signaling pathway. The genes encoding these receptors have been duplicated and then have diversified over evolutionary time, so now there are many members of this gene family.

UNDERSTAND

1. b 2. b 3. c 4. d 5. b 6. d 7. c 8. a

APPLY

1. b 2. c 3. c 4. d 5. d 6. c

SYNTHESIZE

1. All signaling events start with a ligand binding to a receptor. The receptor initiates a chain of events that ultimately leads to a change in cellular behavior. In some cases the change is immediate—for example, the opening of an ion channel. In other cases the change requires more time before it occurs, such as when the MAP kinase pathway becomes activated multiple different kinases become activated and deactivated. Some signals only affect a cell for a short time (the channel example), but other signals can permanently change the cell by changing gene expression, and therefore the number and kind of proteins found in the cell.

2. a. This system involves *both* autocrine and paracrine signaling because Netrin-1 can influence the cells within the crypt that are responsible for its production and the neighboring cells.

   b. The binding of Netrin-1 to its receptor produces the signal for cell growth. This signal would be strongest in the regions of the tissue with the greatest amount of Netrin-1—that is, in the crypts. A concentration gradient of Netrin-1 exists such that the levels of this ligand are lowest at the tips of the villi. Consequently, the greatest amount of cell death would occur at the villi tips.

   c. Tumors occur when cell growth goes on unregulated. In the absence of Netrin-1, the Netrin-1 receptor can trigger cell death—controlling the number of cells that make up the epithelial tissue. Without this mechanism for controlling cell number, tumor formation is more likely.

CHAPTER 10

LEARNING OUTCOME QUESTIONS

10.1 The concerted replication and segregation of chromosomes works well with one small chromosome, but would likely not work as well with many chromosomes.

10.2 No.

10.3 The first irreversible step is the commitment to DNA replication.

10.4 Loss of cohesins would mean that the products of DNA replication would not be kept together. This would make normal mitosis impossible, and thus lead to aneuploid cells and probably be lethal.

10.5 The segregation of chromatids that lose cohesin would be random because they could no longer be held at metaphase attached to opposite poles. This would likely lead to gain and loss of this chromosome in daughter cells due to improper partitioning.

10.6 Tumor suppressor genes are genetically recessive, but proto-oncogenes are dominant. Loss of function for a tumor suppressor gene leads to cancer, whereas inappropriate expression or gain of function lead to cancer with proto-oncogenes.

UNDERSTAND

1. d 2. b 3. b 4. b 5. a 6. c 7. b 8. d

APPLY

1. a 2. c 3. b 4. d 5. c 6. d

SYNTHESIZE

1. If Wee-1 were absent, the cell would have no way to phosphorylate Cdk. If Cdk is not phosphorylated, then it cannot be inhibited. If Cdk is not inhibited, then it will remain active. If Cdk remains active, then it will continue to signal the cell to move through the G_2/M checkpoint, but now in an unregulated manner. The cells would undergo multiple rounds of cell division without the growth associated with G_2. As a consequence, the daughter cells will become smaller and smaller with each division—hence the name of the protein!

2. Growth factor = ligand
   1. Ligand binds to receptor (the growth factor will bind to a growth factor receptor).
   2. A signal is transduced (carried) into the cytoplasm.
   3. A signal cascade is triggered. Multiple intermediate proteins or second messengers will be affected.
   4. A transcription factor will be activated to bind to a specific site on the DNA.
   5. Transcription occurs and the mRNA enters the cytoplasm.
   6. The mRNA is translated and a protein is formed.
   7. The protein functions within the cytoplasm—possibly triggering S phase.

   If you study figure 10.22 you will see a similar pathway for the formation of S phase proteins following receptor–ligand binding by a growth factor. In this diagram various proteins in the signaling pathway become phosphorylated and then dephosphorylated. Ultimately, the Rb protein that regulated the transcription factor E2F becomes phosphorylated. This releases the E2F and allows it to bind to the gene for S phase proteins and cyclins.

3. Proto-oncogenes tend to encode proteins that function in signal transduction pathways that control cell division. When the regulation of these proteins is aberrant, or they are stuck in the "on" state by mutation, it can lead to cancer.

Tumor suppressor genes, on the other hand, tend to be in genes that encode proteins that suppress instead of activate cell division. Thus loss of function for a tumor suppressor gene leads to cancer.

Anaphase I nondisjunction:

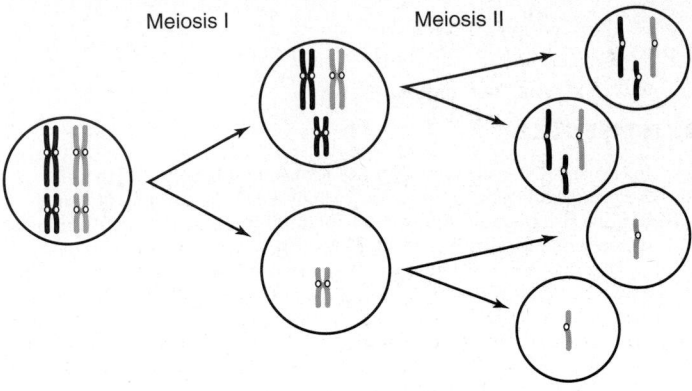

CHAPTER 11

LEARNING OUTCOME QUESTIONS

11.1 Stem cells divide by mitosis to produce one cell that can undergo meiosis and another stem cell.

11.2 No. Keeping sister chromatids together at the first division is key to this reductive division. Homologues segregate at the first division, reducing the number of chromosomes by half.

11.3 An improper disjunction at anaphase I would result in 4 aneuploid gametes: 2 with an extra chromosome and 2 that are missing a chromosome. Nondisjunction at anaphase II would result in 2 normal gametes and 2 aneuploid gametes: 1 with an extra chromosome and 1 missing a chromosome.

11.4 The independent alignment of homologous pairs at metaphase I and the process of crossing over. The first shuffles the genome at the level of entire chromosomes, and the second shuffles the genome at the level of individual chromosomes.

INQUIRY AND DATA ANALYSIS QUESTION

Page 217 Inquiry question: No. At the conclusion of meiosis I each cell has a single copy of each homologue. So, even if the attachment of sister chromatids were lost after a meiosis I division, the results would not be the same as mitosis.

UNDERSTAND

1. c 2. d 3. a 4. b 5. b 6. a 7. b

APPLY

1. c 2. b 3. b 4. d 5. b 6. a

SYNTHESIZE

1. Compare your figure with figure 11.8.
   a. There would be three homologous pairs of chromosomes for an organism with a diploid number of six.
   b. For each pair of homologues, you should now have a maternal and paternal pair.
   c. Many possible arrangements are possible. The key to your image is that it must show the homologues aligned pairwise—not single-file along the metaphase plate. The maternal and paternal homologues *do not* have to align on the same side of the cell. Independent assortment means that the pairs can be mixed.
   d. A diagram of metaphase II would not include the homologous pairs. The pairs have separated during anaphase of meiosis I. Your picture should diagram the haploid number of chromosomes, in this case three, aligned single-file along the metaphase plate. Remember that meiosis II is similar to mitosis.

2. The diploid chromosome number of a mule is 63. The mule receives 32 chromosomes from its horse parent (diploid 64: haploid 32) and another 31 chromosomes from its donkey parent (diploid 61: haploid 31). 32 + 31 = 63. The haploid number for the mule would be one half the diploid number 63 ÷ 2 = 31.5. Can there be a 0.5 chromosome? Even if the horse and donkey chromosomes can pair (no guarantee of that), there will be one chromosome without a partner. This will lead to aneuploid gametes that are not viable.

3. Independent assortment involves the random distribution of maternal versus paternal homologues into the daughter cells produced during meiosis I. The number of possible gametes is equal to 2^n, where n is the haploid number of chromosomes. Crossing over involves the physical exchange of genetic material between homologous chromosomes, creating new combinations of genes on a single chromosome. Crossing over is a relatively rare event that affects large blocks of genetic material, so independent assortment likely has the greatest influence on genetic diversity.

4. a. Nondisjunction occurs at the point when the chromosomes are being pulled to opposite poles. This occurs during anaphase.
   b. Use an image like figure 11.8 and illustrate nondisjunction at anaphase I versus anaphase II

CHAPTER 12

LEARNING OUTCOME QUESTIONS

12.1 Both had an effect, but the approach is probably the most important. In theory, his approach would have worked for any plant, or even animal, he chose. In practice, the ease of both cross- and self-fertilization was helpful.

12.2 1/3 of tall F_2 plants are true-breeding.

12.3 The events of meiosis I are much more important in explaining Mendel's laws. During anaphase I homologues separate and are thus segregated, and the alignment of different homologous pairs at metaphase I is independent.

12.4 The cross is $Aa\ Bb \times aa\ Bb$ and the probability for $(A\_ B\_) = (3/4)\ (1/2) = 3/8$.

12.5 1:1:1:1 dom-dom:dom-rec:rec-dom:rec-rec

12.6 6/16.

INQUIRY AND DATA ANALYSIS QUESTIONS

Page 223 Inquiry question: The ability to control whether the plants self-fertilized or cross-fertilized was of paramount importance in Mendel's studies. Results due to cross-fertilization would have had confounding influences on the predicted number of offspring with a particular phenotype.

Page 225 Data analysis: If the purple F_1 were backcrossed to the white parent, the phenotypic ratio would be 1 purple:1 white, and the genotypic ratio would be 1 non-true-breeding dominant (heterozygous):1 true-breeding recessive (homozygous recessive).

Page 227 Data analysis: Yes. The affected females each had one unaffected parent and thus are heterozygous. This means that the probability of affected offspring would be 50%.

Page 228 Inquiry question: Genetic defects that remain hidden or dormant as heterozygotes in the recessive state are more likely to be revealed in homozygous state among closely related individuals.

Page 231 Data analysis: The probability of being purple-flowered, round, and yellow is the same as the probability of being dominant for all three traits. The probability for the dominant phenotype from a cross of two heterozygotes is 3/4, so the probably of being dominant for all three traits in this cross would be (3/4)(3/4)(3/4) = 27/64.

Page 232 Data analysis:

TABLE 12.2	Dihybrid Testcross	
Actual Genotype	**Results of Testcross**	
	Trait A	**Trait B**
AABB	Trait A breeds true	Trait B breeds true
AaBB	——	Trait B breeds true
AABb	Trait A breeds true	——
AaBb	——	——

Page 233 Data analysis: Cuenot would have observed 2 yellow mice:1 wild type. This is because the homozygous yellow mice die, leaving 2 heterozygotes to 1 homozygous recessive.

Page 235 Inquiry question: Almost certainly, differences in major phenotypic traits of twins would be due to environmental factors such as diet.

Page 236 Data analysis: This is a case of epistasis, in which the albino gene obscures the effects of the black/brown locus. The offspring that are *B_ aa* (where the albino locus is designated by *a*) will appear the same as *bb aa* offspring. Thus the overall ratio is 9 black:3 brown:3+1 albino.

UNDERSTAND

1. b 2. c 3. c 4. c 5. c 6. d

APPLY

1. d 2. d 3. b 4. a 5. c 6. d

SYNTHESIZE

1. The approach to solving this type of problem is to identify the possible gametes. Separate the possible gamete combinations into the boxes along the top and side. Fill in the Punnett square by combining alleles from each parent.

   a. A monohybrid cross between individuals with the genotype *Aa* and *A*

	A	*a*
A	*AA*	*Aa*
a	*aA*	*aa*

   Phenotypic ratio: 3 dominant to 1 recessive

   b. A dihybrid cross between two individuals with the genotype *AaBb*

	AB	*Ab*	*aB*	*ab*
AB	*AABB*	*AABb*	*AaBB*	*AaBb*
Ab	*AAbB*	*AAbb*	*AabB*	*Aabb*
aB	*aABB*	*aABb*	*aaBB*	*aaBb*
ab	*aAbB*	*aAbb*	*aabB*	*aabb*

   Phenotypic ratio: 9 dominant dominant to 3 dominant recessive to 3 recessive dominant to 1 recessive recessive

   Using the product rule: $\text{Prob}(A_\ B_) = (3/4)(3/4) = 9/16$

   $\text{Prob}(A_\ bb) = (3/4)(3/4) = 3/16$

   $\text{Prob}(aa\ B_) = (1/4)(3/4) = 3/16$

   $\text{Prob}(aa\ bb) = (1/4)(1/4) = 1/16$

   c. A dihybrid cross between individuals with the genotype *AaBb* and *aabb*

	AB	*Ab*	*aB*	*ab*
ab	*aAbB*	*aAbb*	*aabB*	*aabb*

   Using the product rule: $\text{Prob}(A_\ B_) = (1/4)(1) = 1/16$

   $\text{Prob}(A_\ bb) = (1/4)(1) = 1/16$

   $\text{Prob}(aa\ B_) = (1/4)(1) = 1/16$

   $\text{Prob}(aa\ bb) = (1/4)(1) = 1/16$

2. The segregation of different alleles for any gene occurs due to the pairing of homologous chromosomes, and the subsequent separation of these homologues during anaphase I. The independent assortment of traits, more accurately the independent segregation of different allele pairs, is due to the independent alignment of chromosomes during metaphase I of meiosis.

3. There seems to be the loss of a genotype since there are only 3 possible outcomes (2 yellow and 1 black). A yellow gene that has a dominant effect on coat color, but also causes lethality when homozygous, could explain the observations. Therefore, a yellow mouse is heterozygous, and crossing two yellow mice yields 1 homozygous yellow (dead):2 heterozygous (appears yellow):1 black. You could test this by crossing the yellow to homozygous black. You should get 1 yellow:1 black, and all black offspring should be true-breeding, and all yellow should behave as above.

4. Two genes are involved, one of which is epistatic to the other. At one gene, there are two alleles: black and brown; at the other gene, there are two alleles: albino and colored. The albino gene is epistatic to the brown gene so that when the animal is homozygous recessive for albino, it is albino regardless of whether it is black or brown at the other locus. This leads to the 4 albino in a Mendelian kind of crossing scheme.

Chapter 13

LEARNING OUTCOME QUESTIONS

13.1 Females would be all wild type; males would be all white-eyed.

13.2 Yes, should be viable and appear female.

13.3 The mt^- DNA could be degraded by a nuclease similar to how bacteria deal with invading viruses. Alternatively, the mt^--containing mitochondria could be excluded from the zygote.

13.4 No, not by genetic crosses.

13.5 Yes. First-division nondisjunction yields four aneuploid gametes, but second-division yields only two aneuploid gametes.

INQUIRY AND DATA ANALYSIS QUESTIONS

Page 244 Inquiry question: There would probably be very little if any recombination, so the expected assortment ratios would have been skewed from the expected 9:3:3:1.

Page 247 Data analysis: Instead of equal proportions of four types of gametes, you would get 45% of each of the two parental types and only 5% of each of the recombinant types. How this would affect a dihybrid cross would depend on the original parents, but it would clearly skew the results such that the 9:3:3:1 phenotypic ratio would be completely obscured. It would have been impossible to conclude that the two loci were behaving independently, as in fact, they are not.

Page 251 Data analysis: What has changed is the mother's age. The older the woman, the higher the risk she has of nondisjunction during meiosis. Thus, she also has a much greater risk of producing a child with Down syndrome.

Page 251 Inquiry question: Nondisjunction produces an XX egg that is fertilized by a Y sperm. A normal X egg is fertilized by an XY sperm produced by nondisjunction. Note that the nondisjunction event that produces the XX egg could be either MI or MII, but the event that produces the XY sperm would have to be MI.

Page 253 Inquiry question: Advanced maternal age, a previous child with birth defects, or a family history of birth defects.

UNDERSTAND

1. c 2. d 3. d 4. a 5. c 6. c 7. c

APPLY

1. c 2. b 3. c 4. b 5. d 6. b

SYNTHESIZE

1. Theoretically, 25% of the children from this cross will be color blind. All of the color blind children will be male, and 50% of the males will be color blind.

2. Parents of heterozygous plant were green wrinkled × yellow round

   Frequency of recombinants is 36+29/1300 = 0.05

   Map distance = 5 cM

3. Male calico cats are very rare. The coloration that is associated with calico cats is the product of X inactivation. X inactivation only occurs in females as a response to dosage levels of the X-linked genes. The only way to get a male calico is to be heterozygous for the color gene and to be the equivalent of a Klinefelter's male (*XXY*).

Chapter 14

LEARNING OUTCOME QUESTIONS

14.1 The 20 different amino acid building blocks offers chemical complexity. This appears to offer informational complexity as well.

14.2 The proper tautomeric forms are necessary for proper base-pairing, which is critical to DNA structure.

14.3 Prior to replication in light N (i.e., ^{14}N) isotope there would be only one band. After one round of replication, there would be two bands with denatured DNA: one heavy and one light.

14.4 The 5′ to 3′ activity is used to remove RNA primers. The 3′ to 5′ activity is used to remove mispaired bases (proofreading).

14.5 A shortening of chromosome ends would eventually affect DNA that encodes important functions.

14.6 No. The number of DNA-damaging agents, in addition to replication errors, would cause lethal damage (this has been tested in yeast).

INQUIRY AND DATA ANALYSIS QUESTIONS

Page 262 Data analysis: The Watson–Crick helical structure fits the measurements made from Franklin's X-ray diffraction pictures. The base-pairing explains why Chargaff observed the regularities in quantities of G = C and A = T, and the base-pairing depends on the proper tautomeric forms of the bases. Taken together, the model rationalizes all of these data.

Page 264 Data analysis: If the Meselson–Stahl experiment was allowed to run for another round, the pattern of bands would be the same as the second round (light molecules and hybrid molecules), but the light, ^{14}N, band would be darker. This is because the products from the ^{14}N alone would remain ^{14}N, and the products from the hybrid molecules ($^{15}N/^{14}N$) would produce one hybrid molecule and one all-^{14}N molecule.

Page 265 Inquiry question: The covalent bonds create a strong backbone for the molecule, making it difficult to disrupt. Individual hydrogen bonds are more easily broken, allowing enzymes to separate the two strands without disrupting the inherent structure of the molecule.

Page 269 Inquiry question: DNA ligase is important in connecting Okazaki fragments during DNA replication. Without it, the lagging strand would not be complete.

Page 273 Inquiry question: The linear structure of chromosomes creates the end problem discussed in the text. It is impossible to finish the ends of linear chromosomes using unidirectional polymerases that require RNA primers. The size of eukaryotic genomes also means that the time necessary to replicate the genome is much greater than in prokaryotes with smaller genomes. Thus the use of multiple origins of replication.

Page 275 Inquiry question: Cells have a variety of DNA repair pathways that allow them to restore damaged DNA to its normal constitution. If DNA repair pathways are compromised, the cell will have a higher mutation rate. This can lead to higher rates of cancer in a multicellular organism such as humans.

UNDERSTAND

1. d   2. a   3. c   4. a   5. c   6. b   7. b

APPLY

1. c   2. b   3. c   4. c   5. a   6. b   7. d   8. c

SYNTHESIZE

1. a. If both bacteria are heat-killed, then the transfer of DNA will have no effect since pathogenicity requires the production of proteins encoded by the DNA. Protein synthesis will not occur in a dead cell.
   b. The nonpathogenic cells will be transformed to pathogenic cells. Loss of proteins will not alter DNA.
   c. The nonpathogenic cells remain nonpathogenic. If the DNA is digested, it will not be transferred and no transformation will occur.

2. The region could be an origin of replication. Origins of replication are adenine- and thymine-rich regions since only these nucleotides form two hydrogen bonds versus the three hydrogen bonds formed between guanine and cytosine, making it easier to separate the two strands of DNA.

    The RNA primer sequences would be 5′-ACUAUUGCUUUAUAA-3′. The sequence is antiparallel to the DNA sequence (review figure 14.16), meaning that the 5′ end of the RNA is matching up with the 3′ end of the DNA. It is also important to remember that in RNA the thymine nucleotide is replaced by uracil (U). Therefore, the adenine in DNA will form a complementary base-pair with uracil.

3. a. *DNA gyrase* functions to relieve torsional strain on the DNA. If DNA gyrase were not functioning, the DNA molecule would undergo supercoiling, causing the DNA to wind up on itself, preventing the continued binding of the polymerases necessary for replication.
   b. *DNA polymerase III* is the primary polymerase involved in the addition of new nucleotides to the growing polymer and in the formation of the phosphodiester bonds that make up the sugar–phosphate backbone. If this enzyme were not functioning, then no new DNA strand would be synthesized and there would be no replication.
   c. *DNA ligase* is involved in the formation of phosphodiester bonds between Okazaki fragments. If this enzyme were not functioning, then the fragments would remain disconnected and would be more susceptible to digestion by nucleases.
   d. *DNA polymerase I* functions to remove and replace the RNA primers that are required for DNA polymerase III function. If DNA polymerase I were not available, then the RNA primers would remain and the replicated DNA would become a mix of DNA and RNA.

CHAPTER 15

LEARNING OUTCOME QUESTIONS

15.1 There is no molecular basis for recognition between amino acids and nucleotides. The tRNA is able to interact with nucleic acid by base-pairing, and an enzyme can covalently attach amino acids to it.

15.2 There would be no specificity to the genetic code. Each codon must specify a single amino acid, although amino acids can have more than one codon.

15.3 Transcription/translation coupling cannot exist in eukaryotes where the two processes are separated in both space and time.

15.4 No. This is a result of the evolutionary history of eukaryotes but is not necessitated by genome complexity.

15.5 Alternative splicing offers flexibility in coding information. One gene can encode multiple proteins.

15.6 This tRNA would be able to "read" STOP codons. This could allow nonsense mutations to be viable, but would cause problems making longer than normal proteins. Most bacterial genes actually have more than one STOP at the end of the gene.

15.7 Attaching amino acids to tRNAs, bringing charged tRNAs to the ribosome, and ribosome translocation all require energy.

15.9 No. It depends on where the breakpoints are that created the inversion, or duplication. For duplications it also depends on the genes that are duplicated.

INQUIRY AND DATA ANALYSIS QUESTIONS

Page 280 Data analysis: This double-mutant strain would only grow on media supplemented with all of the intermediates. This is because ArgE mutants are blocked at the first step. In general, these kinds of double mutants allow you to determine which gene occurs earlier in a pathway: the double mutant will look like an organism with a mutation at the earliest point in the pathway.

Page 281 Inquiry question: One would expect higher amounts of error in transcription over DNA replication. Proofreading is important in DNA replication because errors in DNA replication will be passed on to offspring as mutations. However, RNAs have very short life spans in the cytoplasm, and therefore mistakes are not permanent.

Page 284 Inquiry question: The very strong similarity among organisms indicates a common ancestry of the code.

Page 285 Inquiry question: The promoter acts as a binding site for RNA polymerase. The structure of the promoter provides information both about where to bind and the direction of transcription. If the two sites were identical, the polymerase would need some other cue for the direction of transcription.

Page 289 Data analysis: Splicing can produce multiple transcripts from the same gene. This is shown below with the exons shown in different colors.

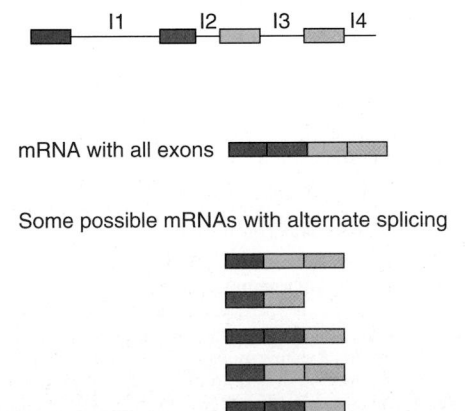

Page 297 Inquiry question: Wobble not only explains the number of tRNAs that are observed due to the increased flexibility in the 5′ position, it also accounts for the degeneracy that is observed in the genetic code. The degenerate base is the one in the wobble position.

UNDERSTAND

1. d   2. c   3. d   4. b   5. c   6. b   7. c

1. b   2. c   3. b   4. d   5. a   6. b   7. b

SYNTHESIZE

1. The predicted sequence of the mRNA for this gene is

   5′–GCAAUGGGCUCGGCAUGCUAAUCC–3′

   The predicted amino acid sequence of the protein is

   5′–GCA AUG GGC UCG GCA UGC UAA UCC–3′

   Met-Gly-Ser-Ala-Cys-STOP

2. A frameshift essentially turns the sequence of bases into a "random" sequence. If you consider the genetic code, 3 of the 64 codons are STOP, so the probability of hitting a STOP in a random sequence is 3/64, or about 1 in every 20 codons.

3. a. mRNA = 5′–GCA AUG GGC UCG GCA UUG CUA AUC C–3′
      The amino acid sequence would then be:
      Met-Gly-Ser-Ala-Leu-Leu-Iso-.
      There is no STOP codon. This is an example of a frameshift mutation. The addition of a nucleotide alters the "reading frame," resulting in a change in the type and number of amino acids in this protein.
   b. mRNA = 5′–GCA AUG GGC UAG GCA UGC UAA UCC–3′
      The amino acid sequence would then be: Met-Gly-STOP.
      This is an example of a nonsense mutation. A single nucleotide change has resulted in the early termination of protein synthesis by altering the codon for Ser into a STOP codon.
   c. mRNA = 5′–GCA AUG GGC UCG GCA AGC UAA UCC –3′
      The amino acid sequence would then be: Met-Gly-Ser-Ala-Ser-STOP.
      This base substitution has affected the codon that would normally encode Cys (UGC) and resulted in the addition of Ser (AGC).

4. The split genes of eukaryotes offers the opportunity to control the splicing process, which does not exist in prokaryotes. This is also true for polyadenylation in eukaryotes. In prokaryotes, transcription/translation coupling offers the opportunity for the process of translation to have an effect on transcription.

Chapter 16

LEARNING OUTCOME QUESTIONS

16.1 The control of gene expression would be more like that in humans (fellow eukaryotes) than in *E. coli*.

16.2 The two helices both interact with DNA, so the spacing between the helices is important for both to be able to bind to DNA.

16.3 The operon would be on all of the time (constitutive expression).

16.4 The loss of a general transcription factor would likely be lethal as it would affect all transcription. The loss of a specific factor would affect only those genes controlled by the factor.

16.5 These genes are necessary for the ordinary functions of the cell. That is, the role of these genes is in ordinary housekeeping and not in any special functions.

16.6 RNA interference offers a way to specifically affect gene expression using drugs made of siRNAs.

16.7 Because there are many proteins in a cell doing a variety of functions, uncontrolled degradation of proteins would be devastating to the cell.

INQUIRY AND DATA ANALYSIS QUESTIONS

Page 308 Inquiry question: The presence of more than one gene in the operon allows for increased control over the elements of the pathway and therefore the product. A single regulatory system can regulate several adjacent genes.

Page 309 Data analysis: A mutation that prevents the repressor from binding to DNA would express the genes for lactose metabolism all the time (constitutive expression). A mutation that prevents the inducer from binding to the repressor would bind to DNA all the time, and thus never express the genes for lactose metabolism (uninducible).

Page 315 Inquiry question: Regulation occurs when various genes have the same regulatory sequences, which bind the same proteins.

Page 324 Inquiry question: Ubiquitin is added to proteins that need to be removed because they are nonfunctional or those that are degraded as part of a normal cellular cycle.

UNDERSTAND

1. c   2. d   3. a   4. c   5. b   6. c   7. b

APPLY

1. c   2. c   3. b   4. d   5. c   6. a   7. c

SYNTHESIZE

1. Mutations that affect binding sites for proteins on DNA will control the expression of genes covalently linked to them. Introducing a wild-type binding site on a plasmid will not affect this. We call this being *cis*-dominant. Mutations in proteins that bind to DNA would be recessive to a wild-type gene introduced on a plasmid.

2. Negative control of transcription occurs when the ability to initiate transcription is reduced. Positive control occurs when the ability to initiate transcription is enhanced. The *lac* operon is regulated by the presence or absence of lactose. The proteins encoded within the operon are specific to the catabolism (breakdown) of lactose. For this reason, operon expression is only required when there is lactose in the environment. Allolactose is formed when lactose is present in the cell. The allolactose binds to a repressor protein, altering its conformation and allowing RNA polymerase to bind. In addition to the role of lactose, there is also a role for the activator protein CAP in regulation of *lac*. When cAMP levels are high then CAP can bind to DNA and make it easier for RNA polymerase to bind to the promoter. The *lac* operon is an example of both positive and negative control.

   The *trp* operon encodes protein manufacture of tryptophan in a cell. This operon must be expressed when cellular levels of tryptophan are low. Conversely, when tryptophan is available in the cell, there is no need to transcribe the operon. The tryptophan repressor must bind tryptophan before it can take on the right shape to bind to the operator. This is an example of negative control.

3. Forms that control gene expression that are unique to eukaryotes include alternative splicing, control of chromatin structure, control of transport of mRNA from the nucleus to the cytoplasm, control of translation by small RNAs, and control of protein levels by ubiquitin-directed destruction. Of these, most are obviously part of the unique features of eukaryotic cells. The only mechanisms that could work in prokaryotes would be translational control by small RNAs and controlled destruction of proteins.

4. Mutation is a permanent change in the DNA. Regulation is a short-term change controlled by the cell. Like mutations, regulation can alter the number of proteins in a cell, change the size of a protein, or eliminate the protein altogether. The key difference is that gene regulation can be reversed in response to changes in the cell's environment. Mutations do not allow for this kind of rapid response.

Chapter 17

LEARNING OUTCOME QUESTIONS

17.1 **#1** Recombinant DNA makes it possible to introduce into a different organism a single gene from the same or a very different species. Gene combinations that are impossible via conventional breeding (e.g., inserting a bacterial gene into a plant) can be created. Breeding can be more targeted and quicker.

#2 *Eco*RI is a restriction enzyme that can be used to cut DNA at specific places. Ligase is used to "glue" together pieces of DNA that have been cut with the same restriction enzyme. The two enzymes make it possible to add foreign DNA into an *E. coli* plasmid.

17.2 A restriction digest DNA fragment is combined with a plasmid that has been cut with the same restriction enzyme. Ligase is used to insert the fragment into the plasmid, which can then be inserted into a bacterial cell, a process called transformation. The cell is cultured and divides many times. The plasmids can then be isolated, and, if desired, the many copies of the fragment can be isolated with a restriction digest and could again be separated using gel electrophoresis.

17.3 Both PCR and DNA replication result in new copies of the DNA made by separating the original strands and using a primer to initiate replication, followed by the addition of dNTPs with the help of a DNA polymerase. In DNA replication, an enzyme called helicase separates the two strands, whereas in PCR, high temperature (95°C) is used to separate the strands. DNA replication uses an RNA primer made by DNA primase, but PCR relies on synthetic DNA primers. The DNA polymerase in most organisms is heat-sensitive. In PCR, a heat-stable DNA polymerase from an archaean that is extremely heat-tolerant is used. Multiple rounds of PCR allow for an exponential increase in copies of the DNA, whereas in most cases, DNA replication results in one new copy of each chromosome before a cell divides.

17.4 A cDNA library is constructed from mRNA. Unlike the gene itself, cDNA does not include the introns or regulatory elements.

17.5 Isolate DNA from cells from you and your neighbor in separate tubes. Digest both DNA samples with the restriction endonucleases and separate the fragments using gel electrophoresis. "Blot" the gel onto nitrocellulose paper and incubate the paper with probes for the relevant RFLPs. Compare the patterns of the RFLP fragments on the two different "blots."

17.6 Your flowchart should start with the DNA sequence for an influenza coat protein, likely a restriction digest fragment. Next the DNA fragment is transformed into a plant cell, and a plant is regenerated and tested. Generations of plants can be bred from a successfully transformed plant. To finish the vaccine, a process must be established to isolate the influenza coat protein from the plant tissue.

17.7 GM plants with pest resistance are created by introducing a single gene into a plant using recombinant technologies. MAB allows traits to be linked to markers within the genome without knowing the sequence of specific genes, thus more traits can be bred into the crop plant. Furthermore, many traits are the result of multiple genes that can all be followed by markers. With MAB, a set of regions within the genome can all be followed and simultaneously bred into the crop plant.

INQUIRY AND DATA ANALYSIS QUESTIONS

Page 331 Inquiry question: A bacterial artificial chromosome or a yeast artificial chromosome would be the best way to go because only a plasmid vector can stably hold up to 10 kb.

Page 334 Data analysis: Six million clones would cover all the DNA sequences in the genome, but there would be no overlapping sequences, presenting a problem if you wanted to use the overlaps to sort out the sequence in which all the clones were found in the genome.

Page 335 Inquiry question: No, cDNA is created using mRNA as a template; therefore, intron sequences would not be expressed using the cDNA.

Page 345 Data analysis: In comparing the weights of the transgenic salmon with the wild salmon, at every day after feeding, the transgenic salmon weighed more than the wild salmon. At the end of the experiment the transgenic salmon's size was 150% of the wild salmon.

UNDERSTAND

1. b   2. b   3. d   4. d   5. c   6. b   7. d   8. a

APPLY

1. d   2. c   3. d

SYNTHESIZE

1. Genes coding for each of the subunits would need to be inserted into different plasmids that are integrated into different bacteria. The cultures would need to be grown separately, and the different protein subunits would then need to be isolated and purified. If the subunits can self-assemble in vitro, then the protein could be functional. It could be difficult to establish just the right conditions for the assembly of the multiple subunits.

2. Your proposal should explain how you will cross the nematode-resistant Chinese soybean and the otherwise robust soybean and isolate DNA from all the offspring in the seedling stage. The STRs will be amplified, and only seedlings that have the STRs associated with nematode resistance will be grown to adulthood and used in the next generation of the breeding study.

CHAPTER 18

LEARNING OUTCOME QUESTIONS

18.1 Banding sites on karyotypes depend on dyes binding to the condensed DNA that is wrapped around protein. The dyes bind to some regions, but not all, and are therefore not evenly spaced along the genome in the way that sequential base-pairs are evenly spaced.

18.2 In DNA replication both strands are copied, but in sequencing only one of the strands is copied. Both replication and sequencing require a primer to start, although in replication the primer is RNA and in sequencing a DNA primer is used. In DNA replication, an entire strand is replicated. In sequencing, the addition of another base-pair is either temporarily or permanently stopped in order to determine the sequence in which base-pairs are added. Sequencing uses labeled dNTPs, whereas DNA replication uses dNTPs without labels.

18.3 One possibility is that transposable elements can move within the genome and create new genetic variability, subject to natural selection.

18.4 The sequences may be very similar, but they could be organized in different syntenic blocks. For example, a shared length of DNA sequence could be found on two different chromosomes in Neandertals and humans.

18.5 From the transcriptome, it is possible to predict the proteins that may be translated and available for use in part of an organism at a specific time in development.

18.6 The promoter region. In the presence of the biotoxin, it would be desirable for the bioluminescent gene to be expressed. One way to regulate this is to have the biotoxin bind to the promoter and support transcription.

INQUIRY AND DATA ANALYSIS QUESTIONS

Page 350 Inquiry question: There is one copy of *bcr* (see with green probe) and one copy of *abl* (seen with red probe). The other *bcr* and *abl* genes are fused, and the yellow color is the result of red plus green fluorescence combined.

Page 351 Data analysis: Clones A and C contain the necessary information to construct the physical map. Clones A, B, and D would also work but that isn't the fewest number of clones.

Page 356 Data analysis: If there are approximately 100,000 alternative splicing events and most genes have multiple introns (only an estimate), then each gene would have about 100,000 splice events divided by 25,000 genes, or 4 splice events per gene. Keep in mind that this is an approximation, and there is likely greater variation among genes.

Page 356 Inquiry question: Count the number of open reading frames, which consist of translation start sites (ATG in the DNA) and enough base-pairs to encode a protein before a STOP codon. This might be an overestimate because not all ATG sequences are part of coding DNA sequences or may not be in the actual reading frame.

Page 358 Inquiry question: Repetitive elements are one of the main obstacles to assembling the DNA sequences in proper order because it is difficult to determine which sequences are overlapping.

Page 365 Inquiry question: Proteins exhibit posttranslational modification and the formation of protein complexes. Additionally, a single gene can code for multiple proteins using alternative splicing.

Page 366 Inquiry question: A proteome is all the proteins coded for by the genome, and the transcriptome is all the RNA present in a cell or tissue at a specific time.

Page 369 Inquiry question: You may be able to take advantage of synteny between the rice and corn genome (see figure 18.10). Let's assume that a drought-tolerance gene has already been identified and mapped in rice. Using what is known about synteny between the rice and corn genomes, you could find the region of the corn genome that corresponds to the rice drought-tolerance gene. This would narrow down the region of the corn genome that you might want to sequence to find your gene. A subsequent step might be to modify the corn gene that corresponds to the rice gene to see if you can increase drought tolerance.

UNDERSTAND

1. b   2. a   3. c   4. d   5. b   6. c   7. b   8. d

APPLY

1. b   2. a   3. d   4. b   5. c   6. d   7. d

SYNTHESIZE

1. The STSs represent unique sequences in the genome. They can be used to align the clones into one contiguous sequence of the genome based on the presence or absence of an STS in a clone. The contig, with aligned clones, would look like this:

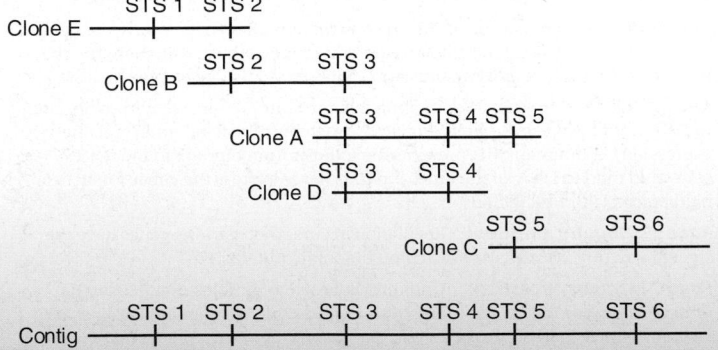

2. The anthrax genome has been sequenced. Investigators would look for differences in the genome between existing natural strains and those collected from a suspected outbreak. The genome of an infectious agent can be modified, or "weaponized," to make it more deadly. Also, single-nucleotide polymorphisms could be used to identify the source of the anthrax. In the case of the Florida anthrax outbreak it was determined that the source was a research laboratory.

CHAPTER 19

LEARNING OUTCOME QUESTIONS

19.2 The early cell divisions are very rapid and do not involve an increase in size between divisions. Interphase is greatly reduced allowing very fast cell divisions.

19.3 This requires experimentation to isolate cell from contact, which would prevent induction, or to follow a particular cell's lineage.

19.4 The nucleus must be reprogrammed. What this means exactly on the molecular level is not clear, but probably involves changes in chromatin structure and methylation patterns.

19.5 Homeotic genes seem to have arisen very early in the evolutionary history of bilaterians. These have been duplicated, and they have diversified with increasing morphological complexity.

19.6 Cell death can be a patterning mechanism. Your fingers were sculpted from a paddle-like structure by cell death.

INQUIRY AND DATA ANALYSIS QUESTIONS

Page 378 Data analysis: If you use a reagent to block FGF signaling, you would get only muscle precursor cells and nerve cord precursor cells. If you had a reagent that causes constitutive signaling, then you would get only mesenchyme precursor cells and notochord precursor cells.

Page 380 Data analysis: The difference in pigmentation pattern acts as a marker for the nuclear and cytoplasmic donors. Since the difference is genetic, the phenotype of the resulting sheep tells you which nucleus gave rise to this individual. Thus Dolly resembled the nuclear donor and not the cytoplasmic donor or the surrogate mother.

UNDERSTAND

1. b   2. d   3. c   4. d   5. b   6. c   7. b

APPLY

1. d   2. a   3. b   4. a   5. c   6. c   7. c

SYNTHESIZE

1. The horizontal lines of the fate map represent cell divisions. Starting with the egg, four cell divisions are required to establish a population of cells that will become nervous tissue. It takes another eight to nine divisions to produce the final number of cells that will make up the nervous system of the worm. It takes seven to eight rounds of cell division to generate the population of cells that will become the gonads. Once established, another seven to eight cell divisions are required to produce the actual gonad cells.

2. Not every cell in a developing embryo will survive. The process of apoptosis is responsible for eliminating cells from the embryo. In *C. elegans*, the process of apoptosis is regulated by three genes: *ced-3, ced-4,* and *ced-9.* Both *ced-3 and ced-4* encode proteases, enzymes that degrade proteins. Interestingly, the *ced-3* protease functions to activate gene expression of the *ced-4* protease. Together, these proteases will destroy the cell from the inside-out. The *ced-9* gene functions to repress the activity of the protease-encoding genes, thereby preventing apoptosis.

3. a. N-cadherin plays a specific role in differentiating cells of the nervous system from ectodermal cells. Ectodermal cells express E-cadherin, but neural cells express N-cadherin. The difference in cell-surface cadherins means that the neural cells lose their contact with the surrounding ectodermal cells and establish new contacts with other neural cells. In the absence of N-cadherin, the nervous system would not form. If you assume that E-cadherin expression is also lost (as would occur normally in development), then these cells would lose all cell–cell contacts and would probably undergo apoptosis.

   b. Integrins mediate the connection between a cell and its surrounding environment, the extracellular matrix (ECM). The loss of integrins would result in the loss of cell adhesion to the ECM. These cells would not be able to move, and, therefore, gastrulation and other developmental processes would be disrupted.

   c. Integrins function by linking the cell's cytoskeleton to the ECM. This connection is critical for cell movement. The deletion of the cytoplasmic domain of the integrin would not affect the ability of integrin to attach to the ECM, but it would prevent the cytoskeleton from getting a "grip." This deletion would likely result in a disruption of development similar to the complete loss of integrin.

4. Adult cells from the patient would be cultured with factors that reprogram the nucleus into pluripotent cells. These cells would then be grown in culture with factors necessary to induce differentiation into a specific cell type that could be transplanted into the patient. This would be easiest for tissue like a liver that regenerates, but could in theory be used for a variety of cell types.

CHAPTER 20

LEARNING OUTCOME QUESTIONS

20.1 Natural selection occurs when some individuals are better suited to their environment than others. These individuals live longer and reproduce more, leaving more offspring with the traits that enabled their parents to thrive. In essence, genetic variation within a population provides the raw material on which natural selection can act thereby leading to evolution.

20.2 #1 To determine if a population is in Hardy–Weinberg equilibrium, it is first necessary to determine the actual allele frequencies, which can be calculated based on the genotype frequencies. After assigning variables p and q to the allele frequencies, we then use the Hardy–Weinberg equation, $p^2 + 2pq + q^2 = 1$ to determine the expected genotype frequencies. If the actual and expected genotype frequencies are the same (or, at least not significantly different), it is safe to say that the population is in Hardy–Weinberg equilibrium.

   #2 You would conclude that one or more of the five evolutionary agents were acting to cause the lack of equilibrium. The next step would be to design studies to test hypotheses about which assumption is not being met.

20.3 There are five mechanisms of evolution: natural selection, mutation, gene flow (migration), genetic drift, and nonrandom mating. Any of these mechanisms can alter allele frequencies within a population, although usually a change in allele frequency results from more than one mechanism working in concert (e.g., mutation can introduce a beneficial new allele into the population, and natural selection will select for that allele such that its frequency increases over the course of two or more generations). Natural selection, the first mechanism and probably the most influential in bringing about evolutionary change, is also the only one to produce adaptive change, that is, change that results in the population being better adapted to its environment. Mutation is the only way in which new alleles can be introduced—it is the ultimate source of all variation. Because it is a relatively rare event, mutation by itself is not a strong agent of allele frequency change; however, in concert with other mechanisms, especially natural selection, it can drastically change the allele frequencies in a population. Gene flow can introduce new alleles into a population from another population of the same species, thus changing the allele frequency within both the recipient and donor populations. Genetic drift is the random, chance factor of evolution—although the results of genetic drift can be negligible in a large population, small populations can undergo drastic changes in allele frequency due to this agent. Finally, nonrandom mating results in populations that vary from Hardy-Weinberg equilibrium not by changing allele frequencies but by changing genotype frequencies—nonrandom mating reduces the proportion of heterozygotes in a population.

20.4 Reproductive success relative to other individuals within an organism's population is referred to as that organism's fitness. Its fitness is determined by its longevity, mating frequency, and the number of offspring it produces for each mating. None of these factors is always the most important in determining reproductive success—instead it is the cumulative effects of all three factors that determine an individual's reproductive success. For example, an individual that has a very long life span but mates only infrequently might have lower fitness than a conspecific that lives only half as long but mates more frequently and with greater success. As seen with the water strider example in this section, traits that are favored for one component of fitness, say, for example, longevity, may be disadvantageous for other components of fitness, say, lifetime fecundity.

20.5 #1 In a population wherein heterozygotes had the lowest fitness, natural selection should favor both homozygous forms. This would result in disruptive selection and a bimodal distribution of traits within the population. Over enough time, it could lead to a speciation event.

   #2 Negative frequency-dependent selection occurs when rare alleles have a fitness advantage over common alleles. As a result, selection is maintained because once an allele becomes rare, selection operates to increase its frequency. Oscillating selection refers to the situation in which the environment changes, first favoring one allele, then another. Over the long term, negative frequency-dependent selection is more likely to preserve variation, because rare alleles are always favored. By contrast, if the environment does not change (oscillate) for an unusually long period of time, the disadvantaged allele may disappear from the population.

20.6 Directional selection occurs when one phenotype has an adaptive advantage over other phenotypes in the population, regardless of its relative frequency within the population. Frequency-dependent selection, on the other hand, results when either a common (positive frequency-dependent selection) or rare (negative frequency-dependent selection) has a selective advantage simply by virtue of its commonality or rarity. In other words, if a mutation introduces a novel allele into a population, directional selection may result in evolution because the allele is advantageous, not because it is rare.

20.7 Background color matching is a form of camouflage used by many species to avoid predation; however, this example of natural selection runs counter to sexual selection—males want to be inconspicuous to predators but attractive to potential mates. For example, to test the effects of predation on background color matching in a species of butterfly, one might raise captive populations of butterflies with a normal variation in coloration. After a few generations, add natural predators to half of the enclosures. After several generations, one would expect the butterflies in the predatory environment to have a high degree of background color matching in order to avoid predation, while the nonpredatory environment would have promoted brightly colored individuals where color would correlate with mating success.

20.8 The dynamics among the different evolutionary mechanisms are very intricate, and it is often difficult, if not impossible, to discern in which direction each process is operating within a population—it is much easier to simply see the final cumulative effects of the various agents of evolutionary change. However, in some cases more than one evolutionary process will operate in the same direction, with the resulting population changing, or evolving, more rapidly than it would have under only one evolutionary mechanism. For example, mutation may introduce a beneficial allele into a population; gene flow could then spread the new allele to other populations. Natural selection will favor this allele within each population, resulting in relatively rapid evolutionary adaptation of a novel phenotype.

20.9 Pleiotropic effects occur with many genes; in other words, a single gene has multiple effects on the phenotype of the individual. Whereas natural selection might favor a particular aspect of the pleiotropic gene, it might select against another aspect of the same gene; thus, pleiotropy often limits the degree to which a phenotype can be altered by natural selection. Epistasis occurs when the expression of one gene is controlled or altered by the existence or expression of another gene. Thus, the outcome of natural selection will depend not just on the genotype of one gene, but the other genotype as well.

INQUIRY AND DATA ANALYSIS QUESTIONS

Page 399 Data analysis: In the example of figure 20.3, the frequency of the recessive white genotype is 0.16. The remaining 84 cats (out of 100) in the population are homozygous or heterozygous black. If the 16 white cats died, they will not contribute recessive white genes to the next generation. Only heterozygous black cats will produce white kittens in a 3:1 ratio of black to white. Homozygous × homozygous black and homozygous × heterozygous black cats will have all black kittens. Since there are 36 homozygous black cats and 48 heterozygous black cats, with a new total of 84 cats, the new frequency of homozygous black cats is 36/84, or 43%, with the heterozygous black cats now comprising 57% of the population. If p^2 = 0.43, then p = 0.65 (approximately), then $1 - p = q$, and q = 0.35. The frequency of white kittens in the next generation, q^2, is 0.12, or 12%.

Page 402 Inquiry question: Rare alleles are more likely to be lost in a population bottleneck specifically because they are rare; the chance that a small population will not include an individual that has that allele is relatively high. For this reason, rare alleles are at risk of being lost. However, if by chance an individual with a rare allele is in the small population, the allele's frequency becomes much greater because of the small number of individuals in the population.

Page 405 Inquiry question: Differential predation might favor brown toads over green toads, green toads might be more susceptible to disease, or green toads might be less able to tolerate variations in climate, among other possibilities.

Page 406 Inquiry question: Since the intermediate-sized water strider has the highest level of fitness, we would expect that the intermediate size would become more prevalent in the population. If the number of eggs laid per day was not affected by body size, the small water striders would be favored because of their tendency to live longer than their larger counterparts.

Page 406 Data analysis: 12-mm striders lay about 2 eggs per day and live about 32 days, for an expected total egg number of 64; the corresponding numbers for 15-mm water striders are 6 and 18, respectively, for a total of 108.

Page 407 Data analysis: In frequency-dependent selection, the fitness of a phenotype is related to its frequency. Thus, for each color type, fitness should be related to frequency, and the color types should not exhibit consistent differences in fitness. By contrast, directional selection is not frequency-dependent; color types should differ in fitness, and those differences should not be affected by frequency.

Page 410 (figure 20.15) Inquiry question: The proportion of flies moving toward light (positive phototropism) would again begin to increase in successive generations.

Page 410 (figure 20.16) Inquiry question: The distribution of birth weights in the human population would expand somewhat to include more babies of higher and lower birth weights.

Page 410 Data analysis: Rates of infant mortality are highest at very low and very high birth weights, and lowest at intermediate birth weights. However, many more babies are born with intermediate birth weights than with extreme births. For example, there are 40 times more individuals born with a birth weight of 6.5 pounds than 2 pounds. As a result, even though the mortality rate is 8 times higher for 2-pound babies than 6.5-pound babies, the absolute number of deaths would be higher for the intermediate-weight size. Overall, the relationship between number of deaths and birth weight would show a peak at intermediate birth weights.

Page 412 Inquiry question: Guppy predators evidently locate their prey using visual cues. The more colorful the guppy, the more likely it is to be seen and thus the more likely it will become prey.

Page 413 Data analysis: On the right-hand side, the index declines with distance from the mine, which suggests that whatever process is operating, its effect is negatively related to distance from the mine. For a given distance from the mine in both directions, index values are higher on the right side, suggesting that the mechanism is stronger on that side. Dispersal of tolerance alleles driven by wind patterns can explain both results: why the allele frequency tails off with distance from the mine, and why the levels are higher on the downwind side of the mine.

Page 414 Inquiry question: Thoroughbred horse breeders have been using selective breeding for certain traits over many decades, effectively removing variation from the population of thoroughbred horses. Unless mutation produces a faster horse, it remains unlikely that winning speeds will improve.

UNDERSTAND

1. a   2. b   3. d   4. a   5. d   6. a   7. d

APPLY

1. d   2. d   3. d

SYNTHESIZE

1. The results depend on coloration of guppies increasing their conspicuousness to predators such that an individual's probability of survival is lower than if it was a drab morph. In the laboratory it may be possible to conduct trials in simulated environments; we would predict, based on the hypothesis of predation, that the predator would capture more of the colorful morph than the drab morph when given access to both. Design of the simulated environment would obviously be critical, but results from such an experiment, if successful, would be a powerful addition to the work already accomplished.

2. On the large lava flows, where the background is almost entirely black, those individuals with black coloration within a population will have a selective advantage because they will be more cryptic to predators. On the other hand, on small flows, which are disrupted by light sand and green plants, dark individuals would be at an adaptive disadvantage for the same reason. You can read more about this in chapter 21; the black peppered moths had an advantage on the trees lacking lichen, but a disadvantage on lichen-covered trees.

3. Ultimately, genetic variation is produced by the process of mutation. However, compared with the speed at which natural selection can reduce variation in traits that are closely related to fitness, mutation alone cannot account for the persistence of genetic variation in traits that are under strong selection. Other processes can account for the observation that genetic variation can persist under strong selection. They include gene flow. Populations are often distributed along environmental gradients of some type. To the extent that different environments favor slightly different variants of phenotypes that have a genetic basis, gene flow among areas in the habitat gradient can introduce new genetic variation or help maintain existing variation. Similarly, just as populations frequently encounter different

selective environments across their range (think of the guppies living above and below the waterfalls in Trinidad), a single population also encounters variation in selective environments across time (oscillating selection). Traits favored this year may not be the same as those favored next year, leading to a switching of natural selection and the maintenance of genetic variation.

CHAPTER 21

LEARNING OUTCOME QUESTIONS

21.1 No. If eating hard seeds caused individuals to develop bigger beaks, then the phenotype is a result of the environment, not the genotype. Natural selection can only act upon those traits with a genetic component. Just as a body builder develops large muscles in his or her lifetime but does not have well-muscled offspring, birds that develop large beaks in their lifetime will not necessarily have offspring with larger beaks.

21.2 An experimental design to test this hypothesis could be as simple as producing enclosures for the moths and placing equal numbers of both morphs into each enclosure and then presenting predatory birds to each enclosure. One enclosure could be used as a control. One enclosure would have a dark background, and the other would have a light background. After several generations, measuring the phenotype frequency of the moths should reveal very clear trends—the enclosure with the dark background should consist of mostly dark moths, the enclosure with the light background mostly light moths, and the neutral enclosure should have an approximately equal ratio of light to dark moths.

21.3 If the trait that is being artificially selected for is due to the environment rather than underlying genotype, then the individuals selected that have that trait will not necessarily pass it on to their offspring.

21.4 The major selective agent in most cases of natural selection is the environment; thus, climatic changes, major continental shifts, and other major geological changes would result in dramatic changes in selective pressure; during these times the rate and direction of evolutionary change would likely be affected in many, if not most, species. On the other hand, during periods of relative environmental stability, the selective pressure does not change and we would not expect to see many major evolutionary events.

21.5 The only other explanation that could be used to explain homologous characteristics and vestigial structures could be mutation. Especially in the case of vestigial structures, if one resulted from a mutation that had pleiotropic effects, and the other effects of the genetic anomaly were selected for, then the vestigial structure would also be selected for, much like a rider on a Congressional bill.

21.6 Convergence occurs when distantly related species experience similar environmental pressures and respond, through natural selection, in similar ways. For example, penguins (birds), sharks (fish), sea lions (mammals), and even the extinct ichthyosaur (reptile) all exhibit the fusiform shape. Each of these animals has similar environmental pressures in that they are all aquatic predators and need to be able to move swiftly and agilely through the water. Clearly their most recent common ancestor does not have the fusiform body shape; thus the similarities are due to convergence (environment) rather than homology (ancestry). However, similar environmental pressures will not always result in convergent evolution. Most importantly, in order for a trait to appear for the first time in a lineage, there must have been a mutation; however, mutations are rare events, and even rarer is a beneficial mutation. There may also be other species that already occupy a particular niche; in these cases it would be unlikely that natural selection would favor traits that would increase the competition between two species.

21.7 It is really neither a hypothesis nor a theory. Theories are the building blocks of scientific knowledge; they have withstood the most rigorous testing and review. Hypotheses, on the other hand, are tentative answers to a question. Unfortunately, a good hypothesis must be testable and falsifiable, and stating that humans came from Mars is not realistically testable or falsifiable; thus, it is, in the realm of biological science, a nonsense statement.

INQUIRY AND DATA ANALYSIS QUESTIONS

Page 419 Inquiry question: Assuming that the size of seeds that parents fed to offspring was unrelated to parent adult size, then no relationship would exist between parent and offspring beak depth—some offspring, by chance, would be fed more small seeds and would develop shallower beaks, whereas others would be fed larger seeds and develop deeper beaks. This result would indicate that beak depth was not a genetically determined trait. On the other hand, if parents with deeper beaks fed their offspring larger seeds, and if seed size determined beak depth, then the same relationship pictured in figure 21.2b would result. In this case, genes and environment would be correlated, and researchers wouldn't know which factor was responsible for beak depth. To distinguish between the two,

researchers bring nestling birds into the laboratory and feed them different-sized seeds to determine if diet determined beak depth (this would be difficult in the Galápagos, though, where Darwin's finches are strictly protected).

Page 419 Data analysis: The parents' mean beak depth would be 9 mm, so we would expect the offspring to have a beak depth of approximately 9 mm. The figure does not allow us to tell whether the sex of the parent matters; however, the fact that there is relatively little scatter around the line (i.e., most of the points lie close to the regression line) suggests that the sex of the parents (i.e., which parent is larger) doesn't have a substantial effect.

Page 421 Inquiry question: Such a parallel trend would suggest that similar processes are operating in both localities. Thus, we would conduct a study to identify similarities. In this case, both areas have experienced coincident reductions in air pollution, which most likely is the cause of the parallel evolutionary trends.

Page 422 Inquiry question: Assuming that small and large individuals would breed with each other, then middle-sized offspring would still be born (the result of matings between small and large flies). Nonetheless, there would also be many small and large individuals (the result of small × small and large × large matings). Thus, the frequency distribution of body sizes would be much broader than the distributions in the figures.

Page 426 Inquiry question: This evolutionary decrease could occur for many reasons. For example, maybe *Nannippus* adapted to forested habitats and thus selection favored smaller size, as it had in the ancestral horses before horses moved into open, grassland habitats. Another possibility is that many species of horses were present at that time, and different-sized horses ate different types of food. By evolving small size, *Nannippus* may have been able to eat a type of food not eaten by the others.

UNDERSTAND

1. d   2. b   3. b   4. a   5. b   6. b   7. b

APPLY

1. a   2. d   3. d

SYNTHESIZE

1. Briefly, they are:
   a. There must be variation among individuals within a population.
   b. Variation among individuals must be related to differences among individuals in their success in producing offspring over their lifetime.
   c. Variation related to lifetime reproductive success must have a genetic (heritable) basis.

2. Figure 21.2a shows in an indirect way that beak depth varies from year to year. Presumably this is a function of variation among individuals in beak size. However, the most important point of figure 21.2a is that it shows the result of selection. That is, if the three conditions hold, we might expect to see average beak depth change accordingly as precipitation varies from year to year. Figure 21.2b is more directly relevant to the conditions noted for natural selection to occur. The figure shows that beak size varies among individuals, *and* that it tends to be inherited.

3. The relationship would be given by a cloud of points with no obvious linear trend in any direction different from a zero slope. In other words, it would be a horizontal line through an approximately circular cloud of points. Such data would suggest that whether a parent(s) has a large or small beak has no bearing on the beak size of its offspring.

4. The direction of evolution would reverse, and the two populations would evolve to be more similar in bristle number, until the differences disappeared entirely. However, the rate at which evolution occurred probably would be slower than in the first part of the experiment. Because selection had initially favored low and high bristle numbers in the two populations at the expense of intermediate bristle numbers, some of the genetic variation for intermediate numbers probably would have been lost from the population. As a result, the rate at which intermediate numbers re-evolved would probably be slower because there was less appropriate genetic variation available.

5. The evolution of horses was not a linear event; instead it occurred over 55 million years and included descendants of 34 different genera. By examining the fossil record, one can see that horse evolution did not occur gradually and steadily; instead several major evolutionary events occurred in response to drastic changes in environmental pressures. The fossil record of horse evolution is remarkably detailed, and shows that although there have been trends toward certain characteristics, change has not been fluid and constant

over time, nor has it been entirely consistent across all of the horse lineages. For example, some lineages experienced rapid increases in body size over relatively short periods of geological time, but other lineages actually saw decreases in body size.

CHAPTER 22

LEARNING OUTCOME QUESTIONS

22.1 **#1** The biological species concept states that different species are capable of mating and producing viable, fertile offspring. If sympatric species are unable to do so, they will remain reproductively isolated and thus distinct species. Along the same lines, gene flow between populations of the same species allow for homogenization of the two populations such that they remain the same species.

#2 The ecological species concept states that sympatric species are adapted to use different parts of the environment, and thus hybrids between them would not be well adapted to either habitat and thus would not survive. Even if they did survive and reproduce, genes from one species that made their way into the other species' gene pool would likely be eliminated by natural selection. The concept would explain the connection of geographic populations of a species. As a result, these populations occupy similar parts of the environment and thus experience similar selective pressures.

22.2 In order for reinforcement to occur and complete the process of speciation, two populations must have some reproductive barriers in place prior to sympatry. In the absence of this initial reproductive isolation we would expect rapid exchange of genes and thus homogenization resulting from gene flow. On the other hand, if two populations are already somewhat reproductively isolated (due to hybrid infertility or a prezygotic barrier such as behavioral isolation), then we would expect natural selection to continue improving the fitness of the nonhybrid offspring, eventually resulting in speciation.

22.3 Reproductive isolation that occurs due to different environments is a factor of natural selection; the environmental pressure favors individuals best suited for that environment. As isolated populations continue to develop, they accumulate differences due to natural selection that eventually result in two populations so different that they are reproductively isolated. Reinforcement, on the other hand, is a process that specifically relates to reproductive isolation. It occurs when natural selection favors nonhybrids because of hybrid infertility or are simply less fit than their parents. In this way, populations that may have been only partly reproductively isolated become completely reproductively isolated.

22.4 Polyploidy occurs instantaneously; in a single generation, the offspring of two different parental species may be reproductively isolated; however, if it is capable of self-fertilization, then it is, according to the biological species concept, a new species. Disruptive selection, on the other hand, requires many generations as reproductive barriers between the two populations must evolve and be reinforced before the two would be considered separate species.

22.5 In the archipelago model, adaptive radiation occurs as each individual island population adapts to its different environmental pressures. On the other hand, in sympatric speciation resulting from disruptive selection, the traits selected are not necessarily best suited for a novel environment but are best able to reduce competition with other individuals. It is in the latter scenario wherein adaptive radiation due to a key innovation is most likely to occur.

22.6 It depends on what species concept you are using to define a given species. Certainly evolutionary change can be punctuated, but in times of changing environmental pressures we would expect adaptation to occur. The adaptations, however, do not necessarily have to lead to the splitting of a species—instead one species could simply adapt in accordance with the environmental changes to which it is subjected. This would be an example of nonbranching, as opposed to branching, evolution; but again, whether the end-result organism is a different species from its ancestral organism that preceded the punctuated event is subject to interpretation.

INQUIRY AND DATA ANALYSIS QUESTIONS

Page 448 Inquiry question: Speciation can occur under allopatric conditions because isolated populations are more likely to diverge over time due to drift or selection. Adaptive radiation tends to occur in places inhabited by only a few other species or where many resources in a habitat are unused. Different environmental conditions typical of adaptive radiation tend to favor certain traits within a population. Allopatric conditions would then generally favor adaptive radiation.

In character displacement, natural selection in each species favors individuals able to use resources not used by the other species. Two species might have evolved from two populations of the same species located in the same environment (sympatric species). Individuals at the extremes of each population are able to use resources not used by the other group. Competition for a resource would

be reduced for these individuals, possibly favoring their survival and leading to selection for the tendency to use the new resource. Character displacement tends to compliment sympatric speciation.

Page 452 Inquiry question: If one area experiences an unfavorable change in climate, a mobile species can move to another area where the climate was like it was before the change. With little environmental change to drive natural selection within that species, stasis would be favored.

UNDERSTAND

1. a 2. c 3. a 4. b 5. a 6. d 7. d 8. b 9. b 10. a

APPLY

1. a 2. d 3. b 4. b

SYNTHESIZE

1. If hybrids between two species have reduced viability or fertility, then natural selection will favor any trait that prevents hybrid matings. This way individuals don't waste time, energy, or resources on such matings and will have greater fitness if they instead spend the time, energy, and resources on mating with members of their own species. For this reason, natural selection will favor any trait that decreases the probability of hybridization. By contrast, once hybridization has occurred, the time, energy, and resources have already been expended. Thus, there is no reason that less fit hybrids would be favored over more fit ones. The only exception is for species that invest considerable time and energy in incubating eggs and rearing the young; for those species, selection may favor reduced viability of hybrids because parents of such individuals will not waste further time and energy on them.

2. The biological species concept, despite its limitations, reveals the continuum of biological processes and the complexity and dynamics of organic evolution. At the very least, the biological species concept provides a mechanism for biologists to communicate about taxa and know that they are talking about the same thing! Perhaps even more significantly, discussion and debate about the meaning of "species" fuels a deeper understanding about biology and evolution in general. It is unlikely that we will ever have a single unifying concept of species, given the vast diversity of life, both extinct and extant.

3. The principle is the same as in character displacement. In sympatry, individuals of the two species that look alike may mate with each other. If the species are not completely interfertile, then individuals hybridizing will be at a selective disadvantage. If a trait appears in one species that allows that species to more easily recognize members of its own species and thus avoid hybridization, then individuals bearing that trait will have higher fitness and that trait will spread through the population.

4. The two species would be expected to have more similar morphology when they are found alone (allopatry) than when they are found together (sympatry), assuming that food resources were the same from one island to the next. This would be the result of character displacement expected under a hypothesis of competition for food when the two species occur in sympatry. A species pair that is more distantly related might not be expected to show the pattern of character displacement since they show greater differences in morphology (and presumably in ecology and behavior as well), which should reduce the potential for competition to drive character divergence.

CHAPTER 23

LEARNING OUTCOME QUESTIONS

23.1 Because of convergent evolution; two distantly related species subjected to the same environmental pressures may be more phenotypically similar than two species with different environmental pressures but a more recent common ancestor. Other reasons for the possible dissimilarity between closely related species include oscillating selection and rapid adaptive radiations in which species rapidly adapt to a new available niche.

23.2 **#1** In some cases wherein characters diverge rapidly relative to the frequency of speciation, it can be difficult to construct a phylogeny using cladistics because the most parsimonious phylogeny may not be the most accurate. In most cases, however, cladistics is a very useful tool for inferring phylogenetic relationships among groups of organisms.

#2 Only shared derived characters indicate that two or more taxa are descended from an ancestor that was not an ancestor of taxa that do not possess the character state in question. Thus, taxa possessing the state are more closely related to each other than they are to taxa without the trait. As the text discusses, this assumption is usually correct, but not when rates of homoplasy are high.

23.3 Yes, in some instances this is possible. For example, assume two populations of a species become geographically isolated from one another in similar environments, and each population diverges and speciation occurs, with one group retaining its ancestral traits and the other deriving new traits. The ancestral group in each population may be part of the same biological species but would be considered polyphyletic because to include their common ancestor would also necessitate including the other, more-derived species (which may have diverged enough to be reproductively isolated).

23.4 Not necessarily; it is possible that the character changed since the common ancestor and is present in each group due to convergence. Although the most recent common ancestor possessing the character is the most parsimonious, and thus the most likely, explanation, it is possible, especially for small clades, that similar environmental pressures resulted in the emergence of the same character state repeatedly during the course of the clade's evolution.

23.5 Hypothetically it is possible; however, the viral analyses and phylogenetic analyses have provided strong evidence that HIV emergence was the other way around; it began as a simian disease and mutated to a human form, and this has occurred several times.

INQUIRY AND DATA ANALYSIS QUESTIONS

Page 459 Data analysis: The data matrix would be similar to this one:

Organisms	Traits		
	Hair	Amniotic membrane	Tail
Salamander	0	0	1
Frog	0	0	0
Lizard	0	1	1
Tiger	1	1	1
Gorilla	1	1	0
Human	1	1	0

Page 460 Inquiry question: In parsimony analyses of phylogenies, the least complex explanation is favored. High rates of evolutionary change and few character states complicate matters. High rates of evolutionary change, such as occur when mutations arise in noncoding portions of DNA, can be misleading when constructing phylogenies. Mutations arising in noncoding DNA are not eliminated by natural selection in the same manner as mutations in coding (functional) DNA. Also, evolution of new character states can be very high in nonfunctional DNA, which can lead to genetic drift. Since DNA has only four nucleotides (four character states), it is highly likely that two species could evolve the same derived character at a particular base position. This leads to a violation of the assumptions of parsimony—that the fewest evolutionary events lead to the best hypothesis of phylogenetic relationships—and resulting phylogenies are inaccurate.

Page 461 Inquiry question: The only other hypothesis is that the most recent common ancestor of birds and bats was also winged. Of course, this scenario is much less parsimonious (and thus much more unlikely) than the convergence hypothesis, especially given the vast number of reptiles and mammals without wings. Most phylogenies are constructed based on the rule of parsimony; in the absence of fossil evidence of other winged animals and molecular data supporting a closer relationship between birds and bats than previously thought, there is no way to test the hypothesis that bird and bat wings are homologous rather than analogous.

Page 467 Inquiry question: One possibility would be to closely examine the patterns of development in different species. In part *a*, all of the species in the blue box with direct development would have inherited that character state from a common ancestor. Consequently, we might expect that the developmental patterns would be similar. By contrast, in part *b* direct development evolved many different times in those species. In the latter case, it is possible that direct development could have evolved in many different ways. Thus, if you compared how development occurs in these species and found that there were different patterns of development in different species, then the hypothesis in part *b* might seem more likely. However, even if all the species inherited direct development from their common ancestor (part *a*), species could diverge evolutionarily, so it is always possible that differences you found might have arisen after divergence from a common ancestor. For this reason, definitive conclusions are always difficult in situations such as this.

Page 470 Inquiry question: If the victim had contracted HIV from a source other than the patient, the most recent common ancestor of the two strains would be much more distant. As it is, the phylogeny shows that the victim and patient strains share a relatively recent ancestor, and that the victim's strain is derived from the patient's strain.

UNDERSTAND

1. d   2. b   3. a   4. b   5. a   6. d   7. d   8. b   9. c

APPLY

1. c   2. d   3. a

SYNTHESIZE

1. Naming of groups can vary; names provided here are just examples. Jaws—shark, salamander, lizard, tiger, gorilla, human (jawed vertebrates); lungs—salamander, lizard, tiger, gorilla, human (terrestrial tetrapods); amniotic membrane—lizard, tiger, gorilla, human (amniote tetrapods); hair—tiger, gorilla, human (mammals); no tail—gorilla, human (humanoid primate); bipedal—human (human).

2. It would seem to be somewhat of a conundrum, or potentially circular chase; choosing a closely related species as an outgroup when we do not even know the relationships of the species of interest. One way of guarding against a poor choice for an outgroup is to choose several species as outgroups and examine how the phylogenetic hypothesis for the group of interest changes as a consequence of using different outgroups. If the choice of outgroup makes little difference, then that might increase one's confidence in the phylogenetic hypotheses for the species of interest. On the other hand, if the choice makes a big difference (different phylogenetic hypotheses result when choosing different outgroups), that might at least lead to the conclusion that we cannot be confident in inferring a robust phylogenetic hypothesis for the group of interest without collecting more data.

3. Recognizing that birds are reptiles potentially provides insight to the biology of both birds and reptiles. For example, some characteristics of birds are clearly of reptilian origin, such as feathers (modified scales), nasal salt-secreting glands, and strategies of osmoregulation/excretion (excreting nitrogenous waste products as uric acid) representing ancestral traits that continue to serve birds well in their environments. On the other hand, some differences from other reptiles (again, feathers) seem to have such profound significance biologically, that they overwhelm similarities visible in shared ancestral characteristics. For example, no extant nonavian reptiles can fly or are endothermic, and these two traits have created a fundamental distinction in the minds of many biologists. Indeed, many vertebrate biologists prefer to continue to distinguish birds from reptiles rather than emphasize their similarities even though they recognize the power of cladistic analysis in helping to shape classification. Ultimately, it may be nothing much more substantial than habit which drives the preference of some biologists to traditional classification schemes.

4. In fact, such evolutionary transitions (the loss of the larval mode, and the re-evolution of a larval mode from direct development) are treated with equal weight under the simplest form of parsimony. However, if it is known from independent methods (e.g., developmental biology) that one kind of change is less likely than another (loss vs. a reversal), these should and can be taken into account in various ways. The simplest way might be to assign weights based on likelihoods; two transitions from larval development to direct development is equal to one reversal from direct development back to a larval mode. In fact, such methods exist, and they are similar in spirit to the statistical approaches used to build specific models of evolutionary change rather than rely on simple parsimony.

5. The structures are both homologous, as forelimbs, and convergent, as wings. In other words, the most recent common ancestor of birds, pterosaurs, and bats had a forelimb similar in morphology to that which these organisms possess—it has similar bones and articulations. Thus, the forelimb itself among these organisms is homologous. The wing, however, is clearly convergent; the most recent common ancestor surely did not have wings (or all other mammals and reptiles would have had to have lost the wing, which violates the rule of parsimony). The wing of flying insects is purely convergent with the vertebrate wing, as the forelimb of the insect is not homologous with the vertebrate forelimb.

6. The biological species concept focuses on processes, in particular those that result in the evolution of a population to the degree that it becomes reproductively isolated from its ancestral population. The process of speciation as utilized by the biological species concept occurs through the interrelatedness of evolutionary mechanisms such as natural selection,

mutation, and genetic drift. On the other hand, the phylogenetic species concept focuses not on process but on history, on the evolutionary patterns that led to the divergence between populations. Neither species concept is more right or more wrong; species concepts are, by their very nature, subjective and potentially controversial.

CHAPTER 24

LEARNING OUTCOME QUESTIONS

24.1 There should be a high degree of similarity between the two genomes because they are relatively closely related. There could be differences in the relative amounts of noncoding DNA. Genes that are necessary for bony skeletal development might be found in the bony fish. The cartilaginous fish might lack those genes or have substantial sequences in the genes needed for skeletal development in bony fish.

24.2 The additional DNA would likely be noncoding DNA that might represent large expanses of retrotransposon DNA that duplicated and transposed multiple times.

24.3 Compare the sequence of the pseudogene with other species. If, for example, it is a pseudogene of an olfactory gene that is found in mice or chimps, the sequences will be much more similar than in a more distantly related species. If horizontal gene transfer explains the origins of the gene, there may not be a very similar gene in closely related species. You might use the BLAST algorithm discussed in chapter 18 to identify similar sequences and then construct a phylogenetic tree to compare the relationships among the different species.

24.4 A SNP can change a single amino acid in the coded peptide. If the new R group is very different, the protein may fold in a different way and not function effectively. SNPs in the *FOXP2* gene may, in part, explain why humans have speech and chimps do not.

Other examples that you may remember from earlier in the text include cystic fibrosis and sickle cell anemia.

24.5 An effective drug might bind only to the region of the pathogen protein that is distinct from the human protein. The drug could render the pathogen protein ineffective without making the human ill. If the seven amino acids that differ are scattered throughout the genome, they might have a minimal effect on the protein, and it would be difficult to develop a drug that could detect small differences. It's possible that the drug could inadvertently affect other areas of the protein as well.

INQUIRY AND DATA ANALYSIS QUESTIONS

Page 476 Data analysis: Rice has 43,000 genes in 430 Mb of DNA (100 genes/Mb), whereas humans have 22,698 genes in 2500 Mb (9 genes/Mb). Rice has over 10 times as many genes as humans per Mb. The organisms closest to humans in terms of number of genes/Mb would be the duck-billed platypus with 8.4 genes/Mb and next would be the domestic cow with 7.6 genes/Mb.

Page 478 Inquiry question: Meiosis in a 3*n* cell would be impossible because three sets of chromosomes cannot be divided equally between two cells. In a 3n cell, all three homologous chromosomes would pair in prophase I, then align during anaphase I. As the homologous chromosomes separate, two of a triplet might go to one cell and the third chromosome would go to the other cell. The same would be true for each set of homologues. Daughter cells would have an unpredictable number of chromosomes.

Page 479 Inquiry question: Polyploidization seems to induce the elimination of duplicated genes. Duplicate genes code for the same gene product. It is reasonable that duplicate genes would be eliminated to decrease the redundancy arising from the translation of several copies of the same gene.

Page 485 Inquiry question: Ape and human genomes show very different patterns of gene transcription activity, even though genes encoding proteins are over 99% similar between chimps and humans. Different genes would be transcribed when comparing apes with humans, and the levels of transcription would vary widely.

UNDERSTAND

1. c 2. d 3. d 4. b 5. b 6. a

APPLY

1. a 2. d 3. d 4. a

SYNTHESIZE

1. The two amino acid difference between the FOXP2 protein in humans and closely related primates must alter the way the protein functions in the brain.

The protein affects motor function in the brain, allowing coordination of larynx, mouth and brain for speech in humans. For example, if the protein affects transcription, there could be differences in the genes that are regulated by FOXP2 in humans and chimps.

2. Human and chimp DNA is close to 99% similar, yet our phenotypes are conspicuously different in many ways. This suggests that a catalog of genes is just the first step to identifying the mechanisms underlying genetically influenced diseases like cancer or cystic fibrosis. Clearly, gene expression, which might involve the actions of multiple noncoding segments of the DNA and other potentially complex regulatory mechanics, are important sources of how phenotypes are formed, and it is likely that many genetically determined diseases result from such complex underlying mechanisms, making the gene identification of genomics just the first step; a necessary but not nearly sufficient strategy. What complete genomes do offer is a starting point to correlating sequence differences among humans with genetic disease, as well as the opportunity to examine how multiple genes and regulatory sequences interact to cause disease.

3. Phylogenetic analysis usually assumes that most genetic and phenotypic variation arises from descent with modification (vertical inheritance). If genetic and phenotypic characteristics can be passed horizontally (i.e., not vertically through genetic lineages), then using patterns of shared character variation to infer genealogical relationships will be subject to potentially significant error. We might expect that organisms with higher rates of HGT will have phylogenetic hypotheses that are less reliable or at least are not resolved as a neatly branching tree.

CHAPTER 25

LEARNING OUTCOME QUESTIONS

25.1 A change in the promoter of a gene necessary for wing development might lead to the repression of wing development in a second segment of a fly in a species that has double wings.

25.2 Yes, although this is not the only explanation. The coding regions could be identical, but the promoter or other regulatory regions could have been altered by mutation, leading to altered patterns of gene expression. To test this hypothesis, the *Eda* gene should be sequenced in both fish and compared.

25.3 The pectoral fins are homoplastic because sharks and whales are only distantly related and pectoral fins are not found in whales' more recent ancestors.

25.4 There is no need for eyes in the dark. Perhaps the fish expend less energy when eyes are not produced and that offered a selective advantage in cavefish. In a habitat with light, a mutation that resulted in a functional *Pax6* would likely be selected for, and over time more of the fish would have eyes. Keep in mind that the probability of a mutation restoring *Pax6* function is very low, but real.

INQUIRY AND DATA ANALYSIS QUESTIONS

Page 495 Inquiry question: Because there is a stop codon located in the middle of the *CAL* (cauliflower) gene-coding sequence, the wild-type function of *CAL* must be concerned with producing flowers. Unlike the ancestral *Brassica oleracea,* cauliflower and broccoli keep branching instead of producing a flower. One interpretation is that a meristem lacking CAL protein is delayed in producing flower parts and instead keeps producing branches, resulting in the broccoli and cauliflower heads. Cauliflower and broccoli heads differ in terms of how advanced floral development is (only meristems in cauliflower and early floral buds in broccoli). This difference could be the result of a later evolutionary event. Additional evolutionary events possibly include large flower heads, unusual head coloration, protective leaves covering flower heads, or head size variants, among other possibilities.

Page 499 Inquiry question: Functional analysis involves the use of a variety of experiments designed to test the function of a specific gene in different species. By mixing and matching parts of the *AP3* and *PI* genes and introducing them into *ap3* mutant plants, it was found that the C terminus sequence of the AP3 protein is essential for specifying petal function. Without the 3 region of the *AP3* gene, the *Arabidopsis* plant cannot make petals.

UNDERSTAND

1. c 2. b 3. a 4. a 5. b 6. d 7. a 8. c 9. c 10. d

APPLY

1. c 2. a 3. d 4. b

SYNTHESIZE

1. Your supervisor might still be correct. The mutation could be in the regulatory region and the *pitx1* gene could be transcribed in cells in different tissues in the two species. You could test this by testing whether or not you are able to obtain the RNA from hindlimbs and the pelvis of both species at different times in development.

2. Development is a highly conserved and constrained process; small perturbations can have drastic consequences, and most of these are negative. Given the thousands or hundreds of thousands of variables that can change in even a simple developmental pathway, most perturbations lead to negative outcomes. Over millions of years, some of these changes will arise under the right circumstances to produce a benefit. In this way, developmental perturbations are not different from what we know about mutations in general. Beneficial mutations are rare, but with enough time they will emerge and spread under specific circumstances.

   Not all mutations provide a selective advantage. For example, reduced body armor increases the fitness of fish in fresh water, but it was not selected for in a marine environment where the armor was important for protection from predators. The new trait can persist at low levels for a very long time until a change in environmental conditions results in an increase in fitness for individuals exhibiting the trait.

3. The latter view represents our current understanding. There are many examples of small gene families (e.g., *Hox*, *MADS*) whose apparent role in generating phenotypic diversity among major groupings of organisms is in altering the expression of other genes. Alterations in timing (heterochrony) or spatial pattern of expression (homeosis) can lead to shifts in developmental events, giving rise to new phenotypes. Many examples are presented in the chapter, such as the developmental variants of two species of sea urchins, one with a normal larval phase, and another with direct development. In this case the two species do not have different sets of developmental genes; rather the expression of those genes differ. Another example that makes the same point is the evolution of an image-forming eye. Recent studies suggest, in contrast to the view that eyes across the animal kingdom evolved independently multiple times, that image-forming eyes from very distantly related taxa (e.g., insects and vertebrates) may trace back to the common origin of the *Pax6* gene. If that view is correct, then genes controlling major developmental patterns would seem to be highly conserved across long periods of time, with expression being the major form of variation.

4. Unless the *Pax6* gene was derived multiple times, it is difficult to hypothesize multiple origins of eyes. Pax6 initiates eye development in many species. The variation in eyes among animals is a result of which genes are expressed and when after Pax6 initiates eye development.

5. Maize relies on *paleoAP3* and *PI* for flower development, whereas tomato has three genes because of a duplication of *paleoAP3*. This duplication event in the ancestor of tomato, but not maize, is correlated with independent petal origin.

6. The direct developing sea urchin has an ancestor that had one or more mutations in genes that were needed to regulate the expression of other genes needed for larval stage development. When those genes were not expressed, there was no larval development and the genes necessary for adult development were expressed.

CHAPTER 26

LEARNING OUTCOME QUESTIONS

26.1 The Phanerozoic eon includes the greatest diversity of life and would be the most informative in terms of fossil record reflecting diverse life-forms. Earlier eons have a more limited fossil record, with primarily unicellular organisms.

26.2 Amino acids, including glycine and alanine, nitrogenous bases like adenine that are found in DNA and RNA, lipids, and simple carbon compounds like CH_2O might be found. If the early life was based on RNA, then ribonucleotides rather than deoxyribonucleotides would be found. Early RNA sequences would have had both enzymatic and information functions. Once the machinery was in place for RNA to encode proteins, amino acid sequences would become less random and gain functions in catalyzing metabolism as well as structural roles and regulating nucleic acid replication, transcription, and translation. Lipids would form early cell membranes, and metabolic pathways would continue to evolve within the confines of the cell membrane. Nucleic acid replication and cell division would evolve early on also.

26.3 Electron microscopy revealing organic walls and vesicles would provide good evidence, combined with spectroscopic analysis indicating the presence of complex carbon molecules.

26.4 Weathering is proposed as a key event associated with both glaciations. Weathering in the late Proterozoic was caused by warm, moist air and by shifts in plate tectonics that increased surface area for weathering. In the Ordovician period, land was being colonized for the first time. Weathering of rocks because of atmospheric changes is not sufficient to explain the CO_2 shift leading to glaciation. Rather, early land plants secreted an organic acid that weathered rocks, releasing phosphorous that ran into the oceans. In the oceans, phosphorous triggered algal blooms that pulled down CO_2 from the atmosphere, triggering a rapid drop in temperature.

26.5 Major pre-Cambrian evolutionary innovations that set the stage for rapid radiation included the evolution of eukaryotic cells with chloroplasts and mitochondria, and meiosis that allowed for sexual reproduction. The new gene combinations created by meiosis provided a set of phenotypes for natural selection to act upon.

INQUIRY AND DATA ANALYSIS QUESTION

Page 510 Data analysis: The graph would look like figure 26.4 because a plot of radioactive decay is the same, regardless of the parent isotope. It will take 7 half-lives. You can calculate this by dividing by 2 for each half-life (1/2, 1/4, 1/8, 1/6, 1/32, 1/64, 1/128). After you get past 1/100, which is 1%, you have the correct half-life. You will come very close to 0%, but there will always be a tiny, although immeasurable, amount of parent isotope since you keep halving the amount, not eliminating it.

UNDERSTAND

1. b   2. d   3. c The correct statement would read, "Brown algae gained chloroplasts by engulfing red algae (endosymbiosis)."   4. c   5. b

APPLY

1. b   2. c   3. d   4. c   5. c

SYNTHESIZE

1. Early land plants lacked roots and had substantially less biomass than the later vascular plants. Alone they could not have sequestered sufficient CO_2 to trigger glaciation. However, just like the vascular plants, they secreted organic acid that enhanced weathering of rocks, leading to the release of phosphorous. The excess phosphorous washed into the oceans. As an essential plant nutrient, the excess phosphorous triggered extensive algal blooms, which did sequester sufficient CO_2, coupled with the land plants, to trigger a glaciation.

2. The collision of tectonic plates could have multiple effects on evolution of life. When the plates collide, less surface area is exposed to water, which could reduce weathering. Reduced weathering would pull less CO_2 out of the atmosphere, which could affect climate. When plates collide, organisms that were previously separated come together. This potentially affects interbreeding and also changes selective pressures affecting evolution.

3. The evolution of meiosis provided a much greater range of genetic diversity for natural selection to act upon. The evolution of multicellularity immediately preceded the Cambrian explosion and was likely a contributing factor to the rapid diversification when ancestors of almost every group of animals evolved. The climate must also have been more hospitable to life during the Cambrian, a period following a global ice age.

CHAPTER 27

LEARNING OUTCOME QUESTIONS

27.1 Viruses use cellular machinery for replication. They do not make all of the proteins necessary for complete replication.

27.2 A prophage carrying such a mutation could not be induced to undergo the lytic cycle.

27.3 This therapy, at present, does not remove all detectable viruses, so it cannot be considered a true cure.

27.4 In addition to a high mutation rate, the influenza genome consists of multiple RNA segments that can recombine during infection. This causes the main antigens for the immune system to shift rapidly.

27.5 Prions carry information in their three-dimensional structure. This 3-D information is different from the essentially one-dimensional genetic information in DNA.

UNDERSTAND

1. c   2. b   3. c   4. d   5. b   6. d   7. b

APPLY

1. c 2. b 3. c 4. d 5. b 6. c 7. c 8. a

SYNTHESIZE

1. A set of genes that are involved in the response to DNA damage are normally induced by the same system. The protein involved destroys a repressor that keeps DNA repair genes unexpressed. Lambda has evolved to use this system to its advantage.

2. Since viruses require the replication machinery of a host cell to replicate, it is unlikely that they existed before the origin of the first cells.

3. This is a complex situation. The relevant factors include the high mutation rate of the virus and the fact that the virus targets the very cells that mount an immune response. The influenza virus also requires a new vaccine every year due to rapid changes in the virus. The smallpox virus was a DNA virus that had antigenic determinants that did not change rapidly, thus making a vaccine possible.

4. Emerging viruses are those that jump species and thus are new to humans. Recent examples include SARS and Ebola.

5. If excision of the λ prophage is imprecise, then the phage produced will carry *E. coli* genes adjacent to the integration site.

CHAPTER 28

LEARNING OUTCOME QUESTIONS

28.1 Archaea have ether-linked instead of ester-linked phospholipids; their cell wall is made of unique material.

28.2 Compare their DNA. The many metabolic tests we have used for years have been supplanted by DNA analysis.

28.3 Transfer of genetic information in bacteria is directional: from donor to recipient and does not involve fusion of gametes.

28.4 Prokaryotes do not have a lot of morphological features, but do have diverse metabolic functions.

28.5 Pathogens tend to evolve to be less virulent. If they are too good at killing, their lifestyle becomes an evolutionary dead end.

28.6 Rotating a crop that has a symbiotic association with nitrogen-fixing bacteria will return nitrogen to the soil depleted by other plants.

INQUIRY AND DATA ANALYSIS QUESTIONS

Page 549 Data analysis: An imprecise excision of the F plasmid produces a so-called F′ plasmid that carries *E. coli* DNA in addition to the normal plasmid DNA. If this is introduced into a new F⁻ cell, it will then produce a new cell that is "diploid" for the region carried on the F′ plasmid. These so-called partial diploids have many genetic uses.

Page 556 Inquiry question: The simplest explanation is that the two STDs are occurring in different populations, and one population has rising levels of sexual activity, while the other has falling levels. However, the rise in incidence of an STD can reflect many parameters other than level of sexual activity. The virulence or infectivity of one or both disease agents may be changing, for example, or some aspect of exposed people may be changing in such a way as to alter susceptibility. Only a thorough public health study can sort this out.

UNDERSTAND

1. b 2. a 3. c 4. c 5. d 6. a 7. b

APPLY

1. c 2. b 3. b 4. c 5. d 6. b 7. a

SYNTHESIZE

1. The study of carbon signatures in rocks using isotopic data assumes that ancient carbon fixation involves one of two pathways that each show a bias toward incorporation of carbon-12. If this bias were not present, it is not possible to infer early carbon fixation by this pathway. This pathway could have arisen even earlier and we would have no way to detect it.

2. The heat killing of the virulent S strain of *Streptococcus* released the genome of the virulent smooth strain into the environment. These strains of *Streptococcus* bacteria are capable of natural transformation. At least some of the rough-strain cells took up smooth-strain genes that encoded the polysaccharide coat from the environment. These genes entered into the rough-strain genome by recombination, and then were expressed. These transformed cells were now smooth bacteria.

3. Use of multiple antibiotics is not a bad idea if all of the bacteria are killed. In the case of some persistent infections, this is an effective strategy. However, it does provide very strong selective pressure for rare genetic events that produce multiple resistances in a single bacterial species. For this reason, it is not a good idea for it to be the normal practice. The more bacteria that undergo this selection for multiple resistance, the more likely it will arise. This is helped by patients not taking their entire course of antibiotic because bacteria may survive by chance and proliferate, with each generation providing the opportunity for new mutations. This is also complicated by the horizontal transfer of resistance via resistance plasmids, and by the existence of transposable genetic elements that can move genes from one piece of DNA to another.

4. Most species on the planet are incapable of fixing nitrogen without the assistance of bacteria. Without nitrogen, amino acids and other compounds cannot be synthesized. Thus a loss of the nitrogen-fixing bacteria due to increased UV radiation levels would reduce the ability of plants to grow, severely limiting the food sources of the animals.

CHAPTER 29

LEARNING OUTCOME QUESTIONS

29.1 Mitochondria and chloroplasts contain their own DNA. Mitochondrial genes are transcribed within the mitochondrion, using mitochondrial ribosomes that are smaller than those of eukaryotic cells and quite similar to bacterial ribosomes. Antibiotics that inhibit protein translation in bacteria also inhibit protein translation in mitochondria. Also, both chloroplasts and mitochondria divide using binary fission like bacteria.

29.2 Meiosis makes sexual reproduction possible. Each generation, new combinations of alleles provide much greater genetic diversity for evolution to act on than is available with asexual reproduction. Different gene combinations were advantageous in different settings, and species began to diverge as natural selection acted on the genetic diversity.

29.3 **#1** Undulating membranes would be effective on surfaces with curvature that may not always be smooth, such as intestinal walls.
#2 Contractile vacuoles collect and remove excess water from within the *Euglena*.

29.4 **#1** The *Plasmodium* often becomes resistant to new poisons and drugs. And, because the *Plasmodium* has multiple hosts, a drug for humans wouldn't eradicate the *Plasmodium* in other stages of its life cycle in the mosquito.
#2 Just based on outward appearances, the sporophyte of brown algae forms a larger bladelike structure, whereas the gametophyte is a small filamentous structure.

29.5 **#1** Both the red and green algae obtained their chloroplasts through endosymbiosis, possibly of the same lineage of photosynthetic bacteria. The red and green algae had diverged before the endosymbiotic events, and the history recorded in their nuclear DNA is a different evolutionary history than that recorded in the plastids derived through endosymbiosis.
#2 The lack of water is the major barrier for sperm that move through water to reach the egg. It is more difficult for sperm to reach the egg on land.

29.6 Construct phylogenies with other traits, including DNA sequences. Then map the presence of amoeboid locomotion with pseudopods on to the other phylogenies. If the pseudopod trait is not clustered, it is likely that it evolved independently multiple times and is thus not a good trait to use in reconstructing protist phylogenies.

29.7 It is unlikely that cellular and plasmodial slime molds are closely related. They both appear in the last section of this chapter because they have yet to be assigned to clades. The substantial differences in their cell biology are inconsistent with a close phylogenetic relationship.

29.8 Comparative genomic studies of choanoflagellates and sponges would be helpful. Considering the similarities among a broader range of genes than just the conserved tyrosine kinase receptor would provide additional evidence.

INQUIRY AND DATA ANALYSIS QUESTIONS

Page 564 Inquiry question: Red and green algae obtained chloroplasts by engulfing photosynthetic bacteria by primary endosymbiosis; chloroplasts in these cells have two membranes. Brown algae obtained chloroplasts by engulfing cells of red algae through secondary endosymbiosis; chloroplasts in cells of brown algae have four membranes. Counting the number of cell membranes of chloroplasts indicates primary or secondary endosymbiosis.

Page 576 Data analysis: Your phylogeny should show a common ancestor for all three algae. The brown alga should branch from a common ancestor of the red and green algae. (See the phylogeny on p. 575 of your text and trace the three algae).

Page 577 Inquiry question: No, they are formed by mitosis. Meiosis produces haploid spores, which divide mitotically to produce multicellular gametophytes. Gametes are produced from haploid gametophyte cells by mitosis.

UNDERSTAND

1. b 2. a 3. b 4. c 5. d 6. b 7. a 8. c 9. b, c 10. a, d 11. d 12. a

APPLY

1. d 2. a 3. a

SYNTHESIZE

1. Cellular and plasmodial slime molds both exhibit group behavior and can produce mobile slime mold masses. However, these two groups are very distantly related phylogenetically.

2. The development of a vaccine, though challenging, will be the most promising in the long run. It is difficult to eradicate all the mosquito vectors, and many eradication methods can be harmful to the environment. Treatments to kill the parasites are also difficult because the parasite is likely to become resistant to each new poison or drug. A vaccine would provide long-term protection without the need to use harmful pesticides or drugs for which drug resistance is a real possibility.

3. For the first experiment, plate the cellular slime molds on a plate that has no bacteria. Spot cyclic AMP and designated places on the plate and determine if the bacteria aggregate around the cAMP.
   For the second experiment, repeat the first experiment using plates that have a uniform coating of bacteria as well as plates with no bacteria. If the cellular slime molds aggregate on both plates, resource scarcity is not an issue. If the cells aggregate only in the absence of bacteria, you can conclude that the attraction to cAMP occurs only under starvation conditions.

CHAPTER 30

LEARNING OUTCOME QUESTIONS

30.1 #1 Make sections and examine them under the microscope to look for tracheids. Only the tracheophytes will have tracheids.
    #2 Gametes in plants are produced by mitosis. Human gametes are produced directly by meiosis.

30.2 Mosses are extremely desiccation-tolerant and can withstand the lack of water. Also, freezing temperatures at the poles are less damaging when mosses have a lower water content.

30.3 The sporophyte generation has evolved to be the larger generation, and therefore an effective means of transporting water and nutrients over greater distances would be advantageous.

30.4 Substantial climate change occurred during that time period. Glaciers had spread, then melted and retreated. The resulting drier climates could have contributed to the extinction of large club mosses. Refer to chapter 26 for more information on changes in Earth's climate over geological time.

30.5 The silica can increase the strength of the hollow-tubed stems and would also deter herbivores.

INQUIRY AND DATA ANALYSIS QUESTIONS

Page 591 Inquiry question: Tracheophytes developed vascular tissue, enabling them to have efficient water- and food-conducting systems. Vascular tissue allowed tracheophytes to grow larger, possibly then enabling them to outcompete smaller, nonvascular land plants. A protective cuticle and stomata that can close during dry conditions also conferred a selective advantage.

Page 592 Data analysis: Tracheophytes first appeared 40 million years after land plants first began diverging. Another 40 million years passed before true leaves appeared.

UNDERSTAND

1. d 2. d 3. c 4. c 5. a 6. d

APPLY

1. d 2. b 3. c 4. d 5. d 6. b

SYNTHESIZE

1. Moss has a dominant gametophyte generation, whereas lycophytes have a dominant sporophyte generation. Perhaps a comparison of the two genomes would provide insight into the genomic differences associated with the evolutionary shift from dominant gametophyte to dominant sporophyte.

2. Because the sporophyte generation of the fern is much larger than the sporophyte generation of the moss, it is expected that the fern sporophyte would have more mitotic divisions in order for the plant to be larger.

3. Moss would face the greater challenge. Without vascular tissue, it would be difficult to move water and nutrients efficiently over 10 m of height. Although the fern gametophyte would be small and on the ground where water would be available for sperm to travel a short distance to fertilize eggs, getting egg and sperm together at the tops of two different moss gametophytes would be challenging at a height of 10 m.

CHAPTER 31

LEARNING OUTCOME QUESTIONS

31.1 The pollen tube grows toward the egg, carrying the sperm within the pollen tube; therefore water is not needed for fertilization.

31.2 The ovule rests, exposed on the scale (a modified leaf), which allows easier access to the ovule through wind pollination.

31.3 Any number of vectors that could acquire the moss gene and also affect the flowering plant could account for horizontal gene transfer. Pathogens including viruses and bacteria, as well as herbivores could potentially facilitate horizontal gene transfer.

31.4 We'd expect a dormancy that required a period of chilling to be broken and/or time in the ground for the seed coat to be weakened by microbes in the soil over the winter so the seedling can force its way out of the seed.

31.5 Fleshy fruits are more likely to encourage animals to eat them. Different colors will attract different animals. For example, birds are attracted to red berries.

INQUIRY AND DATA ANALYSIS QUESTIONS

Page 606 Inquiry question: Comparisons of a single gene could result in an inaccurate phylogenetic tree because it fails to take into account the effects of horizontal gene transfer. For example, the clade of *Amborella trichopoda* is a sister clade to all other flowering plants, but roughly 2/3 of its mitochondrial genes are present due to horizontal gene transfer from other land plants, including more distantly related mosses.

Page 608 Inquiry question: To determine if a moss gene had a function you would employ functional analysis, using a variety of experiments, to test for possible functions of the moss gene in *Amborella*. For example, you could create a mutation in the moss gene and see if there were any phenotypic differences in *Amborella* plants homozygous for the mutant gene. You could place the moss gene promoter in front of a reporter gene and visualize where, if any place, in the plant the gene was expressed.

Page 611 Inquiry question: Endosperm provides nutrients for the developing embryo in most flowering plants. The embryo cannot derive nutrition from soil prior to root development; therefore, without endosperm, the embryo is unlikely to survive, and fitness would be zero.

Page 611 Data analysis: (640 μm – 80 μm)/(8 hr – 1 hr) = 80 μm/hr. This matches the value in the data table. Differences could be the result of reading the graph incorrectly or of rounding in doing the calculation.

Page 612 Inquiry question: The prior sporophyte generation (tan) should be labeled $2n$. The degenerating gametophyte generation (purple) should be labeled $1n$. The next sporophyte generation (blue) should be labeled $2n$.

UNDERSTAND

1. a 2. c 3. b 4. d 5. a 6. d 7. d 8. b 9. a

APPLY

1. c 2. a 3. b 4. a 5. a 6. d

SYNTHESIZE

1. Answers to this question may vary. However, gymnosperms are defined as "naked" seed plants. Therefore, an ovule that is not completely protected by sporophyte tissue would be characteristic of a gymnosperm. To be classified as an angiosperm, evidence of flower structures and double fertilization are

key characteristics, although double fertilization has been observed in some gnetophytes.

2. The purpose of pollination is to bring together the male and female gametes for sexual reproduction. Sexual reproduction is designed to increase the genetic variability of a species. If a plant allows self-pollination, then the amount of genetic diversity will be reduced, but this is a better alternative than not reproducing at all. This would be especially useful in species in which the individuals are widely dispersed.

3. The benefit is that by developing a relationship with a specific pollinator, the plant species increases the chance that its pollen will be brought to another member of its species for pollination. If the pollinator is a generalist, then the pollinator might not travel to another member of the same species, and pollination would not occur. The drawback is that if something happens to the pollinator (extinction or drop in population size), then the plant species would be left with either a reduced or nonexistent means of pollination.

4. The seeds may need to be chilled before they can germinate. You can store them in the refrigerator for several weeks or months and try again. The surface of the seed may need to be scarified (damaged) before it can germinate. Usually this would happen from the effects of weather or if the seed goes through the digestive track of an animal where the seed coat is weakened by acid in the gut of the animal. You could substitute for natural scarification by rubbing your seeds on sandpaper before germinating them. It is possible that your seed needs to be exposed to light or received insufficient water when you first planted it. You may need to soak your seed in water for a bit to imbibe it. Exposing the imbibed seed to sunlight might also increase the chances of germination.

CHAPTER 32

LEARNING OUTCOME QUESTIONS

32.1 **#1** In fungi, mitosis results in duplicated nuclei, but the nuclei remain within a single cell. This lack of cell division following mitosis is very unusual in animals.
  #2 Hyphae are protected by chitin, which is not digested by fungal enzymes.

32.2 Microsporidians lack mitochondria, which are found in *Plasmodium*.

32.3 Blastocladiomycetes are free-living and have mitochondria. Microsporidians are obligate parasites and lack mitochondria.

32.4 Zygospores are more likely to be produced when environmental conditions are not favorable. Sexual reproduction increases the chances of offspring with new combinations of genes that will have an advantage in a changing environment. Also, the zygospore can stay dormant until conditions improve.

32.5 Glomeromycetes depend on the host plant for carbohydrates.

32.6 A dikaryotic cell has two nuclei, each with a single set of chromosomes. A diploid cell has a single nucleus with two sets of chromosomes.

32.7 Preventing the spread of the fungal infection using fungicides and good cultivation practices could help. If farmworkers must tend to infected fields, masks that filter out the spores could protect the workers.

32.8 The fungi that ants consumed may have originally been growing on leaves. Over evolutionary time, mutations that altered ant behavior so the ants would bring leaves to a stash of fungi would have been favored, and the tripartite symbiosis evolved.

32.9 Wind can spread spores over large distances, resulting in the spread of fungal disease.

INQUIRY AND DATA ANALYSIS QUESTION

Page 62 Data analysis: The first fungus had 1.31×10^8 spores/1.1 cm^2 = 1.2×10^8 spores/cm^2 versus the second with 6.87×10^9 spores/8.37 cm^2 = 8.2×10^8 spores/cm^2. The second fungus produced a little over 6 times as many spores per cm^2 as the first fungus.

UNDERSTAND

1. c 2. d 3. a 4. d 5. b 6. d 7. d

APPLY

1. d 2. b 3. a 4. c 5. d

SYNTHESIZE

1. Fungi possess cell walls. Although the composition of these cell walls differs from that of the plants, cell walls are completely absent in animals. Fungi are

also immobile (except for chytrids), and mobility is a key characteristic of the animals.

2. The mycorrhizal relationships between the fungi and plants allow plants to make use of nutrient-poor soil. Without the colonization of land by plants, it is unlikely that animals would have diversified to the level they have achieved today. Lichens are important organisms in the colonization of land. Early landmasses would have been composed primarily of barren rock, with little or no soil for plant colonization. As lichens colonize an area they begin the process of soil formation, which allows other plants to grow.

3. Antibiotics are designed to combat prokaryotic organisms, and fungi are eukaryotic. In addition, fungi possess a cell wall that has a different chemical constitution (chitin) from that of prokaryotes.

4. The fungicide will increase the germination success of your seeds by preventing fungi from harming the seed and emerging seedling. However, because of the close phylogenetic relationship between fungi and animals, a compound that is lethal to fungi may also be harmful to humans. You would not want to get the fungicide on your hands.

CHAPTER 33

LEARNING OUTCOME QUESTIONS

33.1 The rules of parsimony state that the simplest phylogeny is most likely the true phylogeny. As there are living organisms that are both multicellular and unicellular, it stands to reason that the first organisms were unicellular, and multicellularity followed. Animals are also all heterotrophs; if they were the first type of life to have evolved, there would not have been any autotrophs on which they could feed.

33.2 Cephalization, the concentration of nervous tissue in a distinct head region, is intrinsically connected to the onset of bilateral symmetry. Bilateral symmetry promotes the development of a central nerve center, which in turn favors the nervous tissue concentration in the head. In addition, the onset of both cephalization and bilateral symmetry allows for the marriage of directional movement (bilateral symmetry) and the presence of sensory organs facing the direction in which the animal is moving (cephalization).

33.3 This allows systematists to classify animals based solely on derived characteristics. Using features that have only evolved once implies that the species that have that characteristic are more closely related to each other than they are to species that do not have the characteristic.

33.4 The cells of a truly colonial organism, such as a colonial protist, are all structurally and functionally identical; however, sponge cells are differentiated and these cells coordinate to perform functions required by the whole organism. Unlike all other animals, however, sponges do appear much like colonial organisms in that they are not composed of true tissues, and the cells are capable of differentiating from one type to another.

33.5 Two key morphological characters are the number of body layers, making an animal diploblastic or triploblastic, and the type of symmetry. Ctenophores were traditionally thought to be diploblastic, which is the ancestral condition. However, recent studies have suggested that they are, in fact, triploblastic, which is a synapomorphy of bilaterians. In addition, ctenophores exhibit a modified type of radial symmetry, which suggests that this character may not link them with Cnidaria.

33.6 Yes, the coelom has been lost in several clades of protostomes.

INQUIRY AND DATA ANALYSIS QUESTION

Page 646 Inquiry question: None of these characters is a completely reliable indicator of phylogenetic relationships. The coelom only evolved once and is a synapomorphy for the clade comprising protostomes and deuterostomes. Thus, all species with a coelom belong to this clade, but, a number of members of this clade have evolved a pseudocoelom or have become acoelomic. As a result, some species with a coelom are more closely related to species with a pseudocoelom or no coelom than they are to other species with a coelom. A pseudocoelom has evolved convergently in several clades. Thus, the possession of a pseudocoelom is not an indicator of phylogenetic relationships among major animal clades. The acoelomic condition is ancestral for animals, but also re-evolved within Bilateria, so this character state is also not a reliable phylogenetic trait.

UNDERSTAND

1. c 2. b 3. a 4. d 5. a 6. b 7. d 8. a 9. a 10. d 11. a
12. b 13. d

APPLY

1. d   2. b   3. Determinate development indicates that it is a protostome and the fact that it molts places it within the Ecdysozoa. The presence of jointed appendages makes it an arthropod. 4. d

SYNTHESIZE

1. The tree should contain Platyhelminthes and Nemetera on one branch, a second branch should contain nematodes, and a third branch should contain the annelids and the hemichordates. This does not coincide with the information in figure 33.5. Therefore, some of the different types of body cavities have evolved multiple times, and the body cavities are not good characteristics from which to infer phylogenetic relationships.

2. Answers may vary depending on the classification used. Many students will place the Echinoderms near the Cnidaria due to radial symmetry; others will place them closer to the annelids.

CHAPTER 34

LEARNING OUTCOME QUESTIONS

34.1   Molting is a synapomorphy of ecdysozoans, and spiral cleavage of spiralians.

34.2   Tapeworms are parasitic platyhelminthes that live in the digestive system of their host. Tapeworms have a scolex, or head, with hooks for attaching to the wall of their host's digestive system. Another way in which the anatomy of a tapeworm relates to its way of life is its dorsoventrally flattened body and corresponding lack of a digestive system. Tapeworms live in their food; as such they absorb their nutrients directly through the body wall, and their flat bodies facilitate this form of nutrient delivery.

34.3   The cilia both sweep food into the rotifer's mouth and provide the propulsion for locomotion.

34.4   In most invertebrates with a coelom, an important function is internal support by hydrostatic pressure; however, this is unnecessary in most mollusks because the shell serves this purpose.

34.5   With a flow-through digestive tract, food moves in only one direction. This allows for specialization within the tract; sections may be specialized for mechanical and chemical digestion, some for storage, and yet others for absorption. Overall, the specialization yields greater efficiency than does a gastrovascular cavity.

34.6   The main advantage is coordination. A nervous system that serves the entire body allows for coordinated movement and coordinated physiological activities such as reproduction and excretion, even if those systems themselves are segmented. Likewise, a body-wide circulatory system enables efficient oxygen delivery to all of the body cells regardless of the nature of the organism's individual segments.

34.7   Lophophorates are sessile suspension-feeding animals. Much of their body also remains submerged in the ocean floor. Thus, a traditional tubular digestive system would require either the mouth or the anus to be inaccessible to the water column—meaning the animal either could not feed or would have to excrete waste into a closed environment. The U-shaped gut allows them to both acquire nutrients from and excrete waste into their environment.

34.8   *Ascaris lumbricoides,* the intestinal roundworm, infects humans when the human swallows food or water contaminated with roundworm eggs. The most effective ways of preventing the spread of intestinal roundworms is to increase sanitation (especially in food handling), to promote education about the infection process, and to cease using human feces as fertilizer. Not surprisingly, infection by these parasites is most common in areas without modern plumbing.

34.9   One of the defining features of the arthropods is the presence of a chitinous exoskeleton. As arthropods increase in size, the exoskeleton must increase in thickness disproportionately, in order to bear the pull of the animal's muscles. This puts a limit on the size a terrestrial arthropod can reach because the increased bulk of the exoskeleton would prohibit the animal's ability to move. Water is denser than air and thus provides more support; for this reason aquatic arthropods are able to be larger than terrestrial arthropods.

UNDERSTAND

1. d   2. c   3. b   4. a   5. d   6. d   7. d   8. b   9. c   10. d   11. d

APPLY

1. d   2. b   Lobsters, centipedes, and nematods are all ecdysozoans, but centipedes and lobsters have segmented bodies with appendages, which makes them more closely related.   3. a

SYNTHESIZE

1. Clams and scallops are bivalves, which are filter feeders that siphon large amounts of water through their bodies to obtain food. They act as natural pollution-control systems for bays and estuaries. A loss of bivalves (from overfishing, predation, or toxic chemicals) would upset the aquatic ecosystem and allow pollution levels to rise.

2. Chitin is an example of convergent evolution since these organisms do not share a common chitin-equipped ancestor. Chitin is often used in structures that need to withstand the rigors of stress (chaetae, exoskeletons, zoecium, etc.).

3. Since the population size of a parasitic species may be very small (just a few individuals), possessing both male and female reproductive structures would allow the benefits of sexual reproduction.

4. Answers may vary. However, it is known that the tapeworm is not the ancestral form of platyhelminthes; instead it has lost its digestive tract due to its role as an intestinal parasite. As an intestinal parasite, the tapeworm relies on the digestive system of its host to break down nutrients into their building blocks for absorption.

CHAPTER 35

LEARNING OUTCOME QUESTIONS

35.1   Echinoderms are members of the Bilateria, a clade that also includes chordates and protostomes (see figure 33.4). A synapomorphy of the Bilateria is the evolution of bilateral symmetry. Because all bilaterians are bilaterally symmetrical, the evolution of pentaradial symmetry must have occurred at the base of the Echinodermata from a bilaterally symmetrical ancestor.

35.2   Chordates have a truly internal skeleton (an endoskeleton), as opposed to the endoskeleton on echinoderms, which is functionally similar to the exoskeleton of arthropods. Whereas an echinoderm uses tube feet attached to an internal water-vascular system for locomotion, a chordate has muscular attachments to its endoskeleton. Finally, chordates have a suite of four characteristics that are unique to the phylum: a nerve chord, a notochord, pharyngeal slits, and a postanal tail.

35.3   Although mature and immature lancelets are similar in form, the tadpole-like tunicate larvae are markedly different from the sessile, vaselike adult form. Both tunicates and lancelets are chordates, but they differ from vertebrates in that they do not have vertebrae or internal bony skeletons.

35.4   The functions of an exoskeleton include protection and locomotion—arthropod exoskeletons, for example, provide a fulcrum to which the animals' muscles attach. In order to resist the pull of increasingly large muscles, the exoskeleton must dramatically increase in thickness as the animal grows larger. There is thus a limit on the size of an organism with an exoskeleton—if it gets too large it will be unable to move due to the weight and heft of its exoskeleton.

35.5   Lobe-finned fish are able to move their fins independently, whereas ray-finned fish must move their fins simultaneously. This ability to "walk" with their fins indicates that lobe-finned fish are most certainly the ancestors of amphibians.

35.6   #1 The challenges of moving onto land were plentiful for the amphibians. First, amphibians needed to be able to support their body weight and locomote on land; this challenge was overcome by the evolution of legs. Second, amphibians needed to be able to exchange oxygen with the atmosphere; this was accomplished by the evolution of more efficient lungs than their lungfish ancestors as well as cutaneous respiration. Third, since movement on land requires more energy than movement in the water, amphibians needed a more efficient oxygen delivery system to supply their larger muscles; this was accomplished by the evolution of double-loop circulation and a partially divided heart. Finally, the first amphibians needed to develop a way of staying hydrated in a nonaquatic environment, and these early amphibians developed leathery skin that helped prevent desiccation.

    #2 Extant amphibian orders are monophyletic. However, reptiles evolved from a type of amphibian, now extinct. So, if one includes all amphibians, extinct and extant, then amphibians are paraphyletic. But because those amphibian groups more closely related to reptiles have gone extinct, extant amphibians are monophyletic.

35.7   Amphibians remain tied to the water for their reproduction; their eggs are jelly-like and if laid on the land will quickly desiccate. Reptile eggs, on the other hand, are amniotic eggs—they are watertight and contain a yolk, which nourishes the developing embryo, and a series of four protective and nutritive membranes.

35.8   Two primary traits are shared between birds and reptiles. First, both lay amniotic eggs. Second, they both possess scales (which cover the entire reptile body but solely the legs and feet of birds). Birds also share characteristics (e.g., a four-chambered heart) only with one group of reptiles—the crocodilians.

35.9 The most striking convergence between birds and mammals is endothermy, the ability to regulate body temperature internally. Less striking is flight; found in most birds and only one mammal, the ability to fly is another example of convergent evolution.

35.10 Only the hominids constitute a monophyletic group. Prosimians, monkeys, and apes are all paraphyletic—they include the common ancestor but not all descendants: the clade that prosimians share with the common prosimian ancestor excludes all anthropoids, the clade that monkeys share with the common monkey ancestor excludes hominoids, and the clade that apes share with the common ape ancestor excludes hominids.

INQUIRY AND DATA ANALYSIS QUESTIONS

Page 698 Inquiry question: Fish are paraphyletic because the clade comprising all other vertebrates arose from within fishes and is the sister taxon to Sarcopterygii. Reptiles are paraphyletic because birds arose from within and are sister taxon to—among living reptiles—crocodilians. If one considers extinct amphibians, then Amphibia also is paraphyletic, because some extinct amphibian taxa are more closely related to amniotes (the clade comprising reptiles, birds, and mammals) than they are to other amphibians.

Page 717 Inquiry question: The placenta is very similar to reptile and bird eggs, sharing a number of features: chorion, amnion, and yolk sac, and the umbilical cord is derived from the allantois. The reason is that the placenta is the structure that evolved in placental mammals from the egg of their egg-laying ancestors.

UNDERSTAND

1. a   2. c   3. c   4. a   5. c   6. a   7. d   8. a   9. d

APPLY

1. c   2. c   3. b

SYNTHESIZE

1. Increased insulation would have allowed birds to become endothermic and thus to be active at times that ectothermic species could not be active. High body temperature may also allow flight muscles to function more efficiently.

2. Birds evolved from one type of dinosaurs. Thus, in phylogenetic terms, birds are a type of dinosaur.

3. Like the evolution of modern-day horses, the evolution of hominids was not a straight and steady progression to today's *Homo sapiens*. Hominid evolution started with an initial radiation of numerous species. From this group, the evolutionary trend was toward increasing size, similar to what is seen in the evolution of horses. However, like in horse evolution, there are examples of evolutionary decreases in body size as seen in *Homo floresiensis*. Hominid evolution also reveals the coexistence of related species, as seen with *Homo neaderthalensis* and *Homo sapiens*. Hominid evolution, like horse evolution, was not a straight and steady progression to the animal that exists today.

CHAPTER 36
LEARNING OUTCOME QUESTIONS

36.1 Primary growth contributes to the increase in plant height, as well as branching, with shoot and root apical meristems as the source of the growth. Secondary growth makes substantial contributions to the increase in girth of the plant, allowing for a much larger sporophyte generation, with lateral meristems contributing to secondary growth.

36.2 Vessels transport water and are part of the xylem. The cells are dead and only the walls remain. Cylinders of stacked vessels move water from the roots to the leaves of plants. Sieve-tube members are part of the phloem and transport nutrients. Sieve-tube members are living cells, but they lack a nucleus. They rely on neighboring companion cells to carry out some metabolic functions. Like vessels, sieve-tube members are stacked to form a cylinder.

36.3 If abundant water were available, a mutant plant with decreased numbers of root hairs could survive.

36.4 The axillary buds in the axils of the leaves would grow in the absence of the shoot tip. These shoots would eventually flower and reproduce.

36.5 Unlike eudicot leaves, the mesophyll of monocot leaves is not divided into palisade and spongy layers. Instead, cells surrounding the vascular tissues are specialized for carbon fixation. This anatomical variation allows carbon fixation to occur in a part of the leaf where the oxygen concentration is lower, increasing the efficiency of photosynthesis.

INQUIRY AND DATA ANALYSIS QUESTIONS

Page 736 Inquiry question: Three dermal tissue traits that are adaptive for a terrestrial lifestyle include guard cells, trichomes, and root hairs. Guard cells flank an epidermal opening called a stoma and regulate its opening and closing. Stomata are closed when water is scarce, thus conserving water. Trichomes are hairlike outgrowths of the epidermis of stems, leaves, and reproductive organs. Trichomes help to cool leaf surfaces and reduce evaporation from stomata. Root hairs are epidermal extensions of certain cells in young roots and greatly increase the surface area for absorption.

Page 741 Inquiry question: Each time a meristem cell divides, one daughter cell contributes to new tissue and the other continues as a meristem cell. Each cell would have to divide 10 times for every 100 root cap cells that need to be replaced.

UNDERSTAND

1. d   2. d   3. c   4. b   5. b   6. c   7. a   8. b   9. b   10. b   11. d

APPLY

1. b   2. c   3. d   4. c   5. d   6. d   7. c   8. a   9. b   10. a

SYNTHESIZE

1. Roots lack leaves with axillary buds at nodes, although there may be lateral roots that originated from deep within the root. The vascular tissue would have a different pattern in roots and stems. If there is a vascular stele at the core with a pericycle surrounded by a Casparian strip, you are looking at a root.

2. Lenticels increase gas exchange. In wet soil, the opportunity for gas exchange decreases. Lenticels could compensate for decreased gas exchange, which would be adaptive.

3. There are many ways to answer this question. Root modifications could result in increased surface area for absorption under dry conditions, leaf modifications could better balance the loss of water through stomata by optimizing the level of CO_2 relative to O_2 in the leaf. A shoot modification could lead to increased nutrient storage.

4. Whenever water is scarce, the corn plant will wilt. Without the large cross-section vessels, there are insufficient conduits for water in the plant.

5. The plant would grow quickly, use minimal water, have leaf anatomy and physiology with optimized photosynthesis, and have a modified root or shoot that can be used as a food source for humans.

CHAPTER 37
LEARNING OUTCOME QUESTIONS

37.1 Physical pressures include gravity and transpiration, as well as turgor pressure as an expanding cell presses against its cell wall. Increased turgor pressure and other physical pressures are associated with increases in water potential. Solute concentration determines whether water enters or leaves a cell via osmosis. The smallest amount of pressure on the side of the cell membrane with the greater solute concentration that is necessary to stop osmosis is the solute potential. Water potential is the sum of the pressure from physical forces and from the solute potential.

37.2 The apoplastic route. The Casparian strip blocks water movement through the cell walls of the endodermal cells.

37.3 The driving force for transpiration is the gradient between 100% humidity inside the leaf and the external humidity. When the external humidity is low, the rate of transpiration is high, limited primarily by the amount of water available for uptake through the root system. Guttation occurs when transpiration is low, but ions continue to move into the root because of water potential differences. In turn, more water enters the root and can push the liquid up. This push is in contrast to the pull of transpiration.

37.4 Blue light triggers proton transport, creating a proton gradient that drives the opening of the K$^+$ channels. This leads to an influx of water, an increase in turgor in the guard cells, and the opening of the stomata. Increasing temperatures could inhibit the opening of the stomata as well as dry conditions.

37.5 Drought tolerance: Deeply embedded stomata decrease water loss through transpiration because water tension is altered in the crypts.

Flooding tolerance: Pneumatophores are modified roots that grow above flood waters, allowing for oxygen exchange, which is limited for roots under water.

Salt tolerance: Production of high levels of organic molecules in the roots results in uptake of water, even in saline soil, by changing the water potential inside the root.

37.6 Phloem transport can be bidirectional, whereas xylem transport is always from the base to the tip of the plant. Xylem transport depends on transpiration to move the water and dissolved minerals upwards. Phloem transport relies on active transport for loading and unloading at the source and sink, respectively.

INQUIRY AND DATA ANALYSIS QUESTIONS

Page 759 Data analysis: Before equilibrium, the solute potential of the solution is –0.5 MPa, and that of the cell is –0.2 MPa. Since the solution contains more solute than does the cell, water will leave the cell to the point that the cell is plasmolyzed. Initial turgor pressure (Ψ_p) of the cell = 0.05 MPa, and that of the solution is 0 MPa. At equilibrium, both the solution and the cell will have the same Ψ_w. $\Psi_{cell} = -0.2$ MPa + 0.5 MPa = 0.3 MPa before equilibrium is reached. At equilibrium, $\Psi_{cell} = \Psi_{solution} = -0.5$ MPa, thus $\Psi_{w\,cell} = -0.5$ MPa. At equilibrium, the plasmolyzed cell $\Psi_p = 0$ MPa. Finally, using the relationship $\Psi_{W(cell)} = \Psi_p + \Psi_s$ and $\Psi_{W(cell)} = -0.5$ MPa, $\Psi_{P(cell)} = 0$ MPa, then $\Psi_{s(cell)} = -0.5$ MPa.

Page 761 Inquiry question: The fastest route for water movement through cells has the least hindrance, and thus is the symplast route. The route that exerts the most control over what substances enter and leave the cell is the transmembrane route, which is then the best route for moving nutrients into the plant.

Page 763 Data analysis: **#1** The volume that can move through a vessel or tracheid is proportional to r^4. If a mutation increases the radius, r, of a xylem vessel threefold, then the movement of water through the vessel would increase 81-fold ($r^4 = 3^4 = 81$). A plant with larger-diameter vessels can move much more water up its stems.

#2 The volume that can move through a vessel or tracheid is proportional to r^4. $500^4/80^4 = 1525$ times faster. In the context of a plant, temperature, humidity, and whether or not the guard cells are open affects the rate of transpiration. Structural considerations, including cavitation, can also affect the rate of transpiration.

Page 769 Data analysis: Substances in phloem can move at a rate of 50 to 100 cm/hr. Sucrose could move 25 cm in 15 to 30 min. 25 cm/(50 cm/h) = 0.5 hr = 30 min. 25 cm/(100 cm/hr) = 0.25 hr = 15 min.

UNDERSTAND

1. a 2. c 3. d 4. a 5. c 6. d 7. b 8. a 9. d 10. b

APPLY

1. b 2. d 3. b 4. a 5. b

SYNTHESIZE

1. The solute concentration outside the root cells is greater than inside the cells. Thus the solute potential is more negative outside the cell, and water moves out of the root cells and into the soil. Without access to water, your plant wilts.

2. Look for wilty plants since the rate of water movement across the membrane would decrease in the aquaporin mutants.

3. At the level of membrane transport, plants and animals are very similar. Plant cell walls allow plant cells to take up more water than most animal cells, which rupture without the supportive walls. At the level of epidermal cells there is substantial variation among animals. Amphibians exchange water across the skin. Plants have waterproof epidermal tissue but lose water through stomata. Humans sweat, but dogs do not. Some animals have adaptations for living in aquatic or high-saline environment, as do plants. Vascular plants move vast amounts of water through the plant body via the xylem, using evaporation to fuel the transport. Animals with closed circulatory systems can move water throughout the organism and also excrete excess water through the urinary system, which is responsible for osmoregulation.

4. The rate of transpiration is greater during the day than the night. Since water loss first occurs in the upper part of the tree where more leaves with stomata are located, the decrease in water volume in the xylem would first be observed in the upper portion, followed by the lower portion of the tree.

5. Spring year 1—The new carrot seedling undergoes photosynthesis in developing leaves and the sucrose moves toward the growing tip.

   Summer year 1—The developing leaves are sources of carbohydrate, which now moves to the developing root and also the growing young tip.

   Fall year 1—The carrot root is now the sink for all carbohydrates produced by the shoot.

Spring year 2—Stored carbohydrate in the root begins to move upward into the shoot.

Summer year 2—The shoot is flowering, and the developing flowers are the primary sink for carbohydrates from the root and also from photosynthesis in the leaves.

Fall year 2—Seeds are developing and they are the primary sink. The root reserves have been utilized, and any remaining carbohydrates from photosynthesis are transported to developing seeds.

Chapter 38

LEARNING OUTCOME QUESTIONS

38.1 The water potential around the root decreases, and water no longer moves into the root. Transpiration stops, and the plant has insufficient water to maintain life functions.

38.2 Magnesium is found in the center of the chlorophyll molecule. Without sufficient magnesium, chlorophyll deficiencies will result in decreased photosynthesis and decreased yield per acre.

38.3 *Rhizobium* and legume roots have a symbiotic relationship in which the bacteria enter the root through an infection thread, differentiate, and produce NH_3 for the plant in exchange for carbohydrates. Although nitrogen-fixing bacteria are generally limited to the legumes, more than 90% of vascular plants have a symbiotic relationship with mycorrhizal fungi that extend the surface area of the roots and enhance phosphate transfer. Mycorrhizal symbiosis existed first, and some of the signaling pathways in this symbiosis are also found in *Rhizobium* and legume symbioses.

38.4 **#1** As the amount of nitrogen, relative to carbon, decreased in an organism, less protein is made. For example, a plant may be the same size under high- and low-CO_2 atmospheres, but the relative amount of protein is lower under the high-CO_2 conditions. For herbivores, including humans, more plant material would need to be consumed to obtain the same amount of protein under higher versus lower concentrations of CO_2.

#2 Increasing the amount of available nitrogen in the soil is one strategy. This can be accomplished by using chemically produced ammonia for fertilizing, intercropping with nitrogen-fixing legumes, or using organic matter rich in nitrogen for enriching the soil. Efforts to reduce the relative amounts of atmospheric CO_2 would also be helpful.

38.5 Large poplar trees that are not palatable to animals offer a partial solution. Fencing in areas that are undergoing phytoremediation is another possibility, but it would be difficult to isolate all animals, especially birds. Plants that naturally deter herbivores with secondary compounds, including some mustard species (*Brassica* species) could be effective for phytoremediation.

INQUIRY AND DATA ANALYSIS QUESTIONS

Page 778 Inquiry question: CO_2 from the atmosphere and H_2O from the soil are incorporated into carbohydrates during photosynthesis to create much of the mass of the tree. The largest percentage of the mass comes from atmospheric CO_2.

Page 784 Inquiry question: At low and high temperature extremes, enzymes involved in plant respiration are denatured. Plants tend to acclimate to slower long-term changes in temperature, and rates of respiration are able to adjust. Short-term, more dramatic, changes might slow or halt respiration, especially if a temperature change is large enough to cause enzymes to denature.

UNDERSTAND

1. b 2. a 3. c 4. d 5. a 6. b 7. c 8. a

APPLY

1. c 2. a 3. a 4. d

5. a. For the micronutrient problems you would also use the estimate of 1000 kg of potatoes. The conversion you need to use is that 1 ppm is the same as 1 mg/kg. 4 ppm of copper is the same as 4 mg/kg. Multiply this by 1000 kg of potato, and you have 4000 mg or 4 grams of copper up to 30,000 mg or 33 grams.

   b. 15 ppm of zinc is the same as 15 mg/kg. Multiply this by 1000 kg of potato = 15,000 mg or 15 grams up to 100,000 mg or 100 grams.

   c. For potassium, you calculate 0.5% of 1000, which is 5 kg. You would do the same type of calculation for 6%, which is 60 kg.

   d. 25 ppm of iron is the same as 25 mg/kg. Multiply by 1000 kg of potato = 25,000 mg or 25 grams up to 300 grams.

SYNTHESIZE

1. Bacteria that are important for nitrogen fixation could be destroyed. Other microorganisms that make nutrients available to plants could also be destroyed.

2. Grow the tomatoes hydroponically in a complete nutrient solution minus boron and complete nutrient solution with varying concentrations of boron. Compare the coloration of the leaves, the rate of growth (number of new leaves per unit time), and number and size of fruits produced on plants in each treatment group. It would also be helpful to compare the dry weights of plants from each treatment group at the end of the study.

3. Other inputs include both the macronutrients and micronutrients. Nitrogen, potassium, and phosphorous are common macronutrients in fertilizers, and their levels in the circulating water, along with micronutrients, need to be monitored. CO_2 levels in the air also could be monitored.

CHAPTER 39

LEARNING OUTCOME QUESTIONS

39.1 The lipid-based compounds help to create a water-impermeable layer on the leaves.

39.2 A drug prepared from a whole plant or plant tissue would contain a number of different compounds, in addition to the active ingredient. Chemically synthesized or purified substances contain one or more known substances in known quantities.

39.3 It is unlikely that wasps will kill all the caterpillars. When attacked by a caterpillar, the plant releases a volatile substance that attracts the wasp. But, the wasp has to be within the vicinity of the signal when the plant releases the signal. As a result, some caterpillars will escape detection by wasps.

39.4 The local death of cells creates a barrier between the pathogen and the rest of the plant.

INQUIRY AND DATA ANALYSIS QUESTION

Page 795 Inquiry question: Ricin functions as a ribosome-binding protein that limits translation. A very small quantity of ricin was injected into Markov's thigh from the modified tip of his assassin's umbrella. Without translation of proteins in cells, enzymes and other gene products are no longer produced, causing the victim's metabolism to shut down and leading to death.

UNDERSTAND

1. d 2. b 3. b 4. d 5. c 6. a 7. c 8. a 9. d

APPLY

1. c 2. d 3. d 4. b 5. c 6. a

SYNTHESIZE

1. Humans learn quickly, so plants with toxins that made people ill would not become a dietary mainstay. If there was variation in the levels of toxin in the same species in different areas, humans would likely have continued to harvest plants from the area where plants had reduced toxin levels. As domestication continued, seeds would be collected from the plants with reduced toxin levels and grown the following year.

2. For parasitoid wasps to effectively control caterpillars, sufficiently large populations of wasps would need to be maintained in the area where the infestation occurred. As wasps migrate away from the area, new wasps would need to be introduced. The density of wasps is critical because the wasp has to be in the vicinity of the plant being attacked by the caterpillar when the plant releases its volatile chemical signal. Maintaining sufficient density is a major barrier to success.

3. If a plant is flowering or has fruits developing, the systemin will move toward the fruit or flowers, providing protection for the developing seed. If the plant is a biennial, such as a carrot plant, in its first year of growth, systemin will likely be diverted to the root or other storage organ that will reserve food stores for the plant for the following year.

CHAPTER 40

LEARNING OUTCOME QUESTIONS

40.1 Chlorophyll is essential for photosynthesis. Phytochromes regulate plant growth and development using light as a signal. Phytochrome-mediated responses align the plant with the light environment so photosynthesis is maximized, which is advantageous for the plant.

40.2 The plant would not have normal gravitropic responses. Other environmental signals, including light, would determine the direction of plant growth.

40.3 Folding leaves can startle an herbivore that lands on the plant. The herbivore departs, and the plant is protected.

40.4 During the winter months the leaves would cease photosynthesis except on a few warm days. If the weather warmed briefly, water would move into the leaves and photosynthesis would begin. Unfortunately, the minute the temperature dropped, the leaves would freeze and be permanently damaged. Come the spring, the leaves would not be able to function, and the tree would die. It is to the tree's advantage to shed its leaves and grow new, viable leaves in the spring when the danger of freezing is past.

40.5 Gibberellins and brassinosteroids are likely candidates. Applying brassinosteroids or gibberellins to mutant plants grown under low light intensity would be a good initial experiment. If the plants became taller than the wild-type plants, then you would have a candidate hormone. No response does not indicate that neither hormone pathway is altered. It could be that the mutation is in the receptor for the hormone. One way to determine which hormone receptor could be mutated would be to make the mutant transgenic with a wild-type receptor for either gibberellins or brassinosteroids.

INQUIRY AND DATA ANALYSIS QUESTIONS

Page 804 Inquiry question: A number of red-light–mediated responses are linked to phytochrome action alone, including seed germination, shoot elongation, and plant spacing. Only some of the red-light–mediated responses leading to gene expression depend on the action of protein kinases. When phytochrome converts to the Pfr form, a protein kinase triggers phosphorylation that, in turn, initiates a signaling cascade that triggers the translation of certain light-regulated genes. Not all red-light–mediated responses are disrupted in a plant with a mutation in the protein kinase domain of phytochrome.

Page 805 Data analysis: Red light is absorbed by the leaves at a greater rate than far-red light as the sunlight passes through the canopy. Plant-signaling pathways integrate information about the relative amounts of red and far-red light via phytochrome to regulate plant growth and development. For example, seed germination is inhibited as the relative amount of far-red light increases. Thus, below the shady canopy, seed germination would be less likely to occur. With reduced sunlight, the seedlings would have had less chance of survival because of reduced photosynthesis. Alternatively, the information about reduced sunlight could signal rapid stem elongation to get the plant closer to sufficient sunlight for photosynthesis.

Page 807 Inquiry question: Auxin is involved in the phototropic growth responses of plants, including the bending of stems and leaves toward light. Auxin increases the plasticity of plant cells and signals their elongation. The highest concentration of auxin would most likely occur at the tips of stems where sun exposure is maximal.

Page 815 Inquiry question: A chemical substance, such as the hormone auxin, could trigger the elongation of cells on the shaded side of a stem, causing the stem to bend toward the light.

Page 816 Data analysis: A protractor would be a simple tool for determining the angle of the bend relative to either a vertical or horizontal axis.

UNDERSTAND

1. c 2. a 3. d 4. d 5. b 6. d

APPLY

1. b 2. a 3. b 4. c 5. c 6. b 7. d

SYNTHESIZE

1. You are observing etiolation. Etiolation is an energy conservation strategy to help plants growing in the dark to reach the light before they die. They don't green up until light becomes available, and they divert energy to internode elongation. This strategy is useful for potato shoots. The sprouts will be long so they can get to the surface more quickly. They will remain white until exposed to sunlight which will signal the production of chlorophyll.

2. Tropism refers to the growth of an organism in response to an environmental signal such as light. Taxis refers to the movement of an organism in response to an environmental signal. Since plants cannot move, they will not exhibit taxis, but they do exhibit tropisms.

3. Auxin accumulates on the lower side of a stem in a gravitropism, resulting in elongation of cells on the lower side. If auxin or vesicles containing auxin responded to a gravitational field, it would be possible to have a gravitropic response without amyloplasts.

4. Farmers are causing a thigmotropic response. In response to touch, the internodes of the seedlings will increase in diameter. The larger stems will be more resistant to wind and rain once they are moved to the field. The seedlings will be less likely to snap once they are moved to the more challenging environment.

CHAPTER 41

LEARNING OUTCOME QUESTIONS

41.1 Without flowering in angiosperms, sexual reproduction is impossible and the fitness of a plant drops to zero.

41.2 Set up an experiment in a controlled growth chamber with a day–night light regimen that promotes flowering. Then interrupt the night length with a brief exposure to light. If day length is the determining factor, the brief flash of light will not affect flowering. If night length is the determining factor, the light flash may affect the outcome. For example, if the plant requires a long night, interrupting the night will prevent flowering. If the plant flowers whether or not you interrupt the night with light, it may be a short-night plant. In that case, you would want to set up a second experiment where you lengthen the night length. That should prevent the plant from flowering.

41.3 Flowers can attract pollinators, enhancing the probability of reproduction.

41.4 No, because the gametes are formed by meiosis, which allows for new combinations of alleles to combine. You may want to review Mendel's law of independent assortment.

41.5 In gymnosperms, the megagametophyte is the nutritional source for the embryo, in contrast with the double-fertilization event in angiosperms that produces the endosperm and the embryo.

41.6 Retaining the seed in the ground might provide greater stability for the seedling until its root system is established.

41.7 When conditions are uniform and the plant is well adapted to those constant conditions, genetic variation would not be advantageous. Rather, vegetative reproduction would ensure that the genotypes that are well adapted to the current conditions are maintained.

41.8 A biennial life cycle allows an organism to store up substantial reserves to be used to support reproduction during the second season. The downside to this strategy is that the plant might not survive the winter between the two growing seasons, and its fitness would therefore be reduced to zero.

INQUIRY AND DATA ANALYSIS QUESTIONS

Page 833 Inquiry question: Strict levels of *CONSTANS (CO)* gene protein are maintained according to the circadian clock. Phytochrome, the pigment that perceives photoperiod, regulates the transcription of *CO*. By examining posttranslational regulation of *CO*, it might be possible to determine whether protein levels are modulated by means other than transcription. An additional level of control might be needed to ensure that the activation of floral meristem genes coincides with the activation of genes that code for individual flower organs.

Page 834 Inquiry question: Flower production employs up to four genetically regulated pathways. These pathways ensure that the plant flowers when it has reached adult size, when temperature and light levels are optimal, and when nutrition is sufficient to support flowering. All of these factors combine to ensure the success of flowering and the subsequent survival of the plant species.

Page 835 Inquiry question: Once vernalization occurred and nutrition was optimal, flowering could occur in the absence of flower-repressing genes, even if the plant had not achieved adult size. Thus flowering might occur earlier than normal.

Page 849 Data analysis: Seven cell divisions. One cell divides to produce 2 cells, which divide to produce 4, which divide to produce 8, which proceed through four more divisions to produce a total of 128.

Page 849 Inquiry question: #1 Because it is mutant for monopteros, it does not make a functional MONOPTEROS protein, which is a transcription factor needed for transcribing genes required for root development.

#2 If MONOPTEROS could not be repressed, root development would occur whenever there was auxin available. Roots might develop in unexpected places on the plant.

UNDERSTAND

1. a   2. c   3. a   4. d   5. d   6. c   7. d   8. c   9. a   10. d   11. a   12. c   13. c   14. a   15. b   16. c   17. a

APPLY

1. b   2. c   3. b   4. a   5. b   6. c   7. c   8. b   9. c

SYNTHESIZE

1. Poinsettias are short-day plants. The lights from the cars on the new highway interrupt the long night and prevent flowering.

2. Spinach is a long-day plant, and you want to harvest the vegetative, not the reproductive, parts of the plant. Spinach will flower during the summer as the days get longer, but only leaves will be produced during the spring. If you grow and harvest your spinach in the early spring, you will be able to harvest the leaves before the plant flowers and begins to senesce.

3. Cross-pollination increases the genetic diversity of the next generation. However, self-pollination is better than no pollination. The floral morphology of columbine favors cross-pollination, but self-pollination is a backup option. Should this backup option be utilized, there is still one more opportunity for cross-pollination to override self-pollination because the pollen tube from the other plant can still grow through the style more rapidly than the pollen tube from the same plant.

4. Potatoes grown from true seed take longer to produce new potatoes than potatoes grown from tubers. Seeds are easier to store between growing seasons and require much less storage space than whole potatoes. The seed-grown potatoes will have greater genetic diversity than the asexually propagated potato tubers. If environmental conditions vary from year to year, the seed-grown potatoes may have a better yield because different plants will have an advantage under different environmental conditions. The tuber-grown potatoes will be identical. If conditions are optimal for that genotype, the tuber-grown potatoes will outperform the more-variable seed-grown potatoes. But, if conditions are not optimal for the asexually propagated potatoes, the seed-grown potatoes may have the higher yield.

5. Place *Fucus* zygotes on a screen and shine a light from the bottom. If light is more important, the rhizoid will form toward the light, even though that is the opposite direction gravity would dictate. If gravity is more important, the rhizoid will form away from the direction of the light.

CHAPTER 42

LEARNING OUTCOME QUESTIONS

42.1 Organs may be made of multiple tissue types. For example, the heart contains muscle, connective tissue, and epithelial tissue.

42.2 The epithelium in glandular tissue produces secretions, the epithelium has microvilli on the apical surface that increase surface area for absorption.

42.3 Blood is a form of connective tissue because it contains abundant extracellular material: the plasma.

42.4 The function of heart cell requires their being electrically connected. The gap junctions allow the flow of ions between cells.

42.5 Neurons may be a meter long, but this is a very thin projection that still can allow diffusion of materials along its length. They do require specialized transport along microtubules to move proteins from the cell body to the synapse.

42.6 The organ systems may overlap. Consider the respiratory and circulatory systems. These systems are interdependent.

42.7 Yes.

42.8 The distinction should be between the ability to generate metabolic heat to modulate temperature, and the lack of that ability. Thus ectotherms and endotherms have replaced cold-blooded and warm-blooded.

INQUIRY AND DATA ANALYSIS QUESTIONS

Page 879 Inquiry question: After 2 min of shivering, the thoracic muscles have warmed up enough to engage in full contractions. The muscle contractions that allow the full range of motion of the wings utilize kinetic energy in the movement of the wings, rather than releasing the energy as heat, which occurred in the shivering response.

Page 881 Data analysis: Small mammals, with a proportionately larger surface area, dissipate heat readily, which is helpful in a warm environment, but detrimental in a cold environment. In cold conditions, small mammals must seek shelter or have adaptations, such as insulating hair, to maintain body temperature. Because of a greater volume and proportionately less surface area, large mammals are better adapted to cold environments since it takes much longer for them to lose body heat. Hot environments pose a greater challenge for them for the same reason.

Page 882 Data analysis: Pyrogens target the integrating center in the hypothalamus. They act by increasing the "set point," thus leading to a consistently higher body temperature.

UNDERSTAND

1. a   2. c   3. c   4. d   5. a   6. d   7. b   8. b   9. b

APPLY

1. b   2. c   3. c   4. d   5. a   6. c

SYNTHESIZE

1. Yes, both the gut and the skin include epithelial tissue. A disease that affects epithelial cells could affect both the digestive system and the skin. For example, cystic fibrosis affects the ion transport system in epithelial membranes. It is manifested in the lungs, gut, and sweat glands.

2. The nervous system and endocrine system are involved in regulation and maintenance. They function in sensing the internal and external environments and cause changes in the body to respond to changes to maintain homeostasis. The digestive, circulatory, and respiratory systems are also involved in regulation and maintenance. They are grouped together because they all provide necessary nutrients for the body. The digestive system is responsible for the acquisition of nutrients from food; the respiratory system provides oxygen and removes waste (carbon monoxide). The circulatory system transports nutrients to the cells of the body and removes metabolic wastes.

3. Hunger is a negative feedback stimulus. Hunger stimulates an individual to eat, which in turn causes a feeling of fullness that removes hunger. Hunger is the stimulus; eating is the response that removes the stimulus.

4. The internal environment is constantly changing. As you move through your day, muscle activity raises your body temperature, but when you sit down to eat or rest, your temperature cools. The body must constantly adjust to changes in activity or the environment.

CHAPTER 43

LEARNING OUTCOME QUESTIONS

43.1   The somatic nervous system is under conscious control.

43.2   A positive current inward (influx of Na^+) depolarizes the membrane, and a positive current outward (efflux of K^+) repolarizes the membrane.

43.3   Tobacco contains the compound nicotine, which can bind some acetylcholine receptors. This leads to the classic symptoms of addiction due to underlying habituation involving changes to receptor numbers and responses.

43.4   Reflex arcs allow you to respond to a stimulus that is damaging before the information actually arrives at your brain.

43.5   These two systems work in opposition. This may seem counterintuitive, but it is the basis for much of homeostasis.

INQUIRY AND DATA ANALYSIS QUESTIONS

Page 892 Data analysis: The sum of all three is shown in the picture. The other possible pairwise sums of potentials are shown in the following graphs.

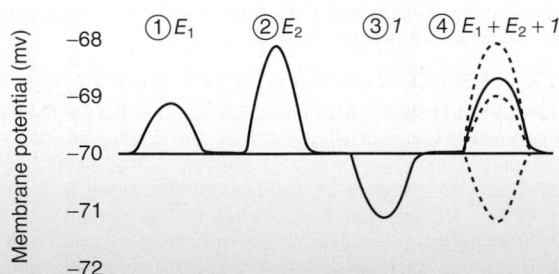

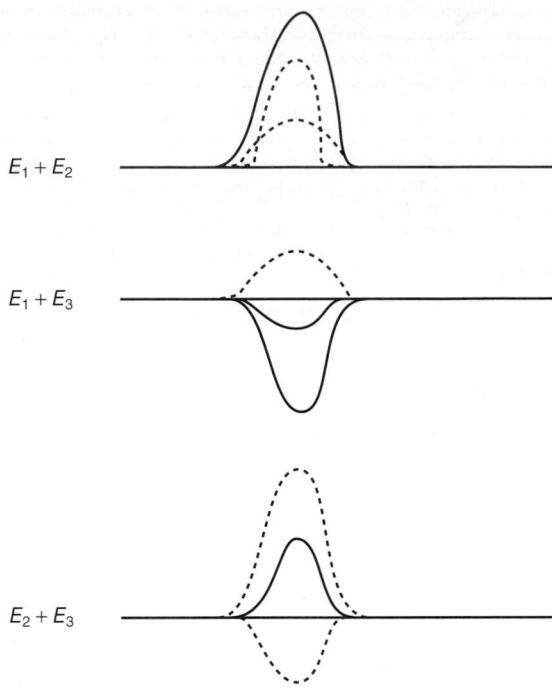

Page 897 Data analysis   The excitatory and inhibitory potentials shown in the figure would sum to a membrane potential close to, or at, the resting potential.

UNDERSTAND

1. a   2. c   3. a   4. a   5. d   6. d   7. a

APPLY

1. d   2. b   3. c   4. a   5. a   6. c   7. d

SYNTHESIZE

1. TEA blocks K^+ channels so that they will not permit the passage of K^+ out of the cell, thereby preventing the cell from returning to the resting potential. Voltage-gated Na^+ channels would still be functional and Na^+ would still flow into the cell, but there would be no repolarization. Na^+ would continue to flow into the cell until an electrochemical equilibrium was reached for Na^+, which is +60 mV. After the membrane potential reached +60 mV, there would be no net movement of Na^+, but the membrane would also not be able to repolarize back to the resting membrane potential. The neuron would no longer be able to function.

   The effects on the postsynaptic cell would be somewhat similar if TEA were applied to the presynaptic cell. The presynaptic cell would depolarize and would continue to release neurotransmitter until it had exhausted its store of synaptic vesicles. As a result, the postsynaptic cell would be bombarded with neurotransmitters and would be stimulated continuously until the stores of presynaptic neurotransmitter were depleted. The postsynaptic cell would, however, recover, being able to repolarize its membrane, and would return to the resting membrane potential.

2. Rising: Na^+ gates open, K^+ closed

   Falling: Na^+ inactivation gate closed, K^+ open

   Undershoot: Na^+ activation gate closed, inactivation gate open, K^+ gate closing.

3. Action potential arrives at the end of the axon.

   Ca^{2+} channels open.

   Ca^{2+} causes synaptic vesicles to fuse with the axon membrane at the synapse.

   Synaptic vesicles release their neurotransmitter.

   Neurotransmitter molecules diffuse across synaptic cleft.

   Postsynaptic receptor proteins bind neurotransmitter.

   Postsynaptic membrane depolarizes.

   If this were an inhibitory synapse, the binding of receptor protein and neurotransmitter would cause the postsynaptic membrane to hyperpolarize.

4. Cells exposed to a stimulus repeatedly may lose their ability to respond. This is known as habituation. Karen's postsynaptic cells may have decreased the number of receptor proteins they produce because the stimulatory signal is so abundant. The result is that it now takes more stimuli to achieve the same result.

CHAPTER 44

LEARNING OUTCOME QUESTIONS

44.1 When the log values of the intensity of the stimulus and the frequency of the resulting action potentials are plotted against each other, a straight line results; this is referred to as a logarithmic relationship.

44.2 Proprioceptors detect the stretching of muscles and subsequently relay information about the relative position and movement of different parts of the organism's body to the central nervous system. This knowledge is critical for the central nervous system; it must be able to respond to these data by signaling the appropriate muscular responses, allowing for balance, coordinated locomotion, and reflexive responses.

44.3 The lateral line system supplements the sense of hearing in fish and amphibian larvae by allowing the organism to detect minute changes in the pressure and vibrations of its environment. This is facilitated by the density of water; without an aquatic environment, the adult, terrestrial amphibian will no longer be able to make use of this system. On land, sound waves are more easily detectable by the sense of hearing than are vibratory or pressure waves by the similar structures of a lateral line.

44.4 Many insects, such as the housefly, have chemoreceptors on their feet with which they can detect the presence of edible materials as they move through their environment. These insects can thus "taste" what they are walking on, and when they encounter an edible substrate they can then descend their proboscis and consume the food.

44.5 Individuals with complete red-green color blindness (those who have no red cones or no green cones, as opposed to those who lack only some red or green cones) would be highly unlikely to be able to learn to distinguish these colors. In order for an individual to perceive the colors in the red-green area of the spectrum, both red and green cones are required; without both cones there is no reference point by which individuals could compare the signals between the retina and the brain. If there are other cues available, such as color saturation or object shape and size, individuals with a less severe form of color blindness may be able to learn to distinguish these colors. In the absence of other references, however, it would be very difficult for individuals with even partial red-green color blindness to distinguish these colors.

44.6 The body temperature of an ectothermic organism is not necessarily the same as the ambient temperature; for example, other reptiles may bask in the sun, wherein on a chilly day the sun may warm the animal's body temperature above the ambient temperature. In this situation, the heat-sensing organs of, for example, a pit viper would still be effective in hunting because it would be able to distinguish the differences between the environment and its ectothermic prey.

INQUIRY AND DATA ANALYSIS QUESTIONS

Page 919 Inquiry question: As the injured fish thrashed around, it would produce vibrations—rapid changes in the pressure of the water. The lateral line system in fish consists of canals filled with sensory cells that send a signal to the brain in response to the changes in water pressure.

Page 926 Inquiry question: Both taste and smell utilize chemoreceptors as sensory receptors, wherein the binding of specific proteins to the receptor induces an action potential, which is sent as a sensory signal to the brain. The chemicals detected by both systems must first be dissolved in extracellular fluid before they can be detected. One major difference between the two systems is that the olfactory system does not route a signal through the thalamus; instead, action potentials are routed directly to the olfactory cortex. Another difference is that olfactory receptors occur in larger numbers—tens of millions, as opposed to tens or hundreds of thousands for taste receptors.

Page 930 Inquiry question: Additional kinds of cones can enhance color vision in two ways. First, they may be sensitive to regions of the color spectrum not covered by other cones, such as those that detect ultraviolet light in some animals. In addition, they may allow enhanced discrimination of color even in regions to which other cones are sensitive. Consider figure 45.19 and the color yellow. Humans have three cones, each of which is sensitive to a varying extent to spectra with a wavelength of approximately 520 nm. The human brain integrates the response of the three cones and interprets the color as yellow. Imagine, however, if we had a fourth cone whose peak sensitivity was in the region of 520 nm. Output

from that cone would provide additional discrimination not possible based on the other three cones, and thus would allow an individual to identify shades of color not distinguishable by individuals lacking that cone.

Page 931 Data analysis: Ultraviolet light has shorter wavelengths than the blue light in figure 44.19, and so the peak absorption would fall to the left of the blue cone's peak absorption of 420 nm. According to figure 8.4, the UV portion of the electromagnetic spectrum is between 10 nm and 400 nm, which is the end of the visual spectrum. The peak would fall in that range, probably closer to the 400 nm to be considered "near-ultraviolet."

UNDERSTAND

1. d   2. b   3. d   4. b   5. a   6. c   7. b   8. d   9. a

APPLY

1. c   2. a   3. c   4. b

SYNTHESIZE

1. When blood pH becomes acidic, chemoreceptors in the circulatory and the nervous systems notify the brain and the body responds by increasing the breathing rate. This causes an increase in the release of carbon dioxide through the lungs. Decreased carbon dioxide levels in the blood cause a decrease in carbonic acid, which, in turn, causes the pH to rise.

2. In order to reach the retina and generate action potentials on the optic nerve, light must first pass through the ganglion and bipolar cells to reach the rods and cones that synapse with the bipolar cells. The bipolar cells then synapse with ganglion cells. These in turn send action potentials to the brain. Because the retina comprises three layers, with the rods and cones located farthest from the pupil, light must travel to the deepest level to set off reactions that move up through the more superficial levels and result in optic signals.

3. Without gravity to force the otoliths down toward the hair cells, the otolith organ will not function properly. The otolith membrane would not rest on the hair cells and would not move in response to movement of the body parallel or perpendicular to the pull of gravity. Consequently, the hair cell would not bend and so would not produce receptor potentials. Because the astronauts can see, they would have an impression of motion—they can see themselves move in relation to objects around them—but with their eyes closed, they would not know if they were moving in relation to their surroundings. Because their proprioceptors would still function, they would be able to sense when they moved their arms or legs, but they would not have the sensation of their enter body moving through space.

   The semicircular canals would not function equally well in zero-gravity conditions. Although the fluid in the semicircular canals is still able to move around, some sensation of angular movement would most likely occur, but the full function of the semicircular canals requires the force of gravity to aid in the directional movement of the fluid in the canals.

CHAPTER 45

LEARNING OUTCOME QUESTIONS

45.1 Neurotransmitters are released at a synapse and act on the postsynaptic membrane. Hormones enter the circulatory system and are thus delivered to the entire body.

45.2 The response of a particular tissue depends first on the receptors on its surface, and second on the response pathways active in a cell. Different receptor subtypes can bind the same hormone, and the same receptor can stimulate different response pathways.

45.3 This might lower the amount of GH in circulation. As a treatment, it may have unwanted side effects.

45.4 With two hormones that have antagonistic effects, the body can maintain a fine-tuned level of blood sugar.

45.5 Reducing blood volume should also reduce blood pressure.

INQUIRY AND DATA ANALYSIS QUESTIONS

Page 952 Data analysis: If hypophysectomy is performed on a very early tadpole, the entire signaling axis is disrupted and metamorphosis is blocked. If the surgery is performed later, TSH signaling has already begun, so metamorphosis is not prevented because the pituitary hormones are no longer needed.

Page 956 Data analysis: The use of juvenile hormone as an insecticide is not lethal, but it does prevent metamorphosis and thus prevents reaching sexual maturity, effectively sterilizing the insect. For an insect whose larval form does damage, preventing the next generation will not stop damage from the present generation and, in fact, may increase it.

UNDERSTAND

1. b   2. d   3. c   4. b   5. d   6. b   7. a

APPLY

1. a   2. c   3. c   4. b   5. b   6. b   7. b

SYNTHESIZE

1. If the target cell for a common hormone or paracrine regulator becomes cancerous, it may become hypersensitive to the messenger. This may in turn cause overproduction of cells, which would result in tumor formation. By blocking the production of the hormone specific to that tissue (for example, sex steroids in breast or prostate tissue), it would be possible to slow the growth rate and decrease the size of the tumor.

2. The same hormone can affect two different organs in different ways because the second messengers triggered by the hormone have different targets inside the cell, owing to the cell's different functions. Epinephrine affects the cells of the heart by increasing metabolism so that their contractions are faster and stronger. However, liver cells do not contract and so the second messenger in liver cells triggers the conversion of glycogen into glucose. That is why hormones are so valuable but also economical to the body. One hormone can be produced, one receptor can be made, and one second-messenger system can be used, but there can be two different targets inside the cell.

3. With hormones such as thyroxine, whose effects are slower and have a broader range of activity, a negative feedback system using one hormone adequately controls the system. However, for certain parameters that have a very narrow range and change constantly within that range, a regulatory system that uses up-and-down regulation is desirable. Too much or too little Ca^{2+} or glucose in the blood can have devastating effects on the body, and so those levels must be controlled within a very narrow range. Relying on negative feedback loops would restrict the quick "on" and "off" responses needed to keep the parameters within a very narrow range.

CHAPTER 46

LEARNING OUTCOME QUESTIONS

46.1   Terrestrial invertebrates experience three limitations due to an exoskeleton. First, animals with an exoskeleton can only grow by shedding, or molting, the exoskeleton, leaving them vulnerable to predation. Second, muscles that act on the exoskeleton cannot strengthen and grow because they are confined within a defined space. Finally, the exoskeleton, in concert with the respiratory system of many terrestrial invertebrates, limits the size to which these animals can grow. In order for the exoskeleton of a terrestrial animal to be strong, it has to have a sufficient surface area, and thus it has to increase in thickness as the animal gets larger. The weight of a thicker exoskeleton would impose debilitating constraints on the animal's ability to move.

46.2   Vitamin D is important for the absorption of dietary calcium as well as the deposition of calcium phosphate in bone. Children undergo a great deal of skeletal growth and development; without sufficient calcium deposition, their bones can become soft and pliable, leading to a condition known as rickets, which causes a bending or bowing of the lower limbs. In the elderly, bone remodeling without adequate mineralization of the bony tissue can lead to brittle bones, a condition known as osteoporosis.

46.3   First, unlike the chitinous exoskeleton, a bony endoskeleton is made of living tissue; thus, the endoskeleton can grow along with the organism. Second, because the muscles that act on the bony endoskeleton are not confined within a rigid structure, they are able to strengthen and grow with increased use. Finally, the size limitations imposed by a heavy exoskeleton that covers the entire organism are overcome by the internal bony skeleton, which can support a greater size and weight without itself becoming too cumbersome.

46.4   Slow-twitch fibers are found primarily in muscles adapted for endurance rather than strength and power. Myoglobin provides oxygen to the muscles for the aerobic respiration of glucose, thus providing a higher ATP yield than anaerobic respiration. Increased mitochondria also increases the ATP productivity of the muscle by increasing availability of cellular respiration and thus allows for sustained aerobic activity.

46.5   Locomotion via alternation of legs requires a greater degree of nervous system coordination and balance; the animal needs to constantly monitor its center of gravity in order to maintain stability. In addition, a series of leaps will cover more ground per unit time and energy expenditure than will movement by alternation of legs.

INQUIRY AND DATA ANALYSIS QUESTIONS

Page 969 Inquiry question: The idea is very similar; both quadrupeds and insects such as grasshoppers have flexors and extensors that exert antagonistic control over many of their muscles. The main difference is in structure rather than function—in a grasshopper, the muscles are covered by the skeletal elements, whereas in organisms with an endoskeleton, the muscles overlie the bony skeleton.

Page 974 (figure 46.18) Inquiry question: Increasing the frequency of stimulation to a maximum rate will yield the maximum amplitude of a summated muscle contraction. The strength of a contraction increases because little or no relaxation time occurs between successive twitches.

Page 974 (figure 46.19) Inquiry question: A rough estimate of the composition of the calf muscle could be obtained by measuring the amount of time the calf muscle takes to reach maximum tension and compare that amount with the contraction speed of muscles of known fiber composition. Alternatively, a small sample of muscle could be extracted and examined for histological differences in fiber composition.

UNDERSTAND

1. d   2. b   3. d   4. a   5. c   6. a   7. b   8. a

APPLY

1. c   2. b   3. d   4. b   5. b   6. b

SYNTHESIZE

1. Although a hydrostatic skeleton might have advantages in terms of ease of transport and flexibility of movement, the exoskeleton would probably do a better job at protecting the delicate instruments within. This agrees with our observations of these support systems on Earth. Worms and marine invertebrates use hydrostatic skeletons, although arthropods ("hard bodies") use an exoskeleton. Worms are very flexible, but easily crushed.

2. The first 90 seconds of muscle activity are anaerobic in which the cells utilize quick sources of energy (creatine phosphate, lactic acid fermentation) to generate ATP. After that, the respiratory and circulatory systems will catch up and begin delivering more oxygen to the muscles, which allows them to use aerobic respiration, a much more efficient method of generating ATP from glucose.

3. If acetylcholinesterase is inhibited, acetylcholine will continue to stimulate muscles to contract. As a result, muscle twitching, and eventually paralysis, will occur. In March 1995, canisters of Sarin were released into a subway system in Tokyo. Twelve people were killed and hundreds injured.

4. Natural selection is not goal-oriented. In other words, evolution does not anticipate environmental pressures, and the structures that result from evolution by natural selection are those most well suited to the previous generations' environment. Since vertebral wing development occurred several times during evolution, it is probable that the animals in question— birds, pterosaurs, and bats—all encountered different evolutionary pressures during wing evolution.

CHAPTER 47

LEARNING OUTCOME QUESTIONS

47.1   The cells and tissues of a one-way digestive system are specialized such that ingestion, digestion, and elimination can happen concurrently, making food processing and energy utilization more efficient. With a gastrovascular cavity, however, all of the cells are exposed to all aspects of digestion.

47.2   Herbivores and omnivores chew their food, using the flat surfaces of their molars to break it into small components and to introduce saliva, which begins the digestive process. Carnivores tear off their food with their sharp teeth (canines and carnasial premolars) and swallow whole chunks of flesh, providing little time or opportunity for digestion to occur in the mouth.

47.3   A chicken sandwich includes carbohydrate (bread), protein (chicken), and fat (mayonnaise). The breakdown of carbohydrates begins with salivary amylase in the mouth. The breakdown of proteins begins in the stomach with pepsinogen, and the emulsification of fats begins in the duodenum with the introduction of bile. Therefore, the chicken begins its breakdown in the stomach.

47.4   Fats are broken down, by emulsification, into fatty acids and monoglycerides, both of which are nonpolar molecules. Nonpolar molecules are able to enter the epithelial cells by simple diffusion.

47.5   The success of any mutation depends on the selective pressures to which that species is subjected. Thus, if two different species are subjected to similar environmental conditions and undergo the same mutation, then, yes, the mutation should be similarly successful. If two species undergo the same mutation but are under different selective pressures, then the mutation may not be successful in both species.

47.6   The sight, taste, and, yes, smell of food are the triggers the digestive system needs to release digestive enzymes and hormones. The saliva and gastric secretions that are required for proper digestion and are triggered by the sense of smell would be affected by anosmia.

47.7   Any ingested compounds that might be dangerous are metabolized first by the liver, thus reducing the risk to the rest of the body.

47.8   Even with normal leptin levels, individuals with reduced sensitivity in the brain to the signaling molecule may still become obese.

INQUIRY AND DATA ANALYSIS QUESTIONS

Page 984 Inquiry question: Recall the old expression: a jack of all trades is master of none. Omnivores are proficient at eating many different types of food, but they're not specialized for any particular type, and thus are less efficient. However, herbivores, with teeth specialized for cutting and grinding, are more effective at eating plant material, and carnivores, with teeth specialized for slicing and tearing, are more efficient at eating muscle tissue. Thus, if a species' diet is solely one type, it is better off having teeth specialized for that purpose.

Page 985 Inquiry question: If the epiglottis does not properly seal off the larynx and trachea, food can accidentally become lodged in the airway, causing choking.

Page 986 Inquiry question: The digestive system secretes a mucus layer that helps to protect the delicate tissues of the alimentary canal from the acidic secretions of the stomach.

Page 988 Data analysis: One meter equals 3.28 feet. The area of a square is length squared. Consider a square with sides 1 meter long. The area is $1 \times 1 = 1$ square meter (1 m^2). But because a meter is 3.28 feet, the area of this square in feet is $3.28 \times 3.28 = 10.76$. Hence, $1 \text{ m}^2 = 10.76 \text{ feet}^2$. Therefore the exact area of the small intestine that is 300 m^2 = 3228 feet^2.

Page 992 Data analysis: The amino acid sequences for lysozyme evolved convergently among ruminants and langur monkeys. Thus, if a phylogeny was constructed using solely the lysozyme molecular data, these species—ruminants and langur monkeys—would be adjacent to each other on the phylogenetic tree.

Page 997 Inquiry question: GIP and CCK send inhibitory signals to the hypothalamus when food is ingested. If the hypothalamus sensors did not work properly, leptin levels would increase; increased leptin levels would result in a loss of appetite.

UNDERSTAND

1. b   2. c   3. c   4. b   5. d   6. a   7. c

APPLY

1. d   2. b   3. c   4. c

SYNTHESIZE

1. Birds feed their young with food they acquire from the environment. The adult bird consumes the food but stores it in her crop. When she returns to the nest, she regurgitates the food into the mouths of the fledglings. Mammals on the other hand feed their young with milk that is produced in the mother's mammary glands. Their young feed by latching onto the mother's nipples and sucking the milk. Mammals have no need for a crop in their digestive system because they feed their young a liquid, which does not require the grinding provided by the crop in birds.

2. Leptin is produced by the adipose cells and serves as a signal for feeding behavior. Since low blood leptin levels signal the brain to initiate feeding, a treatment for obesity would need to raise leptin levels, thereby decreasing appetite.

3. The liver plays many important roles in maintaining homeostasis. Two of those roles are detoxifying drugs and chemicals and producing plasma

proteins. A drop in plasma protein levels is indicative of liver disease, which in turn could be caused by abuse of alcohol or other drugs.

4. The selective pressures that guide the adaptation of mutated alleles within a population were the same in these two groups of organisms. Both ruminants and langur monkeys eat tough, fibrous plant materials, which are broken down by intestinal bacteria. The ruminants and langurs then absorb the nutrients from the cellulose by digesting those bacteria; this is accomplished through the use of these adapted lysozymes. Normal lysozymes, found in saliva and other secretions, work in a relatively neutral pH environment. These intestinal lysozymes, however, needed to adapt to an acidic environment, which explains the level of convergence.

5. Whereas mammalian dentition is adapted to processing different food types, birds are able to process different types of food by breaking up food particles in the gizzard. Bird diets are comparably diverse to mammalian diets; some birds are carnivores, others are insectivores or frugivores, still others omnivores.

CHAPTER 48

LEARNING OUTCOME QUESTIONS

48.1   Fick's Law states that the rate of diffusion (R) can be increased by increasing the surface area of a respiratory surface, increasing the concentration difference between respiratory gases, and decreasing the distance the gases must diffuse: $R = \dfrac{DA \, \Delta p}{d}$. Continually beating cilia increase the concentration difference (Δp).

48.2   Countercurrent flow systems maximize the oxygenation of the blood by increasing Δp; therefore, maintaining a higher oxygen concentration in the water than in the blood throughout the entire diffusion pathway is required. This process is enhanced by allowing water to flow in only one direction, counter to the blood flow. The lamellae, found within a fish's gill filaments, achieves this.

48.3   Birds have a more efficient respiratory system than other terrestrial vertebrates. Birds that live or fly at high altitudes are subjected to lower oxygen partial pressure and thus have evolved a respiratory system that is capable of maximizing the diffusion and retention of oxygen in the lungs. In addition, efficient oxygen exchange is crucial during flight; flying is more energetically taxing than most forms of locomotion, and without efficient oxygen exchange birds would be unable to fly even short distances safely.

48.4   There are both structural and functional differences in bird and mammalian respiration. Both mammals and birds have lungs, but only birds also have air sacs, which they use to move air in and out of the respiratory system, whereas only mammals have a muscular diaphragm used to move air in and out of the lungs. Mammalian lungs are pliable, and gas exchange occurs within small closed-ended sacs called alveoli in the mammalian lung. In contrast, bird lungs are rigid, and gas exchange occurs in the unidirectional parabronchi. In addition, because air flow in mammals is bidirectional, oxygenated and deoxygenated air are mixed, but the unidirectional air flow in birds increases the purity of the oxygen entering the capillaries. Mammalian respiration is less efficient than avian respiration; birds transfer more oxygen with each breath than do mammals. Finally, mammals only have one respiratory cycle, whereas birds have two complete cycles.

48.5   Most oxygen is transported in the blood bound to hemoglobin (forming oxyhemoglobin), whereas only a small percentage is dissolved in the plasma. Carbon dioxide, on the other hand, is predominantly transported as bicarbonate (having first been combined with water to form carbonic acid and then dissociated into bicarbonate and hydrogen ions). Carbon dioxide is also transported dissolved in the plasma and bound to hemoglobin.

INQUIRY AND DATA ANALYSIS QUESTIONS

Page 1002 Data analysis: The rate is a function of area (A) multiplied by differences in pressure (Δp). Thus, increases in either directly increase the rate, so whichever increases to a greater extent would have a greater effect on increasing the rate of diffusion.

Page 1005 Data analysis: Exchange is greatest when the difference in O_2 concentration is greatest. So, for the countercurrent exchange example, the difference is about the same throughout, so exchange would be about the same along the length of the capillaries, although the greatest difference occurs at the top, where the difference is 15% (100% – 85%). In concurrent exchange, the greatest difference is at the bottom, where it is 80% (90% – 10%).

Page 1012 Inquiry question: Fick's Law of Diffusion states that for a dissolved gas, the rate of diffusion is directly proportional to the pressure difference between the two sides of the membrane and to the area over which the diffusion occurs.

In emphysema, alveolar walls break down and alveoli increase in size, effectively reducing the surface area for gas exchange. Emphysema thus reduces the diffusion of gases.

Page 1013 Data analysis: The difference in oxygen content between arteries and veins during rest and exercise shows how much oxygen was unloaded to the tissues.

Page 1014 Data analysis: Not really. A healthy individual still has a substantial oxygen reserve in the blood even after intense exercise.

Page 1014 (figure 48.16) Data analysis: Oxyhemoglobin saturation decreases with increasing temperature, and so for any given P_{O_2}, oxygen will unload at higher temperatures, resulting in lower percent saturation and will remain bound at lower temperatures, resulting in higher percent saturation.

Page 1014 Inquiry question: It increases it. At any pH or temperature, the percentage of O_2 saturation falls (e.g., more O_2 is delivered to tissues) as pressure increases.

UNDERSTAND

1. c 2. d 3. d 4. d 5. b 6. a 7. d 8. c

APPLY

1. d 2. c 3. a 4. c 5. a 6. c

SYNTHESIZE

1. Fish gills have not only a large respiratory surface area but also a countercurrent flow system, which maintains an oxygen concentration gradient throughout the entire exchange pathway, thus providing the most efficient system for the oxygenation of blood. Amphibian respiratory systems are not very efficient. They practice positive-pressure breathing. Bird lungs are quite effective, in that they have a large surface area and one-way air flow; mammals, on the other hand, have only a large surface area but no mechanism to ensure the maintenance of a strong concentration gradient.

2. During exercise, cellular respiration increases the amount of carbon dioxide released, thus decreasing the pH of the blood. In addition, the increased cellular respiration increases temperature, as heat is released during glucose metabolism. Decreased pH and increased temperature both facilitate an approximately 20% increase in oxygen unloading in the peripheral tissues.

3. Unicellular prokaryotic organisms, protists, and many invertebrates are small enough that gas exchange can occur over the body surface directly from the environment. Only larger organisms, for which most cells are not in direct contact with the environment with which gases must be exchanged, require specialized structures for gas exchange.

CHAPTER 49

LEARNING OUTCOME QUESTIONS

49.1 Following an injury to a vessel, vasoconstriction is followed by the accumulation of platelets at the site of injury and the subsequent formation of a platelet plug. This triggers a positive feedback enzyme cascade, attracting more platelets, clotting factors, and other chemicals, each of which continually attracts additional clotting molecules until the clot is formed. The enzyme cascade also causes fibrinogen to come out of solution as fibrin, forming a fibrin clot that will eventually replace the platelet plug.

49.2 When the insect heart contracts, it forces hemolymph out through the vessels and into the body cavities. When it relaxes, the resulting negative pressure gradient, combined with muscular contractions in the body, draws the blood back to the heart.

49.3 The primary advantage of having two ventricles rather than one is that oxygenated blood is separated from deoxygenated blood. In fish and amphibians, oxygenated and deoxygenated blood mix, leading to less oxygen being delivered to the body's cells.

49.4 The delay following auricular contraction allows the atrioventricular valves to close prior to ventricular contraction. Without that delay, the contraction of the ventricles would force blood back up through the valves into the atria.

49.5 During systemic gas exchange, only about 90% of the fluid that diffuses out of the capillaries returns to the blood vessels; the rest moves into the lymphatic vessels, which then return the fluid to the circulatory system via the left and right subclavian veins.

49.6 Breathing rate is regulated to ensure ample oxygen is available to the body. However, the heart rate must be regulated to ensure efficient delivery of the available oxygen to the body cells and tissues. For example, during exertion, respiratory rates will increase in order to increase oxygenation and allow for increased aerobic cellular respiration. But simply increasing the oxygen availability is not enough—the heart rate must also increase so that the additional oxygen can be quickly delivered to the muscles undergoing cellular respiration.

INQUIRY DATA ANALYSIS QUESTIONS

Page 1020 Inquiry question: It depends what organisms you are studying. In mammals, the nucleus is ejected during maturation of the erythrocyte, and thus nuclear DNA cannot be obtained. In other vertebrates, this is not true, and thus nuclear DNA is present in mature red blood cells.

Page 1021 Inquiry question: Erythropoietin is a hormone that stimulates the production of erythrocytes from the myeloid stem cells. If more erythrocytes are produced, the oxygen-carrying capacity of the blood is increased. This could potentially enhance athletic performance and is why erythropoietin is banned from use during the Olympics and other sporting events.

Page 1030 Data analysis: The difference between the two is 15-fold (60/4). Because resistance increases to the fourth power in inverse proportion, the difference in resistance would be 15 to the fourth power, or 15^4, which equals 50,625.

Page 1034 Data analysis: Output equals the number of beats per minute multiplied by volume per beat (output = beats × volume). At rest, output is 5 L/min, so to increase it to 25 L/min requires a fivefold increase. In theory, this could be accomplished by increasing the number of heartbeats fivefold, to 360/min, while keeping volume the same, but this is not physically possible. Similarly, one could keep the number of beats constant but increase the volume fivefold, to 350 mL/beat, but this is also impossible. In reality, both heartbeat number and volume must increase by amounts that, when multiplied together, equal 5. So, for example, if heartbeat doubled to 144 beats/min and volume increased by 2.5-fold to 175 mL/beat, then the total increase (beat × volume) would be a fivefold increase.

UNDERSTAND

1. a 2. b 3. c 4. a 5. c 6. d 7. c 8. a

APPLY

1. a 2. b 3. c 4. a

SYNTHESIZE

1. Antidiuretic hormone (ADH) is secreted by the posterior pituitary but its target cells are in the kidney. In response to the presence of ADH, the kidneys increase the amount of water reabsorbed. This water eventually returns to the plasma where it causes an increase in volume and subsequent increase in blood pressure. Another hormone, aldosterone, also causes an increase in blood pressure by causing the kidney to retain Na^+, which sets up a concentration gradient that also pulls water back into the blood.

2. Blood includes plasma (composed primarily of water with dissolved proteins) and formed elements (red blood cells, white blood cells, and platelets). Lymph is composed of interstitial fluid and is found only within the lymphatic vessels and organs. Both blood and lymph are found in organisms with closed circulatory systems. Hemolymph is both the circulating fluid and the interstitial fluid found in organisms with open circulatory systems.

3. Many argue that the evolution of endothermy was less an adaptation to maintain a constant internal temperature and more an adaptation to function in environments low in oxygen. If this is the case, then yes, it makes sense that the evolution of the four-chambered heart, an adaptation that increases the availability of oxygen in the body tissues and which would be highly beneficial in an oxygen-poor environment, and the evolution of endothermy were related. These two adaptations can also be looked at as related in that the more efficient heart would be able to provide the oxygen necessary for the increased metabolic activity that accompanies endothermy.

4. The SA node acts as a natural pacemaker. If it is malfunctioning, one would expect a slow or irregular heartbeat or irregular electrical activity between the atria and the ventricles.

CHAPTER 50

LEARNING OUTCOME QUESTIONS

50.1 Water moves toward regions of higher osmolarity.

50.2 Nitrogenous waste results from the degradation of old proteins and it is a problem because it is toxic in the body.

50.3 They are both involved in water conservation.

50.4 This may have arisen independently in both the mammalian and avian lineages or have been lost from the reptilian lineage.

50.5 This would increase the osmolarity within the tubule system and thus should decrease the reabsorption of water, leading to water loss.

50.6 Blocking aquaporin channels would prevent reabsorption of water from the collecting duct.

INQUIRY AND DATA ANALYSIS QUESTION

Page 1043 Data analysis: An increase of any solute in the blood will reduce the amount of water that is reabsorbed.

UNDERSTAND

1. d   2. a   3. c   4. c   5. d   6. d   7. d

APPLY

1. d   2. c   3. b   4. b   5. c   6. b

SYNTHESIZE

1. a. Antidiuretic hormone (ADH) is produced in the hypothalamus and is secreted by the posterior pituitary. ADH targets the collecting duct of the nephron and stimulates the reabsorption of water from the urine by increasing the permeability of water in the walls of the duct. The primary stimulus for ADH secretion is an increase in the osmolarity of blood.

   b. Aldosterone is produced and secreted by the adrenal cortex in response to a drop in blood Na^+ concentration. Aldosterone stimulates the distal convoluted tubules to reabsorb Na^+, thus decreasing the excretion of Na^+ in the urine. The reabsorption of Na^+ is followed by the reabsorption of both Cl^- and water, and so aldosterone has the net effect of retaining both salt and water. Aldosterone secretion, however, is not stimulated by a decrease in blood osmolarity, but rather by a decrease in blood volume. A group of cells located at the base of the glomerulus, called the juxtaglomerular apparatus, detects drops in blood volume that then stimulates the renin–angiotensin–aldosterone system.

   c. Atrial natriuretic hormone (ANH) is produced and secreted by the right atrium of the heart, in response to an increase in blood volume. The secretion of ANH results in the reduction of aldosterone secretion. With the secretion of ANH, the distal convoluted tubules reduce the amount of Na^+ that is reabsorbed, and likewise reduce the amount of Cl^- and water that is reabsorbed. The final result is a reduction in blood volume.

     John's normal renal blood flow rate would be 21% of cardiac output, or 7.2 L/min × 0.21 = 1.5 L/min. If John's kidneys are not affected by his circulatory condition, his renal blood flow rate should be about 1.5 L/min.

CHAPTER 51

LEARNING OUTCOME QUESTIONS

51.1 No, innate immunity shows some specificity for classes of molecules common to pathogens.

51.2 Hematopoietic stem cells.

51.3 T-cell receptors are rearranged to generate a large number of different receptors with specific binding abilities. Toll-like receptors are not rearranged and recognize specific classes of molecules, not specific molecules.

51.4 Ig receptors are rearranged to generate many different specificities. TLR innate receptors are not rearranged and bind to specific classes of molecules.

51.5 Allergies are a case of the immune system overreacting, whereas autoimmune disorders involve the immune system being compromised.

51.6 Diagnostic kits use monoclonal antibodies because they are developed against a single specific epitope of an antigen. They are also more efficient to produce because they use cells that can be grown in culture and do not require a more complex process of immunizing animals, bleeding them, and then isolating the antibodies from their sera.

51.7 The main difference between polio and influenza is the rate at which the viruses can change. The poliovirus is a RNA virus with a genome that consists of a single RNA. The viral surface proteins do not change rapidly, allowing immunity via a vaccine. Influenza is an RNA virus with a high mutation rate, which means that surface proteins change rapidly. Influenza has a genome that consists of multiple RNAs, which allows recombination of the different viral RNAs during infection with different strains.

INQUIRY AND DATA ANALYSIS QUESTIONS

Page 1059 Inquiry question: The viruses would be liberated into the body where they could infect numerous additional cells.

Page 1063 Inquiry question: The antigenic properties of the two viruses must be similar enough that immunity to cowpox also enables protection against smallpox.

Page 1074 Inquiry question: The common structure and mechanism of formation of B-cell immunoglobulins (Igs) and T-cell receptors (TCRs) suggests a common ancestral form of adaptive immunity gave rise to the two cell lines existing today.

Page 1078 Data analysis: A high level of hCG in a urine sample will block the binding of the antibody to hCG-coated particles, thus preventing any agglutination.

Page 1079 Inquiry question: Influenza frequently alters its surface antigens, making it impossible to produce a vaccine with a long-term effect. Smallpox virus has a considerably more stable structure.

UNDERSTAND

1. b   2. a   3. b   4. c   5. c   6. b   7. c

APPLY

1. d   2. c   3. c   4. b, a, d, c   5. c   6. b   7. a

SYNTHESIZE

1. It would be difficult to advertise this lotion as immune enhancing. The skin serves as a barrier to infection because it is oily and acidic. Applying a lotion that is watery and alkaline will dilute the protective effects of the skin secretions, thereby inhibiting the immune functions. Perhaps it is time to look for another job.

2. The scratch has caused an inflammatory response. Although it is very likely that some pathogens entered her body through the broken skin, the response is actually generated by the injury to her tissue. The redness is a result of the increased dilation of blood vessels caused by the release of histamine. This also increases the temperature of the skin by bringing warm blood closer to the surface. Leakage of fluid from the vessels causes swelling in the area of the injury, which can cause pressure on the pain sensors in the skin. All of these serve to draw defensive cells and molecules to the injury site, thereby helping to defend her against infection.

3. This could be done a number of ways. However, one method would be to show that viral genetic material never appears within the cells of those who claim immunity. Another method would involve testing for the presence of interferon, which is released by cells in response to viral infection.

4. These data imply that innate immunity is a very ancient defense mechanism. The presence of these proteins in cnidarians indicates that they arose soon after multicellularity.

CHAPTER 52

LEARNING OUTCOME QUESTIONS

52.1 Genetic sex determination essentially guarantees equal sex ratios; when sex ratios are not equal, the predominant sex is selected against because those individuals have more competition for mates. Temperature-dependent sex determination can result in skewed sex ratios in which one sex or the other is selected against. Genetic sex determination, on the other hand, can provide much greater stability within the population, and consequently the genetic characteristics that provide that stability are selected for.

52.2 Estrous cycles occur in most mammals, and most mammalian species have relatively complex social organizations and mating behaviors. The cycling of sexual receptivity allows for these complex mating systems. Specifically, in social groups where male infanticide is a danger, synchronized estrous among females may be selected for as it would eliminate the ability of the male to quickly impregnate the group females. Physiologically, estrous cycles result in the maturation of the egg accompanying the hormones that promote sexual receptivity.

52.3 In mating systems where males compete for mates, sperm competition, a form of sexual selection, is very common. In these social groups, multiple males may mate with a given female, and thus those individuals who produced the highest number of sperm would have a reproductive advantage—a higher likelihood of siring the offspring.

52.4 The answer varies depending on the circumstances. In a species that is very *r*-selected, in other words, one that reproduces early in life and often but does not invest much in the form of parental care, multiple offspring per pregnancy would definitely be favored by natural selection. In *K*-selected species where parental care is very high, on the other hand, single births might be favored because the likelihood of offspring survival is greater if the parental resources are not divided among the offspring.

52.5 The birth control pill works by hormonally controlling the ovulation cycle in women. By releasing progesterone continuously the pill prevents ovulation. Ovulation is a cyclical event and under hormonal control, thus it is relatively easy for the process to be controlled artificially. In addition, the female birth control pill only has to halt the release of a single ovum. An analogous male birth control pill, on the other hand, would have to completely cease sperm production (and men produce millions of sperm each day), and such hormonal upheaval in the male could lead to infertility or other intolerable side effects.

INQUIRY AND DATA ANALYSIS QUESTIONS

Page 1085 Inquiry question: The ultimate goal of any organism is to maximize its relative fitness. Small females are able to reproduce but once they become very large they would be better able to maximize their reproductive success by becoming male, especially in groups where only a few males mate with all of the females. Protandry might evolve in species where the supply of mates is limited and there is relatively little space; a male of such a species in close proximity to another male would have higher reproductive success by becoming female and mating with the available male than by waiting for a female (and then having to compete for her with the other male).

Page 1088 Inquiry question: The evolutionary progression from oviparity to viviparity is a complex process; requiring the development of a placenta or comparable structure. Once a complex structure evolves, it is rare for an evolutionary reversal to occur. Perhaps more importantly, viviparity has several advantages over oviparity, especially in cold environments where eggs are vulnerable to mortality due to cold weather (and predation). In aquatic reptiles such as sea snakes, viviparity allows the female to remain at sea and avoid coming ashore, where both she and her eggs would be exposed to predators.

Page 1094 Inquiry question: Under normal circumstances, the testes produce hormones, testosterone and inhibin, which exert negative feedback inhibition on the hormones produced and secreted by the anterior pituitary (luteinizing hormone and follicle-stimulating hormone). Following castration, testosterone and inhibin are no longer produced, and thus the brain will overproduce LH and FSH. For this reason, hormone therapy is usually prescribed following castration.

Page 1095 Data analysis: In the menstrual cycle FSH exceeds the levels of LH for only a short period at the end of one cycle and continuing to the beginning of the next (approx. days 27–28 and 1–2, respectively).

UNDERSTAND

1. c   2. d   3. d   4. d   5. c   6. c   7. a   8. c   9. a

APPLY

1. a   2. b   3. d   4. a

SYNTHESIZE

1. A mutation that makes *SRY* nonfunctional would mean that the embryo would lack the signal to form male structures during development. Therefore, the embryo would have female genitalia at birth.

2. Amphibians and fish that rely on external fertilization also have access to water. Lizards, birds, and mammals have adaptations that allow them to reproduce away from a watery environment. These adaptations include eggs that have protective shells or internal development, or both.

3. FSH and LH are produced by the anterior pituitary in both males and females. In both cases they play roles in the production of sex hormones and gametogenesis. However, FSH stimulates spermatogenesis in males and oogenesis in females, whereas LH promotes the production of testosterone in males and estradiol in females.

4. It could indeed work. The hormone hCG is produced by the zygote to prevent menstruation, which would in turn prevent implantation in the uterine lining. Blocking the hormone receptors would prevent implantation and therefore pregnancy.

5. Parthenogenic species reproduce from gametes that remain diploid. Sperm are haploid, whereas eggs do not complete meiosis (becoming haploid) until

after fertilization. Therefore, only eggs could develop without DNA from an outside source. In addition, only eggs have the cellular structures needed for development. Therefore only females can undergo parthenogenesis.

CHAPTER 53

LEARNING OUTCOME QUESTIONS

53.1 Ca^{2+} ions act as second messengers and bring about changes in protein activity that result in blocking polyspermy and increasing the rate of protein synthesis within the egg.

53.2 In a mammal, the cells at the four-cell stage are still uncommitted, so separating them will still allow for normal development. In frogs, on the other hand, yolk distribution results in displaced cleavage; thus, at the four-cell stage the cells do not each contain a nucleus, which contains the genetic information required for normal development.

53.3 The cellular behaviors necessary for gastrulation differ across organisms; however, some processes are necessary for any gastrulation to occur. Specifically, cells must rearrange and migrate throughout the developing embryo.

53.4 No—neural crest cell fate is determined by its migratory pathway.

53.5 Marginal zone cells in both the ventral and dorsal regions express bone morphogenetic protein 4 (BMP4). The fate of these cells is determined by the number of receptors on the cell membrane to bind to BMP4; greater BMP4 binding will induce a ventral mesodermal fate. The organizer cells, which previously were thought to activate dorsal development, have been found to actually inhibit ventral development by secreting one of many proteins that block the BMP4 receptors on the dorsal cells.

53.6 Most of the differentiation of the embryo, in which the initial structure formation occurs, happens during the first trimester; the second and third trimesters are primarily times of growth and organ maturation, rather than the actual development and differentiation of structures. Thus, teratogens are most potent during this time of rapid organogenesis.

INQUIRY AND DATA ANALYSIS QUESTIONS

Page 1120 Data analysis: Since both mutants cause a loss of pigment cells, and one is a signaling molecule, the other is likely to be the receptor for the signaling molecule. In the mouse system, the signaling molecule is called Steel and the receptor is called c-Kit (an RTK, see chapter 9 for details).

Page 1127 Inquiry question: High levels of estradiol and progesterone in the absence of pregnancy would still affect the body in the same way. High levels of both hormones would inhibit the release of FSH and LH, thereby preventing ovulation. This is how birth control pills work. The pills contain synthetic forms of either both estradiol and progesterone or just progesterone. The high levels of these hormones in the pill trick the body into thinking that it is pregnant and so the body does not ovulate.

UNDERSTAND

1. d   2. b   3. d   4. c   5. d   6. b   7. a

APPLY

1. c   2. b   3. a   4. d   5. b   6. c

SYNTHESIZE

1. By starting with a series of embryos at various stages, you could try removing cells at each stage. Embryos that failed to compensate for the removal (evidenced by missing structures at maturity) would be those that lost cells after they had become committed; that is, when their fate has been determined.

2. Homeoboxes are sequences of conserved genes that play crucial roles in development of both mammals (Fifi) and *Drosophila* (the fruit fly). In fact, we know that they are more similar than dissimilar; research has demonstrated that both groups use the same transcription factors during organogenesis. The major difference between them is in the genes that are transcribed. Homeoboxes in mammals turn on genes that cause the development of mammalian structures, and those in insects would generate insect structures.

3. After fertilization, the zygote produces hCG, which inhibits menstruation and maintains the corpus luteum. At 10 weeks' gestation, the placenta stops releasing hCG, but it continues to release estradiol and progesterone, which maintain the uterine lining and inhibit the pituitary production of FSH and LH. Without FSH and LH, no ovulation and no menstruation occur.

4. Spemann and Mangold removed cells from the dorsal lip of one amphibian embryo and transplanted them to a different location on a second embryo. The transplanted cells caused cells that would normally form skin and belly to instead form somites and the structures associated with the dorsal area. Because of this and because the secondary dorsal structures contained both host and transplanted cells, Spemann and Mangold concluded that the transplanted cells acted as organizers for dorsal development.

CHAPTER 54

LEARNING OUTCOME QUESTIONS

54.1 Just as with morphological characteristics that enhance an individual's fitness, behavioral characteristics can also affect an individual's survivability and reproductive success. Understanding the evolutionary origins of many behaviors allows biologists insights into animal behavior, including that of humans.

54.2 A male songbird injected with testosterone prior to the usual mating season would likely begin singing prior to the usual mating season. However, since female mating behavior is largely controlled by hormones (estrogen) as well, most likely that male will not have increased fitness (and may actually have decreased fitness, if the singing stops before the females are ready to mate, or if the energetic expenditure from singing for two additional weeks is compensated by reduced sperm production).

54.3 The genetic control over pair-bonding in prairie voles has been fairly well established. The fact that males sometimes seek extra-pair copulations indicates that the formation of pair-bonds is under not only genetic control but also behavioral control.

54.4 In species in which males travel farther from the nest (and thus have larger range sizes), there should be significant sex differences in spatial memory. However, in species without sexual differences in range sizes, males and females should not express sex differences in spatial memory. To test the hypothesis you could perform maze tests on males and females of species with sex differences in range size as well as those species without range size differences between the sexes. (NOTE: Such experiments have been performed and do support the hypothesis that there is a significant correlation between range size and spatial memory, so in species with sex differences in range size there are indeed sex differences in spatial memory. See C.M. Jones, V.A. Braithwaite, S.D. Healy SD. The Evolution of Sex Differences in Spatial Ability. *Behav Neurosci* 2003;117(3): 403–411.

54.5 Although there may be a link between IQ and genes in humans, there is most certainly also an environmental component to IQ. The danger of assigning a genetic correlation to IQ lies in the prospect of selective "breeding" and the emergence of "designer babies."

54.6 One experiment that has been implemented in testing counting ability among different primate and bird species is to present the animal with a number and have him match the target number to one of several arrays containing that number of objects. In another experiment, the animal may be asked to select the appropriate number of individual items within an array of items that equals the target number.

54.7 Butterflies and birds have extremely different anatomy and physiology and thus most likely use very different navigation systems. Birds generally migrate bidirectionally; moving south during the cold months and back north during the warmer months in the northern hemisphere. Usually, then, migrations are multigenerational events and it could be argued that younger birds can learn migratory routes from older generations. Butterflies, on the other hand, fly south to breed and die. Their offspring must then fly north having never been there before.

54.8 In addition to chemical reproductive barriers, many species also employ behavioral and morphological reproductive barriers, such that even if a female moth is attracted by the pheromones of a male of another species, the two may be behaviorally or anatomically incompatible.

54.9 The benefits of territorial behavior must outweigh the potential costs, which may include physical danger due to conflict, energy expenditure, and the loss of foraging or mating time. In a flower that is infrequently encountered, the honeycreeper would lose more energy defending the resource than it could gain by utilizing it. On the other hand, there is usually low competitive pressure for highly abundant resources, thus the bird would expend unnecessary energy defending a resource to which its access is not limited.

54.10 The males should exhibit mate choice because they are the sex with the greater parental investment and energy expenditure; thus, like females of most species, they should be the "choosier" sex.

54.11 Generally, reciprocal behaviors are low-cost, whereas behaviors due to kin selection may be low- or high-cost. Protecting infants from a predator is definitely a high-risk/potentially high-cost behavior; thus it would seem that the behavior is due to kin selection. The only way to truly test this hypothesis, however, is to conduct genetic tests or, in a particularly well-studied population, consult a pedigree.

54.12 **#1** Living in a group is associated with both costs and benefits. The primary cost is increased competition for resources, and the primary benefit is protection from predation.

#2 Altruism toward kin is considered selfish because helping individuals closely related to you will directly affect your inclusive fitness.

#3 Most armies more closely resemble insect societies than vertebrate societies. Insect societies consist of multitudes of individuals congregated for the purpose of supporting and defending a select few individuals. One could think of these few protected and revered individuals as the society the army is charged with protecting. These insect societies, like human armies, are composed of individuals each "assigned" to a particular task. Most vertebrate societies, on the other hand, are less altruistic and express increased competition and aggression between group members. In short, vertebrate societies are composed of individuals whose primary concern is usually their own fitness, whereas insect societies are composed of individuals whose primary concern is the colony itself.

INQUIRY AND DATA ANALYSIS QUESTIONS

Page 1135 Inquiry question: Selection for learning ability would cease, and thus change from one generation to the next in maze-learning ability; would only result from random genetic drift.

Page 1135 Data analysis: In populations with a symmetrical distribution with a peak in the middle (a "bell curve" also called a "normal" curve is an example), the mean value usually corresponds to the peak of the curve. In somewhat skewed distributions, such as those in the seventh generation, the mean usually lies a little bit away from the peak, toward the side of the long tail. In this case, the mean of the parental population is about 64, and that of the fast rat population in generation 7 is around 39, so the difference would be approximately 25. This is an example of directional selection (see chapter 21).

Page 1136 Inquiry question: Normal *fosB* alleles produce a protein that in turn affects enzymes that affect the brain. Ultimately, these enzymes trigger maternal behavior. In the absence of the enzymes, normal maternal behavior does not occur.

Page 1137 Data analysis: This experiment tests the effect on rodent pair-bonding behavior of injecting them with the neuropeptide vasopressin. In the control (normal) treatment, vasopressin is not injected (sometimes scientists inject saline (= salt water) solutions, which have no physiological effect but control for the effect of giving an injection. In the control treatment (red bars), wild-type mice exhibit the highest level of pair-bonding behavior, and montane voles exhibit the lowest.

Page 1140 Inquiry question: Peter Marler's experiments addressed this question and determined that both instinct and learning are instrumental in song development in birds.

Page 1148 Inquiry question: Many factors affect the behavior of an animal other than its attempts to maximize energy intake. For example, avoiding predation is also important. Thus, it may be that larger prey require more time to subdue and ingest, thus making the crabs more vulnerable to predators. Hence, the crabs may trade off decreased energy gain for decreased vulnerability for predators. Many other similar explanations are possible.

Page 1148 Data analysis: Assuming that it takes no time to subdue or eat the prey, the raccoon can get 20 J/hr eating frogs (one every 30 min) and 10 J/hr eating crickets (5 crickets/hr—one every 12 minutes—multiplied by 2 J/cricket). The raccoon should search in the stream for frogs.

Follow-up Question: In this scenario, it takes 45 min to find and eat a frog, which means that, on average, a raccoon can eat 1 1/3 frogs/hr. Multiplied by 10 J/frog, results in an average of 13.33 J/hr. By contrast, it takes 15 min to find and devour a cricket, so four can be eaten per hour, leading to a rate of 8 J/hr. Frogs are still favored, but less so.

Page 1150 Inquiry question: This question is the subject of much current research. Ideas include the possibility that males with longer tails (and therefore more spots) are in better condition (because males in poor condition couldn't survive the disadvantage imposed by the tail). The advantage to a female mating with a male in better condition might be either that the male is less likely to be parasitized and thus less likely to pass that parasite on to the female, or the male may have better genes, which in turn would be passed on to the offspring. Another possibility is the visual system for some reason is better able to detect males with long tails, and thus long-tailed males are preferred by females simply because the longer tails are more easily detected and responded to.

Page 1150 Data analysis: To get the predicted number of mates, find 155 on the *x*-axis and then see the corresponding *y*-axis value where the regression line hits 155 on the *x*-axis. You can visualize this by drawing a vertical line from 155 on the *x*-axis, and then drawing a horizontal line from the intersection of the vertical line and the regression line. Where the horizontal line crosses the *y*-axis is the predicted value, which is approximately 4.

Regression lines are calculated by determining the line that best fits the data. The "best fit" line is the one that minimizes the squared vertical distance of each point from the line. To test this, measure the vertical distance from each point to the regression line, square it, and then add them all together. Then draw a different line and do the same thing. The sum of the squared distances in the new line will be greater.

Note, however, that most points don't fall on the line. That means there is variation in the population, and all of that variation is not accounted for by the regression line. As a result, the regression line can only give a predicted value for a male with a particular number of spots, but in reality, most males will deviate from that line by some unpredictable amount. As a result, you wouldn't expect multiple males with the same number of eyespots to have the same number of mates.

Page 1151 Inquiry question: Yes, the larger the male, the larger the prenuptial gift, which provides energy that the female converts into egg production.

Page 1155 Data analysis: Your aunt is the sister of one of your parents. Siblings share 1/2 of their genes, so your aunt is 1/2 genetically related to your mother or father. You also are 1/2 related to your mother or father. Thus, of the 1/2 of your parent's genes that your aunt has, you are likely to have 1/2 of them. 1/2 of 1/2 = 1/4.

Page 1156 Data analysis: Males get all of their genes from their mother. Because the mother has two copies of each gene, on average, two brothers would be related by 1/2. Because males are haploid, however, their offspring get all of their genes. As a result, cousins that are daughters of brothers would also be related by 1/2 in the genes they get from their father. Because their mothers are unrelated, they would share no genes from their mother, and thus would be related by 1/4. Alternatively, if the cousins were daughters of sisters, the two sisters would share 3/4 of their genes, so the two cousins would share 3/8 of their genes. Because the males were unrelated, the cousins would be 3/16 related.

UNDERSTAND

1. b 2. a 3. a 4. c 5. a 6. c 7. d 8. a 9. b 10. a 11. c 12. d

APPLY

1. Presumably, the model is basic, taking into account only size and energetic value of mussels. However, it may be that larger mussels are in places where shore crabs would be exposed to higher levels of predation or greater physiological stress. Similarly, it could be that the model underestimated time costs or energy returns as a function of mussel size. In the case of large mussels being in a place where shore crabs are exposed to costs not considered by the model, one could test the hypothesis in several ways. First, how are the sizes of mussels distributed in space? If they are completely interspersed that would tend to reject the hypothesis. Alternatively, if the mussels were differentially distributed such that the hypothesis was reasonable, mussels could be experimentally relocated (change their distribution in space) and the diets would be expected to shift to match more closely the situation predicted by the model.

2. The four new pairs may have been living in surrounding habitat that was of lower quality, or they may have been individuals that could not compete for a limited number of suitable territories for breeding. Often, the best territories are won by the most aggressive or largest or otherwise best competitors, meaning that the new territory holders would likely have been less fierce competitors. If new residents were weaker competitors (due to aggression or body size), then the birds not removed would have been able to expand their territories to acquire even more critical resources.

3. The key here is that if the tail feathers are a handicap, then by reducing the handicap in these males should enhance their survival compared with males with naturally shorter tail feathers. The logic is simple. If the male with long tail feathers is superior such that it can survive the negative effect of the long tail feathers, then that superior phenotype should be "exposed" with the removal or reduction of the tail feathers. Various aspects of performance could be measured since it is thought that the tail feathers hinder flying. Can males with shorter tails fly faster? Can males with shorter tails turn better? Ultimately, whether males with shorter tails survive better than males with unmanipulated tails can be measured.

4. Both reciprocity and kin selection explain the evolution of altruistic acts by examining the hidden benefits of the behavior. In both cases, altruism actually *benefits* the individual performing the act in terms of its fitness effects. If it didn't, it would be very hard to explain how such behavior could be maintained because actions that reduce the fitness of an individual should be selected against. Definition of the behavior reflects the apparent paradox of the behavior because it focuses on the cost and not the benefit that also accrues to the actor.

SYNTHESIZE

1. The best experiment for determining whether predatory avoidance of certain coloration patterns would involve rearing a predator without an opportunity to learn avoidance and subsequently presenting the predator with prey with different patterns. If the predator avoids the black and yellow coloration more frequently than expected, the avoidance is most likely innate. If the predator does not express any preference but after injury from a prey with the specific coloration does begin to express a preference, then the avoidance is most likely learned. In this case, the learning is operant conditioning; the predator has learned to associate the coloration with pain and thus subsequently avoids prey with that coloration. To measure the adaptive significance of black and yellow coloration, both poisonous (or stinging) and harmless prey species with the coloration and without the coloration pattern could be presented to predators; if predators avoid both the harmful and the harmless prey, the coloration is evolutionary significant.

2. In many cases the organisms in question are unavailable for or unrealistic to study in a laboratory setting. Model organisms allow behavioral geneticists to overcome this obstacle by determining general patterns and then applying these patterns and findings to other, similar organisms. The primary disadvantage of the model system is, of course, the vast differences that are usually found between groups of taxa; however, when applying general principles, in particular those of genetic behavioral regulation, the benefits of using a model outweigh the costs. Phylogenetic analysis is the best way to determine the scale of applicability when using model organisms.

3. Extra-pair copulations and mating with males that are outside a female's territory are, by and large, more beneficial than costly to the female. By mating with males outside her territory, she reduces the likelihood that a male challenging the owner of her territory would target her offspring; males of many species are infanticidal but would not likely attack infants that could be their own. Historical data have actually shown that in many cases, females are more attracted to infanticidal males if those males win territory prior to their infanticidal behavior.

CHAPTER 55

LEARNING OUTCOME QUESTIONS

55.1 It depends on the type of species in question. Conformers are able to adapt to their environment by adjusting their body temperature and making other physiological adjustments. Over a longer period of time, individuals within a nonconforming species might not adjust to the changing environment, but we would expect the population as a whole to adapt due to natural selection.

55.2 If the populations in question comprised source-sink metapopulations, then the lack of immigration into the sink populations would, most likely, eventually result in the extinction of those populations. The source populations would likely then increase their geographic ranges.

55.3 It depends on the initial sizes of the populations in question; a small population with a high survivorship rate will not necessarily grow faster than a large population with a lower survivorship rate.

55.4 A species with high levels of predation would likely exhibit an earlier age at first reproduction and shorter interbirth intervals in order to maximize its fitness under the selective pressure of the predation. On the other hand, species with few predators have the luxury of waiting until they are more mature before reproducing and can increase the interbirth interval (and thus invest more in each offspring) because their risk of early mortality is decreased.

55.5 Many different factors might affect the carrying capacity of a population. For example, climate changes, even on a relatively small scale, could have large effects on carrying capacity by altering the available water and vegetation, as well as the phenology and distribution of the vegetation. Regardless of the type of change in the environment, however, most populations will move toward carrying capacity; thus, if the carrying capacity is lowered, the population should decrease, and if the carrying capacity is raised, the population should increase.

55.6 A given population can experience both positive and negative density-dependent effects, but not at the same time. Negative density-dependent effects, such as low food availability or high predation pressure, would decrease the population size. On the other hand, positive density-dependent effects, such as is seen with the Allee effect, results in a rapid increase in population size. Since a population cannot both increase and decrease at the same time, the two cannot occur concurrently. However, the selective pressures on a population are on a positive–negative continuum, and the forces shaping population size can not only vary in intensity but can also change direction from negative to positive or positive to negative.

55.7 The two are closely tied together, and both are extremely important if the human population is not to exceed the Earth's carrying capacity. As population growth increases, the human population approaches the planet's carrying capacity; as consumption increases, the carrying capacity is lowered—thus, both trends must be reversed.

INQUIRY AND DATA ANALYSIS QUESTIONS

Page 1164 Inquiry question: Very possibly. How fast a lizard runs is a function of its body temperature. Researchers have shown that lizards in shaded habitats have lower temperatures and thus lower maximal running speeds. In such circumstances, lizards often adopt alternative escape tactics that rely less on rapidly running away from potential predators.

Page 1164 Data analysis: Yes. A large value for the slope (at the extreme, 1.0) indicates that as air temperature increases, body temperature also increases. This indicates that the lizard is conforming to environmental conditions; as the air gets hotter (or colder), so does the lizard. In contrast, a low slope (at the extreme, 0.0) indicates that the lizard's body temperature is not affected by air temperature. This could result if the lizard is extremely effective at regulating its body temperature by moving into and out of the shade.

Page 1169 Inquiry question: Because of their shorter generation times, smaller species tend to reproduce more quickly, and thus would be able to respond more quickly to increased resources in the environment.

Page 1169 Data analysis: In general terms, you could make a reasonable prediction: the bigger the animal, the longer the life span. But it's important to realize that a lot of variation still exists, independent of body size. Notice that the axes of this graph are not arithmetic; each interval covers a larger span than the one before it (e.g., the time in the intervals jumps from one day to one week to one month to one year). If you look carefully at the data, you will see that some species are very similar in body size, but differ greatly in life span. Compare, for example, the mouse and the newt, which are about the same size, but the mouse lives for months and the newt for years.

Page 1170 Data analysis: No. The age classes that produce the most seeds are the two oldest, but so few individuals survive to that age that they make a very minor contribution to the overall reproduction of the population. Instead, you might focus on the Age 2 cohort, as their combination of survivors multiplied by seeds per survivor made the greatest contribution to reproduction. One circumstance in which your answer might be different is if the oldest individuals produced a huge number of seeds per individual. In that case, even though there would only be a few old individuals, they still might make the most important contribution to the population. Such a phenomenon occurs in some fish and other organisms, in which the oldest individuals are also the largest, and the number of eggs produced increases exponentially with body size.

Page 1170 Inquiry question: Based on the survivorship curve of meadow grass, the older the plant, the less likely it is to survive. It would be best to choose a plant that is very young to ensure the longest survival as a house plant. A survivorship curve that is shaped like a type I curve, in which most individuals survive to an old age and then die would also lead you to select a younger plant. A type III survivorship curve, in which only a few individuals manage to survive to an older age, would suggest the selection of a middle-aged plant that had survived the early stages of life since it would also be more likely to survive to old age.

Page 1171 Data analysis: The effect seems to be about equal. Adding two eggs reduces clutch size the following year approximately 0.8; taking away two eggs increases clutch size by almost exactly the same amount! Note that the one outlying point in the figure is what happens when one egg is added. Most likely, this is a random fluctuation, the sort of noise that one encounters in any study of natural populations. Perhaps, for example, just by chance those nests that had an egg added were situated in trees that were especially near food sources, so that the effect of adding an egg was not as great as it might have been otherwise. This sort of natural variation can complicate research on natural populations.

Page 1172 Data analysis: Many factors could affect the amount of energy required, but one obvious one is the total amount of offspring mass produced; two

chicks 3/4 the size of one larger chick probably require more food than the singleton. In this example, you can see that although the average nestling size decreases with increasing clutch size, the amount of the increase is relatively slight. Each nestling in the clutch of 13 still weights about 90% of the weight of the singleton. Hence, parents raising 13 such birds must have put in a lot more effort than the parents of the lone chick in the smallest clutch.

Page 1172 Inquiry question: It depends on the situation. If only large individuals are likely to reproduce (as is the case in some territorial species, in which only large males can hold a territory), then a few large offspring would be favored; alternatively, if body size does not affect survival or reproduction, then producing as many offspring as possible would maximize the representation of an individual's genes in subsequent generations. In many cases, intermediate values are favored by natural selection.

Page 1173 Data analysis: The first population would double in population each generation (2 adults producing 4 offspring), whereas the second population would increase by 2.5 each generation. After one generation, the second species would be 25% greater in population size (2.5/2). After two generations, the second species would have a population of 12.5 (don't ask what half a mouse looks like) and the first species, 8, a difference of 1.56. After 10 generations, the second species would have a population of 19,073, the first species, 1024, a difference of 1863%! In exponential growth, small differences in rate, compounded yearly, can add up to major differences in total numbers.

Page 1174 Inquiry question: Because when the population is below carrying capacity, the population increases in size. As it approaches the carrying capacity, growth rate slows down either from increased death rates, or decreased birth rates, or both, becoming zero as the population hits the carrying capacity. Similarly, populations well above the carrying capacity will experience large decreases in growth rate, resulting either from low birth rates or high death rates, that also approach zero as the population hits the carrying capacity.

Page 1175 (figure 55.19) Inquiry question: There are many possible reasons. Perhaps resources become limited, so that females are not able to produce as many offspring. Another possibility is that space is limited so that, at higher populations, individuals spend more time in interactions with other individuals and squander energy that otherwise could be invested in producing and raising more young.

Page 1175 (figure 55.20) Inquiry question: The answer depends on whether food is the factor regulating population size. If it is, then the number of young produced at a given population size would increase and the juvenile mortality rate would decrease. However, if other factors, such as the availability of water or predators, regulated population size, then food supplementation might have no effect.

Page 1177 Inquiry question: If hare population levels were kept high, then we would expect lynx populations to stay high as well because lynx populations respond to food availability. If lynx populations were maintained at a high level, we would expect hare populations to remain low because increased reproduction of hares would lead to increased food for the lynxes.

Page 1178 Inquiry question: If human populations are regulated by density-dependent factors, then as the population approaches the carrying capacity, either birth rates will decrease or death rates will increase, or both. If populations are regulated by density-independent factors, and if environmental conditions change, then either both rates will decline, death rates will increase, or both.

Page 1179 Inquiry question: The answer depends on whether age-specific birth and death rates stay unchanged. If they do, then the Swedish distribution would remain about the same. By contrast, because birth rates are far outstripping death rates, the Kenyan distribution will become increasingly unbalanced as the bulge of young individuals enter their reproductive years and start producing even more offspring.

Page 1181 Inquiry question: Both are important causes, and the relative importance of the two depends on which resource we are discussing. One thing is clear: The world cannot support its current population size if everyone lived at the level of resource consumption of people in the United States.

UNDERSTAND

1. b   2. c   3. a   4. b   5. d   6. b   7. c

APPLY

1. d   2. c   3. b   4. c

SYNTHESIZE

1. The genetic makeup of isolated populations will change over time based on the basic mechanisms of evolutionary change; that is, natural selection,

mutation, assortative mating, and drift. These same processes affect the genetic makeup of populations in a metapopulation, but the outcomes are likely to be much more complicated. For example, if immigration between a source and a sink population is very high, then local selection in a sink population may be swamped by the regular flow of individuals carrying alleles of lower fitness from a source population where natural selection may not be acting against those alleles; divergence might be slowed or even stopped under some circumstances. On the other hand, if sinks go through repeated population declines such that they often are made up of a very small number of individuals, then they may lose considerable genetic diversity due to drift. If immigration from source populations is greater than zero but not large, these small populations might begin to diverge substantially from other populations in the metapopulation due to drift. The difference is that in the metapopulation, such populations might actually be able to persist and diverge, rather than just going extinct due to small numbers of individuals and no ability to be rescued by neighboring sources.

2. The probability that an animal lives to the next year should decline with age (note that in figure 55.11, all the curves decrease with age), so the cost of reproduction for an old animal would, all else being equal, be lower than for a young animal. The reason is that the cost of reproduction is measured by changes in fitness. Imagine a very old animal that has almost no chance in surviving to another reproductive event; it should spend all its effort on a current reproductive effort since its future success is likely to be zero anyway.

3. If offspring size does not affect offspring quality, then it is in the parent's interest to produce absolutely as many small offspring as possible. In doing so, it would be maximizing its fitness by increasing the number of related individuals in the next generation. This would shift the curve to the right side of the *x*-axis with larger clutch sizes.

4. By increasing the mean generation time (increasing the age at which an individual can begin reproducing; age at first reproduction), keeping all else equal, one would expect that the population growth rate would be reduced. That comes simply from the fact of reducing the number of individuals that are producing offspring in the adult age classes; lower population birth rates would lead to a reduced population growth rate. As to which would have a larger influence, that is hard to say. If the change in generation time (increased age at first reproduction) had an overall larger effect on the total number of offspring an individual female had than a reduced fecundity at any age, then population growth rate would probably be more sensitive to the change in generation time. Under different scenarios, the comparison of these two effects could become more complicated, however. Suffice it to say that population growth control can come from more than one source: fecundity and age at first reproduction.

CHAPTER 56

LEARNING OUTCOME QUESTIONS

56.1 The answer depends on the habitat of the community in question. Some habitats are more hospitable to animals, and others to plants. The abundance of plants and animals in most habitats is also closely tied; thus the variation in abundance of one would affect the variation in abundance of the other.

56.2 It depends on whether we are talking about fundamental niche or realized niche. Two species can certainly have identical fundamental niches and coexist indefinitely, because they could develop different realized niches within the fundamental niche. In order for two species with identical realized niches to coexist indefinitely, the resources within the niche must not be limited.

56.3 This is an example of Batesian mimicry, in which a nonpoisonous species evolves coloration similar to that of a poisonous species.

56.4 In an ecosystem with limited resources and multiple prey species, one prey species could outcompete another to extinction in the absence of a predator. In the presence of the predator, however, the prey species that would have otherwise been driven to extinction by competitive exclusion is able to persist in the community. The predators that lower the likelihood of competitive exclusion are known as keystone predators.

56.5 Selective harvesting of individual trees would be preferable from a community point of view. According to the intermediate disturbance hypothesis, moderate degrees of disturbance, as occur in selective harvesting, increase species richness and biodiversity more than severe disturbances, such as from clear-cutting.

INQUIRY AND DATA ANALYSIS QUESTIONS

Page 1187 (figure 56.2) Inquiry question: The different soil types require very different adaptations, and thus different species are adapted to each soil type.

Page 1187 Data analysis: The ecotone, the area that separates the normal soil from the serpentine soil, is approximately 7–8 m, so the normal and serpentine soils are approximately 7–8 m apart.

Page 1187 (figure 56.3) Inquiry question: The sharp transition between the communities occurs because two different habitats are in close contact with each other.

Page 1189 Inquiry question: Yes. Both species reach higher population densities when they are by themselves than when they are in the presence of each other. The most likely explanation—although there are others—is that they compete for resources, limiting the carrying capacity each species experiences in the presence of the others.

Page 1191 (figure 56.7) Inquiry question: Presumably the resource distribution is the same on the two islands, thus favoring the same intermediate beak size when there is only one species on an island

Page 1191 (figure 56.8) Inquiry question: The kangaroo rats competed with all the other rodent species for resources, keeping the size of other rodent populations smaller. In the absence of competition when the kangaroo rats were removed, more resources were available, which allowed the other rodent populations to increase in size.

Page 1192 Inquiry question: This could be accomplished in a variety of ways. One option would be to provide refuges to give some *Paramecium* a way of escaping the predators. Another option would be to include predators of the *Didinium*, which would limit their populations (see chapter 57).

Page 1201 Data analysis: When rodents are removed, more seeds are available to the ants, so the number of ant colonies initially increases. However, through time, the greater number of large seeds leads to an increase in plants that produce large seeds. The increase in these plants leads to a decrease in the number of plants that produce small seeds, which are more useful to ants. As the number of small seeds declines, the number of ant colonies declines.

Page 1201 Inquiry question: By removing the kangaroo rats from the experimental enclosures and measuring the effects on both plants and ants. At first, the number of small seeds available to ants increases due to the absence of rodents. However, over time, plants that produce large seeds outcompete plants that produce small seeds, and thus fewer small seeds are produced and available to ants; hence, ant populations decline.

UNDERSTAND

1. b 2. a 3. d 4. a 5. b 6. d 7. c 8. c

APPLY

1. d 2. b 3. d 4. d 5. d

SYNTHESIZE

1. Experiments are useful means to test hypotheses about ecological limitations, but they are generally limited to rapidly reproducing species that occur in relatively small areas. Alternative means of studying species' interactions include detailed studies of the mechanisms by which species might interact; sometimes, for long-lived species, instead of monitoring changes in population size, which may take a very long time, other indices can be measured, such as growth or reproductive rate. Another means of assessing interspecific interactions is to study one species in different areas, in only some of which a second species occurs. Such studies must be interpreted cautiously, however, because there may be many important differences between the areas in addition to the difference in the presence or absence of the second species.

2. Adding differentially preferred prey species might have the same effect as putting in a refuge for prey in the single-species system. One way to think about it is that if a highly preferred species becomes rare due to removal by the predator, then a predator might switch to a less desirable species, even if it doesn't taste as good or is harder to catch, simply because it is still provides a better return than chasing after a very rare preferred species. Although the predator has switched, there might be enough time for the preferred species to rebound. All of these dynamics depend on the time it takes a predator to reduce the population size of its prey relative to the time it takes for those prey populations to rebound once the predator pressure is removed.

3. Although the mechanism might be known in this system, hidden interactions might affect interpretations in many ways because ecological systems are complex. For example, what if some other activity of the rodents besides their reduction of large seeds leading to an increase in the number of small seeds was responsible for the positive effect of rodents on ants? One way to test the

specific mechanism would be to increase the abundance of small seeds experimentally independent of any manipulation of rodents. Under the current hypothesis, an increase in ant population size would be expected and should be sustained, unlike the initial increase followed by a decrease seen when rodents are removed.

4. By itself, the pattern shown in figure 56.7 suggests character displacement, but alternative hypotheses are possible. For example, what if the distribution of seeds available on the two islands where the species are found alone is different from that seen where they are found in sympatry? If there were no large and small seeds seen on Los Hermanos or Daphne, just medium-sized ones, then it would be hard to conclude that the bill size on San Cristóbal has diverged relative to that on the other islands just due to competition. This is a general criticism of inferring the process of character displacement with just comparing the size distributions in allopatry and sympatry. In this case, however, the Galápagos system has been very well studied. It has been established that the size distribution of seeds available is not measurably different. Furthermore, natural selection–induced changes seen in the bill size of birds on a single island, in response to drought-induced changes in seed size lend further support to the role of competition in establishing and maintaining these patterns.

5. It is possible, because the definition of an ecosystem depends on scale. In some ecosystems, there may be other, smaller ecosystems operating within it. For example, within a rainforest ecosystem, there are small aquatic ecosystems, ecosystems within the soil, ecosystems on an individual tree. Research seems to indicate that most species behave individualistically, but in some instances groups of species do depend on one another and do function holistically. We would expect this kind of dual-community structure especially in areas of overlap between distinct ecosystems, where ecotones exist.

CHAPTER 57

LEARNING OUTCOME QUESTIONS

57.1 Yes, fertilization with natural materials such as manure is less disruptive to the ecosystem than is chemical fertilization. Many chemical fertilizers, for example, contain higher levels of phosphates than does manure, and thus chemical fertilization has disrupted the natural global phosphorus cycle.

57.2 Both matter and energy flow through ecosystems by changing form, but neither can be created or destroyed. Both matter and energy also flow through the trophic levels within an ecosystem. The flow of matter such as carbon atoms is more complex and multileveled than is energy flow, largely because it is truly a cycle. The atoms in the carbon cycle truly cycle through the ecosystem, with no clear beginning or end. The carbon is changed during the process of cycling from a solid to a gaseous state and back again. On the other hand, energy flow is unidirectional. The ultimate source of the energy in an ecosystem is the Sun. The solar energy is captured by the primary producers at the first trophic level and is changed in form from solar to chemical energy. The chemical energy is transferred from one trophic level to another, until only heat, low-quality energy, remains.

57.3 Yes, there are certainly situations in ecosystems in which the top predators in one trophic chain affect the lower trophic levels, and within the same ecosystem the primary producers affect the higher trophic levels within another trophic chain.

57.4 It depends on whether the amount of sunlight captured by the primary producers was affected. Currently, only approximately 1% of the solar energy in Earth's atmosphere is captured by primary producers for photosynthesis. If less sunlight reached Earth's surface, but a correlating increase in energy capture accompanied the decrease in sunlight, then the primary productivity should not be affected.

57.5 The equilibrium model of island biogeography describes the relationship between species richness and not only island size but also distance from the mainland. A small island closer to the mainland would be expected to have more species than would a larger island that is farther from the mainland.

INQUIRY AND DATA ANALYSIS QUESTIONS

Page 1213 Data analysis: The nitrogen concentration in the runoff from the undisturbed area was between 1 and 2 mg/L, whereas the nitrogen concentration in the runoff in the deforested area varied between about 35 and 80 mg/L. This was at least a 17-fold increase (from 2 to 35) and even more if you consider the maximal values after deforestation. Note that the figure doesn't seem to reflect such a large change, but notice the break in the y-axis from 4 to 40 mg/L. This reflects a change in the scale of the y-axis and is sometimes done in scientific figures to save space when two sets of data vary by a large amount. It is understood that the data within that area continued the progression trend shown.

Page 1217 Data analysis: No. The two highest habitats for world NPP are habitats that differ greatly in NPP per unit area (tropical rainforest, high; open ocean, low). Overall, you can see that the habitats are arranged by NPP per unit area, with highest at the top, but when you look at world NPP, there is no general trend from top to bottom.

Page 1218 Inquiry question: At each link in the food chain, only a small fraction of the energy at one level is converted into mass of organisms at the next level. Much energy is dissipated as heat or excreted.

Page 1219 Inquiry question: In the inverted pyramid, the primary producers reproduce quickly and are eaten quickly, so that at any given time, a small population of primary producers exist relative to the heterotroph population.

Page 1220 (figure 57.14) Inquiry question: Because the trout eat the invertebrates, which graze the algae. With fewer grazers, there is more algae.

Page 1220 (figure 57.15) Inquiry question: The food chain has four levels in this experiment and only three in the previous one. The snakes might reduce the number of fish, which would allow an increase in damselflies, which would reduce the number of chironomids and increase the algae. In other words, lower levels of the food chain would be identical for the "snake and fish" and "no fish and no snake" treatments. Both would differ from the enclosures with only fish.

Page 1222 (figure 57.17) Inquiry question: Herbivores consume much of the algal biomass even as primary productivity increases. Increases in primary productivity can lead to increased herbivore populations. The additional herbivores crop the biomass of the algae even while primary productivity increases.

Page 1222 Data analysis: The angle of the line is called the slope. It indicates how the two variables are positively related. A large slope (greater angle) indicates that as the x-variable increases, so does the y-variable, as in the top and bottom diagrams. A low angle (as in the middle diagram) indicates no relationship: as the x-variable increases, the y-variable changes little. Of course, a negative slope, not indicated here, would indicate a negative relationship, as x increases, y decreases.

Page 1222 (figure 57.18) Inquiry question: Secondary carnivores would keep the primary carnivore biomass at a relatively constant, low level, which would allow herbivore biomass to increase with increasing light level, which would keep primary producer biomass at a low and constant level. In other words, the increased productivity at higher light levels would flow up the food chain. This situation is an extension of figure 57.17 to the addition of a fourth food chain level.

Page 1223 Data analysis: As figure 57.19b illustrates, there is a general declining relationship between number of species in a plot and number of invading species (what scientists called a "statistically significant" relationship). Nonetheless, as the figure illustrates, there is quite a lot of scatter in the data. For example, when there are 24 species in a plot, the number of invaders in different plots ranged from 1 to a very large number. So, in answer to the question, there is an overall relationship, but the ability to make an accurate prediction for any given plot is relatively low because there is so much variation. One would be correct to expect that in general, plots with more species would have invaders, but it is very possible that if one randomly chose one plot with many species and one with few, the opposite outcome might occur.

Page 1224 Inquiry question: (a) Perhaps because an intermediate number of predators is enough to keep numbers of superior competitors down. (b) Perhaps because more habitats are available and thus more different ways of surviving in the environment. (c) Hard to say. Possibly more-stable environments permit greater specialization, thus permitting coexistence of more species.

UNDERSTAND

1. d   2. d   3. b   4. d   5. a   6. a   7. b   8. a   9. a   10. c

APPLY

1. a   2. d   3. d

SYNTHESIZE

1. Because the length of food chains appears to be ultimately limited by the amount of energy entering a system, and the characteristic loss of usable energy (about 90%) as energy is transferred to each higher level, it would be reasonable to expect that the ectotherm-dominated food chains would be longer than the endotherm-dominated chains. In fact, there is some indirect evidence for this from real food chains, and it is also predicted by some advanced ecological models. However, it is difficult to determine whether, in reality, this is the case because all of the complex factors that determine food chain length and structure. Moreover, many practical difficulties are associated with measuring actual food chain length in natural systems.

2. It is critical to distinguish, as this chapter points out, between energy and mass transfer in trophic dynamics of ecosystems. The standing biomass of phytoplankton is not necessarily a reliable measure of the energy contained in the trophic level. If phytoplankton are eaten as quickly as they are produced, they may contribute a tremendous amount of energy, which can never be directly measured by a static biomass sample. The standing crop, therefore, is an incomplete measure of the productivity of the trophic level.

3. As figure 57.17 suggests, trophic structure and dynamics are interrelated and are primary determinants of ecosystem characteristics and behavior. For example, if a particularly abundant herbivore is threatened, energy that is abundant at the level of primary productivity in an ecosystem may be relatively unavailable to higher trophic levels (e.g., carnivores). That is, the herbivores are an important link in transducing energy through an ecosystem. Cascading effects, whether they are driven from the bottom up or from the top down are a characteristic of energy transfer in ecosystems, and that translates into the reality that effects on any particular species are unlikely to be limited to that species itself.

4. There are many ways to answer this question, but the obvious place to start is to think about the many ways plant structural diversity potentially affects animals that are not eating the plants directly. For example, plants may provide shelter, refuges, food for prey, substrate for nesting, among other things. Therefore, increasing complexity might increase the ability of lizards to partition the habitat in more ways, allow more species to escape their predators or seek refuge from harsh physical factors (e.g., cold or hot temperatures), provide a greater substrate for potential prey in terms of food resources, for instance. If we want to know the exact mechanisms for the relationship, we would need to conduct experiments to test specific hypotheses. For example, if we hypothesize that some species require greater structural complexity in order to persist in a particular habitat, we could modify the habitat (reduce plant structure) and test whether species originally present were reduced in numbers or became unable to persist.

CHAPTER 58

LEARNING OUTCOME QUESTIONS

58.1 If the Earth rotated in the opposite direction, the Coriolis effect would be reversed. In other words, winds descending between 30° north or 30° south and the equator would still be moving more slowly than the underlying surface so it would be deflected; however, they would be deflected to the left in the northern hemisphere and to the right in the southern hemisphere. The pattern would be reversed between 30° and 60° because the winds would be moving more rapidly than the underlying surface and would thus be deflected again in the opposite directions from normal—to the left in the northern hemisphere and to the right in the southern hemisphere. All of this would result in Trade Winds that blew from west to east and "Westerlies" that were actually "Easterlies," blowing east to west.

58.2 As with elevation, latitude is a primary determinant of climate and precipitation, which together largely determine the vegetational structure of a particular area, which in turn defines biomes.

58.3 The spring and fall turnovers that occur in freshwater lakes found in temperate climates result in the oxygen-poor water near the bottom of the lake getting remixed with the oxygen-rich water near the top of the lake, essentially eliminating, at least temporarily, the thermocline layer. In the tropics, there is less temperature fluctuation; thus the thermocline layer is more permanent and the oxygen depletion (and resulting paucity of animal life) is sustained.

58.4 Regions affected by the ENSO, or El Niño Southern Oscillation events, experience cyclical warming events in the waters around the coastline. The warmed water lowers the primary productivity, which stresses and subsequently decreases the populations of fish, seabirds, and sea mammals.

58.5 CFCs, or chlorofluorocarbons, are an example of point-source pollution. CFCs and other types of point-source pollutants are, in general, easier to combat because their sources are more easily identified and thus the pollutants more easily eliminated.

58.6 Global climate change and ozone depletion may be interconnected. However, although climate change and ozone depletion are both global environmental concerns due to the impact each has on human health, the environment, economics, and politics, there are some different approaches to combating and understanding each dilemma. Ozone depletion results in an increase in the ultraviolet radiation reaching the Earth's surface. Global climate change, on the other hand, results in long-term changes in sea level, ice flow, and storm activity.

INQUIRY AND DATA ANALYSIS QUESTIONS

Page 1231 Data analysis: Seasonal variation is positively related to the distance from the equator because the farther north (or south), the stronger the Sun's rays are in the winter and the weaker they are in the summer. The effect is much stronger in the northern hemisphere for complicated reasons related to the differing amounts of landmass in the two hemispheres and other factors.

Page 1232 Inquiry question: Because of the tilt of the Earth's axis and the spherical shape of the planet, the light (and heat) from the Sun hits the equator and nearby latitudes more directly than it does at the poles.

Page 1236 Inquiry question: Increased precipitation and temperature allows for the sustainability of a larger variety and biomass of vegetation, and primary productivity is a measure of the rate at which plants convert solar energy into chemical energy.

Page 1236 Data analysis: Figure 58.8 indicates that biomes with approximately 125 cm of precipitation/year vary tremendously in mean annual temperature, from hot savannas to cold taigas. Figure 58.9*b* indicates that productivity is a function of temperature. Hence the variation in figure 58.9*a* results from the different temperatures of ecosytems with intermediate amounts of precipitation.

UNDERSTAND

1. d   2. b   3. a   4. c   5. c   6. c   7. a   8. c

APPLY

1. d   2. b   3. c   4. b

SYNTHESIZE

1. The Earth is tilted on its axis such that regions away from the equator receive less incident solar radiation per unit surface area (because the angle of incidence is oblique). The northern and southern hemispheres alternate between angling toward versus away from the Sun on the Earth's annual orbit. These two facts mean that the annual mean temperature declines as you move away from the equator, and that variation in the mean temperatures of the northern and southern hemispheres is complementary to each other; when one is hot, the other is cold.

2. Energy absorbed by the Earth is maximized at the equator because of the angle of incidence. Because there are large expanses of ocean at the equator, warmed air picks up moisture and rises. As it rises, equatorial air, now saturated with moisture, cools and releases rain, the air falling back to Earth's surface displaced north and south to approximately 30°. The air, warming as it descends, absorbs moisture from the land and vegetation below, resulting in desiccation in the latitudes around 30°.

3. Even though global climate changes have occurred in the past, conservation biologists are concerned about the current warming trend for two reasons. First, the warming rate is rapid, thus the selective pressures on the most vulnerable organisms may be too strong for the species to adapt. Second, the natural areas that covered most of the globe during past climatic changes are now in much more limited, restricted areas, thus greatly impeding the ability of organisms to migrate to more suitable habitats.

CHAPTER 59

LEARNING OUTCOME QUESTIONS

59.1 Unfortunately, most of the Earth's biodiversity hotspots are also areas of the greatest human population growth; human population growth is accompanied by increased resource utilization and exploitation.

59.2 I would tell the shrimp farmers that if they were to shut down the shrimp farm and remediate the natural mangrove swamp on which their property sits, other, more economically lucrative businesses could be developed, such as timber, charcoal production, and offshore fishing.

59.3 Absolutely. The hope of conservation biologists is that even if a species is endangered to the brink of extinction due to habitat degradation, the habitat may someday be restored. The endangered species can be bred in captivity (which also allows for the maintenance of genetic diversity within the species) and either reintroduced to a restored habitat or even introduced to another suitable habitat.

59.4 It depends on the reason for the degradation of the habitat in the first place, but yes, in some cases, habitat restoration can approach a pristine state. For example, the Nashau River in New England was heavily polluted, but habitat restoration efforts returned it to a relatively pristine state. However, because habitat degradation affects so many species within the ecosystem, and the depth and complexity of the trophic relationships within the ecosystem are difficult if not impossible to fully understand, restoration is rarely if ever truly pristine.

INQUIRY AND DATA ANALYSIS QUESTIONS

Page 1260 Inquiry question: Many factors affect human population trends, including resource availability, governmental support for settlement in new areas or for protecting natural areas, and the extent to which governments attempt to manage population growth.

Page 1260 Data analysis: Some hotspots that don't have particularly high population densities today, but are growing quickly are the Chocó, tropical Andes, Madagascar, and Indian Ocean Islands. The Brazilian Cerrado has a very high population growth rate, but because densities there are currently very low, this area is of less immediate concern.

Page 1262 Data analysis: The mangroves provide many economic services. For example, without them, fisheries become less productive and storm damage increases. However, because the people who benefit from these services do not own the mangroves, governmental action is needed to ensure that the value of what economists call "common goods" is protected.

Page 1266 Data analysis: Examination of the figure indicates that for every 90% reduction in area, extinction rate approximately doubles. For example, at an area of 100 (10^2) km², extinction rate is approximately 0.08, but at 10 km², it increases to approximately 0.15, and at 1 km², it drops to about 0.23.

Page 1268 Data analysis: Reproductive success would decline through time, perhaps with the same slope, but negative instead of positive, as shown in the figure.

Page 1269 Inquiry question: As discussed in this chapter, populations that are small face many problems that can reinforce one another and eventually cause extinction.

Pate 1269 Data analysis: Yes. Additional species were hunted when the number of the "catch" in the number of whales already being hunted started to decline. Thus, the number of sperm and sei whales increases markedly at the same time that the number of fin whales decreases (and note that the number of fin whales caught soared when blue whales declined). Then, when sei and sperm whale numbers begin to decline, minke whaling accelerated.

Page 1274 Inquiry question: As we discussed in chapter 21, allele frequencies change randomly in a process called genetic drift. The smaller the population size, the greater these random fluctuations will be. Thus, small populations are particularly prone to one allele being lost from a population due to these random changes.

UNDERSTAND

1. a   2. c   3. d   4. d   5. b   6. c

APPLY

1. d   2. d   3. a   4. d

SYNTHESIZE

1. Although it is true that extinction is a natural part of the existence of a species, several pieces of evidence suggest that current rates of extinction are considerably elevated over the natural background level and the disappearance is associated with human activities (which many of the most pronounced extinction events in the history of the Earth were not). It is important to appreciate the length of time over which the estimate of 99% is made. The history of life on Earth extends back billions of years. Certainly, clear patterns of the emergence and extinction of species in the fossil record extend back many hundreds of millions of years. Since the average time of species' existence is short relative to the great expanse of time over which we can estimate the percentage of species that have disappeared, the perception might be that extinction rates have always been high, when in fact the high number is driven by the great expanse of time of measurement. We have very good evidence that modern extinction rates (over human history) are considerably elevated above background levels. Furthermore, the circumstances of the extinctions may be very different because they are also associated with habitat and resource removal; thus potentially limiting the natural processes that replace extinct species.

2. The problem is not unique and not new. It represents a classic conflict that is the basic source of societal laws and regulations, especially in the management of resources. For example, whether or not to place air pollution scrubbers on the smoke stacks of coal-fired power plants is precisely the same issue. In this case, it is not ecosystem conversion, per se, but the fact that the businesses that run the power plants benefit from their operation, but the public "owns" and relies on the atmosphere is a conflict between public and private interests. Some of the ways to navigate the dilemma is for society to create regulations to protect the public interest. The problem is difficult and clearly does not depend solely on economic valuation of the costs and benefits because there can be considerable debate about those estimates. One only has to look at the global climate change problem to suggest how hard it will be to make progress in an expedient manner.

3. This is not a trivial undertaking, which is why, since the first concerns were raised in the late 1980s, it has taken nearly 15 years to collect evidence showing a decline is likely. Although progress has been made on identifying potential causes, much work remains to be done. Many amphibians are secretive, relatively long-lived, and subject to extreme population fluctuations. Given those facts about their biology, documenting population fluctuations (conducting censuses of the number of individuals in populations) for long periods of time is the only way to ultimately establish the likely fate of populations, and that process is time-consuming and costly.

4. Within an ecosystem, every species is dependent on and depended on by any number of other species. Even the smallest organisms, bacteria, are often specific about the species they feed on, live within, parasitize, and so on. So, the extinction of a single species anywhere in the ecosystem will affect not only the organisms it directly feeds on and that directly feed on it, but also those related more distantly. In the simplest terms, if, for example, a species of rodent goes extinct, the insects and vegetation on which it feeds would no longer be under the same predation pressure and thus could grow out of control, outcompeting other species and leading to their demise. In addition, the predators of the rodent would have to find other prey, which would result in competition with those species' predators. And so on, and so on. The effects could be catastrophic to the entire ecosystem. By looking at the trophic chains in which a particular organism is involved, you could predict the effects its extinction would have on other species.

5. Population size is not necessarily a direct cause of extinction, but it certainly is an indirect cause. Smaller populations have a number of problems that themselves can lead directly to extinction, such as loss of diversity (and thus increased susceptibility to pathogens) and greater vulnerability to natural catastrophes.

Glossary

A

ABO blood group A set of four phenotypes produced by different combinations of three alleles at a single locus; blood types are A, B, AB, and O, depending on which alleles are expressed as antigens on the red blood cell surface.

abscission In vascular plants, the dropping of leaves, flowers, fruits, or stems at the end of the growing season, as the result of the formation of a layer of specialized cells (the abscission zone) and the action of a hormone (ethylene).

absorption spectrum The relationship of absorbance vs. wavelength for a pigment molecule. This indicates which wavelengths are absorbed maximally by a pigment. For example, chlorophyll *a* absorbs most strongly in the violet-blue and red regions of the visible light spectrum.

acceptor stem The 3′ end of a tRNA molecule; the portion that amino acids become attached to during the tRNA charging reaction.

accessory pigment A secondary light-absorbing pigment used in photosynthesis, including chlorophyll *b* and the carotenoids, that complement the absorption spectrum of chlorophyll *a*.

aceolomate An animal, such as a flatworm, having a body plan that has no body cavity; the space between mesoderm and endoderm is filled with cells and organic materials.

acetyl-CoA The product of the transition reaction between glycolysis and the Krebs cycle. Pyruvate is oxidized to acetyl-CoA by NAD^+, also producing CO_2, and NADH.

achiasmate segregation The lining up and subsequent separation of homologues during meiosis I without the formation of chiasmata between homologues; found in *Drosophila* males and some other species.

acid Any substance that dissociates in water to increase the hydrogen ion (H^+) concentration and thus lower the pH.

actin One of the two major proteins that make up vertebrate muscle; the other is myosin.

action potential A transient, all-or-none reversal of the electric potential across a membrane; in neurons, an action potential initiates transmission of a nerve impulse.

action spectrum A measure of the efficiency of different wavelengths of light for photosynthesis. In plants it corresponds to the absorption spectrum of chlorophylls.

activation energy The energy that must be processed by a molecule in order for it to undergo a specific chemical reaction.

active site The region of an enzyme surface to which a specific set of substrates binds, lowering the activation energy required for a particular chemical reaction and so facilitating it.

active transport The pumping of individual ions or other molecules across a cellular membrane from a region of lower concentration to one of higher concentration (i.e., against a concentration gradient); this transport process requires energy, which is typically supplied by the expenditure of ATP.

adaptation A peculiarity of structure, physiology, or behavior that promotes the likelihood of an organism's survival and reproduction in a particular environment.

adapter protein Any of a class of proteins that acts as a link between a receptor and other proteins to initiate signal transduction.

adaptive radiation The evolution of several divergent forms from a primitive and unspecialized ancestor.

adenosine triphosphate (ATP) A nucleotide consisting of adenine, ribose sugar, and three phosphate groups; ATP is the energy currency of cellular metabolism in all organisms.

adherins junction An anchoring junction that connects the actin filaments of one cell with those of adjacent cells or with the extracellular matrix.

ATP synthase The enzyme responsible for producing ATP in oxidative phosphorylation; it uses the energy from a proton gradient to catalyze the reaction $ADP + P_i \longrightarrow ATP$.

adenylyl cyclase An enzyme that produces large amounts of cAMP from ATP; the cAMP acts as a second messenger in a target cell.

adhesion The tendency of water to cling to other polar compounds due to hydrogen bonding.

adipose cells Fat cells, found in loose connective tissue, usually in large groups that form adipose tissue. Each adipose cell can store a droplet of fat (triacylglyceride).

adventitious Referring to a structure arising from an unusual place, such as stems from roots or roots from stems.

aerenchyma In plants, loose parenchymal tissue with large air spaces in it; often found in plants that grow in water.

aerobic Requiring free oxygen; any biological process that can occur in the presence of gaseous oxygen.

aerobic respiration The process that results in the complete oxidation of glucose using oxygen as the final electron acceptor. Oxygen acts as the final electron acceptor for an electron transport chain that produces a proton gradient for the chemiosmotic synthesis of ATP.

aleurone In plants, the outer layer of the endosperm in a seed; on germination, the aleurone produces α-amylase that breaks down the carbohydrates of the endosperm to nourish the embryo.

alga, pl. **algae** A unicellular or simple multicellular photosynthetic organism lacking multicellular sex organs.

allantois A membrane of the amniotic egg that functions in respiration and excretion in birds and reptiles and plays an important role in the development of the placenta in most mammals.

allele One of two or more alternative states of a gene.

allele frequency A measure of the occurrence of an allele in a population, expressed as proportion of the entire population, for example, an occurrence of 0.84 (84%).

allometric growth A pattern of growth in which different components grow at different rates.

allelopathy The release of a substance from the roots of one plant that block the germination of nearby seeds or inhibits the growth of a neighboring plant.

allopatric speciation The differentiation of geographically isolated populations into distinct species.

allopolyploid A polyploid organism that contains the genomes of two or more different species.

allosteric activator A substance that binds to an enzyme's allosteric site and keeps the enzyme in its active configuration.

allosteric inhibitor A noncompetitive inhibitor that binds to an enzyme's allosteric site and prevents the enzyme from changing to its active configuration.

allosteric site A part of an enzyme, away from its active site, that serves as an on/off switch for the function of the enzyme.

alpha (α) helix A form of secondary structure in proteins where the polypeptide chain is wound into a spiral due to interactions between amino and carboxyl groups in the peptide backbone.

alternation of generations A reproductive cycle in which a haploid (*n*) phase (the gametophyte), gives rise to gametes, which, after fusion to form a zygote, germinate to produce a diploid (*2n*) phase (the sporophyte). Spores produced by meiotic division from the sporophyte give rise to new gametophytes, completing the cycle.

alternative splicing In eukaryotes, the production of different mRNAs from a single primary transcript by including different sets of exons.

altruism Self-sacrifice for the benefit of others; in formal terms, the behavior that increases the fitness of the recipient while reducing the fitness of the altruistic individual.

alveolus, pl. **alveoli** One of many small, thin-walled air sacs within the lungs in which the bronchioles terminate.

amino acid The subunit structure from which proteins are produced, consisting of a central carbon atom with a carboxyl group (—COOH), an amino group (—NH₂), a hydrogen, and a side group (*R* group); only the side group differs from one amino acid to another.

aminoacyl-tRNA synthetase Any of a group of enzymes that attach specific amino acids to the correct tRNA during the tRNA-charging reaction. Each of the 20 amino acids has a corresponding enzyme.

amniocentesis Indirect examination of a fetus by tests on cell cultures grown from fetal cells obtained from a sample of the amniotic fluid or tests on the fluid itself.

amnion The innermost of the extraembryonic membranes; the amnion forms a fluid-filled sac around the embryo in amniotic eggs.

amniote A vertebrate that produces an egg surrounded by four membranes, one of which is the amnion; amniote groups are the reptiles, birds, and mammals.

amniotic egg An egg that is isolated and protected from the environment by a more or less impervious shell during the period of its development and that is completely self-sufficient, requiring only oxygen.

ampulla In echinoderms, a muscular sac at the base of a tube foot that contracts to extend the tube foot.

amyloplast A plant organelle called a plastid that specializes in storing starch.

anabolism The biosynthetic or constructive part of metabolism; those chemical reactions involved in biosynthesis.

anaerobic Any process that can occur without oxygen, such as anaerobic fermentation or H₂S photosynthesis.

anaerobic respiration The use of electron transport to generate a proton gradient for chemiosmotic synthesis of ATP using a final electron acceptor other than oxygen.

analogous Structures that are similar in function but different in evolutionary origin, such as the wing of a bat and the wing of a butterfly.

anaphase In mitosis and meiosis II, the stage initiated by the separation of sister chromatids, during which the daughter chromosomes move to opposite poles of the cell; in meiosis I, marked by separation of replicated homologous chromosomes.

anaphase-promoting complex (APC) A protein complex that triggers anaphase; it initiates a series of reactions that ultimately degrades cohesin, the protein complex that holds the sister chromatids together. The sister chromatids are then released and move toward opposite poles in the cell.

anchoring junction A type of cell junction that mechanically attaches the cytoskeleton of a cell to the cytoskeletons of adjacent cells or to the extracellular matrix.

androecium The floral whorl that comprises the stamens.

aneuploidy The condition in an organism whose cells have lost or gained a chromosome; Down syndrome, which results from an extra copy of human chromosome 21, is an example of aneuploidy in humans.

angiosperms The flowering plants, one of five phyla of seed plants. In angiosperms, the ovules at the time of pollination are completely enclosed by tissues.

animal pole In fish and other aquatic vertebrates with asymmetrical yolk distribution in their eggs, the hemisphere of the blastula comprising cells relatively poor in yolk.

anion A negatively charged ion.

annotation In genomics, the process of identifying and making note of "landmarks" in a DNA sequence to assist with recognition of coding and transcribed regions.

anonymous markers Genetic markers in a genome that do not cause a detectable phenotype, but that can be detected using molecular techniques.

antenna complex A complex of hundreds of pigment molecules in a photosystem that collects photons and feeds the light energy to a reaction center.

anther In angiosperm flowers, the pollen-bearing portion of a stamen.

antheridium, pl. antheridia A sperm-producing organ.

anthropoid Any member of the mammalian group consisting of monkeys, apes, and humans.

antibody A protein called immunoglobulin that is produced by lymphocytes in response to a foreign substance (antigen) and released into the bloodstream.

anticodon The three-nucleotide sequence at the end of a transfer RNA molecule that is complementary to, and base-pairs with, an amino-acid–specifying codon in messenger RNA.

antigen A foreign substance, usually a protein or polysaccharide, that stimulates an immune response.

antiporter A carrier protein in a cell's membrane that transports two molecules in opposite directions across the membrane.

anus The terminal opening of the gut; the solid residues of digestion are eliminated through the anus.

aorta (Gr. *aeirein,* to lift) The major artery of vertebrate systemic blood circulation; in mammals, carries oxygenated blood away from the heart to all regions of the body except the lungs.

apical meristem In vascular plants, the growing point at the tip of the root or stem.

apoplast route In plant roots, the pathway for movement of water and minerals that leads through cell walls and between cells.

apoptosis A process of programmed cell death, in which dying cells shrivel and shrink; used in all animal cell development to produce planned and orderly elimination of cells not destined to be present in the final tissue.

aposematic coloration An ecological strategy of some organisms that "advertise" their poisonous nature by the use of bright colors.

aquaporin A membrane channel that allows water to cross the membrane more easily than by diffusion through the membrane.

aquifers Permeable, saturated, underground layers of rock, sand, and gravel, which serve as reservoirs for groundwater.

archegonium, pl. archegonia The multicellular egg-producing organ in bryophytes and some vascular plants.

archenteron The principal cavity of a vertebrate embryo in the gastrula stage; lined with endoderm, it opens up to the outside and represents the future digestive cavity.

arteriole A smaller artery, leading from the arteries to the capillaries.

artificial selection Change in the genetic structure of populations due to selective breeding by humans. Many domestic animal breeds and crop varieties have been produced through artificial selection.

ascomycetes A large group comprising part of the "true fungi." They are characterized by separate hyphae, asexually produced conidiospores, and sexually produced ascospores within asci.

ascus, pl. asci A specialized cell, characteristic of the ascomycetes, in which two haploid nuclei fuse to produce a zygote that divides immediately by meiosis; at maturity, an ascus contains ascospores.

asexual reproduction The process by which an individual inherits all of its chromosomes from a single parent, thus being genetically identical to that parent; cell division is by mitosis only.

A site In a ribosome, the aminoacyl site, which binds to the tRNA carrying the next amino acid to be added to a polypeptide chain.

assembly The phase of a virus's reproductive cycle during which the newly made components are assembled into viral particles.

assortative mating A type of nonrandom mating in which phenotypically similar individuals mate more frequently.

aster In animal cell mitosis, a radial array of microtubules extending from the centrioles toward the plasma membrane, possibly serving to brace the centrioles for retraction of the spindle.

atom The smallest unit of an element that contains all the characteristics of that element. Atoms are the building blocks of matter.

atrial peptide Any of a group of small polypeptide hormones that may be useful in treatment of high blood pressure and kidney failure; produced by cells in the atria of the heart.

atrioventricular (AV) node A slender connection of cardiac muscle cells that receives the heartbeat impulses from the sinoatrial node and conducts them by way of the bundle of His.

atrium An antechamber; in the heart, a thin-walled chamber that receives venous blood and passes it on to the thick-walled ventricle; in the ear, the tympanic cavity.

autonomic nervous system The involuntary neurons and ganglia of the peripheral nervous system of vertebrates; regulates the heart, glands, visceral organs, and smooth muscle.

autopolyploid A polyploid organism that contains a duplicated genome of the same species; may result from a meiotic error.

autosome Any eukaryotic chromosome that is not a sex chromosome; autosomes are present in the same number and kind in both males and females of the species.

autotroph An organism able to build all the complex organic molecules that it requires as its own food source, using only simple inorganic compounds.

auxin (Gr. *auxein*, to increase) A plant hormone that controls cell elongation, among other effects.

auxotroph A mutation, or the organism that carries it, that affects a biochemical pathway causing a nutritional requirement.

avirulent pathogen Any type of normally pathogenic organism or virus that utilizes host resources but does not cause extensive damage or death.

axil In plants, the angle between a leaf's petiole and the stem to which it is attached.

axillary bud In plants, a bud found in the axil of a stem and leaf; an axillary bud may develop into a new shoot or may become a flower.

axon A process extending out from a neuron that conducts impulses away from the cell body.

B

b6–f **complex** *See* cytochrome *b*6–*f* complex.

bacteriophage A virus that infects bacterial cells; also called a *phage*.

Barr body A deeply staining structure, seen in the interphase nucleus of a cell of an individual with more than one X chromosome, that is a condensed and inactivated X. Only one X remains active in each cell after early embryogenesis.

basal body A self-reproducing, cylindrical, cytoplasmic organelle composed of nine triplets of microtubules from which the flagella or cilia arise.

base Any substance that dissociates in water to absorb and therefore decrease the hydrogen ion (H⁺) concentration and thus raise the pH.

base-pair A complementary pair of nucleotide bases, consisting of a purine and a pyrimidine.

basidium, pl. **basidia** A specialized reproductive cell of the basidiomycetes, often club-shaped, in which nuclear fusion and meiosis occur.

basophil A leukocyte containing granules that rupture and release chemicals that enhance the inflammatory response. Important in causing allergic responses.

Batesian mimicry A survival strategy in which a palatable or nontoxic organism resembles another kind of organism that is distasteful or toxic. Both species exhibit warning coloration.

B cell A type of lymphocyte that, when confronted with a suitable antigen, is capable of secreting a specific antibody protein.

behavioral ecology The study of how natural selection shapes behavior.

biennial A plant that normally requires two growing seasons to complete its life cycle. Biennials flower in the second year of their lives.

bilateral symmetry A single plane divides an organism into two structural halves that are mirror images of each other.

bile salts A solution of organic salts that is secreted by the vertebrate liver and temporarily stored in the gallbladder; emulsifies fats in the small intestine.

binary fission Asexual reproduction by division of one cell or body into two equal or nearly equal parts.

binomial distribution The distribution of phenotypes seen among the progeny of a cross in which there are only two alternative alleles.

binomial name The scientific name of a species that consists of two parts, the genus name and the specific species name, for example, *Apis mellifera*.

biochemical pathway A sequence of chemical reactions in which the product of one reaction becomes the substrate of the next reaction. The Krebs cycle is a biochemical pathway.

biodiversity The number of species and their range of behavioral, ecological, physiological, and other adaptations, in an area.

bioenergetics The analysis of how energy powers the activities of living systems.

biofilm A complex bacterial community comprising different species; plaque on teeth is a biofilm.

biogeography The study of the geographic distribution of species.

biological community All the populations of different species living together in one place; for example, all populations that inhabit a mountain meadow.

biological species concept (BSC) The concept that defines species as groups of populations that have the potential to interbreed and that are reproductively isolated from other groups.

biomass The total mass of all the living organisms in a given population, area, or other unit being measured.

biome One of the major terrestrial ecosystems, characterized by climatic and soil conditions; the largest ecological unit.

bipolar cell A specialized type of neuron connecting cone cells to ganglion cells in the visual system. Bipolar cells receive a hyperpolarized stimulus from the cone cell and then transmit a depolarization stimulus to the ganglion cell.

biramous Two-branched; describes the appendages of crustaceans.

blade The broad, expanded part of a leaf; also called the lamina.

blastocoel The central cavity of the blastula stage of vertebrate embryos.

blastodisc In the development of birds, a disclike area on the surface of a large, yolky egg that undergoes cleavage and gives rise to the embryo.

blastomere One of the cells of a blastula.

blastopore In vertebrate development, the opening that connects the archenteron cavity of a gastrula stage embryo with the outside.

blastula In vertebrates, an early embryonic stage consisting of a hollow, fluid-filled ball of cells one layer thick; a vertebrate embryo after cleavage and before gastrulation.

Bohr effect The release of oxygen by hemoglobin molecules in response to elevated ambient levels of CO_2.

bottleneck effect A loss of genetic variability that occurs when a population is reduced drastically in size.

Bowman's capsule In the vertebrate kidney, the bulbous unit of the nephron, which surrounds the glomerulus.

β-oxidation The oxygen-dependent reactions where 2-carbon units of fatty acids are cleaved and combined with CoA to produce acetyl-CoA, which then enters the Krebs cycle. This occurs cyclically until the entire fatty acid is oxidized.

β sheet A form of secondary structure in proteins where the polypeptide folds back on itself one or more times to form a planar structure stabilized by hydrogen bonding between amino and carboxyl groups in the peptide backbone. Also known as a β-pleated sheet.

book lung In some spiders, a unique respiratory system consisting of leaflike plates within a chamber over which gas exchange occurs.

bronchus, pl. **bronchi** One of a pair of respiratory tubes branching from the lower end of the trachea (windpipe) into either lung.

bud An asexually produced outgrowth that develops into a new individual. In plants, an embryonic shoot, often protected by young leaves; buds may give rise to branch shoots.

buffer A substance that resists changes in pH. It releases hydrogen ions (H⁺) when a base is added and absorbs H⁺ when an acid is added.

C

C₃ photosynthesis The main cycle of the dark reactions of photosynthesis, in which CO_2 binds to ribulose 1,5-bisphosphate (RuBP) to form two 3-carbon phosphoglycerate (PGA) molecules.

C₄ photosynthesis A process of CO_2 fixation in photosynthesis by which the first product is the 4-carbon oxaloacetate molecule.

cadherin One of a large group of transmembrane proteins that contain a Ca²⁺-mediated binding between cells; these proteins are responsible for cell-to-cell adhesion between cells of the same type.

callus Undifferentiated tissue; a term used in tissue culture, grafting, and wound healing.

Calvin cycle The dark reactions of C₃ photosynthesis; also called the Calvin–Benson cycle.

calyx The sepals collectively; the outermost flower whorl.

CAM plant Plants that use C₄ carbon fixation at night, then use the stored malate to generate CO_2 during the day to minimize dessication.

Cambrian explosion The huge increase in animal diversity that occurred at the beginning of the Cambrian period.

cAMP response protein (CRP) *See* catabolite activator protein (CAP)

cancer The unrestrained growth and division of cells; it results from a failure of cell division control.

capillary The smallest of the blood vessels; the very thin walls of capillaries are permeable to many molecules, and exchanges between blood and the tissues occur across them; the vessels that connect arteries with veins.

capsid The outermost protein covering of a virus.

capsule In bacteria, a gelatinous layer surrounding the cell wall.

carapace (Fr. from Sp. *carapacho*, shell) Shieldlike plate covering the cephalothorax of decapod crustaceans; the dorsal part of the shell of a turtle.

carbohydrate An organic compound consisting of a chain or ring of carbon atoms to which hydrogen and oxygen atoms are attached in a ratio of approximately 2:1; having the generalized formula $(CH_2O)_n$; carbohydrates include sugars, starch, glycogen, and cellulose.

carbon fixation The conversion of CO_2 into organic compounds during photosynthesis; the first stage of the dark reactions of photosynthesis, in which carbon dioxide from the air is combined with ribulose 1,5-bisphosphate.

carotenoid Any of a group of accessory pigments found in plants; in addition to absorbing light energy, these pigments act as antioxidants, scavenging potentially damaging free radicals.

carpel A leaflike organ in angiosperms that encloses one or more ovules.

carrier protein A membrane protein that binds to a specific molecule that cannot cross the membrane and allows passage through the membrane.

carrying capacity The maximum population size that a habitat can support.

cartilage A connective tissue in skeletons of vertebrates. Cartilage forms much of the skeleton of embryos, very young vertebrates, and some adult vertebrates, such as sharks and their relatives.

Casparian strip In plants, a band that encircles the cell wall of root endodermal cells. Adjacent cells' strips connect, forming a layer through which water cannot pass; therefore, all water entering roots must pass through cell membranes and cytoplasm.

catabolism In a cell, those metabolic reactions that result in the breakdown of complex molecules into simpler compounds, often with the release of energy.

catabolite activator protein (CAP) A protein that, when bound to cAMP, can bind to DNA and activate transcription. The level of cAMP is inversely related to the level of glucose, and CAP/ cAMP in *E. coli* activates the *lac* (lactose) operon. Also called *cAMP response protein* (*CRP*).

catalysis The process by which chemical subunits of larger organic molecules are held and positioned by enzymes that stress their chemical bonds, leading to the disassembly of the larger molecule into its subunits, often with the release of energy.

cation A positively charged ion.

cavitation In plants and animals, the blockage of a vessel by an air bubble that breaks the cohesion of the solution in the vessel; in animals more often called embolism.

$CD4^+$ cell A subtype of helper T cell that is identified by the presence of the CD4 protein on its surface. This cell type is targeted by the HIV virus that causes AIDS.

cecum In vertebrates, a blind pouch at the beginning of the large intestine.

cell cycle The repeating sequence of growth and division through which cells pass each generation.

cell determination The molecular "decision" process by which a cell becomes destined for a particular developmental pathway. This occurs before overt differentiation and can be a stepwise process.

cell-mediated immunity Arm of the adaptive immune system mediated by T cells, which includes cytotoxic cells and cells that assist the rest of the immune system.

cell plate The structure that forms at the equator of the spindle during early telophase in the dividing cells of plants and a few green algae.

cell-surface marker A glycoprotein or glycolipid on the outer surface of a cell's membrane that acts as an identifier; different cell types carry different markers.

cell-surface receptor A cell surface protein that binds a signal molecule and converts the extracellular signal into an intracellular one.

cellular blastoderm In insect embryonic development, the stage during which the nuclei of the syncitial blastoderm become separate cells through membrane formation.

cellular respiration The metabolic harvesting of energy by oxidation, ultimately dependent on molecular oxygen; carried out by the Krebs cycle and oxidative phosphorylation.

cellulose The chief constituent of the cell wall in all green plants, some algae, and a few other organisms; an insoluble complex carbohydrate formed of microfibrils of glucose molecules.

cell wall The rigid, outermost layer of the cells of plants, some protists, and most bacteria; the cell wall surrounds the plasma membrane.

central nervous system (CNS) That portion of the nervous system where most association occurs; in vertebrates, it is composed of the brain and spinal cord; in invertebrates, it usually consists of one or more cords of nervous tissue, together with their associated ganglia.

central vacuole A large, membrane-bounded sac found in plant cells that stores proteins, pigments, and waste materials, and is involved in water balance.

centriole A cytoplasmic organelle located outside the nuclear membrane, identical in structure to a basal body; found in animal cells and in the flagellated cells of other groups; divides and organizes spindle fibers during mitosis and meiosis.

centromere A visible point of constriction on a chromosome that contains repeated DNA sequences that bind specific proteins. These proteins make up the kinetochore to which microtubules attach during cell division.

cephalization The evolution of a head and brain area in the anterior end of animals; thought to be a consequence of bilateral symmetry.

cerebellum The hindbrain region of the vertebrate brain that lies above the medulla (brainstem) and behind the forebrain; it integrates information about body position and motion, coordinates muscular activities, and maintains equilibrium.

cerebral cortex The thin surface layer of neurons and glial cells covering the cerebrum; well developed only in mammals, and particularly prominent in humans. The cerebral cortex is the seat of conscious sensations and voluntary muscular activity.

cerebrum The portion of the vertebrate brain (the forebrain) that occupies the upper part of the skull, consisting of two cerebral hemispheres united by the corpus callosum. It is the primary association center of the brain. It coordinates and processes sensory input and coordinates motor responses.

chaetae Bristles of chitin on each body segment that help anchor annelid worms during locomotion.

channel protein (ion channel) A transmembrane protein with a hydrophilic interior that provides an aqueous channel allowing diffusion of species that cannot cross the membrane. Usually allows passage of specific ions such as K^+, Na^+, or Ca^{2+} across the membrane.

chaperone protein A class of enzymes that help proteins fold into the correct configuration and can refold proteins that have been misfolded or denatured.

character displacement A process in which natural selection favors individuals in a species that use resources not used by other species. This results in evolutionary change leading to species dissimilar in resource use.

character state In cladistics, one of two or more distinguishable forms of a character, such as the presence or absence of teeth in amniote vertebrates.

charging reaction The reaction by which an aminoacyl-tRNA synthetase attaches a specific amino acid to the correct tRNA using energy from ATP.

chelicera, pl. **chelicerae** The first pair of appendages in horseshoe crabs, sea spiders, and arachnids—the chelicerates, a group of arthropods. Chelicerae usually take the form of pincers or fangs.

chemical synapse A close association that allows chemical communication between neurons. A chemical signal (neurotransmitter) released by the first neuron binds to receptors in the membrane of the second neurons.

chemiosmosis The mechanism by which ATP is generated in mitochondria and chloroplasts; energetic electrons excited by light (in chloroplasts) or extracted by oxidation in the Krebs cycle (in mitochondria) are used to drive proton pumps, creating a proton concentration gradient; when protons subsequently flow back across the membrane, they pass through channels that couple their movement to the synthesis of ATP.

chiasma An X-shaped figure that can be seen in the light microscope during meiosis; evidence of crossing over, where two chromatids have exchanged parts; chiasmata move to the ends of the chromosome arms as the homologues separate.

chitin A tough, resistant, nitrogen-containing polysaccharide that forms the cell walls of certain fungi, the exoskeleton of arthropods, and the epidermal cuticle of other surface structures of certain other invertebrates.

chlorophyll The primary type of light-absorbing pigment in photosynthesis. Chlorophyll *a* absorbs light in the violet-blue and the red ranges of the visible light spectrum; chlorophyll *b* is an accessory pigment to chlorophyll *a*, absorbing light in the blue and red-orange ranges. Neither pigment absorbs light in the green range, 500–600 nm.

chloroplast A cell-like organelle present in algae and plants that contains chlorophyll (and usually other pigments) and carries out photosynthesis.

choanocyte A specialized flagellated cell found in sponges; choanocytes line the body interior.

chorion The outer member of the double membrane that surrounds the embryo of reptiles, birds, and mammals; in placental mammals, it contributes to the structure of the placenta.

chorionic villi sampling A technique in which fetal cells are sampled from the chorion of the placenta rather than from the amniotic fluid; this less invasive technique can be used earlier in pregnancy than amniocentesis.

chromatid One of the two daughter strands of a duplicated chromosome that is joined by a single centromere.

chromatin The complex of DNA and proteins of which eukaryotic chromosomes are composed; chromatin is highly uncoiled and diffuse in interphase nuclei, condensing to form the visible chromosomes in prophase.

chromatin-remodeling complex A large protein complex that has been found to modify histones and DNA and that can change the structure of chromatin, moving or transferring nucleosomes.

chromosomal mutation Any mutation that affects chromosome structure.

chromosome The vehicle by which hereditary information is physically transmitted from one generation to the next; in a bacterium, the chromosome consists of a single naked circle of DNA; in eukaryotes, each chromosome consists of a single linear DNA molecule and associated proteins.

chromosomal theory of inheritance The theory stating that hereditary traits are carried on chromosomes.

cilium A short cellular projection from the surface of a eukaryotic cell, having the same internal structure of microtubules in a 9 + 2 arrangement as seen in a flagellum.

circadian rhythm An endogenous cyclical rhythm that oscillates on a daily (24-hour) basis.

circulatory system A network of vessels in coelomate animals that carries fluids to and from different areas of the body.

cisterna A small collecting vessel that pinches off from the end of a Golgi body to form a transport vesicle that moves materials through the cytoplasm.

cisternal space The inner region of a membrane-bounded structure. Usually used to describe the interior of the endoplasmic reticulum; also called the *lumen*.

clade A taxonomic group composed of an ancestor and all its descendents.

cladistics A taxonomic technique used for creating hierarchies of organisms that represent true phylogenetic relationship and descent.

class A taxonomic category between phyla and orders. A class contains one or more orders, and belongs to a particular phylum.

classical conditioning The repeated presentation of a stimulus in association with a response that causes the brain to form an association between the stimulus and the response, even if they have never been associated before.

clathrin A protein located just inside the plasma membrane in eukaryotic cells, in indentations called clathrin-coated pits.

cleavage In vertebrates, a rapid series of successive cell divisions of a fertilized egg, forming a hollow sphere of cells, the blastula.

cleavage furrow The constriction that forms during cytokinesis in animal cells that is responsible for dividing the cell into two daughter cells.

climax vegetation Vegetation encountered in a self-perpetuating community of plants that has proceeded through all the stages of succession and stabilized.

cloaca In some animals, the common exit chamber from the digestive, reproductive, and urinary system; in others, the cloaca may also serve as a respiratory duct.

clone-by-clone sequencing A method of genome sequencing in which a physical map is constructed first, followed by sequencing of fragments and identifying overlap regions.

clonal selection Amplification of a clone of immune cells initiated by antigen recognition.

cloning Producing a cell line or culture all of whose members contain identical copies of a particular nucleotide sequence; an essential element in genetic engineering.

closed circulatory system A circulatory system in which the blood is physically separated from other body fluids.

coacervate A spherical aggregation of lipid molecules in water, held together by hydrophobic forces.

coactivator A protein that functions to link transcriptional activators to the transcription complex consisting of RNA polymerase II and general transcription factors.

cochlea In terrestrial vertebrates, a tubular cavity of the inner ear containing the essential organs for hearing.

coding strand The strand of a DNA duplex that is the same as the RNA encoded by a gene. This strand is not used as a template in transcription, it is complementary to the template.

codominance Describes a case in which two or more alleles of a gene are each dominant to other alleles but not to each other. The phenotype of a heterozygote for codominant alleles exhibit characteristics of each of the homozygous forms. For example, in human blood types, a cross between an AA individual and a BB individual yields AB individuals.

codon The basic unit of the genetic code; a sequence of three adjacent nucleotides in DNA or mRNA that codes for one amino acid.

coelom In animals, a fluid-filled body cavity that develops entirely within the mesoderm.

coenzyme A nonprotein organic molecule such as NAD that plays an accessory role in enzyme-catalyzed processes, often by acting as a donor or acceptor of electrons.

coevolution The simultaneous development of adaptations in two or more populations, species, or other categories that interact so closely that each is a strong selective force on the other.

cofactor One or more nonprotein components required by enzymes in order to function; many cofactors are metal ions, others are organic coenzymes.

cohesin A protein complex that holds sister chromatids together during cell division. The loss of cohesins at the centromere allows the anaphase movement of chromosomes.

collenchyma cell In plants, the cells that form a supporting tissue called collenchyma; often found in regions of primary growth in stems and in some leaves.

colloblast A specialized type of cell found in members of the animal phylum Ctenophora (comb jellies) that bursts on contact with zooplankton, releasing an adhesive substance to help capture this prey.

colonial flagellate hypothesis The proposal first put forth by Haeckel that metazoans descended from colonial protists; supported by the similarity of sponges to choanoflagellate protists.

commensalism A relationship in which one individual lives close to or on another and benefits, and the host is unaffected; a kind of symbiosis.

community All of the species inhabiting a common environment and interacting with one another.

companion cell A specialized parenchyma cell that is associated with each sieve-tube member in the phloem of a plant.

competitive exclusion The hypothesis that two species with identical ecological requirements cannot exist in the same locality indefinitely, and that the more efficient of the two in utilizing the available scarce resources will exclude the other; also known as Gause's principle.

competitive inhibitor An inhibitor that binds to the same active site as an enzyme's substrate, thereby competing with the substrate.

complementary Describes genetic information in which each nucleotide base has a complementary partner with which it forms a base-pair.

complementary DNA (cDNA) A DNA copy of an mRNA transcript; produced by the action of the enzyme reverse transcriptase.

complement system The chemical defense of a vertebrate body that consists of a battery of proteins that become activated by the walls of bacteria and fungi.

complete digestive system A digestive system that has both a mouth and an anus, allowing unidirectional flow of ingested food.

compound eye An organ of sight in many arthropods composed of many independent visual units called ommatidia.

concentration gradient A difference in concentration of a substance from one location to another, often across a membrane.

condensin A protein complex involved in condensation of chromosomes during mitosis and meiosis.

cone (1) In plants, the reproductive structure of a conifer. (2) In vertebrates, a type of light-sensitive neuron in the retina concerned with the perception of color and with the most acute discrimination of detail.

conidia An asexually produced fungal spore.

conjugation Temporary union of two unicellular organisms, during which genetic material is transferred from one cell to the other; occurs in bacteria, protists, and certain algae and fungi.

consensus sequence In genome sequencing, the overall sequence that is consistent with the sequences of individual fragments; computer programs are used to compare sequences and generate a consensus sequence.

conservation of synteny The preservation over evolutionary time of arrangements of DNA segments in related species.

contig A contiguous segment of DNA assembled by analyzing sequence overlaps from smaller fragments.

continuous variation Variation in a trait that occurs along a continuum, such as the trait of height in human beings; often occurs when a trait is determined by more than one gene.

contractile vacuole In protists and some animals, a clear fluid-filled vacuole that takes up water from within the cell and then contracts, releasing it to the outside through a pore in a cyclical manner; functions primarily in osmoregulation and excretion.

conus arteriosus The anteriormost chamber of the embryonic heart in vertebrate animals.

convergent evolution The independent development of similar structures in organisms that are not directly related; often found in organisms living in similar environments.

cork cambium The lateral meristem that forms the periderm, producing cork (phellem) toward the surface (outside) of the plant and phelloderm toward the inside.

cornea The transparent outer layer of the vertebrate eye.

corolla The petals, collectively; usually the conspicuously colored flower whorl.

corpus callosum The band of nerve fibers that connects the two hemispheres of the cerebrum in humans and other primates.

corpus luteum A structure that develops from a ruptured follicle in the ovary after ovulation.

cortex The outer layer of a structure; in animals, the outer, as opposed to the inner, part of an organ; in vascular plants, the primary ground tissue of a stem or root.

cotyledon A seed leaf that generally stores food in dicots or absorbs it in monocots, providing nourishment used during seed germination.

crassulacean acid metabolism (CAM) A mode of carbon dioxide fixation by which CO_2 enters open leaf stomata at night and is used in photosynthesis during the day, when stomata are closed to prevent water loss.

crista A folded extension of the inner membrane of a mitochondrion. Mitochondria contain numerous cristae.

cross-current flow In bird lungs, the latticework of capillaries arranged across the air flow, at a 90° angle.

crossing over In meiosis, the exchange of corresponding chromatid segments between homologous chromosomes; responsible for genetic recombination between homologous chromosomes.

ctenidia Respiratory gills of mollusks; they consist of a system of filamentous projections of the mantle that are rich in blood vessels.

cuticle A waxy or fatty, noncellular layer (formed of a substance called cutin) on the outer wall of epidermal cells.

cutin In plants, a fatty layer produced by the epidermis that forms the cuticle on the outside surface.

cyanobacteria A group of photosynthetic bacteria, sometimes called the "blue-green algae," that contain the chlorophyll pigments most abundant in plants and algae, as well as other pigments.

cyclic AMP (cAMP) A form of adenosine monophosphate (AMP) in which the atoms of the phosphate group form a ring; found in almost all organisms, cAMP functions as an intracellular second messenger that regulates a diverse array of metabolic activities.

cyclic photophosphorylation Reactions that begin with the absorption of light by reaction center chlorophyll that excites an electron. The excited electron returns to the photosystem, generating ATP by chemiosmosis in the process. This is found in the single bacterial photosystem, and can occur in plants in photosystem I.

cyclin Any of a number of proteins that are produced in synchrony with the cell cycle and combine with certain protein kinases, the cyclin-dependent kinases, at certain points during cell division.

cyclin-dependent kinase (Cdk) Any of a group of protein kinase enzymes that control progress through the cell cycle. These enzymes are only active when complexed with cyclin. The cdc2 protein, produced by the *cdc2* gene, was the first Cdk enzyme discovered.

cytochrome Any of several iron-containing protein pigments that serve as electron carriers in transport chains of photosynthesis and cellular respiration.

cytochrome *b6–f* complex A proton pump found in the thylakoid membrane. This complex uses energy from excited electrons to pump protons from the stroma into the thylakoid compartment.

cytokinesis Division of the cytoplasm of a cell after nuclear division.

cytokine Signaling molecules secreted by immune cells that affect other immune cells.

cytoplasm The material within a cell, excluding the nucleus; the protoplasm.

cytoskeleton A network of protein microfilaments and microtubules within the cytoplasm of a eukaryotic cell that maintains the shape of the cell, anchors its organelles, and is involved in animal cell motility.

cytosol The fluid portion of the cytoplasm; it contains dissolved organic molecules and ions.

cytotoxic T cell A special T cell activated during cell-mediated immune response that recognizes and destroys infected body cells.

D

deamination The removal of an amino group; part of the degradation of proteins into compounds that can enter the Krebs cycle.

deductive reasoning The logical application of general principles to predict a specific result. In science, deductive reasoning is used to test the validity of general ideas.

dehydration reaction A type of chemical reaction in which two molecules join to form one larger molecule, simultaneously splitting out a molecule of water; one molecule is stripped of a hydrogen atom, and another is stripped of a hydroxyl group (—OH), resulting in the joining of the two molecules, while the H and —OH released may combine to form a water molecule.

dehydrogenation Chemical reaction involving the loss of a hydrogen atom. This is an oxidation that combines loss of an electron with loss of a proton.

deletion A mutation in which a portion of a chromosome is lost; if too much information is lost, the deletion can be fatal.

demography The properties of the rate of growth and the age structure of populations.

denaturation The loss of the native configuration of a protein or nucleic acid as a result of excessive heat, extremes of pH, chemical modification, or changes in solvent ionic strength or polarity that disrupt hydrophobic interactions; usually accompanied by loss of biological activity.

dendrite A process extending from the cell body of a neuron, typically branched, that conducts impulses toward the cell body.

deoxyribonucleic acid (DNA) The genetic material of all organisms; composed of two complementary chains of nucleotides wound in a double helix.

dephosphorylation The removal of a phosphate group, usually by a phosphatase enzyme. Many proteins can be activated or inactivated by dephosphorylation.

depolarization The movement of ions across a plasma membrane that locally wipes out an electrical potential difference.

derived character A characteristic used in taxonomic analysis representing a departure from the primitive form.

dermal tissue In multicellular organisms, a type of tissue that forms the outer layer of the body and is in contact with the environment; it has a protective function.

desmosome A type of anchoring junction that links adjacent cells by connecting their cytoskeletons with cadherin proteins.

derepression Seen in anabolic operons where the operon that encodes the enzymes for a biochemical pathway is repressed in the presence of the end product of the pathway and derepressed in the absence of the end product. This allows production of the enzymes only when they are necessary.

determinate development A type of development in animals in which each embryonic cell has a predetermined fate in terms of what kind of tissue it will form in the adult.

deuterostome Any member of a grouping of bilaterally symmetrical animals in which the anus develops first and the mouth second; echinoderms and vertebrates are deuterostome animals.

diacylglycerol (DAG) A second messenger that is released, along with inositol-1,4,5-trisphosphate (IP_3), when phospholipase C cleaves PIP_2. DAG can have a variety of cellular effects through activation of protein kinases.

diaphragm (1) In mammals, a sheet of muscle tissue that separates the abdominal and thoracic cavities and functions in breathing. (2) A contraceptive device used to block the entrance to the uterus temporarily and thus prevent sperm from entering during sexual intercourse.

diapsid Any of a group of reptiles that have two pairs of temporal openings in the skull, one lateral and one more dorsal; one lineage of this group gave rise to dinosaurs, modern reptiles, and birds.

diastolic pressure In the measurement of human blood pressure, the minimum pressure between heartbeats (repolarization of the ventricles). *Compare with* systolic pressure.

dicer An enzyme that generates small RNA molecules in a cell by chopping up double-stranded RNAs; dicer produces miRNAs and siRNAs.

dicot Short for dicotyledon; a class of flowering plants generally characterized as having two cotyledons, net-veined leaves, and flower parts usually in fours or fives.

dideoxynucleotide A nucleotide lacking —OH groups at both the 2′ and 3′ positions; used as a chain terminator in the enzymatic sequencing of DNA.

differentiation A developmental process by which a relatively unspecialized cell undergoes a progressive change to a more specialized form or function.

diffusion The net movement of dissolved molecules or other particles from a region where they are more concentrated to a region where they are less concentrated.

dihybrid An individual heterozygous at two different loci; for example *A/a B/b*.

dihybrid cross A single genetic cross involving two different traits, such as flower color and plant height.

dikaryotic In fungi, having pairs of nuclei within each cell.

dioecious Having the male and female elements on different individuals.

diploid Having two sets of chromosomes (2*n*); in animals, twice the number characteristic of gametes; in plants, the chromosome number characteristic of the sporophyte generation; in contrast to haploid (*n*).

directional selection A form of selection in which selection acts to eliminate one extreme from an array of phenotypes.

disaccharide A carbohydrate formed of two simple sugar molecules bonded covalently.

disruptive selection A form of selection in which selection acts to eliminate rather than favor the intermediate type.

dissociation In proteins, the reversible separation of protein subunits from a quaternary structure without altering their tertiary structure. Also refers to the dissolving of ionic compounds in water.

disassortative mating A type of nonrandom mating in which phenotypically different individuals mate more frequently.

diurnal Active during the day.

DNA-binding motif A region found in a regulatory protein that is capable of binding to a specific base sequence in DNA; a critical part of the protein's DNA-binding domain.

DNA fingerprinting An identification technique that makes use of a variety of molecular techniques to identify differences in the DNA of individuals.

DNA gyrase A topoisomerase involved in DNA replication; it relieves the torsional strain caused by unwinding the DNA strands.

DNA library A collection of DNAs in a vector (a plasmid, phage, or artificial chromosome) that taken together represent a complex mixture of DNAs, such as the entire genome, or the cDNAs made from all of the mRNA in a specific cell type.

DNA ligase The enzyme responsible for formation of phosphodiester bonds between adjacent nucleotides in DNA.

DNA microarray An array of DNA fragments on a microscope slide or silicon chip, used in hybridization experiments with labeled mRNA or DNA to identify active and inactive genes, or the presence or absence of particular sequences.

DNA polymerase A class of enzymes that all synthesize DNA from a preexisting template. All synthesize only in the 5′-to-3′ direction, and require a primer to extend.

DNA vaccine A type of vaccine that uses DNA from a virus or bacterium that stimulates the cellular immune response.

domain (1) A distinct modular region of a protein that serves a particular function in the action of the protein, such as a regulatory domain or a DNA-binding domain. (2) In taxonomy, the level higher than kingdom. The three domains currently recognized are Bacteria, Archaea, and Eukarya.

Domain Archaea In the three-domain system of taxonomy, the group that contains only the Archaea, a highly diverse group of unicellular prokaryotes.

Domain Bacteria In the three-domain system of taxonomy, the group that contains only the Bacteria, a vast group of unicellular prokaryotes.

Domain Eukarya In the three-domain system of taxonomy, the group that contains eukaryotic organisms including protists, fungi, plants, and animals.

dominant An allele that is expressed when present in either the heterozygous or the homozygous condition.

dosage compensation A phenomenon by which the expression of genes carried on sex chromosomes is kept the same in males and females, despite a different number of sex chromosomes. In mammals, inactivation of one of the X chromosomes in female cells accomplishes dosage compensation.

double fertilization The fusion of the egg and sperm (resulting in a 2*n* fertilized egg, the zygote) and the simultaneous fusion of the second male gamete with the polar nuclei (resulting in a primary endosperm nucleus, which is often triploid, 3*n*); a unique characteristic of all angiosperms.

double helix The structure of DNA, in which two complementary polynucleotide strands coil around a common helical axis.

duodenum In vertebrates, the upper portion of the small intestine.

duplication A mutation in which a portion of a chromosome is duplicated; if the duplicated region does not lie within a gene, the duplication may have no effect.

E

ecdysis Shedding of outer, cuticular layer; molting, as in insects or crustaceans.

ecdysone Molting hormone of arthropods, which triggers when ecdysis occurs.

ecology The study of interactions of organisms with one another and with their physical environment.

ecosystem A major interacting system that includes organisms and their nonliving environment.

ecotype A locally adapted variant of an organism; differing genetically from other ecotypes.

ectoderm One of the three embryonic germ layers of early vertebrate embryos; ectoderm gives rise to the outer epithelium of the body (skin, hair, nails) and to the nerve tissue, including the sense organs, brain, and spinal cord.

ectomycorrhizae Externally developing mycorrhizae that do not penetrate the cells they surround.

ectotherms Animals such as reptiles, fish, or amphibians, whose body temperature is regulated by their behavior or by their surroundings.

electronegativity A property of atomic nuclei that refers to the affinity of the nuclei for valence electrons; a nucleus that is more electronegative has a greater pull on electrons than one that is less electronegative.

electron transport chain The passage of energetic electrons through a series of membrane-associated electron-carrier molecules to proton pumps embedded within mitochondrial or chloroplast membranes. *See* chemiosmosis.

elongation factor (Ef-Tu) In protein synthesis in *E. coli*, a factor that binds to GTP and to a charged tRNA to accomplish binding of the charged tRNA to the A site of the ribosome, so that elongation of the polypeptide chain can occur.

embryo A multicellular developmental stage that follows cell division of the zygote.

embryonic stem cell (ES cell) A stem cell derived from an early embryo that can develop into different adult tissues and give rise to an adult organism when injected into a blastocyst.

emergent properties Novel properties arising from the way in which components interact. Emergent properties often cannot be deduced solely from knowledge of the individual components.

emerging virus Any virus that originates in one organism but then passes to another; usually refers to transmission to humans.

endergonic Describes a chemical reaction in which the products contain more energy than the reactants, so that free energy must be put into the reaction from an outside source to allow it to proceed.

endocrine gland Ductless gland that secretes hormones into the extracellular spaces, from which they diffuse into the circulatory system.

endocytosis The uptake of material into cells by inclusion within an invagination of the plasma membrane; the uptake of solid material is phagocytosis, and that of dissolved material is pinocytosis.

endoderm One of the three embryonic germ layers of early vertebrate embryos, destined to give rise to the epithelium that lines internal structures and most of the digestive and respiratory tracts.

endodermis In vascular plants, a layer of cells forming the innermost layer of the cortex in roots and some stems.

endomembrane system A system of connected membranous compartments found in eukaryotic cells.

endometrium The lining of the uterus in mammals; thickens in response to secretion of estrogens and progesterone and is sloughed off in menstruation.

endomycorrhizae Mycorrhizae that develop within cells.

endonuclease An enzyme capable of cleaving phosphodiester bonds between nucleotides located internally in a DNA strand.

endoplasmic reticulum (ER) Internal membrane system that forms a netlike array of channels and interconnections within the cytoplasm of eukaryotic cells. The ER is divided into rough (RER) and smooth (SER) compartments.

endorphin One of a group of small neuropeptides produced by the vertebrate brain; like morphine, endorphins modulate pain perception.

endosperm A storage tissue characteristic of the seeds of angiosperms, which develops from the union of a male nucleus and the polar nuclei of the embryo sac. The endosperm is digested by the growing sporophyte either before maturation of the seed or during its germination.

endospore A highly resistant, thick-walled bacterial spore that can survive harsh environmental stress, such as heat or dessication, and then germinate when conditions become favorable.

endosymbiosis Theory that proposes that eukaryotic cells evolved from a symbiosis between different species of prokaryotes.

endotherm An animal capable of maintaining a constant body temperature. *See* homeotherm.

energy level A discrete level, or quantum, of energy that an electron in an atom possesses. To change energy levels, an electron must absorb or release energy.

enhancer A site of regulatory protein binding on the DNA molecule distant from the promoter and start site for a gene's transcription.

enthalpy In a chemical reaction, the energy contained in the chemical bonds of the molecule, symbolized as *H;* in a cellular reaction, the free energy is equal to the enthalpy of the reactant molecules in the reaction.

entropy A measure of the randomness or disorder of a system; a measure of how much energy in a system has become so dispersed (usually as evenly distributed heat) that it is no longer available to do work.

enzyme A protein that is capable of speeding up specific chemical reactions by lowering the required activation energy.

enzyme–substrate complex The complex formed when an enzyme binds with its substrate. This complex often has an altered configuration compared with the nonbound enzyme.

epicotyl The region just above where the cotyledons are attached.

epidermal cell In plants, a cell that collectively forms the outermost layer of the primary plant body; includes specialized cells such as trichomes and guard cells.

epidermis The outermost layers of cells; in plants, the exterior primary tissue of leaves, young stems, and roots; in vertebrates, the nonvascular external layer of skin, of ectodermal origin; in invertebrates, a single layer of ectodermal epithelium.

epididymis A sperm storage vessel; a coiled part of the sperm duct that lies near the testis.

epistasis Interaction between two nonallelic genes in which one of them modifies the phenotypic expression of the other.

epithelium In animals, a type of tissue that covers an exposed surface or lines a tube or cavity.

equilibrium A stable condition; the point at which a chemical reaction proceeds as rapidly in the reverse direction as it does in the forward direction, so that there is no further net change in the concentrations of products or reactants. In ecology, a stable condition that resists change and fairly quickly returns to its original state if disturbed by humans or natural events.

erythrocyte Red blood cell, the carrier of hemoglobin.

erythropoiesis The manufacture of blood cells in the bone marrow.

E site In a ribosome, the exit site that binds to the tRNA that carried the previous amino acid added to the polypeptide chain.

estrus The period of maximum female sexual receptivity, associated with ovulation of the egg.

ethology The study of patterns of animal behavior in nature.

euchromatin That portion of a eukaryotic chromosome that is transcribed into mRNA; contains active genes that are not tightly condensed during interphase.

eukaryote A cell characterized by membrane-bounded organelles, most notably the nucleus, and one that possesses chromosomes whose DNA is associated with proteins; an organism composed of such cells.

eutherian A placental mammal.

eutrophic Refers to a lake in which an abundant supply of minerals and organic matter exists.

evolution Genetic change in a population of organisms; in general, evolution leads to progressive change from simple to complex.

excision repair A nonspecific mechanism to repair damage to DNA during synthesis. The damaged or mismatched region is excised, and DNA polymerase replaces the region removed.

exergonic Describes a chemical reaction in which the products contain less free energy than the reactants, so that free energy is released in the reaction.

exhalant siphon In bivalve mollusks, the siphon through which outgoing water leaves the body.

exocrine gland A type of gland that releases its secretion through a duct, such as a digestive gland or a sweat gland.

exocytosis A type of bulk transport out of cells in which a vacuole fuses with the plasma membrane, discharging the vacuole's contents to the outside.

exon A segment of DNA that is both transcribed into RNA and translated into protein. *See* intron.

exonuclease An enzyme capable of cutting phosphodiester bonds between nucleotides located at an end of a DNA strand. This allows sequential removal of nucleotides from the end of DNA.

exoskeleton An external skeleton, as in arthropods.

experiment A test of one or more hypotheses. Hypotheses make contrasting predictions that can be tested experimentally in control and test experiments where a single variable is altered.

expressed sequence tag (EST) A short sequence of a cDNA that unambiguously identifies the cDNA.

expression vector A type of vector (plasmid or phage) that contains the sequences necessary to drive expression of inserted DNA in a specific cell type.

exteroceptor A receptor that is excited by stimuli from the external world.

extremophile An archaean organism that lives in extreme environments; different archaean species may live in hot springs (thermophiles), highly saline environments (halophiles), highly acidic or basic environments, or under high pressure at the bottom of oceans.

F

5′ cap In eukaryotes, a structure added to the 5′ end of an mRNA consisting of methylated GTP attached by a 5′ to 5′ bond. The cap protects this end from degradation and is involved in the initiation of translation.

facilitated diffusion Carrier-assisted diffusion of molecules across a cellular membrane through specific channels from a region of higher concentration to one of lower concentration; the process is driven by the concentration gradient and does not require cellular energy from ATP.

family A taxonomic grouping of similar species above the level of genus.

fat A molecule composed of glycerol and three fatty acid molecules.

feedback inhibition Control mechanism whereby an increase in the concentration of some molecules inhibits the synthesis of that molecule.

fermentation The enzyme-catalyzed extraction of energy from organic compounds without the involvement of oxygen.

fertilization The fusion of two haploid gamete nuclei to form a diploid zygote nucleus.

fibroblast A flat, irregularly branching cell of connective tissue that secretes structurally strong proteins into the matrix between the cells.

first filial (F$_1$) generation The offspring resulting from a cross between a parental generation (P); in experimental crosses, these parents usually have different phenotypes.

First Law of Thermodynamics Energy cannot be created or destroyed, but can only undergo conversion from one form to another; thus, the amount of energy in the universe is unchangeable.

fitness The genetic contribution of an individual to succeeding generations. Relative fitness refers to the fitness of an individual relative to other individuals in a population.

fixed action pattern A stereotyped animal behavior response, thought by ethologists to be based on programmed neural circuits.

flagellin The protein composing bacterial flagella, which allow a cell to move through an aqueous environment.

flagellum A long, threadlike structure protruding from the surface of a cell and used in locomotion.

flame cell A specialized cell found in the network of tubules inside flatworms that assists in water regulation and some waste excretion.

flavin adenine dinucleotide (FAD, FADH$_2$) A cofactor that acts as a soluble (not membrane-bound) electron carrier (can be reversibly oxidized and reduced).

fluorescent in situ hybridization (FISH) A cytological method used to find specific DNA sequences on chromosomes with a specific fluorescently labeled probe.

food security Having access to sufficient, safe food to avoid malnutrition and starvation; a global human issue.

foraging behavior A collective term for the many complex, evolved behaviors that influence what an animal eats and how the food is obtained.

founder effect The effect by which rare alleles and combinations of alleles may be enhanced in new populations.

fovea A small depression in the center of the retina with a high concentration of cones; the area of sharpest vision.

frameshift mutation A mutation in which a base is added or deleted from the DNA sequence. These changes alter the reading frame downstream of the mutation.

free energy Energy available to do work.

free radical An ionized atom with one or more unpaired electrons, resulting from electrons that have been energized by ionizing radiation being ejected from the atom; free radicals react violently with other molecules, such as DNA, causing damage by mutation.

frequency-dependent selection A type of selection that depends on how frequently or infrequently a phenotype occurs in a population.

fruit In angiosperms, a mature, ripened ovary (or group of ovaries), containing the seeds.

functional genomics The study of the function of genes and their products, beyond simply ascertaining gene sequences.

functional group A molecular group attached to a hydrocarbon that confers chemical properties or reactivities. Examples include hydroxyl (—OH), carboxylic acid (—COOH) and amino groups (—NH$_2$).

fundamental niche Also referred to as the hypothetical niche, this is the entire niche an organism could fill if there were no other interacting factors (such as competition or predation).

G

G$_0$ phase The stage of the cell cycle occupied by cells that are not actively dividing.

G$_1$ phase The phase of the cell cycle after cytokinesis and before DNA replication called the first "gap" phase. This phase is the primary growth phase of a cell.

G$_1$/S checkpoint The primary control point at which a cell "decides" whether or not to divide. Also called START and the restriction point.

G$_2$ phase The phase of the cell cycle between DNA replication and mitosis called the second "gap" phase. During this phase, the cell prepares for mitosis.

G$_2$/M checkpoint The second cell-division control point, at which division can be delayed if DNA has not been properly replicated or is damaged.

gametangium, pl. **gametangia** A cell or organ in which gametes are formed.

gamete A haploid reproductive cell.

gametocytes Cells in the malarial sporozoite life cycle capable of giving rise to gametes when in the correct host.

gametophyte In plants, the haploid (*n*), gamete-producing generation, which alternates with the diploid (2*n*) sporophyte.

ganglion, pl. **ganglia** An aggregation of nerve cell bodies; in invertebrates, ganglia are the integrative centers; in vertebrates, the term is restricted to aggregations of nerve cell bodies located outside the central nervous system.

gap gene Any of certain genes in *Drosophila* development that divide the embryo into large blocks in the process of segmentation; *hunchback* is a gap gene.

gap junction A junction between adjacent animal cells that allows the passage of materials between the cells.

gastrodermis In eumetazoan animals, the layer of digestive tissue that develops from the endoderm.

gastrula In vertebrates, the embryonic stage in which the blastula with its single layer of cells turns into a three-layered embryo made up of ectoderm, mesoderm, and endoderm.

gastrulation Developmental process that converts blastula into embryo with three embryonic germ layers: endoderm, mesoderm, and ectoderm. Involves massive cell migration to convert the hollow structure into a three-layered structure.

gene The basic unit of heredity; a sequence of DNA nucleotides on a chromosome that encodes a protein, tRNA, or rRNA molecule, or regulates the transcription of such a sequence.

gene conversion Alteration of one homologous chromosome by the cell's error-detection and repair system to make it resemble the sequence on the other homologue.

gene expression The conversion of the genotype into the phenotype; the process by which DNA is transcribed into RNA, which is then translated into a protein product.

gene pool All the alleles present in a species.

gene-for-gene hypothesis A plant defense mechanism in which a specific protein encoded by a viral, bacterial, or fungal pathogen binds to a protein encoded by a plant gene and triggers a defense response in the plant.

general transcription factor Any of a group of transcription factors that are required for formation of an initiation complex by RNA polymerase II at a promoter. This allows a basal level that can be increased by the action of specific factors.

generalized transduction A form of gene transfer in prokaryotes in which any gene can be transferred between cells. This uses a lytic bacteriophage as a carrier where the virion is accidentally packaged with host DNA.

genetic counseling The process of evaluating the risk of genetic defects occurring in offspring, testing for these defects in unborn children, and providing the parents with information about these risks and conditions.

genetic drift Random fluctuation in allele frequencies over time by chance.

genetic map An abstract map that places the relative location of genes on a chromosome based on recombination frequency.

genome The entire DNA sequence of an organism.

genomic imprinting Describes an exception to Mendelian genetics in some mammals in which the phenotype caused by an allele is exhibited when the allele comes from one parent, but not from the other.

genomic library A DNA library that contains a representation of the entire genome of an organism.

genomics The study of genomes as opposed to individual genes.

genotype The genetic constitution underlying a single trait or set of traits.

genotype frequency A measure of the occurrence of a genotype in a population, expressed as a proportion of the entire population, for example, an occurrence of 0.25 (25%) for a homozygous recessive genotype.

genus, pl. **genera** A taxonomic group that ranks below a family and above a species.

germination The resumption of growth and development by a spore or seed.

germ layers The three cell layers formed at gastrulation of the embryo that foreshadow the future organization of tissues; the layers, from the outside inward, are the ectoderm, the mesoderm, and the endoderm.

germ-line cells During zygote development, cells that are set aside from the somatic cells and that will eventually undergo meiosis to produce gametes.

gill (1) In aquatic animals, a respiratory organ, usually a thin-walled projection from some part of the external body surface, endowed with a rich capillary bed and having a large surface area. (2) In basidiomycete fungi, the plates on the underside of the cap.

globular protein Proteins with a compact tertiary structure with hydrophobic amino acids mainly in the interior.

glomerular filtrate The fluid that passes out of the capillaries of each glomerulus.

glomerulus A cluster of capillaries enclosed by Bowman's capsule.

glucagon A vertebrate hormone produced in the pancreas that acts to initiate the breakdown of glycogen to glucose subunits.

gluconeogenesis The synthesis of glucose from noncarbohydrates (such as proteins or fats).

glucose A common six-carbon sugar ($C_6H_{12}O_6$); the most common monosaccharide in most organisms.

glucose repression In *E. coli*, the preferential use of glucose even when other sugars are present; transcription of mRNA encoding the enzymes for utilizing the other sugars does not occur.

glycocalyx A "sugar coating" on the surface of a cell resulting from the presence of polysaccharides on glycolipids and glycoproteins embedded in the outer layer of the plasma membrane.

glycogen Animal starch; a complex branched polysaccharide that serves as a food reserve in animals, bacteria, and fungi.

glycolipid Lipid molecule modified within the Golgi complex by having a short sugar chain (polysaccharide) attached.

glycolysis The anaerobic breakdown of glucose; this enzyme-catalyzed process yields two molecules of pyruvate with a net of two molecules of ATP.

glycoprotein Protein molecule modified within the Golgi complex by having a short sugar chain (polysaccharide) attached.

glyoxysome A small cellular organelle or microbody containing enzymes necessary for conversion of fats into carbohydrates.

glyphosate A biodegradable herbicide that works by inhibiting EPSP synthetase, a plant enzyme that makes aromatic amino acids; genetic engineering has allowed crop species to be created that are resistant to glyphosate.

Golgi apparatus (Golgi body) A collection of flattened stacks of membranes in the cytoplasm of eukaryotic cells; functions in collection, packaging, and distribution of molecules synthesized in the cell.

G protein A protein that binds guanosine triphosphate (GTP) and assists in the function of cell-surface receptors. When the receptor binds its signal molecule, the G protein binds GTP and is activated to start a chain of events within the cell.

G protein-coupled receptor (GPCR) A receptor that acts through a heterotrimeric (three component) G protein to activate effector proteins. The effector proteins then function as enzymes to produce second messengers such as cAMP or IP_3.

gradualism The view that species change very slowly in ways that may be imperceptible from one generation to the next but that accumulate and lead to major changes over thousands or millions of years.

Gram stain Staining technique that divides bacteria into gram-negative or gram-positive based on retention of a violet dye. Differences in staining are due to cell wall construction.

granum (pl. grana) A stacked column of flattened, interconnected disks (thylakoids) that are part of the thylakoid membrane system in chloroplasts.

gravitropism Growth response to gravity in plants; formerly called geotropism.

ground meristem The primary meristem, or meristematic tissue, that gives rise to the plant body (except for the epidermis and vascular tissues).

ground tissue In plants, a type of tissue that performs many functions, including support, storage, secretion, and photosynthesis; may consist of many cell types.

growth factor Any of a number of proteins that bind to membrane receptors and initiate intracellular signaling systems that result in cell growth and division.

guard cell In plants, one of a pair of sausage-shaped cells flanking a stoma; the guard cells open and close the stomata.

guttation The exudation of liquid water from leaves due to root pressure.

gymnosperm A seed plant with seeds not enclosed in an ovary; conifers are gymnosperms.

gynoecium The aggregate of carpels in the flower of a seed plant.

H

habitat The environment of an organism; the place where it is usually found.

habituation A form of learning; a diminishing response to a repeated stimulus.

halophyte A plant that is salt-tolerant.

haplodiploidy A phenomenon occurring in certain organisms such as wasps, wherein both haploid (male) and diploid (female) individuals are encountered.

haploid Having only one set of chromosomes (n), in contrast to diploid ($2n$).

haplotype A region of a chromosome that is usually inherited intact, that is, it does not undergo recombination. These are identified based on analysis of SNPs.

Hardy-Weinberg equilibrium A mathematical description of the fact that allele and genotype frequencies remain constant in a random-mating population in the absence of inbreeding, selection, or other evolutionary forces; usually stated: if the frequency of allele a is p and the frequency of allele b is q, then the genotype frequencies after one generation of random mating will always be $p_2 + 2pq + q_2 = 1$.

Haversian canal Narrow channels that run parallel to the length of a bone and contain blood vessels and nerve cells.

heat A measure of the random motion of molecules; the greater the heat, the greater the motion. Heat is one form of kinetic energy.

heat of vaporization The amount of energy required to change 1 g of a substance from a liquid to a gas.

heavy metal Any of the metallic elements with high atomic numbers, such as arsenic, cadmium, lead, etc. Many heavy metals are toxic to animals even in small amounts.

helicase Any of a group of enzymes that unwind the two DNA strands in the double helix to facilitate DNA replication.

helix-turn-helix motif A common DNA-binding motif found in regulatory proteins; it consists of two α-helices linked by a nonhelical segment (the "turn").

helper T cell A class of white blood cells that initiates both the cell-mediated immune response and the humoral immune response; helper T cells are the targets of the AIDS virus (HIV).

hemoglobin A globular protein in vertebrate red blood cells and in the plasma of many invertebrates that carries oxygen and carbon dioxide.

hemopoietic stem cell The cells in bone marrow where blood cells are formed.

hermaphroditism Condition in which an organism has both male and female functional reproductive organs.

heterochromatin The portion of a eukaryotic chromosome that is not transcribed into RNA; remains condensed in interphase and stains intensely in histological preparations.

heterochrony An alteration in the timing of developmental events due to a genetic change; for example, a mutation that delays flowering in plants.

heterokaryotic In fungi, having two or more genetically distinct types of nuclei within the same mycelium.

heterosporous In vascular plants, having spores of two kinds, namely, microspores and megaspores.

heterotroph An organism that cannot derive energy from photosynthesis or inorganic chemicals, and so must feed on other plants and animals, obtaining chemical energy by degrading their organic molecules.

heterozygote advantage The situation in which individuals heterozygous for a trait have a selective advantage over those who are homozygous; an example is sickle cell anemia.

heterozygous Having two different alleles of the same gene; the term is usually applied to one or more specific loci, as in "heterozygous with respect to the *W* locus" (that is, the genotype is *W/w*).

Hfr cell An *E. coli* cell that has a high frequency of recombination due to integration of an F plasmid into its genome.

histone One of a group of relatively small, very basic polypeptides, rich in arginine and lysine, forming the core of nucleosomes around which DNA is wrapped in the first stage of chromosome condensation.

histone protein Any of eight proteins with an overall positive charge that associate in a complex. The DNA duplex coils around a core of eight histone proteins, held by its negatively charged phosphate groups, forming a nucleosome.

holoblastic cleavage Process in vertebrate embryos in which the cleavage divisions all occur at the same rate, yielding a uniform cell size in the blastula.

homeobox A sequence of 180 nucleotides located in homeotic genes that produces a 60-amino-acid peptide sequence (the homeodomain) active in transcription factors.

homeodomain motif A special class of helix-turn-helix motifs found in regulatory proteins that control development in eukaryotes.

homeosis A change in the normal spatial pattern of gene expression that can result in homeotic mutants where a wild-type structure develops in the wrong place in or on the organism.

homeostasis The maintenance of a relatively stable internal physiological environment in an organism; usually involves some form of feedback self-regulation.

homeotherm An organism, such as a bird or mammal, capable of maintaining a stable body temperature independent of the environmental temperature. *See* endotherm.

homeotic gene One of a series of "master switch" genes that determine the form of segments developing in the embryo.

hominid Any primate in the human family, Hominidae. *Homo sapiens* is the only living representative.

hominoid Collectively, hominids and apes; the monkeys and hominoids constitute the anthropoid primates.

homokaryotic In fungi, having nuclei with the same genetic makeup within a mycelium.

homologue One of a pair of chromosomes of the same kind located in a diploid cell; one copy of each pair of homologues comes from each gamete that formed the zygote.

homologous (1) Refers to similar structures that have the same evolutionary origin. (2) Refers to a pair of the same kind of chromosome in a diploid cell.

homoplasy In cladistics, a shared character state that has not been inherited from a common ancestor exhibiting that state; may result from convergent evolution or evolutionary reversal. The wings of birds and of bats, which are convergent structures, are examples.

homosporous In some plants, production of only one type of spore rather than differentiated types. *Compare with* heterosporous.

homozygous Being a homozygote, having two identical alleles of the same gene; the term is usually applied to one or more specific loci, as in "homozygous with respect to the *W* locus" (i.e., the genotype is *W/W* or *w/w*).

horizontal gene transfer (HGT) The passing of genes laterally between species; more prevalent very early in the history of life.

hormone A molecule, usually a peptide or steroid, that is produced in one part of an organism and triggers a specific cellular reaction in target tissues and organs some distance away.

host range The range of organisms that can be infected by a particular virus.

Hox **gene** A group of homeobox-containing genes that control developmental events, usually found organized into clusters of genes. These genes have been conserved in many different multicellular animals, both invertebrates and vertebrates, although the number of clusters changes in lineages, leading to four clusters in vertebrates.

humoral immunity Arm of the adaptive immune system involving B cells that produce soluble antibodies specific for foreign antigens.

humus Partly decayed organic material found in topsoil.

hybridization The mating of unlike parents.

hydration shell A "cloud" of water molecules surrounding a dissolved substance, such as sucrose or Na$^+$ and Cl$^-$ ions.

hydrogen bond A weak association formed with hydrogen in polar covalent bonds. The partially positive hydrogen is attracted to partially negative atoms in polar covalent bonds. In water, oxygen and hydrogen in different water molecules form hydrogen bonds.

hydrolysis reaction A reaction that breaks a bond by the addition of water. This is the reverse of dehydration, a reaction that joins molecules with the loss of water.

hydrophilic Literally translates as "water-loving" and describes substances that are soluble in water. These must be either polar or charged (ions).

hydrophobic Literally translates as "water-fearing" and describes nonpolar substances that are not soluble in water. Nonpolar molecules in water associate with each other and form droplets.

hydrophobic exclusion The tendency of nonpolar molecules to aggregate together when placed in water. Exclusion refers to the action of water in forcing these molecules together.

hydrostatic skeleton The skeleton of most soft-bodied invertebrates that have neither an internal nor an external skeleton. They use the relative incompressibility of the water within their bodies as a kind of skeleton.

hyperosmotic The condition in which a (hyperosmotic) solution has a higher osmotic concentration than that of a second solution. *Compare with* hypoosmotic.

hyperpolarization Above-normal negativity of a cell membrane during its resting potential.

hypersensitive response Plants respond to pathogens by selectively killing plant cells to block the spread of the pathogen.

hypertonic A solution with a higher concentration of solutes than the cell. A cell in a hypertonic solution tends to lose water by osmosis.

hypha, pl. **hyphae** A filament of a fungus or oomycete; collectively, the hyphae constitute the mycelium.

hypocotyl The region immediately below where the cotyledons are attached.

hypoosmotic The condition in which a (hypoosmotic) solution has a lower osmotic concentration than that of a second solution. *Compare with* hyperosmotic.

hypothalamus A region of the vertebrate brain just below the cerebral hemispheres, under the thalamus; a center of the autonomic nervous system, responsible for the integration and correlation of many neural and endocrine functions.

hypotonic A solution with a lower concentration of solutes than the cell. A cell in a hypotonic solution tends to take in water by osmosis.

I

icosahedron A structure consisting of 20 equilateral triangular facets; this is commonly seen in viruses and forms one kind of viral capsid.

imaginal disk One of about a dozen groups of cells set aside in the abdomen of a larval insect and committed to forming key parts of the adult insect's body.

immune response In vertebrates, a defensive reaction of the body to invasion by a foreign substance or organism. *See* antibody and B cell.

immunoglobulin An antibody molecule.

immunological tolerance Process where immune system learns to not react to self-antigens.

in vitro mutagenesis The ability to create mutations at any site in a cloned gene to examine the mutations' effects on function.

inbreeding The breeding of genetically related plants or animals; inbreeding tends to increase homozygosity.

inclusive fitness Describes the sum of the number of genes directly passed on in an individual's offspring and those genes passed on indirectly by kin (other than offspring) whose existence results from the benefit of the individual's altruism.

incomplete dominance Describes a case in which two or more alleles of a gene do not display clear dominance. The phenotype of a heterozygote is intermediate between the homozygous forms. For example, crossing red-flowered with white-flowered four o'clocks yields pink heterozygotes.

independent assortment In a dihybrid cross, describes the random assortment of alleles for each of the genes. For genes on different chromosomes this results from the random orientations of different homologous pairs during metaphase I of meiosis. For genes on the same chromosome, this occurs when the two loci are far enough apart for roughly equal numbers of odd- and even-numbered multiple crossover events.

indeterminate development A type of development in animals in which the first few embryonic cells are identical daughter cells, any one of which could develop separately into a complete organism; their fate is indeterminate.

inducer exclusion Part of the mechanism of glucose repression in *E. coli* in which the presence of glucose prevents the entry of lactose such that the *lac* operon cannot be induced.

induction (1) Production of enzymes in response to a substrate; a mechanism by which binding of an inducer to a repressor allows transcription of an operon. This is seen in catabolic operons and results in production of enzymes to degrade a compound only when it is available. (2) In embryonic development, the process by which the development of a cell is influenced by interaction with an adjacent cell.

inductive reasoning The logical application of specific observations to make a generalization. In science, inductive reasoning is used to formulate testable hypotheses.

industrial melanism Phrase used to describe the evolutionary process in which initially light-colored organisms become dark as a result of natural selection.

inflammatory response A generalized nonspecific response to infection that acts to clear an infected area of infecting microbes and dead tissue cells so that tissue repair can begin.

inhalant siphon In bivalve mollusks, the siphon through which incoming water enters the body.

inheritance of acquired characteristics Also known as Lamarckism; the theory, now discounted, that individuals genetically pass on to their offspring physical and behavioral changes developed during the individuals' own lifetime.

inhibitor A substance that binds to an enzyme and decreases its activity.

initiation factor One of several proteins involved in the formation of an initiation complex in prokaryote polypeptide synthesis.

initiator tRNA A tRNA molecule involved in the beginning of translation. In prokaryotes, the initiator tRNA is charged with *N*-formylmethionine (tRNA^{fMet}); in eukaryotes, the tRNA is charged simply with methionine.

inorganic phosphate A phosphate molecule that is not a part of an organic molecule; inorganic phosphate groups are added and removed in the formation and breakdown of ATP and in many other cellular reactions.

inositol-1,4,5-trisphosphate (IP$_3$) Second messenger produced by the cleavage of phosphatidylinositol-4,5-bisphosphate.

insertional inactivation Destruction of a gene's function by the insertion of a transposon.

instar A larval developmental stage in insects.

integrin Any of a group of cell-surface proteins involved in adhesion of cells to substrates.

Critical to migrating cells moving through the cell matrix in tissues such as connective tissue.

intercalary meristem A type of meristem that arises in stem internodes in some plants, such as corn and horsetails; responsible for elongation of the internodes.

interferon In vertebrates, a protein produced in virus-infected cells that inhibits viral multiplication.

intermembrane space The outer compartment of a mitochondrion that lies between the two membranes.

interneuron (association neuron) A nerve cell found only in the middle of the spinal cord that acts as a functional link between sensory neurons and motor neurons.

internode In plants, the region of a stem between two successive nodes.

interoceptor A receptor that senses information related to the body itself, its internal condition, and its position.

interphase The period between two mitotic or meiotic divisions in which a cell grows and its DNA replicates; includes G_1, S, and G_2 phases.

intracellular receptor A signal receptor that binds a ligand inside a cell, such as the receptors for NO, steroid hormones, vitamin D, and thyroid hormones.

intron Portion of mRNA as transcribed from eukaryotic DNA that is removed by enzymes before the mature mRNA is translated into protein. *See* exon.

inversion A reversal in order of a segment of a chromosome; also, to turn inside out, as in embryogenesis of sponges or discharge of a nematocyst.

ionizing radiation High-energy radiation that is highly mutagenic, producing free radicals that react with DNA; includes X-rays and γ-rays.

isomer One of a group of molecules identical in atomic composition but differing in structural arrangement; for example, glucose and fructose.

isosmotic The condition in which the osmotic concentrations of two solutions are equal, so that no net water movement occurs between them by osmosis.

isotonic A solution having the same concentration of solutes as the cell. A cell in an isotonic solution takes in and loses the same amount of water.

isotope Different forms of the same element with the same number of protons but different numbers of neutrons.

J

jasmonic acid An organic molecule that is part of a plant's wound response; it signals the production of a proteinase inhibitor.

K

karyotype The morphology of the chromosomes of an organism as viewed with a light microscope.

keratin A tough, fibrous protein formed in epidermal tissues and modified into skin, feathers, hair, and hard structures such as horns and nails.

key innovation A newly evolved trait in a species that allows members to use resources or

other aspects of the environment that were previously inaccessible.

kidney In vertebrates, the organ that filters the blood to remove nitrogenous wastes and regulates the balance of water and solutes in blood plasma.

kilocalorie Unit describing the amount of heat required to raise the temperature of a kilogram of water by 1°C; sometimes called a Calorie, equivalent to 1000 calories.

kinase cascade A series of protein kinases that phosphorylate each other in succession; a kinase cascade can amplify signals during the signal transduction process.

kinesis Changes in activity level in an animal that are dependent on stimulus intensity. *See* kinetic energy.

kinetic energy The energy of motion.

kinetochore Disk-shaped protein structure within the centromere to which the spindle fibers attach during mitosis or meiosis. *See* centromere.

kingdom The second highest commonly used taxonomic category.

kin selection Selection favoring relatives; an increase in the frequency of related individuals (kin) in a population, leading to an increase in the relative frequency in the population of those alleles shared by members of the kin group.

knockout mice Mice in which a known gene is inactivated ("knocked out") using recombinant DNA and ES cells.

Krebs cycle Another name for the citric acid cycle; also called the tricarboxylic acid (TCA) cycle.

L

labrum The upper lip of insects and crustaceans situated above or in front of the mandibles.

lac **operon** In *E. coli*, the operon containing genes that encode the enzymes to metabolize lactose.

lagging strand The DNA strand that must be synthesized discontinuously because of the 5′-to-3′ directionality of DNA polymerase during replication, and the antiparallel nature of DNA. Compare *leading strand*.

larva A developmental stage that is unlike the adult found in organisms that undergo metamorphosis. Embryos develop into larvae that produce the adult form by metamorphosis.

larynx The voice box; a cartilaginous organ that lies between the pharynx and trachea and is responsible for sound production in vertebrates.

lateral line system A sensory system encountered in fish, through which mechanoreceptors in a line down the side of the fish are sensitive to motion.

lateral meristems In vascular plants, the meristems that give rise to secondary tissue; the vascular cambium and cork cambium.

Law of Independent Assortment Mendel's second law of heredity, stating that genes located on nonhomologous chromosomes assort independently of one another.

Law of Segregation Mendel's first law of heredity, stating that alternative alleles for the same gene segregate from each other in production of gametes.

leading strand The DNA strand that can be synthesized continuously from the origin of replication. Compare *lagging strand*.

leaf primordium, pl. primordia A lateral outgrowth from the apical meristem that will eventually become a leaf.

lenticels Spongy areas in the cork surfaces of stem, roots, and other plant parts that allow interchange of gases between internal tissues and the atmosphere through the periderm.

leucine zipper motif A motif in regulatory proteins in which two different protein subunits associate to form a single DNA-binding site; the proteins are connected by an association between hydrophobic regions containing leucines (the "zipper").

leucoplast In plant cells, a colorless plastid in which starch grains are stored; usually found in cells not exposed to light.

leukocyte A white blood cell; a diverse array of nonhemoglobin-containing blood cells, including phagocytic macrophages and antibody-producing lymphocytes.

lichen Symbiotic association between a fungus and a photosynthetic organism such as a green alga or cyanobacterium.

ligand A signaling molecule that binds to a specific receptor protein, initiating signal transduction in cells.

light-dependent reactions In photosynthesis, the reactions in which light energy is captured and used in production of ATP and NADPH. In plants this involves the action of two linked photosystems.

light-independent reactions In photosynthesis, the reactions of the Calvin cycle in which ATP and NADPH from the light-dependent reactions are used to reduce CO_2 and produce organic compounds such as glucose. This involves the process of carbon fixation, or the conversion of inorganic carbon (CO_2) to organic carbon (ultimately carbohydrates).

lignin A highly branched polymer that makes plant cell walls more rigid; an important component of wood.

limbic system The hypothalamus, together with the network of neurons that link the hypothalamus to some areas of the cerebral cortex. Responsible for many of the most deep-seated drives and emotions of vertebrates, including pain, anger, sex, hunger, thirst, and pleasure.

linked genes Genes that are physically close together and therefore tend to segregate together; recombination occurring between linked genes can be used to produce a map of genetic distance for a chromosome.

linkage disequilibrium Association of alleles for 2 or more loci in a population that is higher than expected by chance.

lipase An enzyme that catalyzes the hydrolysis of fats.

lipid A nonpolar hydrophobic organic molecule that is insoluble in water (which is polar) but dissolves readily in nonpolar organic solvents; includes fats, oils, waxes, steroids, phospholipids, and carotenoids.

lipid bilayer The structure of a cellular membrane, in which two layers of phospholipids spontaneously align so that the hydrophilic head groups are exposed to water, while the hydrophobic fatty acid tails are pointed toward the center of the membrane.

lipopolysaccharide A lipid with a polysaccharide molecule attached; found in the outer membrane layer of gram-negative bacteria; the outer membrane layer protects the cell wall from antibiotic attack.

locus The position on a chromosome where a gene is located.

long interspersed element (LINE) Any of a type of large transposable element found in humans and other primates that contains all the biochemical machinery needed for transposition.

long terminal repeat (LTR) A particular type of retrotransposon that has repeated elements at its ends. These elements make up 8% of the human genome.

loop of Henle In the kidney of birds and mammals, a hairpin-shaped portion of the renal tubule in which water and salt are reabsorbed from the glomerular filtrate by diffusion.

lophophore A horseshoe-shaped crown of ciliated tentacles that surrounds the mouth of certain spiralian animals; seen in the phyla Brachiopoda and Bryozoa.

lumen A term for any bounded opening; for example, the cisternal space of the endoplasmic reticulum of eukaryotic cells, the passage through which blood flows inside a blood vessel, and the passage through which material moves inside the intestine during digestion.

luteal phase The second phase of the female reproductive cycle, during which the mature eggs are released into the fallopian tubes, a process called ovulation.

lymph In animals, a colorless fluid derived from blood by filtration through capillary walls in the tissues.

lymphatic system In animals, an open vascular system that reclaims water that has entered interstitial regions from the bloodstream (lymph); includes the lymph nodes, spleen, thymus, and tonsils.

lymphocyte A type of white blood cell. Lymphocytes are responsible for the immune response; there are two principal classes: B cells and T cells.

lymphokine A regulatory molecule that is secreted by lymphocytes. In the immune response, lymphokines secreted by helper T cells unleash the cell-mediated immune response.

lysis Disintegration of a cell by rupture of its plasma membrane.

lysogenic cycle A viral cycle in which the viral DNA becomes integrated into the host chromosome and is replicated during cell reproduction. Results in vertical rather than horizontal transmission.

lysosome A membrane-bounded vesicle containing digestive enzymes that is produced by the Golgi apparatus in eukaryotic cells.

lytic cycle A viral cycle in which the host cell is killed (lysed) by the virus after viral duplication to release viral particles.

M

macroevolution The creation of new species and the extinction of old ones.

macromolecule An extremely large biological molecule; refers specifically to proteins, nucleic acids, polysaccharides, lipids, and complexes of these.

macronutrients Inorganic chemical elements required in large amounts for plant growth, such as nitrogen, potassium, calcium, phosphorus, magnesium, and sulfur.

macrophage A large phagocytic cell that is able to engulf and digest cellular debris and invading bacteria.

madreporite A sievelike plate on the surface of echinoderms through which water enters the water–vascular system.

MADS **box gene** Any of a family of genes identified by possessing shared motifs that are the predominant homeotic genes of plants; a small number of *MADS* box genes are also found in animals.

major groove The larger of the two grooves in a DNA helix, where the paired nucleotides' hydrogen bonds are accessible; regulatory proteins can recognize and bind to regions in the major groove.

major histocompatibility complex (MHC) A set of protein cell-surface markers anchored in the plasma membrane, which the immune system uses to identify "self." All the cells of a given individual have the same "self" marker, called an MHC protein.

Malpighian tubules Blind tubules opening into the hindgut of terrestrial arthropods; they function as excretory organs.

mandibles In crustaceans, insects, and myriapods, the appendages immediately posterior to the antennae; used to seize, hold, bite, or chew food.

mantle The soft, outermost layer of the body wall in mollusks; the mantle secretes the shell.

map unit Each 1% of recombination frequency between two genetic loci; the unit is termed a centimorgan (cM) or simply a map unit (m.u.).

marsupial A mammal in which the young are born early in their development, sometimes as soon as eight days after fertilization, and are retained in a pouch.

mass extinction A relatively sudden, sharp decline in the number of species; for example, the extinction at the end of the Cretaceous period in which the dinosaurs and a variety of other organisms disappeared.

mass flow hypothesis The overall process by which materials move in the phloem of plants.

mast cells Leukocytes with granules containing molecules that initiate inflammation.

maternal inheritance A mode of uniparental inheritance from the female parent; for example, in humans mitochondria and their genomes are inherited from the mother.

matrix In mitochondria, the solution in the interior space surrounded by the cristae that contains the enzymes and other molecules involved in oxidative respiration; more generally, that part of a tissue within which an organ or process is embedded.

medusa A free-floating, often umbrella-shaped body form found in cnidarian animals, such as jellyfish.

megapascal (MPa) A unit of measure used for pressure in water potential.

megaphyll In plants, a leaf that has several to many veins connecting it to the vascular cylinder of the stem; most plants have megaphylls.

mesoglea A layer of gelatinous material found between the epidermis and gastrodermis of eumetazoans; it contains the muscles in most of these animals.

mesohyl A gelatinous, protein-rich matrix found between the choanocyte layer and the epithelial layer of the body of a sponge; various types of amoeboid cells may occur in the mesohyl.

metacercaria An encysted form of a larval liver fluke, found in muscle tissue of an infected animal; if the muscle is eaten, cysts dissolve in the digestive tract, releasing the flukes into the body of the new host.

methylation The addition of a methyl group to bases (primarily cytosine) in DNA. Cytosine methylation is correlated with DNA that is not expressed.

meiosis I The first round of cell division in meiosis; it is referred to as a "reduction division" because homologous chromosomes separate, and the daughter cells have only the haploid number of chromosomes.

meiosis II The second round of division in meiosis, during which the two haploid cells from meiosis I undergo a mitosis-like division without DNA replication to produce four haploid daughter cells.

membrane receptor A signal receptor present as an integral protein in the cell membrane, such as GPCRs, chemically gated ion channels in neurons, and RTKs.

Mendelian ratio The characteristic dominant-to-recessive phenotypic ratios that Mendel observed in his genetics experiments. For example, the F_2 generation in a monohybrid cross shows a ratio of 3:1; the F_2 generation in a dihybrid cross shows a ratio of 9:3:3:1.

menstruation Periodic sloughing off of the blood-enriched lining of the uterus when pregnancy does not occur.

meristem Undifferentiated plant tissue from which new cells arise.

meroblastic cleavage A type of cleavage in the eggs of reptiles, birds, and some fish. Occurs only on the blastodisc.

mesoderm One of the three embryonic germ layers that form in the gastrula; gives rise to muscle, bone and other connective tissue, the peritoneum, the circulatory system, and most of the excretory and reproductive systems.

mesophyll The photosynthetic parenchyma of a leaf, located within the epidermis.

messenger RNA (mRNA) The RNA transcribed from structural genes; RNA molecules complementary to a portion of one strand of DNA, which are translated by the ribosomes to form protein.

metabolism The sum of all chemical processes occurring within a living cell or organism.

metamorphosis Process in which a marked change in form takes place during postembryonic development as, for example, from tadpole to frog.

metaphase The stage of mitosis or meiosis during which microtubules become organized into a spindle and the chromosomes come to lie in the spindle's equatorial plane.

metastasis The process by which cancer cells move from their point of origin to other locations in the body; also, a population of cancer cells in a secondary location, the result of movement from the primary tumor.

methanogens Obligate, anaerobic archaebacteria that produce methane.

microarray DNA sequences are placed on a microscope slide or chip with a robot. The microarray can then be probed with RNA from specific tissues to identify expressed DNA.

microbody A cellular organelle bounded by a single membrane and containing a variety of enzymes; generally derived from endoplasmic reticulum; includes peroxisomes and glyoxysomes.

microevolution Refers to the evolutionary process itself. Evolution within a species. Also called adaptation.

micronutrient A mineral required in only minute amounts for plant growth, such as iron, chlorine, copper, manganese, zinc, molybdenum, and boron.

microphyll In plants, a leaf that has only one vein connecting it to the vascular cylinder of the stem; the club mosses in particular have microphylls.

micropyle In the ovules of seed plants, an opening in the integuments through which the pollen tube usually enters.

micro-RNA (miRNA) A class of RNAs that are very short and only recently could be detected. *See also* small interfering RNAs (siRNAs).

microtubule In eukaryotic cells, a long, hollow protein cylinder, composed of the protein tubulin; these influence cell shape, move the chromosomes in cell division, and provide the functional internal structure of cilia and flagella.

microvillus Cytoplasmic projection from epithelial cells; microvilli greatly increase the surface area of the small intestine.

middle lamella The layer of intercellular material, rich in pectic compounds, that cements together the primary walls of adjacent plant cells.

mimicry The resemblance in form, color, or behavior of certain organisms (mimics) to other more powerful or more protected ones (models).

miracidium The ciliated first-stage larva inside the egg of the liver fluke; eggs are passed in feces, and if they reach water they may be eaten by a host snail in which they continue their life cycle.

missense mutation A base substitution mutation that results in the alteration of a single amino acid.

mitochondrion The organelle called the powerhouse of the cell. Consists of an outer membrane, an elaborate inner membrane that supports electron transport and chemiosmotic synthesis of ATP, and a soluble matrix containing Krebs cycle enzymes.

mitogen-activated protein (MAP) kinase Any of a class of protein kinases that activate transcription factors to alter gene expression. A mitogen is any molecule that stimulates cell division. MAP kinases are activated by kinase cascades.

mitosis Somatic cell division; nuclear division in which the duplicated chromosomes separate to form two genetically identical daughter nuclei.

molar concentration Concentration expressed as moles of a substance in 1 L of pure water.

mole The weight of a substance in grams that corresponds to the atomic masses of all the component atoms in a molecule of that substance. One mole of a compound always contains 6.023×10^{23} molecules.

molecular clock method In evolutionary theory, the method in which the rate of evolution of a molecule is constant through time.

molecular cloning The isolation and amplification of a specific sequence of DNA.

monocot Short for monocotyledon; flowering plant in which the embryos have only one cotyledon, the floral parts are generally in threes, and the leaves typically are parallel-veined.

monocyte A type of leukocyte that becomes a phagocytic cell (macrophage) after moving into tissues.

monoecious A plant in which the staminate and pistillate flowers are separate, but borne on the same individual.

monomer The smallest chemical subunit of a polymer. The monosaccharide α-glucose is the monomer found in plant starch, a polysaccharide.

monophyletic In phylogenetic classification, a group that includes the most recent common ancestor of the group and all its descendants. A clade is a monophyletic group.

monosaccharide A simple sugar that cannot be decomposed into smaller sugar molecules.

monosomic Describes the condition in which a chromosome has been lost due to nondisjunction during meiosis, producing a diploid embryo with only one of these autosomes.

monotreme An egg-laying mammal.

morphogen A signal molecule produced by an embryonic organizer region that informs surrounding cells of their distance from the organizer, thus determining relative positions of cells during development.

morphogenesis The development of an organism's body form, namely its organs and anatomical features; it may involve apoptosis as well as cell division, differentiation, and changes in cell shape.

morphology The form and structure of an organism.

morula Solid ball of cells in the early stage of embryonic development.

mosaic development A pattern of embryonic development in which initial cells produced by cleavage divisions contain different developmental signals (determinants) from the egg, setting the individual cells on different developmental paths.

motif A substructure in proteins that confers function and can be found in multiple proteins. One example is the helix-turn-helix motif found in a number of proteins that is used to bind to DNA.

motor (efferent) neuron Neuron that transmits nerve impulses from the central nervous system to an effector, which is typically a muscle or gland.

M phase The phase of cell division during which chromosomes are separated. The spindle assembles, binds to the chromosomes, and moves the sister chromatids apart.

M phase-promoting factor (MPF) A Cdk enzyme active at the G_2/M checkpoint.

Müllerian mimicry A phenomenon in which two or more unrelated but protected species resemble one another, thus achieving a kind of group defense.

multidrug-resistant (MDR) strain Any bacterial strain that has become resistant to more than one antibiotic drug; MDR *Staphylococcus* strains, for example, are responsible for many infection deaths.

multienzyme complex An assembly consisting of several enzymes catalyzing different steps in a sequence of reactions. Close proximity of these related enzymes speeds the overall process, making it more efficient.

multigene family A collection of related genes on a single chromosome or on different chromosomes.

muscle fiber A long, cylindrical, multinucleated cell containing numerous myofibrils, which is capable of contraction when stimulated.

mutagen An agent that induces changes in DNA (mutations); includes physical agents that damage DNA and chemicals that alter DNA bases.

mutation A permanent change in a cell's DNA; includes changes in nucleotide sequence, alteration of gene position, gene loss or duplication, and insertion of foreign sequences.

mutualism A symbiotic association in which two (or more) organisms live together, and both members benefit.

mycelium, pl. **mycelia** In fungi, a mass of hyphae.

mycorrhiza, pl. **mycorrhizae** A symbiotic association between fungi and the roots of a plant.

myelin sheath A fatty layer surrounding the long axons of motor neurons in the peripheral nervous system of vertebrates.

myofilament A contractile microfilament, composed largely of actin and myosin, within muscle.

myosin One of the two protein components of microfilaments (the other is actin); a principal component of vertebrate muscle.

N

natural killer cell A cell that does not kill invading microbes, but rather, the cells infected by them.

natural selection The differential reproduction of genotypes; caused by factors in the environment; leads to evolutionary change.

nauplius A larval form characteristic of crustaceans.

negative control A type of control at the level of DNA transcription initiation in which the frequency of initiation is decreased; repressor proteins mediate negative control.

negative feedback A homeostatic control mechanism whereby an increase in some substance or activity inhibits the process leading to the increase; also known as feedback inhibition.

nematocyst A harpoonlike structure found in the cnidocytes of animals in the phylum Cnidaria, which includes the jellyfish among other groups; the nematocyst, when released, stings and helps capture prey.

nephridium, pl. **nephridia** In invertebrates, a tubular excretory structure.

nephrid organ A filtration system of many freshwater invertebrates in which water and waste pass from the body across the membrane into a collecting organ, from which they are expelled to the outside through a pore.

nephron Functional unit of the vertebrate kidney; one of numerous tubules involved in filtration and selective reabsorption of blood; each nephron consists of a Bowman's capsule, an enclosed glomerulus, and a long attached tubule; in humans, called a renal tubule.

nephrostome The funnel-shaped opening that leads to the nephridium, which is the excretory organ of mollusks.

nerve A group or bundle of nerve fibers (axons) with accompanying neurological cells, held together by connective tissue; located in the peripheral nervous system.

nerve cord One of the distinguishing features of chordates, running lengthwise just beneath the embryo's dorsal surface; in vertebrates, differentiates into the brain and spinal cord.

neural crest A special strip of cells that develops just before the neural groove closes over to form the neural tube in embryonic development.

neural groove The long groove formed along the long axis of the embryo by a layer of ectodermal cells.

neural tube The dorsal tube, formed from the neural plate, that differentiates into the brain and spinal cord.

neuroglia Nonconducting nerve cells that are intimately associated with neurons and appear to provide nutritional support.

neuromuscular junction The structure formed when the tips of axons contact (innervate) a muscle fiber.

neuron A nerve cell specialized for signal transmission; includes cell body, dendrites, and axon.

neurotransmitter A chemical released at the axon terminal of a neuron that travels across the synaptic cleft, binds a specific receptor on the far side, and depending on the nature of the receptor, depolarizes or hyperpolarizes a second neuron or a muscle or gland cell.

neurulation A process in early embryonic development by which a dorsal band of ectoderm thickens and rolls into the neural tube.

neutrophil An abundant type of granulocyte capable of engulfing microorganisms and other foreign particles; neutrophils comprise about 50–70% of the total number of white blood cells.

niche The role played by a particular species in its environment.

nicotinamide adenine dinucleotide (NAD) A molecule that becomes reduced (to NADH) as it carries high-energy electrons from oxidized molecules and delivers them to ATP-producing pathways in the cell.

NADH dehydrogenase An enzyme located on the inner mitochondrial membrane that catalyzes the oxidation by NAD^+ of pyruvate to acetyl-CoA. This reaction links glycolysis and the Krebs cycle.

nitrification The oxidization of ammonia or nitrite to produce nitrate, the form of nitrogen taken up by plants; some bacteria are capable of nitrification.

nociceptor A naked dendrite that acts as a receptor in response to a pain stimulus.

nocturnal Active primarily at night.

node The part of a plant stem where one or more leaves are attached. *See* internode.

node of Ranvier A gap formed at the point where two Schwann cells meet and where the axon is in direct contact with the surrounding intercellular fluid.

nodule In plants, a specialized tissue that surrounds and houses beneficial bacteria, such as root nodules of legumes that contain nitrogen-fixing bacteria.

nonassociative learning A learned behavior that does not require an animal to form an association between two stimuli, or between a stimulus and a response.

noncompetitive inhibitor An inhibitor that binds to a location other than the active site of an enzyme, changing the enzyme's shape so that it cannot bind the substrate.

noncyclic photophosphorylation The set of light-dependent reactions of the two plant photosystems, in which excited electrons are shuttled between the two photosystems, producing a proton gradient that is used for the chemiosmotic synthesis of ATP. The electrons are used to reduce NADP to NADPH. Lost electrons are replaced by the oxidation of water producing O_2.

nondisjunction The failure of homologues or sister chromatids to separate during mitosis or meiosis, resulting in an aneuploid cell or gamete.

nonextreme archaea Archaean groups that are not extremophiles, living in more moderate environments on Earth today.

nonpolar Said of a covalent bond that involves equal sharing of electrons. Can also refer to a compound held together by nonpolar covalent bonds.

nonsense codon One of three codons (UAA, UAG, and UGA) that are not recognized by tRNAs, thus serving as "stop" signals in the mRNA message and terminating translation.

nonsense mutation A base substitution in which a codon is changed into a stop codon. The protein is truncated because of premature termination.

Northern blot A blotting technique used to identify a specific mRNA sequence in a complex mixture. *See* Southern blot.

notochord In chordates, a dorsal rod of cartilage that runs the length of the body and forms the primitive axial skeleton in the embryos of all chordates.

nucellus Tissue composing the chief pair of young ovules, in which the embryo sac develops; equivalent to a megasporangium.

nuclear envelope The bounding structure of the eukaryotic nucleus. Composed of two phospholipid bilayers with the outer one connected to the endoplasmic reticulum.

nuclear pore One of a multitude of tiny but complex openings in the nuclear envelope that allow selective passage of proteins and nucleic acids into and out of the nucleus.

nuclear receptor Intracellular receptors are found in both the cytoplasm and the nucleus. The site of action of the hormone–receptor complex is in the nucleus where they modify gene expression.

nucleic acid A nucleotide polymer; chief types are deoxyribonucleic acid (DNA), which is double-stranded, and ribonucleic acid (RNA), which is typically single-stranded.

nucleoid The area of a prokaryotic cell, usually near the center, that contains the genome in the form of DNA compacted with protein.

nucleolus In eukaryotes, the site of rRNA synthesis; a spherical body composed chiefly of rRNA in the process of being transcribed from multiple copies of rRNA genes.

nucleosome A complex consisting of a DNA duplex wound around a core of eight histone proteins.

nucleotide A single unit of nucleic acid, composed of a phosphate, a five-carbon sugar (either ribose or deoxyribose), and a purine or a pyrimidine.

nucleus In atoms, the central core, containing positively charged protons and (in all but hydrogen) electrically neutral neutrons; in eukaryotic cells, the membranous organelle that houses the chromosomal DNA; in the central nervous system, a cluster of nerve cell bodies.

nutritional mutation A mutation affecting a synthetic pathway for a vital compound, such as an amino acid or vitamin; microorganisms with a nutritional mutation must be grown on medium that supplies the missing nutrient.

O

ocellus, pl. **ocelli** A simple light receptor common among invertebrates.

octet rule Rule to describe patterns of chemical bonding in main group elements that require a total of eight electrons to complete their outer electron shell.

Okazaki fragment A short segment of DNA produced by discontinuous replication elongating in the 5′-to-3′ direction away from the replication.

olfaction The function of smelling.

ommatidium, pl. **ommatidia** The visual unit in the compound eye of arthropods; contains light-sensitive cells and a lens able to form an image.

oncogene A mutant form of a growth-regulating gene that is inappropriately "on," causing unrestrained cell growth and division.

oocyst The zygote in a sporozoan life cycle. It is surrounded by a tough cyst to prevent dehydration or other damage.

open circulatory system A circulatory system in which the blood flows into sinuses in which it mixes with body fluid and then reenters the vessels in another location.

open reading frame (ORF) A region of DNA that encodes a sequence of amino acids with no stop codons in the reading frame.

operant conditioning A learning mechanism in which the reward follows only after the correct behavioral response.

operator A regulatory site on DNA to which a repressor can bind to prevent or decrease initiation of transcription.

operculum A flat, bony, external protective covering over the gill chamber in fish.

operon A cluster of adjacent structural genes transcribed as a unit into a single mRNA molecule.

opisthosoma The posterior portion of the body of an arachnid.

oral surface The surface on which the mouth is found; used as a reference when describing the body structure of echinoderms because of their adult radial symmetry.

orbital A region around the nucleus of an atom with a high probability of containing an electron. The position of electrons can only be described by these probability distributions.

order A category of classification above the level of family and below that of class.

organ A body structure composed of several different tissues grouped in a structural and functional unit.

organelle Specialized part of a cell; literally, a small cytoplasmic organ.

orthologues Genes that reflect the conservation of a single gene found in an ancestor.

oscillating selection The situation in which selection alternately favors one phenotype at one time, and a different phenotype at a another time, for example, during drought conditions versus during wet conditions.

osculum A specialized, larger pore in sponges through which filtered water is forced to the outside of the body.

osmoconformer An animal that maintains the osmotic concentration of its body fluids at about the same level as that of the medium in which it is living.

osmosis The diffusion of water across a selectively permeable membrane (a membrane that permits the free passage of water but prevents or retards the passage of a solute); in the absence of differences in pressure or volume, the net movement of water is from the side containing a lower concentration of solute to the side containing a higher concentration.

osmotic concentration The property of a solution that takes into account all dissolved solutes in the solution; if two solutions with different osmotic concentrations are separated by a water-permeable membrane, water will move from the solution with lower osmotic concentration to the solution with higher osmotic concentration.

osmotic pressure The potential pressure developed by a solution separated from pure water by a differentially permeable membrane. The higher the solute concentration, the greater the osmotic potential of the solution; also called *osmotic potential.*

ossicle Any of a number of movable or fixed calcium-rich plates that collectively make up the endoskeleton of echinoderms.

osteoblast A bone-forming cell.

osteocyte A mature osteoblast.

outcrossing Breeding with individuals other than oneself or one's close relatives.

ovary (1) In animals, the organ in which eggs are produced. (2) In flowering plants, the enlarged basal portion of a carpel that contains the ovule(s); the ovary matures to become the fruit.

oviduct In vertebrates, the passageway through which ova (eggs) travel from the ovary to the uterus.

oviparity Refers to a type of reproduction in which the eggs are developed after leaving the body of the mother, as in reptiles.

ovoviviparity Refers to a type of reproduction in which young hatch from eggs that are retained in the mother's uterus.

ovulation In animals, the release of an egg or eggs from the ovary.

ovum, pl. ova The egg cell; female gamete.

oxidation Loss of an electron by an atom or molecule; in metabolism, often associated with a gain of oxygen or a loss of hydrogen.

oxidation–reduction reaction A type of paired reaction in living systems in which electrons lost from one atom (oxidation) are gained by another atom (reduction). Termed a *redox reaction* for short.

oxidative phosphorylation Synthesis of ATP by ATP synthase using energy from a proton gradient. The proton gradient is generated by electron transport, which requires oxygen.

oxygen debt The amount of oxygen required to convert the lactic acid generated in the muscles during exercise back into glucose.

oxytocin A hormone of the posterior pituitary gland that affects uterine contractions during childbirth and stimulates lactation.

ozone O_3, a stratospheric layer of the Earth's atmosphere responsible for filtering out ultraviolet radiation supplied by the Sun.

P

***p53* gene** The gene that produces the p53 protein that monitors DNA integrity and halts cell division if DNA damage is detected. Many types of cancer are associated with a damaged or absent *p53* gene.

pacemaker A patch of excitatory tissue in the vertebrate heart that initiates the heartbeat.

pair-rule gene Any of certain genes in *Drosophila* development controlled by the gap genes that are expressed in stripes that subdivide the embryo in the process of segmentation.

paleopolyploid An ancient polyploid organism used in analysis of polyploidy events in the study of a species' genome evolution.

palisade parenchyma In plant leaves, the columnar, chloroplast-containing parenchyma cells of the mesophyll. Also called *palisade cells.*

panspermia The hypothesis that meteors or cosmic dust may have brought significant amounts of complex organic molecules to Earth, kicking off the evolution of life.

papilla A small projection of tissue.

paracrine A type of chemical signaling between cells in which the effects are local and short-lived.

paralogues Two genes within an organism that arose from the duplication of one gene in an ancestor.

paraphyletic In phylogenetic classification, a group that includes the most recent common ancestor of the group, but not all its descendants.

parapodia One of the paired lateral processes on each side of most segments in polychaete annelids.

parasexuality In certain fungi, the fusion and segregation of heterokaryotic haploid nuclei to produce recombinant nuclei.

parasitism A living arrangement in which an organism lives on or in an organism of a different species and derives nutrients from it.

parenchyma cell The most common type of plant cell; characterized by large vacuoles, thin walls, and functional nuclei.

parthenogenesis The development of an egg without fertilization, as in aphids, bees, ants, and some lizards.

partial diploid (merodiploid) Describes an *E. coli* cell that carries an F′ plasmid with host genes. This makes the cell diploid for the genes carried by the F′ plasmid.

partial pressure The components of each individual gas—such as nitrogen, oxygen, and carbon dioxide—that together constitute the total air pressure.

passive transport The movement of substances across a cell's membrane without the expenditure of energy.

pedigree A consistent graphic representation of matings and offspring over multiple generations for a particular genetic trait, such as albinism or hemophilia.

pedipalps A pair of specialized appendages found in arachnids; in male spiders, these are specialized as copulatory organs, whereas in scorpions they are large pincers.

pelagic Free-swimming, usually in open water.

pellicle A tough, flexible covering in ciliates and euglenoids.

pentaradial symmetry The five-part radial symmetry characteristic of adult echinoderms.

peptide bond The type of bond that links amino acids together in proteins through a dehydration reaction.

peptidoglycan A component of the cell wall of bacteria, consisting of carbohydrate polymers linked by protein cross-bridges.

peptidyl transferase In translation, the enzyme responsible for catalyzing the formation of a peptide bond between each new amino acid and the previous amino acid in a growing polypeptide chain.

perianth In flowering plants, the petals and sepals taken together.

pericycle In vascular plants, one or more cell layers surrounding the vascular tissues of the root, bounded externally by the endodermis and internally by the phloem.

periderm Outer protective tissue in vascular plants that is produced by the cork cambium and functionally replaces epidermis when it is destroyed during secondary growth; the periderm includes the cork, cork cambium, and phelloderm.

peristalsis In animals, a series of alternating contracting and relaxing muscle movements along the length of a tube such as the oviduct or alimentary canal that tend to force material such as an egg cell or food through the tube.

peroxisome A microbody that plays an important role in the breakdown of highly oxidative hydrogen peroxide by catalase.

petal A flower part, usually conspicuously colored; one of the units of the corolla.

petiole The stalk of a leaf.

phage conversion The phenomenon by which DNA from a virus, incorporated into a host cell's genome, alters the host cell's function in a significant way; for example, the conversion of *Vibrio cholerae* bacteria into a pathogenic form that releases cholera toxin.

phage lambda (λ) A well-known bacteriophage that has been widely used in genetic studies and is often a vector for DNA libraries.

phagocyte Any cell that engulfs and devours microorganisms or other particles.

phagocytosis Endocytosis of a solid particle; the plasma membrane folds inward around the particle (which may be another cell) and engulfs it to form a vacuole.

pharyngeal pouches In chordates, embryonic regions that become pharyngeal slits in aquatic and marine chordates and vertebrates, but do not develop openings to the outside in terrestrial vertebrates.

pharyngeal slits One of the distinguishing features of chordates; a group of openings on each side of the anterior region that form a passageway from the pharynx and esophagus to the external environment.

pharynx A muscular structure lying posterior to the mouth in many animals; aids in propelling food into the digestive tract.

phenotype The realized expression of the genotype; the physical appearance or functional expression of a trait.

pheromone Chemical substance released by one organism that influences the behavior or physiological processes of another organism of the same species. Pheromones serve as sex attractants, as trail markers, and as alarm signals.

phloem In vascular plants, a food-conducting tissue basically composed of sieve elements, various kinds of parenchyma cells, fibers, and sclereids.

phoronid Any of a group of lophophorate invertebrates, now classified in the phylum Brachiopoda, that burrows into soft underwater substrates and secretes a chitinous tube in which it lives out its life; it extends its lophophore tentacles to feed on drifting food particles.

phosphatase Any of a number of enzymes that removes a phosphate group from a protein, reversing the action of a kinase.

phosphodiester bond The linkage between two sugars in the backbone of a nucleic acid molecule; the phosphate group connects the pentose sugars through a pair of ester bonds.

phospholipid Similar in structure to a fat, but having only two fatty acids attached to the glycerol backbone, with the third space linked to a phosphorylated molecule; contains a polar hydrophilic "head" end (phosphate group) and a nonpolar hydrophobic "tail" end (fatty acids).

phospholipid bilayer The main component of cell membranes; phospholipids naturally associate in a bilayer with hydrophobic fatty acids oriented to the inside and hydrophilic phosphate groups facing outward on both sides.

phosphorylation Chemical reaction resulting in the addition of a phosphate group to an organic molecule. Phosphorylation of ADP yields ATP. Many proteins are also activated or inactivated by phosphorylation.

photic zone The area in an aquatic habitat that receives sufficient light for photosynthesis to occur and net primary productivity to be positive.

photoelectric effect The ability of a beam of light to excite electrons, creating an electrical current.

photon A particle of light having a discrete amount of energy. The wave concept of light explains the different colors of the spectrum, whereas the particle concept of light explains the energy transfers during photosynthesis.

photoperiodism The tendency of biological reactions to respond to the duration and timing of day and night; a mechanism for measuring seasonal time.

photoreceptor A light-sensitive sensory cell.

photorespiration Action of the enzyme rubisco, which catalyzes the oxidization of RuBP, releasing CO_2; this reverses carbon fixation and can reduce the yield of photosynthesis.

photosystem An organized complex of chlorophyll, other pigments, and proteins that traps light energy as excited electrons. Plants have two linked photosystems in the thylakoid membrane of chloroplasts. Photosystem II passes an excited electron through an electron transport chain to photosystem I to replace an excited electron passed to NADPH. The electron lost from photosystem II is replaced by the oxidation of water.

phototropism In plants, a growth response to a light stimulus.

pH scale A scale used to measure acidity and basicity. Defined as the negative log of H^+ concentration. Ranges from 0 to 14. A value of 7 is neutral; below 7 is acidic and above 7 is basic.

phycobiloprotein A type of accessory pigment found in cyanobacteria and some algae. Complexes of phycobiloprotein are able to absorb light energy in the green range.

phycologist A scientist who studies algae.

phyllotaxy In plants, a spiral pattern of leaf arrangement on a stem in which sequential leaves are at a 137.5° angle to one another, an angle related to the golden mean.

phylogenetic species concept (PSC) The concept that defines species on the basis of their phylogenetic relationships.

phylogenetic tree A pattern of descent generated by analysis of similarities and differences among organisms. Modern gene-sequencing techniques have produced phylogenetic trees showing the evolutionary history of individual genes.

phylogeny The evolutionary history of an organism, including which species are closely related and in what order related species evolved; often represented in the form of an evolutionary tree.

phylum, pl. phyla A major category, between kingdom and class, of taxonomic classifications.

physical map A map of the DNA sequence of a chromosome or genome based on actual landmarks within the DNA.

phytochrome A plant pigment that is associated with the absorption of light; photoreceptor for red to far-red light.

phytoestrogen One of a number of secondary metabolites in some plants that are structurally and functionally similar to the animal hormone estrogen.

phytoremediation The process that uses plants to remove contamination from soil or water.

pigment A molecule that absorbs light.

pilus, pl. pili Extensions of a bacterial cell enabling it to transfer genetic materials from one individual to another or to adhere to substrates.

pinocytosis The process of fluid uptake by endocytosis in a cell.

pistil Central organ of flowers, typically consisting of ovary, style, and stigma; a pistil may consist of one or more fused carpels and is more technically and better known as the gynoecium.

pith The ground tissue occupying the center of the stem or root within the vascular cylinder.

pituitary gland Endocrine gland at the base of the hypothalamus composed of anterior and posterior lobes. Pituitary hormones affect a wide variety of processes in vertebrates.

placenta, pl. placentae (1) In flowering plants, the part of the ovary wall to which the ovules or seeds are attached. (2) In mammals, a tissue formed in part from the inner lining of the uterus and in part from other membranes, through which the embryo (later the fetus) is nourished while in the uterus and through which wastes are carried away.

plankton Free-floating, mostly microscopic, aquatic organisms.

plant receptor kinase Any of a group of plant membrane receptors that, when activated by binding ligand, have kinase enzymatic activity. These receptors phosphorylate serine or threonine, unlike RTKs in animals that phosphorylate tyrosine.

planula A ciliated, free-swimming larva produced by the medusae of cnidarian animals.

plasma The fluid of vertebrate blood; contains dissolved salts, metabolic wastes, hormones, and a variety of proteins, including antibodies and albumin; blood minus the blood cells.

plasma cell An antibody-producing cell resulting from the multiplication and differentiation of a B lymphocyte that has interacted with an antigen.

plasma membrane The membrane surrounding the cytoplasm of a cell; consists of a single phospholipid bilayer with embedded proteins.

plasmid A small fragment of extrachromosomal DNA, usually circular, that replicates independently of the main chromosome, although it may have been derived from it.

plasmodesmata In plants, cytoplasmic connections between adjacent cells.

plasmodium Stage in the life cycle of myxomycetes (plasmodial slime molds); a multinucleate mass of protoplasm surrounded by a membrane.

plasmolysis The shrinking of a plant cell in a hypertonic solution such that it pulls away from the cell wall.

plastid An organelle in the cells of photosynthetic eukaryotes that is the site of photosynthesis and, in plants and green algae, of starch storage.

platelet In mammals, a fragment of a white blood cell that circulates in the blood and functions in the formation of blood clots at sites of injury.

pleiotropy Condition in which an individual allele has more than one effect on production of the phenotype.

plesiomorphy In cladistics, another term for an ancestral character state.

plumule The epicotyl of a plant with its two young leaves.

point mutation An alteration of one nucleotide in a chromosomal DNA molecule.

polar body Minute, nonfunctioning cell produced during the meiotic divisions leading to gamete formation in vertebrates.

polar covalent bond A covalent bond in which electrons are shared unequally due to differences in electronegativity of the atoms involved. One atom has a partial negative charge and the other a partial positive charge, even though the molecule is electrically neutral overall.

polarity (1) Refers to unequal charge distribution in a molecule such as water, which has a positive region and a negative region although it is neutral overall. (2) Refers to axial differences in a developing embryo that result in anterior–posterior and dorsal–ventral axes in a bilaterally symmetrical animal.

polarize In cladistics, to determine whether character states are ancestral or derived.

pollen tube A tube formed after germination of the pollen grain; carries the male gametes into the ovule.

pollination The transfer of pollen from an anther to a stigma.

polyandry The condition in which a female mates with more than one male.

polyclonal antibody An antibody response in which an antigen elicits many different antibodies, each fitting a different portion of the antigen surface.

polygenic inheritance Describes a mode of inheritance in which more than one gene affects a trait, such as height in human beings; polygenic inheritance may produce a continuous range of phenotypic values, rather than discrete either–or values.

polygyny A mating choice in which a male mates with more than one female.

polymer A molecule composed of many similar or identical molecular subunits; starch is a polymer of glucose.

polymerase chain reaction (PCR) A process by which DNA polymerase is used to copy a sequence of interest repeatedly, making millions of copies of the same DNA.

polymorphism The presence in a population of more than one allele of a gene at a frequency greater than that of newly arising mutations.

polyp A typically sessile, cylindrical body form found in cnidarian animals, such as hydras.

polypeptide A molecule consisting of many joined amino acids; not usually as complex as a protein.

polyphyletic In phylogenetic classification, a group that does not include the most recent common ancestor of all members of the group.

polyploidy Condition in which one or more entire sets of chromosomes is added to the diploid genome.

polysaccharide A carbohydrate composed of many monosaccharide sugar subunits linked together in a long chain; examples are glycogen, starch, and cellulose.

polyunsaturated fat A fat molecule having at least two double bonds between adjacent carbons in one or more of the fatty acid chains.

population Any group of individuals, usually of a single species, occupying a given area at the same time.

population genetics The study of the properties of genes in populations.

positive control A type of control at the level of DNA transcription initiation in which the frequency of initiation is increased; activator proteins mediate positive control.

posttranscriptional control A mechanism of control over gene expression that operates after the transcription of mRNA is complete.

postzygotic isolating mechanism A type of reproductive isolation in which zygotes are produced but are unable to develop into reproducing adults; these mechanisms may range from inviability of zygotes or embryos to adults that are sterile.

potential energy Energy that is not being used, but could be; energy in a potentially usable form; often called "energy of position."

precapillary sphincter A ring of muscle that guards each capillary loop and that, when closed, blocks flow through the capillary.

pre-mRNA splicing In eukaryotes, the process by which introns are removed from the primary transcript to produce mature mRNA; pre-mRNA splicing occurs in the nucleus.

pressure potential In plants, the turgor pressure resulting from pressure against the cell wall.

prezygotic isolating mechanism A type of reproductive isolation in which the formation of a zygote is prevented; these mechanisms may range from physical separation in different habitats to gametic in which gametes are incapable of fusing.

primary endosperm nucleus In flowering plants, the result of the fusion of a sperm nucleus and the (usually) two polar nuclei.

primary growth In vascular plants, growth originating in the apical meristems of shoots and roots; results in an increase in length.

primary immune response The first response of an immune system to a foreign antigen.

If the system is challenged again with the same antigen, the memory cells created during the primary response will respond more quickly.

primary induction Inductions between the three primary tissue types: mesoderm and endoderm.

primary meristem Any of the three meristems produced by the apical meristem; primary meristems give rise to the dermal, vascular, and ground tissues.

primary nondisjunction Failure of chromosomes to separate properly at meiosis I.

primary phloem The cells involved in food conduction in plants.

primary plant body The part of a plant consisting of young, soft shoots and roots derived from apical meristem tissues.

primary productivity The amount of energy produced by photosynthetic organisms in a community.

primary structure The specific amino acid sequence of a protein.

primary tissues Tissues that make up the primary plant body.

primary transcript The initial mRNA molecule copied from a gene by RNA polymerase, containing a faithful copy of the entire gene, including introns as well as exons.

primary wall In plants, the wall layer deposited during the period of cell expansion.

primase The enzyme that synthesizes the RNA primers required by DNA polymerases.

primate Monkeys and apes (including humans).

primitive streak In the early embryos of birds, reptiles, and mammals, a dorsal, longitudinal strip of ectoderm and mesoderm that is equivalent to the blastopore in other forms.

primordium In plants, a bulge on the young shoot produced by the apical meristem; primordia can differentiate into leaves, other shoots, or flowers.

principle of parsimony Principle stating that scientists should favor the hypothesis that requires the fewest assumptions.

prions Infectious proteinaceous particles.

procambium In vascular plants, a primary meristematic tissue that gives rise to primary vascular tissues.

product rule *See* rule of multiplication.

proglottid A repeated body segment in tapeworms that contains both male and female reproductive organs; proglottids eventually form eggs and embryos, which leave the host's body in feces.

prokaryote A bacterium; a cell lacking a membrane-bounded nucleus or membrane-bounded organelles.

prometaphase The transitional phase between prophase and metaphase during which the spindle attaches to the kinetochores of sister chromatids.

promoter A DNA sequence that provides a recognition and attachment site for RNA polymerase to begin the process of gene transcription; it is located upstream from the transcription start site.

prophase The phase of cell division that begins when the condensed chromosomes become visible and ends when the nuclear envelope breaks down. The assembly of the spindle takes place during prophase.

proprioceptor In vertebrates, a sensory receptor that senses the body's position and movements.

prosimian Any member of the mammalian group that is a sister group to the anthropoids; prosimian means "before monkeys." Members include the lemurs, lorises, and tarsiers.

prosoma The anterior portion of the body of an arachnid, which bears all the appendages.

prostaglandins A group of modified fatty acids that function as chemical messengers.

prostate gland In male mammals, a mass of glandular tissue at the base of the urethra that secretes an alkaline fluid that has a stimulating effect on the sperm as they are released.

protease An enzyme that degrades proteins by breaking peptide bonds; in cells, proteases are often compartmentalized into vesicles such as lysosomes.

proteasome A large, cylindrical cellular organelle that degrades proteins marked with ubiquitin.

protein A chain of amino acids joined by peptide bonds.

protein kinase An enzyme that adds phosphate groups to proteins, changing their activity.

protein microarray An array of proteins on a microscope slide or silicon chip. The array may be used with a variety of probes, including antibodies, to analyze the presence or absence of specific proteins in a complex mixture.

proteome All the proteins coded for by a particular genome.

proteomics The study of the proteomes of organisms. This is related to functional genomics as the proteome is responsible for much of the function encoded by a genome.

protoderm The primary meristem that gives rise to the dermal tissue.

proton pump A protein channel in a membrane of the cell that expends energy to transport protons against a concentration gradient; involved in the chemiosmotic generation of ATP.

proto-oncogene A normal cellular gene that can act as an oncogene when mutated.

protostome Any member of a grouping of bilaterally symmetrical animals in which the mouth develops first and the anus second; flatworms, nematodes, mollusks, annelids, and arthropods are protostomes.

pseudocoel A body cavity located between the endoderm and mesoderm.

pseudogene A copy of a gene that is not transcribed.

pseudomurien A component of the cell wall of archaea; it is similar to peptidoglycan in structure and function but contains different components.

pseudopod A nonpermanent cytoplasmic extension of the cell body.

P site In a ribosome, the peptidyl site that binds to the tRNA attached to the growing polypeptide chain.

punctuated equilibrium A hypothesis about the mechanism of evolutionary change proposing that long periods of little or no change are punctuated by periods of rapid evolution.

Punnett square A diagrammatic way of showing the possible genotypes and phenotypes of genetic crosses.

pupa A developmental stage of some insects in which the organism is nonfeeding, immotile, and sometimes encapsulated or in a cocoon; the pupal stage occurs between the larval and adult phases.

purine The larger of the two general kinds of nucleotide base found in DNA and RNA; a nitrogenous base with a double-ring structure, such as adenine or guanine.

pyrimidine The smaller of two general kinds of nucleotide base found in DNA and RNA; a nitrogenous base with a single-ring structure, such as cytosine, thymine, or uracil.

pyruvate A three-carbon molecule that is the end product of glycolysis; each glucose molecule yields two pyruvate molecules.

Q

quantitative trait A trait that is determined by the effects of more than one gene; such a trait usually exhibits continuous variation rather than discrete either–or values.

quaternary structure The structural level of a protein composed of more than one polypeptide chain, each of which has its own tertiary structure; the individual chains are called subunits.

R

radial canal Any of five canals that connect to the ring canal of an echinoderm's water–vascular system.

radial cleavage The embryonic cleavage pattern of deuterostome animals in which cells divide parallel to and at right angles to the polar axis of the embryo.

radial symmetry A type of structural symmetry with a circular plan, such that dividing the body or structure through the midpoint in any direction yields two identical sections.

radicle The part of the plant embryo that develops into the root.

radioactive isotope An isotope that is unstable and undergoes radioactive decay, releasing energy.

radioactivity The emission of nuclear particles and rays by unstable atoms as they decay into more stable forms.

radula Rasping tongue found in most mollusks.

reaction center A transmembrane protein complex in a photosystem that receives energy from the antenna complex exciting an electron that is passed to an acceptor molecule.

reading frame The correct succession of nucleotides in triplet codons that specify amino acids on translation. The reading frame is established by the first codon in the sequence as there are no spaces in the genetic code.

realized niche The actual niche occupied by an organism when all biotic and abiotic interactions are taken into account.

receptor-mediated endocytosis Process by which specific macromolecules are transported into eukaryotic cells at clathrin-coated pits, after binding to specific cell-surface receptors.

receptor protein A highly specific cell-surface receptor embedded in a cell membrane that responds only to a specific messenger molecule.

receptor tyrosine kinase (RTK) A diverse group of membrane receptors that when activated have kinase enzymatic activity. Specifically, they phosphorylate proteins on tyrosine. Their activation can lead to diverse cellular responses.

recessive An allele that is only expressed when present in the homozygous condition, but being "hidden" by the expression of a dominant allele in the heterozygous condition.

redia A secondary, nonciliated larva produced in the sporocysts of liver flukes.

regulatory protein Any of a group of proteins that modulates the ability of RNA polymerase to bind to a promoter and begin DNA transcription.

replicon An origin of DNA replication and the DNA whose replication is controlled by this origin. In prokaryotic replication, the chromosome plus the origin consist of a single replicon; eukaryotic chromosomes consist of multiple replicons.

replisome The macromolecular assembly of enzymes involved in DNA replication; analogous to the ribosome in protein synthesis.

reciprocal altruism Performance of an altruistic act with the expectation that the favor will be returned. A key and very controversial assumption of many theories dealing with the evolution of social behavior. *See* altruism.

reciprocal cross A genetic cross involving a single trait in which the sex of the parents is reversed; for example, if pollen from a white-flowered plant is used to fertilize a purple-flowered plant, the reciprocal cross would be pollen from a purple-flowered plant used to fertilize a white-flowered plant.

reciprocal recombination A mechanism of genetic recombination that occurs only in eukaryotic organisms, in which two chromosomes trade segments; can occur between nonhomologous chromosomes as well as the more usual exchange between homologous chromosomes in meiosis.

recombinant DNA Fragments of DNA from two different species, such as a bacterium and a mammal, spliced together in the laboratory into a single molecule.

recombination frequency The value obtained by dividing the number of recombinant progeny by the total progeny in a genetic cross. This value is converted into a percentage, and each 1% is termed a map unit.

reduction The gain of an electron by an atom, often with an associated proton.

reflex In the nervous system, a motor response subject to little associative modification; a reflex is among the simplest neural pathways, involving only a sensory neuron, sometimes (but not always) an interneuron, and one or more motor neurons.

reflex arc The nerve path in the body that leads from stimulus to reflex action.

refractory period The recovery period after membrane depolarization during which the membrane is unable to respond to additional stimulation.

reinforcement In speciation, the process by which partial reproductive isolation between populations is increased by selection against mating between members of the two populations, eventually resulting in complete reproductive isolation.

replica plating A method of transferring bacterial colonies from one plate to another to make a copy of the original plate; an impression of colonies growing on a Petri plate is made on a velvet surface, which is then used to transfer the colonies to plates containing different media, such that auxotrophs can be identified.

replication fork The Y-shaped end of a growing replication bubble in a DNA molecule undergoing replication.

repolarization Return of the ions in a nerve to their resting potential distribution following depolarization.

repression In general, control of gene expression by preventing transcription. Specifically, in bacteria such as *E. coli* this is mediated by repressor proteins. In anabolic operons, repressors bind DNA in the absence of corepressors to repress an operon.

repressor A protein that regulates DNA transcription by preventing RNA polymerase from attaching to the promoter and transcribing the structural gene. *See* operator.

reproductive isolating mechanism Any barrier that prevents genetic exchange between species.

residual volume The amount of air remaining in the lungs after the maximum amount of air has been exhaled.

resting membrane potential The charge difference (difference in electric potential) that exists across a neuron at rest (about 70 mV).

restriction endonuclease An enzyme that cleaves a DNA duplex molecule at a particular base sequence, usually within or near a palindromic sequence; also called a restriction enzyme.

restriction fragment length polymorphism (RFLP) Restriction enzymes recognize very specific DNA sequences. Alleles of the same gene or surrounding sequences may have base-pair differences, so that DNA near one allele is cut into a different-length fragment than DNA near the other allele. These different fragments separate based on size on electrophoresis gels.

retina The photosensitive layer of the vertebrate eye; contains several layers of neurons and light receptors (rods and cones); receives the image formed by the lens and transmits it to the brain via the optic nerve.

retinoblastoma susceptibility gene (*Rb*) A gene that, when mutated, predisposes individuals to a rare form of cancer of the retina; one of the first tumor-suppressor genes discovered.

retrovirus An RNA virus. When a retrovirus enters a cell, a viral enzyme (reverse transcriptase) transcribes viral RNA into duplex DNA, which the cell's machinery replicates and transcribes as if it were its own.

reverse genetics An approach by which a researcher uses a cloned gene of unknown function, creates a mutation, and introduces the mutant gene back into the organism to assess the effect of the mutation.

reverse transcriptase A viral enzyme found in retroviruses that is capable of converting their RNA genome into a DNA copy.

Rh blood group A set of cell-surface markers (antigens) on the surface of red blood cells in humans and rhesus monkeys (for which it is named); although there are several alleles, they are grouped into two main types: Rh-positive and Rh-negative.

rhizome In vascular plants, a more or less horizontal underground stem; may be enlarged for storage or may function in vegetative reproduction.

rhynchocoel A true coelomic cavity in ribbonworms that serves as a hydraulic power source for extending the proboscis.

ribonucleic acid (RNA) A class of nucleic acids characterized by the presence of the sugar ribose and the pyrimidine uracil; includes mRNA, tRNA, and rRNA.

ribosomal RNA (rRNA) A class of RNA molecules found, together with characteristic proteins, in ribosomes; transcribed from the DNA of the nucleolus.

ribosome The molecular machine that carries out protein synthesis; the most complicated aggregation of proteins in a cell, also containing three different rRNA molecules.

ribosome-binding sequence (RBS) In prokaryotes, a conserved sequence at the 5′ end of mRNA that is complementary to the 3′ end of a small subunit rRNA and helps to position the ribosome during initiation.

ribozyme An RNA molecule that can behave as an enzyme, sometimes catalyzing its own assembly; rRNA also acts as a ribozyme in the polymerization of amino acids to form protein.

ribulose 1,5-bisphosphate (RuBP) In the Calvin cycle, the five-carbon sugar to which CO_2 is attached, accomplishing carbon fixation. This reaction is catalyzed by the enzyme rubisco.

ribulose bisphosphate carboxylase/oxygenase (rubisco) The four-subunit enzyme in the chloroplast that catalyzes the carbon fixation reaction joining CO_2 to RuBP.

RNA interference A type of gene silencing in which the mRNA transcript is prevented from being translated; small interfering RNAs (siRNAs) have been found to bind to mRNA and target its degradation prior to its translation.

RNA polymerase An enzyme that catalyzes the assembly of an mRNA molecule, the sequence of which is complementary to a DNA molecule used as a template. *See* transcription.

RNA primer In DNA replication, a sequence of about 10 RNA nucleotides complementary to unwound DNA that attaches at a replication fork; the DNA polymerase uses the RNA primer as a starting point for addition of DNA nucleotides to form the new DNA strand; the RNA primer is later removed and replaced by DNA nucleotides.

RNA splicing A nuclear process by which intron sequences of a primary mRNA transcript are cut out and the exon sequences spliced together to give the correct linkages of genetic information that will be used in protein construction.

rod Light-sensitive nerve cell found in the vertebrate retina; sensitive to very dim light; responsible for "night vision."

root The usually descending axis of a plant, normally below ground, which anchors the plant and serves as the major point of entry for water and minerals.

root cap In plants, a tissue structure at the growing tips of roots that protects the root apical meristem as the root pushes through the soil; cells of the root cap are continually lost and replaced.

root hair In plants, a tubular extension from an epidermal cell located just behind the root tip; root hairs greatly increase the surface area for absorption.

root pressure In plants, pressure exerted by water in the roots in response to a solute potential in the absence of transpiration; often occurs at night. Root pressure can result in guttation, excretion of water from cells of leaves as dew.

root system In plants, the portion of the plant body that anchors the plant and absorbs ions and water.

R plasmid A resistance plasmid; a conjugative plasmid that picks up antibiotic resistance genes and can therefore transfer resistance from one bacterium to another.

rule of addition The rule stating that for two independent events, the probability of either event occurring is the sum of the individual probabilities.

rule of multiplication The rule stating that for two independent events, the probability of both events occurring is the product of the individual probabilities.

rumen An "extra stomach" in cows and related mammals wherein digestion of cellulose occurs and from which partially digested material can be ejected back into the mouth.

S

salicylic acid In plants, an organic molecule that is a long-distance signal in systemic acquired resistance.

saltatory conduction A very fast form of nerve impulse conduction in which the impulses leap from node to node over insulated portions.

saprobes Heterotrophic organisms that digest their food externally (e.g., most fungi).

sarcolemma The specialized cell membrane in a muscle cell.

sarcomere Fundamental unit of contraction in skeletal muscle; repeating bands of actin and myosin that appear between two Z lines.

sarcoplasmic reticulum The endoplasmic reticulum of a muscle cell. A sleeve of membrane that wraps around each myofilament.

satellite DNA A nontranscribed region of the chromosome with a distinctive base composition; a short nucleotide sequence repeated tandemly many thousands of times.

saturated fat A fat composed of fatty acids in which all the internal carbon atoms contain the maximum possible number of hydrogen atoms.

Schwann cells The supporting cells associated with projecting axons, along with all the other nerve cells that make up the peripheral nervous system.

sclereid In vascular plants, a sclerenchyma cell with a thick, lignified, secondary wall having many pits; not elongate like a fiber.

sclerenchyma cell Tough, thick-walled cells that strengthen plant tissues.

scolex The attachment organ at the anterior end of a tapeworm.

scrotum The pouch that contains the testes in most mammals.

scutellum The modified cotyledon in cereal grains.

second filial (F₂) generation The offspring resulting from a cross between members of the first filial (F₁) generation.

secondary cell wall In plants, the innermost layer of the cell wall. Secondary walls have a highly organized microfibrillar structure and are often impregnated with lignin.

secondary growth In vascular plants, an increase in stem and root diameter made possible by cell division of the lateral meristems.

secondary immune response The swifter response of the body the second time it is invaded by the same pathogen because of the presence of memory cells, which quickly become antibody-producing plasma cells.

secondary induction An induction between tissues that have already differentiated.

secondary metabolite A molecule not directly involved in growth, development, or reproduction of an organism; in plants these molecules, which include nicotine, caffeine, tannins, and menthols, can discourage herbivores.

secondary plant body The part of a plant consisting of secondary tissues from lateral meristem tissues; the older trunk, branches, and roots of woody plants.

secondary structure In a protein, hydrogen-bonding interactions between —CO and —NH groups of the primary structure.

secondary tissue Any tissue formed from lateral meristems in trees and shrubs.

Second Law of Thermodynamics A statement concerning the transformation of potential energy into heat; it says that disorder (entropy) is continually increasing in the universe as energy changes occur, so disorder is more likely than order.

second messenger A small molecule or ion that carries the message from a receptor on the target cell surface into the cytoplasm.

seed bank Ungerminated seeds in the soil of an area. Regeneration of plants after events such as fire often depends on the presence of a seed bank.

seed coat In plants, the outer layers of the ovule, which become a relatively impermeable barrier to protect the dormant embryo and stored food.

segment polarity gene Any of certain genes in *Drosophila* development that are expressed in stripes that subdivide the stripes created by the pair-rule genes in the process of segmentation.

segmentation The division of the developing animal body into repeated units; segmentation allows for redundant systems and more efficient locomotion.

segmentation gene Any of the three classes of genes that control development of the segmented body plan of insects; includes the gap genes, pair-rule genes, and segment polarity genes.

segregation The process by which alternative forms of traits are expressed in offspring rather than blending each trait of the parents in the offspring.

selection The process by which some organisms leave more offspring than competing ones, and their genetic traits tend to appear in greater proportions among members of succeeding generations than the traits of those individuals that leave fewer offspring.

selectively permeable Condition in which a membrane is permeable to some substances but not to others.

self-fertilization The union of egg and sperm produced by a single hermaphroditic organism.

semen In reptiles and mammals, sperm-bearing fluid expelled from the penis during male orgasm.

semicircular canal Any of three fluid-filled canals in the inner ear that help to maintain balance.

semiconservative replication DNA replication in which each strand of the original duplex serves as the template for construction of a totally new complementary strand, so the original duplex is partially conserved in each of the two new DNA molecules.

senescent Aged, or in the process of aging.

sensory (afferent) neuron A neuron that transmits nerve impulses from a sensory receptor to the central nervous system or central ganglion.

sensory setae In insects, bristles attached to the nervous system that are sensitive mechanical and chemical stimulation; most abundant on antennae and legs.

sepal A member of the outermost floral whorl of a flowering plant.

septation In prokaryotic cell division, the formation of a septum where new cell membrane and cell wall is formed to separate the two daughter cells.

septum, pl. **septa** A wall between two cavities.

sequence-tagged site (STS) A small stretch of DNA that is unique in a genome, that is, it occurs only once; useful as a physical marker on genomic maps.

seta, pl. **setae** (L., bristle) In an annelid, bristles of chitin that help anchor the worm during locomotion or when it is in its burrow.

severe acute respiratory syndrome (SARS) A respiratory infection with an 8% mortality rate that is caused by a coronavirus.

sex chromosome A chromosome that is related to sex; in humans, the sex chromosomes are the X and Y chromosomes.

sex-linked A trait determined by a gene carried on the X chromosome and absent on the Y chromosome.

Sexual dimorphism Morphological differences between the sexes of a species.

sexual reproduction The process of producing offspring through an alternation of fertilization (producing diploid cells) and meiotic reduction in chromosome number (producing haploid cells).

sexual selection A type of differential reproduction that results from variable success in obtaining mates.

shared derived character In cladistics, character states that are shared by species and that are different from the ancestral character state.

shoot In vascular plants, the aboveground portions, such as the stem and leaves.

short interspersed element (SINE) Any of a type of retrotransposon found in humans and other primates that does not contain the biochemical machinery needed for transposition; half a million copies of a SINE element called Alu is nested in the LINEs of the human genome.

shotgun sequencing The method of DNA sequencing in which the DNA is randomly cut into small fragments, and the fragments cloned and sequenced. A computer is then used to assemble a final sequence.

sieve cell In the phloem of vascular plants, a long, slender element with relatively unspecialized sieve areas and with tapering end walls that lack sieve plates.

signal recognition particle (SRP) In eukaryotes, a cytoplasmic complex of proteins that recognizes and binds to the signal sequence of a polypeptide, and then docks with a receptor that forms a channel in the ER membrane. In this way the polypeptide is released into the lumen of the ER.

signal transduction The events that occur within a cell on receipt of a signal, ligand binding to a receptor protein. Signal transduction pathways produce the cellular response to a signaling molecule.

simple sequence repeat (SSR) A one- to three-nucleotide sequence such as CA or CCG that is repeated thousands of times.

single-nucleotide polymorphism (SNP) A site present in at least 1% of the population at which individuals differ by a single nucleotide. These can be used as genetic markers to map unknown genes or traits.

sinus A cavity or space in tissues or in bone.

sister chromatid One of two identical copies of each chromosome, still linked at the centromere, produced as the chromosomes duplicate for mitotic division; similarly, one of two identical copies of each homologous chromosome present in a tetrad at meiosis.

small interfering RNAs (siRNAs) A class of micro-RNAs that appear to be involved in control of gene transcription and that play a role in protecting cells from viral attack.

small nuclear ribonucleoprotein particles (snRNP) In eukaryotes, a complex composed of snRNA and protein that clusters together with other snRNPs to form the spliceosome, which removes introns from the primary transcript.

small nuclear RNA (snRNA) In eukaryotes, a small RNA sequence that, as part of a small nuclear ribonucleoprotein complex, facilitates recognition and excision of introns by base-pairing with the 5′ end of an intron or at a branch site of the same intron.

sodium–potassium pump Transmembrane channels engaged in the active (ATP-driven) transport of Na⁺, exchanging them for K⁺, where both ions are being moved against their respective concentration gradients; maintains the resting membrane potential of neurons and other cells.

solute A molecule dissolved in some solution; as a general rule, solutes dissolve only in solutions of similar polarity; for example, glucose (polar) dissolves in (forms hydrogen bonds with) water (also polar), but not in vegetable oil (nonpolar).

solute potential The amount of osmotic pressure arising from the presence of a solute or solutes in water; measure by counterbalancing the pressure until osmosis stops.

solvent The medium in which one or more solutes is dissolved.

somatic cell Any of the cells of a multicellular organism except those that are destined to form gametes (germ-line cells).

somatic cell nuclear transfer (SCNT) The transfer of the nucleus of a somatic cell into an enucleated egg cell that then undergoes development. Can be used to make ES cells and to create cloned animals.

somatic mutation A change in genetic information (mutation) occurring in one of the somatic cells of a multicellular organism, not passed from one generation to the next.

somatic nervous system In vertebrates, the neurons of the peripheral nervous system that control skeletal muscle.

somite One of the blocks, or segments, of tissue into which the mesoderm is divided during differentiation of the vertebrate embryo.

Southern blot A technique in which DNA fragments are separated by gel electrophoresis, denatured into single-stranded DNA, and then "blotted" onto a sheet of filter paper; the filter is then incubated with a labeled probe to locate DNA sequences of interest.

S phase The phase of the cell cycle during which DNA replication occurs.

specialized transduction The transfer of only a few specific genes into a bacterium, using a lysogenic bacteriophage as a carrier.

speciation The process by which new species arise, either by transformation of one species into another, or by the splitting of one ancestral species into two descendant species.

species, pl. **species** A kind of organism; species are designated by binomial names written in italics.

specific heat The amount of heat that must be absorbed or lost by 1 g of a substance to raise or lower its temperature 1°C.

specific transcription factor Any of a great number of transcription factors that act in a time- or tissue-dependent manner to increase DNA transcription above the basal level.

spectrin A scaffold of proteins that links plasma membrane proteins to actin filaments in the cytoplasm of red blood cells, producing their characteristic biconcave shape.

spermatid In animals, each of four haploid (*n*) cells that result from the meiotic divisions of a spermatocyte; each spermatid differentiates into a sperm cell.

spermatozoa The male gamete, usually smaller than the female gamete, and usually motile.

sphincter In vertebrate animals, a ring-shaped muscle capable of closing a tubular opening by constriction (e.g., between stomach and small intestine or between anus and exterior).

spicule Any of a number of minute needles of silica or calcium carbonate made in the mesohyl by some kinds of sponges as a structural component.

spindle The structure composed of microtubules radiating from the poles of the dividing cell that will ultimately guide the sister chromatids to the two poles.

spindle apparatus The assembly that carries out the separation of chromosomes during cell division; composed of microtubules (spindle fibers) and assembled during prophase at the equator of the dividing cell.

spindle checkpoint The third cell-division checkpoint, at which all chromosomes must be attached to the spindle. Passage through this checkpoint commits the cell to anaphase.

spinnerets Organs at the posterior end of a spider's abdomen that secrete a fluid protein that becomes silk.

spiracle External opening of a trachea in arthropods.

spiral cleavage The embryonic cleavage pattern of some protostome animals in which cells divide at an angle oblique to the polar axis of the embryo; a line drawn through the sequence of dividing cells forms a spiral.

spiralian A member of a group of invertebrate animals; many groups exhibit spiral cleavage. Mollusks, annelids, and flatworms are examples of spiralians.

spliceosome In eukaryotes, a complex composed of multiple snRNPs and other associated proteins that is responsible for excision of introns and joining of exons to convert the primary transcript into the mature mRNA.

spongin A tough protein made by many kinds of sponges as a structural component within the mesohyl.

spongy parenchyma A leaf tissue composed of loosely arranged, chloroplast-bearing cells. *See* palisade parenchyma.

sporangium, pl. **sporangia** A structure in which spores are produced.

spore A haploid reproductive cell, usually unicellular, capable of developing into an adult without fusion with another cell.

sporophyte The spore-producing, diploid (2*n*) phase in the life cycle of a plant having alternation of generations.

stabilizing selection A form of selection in which selection acts to eliminate both extremes from a range of phenotypes.

stamen The organ of a flower that produces the pollen; usually consists of anther and filament; collectively, the stamens make up the androecium.

starch An insoluble polymer of glucose; the chief food storage substance of plants.

start codon The AUG triplet, which indicates the site of the beginning of mRNA translation; this codon also codes for the amino acid methionine.

stasis A period of time during which little evolutionary change occurs.

statocyst Sensory receptor sensitive to gravity and motion.

stele The central vascular cylinder of stems and roots.

stem cell A relatively undifferentiated cell in animal tissue that can divide to produce more differentiated tissue cells.

stereoscopic vision Ability to perceive a single, three-dimensional image from the simultaneous but slightly divergent two-dimensional images delivered to the brain by each eye.

stigma (1) In angiosperm flowers, the region of a carpel that serves as a receptive surface for pollen grains. (2) Light-sensitive eyespot of some algae.

stipules Leaflike appendages that occur at the base of some flowering plant leaves or stems.

stolon A stem that grows horizontally along the ground surface and may form adventitious roots, such as runners of the strawberry plant.

stoma, pl. **stomata** In plants, a minute opening bordered by guard cells in the epidermis of leaves and stems; water passes out of a plant mainly through the stomata.

stop codon Any of the three codons UAA, UAG, and UGA, that indicate the point at which mRNA translation is to be terminated.

stratify To hold plant seeds at a cold temperature for a certain period of time; seeds of many plants will not germinate without exposure to cold and subsequent warming.

stratum corneum The outer layer of the epidermis of the skin of the vertebrate body.

striated muscle Skeletal voluntary muscle and cardiac muscle.

stroma In chloroplasts, the semiliquid substance that surrounds the thylakoid system and that contains the enzymes needed to assemble organic molecules from CO_2.

stromatolite A fossilized mat of ancient bacteria formed as long as 2 BYA, in which the bacterial remains individually resemble some modern-day bacteria.

style In flowers, the slender column of tissue that arises from the top of the ovary and through which the pollen tube grows.

stylet A piercing organ, usually a mouthpart, in some species of invertebrates.

suberin In plants, a fatty acid chain that forms the impermeable barrier in the Casparian strip of root endoderm.

subspecies A geographically defined population or group of populations within a single species that has distinctive characteristics.

substrate (1) The foundation to which an organism is attached. (2) A molecule on which an enzyme acts.

subunit vaccine A type of vaccine created by using a subunit of a viral protein coat to elicit an immune response; may be useful in preventing viral diseases such as hepatitis B.

succession In ecology, the slow, orderly progression of changes in community composition that takes place through time.

summation Repetitive activation of the motor neuron resulting in maximum sustained contraction of a muscle.

supercoiling The coiling in space of double-stranded DNA molecules due to torsional strain, such as occurs when the helix is unwound.

surface tension A tautness of the surface of a liquid, caused by the cohesion of the molecules of liquid. Water has an extremely high surface tension.

surface area-to-volume ratio Relationship of the surface area of a structure, such as a cell, to the volume it contains.

suspensor In gymnosperms and angiosperms, the suspensor develops from one of the first two cells of a dividing zygote; the suspensor of an angiosperm is a nutrient conduit from maternal tissue to the embryo. In gymnosperms the suspensor positions the embryo closer to stored food reserves.

swim bladder An organ encountered only in the bony fish that helps the fish regulate its buoyancy by increasing or decreasing the amount of gas in the bladder via the esophagus or a specialized network of capillaries.

swimmerets In lobsters and crayfish, appendages that occur in lines along the ventral surface of the abdomen and are used in swimming and reproduction.

symbiosis The condition in which two or more dissimilar organisms live together in close association; includes parasitism (harmful to one of the organisms), commensalism (beneficial to one, of no significance to the other), and mutualism (advantageous to both).

sympatric speciation The differentiation of populations within a common geographic area into species.

symplast route In plant roots, the pathway for movement of water and minerals within the cell cytoplasm that leads through plasmodesmata that connect cells.

symplesiomorphy In cladistics, another term for a shared ancestral character state.

symporter A carrier protein in a cell's membrane that transports two molecules or ions in the same direction across the membrane.

synapomorphy In systematics, a derived character that is shared by clade members.

synapse A junction between a neuron and another neuron or muscle cell; the two cells do not touch, the gap being bridged by neurotransmitter molecules.

synapsid Any of an early group of reptiles that had a pair of temporal openings in the skull behind the eye sockets; jaw muscles attached to these openings. Early ancestors of mammals belonged to this group.

synapsis The point-by-point alignment (pairing) of homologous chromosomes that occurs before the first meiotic division; crossing over takes place during synapsis.

synaptic cleft The space between two adjacent neurons.

synaptic vesicle A vesicle of a neurotransmitter produced by the axon terminal of a nerve. The filled vesicle migrates to the presynaptic membrane, fuses with it, and releases the neurotransmitter into the synaptic cleft.

synaptonemal complex A protein lattice that forms between two homologous chromosomes in prophase I of meiosis, holding the replicated chromosomes in precise register with each other so that base-pairs can form between nonsister chromatids for crossing over that is usually exact within a gene sequence.

syncytial blastoderm A structure composed of a single large cytoplasm containing about 4000 nuclei in embryonic development of insects such as *Drosophila*.

syngamy The process by which two haploid cells (gametes) fuse to form a diploid zygote; fertilization.

synthetic polyploidy A polyploidy organism created by crossing organisms most closely related to an ancestral species and then manipulating the offspring.

systematics The reconstruction and study of evolutionary relationships.

systemic acquired resistance (SAR) In plants, a longer-term response to a pathogen or pest attack that can last days to weeks and allow the plant to respond quickly to later attacks by a range of pathogens.

systemin In plants, an 18-amino-acid peptide that is produced by damaged or injured leaves that leads to the wound response.

systolic pressure A measurement of how hard the heart is contracting. When measured during a blood pressure reading, ventricular systole (contraction) is what is being monitored.

T

3′ poly-A tail In eukaryotes, a series of 1–200 adenine residues added to the 3′ end of an mRNA; the tail appears to enhance the stability of the mRNA by protecting it from degradation.

T box A transcription factor protein domain that has been conserved, although with differing developmental effects, in invertebrates and chordates.

tagma, pl. **tagmata** A compound body section of an arthropod resulting from embryonic fusion of two or more segments; for example, head, thorax, abdomen.

Taq polymerase A DNA polymerase isolated from the thermophilic bacterium *Thermus aquaticus* (Taq); this polymerase is functional at higher temperatures, and is used in PCR amplification of DNA.

TATA box In eukaryotes, a sequence located upstream of the transcription start site. The TATA box is one element of eukaryotic core promoters for RNA polymerase II.

taxis, pl. **taxes** An orientation movement by a (usually) simple organism in response to an environmental stimulus.

taxonomy The science of classifying living things. By agreement among taxonomists, no two organisms can have the same name, and all names are expressed in Latin.

T cell A type of lymphocyte involved in cell-mediated immunity and interactions with B cells; the "T" refers to the fact that T cells are produced in the thymus.

telencephalon The most anterior portion of the brain, including the cerebrum and associated structures.

telomerase An enzyme that synthesizes telomeres on eukaryotic chromosomes using an internal RNA template.

telomere A specialized nontranscribed structure that caps each end of a chromosome.

telophase The phase of cell division during which the spindle breaks down, the nuclear envelope of each daughter cell forms, and the chromosomes uncoil and become diffuse.

telson The tail spine of lobsters and crayfish.

temperate (lysogenic) phage A virus that is capable of incorporating its DNA into the host cell's DNA, where it remains for an indeterminate length of time and is replicated as the cell's DNA replicates.

template strand The DNA strand that is used as a template in transcription. This strand is copied to produce a complementary mRNA transcript.

tendon (Gr. *tendon,* stretch) A strap of cartilage that attaches muscle to bone.

tensile strength A measure of the cohesiveness of a substance; its resistance to being broken apart. Water in narrow plant vessels has tensile strength that helps keep the water column continuous.

tertiary structure The folded shape of a protein, produced by hydrophobic interactions with water, ionic and covalent bonding between side chains of different amino acids, and van der Waal's forces; may be changed by denaturation so that the protein becomes inactive.

testcross A mating between a phenotypically dominant individual of unknown genotype and a homozygous "tester," done to determine whether the phenotypically dominant individual is homozygous or heterozygous for the relevant gene.

testis, pl. **testes** In mammals, the sperm-producing organ.

tetanus Sustained forceful muscle contraction with no relaxation.

thalamus That part of the vertebrate forebrain just posterior to the cerebrum; governs the flow of information from all other parts of the nervous system to the cerebrum.

therapeutic cloning The use of somatic cell nuclear transfer to create stem cells from a single individual that may be reimplanted in that individual to replace damaged cells, such as in a skin graft.

thermodynamics The study of transformations of energy, using heat as the most convenient form of measurement of energy.

thermogenesis Generation of internal heat by endothermic animals to modulate temperature.

thigmotropism In plants, unequal growth in some structure that comes about as a result of physical contact with an object.

threshold The minimum amount of stimulus required for a nerve to fire (depolarize).

thylakoid In chloroplasts, a complex, organized internal membrane composed of flattened disks, which contain the photosystems involved in the light-dependent reactions of photosynthesis.

Ti (tumor-inducing) plasmid A plasmid found in the plant bacterium *Agrobacterium tumefaciens* that has been extensively used to introduce recombinant DNA into broadleaf plants. Recent modifications have allowed its use with cereal grains as well.

tight junction Region of actual fusion of plasma membranes between two adjacent animal cells that prevents materials from leaking through the tissue.

tissue A group of similar cells organized into a structural and functional unit.

tissue plasminogen activator (TPA) A human protein that causes blood clots to dissolve; if used within 3 hours of an ischemic stroke, TPA may prevent disability.

tissue-specific stem cell A stem cell that is capable of developing into the cells of a certain tissue, such as muscle or epithelium; these cells persist even in adults.

tissue system In plants, any of the three types of tissue; called a system because the tissue extends throughout the roots and shoots.

tissue tropism The affinity of a virus for certain cells within a multicellular host; for example, hepatitis B virus targets liver cells.

tonoplast The membrane surrounding the central vacuole in plant cells that contains water channels; helps maintain the cell's osmotic balance.

topoisomerase Any of a class of enzymes that can change the topological state of DNA to relieve torsion caused by unwinding.

torsion The process in embryonic development of gastropods by which the mantle cavity and anus move from a posterior location to the front of the body, closer to the location of the mouth.

totipotent A cell that possesses the full genetic potential of the organism.

trachea, pl. tracheae A tube for breathing; in terrestrial vertebrates, the windpipe that carries air between the larynx and bronchi (which leads to the lungs); in insects and some other terrestrial arthropods, a system of chitin-lined air ducts.

tracheids In plant xylem, dead cells that taper at the ends and overlap one another.

tracheole The smallest branches of the respiratory system of terrestrial arthropods; tracheoles convey air from the tracheae, which connect to the outside of the body at spiracles.

trait In genetics, a characteristic that has alternative forms, such as purple or white flower color in pea plants or different blood type in humans.

transcription The enzyme-catalyzed assembly of an RNA molecule complementary to a strand of DNA.

transcription complex The complex of RNA polymerase II plus necessary activators, coactivators, transcription factors, and other factors that are engaged in actively transcribing DNA.

transcription factor One of a set of proteins required for RNA polymerase to bind to a eukaryotic promoter region, become stabilized, and begin the transcription process.

transcription bubble The region containing the RNA polymerase, the DNA template, and the RNA transcript, so called because of the locally unwound "bubble" of DNA.

transcription unit The region of DNA between a promoter and a terminator.

transcriptome All the RNA present in a cell or tissue at a given time.

transfection The transformation of eukaryotic cells in culture.

transfer RNA (tRNA) A class of small RNAs (about 80 nucleotides) with two functional sites; at one site, an "activating enzyme" adds a specific amino acid, while the other site carries the nucleotide triplet (anticodon) specific for that amino acid.

transformation The uptake of DNA directly from the environment; a natural process in some bacterial species.

transgenic organism An organism into which a gene has been introduced without conventional breeding, that is, through genetic engineering techniques.

translation The assembly of a protein on the ribosomes, using mRNA to specify the order of amino acids.

translation repressor protein One of a number of proteins that prevents translation of mRNA by binding to the beginning of the transcript and preventing its attachment to a ribosome.

translocation (1) In plants, the long-distance transport of soluble food molecules (mostly sucrose), which occurs primarily in the sieve tubes of phloem tissue. (2) In genetics, the interchange of chromosome segments between nonhomologous chromosomes.

transmembrane domain Hydrophobic region of a transmembrane protein that anchors it in the membrane. Often composed of α-helices, but sometimes utilizing β-pleated sheets to form a barrel-shaped pore.

transmembrane route In plant roots, the pathway for movement of water and minerals that crosses the cell membrane and also the membrane of vacuoles inside the cell.

transpiration The loss of water vapor by plant parts; most transpiration occurs through the stomata.

transposable elements Segments of DNA that are able to move from one location on a chromosome to another. Also termed *transposons* or *mobile genetic elements*.

transposition Type of genetic recombination in which transposable elements (transposons) move from one site in the DNA sequence to another, apparently randomly.

transposon DNA sequence capable of transposition.

trichome In plants, a hairlike outgrowth from an epidermal cell; glandular trichomes secrete oils or other substances that deter insects.

triglyceride (triacylglycerol) An individual fat molecule, composed of a glycerol and three fatty acids.

triploid Possessing three sets of chromosomes.

trisomic Describes the condition in which an additional chromosome has been gained due to nondisjunction during meiosis, and the diploid embryo therefore has three of these autosomes. In humans, trisomic individuals may survive if the autosome is small; Down syndrome individuals are trisomic for chromosome 21.

trochophore A specialized type of free-living larva found in lophotrochozoans.

trophic level A step in the movement of energy through an ecosystem.

trophoblast In vertebrate embryos, the outer ectodermal layer of the blastodermic vesicle; in mammals, it is part of the chorion and attaches to the uterine wall.

tropism Response to an external stimulus.

tropomyosin Low-molecular-weight protein surrounding the actin filaments of striated muscle.

troponin Complex of globular proteins positioned at intervals along the actin filament of skeletal muscle; thought to serve as a calcium-dependent "switch" in muscle contraction.

***trp* operon** In *E. coli*, the operon containing genes that code for enzymes that synthesize tryptophan.

true-breeding Said of a breed or variety of organism in which offspring are uniform and consistent from one generation to the next. This is due to the genotypes that determine relevant traits being homozygous.

tube foot In echinoderms, a flexible, external extension of the water–vascular system that is capable of attaching to a surface through suction.

tubulin Globular protein subunit forming the hollow cylinder of microtubules.

tumor-suppressor gene A gene that normally functions to inhibit cell division; mutated forms can lead to the unrestrained cell division of cancer, but only when both copies of the gene are mutant.

turgor pressure The internal pressure inside a plant cell, resulting from osmotic intake of water, that presses its cell membrane tightly against the cell wall, making the cell rigid. Also known as *hydrostatic pressure*.

tympanum In some groups of insects, a thin membrane associated with the tracheal air sacs that functions as a sound receptor; paired on each side of the abdomen.

U

ubiquitin A 76-amino-acid protein that virtually all eukaryotic cells attach as a marker to proteins that are to be degraded.

unequal crossing over A process by which a crossover in a small region of misalignment at synapsis causes two homologous chromosomes to exchange segments of unequal length.

uniporter A carrier protein in a cell's membrane that transports only a single type of molecule or ion.

uniramous Single-branched; describes the appendages of insects.

unsaturated fat A fat molecule in which one or more of the fatty acids contain fewer than the maximum number of hydrogens attached to their carbons.

urea An organic molecule formed in the vertebrate liver; the principal form of disposal of nitrogenous wastes by mammals.

urethra The tube carrying urine from the bladder to the exterior of mammals.

uric acid Insoluble nitrogenous waste products produced largely by reptiles, birds, and insects.

urine The liquid waste filtered from the blood by the kidney and stored in the bladder pending elimination through the urethra.

uropod One of a group of flattened appendages at the end of the abdomen of lobsters and crayfish that collectively act as a tail for a rapid burst of speed.

uterus In mammals, a chamber in which the developing embryo is contained and nurtured during pregnancy.

V

vacuole A membrane-bounded sac in the cytoplasm of some cells, used for storage or digestion purposes in different kinds of cells; plant cells often contain a large central vacuole that stores water, proteins, and waste materials.

valence electron An electron in the outermost energy level of an atom.

variable A factor that influences a process, outcome, or observation. In experiments, scientists attempt to isolate variables to test hypotheses.

vascular cambium In vascular plants, a cylindrical sheath of meristematic cells, the division of which produces secondary phloem outwardly and secondary xylem inwardly; the activity of the vascular cambium increases stem or root diameter.

vascular tissue Containing or concerning vessels that conduct fluid.

vas deferens In mammals, the tube carrying sperm from the testes to the urethra.

vasopressin A posterior pituitary hormone that regulates the kidney's retention of water.

vector In molecular biology, a plasmid, phage or artificial chromosome that allows propagation of recombinant DNA in a host cell into which it is introduced.

vegetal pole The hemisphere of the zygote comprising cells rich in yolk.

vein (1) In plants, a vascular bundle forming a part of the framework of the conducting and supporting tissue of a stem or leaf. (2) In animals, a blood vessel carrying blood from the tissues to the heart.

veliger The second larval stage of mollusks following the trochophore stage, during which the beginning of a foot, shell, and mantle can be seen.

ventricle A muscular chamber of the heart that receives blood from an atrium and pumps blood out to either the lungs or the body tissues.

vertebrate A chordate with a spinal column; in vertebrates, the notochord develops into the vertebral column composed of a series of vertebrae that enclose and protect the dorsal nerve cord.

vertical gene transfer (VGT) The passing of genes from one generation to the next within a species.

vesicle A small intracellular, membrane-bounded sac in which various substances are transported or stored.

vessel element In vascular plants, a typically elongated cell, dead at maturity, which conducts water and solutes in the xylem.

vestibular apparatus The complicated sensory apparatus of the inner ear that provides for balance and orientation of the head in vertebrates.

vestigial structure A morphological feature that has no apparent current function and is thought to be an evolutionary relic; for example, the vestigial hip bones of boa constrictors.

villus, pl. villi In vertebrates, one of the minute, fingerlike projections lining the small intestine that serve to increase the absorptive surface area of the intestine.

virion A single virus particle.

viroid Any of a group of small, naked RNA molecules that are capable of causing plant diseases, presumably by disrupting chromosome integrity.

virus Any of a group of complex biochemical entities consisting of genetic material wrapped in protein; viruses can reproduce only within living host cells and are thus not considered organisms.

visceral mass Internal organs in the body cavity of an animal.

vitamin An organic substance that cannot be synthesized by a particular organism but is required in small amounts for normal metabolic function.

viviparity Refers to reproduction in which eggs develop within the mother's body and young are born free-living.

voltage-gated ion channel A transmembrane pathway for an ion that is opened or closed by a change in the voltage, or charge difference, across the plasma membrane.

W

water potential The potential energy of water molecules. Regardless of the reason (e.g., gravity, pressure, concentration of solute particles) for the water potential, water moves from a region where water potential is greater to a region where water potential is lower.

water–vascular system A fluid-filled hydraulic system found only in echinoderms that provides body support and a unique type of locomotion via extensions called tube feet.

Western blot A blotting technique used to identify specific protein sequences in a complex mixture. *See* Southern blot.

wild type In genetics, the phenotype or genotype that is characteristic of the majority of individuals of a species in a natural environment.

wobble pairing Refers to flexibility in the pairing between the base at the 5′ end of a tRNA anticodon and the base at the 3′ end of an mRNA codon. This flexibility allows a single tRNA to read more than one mRNA codon.

wound response In plants, a signaling pathway initiated by leaf damage, such as being chewed by a herbivore, and lead to the production of proteinase inhibitors that give herbivores indigestion.

X

X chromosome One of two sex chromosomes; in mammals and in *Drosophila*, female individuals have two X chromosomes.

xylem In vascular plants, a specialized tissue, composed primarily of elongate, thick-walled conducting cells, which transports water and solutes through the plant body.

Y

Y chromosome One of two sex chromosomes; in mammals and in *Drosophila*, male individuals have a Y chromosome and an X chromosome; the Y determines maleness.

yolk plug A plug occurring in the blastopore of amphibians during formation of the archenteron in embryological development.

yolk sac The membrane that surrounds the yolk of an egg and connects the yolk, a rich food supply, to the embryo via blood vessels.

Z

zinc finger motif A type of DNA-binding motif in regulatory proteins that incorporates zinc atoms in its structure.

zona pellucida An outer membrane that encases a mammalian egg.

zone of cell division In plants, the part of the young root that includes the root apical meristem and the cells just posterior to it; cells in this zone divide every 12–36 hr.

zone of elongation In plants, the part of the young root that lies just posterior to the zone of cell division; cells in this zone elongate, causing the root to lengthen.

zone of maturation In plants, the part of the root that lies posterior to the zone of elongation; cells in this zone differentiate into specific cell types.

zoospore A motile spore.

zooxanthellae Symbiotic photosynthetic protists in the tissues of corals.

zygomycetes A type of fungus whose chief characteristic is the production of sexual structures called zygosporangia, which result from the fusion of two of its simple reproductive organs.

zygote The diploid ($2n$) cell resulting from the fusion of male and female gametes (fertilization).

Photo Credits

Chapter 1
Opener: © Soames Summerhays/Natural Visions; 1.1(organelle): © Dr. Donald Fawcett & Porter/Visuals Unlimited; 1.1(cell): © Steve Gschmeissner/Getty Images; 1.1(tissue): © Ed Reschke; 1.1(organism): © Russell Illig/Getty Images RF; 1.1(population): © George Ostertag/agefotostock; 1.1(species): © PhotoDisc/Volume 44 RF; 1.1(community): © Ryan McGinnis/Alamy; 1.1(ecosystem): © Robert and Jean Pollock; 1.1(biosphere): NASA; 1.5: © Huntington Library/SuperStock; 1.11a: © Dennis Kunkel/Phototake; 1.11b: © Karl E. Deckart/Phototake; 1.12(plantae left): © Alan L. Detrick/Photo Researchers, Inc.; 1.12(plantae middle): © David M. Dennis/Animals Animals; 1.12(plantae right): © Corbis/Volume 46 RF; 1.12(fungi left): © Royalty-Free/Corbis; 1.12(fungi middle): © Mediscan/Corbis; 1.12(fungi right): © PhotoDisc BS/Volume 15 RF; 1.12(animalia left): © Royalty-Free/Corbis; 1.12(animalia middle): © Tom Brakefield/Corbis; 1.12(animalia right): © PhotoDisc/Volume 44 RF; 1.12(protista left): © Corbis/Volume 64 RF; 1.12(protista middle): © Tom Adams/Visuals Unlimited; 1.12(protista right): © Douglas P. Wilson/Frank Lane Picture Agency/Corbis; 1.12(archaea left): © Ralph Robinson/Visuals Unlimited; 1.12(archaea right): © Kari Lounatman/Photo Researchers, Inc.; 1.12(bacteria left): © Dwight R. Kuhn; 1.12(bacteria right): © Alfred Pasieka/SPL/Photo Researchers, Inc.; pp.15-16: © Soames Summerhays/Natural Visions.

Chapter 2
Opener: Courtesy of IBM Zurich Research Laboratory. Unauthorized use not permitted; 2.2: Image Courtesy of Bruker Corporation; 2.10a: © Glen Allison/Getty Images RF; 2.10b: © PhotoLink/Getty Images RF; 2.10c: © Jeff Vanuga/Corbis; 2.13: © Hermann Eisenbeiss/National Audubon Society Collection/Photo Researchers, Inc.; pp. 31-32: Courtesy of IBM Zurich Research Laboratory. Unauthorized use not permitted.

Chapter 3
Opener: © Deco/Alamy; 3.10b: © Asa Thoresen/Photo Researchers, Inc.; 3.10c: © J. Carson/Custom Medical Stock Photo; 3.11b: © Science VU/Visuals Unlimited; 3.12: © Scott Johnson/Animals Animals; 3.13a: © Driscoll, Youngquist & Baldeschwieler, Caltech/SPL/Photo Researchers, Inc.; 3.13b: © M. Freeman/PhotoLink/Getty Images RF; pp. 56-57: © Deco/Alamy.

Chapter 4
Opener: © Dr. Gopal Murti/Photo Researchers, Inc.; p. 62(bright-field microscope): © David M. Phillips/Visuals Unlimited; p. 62(dark-field microscope): © Mike Abbey/Visuals Unlimited; p. 62(phase-contrast microscope): © David M. Phillips/Visuals Unlimited; p. 62(differential-interference-contrast microscope): © Mike Abbey/Visuals Unlimited; p. 62(fluorescence microscope): © Dr. Torsten Wittmann/Photo Researchers, Inc.; p. 62(confocal microscope): © Med. Mic. Sciences, Cardiff Uni./Wellcome Images; p. 62(transmission electron microscope): © Microworks/Phototake; p. 62 (scanning electron microscope): © Stanley Flegler/Visuals Unlimited; p. 62(bottom right): © Dr. Donald Fawcett/Visuals Unlimited; 4.3: © Phototake; 4.4: Courtesy of E.H. Newcomb & T.D. Pugh, University of Wisconsin; 4.5a: © Eye of Science/Photo Researchers, Inc.; 4.8b: © Dr. Richard Kessel & Dr. Gene Shih/Visuals Unlimited; 4.8c: © John T. Hansen, Ph.D./Phototake; 4.8d: © Dr. Ueli Aebi; 4.10(inset): © Dr. Donald Fawcett & R. Bolender/Visuals Unlimited; 4.11(inset): © Dennis Kunkel/Phototake; 4.14(inset): From S.E. Frederick and E.H. Newcomb, "Microbody-like organelles in leaf cells," *Science*, 163:1353-5. © 21 March 1969. Reprinted with permission from AAAS; 4.15(inset): © Henry Aldrich/Visuals Unlimited; 4.16(inset): © Dr. Donald Fawcett & Dr. Porter/Visuals Unlimited; 4.17(inset): © Dr. Jeremy Burgess/Photo Researchers, Inc.; 4.23(top & bottom insets): © William Dentler, University of Kansas; 4.24a-b: © SPL/Photo Researchers, Inc.; 4.25: © Biophoto Associates/Photo Researchers, Inc.; 4.28a: Courtesy of Daniel Goodenough; 4.28b: © Dr. Donald Fawcett/Visuals Unlimited; 4.28c: © Dr. Donald Fawcett/D. Albertini/Visuals Unlimited; pp. 85-86: © Dr. Gopal Murti/Photo Researchers, Inc.

Chapter 5
Opener: © Dr. Gopal Murti/SPL/Photo Researchers, Inc.; 5.2b: © Whitney L. Stutts, University of Florida; p. 91: © Dr. Donald Fawcett/Photo Researchers, Inc.; 5.4 (4): © Dr. Donald Fawcett/Visuals Unlimited; 5.12: © David M. Phillips/Visuals Unlimited; 5.15a: CDC/Dr. Edwin P. Ewing, Jr.; 5.15b: © BCC Microimaging, Inc. Reproduced with permission; 5.15c (top-bottom): Reproduced with permission from M.M. Perry and A.B. Gilbert, "Yolk transport in the ovarian follicle of the hen (*Gallus domesticus*): lipoprotein-like particles at the periphery of the oocyte in the rapid growth phase," *Journal of Cell Science*, 39:257-72, October 1979. © The Company of Biologists; 5.16b: © Dr. Brigit Satir; pp. 105-106: © Dr. Gopal Murti/SPL/Photo Researchers, Inc.

Chapter 6
Opener: © Robert Caputo/Aurora Photos; 6.3: © Jill Braaten; 6.11b: © Professor Emeritus Lester J. Reed, University of Texas at Austin; pp. 119-120: © Robert Caputo/Aurora Photos.

Chapter 7
Opener: © Creatas/PunchStock RF; 7.18a: © Wolfgang Baumeister/Photo Researchers, Inc.; 7.18b: NPS Photo; pp. 144-145: © Creatas/PunchStock RF.

Chapter 8
Opener: © Royalty-Free/Corbis; 8.1(middle right): Courtesy Dr. Kenneth Miller, Brown University; 8.8: © Eric Soder/pixsource.com; 8.20: © Dr. Jeremy Burgess/Photo Researchers, Inc.; 8.22a: © John Shaw/Photo Researchers, Inc.; 8.22b: © Joseph Nettis/National Audubon Society Collection/Photo Researchers, Inc.; 8.24(inset): © 2011 Jessica Solomatenko/Getty Images RF; pp. 166-167: © Royalty-Free/Corbis.

Chapter 9
Opener & pp. 183-184: © RMF/Scientifica/Visuals Unlimited.

Chapter 10
Opener: © Stem Jems/Photo Researchers, Inc.; 10.2: Courtesy of William Margolin; 10.4: © Biophoto Associates/Photo Researchers, Inc.; 10.6: © CNRI/Photo Researchers, Inc.; 10.10: Image courtesy of S. Hauf and J-M. Peters, IMP, Vienna, Austria; 10.11-10.12: © Andrew S. Bajer, University of Oregon; 10.13: © Dr. Jeremy Pickett-Heaps; 10.14a: © David M. Phillips/Visuals Unlimited; 10.14b: © Guenter Albrecht-Buehler, Northwestern University, Chicago; 10.15(top): © E.H. Newcomb & W.P. Wergin/Biological Photo Service; pp. 205-206: © Stem Jems/Photo Researchers, Inc.

Chapter 11
Opener: © Science VU/L. Maziarski/Visuals Unlimited; 11.3b: Reprinted with permission from the *Annual Review of Genetics*, Volume 6 © 1972 by Annual Reviews, www.annualreviews.org; 11.6: © Clare A. Hasenkampf/Biological Photo Service; pp. 218-219: © Science VU/L. Maziarski/Visuals Unlimited.

Chapter 12
Opener: © Corbis RF; 12.1: © Norbert Schaefer/Corbis; 12.2: © David Sieren/Visuals Unlimited; 12.3: © Leslie Holzer/Photo Researchers, Inc.; 12.11(top): From Albert F. Blakeslee, "CORN AND MEN: The Interacting Influence of Heredity and Environment—Movements for Betterment of Men, or Corn, or Any Other Living Thing, One-sided Unless They Take Both Factors into Account," *Journal of Heredity*, 1914, 5:511-8, by permission of Oxford University Press; 12.14: © DK Limited/Corbis; pp. 237-238: © Corbis RF.

Chapter 13
Opener: © Adrian T. Sumner/Photo Researchers, Inc.; 13.1: © Cabisco/Phototake; p.

Collection/Photo Researchers, Inc.; 54.12a: © Linda Koebner/Bruce Coleman Inc./Photoshot; 54.12b: © Tom & Pat Leeson/Photo Researchers, Inc.; 54.13a-c: © SuperStock; 54.14: Courtesy of Bernd Heinrich; 54.15b: © Roy Toft/Getty Images; 54.15c: © George Lepp/Getty Images; 54.18(inset): © Dwight R. Kuhn; 54.21: © Tom Leeson; 54.22b: © Scott Camazine/Photo Researchers, Inc.; 54.23a: © Gerald Cubitt; 54.24: © Nina Leen, Life Magazine/Time, Inc./Getty Images; 54.27(left): © Peter Steyn/Getty Images; 54.27(right): © Gerald C. Kelley/Photo Researchers, Inc.; 54.28a: © Bruce Beehler/Photo Researchers, Inc.; 54.28b: © B. Chudleigh/VIREO; 54.30: © Cathy & Gordon ILLG; 54.31a: © Michael & Patricia Fogden/Minden Pictures/National Geographic Image Collection; 54.32a: © Dr. Douglas Ross; 54.33: © Nick Gordon/ardea.com; 54.35: © Steve Hopkin/Getty Images; 54.36: © Heinrich van den Berg/Getty Images; 54.37: © Mark Moffett/Minden Pictures; 54.38: © Nigel Dennis/National Audubon Society Collection/Photo Researchers, Inc.; pp. 1159-1160: © K. Ammann/Bruce Coleman Inc./Photoshot.

Chapter 55

Opener: © PhotoDisc/Volume 44 RF; 55.1: © Michael Fogden/Animals Animals; 55.3(inset): © Melissa Losos; 55.4: © Stone Nature Photography/Alamy; 55.6(inset): © Duncan Usher/ardea.com; 55.13a: © Christian Kerihuel; 55.15: © Barry Sinervo; 55.21(left): © Juan Medina/Reuters/Corbis; 55.21 (right): © Juniors Tierbildarchiv/Photoshot; pp. 1182-1183: © PhotoDisc/Volume 44 RF.

Chapter 56

Opener: © Corbis RF; 56.1: © Daryl & Sharna Balfour/Okopia/Photo Researchers, Inc.; 56.6a-d: © Jonathan Losos; 56.10a: © Edward S. Ross; 56.10b: © Raymond Mendez/Animals Animals; 56.11a-b: © Lincoln P. Brower; 56.12: © Michael & Patricia Fogden/Corbis; 56.13: © Milton Tierney/Visuals Unlimited; 56.15: © Merlin D. Tuttle/Bat Conservation International; 56.16: © Alex Wild/Visuals Unlimited; 56.17: © Charles T. Bryson, USDA Agricultural Research Service, Bugwood.org; 56.19: © Eastcott/Momatiuk/The Image Works; 56.20: © PhotoDisc/Volume 44 RF; 56.21a: © F. Stuart Westmorland/Photo Researchers, Inc.; 56.21b: © Ann Rosenfeld/Animals Animals; 56.23: © David Hosking/National Audubon Society Collection/Photo Researchers, Inc.; 56.24b-d: © Tom Bean; 56.25a-b: © Studio Carlo Dani/Animals Animals; 56.26: © Educational Images Ltd., Elmira, NY, USA. Used by Permission; pp. 1204-1205: © Corbis RF.

Chapter 57

Opener: © PhotoDisc/Getty Images RF; 57.3: © Worldwide Picture Library/Alamy; 57.6 (bottom): Jeff Schmaltz, MODIS Rapid Response Team, NASA/GSFC; 57.7a: U.S. Forest Service; 57.19a: © Layne Kennedy/Corbis; pp. 1227-1228: © PhotoDisc/Getty Images RF.

Chapter 58

Opener: GSFC/NASA; 58.10: © Andoni Canela/agefotostock; 58.11: © Image Plan/Corbis RF; 58.14a: © Art Wolfe/Photo Researchers, Inc.; 58.14b: © Bill Banaszowski/Visuals Unlimited; 58.16: Provided by the SeaWiFS Project, NASA/Goddard Space Flight Center, and ORBIMAGE; 58.17: © Digital Vision/Getty Images RF; 58.19a: © Jim Church; 58.19b: © Ralph White/Corbis; 58.19b(inset): NOAA Pacific Marine Environmental Laboratory's Vents Program; 58.22a: © Environmental Images/agefotostock; 58.22b: © Frans Lanting/Corbis; 58.23: © Gilbert S. Grant/National Audubon Society Collection/Photo Researchers, Inc.; 58.25a: NASA/Goddard Space Flight Center Scientific Visualization Studio; 58.27: NASA Goddard Institute for Space Studies; 58.29(top): © Dr. Bruno Messerli; 58.29(bottom): © Ulrich Doering/agefotostock; pp. 1253-1255: GSFC/NASA.

Chapter 59

Opener: © Norbert Rosing/National Geographic Image Collection; 59.1a: Courtesy of Rhoda Knight Kalt & DN-Images; 59.3: © Robert Harding Images/Masterfile; 59.4(1): © Frank Krahmer/Masterfile; 59.4(2): © Kevin Schafer/Corbis; 59.4(3): © Heather Angel/Natural Visions; 59.4(4): © NHPA/Photoshot; 59.6a: © Edward S. Ross; 59.6b: © Inga Spence/Visuals Unlimited; 59.7a: © Juan Carlos Muñoz/agefotostock; 59.7b: © Andoni Canela/agefotostock; 59.9: © Michael & Patricia Fogden/Corbis; 59.10(1): © Brian Rogers/Natural Visions; 59.10(2): © David M. Dennis/Animals Animals; 59.10(3): © Craig K. Lorenz/Photo Researchers, Inc.; 59.10(4): © David A. Northcott/Corbis; 59.14(left): © Mark Moffett/Minden Pictures/Corbis; 59.14(right): © Dr. Morley Read/Photo Researchers, Inc.; 59.15: © John Gerlach/Animals Animals; 59.17: © Peter Yates/SPL/Photo Researchers, Inc.; 59.18(left): © Jack Jeffrey; 59.18(right): © Jack Jeffrey/PhotoReseourceHawaii.com; 59.19: © Tom McHugh/Photo Researchers, Inc.; 59.21: © Merlin D. Tuttle/Bat Conservation International; 59.22a: ANSP © Steven Holt/stockpix.com; 59.22b: U.S. Fish and Wildlife Service; 59.24: © William Weber/Visuals Unlimited; 59.25a-b: © University of Wisconsin-Madison Arboretum; 59.27: © Studio Carlo Dani/Animals Animals; pp. 1278-1279: © Norbert Rosing/National Geographic Image Collection.

Boldface page numbers correspond with **boldface terms** in the text. Page numbers followed by an "f" indicate figures; page numbers followed by a "t" indicate tabular material.